List of Families for A CALIFORNIA FLORA

A CALIFORNIA FLORA

A CALIFORNIA FLORA

PHILIP A. MUNZ

IN COLLABORATION WITH

DAVID D. KECK

WITH

SUPPLEMENT

BY
PHILIP A. MUNZ

UNIVERSITY OF CALIFORNIA PRESS

BERKELEY   LOS ANGELES   LONDON

A CALIFORNIA FLORA:

UNIVERSITY OF CALIFORNIA PRESS
BERKELEY AND LOS ANGELES, CALIFORNIA
UNIVERSITY OF CALIFORNIA PRESS, LTD.
LONDON, ENGLAND
© 1959 BY THE REGENTS OF THE UNIVERSITY OF CALIFORNIA
LIBRARY OF CONGRESS CATALOG CARD NUMBER: 59-5146

SUPPLEMENT:

COPYRIGHT © 1968, BY
THE REGENTS OF THE UNIVERSITY OF CALIFORNIA
COMBINED EDITION, 1973
ISBN: 0-520-02405-2
Second printing, 1975

To the Memory of

ALLEN L. CHICKERING

this book is affectionately dedicated. As
chairman of the Board of Trustees of the
Rancho Santa Ana Botanic Garden from its
beginning to 1958, Mr. Chickering continuously and devotedly served the institution.
His liberal policy was an important factor
in its development and in making this flora
possible.

Contents

A CALIFORNIA FLORA

Introduction

SIZE AND TOPOGRAPHY

(Atwood, W. W., The physiographic provinces of North America, pp. 439–518, 1940. Reed, R. D., Geology of California. Am. Assn. Petroleum Geologists, Tulsa, Okla., pp. 1–355, 1933. Sprague, M., in U.S. Dept. Agric. Yearbook Climate and man, pp. 795–797, 1941.)

Third largest state in the Union, California has an area of 158,297 square miles. It is larger than New York, Ohio, and all the New England states combined. By virtue of size alone California should have many plant species, and with its exceedingly diverse topography, it would be expected to maintain a remarkably large and interesting flora. No other state has such a range of physical conditions. Elevations range from 276 feet below sea level in Death Valley to 14,495 above at the top of Mount Whitney; within a few miles of each other are the lowest and highest points in the country.

In general, California consists of two great series of mountain ranges: an outer, the Coast Ranges, and an inner, the Sierra Nevada plus the southern end of the Cascade Range, including Mount Lassen and Mount Shasta. Between these two axes lies the Great Central Valley whose Sacramento and San Joaquin river systems drain through the Golden Gate. The Sierra Nevada consists primarily of an immense granitic block 400 miles long and 50 to 80 miles wide, and extends from Plumas County to Kern County. Its steep eastern face arises abruptly from 5000 to 10,000 feet above the alluvial-filled basin to the east, whereas the west slope is a more gradual tilted plateau scored by deep canyons containing a series of sizable rivers. The Sierra Nevada is notable in the United States for its display of cirques, moraines, lakes, and glacial valleys. The Cascade series, on the other hand, is volcanic; the southern section, extending from Mount Lassen in California to the Columbia River gorge, has at least 120 volcanoes.

1

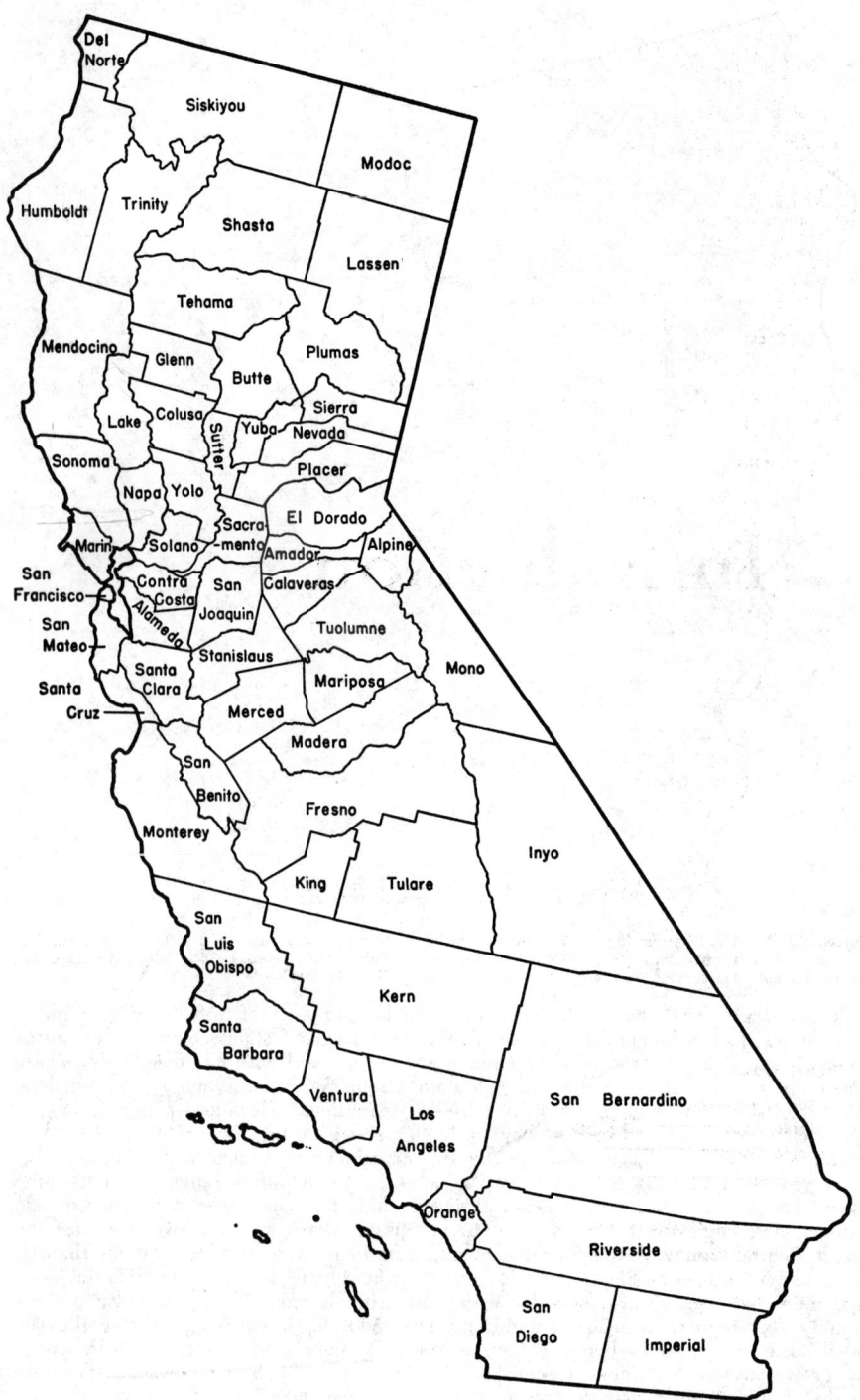

County map of California.

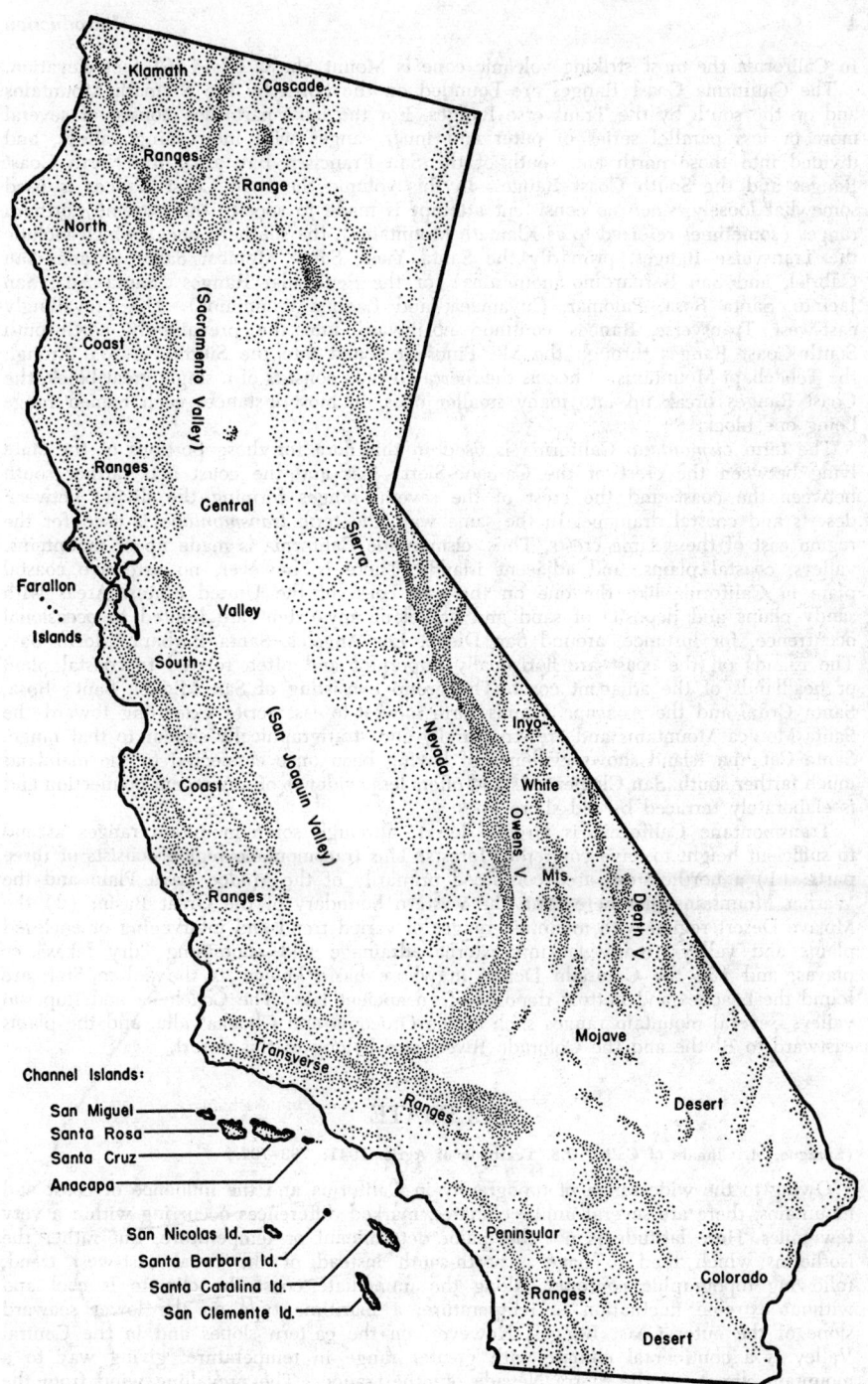

Klamath

Cascade

Ranges

Range

North

Coast

(Sacramento Valley)

Ranges

Central

Sierra

Farallon

Valley

Islands

South

Nevada

Coast

(San Joaquin Valley)

Inyo

Santa

White

Owens V.

Mts.

Ranges

Death V.

Mojave

Transverse

Desert

Ranges

Channel Islands:

San Miguel

Santa Rosa

Santa Cruz

Anacapa

San Nicolas Id.

Peninsular

Santa Barbara Id.

Colorado

Santa Catalina Id.

Ranges

San Clemente Id.

Desert

Topographic map of California.

In California the most striking volcanic cone is Mount Shasta of 14,161 feet elevation.

The California Coast Ranges are bounded on the north by the Klamath Mountains and on the south by the Transverse Ranges. For the most part they consist of several more or less parallel series of outer and inner ranges with intervening valleys, and divided into those north and south of the San Francisco Bay area, the North Coast Ranges and the South Coast Ranges. In this volume the term Coast Ranges is used somewhat loosely, since no consistent attempt is made to specify the extreme northern ranges (sometimes referred to as Klamath Mountains), the southern ones (which include the Transverse Ranges, primarily the Santa Ynez, Santa Monica, Santa Susana, San Gabriel, and San Bernardino mountains), or the Peninsular Ranges (Santa Ana, San Jacinto, Santa Rosa, Palomar, Cuyamaca, and Laguna mountains). The prevailingly east-west Transverse Ranges continue northward into the prevailingly north-south South Coast Ranges through the Mt. Pinos area and into the Sierra Nevada through the Tehachapi Mountains. Whereas the Sierra Nevada consists of a single great block, the Coast Ranges break up into many smaller ones, in most instances each named range being one block.

The term *cismontane* California is used in this flora for those portions of the state lying between the crest of the Cascade-Sierra axis and the coast and farther south between the coast and the crest of the several ranges forming the divide between deserts and coastal drainage. In the same way, the term *transmontane* is used for the region east of these same crests. Thus, cismontane California is made up of mountains, valleys, coastal plains, and adjacent islands. There is, however, no extensive coastal plain in California like the one on the east coast of the United States. Areas with sandy plains and deposits of sand and gravel of any extent are limited to occasional occurrence, for instance, around San Diego, Los Angeles, Santa Barbara, Morro Bay. The islands off the coast are floristically important and often related to coastal plain or headlands of the adjacent coast. The group consisting of San Miguel, Santa Rosa, Santa Cruz, and the Anacapa islands form a west-to-east series extending toward the Santa Monica Mountains and are structurally and stratigraphically similar to that range. Santa Catalina Island shows evidence of having been once connected to the mainland much farther south. San Clemente Island offers less evidence of any recent connection and is elaborately terraced by old shore lines.

Transmontane California is largely desert, although some mountain ranges ascend to sufficient height to have coniferous forests. This transmontane region consists of three parts: (1) a northeastern area composed primarily of the Modoc Lava Plain and the Warner Mountains that here form the western boundary of the Great Basin; (2) the Mojave Desert region with mountain ranges of varied trend and intervening or enclosed plains and valleys, many lacking external drainage and containing "dry lakes" or playas; and (3) the Colorado Desert in whose basin known as the Salton Sink are found the beaches and bottom deposits of an ancient lake. The Coachella and Imperial valleys, several mountain ranges such as the Orocopia and Chuckawalla, and the plains eastward to Blythe and the Colorado River, also belong to this desert.

CLIMATE

(Sprague, M. Climate of Calif., U.S. Yearbook of Agric. 1941: 783–797.)

Owing to the wide range of topography in California and the influence of coast and mountains, there are several unlike climates, marked differences occurring within a very few miles. Here latitude is not the major determinant of temperature, but rather the isotherms which tend to have a north-south instead of the usual east-west trend, following topographic contours. Along the immediate coast the climate is cool and without extreme fluctuation in temperature, a maritime type on the lower seaward slope of the outer Coast Ranges. However, on the eastern slopes and in the Central Valley is a continental climate with greater range in temperature, giving way to a mountain climate in the Sierra Nevada or other ranges. The prevailing wind from the west moderates the climate and increases precipitation. Most of the state is characterized by a wet and a dry season, the former falling usually in the cooler months. The eastern and western slopes of the Coast Ranges and the Sierra Nevada as a whole rise to high enough elevations for snow, in fact for freezing temperatures in all months of the year. Near the coast there may be 365 days frost-free each year, but above 6000 feet less than 100. The amount of precipitation is influenced by the distance from the ocean, by the elevation, shape, and steepness of the mountain slopes and their direction in relation to the moisture-bearing winds. In general, precipitation increases from south to north and is heavier on southern and western slopes. Average annual precipitation varies from less than 2 inches in Death Valley to over 109 inches in parts of Del Norte County. Some desert areas may have entire years without moisture, while a station in Del Norte County has recorded as much as 153.54 inches. Mountain stations sometimes receive 10 inches in 24 hours, particularly in southern California, and torrential floods may ensue. Snowfall may range from a trace or none near the southern coast to 449 inches in the Sierra Nevada.

An important climatic factor for the vegetation is fog, which is most frequent in coastal and neighboring foothill districts. It increases generally with latitude and altitude. In summer it is more frequent than in winter along the coast and on windward slopes. In winter it may be very important in interior areas, for instance the Central Valley.

GEOLOGICAL HISTORY
By Daniel I. Axelrod

Early in the Cenozoic, which commenced approximately 75 million years ago, the vegetation of North America comprised three geofloras (a geoflora is a major vegetation unit that has maintained its essential identity through time and space). The southern half of the continent was occupied by the broad-leaved evergreen Neotropical-Tertiary Geoflora, the northern half by the temperate Arcto-Tertiary Geoflora of mixed deciduous hardwoods and conifers, and between them, centered in the area of southwestern North America, the sclerophyllous and microphyllous Madro-Tertiary Geoflora was making its initial appearance.

The **Neotropical-Tertiary Geoflora** migrated northward during the Eocene, overlapping localities where warm-temperate Cretaceous floras have been recorded. This northward migration carried it to southeastern Alaska on the coast and to near the Canadian border in the interior, regions where it is recorded in association with the temperate Arcto-Tertiary Geoflora which occupied higher cooler latitudes. The Neotropical-Tertiary Geoflora was dominated by broad-leaved evergreens of tropical families, distributed in genera such as Chrysophyllum, Cinnamomum, Cycas, Dioon, Ficus, Lonchocarpus, Mallotus, Meliosma, Persea, Petrea, Sabal, Tetracera, and a host of others which now find their nearest counterparts in the subtropical forests from southern Mexico to Panama, as well as in southern Asia, regions where rainfall is at least 80 inches annually and where climate is uniformly warm through the year. The close resemblance between the subtropical Eocene forests of eastern Oregon and those of the western coastal area clearly indicates that the Cascade Mountains were not yet in existence to form a topographic-climatic barrier at that time. Likewise, the similarity between the subtropical forests of coastal southern California and of New Mexico indicates that climate was relatively uniform across the area, that plant migration was readily possible over the region, and that a high cordillera did not exist in the interior. Clearly, the desert flora that now occupies the intervening region was not then in existence.

In response to the trend toward a cooler and drier climate following the Eocene, the Neotropical-Tertiary Geoflora was gradually restricted southward and coastward. Remnants of its subtropical forests persisted in the warm coastal valleys of California and Oregon into the Miocene, and a few of its relicts, notably Persea and Sabal, survived into the later Pliocene in California, disappearing at the close of the epoch as summer rains were eliminated and temperatures were lowered. No plants in California today represent direct descendants of this humid subtropical forest. However, a number of California genera and species were derived from it secondarily. They comprise subtropic to warm-temperate groups that became adapted to expanding dry subtropical and warm temperate areas over southwestern North America following the Early Eocene, and represent part of the Madro-Tertiary Geoflora.

The **Arcto-Tertiary Geoflora** had an holarctic distribution at high and high-middle latitudes during the Early Tertiary. In western North America it comprised three principal elements:

(1) Its *West American Element* included fossil species closely related to present-day dominants of:

(a) the Coast Forest, with Abies, Acer, Alnus, Chamaecyparis, Cornus, Gaultheria, Lithocarpus, Picea, Populus, Pseudotsuga, Sequoia, Thuja, Vaccinium, and others;

(b) the Sierran Forest, with Abies, Acer, Alnus, Cornus, Fraxinus, Pinus, Picea, Pseudotsuga, Quercus, and others;

(c) The Border-Redwood Forest, with Acer, Alnus, Castanopsis, Platanus, Populus, and Quercus.

(2) The *East American Element* included plants related to those now in the mixed deciduous hardwood forests of summer-wet eastern North America, such as Carpinus, Carya, Castanea, Fagus, Hamamelis, Liquidambar, Nyssa, Taxodium, Ulmus, etc. It also included species of genera of holarctic occurrence, such as Acer, Betula, Cornus, Fraxinus, Populus, Prunus, Salix, and others whose fossil species are closely allied to those now in the eastern part of this continent.

(3) The *East Asian Element* of mixed deciduous forest species comprised such genera as *Ailanthus, Alangium, Cercidiphyllum, Ginkgo, Glyptostrobus, Metasequoia, Pterocarya,* which are no longer native to North America; species of *Carpinus, Carya, Castanea, Fagus, Gordonia, Liquidambar, Lindera,* and *Ulmus*—genera which occur also in eastern North America; and species of genera of holarctic occurrence that occur also in both western and eastern North America, notably species of *Acer, Betula, Cornus, Fraxinus, Juglans, Picea, Pinus, Prunus, Quercus, Rhododendron, Rosa, Salix, Thuja,* and *Vaccinium.*

Species of these elements regularly were admixed in a forest of generalized floristic composition, though the geoflora was in no sense homogeneous over the area of its occurrence. It comprised several forest types reflecting climatic differences imposed by position with respect to the coast, mountains, and altitude. Diverse edaphic conditions also controlled the forest dominants, notably in the taxodioid genera—Glyptostrobus, Metasequoia, Sequoia, and Taxodium. In response to the gradual development of increasingly more emergent continents following the Early Tertiary and to the accompanying trend toward lowered temperature, this geoflora had gradually migrated southward from near the Canadian border, where it occurred in the Paleocene, to central Nevada and northern California by the Middle Oligocene. With increasing aridity during the Miocene and Early Pliocene, it was confined gradually to moister upland sites over the interior and to the humid coastal sector. With the accompanying reduction in summer rain, species of the East American and East Asian Elements rapidly were reduced in numbers during the Late Tertiary, and only a few (Castanea, Pterocarya, Trapa, Ulmus) lingered on into the Late Pliocene in the mild coastal strip from central California northward, becoming extinct there in the early part of the Pleistocene.

The surviving *West American Element,* including species similar to those now dominating our western conifer forests, is thus a segregate of the Arcto-Tertiary Geoflora. In the far West its species became adapted to a new climate—one typified by winter rain and summer drought. Its major communities were differentiated chiefly in the later Pliocene when more diverse climates developed over the area in response to important topographic changes—chiefly the elevation of the Sierra Nevada, the Coast Ranges, and the Transverse and Peninsular ranges. Available paleobotanical data suggest that in California:

(1) *a.* Members of the Sierran forest, together with their Arcto-Tertiary associates, were commencing to invade the Sierra Nevada in the Late Miocene. However, zonal differentiation of the major communities—Yellow Pine Forest, Fir Forest, Subalpine Forest—took place chiefly in the Late Pliocene and Quaternary as the mountain block was elevated to its present heights by warping and faulting.

b. Species of the Sierran forest apparently invaded the higher North Coast Ranges chiefly in the late Tertiary, and probably only during the later Pliocene when the region had been lifted almost to its present altitude.

 c. The Sierran forest probably did not occupy southern California until the Middle Pleistocene, at which time the mountains attained heights sufficient to support it.

 (2) Members of the Coast Forest (Sequoia, Chamaecyparis, Thuja, Abies, etc.), which occupied the coastward slopes of the northern Sierra Nevada by the close of the Miocene, gradually shifted to a more coastal position during the Pliocene as aridity increased. A few of its species, such as *Castanopsis chrysophylla, Taxus brevifolia, Vaccinium parvifolium,* still persist in scarcely modified form in moist relict sites in the northern Sierra.

 (3) The Border-Redwood (oak-madrone) Forest represents an ecotone (transitional area) between the Arcto-Tertiary and Madro-Tertiary Geofloras, and first became established in central California during the Pliocene.

 The **Madro-Tertiary Geoflora,** named for the Sierra Madre of northern Mexico where many of its relicts now survive, had an origin in southwestern North America. In contrast to the Arcto-Tertiary Geoflora, it chiefly comprised small-leaved drought-deciduous sclerophyllous plants. They are principally dry-tropical to warm-temperate types that apparently were derived from more humid tropical alliances of the Neotropical-Tertiary Geoflora that became adapted to the expanding dry climate. Some of them, however, appear to represent members of Arcto-Tertiary groups that also developed in response to increasing aridity in the middle and later Tertiary. Fossil plants apparently ancestral to Madro-Tertiary species have been discovered in the Late Cretaceous and Paleocene floras of southwestern North America. The geoflora had come into existence by the Middle Eocene and spread over the southwestern part of the continent during succeeding epochs as dry climates expanded. Several of the present major vegetation types were represented, including oak-conifer woodland, chaparral, thorn scrub, desert grassland, and sage. Part of the desert flora as well has been derived from it.

 Woodland vegetation comprised three elements:

 (1) *California Woodland Element,* with species of Arbutus, Juglans, Pinus (digger pine, pinyon), Quercus (live oaks), Lyonothamnus, Populus, Umbellularia, and their common associates. It included close ancestors of woodland plants now in California which form several different woodland associations, such as the insular woodland, digger pine woodland, and oak-walnut woodland.

 (2) *Sierra Madrean Woodland Element,* with species of Arbutus, Cupressus, Persea, Robinia, Populus, Sapindus, Ungnadia, Ilex, Quercus, and Vauquelinia, whose nearest relatives survive now in the summer-wet area from Arizona to west Texas and southward into Mexico.

 (3) *Conifer Woodland Element,* with pinyon and juniper, and their associates of Amelanchier, Prunus (Emplectocladus), Cercocarpus, Holodiscus, Purshia, Symphoricarpos. It included species related to those now making up the conifer woodland from eastern California to the Rocky Mountains.

 These elements were admixed in a woodland of generalized floristic composition, which showed regional differences in response to climate as governed chiefly by latitude, altitude, position with respect to the coast, and age. As summer rains were successively reduced during the later Pliocene, Sierra Madrean species were largely eliminated from the far West, and the surviving California and Conifer woodland elements persisted and became adapted to a region of summer drought and winter rain. Paleobotanical data suggest the following history for the surviving California woodland communities:

 (1) Conifer (pinyon-juniper) woodland represents a community that adapted to the cold-semiarid climate of the desert slopes after California oak woodland species were eliminated from the region owing to lowered temperature at the close of the Tertiary.

 (2) Differentiation of the lowland floras of central and southern California is an event of the Late Pliocene and Early Pleistocene, for relatives of present southern California species (*Juglans californica, Lyonothamnus floribundus, Quercus Palmeri, Q. tomentella, Rhus laurina, R. ovata*) ranged into central California during the Middle Pliocene. They were all but eliminated there in the Late Pliocene and Early Pleistocene as temperatures were lowered, though a few still have relict outposts in the region.

 (3) Digger pine woodland appears to have recently become adapted to the colder

winters of central California, for its Pliocene counterpart ranged into southern California where it lived close to thorn scrub.

(4) The Insular (Catalinan) woodland had almost disappeared from the mainland by the end of the Pliocene as temperatures were lowered, and now survives in a maritime climate of equable temperature in which frosts are largely unknown.

(5) The cismontane flora of southern California was differentiated into an interior and a coastal sector in the Quaternary as the Puente Hills–Santa Ana Mountains barrier was elevated. The development of the more continental climate over the interior eliminated most of the coastward slope species, although a few (*Rhus laurina, R. integrifolia, Juglans californica*) have relict occurrences in the interior.

Chaparral of the Madro-Tertiary Geoflora included a California Chaparral Element, with fossil species of Arctostaphylos, Ceanothus, Cercocarpus, Dendromecon, Garrya, Heteromeles, Prunus, Quercus, Rhamnus, and Rhus closely similar to those now dominating the California communities. Mixed with them were species of the Southwestern Chaparral Element, including plants of many of these same genera whose nearest relatives are in the region from Arizona eastward and southward into Mexico. They apparently disappeared from California in the later Pliocene as summer rains ceased. The occurrence of close fossil relatives of present-day southern California species in central California, as well as in Nevada and Oregon, in the Pliocene, shows that the modern areas of chaparral are of recent derivation. Furthermore, in southern California ancestors of present coastal types ranged into the interior during the Pliocene, but apparently disappeared in the Late Pliocene when the climate became more continental.

Whereas the Madro-Tertiary woodland and chaparral ranged widely over the lowlands of the western United States in the Late Tertiary, thorn forest vegetation of the Madro-Tertiary Geoflora was confined chiefly to the warmer region of southern California and areas to the south and east. It included species of Acacia, Bursera, Colubrina, Dodonaea, Erythea, Ficus, Karwinskia, Lysiloma, Pithecolobium, Prosopis, Randia, and others whose nearest relatives now comprise arid subtropical scrub vegetation in northeastern and northwestern Mexico. Lowered winter temperature and the cessation of summer rain eliminated the community from California in the later Pliocene, though some of its species (for instance, Prosopis, Acacia, Condalia) have become adapted to the desert of southeastern Californiᵃ Arizona, and northern Mexico.

With the rapid expansion of dry climates in the Middle Pliocene, woodland and chaparral for the most part vanished from the lowlands of the present central and southern desert region and from the forest of the northern sector. In the later Pliocene and Pleistocene the rapidly rising Cascades, Sierra Nevada, and the Peninsular Ranges brought a drier climate to their lee, and a desert flora came into existence. Its species derived chiefly from those represented in the geofloras that dominated the area earlier in the Pliocene. The colder Great Basin and Mojave deserts received increments from the surviving Arcto-Tertiary Geoflora, the Sonoran and Mojave from the Madro-Tertiary.

Selected References

Axelrod, D. I., A Miocene flora from the western border of the Mohave Desert, Carnegie Inst. Wash. Pub. 516, pp. 1–129, 1939.

————, Late Tertiary floras from the Great Basin and border areas, Bull. Torrey Bot. Club, vol. 67, pp. 477–487, 1940.

————, Climate and evolution in western North America during Middle Pliocene time, Evol., vol. 2, pp. 127–144, 1948.

————, Evolution of desert vegetation in western North America, Carnegie Inst. Wash. Pub. 590, pp. 215–306, 1950.

————, A theory of angiosperm evolution, Evol., vol. 6, pp. 29–60, 1952.

————, Mio-Pliocene floras from west-central Nevada, Univ. Calif. Publ. Geol. Sci., vol. 33, pp. 1–322, 1956.

————, Late Tertiary floras and the Sierra Nevadan uplift, Bull. Geol. Soc. Amer., vol. 68, pp. 19–45, 1957.

————, Evolution of the Madro-Tertiary Geoflora, Bot. Rev., vol. 24, pp. 433–509, 1958.

Chaney, R. W., Succession and distribution of Cenozoic floras around the northern Pacific basin, *in* Essays in Geobotany in honor of William Albert Setchell, Univ. Calif. Press, 1936.

————, Paleoecological interpretations of Cenozoic plants in western North America, Bot. Rev., vol. 4, pp. 371–396, 1938.

————, Tertiary centers and migration routes, Ecol. Monogr., vol. 17, pp. 139–148, 1947.

————, and E. I. Sanborn, Goshen flora of west-central Oregon, Carnegie Inst. Wash. Pub. 439, pp. 1–103, 1933.

————, C. Condit, and D. I. Axelrod, Pliocene floras of California and Oregon, Carnegie Inst. Wash. Pub. 553, pp. 1–407, 1944.

————, and D. I. Axelrod, Miocene forests of the Columbia Plateau, Carnegie Inst. Wash. Pub. 617, 1960.

Jahns, R. J., ed., Geology of southern California, Calif. State Div. of Mines Bull. 170, 1954.

MacGinitie, H. D., A Middle Eocene flora from the central Sierra Nevada, Carnegie Inst. Wash. Pub. 534, pp. 1–173, 1941.

RECENT GEOLOGICAL HISTORY AND THE VEGETATION

As described in the preceding section by Daniel I. Axelrod, much of the vegetation of western North America has come about by climatic changes involving moisture and temperature. With an early Eocene rainfall of possibly 80 inches or more for much of the West, the change progressed to a Miocene possibility of about 12 to 25 inches on the western border of the Mojave Desert and a flora new to the region developed. In general, this drier area arose east of the Cascade-Sierran axis, but summer dryness throughout what is now California became important and the sclerophyllous types of woody plants increased. Thus a flora arose, small-leaved drought-resistant and drought-deciduous in comparison to the early Tertiary floras. By Pliocene, forest and woodland had to move toward the coast or into the mountains. The lower elevations of what is now desert became dominated by Scrub and Grassland. Desert species evolved slowly from floras already present, the colder deserts getting some from Arcto-Tertiary sources, the warmer from Madro-Tertiary. Thus, *Artemisia tridentata, Sarcobatus, Atriplex, Eurotia,* and other Eurasian genera of the present Great Basin Desert exemplify the former; genera like *Larrea, Eriogonum, Grayia,* and *Tetradymia* seem to be of southern origin.

During the Pleistocene temperate vegetation showed farther advance southward than during any other period of the Cenozoic, correlated undoubtedly with the glaciation that left its marks on the California mountains. The Redwood Forest has been recorded as far south as Carpenteria and Santa Cruz Island, perhaps 200 miles south of its present limits. Glaciation in California may have meant not only a climate cooler than at present, but more moist. Much evidence for this exists not only throughout the Sierra Nevada, where signs of past glaciation are prevalent and some minor glaciers still persist, but also at lower elevations. The former lakes, the former greater depths of present lakes, and the no-longer-functional stream beds of the Mojave Desert further verify the theory. According to Flint (Glacial geology and the Pleistocene epoch, John Wiley & Sons, 1947) Mono Lake had attained a depth of nearly 900 feet, Owens Lake was 250 feet higher and overflowed into the basins of China and Searles lakes, and water from Searles Basin overflowed into Panamint Basin to form a lake 930 feet deep, which in turn apparently spilled into Death Valley producing a lake 90 miles long and 600 feet deep. Other basins, as in the San Joaquin Valley, had lake stages correlated with glacial stages. Southward migrations during such colder and wetter periods, with subsequent and perhaps intermediate northward migrations, have undoubtedly had a great influence on both the evolution and distribution of our present vegetation. Extensive migration must certainly have brought into the same area species not previously sympatric, permitting much hybridization. Movement was not only northward and southward but from one altitude to another. Many of our rarer montane species of southern California persist as relicts which have climbed to higher elevations with warmer, drier climate. Undoubtedly some of the desert species (like various Boutelouas and other grasses) known from occasional collections in California are relicts disappearing because of a possible decrease in the summer showers. The persistence of over 300 species of Sierran plants in the pine belt of southern California can best be explained by extensive southward migration in the Glacial Period.

The student of California botany is constantly impressed by the influence of summer dryness, even though the annual precipitation may be fairly high. What else could explain the absence of so many alpine plants from the Sierra Nevada, species often circumpolar in distribution and common far south in the Rocky Mountains? This lack of many widely distributed alpine species has been compensated for by evolution of high altitude species in genera common in the California flora of somewhat lower elevation; hence our alpine flora has many endemics. Furthermore, a region so broken up into separate mountain ranges and low-lying valleys, with the many ecological niches often quite isolated from each other, favors endemism. It is not surprising, then, to find in California many exceedingly local species and subspecies, therefore a flora extremely rich in number of forms and of many very large genera. Take as examples Arctostaphylos, Astragalus, Calochortus, Ceanothus, Eriogonum, Lupinus, Penstemon, and Trifolium. Some follow one pattern in their evolution, others another, but all have undergone rather rapid and highly polymorphic differentiation.

CALIFORNIA PLANT COMMUNITIES

(Munz, Philip A., and David D. Keck, California plant communities, El Aliso 2: 87–105, 1949, and 2: 199–202, 1950.)

For discussion of habitat and distribution of the species treated in this book it seemed desirable to work out a classification of Plant Communities which would indicate more definitely the associated species than does the Merriam Life Zone system. To that end a system proposed has now been tried for several years and proves quite satisfactory. It involves the use of twenty-nine Communities that can be grouped into *vegetation types* and these in turn into *biotic provinces*.

BIOTIC PROVINCES

In a discussion of the California flora the biotic provinces are well worth considering as they reveal to some extent the relationship of the flora from one area to another both within and without the state. Dice (p. 3 of The biotic provinces of N. Am. [Univ. Mich. Press, 1943], pp. 1–73) says that a biotic province "covers a considerable and continuous geographic area and is characterized by the occurrence of one or more important ecological associations that differ, at least in proportional area covered, from the associations of adjacent provinces. In general, biotic provinces are characterized also by peculiarities of vegetation type, ecological climax, flora, fauna, climate, physiography, and soil." The provinces are the result of interaction of past and present forces, and each includes several vegetation types and Plant Communities. For California we recognize five biotic provinces:

(A) **Oregonian,** the "Vancouveran" proper of Van Dyke (The distribution of insects in western North America. Ann. Entom. Soc. Am. 12: 1–12, 1919). We accept the name "Oregonian," since it has priority, having been proposed as early as 1859 by J. G. Cooper, because it is better known than Van Dyke's name, and also because this province differs considerably between California, on the one hand, and Vancouver Island, whence it drew its name, on the other. It applies to the cool moist coastal strip extending southward to San Francisco Bay and small elements in Monterey County. Much the same as the "Redwood Transition Zone" of Jepson, it is the southern limit for many species of north coastal distribution.

(B) **Californian,** those portions of California west of the Sierra Nevada and the southern mountains, thus including the interior valleys and their surrounding hills in the central and northern parts of the state, as well as the southern coastal area and Coast Ranges south of San Francisco Bay. It reaches its southern limit in northern Baja California. There are much endemism, considerable Mexican influence, and a marked similarity in some genera and species to the temperate parts of Chile and Peru. Although the climate is highly diversified, the rains fall almost entirely in the winter and the summer season is long and dry.

(C) **Sierran,** the great montane area that runs interruptedly from southern Oregon through Mount Shasta, the Sierra Nevada, Tehachapi Mountains, San Gabriel, San Bernardino, San Jacinto, and Cuyamaca ranges to the San Pedro Martir of Baja California. It begins with the yellow pine belt and extends to the summits of the mountains. In its lower and middle elevations is a large floral element derived from the surrounding lowlands. There is also much endemism. Only at the higher altitudes are found the widespread boreal species that occur with greater frequency in such other ranges as the Cascades and the Rocky Mountains. This is a region of winter snow and some summer rain.

(D) **Nevadan,** largely the "Great Basin" of Van Dyke and the "Artemisian" of Dice, occurring east of the Cascade-Sierran axis from Owens Valley northward. Some elements extend into Siskiyou County. The lower plains are covered with *Artemisia tridentata,* but there are interrupted mountain ranges with some forest and woodland. Historically the general affinity is to the east and south. The winters are cold, the summers hot. Precipitation is relatively light, usually in the form of winter snow.

(E) **Southern Desert,** called "Sonoran" by Van Dyke, "Sonoran" and "Mohavian" by Dice, and "Mojave Desert" and "Sonoran Desert" by Shreve. Since the term "Sonoran" has had so different a meaning in the Merriam system of life zones, we are not employing it here. And, although we recognize that the Mojave and Colorado deserts differ in many respects, we are considering the two together because the Creosote Bush Scrub occupies the largest single area in both. There is decided affinity with the flora to the southeast. The deserts are known not only for their dryness, but also for the diurnal and seasonal temperature extremes.

VEGETATION TYPES

The vegetation of an area can be, and almost automatically is, thought of in terms of physiognomy, or the structural units into which it can be divided. The major vegetation types—grassland, chaparral, woodland, coniferous forest, marsh, scrub, and the like—are fairly obvious. Such types are often referred to as "plant formations" or "climax formations," each of which is the "product of the complex of climatic factors effective in a region" (p. 224 of H. J. Oosting's The study of plant communities [W. H. Freeman & Co., 1948], pp. 1–389). We recognize for California the following major vegetation types: I. Strand; II. Salt Marsh; III. Freshwater Marsh; IV. Scrub; V. Coniferous Forest; VI. Mixed Evergreen Forest; VII. Woodland-Savanna; VIII. Chaparral; IX. Grassland; X. Alpine Fell-fields; and XI. Desert Woodland. These names that are not self-explanatory will be made plain by the discussion under "Communities."

These vegetation types, such as grassland, are not necessarily uniform. For instance, the grasslands of the Great Plains east of the Rocky Mountains, with winter snow but maximum precipitation in summer, are quite different from the grasslands of California, which receive all their precipitation in winter and in the form of rain. The grasslands of southeastern Washington, with major precipitation in winter but some in summer too, and with cooler year-round temperatures, differ from either. In each of these three areas the climatic conditions favor a grassland climax, but as they are not alike, the constituent species of grasses and other herbs also differ. It is to such different phases of the vegetation type that the name *Plant Community* is applied in this book.

PLANT COMMUNITIES

Oosting (1948) defines a *community* as "an aggregation of living organisms having mutual relationships among themselves and to their environment." Such a term can be more or less inclusive and agrees with the action of the Third International Botanical Congress at Brussels in 1910 in adopting the word *community* "to cover ecological units of every degree." We are using the term Plant Community for each regional element of the vegetation that is characterized by the presence of certain dominant species. In other words, the community is floristically determined. A vegetation type may consist of one to several communities.

The great majority of California plant communities have a climatic rather than a purely edaphic basis. Most of these can readily be divided into fairly distinct smaller groups. Take, for example, the Yellow Pine Forest. It is not at all a uniform community, but contains the plants of wet meadows, lake shores and stream banks, rocky outcrops, as well as open and pine-covered benches and slopes. In this book such minor subdivisions are not named. The few communities that have an edaphic basis include the strand, marsh, and saline groups. In these instances the vegetation type and community are commensurate.

In California's diversified topography and its exceedingly complex distribution of climates, the communities naturally must often be geographically discontinuous, with parts of one occurring as islands within another. A north-facing slope, for example, may differ from one facing south; the former may have oak-woodland, the latter, chaparral. Thus a community may be much dissected, especially near its margins, and may dovetail between adjacent communities so the lines of contact are irregular and transitional areas numerous, at times even rather extensive. The situation in some places is so complex as to be open to diverse interpretations and various observers are liable to disagree on treatment.

The present classification, an essentially practical one for our purposes, defines twenty-nine *communities* within eleven *vegetation types* (see Table I) and five *biotic provinces*. The individual communities were arrived at by recognizing each that was rather easily distinguishable from the others by having a number of species more or less restricted to it, including at least a few dominants or characteristic indicator species. This system is used for a specific purpose in this book and does not necessarily coincide with one designed for purely ecological considerations.

Many species of widespread distribution occupy a number of climatic zones and have correspondingly a number of climatic races or ecotypes, each fitting into its own environmental niche. In the classification herewith proposed, it is hoped that pointing out the various communities in which the species occur will give some indication of the number of ecotypes to be expected. The fact that one species may be restricted to a single community and another may occupy several, in which quite different conditions prevail, should be suggestive of ecotypic constitution and of the location of various ecotypes, if they do exist.

Following are short discussions of the twenty-nine Plant Communities, listing first the indicator species, then distribution and type of area covered, pertinent climatic data, and giving for most a brief outline of the appearance of the vegetation. The most significant indicator-species of the various communities are those most closely restricted in their ecological requirements—that is, those species of only one or very few ecotypes. Tree indicators, when present, are stressed above the herbs because they have more effect on the microhabitat or microclimate, and they are more widely known and more readily identified.

1. Coastal Strand

Artemisia pycnocephala, Franseria Chamissonis, Lathyrus littoralis, Lupinus arboreus, L. Chamissonis, Abronia maritima, A. umbellata, Oenothera cheiranthifolia, Atriplex leucophylla, Fragaria chiloensis, Poa Douglasii, Haplopappus ericoides, Mesembryanthemum nodiflorum, M. crystallinum, Convolvulus Soldanella.

Sandy beaches and dunes scattered along the entire coast.

Annual rainfall 15 to 70 inches, with much fog and wind; growing season 12 months, with 350 to 365 frost-free days; small seasonal and diurnal fluctuations in temperature; mean summer maxima 61°–72°, mean winter minima 39°–47° F.

Vegetation low or prostrate, often succulent, late flowering. The constitution of this community varies considerably from north to south, some species reaching their southern limit at Cape Mendocino, some at Monterey Peninsula, and some at Point Conception. A number of others, however, exemplify the continuity of the community by extending the entire length of the state and beyond.

2. Coastal Salt Marsh

Salicornia virginica, S. subterminalis, Suaeda californica and var. pubescens, Distichlis spicata, Spartina leiantha, Limonium californicum, Frankenia grandifolia, Triglochin maritima.

Salt marshes along the coast, from sea level to 10 feet.

Average rainfall 15 to 40 inches; growing season 12 months, with 330 to 365 frost-free days; small seasonal and diurnal fluctuations in temperature; temperature range about as in Coastal Strand.

Most extensive on tidelands.

TABLE I

MAJOR VEGETATION TYPES AND PLANT COMMUNITIES OF CALIFORNIA

Vegetation Type	Plant Community
I. Strand	1. Coastal Strand
II. Salt Marsh	2. Coastal Salt Marsh
III. Freshwater Marsh	3. Freshwater Marsh
IV. Scrub	4. Northern Coastal Scrub
	5. Coastal Sage Scrub
	6. Sagebrush Scrub
	7. Shadscale Scrub
	8. Creosote Bush Scrub
	9. Alkali Sink
V. Coniferous Forest	10. North Coastal Coniferous Forest
	11. Closed-cone Pine Forest
	12. Redwood Forest
	13. Douglas-Fir Forest
	14. Yellow Pine Forest
	15. Red Fir Forest
	16. Lodgepole Forest
	17. Subalpine Forest
	18. Bristle-cone Pine Forest
VI. Mixed Evergreen Forest	19. Mixed Evergreen Forest
VII. Woodland-Savanna	20. Northern Oak Woodland
	21. Southern Oak Woodland
	22. Foothill Woodland
VIII. Chaparral	23. Chaparral
IX. Grassland	24. Coastal Prairie
	25. Valley Grassland
X. Alpine Fell-fields	26. Alpine Fell-fields
XI. Desert Woodland	27. Northern Juniper Woodland
	28. Pinyon-Juniper Woodland
	29. Joshua Tree Woodland

3. Freshwater Marsh

Scirpus Olneyi, S. validus, S. acutus, S. californicus, Typha latifolia, T. domingensis (*T. angustifolia*), *Heleocharis palustris, Carex senta, C. obnupta.*

Marshes of interior valleys such as near Tulare Lake, river-bottom lagoons, and near coast back of immediate salty areas, from sea level to about 500 feet.

Climatic conditions variable, but growing season long and physical conditions relatively constant.

4. Northern Coastal Scrub

Baccharis pilularis, Mimulus aurantiacus, Castilleja latifolia, Rubus vitifolius, Lupinus variicolor, Heracleum lanatum, Eriophyllum staechadifolium, Gaultheria Shallon, Anaphalis margaritacea, Artemisia Suksdorfii, Erigeron glaucus.

Narrow coastal strip from southern Oregon to San Mateo County and from Pacific Grove to Point Sur, lying between the Coastal Strand and the Redwood Forest at elevations mostly below 500 feet.

Annual rainfall 25 to 75 inches, with much fog and wind; growing season 10 to 12 months, with 300 to 350 frost-free days; little fluctuation in temperature, mean summer maxima 63°–75°, mean winter minima 35°–40° F.

Rather low plants rarely over 6 feet in height, sometimes dense, but often with extensive areas of grass (*Danthonia californica, Deschampsia caespitosa* ssp. *holciformis, Calamagrostis nutkaensis, Holcus lanatus,* etc.) between.

5. Coastal Sage Scrub

Artemisia californica, Salvia apiana, S. mellifera, S. leucophylla, Eriogonum fascicu-

latum, Rhus integrifolia, Encelia californica, Horkelia cuneata, Haplopappus squarrosus, H. venetus, Eriophyllum confertiflorum.

Usually dry rocky or gravelly slopes, South Coast Ranges to Baja California, mostly below 3000 feet and below the Chaparral.

Annual rainfall 10 to 20 inches; growing season 8 to 12 months, with 230 to 350 frost-free days; mean summer maximum temperatures 68°–90°, mean winter minima 37°–48° F.

Plants half-shrubs, 1 to 5 feet tall or somewhat woodier and larger, forming a more open community than Chaparral.

6. Sagebrush Scrub

Artemisia tridentata, A. arbuscula ssp. *nova, A. cana, Coleogyne ramosissima, Chrysothamnus nauseosus* sspp. *speciosus* and *mohavensis, C. viscidiflorus, Atriplex confertifolia, A. canescens, Tetradymia spinosa, Purshia tridentata, P. glandulosa.*

Deep pervious soil along the east base of the Sierra Nevada from Modoc County south to the San Bernardino Mountains, mostly at elevations of 4000 to 7500 feet; occasional in Siskiyou and San Diego counties.

Average precipitation 8 to 15 inches mostly as winter snow; growing season 3.5 to 6 months, with 70 to 130 frost-free days; mean summer maximum temperatures 83°–95°, mean winter minima 8°–27° F.

Low silvery-gray shrubs 2 to 7 feet tall, interspersed with greener plants.

7. Shadscale Scrub

Atriplex confertifolia, Grayia spinosa, Eurotia lanata, Kochia californica, Artemisia spinescens, Menodora spinescens, Gutierrezia Sarothrae, Coleogyne ramosissima.

In heavy soil, often with underlying hardpan, of mesas and flats at 3000 to 6000 feet, about the Mojave Desert, Owens Valley, etc.

Average rainfall 3 to 7 inches; growing season limited by water; frost-free days 150 to 250; temperatures similar to those in Joshua Tree Woodland.

Plants largely 1 to 1.5 feet tall, shallow-rooted, and covering large monotonous areas between Creosote Bush Scrub and Joshua Tree Woodland.

8. Creosote Bush Scrub

Larrea divaricata, Franseria dumosa, Fouquieria splendens, Dalea californica. D. Schottii, D. spinosa, Lycium brevipes, L. Andersonii, Hymenoclea Salsola, Encelia farinosa, E. frutescens, Sphaeralcea ambigua, Baccharis sergiloides, Echinocereus Engelmannii, Opuntia Bigelovii, O. echinocarpa, O. basilaris; Prosopis juliflora var. *glandulosa, Olneya Tesota, Pluchea sericea,* and *Chilopsis linearis* along the watercourses.

Well-drained soil of slopes, fans, and valleys, usually below 3500 feet, in deserts from southern end of Owens Valley to Mexico.

Average rainfall mostly 2 to 8 inches, some as summer showers; frost-free days 180 to 345; highly variable seasonal and diurnal temperatures, mean summer maxima 100°–110°, means winter minima 30°–42° F.

Shrubs 2 to 10 feet tall, widely spaced, largely dormant between rainy periods.

9. Alkali Sink

Atriplex polycarpa, A. lentiformis, A. Breweri, A. spinifera, A. Parryi, Sarcobatus vermiculatus, Allenrolfea occidentalis, Suaeda Torreyana var. *ramosissima, Salicornia virginica, Frankenia grandifolia* var. *campestris.*

Poorly drained alkaline flats and playas in floor of Great Central Valley and of arid regions east of the Sierra Nevada, and in such sinks as Panamint and Death valleys, mostly at less than 4000 feet elevation.

Average rainfall 1.5 to 7 inches; frost-free days 200 to 335; highly variable seasonal and diurnal temperatures, mean summer maxima 106°–116°, mean winter minima 28°–37° F.

Low scattered gray or fleshy halophytes where there is poor or no drainage, as about dry lakes; under this community are grouped several associations that are perhaps more distinct and cover larger areas in the deserts of Nevada and Utah.

10. North Coastal Coniferous Forest

Thuja plicata, Tsuga heterophylla, Picea sitchensis, Pseudotsuga Menziesii, Abies grandis, Chamaecyparis Lawsoniana, Rhamnus Purshiana, Acer circinatum.

Outer North Coast Range, Mendocino County northward, from near sea level up to 1000 feet or more; in occasional restricted patches as far south as Sonoma County.

Average rainfall 40 to 110 inches, with frequent dense fogs; growing season ε to 12 months, with 225 to 360 frost-free days; temperature mild and equable; mean summer maxima 62°–70°, mean winter minima 38°–42° F.

Trees 150 to 200 feet tall or more; the forest dense and continuous, often with much undergrowth. Of increasing importance northward through Oregon and Washington.

11. Closed-cone Pine Forest

Pinus muricata, P. contorta, P. radiata, P. remorata, Cupressus macrocarpa, C. pygmaea, C. Goveniana.

Interrupted forest from Mendocino plains southward near the immediate coast to Santa Barbara County, from near sea level to 1200 feet. Northward it is on the seaward side of the redwoods in barren soils.

Average rainfall 20 to 60 inches, much fog; growing season 9 to 12 months, with 270 to 360 frost-free days; climate cool with temperatures comparable with those in the Redwood Forest.

Trees 30 to nearly 100 feet tall, in a relatively dense forest.

12. Redwood Forest

Sequoia sempervirens, Pseudotsuga Menziesii, Myrica californica, Lithocarpus densiflora, Vaccinium ovatum, Gaultheria Shallon, Rhododendron californicum, Oxalis oregona, Vancouveria parviflora, Polystichum munitum, Whipplea modesta.

Seaward slopes of outer Coast Ranges, 10 to 2000 feet (even to 3000 feet in Santa Lucia Mountains), from Del Norte County and adjacent Oregon to Santa Cruz County, with outliers along the coast of central Monterey County.

Average rainfall 35 to 100 inches, with dense dripping fog in dry season; growing season 6 to 12 months, with 200 to 350 frost-free days; not much change in temperature diurnally or seasonally, the mean summer maxima 68°–84°, the mean winter minima 33°–40° F.

Trees very tall, even to 350 feet, in a heavy, dense forest.

13. Douglas-Fir Forest

Pseudotsuga Menziesii, Lithocarpus densiflora, Arbutus Menziesii, Castanopsis chrysophylla, Pinus Lambertiana.

North Coast Ranges from Mendocino County northward, scattered remnants southward to Sonoma and Marin counties, mostly east of the Redwood Forest and to elevations of 4500 feet, but in some places reaching almost to the coast.

Climatic data much as for Mixed Evergreen Forest.

Trees to 200 feet high, in dense forests often of pure stands of *Pseudotsuga*. Apparently best developed on east and north slopes in California. Common northward to British Columbia.

14. Yellow Pine Forest

Pinus ponderosa, P. Lambertiana, Libocedrus decurrens, Abies concolor, Pseudotsuga Menziesii, Quercus Kelloggii, Ribes nevadense, R. Roezlii, Rubus parviflorus, Chamaebatia foliolosa, Arctostaphylos patula, A. Mariposa, Ceanothus integerrimus.

North Coast Ranges, 3000 to 6000 feet; northern California, 1200 to 5500 feet; Sierra Nevada, 2000 to 6500 or 7000 feet; southern California, 5000 to 8000 feet.

Average precipitation 25 to 80 inches, partly as snow; growing season 4 to 7 months, with 90 to 210 frost-free days; mean summer maximum temperatures 80°–93°, mean winter minima 22°–34° F.

Trees 75 to 200 feet tall, in extensive continuous forests.

15. Red Fir Forest

Abies magnifica, Pinus Murrayana, P. monticola, P. Jeffreyi, Castanopsis sempervirens, Ceanothus cordulatus, Ipomopsis aggregata, Populus tremuloides.

Above 6000 feet in North Coast Ranges; northern California, 5500 to 7500 feet; Sierra Nevada, 6000 to 9000 feet; southern California, 8000 to about 9500 feet.

Average precipitation 35 to 65 inches, with heavy winter snow; growing season 3 to 4.5 months, with 40 to 70 frost-free days; mean summer maximum temperatures 73°–85°, mean winter minima 16°–26° F.

Trees to 100 feet tall or more, in dense forests.

16. Lodgepole Forest

Pinus Murrayana, Tsuga Mertensiana, Artemisia Rothrockii, Potentilla Breweri, Castilleja Culbertsonii, Pedicularis attolens, Haplopappus apargioides, Senecio lugens.

Northernmost California to central Sierra Nevada, where it grows from about 8300 to 9500 feet.

Average precipitation about 30 to 60 inches, mostly as snow; growing season 9 to 14 weeks, with frost-free days as many as 40; mean summer maximum temperatures 67°–75°, mean winter minima 10°–18° F.

Trees to 50 or 60 feet tall, in rather open forest with extensive meadows scattered through it.

17. Subalpine Forest

Pinus albicaulis, P. Balfouriana, P. flexilis, P. Murrayana, Tsuga Mertensiana, Salix petrophila, Eriogonum incanum, Ribes cereum, R. montigenum, Aquilegia pubescens, Sedum obtusatum, Potentilla fruticosa, Cassiope Mertensiana, Phyllodoce Breweri, Penstemon heterodoxus.

The most boreal forest in California; in northern California from about 8000 to 9500 feet; Sierra Nevada, 9500 to 11,000 feet; poorly represented in southern California and above 9500 feet.

Average precipitation about 30 to 50 inches, dropping as low as 15 inches on the east side of the crest, mostly as snow, with heavy snow cover in winter; growing season 7 to 9 weeks, and killing frost possible in every month; mean summer maximum temperatures probably not over 65° F., winter minima unknown.

Trees from elfin wood (Krummholz) to 40 feet tall or more, usually rather scattered.

18. Bristle-cone Pine Forest

Pinus aristata, P. flexilis, Cercocarpus ledifolius, Haplopappus Gilmani, Chrysothamnus gramineus, C. viscidiflorus ssp. *pumilus, C. Parryi* ssp. *asper, Artemisia arbuscula* ssp. *nova, Phacelia frigida, Lomatium inyoense, Cryptantha Roosiorum, C. Hoffmannii, Astragalus platytropis, Cymopterus cinerarius, Penstemon scapoides, Stipa pinetorum.*

Inyo-White Mountains, Panamint Mountains, Funeral and Grapevine mountains of Mono and Inyo counties, at 8500 to 11,500 feet, occasionally as low as 7200 feet.

Precipitation data available for a three-year period only and ranged from 10 to 22 inches with average of 15. Snowfall averaged 129 inches. Frost-free days from 50 to 90; mean summer maximum temperatures 54°–66°, mean winter minima 3°–21°.

Open forest of trees 15 to 40 feet high, on more or less brushy and rocky slopes.

19. Mixed Evergreen Forest

Lithocarpus densiflora, Arbutus Menziesii, Pseudotsuga Menziesii, Castanopsis chrysophylla, Umbellularia californica, Acer macrophyllum, Quercus chrysolepis, Q. Kelloggii, Corylus californica, Cornus Nuttallii, Ceanothus Parryi, C. thyrsiflorus.

Along inner edge of the Redwood Forest and on higher hills within it, mostly in the North Coast Ranges, but as far south as the Santa Cruz Mountains and the north side of the Santa Lucia Mountains, at elevations of 200 to 2500 feet.

Average rainfall 25 to 65 inches, with some fog; growing season 7 to 11 months, with 200 to 300 frost-free days; mean summer maximum temperatures 75°–95° F., mean winter minima 29°–39° F.

Trees to 100 feet or more tall, in rather close stands, often with brush beneath and with grassland islands referable to Coastal Prairie. Many members of this community, which fraternize with the Redwood, also enter the Yellow Pine Forest as important constituents, even accompanying it from the Coast Ranges to Mount Shasta and well southward along the Sierra Nevada.

20. Northern Oak Woodland

Quercus Garryana, Q. Kelloggii, Q. chrysolepis, Q. Wislizenii, Acer macrophyllum, Aesculus californica, Arctostaphylos Manzanita.

North Coast Ranges from Humboldt and Trinity counties as far south as Napa County and inland from the Redwood Forest to the Yolla Bolly Mountains, ascending to 3000 or even 5000 feet.

Average rainfall 25 to 40 inches; growing season 6 to 9 months, with 180 to 265 frost-free days; mean summer maximum temperatures 80°–94° F., and mean winter minima 31°–38° F.

Trees 25 to 75 feet tall, in rather open woodland with little undergrowth.

21. Southern Oak Woodland

Quercus agrifolia, Q. Engelmannii, Juglans californica, Rhus integrifolia, R. ovata, R. trilobata.

Valleys of interior southern California from Los Angeles County to San Diego County and ascending to about 5000 feet at Vandeventer Flat in the San Jacinto Mountains.

Average rainfall 15 to 25 inches, often of torrential type with rapid runoff; growing season 7 to 10 months, with 200 to 350 frost-free days; mean summer maximum temperatures 84°–92°, mean winter minima 32°–44° F.

Trees 20 to 60 feet tall, with grassland or few soft shrubs between them.

22. Foothill Woodland

Pinus Sabiniana, P. Coulteri in upper parts, *Quercus Douglasii, Q. chrysolepis, Q. agrifolia, Q. Wislizenii, Q. lobata, Umbellularia californica, Aesculus californica, Rhamnus californica* ssp. *tomentella, Ceanothus cuneatus, Cercis occidentalis, Ribes quercetorum, Eriodictyon californicum.*

Foothills and valley borders, 400 to 3000 feet, fingering upward on warm slopes to 5000 feet; inner Coast Ranges, Trinity County to Santa Barbara County; western foothills of the Sierra Nevada, reaching southern limit in northwestern Los Angeles County.

Average rainfall 15 to 40 inches, little or no fog; growing season 6 to 10 months, with 175 to 310 frost-free days; hot dry summers, with mean maximum temperatures 75°–96°, and mean winter minima 29°–42° F.

Trees 15 to 70 feet tall, in dense or open woodland, with scattered brush and grassland between the trees. This composite community contains both the oak parklands of the valley floors and the digger pine woodland of the surrounding slopes.

23. Chaparral

Adenostoma fasciculatum, Heteromeles arbutifolia, Rhamnus californica, R. crocea, Quercus dumosa, Cercocarpus betuloides, Yucca Whipplei, Fremontia californica, Prunus ilicifolia, Ceanothus spp., *Arctostaphylos* spp., *Pickeringia montana, Trichostema lanatum.*

Dry slopes and ridges in Coast Ranges from Shasta County south, and below the Yellow Pine Forest on the western slopes of the Sierra Nevada and more southern mountains. Rocky, gravelly, or fairly heavy soils.

Average rainfall 14 to 25 inches; hot dry summers and cool but not cold winters; growing season 8 to 12 months, with 250 to 360 frost-free days; mean summer maximum temperatures 82°–94°, mean winter minima 29°–45° F.

A broad-leaved sclerophyll type of vegetation, 3 to 6 or 10 feet high and dense, often nearly impenetrable. Subject to fire, following which many of the shrubs tend to stump-sprout.

24. Coastal Prairie

Festuca idahoensis, Danthonia californica, Calamagrostis nutkaensis, Deschampsia caespitosa ssp. *holciformis, Holcus lanatus, Pteridium aquilinum* var. *pubescens, Carex tumulicola, Brodiaea pulchella, Iris Douglasiana, Sisyrinchium bellum, Calochortus luteus, Ranunculus californicus, Lupinus formosus, L. variicolor, Sanicula arctopoides, Chrysopsis Bolanderi, Grindelia hirsutula.*

Open temperate hill-grasslands or glades or bald hills; west slopes of outer and middle Coast Ranges from Mendocino and Trinity counties northward and as scattered patches south to San Francisco Bay. Occurring mostly below 4000 feet.

Climatic data much as for Northern Oak Woodland.

Originally bunch grasses with various flowering herbs; now partly superseded by annual introduced weedy grasses. Sometimes divided into a coastal strip, where there is intergradation with our Northern Coastal Scrub, and hill-prairie or the open hill-grasslands. Since both occur fairly near the coast and have much the same species, we are keeping these "temperate grasslands" of northern affinities as one community, separate from the more interior Valley Grassland community of more southern relationships.

25. Valley Grassland

Originally with various bunch grasses such as *Stipa pulchra, S. cernua, Poa scabrella,* and *Aristida divaricata;* now because of overgrazing largely replaced by annual species of *Bromus, Festuca, Avena,* etc.

Great Central Valley and low hot valleys of inner Coast Ranges, such as Salinas and San Benito valleys, Antelope Valley; ascending to about 4000 feet in Tehachapi Mountains and eastern San Diego County; along the coast from San Luis Obispo County south.

Average rainfall 6 to 20 inches; growing season 7 to 11 months, with 205 to 325 frost-free days; mean maximum summer temperatures 88°–102°, mean winter minima 32°–38° F.

Subtropical type of open treeless grassland, with winter rain and hot dry summers; rich display of flowers in wet springs. Local habitats, such as "hog wallows," with distinctive floras.

26. Alpine Fell-fields

Carex Helleri, C. Breweri, Festuca brachyphylla, Poa rupicola, P. Suksdorfii, Luzula spicata, Eriogonum ovalifolium, Oxyria digyna, Draba densifolia, D. Breweri, D. Lemmonii, D. oligosperma, Phoenicaulis eurycarpa, Ivesia Shockleyi, Potentilla diversifolia, Astragalus tegetarius, Epilobium obcordatum, Podistera nevadensis, Polemonium eximium, Cryptantha nubigena, Castilleja nana, Penstemon Davidsonii, Haplopappus Macronema, Hulsea algida.

Above tree growth; northern California mostly above 9500 feet; Sierra Nevada mostly above 10,500 feet; San Bernardino and San Jacinto mountains with bare suggestion on highest peaks.

Average precipitation about 25 to 35 inches, predominantly as snow; swept by gales in winter with deep drifts of snow accumulating locally; growing season 4 to 7 weeks and killing frost possible at any time; intense illumination; mean summer maximum temperatures probably not over 55°–60° F., winter minima unknown.

Almost entirely perennial herbs, scattered or forming low turf, or among rocks; many cushion plants.

27. Northern Juniper Woodland

Juniperus occidentalis, Pinus Jeffreyi, P. monophylla, Artemisia tridentata, Penstemon speciosus.

Great Basin Plateau to the base of the Sierra Nevada from Modoc County to southern Mono County, 4200 to 5600 feet in the north, 6000 to 7000 feet in the south.

Average precipitation 10 to 30 inches, largely as snow; growing season 2 to 5 months, with 70 to 140 frost-free days; mean summer maximum temperatures 82°–89°, mean winter minima 10°–20° F.

Open forest of trees 10 to 60 feet tall, on brush-covered slopes and flats.

28. Pinyon-Juniper Woodland

Pinus monophylla, Juniperus californica or *J. osteosperma* (*J. utahensis*), *Quercus turbinella, Purshia glandulosa, Cowania Stansburiana, Fallugia paradoxa, Cercocarpus ledifolius, Yucca schidigera, Y. baccata.*

East base of Sierra Nevada, White-Inyo ranges southward through higher mountains of Mojave Desert, mostly at elevations of 5000 to 8000 feet, and between Yellow Pine Forest and Joshua Tree Woodland or Sagebrush Scrub.

Average precipitation 12 to 20 inches, with some snow and some summer showers; growing season 5 to 8 months, with 150 to 250 frost-free days; mean summer maximum temperatures about 88°–95°, mean winter minima about 20°–30° F.

Trees 10 to 30 feet tall, in open stands with shrubs between.

29. Joshua Tree Woodland

Yucca brevifolia and var. *Jaegeriana, Juniperus californica* or *J. osteosperma* (*J. utahensis*), *Salazaria mexicana, Lycium Andersonii, L. Cooperi, Eriogonum fasciculatum* var. *polifolium, Tetradymia axillaris.*

Well-drained mesas and slopes 2500 to 4000 feet or higher, from southern Owens Valley to Little San Bernardino Mountains and southern Nevada and Utah.

Average rainfall about 6 to 15 inches, with summer showers; growing season on the deserts limited by water rather than by temperature; frost-free days 200 to 250; mean summer maximum temperatures 95°–100°, mean winter minima 22°–32° F.

Trees 10 to 30 feet high, scattered, with shrubs and herbs between.

ACKNOWLEDGMENTS

It is impossible to give due credit to all those who have so generously helped in one way or another in the preparation of this Flora. We extend to those persons specifically named and to the many we do not name our real gratitude for assistance given. Among those who should be particularly thanked are: John Thomas Howell for his treatment of the genus Carex and for much help in many groups like Eriogonum, Arctostaphylos,

Lupinus, Cirsium, etc.; Rimo C. Bacigalupi for much assistance in Castilleja, Mimulus, Collinsia, etc.; Daniel I. Axelrod for his account of Geological History; Roxana S. Ferris for aid with Galium, various Compositae, and other groups; Verne E. Grant for the preparation of the manuscript on Cyperaceae exclusive of Carex and for much help with Polemoniaceae; and Rupert C. Barneby for the treatment of Astragalus and Oxytropis. We are indebted for help also to Alva R. Grant in Gilia, F. Harlan Lewis and Peter H. Raven in various groups, Mildred E. Mathias and Lincoln Constance in Umbelliferae, David B. Dunn for great aid in Lupinus, Frederick J. Hermann in Juncus, Henry J. Thompson in Dodecatheon, Lawrence R. Heckard in Phacelia, A. M. Vollmer and Ira L. Wiggins in Lilium, Marion Ownbey for much new information and many corrections in Allium, Reid Moran in Dudleya, Milo S. Baker in Viola, Arthur J. Cronquist in Compositae and some other Gamopetalae, Sherwin Carlquist in Eriophyllum and other Compositae, Douglas M. Post in Frasera, Lee W. Lenz in Iris, Gerald L. Ownbey for the manuscript on Argemone, and George B. Rossbach in Erysimum. Percy C. Everett and Edward K. Balls have given much information learned from living plants; George Garrettson and Ethelbert Johnson have helped with specimens of unusual weeds. Ralph D. Cornell has contributed the Kodachrome for the frontispiece. John C. Roos has added many records and given us much information. Richard J. Shaw and Stephen S. Tillett have prepared the drawings, the former those of pteridophytes and many monocotyledons, the latter all the others. Gloria Campbell and Gladys Boggess have typed most of the manuscript.

David D. Keck has worked out the preliminary general sequence of families used in this Flora, as well as the Glossary and List of Authors and their abbreviations. He prepared the treatment of Rumex, of Poa, and of about the first half of the family of Compositae. He has also done a great deal of editing and reading of proof.

ABBREVIATIONS

acc. – according to
adv. – adventive
Afr. – Africa
ak., aks. – akene, akenes
Alta. – Alberta
Am. – America, American
Ariz. – Arizona
Ark. – Arkansas
auth. – authors
B.C. – British Columbia
ca. – circa (about or approximately)
Calif. – California
caps. – capsule
cent. – central
cm. – centimeter, centimeters
co., cos. – county, counties
Colo. – Colorado
cult. – cultivated, cultivation
diam. – diameter
dm. – decimeter, decimeters
e. – east, eastern, eastward
elev., elevs. – elevation, elevations
Eu. – Europe
F. – forest
f. – forma
fil., fils. – filament, filaments
fl., fls. – flower, flowers, floral
Fla. – Florida
fld. – flowered
fr., frs. – fruit, fruits
ft. – foot, feet

hemis. – hemisphere
id., ids. – island, islands
Ida. – Idaho
Ill. – Illinois
Ind. – Indiana
infl., infls. – inflorescence, inflorescences
introd. – introduction, introduced
invol., invols. – involucre, involucres
Kans. – Kansas
Ky. – Kentucky
La. – Louisiana
L. Calif. – Lower California
lf. – leaf
lft., lfts. – leaflet, leaflets
loc., locs. – locality, localities
lvd. – leaved
lvs. – leaves
m. – meter, meters
Man. – Manitoba
Medit. – Mediterranean
Mex. – Mexico, Mexican
Mich. – Michigan
Minn. – Minnesota
Miss. – Mississippi
mm. – millimeter, millimeters
Mont. – Montana
mt., mts. – mountain, mountains
n. – north, northern, northward
natur. – naturalized
N. Dak. – North Dakota
ne. – northeast, northeastern

n. Mex. – northern Mexico
Nebr. – Nebraska
Nev. – Nevada
New Mex. – New Mexico
Nfld. – Newfoundland
no. – number
nw. – northwest, northwestern
Okla. – Oklahoma
Ore. – Oregon
R. – river
ref., refs. – reference, references
s. – south, southern, southward
Sask. – Saskatchewan
S. Dak. – South Dakota
se. – southeast, southeastern
segm., segms. – segment, segments
Son. – Sonora

sp., spp. – species (singular and plural)
ssp., sspp. – subspecies (singular and plural)
sw. – southwest, southwestern
temp. – temperate
Tex. – Texas
trop. – tropical, tropics
U.S. – United States
V. – valley
Va. – Virginia
var., vars. – variety, varieties
w. – west, western, westward
Wash. – Washington
wd., wds. – woodland, woodlands
W.I. – West Indies
Wis. – Wisconsin
Wyo. – Wyoming

SIGNS USED

± – more or less
♂ – staminate
♀ – pistillate
× – sign of hybrid, used before the species name

NOTE TO THE READER

The question will naturally occur to the reader of this book as to the authors' position regarding use of subspecies and varieties, as no consistent policy seems to run through its pages. While in our own monographic work we tend to the use of subspecies, we feel that such monographic treatment of various groups of plants is now so uneven that we are in no position to make the hundreds of new combinations necessary for consistency. We have therefore attempted to avail ourselves of as much of the literature as possible with as little change as possible. Thus, in one genus we have used subspecies, in another varieties.

It is impossible to create a volume of this sort and not yearn for additional time for the study of each group covered. One knows that he can never make available such a manual if he takes the time that he should like for careful work on one species-complex after another. As yet, experimental studies have been completed in relatively few of our large and difficult genera. Even there different workers do not entirely agree as to classification. As in all manuals, then, this book is uneven in its merit, varying from group to group.

We hope that the use of the Plant Community concept may be of real value. Often a species is shown to appear in quite different environments, for example, in the Coast Ranges and in the Sierran foothills. This type of information may point out the possibility of different ecotypes within a species or subspecies in a way that may not have been evident in earlier manuals and may suggest to geneticists and evolutionists possible groups for investigation.

Divisions

KEY TO DIVISIONS

A. Plants without seeds or fls., reproducing by 1-celled spores borne in sporangia.
    B. Plants not fernlike and not free-floating, the lvs. mostly minute or ± scalelike (except in *Isoetes*), 1-veined.
        C. Stems not jointed; lvs. not whorled and not forming a sheath at the node
                                                          1. *Lepidophyta*   p. 21
        CC. Stems of hollow joints; lvs. whorled and forming a sheath at the solid nodes
                                                          2. *Calamophyta*   p. 26
    BB. Plants fernlike and with large lvs. (fronds) or free-floating aquatics and with small or scalelike overlapping lvs.; mostly elaborately veined . . . . 3. *Pterophyta*   p. 28
AA. Plants with seeds usually produced by cones or by fls.
    B. Seeds not inclosed in ripened pistils, but naked and usually borne on the surface of a scale; the crowded or overlapping scales usually forming a cone or strobilus; plants not bearing typical fls. . . . . 4. *Coniferophyta*   p. 47
    BB. Seeds inclosed in ripened pistils; plants producing true fls. . . . . 5. *Anthophyta*   p. 67

Division I. LEPIDOPHYTA (Fig. 1)

Class 1. LYCOPODINAE

Plant-body a sporophyte with slender stems and roots and mostly small spiral or paired or whorled 1-veined lvs. (microphylls). Sporangium solitary on the adaxial face and near the base of the sporangium-bearing lf. (sporophyll); sporophylls mostly segregated from the sterile foliage lvs. and aggregated into ± loose strobili. Spores developing into minute or microscopic gametophytes producing antheridia and archegonia, the ♂ and ♀ sex-organs. Fusion of gametes results in a new sporophyte.

21

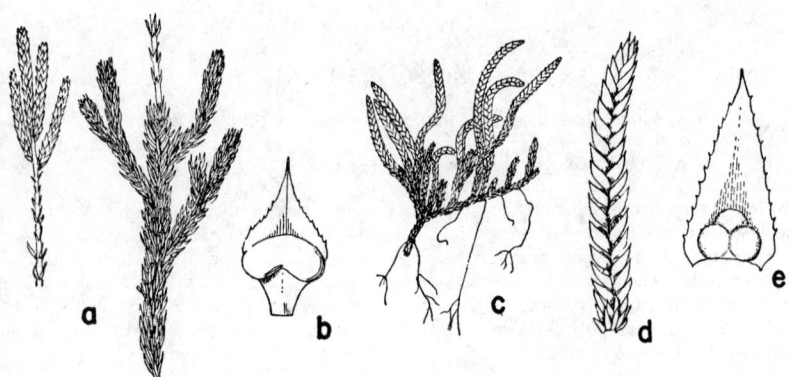

Fig. 1. LEPIDOPHYTA. *Lycopodium clavatum: a,* habit, × ½, with scalelike lvs. and, to the left 4 strobili with overlapping sporophylls; *b,* sporophyll, × 9, with adaxial sporangium. *Selaginella: c,* habit, × ½, with strobili above; *d,* strobilus, with microsporophylls above (resembling *b*) and megasporophylls below; *e,* megasporophyll with adaxial megasporangium bulging with contained tetrad of megaspores.

KEY TO ORDERS

A.   Stems above ground, ± elongate and branched, covered with many minute overlapping
     scalelike to lanceolate lvs. less than 1 cm. long.
  B.   Sterile lvs. without a ligule and 6–8 mm. long; spores all alike (homosporous)
                                                               1. *Lycopodiales*   p. 22
  BB. Sterile lvs. with a minute transverse ligule near the base; lvs. less than 5 mm. long;
     spores of 2 kinds (heterosporous) .... 2. *Selaginellales*   p. 23
AA. Stems underground, cormlike, bearing tufted lvs. 2.5–25 cm. long ...... 3. *Isoetales*   p. 25

Order 1. LYCOPODIALES

Characters as given for the family *Lycopodiaceae*

1. Lycopòdiàceae. CLUB-MOSS FAMILY

Low, usually coarsely mosslike, with elongate simple to much-branched stems covered with small simple 1-nerved evergreen mostly entire lvs. imbricated or crowded into 4–16 ranks. Sporangia uniform, solitary in axils of foliage lvs. or of ± reduced aggregated lvs. (sporophylls) that may form conelike clusters. Spores small, yellow, numerous, all of 1 kind. Two genera.

1. Lycopòdium L. CLUB-MOSS

Characters as given for the family. Ca. 200 spp., widely distributed. (Greek, *lukus,* wolf, and *pous,* foot, from a fancied resemblance.)

1. **L. clavàtum** L. Stem creeping, much-elongated, forking, wiry, with ascending branches 0.5–4 dm. high; foliage-lvs. linear-subulate, bristle-tipped, entire or denticulate, 6–8 mm. long; those of the peduncles shorter, more scattered; penduncles 3–15 cm. long; cones 1–4, narrowly cylindric, 3–10 cm. long; sporophylls deltoid-ovate, straw-colored, ca. 2.5 mm. long, bristle-tipped, the margins erose-fimbriate.—Forming dense masses on trees and banks, at ca. 500 ft.; Douglas-Fir F.; Humboldt Co.; to Alaska, Atlantic Coast; Eurasia. The spores of this and other large spp. are the source of Lycopodium powder.

2. **L. inundatum** L., a species with sporophylls scarcely different from the foliage leaves, occurs at Humboldt Bay.

Order 2. SELAGINELLALES

Characters as given for the family *Selaginellaceae*

1. Selaginellàceae

Low leafy terrestrial plants, freely branched, with numerous very small, usually imbricate lvs. that are spirally arranged and all alike or in 4 longitudinal rows, and of 2 kinds. Sporangia in terminal quadrangular sessile spikes or strobili of slightly modified lvs. (sporophylls), each strobilus containing both microsporophylls and macrosporophylls, the sporangia solitary, adaxial near the base of the sporophylls; the macrosporangia with 1–4 rather large macrospores, the microsporangia with many minute reddish or orange microspores. One genus.

1. Selaginélla Beauv. LITTLE CLUB-MOSS. SPIKE-MOSS

Characters of the family. Ca. 600 spp., widely distributed, especially in the tropics. (Name, a diminutive of *Selago*, a classical name for some sp. of Lycopodium.)

(Tryon, R. M. S. rupestris and its allies. Ann. Mo. Bot. Gard. 42: 1–99, 1955.)
A.   Lvs. in 4 ranks, those of the 2 upper rows rhombic-ovate, of the 2 lateral rows oval-oblong. N. Calif. . . . . 1. *S. Douglasii*
AA. Lvs. arranged radially in many ranks, lanceolate to deltoid-subulate.
   B.   Stems erect or ascending, rooting only at or near the base . . . . 2. *S. Bigelovii*
   BB. Stems prostrate to decumbent or pendent, rooting at or near the apex.
      C.   Lvs. on the underside of main stems (between the first and third branches back from the apex of the stem) not decurrent.
         D.   The lvs. lacking a terminal seta. About San Diego . . . . 3. *S. cinerascens*
         DD. The lvs. ending in a terminal seta.
            E.   Branches somewhat dorsiventral, the lvs. of the under-ranks largest, obliquely imbricate. Mts. from Mt. Shasta to Tulare Co. . . . . 4. *S. Hanseni*
            EE. Branches not at all dorsiventral, the lvs. uniform, equally ascending or appressed-imbricate on all sides. From Marin Co. n. . . . . 5. *S. Wallacei*
      CC. Lvs. on the underside of the main stems strongly decurrent.
         D.   Lvs. lacking a terminal seta in maturity; plants strongly dorsiventral. Desert
                                                         11. *S. eremophila*
         DD. Lvs. with a terminal seta; stems radially symmetrical to dorsiventral.
            E.   Stems 15–80 cm. long, trailing or pendent from trees; lvs. wrinkled, long-decurrent. Near coast from Humboldt Co. n. . . . . 6. *S. oregana*
            EE. Stems mostly shorter, stiffer, terrestrial; lvs. not wrinkled.
               F.   Setae from ⅓ to ⅔ the length of the blade of the lf.
                   G.   Under lvs. mostly definitely longer than the upper on the same part of the stem; dry leafy stems not readily fragmenting. Extreme n. Calif. . . . . 7. *S. densa*
                   GG. Under lvs. of stems ca. as long as upper lvs. of same part of stem; dry leafy stems readily fragmenting. Mts. of s. Calif.
                                                   10. *S. asprella*
            FF. Setae less than ⅓ the length of the lf.-blades.
                 G.   The setae yellowish, smooth; cilia 2–7 on each side of lvs. or wanting. Pine belt from San Jacinto Mts. to Sierra Nevada
                                                       8. *S. Watsoni*
                 GG. The setae white, ± scabrous; cilia more numerous. Panamint and Providence mts. . . . . 9. *S. leucobryoides*

1. **S. Douglásii** (Hook & Grev.) Spring. [*Lycopodium ovalifolium* Hook. & Grev., not Desv. *L. D.* Hook. & Grev.] Stems prostrate, creeping, 8–40 cm. long, rooting throughout; main branches alternate, with flat assurgent divisions, leafy throughout; lateral lvs. distant, spreading, 2–3 mm. long, those of the upper plane smaller, pointed forward; strobili many, sharply quadrangular, 0.5–2 cm. long; sporophylls closely imbricate, cordate-acuminate, sharply keeled.—Moist shaded rocks and banks; reported from n. Calif.; Ore. to B.C., Ida.

2. **S. Bigelòvii** Underw. Stems slender, shortly branched, ascending to erect, 5–20 cm. long, from widely creeping rhizomes; lvs. appressed-imbricate, gray-green, the blade 1.2–1.8 mm. long, ciliate, narrow-lanceolate, with a short white seta; strobili 4–15 mm. long, erect at the tips of short lateral branches, the sporophylls ovate, ca. 2 mm.

long.—Dry rocky banks below 6000 ft.; Chaparral, Coastal Sage Scrub, Foothill Wd., etc.; Coast Ranges from Sonoma Co. s. to n. L. Calif., s. Sierra Nevada of Tulare and Kern cos.; along w. edge of deserts; Santa Barbara Ids.

3. **S. cineráscens** A. A. Eat. [*S. bryoides* Underw.] Plants forming a close ashen carpet on the ground; stems wiry, 5–12 cm. long, branched; lvs. appressed-imbricate, ashy-gray, linear to broadly subulate, ciliate, not setigerous, 1.2–1.5 mm. long, with deep dorsal groove; strobili 2–4 mm. long; sporophylls broadly cordate, short-acuminate, ciliate.— Dry slopes and mesas; Coastal Sage Scrub, Chaparral; sw. San Diego Co.; adjacent L. Calif.

4. **S. Hánseni** Hieron. [*S. Bolanderi* Hieron. *S. rupestris* vars. *B.* and *H.* Jeps.] Stems prostrate, 5–25 cm. long, pinnately branched, strongly dorsiventral; lvs. of upper side lanceolate to ligulate, 2–2.5 mm. long, of lower side somewhat longer, all ± glaucous, ciliate, the seta whitish-hyaline, nearly smooth; strobili 5–9 mm. long, the sporophylls deltoid-ovate, acuminate, 2–2.8 mm. long, many-ciliate.—On dry rocky slopes and banks, 1000–5000 ft.; Foothill Wd., Yellow Pine F.; w. base of Sierra Nevada from Tulare Co. n. to Mt. Shasta; Santa Lucia Mts.

5. **S. Wallàcei** Hieron. [*S. rupestris* var. *W.* Frye.] Loosely cespitose, with prostrate stems 5–15 cm. long and numerous ascending branches 1–4 cm. long, freely branched; lvs. chartaceous, subglaucous, linear-oblong, obtusish, 1.5–3 mm. long, ciliate and with a terminal seta; strobili many, 1–3 cm. long, curved; sporophylls ovate-deltoid, 2–2.8 mm. long, abruptly setigerous, ciliate or not.—Damp shaded places to dry rocky slopes, up to 5000 ft.; N. Coastal Scrub, Mixed Evergreen F., N. Oak Wd., etc., Yellow Pine F.; Coast Ranges from Marin Co. n. to Humboldt and Siskiyou cos.; to B.C., Mont.

6. **S. oregàna** D. C. Eat. [*S. struthioides* of Calif. refs.] Stems slender, widely trailing or pendent, 1.5–9 dm. long, the branches long, remote, strongly curled in the dormant state, or in terrestrial forms making irregular mats with intricate branches; upper and under lvs. subequal, bright green, loosely imbricate, decurrent, narrowly lance-attenuate, 2.4–3.4 mm. long, with a short yellowish green seta; strobili many, 1.5–3 cm. long, curved; sporophylls ovate, long-acuminate, ca. 2 mm. long, few-ciliate.—On fallen trunks, damp shaded rocks and banks, etc. below 1000 ft.; Redwood F., Mixed Evergreen F.; Humboldt and Del Norte cos.; to w. Wash.

7. **S. dénsa** Rydb. var. **scopulòrum** (Maxon) Tryon. [*S. s. Maxon.*] Prostrate, short-creeping, the stems 2–6 cm. long, pinnately branched, forming large mats; lvs. appressed-imbricate, chartaceous, 2.4–3.2 mm. long, subglaucous, linear to lance-subulate, ± obtuse, but ending in a short seta and few-ciliate; strobili many, 1–2.5 cm. long, slender, sub-erect; the sporophylls deeply concave, broadly ovate, long-acuminate, ciliate.—Open rocky places, 5000–7000 ft.; Red Fir F., Lodgepole F.; Siskiyou Co.; to B.C., Mont., w. Tex., Ariz.

8. **S. Wátsoni** Underw. Cespitose, the stems prostrate, short, 2–8 cm. long, with numerous branches 1–2 cm. long, ascending; lvs. densely appressed-imbricate, thick, linear-oblong, ca. 2–2.4 mm. long, with broad dorsal groove and stout smooth seta, with few or no cilia; strobili 1–2.5 cm. long, sharply quadrangular, with ovate sporophylls ciliated on basal half.—Locally frequent in dry rocky places, 7500–14,000 ft.; Lodgepole F., Sub-alpine F., Bristle-cone Pine F., Alpine Fell-fields; Santa Rosa Mts. and San Jacinto Mts. n. through Sierra Nevada to Nevada Co., White Mts.; to Ore., Mont., Utah.

9. **S. leucobryoìdes** Maxon. Plants cushionlike, the stems 1–2 cm. long, readily fragmenting when dry, with thick short erect branches; lvs. uniform, appressed-imbricate, glaucous, linear-subulate, 2.5–3.2 mm. long, short-setigerous, with 8–16 cilia on each side; strobili many, 5–10 mm. long, erect; sporophylls rigidly appressed-imbricate, deltoid-ovate, ciliated and toothed.—Rock-crevices and slopes, at 2000–7500 ft.; Shad-scale Scrub to Pinyon-Juniper Wd.; Panamint and Providence mts., Mojave Desert.

10. **S. asprélla** Maxon. Stems loosely matted, 3–6 cm. long, creeping, with laxly ascending branches 1–2 cm. long; lvs. ascending, somewhat imbricate, deltoid-subulate, glaucous, white-margined, eciliate or ciliate near the apex, 2.8–3.2 mm. long, the setae 0.7–0.9 mm. long, white-hyaline, serrulate, forming conspicuous tufts at ends of branches; strobili lax, 1–2 cm. long, arcuate, the sporophylls 2.5–3 mm. long, long-acuminate.— Dry rocky places, 5500–8600 ft.; Yellow Pine F., Red Fir F., Lodgepole F.; Laguna Mts. (San Diego Co.), San Jacinto, San Bernardino and San Gabriel mts., Santa Ana Mts.

11. **S. eremóphila** Maxon. [*S. Parishii* of Calif. references.] Plants prostrate, dorsi-

ventral, the stems 5–12 cm. long, with spreading branches; lvs. crowded, broadly lanceolate, those of lower side 2 mm. long, thin, acutish, not setigerous when mature, with ca. 25 cilia on each side, lvs. of upper side subvertical, 1–1.4 mm. long, with 6–12 cilia on each side; strobili many, curved, 6–10 mm. long, with 12–18 cilia on each side.—Sheltered places among rocks, below 3000 ft.; Creosote Bush Scrub; canyons along w. edge of Colo. Desert, Chuckawalla Mts.; Ariz., L. Calif.

Order 3. ISOETALES

Characters as given for the family *Isoetaceae*

1. Isoetàceae. Quillwort Family

Small aquatic, amphibious, or terrestrial herbs, with a 2–3-lobed corm crowned with many sedgelike elongate-subulate lvs. with large basal sporangia. Sporangia solitary, on upper surface of lf.-base, of 2 kinds: micro- and mega-sporangia, the former producing small spores each of which forms an antheridium, the latter with large spores forming female gametophytes. One genus of world-wide distribution.

1. Isòetes L. Quillwort

Corm fleshy, bearing roots below. Lvs. with an upper sterile part that is septate, with 4 longitudinal series of air-spaces and with or without peripheral strands, with or without stomates. Lvs. with a fertile lower portion bearing a large plano-convex imbedded sporangium partly covered apically by a thin membrane (velum) above which is a small triangular structure (ligule). Co. 60 spp., cosmopolitan. (Greek, *isos*, ever and *etas*, green, of uncertain application.)

(Pfeiffer, Norma E. Monograph of the Isoetaceae. Ann. Mo. Bot. Gard. 9: 79–232, 1922.)
A. Plants mostly submersed; peripheral strands lacking in lvs.
    B. Stomates lacking; megaspores with irregular crests on basal faces . . . . 1. *I. occidentalis*
    BB. Stomates present, rather few; megaspores not crested on basal faces.
        C. Megaspores tubercled. San Bernardino Mts., Sierra Nevada and north . . 2. *I. Bolanderi*
        CC. Megaspores spinose. Trinity and Siskiyou cos. . . . . 3. *I. muricata*
AA. Plants amphibious or terrestrial; peripheral strands often present.
    B. Corm 3-lobed; velum complete.
        C. Peripheral strands 3; megaspores frosted, mostly tubercled; lvs. 3–20 cm. long
            4. *I. Nuttallii*
        CC. Peripheral strands 0; megaspores usually glossy, mostly smooth; lvs. 3–6 cm. long
            5. *I. Orcuttii*
    BB. Corm 2-lobed, velum ⅓ complete; peripheral strands 4–12; lvs. 5–28 cm. long
        6. *I. Howellii*

1. **I. occidentàlis** Henders. [*I. lacustris* var. *paupercula* Engelm.] Submersed; stem 2-lobed; lvs. commonly 9–30(–60), dark green, rigid, 5–20 cm. long, without peripheral strands or stomates; ligule short-triangular; sporangium almost round, 5–6 mm. long, the velum covering ca. ⅓; megaspores cream-color, ca. 0.5–0.6 mm. in diam., marked with low irregular crests forming a network on the basal face.—Mt. lakes at 6000–11,000 ft.; Red Fir F. to Subalpine F.; Sierra Nevada to Siskiyou Co.; to Ida., Wyo., Colo.

2. **I. Bolánderi** Engelm. [*I. californica* Engelm. *I. B.* var. *Parryi* Engelm. *I. B.* var. *Sonnei* Henders.] Submersed; corm deeply 2-lobed; lvs. 6–25, tapering to a very fine point, bright green, soft, the lvs. 6–15(–25) cm. long, with some stomates but usually no peripheral strands; ligule small, cordate; sporangia 3–4 mm. long, covered ca. ⅓ by the velum; megaspores ca. 0.3–0.4 mm. in diam., with low tubercles.—Mt. lakes and ponds, 5000–11,000 ft.; Montane Coniferous F.; San Bernardino Mts., Sierra Nevada; to Wash., Wyo., Ariz.

Var. **pygmaèa** (Engelm.) Clute. [*I. p.* Engelm.] Lvs. 2–3 cm. long.—High mt. lakes and streams; s. Sierra Nevada.

3. **I. muricàta** Durieu var. **hespéria** Reed. [*I. Braunii* of Calif. refs., not Durieu.] Submersed, or in dry seasons emersed, the corm 2-lobed; lvs. 10–18, straight or recurved,

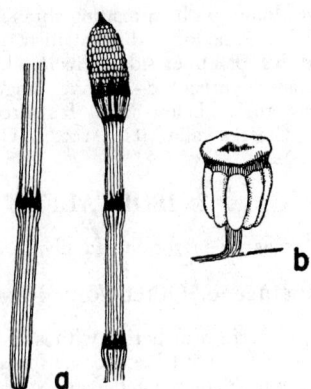

Fig. 2. CALAMOPHYTA. *Equisetum: a,* habit, × ½, with hollow ridged internodes and nodal whorls of scalelike sheathing lvs. and terminal cone of whorls of peltate structures; *b,* peltate structure enlarged, showing circle of sporangia.

firm, 5–10(–13) cm. long, with few stomates toward the tip and no peripheral strands; ligule deltoid; velum covering ca. ¼ of the sporangium; megaspores 0.4–0.5 mm. in diam., with rather long fine spinules.—In lakes at ca. 8000 ft., Subalpine F.; head of Trinity R., n. Calif.; to B.C., Ida., Colo.

4. **I. Nuttállii** A. Br. ex Engelm. Terrestrial; corm 3-lobed; lvs. 5–75, spreading, 2.5–20 cm. long, averaging 8 cm., usually with a characteristic twist, many stomates and 3 peripheral strands; ligule small, triangular; velum completely covering the 4–7 mm. sporangia; megaspores mostly grayish, 0.26–0.56 mm. in diam., mostly frosted and ± tuberculate.—Mud near vernal pools, etc. below 9000 ft.; Chaparral to Lodgepole F.; mesas near San Diego, n. through Coast Ranges and Sierra Nevada to Wash.

5. **I. Orcúttii** A. A. Eat. [*I. Nuttallii* var. *O.* Clute.] Corm 3-lobed; lvs. 3–25, spreading, 3–6 cm. long, with many stomates and no peripheral strands; velum complete; megaspores gray (or brown when wet), 0.2–0.4 mm. in diam., smooth, glossy, sometimes ± tubercled.—In water of vernal pools, at low elevs.; Chaparral, V. Grassland, etc.; mesas near San Diego to Contra Costa and Sacramento cos.

6. **I. Howéllii** Engelm. [*I. melanopoda* var. *californica* A. A. Eat.] Amphibious; corm 2-lobed; lvs. mostly 10–30, bright green, spreading, 5–28 cm. long, with many stomates and usually 4 peripheral strands; ligule narrow, elongate-triangular; velum covering ⅓ of the sporangium; megaspores white (tan when wet), 0.25–0.6 mm. in diam., usually with tubercles and anastamosing and distinct crests.—In water and on mud, below 9000 ft.; many Plant Communities; mesas near San Diego to San Bernardino Mts., Coast Ranges, Sierra Nevada, to Del Norte and Modoc cos.; to Wash., Mont.

Division II. CALAMOPHYTA (Fig. 2)

Class I. EQUISETINAE

With a single living order (Equisetales) and family *Equisetaceae*. Characters as given for that family.

1. Equisetàceae. Horsetail Family

Rushlike, often branching plants, with perennial creeping branching rhizomes rooted at the nodes. Aerial stems perennial or annual, cylindrical, fluted, simple or with whorled branches at the solid sheathed nodes, the internodes generally hollow. Sheaths at the nodes made up of the united minute whorled lvs. which may be free at the tips. Surfaces of the stems overlaid with silica; stomates arranged in bands or rows in the grooves. Strobili or cones terminal, formed of whorls of stalked peltate structures around a cent. axis, each with a circle of sporangia beneath. Spores uniform, green, the outer layer forming ± hygroscopic bands. A single genus.

1. Equisètum L. Horsetail. Scouring-Rush

Characters of the family. Ca. 25 spp., widely distributed. (Latin, *equus*, horse, and *seta*, bristle.)

Aerial stems all green, not dimorphous; stomates in regular rows.
 The aerial stems annual, not living from year to year; cones blunt, not rigidly apiculate.
  The main stems rough with crossbands of silica, and with basal clusters of small branching sterile shoots . . . . 1. *E. Funstoni*
  The main stems smooth, with or without crossbands of silica; lacking basal clusters.
                             2. *E. kansanum*
 The aerial stems perennial; cones rigidly pointed.
  Sheaths much longer than broad, ± funnel-shaped, green, normally with a narrow black band at the top . . . . 3. *E. laevigatum*
  Sheaths nearly or quite as broad as long, cylindrical, usually ashy with 2 black bands
                             4. *E. hyemale*
Aerial stems of 2 kinds, the fertile ones not green or branched, the sterile green, with many slender branches; stomates scattered.
 Sterile stems stout, generally over 3 mm. thick and 6–12 or more dm. tall; sheaths with 20–30 teeth . . . . 5. *E. Telmateia*
 Sterile stems slender, generally less than 3 mm. thick and less than 6 dm. tall; sheaths with 8–12 teeth . . . . 6. *E. arvense*

1. **E. Fúnstoni** A. A. Eat. [*E. fontinale* Copel.] Aerial stems annual, the main ones erect, very rough with sharply projecting transverse bands of silica and 20–30-ridged; sheaths elongate, green with the persistent bases of the teeth strongly incurved with age; tuft of sterile or fertile branched spreading slender shoots at base of main stems (in some forms, only these basal shoots develop); cones slender, 12–28 mm. long, blunt.—Common in moist places below 8000 ft.; many Plant Communities; from San Diego Co. to Humboldt, Butte, and Siskiyou cos.; occasional on the desert, Catalina Id.; n. L. Calif. A number of growth forms have been named.

2. **E. kansánum** J. H. Schaffn. Aerial stems annual, single or clustered, erect, 3–8 dm. high, usually without branches or basal tuft, light green, smooth or slightly roughened, 15–30-ridged; sheaths elongate, green, dilated upward, the teeth largely deciduous, the rim of the sheath not so incurved as in *E. Funstoni;* cones 1–3 cm. long, oblong-ovoid, obtuse.—Occasional in moist places below 6500 ft.; several Plant Communities; cismontane Calif. inland to Santa Rosa Mts. (Riverside Co.), Owens V.; to B.C., Ontario, Mo., New Mex.

3. **E. laevigátum** A. Br. [*E. hyemale* var. *intermedium* A. A. Eat.] Aerial stems persisting for 2 years, erect, tufted, rather simple, 3–8 dm. high, 20–30-ridged, smoothish or rough, the ridges with crossbands of silica; sheaths longer than broad, dilated upward, the lower with basal and apical black rings; teeth often deciduous; cones 1–2 cm. long, sharply apiculate.—Occasional in wet places below 6500 ft.; many Plant Communities; cismontane Calif. to Lassen and Modoc cos.; B.C., N.Y., Tex., L. Calif.

4. **E. hymàle** L. var. **robústum** (A. Br.) A. A. Eat. [*E. r.* A. Br. *E. praealtum* Raf.] Aerial stems persisting for several years, erect, rigid, simple or at length branched, 0.5–2 mm. long, 4–12 mm. thick, with 16–48 ridges, these very rough with a row of transverse bands of silica; sheaths but little longer than broad, cylindrical, ashy with black bands at both ends, teeth quite persistent; lvs. 3-keeled, the cent. keel rarely grooved; cones oval, 1–2.5 cm. long, sharply apiculate.—Occasional in colonies below 8500 ft.; many Plant Communities; cismontane and montane Calif.; to B.C., Quebec and most of U.S.

Var. **califórnicum** Milde. Ridges usually with 2 distinct rows of transverse bands of silica; lvs. 4-keeled, the cent. groove narrow, but evident.—Occasional in moist places; Santa Cruz Id., Santa Clara Co. n. to Butte Co.; to Alaska, Nev., New Mex.

5. **E. Telmatèia** Ehrh. var. **Bráunii** Milde. [*E. maximum* Lam.] Giant Horsetail. Fertile stems erect, short-lived. 2–6 dm. high, whitish or brownish, smooth, with loose membranous sheaths 2–5 cm. long, and with 20–30 teeth; sterile stems 5–25 dm. tall,

5–20 mm. thick, pale green, 20–40-ridged, the sheaths cylindrical, the teeth broadly hyaline-margined; branches solid, in dense whorls; cones stout-pedunculate, 4–8 cm. long.—Occasional in swampy places and along streams, below 4500 ft.; several Plant Communities; cismontane Calif.; to B.C., also in Mich.

6. **E. arvénse** L. COMMON HORSETAIL. Fertile stems 5–25 cm. high, flesh-colored, soon dying, the pale sheaths with 8–12 lanceolate brownish teeth; sterile stems 10–60 cm. high, green, 6–14-ridged, roughish, the sheaths with hyaline-margined teeth; branches slender, in dense whorls; cones lance-ovoid, 2–3 cm. long.—Wet places below 10,000 ft.; many Plant Communities; cismontane Calif., Inyo Co. to Modoc Co.; to Alaska, Labrador, Nfld., most of the U.S., Eurasia. Many growth forms have been named.

Division III. PTEROPHYTA. FERNS

Class I. FILICINAE

Terrestrial sometimes epiphytic or aquatic plants, usually with evident alternation of generations. The sporophyte generation, the fern-plant, usually differentiated into stems, roots and lvs. In the homosporous ferns all or certain lvs. (sporophylls) bear the minute sporangia which produce the 1-celled spores that develop into the small usually independent gametophyte generation. This forms antheridia and archegonia and fertilization results in a new sporophyte. Heterosporous ferns aquatic and floating, or on mud; producing the sporangia in special structures (sporocarps) that contain micro- and macro-sporangia and result in microscopic ♂ and ♀ gametophytes.

KEY TO ORDERS

A. Plants fernlike, terrestrial, producing 1 kind of spore in sporangia.
  B. Sporangia large, globular, without an annulus, borne in a stalked spike or panicle from the base of the green blade which appears lateral; rootstock almost none; frond usually 1, seeming to arise from a cluster of fleshy roots. (Eusporangiatae)
      1. *Ophioglossales* p. 28
  BB. Sporangia minute, stalked, with an annulus of thick-walled cells on one side and borne on the underside of the frond; rootstock developed; fronds more than 1; roots not fleshy. (Leptosporangiatae) .. .. 2. *Filicales* p. 30
AA. Plants not particularly fernlike, but floating aquatics or creeping on mud, producing 2 kinds of spores in round to ovoid bony sporocarps near the base of the lf.-stalks or on the underside of the branches.
  B. Lvs. circinate, petioled, the blades sometimes lacking; plants mostly creeping on and rooting in the mud   .... 3. *Marsileales* p. 46
  BB. Lvs. not circinate, sessile, overlapping; plants mostly floating on water
      4. *Salviniales* p. 47

Order 1. OPHIOGLOSSALES

Characters as given for the family *Ophioglossaceae*

1. Ophioglossaceae. ADDER'S-TONGUE FAMILY

Perennial herbs, with a short fleshy rhizome, ± fleshy fibrous roots and 1–several lvs. with a bud containing the undeveloped succeeding lvs. commonly enclosed in the sheathing base of the stalk of the last expanded lf. Lvs. erect (in ours), simple to variously compound, commonly with a sterile blade near the base of which arises a stalked sporebearing spike or panicle. Sporangia naked, developed partly from sub-epidermal tissue (eusporangiate). Spores all alike (plants homosporous), yellowish. Gametophytes tuberlike, underground. Three genera, widely distributed.

Sterile lf.-blade simple; fertile portion a spike; venation reticulate   .... 1. *Ophioglossum* p. 29
Sterile lf.-blade lobed or divided; fertile portion usually a panicle; venation dichotomous, open .... 2. *Botrychium* p. 29

1. Ophioglóssum L. Adder's-Tongue Fern

Rootstock erect, fleshy, short; the bud for the following year by the side of the base of the naked stalk; sterile portion of lf.-blades fleshy, simple; fertile portion a spike with 2 rows of connate sporangia. Perhaps 40 spp., widely distributed in both hemis. (Greek, *ophis*, snake, and *glossa*, tongue.)

(Clausen, R. T., Mem. Torr. Bot. Club 19[2]: 111–164, 1938.)

Lvs. elliptical or oblong-elliptical, usually with 8–20 parallel veins passing down through the base of the blade which is 2.5–12 cm. long. Siskiyou Co. .... 1. *O. vulgatum*
Lvs. lanceolate, the principal parallel veins 3–7, blade 1.5–5 cm. long. Amador and San Diego cos.
2. *O. californicum*

1. **O. vulgàtum** L. Fronds usually solitary, 10–40 cm. high; sterile blade flat, 1–5 cm. broad, rounded to obtuse at apex; actual fertile portion of spike 2–4 cm. long with 10–30(–50) pairs of sporangia.—Open swamp near Sisson, Siskiyou Co., *Blasdale* in 1894; to Alaska, Quebec, Gulf States; Eurasia. July.

2. **O. califórnicum** Prantl. [*O. lusitanicum* ssp. *c.* Clausen.] Plant 3–13 cm. high, the sterile blade ± fleshy, 0.5–1.5 cm. wide, tending to be acute; fertile part of spike 0.5–2 cm. long, with 8–15 pairs of sporangia.—Vernal pools; Ione, Amador Co., Monterey, and San Diego; to L. Calif., Mex. Dec.–April.

2. Botrýchium Sw. Grape Fern. Moonwort

Rootstock small, erect; bud for following year near base of naked stalk. Fronds 1–3; sterile portion erect or bent down in vernation, stalked or sessile when developed, 1–4 times pinnate; fertile portion a spike to a panicle, the large globose sporangia distinct, borne in 2 rows on the ultimate divisions. Ca. 25 spp., mostly in temp. regions of both hemis. (Greek, *botrus*, a cluster, referring to the groups of sporangia.)

(Clausen, R. T., Mem. Torr. Bot. Club 19: 22–107, 1938.)

Sterile lf.-blades 10–40 cm. long, ternately decompound; buds hairy .... 1. *B. multifidum*
Sterile lf.-blades 1–12 cm. long, pinnately or palmately divided; buds glabrous.
   Common stalk of lf. ca. ½ to entirely underground; sterile blade erect in vernation (to be sought in the bud for next season) .... 2. *B. simplex*
   Common stalk of lf. almost entirely above ground; sterile blade with the apex decurved in vernation .... 3. *B. Lunaria*

1. **B. multífidum** (Gmel.) Rupr. ssp. **silaifòlium** (Presl) Clausen. [*B. s.* Presl. *B. occidentale* Underw. *B. californicum* Underw. *B. s.* var. *c.* Jeps. *B. m.* ssp. *c.* Clausen.] Plant fleshy, rather large, 1–4.5 dm. high, the sterile blade coriaceous to membranous, 3–25 cm. long, 7–35 cm. broad, ternately decompound into ovate to obovate or oblong ultimate divisions 4–20 mm. long; fertile spike 1.5–6 dm. high, decompound, diffuse, the stalk 1–3 dm. long; bud hairy, both the blade and fertile stalks of next year's frond completely deflexed in vernation.—Moist meadows and borders of woods, below 8000 ft.; Red Fir F., N. Coastal Coniferous F., etc.; along the coast and in the Coast Ranges from Monterey n., Sierra Nevada from Tulare Co. n., to Modoc, Siskiyou, and Del Norte cos.; to B.C., New Brunswick, N.J.

Plants stout, fleshy, 1–2 dm. high, with the sterile blade 6–9 cm. long, the stalk of the fertile spike less than 1 dm. long, are the ssp. **Còulteri** (Underw.) Clausen. [*B. C.* Underw.] Intergrading with ssp. *silaifòlium* and ranging to Wash., Mont., Colo. More common in Calif. at elevs. from 7400–11,200 ft., from San Bernardino Mts., Mt. Pinos, Sierra Nevada, White Mts.; to Wash., Colo., Wyo., also in New England and n. Eu. is var. *compósitum* (Lasch) Milde [*B. Kannenbergii* var. *c.* Lasch] with the sterile blade ternate, the 2 basal divisions being again pinnate.

2. **B. símplex** E. Hitchc. Plants slender, 3–20 cm. high; sterile blade simple, entire or once pinnate, inserted either basally or suprabasally, distinctly stalked, 1–5 cm. long, 0.5–2.5 cm. wide, the segms. cuneate or flabelliform; fertile spike 0.3–5 cm. long on a stalk 0.5–8 cm. long; sporangia 0.8–1.2 mm. in diam.—Open meadows and damp

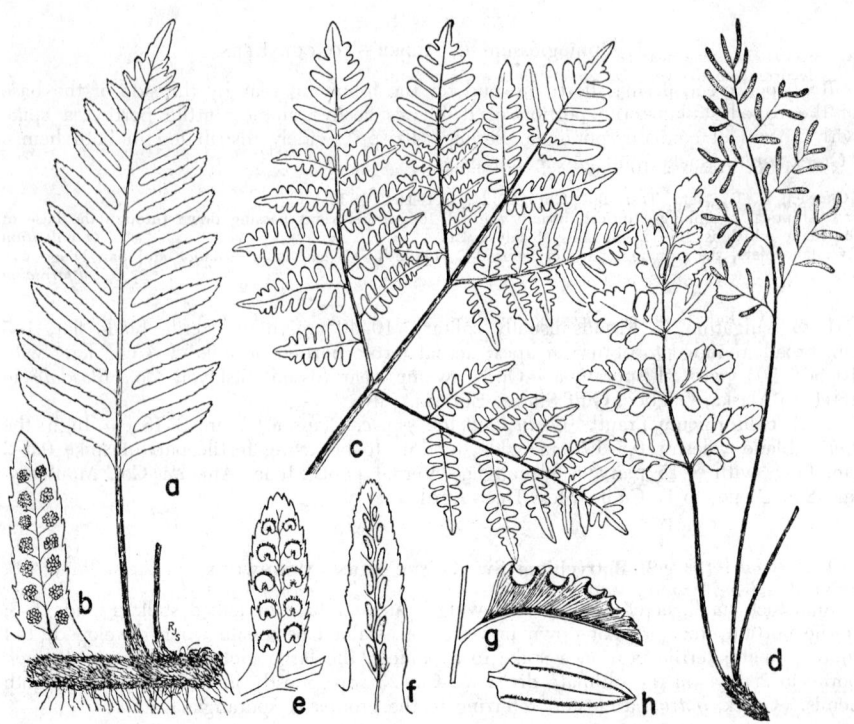

Fig. 3. FILICALES. *Polypodium: a,* habit, × ½, pinnatifid lf. and creeping rhizome; *b,* lf.-segm. from beneath with naked sori of sporangia. *Pteridium: c,* small frond, × ½, twice-pinnate, then pinnatifid. *Cryptogramma: d,* habit, × ½, to show sterile frond with broad pinnules and fertile frond with narrower ones with revolute margins. *Dryopteris: e,* pinnule with each sorus covered with true indusium (reniform). *Woodwardia: f,* pinnule with elongate indusia. *Adiantum: g,* pinnule with false indusium (portion of lf.-margin folded over). *Pellaea: h,* pinnule with false indusia formed by whole lf.-margin.

places, 5000–8500 ft.; Red Fir F., Lodgepole F.; San Bernardino Mts., Sierra Nevada, Siskiyou Co.; to B.C., Nfld., Pa., New Mex.; Eurasia.

3. **B. Lunària** (L.) Sw. [*Osmunda* L. L.] MOONWORT. Plants rather stout, 4–25 cm. high; fronds erect, the common stalk 2–13 cm. long, sterile blade commonly sessile, coriaceous, 1–6 cm. long, ca. half as wide, pinnately divided, with flabellate often overlapping lobes; fertile stalk 1–6 cm. long, the fruiting spike 0.5–8 cm. long; sporangia 0.5–1 mm. in diam.; $n = 45$ (Manton, 1950).—Meadows and moist grassy places; reported from San Bernardino Mts. and Modoc Co., but the usual form in Calif. is:

Var. **minganénse** (Victorin) Dole. [*B. m.* Victorin.] Sterile blade more than twice as long as wide, the segms. usually remote from each other.—At 7000–10,300 ft.; Red Fir F. to Subalpine F.; San Bernardino Mts., e. San Gabriel Mts., Sierra Nevada of Tulare Co., Warner Mts.; to Alaska, e. Canada., Mich., Wis.

Order 2. FILICALES (Fig. 3)

Herbaceous to arborescent ferns, the stem varying from a creeping rhizome to an erect trunk. Lvs. of circinate vernation, several to many, sometimes differentiated into sterile and fertile types or portions. Sporangia on the underside of the blades, scattered or in separate clusters (sori). Sporangium leptosporangiate (developed from a single epidermal cell), with the wall 1-cell thick and bearing an annulus. Gametophyte mostly a minute separate thallus. Several families and many genera (perhaps 300), with several thousand spp.

KEY TO FAMILIES

A. Sporangia borne near or at the margin of the lvs. (i.e. near or at the apex of veins) or decurrent on the veins or completely covering them.
B. Fronds (sterile and fertile) all alike or nearly so .... 1. *Pteridaceae* p. 31
BB. Fronds dimorphous, the fertile blades with contracted linear segms.
    C. Fronds 1–3 dm. long, ± ovate. High montane.
        1. *Pteridaceae* (*Cryptogramma*) p. 36
    CC. Fronds 2–10 dm. long, with linear to sublanceolate blades. N. Coast
        3. *Blechnaceae* (*Blechnum*) p. 44
AA. Sporangia not marginal, but borne in separate sori away from the margins.
B. Sori round to oval.
    C. Indusia wanting, the sori naked.
        D. Stipes jointed to the rhizome; blades pinnatifid ...... 5. *Polypodiaceae* p. 45
        DD. Stipes continuous with the rhizome; blades 2-pinnate, then pinnatifid
            2. *Aspidiaceae* (*Athyrium*) p. 43
    CC. Indusia present, although sometimes soon concealed by the maturing sporangia; stipes not jointed to the rhizome .... 2. *Aspidiaceae* p. 39
BB. Sori oblong, linear to lunate, or horseshoe-shaped.
    C. Fronds 6–20 dm. tall; indusium often ± curved.
        D. Venation partly areolate; sori 2–6 mm. long
            3. *Blechnaceae* (*Woodwardia*) p. 44
        DD. Venation entirely open; sori to ca. 1 mm. long
            2. *Aspidiaceae* (*Athyrium*) p. 43
    CC. Fronds 0.3–2.5 dm. tall, pinnate; indusium straight .... 4. *Aspleniaceae* p. 44

1. Pteridàceae

Terrestrial ferns with a creeping rhizome or ascending to erect stem, the indument of hairs or paleae. Fronds pinnate in plan, decompound to simple and entire, not articulate to the rhizome. Sori typically marginal and protected by an indusium opening toward the margin, or by a reflexed margin, or elongate along the veins and without an indusium, or covering the whole fertile surface. Annulus sometimes oblique and uninterrupted, or mostly longitudinal and interrupted, the sporangium opening through a definite stomium. Spores almost always tetrahedral. Ca. 63 genera of temp. and trop. regions.

A. Sporangia following the veins throughout, hence sporangia on back of fronds. .. 8. *Pityrogramma*
AA. Sporangia borne at or near the apex of the veins, hence near margin of fronds.
B. Fronds dimorphous, the fertile pinnae very narrow. High montane ...... 6. *Cryptogramma*
BB. Fronds uniform or nearly so (sterile and fertile pinnae almost alike).
    C. Plants large, coarse, the fronds mostly 3–15 dm. high, stipes light-colored, not brown; indusium double, the inner minute, concealed .... 1. *Pteridium*
    CC. Plants mainly small, usually less than 3 dm. high; stipes mostly brown or purplish; indusium single or none.
        D. Reflexed lf.-margin not continuous, appearing as separate large indusia; ultimate segms. of frond at least 1 cm. broad; maidenhair ferns .... 9. *Adiantum*
        DD. Reflexed lf.-margin continuous, or if discontinuous, the ultimate segms. of the frond 1–5 mm. broad.
            E. Foliage nearly or quite glabrous, the inrolled margins usually quite continuous.
                F. Lf.-blades deltoid, not much longer than broad.
                    G. Sori solitary, with separate short round-lunate indusia; ultimate lf.-segms. 3–7 mm. long .... 4. *Aspidotis*
                    GG. Sori contiguous, with a common narrowly linear indusium formed by the inrolled lf.-margin.
                          H. Pinnules lance-linear, 5–15 mm. long, mucronate. From San Luis Obispo and Tulare cos. north .... 7. *Onychium*
                          HH. Pinnules oval or elliptical, 3–17 mm. long, mostly retuse. Ore. to L. Calif. .... 5. *Pellaea andromedaefolia*
                FF. Lf.-blades narrower, much longer than broad .... 5. *Pellaea*
        EE. Foliage mostly hairy or scaly, if subglabrous, then with margins not inrolled or the inrolling discontinuous; lf.-blades much longer than broad.
            F. Fronds lanceolate to lance-oblong, not waxy-granular ... 2. *Cheilanthes*
            FF. Fronds deltoid-pentagonal, waxy-granular beneath .... 3. *Aleuritopteris*

1. Pterídium Scop. BRAKE or BRACKEN

Coarse ferns with long-creeping branched underground rhizomes clothed with hairs. Fronds stout, erect to reclining, with stout stipes having a feltlike covering near the base. Blade pinnately compound, coriaceous, ± densely hairy, triangular to elongate, the ultimate segms. entire to lobed, the veins' free except for a marginal strand. Sorus continuous along the lf.-margin, borne on the connecting vein, with a double indusium, the outer (false) formed by the reflexed margin, the inner (true) developed or obsolescent, nearly concealed by the slender-stalked sporangia; annulus of ca. 13 thickened cells; spores tetrahedral or globose-tetrahedral, smooth. Ca. 6–8 widely distributed spp. (Diminutive of *Pteris*, a fern genus.)

1. **P. aquilìnum** (L.) Kuhn var. **pubescens** Underw. [*P. a.* var. *lanuginosum* (Bong.) Fern. *Pteris a.* var. *l.* Bong.] Lvs. erect or ascending 3–15 dm. tall; stipes straw-colored; blades 2–10 or more dm. long, usually 3 times pinnate in lower part; segms. linear-oblong, pubescent to tomentose beneath, sometimes hairy above, mostly 0.5–2.5 cm. long, entire or ± sinuate; indusia narrow, villous; *n* = 52 (Manton, 1950).—In moist places at lower elevs., common groundcover in forests at higher ones, up to 10,000 ft.; many Plant Communities from Subalpine F. to Coastal Sage Scrub, etc.; coastal strand to high mts., widely distributed in Calif.; to Alaska, S. Dak., nw. Mex., w. Tex., scatteringly to e. Canada.

2. Cheilánthes Sw.

Mostly small xerophilous ferns with short-creeping or ± erect scaly rhizomes; fronds pinnate to decompound, not usually deltoid, hairy or scaly, rarely smooth, with veins ending free. Sori marginal, on the tips of the veins; indusium formed by the ± modified reflexed margin, typically discrete, but more often ± confluent, often obsolescent or wanting. Sporangia with annulus of 14–24 thickened cells and many-celled stomium. Spores globose-tetrahedral, smooth, granulose or sometimes corrugated. Ca. 180 spp. of warmer parts of world. (Greek, *cheilos*, margin, and *anthus*, fl.)

Sori naked, the lf-margin scarcely if at all inrolled. (*Notholaena*)
  Fronds not at all hairy or scaly; the blade lanceolate with remote pinnae . . . . 1. *C. Jonesii*
  Fronds hairy or scaly.
    The fronds scaly, once pinnate or bipinnatifid . . . . 2. *C. sinuata*
    The fronds woolly, bipinnate or tripinnatifid.
      Lvs. averaging 10–15 cm. long; tomentum close; plant not at all viscid. Cismontane
                                                      3. *C. Newberryi*
      Lvs. averaging 7–12 cm. long; tomentum very loose; plant ± viscid. Deserts . . . . . . 4. *C. Parryi*
Sori covered with inrolled lf.-margin. (*Cheilanthes*)
  The sori 1–2 on each lobe, with separate indusia; segms. flattish.
    Fronds minutely viscid-glandular, inconspicuously hairy at base of stipe only. Deserts
                                                       5. *C. viscida*
    Fronds with spreading whitish septate hairs. Cismontane . . . . 6. *C. Cooperae*
  The sori several–many on each lobe or segm., with a common indusium; segms. beadlike.
    Fronds without scales or coarse fibers. E. Mojave Desert . . . . 7. *C. Feei*
    Fronds with scales or fibers.
      Segms. hairy on upper surface. San Jacinto Mts.
        Rachis with scales; pinnules loosely woolly beneath . . . . 8. *C. Parishii*
        Rachis with coarse fibrils; pinnules densely woolly beneath . . . . 9. *C. fibrillosa*
      Segms. without hairs on upper surface.
        Blades usually twice pinnate; segms. mostly oblong, densely tomentose beneath. From Marin
        and Mariposa cos. n. . . . . 10. *C. gracillima*
        Blades 3–4 times pinnate; segms. mostly roundish to oval, imbricate-paleaceous beneath.
          Fronds few, 5–10 mm. apart; segms. mostly subcordate-orbicular, fertile nearly to the
          base, the subentire margin closely revolute. Cismontane from Santa Barbara Co. to n. L.
          Calif. . . . . 11. *C. Clevelandii*
          Fronds many, close; segms. oval or irregularly roundish, fertile distally, the ± crenate
          margins deeply recurved.
            Segms. with a few minute pale stellate scales above; scales beneath small, numerous,
            dark, many reduced to entangled cilia. Santa Clara, Sonoma, and Mendocino cos
                                             12. *C. intertexta*
            Segms. devoid of scales above; scales beneath large, much exceeding the segms., white
            to light castaneous.
              Rhizome-scales narrow, rigid, with strongly sclerotic walls; scales of blade with deeply
              cordate base with overlapping basal lobes. San Gabriel, Santa Ana, and Laguna mts.
              to Ariz. . . . . 13. *C. Covillei*
              Rhizome-scales broader, thinner, pale brown or only partly sclerotic-walled; scales of
              blade rounded at base or merely subcordate. E. Mojave Desert to Ariz.
                                               14. *C. Wootonii*

1. **C. Jònesii** (Maxon) Munz. [*Notholaena J.* Maxon.] Tufted, with short oblique light brown paleaceous rhizome, the scales linear, long-attenuate; fronds few to

many, 3–15 cm. long, with curved red-brown stipes; blades 2–8 cm. long, mostly twice-pinnate, the 3–7 pairs of pinnae distant; pinnules entire or crenately lobed, rounded, 2–5 mm. long, glabrous, not at all pulverulent; sporangia in a submarginal band.— Crevices of limestone cliffs, 3500–6000 ft.; Joshua Tree Wd., Pinyon-Juniper Wd., etc.; scattered locations Mojave Desert and Palm Springs region, to White Mts., upper Tule R., Tulare Co. and Santa Barbara Co.; sw. Utah, Ariz.

2. **C. sinuàta** (Lag.) Domin var. **cochisénsis** (Goodd.) Munz. [*Notholaena c.* Goodd. *N. s.* var. *c.* Weath.] Erect plants 1–3 dm. high, the rootstock short, woody with reddish, linear, sparsely ciliate scales; fronds tufted, simply pinnate, grayish-green above, grayish-brown and scaly beneath, the pinnae subquadrate, 5–7(–10) mm. long, almost as broad, with 1–2(–3) pairs of lobes.—Dry limestone slopes and crevices at 3200–5500 ft.; Joshua Tree Wd., Pinyon-Juniper Wd.; Providence and Clark mts., e. San Bernardino Co.; to Ariz., n. Mex.

3. **C. Newbérryi** (D. C. Eat.) Domin. [*Notholaena N.* D. C. Eat.] COTTON-FERN. Rhizomes slender, branched, creeping, appressed-scaly; fronds erect, clustered, 8–22 cm. long, the stipes 4–14 cm. long, wiry, purplish-brown; blades narrow-oblong to oblong-lanceolate, 3 times pinnate, finely and densely tawny-tomentose (or at first white) beneath, whitish-tomentulose above; pinnae ca. 10 pairs; pinnules crowded, round-obovate, minute; sporangia mostly exposed in age.—Common on dry rocky slopes and walls below 2500 ft.; Chaparral, Coastal Sage Scrub; Ventura Co. and s. base of San Gabriel Mts. to San Bernardino and San Diego cos., San Clemente Id.

4. **C. Párryi** (D. C. Eat.) Domin. [*Notholaena P.* D. C. Eat.) PARRY CLOAK FERN. Rhizomes short, thick, tufted, densely paleaceous with brownish scales 4–6 mm. long and entire; fronds many, clustered, 5–20 cm. long; stipes purplish-brown, slender, hirsute, 3–8 cm. long; blades linear- to ovate-oblong, 3–8 cm. long, 3 times pinnate, pinnae 5–9 pairs, the ultimate segms. round to ovate-oblong, 2–3 mm. long, closely enveloped in dense light gray or brownish wool; sporangia dark, partly evident in age.— Common, under overhanging rocks, etc., below 7000 ft.; Creosote Bush Scrub to Pinyon-Juniper Wd.; deserts from White Mts. s. to Colo. Desert; to Utah, Ariz.

5. **C. víscida** Davenp. Rhizome thick, with subglobose divisions and red-brown lanceolate scales; fronds tufted, 1–2 dm. long; stipes 4–10 cm. long, slender, brittle, dark brown, viscid-glandular; rachises glandular and hairy; blades linear-oblong, twice-pinnate, the ultimate segms. incised or toothed, minute, green, not hairy, viscid-glandular; sori solitary on the ultimate lobes.—Occasional about rocks below 4000 ft.; Creosote Bush Scrub; w. edge of Colo. Desert, Twentynine Palms, Granite Mts. near Victorville, Darwin, Panamint Mts.; L. Calif.

6. **C. Coóperae** D. C. Eat. Rhizome short, thick, with tufted scales 4–7 mm. long; fronds tufted, 5–20 cm. long; stipes 1–7 cm. long, dark brown, with flattish, gland-tipped hairs; blades narrow- to deltoid-oblong, 4–15 cm. long, twice pinnate, the pinnules entire to lobed, 4–6 mm. long, grayish-green, with many flattish septate gland-tipped hairs; sori 1 or 2 to a lobe.—Occasional in limestone clefts below 2000 ft.; Coastal Sage Scrub, Chaparral, Foothill Wd.; Slover Mt. near Colton, Piru, Santa Inez Mts., foothills of Sierra Nevada from Mariposa to Eldorado and Shasta cos.

7. **C. Feèi** T. Moore. [*Myriopteris gracilis* Fée. *C. g.* Riehl, not Kaulf.] Rhizome with several short scaly branches bearing tufted linear scales 5–7 mm. long; fronds many, tufted, 8–25 cm. long; stipes 3–12 cm. long, slender, brown, ± pilose; blades lance-oblong to narrow-ovate, thinly villous and beneath with pale brown tomentum, usually thrice-pinnate, the ultimate segms. gray-green, oval to rounded, simple or lobed, the margins narrowly recurved.—Limestone crevices, 4000–6500 ft.; Pinyon-Juniper Wd., Joshua Tree Wd.; Providence, New York, Clark, and Inyo-White mts. of Mojave Desert; to B.C., Tex., n. Mex.

8. **C. Paríshii** Davenp. Rhizome short, with tufted scales 3–4 mm. long; fronds 5–15 cm. long, the stipes half that length, dark brown, fibrillose-paleaceous; blades 3–4 times pinnate, lance-oblong, villous above and beneath with long flexuous hyaline hairs, paleaceous on rachises beneath with light brown elongate scales; ultimate lf.-segms. minute, with ± recurved margins, gray-green, minute.—About rocks, Creosote Bush Scrub; Andreas Canyon near Palm Springs, Riverside Co.

9. **C. fibrillòsa** Davenp. ex Underw. [*C. lanuginosa* var. *f.* Davenp.] Much like *C. Parishii*, the rachis fibrillose, not scaly; blades thrice pinnate, the segms. woolly beneath,

subglabrate in age.—Known from a single collection in Chaparral; San Jacinto Canyon, Riverside Co.

10. **C. gracíllima** D. C. Eat. in Torr. LACE FERN. Rhizomes tufted, with many short branches and densely paleaceous with light-brown narrow scales 2–3 mm. long; lvs. many, 1–2.5 dm. long, the stipes 0.5–1.5 dm. long, dark brown, soon naked; blades linear or lanceolate, mostly twice-pinnate into dull or yellowish green oblong segms., ± oblique and close or remote with a few minute scales above, densely cinnamon-brown-tomentose beneath, the edges deeply recurved.—Among rocks at 2500–9000 ft.; Yellow Pine F. to Lodgepole F.; Coast Ranges from Marin Co. n., Sierra Nevada from Tulare Co. n.; to B.C., w. Mont.

11. **C. Clevelándii** D. C. Eat. Rhizome slender, creeping, woody, brown-scaly with glossy scales 2.5–3 mm. long; fronds scattered, 1–4 dm. tall; stipes stout, 7–24 cm. long, light brown, with a few linear pale scales; blades lance-linear to deltoid-oblong, 3–4 times pinnate, the rachises and segms. densely paleaceous beneath with small imbricate brownish scales, glabrous and green above, ultimate segms. minute, ± rounded, with revolute border.—Frequent in rocky places below 5000 ft.; Chaparral, Coastal Sage Scrub; cismontane San Diego Co. to Banning; near Santa Barbara; Santa Cruz and Santa Rosa ids.; n. L. Calif.

12. **C. intertéxta** (Maxon) Maxon in Abrams. [*C. Covillei* var. *i.* Maxon.] Rhizomes short-creeping, subnodose; appressed-paleaceous with linear-lanceolate brownish scales ca. 2 mm. long; fronds many, close, 8–25 cm. high; stipes 5–18 cm. long, purplish-brown, wiry, ± scattered-paleaceous at least when young; blades ovate-deltoid to ± oblong, thrice pinnate, the ultimate segms. with a few minute pale stellate scales above, beneath thickly covered with bright castaneous imbricate ± ciliate scales, beadlike, 1–1.5 mm. in diam. and with strongly recurved margins.—Occasional, rock-crevices, mostly between 2500 and 9000 ft.; Mixed Evergreen F. to Lodgepole F.; Coast Ranges from Santa Clara Co. to Mendocino Co., Sierra Nevada from Tulare Co. n., to Siskiyou Co. Doubtfully distinct from the next.

13. **C. Covíllei** Maxon. Rhizome short, creeping, appressed-paleaceous with linear to lanceolate scales 1.5–2.5 mm. long, brown or darker; fronds many, 1–3 dm. long; stipes 5–15 cm. long, brown to dark purplish, with small linear scales; blades oblong or ± deltoid, thrice pinnate, the rachises and under surface of pinnules imbricate-paleaceous with cordate scales exceeding the ± oval or roundish beadlike green segms. that are naked and glabrous above.—Common in rocky places, 1500–9000 ft.; Chaparral, Yellow Pine F., Pinyon-Juniper Wd., Joshua Tree Wd., etc.; deserts and cismontane s. Calif. to San Luis Obispo Co., Tulare Co.; to Nev., Ariz., L. Calif.

14. **C. Wootònii** Maxon. Rhizome creeping, slender, pale brown, with loosely imbricate oblong-ovate to lance-oblong scales 2–3 mm. long; fronds several, 0.5–3 cm. apart, 1–3 dm. long; stipe slender, 5–18 cm. long, castaneous; blades oblong, thrice pinnate, the rachises and under surface imbricately paleaceous with brownish scales scarcely cordate at base, narrow-ovate, conspicuously long-ciliate especially toward base. —Dry rocky places, 4000–8000 ft.; Joshua Tree Wd., Pinyon-Juniper Wd.; Panamint, Inyo-White, Providence, and New York mts., e. Mojave Desert; to New Mex.

3. Aleuritópteris Fée

Small terrestrial ferns with short ascending rhizomes clothed with black linear-setaceous paleae. Stipes crowded, black and polished, naked or with brown paleae. Blades deltoid and bipinnatifid to ovate with large bipinnatifid basal pinnae, not finely dissected, ± densely white- or yellow-waxy beneath, the veins free. Sori marginal, on vein tips, not laterally confluent but commonly in contact, protected by scarious reflexed margins which are individual and separate or ± continuous. Sporangia few and large, the annulus of 16–32 thickened cells, stomium large. Spores round or ± tetrahedral, dark, usually granular, rarely reticulate-spinulose. Ca. 15 spp., mostly N. Temp. (Greek, *aleuron*, flour, and *pteris*, fern, because of the waxy mealy under lf.-surface.)

1. **A. cretàcea** (Liebm.) Fourn. [*Notholaena c.* Liebm. *N. californica* D. C. Eat.] Tufted plants, the rhizomes nodose-multicipital, densely paleaceous; scales rigidly acicular, 3.5–4.5 mm. long, dark reddish-brown to blackish, denticulate; fronds 4–12 cm. high, few to many; stipes brown; blades deltoid-pentagonal, 2–5 cm. long, 2–4

cm. wide, 3-pinnate on the large basal pinnae, the pinnae and segms. close, the segms. with incurved margins, oblong, obtuse, greenish above, yellowish- or whitish-waxy beneath, the sporangia not concealed.—About rocks and cliffs below 2500 ft.; Coastal Sage Scrub, Creosote Bush Scrub; mts. from Victorville, Colton, Palm Springs, Yaqui Well and Beale's Well (Colo. Desert); to n. L. Calif., Ariz.

Ssp. **nigréscens** (Ewan) Munz. [*Notholaena californica* ssp. *n.* Ewan.] Fronds 10–15 cm. high, the stipes almost black, 6–10 cm. long; blades 3.5–4.5 cm. wide, 4–6 cm. long, the pinnae more distant, dark olivaceous or ferrous.—Canyons between 750 and 2000 ft.; Chaparral; s. base of San Gabriel Mts., Los Angeles Co.

4. Aspidòtis Nutt. ex Copel.

Small ferns with short-creeping rhizome clothed with blackish linear-setaceous paleae. Fronds approximate, glabrous, subcoriaceous; stipe brown, polished; blade deltoid-ovate, pinnately decompound, finely dissected with acute pinnules and teeth. Sori small, in the sinuses of the teeth, with a scarious reflexed indusium. Sporangia few; annulus of 20 thickened cells; spores globose-tetrahedral, smooth or muriculate. Two spp., of Calif. and Mex. (Greek, *aspis,* shield, i.e., having a shield.)

1. **A. califórnica** Nutt. ex Copel. [*Hypolepis c.* Hook. *Cheilanthes c.* Mett.] CALI-FORNIA LACE FERN. Rhizome-scales 2–2.5 mm. long; fronds 0.5–3.5 dm. long, many; stipes dark brown, glossy, wiry, 0.5–2.5 dm. long, nearly naked; blades ± 4 times pinnate, the segms. bright green, glabrous, linear to elliptic, oblique, decurrent, acute, 2–3 mm. long, the sori solitary at enlarged vein-tips, with round-lunate ample false indusium adherent to slender saccate marginal tooth at each side.—Common on dry shaded slopes and cliffs below 2500 ft.; many Plant Communities; Coast Ranges from San Diego to Humboldt cos., Sierra Nevada n. to Butte Co.; Santa Barbara Ids.; n. L. Calif.

5. Pellaèa Link. CLIFF-BRAKE

Ours mostly rather small ferns with short or elongate rhizomes and erect glabrous persistent lvs. Blades 1–4 times pinnate, the rachises usually dark, lustrous; segms. round-oval to linear, small to larger, ± articulate; veins free, not thickened at the tips of the branches. Sori terminal and subterminal, rounded or oblong, often confluent in a submarginal band, ± concealed by the strongly revolute indusiform margin. Ca. 80 spp., mostly temp. (Greek, *pellos,* dusky, referring to the stipes.)

(Tryon, Alice F. A revision of the fern genus Pellaea section Pellaea. Ann. Mo. Bot. Gard. 44: 125–193, 1957.)

1. **P. compácta** (Davenp.) Maxon. [*P. Wrightiana c.* Davenp. *P. W.* var. *californica* Lemmon.] Rhizome woody, stout, nodose, with tufted dark or gray-brown linear scales 5–7 mm. long; fronds mostly fertile, many, clustered, 2–3.5 dm. long, stiff; stipes dark brown, 1–1.3 dm. long, naked; blades narrow-oblong, 2.5–4.5 cm. wide, twice-pinnate, the ultimate segms. gray-green, 5–7 pairs for each pinna, subsessile, 3–4 mm. long, the fertile ones strongly revolute, transversely wrinkled; sporangia almost concealed in the incurved border.—Dry stony slopes, 4500–8500 ft.; Montane Coniferous F. from

Santa Rosa Mts. to Mt. Pinos, and in Sierra Nevada to Placer Co.; in Pinyon-Juniper Wd., mts. of Mojave Desert.

2. **P. brachýptera** (T. Moore) Baker. [*Platyloma b.* T. Moore.] The woody horizontal rhizome thick, nodose, with brown numerous acicular scales 5–8 mm. long; fronds clustered, 1.5–4 dm. long; stipes stout, purplish-brown, at least as long as blades, naked; blades 1.5–4 cm. wide, twice-pinnate, the ultimate segms. dull or gray-green, 5–11 on a pinna, spreading, usually close, linear, 6–17 mm. long, revolute; sporangia concealed.—Dry rocky places, 3000–8000 ft.; Montane Coniferous F.; Plumas and Lake cos. n.; to sw. Ore.

3. **P. mucronàta** (D. C. Eat.) D. C. Eat. [*Allosorus m.* D. C. Eat. *P. ornithopus* Hook.] BIRD's-FOOT FERN. Rhizome thick, woody, with closely tufted castaneous scales 5–10 mm. long and having denticulate edges; fronds mostly fertile, 1.5–5 dm. long, stiff; stipes 0.4–2 dm. long, purplish-brown, dullish; blades 1–3 dm. long, 2–3 times pinnate, with distant pinnae at nearly right angles to the rachis divided into 6–20 pairs of usually ternate pinnules, or these with as many as 11 linear-oblong to elliptical gray-green segms. 2–6 mm. long, wrinkled, revolute to the middle; sporangia mostly hidden.—Dry rocky slopes, mostly below 6000 ft.; many Plant Communities, but especially Coastal Sage Scrub, Chaparral, Foothill Wd., etc.; nearly throughout cismontane Calif., sparingly on the desert; n. L. Calif.

Var. **califòrnica** (Lemmon) M. & J. [*P. Wrightiana* var. *c.* Lemmon. *Platyloma bella* T. Moore. *Pellaea b.* Baker. *P. W.* var. *compacta* Davenp. *P. c.* Maxon.] Pinnae ascending at acute angles to the rachis, becoming imbricate toward the apical portion of the frond; pinnules less than 10 pairs per pinna, usually entire.—Mostly above 6000 ft.; Montane Coniferous F.; cent. Sierra Nevada and mts. of s. Calif.

4. **P. andromedaefòlia** (Kaulf.) Fée. [*Pteris a.* Kaulf. *Pellaea a.* var. *rubens* D. C. Eat. *P. rafaelensis* Mox.] COFFEE FERN. Rhizome slender, wide-creeping, with imbricated narrow hair-pointed scales 1.5–3.5 mm. long; fronds distichous, 1.5–7 dm. long, the stipes 0.5–4 dm. long, glaucous, straw-colored; blades 0.5–2 dm. broad, 2–4 times pinnate, with glabrous or glandular-pubescent rachises and alternate or subopposite pinnae; pinnules remote, oblong, obtuse or retuse, the sterile ones flat, the fertile strongly revolute, all ± reddish; sporangia partly hidden.—Dry rather stony places, mostly below 4000 ft.; many Plant Communities; cismontane s. Calif. to Mendocino and Butte cos.; n. L. Calif., Channel Ids. A slightly pubescent form about San Diego has been called var. *pubéscens* Baker.

5. **P. Bridgèsii** Hook. Rhizome short-creeping, with tufted brown scales 5–7 mm. long and linear-attenuate; fronds tufted, 1–3.5 dm. long, stiff, the stipes 0.4–2 dm. long, castaneous, shining; blades narrow, gray-green, 1–2.5 cm. wide, pinnate with 5–16 pairs of sessile cordate-oblong to broadly oval usually conduplicate falcate pinnae and a terminal one; sterile pinnae roundish; sporangia decurrent in a broad intramarginal band not concealed by the wrinkled or plane border.—Dry exposed rocky places, at 6000–11,000 ft.; Montane Coniferous F.; Sierra Nevada from Tulare Co. to Sierra Co.; Ida.

6. **P. Brèweri** D. C. Eat. With habit of *P. Bridgesii*, the rhizome-scales twisted, 7–10 mm. long; fronds 0.5–2 dm. long; stipes 3–10 cm. long, glossy, bright brown, transversely corrugate; blades 1.5–3.5 cm. broad, with 6–12 pairs of pinnae that are mostly 2-parted with the upper lobe larger; sporangia terminal and subterminal, confluent, nearly hidden by the reflexed margin.—Exposed dry rocky places, 7000–12,000 ft.; Montane Coniferous F.; Sierra Nevada from Tulare Co. n., to Siskiyou Co., also in White, Panamint, and San Bernardino mts.; to Wash., Wyo., Ida., Utah.

6. Cryptográmma R. Br. in Richards.

Rather small ferns with stout ascending rhizomes clothed with thin brown paleae. Fronds crowded, glabrous, green, dimorphous, the fertile usually having narrower and longer pinnules than the sterile. Blades 2–3 times pinnate; veins free. Sorus submarginal, covering the branches of forked veins, protected by a continuous reflexed margin. Annulus of 20–24 thickened cells; spores tetrahedral or exceptionally bilateral, hyaline, tuberculate. Five spp.: 1 Chilean, 2 Eurasian, 2 N. Am. (Greek, *cryptos*, hidden, and *gramme*, line, because of the concealed sori.)

1. C. acrostichoìdes R. Br. in Richards. [*C. crispa* ssp. *a.* Hult.] ROCK-BRAKE. PARSLEY FERN. Rhizomes in tufts, the scales lance-ovate, rusty to dark brown, or striped, ca. 4–5 mm. long; fronds many, closely clustered, 1–3 dm. high; stipes of fertile lvs. straw-colored, of sterile green; sterile blades 3–12 cm. long, 2–3 times pinnate, ovate to lance-ovate, the pinnae few, close, with ultimate segms. crowded, ovate to oblong; fertile blades simpler, the ultimate segms. linear-oblong, 6–12 mm. long, ca. 2 mm. wide, with revolute margins.—Rocky ledges, cliffs, etc., 6000–11,000 ft. (or lower in N. Coast Ranges); upper Montane Coniferous F.; San Jacinto and San Bernardino mts., White Mts., Sierra Nevada, to Modoc and Humboldt cos.; to Alaska, Labrador, New Mex.

7. Onýchium Kaulf.

Small to middle-sized ferns with creeping rhizome or short and compact, paleate. Fronds tripinnate or more compound, with ultimate pinnules glabrous, small, narrow, herbaceous or ± coriaceous, the veins free except for a fertile commissure connecting the tips. Sori continuous along the margins, protected by a scarious introrse marginal or submarginal indusium, so broad that the 2 meet at the midrib. Annulus of ca. 20 thickened cells; spores tetrahedral, hyaline, typically with a thick ribbed or tuberculate epispore. Several spp. (Greek, *onychion*, little claw.)

1. O. dénsum Brack. in Wilkes. [*Pellaea densa* Hook. *Cheilanthes siliquosa* Maxon.] CLIFF-BRAKE. Cespitose with oblique divisions to the rhizome and acicular brownish subentire shining scales; fronds many, crowded, 0.5–3 dm. high; the stipes 0.5–2 dm. long, brownish, flexuous; blades mostly fertile, broadly ovate, 2–8 cm. long, 3 times pinnate below, glabrous, the pinnae few, close, oblique; ultimate segms. linear, 5–15 mm. long, 1–1.5 mm. wide, mucronate, the margins revolute, with a continuous erose indusium; sterile fronds few, the segms. broader.—Rocky places, cliffs, etc., at 6000–8500 ft. in Sierra Nevada, 1000–6500 in Coast Ranges; N. Oak Wd., Douglas-Fir F., Yellow Pine F., Red Fir F., etc.; San Luis Obispo and Kern cos. to B.C., Quebec, Mont., Wyo.

8. Pityrográmma Link

Small to medium-sized ferns with short-creeping to oblique rhizomes covered with dark rigid lance-linear paleae. fronds crowded, uniform, the stipe scaly at base, dark, polished; blades 1–3 times pinnate, linear to ovate, herbaceous to subcoriaceous, ± densely waxy-powdered beneath, sometimes glandular above, usually without scales. Sori borne along the veins, confluent, non-indusiate. Annulus of 20–24 thickened cells; spores globose-tetrahedral, dark, irregularly reticulate-ribbed. Perhaps 40 spp., mostly of trop. Am. (Greek, *pityron*, bran, and *gramme*, line, because of the furfuraceous sori.)

(Weatherby, C. A. Varieties of P. triangularis. Rhodora 22: 113–120, 1920.)

1. P. triangulàris (Kaulf.) Maxon. [*Gymnogramme t.* Kaulf. *Gymnopteris t.* Underw. *Ceropteris t.* Underw.] GOLDENBACK FERN. Rhizome stout, ascending or short-creeping, with brownish or blackish scales ca. 2–2.5 mm. long; fronds many, 1–4 dm. high, the stipes red-brown when young, darker in age, twice as long as the blades, these deltoid-pentagonal, 6–18 cm. long, almost as broad, pinnate, or the basal pinnae again pinnate; pinnules oblong, lobed to subentire, glabrous above, coriaceous, yellow-powdery beneath; sori ± covering the backs of the pinnules, following the veins; spores round to deltoid in outline.—Common in ± rocky shaded places, below 5000 ft.; many Plant Communities; most of cismontane Calif. from n. L. Calif. to B.C.

Upper surface of the ± coriaceous blade glabrous; indument bright to pale yellow; basal segms. of the lowest pinnae mostly elongate and pinnatifid .... *P. triangularis*
Upper surface of blade glandular or viscid or both; indument beneath mostly white, rarely none.
  The upper surface of the ± coriaceous blade viscid, often also with yellowish stalked glands; lower basal segms. of the lowest pinnae mostly undulate-crenate, not pinnatifid. Coastal San Diego Co. to Orange Co., Ids. .... *Var. viscosa*
  The upper surface of the often rather thin blade glandular only, not viscid; lower basal segms. of the lowest pinnae usually elongate, dilated, deeply pinnatifid.
    Stipes mostly blackish, glandular, white-farinose above and near base, not very lustrous; blade

thin, soft, usually grayish above with whitish glands; spores round to deltoid in outline. Foothills
from Butte to Tulare cos. . . . . *Var. pallida*
Stipes mostly red-brown, subglabrous; blade subcoriaceous, with scattered yellowish glands above;
spores trilobate in outline. Deserts . . . . *Var. Maxoni*

Var. viscòsa (Nutt. ex D. C. Eat.) Weath. [*Gymnogramme t.* var. *v.* Nutt. ex D. C.
Eat. *Ceropteris v.* Underw. *P. v.* Maxon.] SILVERBACK FERN. Stipes red-brown; blades
viscid above, with white powder beneath.—Dry slopes; Coastal Sage Scrub, Chaparral;
w. San Diego Co. to Laguna Beach, Orange Co.; Catalina, San Clemente, Santa Cruz,
Santa Rosa ids.
Var. pállida Weath. Stipes blackish, glandular and white-farinose above, white-
powdery beneath.—Occasional, Foothill Wd., etc., foothills bordering the Cent. V.
from Butte to Kern and Santa Clara cos.
Var. Máxoni Weath. Stipes red-brown, glossy; blades nearly 3-pinnate below, sparsely
yellowish-glandular above, light yellow or whitish-powdery beneath.—Rocky canyons,
etc.; Creosote Bush Scrub to Pinyon-Juniper Wd.; e. Mojave Desert, Colo. Desert; to
Ariz., Son., L. Calif.

9. Adiántum L. MAIDENHAIR

Delicate ferns with long-creeping or short and ascending, paleate rhizomes, the
scales usually brown to black, narrow. Fronds distichous or in several ranks, ascending
to drooping, with firm dark usually shining stipes; blades simple, 1–3 times pinnate or
decompound, mostly broad, mostly firm-herbaceous, the veins free or rarely anastomosing.
Sori appearing marginal, the sporangia borne along and sometimes between the ends
of the free, forking veins, on the lower side of the reflexed indusiform marginal lobes
of the pinnules. Sporangia with an annulus of ca. 18 thickened cells; spores tetrahedral,
dark, smooth. Ca. 200 spp., largely of trop. Am. (Greek, *a*, without, and *diaine*, un-
wetted, referring to shedding of rain drops.)

Blade at least as wide as long, divided at base into 2 equal parts, each with several pinnate branches
1. *A. pedatum*
Blade much longer than wide, not forked, but with a continuous main rachis.
Indusia nearly continuous, becoming 8 mm. long; segms. with ± rounded, scarcely lobed margin
and regular outline . . . . 2. *A. Jordani*
Indusia distinct, ca. 2 mm. long; segms. wedge-shaped, deeply lobed and with irregular outline
3. *A. Capillus-Veneris*

1. **A. pedàtum** L. var. **aleùticum** Rupr. FIVE-FINGER FERN. Rhizome rather short,
thick, with brown lance-oblong to deltoid scales; fronds close, erect, 2–8 dm. high,
the stipe stout, dark, chaffy at base; blades mostly reniform-orbicular in outline, 1–5
dm. broad, forked at base, the ± recurved branches bearing on the outer side several
slender spreading pinnate divisions 1–4 dm. long; pinnules close, short-stalked, numerous,
oblong to deltoid-oblong, the lower margin entire, the upper deeply cleft; sori linear
to oblong-lunate, solitary on the truncate lobes.—Moist shaded rock-crevices and in
swampy woods and canyons, from sea level to 10,700 ft.; Redwood F., Mixed Evergreen F., Chaparral, Montane Coniferous F., etc.; San Bernardino and San Gabriel
mts. through the Coast Ranges and Sierra Nevada, Santa Cruz Id.; to Alaska, Utah,
Quebec.
2. **A. Jórdanii** K. Mull. [*A. emarginatum* D. C. Eat.] CALIF. MAIDENHAIR. Rhizome
creeping, slender, rather densely paleaceous with dark brown rigid attenuate scales;
fronds several, rather close, 2–5 dm. long, the stipes almost equal to blades, dark
brown; blades ovate, 2–3 times pinnate at base, 1-pinnate near tip, the pinnules
rounded, 6–25 mm. broad, entire below, shallowly 2–5-lobed at the outer edge, the
sori close, 4–8 mm. long.—Damp shaded banks at base of rocks and trees, mostly below
3500 ft.; several Plant Communities; Coast Ranges from San Diego Co. n. to Ore.,
occasional in foothills of Sierra Nevada, on ids. off s. Calif. coast.
3. **A. Capíllus-Véneris** L. [*A. modestum* Underw.] VENUS-HAIR FERN. Rhizome
creeping, slender; the scales thin, light brown, lance-linear, entire; fronds ± spaced,
ascending to pendent, 2–7 dm. long, the stipes slender, almost black, to ca. as long as
blades; blades 1–4 dm. long, 1–2 dm. wide, 2–3 times pinnate at base, the upper
third once pinnate, pinnules stalked, obovate to rhombic, etc., 5–30 mm. long, cuneate

at base, the outer edge lobed or incised, with toothed margins; sori mostly oblong-lunate, solitary on the lobes, 1–2 mm. long; $n = 30$ (Manton, 1950.)—Calcareous seeps on rocky walls, etc., mostly below 4000 ft.; many Plant Communities; widely scattered in cismontane Calif., especially in s. part of state; occasional on desert; Santa Cruz Id., Catalina Id.; warmer regions in both hemis.

2. Aspidiàceae

Mostly terrestrial ferns, with creeping ascending or erect rhizome, rarely forming a short trunk or scandent, paleate. Stipe rarely articulate; fronds pinnate in plan, simple to decompound, mostly uniform. Sori typically away from the margin, rarely marginal, typically round, sometimes elongate, or the sporangia extending indefinitely along the veins and even over the surface; indusium usually present, fixed beneath the sorus, and opening around the margin, but sometimes peltate, or opening over the sorus, or elongate, sometimes wanting. Annulus longitudinal, of 10–40 thickened cells, interrupted by the pedicel. Spores bilateral, epispore usually present, often conspicuous. Ca. 66 genera, largely trop.

A.   Indusia wanting, the sorus roundish; ultimate segms. of lvs. very narrow. Alpine ... **6. *Athyrium***
AA. Indusia present, peltate or reniform.
    B.   Indusium round, centrally attached; lvs. once pinnate, the basal pinnae sometimes pinnate; texture coriaceous .... **2. *Polystichum***
    BB. Indusium not centrally attached; lvs. at least bipinnate throughout; texture not coriaceous.
        C.   Stipes slender, less than 1.5 mm. in diam.; blade of frond 1–1.5 dm. long and 5–8 cm. wide.
            D.   Indusium under the sorus, with stellate divisions .... **1. *Woodsia***
            DD. Indusium hoodlike, fixed on one side, with a broad base .... **5. *Cystopteris***
        CC. Stipes coarser, 2–4 mm. in diam.; blade of frond 2.5–5 or more dm. long and 1–2 dm. wide.
            D.   Indusium distinctly reniform and quite circular in outline, attached along the sinus.
                E.   Blades with distinctly toothed segms., if only 1–2 times pinnate, sometimes 3 times pinnate; veins 2-several times forked .... **3. *Dryopteris***
                EE. Blades with entire or subentire segms.; veins simple or once forked
                                        4. *Lastrea*
            DD. Indusium merely curved, elongate rather than round, attached along the inner side .... **6. *Athyrium***

1. Woódsia R. Br.

Rhizome erect, clothed with broad thin paleae. Fronds pinnate or bipinnate, hairy, sometimes also scaly, or glabrescent, numerous, densely clustered; veins free. Sori roundish, separate, or ± confluent in age, the indusia basal, breaking irregularly into lacerate divisions, or stellate, the filiform divisions often concealed by sporangia. Sporangia small, globose, the annulus longitudinal, of 18–20 thickened cells; spores ± reticulate. Ca. 40 spp., best developed in China. (Named for J. *Woods*, 1776–1864, English botanist.)

Blades glabrous; indusia divided into hairlike segms.; fronds 1–2 dm. high .... **1. *W. oregana***
Blades glandular-puberulent; indusia divided into lanceolate segms.; fronds 1–4 dm. high
                                                      2. *W. scopulina*

1. **W. oregàna** D. C. Eat. Rhizomes short-creeping, rather slender, tufted, densely paleaceous with pale brown often dark-striped scales; fronds many, 0.5–2.5 dm. high; stipes 2–12 cm. long, straw-colored with a brownish base; blades lance-oblong to linear, 5–12 cm. long, glabrous, bright green, delicately herbaceous, with 6–12 pairs of pinnatifid deltoid-oblong pinnae with toothed lobes; sori submarginal.—Rare in dry rocky places, 4000–11,000 ft.; Pinyon-Juniper Wd., Yellow Pine F. to Subalpine F.; Santa Rosa and San Jacinto mts. (Riverside Co.), higher mts. of Mojave Desert, Sierra Nevada, to Modoc Co.; to B.C., S. Dak., New Mex., also Quebec.

2. **W. scopulìna** D. C. Eat. Similar to no. 1, larger, the blades with numerous light green glandular-puberulent pinnae and few to many flat septate hairs; indusial segms. lanceolate.—Occasional, in exposed rocky places, 4000–12,000 ft.; San Bernardino Mts., Sierra Nevada, White Mts.; to Alaska, Quebec, Colo., N. Car.

2. Polýstichum Roth

Rather coarse ferns with mostly short ascending paleate rhizomes, the scales mostly lacerate. Fronds several, rigidly ascending or recurved; stipes paleate; blades uniform or dimorphous, pinnate to decompound, the ultimate divisions or teeth usually mucronate, the segms. of harsh texture; veins free. Sori on the veins, round, the indusium peltate or rarely none. Annulus usually of 18 or more cells. Spores bilateral, oblong to roundish, usually tuberculate or echinulate. Ca. 175 spp., mainly of temp. and boreal regions. (Greek, *polus*, many, and *stichos*, row, some spp. having many rows of sori.)

A. Fronds once pinnate, the pinnae variously serrate or incised, never lobed or pinnatifid.
  B. Stipes 1–6 cm. long; pinnae mostly lance-oblong, or the lowermost deltoid, serrate-dentate, the teeth spreading. Plumas and Siskiyou cos. .... 1. *P. Lonchitis*
  BB. Stipes 5–50 cm. long; pinnae lance-linear, incised to biserrate, the teeth incurved. Widely distributed .... 2. *P. munitum*
AA. Fronds with at least the lower pinnae pinnately lobed or divided.
  B. Teeth of the frond-segms. not pungent or spine-tipped. Siskiyou and Trinity cos.
                                                                    3. *P. Lemmonii*
  BB. Teeth of the frond-segms. pungent or spine-tipped.
    C. Pinnae lobed or divided, but not completely pinnate.
      D. Pinnae ± deltoid, mostly with not more than 6 lobes on a side. Pine belt
                                                                    4. *P. scopulinum*
      DD. Pinnae broadly oblong, largely with 12 or more lobes on a side. Below the pine belt .... 5. *P. californicum*
    CC. Pinnae largely completely pinnate, the lowest pinnules again cleft or divided
                                                                    6. *P. Dudleyi*

1. **P. Lonchìtis** (L.) Roth. [*Polypodium* L. L.] HOLLY FERN. Rhizome large, woody, ± erect, with large ovate rusty-brown scales; fronds clustered, persistent, rigid, 1.5–6 dm. long, lance-linear; the stipes thick, straw-colored, paleaceous; blades 1–5.5 dm. long, 2–7 cm. wide, once-pinnate, with stout scaly rachis; pinnae many, close, oblong, falcate, acute, cuspidate-serrulate, the basal ones deltoid; sori large, contiguous, in 2 rows on the pinnae, nearly medial; indusia entire; $n = 41$ (Manton, 1950).—Shaded rocky places, at 5000–7000 ft.; Red Fir F., Subalpine F.; Siskiyou and Plumas cos.; to Alaska, Nova Scotia, Rocky Mts.; Eurasia.

2. **P. munìtum** (Kaulf.) Presl. [*Aspidium m.* Kaulf.] SWORD FERN. Coarse evergreen fern from strong woody suberect very scaly rhizomes; fronds many, sometimes 75–100, rigidly ascending in heavy crowns or clumps, 6–14 dm. high; stipes stout, 1–3(–6) dm. long, conspicuously paleaceous with large lanceolate chestnut-brown scales mixed with shorter lance-linear ciliate ones; blades pinnate, lanceolate, rather short-acuminate to slender-subcaudate at tip, dark lustrous green above, paler beneath, 3–6(–10) dm. long, 9–16(–25) cm. broad; pinnae densely and evenly placed, very many, narrow-lanceolate, auriculate, straight or falcate, pungently toothed or incised, the teeth short, firm, bristle-tipped; sori 1–1.5 mm. broad, usually in 2 rows, submarginal; indusia subrotund, fringed, irregularly and tardily deciduous.—Common in damp woods at mostly below 2500 ft.; Redwood F., N. Coniferous F., Douglas-Fir F.; etc.; near the coast from Monterey Co. to Del Norte Co.; Santa Cruz Id.; to Alaska, Mont.

Var. **ímbricans** (D. C. Eat.) Maxon. [*Aspidium m.* var. *i.* D. C. Eat.] Fronds 3–5 dm. long, the stipes with a basal tuft of scales; pinnae crowded, obliquely imbricate, mostly 2–3 cm. long.—With the sp. and sspp. from San Diego Co. to B.C.

SSD. **cúrtum** Ewan. Rhizomes less robust, shorter, very chaffy; fronds fewer, usually 15–30, 4–8 dm. high, the more slender stipes 2–3 dm. long, densely paleaceous near base, sparingly so above; rachises finely and inconspicuously paleaceous; blades gradually tapering to tip, 2.5–4(–5) dm. long, 6–9 cm. wide, usually with lance-oblong pinnae.—Canyon-slopes, 1500–8600 ft.; Chaparral to Montane Coniferous F.; Santa Lucia Mts. to mts. of San Diego Co.

Ssp. **nudàtum** (D. C. Eat.) Ewan. [*Aspidium m.* var. *n.* D. C. Eat. *P. m.* var. *n.* Gilbert.] Rhizome slender; fronds few, almost scaleless, the slender stipes 1–2 dm. long; blades lance-ovate, 2–3 dm. long, 7–9 cm. wide.—Occasional in rock-crevices, 4500–6700 ft.; Yellow Pine F., Red Fir F.; Sierra Nevada from Plumas to Tulare cos.

3. **P. Lemmònii** Underw. [*P. mohrioides* var. *L.* Fern.] SHASTA FERN. Rhizomes tufted, stout, thickly covered with bases of old stipes; fronds ascending, 1–4 dm. long;

stipes 0.3–1.5 dm. long, straw-colored, with a very paleaceous base, the scales scattered also above and on the rachis, mostly large, lanceolate to lance-ovate, brown; blades linear to narrowly lance-oblong, 1–3 dm. long, 1.5–5 cm. wide, pinnate then pinnatifid; pinnae many, ± imbricate, deltoid-oblong to -ovate, pinnately divided at base, deeply lobed or divided nearly to the tip into close ± ovate divisions, crenate or crenately lobed; sori on the apical pinnae, 1–several on a segm.; indusia large, thin, erose-dentate.—Moist granitic places among rocks, 4500–6500 ft.; Red Fir F., Subalpine F.; mts. of Siskiyou and Trinity cos.; to Wash.

4. **P. scopulìnum** (D. C. Eat.) Maxon. [*Aspidium aculeatum* var. *s.* D. C. Eat. *P. mohrioides* var. *s.* Fern.] Rhizome stout, erect or decumbent, paleaceous with light brown linear to oblong-ovate denticulate scales; fronds 6–10, erect-spreading, 1.5–4(–6) dm. long; stipes 0.3–1.4 dm. long, stout, grooved, paleaceous especially toward the base; blades linear to lance-oblong, coriaceous, fibrillose-paleaceous beneath, 1–3 dm. long, 2.5–6 cm. wide, pinnate; pinnae many, deltoid-ovate to -oblong, ± pinnately divided at the base, the lobes and teeth oblique, pungent; sori many, large, close, near the middle, in 2 confluent rows; indusia erose-dentate.—Dry crevices and rocky places, 5000–10,500 ft.; Montane Coniferous F.; San Gabriel and San Bernardino mts., n. Sierra Nevada to Siskiyou, Trinity and Glenn cos.; to Wash., Ida., Utah, Quebec.

5. **P. califórnicum** (D. C. Eat.) Underw. [*Aspidium c.* D. C. Eat.] Rhizome stout, suberect, copiously paleaceous with large dark brown attenuate scales; fronds several, ascending, 3–11 dm. long; stipes 0.5–3.5 dm. long, paleaceous at base; blades linear-oblong to -lanceolate, 2–7.5 dm. long, 0.5–2 dm. wide, subcoriaceous, fibrillose beneath, minutely paleaceous on rachis; pinnae many, linear from a broader base, obliquely large, several pairs on a segm.; indusia erose-ciliate.—Creek banks, canyons, etc. below 1000 ft.; Redwood F., Mixed Evergreen F.; near the coast from Santa Cruz and Santa Clara cos. to Mendocino Co. pinnatifid to incised, with elliptical decurrent close broadly joined lobes, aristate; sori

6. **P. Dúdleyi** Maxon. [*P. aculeatum* var. *D.* Jeps.] Resembling *P. californicum*, the rhizome decumbent, paleaceous with linear to ovate thin denticulate-ciliate scales; stipes paleaceous; blades twice pinnate, lance-oblong to -ovate, attenuate, 2.5–7.5 dm. long, 0.8–2.5 dm. wide; pinnae mostly oblique, contiguous, linear to lance-oblong, filiform-paleaceous, especially beneath; pinnules obliquely ovate to ovate-oblong, auriculate, the upper basal cleft or divided, the others serrate to incised, with short-awned incurved teeth.—Rocky canyons; Redwood F., Mixed Evergreen F., etc.; near the coast from n. San Luis Obispo Co. to Marin Co.

3. Dryópteris Adans. WOOD FERN. SHIELD FERN

Mainly woodland ferns, of moderate or large size; rhizome short and stout, ascending to erect or creeping, with broad paleae that are entire or glandular-margined. Fronds borne singly or in a crown, the elongate stipes commonly scaly. Blades bipinnatifid to decompound, uniform, firm in texture, mostly glabrous; veins free, forked. Sori normally on the veins, round, indusiate or not, the indusia round-reniform and attached by the inner end of the sinus. Annulus of 14 or more cells. Spores bilateral, tuberculate or echinulate. A genus of ca. 150 spp., mostly trop. (Greek, *drys*, oak, and *pteris*, fern.)

Blades essentially three times pinnate . . . . 1. *D. dilatata*
Blades 1–2 times pinnate.
    Pinnae sessile, lance-oblong, the lower basal pinnule with a semicordate base, this overlying the
    primary rachis; veinlets all ending in salient spinelike teeth. Common . . . . 2. *D. arguta*
    Pinnae mostly short-stalked, deltoid-lanceolate, the basal pinnules symmetrical, not semicordate,
    not overlying the primary rachis; veinlets usually ending in curved teeth. San Bernardino Mts.
    . . . . 3. *D. Filix-mas*

1. **D. dilatàta** (Hoffm.) Gray. [*Polypodium d.* Hoffm. *D. spinulosa* var. *d.* Underw. *Aspidium s.* var. *d.* Link.] Rhizome stout, woody, creeping or ascending, chaffy with brownish scales; fronds 3–10 dm. long, spreading, borne in a crown; stipes stout, dark, 1.5–4.5 dm long, paleaceous; blades ovate to deltoid or ovate-oblong, 1.5–9 dm. long, 1.4 dm. broad, nearly or quite 3 times pinnate; pinnae lanceolate or the lower pairs deltoid; pinnules oblong-ovate, their segms. herbaceous, mucronate-dentate; sori mostly subterminal; indusia glabrous or sparsely glandular; $n = 82$ (Manton, 1950).—In dense

woods, on decaying logs, etc., below 1500 ft.; Redwood F., Mixed Evergreen F.; near coast from San Mateo Co. to Del Norte Co.; to Alaska, Atlantic Coast; Eurasia.

2. **D. argùta** (Kaulf.) Watt. [*Aspidium a.* Kaulf. *D. rigida* var. *a.* Underw.] Rhizomes stout, short-creeping, woody, with thin attenuate bright brown scales; fronds several, close, erect, 3–8 dm. long; stipes stout, scaly, shorter than blades; blades lance-ovate to oblong, acuminate, 2.5–6 dm. long, 1–3 dm. broad, mostly twice-pinnate; pinnules lance-oblong, subcoriaceous, rounded-obtuse, serrate to incise, the teeth often spinelike; sori in 2 rows, large, close; indusia firm with a deep narrow sinus, the margins glandulose.— Common on shaded slopes and in open woods, mostly below 5000 ft.; many Plant Communities; cismontane Calif. from San Diego Co. to Wash.; Santa Barbara Ids.

3. **D. Fìlix-más** (L.) Schott. [*Polypodium F.-m.* L. *Aspidium F.-m.* Sw.] MALE FERN. Fronds 3–10 dm. long; stipes scaly; blades 2.5–8 dm. long, nearly twice-pinnate, the basal pinnules not subcordate, marginal teeth curved, not spinelike; $n = 82$ (Manton, 1950).—Known from a single collection in 1882 at 8000 ft.; Holcomb V., San Bernardino Mts.; Ariz. to B.C., Atlantic Coast; Eurasia.

4. Lastrèa Bory

Moderate-sized ferns; the rhizome short- or long-creeping, or ascending to erect, paleate, the paleae rarely dense, often pubescent. Lf-blades typically bipinnatifid and usually narrowed toward both ends, rarely more compound, often hairy with simple unicellular hairs; veins free, usually simple and reaching the margin. Sori on lower surface on veins (rarely terminal) small, round or rarely elongate. Indusium round-reniform if present. Spores bilateral. Ca. 500 spp., quite cosmopolitan. (Meaning unknown.)

Blades oblong to ovate-oblong, not narrowed at base, distinctly puberulent beneath. Wet places in canyons, s. Calif. . . . . 1. *L. augescens*
Blades lanceolate or oblanceolate, distinctly narrowed toward base, glabrous or slightly hairy beneath. From Mariposa and Trinity cos. n. . . . . 2. *L. oregana*

1. **L. augéscens** (Link) J. Sm. [*Aspidium a.* Link. *Thelypteris a.* M. & J. *T. Feei* and *T. puberula* of Calif. refs.] Rhizome woody, slender, long-creeping, with few apical rusty linear-attenuate short-hirsute scales; fronds few, 3–10 dm. high; stipes stout, straw-colored, naked; blades broadly oblong, 3–7 dm. long, 1.5–4 dm. wide, pinnate-pinnatifid, freely puberulous beneath; pinnae spreading, linear-attenuate, cut to ca. 2–3 mm. from the stout midrib; segms. oblique, narrow-oblong, subfalcate, acuminate, subentire; veins 8–11 pairs, simple, the 2 lower pairs running to the hyaline sinus; sori small, close, supramedial; indusia firm, pilose.—Occasional in wet shaded canyons below 3000 ft.; Chaparral, Creosote Bush Scrub; about Santa Barbara, s. face of San Gabriel Mts., e. base of San Jacinto Mts.; to L. Calif. and Mex.

2. **L. oregàna** (C. Chr.) Copel. [*Dryopteris o.* C. Chr. *Aspidium nevadense* D. C. Eat., not Boiss.] Rhizome slender, creeping, apically covered with old stipe-bases; scales ovate, yellow-brown, thin, concave, slightly dentate; fronds few, in close hibernating crown, 5–9 dm. long; stipes slender, straw-colored, almost bare; blades elliptic-lanceolate, acuminate, 4–6 dm. long, 1–1.5 dm. broad, narrow-based, pinnate-pinnatifid; pinnae sessile, the basal usually remote, small, the large ones close, linear; segms. close, oblong, entire or ± crenate, minutely resinous-glandular and ± ciliate; sori small, submarginal; indusia long-ciliate, glandular.—Along streams, 3000–5000 ft.; Montane Coniferous F.; Sierra Nevada from Tuolumne Co. n., Coast Ranges from Trinity Co. n.; to w. Ore.

5. Cystópteris Bernh.

Rather delicate ferns with slender creeping rhizomes. Fronds erect or ± spreading, the stipes slender, not jointed to rhizome; blades 1–4-pinnate, delicate, the fertile commonly less leafy and longer-stalked than the sterile. Sori roundish, on the veins, separate; indusium attached to the base of the receptacle and partly under the sorus, soon pushed back by the sporangia and partly hidden, withering, the sori in age seemingly naked. Sporangium roundish, on a slender pedicel, the longitudinal annulus of 14–16 thickened cells; spores reniform, smooth or muriculate. A small genus, widely distributed. (Greek, *cystis*, bladder, and *pteris*, fern.)

1. **C. frágilis** (L.) Bernh. [*Polypodium f. L. Filix f.* Gilib.] BRITTLE FERN. Rhizome with thin ovate acuminate scales toward the tip; fronds few to several; stipes slender, brittle, smooth, stramineous or greenish, mostly 0.5–2 dm. long; blades variable, mostly lance-oblong to -ovate, acuminate, 1–2.5 dm. long, nearly or quite 2-pinnate, thin; pinnules decurrent on the margined rachis; veinlets mostly excurrent to the marginal teeth; sori small; indusia roundish or ovate-acuminate, deeply convex; $n = 84$ (Manton, 1950). —Mostly moist ± rocky places, frequently ± shaded; mostly above 3500 ft. in s. Calif., lower in n., ascending to 12,000 ft.; many Plant Communities; montane and cismontane Calif., occasional in desert mts.; to Alaska, Atlantic Coast, Ga., Ala., trop. Am.; Eurasia.

6. Athýrium Roth. LADY FERN

Medium-sized to large upright ferns, the rhizome commonly erect, or long-creeping, with membranous thin-walled scales. Roots mostly stout, black. Fronds usually large; the stipe short to long. Blades usually pinnately decompound, sometimes pinnate or simple, herbaceous or coriaceous, mostly glabrous except on the axes; veins typically free. Sori on the surface, typically elongate along one or both sides of the veins, the indusium typically curved across the distal end of the sorus, sometimes wanting. Sporangia with slender pedicels, the annulus of 12–20, commonly 16, thickened cells. Spores bilateral. Ca. 600 spp., widely distributed. (Greek, *a*, without, and *thurium*, shield.)

Indusium mostly crescent-shaped, broadly hooked or horseshoe-shaped; pinnules mostly close
.... 1. *A. Filix-femina*
Indusium lacking; pinnules mostly distant .... 2. *A. alpestre*

1. **A. Fìlix-fémina** (L.) Roth var. **califórnicum** Butters. Rhizome erect or ascending, stout, with dark brown lanceolate scales to 1 cm. long; fronds 6–15 dm. high, erect-arching, mostly 1–2 dm. broad; stipes short, straw-colored, paleaceous at the dark base; blades 2–3 times pinnate, the pinnae lance-oblong, acute or acuminate, sessile, ascending-spreading, puberulent underneath along the rachises, not paleaceous; pinnules lance-oblong, crenate-dentate to pinnatifid, largely 0.5–2 cm. long; sori oblong to lunate, usually less than 1 mm. long; indusia toothed or ciliate on the free edge.—Along streams and in meadows, mostly 4000–9500 ft.; Yellow Pine F. to Subalpine F.; San Jacinto and San Bernardino mts., Sierra Nevada, N. Coast Ranges from Yolla Bolly Mts. n., to Modoc and Siskiyou cos.; to Ida., Colo., n. Mex.

Var. **sitchénse** Rupr. [*A. F.* ssp. *cyclosorum* (Rupr.) C. Chr.] Blades mostly wider, the pinnae 1–2 dm. long, spreading at right angles to the main rachis and with scattering scales on underside of their rachises, not puberulent; pinnules largely 2–3 cm. long.—Swampy places at low elevs. near the coast; Mixed Evergreen F., Redwood F., etc.; San Luis Obispo Co. to Alaska, Ida., Quebec; Eurasia; Santa Cruz Id.

2. **A. alpéstre** (Hoppe) Rylands var. **americànum** Butters. [*A. americanum* Maxon. *Phegopteris alpestris* var. *americana* Jeps.] Rhizomes erect or decumbent, branched, making massive rounded tufts; scales many, brown, thin, to ca. 1 cm. long; fronds clustered, 2–9 dm. long; stipes short, straw-colored from a darker base, sparsely chaffy; blades lance-oblong or narrower, 2–6 dm. long, 0.4–2.5 dm. wide, usually twice pinnate, then pinnatifid; pinnae ascending-spreading, acuminate; pinnules delicate, stalked, separated, ± oblong, incised; sori many, round, without indusia.—Meadows, and moist places, 5500–11,500 ft.; Red Fir F. to Subalpine F.; Sierra Nevada, White Mts., n. to Siskiyou Co.; to Alaska, Colo.

3. Blechnàceae

Terrestrial ferns, with creeping or erect paleaceous rhizomes. Stipes not articulate. Fronds commonly large and coarse, mostly pinnate or pinnatifid, sometimes more compound. Veinlets branching and anastamosing. Sori on the secondary veins, discrete or united. Indusium open on the costal side, rarely wanting. Sporangia large; annulus longitudinal, interrupted. Spores bilateral, usually without epispore. Eight genera.

Blades pinnate into entire pinnae; sterile and fertile fronds dissimilar .... 1. *Blechnum*
Blades pinnate into lobed or pinnatifid pinnae; fronds all alike .... 2. *Woodwardia*

1. Bléchnum L. Deer Fern

Rhizome mostly stout, ascending to erect, scaly with narrow dark paleae. Fronds cespitose or imbricate, dimorphous, usually pinnate and coriaceous, glabrous, the margins entire or serrate; sterile veins free. Sori elongate, parallel to the midrib, one on each side, normally continuous; indusium intramarginal, opening toward the midrib, entire to lacerate. Sporangia usually crowded, rather large; annulus of 14–28, largely 20, cells. Spores bilateral, subglobose to subreniform, usually smooth. Over 200 spp., mostly of S. Hemis. (Greek, *blechnon*, a kind of fern.)

1. **B. Spìcant** (L.) Roth. [*Osmunda* S. L. *Lomaria* S. Desv. *Struthiopteris* S. Weis.] Rhizome short-creeping, 1–2 cm. thick, woody, covered with brownish linear to lance-linear scales 5–10 mm. long; sterile fronds many, erect, 2–10 dm. long, evergreen; stipes 2–30 cm. long, brownish; blades linear to ± lanceolate, 1–8 dm. long, 2–9 cm. broad, pinnatifid to pinnate, the segms. many, linear-oblong to linear, entire to crenulate; fertile fronds few, cent., 4–15 dm. long; the stipes long; pinnae distant, mostly linear from a broader base, the back nearly covered by the sporangia; $n = 34$ (Manton, 1950).—Wet sheltered places; Redwood F., Mixed Evergreen F.; Santa Cruz Co. to Del Norte Co.; to Alaska; Eurasia.

2. Woodwárdia Sm. Chain Fern

Coarse rather large ferns with stout erect to short-creeping rhizomes. Fronds several to many in a crown, firm, typically bipinnatifid, with entire or serrulate margins; veinlets anastomosing. Sori borne on the outer horizontal veins of a continuous series of elongate areoles, the indusium elongate, arched. Annulus of 18–24 cells. Spores bilateral, smooth or flocculose. Ca. 5 spp. of N. Am. and Eurasia. (Named for T. J. *Woodward*, English botanist.)

1. **W. fimbriàta** Sm. in Rees. [*W. Chamissoi* Brack. *W. radicans* var. *americana* Hook.] Rhizome woody, the scales lance-attenuate, 1–3 cm. long, glossy, bright brown, entire; fronds almost erect, in a circle, 1–2 m. high; stipes short, straw-colored from a brown base; blades 2–5 dm. wide, ± oblong, pinnate, narrow at base; pinnae deeply pinnatifid, 1–2.5 dm. long, the segms. lanceolate, spinulose-serrate, firm-herbaceous; indusia almost straight, 1.5–6 mm. long.—Springy and boggy places in canyons below 5000 (8000) ft.; many Plant Communities; Santa Cruz Id.; occasional along the desert edge, cismontane Calif. to B.C., Ariz.

4. Aspleniàceae

Terrestrial, sometimes epiphytic ferns with creeping or suberect paleaceous rhizomes. Stipes not articulate. Fronds simple to decompound, small to large, mostly firm in texture; veins forking, free or anastomosing. Sori elongate along the veinlets, with elongate indusia attached to the veinlets. Annulus longitudinal, incomplete, commonly of ca. 20 cells. Spores bilateral. Ca 9 genera.

1. Asplènium L. Spleenwort

Ours smallish ferns, with erect or ascending rhizomes and rigid scales. Fronds pin-nate. Stipes and rachises dark. Veins free. Sori oblong or linear, straight, oblique to midrib. Indusium fixed lengthwise by one edge. Ca. 700 spp., cosmopolitan. (Greek, *a*, without, and *splen*, spleen, once used medicinally.)

Stipe and rachis dark chestnut or purplish-brown throughout. S. Calif. . . . . 1. *A. vespertinum*
Stipe red-brown below, the upper part and rachis green. Sierra Nevada . . . . 2. *A. viride*

1. **A. vespertìnum** Maxon. [*A. Trichomanes* var. v. Jeps.] Tufted evergreen with ascending or erect rhizome 1–2 cm. long, with linear-acicular scales 2–2.5 mm. long, dark brown; fronds numerous, 5–28 cm. long; stipes purplish-brown, shining; blades linear or lance-oblong, 1–2.5 cm. wide, the rachis fibrillose; pinnae 20–30 pairs, oblong to linear-oblong, deeply crenate, 6–10 mm. long; sori 4–12 on a pinna, with narrow firm

crenulate indusium.—Moist shaded rocky places below 3000 ft.; Chaparral, Coastal Sage Scrub, S. Oak Wd.; from s. face of San Gabriel Mts. to San Diego Co.

2. **A. víride** Huds. Stipes red-brown at base, green in upper part as is the rachis; fronds 3–20 cm. long; blades oblong-linear to narrowly linear-lanceolate, 0.5–1.5 cm. broad, with 5–25 pairs of pinnae, the lower broadly rounded-deltoid, the upper rhombic to obliquely ovate; sori 2–4 pairs, remote from margin, soon confluent and concealing the delicate subentire indusia.—Crevices in north-facing cliff, South Butte, n. of Sierra City, Sierra Co. at 7500 ft.; to Alaska, Nfld., Vermont; Eurasia.

5. Polypodiàceae

Epiphytic, sometimes terrestrial ferns with creeping or ascending rhizome bearing broad to setiform scales. Stipes usually articulate. Fronds usually simple to pinnate, usually firm in texture, glabrous to hairy or scaly; venation free or usually reticulate. Indusia lacking, the sori typically round, sometimes elongate along the veins and sporangia sometimes spread over the laminar surface. Annulus usually of 12–14 cells; stomium well developed. Spores bilateral, sometimes tetrahedral, with or without a thin epispore. Ca. 65 genera.

1. Polypòdium L. POLYPODY

Epiphytes, rarely terrestrial, with creeping paleaceous rhizome. Fronds with articulate stipes, uniform, pinnatifid to compound, glabrous or paleate, rarely pubescent. Sori on back of frond, terminal or nearly so on the lowest free veinlet included in an areole; round to elliptical. Ca. 75 spp., mostly of N. Hemis. (Greek, *polus*, many, and *pous*, foot, some spp. having many knoblike places on the rhizome.)

Blades coriaceous or leathery; rhizome glaucous; sori 3–4 mm. broad. Coastal . . . . 1. *P. Scouleri*
Blades not leathery; rhizome not glaucous; sori 1–3 mm. broad.
  Pinnae usually less than 2.5 cm. long, rounded at tips, with entire or crenate margins. At 5000–
  8500 ft. . . . . 2. *P. vulgare*
  Pinnae usually more than 3 cm. long, pointed, with serrate margins. Below 4000 ft.
    Segms. oblong, acute or obtuse; veins ± areolate, mostly opaque. Widely distributed
                                                      3. *P. californicum*
    Segms. deltoid-linear, attenuate-acute; veins free, mostly translucent. Coastal . . 4. *P. Glycyrrhiza*

1. **P. Scoùleri** Hook. & Grev. [*P. pachyphyllum* D. C. Eat. *P. carnosum* Kell.] Rhizome woody, creeping, 6–10 mm. thick, loosely chaffy, white-pruinose and naked in age, the scales dark brown, ca. 1 cm. long, denticulate; fronds few, 1.5–7 dm. long; stipes stout, naked, shorter than the blade; blades deltoid-ovate, ca. 1–4 dm. long, pinnatisect into 2–14 pairs of linear to narrowly oblong obtuse spreading coriaceous crenate segms. 8–20 mm. broad; midribs scaly beneath; veins joined in a series of areoles; sori crowded against the midrib, mostly on the upper segms.—Mossy logs, cliffs and slopes below 1500 ft.; N. Coastal Scrub, Coastal Strand, Redwood F., Mixed Evergreen F., etc.; Santa Cruz Id.; coast from Santa Cruz Co. to Del Norte Co.; to B.C.

2. **P. vulgàre** L. var. **columbiànum** Gilbert. [*P. hesperium* Maxon.] Rhizome creeping, firm, ca. 5 mm. thick, densely paleaceous with ovate acuminate scales 3–5 mm. long; fronds rather close, mostly 1–2 dm. long; stipes ca. as long as blade, straw-colored, naked; blades deltoid-oblong to linear-oblong, pinnatifid nearly to the naked rachis; segms. narrow-oblong to oval, rounded-obtuse, crenate to crenate-serrulate; veins mostly twice forked, translucent; sori round-oval, medial. Rare, rock-ledges and crevices, 5000–8500 ft.; Yellow Pine F., Red Fir F.; San Jacinto Mts., San Bernardino Mts., Sierra Nevada; to Alaska, S. Dak., New Mex., L. Calif.

3. **P. califórnicum** Kaulf. [*P. intermedium* H. & A. *P. vulgare* var. *i.* Fern.] Rhizome creeping, 5–10 mm. thick, with deciduous deltoid-ovate rusty brown scales 3–7 mm. long; fronds not evergreen, 1–3.5 dm. high; stipes stout, straw-colored, naked, mostly shorter than the blades; blades oblong to narrowly ovate, pinnatifid nearly to the rachis, mostly 0.5–3 dm. long, 0.5–1.5 dm. wide, the segms. membranous, linear-oblong, acutish to obtuse, 3–7 cm. long, veins mostly dark, opaque, 3–5 times forked, ± casually joined and often forming an irregular series of areoles; sori oval, slightly inframedial.—Common on rocky ledges and moist banks below 4000 ft.; Chaparral, Coastal Sage Scrub,

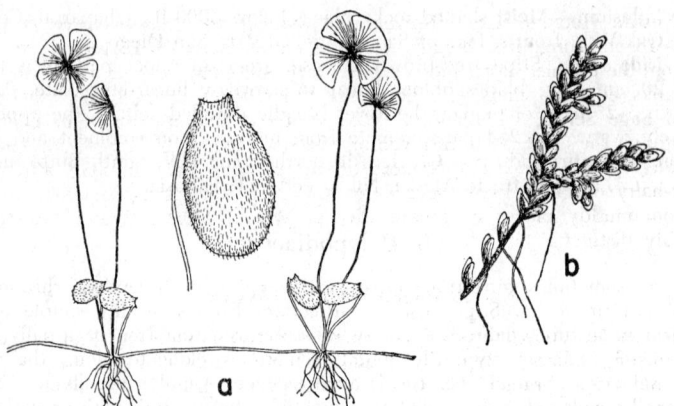

Fig. 4. MARSILEALES. *Marsilea: a,* habit, × ½, creeping plants with erect petioled 4-foliolate blades and hard bean-shaped sporocarps.
SALVINIALES *Azolla: b,* small floating plant, × 2, showing imbricate 2-lobed lvs. and rootlets beneath.

Foothill Wd., Mixed Evergreen F., etc.; cismontane s. Calif. through the Coast Ranges and foothills of Sierra Nevada to Humboldt and Butte cos.; Santa Barbara Ids.; L. Calif. Unusually coriaceous specimens from exposed sea bluffs, rocks, etc., are var. *Kaulfùssii* D. C. Eat.
   4. **P. Glycyrrhìza** D. C. Eat. [*P. vulgare* var. *occidentale* Hook. *P. falcatum* Kell.] LICORICE FERN. Resembling no. 3, but rhizomes 3–5 mm. thick, their scales 5–9 mm. long; lf.-blades mostly lanceolate, the segms. linear-attenuate, ± falcate, attenuate-acute at tips; veins free, mostly 3-forked and translucent.—On rocks, mossy tree trunks, logs, etc., below 2000 ft.; Redwood F., Mixed Evergreen F., Yellow Pine F.; near the coast from Monterey Co. to Alaska.

Order 3. MARSILEALES (Fig. 4)

Characters as given for the family *Marsileaceae*

1. Marsileàceae. MARSILEA FAMILY

Small perennial herbs with slender creeping rhizomes rooting in the mud. Lvs. petioled and with 2- to 4-foliolate blades, or filiform and bladeless. Sori borne within hard round to bean-shaped sporocarps which arise from the base of the petioles or near them and bear microspores that produce antheridia and megaspores that produce female gametophytes. Three genera.

Lf.-blades present; plants hairy .... 1. *Marsilea*
Lf.-blades absent; plants glabrous .... 2. *Pilularia*

1. Marsílea L.

Plants on mud or in shallow water and with floating lvs. Lfts. 4. Sporocarps ovoid or bean-shaped, usually with 2 teeth near the base, 2-chambered vertically, and with many transverse partitions, each compartment being 1 sorus and containing both microsporangia and megasporangia. Ca. 70 spp., in most parts of the world. (Named for A. *Marsigli,* Italian botanist of 18th century.)

The upper tooth of the sporocarp sharp, conspicuous; each sorus with 12–20 megasporangia
                                                                            1. *M. vestita*
The upper tooth of the sporocarp obsolete or a rounded tubercle; each sorus with 6–9 megasporangia
                                                                            2. *M. oligospora*

1. **M. vestìta** Hook. & Grev. Rhizomes long-creeping, densely hairy at the nodes; petioles 2–18 cm. long; lfts. broadly cuneate, entire, hairy, 5–15 mm. long; peduncles distinct from the petiole, short; sporocarps solitary, 4–8 mm. long, at first densely hairy.— Muddy banks, edge of ponds, etc., especially about vernal pools, below 7000 ft.; mostly in Coastal Sage Scrub, V. Grassland, etc.; scattered stations from San Diego Co. n., largely through the Cent. V. to Siskiyou and Modoc cos.; to B.C., S. Dak., Tex.

2. **M. oligóspora** Goodd. Plants more tufted; petioles 3–8 cm. long; lfts. 5–10 mm. long, soft-hairy or glabrescent; sporocarps 4–6 mm. long, strigose when young.—Reported from muddy places in Tulare Co.; to Wash., Mont., Wyo. These plants are questionably distinct.

2. Pilulària L. PILL-WORT

Rhizome slender, short-creeping; lvs. minute, filiform, bladeless. Sporocarps globose, short-peduncled, axillary, longitudinally 2–4-loculed, each locule a sorus with microsporangia and megasporangia. Ca. 6 widely distributed spp. (Latin, *pilula*, a little ball, referring to the sporocarps.)

1. **P. americàna** A. Br. Lvs. 2–4(–5) cm. long, 1-more at a node, filiform; sporocarps ca. 2 mm. in diam.—Occasional in heavy soil, largely of vernal pools, below 5500 ft.; V. Grassland, etc.; San Diego region to Siskiyou and Modoc cos.; Ore.; also in Ark.

Order 4. SALVINIALES (Fig. 4)

Characters as given for the family *Salviniaceae*

1. Salviniàceae. SALVINIA FAMILY

Floating aquatic plants, with ± elongate and sometimes branching axis bearing apparently 2-ranked lvs. Sporocarps soft and thin-walled, 1-loculed, each containing a cent. receptacle bearing microsporangia or megasporangia. Two genera.

1. Azólla Lam.

Small mosslike plants with pinnately branched stems covered with minute imbricate 2-lobed lvs. producing rootlets beneath. Sporocarps of 2 kinds, borne in pairs beneath the stem, the small ones ovoid and bearing 1 megaspore, the larger globose and containing masses of microspores. Six spp., widely distributed. (Greek name of doubtful origin.)

1. **A. filículoìdes** Lam. Plants compact, often reddish, 1–2.5 cm. long, easily breaking apart; lvs. ovate, 1 mm. long, deeply 2-lobed, the dorsal lobe papillate-hairy, the ventral submersed and forming the sporocarp.—Frequent in sluggish water at low elevs., cismontane Calif., Modoc Co., Mojave R.; to Wash., Ariz., Mex., S. Am. Plants from cent. and n. Calif., smaller (1–3 cm. across instead of 2–6 cm.) and with septate glochidia on the microsporangial massulae, have been referred to *A. mexicana* Presl.

Division IV. CONIFEROPHYTA. CONE-BEARING PLANTS

Woody plants with scaly to needlelike or broader simple lvs. Monoecious or occasionally dioecious, the ♂ structures (microsporophylls) usually in short-lived strobili and bearing the sporangia on their abaxial surface. Ovulate structures in cones or not, the ovule or ovules borne terminally and singly, or more commonly adaxially on conescales, hence no development of a stigma. Seeds thus not inclosed in a seed-vessel, but released when the cone-scales pull apart. An ancient division with some extinct fossil groups and ca. 500 living spp.

KEY TO ORDERS

A.   Stems not jointed; lvs. needlelike or linear, sometimes scalelike and then closely overlapping; plants mostly with resin. Mostly trees or arborescent.

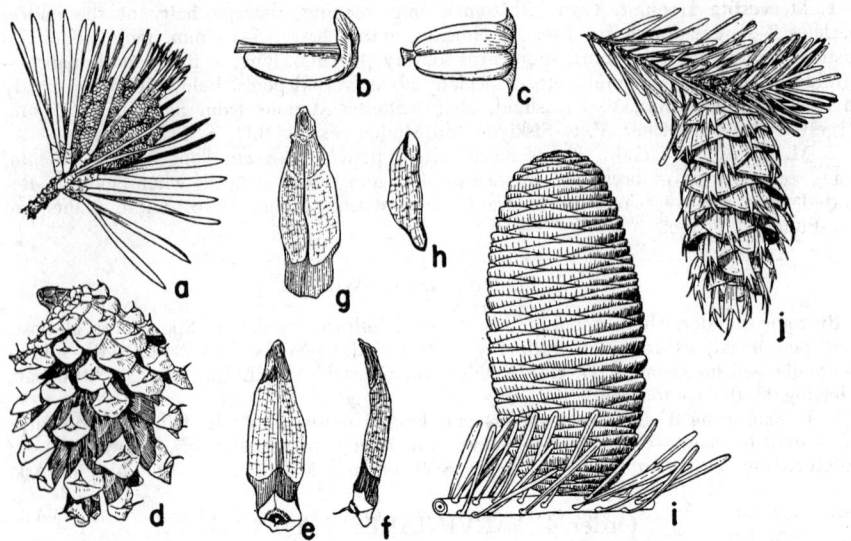

Fig. 5. PINACEAE. *Pinus Murrayana: a,* twig, × ½, with fascicles of lvs. (needles) and cluster of ♂ cones; *b* and *c,* lateral and abaxial views, respectively, of microsporophyll (stamen), × 5, with 2 microsporangia (pollen-sacs); *d,* ♀ cone, × ¾; *e* and *f,* ovuliferous scales, × 1, under (abaxial) and lateral views, respectively, with thickened exposed part (apophysis) ending in an elev. (umbo); *g,* upper (adaxial) side of ovuliferous scale with umbo out of sight; *h,* seed with long wing. *Abies concolor: i,* twig, × ½, with solitary needles and erect ♀ cone. *Pseudotsuga Menziesii: j,* twig, × ½, with solitary needles and pendent ♀ cone with prominent 3-toothed papery bracts subtending the ovuliferous scales.

B.   Fr. a dry cone (or in *Juniperus* ± berrylike) of several to many scales, each scale
         bearing 1 to several seeds; lvs. needlelike to scalelike . . . . 1. *Coniferales*   p. 48
BB. Fr. solitary, berrylike or drupelike, 1-seeded; lvs. linear, in flat sprays. .   2. *Taxales*   p. 64
AA. Stems jointed; lvs. scalelike in 2's or 3's in widely separated whorls; plants without
         resin; cones small, with thin scales. Desert shrubs. . . . . 3. *Ephedrales*   p. 65

Order 1. CONIFERALES

Ovulate cone usually dry, ± woody, fleshy in *Juniperus* but the constituent ± fused ovuliferous scales evident even there. Several families.

KEY TO FAMILIES

Lvs. needlelike or linear, not scalelike (see also a shrubby sp. of *Juniperus*).
   Cone-scales overlapping, each bearing 2 seeds . . . . 1. *Pinaceae*   p. 48
   Cone-scales not overlapping, each bearing 5–7 seeds . . . . 2. *Taxodiaceae* (*Sequoia*)   p. 57
Lvs. scalelike, mostly decurrent, thickly clothing the branches.
   Cones 5–8 cm. long, the seeds ca. 12 mm. long, wing-margined
                                                      2. *Taxodiaceae* (*Sequoiadendron*)   p. 58
   Cones 1–2.5 cm. long, seeds often smaller, winged or not . . . . 3. *Cupressaceae*   p. 58

1. Pinàceae. PINE FAMILY (Fig. 5)

Resinous, mostly evergreen trees, sometimes shrubs, with linear, needlelike, spirally arranged, solitary or fascicled lvs. (in the latter case the fascicles borne in the axils of scalelike lvs.). Stamens and ovules in different cones on the same tree. Staminate cones with many spirally arranged stamens, each having 2 pollen-sacs beneath, the cones drying and soon deciduous after anthesis. Ovulate cones with spirally arranged scales, each of which is subtended by a bract, the cones maturing in 2 or 3 years and becoming ± woody; ovules naked, 2 at base of each scale on upper side, developing into seeds usually bearing a terminal wing made of tissue from surface parts of the ovuliferous scale. Ca. 9 genera and 200 spp.

Cones erect on the branch, the scales falling separately at maturity; lvs. solitary, blunt, flat in cross-section; terminal buds blunt, enveloped in resin .... 1. *Abies*
Cones reflexed or pendulous, the scales persistent; lvs. solitary or fascicled, acute to sharp at apex; terminal buds mostly pointed, not enveloped in resin.
  Lvs. of 2 kinds: persistent needle-lvs. in fascicles of 1–5 and deciduous scale-lvs.; cones maturing after the first year, the bracts subtending the scales minute .... 2. *Pinus*
  Lvs. of 1 kind, linear; cones maturing the first year, the bracts obvious.
    Branchlets roughened by persistent lf.-bases; bracts of cone-scales not exserted.
      Lf. short-petioled, flattened, stomatiferous below .... 3. *Tsuga*
      Lf. sessile, usually ± 4-sided, if flattened not stomatiferous below .... 4. *Picea*
    Branchlets not roughened; bracts of cone-scales conspicuously exserted .... 5. *Pseudotsuga*

1. Àbies Mill. FIR

Pyramidal evergreen trees, with bark of old trees thick and furrowed. Lvs. linear to lance-linear, entire, sessile, flattened and usually grooved above, keeled beneath and with 2 pale stomatic bands, spirally arranged but often appearing 2-ranked, or more frequently curved upward; abscission-scar smooth. Cones from axillary winter buds, the ♂ borne on the under side of the rather stiff branches of upper half of tree. Ovulate cones on topmost branches, erect, with numerous 2-ovuled, imbricated scales which are longer or shorter than the subtending bracts, stipitate at base, rather thin and falling away separately on the tree. Seeds with large thin wing. Ca. 40 spp. of N. Hemis. (Classical Latin name.)

Bracts subtending cone-scales with elongate bristlelike tips much exceeding the scales; lvs. alike all over the tree; bark mostly smoothish. Santa Lucia Mts. .... 1. *A. bracteata*
Bracts subtending cone-scales shorter than scales or, if exserted, lacking long bristle-tips; lvs. of lower and uppermost branches ± different.
  Cones mostly 6–12 cm. long; bracts ca. half as long as scales; lvs. flat.
    Lvs. dark green above with stomata only on the silvery white lower surface, notched at apex; bark whitish. N. Coast Ranges, at low elevs. .... 2. *A. grandis*
    Lvs. blue-green with stomata on both sides, mostly acute at apex; bark grayish. Montane
                                                                  3. *A. concolor*
  Cones mostly 10–20 cm. long; bracts at least ¾ as long as scales; lvs. ± 4-sided.
    Bracts of cone not exserted (except in var. *shastensis*); lvs. almost equally 4-sided. Sierra Nevada and N. Coast Ranges .... 4. *A. magnifica*
    Bracts of cone much exserted and reflexed; lvs. of lower branches somewhat flattened. Siskiyou Mts.
                                                                        5. *A. procera*

1. **A. bracteàta** D. Don ex Poiteau [*Pinus b.* D. Don. *P. venusta* Dougl. *A. v.* K. Koch.] SANTA LUCIA FIR. Tree, 10–30(–50) m. tall, with a narrow spirelike crown; bark of old trunks slightly fissured, broken into appressed scales; branches declined or drooping; branchlets glabrous, glaucous when young; lvs. flat, stiff, sharp-pointed, 3–5.5 cm. long, dark green above, white-banded with stomata beneath; ♂ catkins pale yellow; ♀ cones round-ovoid, 6–10 cm. long, on stout peduncles; bracts much exserted, ending in awn-like tips 2–3 cm. long; seeds 6 mm. long, red-brown.—Rocky slopes and canyons, from ca. 2000–4500 ft.; Mixed Evergreen F., Yellow Pine F.; Santa Lucia Mts., Monterey Co. May.

2. **A. grándis** (Dougl.) Lindl. [*Pinus g.* Dougl. ex G. Don in Lamb.] GIANT FIR. Tree 12–90 m. tall, with brownish bark and long drooping branches; branchlets glabrous to minutely pubescent; lvs. 3–5 cm. long, rounded and bifid at apex, flat, thin, deeply grooved, forming flat sprays, shining dark green above with white bands beneath; cones long-oblong in outline, 5–10 cm. long, bright green, the bracts hidden, with a short awl-like point on the roundish tip; scales broadly fan-shaped, puberulent; seeds pale brown, 8 mm. long, the wings ca. twice as long.—Low hills and valleys, near the coast; Red-wood F., N. Coastal Coniferous F., Mixed Evergreen F.; Sonoma to Del Norte cos.; to B.C., Mont. April–May. Coarse, not durable wood used for boxing.

3. **A. cóncolor** (Gord. & Glend.) Lindl. [*Picea c.* Gord. & Glend. *P. Lowiana* Gord. *A. c.* var. *L.* Lemmon.] WHITE FIR. Tree 15–70 m. tall, with rather narrow spirelike crown and short stiff branches; branchlets nearly glabrous; older bark gray, furrowed; lvs. 3–6 cm. long, acute or rounded at tip, irregularly arranged, ± erect on the branches, bluish-green and stomatiferous above, with pale stomatiferous bands beneath separated by a median keel; ♂ cones usually reddish; ♀ cones oblong-cylindric, 7–12 cm. long,

greenish or purplish becoming brown, the bracts thin, scarcely half as long as scales and usually with a short slender point on the roundish end; seeds ca. 10–12 mm. long, dark brown; $n = 12$ (Sax & Sax, 1933).—Common on dry slopes and rocky places, 3000–10,000 ft.; mostly Yellow Pine F., occasional to Lodgepole F.; mts. of San Diego Co. n. through the Sierra Nevada and N. Coast Ranges, to Del Norte, Siskiyou and Modoc cos.; occasional in desert ranges like Kingston and Clark; to Ore., s. Rocky Mts., L. Calif. May–June. Wood of second-grade quality, used for packing-cases, etc.

4. **A. magnífica** A. Murr. [*A. nobilis* var. *m.* Kell.] RED FIR. Tree 20 to 60 m. high, the branches comparatively short, in horizontal sprays; old bark dark red, roughly fissured; branchlets puberulent the first season, later glabrous and silvery-gray; lvs. 2.0–3.5 cm. long, obtuse and entire at tip, ribbed on both sides and almost equally 4-sided with stomates on each surface, at first glaucous, later blue-green, those on upper branches curved upward and acute, on lower 2-ranked; ♂ cone deep purple-red; ♀ cones oblong-cylindric, 15–20 cm. long, the scales stalked, broad, with upturned edges, the bracts ⅔ as long as scales, abruptly contracted to a narrow tip; seed deep brown, 1.2–1.8 cm. long.— Dry slopes and ridges, 5000–9000 ft.; Red Fir F., etc.; Sierra Nevada from Kern Co. n., Coast Ranges from Lake Co. n.; to s. Ore. June. Wood light, soft, fairly durable, light red-brown, used largely for fuel and coarse lumber. Trunks brittle and windfalls frequent.

Var. **shasténsis** Lemmon. [*A. s.* Lemmon.] SHASTA FIR. Bracts of ♀ cones well exserted and yellowish, ± broadened at apex.—With the sp. and sometimes replacing it.

5. **A. prócera** Rehd. [*Pinus nobilis* Dougl. *A. n.* Lindl., not A. Dietr.] NOBLE FIR. Tree 30–70 m. tall, with short rigid branches; bark red-brown, deeply fissured; branchlets with fine reddish pubescence; lvs. 2.5–3.5 cm. long, entire or emarginate at apex, blue-green, stomatiferous on all 4 surfaces; cones oblong-cylindric, 10–20 cm. long, green becoming purplish-brown, the bracts exserted, reflexed so as almost to conceal the scales; seeds pale red-brown, ca. 12 mm. long.—At 3000–5000 ft.; Red Fir F.; Siskiyou Mts.; to Wash. The wood fairly hard and strong, of use for interior finish and packing cases.

2. Pìnus L. PINE

Evergreen trees, rarely shrubs, with furrowed or scaly bark. Lvs. needlelike, borne in clusters of 1–several on deciduous spurs in the axils of scalelike primary lvs. and enclosed in the bud by numerous scales lengthening and forming a sheath at the base of each cluster. Staminate catkins or cones of many overlapping stamens each with 2 pollen-sacs on the lower side, the cones early deciduous. Ovulate cones becoming woody, composed of many ovuliferous scales with 2 ovules on each upper (adaxial) surface and maturing at the end of the second or third season. Each scale having a ± thickened exposed part (the apophysis) with a terminal or dorsal brown protuberance or elevation (umbo). Seeds usually obovoid, usually winged at tip; cotyledons 3–18. Ca. 80 spp. of N. Hemis. (Classical Latin name.)

(Shaw, G. R. The genus Pinus. Pub. Arnold Arboretum No. 5: 1–96, 1914.)

A. Sheaths of the lf.-clusters deciduous in the first year; base of the bracts of the lf.-bearing spurs not decurrent; needles with 1 cent. vascular strand. (Subgenus Haploxylon.)

    B. Lvs. 5 in a cluster.

        C. The cones with terminal unarmed umbos (elevs.) on the scales.

            D. The cones remaining closed, 2–7 cm. long; cone-scales thick at tip, not closely overlapping; needles 3.5–5.5 cm. long; seeds wingless. Subalpine .. 1. *P. albicaulis*

           DD. The cones opening at maturity, mostly 10–40 cm. long; cone-scales thinner at tip, overlapping; needles 3–10 cm. long; seeds winged.

                E. Cones with short stalks or almost none, the cone-body 8–25 cm. long; needles 2.5–7 cm. long. Mts. bordering the deserts .... 2. *P. flexilis*

               EE. Cones with long well developed stalks; needles 5–10 cm. long. General montane.

                    F. Lvs. sharp-pointed, with conspicuous white lines on the back; cones 25–40 cm. long; seed-wings rounded at apex. Mostly below 7000 ft.

                                  3. *P. Lambertiana*

                  FF. Lvs. obtuse at apex, without conspicuous white lines on back; cones 10–20 cm. long; seed-wings acute. Between 6000 and 9000 ft.

                                    4. *P. monticola*

        CC. The cones with dorsal umbos armed with slender prickles.

            D. Cones 7–9 cm. long, their scales armed with long slender incurved prickles; needles 25–35 mm. long. Inyo-White and Panamint ranges .... 8. *P. aristata*

DD. Cones 3.5–7.5 cm. long, their scales armed with minute incurved prickles; needles
    mostly 2–2.5 cm. long. Sierra Nevada and N. Coast Ranges .... 9. *P. Balfouriana*
BB. Lvs. 1–4 in a cluster.
  C. The lvs. commonly in 1's or 2's and more than 1 mm. wide, ± grayish green, not very
    glaucous.
    D. Lvs. mostly single, terete. Common in desert mts. .... 5. *P. monophylla*
    DD. Lvs. mostly in 2's, semiterete, deeply channeled. Rare, in Little San Bernardino
      and New York mts. .... 6. *P. edulis*
  CC. The lvs. commonly in 4's, ± glaucous, scarcely 1 mm. wide. W. edge of Colo. Desert
                                           7. *P. quadrifolia*
AA. Sheaths of the lf-clusters persistent; base of the spur-bracts decurrent; needles with 2 vascular
  strands. (Subgenus *Diploxylon*.)
  B. Lvs. mostly 2 in a cluster; cones 3–7 cm. long.
    C. The cones lateral, 5–8 cm. long, asymmetrical, persistent and remaining closed for some
      years, with stout prickles; needles 10–15 cm. long. Coastal.
      D. Scale-tips of young ♀ cones reflexed; mature cones with tips of scales produced
        into a strongly recurved hook; cones reflexed. Mainland .... 10. *P. muricata*
      DD. Scale-tips of young ♀ cones erect; mature cones with tips of scales plane or slightly
        rounded; cones at right angles to the branch. Insular .... 11. *P. remorata*
    CC. The cones subterminal, 3–5 cm. long, almost symmetrical, opening when mature, decid-
      uous; needles 3–5 cm. long.
      D. Bark fairly thick, furrowed on older trunks; lvs. dark green. N. coastal
                                          12. *P. contorta*
      DD. Bark thin, scaly even on old trunks; lvs. yellow-green. Subalpine
                                        13. *P. Murrayana*
  BB. Lvs. more than 2 in a cluster; cones mostly larger.
    C. Needles 5 in a cluster, gray-green, 20–30 cm. long; cones 10–15 cm. long. S. coast
                                        14. *P. Torreyana*
    CC. Needles mostly 3 in a cluster.
      D. Cones subterminal, symmetrical, opening at maturity, usually deciduous above the
        basal scales persistent on the branch.
        E. Lvs. yellow-green; branchlets not glaucous; cones 7–15 cm. long
                                      15. *P. ponderosa*
        EE. Lvs. dull gray-green; branchlets glaucous; cones 15–35 cm. long
                                        16. *P. Jeffreyi*
      DD. Cones lateral.
        E. The cones asymmetrical (the outer scales much enlarged), remaining closed
        and attached to the branches for years; needles 8–17 cm. long.
          F. Outer cone-scales with rounded apex, the cones with minute prickles;
          needles deep green. Coastal .... 17. *P. radiata*
         FF. Outer cone-scales with prominent knoblike tips, the cones with stout
         prickles; needles yellowish green. Montane .... 18. *P. attenuata*
        EE. Cones ± symmetrical, opening at maturity, deciduous; needles 15–30 cm. long.
         F. The cones oblong-ovoid, the seeds longer than their wings; lvs. gray-
          green, drooping .... 19. *P. Sabiniana*
         FF. The cones oblong-conic, the seeds shorter than their wings; lvs. blue-green,
         erect .... 20. *P. Coulteri*

1. **P. albicáulis** Engelm. [*P. flexilis* var. *a.* Engelm. *Apinus a.* Rydb. *P. shasta* Carr.]
WHITEBARK PINE. From low and dwarfish, even prostrate, to sometimes 15 m. tall, with
twisted or crooked trunk; bark thin, broken by narrow fissures in thin brownish or whitish scales; branchlets stout, red-brown to almost orange, mostly puberulous when
young; lvs. in 5's, stout, 3.5–7 cm. long, rigid, dark green, densely clothing the twigs,
with 1–3 rows of dorsal stomates; ♂ catkins ca. 1 cm. long, often ± red-purple; ♀ cones
ovoid to subglobose, 3.5–7.5 cm. long, purple, remaining closed at maturity, the scales
thickened, acute, often armed with stout pointed umbos; seeds wingless, acute, 8–12 mm.
long, dark chestnut-brown.—Dry rocky places, 8000–12,000 ft.; Subalpine F.; Sierra
Nevada from Tulare Co. n., to Modoc and Siskiyou cos.; to B.C., Alta., Wyo. July–Aug.
Wood light, soft, close-grained, brittle. Seeds large, sweet.

2. **P. fléxilis** James. [*Apinus f.* Rydb.] LIMBER PINE. Tree 10–20 m. high, with a short
trunk and ultimately broad crown; bark at first smooth and grayish white, later dark
and broken into brown plates covered with thin scales; branchlets stout, orange-green,
finely pubescent at first; lvs. in 5's, stout, erect, stiff, dark green, with 1–4 rows of
stomates on each side; ♂ catkins reddish, ca. 1 cm. long; ♀ cones subcylindric, 8–25 cm.
long, light brown with ± of a purplish tinge, the scales thickened and often incurved at
apex; seeds dark red-brown, mottled with black, 10–12 mm. long, compressed, the
narrow wings usually persistent on the scale; *n* = 12 (Sax & Sax, 1933).—Dry slopes
and ridges, 7500–11,600 ft.; Lodgepole F., Subalpine F., Bristlecone Pine F.; Warner
Mts., e. slope of Sierra Nevada from Mono Pass s., mts. of s. Calif. to San Jacinto and

Santa Rosa mts., Panamint, Inyo-White ranges; to se. B.C. and Rocky Mts. to Mex. July–Aug. Wood light, soft, close-grained, only occasionally made into lumber.

3. **P. Lambertiàna** Dougl. [*Strobus L.* Mold.] SUGAR PINE. The largest of all pines, 20–75 m. tall, the crown open and narrow when young, later flat-topped, the older branches well spaced, wide-spreading; bark smooth, dark green when young, later deeply and irregularly divided into platelike ridges with loose reddish-brown scales; branchlets stout, pubescent when young; lvs. in 5's, slender, 7–10 cm. long, with several rows of stomates on both sides; ♂ catkins yellow, 8–10 mm. long; ♀ cones cylindric, 25–45 cm. long, on stalks 5–8 cm. long, the scales often 4 cm. wide, with thin tips and terminal scarlike umbos; seeds dark, ca. 9–12 mm. long, the wings twice as long; $2n = 24$ (Mehra & Khoshoo, 1948).—Common forest tree, 2500–9000 ft.; Yellow Pine F., Red Fir F.; mts. from s. Calif. to Ore., L. Calif. May–June. Wood light, soft, straight-grained, of high commercial value. Considered by many to be the most beautiful pine.

4. **P. monticola** Dougl. [*P. Strobus* var. *m.* Nutt. *P. m.* var. *minima* Lemmon. *Strobus monticola* Rydb.] WESTERN WHITE PINE. SILVER PINE. Forest tree 15–50 m. tall, with a slender crown in thick stands or more open and spreading when old and exposed; bark thin and light gray on young stems, divided into squarish brownish small plates in age; branchlets stout, brownish, at first puberulent; lvs. in 5's, mostly obtuse, 5–10 cm. long, blue-green and glaucous, with white bands of stomates on cent. side; ♂ catkins yellow, ca. 8 mm. long; ♀ cones cylindric, 10–20 cm. long, light brown, the scales thin, smooth, with scarlike umbo; seeds brown, mottled with black, ca. 8 mm. long, the wings 3 times as long.—Scattered, between 2000 and 9800 ft.; Red Fir F. to Subalpine F.; Sierra Nevada, to Del Norte, Siskiyou and Modoc cos.; to B.C., Mont. June–July. Wood light, soft, straight-grained, sometimes used in construction and interior finish.

5. **P. monophýlla** Torr. & Frém. [*P. Fremontiana* Endl. *P. edulis* var. *m.* Torr. *P. cembroides* var. *m.* Voss. *Caryopitys m.* Rydb.] ONE-LVD. PINYON. A low tree 5–15 m. tall, usually with divided trunk, ± flat-topped in age; bark with narrow flat ridges; branchlets glandular-puberulent at first; lvs. mostly in 1's, rigid, incurved, pale gray-green, 2.5–3.5 cm. long; ♂ catkins ca. 6 mm. long; ♀ cones subglobose to broadly ovoid, 3.5–5.5 cm. long, brown, the scales 4-sided, knobbed at tip; seeds ca. 15 mm. long, the wings somewhat shorter.—Dry rocky slopes and ridges, at 3500–9000 ft.; Pinyon-Juniper Wd., Foothill Wd.; e. slope of Sierra Nevada from Mono Co. s. and in mts. in and bordering Mojave Desert; head of San Joaquin V. and at scattered stations on w. base of Sierra Nevada to Tuolumne Co.; to Utah, Ariz., L. Calif. May. Wood light, soft, brittle, used largely for fuel and charcoal. Seeds important for food.

6. **P. édulis** Engelm. [*Caryopitys e.* Small. *P. cembroides* var. *e.* Voss.] Differing from *P. monophylla* chiefly in its 2- or 3-lvd. clusters and perhaps scarcely specifically distinct.—At 4200–7000 ft.; Pinyon-Juniper Wd. with *P. monophylla*; Little San Bernardino Mts., New York Mts.; to Wyo., Tex., Ariz., n. Mex. Seeds important food.

7. **P. quadrifòlia** Parl. ex Sudw. [*P. Parryana* Engelm., not Gord. *P. cembroides* var. *P.* Voss.] FOUR-LVD. PINYON. Much like the 2 preceding in habit, the branchlets pubescent when young; lvs. mostly 4, pale blue-green on back, with conspicuous rows of stomates on inner surfaces; seed 15 mm. long, the wings ca. 3 mm. long.—Dry slopes, 2500–5500 ft.; Pinyon-Juniper Wd.; w. edge of Colo. Desert from Santa Rosa and San Jacinto mts. to L. Calif. May. Seeds edible.

8. **P. aristàta** Engelm. [*P. Balfouriana* var. *a.* Engelm.] BRISTLECONE PINE. Bushy, heavy trees, 6–16 m. tall, sometimes with very thick trunks; bark thin, red-brown, with flat irregular ridges and appressed scales; branchlets light orange-color, glabrous or ± puberulous at first; lvs. in 5's, rather slender, 2.5–3.5 cm. long, densely clothing the younger parts of the long branchlets, whitish on inner surfaces; ♂ catkins 10–12 mm. long; ♀ cones almost sessile, ovoid, 7–9 cm. long, dark purplish-brown, the scales thickened and ridged, with a slender incurved prickle ca. 6 mm. long; seeds light brown, with darker mottling, ca. 8 mm. long, the wings slightly longer.—In open, often almost pure stands on dry rocky slopes at 7500–11,500 ft.; Bristlecone Pine F.; Inyo-White ranges, Panamint, Funeral, and Grapevine mts.; to Colo., New Mex. June–July.

9. **P. Balfouriàna** Grev. & Balf. in A. Murr. FOXTAIL PINE. Low tree 6–15 m. tall, rather stout, with short spreading branches; bark thin, smooth, white to brown, with broad flat ridges; branchlets long, ± puberulous at first, densely clothed toward ends with crowded lvs.; lvs. stout, stiff, in 5's, dark blue-green on back, whitish on inner sur-

faces, with many rows of stomates; ♂ catkins 10–12 mm. long; ♀ cones ovoid, sessile, dark red-brown, 3.5–7.5 cm. long, the scales thick and ridged, ± 4-sided, with minute incurved prickles; seeds pale, mottled, ca. 8 mm. long, and wings ca. 3 times as long.—Dry rocky slopes and ridges, 6000–11,500 ft.; Subalpine F.; Sierra Nevada in Tulare and Inyo cos., mts. of Trinity and Siskiyou cos. July. Wood light, soft, brittle.

10. **P. muricàta** D. Don. [*P. Edgariana* Hartw.] BISHOP PINE. Small tree 15–25 m. tall, with thick spreading branches; older bark thick, deeply furrowed and ridged, with dark purplish-brown scales; branchlets brownish, rather slender, glabrous; lvs. 2, in crowded clusters, rigid, dark yellow-green, 10–15 cm. long, persisting 2–3 years only; ♂ catkins ca. 8 mm. long; ♀ cones sessile, usually whorled, deflexed, 5–7 cm. long, asymmetrical with much enlarged outer scales prolonged into prominent knobs with stout recurved prickles; cones persisting closed for many years; seeds ca. 6 mm. long, nearly triangular, with a dark rough shell, the wings 3–4 times as long.—Low hills and flats; Closed-cone Pine F.; near the coast from Humboldt to Santa Barbara cos., Santa Cruz Id. March–April. The wood is light, strong, hard, coarse-grained; occasionally used for lumber.

11. **P. remoràta** Mason. SANTA CRUZ ID. PINE. Much like *P. muricata* and doubtfully distinct, the lvs. dark green; cones ovoid, more symmetrical, scarcely or not deflexed, 5–8 cm. long, each scale plane or slightly rounded, armed with a minute often deciduous prickle.—Associated with *P. muricata* on Santa Cruz Id. and sparingly so near La Purisima Mission, Santa Barbara Co. It has been stated by H. L. Mason that this sp. has been distinct since early Pleistocene and is now a remnant of a past flora.

12. **P. contòrta** Dougl. ex Loud. [*P. Boursieri* Carr. *P. tenuis* Lemmon.] BEACH PINE. Low tree, 3–10(–16) m. tall, with short trunk and broad compact or open crown; older bark deeply and irregularly fissured into small oblong plates with dark red-brown scales; young branchlets stout, light orange at first, later dark brown to almost black; lvs. in 2's, dark green, 3–5 cm. long, densely clothing the branches and lasting mostly 2–3 years; ♂ catkins ca. 8 mm. long, orange-red; ♀ cones ovoid to subcylindric, somewhat oblique at base, horizontal or declined, often clustered, 2–5 cm. long, with thin slightly concave scales, armed with slender, ± recurved, often deciduous prickles, chestnut-brown, mostly opening at maturity, mostly long-persistent; seeds ca. 4 mm. long, dark red-brown, the wings 2–3 times as long; $2n = 24$ (Langlet, 1934).—Coastal Strand, Closed-cone Pine F.; coast from Mendocino Co. n.; to Alaska. May. The wood is light, hard, strong, brittle, coarse-grained, occasionally used as fuel. A canelike dwarf, 6–15 dm. high, with very small cones, and growing in pine barrens in Mendocino Co. has been called var. **Bolánderi** (Parl.) Vasey. [*P. B.* Parl.]

13. **P. Murrayàna** Grev. & Balf. in A. Murr. [*P. contorta* var. M. Engelm. *P. c.* var. *latifolia* Engelm.] LODGEPOLE PINE. TAMARACK PINE. A tree 15–40 m. tall, slender and straight when in close stands, heavier and more open and branched in older and spaced trees; bark very thin, close and firm, light brown, covered with small thin scales; branchlets slender, light orange when young; lvs. in 2's, yellow-green, 3–6 cm. long; cones much as in *P. contorta*, less oblique and tending to be less persistent.—Often forming dense stands, especially after fire, on dry slopes and even invading moist meadows, at 5000–11,000 ft.; Lodgepole F., Subalpine F.; from San Jacinto Mts. n., through the Sierra Nevada to Siskiyou Mts., Mt. Shasta, etc.; to Alaska, Rocky Mts., L. Calif. July. Wood light, pitchy, soft, close, straight-grained, used (largely outside of Calif.) for railroad ties, mine-timbers, etc. Variable in length and width of the needles.

14. **P. Torreyàna** Parry ex Carr. [*P. lophosperma* Lindl.] TORREY PINE. Tree usually ca. 10–15 m. high in nature with broad open crown, but in cult. much taller and more symmetrical; bark 2–3 cm. thick, red-brown, with irregular broad ridges; branchlets stout, at first green, then purplish; lvs. forming large terminal tufts, borne in 5's, dull green, stout, 20–30 cm. long, with many rows of stomates on all 3 surfaces; ♂ catkins yellowish, 2–3.5 cm. long, in dense masses; ♀ cones on elongate stalks, broadly ovoid, 10–15 cm. long, chestnut-brown, the scales thickened apically into a low pyramidal knob with a minute prickle; seeds oval, ± angled, 2–2.5 cm. long, hard-shelled, brown, ± mottled, nearly surrounded by the cm.-long wings.—Restricted endemics, on dry slopes below 500 ft.; near Del Mar on the San Diego Co. coast and Santa Rosa Id. Jan.–March. Wood light, soft, coarse-grained; now cult. in New Zealand, Kenya, etc.

15. **P. ponderòsa** Dougl. ex P. & C. Lawson. [*P. Benthamiana* Hartw.] YELLOW PINE.

PONDEROSA PINE. Long-lived tree 15–70 m. high, the branches short, often pendulous, generally turned up at ends; bark on old trees often 6–10 cm. thick and separated into broad yellow-brown to reddish flat plates with small soft concave scales whose smooth inner surface is sulphur-yellow and with small dark resin-pits throughout; odor in the furrows resinous; branchlets of season shining green, or of previous year brownish, not at all glaucous; lvs. glossy, yellow-green, in 3's, lasting ca. 3–5 years, 12–25 cm. long, ± scabrous on margins, the rows of stomates scarcely distinguishable; ♂ catkins 2–3 cm. long, in short dense clusters; ♀ cones subterminal, ± horizontal, short-stalked, oval, mostly 7–15 cm. long, the spreading scales appearing ± slender and widely spaced, the prickles short, protruding outward from the umbos, prickly when clasped; seeds ovoid, acute, 6–7 mm. long, dark purple, ± mottled, the wings 4–5 times as long, oblique at apex; *n* = 12 (Sax & Sax, 1933).—Forming large parklike forests at 2000–8500 ft.; largely Yellow Pine F.; mts. from San Diego Co. to N. Coast Ranges, Sierra Nevada and Modoc Co.; to L. Calif., B.C., etc. and in somewhat different form to Rocky Mts. May–June. The wood is hard, strong, comparatively fine-grained and much used for many kinds of construction.

16. **P. Jéffreyi** Grev. & Balf. in A. Murr. [*P. ponderosa* var. *J.* Balf. ex Vasey. *P. deflexa* Torr.] JEFFREY PINE. Tree 20–60 m. tall, with a longer more symmetrical crown and the bark tending to be darker than in *P. ponderosa*, in narrower or sometimes as broad plates, the inner surface of the scales creamy pinkish- or chocolate-brown, the scales harder, resin-pits lacking; odor in the deep furrows rather strong, pleasant, sweet, ± vanilla-like; branchlets of season very glaucous or pruinose; lvs. in 3's, ± blue-green, the needles 12–28 cm. long, persisting 5–8 years, scabrous on edges, with plainly marked rows of stomates; ♂ cones 20–35 mm. long; ♀ cones subterminal, short-stalked, long-oval, 15–25 cm. long, the scales numerous, stout, closely compacted, almost vertical to the cone-axis, with long prickles mostly deflexed, the points seldom protruding outward and scarcely perceptible when clasped in the hand; seeds 10–13 mm. long, mottled, the wings to 3 cm. long; *n* = 12 (Sax & Sax, 1933).—Dry slopes mostly 6000–9000(–10,600) ft.; Yellow Pine F., Red Fir F.; Sierra Nevada, especially on the e. slope, to n. L. Calif., and to Yolla Bolly and Siskiyou mts., etc.; s. Ore. May–June. Often forming almost unmixed forests, at other times hybridizing freely with *P. ponderosa*, less frequently with *P. Coulteri*.

17. **P. radiàta** D. Don. [*P. insignis* Dougl. ex Loud. *P. californica* H. & A.] MONTEREY PINE. Symmetrical tree or flat-topped in age, 15–25(–35) m. high; bark dark-brown, with narrow ridges; branchlets ± orange, at first glaucous, later dark red-brown; lvs. in 3's, deep glossy green, 8–15 cm. long, slender, persisting ca. 3 years; ♂ catkins yellow, ca. 12 mm. long; ♀ cones asymmetrical-ovoid, short-stalked, reflexed, 7–15 cm. long, brown, shining, bluntly pointed, lower scales on outer side of cone thickened and mammillate-knobbed, armed with slender prickles that mostly wear off, the cones remaining closed and persisting for years; seeds dark, rough, 6–7 mm. long, the wings 12–18 mm. long, longitudinally brown-striped; 2*n* = 24 (Mehra & Khoshoo, 1948).— Endemic on dry bluffs and slopes below 1000 ft.; Closed-cone Pine F.; Año Nuevo Point in Santa Cruz Co., Monterey Peninsula, and Cambria in San Luis Obispo Co. Cones from the different groves ± dissimilar. April. Wood light, soft, close-grained, not strong. An important cult. timber tree in New Zealand, Australia, Afr., etc.

18. **P. attenuàta** Lemmon. [*P. tuberculata* Gord. *P. californica* Hartw., not H. & A.] KNOBCONE PINE. Tree 2–12(–15) m. high, forming in age a straggling crown; old bark thin, with low ridges and loose scales; branchlets pale brown; lvs. in 3's, slender, stiff, pale yellow-green, 8–17 cm. long, with evident rows of stomates; ♂ cones ca. 12 mm. long; ♀ cones short-stalked, often whorled, remaining closed and persisting for years, light brown, narrow-ovoid, asymmetrical, the scales on the outer side enlarged into prominent knobs with thick flattened incurved prickles; seeds dark, ellipsoid, compressed, 5–7 mm. long, the narrow wings 25–35 mm. long.—Dry barren or rocky places, below 4000 ft.; Douglas-Fir F., Closed-cone Pine F., Foothill Wd., Chaparral; widely scattered from Santa Ana Mts., Orange Co., San Bernardino Mts., Santa Lucia Mts., Santa Cruz Mts., Moraga Ridge, Mt. St. Helena, etc., to Del Norte and Siskiyou cos., s. in Sierra Nevada to Mariposa Co.; s. Ore. March–May. Wood light, soft, not strong, coarse-grained. P. × **attenuradiata** Stockw. & Right., name given to the hybrid between *P. attenuata* and *P. radiata* occurring near Swanton, Santa Cruz Co.

19. **P. Sabiniàna** Dougl. Digger Pine. Tree 12–18(–30) m. high, the trunk dividing and supporting an open sparsely leafy crown; bark dark brown, often tinged purple, with low broad ridges and large brown scales; branchlets stout, at first pale and glaucous, later darker; lvs. in 3's, stout, gray-green, 20–30 cm. long, with many bands of stomates and lasting 3–4 years; ♂ catkins 3–4 cm. long; ♀ cones long-stalked, deflexed, broadly ovoid to subglobose, mostly 15–25 cm. long, light- to chocolate-brown, remaining on trees for some years after shedding seeds, the scales ending in stout flattened downwardly projecting hooks; seeds blackish-brown, 20–24 mm. long, with very short wings.—Dry slopes and ridges, below 4500 ft.; Foothill Wd.; hills bordering Cent. V., inner Coast Ranges, nearing the coast in Santa Lucia Mts., s. to n. Los Angeles Co. April–May. Wood light, soft, not strong, close-grained, brittle, of little use. Seeds sweet and edible.

20. **P. Coùlteri** D. Don. [*P. macrocarpa* Lindl. *P. C.* var. *diabloensis* Lemmon.] Coulter Pine. Tree 12–25 m. high with broad pyramidal or asymmetrical crown; bark blackish-brown, deeply broad-ridged with thin scales; branchlets stout, rough, at first dark orange-brown, later almost black; lvs. in 3's, stiff, stout, deep blue-green, 15–30 cm. long, with many stomatal bands, persisting for 3–4 years; ♂ catkins ca. 2.5 cm. long; ♀ cones oblong-ovoid, 20–30 cm. long, buff, hanging on the trees for several years after releasing the seeds; scales with prominent stout incurved flattened spurs; seeds ellipsoid, dark chestnut-brown, 12–18 mm. long, the wings 25–30 mm. long.— Dry rocky slopes, 1000–7000 ft.; Yellow Pine F., Foothill Wd.; inner Coast Ranges from Contra Costa Co. s. to mts. of s. Calif., n. L. Calif. May–June. Wood light, soft, not strong, brittle, coarse-grained, used for second-class lumber. Seeds were used by Indians.

3. Pìcea Dietr. Spruce

Pyramidal trees with tapering trunks, thin scaly bark, and slender whorled branches. Lvs. linear, spirally arranged, projecting from the branchlet in all directions or sometimes appearing 2-ranked by twisting, persistent for several years, mostly 4-sided and bearing stomates on all sides, sometimes flattened and stomatiferous on upper side. Staminate cones of spirally arranged stamens, enlarged and scalelike at apex, each with 2 pollen-sacs. Ovulate cones pendent, maturing the first autumn, oblong-cylindric to ovoid, mostly scattered over upper half of tree, the scales thin, exceeding the bracts and persistent on the cone-axis. Seeds 2 on each scale, with long wings and 4–15 cotyledons. Ca. 40 spp. of cooler parts of N. Temp. (Ancient Latin name.)

Lvs. 4-sided, with stomates on all sides. Siskiyou and Shasta cos. . . . . 1. *P. Engelmannii*
Lvs. flattened, mostly with stomates on upper side.
  Branchlets pubescent; lvs. obtuse; cone-scales entire. Mts., Siskiyou Co. . . . . 2. *P. Breweriana*
  Branchlets glabrous; lvs. acute or acuminate; cone-scales toothed above the middle. Coast from Mendocino Co. n. . . . . 3. *P. sitchensis*

1. **P. Engelmánnii** Parry ex Engelm. Engelmann Spruce. Tree 15–40 m. tall, with spreading branches in regular whorls and forming a narrow compact head; bark broken into large loose brownish-gray scales; branchlets slender, pubescent when young; lvs. ill-scented, soft, flexible, 25–30 mm. long, acute, glaucous when young, later dark blue-green; cones sessile or nearly so, oblong-cylindric, ca. 5–6 cm. long, light chestnut-brown, deciduous after shedding seeds, the scales thin, flexible, erose-dentate to subentire; seeds very dark, ca. 3 mm. long.—Moist slopes and canyons, 4500–4700 ft.; Red Fir F.; Clarke Creek, Shasta Co., Sugar Creek (tributary to Scott R.), Siskiyou Co.; to B.C., Alta. to New Mex. N. of our area an important forest tree with light soft wood used for construction, fuel, etc.

2. **P. Breweriàna** Wats. Weeping Spruce. Tree to 30 m. tall, the lower branches with slender pendulous branchlets that are finely pubescent for 2 years; bark red-brown, broken into long thin appressed scales; lvs. spreading from all sides of branchlets, 2–2.5 cm. long, obtuse, dark green and rounded on lower surface, flat and paler on upper side with longitudinal bands of stomata; ♂ catkins dark purple; ♀ cones stalked, oblong, 7–10 cm. long, the scales thin, flat, rounded and entire at tips; bracts oblong, acute, ca. ¼ as long as scales; seeds dark brown, 3 mm. long.—Cold hollows and n.

slopes, 4600–7500 ft.; Red Fir F.; mts. of Del Norte, Trinity, and Siskiyou cos.; sw. Ore. Wood soft, heavy, close-grained.

3. **P. sitchénsis** (Bong.) Carr. [*Pinus s.* Bong. *P. Menziesii* Dougl.] SITKA SPRUCE. A tree to 35 m. tall, with wide spreading rigid branches and drooping branchlets; bark dark red-brown with thin loose scales; lvs. 1.5–2.5 cm. long, spreading from all sides of the branchlets, pointed, rounded on the green lower surface, flat and white above with 2 longitudinal bands of stomates; ♂ cones dark red; ♀ cones long-oblong, dull brown, 5–10 cm. long, the scales oblong, rounded and denticulate above; bracts half as long as scales, lanceolate, toothed; seeds red-brown, ca. 3 mm. long; $2n = 24$ (Thomas, 1954).—Moist, even swampy places below 1200 ft.; Coastal Strand, N. Coastal Coniferous F., Redwood F., etc.; near the coast from Mendocino Co. n.; to Alaska. April–May. Wood light, soft, not strong, made into lumber for interior finish, fencing, etc.

4. Tsùga Carr. HEMLOCK

Pyramidal evergreen trees with deeply furrowed astringent bark, nodding leading shoots, slender scattered horizontal or pendulous branches with slender ultimate branchlets forming graceful pendent masses of foliage. Lvs. flat or angular, spirally disposed, but appearing ± 2-ranked by twisting, stomate-bearing on both surfaces. Staminate catkins in axils of lvs. of previous year, globose. Ovulate cones terminal, oblong-cylindric to oblong-ovoid, usually pendulous, subsessile, green or purplish, to brown, with roundish concave thin persistent scales and minute bracts. Seeds light brown, ovoid-oblong, compressed, almost surrounded by the longer obovate-oblong wings; cotyledons 3–6. Ca. 10 spp. of temp. N. Am. and Asia. (The Japanese name.)

Lvs. spirally arranged around the branchlets, rounded above, with whitish stomate-bearing lines on both sides; cones 3–7.5 cm. long. High montane .... 1. *T. Mertensiana*
Lvs. 2-ranked, flattish, with whitish bands below; cones 2–2.5 cm. long. Outer Coast Ranges from Sonoma Co. n. .... 2. *T. heterophylla*

1. **T. Mertensiàna** (Bong.) Carr. [*Pinus M.* Bong. *Hesperopeuce M.* Rydb. *T. Roezlii* Carr. *T. crassifolia* Flous.] MOUNTAIN HEMLOCK. Forest tree 20–45 m. tall with a slender pyramidal crown, the old bark deeply divided into rounded ridges with red-brown scales; branchlets slender, light reddish-brown, densely pubescent for 2–3 years; lvs. ± curved, 1.5–2 cm. long; ♂ catkins bluish; ♀ cones sessile, oblong-cylindric, narrowed at both ends, 2.5–7.5 cm. long, the scales thin, ca. as broad as long, with cuneate base; bracts ¼ as long, sharp-pointed at tip, cuneate at base; seeds 3 mm. long, the wings ca. 4 times as long.—Wooded slopes, 6000–11,000 ft.; largely Subalpine F.; Sierra Nevada from Fresno Co. n., to Trinity and Siskiyou cos.; to Alaska, Mont. Wood light, soft, not strong, close-grained, occasionally used for lumber.

2. **T. heterophýlla** (Raf.) Sarg. [*Abies h.* Raf. *A. Bridgesii* Kell.] WESTERN HEMLOCK. Forest tree to 60 m. high, with narrow pyramidal crown; old bark brown with closely appressed scales; branchlets pale yellow-brown, pubescent; lvs. 0.6–2 cm. long, distinctly grooved; ♂ catkins yellow; ♀ cones oblong-oval, 2–2.5 cm. long, the scales puberulent, longer than broad, bracts ¼ as long as scales, abruptly rounded at tip, abruptly contracted at base; seeds 3 mm. long, the wings 2–3 times as long.—Wooded slopes below 2000 ft.; Mixed Evergreen F., Closed-cone Pine F., N. Coastal Coniferous F.; Coast Ranges from Sonoma Co. n.; to Alaska, Ida., Mont. Bark of considerable use for tanning; wood more durable than of other Am. hemlocks and used for construction.

5. Pseudotsùga Carr.

Pyramidal evergreen trees with long whorled branches, drooping branchlets and thick deeply furrowed bark. Winter-buds pointed, not enclosed in resin. Lvs. flat, petiolate, linear, appearing 2-ranked because of twisting, grooved and dark green above, paler beneath with several rows of stomates, persisting for 6–8 years. Staminate catkins axillary on the under side of the branchlets. Ovulate cones terminal or nearly so, of many concave spreading rounded scales, each subtended by a longer prominent, mostly 3-toothed bract, the cones maturing the first autumn. Seeds oblong, triangular, shorter than the dark membranous wings; cotyledons 6–12. Ca. 5 spp. of w. N. Am. and e. Asia. (Greek, *pseudos*, false, and *tsuga*, hemlock.)

Cones 5–8 cm. long, the bracts conspicuously exserted; lvs. usually rounded at apex. From Fresno and Monterey cos. n. . . . . 1. *P. Menziesii*
Cones 10–16 cm. long, the bracts slightly exserted; lvs. acuminate. From Santa Barbara and Kern cos. s. . . . . 2. *P. macrocarpa*

1. **P. Menzièsii** (Mirb.) Franco. [*Abies M.* Mirb. *Pinus taxifolia* Lamb., not Salisb. *Abies t.* Poir., not Desf. *Pseudotsuga t.* Britton. *Pinus Douglasii* Sab. *Pseudotsuga D.* Carr. *Abies mucronata* Raf. *P. m.* Sudw.] DOUGLAS-FIR. Forest tree to 70 m. tall, with a narrow or broad pyramidal crown, the slender crowded branches with long pendulous lateral branches; bark becoming deeply fissured, with rather dark broad ridges; branchlets slender, pubescent for 3–4 years, brownish; lvs. 2–3 cm. long, obtuse, dark yellow-green, sometimes more bluish, persisting for ca. 8 years; ♂ catkins yellow, with red tinge; ♀ cones pendent, with rounded slightly concave flexible puberulent scales and longer bracts ca. 6 mm. wide; seeds ca. 6 mm. long, wings slightly more; $n = 13$ (Sax & Sax, 1933).—Moist slopes largely below 5000 ft.; Redwood F., Douglas-Fir F., Yellow Pine F., N. Coastal Coniferous F., Mixed Evergreen F.; Coast Ranges from Monterey Co. n., Sierra Nevada from Fresno Co. n.; to B.C., and w. Nev. Represented in the Rocky Mts. by var. **glàuca** (Beissn.) Franco. The most important lumber tree of N. Am., known in the trade as "Oregon Pine."

2. **P. macrocárpa** (Vasey) Mayr. [*Abies m.* Vasey. *P. Douglasii* var. *m.* Engelm. *P. californica* Flous.] BIG-CONE SPRUCE. Tree 12–20 m. tall, the lower branches elongated and with slender often pendulous branchlets; bark divided into broad rounded ridges; lvs. blue-green, 2–3 cm. long, pointed; ♂ catkins pale yellow; ♀ cones 10–15 cm. long, short-stalked, the scales thick and rigid, the bracts ca. 8 mm. wide; seeds 12 mm. long, equal to the wings.—Dry slopes and canyons, 2000–6000 ft.; Chaparral, Yellow Pine F.; mts. from Santa Barbara Co. and Tehachapi and San Emigdio ranges to San Diego Co.

2. Taxodiàceae. TAXODIUM FAMILY

Evergreen or deciduous trees, monoecious. Lvs. spirally arranged, needle- or scalelike, decurrent. Staminate catkins terminal or axillary, of spirally arranged stamens with 2–9 pollen-sacs on the underside. Ovulate cones terminal, woody, of numerous thickened wide-spreading scales spirally arranged, bearing 2–9 ovules, and without distinct bracts. Seeds with small winglike borders. Ca. 10 genera and 16 spp., mostly in N. Hemis. Sometimes included in Pinaceae.

Lvs. of 2 kinds, the most linear and spreading in 2 ranks; buds scaly; cones maturing in 1 season. Near the coast . . . . 1. *Sequoia*
Lvs. all scalelike, appressed; buds naked; cones maturing in second season. Sierra Nevada.
2. *Sequoiadendron*

1. Sequòia Endl. REDWOOD

Tall evergreen trees, with columnar trunks and horizontal or drooping branches. Vegetative reproduction at base of old trunks and stumps. Buds scaly. Lvs. dimorphic, those on the vigorous terminal shoots small, scalelike, those of other branches ± falcate, lanceolate, petioled, ± revolute, obscurely keeled above and with 2 narrow bands of stomates, glaucous and stomatiferous beneath. Staminate cones stipitate. Ovulate cones maturing the first season, becoming brown and shedding the seeds at maturity, the axis slender; scales 15–20, obliquely shield-shaped, easily broken off, terminated by a flattened spine. Ovules 3–7, erect, in a single arched row near the margin of the scales. Seeds mostly 2–5 on a scale, each with 2 spongy wings not as broad as the body of the seed. One sp. (*Sequoyah*, A Cherokee half-breed, 1770–1843).

1. **S. sempervìrens** (D. Don) Endl. [*Taxodium s.* D. Don in Lamb. *Steinhauera s.* Voss. *S. religiosa* Presl.] From 50–80(–100) m. high, the trunk strongly buttressed at the base, slightly tapering upward and reaching a diam. of over 6 m.; bark 1.5–3 dm. thick, red, spongy-fibrous, broadly ridged; branchlets slender, spreading, forming flat sprays with lvs. 1.2–2.5 cm. long, sharp-pointed, dark green and shining above; ♂ cones ovoid, green; ♀ cones broadly oblong, 2.0–3.5 cm. long, red-brown; seeds lance-oblong, light brown, 5–6 mm. long, narrowly winged on the sides; $n = $ ca. 33 (Stebbins, 1948). —Flats and slopes mostly below 2000 ft. and in the coastal fog belt; Redwood F.; from Santa Lucia Mts. to sw. Ore. March. Wood reddish, light, soft, straight-grained, easily

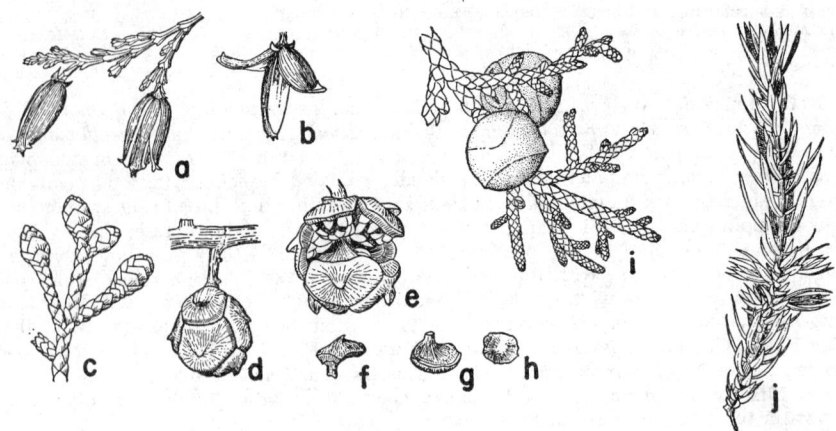

Fig. 6. CUPRESSACEAE. *Libocedrus decurrens: a,* twig, × ½, with scalelike lvs. and ♀ cones; *b,* cone, × ½, with ovuliferous scales pulling apart. *Cupressus Forbesii: c,* twig, × 1, with imbricate scalelike lvs. and ♂ cones; *d,* ♀ cone, × ½, with peltate ovuliferous scales, each bearing cent. umbo; *e,* cone with scales pulling apart to reveal seeds; *f* and *g,* ovuliferous scales, × ½; *h,* seed, × 1. *Juniperus californica: i,* twig, × 1, with scalelike lvs. and ♀ cones formed by union of fleshy scales. *J. communis* var. *saxatilis: j,* twig, × 1, with needlelike lvs.

split, decay-resisting and very important for lumber. Among the world's largest trees, being the tallest known, reaching a height of 340 ft., and often attaining an age of 1500 years.

2. Sequoiadéndron Buchh. Giant-Sequoia. Big Tree

Massive tree, evergreen, tall, with thick bark divided into broad rounded ridges separating into loose light cinnamon-red flat fibers. Lvs. all small, scalelike, sessile. No vegetative reproduction from base. Buds naked. Staminate cones sessile. Ovulate cones remaining green and attached to the tree for many years, the axis stout and woody, with 25–40 wedge-shaped scales that are not easily broken off, terminating in a long terete spine ± persistent, and with 3–12 or more ovules in a double crescentic row. Seeds 3–9 maturing the second season, with 2 thin lateral wings broader than the seed-body. One sp. (*Sequoia* and *dendron,* tree.)

1. **S. gigantèum** (Lindl.) Buchh. [*Wellingtonia* g. Lindl. *Americus* g. Anon. *Sequoia* g. Dcne., not Endl. S. *Wellingtonia* Seem. *Taxodium* g. Kell. & Behr. *Sequoia Washingtoniana* Sudw.] Trunks strongly buttressed, 50–80(–100) m. tall and 5–12 m. in diam., often unbranched for 50 m., then with an open crown in age of a few giant upturned branches, or with a more compact conical crown when younger; bark 3–6 dm. thick; lvs. decurrent, thickly set, appressed or with spreading tips, 3–6 mm. long, or to 12 mm. on leading shoots, blue-green; ♂ cones ca. 6 mm. long; ♀ oblong-ovoid, 5–8 cm. long, red-brown, the scales abruptly dilated into grooved disks; seeds light brown, ca. 6 mm. long, the wings broader than the body; $n = 11$ (Jensen & Levan, 1941).— At 4600–8400 ft. in isolated groves of few trees in the n. to very many in the s.; Yellow Pine F., Red Fir F.; w. slope of Sierra Nevada from Placer to Tulare cos. Wood light, soft, not strong, brittle, used for construction, shingles, etc. The most massive and perhaps the oldest living trees, possibly reaching an age of 3500 years.

3. Cupressàceae. Cypress Family (Fig. 6)

Monoecious or dioecious evergreen trees or shrubs, with opposite or whorled scalelike or sometimes linear lvs. that are decurrent and thickly clothing the branchlets or sometimes jointed at base. Staminate cones small, terminal on the branchlets or axillary; stamens with mostly 3–6 pollen-sacs attached to the lower half of the thin shieldlike expanded portion. Ovuliferous scales paired or whorled with 1–many erect ovules near the base. Cones woody or fleshy, the scales shield-shaped or imbricate; seeds often angled or winged; cotyledons 2–several. Ca. 20 genera in both hemis.

Fr. a woody cone; lvs. scalelike, decussate.
  Cones and their scales oblong, imbricated or valvate, maturing the first year; seeds 2 to each scale.
    Cone-scales 6, only the middle pair fertile; lvs. appearing in whorls of 4; seeds unequally 2-winged .... 1. *Libocedrus*
    Cone-scales 8–12, the 2–3 middle pairs fertile; lvs. obviously paired; seeds equally 2-winged
                                                                   2. *Thuja*
  Cones subglobose, their scales shield- or wedge-shaped, maturing in 1 or 2 years; seeds few to many on each scale.
    Cones maturing the second year; seeds many to each scale; branchlets mostly not flattened
                                                                    3. *Cupressus*
    Cones maturing the first year; seeds few to each scale; branchlets forming flat sprays
                                                            4. *Chamaecyparis*
Fr. berrylike, formed by the fusion of the scales; ovules 1–2; lvs. in 2's or 3's .... 5. *Juniperus*

1. Libocèdrus Endl.

Evergreen trees with distichous strongly compressed branchlets forming flat sprays. Lvs. scalelike, 4-ranked, closely and distinctly imbricate, decussate, decurrent, the pairs of ca. equal length, the dorsi-ventral narrow, the lateral keeled, overlapping the dorsi-ventral laterally. Monoecious, the ♂ cones oblong, of 6–16 decussate scales each with 4 pollen-sacs on the under side. Ovulate cones oblong, ± truncate, maturing the first year, the scales in 3 pairs, woody, imbricate, mucronulate on the back near the tip, the lower pair small, sterile, the middle pair 3–6 times longer, fertile, erect, oblong, the upper pair linear, sterile, connate into a flat woody plate. Seeds 2 on each fertile scale, compressed, with 2 unequal lateral oblong wings, the larger almost as long as the scales, the smaller ca. half as long; cotyledons 2. Ca. 12 spp., widely scattered. (Greek, *libas*, resin, and *kedros*, cedar.)

    1. **L. decúrrens** Torr. [*Thuja d.* Voss. *Heyderia d.* K. Koch.] INCENSE-CEDAR. Forest tree 25–35(–50) m. high, evergreen, aromatic, with a straight conical trunk from a broad base, the lower branches curved downward, the upper erect; crown conical; bark 1–2.5 cm. thick, cinnamon-brown, fibrous; branchlets flattened, often vertically placed; lvs. light green, 3–10 mm. long; ♂ catkins yellow, 5–6 mm. long; ♀ cones pendulous, oblong, 2–2.5 cm. long; seeds 8–10 mm. long.—Mt. slopes and canyons, 2400–8200 ft.; Mixed Evergreen F., Yellow Pine F.; mts. from n. L. Calif. to Ore., w. Nev. April–May. Wood light, soft, close-grained, durable in the ground, usually with cavities due to dry rot; used for shingles, posts, lead pencils, railroad ties.

2. Thùja L. ARBORVITAE

Aromatic evergreen trees with pyramidal heads and slender erect or spreading branches. Branchlets pendulous, flattened, forming frondlike horizontal sprays. Buds naked. Bark thin, scaly. Lvs. scalelike, decussate, acute, stomatiferous on back, in 4 longitudinal series, the lateral pair compressed and prominently keeled, nearly concealing the other pair. Monoecious, the cones terminal and solitary; ♂ ovoid, the scales 4–6, peltate, bearing 2–4 subglobose pollen-sacs; ♀ maturing the first autumn, ovoid-oblong, the scales thin, flexible, oblong, acute, in 4–6 pairs, mucronate at apex, the 2–3 middle pairs fertile. Seeds mostly 2, ovoid, compressed, acute, light brown, with broad lateral wings distinct at apex; cotyledons 2. Six spp., in N. Am. and e. Asia. (Ancient Greek name for a resinous tree.)

    1. **T. plicàta** Donn ex D. Don in Lamb. [*T. gigantea* Nutt. *T. Menziesii* Carr.] GIANT-CEDAR. CANOE-CEDAR. Forest tree to 70 m. tall, buttressed at base, tapering upward into a simple or sometimes divided apex, the base to 5 m. in diam.; branchlets bright green, compressed, yellow-green at first, later brownish, usually falling after second year; old bark 1–2 cm. thick, cinnamon-brown, ridged and fibrous; lvs. ovate, bright green, shining, those of leading shoots often glandular, long-pointed, to 6 mm. long, of other shoots ca. 3 mm. long, obscurely gland-pitted; ♂ cones brownish, ca. 2 mm. long; ♀ cones reflexed, 10–12 mm. long, clustered, the scales leathery, mucronate; seeds ca. 6 mm. long, the wings divergent at apex and slightly longer than seed-body;

$n = 11$ (Sax & Sax, 1932).—Moist places below 2000 ft.; N. Coastal Coniferous F.; outer Coast Ranges from Mendocino Co. n.; to Alaska, w. Mont. Wood light, soft, not strong, coarse, easily split, used for interior finish, important for shingles.

3. Cupréssus L. CYPRESS

Evergreen trees, sometimes shrubby, ± resinous, with stout erect or horizontal branches. Buds naked. Branchlets slender, 4-angled. Old bark fibrous or exfoliating in plates, ± brown. Lvs. opposite, small, of 3 sorts: linear juvenile, prickly, ± divaricate; normal appressed, acute but blunt, seldom over 2 mm. long, often ridged or keeled and bearing a gland on back; and those on vigorous shoots elongate, to 20 mm., with spreading acute tip. Staminate and ♀ cones on separate branchlets of same tree, the former mostly 3–5 mm. long, 4-sided to subcylindrical, with 10–20 triangular opposite scales, each with 3–10 pollen-sacs and ± ciliate upper margins. Ovulate cones solitary at tips of short branchlets, numerous and appearing clustered, maturing at end of second season, dull brown or gray, globose to oblong, with much thickened peltate scales with cent. boss or umbo, separating at maturity or remaining closed for years. Seeds many on all but upper and lower pairs of scales, often over 100 per cone, irregular in shape, with narrow thin wing; cotyledons 3–5. Perhaps 25–30 spp. of w. N. Am., s. Eu., w. Asia. Widely cult. (Classical name of the cypress.)

(Wolf, C. B. Taxonomic and distributional studies of the New World Cypresses. El Aliso 1: 1–250, 1948.)

A. Dorsal surface of mature lvs. usually with a gland or pit; (see also *C. Sargentii*); foliage gray or glaucous gray-green.
    B. Branchlets forming flat sprays, i.e., arranged in one plane; bark of trunk rough and fibrous. Amador and Sonoma cos. to Butte and Shasta cos. . . . . 1. *C. Macnabiana*
    BB. Branchlets not forming flat sprays; bark fibrous or exfoliating in plates.
        C. Branchlets slender, less than 1.3 mm. in diam.; crown open; ♂ cones mostly 2–3 mm. long, usually with 8 scales; ♀ cones mostly 10–20 mm. in diam., seeds mostly 3 mm. long. Siskiyou Co. . . . . 2. *C. Bakeri*
        CC. Branchlets usually more than 1.3 mm. in diam.; crown compact; ♂ cones mostly more than 3 mm. long, usually of 10–20 scales; ♀ cones mostly 25–30 mm. in diam.
            D. Bark of older trunks gray-brown, fibrous, not exfoliating (of younger branches smooth, cherry-red or brown); mature cones mostly longer than thick, the umbos not very prominent; seeds ca. 4–5 mm. long. Kern Co. . . . . 3. *C. nevadensis*
            DD. Bark of older trunks thin, cherry-red, smooth, exfoliating; mature cones globose, the umbos 3–4 mm. high; seeds 5–8 mm. long. San Diego Co.
                                                            4. *C. Stephensonii*
AA. Dorsal surface of mature lvs. usually lacking a gland or pit; foliage bright green, dark green or dusty green, not gray.
    B. Trunks and lower branches with a cherry-red or mahogany-brown smooth polished bark, exfoliating in thin plates; foliage rather bright green; ♂ cones with 10–14 scales. Orange and San Diego cos. . . . . 5. *C. Forbesii*
    BB. Trunks and older branches with fibrous, nonexfoliating bark or if smooth on younger branches not cherry-red or mahogany-brown.
        C. Foliage dull, dusty green; branchlets rather thick and harsh; lvs. sometimes with dorsal pits; seeds glaucous. Santa Barbara to Mendocino cos. . . . . 10. *C. Sargentii*
        CC. Foliage bright green or dark green; pits mostly absent on lvs.; seeds sometimes glaucous.
            D. Branchlets rather thick and harsh; ♂ cones with 6–10 pollen sacs on the cent. scales; ♀ cones mostly longer than thick, 25–35 mm. long; seeds shiny, dark brown, 4–6 mm. long. Monterey Co. . . . . 6. *C. macrocarpa*
            DD. Branchlets slender, not harsh; cent. scales of ♂ cones with not more than 6 pollen sacs; ♀ cones subglobose, usually not more than 20 mm. in diam.
                E. Tree with long slender whiplike leader, to 30 m. tall (in sterile soil dwarfed and fertile when 1 m. tall); foliage deep dark dull blackish-green; seeds shiny black to dull brown. Mendocino Co. . . . . 7. *C. pygmaea*
                EE. Tree with compact pyramidal crown, rarely over 10 m. tall; foliage light bright green or slightly yellow.
                    F. Ovulate cones usually 10–15 mm. long; seeds dull black or glaucous, with inconspicuous hilum. Monterey Co. . . . . 8. *C. Goveniana*
                    FF. Ovulate cones usually over 15 mm. long; seeds dull brown, usually slightly glaucous, with conspicuous hilum. Santa Cruz Mts.
                                              9. *C. Abramsiana*

1. C. Macnabiàna A. Murr. [*C. attenuata* Gord. & Glend.] MCNAB CYPRESS. Small tree or shrublike, to ca. 10 m. tall, with a branched crown broader than high; bark gray, furrowed, fibrous; branchlets ca. 1 mm. thick, forming flat sprays; foliage fragrant, dull gray-green, the lvs. on vigorous shoots 5–10 mm. long, on others 1.5 mm. long, blunt, conspicuously pitted with an active glandular pit; ♂ cones 2–3 mm. long, usually of 8 scales each with 3–5 pollen-sacs; ♀ cones brownish or gray, unopened for many years, 2–2.5 cm. long, with conical umbos to 2 or 4 mm. long; seeds 75–105, irregular, 2–4.5 mm. long, brownish, with 3–5 cotyledons.—Dry flats and slopes, largely at 1000–2600 ft.; Foothill Wd., Chaparral; foothills of Sierra Nevada from Amador to Butte cos., inner Coast Ranges from Shasta to Sonoma cos.

2. **C. Bàkeri** Jeps. [*C. Macnabiana* var. *B.* Jeps.) MODOC CYPRESS. Slender tree 10–15 m. high, with a narrow crown and red-brown or cherrylike bark, only partially exfoliating and becoming grayish in extreme age; branchlets ± evenly disposed on all sides of the branches, rather slender, mostly less than 12 mm. long, 0.5–1 mm. thick; lvs. ca. 2 mm. long, subacute, slightly ridged or rounded on back, abruptly constricted at tip, with conspicuous active dorsal gland; foliage gray-green; lvs. on vigorous shoots 3–10 mm. long; ♂ cones 2–3 mm. long, subglobose; ♀ cones to ca. 12 mm. long, less in diam., gray, slightly warty, the upper umbos rather conspicuous; seeds 50–85, light tan, 3–4 mm. long, the wing almost 1 mm. broad; cotyledons 3–4.—On dry serpentine and volcanic soil, at 3800–6000 ft.; Chaparral, Yellow Pine F.; Burney Springs area in Shasta Co., extreme se. Siskiyou Co. near Timbered Crater.

Ssp. **Matthèwsii** C. B. Wolf. SISKIYOU CYPRESS. Trees largely 15–30 m. high; foliage light gray-green; branchlets mostly over 12 mm. long, 1–1.3 mm. thick; ♀ cones 15–23 mm. in diam., light gray-brown, rather warty, the umbos rather inconspicuous.—Largely 4000–6000 ft.; Yellow Pine F.; Siskiyou Mts., Goose Nest Mt., Siskiyou Co.

3. **C. nevadénsis** Abrams. [*C. Macnabiana* ssp. *n.* Abrams.] PIUTE CYPRESS. Erect tree to ca. 10 m. high with equally broad pyramidal crown; old bark gray-brown, fibrous, 1–2 cm. thick; younger bark somewhat cherry-red or brownish, occasionally exfoliating; foliage soft gray-green, glaucous; branchlets 1–1.5 mm. thick, 4-sided; lvs. ca. 2 mm. long, sharply acute, with active dorsal gland, fimbriate-ciliate on their edges; lvs. on vigorous branches 8–12 mm. long; ♂ cones 3–5 mm. long, of 10–16 scales, each with 5–6 pollen-sacs; ♀ cones 20–30 mm. long, mostly longer than thick, of 6–8 scales with somewhat conical umbos; seeds light tan, 3–5 mm. long, the winged margin rather conspicuous; cotyledons 4–5.—Dry slopes at 4000–6000 ft.; Foothill Wd.; Piute Mts., Kern Co. Feb.–March.

4. **C. Stephensònii** C. B. Wolf. CUYAMACA CYPRESS. Erect or ± spreading tree 10–16 m. high, the bark thin, cherry-red, smooth, exfoliating; branchlets rather thick and stiff, 1.5–2 mm. thick, ± 4-sided; foliage blue-gray or gray-green; lvs. ca. 1 mm. long, acute, with active dorsal pit-gland; ♂ cones 2–4 mm. long, usually with 10–12 scales, each with 3–5 pollen-sacs; ♀ cones with 6–8 scales, these with conspicuous umbo 3–4 mm. high, the cones to ca. 25 mm. in diam.; seeds 100–125, dark brown, 5–8 mm. long, the wings almost as broad as the body; cotyledons 3–4.—Dry slopes at 3000–4000 ft.; Chaparral; headwaters of King Creek, sw. slope of Cuyamaca Peak, e. San Diego Co.

5. **C. Fórbesii** Jeps. TECATE CYPRESS. Small tree, usually less than 10 m. high, with an irregular spreading crown and exfoliating mahogany-brown or cherry-red bark; branchlets 1–1.4 mm. thick; foliage light rich green to ± dull green; lvs. 1.5 mm. long, acute, rounded or ridged on back, with an inconspicuous pit; ♂ cones 3–4 mm. long, mostly with 12–14 scales each with 3–4 pollen-sacs; ♀ cones dull brown or gray, globose, to 30 mm. in diam., the umbos inconspicuous, to 4 or 5 mm. high, the cones not opening for many years; seeds rich dark brown, 5–6 mm. long, the thin wing nearly 1 mm. broad; cotyledons 3–6.—Dry slopes, 1500–5000 ft.; Chaparral; Claymine and Gypsum canyons, Santa Ana Mts. in Orange Co., Guatay Mt., Otay Mt. and Mt. Tecate, San Diego Co.

6. **C. macrocárpa** Hartw. ex Gord. [*C. Hartwegii* Carr.] MONTEREY CYPRESS. Tree to 20 or 25 m. high, at first symmetrical and with cent. leader, or in old trees with broad flat-topped often asymmetrical crown; bark at first rich-brown, later ashy-gray, with flat connected ridges separating into thick persistent scales; branchlets 2–3 mm. thick; foliage rich bright green; lvs. scarcely 2 mm. long, acute, with faint depressed spot on back, or 5–10 mm. long on vigorous shoots; ♂ cones 6 mm. long, of 12–14 scales each with 6–8 pollen-sacs; ♀ cones globose or slightly elongate, 25–35 mm. long, the scales 8–12 with inconspicuous umbos; cones closed for some years; seeds dark brown, 5–6 mm. long, the wing 1–2 mm. wide; cotyledons usually 4.—Exposed headland and dry places; Closed-cone Pine F.; Monterey Peninsula.

7. **C. pygmaèa** (Lemmon) Sarg. [*C. Goveniana* var. *parva* Lemmon. *C. G.* var.

pigmaea Lemmon.] Pygmy Cypress. Mendocino Cypress. Dwarf trees ca. 1 m. high
when in sterile soil, or 10–50 m. tall in better soils and with a straight cent. trunk with
gray-brown shreddy fibrous bark; branchlets 1–1.5 mm. in diam.; foliage dark, dull
green; ♂ cones 3–4 mm. long, of 12–14 scales; ♀ cones subglobose, ca. 15 mm. in
diam., with inconspicuous umbos on the 8–10 scales; seeds ca. 130, flattish, irregular,
3–4 mm. long, with stiff thin narrow margin; cotyledons 3–4.—Flats and slopes below
1500 ft.; Closed-cone Pine F., Mixed Evergreen F., N. Coastal Coniferous F., etc.;
back of Anchor Bay and near Ft. Bragg, Mendocino Co.

8. **C. Goveniàna** Gord. [*C. californica* Carr.] Gowen Cypress. Small tree 5–7 m.
high, with a spread of 2–4 m., rather sparsely branched; bark smooth brown to gray,
becoming rough and fibrous on old trees; branchlets 1–1.5 mm. thick; foliage light
rich green, the lvs. 1–2 mm. long, bluntish, mostly without visible dorsal pits; ♂ cones
3–4 mm. long, of 12–14 scales with ± fimbriate margins and 3–several pollen-sacs;
♀ cones brown to gray-brown, to dull gray, 10–15 mm. long, subglobose, remaining
closed for years; seeds 90–110, dull dark brown to almost black, 3–4 mm. long, narrowly
margined; cotyledons 3–5.—Dry slopes below 1000 ft.; Closed-cone Pine F.; Huckleberry Hill and San Jose Creek, Monterey Co.

9. **C. Abramsiàna** C. B. Wolf. Santa Cruz Cypress. Erect densely branched tree to
10 m. high, with symmetrical pyramidal crown and branches nearly to the ground;
bark gray, rather thin, broken into vertical strips or plates, or fibrous and shreddy;
foliage rich light bright green; branchlets 1–1.5 mm. thick; lvs. ca. 1.5 mm. long,
acutish, with a closed dorsal pit; ♂ cones 3–4 mm. long, usually with ca. 12 scales with
4–6 pollen-sacs; ♀ cones subglobose or ± elongate, 20–30 mm. long, usually with ca.
8 scales with broad cent. humps; seeds ca. 62, angular, 3–5 mm. long, with a narrow
hard wing and 3–4 cotyledons.—Dry slopes at ca. 1600–2500 ft.; Yellow Pine F.,
Closed-cone Pine F.; Bonnie Doon and Eagle Rock, Santa Cruz Mts.

10. **C. Sargéntii** Jeps. [*C. S.* var. *Duttonii* Jeps.] Sargent Cypress. Slender or bushy
tree 10–15(–25) m. tall, the main trunk usually straight, the crown narrow or broad;
bark gray or dark brown to almost black, thick and fibrous; branchlets ca. 2 mm. thick;
foliage dull green; lvs. almost 2 mm. long, blunt, often with a dorsal pit in older lvs.;
♂ cones 3–4 mm. long, usually with ca. 10–12 scales, each with 3–4 pollen-sacs; ♀ cones
mostly 20–25 mm. long, subglobose to ± elongate, the scales mostly 6–8, with inconspicuous umbos; seeds 100 or more, 4–6 mm. long, dark brown, glaucous, the thin
wing 0.5–1 mm. broad; cotyledons mostly 3–4.—Dry slopes, sometimes on serpentine,
at ca. 700–3000 ft.; Chaparral, Closed-cone Pine F., Yellow Pine F.; Coast Ranges
from Zaca Peak, Santa Barbara Co. to Red Mt., n. Mendocino Co.

4. Chamaecýparis Spach. False-Cypress

Ours tall pyramidal trees, with thin scaly or thicker furrowed bark and nodding
leading shoots. Branches spreading; branchlets flattened or ultimately terete, 2-ranked,
in a horizontal plane. Lvs. scalelike, opposite in pairs, densely clothing the branchlets,
needle-shaped in juvenile state. Staminate and ♀ cones on different branchlets, the
former oblong, yellow or red, of numerous decussate stamens, each with 2 globose
pollen-sacs. Ovulate cones globose, erect, the scales peltate, 6–12, each with 2–5 erect
ovules. Seeds 1–5, oblong-ovoid, with 2 prominent wings; cotyledons 2. Ca. 6 spp., of
N. Am. and e. Asia. (Greek, *chamai*, dwarf, and *kuparissos*, cypress.)

Bark often 15–20 cm. thick, divided into broad rounded ridges; branchlets slender, flattened; lvs.
conspicuously glandular on the back .... 1. *C. Lawsoniana*
Bark ca. 1–2 cm. thick, with flat scaly ridges; branchlets scarcely flattened; lvs. not or obscurely
glandular on back .... 2. *C. nootkatensis*

1. **C. Lawsoniàna** (A. Murr.) Parl. [*Cupressus L.* A. Murr. *C. fragrans* Kell.] Lawson-
Cypress. Port-Orford-Cedar. Tree 25–60 m. high, with an enlarged base, tall trunk
and spire-like head of small horizontal or pendulous branches bearing slender flattened
branchlets; bark dark red-brown, smooth on young trees; lvs. bright green above, glaucous
below, ca. 1.5 mm. long, up to 6 mm. on vigorous leading shoots; ♂ cones oblong, the
stamens ca. 12, with red expanded part; ♀ cones numerous on upper branches, globose,
8 mm. in diam., reddish-brown, ± glaucous, the 8–10 scales with a short conical projection at the apex; seeds ovate, slightly flattened, 3–4 mm. long, narrowly wing-margined

on each side; $n = 11$ (Sax & Sax, 1933).—Moist slopes and canyons, often on serpentine, below 4800 ft.; N. Coastal Coniferous F., Mixed Evergreen F., to Yellow Pine F., etc.; w. Shasta to Humboldt, Del Norte, and Siskiyou cos.; w. Ore. Wood light, hard, strong, close-grained, the lumber used for interior finish, shipbuilding, matches, etc.

2. **C. nootkaténsis** (D. Don) Spach. [*Cupressus n.* D. Don in Lamb.] ALASKA-CEDAR. Differing from the preceding in its less flattened branchlets; thinner bark; less glandular, more blue-green lvs.; yellow stamens; cones 12 mm. in diam., with 4–6 scales.—Red Fir F.; known from 2 localities in Siskiyou Co.: ne. slope of Mt. Emily and Little Grayback, at ca. 5000 ft.; n. to Alaska.

5. Juníperus L. JUNIPER

Aromatic evergreen trees or shrubs with thin shredding bark and opposite or whorled (in 3's) lvs. that are needlelike and spreading or scalelike and appressed. Plants dioecious or sometimes monoecious. Staminate cones minute, solitary or clustered, with ovate or peltate scales bearing 3–6 pollen-sacs. Ovulate cones ovoid to subglobose, berrylike by the union of the 2–6 fleshy scales, some or all of which bear 1–3 erect ovules. Seeds wingless, ovoid, terete or angled, brown, with a large 2-lobed hilum; cotyledons 2–6. Ca. 60 spp. of N. Hemis. (Ancient classical name.)

Lvs. jointed at the base, not decurrent, subulate; ♂ catkins axillary; usually prostrate shrub
1. *J. communis*
Lvs. decurrent on the branches, mostly scalelike; ♂ cones terminal; erect shrub or tree.
Fr. red or reddish-brown, with sweet dry pulp; cotyledons 4–6; shrubby or small tree, mostly 2–6 m. tall.
Lvs. mostly in 2's, glandular-pitted on the back, rounded at apex. W. edge of deserts to S. Coast Ranges and Sierra Nevada .... 2. *J. californica*
Lvs. mostly in 3's, not glandular-pitted on back, acute. Inyo-White Mts. to mts. of e. San Bernardino Co. .... 3. *J. osteosperma*
Fr. blue or blue-black, with resinous pulp; tree 5–20 m. tall. Mts. from s. Calif. and N. Coast Ranges .... 4. *J. occidentalis*

1. **J. communis** L. var. **saxátilis** Pall. [*J. c.* var. *montana* Ait. *J. sibirica* Burgsd.] DWARF JUNIPER. Low or prostrate shrub to ca. 3–10 dm. tall and 1–2 or more m. across; lvs. rigid, linear-lanceolate, pungently acute, 6–12 mm. long, dark green, shining and strongly convex beneath with a broad white band of stomates; ♂ cones 3–6 mm. long, the stamens with a short subulate point; berries globose, 7–9 mm. in diam., bright blue, with a white bloom; seeds 1–3, ovoid, acute, angled, ca. 4–5 mm. long.—Stony or wooded slopes at 6400–11,000 ft.; mostly in Lodgepole F., Subalpine F., Sierra Nevada from Mono Pass n., to Mt. Shasta; at 300–5000 ft. in shrubby places, Yellow Pine F., Douglas-Fir F., etc.; Del Norte and Siskiyou cos.; to Alaska, Greenland; Eurasia.

2. **J. califórnica** Carr. [*Sabina c.* Antoine.] CALIFORNIA JUNIPER. Arborescent shrub 1–4 m. high, with stout irregular stems and a broad erect but fairly open head, rarely a small tree to 10 m. tall; bark ashy-gray; branchlets stiff; lvs. scalelike, closely appressed, slightly keeled and conspicuously glandular-pitted on back, bluntly pointed, cartilaginously fringed on margins, 3–4 mm. long; berries at first bluish with a dense bloom, later reddish-brown beneath the bloom, oblong-ovoid, 12–18 mm. long, nearly smooth; seeds 1–2, sharp-pointed, angled, 6–9 mm. long, light brown and shining except for the dull yellowish base; cotyledons 4–6.—Dry slopes and flats mostly below 5000 ft.; Pinyon-Juniper Wd., Joshua Tree Wd.; desert slopes from w. edge of Colo. Desert and Joshua Tree National Monument to Kern Co.; interior cismontane s. Calif., to w. slope of s. Sierra Nevada (largely in Foothill Wd.) and inner Coast Ranges n. to Tehama Co.; approaching the coast in Santa Lucia Mts. Jan.–March.

3. **J. osteospérma** (Torr.) Little. [*J. tetragona* var. *o.* Torr. *Sabina o.* Antoine. *J. californica* var. *utahensis* Engelm. *J. u.* Lemmon.] UTAH JUNIPER. Arborescent shrub or small tree, mostly 3–6 m. tall, forming rounded clumps or crowns; branchlets stiff; lvs. without glands or inconspicuously glandular, acute to acuminate, 2–3 mm. long; fr. 6–9 mm. long, rounded or oblong, red-brown beneath the bloom when mature; seeds 1–2, ovoid, strongly angled, ca. 7 mm. long; cotyledons 4–6.—Dry slopes and flats, 4800–8500 ft.; Pinyon-Juniper Wd.; mts. of e. Mojave Desert to Bridgeport region, Mono Co.; to sw. Ida., sw. Wyo., and w. New Mex.

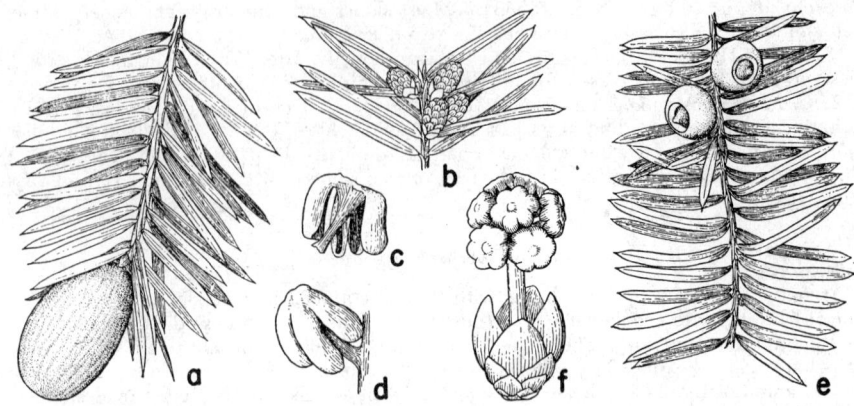

Fig. 7. TAXACEAE. *Torreya californica: a*, twig, × ½, with lvs. and terminal drupelike fruit; *b*, ♂
cones, × ½; *c* and *d*, stamens, × 5, with several pendent sporangia. *Taxus brevifolia: e*, twig, × 1,
with 2 mature ovules, each surrounded by fleshy aril; *f*, ♂ cone, × 4, with stalked head of stamens.

4. **J. occidentàlis** Hook. [*Sabina o*. Antoine.] WESTERN JUNIPER. Tree 5–20 m. tall,
with a well defined trunk to 1(–2) m. thick and with cinnamon-brown shreddy bark;
branches often large and spreading; lvs. in 2's or 3's, closely appressed, rounded and
gland-pitted on the back, denticulately fringed, gray-green, 3 mm. long; berries rounded
to oblong-ovoid, 6–8 mm. long, blue-black at maturity, very glaucous; seeds 2–3, ovoid,
acute, rounded and grooved or pitted on the back, ca. 6 mm. long; cotyledons 2.—
Dry slopes and flats, 3000–10,500 ft.; N. Juniper Wd., Montane Coniferous F.; San
Bernardino Mts. n. through the Sierra Nevada, to Modoc and Siskiyou cos., Yolla Bolly
Mts.; to Wash., Ida., w. Nev. Old trees often grow from cracks in the solid granite
of the higher mts. and become picturesque specimens. Others in favorable conditions
may attain a diam. of 5–6 m. and an estimated age of 3000 years.

Order 2. TAXALES

Characters as given for the family *Taxaceae*

1. Taxàceae. YEW FAMILY (Fig. 7)

Evergreen trees or shrubs with scaly or fissured bark and linear to lanceolate, stiff,
entire, spirally arranged lvs. that usually appear 2-ranked because of the twisted
flattened petioles. Stamens and ovules borne on different trees, the former solitary or in
small spikes in lf.-axils; sporangia 2–8, pendent, surrounded by bud-scales. Ovule solitary,
terminal on a short axillary branch. Seed bony, loosely surrounded in a fleshy cup or
aril or quite enveloped in it and appearing drupelike; cotyledons 2. Three genera and
ca. 15 spp.

Fr. red, berrylike, open at apex; lvs. 1–2 cm. long .... 1. *Taxus*
Fr. green or purplish, closed at apex; lvs. 2.5–6 cm. long .... 2. *Torreya*

1. Táxus L. YEW

Bark scaly; branches usually horizontal or drooping. Lvs. flat, keeled on upper surface, bearing stomates and paler beneath, mostly bluntish at apex. Stamens in globose
stalked heads, 4–8, exposed at anthesis by opening of the surrounding scales. Ovule
on a circular disk which becomes the aril in fr. Seeds ripe the first autumn. Ca. 8 spp.
of N. Hemis., sometimes treated as 1 sp. with several sspp. Used ornamentally and
for the elastic durable wood. (Classical Latin name, probably from Greek *toxon*, a
bow, the wood being used for bows.)

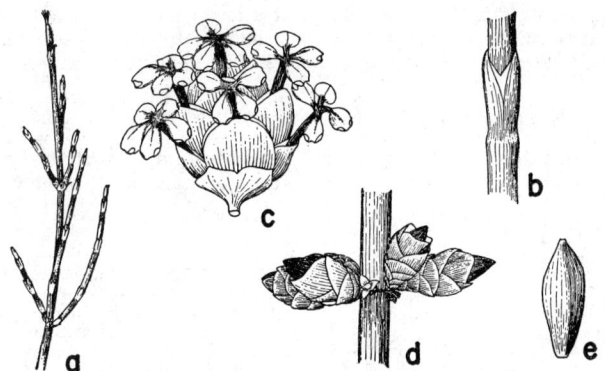

Fig. 8. EPHEDRACEAE. *Ephedra californica:* a, twig, × ½, with scalelike lvs.; b, detail of lvs., × 3; c, ♂ cone, × 4, with stamens united into columns; d, ♀ cone, × 1½, cent. ovule projecting from bracts; e, seed, × 2.

1. T. brevifòlia Nutt. [*T. baccata* ssp. *b.* Pilg. *T. Boursieri* Carr. *T. Lindleyana* A. Murr.] Tree 5–25 m. high, the bark ca. 5–6 mm. thick, with small dark red-brown scales and horizontal, ± drooping branches; lvs. ca. 12–18 mm. long, 1–2 mm. wide, abruptly short-pointed, deep yellow-green above, paler beneath, forming flat sprays; fr. ovoid ca. 1 cm. long; seed 2–4-angled, 5–6 mm. long.—Scattered in damp, ± shaded places like canyons, below 7000 ft.; Mixed Evergreen F., Douglas-Fir F., Yellow Pine F., Red Fir F., etc.; Coast Ranges from Santa Cruz Mts. and Marin Co. n., w. slope of the Sierra Nevada from Tulare Co. n.; to Alaska, Mont. April–May. Wood heavy, strong, hard, red, used by Indians for paddles, spear handles, bows, etc.

2. Tórreya Arn.

Bark fissured; branches whorled, with subopposite branchlets. Lvs. linear to lance-linear, spiny-pointed, with 2 narrow glaucous bands beneath, the midrib not keeled above. Stamens in whorls of 6–8, surrounded by bud-scales at base, each whorl consisting of 4 stamens with 4 pollen-sacs. Ovule completely covered by the fleshy coat and drupelike, the seed with a thick outer woody coat and ruminate albumen; embryo small. Six spp. of N. Am. and Asia. (Named for John *Torrey,* 1796–1873, New York botanist.)

1. T. califórnica Torr. [*Tumion c.* Greene. *Torreya myristica* Hook. f.] CALIFORNIA-NUTMEG. Evergreen tree 5–30 m. tall, with slender spreading branches and stiff, dark green lvs. 2.5–7 cm. long, 2–3 mm. wide; fr. ovoid-oblong, 2.5–3.5 cm. long, green with purplish markings; seeds obovoid-ellipsoid, ca. 2.5–3 cm. long.—Cool shaded slopes and canyons, below 4500 ft.; Mixed Evergreen F., Douglas-Fir F., Yellow Pine F., etc.; Coast Ranges from Santa Cruz to Mendocino cos., w. slope of Sierra Nevada from Tulare to Tehama cos. April–May. Wood light, soft, yellow, used occasionally for fence posts.

Order 3. EPHEDRALES

Characters of the family *Ephedraceae*

1. Ephedràceae. EPHEDRA FAMILY (Fig. 8)

Shrubs or small trees with jointed grooved green stems and scalelike opposite or whorled lvs. Plants usually dioecious, the axillary cones with basal decussate scalelike bracts. Staminate cones with 2–8 stamens free or united into a column and each with 1–8 terminal usually bilocular pollen-sacs. Ovulate cones with 1–3 cent. ovules enclosed in an urn-shaped envelope formed by union of a lower pair of bracts to form an outer integument and of an upper pair to form an elongate tubular inner integument. Seed enclosed in the indurate envelope; cotyledons 2. One genus of arid parts of N. Hemis.

1. Ephèdra L. EPHEDRA. MORMON or MEXICAN TEA

Broomlike shrubs, the branches somewhat resembling an Equisetum. Perhaps 40 spp. of N. and S. Am. and Eurasia. (Ancient name used by Pliny for the horsetails.)

(Cutler, H. C. Monograph of the N. Am. spp. of the genus Ephedra. Ann. Mo. Bot. Gard. 26: 373–424, 1939.)
A.   Lvs. and bracts in 3's; cones sessile.
    B.   Lvs. persisting, becoming shredded and gray in age, 5–13 mm. long; tips of twigs spinose
                                                                     1. *E. trifurca*
    BB. Lvs. falling off in age or remaining firm, 2–6 mm. long; twig-tips not spinose.
        C.   Seeds brown, smooth, subglobose; young stems yellow-green, almost smooth, except for the longitudinal furrows. Widely distributed . . . . 2. *E. californica*
        CC. Seeds cream to light brown, rough, angular, less than half as thick as long; young stems pale gray-green, slightly roughened and with longitudinal furrows. Death V. region . . . . 3. *E. funerea*
AA. Lvs. and bracts in 2's; cones often peduncled.
    B.   Seeds solitary, or if more than 1, light in color.
        C.   The seeds smooth, brown to chestnut in color; lf.-bases brown and persistent; stems usually rough . . . . 4. *E. aspera*
        CC. The seeds furrowed or scabrous, light brown to gray-green; lf.-bases gray and deciduous; stem usually smooth . . . . 5. *E. fasciculata*
    BB. Seeds paired, brown to almost black.
        C.   Lf.-bases brown; seeds lightly furrowed longitudinally; branchlets green, not glaucous, 1–1.5 mm. in diam., numerous, tending to be erect and ± parallel . . . . 6. *E. viridis*
        CC. Lf.-bases gray; seeds smooth; branchlets gray-green, ± glaucous, ca. 2 mm. in diam., fewer and divergent . . . . 7. *E. nevadensis*

1. **E. trifúrca** Torr. in Emory. Erect, 0.5–2 m. tall, with rigid terete branchlets to 3.5 mm. thick and internodes 3–9 cm. long, at first pale green almost smooth, later yellow, then gray-green; lvs. in 3's, spinosely tipped, their sheath becoming fibrous, grayish, persistent; ♂ cones 1-several, obovoid, 6–9 mm. long, the bracts in 3's, membranaceous, red-brown; staminal column 4–5 mm. long with 4–5 anthers; ♀ cones obovoid, 10–14 mm. long, the bracts in 6–9 whorls of 3, orbicular, 8–12 mm. long, translucent except for the red-brown center; seed 1, sometimes 2–3, light brown, smooth, 9–14 mm. long.—Dry sandy and rocky places, below 2000 ft.; Creosote Bush Scrub; Colo. Desert; to w. Tex., nw. Mex., L. Calif. March–April.

2. **E. califórnica** Wats. Erect or spreading, 0.3–1 m. high, with semiflexible to rigid branches to 4 mm. thick and internodes 3–6 cm. long, at first yellow-green, glaucous, almost smooth; lvs. in 3's, 2–6 mm. long, with a green-brown dorsi-median thickening, the hard brown sheath subpersistent; ♂ cones 1-several, ovoid, 6–7.5 mm. long, the bracts in 3's, in 8–12 whorls, ovate, 2.5–3 mm. long, membranaceous, light yellow with hyaline margins; staminal column 3–5 mm. long, with 3–7 subsessile anthers; ♀ cones ovate, 7–10 mm. long, the bracts in 4–6 whorls, round, 5–7 mm. long, pale yellow-translucent except for the orange- or green-yellow center and base; seed mostly 1, subglobose, 7–9 mm. long, brown to chestnut.—Dry slopes and fans largely below 3000 ft.; Creosote Bush Scrub, both deserts; in Chaparral, San Diego Co.; and V. Grassland along inner Coast Ranges n. to Merced Co.; L. Calif. March–April.

3. **E. funèrea** Cov. & Mort. [*E. californica* var. *f.* L. Benson.] Erect, 0.3–1.5 m. high, with stiff hard branches to 3.5 mm. thick and internodes 2–6 cm. long, pale, gray-green, glaucous, slightly roughened; lvs. in 3's, 3–6 mm. long; ♂ cones elongate-elliptic, 5–8 mm. long, with 6–9 whorls of 3 bracts, these ovate, 3–4 mm. long, membranaceous, yellowish; ♀ cones lance-obovate, 8–13 mm. long, the bracts in 6–9 whorls, obovate, 4–8 mm. long, yellow-translucent except for the green cent. part; seeds mostly 1, tetragonal, pale green to light brown, 6–9 mm. long, smooth to scabrous.—At ca. 2000–5000 ft.; Creosote Bush Scrub; Death V. region; sw. Nev. March–April.

4. **E. áspera** Engelm. ex Wats. [*E. nevadensis* var. *a.* L. Benson.] Erect, 0.3–1.2 m. high, with rigid terete branchlets to 3 mm. thick, with internodes 1–5 cm. long, pale to dark green, mostly scabrous, becoming yellow; lvs. opposite, 1.2–5 mm. long, the basal sheath subpersistent; ♂ cones mostly paired, obovoid, 4–7 mm. long, the bracts in

6–10 pairs, obovate, 3 mm. long, membranaceous, yellow to red-brown; staminal column 4–5 mm. long, with 4–6 subsessile anthers; ♀ cones ovoid, 6–10 mm. long, with 5–7 pairs of bracts, these round, 2–5 mm. long, thickened, red-brown with membranaceous margins; seed 1, smooth to slightly rough, light brown to chestnut, 5–8 mm. long.—Occasional on dry rocky slopes below 5000 ft.; Creosote Bush Scrub, Joshua Tree Wd.; Mojave and Colo. deserts; to Tex., n. Mex. March–April.

5. **E. fasciculàta** A. Nels. Low, often prostrate, 0.3–1 m. high, the branches flexuous, up to 3.5 mm. thick with internodes 1–5 cm. long, pale green, smooth to ± roughened, becoming yellowed; lvs. opposite, 1–3 mm. long, with a hyaline subpersistent white sheath; ♂ cones 2-several, narrow-ellipsoid, 4–8 mm. long, sessile, with 4–8 pairs of obovate bracts 2–3 mm. long, membranaceous, light yellow; staminal column 3–9 mm. long, with 6–10 sessile or short-stipitate anthers; ♀ cones ellipsoidal, 6–13 mm. long, with 4–7 whorls of elliptic bracts 3–7 mm. long with slightly thickened light brown to green backs and hyaline margins; seed usually 1, longitudinally furrowed, light brown, 8–13 mm. long.—Occasional from sandy places in Creosote Bush Scrub, as at Kelso and Palm Springs; to Ariz.

Var. **Clòkeyi** (Cutler) Clokey. [*E. C. Cutler.*] Erect; ♀ cones obovoid, 6–10 mm. long; seed 5–8 mm. long.—Dry slopes and washes; Creosote Bush Scrub; both deserts.

6. **E. víridis** Cov. [*E. nevadensis* var. *v.* Jones.] Erect, 0.5–1.5 m. high, with very numerous broomlike yellow-green scabrous slender branchlets; lvs. opposite, 1.5–4 mm. long, setaceously tipped from a dorsi-median thickening, deciduous and leaving a thickened persistent brown base; ♂ cones 2 or more, obovoid, 5–7 mm. long, with 6–10 pairs of bracts 2–4 mm. long, membranaceous, light yellow; staminal column 2–4 mm. long, with 5–8 sessile anthers; ♀ cones obovoid, 6–10 mm. long, with 4–8 pairs of ovate bracts 4–7 mm. long; seeds paired, brown, trigonal, smooth, 5–8 mm. long.—Frequent on dry rocky slopes, canyon walls, etc., at 3000–7500 ft.; Creosote Bush Scrub to Pinyon-Juniper Wd.; deserts from w. edge of Colo. Desert through Mojave Desert along the e. slope of the Sierra Nevada to Mono and Lassen cos.; to w. Colo., Utah, Ariz. April–June.

7. **E. nevadénsis** Wats. Spreading-erect, to ca. 1.2 m. high, the young stems pale green, glaucous, almost smooth, later yellow or gray; lvs. in 2's, 2–4(–7) mm. long, with a dorsi-median thickening, splitting and falling off, leaving gray bases; ♂ cones 1-several, ellipsoid, 4–8 mm. long, with 5–9 pairs of obovate bracts 3–4 mm. long, membranaceous, yellow to light brown; staminal column 3–5 mm. long, with 6–9 subsessile anthers; ♀ cones roundish, 5–11 mm. long, pedunculate, with 3–5 pairs of bracts 4–8 mm. long, light brown to yellow-green; seeds paired, smooth, brown, 6–9 mm. long.—Common on dry slopes and hills mostly below 4500 ft.; Creosote Bush Scrub, Joshua Tree Wd.; Colo. and Mojave deserts to Owens V.; to Utah, Ariz. March–April.

Division V. **ANTHOPHYTA**. Flowering Plants

Trees, shrubs or herbs; with true fls., consisting of a cent. axis (receptacle) bearing typically 4 series of appendages: calyx or sepals, corolla or petals, stamens and carpels (pistil). Any one or more of these series may be lacking or may have fusion between members of the same series (coalescence) or of neighboring series (adnation). Stamens typically of an elongate basal part (fil.) and a slightly expanded distal part (anther) with mostly 4 parallel microsporangia. Pistil of 1 or more fused carpels and consisting typically of a basal ovary containing the ovules, a ± elongate style and a terminal stigma for receiving the pollen-grains. These send pollen-tubes to the ovules and fertilization occurs. The matured ovules or seeds are thus inclosed in a fr. (ripened ovary with any accessory parts). Ca. 250,000 spp. and the dominant land-plants of today. As here used the term Anthophyta is the equivalent of the more common Angiospermae.

KEY TO CLASSES

Class 1. DICOTYLEDONEAE

Stem exogenous, with pith, xylem, cambium, phloem and cortex, and if of more than short duration, increasing in diam. by annual rings formed by the cambium. Lvs. net-veined, usually pinnately or palmately so. Fls. mostly with parts in 4's or 5's. Embryo with 2 cotyledons in most cases, the first lvs. of the germinating plantlet opposite. Ca. ¾ of the flowering plants belong here.

KEY TO FAMILIES

Group 1. WOODY PLANTS WITHOUT PETALS, OR PETALS NOT EVIDENT

FF. Fls. smaller.
  G. Plants small prostrate shrubs of the Del Norte Co. coast; ovary 6–9 loculed; stigmas 6–9-parted .... *Empetraceae* p. 432
  GG. Plants not as above.
    H. Ovary 3–4-loculed; styles 3–4.
      I. Fr. a lobed caps.; lvs. often alternate
        Euphorbiaceae p. 159
      II. Fr. a smooth cylindrical caps.; lvs. opposite
        Buxaceae p. 985
    HH. Ovary 1-celled (or if 2–3-celled, the style single); styles 1–3.
      I. Fr. an ak. or utricle.
        J. Styles 2–3; ovary not surrounded by a floral tube.
          K. Fls. subtended by an invol. of ± united bracts; ak. triangular or lenticular
            Polygonaceae p. 320
          KK. Fls. without an invol. of bracts; fr. mostly a utricle or depressed-globose ak.
            Chenopodiaceae p. 366
        JJ. Style 1, ± plumose in fr.; ovary surrounded by floral tube
          Rosaceae (Cercocarpus, Coleogyne) p. 754
      II. Fr. a samara, drupe or berry.
        J. Lvs. opposite.
          K. Plants with silvery-scurfy branchlets and lvs.; fr. red, fleshy ..... *Elaeagnaceae* p. 987
          KK. Plants glabrous; fr. a samara or blackish drupe .... *Oleaceae* p. 446
        JJ. Lvs. alternate; fr. a drupe or berrylike
          Rhamnaceae p. 970

Group 2. HERBACEOUS PLANTS WITHOUT PETALS, OR PETALS NOT EVIDENT

A. Aquatic plants growing ± submerged, or, as water dries up, on wet mud.
  B. Lvs. largely dissected, whorled.
    C. Ovary superior; fr. an ak.; lvs. 3-times forked .... *Ceratophyllaceae* p. 113
    CC. Ovary inferior; fr. a nutlet or splitting into 4 nutlets; lvs. ± pinnatifid
      Haloragaceae p. 961
  BB. Lvs. entire, opposite, often crowded into terminal rosettes.
    C. Ovary superior, splitting into 4 parts when ripe; lvs. 3–8 mm. long
      Callitrichaceae p. 963
    CC. Ovary inferior, forming a caps.; lvs. 15–25 mm. long
      Onagraceae (Ludwigia) p. 925
AA. Land plants, sometimes growing in wet places.
  B. Ovary superior, free from the calyx, although sometimes surrounded by it.
    C. Perianth lacking entirely.
      D. Fls. perfect, borne in a spike.
        E. Lvs. 3-foliolate; spike without an invol. ... *Berberidaceae* (Achlys) p. 111
        EE. Lvs. entire; spike subtended by a conspicuous invol. .. *Saururaceae* p. 114
      DD. Fls. imperfect, either ♂ or ♀.
        E. Fls. borne in clusters often surrounded by an invol. resembling a perianth; caps. 3-lobed; sap often milky; plants monoecious
          Euphorbiaceae p. 159
        EE. Fls. borne in catkins disposed in terminal spikes; plants dioecious. Seashore .... *Batidaceae* p. 396
    CC. Perianth present as a calyx (often petaloid).
      D. Pistils more than 1 and distinct; stamens 10-many ..... *Ranunculaceae* p. 78
      DD. Pistil 1.
        E. Fls. perigynous, the ovary inclosed in or seated in a floral tube.
          F. Stipules present; lvs. alternate; fr. an ak. .... *Rosaceae* p. 754
          FF. Stipules none; lvs. opposite.
            G. Stamens many; fr. circumscissile
              Aizocaceae (Sesuvium) p. 308
            GG. Stamens 3–5; fr. an ak. .... *Nyctaginaceae* p. 388
        EE. Fls. hypogynous, the ovary not so inclosed.
          F. Style and stigma single.
            G. Calyx not tubular.
              H. Lvs. subulate, squarrose-spreading, 3–6 mm. long
                Caryophyllaceae (Loeflingia) p. 285
              HH. Lvs. ovate, the blades 10–20 mm. long.
                I. Calyx 5–6-parted; stamens 6–7
                  Euphorbiaceae (Eremocarpus) p. 162
                II. Calyx largely 4-parted; stamens 4 ... *Urticaceae* p. 920
            GG. Calyx tubular, corollalike, subtended by bracts often forming a calyxlike invol. .... *Nyctaginaceae* p. 388

FF. Styles and stigmas more than 1.
    G.   Lvs. mostly deeply palmately 5–7-lobed or with 5–7 lfts.;
          fls. imperfect .... *Moraceae*   p. 919
    GG. Lvs. mostly entire or shallowly lobed; fls. perfect or imper-
          fect.
        H.   Ovary 1-loculed (except sometimes in *Silene* which
            has a tubular gamosepalous calyx).
          I. Ovule or seed solitary; fr. mostly an ak. or utri-
             cle; lvs. often alternate.
            J. Fls. borne in a tubular to campanulate invol.
                           Polygonaceae   p. 320
            JJ. Fls. not borne in an invol.
             K.   Lvs. with evident stipular sheath above
                 each node ....*Polygonaceae*   p. 320
             KK. Lvs. lacking such stipular sheath.
                L.   Calyx mostly 6-cleft; stamens 3, 6,
                   or 9 .... *Polygonaceae*   p. 320
                LL. Calyx-lobes or sepals 1, 4 or 5.
                  M.   Bracts subtending fls. not scar-
                     ious; plants mostly mealy,
                     scurfy, or fleshy
                         Chenopodiaceae   p. 366
                  MM. Bracts subtending fls. scari-
                    ous; plants not mealy, scurfy,
                    or fleshy ... *Amaranthaceae*   p. 384
          II. Ovules or seeds more than 1; fr. a caps.; lvs.
            opposite.
            J. Lvs. of each pair very unequal .. *Aizoaceae*   p. 306
            JJ. Lvs. of each pair normally ca. equal
                             Caryophyllaceae   p. 273
    HH. Ovary more than 1-loculed.
        I. Fr. berrylike, ca. 10-lobed, made up of fleshy
          coalesced carpels .... *Phytolaccaceae*   p. 388
        II. Fr. a caps. of 2–5 fused carpels.
          J. Lvs. whorled or opposite ...... *Aizoaceae*   p. 306
          JJ. Lvs. alternate .... *Euphorbiaceae*   p. 159
BB. Ovary ± inferior, partly or wholly adnate to the floral tube.
  C.   Fls. imperfect; stamens 8-many; fr. a 1-loculed caps.
    D.   Lvs. divided; large green herb .... *Datiscaceae*   p. 177
    DD. Lvs. reduced to minute bracts; very small parasite on Dalea
                                Rafflesiaceae   p. 966
  CC. Fls. perfect.
    D.   Calyx 3-lobed, the lobes 1–5 cm. long ..... *Aristolochiaceae* (Asarum)   p. 965
    DD. Calyx 4–5-lobed, shorter.
        E. Fr. a bony nut, indehiscent; lvs. rhombic-ovate, entire
                          Aizoaceae (Tetragonia)   p. 308
        EE. Fr. a caps.
          F. Lvs. entire, oblong, glabrous; stamens 3–6. Root-parasite
                           Santalaceae   p. 988
          FF. Lvs. toothed or crenate to lobed, roundish; stamens 4, 5, or 8
                           Saxifragaceae   p. 730

Group 3. PLANTS WITH SEPARATE PETALS; STAMENS NUMEROUS, MORE THAN TWICE AS MANY AS PETALS
A Ovary superior.
  B.   Aquatic plants with floating lf.-blades .... *Nymphaeaceae*   p. 112
  BB. Land plants, if in wet places, not having floating lvs.
    C.   Plants woody, i.e., shrubs or trees.
      D.   Plants with evident stipules.
        E. Shrubs with recurved clawlike thorns and pinnate lvs.; pistil 1, form-
          ing a legume   . . . . *Leguminosae* (Acacia)   p. 797
        EE. Shrubs not both thorny and with pinnate lvs.
          F. Stamens united into a tube around the pistil; lvs. rounded, often
            stellate-pubescent .... *Malvaceae*   p. 116
          FF. Stamens not united into a tube; lvs. various, not stellate-
            pubescent .... *Rosaceae*   p. 754
      DD. Plants without stipules.
        E. Desert plants with conspicuously thorny branches
                    Simarubaceae (Holacantha)   p. 993
        EE. Cismontane or desert plants, not thorny.
          F. Fr. not of separate follicles; calyx not of 5 persistent lobes.
            G.   Pistils many, concealed in a hollow receptacle; petals
                many; lvs. opposite .... *Calycanthaceae*   p. 76
            GG. Pistil 1; petals 4–6; lvs. alternate ...... *Papaveraceae*   p. 192
          FF. Fr. of 2–9 separate follicles; calyx of 5 persistent lobes
                            Crossosomataceae   p. 795

CC. Plants herbaceous, or woody at base only, not definite shrubs.
  D.   Sepals 2 (joined into a pointed cap in *Eschscholzia*).
      E.   Plants fleshy; sepals persistent ... . *Portulacaceae* p. 295
      EE. Plants not fleshy; sepals caducous at anthesis .... *Papaveraceae* p. 192
  DD. Sepals normally more than 2.
      E.   Stamens united into a tube around the pistil; lvs. rounded
                                                    Malvaceae p. 116
      EE. Stamens not so united.
          F.   Maturing ovary at summit exposing the contained seeds; fls.
              irregular .... *Resedaceae* p. 204
          FF. Maturing ovary closed at summit; fls. mostly regular.
             G.   Plants insectivorous, the lvs. modified with pitcherlike
                petioles .... *Sarraceniaceae* p. 137
             GG. Plants not insectivorous, the lvs. not pitcherlike.
                H.   Stamens hypogynous.
                    I.   Lvs. opposite.
                      J.   Lf.-blades punctate with pellucid dots; sta-
                          mens usually in 3 or 5 bundles
                                      Hypericaceae p. 191
                    JJ. Lf.-blades not so punctate; stamens separate.
                                        Cistaceae p. 173
                II. Lvs. alternate.
                      J.   Lf.-blades compound with 3–5 lfts. and a
                        strong disagreeable odor; ovary short-stipitate
                              Capparidaceae (Polanisia) p. 207
                    JJ. Lf.-blades not as above; ovary sessile.
                        K.   Ovary 3-loculed; stamens united in ranks
                          of 5 .... *Euphorbiaceae* (Ditaxis) p. 162
                      KK. Ovary or ovaries 1-loculed; stamens spi-
                        rally arranged.
                          L.   Sepals persistent; anthers maturing
                            in centrifugal sequence
                                Paeoniaceae p. 77
                        LL. Sepals deciduous; anthers maturing
                          centripetally .. . *Ranunculaceae* p. 78
             HH. Stamens perigynous; stipules usually well developed
                                      Rosaceae p. 754

AA. Ovary at least partly inferior.
  B.   Ovary only partly inferior.
      C.   Fr. a dry caps. .... *Saxifragaceae* p. 730
      CC. Fr. a fleshy pome .... *Rosaceae* p. 754
  BB. Ovary wholly inferior.
      C.   Petals numerous (technically may be modified stamens); plant succulent.
         D.   Plants not spiny; lvs. well developed   *Aizoaceae* (Mesembryanthemum) p. 308
         DD. Plants spiny, with lvs. much reduced and caducous, or wanting
                                              Cactaceae p. 309
      CC. Petals few; plants not succulent.
         D.   Trees or shrubs; fr. fleshy .... *Rosaceae* p. 754
         DD. Herbs; fr. a caps. .... *Loasaceae* p. 177

Group 4. PLANTS WITH SEPARATE PETALS; STAMENS NOT MORE THAN TWICE AS MANY AS PETALS

A. Pistils more than 1, nearly or quite separate.
  B.   Plants fleshy, at least the lvs. so; lvs. simple .... *Crassulaceae* p. 718
  BB. Plants not succulent.
      C.   Insertion of stamens on the floral tube, perigynous.
         D.   Stipules usually absent; pistils mostly fewer than sepals .. *Saxifragaceae* p. 730
         DD. Stipules usually present; pistils mostly as many as sepals   *Rosaceae* p. 754
      CC. Insertion of stamens hypogynous; small herbs with long taillike spike of pistils
                                      Ranunculaceae (Myosurus) p. 79
AA. Pistil 1, of 1 or more ± united carpels.
  B.   Plants climbing by means of tendrils; lvs. palmately veined.
      C.   Fr. a berry, the frs. borne in clusters. Grapes   *Vitaceae* p. 969
      CC. Fr. a pepo, the frs. borne singly. Gourds, melons, etc. ...... *Cucurbitaceae* p. 1058
  BB. Plants not climbing by means of tendrils.
      C.   Styles 2–5, separate to near base.
         D.   Plants definitely woody, well developed shrubs or trees.
           E.   Lvs. small, scalelike, appressed; fls. minute, in large clusters
                                            Tamaricaceae p. 174
          EE. Lvs. well developed.
             F.   Ovary superior.
               G.   Fr. a samara; lvs. palmately veined, opposite .. *Aceraceae* p. 995
               GG. Fr. not a samara; lvs. pinnately veined.
                  H.   Lvs. alternate; fr. a 1-seeded ± dry drupe
                                         Anacardiaceae p. 997

HH. Lvs. opposite; fr. a several-seeded caps.

Staphyleaceae p. 986

FF. Ovary appearing ± inferior.

  G. Lvs. alternate, with stipules.

    H. Fr. a pome, the dry pistils technically superior and surrounded by a fleshy floral tube; lvs. ± oblong, pinnately veined . . . . *Rosaceae* (Heteromeles) p. 794

    HH. Fr. a berry, inferior; lvs. mostly palmately lobed

Saxifragaceae (Ribes) p. 745

  GG. Lvs. opposite, without stipules

Saxifragaceae (Jamesia, Whipplea) p. 730

DD. Plants herbaceous.

  E. The plant a submerged aquatic or on wet mud flats and rooting at nodes.

    F. Lvs. ± finely dissected . . . . *Haloragaceae* (Myriophyllum) p. 961

    FF. Lvs. entire, opposite . . . . *Elatinaceae* p. 271

EE. The plant a normal terrestrial plant.

  F. Ovary superior.

    G. Lvs. modified for catching insects, with numerous stout glandular hairs on upper surface . . . . *Droseraceae* p. 138

    GG. Lvs. not insectivorous.

      H. Lvs. compound, with 3 lfts.; stamens 10

Oxalidaceae p. 146

      HH. Lvs. simple.

        I. The lvs. largely basal.

          J. Plants fleshy, not maritime . . *Portulacaceae* p. 295

          JJ. Plants not fleshy, maritime

Plumbaginaceae p. 409

        II. The lvs. mostly cauline. – on stem

          J. Lvs. for most part alternate.

            K. Sepals 2; plants ± fleshy; stamens 1–3

Portulacaceae p. 295

            KK. Sepals 3-several; plants not fleshy.

              L. Stamens 10; plants ± gray-pubescent . . . *Euphorbiaceae* (Ditaxis) p. 162

              LL. Stamens 3–8; plants not gray-pubescent.

                M. Fls. regular, 5-merous; caps. closed until mature

Caryophyllaceae p. 273

                MM. Fls. irregular, 2–7-merous; caps. open at top before maturity . . . . *Resedaceae* p. 204

          JJ. Lvs. opposite.

            K. Stamens 5 or 10; caps. 3–10-valved or -toothed; placenta largely cent.

Caryophyllaceae p. 273

            KK. Stamens 4–7; caps. 2–4-valved; placenta basal but parietal . . . . . *Frankeniaceae* p. 176

  F. Ovary ± inferior.

    G. Ovules several in each cavity of the ovary; fr. a caps. or many-seeded berry . . . . *Saxifragaceae* p. 730

    GG. Ovules solitary in each cavity of the ovary; fls. in umbels.

      H. Fr. a berry; lvs. 3–15 dm. long . . . . *Araliaceae* p. 999

      HH. Fr. dry, splitting into 2 carpels; lvs. mostly smaller

Umbelliferae p. 1000

CC. Style 1, sometimes ± divided toward apex.

  D. Ovary inferior.

    E. Plants well developed shrubs or arborescent; fr. fleshy.

      F. Fls. racemose or solitary; lvs. alternate; stamens mostly 5.

Saxifragaceae (Ribes) p. 745

      FF. Fls. in heads, cymes or umbels; lvs. mostly opposite; stamens 4

Cornaceae p. 1034

    EE. Plants herbaceous; fr. dry; fls. mostly 4-merous . . . . . . *Onagraceae* p. 923

  DD. Ovary superior, but sometimes appearing inferior because inclosed in but not adnate to floral tube.

    E. Plants well developed shrubs or trees.

      F. Fls. irregular, the petals not all alike.

        G. Petals 3 (2 forming a pair, the 3d hooded above); sepals 5, the lateral petaloid . . . . *Polygalaceae* p. 155

        GG. Petals 4–5; calyx not petaloid.

          H. Ovary 3-loculed; lvs. opposite, palmately compound; fls. not papilionaceous, white in a large thyrse; fr. a 1-seeded caps. . . . . *Hippocastanaceae* p. 994

          HH. Ovary 1-loculed (sometimes divided into 2 parts by a false partition).

I. Fls. papilionaceous; fr. a legume; lvs. mostly
compound; stamens more than 4 .. *Leguminosae*   p. 795
II. Fls. not papilionaceous; fr. a spiny 1-seeded pod;
lvs. simple; stamens 4 .... *Krameriaceae*   p. 897
FF. Fls. regular, the petals essentially alike.
   G.   Lvs. compound, consisting of 2 or more lfts.
     H.   The lvs. twice pinnate, 3–6 dm. long; fls. purplish.
Introd. tree .... *Meliaceae*   p. 999
     HH. The lvs. once divided, shorter.
       I.   Fr. a samara.
         J.   Wing of fr. terminal; lfts. 3–9; lvs. opposite
Oleaceae (Fraxinus)   p. 447
         JJ. Wing of fr. all way around; lfts. 3; lvs. al-
ternate .... *Rutaceae* (Ptelea)   p. 992
       II. Fr. not a samara.
         J.   The fr. fleshy; petals 3 plus 3; lfts. mostly
more than 3 .... *Berberidaceae*   p. 107
         JJ. The fr. dry; petals in 1 series.
           K.   Ovary 1-loculed, becoming a legume;
stamens mostly 10 ..... *Leguminosae*   p. 795
           KK. Ovary more than 1-loculed.
             L.   Lfts. 13–25; fls. 5-merous
Burseraceae   p. 996
             LL. Lfts. 2–3.
               M.   Sepals and petals 5; fr. not
inflated .... *Zygophyllaceae*   p. 157
               MM. Sepals and petals 4; fr. in-
flated
Capparidaceae (Isomeris)   p. 206
   GG. Lvs. simple, but sometimes deeply divided.
     H.   Fr. stipitate, 2-locular; fls. yellow
Cruciferae (Stanleya)   p. 213
     HH. Fr. sessile.
       I.   The fr. a legume, oblong, flat; stamens 10; fls.
red-purple .... *Leguminosae* (Cercis)   p. 799
       II. The fr. not a legume.
         J.   Trees with large palmately lobed lvs.; fls. im-
perfect, reduced, in spherical clusters
Platanaceae   p. 897
         JJ. Shrubs with pinnately veined lvs.
           K.   Stamens 4–5 (sometimes 6–10 in *For-
sellesia*, a small spiny desert shrub).
             L.   The stamens opposite the petals
Rhamnaceae   p. 970
             LL. The stamens alternate with the pet-
als, or more numerous
Celastraceae   p. 966
           KK. Stamens 10.
             L.   Lvs. mostly 3–6 cm. long. Boggy
places ...... *Ericaceae* (Ledum)   p. 412
             LL. Lvs. not more than 1 cm. long. Dry
places ... *Rosaceae* (Adenostoma)   p. 778
EE. Plants herbaceous.
   F.   Petals 6 in 2 circles; lvs. 1–2 times ternately compound, basal
Berberidaceae (Vancouveria)   p. 111
   FF. Petals not as above.
     G.   Sepals 2 or 3.
       H.   Plants fleshy, succulent; sepals 2, persistent
Portulacaceae   p. 295
       HH. Plants not succulent.
         I.   Sepals caducous, 2 or 3; stamens 6–12; lvs. en-
tire ....:...... *Papaveraceae*   p. 192
         II. Sepals not early caducous; stamens 6; lvs. dis-
sected.
           J.   Stamens in 2 sets of 3 each; petals 4, in 2
dissimilar pairs .... *Fumariaceae*   p. 202
           JJ. Stamens in 1 set; petals 3, alike
Limnanthaceae   p. 149
   GG. Sepals 4 or 5, sometimes more.
     H.   Fls. nearly or quite regular.
       I.   Lvs. compound.
         J.   Fr. 1-loculed, forming a legume; lvs. pinnate
Leguminosae (Cassia)   p. 800
         JJ. Fr. 2–5-loculed; fr. not a legume.
           K.   Lfts. 3; lvs. alternate.

Group 5. Plants with petals ± connate

E. Pistils 2 (ovaries distinct, but styles or stigmas united); plants with milky juice.
    F. Styles united; stamens connivent around the stigma
                                                      Apocynaceae p. 449
    FF. Styles distinct below; stamens monadelphous and adnate to the stylar column .... *Asclepiadaceae* p. 452
EE. Pistil 1.
    F. Ovary 1-loculed, 1-ovuled; style and stigma 1; fr. dry, hard; the seemingly colored corolla really a calyx inside a 1-fld. invol.
                                     Nyctaginaceae p. 388
    FF. Ovary and fr. not as above.
        G. Stamens as many as the corolla-lobes and opposite them, or with an additional series of staminodia; placentae cent. or basal.
            H. Fr. a several-seeded caps.; style 1
                                   Primulaceae p. 398
            HH. Fr. 1-seeded; styles 5 .... *Plumbaginaceae* p. 409
        GG. Stamens as many as or fewer than the corolla-lobes and alternate with them.
            H. Corolla small, dry-scarious, veinless; caps. opening by a lid; lvs. usually basal; stamens 2 or 4
                                   Plantaginaceae p. 405
            HH. Corolla not dry-scarious, veiny; caps. usually not as above.
                I. Ovary 4-loculed, commonly 4-lobed, each lobe forming a nutlet when mature (though some may abort); infl. usually a scorpioid cyme
                                     Boraginaceae p. 552
                II. Ovary 1-, 2-, or 3-loculed.
                    J. Stamens 2, opposite each other; shrub; caps. circumscissile ...... *Oleaceae* (Menodora) p. 448
                    JJ. Stamens 4–7; fr. not opening by a lid.
                        K. Anthers opening by terminal pores; fls. large, white or yellowish
                              Ericaceae (Rhododendron) p. 412
                      KK. Anthers opening by longitudinal slits.
                          L. Style 3-cleft; ovary 3-loculed; caps. 3-valved .... *Polemoniaceae* p. 468
                          LL. Style not 3-cleft; ovary 1–2-loculed.
                              M. Calyx 4–5-toothed or -cleft; style 1, entire.
                                  N. Ovary 1-loculed; stigmas 2; lvs. opposite
                                         Gentianaceae p. 437
                                NN. Ovary 2-loculed; stigma usually 1; lvs. alternate
                                       Solanaceae p. 590
                          MM. Calyx of 5 distinct sepals, or sepals united only at very base; styles 2 or 1, usually partly divided. (Go to N.)
N. Plants twining or trailing; infl. not coiled; corolla plaited in bud
                                       Convolvulaceae p. 457
NN. Plants erect or diffuse; infl. cymose, often coiled; corolla not plaited in bud .... *Hydrophyllaceae* p. 515
DD. Corolla ± irregular.
    E. Fr. of 2–4 nutlets; lvs. opposite.
        F. Ovary not lobed; style apical, entire .... *Verbenaceae* p. 686
        FF. Ovary 4-lobed; style arising between the ovary-lobes, cleft at apex .... *Labiatae* p. 689
    EE. Fr. a caps.
        G. Ovary 1-loculed.
            H. Plants aquatic, often with dissected lvs.; corolla spurred; stamens 2 .... *Lentibulariaceae* p. 680
            HH. Plants terrestrial, the lvs. not dissected; corolla gibbous; stamens 4 .... *Martyniaceae* p. 679
        GG. Ovary 2-loculed.
            H. Seeds not winged; placentae axile.
                I. Calyx with a pair of bractlets at base; desert shrub with corolla red; stamens 2 .. *Acanthaceae* p. 685
                II. Calyx without bractlets; herbs or shrubs; stamens 5, 4, or rarely 2 .... *Scrophulariaceae* p. 603
            HH. Seeds winged; placentae parietal; caps. linear, 15–25 cm. long. Desert shrub .... *Bignoniaceae* p. 678

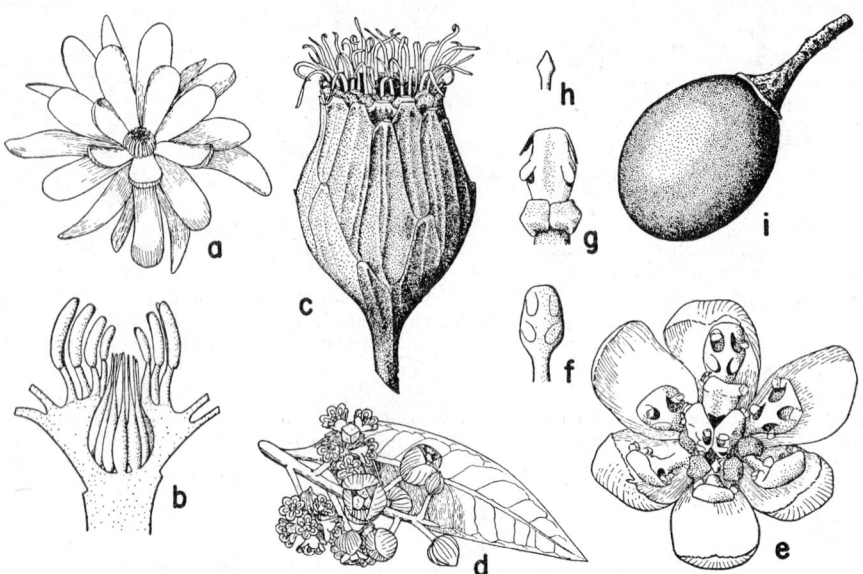

Fig. 9. CALYCANTHACEAE. *Calycanthus occidentalis: a,* fl., × ½, perianth-parts imbricated; *b,* fl. in section, × 2½, cent. free carpels in hollow receptacle with numerous stamens and some perianth-bases; *c,* fr., × 1, matured receptacle.

LAURACEAE. *Umbellularia californica: d,* lf. and infl., × ½; *e,* fl., × 5, with 2 series each of sepals and stamens; *f,* × 5, outer stamen; *g,* inner stamen, × 5, with basal glands and with anther-flaps; *h,* staminodium, × 5; *i,* drupe, × 1.

AA. Ovary inferior or partly so.
  B. Stamens more than 5.
    C. Anthers opening by terminal pores or chinks; fr. fleshy .... *Ericaceae* p. 410
    CC. Anthers opening by longitudinal slits; fr. dry .... *Styracaceae* p. 397
  BB. Stamens 5 or fewer.
    C. Stamens distinct.
      D. Lvs. alternate; fls. regular (± irregular in *Nemacladus* & *Parishella*); stamens 5 .... *Campanulaceae* p. 1061
      DD. Lvs. opposite or whorled.
        E. Stamens 1–3; fls. irregular; fr. 1-seeded .... *Valerianaceae* p. 1052
        EE. Stamens 4–5, rarely 2; fls. often regular.
          F. Ovary 1-loculed; fls. in short spikes or involucral heads; fr. an ak. Introd. herbs .... *Dipsacaceae* p. 1057
          FF. Ovary 2–5-loculed; fls. not so arranged; fr. usually not a solitary ak.
            G. Herbs, or if shrubs, with narrow ± whorled lvs. or with fls. in heads; ovary 2-loculed; lvs. opposite and stipulate or whorled and estipulate .... *Rubiaceae* p. 1037
            GG. Shrubs, or if herbs, with broad lvs.; ovary mostly 3–5-loculed; lvs. opposite or perfoliate, neither whorled nor with true stipules .... *Caprifoliaceae* p. 1046
    CC. Stamens united by the anthers.
      D. Plants bearing tendrils; lvs. palmate; stamens 5, with 2 pairs united and appearing to be 3 .... *Cucurbitaceae* p. 1058
      DD. Plants not tendril-bearing.
        E. Fls. not in involucrate heads; stamens free from the corolla *Campanulaceae* p. 1061
        EE. Fls. in involucrate heads; stamens adnate to the corolla *Compositae* p. 1073

1. Calycanthàceae. CALYCANTHUS FAMILY (Fig. 9)

Aromatic deciduous or evergreen shrubs with lvs. opposite, entire, short-petioled, exstipulate. Fls. rather large, solitary, regular, bisexual, on lateral leafy branches. Sepals and petals similar, imbricated in several series. Stamens many, inserted on the receptacle, the inner sterile; fils. short. Pistils many, distinct, inserted on inner face of the hollow receptacle, each 1-loculed and 1–2-ovuled; style filiform. Fr. consisting of many 1-seeded aks. completely enclosed by the (in ours) ovoid receptacle. Seeds erect, without endosperm; cotyledons foliaceous, convolute. Two genera and ca. 6 spp., native in N. Am. and e. Asia.

1. Calycánthus L. Spice-Bush. Sweet-Shrub

Deciduous. Fls. reddish-brown. Stamens many, inserted in several rows. Four spp. of N. Am. shrubs, often grown for the fragrance. (Greek, *kalyx,* calyx, and *anthos,* fl., referring to the colored calyx.)

1. C. occidentàlis H. & A. [*Butneria o.* Greene.] Erect usually rounded shrub, 1–3(–6) m. tall and often of greater diam., pleasantly aromatic when bruised; lvs. ovate to lance-oblong, 5–12(–17) cm. long, 2–7(–8.5) cm. wide, acute at apex, usually rounded sometimes cordate at base, somewhat scabrous above, ± pubescent beneath, firm, very short-petioled; fls. solitary, terminal, short-peduncled; sepals and petals linear-spatulate, 2–6 cm. long, rather fleshy, pubescent; sterile fils. villous; fr. ovoid, cuplike, 2.5–4.5 cm. long, conspicuously veined; aks. many, oblong-oblanceolate, 8–12 mm. long, velvety-villous; $n = 11$ (Cave, 1949).—Moist places, canyon streams and about ponds, below 4000 ft.; Foothill Wd., Yellow Pine F.; N. Coast Ranges, Napa Co. to Trinity Co. and w. base of Sierra Nevada, Tulare Co. to Shasta Co. April–Aug.

2. Lauràceae. Laurel Family (Fig. 9)

Aromatic trees and shrubs, usually evergreen. Lvs. alternate, rarely opposite, simple, exstipulate, mostly leathery, punctate. Fls. perfect or imperfect, yellow or greenish, regular. Calyx usually 6-parted, the segms. in 2 series. Petals 0. Stamens in 3–4 whorls of 3 each, some frequently reduced to staminodia; anthers 2–4-celled, opening by uplifting valves. Ovary usually superior and free, 1-celled, 1-ovuled, with a single style. Fr. a berry or drupe, indehiscent. Ca. 40 genera and 1000 spp., widely distributed in trop., less so in temp. regions. Some grown for ornament, some (avocado, cinnamon, and camphor) of economic importance.

1. Umbellulària Nutt. California-Bay. California-Laurel

Evergreen pungently aromatic trees. Lvs. alternate, entire, coriaceous. Fls. bisexual, in simple peduncled umbels in upper axils. Sepals 6, deciduous. Stamens 9, the 3 inner with 2 stalked orange glands at base. Anthers 4-celled. Fr. a drupe. One sp. (Latin, *umbellula,* a small umbel.)

1. U. califórnica (H. & A.) Nutt. [*Tetranthera c.* H. & A. *Oreodaphne c.* Nees.] Tree with broad crown, and to 30(–45) m. high, or an erect shrub in dryer places; bark greenish- to reddish-brown; lvs. oblong to oblong-lanceolate, 3–8(–10) cm. long, 1.5–3 cm. wide, short-petioled, obtusely acuminate, shining, glabrous to somewhat pubescent, yellowish-green; fls. 6–10 on a peduncle, yellow-green; sepals 6–8 mm. long, oblong-ovate; drupe usually solitary, rounded-ovoid, 2–2.5 cm. long, greenish, becoming dark purple when ripe; stone light brown, ellipsoid, smooth; $2n = 24$ (Bambacioni, 1941).—Common, canyons and valleys, below 5000 ft.; Chaparral, Foothill Wd., lower Yellow Pine F., Mixed Evergreen F., Redwood F.; Coast Ranges and Sierra Nevada, San Diego Co. to nw. Calif. and sw. Ore., where it reaches its greatest size and is known as Myrtle. The wood is hard, strong, and takes a high polish; used for turning. Dec.–May.

Var. fresnénsis Eastw. Branches of infl. and lower surface of lvs. finely tomentulose.—Fresno Co.

3. Paeoniàceae. Paeony Family

Large shrubs or herbs with spirally arranged estipulate lvs. Fls. large, terminal, usually solitary, perfect, actinomorphic, hypogynous. Calyx mostly of 5 free sepals. Corolla of 5–10(–13) large free petals. Stamens many, maturing in centrifugal sequence, often united below into a ring. Carpels 2–5, free, with fleshy walls and mounted on a fleshy disk. Ovules several, each with 2 integuments of which the outer projects beyond the inner. Fr. of 2–5 large follicles each with several seeds. A single genus. Differing from *Ranunculaceae*, in which usually included, in having persistent sepals and stamens maturing in centrifugal sequence.

1. Paeònia L. PEONY

Ours perennial herbs with fascicled fleshy roots. Lvs. basal and cauline, alternate, large, ternately divided. Fls. large, solitary, terminal. Sepals 5 or 6, persistent, green. Petals mostly 5 or 6, with the numerous stamens borne on a fleshy disk adnate to the base of the calyx. Pistils 2–5, free, becoming large fleshy dehiscent follicles, with several large seeds. Ca. 30 spp. of the N. Hemis., largely Asian, many of horticultural value. (Greek, *Paeon*, the physician of the gods.)

Stems 2–4 dm. high, 5–8-lvd.; lvs. glaucous, their primary divisions with distinct petiolules below the abruptly contracted bases; petals round, shorter than inner sepals . . . . 1. *P. Brownii*
Stems 3.5–7.5 dm. high, 7–12-lvd.; lvs. green, their primary divisions subsessile, cuneate at base; petals elliptic, longer than inner sepals . . . . 2. *P. californica*

1. **P. Brownii** Dougl. ex Hook. Somewhat fleshy, with deep heavy roots; stems one to several, 2–4 dm. high, mostly simple; lvs. 5–8 per stem, glaucous, biternate, usually somewhat fleshy, the primary segms. abruptly contracted at base, petiolulate, mostly 3–6 cm. long, 2–5 cm. wide, the ultimate lobes elliptic and usually obtuse; petioles mostly 2–7 cm. long; fls. solitary, subglobose, on recurved leafy peduncles; sepals concave, roundish; petals rounded, 8–13 mm. long, mostly slightly wider, maroon to bronze in center, yellowish or greenish at margin; fils. 3–5 mm. long; anthers 2–4 mm. long; follicles 2–4 cm. long; seeds blackish, cylindrical, 10–12 mm. long; $n = 5$ (Stebbins, 1938).—Dry slopes, 3000–7300 ft.; Sagebrush Scrub, Yellow Pine F., Chaparral; Coast Ranges and Sierra Nevada from Santa Clara and Tuolumne cos. n.; to B.C., Wyo., Nev. April–May.

2. **P. califórnica** Nutt. ex T. & G. [*P. Brownii* var. *c.* Lynch, *P. B.* ssp. *c.* Abrams.] Stems 5–30, mostly branched, 3.5–7 dm. high; lvs. 7–12 per stem, green, thin, easily wilting, the primary segms. cuneate at base, sessile or with short winged petiolules, mostly 3–7 cm. long, 1–4 cm. wide, the ultimate lobes lanceolate or narrow-elliptic, usually acute; petals elliptic, 15–25 mm. long, 11–18 mm. wide, deep blackish-red at center, pink at margin; fils. 5–8 mm. long; anthers 3–7 mm. long; follicles 3–4 cm. long; seeds blackish, subterete, slightly curved, 1.3–2 cm. long, 5–8 mm. thick; $n = 5$ (Stebbins, 1938).—Brushy places, below 4000 ft.; Chaparral, Coastal Sage Scrub; Coast Ranges from Monterey Co. to San Diego Co. Jan.–March.

4. Ranunculàceae. CROWFOOT FAMILY (Fig. 10)

Herbs, sometimes small shrubs or woody climbers. Lvs. alternate or opposite or basal, the blades simple to compound; petioles with dilated base, exstipulate. Fls. with the parts usually all present, free and distinct, but exceptions are many: petals may be lacking and sepals petaloid; sepals and petals may be spurred; pistils may be somewhat united. Sepals mostly 3–15. Petals usually the same. Stamens usually many, hypogynous, maturing in centripetal sequence. Pistils few or many; ovary superior, 1-celled, with 1-many ovules and basal or parietal placentae. Fr. an ak. or follicle, sometimes a berry or caps. Ca. 35 genera and perhaps 1500 spp., mostly of N. Temp. and Arctic regions.

Pistils few- to several-ovuled, becoming follicles or berries when ripe.
  Fls. irregular.
    Upper sepal spurred; petals 4, the upper pair spurred and included in the spurred sepal
                                                                                4. *Delphinium*
    Upper sepal hooded; petals 2, clawed, included in the hooded sepal . . . . 5. *Aconitum*
  Fls. regular.
    Petals produced backwards into long spurs . . . . 10. *Aquilegia*
    Petals not spurred.
      The petals 6–25 mm. long . . . . 13. *Coptis*
      The petals none or not over 3 mm. long (sepals petaloid).
        Sepals 12–16 mm. long; lvs. undivided . . . . 1. *Caltha*

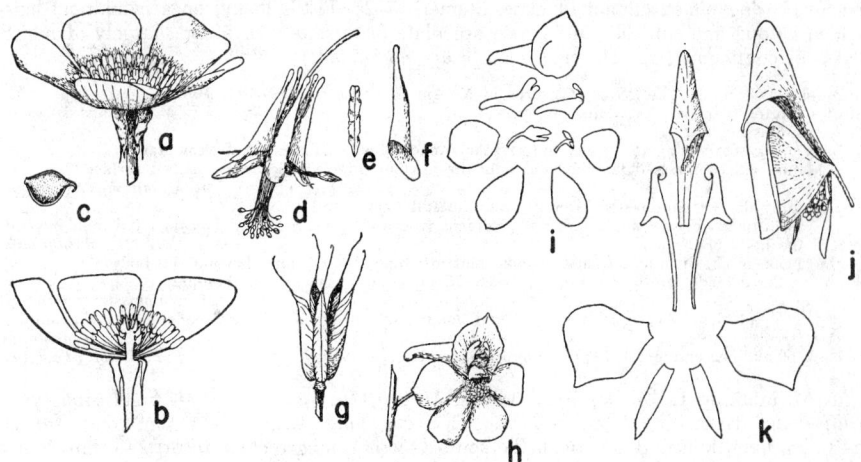

Fig. 10. RANUNCULACEAE. *Ranunculus repens: a*, fl., × 1, with 5 reflexed sepals, spreading petals, numerous stamens; *b*, fl. in section, with cent. receptacle and aks.; *c*, ak., × 6. *Aquilegia formosa: d*, fl., × 1, sepals spreading and between the erect spurred petals; *e*, staminodium; *f*, petal with short lamina and tubular spur; *g*, follicles, × 1. *Delphinium Nuttallianum: h*, fl., × 1, with 5 outer sepals (the uppermost spurred) and 4 inner petals, the perianth pulled apart in *i*. *Aconitum columbianum: j*, fl. from side with upper sepal or helmet, 2 lateral sepals and pendent linear lower sepals; *k*, perianth pulled apart, with 5 sepals and 2 petals which grow up into the helmet.

Sepals 2–9 mm. long; lvs. compound.
    Fls. solitary, cymose or panicled, 4–9 mm. long .... 9. *Isopyrum*
    Fls. racemose, 2–3 mm. long .... 3. *Actaea*
Pistils 1-ovuled, becoming aks.
  Stem-lvs. opposite or whorled; petals none.
    The stem-lvs. forming a single whorl; herbs .... 7. *Anemone*
    The stem-lvs. many, opposite; woody vines .... 8. *Clematis*
  Stem-lvs. alternate or none.
    Petals present (rarely absent in *Ranunculus*).
      Sepals spurred; small annuals with basal linear lvs. .... 2. *Myosurus*
      Sepals not spurred; lvs. usually cauline as well as basal .... 6. *Ranunculus*
    Petals absent (sepals may be petaloid).
      Lvs. deeply lobed; but not decompound; fls. bisexual .... 12. *Trautvetteria*
      Lvs. decompound; fls. usually unisexual .... 11. *Thalictrum*

1. Cáltha L. Marsh-Marigold

Fleshy perennial herbs from short vertical rootstocks with fascicled fibrous roots. Lvs. large, simple, mostly basal. Fls. 1-few per stem. Sepals 5–9, petaloid. Petals 0. Stamens many; fils. short. Pistils several to many, sessile, becoming many-seeded follicles. Ca. 15 spp. of wet places in the Arctic and N. Temp. regions. (*Caltha*, Latin name for marigold.)

1. **C. Howéllii** (Huth) Greene. [*C. biflora* var. *H.* Huth. *C. biflora* auth., not DC.] Lvs. round-reniform, all basal, obscurely repand-crenate, the blades 3–10 cm. wide, the petioles 5–30 cm. long, fleshy; fl. solitary on a bractless scape which exceeds the lvs.; sepals white, 6–10, oblong, 12–16 mm. long; stamens ca. 6 mm. long; follicles stipitate, 10–15 mm. long; seeds brownish, narrowed at both ends, ca. 2 mm. long.—Marshy and boggy places, 4500–10,500 ft.; Montane Coniferous F.; Sierra Nevada from Tulare Co. n. and in N. Coast Ranges; s. Ore. May–July.

2. Myosùrus L. Mouse-Tail

Small tufted annuals with fibrous roots. Lvs. basal, entire, linear or linear-spatulate. Fls. minute, greenish-yellow to whitish, solitary on naked scapes. Sepals 5, sometimes 6–7, spurred at base. Petals of same number if present, greenish-yellow, each with a nectar-bearing pit at summit of claw. Stamens 5–25. Pistils many, on a cylindrical axis which is long and spikelike in fr. Aks. apiculate or aristate. Ca. 5 spp., largely of moist flats, all continents. (Greek, *mus*, mouse, and *oura*, tail.)

(Campbell, Gloria R. The genus Myosurus in N. Am. El Aliso 2: 389–403, 1952.)
Scapes shorter than or rarely equaling lvs. .... 1. *M. minimus* ssp. *apus*
Scapes surpassing lvs.
Back of ak. (excepting the beak) longer than broad. Mostly montane and cismontane.
Mature ak. compressed laterally and only the median ridge visible on the dorsal surface
                                                                1. *M. minimus* ssp. *major*
Mature ak. not compressed laterally, but the full back visible.
Beak of ak. closely appressed, not divergent from the spike, extended less than 0.5 mm. beyond
the body of the ak. .... 1. *M. minimus*
Beak of ak. divergent from the spike and extended 0.5–1.5 mm. beyond the body.
Sepal-blade faintly 3-nerved; spike 6–25 mm. long; back of ak. rhombic
                                                                1. *M. minimus* ssp. *montanus*
Sepal-blade 1-nerved; spike 3–9 mm. long; back of ak. elliptic or broader below the middle
                                                                2. *M. aristatus*
Back of ak. (excepting the beak) roundish in outline. E. Mojave Desert .... 3. *M. cupulatus*

1. **M. mínimus** L. [*M. lepturus* Howell. *M. aristatus* var. *l.* Jeps. *M. breviscapus* var. *californicus* Huth.] Lvs. linear-filiform, 2–8 cm. long; scapes 3–15 cm. long; sepals 2–3 mm. long, faintly 3–5-nerved, the spurs 1–3 mm. long; petals linear, 2–3 mm. long; carpel-spike 1–5 cm. long, 2–3 mm. thick at base; aks. rhombic on back, 1–2.5 mm. long, with closely appressed erect beaks 0.25 mm. long; seed green, ovoid to ellipsoid, 0.66–1 mm. long, mostly 2–3 times as long as thick.—Moist places, such as vernal pools, generally below 2500 ft. (to 4800 in San Jacinto Mts.); mostly V. Grassland, Chaparral, N. Oak Wd.; San Diego Co. to Siskiyou and Modoc cos.; to B.C., Atlantic Coast, Eurasia, etc. March–June.

KEY TO SUBSPECIES AND VARIETIES

Scapes surpassing the lvs.
Mature ak. not compressed laterally, 0.5–1.5 mm. across, the full back of the ak. visible on surface
of spike.
Beak of ak. extended less than 0.5 mm. beyond the body, not divergent from spike.
Mature ak. truncatish at base in side view, lateral ridges lacking on back, ak. usually over
1.5 mm. long; spike mostly 2–3 mm. thick at base .... *M. minimus*
Mature ak. rounded at base in side view, lateral ridges present on back, ak. usually less than
1.5 mm. long; spike 1–1.5 mm. thick at base .... var. *filiformis*
Beak of ak. extended 0.5–1.5 mm. beyond body, divergent .... ssp. *montanus*
Mature ak. compressed laterally, 0.3–0.6 mm. across, only the median ridge visible on spike surface
                                                                ssp. *major*
Scapes shorter than or rarely equaling the lvs.
Mature aks. not compressed laterally, 0.7–1.5 mm. long, the beak extended 0.25 mm. beyond the
body. S. Calif. .... ssp. *apus*
Mature aks. compressed laterally, 1.8–2.3 mm. long, the beak extended 0.5–1.2 mm. beyond the
body. Cent. and n. Calif. .... ssp. *apus* var. *sessiliflorus*

Var. **filifórmis** Greene. Very slender, the spike ca. 1–1.5 mm. thick at base; seed less than twice as long as thick.—Vernal pools, below 3000 ft.; V. Grassland, Chaparral, N. Oak Wd.; San Diego, inner Coast Ranges and borders of Cent. V., from Kern Co. to Mendocino Co. March–May.

Ssp. **montànus** Campb. Spike ca. 2 mm. thick at base; beak of ak. extended 0.5–1 mm. beyond the body, divergent; seed 0.8–1.7 mm. long, 2–4 times longer than wide.—Wet places, 5000–10,000 ft.; Montane Coniferous F.; San Bernardino Mts., Sierra Nevada from Tulare Co. to Plumas Co., Trinity Co.; to Canada, Rocky Mts. July–Aug.

Ssp. **màjor** (Greene) Campb. [*M. m.* Greene. *M. minimus* var. *m.* Davis.] Lvs. 2–8 cm. long; scapes 5–16 cm. long; spike 2–3 mm. thick at base; ak. narrow, compressed laterally; seed 1–1.25 mm. long.—Near Edgewood, Siskiyou Co.; to Wash.

Ssp. **àpus** (Greene) Campb. [*M. minimus* var. *a.* Greene.] Lvs. 2–9 cm. long; scapes shorter or barely as long; spike 2–3 mm. thick at base; back of aks. rhombic, the beak extended 0.25 mm. beyond the body; seed 0.6–0.8 mm. long.—Vernal pools, below 1500 ft.; V. Grassland, Coastal Sage Scrub; San Diego to w. Riverside Co. April–May.
Var. **sessiflòrus** (Huth) Campb. [*M. aristatus* var. *s.* Huth. *M. sessilis* Wats. *M. alopecuroides* Greene.] Lvs. mostly 1–5 cm. longer than the scapes; mature aks. com-

pressed laterally, the beak extended 0.5–1.25 mm. beyond the body.—V. Grassland; Cent. V. and adjacent Coast Ranges, Fresno Co. to Glenn Co. March–April.

2. **M. aristàtus** Benth. in Hook. [*M. apetalus* auth., not Gay.] Lvs. narrow-linear to linear-spatulate, 1–7 cm. long; scapes 2–5(–10) cm. high, slender; sepals linear-oblong, 1–2 mm. long, the spurs almost as long; spikes 5–10 mm. long; aks. elliptic on back, keeled and with a marginal nerve on each side of keel; beak ca. as long as body and divergent.—Moist places, 4500–7000 ft.; Sagebrush Scrub, N. Juniper Wd., Montane Coniferous F.; Doble (San Bernardino Mts.), Fallen Leaf Lake (Eldorado Co.), and Modoc Co.; to B.C., Rocky Mts. May–July.

3. **M. cupulàtus** Wats. Lvs. filiform-linear, 1–6 cm. long; scapes 2–10 cm. high, slender; sepals 1.5–3 mm. long, the spurs ca. as long; spikes 5–20(–40) mm. long, 2–3 mm. thick at base; aks. rounded, dorsally cupulate, the beak somewhat spreading, ca. 0.5 mm. long.—Dry limestone slopes, 3700–5500 ft.; Joshua Tree Wd., Pinyon-Juniper Wd.; Little San Bernardino, Providence, New York, and Kingston mts.; to New Mex. April–May.

3. Actaèa L. BANEBERRY

Erect perennial herbs from short branching rootstocks. Stems tall, with 1–3 large 2–3-times ternate lvs. Fls. small, white, in terminal racemes. Sepals 4–5, petaloid, caducous. Petals 4–5, small, flat, spatulate, with slender claws. Stamens many. Pistil 1, with a broad sessile 2-lobed stigma; ovary 1-celled with 2 parietal placentae and several to many ovules. Fr. a berry; seeds smooth, lens-shaped, in 2 horizontal rows. Ca. 6 spp. of rich woods of N. Temp. Zone. (Greek name of the elder.)

1. **A. rùbra** (Ait.) Willd. ssp. **argùta** (Nutt.) Hult. [*A. a.* Nutt. ex T. & G. *A. spicata* var. *a.* Torr.] Stems 1–several, 2–6(–8) dm. tall, branching above, sparsely pubescent; lvs. all cauline, the lowest 1–2 dm. wide, with thin mostly ovate incised and serrate lfts. 2.5–6 cm. long and petioles to ca. 1 dm. long; racemes 1–5, dense in fl., 6–10 cm. long in fr.; pedicels 4–8 mm. long; sepals and petals 2–3 mm. long, falling away; stamens 5 mm. long; berries red or white, oblong-ovoid to rounded, shining, 6–8 mm. long; seeds dark brown, compressed, pitted, ca. 3 mm. long; $n = 8$ (Rodriguez, 1949).—Rich moist woods, below 10,000 ft.; Montane Coniferous F., Mixed Evergreen F., Redwood F.; Sierra Nevada s. to Tulare Co., San Bernardino Mts., Coast Ranges s. to San Luis Obispo Co.; to Alaska, Rocky Mts. May–June.

4. Delphínium L. LARKSPUR

Erect branching herbs, ours mostly perennial from a caudex, with fibrous or fleshy roots. Lvs. palmately lobed or divided. Fls. irregular, in showy spikes or racemes which may be paniculate. Sepals 5, blue, white, yellow, or red, the posterior sepal prolonged into a spur into which project the nectary-bearing spurs of the 2 upper petals. Petals 4, in unequal pairs, the 2 lateral small and short-clawed to obsolete. Stamens many, usually included. Pistils 1–5, sessile, maturing into many-seeded dehiscent follicles. Seeds obpyramidal, the angles winged or wingless, sometimes inclosed in a loose papery membrane. Ca. 250 spp. of the N. Temp. Zone, some important in hort., and some as stock-poisoning plants. The poisoning of cattle from eating delphinium plants is known as delphinosis; in Calif. delphinosis is responsible for more cattle losses than any other plant poisoning. *D. Menziesii* is probably the most destructive species in Calif. because of its occurrence in such enormous masses. Larkspurs are poisonous for cattle and horses, but usually not for sheep. (Latin, *delphinus,* dolphin, because of fl.-shape in some spp.)

ʃEwan, J. A synopsis of the N. Am. spp. of Delphinium. Univ. Colo. Phys. & Biol. Studies 2 [2]: 55–244, 1945. Lewis, H., & C. Epling. A taxonomic study of Californian delphiniums. Brittonia 8: 1–22, 1954.)

A. Plants annual; pistil 1. Garden escape .... 1. *D. Ajacis*
AA. Plants perennial; pistils 3. Natives.
  B. Fls. red or yellow to yellowish-white.
    C. Lvs. divided into linear or narrowly lanceolate primary divisions; fls. usually scarlet
                                                   30. *D. cardinale*
  CC. Lvs. divided into broad primary divisions.

  D. Fls. rose-pink.
    E. Stems hollow, 7–10 dm. high; sepals acuminate . . . . 2. *D. Purpusii*
    EE. Stems not hollow, 2–4 dm. high; sepals rounded
                       6. *D. gracilentum* f. *versicolor*
  DD. Fls. not rose-pink; blade of lower petal bifid.
    E. The fls. orange- or dull-red; follicles divergent . . . . 13. *D. nudicaule*
    EE. The fls. cream or yellowish; follicles erect . . . . 14. *D. luteum*
BB. Fls. blue to purple, sometimes bluish-white.
  C. Plants scapose, with fleshy basal trifid lvs., the divisions entire to toothed. Inner N.
    Coast Ranges . . . . 3. *D. uliginosum*
  CC. Plants not both scapose and with simply tripartite lvs.
    D. The stems very slender at base and easily separated from the tuberous roots.
      E. Stems and lvs. pubescent.
        F. Plants 1–3 dm. high; follicles 7–10 mm. long.
          G. Sepals 12–16 mm. long; follicles glabrous. Santa Cruz Co. n.
                           4. *D. decorum*
          GG. Sepals 9–10 mm. long; follicles thinly hairy. S. Sierra Nevada
                           7. *D. pratense*
        FF. Plants 2–7 dm. high: follicles 10–14 mm. long, hairy. Mendocino Co. n.
                         8. *D. Menziesii*
      EE. Stems and lvs. mostly glabrous, except sometimes in infl.
        F. Sepals glabrous without; follicles glabrous.
          G. The sepals 11–15 mm. long; follicles 10–14 mm. long. Coast
            Ranges, Lake and Colusa cos. to L. Calif.; Sierra Nevada from
            Butte to Eldorado cos. . . . . 5. *D. patens*
          GG. The sepals 5–10 mm. long.
            H. Each upper petal with a subapical lower lobe, the tip thus
              mitten-shaped. Pine belt, Mariposa Co. n.
                       10. *D. depauperatum*
            HH. Each upper petal not so lobed.
              I. Fls. mostly 15–20, dark; petioles 10–20 cm. long. At
                1000–5000 ft., Butte Co. to Mariposa Co.
                       6. *D. gracilentum*
              II. Fls. mostly 6–10, pale; petioles 3–6 cm. long. At 6000–
                8000 ft., Fresno and Tulare cos. . . . . 7. *D. pratense*
        FF. Sepals pubescent to hairy.
          G. Follicles and infl. glandular-pubescent.
            H. Sepals pale lavender; follicles ca. 1 cm. long. Sutter Co. to
              Kern Co. . . . . 5. *D. patens* ssp. *Greenei*
            HH. Sepals blue; follicles 12–17 mm. long. Modoc Co.
                       11. *D. diversifolium*
          GG. Follicles and infl. glabrous to pubescent, but not glandular.
            H. Lvs. chiefly cauline.
              I. Follicles thinly hairy; sepals 12–16 mm. long. Dry places,
                Modoc and Siskiyou cos. to Mariposa Co.
                       9. *D. Nuttallianum*
              II. Follicles glabrous; sepals 8–10 mm. long. Damp places,
                Mariposa to Siskiyou cos. . . . . 10. *D. depauperatum*
            HH. Lvs. basal; follicles glabrous. Wet places, Del Norte Co. to
              Siskiyou and Eldorado cos. . . . . 12. *D. Sonnei*
  DD. The stem not attenuate at base and rather firmly attached to woody or fibrous (not
    tuberous) roots.
    E. Plants mostly 1–2 m. tall, leafy up to the dense, many-fld. raceme.
      F. Sepals 15–25 mm. long; follicles 15–25 mm. long; sinus of lower petals
        4–5 mm. deep . . . . 17. *D. trolliifolium*
      FF. Sepals 6–12 mm. long; follicles 9–18 mm. long; sinus of lower petals
        1–2.5 mm. deep.
        G. Fls. whitish or yellowish with some green or lavender; spur 6–8 mm.
          long. Coast Ranges, Contra Costa Co. to Santa Clara Co.
                       19. *D. californicum*
        GG. Fls. deep blue to violet-purple; spur 8–13 mm. long.
          H. Stems glabrous and glaucous to the infl.; primary divisions of
            lvs. broad, cuneate, laciniately toothed; sinus of lower petals
            2–2.5 mm. deep. Siskiyou Co. to Madera Co.; Los Angeles
            and San Bernardino cos. . . . . 15. *D. glaucum*
          HH. Stems puberulent above; primary divisions of lvs. divided al-
            most to base into linear lobes; sinus of lower petals 1 mm.
            deep. Modoc Co. . . . . 20. *D. stachydeum*
    EE. Plants mostly much less than 1 m. tall; at least the uppermost lvs. beneath
      the infl. reduced to leafy bracts.
      F. Fls. strongly bicolored, the darker sepals contrasting markedly with
        the light-colored petals.
        G. Raceme hirsutulous; sepals blue to violet; sinus of upper petals
          closed, 3–4 mm. deep. Coastal San Luis Obispo and Santa Barbara
         cos. . . . . 26. *D. Parryi* ssp. *Blochmanae*

GG. Raceme glabrous; sepals light blue; sinus of upper petals open, 1–3 mm. deep. Interior, from Glenn and Butte cos. to Kern Co.
    27. *D. recurvatum*

FF. Fls. not conspicuously bicolored.
  G. The fls. deep blue to purple.
    H. Stems essentially glabrous; lvs. mostly green at anthesis.
      I. Lvs. glabrous, pentagonal in outline, with broadly cuneate approximate divisions.
        J. Lf.-blades mostly 4–6 cm. wide, with acute teeth; petioles 15–20 cm. long; follicles pubescent. High montane .... 16. *D. polycladon*
        JJ. Lf.-blades mostly 6.5–7.5 cm. wide, with crenate teeth; petioles 12–15 cm. long; follicles glabrous. Sonoma and Marin cos. .... 18. *D. Bakeri*
      II. Lvs. pubescent to hairy, the divisions not as above.
        J. Sepals 8–24 mm. long.
          K. Petioles mostly spreading-hairy. Mostly Cent. V. and borders .... 29. *D. variegatum*
          KK. Petioles glabrous to puberulent.
            L. Follicles 14–28 mm. long; lf.-blades 1–3 cm. wide, rounded in outline, with oblong segms. Mono Co. n. .... 31. *D. Andersonii*
            LL. Follicles 6–10 mm. long; lf.-blades mostly 3–6 cm. wide, with angular outline. W. edge of Colo. Desert
              24. *D. Parishii* ssp. *subglobosum*
        JJ. Sepals 5–7 mm. long. Desert slopes, San Bernardino Mts. .... 24. *D. Parishii*
    HH. Stems pubescent to hairy; lvs. usually withered at anthesis.
      I. Petioles with long spreading hairs.
        J. Midveins of lvs. mostly rusty-brownish; seeds with white-margined angles .... 23. *D. hesperium*
        JJ. Midveins not rusty-brownish; seeds scaly-echinate
          22. *D. Hanseni*
      II. Petioles strigose to puberulent.
        J. Lf.-blades with prominent rusty veins, the primary divisions of lvs. at least 3 mm. wide. Humboldt Co. to Kern Co. .... 23. *D. hesperium*
        JJ. Lf.-blades not so veined, the primary divisions narrower (except sometimes in var. *maritimum*). Santa Barbara Co. to L. Calif. .... 26. *D. Parryi*

GG. The fls. pale blue to whitish.
  H. Stems essentially glabrous.
    I. Fls. white or pinkish; follicles ca. 20 mm. long, hairy. Cent. V. .... 21. *D. gypsophilum*
    II. Fls. light blue to lavender; follicles 10–14 mm. long.
      J. Petioles conspicuously spreading-hairy; sepals 12–16 mm. long. San Luis Obispo Co. ... 29. *D. variegatum*
      JJ. Petioles glabrous to short-pubescent; sepals 8–12 mm. long. Deserts and desert-borders ..... 24. *D. Parishii*
  HH. Stems puberulent to hairy.
    I. Sepals pale purple, 6–8 mm. long; follicles glabrous. Mts. of Riverside and San Diego cos.
      23. *D. hesperium* ssp. *cuyamacae*
    II. Sepals white to pink or greenish, 9–10 mm. long; follicles puberulent. From Kern Co. n.
      J. Lvs. well distributed along stem; fls. white to pink, rather crowded .... 23. *D. hesperium* ssp. *pallescens*
      JJ. Lvs. largely basal; fls. greenish-white to bluish-purple, not crowded .... 22. *D. Hanseni*

1. **D. Ajàcis** L. [*D. Gayanum* Wilmott.] Rocket Larkspur. Annual, pubescent, 3–6 dm. tall, few-branched; basal and lower lvs. long-petioled, the blades divided to the base into 3 parts, these dissected into linear segms.; upper lvs. sessile; pedicels ascending, 2–4 cm. long, with subulate bracts; fls. blue or violet to rose, pink or white, the spur upwardly curved, ca. 2 cm. long; pistil 1; follicle pubescent; seeds transversely thin-ridged; n = 8 (Gregory, 1941).—Occasional garden escape; native of Eu. May–July.

2. **D. Purpùsii** Bdg. [*D. roseum* Heller.] Perennial, from a deep-seated woody-fibrous rootstock; stems hollow, 7–10 dm. high, simple or nearly so, thinly hirsute and glandular above; lvs. brittle, mostly lower-cauline, with lower petioles 10–15 cm. long, blades 7–9 cm. wide, ca. equally 5-palmatifid, with narrow sinuses and broad divisions, the teeth abruptly acute, pubescent on margins and veins beneath; racemes rather

loose, 5–14-fld.; sepals rose-pink, narrow-ovate, acuminate, 7–10 mm. long, the spur 13–15 mm. long; upper petals whitish, 7–8 mm. long, the lower rose, split nearly to the middle; follicles 18–23 mm. long, oblong, erect, glabrous; seeds 2–3 mm. long, black, the angles winged with a white loose cellular coat; $n = 8$ (Lewis et al., 1951).— Dry rocky slopes, below 4000 ft.; Chaparral, Foothill Wd., Kern Co. April–May.

3. **D. uliginòsum** Curran. [*D. decorum* var. *u.* Huth.] Roots clustered, fusiform; plants scapose, 3–5 dm. high, fine-hairy under a lens; basal lvs. 2–4 cm. wide, cuneate-orbicular, 3-cleft, the primary divisions entire to 3–5-toothed or crenately lobed, the ultimate segms. rounded or barely acute; petioles 4–8 cm. long; racemes interrupted, 3–12-fld.; sepals rich bright blue, ovate to obovate, 10–15 mm. long, rounded to acute, with median puberulent band; spur slender, often curved upward, 12–15 mm. long; upper petals whitish, crisped at the oblique violet apex; lower petals violet, ovate, rounded, white-ciliate, bifid, the sinus narrow, 2–4 mm. deep; follicles erect, slender, 1 cm. long; seeds 1 mm. long, black, muriculate, weakly angled with little or no development of membranous wings; $2n = 16$ (Lewis *et al.*, 1951).—Wet heavy soil, near streams and swamps in serpentine areas; Chaparral, Foothill Wd.; inner N. Coast Ranges, Napa, Lake, and Colusa cos. May–June.

4. **D. decòrum** F. & M. [*D. d.* var. *racemosum* Eastw.] Roots fleshy, in a small cluster; stems 1-several, simple or few-branched, erect, 1–2(–3) dm. high, pubescent; lvs. mostly near the base, tripartite, 3–4 cm. wide, cent. division entire or trifid, the lateral 1–2-bifid, somewhat pubescent, fleshy; cauline lvs. 1–2, reduced; fls. 2–5(–8), villous; sepals blue-purple, oblong-ovate, obtuse or acute, 12–16 mm. long, 7–8 mm. wide, with median pubescent band; upper petals oblique, whitish, with distinct lateral lobe; lower petals ovate, bluish, bifid, with open sinus 1.5–2.5 mm. deep; follicles erect, glabrous, ca. 1 cm. long; seeds reticulate-roughened, margined at the angles.—Open grassy slopes, below 2000 ft.; N. Coastal Scrub, Coastal Prairie; near the coast, Mendocino Co. to Santa Cruz Co. March–May.

Ssp. **Tràcyi** Ewan. [*D. antoninum* Eastw.] Stems often branched at base, hollow below, thinly hirsute in infl.; lvs. few, the lateral primary segms. deeply palmatisect; raceme loosely irregular, with elongate lower branches; $n = 8$ (Lewis *et al.*, 1951).— Grassy slopes and wooded hills, 2000–7000 ft.; Yellow Pine F., Mixed Evergreen F., and Coastal Prairie; N. Coast Ranges, Lake Co. to Humboldt Co.; s. Ore. May–June.

5. **D. pàtens** Benth. [*D. decorum* var. *p.* Gray. *D. d.* var. *sonomensis* Eastw.] Roots tuberiform, in a globose cluster; stems simple or branched, 2–4(–5) dm. tall, fistulous, mostly glabrous; lvs. few, mostly near the base, 4–9 cm. wide, tripartite, the cent. part trifid, the lateral bifid, ultimate lobes oblong, obtusish; lower petioles 5–15 cm. long; raceme simple or branched, open, 6–15-fld.; sepals mostly dark blue, ovate to ovate-oblong, acute, 10–13 mm. long, 5–6(–8) mm. wide, mostly glabrous; spur rather stout, 8–11 mm. long; upper petals oblique, whitish or cream, with blue lines; lower petals ovate, whitish with blue lines to bluish, with cent. mass of hairs, the sinus 2 mm. deep, somewhat open; follicles oblong, erect, 10–14 mm. long, glabrous; seeds prismatic-oblong, dark brown, rugulose, ca. 2 mm. long, with narrow pale wing-margins; $2n = 16$ (Lewis *et al.*, 1951).—Borders of thickets and in woods, below 3500 ft.; Chaparral, Foothill Wd.; Coast Ranges, Lake and Colusa cos. to n. Santa Barbara Co., and base Sierra Nevada, Butte Co. to Eldorado Co. March–May.

Ssp. **Greènei** (Eastw.) Ewàn. [*D. G.* Eastw.] Stems 2–4.5 dm. tall, glandular-pubescent in infl.; lvs. tripartite, 3–5 cm. wide; racemes 3–12-fld.; sepals pale lavender, sparingly glandular-pubescent; follicles glandular-pubescent, ca. 1 cm. long.—Wooded canyons below 4500 ft.; Foothill Wd.; Kern Co. to Sutter Co. May.

Ssp. **montànum** (Munz) Ewan. [*D. Parryi* var. *m.* Munz.] Rather stout, 2.5–3.5 dm. tall, glabrous; lvs. mostly basal and with main divisions shallowly lobed, a few lvs. cauline and deeply palmatisect; racemes compact, 6–12-fld.—At 5000–7500 ft.; Yellow Pine F.; Ventura Co. to San Bernardino Co. May–June.

Ssp. **hepaticoìdeum** Ewan. Stems slender, procumbent, 2–5(–9) dm. long, glabrous; lvs. near base, long-petioled, 4–7(–9) cm. wide, with 3 broadly obovate primary divisions, the cent. shallowly 3-lobed, the lateral 2-lobed; racemes lax, with long pedicels; sepals 13–15 mm. long, glabrous; follicles 13–14 mm. long.—Shaded canyons, mostly below 5000 ft.; Chaparral, S. Oak Wd.; Santa Barbara Co. to San Diego Co.; L. Calif. April–May.

6. **D. graciléntum** Greene. [*D. decorum* var. *g.* Davis.] Roots shallow, tuberiform; stem slender, often procumbent, 2–4 dm. long, simple, glabrous except for glandular-pubescent infl., ± glaucous; lvs. few, mostly basal, thin, equally 5-fid, 4–5(–10) cm. wide, the primary divisions with oblong obtusish ultimate lobes, pubescent especially beneath; petioles 1–2 dm. long; racemes lax, narrow, (8–)15–20-fld.; pedicels filiform, spreading-ascending; sepals dark deep blue, ovate, rounded, 6–10 mm. long, subglabrous or with some median hair; spur straightish, 10–12 mm. long; upper petals white, narrow-oblique; lower petals blue, round-ovate, with an open sinus 2 mm. deep; follicles glabrous, somewhat spreading, 8–10(–12) mm. long; seeds obpyramidal, 2 mm. long, muriculate, black, the truncate summit with white membranous rim, angles wingless; $2n = 16$ (Lewis *et al.*, 1951).—Partial shade, mostly damp places, 1000–8000 ft.; Foothill Wd., Yellow Pine F.; w. base of Sierra Nevada, Butte Co. to Mariposa Co. April–July. F. *versicolor* Ewan. Fls. pale pink.—With the sp. and s. to Kern Co. May.

7. **D. praténse** Eastw. Root small, tuberiform; stems glabrous below, ± pubescent above, 2.5–3 dm. tall, simple or branched, slender; lvs. few, 2.5–4(–5) cm. wide, ± pubescent, 5-fid, the primary divisions with few long, linear-oblong lobes; petioles 3–6 cm. long; racemes laxly 6–10-fld., with conspicuous bracts and hirsute pedicels 1.5–5 cm. long; sepals narrow-ovate, violet-blue, subglabrous, 9–10 mm. long, 4–6 mm. wide; spur slender, 1 cm. long; upper petals whitish, narrow-oblique; lower petals pale, deltoid-ovate, the open sinus 1–2 mm. deep; follicles thinly hairy, somewhat spreading, less than 1 cm. long; seeds brown, rugulose, with terminal membranous crown.—Meadows and open woods, 6000–8500 ft.; Montane Coniferous F.; Fresno and Tulare cos. June–July.

8. **D. Menzièsii** DC. Tubers in small shallow globose cluster; stems 2–5(–7) dm. high, soft-pubescent with spreading white hairs; lvs. both cauline and radical, reduced upward to leafy bracts, the blades round-pentagonal, 3–6 cm. wide, pubescent, palmatifid into broad approximate cuneate divisions, these shallowly to deeply lobed, the ultimate segms. blunt to short-acute, mucronulate; lower petioles 5–8(–15) cm. long; racemes 3–10-fld., short, hairy; pedicels spreading; sepals deep rich blue, oblong-ovate, 10–15 mm. long, 9–11 mm. wide, apiculate, hairy; upper petals pale, rhombic; lower petals rounded, dark blue, sometimes white-lined, 7–9 mm. wide, retuse to 2-lobed; follicles somewhat spreading, 10–14 mm. long, white-hairy; seeds few, oblong-prismatic, brownish, 1.5 mm. long, narrowly wing-margined.—Open places above the ocean; N. Coastal Scrub; Mendocino Co. n.; to B.C. March–May.

9. **D. Nuttalliànum** Pritz. ex Walp. [*D. pauciflorum* Nutt. ex T. & G., not D. Don.] Roots tuberiform, in a small cluster; stems slender, subglabrous below, pubescent above, glandular or not, 1–3(–4) dm. tall; lvs. few, chiefly cauline, lowest withering early, rounded in outline, 3–5 cm. wide, with 3 or 5 approximate primary divisions, these subsimple to palmatisect into oblong obtuse lobes, nearly or quite glabrous, petioles 5–7 cm. long; upper lvs. 3-parted, bractlike; racemes few-fld., crowded or open; fls. nodding, bright blue to purplish; pedicels ascending, slender; bracts leaflike to linear; sepals oblong-ovate, obtuse or acute, 8–12 mm. long, 5–7 mm. wide, loosely hairy to pubescent; spur slender, 10–15 mm. long; upper petals whitish, short-oblique; lower petals narrow-ovate, translucent, the narrow sinus 1–2 mm. deep; follicles erect or flaring, 1–1.8 cm. long, thinly hirsute; seeds quadrate-oblong, dark, 1–1.5 mm. long, glabrous, with narrow wing-angles; $2n = 16$ (Lewis *et al.*, 1951).—Grassy and brushy places and open woods, 5000–10,000 ft.; Sagebrush Scrub, Chaparral, Yellow Pine F., Lodgepole F.; Modoc and Siskiyou cos. to Mariposa Co. May–July.

10. **D. depauperàtum** Nutt. ex T. & G. [*D. pauciflorum* var. *d.* Gray.] Rootstock slender, vertical or with cluster of branches; stems very slender, weak, subscapose, 0.6–3 dm. long, glabrous; lvs. few, near the base, angulate-rounded, subglabrous, 2–5 cm. wide, palmatifid into 3 or 5 simple divisions or these with lance-oblong lobes; fls. few; pedicels slender, divaricate; bracts entire to lobed; sepals bright dark blue, mostly glabrous, oblong-ovate, 5–8(–10) mm. long, 3.5–5 mm. wide; spur slender, straight, 9–15 mm. long; upper petals narrow, white-edged, oblique; lower petals round-ovate, with cent. cluster of hairs, the sinus ca. 2 mm. deep; follicles ovoid, erect, glabrous, 8–12 mm. long; seeds few, 2–2.5 mm. long, dark, wrinkled but shining.—Damp thickets and edge of woods, 4000–9500 ft.; Sagebrush Scrub, Yellow Pine F., Red Fir F., Lodgepole F.; Siskiyou Co. to Lassen and Mariposa cos.; to B.C., Alta., Nev. May–July.

11. **D. diversifòlium** Greene ssp. **harneyénse** Ewan. Root slender, fusiform; stems erect, slender to stoutish, 3–4.5 dm. tall, hirsutulous to glabrate below, glandular-hairy in infl.; lvs. largely basal, 3–6 cm. wide, glabrous, deeply trifid, the primary divisions toothed to incised; petioles rather long; fls. several; sepals blue, oblong-ovate, 10–13 mm. long, acute, with broad band of short hairs; spur slender, 14–17 mm. long; upper petals often pale, narrow-oblong, emarginate to bifid; lower petals quadrate, with cluster of white hairs, bifid, the sinus open, 2–2.5 mm. deep; follicles straight, 12–17 mm. long, glandular-pubescent; seeds quadrate-rounded, 1.5 mm. long, dark, with narrow white-winged angles.—Moist banks and meadows; Sagebrush Scrub, Yellow Pine F.; Goose Lake V., Sierra V., Modoc Co.; to Ore., Ida. May–July.

12. **D. Sónnei** Greene. [*D. decorum* var. *nevadense* Wats. *D. pauciflorum* var. *n.* Gray. *D. p.* var. *S.* Smiley.] Roots tuberiform or rootstock fusiform; stems erect or ascending, glabrous, 1.5–3 dm. tall, simple or few-branched; lvs. mostly 4–6 in basal cluster, fleshy, 3–9 cm. wide, round, 5-palmatifid, the primary divisions pinnatifid into oblong obtuse glabrous segms.; petioles 5–15 cm. long; cauline lvs. 1–2, reduced; racemes open, 7–12-fld.; pedicels spreading; bracts linear; sepals oblong-ovate, purplish-blue, 10–14 mm. long, 5–6 mm. wide, sparingly hairy; spur nearly straight, 11–15 mm. long; upper petals whitish, often with blue veins; lower petals oblong-ovate, crisped, with cent. mass of hairs, the sinus narrow, 2 mm. deep; follicles erect or divergent, glabrous, 12–15 mm. long; seeds quadrate-ovoid, 1–1.5 mm. long, black, shining, narrowly wing-angled.—Moist places, such as swales and seeps on banks, openings in woods, 1000–7000 ft.; Mixed Evergreen F., Yellow Pine F.; Del Norte and Siskiyou cos. to Eldorado Co.; Nev., Ore. June–July.

13. **D. nudicaùle** T. & G. [*D. sarcophyllum* H. & A. *D. decorum* var. *n.* Huth. *D. n.* var. *elatius* W. Thomps. *D. armeniacum* Heller. *Delphinastrum n.* Nieuwl.] Rootstock elongate, thin, fibrous, with fascicled roots; stems erect or ascending, glabrous, often glaucous, 2–6(–8) dm. high; lvs. basal or nearly so to largely cauline, 3–10 cm. wide, 3–5-fid, the cent. division broadly cuneate-obovate, the ultimate lobes shallow, rounded, nipple-mucronate, glabrous to sparsely pubescent; petioles 5–15 cm. long; racemes few-fld., open, with long ascending pedicels; fls. cornucopia-shaped; sepals orange- or dull-red, sometimes yellow, ovate, 10–12 mm. long, acute, ciliolate, glabrous or nearly so; spur 15–20 mm. long; upper petals yellow with red tips, obliquely ovate, bidentate and ciliolate at tip; lower petals scarcely widened toward summit, bifid, the sinus narrow, 2–3 mm. deep; follicles divergent, glabrous, 1.5–2 cm. long; seeds 2–3 mm. long, brown to dark, rugulose, ovoid to obpyramidal, wing-margined; $2n = 16$ (Lewis *et al.*, 1951).—Dry slopes among shrubs and in woods, below 6500 ft.; Chaparral, Foothill Wd., Mixed Evergreen F., Yellow Pine F.; Coast Ranges s. to Monterey Co., Sierra Nevada s. to Plumas and Butte cos., Mariposa Co.; s. Ore. March–June. Hybrids between this sp. and the *D. decorum* complex range in color from coral and yellow to lavender and purple.

14. **D. lùteum** Heller. [*D. nudicaule* var. *l.* Jeps.] Roots deep-seated, thick, vertical; stems several, fleshy, pubescent at base and in infl., otherwise glabrous, 2–4 dm. tall; lvs. mostly near base, few on stem, 3–6 cm. wide, 3–5-parted, the divisions again lobed or toothed, obtuse to acute, ± pubescent, with petioles 10–20 cm. long; racemes open, 2–12-fld.; sepals yellowish, tipped purplish, broadly ovate, 15–16 mm. long, 9–10 mm. wide, pubescent; spur 14–17 mm. long; upper petals narrow, not lobed; lower petals oblong-ovate, the sinus closed, 3–4 mm. deep; follicles erect, glabrous, 14–16 mm. long; seeds oblong-ovoid, ca. 2 mm. long, dark with narrow white margin.—Open places on sea bluffs; N. Coastal Scrub; Bodega region, Sonoma Co. March–May.

15. **D. glaùcum** Wats. [*D. scopulorum* var. *g.* Gray. *Delphinastrum g.* Nieuwl.] Caudex stout, woody; stems coarse, glabrous, glaucous, fistulous, leafy, 1–2(–2.5) m. tall; lvs. 8–15(–20) cm. broad, palmatifid into broad cuneate acutely incised and toothed segms., glabrous, or pubescent beneath; petioles 4–10 cm. long; racemes 1–4 dm. long, many-fld., glabrous to somewhat pubescent, sometimes branched at base; pedicels ascending, 1–4 cm. long; lower bracts leafy, upper subulate; sepals light to dark violet-purple, rhombic-ovate, abruptly acute, 8–12 mm. long, 5–7 mm. wide, puberulent on back; spur rather stout, 8–10 mm. long; upper petals narrow, oblong, notched, purple-tipped; lower petals rhombic-ovate, the sinus nearly closed, 2–2.5 mm. deep; follicles erect, glabrous or puberulent, 10–18 mm. long; seeds ovoid, 3 mm. long,

straw-colored, wing-angled; $2n = 16$ (Lewis *et al.*, 1951).—Wet meadows and near streams, 5000–10,600 ft.; Montane Coniferous F.; Siskiyou Co. to Madera Co., San Gabriel and San Bernardino mts.; to Alaska, Rocky Mts. July–Sept.

16. **D. polyclàdon** Eastw. [*D. luporum* Greene. *D. scopulorum* var. *l.* Jeps.] Rootstock slender, vertical; stems several, simple, glabrous, mostly 5–8 dm. high; lvs. largely basal, round-reniform in outline, 3–8(–10) cm. wide, palmatifid into cuneate divisions, these mostly 3-lobed, glabrous; petioles 5–18 cm. long; racemes lax, 1–2 dm. long, few-fld., simple or with basal branches; pedicels 1–5 cm. long, ascending, glabrous to villous; sepals dark blue to blue-purple, broadly ovate, rounded, apiculate, glabrous to villous, 9–14 mm. long; spur slender, 12–17 mm. long; upper petals pale, short-oblique; lower petals ovate, hairy, the open sinus 2 mm. deep; follicles pubescent, 12–18 mm. long; seeds 1.5 mm. long, smoky, narrowly wing-angled; $2n = 16$ (Lewis *et al.*, 1951).— Among rocks and willows along creeks and in meadows, 7500–11,150 ft.; Lodgepole F. to Alpine Fell-fields; Sierra Nevada from Eldorado Co. to Tulare Co. July–Sept.

17. **D. trolliifòlium** Gray. [*D. exaltatum* var. *t.* Huth. *Delphinastrum t.* Nieuwl.] Rootstock deep, vertical, woody; stems stout 6–15(–18) dm. high, leafy throughout, glabrous or ± pubescent; lower lvs. 10–15 cm. wide, rounded in outline, 5–7-parted, the primary divisions approximate, cuneate-obovate, 3-cleft and laciniately lobed, glabrous or nearly so, the petioles to 25 cm. long, the upper cauline smaller, with primary divisions less approximate and with few coarse oblong-ovate lobes; racemes 1–3 dm. long, lax, mostly 15–30-fld.; pedicels divergent, 1–5 cm. long, ± villous; sepals violet-purple, ovate-lanceolate to elliptic, 15–25 mm. long, obtuse to acute, somewhat appressed-villous on backs; spur stout, somewhat recurved at tip, 15–20 mm. long; upper petals whitish, oblique, narrow, the lower purplish, oblong, villous, the sinus open, 4–5 mm. deep; follicles usually arcuate, 15–25 mm. long, glabrous; seeds compressed-spheroidal, dark, 1–1.5 mm. long, narrowly winged on angles; $2n = 16$ (Lewis *et al.*, 1951).—Moist partially shaded places, below 3500 ft.; Mixed Evergreen F.; Humboldt Co.; to w. Ore. April–July.

18. **D. Bàkeri** Ewan. Root-cluster fleshy; stems erect, subglabrous, 5–6 dm. high, leafy throughout; lvs. shallowly 5-parted, 6–7.5 cm. wide, glabrous, the primary divisions broadly cuneate, crenate with deep apiculate teeth; petioles to 15 cm. long; racemes rather loosely 5–15-fld., pedicels spreading-ascending, 8–20 mm. long; sepals dark blue or purplish, lance-ovate, acute, 11–13 mm. long, glabrous; spur slender, ca. as long as sepals; upper petals white, oblique; lower petals oblong, villous, blue-purple, the sinus 1–1.5 mm. deep; follicles suberect, glabrous.—Low brush and fence rows, below 600 ft.; V. Grassland; Coleman V., Sonoma Co. and Tomales, Marin Co. April–May.

19. **D. califórnicum** T. & G. [*D. exaltatum* var. *c.* Huth. *D. c.* var. *laxiusculum* Huth.] Roots thick, woody, branched; stems simple, stout, hollow, 5–20 dm. high, leafy, pubescent below, subglabrous to thinly hairy in infl.; lvs. thin, 5–10(–15) cm. wide, 5–7-palmatifid (upper 3–5), the divisions cuneate, incisely lobed and toothed; petioles 10–15 cm. long; racemes dense, 3–5 dm. long, sometimes branched below; pedicels ascending, 1–4 cm. long; fls. dull bluish or purplish; sepals whitish, tinged green or lavender, ovate, blunt, pubescent, 6–10 mm. long; spur stout, arched, 6–8 mm. long; upper petals beaked, villous; lower petals hairy, the sinus open, ca. 1 mm. deep; follicles erect, glabrous, 9–13 mm. long; seeds prismatic, 2–2.5 mm. long, dark, narrowly wing-angled; $2n = 16$ (Lewis *et al.*, 1951).—Moist, partly shaded places, ravines and slopes, below 2000 ft.; Mixed Evergreen F.; slopes of outer Coast Ranges, San Francisco Co. to Monterey Co. May–June. F. *longipìlis* Ewan. Infl. with long spreading or curling hairs.— Monterey to Cambria, San Luis Obispo Co. May–June.

Ssp. **intèrius** (Eastw.) Ewan. [*D. c.* var. *i.* Eastw.] Plant 1.5–3 m. tall, paniculately branched above; racemes interrupted; pedicels and axis of infl. glabrous; fls. sordid, yellowish.—Wet places; Foothill Wd.; inner Coast Range, Contra Costa Co. to Santa Clara Co. April–June.

20. **D. stachýdeum** (Gray) Tides. [*D. scopulorum* var. *s.* Gray.] Roots woody, vertical, with large crown; stems hollow, 6–20 dm. high, simple, rather equably leafy, puberulent above; lvs. green at anthesis, the mid-cauline 6–10 cm. wide, palmatisect to base, the primary divisions deeply divided into linear, mostly oblong segms., acute, somewhat pubescent especially beneath; petioles 10–16 cm. long; racemes many-fld., dense-spicate, 2.5–4 cm. wide; pedicels ascending, 1–2 cm. long; sepals a lively blue, oblong-ovate,

7–9 mm. long, puberulent; spur slender, 11–13 mm. long; upper petals bluish only at tips; lower petals rounded, darker blue, bearded, the sinus closed, 1 mm. deep; follicles 10–12 mm. long, hirsutulous; seeds brownish, 2.5 mm. long, winged on angles.—Dry rocky ridge, 8000 ft.; Lodgepole F.; Warner Mts., Modoc Co.; to Ore., Ida. July–Aug.

21. **D. gypsóphilum** Ewan. Rootstock stout, branched, with fibrous roots; stems erect, green, glabrous except for some appressed pubescence in the raceme, 4–10(–12) dm. high; lvs. withering early, well distributed up the stem, 3–5 cm. wide, trifid into few-lobed divisions, the lobes glabrous or nearly so, linear, mucronate; fls. 15-many, rather crowded; pedicels 1–2 cm. long, ascending; sepals clear white, ovate, acute to obtuse, puberulent, not recurved at tips, 15–20 mm. long; spur slender, upcurved, 10–14 mm. long; upper petals whitish, rounded, short-oblique; lower petals rounded, bearded, the open sinus 1–1.8 mm. deep; follicles 18–22 mm. long, sparsely hairy; seeds trigonous-ovoid, 2–2.5 mm. long, dark with broad white wings; $2n = 16$ or 32 (Lewis *et al.*, 1951).—Open slopes and fields, below 3000 ft.; Alkali Sink, V. Grassland; San Joaquin Co. to Kern Co. April–May.

Ssp. **parviflòrum** Lewis & Epl. Sepals dingy white, 11–13 mm. long.—Foothill Wd.; Monterey and San Luis Obispo cos.

22. **D. Hánseni** (Greene) Greene. [*D. hesperium* var. *H.* Greene.] Taproot short, slender; stems 4–9 dm. tall, greenish, appressed-pubescent, somewhat pilose at base; lvs. withering early, largely basal, the lower rhombic, hairy, 4–9 cm. wide, shallowly palmatifid, the upper smaller, palmatisect into narrow divisions; petioles ascending, ca. twice as long as blade; raceme rather compact, spicate, many-fld.; pedicels suberect, 1–2 cm. long; sepals dark purple to bluish or reddish, sometimes whitish, oblong, obtuse, 6–8 mm. long, pubescent; spur slender, curving, 7–10 mm. long; upper petals pale, narrow, oblique, striped; lower petals darker, rounded, bearded, the sinus open, 2–3 mm. deep; follicles erect, 10–14 mm. long, hairy; seeds obpyramidal, white, scaly-echinate; $2n = 16$ or 32 (Lewis *et al.*, 1951).—Open grassy places and clearings, below 3700 ft.; Foothill Wd.; base of Sierra Nevada from Butte Co. to Fresno Co. April–May. The following are poorly marked and have been proposed: (1.) Ssp. **arcuàtum** (Greene) Ewan. [*D. H.* var. *a.* Greene.] Slender, straggling or procumbent, 4–8 dm. tall; lvs. mostly basal, green at anthesis, glabrous except for long spreading hairs on veins and margins; racemes loose; fls. greenish-white with reddish umbo; sepals 9–10 mm. long.— Rocky and sandy places, commonly among shrubs and boulders, 1500–4000 ft.; Foothill Wd., Yellow Pine F.; Mariposa Co. to Kern Co. May–July. (2.) Ssp. **kernénse** (A. Davids.) Ewan. [*D. H.* var. *k.* A. Davids.] Stems stout, 2–4 dm. high; lvs. in close basal tuft, white-hairy; fls. crowded, pale bluish white; sepals 7–10 mm. long, recurved.— Grassy slopes among shrubs, 1000–3300 ft.; Chaparral, Foothill Wd.; Greenhorn and Tehachapi mts., Kern Co. May–June.

23. **D. hespérium** Gray. Rootstock stout, elongate or with cluster of fibrous roots; stems erect, simple, rather slender, puberulent throughout, 3–6(–9) dm. high; lvs. tending to wither early, well distributed along the stem, the blades mostly 2–4 cm. wide, 3–5-palmatifid, the primary divisions lobed with oblong apiculate segms., pubescent on both surfaces; petioles mostly 3–6 cm. long; racemes strict, densely-fld., 5–20 cm. long; pedicels mostly 5–15 mm. long; sepals dark blue or blue-purple, oblong-ovate, obtuse or rounded, 8–12 mm. long, with puberulent band; spur nearly straight, 9–15 mm. long; upper petals clavate-oblique, white-edged; lower petals rounded, bearded, the sinus narrow, 1–3 mm. deep; follicles puberulent, straight, 10–12 mm. long; seeds ovoid, dark, ca. 1 mm. long, the angles white-margined; $2n = 16$ (Lewis *et al.*, 1951).—Grassy slopes and ridges, below 2500 ft.; Mixed Evergreen F., Foothill Wd.; inner Coast Ranges, Humboldt and Butte cos. to Santa Clara Co. April–June. Forma *hirsùtum* Ewan. Stems, petioles, and pedicels white-hirsute.—Below 3000 ft.; Foothill Wd.; Mendocino and Butte cos. to Tulare Co. May–June.

Ssp. **palléscens** (Ewan) Lewis & Epl. [*D. h.* forma *p.* Ewan.] Plants 4–10 dm. tall; fls. white to pink.—Tehama Co. to Santa Clara and Kern cos. May–July.

Ssp. **cuyamácae** (Abrams) Lewis & Epl. [*D. c.* Abrams.] Fls. pale blue to pale violet; sepals 6–8 mm. long; follicles glabrous.—Grassy meadows, 4000–5000 ft.; lower edge of Yellow Pine F.; s. San Jacinto Mts. to Palomar and Cuyamaca mts. May–July.

24. **D. Paríshii** Gray. [*D. amabile* Tides. *D. a.* ssp. *Clarianum* Ewan.] Perennial from a root-crown with several woody roots; stems erect, 1.5–6 dm. tall, 1–several, often hollow, glabrous to pubescent, ± glaucous; lvs. largely basal or well distributed up the

lower stem, often withered at anthesis, deltoid in outline, 2–8 cm. wide, the primary divisions cuneate with linear to oblong lobes 2–5 mm. wide, subglabrous to pubescent (especially beneath); petioles 3–15 cm. long; racemes compact to open, ca. 5–25-fld.; pedicels stout, 0.5–4 cm. long; sepals ovate, lavender- to light- or azure-blue, obtusish, mostly pubescent, 8–12 mm. long; spur curved or straight, 8–13 mm. long; upper petals whitish, notched; lower petals bluish or violet, mostly hairy, the sinus 1.5–2.5 mm. deep, usually open; follicles glabrous to puberulent, shining, erect, 8–14 mm. long; seeds dark, 2–4 mm. long, with copious loose white pellicle; $2n = 16$ (Lewis *et al.*, 1951).—Gravelly benches and washes, below 7500 ft.; Creosote Bush Scrub, Joshua Tree Wd., Pinyon-Juniper Wd.; Mojave Desert from Mono Co. s. to w. end of Coachella V., Riverside Co. April–June.

Ssp. **subglobòsum** (Wiggins) Lewis & Epl. [*D. s.* Wiggins. *D. Parryi* var. *s.* Munz. *D. collinum* Ewan.] Stems slender, not at all fistulous; fls. dark blue-purple; sepals 8–10 mm. long; spur 10–13 mm. long, the tip curved downward; upper petals bluish to whitish, notched; lower petals rounded, villous, bright blue, the sinus ca. 2.5 mm. deep; follicles sparsely hairy; seeds 1.5–2 mm. long.—Dry stony fans and slopes, below 5000 ft.; Creosote Bush Scrub, Chaparral, Pinyon-Juniper Wd.; w. edge of Colo. Desert, from Santa Rosa Mts. to L. Calif. March–May.

Ssp. **purpurèum** Lewis & Epl. Stems largely fistulous, puberulent, glaucous; lvs. dissected into narrow lobes which are conspicuously pubescent with straight or curled hairs; sepals pinkish-purple.—Pinyon-Juniper Wd.; Mt. Pinos region, Cuyama V., Kern and Ventura cos. May–June.

25. **D. inópinum** (Jeps.) Lewis & Epl. [*D. Parishii* var. *i.* Jeps. *D. P.* var. *pallidum* Munz. *D. amabile* ssp. *p.* Ewan.] Much like *D. Parishii;* lvs. mostly basal, generally glabrous, ± fleshy; fls. small, cupped, white; sepals 6–7 mm. long, somewhat recurved at tips; spur 7–10 mm. long; follicles 7–9 mm. long, glabrous; $2n = 16$ (Lewis *et al.*, 1951).—Open places, 5000–8000 ft.; Yellow Pine F.; Upper Kern River Canyon, Mt. Pinos, Kern Co. June–July.

26. **D. Párryi** Gray. [*D. P.* var. *maritimum* A. Davids. *D. P.* ssp. *Eastwoodae* Ewan.] Caudex with woody deep-seated roots; stems slender, usually simple, usually puberulent, 3–9 dm. tall; lvs. chiefly cauline, the lower withering early, main lf.-blades 2–8 cm. wide, 3–5-parted, then further divided into linear lobes, puberulent; petioles to ca. 1 dm. long; racemes 6–18 cm. long, usually loose, few–many-fld.; pedicels ascending, 1.5–3 cm. long; bracts filiform; sepals rotate, deep purplish-blue, ovate, rounded to acutish, 10–15 mm. long, puberulent; spur nearly straight, ca. as long as sepals; upper petals exserted, oblique, whitish; lower petals rounded, purplish-blue, floccose, the sinus rather open, 3 mm. deep; follicles puberulent, 10–15 mm. long; seeds ovoid, 1–2 mm. long, dark, nearly wingless to almost invested by white wings; $2n = 16$ (Lewis *et al.*, 1951).—Common, open slopes and mesas, below 6500 ft.; V. Grassland, Coastal Sage Scrub, Chaparral, S. Oak Wd., Yellow Pine F.; Santa Barbara Co. to San Diego Co.; Santa Barbara Ids.; L. Calif. April–May. Plants with lvs. divided into segms. 5–7 mm. wide and from coastal bluffs and beaches of Catalina Id. and Orange and San Diego cos. to L. Calif. are f. *maritimum* (A. Davids.) Ewan. Those with linear lf.-segms. that tend to inroll and with lighter blue fls., growing on serpentine in San Luis Obispo Co., have been named ssp. *Eastwoòdae* Ewan.

Ssp. **seditiòsum** (Jeps.) Ewan. [*D. hesperium* var. *s.* Jeps.] Leafy throughout, the lvs. dissected into linear-filiform lobes, commonly involute, hirsutulous; bracts conspicuous; fls. somewhat cupped, in a dense infl.; sepals dark blue, 8–12 mm. long; upper petals white, the lower with closed sinus 1–2 mm. deep; follicles 9–13 mm. long, puberulent to glabrous; seeds 1.5–2.5 mm. long; $n = 8$ (Lewis *et al.*, 1951).—Grassy slopes; Foothill Wd., Chaparral; inner S. Coast Ranges, Santa Clara Co. to Santa Barbara Co. April–May.

Ssp. **Blóchmanae** (Greene) Lewis & Epl. [*D. B.* Greene. *D. variegatum* var. *B.* Davis. *D. Parryi* var. *B.* Jeps.] Fls. larger; sepals indigo to violet, ovate to oblong, sparsely strigulose, 15–18 mm. long; spur 11–13 mm. long; upper petals white, blunt, oblique-oblong; lower petals lavender, rounded, 7–8 mm. wide, bearded, the sinus closed, 3–4 mm. deep; seeds ovoid, smooth, dark, 1 mm. long, with narrow white wing-margins; $2n = 16$ (Lewis *et al.*, 1951).—Sand of fixed dunes; Chaparral; Nipomo Mesa, San Luis Obispo Co. to near Lompoc, Santa Barbara Co. April–May.

27. **D. recurvàtum** Greene. [*D. hesperium* var. *r.* Davis.] Roots shallow, heavy, woody-

fibrous; stems reddish or purplish, erect, 2–6(–9) dm. tall, glabrous or puberulent below, glabrous in infl.; lvs. several, largely cauline, 1.5–3 cm. wide, palmatifid into few-parted divisions, the ultimate segms. hairy beneath, blunt, mucronate; petioles ascending-erect, 3–7 cm. long; fls. 15–24; pedicels 1–3 cm. long, ascending; sepals light blue, oblong-ovate, blunt, incurved at tips, sparsely strigulose, 10–16 mm. long; spur straightish, 10–14 mm. long; upper petals white or cream, conspicuous; lower petals whitish to pale-blue, bearded, deltoid-ovate, the sinus open, 1–3 mm. deep; follicles 9–12 mm. long, thinly hairy; seeds light-colored, 1 mm. long, broadly white-winged; $2n = 16$ (Lewis *et al.*, 1951).—Subalkaline soil of brushy or open places; Alkali Sink, V. Grassland; Glenn and Butte cos., Contra Costa Co. to Kern Co. March–May.

28. **D. umbraculòrum** Lewis & Epl. Perennial with shallow fibrous fleshy roots; stems firmly attached, simple or branched near infl., 4–8 dm. tall, glaucous and glabrous or ± puberulent, strongly anthocyanous; lvs. sparsely puberulent to glabrous, the basal blades with few broad rounded lobes, usually 5, these in turn shallowly 3-lobed; fls. ca. 10–25; lateral sepals 8–14 mm. long, 4–7 mm. broad, rotate, lavender to deep blue-purple with darker veins; petals of same color or upper lighter; follicles 10–15 mm. long, puberulent; seeds with loose smooth membranous not inflated white to brownish coat, sometimes flecked with black; $n = 8$ (Lewis & Epling, 1954).—Usually shaded places; largely Foothill Wd.; Monterey and San Luis Obispo cos. May–June.

29. **D. variegàtum** T. & G. [*D. grandiflorum* var. *v.* H. & A.] Roots clustered, thickened; stems simple, 2.5–6 dm. tall, subglabrous to hairy; lvs. subbasal to mostly cauline, the blades 1–5 cm. wide, tripartite, the ultimate segms. linear to short and rounded, ± involute, puberulent to hirsute; racemes rather few-fld.; sepals light blue to rich blue-purple, 12–17 mm. long, hirsutulous; spur straightish, 10–15 mm. long; upper petals yellow or whitish, the lower rounded, 6–8 mm. wide, with closed sinus 1–1.5 mm. deep; follicles short-oblong, erect, ca. 1 cm. long; seeds quadrate, smoky, 1–2 mm. long, with broad pale wings; $2n = 16$ or 32 (Lewis *et al.*, 1951).—Open grassy hills and woods; V. Grassland, Foothill Wd.; San Luis Obispo and Tulare cos to Tehama Co. April–May. Variable and with hybrids with other spp. Ewan recognized the hairier plants from outside the upper Salinas V. as ssp. *apiculàtum* (Greene) Ewan [*D. a.* Greene], and plants with few lvs. and sepals 8–10 mm. long, incurved at tips, occurring from Napa and Lake cos. to Tehama Co., as f. *Emíliae* (Greene) Ewan [*D. E.* Greene]. Plants with subglabrous stems, few reduced lvs., 2–6 fls., and sepals 13–16 mm. long, growing from Placer Co. to Fresno Co., are f. *subnùdum* (Eastw.) Ewan [*D. s.* Eastw.]. A large-fld. variant is f. *supérbum* Ewan, with royal purple sepals 16–24 mm. long, San Mateo Co. to Butte and Tehama cos.

30. **D. cardinàle** Hook. [*D. coccineum* Torr. *D. cardinale* var. *angustifolium* Huth.] SCARLET LARKSPUR. Roots deep, thickened, woody; stems erect, simple or branched above, 1–2 m. tall, hollow, mostly puberulent throughout; basal lvs. withered at anthesis, 5–20 cm. wide, 5-parted, the primary divisions cuneate, shallowly to deeply 3-lobed, petioles 1–2.5 dm. long; cauline lvs. 5–7-parted, the primary divisions deeply divided into divergent linear to oblanceolate segms., subglabrous or thinly hairy on the veins; fls. in open racemes or panicles; bracts linear, simple or palmate; pedicels ascending, mostly 2–6 cm. long; sepals scarlet, rarely yellow, ovate, obtuse, acutish, mostly glabrous, 10–16 mm. long; spur stout, 15–20 mm. long; upper petals exserted, yellow with scarlet tips, narrow-oblong, with narrow upper and wider lower lobes; lower petals narrow-oblong, the sinus closed, 1 mm. deep; follicles erect, glabrous, 1–1.5 cm. long; seeds quadrangular, 2–3 mm. long, dark with sharp angles; $2n = 16$ (Lewis *et al.*, 1951).—Dry openings in brush and woods, below 5000 ft.; Coastal Sage Scrub, Chaparral, Foothill Wd.; Monterey Co. to San Diego Co.; L. Calif. May–July.

31. **D. Andersònii** Gray. [*D. tricorne* var. A. Huth. *Delphinastrum* A. Nieuwl.] Roots branched, woody-fibrous; stems simple, 2–6 dm. tall, hollow, glabrous; lvs. basal, green at anthesis, rounded, 1–3 cm. wide, the primary divisions approximate, with oblong mucronate somewhat fleshy glabrate or pubescent segms.; petioles 4–10 cm. long; racemes mostly 5–12-fld., with ascending glabrous pedicels 1–6 cm. long; bracts linear, pubescent; sepals rich blue to purplish, oblong-ovate, obtusish, 10–14 mm. long, somewhat villous; spur thick, straightish, 10–12 mm. long; upper petals whitish, emarginate; lower petals rounded, the sinus open, 2.5–4 mm. deep; follicles 14–24 mm. long, somewhat pubescent; seeds black, 2.5 mm. long, white-winged on the angles; $2n = 16$

(Lewis *et al.*, 1951).—Sandy and volcanic soil, among shrubs and pines, 5000–7500 ft.; Sagebrush Scrub, Yellow Pine F.; Lassen Co. to Mono Co.; Nev., Ida. May–June.

Ssp. **cognàtum** (Greene) Ewan. [*D. c.* Greene. *D. bicolor* var. *c.* Davis.] Stems hardly hollow; lvs. cauline and basal, withering at anthesis, not fleshy; spur longer than sepals; upper petals with purple lines; follicles 20–28 mm. long.—Volcanic soil; Sagebrush Scrub; Siskiyou Co. to Modoc and Lassen cos.; Ore. to Utah. April–June.

5. Aconìtum L. ACONITE. MONKSHOOD

Perennial herbs from ± tuberous roots. Stems erect, ascending, to reclining, sometimes climbing on other plants. Lvs. palmately lobed, usually well distributed. Fls. showy, irregular, in racemes or panicles. Sepals 5, the upper (helmet or hood) arched, the 2 lateral round to reniform, the 2 lower narrower and less showy. Petals 2–5, the 2 upper concealed in the helmet, clawed at base and bearing at the summit a capitate to coiled glandlike nectariferous part (spur) and an expanded somewhat pendent liplike part (lamina). Stamens many; fils. expanded at base. Pistils 3–5, sessile, free, becoming follicles. Seeds many, 3-cornered, sometimes winged. A genus of perhaps 100 spp. of Temp. and cooler N. Temp. Zone; some grown for ornament, others sought as source of drugs. (Ancient Greek name.)

1. **A. columbiànum** Nutt. Tuber 1–3 cm. long; stems mostly erect and stout, sometimes weak and reclining or climbing, 5–20 dm. tall, glabrous below, spreading-pubescent and even viscid in infl.; lvs. thin, 5–12(–15) cm. wide, deeply 3–5-cleft, the primary divisions rhombic-cuneate at base, then laciniately toothed or cleft, subglabrous to finely pubescent; lower petioles longer than blades; upper lvs. reduced; infl. rather lax, 1–6 dm. long; lower bracts tripartite, foliose, the upper linear; lower pedicels elongate; fls. purplish-blue, sometimes pale, villous, 3–4 cm. high; hood 15–25 mm. high, mostly ca. 5–8 mm. wide above, rather straight in front profile, with a slender beak 6–8 mm. long; lateral sepals 14–16 mm. long, the lower 10–12 mm. long; pistils 3, 12–18 mm. long in fr.; seeds lamellate, 3–4 mm. long.—Moist places as about meadows, especially in willow thickets, 4000–8000 ft.; Montane Coniferous F.; Sierra Nevada, and N. Coast Ranges s. to Trinity Co. July–Aug. Exceedingly variable and in need of study to determine the true status of the variants. For Calif. they may be considered as follows:

A. Plants with bulbils in some or many of the lf.-axils.
B. Cauline lf.-blades 7–12(–18) cm. wide the primary middle segm. 1.5–3 cm. wide below where it is lobed. Cent. Sierra Nevada .... 1a. *A. Hanseni*
BB. Cauline lf.-blades 4–7(–10) cm. wide, the primary middle segm. 0.8–1.5(–2.5) cm. wide below where it is lobed. Siskiyou Co. .... 1b. *A. viviparum*
AA. Plants not bearing bulbils in lf-axils.
B. Upper part of plant climbing or with slender twining or sinuous branches. Humboldt Co. to Glenn and Plumas cos. .... 1d. *A. geranioides*
BB. Upper part of plant not climbing or sinuous; plant erect.
C. Primary segms. of main cauline lvs. dissected (divided more than halfway from margin to midrib) into lance-linear lobes. Klamath Lake region .. .. 1c. *A. Leibergii*
CC. Primary segms. of main cauline lvs. coarsely dentate, the undivided cuneate-rhombic part wider than the teeth are deep. Widespread .... 1e. *A. columbianum*

1a. **A. Hánseni** Greene. Plants 6–12 dm. high; many lvs. and even bracts bulbiferous; helmet 15–20 mm. high; beak descending, 7–10 mm. long.—Eldorado and Mariposa cos.

1b. **A. vivíparum** Greene. Plants 3–6(–9) dm. high; upper lvs. bulbiferous; sometimes fls. replaced by bulblets; helmet 15–20 mm. high; beak horizontal, 6–8 mm. long.—Del Norte and Siskiyou cos.; adjacent Ore.

1c. **A. Leibérgii** Greene. Like *A. viviparum*, but not bulbiferous, and lvs. with narrowly instead of broadly cuneate divisions.—Humboldt and Trinity cos.; sw. Ore.

1d. **A. geranioides** Greene. Plants 7–20 dm. tall, often leaning, clambering, or reclining, the upper parts even twining; primary segms. of lvs. cuneate-obovate, deeply 3-lobed, then coarsely toothed; helmet 13–17 mm. high; beak horizontal, 5–8 mm. long.—Humboldt Co. to Glenn, Siskiyou, and Plumas cos.; s. Ore.

1e. **A. columbiànum** Nutt. in T. & G. [*A. Helleri* Greene. *A. cheirophyllum* Greene.] Plants 6–14 dm. high, erect; primary segms. of lvs. rhombic-lanceolate to -ovate, shallowly 3-lobed above the middle; helmet 15–24 mm. high; beak descending, 6–8 mm. long.—Modoc Co. to Tulare and Inyo cos.; to B.C. and Rocky Mts.

6. Ranúnculus L. Buttercup. Crowfoot

Annual and perennial herbs. Root systems from fibrous to fleshy to tuberous. Lvs. alternate, entire to compound, often largely basal. Fls. solitary on ends of stem or the many branches, or sometimes ± panicled or corymbose, yellow or white. Sepals usually 5; petals mostly 5, sometimes fewer or more, each with a nectary at the base. Stamens mostly rather many, short. Pistils 5 to many, the ovary 1-loculed and 1-ovuled, ripening into a small hard ak. with a persistent beak (style). Frs. mostly clustered on the receptacle to form a headlike group. Perhaps 250 spp., widely dispersed in temp. and cold countries, often in damp or wet places; some of hort. value. (Latin, little frog, because of the moist habitat for many of the spp.)

(Benson, L. A treatise on the N. Am. Ranunculi. Am. Midl. Nat. 40: 1–261, 1948. Drew, W. B. The N. Am. representatives of Ranunculus, section Batrachium. Rhodora 38: 1–47, 1936.)
A.   Petals white; aquatics with the submersed lvs. dissected; aks. transversely ridged.
    B.   Receptacle glabrous; style in anthesis 2–3 times as long as ovary .... 29. *R. Lobbii*
    BB. Receptacle with short hairs; style in anthesis not longer than ovary.
        C.   Pedicels not recurved at fruiting time; submersed lvs. usually petioled and ca. as long
             as internodes .... 30. *R. aquatilis*
        CC. Pedicels recurved at bases at fruiting time; submersed lvs. usually sessile and much
             shorter than internodes .... 31. *R. subrigidus*
AA. Petals yellow or red (rarely white in age); plants mostly terrestrial; aks. not transversely ridged.
    B.   Petals red, ca. 1.5 cm. long; aks. 7–14 mm. long .... 28. *R. Andersonii*
    BB. Petals yellow, usually shorter; aks. mostly 1–5 mm. long.
        C.   Plants annual.
            D.   Lvs. entire or shallowly toothed; stems reclined, rooting at lower nodes.
                E.   Sepals and petals 1–1.5 mm. long; upper stem-lvs. sessile, ± lanceolate
                       22. *R. pusillus*
                EE. Sepals and petals 2–2.5 mm. long; upper stem-lvs. petioled, ovate
                       23. *R. alveolatus*
            DD. Lvs. 3(or 3–5)-parted or -foliolate; stems ascending, not usually rooting at nodes.
                E.   Aks. covered with spines or papillae or hooked hairs.
                    F.   Petals 5–8 mm. long; aks. 5–6 mm. long.
                        G.   Aks. 10–20, not spiny on margins; stems glabrous.
                             13. *R. muricatus*
                      GG. Aks. 5, with long spines on margins; stems thinly hirsute
                             14. *R. arvensis*
                    FF. Petals 1–3 mm. long (or none); aks. 1.5–2 mm. long.
                      G.   Ak.-beak ca. 0.5 mm. long; basal lvs. 2–2.5 cm. broad.
                        H.   The petals usually 1, 2, or 0; ak.-papillae produced into
                           hooked bristles. Common native .... 10. *R. hebecarpus*
                        HH. The petals 5; ak.-papillae ending in hooked spines. Sparingly
                           natur. .... 11. *R. parviflorus*
                      GG. Ak.-beak 2 mm. long; basal lvs. 4–14 cm. broad .. 7. *R. uncinatus*
                EE. Aks. smooth, at least not papillate or spiny.
                    F.   Beak of ak. well developed, 1–2 mm. long. Plants terrestrial.
                      G.   Stamens 10–15; receptacle glabrous, mostly 1–2 mm. long in fr.
                             7. *R. uncinatus*
                      GG. Stamens 15–35; receptacle hispid. 4–5 mm. long in fr.
                             8. *R. Macounii*
                  FF. Beak of ak. practically lacking. Plants palustrine or aquatic
                           24. *R. sceleratus*
    CC. Plants perennial.
        D.   The plants, at least the stems, hirsute or evidently pubescent.
            E.   Plants with runners, which root at nodes ... . 1. *R. repens*
            EE. Plants lacking runners, and not rooting at nodes.
                F.   Basal lvs. entire, lanceolate; ak.-beak not recurved
                       17. *R. alismaefolius* var. *Lemmonii*
                FF. Basal lvs. 3–7-lobed or -foliolate, broader.
                    G.   Petals 2.5–6 mm. long.
                      H.   Receptacle glabrous; sepals yellow-green. Widely distributed
                             7. *R. uncinatus*
                      HH. Receptacle hispid; sepals often tinged purple. Modoc Co.
                             8. *R. Macounii*
                GG. Petals 7–18 mm. long.
                    H.   Receptacle pubescent or hispid.
                      I.   Sepals spreading at base; stem-base forming a bulbous
                         thickening 10–13 mm. in diam. Sparingly natur.
                         2. *R. bulbosus*
                    II. Sepals reflexed from base; stem-base not bulbous.
                      J.   Petals 9–16 .... 5. *R. californicus*

JJ. Petals 5.
  K.   Aks. smooth, the beaks straight, ca. 4 mm. long.
       Widespread native .... 9. *R. orthorhynchus*
  KK. Aks. usually papillate, the beaks curved at tip,
       0.3 mm. long. Sparingly natur. . . 12. *R. Sardous*
HH. Receptacle glabrous.
  I.   Sepals spreading; petal-width at least ⅔ the length. Spar-
       ingly natur. .... 3. *R. acris*
  II. Sepals reflexed; petal-width less than ⅔ the length. Com-
       mon natives.
    J.   Body of ak. mostly 4–5 mm. long, the beak with broad,
         thin base 0.6–1 mm. wide; petals 5–10
                                                    6. *R. canus*
    JJ. Body of ak. mostly 2–3 mm. long, the beak with base
         less than 0.5 mm. wide.
      K.   Petals mostly 5–6, mostly 1–2 times as long as
           broad .... 4. *R. occidentalis*
      KK. Petals mostly 8–15, mostly 2–3 times as long as
           broad .... 5. *R. californicus*
DD. The plants glabrous or nearly so.
  E.   Lower lvs. 3–7-parted or -foliate or even dissected.
    F.   Stems 0.4–1.5 dm. long; ak.-beaks straight.
      G.   Head of aks. ovoid, 4–7 mm. in diam.; basal lf.-blades 2.5–4 cm.
           broad .... 15. *R. Eschscholtzii*
      GG. Head of aks. globose, 10–20 mm. in diam.; basal lf.-blades 1–1.8
           cm. broad .... 16. *R. glaberrimus*
    FF. Stems 1.5–9 dm. long; ak.-beaks mostly curved.
      G.   Ak.-body 5–6 mm. long, covered with stout curved spines; recepta-
           cle hispid. Sparingly natur. .... 13. *R. muricatus*
      GG. Ak.-body 2–3.5 mm. long, not spiny.
        H.   Receptacle glabrous.
          I.   Petals 6–15 mm. long.
            J.   Petals mostly 5–6 .... 4. *R. occidentalis*
            JJ. Petals mostly 7–16 .... 5. *R. californicus*
          II. Petals 2.5–4 mm. long .... 7. *R. uncinatus*
        HH. Receptacle hispid or hairy.
          I.   Stems not rooting at nodes; plants terrestrial; lvs. not dis-
               sected.
            J.   Petals 3–5 mm. long .... 8. *R. Macounii*
            JJ. Petals 10–17 mm. long
                                   9. *R. orthorhynchus* var. *Bloomeri*
          II. Stems rooting at nodes, floating or reclining; lvs. dissected
                                                    25. *R. flabellaris*
  EE. Lower lvs. entire or shallowly lobed.
    F.   Stems not rooting at the nodes.
      G.   Head of aks, 10–20 mm. in diam.; aks. usually pubescent
                                                    16. *R. glaberrimus*
      GG. Head of aks. 3–8 mm. in diam.; aks. glabrous.
        H.   Sepals greenish, 1.5–5 mm. long.
          I.   Petals ca. 10 mm. long; sepals 3–5 mm. long; lvs. 0.5–
               3 cm. broad .... 17. *R. alismaefolius*
          II. Petals 3–6 mm. long; sepals 1.5–2.5 mm. long; lvs. 3–5
               cm. broad .... 21. *R. Populago*
        HH. Sepals white, 7–10 mm. long; petals 4–5 mm. long
                                                    27. *R. hystriculus*
    FF. Stems rooting at the nodes.
      G.   The stems not differentiated into stolons and scapes; receptacle
           glabrous, 1–2 mm. long in fr.
        H.   Lower lvs. 1.5–12 mm. wide; sepals 3–5 mm. long
                                                    18. *R. Flammula*
        HH. Lower lvs. 10–28 mm. wide; sepals 2–3 mm. long.
          I.   Roots not fusiform-thickened at base. Once collected in
               Owens V. .... 19. *R. hydrocharoides*
          II. Roots fusiform-thickened at base. Siskiyou C.
                                                    20. *R. Gormanii*
      GG. The stems differentiated into stolons and scapes; receptacle hairy,
           cylindroid, 4–7 mm. long in fr. .... 26. *R. Cymbalaria*

1. **R. rèpens** L. Perennial with runners 2–8 dm. long; roots filiform; fl.-stems ascending, branched, hairy, 3–6 dm. long; lvs. ternately compound, 2–10 cm. wide, the lfts.
lobed and toothed, ± hairy; petioles 3–15 cm. long, mostly hairy; stem-lvs. alternate,
petioled; pedicels 2–12 cm. long; sepals 5, greenish, 5–7 mm. long, deciduous; petals
5, yellow, cuneate-obovate, 7–13 mm. long, the nectary-scale glabrous; stamens 50–80;
aks. 20–25, in subglobose head, each obovoid-discoid, 2.5 mm. long, glabrous, the beak

stout, recurved, 1 mm. long; receptacle mostly pubescent; $2n = 32$ (Neves, 1944).—Natur. about lawns and low places, chiefly from Monterey and Fresno cos. n.; to Alaska, Atlantic Coast; native of Old World. May–Aug.

Var. **pleniflòrus** Fern. Less stoloniferous; lfts. less cuneate; petals numerous, the fls. hemispherical.—Common in gardens and occasionally escaping. April–June.

2. **R. bulbòsus** L. Perennial, with tuberous base 10–13 mm. thick; stems erect, 3–7 dm. long, strigose; lf.-blades ovate in outline, 1.5–3 cm. broad, the 3 lfts. lobed and toothed; petioles 5–14 cm. long; upper cauline lvs. sessile; pedicels 2–10 cm. long; sepals 5, greenish-yellow, reflexed at middle, 7 mm. long; petals 5, yellow, orbicular-obovate, 10–14 mm. long; the nectary-scale glabrous; stamens 35–60; aks. 12–30, in globose-ovoid head, each 2.8 mm. long, glabrous, the beak 0.4 mm. long, recurved; receptacle pubescent; $2n = 16$ (Neves, 1944).—Natur. in Humboldt Co.; to Wash., Atlantic Coast; native of Old World. May–June.

3. **R. ácris** L. Perennial, mostly hirsute, with stout roots; stems several, erect, 5–10 dm. high, branched above; lf.-blades pentagonal in outline, 4–8 cm. wide, 3-parted, then lobed, appressed-hirsute; petioles 5–17 cm. long, hairy; upper cauline lvs. sessile; pedicels 2–10 cm. long; sepals 5, greenish, spreading, 4–7 mm. long; petals 5, yellow, cuneate-obovate, 8–14 mm. long, the nectary-scale glabrous; stamens 40–80; aks. 25–40, in globose head, each obovoid-discoid, 2–2.5 mm. long, glabrous, the beak ca. 0.5 mm. long, recurved; receptacle glabrous; $2n = 28, 56$ (Löve & Löve, 1944).—Natur. in moist places, Humboldt Co.; to Wash., Atlantic Coast; native of Old World. June–Aug.

✓4. **R. occidentàlis** Nutt. var. **Eisénii** (Kell.) Gray. [*R. E.* Kell.] Perennial, with slender roots; stems erect, 3–7 dm. high, 2–5 mm. thick, branched above, mostly hirsute; basal lvs. fan-shaped or semicircular in outline, 2–5 cm. wide, 3-parted, the lobes cuneate, again lobed, strigose, the ultimate teeth acutish; petioles 3–12 cm. long, hirsute; pedicels 2–10 cm. long; sepals 5, greenish-yellow, reflexed at middle, 4–8 mm. long, promptly deciduous; petals usually 5, yellow, elliptic, 7–12 mm. long, 4–8 mm. wide; stamens 25–50; aks. mostly 8–15, in hemispheric cluster, each discoid to discoid-obovoid, 2–3.5 mm. long, glabrous to strigose, the beak 0.5–1 mm. long, usually recurved, extending from the margin of the body; receptacle glabrous.—Vernally moist ground, mostly 300–6000 ft.; Foothill Wd., Yellow Pine F.; Modoc Co., inner N. Coast Ranges, Humboldt and Siskiyou cos. to Napa Co., Sierra Nevada foothills to Tehachapi Mts.; s. Ore. March–May.

KEY TO CALIF. VARIETIES

Ak.-beak 0.5–1.3 mm. long, usually curved; stems 3–7 dm. long. Widely distributed.
  Stems erect or suberect.
    Lvs. 2–5 cm. wide, the parts cuneate, with acutish ultimate lobes; petals 4–8 mm. wide; stems
    2–5 mm. thick .... var. *Eisenii*
    Lvs. 1.5–3 cm. wide, the parts oblong or narrow-cuneate, with sharply acute ultimate lobes;
    petals 2.5–3.5 mm. wide; stems 1–2 mm. thick .... var. *Rattanii*
  Stems flexuous, mostly reclining; lvs. with lanceolate main divisions; petals 1.5–3 mm. wide
                                        var. *ultramontanus*
Ak.-beak 1.5–2 mm. long, straight except at hooked tip; stems 2–4 dm. long. Siskiyou–Modoc cos.
  Lvs. with lanceolate divisions. Plants from elevs. of 4000–7000 ft. .... var. *dissectus*
  Lvs. with cuneate divisions. Plants from 1000–4000 ft. .... var. *Howellii*

Var. **Rattánii** Gray. [*R. R.* Howell.] Stems erect, 3–5 dm. long, 1–2 mm. thick; lvs. 1.5–3 cm. wide, 3-parted into narrow-cuneate divisions, the ultimate lobes sharply acute; petals mostly 5–8 mm. long, 2.5–3.5 mm. wide; aks. and beaks as above.—Vernally moist places, below 3500 ft.; Coastal Prairie; outer Coast Ranges, s. to Lake Co.; w. Ore. April–June.

Var. **ultramontànus** Greene. [*R. u.* Heller. *R. alceus* Greene. *R. o.* var. *a.* Jeps.] Stems flexuous, usually reclining, 3–6 dm. long, 1–2 mm. thick; lvs. 2–6 cm. broad, 3-parted or with 3–5 lfts., the divisions lanceolate to cuneate, lobed, ultimate lobes lance-ovate, acutish; petals 6–8 mm. long, 1.5–3 mm. wide; ak.-body 2.5–3 mm. long, the beak 0.7–1.3 mm. long, produced from the apex of the body.—Meadows and moist places, 4000–6500 ft.; upper Foothill Wd., Montane Coniferous F.; N. Coast Ranges s. to Lake Co., Sierra Nevada s. to Tuolumne and Inyo cos.; w. Nev., cent. Ore. June–July.

Var. **disséctus** Henders. [*R. ciliosus* Howell. *R. marmorarius* Jeps. & Tracy.] Stems erect or reclining, 2–3 dm. long, 1.5–2.5 mm. thick; lvs. thin, 3-parted, the divisions lanceolate and simple or parted, the ultimate divisions somewhat lanceolate; petals 5–10 mm. long, 4–6 mm. broad; ak.-body 2.5–3 mm. long, the beak 1.5–2 mm. long, straight, with terminal hook, prolonged from margin of body.—Meadows and open places, 4000–7000 ft.; Montane Coniferous F.; Siskiyou and Modoc cos.; s. Ore. June–July.

Var. Howéllii Greene. [*R. H.* Greene.] Stems suberect, 2–4 dm. long, 1–3 mm. thick; lvs. 2–4.5 cm. broad, 3-parted, the lobes cuneate and again lobed, the ultimate lobes cuneate; petals 7–12(–18) mm. long, 3–6(–8) mm. wide; ak.-body 3 mm. long, the beak 2 mm. long, straight, hooked only at tip, prolonging the margin of the body.—Wooded hills, 1000–4000 ft.; Foothill Wd., Yellow Pine F.; Siskiyou Co.; sw. Ore. March–May.

5. **R. califórnicus** Benth. [*R. dissectus* H. & A. *R. c.* var. *latilobus* Gray. *R. l.* Parish.] Perennial, with slender roots; stems erect, 3–7 dm. high, 1.5–5 mm. in diam., hirsute to subglabrous, branched above; basal lvs. 2–7 cm. broad, long-ovate or orbicular, 3–5-foliolate or 3-parted, the lfts. or lobes cuneate, lobed, mostly hirsute, the ultimate lobes acute; petioles 5–25 cm. long; sepals greenish-yellow, reflexed, 4–8 mm. long; petals mostly 9–16, yellow, 8–15 mm. long, 3–5 mm. broad; stamens 30–60; aks. 5–35, in a rounded head, each obovoid-discoid, 2–2.5 mm. long, the beaks stout, recurved, 0.4–0.8 mm. long; $2n = 28$ (Coonen, 1939).—Vernally moist slopes and meadows, below 3000 ft.; N. Coastal Scrub, Foothill Wd., N. Oak Wd., Mixed Evergreen F., V. Grassland; Coast Ranges throughout Calif., Sierra Nevada foothills, Amador, Calaveras, and Madera cos.; s. Ore., L. Calif. Feb.–May.

KEY TO VARIETIES

Stems prostrate, 1–2.5 dm. long; lvs. with ultimate lobes cuneate, rounded or obtuse. Coastal bluffs
          var. *cuneatus*
Stems mostly erect, 2.5–7 dm. high; lvs. with ultimate lobes narrower and acute.
  The stems mostly hirsute.
    Petals mostly 8–15 mm. long; lvs. mostly 3–5-foliolate. Below 2000 ft. .... *R. californicus*
    Petals mostly 6–8 mm. long; lvs. 3-parted. S. Montane, above 4500 feet .... var. *austromontanus*
  The stems nearly or quite glabrous.
    Aks. 2.5 mm. long; divisions of lvs. broad, shallowly lobed. Inner edge of Redwood Belt
          var. *gratus*
    Aks. 1–2 mm. long; divisions of lvs. deeply lobed into narrow segms. Cent. V. ... var. *rugulosus*

Var. gràtus Jeps. Subglabrous; lvs. 3-lobed, the divisions broadly cuneate, shallowly lobed, thinly pubescent; petals mostly 5–9, 3–10 mm. long, 2–5 mm. broad; aks. nearly discoid, 2–2.5 mm. long, the beaks 1–1.5 mm. long.—Canyons and n. slopes; inner edge of Redwood F.; Humboldt Co. to Monterey Co. Feb.–May.

Var. cuneàtus Greene. [*R. californicus* var. *crassifolius* Greene.] Stems hirsute to pubescent, prostrate, 1–2.5 dm. long; lvs. ± cordate, 3-lobed, the lobes cuneate and divided into ultimate rounded or obtuse lobes; aks. as in the sp.—Coastal bluffs and hills; Coastal Prairie, N. Coastal Scrub; Del Norte Co. to Monterey Co.; Santa Cruz and San Miguel ids.; w. Ore. Feb.–May.

Var. rugulòsus (Greene) L. Benson. [*R. r.* Greene.] Stems nearly glabrous, 3–6 dm. high; lvs. 3-parted or 3–5-foliolate, ovate in outline, 2–8 cm. wide, again lobed into oblanceolate or narrowly cuneate segms., sparsely pubescent; petals 7–12, 7–11 mm. long, 3 mm. broad; aks. obovoid, 1–2 mm. long, the beak 0.8 mm. long.—Moist stream banks, below 3000 ft.; V. Grassland, Foothill Wd.; Colusa and Sutter cos. to Tulare Co. March–May.

Var. austromontànus L. Benson. [*R. c.* var. *ludovicianus* auth., not *R. l.* Greene.] Densely pilose-hirsute; lvs. fan-shaped, the segms. cuneate to oblanceolate; petals 7–12, 6–8 mm. long, 2–4(–5) mm. wide; aks. obovoid-discoid, 2–2.5(–3) mm. long, the beak 0.5 mm. long.—Meadows and stream banks, 4500–7500 ft.; mostly Montane Coniferous F.; San Bernardino Mts. to San Diego Co. May–July.

6. **R. cànus** Benth. [*R. occidentalis* var. *c.* Gray. *R. californicus* var. *c.* Brew. & Wats. *R. canus* var. *Blankinshipii* Rob. *R. B.* and *R. longilobus* Heller.] Perennial, with slender roots; stems erect, hirsute, branched above, 4–9 dm. high, 3–7 mm. thick; lvs. ovate in outline, 5–8 cm. broad, 3-parted or 3(5)-foliolate, each part cuneate, not much divided, with ultimate lobes acute, strigose on both surfaces; petioles to 10 or 12 cm. long, often pilose; sepals greenish-yellow, reflexed at middle, deciduous, 5–7 mm. long; petals 5–10, yellow, 11–14 mm. long, 5–7 mm. broad; stamens 40–70; aks. 12–30,

in a rounded head, each discoid, 4–5 mm. long, the beak deltoid, 0.5–1 mm. long, recurved at tip; receptacle glabrous.—Heavy soil, open places, below 1000 ft.; V. Grassland, lower edge of Foothill Wd.; in and about Cent. V., Tehama and Butte cos. to Contra Costa and Tulare cos. Feb.–April.

KEY TO VARIETIES

Petals 5–10, 5–7 mm. wide; basal lvs. 5–8 cm. broad. N. of Kern Co.
  Ultimate segms. of basal lvs. 4–10 mm. broad, hairy but greenish on both surfaces . . . . . . *R. canus*
  Ultimate segms. of basal lvs. 2–4 mm. broad, gray-hairy on lower surface only, sparsely hairy and
  green above . . . . var. *laetus*
Petals 10–23, 3–5 mm. wide; basal lvs. 2–5 cm. broad. Kern and San Bernardino cos.
                                                               var. *ludovicianus*

Var. laètus (Greene) L. Benson. [*R. californicus* var. *l.* and var. *canescens* Greene. *R. canus* var. *hesperoxys* Davis.] Lfts. of basal lvs. dissected into lance-linear segms. 2–4(–8) mm. broad, appressed-silky beneath, green above.—Very heavy soil, n. slopes, foothills; upper V. Grassland, Foothill Wd.; Sutter Co. to Contra Costa Co. and in lower San Joaquin V. March–April.

Var. ludoviciànus (Greene) L. Benson. [*R. l.* Greene, *R. californicus* var. *l.* Davis.] Stems very hairy, 2.5–4.5 dm. high, 3–5 mm. thick; lvs. 2–5 cm. wide, 3-parted into cuneate, appressed-hirsute lobes; petals 10–23, 10–13 mm. long, 3–5 mm. broad.— Vernally moist ground, 3300–7500 ft.; Foothill Wd., Yellow Pine F.; Kern and San Bernardino cos. March–May.

7. **R. uncinàtus** D. Don in G. Don. [*R. tenellus* Nutt. *R. Nelsonii* var. *t.* Gray. *R. occidentalis* var. *t.* Gray. *R. Bongardii* var. *t.* Greene. *R. Nelsonii* var. *glabriusculus* Holz. *R. Douglasii* Howell. *R. arcuatus* Heller.] Annual or perennial, glabrous to sparsely hirsute with white hairs, 2–4.5 dm. high; basal lvs. cordate-reniform in outline, 2–9 cm. broad, 3-parted, the parts lobed, glabrous, with ultimate segms. mostly obtuse; petioles 5–20(–25) cm. long, glabrous to hirsute; cauline lvs. often larger than basal; pedicels to 2 cm. long in fl., twice that in fr.; sepals yellowish-green, reflexed, 3 mm. long; petals 5, yellow, 2.5–3(–6) mm. long, 1–2(–3) mm. broad; stamens 10–15; aks. 5–30, in a round head, the body discoid, obovoid or ellipsoid, 2–2.5 mm. long, glabrous, the beak 1–1.5 mm. long, recurved, hooked at tip; receptacle glabrous.— Occasional, moist shaded places, below 8000 ft.; Redwood F., Coast Ranges s. to Sonoma Co.; Montane Coniferous F., Sierra Nevada and San Bernardino Mts.; to Alaska, Rocky Mts. May–July.

Var. parviflòrus (Torr.) L. Benson. [*R. occidentalis* var. *p.* Torr. *R. o.* var. *Lyallii* Gray. *R. Bongardii* Greene. *R. tenellus* var. *L.* Rob. *R. Greenei* Howell. *R. Bongardii* var. *G.* Piper.] Stems with stiff reddish-brown hairs; lvs. with ultimate segms. acute, appressed-hispidulous; cauline lvs. usually smaller than basal; aks. hispid, the beak 2 mm. long.—Shaded places below 3000 ft.; Redwood F., N. Coast Ranges s. to Marin Co.; less frequent up to 7000 ft., Montane Coniferous F., Sierra Nevada and San Bernardino Mts.; to Alaska, Rocky Mts. May–July.

8. **R. Macoùnii** Britton. [*R. hispidus* var. *oreganus* Gray. *R. o.* Howell. *R. rudis* Greene. *R. rivularis* Rydb.] Annual or perennial; stems reclining to suberect, 3–9 dm. long, densely hirsute to glabrous; lower lvs. deltoid in outline, 5–13 cm. wide, 3-parted or 3-foliolate, the main divisions parted, then lobed, pubescent to glabrous; petioles 5–20 cm. long, hirsute to glabrous; sepals yellowish, often tinged purple, reflexed, 4–6 mm. long; petals 5, yellow, obovate, 3–6 mm. long, 2.5–5 mm. wide; stamens 15–35; aks. 20–50, in an ovoid head, each obovoid, 2–3 mm. long, glabrous, the beak subdeltoid, 1–1.2 mm. long; receptacle hispid.—Marshy and wet places, ca. 5000 ft.; Sagebrush Scrub, N. Juniper Wd.; Goose Lake, Modoc Co.; to Alaska, Labrador, Nebr., New Mex. June–July.

9. **R. orthorhýnchus** Hook. Perennial, the roots somewhat fleshy; stems hirsute, 1.5–5 dm. high, 1.5–3 mm. thick; basal lvs. 3–8 cm. wide, pinnate, with 3–7 lfts., these twice lobed into linear or cuneate acute divisions; petioles 3–15 cm. long, hairy to glabrous; pedicels strigose to glabrous; sepals pilose, greenish-yellow, somewhat reflexed, 6–8 mm. long; petals 5, yellow, 8–19 mm. long, 4–7 mm. broad, truncate; stamens 20–40; aks. 12–20, ellipsoid, 3 mm. long, keeled, with slender, straight beaks ca. 4 mm. long; receptacle 2–3 mm. long in fr., hispidulous.—Meadows, below 6500 ft.; Redwood F.,

Montane Coniferous F.; N. Coast Ranges s. to Glenn Co., Sierra Nevada to Tuolumne Co.; Modoc Co.; to B.C. May–July.

KEY TO VARIETIES

Stems 6–12 dm. long, often 7–9 mm. thick; receptacle 5–9 mm. long in fr. .... var. *platyphyllus*
Stems 1.5–5 dm. long, mostly 2–4 mm. thick; receptacle 2–4 mm. long in fr.
 Sepals glabrous; petals 5–8, emarginate .... var. *Bloomeri*
 Sepals pilose; petals 5, truncate.
  Divisions of basal lvs. longer than wide, deeply and acutely lobed. N. Coast Ranges and cent. Sierra Nevada .... *R. orthorhynchus*
  Divisions of basal lvs. not longer than wide, shallowly and obtusely lobed. Sierra Nevada s. of Yosemite .... var. *Hallii*

Var. **platyphýllus** Gray. [*R. p.* Piper. *R. maximus* Greene. *R. o.* var. *m.* Jeps. *R. macranthus* Brew. & Wats., not Scheele.] Stems 4–12 dm. high, 4–9 mm. thick, hirsute; lvs. as in the sp. but larger; sepals pilose; petals 5, 10–15 mm. long, 5–10 mm. broad, truncate; aks. 20–35, the beaks 2.5–3 mm. long; receptacle 5–9 mm. long in fr., hispidulous.—Meadows, below 7000 ft.; many Communities, coast to mts.; N. Coast Ranges s. to Marin Co., Alameda Co., and inland to San Joaquin Co.; to Alaska, Mont., Wyo. May–July.

Var. **Bloòmeri** (Wats.) L. Benson. [*R. B.* Wats.] Glabrous or sparingly hairy; stems 1.5–5 dm. high, succulent; lvs. simple or usually 3-foliolate, cordate-ovate in outline, the lfts. rounded, shallowly lobed, subglabrous; sepals glabrous; petals 5–8, 10–17 mm. long, 4–10 mm. wide, emarginate; aks. ca. 3.5 mm. long, the beaks 2.5 mm. long.—Wet heavy soil in meadows and near ditches; N. Oak Wd., Foothill Wd.; Mendocino Co. to Santa Clara Co.; Lake Co., Ore. March–May.

Var. **Hàllii** Jeps. Like the sp., but ultimate divisions of lvs. broad as long, shallowly and obtusely lobed; beak of aks. more distinctly a prolongation of the marginal keel of the body.—Meadows, 4000–7000 ft.; Montane Coniferous F.; Sierra Nevada from Tuolumne Co. to Tulare Co. June–July.

10. **R. hebecárpus** H. & A. [*R. h.* var. *pusillus* Brew. & Wats.] Annuals, with filiform roots; stems erect, slender, pubescent, 1–3 dm. high, freely branched throughout; basal lvs. cordate-reniform in outline, 1–2 cm. wide, 3-parted and then lobed, the ultimate segms. oblong-ovate, acute, appressed-hairy; petioles 3–9 cm. long; pedicels 1–16 mm. long; sepals yellow, 1.5 mm. long, scarious-margined, pubescent; petals 0–5, yellow, spatulate, 1.5 mm. long; stamens 8–15; aks. 4–10 in a round head, each 2 mm. long, papillate with hooked bristles, the beak stout, 0.6 mm. long, hooked at tip; receptacle glabrous.—Shaded places, below 3000 ft.; many Communities, particularly Foothill Wd., Chaparral; from Siskiyou and Modoc cos. throughout foothill region w. of Sierra Nevada; to Wash., Ida., L. Calif. March–May.

11. **R. parviflòrus** L. Annuals with slender roots; stems erect, 1–3 dm. long, slender, branched, hirsute; lvs. reniform, 2–2.5 cm. broad, 3-parted, then lobed, the ultimate segms. acute, hirsute; petioles 3–6 cm. long; pedicels to 1.5 cm. long; sepals greenish-yellow, spreading, 1 mm. long; petals 5, yellow, elliptic, 1–2 mm. long; stamens ca. 10; aks. 10–20 in a globose head, each obovoid, 1.5 mm. long, with reddish-brown papillae produced into slender hooks, the beak deltoid, recurved, 0.5 mm. long; receptacle glabrous.—Natur. in waste places, Humboldt and Monterey cos.; Atlantic Coast; native of Old World. June–Aug.

12. **R. Sardòus** Crantz. [*R. parvulus* L.] Perennial with filiform roots; stems suberect, 1–5 dm. long, slender, branching, hirsute; lvs. broadly cordate in outline, 2–2.5 cm. wide, 3-foliolate, the lfts. parted and lobed with deltoid pubescent ultimate segms.; petioles 3–16 cm. long, hirsute; pedicels 3–6 cm. long; sepals greenish-yellow, reflexed, 3–5 mm. long, pilose; petals 5, yellow, 8–9 mm. long, 5–7 mm. broad; stamens 25–50; aks. 12–25 in a rounded head, each subcircular, 2–3 mm. long, papillate to glabrous, the beak deltoid, 0.3 mm. long, curved at tip; receptacle hairy; $2n = 48$ (Neves, 1944). —Natur. in waste places, Humboldt Co.; Ore., Atlantic Coast; native of Old World. May–July.

13. **R. muricàtus** L. Glabrous annuals or perennials with fairly coarse roots; stems reclining to erect, 2–5 dm. long, 2–5 mm. thick, branched; lvs. rounded to reniform in outline, 2–6 cm. broad, 3-parted, the parts shallowly lobed with rounded-acute

segms.; petioles 4–15 cm. long; pedicels 1–5 cm. long; sepals greenish, spreading, 4–7 mm. long, with a few bristly hairs; petals 5, yellow, obovate, 5–8 mm. long, 3–4 mm. wide; stamens few; aks. 10–20, in a globose cluster; each 5.5 mm. long, obovate, with many stout curved spines, the beak 2–2.5 mm. long, curved; receptacle hispid; $2n = 48$ (Neves, 1944).—Natur. in moist places, low elevs., Del Norte, Shasta, and Butte cos. to San Mateo and Mariposa cos.; to Wash., e. U.S.; native of Old World. May–June.

14. **R. arvénsis** L. Annual with fibrous roots; stems erect, 1.5–5 dm. high, slender, simple to branched, thinly hirsute; lvs. 1.5–4.5 cm. wide, cuneate and apically toothed to 3-parted with oblanceolate apically toothed lobes, glabrous to strigose; petioles 1–5 cm. long; pedicels 1–5 cm. long, pubescent; sepals greenish-yellow, spreading, elliptic, strigose, 4–7 mm. long; petals 5, yellow, obovate, 6–8 mm. long, 3–3.5 mm. wide; stamens 10–15; aks. 5, whorled, obovate, 5 mm. long, spiny, the beak stout, curved, 2.5–3 mm. long; receptacle somewhat pubescent.—Natur. in waste places, Mendocino, El Dorado, and Mariposa cos.; Ore., Ida., Utah, Atlantic Coast; native of Eu. May–June.

15. **R. Eschschòltzii** Schlecht. Glabrous perennials with caudex 1–2 cm. long, 3–6 mm. thick, simple, bearing slender roots; stems erect or decumbent, 4–15 cm. long, scapose; lvs. rounded in outline, 1.5–3 cm. wide, deeply 3-parted, the middle lobe 3-lobed or entire, the lateral parted, the ultimate segms. and sinuses rounded; petioles 2–6 cm. long; stipular lf.-bases 1–2 cm. long, disintegrating annually; pedicels 1–10 cm. long; sepals yellow, tinged lavender, glabrous or sparsely hairy, 4–8 mm. long; petals 5, yellow, cuneate-obovate, 7–11 mm. long, 5–10 mm. broad; stamens 20–40; aks. in ovoid head, each oblong-obovoid, glabrous, ca. 1.5 mm. long, the beak slender, rather straight, ca. 1 mm. long; receptacle glabrous.—Meadows and about rocks, 8000–11,500 ft.; Subalpine F., Alpine Fell-fields; Siskiyou and Trinity cos., Sierra Nevada s. to Tulare and Inyo cos.; to Alaska and Rocky Mts. July–Aug.

KEY TO VARIETIES

Ultimate segms. and sinuses of basal lvs. rounded or obtuse. Widely distributed.
    Caudex 1–2 cm. long, unbranched, 3–6 mm. thick; scarious stipular lf.-bases disintegrating annually
                                                                           R. Eschscholtzii
    Caudex 1.5–5 cm. long, usually branched, 5–12 mm. thick; scarious stipular lf.-bases persistent
                                                                            var. *oxynotus*
Ultimate segms. and sinuses of basal lvs. sharply acute. Siskiyou Co. . . . . var. *Suksdorfii*

Var. **Suksdórfii** (Gray) L. Benson. [*R. S.* Gray.] Sinuses of lvs. and lobes sharply acute, the basal lvs. deeply parted and the middle lobe again 3-lobed.—Marble Mts., Siskiyou Co.; to Wash., Mont., Wyo. July–Aug.

Var. **oxynòtus** (Gray) Jeps. [*R. o.* Gray.] Caudex 1.5–5 cm. long, 5–12 mm. thick, usually branched; lf.-sinuses and -lobes rounded or obtuse, the middle lobe of basal lvs. usually entire; stipular lf.-bases thickened, persistent for 2–3 years.—At 9000–13,300 ft.; Subalpine F., Alpine Fell-fields; Modoc Co., Sierra Nevada from Sierra Co. to Tulare Co., Sweetwater Mts., Mono Co., White Mts., Inyo Co., San Bernardino and San Jacinto mts.; Nev. July–Aug.

16. **R. glabérrimus** Hook. [*R. Austinae* Greene.] Glabrous perennials, with fleshy roots 2–3 mm. thick; stems prostrate or ascending, 5–15 cm. long, 1–2 mm. thick, 1–6-fld.; blades of basal lvs. rounded to obovate, 2–3 cm. long, 1–2 cm. wide, entire or 3–5-lobed at apices, rarely dissected into narrow divisions, rounded or tapering at base, rounded at apex; petioles 3–9 cm. long; pedicels 1–10 cm. long; sepals tinged lavender, spreading, slightly pubescent, 5–8 mm. long; petals usually 5, yellow, or white in age, obovate, 6–15 mm. long, 5–10 mm. wide, truncate or 2-lobed; stamens 40–60; aks. 75–150, in a large globose head, each obovoid, 2 mm. long, usually pubescent, the beak 0.6 mm. long, not recurved; receptacle glabrous.—Sandy places, ca. 5000 ft.; Sagebrush Scrub; s. to Plumas and Siskiyou cos.; to B.C., Mont., S. Dak. April–May.

Var. **ellípticus** (Greene) Greene. [*R. e.* Greene.] Basal lvs. entire, elliptic to oblanceolate, 3–5 cm. long, tapering into petioles 3–5 cm. long.—Meadows; Montane Coniferous F.; Mono and Nevada cos. to Modoc Co.; to B.C., N. and S. Dak., Nebr., New Mex. April–June.

17. **R. alismaefòlius** Geyer ex Benth. [*R. Bolanderi* Greene.] Glabrous perennial with 15–30 gradually tapering roots; stems erect, 3–8 dm. long, mostly 4–6 mm. thick, branched above, fistulous, several- to many-fld.; basal lvs. simple, lanceolate, 1–3 cm.

broad, entire or serrulate, tapering into a broad petiole 5–15 cm. long; pedicels to 10 cm. long; sepals yellowish-green, spreading, glabrous to pubescent, 3–5 mm. long; petals 5, yellow, obovate, ca. 10 mm. long, 5 mm. wide; stamens 30–80; aks. 30–50, in a rounded head, glabrous, cuneate-obovoid, 1.5–2.5 mm. long, the beak 0.6–0.9 mm. long, not recurved; receptacle 4–5 mm. long in fr., glabrous.—Muddy banks and ditches, below 5000 ft.; Mixed Evergreen F., Yellow Pine F.; N. Coast Ranges s. to Mendocino Co., Sierra Nevada s. to Sierra Co.; to B.C., Ida., Mont. May–June.

KEY TO CALIFORNIA VARIETIES

Receptacle 35 mm. long in fr.; lvs. 1–3 cm. broad; stems 3–8 dm. long, erect .... *R. alismaefolius*
Receptable 1–2 mm. long (except in *Lemmonii*); lvs. 0.5–1 cm. broad; stems 1–3 dm. long, suberect to reclining.
  Stems glabrous; petals 6–8 mm. long.
    Basal lf.-blades 3–7 cm. long; roots not tuberous-thickened at bases .... var. *Hartwegii*
    Basal lf.-blades 2–4 cm. long; roots somewhat tuberous-thickened at bases ...... var. *alismellus*
  Stems commonly pilose; petals mostly ca. 10 mm. long .... var. *Lemmonii*

Var. **Hartwégii** (Greene) Jeps. [*R. H.* Greene.] Glabrous; stems suberect or reclining below, 2–4 dm. long, 1.5–2.5 mm. thick; basal lvs. lanceolate, entire, 5–10 mm. wide, tapering into the petiole; petals 6–8 mm. long; aks. 20–30, 2 mm. long, glabrous; receptacle 1–1.5 mm. long in fr.—Meadows, 4500–7500 ft.; Montane Coniferous F.; N. Coast Ranges s. to n. Lake Co., Sierra Nevada s. to Fresno Co.; to Wash., Ida., Nev. May–June.

Var **alisméllus** Gray. [*R. a.* Greene.] Glabrous; stems ascending, 1–3 dm. high, mostly 1–1.5 mm. thick; blades of basal lvs. lance-ovate, 2–4 cm. long, 4–10 mm. wide, abruptly narrowed at base; petals ca. 6 mm. long; aks. 10–30, glabrous, 1.5 mm. long; receptacle ca. 1 mm. long.—Meadows and wet banks, 4500–6500 ft.; Montane Coniferous F.; N. Coast Ranges s. to Glenn Co., Sierra Nevada (up to 12,000 ft.), San Bernardino and San Jacinto mts.; to Wash., Mont., L. Calif. June–July.

Var. **Lemmònii** (Gray) L. Benson. [*R. L.* Gray.] Commonly pilose; stems decumbent, 1.5–3 dm. long, 1.5–3 mm. thick; basal lf.-blades lanceolate, 3–9 cm. long, mostly 5–10 mm. wide, tapering at base; petals 8–16 mm. long; aks. ca. 20, mostly pubescent, 2 mm. long; receptacle 2–5 mm. long in fr.—Meadows, 5000–6500 ft.; Sagebrush Scrub, N. Juniper Wd.; Modoc Co. to Nevada Co. May–June.

18. **R. Flámmula** L. var. **ovàlis** (Bigel.) L. Benson. [*R. filiformis* var. *o.* Bigel. *R. reptans* var. *o.* T. & G. *R. F.* var. *intermedius* and *R. i.* as to Calif. refs.] Almost glabrous perennials with filiform roots; stems reclining, usually rooting at nodes, slender, 1–4 dm. long; lvs. often fascicled at rooting nodes, simple, entire, linear-spatulate to oblanceolate, 1.5–5 cm. long, 1.5–7 mm. wide; petioles 1–6 cm. long, not sheathing the stems; pedicels from any node, 1–8 cm. long; sepals greenish-yellow, spreading or somewhat reflexed, 3–5 mm. long, glabrous to strigose; petals 5 or 10, yellow, obovate, 3–6 mm. long; stamens 20–30; aks. 10–25, in a rounded head, each 1–1.5 mm. long, with very short recurved beak; receptacle glabrous.—Muddy and marshy places; Coastal Prairie, Redwood F., Coast Ranges s. to Marin Co.; Montane Coniferous F. below 7500 ft., Sierra Nevada s. to Fresno Co., San Bernardino Mts.; to Alaska, Labrador, Pa., New Mex. July–Aug.

Var. **samolifòlius** (Greene) L. Benson. [*R. s.* Greene. *R. reptans* var. *s.* L. Benson.] A poorly marked var. with lvs. ovate to broadly oblanceolate, 10–12 mm. wide, usually sessile and sheathing the stem.—Meadows, 4500–6500 ft.; Montane Coniferous F.; Siskiyou Co. to Plumas Co.; Ore.

19. **R. hydrocharoìdes** Gray. Glabrous perennials with slender roots; stems procumbent or floating or suberect, rooting at lower nodes, 1–2.5 dm. long, 1.5–4 mm. thick, fistulous; lf.-blades entire, ovate or elliptic to suborbicular, 1–2.8 cm. wide, on petioles 2–8 cm. long; pedicels mostly 2–7 cm. long; sepals 5, greenish-yellow, spreading, ovate, 2–3 mm. long; petals 5–7, light yellow, 3–5 mm. long, 2–3 mm. wide; stamens 10–40; aks. 10–25, in a rounded head, each obovoid, 1.3 mm. long, the beak rather straight.—Marshes and streams; Owens V., *Kellogg in 1874;* Ariz., New Mex., L. Calif. to Chihuahua. July.

20. **R. Gormánii** Greene. [*R. reptans* var. *G.* Davis.] Glabrous perennials, with slender roots thickened at base; stems rooting at 1–3 nodes, 5–12 cm. long, slender, 1–2-fld.;

lf.-blades entire, ovate, 2–3 cm. long, 1.5–2 cm. wide, on petioles 2–4 cm. long; pedicels 3–5 cm. long; sepals yellow, spreading, narrow-obovate, 2–3 mm. long, glabrous; petals 5, light yellow, 5–6 mm. long, 2–3 mm. wide; stamens 10–20; aks. 6–15, flat-obovoid, 1.5 mm. long, the beak stout, 0.6 mm. long, recurved at tip; receptacle glabrous.—Mt. meadows, 5000–6000 ft.; Subalpine F.; Del Norte, Humboldt and Siskiyou cos.; Ore. June–July.

21. **R. Populàgo** Greene. [*R. alismellus* var. *P*. Davis. *R. Cusickii* Jones.] Glabrous perennials, with roots slightly thickened at base; stems declined or erect, not rooting, 1–4 dm. long, 6–10-fld.; basal lf.-blades round-reniform to ovate, 3–5 cm. long and broad, usually denticulate, cordate or truncate at base, rounded to acute at apex; petioles 4–12 cm. long; pedicels 1–5 cm. long; sepals green, spreading, 1.5–2.5 mm. long, 1–1.5 mm. wide, glabrous; petals 5, light yellow, obovate, 3–6 mm. long, 1.5–3 mm. broad; stamens 10–30; aks. 10–25, in a hemispheric head, each flat-obovoid, 1.5 mm. long, the beak stout, 0.6 mm. long, bent; receptacle glabrous or pubescent.—Meadows and boggy places, 5000–6000 ft.; Red Fir F., Subalpine F.; from Siskiyou and Butte cos.; to Wash., Ida. June–July.

22. **R. pusíllus** Poir. [*R. p.* var. *Lindheimeri* Gray. *R. Biolettii* Greene.] Glabrous annual, with filiform roots; stems reclining, usually rooting at lowest nodes, 1–4 dm. long, slender, branched; lower lvs. simple, oblong to ovate, 6–40 mm. long, 5–15 mm. wide, entire or shallowly toothed, on petioles 1–6 cm. long; pedicels 0.5–5 cm. long; sepals greenish-yellow, spreading, 1–1.5 mm. long; petals 1–3, yellow, obovate, 1–1.5 mm. long, truncate; stamens 5–10; aks. 15–50, in a hemispheric head, each oblong-obovoid, 1 mm. long, with very minute beak; receptacle glabrous.—Shallow water and marshy places, low elevs.; N. Oak Wd. and coastal Communities; outer Coast Ranges, Humboldt Co. to Napa Co. and Santa Cruz Mts.; Mo. to N.Y., Fla., Tex. April–May.

23. **R. alveolàtus** Carter in L. Benson & Carter. Annual, glabrous or sparsely pubescent; roots filiform; stems decumbent, rooting at lower nodes, 1–3 dm. long, slender; lf.-blades simple, ovate to lance-ovate, 6–20 mm. long, 4–12 mm. wide, entire or slightly toothed, with petioles to 8 cm. long; pedicels 0.5–4 cm. long; sepals 3, ovate, 2–2.5 mm. long, membranous near base, rather persistent; petals 2–3, yellow, 2–2.5 mm. long; stamens 4–5; aks. 15–25 in an ovoid head, each flat-ovoid, 1.5 mm. long, the beak obscure; receptacle glabrous.—Wet places, below 3300 ft.; Foothill Wd., V. Grassland; foothills of Sierra Nevada from Calaveras Co. to Placer Co. April–May.

24. **R. sceleràtus** L. Glabrous or rarely hirsute annuals, with fleshy roots; stems erect, rarely rooting, 1–8 dm. high, 2–15 mm. thick, freely branched, inflated; lower lvs. reniform in outline, mostly 2–6 cm. broad, deeply 3-parted, these divisions then somewhat lobed; petioles mostly 3–12 cm. long; cauline lvs. with oblanceolate primary divisions and apically toothed; pedicels 1–3 cm. long; sepals greenish-yellow, spreading, persistent, 2–3 mm. long; petals 5, light yellow, 2–4 mm. long; stamens 10–25; aks. obovoid, ca. 1 mm. long, with minute transverse ridges, the beak almost lacking; receptacle usually pubescent; 2n = 32 (Gregory, 1939).—Lake borders, marshes, etc.; sparingly natur., as near Stockton; Ore., Wash., e. U.S.; native of Old World. June–Sept.

Var. **multífidus** Nutt. in T. & G. [*R. eremogenes* Greene. *R. s.* ssp. *m.* Hult.] Basal lvs. with primary parts again deeply parted; cauline lvs. with narrowly cuneate divisions apically lobed; ak. not transversely ridged.—Wet places, below 6500 ft.; Sagebrush Scrub, N. Juniper Wd.; Siskiyou and Modoc cos.; to Alaska, Minn., New Mex. May–Aug.

25. **R. flabellàris** Raf. in Bigel. [*R. multifidus* Pursh, not Forsk.] Glabrous rarely hairy perennials with filiform roots; stems floating or reclining, rooting at lower nodes, 3–7 dm. long, 2–4 mm. thick, branched; lvs. all cauline, triternately dissected, 2–12 cm. wide, with many ribbonlike divisions 1–2 mm. wide in aquatic forms, or parted and then lobed in palustrine forms; petioles very short; pedicels 1–5 cm. long; sepals greenish-yellow, spreading, 5–8 mm. long, early deciduous; petals 5–8, yellow, obovate, 7–15 mm. long; stamens 50–80; aks. 50–75, in an ovoid head, each obovoid, 2 mm. long, smooth, the beak flat, 1.5 mm. long; receptacle hairy.—Mud or shallow water, below 6000 ft.; Montane Coniferous F.; Humboldt, Mendocino, Plumas, and Modoc cos.; to B.C., Quebec, N.J., Va., La. June–Aug.

26. **R. Cymbalària** Pursh var. **saximontànus** Fern. Glabrous or sparingly hairy perennials with slender roots, stoloniferous; stems ca. 1 mm. thick, several dm. long; basal

lvs. simple, cordate, ovate or even reniform, 1–4 cm. long, 1–3 cm. broad, mostly crenate, on petioles 2–5 cm. long; scapes erect, 5–30 cm. high, 1–several-fld.; sepals greenish-yellow, spreading, 4–8 mm. long, promptly deciduous; petals 5–12, yellow, obovate, 4–8 mm. long; stamens 20–35; aks. many, in an elongate head, each cuneate-oblong, 1.5 mm. long, the beak 0.3 mm. long, not curved; receptacle hairy.—Muddy places, often somewhat alkaline, below 9700 ft.; many Plant Communities; Humboldt Co., Sierra Nevada from Mono and Tuolumne cos. to Kern Co., Mojave Desert, San Bernardino Mts., Santa Ana River system; to B.C., Man., Nebr., Tex., L. Calif. June–Aug.

27. **R. hystrículus** Gray. [*Kumlienia h.* Greene.] Glabrous perennials with somewhat fleshy roots and short caudex; scapes erect, 1.5–4 dm. high; basal lf.-blades simple, rounded, 2–4 cm. long, 3–5 cm. broad, shallowly 3-lobed, the lobes again shallowly lobed with rounded ultimate divisions; petioles 3–9 cm. long; sepals white, spreading, glabrous, deciduous, 7–10 mm. long; petals 8–12, yellowish or greenish, 4–5 mm. long; aks. 25–30, in a rounded head, each oblong, 3 mm. long, the beak slender, 1.3 mm. long, hooked at tip; receptacle glabrous.—Wet places among rocks and near streams, 3300–6000 ft.; mostly Yellow Pine F.; Sierra Nevada from Butte Co. to Tulare Co. April–June.

28. **R. Andersònii** Gray. [*Oxygraphis A.* Freyn. *Beckwithia A.* Jeps. *B. Austinae* Jeps.] Glabrous perennial with short caudex and slender roots; scapes erect, 1–2 dm. high, 1–2-fld.; lvs. cordate or rounded in outline, 1–3 cm. long, 2–4 cm. broad, 3-foliolate, the lfts. dissected into ovate to linear segms. 7–8 mm. long, 1–3 mm. wide; petioles 3–6 cm. long; sepals reddish, thin, spreading, 7–8 mm. long, persistent; petals 5, pink to red, cuneate to round, 1.5 cm. long; stamens 75–100; aks. 15–25 in a round cluster, each utricular, obovoid, 7–14 mm. long, with very short beak; receptacle pubescent.—Dry rocky slopes, 3000–7500 ft.; Sagebrush Scrub, Pinyon-Juniper Wd.; Panamint Mts. n. to Lassen and Modoc cos.; Ore., Ida., Nev. June.

29. **R. Lóbbii** (Hiern) Gray. [*R. Hydrocharis* f. *L.* Hiern. *R. hederaceus* var. *L.* Brew. & Wats. *R. aquatilis* var. *L.* Wats. *Batrachium L.* Howell.] Glabrous aquatic annuals; stems floating, rooting at lower nodes, 1–10 dm. long, branched; submersed lvs. 1–4 cm. long, 3–5 cm. wide, dissected into filiform divisions; upper lvs. floating, 5–8 mm. long, 3–10 mm. broad, deeply parted into 3 elliptic, very shallowly lobed parts; petioles 1–3 cm. long; pedicels 0.5–2 cm. long; sepals light green, 2–3 mm. long; petals 5, white, 4–6 mm. long; stamens 5–15; aks. 4–6, obovoid, 2.5 mm. long, the beak very short; receptacle glabrous.—Shallow vernal ponds, low elevs.; Redwood F., Mixed Evergreen F., N. Oak Wd.; Sonoma and Lake cos. to Santa Clara and Alameda cos. Feb.–April.

30. **R. aquátilis** L. var. **hispídulus** E. Drew. [*R. a.* f. *heterophyllus* Gray. *R. Grayanus* Freyn. *Batrachium G.* Rydb. *R. trichophyllus* var. *hispidulus* W. Drew.] Glabrous or hispidulous aquatic perennials with submersed stems 2–6(–15) dm. long, 1–2 mm. thick; submersed lvs. 2–4 cm. long, 3–5 cm. wide, usually shorter than internodes, dissected into filiform divisions; floating lvs. 1–10, simple, 1–2 cm. wide, 3-lobed or -parted, the lobes again forked or parted; pedicels 1–3 cm. long; sepals light green, spreading, 2–3 mm. long; petals 5, white, 4–6 mm. long; stamens 10–25; aks. 15–25, glabrous, 1–1.5 mm. long, with minute beaks; receptacle with tufted hairs.—Ponds, ditches, and slow streams, sea level to ca. 6000 ft.; many Plant Communities; Coast Ranges s. to Monterey Co., inland to Lake, San Joaquin, and Sacramento cos., Siskiyou and Modoc cos. s. to Mariposa Co., Cuyamaca Lake, San Diego Co.; to Alaska, Utah, Wyo. April–July.

Var. **Harrísii** L. Benson. Stems 2–5 mm. thick; lvs. all submersed, 2.5–5 cm. long; petals ca. 8 mm. long; aks. 2.5 mm. long, sparsely hispidulous; receptacle hispidulous.—Ponds, 2000–3000 ft.; Humboldt Co. to Shasta Co.; e. Ore. July–Aug.

Var. **capillàceus** (Thuill.) DC. [*R. c.* Thuill. *R. trichophyllus* Chaix. *Batrachium t.* F. Schultz. *R. a.* var. *t.* Gray. *R. a.* var. *brachypus* H. & A. *B. Bakeri* and *pedunculare* Greene.] Stems 1–2.5 mm. in diam.; lvs. all submersed and dissected; petals 4–8 mm. long; aks. glabrous or nearly so, 1–1.5 mm. long.—Ponds and slow streams, below 10,000 ft.; many Plant Communities; widely distributed in Calif., scarce near coast; to Alaska, Atlantic Coast, Mex.; Old World. April–Aug.

31. **R. subrígidus** W. Drew. [*R. circinatus* var. *subrigidus* L. Benson.] Glabrous aquatic

perennials with submersed stems 2–6 dm. long; lvs. all submersed, finely dissected into filiform divisions, 1–2 cm. long, 1.5–3 cm. broad, much shorter than the internodes; pedicels 2–5 cm. long, recurved in fr.; sepals light green, spreading, 3–5 mm. long; petals 5, white, 5–9 mm. long; stamens 5–10; aks, 30–45(–80), in a globose head, each obovoid, 1–1.5 mm. long, the beak very short; receptacle hispidulous.—Uncommon, ponds and slow streams; various Communities; San Francisco, Lassen, Modoc, and Siskiyou cos.; to B.C., Quebec, Mass., Mex. May–July.

7. Anemòne L. ANÉMONE. WINDFLOWER

Perennial herbs, low to fairly tall. Lvs. mostly radical, lobed, divided or dissected, sometimes compound; stem-lvs. 2 or 3 together forming an invol. subtending the fl. or remote from it. Fls. mostly showy, apetalous, the sepals colored or petaloid, 4–20. Stamens many, shorter than the sepals. Pistils many; ovary 1-ovuled. Fr. tailed or appendaged aks., ribless, flattened, forming a head. Styles short. Ca. 100 spp., mostly of the N. Temp. Zone; a few of hort. importance. (Ancient Greek name, from *anemos*, wind.)

Styles becoming 20–35 mm. long and plumose in fr.; sepals 2–3 cm. long. (*Pulsatilla*)
                                                                        1. A. occidentalis
Styles less than 5 mm. long, not plumose; sepals mostly less than 2 cm. long. (*Euanemone*)
  Lvs. dissected into narrow almost linear segms.; aks. ± woolly.
    Stems from the crown of a woody taproot. N. Calif.
      Styles 1–1.3 mm. long; plant mostly 2–5 dm. high .... 2. A. multifida
      Styles 1.5–3 mm. long; plant mostly lower .... 3. A. Drummondii
    Stems from a fusiform tuber. Deserts .... 4. A. tuberosa
  Lvs. not dissected into linear segms.; aks. not woolly; stems from horizontal rootstocks.
    Involucral lvs. simple, sessile .... 5. A. deltoidea
    Involucral lvs. 3-foliolate, petioled .... 6. A. quinquefolia

1. **A. occidentàlis** Wats. [*Pulsatilla o.* Freyn.] PASQUE FLOWER. Caudex stout, with thick vertical root; stems 1–several, 1–6 dm. high, long-villous when young, often glabrate; basal lvs. few, 4–8 cm. wide, ternate, the silky-villous divisions twice pinnately dissected into linear or lance-linear segms.; petioles 3–10 cm. long, villous; involucral lvs. similar, sessile or nearly so; fls. solitary; sepals 5–8, white or purplish, villous without, oblong to ovate, 2–3 cm. long; heads large in fr. because of the conspicuous, plumose styles; aks. villous, reflexed in age; styles 2–3.5 cm. long.—Dry rocky slopes, 5500–10,000 ft.; Montane Coniferous F.; Sierra Nevada from Tulare Co. n., to Shasta, Siskiyou and Trinity cos.; to B.C., Mont. July–Aug.

2. **A. multifida** Poir. [*A. m.* var. *globosa* T. & G. *A. g.* Nutt. ex Pritz.] Plants from stout rootstocks; stems 1–several, erect or ascending, 1–5 dm. high, soft-villous; basal lvs. many, the petioles 3–12 cm. long, villous; blades 3–10 cm. broad, 2–3-parted, with ultimate linear to lance-linear divisions, nearly glabrous above; involucral lvs. short-petioled part way up the stem; peduncles 1–3; sepals greenish yellow to pinkish or purplish, ovate, 7–11 mm. long, villous without; aks. silky-villous, in a rounded head 8–10 mm. high; styles 1–1.3 mm. long; *n* = 16 (Moffett, 1932).—Open fields to rocky places, 5000–7000 ft.; Montane Coniferous F.; Siskiyou Co.; to Alaska, Sask., Quebec; Chile. June–July.

3. **A. Drummóndii** Wats. [*A. baldensis* auth., not L. *A. californica* Eastw.] Plants with a stout root-crown; stems 1 to several, 1–3 dm. high, villous; basal lvs. several, 2–5 cm. wide, villous, 3–4-fid, the ultimate divisions linear; petioles 2–10 cm. long, villous; involucral lvs. sessile or nearly so; peduncles usually solitary, 3–10(–14) cm. long; sepals 5–8, ovate, 8–16 mm. long, white, tinged blue, villous without; aks. in rounded head, densely woolly; styles slender, 1.5–3 mm. long.—Talus and gravelly or rocky slopes, 5000–10,600 ft.; Montane Coniferous F.; Sierra Nevada from Inyo Co. n., N. Coast Ranges from Trinity Co. n.; to Alaska, Alta., Ida. May–Aug.

4. **A. tuberòsa** Rydb. [*A. sphenophylla* Cov., not Poepp.] Root tuberous; stems 1–3 dm. tall, subglabrous; lvs. few, 3–5 cm. wide, twice ternate, subglabrous, the divisions cuneate, ternately cleft; involucral bracts similar but short-petioled; peduncles 1–2, strigose; sepals rose, linear-oblong, 10–14 mm. long; aks. densely woolly, in an ellipsoid head; styles filiform, 1.5 mm. long.—Dry rocky slopes, 3000–5000 ft.; Joshua Tree Wd., Pinyon-Juniper Wd.; w. edge of Colo. Desert, e. Mojave Desert; to Utah, New Mex. April–May.

5. **A. deltoìdea** Hook. Rootstocks slender, creeping; stems slender, 1–3 dm. high, sub-glabrous or with scattered hairs; basal lvs. mostly solitary, 3-foliolate, the lfts. ovate, crenate-dentate, 3–6 cm. long, scattered-hairy beneath on veins; petioles 10–15 cm. long, slender, pilose; the 3 involucral lvs. like one basal, but subsessile; peduncles 1–2, solitary, glabrous or few-haired; sepals white, commonly 5, ovate to obovate, 1.5–2.5 cm. long; aks. hirsute below, glabrous above; styles less than 0.5 mm. long.—Dry forest floor, below 5500 ft.; Yellow Pine F., Mixed Evergreen F.; Mendocino, Humboldt and Siski-you cos.; to B.C. April–July.

6. **A. quinquefòlia** L. var. **Gràyi** (Behr & Kell.) Jeps. [*A. G.* Behr & Kell. *A. nemorosa* var. *G.* Greene.] Rootstocks rather slender, horizontal; stems slender, 1–3 dm. high, glabrous; basal lf. simple, trifid, long-petioled, usually lacking at anthesis; involucral lvs. on petioles 1–4 cm. long, 3-foliolate, the lfts. obovate, crenately toothed or incised above the cuneate base, 2–6 cm. long, sparsely strigose above and beneath; peduncle 1, strigose; sepals mostly 5, white, sometimes tinged purple, elliptic-obovate, 8–15 mm. long; stamens, ca. 25–40; fruiting heads nodding; aks. pubescent; styles scarcely 1 mm. long.—Moist wooded slopes near the coast, below 2000 ft.; Redwood F., Mixed Evergreen F.; Santa Cruz Mts. to Sonoma Co. April–May.

Var. **oregàna** (Gray) Rob. [*A. o.* Gray. *A. adamsiana* Eastw.] Basal lf. usually present at anthesis; sepals 10–20 mm. long, mostly blue or pink; stamens ca. 40–50.—Moist woods, below 3100 ft.; Yellow Pine F.; Siskiyou Mts.; to Wash. May.

Var. **mìnor** (Eastw.) Munz. [*A. adamsiana* var. *m.* Eastw. *A. oligantha* Eastw.] Basal lf. usually lacking at anthesis; lfts. of invol. 1–2.5 cm. long, broadly rhombic-ovate, rounded at apex; sepals white to rose, 4–10 mm. long; stamens 20–30; aks. sparingly silky.—Shaded places, forest floor, below 5000 ft.; largely Yellow Pine F.; Del Norte Co. to Plumas Co. April–May.

8. Clématis L. CLEMATIS. VIRGINS-BOWER

Half-woody vines which climb by clasping or twining petioles, sometimes erect perennial herbs. Lvs. opposite, entire to pinnately compound. Fls. small to large, solitary to paniculate, sometimes individually showy, sometimes as a mass, on axillary peduncles, perfect to imperfect (plants dioecious or polygamo-dioecious). Sepals 4–5, valvate in bud. Petals 0, or sometimes represented by staminoid structures. Stamens many. Pistils many, 1-ovuled, forming 1-seeded, mostly tailed aks. in a head. Over 200 spp., mostly of temp. regions; many of hort. importance. (Ancient Greek name for some climbing plant.)

Fls. many in cymose panicles; lfts. 5–7 .... 1. *C. ligusticifolia*
Fls. 1–3 on a peduncle; lfts. 3–5.
    Sepals 1.5–2.5 cm. long, pubescent above; aks. pubescent .... 2. *C. lasiantha*
    Sepals mostly 0.8–1.2 cm. long, glabrous above; aks. glabrous .... 3. *C. pauciflora*

1. **C. ligusticifòlia** Nutt. in T. & G. Climbing, 4–6(–15) m. high; branchlets almost glabrous except in infl.; lvs. pinnately 5–7-foliolate; lfts. lanceolate to lance-ovate (or ovate), rounded to subcuneate at base, mostly acuminate at apex, subentire to 3-lobed or coarsely toothed from ca. the middle, 2–8 cm. long, somewhat strigose especially beneath; petioles commonly 3–7 cm. long, the petiolules 1–3 cm.; peduncles densely strigose, 3–10 cm. long; fls. few to many; sepals white, oblong-oblanceolate, densely woolly-silky within and without, ca. 1 cm. long; fils. somewhat dilated; anthers of ♀ fls. sterile; aks. pubescent, the styles in fr. plumose, 2–4 cm. long; *n* = 8 (Meurman & Th., 1939).—Climbing over bushes and in trees, largely along streams and in moist places, below 7000 ft.; many Plant Communities; Coast Ranges and Sierra Nevada; L. Calif. to B.C. and Rocky Mts. March–Aug. Infusions of this vine were used by early settlers for sores and cuts on horses and by Indians for colds and sore throats. A variable sp., some plants from s. Calif. tending to have silky-canescent lvs., are var. *califórnica* Wats. [*C. biflora* Eastw.], and those from the White Mts. to e. Wash., with lfts. broad, subglabrous, and cordate at base, are near var. *brevifòlia* Nutt. [*C. b.* Howell.]

2. **C. lasiántha** Nutt. in T. & G. Woody climber to 4 or 5 m. high; branchlets woolly-pubescent; lfts. mostly 3, mostly broad-ovate, 2–5 cm. long, rounded to subcordate at base, coarsely toothed to 3-lobed, with rounded teeth, strigose-pubescent especially beneath; petioles commonly 2–5 cm. long, the petiolules 0.3–1 cm.; peduncles mostly 1-fld., 4–12 cm. long; sepals broadly oblong, silky-woolly without and within, 1.5–2.5 cm. long;

fils. somewhat flattened; aks. pubescent, the styles in fr. plumose, 2.5–3 cm. long; $n = 8$ (Meurman & Th., 1939).—Clambering over shrubs and in low trees, in canyons and near streams, below 6000 ft.; largely Chaparral, sometimes other Communities as Yellow Pine F.; Coast Ranges and Sierra Nevada foothills, Trinity and Shasta cos. to L. Calif. March–June.

3. **C. pauciflòra** Nutt. in T. & G. [*C. parviflora* Nutt.] Woody climber, 2–4 m. high; branchlets rather sparsely hairy; lfts. 3–5, round-ovate, toothed or lobed, mostly rounded at base and acute at apex, 1–2 cm. long, glabrous to sparsely strigose; petioles mostly 1–3 cm. long, the petiolules 0.2–1 cm.; peduncles 1–3-fld., mostly 2–3 cm. long; sepals white, oblong, 8–12(–15) mm. long, thin, woolly-pubescent without, glabrous within; aks. glabrous; styles plumose, in fr. 2–4 cm. long.—Clambering over shrubs, rocky slopes, and canyons, below 4000 ft.; Chaparral; Los Angeles Co. s., inland to Little San Bernardino Mts.; L. Calif. Jan.–April. Some plants from Santa Cruz Id. are apparently segregates of a cross between *C. lasiantha* and this sp., having sepals as long as in the former, but glabrous within, and aks. sometimes hairy, sometimes glabrous.

9. Isopỳrum L.

Low slender glabrous perennial herbs, ours with fleshy fibrous roots. Lvs. 2–3-ternate. Fls. mostly whitish, solitary or in cymes or panicles. Sepals 5–6, petaloid. Petals 0 in ours. Stamens many; fils. enlarged upward. Pistils few to many, sessile or stipitate, several-ovuled, becoming follicles. Seeds smooth. Ca. 25 spp., of N. Temp. Zone. (*Isopyron*, the Greek name for a sp. of *Fumaria*.)

Sepals 7–9 mm. long; stamens more than 20 .... 1. *I. occidentale*
Sepals 4–5 mm. long; stamens ca. 10 .... 2. *I. stipitatum*

1. **I. occidentàle** H. & A. [*I. o.* var. *coloratum* Greene. *Enemion o.* Drum. & Hutch.] Stems 1–several, slender, 10–25 cm. high, branched above; basal lf. long-petioled (5–10 cm.), with cuneate 2–3-lobed lfts. 1–2 cm. long, glaucous beneath; cauline lvs. reduced, few; sepals white, occasionally pink or purplish, oblong-obovate, 7–9 mm. long; stamens ca. 5 mm. long; follicles 8–10 mm. long, compressed, sessile; seeds several.—Infrequent and local, shaded places as among shrubs, below 5000 ft.; Chaparral, Foothill Wd., Yellow Pine F.; inner Coast Ranges from Los Angeles Co. (Bouquet Canyon) to Santa Clara and Napa cos., w. base of Sierra Nevada, Butte Co. to Kern Co. March–April.

2. **I. stipitàtum** Gray. [*I. Clarkei* Kell. *Enemion s.* Drum. & Hutch.] Tufted, 5–12 cm. high; basal lvs. ca. as high, the lfts. entire or divided into oblong or oblance-oblong parts, glaucous on both sides, 4–6 mm. long; stem-lvs. few, reduced; sepals whitish, oblance-oblong; stamens ca. 3 mm. long; follicles 6–7 mm. long, stipitate; seeds 3–5.—Brushy and wooded slopes, 2000–4500 ft.; Chaparral and Foothill Wd.; Coast Ranges, Santa Clara and Alameda cos., Mendocino Co. to Siskiyou and Modoc cos.; s. Ore. Feb.–April.

10. Aquilègia L. COLUMBINE

Perennial herbs from a thick caudex. Stems usually several, erect, branched. Lvs. largely basal, 2–3-ternate; the cauline gradually reduced up the stem. Fls. terminating the branches, showy, pendent or erect. Sepals 5, alike, colored. Petals 5, usually with a broad lamina projected to the front and a long hollow nectariferous spur projecting backward. Stamens many, with filiform fils.; innermost stamens represented by a sheath of flattened erect staminodia. Pistils usually 5, free, sessile, with terminal styles. Follicles separate, erect, reticulate-veined. Seeds black, shining, many, narrow-obovoid. Ca. 65 spp. of the N. Temp. Zone; many grown for their fls. (Derivation uncertain, possibly from Latin, *aquila*, eagle, because the spurs suggest claws, or *aquilegus*, water-drawer, because many grow in moist places.)

(Munz, P. A. The cultivated and wild columbines. Gentes Herb. 7: 1–150, 1946.)
Fls. erect or nearly so, cream to yellow or pink. Subalpine .... 1. *A. pubescens*
Fls. nodding, red and yellow. Mostly at lower elevs.
   Basal lvs. biternate; lfts. 2–5 cm. long; laminae of petals 1–6 mm. long .... 2. *A. formosa*
   Basal lvs. mostly triternate.

Lfts. grayish, 0.5–3 cm. long, with rounded lobes; laminae 3–5 mm. long; plants not markedly
viscid . . . . 3. A. *Shockleyi*
Lfts. green, 1–4 cm. long, with pointed tips to lobes; laminae lacking; plants viscid
4. A. *eximia*

1. **A. pubéscens** Cov. Stems tufted, 2–4.5 dm. high, glandular-pubescent above or
throughout; basal lvs. ternate to biternate, thickish, pale green above, sometimes
glaucous beneath, glabrous to pubescent on both surfaces; petioles 5–20 cm. long; lfts.
crowded, rounded-obovate, deeply cleft, 1–2(–3) cm. wide, the divisions few-toothed or
-lobed; stem-lvs. few, reduced upward; pedicels short; fls. glandular-puberulent; sepals
spreading, lance-oblong to ovate, acutish, 1.5–2 cm. long, 5–9 mm. wide; petals
oblong, rounded to retuse at apex, 8–12 mm. long; spurs straight to spreading,
2.5–4 cm. long; stamens 3–6 mm. longer than laminae; follicles pubescent, 2–2.5
cm. long, the pubescent styles an additional 10–12 mm.; seeds 1.5 mm. long.—
Rocky places and talus, 9000–12,000 ft.; Subalpine F., Alpine Fell-fields; Sierra Nevada,
Tuolumne Co. to Tulare and Inyo cos. June –Aug. Hybridizing rather freely with *A.
formosa* var. *pauciflora*.

2. **A. formòsa** Fisch. in DC. [*A. canadensis* var. *f.* Wats. *A. emarginata* Eastw.] Stems
mostly 5–10 dm. high, glaucous, mostly glabrous below, openly branched and glandular-
pubescent above; basal lvs., biternate, thin, green above, glaucous beneath, glabrous
to pubescent; petioles 1–2 dm. long; lfts. cuneate-obovate to suborbicular, mostly 2–4
cm. long, cleft to about the middle, then variously lobed and with rounded teeth; lvs. reduced up the stem; fls. pendent, ± pubescent; sepals red, ovate-lanceolate, wide-spreading
to reflexed, 15–25 mm. long, 5–10 mm. wide, acute to acuminate; laminae yellow,
rounded to truncate or emarginate, 3–5 mm. long; spurs red, 10–20 mm. long; stamens
10–15 mm. longer than laminae; follicles glandular-pubescent, 15–25 mm. long, the
styles an additional 10–15 mm.; seeds ca. 2 mm. long.—Moist woods, 4000–9000 ft.;
Montane Coniferous F., N. Juniper Wd.; Mono and Fresno cos. to Modoc, Siskiyou
and Del Norte cos.; to Alaska, Mont., Utah. June–Aug.

Forma **anómala** J. T. Howell. Laminae ca. 5 mm. long; spurs nearly lacking.—Montane;
Siskiyou and Plumas cos. July.

Var. **truncàta** (F. & M.) Baker. [*A. t.* F. & M. *A. f.* ssp. *t.* Pays. *A. californica* Hartw.
ex Lindl.] Stems glabrous or sparingly pubescent, 5–10 dm. high; sepals 10–20 mm.
long; laminae 1–2 mm. long; $2n = 14$ (Skalinska, 1931).—Common in moist places,
below 8000 ft.; most Plant Communities; Sierra Nevada and Coast Ranges s. to Riverside
Co.; s. Ore., w. Nev. April–Aug.

Var. **hypolàsia** (Greene) Munz. [*A. h.* Greene.] Stems viscid-pubescent, 5–10 dm.
high; sepals 10–20 mm. long; laminae 1–2 mm. long.—Moist places, below 7000 ft.;
Chaparral, Montane Coniferous F.; Los Angeles Co. s.; L. Calif. May–Aug.

Var. **pauciflòra** (Greene) Boothman. [*A. p.* Greene. *A. f.* ssp. *p.* Pays. *A. truncata* var.
p. Jeps.] Lvs. mostly basal; stems simple, scapelike, 1.5–3 dm. high; laminae 2–3 mm.
long.—Along streams and in moist places, 5000–10,500 ft.; Montane Coniferous F.;
Sierra Nevada, Shasta Co. to Kern Co., San Bernardino Co. June–Aug.

3. **A. Shóckleyi** Eastw. [*A. formosa* sspp. *dissecta* and *caelifax* Pays. *A. f.* var. *c.*
Munz. *A. mohavensis* Munz.] Stems 4–8 dm. high, glabrous and glaucous except for the
branched, glandular-pubescent infl.; basal lvs. mostly triternate, pale green above, more
glaucous beneath, glabrous or somewhat pilose, somewhat glutinous; petioles 5–20 cm.
long; lfts. cuneate-obovate to suborbicular, 5–30 mm. long, cleft to middle or deeper,
the principal segms. 2–3-lobed; cauline lvs. gradually reduced upward; fls. nodding,
glandular-pubescent; sepals red or with some green or yellow, spreading or reflexed,
lanceolate to elliptic, 10–20 mm. long, 4–8 mm. wide, acutish; laminae yellow, rounded-
truncate, 3–5 mm. long; spurs yellow-red to red, 12–25 mm. long; stamens 10–16 mm.
longer than laminae; follicles glandular-puberulent, 17–25 mm. long, the styles ca.
12 mm. more; seeds almost 2 mm. long.—Springy places, 5000–7700 ft.; Pinyon-Juniper
Wd.; Panamint, Argus, Clark, and New York mts., Mojave Desert; Nev. June–July.

4. **A. exímia** Van Houtte ex Planch. [*A. Tracyi* Jeps. *A. adiantoides* Greene. *A. fontinalis* J. T. Howell.] Plant densely glandular-pubescent throughout, 5–10 dm. high;
basal lvs. triternate; petioles 1–2 dm. long; lfts. broadly cuneate-obovate to suborbicular,
1–4 cm. long, greenish above, glaucous beneath, cleft to near the middle, then lobed
and with pointed teeth; fls. nodding, viscid-puberulent; sepals spreading or reflexed,

reddish, ovate-lanceolate, 15–25 mm. long, 5–10 mm. wide, subacuminate; laminae obsolete; spurs scarlet with yellow near the orifice, 18–30 mm. long, 8–10 mm. wide; stamens 15–25 mm. long; follicles 15–25 mm. long, the styles almost as long; seeds ca. 2 mm. long.—Springy places, often on serpentine, below 6000 ft.; Foothill Wd., Mixed Evergreen F., Chaparral; Coast Ranges, Mendocino Co. to Ventura Co. May–Aug.

11. Thalíctrum L. MEADOW-RUE

Erect perennial herbs from a short caudex. Lvs. alternate, ternately decompound, ample, the amplexicaul petioles dilated at base. Fls. often unisexual, paniculate, subcorymbose, or rarely racemose, numerous. Sepals 4(–7), fugacious, greenish or petaloid. Petals none. Stamens free, numerous, exceeding the sepals at anthesis, the fils. often colored, erect or pendent. Carpels several on a small receptacle, with a single pendulous ovule, the aks. often compressed, sometimes inflated, longitudinally ribbed or veined. A genus of ca. 120 spp., most abundant in temp. N. Am. and Eurasia. (Greek, *thalictron,* name given to the Meadow-Rue by Dioscorides.)

(Boivin, Bernard. American Thalictra and their Old World allies. Contr. Gray Herb. 142, 1944.)
A. Fls. perfect.
    B. Stem scapose; fls. racemose; fils. filiform; anthers linear-oblong, apiculate; stigma deltoid-winged .... 1. *T. alpinum*
    BB. Stem leafy; fls. paniculate; fils. linear-spatulate; anthers ovate or oblong, obtuse; stigma linear
                                                              2. *T. sparsiflorum*
AA. Fls. unisexual, paniculate; fils. filiform; anthers linear-oblong, apiculate.
    B. Dry aks. distinctly compressed, much broader than thick; fils. 4–6(–7) mm. long.
        C. Carpels and under surface of upper lvs. glandular-puberulent, the herbage green; aks. ovate-lanceolate, strongly 3–5-nerved, not obviously reflexed or inflated .. 3. *T. Fendleri*
        CC. Carpels and lvs. entirely glabrous, the herbage glaucous; aks. obovate, 1-nerved, or the several veins ± reticulate, ± reflexed and when fresh turgid .... 4. *T. polycarpum*
    BB. Dry aks. not much compressed, almost as thick as broad; fils. 6–9 mm. long
                                                                              5. *T. occidentale*

1. **T. alpìnum** L. [*T. monoense* Greene.] Cespitose or somewhat stoloniferous, glabrous throughout, 1–2.5 dm. high, the scapose stem simple; lvs. basal, few (1–6), 2–4-ternate, 2–5 cm. long, the flabellate lfts. thick, strongly veined, glaucous, dull on both surfaces, the margin revolute; raceme nodding, the peduncles recurving after anthesis; sepals 1.5–2.3 mm. long; stamens 8–15; carpels 3–6; aks. obliquely oblanceolate to obovate, ca. 3 mm. long, 5–6-ribbed on each side; $n = 7$ (Kuhn, 1930).—In moist meadows and bogs, stream borders or lake margins, 10,500–12,000 ft.; Alpine Fell-fields; rare in Calif., in the White Mts., Mono Co., and Rock Creek Lake Basin, Inyo Co.; e. to Rocky Mts., whence n. to the Arctic; circumpolar and most abundant in the belt of the Arctic Circle. The w. Am. material, with dull, light-colored lfts., is sometimes distinguished from the far northern, with more lustrous darker lfts., as the var. *hebètum* Boivin. June–Aug.

2. **T. sparsiflòrum** Turcz. [*T. s.* var. *nevadense* Boivin.] Erect, slender, branching above, 3–12 dm. high, minutely glandular-puberulent throughout, especially on the carpels and the under surface of the lvs.; lvs. 2–4-ternate, the upper subsessile, the lfts. 1–2 cm. long, rotund or cordate at base; panicle foliose; sepals 3–4 mm. long; aks. 4–12, semi-obovate (scimitar-shaped), strongly compressed, the dorsal angle straight, beaked and stipitate, the 4–5 lateral nerves curving upward; $n = 21$ (Kuhn, 1928).—Moist stream banks and bogs, often in willow thickets, 5000–10,900 ft.; Red Fir F. to Subalpine F.; Mt. Lassen s. throughout the Sierra Nevada, White, San Bernardino, and San Jacinto mts.; rare in Ore., but frequent through the Rocky Mts. to Alaska; e. Siberia. Not common in the cent. Sierra Nevada and s. Calif. July–Aug.

3. **T. Féndleri** Engelm. ex Gray. [*T. F.* var. *platycarpum* Trel. *T. p.* Greene. *T. hesperium* Greene. *T. polycarpum* Wats. var. *h.* Jeps. *T. F.* var. *h.* Jeps.] Habit of *sparsiflorum,* 6–15 dm. high, minutely glandular-puberulent above, especially on the carpels, pedicels, and under surface of the lvs., not glaucous; lvs. 2–4-ternate, the lfts. roughly 3-lobed and crenate; panicle open, foliose; sepals narrowly to broadly elliptic, 2–3 mm. long, scarious; aks. oblanceolate to obovate, ± ventricose and compressed, subsessile, 5–6 mm. long, 3–4-ribbed on each side, the nerves rarely branched but not anastomosing; $n = 14$ (Langlet, 1927; Kuhn, 1930; Clausen, 1940), 28, and 35 (Clausen, 1940).—Moist soil near streams and meadows, often in shade and thickets, occasional on dry slopes, 4000–

10,000 ft.; Montane Coniferous F.; Warner Mts., Modoc Co., s. throughout the Sierra Nevada to the Laguna Mts., San Diego Co., locally near the coast in n. Monterey Co.; to cent. Ore., Wyo., Tex., n. L. Calif. Closely related to the next. May–Aug.

4. **T. polycárpum** (Torr.) Wats. [*T. Fendleri* var. *p.* Torr. *T. caesium, bernardinum, coreospermum, lentiginosum, ametrum, mendocinum, latiusculum,* and *magarum* Greene. *T. p.* var. *caesium* Jeps.] More robust, 6–18 dm. high, glabrous throughout and ± glaucous; aks. many, in globular heads, vesicular-inflated when fresh, subglobose to obovoid, obliquely ventricose, subsessile, 4–7 mm. long, often with 1, sometimes 2, longitudinal wavy nerves on each side with branched anastomosing nervelets; otherwise as in *Fendleri.*—Frequent, mostly below 2500 ft.; Mixed Evergreen F., N. Oak, Foothill, and S. Oak wds., Coast Ranges from Humboldt Co. to San Diego Co., occasional in the Sierran foothills from Butte Co. to Kern Co.; n. to Columbia R. March–June. This sp. and *Fendleri* comprise a polyploid complex in which reproduction is presumably in large part apomictic. From this the polymorphy is explained together with the occurrence of recombination types in the Coast Ranges, such as glaucous types otherwise referable to *Fendleri,* and slightly glandular types otherwise referable to *polycarpum.*

5. **T. occidentàle** Gray var. **palousénse** St. John. [*T. heterophyllum* Nutt., not Lejeune.] Plants 2–8 dm. tall, glabrous; lvs. 3–4-ternate; lfts. thin, round to obovate-cuneate, 3-lobed; infl. 5–20 cm. long, the peduncles divaricate, 1–3.5 cm. long; sepals 3–5 mm. long in ♂ fls., 1.5–2.5 mm. in ♀; fils. purplish, 6–9 mm. long; aks. reflexed in maturity, narrow-fusiform, not compressed, 4–7 mm. long, 2–3 mm. wide, 8–12-ribbed, with shallow nonreticulate intervals; stipe 1.2–1.5 mm. long.—Thickets along streams, at ca. 5300 ft.; Yellow Pine F.; Modoc Co.; to B.C. and n. Rocky Mts. May–July.

12. Trautvettèria F. & M.

Large perennial herbs, with slender underground rootstocks and fascicled roots. Stems slender, erect, branching above and forming corymbose cymes. Lvs. large, palmately lobed, the basal long-petioled, the cauline smaller and short-petioled to sessile. Sepals 3–5, strongly concave, greenish-white, early deciduous. Petals 0. Stamens many, conspicuous, the fils. white, clavate. Pistils many, forming inflated akenes with short recurved styles. A single sp. (E. R. *Trautvetter,* 19th-century Russian botanist.)

1. **T. grándis** Nutt. ex T. & M. [*T. fimbriata* and *rotundata* Greene.] Glabrous or nearly so, 5–10 dm. high; lower lvs. 1–2 dm. wide, deeply 5–11-lobed, the lobes acute, irregularly and sharply deeply toothed; sepals 3–6 mm. long; fils. 7–10 mm. long; aks. glabrous, 3–5 mm. long.—Swamps and along streams, 4000–5000 ft.; Red Fir F., Subalpine F.; s. to Trinity and Placer cos.; to B.C., Mont., New Mex., Ariz. July–Aug.

13. Cóptis Salisb. GOLDTHREAD

Low perennial herbs with slender yellow rootstocks and scaly stolons. Lvs. basal, divided or compound. Stems scapose, with 1–3 small white fls. Sepals 5–7, petaloid. Petals 5–7, club-shaped. Stamens many. Pistils 3–7, stipitate, several-ovuled, becoming divergent follicles. Seeds several, smooth, shining. Ca. 10 spp. of cooler N Temp. Zone. (Greek, *koptein,* to cut, because of the divided lvs.)

1. **C. laciniàta** Gray. Lvs. trifoliolate, each lft. ovate, 2–6 cm. long, shining above, glabrous or puberulent on veins beneath, 3–5-parted or -cleft; petioles 4–12 cm. long; scapes 6–15 cm. high, 2–3-fld.; sepals greenish-white, linear, 8–10 mm. long; petals somewhat shorter; stamens 2 mm. long; stipes 5–7 mm. long; follicles 10–12 mm. long, open apically.—Wet cliffs and moist places; Redwood F., Douglas-Fir F.; Mendocino Co. n.; to Wash. March–April.

5. Berberidàceae. BARBERRY FAMILY (Fig. 11)

Herbs or shrubs. Lvs. alternate or basal, simple or compound. Fls. bisexual, regular, solitary or racemed. Sepals and petals hypogynous, usually imbricated in 2 or more series, usually similar and in sets of 3's. Stamens hypogynous, as many as petals and opposite them, sometimes twice as many; anthers usually opening by 2 uplifting valves.

Fig. 11. BERBERIDACEAE. *Berberis: a,* pinnate lf. and fl. with outer and inner series of sepals and 6 petals, 6 stamens and cent. pistil; *b,* petal with opposite stamen showing 2 uplifting valves on anther.

Pistil of 1 carpel; style short or lacking; ovary 1-celled; ovules 2-many, basal or parietal. Fr. a berry or caps. Ca. 10 genera and 200 spp., of temp. regions, largely of the N. Hemis.

Plants woody, with spiny-toothed lvs. .... 1. *Berberis*
Plants herbaceous, with basal nonspiny lvs.
  Fls. lacking sepals and petals; lfts. sessile .... 2. *Achlys*
  Fls. with sepals and petals; lfts. petiolulate .... 3. *Vancouveria*

1. Bérberis L. BARBERRY

Deciduous, or in ours evergreen, spiny or unarmed shrubs, often with underground rootstocklike branches so as to form large patches. Wood and inner bark yellow. Lvs. simple or in ours once-pinnate, alternate, with lfts. 3 to many, prickly, glabrous. Fls. perfect, yellow, in drooping racemes which are fascicled or solitary. Sepals 6, in 2 series, petallike, falling early, subtended by 3 bractlets. Petals 6, in 2 series, usually with a pair of glands near the base. Stamens 6, opposite the petals. Ovary superior, 1-celled, consisting of 1 carpel; stigma sessile. Fr. a few-seeded berry, sometimes becoming dry. As here treated including *Mahonia* (unarmed and with pinnate lvs.). A genus of over 200 spp.; N. and S. Am., Eurasia, N. Afr. Many cult. as ornamentals. (*Berberis,* an Arabic name.)

(Abrams, L. R. The Mahonias of the Pacific states. Phytologia 1: 89–94, 1934.)
A. Bud-scales persistent, mostly 20–40 mm. long; lfts. mostly 11–23; fils. without teeth
                                                            1. *B. nervosa*
AA. Bud-scales deciduous, 2–5 mm. long; lfts. mostly 3–11; fils. with a pair of recurved teeth near
    the apex.
    B. Racemes densely many-fld., the floral bracts deltoid-ovate, obtuse or acute; lfts. mostly 2–6
        cm. wide.
        C. Lfts. glossy-green above and beneath, the lower surface not microscopically papillate.
            D. The lfts. crowded on the rachis, often overlapping, ovate, mostly obtuse, the basal
                lfts. near base of petiole .... 2. *B. pinnata*
           DD. The lfts. usually not crowded or overlapping, often lance-ovate and acute, the
               basal lfts. usually 2–4 cm. from base of petiole ... . 3. *B. Aquifolium*
        CC. Lfts. dull beneath, the lower surface microscopically papillate.
            D. The lfts. bright green and glossy above.
               E. Each margin of lft. usually with 7–11 teeth tipped with slender spines. Coast
                  Ranges .... 4. *B. Piperiana*
              EE. Each margin of lft. usually with 12–16 teeth tipped with bristles. Sierra Ne-
                  vada .... 5. *B. Sonnei*
           DD. The lfts. dull or gray-green above.
               E. Each margin of lft. with usually 12 or more bristle-tipped teeth; stems 1–2
                  dm. tall. N. Calif. .... 6. *B. repens*
              EE. Each margin of lft. with usually 5–9 stouter spines; stems 1.5–18 dm. tall.
                  Widespread.
                    F. Lfts. mostly not cordate at base, oblong-ovate, mostly with 3–10 teeth
                      on each edge.
                      G. Lvs. plane, blue-green above, the marginal teeth medium-sized, not
                        at right angles to lf.-surface .... 7. *B. pumila*
                    GG. Lvs. strongly crisped, gray-green above, the marginal teeth very
                        stout and spinose, almost vertical to the lf.-surface .. 8. *B. Dictyota*
                FF. Lfts. cordate at base, suborbicular, with 8–20 teeth on each edge
                                          9. *B. amplectens*
    BB. Racemes loosely 3–9-fld., the floral bracts mostly lanceolate, subacuminate; lfts. mostly 1–2
        cm. wide.

C. Lfts. stiff-rigid, somewhat folded along the midribs, the 2 sides elevated and strongly crisped with some of the stout marginal spines almost vertical to the lf.-surfaces.
  D. Berries dry, inflated, 10–15 mm. in diam.; seeds almost black, 4–5 mm. long; terminal lft. rarely more than twice as long as wide. Cushenbury Springs, Mojave Desert .... 10. *B. Fremontii*
  DD. Berries juicy, not inflated, ca. 6–8 mm. in diam.; seeds red-brown, ca. 3 mm. long.
    E. Terminal lft. mostly 2–5 times as long as wide; berries plum-colored. E, Mojave Desert .... 11. *B. haematocarpa*
    EE. Terminal lft. mostly not more than twice as long as wide; berries yellowish-red. Se. San Diego Co. .... 12. *B. Higginsae*
  CC. Lfts. less rigid, almost plane, the bristlelike spines almost in the plane of the lf.-surfaces; berries red .... 13. *B. Nevinii*

1. **B. nervòsa** Pursh. [*Mahonia n.* Nutt. *Odostemon n.* Rydb. *M. glumosa* DC.] Stems simple, 3–6(–9) dm. high, from long roostocks; scales of terminal bud glumaceous, almost spiny when dry, persistent, brownish, mostly 2–4 cm. long; lvs. in terminal tuft, 25–45 cm. long; lfts. 7–21, ovate to lance-ovate, acute, glossy-green on both surfaces, 3–8 cm. long, 2–4 cm. wide, the margins serrulate with ca. 6–12 bristlelike teeth on each side; petioles commonly 3–7 cm. long; racemes erect, 7–20 cm. long; pedicels 5–8 mm. long; inner sepals 7 mm. long; berries blue with a gray bloom, 8–10 mm. long.—Shaded slopes, below 6000 ft.; Redwood F., N. Coastal Coniferous F., Douglas-Fir F.; Coast Ranges, Monterey and San Benito cos. to Del Norte, Shasta, and Siskiyou cos.; to B.C., Ida. April–June.

2. **B. pinnàta** Lag. [*Mahonia p.* Fedde. *M. fascicularis* DC. *Odostemon f.* Abrams.] Stems branched, erect, 3–16 dm. high; bud-scales deciduous; lvs. 5–12 cm. long; lfts. 5–9(–17), crowded on the rachis, usually overlapping, ovate to oblong, 2.5–5 cm. long, 2–3.5 cm. wide, glossy-green above, paler but glossy beneath, rather thin, rather deeply and sinuately spinulose–dentate with 10–20 teeth on each side, the lowest lfts. near base of petiole; racemes fascicled, 3–6 cm. long; inner sepals to 6 mm. long; berries ca. 6 mm. long, blue-glaucous.—Rocky exposed places, on wooded slopes, below 4000 ft.; mostly in Douglas-Fir F., Redwood F., and Mixed Evergreen F.; Coast Ranges, s. Ore. to San Gabriel Mts., Los Angeles Co., mts. of San Diego Co.; L. Calif. March–M y.
Ssp. **insulàris** Munz. Stems 2–6(–8) m. tall, leaning on trees for support; lfts. subentire or shallowly dentate, oblong, thin, 2–9 cm. long, 1.5–5 cm. wide.—Among trees; Closed-cone Pine F.; Santa Cruz Id. and Santa Rosa Id. March.

3. **B. Aquifòlium** Pursh. [*Mahonia A.* Nutt. *Odostemon A.* Rydb.] OREGON-GRAPE. Stems erect, branched, 1–2(–3) m. tall; lvs. 10–25 cm. long; lfts. 5–9(–11), oblong-ovate to oblong-lanceolate, 2.5–6 cm. long, 2–4 cm. wide, coriaceous but rather thin, glossy-green above and below, shallowly sinuate-dentate, with 10–20 slender spines on each edge, the basal lfts. usually 2–4 cm. from base of petiole; racemes 3–7 cm. long, clustered; inner sepals 7–8 mm. long; fr. ellipsoid, blue and glaucous, 8–9 mm. long; $n = 14$ (Dermen, 1931).—Wooded slopes, below 7000 ft.; Douglas-Fir F., Red Fir F.; Humboldt and Trinity cos. to Modoc Co.; to B.C., Ida. March–May.

4. **B. Piperiàna** (Abrams) McMinn. [*Mahonia P.* Abrams.] Stems erect, branched or simple, 2–6(–8) dm. tall; lvs. 10–20 cm. long; lfts. 5–9, ovate, 2.5–6 cm. long, 1.5–4 cm. wide, glossy-green above, dull and gray-green beneath, usually spinose-dentate with 7–10(–15) teeth on each edge, the lowest lfts. 1.5–3 cm. from base of petiole; racemes fascicled, 3–7 cm. long; inner sepals 6–7 mm. long; berries ellipsoid-ovoid, blue-black, ca. 6 mm. long.—Dry open or wooded slopes, at least sometimes on serpentine, 3000–5000 ft.; Mixed Evergreen F., Douglas-Fir F., Yellow Pine F., Coast Ranges from Lake Co. to Del Norte and Siskiyou cos.; less common in Chaparral and Yellow Pine F., San Gabriel Mts. to San Diego Co. and L. Calif.; s. Ore. March–June.

5. **B. Sónnei** (Abrams) McMinn. [*Mahonia S.* Abrams.] Stems 2.5–6 dm. high; lvs. 10–25 cm. long; lfts. 5, oblong-ovate, 4–8 cm. long, 2.5–3.5 cm. wide, glossy-green above, dull and paler beneath, with 12–20 bristlelike teeth on each margin, the lowest lfts. 2–3 cm. from base of petiole; racemes densely fld., 4–7 cm. long; berry ovoid, blue-black, ca. 6 mm. long.—Rocky banks, 7000 ft.; Montane Coniferous F.; cent. Sierra Nevada. April–May.

6. **B. rèpens** Lindl. [*Mahonia r.* G. Don. *Odostemon r.* Ckll.] Stems 1–2 dm. high, from underground creeping stolonlike bases; lvs. 7–25 cm. long; lfts. mostly 5, round-ovate to oblong-ovate, 3–9 cm. long, 1.5–4 cm. wide, quite plane, dull green above, dull and paler beneath, with 8–20 bristle-tipped teeth on each margin; racemes densely-fld.,

3–8 cm. long; inner sepals ca. 5 mm. long; berries ovoid to oblong, blue-glaucous, 5–8 mm. long; $n = 14$ (Dermen, 1931).—Dry open woods; Yellow Pine F.; Inyo Co., Modoc Co.; to B.C., Black Hills and New Mex. April–June.

7. **B. pùmila** Greene. [*Mahonia p.* Fedde. *Odostemon p.* Heller.] Stems ascending or erect, 1.5–4 dm. high, simple or branched; lvs. 8–15 cm. long; lfts. mostly 5–9, broadly ovate to oblong-ovate, 4–6 cm. long, 2–4.5 cm. wide, blunt at apex, dull and strongly reticulate-veined above, paler (glaucous) and dull beneath, plane or slightly crisped, with 3–10 medium-sized spiny teeth on each margin, the lowest lfts. 1–4 cm. from base of petiole; racemes dense, fascicled, 2–5 cm. long; inner sepals 7–8 mm. long; berries oblong-ovoid, blue-black, 5–8 mm. long.—Rocky outcrops and clay slopes, in sun or partial shade, 1000–4000 ft.; Yellow Pine F.; N. Coast Ranges s. to Lake Co., Sierra Nevada s. to Mariposa Co.; s. Ore. March–May.

8. **B. Dictyòta** Jeps. [*Mahonia D.* Fedde. *Odostemon D.* Ckll. *B. californica* Jeps.] Stems erect, few-branched, 3–18 dm. high; lvs. 5–10 cm. long; lfts. mostly 5 (3 or 7), broadly oblong-ovate, blunt at apex, obtuse to subtruncate at base, 3–7 cm. long, 2–7 cm. wide, stiff-coriaceous, dull above, paler (glaucous) and dull beneath, prominently veined, the margins strongly crisped and with 3–8 stout spinose teeth on each side, the lowest lfts. 0.5–3 cm. from base of petioles; racemes fascicled, rather open, mostly 3–9 cm. long; inner sepals 5–7 mm. long; berries ovoid, blue-black, 6–7 mm. long.—Dry rocky places, ca. 2000–6000 ft.; Chaparral, Foothill Wd., Yellow Pine F.; inner Coast Ranges and mts. at head of Sacramento V., foothills of Sierra Nevada, Marysville Buttes, to mts. of e. San Diego Co. April–May. Exceedingly variable.

9. **B. ampléctens** (Eastw.) Wheeler. [*Mahonia a.* Eastw.] Stems 1.5–6 dm. high; lvs. 8–15 cm. long; lfts. 5–7(3), suborbicular, ± closed-cordate at base, 3–8 cm. long, 2–8 cm. wide, veiny, stiff-coriaceous, dull green above, glaucous, paler and dull beneath, fairly plane, but with 8–20 spinose teeth on each side, the basal lfts. 2–4 cm. from base of petiole; racemes fascicled, rather lax, 3–6 cm. long; inner sepals ca. 5 mm. long.— Rocky slopes, 3500–6000 ft.; Chaparral, Yellow Pine F.; Santa Rosa Mts., Riverside Co. to Laguna Mts., San Diego Co. April–May.

10. **B. Fremóntii** Torr. [*Mahonia F.* Fedde. *Odostemon F.* Rydb.] Erect shrub 1–2(–4) m. tall with stiff erect branches; lvs. glaucous, blue-green and dull above, paler beneath, stiff-rigid, ± plainly reticulate; lfts. 3–7, strongly crisped, some teeth almost perpendicular to the plane of the upper surface, others to that of the lower, with mostly 3–5 stoutish spines (2–3 mm. long) on each edge, the basal lfts. 0.5–1.5 cm. from the base of the petiole; terminal lft. rhombic-ovate to pointed-ovate in outline, mostly 1.5–2.5 cm. long, 1–2 cm. wide; lateral lfts. ± angled-squarish in outline, frequently ± asymmetrical, 1–1.5 cm. long, almost as wide; racemes somewhat clustered, 3–9-fld., 2–3.5 cm. long, rather open; fls. ca. 6 mm. long; mature berries light yellow to red, dry, ± inflated and spongy, 10–15 mm. in diam.; seeds almost black, flat, 4–5 mm. long.—Rare, dry rocky places, 3000–5000 ft.; Joshua Tree Wd., Pinyon-Juniper Wd.; Cushenbury Springs, Mojave Desert; to Utah, Colo., Ariz., April–June.

11. **B. haematocárpa** Woot. [*B. Nevinii* var. *h.* L. Benson.] Much like *B. Fremontii* in habit and size; foliage glaucous, greener; lvs. compact, crowded, stiff-rigid, strongly crisped and spine-toothed; terminal lft. ± lance-ovate, mostly 3–6(–10) cm. long, 0.7–1.5 cm. wide, long-acuminate; lateral lfts. ovate to lance-ovate, ± pointed, 1.5–3 cm. long, 1–1.5 cm. wide; fls. ca. 4.5–5 mm. long; berries juicy, purplish-red, with a bloom, ca. 7–8 mm. in diam.; seeds dark red-brown, plump, ca. 3 mm. long.—Dry rocky places, 4500–5500 ft.; Pinyon-Juniper Wd.; Barnwell (New York Mts.) and Old-Dad-Granite Mts., e. Mojave Desert; to w. Tex., n. Mex. May–June.

12. **B. Hígginsae** Munz. With much the habit of the 2 preceding spp., stiff, 1–2.5 m. tall, glaucous; lvs. rigid, strongly crisped, spine-toothed, with spines 2–3.5 mm. long; lfts. 1–7, the terminal squarish-ovate to lance-ovate, 1.5–3.5 cm. long, 1–2 cm. wide; lateral lfts. ± squarish-ovate, 1–2.5(–3) cm. long, 1–2 cm. wide; racemes ca. 3.5–6 cm. long; fls. yellow, ca. 5 mm. long; berries yellowish-red, ca. 6–7 mm. in diam., drying dark; seeds dark red-brown, plump, 2.7–3.2 mm. long.—Dry rocky points and slopes, ca. 3000–4500 ft.; Chaparral, Pinyon-Juniper Wd.; se. San Diego Co. (Boulevard, Dubber) and adjacent L. Calif. May.

13. **B. Nevínii** Gray. [*Mahonia N.* Fedde. *Odostemon N.* Abrams.] Large rounded shrub, 1–4 m. tall, with stiff branched stems; lvs. 4–8 cm. long; lfts. mostly 3–5, lanceo-

late or the lateral lance-ovate, acuminate, somewhat coriaceous, 2–4 cm. long, 0.6–1 cm. wide, blue-green and rather dull on both surfaces, paler beneath, rather plane, spinulose-serrate with 5–16 bristlelike teeth on each side ca. 1 mm. long, the basal lfts. 0.5–1 cm. from the base of the petiole; racemes loosely fld., 2.5–5 cm. long; inner sepals 3–4 mm. long; berries juicy, yellowish-red to red, rounded, 6–8 mm. in diam.; seeds plump, brownish, 3.5–4 mm. long.—Sandy and gravelly places below 2000 ft.; Coastal Sage Scrub, Chaparral; San Fernando V. and Arroyo Seco near Pasadena, in both of which places now largely extinct; San Timoteo Canyon near Redlands (San Bernardino Co.), Dripping Springs near Aguanga (Riverside Co.), San Felipe Wash (e. San Diego Co.). March–April.

2. Áchlys DC. Deer-Foot. Vanilla-Leaf

Perennial herbs with branching scaly creeping rootstocks. Lvs. long-petioled, trifoliate, large. Fls. perfect, bractless, in a short dense scapose spike. Sepals 0. Petals 0. Stamens 6–13; fils. long, the outer dilated upward. Ovary 1-ovuled; stigma broad, sessile. Fr. becoming dry, indehiscent, broadly lunate. Two spp., 1 in Japan. (*Achlus,* Greek god of night.)

1. **A. triphýlla** (Sm.) DC. [*Leontice t.* Sm.] Glabrous; scapes solitary, slender, 25–50 cm. high; lf. solitary, suborbicular, ca. as long as or slightly shorter than scape; lfts. broadly fan-shaped, coarsely sinuate-dentate, 5–10 cm. long; spike 2.5–5 cm. long; fr. reddish, 3–4.5 mm. long.—Moist upland forests, below 5000 ft.; Douglas-Fir F., Yellow Pine F., Mixed Evergreen F., Redwood F.; N. Coast Ranges s. to Mendocino Co.; to B.C. April–June.

3. Vancouvèria Morr. & Dec.

Perennial fernlike herbs. Rhizome creeping, elongated, slender, scaly. Lvs. basal, usually biternate, with united sheathing stipules and cordate blunt or indented often undulate-margined and usually obscurely 3-lobed lfts. Stem scapose, sometimes unifoliate. Fls. few or many, in a simple or compound glabrous or glandular infl., trimerous, regular, nodding, small. Outer sepals 6–9, small, unequal, early deciduous; inner sepals 6, petaloid, spatulate, reflexed, larger. Petals 6, ligulate, reflexed, bearing a nectary at apex. Stamens 6, connivent, erect, with anthers ca. as long as fils., dehiscing by 2 oblong valves. Ovary and style 1; stigma cup-shaped; ovules 2–10, on the ventral suture. Follicle 2-valved, splitting to base. Seeds conspicuously arillate. A genus of 3 spp., all found in Calif. and limited to the Pacific States.

(Stearn, W. T. Epimedium and Vancouveria, a monograph. Jour. Linn. Soc. Bot. 51: 409–535, 1938.)

Lvs. deciduous; lfts. membranaceous throughout; pedicels glabrous; fls. white . . . . 1. *V. hexandra*
Lvs. persistent; lfts. thickened, with coriaceous margin; pedicels glandular.
  Fls. yellow, 12–15 mm. long; outer sepals, stamens, and ovary glandular-pubescent

                                                            2. *V. chrysantha*
  Fls. white, 7–8 mm. long; outer sepals, stamens, and ovary glabrous . . . . 3. *V. planipetala*

1. **V. hexándra** (Hook.) Morr. & Dec. [*Epimedium h.* Hook. **V. parvifolia** Greene.] Plant 1–4 dm. high, the scapose stem glabrous above, pilose at base, usually exceeding the lvs.; lvs. normally basal, bi- or tri-ternate, sparingly pubescent, perishing in autumn; lfts. narrowly to broadly ovate, 2–6 cm. long, membranaceous, cordate at base (with open sinus), usually 3-lobed above the middle, at least the mid-lobe rounded and in-dented, bright green above, glaucescent beneath; panicle loose, glabrous, 5–30-fld., the lower peduncles mostly 3-fld. and up to 12 cm. long, the upper 1-fld.; fls. white, 10–14 mm. long; outer sepals unequal, the outermost very short, the inner up to 5 mm. long; inner sepals spatulate, 6.5–9 mm. long, 3–4 mm. wide; petals long-clawed, shorter than the inner sepals, expanded and folded over at tip to form a quadrate nectary 2 mm. wide; stamens 4 mm. long, like the sepals sparingly dotted with short red-glandular hairs; ovary very glandular; follicles 1.5 cm. long, 1–6-seeded, the 3-mm.-long black seeds nearly covered by the aril; *n* = 6 (Langlet, 1928).—Deep shade, 50–4500 ft.; Redwood F., Douglas-Fir F.; Mendocino Co. to Del Norte and w. Siskiyou cos.; along the coast of Wash. May–July.

2. **V. chrysántha** Greene. [*V. hexandra* var. *aurea* Rattan. *V. h.* var. *c.* Greene. *Epⁱ*

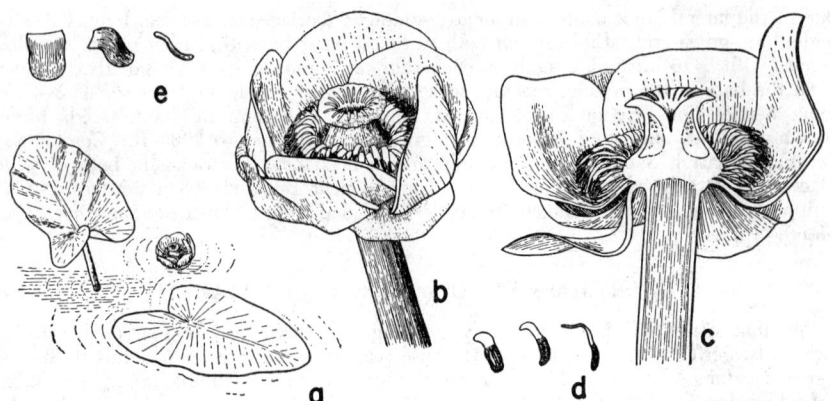

Fig. 12. NYMPHAEACEAE. *Nuphar polysepalum:* a, habit; b, fl., × ½, and c, in section, with cent. several-loculed pistil, numerous stamens and then petals and sepals; d, stamens; e, transition from stamens to petals.

medium c. Kom.] Plant 2–4 dm. high; lvs. basal, biternate or 3- or 5-foliolate, with pilose petioles, persistent through the winter; lfts. ovate or broader, to 4 cm. long and 4 cm. broad, subcoriaceous, the margin cartilaginously thickened and crisped, obscurely 3-lobed, dark green and glabrous above, glaucescent and pubescent beneath; panicle loose, very glandular, 4–15-fld., usually subracemose; fls. yellow, 10–14 mm. long, similar to those of *hexandra.*—Rather open situations and in thickets, 500–4000 ft.; Mixed Evergreen F., Yellow Pine F.; w. end of the Siskiyou Mts., in Del Norte and Siskiyou cos. and sw. Ore. June.

3. **V. planipétala** Calloni. [*Epimedium p.* Citerne. *E. hexandrum* f. *p.* Himmelb. *V. parviflora* Greene. *V. chrysantha* var. *parviflora* Jeps. *E. parviflorum* Kom. *V. crispa, Vaseyi,* and *concolor* Greene.] INSIDE-OUT FLOWER. REDWOOD-IVY. Plant 1.5–5 dm. high; lvs. basal, usually biternate, sometimes ternately compound with the divisions 5-foliolate, persistent through the winter; lfts. broadly ovate, to 4 cm. long, subcoriaceous, the margin cartilaginously thickened and ± crisped, cordate at base, 3-lobed, dark glossy green and glabrous above, dull and glaucescent beneath with short sparse hairs; panicle loose, glandular, 25–50-fld., rarely subracemose; fls. white or lavender-tinged, 6–8 mm. long; outer sepals unequal, up to 3 mm. long; inner sepals spatulate, erose-margined, 4 mm. long, 2 mm. wide; petals shorter than the inner sepals, oblanceolate, flat, ± 3-lobed but not saccate at the nectar-bearing tip; stamens 2 mm. long; ovary, like the sepals and stamens, glabrous; follicle 4–6 mm. long, 1–3-seeded.—Shade of woods, 100–2000 ft.; Redwood F.; Santa Lucia Mts., Monterey Co. to sw. Ore. May–June.

6. Nymphaeàceae. WATER-LILY FAMILY (Fig. 12)

Aquatic perennial acaulescent herbs, often with large horizontal rootstocks. Lvs. mostly large, simple, rising directly from the rootstock, floating or submersed. Fls. axillary, solitary, peduncled, usually bisexual. Sepals commonly 4 or 3 (5), sometimes many. Petals 3 to many, sometimes stamenlike. Stamens 3–many, hypogynous; fils. usually broad. Carpels 2–many, distinct or connate or immersed in a spongy receptacle with lamellate placentation. Fr. formed of the separate carpels or usually of the many connate carpels, or of the combined carpels and receptacle and opening by holes at the top. Seeds usually with pulpy aril. Ca. 8 genera and 80 spp. of fresh water and widely distributed; many highly ornamental.

Lvs. peltate, lacking sinus on one side; sepals and petals 3 each; carpels separate . . . . . . 1. *Brasenia*
Lvs. with sinus reaching to summit of petiole; sepals and petals more numerous; carpels united
2. *Nuphar*

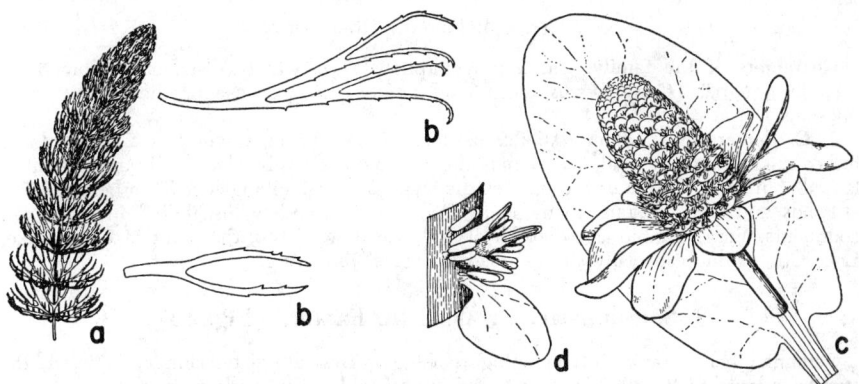

Fig. 13. CERATOPHYLLACEAE. *Ceratophyllum demersum: a,* habit, × ½; *b,* 2 types of lvs., × 2. SAURURACEAE. *Anemopsis californica: c,* basal lf. and flowering spike, × 1, with basal white invol. and small white bracts subtending fls.; *d,* single fl., × 5, with 6 stamens and 3 carpels and subtended by bract.

1. Brasènia Schreb. WATER-SHIELD

Creeping rootstock with slender branching stems. Lvs. entire, peltate, floating, long-petioled. Fls. small, purple; sepals and petals similar, mostly 3, persistent. Stamens 12–18; fils. filiform. Carpels 4–18, separate; ovules 1–2. Fr. leathery, indehiscent, 1–2-seeded. One sp. of N. Am., Asia, Afr., Australia.

1. **B. Schrèberi** J. F. Gmel. [*B. peltata* Pursh.] Lf.-blades oval to elliptic, 5–15 cm. long, green above, often purple beneath; sepals and petals 10–15 mm. long, linear-oblong; frs. oblong, 6–8 mm. long.—Ponds, ditches, and slow streams, below 7000 ft.; various Plant Communities; Kern and Lake cos. n.; through most of temp. N. Am. June–Aug.

2. Nùphar Sm. COW-LILY. YELLOW POND-LILY

Rootstocks stout, creeping. Lvs. floating or erect, orbicular to lanceolate, entire, with a deep sinus; some submerged lvs. often present, thin, delicate. Fls. yellow to purplish, usually standing above the water. Sepals 5–12, round, thick, yellow at least inside. Petals and stamens similar, numerous, inserted on a receptacle below the ovary. Pistil several-celled, with disklike several-rayed stigma. Fr. usually ripening above water; seeds many. Ca. 25 spp. of wide distribution in N. Hemis. (*Nuphar,* the Arabic name.)

1. **N. polysépalum** Engelm. [*Nymphaea p.* Greene. *Nymphozanthus p.* Fern.] Lvs. long-petioled, the blades deeply cordate, broadly oval, 1–4 dm. long, pinnately veined, floating or emergent; sepals 7–9, yellow, often tinged red, rounded and concave, the inner to 5 cm. long; petals narrow-cuneate, largely hidden by the stamens; anthers reddish; stigma-rays 15–25; fr. ovoid to subglobose, 3–3.5 cm. thick.—Ponds and slow streams, below 7500 ft.; many Plant Communities; Coast Ranges s. to San Luis Obispo Co., Sierra Nevada s. to Mariposa Co.; to Alaska, S. Dak., Colo. April–Sept.

7. Ceratophyllàceae. HORNWORT FAMILY (Fig. 13)

Aquatic herbs, submerged, with slender branching stems. Lvs. whorled, dissected into filiform, stiffish, serrulate divisions. Plants monoecious, the ♂ and ♀ fls. usually on separate nodes, sessile, without perianth, but surrounded by a calyxlike 8–12-cleft invol. Stamens distinct, 8–18, with subsessile anthers. Pistil usually 1; ovary 1-celled, 1-ovuled, with persistent style. Fr. a spinose or tuberculate ak. A single genus.

1. Ceratophýllum L. Hornwort

Characters of the family. Ca. 3 polymorphous spp., widely distributed. Sometimes used in aquaria. (Greek, *keras*, horn, and *phullon*, lf., because of the narrow rigid divisions.)

1. **C. demérsum** L. Stems 0.5–2.5 m. long; lvs. 5–12 in a whorl, 1–2.5 cm. long, forked 2–3 times; fr. oval, ca. 5 mm. long, smooth or tubercled, with or without a lateral spur on each side of the base of the style; $2n = 24$ (Langlet & Söderberg, 1927). —Ponds and slow streams; many Plant Communities; widely distributed in Calif.; all continents. Very variable as to length of and crowding of lvs., the plant in Clear Lake, Lake Co., having especially short, broad lf.-segms. June–Aug.

8. Saururàceae. Lizard-Tail Family (Fig. 13)

Perennial herbs; ours with creeping rootstocks. Lvs. alternate, simple, petioled, the stipules adnate to the petiole. Stems nodose, scapelike. Fls. perfect, in dense spikes or racemes; bracts conspicuous. Perianth absent. Stamens mostly 6 or 8, free or adnate to ovary at base or epigynous; anthers 2-celled. Pistil of 3–4 free or connate carpels; ovules 2–4 in each free carpel, or 6–8 on each parietal placenta, if carpels united. Fr. a caps. or berry. Seeds globose or ovoid, with copious perisperm. Three genera and 4 spp. of N. Am. and Asia.

1. Anemópsis Hook. Yerba Mansa

Stems nodose, scapelike, stoloniferous from thick creeping rootstocks. Lvs. mostly basal, minutely punctate. Spike conical, with basal, persistent, white invol. of several bracts; each fl. (except the lowest) subtended by a small white bract. Ovary sunk in the rachis of the spike; stigmas 3–4. Fr. a caps. One sp., formerly used medicinally for diseases of skin and blood. (Greek, *anemone*, and *opsis*, anemonelike.)

1. **A. califórnica** Hook. [*A. Bolanderi* A. DC. *Houttuynia c.* Benth. & Hook.] Stems 1–5 dm. high, woolly-pubescent, with a broadly ovate clasping lf. above the middle and 1–3 small lvs. in this axil; basal lvs. elliptic-oblong, 4–18 cm. long, on equally long petioles, entire, cordate at base; spikes 1–4 cm. long; involucral bracts white or reddish beneath, 1–3 cm. long; floral bracts 5–6 mm. long; ovules 6–10 on each placenta. —Common in wet, especially somewhat alkaline, places, below 6500 ft.; many Plant Communities below Yellow Pine F.; Sacramento V., and Santa Clara and Inyo cos. s. to L. Calif.; Nev. to Tex., Mex. March–Sept.

9. Sterculiàceae. Cacao Family (Fig. 14)

Trees, shrubs, or herbs, chiefly with stellate pubescence. Lvs. alternate, simple or rarely compound, stipulate. Fls. small or large, perfect, regular or nearly so. Calyx usually 5-lobed. Petals 5 or 0, free or united with stamen tube. Fertile stamens 5, the fils. ± united below, with staminodia sometimes also present. Fr. a 1–5-celled caps. A family of ca. 60 genera and 700 spp., usually of warmer regions.

Fls. yellow, 3–8 cm. across; shrub 1–6 m. tall; lvs. usually lobed .... 1. *Fremontia*
Fls. brownish, 0.4–0.7 cm. across; shrub 1–3 dm. tall; lvs. not lobed .... 2. *Ayenia*

1. Fremóntia Torr. Fremontia. Flannel Bush

Large shrubs or small trees, evergreen but losing many lvs. during dry season, stellate-pubescent and with mucilaginous inner bark. Lvs. simple, alternate, subentire to 3-, 5-, or 7-lobed. Fls. mostly bisexual, large, showy, regular, solitary, opposite the lvs. on the younger branches or on short lateral spurs. Calyx subtended by involucel of ca. 3 lance-subulate bractlets, petaloid, open-campanulate, 5-lobed to below the middle, externally stellate-pubescent, internally with a pit at the base of each lobe. Petals 0. Stamens 5, alternate with the sepals, joined by their fils. for about half their length. Ovary superior, 4- or 5-celled, surrounded by the base of the fil.-tube; style filiform,

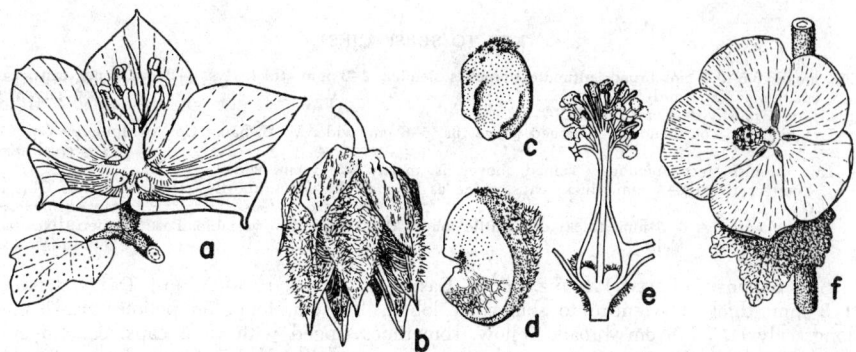

Fig. 14. STERCULIACEAE. *Fremontia: a,* fl., × 1, with 5-lobed petaloid calyx, 5 united stamens with alternating pits at base, cent. style; *b,* 5-valved caps., × 1, subtended by persistent withered calyx.
MALVACEAE. *Sidalcea: c,* carpel, × 4. *Malacothamnus; d,* carpel, × 5; *e,* longitudinal section of pistil and column of stamens, × 2½, showing base of attached petal to right and base of calyx; *f,* fl., × 1, with 5 petals.

exserted beyond the stamens. Fr. a densely bristly-hairy caps., 4- or 5-valved, dehiscent from apex, persisting for months. Seeds dark, 2 or 3 in each locule. As here treated, a genus of 2 spp., of considerable horticultural value. (Named for John C. *Frémont,* early explorer of the West.)

(Harvey, M. A revision of the genus Fremontia. Madroño 7: 100–110, 1943.)
Lvs. palmately 5–7-veined from base; pits at base of calyx-lobes lacking long hairs; fls. 6–9 cm. across, rather open-campanulate, borne on main twigs; seeds shining, without caruncle ...... 1. *F. mexicana*
Lvs. 1–3-veined at base; pits at base of calyx-lobes usually with long hairs; fls. 3–6 cm. wide, flat, borne on lateral branchlets; seeds dull, with large terminal caruncle .... 2. *F. californica*

1. **F. mexicàna** (A. Davids.) Macbr. [*Fremontodendron m.* A. Davids. *Fremontia californica* var. *m.* Jeps.] Tall, stiff, 2–6 m. high and ca. as thick; branches with dense stellate tomentum at first yellowish, later dark; lvs. thickish, rounded in outline, 2.5–7 cm. wide, cordate at base, with 5–7 main veins from base, shallowly lobed, dark green and sparsely stellate-pubescent above, densely tawny-tomentose beneath; petioles stout, mostly 2–4 cm. long; calyx somewhat shallowly campanulate, 6–9 cm. across, orange, becoming somewhat reddish at base on outside, somewhat puberulent within the basal pits; calyx conical, acuminate, mostly 3–4 cm. long, ca. half as thick; seeds black, shiny, ellipsoid-ovoid, not conspicuously carunculate, 4–5 mm. long.—Local in dry canyons, ca. 1500 ft.; Chaparral, S. Oak Wd.; Otay Mt., Jamul Mt., San Diego Co.; adjacent L. Calif. March–June. The fls. are among the lvs., hence do not make showy masses; they continue to come into bloom over some months.

2. **F. califórnica** Torr. [*Cheiranthodendron c.* Baill. *Fremontodendron c.* Cov.] Open spreading shrub 1.5–4(–7) m. tall; lvs. and fls. mostly on short lateral spurlike branch-lets; lvs. round- to elliptic-ovate, ± 3-lobed, dull green and sparsely stellate-pubescent above, densely tawny-stellate beneath, 1–3(–7) cm. long; petioles 1–3(–4.5) cm. long; calyx clear yellow, flat, 3.5–6 cm. in diam., the basal glands usually with long hairs; caps. ovoid, acute, 2.5–3.5 cm. long, almost as thick; seeds dark, dull, ovoid, 3–4 mm. long, with a subterminal easily removed brownish caruncle 1–1.5 mm. long; $n = 20$ (Lenz, 1950).—Dry mostly granitic slopes, 3000–6000 ft.; Chaparral, Yellow Pine F., Pinyon-Juniper Wd.; w. base Sierra Nevada from s. Shasta Co. to Kern Co., s. through the mts. to San Diego Co.; Ariz. Mostly May–June. The fls. are in great showy masses and tend to bloom out at one time. Some plants in the n. Sierran foothills with bright green upper surface to the lvs. and more whitish pubescence on lower surface have been described as var. *víridis* M. Harvey. Tehama Co. to Butte Co. In Tulare and Kern cos. some plants with subentire lvs. and petioles ⅓–½ as long as the blades have been named var. *íntegra* M. Harvey. In e. San Diego Co. some plants have subentire lvs. and petioles more than half the length of the blade; they are var. *diegénsis* M. Harvey.

KEY TO SUBSPECIES

Fls. small, 2.5–3.5 cm. broad; ultimate branches slender, 2–3 mm. thick; lvs. 1–2 cm. long; caps. ca.
1.5 cm. long. Napa and Lake cos. .... ssp. *napensis*
Fls. larger, mostly 4–6 cm. broad.
    Lvs. thickish, conspicuously 3-veined above; fls. 5–6 cm. wide. W. Tehama Co. to Monterey Co.
                                                                                                ssp. *crassifolia*
    Lvs. thinnish, inconspicuously veined above; fls. mostly 3.5–5 cm. wide.
        Younger twigs 2–3 mm. thick; caps. twice as long as thick; lvs. mostly entire. San Luis Obispo
        Co. .... ssp. *obispoensis*
        Younger twigs 3–5 mm. thick; caps. little longer than thick; lvs. variable. Base of Sierra Nevada
        and mts. of s. Calif. .... *F. californica*

Ssp. **napénsis** (Eastw.) Munz. [*F. n.* Eastw. *F. c.* var. *n.* McMinn.] Twigs slender,
1–3 mm. thick; lvs. entire to somewhat lobed, 1–2 cm. long, on petioles 0.2–1 cm.
long; calyx 2.5–3.5 cm. broad, yellow, sometimes tinged with rose; caps. ca. 1.5 cm.
long and thick; seed ca. 3 mm. with caruncle an additional 1.5–2 mm. long.—Brushy
slopes, ca. 1500–1800 ft.; Chaparral; Napa, Lake, and Yolo cos. May.
    Ssp. **obispoénsis** (Eastw.) Munz. [*F. o.* Eastw.] Twigs 2–3 mm. thick; lvs. coriaceous,
subentire, 1.5–2.5 cm. long; petiole ca. half as long; calyx 5–6 cm. in diam.; caps.
acuminate, 2.5–3 cm. long, 1 cm. in diam.—Canyon-slopes; Chaparral; San Luis
Obispo Co. and n. Santa Barbara Co. May.
    Ssp. **crassifòlia** (Eastw.) Abrams. [*F. c.* Eastw.] Twigs heavy, 3–5 mm. thick; lvs.
heavy, 3-lobed, 2.5–4.5 cm. long; petioles ¼–½ as long; calyx ca. 6 cm. wide; caps.
2.5–3.5 cm. long, 1.5 cm. in diam.—Chaparral; Coast Ranges from Tehama Co. to
Monterey Co. April–May.

2. Ayènia Loefl.

Herbs or ours a subshrub. Lvs. serrate-dentate. Fls. small, pedicellate, in axillary
clusters. Calyx 5-lobed. Petals 5, long-clawed, hooded, the hoods inflexed, adnate to
the calyx-tube and covering the anthers. Anthers 3-celled, solitary in sinuses of stamen-
tube and alternating with truncate staminodia. Caps. stipitate, 5-celled, splitting into
5 one-seeded bivalvate carpels. Seeds in our narrowly obovoid, dark, very irregularly
and deeply pitted-rugose. Ca. 15 spp. of warmer parts of New World. (Named for the
Duc d'*Ayen*.)
    1. **A. califórnica** Jeps. [*A. pusilla* auth., not L. *A. compacta* auth., not Rose.] Low
shrub, with several stems slender, 1–2(–3) dm. tall, strigulose; lvs. ovate to oblong-
ovate, dentate-serrate, 6–10 mm. long, on petioles ca. half as long; fls. brownish, 2 mm.
long; caps. ca. 3 mm. long, finely pubescent and with short blunt tubercles; seed
almost 2 mm. long.—Occasional in dry rocky canyons, below 1500 ft.; Creosote Bush
Scrub; w. edge of Colo. Desert and Eagle Mts. March–April.

10. Malvàceae. MALLOW FAMILY (Fig. 14)

Herbs or soft-woody shrubs or even trees, usually with stellate or branched pubescence.
Lvs. alternate, palmately-ribbed and usually -lobed, with small deciduous stipules. Fls.
regular and bisexual, occasionally unisexual, mostly 5-merous, white, yellow, pink or
red to purple. Calyx 5-lobed, often subtended by calyxlike bracts forming an involucel.
Petals 5, convolute. Stamens many, hypogynous, cohering into a tube or column around
the styles and often adnate to the petals. Pistil of several united carpels, the ovary
superior, rarely 1-celled, mostly with as many cells as styles or stigmas; styles mostly
united below. Fr. a loculicidal caps. or sometimes berrylike or more often separating
into carpels which are aks. or follicles. Ca. 50 genera and 1000 spp. of temp. and trop.
regions, many important economically for their fibers (cotton), food products, and as
ornamentals.

A. Fr. a caps., the carpels not separating from each other or the axis; fls. solitary in upper axils;
    ovules several in each carpel .... 14. *Hibiscus*
AA. Fr. of several carpels, usually separating from each other and the axis when mature.
    B. Style-branches terminating in capitate or truncate stigmas.
        C. Involucel of bractlets below calyx wanting.

D. Carpels not sharply differentiated into a reticulate basal and a wider smooth apical portion; mostly herbaceous.
  E. Petals yellow or orange; carpels 2–9-seeded .... 1. *Abutilon*
  EE. Petals blue-violet to lavender; carpels 1-seeded .... 12. *Anoda*
 DD. Carpels sharply differentiated into a reticulate basal and a wider winged smooth apical portion; rather woody .... 2. *Horsfordia*
CC. Involucel of 1 to several bractlets below calyx present.
  D. Carpels sharply differentiated into a reticulate basal and a wider winged smooth apical portion; plants herbaceous or suffruticose .... 3. *Sphaeralcea*
  DD. Carpels not so differentiated.
    E. Ovules 2 or more per carpel; carpels hirsute.
      F. Plant a tall erect herb; carpels 8–10 mm. high, oblong, 1-celled
                                                              4. *Iliamna*
      FF. Plant a low spreading herb; carpels 4 mm. high, reniform, 2-celled
                                                              7. *Modiola*
    EE. Ovules solitary; carpels not hirsute.
      F. Plants not lepidote (beset with small scurfy scales); fls. white, pink, lavender, or purplish.
        G. Carpels indehiscent, rugose on the sides; our species annual
                                                              5. *Malvastrum*
        GG. Carpels splitting into halves at maturity, not rugose; shrubby
                                                              6. *Malacothamnus*
      FF. Plants lepidote; petals yellowish .... 11. *Sida*
BB. Style-branches filiform, longitudinally stigmatic on the inner side.
  C. Involucel of 3–9 connate bractlets.
    D. Axis of fr. surpassing the carpels, forming a cone or projection in the center
                                                              8. *Lavatera*
    DD. Axis of fr. not extending above the ring of carpels .... 9. *Althaea*
  CC. Involucel of 1–3 distinct bractlets, or lacking.
    D. Bractlets of involucel 3; weedy introd. annuals .... 10. *Malva*
    DD. Bractlets 0 or 1; native annuals or perennials .... 13. *Sidalcea*

1. Abùtilon Mill. INDIAN-MALLOW. FLOWERING-MAPLE

Perennial, herbaceous or shrubby, stellate-canescent or -tomentose, or hirsute with simple hairs. Lvs. alternate, petioled, crenate or dentate, cordate at base, not or obscurely lobed. Fls. solitary and axillary or in leafy panicles, often drooping, without involucel, the calyx 5-cleft and sometimes bright-colored. Corolla trumpet-shaped or bell-shaped, commonly orange or yellow; staminal column anther-bearing at apex. Fr. truncate-cylindric or subglobose, the carpels smooth-sided, dehiscent nearly to base when ripe; ovules 2 or more per carpel. Ca. 100 spp. in many warm regions, some cult. as ornamentals. (Name of Arabic origin.)

Carpels not conspicuously inflated or thin-walled; fr. not globose.
  Plants annual; lvs. often 10–20 cm. wide; carpels more than 10, with long divergent awns.
                                                              1. *A. Theophrasti*
  Plants perennial; lvs. 1–5 cm. wide; carpels usually fewer.
    Carpels seldom more than 5; petals pink or red, 4–6 mm. long .... 2. *A. parvulum*
    Carpels usually 7 or more; petals orange, 15–20 mm. long .... 3. *A. Palmeri*
Carpels conspicuously inflated, with thin walls; fr. globose .... 4. *A. crispum*

1. **A. Theophrásti** Medic. VELVET LEAF. Plant 1–2 m. tall; lvs. ovate-orbicular, cordate, velvety; fls. yellow, 6–8 mm. long.—Occasional escape (Riverside, Santa Ana R. Canyon, San Diego); native of s. Asia. July–Aug.

2. **A. párvulum** Gray. Cespitose perennial; stems slender, 1–4 dm. tall, cinereous-stellate, diffusely branched; lvs. ovate-cordate, dentate, 1–2.5 cm. wide, the petioles 5–10 mm. long; calyx 3–4 mm. long, reflexed in fr.; petals 4–6 mm. long; fr. 7–8 mm. long, stellate-puberulent.—Arid rocky slopes, 3000–4000 ft.; Shadscale Scrub; Providence Mts.; to Colo., Ariz., Mex. April–May.

3. **A. Pálmeri** Gray. Woody at base, 7–8 dm. high, stellate-villous; lvs. velvety, round-cordate, 3–5 cm. wide, obscurely 3-lobed; petioles 3–5 cm. long; calyx 9–12 mm. long, the lobes round-ovate, acuminate; petals rich orange, ca. 2 cm. long; fr. very villous, ca. 10 mm. long.—Dry slopes, below 2500 ft.; Creosote Bush Scrub; base of Laguna Mts., e. San Diego Co., Chuckwalla Mts., Riverside Co.; to Ariz., Son., L. Calif. April–May.

4. **A. crispum** (L.) Sweet. [*Sida c.* L. *Gayoides c.* Small.] Annual, 3–7 dm. high, with slender branches; lvs. ovate, cordate, often prominently veined beneath, the blades 1–5 cm. long; pedicels filiform, 2–5 cm. long, often deflexed at the joint; petals pale yellow, 7–9 mm. long; fr. globose; carpels conspicuously inflated, muticous, 6–8 mm.

long, with thin membranaceous walls, hispid with rather long simple hairs; $n = 7$ (Skovsted, 1935).—Creosote Bush Scrub; Mt. Springs Grade, e. San Diego Co.; Ariz. to trop. Am.

2. Horsfórdia Gray

Erect, sparingly branched, rather woody, densely stellate-canescent or -tomentose, somewhat yellowish. Lvs. cordate to lanceolate, thickish, somewhat denticulate or crenulate, truncate or subcordate at base. Fls. axillary, 1 to few, peduncied, often in leafy panicles; bractlets none. Calyx 5-lobed. Carpels 8–12, disjoined at maturity, the apical part scarious, winged, the basal portion firm and reticulate. Seeds of the 2 parts of the carpel unlike. Spp. 4, sw. U.S. and adjacent Mex. (F.H. *Horsford*, a New England botanist.)

Petals pink (drying bluish), 10–15 mm. long; plant dull green .... 1. *H. alata*
Petals yellow, ca. 8 mm. long; plant vivid yellow-green .... 2. *H. Newberryi*

1. **H. alàta** (Wats.) Gray. [*Sida a.* Wats.] Stems 1–3 m. tall; lvs. subcordate, ovate-lanceolate, 2–9 cm. long; pedicels 8–16 mm. long; calyx-lobes ovate, acuminate; petals obovate, hairy at base, 15–18 mm. long; carpels 10–12, becoming 8 mm. long, the upper part 2-winged and 3 times the length of the lower part; seeds solitary.—Rocky canyons and sandy washes, ca. 500 ft.; Creosote Bush Scrub; Coral Reef Ranch, Coachella V., Riverside Co.; to Ariz., Son., L. Calif. March–April, Nov.–Dec.

2. **H. Newbérryi** (Wats.) Gray. [*Abutilon N.* Wats.] Stems 1–2.5 m. tall, becoming woody but flowering the first year; lvs. cordate, ± ovate, 3–9 cm. long; pedicels 5–12 mm. long; calyx-lobes subdeltoid, acute; petals rounded, 7–8 mm. long; carpels 8–9, ca. 6 mm. long in fr., the wings not much longer than lower part; seeds 2–3 per carpel.—Dry rocky places, below 2500 ft.; Creosote Bush Scrub; w. Colo. Desert; to Ariz., Son., L. Calif. March–April, Nov.–Dec.

3. Sphaerálcea St. Hil. GLOBEMALLOW

Perennial herbs or suffruticose plants, usually closely stellate-pubescent. Lvs. fairly thick, shallowly dentate to pedately dissected. Infl. racemose or paniculate. Involucel of 2 or 3 bractlets, usually deciduous soon after anthesis. Calyx 5-lobed. Petals 5, usually red (grenadine), sometimes pink or lavender. Stamens in one series. Stigmas capitate. Carpels 5 or more, 1–3-ovuled, 1-celled, relatively thick, less than 8 mm. high, differentiated into a dehiscent unreticulate apical portion and an indehiscent reticulate basal portion. Seeds reniform, usually pubescent. Perhaps 60 spp. of warmer parts of New World, a few S. African. (Greek, *sphaera*, globe, and *alkea*, mallow, because of the spherical fr.)

(Kearney, T. H. The North American species of Sphaeralcea subgenus Eusphaeralcea. Univ. Calif. Pub. Bot. 19: 1–128, 1935.)
A. Plants annual or biennial; petals orange; the reticulate basal part forming ⅔ or more of the carpel and conspicuously wider than the unreticulate apical part.
    B. Plant yellowish-canescent; lvs. thick and firm, crenulate; carpels deeply notched
                                                                  1. *S. Orcuttii*
    BB. Plant grayish-pubescent; lvs. thin and soft, coarsely crenate; carpels shallowly notched
                                                                    2. *S. Coulteri*
AA. Plants perennial; petals red (grenadine) to white, pink or lavender; the reticulate part forming less than ⅔ of the carpel and not conspicuously wider than the unreticulate part.
    B. Lvs. not more than ⅓ as wide as long; pubescence yellowish .... 6. *S. angustifolia*
    BB. Lvs. at least half as wide as long.
        C. The lvs. deeply cleft or divided; infl. narrow, many-fld. Lassen Co.
                                                        9. *S. grossulariaefolia*
        CC. The lvs. not deeply cleft.
            D. Calyx at anthesis 11–20 mm. long.
                E. Reticulate part of carpel rugose or muricate on back, the reticulations prominent and coarse. Widely distributed .... 4. *S. ambigua*
                EE. Reticulate part of carpel smooth or nearly so on back, the reticulations less prominent and finer. Panimint Mts. .... 5. *S. Rusbyi*
            DD. Calyx at anthesis 4–10 mm. long.
                E. Reticulate part of carpel rugose or muricate on back. Mostly s. Calif.
                    F. Fr. truncate-conical; lvs. ca. as wide as long; infl. usually an open panicle
                                                            4. *S. ambigua*

FF. Fr. hemispherical or nearly so; lvs. definitely longer than wide; infl. narrow .... 3. *S. Emoryi*
   EE. Reticulate part of carpel smooth or nearly so on back. Cent. Calif.
      F. Pubescence dense, grayish; lf.-blades finely crenate; carpels broadly ovate, acutish. Inyo Co. .... 7. *S. parvifolia*
      FF. Pubescence sparse and plant bright green; lf.-blades coarsely crenate or dentate; carpels suborbicular, very obtuse. Placer Co. .. 8. *S. Munroana*

1. **S. Orcúttii** Rose. Annual or biennial, with large taproot, densely yellowish-canescent, erect, 6–9(–12) dm. tall, stout; lvs. thick, firm, deltoid-ovate, shallowly 3-lobed near base, usually subcordate, crenulate, 3–5 cm. long, ⅗–⅘ as wide; petioles 1–2.5 cm. long; infl. long, narrow, many-fld.; bractlets linear, 2–3 mm. long; calyx at anthesis 4–7 mm. high, the lobes lance-ovate, acuminate; petals ± orange, 8–12 mm. long; fr. hemispherical; carpels 12–17, with thin scarious walls, 2.5–3 mm. high, deeply notched, the indehiscent part forming ¾–⅘ of carpel and much wider than the dehiscent part, prominently reticulate, fenestrate; seeds 1, sometimes 2; *n* = 5 (Webber, 1936).—Fairly frequent in dry sandy semialkaline places, at low elevs.; Creosote Bush Scrub; Imperial Co.; Ariz., L. Calif. Feb.–Aug.

2. **S. Coùlteri** (Wats.) Gray. [*Malvastrum C.* Wats.] Annual, with slender taproot, rather sparsely pubescent, erect, 2–10(–15) dm. high, rather slender; lvs. thin, soft, broadly ovate to suborbicular, truncate to cordate at base, scarcely lobed to deeply 3–5-lobed, coarsely crenate, 1–3 cm. long; petioles 1–4 cm. long; infl. thyrsoid, few-to many-fld.; bractlets linear, 2–3 mm. long; calyx at anthesis 5–7 mm. long, the lobes lanceolate, acuminate; petals ± orange, 8–15 mm. long; fr. hemispheric; carpels 14–22, thin-walled, 2–2.5 mm. high, reniform, shallowly notched, the reticulate part forming ⅔–¾ of carpel, fenestrate; seed 1, glabrous or sparsely pubescent; *n* = 5 (Webber, 1936).—Rare, dry sandy places, low elevs.; Creosote Bush Scrub; Imperial Co.; to Ariz. and Sinaloa. March–April.

3. **S. Emòryi** Torr. in Gray. [*S. angustifolia* var. *gavisa* Jeps.] Perennial, with a stout woody crown, several stems 3–10(–12) dm. tall, grayish-canescent; lvs. thickish, ovate-oblong, usually subcordate and somewhat rounded-angulate at base, crenulate to dentate, 2–9 cm. long, ½–¾ as wide; petioles 1–6 cm. long; infl. a many-fld. usually narrow thyrse, leafy; bractlets linear, 1–2 mm. long; calyx at anthesis 5–10 mm. long, the lobes deltoid-ovate to lanceolate, acute to acuminate; petals grenadine, sometimes pink or lavender, 10–20 mm. long; fr. truncate-conical; carpels 11–16, with chartaceous walls, 3.5–6 mm. high, rather deeply notched, with prominent ventral beak, the reticulate part forming ⅓–½ of carpel, usually rugose-tuberculate on back; seeds usually 2, usually pubescent; *n* = 15, 25 (Webber, 1936).—Sandy or loamy places along roads and in fields, below 2000 ft.; Creosote Bush Scrub; desert parts of Riverside and Imperial cos.; Nev., Ariz., L. Calif. March–May. Exceedingly variable.

KEY TO SUBSPECIES

Lf.-blades ⅓–¾ as wide as long, cleft near base, 3–9 cm. long.
  Lvs. ½–¾ as wide as long, mostly cordate at base; carpels rather thick-walled and coarsely reticulate.
    Stems and lvs. densely pubescent, grayish-canescent; lvs. thickish, scarcely lobed ..... *S. Emoryi*
    Stems and lvs. less pubescent, greener; lvs. thinnish, definitely lobed .... ssp. *variabilis*
  Lvs. ⅓–½ as wide as long, at most subcordate; carpels thin-walled, finely reticulate
                                  ssp. *nevadensis*
Lf.-blades ca. as wide as long, cleft far above base, 2–5 cm. long .... ssp. *arida*

Ssp. **variábilis** (Ckll.) Kearn. [*S. v.* Ckll. *S. Fendleri* var. *californica* Parish. *S. F.* var. *v.* Ckll.] Herbage greener, more sparsely pubescent, lvs. thinner, 3-cleft, with broad to narrow lateral lobes and longer ± pinnatifid mid-lobe; *n* = 10 or 15 (Webber, 1936). —Dry places in fields and along roads, below 3000 ft.; Coastal Sage Scrub, Creosote Bush Scrub; Upland, Colton, Redlands to s. Mojave and n. Colo. deserts; Ariz. March–May.

Ssp. **nevadénsis** Kearn. Lvs. ⅓–½ as wide as long; carpels thin-walled, finely reticulate. —Creosote Bush Scrub; e. Riverside Co. to e. Inyo Co.; Nev., Ariz. March–May.

Ssp. **árida** (Rose) Kearn. [*S. a.* Rose.] Lvs. not more than 5 cm. long, almost as wide; carpels 8–12, thin-walled, finely reticulate; *n* = 10 (Webber, 1936).—Creosote Bush Scrub; e. Imperial Co.; to Nev., Son., Sinaloa. April.

4. **S. ambígua** Gray. [*S. a.* var. *Keckii* Munz.] DESERT-HOLLYHOCK. DESERT-MALLOW.

Perennial, with thick woody crown and woody lower stems, whitish- or yellowish-canescent; stems few to many, erect or ascending, 5–10 dm. high; lvs. thickish, ovate to suborbicular, ± cordate at base, blunt at apex, crenulate to crenate, 3-lobed near middle, 2–6 cm. long and ca. as wide; infl. usually open-paniculate with few to many fls.; bractlets 3–5 mm. long; calyx at anthesis 6–20 mm. high, the lobes lanceolate, acuminate; petals grenadine, 15–35 mm. long; fr. hemispherical; carpels 12–16, with thickish chartaceous walls, 3.5–6 mm. high, helmet-shaped, rather deeply notched, usually rugose and muricate dorsally, the indehiscent part ca. ⅓ of carpel, coarse-reticulate; seeds usually 2, pubescent; $n = 10$ or 15 (Webber, 1936).—Dry rocky slopes and canyons, mostly below 4000 ft.; Creosote Bush Scrub, Joshua Tree Wd.; throughout the Calif. deserts; to Utah, Son., L. Calif. March–June. Exceedingly variable.

KEY TO SUBSPECIES

Stems usually woody above base and more than 5 dm. high; infl. usually an open long-branched panicle.
 Petals grenadine .... *S. ambigua*
 Petals pink or lavender, often drying violet .... ssp. *rosacea*
Stems usually herbaceous above the crown and less than 5 dm. high; infl. racemiform or narrow-thyrsoid.
 Pubescence of stems usually whitish; carpels coarsely reticulate; lvs. thin, not conspicuously rugose-veined .... ssp. *monticola*
 Pubescence of stems usually yellowish; carpels finely reticulate; lvs. thickish, conspicuously rugose-veined .... ssp. *rugosa*

Ssp. **rosàcea** (M. & J.) Kearn. [*S. r.* M. & J. *S. purpurea* Parish ex Jeps.] Stems woody at base, white-pubescent; petals pale purplish-pink, drying rose-violet; anthers usually purple; $n = 15$ (Webber, 1936).—Below 3600 ft.; Creosote Bush Scrub; w. edge of Colo. Desert; Ariz., L. Calif. March–May.

Ssp. **montícola** Kearn. [*S. pulchella* Jeps. *S. a.* var. *aculeata* Jeps.] Stems herbaceous, whitish-pubescent; lvs. 1.5–3 cm. long; infl. narrow, few-fld.; petals 15–20 mm. long; carpels less rugose dorsally, the coarsely reticulate portion small; $n = 10$ (Webber, 1936).—Dry rocky slopes, mostly 4000–7000 ft.; Pinyon-Juniper Wd.; Mono Co. to San Bernardino Co.; Nev. April–June.

Ssp. **rugòsa** Kearn. Stems herbaceous, yellowish-pubescent; infl. narrow, thyrsoid, interrupted; fls. many, 12–15 mm. long; carpels finely reticulate; $n = 5$ (Webber, 1936).—Dry slopes, 3000–6000 ft.; Joshua Tree Wd., Pinyon-Juniper Wd., Yellow Pine F.; Victorville (Mojave Desert), San Jacinto Mts. to L. Calif. Occasional lower, as at San Jacinto. April–June.

5. **S. Rúsbyi** Gray ssp. **eremícola** (Jeps.) Kearn. [*S. e.* Jeps.] Perennial, with thick woody crown, green and sparsely pubescent, 3–6 dm. high; lvs. thin, round-cordate, 4–20 mm. long, 3–5-parted, the lobes then cleft and toothed; fls. few in a loose panicle; calyx woolly, 11–14 mm. long; petals apricot, 14–16 mm. long; carpels 10–12, with chartaceous walls, 4–5 mm. high, shallowly and broadly notched, the indehiscent part faintly reticulate; seeds 2; $n = 5$ (Webber, 1936).—At 4000–4400 ft.; Creosote Bush Scrub; Emigrant Canyon, Panamint Mts., Inyo Co. May–June.

6. **S. angustifòlia** (Cav.) G. Don ssp. **cuspidàta** (Gray) Kearn. [*S. c.* Britton.] Perennial with thick woody crown, densely canescent, erect, 8–18 dm. high; lvs. thickish, prominently veined, often revolute, linear-lanceolate, 2–7 cm. long, usually ca. ⅕ as wide, subhastately lobed basally; petioles 0.5–2.5 cm. long; infl. long, narrow, interrupted, many-fld., leafy; calyx at anthesis 5–9 mm. long, the lobes lanceolate; petals grenadine to almost pink, 7–12 mm. long; fr. truncate-conical; carpels 10–16, with chartaceous walls, 4–6 mm. high, shallowly notched, the indehiscent portion prominently reticulate; seeds 1–3, usually glabrous; $n = 5$ or 10 (Webber, 1936).—Reported from near Indio and Hayfields (Riverside Co.) and Los Angeles; Kans. to Mex., Ariz. Summer.

7. **S. parvifòlia** A. Nels. Perennial, grayish- or whitish- canescent; stems several, 6–10 dm. high; lvs. thickish, prominently veined beneath, broadly ovate to suborbicular, 1.5–4 cm. long, not lobed to shallowly 3-lobed near the middle with broad rounded lobes; petioles 1–3 cm. long; infl. many-fld., narrowly thyrsoid-glomerate; calyx densely pubescent, 4–8 mm. long at anthesis, the lobes lance-ovate, short-acuminate; petals grenadine, 8–18 mm. long; carpels 9–12, with chartaceous walls, 3–5 mm. high,

broadly notched, the indehiscent part ca. ¼ of carpel, finely rather faintly reticulate; seeds usually 2; $n = 5$ or 10 (Webber, 1936).—Dry slopes, 5000–7000 ft.; Pinyon-Juniper Wd.; Inyo Co. (Bishop Creek and White Mts.); to Colo., New Mex. June–July.

8. **S. Munroàna** (Dougl.) Spach. [*Malva M.* Dougl. *Malvastrum M.* Gray.] Perennial with woody crown and herbaceous slender stems 4–9 dm. high, sparsely pubescent to canescent; lvs. thin, broadly ovate, subcordate to subtruncate at base, rounded to obtuse at apex, shallowly 3–5-lobed, crenate to dentate, 1.5–5 cm. long; petioles 2–4 cm. long; infl. many-fld., narrowly thyrsoid-glomerate; calyx at anthesis 4–9 mm. high, the lobes mostly deltoid-ovate, acute; petals grenadine, 10–18 mm. long; fr. hemispherical; carpels 10–12, with chartaceous walls, 3–3.5 mm. high, shallowly notched, the indehiscent part forming ⅓–⅗ of carpel, finely reticulate; seeds mostly 1.—Dry places, Squaw Creek, Placer Co.; to B.C., Mont., Wyo. May–June.

9. **S. grossulariaefòlia** (H. & A.) Rydb. [*Sida g.* H. & A. *Malvastrum g.* Gray.] Perennial, densely whitish-canescent, the stems few, 7–10 dm. long; lvs. rather prominently veined beneath, deltoid or broadly ovate, ± cordate, pedately 5-cleft into cuneate-obovate divisions, then again cleft or parted and coarsely few-toothed, 1.5–4 cm. long, ca. as wide; petioles 1–3.5 cm. long; infl. usually many-fld., narrowly thyrsoid-glomerate; calyx at anthesis 5–8 mm. high, the lobes lance-ovate; petals grenadine, 8–20 mm. long; fr. hemispherical; carpels ca. 12, with chartaceous walls, ca. 3 mm. high, shallowly notched, the indehiscent part ca. half the carpel, finely reticulate; seed usually 1.—Dry places in volcanic soil, Hot Springs Peak, Lassen Co.; to Wash. and Utah. May–June.

4. Iliámna Greene

Plants herbaceous above the woody caudex, sparsely stellate-pubescent. Stems tall, leafy. Lvs. large, aceriform, thin, rather shallowly 3–7-cleft with broad triangular lobes. Fls. large, in long interrupted thyrsoid panicles. Involucel of 3 narrow persistent bractlets. Sepals connate at base. Petals white, pink, or rose-purple, villous along margin of claws. Stamineal column hirsute. Stigmas capitate. Carpels oblong, thin-walled, rounded at apex, smooth and glabrous on sides, bordered toward back with narrow band of stellate hairs and densely hirsute on back with coarse erect simple hairs 1–4 mm. long, dehiscent from apex to base on back and ca. ⅔ down the ventral edge, remaining attached to axis by threads. Seeds ± reniform, dark brown, usually 3. Seven spp. mostly of w. N. Am. (Name of Greek derivation, but significance uncertain.)

(Wiggins, I. L. A resurrection and revision of the genus Iliamna Greene. Contr. Dudley Herb. 1: 213–229, 1936.)

Lvs. deeply 5–7-lobed, 5–20 cm. broad; plant 15–20 dm. high .... 1. *I. latibracteata*
Lvs. (at least the upper) shallowly 3-lobed, 1–4 cm. broad; plant mostly 3–10 dm. high
2. *I. Bakeri*

1. **I. latibracteàta** Wiggins. [*Sphaeralcea rivularis* var. *cismontana* Jeps. *S. acerifolia* auth., not Nutt.] Plant 1.5–2 m. high, stellate-pubescent; lvs. suborbicular in outline, cordate, 8–25 cm. wide, green above, canescent beneath, the lobes lance-ovate, with broad teeth; petioles 5–14 cm. long; fls. in crowded spicate clusters; bractlets lanceolate, 10–14 mm. long; calyx 8–10 mm. long at anthesis, the triangular-ovate lobes 5–8 mm. long; petals 2.5–3 cm. long, rose to lavender with purple base; carpels 10–14, oblong, 8–10 mm. high, with conspicuous bristles on back, 2–3-seeded; seeds 2 mm. high, puberulent.—Moist often shaded places, 200–2500 ft.; Redwood F.; Humboldt Co. to sw. Ore. June–July.

2. **I. Bàkeri** (Jeps.) Wiggins. [*Sphaeralcea B.* Jeps.] Plant 3–7(–10) dm. high, harshly stellate-pubescent throughout; lvs. suborbicular to cuneate-obovate, the lower truncatish at base, 3–5-lobed, irregularly serrate, 1.5–4.5(–7) cm. long; petioles 1–2.5(–6) cm. long; fls. solitary or in 2–3-fld. axillary clusters; bractlets linear-oblong, 6–12 mm. long; calyx 9–16 mm. long at anthesis, the lobes ovate, 4–10 mm. long, acuminate; petals lavender to lavender-pink, 15–30 mm. long; carpels 8–10 mm. high, bristly on back, 3–4-seeded; seeds 2 mm. long, puberulent.—Volcanic loam and lava beds, 3000–5000 ft.; N. Juniper Wd.; Shasta, Modoc, and Siskiyou cos.; s. Ore. July–Aug.

5. Malvástrum Gray

Plants annual or perennial, herbaceous or somewhat woody, sparsely to densely stellate-pubescent or hispid. Lvs. ovate to orbicular, crenate or palmately cleft. Fls. axillary or in terminal bracted infl. Calyx subtended by an involucel of narrow bractlets. Sepals somewhat united at base. Petals white to purple, or yellow. Stamens many, united into a tube. Carpels few, indehiscent, compressed or somewhat turgid, rugose on the sides, sometimes tuberculate on the back, 1-ovuled, not sharply differentiated apically and basally. A genus of perhaps 50 spp. of Am. and S. Afr. (A name derived from *malva*, mallow.)

Lvs. crenate; petals 2–3 cm. long, each with a dark spot .... 1. *M. rotundifolium*
Lvs. 5–7-lobed; petals not spotted.
   Petals 5–6 mm. long; calyx 3–5 mm. long .... 2. *M. exile*
   Petals 10–25 mm. long; calyx 7–12 mm. long.
     The petals white to pale lavender, 10–13 mm. long; calyx-lobes tapering gradually to a slender
     tip .... 3. *M. kernense*
     The petals purplish, 12–30 mm. long; calyx-lobes abruptly acuminate .... 4. *M. Parryi*

1. **M. rotundifòlium** Gray. [*Eremalche r.* Greene. *Sphaeralcea r.* Jeps.] DESERT FIVESPOT. Erect annual, 1–4(–6) dm. high, simple or branched, the stems and petioles hispid with long mostly simple hairs; lvs. remote, suborbicular, cordate, coarsely crenate, 2–5 cm. wide, on petioles 2–10 cm. long; fls. in racemose corymbs; bractlets filiform, 6–10 mm. long; calyx 10–15 mm. long, stellate-hispid, the lobes ovate, acuminate; petals 2–3 cm. long, rose-pink to lilac, drying purplish, each with conspicuous dark spot, the corolla subglobose when fresh; carpels many, thin, flat, black at maturity, reticulate near the edges, 2.5–3 mm. high.—Frequent in washes and on mesas, below 3800 ft.; Creosote Bush Scrub; Mojave and Colo. deserts; Ariz., Nev. March–May.

2. **M. éxile** Gray. [*Malveopsis e.* Kuntze. *Eremalche e.* Greene. *Sphaeralcea e.* Jeps.] Annual, decumbent or prostrate, with several stems from base, these 1–4 dm. long, stellate-pubescent; lvs. suborbicular, 8–18 mm. wide, palmately 3–5-cleft with rounded lobes; petioles slender, 1–3.5 cm. long; bractlets slender, 3–4 mm. long; calyx 3–5 mm. long; petals whitish to pink or lavender, 5–6 mm. long; carpels not more than 15, rather turgid, transversely wrinkled, grayish, 1–1.5 mm. long.—Common in dry open places, below 5000 ft.; Creosote Bush Scrub; Mojave and Colo. deserts; to Utah, Ariz.; has been reported from near Riverside. March–May.

3. **M. kernénse** (C. B. Wolf) Munz. [*Eremalche k.* C. B. Wolf.] Erect or prostrate annual, usually branched from base, the stems 1–2 dm. long, with scattered stellate hairs; lvs. suborbicular, 1–3 cm. long, 3–5-cleft, the lobes crenate; petioles 1–4 cm. long; bractlets filiform, 3–4 mm. long; calyx 8–10 mm. long, the lobes 2–3 mm. wide, tapering gradually to a slender tip; petals whitish to lavender, 10–13 mm. long; carpels 8–13, grayish or light brown, rather turgid, transversely corrugate on back, 2 mm. long.— Dry open clay flats, 600–900 ft.; Shadscale Scrub; Temblor V., near McKittrick, Kern Co. April–May.

4. **M. Párryi** Greene. [*Eremalche P.* Greene. *Sphaeralcea P.* Jeps.] Annual, with several stems from near the base, or ± simple, erect or ascending, subglabrous to somewhat stellate-pubescent; lvs. twice-lobed or -cleft, 2–4 cm. wide; petioles 1–5 cm. long, slender; fls. pedicelled, in loose terminal corymbs; bractlets linear, 9–11 mm. long; calyx loosely stellate-pubescent, 10–14 mm. long, the lobes ovate, 4–5 mm. wide, abruptly acuminate; petals pinkish lavender to purplish, darker on drying, mostly 1.5–3 cm. long; carpels turgid, ca. 1.5 mm. long, rugose-reticulate, somewhat brownish. —Dry flats and hills, below 4500 ft.; Pinyon-Juniper Wd., Foothill Wd., V. Grassland; largely about the San Joaquin V., n. Ventura Co. to Alameda Co. March–May.

6. Malacothámnus Greene

Shrubs, sometimes suffrutescent, with long usually flexuous branches, ± densely stellate-tomentose or -pubescent. Lvs. petioled, ± evidently and palmately 3–5-lobed. Fls. pink or lavender, 1 to several at a node, and in axillary ± capitate clusters forming elongate interrupted spikes or in open panicles. Calyx subtended by an involucel of bractlets.

Sepals 5, united at base. Petals 5, distinct, but often joined at base to the stamen-tube. Stamens many, the fils. united into a tube. Carpels several, completely dehiscent (splitting into halves at maturity), muticous, unappendaged, not rugose, 1-ovuled. A genus of 20 spp. of sw. N. Am.

(Kearney, T. H. The genus Malacothamnus. Leafl. W. Bot. 6: 113–140, 1951.)
A.   Fls. in dense terminal heads, these with conspicuous invols. of membranous or leafy bracts ca. as long as calyces. Monterey and San Luis Obispo cos. .... 1. *M. Palmeri*
AA. Fls. not in dense heads, or, if so, these not conspicuously involucrate or terminal.
     B.   Calyx-lobes mostly 4–8 mm. wide at base; bractlets subtending the calyx broadly lanceolate to ovate; lvs. ± cordate at base. San Benito, Fresno, and Monterey cos. .. 2. *M. aboriginum*
     BB. Calyx-lobes mostly not more than 3 mm. wide at base; bractlets narrowly lanceolate to filiform.
        C.   Calyx conspicuously and densely white-lanate, the hairs ± concealing the calyx-lobes.
            D.   Stems closely white-tomentose with very short hairs; lvs. mostly truncate or sub-cuneate at base, finely crenate-dentate; bractlets ca. as long as calyx-tube.
               E.   Infl. open-paniculate, the lower branches usually few-fld., the fls. with long pedicels. San Luis Obispo Co. .... 3. *M. niveus*
               EE. Infl. thyrsoid-glomerate, the lower branches short, with many subsessile fls. Tehama Co. to Yolo Co. .... 4. *M. Helleri*
            DD. Stems more loosely pubescent with longer hairs; lvs. mostly cordate at base, coarsely crenate; bractlets ca. as long as calyx.
               E.   Lvs. thick, velvety-tomentose. Bordering Cent. V. .... 5. *M. Fremontii*
               EE. Lvs. rather thin, pubescent (but not densely so). S. Mojave Desert
                                                     6. *M. orbiculatus*
        CC. Calyx not conspicuously and densely lanate, the pubescence sparser or looser or shorter, not concealing the calyx-lobes, or if the calyx sometimes rather woolly, then the infl. very narrow and few-fld.
            D.   Hairs of calyx long, simple to few-branched (the arms to 2 mm. or more long).
               E.   Stems conspicuously shaggy-tomentose with grayish hairs; lvs. angulately 3–5-lobed, bicolored, soft-tomentose beneath. San Clemente Id.
                                                 7. *M. clementinus*
               EE. Stems closely pubescent with yellowish hairs; lvs. simple or with rounded lobes, not strongly bicolored or tomentose beneath. Mainland
                                                 8. *M. densiflorus*
            DD. Hairs of calyx short (to 1 mm. long) and many-armed.
               E.   Infl. open-paniculate, rather few-fld., with long slender rather flexuous ascend-ing-spreading branches.
                  F.   Calyx 9–11 mm. long, the lobes 2–3 times as long as the tube; bractlets broadly subulate, thick, whitish. Monterey Co. .... 9. *M. Abbottii*
                  FF. Calyx 6–7 mm. long, the lobes less than twice the tube; bractlets nar-rowly subulate, thin, dark. San Luis Obispo Co. .... 10. *M. gracilis*
               EE. Infl. contracted, or if open, the branches strictly ascending.
                  F.   Stems loosely pubescent.
                     G.   Lvs. deeply 3–5-lobed; petioles and lf.-veins very stout; bractlets not more than ⅓ the calyx. San Fernando V. ...... 11. *M. Davidsonii*
                     GG. Lvs. simple or shallowly lobed; petioles and lf.-veins slender; bract-lets usually at least half as long as calyx.
                        H.   The lvs. cordate at base; calyx-lobes ovate, 3–7 mm. long. Desert borders, Inyo Co. to San Bernardino Co.
                                                   6. *M. orbiculatus*
                      HH. The lvs. mostly truncate at base.
                        I.   Calyx 9–12 mm. long, the lobes lance-ovate, 6–8 mm. long. Fresno Co. to San Gabriel Mts.
                                             12. *M. marrubioides*
                        II. Calyx 5–8 mm. long, the lobes deltoid-ovate, ca. 3 mm. long. San Mateo, Santa Clara, and Santa Cruz cos.
                                             13. *M. arcuatus*
                 FF. Stems appressed-pubescent, or if pubescence loose, then the calyx-lobes abruptly acuminate.
                   G.   Calyx 4–6 mm. long.
                     H.   Stems striate-angular; lvs. deeply lobed; petals 10–12 mm. long. Mendocino Co. .... 14. *M. mendocinensis*
                     HH. Stems not striate-angular; lvs. shallowly lobed; petals 13–15 mm. long. Contra Costa Co. to Santa Clara Co.
                                             15. *M. Hallii*
                 GG. Calyx 6–12 mm. long.
                   H.   Calyx-lobes abruptly acuminate
                                   8. *M. densiflorus* var. *viscidus*
                   HH. Calyx-lobes acutish.
                      I.   Lvs. thickish, truncate or cuneate at base, not or scarcely lobed.
                        J.   Infl. subracemose, few-fld., the fls. 1–3 at each node; pubescence whitish. San Luis Obispo Co.
                                         16. *M. Jonesii*

JJ. Infl. open-paniculate, the fls. 3 or more at a node;
pubescence yellowish. San Bernardino
17. M. *Parishii*
II. Lvs. thin, mostly cordate at base. S. Calif.
18. M. *fasciculatus*

1. **M. Pálmeri** (Wats.) Greene. [*Malvastrum P.* Wats. *Sphaeralcea P.* Jeps.] Suffrutescent, 1.5–2.5 m. tall, 2–4 m. across, with coarse leafy stems; lf.-blades broadly ovate to almost round, 2–7 cm. wide, slightly longer, dark green, stellate-pubescent above and beneath, ± 3-lobed, crenate-dentate; petioles 1–3 cm. long; fls. sessile in a terminal headlike cluster or with 1–2 axillary clusters below, these clusters dense, 3–5 cm. wide, with leafy ± lobed oblong-lanceolate bracts 1.5–2 cm. long; bractlets linear to ovate; calyx 14–18 mm. long, the lobes lance-ovate, hirsute, 12–14 mm. long, acuminate; petals rose, 20–30 mm. long, hairy near base; carpels densely stellate, ca. 4 mm. high; seed brown, ca. 2.5 mm. long.—Dry rocky slopes, below 1200 ft.; Chaparral; foothills of Santa Lucia Mts., San Luis Obispo Co. May–July.

Var. **involucràtus** (Rob.) Kearn. [*Malvastrum i.* Rob. *M. P.* var. *i.* McMinn.] Lvs. soon glabrous above; outer involucral bracts not lobed, oblong to broadly ovate; calyx 10–14 mm. long; petals 14–20 mm. long.—Foothill Wd.; Jolon, near King City, and Carmel V., Monterey Co. May–Aug. Plants from Arroyo Seco road at 3600 ft., intermediate between the sp. and the var., have been described as var. *lucianus* Kearn.

2. **M. aboríginum** (Rob.) Greene. [*Malvastrum a.* Rob. *Sphaeralcea a.* Jeps.] Coarse, woody below, 1–3 m. tall, densely felted-tomentose on stems and twigs; lf.-blades graygreen, broadly ovate, 3–6 cm. long, 3–5-lobed, crenate-dentate, cordate at base, stellate-pubescent; petioles 1–3.5 cm. long; fls. sessile in headlike clusters in upper axils, forming an elongate ± naked infl.; bractlets 3, ovate, 6–8 mm. long; calyx strongly angled in bud, 8–10 mm. long, the lobes 4–6 mm. long and somewhat wider; petals rose, 12–14 mm. long; carpels ca. 3 mm. high; seed brownish, ca. 1.5 mm. long.—Dry hills, 700–1600 ft.; Foothill Wd.; away from the coast, S. Coast Range, San Benito, Fresno, and Monterey cos. May–Sept.

3. **M. níveus** (Eastw.) Kearn. [*Malvastrum n.* Eastw. *M. Fremontii* var. *n.* McMinn. *M. fragrans* Eastw., not Harv. & Gray.] Erect, 1–2 m. tall, loosely branched, densely and closely white-tomentose with very short hairs and slender branches; lvs. rounded-deltoid, scarcely or obscurely lobed, ± truncate at base, crenate to crenate-dentate, 1.5–4 cm. wide, ca. as long, white-tomentose, on petioles 7–14 mm. long; infl. open-paniculate, the lower branches slender, usually elongate, few-fld., many of the fls. long-pedicelled; bractlets linear, dark, 3–4 mm. long; calyx 7–10 mm. long, white-woolly, the lobes deltoid, evident, longer than wide; petals 15–20 mm. long, lavender-pink; carpels 2.5–3 mm. high, deeply incised, broadly stalked; seeds papillate-stellulate.—At 1200–2600 ft.; Chaparral; San Luis Obispo Co. May–July.

4. **M. Hélleri** (Eastw.) Kearn. [*Malvastrum H.* Eastw. *M. Fremontii* var. *H.* McMinn. *Sphaeralcea F.* var. *exfibulosa* Jeps.] Erect shrub, 1–1.5 m. high, closely and densely white-tomentose; lvs. ovate to suborbicular, conspicuously veined, crenate, truncate at base, 2–6 cm. long and wide, whitish stellate-pubescent beneath, greener above; petioles 1–4 cm. long; infl. thyrsoid-glomerate, or the lower branches elongate, many-fld., the fls. subsessile; bractlets linear, dark, 4–5 mm. long; calyx 5–7 mm. long, white-woolly, the lobes ovate, obtuse, obscured by the wool; petals 10–12 mm. long, rose; carpels 2.5 mm. high, shallowly incised; seeds obscurely stellulate.—Gravelly washes, inner N. Coast Ranges, Tehama Co. to Yolo Co. June–Aug.

5. **M. Fremóntii** (Torr. ex Gray) Greene. [*Malvastrum F.* Torr. ex Gray. *Sphaeralcea F.* Jeps.] Woody below, erect, 1–2 m. high, densely white-tomentose; lf.-blades thick, round-ovate, scarcely to plainly but shallowly 5–7-lobed, crenate, 3–10 cm. long, soft-tomentose on both surfaces, rounded to somewhat cordate at base; petioles 1–3.5 cm. long; fls. pink or rose, in headlike clusters in the axils on upper branchlets; bractlets 3, linear-filiform, dark, ca. as long as calyx; calyx subglobose in bud, densely white-woolly, ca. 7 mm. long, the segms. hidden except for the short pointed tips; petals 12–20 mm. long; carpels glabrous except for the summit, ca. 3 mm. high; seeds 1.5 mm. long, dark gray; $n = 17$ (Webber, 1936).—Occasional, on dry open slopes, especially on burns, 200–2500 ft.; Foothill Wd., Chaparral; w. base of Sierra Nevada, Tulare Co. to Amador Co., inner N. Coast Ranges, Tehama Co. and Contra Costa Co. April–Sept.

Ssp. **cercóphorus** (Rob.) Munz. [*Malvastrum F.* var. *c.* Rob. *Sphaeralcea F.* var. *c.* Jeps. *Malvastrum Howellii* Eastw. and var. *cordatum* Eastw. *Malacothamnus H.* Kearn.] Calyx ca. 10 mm. long, the segms. attenuate-caudate.—Below 3500 ft., inner Coast Ranges, Contra Costa to Santa Clara cos., foothills of Sierra Nevada, Madera and Calaveras cos. May–Sept.

6. **M. orbiculàtus** (Greene) Greene. [*Malvastrum o.* Greene. *M. Fremontii* var. *o.* Jtn. *Sphaeralcea o.* Jeps.] Suffruticose, 5–20 dm. high, with mostly stout simple branches densely soft-stellate-tomentose; lvs. rounded, thinnish, 3–7 cm. wide, coarsely crenate, sometimes 3–5-lobed, greener above than beneath; petioles mostly 1–2 cm. long; fls. rose, few in sessile axillary clusters forming long interrupted spicate panicles; bractlets linear, shorter than calyx; calyx ± hispid-stellate to densely lanate, 7–10 mm. long, the segms. ovate, 3–7 mm. long, acute to acuminate; petals 12–16 mm. long; carpels pubescent on upper half, 2–2.5 mm. high; seeds 1.5 mm. long; *n* = 17 (Webber, 1936).— Dry stony slopes, 3000–8500 ft.; Pinyon-Juniper Wd., Yellow Pine F.; borders of Mojave Desert from e. slope of Sierra Nevada, Inyo Co. to n. slope of San Bernardino Mts. June–Oct.

7. **M. clementìnus** (M. & J.) Kearn. [*Malvastrum c.* M. & J. *Sphaeralcea orbiculata* var. *c.* Jeps.] Rounded subshrub to 1 m. tall, with numerous shaggy, stellate-tomentose branches; lf.-blades 3–5-lobed, deeply cordate at base, broadly ovate, 3–5 cm. wide, greenish above, whiter and soft-stellate beneath; petioles 1–1.5 cm. long; fls. many, subsessile and densely glomerate in uppermost axils forming interrupted spikes 1–2 dm. long; bractlets filiform, almost as long as calyx; calyx 7–8 mm. high, loosely stellate-tomentose, the lobes broadly lanceolate, acute, 4 mm. long; petals pink, ca. 13 mm. long, oblong-obovate; carpels stellate-tomentose at summit, 2.5–3 mm. high; seeds almost 2 mm. long.—Rocky canyon-walls; Coastal Sage Scrub; Lemon Tank, San Clemente Id. April.

8. **M. densiflòrus** (Wats.) Greene. [*Malvastrum d.* Wats. *Sphaeralcea d.* Jeps.] Erect, 1–2 m. tall, with yellowish close scurfy pubescence; lvs. rounded-ovate, simple or shallowly rounded-lobed, truncate to cordate at base, 2–5 cm. wide, green, sparsely pubescent beneath; petioles 5–20 mm. long; infl. an interrupted spike of dense sessile heads, ± glandular; bractlets filiform, viscid, 7–16 mm. long; calyx 10–14 mm. long, loosely hirsute with long subsimple or stellate hairs or subglabrous, the lobes lance-ovate, 5–12 mm. long; petals rose-pink, 10–15 mm. long; carpels 2.2–2.8 mm. high, subsessile, shallowly incised.—Dry slopes, mostly 1000–4000 ft.; Chaparral; Santa Ana Mts. to Palm Springs (Riverside Co.) and Cuyamaca Mts. (San Diego Co.). April–July.

Var. **víscidus** (Abrams) Kearn. [*Malvastrum v.* Abrams. *M. d.* var. *v.* Estes. *Sphaeralcea d.* var. *v.* Jeps. *Malvastrum marrubioides* var. *v.* McMinn.] Bractlets 4–6 mm. long; calyx-lobes 3–7 mm. long, broadly ovate; *n* = 17 (Webber, 1936).—San Diego Co.; L. Calif. March–June.

9. **M. Ábbottii** (Eastw.) Kearn. [*Malvastrum A.* Eastw.] Erect shrub, 1–2 m. tall, white-tomentose; lf.-blades broadly ovate, 2–5 cm. long, truncate at base, gray and soft-tomentose, crenate; petioles 1–2.5 cm. long; infl. a panicle with ascending branches; bractlets linear-lanceolate, thick, whitish, 5 mm. long; calyx 9–12 mm. long, finely white-tomentose, the lobes 5–6 mm. long, long-acuminate; petals rose, 18–20 mm. long. —Among willows, Salinas R., Monterey Co. June–Oct.

10. **M. grácilis** (Eastw.) Kearn. [*Malvastrum g.* Eastw.] Erect slender shrub, 1–2 m. tall; twigs very slender, whitish with dense stellate tomentum; lf.-blades ovate, often 3-lobed, 1–5 cm. long, truncate or rounded at base, crenate, densely pale-tomentose; petioles 0.5–2 cm. long; infl. paniculate, with very slender branches; bractlets linear, purplish, 3–4 mm. long; calyx 6–7 mm. long, reddish-purple, somewhat glandular and tomentose, the lobes gradually acuminate; petals rose, 12–15 mm. long; carpels stellate-tomentose at apex, 3–4 mm. high.—Rocky dry slopes; Chaparral; San Luis Obispo Co. June–Oct.

11. **M. Davidsònii** (Rob.) Greene. [*Malvastrum D.* Rob. *Sphaeralcea D.* Jeps.] Erect coarse shrub, 2–5 m. high; branches stout, densely shaggy with stellate tomentum; lf.-blades thick, round-cordate, 2.5–10 cm. wide, 5-angled or shallowly 3–5-lobed, densely stellate-tomentose; petioles 1.5–3 cm. long; fls. numerous, in short racemes forming panicles 2.5–4.5 dm. long; bractlets 3 mm. long; calyx densely stellate-tomentose, 5–8 mm. long, the lobes ovate, acute, 2–4 mm. long, obscured by the tomentum; petals

pink or rose, 12–15 mm. long; carpels tomentulose at summit, ca. 3 mm. high, incised; seeds dark; $n = 17$ (Webber, 1936).—Sandy washes and flats, 1000–1200 ft.; Coastal Sage Scrub; San Fernando V., Los Angeles Co. June–Sept.

12. **M. marrubioìdes** (Dur. & Hilg.) Greene. [*Malvastrum m.* Dur. & Hilg. *M. gabrielense* M. & J. *Sphaeralcea densiflora* var. *g.* Jeps.] Shrubby below, 6–20 dm. high, with erect slender branches and close stellate pubescence; lvs. rounded, obscurely lobed, pale green, 1.5–4.5 cm. wide; petioles 1–2.5 cm. long; fls. few in glomerules at upper axils, forming a spicate or subpaniculate scarcely glandular infl.; bractlets subulate, 7–12 mm. long, lanceolate; calyx 10–12 mm. long, the lobes ovate-lanceolate, 6–8 mm. long, densely stellate-tomentose; petals pink, 16–18 mm. long; carpels stellate-pubescent at apex, 2.5–3.5 mm. high, shallowly incised.—Dry stony slopes, 1500–7000 ft.; Chaparral, Foothill Wd.; Sierra Nevada foothills (Fresno Co.); mts. at w. end of Mojave Desert, Mt. Pinos s. to San Gabriel Mts. June–Aug.

13. **M. arcuàtus** (Greene) Greene. [*Malveopsis a.* Greene. *Malvastrum a.* Rob. *Sphaeralcea a.* Arthur.] Erect, woody below, 6–18 dm. tall, the stems densely white-tomentose; lf.-blades thickish, rugose, ovate to rhombic-ovate, 2–5 cm. long, not or slightly lobed, crenate-dentate, greenish above, densely canescent-tomentose beneath; petioles 5–30 mm. long; fls. in dense headlike clusters sessile in upper axils and forming interrupted spikes; bractlets linear, slightly shorter than calyx; calyx whitish or somewhat rusty, soft-tomentose, 6–8 mm. long, the lobes deltoid-ovate, ca. 3 mm. long, acute; petals rose, 12–18 mm. long; carpels tomentose above, 3 mm. long; seeds grayish, 1.5 mm. long.—Brushy canyons, below 1000 ft.; Chaparral; e. slope of Coast Range, San Mateo, Santa Clara and Santa Cruz cos. April–July.

14. **M. mendocinénsis** (Eastw.) Kearn. [*Malvastrum m.* Eastw.] Shrub 1–2 m. high, erect, white-tomentose; branches slender, striate-angular; lvs. ovate, obtusely 3–5-lobed, cordate at base, stellate-scabrid above, stellate-tomentose and paler beneath, 5–7 cm. long; petioles 2–3 cm. long; infl. contracted, the branches short; bracteoles ca. 1 mm. long; calyx 4–6 mm. long, with short, obtuse, deltoid lobes; petals rose, ca. 12 mm. long; carpels ca. 2.2 mm. high, stellate-tomentose, sessile.—Open banks at ca. 700 ft.; N. Oak Wd.; 5 mi. sw. of Ukiah, Mendocino Co. May–June.

15. **M. Hállii** (Eastw.) Kearn. [*Malvastrum H.* Eastw. *Sphaeralcea fasciculata* var. *Elmeri* Jeps.] Shrub 1–2 m. tall, with close, short-rayed pubescence; lvs. suborbicular, shallowly 3–5-lobed, usually subcordate, 1.5–6 cm. wide, greenish and subglabrous above, whiter and more densely stellate beneath; petioles 1–2.5 cm. long; infl. contracted, the branches usually short; calyx ca. 5 mm. long, the lobes deltoid, acute, 2 mm. long; petals rose, 13–15 mm. long; carpels 2.2–3 mm. long, subsessile.—Stony slopes; Chaparral; inner Coast Range, Mt. Diablo, Contra Costa Co., Pacheco Pass, etc., Santa Clara Co. May–July.

16. **M. Jònesii** (Munz) Kearn. [*Malvastrum J.* Munz. *Sphaeralcea fasciculata* var. *J.* Jeps. *Malvastrum Dudleyi* Eastw.] Erect shrub, 6–12 dm. high, with many slender branches, covered with dense soft whitish stellate tomentum; lf.-blades suborbicular, firm, not or obscurely 3–5-lobed, coarsely and irregularly crenate-dentate, 1–2.5 cm. long, rounded to cuneate at base, pale and closely velvety-pubescent; petioles 1–2 cm. long; fls. solitary or 2–3 in upper axils, forming an interrupted spicate infl.; bractlets subulate, 3–4 mm. long; calyx 6–8 mm. long, loosely stellate-tomentose, the lobes ovate, acute, ca. 5 mm. long; petals pink or rose, 12–14 mm. long; carpels 3 mm. high, tomentose at apex; seeds dark.—Dry slopes; Chaparral, Foothill Wd.; near Paso Robles and Santa Lucia Mts., San Luis Obispo Co. May–July.

17. **M. Paríshii** (Eastw.) Kearn. [*Malvastrum P.* Eastw.] Erect shrub, white-tomentose; lvs. thin, rhombic-ovate, distinctly but shallowly 3-lobed, cuneate at base, crenate, ca. 5 cm. long, 4 cm. wide, green above, densely white-tomentose beneath; petioles ca. 1 cm. long; infl. open-paniculate with elongate lower branches and usually 3 or more fls. per node; bracteoles filiform, ca. 2 mm. long; calyx 8–9 mm. long, the lobes ovate, acute, nerved, 5 mm. long; petals rose, ca. 12 mm. long; carpels 3 mm. high, conspicuously stalked.—Near San Bernardino, 1000–1500 ft. June–July.

18. **M. fasciculàtus** (Nutt.) Greene. [*Malva f.* Nutt. ex T. & G. *Malveopsis f.* Kuntze. *Malvastrum f.* Greene. *M. Thurberi* Gray. *Sphaeralcea f.* Arthur.] Shrub 1–5 m. tall, with long slender wandlike branches covered with short soft tomentum; lf.-blades round-ovate, mostly shallowly lobed, 2–4(–6) cm. wide, densely canescent beneath,

less so and somewhat greener above, crenate or crenate-dentate; petioles 0.5–1 cm. long; fls. in sessile or nearly sessile headlike clusters forming an interrupted spicate infl.; bractlets linear, 2–4 mm. long; calyx 6–8 mm. long, stellate-pubescent, the lobes deltoid, acute to obtuse; petals pink, 12–18 mm. long; carpels stellate-pubescent at summit, 2.5–3 mm. high; seeds ca. 1.5 mm. long; $n = 17$ (Webber, 1936).—Common on dry slopes and canyon-sides, below 2500 ft.; Chaparral, Coastal Sage Scrub; s. Riverside Co. to L. Calif. April–July.

KEY TO VARIETIES

Infl. of interrupted spike of headlike clusters, or with lower branches sometimes to 6 cm. long; calyx loosely pubescent with long hairs.
    Lvs. small, to 4 cm. long, not conspicuously bicolored. S. Riverside Co. to L. Calif. . . . *M. fasciculatus*
    Lvs. larger, to 7 cm. long, quite conspicuously bicolored. Catalina Id. . . . . var. *catalinensis*
Infl. open-paniculate, the lower branches often long and loosely fld.; calyx closely pubescent.
    Lvs. essentially concolored, 2–4 cm. wide. Coastal Santa Barbara and Ventura cos. . . . var. *Nuttallii*
    Lvs. bicolored, paler beneath.
        Lf.-blades 3–7 cm. wide; panicle many-branched, the ultimate divisions rigid, not racemose.
        Santa Cruz Id. . . . . var. *nesioticus*
        Lf.-blades 2–4 cm. wide; panicle few-branched, the ultimate divisions elongate, commonly racemose. San Bernardino Co. to Riverside, Orange and Ventura cos. . . . . var. *laxiflorus*

Var. **laxiflòrus** (Gray) Kearn. [*Malvastrum Thurberi* var. *l.* Gray. *M. splendidum* Kell. *Malveopsis s.* Kuntze. *Malacothamnus f. s.* Abrams. *Malvastrum f.* var. *l.* M. & J. *Sphaeralcea f.* var. *l.* Jeps.] Lvs. somewhat bicolored; infl. branched and paniculate; carpels 2–3 mm. high; $n = 17$ (Webber, 1936).—Below 5500 ft., Orange and n. Riverside cos. to Cajon Pass, San Bernardino Co., to s. Ventura Co. May–Aug.

Var. **Nuttállii** (Abrams) Kearn. [*Malacothamnus N.* Abrams. *Malvastrum N.* Davids. & Mox. *Sphaeralcea f.* var. *N.* Jeps. *Malvastrum f.* var. *N.* McMinn.] Lvs. closely hoary-stellate on both surfaces; panicle elongate, few-branched; carpels 3–5 mm. high; $n = 17$ (Webber, 1936).—Chaparral; Ventura and Santa Barbara cos. June–Aug.

Var. **nesióticus** (Rob.) Kearn. [*Malvastrum n.* Rob. *M. f.* var. *n.* McMinn. *Malacothamnus n.* Abrams. *Sphaeralcea n.* Jeps. *S. f.* var. *n.* Jeps.] Lvs. thinnish, 5–7 cm. long, cordate, subglabrate above, deeply lobed; panicles open, terminal, quite leafless; calyx finely canescent; carpels ca. 4 mm. high.—Steep canyons, Santa Cruz Id. June–July.

Var. **catalinénsis** (Eastw.) Kearn. [*Malvastrum c.* Eastw. *M. f.* var. *c.* McMinn.] Lvs. thin, subglabrous above, deeply lobed, 5–7 cm. long; fls. in congested clusters in short-branched panicle; calyx stellate-tomentose; carpels 3.2–3.8 mm. high.—Catalina Id. Closely matched by plants from Santa Monica Mts. May.–Aug.

7. Modìola Moench

Low diffuse perennial herb. Lvs. rounded, coarsely crenate to palmately incised. Fls. small, purplish to orange, solitary in axils, each with involucel of 3 bractlets. Calyx in ours enlarging in fr., the 5 lobes ovate, hirsute, acute. Petals obovate. Fr. depressed, of 14–25 thin-coriaceous reniform carpels, each septate between the 2 seeds. Small genus of Am. and Afr. (Latin, *modiolus*, the nave of a wheel, because of the fr.)

1. **M. caroliniàna** (L.) G. Don. [*Malva c.* L.] Stems spreading, pubescent, 3–6 dm. long; lvs. 2–5 cm. wide, on slender petioles; calyx 5–6 mm. long at anthesis, slightly enlarged in fr.; petals dull red, 5–8 mm. long; carpels hirsute, ca. 4 mm. high, beaked at outer edge of top.—Rather widely natur. in Calif. at lower elevs.; native of trop. Am. April–Sept.

8. Lavatèra L. TREE-MALLOW

Ours erect to arborescent herbs or shrubs. Lvs. angled or lobed, maplelike in ours, long-petioled. Fls. showy, axillary or in terminal racemes, each with a 3-lobed involucel in ours. Calyx 5-lobed. Petals reflexed after anthesis, clawed, emarginate or truncate. Fr. depressed, the carpels in ours 5–8, smooth, 1-seeded. Ca. 25 spp. of Medit. region to Asia, Australia, Canary Ids., ids. off Calif. and L. Calif. (*Lavater*, two Swiss brothers, of the time of Tournefort.)

Involucel-lobes lanceolate; petals 2.5–4.5 cm. long .... 1. *L. assurgentiflora*
Involucel-lobes rounded-ovate; petals 1–2 cm. long.
   Involucel surpassing calyx; lvs. densely soft-downy .... 2. *L. arborea*
   Involucel shorter than calyx; lvs. sparsely pubescent .... 3. *L. cretica*

   1. **L. assurgentiflòra** Kell. [*Saviniona a.* Greene. S. *clementina, reticulata, dendroidea* and *suspensa* Greene.] MALVA ROSA. Erect, shrubby, 1–4 m. tall, with thick glabrous to pubescent twigs; lvs. 5–15 cm. wide, 5–7-lobed, the lobes ovate-triangular, coarsely and irregularly toothed, pale beneath; petioles 5–15 cm. long; bractlets ca. 3, lance-ovate, 5–9 mm. long; calyx densely stellate-pubescent, 12–15 mm. long; petals rose with darker veins, 2.5–4.5 cm. long, with long narrow glabrous claws and a pair of dense hairy tufts at base; carpels woody, ca. 6 mm. high, rounded on back, triangular in cross section, glabrous to pubescent.—Variable, native of sandy flats and rocky places; Coastal Sage Scrub; Santa Barbara Ids., cult. and escaped along mainland coast. March–Nov.
   2. **L. arbòrea** L. Shrubby, becoming 1–3 m. tall; lvs. 5–20 cm. broad, softly stellate-pubescent on both sides, shallowly and unequally 5–9-lobed, crenate; petioles 3–8 cm. long; fls. many, in short leafy racemes or axillary clusters, short-pedicelled; involucel exceeding calyx, with broad rounded lobes; calyx ca. 4 mm. long at anthesis; petals pale purple-red with dark purple veins at base, 1.5–2 cm. long; carpels 7–9, subglabrous, ± reticulate; $n = 22$ (Skovsted, 1935).—Natur. on bluffs and dunes near the coast, Mendocino Co. to San Mateo Co.; introd. from Eu. June–July.
   3. **L. crètica** L. Winter annual, 1–3 m. tall, sparsely pubescent; lower lvs. suborbicular, the upper shallowly and broadly 5-lobed, crenate, truncate to subcordate at base, 4–10 cm. wide, on longer petioles; involucel shorter than calyx, with broad rounded lobes; calyx ca. 4 mm. long at anthesis, later much enlarged and surrounding the fr.; petals pinkish to lilac, 10–16 mm. long; fr. depressed, of 7–10 glabrous or puberulent relatively smooth carpels; $n =$ ca. 56 (Skovsted, 1935).—Natur. on San Francisco Peninsula and Marin Co. coast; native of Medit. region. May–July.

9. Althaèa L. HOLLYHOCK

   Biennial to perennial herbs, tall and with leafy stems. Lvs. lobed or parted. Fls. solitary or racemose, axillary, usually toward the top of the stem. Involucel of 6–9 bractlets. Otherwise like *Malva*. A genus of ca. 15 spp. of Old World temp. region. (Greek, *althaino*, to cure.)
   1. **A. ròsea** (L.) Cav. [*Alcaea r.* L.] HOLLYHOCK. Mostly biennial, 2–3 m. tall, the stems strict, hairy; lvs. round-cordate, 5–7-lobed or -angled, commonly 1–2 dm. wide, crenate, long-petioled; fls. 6–10 cm. wide, subsessile in elongate, terminal spicate raceme, of many colors.—Occasional escape from cult.; native of China. Summer.

10. Málva L. MALLOW. CHEESES

   Annual or biennial herbs, sparsely pubescent or glabrate. Lvs. alternate, orbicular or reniform, ± lobed or dissected. Fls. solitary or clustered in lf.-axils. Involucel of 3 bractlets like an outer calyx. Calyx 5-cleft into broad lobes. Petals 5, emarginate, white to pink or purplish. Styles numerous, stigmatic down the inner side. Fr. depressed, disklike, separating at maturity into the many 1-seeded compressed reniform indehiscent carpels. Ca. 30 spp. of the Old World; ours natur. (Latin name from Greek, *malache*, referring to the emollient lvs.)

Bractlets of involucel ovate to oblong.
   Petals 2–4 times the length of calyx; corolla 2–3 cm. wide. .... 1. *M. sylvestris*
   Petals 1–2 times the length of calyx; corolla 1–1.5 cm. wide .... 2. *M. nicaeensis*
Bractlets linear to nearly so.
   Petals twice the length of calyx; carpels quite smooth on back, rounded at margins.
      Stems procumbent; lvs. obscurely lobed; fls. clearly pedicelled .... 3. *M. neglecta*
      Stems erect; lvs. definitely lobed; fls. subsessile .... 4. *M. verticillata*
   Petals scarcely longer than calyx; carpels rugose-reticulate on back, sharp-margined.
      Claws of petals glabrous; calyx enlarged in fr. and widely spreading, veiny-reticulate
                                                      5. *M. parviflora*
      Claws of petals bearded; calyx barely enlarged and mostly closed over fr., scarcely reticulate
                                                      6. *M. rotundifolia*

   1. **M. sylvéstris** L. HIGH MALLOW. Biennial, ascending, 3–10 dm. tall, branched, rough-hairy; lvs. 3–7 cm. wide, round-cordate or reniform, with 5–7 triangular sharply

pointed lobes, long-petioled; fls. 2–6 in the axils; involucels 4–5 mm. long; calyx ca. 6 mm. long, the lobes short and broad; petals rose-purple with deeper veins, 2–2.5 cm. long, deeply notched; carpels ca. 10, wrinkled-veiny on back, sharp-edged, glabrous or nearly so; $2n = 42$ (Skovsted, 1935).—Reported as garden escape, but our plant apparently the:

Ssp. **mauritiàna** (L.) Boiss. [*M. m.* L.] Plant glabrous or nearly so; lf.-lobes shallowly round-lobed; petals deeper-colored.—Occasionally natur. as at Healdsburg, Ventura, etc.; native of Eurasia.

2. **M. nicaeénsis** All. [*M. borealis* auth., not Wallm.] Annual, ascending, 2–6 dm. high, pubescent; lvs. rounded-reniform, shallowly 5–7-lobed, crenate, 3–10(–12) cm. wide, long-petioled; bractlets lance-ovate, 4–5 mm. long; calyx 4–5 mm. long at anthesis, becoming enlarged, veiny and closed over fr.; petals pinkish to blue-violet, 10–12 mm. long; carpels 7–9, rugose-reticulate, sharp-edged, glabrous to pubescent; $n = 21$ (Skovsted, 1935).—Occasional weed in waste places; as about towns; introd. from Eurasia. March–Sept.

3. **M. neglécta** Wallr. [*M. rotundifolia* many auth.] Annual or biennial; stems procumbent, somewhat pubescent, 2–5 dm. long; lvs. round-cordate, crenate, not to obscurely 5–7-lobed, 2–6 cm. wide; petioles commonly 5–20 cm. long; fls. fascicled in axils, commonly pedicelled; bractlets linear, 3–5 mm. long; calyx ca. 4–6 mm. long at anthesis, the lobes acuminate; petals pale lilac or white, 8–13 mm. long; carpels ca. 15, puberulent, rounded but not reticulate on back, with rounded margins; $2n = 42$ (Skovsted, 1935).—Weed in waste places and about gardens, n. and cent. Calif.; native of Eurasia. May–Oct.

4. **M. verticillàta** L. var. **críspa** L. [*M. c.* L.] CURLED MALLOW. Erect smoothish annual, 6–18 dm. high; lvs. crisped, 5–7-lobed, crenate, long-petioled; fls. sessile or nearly so, white to purple, crowded in axils; bractlets narrow; calyx ca. 4–5 mm. long, the lobes acuminate; petals 8–10 mm. long; carpels glabrous, smoothish or obscurely reticulate, the edges rounded; $n = $ ca. 56 (Skovsted, 1935).—Reported from Mill V., Marin Co.; native of Old World.

5. **M. parviflòra** L. CHEESEWEED. Annual, erect, branched, 3–8 dm. high, sparsely stellate-pubescent to almost glabrous; lvs. roundish, somewhat angulate-lobed, 2–8 cm. wide, on somewhat longer petioles; fls. axillary, short-pedicelled; bractlets linear, 1–2 mm. long; calyx ca. 3 mm. long at anthesis, enlarged, reticulate and wide-spreading in fr.; petals whitish to pinkish, 4–5 mm. long; carpels ca. 11, rugose-reticulate, thin-margined, denticulate on angles, pubescent on back in our form; $2n = 42$ (Skovsted, 1941).—Common weed in orchards, waste places, almost throughout Calif.; native of Eurasia. Can be found in bloom most of the year.

6. **M. rotundifòlia** L. Similar to *M. parviflora*, but petals with bearded claws; calyx scarcely enlarged or reticulate and tending to close over fr.; carpels mostly entire on angles; $n = 21$ (Skovsted, 1935).—Not certainly found in Calif.; native of Eu.

11. Sìda L.

Ours low canescent stellate or cinereous-puberulent herbs. Lvs. crenate, serrate or lobed. Fls. axillary, solitary or in small cymules. Involucel usually lacking, in ours mostly of 1–3 linear deciduous bractlets. Calyx 5-lobed. Petals 5. Carpels 5–10, 1-seeded, indehiscent or dehiscent only part way from apex, ± rugose and often reticulate on the sides. Ca. 150 spp. of warmer parts of world. (Unexplained Greek name of some plant.)

Plant perennial; petals 10–12 mm. long .... 1. *S. hederacea*
Plant annual; petals 4–8 mm. long .... 2. *S. rhombifolia*

1. **S. hederàcea** (Dougl.) Torr. [*Malva h.* Dougl. *M. californica* Presl. *Disella h.* Greene.] ALKALI-MALLOW. Stems from elongate rootstocks, decumbent or prostrate, whitish-stellate, 1–4 dm. long; lvs. round-reniform to broadly deltoid, dentate, rounded at apex, 1.5–4.5 cm. wide, on petioles 1–3 cm. long; calyx 6–7 mm. long; petals yellowish, 10–12 mm. long; carpels 6–10, indehiscent, reticulate on sides; $2n = 22$ (Heiser & Whitaker, 1948).—Moist, ± saline places, below 6000 ft.; many Plant Communities; widely distributed in Calif.; to Wash., Okla., Mex. May–Oct.

2. **S. rhombifòlia** L. Annual, to 1 m. high; lvs. subsessile to short-petioled, rhombic-

oblong to oblanceolate, cinereous-puberulent beneath, serrulate; peduncles ± elongate; calyx cinereous, 5–10-nerved at base; petals pale yellow, often red at base, 4–8 mm. long.—Weed in cotton field in Madera Co.; natur. from trop.

12. Anòda Cav.

Annuals; sparsely hirsute to puberulent or tomentose. Lvs., especially the upper, often hastate. Involucel lacking. Calyx 5-cleft. Petals bluish to lavender. Styles 5–20, tipped by capitate stigmas. Fr. depressed, hemispheric or disklike; carpels usually umbonate or spurred on the back, with lateral walls fragile and usually breaking up before maturity, the inner layer making a sacklike envelope about the seed or becoming closely adherent to the seed. Ca. 10 spp., mostly Mex. and S. Am. (A colloquial name from Ceylon.)

1. **A. cristàta** (L.) Schlecht. var. **digitàta** (Gray) Hochr. [*A. arizonica* var. *d.* Gray.] Hairy, to 1 m. high; lvs. narrow-ovate in outline, angulate-lobed, often hastate, truncate at base, 2–6 cm. long, on somewhat shorter petioles; fls. axillary, purple; calyx becoming flat, with 5 long-acuminate lobes, in fr. ca. 2–2.5 cm. across; petals 2–2.5 cm. long; carpels dark green, 9–20, hispid, their backs separated by pale bands; seeds pubescent.— Reported as garden weed at Placerville and near Stockton; Ariz., w. Tex., Mex. June–Sept.

13. Sidálcea Gray. Checker

Annual or perennial herbs. Lvs. rounded, frequently palmately or pedately parted or lobed, with small stipules, petioled. Fls. in terminal racemes or spikes, perfect or with abortive anthers, these ♀ fls. small. Involucel lacking or of 1 bractlet. Calyx 5-lobed. Petals 5, emarginate or truncate, purple to rose pink, or white. Stamen-tube pubescent, with double series of fils., the outer series of phalanges (sets of united fils.) often distinctly below the inner. Carpels 5–9, beakless or beaked, 1-seeded, dehiscent. Ca. 22 spp. of w. N. Am., used somewhat in horticulture. (*Sida* and *Alcea*, 2 malvaceous genera.)

(Hitchcock, C. L. A study of the perennial spp. of Sidalcea. Univ. Wash. Pub. Biol. 18: 1–79, 1957.)
- A.   Plants annual.
     B.   Upper stipules divided into linear segms.; bracts at least as long as calyx .. 1. *S. diploscypha*
     BB. Upper stipules mostly simple; bracts much shorter than calyx.
          C.   Divisions of upper cauline lvs. cuneate-obovate, 2–5-toothed at apex; stems hirsute-pubescent throughout; fls. few. Tulare Co. .... 2. *S. Keckii*
          CC. Divisions of upper cauline lvs. sublinear, simple, rarely 2-toothed; stems mostly subglabrous below.
               D.   Outer fils. divided above into small sets; plants glabrate or minutely puberulent; carpels glabrous .... 3. *S. Hartwegii*
               DD. Outer fils. united into a continuous series.
                    E.   Carpels rugose-reticulate on back, pubescent; fls. many, in dense infl.; plants densely hirsute above .... 4. *S. hirsuta*
                    EE. Carpels longitudinally grooved on back, glabrous; fls. few to many, in open infl.; plants glabrate or sparsely pubescent .... 5. *S. calycosa*
- AA. Plants perennial.
     B.   Lvs. not deeply parted or divided; staminal column not conspicuously double.
          C.   Fls. rose-purple; lvs. fan-shaped, scarcely or not lobed; infl. of elongate spicate racemes
                                                                                     17. *S. Hickmanii*
          CC. Fls. whitish; lvs. vitiform, shallowly lobed; infl. of oblong spikes . 18. *S. malachroides*
     BB. Lvs. (at least the upper) usually deeply parted or divided; staminal column conspicuously double, the outer phalanges narrow and cleft.
          C.   Plants with enlarged rather fleshy taproots or fascicled roots, not at all rhizomatous; stems usually hirsute toward base; calyx mostly hirsute (at least in part) with pustulose hairs; racemes mostly elongate and loosely fld.; pedicels slender and at least as long as the calyx; carpels smooth to lightly reticulate on the sides .... 16. *S. neomexicana*
          CC. Plants usually with tough and fibrous rather than fleshy roots, often rhizomatous; stems often stellate-pubescent at base; calyx seldom with pustulose hairs; racemes often spikelike and very closely fld.; carpels frequently conspicuously reticulate-alveolate.
               D.   Carpels smooth; plants without rootstocks; stems leafless or nearly so except at their softly hirsute bases; racemes spikelike, but usually rather open and loosely fld.; pedicels 1–2(–3) mm. long; lvs. 3(5)-lobate, the cauline usually ternately dissected into linear or narrowly oblong segms. .... 14. *S. pedata*
               DD. Carpels usually roughened or plant not otherwise as above.
                    E.   Rootstocks lacking, the plants from taproots and usually a branching crown, often with decumbent branches but these not freely rooting; herbage often very conspicuously glaucous.

    F. Roots fleshy, simple to fascicled; stems finely stellate to more coarsely hairy with simple forked 4-rayed hairs; lvs. fleshy, glaucous, densely pubescent with 2- and 4-rayed to stellate hairs; racemes elongate, loosely many-fld.; pedicels 2–8 mm. long; calyx 5–8 mm. long, finely stellate. Alkaline places .... 15. *S. Covillei*

  FF. Roots not fleshy, or if so, calyx not uniformly finely stellate, or stems coarsely hairy, or racemes more spikelike or more closely fld.

    G. Plants very glaucous; lower part of stems with small appressed 7–10-rayed hairs; racemes open, loosely 3–12(–15)-fld.; pedicels slender, 3–10 mm. long; lvs. finely stellate dorsally, mostly with 5 main lobes; calyx finely rather uniformly stellate.

      H. Lvs. usually 5-lobed, the 2 lower lobes usually not cleft more than halfway to base, all usually 3-toothed to entire; plants glaucous but not grayish .... 7. *S. glaucescens*

     HH. Lvs. apparently 7-lobed, the first 2 of the 5 primary segms. usually cleft more than half their length, the lobes from deeply toothed to laciniately cleft or pedately parted; plants grayish-glaucous .... 8. *S. multifida*

   GG. Plants usually not particularly glaucous, or stems not finely appressed-stellate as above, or racemes many-fld. and often spikelike.

      H. Carpels usually prominently reticulate-pitted; racemes not spikelike, the pedicels mostly 3–8 mm. long; petals frequently prominently white-lined .... 6. *S. malvaeflora*

     HH. Carpels mostly smooth to lightly reticulate; racemes usually spikelike at anthesis, often greatly congested; pedicels mostly 1–3(–5) mm. long; petals not prominently white-lined.

        I. Calyx somewhat accrescent, the lobes more ovate-lanceolate than lanceolate; basal portion of stems finely hairy, not long-hirsute; pedicels 1–2 mm. long; carpels lightly reticulate on back, the sides distinctly reticulate-alveolate
                                 12. *S. setosa*

        II. Calyx usually not accrescent, the lobes lanceolate; basal portion of stem usually coarsely stellate to long-hirsute; pedicels 1–5 mm. long; carpels nearly smooth
                                   13. *S. oregana*

 EE. Rootstocks present although often short and thick; stems often trailing and freely rooting; plants not particularly glaucous.

    F. Carpels narrowly wing-margined, lightly reticulate, usually glabrous; calyx uniformly finely stellate, 10–15 mm. long, the lobes conspicuously 3-nerved; petals pale pink, often drying to yellowish, mostly 2–3.5 cm. long; racemes much elongate, loosely many-fld.; stems finely and densely stellate at base .... 9. *S. robusta*

  FF. Carpels not wing-margined, sparsely puberulent to stellate on back; calyx often with mixed pubescence; stems seldom uniformly and finely stellate at base.

    G. Rootstocks short and thick; racemes spikelike, closely many-fld.; pedicels but 1–2 mm. long; calyx 6–8 mm. long at anthesis, accrescent, finely stellate and conspicuously bristly with longer coarser hairs; petals 8–15 mm. long; carpels ca. 2.5 mm. long, reticulate-alveolate on sides and edges, smoothish on back .... 12. *S. setosa*

   GG. Rootstocks usually well developed or racemes elongate and loosely fld. and pedicels well over 2 mm. long; calyx often uniformly stellate; petals and carpels various.

      H. Racemes spikelike; fls. many, crowded or interruptedly glomerate; pedicels 1–3 mm. long; calyx 5–9 mm. long, finely stellate and with mixture of slender soft hairs as much as 2 mm. long; petals 5–15 mm. long; carpels ca. 2.5 mm. long, glabrous to finely stellate; stems long-hairy toward base; rootstocks usually well developed .... 11. *S. ranunculacea*

     HH. Racemes various but usually elongate and loosely fld.; calyx 5–12 mm. long, often uniformly stellate; petals at least 10 mm. long; carpels 3–4 mm. long; rootstocks sometimes imperfectly developed.

        I. Plants from very widespread rootstocks or the stems trailing and freely rooting, 2–5 dm. tall, mostly abundantly hirsute at base with slender hairs as much as 3 mm. long; lvs. usually hirsute on both surfaces, rarely stellate; calyx finely stellate but margins and back of lobes usually with mixture of longer coarser stellae; carpels ca. 3 mm. long, densely bristly-stellate on back .... 10. *S. reptans*

        II. Plants various, but if widely rhizomatous or with trailing and rooting branches, then either the lvs. or the lower portion of the stems usually stellate; calyx often uniformly stellate; carpels 3–4 mm. long, usually not stellate on back
                                6. *S. malvaeflora*

1. **S. diploscýpha** (T. & G.) Gray. [*Sida d.* T. & G. *Sidalcea d.* var. *minor* Gray. *S. secundiflora* Greene.] Annual, erect, simple or branched, 2–6 dm. tall, pilose-hirsute throughout, also with short stellate pubescence; basal lvs. rounded, 1–2.5 cm. wide, long-petioled, crenate, dying early, the cauline 2–6 cm. wide, rather deeply parted into lobed segms.; infl. few-fld., with conspicuous palmately parted bracts, these 5–14 mm. long and with filiform divisions; calyx 10–14 mm. long, the lobes lanceolate-subulate; petals 1.5–3 cm. long, purple to pink, often with dark purplish spot; fr. depressed, the carpels subreniform, glabrous, rugulose, ca. 2.5 mm. high.—Open hillsides and plains, below 3000 ft.; V. Grassland, Coastal Prairie, N. Oak Wd., Foothill Wd.; Coast Ranges (Humboldt Co. to San Luis Obispo Co.), Sacramento V., foothills of Sierra Nevada (Shasta Co. to Mariposa Co.). April–June.

2. **S. Kéckii** Wiggins. Slender annual, 1.5–3.5 dm. tall, hirsute and somewhat stellate-pubescent; lower lvs. 1.5–2.5 cm. wide, shallowly 7–9-lobed, the lobes 3–5-toothed, upper lvs. more deeply cuneate-lobed; petioles 2–4.5 cm. long; fls. few; bracts 3–7 mm. long, bifid; calyx 8–11 mm. long, the lobes linear-lanceolate; petals rose, 10–20 mm. long; carpels 3–4 mm. high, favose-reticulate, glabrous or with 1–3 hairs at apex.— Grassy slopes at ca. 1200 ft.; White R., near Glenville, Tulare Co. April.

3. **S. Hartwégii** Gray ex Benth. [*S. tenella* Greene. *S. H.* var. *t.* Gray.] Annual, simple or few-branched, 1.5–3 dm. high, glabrescent or minutely stellate-pubescent (especially in infl.); lvs. 1.5–4(–6) cm. wide, the basal lobed, the cauline pedately 5–7-divided into linear entire to trifid divisions; petioles 0.6–3 cm. long; fls. few; bracts 2–3 mm. long, bidentate to bifid; calyx 5–10 mm. long, the lobes long-acuminate; petals rose-purple, 12–16 mm. long; carpels glabrous, alveolate-reticulate, ca. 2.5 mm. high.—Open clay flats to rocky places among trees, below 3500 ft.; V. Grassland, N. Oak Wd., Foothill Wd.; Coast Ranges (Mendocino Co. to Napa Co.), Sacramento V., foothills of Sierra Nevada (Shasta Co. to Mariposa Co.). March–June.

4. **S. hirsùta** Gray. [*S. delphinifolia* Gray, not Nutt.] Stout erect simple or few-branched annual, mostly 3–6(–8) dm. high, glabrate below, soft-hirsute above; basal lvs. crenately lobed, cauline 3–8 cm. wide, 7–9-parted into linear segms.; infl. densely spicate, many-fld.; bracts bifid, 4–8 mm. long; calyx tawny-hirsute, 8–10 mm. long, the base somewhat chartaceous, the lobes long-acuminate; petals rose-purple, 15–25 mm. long; carpels stellate-pubescent, favose-reticulate, 3–4 mm. high, with a bristly beak.— Low wet places which soon become dry, below 1000 ft.; V. Grassland, Foothill Wd.; Shasta Co. to Merced Co., e. Mendocino Co. April–May.

5. **S. calycòsa** Jones. [*S. sulcata* Curran.] Erect slender annual, sparingly branched, glabrous or sparingly pubescent, 3–8 dm. high, the stems sometimes prostrate at base and rooting along the internodes; basal lvs. rounded, crenate or shallowly lobed, cauline 2–5 cm. wide, 5–11-parted into oblanceolate or cuneate segms., these entire or toothed; fls. in open peduncled racemes; bracts 2–5 mm. long, bifid into lanceolate lobes; calyx thinly hirsute, 4–7 mm. long, ± scarious in age, the lobes lance-ovate, acuminate; petals light purple, 12–20 mm. long; carpels striate-ridged on back, reticulate laterally, glabrous, ca. 2.5 mm. high.—Wet places, below 3500 ft.; Foothill Wd., N. Oak Wd.; Mendocino Co. to Marin and Napa cos., and Shasta Co. to Tulare Co. April–July.

Ssp. **rhizómata** (Jeps.) Munz. [*S. r.* Jeps.] Perennial with long-creeping rooting stoloniferous base, the stems succulent, erect or ascending, 3–5 dm. high, glabrous or sparingly hirsute above; basal lvs. 3–10 cm. wide, shallowly incised, the cauline 7–11-divided into broadly cuneate divisions; infl. spicate; bracts 8–12 mm. long, deeply 2-lobed, the lobes ovate; calyx 6–12 mm. long, the lobes ovate, acuminate; petals light purple, 20–25 mm. long; carpels striate-grooved on back, reticulate on sides, 4.5 mm. high, with slender deciduous beaks.—Among tussocks of sedge and rush; Coastal Salt Marsh; Point Reyes Peninsula, Marin Co. May–July.

6. **S. malvaeflòra** (DC.) Gray ex Benth. [*Sida m.* DC. *S. delphinifolia* Nutt. *Sidalcea humilis* Gray. *S. scabra* Greene. *S. rostrata* Eastw.] Perennial with rather widely spreading rootstocks from heavy root; stems 1.5–6 dm. tall, coarsely pubescent below with simple to 4-rayed spreading to appressed hairs, more stellate upward; lvs. mostly 2–6 cm. broad, often markedly fleshy, usually simply to bifurcately hirsute beneath and cruciately stellate above, long-petioled, the basal from roundish to ± reniform, shallowly 7–9-lobed and coarsely crenate, the cauline similar or more deeply lobed, the floral often divided to base; racemes mostly simple, congested to elongate and open;

pedicels 3–20 mm. long; calyx 8–12 mm. long, finely appressed-stellate and with longer coarse 2–4-rayed hairs; petals 1–2.5 cm. long, usually white-veined; carpels ca. 4 mm. long, sparsely glandular-puberulent, coarsely reticulate-faveolate, short-beaked; $2n = 40$, 60 (Kruckeberg, 1957).—Frequent on open grassy slopes and mesas, low elevs.; Coastal Prairie, Mixed Evergreen F., etc.; Mendocino Co. to Santa Barbara Co.

KEY TO SUBSPECIES

A. Lvs. chiefly basal, the cauline few, not dissected into linear segms.; infl. open, elongate; rootstocks usually poorly developed; carpels 2.5–3 mm. long, lightly reticulate-alveolate. Kern and e. Santa Barbara cos. to L. Calif. . . . . ssp. *sparsifolia*
AA. Lvs. usually also on the stem; infl. typically few-fld., more congested; rootstocks usually well developed; carpels often over 4 mm. long, usually prominently reticulate-alveolate.
    B. Plants usually purplish, glabrous to sparsely hirsute at base; stipules purplish; basal lvs. cuneate to reniform, fleshy, crenate but scarcely lobed; lower pedicels to 2 and 3 cm. long; calyx purplish, moderately pubescent with small stellae and longer tangled to straight hairs. Sonoma Co. to s. Mendocino Co. . . . . ssp. *purpurea*
    BB. Plants more usually not purplish at base or on stipules and calyx, rather densely hairy at base; basal lvs. mostly cordate to reniform, often lobed; pedicels often short and stout; calyx usually rather densely hairy.
        C. Cauline lvs. usually dissected into linear segms.; racemes open, loosely many-fld., often over 1.5 dm. long; calyx 7–11 mm. long, finely stellate, ± bristly with coarser longer hairs or finely stellate overall. San Luis Obispo Co. to Sonoma Co. . . . . . ssp. *laciniata*
        CC. Cauline lvs. not dissected into linear segms.; racemes mostly short, rather condensed, usually not much over 1 dm. long; calyx variously pubescent.
            D. Stems abundantly coarsely harsh-hirsute near base with simple or forked hairs; racemes loosely 5–20-fld.; calyx prominently nerved, densely and finely bristly-stellate; carpels ca. 4 mm. long, deeply reticulate-alveolate; corolla pale pink. S. Siskiyou Co. to s. Shasta Co. . . . . ssp. *celata*
            DD. Stems more usually stellate; racemes often short and congested; calyx frequently with a scattering of longer and coarser hairs among the stellae; carpels usually less than 4 mm. long; corolla pinkish-rose.
                E. Stems hairy throughout with 4(or more)-rayed spreading soft stellae; lvs. stellate on both surfaces or with a mixture of forked hairs ventrally; calyx grayish with uniform coarse many-rayed stellae. Coastal mts. of Santa Barbara and Ventura cos. . . . . ssp. *californica*
                EE. Stems usually partially hirsute below, or the lvs. not stellate ventrally, or the calyx finely stellate.
                    F. Stems finely stellate throughout or glabrous above; infl. elongate and loosely 2–15-fld.; calyx densely, uniformly and finely stellate. N. Coast Ranges from Trinity Co. n. . . . . ssp. *nana*
                    FF. Stems usually coarsely hirsute to stellate; infl. often short and congested or with more than 15 fls.; calyx usually not uniformly finely stellate.
                        G. Carpels frequently stellate, 3–3.5 mm. long; calyx mostly ± uniformly finely stellate but often with a scattering of somewhat longer stellae on the lobes.
                          H. Stems glabrous to sparsely hirsute or stellate below, glabrous and very slender above, brittle when fresh; carpels ca. 3.5 mm. long; fls. usually less than 15. Siskiyou Mts.
                                    ssp. *elegans*
                          HH. Stems usually rough-pubescent with stellae and simple hairs, or if with one kind only, not particularly slender or brittle above; carpels usually ca. 3 mm. long; fls. mostly more than 15. Humboldt and Siskiyou cos. n. . . . . ssp. *asprella*
                      GG. Carpels not stellate, often over 3.5 mm. long; calyx usually rather sparsely finely appressed-stellate but with mixture of much longer coarser forked to 4-rayed hairs on the lobes.
                          H. Lvs. finely stellate on dorsal surfaces; stems mostly hirsute with slender simple hairs; carpels ca. 3 mm. long; calyx rather sparsely finely stellate and somewhat hirsute with longer simple to 4-rayed hairs; racemes open and few-fld. San Bernardino Mts. . . . . ssp. *dolosa*
                          HH. Lvs. coarsely stellate dorsally; stems usually coarsely hirsute (to stellate); carpels 3.5–4 mm. long; calyx coarsely stellate, often ± lanate at base.
                              I. Calyx without longer tangled hairs at base; lower portions of stems coarsely pubescent with forked to cruciate hairs; fls. many, closely arranged; stems elongate, trailing, rooting freely. Humboldt Co. n. . . . . ssp. *patula*
                              II. Calyx often with longer ± tangled hairs at base; lower parts of stems coarsely hirsute to cruciately hairy; racemes often open, elongate; stems less elongate, rhizomelike at base. Near coast, Mendocino Co. to Los Angeles Co.
                                        S. malvacflora

Ssp. **sparsifòlia** C. L. Hitchc. In ± moist places; several Plant Communities; from Santa Barbara and Kern cos. s. Hitchcock recognizes 4 vars. largely separable on pubescence and lf.-size.

Ssp. **purpùrea** C. L. Hitchc. Mixed Evergreen F., etc.; near the coast from n. Sonoma Co. to s. Mendocino Co.

Ssp. **laciniàta** C. L. Hitchc. Valleys in Coast Ranges; V. Grassland, Foothill Wd., etc.; Sonoma Co. to San Luis Obispo and San Benito cos. Hitchcock has 2 vars. based on calyx-pubescence.

Ssp. **celàta** (Jeps.) C. L. Hitchc. [*S. m.* var. *c.* Jeps.] Mixed Evergreen F., Yellow Pine F., etc.; Siskiyou, Trinity and Shasta cos.

Ssp. **califórnica** (Nutt.) C. L. Hitchc. [*Sida c.* Nutt. in T. & G. *Sidalcea c.* Gray.] Coastal Sage Scrub, etc.; Santa Barbara and Ventura cos.

Ssp. **nàna** (Jeps.) C. L. Hitchc. [*S. reptans* var. *n.* Jeps.] Mixed Evergreen F., etc.; Trinity and Siskiyou cos.; sw. Ore.

Ssp. **élegans** (Greene) C. L. Hitchc. [*S. e.* Greene.] Mixed Evergreen F.; Del Norte and Siskiyou cos.; sw. Ore.

Ssp. **asprélla** (Greene) C. L. Hitchc. [*S. a.* Greene. *S. m.* var. *a.* Jeps.] Foothill Wd., Yellow Pine F., etc.; Tuolumne Co. to Plumas and Butte cos., Trinity and Siskiyou cos.; s. Ore.

Ssp. **dolòsa** C. L. Hitchc. Yellow Pine F., etc.; San Bernardino Mts.

Ssp. **pátula** C. L. Hitchc. Redwood F., etc.; Humboldt and Del Norte cos.; s. Ore.

7. **S. glaucéscens** Greene. [*S. montana* Congd.] Stems slender, from a woody root-crown, procumbent to erect, glaucous, subglabrous or minutely stellate-puberulent, 3–7 dm. long; lvs. all much alike, deeply 5–7-lobed or -parted, the divisions linear to oblong or narrow-cuneate, entire or 3–5-lobed; infl. a slender lax raceme, ± stellate-pubescent; bracts mostly bifid, 2–3 mm. long, the divisions linear-lanceolate; calyx 5–7 mm. long, the lobes lanceolate, acuminate, broader at the base and veiny in fr.; petals pink to rose, 8–20 mm. long; fr. depressed, the carpels reticulate and dorsally grooved, ca. 2.5 mm. high, glabrous; $2n = 40$ (Kruckeberg, 1957).—Dry grassy places or open woods, 3000–11,000 ft.; Montane Coniferous F.; Sierra Nevada from Tulare Co. n., to Modoc and Siskiyou cos.; w. Nev. May–July.

8. **S. multífida** Greene. Very glaucous, 1–5 dm. tall, from a stout woody crown; stems many, sparsely to densely hairy with small many-rayed stellae; lvs. 1.5–4 cm. broad, chiefly basal, mostly deeply 5-lobed, the lower lobes again divided, all 7 segms. narrow, laciniately cleft to parted; racemes open, 3–9-fld., glabrous to finely stellate; pedicels slender, 3–10 mm. long; calyx uniformly finely stellate, 7–10 mm. long; petals 1–2.5 cm. long; carpels 6–7, 3.5–4 mm. long, reticulate-pitted on sides and back, glandular-puberulent on back; $2n = 20$ (Kruckeberg, 1957).—Sagebrush Scrub, Yellow Pine F.; n. Mono and se. Tulare cos.; Nev. May–July.

9. **S. robústa** Heller ex Roush. [*S. asprella* var. *r.* Jeps.] Plant 5–12 dm. tall, with well developed rhizomes; stems densely and finely pubescent with spreading basal stellae, glabrous above and into infl., glaucous; lvs. mostly on lower third of stem, 3–8 cm. broad, the basal shallowly 5–7-lobed, coarsely crenate, the lower cauline cleft halfway or more into oblong-obovate coarsely dentate lobes; racemes to 4 dm. long, open, lax; calyx 10–15 mm. long, uniformly and finely densely stellate, the lobes 3-nerved; petals mostly 2–3.5 cm. long; carpels glabrous to sparsely glandular-puberulent, 3–3.5 mm. long, finely reticulate-pitted on sides, less so on back; $2n = 20$ (Kruckeberg, 1957).—Foothill Wd., Chaparral; near Chico, Butte Co. April–May.

10. **S. réptans** Greene. [*S. spicata* var. *r.* Jeps. *S. favosa* Congd.] Stems 1 to few, from running rootstocks, decumbent at base and rooting at nodes, 2–5 dm. high, slender, hirsute; lvs. mostly from lower stem, sparsely hairy, coarsely dentate to lobed or incised, 2–7(–10) cm. wide, the segms. lobed; infl. few-fld., racemose; bracts linear to oblong, simple or bidentate, 4–7 mm. long; calyx 8–10 mm. long, the lobes ovate to lance-ovate, acute to acuminate, stellate-pubescent and with longer simple hairs on margin; petals purple, to deep pink, 12–18 mm. long; carpels reticulate-favose, pubescent, beaked, ca. 3 mm. long; $2n = 20$ (Kruckeberg, 1957).—Wet meadows, 4000–7600 ft.; Montane Coniferous F.; Amador Co. to Tulare Co. July–Aug.

11. **S. ranunculàcea** Greene. [*S. spicata* var. *r.* Roush. *S. reptans* var. *r.* Jeps. *S. interrupta* Greene.] Wide-spreading from slender rootstocks, 2–5 dm. tall, long-hairy at base

with simple to 4-rayed hairs; lvs. rather fleshy, 2.5–6 cm. broad, soft but coarsely stellate above and beneath, the basal shallowly 5-lobed, the middle and upper more deeply 5–7-parted; racemes many-fld., spikelike; pedicels mostly 1–3 mm. long; calyx 5–9 mm. long, finely stellate and long-hirsute; petals pink-magenta, 5–15 mm. long; carpels ca. 2.5 mm. long, lightly reticulate on sides, weakly stellate to glabrous on back; $2n = 20$ (Kruckeberg, 1957).—Wet meadows and banks, 6500–9000 ft.; Montane Coniferous F.; Tulare and Kern cos. July–Aug.

12. **S. setòsa** C. L. Hitchc. Perennial from a thick heavy root and short rootstocks; stems 5–10 dm. tall, pubescent at base with fine soft stellae and some longer simple or forked hairs, finely stellate above; lvs. stellate above and beneath, the basal long-petioled, 5–10 cm. broad, shallowly 5–9-lobed, the lobes cuneate-obovate, coarsely toothed, upper divided into 5–9 entire to laciniate segms.; infl. usually compound, the racemes spicate, many-fld., 3–7 cm. long; pedicels largely 1–2 mm. long; calyx 5–8 mm. long, accrescent and to 10 mm. in fr., finely stellate but conspicuously bristly; petals pinkish-lavender, 5–15 mm. long; carpels ca. 2.5 mm. long, sparsely glandular-puberulent and lightly reticulate on back, reticulate-alveolate on sides; $2n = 40$, 60 (Kruckeberg, 1957).—At Edgewood, Siskiyou Co.; s. Ore. June.

13. **S. oregàna** (Nutt.) Gray. [*Sida o.* Nutt. in T. & G.] Perennial with a heavy tap-root; stems many, glabrous to sparsely hairy with large appressed 2–4-rayed stellae, usually rather conspicuously glaucous; lvs. 3–10 cm. broad, stellate above, hirsute to stellate beneath, the basal shallowly 5–7-lobed and coarsely crenate, the cauline more deeply segmented usually with 7 coarsely few-toothed to lacerate lobes; racemes mostly simple, loosely to closely fld., up to 3 dm. long in fr.; pedicels 2–10 mm. long; calyx 5–8 mm. long, uniformly and densely finely stellate; petals 10–20 mm. long, light to deep pink; carpels ca. 3 mm. long, usually reticulate-alveolate on sides, rugose-roughened on back.—Sagebrush Scrub, etc.; Shasta Co.; to Ore., Utah, Wyo., Nev. June–July. Plants with stem-base more pubescent are var. *màxima* (Peck) C. L. Hitchc. Siskiyou Co. to Lassen and Modoc cos.; to Wash.

KEY TO SUBSPECIES

Stems usually stellate near the base; infl. open at least after anthesis; calyx mostly over 6 mm. long
                                          S. oregana
Stems usually hirsute at base with simple hairs; infl. densely crowded, spicate; calyx ca. 5 mm. long at anthesis.
 Plants usually over 9 dm. tall, somewhat rhizomatous; carpels mostly over 2.5 mm. long; calyx conspicuously accrescent in fr.
  Calyx densely hirsute as well as stellate, with slender hairs 1.5–2.5 mm. long; calyx in fr. ca. 2 times as long as in anthesis . . . . spp. *eximia*
  Calyx stellate only, or but sparsely hirsute; calyx in fr. ca. 1.5 times as long as at anthesis
                                        ssp. *valida*
 Plants mostly less than 9 dm. tall, usually not rhizomatous; carpels seldom over 2.5 mm. long; calyx not conspicuously accrescent.
  Infl. very much compounded; racemes few-fld., at anthesis usually less than 2.5 cm. long; carpels very sparsely glandular-puberulent to glabrous, ca. 2.2 mm. long; rhizomes often developed
                                   ssp. *hydrophila*
  Infl. simple to compound, many-fld., usually well over 2.5 cm. long at anthesis; carpels moderately glandular-puberulent on back, otherwise variable; rhizomes lacking . . . . ssp. *spicata*

Ssp. **exímia** (Greene) C. L. Hitchc. [*S. e.* Greene.] Stems mostly 9–12 dm. tall, conspicuously hirsute at base.—Wet meadows, below 3500 ft.; Redwood F., Mixed Evergreen F.; Mendocino Co. to Siskiyou and Humboldt cos.; s. Ore. June–July.

Ssp. **válida** (Greene) C. L. Hitchc. [*S. v.* Greene. *S. spicata* ssp. *v.* Wiggins.] Plants mostly over 10 dm. high; $2n = 20$ (Kruckeberg, 1957).—Edge of Freshwater Marsh; Kenwood, Sonoma Co.

Ssp. **hydróphila** (Heller) C. L. Hitchc. [*S. h.* Heller.] Stems 3–9 dm. high; $2n = 20$ (Kruckeberg, 1957).—Mixed Evergreen F., Yellow Pine F.; Lake and Napa cos. July–Aug.

Ssp. **spicàta** (Regel) C. L. Hitchc. [*Callirhoe s.* Regel. *S. s.* Greene. *S. o.* var. *s.* Jeps.] Plants 3–8 dm. tall; $2n = 20$, 40 (Kruckeberg, 1957).—Montane Coniferous F.; Sierra Nevada to Siskiyou, Trinity and Modoc cos.; Ore. June–Aug.

14. **S. pedàta** Gray. [*S. spicata* var. *p.* Jeps.] Stems few to several from the root-crown, ascending, simple, subscapose, 1–3 dm. tall, glabrous to hirsute, reddish; lvs.

many, 2–4 cm. wide, pedately 5–7-parted or -divided, the divisions narrowly cuneate, 3-lobed, the ultimate segms. linear to oblong; infl. a many-fld., at length elongate, spicate, stellate-pubescent raceme; bracts 3–5 mm. long, simple to bifid; calyx 4–5 mm. long, the lobes lanceolate, widening in fr.; petals rose-purple, 8–12 mm. long; carpels 5, rounded, smooth, ca. 3 mm. high; $2n = 20$ (Kruckeberg, 1957).—Wet meadows, 6500–7500 ft.; Montane Coniferous F.; San Bernardino Mts. May–July.

15. **S. Covíllei** Greene. [*S. malvaeflora* var. *C.* Roush.] Plants from fleshy roots without rootstocks; stems several, 2–6 dm. tall, rather finely pubescent with 5–9-rayed stellae to more coarsely hairy with 2–4-rayed or simple hairs; lvs. chiefly at base, fleshy, glaucous, with 2–4-rayed hairs to stellate, the basal usually shallowly 5–7-lobed and sparingly coarsely crenate to lobed half their width, cauline divided more deeply; infl. simple to compound, elongate, loosely many-fld., finely stellate; pedicels 2–8 mm. long; calyx 5–8 mm. long, finely and densely stellate; petals pinkish-lavender, 10–15 mm. long; carpels ca. 2.5 mm. long, rather prominently reticulate but not pitted on sides, less so on back; $2n = 20$ (Kruckeberg, 1957).—Alkaline meadows; Shadscale Scrub; Owens V. May–June.

16. **S. neomexicàna** Gray ssp. **Thúrberi** (Rob. ex Gray) C. L. Hitchc. [*S. parviflora* var. *T.* Rob. *S. p.* Greene. *S. nitrophila* Parish.] Stems 2–9 dm. tall from fleshy fusiform simple to clustered roots, hirsute to stellate below or glabrous, frequently ± stellate above; lvs. 1.5–4.5 cm. wide, the lower 5–9-lobed or -parted into lobed or toothed divisions, the upper into linear entire or lobed divisions; infl. slender-racemose, many-fld., stellate-pubescent; calyx 5–7 mm. long, the lobes lance-acuminate, moderately stellate-pubescent and with some longer hairs; petals rose, mostly 6–12 mm. long; carpels lightly and coarsely reticulate on the sides, usually smooth on back, ca. 2 mm. long; $2n = 20$ (Kruckeberg, 1957).—Alkaline usually wet places; Coastal Sage Scrub, Chaparral, Creosote Bush Scrub; Los Angeles, Orange, Riverside and San Bernardino cos.; to New Mex. April–June.

17. **S. Hickmánii** Greene. Perennial from thick woody root-crown; stems 4–8 dm. tall, grayish throughout with harsh coarse stellae, leafy; lvs. 1–6 cm. broad, the basal reniform-flabelliform, very shallowly lobed, coarsely crenate-dentate, the upper cauline similar or more deeply lobed; infl. usually compound, often ± glomerate, the lowest fls. often in axils of reduced foliage lvs.; bracts mostly 2–5 mm. long; pedicels mostly 1–3 mm. long; bractlets below calyx 2–7 mm. long; calyx densely stellate, 6–10 mm. long; petals pale pinkish-lavender, 5–18 mm. long; carpels glabrous, 2–2.5 mm. long, dorsally ridged and shallowly corrugate-reticulate on sides and across back.—Dry ridges, 500–2700 ft.; Chaparral; Santa Lucia Mts. June–July.

KEY TO SUBSPECIES

Bracts linear to linear-lanceolate or oblong, from not much longer than pedicels to half the length of the calyx; bracteoles below calyx usually not much more than ⅔ the length of the calyx.
  Calyx densely stellate, grayish, the stellae of the sepal-margins much longer than those on the backs. Santa Lucia Mts. . . . . S. Hickmanii
  Calyx more sparsely stellate, the stellae uniform. Marin Co. . . . . ssp. *viridis*
Bracts broader, more nearly lanceolate, they and the calyx-bracteoles usually ca. equaling the tips of the sepals.
  Cauline lvs. divided nearly to base. San Luis Obispo Co. . . . . ssp. *anomala*
  Cauline lvs. shallowly lobed, the lobes not over ¼ the width of the lf. Santa Barbara and San Bernardino cos. . . . . ssp. *Parishii*

Ssp. **víridis** C. L. Hitchc. Lvs. greenish rather than grayish; bracts 2–5 mm. long; bractlets 2–4 mm. long.—Chaparral; Big Carson Ridge, Marin Co. June.

Ssp. **anómala** C. L. Hitchc. Upper and middle stem-lvs. lobed nearly to the petiole into 5–7 cuneate-obovate crenate-dentate segms.; infl. much compounded; calyx 10–12 mm. long; petals 12–15 mm. long.—Two and a quarter miles nw. of Cuesta, San Luis Obispo Co. May.

Ssp. **Paríshii** (Rob.) C. L. Hitchc. [*S. H.* var. *P.* Rob. *S. P.* Rob. ex Davids. & Mox.] Coarsely grayish stellate throughout; bracts 6–10 mm. long; bractlets of calyx ca. as long as calyx; lvs. shallowly lobed.—Local on dry slopes, 4500–7000 ft.; Chaparral, Yellow Pine F.; San Bernardino Mts., Mission Pine, Santa Barbara Co. June–Aug.

18. **S. malachroìdes** (H. & A.) Gray. [*Malva m.* H. & A. *Hesperalcea m.* Greene. *S.*

vitifolia Gray.] Suffrutescent perennial, 6–15 dm. high, erect, stout, leafy-branched above, stellate-hispidulous to hirsute; lvs. grapelike, 2–8(–20) cm. wide, palmately but shallowly lobed, the lobes dentate; infl. densely spicate, stellate-hirsute; bracts purplish, linear, 4–6 mm. long, simple or bifid; calyx membranous in age, ca. 5 mm. long, the lobes lance-ovate, acute; petals white, ca. 1 cm. long; carpels 7–9, half dehiscent, glabrous, smooth, ca. 3 mm. high; $2n = 20$ (Kruckeberg, 1957). Along the coast, especially in disturbed areas, below 2000 ft.; Redwood F., Mixed Evergreen F.; Monterey Co.; s. Ore. May–July.

14. Hibíscus L. Rose-Mallow

Herbs or shrubs. Lvs. palmately veined, lobed or parted. Fls. bisexual, 5-merous, mostly bell-shaped, axillary or paniculate, often large and showy. Involucel of few to several slender to broad bractlets. Calyx 5-toothed or -cleft. Corolla of 5 petals. Staminal column anther-bearing below the naked truncate 5-toothed summit. Style-branches 5. Ovary 5-celled, with 2-many ovules in each cell; fr. a 5-valved loculicidal caps. Seeds several in each locule, long-hairy. A genus of ca. 200 spp. of warmer regions around the world; many used for ornament, some for food and fibers. (Ancient Greek and Latin name for some large mallow.)

A. Plants low annuals, 1–5 dm. high, with 3–5-parted lvs. and bladdery-inflated calyces
                           1. *H. Trionum*
AA. Plants taller or woody at base, with toothed (not parted) lvs. and calyces not bladdery-inflated.
 B. Lvs. cordate, 6–10 cm. wide; corolla 6–10 cm. long .... 2. *H. californicus*
 BB. Lvs. ovate, 1–2.5 cm. wide; corolla 1–2.5 cm. long .... 3. *H. denudatus*

1. **H. Triònum** L. Flower-of-an-Hour. Hairy annual, 3–6 dm. high, some branches prostrate; lvs. 2–3 cm. wide, 3–5-parted, the divisions coarsely toothed; fls. axillary, solitary; bractlets of involucel linear; calyx inflated, setose, papery, 5-winged, with many dark nerves, becoming 2.5 cm. long in fr.; petals ca. 2 cm. long, sulphur-yellow with blackish basal spot; caps. 15–18 mm. long; seeds sparsely hairy, reniform, ca. 2 mm. long.—Casual weed in waste places and about gardens; Trinity Co., Stockton, Riverside Co., Orange Co.; native of cent. Afr. Aug.–Sept.

2. **H. califórnicus** Kell. Annual, with stout erect stems 1–2 m. tall, densely stellate-pubescent; lvs. bicolored, soft-stellate, cordate, ovate, crenate-dentate, acuminate, 6–10 cm. long, almost as wide, long-petioled; peduncles subterminal, 5–9 cm. long, jointed near middle and united at base with the lf.-petiole; bractlets of involucel 3–4 cm. long; calyx 3–4 cm. long, cleft to middle; petals white or pinkish, with deep crimson center, 6–10 cm. long; caps. 2.5–3 cm. long; seed round, glabrous, minutely papillate, ca. 2.5 mm. long.—Moist banks; Freshwater Marsh; lower Sacramento and San Joaquin rivers, Contra Costa and San Joaquin cos. to Butte Co. Aug.–Sept.

3. **H. denudàtus** Benth. Tufted suffrutescent perennial 3–7 dm. tall, yellowish with a dense stellate tomentum; lvs. ovate, 1–2.5 cm. long, serrulate, short-petioled; fls. in upper axils, short-peduncled; bractlets of involucel 2–5 mm. long, setaceous; calyx 10–14 mm. long, cleft nearly to base; petals pinkish-lavender to whitish, often with some red near base, 15–22 mm. long; caps. 5–7 mm. long, dehiscent to base; seeds dark, ca. 2.5 mm. long, surrounded by much longer silky hairs.—Rocky slopes and canyons, below 2500 ft.; Creosote Bush Scrub; Colo. Desert; to Tex., n. Mex. Feb.–May.

11. Sarraceniàceae. Pitcher-Plant Family (Fig. 15)

Insectivorous herbs of swamps and bogs, with short rootstocks. Lvs. radical, tubular or pitcher-shaped with small laminae. Fls. scapose, solitary or few in racemes, nodding, bisexual. Sepals 4–5, free, hypogynous, imbricate, often colored, persistent. Petals 5, free, imbricate, sometimes absent. Stamens many, free, hypogynous; anthers 2-celled, opening lengthwise, versatile. Pistil 1, free, 3–5-celled, the placentae axile; style simple, often peltately expanded at the apex. Fr. a loculicidal caps. Seeds many, small, reticulate, with fleshy endosperm and minute embryo. An Am. family of ca. 3 genera and 10 spp.; *Sarracenia*, the common pitcher-plant of e. N. Am., having 8.

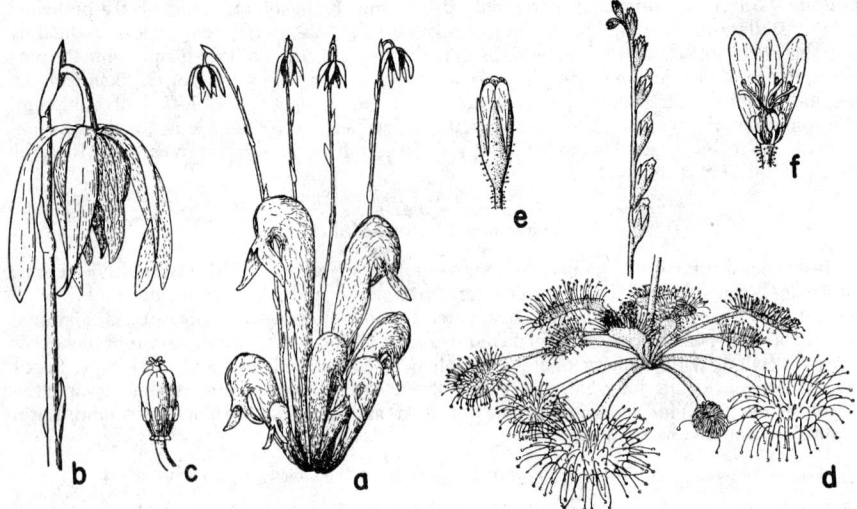

Fig. 15. SARRACENIACEAE. *Darlingtonia californica: a,* habit, × ⅛, showing arched pitcherlike or tubular lvs. and fl.-stalks; *b,* fl., × ½, with 5 sepals, 5 petals; *c,* fl. with perianth removed to show 12 stamens and 5-loculed pistil.

DROSERACEAE. *Drosera rotundifolia: d,* habit, × 1, with basal insectivorous lvs. and scapose infl.; *e,* 5-merous fl., × 2½; *f,* fl. in section with petals, stamens and 4 2-parted styles.

1. Darlingtònia Torr. CALIFORNIA PITCHER-PLANT

Sepals 5. Petals 5. Stamens 12–15, in 1 series; fils. subulate; anthers with uneven sacs and twisted so that the smaller sac is next to the ovary. Seeds broadly clavate, covered with soft spinelike processes. One sp. (Wm. *Darlington,* 1782–1863, Am. botanist.)

1. **D. califórnica** Torr. [*Chrysamphora c.* Greene.] Lvs. greenish-yellow, 2–6 dm. high, conspicuously veined, gradually enlarged upward into a rounded dotted hood with opening underneath the hood and an apical 2-lobed appendage; scapes equaling or exceeding lvs., with several yellow scales and 1 fl.; sepals yellow-green with purplish lines, oblong to oblanceolate, 4–6 cm. long; petals dark purple, veined, 2–4 cm. long; stamens ca. 6 mm. long; caps. obovoid, 2.5–4.5 cm. long; seeds light reddish-brown, ca. 2 mm. long; $n = 15$ (Bell, 1949).—Marshy and boggy places, 300–6000 ft.; Redwood F., Douglas-Fir F., Yellow Pine F., Red Fir F.; Del Norte, Trinity, Siskiyou, Plumas and Nevada cos.; Ore. April–June.

12. Droseràceae. SUNDEW FAMILY (Fig. 15)

Insectivorous herbs of bogs or marshes. Roots weak. Lvs. usually in a basal rosette, circinate in the bud, mostly bearing tentacular gland-tipped sensitive hairs. Fls. perfect, hypogynous. Sepals persistent. Petals free, delicate, fugacious. Styles 2–5, with parietal placentation and numerous anatropous ovules. Caps. loculicidal. Seeds with fleshy endosperm, straight embryo, and short cotyledons.—Four genera, 3 of which are monotypic: *Dionaea,* the Venus' Fly-trap of N. Carolina; *Drosophyllum,* of the Hibernian Peninsula and adjacent Morocco; *Aldrovanda,* an aquatic restricted to the Old World; and *Drosera,* a rather large genus distributed to all continents.

1. Drósera L. SUNDEW. DEW PLANT

Mostly perennial herbs of brownish or reddish hue, the sensitive hairs of the basal lvs. secreting a glutinous fluid by which insects are entrapped. Fls. 5-merous, rarely 4- or

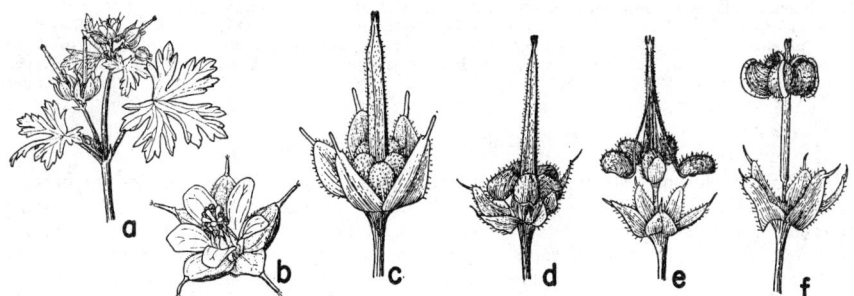

Fig. 16. GERANIACEAE. *Geranium: a,* tip of branch with lvs. and fl.-clusters, × ¾; *b,* fl., × 2½, with 5 sepals, 5 petals, 10 stamens, and 5 styles; *c–f,* a series to show the 5 carpels separating and coiling away from the cent. axis, × 2.

8-merous, but the pistils 3–5, borne in scorpioid cymes on a scape (in ours). Anthers extrorse, versatile. Ovary 1-celled. Styles 2–5 (usually 3), 2- to 5-partite. Caps. usually 3-valved. Ca. 85 spp., more than 50 of them in Australia and New Zealand. (Greek, *droseros,* dewy, from the dewlike glands tipping the hairs of the lvs.)

Lf.-blades suborbicular, broader than long; seeds finely striate .... 1. *D. rotundifolia*
Lf.-blades elongate-spatulate; seeds areolate-striate .... 2. *D. anglica*

1. **D. rotundifòlia** L. Lvs. in a spreading basal rosette, the petiole 1.5–5 cm. long, flat, pilose along the upper surface, the blade 4–10 mm. long, broader than long and abruptly tapering to the petiole, the upper surface clothed with tentacular gland-tipped reddish hairs that are longest on the margin, the stipules 4–6 mm. long, adnate, membranous, fimbriate; scape glabrous, rarely forked, 5–25 cm. high, 2–15-fld., the short-pedicelled fls. not in the axils of the minute bracts; sepals oblong, united at base, shorter than the 5–6 mm. long white or pinkish ephemeral petals; seeds fusiform, 1–1.5 mm. long, the testa loose; *n* = 10 (Rosenberg, 1899).—In cold swamps and sphagnum bogs, often terrestrial, below 8000 ft., sometimes associated with *Darlingtonia,* Montane Coniferous F., intermittent through the Sierra Nevada, Tulare Co., n. to Mt. Shasta; near sea level to 4000 ft., Mixed Evergreen F., Redwood F., Coast Ranges, Sonoma Co. to Del Norte Co.; to Alaska, Mont., Great Lakes to the Atlantic; Eurasia. July–Aug.

2. **D. ánglica** Huds. [*D. longifolia* of w. manuals.] Lvs. erect in the rosette, the petiole 3–7 cm. long, glabrous or minutely glandular, the blade 15–35 mm. long, 3–5 mm. wide, gradually narrowed to the petiole, the tentacular hairs and stipules as in *rotundifolia;* scape 1–8-fld., otherwise it and the fls. as in *rotundifolia; n* = 20 (Rosenberg, 1903).—Rare, in cold bogs, preferring wetter places than *rotundifolia,* at 4300–6000 ft.; Yellow Pine F.; Plumas and Sierra Cos.; to Alaska, Yellowstone Park, also Great Lakes and e. Canada; more abundant in n. Eurasia; Hawaiian Ids. July–Aug.

The hybrid between *D. rotundifolia* and *D. anglica* has been found several times in bogs where the two grow together. It is of particular interest in biological literature because it was the first interspecific hybrid to be studied cytologically (O. Rosenberg, 1903).

13. Geraniàceae. GERANIUM FAMILY (Fig. 16)

Annual or perennial herbs or somewhat woody. Lvs. opposite or alternate, simple or compound, ours with stipules. Fls. bisexual, 5-merous, hypogynous. Sepals imbricate in bud, persistent. Petals usually imbricate. Disk often with 5 glands alternating with the petals. Stamens as many, or 2 or 3 times as many, as the petals, some of them frequently sterile; fils. tending to be basally somewhat connate. Ovary deeply lobed, the carpels 2-ovuled, 1-seeded, with axile placentation. Styles long, adnate to the elongate axis, separating and coiling when mature. Eleven genera and ca. 650 spp. of temp. and subtrop. regions.

Fls. without nectar-spur, and with glands alternating with the petals; petals alike.
    Lvs. palmately veined or divided; stamens all bearing anthers .... 1. *Geranium*
    Lvs. pinnately veined or divided; stamens having outer fils. without anthers .... 2. *Erodium*
Fls. with nectar-spur at base of calyx (discovered by sectioning pedicel) and without glands; upper
    petals usually larger .... 3. *Pelargonium*

1. Geràníum L. CRANESBILL

Herbs with forking stems and swollen nodes. Lvs. palmately lobed or parted. Peduncles 1–3(–7)-fld., axillary, often in subcymose arrangement. Fls. regular, bisexual, 5-merous. Sepals imbricated. Petals hypogynous. Stamens 10 (rarely 5); fils. sometimes somewhat connate at base; anthers all well formed, the 5 longer alternate with the petals and with basal glands. Style-column usually beaked, the styles in fr. nearly glabrous inside. Carpels turgid, permanently attached to the styles. Seeds smooth or pitted. Ca. 250 spp. of temp. regions around the world, some weedy annuals, a few of the perennials cult. (Greek, *geranos*, a crane, from the beaklike fr.)

(Fernald, M. L. G. carolinianum and allies of northeastern N. Am. Rhodora 37: 295–301, 1935.
Jones, G. N., and F. F. A revision of the perennial species of Geranium of the U.S. and Canada.
Rhodora 45: 5–26, 32–53, 1943.)
A. Petals 3–8(–10) mm. long; mature style-column less than 2 cm. long; annuals or perennials,
    mostly introd.
    B. Sepals prominently awned or subulate-tipped; seeds reticulate.
        C. Plants perennial from a carrotlike root; no hairs with gland-tips.
            D. Stems with retrorse-spreading stiff shining hairs; petals scarcely longer than sepals;
               seeds coarsely reticulate .... 1. *G. pilosum*
           DD. Stems with retrorse appressed dull hairs; petals ca. twice as long as sepals; seeds
               finely reticulate .... 2. *G. retrorsum*
        CC. Plants annual; some hairs of pedicels gland-tipped.
            D. Fruiting pedicels much longer than calyx; beak of mature fr. 3–6 mm. long
                                                        3. *G. Bicknellii*
           DD. Fruiting pedicels scarcely or not longer than calyx; beak of mature fr. 1–2 mm. long.
              E. Carpel-bodies with short stiff subequal spreading hairs; lobes of upper lvs.
                 acute; pits of seed square or rounded .... 4. *G. dissectum*
              EE. Carpel-bodies with long ± unequal ascending hairs; lobes of lvs. obtuse; pits
                 of seed ± elongate .... 5. *G. carolinianum*
    BB. Sepals awnless, at most with callous tips; seeds smooth or minutely granular.
        C. Carpel-bodies finely pubescent, not cross-wrinkled; style-column beakless.
            D. Petals 7–8 mm. long; plant perennial; carpel-bodies puberulent; seeds reticulate
                                           6. *G. pyrenaicum*
            DD. Petals 4–5 mm. long; plant annual or biennial; carpel-bodies strigose-pubescent;
               seeds smooth .... 7. *G. pusillum*
         CC. Carpel-bodies glabrous, mostly conspicuously cross-wrinkled; style-column with filiform
            beak 1–2 mm. long ....8. *G. molle*
AA. Petals 10–20 mm. long; mature style-column 2–5 cm. long; native perennials.
    B. Petals glabrous except at the ciliate base; free style-branches 1–2.5 mm. long; pedicels erect
        in fr. Nw. Calif. .... 9. *G. oreganum*
    BB. Petals pilose on inner surface for ¼–½ their length; free style-branches 3–9 mm. long;
        fruiting pedicels spreading or reflexed and then bent upward.
        C. The petals pilose ¼–⅓ their length. N. Calif.
            D. Petioles of basal lvs. and the lower internode of stem copiously glandular-villous and
               -pubescent. Siskiyou Co. .... 10. *G. viscosissimum*
            DD. Petioles of basal lvs. and the lower internode strigose or subglabrous, not glandular.
               Modoc Co. .... 11. *G. nervosum*
         CC. The petals pilose about half their length. Mostly cent. and s. Calif. (except no. 12).
            D. Lf.-lobes abruptly attenuate, the caudate tips 4–6 mm. long; style-branches 5 mm.
               long; pedicels with yellowish glands at tips of hairs. Modoc Co.
                                            12. *G. attenuilobum*
            DD. Lf.-lobes acute to acuminate or obtusish, not caudate-attenuate.
               E. Style-branches 3–4 mm. long; petals white or whitish; pedicels usually with
                 purplish glands. Sierra Nevada to s. Calif. .... 13. *G. Richardsonii*
              EE. Style-branches 4.5–9 mm. long; petals pink or lavender.
                   F. Stems subglabrous to finely retrorse-pubescent; style-branches 4.5–5.5
                     mm. long; pedicels with short gland-tipped hairs .... 14. *G. concinnum*
                 FF. Stems villous; style-branches 6–9 mm. long; pedicels glandular-villosulous
                     and often with longer glandless hairs .... 15. *G. californicum*

    1. **G. pilòsum** Forst. f. Perennial, with a thick taproot, caudex branched; stems slender, branched, 3–5 dm. long, with retrorse-spreading hispid translucent shining hairs; lvs. 1.5–3.5 cm. wide, incisely 3–5-parted, the cuneate divisions again obtusely 3–5-lobed, hairy above and beneath; petioles 2–6 cm. long (or the lower longer), with

retrorse-spreading hairs; peduncles usually 2-fld., 0.5–3(–5) cm. long; pedicels ca. 1–2 cm. long; sepals awn-tipped, ca. 4–5 mm. long; petals broad-obovate, bright reddish-lilac, 6–7 mm. long; style-column ca. 15 mm. long, minutely pubescent, not beaked; carpel-bodies dark, sparsely hairy, ca. 3 mm. long; seeds coarsely reticulate.—Occasionally natur. in moist partly shaded places at low elevs. near the coast; N. Coastal Scrub, Mixed Evergreen F.; Humboldt Co. to Alameda Co.; native of Australasia. May–June.

2. **G. retrórsum** L'Hér. [*G. pilosum* var. *r.* Jeps.] Like *G. pilosum* but with hairs of stems and whole plant appressed, retrorse, whitish, dull; peduncles 3–30 mm. long; pedicels 4–16 mm. long; sepals 3–4 mm. long; petals 5–6 mm. long; style-column ca. 1.5 cm. long; carpel-bodies ca. 2.5 mm. long, strigose; seeds finely reticulate.—Occasionally natur. in grassy places along roads or edge of brush, near the coast; N. Coastal Scrub, Mixed Evergreen F.; Humboldt Co. to Ventura Co., Palomar Mts.; native of Australasia. May–Sept.

3. **G. Bicknéllii** Britton var. **lóngipes** (Wats.) Fern. [*G. carolinianum* var. *l.* Wats. *G. l.* Goodd.] Annual, with divaricate slender branches, 1–5 dm. high, with short stiff white hairs and in the upper parts also gland-tipped hairs; lvs. mostly 2–5 cm. wide, 5-parted, the cuneate-obovate divisions cleft into oblong acute lobes, stiff-strigose; petioles slender and except for the uppermost much longer than the blades; peduncles 3–10 cm. long, slender, mostly 2-fld.; pedicels filiform, 1.5–3 cm. long; sepals 5–6 mm. long, slender-awned; petals pale purple, 5–7 mm. long; style-column ca. 15–18 mm. long, including the 3–6 mm. long beak; carpel-bodies 2.5–3 mm. long with loosely appressed stiff hairs; seeds dark, oblong, rather finely reticulate.—Open and disturbed places, 2000–4000 ft.; Mixed Evergreen F., Yellow Pine F.; Marin Co. to Humboldt and Siskiyou cos.; to B.C., Ont., Colo. July–Aug.

4. **G. disséctum** L. [*G. laxum* Hanks.] Annual, usually freely branched, 3–6 dm. high, retrorsely pubescent; lvs. 3–6 cm. broad, deeply 5-parted, the divisions again divided into broadly linear acute segms., stiff-pubescent especially along the veins beneath; petioles very long, except on uppermost lvs.; peduncles mostly 2-fld., 5–50 mm. long; pedicels mostly 8–15 mm. in fr., glandular-pubescent; sepals awn-tipped, 5–7 mm. long; petals rose-purple, equaling sepals; style-column ca. 12 mm. long including the 2 mm. beak; carpel-bodies 2–2.5 mm. long, the hairs subequal, stiff, spreading; seeds subglobose, brown, strongly reticulate, with subuniform, square or rounded, unequally thick-walled pits; $2n = 22$ (Löve & Löve, 1944).—Waste and open places or borders of brush, mostly below 3000 ft.; widely distributed and in many Plant Communities; to Atlantic Coast. Natur. from Eu. March–June.

5. **G. caroliniànum** L. Annual, branched, erect or ascending, 2–4 dm. high, densely retrorse-hirsute with short subuniform hairs; lvs. 3–6 cm. broad, 5–7-parted, the cuneate divisions ± cleft into linear or oblong mostly obtuse lobes; lower petioles very long, upper short; peduncles mostly 2-fld., 1–3 cm. long, solitary or loosely aggregated into terminal corymbs; pedicels mostly 2–7 mm. long, with some gland-tipped hairs among the others; sepals 5–7 mm. long, 3-veined, awn-tipped; petals pink or paler, ca. as long as sepals; style-column 10–14 mm. long including the 2 mm. beak; carpel-bodies 2.5–3 mm. long, with fine short hairs and some long ascending ones 1 mm. long; seeds oblong, reticulate, with 25–35 rows of somewhat elongate areolae; $n = 26$ (Shaw, 1952).—Frequent in grassy and shaded places, below 5000 ft.; many Plant Communities; from the coast to the Sierra Nevada and the length of the state; to B.C. and Atlantic Coast. April–July. Plants from Berry Canyon and Little Chico (Butte Co.) seem to approach *G. sphaerospermum* Fern. in their broad 5-veined sepals and rounder seeds with more numerous rows of pits.

6. **G. pyrenàicum** Burm. f. Perennial, with short scaly caudex; stems retrorsely villous; lvs. 5–7-cleft, the divisions with 3–5 broad obtuse oblong lobes; peduncles 1–4 cm. long; pedicels glandular-puberulent, 2–3 cm. long; sepals callous-tipped, 5–8 mm. long; petals purple, 8–15 mm. long; style-column ca. 12–14 mm. long, scarcely beaked; carpel-bodies ca. 3 mm. long, puberulent, not wrinkled; seeds minutely granular; $2n = 28$ (Warburg, 1938).—Reported from San Bernardino Mts.; native of Eu. June–July.

7. **G. pusíllum** Burm. f. Annual; stems branched, 1.5–5 dm. long, decumbent or prostrate, puberulent; lvs. 1.5–5 cm. broad, 5–7 parted, the divisions toothed or lobed into obtuse segms.; peduncles 2-fld., 5–15 mm. long; pedicels very slender, glandular-

puberulent, ca. as long as peduncles; sepals not awned, ca. 3–4 mm. long; petals pinkish to violet, 4–5 mm. long; style-column 8–9 mm. long, scarcely beaked; carpel-bodies 2 mm. long, strigose, not wrinkled; seeds smooth, brown, ca. 1 mm. long; $n = 13$ (Shaw, 1952).—Occasional weed, as about lawns and waste places, n. coastal Calif., Susanville, Lassen Co.; to B.C. and Atlantic Coast; native of Eu. June–Sept.

8. **G. mólle** L. Annual, or biennial and with thickened root; stems branched, decumbent, 1–4 dm. long, villous-pubescent; lvs. 2–6 cm. broad, mostly 5–7-cleft the broad cuneate segms. toothed or lobed; petioles pilose; peduncles very slender, 2-fld., 1–3 cm. long; pedicels 5–20 mm. long, filiform, glandular-pubescent; sepals 3–4 mm. long, awnless; petals rose-purple, 3–5 mm. long; style-column 8–12 mm. long including the filiform 1–2 mm. beak; carpel-bodies glabrous, usually conspicuously cross-wrinkled, ca. 2 mm. long; seeds smooth, brown, oblong, 1 mm. long; $n = 13$ (Warburg, 1938). —Occasional in grassy, brushy or wooded places or about lawns, over most of the coastal part of state; to B.C. and Atlantic Coast; native of Eu. May–July.

9. **G. oregànum** Howell. [*G. incisum* Nutt., not Andr.] Perennial; stem 1, glabrous to pilose and ± glandular, 4–8 dm. high; lower petioles glabrous to retrorsely pubescent, 1.5–4 dm. long; blades 10–12 cm. wide, sparsely strigose, 5–7-lobed, the divisions cuneate and deeply irregularly toothed; peduncles 1–15 cm. long, these and pedicels glandular-villous with purple-tipped hairs; pedicels 2, erect in fr., 1–3 cm. long; sepals 7–11 mm. long, awned; petals deep rose-purple, glabrous except at the ciliate base, 13–17 mm. long; style-column 3–5 cm. long including the 3–4 mm. beak; style-branches 2 mm. long; carpel-bodies 6–8 mm. long, hispidulous to glandular-pubescent; seeds reticulate, 3–3.5 mm. long.—Meadows, thickets and woods, at 2500–5000 ft.; largely Yellow Pine F.; Humboldt and Siskiyou cos.; to Wash. June–July.

10. **G. viscosíssimum** F. & M. [*G. incisum* in part, auth., not Andr.] Perennial with simple caudex and 1 to few stems, 3–8 dm. tall, the lower internodes glandular-villous or pilose, and usually also short glandular-pubescent; lower petioles 1–3 dm. long, glandular-pubescent; blades 6–12 cm. wide, pubescent, 5–7-parted, the rhombic to obovate segms. deeply incised with lance-oblong acute to acuminate lobes; upper lvs. smaller; peduncles 1.5–6 cm. long, these and the pedicels and calyces densely glandular-villous; pedicels in fr. 1.5–3 cm. long, reflexed and bent upward; sepals 8–12 mm. long, awn-tipped; petals pink to rose-purple, dark-veined, 1.2–2 cm. long, pilose ca. ⅓ their length; style-column 2.5–3 cm. long, including the 4–5 mm. beak; style-branches 4–5 mm. long; carpel-bodies 5–6 mm. long, glandular-pubescent, with hair-tuft at base, glabrate at apex; seeds reticulate, 3–4 mm. long.—Open woods and meadows; Yellow Pine F.; Siskiyou Co. (Quartz V.); to B.C. and Sask. May–July.

11. **G. nervòsum** Rydb. [*G. strigosius* St. John. *G. incisum* auth., in part. *G. strigosum* Rydb., not Burm. f.] Perennial; stems mostly solitary, 3–7 dm. high, the lower internode mostly strigose or retrorse-pubescent; lower petioles 0.5–2.5 dm. long, puberulent to strigose; blades 3–10 cm. wide, strigulose or finely pilose, deeply 5–7-parted, the divisions obovate to rhombic with incised lobes and acute tips; cauline lvs. smaller; peduncles 0.5–3 cm. long, minutely pilose or glandular; pedicels 2(–3–4), glandular-pubescent, 1–5 cm. long, in fr. reflexed and bent upward; sepals 8–10 mm. long, awned; petals 1.5–2 cm. long, rose-purple, veined, pilose usually less than ⅓ their length; style-column 3–3.5 cm. long including the 4 mm. beak; style-branches 4–5 mm. long; carpel-bodies 6 mm. long, puberulent to glandular-pubescent; seeds reticulate, 3.5–4 mm. long; $n = 26$ (Shaw, 1952).—Grassy places, creek-bottoms, stream-banks, etc., 5000–6000 ft.; Yellow Pine F., N. Juniper Wd.; Modoc Co.; to B.C., Alta. May–July.

12. **G. attenuilòbum** Jones & Jones. Perennial; stem glandular-puberulent; cauline lvs. 5-parted, 2.5–6 cm. broad, pilosulous above, paler and glandular along the veins beneath, the ultimate lobes abruptly attenuate with caudate tips 4–6 mm. long; peduncles 3–9 cm. long, glandular-puberulent; pedicels 1–5 cm. long; sepals 8–11 mm. long, the awns 2.5–3 mm. long; petals 12–16 mm. long, pilose ½ their length; style-branches 5 mm. long.—At 5000–6000 ft.; Yellow Pine F.; Warner Mts., Modoc Co. June–July. An uncertain entity based on inadequate material.

13. **G. Richardsònii** Fisch. & Trautv. Perennial; stems 1 to few, 3–9 dm. high, glabrous or sparsely pubescent; lower petioles 0.5–3 dm. long, glabrous or sparsely retrorse-pubescent; blades 3–15 cm. broad, 5–7-parted, the segms. rhombic, sparsely strigose above and on veins beneath, with acute to acuminate lobes; upper lvs. reduced; peduncles 2–12 cm. long, they and the pedicels glandular-pilose, the hairs tipped with

purplish glands; pedicels paired, slender, 1–2 cm. long, reflexed and bent upward in fr.; sepals 6–12 mm. long, awn-tipped; petals 10–18 mm. long, white or pinkish, with purple veins, pilose inside ca. ½ their length; style-column 2–2.5 cm. long including the 2 mm. beak; style-branches 3–5 mm. long; carpel-bodies 3–4 mm. long, sparingly pubescent and with some longer stiff hairs; seed coarsely reticulate, 2.5–3.5 mm. long; *n* = 26 (Shaw, 1952).—Moist places as about meadows, 4000–9000 ft.; Yellow Pine F., Red Fir F., Lodgepole F.; Sierra Nevada (from Nevada Co. s.) to San Jacinto Mts.; to e. B.C., Sask., New Mex. July–Aug.

14. **G. concínnum** Jones & Jones. Perennial; stems 1–few, slender, 1–4 dm. tall, retrorse-pubescent to subglabrous; basal petioles 1–2 dm. long, retrorse-pubescent, the blades 2–6 cm. broad, 5–7-parted, the segms. rhombic, strigulose above and beneath, deeply incised with lanceolate lobes; peduncles 5–11 cm. long, finely glandular-puberulent; pedicels 2.5–11 cm. long, in fr. erect or ascending, glandular-pubescent, the glands yellowish; sepals 6–8 mm. long, awned; petals lavender to pink, 10–15 mm. long, pilose ½–¾ their length; style-column 2–2.5 cm. long including the 3–5 mm. beak; style-branches 4–5.5 mm. long; carpel-bodies 4–5 mm. long, pubescent; seeds minutely reticulate, ca. 3 mm. long.—Moist places, 7000–8000 ft.; Yellow Pine F., Red Fir F.; mts. from Tulare Co. to San Bernardino Co. June–Aug.

15. **G. califórnicum** Jones & Jones. [*G. leucanthum* Small, not Griseb. *G. caespitosum* auth., not James.] Perennial; stems 1–3, sparsely villous, 2–6 dm. high, glandular-pilose above; lower petioles 5–25 cm. long, sparsely villous, the blades 3–8 cm. wide, usually 5-parted, the divisions rhombic or cuneate, strigulose above, soft-hirsute beneath especially along the veins, with acute ultimate lobes; peduncles 2–8 cm. long; pedicels 2–3, mostly 3–12 cm. long, glandular-villous and with longer eglandular hairs; sepals 6–9 mm. long, awned; petals rose-pink to white, dark-veined, 12–16 mm. long, pilose ca. half their length; style-column 2–2.5 cm. long, including the 3–4 mm. beak; style-branches 6–9 mm. long; carpel-bodies 4–5 mm. long, sparsely glandular-hairy; seeds faintly reticulate, ca. 3 mm. long.—Damp woods and meadows, 4000–8000 ft.; Yellow Pine F. to Lodgepole F.; Tuolumne Co. to San Diego Co. June–July.

2. Eròdium L'Hér. STORKSBILL. FILAREE. CLOCKS

Ours annual herbs. Lvs. usually at first forming close rosette on ground, pinnately veined, simple to pinnate, opposite, with 1 interpetiolar stipule on one side and 2 on the other. Pedicels commonly retrocurved in fr. Sepals 5. Petals 5. Stamens 5, alternating with 5 scalelike staminodia. Style-column very elongate, the styles bearded inside, spirally coiled when freed from the central axis. Carpel-bodies narrow, spindle-shaped, indehiscent. Ca. 60 spp., widespread in temp. and semitrop. regions, some of importance for forage. (Greek, *erodios*, a heron, because of the long beak on the fr.)

(Dayton, W. A. Alfileria seed. Rhodora 39: 233–235, 1937. Knuth, R. Geraniaceae *in* Engler, A., Das Pflanzenreich, IV, 129: 221–290, 1912.)

A. Lvs. ± cordate at base, usually 3–more-lobed.
  B. Petals unequal, much exceeding sepals; pedicels strigulose, not glandular. Deserts
                                                                   1. *E. texanum*
  BB. Petals equal, rarely longer than sepals. Cismontane.
    C. Style-column 2–2.5 cm. long; sepals ca. 5 mm. long; pedicels ± glandular.
                                                    2. *E. malacoides*
    CC. Style-column 4–6 cm. long; sepals 6–10 mm. long.
      D. Lvs. shallowly 5–7-lobed; pedicels glandular-pilose; carpel-bodies 8–10 mm. long .... 3. *E. macrophyllum*
      DD. Lvs. deeply 3–5-parted; pedicels retrorse-pubescent, not glandular; carpel-bodies 5–6 mm. long .... 4. *E. cygnorum*
AA. Lvs. not cordate at base.
  B. The lvs. simple, often lobed or divided; style-column 5–12 cm. long.
    C. Style-column 5–9 cm. long; concavities at top of fr. subtended by a single fold, the upper part of the carpel-body ± pubescent; sepals with a short green tip
                                            5. *E. obtusiplicatum*
    CC. Style-column 9–12 cm. long; concavities at top of fr. subtended by 2 folds, the upper part of the carpel-body glabrous; sepals with a prominent reddish mucro .. 6. *E. Botrys*
  BB. The lvs. pinnate; style-column 2.5–4 cm. long.
    C. Lfts. broad, coarsely toothed or serrate; sepal-tips not setose; claws of petals glabrous
                                               7. *E. moschatum*
    CC. Lfts. pinnately lobed or divided; sepal-tips setose; claws of petals ciliate
                                                 8. *E. cicutarium*

1. **E. texànum** Gray. Stems few to several, decumbent, 1–4 dm. long, puberulent, not glandular, the peduncles and pedicels canescent-strigulose; lf.-blades ovate to rounded in outline, 1–3 cm. long, cordate, with 3 rounded lobes; stipules lanceolate to deltoid; peduncles 1–3-fld., mostly 1–3 cm. long; pedicels 3–10 mm. long; sepals 5–8 mm. long, silvery with purple veins, mucronate; petals purple, twice as long as sepals; style-column 4–6.5 cm. long; carpel-bodies 7–8 mm. long, sparsely pubescent, not truncate at apex, the spirally coiled part with ca. 4 turns when mature; $2n = 20$ (Baker, 1954).—Dry sandy or gravelly places, below 3500 ft.; Creosote Bush Scrub; Colo. and e. Mojave deserts, Jurupa Hills, near Riverside; to Tex., L. Calif. March–May.

2. **E. malacoìdes** (L.) Willd. [*Geranium m.* L.] Stems 1–4 dm. long, pilose; lf.-blades cordate-ovate, 2–4 cm. long, pilose, entire to shallowly lobed, crenate to dentate; stipules lanceolate; peduncles 2–8 cm. long, 2–8-fld., densely glandular-pilose; sepals ca. 4–5 mm. long; petals ca. as long; style-column 2–3 cm. long; carpel-bodies small, short-hirsute, the apical concavities subtended by a single fold.—Reported from Berkeley, Mt. Diablo, and the Sacramento V.; native of Medit. region.

3. **E. macrophýllum** H. & A. Stem proper very short, the lvs. and peduncles sub-basal, the latter 1–3 dm. high; plant puberulent and ± glandular-puberulent especially above; lf.-blades reniform-cordate, 2–5 cm. wide, crenate, usually shallowly lobed, the petioles 3–12 cm. long; stipules ovate; peduncles 2–3(–6)-fld.; pedicels 5–20 mm. long; sepals 8–10 mm. long; petals white or sometimes rose-red or purple, 10–16 mm. long; style-column 4–5 cm. long; carpel-bodies truncate, 8–10 mm. long, pubescent, the spirally coiled part with 2–3 turns.—Open fields and grassy places, below 3500 ft.; V. Grassland, Foothill Wd.; inner Coast Ranges from Tehama Co. to Kern Co., nearer the coast s. to San Dimas, Santa Monica Mts., Santa Cruz Id.; L. Calif. March–May. Plants from S. Coast Ranges, with more glandular pubescence, 5–6-fld. umbels, and red petals have been designated as var. *califórnicum* (Greene) Jeps. [*E. c.* Greene.]

4. **E. cygnòrum** Nees. Stems 1–3 dm. long, decumbent to ascending, pilose; lf.-blades ovate in outline, 2–4 cm. long, 3–5-parted into cuneate lobed divisions; stipules ovate; peduncles 1–10 cm. long, commonly 3–5-fld.; pedicels filiform, 1–2.5 cm. long, retrorse-pubescent; sepals 6–7 mm. long, mucronate; petals 7–8 mm. long, blue; style-column 5–6 cm. long; carpel-bodies stiff-pubescent, 5–6 mm. long, the apical concavities subtended by a single fold.—Reported from San Diego, Corona; native of Australia.

5. **E. obtusiplicàtum** (Maire, Weiller & Wilcz.) J. T. Howell. [*E. Botrys* var. *o.* Maire, Weiller & Wilcz. *E. Botrys* f. *montanum* Brumhard.] Stems ascending to erect, 1–4 dm. long, hirsute and somewhat glandular; lf.-blades oblong-ovate, 2–8 cm. long, the lower with 3–7 rounded shallow lobes, the upper more deeply pinnatifid, pilose especially on veins and margins; stipules ovate; peduncles 1–10 cm. long, glandular-pubescent; pedicels 1–5, glandular-pubescent, 0.5–1.5 cm. long; sepals 7–8 mm. long, to 11 mm. in fr., scarcely mucronate; petals lavender, 8–11 mm. long; style-column 5.5–8.5 cm. long, the spirally coiled part with numerous turns; carpel-bodies 6–8 mm. long, with short stiff spreading hairs, the apex pubescent and its concavities surrounded by one fold; $2n = 40$ (Baker, 1954).—Locally common in open grassy places, at low elevs. especially in V. Grassland and Foothill Wd., also in other communities; Humboldt Co. to San Diego Co.; Ore.; natur. from N. Afr. April–Aug.

6. **E. Bòtrys** (Cav.) Bertol. [*Geranium B.* Cav.] Stems semiprostrate to suberect, 1–9 dm. long, retrocurved-hirsute; lf.-blades ovate to oblong-ovate, 3–8 cm. long, shallowly to deeply lobed or pinnatifid, setose-pilose on veins and margins; stipules ovate; peduncles 2–20 cm. long, glandular-pubescent; pedicels 1–4, glandular-pubescent, 1.5–2.5 cm. long; sepals 7–8 mm. long, 13–15 mm. in fr., with prominent reddish mucro; petals ca. 1.5 cm. long, lavender; style-column 9.5–12.5 cm. long, the spirally coiled part with numerous turns; carpel-bodies 8–10 mm. long, with short stiff spreading hairs, the apex subglabrous and its concavities surrounded by 2 folds; $n = 20$ (Heiser & Whitaker, 1948).—Grassy places at low elevs. almost throughout Calif. except in the desert; with, but often less common than, *E. obtusiplicatum;* native of Medit. region. March–May.

7. **E. moschàtum** (L.) L'Hér. [*Geranium m.* L.] Stems rather fleshy, decumbent to ascending, 1–6 dm. long, glandular-pubescent; lvs. 6–40 cm. long (including petioles),

pinnate, the lfts. ovate to oblong-ovate, 1–3 cm. long, serrate to cleft, the terminal 3–5-parted; stipules rounded-ovate; peduncles 6–20 cm. long, glandular-pubescent; pedicels 6–13, glandular-pubescent and with subappressed nonglandular hairs, ca. 6–20 mm. long; sepals 6–7 mm. long, mucronate, but mostly without terminal setae; petals rose-violet, not spotted, somewhat longer; style-column 2–4 cm. long, the coiled part: with several turns; carpel-bodies 4–5 mm. long, stiff-pubescent, the apical concavities glabrous, oblong, subtended by a concentric fold; $2n = 20$ (Gauger, 1937).—Common, especially in loams and heavy soils at low elevs., throughout cismontane Calif.; natur. from Medit. region. Feb.–May.

✻ 8. **E. cicutàrium** (L.) L'Hér. [*Geranium c.* L.] Stems slender, decumbent, 1–5 dm. long, strigulose and glandular-pubescent; lvs. commonly 3–10 cm. long, pinnate, the lfts. incisely pinnatifid; stipules lanceolate; peduncles 5–15 cm. long, glandular-pubescent, slender; pedicels 2–10, glandular-pubescent, 8–18 mm. long; sepals 3–5 mm. long, short-mucronate and with 1–2 white bristles; petals rose-lavender, 5–7 mm. long, ciliate at the base, 2-spotted; style-column 2–4 cm. long, the coiled parts with several turns; carpel-bodies 4–5 mm. long, stiff-pubescent, the apical concavities glabrous, circular, without a subtending fold; $2n = 40$ (Andreas, 1947).—Common everywhere in Calif., in open cult. and dry places below 6000 ft.; natur. very early from the Medit. region. Feb.–May.

3. Pelargònium L'Hér. GERANIUM

Annual or perennial herbs, sometimes woody, often succulent, often strong-smelling. Lvs. alternate or opposite, palmately or pinnately veined, lobed or dissected. Peduncles axillary, umbellately 2–many-fld. Fls. of many colors, irregular. Sepals 5, imbricate, the posterior with a nectar-spur that is joined to the pedicel for much of the latter's length. Petals 5, the 2 upper mostly larger and more prominently colored. Stamens 10, somewhat connate at the base, part of them without anthers. Fr. of 5 carpels, the style-column elongate, the styles pubescent within and spirally coiled when free from the cent. axis. Nearly 250 spp., mostly from S. Afr., many grown for ornament and their contained oils. (Greek, *pelargos*, stork, from the bill-shaped fr.)

A number of spp. have been reported as escapes from cult. in Calif., and apparently mostly do not persist long. They are to be sought along the coast and in waste places and about city dumps. The following key is presented to help in identifying such waifs. Many of the cult. plants are much modified from the original wild spp., hence difficult to determine.

A. Plant weak and trailing or climbing; lvs. quite smooth and shining . . . . 1. *P. peltatum*
AA. Plant erect or sometimes spreading; lvs. variously pubescent or hairy.
    B. Stems fleshy, at least when young; lvs. alternate, scarcely if at all lobed.
        C. Lvs. uniformly green, without color zone inside the margin; sepals 7–8 mm. long;
             pedicels proper (below spur) and calyx copiously hirsute . . . . 2. *P. inquinans*
        CC. Lvs. with a broad color zone inside the margin; sepals 9–11 mm. long; pedicels and
             calyx glandular-pubescent.
            D. Stems usually solitary, 2–4 dm. high; stipules pointed, longer than broad
                                                     3. *P. zonale*
            DD. Stems several, usually much higher; stipules rounded, broader than long
                                                   4. *P. hortorum*
    BB. Stems not fleshy.
        C. Plants annual or perennial herbs; petals 3–5 mm. long; lvs. sometimes opposite, shal-
            lowly lobed.
            D. Calyx-spur 1–1.5 mm. long; stems spreading-pubescent . . . . 5. *P. inodorum*
            DD. Calyx-spur 4–15 mm. long; stems almost glabrous . . . . 6. *P. grossularioides*
        CC. Plants shrubby; petals 10–40 mm. long; lvs. alternate.
            D. Lvs. angled or obscurely lobed, sharp-toothed; petals commonly 3–4 cm. long;
               fls. on well-formed pedicels . . . . 7. *P. domesticum*
            DD. Lvs. obviously palmately lobed; petals 1.2–2 cm. long; fls. subsessile.
                E. Lobes of lvs. pinnatifid into divisions ca. 6 mm. wide with dentate margins;
                   petals rose or pink, veined purple, ca. 12 mm. long . . . . . . 8. *P. graveolens*
                EE. Lobes of lvs. ca. 2–2.5 cm. wide, shallow and obtuse.
                    F. Plants diffuse or procumbent; lvs. lobed more than halfway to midrib;
                       petals rose, veined purple, ca. 2 cm. long. Rose-scented
                                                  9. *P. capitatum*
                  FF. Plants erect; lvs. with more shallow lobes; petals 12–15 mm. long, rose
                     or pink with darker veins . . . . 10. *P. vitifolium*

1. **P. peltàtum** (L.) Ait. [*Geranium p.* L.] IVY GERANIUM. Lvs. alternate, somewhat succulent, shining; fls. commonly rose, white or lavender, the upper petals blotched and striped; $2n = 36$ (Gauger, 1937).—Very successful in cult. and to be expected as escape on the immediate coast. S. Afr.

2. **P. inquìnans** (L.) Ait. [*Geranium i.* L.] Lvs. alternate, ca. 7-lobed to ⅕ or ⅙ way to the base, crenate-dentate; calyx-spur 2–3 cm. long; petals 15–20 mm. long; $2n = 18$ (Takagi, 1928).—Reported as occasional escape. S. Afr.

3. **P. zonàle** (L.) Ait. [*Geranium z.* L.] HORSESHOE GERANIUM. Sparsely branched, ± pilose; lvs. alternate, round-cordate, crenate-dentate, obscurely many-lobed, mostly with a dark horseshoe band; stipules ovate, pointed; calyx-spur 2.5–3.5 cm. long; fls. red to pink.—Reported from Oceanside, Monterey, San Francisco. S. Afr.

4. **P. × hortòrum** Bailey. FISH GERANIUM. Freely branched; with fishy odor; lvs. alternate, round to reniform, somewhat scalloped, crenate-toothed, mostly with a color band; stipules broad, rounded; calyx-spur 2–3 cm. long; fls. of many colors.—The common bedding geranium, of horticultural origin; occasional escape near cult. areas.

5. **P. inodòrum** Willd. [*P. clandestinum* L'Hér.] Spreading-pubescent annual, ascending to erect, 1.5–5 dm. high; lvs. 2–5 cm. wide, crenately 3–5-lobed; fls. 4–10; pedicels 4–6 mm. long; sepals ca. 4 mm. long; petals 4–6 mm. long, pink to rose with darker veins.—Once reported as weed in Santa Ana; native of Australia.

6. **P. grossularioìdes** (L.) Ait. [*Geranium g.* L. *P. anceps* L'Hér.] Subglabrous annual or perennial, suberect, 3–6 dm. high, with pungent turpentinelike odor; lvs. 2–5 cm. wide, shallowly several-lobed, crenate-dentate; fls. 3–10; pedicels 2–5 mm. long; sepals ca. 4 mm. long; petals red, 6 mm. long.—Occasional in waste ground, Mendocino Co. to San Francisco Bay, also Fallbrook; S. Afr. April–July.

7. **P. × domésticum** Bailey. MARTHA WASHINGTON GERANIUM. PELARGONIUM. Erect, woody, 5–15 dm. high, not succulent, soft-hairy; lvs. alternate, obscurely lobed, dentate; stipules acuminate; fls. many, showy, the upper petals with dark blotches.—Of horticultural origin.

8. **P. gravèolens** L'Hér. ROSE GERANIUM. Woody, hairy, to 1 m. high; lvs. deeply 5–7-lobed, the divisions again lobed and toothed, fragrant; stipules ovate, abruptly acuminate; fls. 5–10, subsessile; calyx-spur 6–9 mm. long; petals ca. 12 mm. long.—The commonest rose geranium in cult.; S. Afr.

9. **P. capitàtum** (L.) Ait. [*Geranium c.* L.] Stems spreading or trailing, densely white-hairy, woody, to 1 m. long; lvs. 3–5-lobed, rose-scented; stipules ovate, subacuminate; fls. 9–20, subsessile; calyx-spur 4–5 mm. long; petals ca. 2 cm. long.—Once reported from s. Calif.; S. Afr.

10. **P. vitifòlium** (L.) Ait. [*Geranium v.* L.] Woody, to 1 m. high, grayish-hirsute; lvs. with ca. 3 main shallow rounded lobes, crenate-dentate; stipules ovate, acute; fls. 7–16, subsessile; calyx-spur 3–5 mm. long; petals 13–15 mm. long.—Well established near Pigeon Point, San Francisco, San Mateo Co., Ft. Bragg etc.; S. Afr.

14. Oxalidàceae. OXALIS or WOOD-SORREL FAMILY (Fig. 17)

Annual or perennial herbs to shrubs or trees, with acid sap. Lvs. usually compound, either digitate or pinnate, alternate or the basal opposite. Fls. bisexual, regular; sepals and petals 5, sometimes somewhat united at base. Stamens 10, hypogynous, the outer 5 opposite the petals; fils. joined near base and at least some with basal glandular appendages. Styles 5, separate; stigmas capitate or somewhat bifid. Ovary superior, 5-celled, 2- to many-ovuled. Fr. a caps. or berrylike. Ca. 10 genera and over 500 spp. widely spread in temp. and trop. regions.

1. Óxalis L. WOOD-SORREL

Caulescent or acaulescent, annual or perennial, often with bulbous or tuberous underground parts or creeping rootstocks. Lfts. 3 or more, sensitive to light (folding at night). Fls. 1 to several in umbellike cymes on axillary peduncles, yellow, white, pink to red or violet. Stamens 5 long and 5 short. Styles 5. Fr. a loculicidal caps. Nearly 500 spp., many of horticultural value, some persistent weeds, a few grown for the edible bulbous parts. (Greek, *oxus*, sour.)

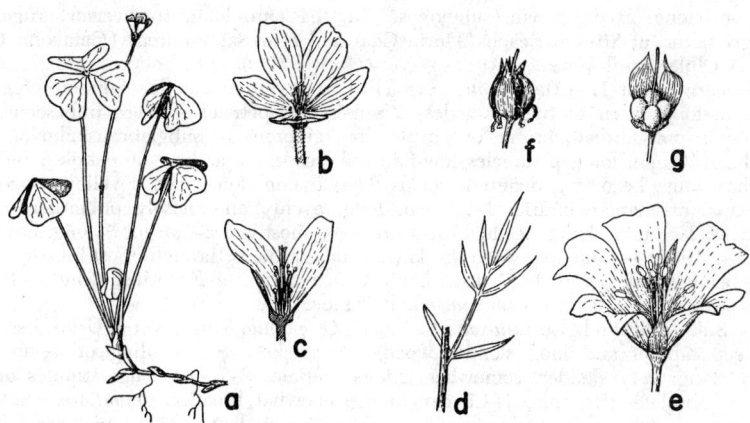

Fig. 17. OXALIDACEAE. *Oxalis: a,* habit, × ¼; *b,* fl., × 1; *c,* fl. in section, with 5 cent. styles, 5 long and 5 short stamens.

LIMNANTHACEAE. *Limnanthes Douglasii: d,* pinnately dissected lf., × 1; *e,* fl., × 2; *f,* old fl. with persistent sepals; *g,* the 4 nutlets about a common gynobasic style, × 1½.

(Wiegand, K. M. Oxalis corniculata and its relatives in N. Am. Rhodora 27: 113–124, 133–139, 1925. Small, J. K., N. Am. Fl. 25[1]: 25–57, 1907. Knuth, R., *in* Das Pflanzenreich, IV, 130: 43–389, 1930.)

A. Petals yellow.
  B. The plants acaulescent, with underground bulblets; petals ca. 2 cm. long
                                                    1. *O. Pes-caprae*
    BB. The plants caulescent, without bulblets; petals 8–16 mm. long.
      C. Plants annual; lf.-bearing stems 3–7 cm. high .... 2. *O. laxa*
      CC. Plants perennial; lf.-bearing stems 1–3 dm. long.
        D. Stems creeping from a slender taproot and rooting at the nodes. Garden weed
                                               3. *O. corniculata*
          DD. Stems erect or decumbent, not rooting at nodes. Native plants.
            E. Petals 12–16 mm. long; seeds 2–2.4 mm. long; lfts. 13–26 mm. broad. Humboldt Co. .... 4. *O. Suksdorfii*
            EE. Petals 8–10 mm. long; seeds 1.2–1.6 mm. long; lfts. 7–16 mm. broad.
              F. Pedicels spreading-pubescent; styles 1–2.5 mm. long; stipules usually broad. Mendocino Co. to San Diego .... 5. *O. pilosa*
              FF. Pedicels strigulose to glabrate; styles 2.5–3.5 mm. long; stipules narrow or almost lacking. Santa Barbara to L. Calif. .... 6. *O. californica*
AA. Petals white to red or violet.
  B. Lvs. sessile, cauline; lfts. linear to oblong; petals violet .... 7. *O. hirta*
  BB. Lvs. petioled, basal or subbasal; lfts. obcordate.
    C. Plants with slender scaly rootstocks.
      D. Fls. solitary; caps. round-ovoid .... 8. *O. oregana*
      DD. Fls. 3–6, umbellate; caps. linear .... 9. *O. trilliifolia*
    CC. Plants with woody crown and root-tubers or with scaly bulbs, acaulescent.
      D. Pedicels and calyx pubescent; plant from a thickened and tuberiform root
                                              10. *O. rubra*
      DD. Pedicels and calyx glabrous; plant from a scaly bulb .... 11. *O. Martiana*

1. **O. Pes-cáprae** L. [*O. cernua* Thunb. *Bulboxalis c.* Small.] BERMUDA-BUTTERCUP. Perennial from a deep rootstock with scaly bulbs, acaulescent; lvs. on glabrous petioles 1–2 dm. long; lfts. obcordate-bilobed, 1–2.5 cm. long, as wide or wider, green and mostly glabrous above, glaucous and thinly hairy beneath; peduncles 1.5–3 dm. high, glabrous; pedicels 3–10 or more, at first cernuous, then erect, pubescent, 1–2 cm. long; sepals lanceolate, 5–7 mm. long; petals deep yellow, 1.5–2.5 cm. long; caps. 5–7 mm. long.—Becoming abundantly natur. in orchards, fields and waste places at low elevs.; Fla., W.I.; native of S. Afr. Nov.–March.

2. **O. láxa** H. & A. Annual; stems 2–7 cm. high, very densely leafy; petioles 3–10(–20) cm. long, slender, pubescent; lfts. obcordate, pubescent and green above and beneath, 1–2 cm. long, almost as wide; peduncles 6–20 cm. long, slender, frequently 2-forked and with 3–6 fls. on each fork; pedicels pubescent, somewhat glandular, 5–12 mm. long; sepals glandular-pubescent, 1.5–3 mm. long; petals yellow, ca. as long; caps.

ca. 3 mm. long; seeds brown, subglobose, ca. 0.4 mm. long, transversely rugose.—Sparingly natur. at Stinson Beach (Marin Co.) and near San Andreas (Calaveras Co.); native of Chile. April–June.

3. **O. corniculàta** L. [*Xanthoxalis c.* Small.] Stems from slender taproot, creeping, 0.5–3 dm. long, often rooting at nodes, ± pubescent; petioles slender, pubescent, 1–5 cm. long; stipules broad, brown or purple; lfts. glabrous or subglabrous above, hairy beneath, 6–10 mm. long; peduncles few-fld., pubescent, ca. as long as petioles; pedicels 4–20 mm. long, becoming deflexed; sepals 2.5–4.5 mm. long; petals yellow, 4–8 mm. long; caps. prismatic-cylindric, 1–2.5 cm. long, evenly and closely puberulent; beak and styles 1–3 mm. long; seeds brown, rugose, mostly 1.2–1.5 mm. long; $2n = 24$ (Rutland, 1941).—Common weed in lawns and gardens through most of the state, having been introd. before 1837; to Atlantic Coast; native of Eu. Most months of year. A form with purple lvs. is var. *atropurpùrea* Planch.

4. **O. Suksdórfii** Trel. [*Xanthoxalis S.* Small. *O. pumila* Nutt., not d'Urv.] Perennial with deep taproot and long slender woody rootstocks; stems trailing or decumbent, 1–3 dm. long, very slender, somewhat villous; petioles 2–7 cm. long; stipules oblong or almost obsolete; lfts. thin, 1–1.5 cm. long, somewhat broader, obcordate, scattered-hairy above and beneath; peduncles ca. as long as petioles, 1–2-fld.; pedicels filiform, 1–3 cm. long, strigose; sepals 4.5–6 mm. long; petals yellow, 1.3–1.6 cm. long; caps. ovoid, 7–10 mm. long, puberulent; styles 5–6 mm. long.—Damp shaded woods, below 2000 ft.; Redwood F., Mixed Evergreen F.; Humboldt and Del Norte cos.; to Wash. May–July.

5. **O. pilòsa** Nutt. [*O. Wrightii* var. *p.* Wieg. and var. *subpilosa* Wieg. *Xanthoxalis p.* Small.] Perennial with stout woody taproot; plants loosely cespitose, the stems slender, decumbent, 1–4 dm. long, spreading-pubescent; petioles 2–7 cm. long, pubescent; stipules wide, well developed; lfts. 5–15 mm. long, ca. as wide, obcordate, hairy beneath and ± so above; peduncles axillary, 1–1.5 times as long as petioles, spreading-pubescent, 1–3-fld.; pedicels 1.5–3 cm. long; sepals 4.5–6 mm. long; petals yellow, 8–12 mm. long; caps. cylindrical, 12–18 mm. long, closely puberulent; styles mostly 1–2 mm. long; seeds 1.2–1.5 mm. long, brown, transversely corrugate.—Occasional, open hills and brushy hillsides mostly near the immediate coast; Chaparral, Coastal Scrub; San Diego Co. (Pala) to Mendocino Co.; Santa Cruz Id. Most of the year. Doubtfully distinct from the Mexican sp. *O. albicans* HBK.

6. **O. califórnica** (Abrams) Knuth. [*Xanthoxalis c.* Abrams.] Much like *O. pilosa,* but the pubescence appressed; stipules narrow or almost obsolete; styles 2.5–3.5 mm. long; seeds 1.5–1.6 mm. long.—Common on dry brushy, stony slopes, below 2000 ft.; Coastal Sage Scrub, Chaparral; Santa Barbara to L. Calif.; Santa Cruz and Catalina ids. March–May.

7. **O. hírta** L. Perennial, with bulbs; stems erect or decumbent, branched, pubescent, leafy, 3–30 cm. long; lvs. subsessile; lfts. oblong-cuneate, lanceolate or linear-cuneiform, obtuse to emarginate, 5–15 mm. long, 1.5–3 mm. wide, glabrous above, hairy beneath; peduncles axillary, 2–6 times as long as lvs., hairy; sepals lanceolate, hairy, 4–5 mm. long; petals 15–23 mm. long, the claws yellow, blades violet; $2n = 30$ (Yamashita, 1935).—Reported from Montebello (Los Angeles Co.) and Salinas (Monterey Co.); native of S. Afr.

8. **O. oregàna** Nutt. [*O. o.* var. *Tracyi* Jeps.] REDWOOD-SORREL. Rootstocks branching, wiry, scaly; plant acaulescent above ground; lvs. many, the petioles ± rusty-villous, 5–17 cm. long; lfts. broadly obcordate, 1–3 cm. long, somewhat wider, glabrous and green above, glaucous and long-hairy beneath; peduncles mostly shorter than lvs., 1-fld., villous; pedicels 0.5–2.5 cm. long; sepals lance-oblong, 4–8 mm. long, pubescent; petals white or pinkish, often veined purple, oblong-obovate, 8–20 mm. long; styles pubescent, 2.5–3 mm. long; caps. round-ovoid, 7–8 mm. long, subglabrous.—Primarily in shade; Redwood F., Douglas-Fir F.; Monterey Co. n.; to Wash. April–Sept. Forma *Smalliàna* (Knuth) Munz. [*O. S.* Knuth. *O. Smallii* in error. *O. macra* Small, not Schlechter.] Fls. 2–2.5 cm. long, deep rose-purple.—With the sp., Monterey Co. to Del Norte Co.

9. **O. trilliifòlia** Hook. [*Hesperoxalis t.* Small.] Rootstocks vertical, fleshy-scaly; plant above ground acaulescent; petioles glabrous, 1–3 dm. long; lfts. ± obcordate, 2–6 cm. long, as wide or wider, glabrous above, sparsely hairy beneath; peduncles glabrous, ca.

as long as lvs., 3–15-fld.; pedicels 1–2.5 cm. long, somewhat pilose; fls. pendent; sepals 5–6 mm. long, lance-oblong; petals white to pinkish, 8–12 mm. long; caps. linear, erect, 1.5–2.5 cm. long, long-beaked, glabrous; styles ca. 2 mm. long; seeds oblong, brown, 2–2.5 mm. long, smooth.—Moist forest floor, below 6000 ft.; Red Fir F., Douglas-Fir F.; Humboldt and Trinity cos.; to Wash. May–Sept.

10. **O. rùbra** St. Hil. Perennial from a thick tuberlike woody root; plant acaulescent; lvs. many, with slender subglabrate petioles 1.5–2.5 dm. long; lfts. obcordate, papillose with dark spots, ± appressed-hairy on both surfaces but especially beneath, 1.5–2 cm. long and wide; peduncles surpassing lvs., strigose above, ca. 6–12-fld.; pedicels 12–25 mm. long, strigose; sepals ca. 4 mm. long, with apical orange spots; petals rose with darker veins, sometimes lilac or whitish, ca. 11–12 mm. long, somewhat strigose without; caps. oblong, puberulent at apex; seeds brown, ca. 2 mm. long; $2n = 42$ (Heitz, 1927).—Locally natur. near towns, Humboldt Co., Marin Co., etc.; native of S. Am. May–July.

11. **O. Martiàna** Zucc. [*Ionoxalis M.* Small.] Perennial, with scaly bulbs, acaulescent; petioles 1–3 dm. high, villous; lfts. suborbicular, 2.5–6 cm. wide, deep green and scattered-hairy above, paler, glandular-punctate and more pubescent beneath; peduncles slightly exceeding petioles, villous, many-fld.; pedicels 1–3 cm. long, strigose; sepals linear-oblong, 5–6 mm. long, each with 2 apical tubercles; petals violet or red-purple, 12–18 mm. long; caps. oblong, ca. 1 cm. long.—Reported by J. T. Howell as established in park at Ross (Marin Co.); native of trop. Am.

✓15. Limnanthàceae. False Mermaid Family (Fig. 17)

Low tender glabrous annual herbs of moist places. Lvs. alternate, pinnately dissected, exstipulate. Fls. bisexual, regular, solitary, axillary, 3–6-merous, subperigynous. Sepals valvate, persistent. Petals contorted. Stamens 6 or 10, free, some with a gland at the base. Carpels 3 or 5, nearly distinct, connected by a common gynobasic style; ovules solitary in each carpel. Fr. of semidrupaceous free ± tuberculate nutlets. Seeds erect, without endosperm; embryo straight. Two genera, N. Am.

Fls. commonly 5-merous; petals longer than sepals .... 1. *Limnanthes*
Fls. 3-merous; petals shorter than sepals .... 2. *Floerkea*

1. Limnánthes R. Br. Meadow-Foam

Lvs. usually pinnate, then pinnatifid. Sepals and petals usually 5, sometimes 4 or 6. Petals with a U-shaped band of hairs on the claw. Stigmas 5, capitate. Seven spp. (Greek, *limne*, marsh, and *anthos*, fl., because of habitat.)

(Mason, C. T. A systematic study of the genus Limnanthes. Univ. Calif. Pub. Bot. 25: 455–512, 1952.)
A. Sepals with long white fine somewhat curly hairs at least within, sometimes also without; lvs. ±
    pilose; petals scarcely if at all emarginate.
    B. Petals obviously exceeding sepals.
        C. Corolla bowl-shaped in outline; stamens 5–6 mm. long .... 1. *L. alba*
        CC. Corolla bell-shaped in outline; stamens 3 mm. long .... 5. *L. montana*
    BB. Petals scarcely if at all longer than sepals .... 2. *L. floccosa*
AA. Sepals mostly glabrous, sometimes slightly coarse-ciliate; lvs. glabrous; petals often emarginate.
    ✓B. Stamens 5–8 mm. long .... 3. *L. Douglasii*
    BB. Stamens 2.5–4 mm. long.
        C. Corolla bell-shaped in outline; nutlets smooth except for the apical triangular scales.
            D. Petals 10–14 mm. long, with brown-purple veins, the claws with 2 rows of hairs.
                Amador Co. to Tuolumne Co. .... 4. *L. striata*
            DD. Petals 7–10 mm. long, not colored-veined, glabrous. Tulare Co. .. 5. *L. montana*
        CC. Corolla bowl-shaped, petals not with colored veins, the claws lacking lines of hair;
            nutlets densely beset with low pyramidal tubercles.
            D. Stems suberect; lfts. ovate, entire; nutlets 2–3 mm. long. Mendocino Co.
                                                                                6. *L. Bakeri*
            DD. Stems with wide-spreading branches; lfts. pinnatifid into 3–5 oblanceolate segms.;
                nutlets 4 mm. long. San Diego Co. .... 7. *L. gracilis*

1. **L. álba** Hartw. in Benth. [*Floerkea a.* Greene. *L. Douglasii* var. *a.* Rattan. *L. a.* var. *detonsa* Jeps.] Simple or branched from base, 1–3 dm. high, mostly erect; lvs. 3–10 cm. long (including petioles), long-villous, once to twice pinnately dissected into linear or lanceolate divisions; pedicels spreading-ascending, slender, 2–8 cm. long;

calyx with long somewhat curly white hairs within and without; sepals lance-ovate to lanceolate, subacuminate, 5–8 mm. long, or longer in fr.; corolla bowl-shaped; petals white, sometimes rose-pink at apex in age, cuneate-obovate, truncate or very slightly emarginate, 10–14 mm. long; stamens 5–6 mm. long; nutlets obovoid, dark brown, 3 mm. high, with low pyramidal tubercles near apex; *n* = 5 (Mason, 1952).—Vernal pools or low moist open places, 50–4000 ft.; V. Grassland, Foothill Wd., Yellow Pine F.; Butte Co. to Merced Co. April–June. A form about Chico, Butte Co., has petals scarcely if any longer than the sepals.

Var. versícolor (Greene) C. T. Mason. [*Floerkea v.* Greene. *L. v.* Rydb.] Lvs. nearly or quite glabrous; sepals hairy only within or not at all.—From ca. 1000–5300 ft.; Foothill Wd., Yellow Pine F.; Shasta Co. to Tuolumne Co. May–June.

2. **L. floccòsa** Howell. Simple or branched from the base, 5–20 cm. high, erect or recurved, glabrous or somewhat pilose above; lvs. 3–7 cm. long, glabrous to pilose, 5–7-pinnate, the pinnae simple to 3-parted into lance-elliptic segms.; pedicels mostly 2–6 cm. long; sepals narrow-ovate, acuminate, 6–10 mm. long, loosely long-woolly without, densely so within; petals white or drying pink, obovate, rounded at apex, shorter than to ca. as long as sepals; stamens ca. 4 mm. long; nutlets 4–5 mm. high, brownish, with light-colored, pyramidal, granular tubercles on upper part; *n* = 5 (Mason, 1952).—Clay depressions, mostly below 1000 ft.; V. Grassland; Butte, Siskiyou and Shasta cos.; Jackson Co., Ore. March–May.

3. **L. Douglasii** R. Br. [*Floerkea D.* Baill. *L. sulphurea* Loud.] Mostly branched at base, glabrous, ascending to almost erect, 1–4 dm. high; lvs. 5–12 cm. long, glabrous, pinnately divided, the divisions 5–11 and entire to incisely toothed or 3–5-lobed; pedicels becoming 5–10 cm. long; sepals lanceolate, 6–10 mm. long, glabrous, sharply acute; fls. bowl-shaped; petals yellow with white tips, cuneate-obovate or -oblong, 8–16 mm. long, rather deeply notched at apex, somewhat long-hairy within and with a U-shaped band of shorter hairs at base; stamens 5–8 mm. long; nutlets dark, ca. 2.5–4 mm. high, smooth to somewhat triangular-tuberculate; *n* = 5 (Mason, 1952).—Moist places, such as vernal pools, seepages, etc., mostly below 3000 ft.; V. Grassland, Foothill Wd.; Coast Ranges, San Benito Co. to s. Ore. March–May.

KEY TO VARIETIES

Nutlets smooth, wrinkled or with scattered tubercles, if covered with tubercles then the petals colored; lfts. ovate.
    Living petals with some yellow.
        Petals yellow with white tips .... *L. Douglasii*
        Petals yellow .... var. *sulphurea*
    Living petals white .... var. *nivea*
Nutlets with high prominent ridges; petals white with rose veins; lfts. sublinear .... var. *rosea*

Var. sulphùrea C. T. Mason. Lfts. broadly obovate, 7–13, incisely toothed or lobed; petals 12–18 mm. long, cuneate to obovate, yellow, emarginate; nutlets 3.5–5 mm. long, smooth or wrinkled, often with a crown of tubercles.—N. Coastal Scrub; Point Reyes Peninsula, Marin Co.

Var. nívea C. T. Mason. Lfts. 5–11, incisely toothed, often 3–5-lobed; petals cuneate, emarginate, white, often with prominent dark purple veins; nutlets 3–4 mm. long, smooth or wrinkled, occasionally with a crown of tubercles.—Coast Ranges from San Luis Obispo Co. to Humboldt Co.

Var. ròsea (Hartw. in Benth.) C. T. Mason. [*L. r.* Hartw. *Floerkea r.* Greene.] Lfts. 7–11, lobed or parted with linear segms.; petals cuneate to obovate, emarginate to obcordate, white with pink veins or white with cream base and aging rose pink; nutlets 4–5 mm. long, with high prominent ridges or tubercles, often appearing white.—Moist spots, below 1500 ft.; V. Grassland, Foothill Wd.; Coast Ranges to Sierra Nevada foothills and as far s. as Madera Co. March–May.

4. **L. striàta** Jeps. Stems several, ascending, 1–3 dm. high, glabrous; lvs. like those of *L. Douglasii;* pedicels 5–20 cm. long, slender; sepals lance-linear, glabrous, 5–7 mm. long; corolla campanulate; petals white with green-yellow base and striate with brown-purple lines, oblance-obovate, scarcely emarginate, 10–14 mm. long, the claw with 2 rows of hairs; stamens 3–4 mm. long; nutlets red-brown, barely 2 mm. high, rather

smooth except for short triangular tubercles at the summit; $n = 5$ (Mason, 1952).—Open moist ground, 600–1700 ft.; Foothill Wd.; Eldorado Co. to Mariposa Co. March–May.

5. **L. montàna** Jeps. Stems slender, usually several, ascending, 1–2 dm. high; lvs. 4–9 cm. long, pinnate, then pinnatifid into 3–5 oblanceolate lobes, sometimes hairy; pedicels 3–7 cm. long, slender; sepals ± loosely hairy, lanceolate, 4–5 mm. long; corolla campanulate; petals white, spatulate-obovate, rounded-truncate, 7–10 mm. long, glabrous; stamens ca. 2.5 mm. long; nutlets scarcely 2 mm. high, acutely tubercled near summit; $n = 5$ (Mason, 1952).—Seeps and moist places, 1000–5500 ft.; Foothill Wd., Yellow Pine F.; Mariposa Co. to Tulare Co. March–May.

6. **L. Bàkeri** J. T. Howell. Stems 1–few, suberect or ascending, 1–3 dm. high; lvs. 5–15 cm. long, 5-foliolate, glabrous, the lfts. ovate to elliptic, obtuse, entire; pedicels 3–8 cm. long; sepals lance-ovate, 5–8 mm. long, glabrous, acuminate; corolla bowl-shaped; petals white with some yellow at base, 6–9 mm. long, cuneate-oblanceolate, scarcely or not emarginate, with some scattered long hairs within, but none at base; stamens 2–3 mm. long; nutlets brownish, 2–3 mm. long, densely beset with low tubercles at the apex; $n = 5$ (Mason, 1952).—Moist places in fields; Foothill Wd.; n. of Willits, Mendocino Co. April–May.

7. **L. grácilis** Howell var. **Paríshii** (Jeps.) C. T. Mason. [*L. versicolor* var. *P.* Jeps.] Stems mostly with widely divaricate branches from base, 1–2 dm. long, glabrous; lvs. 2–6 cm. long, with ca. 5 pinnae, these pinnatifid into 3–5 oblanceolate segms., glabrous; pedicels 2–8 cm. long; sepals lance-ovate, glabrous or with a few long hairs, acuminate, 5–8 mm. long; fls. bowl-shaped; petals white, aging pink, obovate, emarginate, 8–10 mm. long, glabrous; stamens 3–3.5 mm. long; nutlets brown, 4 mm. long, with low pyramidal tubercles over most of surface; $n = 5$ (Mason, 1952).—Moist lake shores and wet places, 4500–5000 ft.; Yellow Pine F.; Cuyamaca and Laguna mts., San Diego Co. April–May.

2. Floérkea Willd.

Like *Limnanthes*, but lvs. only once-pinnate. Fls. 3-merous. Petals white, ca. half as long as sepals. Stamens 6. Stigmas 3. One sp. of N. Am. (H. G. *Floerke*, 1764–1835, German botanist.)

1. **F. proserpinacoìdes** Willd. [*F. occidentalis* Rydb.] Slender glabrous annual, 3–20 cm. high, simple or branched; lfts. 3–7, oblong to narrow-oblanceolate, 4–15 mm. long; pedicels 1–3 cm. long, recurved-arcuate in fr.; sepals 2–4 mm. long; nutlets 2–3, rounded, 2–3.5 mm. long, with small pyramidal tubercles.—Places moist in early season, 5000–9500 ft.; Sagebrush Scrub, Red Fir F., Lodgepole F.; Tehama, Siskiyou and Modoc cos. to Placer Co., Tulare Co.; to Wash. and Atlantic Coast. May–July.

16. Tropaeolàceae. Tropaeolum Family (Fig. 18)

Succulent prostrate or vinelike herbs, often climbing by coiling petioles; annual or perennial. Roots sometimes tuberous. Lvs. alternate, digitately angled or peltate, sometimes lobed. Fls. solitary, axillary, very irregular, usually showy. Sepals 5, one produced into a spur. Petals 5, rarely fewer, clawed, the 2 upper unlike the others. Stamens 8, free, unequal. Ovary superior, 3-lobed, 3-celled; cells 1-ovuled; placentation axile. Style 1, apical; stigmas 3, linear. Fr. of 3 indehiscent 1-seeded carpels which separate from cent. axis when mature. Single genus with ca. 50 spp. of Latin Am.

1. Tropaèolum L. Nasturtium

Characters of family. (Greek, *trophy*, from the shieldlike lvs.)

1. **T. màjus** L. Garden Nasturtium. Annual or of longer duration, dwarf or climbing, glabrous; lvs. peltate, 4–15 cm. wide, ca. 9-nerved; fls. yellow, red, etc., 3–6 cm. wide; spur 2–4 cm. long; $2n = 28$ (Warburg, 1938).—Occasionally escaping, especially along the coast, and persisting in waste places, dumps, etc.; native of S. Am. Fls. much of year.

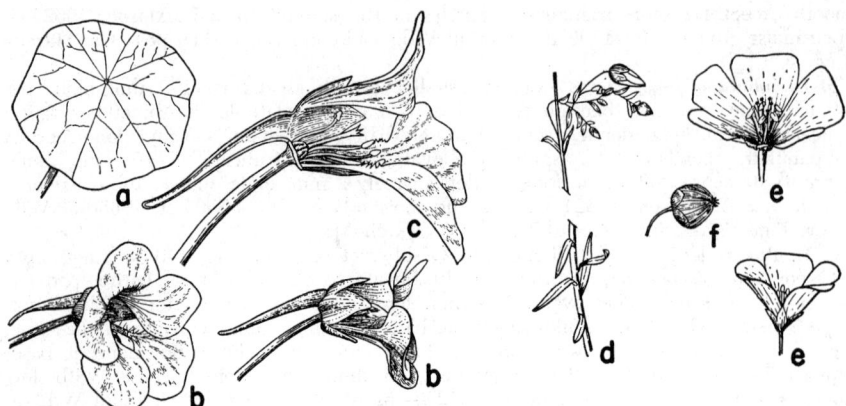

Fig. 18. TROPAEOLACEAE. *Tropaeolum majus: a,* peltate lf., × ½; *b* and *b,* 2 views of fl., × ½, with 5 sepals (1 spurred), 5 clawed petals; *c,* fl. in section, with cent. pistil and 3-lobed ovary, free stamens, and spur from 1 sepal.
LINACEAE. *Linum perenne* ssp. *Lewisii: d,* habit, × ½; *e* and *e,* internal and external views of fl. with 5 styles and 5 stamens; *f,* caps., × 1, surrounded by calyx.

17. Linàceae. FLAX FAMILY (Fig. 18)

Herbs or shrubs. Lvs. simple, alternate or opposite; stipules small or lacking. Fls. bisexual, regular. Sepals 5 (rarely 4), free or partially united, imbricate. Petals 5 (or 4) and alternate with sepals, convolute. Stamens 5 (or 4) and alternate with petals; fils. connate at base, often alternating with small staminodia; anthers introrse, 2-celled. Ovary superior, 2–5-celled or seemingly 4–10-celled by false septa; ovules 2 in each cell; styles as many as ovary-cells, ± free, filiform, with simple subcapitate stigmas. Fr. a caps., often septicidal; seeds compressed, shining, with straight embryo and flat cotyledons. Ca. 14 genera and 150 spp., widely distributed, a few important as ornamentals or for flax, linseed oil, etc.

1. Lìnum L. FLAX

Annual or perennial herbs, with tough fibrous cortex. Lvs. sessile, entire, without stipules or with stipular glands. Fls. in terminal or axillary racemes, corymbs or cymes, red, yellow, blue or white. Sepals, petals, stamens 5, the petals fugacious. Ovary 5- or 10-celled. Caps. dehiscent or indehiscent. Seeds mucilaginous. Ninety or more spp. of temp. and warm regions. (*Linum,* Latin for flax.)

(Small, J. K., N. Am. Fl. 25: 67–87, 1907.)
A. Petals blue, 10–20 mm. long; styles 5; caps. 5–10 mm. long.
    B. Stigmas elongate. Introd. plants.
        C. Sepals not longer than caps.; fls. blue.
            D. Petals ca. 1.5 cm. long; fr. 7–10 mm. long; lvs. narrowly lanceolate; seeds 4 mm.
                long, brown, shiny but somewhat roughened .... 1. *L. usitatissimum*
           DD. Petals ca. 1 cm. long; fr. 5 mm. long; lvs. linear; seeds 3 mm. long, brownish-
                green, shining and smooth .... 2. *L. angustifolium*
        CC. Sepals longer than caps.; fls. red .... 3. *L. grandiflorum*
    BB. Stigmas nearly as broad as long; petals mostly 1.5–2 cm. long. Native perennial
                                                    4. *L. perenne*
AA. Petals white, rose or yellow, 2–8 mm. long (–15 in *puberulum*); styles 2–3, sometimes quite
    united; caps. 2–4 mm. long.
    B. Styles united for some distance; fls. yellow.
        C. Styles 2, united to below middle; petals ca. 3 mm. long. Coast Ranges . 5. *L. digynum*
        CC. Styles 5, united to near apex; petals 12–15 mm. long. Deserts ...... 6. *L. puberulum*
    BB. Styles free; fls. rose to white or yellow.
        C. Lvs. and bracts with marginal stipitate glands.
            D. Pedicels 2–4 mm. long; sepals acuminate; lvs. ovate. Lake and Colusa cos.
                                          7. *L. drymarioides*

DD. Pedicels mostly 5–10 mm. long; sepals acute; lvs. linear-lanceolate. Mendocino and Lake cos. .... 8. *L. adenophyllum*
CC. Lvs. and bracts entire.
  D. Pedicels mostly 2–5 mm. long; stipule-glands prominent.
    E. Petals yellow; outer sepals evidently toothed. Vaca Mts. to Mt. Diablo 9. *L. Breweri*
    EE. Petals rose to white; outer sepals with few or no teeth.
      F. Sepals pubescent; petals 6–8 mm. long. Coastal, Marin Co. to San Francisco Peninsula .... 10. *L. congestum*
      FF. Sepals glabrous; petals 4–6 mm. long. Interior, Butte and Glenn cos. to San Benito Co. .... 11. *L. californicum*
  DD. Pedicels, at least the lower, 8–20 mm. long; stipule-glands inconspicuous.
    E. Petals 6–7 mm. long, pink to white. Humboldt to Sonoma, Napa, and Santa Clara cos. .... 12. *L. spergulinum*
    EE. Petals mostly 3–4 mm. long.
      F. Fls. pink to white. Widely distributed .... 13. *L. micranthum*
      FF. Fls. yellow. Lake, Napa, and Santa Clara cos.
        G. Styles 3; secondary branches of infl. mostly alternate; stamen-cup at base of fils. 5-toothed .... 14. *L. Clevelandii*
        GG. Styles 2; secondary branches of infl. mostly opposite; stamen-cup 10-toothed .... 15. *L. bicarpellatum*

1. **L. usitatíssimum** L. COMMON FLAX. Slender-branched annual, 3–9 dm. high, glabrous, with stems somewhat angled above; lvs. ± lanceolate, erect or ascending, 3-nerved, 1–3.5 cm. long; pedicels erect, 2–2.5 cm. long; sepals acuminate, unequal, 6–9 mm. long; petals blue, occasionally white, 1–1.5 cm. long; styles nearly distinct, with clavate stigmas; caps. 7–10 mm. high, round-ovoid; seeds ca. 4 mm. long; $2n = 30$ (Ray, 1944), 32 (Kostoff, 1940).—Cult. as a field crop and escaping as a casual weed in waste places and along roads; introd. from Eu. Feb.–May.

⌐2. **L. angustifòlium** Huds. Short-lived perennial, branched from base, the stems ascending, 2–10 dm. high, glabrous, densely lfy.; lvs. linear, acuminate, 5–25 mm. long; pedicels mostly 5–18 mm. long; sepals mucronate, 4–6 mm. long; petals blue, 8–10 mm. long; styles distinct; caps. ca. 4 mm. high, subglobose; seeds ca. 2.5 mm. long.— Becoming locally common on grassy hills and in valleys, particularly in Grassland and Wd. communities near the coast, San Mateo Co. to s. Ore.; introd. from Medit. region. April–Aug.

3. **L. grandiflòrum** Desf. Annual, 2–5 dm. high, much branched; main lvs. lanceolate, glaucous, 1–2 cm. long; pedicels 2–7 cm. long, slender; sepals 7–8 mm. long, lanceolate, ciliated with rather stiff hairs; petals rich red, 1.5–2 cm. long; styles 5, with linear stigmas; caps. shorter than sepals.—Grown in fl.-gardens and escaping occasionally, as in Contra Costa Co. and Santa Barbara; native of N. Afr. April–June.

4. **L. perénne** L. ssp. **Lewísii** (Pursh) Hult. [*L. L.* Pursh. *L. p.* var. *L.* Eat. & Wright. *L. L.* var. *alpicola* Jeps. *L. decurrens* Kell. *L. Lyallanum* Alef.] Glabrous perennial, usually with several densely lfy. stems from a branched caudex, 1.5–7.5 dm. high; lvs. linear to lance-linear, acute, ascending-erect, 1–2 cm. long; fls. in lfy. 1-sided racemes, blue, rarely white; pedicels mostly 1–3 cm. long; sepals ovate, 4–6 mm. long, obtuse to short-mucronate, with smooth hyaline margins; petals 1–1.5 cm. long; styles 5, with capitate stigmas; caps. round-ovoid, acute, 5–8 mm. long, the septa ciliate; seeds brown, shining, 3.5–4.5 mm. long.—Dry slopes and ridges, mostly 4000–11,000 ft.; Montane Coniferous F., Pinyon-Juniper Wd.; mts. from Siskiyou Co. through the Sierra Nevada to L. Calif. Apparently also in Foothill Wd., S. Coast Ranges, San Benito Co. to San Luis Obispo Co.; to Alaska, James Bay, Tex., n. Mex. May–Sept. A variable plant in need of study and differing from the Eurasian *perenne* chiefly by its greater stature and fewer branches in the infl.

5. **L. dígynum** Gray. [*Cathartolinum d.* Small.] Annual, mostly 6–20 cm. high; stem slender, frequently simple below, forked above, glabrous; lvs. mostly opposite, erect or ascending, elliptic to oblong, obtuse to acutish, 5–16 mm. long, the lower entire, the uppermost sharply serrate; pedicels to 1 or 2 mm. long, subtended by leafy bracts; sepals unequal, obtuse, 2–3 mm. long, gland-toothed; petals yellow, not appendaged, 3 mm. long; styles 2, partly united, scarcely 1 mm. long, with capitate stigmas; caps. ca. 2 mm. long; seeds lance-obovoid, brown, ca. 1 mm. long; $n = 8$ (Raven, 1958).—Moist grassy meadows, 3500–4700 ft.; Yellow Pine F.; Sierra Nevada, Fresno Co. n.; to Wash. June–July.

6. **L. pubérulum** (Engelm.) Heller. [*L. rigidum* var. *p.* Engelm. *Cathartolinum p.* Small.] Short-lived pale green perennial, 1–3 dm. tall, branched from base, puberulent throughout, the stems striate-angled; lvs. crowded below, linear to subulate, 5–10 mm. long, the upper lvs. and bracts gland-toothed; pedicels in open bracteate cymes; sepals unequal, lanceolate to lance-ovate, acuminate, gland-toothed, 4–6 mm. long; petals coppery-orange to yellow, with reddish base, 10–15 mm. long; styles 5, united to near apex; caps. somewhat ovoid, ca. 4 mm. long; seeds flat, ca. 2.5 mm. long.—Dry slopes and ridges, 4500–6000 ft.; Pinyon-Juniper Wd.; e. Mojave Desert (New York Mts., Clark Mt.); to Colo., Tex. May–July.

7. **L. drymarioìdes** Curran. [*Hesperolinon d.* Small.] Annual, dichotomously branched from base or just above, 1–2.5 dm. high; stems slender, dark, pubescent; lvs. alternate, ovate, 4–10 mm. long, gland-toothed, acute; infl. several times forked; pedicels 2–6 mm. long; sepals lanceolate, subacuminate, somewhat gland-serrate, 2.5–3.5 mm. long, pubescent; petals rose with darker veins, 4–5 mm. long, with minute cent. scale and deflexed lateral scales; styles 3, free, subclavate, ca. 1 mm. long; caps. ca. 2 mm. long; seeds brown, flat, slightly more than 1 mm. long.—Rare on dry slopes, ca. 2500 ft.; Foothill Wd.; Lake and w. Colusa cos. July–Aug.

8. **L. adenophýllum** Gray. [*Hesperolinon a.* Small.] Annual, mostly simple below and repeatedly forked above, 1–3 dm. high, ± pubescent; lvs. mostly alternate, linear-lanceolate, 5–20 mm. long, stipitate-glandular on margins; pedicels filiform, 2–14 mm. long; sepals lance-oblong, 2.5–3.5 mm. long, subacuminate, sparingly gland-toothed; petals pale yellow, 4–5 mm. long, the cent. scale oblong, the lateral appendages pubescent; styles 3, free, filiform, 1.5 mm. long; caps. ca. 2 mm. long; seeds ca. 1.5 mm. long.—Dry brushy hills and woods, at least partly on serpentine, 1500–4500 ft.; Chaparral, N. Oak Wd.; Mendocino and Lake cos. June–July.

9. **L. Brèweri** Gray. [*Hesperolinon B.* Small.] Annual, 1–3.5 dm. tall, dichotomously branched above the base; stems slender, reddish, glabrous except above some of the nodes; lvs. opposite or alternate, linear to filiform, 1–2 cm. long; pedicels 1–4 mm. long; sepals broadly lanceolate, acuminate, 3–4 mm. long, sparingly gland-toothed; petals bright yellow, 6–7 mm. long, with an oblong cent. scale and small pubescent lateral scales; styles 3, free, ca. 3 mm. long; caps. round-ovoid, ca. 2.5 mm. long; seeds narrow, ca. 2 mm. long.—Grassy or brushy slopes, mostly partly shaded, at least partly on serpentine, 400–3300 ft.; Chaparral, Foothill Wd.; inner Coast Ranges, Vaca Mts. to Mt. Diablo. May–July.

10. **L. congéstum** Gray. [*L. californicum* var. *c.* Jeps. *Hesperolinon c.* Small.] Annual, 1–4 dm. tall, corymbosely branched near middle, the stems angled, ± pubescent above the nodes; lvs. linear or nearly so, sometimes pubescent, 1–2.5 cm. long; pedicels 1–8 mm. long; sepals lance-ovate, pubescent, acuminate, 3–4.5 mm. long, the inner somewhat gland-toothed; petals rose to whitish, 6–7 mm. long, with narrow cent. scale and shorter lateral ones; styles 3, ca. 2 mm. long; caps. short-ovoid, ca. 2.5 mm. long; seeds narrow, ca. 1.5 mm. long.—Dry slopes on serpentine; N. Coastal Scrub, Coastal Prairie; Tiburon Peninsula, Marin Co. to Crystal Springs, Santa Mateo Co. May–June.

11. **L. califórnicum** Benth. [*Hesperolinon c.* Small. *L. c.* var. *confertum* Gray ex Trel.] Annual, 1–5 dm. tall, paniculately branched above, the branches angled-striate, mostly glabrous; lvs. linear, 1–2.5 cm. long; pedicels 1–6 mm. long; sepals lanceolate, glabrous, acuminate, 3–5 mm. long, the inner somewhat gland-toothed; petals pinkish to whitish, 4–6 mm. long; styles 3, ca. 4–5 mm. long; caps. ovoid, ca. 3 mm. long; seeds narrow, almost 2 mm. long.—Open grassy or rocky places on serpentine or shale, below 2000 ft.; Chaparral, V. Grassland, Foothill Wd.; inner Coast Ranges and adjacent plains, Glenn and Butte cos. to San Carlos Range. May–June.

12. **L. spergulìnum** Gray. [*Hesperolinon s.* Small.] Annual, repeatedly dichotomously branched beginning near base, glabrous or with some pubescence above the nodes, 1–5 dm. high; lvs. linear, 1–2.5 cm. long; pedicels capillary, 6–20 mm. long; sepals ovate, acutish, 1.5 mm. long, especially the inner with some gland-teeth; petals pinkish or white, 6–7 mm. long, the middle appendage large, erect, hairy within, the lateral small, hairy, toothlike; styles 3, ca. 4 mm. long; caps. ovoid, 1–2 mm. long.—Dry slopes, away from the immediate coast, 800–3000 ft.; Chaparral, N. Oak Wd., Foothill Wd.; Coast Ranges from Humboldt and Lake cos. to Sonoma and Napa cos., Mt. Hamilton Range and San Carlos Range? June–July.

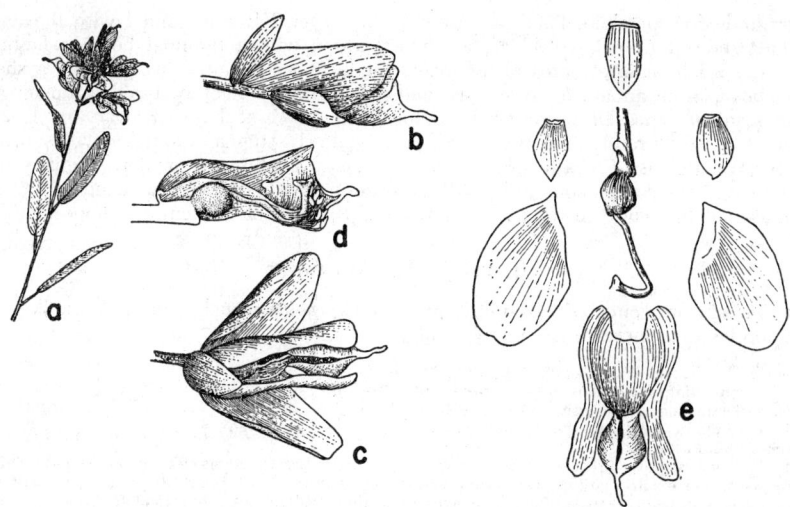

Fig. 19. POLYGALACEAE. *Polygala cornuta: a,* habit, × ½; *b,* fl. from side, × 2½; *c,* fl. from above and *d,* in section; *e,* fl. pulled apart, to show 5 sepals (the 2 inner being the enlarged lateral wings) and 3 united petals below (2 of them lateral and 1 keel-shaped).

13. **L. micránthum** Gray. [*Hesperolinon m.* Small.] Annual, freely branched at or usually above the base, 1–4 dm. high, the branches very slender, minutely pubescent above the nodes; lvs. linear, acutish, 1–2.5 cm. long; pedicels 5–16 mm. long, filiform; sepals lanceolate to somewhat oblong, 1–2 mm. long with some glandular teeth; petals pale pink to white, mostly 2–3 mm. long, the cent. scale ciliate, the lateral often almost obsolete; styles 3, scarcely 1 mm. long; caps. 1–2 mm. long; seeds brown, narrow, with dark spots, 1–1.5 mm. long.—Often on serpentine, open slopes and ridges, mostly 1000–5500 ft.; Chaparral, Foothill Wd., Yellow Pine F.; interior San Diego Co., w. Riverside Co., Santa Monica Mts., w. edge Mojave Desert, Santa Ynez Mts. through Coast Ranges to Ore., w. Sierra Nevada to Modoc and Siskiyou cos. May–July.

14. **L. Clevelándii** Greene. [*Hesperolinon C.* Small.] Annual 1–3 dm. tall, repeatedly dichotomous usually from about the middle, slightly pubescent above the nodes; lvs. linear or somewhat wider, 5–20 mm. long; pedicels filiform, mostly 3–15 mm. long; sepals lance-oblong, ca. 2 mm. long, with some glandular teeth; petals yellow, 3–4 mm. long, cent. scale erect, the lateral toothlike; stamens forming a 5-toothed cup at base of fils.; styles 3, ca. 1 mm. long; caps. 1.5–2 mm. long; seeds narrow, brown, ca. 1 mm. long.—On serpentine of rocky slopes, 1000–3500 ft.; Chaparral; Lake and adjacent Napa and Mendocino cos., Red Mts. on Stanislaus–Santa Clara Co. line. May–July.

15. **L. bicarpellàtum** H. K. Sharsm. Annual, 8–30 cm. tall, puberulous above nodes, branched mostly from near middle, the secondary branches mostly opposite; lvs. linear, 15–20 mm. long; pedicels 2–12 mm. long; sepals 2–3 mm. long, lanceolate, somewhat gland-toothed; petals yellow, 3–4 mm. long, the cent. appendage hairy, the lateral thickish; stamen-cup at base of fls. 10-toothed; styles 2, 3 mm. long; caps. ca. 1.5 mm. long; seeds narrow, brown, slightly more than 1 mm. long.—Clay and serpentine, brushy slopes and open woods, ca. 1200–1500 ft.; Chaparral, Foothill Wd.; s. Lake and Napa cos. May–June.

18. Polýgalàceae. Milkwort Family (Fig. 19)

Herbs, shrubs or trees. Lvs. alternate, rarely opposite or whorled, simple; stipules 0 or small glands; petioles short. Fls. bisexual, irregular, usually racemose or spicate, each subtended by a bract and 2 bractlets. Sepals 5, free, imbricate, the 2 inner (wings) larger and often petaloid. Petals 3, rarely 5, the upper 2 free or united with the lower-most or anterior (keel) which is boat-shaped, often with a terminal beak or fimbriate crest, the 2 lateral rarely present. Stamens usually 8, with united fils. forming a sheath split above, often adnate to the petals; anthers usually confluently 1-celled, opening by a subterminal pore. Disc present or reduced to gland at base of ovary or lacking. Pistil of 1 or 2, rarely 3–5, united carpels; style 1; stigma 2-lobed. Ovules usually solitary, pendulous. Fr. a caps., drupe or samara, dehiscent or not. Seeds usually pubescent, with conspicuous aril, with endosperm. With ca. 10 genera and 1000 spp. from around the world in trop. and temp. regions; some of horticultural importance.

1. Polýgala L. Milkwort

A widespread genus of 500–600 spp. (Greek, *polus*, much, and *gala*, milk; some spp. supposed to increase the flow of milk.)

(Blake, S. F., N. Am. Fl. 25: 305–370, 1924.)
Branches spreading-puberulous or -pubescent, spine-tipped.
    Fls. 4–5 mm. long, yellowish; plants 12–90 cm. high .... 1. *P. acanthoclada*
    Fls. 10–11 mm. long, pink-purple and yellow; plants 5–15 cm. high .... 2. *P. subspinosa*
Branches glabrous or strigulose, not spine-tipped.
    Fls. of 2 kinds, those in racemes at base of plant lacking petals, others with petals; sepals glabrous
        except for the ciliate margins; caps. thin-walled, reticulate .... 3. *P. californica*
    Fls. all having petals; sepals often pubescent; caps. thick-walled, scarcely reticulate .. 4. *P. cornuta*

1. **P. acánthoclàda** Gray. Spiny cinereous much branched shrub, 4–7 dm. high, the pubescence short, white, spreading; lvs. oblanceolate to linear-spatulate, 8–15 mm. long, short-petioled, puberulous; fls. yellowish; outer sepals 2–3 mm. long, the wings 4–5 mm. long; petals ca. 3 mm. long, slightly purplish at apex; caps. orbicular-truncate, notched, ca. 4 mm. long; seed straw-colored, somewhat pubescent, cylindro-obovoid, carunculate, ca. 2–2.5 mm. long.—Dry stony slopes and mesas, 2500–6000 ft.; Shadscale Scrub, Joshua Tree Wd., Pinyon-Juniper Wd.; e. Mojave Desert (Eagle, New York, and Shadow mts.); to Ariz., Colo. May–Aug.

2. **P. subspinòsa** Wats. [*P. lasseniana* Heller.] Much-branched tufted undershrub, 5–15 cm. high, pale green, densely puberulous, the branches spine-tipped; lvs. short-petioled, obovate to elliptic-oblanceolate, 1–2 cm. long, sparsely ciliolate, ± pubescent; fls. 6–9, pink-purple and yellow; sepals elliptic, hyaline, the outer puberulent along their middle, 4–6 mm. long, the wings ca. 10 mm. long, glabrous; petals ca. 8 mm. long, the keel 9–10.5 mm. long, with a blunt entire straight beak 1.5–2 mm. long; caps. oval, reticulate, pubescent especially on margin, 7–8 mm. long; seed ellipsoid, pubescent, curved-rostrate at base, 4 mm. long, the aril ca. 1.5 mm. long.—Gravelly wash, 4500 ft.; Sagebrush Scrub; near Secret V., Lassen Co.; to Utah, New Mex., Ariz. June–July.

Var. **heterorhýncha** Barneby. Lvs. sessile, elliptic; keel ca. 13 mm. long, the beak 2–3 mm. long, not notched on under side.—Alkaline calcareous hills, 3000–4000 ft.; Shadscale Scrub; e. Inyo Co., Calif. and Nye Co., Nev. April–May.

3. **P. califórnica** Nutt. [*P. cucullata* Benth.] Stems many, erect or spreading, slender, 3–35 cm. high, woody at base, somewhat incurved-puberulous, with short racemes of cleistogamous fls. near base; lvs. broadly lance-oblong to elliptic, 1–4 cm. long, somewhat strigulose along veins, mostly obtuse or rounded, distinctly short-petioled; terminal racemes loosely 3–10-fld.; bracts linear, deciduous; pedicels 2.5–4.5 mm. long; fls. bright rose to paler; outer sepals ciliolate, oblong to oblong-obovate, 5–6 mm. long, the wings obovate or subspatulate, emarginate, 10–11.5 mm. long, glabrous except for ciliolate upper margin near base; keel ca. 12 mm. long, ciliate near base of saccate part, the beak blunt, minutely papillose, ca. 3 mm. long; caps. suborbicular, reticulate, ca. 7 mm. long; seed-body obovoid, pilose, ca. 3 mm. long, the aril almost 3 mm. long, rounded and almost concealed by the 2 broad lobes covering the end of the seed.— Rocky ridges and slopes in brush or open woods, below 3000 ft.; Chaparral, Closed-cone Pine F., Redwood F.; Santa Cruz Id., outer Coast Ranges from n. San Luis Obispo Co. to s. Ore. March–July.

4. **P. cornùta** Kell. Stems many, slender, 6–9 dm. high, subglabrous to sparsely incurved-puberulous, woody below; lvs. ovate to almost linear, ciliolate, strigulose on

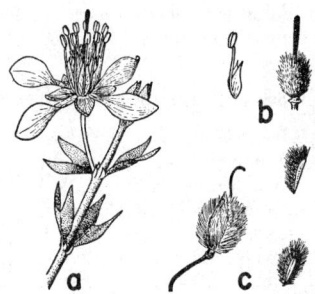

Fig. 20. ZYGOPHYLLACEAE. *Larrea divaricata: a,* habit, × 1, with lvs. divided into 2 lfts., fl. 5-merous, petals clawed, stamens 10, pistil 1; *b,* stamen and pistil; *c,* fr. dehiscing into 5 1-seeded carpels.

veins, 2–4 cm. long, mostly obtuse or rounded, short-petioled; fls. 4–20, yellowish to greenish-white, sometimes with dull plum color; pedicels 2–4 mm. long; outer sepals ovate, ca. 4 mm. long, densely puberulent, the wings obovate, ca. 10 mm. long, densely puberulent; keel ciliolate at base, 11–12 mm. long, including the straight, 1.5–2 mm. long beak; caps. orbicular-ovate, firm-walled, ciliolate, ca. 8 mm. long; seed-body ellipsoid-obovoid, pilose, 3.5–4 mm. long, the aril 3.5–4 mm. high, pilose, the 2 lateral lobes appressed.—Rocky or gravelly slopes, mostly 1000–5000 ft.; N. Coastal Coniferous F., Yellow Pine F., Douglas-Fir F.; Humboldt, Trinity, and Siskiyou cos. to the Sierra Nevada and s. to Fresno Co. June–Aug.

Var. **Pollárdii** Munz. Fls. greenish-white; wings ca. 7–8 mm. long, densely puberulent; keel ca. 9–10 mm. long, including the 1–1.5 mm. long beak.—Shaded canyons, below 2000 ft.; Chaparral, S. Oak Wd.; Santa Barbara and Ventura cos. May–July.

Var. **Físhiae** (Parry) Jeps. [*P. F.* Parry.] Fls. red-purple at least in age; wings 7–8 mm. long, glabrous except for the ciliolate margins; keel 8–9 mm. long including the 1 mm. beak.—Shaded rocky places in canyons, below 3000 ft.; S. Oak Wd.; Santa Monica Mts., Mt. Wilson, Santa Ana Mts., etc.; L. Calif. June–Aug.

19. Zygophyllàceae. Caltrop Family (Fig. 20)

Herbs or shrubs, often with branches jointed at nodes. Lvs. opposite or alternate, pinnate or 2–3-foliolate, not gland-dotted; stipules paired, persistent, often spinescent. Fls. bisexual, regular, 1 or 2 in the axils of the stipules. Stipules 5, rarely 4, usually free and imbricate. Petals 5, rarely 4 or 0, free, usually imbricate or contorted. Disk usually present. Stamens free, essentially hypogynous, usually twice as many as petals; anthers 2-celled, opening lengthwise. Pistil usually of 4 or 5 united carpels, or twice as many; ovary mostly superior, sessile; style simple; stigmas 1 or 5; ovules 2 or more in each cell. Fr. various, forming a caps. or splitting into 5–12 indehiscent nutlets. Seeds mostly with some endosperm; embryo straight with flat cotyledons.

Stipules spiny; lvs. 3-foliolate; fls. purplish .... 1. *Fagonia*
Stipules not spiny; lvs. 2–many-pinnate; fls. yellow or orange.
    Lfts. 1 pair; stamens with scalelike appendages.
        Plant a well-developed shrub; fr. hairy, globose. Common desert native .... 2. *Larrea*
        Plant suffrutescent; fr. glabrous, elongate. Rare waif .... 3. *Zygophyllum*
    Lfts. 4 or more pairs; stamens not appendaged.
        Fr. flat, radiate, breaking up into 5 very spiny nutlets .... 4. *Tribulus*
        Fr. hemispheric or higher, not radiate, breaking up into 8–12, merely tubercled nutlets
                                                    5. *Kallstroemia*

1. Fagònia L.

Ours low suffrutescent branching diffuse spreading plants. Lvs. in our digitately 3-foliolate; stipules spiny. Fls. solitary, purplish, small, 5-merous. Petals clawed, deciduous. Ovary with 5 two-ovuled cells. Fr. pyramidal-ovoid, deeply 5-angled, each

carpel 1-seeded and ventrally dehiscent. Seeds erect, compressed, broadly oblong. Ca.
18 spp. of Medit. region, sw. Afr., Chile, sw. N. Am. (G. C. *Fagon*, French botanist of
17th century.)

(Johnston, I. M., Proc. Calif. Acad. Sci. [IV], 12: 1049–1052, 1924.)

1. **F. califórnica** Benth. ssp. **laèvis** (Standl.) Wiggins. [*F. l.* Standl. *F. chilensis* var.
l. Jtn.] Compact, intricately branched, 1–4 dm. high, the stems somewhat angled,
smooth or minutely scabrous above; stipules subulate, 2–3(–5) mm. long, spreading; lfts.
lanceolate, 3–10 mm. long, 1–3 mm. wide, cuspidate; fls. in open cymes; sepals ca. 2
mm. long; petals 5–8 mm. long; fr. deflexed, 4–5 mm. long, reticulate, pubescent, with
beak 1–2 mm. long; seeds subovate, brown, mottled, ca. 2 mm. long.—Common on
dry often rocky slopes, mostly below 2000 ft.; Creosote Bush Scrub; through most of
Colo. Desert, occasional on s. Mojave Desert (Sheephole Mts.), sw. San Diego Co. as at
Otay Dam; to Ariz., Son., L. Calif. March–May, Nov.–Jan.
    Var. **glutinòsa** Vail. [*F. chilensis* var. *g.* Jtn. *F. viscosa* Rydb. *F. californica* var. *Bar-
clayana* of Jeps., not Benth.] Stipules 4–12 mm. long; lfts. 1–2 cm. long, 3–9 mm. wide;
upper stems glandular.—Common, Indio to Orocopia Mts., less so s.; to Son., L. Calif.
Same months.

2. Lárrea Cav. CREOSOTE BUSH

Evergreen strong-scented resinous shrubs. Lvs. opposite, with 2 divaricate sessile
asymmetrical olive-green lfts. Fls. solitary, yellow. Sepals 5, unequal, imbricate, deciduous. Petals 5, clawed, oblong-spatulate. Stamens 10, on small 10-lobed disk. Pistil of 5
united carpels; style slender, with 5 stigmas, the ovary-cells ca. 6-ovuled. Fr. globose,
hairy, separating at maturity into 5 indehiscent 1-seeded carpels. Perhaps 3–4 spp. of
the warmer dry parts of New World. (Named for J. A. de *Larrea*, Spaniard interested
in science.)

1. **L. divaricàta** Cav. [*Zygophyllum tridentatum* Moç. & Ses. ex DC. *L. t.* Cov. *L.
glutinosa* Engelm. *Covillea g.* Rydb. *C. t.* Vail. *L. t.* var. *g.* Jeps. *Schroeterella d., g.,* and
t. Briq. *Neoschroetera d., g.,* and *t.* Briq.] Branches 1–3(–4) m. tall, brittle, grayish,
irregular, with dark glandular bands at the nodes, densely leafy toward tips; stipules
brown, persistent, ca. 1 mm. long; lfts. obliquely lance-ovate, 5–10 mm. long, indistinctly 3-veined, entire; sepals round-ovate, strigose, 5–7 mm. long; petals 5–8 mm.
long, partly twisted like vanes of a windmill; fr. 4–5 mm. long, rusty long-villous, the
style slender, persistent, 5 mm. long; $2n = 52$, 104 (Covas, 1949).—The dominant
shrub over great areas of desert from Inyo Co. s., on dry slopes and plains up to ca.
5000 ft.; Creosote Bush Scrub; occasional in interior cismontane valleys as at Poso
Creek in Kern Co. and Aguanga and Hemet in Riverside Co.; to Utah, Tex., Mex., S.
Am. April–May. A sp. of economic value to the Indians for its antiseptic properties, medicinal use, as fuel, etc. It is a question whether N. and S. Am. materials are conspecific,
but the amount of variation in each makes separation difficult; the chromosome count
reported for S. Am. is $2n = 26$ (Covas & Schnack, 1946).

3. Zygophýllum L. BEAN-CAPER

Much branched herbs or subshrubs with fleshy branches. Lvs. stipulate, fleshy, opposite, composed in ours of 1 pair of lfts. Fls. axillary, solitary. Sepals 4–5, Petals 4–5,
clawed. Disk fleshy. Stamens 8–10, inserted at base of disk, each with a scale at base.
Ovary sessile, angular, 4–5-celled, tapering into a style. Fr. angular or winged, indehiscent or dehiscent; each cell 1-seeded. Genus of ca. 70 spp. of Medit. region, S. Afr.,
Asia, Australasia. (Greek, *zygon*, yoke and *phyllon*, lf., because of the paired lfts.)

1. **Z. Fabàgo** L. var. **brachycárpum** Boiss. Herbaceous or suffrutescent, with deep
strong root and erect branching tops, 3–6 dm. high, glabrous; lfts. obovate, 1–4 cm.
long; fls. copper and yellow; sepals 5, oblong, 6–7 mm. long, rounded at apex; petals five,
7–8 mm. long; pedicel reflexed in fr.; capsule 5-angled, ovate-oblong, 10–12 mm. long;
seeds gray-brown, obovate, 2–3 mm. long.—Established near Rosamond, Muroc, etc.,
w. Mojave Desert; reported also from Hamlin, Stanislaus Co.; introd. from Old World.
Summer.

4. Tríbulus L. CALTROP. PUNCTURE-VINE

Pubescent herbs with weak often prostrate stems. Lvs. even-pinnate; stipules membranaceous. Fls. solitary, axillary, 5(or 4)-merous. Fils. 10 (8), slender, unappendaged. Carpels as many as petals, surrounded at base by an annular 10-lobed disk; styles united into short stout column; ovules 3–5 per cell. Fr. depressed, spinose, separating into 5 bony indehiscent, 3–5-seeded carpels. Ca. 12 spp., widely distributed, especially in trop. (Latin name of *caltrop*, the shape of which is suggested by the 3-pronged fr.)

1. **T. terréstris** L. Annual, trailing, branched from base, the stems 2–10 dm. long; lfts. 4–7 pairs, oblong or elliptic, 5–12 mm. long; petals yellow, obovate, 3–4 mm. long; carpels crested and armed with 2–4 spreading spines 4–6 mm. long; $2n = 24$ (Negodi, 1939), 48 (Schnack & Covas, 1947).—Natur. in waste places, along roadsides, etc. through much of Calif. below 5000 ft., even on the deserts; to Atlantic Coast; from Old World. April–Oct. A serious pest.

5. Kallstroèmia Scop.

Annual herbs much like *Tribulus,* but with fr. dehiscing into twice as many indehiscent 1-seeded tuberculate, not spiny, nutlets and leaving a ± persistent cent. axis. Perhaps a dozen spp. of New World and Australia. (In honor of *Kallstroem,* obscure contemporary of Scopoli.)

(Vail, A. M., and P. A. Rydberg., N. Am. Fl. 25: 110–114, 1910.)
Petals 4–8(–12) mm. long, orange-yellow, fading whitish; beak of fr. 2–6 mm. long, strongly conic at base.
  Beak of fr. 2–3 mm. long, shorter than nutlets; petals 3–5 mm. long . . . . 1. *K. califórnica*
  Beak of fr. 4–6 mm. long, mostly longer than nutlets; petals 6–12 mm. long . . . . . . 2. *K. parviflora*
Petals 15–30 mm. long, orange; beak of fr. 8–11 mm. long, slender, scarcely enlarged at base
                                                        3. *K. grandiflora*

1. **K. califórnica** (Wats.) Vail. [*Tribulus c.* Wats.] Stems decumbent, branched, 1–6.5 dm. long, whitish pubescent; stipules ca. 2 mm. long; lfts. 5–7 pairs, elliptic, 4–8 mm. long; pedicels 1–3 cm. long in fr.; sepals deciduous, 3–4 mm. long; petals 3–5 mm. long; fr. strigulose, ovoid-globose, ca. 3 mm. long, with sharp tubercles on back of carpels, the beak glabrous, slender, shorter than carpels.—Occasional or common in some seasons, in sandy places, below 2500 ft.; Creosote Bush Scrub; Mojave Desert (Cronise V., Twentynine Palms) to Imperial V.; to Ariz., Mex. Aug.–Oct.

2. **K. parviflòra** Nort. Much like *K. californica,* but stipules 6–7 mm. long; lfts. 3–5 pairs; pedicels more conspicuously enlarged upward; sepals more persistent, 5 mm. long; petals 6–12 mm. long; fr. 3–4 mm. long, with rounded tubercles, the beak 4–6 mm. long.—Sandy slopes, at 5600 ft.; Joshua Tree Wd.; Clark Mt., e. Mojave Desert, Warners Hot Springs, e. San Diego Co.; to Miss., Mex. Aug.–Oct.

3. **K. grandiflòra** Torr. ex Gray. Stems suberect, rather diffusely branched, 2–7 dm. long, pubescent, somewhat hirsute; stipules 5–7 mm. long; lfts. 5–9 pairs, oblong, 10–25 mm. long; pedicels 2–6 cm. long in fr., thickened upward; sepals linear, hirsute, 8–15 mm. long, persistent; petals 12–30 mm. long; fr. 4–5 mm. long, tuberculate, the beak mostly 8–10 mm. long.—Sandy soil, at 900 ft.; Creosote Bush Scrub; near Desert Center, n. Colo. Desert; Ariz. to Tex. and Mex. Sept.–Oct.

20. Euphorbiàceae. SPURGE FAMILY (Fig. 21)

Herbs, shrubs or trees usually with a milky acrid sap; some succulent and cactuslike, others of very different habit. Lvs. simple, alternate, opposite or whorled, usually stipulate, sometimes with glands at base. Plants monoecious or dioecious, the infl. variable; the fls. sometimes in a *cyathium,* i.e., the naked ♀ fl. surrounded by ♂ fls., each a single stamen jointed at union of pedicel to peduncle, and all surrounded by an invol. Calyx and corolla present or absent, frequently different in ♂ and ♀ fls., the parts free or rarely united. Disk present or reduced to glands. Stamens as many or twice as many as sepals, or more numerous, or 1. Ovary superior, usually 3-celled, with 1 or sometimes 2 pendulous ovules in a cell; styles free or united. Fr. usually a 3-lobed caps.,

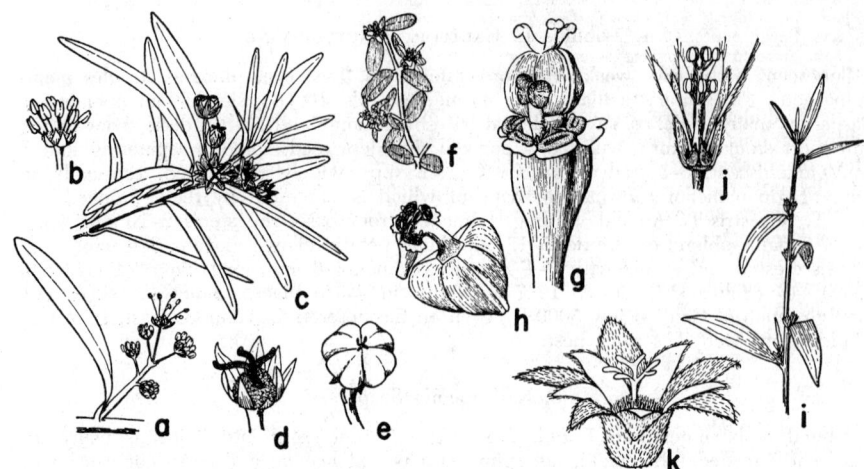

Fig. 21. EUPHORBIACEAE. *Tetracoccus dioicus: a, ♂* fls., × 1; *b,* single ♂ fl., × 2, with stamens and sepals; *c,* ♀ branch, × 1; *d,* ♀ fl. with sepals and pistil, × 2; *e,* 4-loculed ovary, × 1. *Euphorbia: f,* branch, × 1; *g,* cupulate cyathium, × 15, with circle of glands at summit, each bearing a flaring petaloid appendage, and with ♂ fl. consisting of 1 stamen jointed on its pedicel, and a ♀ fl. or 3-loculed pistil; *h,* more mature ovary farther exserted from cyathium, × 7. *Ditaxis lanceolata: i,* habit, × ½; *j,* ♂ fl., × 4, with sepals, slightly longer petals, and 2 series of stamens; *k,* × 4, ♀ fl., with sepals, petals, and pistil.

the lobes or carpels separating elastically from a persistent axis and 2-valved. Seed anatropous; embryo straight; cotyledons flat. Ca. 280 genera and more than 8000 spp., widely distributed. Some furnish food, some valuable oils, some are of ornamental and medical use.

A. Fls. with a calyx, not in the base of an invol. (cyathium).
   B. Plants definitely woody.
      C. Lvs. palmately lobed, 1–5 dm. long .... 7. *Ricinus*
      CC. Lvs. not palmately lobed, less than 0.5 dm. long.
         D. Fls. in umbellate axillary clusters; lvs. usually entire .... 1. *Tetracoccus*
         DD. Fls. in racemes or spikes; lvs. dentate.
            E. Lvs. 6–10 mm. long, gray at least beneath, 1-veined. Deserts .. 5. *Bernardia*
            EE. Lvs. 10–30 mm. long, green, often 3–5-veined from base. Coastal drainage
                                                      6. *Acalypha*
   BB. Plants herbaceous, at most slightly woody at base.
      C. The plants armed with stiff stinging hairs.
         D. Herbage gray-stellate as well as hispid. Cismontane annual ..... 3. *Eremocarpus*
         DD. Herbage green, hispid only, not stellate. Desert perennial .... 8. *Tragia*
      CC. The plants not armed with stinging hairs.
         D. Herbage densely stellate-scurfy .... 2. *Croton*
         DD. Herbage not stellate-pubescent.
            E. Lvs. alternate; styles 3.
               F. Petals 5, straw-colored, most evident in ♂ fls.; styles 2-cleft
                                                    4. *Ditaxis*
               FF. Petals lacking; styles entire .... 9. *Stillingia*
            EE. Lvs. opposite; styles 2 .... 10. *Mercurialis*
AA. Fls. lacking a calyx, included in a cup-shaped and calyxlike invol. (cyathium) which surrounds
    several ♂ fls. and 1 ♀ fl. with a 3-lobed pistil .... 11. *Euphorbia*

1. Tetracóccus Engelm.

Shrubs with alternate or subopposite entire or toothed lvs. Plants usually dioecious, the fls. small, apetalous; the ♂ in cymose racemes or umbellate clusters with 4–6–8-parted calyx and 4–6–9 stamens. Pistillate fls. solitary with 6–12-parted calyx. Ovary 3–4-celled; styles 3–4, distinct. Ovules 2 in each locule. Seeds usually solitary, strophiolate. A genus of ca. 4 spp. of Calif., Ariz., L. Calif. (Greek, *tetra,* four, and *kokkos,* fr., because of the 4-lobed caps. in original sp.)

(Dressler, R. L., The genus Tetracoccus, Rhodora 56: 45–60, 1954.)
Plants not spiny; lvs. opposite or subopposite; caps. 4-celled.
Lvs. linear, entire. Chaparral of coastal San Diego Co. . . . . 1. *T. dioicus*
Lvs. ovate to lance-ovate, toothed. Death V. region . . . . 2. *T. ilicifolius*
Plants spiny; lvs. alternate; caps. 3-celled. N. edge of Colo. Desert . . . . 3. *T. Hallii*

1. **T. dioìcus** Parry. Erect spreading shrub 0.5–1.5 m. high, the young branches reddish, glabrous, slender; lvs. linear, 2–3 cm. long, entire, on petioles 1–2 mm. long; ♂ fls. few, on short axillary pedicels, the calyx 6–10-parted, ca. 1 mm. long; fils. pubescent; ♀ fl. glabrous, the calyx 3–5 mm. long; caps. depressed-globose, 4-lobed, 8-grooved, ca. 8 mm. long, somewhat thicker; seeds red-brown, shining, asymmetrically oblanceolate in outline, somewhat flat, ca. 5 mm. long, the pale basal strophiole ca. 1 mm. long, somewhat 2-lobed.—Dry stony slopes, below 2500 ft.; Chaparral; San Juan Camp on Ortega Highway, Orange Co., through coastal San Diego Co., inland to Jacumba; L. Calif. April–May.
2. **T. ilicifòlius** Cov. & Gilman. Erect sparsely branched shrub, 3–15 dm. high; lvs. opposite, oblong-ovate to lance-ovate, 1–3 cm. long, 0.6–1.2 cm. wide, thick, pinnately veined, 3–8-toothed on each side, villous when young, glabrate, the petioles 1–2 mm. long; ♂ fls. in branched pubescent clusters 2–3 cm. long, the calyx villous, ca. 1 mm. long, red-tipped; ♀ fls. villous, 4–5 mm. long; caps. 4-celled, oblong-orbicular, 5–6 mm. long, usually with 2 seeds per cell. Dry slopes, 4000–5500 ft., in Inyo Co. (Falls Canyon, Grapevine Mts. and Death V. Canyon and Tetracoccus Peak, Panamint Mts.) May–June.
3. **T. Hállii** Bdg. [*Halliophytum H.* Jtn. *H. fasciculatum* var. *H.* McMinn. *Securinega f.* var. *H.* Jeps.] Erect shrub 5–20 dm. high, with stiff divaricate spinescent branches; lvs. entire, pubescent when young, oblanceolate, 3–9 mm. long, borne in small fascicles on short spurlike branchlets; ♂ fls. in axillary umbels, the sepals 4–6, 1–2 mm. long; ♀ on short pedicels; caps. 3-celled, oblong-globose, pubescent, 3–12 mm. long, 3-seeded; seeds smooth, ± ovoid, 4–7 mm. long, wrinkled ventrally, with a thin rudimentary caruncle; $2n = 24$ (Perry, 1943).—Dry rocky slopes and benches below 3600 ft.; Creosote Bush Scrub; Eagle, Cottonwood and Chuckawalla mts. of Riverside Co. to near Needles and Ivanpah, San Bernardino Co., etc.; sw. Ariz. March–May.

2. Cròton L. CROTON

Shrubs or herbs (ours perennial stellate-scurfy herbs). Lvs. alternate, petioled, simple; stipules obsolete. Monoecious or in ours dioecious; infl. mostly racemose. Staminate fls. in ours in terminal racemes, with 5(4–6)-parted calyx, the petals present or absent, the glands or lobes of disk alternate with petals; receptacle usually hairy; stamens 5 or more, the anthers inflexed in bud. Pistillate fls. mainly solitary or in ours racemose, below the ♂, the calyx 5–10-cleft; petals 0 or minute. Ovary 3(rarely 2–4)-locular, each cell 1-ovuled; styles as many as locules, 1–3-cleft. Caps. usually 3-lobed, globose, 3- or 1-seeded. Seeds carunculate, smooth, shining. Perhaps 600 spp. of warm or hot regions, all continents but Eu. (Greek, *kroton*, a tick, the old name of Castor-bean because of appearance of the seeds.) Malodorous and ± poisonous plants, an Asiatic sp. yielding oil of croton and *C. californicus* used by the Indians for stupefying fish.

(Ferguson, A. M. Crotons of the U.S. Rep. Mo. Bot. Gard. 12: 33–73, 1901.)
Caps. 10–11 mm. high; seeds 7–8 mm. long; lvs. 2–8 mm. wide. Sand dunes w. of Yuma
1. *C. Wigginsii*
Caps. 4–7 mm. high; seeds 3–5.5 mm. long; lvs. 5–20 mm. wide.
Lf.-blades mostly 2–4.5 cm. long; petioles mostly 0.5–3.5 cm. long.
Lvs. mostly 8–20 mm. wide; petioles 1–3.5 cm. long. Cismontane . . . . 2. *C. californicus*
Lvs. mostly 5–10 mm. wide; petioles 0.5–1 cm. long. Coastal s. Calif. . . . . var. *tenuis*
Lf.-blades mostly 0.5–2 cm. long; petioles 0.3–0.7 cm. long. Deserts . . . . var. *mohavensis*

1. **C. Wìgginsii** Wheeler. [*C. arenicola* Rose & Standl., not Small.] Shrub, 5–8 dm. high, much-branched, densely silvery-stellate throughout; lvs. linear to lanceolate, 20–65 mm. long, 2–8(–15) mm. wide; petioles 5–14 mm long; plants dioecious; ♂ fls. in racemes 15–30 mm. long, the calyx-lobes ovate, obtuse, ca. 2 mm. long; ♀ raceme ca. 2.5–3 cm. long, the calyx-lobes obtuse, ovate; caps. 9–11 mm. high, densely stellate; seeds oval or oblong, 7–8 mm. long, variegated with brown and gray, the caruncle

small, stipitate.—Creosote Bush Scrub; sand dunes w. of Yuma; to Son. March–May. Doubtfully distinct from:

2. **C. califórnicus** Muell.-Arg. [*C. c.* var. *major* Wats. *Oxydectes c.* Kuntze. *Hendecandra procumbens* Esch.] Erect or spreading perennial, often suffruticose, branched, 2–10 dm. high, ± stellate-hoary; petioles slender, 1–3.5 cm. long; lf.-blades oblong-elliptic, 1.5–4 cm. long, 1–2 cm. wide, greener above than beneath, entire; ♂ fls. corymbose at anthesis, the racemes becoming 10–15 mm. long, the fls. dropping off, the calyx 1–2 mm. long; ♀ racemes short, few-fld.; ♀ calyx 2–3 mm. long; styles 2–2.5 mm. long; caps. 5–7 mm. long; seeds rounded-obovoid, 3–4 mm. long, black or mottled, the caruncle triangular or reniform.—Locally frequent, mostly in dry sandy places, as beaches and washes, below 4000 ft.; Coastal Sage Scrub, Coastal Strand, Chaparral; Antioch and San Francisco to L. Calif. March–Oct. Variable and partly replaced locally by the poorly defined:

Var. **ténuis** (Wats.) Ferg. [*C. t.* Wats.] Stems very slender; lvs. generally less than 1 cm. wide, hoarier, on petioles less than 1 cm. long.—Sandy places, such as dunes and beaches; Coastal Strand, Coastal Sage Scrub; Ventura Co. to L. Calif.

Var. **mohavénsis** Ferg. Lvs. pale olive-green, 1–2 cm. long; petioles less than 1 cm. long.—Sandy places; Creosote Bush Scrub; deserts; Ariz.

3. Eremocárpus Benth. Turkey-Mullein. Dove Weed

Low broad gray heavy-scented annual with stellate pubescence and longer stinging hairs. Lvs. alternate, entire, 3-nerved. Monoecious. Staminate fls. in terminal cymes, with 5–6-parted calyx, no corolla, 6–7 stamens on hairy receptacle. Pistillate fls. 1–3 in lower axils, without calyx or corolla. Ovary with 4–5 small glands at base, 1 locule; style undivided. Ovule 1. Caps. 2-valved, 1-seeded. Seed gray, smooth, shining, ecarunculate. One sp. of w. U.S. (Greek, *eremos*, solitary, and *karpos*, fr.)

1. **E. setígerus** (Hook.) Benth. (*Croton s.* Hook. *Piscaria s.* Piper.] Stems dichotomously branched from base, forming dense rounded masses 3–20 cm. high and 5–80 cm. across; lvs. ovate to suborbicular, 1–6 cm. long, on petioles ca. as long; ♂ fls. pedicelled, the calyx ca. 2 mm. long; stamens exserted; pistil pubescent; caps. 4 mm. long; seeds dark, somewhat variegated, 3–4 mm. long, somewhat ridged; $2n = 20$ (Heiser & Whitaker, 1948).—Common in dry open places, in sandy or heavy soil, mostly below 2500 ft., sometimes higher; Coastal Sage Scrub, V. Grassland, Foothill Wd., Oak Wd.; through most of cismontane Calif. especially away from the immediate coast, occasional at edge of desert; to Wash. May–Oct. Most growth after the rainy season. Seeds much eaten by doves and quail. Plants used by the Indians to stupefy fish.

4. Ditáxis Vahl.

Herbs or shrubs, mostly with 2-branched hairs (the branches appressed and extending in opposite directions). Lvs. simple, alternate, with small stipules. Mostly monoecious, rarely dioecious. Fls. in short or reduced axillary few-fld. usually bracteate clusters. Calyx 5-parted. Petals 5, reduced or wanting in ♀ fls. Stamens 8–12, usually 10, in 2 series. Ovary 3-celled, 3-ovuled; styles 3, bifid. Caps. 3-lobed, each locule 2-valved. Seeds subglobose, ecarunculate, reticulate or pitted. Ca. 45 spp. of temp. and warmer parts of New World. (Greek, *dis*, two, and *taxis*, rank, the stamens in 2 whorls.)

(Pax, F., *in* Das Pflanzenreich, IV, 147 [VI]: 51–77, 1912.)

A. Plants glabrous .... 1. *D. californica*
AA. Plants strigose to pubescent.
    B. Gland-tipped teeth present on lvs. and bracts .... 2. *D. adenophora*
    BB. Gland-tipped teeth mostly lacking.
        C. Pistillate sepals white-margined; plants herbaceous.
            D. Lvs. mostly broadly cuneate-spatulate with truncate coarsely toothed apices; depressions on seeds not with radiating ridges .... 3. *D. serrata*
            DD. Lvs. oblanceolate, acute, serrulate or entire; depressions on seeds with minute radiating ridges .... 4. *D. neomexicana*
        CC. Pistillate sepals not white-margined; lvs. linear-lanceolate, entire; plants woody at base
                                                5. *D. lanceolata*

1. **D. califórnica** (Bdg.) Pax & K. Hoffm. [*Argythamnia c.* Bdg.] Perennial, becoming woody near base, glabrous throughout; the stems several, branched, 2–4 dm. long;

petioles 2–10 mm. long; stipules filiform, 2 mm. long; lf.-blades obovate to lance-oblong, green, mostly minutely serrulate, 1–3.5 cm. long, abruptly acute, conspicuously veined; racemes very short, few-fld., with ovate hyaline bracts 1.5–2 mm. long; ♂ fls. 2–4, ♀ 1; sepals of ♂ fls. lanceolate, 2.5 mm. long, recurved, hyaline-margined; petals white, hyaline, 2.5 mm. long; stamens 5 + 5; ♀ fl. with lance-linear sepals ca. 3.5 mm. long, petals shorter; styles free, bifid; caps. ca. 3 mm. long; seeds brownish, 1.5 mm. long, round-obovoid, reticulate with low ridges and faintly lined between.—Infrequent in sandy washes and on canyon-floors, 400–3000 ft.; Creosote Bush Scrub; Santa Rosa Mts. to s. side of Eagle Mts., Riverside Co. March–May, Oct.–Dec.

2. **D. adenóphora** (Gray) Pax & K. Hoffm. [*Argythamnia a.* Gray. *A. Clariana* Jeps.] Perennial, 2.5–4 dm. high, freely branched, rather thinly hairy; petioles 3–8 mm. long; stipules linear, ca. 2 mm. long; lf.-blades lanceolate to ovate, 1.5–3.5 cm. long, 15–18 mm. wide, acute, glandular-serrate; bracts lanceolate, 5 mm. long, glandular-serrate; racemes short, with 1 ♀ and 3–4 ♂ fls.; ♂ fl. with lanceolate nonglandular obscurely margined sepals ca. 3 mm. long, the petals pinkish, ca. 4 mm. long; stamens 10, in 2 whorls; ♀ fl. with lance-ovate glandular-serrulate sepals ca. 3 mm. long and white petals ca. 4 mm. long; ovary hairy; styles bifid; caps. ca. 5 mm. thick; seeds trigonous-ovoid, papillose, pitted-rugose, ca. 3 mm. long.—Rare, sandy flats, below 500 ft.; Creosote Bush Scrub; Coachella V. (Riverside Co.); Ariz., Son. Dec.–March.

3. **D. serràta** (Torr.) Heller. [*Aphora s.* Torr. *Argythamnia s.* Muell.-Arg.] Bushy annual to perennial, with several stems 1–3(–4) dm. high, densely strigose; petioles 2–5 mm. long; lvs. ovate to obovate, 1–2.5 cm. long, 7–12 mm. wide, usually serrulate, plainly veined, obtusish; racemes with 1–2 ♀ and 2–4 ♂ fls.; bracts lanceolate, 1–1.5 mm. long; ♂ fls. with calyx-lobes lance-linear, acute, scarcely 2 mm. long, hyaline-margined; petals slightly longer, oblanceolate, acuminate; stamens 10, in 2 whorls; ♀ fl. with calyx ca. 3 mm. long, the petals somewhat shorter; styles 3, bifid; caps. 4 mm. thick; seeds subglobose, 2 mm. long, grayish or brownish, reticulate, and with lines of very minute pits.—Occasional in dry rocky and sandy places, below 2500 ft.; Creosote Bush Scrub; e. Mojave Desert (Newberry Springs, Bagdad, Kelso, Needles) and Colo. Desert; to Tex., Mex. April–Nov.

4. **D. neomexicàna** (Muell.-Arg.) Heller. [*Argythamnia n.* Muell.-Arg.] Annual to perennial, not woody, 1–3.5 dm. high; lvs. lanceolate to oblanceolate, 1–2.5 cm. long, 6–11 mm. wide, acute, serrulate or entire, strigose with long setalike hairs; fls. few, congested in axils; ♂ fls. 1.5–2 mm. long, the petals exceeding the sepals; ♀ sepals linear-lanceolate, 3.5–5 mm. long, conspicuously white-margined; ♀ petals 1.5–2.5 mm. long, lance-ovate, somewhat hairy on back; styles united only at base, with narrow stigma-lobes; seeds ovoid, brownish, shallowly faveolate, the depressions marked with minute radiating ridges.—Dry slopes; Creosote Bush Scrub; Mojave Desert to L. Calif., Tex., Son. March–Dec.

5. **D. lanceolàta** (Benth.) Pax & K. Hoffm. [*Serophyton l.* Benth. *Argythamnia l.* Muell.-Arg. *A. sericophylla* Gray. *D. s.* Heller.] Somewhat woody below, freely branched, brittle, silvery-strigose on younger twigs, 2–4.5 dm. high; petioles 1–3 mm. long; stipules ca. 1 mm. long; lf.-blades lance-linear to -ovate, acute, 1–3 cm. long, 8–10 mm. wide, entire, greenish-glabrate; racemes 6–8 mm. long, with 1 ♀ and few ♂ fls.; bracts ovate, 1 mm. long; ♂ fls. with calyx-lobes lanceolate, ca. 4 mm. long, strigose; petals hairy, slightly longer; stamens 10, in 2 whorls; ♀ fl. with lance-ovate calyx-lobes 4–5 mm. long; petals hairy, slightly shorter; ovary hispid; styles hirsute, bifid; caps. ca. 5 mm. thick; seeds brownish-gray, globose-ovoid, ca. 2 mm. long, with shallow circular pits having radiating lines.—Frequent, rocky slopes and canyons, below 2000 ft.; Creosote Bush Scrub; Colo. Desert; Ariz., L. Calif. March–May.

5. Bernárdia Houst. ex P. Br.

Ours a deciduous shrub. Lvs. alternate, petiolate, serrate, stipulate. Ours monoecious, some spp. dioecious. Staminate fls. 2 or more in axils of bracts of small racemiform panicles; calyx-lobes 3 or 4; stamens 3–6, the fils. distinct. Pistillate fl. solitary, terminal; sepals 5; ovary 3-celled, 3-ovuled; styles 3, short, bifid. Seeds ecarunculate. Ca. 20 subtrop. or trop. Am. spp. (P. F. *Bernard*, 1749–1825, French botanist.)

1. **B. incàna** Mort. [*B. myricaefolia* auth., not Wats.] Many-stemmed, 1–2 m. high,

irregularly branched, with close grayish stellate pubescence; lvs. oblong-ovate, 8–16 mm. long, crenate, thickish, obtuse to rounded at apex; petioles 1–2 mm. long; ♂ infl. axillary, the calyx ca. 1 mm. long; ♀ calyx ca. 1 mm. long; caps. 8–11 mm. thick; seeds subglobose, light brown, mottled, smoothish, 4–5 mm. long.—Occasional in dry rocky canyons, below 4100 ft.; Creosote Bush Scrub; Colo. Desert (Pinyon Wells and Eagle Mts. s.); Ariz. April–May, Oct.–Nov.

6. Acalypha L.

Herbs or shrubs. Lvs. alternate, simple, petioled, stipulate. Monoecious. Infl. of terminal or axillary spikes or racemes, ♂, mixed, or ♀. Petals lacking. Staminate fls. several to many in axil of each bract, pedicelled; sepals 4; stamens 6–8, distinct, on a raised central receptacle. Pistillate fls. sessile, 1–2 in each bract-axil; calyx 3–8-lobed. Carpels 3 (2), 3(2)-ovuled; styles distinct, much dissected. Caps. usually 3-celled. Seeds small, ovoid, with small caruncles. Ca. 250 spp., mostly of trop. of both hemis., often grown for their foliage. (Greek, *akalephes*, nettle.)

1. **A. califórnica** Benth. CALIFORNIA COPPERLEAF. Rounded shrub with many slender reddish twigs, 3–12 dm. high, puberulent; lvs. ovate, sometimes cordate, obtuse, finely crenate-dentate, 1–2 cm. long, tinged reddish; petioles 3–15 mm. long; stipules linear, 3–5 mm. long; ♂ spikes reddish, 1.5–3 cm. long, on short peduncles; calyx ca. 0.5 mm. long, pubescent; ♀ spikes mostly shorter, with conspicuous glandular-pubescent bracts; calyx ca. 1 mm. long; caps. ca. 3 mm. across; seeds broadly obovoid, grayish-brown, minutely cellular-pitted, ca. 1.3 mm. long, with flat rhombic caruncle.—Locally frequent, dry granite slopes, 700–4000 ft.; Chaparral, S. Oak Wd.; cismontane San Diego Co. to desert edge; L. Calif. Jan.–June.

7. Rícinus L. CASTOR-BEAN

Monoecious arborescent shrub or in colder places grown as annual; glabrous, glaucous. Lvs. large, alternate, peltate, palmately 5–11-lobed, the lobes serrate; petioles with conspicuous glands. Fls. in racemose or panicled clusters, ♀ above the ♂. Calyx 3–5-parted. Petals 0. Disk 0. Stamens many, the fils. much-branched. Ovary 3-celled, 3-ovuled; styles 3, bifid, plumose, red. Caps. soft-spiny, with 3 two-valved carpels. Seeds glabrous, carunculate, variously marked and colored. One variable sp. with many horticultural forms. (Named for the Medit. sheep-tick, *Ricinus*, because the seed looks like a tick.)

1. **R. commùnis** L. With us a shrub 1–3 m. tall; lvs. 1–4 dm. broad; caps. 10–20 mm. thick; seeds compressed oblong-ellipsoid, 10–14 mm. long, lustrous-silvery and brown, ± mottled, the caruncle 3–4 mm. wide, flattish; $2n = 20$ (Hagerup, 1932).—Frequent as an escape in waste places, especially in s. Calif.; native of Asia and Afr. Most of the year.

8. Tràgia L.

Perennial herbs, sometimes twining, with stinging hairs. Lvs. mostly alternate, stipulate, petiolate, simple or compound, serrate. Monoecious; fls. in bracteate racemes, these terminal or opposite the lvs.; ♂ fls. above, the ♀ few and at the base, all with small bracts and apetalous. Staminate calyx 3(5)-parted; stamens mostly 3–5; fils. short. Pistillate calyx 3–8-parted; style 3-cleft, the branches simple. Caps. 3-celled, bristly, 3-seeded, separating in 2-valved carpels. Seeds not carunculate. Ca. 100 spp., widely distributed in trop. and subtrop. of all continents but Eu. (*Tragus*, Latin name of Hieronymus Bock, 1498–1554, German herbalist.)

1. **T. stylàris** Muell.-Arg. [*T. ramosa* Torr.] Perennial herb with woody caudex; stems several, slender, 1–3 dm. high; lvs. light green, lance-ovate to oblong-lanceolate, 1–2 cm. long, coarsely and sharply serrate; ♂ fls. 2–4, with 4–5-lobed calyx ca. 1 mm. long; stamens mostly 4–5; ♀ fls. solitary; the calyx 5-lobed, 1.5–2 mm. long; caps. depressed, 6–8 mm. across; seeds globose, brown, smooth, 2 mm. in diam.—Occasional on dry, rocky slopes, 3000–5500 ft.; Shadscale Scrub, Pinyon-Juniper Wd.; Providence and Clark mts., e. Mojave Desert; to Colo., Tex. Apr.–May.

9. Stillíngia Garden ex L.

Glabrous monoecious herbs. Lvs. alternate, mostly 2-glandular at base. Fls. bracteolate, apetalous, in terminal or axillary spikes which are ♀ at base. Staminate fls. with 2-lobed calyx and 2 stamens. Pistillate fls. 1–6, the calyx 3-parted or lacking. Ovary 3-celled, 3-ovuled; styles 3, filiform, simple. Caps. 3-lobed, separating into 2-valved carpels. Seeds ovoid or ovoid-ellipsoid, small, the caruncle minute or lacking. Ca. 15 spp. of warm parts of New World and Pacific and Indian oceans. (Named for B. *Stillingfleet,* 1702–1771, English naturalist.)

(Rogers, D. J. A revision of Stillingia in the New World. Ann. Mo. Bot. Gard. 38: 207–259, 1951.)
Lvs. ovate, sharply serrate, 3-nerved; mostly annual .... 1. *S. spinulosa*
Lvs. linear, entire or nearly so, 1-nerved; perennials.
  Plant forming a rounded leafy bush; infl. not much higher than foliage; lvs. tending to have few slender teeth near base; caps. 3–3.5 mm. high .... 2. *S. paucidentata*
  Plant with fewer erect sparsely leafy stems; infl. overtopping foliage; lvs. entire or serrulate at apex; caps. 2.5 mm. high .... 3. *S. linearifolia*

1. **S. spinulòsa** Torr. in Emory. [*Sapium annuum* Torr. *Stillingia a.* Muell-Arg.] Tufted annual or perennial, 5–20 cm. high, much-branched, densely lfy.; lvs. rhombic-ovate, spinulose-serrate, 3-nerved, 1–3 cm. long, acuminate, narrowed into short petiole; spikes mostly axillary, 5–12 mm. long, with trumpet-shaped glands at base of bracts; ♂ calyx ca. 0.5 mm. long; ♀ calyx ca. 1 mm. long; caps. 4–5 mm. high; styles ca. 3 mm. long; seeds cylindro-ovoid, grayish, with some brown mottling, ca. 3 mm. long, with numerous fine somewhat wavy longitudinal lines; caruncle obsolete.—Frequent in dry sandy places, below 3000 ft.; Creosote Bush Scrub; Mojave and Colo. deserts; Nev., Ariz. March–May.

2. **S. paucidentàta** Wats. [*S. linearifolia* var. *p.* Jeps.] Perennial, 2–4.5 dm. high, the stems freely branched above; lvs. linear, or the lower lanceolate, 3–5 cm. long, mostly 1–5 mm. wide, with 2–3 setaceous teeth on each side near base, short-petioled; spikes numerous, at tips of branches, 2–6(–10) cm. long, 3–4 mm. thick, with subcrateriform glands; ♂ fls. many, with calyx ca. 1.5 mm. and stamens ca. 3 mm. long; ♀ calyx ca. 2 mm. long; caps. 3–4 mm. long; styles 2 mm. long; seeds obovoid, gray, smooth, 2.5 mm. long.—Locally frequent, sandy open flats and slopes, 2000–4000 ft.; Creosote Bush Scrub; w. half of Mojave Desert; below sea-level near Mecca; w. Ariz. April–June.

3. **S. linearifòlia** Wats. [*S. gymnogyna* Pax & K. Hoffm.] Perennial, 3–8 dm. tall, the stems slender, loosely branched; lvs. linear, usually quite entire, 1–4 cm. long, 1–2 mm. wide; spikes terminal, slender, 1–2 mm. thick in ♂ part, lax, 2–5 cm. long; ♂ calyx 0.6 mm. long; stamens 1.5 mm. long; ♀ calyx almost obsolete; styles ca. 1 mm. long; caps. 2.5 mm. high; seeds gray, asymmetrically obovoid, smoothish, ca. 2 mm. long.—Occasional in washes and rocky places, below 3500 ft.; Creosote Bush Scrub; Colo. and Mojave deserts; more frequent on dry slopes and plains, interior cismontane valleys below 3000 ft. (5000 ft. at Idyllwild); Coastal Sage Scrub, Chaparral; San Bernardino Co. to San Diego Co., also as a waif in Orange Co.; to Ariz., Mex. March–May.

10. Mercuriàlis L. MERCURY

Monoecious or dioecious herbs. Lvs. opposite, pinnately veined. Fls. apetalous, small, green, in interrupted axillary spikes. Calyx 3-parted. Stamens 8–20. Ovary usually 2-celled, 2-ovuled; styles 2. Caps. 2-celled. Ca. 7 spp. of Eurasia and N. Afr. (A plant used by Pliny and referring to the god *Mercury.*)

1. **M. ánnua** L. Dioecious annual, glabrous, erect, 1.5–3 dm. high, openly branched; lvs. ovate to lance-ovate, serrate, 2–5 cm. long, subacuminate, on petioles 0.5–2.5 cm. long; ♂ spikes interrupted, pedunculate, the calyces ca. 1 mm. long, globose in bud; ♀ fls. 2–3 in axils, the calyces ca. 1 mm. long; caps. hispid, 2.5 mm. long; styles ca. 1 mm. long; seeds globose-obovoid, dark, shining, shallowly pitted, 1.5 mm. long, with whitish flat subtriangular caruncle; $2n = 16, 32$ (Ehrenberg, 1945).—Locally natur. in cult. fields, San Mateo Co.; e. U.S.; native of Eu. March–Sept.

11. Euphórbia L. SPURGE

Monoecious herbs or shrubs with milky acrid juice. Stems leafy and slender to almost leafless and fleshy, unarmed or armed. Lvs. simple, alternate or opposite, entire or dentate, stipulate or estipulate. Fls. in a cyathium (cupulate invol. resembling a calyx or corolla with united lobes) with 1–5 nectariferous glands on its margin alternating with the lobes and often with petaloid appendages from beneath the glands. Staminate fls. in 5 fascicles in the cyathium, 1–several per fascicle and each a single stamen jointed on its pedicel and usually subtended by a minute bract. Pistillate fl. solitary in center of cyathium, becoming exserted, consisting of a 3-celled, 3-ovuled ovary sometimes subtended by 3 small scales; styles 3, each usually 2-cleft. Caps. usually nodding, separating into 3 two-valved carpels. Seeds often carunculate. A diversified genus of perhaps 1000 spp., mostly temp., many cactuslike or of otherwise highly modified habit. (*Euphorbus*, physician of Numidia.)

(Norton, J. B. S. N. Am. Spp. of Euphorbia section Tithymalus. Ann. Rep. Mo. Bot. Gard. 11: 85–144, 1900. Wheeler, L. C. Euphorbia subgenus Chamaesyce in Canada and the U.S. exclusive of S. Florida. Rhodora 43: 97–154, 168–205, 223–286, 1941.)

A. Glands of the cyathium without petaloid appendages; lvs. essentially symmetrical.
  B. Lvs. linear; cyathia in terminal headlike clusters, their glands cup-shaped in ours, concealed by the inflexed segms. of the invol. (Subgenus Poinsettia) . . . . 1. *E. eriantha*
  BB. Lvs. (at least in the infl.) ovate to rounded; cyathia solitary or in cymes, their glands not cup-shaped. (Subgenus Esula.)
    C. Glands elliptic or transversely oval; lvs. serrulate.
      D. Floral lvs. broad at base; caps. with fleshy ellipsoid-lenticular tubercles
                                                                2. *E. spathulata*
      DD. Floral lvs. narrowed at base; caps. smooth; seeds subglobose-ovoid
                                                                3. *E. Helioscopia*
    CC. Glands crescent-shaped or 2-horned; lvs. largely entire.
      D. Stem-lvs. opposite, strongly decussate; caps. 8–12 mm. high . . . . . 8. *E. Lathyris*
      DD. Stem-lvs. alternate; caps. 2–4 mm. high.
        E. Plants perennial with widely creeping rootstocks; seeds smooth; caps. warty
                                                                4. *E. Esula*
        EE. Plants annual or cespitose perennials; seeds pitted or ridged; caps. smooth or keeled, not warty.
          F. Floral lvs. lance-linear, much like the cauline; plants annual
                                                                5. *E. exigua*
          FF. Floral lvs. lance-ovate or wider, mostly unlike the cauline.
            G. Plants annual.
              H. Caps. with thin keels; stem lvs. petioled; seeds with deep pits . . . . 6. *E. Peplus*
              HH. Caps. smooth; stem lvs. subsessile; seeds vermiculate-ridged to almost smooth . . . . 7. *E. crenulata*
            GG. Plants perennial.
              H. Glands hornless, lacerate; stem lvs. elliptic, tapering to acute tips . . . . 10. *E. incisa*
              HH. Glands horned, not lacerate; stem lvs. oblong to suborbicular
                                                                9. *E. Palmeri*
AA. Glands of the cyathium with petaloid appendages or, if these wanting, the lvs. all opposite and with inequilateral bases.
  B. Lvs. alternate, their bases symmetrical; plant a shrub 4–10 dm. tall. (Subgenus Agaloma)
                                                                11. *E. misera*
  BB. Lvs. all opposite, their bases usually strongly inequilateral; plants herbaceous or suffrutescent, mostly prostrate or low. (Subgenus Chamaesyce)
    C. Ovary and caps. hairy.
      D. Perennials.
        E. Invols. or cyathia urceolate; ♂ fls. up to 12; appendages entire or crenate
                                                                29. *E. arizonica*
        EE. Invols. or cyathia not urceolate; ♂ fls. 15–60.
          F. Seeds scarcely angled, narrow-ovoid, encircled by 4–5 rounded ridges. Colo. Desert . . . . 17. *E. pediculifera*
          FF. Seeds quadrangular, smooth to slightly wrinkled.
            G. Herbage with short straight spreading hairs
                                                                20. *E. polycarpa* var. *hirtella*
            GG. Herbage with appressed long and weak or matted hairs.
              H. Appendages wider than the glands, with short spreading hairs beneath and on the margins. Desert of Inyo and Kern cos.
                                                                18. *E. vallis-mortae*
              HH. Appendages wider than to narrower than glands, glabrous or rarely with few short hairs beneath. Los Angeles to San Diego and Colo. Desert . . . . 19. *E. melanadenia*

DD. Annuals.
  E. Staminate fls. 40–60 per cyathium; caps. 2.5 mm. in diam. Glenn Co.
                                                      13. *E. ocellata* var. *Rattanii*
  EE. Staminate fls. up to 15; caps. less than 2 mm. in diam.
    F. Invols. urceolate; appendages deeply 3–4-parted ...... 30. *E. setiloba*
    FF. Invols. obconic to campanulate.
      G. Glands without appendages or with mere rudiments; seeds smooth;
          lvs. not more than 8 mm. long .... 22. *E. micromera*
      GG. Glands appendaged; seeds granular or ridged.
        H. Caps. 1.5–1.9 mm. long; seeds 1–1.4 mm. long, not trans-
          versely ridged; appendages narrower than glands
                                26. *E. serpyllifolia* var. *hirtula*
        HH. Caps. 1–1.4 mm. long; seeds 0.9–1 mm. long, transversely
           ridged.
          I. Seeds with low rounded transverse ridges not whitened
            on summit; caps. strigose; lvs. 4–17 mm. long; ap-
            pendages narrower than glands .... 31. *E. supina*
          II. Seeds with narrow sharp transverse ridges whitened on
            summit; caps. with spreading hairs; lvs. mostly 4–8 mm.
            long; appendages 1–2 times as wide as glands
                                        32. *E. prostrata*
CC. Ovary and caps. glabrous.
  D. Stipules united into a broad white membranous scale ...... 23. *E. albomarginata*
  DD. Stipules not so united.
    E. Plants annual.
      F. Lvs. linear, symmetric, glabrous; plants erect. E. Mojave Desert.
        G. Caps. roundly 3-lobed, 2 mm. long; seeds 1.8 mm. long, smooth
                                            14. *E. Parryi*
        GG. Caps. sharply 3-angled, 1.3 mm. long; seeds 1–1.3 mm. long,
          with transverse rounded ridges .... 15. *E. revoluta*
      FF. Lvs. lanceolate to ovate or rounded, not symmetric at base.
        G. Margins of lvs. entire or nearly so.
          H. Glands radially elongate; seeds rounded on back, the face
            nearly flat .... 12. *E. platysperma*
          HH. Glands strictly discoid or transversely oblong; seeds ovoid or
            quadrangular, the face not flat.
            I. Glands with wide appendages; seeds with 4–6 trans-
              verse rounded ridges. Colo. Desert. .. 28. *E. Abramsiana*
            II. Glands not appendaged; seeds not thus transversely
              ridged.
              J. Caps. 2–2.3 mm. long; seeds 1.1–1.4 mm. thick; ♂
                fls. 40–60 .... 13. *E. ocellata*
              JJ. Caps. 1.3 mm. long; seeds 0.5 mm. thick; ♂ fls. 2–5
                                  22. *E. micromera*
        GG. Margins of lvs. toothed, at least toward apex.
          H. Styles entire; appendages of glands parted in 3–5 ligule-
            like structures 1 mm. long. Cent. V. ..... 25. *E. Hooveri*
          HH. Styles divided usually to ⅓–½ their length; appendages
            entire to slightly lobed.
            I. Stems mostly erect; lvs. oblong to oblong-lanceolate,
              8–35 mm. long; seeds with finely rippled surface
                                    16. *E. maculata*
            II. Stems prostrate or nearly so.
              J. Plants essentially glabrous.
                K. Seeds not transversely ridged, 1–1.4 mm. long;
                  appendages not as wide as glands
                              26. *E. serpyllifolia*
                KK. Seeds with 3–4 transverse ridges and 0.7–0.9
                  mm. long; appendages 1–1.5 times as wide as
                  glands .... 27. *E. glyptosperma*
              JJ. Plants pubescent; seeds with 4–6 transverse ridges
                and 0.6–0.7 mm. long .... 28. *E. Abramsiana*
    EE. Plants perennial.
      F. Glands discoid, without appendages; caps. 1.7 mm. long; lvs. 2–4 mm.
        long .... 21. *E. Parishii*
      FF. Glands appendaged; lvs. mostly 3–10 mm. long.
        G. Caps. 1.1–1.3 mm. long; ventral stipules united .. 20. *E. polycarpa*
        GG. Caps. 2.3–2.5 mm. long; stipules distinct ...... 24. *E. Fendleri*

1. **E. eriántha** Benth. [*Poinsettia e.* Rose & Standl.] Green erect annual, 1.5–5 dm. tall, freely branched especially above the base, glabrous except at the strigose tips; lvs. narrow-linear, sparsely strigose, 2–5 cm. long, entire, short-petioled, the uppermost forming whorls near the fl.-clusters; invols. (cyathia) 1–4 at end of each branch, white-strigulose, ca. 2 mm. high; glands thin, 3–5, concealed by the inflexed segms. of the

invol.; styles entire; caps. strigulose, 4–5 mm. long; seeds quadrate-oblong, whitish, compressed, coarsely wrinkled, 3.5–4 mm. long, with a stipitate reniform caruncle.— Infrequent, rocky canyons and mesas, below 3000 ft.; Creosote Bush Scrub; Colo. Desert from Eagle Mts. and Andreas Canyon s.; to Tex., Mex. March–April.

2. **E. spathulàta** Lam. [*E. dictyosperma* F. & M. *Tithymalus d.* Heller.] Annual, erect, 2–4.5 dm. high, simple or branched, glabrous; lower lvs. alternate, oblong- or obovate-spatulate, serrulate, 1–3 cm. long, subsessile; floral lvs. opposite, broadly ovate, subcordate or truncate-based, 6–12 mm. long; infl. umbellike, the rays forked; invols. companulate, ca. 1 mm. high; glands small, sessile, yellow, transversely oval; styles bifid; caps. subglobose, 2–3 mm. long, warty-tuberculate; seeds finely reticulate, ellipsoid-lenticular, 1.3–1.5 mm. long, yellowish brown.—Occasional, open and disturbed places on dry slopes and flats, below 4000 ft.; many Plant Communities; throughout cismontane. Calif.; to Wash., Minn., Ala., Mex.; S. Am. March–June.

3. **E. Helioscòpia** L. [*Tithymalus H.* Hill.] WARTWEED. Annual, stems ascending, 1–5 dm. high, rather stout, with few long scattered hairs above; lvs. all obovate, finely serrate, obtuse or retuse, subsessile, 1–2.5 cm. long; umbel usually 5-rayed, the branches then in 3's, ultimately 2's; floral lvs. narrowed to base; glands round to elliptical, yellow, stalked; caps. smooth, subglobose, 2.5–3 mm. high; seeds round-obovoid, 2 mm. long, dark, with coarse honeycomb-like reticulations and yellow reniform caruncle; $2n = 42$ (Perry, 1943).—Natur. as weed near El Monte (Los Angeles Co.), San Francisco, and Elk, Mendocino Co.; from Eu. April–July.

4. **E. Ésula** L. [*Tithymalus E.* Hill. *E. virgata* Waldst. & Kit.] LEAFY SPURGE. Perennial, 3–8 dm. high, with slender creeping rootstocks; cauline lvs. broadly linear to narrowly lance-oblong, 2–2.5 cm. long; floral lvs. broadly ovate, 1–1.3 cm. long; invols. 2.5 mm. high; glands strongly 2-horned; caps. warty, ca. 5 mm. high; seeds ellipsoid-ovoid, yellow-brown, ca. 2 mm. long, with yellow, emarginate caruncle.—Reported from Modoc, Lassen and Siskiyou cos.; e. U.S.; natur. from Eu. June–July.

5. **E. exígua** L. [*Tithymalus e.* Hill.] Annual, erect to depressed, 0.5–4 dm. tall, branched from near middle; lvs. linear, 5–25 mm. long, sessile; floral lvs. lanceolate, connate on one side, 6–12 mm. long; cyathia 1 mm. high, the glands 2-horned, caps. ovoid, ca. 2 mm. long, smooth; seeds quadrangular, ovoid, 1.2–1.5 mm. long, tuberculate, the caruncle small, conical.—Reported long ago from Santa Clara Co.; e. U.S.; native of Eu.

6. **E. Péplus** L. [*Tithymalus P.* Hill.] PETTY SPURGE. Glabrous erect annual, 1–3.5 dm. high, simple or branched from below; cauline lvs. obovate to roundish, obtuse to retuse, 1–2.5 cm. long, distinctly petioled, entire; floral lvs. ovate, obtuse, 0.6–1.5 cm. long; cyathia 1.5 mm. high, campanulate; glands large, yellow, with spreading narrow horns; caps. 2 mm. long, each carpel with a broad thin dorsal keel; seeds oblong, 1.2 mm. long, cellular-punctate, ashy, subhexagonal, the 4 outer faces each with 3–4 rounded pits, the 2 inner with a longitudinal furrow, no caruncle; $2n = 16$ (Perry, 1943).—Frequent in moist places as garden weed, mostly about towns; Atlantic Coast; from Eu. Feb.–Aug.

7. **E. crenulàta** Engelm. [*E. leptocera* Engelm. *Tithymalus c.* Heller. *E. c.* var. *franciscana* Nort. *T. f.* Heller. *E. Nortoniana* A. Nels.] Annual or of longer duration, glabrous; stems 1 or more, simple or branched below and above, 2–6 dm. high; cauline lvs. obovate to spatulate, obtuse, entire, 1.5–3.5 cm. long, subsessile; floral lvs. opposite or in 3's, subcordate, deltoid to rhombic-ovate, 5–15 mm. long; invols. ca. 2 mm. high, with denticulate lobes; glands crescent-shaped, the horns slender, sometimes cleft; caps. smooth, ca. 3 mm. long; seeds ash-colored, oblong-obovoid, irregularly vermiculate-ridged to almost smooth, 2–2.5 mm. long, the yellowish caruncle reniform, somewhat elevated.—Common and widespread in dry places, below 5000 ft.; most Plant Communities from Closed-cone Pine F. and Coastal Scrub to Yellow Pine F. in the Sierra Nevada; Ore. March–Aug.

8. **E. Láthyris** L. [*Tithymalus L.* Hill.] CAPER SPURGE. Annual or biennial, coarse, 3–10 dm. high, glabrous, erect, usually simple below, glaucous; cauline lvs. opposite, strongly decussate, 5–14 cm. long, oblong-lanceolate with cordate clasping base; floral lvs. ovate, cordate, 2–6 cm. long; cyathia ca. 4 mm. long; glands crescent-shaped, strongly 2-horned; caps. 8–12 mm. long; seeds oblong-ovoid, brown with darker spots, cellular-punctate and shallowly reticulate, 4–5 mm. long, the caruncle yellow, helmet-

like.—Occasional weed in waste places, roadsides, etc., Humboldt and Siskiyou cos. to San Diego; to Atlantic Coast; from Eu. Most of year.

9. **E. Pálmeri** Engelm. [*Tithymalus P.* Abrams.] Glabrous somewhat glaucous perennial with several ascending to erect stems from a heavy root-crown, 1–3.5 dm. high, these fairly simple below the few-rayed summit; cauline lvs. obovate, rounded or obtuse at apex, sessile, 0.5–2 cm. long, whorled and larger at base of umbel; floral lvs. round-ovate to subreniform; invols. ca. 3 mm. long, whitish, with rounded entire ciliate lobes; glands crenate, slightly 2-horned; caps. ovoid, 4 mm. long; seeds oblong-ovoid, ashy, 2.5 mm. long, shallowly reticulate-ridged, the brownish caruncle subconic, easily removed.—Common on dry slopes and benches, 4000–9000 ft.; Yellow Pine F., Red Fir F.; Mt. Pinos to Laguna Mts.; to Utah, Ariz.

10. **E. incìsa** Engelm. [*E. schizoloba* Engelm. *Tithymalus s.* Nort.] With habit of *E. Palmeri,* but stems more slender; cauline lvs. ovate to elliptic, 6–14 mm. long, mucronate; floral lvs. ovate to broadly subcordate; invols. whitish, 2–3 mm. high; glands broad, irregularly toothed, hornless; caps. oblong-ovoid, ca. 4 mm. long; seeds oblong-cylindric, 2–2.5 mm. long, irregularly reticulate with low flat ridges, the caruncle small, conical.— Gravelly or rocky slopes, mostly 3000–5000 ft.; Creosote Bush Scrub, Pinyon-Juniper Wd.; e. Mojave Desert (Panamint Mts. to Sheephole and Old Woman mts., and e.); Nev., Ariz. March–May.

11. **E. mísera** Benth. [*Trichosterigma m.* Kl. & Gke.] Irregularly branched shrub 3–9 dm. tall, with grayish twigs and puberulent young growth; lvs. mostly fascicled, round-ovate, entire, 4–15 mm. long, on short slender petioles; stipules fimbriate; cyathia solitary, terminal, 2–3 mm. long, with short inflexed lobes; glands purple with whitish crenulate appendages; caps. 4–5 mm. long, rather smooth; seeds round-ovoid, 2.5 mm. long, slightly pitted or reticulate-wrinkled.—Occasional on sea bluffs; Coastal Sage Scrub; Corona del Mar to San Diego, Catalina Id., and in Creosote Bush Scrub at Whitewater, Colo. Desert; L. Calif. Jan.–Aug.

12. **E. platyspérma** Engelm. [*E. eremica* Jeps.] Glabrous annual, the stems prostrate, 1–2.5 dm. long; lvs. oblong to obovate, 5–10 mm. long; stipules 1.5–2 mm. long, mostly distinct, with 2–3 divisions; cyathia solitary at the nodes, shallow-campanulate, 1.5–2 mm. in diam.; glands slightly radially elongate, 1 mm. wide, the margin sometimes produced into 2 short rounded lobes, not appendaged; ♂ fls. ca. 50; caps. glabrous, round-ovoid, ca. 4 mm. long; seeds white, microreticulate, 2.5–3 mm. long, broadly oblong, the back rounded, smooth, the face with 2 smooth flat slightly concave facets separated by the raphe.—Rare, sandy soil; Creosote Bush Scrub; near Thousand Palms in Coachella V., Yuma. May.

13. **E. ocellàta** Dur. & Hilg. [*Chamaesyce o.* Millsp. *C. sulfurea* Millsp. *E. o.* var. *s.* Jeps.] Prostrate glabrous annual, the stems 1–2 dm. long; median lvs. ovate-deltoid-falcate, 4–10 mm. long, blunt or mucronulate; stipules mostly distinct, filiform or wider, entire or parted; cyathia solitary at the nodes, turbinate to campanulate, 1.5–2 mm long; glands not appendaged, discoid or slightly elongate radially, ca. 0.6 mm. wide; ♂ fls. 40–60; caps. glabrous, subglobose, 2–2.3 mm. long; seeds round-ovoid, whitish, 1.3–1.6 mm. long, smooth to rugose, cellular-punctate.—Dry sandy open places, below 1500 ft.; V. Grassland, Foothill Wd., Coastal Sage Scrub; Shasta Co. through interior valleys to Kern Co., Colton, San Bernardino Co. May–Oct.

Var. **Rattánii** (Wats.) Wheeler. [*E. R.* Wats. *Chamaesyce R.* Millsp.] Plant short-pubescent throughout; glands often with narrow white appendages; seeds somewhat quadrangular.—Sandy or gravelly places, ca. 250–350 ft.; V. Grassland; lower Stony Creek drainage, Glenn Co.

Var. **arenícola** (Parish) Jeps. [*E. a.* Parish. *Chamaesyce a.* Millsp.] Glabrous; lvs. lance-ovate, not or scarcely falcate, 6–15 mm. long, acute; seeds ovoid or slightly angled on back and sides, very smooth.—Occasional, sandy places, below 2500 ft.; Creosote Bush Scrub; Mojave Desert, Searle's Lake s. and e.; to Utah, Ariz. May–Sept.

14. **E. Párryi** Engelm. [*Chamaesyce P.* Rydb.] Glabrous annual, prostrate to ascending, branched from base, the branches 0.6–3.0 dm. long; lvs. linear, entire, 1–3 cm. long; stipules linear, distinct, entire or parted; cyathia on long peduncles, cupuliform, 1.5–1.7 mm. in diam.; glands transversely oval, ca. 0.4 mm. long, with narrow white entire appendages; caps. ca. 2 mm. long, with rounded lobes, glabrous; seeds triangular-

ovoid, 1.8 mm. long, mottled brown and white.—Sand dunes, ca. 2300 ft.; Creosote Bush Scrub; Kelso, Mojave Desert; to Colo., Tex., Chihuahua. May–June.

15. **E. revolùta** Englem. [*Chamaesyce r.* Small.] Glabrous erect ± purplish annual, 0.5–2.0 dm. tall, diffusely branched from below; lvs. linear, entire, revolute, 1–2.5 cm. long; stipules entire, distinct; cyathia solitary in the forks, broadly obconic, ca. 1 mm. in diam.; glands round, 0.2–0.3 mm. wide, with appendages barely evident to radially elongate; caps. glabrous, 1.4 mm. long, sharply angled; seeds triangular-pyramidal to sharply 4-angled, 1–1.3 mm. long, whitish, with 2 transverse rounded ridges.—Rocky slope, at 4000 ft.; Creosote Bush Scrub; n. slope of Clark Mt., e. Mojave Desert; to Colo., New Mex., Chihuahua. Aug.–Sept.

16. **E. maculàta** L. [*Chamaesyce m.* Small. *E. nutans* Lag. *E. Preslii* Guss.] Erect or ascending annual, 1–9 dm. high, simple or mostly branched, glabrate except for the crisp-pubescent young tips; lvs. oblong, 1–3 cm. long, serrulate, very short-petioled; stipules mostly united, broadly triangular-acuminate, ciliate; cyathia solitary or clustered, obconic, 0.7–1 mm. across; glands stipitate, round to transversely elliptical, with rudimentary appendages; ♂ fls. 5–11; caps. 2–2.3 mm. long, glabrous, rather acutely lobed; seeds grayish to brown, ovoid-quadrangular, 1.1–1.6 mm. long, with finely rippled facets; $2n = 28$ (Perry, 1943).—Occasional weed in cismontane Calif.; native, Atlantic Coast to Miss. V. and S. Am. April–Oct.

17. **E. pediculífera** Engelm. [*Chamaesyce p.* Rose & Standl.] Prostrate to ascending perennial from stout taproot, usually gray-strigulose throughout; stems numerous, forked, 1–4 dm. long; lvs. oblique-ovate to oblong, 4–20 mm. long, entire; stipules minute, triangular, those on lower side of stem united; cyathia mostly solitary at the nodes, campanulate, 1.5–2 mm. long; glands transversely oblong, dark red-purple, with appendages from obsolete to 2 mm. wide; ♂ fls. 22–25; caps. strigulose, 2 mm. in diam., 2 mm. long, obtusely lobed; seeds white, narrow-ovoid, 1–1.3 mm. long, encircled by 4–5 rounded ridges.—Infrequent, dry slopes and washes, below 1500 ft.; Creosote Bush Scrub; Colo. Desert (Vallecitos, Midway Well, 20 mi. ne. of Ogilby); to Ariz., L. Calif., Sinaloa. Jan.–April.

18. **E. vallis-mórtae** (Millsp.) J. T. Howell. [*Chamaesyce v.* Millsp.] Perennial, much-branched, forming a dense rounded plant 5–15 cm. high, hoary-tomentose throughout; lvs. suborbicular to oblong-ovate, 4–8 mm. long, entire; stipules on lower side of stem united, on upper side the upper distinct, filiform, hairy; cyathia solitary, campanulate, ca. 2 mm. across; glands transversely oblong, mostly reddish, to 1 mm. long, the white appendages as wide as glands, entire or crenulate; ♂ fls. 17–22; caps. tomentose, 3-angled, 2 mm. long; seeds sharply 4-angled, 1.4–1.7 mm. long, white, the facets nearly smooth.—Dry sandy places, below 4000 ft.; Creosote Bush Scrub; w. Mojave Desert from Owens Lake to Red Rock Canyon. May–Oct.

19. **E. melanadènia** Torr. [*Chamaesyce m.* Millsp. *E. polycarpa* var. *vestita* Wats. *E. p.* var. *appendiculata* Munz. *E. cinerascens* var. *a.* Engelm. *Chamaesyce aureola* Millsp.] Perennial with usually ascending forking stems 8–20 cm. long, hoary-tomentose throughout; lvs. ovate, 2–8 mm. long, entire; stipules on lower side of stem mostly united, linear, hairy, the upper distinct; cyathia solitary, open-campanulate, 1.2–1.5 mm. in diam.; glands transversely oblong, dark red, usually with conspicuous white appendages; ♂ fls. 15–20; caps. tomentose, sharply angled, ovoid, 1.5–1.7 mm. long; seeds white, quadrangular, 1.2–1.5 mm. long, the facets smooth or slightly wrinkled.—Dry stony slopes or flats, below 4000 ft.; Chaparral; Santa Monica Mts., San Gabriel Mts., e. San Diego Co.; Ariz., L. Calif. Dec.–May.

20. **E. polycárpa** Benth. [*Chamaesyce p.* Millsp.] Mostly glabrous prostrate or ascending perennial from slender taproot, the stems 5–25 cm. long; lvs. almost round to oblong, 2–8 mm. long, entire; stipules lanceolate, mostly ciliate, those on ventral side of stem united; invols. solitary at nodes, campanulate, 1–1.5 mm. wide; glands transversely oblong, maroon, 0.5–0.7 mm. long, the appendages broader than the glands; ♂ fls. 15–32; caps. glabrous, rounded, sharply 3-angled, 1.1–1.3 mm. in diam.; seeds rather sharply quadrangular, whitish, 1–1.3 mm. long, the facets smoothish.—Dry slopes and washes, below 3000 ft.; Coastal Sage Scrub, Chaparral; Ventura Co. to San Diego; Creosote Bush Scrub, Mojave and Colo. deserts; to Nev., Ariz., Son., L. Calif. Most of the year. Intergrading freely with:

Var. **hirtélla** Boiss. [*Chamaesyce p.* var. *h.* Millsp. *C. tonsita* Millsp.] Herbage pubescent; appendages narrower than glands.—Sandy places; Creosote Bush Scrub; s. Mojave Desert, Colo. Desert; s. Nev. to Son., L. Calif. Most months of year.

21. **E. Paríshii** Greene. [*Chamaesyce P.* Millsp. *E. polycarpa* var. *P.* Jeps. *E. patellifera* J. T. Howell.] Glabrous perennial, forming prostrate mats; stems 1–2.5 dm. long; lvs. ovate, entire, 2–4 mm. long; stipules broadly linear, entire, ciliate, those on lower surface of stem somewhat united; cyathia solitary, campanulate, 1–1.2 mm. wide; glands discoid, unappendaged, 0.5 mm. wide, yellow or reddish; ♂ fls. 40–50; caps. glabrous, sharply 3-angled, 1.7 mm. long, rounded; seeds white, quadrangular, 1.5 mm. long, the facets faintly wrinkled.—Infrequent, washes and fans, below 3000 ft.; Creosote Bush Scrub; both deserts from Death V. s. to e. San Diego Co.; Nev. April–Oct.

22. **E. micrómera** Boiss. [*E. polycarpa* var. *m.* Millsp. *Chamaesyce m.* Woot. & Standl. *E. pseudoserpyllifolia* Millsp. f. *villosa* J. T. Howell. *E. setiloba* var. *nodulosa* Jeps.] Prostrate annual, glabrous or pubescent, the stems 1–2.5 dm. long; lvs. ovate or oblong, 2–7 mm. long, entire; stipules triangular, ciliate, those of lower side of stem distinct, of upper often united; invols. short-campanulate, ca. 0.9 mm. in diam.; glands discoid, pink or red, mostly unappendaged, 0.1–0.15 mm. wide; ♂ fls. 2–5; caps. glabrous to pubescent, subglobose, 1.3 mm. long; seeds sharply quadrangular, whitish, 1.1–1.3 mm. long, the facets smooth or faintly wrinkled, convex.—Sandy places, below 3000 ft.; Creosote Bush Scrub; deserts from Inyo Co. to San Diego Co.; to Utah, Tex., Mex.; Peru. Mostly Sept.–Dec., also April–June.

23. **E. albomargináta** T. & G. [*Chamaesyce a.* Small.] RATTLESNAKE WEED. Glabrous prostrate perennial, the stems 5–25 cm. long; lvs. rounded to oblong, 3–8 mm. long, with thin whitish entire margin; stipules united into a conspicuous white membranous triangular scale, entire or somewhat lacerate; cyathia solitary at nodes, 1.5–2 mm. in diam., campanulate to obconic; glands transversely oblong, concave, mostly maroon, 0.5–1 mm. long, the appendages conspicuous, white, entire or subcrenate; ♂ fls. 15–30; caps. glabrous, ovoid, sharply 3-angled, 1.7–2.3 mm. long; seeds rounded-quadrangular, oblong, 1.3–1.7 mm. long, opaque-white, the facets smooth.—Common on dry slopes and in fields, below 4000 ft.; Coastal Sage Scrub, Chaparral, Los Angeles Co. to San Diego Co.; V. Grassland in Tulare and Kern cos.; and up to 7500 ft., Creosote Bush Scrub, Pinyon-Juniper Wd., Joshua Tree Wd., deserts from Inyo Co. to Imperial Co.; to Utah, Tex., Mex. April–Nov.

24. **E. Féndleri** T. & G. [*Chamaesyce F.* Small.] Glabrous perennial from a deep-set taproot, the stems decumbent to erect, 5–15 cm. long; lvs. entire, round-ovate to lance-ovate, 3–11 mm. long; stipules distinct, linear, mostly entire and glabrous; cyathia solitary at nodes, campanulate to turbinate, 1.3–1.7 mm. in diam.; glands reddish, 2–4 times as long as wide, up to 1 mm. long, the appendages white, crenate, ca. as wide as gland; ♂ fls. 25–35; caps. glabrous, rounded, ca. 2.4 mm. long; seeds quadrangular, 2–2.2 mm. long, white, with smooth front facets and slightly wrinkled back facets.— Occasional on dry slopes, 5000–7500 ft.; Pinyon-Juniper Wd.; White Mts. (Inyo Co.), New York and Clark mts. (San Bernardino Co.); to Nebr., Tex., Son. May–Oct.

25. **E. Hoòveri** Wheeler. Prostrate or decumbent glabrous annual, the stems 5–20 cm. long; lvs. round-cordate to -reniform, papillate, 2–5 mm. long, spinulose-serrate; stipules united, lacerate; cyathia solitary, campanulate, 1.7–2 mm. in diam.; glands transversely oval, red, aging olive, ca. 0.5 mm. long, the appendages white, parted into 3–5 linear lobes ca. 1 mm. long; ♂ fls. 30–35; caps. glabrous, subglobose, roundly lobed, 1.6–1.9 mm. long; seeds ovoid-quadrangular, 1.4–1.6 mm. long, white, the back semi-circular, the facets with low irregular ridges.—Dried mud flats; V. Grassland; Cent. V. (near Vina, Tehama Co., Visalia and Yettem, Tulare Co.). July.

26. **E. serpýllifòlia** Pers. [*Chamaesyce s.* Small. *E. occidentalis* E. Drew. *E. s.* var. *o.* Jeps. *E. s.* var. *rugulosa* Engelm. *E. r.* Greene.] Glabrous annual, prostrate or erect, the stems 5–35 cm. long; lvs. ovate, oblong, or obovate, 3–14 mm. long, usually serrulate at least toward apex; stipules distinct, linear, entire or few-parted, mostly glabrous; cyathia solitary, 0.8–1.2 mm. in diam., campanulate; glands transversely oblong, 0.2–0.5 mm. long, with narrow white entire to subdentate appendages; ♂ fls. 5–18; caps. glabrous, sharply 3-angled, 1.5–1.9 mm. long; seeds acutely quadrangular, turgid-ovoid, 1–1.4 mm. long, the facets smooth to rugulose, white to brown.—Common, in dry disturbed places, below 7000 ft.; many Plant Communities; through most of Calif.; to B.C., Mich., Ia., Mex. Mostly Aug.–Oct.

Var. **hírtula** (Engelm.) Wheeler. [*E. h.* Engelm. *Chamaesyce h.* Millsp.] Plants ±
villous; lvs. 3–7 mm. long.—Dry places, mostly 4000–8000 ft., Yellow Pine F., Mariposa
Co. to San Diego Co.; lower and in Chaparral and Foothill Wd., Monterey Co.
June–Sept.

27. **E. glyptospérma** Engelm. [*Chamaesyce g.* Small.] Glabrous annual, stems mostly
prostrate, sometimes ascending, 5–25 cm. long; lvs. oblong, often subfalcate, 3–15
mm. long, serrulate at least toward apex; stipules subulate, long-attenuate, with linear
divisions; cyathia solitary, obconic, 0.6–0.9 mm. in diam.; glands transversely elliptic
to oblong, 0.2–0.4 mm. long, with white appendages 1–1.5 times as wide as glands
and ± crenulate; ♂ fls. 1–5; caps. glabrous, sharply 3-angled, 1.5–1.7 mm. long; seeds
sharply quadrangular, 1–1.3 mm. long, white to tan, the facets with 3–4 rounded
transverse ridges.—Reported from dry ground along Sacramento R. near Redding,
Shasta Co.; B.C., New Brunswick, Ind., Tex. Sept.

28. **E. Abramsiàna** Wheeler. [*E. pediculifera* var. *A.* Ewan in Jeps.] Prostrate annual,
finely pubescent to subglabrous, the stems 5–25 cm. long; lvs. ovate- to elliptic-oblong,
2–12 mm. long, sometimes serrulate at apex; stipules distinct, 2–5-parted; cyathia
mostly on congested lateral branches; invols. obconic, 0.6 mm. across; glands roundish
to transversely elliptic, ca. 0.2 mm. long, with wide white appendages entire or somewhat 2-lobed; ♂ fls. 3–5; caps. glabrous, round-oblong, ca. 1.5 mm. long; seeds sharply
quadrangular, white, 1–1.4 mm. long, the facets with 4–6 irregular transverse rounded
ridges.—Sandy flats; Creosote Bush Scrub; Imperial V.; to Ariz., Sinaloa. Sept.–Nov.

29. **E. arizónica** Engelm. [*Chamaesyce a.* Arthur.] Perennial from a woody taproot,
prostrate and matted to erect, the stems 1–3 dm. long, finely clavate-hairy; lvs. deltoid-
ovate to ovate-oblong, entire, 2–10 mm. long, reddish; stipules minute; cyathia long-
turbinate, constricted above, ca. 1.5 mm. high; glands red, concave, almost twice as
long as wide, 0.4 mm. long, the appendages oval, pinkish, to 1 mm. long; ♂ fls. mostly
6–7; caps. with spreading hairs, subglobose, ca. 1.5 mm. long; seeds quadrangular,
whitish, 1–1.2 mm. long, the facets with low, often anastomosing ridges.—Rare;
Creosote Bush Scrub; Colo. Desert (Andreas and Palm canyons, Borrego); to Tex., n.
Mex. March–April.

30. **E. setilòba** Engelm. [*Chamaesyce s.* Millsp.] Annual, mostly prostrate, soft-
pubescent; stems 3–15 cm. long; lvs. oblong to oblong-ovate, 2–7 mm. long, entire;
stipules not evident; invols. turbinate, constricted at summit, ca. 1.2 mm. high; glands
red, mostly transversely oblong, 0.1–0.2 mm. long, the appendages white, ca. 1 mm.
long and wide, parted into 3–5 narrow segms.; ♂ fls. 3–7; caps. hairy, globose, sharply
angled, ca. 1.1 mm. in diam.; seeds brownish white, sharply quadrangular, 1 mm. long,
the facets with low irregular wrinkles.—Sandy places, below 5100 ft.; Creosote Bush
Scrub, Joshua Tree Wd., Pinyon-Juniper Wd.; deserts from Inyo Co. to Imperial and
San Diego cos.; to Nev., Tex., n. Mex. Most months of year.

31. **E. supìna** Raf. [*Chamaesyce s.* Mold. *E. maculata* auth., not L.] Annual, prostrate
or ascending, ± villous; stems 1–4.5 dm. long; lvs. elliptic-ovate to oblong, serrulate
to subentire, 4–17 mm. long; stipules subulate, sometimes lacerate; cyathia mostly on
congested lateral branches, obconic, 0.8 mm. in diam.; glands transversely elongate, ca.
0.2 mm. long, the appendages narrow, white, crenulate; ♂ fls. mostly 4–5; caps. strigose,
sharply angled, ca. 1.4 mm. long; seeds quadrangular, ca. 1 mm. long, the facets whitish-
brown and with low transverse ridges.—Weed about waste places, especially near
towns, throughout cismontane Calif.; introd. from e. U.S. May–Oct.

32. **E. prostràta** Ait. [*Chamaesyce p.* Small. *E. Chamaesyce* auth., not L.] Prostrate
to decumbent annual, crisped-hairy to glabrate; stems 5–25 cm. long; lvs. broadly elliptical to elliptic-oblong, 3–11 mm. long, often serrulate; stipules triangular-subulate, short-
hairy, often lacerate, those on ventral side of stem sometimes united; cyathia mostly
on condensed lateral branchlets, 0.6–0.9 mm. in diam., obconic; glands transversely
oval to oblong, 0.2–0.3 mm. long, the appendages white, 1–2 times as wide as gland;
♂ fls. 4; caps. glabrate on faces, hairy on angles, 1–1.4 mm. long; seeds sharply
quadrangular, 1 mm. long, whitish, the facets with low transverse wrinkles.—Weed
in Ojai, Ventura Co. and Mill Valley, Marin Co.; natur. also in e. U.S. from trop. Am.
Aug.–Sept.

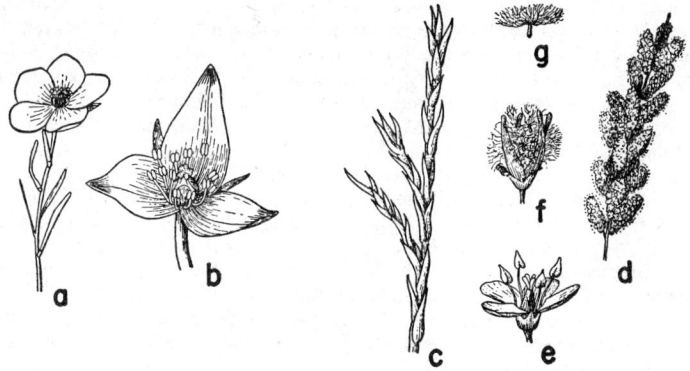

Fig. 22. CISTACEAE. *Helianthemum scoparium:* a, fl., × 1, with 5 petals; b, × 5, the petals removed to show 3 large and 2 small sepals.
TAMARICACEAE. *Tamarix parviflora:* c, leafy branch, × 4; d, infl., × ½; e, 4-merous fl., × 5; f, fl. with dehiscent caps., × 2½; g, seed with coma.

21. Cistàceae. Rock-Rose Family (Fig. 22)

Shrubs or herbs. Lvs. opposite, sometimes alternate, simple, entire; stipules well developed to none. Fls. regular, usually bisexual, solitary or in racemes or panicles. Sepals 5, persistent, 2 wholly external, smaller and bractlike or lacking, the 3 inner persistent. Petals 5, rarely fewer or 0, usually ephemeral. Stamens many. Ovary superior, sessile, 1-celled or falsely 5–10-celled; ovules few to many, on parietal placentae. Style simple; stigma entire or 3–10-lobed. Fr. a 3–5-valved caps.; seeds orthotropous, exalbuminous. Ca. 8 genera and 160 spp., mostly of N. Hemis., some grown as ornamentals.

Valves of caps. 3; lvs. linear to lanceolate. Native subshrubs or herbs .... 1. *Helianthemum*
Valves of caps. 5 or 10; lvs. ovate to oblong. Escaped shrub .... 2. *Cistus*

1. Heliánthemum Mill. Rock-Rose

Herbs or subshrubs. Lvs. in ours largely alternate. Fls. yellow, the broad petals crumpled in the bud. Stigma 3-lobed. Caps. 3-valved. Ca. 120 spp. of Medit. region and N. and S. Am. Our spp. particularly abundant after fires. (Greek, *helios,* the sun, and *anthemon,* fl., since the fls. open only in the sun.)

(Schreiber, B. O. The genus Helianthemum in Calif. Madroño 5: 81–85, 1939.)
Infl. corymbose or short-paniculate, densely glandular. Santa Cruz Id. .... 1. *H. Greenei*
Infl. racemose or paniculate, not glandular.
   Lvs. narrowly linear, green, glabrous to sparsely stellate-pubescent. Widely distributed
                                               2. *H. scoparium*
   Lvs. linear-lanceolate or oblanceolate, densely stellate-pubescent. Amador Co. ... 3. *H. suffrutescens*

1. **H. Greènei** Rob. [*H. occidentale* Greene, not Nym. *Halimium o.* Gross.] Low tufted perennial, 1–2(–3) dm. high, from a woody base, stellate-villous; lvs. linear-lanceolate to oblanceolate, 1.5–3 cm. long, 3–8 mm. wide, the upper linear; infl. strongly glandular-pubescent; outer sepals lanceolate to linear, 3–5 mm. long, the inner thick, ovate, acuminate, 6–7 mm. long; petals 5–9 mm. long; caps. sharply acute, 7–8 mm. long; seeds irregular, subpyramidal, black mottled white, cellular-pitted, ca. 1 mm. long.—Dry slopes and stony ridges; Chaparral; Santa Cruz and San Miguel ids. March–May.

2. **H. scopàrium** Nutt. [*Halimium s.* Gross. *Crocanthemum s.* Millsp. *Linum trisepalum* Kell.] Suffrutescent, with many ascending or spreading stems, 2–3 dm. long, divaricate, ± mottled, but with ultimate twigs erect, greenish, somewhat stellate-pubescent to glabrate; lvs. narrowly linear, somewhat revolute, early deciduous, 1–2.5 cm. long, 1.5–3 mm. wide; infl. a narrow terminal panicle, leafy, few-fld.; outer sepals linear, 2–3 mm. long, the inner ovate, 4–5 mm. long; petals 5–7 mm. long; caps. 3–4 mm. long, ovoid; seeds irregular, somewhat angled, black, cellular-papillose, ca. 1 mm. long.

—Sandy flats and slopes near the coast; Closed-cone Pine F., N. Coastal Scrub; Mendocino Co. to Santa Barbara Co., Santa Cruz and Santa Rosa ids. March–June.

Var. **vulgàre** Jeps. Stems 2–3 dm. high, suberect; lvs. 1–2.5 cm. long, 0.5–1.5 mm. wide; infl. sparsely leafy, narrow, many-fld.; inner sepals 2–3.5 mm. long; petals 4–6 mm. long.—Dry slopes and rocky ridges, below 4000 ft.; Chaparral; Lake Co. to Santa Barbara Co. and Eldorado Co. to Mariposa Co., in s. Calif. near the coast to n. L. Calif.; Catalina Id. March–May.

Var. **Aldersònii** (Greene) Munz. [_H. A._ Greene.] Plants 3–9 dm. tall; lvs. 3–6 cm. long, 2–3.5 mm. wide; infl. open; inner sepals 5–6 mm. long; petals 8–12 mm. long.— Dry sandy or rocky places, below 5000 ft.; chiefly Chaparral; interior s. Calif. from Etiwanda and City Creek, San Bernardino Co. to L. Calif. March–July.

3. **H. suffrutéscens** Schreib. Erect shrub 4–8 dm. high, the twigs woody or herbaceous, with persistent stellate tomentum; lvs. lance-linear to oblanceolate, flat, 1–4 cm. long, 2–8 mm. wide, stellate-pubescent, persistent; panicles leafy; outer sepals 2 mm. long, the inner 4–5 mm. long; petals 6 mm. long; caps. acute.—At ca. 500 ft.; Chaparral; near Bisbee Peak, Amador Co. April–May.

2. Cístus L. ROCK-ROSE

Shrubs. Lvs. opposite, estipulate. Fls. showy in terminal cymes. Sepals 3 or 5; petals 5. Stamens many. Ovary with 5 placentae; style with 5–10-lobed stigma. Caps. 5- or 10-valved. Ca. 20 spp. of Medit. region, several cult. as ornamentals. (Ancient Greek name.)

1. **C. villòsus** L. var. **córsicus** (Lois.) Gross. [_C. c._ Lois.] Erect glandular-pubescent shrub ca. 1 m. high; lvs. ovate to ovate-oblong, 3–7 cm. long, sparsely stellate-pilose above, simple-pilose or glabrate beneath, glandular on both surfaces; fls. 1–4; sepals 5, ovate, acuminate, 13–14 mm. long; petals 20–30 mm. long, mauve or purplish; caps. oblong-globose, ca. 1 cm. long; seeds brownish, shining.—Natur. on openly wooded hills near Black Point, Marin Co.; native of Medit. region.

22. Tamaricàceae. TAMARISK FAMILY (Fig. 22)

Shrubs or small trees. Lvs. alternate, entire, thickish, small, scalelike, exstipulate. Fls. regular, bisexual, small, 4–5-merous. Sepals free or ± united. Petals imbricated, usually withering and persistent. Stamens as many or twice as many as petals, free or united, inserted on a fleshy disk. Ovary 1-celled; styles 3–5; placentae 3–5, parietal. Caps. 3–5-valved; seeds many, usually with terminal tuft of hairs. Genera 4; spp. ca. 100.

1. Támarix L. TAMARISK

Branchlets slender, with minute appressed scaly lvs. Ca. 75 spp., Medit. region to E. Indies and Japan. (_Tamaris_, Spanish river.)

(McClintock, E. Studies in Calif. ornamental plants. 3. The tamarisks. Jour. Calif. Hort. Soc. 12: 76–83, 1951.)

Sepals and petals 5; fls. in spring and summer, borne in racemes on the current year's branches and forming terminal panicles.
  Lvs. and bracts not sheathing; fls. short-pedicelled
    Fils. inserted between the lobes of the disk; petals usually present on frs. ...... 1. _T. pentandra_
    Fils. broadened toward base and confluent with angles of the disk; petals usually deciduous from frs. .... 2. _T. gallica_
  Lvs. and bracts sheathing; fls. sessile .... 3. _T. aphylla_
Sepals and petals 4; fls. in spring before the lvs., borne in lateral racemes on branches of the previous year .... 4. _T. tetrandra_

1. **T. pentándra** Pall. [_T. gallica_ auth., not L.] Glabrous loosely branched shrub or small tree, 1–3(–6) m. tall; lvs. rhombic-ovate, acute or acuminate, keeled, 0.5–3.5 mm. long, the margins scarious; fls. white or pinkish, in slender racemes 2–5 cm. long and grouped in panicles; petals deciduous, 1.5–2 mm. long; stamens 5, the fils. 1.5–2.5 mm. long, attached between lobes of disk; anthers blunt to mucronate; caps. lance-ovoid, 3–4 mm. long; seeds less than 1 mm. long, with white coma; $2n = 24$ (Bowden, 1940).—Well established and widely scattered along water courses and in low places,

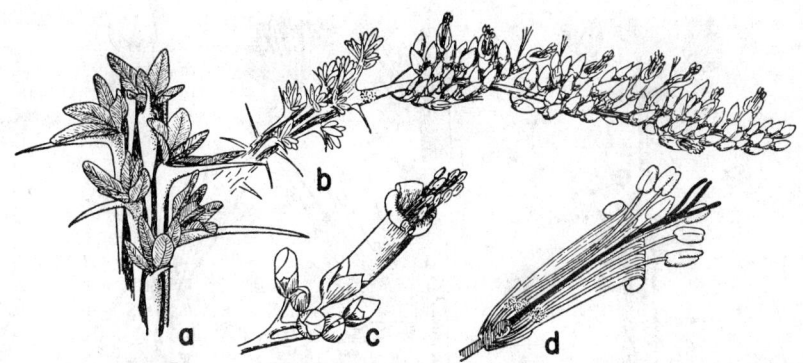

Fig. 23. FOUQUIERIACEAE. *Fouquieria splendens: a,* habit, × ½, with furrowed twig, divaricate spines, fascicled secondary lvs.; *b,* infl., × ⅙; *c,* tubular fl., × 1¼, the imbricated sepals unequal; *d,* fl. in section, × 1½, the stigma 3-lobed.

below 4000 ft., coastal and desert drainage; native Medit. region. April–Aug., but fls. may be found much of year.

2. **T. gállica** L. [*T. anglica* Webb.] Much like *T. pentandra;* sepals 0.5–1.0 mm. long; petals mostly deciduous from mature fr.; fls. attached to the corners of the angled disk; anthers mucronate; $2n = 24$ (Bowden, 1940).—Uncommon in cult. and occasional as an escape; native of Old World. June–Aug.

3. **T. aphýlla** (L.) Karst. [*Thuja a.* L. *Tamarix articulata* Vahl.] ATHEL. Tree 6–10 m. high, with grayish jointed branchlets; lvs. glaucous, sheathing, minute; fls. pinkish-white, sessile, in spicate racemes; stamens 5; $2n = 24$ (Bowden, 1945).—Planted as windbreak in Sacramento, San Joaquin, Coachella, Imperial and other hot valleys and to be expected in low waste places; native Asia and Afr. May–July.

4. **T. tetrándra** Pall. [*T. parviflora* DC.] Shrub or small tree, 3–5 m. tall with many slender branches; lvs. acuminate, ca. 1.5 mm. long; fls. pink, 4-merous, borne in spring in slender racemes 2.5–3.5 cm. long on branches of previous year; stamens 4, the anthers rather long-pointed; caps. ca. 3 mm. long.—Occasional escape in river beds, as Lake, San Benito, Inyo, Kern, and San Diego cos.; native of Medit. region. March–April.

23. **Fouquieriàceae.** OCOTILLO FAMILY (Fig. 23)

Resinous spiny shrubs with erect virgate stems. Primary lvs. soon deciduous (with dry weather), the petioles developing into heavy phyllodial thorns, in the axils of which there later appear fascicles of secondary lvs. Fls. showy, bisexual, in terminal panicles. Sepals 5, imbricated, unequal. Petals 5, hypogynous, connate into a tube, imbricate. Stamens 10 or more, 1–2-seriate, hypogynous, with a basal toothed portion. Disk annular, small. Ovary 1-celled, with 3 parietal septiform placentae, each with about 6 ovules. Fr. a caps. Seeds oblong, compressed, winged, the wing breaking up into hairlike parts. Two genera and 8 spp.; sw. N. Am.

1. **Fouquièria** HBK. OCOTILLO. CANDLEWOOD

Our only genus. Spp. 7; Mex. and sw. U.S. Sometimes planted as hedge-fences. (P. E. *Fouquier,* Parisian medical prof.)

(Shreve, F. Fouquieriaceae, *in* Pflanzenareale 3: 3–4, 1931.)

1. **F. spléndens** Engelm. Stems stout, several to many from the base, stiff, mostly simple and canelike, sometimes few-branched, 2–7 m. tall, gray with darker furrows and stout divaricate spines; lvs. fleshy, oblong-obovate, 1–2.5 cm. long, 1-nerved; panicles 1–2.5 dm. long, dense, many-fld.; sepals 4–7 mm. long, rounded; corolla tubular,

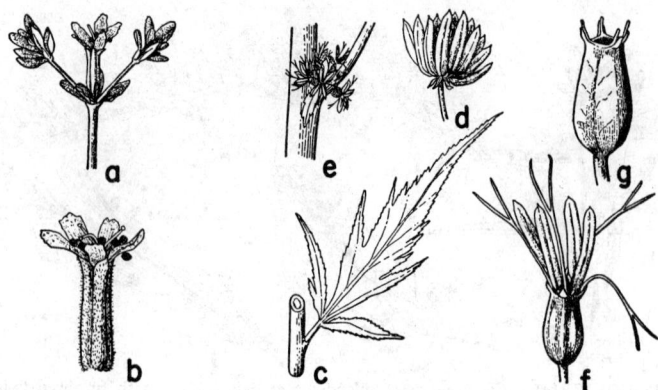

Fig. 24. FRANKENIACEAE. *Frankenia grandifolia: a*, sessile fl. with branches beneath, × 1; *b*, fl., × 2, sepals connate into a tube, petals 5, stamens 6, style 3-cleft.
DATISCACEAE. *Datisca glomerata: c*, lf., × ¼; *d*, ♂ fl., × 2, with many stamens, *e*, ♀ fls., × ½; *f*, ♀ fl., × 2, with 4 stamens and forked styles; *g*, caps., × 2.

scarlet, 2–2.5 cm. long, the lobes rounded, recurved; stamens 10–17, exserted; caps. 1.5 cm. long, 3-valved; seeds white, fringed; $2n = 16$ (Johansen, 1936).—Dry mostly rocky places, below 2500 ft.; Creosote Bush Scrub; se. Mojave Desert and Colo. Desert; to Tex., Mex. March–July. Lvs. soon deciduous, but renewed with both spring and late summer rains.

24. Frankeniàceae. FRANKENIA FAMILY (Fig. 24)

Low perennial herbs or undershrubs. Lvs. opposite, entire, subsessile, exstipulate. Fls. bisexual, regular, small, solitary or in cymes. Sepals 4–6, persistent, connate, tubular. Petals as many, clawed, imbricate, with scalelike appendage within. Stamens usually 6, hypogynous. Ovary superior, 1-celled, with 2–4 parietal placentae and few to many ovules. Fr. a caps., opening by valves. Seeds with endosperm and straight embryo. Four genera and ca. 65 spp., widely distributed in subtrop. and temp. regions.

1. Frankènia L.

Primary lvs. with axillary fascicles. Style 2–3-cleft. Caps. linear, angled. Ca. 50–60 spp. of maritime plants. (J. *Franke,* 1590–1661, the first writer on Swedish plants.)

Lvs. 5–15 mm. long, linear-oblanceolate to obovate, ± expanded; style 3-cleft ...... 1. *F. grandifolia*
Lvs. 2–4 mm. long, revolute and terete; style 2-cleft .... 2. *F. Palmeri*

1. **F. grandifòlia** Cham. & Schlecht. [*Velezia latifolia* Eschs.] Bushy, herbaceous or suffrutescent, 1.5–3 dm. high, glabrous to pubescent or subhirsute; lower lvs. obovate, somewhat revolute, 5–15 mm. long, subsessile, united in pairs by membranous base, upper narrower; calyx narrow-cylindric, 6–7 mm. long, furrowed, with acute teeth; petals pinkish, 2–4 mm. longer than calyx; stamens 4–7, commonly 5; caps. linear, ca. 5 mm. long; seeds brown, sharply subfusiform, ca. 1 mm. long.—Salt marshes, beaches, alkali flats along or near the coast; Coastal Salt Marsh, Coastal Strand; Marin and Solano cos. to L. Calif., also on the ids. June–Oct.

Var. **campéstris** Gray. More tufted; lvs. spatulate, more revolute.—Alkali flats of interior, below 5000 ft.; Alkali Sink; Great V., S. Coast Ranges and Owens V. to Riverside Co.; Nev.

2. **F. Pálmeri** Wats. Shrub, 1–2 dm. high, densely leafy; lvs. oblong-linear, canescent, terete, 2–4 mm. long; calyx 3 mm. long; petals whitish; seeds few.—Coastal Salt Marsh; San Diego; to L. Calif., Son. May–July.

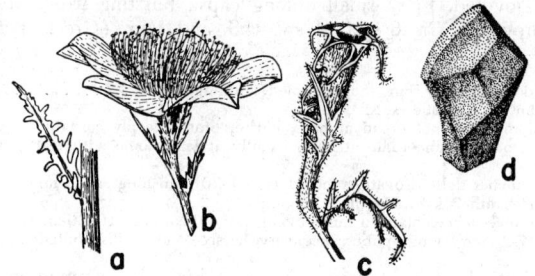

Fig. 25. LOASACEAE. *Mentzelia Lindleyi: a,* small lf., × ½; *b,* fl., × ½, with 5 petals, many stamens; *c,* maturing inferior ovary, × 1, the 5 sepals persistent; *d,* seed, × 5, angled.

25. Datiscàceae. DATISCA FAMILY (Fig. 24)

Herbs or trees. Lvs. alternate, simple or pinnate, exstipulate. Fls. dioecious or rarely bisexual, in spikes or racemes. Staminate fls. with 3–9 calyx-lobes; petals small, 8 or 0; stamens 4–25. Pistillate and perfect fls. with inferior ovary, the stamens reduced to staminodia or developed. Ovary 1-celled, open or closed at apex, the placentae parietal; styles free; ovules many. Caps. opening among the styles; seeds many, minute, with little endosperm and cylindric straight embryo. Small family of n. trop. and subtrop.

1. Datísca L.

Stout perennial glabrous herb. Lvs. pinnately incised. Corolla O. Staminate calyx with 4–9 unequal lobes, 8–12 stamens. Pistillate calyx 3-toothed, sometimes with 2–4 stamens. Two spp., one in Asia. (Meaning unknown.)

1. **D. glomeràta** (Presl) Baill. [*Tricerastes g.* Presl.] Erect, 1 m. or more high, branched; lvs. ovate to lanceolate in outline, acuminate, 1–2 dm. long, the segms. lanceolate, sharply incised-serrate; petioles 2–3 cm. long; fls. several in each axil of a leafy raceme; ♂ calyces 2 mm., ♀ 5–8 mm. long; styles ca. 6 mm. long; caps. ca. 8 mm. long; seeds light brown, subcylindric, ca. 1 mm. long and with ca. 11–12 rows of small pits.—Dry stream beds and washes in mts. and hills, below 6500 ft.; Yellow Pine F., Foothill Wd., Chaparral, Mixed Evergreen F., etc.; Siskiyou and Shasta cos. to San Diego Co., occasional at desert edge; L. Calif. May–July.

26. Loasàceae LOASA FAMILY (Fig. 25)

Herbs or somewhat woody plants, with rough often stinging hairs. Lvs. opposite or (in ours) alternate, entire to variously divided, exstipulate. Fls. bisexual, solitary to cymose or capitate, regular. Floral tube adnate to ovary. Sepals 4–5, rather persistent. Petals 4–5, sessile or clawed. Stamens usually many; fils. free or collected into bundles, filiform or dilated, sometimes petaloid. Ovary inferior, 1–3-celled; style 1; ovules 1–many, parietal. Fr. a caps. Seeds with or without endosperm; embryo straight, linear. Ca. 13 genera, 250 spp., mainly of trop. and temp. Am.

Stamens 5; style entire; seed solitary ....1. *Petalonyx*
Stamens many; style often cleft; seeds few to many.
  The stamens inserted below the petals; placentae 3; style 3-cleft to entire .... 2. *Mentzelia*
  The stamens inserted on the petals; placentae 5.
    Petals whitish, 2.5–4 cm. long, united only at base; style 5-cleft .... 3. *Eucnide*
    Petals yellowish, 1.8 cm. long, united ¾ their length; style entire .... 4. *Sympetaleia*

1. Petalónyx Gray. SANDPAPER PLANT

More or less shrubby herbs, scabrous with short barbed hairs. Lvs. alternate, simple, persistent. Fls. bisexual, small, whitish or yellowish, in terminal heads or short spikes. Sepals 5, deciduous. Petals 5, with long connivent claws. Stamens 5, long-exserted.

Ovary 1-celled, 1-ovuled. Fr. a small oblong caps., bursting irregularly. Seed oblong-ovoid, shining, brownish. Ca. 6 spp. of sw. U.S. and Mex. (Greek, *petalon*, petal, and *onyx*, claw.)

Lvs. 1–2.5 cm. wide, distinctly but short-petioled, shining .... 1. *P. nitidus*
Lvs. usually less than 1 cm. wide, sessile, dull.
   The lvs. greenish, rounded, not broad at base, entire; bracts deeply cordate .... 2. *P. linearis*
   The lvs. grayish, broad to subcordate at base, usually at least some with teeth; bracts subcordate or
   rounded at base.
      Lvs. of main branches deltoid-ovate, subcordate, 10–20 mm. long, 8–13 mm. wide; stamens 5 mm.
      long; pubescence rather soft. Death V. region .... 3. *P. Gilmanii*
      Lvs. of main branches lanceolate to lance-ovate, not subcordate, 10–25 mm. long, 3–6(–10) mm.
      wide; stamens 7–8 mm. long; pubescence very harsh. Widely distributed ...... 4. *P. Thurberi*

1. **P. nítidus** Wats. Woody at base, 1.5–3 dm. high; lvs. shining, round-ovate, 1.5–3 cm. long, coarsely few-toothed on sides; petioles 2–6 mm. long; infl. densely paniculate; bracts ovate, scaberulous, short-ciliate, 6–8 mm. long; sepals sublinear, white, 1.5–2 mm. long; petals whitish, 7–8 mm. long; stamens 9–10 mm. long; caps. ca. 1.5 mm. long.—Uncommon, dry rocky slopes, 3500–6500 ft.; Creosote Bush Scrub, Joshua Tree Wd., Pinyon-Juniper Wd.; mts. of Inyo Co. and San Bernardino Co.; to Utah, Ariz. May–July.

2. **P. lineàris** Greene. Weak rounded bushy shrub, 1.5–4 dm. tall; lvs. lance-oblong to linear, green, scabrellous, entire, 1–2.5 cm. long, 3–6(–8) mm. wide, sessile, rounded at base; infl. congested, terminal; bracts round-ovate, 6–10 mm. long, with deeply cordate base; sepals lance-linear, ca. 1 mm. long; petals white, 2.5–3 mm. long; stamens ca. 4 mm. long; caps. 2 mm. long.—Rocky places in canyons, below 3000 ft.; Creosote Bush Scrub; Colo. Desert from Coachella V. and Whipple Mts. to L. Calif., Ariz. March–May.

3. **P. Gilmánii** Munz. Diffuse rounded cinereous shrub, to ca. 1 m. tall, rather soft pilose-hirsute; lvs. deltoid-ovate, sessile, subcordate, subentire but somewhat wavy, 10–20 mm. long, 8–13 mm. wide, the upper smaller; spikes dense, the bracts subcordate, sessile, 4–6.5 mm. long; sepals lance-linear, 2 mm. long; petals white, 3–4 mm. long; stamens 5 mm. long; caps. ca. 2 mm. long.—Rare, dry washes and slopes, 1500–3000 ft.; Creosote Bush Scrub; Death V. region (Ryan Wash, Ubehebe Crater). May–June, Sept.–Nov.

4. **P. Thúrberi** Gray. Woody at base, low, spreading, grayish, the stems 3–6 dm. long, finely scabrous throughout; lvs. sessile, with broad base, linear- to lance-ovate, 6–25 mm. long, entire or few-toothed; spikes short, dense, the bracts ovate, acuminate, toothed at base, greenish, 5–6 mm. long; sepals 1.5–2 mm. long; petals whitish, 4–5 mm. long; stamens 7–8 mm. long; caps. ca. 2 mm. long.—Frequent in dry sandy or gravelly places, below 4000 ft.; Creosote Bush Scrub; Mojave and Colo. deserts from Inyo Co. to Nev., Ariz., L. Calif. May–July.

2. Mentzèlia L. Blazing-Star

Annual to perennial herbs (or woody), freely branched, covered with various types of rigid tenacious barbed hairs. Lvs. alternate, brittle at least in age, entire, lobed or pinnatifid, sessile or petioled,. scabrous and adhesive to clothing or fur. Fls. bisexual, solitary or cymose. Sepals 5. Petals 5–10, imbricated, mostly free. Stamens many, free or in fascicles opposite the petals, the outer sometimes sterile and petaloid. Ovary inferior, 1-chambered; style simple or 3-cleft at apex; ovules few to many. Caps. cylindric to oblong or turbinate, dehiscent at summit. Seeds prismatic, irregularly angled or flat, with scanty endosperm. Ca. 50 spp. of trop. and subtrop. Am. and w. U.S. (C. *Mentzel*, 1622–1701, German botanist.)

(Darlington, J. A. Monograph of the genus Mentzelia. Ann. Mo. Bot. Gard. 21: 103–226, 1934.)
A. Outer fils. not cleft at apex; petals yellow or orange when fresh; bracts below fls. lacking, or if
    present, not deeply laciniate.
    B. Caps. turbinate or broadly obconic, not more than 3 times as long as thick; seeds horizontal,
       usually lenticular and winged; plants usually biennial or perennial; some hairs glochidiate.
       Section Bartonia)
       C. Petals 5–8 cm. long; sepals 2–4 cm. long .... 1. *M. laevicaulis*
       CC. Petals 0.5–2 cm. long; sepals 0.4–1.2 cm. long.
          D. Fls. solitary, axillary or in upper forks.

E. Plants perennial; sepals 4–5 mm. long; caps. erect . . . . 3. *M. Torreyi*
EE. Plants annual; sepals 6–9 mm. long; caps. reflexed . . . . 4. *M. reflexa*
DD. Fls. not solitary, but cymose or corymbose.
    E. Lvs. sinuate-dentate or nearly entire; caps. 8–10 mm. long.
      F. Cauline lvs. cordate-clasping, oblong; pubescence soft. Deserts above
        3500 ft. . . . . 2. *M. leucophylla*
      FF. Cauline lvs. sessile but not clasping, the upper almost round in outline;
        pubescence harsh. Deserts below 3000 ft. . . . . 5. *M. puberula*
    EE. Lvs. pinnatifid; caps. 10–20 mm. long . . . . 6. *M. multiflora*
BB. Caps. clavate or subcylindric, more than 5 times as long as thick; seeds pendulous, thick,
faceted and angled, not winged; plants annual; hairs not glochidiate. (Section Trachyphytum
or Acrolasia.)
  C. Floral bracts round-ovate to lance-ovate, partly hiding the crowded fls.
    D. Stamens filiform; floral bracts white-membranous . . . . 13. *M. congesta*
    DD. Stamens (the outer) broad; fl. bracts green . . . . 16. *M. micrantha*
  CC. Floral bracts lance-linear or wider, not concealing fls.
    D. Seeds irregularly angled, tuberculate under a lens.
      E. Petals rounded to retuse at apex, 2–20 mm. long; caps. 10–22 mm. long.
        F. The petals yellow, not orange or coppery. Deserts.
          G. Petals 2–4 mm. long; lvs. dentate to entire . . . . 7. *M. albicaulis*
          GG. Petals 5–15 mm. long; lvs. mostly deeply pinnatifid.
            H. Petals 7–10(–12) mm. long . . . . 8. *M. Veatchiana*
            HH. Petals 12–20 mm. long . . . . 9. *M. nitens*
        FF. The petals orange or coppery. Mostly cismontane.
          G. Stems greenish; lvs. entire to pinnatifid; petals 12–20 mm. long
            10. *M. gracilenta*
          GG. Stems whitish; lvs. mostly pectinately pinnatifid; petals 6–10 mm.
            long . . . . 11. *M. pectinata*
      EE. Petals abruptly acuminate to acute with definite mucro and 20–40 mm.
        long; caps. 25–40 mm. long . . . . 12. *M. Lindleyi*
    DD. Seeds cubic-rhomboid, regularly angled with grooves on some angles, almost
    smooth or granular.
      E. Lvs. usually entire, rarely toothed; petals 3–4 mm. long . . . . 14. *M. dispersa*
      EE. Lvs. deeply pinnatifid; petals 6–8 mm. long . . . . 15. *M. affinis*
AA. Outer fils. cleft at the dilated apex, the cent. anther-bearing part subtended by 2 lateral lobes;
petals ochroleucous, satiny; bracts closely subtending fls. and deeply laciniately lobed. (Section
Bicuspidaria)
  B. Floral bracts white-membranous with green margins . . . . 17. *M. involucrata*
  BB. Floral bracts green.
    C. Bracts and upper lvs. subcordate-clasping; caps. erect, 20–24 mm. long. Sw. Colo.
    Desert . . . . 18. *M. hirsutissima*
    CC. Bracts and upper lvs. not clasping; mature caps. reflexed, 12–16 mm. long. Mojave
    Desert . . . . 19. *M. tricuspis*

1. **M. laevicaùlis** (Dougl.) T. & G. [*Bartonia l.* Dougl.] Coarse erect stout biennial,
4–16(–20) dm. high; stems branched, shining white, subglabrous below, stiff-hairy
above; rosette-lvs. oblanceolate, deeply sinuately pinnately lobed, 1–3 dm. long, the
cauline pinnatifid, ovate-lanceolate, sessile, 1–10 cm. long; fls. in terminal clusters of
1–3; floral tube cylindric, 1.5–3 cm. long; sepals lanceolate, acuminate, 2–4 cm.
long, reflexed in fr.; petals light yellow, 5–8 cm. long, lanceolate; stamens many, the
outer 5 with dilated fils.; style 3.5–6 cm. long; caps. subcylindric, 3–4 cm. long; seeds
flat, round-oblong, winged all around, gray-brown, 2–3 mm. long; $n = 11$ (Thompson
& Lewis, 1955).—Dry usually disturbed gravelly and stony places, below 8500 ft.; many
Plant Communities; inner and middle Coast Ranges, Monterey Co. to Siskiyou and
Humboldt cos., Sierra Nevada from Tulare Co. to Plumas Co., Inyo Co. to Modoc Co.,
Mojave Desert, interior cismontane s. Calif.; to Wash., Mont., Utah. June–Oct.

2. **M. leucophýlla** Bdg. [*M. oreophila* J. Darl.] Biennial, 3–4 dm. high; stems soft-
pubescent; basal lvs. oblanc-oblong, 6–8 cm. long, ashy, sinuate-dentate, the cauline
oblong, 2–4 cm. long, cordate-clasping, scabrous-pubescent but hairs scarcely enlarged at base; infl. few-branched; sepals lanceolate, obtuse, 6–10 mm. long; petals pale
yellow, retuse, pubescent at apex, 9–11 mm. long; outer stamens petaloid; caps. 8–10
mm. long, almost as thick; seeds flat, narrowly margined.—Gravelly washes in dry
canyons, 3000–6000 ft.; Creosote Bush Scrub, Joshua Tree Wd.; Argus Mts., e. Inyo
and San Bernardino cos.; sw. Nev. May–June.

3. **M. Tórreyi** Gray. Cespitose perennial, densely branched, 5–15 cm. high; stems
several, white, pubescent and stiff-hairy; lvs. sessile, lanceolate to ovate in outline, 2–5
cm. long, 3–5-cleft into lance-acuminate divisions with revolute margins, the cent. lobe
spine-tipped; fls. solitary, axillary; floral tube ca. 3 mm. long, oblong; sepals linear-
subulate, 4–5 mm. long, not reflexed in fr.; petals 5, pale yellow, oblanceolate, 8–10

mm. long; stamen-fils. all filiform; style 3-cleft; caps. urceolate, 4–6 mm. long; seeds 4–5(–9), oblong, 5-angled, subrugulose, 2 mm. long, dark, shiny.—Mostly dry volcanic soil, 4000–6750 ft.; Pinyon-Juniper Wd., Shadscale Scrub; Mono Co.; to Ida., Nev. June–Aug.

4. **M. refléxa** Cov. Stout diffusely branched annual, 1–2 dm. high, densely short-hirsute; lower lvs. narrow-oblanceolate, short-petioled, soft-pubescent, 5–7 cm. long, the cauline ovate to subhastate, nearly sessile, 2–5 cm. long, irregularly deep-sinuate; fls. solitary in upper forks; fl.-tube obovoid, 7–10 mm. long; sepals narrow-subulate, 6–9 mm. long; petals 8, yellow, almost as long as sepals; fils. 9–15, somewhat dilated, narrowed abruptly near apex, 3–5 mm. long; style stout, 3-cleft; caps. oblong-obovoid, reflexed at maturity; seeds 10–12, irregularly angular, obovate, compressed, muriculate, transversely grooved on both faces, 1.7–2 mm. long.—Dry rocky places, below 4000 ft.; Creosote Bush Scrub; Death V. region to Baker and Kelso, Mojave Desert. March–May.

5. **M. pubérula** J. Darl. Perennial, 1–3 dm. high, openly branched; stems white, scabrous-pubescent; basal lvs. oblong-lanceolate, 6–8 cm. long, short-petioled, scabrous (the hairs subconical, enlarged at base), irregularly and coarsely sinuate-dentate, the cauline round-ovate to ovate-oblong, 1–6 cm. long, sessile or subsessile, but not clasping, coarsely dentate; fls. yellow, at ends of branches; fl.-tube 5–6 mm. long; sepals 5, lanceolate, revolute, persistent, 6–8 mm. long; petals broadly oblanceolate, 8–10 mm. long, obtuse to emarginate; outer fils. petaloid, the inner linear; caps. turbinate, 8–10 mm. long; seeds light brown, slightly punctate, broadly winged, ca. 2 mm. in diam.—Occasional in rocky or gravelly places, below 2500 ft.; Creosote Bush Scrub; from Ord Mts. (Mojave Desert) to Chocolate Mts. (Colo. Desert); Ariz., L. Calif. March–May.

6. **M. multiflòra** (Nutt.) Gray. [*Bartonia m.* Hook. *Nuttallia m.* Greene. *M. longiloba* J. Darl.] Perennial from a stout taproot, 4–8 dm. high, with smooth or slightly scabrous white stems, pubescent upward; lower lvs. short-petioled, the upper sessile, all lanceolate, regularly sinuate-pinnatifid, 2–10 cm. long, rather green, scabrous; fls. corymbose, each branch ending in a cluster of 3–4; fl.-tube 7–10 mm. long; sepals 5, lance-subulate, persistent, often becoming reflexed, 8–12 mm. long; petals 10, oblong-oval, obtuse, 15–20 mm. long, yellow; stamens many, the outer fils. somewhat broadened; style filiform; caps. urceolate, 10–20 mm. long, 5–8 mm. thick, scabrous, rounded below; seeds pale, flat, broadly wing-margined, ca. 2.5 mm. in diam.; $n = 9$ (Thompson & Lewis, 1955).—Dry sandy and gravelly places, below 6500 ft.; Creosote Bush Scrub, Pinyon-Juniper Wd.; deserts, Inyo Co. to Imperial Co.; to Wyo., Tex., Mex. April–June.

7. **M. albicaùlis** Dougl. ex Hook. [*Bartonia a.* Dougl. *Acrolasia a.* Rydb.] Annual, 1–4 dm. high, with slender white shining glabrous to pubescent stems; lvs. sessile, scabrous, the lower narrow-oblanceolate, 3–5 cm. long, dentate to entire, the upper linear to lanceolate, 2–3 cm. long, entire to somewhat lobed; fls. axillary, sessile, the lower solitary, the upper in corymbose clusters of 3; fl.-tube narrow-obconic, 5–6 mm. long; sepals lance-subulate, 2–2.5 mm. long; petals yellow, obovate, 2–6 mm. long, emarginate; stamens shorter than petals; fils filiform; caps. linear-cylindric, 10–16 mm. long, ca. 2 mm. thick; seeds irregularly cubical, angled, finely tuberculate, ca. 0.7 mm. long; $n = 27$ (Thompson & Lewis, 1955).—Common, dry sandy places, below 7000 (10,000) ft.; many Communities from Creosote Bush Scrub to Yellow Pine F.; e. of Cascade-Sierran axis from Colo. Desert to B.C., Nebr., New Mex. Occasional, inner S. Coast Ranges. March–July.

8. **M. Veatchiàna** Kell. [*M. gracilenta* var. *V.* Jeps. *M. albicaulis* var. *V.* Urb. & Gilg.] Resembling *M. albicaulis* in habit and stature; stems white, slender; lvs. mostly deeply pinnatifid; petals yellow, 7–12 mm. long; $n = 18$ (Thompson & Lewis, 1955).—Dry sandy places; Creosote Bush Scrub, etc.; Colo. and Mojave deserts to se. Ore., Utah, Ariz. April–June.

9. **M. nìtens** Greene. [*Acrolasia n.* Rydb. *M. gracilenta* var. *n.* Jeps. *M. Lindleyi* var. *eremophila* Jeps. *M. n.* var. *e.* J. Darl. *M. g.* var. *e.* Jeps.] Stems glabrous, white, shining; lvs. pinnatifid with linear lobes or the uppermost subentire; sepals 4–7 mm. long; petals bright yellow, 12–20 mm. long; $n = 9$ (Thompson & Lewis, 1955).—Occasional, 1500–5000 ft.; Creosote Bush Scrub, Pinyon-Juniper Wd.; Mono Co. to Riverside Co.; to Ariz., Utah. April–June.

10. **M. gracilénta** T. & G. [*M. albicaulis* var. *g.* Wats. *Acrolasia g.* Rydb.] Stems greenish, pubescent; lvs. oblong to lance-linear, entire to pinnatifid; sepals 6–10 mm. long; petals golden to cream, rounded to retuse at apex, 10–20 mm. long; $n = 9$ (Thompson & Lewis, 1955).—Dry open often rocky places, below 3000 ft.; Chaparral, Foothill Wd.; S. Coast Ranges, Monterey Co. to Ventura and Kern cos. March–May.

11. **M. pectinàta** Kell. [*M. albicaulis* var. *p.* Urb. & Gilg. *M. gracilenta* var. *p.* Jeps.] Stems greenish-white, stoutish; lvs. broadly lanceolate, pectinately lobed or pinnatifid; sepals 5–6 mm. long; petals orange above, coppery-red toward base, 10–12 mm. long; $n = 9$ (Thompson & Lewis, 1955).—Occasional, dry places, below 6000 ft.; Yellow Pine F., Sagebrush Scrub, Foothill Wd., etc.; w. base of Sierra Nevada from Plumas Co. to Kern Co., inner S. Coast Ranges from Monterey Co. to San Diego Co.; Santa Cruz Id. April–June.

12. **M. Líndleyi** T. & G. [*Bartonia aurea* Lindl. *M. Bartonia* Steud. *M. a.* Baill., not Nutt.] Hispid annual, 1–6 dm. high, mostly freely branched; lvs. lanceolate to ovate, sessile, slightly clasping at base, 2–17 cm. long, mostly pectinately pinnatifid; fls. axillary and solitary or in terminal clusters of 2–3; sepals 10–17 mm. long, acuminate; petals obovate, abruptly acute with a definite mucro, 20–40 mm. long, 16–33 mm. wide, golden-yellow with orange-red base; stamens 2–3 cm. long, the fils. mostly filiform, or the outer somewhat dilated at base; caps. linear-clavate, 2.5–4 cm. long, hirsute; seeds many, gray-brown, somewhat mottled, irregularly angled, 1–1.5 mm. long, microscopically tessellate; $2n = 36$ (Sugiura, 1936).—Sunny rocky slopes, below 2500 ft.; Coastal Sage Scrub, Foothill Wd.; S. Coast Ranges, Alameda Co. to Santa Clara Co. and w. Stanislaus and Fresno cos. Occasional elsewhere as escape from gardens. April–June.

Ssp. **cròcea** (Kell.) C. B. Wolf. [*M. c.* Kell.] Petals 20–30 mm. long, 10–15 mm. wide, with less pronounced mucro; $n = 18$ (Thompson & Lewis, 1955).—Exposed rocky places, below 4000 ft.; Foothill Wd.; foothills of Sierra Nevada from Tuolumne Co. to Tulare Co. March–May.

13. **M. congésta** (Nutt.) T. & G. [*Trachyphytum c.* Nutt., pro synon. *Acrolasia c.* Rydb.] Annual, 1–4 dm. high, mostly branched, hispidulous; lower lvs. lanceolate, 3–9 cm. long, pinnatifid, the upper reduced, clasping; fls. congested at ends of branches, subtended by conspicuous 3-lobed broadly ovate bracts 6–18 mm. long and white-membranous in center; sepals 2–3 mm. long; petals pale yellow with orange base, 4–5 mm. long; caps. cylindrical, 8–10 mm. long; seeds irregularly round-ovoid, angled, minutely punctate, 1–1.3 mm. long; $n = 9$ (Thompson & Lewis, 1955).—Burns and loose coarse soil, 4000–9000 ft.; Creosote Bush Scrub, Sagebrush Scrub, Pinyon-Juniper Wd., N. Juniper Wd.; Argus Mts. and Panamint Mts., along Sierra Nevada to Ore., Nev.; Yellow Pine F., Palomar Mts., San Diego Co. May–July.

Var. **Davidsoniàna** (Abrams) Macbr. [*Acrolasia D.* Abrams. *M. D.* Abrams.] Stems simple to few-branched; lvs. lance-linear, mostly subentire; sepals ca. 1 mm. long; petals 2–3 mm. long.—Dry gravelly soil, 4200–8100 ft.; Pinyon-Juniper Wd., Yellow Pine F.; Mt. Pinos and Panamint Mts. to San Bernardino Mts.; apparently also in w. Nev. May–July.

14. **M. dispérsa** Wats. [*Acrolasia d.* A. Davids. *M. pinetorum* Heller. *M. d.* var. *obtusa* Jeps. *M. d.* var. *p.* Jeps. *M. d.* var. *latifolia* Macbr. *A. montana* A. Davids. *A. desertorum* A. Davids.] Annual, 1–3(–4) dm. high; stems slender, whitish, somewhat pubescent; lower lvs. lanceolate, entire, 3–10 cm. long, the upper ovate, entire to sinuate-pinnatifid, 1–3 cm. long, all sessile; fls. approximate near ends of branches; sepals 1.5–3 mm. long; petals 3–4 mm. long, yellow with basal orange spot; caps. very narrowly clavate, 15–25 mm. long; seeds rhomboid-cubic, smoothish but somewhat muriculate under magnification, grooved on the angles.—Dry, disturbed, sandy to rocky places, below 6500 (8500) ft.; many Plant Communities; through most of Calif.; to Wash., Rocky Mts. May–Aug.

15. **M. affìnis** Greene. [*Acrolasia a.* Rydb. *A. viridescens* Heller.] Stems somewhat coarser; lvs. ovate-lanceolate, mostly pinnatifid; sepals 4–6 mm. long; petals 6–8 mm. long; $n = 9$ (Thompson & Lewis, 1955).—Open places, below 5000 ft.; V. Grassland, inner S. Coast Ranges and San Joaquin V., Alameda Co. to Kern Co., and s. away from the immediate coast to L. Calif.; Creosote Bush Scrub; Inyo Co. to Imperial Co.; Ariz. April–May.

16. **M. micrántha** (H. & A.) T. & G. [*Bartonia m.* H. & A. *Acrolasia m.* Rydb. *A. m.*

var. **stricta** A. Davids. *M. m.* var. *s.* A. Davids. *A. catalinensis* Millsp.] Annual, 1–4(–8) dm. high; stems branched from base, rather coarse, greenish to whitish, ± pubescent; lower lvs. lanceolate to ovate, 1–15 cm. long, sinuately toothed or serrate, the upper subcordate at base, lanceolate to ovate, sinuate-toothed, 2–8 cm. long, all rough-hispid; fls. congested in compact clusters, scarcely if at all exceeding the broad leafy bracts; sepals 2–3 mm. long, erect; petals pale yellow, 3–4 mm. long; the 5 outer stamens petallike, the inner fils. linear; caps. linear, 7–9 mm. long; seeds prismatic with grooved angles, 1.5 mm. long, minutely punctate; $n = 9$ (Thompson & Lewis, 1955).—Dry disturbed places such as burns, below 5000 ft.; Chaparral; Coast Ranges, Trinity Co. to San Bernardino and San Diego cos.; Catalina and San Clemente ids. April–July.

17. **M. involucràta** Wats. [*Bicuspidaria i.* Rydb.] Hispid erect annual, usually branched from base, 1–4 dm. high, with stout white pubescent stems; lvs. linear to oblong-lanceolate, 4–12(–16) cm. long, coarsely sinuate-dentate, the upper sessile and clasping; fls. solitary, terminal, subtended by a pair of broad white scarious deeply toothed green-tipped bracts 1.5–3 cm. long; sepals 8–14 mm. long; petals pale cream-color, satiny, with reddish veins and base, obovate, abruptly acuminate, 15–30 mm. long; stamens many, each outer fil. with 2 long linear cusps; caps. subcylindric, 15–25 mm. long, ca. 5–6 mm. thick, papery; seeds horizontally flattened, not margined, irregular-oval to round-quadrangular, minutely granular, 3–3.5 mm. long; $n = 9$ (Thompson & Lewis, 1955).—Common in sandy, gravelly, or rocky places, below 4500 ft.; Creosote Bush Scrub; deserts from Inyo Co. to Son. and L. Calif. Jan.–May.

Var. **megalántha** Jtn. Petals 35–65 mm. long.—Chiefly Coachella V., also Twentynine Palms, Eagle Mts., Chuckwalla Mts. Jan.–April.

18. **M. hirsutíssima** Wats. var. **stenophýlla** (Urb. & Gilg.) Jtn. [*M. s.* Urb. & Gilg. *M. Peirsonii* Jeps.] Diffusely branched annual or of longer duration, somewhat hirsute above, 1.5–3 dm. high, the older stems white, rather coarse, glabrate; lvs. ovate to lance-ovate, 2.5–8 cm. long, sessile, acuminate, pinnatifid with acute divaricate segms.; bracts clasping, green, acuminate, ca. 2 cm. long, with narrow divaricate lobes; sepals dark, 12–16 mm. long, lanceolate, long-acuminate; petals pale yellow, 16–22 mm. long, obovate, abruptly acuminate or acute; stamens many, the outer fils. with 2 short lateral lobes and a cent. anther-bearing lobe twice as long as lateral; caps. cylindric, 20–24 mm. long, 4–5 mm. thick; seeds grayish, irregularly oval, ca. 2 mm. long, irregularly rugose, minutely granular-punctate.—Rare, in sandy places, 800–2300 ft.; Creosote Bush Scrub; Mountain Springs region, sw. Colo. Desert; n. L. Calif. April–May.

19. **M. tricúspis** Gray. [*Bicuspidaria t.* Rydb. *Nuttallia t.* Davids. & Mox.] Hispidulous annual 5–20 cm. high, simple or branched; lvs. linear- to ovate-lanceolate, 2–7 cm. long, subentire to sinuate-dentate; fls. solitary, pedicelled, subtended by greenish lance-linear subpinnatifid bracts 1–2 cm. long; sepals 10–13 mm. long, subulate-acuminate; petals pale yellow, narrow-obovate, sharply apiculate, 16–25 mm. long; stamens many, the outer fils. with lateral lobes 2 or more mm. long; caps. oblong-cylindric, reflexed, 12–16 mm. long, 4–5 mm. thick; seeds broadly oblong, rugose with 2–3 deep folds, gray-white, minutely granular-punctate, ca. 1.7 mm. long.—Sandy places, below 2500 ft.; Creosote Bush Scrub; e. Mojave Desert, especially about Needles; to Utah, Ariz. March–May.

Var. **brevicornùta** Jtn. [*Acrolasia tridentata* A. Davids.] Lateral lobes of outer fils. oblong, ca. 0.5 mm. long.—Dry sandy and gravelly places; Creosote Bush Scrub; Mojave Desert from Red Rock Canyon, Kern Co. to Daggett and Newberry Mts., San Bernardino Co. April–May.

3. Eucnìde Zucc. ROCK-NETTLE

Shrubs or herbs armed with stinging barbed hairs. Lvs. alternate, obovate, petioled, toothed. Fls. pedicelled, mostly in terminal bracted cymes. Sepals 5, persistent. Petals 5, deciduous, united at base. Stamens many, the fils. filiform, united below, adnate to petals. Style 5-cleft. Placentae 5; ovules many. Caps. obovoid, 5-valved at top; seeds many, minute, longitudinally striate. Ca. 8 spp. of sw. U.S. and Mex. (Greek, *eu,* true, and *cnide,* sea-nettle).

1. **E. ùrens** (Gray) Parry. [*Mentzelia u.* Gray.] Rounded bush 3–6 dm. high, often much broader, very hispid, the long hairs barbed or not, also finely puberulent, with straw-colored stout often decumbent stems; lf.-blades ovate, coarsely toothed, 2–6

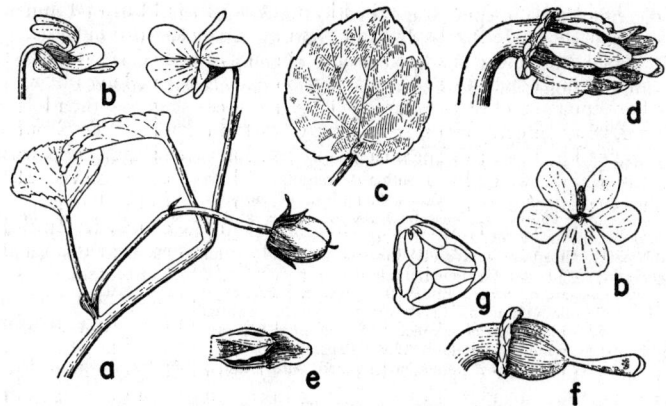

Fig. 26. VIOLACEAE. *Viola sempervirens: a,* habit, × 1; *b* and *b,* fl., × 1, from side and front; *c,* lf., × 1; *d,* stamens after removal of perianth, × 5, with broad fils.; *e,* stamen, × 5, showing broad fil. continued beyond the anther; *f,* pistil, × 5; *g,* cross section of ovary, × 2½, with parietal placentation.

cm. long, on somewhat shorter hispid petioles; pedicels 6–12 mm. long; sepals lanceolate, 15–18 mm. long; petals cream, narrow-obovate, 2.5–4 cm. long; stamens 10–15 mm. long; caps. obovoid, 10–12 mm. long.—Dry rocky places, 2000–4500 ft.; Creosote Bush Scrub; Death V. region, Inyo and Argus mts.; to Utah, Ariz., L. Calif. April–June.

4. Sympetalèia Gray

Bristly annual, the hairs simple and barbed. Lvs. alternate, round, ± cordate, toothed to shallowly lobed, petioled. Fls. solitary, axillary. Sepals 5, persistent. Petals 5, united into a tube with short lobes. Stamens many, inserted on corolla-tube. Anthers 1-celled. Placentae 5; ovules many. Caps. oblong-obovoid, truncate, 5-valved at apex. Seeds minute. Two spp. of sw. U.S. and adjacent Mex. (Greek, *sym,* united, and *petalon,* lf. or petal).

1. **S. rupéstris** (Baill.) Gray. [*Loasella r.* Baill.] Branched, 1–3 dm. high, harshly pubescent; lvs. 2–10 cm. wide, unevenly and coarsely dentate; sepals elliptic, 4–5 mm. long, green; corolla yellow, tubular, 10–12 mm. long, the lobes 2–3 mm. long, green, ovate; caps. 7–8 mm. long, on recurved or contorted pedicels twice as long.— Reported from Painted Gorge, 7 mi. n. of Coyote Wells, sw. Imperial Co.; L. Calif. to Son. Dec.–April.

27. Violàceae. Violet Family (Fig. 26)

Herbs (with us), shrubs or rarely trees. Lvs. alternate or basal, simple, entire to laciniate, with stipules. Fls., in ours, irregular, axillary, nodding, bisexual, the peduncles usually 2-bracted. Sepals 5, free or slightly connate, nearly alike, usually persistent. Petals 5, hypogynous, the lower one usually larger than the others and saccate or spurred. Stamens 5, alternate with petals, the fils. short, broad, continued beyond the anther-locules, often connate into a ring around the ovary. Ovary free, sessile, 1-celled, with 3 (2–5) parietal placentae bearing 1–many ovules; style usually clavate, with the simple stigma turned to one side. Fr. a caps., dehiscent by valves. Seeds large, with hard seed-coat and straight embryo. Ca. 15 genera and 400 spp. of wide distribution.

1. Viòla L. Violet

Annual to perennial herbs. Fls. in most spp. of 2 kinds, those of early season with showy petals and mostly fertile, of later season cleistogamous, apetalous, and often producing seeds. Lower petal spurred, the other 4 an upper usually larger pair and a lateral pair. Stamens 5, enclosing stamineal cavity, the 2 lower with nectarlike appendages projecting into the spur. Caps. ovoid, the 3 valves boat-shaped and with thick rigid keels and thin sides; the latter dry and contract when mature and cause the seeds to be discharged forcibly. Over 300 spp., mostly of temp. zones; many, like the Common Pansy and Garden Violet, of great beauty. (*Viola,* the classical name.)

(Baird, Viola B. Wild violets of N. Am. Univ. Calif. Press, 1941. Baker, M. S. Studies in western violets. A series of papers *in* Madroño and Leafl. W. Bot.)
- A. Stipules almost as large as lvs., leaflike with large terminal green lobe and pinnatisect at base; annual; petals pale yellow. Introd. (Section Melanium.) . . . . 23. *V. arvensis*
- AA. Stipules much smaller and very different from lvs.; perennials.
  - B. Fls. with some yellow, at least at bases of petals and on spur; access to the spur cavity blocked by the much expanded retrorsely bearded thickened head of the style; bearding of lateral petals short and clavate; spur ca. as broad as long. Plants of dry habitats (except *V. glabella*). (Section Chamaemelanium)
    - C. Lvs. not cut or lobed, but with margins entire or serrate-crenate.
      - D. Corolla yellow on face except for dark veining.
        - E. Plants strictly erect with cauline lvs. and fls. crowded at ends of stems which are naked below; caps. acute.
          - F. Upper petals not backed with brown; stipules almost entire, ± scarious 1. *V. glabella*
          - FF. Upper petals backed with brown; stipules deeply toothed, ± green 2. *V. lobata*
        - EE. Plants erect to prostrate, but with lvs. and fls. scattered along the stems; caps. obtuse.
          - F. Stems prostrate, stolonlike; lvs. coriaceous, evergreen, round-cordate. Redwood F. . . . . 3. *V. sempervirens*
          - FF. Stems ascending to erect; at least upper lvs. narrowed.
            - G. Crown of rootstock near the surface; fls. 2 cm. or less across; cleistogamous fls. present in upper axils.
              - H. Plants woolly; caps. woolly.
                - I. Petals yellowish on back. Plumas Co. to Eldorado Co. 6. *V. tomentosa*
                - II. Petals purplish on back. Mono Co. . . . . 7. *V. aurea*
              - HH. Plants pubescent to subglabrous, not truly woolly.
                - I. Caps. glabrous or glabrate; petals not purplish on backs, although the whole corolla may be somewhat purplish in age.
                  - J. Lf.-blades entire, 2–5 cm. long. Placer Co. to s. Ore. 5. *V. Bakeri*
                  - JJ. Lf.-blades ± toothed. Humboldt Co. to Modoc Co. 4. *V. praemorsa*
                - II. Caps. minutely pubescent; petals with some purple on backs.
                  - J. Blades of basal lvs. as long as or longer than wide.
                    - K. Petals 6–10 mm. long; lvs. rather spreading, greenish to grayish, usually tinted purple; upper lvs. subentire or irregularly or few(2–4)-toothed. Yellow Pine F. or higher . . . . . . 8. *V. purpurea*
                    - KK. Petals 10–12 mm. long; lvs. erect, grayish, without purple tinting; upper lvs. rather regularly 5–6-dentate on each side. Below Yellow Pine F. 9. *V. quercetorum*
                  - JJ. Blades of basal lvs. commonly wider than long.
                    - K. Teeth of basal lvs. shallow and very irregular. Mostly n. Calif. . . . . 8. *V. purpurea*
                    - KK. Teeth of basal lvs. deep (2–3 mm.) and regular. Mts. at w. edge of Mojave Desert 7. *V. aurea* ssp. *mohavensis*
            - GG. Rootstocks short (1–2 cm. long), soft; fls. more than 2 cm. across; cleistogamous fls. lacking. Below Yellow Pine F. 10. *V. pedunculata*
      - DD. Corolla white with purple veins or blotches; upper petals red-violet on back.
        - E. Plants pubescent; lvs. mostly cordate at base, rugose . . . . 11. *V. ocellata*
        - EE. Plants glabrous; lvs. truncate or cuneate at base, not rugose . . 12. *V. cuneata*
    - CC. Lvs. deeply cut or lobed.
      - D. Lf.-blades palmately 3–7-lobed, these divisions broad and not dissected; stems naked below . . . . 2. *V. lobata*
      - DD. Lf.-blades dissected into narrow ultimate divisions.
        - E. Lvs. twice palmately divided; petals pale yellow; cleistogamous fls. present. Plants of wooded or brushy slopes . . . . 13. *V. Sheltonii*
        - EE. Lvs. 2–3 times pinnatifid; cleistogamous fls. lacking. Plants of open grassy places.
          - F. Upper petals, as well as others, bright yellow in front, the 2 upper brownish on back . . . . 14. *V. Douglasii*
          - FF. Upper petals purplish.

    G.   Plants minutely pubescent; lateral and lower petals bluish or white with yellow base .... 15. *V. Beckwithii*
    GG.   Plants glabrous; lateral and lower petals yellow or cream-color 16. *V. Hallii*
  BB. Fls. white, blue, or purple, without any yellow; access to spur cavity not blocked by the style (the stigma held above the floor of the lower petal); bearding of lateral petals longer, scarcely clavate; spur longer than broad. Plants of moist places. (Section Nomimium)
    C. Plants lacking lf.-bearing stems (although they may have stolons), the lvs. and peduncles arising from the base or crown.
      D. Petals white, 6–10 mm. long; lvs. rounded to ovate-cordate, mostly less than 2 cm. wide. Widespread in mts. .... 18. *V. Macloskeyi*
      DD. Petals pale violet to blue; lvs. mostly 2.5–5 cm. wide, or if less, not rounded.
        E. Lvs. round to reniform.
          F. Plants not stoloniferous; rhizome thick and fleshy; petals bright violet, 10–14 mm. long. San Jacinto Mts. to Modoc Co. .. 17. *V. nephrophylla*
          FF. Plants stoloniferous, the rhizome cordlike.
            G. Petals pale violet to almost white, 6–10 mm. long; fls. not fragrant; plant glabrous. Native, Mendocino Co. n. .. 20. *V. palustris*
            GG. Petals deep violet, ca. 14 mm. long; fls. very fragrant; plant pubescent. Escape from cult. .... 21. *V. odorata*
        EE. Lvs. lanceolate to lance-ovate; petals nearly white, 10–14 mm. long. Del Norte Co. .... 19. *V. lanceolata*
  CC. Plants with evident leaf-bearing stems; petals violet, 8–12 mm. long. Widespread in mts. 22. *V. adunca*

1. **V. glabélla** Nutt. Bright green glabrous or finely pubescent perennial, from a thick scaly horizontal rootstock; stems rather weak, suberect, 1–3 dm. high; basal lvs. 2–3, reniform-cordate, crenate, abruptly acute, 3–9 cm. wide, usually somewhat shorter, on petioles 5–25 cm. long; cauline lvs. near apex of stem only, similar to basal lvs. or ovate-cordate but smaller and on shorter petioles; stipules thin-membranous, ca. 4–10 mm. long, entire or few-toothed; peduncles ca. 2–8 cm. long; sepals 5–7 mm. long; petals deep yellow, 6–16 mm. long, the lateral and lower with purple veins, the lateral bearded; spur saccate, 1–2 mm. long; cleistogamous fls. ca. 2.5 mm. long; caps. oblong, 7–8 mm. long, abruptly acute; seeds pale brown, shining, ca. 2 mm. long; $n = 12$ (Gershoy, 1934).—Wet shaded places in woods, as along stream banks, up to 8000 ft.; Redwood F., Mixed Evergreen F., Yellow Pine F., Red Fir F.; Sierra Nevada from Tulare Co. n. and Coast Ranges from Monterey Co. n.; to Alaska and n. Rocky Mts. March–July.

2. **V. lobàta** Benth. [*V. sequoiensis* Kell. *V. dactylifera* Greene.] Mostly pubescent perennials, from a short thickish rootstock; stem erect, 1–3(–3.5) dm. high; lvs. mostly at summit of stem, sometimes 1 from rootstock, ovate to wedge-shaped or almost reniform in outline, 2.5–10 cm. wide, usually shorter, palmately 3–7(–9)-cleft or -divided into lance-oblong lobes, the petioles of cauline lvs. mostly 1–5 cm. long; stipules green, 10–15 mm. long, sharply serrate; peduncles ca. as long as lvs.; sepals 6–10 mm. long; petals deep yellow, 8–15 mm. long, the 2 upper purple on back, all or lower 3 with purple-brown veins toward base, the lateral yellow-bearded; spur saccate, 1–2 mm. long; cleistogamous fls. in uppermost axils; caps. ellipso-ovoid, acute, 8–10 mm. long; seeds buff, blotched with brown, shining, 2 mm. long; $n = 6$ (Gershoy, 1934).—Rather dry slopes, in open woods, 1000–6500 ft.; Yellow Pine F., Douglas-Fir F.; Sierra Nevada from Tulare Co. n., Cuyamaca Mts. (San Diego Co.), Santa Lucia Mts., N. Coast Ranges from Napa Co. n.; s. Ore. April–July.

Var. **integrifòlia** Wats. [*V. deltoidea* Greene.] Like the sp. but with the lvs. not lobed, cordate-ovate, irregularly crenate.—With the sp. but in some areas of nw. Calif. and adjacent Ore. quite supplanting it. V. **califórnica** M.S. Baker, with thinner, concolorous lvs., more regularly toothed, and with petals faintly darkened on backs, comes from Humboldt and Trinity cos. and may be a hybrid of the above with V. glabella.

Ssp. **psychòdes** (Greene) Munz. [*V. p.* Greene.] Much like the sp. and intergrading with it, but more glaucous, glabrous; lvs. puncticulate beneath and reticulate; peduncles tending to surpass the lvs.—On serpentine, 1500–6500 ft.; Yellow Pine F., Mixed Evergreen F.; Trinity, Shasta, Butte, Siskiyou, and Del Norte cos.; to sw. Ore. April–July.

3. **V. sempérvirens** Greene. [*V. sarmentosa* Dougl., not Bieb.] EVERGREEN VIOLET. Evergreen subglabrous perennial from short scaly rootstocks, producing stolonlike stems 1–3 dm. long, with scattered lvs. and rooting at the nodes; lvs. round-ovate to slightly cordate, crenate, often rusty beneath, 1.5–3 cm. broad; petioles 3–16 cm. long; stipules ovate-subulate, brown-scarious, 5–7 mm. long; peduncles 5–10 cm. long; sepals 4–6

mm. long; petals lemon-yellow, 8–12 mm. long, the 3 lower faintly purple-veined, the lateral bearded at the base; spur broad, 1–2 mm. long; cleistogamous fls. in axils of stolons; caps. round-ovoid, mottled with purple, ca. 5 mm. long; seeds brown, tinged purple, ca. 2 mm. long; $n = 12$ (Clausen, 1929), 24 (Gershoy, 1934).—Mostly shaded woods, below 3500 ft.; Redwood F., Closed-cone Pine F., Mixed Evergreen F., Douglas-Fir F.; Coast Ranges from Monterey Co. n.; to B.C. Feb.–April.

4. **V. praemórsa** Dougl. [*V. Nuttallii* var. *p.* Wats. *V. N.* ssp. *p.* Piper.] Almost acaulescent perennial or stems later 3–6 cm. long, from a short vertical scaly rootstock; plant rather villous; lf.-blades ovate or oblong-ovate, cuneate or truncate at base, obtusish at apex, ± crenate-dentate, 2.5–5 cm. long, the petioles ½–1½ times as long; stipules lanceolate, entire, membranous; peduncles equaling or exceeding lvs.; sepals ca. 7 mm. long; petals deep yellow, 1–1.5 cm. long, the 3 lower veined brown-purple, the lateral bearded; cleistogamous fls. in upper axils; caps. green to purplish, broadly ovoid, puberulent; seeds buff to light brown, 3 mm. long; $n = 18$ (Clausen, 1949).— Rather dry open woods, 1700–5500 ft.; Yellow Pine F., Douglas-Fir F.; Coast Ranges, Mendocino Co. and Siskiyou Co. n.; to B.C. April–May.

Ssp. **màjor** (Hook.) M.S. Baker. [*V. Nuttallii* var. *m.* Hook. *V. Brooksii* Kell. *V. p.* var. *m.* Peck.] Stems 10–30 cm. long; lvs. 4–8 cm. long on petioles 8–15 cm. long; $n = 24$ (Clausen, 1949).—Meadows and slopes that dry early, 2700–5000 ft.; Yellow Pine F., N. Oak Wd.; Humboldt Co. to Modoc Co.; to e. Wash., Ida. April–June.

Ssp. **oregòna** Baker & Clausen [*V. p.* var. *o.* Peck.] Plants short-pubescent, sub-acaulescent; lvs. lance-ovate to ovate, 2–4 cm. long on petioles 1.5–2 times as long; petals ca. 1 cm. long; $n = 24$ (Clausen, 1949).—Dry, wooded slopes, 2500–8000 ft.; N. Juniper Wd., Yellow Pine F.; Siskiyou and Modoc cos.; Klamath Co., Ore. April–June.

5. **V. Bàkeri** Greene. Perennial from a large woody deep-seated taproot; stems 2–5 cm. long; lvs. lanceolate to narrow-ovate, entire, 2.5–4 cm. long, subglabrous to short-pubescent on margins and veins or the surfaces; petioles as long or longer, mostly retrorse-pubescent; stipules lanceolate, entire; peduncles slender, equal to or shorter than lvs.; sepals 4–5 mm. long; petals light yellow, 8–10 mm. long, the lower 3 with brownish veins, the lateral scaly-puberulent near base; spur ca. 1 mm. long; cleistogamous fls. in upper axils; caps. round-ovoid, glabrous, 5–6 mm. long; seeds brown, 3 mm. long; $n = 24$ (Clausen, 1949).—Forest floor, 4500–8000 ft.; mostly Yellow Pine F. and Red Fir F.; Fresno Co. to Humboldt Co.; Ore. May–July.

Ssp. **grándis** Baker & Clausen. Stems to 15 cm. long; lf.-blades 4–6 cm. long; petioles 5–14 cm. long; caps. ca. 8–9 mm. long; seeds 3.5 mm. long; $n = 24$ (Clausen, 1949).— 7000–8000 ft., Placer and Plumas cos.

6. **V. tomentòsa** Baker & Clausen. Gray loose-woolly perennial, from a large woody deep-seated taproot; stems 3–15 cm. long, prostrate in sun, suberect in shade; radical lvs. 2–5, erect, narrow-ovate to elliptic, entire or nearly so, 1.5–5 cm. long, on petioles 2–5 cm. long; stem-lvs. smaller; stipules lanceolate, 5–15 mm. long; peduncles 1–4 cm. long, the bractlets scarcely evident; sepals 5–6 mm. long; petals 6–7 mm. long, clear yellow, the upper with or without faint darkening on the back, the lateral with 2 short dark lines near base, short-bearded; spur scarcely 1 mm. long; cleistogamous fls. none; caps. woolly, subglobose, 4 mm. long; seeds brown, mottled lighter, 2.5–2.8 mm. long; $n = 6$ (Clausen, 1949).—Local in dry gravelly places, 5000–6500 ft.; Yellow Pine F., Lodgepole F.; Sierra Nevada from Plumas Co. to Eldorado Co. June–Aug.

7. **V. áurea** Kell. [*V. purpurea* var. *a.* M.S. Baker.] Grayish loosely long-pubescent almost tomentose perennial from a deep vertical root, subacaulescent or with stems to ca. 6 cm. high; lf.-blades rounded or subreniform to lance-ovate, 1–3 cm. long, coarsely and rather regularly sinuate-dentate, the teeth with white glandular tips, the lvs. rounded to obtuse at apex, the petioles 2–7 cm. long; stipules hyaline to greenish, lance-linear, 5–10 mm. long, somewhat toothed and ciliate; peduncles slightly surpassing lvs.; sepals 5–6 mm. long; petals deep yellow, 8–12 mm. long, the 2 upper purple-brown on back, the 3 lower with purple-brown veins; spur saccate, 1–2 mm. deep; cleistogamous fls. in upper axils; caps. subglobose, 5–6 mm. long, densely short-villous; seeds brownish, almost 3 mm. long; $n = 6$ (Clausen, 1949).—Dry sandy slopes, 4500–6500 ft.; Pinyon-Juniper Wd., Sagebrush Scrub; Mono Co.; w. Nev. April–June.

Ssp. **mohavénsis** Baker & Clausen. Grayish, with a short rather dense pubescence; stems mostly 5–15 cm. long; cauline lvs. subacute, with rather pointed teeth; seeds ca.

1.7 mm. long; $n = 6$ (Clausen, 1949).—Dry rocky or gravelly places, 3000–7500 ft.; Sagebrush Scrub, Chaparral, Yellow Pine F.; about the w. Mojave Desert from Mono Co. and Tehachapi Mts. and Mt. Pinos to San Bernardino Mts. April–June.

8. **V. purpùrea** Kell., not Stev. [*V. Kelloggii* A. Nels.] Perennial from a strong woody taproot; stems 6–20 cm. high, depressed in sun, suberect in shade, retrorse-pubescent; herbage puberulent; radical lvs. 1–5, rounded, with subcuneate base, purple-tinted, somewhat succulent, glabrate above, irregularly sinuate-dentate, 1.5–3 cm. wide, 1.8–3.5 cm. long, on petioles 4–11 cm. long and with the scarious stipules adnate to petiole, the free tips lanceolate to triangular, 2–3 mm. long; cauline lvs. ovate, rather regularly crenate-serrate, the stipules greenish, unequal, coarsely toothed, subovate, 3–14 mm. long; peduncles exceeding lvs., 3–10 cm. long, the bracts filiform, near the middle; sepals 4–6 mm. long; petals deep lemon-yellow, 8–10 mm. long, the 2 upper purplish on back, the 3 lower with purplish-brown veins and the 2 lateral bearded; spur saccate, 1–2 mm. long; cleistogamous fls. present at upper axils; caps. subglobose, puberulent, 5–6 mm. long; seeds dull dark brown, 2.3 mm. long; $n = 6$ (Clausen, 1949).—Dry slopes, 1800–6000 ft.; Yellow Pine F.; mts. of San Diego Co. n. and along the w. Sierra Nevada to near Mt. Lassen, Mt. Shasta and Siskiyou Mts., in Coast Ranges from Mt. Pinos to San Rafael Mts. and Lake Co. n.; to Ore. April–June.

KEY TO SUBSPECIES

A. Radical lvs. ± cuneate at base.
    B. Foliage green; stems well developed above ground; petals 8–10 mm. long. Forest plants.
        C. Uppermost lvs. ovate, toothed; radical lvs. rounded to obtuse at apex. Yellow Pine F.
                                                                  V. purpurea
        CC. Uppermost lvs. ± lanceolate, often entire; radical lvs. acute. Mostly above Yellow
           Pine F. .... ssp. *mesophyta*
    BB. Foliage grayish; stems mostly buried; petals 6–8 mm. long. Open rocky places
                                                             ssp. *xerophyta*
AA. Radical lvs. subcordate to truncate at base.
    B. Basal lvs. mostly rounded at apex.
        C. Foliage green; seeds not mottled. Plants of forests.
           D. Margins of upper lvs. entire; stems scarcely developed above ground
                                                  ssp. *integrifolia*
           DD. Margins of upper lvs. toothed, or if entire the plants with well developed stems
              above ground .... ssp. *dimorpha*
        CC. Foliage grayish; stems mostly buried; seeds mottled. Open sagebrush flats
                                                         ssp. *geophyta*
    BB. Basal lvs. mostly obtusely pointed; lvs. deeply dentate, almost lobed .... ssp. *atriplicifolia*

Ssp. **mesophỳta** Baker & Clausen. [*V. p.* var. *m.* Peck.] Erect, green, 10–20 cm. high; basal lvs. ovate, on petioles 3–13 cm. long; cauline lvs. lance-ovate, 2–6 cm. long, undulate-denticulate; petals 8–10 mm. long; seeds gray, mottled with brown, 2.5 mm. long; $n = 6$ (Clausen, 1949).—Dry slopes, 5000–10,500 ft.; Yellow Pine F., Red Fir F., Lodgepole F.; San Jacinto Mts. to Plumas Co. May–June.

Ssp. **xerophỳta** Baker & Clausen. [*V. p.* var. *grisea* Jeps.] Gray, densely pubescent, 3–10(–12) cm. high; cauline lvs. mostly lanceolate, 1–3.5 cm. long, irregularly dentate; petals 6–8 mm. long; seeds gray, mottled with brown, 2–3 mm. long; $n = 6$ (Clausen, 1949).—Dry slopes, 5000–11,000 ft.; Yellow Pine F., Red Fir F., Lodgepole F.; San Jacinto Mts. to Sierra Co. May–July. (*V. pinetorum* Greene is based on a type which is apparently a hybrid between sspp. *mesophyta* and *xerophyta*, hence not available as a name.)

Ssp. **integrifòlia** Baker & Clausen. Stems almost buried; first lvs. green above, round, 1–1.8 cm. long, shallowly and irregularly dentate, the base subcordate or truncate, on petioles 3–5 cm. long; cauline lvs. entire, ovate to lance-oblong; petals 6–8 mm. long; seeds dark brown, 2.9 mm. long; $n = 6$ (Clausen, 1949).—Dry slopes of pumice or rock, 6000–9000 ft.; Red Fir F., Lodgepole F.; Inyo and Mariposa cos. to Mt. Lassen, Mt. Shasta, Mendocino and Glenn cos.; s. Ore., w. Nev. May–July.

Ssp. **dimórpha** Baker & Clausen. Green, 3–25 cm. high; early lvs. round, with truncate or subcordate base, irregularly dentate, on petioles 2–8 cm. long; cauline lvs. lance-ovate to lanceolate, 1.5–3 cm. long, repand-denticulate; petals 7–9 mm. long; seeds brownish, 2.5 mm. long; $n = 6$ (Clausen, 1949).—Dry slopes in woods, 4300–8000 ft.; Yellow Pine F., Lodgepole F.; e. slope of Sierra Nevada from Inyo Co. n., Plumas Co. through Lassen and Siskiyou cos. to Deschutes Co., Ore. May–July.

Ssp. **geophỳta** Baker & Clausen. [*V. p.* var. *g.* Peck.] Stems largely buried; herbage ± canescent-pubescent; lower lvs. rounded, purplish, coarsely dentate, 1–3 cm. long, subcordate at base; cauline lvs. ovate to ovate-lanceolate, rather sharply serrate-dentate; petals 6–8 mm. long; seeds mottled gray and brown, 2.4 mm. long; $n = 6$ (Clausen, 1949).—Open sandy flats, 3000–8000 ft.; Sagebrush Scrub; Mono Co. to se. Ore. April–June.

Ssp. **atriplicifòlia** (Greene) Baker & Clausen. [*V. a.* Greene. *V. p.* var. *a.* Peck.] Green, the stems to 5 cm. long above ground; early lvs. round-ovate, with few deep sharp teeth, almost like lobes in some cases; later lvs. triangular, subentire to coarsely irregularly few-toothed, acute; petals 6–9 mm. long; seeds light brown, 2.4 mm. long; $n = 6$ (Clausen, 1949).—Dry often exposed slopes, 5000–10,000 ft.; Sagebrush Scrub, N. Juniper Wd., Lodgepole F.; e. side of Sierra Nevada, Inyo Co. to Modoc Co.; to Wash., Wyo.

9. **V. quercetòrum** Baker & Clausen. Perennial, from a woody taproot, the foliage grayish-green, with little purple, puberulent throughout; stems 5–15 cm. long; lowest lvs. erect, rounded to ovate, obtuse, with base cuneate to subcordate, the margin irregularly sinuate-dentate, 2–5 cm. long, 2–3 cm. wide, on petioles 3–9 cm. long; upper lvs. smaller, more acute; stipules of lower lvs. subscarious, with short free tip, those of upper lvs. green, unequal, entire to toothed, 5–20 mm. long; peduncles 4–13 cm. long, usually exceeding lvs.; sepals 6–8 mm. long; petals yellow, ± darkened on back, 10–12 mm. long, the lateral clavate-bearded; spur short; cleistogamous fls. in upper axils; caps. puberulent, 7–8 mm. long, round-ovoid; seeds dark brown, 2.7 mm. long; $n = 12$ (Clausen, 1949).—Dry grassy or brushy slopes, below 7000 ft.; Chaparral, S. Oak Wd., Foothill Wd., N. Oak Wd., rarely Yellow Pine F.; hills about Cent. V., and inner Coast Ranges to San Diego Co. and s. Ore. March–June.

10. **V. pedunculàta** T. & G. JOHNNY-JUMP-UP. WILD PANSY. Perennial, with several slender stems from a deep short thick spongy rootstock with numerous fleshy roots; stems several, branching, decumbent, puberulent, 1–3.5 dm. long; lvs. bright green, deltoid-ovate to subcordate, coarsely but shallowly crenate, obtuse, 1.5–4 cm. long, ca. as wide, subglabrous to puberulent, the petioles 3–5(–6.5) cm. long; stipules green to hyaline, deeply toothed; peduncles much exceeding lvs., to 1.5 dm. long; sepals 6–8 mm. long; petals orange-yellow, 10–16(–20) mm. long, the 2 upper red-brown on the back, the lateral bearded, the 3 lower veined with dark brown; spur 1–2 mm. long; cleistogamous fls. lacking; caps. glabrous, 8–11 mm. long; seeds dark brown, shining, 2.7 mm. long; $n = 6$ (Clausen, 1949).—Common on grassy slopes, below 2500 ft.; V. Grassland, N. Oak Wd., Coastal Sage Scrub; Coast Ranges from Sonoma and Colusa cos. to L. Calif.; Santa Barbara Ids. Feb.–April.

Ssp. **tenuifòlia** Baker & Clausen. Lvs. deltoid, mostly 1–2 cm. long, less wide, acute; petals yellow, 8–15 mm. long; $n = 6$ (Clausen, 1949).—Rocky uplands; Foothill Wd.; inner Coast Ranges from w. Merced Co. to e. San Luis Obispo Co., w. base of Sierra Nevada (Trimmer Springs, Fresno Co., Springville, Tulare Co.). Feb.–April.

11. **V. ocellàta** T. & G. Pubescent perennial with vertical scaly rootstock; stems slender, suberect, 1–3 dm. high; lower lvs. cordate-ovate, crenate, acute to subacuminate, 2–6 cm. long, on petioles 3–15 cm. long; upper lvs. smaller, less cordate; stipules brown-scarious, somewhat laciniate-ciliate; peduncles mostly 2–4 cm. long, scarcely overtopping lvs.; sepals 6–7 mm. long; petals mostly 8–12 mm. long, white with yellow near the base, at least the 2 upper deep red-violet on back, the laterals with small purple eye-spot near the base and with yellow bearding, the lower with purple veins; spur saccate; cleistogamous fls. in uppermost axils; caps. round-ovoid, minutely puberulent, 6–7 mm. long; seeds brown-purple, ca. 2 mm. long; $n = 6$ (Gershoy, 1934).—Rocky and grassy banks and in thickets, often in serpentine, below 3500 ft.; Redwood F., Chaparral, Mixed Evergreen F., Yellow Pine F.; Coast Ranges from Monterey Co. to Del Norte and Trinity cos., Shasta Co. March–June.

12. **V. cuneàta** Wats. Glabrous perennial from a horizontal scaly rootstock; stems slender, 7–20 cm. high; basal lvs. round-ovate to deltoid, crenulate, subcuneate to truncatish at base, abruptly acute, 1–2.5 cm. long, sometimes wider, purple-veined; petioles 2–7 cm. long; cauline lvs. smaller; stipules greenish, subentire or gland-toothed; peduncles scarcely exceeding lvs.; sepals 4–5 mm. long; petals 8–10 mm. long, white, the 2 upper may have a purplish base, the lateral with purple eye-spot near the base and

clavate hairs, the lower veined purple, all deep red-violet on back; spur yellowish, saccate; cleistogamous fls. in upper axils; caps. subglobose, purplish, ca. 5 mm. in diam.; seeds deep brown-purple, ca. 2 mm. long.—Moist open forests, 1200–5000 ft.; Yellow Pine F., Douglas-Fir F.; Mendocino Co. to s. Ore. and e. to Shasta Co. March–June.

13. **V. Sheltònii** Torr. Glabrous perennial from a short stout rootstock; stems rising only a little above ground; lvs. 2–6(–10) cm. wide, broader than long, dark blue-green, palmately 3-divided, the cuneate-obovate divisions again palmately 3-parted and cleft into linear-oblong lobes, ± ciliolate; petioles 3–10(–15) cm. long; stipules lance-ovate, membranous, the upper somewhat lacerate; peduncles mostly 8–15 cm. long, somewhat overtopping the lvs.; sepals 6–8 mm. long, ciliate; petals 10–15 mm. long, deep lemon-yellow, the 3 lower veined brown-purple, the 2 upper brown-purple on back, the lateral with clavate hairs; spur saccate; cleistogamous fls. at upper axils of the underground slender stems; caps. oblong-ovoid, glabrous, 7–8 mm. long, green with purple streaks; seeds brownish, shining, 2.5 mm. long.—Rich loam, shade of open woods or brushy places, 2500–8000 ft.; Mixed Evergreen F., Yellow Pine F., Chaparral, Red Fir F.; Santa Ana Mts. (Orange Co.), San Gabriel Mts., Sierra Nevada from Tulare Co. to Modoc Co., Coast Ranges from Santa Clara Co. to Siskiyou Co.; to Wash. April–July.

14. **V. Douglásii** Steud. [*V. chrysantha* Hook., not Schrad. *V. pruinosa* Pollard.] More or less pubescent perennial, with a deep short rootstock and a cluster of strong roots; stems 5–10 cm. long, partly underground; lvs. ovate in outline, 2–5 cm. long, bipinnately 3–5-parted into 3–5 cleft linear or oblong segms.; petioles 2–6 cm. long; stipules lanceolate, entire to incised, the upper greenish; peduncles 5–12 cm. long, usually surpassing the lvs.; sepals 6–12 mm. long, ciliate; petals mostly 8–16 mm. long, light golden-yellow with dark brown veins, the 2 upper brown to very dark on back, the lateral yellow-bearded; spur saccate; cleistogamous fls. absent; caps. green, acute, oblong, glabrous, 6–9 mm. long; seeds pale buff, ca. 2.5 mm. long.—Grassy slopes and flats, moist in early season, 3500–7500 ft.; S. Oak Wd., Yellow Pine F.; mts. of s. Calif.; and 100–4000 ft.; Foothill Wd., N. Oak Wd.; Coast Ranges from San Luis Obispo Co. to Siskiyou Co. and foothills of Sierra Nevada; Ore. March–May.

15. **V. Beckwíthii** T. & G. Puberulent to almost glabrous perennial from a deep short rootstock; stems almost underground, clustered; lvs. 2.5–3 cm. long, palmately 3-parted then bipinnately parted into ultimate linear or spatulate often mucronate segms.; petioles 1–5(–6) cm. long; upper stipules lance-oblong, ciliate-lacerate; peduncles 5–7 cm. long, mostly slightly exceeding lvs.; sepals 7–8 mm. long; petals 8–14 mm. long, the 2 upper dark red-violet, the 3 lower lilac with yellow area at base and veined dark violet; lateral petals with yellow beard; spur saccate; cleistogamous fls. lacking; caps. green, oblong-ovoid, glabrous, 7–8 mm. long; seeds buff, shining, ca. 3 mm. long.—Dry gravelly places, often among shrubs, 3000–6000 ft.; Sagebrush Scrub, N. Juniper Wd., Yellow Pine F.; e. of Sierra Nevada from Inyo Co. to Modoc and Siskiyou cos.; to Ore., Ida., Utah. March–May.

Ssp. **glabràta** M. S. Baker. Foliage glabrous; lf.-segms. wider.—Open spots; Yellow Pine F.; Sierra Co. to Lassen Co.

16. **V. Hállii** Gray. Habit of *V. Beckwithii* but more glabrous, the ultimate lf.-segms. often lanceolate; stipules greener, more apt to be entire; upper petals dark red-violet, almost black on back, the 3 lower cream with deep yellow base and dark violet veining; caps. green, streaked with purple; seeds buff, shining, ca. 3.5 mm. long; $n = 30$–36 (Gershoy, 1934).—Open often rocky places, wet in early spring, 1000–6500 ft.; Mixed Evergreen F., Yellow Pine F.; Mendocino Co. to Siskiyou Co.; w. Ore. April–June.

17. **V. nephrophýlla** Greene. Nearly glabrous acaulescent perennial from a short rather thick rootstock with long fibrous roots; lvs. broadly cordate-ovate to subreniform, crenate, 2–7 cm. wide, acute or obtuse, the petioles brownish, 5–15 cm. long; stipules ovate-lanceolate, membranous, somewhat gland-toothed; peduncles 10–25 cm. long; sepals 5–6 mm. long, lance-oblong; petals 12–18 mm. long, deep blue-violet with white basal bearding and the 3 lower with dark veins; spur ca. 3 mm. long, almost as wide; cleistogamous fls. on shorter peduncles, at first erect, later inclined; caps. glabrous, ovoid, 6–9 mm. long; seeds buff, shining, ca. 2 mm. long.—Occasional, shaded cool, wet places, 3500–7000 ft.; mostly Yellow Pine F.; Cuyamaca Mts. of San Diego Co. to San Gabriel Mts., e. slope Sierra Nevada from Inyo Co. to Modoc Co.; to B.C., Nfld., New Mex. May–June.

18. **V. Maclóskeyi** Lloyd. [*V. blanda* auth., not Willd. *V. b.* var. *M.* Jeps. *V. anodonta* Greene. *V. parnassifolia* Greene.] Subglabrous stemless perennial from a slender creeping rootstock, sending out leafy stolons in late season and forming dense patches; lvs. round-reniform to cordate-ovate, thin, subentire to crenulate, 1–2(–3) cm. long; petioles slender, 1–6(–10) cm. long, glabrous to somewhat villous; stipules greenish, subovate, usually entire; peduncles 2–15 cm. long; sepals ovate-lanceolate, 3–4 mm. long, glabrous; petals 6–9 mm. long, white, the 3 lower with purple veins, the lateral sometimes bearded; spur 2–3 mm. long; cleistogamous fls. present; caps. green, ovoid, ca. 6 mm. long; seeds dark, ca. 1 mm. long; $n = 12$ (Clausen, 1936).—Wet banks and meadows, 3500–11,000 ft.; Yellow Pine F. to Subalpine F.; San Jacinto Mts. to Sierra Nevada, N. Coast Ranges from Lake Co. n.; to B.C. May–Aug. A single collection from Spirit Lake, Siskiyou Co. has been referred to ssp. **pállens** (Banks) M. S. Baker, having lvs. with 8 or more crenations on each edge.

19. **V. lanceolàta** ssp. **occidentàlis** (Gray) Russell. [*V. primulifolia* var. *o.* Gray. *V. o.* Howell.] Acaulescent glabrous perennial from stoutish rootstocks that produce late-season slender stolons; lvs. rhombic-ovate, 2–7 cm. long, obscurely crenate, acutish at both ends, on petioles 3–10 cm. long; stipules lance-oblong somewhat serrulate, ca. 4 mm. long; peduncles reddish, 6–12 cm. long, scarcely equaling lvs.; sepals 3–4 mm. long; petals 10–14 mm. long, pure white, the 3 lower with purple veins, the lateral heavily bearded; spur saccate; cleistogamous fls. lacking; caps. green, ellipsoid, 6–7 mm. long; seeds brown to dark, shining, 1.3 mm. long; $n = 12$ (Clausen, 1936).—Rare in marshes or bogs, below 2500 ft.; Mixed Evergreen F.; Del Norte Co. and sw. Ore. April.

20. **V. palústris** L. [*V. Howellii* auth., not Gray.] Glabrous creeping perennial with slender rootstock and late-season runners; lvs. round-cordate to -ovate, slightly crenulate, 2–5 cm. wide, the petioles 4–10 cm. long; stipules green, streaked red, lance-ovate, with scattered hairlike points on the edge; peduncles exceeding length of lvs.; sepals lance-oblong, 4–5 mm. long; petals acute, 8–12 mm. long, lilac to almost white, with some darker veining, the lateral somewhat bearded; spur broad, ca. 3 mm. long; cleistogamous fls. present; caps. green, ovoid, 6–7 mm. long; seeds dark brown, 1.6 mm. long; $n = 24$ (Gershoy, 1934).—Cool wet brushy places, below 500 ft.; N. Coastal Scrub; Mendocino Co. to Alaska, Atlantic Coast, Eurasia. May–July.

21. **V. odoràta** L. ENGLISH VIOLET. Acaulescent perennial spreading by long leafy prostrate stolons, flowering the second year; lvs. broadly cordate-ovate, finely pubescent, 2–5 cm. wide, obtusely serrulate; petioles 5–10 cm. long; stipules lance-ovate, acuminate, sparingly toothed; peduncles commonly 10–15 cm. high; sepals ca. 6–7 mm. long, blunt; petals ca. 15 mm. long, deep violet, with some white at base; fls. fragrant; spur 5–7 mm. long; style bent down at tip like a hook; cleistogamous fls. present; caps. closely pubescent, purple, plump-ovoid, ca. 8 mm. long; seeds cream-color, 3.5–4 mm. long; $n = 10$ (Clausen, 1931).—Common in gardens and occasionally reported as an escape, as at Pinecrest, Tuolumne Co., Mt. Tamalpais; native of Eurasia. Dec.–March.

22. **V. adúnca** Sm. [*V. canina* var. *a.* Gray. *V. filipes* Greene.] Usually a puberulent perennial from a slender rootstock; stems becoming 4–20 cm. long, but sometimes very short at time of flowering; lvs. thickish, round-ovate, somewhat heart-shaped at base, obtuse at apex, obscurely crenate, often brown-dotted beneath in dry specimens, 1–4 cm. long; petioles 1–6 cm. long; stipules linear-lanceolate, acute, mostly lacerate-toothed; peduncles from shorter than to much longer than lvs.; sepals ca. 5 mm. long, lance-linear; petals deep to pale violet, rarely whitish, 8–13 mm. long, the 3 lower white at base and veined purple, the lateral white-bearded; spur 5–12 mm. long, rather broad, obtuse, often hooked at tip; cleistogamous fls. present; caps. buff, narrow-ovoid, 6–8 mm. long; seeds dark brown, 1.5–2 mm. long; $n = 10$ (Gershoy, 1934).—Damp banks and edge of meadows in most Forest Communities of state; at 5000–8000 ft.; mts. of s. Calif.; 3300–8500 ft., Sierra Nevada to Modoc Co.; 25–6000 ft., Coast Ranges from Monterey Co. to Del Norte and Siskiyou cos.; to Alaska, Quebec, Colo. March–July.

Var. **oxycéras** (Wats.) Jeps. [*V. canina* var. *o.* Wats. *V. o.* Greene. *V. uncinulata* Greene.] Stems 1–4 cm. high; plant quite glabrous; stipules often subentire; lvs. truncate at base, rounded at tip; spur slender, pointed at tip; petals often 6–8 mm. long.—Damp shaded places, 9000–11,500 ft.; Subalpine F., Alpine Fell-fields; Tulare Co. to Siskiyou Co.; Ore. June–Aug.

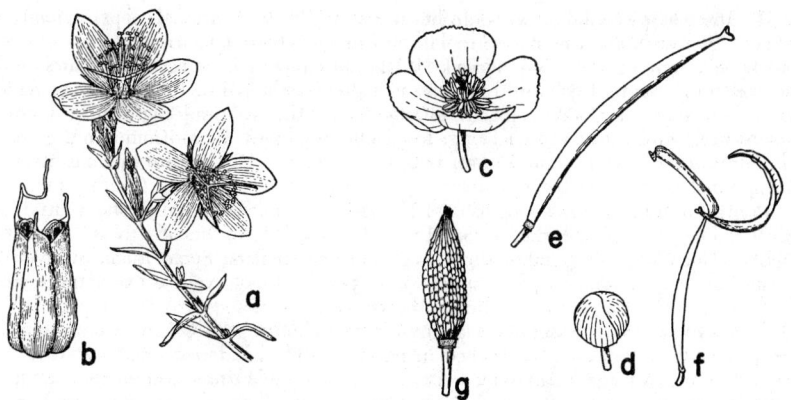

Fig. 27. HYPERICACEAE. *Hypericum concinnum: a,* habit, × ½, petals 5, stamens many, styles 3; *b,* caps., × 3.
PAPAVERACEAE. *Dendromecon: c,* fl., × ½, petals 4, stamens many; *d,* bud, × ½, sepals 2; *e,* caps., × ½, *f,* caps., × ¼, elastically dehiscent from base upward. *Platystemon californicus: g,* fr., × 1, of many carpels, at first united, later moniliform and separate.

23. **V. arvénsis** Murr. Freely branched annual, spreading or decumbent, 1–3 dm. high, harsh-puberulent; lvs. oblong-lanceolate to -ovate, coarsely crenate-serrate, 1.5–3 cm. long, cuneate at base, short-petioled; stipules green, almost as large as lvs., cut into linear segms. and with a large terminal portion; peduncles 5–8 cm. long; sepals lance-linear, acuminate, 6–7 mm. long; petals pale yellow, 7–10 mm. long; caps. subglobose, glabrous, 6–7 mm. long; seeds buff, 1.5–1.7 mm. long; $n = 17$ (Clausen, 1931).—Occasional escape, as at Scott V., Siskiyou Co.; native of Eu. May–July.

28. Hypericàceae. St. John's Wort Family (Fig. 27)

Herbs or shrubs. Lvs. opposite, entire, exstipulate, mostly sessile, glandular-punctate. Fls. perfect, regular, hypogynous. Sepals 5 or 4, green, persistent. Petals 5 or 4, mostly oblique and convolute in the bud. Stamens usually many, distinct or ± united into 3–5 clusters. Ovary superior, 1- or 3–7-celled; ours with 3 styles. Caps. 1-loculed with 2–5 parietal placentae or 3–7-loculed, mostly septicidal. Seeds many, small, anatropous, without endosperm. Ca. 10 genera and 300 spp. of temp. and warm regions.

1. Hypéricum L. (By Usage **Hypéricum**.) St. John's Wort

Plants glabrous. Lvs. several-nerved from base, sessile. Fls. often yellow, solitary or in terminal cymes; petals deciduous. Ca. 200 spp. of N. Hemis., some in cult. (Ancient Greek name.)

Petals not longer than sepals; styles short; caps. 1-celled.
  Stems ascending; lvs. 10–20 mm. long .... 1. *H. mutilum*
  Stems mostly prostrate; lvs. 4–12 mm. long .... 2. *H. anagalloides*
Petals much longer than sepals; styles long; caps. 3-celled.
  Sepals linear-deltoid, acuminate; stems with sterile shoots in lower lf.-axils ...... 3. *H. perforatum*
  Sepals ovate to obovate, acute to obtuse; stems lacking sterile shoots in lower axils.
    Lvs. oblong to ovate, mostly flat; sepals mostly without black dots .... 4. *H. formosum*
    Lvs. linear to lanceolate, mostly folded; sepals with many black marginal dots .. 5. *H. concinnum*

1. **H. mùtilum** L. Perennial with leafy-bracted decumbent base; stems slender, weak, 2–5 dm. high, branched above; lvs. ovate to narrow-oblong, partly clasping, 5-nerved, 1–2 cm. long; fls. in terminal leafy cymes; sepals lance-linear, acute, ca. 4 mm. long; petals yellow, shorter than sepals; stamens 6–12; caps. 2.5–3.5 mm. long, short-ellipsoid. —River and pond banks; Freshwater Marsh; Sacramento R. from Shasta Co. and s., Kelseyville, Lake Co., Prattville, Yosemite V., lower San Joaquin V.; native, Nova Scotia to Tex. Aug.–Sept.

2. **H. anagalloìdes** Cham. & Schlecht. Tinker's Penny. Annual or perennial, procumbent, often in mats, the stems rooting at nodes, 5–15(–20) cm. long; lvs. elliptic to ovate or orbicular, obtuse, 5–7-nerved, 4–12 mm. long; fls. in few-fld. cymes, golden to salmon-color; sepals 2.5–3 mm. long; petals slightly shorter to longer; stamens 15–25; caps. ca. 3 mm. long.—Wet places, 4000–10,000 ft.; Yellow Pine F., Red Fir F.; Sierra Nevada and mts. of s. Calif.; below 5000 ft.; many Plant Communities; coastal marshes to mts. of Coast Ranges, San Luis Obispo Co. to Del Norte and Siskiyou cos.; to B.C., Mont. June–Aug.

3. **H. perforàtum** L. Klamath Weed. Perennial with leafy basal offshoots, the stems simple, tough, 3–10(–12) dm. high, much branched; lvs. linear- to elliptic-oblong, revolute, 1.5–2.5 cm. long, subtending short leafy branchlets; cymes densely fld.; sepals linear-lanceolate, acuminate, 4–5 mm. long; petals orange-yellow, sometimes black-dotted, 8–12 mm. long, twisting after anthesis; stamens many, in 3–5 groups; caps. 7–8 mm. long, narrow-oblong; seeds black, shining, subcylindric, 0.6–0.7 mm. long, reticulate; $2n = 32$ (Noack, 1939).—A bad weed in pastures and abandoned or poorly cult. fields, below 4500 ft., Wynola (San Diego Co.), Sierra Nevada from Tuolumne Co. n. and Coast Ranges from Santa Clara and Santa Cruz cos. n.; especially serious over hundreds of thousands of acres in the nw. cos.; native of Eu. Somewhat poisonous to livestock. June–Sept.

4. **H. formòsum** HBK. var. **Scoùleri** (Hook.) Coult. [*H. S.* Hook.] Perennial with erect stems from running rootstocks, slender, 2–7 dm. high, simple or branched above; lvs. oblong-ovate, obtuse, 1–2.5 cm. long, several-nerved, black-dotted along margin; cymes panicled; sepals ovate, obtuse, 3 mm. long, black-dotted on margin; petals yellow, 7–10 mm. long, obovate; stamens many, in 3 groups; caps. 3-lobed, 6–7 mm. long; seeds brownish, ca. 0.6 mm. long, reticulate-pitted.—Frequent, wet meadows and banks, 4000–7500 ft.; Yellow Pine F.; mts. of s. Calif., Sierra Nevada, occasionally lower in Chaparral; 0–5000 ft.; Redwood F., Douglas-Fir F., Yellow Pine F., Coast Ranges from Monterey Co. n.; to B.C., Mont. June–Aug.

5. **H. concínnum** Benth. Gold Wire. Bushy perennial with a woody crown and numerous wiry stems 1.5–3 dm. high; lvs. linear to lanceolate, acute, usually folded, 2–4 cm. long; cymes rather dense, few-fld.; sepals ovate, abruptly acuminate, 7–8 mm. long, black-dotted at margin; petals yellow, obovate, 12–15 mm. long, black-dotted on margin; stamens many, in 3 groups; caps. 6–7 mm. long; seeds dark, cylindric, shining, 1 mm. long, minutely and shallowly pitted.—Dry brushy slopes, below 3000 ft.; Chaparral, Yellow Pine F.; Sierra Nevada from Mariposa Co. to Shasta Co., N. Coast Ranges from Marin Co. to Mendocino Co. May–July.

29. Papaveràceae. Poppy Family (Fig. 27)

Herbs or shrubs, with milky yellow or colorless sap. Lvs. alternate or sometimes opposite, without stipules. Fls. regular, complete, usually solitary and peduncled. Sepals 2, rarely 3 or 4, sometimes united, caducous. Petals twice as many as sepals. Stamens mostly numerous, hypogynous, distinct. Pistil 1, of 2–several usually united carpels; ovary superior, usually 1-celled, with parietal placentae; style short; stigma simple or lobed. Fr. a caps., usually dehiscent by apical pores or valves; seeds usually many. Ca. 25 genera and 200 spp., widely distributed but most abundant in w. N. Am.

A. Lvs. mainly opposite or whorled.
  B. Stamens many; carpels 6–25, torulose and separating when mature; petals rather persistent
        1. *Platystemon*
  BB. Stamens 6–12; carpels mostly 3, united; petals falling early .... 2. *Meconella*
AA. Lvs. mainly alternate or basal.
  B. Plants shrubby at least at base, mostly 1 m. or more tall.
        C. Fls. white; lvs. lobed .... 3. *Romneya*
        CC. Fls. yellow; lvs. entire .... 5. *Dendromecon*
  BB. Plants herbaceous, usually less than 1 m. tall.
        C. Plant spiny or conspicuously long-hairy.
              D. Herbage long-hairy; lvs. bunched at base of plant .... 4. *Arctomecon*
              DD. Herbage spiny; lvs. well distributed .... 9. *Argemone*
        CC. Plant not spiny or conspicuously long-hairy.
              D. Lvs. multifid into linear segms.

E. Sepals separate, falling as fl. opens. Garden escape ...... 6. *Hunnemannia*
EE. Sepals united into a conical cap which is pushed off as fl. opens. Common
native .... 7. *Eschscholzia*
DD. Lvs. not multifid into linear segms.
E. Caps. long-linear; stigma 2-lobed; upper lvs. clasping. Rare garden escape
8. *Glaucium*
EE. Caps. ovoid to turbinate, obovoid or globose.
F. Plants usually 20–120 cm. high; stamens many. Cismontane.
G. Style slender, short; stigma capitate .... 10. *Stylomecon*
GG. Style lacking; stigma discoid .... 11. *Papaver*
FF. Plants 2–3 cm. high; stamens 6–9. Deserts .... 12. *Canbya*

1. Platystèmon Benth. CREAM CUPS

Low villous annuals with opposite entire lvs. Sepals 3, ovate, tardily deciduous.
Petals 6, white to yellowish, tardily deciduous. Stamens many; fils. flattened; anthers
linear. Carpels 6–25, at first united, separate in fr., each several-ovuled, becoming moniliform and breaking transversely into indehiscent 1-seeded joints. Stigmas linear, free.
Seeds dark, quadrate, variously sculptured. A genus of Calif. and adjacent areas, with
apparently 1 variable sp., for which almost 60 segregates have been proposed. (Greek,
platus, broad, and *stemon*, stamen.)

(Fedde, F., *in* Das Pflanzenreich, IV, 104: 106–131, 1909.)

1. **P. califórnicus** Benth. Plant 1–3 dm. high, with many stems from the base,
soft-pilose, the branches ± decumbent; lvs. largely on lower part of plant, lance-linear,
subsessile, 2–5(–8) cm. long; fls. subscapose, solitary on peduncles 1–2 dm. long;
sepals villous, forming rounded to oblong-ovoid long-hairy buds 6–10 mm. long; petals
usually cream, 8–16(–20) mm. long; carpels forming an erect glabrous or setose oblong-
cylindric head 10–16 mm. long beaked by the persistent styles (an added 4–8 mm.
long); seeds ca. 1 mm. long; *n* = 6 (Sugiura, 1937).—Common in open grassy clay
or sandy places, also on burns, usually below 3000 ft.; V. Grassland, Chaparral, Oak
Wd., etc.; most of cismontane Calif., w. edge of deserts; to Utah, Ariz., L. Calif.
March–May.
Exceedingly variable, the extreme variations may be keyed:

A. Mature fr. nodding. W. San Diego Co., Santa Cruz and Santa Rosa ids. ...... var. *nùtans* Bdg.
AA. Mature fr. erect.
B. Plants low, often less than 1 dm. tall; fr. scarcely torulose.
C. The plants glabrous. San Miguel, San Nicolas, Santa Rosa ids.
var. *ornithòpus* (Greene) Munz
[*P. ornithopus* Greene.]
CC. The plants sparingly short-pilose. Santa Barbara Id. .... var. *ciliàtus* Dunkle
BB. Plants over 1 dm. tall; fr. usually torulose.
C. Petals spreading nearly rotately from the base; young carpels white with stiffish hairs.
W. foothills of Sierra Nevada at 3500–5000 ft.
var. *horrídulus* (Greene) Jeps. [*P. horridulus* Greene.]
CC. Petals cupped-ascending; young carpels glabrous to hairy.
D. The petals cream; plant moderately hairy. General cismontane Calif.
P. califórnicus
DD. The petals yellow; plant excessively long-hairy. Cuyama V. and Tehachapi Mts. s.
along the desert slopes to L. Calif. .... var. *crinìtus* Greene

2. Meconélla Nutt. in T. & G.

Low, even diminutive, slender-stemmed annuals. Lvs. opposite. Sepals 3, rarely 2.
Petals 6, rarely 4, deciduous. Stamens 6–12, sometimes many; fils. not or slightly
widened. Carpels 3, combined into a 1-celled 3-lobed or subterete ovary. Stigmas ovate
to subulate. Caps. 3-valved, dehiscent through the parietal placentae. Seeds numerous,
minute, ovoid to round-reniform, dark, shining. A genus of 3–4 spp. of Pacific N. Am.
(Greek, *mekon*, poppy, and *ella*, diminutive.)

(Fedde, F., Hesperomecon *and* Meconella, *in* Das Pflanzenreich, IV, 104: 100–106, 1909.)
Lvs. all basal; fls. scapose, yellowish .... 1. *M. linearis*
Lvs. basal and cauline; fls. on branches as well as main stems, white.
Stamens in 2 series, 8–16 in number; petals mostly 8–10 mm. long .... 2. *M. californica*
Stamens in 1 series, 4–6; petals 3–5 mm. long .... 3. *M. oregana*

1. **M. lineàris** (Benth.) Nels. & Macbr. [*Platystigma l.* Benth. *Platystemon l.* Curran. *Hesperomecon l.* Greene. *H. filiformis* Fedde. *H. affinis, Platystemon, stricta, angusta* and *luteola* Greene.] Plants 0.5–2.5 dm. high, with stiff spreading hairs; lvs. basal, linear, 2.5–7 cm. long, sessile; fls. scapose; sepals brownish, 4–5 mm. long; petals cream with light yellow base, obovate, 6–15 mm. long; stamens many, with linear to somewhat dilated fils.; caps. smooth, glabrous, somewhat 3-lobed, narrow-obovoid, 10–14 mm. long, with lance-deltoid stigmas an added 2 mm. long; seeds black, shining, reniform-obovoid, ca. 0.4 mm. long.—Sandy grassy slopes or washes below 3300 ft.; V. Grassland, N. Oak Wd., Foothill Wd., Chaparral; Sierra Nevada; from Fresno Co. to Tehachapi Mts., Coast Ranges from Santa Barbara Co. to Ore. March–June. A form with outer petals lemon-yellow and the inner ones cream-colored occurs here and there from Tehachapi to Ft. Ross and has been called var. *pulchélla* (Greene) Jeps. [*Hesperomecon p.* Greene.]

2. **M. califórnica** Torr. [*Platystigma c.* Benth. & Hook. *M. oregana* var. *c.* Jeps. *Platystemon Torreyi* Greene. *M. collina* and *octandra* Greene. *M. oregana* var. *o.* Jeps.] Slender erect glabrous annual, 1–2 dm. high; lvs. entire, the basal spatulate, 1–2.5 cm. long with winged petioles, the cauline oblanceolate to linear, shorter; peduncles 5–7 cm. long; sepals often reddish, 2–4 mm. long; petals white, elliptic to oblong, 5–10 mm. long; stamens 8–12(–16), in 2 series, the fils. mostly linear and longer than anthers; caps. linear, twisted when mature, 15–35 mm. long; seeds obovoid, shining, 0.6 mm. long.—Open rocky slopes, below 3000 ft.; Foothill Wd., Chaparral; San Francisco Bay region, foothills of Sierra Nevada, from Shasta Co. to Kern Co. March–May.

3. **M. oregàna** Nutt. in T. & G. [*Platystigma o.* Wats.] Glabrous, with very slender stems 4–12 cm. high; basal lvs. spatulate to obovate, 6–20 mm. long, short-petioled, the cauline sessile and shorter; peduncles 2–4 cm. long; sepals 2 mm. long, often purplish; petals white, oblanceolate to obovate, 3–5 mm. long; stamens 4–6, the anthers hardly 0.5 mm. long, much shorter than fils.; caps. linear, 15–25 mm. long, erect, often twisted.—Doubtfully in Calif.; Ore. to B.C. April–May.

Var. **denticulàta** (Greene) Jeps. [*M. d.* Greene. *M. kakoethes* Fedde.] Glabrous, 5–30 cm. high; basal lvs. ovate to obovate, 2.5–4 cm. long, petioled; cauline lvs. spatulate to linear, 1–3 cm. long, sometimes denticulate, subsessile; petals 2–4 mm. long; anthers 6, linear, 1 mm. long, almost equal to or longer than fils.; caps. linear, 2–3 cm. long, twisted; seeds reniform-obovoid, black, shining, ca. 0.5 mm. long.—Frequent but inconspicuous, shaded canyons below 3300 ft.; Chaparral, Coastal Sage Scrub; San Diego Co. to Monterey Co.; Santa Cruz Id. March–May.

3. Rómneya Harv. MATILIJA POPPY

Large glabrous bushy suffruticose somewhat glaucous perennial, with colorless bitter juice and creeping underground rootstocks giving rise to large patches from an original plant. Lvs. gray-green, alternate, pinnatifid. Fls. few at summit of stems. Sepals 3. Petals 6, white, very large. Stamens many. Style 0; stigmas 7–12, partly cohering in a ring. Ovary and coriaceous caps. oblong to ovoid, strigose-hispid, with 7–12 placentae, some or most of which meet in the axis and so form partitions. Seeds ovoid, roughened, dull, slightly incurved. Spp. 2, s. Calif. and n. L. Calif.; cult. for the handsome fls. (T. *Romney* Robinson, Irish astronomer, friend of T. Coulter, the discoverer of the genus.)

Sepals and summit of peduncle glabrous; peduncles not conspicuously leafy near summit; lobes of lvs. mostly more than 12 mm. wide .... 1. *R. Coulteri*

Sepals and summit of peduncle setose; peduncles leafy to top; lobes of lvs. mostly less than 10 mm. wide .... 2. *R. trichocalyx*

1. **R. Còulteri** Harv. Stems heavy, 1–2.5 m. tall, branching, leafy; lvs. firm, round-ovate in outline, petioled, 5–20 cm. long, pinnately parted or divided into 3–5 main lanceolate to ovate divisions, these in turn sparingly dentate to 2–3-cleft and mostly 10–20 mm. wide, sparsely spinulose-ciliate; fls. ca. 5–8 per stem; buds glabrous, round-ovoid, somewhat beaked, 2.5–3 cm. long; petals crinkled, 6–10 cm. long; stamens yellow; caps. 3–4 cm. long; seeds dark brown, 1.3–1.5 mm. long, microscopically papillose and with larger roughenings or subreticulate ridges; $2n = 38$ (Bilquez, 1951).

—Dry washes and canyons, below 4000 ft.; Chaparral, Coastal Sage Scrub; away from the immediate coast, Santa Ana Mts. to San Diego Co. May–July.

2. **R. trichocàlyx** Eastw. [*R. Coulteri* var. *t.* Jeps.] Stems more slender; lvs. mostly 3–10 cm. long, the lobes 3–10 mm. wide; peduncles leafy, spreading-bristly below receptacle; buds round, not beaked, 2–2.5 cm. long, appressed-setose; petals 4–8 cm. long; seeds straw color to brown, rather smooth.—Below 3600 ft.; Chaparral, Coastal Sage Scrub; Ventura and San Diego cos. to L. Calif. May–July.

Hybrids between the 2 spp. are in the nursery trade.

4. Arctomècon Torr. & Frém.

Rather low perennial herbs with stout taproot. Lvs. many, long-hirsute, largely basal, cuneate-obovate, mostly toothed at apex. Fls. large, 1–several at ends of long naked stems, nodding in bud. Sepals 2 or 3. Petals 4 or 6. Stamens many. Styles united, short; stigmas united, 3–6, cordate-bilobed. Caps. ovoid to oblong-linear, 3–6-valved. Seeds rather few, oblong-obovoid. Three spp. of sw. U.S. (Greek, *arctos*, a bear, and *mecon*, poppy, because of hairiness.)

1. **A. Merriàmii** Cov. Desert Poppy. Plants 2–4(–5) dm. high, branched at base, glaucous; lvs. 2.5–7.5 cm. long, gradually narrowed at base into petioles ca. as long; peduncles 2–3.5 dm. long, naked, glabrous, one-fld.; sepals 3, villous, 1.5–2 cm. long; petals 6, white, obcordate, 2.5–4 cm. long; caps. narrow-obovoid, 2.5–3.5 cm. long; seeds black, shining, 1.5–2 mm. long, rather coarsely reticulate-ridged.—Rare, loose rocky slopes, 3000–4500 ft.; Creosote Bush Scrub; Death V. region; to Clark Co., Nev. April–May.

A. califórnica Torr. & Frém., with several-fld. stems and yellow petals, occurs just e. of the state line in Clark Co., Nev. and is to be sought in the e. Mojave Desert.

5. Dendromècon Benth. Tree Poppy

Glabrous openly branched evergreen shrubs. Lvs. alternate, simple, entire, coriaceous, yellowish to somewhat grayish green. Fls. solitary, terminal on the short branchlets. Sepals 2. Petals 4, yellow. Stamens numerous, with short fils. Stigmas 2, oblong. Caps. linear, 1-celled, elastically 2-valved from the base upward. Seeds obovoid to rounded, finely pitted, carunculate at the hilum. Here treated as having 2 spp., although 20 have been proposed. (Greek, *dendron*, tree, and *mekon*, poppy.)

Lvs. mostly lanceolate, 3–8 times as long as wide, acuminate at apex, microscopically denticulate on margins; peduncles usually definitely surpassing the lvs.; body of seed 2–2.5 mm. long. Mainland
D. rigida

Lvs. oblong-ovate to elliptic or rounded-oblong, 1.5–3 times as long as wide, mostly rounded to acute at apex, quite entire on margins; peduncles scarcely exceeding lvs.; body of seed 2.5–3 mm. long
2. *D. Harfordii*

1. **D. rígida** Benth. Stiff rounded shrub 1–3(–6) m. high, the main stems with grayish or whitish shredding bark; lvs. lance-linear to lance-oblong, 2.5–10 cm. long, 0.7–2.5 cm. wide, coriaceous, reticulate, subacuminate, scabrous-denticulate under a lens, vertical; petioles usually twisted, 2–8 mm. long; peduncles exceeding lvs.; sepals falling early, round, 8–10 mm. long; petals yellow, obovate to rounded, 2–3 cm. long; caps. linear, arcuate, 5–10 cm. long; seeds brownish olive, 2–2.5 mm. long, finely reticulate, the caruncle pale, subpeltate, an additional 0.5–1 mm. long.—Rather common on dry slopes and in stony washes, below 6000 ft.; Chaparral; Coast Ranges from Sonoma Co. to L. Calif., w. base of Sierra Nevada, Shasta Co. to Tulare Co. April–June (–Aug.). Exceedingly variable.

2. **D. Harfórdii** Kell. [*D. rigida* var. *H.* K. Bdg. *D. densifolia* and *flexilis* Greene.] More or less rounded erect shrub or tree 2–6 m. high, but with some branches spreading or even drooping; lvs. rather deep green, crowded on branches, elliptic to oblong-ovate, usually not more than twice as long as wide, the main ones 3–8 cm. long, usually rounded at tip; peduncles not exceeding lvs; petals 2–4 cm. long; caps. arcuate, 7–10 cm. long.—Brushy slopes; Chaparral; Santa Cruz and Santa Rosa ids. April–July, but with some fls. most of year.

Var. rhamnoìdes (Greene) Munz. [*D. r.* Greene. *D. arborea* Greene.] Lvs. paler green, less crowded, mostly 2–3 times as long as wide, the main ones 6–13 cm. long, acute-mucronulate at tip; *n* = 28 (Lenz, 1950).—Brushy slopes, Santa Catalina and San Clemente ids.

6. Hunnemánnia Sweet. MEXICAN TULIP POPPY

Perennial herb, glabrous, bushy, glaucous, much like *Eschscholzia,* but sepals separate. Receptacle scarcely dilated. Stamens orange. Monotypic genus from Mex. (*J. Hunneman,* English botanist, died in 1839.)

1. **H. fumariifólia** Sweet. Plant 5–6 dm. high, with dissected lvs.; sepals 1.4–1.6 cm. long; petals yellow, 2.5–3 cm. long; caps. linear, ca. 10 cm. long; seeds ovoid, punctate; *n* = 7 (Sugiura, 1936).—Occasional escape from cult., as at Oceano, *Wolf;* native of Mex. April–June.

7. Eschschólzia Cham. in Nees. CALIFORNIA POPPY

Annual or perennial herbs with colorless juice. Lvs. alternate, mostly glabrous, ternately finely dissected. Fls. yellow to red-orange, or in cult. forms with other colors, peduncled. Torus dilated to form a funnel-shaped base for the pistil. Sepals 2, completely united into a cap (calyptra) pushed off by the expanding petals. Petals 4, rarely 6 or 8. Stamens many to as few as 16; fils. short; anthers linear. Ovary cylindric, 1-celled, with 2 placentae; styles short; stigma with 4–6 linear divergent lobes. Caps. elongated, 10-nerved, 2-valved from base toward apex. Seeds subglobose to slightly elongate, reticulate or rough-tubercled or pitted. Of much horticultural interest, many color-forms having been developed. Ca. 8 or 10 spp., from Columbia R. to n. Mex. (Named for Dr. J. F. *Eschscholtz,* 1793–1831, surgeon and naturalist with Russian expeditions to Pacific Coast in 1816 and 1824.)

(Fedde, F., *in* Das Pflanzenreich, IV, 104: 144–202, 1909.)
A.   Torus with only an erect hyaline rim, the outer rim absent or rudimentary; annuals with entire cotyledons.
   B.   Buds nodding, usually with some hairs; herbage ± canescent-hairy. S. Coast Ranges.
                                                                      5. *E. Lemmonii*
   BB. Buds erect, usually glabrous; herbage usually glabrous.
      C.   Stems scapose, the lvs. mostly crowded in a basal tuft.
         D.   Lvs. usually twice ternate into relatively few rather long ultimate lobes; seeds mostly burlike. Cent. V. . . . . 6. *E. Lobbii*
         DD. Lvs. usually 3–4 times ternate; seeds reticulate or pitted.
           E.   Lf.-blades mostly 4 times ternate into very short sharp-pointed lobes; seeds weakly reticulate. S. Mojave and Colo. deserts . . . . 4. *E. Parishii*
           EE. Lf.-blades mostly 3 times ternate into blunt lobes.
              F.   Calyptra long-apiculate; seeds reticulate to almost bur-like. Cent. V. and borders . . . . 1. *E. caespitosa*
              FF. Calyptra acute but not long-apiculate; seeds with deep remote pits. Mojave Desert . . . . 7. *E. glyptosperma*
      CC. Stems leafy.
         D.   Petals 3–6 mm. long. Deserts . . . . 3. *E. minutiflora*
         DD. Petals 6–40 mm. long.
           E.   Lvs. mostly 3 times ternate into blunt lobes. Cent. V. and borders
                                                         1. *E. caespitosa*
           EE. Lvs. mostly 4 times ternate.
              F.   Ultimate lobes of lvs. blunt; calyptra short-apiculate. Insular
                                                         2. *E. elegans*
              FF. Ultimate lobes of lvs. sharp-pointed; calyptra long-apiculate. Desert
                                                         4. *E. Parishii*
AA. Torus with 2 rims, the inner erect and hyaline, the outer spreading; perennials or annuals with 2-cleft cotyledons . . . . 8. *E. californica*

1. **E. caespitòsa** Benth. [*E. tenuifolia* Benth. *E. rhombipetala* Greene. *E. c.* var. *r.* Jeps. *E. hypecoides* Benth. *E. c.* var. *h.* Gray.] Annual, usually with several stems from a tuft of basal lvs., glabrous or with patches of short stiff hairs, somewhat glaucous, 1–3(–4) dm. high; stems scapose to leafy; lvs. dissected into many narrow divisions which are mostly under 1 cm. long; buds erect; torus turbinate, without a spreading outer rim; calyptra ovoid-elliptic, apiculate, 10–18 mm. long; petals bright yellow, 1–2.5 cm. long; caps. 5–8 cm. long, 1.5–2 mm. thick; seeds subglobose to somewhat

longer than thick, 1–1.3 mm. long, gray-brown, with a network of thin low roughened ridges forming rows of ca. 5–6 irregular meshes; $n = 6$ (Smith, 1937).—Dry flats and brushy slopes, below 3500 ft.; V. Grassland, Foothill Wd., Chaparral; about the Great Cent. V., s. at altitudes up to 5000 ft., to San Bernardino and Orange cos. The typical form of the sp. has scapose stems, those with leafy stems constitute the var. *hypecoìdes,* which is the more common form and occurs throughout the range. March–June.

Ssp. **kernénsis** Munz. Stems coarse, leafy; petals deep orange, 2.5–4 cm. long; caps. 2.5–4 mm. thick; seed 1.3–1.5 mm. long, rounded, the ridges fluted, prominent, some seeds almost burlike.—Heavy soil, 1000–2000 ft.; V. Grassland; hills s. of Bakersfield, Tejon Pass region, Kern Co. March–April.

2. **E. élegans** Greene. [*E. Wrigleyana* Millsp. & Nutt.] Stout leafy-stemmed glabrous glaucous annual, 1.5–4 dm. high; lvs. conspicuously glaucous, the blades 1–3 cm. long, very finely dissected; peduncles 1–4 cm. long, erect in bud; torus broadly turbinate, without a spreading outer rim; calyptra ovoid, short-apiculate, 4–10 mm. long; petals yellow, often orange at base, 5–15 mm. long; caps. 4–7 cm. long; seeds round, dark, 0.6–0.7 mm. thick, reticulate with low rough ridges.—Channel Ids. to Guadalupe Id. March–May.

3. **E. minutiflòra** Wats. [*E. m.* var. *rutaefolia* (Greene) Jeps. *E. r.* Greene.] Mostly glabrous somewhat glaucous annual, with a number of ascending stems 1–4.5 dm. long, branched, leafy throughout; lvs. thickish, the blades mostly 1–3 cm. long, finely to coarsely dissected; fls. on short peduncles scattered along the stems; torus short-turbinate, the outer rim erect with hyaline inner edge; calyptra 4–6 mm. long, short-apiculate; petals yellow, 3–6 mm. long; stamens ca. 12; caps. 3–5(–6) cm. long; seeds round-oblong, ca. 1 mm. long, dark brown, the ridges of the reticulum rather firm and forming rows of ca. 8 cells; $2n = 36$ (Lewis & Snow, 1951).—Common in sandy and gravelly places, below 4500 (9000) ft.; Creosote Bush Scrub, Pinyon-Juniper Wd., Joshua Tree Wd.; deserts of Calif. from Mono Co. to L. Calif., Utah, Ariz. Possibly also along w. edge of upper San Joaquin V. March–May.

Var. **darwinénsis** Jones. [*E. m.* var. *Parishii* auth., in part.] Petals ca. 8 mm. long.—N. Mojave Desert.

4. **E. Paríshii** Greene. [*E. minutiflora* var. *P.* Jeps., as to type.] Annual, erect, several-stemmed, glabrous, somewhat glaucous, 2–3.5 dm. high; lvs. largely basal, dissected into linear lobes, the lower blades mostly 2–4 cm. long, the cauline scattered and much reduced; peduncles very slender, short; torus turbinate, the outer margin mostly erect with an inner evident hyaline edge; calyptra ca. 1.5 cm. long, conspicuously apiculate; petals yellow, 1.5–3 cm. long; stamens ca. 24; caps. mostly 5–7 cm. long; seeds dark brown, 1–1.5 mm. long, round to oblong, the ridges of the reticulum very low and indefinite; $2n = 12$ (Lewis and Snow, 1951).—Dry rocky slopes, below 4000 ft.; mostly Creosote Bush Scrub; s. Mojave Desert from Barstow, Sheephole Mts., and Cottonwood Springs to Colo. Desert. March–April.

5. **E. Lemmònii** Greene. [*E. L.* vars. *laxa* and *cuspidata* Greene. *E. eximia, alcicornis* and *asprella* Greene. *E. L.* var. *asprella* Jeps. *E. delitescens* Fedde. *E. urceolata* Eastw.] Annual, ± canescent with short coarse sometimes curved hairs, glaucous; stems branched, ascending or decumbent, 1–3 dm. long; lvs. ± pubescent, glaucous, finely dissected, the blades 1–3.5 cm. long; peduncles 2–15 cm. long, nodding in bud; torus urn-shaped, the inner hyaline margin ± evident; calyptra white-pubescent, 7–15 mm. long, apiculate; petals orange to yellow, 1–3 cm. long; caps. 3–7 cm. long; seeds round-oblong, reticulate.—Dry grassy plains and slopes, below 2500 ft.; V. Grassland; inner S. Coast Ranges, San Benito and Monterey cos. to w. Kern and n. Santa Barbara cos. March–May. Exceedingly variable as to habit (cespitose and scapose to branched and short-peduncled), pubescence, fl.-size, etc. Plants with buds nearly or quite glabrous have been designated as var. *aspréla* (Greene) Jeps. but probably should bear the earlier name var. *láxa* Greene.

6. **E. Lóbbii** Greene. [*E. graminea* Fedde. *E. pulchella* and *unguiculata* Greene.] Tufted subglabrous annual, scarcely if at all glaucous; lvs. largely basal, the blades 1–3 cm. long, dissected into rather few linear segms. to 1 cm. long; stems scapose, 1–3 dm. high; buds erect; torus short-turbinate; calyptra round-ovoid, 6–10 mm. long, acute to short-acuminate, but not long-apiculate; petals yellow, 7–15 mm. long; caps. 3–5 cm. long; seeds brown, round-oblong, ca. 1 mm. in diam., usually ± burlike

because of the papery brown lamellae growing out of the ridges; $n = 6$ (Sugiura, 1940 for *E. pulchella*).—Open grassy slopes and flats, fallow fields, below 2000 ft.; mostly V. Grassland; inner N. Coast Ranges and foothills of Sierra Nevada, s. to Tulare Co. March–April.

7. **E. glyptospérma** Greene. [*E. paupercula* Greene.] Glaucous glabrous tufted annual with many slender erect scapose stems 1–3 dm. high; lvs. basal, of equal length, the blades 1–2 cm. long, very finely dissected; buds erect; torus turbinate, without spreading outer rim; calyptra ovoid to lance-ovoid, 8–12 mm. long, acute but not long-apiculate; petals yellow, 1–2.5 cm. long; caps. 4–7 cm. long; seeds gray, globose, 1–1.3 mm. in diam., with deep rather remote pits; $2n = 14$ (Lewis & Snow, 1951).—Open flats and slopes, up to 5000 ft.; Creosote Bush Scrub, Joshua Tree Wd.; Mojave Desert from Inyo Co. to n. Riverside Co.; to Utah, Ariz. March–May.

8. **E. califórnica** Cham. Annual to perennial, flowering the first year, freely branched, ± glaucous, generally nearly glabrous, the stems becoming 2–6 dm. long and falling over in age; lvs. ternately several times dissected into narrow segms., the blades 2–6 cm. long, the basal lvs. petioled; peduncles 3–15 cm. long; torus with 2 rims, the inner erect and hyaline, the outer spreading and 2–4 mm. wide; calyptra 1–4 cm. long, variable in shape; petals deep orange to light yellow, 2–6 cm. long; caps. 3–8(–10) cm. long; seeds gray-brown, round-oblong, 1.2–1.5 mm. long, reticulate with ca. 8–10 meshes per row; $n = 6$ (Lawrence, 1930).—Common in grassy and open places, up to 6500 ft.; many Plant Communities; most of cismontane Calif. and w. part of Mojave Desert. Feb.–Sept. Exceedingly variable, with over 50 spp. proposed for the state; much in need of study; the following tendencies seem most marked:

(1) The typical plant from dunes and bluffs along the coast, a heavy-rooted glaucous perennial, with smooth broad compact lvs. and yellow fls.; Channel Ids. n. to Mendocino Co.

(2) An inland form or at least away from the very coast; perennial, glaucous, smooth-lvd., less compact, marked by great seasonal variation in fl.-size and color (large and deep orange in early spring, small and light yellow in late season); Columbia R., Wash. through interior Calif. to s. Calif. This is var. *cròcea* (Benth.) Jeps. [*E. c.* Benth.]; perhaps the earliest available varietal name is var. *Douglásii* (Benth.) Gray [*E. D.* Benth.].

(3) No. 2 is largely replaced in the San Joaquin V. and s. Calif. by an annual form: var. *peninsuláris* (Greene) Munz [*E. p.* Greene], for which the earliest varietal name seems to be *E. arvensis* var. *dilatàta* Greene.

(4) A perennial with very gray roughish puberulent lvs. (appearing pitted when dry and under a lens) and prostrate stems; sand dunes, San Miguel Id., Surf to Monterey. This is var. *marítima* (Greene) Jeps. [*E. m.* Greene].

8. Glaùcium Mill. SEA POPPY

Glaucous herbs with saffron-colored sap. Lvs. alternate, pinnatifid or dissected, the basal petioled, the cauline clasping. Fls. solitary, axillary and terminal. Sepals 2. Petals 4, yellow or in some spp. red. Stamens many. Stigma nearly sessile, 2-lobed, the lobes convex, dilated. Caps. long-linear, dehiscent to the base. Seeds ovoid-reniform, with numerous shallow depressions. Ca. 6 spp., mostly Medit. (Greek name referring to glaucousness.)

1. **G. flàvum** Crantz. [*Chelidonium G.* L.] Somewhat papillose-hairy biennial, stout, branched, 6–9 dm. high; lvs. ovate to oblong in outline, 5–15 cm. long, pinnatifid to sinuate-lobed; fls. golden to orange; sepals acute, pilose; petals 2–3 cm. long; caps. 15–20 cm. long; seeds 1.5 mm. broad; $n = 6$ (Sugiura, 1936).—Occasional in waste places; as Elsinore (*Baer*); native of Eu. June–Sept.

9. Argemòne L. PRICKLY POPPY

Text prepared by Gerald B. Ownbey

Caulescent annual or perennial herbs (one shrub) with yellow or orange latex. Stems glaucous, 1–several, erect or ascending, 2–20 dm. tall, cymosely branched, smooth to closely prickly. Lvs. glaucous, apetiolate, oblanceolate to obovate below, becoming elliptical to ovate upward, the uppermost often much reduced and bracteate, the margin dentate, prickly, the lamina unlobed to variously lobed or pinnatifid, the surfaces smooth to closely prickly, hispid or crisped-hispid, the areas over the veins often of a lighter bluish cast than the remaining surfaces. Buds oblong, obovate, elliptical or subspherical, the sepals caducous, normally 3 (2–6), imbricated, smooth to prickly or hispid, each with a subterminal horn tipped by an indurated prickle. Fls. 3–15 cm. in diam. Petals normally 6, in two whorls of 3, narrowly elliptical to obcuneate, obovate or suborbicular, yellow, golden, lavender, or white. Stamens 20–250 or more; fils. filiform, yellow, red, or lavender; anthers linear, extrorse, coiled after dehiscence. Stigma 3–7-lobed. Style essentially absent or to 3 mm. long. Caps. 3–7-carpellate, unilocular, fluted on the sutures, the valves apically splitting away from the placentae for about ⅓ their length, the unopened caps. elliptical, ovate or lanceolate, the surface glabrous to commonly spinescent. Seeds numerous, subspherical, anatropous, apiculate at the micropyle, scrobiculate. Ca. 30 spp. in N. Am., S. Am., Hawaii; 1 sp. introd. in the trop. and subtrop. throughout the world. (Greek, *argema*, cataract of the eye, for which the juice of a poppylike plant of the same name was a supposed remedy.)

Stamens ca. 100–120; latex orange when fresh; caps. ca. 25–30 mm. long, armed with large rather even-sized spines, but not closely so; stems armed with scattered spines; lvs. smooth to sparingly prickly on the veins especially on the midrib beneath; sepal-horns smooth . . . . 1. *A. corymbosa*
Stamens ca. 150–250; latex pale lemon-yellow, rarely more intensely yellow when fresh; caps. ca. 35–55 mm. long, armed with widely spaced spines to closely spinescent and prickly; stems armed with scattered spines to closely prickly; lvs. smooth or very sparingly prickly on the main veins to closely prickly throughout; sepal-horns smooth to closely prickly . . . . 2. *A. munita*

1. **A. corymbòsa** Greene. [*A. intermedia* var. *c.* Eastw.] Perennial with orange latex; stems 5–8 dm. tall, armed with scattered stout perpendicular or retrorse spines; lvs. coriaceous, the basal shallowly lobed, the middle and upper more shallowly lobed to unlobed, the uppermost clasping, the upper surfaces smooth or sparingly prickly on the main veins, the lower surfaces prickly on the main veins; buds sparingly prickly, the sepal-horns smooth; fls. 5–9 cm. in diam.; petals white; stamens ca. 100–120; caps. ovate to lanceolate, 4–5-carpellate, 10–15 mm. wide excluding the armature, 25–30 mm. long including the stigma, stoutly and rather evenly but not closely spinescent, the largest spines ca. 6–7 mm. long; seeds ca. 1.5 mm. long; $n = 28$ (Ownbey, 1957).—Dry slopes and flats, 1400–3500 ft.; Creosote Bush Scrub; Mojave Desert. April–May.

2. **A. munìta** Dur. & Hilg. [*A. platyceras* Calif. refs., not Link & Otto.] Annual or perennial with pale lemon-yellow rarely more intensely yellow latex; stems 6–15 dm. tall, sparingly spinescent to closely prickly; lvs. lobed ca. halfway to midrib below, more shallowly upward, the lobes usually rounded at apex, sometimes angular, the uppermost lvs. often clasping, the upper surfaces of the lvs. sparingly prickly on larger veins and also often between the veins both above and below; buds simply and sparingly to moderately prickly, the sepal-horns scatteringly prickly at base; fls. 5–13 cm. in diam.; petals white; stamens ca. 150–250; caps. elliptical, elliptic-lanceolate or lanceolate, 3–5-carpellate, 9–18 mm. wide excluding the armature, 35–55 mm. long including the stigma, moderately spinescent with spreading spines and with a few lesser spines and prickles, the largest spines ca. 8 mm. long; seeds ca. 1.8–2.6 mm. long; $n = 14$ (Ownbey, 1957).—Occasional, 250–6000 ft.; Chaparral, etc.; Coast Ranges from San Luis Obispo Co. to L. Calif. Aug.

KEY TO SUBSPECIES

A. Stems with ca. 0–80 prickles per sq. cm. 5 cm. below the oldest caps.; lf.-surfaces smooth or sparingly prickly.
   B. Stems with ca. 0–10 prickles per sq. cm. 5 cm. below the oldest caps.; lf.-surfaces usually totally smooth above, and with a few prickles on the main veins beneath; caps. armed with large widely spaced spines . . . . ssp. *robusta*
   BB. Stems more copiously prickly; lf.-surfaces prickly on the veins and usually also sparingly prickly between the main veins; caps. armed with large spines interspersed with uneven-sized smaller ones.
      C. Stems with ca. 10–30 prickles per sq. cm. 5 cm. below the oldest caps.; lf.-surfaces sparingly prickly on the veins and usually between the veins . . . . ssp. *munita*
      CC. Stems with about 50–80 prickles per sq. cm. 5 cm. below the oldest caps.; lf.-surfaces more prickly on and between the veins, but not closely so . ssp. *munita* × ssp. *rotundata*

AA. Stems with about 80–500 prickles per sq. cm. 5 cm. below the oldest caps.; lf.-surfaces moderately to closely prickly.
   B. Stems usually purplish, usually closely armed with uneven-sized prickles of moderate length; buds usually with numerous larger compound prickles interspersed with smaller uneven-sized ones.
      C. Stems with about 80–180 prickles per sq. cm. 5 cm. below the oldest caps.; bud-prickles all simple; less copiously prickly throughout than the following ssp.
                                                  ssp. *munita* × ssp. *rotundata*
      CC. Stems with about 120–500 prickles per sq. cm. 5 cm. below the oldest caps.; largest bud-prickles usually compound; usually very copiously prickly throughout
                                                   ssp. *rotundata*
  BB. Stems usually greenish-white, usually closely armed with very long uneven-sized prickles, i.e., with about 100–300 prickles per sq. cm. 5 cm. below the oldest caps.; buds usually with numerous simple more even-sized prickles .... ssp. *argentea*

Ssp. **robústa** G. Ownbey. Annual; stems 5–16 dm. tall, usually tinged with purple, sparsely prickly; uppermost leaves clasping; buds sparingly prickly, the sepal-horns smooth; caps. armed with scattered mostly subequal often reflexed spines; $n = 14$ (Ownbey, 1957).—Chaparral, 5000–5680 ft.; Santa Ana Mts., Orange Co. July.

Ssp. **rotundàta** (Rydb.) G. Ownbey. [*A. r.* Rydb.] Perennial; stems 4–10 dm. tall, often tinged with purple, usually closely prickly throughout; lf.-surfaces usually closely prickly both above and below; buds usually copiously prickly, the larger prickles often compound, the sepal-horns usually very prickly; caps. usually closely armed with straight or slightly incurved usually simple spines interspersed with smaller to minute spines and prickles, the armature partially obscuring the capsular surface; $n = 14$ (Ownbey, 1957).—Usually at 4000–8500 ft.; Chaparral, Pinyon-Juniper Wd., Montane Coniferous F.; mts. of Mojave Desert, San Bernardino and San Gabriel mts., e. slope of Sierra Nevada to Shasta Co., localized in Lake and Colusa cos.; nw. Ariz., Nev., Utah. June–Sept. Probable hybrids occur with the typical form of the sp.; variable, but intermediate in foliage and armature between the putative parents.—At 1200–3200(–8000) ft.; Coast Ranges, San Diego Co. to Contra Costa Co.

Ssp. **argentèa** G. Ownbey. Perennial; stems 4–8 dm. tall, greenish-white, mostly closely armed with long slender perpendicular prickles; lf.-surfaces usually copiously prickly both above and below; buds copiously armed with simple ascending prickles, the sepal-horns prickly; caps.-armature consisting of numerous uneven-sized spines and prickles, the surface partially obscured; $n = 14$ (Ownbey, 1957).—At 1500–3000 ft.; Creosote Bush Scrub; in and adjacent to the dry desert mt. ranges, Inyo Co. to San Diego and Imperial cos.; s. Nev., w. Ariz. March–May.

10. Stylomècon G. Tayl.

Erect annual, the stem simple or branched with yellow sap. Lvs. alternate, pinnatisect or pinnatifid, the lobes entire or in turn dissected or lobed. Fls. axillary, on elongate peduncles. Sepals 2, deciduous. Petals 4, orange-red with basal purplish spot above the green claw. Stamens many. Caps. turbinate, glabrous, with 4–11 parietal placentae and dehiscing near their apex. Style slender, short, persistent; stigma capitate. Seeds many, reniform, rugose-reticulate. One sp. (Greek, *stylus*, style, and *mekon*, poppy.)

1. **S. heterophýlla** (Benth.) G. Tayl. [*Meconopsis h.* Benth. *M. crassifolia* Benth. *Papaver h.* Greene. *P. h.* var. *c.* Jeps.] Plants 3–6 dm. high, glabrous, or somewhat pilose below; lvs. 2–12 cm. long, including the petioles; peduncles 5–20 cm. long; sepals 4–10 mm. long; petals 1–2 cm. long; caps. clavate-obovoid, 8–15 mm. long, with yellowish ribs; seeds dark, ca. 0.4 mm. long.—Occasional on grassy and brushy slopes, below 4000 ft.; Chaparral, V. Grassland, Foothill Wd., Oak Wd.; Coast Ranges from Lake Co. s., San Joaquin V., foothills of s. Sierra Nevada; to L. Calif.; Channel Ids. Most common s. April–May.

11. Papàver L. POPPY

Annual or perennial herbs with yellowish or milky juice. Lvs. pinnately lobed or divided. Fls. solitary on long peduncles, drooping in the bud. Sepals 2. Petals 4. Stamens many. Ovary and caps. obovoid to subglobose, with 4–20 intruded placentae; caps.

dehiscent by transverse pores under the edge of the sessile round flat or subconical disk formed by the united radiating stigmas. Seeds many, minute, reniform, variously striate and pitted. Ca. 50 spp., mostly of Old World; some well known as garden fls. (Classical name of the *Poppy*.)

(Fedde, F., *in* Das Pflanzenreich, IV, 204: 288–386, 1909.)
Cauline lvs. with broad clasping base; caps. 3–5 cm. long .... 1. *P. somniferum*
Cauline lvs. not clasping but with narrowed base; caps. 1–2 cm. long.
  Peduncles with weak spreading hairs or glabrous; caps.-disk 5–18-rayed.
    Petals 2–3 cm. long; caps. broad-obovoid. Escape from cult., in waste places .... 2. *P. Rhoeas*
    Petals 1–2 cm. long; caps. clavate-turbinate. Native on burns and in woods .. 3. *P. californicum*
  Peduncles with stiff appressed hairs; caps.-disk 4–6-rayed. Introd. spp.
    Petals ca. 2 cm. long; caps. 1.5–1.8 cm. long; lf.-segms. linear-lanceolate ..... 4. *P. Argemone*
    Petals less than 2 cm. long; caps. 1 cm. or less long; lf.-segms. oblong-ovate ..... 5. *P. apulum*

1. **P. somníferum** L. OPIUM POPPY. Glaucous annual, glabrous or slightly hairy, 6–12 dm. high, with rather coarse stems and peduncles; lower lvs. petioled, the upper oblong, clasping, cordate, unequally coarse-toothed to lobed, 1–2 dm. long; sepals 1.5–2 cm. long; petals round, entire, white, pink, red or purple, 4–6 cm. long; caps. globose, 3–5 cm. long, the disk plane; seeds dark, reticulate-favose, 0.6–0.7 mm. long; $n = 11$ (Furusato, 1940).—Common Garden Poppy, reported as occasional escape in waste places; native of Old World. Summer.

2. **P. Rhoèas** L. RED or CORN POPPY. Slender branching annual, 3–8 dm. high, mostly with spreading hairs; lvs. 3–15 cm. long, somewhat clustered about base of peduncles, irregularly pinnatifid and divided with serrate lanceolate segms.; sepals ca. 1 cm. long, setose; petals round, cinnabar-red, or purple or scarlet, or white or white with red margins, sometimes with dark spot, 2–3 cm. long; caps. round to obovoid, 1–2 cm. long, the disk flat; seeds minute, dark; $n = 7$ (Lawrence, 1930).—Common in gardens, reported as occasional escape; native of Old World.

3. **P. califórnicum** Gray. [*P. Lemmonii* Greene.] Slender glabrous or pilose annual, 3–6 dm. tall; lvs. 3–9 cm. long, pinnately divided into oblong or rounded toothed or lobed segms.; sepals 8–10 mm. long, pilose; petals 1–2 cm. long, brick red with basal greenish spot; caps. clavate-turbinate, 1–1.6 cm. long, the disk flat; seeds dark, coarsely rugose-reticulate, 0.4–0.5 mm. long.—Burns and disturbed places, below 2500 ft.; Chaparral and Oak Wd.; Coast Ranges from Marin Co. to San Diego Co. April–May.

4. **P. Argemòne** L. Annual erect or ascending herb, 2–5 dm. high, setulose; lvs. 3–12 cm. long, mostly bipinnatisect, the segms. acute, lance-linear; peduncles coarsely strigose; sepals 5–10 mm. long, strigose to glabrous; petals oblong-obovate, ca. 2 cm. long, pale scarlet, with a dark basal spot; caps. setose, clavate-cylindric, 1.5–2 cm. long, the stigma elevated; seeds dark, foveolate, ca. 0.8 mm. long; $n = 6$ (Beale, 1939). —Reported as occasional waif; native of Old World. May?

5. **P. ápulum** Ten. var. **micránthum** (Boreau) Fedde. [*P. m.* Boreau.] Annual, 2–4 dm. high, branched, hispid at base; lvs. bi- or tri-pinnatifid with oval-oblong segms.; petals pale rose, elliptic-obovate, 1.5–1.8 cm. long; caps. ovoid to elliptic-obovoid, ca. 1 cm. long, with elevated disc.—Reported from Point of Rocks, Antelope Plains, w. Kern Co. (*C. Smith*); native of Eu. March–April.

12. Cánbya Parry

Minute almost acaulescent glabrous tufted annuals. Lvs. alternate, linear-oblong, fleshy, entire. Scapes filiform, 1-fld. Sepals 3, caducous. Petals 6, persistent, obovate. Stamens 6–9. Ovary 1-celled with 3 placentae; style 0; stigmas 3, linear. Caps. ovoid, 3-valved from apex down. Seeds many, elongate-oblong, slightly arcuate. Spp. 2, Ore. to Calif. (W. M. *Canby*, Delaware botanist of 19th century.)

1. **C. cándida** Parry. Plants 2–3 cm. high; lvs. 5–7 mm. long; petals white, 3 mm. long, closing over the caps. after anthesis; caps. 2–2.5 mm. long; seeds shining, dark brown, ca. 0.6 mm. long.—Sandy places, 2000–4000 ft.; Creosote Bush Scrub, Joshua Tree Wd.; w. Mojave Desert from Walker Pass to Victorville region. April–May.

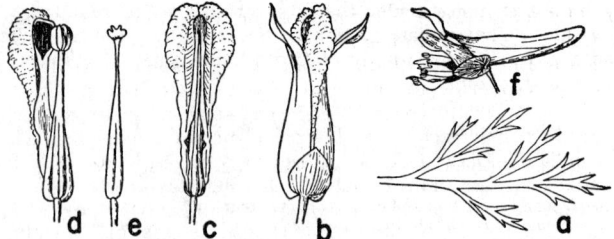

Fig. 28. FUMARIACEAE. *Dicentra chrysantha: a,* part of dissected lf., × ½; *b,* fl., × 1, sepal short, outer 2 petals saccate at base, divaricate at tip; *c,* outer petals removed, the inner 2 clawed below, crested above; *d,* 1 set of 3 stamens; *e,* pistil. *Corydalis Caseana: f,* fl., × 1, with spur from 1 of outer 2 petals.

30. Fumariàceae. FUMITORY FAMILY (Fig. 28)

Herbs with brittle stems and watery juice. Lvs. basal or alternate, usually much dissected, without stipules. Fls. perfect, irregular, hypogynous. Sepals 2, small, deciduous. Petals 4, somewhat connivent, the 2 outer spreading at apex and 1 or both often saccate or spurred at base; the 2 inner smaller and narrower and sometimes coherent to apex. Stamens 4, free and opposite the petals, or 6 and united in 2 sets. Pistil of 2 carpels; style slender; stigma 2-lobed; ovary 1-celled with 2 parietal placentae. Fr. a caps. with 2 valves and several seeds or a 1-seeded nut. Seeds shining, crested or not, the minute embryo in a fleshy endosperm. Five genera and ca. 200 spp., mostly natives of N. Temp. Zone.

Outer petals both saccate or spurred at base .... 1. *Dicentra*
Outer petals with one spurred, the other not.
   Fr. with more than 1 seed, an elongate dehiscent caps.; seed with a crest or aril; plants perennial
                                                  2. *Corydalis*
   Fr. a 1-seeded globose nut; seed not crested; plants annual .... 3. *Fumaria*

1. Dicéntra Bernh. BLEEDING HEART

Perennial herbs. Lvs. basal or basal and cauline, dissected. Fls. solitary or in racemes or panicles, irregular, usually flattened and heart-shaped. Sepals small. Outer 2 petals saccate at base, spreading at apex; inner 2 narrow, clawed, usually cohering above and crested on back. Stamens 6, in 2 sets opposite the outer petals. Caps. several-seeded, elongate, 2-valved. Seeds crested or not. A genus of 15–16 spp. of N. Am. and Asia; of some ornamental value. (Greek, *dis,* twice, and *centron,* spur.)

Stems leafy, 5–16 dm. high; corolla deciduous; fls. erect.
   Fls. golden yellow, 12–15 mm. long, the outer petals spreading or recurved to middle
                                                          1. *D. chrysantha*
   Fls. off-white to cream, 20–25 mm. long, the outer petals spreading only at tips . 2. *D. ochroleuca*
Stems scapose, 0.5–4 dm. high; corolla rather persistent; fls. nodding.
   Plants with heavy creeping rootstocks; fls. several in a compact panicle .... 3. *D. formosa*
   Plants with fascicles of tuberous roots; fls. 1–3.
     Recurving tips of outer petals longer than the body of the petals .... 4. *D. uniflora*
     Recurving tips of outer petals shorter than the body .... 5. *D. pauciflora*

1. D. chrysántha (H. & A.) Walp. [*Diclytra c.* H. & A. *Bikukulla c.* Cov.] GOLDEN

EAR-DROPS. Glaucous, erect, with several coarse stems from stout roots, 5–15 dm. high; lvs. basal and on lower stems, bipinnate, rather stiff, 15–30 cm. long (including petiole), the pinnules 1–3 cm. long and further divided into usually narrow lobes; panicle loose, narrow, 2–5 dm. long, terminal, many-fld.; sepals suborbicular, 3–4 mm. long, caducous; corolla oblong, bright yellow, slightly cordate, 12–15 mm. long, the outer petals saccate below the tip, spreading in upper half; crest of inner petals narrow, crisped; caps. mostly 15–25 mm. long, lance-ovoid; seeds black, subreniform to almost round, densely papillate, 1.5–2 mm. long.—Frequent on burns and in disturbed places, dry slopes, below 5000 (7300) ft.; Chaparral, Yellow Pine F., Foothill Wd.; inner

Coast Ranges, Mendocino Co. s., foothills of Sierra Nevada from Calaveras Co. s., cismontane s. Calif.; L. Calif. April–Sept.

2. **D. ochroleùca** Engelm. [*Diclytra o.* Greene. *Bikukulla o.* Heller.] Habit of *D. chrysantha*, but lvs. 3-pinnate, the pinnules dissected into narrow lobes; sepals ca. 5 mm. long; corolla off-white to cream, 20–25 mm. long, only the purplish tips of the outer petals spreading; inner petals purple-tipped, broad-crested; seeds ca. 1.2–1.4 mm. long, densely papillate.—Occasional in dry disturbed places, below 3000 ft.; Chaparral; Santa Lucia and Santa Ynez mts. to Santa Ana Mts. May–July.

3. **D. formòsa** (Andr.) Walp. [*Fumaria f.* Andr. *Corydalis f.* Pursh. *Diclytra saccata* Nutt. in T. & G. *Bikukulla f.* Cov.] BLEEDING HEART. Scapose plants with a fleshy rootstock; slender stems 2–4.5 dm. high; lvs. basal, long-petioled, 2–5 dm. long, biternately compound, glaucous beneath or also above, the ultimate segms. oblong, 2–5 cm. long, pinnately cleft or incised into divisions 1.5–4 mm. wide; fls. several, ± nodding, in a small panicle; sepals ovate to lanceolate, 3–5 mm. long; corolla rose-purple to whitish, ca. 14–18 mm. long, cordate at base, the outer petals with spreading ovate tips, the inner wing-crested on the back; caps. 14–20 mm. long; seeds subreniform, ca. 1.5 mm. long, shining, black, finely papillate, with a light-colored arillike crest; $2n = 24$ (Kellet, 1937).—Damp ± shaded places, 0–7000 ft.; Redwood F., Yellow Pine F., Red Fir F.; N. Oak Wd., etc.; Coast Ranges from Santa Cruz Co. n., Sierra Nevada from Tulare Co. to Siskiyou Co. March –July. Variable as to dissection of foliage and color of fls.

Ssp. **oregàna** (Eastw.) Munz. [*D. o.* Eastw. *D. formosa f. o.* Van Melle.] Plants 1.5–2.5 dm. high; corolla very shallowly cordate, the outer petals ochroleucous, the inner with rose-colored tips.—Mixed Evergreen F.; state boundary; w. base Siskiyou Mts., Del Norte Co.; adjacent Ore. April–May.

Ssp. **nevadénsis** (Eastw.) Munz. [*D. n.* Eastw.] Lvs. finely dissected, the divisions 1–2 mm. wide; sepals 5–8 mm. long; corolla more rounded at base, 13–15 mm. long, the outer petals ochroleucous to pinkish, the inner white with pale yellow tinge.— Moist places, 7500–10,000 ft.; Lodgepole F., Subalpine F.; Tulare Co. July.

4. **D. uniflòra** Kell. [*Diclytra u.* Greene. *Bikukulla u.* Howell.] STEER'S HEAD. Scapose, from a fascicle of tubers, 3–7 cm. high; lvs. 4–6 cm. long, 1–3, basal, 2–3-ternate into oblong lobes, glaucous beneath; scapes 1-fld.; sepals lance-oblong, 5–6 mm. long; fls. white or pink to lilac, ca. 15 mm. long, cordate at base, the outer petals narrow and recurved to below the middle, the inner purple-tipped, not crested; caps. ovoid, ca. 12 mm. long; seeds rounded, shining, black, obscurely reticulate, crested.—Gravelly or rocky places, 5400–12,000 ft.; Red Fir F., Lodgepole F., Subalpine F., Alpine Fell-fields; Glenn Co., Siskiyou Mts., Sierra Nevada from Fresno Co. n.; to Wash., Utah. May–July.

5. **D. pauciflòra** Wats. [*Diclytra p.* Greene. *Bikukulla p.* Cov.] Much like *D. uniflora*, but the lf.-segms. narrowly linear and more acute; scapes 1–3-fld.; sepals lance-ovate; corolla 18–20 mm. long, deeply cordate, the outer petals reflexed only toward tips, the inner with more abruptly expanded apex.—Gravelly places, 4000–10,000 ft.; Subalpine F., Alpine Fell-fields; Salmon Alps, Scott and Trinity mts., Sierra Nevada of Plumas and Tulare cos. June–July.

2. Corýdalis Medic.

Annual or mostly perennial herbs. Lvs. simple or pinnate, the pinnae 1–2-divided and incised. Infl. a terminal bracteate raceme or panicle. Fls. bilaterally symmetrical; sepals 2, caducous; petals 4, the 2 outer dissimilar with 1 spurred, both ± keeled and hooded at apex, the 2 inner similar, connate at apex, clawed. Stamens 6, in 2 bundles of 3 each. Stigma persistent, flattened, with 4–8 papillary stigmatic surfaces. Caps. many-seeded. Seeds round-reniform, somewhat compressed, smooth or not, with an arillike crest. Perhaps 100 spp. of N. Temp. Zone and S. Afr., a few used as garden plants. (Greek, *korydalis*, name for the crested lark.)

(Ownbey, G. B. Monograph of the N. Am. Spp. of Corydalis. Ann. Mo. Bot. Gard. 34: 187–259, 1947.)

Plant perennial; fls. pink to white with purple tips; spur 12–16 mm. long .... 1. *C. Caseana*
Plant winter annual or biennial; fls. yellow; spur 4–5 mm. long .... 2. *C. aurea*

1. **C. Caseàna** Gray. [*C. Bidwelliae* Wats. *Capnodes C.* Greene. *C. Bidwellianum* Greene.] Glaucous perennial from thickened roots; stems 1–several, stout, 5–10 dm.

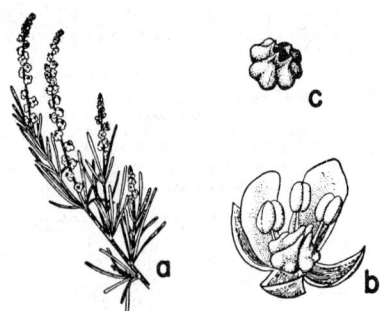

Fig. 29. RESEDACEAE. *Oligomeris linifolia: a,* habit, × ½; *b,* fl., × 12, sepals 4, petals 2, stamens 3, ovary 4-lobed; *c,* caps., × 2½, gaping at summit.

high; lvs. ca. 5, 15–35 cm. long, pinnate, the primary segms. again 1–2-pinnatifid or deeply incised, the ultimate divisions lance-elliptic to ovate, 1–2.5 cm. long, apiculate; infl. a dense narrow panicle, 5–12 cm. long; fls. white or pinkish with petal-tips purple; sepals 2–4 mm. long; spur slender, the hood crested; unspurred outer petal 10–12 mm. long; inner petals 9–10 mm. long; style ca. 3 mm. long; caps. 10–15 mm. long, oblong; seeds shiny, black, minutely papillose, ca. 2.5 mm. in diam.—Moist shade, 4000–9000 ft.; Yellow Pine F., Red Fir F.; Shasta and Lassen cos. to Placer Co. June–Aug.

2. **C. aùrea** Willd. [*Capnodes a.* Kuntze.] Glaucous winter annual or biennial, much branched, leafy, 1–4 dm. high; lvs. bipinnate, the ultimate segms. elliptical, ca. 3–8 mm. long, subapiculate; racemes scarcely if at all exceeding the lvs., rather few-fld.; sepals 1–3 mm. long; fls. yellow, the spurred petal 13–16 mm. long including the 4–5 mm. long spur; inner petals 8–10 mm. long; caps. 18–24(–30) mm. long, linear; seeds shiny, smoothish to obscurely marked, nearly 2 mm. in diam.; $n = 8$ (G. Ownbey, 1951).— Loose open often disturbed places, 5000–7500 ft.; Sagebrush Scrub, Yellow Pine F.; Modoc and Mono cos.; to Alaska, Quebec, Tex. May–Aug.

3. Fumària. L. FUMITORY

Annuals with branching leafy stems. Lvs. decompound and finely dissected. Fls. small, in dense racemes or spikes. Sepals 2, scalelike. Petals 4, one of the outer pair spurred at base, the inner pair narrow, coherent at apex, keeled or crested on back. Stamens in 2 bundles. Style deciduous; ovary subglobose, 1-ovuled. Fr. 1-seeded, indehiscent. Seeds crestless. Ca. 15 spp. of Old World. (Latin, *fumus,* smoke, presumably because of odor of fresh roots.)

Fls. purple, 5–7 mm. long; nutlet depressed-globose .... 1. *F. officinalis*
Fls. cream with inner petals purple at tip, 3–4 mm. long; nutlet apiculate .... 2. *F. parviflora*

1. **F. officinalis** L. Glabrous, diffusely branched, the stems 2–6 dm. long; lvs. 2–6 cm. long (including petiole), finely dissected into segms. ca. 2 mm. wide; racemes narrow, 3–7 cm. long; pedicels 2–4 mm. long; sepals 2–3 mm. long; nutlet ca. 2 mm. in diam.— Occasional in waste places as weed; native of Eu. April–July.

2. **F. parviflòra** Lam. Much like the preceding, but lf.-segms. narrower, channeled; fls. paler, smaller; sepals 0.5–1 mm. long; fr. not depressed at apex; $2n = 28$ (Negodi, 1936).—Occasional as ruderal; native of Eu. March–May.

31. Resedàceae. MIGNONETTE FAMILY (Fig. 29)

Annual or perennial herbs with watery juice, rarely woody. Lvs. alternate, simple or pinnately divided, with small glandlike stipules. Fls. small, perfect, irregular, in racemes or spikes. Calyx 4–7-parted, persistent. Petals small and inconspicuous, 2–7 or 0, often laciniate, clawed. Stamens 3–40, usually in a ± one-sided hypogynous disk; fils. free or united at base; anthers 2-celled, introrse. Ovary of 2–6 free or connate carpels,

1-celled, closed or opening at the top, each carpel with its own stigma. **Fr.** a gaping caps. or a berry; seeds many, subreniform, without endosperm. Ca. 6 genera and 65 spp., mostly of Medit. region.

Petals 4–7; disk cup-shaped; lvs. often pinnatifid .... 1. *Reseda*
Petals 2; disk lacking; lvs. entire .... 2. *Oligomeris*

1. Resèda L. MIGNONETTE

Erect or decumbent herbs, sometimes almost woody at base. Lvs. entire, lobed or pinnatifid. Fls. small, in terminal spikes or racemes. Petals 4–7, cleft, unequal. Disk cup-shaped. Stamens 8–40, attached on one side of the fl. Caps. 3–6-horned or -angled, opening only at the top at maturity. Ca. 50 spp., mostly of Medit. region and about the Red Sea. (Latin, *resedare*, to assuage or calm, because of supposed sedative properties.)

Caps. erect or ascending.
  Lvs. entire; caps. subglobose .... 1. *R. Luteola*
  Lvs. pinnatifid or lobed; caps. oblong.
    Fls. pale yellow; stamens 15–20; lowest petal not cleft .... 2. *R. lutea*
    Fls. greenish-white; stamens 12–15; all petals cleft .... 3. *R. alba*
Caps. pendent, the opening directed downward; lvs. mostly entire or 3-lobed .... 4. *R. odorata*

1. **R. Lutèola** L. DYER'S ROCKET. Biennial, glabrous, erect, 4–8 dm. high, simple or branched; lvs. linear to lanceolate, sessile and narrowed toward base or the lowest petioled, entire, 2–10 cm. long; racemes spicate, many-fld., becoming 2–3.5 dm. long; fls. greenish-yellow; sepals 4; petals 4 or 5, 2–4 mm. long, the lower linear and entire, the upper lobed; stamens 20–30; caps. globose, 4–6 mm. in diam., with 3–4 terminal teeth; seeds round-reniform, almost black, shining, ca. 1 mm. long; $2n = 24, 26, 28$ (Eigsti, 1936).—Reported as escape from San Diego Co., Mendocino Co., etc.; native of Eu. May–Oct.

2. **R. lùtea** L. YELLOW MIGNONETTE. Biennial to perennial, pubescent to glabrous; stems slender, ascending to suberect, 3–8 dm. long; lvs. bipinnatifid to pinnate-parted, 5–10 cm. long, the segms. oblong to linear; racemes narrow, becoming 1–2 dm. long; sepals mostly 6; petals 6, all but the lowest cleft, 3–4 mm. long; stamens 15–20; caps. ovoid-oblong, 7–8 mm. long, usually apically-toothed; seeds obovoid, shining, smooth, 1.4–1.8 mm. long, blackish; $2n = 48$ (Eigsti, 1936).—Sparingly established in waste places, as in Lake, Los Angeles, and San Diego cos.; native of Eu. May–Sept.

3. **R. álba** L. WHITE MIGNONETTE. Biennial to perennial, glabrous, erect, 3–8 dm. tall; lvs. 3–12 cm. long, deeply pinnatifid into lance-oblong segms.; fls. white in dense spikelike racemes; petals 6 (5), 4–5 mm. long, 3-cleft; caps. ovoid-oblong, 10–12 mm. long, usually 4-toothed; seeds subreniform, dark, dull, minutely papillose, ca. 1 mm. long; $2n = 20$ (Eigsti, 1936).—Established in waste places, roadsides, etc., especially in s. Calif., also at Monterey, San Francisco; native of Eu. May–Nov.

4. **R. odoràta** L. GARDEN MIGNONETTE. Annual, branched, spreading and decumbent with age, the stems 1.5–6 dm. long; lvs. spatulate to oblong, entire or few-lobed, 2–7 cm. long; fls. yellowish-white to orange, fragrant, in racemes that become loose; petals finely cleft; anthers golden; caps. obovoid-rounded, 5–7 mm. long, usually 3-horned; seeds oblong-reniform, greenish-yellow, ca. 1.5 mm. long, somewhat round-tuberculate; $2n = 12$ (Oksijuk, 1929).—Occasional escape from gardens; native of n. Afr. Feb.–Sept.

2. Oligómeris Camb.

Low branching herbs. Lvs. linear, entire. Fls. small, greenish, in terminal spikes. Sepals 4. Petals 2, entire or lobed, persistent. Stamens 3–10. Caps. 4-lobed, each lobe sulcate on back. Seeds many. Ca. 5 spp., sw. U.S., Mex., Afr., Asia. (Greek, *oligos*, little, and *meris*, parts.)

1. **O. linifòlia** (Vahl) Macbr. [*Reseda l.* Vahl. *Ellimia ruderalis* Nutt. *O. subulata* Webb.] Annual, erect, glabrous, rather fleshy, 1–3 dm. high; lvs. often fascicled, 1–2.5 cm. long; spikes bracteate, densely fld., 2–10 cm. long; fls. ca. 1.5 mm. long; petals oblong, whitish, obscurely lobed; stamens 3; caps. 1.5 mm. long, 2.5 mm. thick,

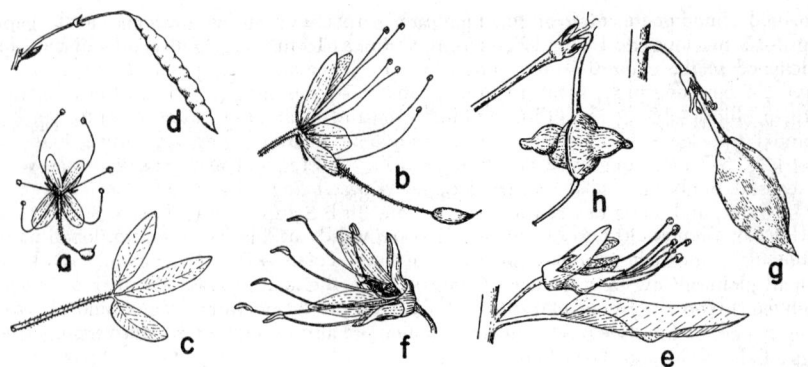

Fig. 30. CAPPARIDACEAE. *Cleome lutea: a,* fl., front view, × 1; *b,* side view, × 2, with 4 sepals, 4 petals, 6 stamens, stipitate ovary; *c,* trifoliolate lf.; *d,* stipitate caps. *Isomeris arborea: e,* fl. in axil of leaflike bract, × 1; *f,* fl., × 1, with 4 sepals, 4 petals, 6 stamens; *g,* caps., × ⅕, stipitate. *Cleomella obtusifolia: h,* fr., × 2½, with stipe and short ovary.

4-toothed; seeds round-reniform, black, shining, ca. 0.5 mm. long.—Common in open often subsaline places, below 3000 ft.; Creosote Bush Scrub, Alkali Sink; deserts; occasional on sea bluffs, alkaline places, Santa Rosa and Santa Catalina ids., Santa Monica to L. Calif.; to Tex., Mex. Feb.–July.

32. Capparidàceae. CAPER FAMILY (Fig. 30)

Herbs or shrubs with ill-smelling foliage and watery sap. Lvs. alternate, usually palmately compound with 3–5(–7) entire lfts. Fls. perfect, regular or irregular, in bracted terminal racemes or solitary. Sepals (in ours) 4, free or united, often persistent. Petals (in ours) 4, not much clawed. Stamens 6 to many, subequal, inserted on the receptacle which is often thickened or produced between the stamens and petals. Ovary superior, of 2 united carpels, usually 1-celled with 2 parietal placentae, few- to many-ovuled. Stipe usually well developed. Fr. a caps. or fleshy. Seeds ± reniform, without endosperm; cotyledons usually coiled. Ca. 35 genera and 450 spp., mostly of warmer regions, one furnishing the capers of commerce, a few grown as ornamentals.

A. Shrubs with conspicuously inflated caps.; petals over 1 cm. long .... 1. *Isomeris*
AA. Herbs with caps. not or scarcely inflated; petals often less than 1 cm. long.
    B. Fls. in dense axillary glomerules; style spinescent in fr. .... 2. *Oxystylis*
    BB. Fls. axillary or racemose, not in glomerules.
        C. Stamens 8–32, purple; stipe scarcely developed. Ne. Calif. .... 3. *Polanisia*
        CC. Stamens 6, mostly yellow.
            D. Fr. didymous, 2-celled, each valve enclosing a seed and falling away with it
                                                            4. *Wislizenia*
            DD. Fr. not didymous, 1-celled, several- to many-seeded.
                E. Caps. 12–45 mm. long, linear to oblong; petals 6–12 mm. long, generally
                    clawed .˙.. 5. *Cleome*
                EE. Caps. 3–6 mm. long, often somewhat wider; petals 1.5–7 mm. long, not
                    clawed .... 6. *Cleomella*

1. Isómeris Nutt. BLADDERPOD

Ill-scented widely branched glaucous puberulent shrubs. Lvs. alternate, trifoliate, petioled. Fls. yellow, in dense terminal bracted racemes. Calyx persistent, 4-cleft. Petals 4, oblong, sessile, the 2 lower more spreading than the 2 upper. Stamens 6, long-exserted. Torus hemispheric. Caps. large, inflated, pendulous, coriaceous, long-stipitate, tardily dehiscent in 2 valves. Seeds few, large, smooth, somewhat obovoid with pointed base. Monotypic. (Greek, *isos,* equal, and *meris,* part.)

1. **I. arbórea** Nutt. [*Cleome I.* Greene.] Rounded, erect, 6–15(–25) dm. high; petioles 0.5–2.0 cm. long; lfts. oblong to elliptic-oblanceolate, 1–3.5 cm. long, 0.3–0.8

cm. wide, mucronate, entire, the uppermost sometimes solitary; racemes mostly becoming 3–15 cm. long, with simple green bracts to ca. 1 cm. long; pedicels 6–12 mm. long, thickened in fr.; calyx 6–8 mm. long, the lobes subacuminate; petals 10–16 mm. long; stipe 1–2 cm. long in fr., stout, recurved; caps. 2.5–5 cm. long, 1–1.5 cm. thick, oblanceolate to elliptic-oblong in outline, gradually narrowed at base, more abruptly so at the pointed tip; seeds smooth, hard, with prominent incurved embryo, brown with some mottling, 6–7 mm. long; $n = 17$ (Billings, 1937).—Frequent in subalkaline places, such as coastal bluffs and hills, stabilized dunes; Coastal Sage Scrub; San Luis Obispo Co. to L. Calif., and along desert washes; Creosote Bush Scrub, Joshua Tree Wd.; w. Mojave and Colo. deserts. Fls. most of year. Exceedingly variable as to foliage and fr. The following ill-defined extremes are to be noted:

Var. **globòsa** Cov. [*I. g.* Heller.] Caps. subglobose, 2–3 cm. long, abruptly narrowed at both ends, the apex not sharp-pointed.—With the typical form in s. Calif. and to w. Fresno and e. Monterey cos., being common in and ascending to 4000 ft. in Liebre Mts., Bakersfield and Tehachapi regions.

Var. **insuláris** Jeps. Caps. almost as thick as long, but gradually narrowed at ends, with sharply pointed apex.—Santa Rosa, Santa Catalina and Coronado ids.

Var. **angustàta** Parish. [*I. a.* Parish.] Caps. narrow, 5–12 mm. thick, strongly attenuate to both ends.—Sandy washes below 4000 ft.; Creosote Bush Scrub; Mojave and Colo. deserts, to coast at San Diego.

2. Oxýstylis Torr. & Frém.

Branching erect minutely scaberulous annual. Lvs. trifoliate, petioled. Fls. in head-like axillary glomerules. Sepals 4, oblong-linear. Petals 4, yellow, oblong-ovate. Stamens 4, exserted, borne on a somewhat fleshy elevated torus. Ovary didymous, borne on a short stout stipe; style elongate, subulate, forming a stiff spine. Fr. of 2 1-seeded rounded faintly reticulate nutlets. One sp. (Greek, *oxus*, sharp, and *stylis*, column or style.)

1. **O. lùtea** Torr. & Frém. Stems yellowish, stout, 3–9 dm. high, flowering from very base; petioles 3–6 cm. long; lfts. green, elliptic, 1–3 cm. long, mucronulate; petals ca. 1.5 mm. long; fruiting pedicels recurved, 3–4 mm. long; stipe 1–2 mm. long; nutlets globose-ovoid, 1.5 mm. in diam.; style 4–7 mm. long.—Alkaline washes and flats, below 2000 ft.; Creosote Bush Scrub, Alkali Sink; Death V. Region to Tecopa and adjacent Nev. March–Oct.

3. Polanísia Raf. CLAMMYWEED

Heavy-scented viscid branching annuals. Lvs. mostly palmately 3–5-foliolate. Fls. in terminal racemes with simple bracts. Sepals 4, deciduous. Petals 4, slender or clawed. Receptacle not elongated, with a gland behind the ovary. Stamens 8–32. Ovary sessile or short-stipitate. Caps. elongate, cylindrical or compressed, erect on spreading pedicels, many-seeded, 2-valved from summit. Seeds rugose or reticulate. Perhaps 20–25 spp. of N. Am. (Greek, *polus*, many, and *anisos*, unequal, referring to stamens.)

1. **P. trachyspérma** T. & G. [*Jacksonia t.* Greene.] Plants 2–8 dm. high, glandular-pubescent; petioles 1.5–4 cm. long; lfts. 3, oblanceolate to narrowly ovate, 1.5–2.5 cm. long; pedicels 1–2 cm. long; sepals lanceolate, purple-tinged, 4–5 mm. long; petals whitish, obovate, 8–12 mm. long, notched at apex, clawed; stamens purple, long-exserted; style 4–6 mm. long; caps. almost sessile, cylindric, 3–5 cm. long; seeds almost round, brown, rough-reticulate, ca. 2 mm. long; $n = 10$ (Raghavan, 1938).—Sandy or gravelly places, 4500–5000 ft.; Sagebrush Scrub; Modoc Co.; to B.C., N. Dak., Ariz., Mex. July–Aug.

4. Wislizènia Engelm.

Erect branched ill-scented annuals. Lvs. mostly trifoliate, with small bristlelike deciduous stipules. Fls. yellow, racemose, inconspicuously bracteate. Sepals and petals 4. Stamens 6, much exserted. Stipe reflexed in fr. Fr. 2-seeded, didymous, each valve closely contracted upon its single seed and falling away with it like a nutlet. Style elongate, persistent. A small genus of sw. U.S. and nw. Mex. (Named for A. *Wislizenius,* early botanical collector in the Southwest.)

1. **W. refrácta** Engelm. [*W. californica* Greene. *W. Palmeri* Gray. *W. divaricata* Greene.] JACKASS-CLOVER. Two to 15 dm. tall, subglabrous; lfts. oblanceolate or wider, 1–3(–4) cm. long, scaberulous, mucronulate, entire; raceme dense at anthesis, later elongate; pedicels commonly 5–10 mm. long; sepals ca. 1.5 mm. long, united at base, denticulate; petals 3 mm. long; stipe 5–10 mm. long; nutlets obovoid to almost round, 2–3 mm. long, reticulate-ridged, ± tubercled at summit; style 4–5 mm. long, hairlike.—Subalkaline soil; Alkali Sink and adjacent areas; Sacramento to Bakersfield (possibly as an immigrant); occasional on deserts; to New Mex., Son., L. Calif. April-Nov. Desert plants sometimes unifoliolate in upper parts.

5. Cleóme L.

Annual herbs to woody plants. Lvs. simple or digitately 3–7-foliolate. Fls. solitary or racemose. Sepals 4, distinct or united at base. Petals 4, entire, cruciate. Stamens 6, rarely 4. Ovary stipitate, with a gland at base; style short or none. Caps. linear to oblong, erect or pendulous. Seeds round-reniform, several to many. Ca. 75 spp., mostly in trop. of Am. and Afr.; some grown for the fls. (Ancient name of some mustardlike plant.)

Sepals united at base; stamens long-exserted.
    Fls. pink or purplish, to white; lfts. 3 .... 1. *C. serrulata*
    Fls. yellow; lfts. largely 5–7 .... 2. *C. lutea*
Sepals separate to base.
    Stamens not exceeding petals; caps. glabrous .... 3. *C. sparsifolia*
    Stamens much exceeding petals; caps. pubescent .... 4. *C. platycarpa*

1. **C. serrulàta** Pursh. [*Peritoma s.* DC.] ROCKY MOUNTAIN BEEPLANT. Erect glabrous to sparsely pubescent annual, 3–10 dm. high, branched, especially above; lower petioles to 5 or 6 cm. long; lfts. lanceolate to oblanceolate, 2–7 cm. long, mostly entire; racemes dense in fl., to 2.5 dm. long in fr.; pedicels slender, 1–1.5 cm. long; calyx ca. 2 mm. long; petals oblong, slightly clawed, 10–12 mm. long; stipe becoming as long as pedicel; caps. linear, 2.5–6.5 cm. long, 4–6 mm. thick, acute, glabrous; $n = 16$ (Rollins, 1939).—Occasional, probably as a waif, s. Calif. and Tulare Co.; native, Sagebrush Scrub; Siskiyou Co.; to e. Wash., Great Plains. May–Aug.

2. **C. lùtea** Hook. Habit of the preceding; lfts. 3–5(–7), oblong or lance-oblong, 1.5–5 cm. long; petals yellow, 6–8 mm. long; pod linear, 2–4 cm. long, on a stipe ca. 1 cm. long; seeds reniform-orbicular, yellowish, ca. 2 mm. long, scattered-tuberculate; $n = $ ca. 16 (Rollins, 1939).—Sandy flats, 3600–5500 ft.; Pinyon-Juniper Wd., Creosote Bush Scrub, Sagebrush Scrub; Inyo and Mono cos., Mountain Pass, e. San Bernardino Co.; to Wash., Nebr., New Mex. May–Aug.

3. **C. sparsifòlia** Wats. Glabrous branched annual, 1–4 dm. high; lvs. few, the petioles 1–4 cm. long; lfts. 3, cuneate to obovate, 5–10 mm. long; racemes few-fld.; pedicels 5–8 mm. long; sepals 4, 1–2 mm. long; petals spatulate, greenish with white margin, 7–8 mm. long; stipe 3–5 mm. long; caps. linear, 1.5–2.5 cm. long; seeds oblong-reniform, compressed, gray with darker mottling, ca. 2 mm. long, cellular-punctate.—Sandy places, 3000–6500 ft.; Creosote Bush Scrub, Pinyon-Juniper Wd.; Mono and Inyo cos.; w. Nev. May–Aug.

4. **C. platycárpa** Torr. Erect annual, viscid-pubescent, 3–7 dm. high; petioles 1–4 cm. long; lfts. 3, oblong to subovate, entire, 1–2.5 cm. long; fruiting raceme 1–4 dm. long, densely leafy-bracted; pedicels 1–1.5 cm. long; petals yellow, oblong-spatulate, 7–8 mm. long; stipe 1–1.8 cm. long; caps. oblong-elliptic, 12–20 mm. long, flat; seeds yellowish, suborbicular, compressed, 2.5 mm. long, microscopically punctate.—Subalkaline clayey soil, 2500–5000 ft.; Sagebrush Scrub; Butte Co. to Siskiyou and Modoc cos.; Nev. to Ore., Ida. May–Aug.

6. Cleomélla DC. STINKWEED

Glabrous to pubescent erect or diffuse annuals. Lvs. mostly trifoliolate, petioled. Fls. yellow, small, mostly racemose or axillary. Sepals 4. Petals 4. Stamens 6. Caps. rhombic, broader than long, small, few-seeded, the valves hemispheric or laterally produced into short horns, readily shed. Seeds reniform-orbicular, somewhat compressed. Ca. 12 spp. of N. Am. (Diminutive of *Cleome.*)

(Payson, E. B. A synoptical revision of the genus Cleomella. Univ. Wyo. Pub. Sci., Bot. 1: 29–46, 1922.)

Stipes much longer than the caps.
  Lvs. and frs. glabrous; sepals not ciliate.
    Lfts. linear-oblong .... 1. *C. plocasperma*
    Lfts. ovate-oblong .... 2. *C. Hillmanii*
  Lvs. and frs. pubescent; sepals ciliate .... 3. *C. obtusifolia*
Stipes shorter than the caps.
  Fls. racemose; pedicels 10–20 mm. long .... 4. *C. parviflora*
  Fls. in axils of ordinary lvs.; pedicels 2–3 mm. long .... 5. *C. brevipes*

1. **C. plocaspérma** Wats. [*C. oocarpa* Gray] Glabrous and often glaucous, erect, usually widely branched, 2–4 dm. high; lfts. oblong-linear, 15–25 mm. long, on usually somewhat shorter petioles; racemes dense at anthesis, later more lax; lower bracts trifoliolate, the upper unifoliolate; pedicels 5–10 mm. long, filiform, divergent-ascending; stipes 4–8 mm. long; calyx glabrous, ca. 1 mm. long; petals pale yellow, oblong, ca. 5 mm. long; stamens exserted; caps. irregularly rhombic-ovoid, ca. 4 mm. long, the valves angular or low-conical; seeds few, straw-colored, sometimes mottled with brown, smooth or microscopically roughened, flattened-obovoid, almost 2 mm. long.—Alkaline flats, ca. 4000 ft.; Sagebrush Scrub; Lassen Co. to Ore., Ida., Utah. May–July.

Var. **mojavénsis** (Pays.) Crum. [*C. m.* Pays. *C. p.* var. *stricta* Crum.] Taller, 3–6 dm. high; pedicels and stipes each ca. 1 cm. long.—Alkaline seeps and flats, 3000–5600 ft.; Creosote Bush Scrub, Alkali Sink; Mojave Desert, Inyo Co. to w. San Bernardino Co. May–Oct. Late-season collections have fruiting racemes to 6 dm. long and are the basis for var. *stricta.*

2. **C. Hillmánii** A. Nels. [*C. longipes* var. *grandiflora* Wats. *C. g.* Cov.] Glabrous, simple to branched, 1–3 dm. high; lvs. 3-foliolate; lfts. oblong-ovate, 1–2 cm. long; pedicels 12–15 mm. long; stipes 8–14 mm. long; petals orange-yellow, 6–7 mm. long; caps. 4–6 mm. long, rhombic; seeds 3–7, smooth.—Dry slopes, 4500–5500 ft.; Sagebrush Scrub; about Reno, Nev., to be expected in adjacent Calif. April–June.

3. **C. obtusifòlia** Torr. & Frém. [*C. o.* vars. *Jonesii* and *florifera* Crum. *C. taurocranos* A. Nels.] Diffusely branched, 8–15 cm. high or with longer trailing stems; stems glabrous; petioles 1–3 cm. long; lfts. obovate to oblong, somewhat fleshy, 8–15 mm. long, subglabrous above, pubescent beneath; stipules fimbriate; fls. from near base of stems; pedicels 6–12 mm. long; calyx ciliate, ca. 1.5 mm. long; petals 5–6 mm. long, yellow; stipe reflexed in fr., ca. 6 mm. long; caps. ca. 4 mm. long, 6–8 mm. broad, the valves pubescent, laterally conical to hornlike; seeds suborbicular, compressed, brown or mottled with gray, smoothish, ca. 1 mm. in diam.—Alkaline flats, below 4000 ft.; Creosote Bush Scrub, Joshua Tree Wd.; Inyo Co. to Colo. Desert; Ariz., Nev. April–Oct. Plants with pubescent stems have been called var. *pubéscens* A. Nels.

4. **C. parviflòra** Gray. [*C. alata* Eastw.] Mostly branched from base, glabrous, 1–3(–4) dm. high; petioles 2–10 mm. long; lfts. linear, 10–25 mm. long; infl. lax, somewhat recurved; pedicels 10–15(–25) mm. long; stipes scarcely 1 mm. long; calyx ca. 0.5 mm. long, the lobes acuminate; petals yellow, 1.5–2 mm. long; caps. ovoid-rhombic, ca. 4 mm. long; seeds several, smooth, stramineous, obovoid, compressed, ca. 1.2 mm. long. —Alkali flats, 2500–6700 ft.; Alkali Sink, Creosote Bush Scrub, Joshua Tree Wd., Pinyon-Juniper Wd.; e. Lassen and Mono cos. to w. San Bernardino Co.; w. Nev. May–Aug.

5. **C. brévipes** Wats. Diffusely branched from base, somewhat scaberulous, 5–30 cm. high; lvs. subsessile; lfts. linear-spatulate, 5–15 mm. long; fls. solitary in axils of most lvs.; pedicels ca. 2–3 mm. long, recurved in fr.; calyx ca. 0.5 mm. long, the lobes long-acuminate; petals ca. 1.5 mm. long; caps. scarcely stipitate, ovoid, 3 mm. long; style ca. 0.4 mm. long; seeds stramineous, flattened-obovoid, shining, 1.6–2 mm. long.—Alkaline seeps, 1400–4000 ft.; Alkali Sink; Mojave Desert (Owens Lake, Keeler, Camp Cady, Tecopa); w. Nev. May–Oct.

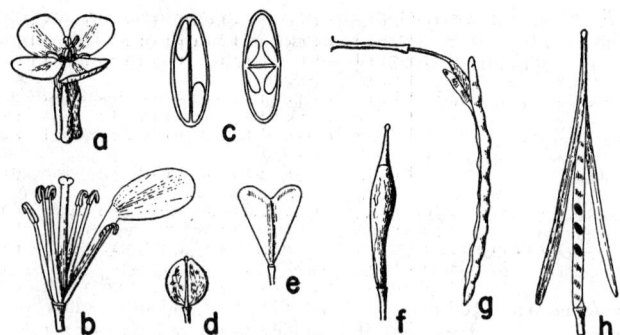

Fig. 31. CRUCIFERAE. *Erysimum: a,* fl., × 1, sepals 4, petals 4; *b,* fl. opened up, petal clawed, stamens 4 long and 2 short (tetradynamous). Diagrams of cross sections of 2 ovaries of Cruciferae, *c,* to show compression parallel to and contrary to septum. *Lepidium: d,* and Capsella: *e,* 2 types of silicles. *Brassica: f,* silique with terminal beak. *Stanleya: g,* stipitate silique. *Erysimum: h,* sessile silique dehiscing by 2 valves from base.

33. Crucíferae. MUSTARD FAMILY (Fig. 31)

Herbs or rarely shrubby plants, with a pungent watery juice. Lvs. alternate, entire, lobed, dissected or pinnate, without stipules. Fls. in terminal racemes or corymbs, rarely solitary and terminal. Sepals 4, erect or somewhat spreading, rather alike, deciduous. Petals 4, commonly clawed, the blades spreading in the form of a cross. Stamens 6 (rarely 4 or 2), two of them inserted lower and shorter. Nectar glands commonly 4. Ovary superior, usually 2-celled by a thin partition stretched between the 2 marginal placentae, from which, when ripe, the valves usually separate. Style 1; stigma 2-lobed or entire. Fr. a 2-celled caps., long and narrow (silique) or short (silicle), sometimes indehiscent and 1-celled, sometimes separating across into 1-seeded joints. Seeds campylotropous, without endosperm, filled with the large embryo which is curved and folded so that the cotyledons have their margins of one side against the radicle (accumbent) or the back of one cotyledon against the radicle (incumbent), or the cotyledons may be folded upon themselves (conduplicate). A large family, widely distributed, with many plants of economic or horticultural value (cabbage, mustard, radish, turnip, rutabaga, cauliflower, stock, etc.).

A. Frs. indehiscent, 1-celled, 1-seeded, thin and flat; pedicels recurved in fr., at least at their tips.
    B. Silicles not winged, with hooked hairs; pubescence of stems of branched hairs . 41. *Athysanus*
    BB. Silicles wing-margined all around, not with hooked hairs; pubescence of stems none or simple.
        C.   Silicles suborbicular, 2–10 mm. long .... 43. *Thysanocarpus*
        CC. Silicles oblong, 12–15 mm. long .... 17. *Isatis*
AA. Frs. dehiscent by valves or breaking transversely into joints, normally containing 2 or more seeds.
    B. Siliques at length breaking transversely into seed-bearing indehiscent joints.
        C.   Frs. transversely 2-jointed, 2-seeded; petals 5–10 mm. long.
            D.   Fls. purplish or whitish. Plants of sea-beaches .... 16. *Cakile*
            DD. Fls. yellow with darker veins. Occasional waif .... 18. *Rapistrum*
        CC. Frs. breaking irregularly into joints, several-seeded; petals 10–20 mm. long.
            D.   Siliques 3–5 mm. thick, on pedicels 10–20 mm. long, the locules 1-seriate
                                           22. *Raphanus*
            DD. Siliques 1.5–2 mm. thick, on pedicels 2–5 mm. long, the locules 2-seriate
                                          52. *Chorispora*
    BB. Siliques or frs. not breaking transversely into seed-bearing joints, but dehiscent by valves.
        C.   Fr. a silique, elongate, several times longer than wide or thick.
            D.   Siliques long-stipitate, the stipe 1–2 cm. long; anthers twisted; sepals spreading
                or recurved at anthesis .... 1. *Stanleya*
            DD. Siliques sessile or with much shorter stipes; anthers usually not twisted; sepals usually erect.
               E.   The siliques compressed contrary to the narrow partition; racemes leafy
                                          31. *Tropidocarpum*
               EE. The siliques not compressed contrary to the partition; racemes leafless.
                  F.   Stigma-lobes situated over the valves; anthers sagittate at base; petals usually linear.

G. Calyx urn-shaped or flask-shaped at anthesis, sessile or subsessile.
  H. Siliques conspicuously beaked, pendent, sessile, strongly compressed; valves of mature fr. remaining attached at tip; seeds flat, winged. Annual .... 5. *Streptanthella*
  HH. Siliques not or inconspicuously beaked, erect to deflexed, the valves separating completely. Annual to perennial.
    I. The siliques flattened; seeds flat, usually winged
        3. *Streptanthus*
    II. The siliques terete or 4-sided; seeds wingless or nearly so
        4. *Caulanthus*
GG. Calyx open at anthesis, not urn-shaped; siliques sometimes short-stipitate, mostly terete or 4-sided when mature .. 2. *Thelypodium*
FF. Stigma-lobes situated over the placentae; anthers mostly not sagittate at base; petals frequently wider than linear.
  G. Basal lvs. forming definite rosettes, from which arise the flowering stems.
    H. Fls. yellow; stems angled .... 23. *Barbarea*
    HH. Fls. white; stems terete.
      I. Lvs. simple, undivided or sinuate to lyrate.
        J. Plants glabrous or with simple hairs .... 46. *Arabis*
        JJ. Plants with forked or stellate hairs.
          K. Siliques tetragonal-cylindric, 1–1.5 cm. long
            13. *Arabidopsis*
          KK. Siliques flat, mostly longer.
            L. Stems leafy; siliques linear .... 46. *Arabis*
            LL. Stems scapose; siliques lanceolate to ovate
              36. *Phoenicaulis*
      II. Lvs. deeply pinnatifid or pinnate.
        J. The lvs. pinnate, with definite lfts.; seeds wingless
          28. *Cardamine*
        JJ. The lvs. pinnatifid; seeds winged .... 47. *Sibara*
  GG. Basal lvs. not in well-formed rosettes.
    H. Frs. with a long indehiscent conical to flattened beak.
      I. Seeds globose, in 1 row in each locule .... 20. *Brassica*
      II. Seeds ellipsoid to ovoid, in 2 rows in each locule.
        J. Petals 15–20 mm. long; siliques 4-angled, the beak flattened .... 19. *Eruca*
        JJ. Petals 8–10 mm. long; siliques terete, the beak slender-conical .... 21. *Diplotaxis*
    HH. Frs. not with indehiscent beak, but dehiscent to tip.
      I. Pubescence simple or wanting.
        J. Siliques strongly flattened.
          K. Plants with fleshy rootstocks; stems leafless in lower half .... 29. *Dentaria*
          KK. Plants not from fleshy rootstocks; stems leafy or leafless above.
            L. Valves of silique not nerved; lvs. usually pinnate .... 28. *Cardamine*
            LL. Valves of silique nerved; lvs. simple
              46. *Arabis*
        JJ. Siliques terete or 4-angled.
          K. Lvs. entire, simple, deeply auriculate-clasping
            53. *Conringia*
          KK. Lvs., at least the lower, pinnate to pinnatifid or deeply lobed.
            L. Plants without creeping bases, mostly annual or biennial; seeds uniseriate (except in *Rorippa*).
              M. Pods somewhat 4-angled; biennial or perennial, of wet places . 23. *Barbarea*
              MM. Pods terete; annuals. (Go to N.)
N. Siliques 2–10 cm. long (if shorter, closely appressed). Weeds of fields, roadsides, etc. .... 12. *Sisymbrium*
NN. Siliques 0.4–1.2 cm. long, divaricate. Wet places .... 24. *Rorippa*
            LL. Plants with creeping base, perennial; seeds biseriate.
              M. Stems rooting at nodes; fls. white
                25. *Nasturtium*
              MM. Stems from underground rhizomes; fls. yellow .... 24. *Rorippa*
      II. Pubescence of forked or stellate hairs.
        J. Lvs. pinnatifid to pinnate.
          K. Plants annual, with finely dissected to bipinnatifid lvs. .... 15. *Descurainia*
          KK. Plants perennial from heavy caudex.

      L. Stems 4–5 dm. high; lvs. subentire to coarsely dentate or lobed. Inyo Co.
                                   13. *Halimolobus*
     LL. Stems 0.5–1 dm. high; lvs. pinnate with linear segms. Lassen Co. .. 45. *Polyctenium*
    JJ. Lvs. entire or nearly so.
      K. Fls. yellow to orange .... 48. *Erysimum*
     KK. Fls. white to rose or purple.
        L. Petals 3–15 mm. long; stigmas not horned on back; epidermal cells of septum of silique not reticulate. Widespread natives
                                   46. *Arabis*
       LL. Petals ca. 20 mm. long; stigmas horned on back; epidermal cells of silique-septum reticulate. Sparingly natur. near seacoast
                                  51. *Matthiola*

CC. Fr. a silicle, 1–2(–3) times longer than wide or thick.
  D. Silicles evidently flattened.
    E. The silicles flattened parallel to the broad partition.
      F. Petals 18–20 mm. long, white to purple .... 27. *Lunaria*
     FF. Petals 1.5–8 mm. long.
        G. Stems scapose, 1-fld.; silicles suborbicular; seeds broadly winged
                                    30. *Idahoa*
       GG. Stems not scapose, with more than 1 fl.; seeds not broadly winged.
          H. Pubescence of appressed 2-branched hairs attached at middle; silicles orbicular with flat margins .... 50. *Lobularia*
         HH. Pubescence stellate or forked, but not as above.
            I. Silicles orbicular, margined; lvs. spatulate, entire. Introd. weed .... 49. *Alyssum*
           II. Silicles ovate to elliptical, oblong or almost linear, not margined.
              J. The silicles soon dehiscent; infl. not secund or more than twice as long as rest of stem; fls. yellow or white; seeds smooth .... 40. *Draba*
             JJ. The silicles very tardily dehiscent; infl. secund, 3–4 times as long as rest of stem; fls. white, 2 mm. long; seed hispidulous .... 42. *Heterodraba*
    EE. The silicles flattened contrary to the narrow partition.
      F. Silicles didymous (twinlike), the 2 locules somewhat rounded with deep notch between both at summit and base.
        G. The silicles 6–18 mm. wide; pubescence stellate ... 33. *Dithyraea*
       GG. The silicles ca. 2 mm. wide; pubescence not stellate . 9. *Coronopus*
     FF. Silicles not didymous.
        G. The silicles inverted-triangular, broad at summit; annuals; fls. ca. 2 mm. long. Introd. weed .... 38. *Capsella*
       GG. The silicles orbicular to oval or cuneate.
         H. Petals 18–20 mm. long. Suffrutescent perennial of Colo. Desert .... 32. *Lyrocarpa*
         HH. Petals 1–5 mm. long.
           I. Cells of the silicles 1-seeded .... 8. *Lepidium*
          II. Cells of the silicles 2–many-seeded.
             J. Silicles notched to subtruncate at apex, 4–18 mm. long; petals 3–5 mm. long .... 11. *Thlaspi*
             JJ. Silicles rounded at apex, 2–4 mm. long; petals 1 mm. long .... 37. *Hutchinsia*

DD. Silicles turgid or inflated, not flattened.
  E. Plants in ± scapose tufts, the lvs. mostly basal.
    F. Lvs. pinnatifid; fls. white, 4–5 mm. long; heavy-rooted perennial. Lassen Co. .... 44. *Smelowskia*
    FF. Lvs. simple, entire or nearly so.
      G. Plant glabrous; lvs. subulate; petals minute. Aquatic perennial
                                    6. *Subularia*
      GG. Plants not glabrous; lvs. not subulate.
        H. Petals white to violet, ca. 1 cm. long. Rare escape
                                  10. *Ionopsidium*
        HH. Petals yellow, or if whitish, much smaller.
          I. Silicles didymous, 10–15 mm. long; petals yellow, 10–12 long. E. Mojave Desert .... 34. *Physaria*
         II. Silicles not didymous, 3–6 mm. long.
           J. Lvs. 20–50 mm. long; silicles subglobose, ± stellate-pubescent .... 35. *Lesquerella*
          JJ. Lvs. 2–12 mm. long; silicles ovoid, with mostly simple hairs .... 40. *Draba*
  EE. Plants with leafy stems.
    F. Fls. white; long-lived perennials.

G. Petals 2–3 mm. long; lvs. 1–6 cm. long; silicles 2.5–4 mm. long
7. *Cardaria*
GG. Petals 5–7 mm. long; lvs. 10–40 cm. long; silicles 4–6 mm. long
26. *Armoracia*
FF. Fls. yellow; annual to perennial.
  G. Valves of the silicles 1-nerved; cauline lvs. sagittate-clasping, not lobed .... 39. *Camelina*
  GG. Valves of the silicles not nerved; cauline lvs. not sagittate-clasping.
    H. Lvs. pinnatifid. Plants of wet places .... 24. *Rorippa*
    HH. Lvs. entire. Plants of dry places .... 35. *Lesquerella*

1. Stánleya Nutt. Prince's Plume

Annual or perennial or even suffruticose, glabrous or pubescent with simple hairs. Stems simple or branched. Radical lvs. forming a basal rosette or not; cauline petioled to sessile and auriculate, pinnately dissected to entire. Infl. racemose. Sepals linear-oblong, spreading or reflexed at anthesis. Petals white to yellow. Fils. subequal, with glandular tissue surrounding base of single stamens but obsolete or only on underside of paired fils. Siliques linear, flattened parallel to septum or subterete, long-stipitate; stigma sessile or nearly so. Seeds oblong, marginless, numerous. Six spp. of w. U.S. (Named for Lord Edward *Stanley*, English ornithologist of 19th century.)

(Rollins, R. C. The cruciferous genus Stanleya. Lloydia 2: 109–127, 1939.)
Middle and upper cauline lvs. petioled. Mono Co. s.
  Lower lvs. pinnate; inner surface of petal-claw villous ....1. *S. pinnata*
  Lower lvs. entire or runcinate; inner surface of petal-claw glabrous .... 2. *S. elata*
Middle and upper cauline lvs. sessile, auriculate-clasping. Lassen Co. .... 3. *S. viridiflora*

1. **S. pinnàta** (Pursh) Britton. [*Cleome p.* Pursh. *S. pinnatifida* Nutt.] Suffrutescent, glabrous to pubescent, the stems simple or usually branched above, glaucous, 4–15 dm. high; truly radical lvs. lacking, the lower cauline usually pinnatifid into lanceolate segms., rarely bipinnate, broadly lanceolate in outline, 5–20 cm. long, petioled, the upper cauline oblanceolate, entire or divided, 3–6 cm. long; infl. becoming 2–6 dm. long in fr., many-fld.; pedicels 6–12 mm. long, glabrous to somewhat pilose; sepals glabrous, linear, dilated at base, 10–15 mm. long; petals yellow, 12–16 mm. long, the sharply differentiated blade oblong to almost oval, 3–6 mm. long, the claw villous, brownish; fils. pilose at base; siliques 3–8 cm. long, linear, subterete, spreading, arcuate to almost straight; stipe 1–2.5 cm. long; seeds brown, oblong, wingless, plump, ca. 2 mm. long, 1 mm. broad; *n* = 12 (Rollins, 1939).—Seleniferous soil, desert slopes and washes, 1000–5000 ft.; Creosote Bush Scrub, Joshua Tree Wd., Pinyon-Juniper Wd.; n. base Santa Rosa Mts. to Cuyama V. and Inyo Co.; e. to N. Dak., Kans., Tex. April–Sept.

Ssp. **inyoénsis** Munz & Roos. Plant a distinct shrub with a basal trunk 4–8 cm. thick; herbage yellow-green throughout; sepals 8–10 mm. long; petals 10 mm. long.—Sandy places, 3000 ft.; Creosote Bush Scrub; se. end of Eureka V., e. base of Inyo Mts.

2. **S. elàta** Jones. Perennial, erect, with 1–several stems simple or branched above, 6–15 dm. high, glabrous, glaucous; lvs. petioled, lance-ovate, 1–2 dm. long, entire or with a few small basal lobes; infl. becoming 1–5 dm. long; pedicels glabrous, 5–10 mm. long; sepals oblong-linear, yellowish, glabrous, 8–12 mm. long; petals yellow to whitish, 8–10 mm. long, the blade only ca. 1 mm. wide; stipes ca. 2 cm. long; siliques subterete, arcuate or spreading, 5–10 cm. long; seeds brown, oblong, wingless, 2 mm. long; *n* = 12 (Rollins, 1939).—Desert washes and slopes, 4500–6500 ft.; Joshua Tree Wd., Pinyon-Juniper Wd.; White Mts. to Panamint and Argus mts.; Nev., Ariz. May–July.

3. **S. viridiflòra** Nutt. Perennial, glabrous, 1-stemmed, 3–12 dm. high; basal lvs. forming rosette, petioled, entire to somewhat divided, obovate to oblanceolate, 1–3 dm. long, the cauline reduced upward to lanceolate sagittate lvs. 2–4 cm. long; infl. 1–5 dm. long; pedicels 4–7 mm. long; sepals linear-oblong, 12–16 mm. long; petals linear-oblong, erose at apex, yellow to almost white, 15–20 mm. long; stipes 15–25 mm. long; siliques subterete, arcuate, 4–7 cm. long; seeds brown, oblong, 2–3 mm. long; *n* = 12 (Rollins, 1939).—Reported by R. C. Barneby from alkaline clay slopes at 4400 ft., Smoke Creek, Lassen Co.; e. Ore. to Wyo., Nev. May–Aug.

2. Thelypòdium Endl.

Annual to perennial herbs with mostly erect simple or branched stems; glabrous to hirsute with simple hairs. Basal lvs. mostly petioled, the cauline petioled to sessile. Infl. usually racemose, at least in fr. Sepals scarcely if at all saccate. Petals linear to oblanceolate or oblong, clawed, plane or crisped, white or cream to purple or lilac. Stamens 6, ± exserted. Siliques stipitate to subsessile, terete or somewhat flattened parallel to the partition, erect to pendent; style short; stigma small, entire to slightly 2-lobed. Seeds oblong, somewhat flattened, not winged. Ca. 16 spp. of w. N. Am. (Greek, *thelus*, female, and *pus*, foot, referring to presence of stipe.)

(Payson, E. B. A monographic study of Thelypodium and its immediate allies. Ann. Mo. Bot. Gard. 9: 233–324, 1922.)
A. Cauline lvs. sagittate or auriculate-clasping at base.
  B. Petals 8–14 mm. long; at least the lower lvs. usually toothed.
    C. Pedicels 1–2 mm. long; plants short-lived perennials. Siskiyou Co. . . 1. *T. brachycarpum*
    CC. Pedicels 3–6 mm. long; plants biennial.
      D. Sepals 3–4 mm. long; infl. dense. Lassen Co. to Inyo Co. . . . . 2. *T. crispum*
      DD. Sepals 6–9 mm. long; infl. lax.
        E. Petals spatulate, twice as long as sepals. Ne. Calif. . . . . .3. *T. Howellii*
        EE. Petals narrowly linear, 1½ times the sepals. San Bernardino Mts.
                           4. *T. stenopetalum*
  BB. Petals ca. 5 mm. long; lvs. all entire; heavy-rooted perennial . . . . .5. *T. flexuosum*
AA. Cauline lvs. not sagittate or amplexicaul.
  B. Plants deep-rooted perennials with woody base; petals purplish, 11–14 mm. long. Inyo Mts.
                           6. *T. Jaegeri*
  BB. Plants annual or biennial; petals whitish or faintly colored.
    C. Fruiting racemes dense, spikelike; plants biennial.
      D. Cauline lvs. toothed or lobed; siliques 3–8 cm. long. N. Calif. . . . 7. *T. laciniatum*
      DD. Cauline lvs. entire; siliques 1.5–2.5 cm. long. About Mojave Desert
                          8. *T. integrifolium*
    CC. Fruiting racemes open; plants annual.
      D. Petals crisped, 9–14 mm. long . . . . .9. *T. flavescens*
      DD. Petals flat, 4–5 mm. long.
        E. Sepals purplish, spreading; pedicels 5–6 mm. long . . . . . . 10. *T. Lemmonii*
        EE. Sepals green or yellowish, erect; pedicels 2–4 mm. long . . 11. *T. lasiophyllum*

1. **T. brachycárpum** Torr. [*Thelypodiopsis b.* O. E. Schulz.] Glabrous short-lived perennial, sometimes somewhat hirsute at base; stems simple or few-branched, ascending, 3–15 dm. high; basal lvs. oblanceolate to spatulate, lyrate-pinnatifid to toothed, 3–6 cm. long, petioled; cauline lvs. 1–4 cm. long, sessile with sagittate base, entire or toothed, acute; racemes dense, 5–20 cm. long; pedicels stout, divergent, 1–2 mm. long; sepals ± crisped, yellow-green to whitish, 4–5 mm. long, lance-linear; petals white, linear, 8–14 mm. long; fils. subequal, the anthers exserted, sagittate at base, apiculate; siliques arcuate-ascending, unequally torulose, 15–30 mm. long, on a stipe 1–1.5 mm. long; seeds brown, oblong-ovate, ca. 1.5 mm. long, faintly reticulate-pitted.—Somewhat alkaline clay soil, 3000–6500 ft.; Sagebrush Scrub, Yellow Pine F.; Siskiyou Co.; adjacent Ore. June–Aug.

2. **T. críspum** Greene. [*Thelypodiopsis c.* O. E. Schulz. *Thelypodium brachycarpum* var. *c.* Jeps.] Winter annual or biennial, usually 1-stemmed, 3–7 dm. high, mostly simple or few-branched above; rosette-lvs. obovate to oblanceolate, subentire to lyrately lobed, short-petioled, 2–5 cm. long, the cauline 1–4 cm. long, linear-sagittate, sessile, acute; racemes dense, becoming 1–3 dm. long; pedicels slender, suberect, 3–5 mm. long; sepals blunt, crisped, white or purplish, 3–4 mm. long; petals white, linear-spatulate, ca. 8 mm. long; fils. subequal, the anthers exserted, sagittate at base, apiculate; stipes ca. 1 mm. long; siliques unequally torulose, divaricate-ascending, 10–25 mm. long; seeds brown, ca. 1 mm. long.—Subalkaline places, as near hot springs, in travertine, etc., 4500–9000 ft.; Pinyon-Juniper Wd., Sagebrush Scrub; Lassen Co. to n. Inyo Co.; w. Nev. June–July.

3. **T. Howéllii** Wats. [*Streptanthus H.* Jones. *Thelypodiopsis H.* O. E. Schulz. *Thelypodium simplex* Greene.] Biennial, glabrous and glaucous or hirsute near base, the stems erect, slender, simple or branched below or above, 3–8 dm. high; basal lvs. in a rosette, oblanceolate, lyrate to entire, petioled, 2–4 cm. long; cauline lvs. entire, acute, lance-linear, suberect, with sagittate base, 1–4 cm. long; infl. lax, pedicels ascending, 4–5 mm. long; sepals oblong, often purplish, scarious-margined, acuminate, ca. 7 mm. long;

petals purple to white, spatulate, crisped, ca. 14 mm. long; stamens slightly exceeding sepals, the anthers sagittate; stipes scarcely developed; siliques erect or ascending, 2–5 cm. long; seeds brown, thick, ca. 1 mm. long.—Alkaline adobe meadows, 4000–5000 ft.; Sagebrush Scrub; Lassen and Modoc cos.; to e. Ore. June–July.

4. **T. stenopétalum** Wats. Biennial, glabrous, glaucous, branched from base, with simple stems 3–5 dm. tall; radical lvs. soon withering, entire or repand, oblanceolate, the cauline oblong-lanceolate, erect, sagittate at base, 1–5 cm. long; infl. lax, 1–2 dm. long; pedicels ascending, 4–6 mm. long; sepals linear, greenish-purple, 6–9 mm. long; petals narrowly linear, whitish, crisped above, 10–14 mm. long; fils. tetradynamous, 8–14 mm. long, the anthers coiled when dry, apiculate, ca. 5 mm. long; siliques sessile, ascending, 4–5 cm. long, slender; seeds oblong, brown, ca. 1 mm. long.—Uncommon, stony slopes, 6500–7000 ft.; Yellow Pine F.; Bear V., San Bernardino Mts. June–July.

5. **T. flexuòsum** Rob. in Gray. Glabrous perennial, with thick caudex; stems 3–5 dm. long, slender, branched, ascending; lower lvs. entire, lanceolate, narrowed into slender petioles, obtuse, 8–16 cm. long, the cauline remote, lance-linear, acuminate, with auriculate base, 2–7 cm. long; infl. lax, 5–10 cm. long; pedicels divergent-ascending, 5–8 mm. long; sepals oblong, hyaline-margined, blunt, 2.5 mm. long; petals pale purplish to white, spatulate, ca. 5 mm. long; fils. scarcely exserted, the anthers 1–2 mm. long, sagittate, not apiculate; siliques subsessile, reticulate, 15–25 mm. long; seeds narrow, brown, ca. 2 mm. long.—Alkaline flats, 3000–5000 ft.; Sagebrush Scrub; Modoc and Siskiyou cos.; se. Ore., nw. Nev. May–June.

6. **T. Jàegeri** Roll. Cespitose deep-rooted perennial, with woody base, many branching stems, glabrous, glaucous, 1–3 dm. high; cauline lvs. slender-petioled, repand to fewlobed, ovate to elliptical, cuneate at base, 2.5–6 cm. long, 1–2.5 cm. wide; infl. lax, 4–10 cm. long; pedicels divaricate, 8–14 mm. long; sepals oblong, purplish, 5–7 mm. long, blunt; petals purple to pink-purple, spatulate, purple-veined, 11–14 mm. long, 3–4 mm. wide; stamens not exserted, the anthers ca. 2.5 mm. long; siliques subsessile, linear, subterete, 3–5 cm. long, divaricate, almost straight; seeds almost 1 mm. long.— Shaded rock-crevices, 6000–8000 ft.; Pinyon-Juniper Wd.; s. end of Inyo Mts., ca. 17–20 miles n. of Darwin, Inyo Co. May–June.

7. **T. laciniàtum** (Hook.) Endl. [*Macropodium l.* Hook. *Pachypodium l.* Nutt.] Glabrous glaucous biennial, with stout often hollow stems, irregularly branched upward, 3–20 dm. high; radical lvs. petioled, deltoid-lanceolate, 1–5 dm. long, irregularly and deeply lobed, the cauline petioled, subpinnatifid to subentire; racemes dense, 1–5 dm. long; pedicels stout, spreading, 3–5 mm. long; sepals pale, acute, lance-oblong, 4–6 mm. long; petals white, sublinear, 7–12 mm. long; fils. 5–12 mm. long, the anthers 2–4 mm. long, apiculate; stipes 2–4 mm. long; siliques spreading or recurved, 3–8(–12) cm. long, ca. 1 mm. wide, somewhat flattened parallel to the septum; seeds brown, ca. 1.2 mm. long, finely cellular-reticulate.—Ledges, crevices and banks, 2000–6000 ft.; Sagebrush Scrub; from Big Pine, Inyo Co. to Yreka, Siskiyou Co.; to Wash., Ida., Nev. May–June.

Var. **milleflòrum** (A. Nels.) Pays. [*T. m.* A. Nels.] Lvs. nearly or quite entire; pedicels curved upward, 3–4 mm. long; stipes 1–2 mm. long.—Sierra V. (*Lemmon in 1875*), Amedee, Honey Lake V.; range of sp.

8. **T. integrifòlium** (Nutt.) Endl. [*Pachypodium i.* Nutt. *T. affine* Greene.] Glabrous biennials, erect, branched above, 1–2 m. tall; basal lvs. oblong-elliptical, subentire, 1–3 dm. long, acute, gradually narrowed into the rather long petiole, the cauline linear to oblanceolate, 2–6 cm. long, subsessile; infl. paniculate, the racemes dense, 5–15 cm. long; pedicels slender to stout, 3–8 mm. long; sepals lance-oblong, greenish with hyaline margins, 4–5 mm. long; petals white to bluish or pinkish, spatulate, 6–9 mm. long; fils. scarcely exserted, the anthers ca. 2 mm. long; stipes ca. 1 mm. long; siliques horizontal-ascending, subterete, torulose, slightly arcuate, 1.5–2.5 cm. long; seeds ca. 1.5 mm. long, oblong, brown, faintly cellular-pitted.—Mostly about alkaline seeps, 3000–7000 ft.; Joshua Tree Wd., Pinyon-Juniper Wd.; w. edge Mojave Desert, n. base of San Bernardino Mts., Antelope V., Tehachapi Mts., White Mts., Mono Lake; to Wash., Nev. July–Oct. Variable as to fl.-color, no. of nectary-glands, etc.

9. **T. flavéscens** (Hook.) Wats. [*Streptanthus f.* Hook. *Guillenia f.* Greene. *Caulanthus f.* Pays. *C. procerus* Wats. *S. p.* Brew. *T. Hookeri* Greene. *T. Greenei* Jeps. *G. H.* Greene, *S. Dudleyi* Eastw. *S. lilacinus* Hoov.] Annual, glabrous and glaucous or sparsely hirsute, the stems simple or branched above, erect, 3–10 dm. high; basal

lvs. petioled, lanceolate to oblanceolate, sinuate to pinnatifid, 5–20 cm. long; cauline lvs. sessile or short-petioled, subentire to toothed; infl. rather lax; pedicels rather stout, ascending, 5–7 mm. long; sepals lanceolate, yellowish with broad hyaline margins, crisped, 5–10 mm. long; petals white to cream, sometimes purple-veined, narrow, crisped, recurved, 8–14 mm. long; fils. included, the anthers greenish, apiculate, 2.5–3.5 mm. long; siliques subsessile, erect to reflexed, subterete, glabrous to somewhat hirsute, 4–8 cm. long; seeds dark brown, oblong, ca. 1.5 mm. long.—Open valleys and slopes, largely on serpentine, below 2000 ft.; V. Grassland; inner Coast Ranges, Solano Co. to San Benito Co. March–April.

10. **T. Lemmònii** Greene. [*Caulanthus anceps* Pays.] Glabrous glaucous annual or somewhat pilose near base, the stems erect, simple or few-branched above, 3–12 dm. high; lower lvs. oblong, repand-dentate to somewhat lobed, 6–15 cm. long, short-petioled, the cauline nearly or quite sessile, lanceolate, subentire to denticulate; racemes rather lax; pedicels slender, ascending to reflexed at maturity, 5–6 mm. long; sepals oblong, blunt, purplish with scarious margins, 3–4 mm. long; petals whitish, with lilac veins, oblanceolate, obtuse, 4–5 mm. long; fils. included, the anthers ca. 1.5 mm. long, not apiculate; siliques erect to reflexed, terete, subsessile, 3–5 cm. long; seeds narrow, brown, ca. 1.5 mm. long.—Open slopes and flats, below 3000 ft.; V. Grassland; inner S. Coast Ranges, San Joaquin Co. to San Luis Obispo and Kern cos. March–April.

11. **T. lasiophýllum** (H. & A.) Greene. [*Turritis l.* H. & A. *Sisymbrium. l.* K. Bdg. *Guillenia l.* Greene. *S. reflexum* Nutt. *S. deflexum* Harv. *Thelypodium neglectum* Jones. *Erysimum retrofractum* Torr.] Annual, usually hirsute at base, erect, simple or branched above, 2–10(–18) dm. high; lvs. all petioled, the lower oblong or oblanceolate, 3–12 cm. long, irregularly pinnatifid with divaricate segms. or only sinuate-dentate, the upper much reduced; infl. much elongate in fr.; pedicels 2–4 mm. long, usually reflexed in age; sepals greenish, oblong, 3 mm. long; petals yellowish-white, spatulate, ca. 6 mm. long; fils. included, the anthers 1–1.5 mm. long, not apiculate; siliques reflexed, terete, sessile or subsessile, straight or somewhat curved, 3–6 cm. long; seeds brown, oblong, faintly reticulate, ca. 1 mm. long.—Frequent on slopes and especially on burns, in washes, canyons, etc., below 5000 ft.; grasslands and open woodlands; Coast Ranges and most of s. Calif. w. of the deserts; Channel Ids.; to Wash., L. Calif. March–June.

KEY TO VARIETIES

Siliques reflexed.
    Lvs. with acute lobes; siliques ± straight. Cismontane .... *T. lasiophyllum*
    Lvs. with more rounded lobes; siliques curved outward. Deserts .... var. *utahense*
Siliques erect or ascending.
    Pedicels 2–3 mm. long; siliques slender, less than 1 mm. thick .... var. *inalienum*
    Pedicels ca. 1 mm. long; siliques stout, ca. 1.5 mm. thick .... var. *rigidum*

Var. **inaliènum** Rob. [*Caulanthus l.* var. *i.* Pays. *Sisymbrium acutangulum* Brew. & Wats. *S. acuticarpum* Jones.] Siliques slender, erect or ascending, less than 1 mm. thick. —Coast Ranges, Solano Co. to San Luis Obispo Co.

Var. **rígidum** (Greene) Rob. [*T. r.* Greene.] Siliques stout, ascending, more than 1 mm. thick; pedicels very short.—Glenn Co. to Contra Costa Co.

Var. **utahénse** (Rydb.) Jeps. [*T. u.* Rydb.] Lvs. thin, glabrous, with rounded lobes; siliques reflexed, usually curved outward.—Common, washes and among bushes; Creosote Bush Scrub, Joshua Tree Wd., Pinyon-Juniper Wd.; Mojave and Colo. deserts; to Utah, Ariz. April–May.

3. **Streptánthus** Nutt.

Annual or perennial herbs, glabrous or pubescent with unbranched hairs. Lvs. entire to lyrate-pinnatifid, the upper usually clasping. Infl. racemose or paniculate. Calyx flask-shaped, closed or nearly so at anthesis, often brightly colored. Petals usually narrow, with crisped or channeled blades. Stamens often in 3 pairs according to length, the longest pair often connate and with reduced anthers. Siliques linear, flattened parallel to the septum, erect, divaricate or pendent, sessile or subsessile. Style usually short. Stigma entire or slightly 2-lobed. Seeds flat, usually winged. Ca. 25 spp. of sw. U.S. (Greek, *streptas*, twisted, and *anthos*, fl., because of the petals.)

A. Plants essentially glabrous.
  B. Infl. without broad leafy bracts.
    C. Usually perennial; sepals usually with short stiff apical hairs; fils. of all stamens distinct.
      D. Lvs. not auriculate-clasping, narrowed at base. Siskiyou Mts. ...... 6. *S. Howellii*
      DD. Lvs., at least of upper stems, auriculate-clasping or at least widened at base.
        E. Petals purple, sometimes with white tips or edges.
          F. The petals 10–14 mm. long; main stem-lvs. not conspicuously crowded.
            G. Style 1–3 mm. long; stems mostly stout, 4–15 dm. high.
              H. Siliques arcuate, divergent, 1.5–2.5 mm. wide; plants biennial.
                  San Bernardino Mts. to L. Calif. .... 1. *S. campestris*
              HH. Siliques almost straight, suberect, 2.5–4 mm. wide; plants
                  perennial. New York and Panamint mts. n. ... 3. *S. cordatus*
            GG. Style absent; stems slender, 1.5–4 dm. high. Mono Co.
                                                                    4. *S. oliganthus*
          FF. The petals 6–8 mm. long; cauline lvs. crowded, longer than internodes.
              Siskiyou Mts. .... 5. *S. barbatus*
        EE. Petals white.
          F. The petals 7–9 mm. long; stigma indistinctly lobed. San Gabriel Mts.
              to Laguna Mts. .... 2. *S. bernardinus*
          FF. The petals 11 mm. long; stigma plainly 2-lobed. N. Calif.
                                                                    7. *S. tortuosus*
    CC. Annual; sepals lacking short stiff terminal hairs; fils. of upper pair of stamens connate.
      D. Siliques ascending to recurved-spreading, 2.5–7 cm. long; sepals about equally
          broad.
        E. Lower lvs. saliently lobed, spatulate-obovate, long-petioled; plants 0.5–1.8
            dm. high. Marin Co. .... 9. *S. batrachopus*
        EE. Lower lvs. toothed or entire, short-petioled; plants mostly 3–7 dm. high.
          F. Basal lvs. oblanceolate to ovate to obovate; petals in dissimilar pairs.
            G. Sepals alike; basal lvs. broadly obovate to ovate; petals 7–9 mm.
                long. Glenn Co. to San Benito Co. .... 10. *S. Breweri*
            GG. Sepals with the 3 upper connivent and the lower one spreading;
                basal lvs. oblanceolate; petals ca. 10 mm. long. Marin Co. to
                Santa Clara Co. .... 12. *S. glandulosus*
          FF. Basal lvs. linear-lanceolate; petals in similar pairs. Mendocino Co. to
              Napa Co. .... 11. *S. barbiger*
      DD. Siliques pendent, 2–3 cm. long; outer sepals much wider than inner. Sierra Nevada
          foothills .... 17. *S. polygaloides*
  BB. Infl. with some conspicuous broad leafy bracts among lower fls.
    C. Middle cauline lvs. oblong to obovate.
      D. Siliques arcuate-spreading. Widespread .... 7. *S. tortuosus*
      DD. Siliques erect. Kings-Kern Divide. .... 8. *S. gracilis*
    CC. Middle cauline lvs. linear or pinnate with linear divisions; siliques reflexed. W. base
        of Sierra Nevada .... 18. *S. diversifolius*
AA. Plants ± setose-hispid especially toward base.
  B. Perennial; fils. distinct. San Gabriel Mts. to Laguna Mts. .... 2. *S. bernardinus*
  BB. Annuals.
    C. Siliques pendent; fils. distinct; petals 12–14 mm. long. Los Angeles Co. to L. Calif.
                                                                16. *S. heterophyllus*
    CC. Siliques mostly erect to spreading; longer fils. united; petals mostly 6–11 mm. long.
      D. Plants less than 2 dm. high.
        E. Lvs. cuneate-obovate; petals 6–8 mm. long; siliques 4–7 cm. long. Mt. Diablo
                                                                14. *S. hispidus*
        EE. Lvs. oblong-orbicular; petals 10 mm. long; siliques 1.5–2 cm. long. Mt.
            Hamilton .... 15. *S. callistus*
      DD. Plants mostly 2–6 dm. high.
        E. Terminal fls. normal or slightly reduced; stigmas entire. Mendocino Co. to
            San Luis Obispo Co. .... 12. *S. glandulosus*
        EE. Terminal fls. sterile, forming a conspicuous dark tuft of numerous elongate
            sepals; stigmas somewhat 2-lobed. San Benito and Monterey cos. to Ventura Co.
                                                                13. *S. insignis*

1. **S. campéstris** Wats. Apparently largely biennial, glabrous, glaucous, the stems few,
stout, 6–15 dm. high, mostly rather simple; basal lvs. obovate or broadly oblanceolate, 5–10
cm. long, thickish, rounded at apex, petioled, dentate with setosely tipped-teeth and
margins ± ciliate; cauline lvs. lance-oblong, auriculate-clasping; racemes lax, becoming
2–3.5 dm. long; pedicels 5–10 mm. long, divergent-ascending, stout in fr.; sepals
purple, narrow-oblong, ca. 8 mm. long, bristle-tipped; petals purple, recurved, ca. 1
cm. long; fils. distinct; siliques ascending-divergent, arcuate, 7–14 cm. long, 1.5–2.5
mm. wide; style 1–2 mm. long; stigma slightly 2-lobed; seeds brown, oblong, 2–3
mm. long, winged.—Occasional, dry rocky slopes, 3000–7000 ft.; Chaparral, Pinyon-
Juniper Wd., Yellow Pine F.; San Bernardino and Little San Bernardino mts. to n. L.
Calif. May–July.

2. **S. bernardìnus** (Greene) Parish. [*Agianthus b.* Greene. *S. campestris* var. *b.* Jtn. *A. jacobaeus* Greene. *S. c.* var. *j.* Jeps.] Glaucous, mostly perennial, with rather slender stems 3–6 dm. high; basal lvs. obovate to broadly oblanceolate, 3–6 cm. long, thickish, petioled, coarsely dentate, sometimes setose-ciliate; cauline lvs. subentire, lanceolate to lance-ovate, 1.5–4 cm. long, auriculate-clasping; racemes lax, becoming 1–2.5 dm. long; pedicels 3–6 mm. long, stout in fr.; sepals broad-oblong, whitish, 5 mm. long, somewhat crisped; petals white, 7–9 mm. long; siliques subsessile, 5–8 cm. long, 1–1.5 mm. wide; style 1–2 mm. long; stigma scarcely 2-lobed; seeds brown, ca. 2 mm. long, winged.— Dry slopes, 4000–7500 ft.; mostly Yellow Pine F.; San Gabriel Mts. to Laguna Mts. June–July.

3. **S. cordàtus** Nutt. [*Euklisia c.* Rydb. *Cartiera c.* Greene. *S. crassifolius* Greene. *S. cordatus* vars. *exiguus* and *Duranii* Jeps.] Short-lived glabrous glaucous perennial, the stems mostly simple, stoutish, 3–8 dm. high; basal lvs. spatulate-obovate, variously dentate, often setose-ciliate, petioled, 2.5–7 cm. long; cauline lvs. broadly oblong, 2–6 cm. long, sagittate at base, entire, rounded at apex, tending to be rather crowded on stem; racemes rather lax, 1–3 dm. long; pedicels ascending, 5–10 mm. long, stout in fr.; sepals greenish or purplish, broadly oblong, 5–8 mm. long with terminal tuft of bristles; petals purple, white-margined, recurved, 10–14 mm. long; stamens distinct; siliques ascending or spreading, 5–8 cm. long, 3–4 mm. wide; style 2–3 mm. long; stigma distinctly 2-lobed; seeds broadly oblong, winged, brown, ca. 2.5 mm. long; $n = 12$ (Rollins, 1939).—Dry stony slopes, 4000–10,000 ft.; Pinyon-Juniper Wd., Yellow Pine F.; New York, Panamint, and White mts., and along e. slope of Sierra Nevada to se. Ore.; to Wyo., Colo., Ariz. May–July.

4. **S. oligánthus** Roll. Perennial; stems 1 or 2 from the base, simple or rarely with 1 or 2 branches above, glabrous, glaucous, 1.5–4 dm. high; basal lvs. lanceolate to oblanceolate, entire, apiculate, long-petioled, 4–8 cm. long, ciliate; upper lvs. sessile, oblong, sagittate, 2–5 cm. long; racemes few-fld.; pedicels 3–6 mm. long; sepals purple, glabrous, oblong, 6–10 mm. long; petals purple-tipped, pale below, lanceolate, crisped, 10–15 mm. long; siliques glabrous, flattened parallel to septum, obtuse at apex, nerved below middle, divaricate to ascending, 4–7 cm. long, 2–3 mm. wide; stigma sessile; seeds flat, winged, suborbicular, ca. 2 mm. broad.—Rocky slopes, 8000–8200 ft.; Red Fir F.; Sweetwater and Masonic mts., Mono Co. June–July.

5. **S. barbàtus** Wats. [*Cartiera b.* Greene.] Glabrous glaucous short-lived perennial, from deep-seated rootstock; stems 1–few, mostly simple, slender, 2–5 dm. high; lvs. many, crowded, longer than internodes, round-oblong with cordate clasping base, subentire, 1–2.5 cm. long; racemes 1–3 dm. long, lax in fr.; pedicels erect or ascending, 4–7 mm. long; sepals purple, obtusish, apically bearded, 5–7 mm. long; petals purplish, 6–8 mm. long; siliques arcuate-spreading, subsessile, 2.5–6.5 cm. long, 2–3 mm. wide; style 1–2 mm. long; stigma entire; seeds dark brown, almost round, ca. 2 mm. long, narrowly wing-margined at ends.—Dry rocky slopes, 2700–5000 ft.; Yellow Pine F.; Trinity and Siskiyou cos. June–Aug.

6. **S. Howéllii** Wats. [*Cartiera H.* Greene.] Glabrous glaucous perennial, 3–7 dm. high; lvs. rather fleshy, the lower petioled, spatulate-obovate, 2–10 cm. long, entire to sinuate-dentate, the upper entire, narrow, oblong-spatulate, not auriculate-clasping; racemes lax, 1–3 dm. long; pedicels ascending, 7–12 mm. long; sepals purplish, oblong, 5–7 mm. long; petals oblong-linear, maroon to purple, 8–10 mm. long; siliques divergent-arcuate, 5–8 cm. long, 2.5–3.5 mm. wide; style barely 1 mm. long; stigma entire; seeds broadly winged, ca. 3 mm. long; $n = 14$ (Kruckeberg, 1957).—Dry serpentine slopes, 2000–3000 ft.; Mixed Evergreen F.; Siskiyou Mts., Del Norte Co.; adjacent Ore. July–Aug.

7. **S. tortuòsus** Kell. [*Pleiocardia t.* Greene. *S. foliosus* Greene. *S. t.* vars. *optatus* and *oblongus* Jeps. *S. sanhedrensis* Eastw.] Glabrous and rather glaucous annual or biennial, 2–10 dm. high, the stems 1 to several, simple or somewhat branched above; lower lvs. spatulate-obovate to -oblanceolate, short-petioled, entire or toothed, 3–8 cm. long; middle cauline lvs. oblong to obovate, toothed to subentire, clasping, 2–9 cm. long; uppermost lvs. in lower part of infl., oblong-ovate, deeply clasping; racemes mostly 5–10 cm. long, rather compact, ± secund; pedicels 3–10(–15) mm. long, ascending; sepals usually purplish, 6–8 mm. long, recurved at the pointed glabrous tips; petals purplish or yellowish-white, usually with purple veins, 10–12 mm. long; siliques 6–12 cm. long,

arcuate-spreading, 1.5–3 mm. wide; style scarcely developed; stigma not lobed; seeds round-oblong, brown, narrowly winged, ca. 2 mm. long.—Exceedingly variable; dry rocky slopes, 1000–6500 ft.; Mixed Evergreen F., Yellow Pine F.; N. Coast Ranges s. to Sonoma Co., Sierra Nevada to Tulare Co. May–Aug.

KEY TO VARIETIES

Sepals 5–8 mm. long; annual or biennial.
  Petals mostly 6–8 mm. long.
    Plants from below 7000 ft., not bushy, 2–10 dm. high; usually with purplish sepals .. *S. tortuosus*
    Plants from above 7000 ft., mostly bushy, 1–2 dm. high.
      Sepals purplish .... var. *orbiculatus*
      Sepals yellowish ....var. *flavescens*
  Petals 8–10 mm. long; sepals yellowish. Nw. Calif. .... var. *pallidus*
Sepals ca. 10 mm. long; perennial .... var. *suffrutescens*

Var. **orbiculàtus** (Greene) Hall. [*S. o.* Greene. *Pleiocardia o.* Greene.] Bushy, mostly 1–2 dm. high; sepals 5–6 mm. long; petals 7–9 mm. long.—Dry slopes, 7000–11,500 ft.; largely Lodgepole F. and Subalpine F.; Sierra Nevada to Mt. Shasta and Humboldt Co., where it occurs considerably lower; s. Ore. June–Sept. Intergrading freely with the sp.

Var. **flavéscens** Jeps. Sepals yellowish, 4–6 mm. long; petals yellow, 6–8 mm. long.— At 10,000–10,500 ft.; Subalpine F.; Farewell Gap, Sawtooth Range, Tulare Co. Aug.

Var. **pállidus** Jeps. Sepals yellowish, 5–8 mm. long; petals pale, purple-veined, 8–10 mm. long.—From 1500–6000 ft.; Mixed Evergreen F., Yellow Pine F.; Trinity and Humboldt cos. May–July.

Var. **suffrutéscens** (Greene) Jeps. [*S. s.* Greene.] Perennial with woody branched caudex; sepals purplish, ca. 1 cm. long.—Hood's Peak, Sonoma Co. and Table Mt., Butte Co. April–May.

8. **S. grácilis** Eastw. [*Pleiocardia g.* Greene.] Slender glabrous glaucous annual, often branched from near base, 1–3 dm. high; lower lvs. rounded to oblanceolate, sinuate-dentate to lobed, slender-petioled, 1–4 cm. long; cauline lvs. entire to lobed, oblong to ovate, 0.5–1.3 cm. long, largely sessile and clasping; racemes 3–12 cm. long with 1 or 2 ovate sessile bracts below; pedicels erect or ascending, 3–6 mm. long; sepals purple, spreading and glabrous at tip, 4–5 mm. long; petals pinkish, 7–8 mm. long, with broad limb; fils. distinct; siliques very short-stipitate, suberect, 3–7 cm. long, 1–1.5 mm. wide; style scarcely evident; stigma round; seeds scarcely winged, ca. 1 mm. long. —Dry slopes of disintegrated granite, 10,000–11,000 ft.; Subalpine F.; Kings-Kern Divide, s. Sierra Nevada. July–Aug.

9. **S. batràchopus** Morrison. Erect simple or branched glabrous glaucous annual, 5–18 cm. high; lower lvs. oblong to spatulate-obovate, saliently lobed, petioled, 1–2.5 cm. long; upper lvs. sessile, auriculate-clasping, linear-lanceolate, entire, ca. 1 cm. long; racemes lax, 5–10 cm. long; pedicels ascending, 2–3 mm. long; sepals green or purple, with white recurved glabrous tips, ca. 4 mm. long; petals white with purple midvein, 6–7 mm. long; stamens of each pair ± connate; siliques arcuate-spreading, 2.5–3 cm. long, ca. 1.5 mm. wide; style ca. 1 mm. long; stigma entire; seeds brown, winged, 2 mm. long.—Serpentine outcrops, below 2000 ft.; Chaparral; Mt. Tamalpais and Carson Ridge, Marin Co. May–June.

10. **S. Brèweri** Gray. [*Pleiocardia B.* Greene.] Glabrous glaucous annual, branching from base, 3–6 dm. high; basal lvs. broadly ovate to obovate, short-petioled, entire to dentate, 3–12 cm. long; cauline lvs. sessile, clasping, the upper entire, lanceolate, much reduced; racemes 1–3 dm. long; pedicels 1–3 mm. long; sepals largely purple, 5–7 mm. long, the tips recurved, not setose; upper petals white or with purple veins, the lower with darker veins, all recurved, 8–10 mm. long; longer fils. connate; siliques erect or ascending, usually somewhat incurved, 3–5 cm. long, 1 mm. wide; stigma sub-sessile, entire; seeds winged only at apex if at all, rounded-oblong, ca. 1.2 mm. long.— Talus and gravel on serpentine slopes, 1500–4000 ft.; Chaparral; inner Coast Ranges, Glenn Co. to San Benito Co. May–July.

Var. **hespéridis** (Jeps.) Jeps. [*S. h.* Jeps. *Pleiocardia h.* Greene.] Stem few-branched; racemes secund; sepals mostly green, 4–6 mm. long; petals 7–8 mm. long; siliques recurved-spreading.—Serpentine; Chaparral; Lake and Napa cos.

11. **S. bárbiger** Greene. [*Mesoreanthus b.* Greene. *M. fallax* and *vimineus* Greene.]

Glabrous glaucous erect annual, branched above, 3–7 dm. high; lvs. sublinear, erect, entire, the lower 6–10 cm. long, the upper very narrow, reduced; racemes lax, 5–25 cm. long; pedicels erect, 2–3 mm. long; sepals green to purplish, hyaline on margins and recurved tips, sometimes with stiff hairs but no apical tuft, 6–7 mm. long; petals ca. 1 cm. long, not crisped, the 2 upper white, the 2 lower with purple midvein; upper fils. connate; siliques spreading, somewhat recurved, sessile, 5–7 cm. long, ca. 1.5 mm. wide; stigma sessile, entire; seeds not winged.—Serpentine ridges, 500–1500 ft.; Chaparral; inner Coast Range, Mendocino Co. to Napa Co. May–July.

12. **S. glandulòsus** Hook. [*S. peramoenus, Mildredae, Biolettii,* and *asper* Greene. *Euclisia g., M., B., a., elatior* and *Bakeri* Greene.] JEWEL-FLOWER. Annual with simple or branched stems, setose-hirsute at least in lower parts, 3–6 dm. high; lower lvs. oblanceolate, coarsely and saliently toothed, petioled, 2–8 cm. long; upper lvs. linear to lanceolate, entire to dentate, auriculate-clasping; racemes lax, 5–30 cm. long; pedicels mostly 4–10 mm. long, ascending; sepals usually purple, 5–7 mm. long, oval, glabrous or nearly so, rounded at apex, the 3 upper connivent, the lower one spreading; petals purple or white with purple veins, recurved-spreading, ca. 1 cm. long, the upper pair commonly longer and darker than the lower; longest fils. connate; siliques ascending-spreading, sessile, 5–9 cm. long, ca. 2 mm. wide, sometimes hispid; stigma sessile, entire; seeds broadly elliptic-oblong, narrowly winged, ca. 1.2 mm. long.—A polymorphic sp. from dry often loose disturbed soil and rocky ridges, below 4000 ft.; N. Oak Wd., Foothill Wd., Chaparral; Coast Ranges from Sonoma and Solano cos. to San Luis Obispo Co. April–May.

KEY TO VARIETIES

Siliques ascending to erect.
  Plants ± hispid; pedicels mostly less than 1 cm. long.
    Sepals purple.
      Upper lvs. lanceolate to linear, entire or somewhat toothed; plants mostly branched above, the branches ascending. Sonoma Co. to San Luis Obispo Co. . . . . S. *glandulosus*
      Upper lvs. lance-oblong, with pairs of salient teeth; plants branched below, the branches wide-spreading. Marin Co. . . . . var. *pulchellus*
    Sepals whitish or pale lavender. Contra Costa and Santa Clara cos. . . . . var. *albidus*
  Plants almost or quite glabrous; pedicels often more than 1 cm. long. Tiburon Peninsula . . var. *niger*
Siliques spreading-recurved to pendulous; sepals mostly yellowish to pale. Mendocino Co. to Marin Co.
                                                              var. *secundus*

Var. **álbidus** (Greene) Jeps. [*S. a.* Greene. *Euclisia a.* Greene.] Nearly glabrous; lvs. shallowly denticulate, lanceolate; sepals tinged with pale rose-lavender, 5–7 mm. long; petals whitish with light purplish veins, crisped, 8–11 mm. long; siliques erect.—Serpentine outcrops, 600–2500 ft.; Chaparral; w. base of Mt. Hamilton Range to Mt. Diablo, Santa Clara and Contra Costa cos. April–June.

Var. **pulchéllus** (Greene) Jeps. [*S. p.* Greene. *Euclisia p.* Greene.] With few widely divaricate branches, setose-hirsute; lvs. oblong-lanceolate, with opposite salient teeth; pedicels setose, 2–4 mm. long; sepals purple, somewhat hispid, 5–6 mm. long; siliques erect, usually setose.—Mostly on serpentine ridges; Chaparral; Mt. Tamalpais and Carson Ridge, Marin Co. May–July.

Var. **nìger** (Greene) Munz. [*S. n.* Greene. *Euclisia n.* Greene. *E. violacea* Greene?] Nearly or quite glabrous throughout; sepals very dark purple, 6–8 mm. long; petals with purple claw and white blade with purple midvein, 10–12 mm. long; siliques erect or ascending, almost straight, 4–7 cm. long.—Serpentine slopes, below 500 ft.; Coastal Prairie; Tiburon Peninsula, Marin Co. May–June.

Var. **secúndus** (Greene) Munz. [*S. s.* and *S. versicolor* Greene. *Euclisia s.* and *v.* Greene.] Infl. strongly secund; sepals usually yellowish, sometimes tinged lavender, ca. 6 mm. long; petals pale, 10–12 mm. long; siliques recurved or pendulous.—Rocky ridges and slides of serpentine or sedimentary rocks; Chaparral, N. Oak Wd.; Coast Range, Mendocino Co. to Marin Co. April–June.

13. **S. insígnis** Jeps. Annual, slender, sparsely hirsute-hispid throughout, usually branched, 1–4.5 dm. high; lvs. deeply lobed, almost pinnatifid, ± lanceolate in outline, 2.5–6 cm. long, sessile, auriculate; racemes lax, with a terminal tuft of dark sterile fls.; pedicels spreading, 2–3 mm. long; sepals dark purple, sparsely short-hispid, 5–6 mm. long; petals narrow, white with dark midvein, crisped, 9–12 mm. long; siliques

flattened, erect, straight, 5–7 cm. long, ca. 2 mm. wide, bristly-hispid; style ca. 1 mm. long; stigma somewhat 2-lobed; seeds round-oblong, winged, 1.5–2 mm. long.—Largely on serpentine, below 4500 ft.; Chaparral, Pinyon-Juniper Wd., Foothill Wd.; inner S. Coast Ranges of San Benito, Fresno and Monterey cos., Santa Ynez Mts. of Ventura Co. April–May.

14. **S. híspidus** Gray. [*Euclisia h.* Greene.] Hirsute-hispid annual 0.5–2 dm. high, compact; lvs. cuneate-obovate, all except lower sessile, with coarse obtuse teeth near rounded apex, 1–3 cm. long; racemes 5–12 cm. long, without or with few terminal sterile fls.; pedicels ascending, 2–4 mm. long; sepals purplish, 5–6 mm. long, mostly densely hispid; petals light purple, with white margins, somewhat crisped, 6–8 mm. long; siliques flat, hispid-hirsute, erect, straight, 4–7 cm. long, 1–2 mm. wide; stigma entire, subsessile; seeds winged, almost 2 mm. long.—Talus or rocky outcrops, 2000–3850 ft.; Chaparral; Mt. Diablo, Contra Costa Co. March–June.

15. **S. callístus** Morrison. Simple or branched, 3–6 cm. high, the branches sub-horizontal, sparingly hispid throughout; lvs. sessile or nearly so, oblong-orbicular, 3–15 mm. long, coarsely dentate, the upper clasping; raceme terminated by sterile fls.; pedicels 1–2 mm. long; sepals green, lance-ovate, keeled, 5 mm. long; petals purple, 10 mm. long, with whitish undulate margins; siliques erect, terete, incurved, 15–20 mm. long, hispid; stigma sessile; seeds rounded, wingless, 1 mm. long.—Talus, ca. 2000 ft.; Foothill Wd.; Arroyo Bayo, Mt. Hamilton, Santa Clara Co. April–May.

16. **S. heterophýllus** Nutt. [*Caulanthus h.* Pays.] Annual, hirsute below, erect, simple or branched, 3–10 dm. high; lvs. linear to lanceolate, pinnatifid with divaricate lobes, 5–10 cm. long, all but the lowest amplexicaul; racemes lax, 2–4 dm. long; pedicels 4–8 mm. long, hirsute, recurved or reflexed; sepals purplish or greenish, 8–9 mm. long, linear-lanceolate, somewhat divergent at tips; petals pale with purple veining, linear, recurved, 12–14 mm. long; fils. distinct; siliques pendent, straight, subcompressed, glabrous, 5–8 cm. long, 1.5–2 mm. wide; style ca. 1 mm. long; stigma slightly 2-lobed; seeds narrowly winged.—Occasional, disturbed places such as burns, below 3500 ft.; Chaparral, Coastal Sage Scrub; Los Angeles Co. to L. Calif. March–May.

17. **S. polygaloìdes** Gray. [*Microsemia p.* Greene.] Glabrous annual; stems slender, branched upwards, 2.5–9 dm. high; lvs. entire, linear, the upper sagittate-clasping; racemes lax, 5–15 cm. long; pedicels 2–3 mm. long, recurved in age; sepals purple to yellowish, 5–6 mm. long, the upper one suborbicular, retuse, the 2 lateral small, the lower one keeled, ovate, with acutish tip; petals linear-oblong, yellow or white with dark veins, the upper erect, the lower curved outwards, ca. 6 mm. long; upper fils. connate; siliques pendent, almost straight, somewhat flattened-quadrangular, 2–3 cm. long, 1 mm. wide; style ca. 1 mm. long; stigma entire; seeds oblong, slightly winged at tip, ca. 1.4 mm. long.—Dry open clay or serpentine slopes, below 3000 ft.; Foothill Wd., Yellow Pine F.; foothills of Sierra Nevada from Butte Co. to Fresno Co. May–June.

18. **S. diversifòlius** Wats. [*Mitophyllum d.* Greene. *S. linearis* and *foliosus* Greene. *Pleiocardia fenestrata* Greene.] Annual, glabrous, erect, slender-stemmed, branched above, 2–5 dm. high; lower lvs. entire, linear-filiform; cauline 1–5 cm. long, linear and entire to somewhat wider and pinnatifid into few linear lobes; upper lvs. and bracts entire, cordate-ovate, long-pointed at apex, clasping at base; racemes lax; pedicels ascending, 2–5 mm. long; sepals yellowish, 5–6 mm. long, with recurved tips; petals yellow to whitish, recurved, 8–9 mm. long; siliques reflexed, straightish, flattened, 4–8 cm. long, 1–2 mm. wide; styles 1–1.5 mm. long; stigma 2-lobed; seeds oblong, winged at least at ends, ca. 1.3 mm. long.—Dry rocky slopes, below 5000 ft.; Yellow Pine F., Foothill Wd.; Amador and Butte cos. to Tulare Co. April–July.

4. Caulánthus Wats.

Mostly annual sometimes perennial herbs, glabrous or pubescent with simple hairs; stems simple or branched. Basal lvs. usually not forming a conspicuous rosette; cauline lvs. short-petioled or sessile. Fls. racemose, purple, yellow, or white. Calyx somewhat flask-shaped, closed at anthesis or nearly so. Petals with narrow often crisped blades. Stamens 6, equal, distinct or sometimes united in pairs. Siliques divaricate, erect or deflexed, terete or only slightly flattened, sessile or nearly so; style usually short; stigma entire or 2-lobed. Seeds wingless or with narrow wings. Ca. 18 spp. of arid w. N. Am. (Greek, *kaulos*, stem, and *anthos*, fl., referring to cauliflower, since some spp. can be used like it.)

(Payson, E. B. A monographic study of Thelypodium and its immediate allies. Ann. Mo. Bot. Gard. 9: 233–324, 1922.)
A. Cauline lvs. sessile and auriculate at base.
  B. Plants glabrous or inconspicuously pubescent.
    C. Stems not conspicuously inflated.
      D. Stigma entire or indistinctly 2-lobed.
        E. Fls. purplish; siliques erect or divaricate, 6–8 cm. long. Mostly s. montane
                  1. *C. amplexicaulis*
        EE. Fls. yellow; siliques reflexed, 2–4.5 cm. long. Deserts .... 2. *C. Cooperi*
      DD. Stigma distinctly 2-lobed.
        E. Siliques 2–4 cm. long; fils. all distinct .... 3. *C. californicus*
        EE. Siliques 5–13 cm. long; at least 1 pair of fils. united .... 4. *C. Coulteri*
    CC. Stems conspicuously inflated; siliques erect; stigma deeply 2-lobed ...... 5. *C. inflatus*
  BB. Plants evidently hirsute or pilose.
    C. Stigma distinctly 2-lobed.
      D. Calyx yellowish; stigma shallowly lobed. Riverside Co. s. .... 6. *C. simulans*
      DD. Calyx purplish; stigma deeply lobed. Los Angeles Co. to Monterey and Madera cos.
                  4. *C. Coulteri*
    CC. Stigma very small, entire. S. San Diego Co. .... 7. *C. stenocarpus*
AA. Cauline lvs. sessile or petioled, but not auriculate.
  B. Calyx densely hispid-hirsute. W. edge of Colo. Desert .... 8. *C. Hallii*
  BB. Calyx glabrous to sparsely pilose.
    C. Plants pilose to hirsute, especially near base. Inyo Co. n. .... 9. *C. pilosus*
    CC. Plants glabrous.
      D. Sepals hairy; stems ± inflated; perennial. E. Mojave Desert .... 10. *C. crassicaulis*
      DD. Sepals glabrous; stems mostly not inflated.
        E. Plants perennial; lvs. usually lyrate; petals purplish. San Gabriel and San Bernardino mts. to Utah .... 11. *C. major*
        EE. Plants annual or perennial; lvs. usually entire; petals greenish. Grapevine and White mts. .... 12. *C. glaucus*

1. **C. amplexicaùlis** Wats. [*Euclisia a.* Greene. *Streptanthus a.* Jeps. *Pleiocardia magna* Greene?] Glabrous glaucous annual, simple or branched, 1–5 dm. high; basal lvs. oblong-oblanceolate, 3–10 cm. long, sinuate-dentate, with broadly winged petiole; cauline lvs. oblong, 1–7 cm. long, deeply amplexicaul, sinuate-dentate to entire, obtuse; racemes lax, few-fld., 5–20 cm. long; pedicels ascending, 8–15 mm. long; sepals purple, somewhat saccate, 5–7 mm. long, the tips recurved; petals purplish, linear, 8–10 mm. long, crisped above; siliques spreading, curved, terete, 6–8 cm. long, ca. 1 mm. thick, subsessile; style 1.5–2 mm. long; stigma entire; $n = 14$ (Kruckeberg, 1957).—Loose dry slopes, 5000–8500 ft.; Yellow Pine F.; San Bernardino and San Gabriel mts., Mt. Pinos; less frequent, Joshua Tree Wd.; Antelope V., w. Mojave Desert. May–July.

2. **C. Coòperi** (Wats.) Pays. [*Thelypodium C.* Wats. *Guillenia C.* Greene.] Glabrous or subglabrous annual, with simple or branched somewhat flexuous slender stems 2–6 dm. long; radical lvs. oblanceolate, sinuate, 2–6 cm. long, with short winged petioles; cauline lvs. oblong to lanceolate, mostly entire, 1–3 cm. long, with clasping base; racemes lax; pedicels stout, recurved, 1–3 mm. long; sepals greenish, 5–6 mm. long; petals yellowish, 7–9 mm. long, somewhat crisped; siliques deflexed, terete, often arcuate, 2–4 cm. long; style 1–2 mm. long; stigma small, somewhat 2-lobed; seeds brown, oblong, not winged, ca. 1.5 mm. long, faintly cellular-reticulate.—Common, often among shrubs, on washes and slopes, below 7000 ft.; Creosote Bush Scrub, Joshua Tree Wd.; both deserts; to Nev., Ariz. March–April.

3. **C. califórnicus** (Wats.) Pays. [*Stanfordia c.* Wats. *Streptanthus c.* Greene.] Glabrous annual, or sparsely pilose near base, erect, branched, 2–5 dm. high; basal lvs. oblanceolate, sinuately lobed to pinnatifid, 3–6 cm. long, obtuse; cauline lvs. oblong to ovate, dentate, amplexicaul, obtuse; racemes lax, secund; pedicels pilose, 5–10 mm. long, often recurved in age; sepals unequal, white-membranous and saccate below, purple-tipped, 6–9 mm. long; petals narrow, whitish, wavy-margined, slightly exceeding sepals; longer fils. sometimes slightly united; siliques ascending or deflexed, straight, slightly compressed, 2–4 cm. long, 2–4 mm. wide; styles 1–1.5 mm. long; stigma deeply 2-lobed; seeds dark, brown, round-oblong, not winged, 1–1.2 mm. long.—Dry plains and slopes, below 3000 ft.; V. Grassland; upper San Joaquin V., Fresno Co. to Kern Co., and e. San Luis Obispo and adjacent Santa Barbara cos. March–April.

4. **C. Còulteri** Wats. [*Streptanthus C.* Greene.] Annual, hirsute-pubescent especially below, erect, the stem simple or branched, 3–8 dm. high; lower lvs. oblong-oblanceolate, sinuate-pinnatifid, 5–10 cm. long, petioled; cauline oblanceolate to oblong to lanceolate, 4–8 cm. long, subentire to sinuate-dentate, mostly amplexicaul; racemes lax, with terminal tufts of purplish buds; pedicels hirsute, reflexed, 3–10 mm. long; sepals purple, later yellowish or greenish, glabrous to hirsute, unequal, scarcely saccate, 8–14 mm. long; petals whitish with purple veins, widely spreading, crisped, 12–18 mm. long; longer fils. united; siliques divergent-descending or pendent to ascending, glabrous, subterete to somewhat flattened, 5–10 cm. long, 2–3 mm. wide; style ca. 1 mm. long; stigma-lobes ca. 1 mm. long; seeds dark brown, oblong, not winged, ca. 2 mm. long, faintly cellular-reticulate.—Dry slopes, below 5000 ft.; V. Grassland, Chaparral, Foothill Wd.; foothills of Sierra Nevada from Madera Co. to Kern Co., San Joaquin V., s. to Elizabeth Lake, Los Angeles Co. March–May.

Var. **Lemmònii** (Wats.) Munz [*C. L.* Wats. *Streptanthus L.* Jeps. *S. Parryi* Greene.] Plant less hirsute; lvs. less deeply dentate; pedicels becoming 10–20 mm. long; siliques erect; stigma-lobes 1.5–3 mm. long.—Grassy slopes, below 1600 ft.; V. Grassland; inner S. Coast Ranges, Alameda Co. to Santa Barbara Co. March–April.

5. **C. inflàtus** Wats. [*Streptanthus i.* Greene.] Squaw-Cabbage. Desert Candle. Glabrous erect annual, sometimes slightly hirsute at base; stem usually simple, conspicuously inflated, 2–7 dm. high; lvs. oblong to ovate, or the lower oblanceolate, all clasping at base, entire or denticulate, 2–7 cm. long; racemes lax below the terminal bud-tuft; pedicels ascending, glabrous to pilose, ca. 3 mm. long; sepals purple in bud, at anthesis white with purple tips, glabrous, subequal, acute, 8–10 mm. long, scarious on margins; petals white, linear, crisped near tip, slightly exceeding sepals; longer fils. coherent; siliques stout, erect or ascending, 5–10 cm. long; stigma subsessile, deeply 2-lobed; seeds oblique-oblong, dark brown, wingless, 2.5–3.5 mm. long.—Common on open flats and among brush, below 5000 ft.; Creosote Bush Scrub, V. Grassland, Joshua Tree Wd.; inner S. Coast Ranges and San Joaquin V. from w. Fresno Co. to Barstow region., w. Mojave Desert. March–May.

6. **C. símulans** Pays. [*Streptanthus s.* Jeps.] Erect branching annual, hirsute near base, 3–4 dm. high; lower lvs. oblong to lanceolate, sparingly hirsute, deeply sinuate-dentate to subentire, 2–6 cm. long; upper cauline lvs. reduced, amplexicaul; racemes lax; pedicels recurved, hairy, 3–5 mm. long; sepals yellowish, sparsely hirsute, 5–6 mm. long; petals whitish, narrow, crisped, recurved at tip, 8–10 mm. long; siliques slender, straight, descending, terete, 4–6 cm. long; style ca. 1 mm. long; stigma 2-lobed; seeds oblong, not winged, ca. 1 mm. long.—Uncommon in rocky places, 2000–5500 ft.; Chaparral, Pinyon-Juniper Wd.; Santa Rosa Mts., Riverside Co. to interior San Diego Co. April–June.

7. **C. stenocárpus** Pays. Slender erect hirsute annual, simple or branched, 3–4 dm. high; cauline lvs. scattered, linear-lanceolate, subentire, sessile, clasping, 1–2 cm. long, the lowermost sinuate-dentate; racemes lax; pedicels hirsute, recurved, 1–3 mm. long; sepals purple, subglabrous, 4 mm. long; petals veined with purple, 6 mm. long, broadly linear; siliques descending, subterete, 2–4.5 cm. long, subglabrous; style 1–2 mm. long; stigma small, entire; seeds not winged.—Dry slopes, especially on burns; Chaparral; San Diego Co. (Harbison Canyon, Dehesa, Bernardo); n. L. Calif. (Ensenada). April–May.

8. **C. Hállii** Pays. [*Streptanthus H.* Jeps.] Annual, simple or branched, sometimes with inflated stems, 2–5(–9) dm. high, somewhat stiff-hairy on lower stems, lvs. and infl.; lower lvs. oblanceolate to oblong, coarsely sinuate-dentate to pinnatifid, 3–10 cm. long, short-petioled; cauline lvs. reduced, not auriculate, mostly coarse-dentate; racemes very lax; pedicels divergent to recurved, 6–15 mm. long; sepals yellowish, hirsute, lanceolate, 6–8 mm. long; petals yellowish-white, 8–9 mm. long, spatulate, not crisped; fils. distinct; siliques terete, glabrous, divaricate, 6–10 cm. long; style 1.5–2 mm. long; stigma deeply 2-lobed; seeds dark brown, oblong, not winged.—Occasional, about washes and dry places, 2000–6000 ft.; Creosote Bush Scrub, Joshua Tree Wd., Pinyon-Juniper Wd.; Little San Bernardino Mts., along w. edge of Colo. Desert to e. San Diego Co. April–May.

9. **C. pilòsus** Wats. [*Streptanthus p.* Jeps.] Biennial or short-lived perennial, sparingly to densely hirsute about base, 3–10 dm. high; lower lvs. runcinate-pinnatifid, 5–12 cm.

long, petioled, oblanceolate in outline; cauline lvs. reduced, less lobed, not auriculate; racemes lax, several dm. long; pedicels ascending, 5–8 mm. long; sepals green to purplish, ± pilose, 6–8 mm. long; petals whitish, 8–10 mm. long, crisped; siliques spreading, often arcuate, 6–12 cm. long; style ca. 1 mm. long; stigma evidently 2-lobed; seeds narrow-oblong, not winged, almost 2 mm. long.—Dry slopes, washes, etc., 4500–9000 ft.; Joshua Tree Wd., Pinyon-Juniper Wd., Shadscale Scrub; from Darwin, Inyo Co. to e. Ore., Ida., Utah. April–July.

10. **C. crassicaùlis** (Torr.) Wats. [*Streptanthus c.* Torr.] Short-lived perennial; stems inflated, glaucous, glabrous, unbranched, 3–10 dm. high; lower lvs. oblanceolate, subentire or sinuate-dentate to runcinate, petioled, 5–15 cm. long; upper lvs. reduced, not clasping; racemes lax, to ca. 4 dm. long; pedicels stout, hirsute, 3–5 mm. long; sepals purplish, color almost concealed by the dense white hairiness, subequal, 10–15 mm. long; petals dark purple with white margins, narrow-oblong, 10–15 mm. long, not crisped; fils. distinct; siliques erect or ascending, rather stout, 10–13 cm. long; stigma subsessile, with 2 broad lobes; seeds narrowly oblong-obovate, brown, ca. 3 mm. long; $n = 12$ (Rollins, 1939).—Uncommon, dry slopes and canyons, 4000–8000 ft.; Pinyon-Juniper Wd.; Clark Mt., Kingston Mts., White Mts.; to Wyo., Utah. April–June.

11. **C. màjor** (Jones) Pays. [*C. crassicaulis* var. *m.* Jones. *Streptanthus m.* Jeps.] Glabrous perennial, somewhat glaucous, with few erect stems 3–8 dm. high, often inflated, simple or few-branched; lower lvs. tufted, oblanceolate in outline, runcinate, or lyrate, 5–18 cm. long, petioled; upper lvs. few, remote, linear to lanceolate, reduced, not auriculate; racemes rather few-fld., lax; pedicels stout, ascending, 3–5 mm. long; sepals glabrous, purple or yellowish with purple tips, 7–9 mm. long; petals purplish, broadly linear, somewhat crisped, ca. 12–13 mm. long; siliques erect or ascending, 8–12 cm. long; stigma subsessile, shallowly 2-lobed; seeds oblong-elliptic, brown, not winged, ca. 2.5 mm. long.—Occasional on dry loose and rocky slopes, 5500–7500 ft.; Joshua Tree Wd., Pinyon-Juniper Wd., Yellow Pine F.; San Gabriel and San Bernardino mts., Grapevine Mts. (Death V.), New York and Providence mts.; to Utah. May–July.

12. **C. glaùcus** Wats. [*Streptanthus g.* Wats.] Perennial, glabrous, glaucous, the stems several, 3–7 dm. high; lower lvs. suborbicular to oblong-ovate, obtuse, 5–12 cm. long, entire or lobed at base of blade only, petioled; cauline lvs. few, narrower, reduced, not amplexicaul; racemes lax, becoming 5–6 dm. long; pedicels erect or ascending, 7–16 mm. long; sepals greenish or purplish, glabrous, 8–10 mm. long; petals greenish, narrow, recurved at tip, 15–17 mm. long; siliques divaricate, frequently arcuate, 5–10 cm. long; stigma subsessile, deeply 2-lobed; seeds oblong-ovate, ca. 1.3 mm. long, not winged.—Dry rocky slopes, 5000–7500 ft.; Pinyon-Juniper Wd.; White Mts. to Grapevine Mts., Inyo Co.; Nev. May–June.

5. Streptanthélla Rydb.

Glabrous annuals with slender usually branched stems. Lvs. entire or shallowly dentate, linear-lanceolate, narrowed at base. Fls. small, in racemes. Sepals with lateral pair saccate at base. Petals narrow, crisped or channeled. Siliques pendent on recurved pedicels, not stipitate, strongly compressed, dehiscent at base but not separating at apex and provided with a conspicuous beak simulating a persistent style. Stigma subentire. Seeds oblong, flat, winged. One sp. (Diminutive of *Streptanthus.*)

1. **S. longiróstris** (Wats.) Rydb. [*Arabis l.* Wats. *Streptanthus l.* Wats. *Thelypodium l.* Jeps. *Guillenia rostrata* Greene.] Plants glaucous, 2–6 dm. tall; lvs. 2–5 cm. long, the upper reduced and entire; pedicels 2–5 mm. long; sepals greenish or tipped with purple, 4–6 mm. long; petals yellowish, linear-spatulate, 5–8 mm. long; siliques 3–6 cm. long; seeds almost 2 mm. long.—Common, especially about shrubs, sandy flats and washes, below 6000 ft.; Creosote Bush Scrub, Sagebrush Scrub, Joshua Tree Wd., Pinyon-Juniper Wd., V. Grassland; inner S. Coast Ranges, Monterey Co. to Kern Co., and deserts from Inyo Co. to L. Calif., e. Wash., Wyo., New Mex. March–May.

Var. derelícta J. T. Howell. Not glaucous; most lvs. pinnatifid into few, narrow, divaricate lobes.—Sandy flats near sea-level; Creosote Bush Scrub; Coachella V. and Desert Center, Riverside Co. to Yuma and Son. Feb.–May.

6. Subulària L. Awlwort

Small stemless aquatic perennials. Lvs. subulate, in basal tufts. Fls. few, minute, on naked scapes. Sepals broadly ovate, greenish. Petals white. Stamens 6, subequal. Style none. Silicles short-stipitate, round-ovoid, the valves turgid, 1-ribbed dorsally. Seeds few in each cell, marginless. Two spp. of N. Hemis. (Latin, *subula*, an awl, referring to lvs.)

1. **S. aquática** L. Glabrous; lvs. 1–3 cm. high; scapes 3–10 cm. long; pedicels 2–5 mm. long; silicles 2–3 mm. long.—Rare, wet banks and shallow water, 7000–10,000 ft.; Lodgepole F., Subalpine F.; Sierra Nevada (East Lake, Tulare Co.; Mono Pass, Dana Meadows, Donner Lake, Lake Tahoe, Webber Lake); to Alaska and Nfld.; Eurasia. July–Aug.

7. Cardària Desv.

Perennial rhizomatous herbs. Stems branched, erect or decumbent, pubescent. Lvs. oblong, dentate, the lower petioled, the upper clasping. Fls. small, in corymbed racemes. Sepals alike at base, scarious-margined. Petals white. Stamens 6; fils. free, toothless. Silicles ovoid, subglobose or cordate, inflated, nearly or quite indehiscent, with thin fenestrate or entire septum; style slender. Seeds 2–4, pendulous, wingless. Four spp. of Old World. (Name from the *cordiform* fr. of first sp. described.)

(Rollins, R. C. On two weedy crucifers. Rhodora 42: 302–306, 1940.)
Silicles glabrous, broader than long, notched at base; style 1–2 mm. long .... 1. *C. Draba*
Silicles pubescent, longer than broad, not notched at base; style 2–3 mm. long ...... 2. *C. pubescens*

1. **C. Dràba** (L.) Desv. [*Lepidium D.* L.] Hoary-Cress. Pubescent or somewhat tomentulose, several-stemmed, 3–4 dm. high, leafy; lvs. 3–6 cm. long, the lower petioled, the upper narrowed to an auriculate base; racemes in a terminal paniculate cluster; pedicels spreading, slender, 6–12 mm. long; fls. 2–3 mm. long; silicles 3–4 mm. long, reniform or depressed-cordate; style slender, ca. 1 mm. long; seeds ovoid or ellipsoid, brown, almost smooth, compressed; $2n = 64$ (Manton, 1932).—Widespread as pernicious weed in fields and waste places at low elevs.; to Wash. and Atlantic Coast; native of Eu. March–June.

Var. **rèpens** (Schrenk) O. E. Schulz. [*Physolepidion r.* Schrenk.] Silicles orbicular, obtuse or acute at base, 4–5 mm. long.—A poorly marked segregate, occasional in Calif.; to Wash., Dak.; natur. from Eu.

2. **C. pubéscens** (C. A. Mey.) Roll. var. **elongàta** Roll. [*Hymenophysa p.* auth.] Whitetop. Minutely pubescent, 1–4 dm. high; lvs. 1–3.5 cm. long, clasping; pedicels slender, 3–8 mm. long; silicles pubescent, subglobose, 2.5–3.5 mm. long; style almost as long; seeds brown, pitted-reticulate, ca. 1.3 mm. long.—Occasional weed in waste places, alfalfa fields, etc.; to Wash., Mich., Pa.; probably from Asia. April–July. Confused with *C. Draba*.

8. Lepídium L. Peppergrass

Annual to suffrutescent perennials, glabrous to hirsute with simple hairs. Lvs. entire to bi- or tri-pinnate, sometimes clasping or perfoliate. Fls. minute, in racemes. Pedicels divaricate, terete to flattened. Sepals often pubescent. Petals white or yellow or absent. Stamens 2, 4, or 6. Silicles round, ovate, elliptic, or obovate, reticulate to smooth, glabrous to hirsute, strongly obcompressed, usually notched or lobed at the more or less winged apex. Style lacking or developed. Seeds 2, flattened. Ca. 130 spp., widely distributed; one grown for salad. (Greek, *lepidion*, a little scale, from shape of pods.)

(Hitchcock, C. L. The genus Lepidium in the U.S. Madroño 3: 265–320, 1936.)
A.   Lvs. (at least some of the cauline) perfoliate or with clasping bases.
    B.   Frs. inflated and with conspicuously winged margin; cauline lvs. sagittate at base
                                      1. *L. campestre*
    BB. Frs. not greatly inflated, not winged; cauline lvs. perfoliate .... 2. *L. perfoliatum*
AA. Lvs. not perfoliate or with clasping base.
    B.   Styles none, or less than 0.3 mm. long, usually shorter than notch of fr.
        C.   Silicles notched at apex but not or slightly winged.
            D.   Pedicels terete or only slightly flattened.

E.   Stems glabrous or nearly so; stem-lvs. 1–4 cm. broad, entire or at most
     dentate; stamens 6 .... 3. *L. latifolium*
EE. Stems pubescent or stem-lvs. less than 1 cm. broad, or both; stamens mostly
     2 or 4.
     F.   Sepals persisting until frs. almost mature; pedicels somewhat narrowly
          wing-margined; cauline lvs. pinnatifid to lobed.
          G.   Silicles ovate, prominently reticulate, with 2 acute winged diver-
               gent apical teeth; pedicels winged .... 4. *L. strictum*
          GG. Silicles oval to obovate, not reticulate, with rounded apex;
               pedicels not winged .... 5. *L. oblongum*
     FF. Sepals deciduous along with petals and stamens or soon after; pedicels
          not wing-margined; cauline lvs. mostly entire.
          G.   Petals wanting or shorter than sepals; cotyledons incumbent in
               the seed (with the back of 1 against the radicle).
               H.   Lower lvs. coarsely toothed; silicles 2.5–3.5 mm. long.
                    I.   Branches bearing many short corymbiform racemes in the
                         lf.-axils as well as longer terminal ones
                                                            6. *L. ramosissimum*
                    II. Branches bearing simple elongate naked racemes
                                                            7. *L. densiflorum*
               HH. Lower lvs. pinnatifid with incised-dentate divergent lobes;
                    silicles 1.5 mm. long .... 8. *L. pinnatifidum*
          GG. Petals equaling or exceeding sepals; cotyledons incumbent or ac-
               cumbent (the edges against the radicle) .... 9. *L. virginicum*
DD. Pedicels strongly flattened.
     E.   Petals lacking; plant freely branched from base, pubescent; stamens 2 or 4
                                                            10. *L. lasiocarpum*
     EE. Petals evident; plant usually simple below, subglabrous; stamens 6
                                                            11. *L. nitidum*
CC. Silicles plainly winged at apex with 2 prominent divergent lobes or teeth.
     D.   Petals lacking or up to 1.3 mm. long; silicles 2.5–4.5 mm. long.
          E.   Pedicels much flattened; silicles pubescent .... 12. *L. dictyotum*
          EE. Pedicels slender; silicles glabrous .... 13. *L. oxycarpum*
     DD. Petals 2–4 mm. long; silicles 5.5–7 mm. long .... 14. *L. latipes*
BB. Styles well developed, at least 0.3 mm. long and exceeding notch of frs.
     C.   Fls. white.
          D.   Plants perennial, somewhat woody at base, glabrous or puberulent.
               E.   Fls. 4 mm. long; silicles 5–6 mm. long .... 16. *L. Fremontii*
               EE. Fls. 2 mm. long; silicles 2 mm. long .... 17. *L. montanum*
          DD. Plants annual or biennial, hirsute to villous .... 19. *L. Thurberi*
     CC. Fls. yellow; plants annual.
          D.   Silicles notched at apex; plants prostrate or nearly so .... 15. *L. flavum*
          DD. Silicles not notched at apex; plants erect ... . 18. *L. Jaredii*

1. **L. campéstre** (L.) R.Br. [*Thlaspi c.* L.] Cow-Cress. Densely short-villous annual
or biennial, simple to branched, 1.5–7 dm. high; lvs. oblanceolate, the basal 4–12 cm.
long, ca. 1 cm. wide, pinnatifid, lyrate to entire, petioled, the cauline oblong-oblanceolate,
denticulate, suberect, overlapping, sagittate-clasping; pedicels slender, slightly flattened,
5–8 mm. long; sepals ca. 1.5 mm. long, villous to subglabrous; petals white or yellowish,
ca. 2 mm. long; silicles winged on margin and apex, oblong-ovate, 5–6 mm. long,
slightly emarginate, concave above, often papillose; style 0.2–0.6 mm. long; seeds dark
brown, ca. 2 mm. long, the cotyledons incumbent; $n = 8$ (Wulff, 1939).—Infrequent
weed about waste and disturbed places, mostly above 3000 ft.; Plumas, Placer, Eldorado, Trinity, and Siskiyou cos.; to Wash., Atlantic Coast; natur. from Eu. May–July.

2. **L. perfoliàtum** L. Shield-Cress. Erect annual, mostly glabrous, branched, 2–5
dm. high; lower lvs. bipinnatifid into linear segms., the middle cauline entire, auriculate,
the upper perfoliate; pedicels spreading, slender, terete, 4–8 mm. long; sepals pilose,
ca. 1 mm. long; petals narrow, slightly longer, yellow; silicles rhombic-ovate, ca. 4
mm. long, minutely notched; style about equal to sinus; seeds elliptic-oblong, almost 2
mm. long, narrow-winged, the cotyledons incumbent; $n = 8$ (Jaretzky, 1929).—
Sparingly but widely natur. in Calif. below 7000 ft.; to Wash., Atlantic Coast; from
Eu. March–June.

3. **L. latifòlium** L. Perennial, 4–10 dm. tall, from widely spreading underground
root system, glabrous or nearly so; lvs. entire to dentate, the basal oblong, 1–3 dm.
long, 5–8 cm. wide, long-petioled, the cauline reduced, the upper subsessile; racemes
many-fld., compound; pedicels terete, slender, 3–4 mm. long; sepals oval, less than 1
mm. long, somewhat pilose; petals white, spatulate, ca. 1.5 mm. long; silicles round-
ovate, somewhat pilose, ca. 2 mm. long, not or scarcely emarginate; stigma almost
sessile; seeds ellipsoid, strongly compressed, not winged; $2n = 24$ (Manton, 1932).—

Occasional weed in waste places, etc. (Stanislaus, San Joaquin, Santa Clara, Orange cos.); Mex., New Eng.; native of Eu. June–Aug.

4. **L. stríctum** (Wats.) Rattan. [*L. oxycarpum* var. *s.* Wats. *L. pubescens* auth., not Desv. *L. reticulatum* Howell, not Thell.] Pubescent annual, erect to spreading with branches 0.5–2 dm. long; lvs. bipinnatifid to laciniately lobed with linear divisions, basal lvs. 3–7 cm. long, glabrous or nearly so; racemes crowded, 3–5 cm. long; pedicels divergent, somewhat flattened and wing-margined, 1–2 mm. long; sepals scarcely 1 mm. long, pilose, persistent; petals minute; silicles ovate to rounded, 2.2–3 mm. long, distinctly reticulate, biconvex, with small wings at apex and open sinus; style 0; seeds ca. 1 mm. long with incumbent cotyledons; $2n = 16$ (Manton, 1932).—Common in hard beaten soil, along paths, etc., and particularly in cent. Calif., to Humboldt and Los Angeles cos.; to Ore., Utah; apparently introd. from S. Am. March–May.

5. **L. oblóngum** Small. [*L. Greenei* Thell. *L. bipinnatifidum* auth., not Desv.] Much branched diffuse annual, 0.5–2 dm. tall, hirtellous to villous; lvs. pinnatifid to laciniately lobed or cleft, the basal lvs. ca. 3 cm. long with lobed pinnae, the cauline smaller, laciniate with central rachis to 4 mm. wide; racemes many, 6–9 cm. long; pedicels somewhat flattened, scarcely wing-margined; sepals slightly over 1 mm. long, not long-persistent; petals 0 or minute; silicles glabrous or sparsely pectinate, flat, oval to oblong-obovate, 2.5–3.5 mm. long, indistinctly reticulate, narrow-winged, the apices rounded with a small v-shaped sinus; stigma subsessile; seeds ca. 1 mm. long with incumbent cotyledons.—Well distributed in s. Calif., occasional to cent. Calif., Channel Ids.; apparently a S. Am. native. March–May.

6. **L. ramosíssimum** A. Nels. Much branched, puberulent, apparently biennial, 1.5–5 dm. tall; lower lvs. few-toothed, the upper linear, entire; infl. with many short corymbiform racemes in lf.-axils as well as the longer terminal racemes; pedicels somewhat wing-margined; sepals ca. 1 mm. long, pubescent; petals linear, shorter; silicles 2.5–3.5 mm. long, elliptic, shallowly notched and winged at apex; stigma sessile; cotyledons incumbent.—Reported from Chollas V., San Diego Co.; Rocky Mts.

7. **L. densiflórum** Schrad. [*L. ruderale* and *L. apetalum* of many auth.] Diffuse annual, 3–5 dm. tall, puberulent to pubescent; lvs. mostly oblanceolate, the basal 4–6(–8) cm. long, coarsely toothed, the divisions also toothed, the cauline entire to somewhat toothed; racemes many, 6–15 cm. long; pedicels scarcely flattened; sepals ca. 1 mm. long, usually somewhat pilose; petals mostly lacking; silicles round-obovate to elliptic-ovate, glabrous to puberulent, ca. 2.5 mm. long, narrowly notched with somewhat winged apex; style obsolete; cotyledons incumbent; $2n = 32$ (Manton, 1932).—Yosemite V.; common e. of Rocky Mts.

KEY TO VARIETIES

Silicles averaging ca. 2.5 mm. long, elliptic-ovate to round-obovate; pedicels slightly flattened
                                                                        L. densiflorum
Silicles averaging 3–3.5 mm. long, oblong-obovate; pedicels rather flat.
  Pedicels flattened chiefly on lower side, not twice as broad as thick.
    Silicles glabrous .... var. *Bourgeauanum*
    Silicles pubescent .... var. *pubicarpum*
  Pedicels flattened on upper and lower sides, about twice as broad as thick .... var. *ramosum*

Var. **Bourgeauànum** (Thell.) C. L. Hitchc. [*L. B.* Thell.] Pedicels flat; silicles oblong, glabrous, ca. 3 mm. long.—Colusa Co., Siskiyou Co.; to Wash., Alaska, Rocky Mts.

Var. **pubicárpum** (A. Nels.) Thell. [*L. p.* A. Nels.] Pedicels not much flattened; silicles pubescent, 3–3.5 mm. long, oblong-obovate; $2n = 32$ (Manton, 1932).—Lassen and Shasta cos. n.; to Wash., Mont., Wyo.

Var. **ramòsum** (A. Nels.) Thell. [*L. r.* A. Nels.] Pedicels flattened, almost twice as broad as thick; silicles ca. 3.5 mm. long, glabrous.—Reported from Barstow, Mojave Desert; to Wyo., New Mex.

8. **L. pinnatífidum** Ledeb. Erect, annual or biennial, subglabrous, branched, leafy, 2–4 dm. high; basal lvs. broadly lanceolate in outline, 5–8 cm. long, pinnatifid with incised-dentate divergent lobes; cauline lvs. smaller, less divided; racemes terminal and axillary; pedicels pubescent, slender, terete, divergent, 3–4 mm. long; sepals lance-ovate to subelliptic, white-margined, setulose, 1 mm. long; petals rudimentary; silicles broadly- to ovate-elliptic, minutely wing-margined, obtusely emarginate, ca. 1.5

mm. long; seeds compressed, ovoid-ellipsoid, somewhat tuberculate, brownish, less than
1 mm. long.—Reported from along a railroad, Oak View, Ventura Co., in 1947 and
from saline soil, Smeltzer, Orange Co., in 1931; native of Old World. May.

9. **L. virgínicum** L. Freely branched annual, 1.5–6 dm. tall, glabrous to minutely
pubescent; basal lvs. incised to pinnate, 5–15 cm. long; cauline ascending, lanceolate
to linear, entire to incised, reduced; racemes numerous, many-fld., elongate; pedicels
slender, terete, usually somewhat longer than frs.; sepals ca. 1 mm. long; petals white,
equaling or exceeding sepals; silicles usually somewhat longer than broad, 2.5–4 mm.
long, scarcely margined, shallowly notched at apex; style almost lacking; seeds 1–1.5
mm. long, with accumbent cotyledons; $2n = 32$ (Manton, 1932).—Occasional introd.
in Calif. from E. Coast; reported from Napa Co.

Cotyledons accumbent .... *L. virginicum*
Cotyledons incumbent to oblique.
  Stems and pedicels pubescent.
    Cauline lvs. simple to incised; plants mostly 3–6 dm. tall .... var. *pubescens*
    Cauline lvs. parted or with narrow lobes; plants 1–2 dm. tall .... var. *Robinsonii*
  Stems and pedicels glabrous .... var. *medium*

Var. **pubéscens** (Greene) Thell. [*L. intermedium* var. *p.* Greene. *L. medium* var. *p.*
Rob. *L. v.* ssp. *texanum* Thell. and *L. t.* auth., not Buckl. *L. bernardinum* Abrams.] The
common form in Calif., widespread in waste places, along roadsides, etc., below 7000 ft.;
many Plant Communities; to B.C. and Rocky Mts. March–Aug.

Var. **Robinsònii** (Thell.) C. L. Hitchc. [*L. R.* Thell. *L. californicum* Nutt.] Occasional
at low elevs.; Chaparral, Coastal Sage Scrub; cismontane s. Calif. from Los Angeles Co.
s., Channel Ids.; L. Calif. Jan.–April.

Var. **mèdium** (Greene) C. L. Hitchc. [*L. m.* Greene. *L. intermedium* Gray, not Rich.]
N. Calif. (Humboldt Co.); to Wash., Wyo., Okla.

10. **L. lasiocárpum** Nutt. Branched spreading annual, the branches 0.5–2.5 dm.
long, hirsute-hispid; lvs. linear to oblanceolate, toothed to pinnatifid, 1–2 cm. long or
the lower to 6 cm. long and petioled; racemes 3–8 cm. long; pedicels distinctly flattened
on both surfaces, 2–5 mm. long, usually pubescent on lower side; sepals ca. 1 mm.
long; petals narrow, usually shorter than sepals, sometimes lacking; silicles subglabrous
to hispid, suborbicular to somewhat longer than wide, 3–4.5 mm. wide, finely reticulate,
slightly winged at apex; style almost or quite lacking; seeds ca. 1.2 mm. long, with
incumbent cotyledons.—Common on grassy slopes and sandy flats, below 5000 ft.;
Creosote Bush Scrub, Shadscale Scrub, deserts from Inyo Co. to Imperial Co., less
frequent in cismontane s. Calif.; V. Grassland, Coastal Sage Scrub, Santa Barbara to
L. Calif.; Coastal Strand, Marin Co.; to Utah, Ariz. Feb.–May. Intergrading with:

Var. **georgìnum** (Rydb.) C. L. Hitchc. [*L. g.* Rydb. *L. lasiocarpum* ssp. *g.* Thell.]
Plants fewer-branched, less harshly pubescent; pedicels glabrous on lower side.—
Occasional on deserts; to Utah, Ariz. April–May.

11. **L. nítidum** Nutt. [*L. leiocarpum* H. & A. *L. n.* var. *insigne* Greene.] Annual,
usually erect and simple at base, sometimes with spreading branches from base; stems
glabrous to moderately pubescent, 0.5–4 dm. long; lower lvs. 3–10 cm. long, pinnately
parted into narrow segms., the cauline smaller, pinnatifid to entire; racemes rather lax
in fr.; pedicels densely puberulent, very much flattened; sepals ovate, ca. 1 mm. long;
petals spatulate, 0.5–1.5 mm. long; silicles ovate to suborbicular, convex below, somewhat concave above, glabrous, 3.5–6 mm. long, without divergent apices, the margins
upturned; stigma subsessile; seeds ca. 2 mm. long, with incumbent cotyledons.—Common
on open places, below 3000 ft., throughout the state except on the desert; Chaparral,
V. Grassland, Coastal Sage Scrub, etc.; to Wash. and L. Calif. Feb.–May.

Var. **Howéllii** C. L. Hitchc. Stems densely pubescent; silicles minutely pubescent on
margins.—Creosote Bush Scrub; w. Mojave Desert.

Var. **oregànum** (Howell) C. L. Hitchc. [*L. o.* Howell. *L. strictum* var. *o.* Rob.] Stems
glabrous or inconspicuously pubescent; silicles with distinct divergent apices prolonged
beyond the general oval contour.—Occasional in Coast Ranges, San Luis Obispo Co.
to Siskiyou Co.; s. Ore. Possibly a hybrid of *L. dictyotum* and *L. nitidum*.

12. **L. dictyòtum** Gray. [*L. d.* var. *macrocarpum* Thell. *L. acutidens* var. *microcarpum*
Thell.] Low pubescent annual, the branches decumbent to ascending, 0.2–2 dm. long;
lower lvs. usually pinnatifid with 2–5 pairs of linear lobes, these sometimes cleft;

cauline lvs. mostly entire, linear; racemes many-fld., usually rather compact; pedicels flattened, 1.5–3.5 mm. long, somewhat pubescent; sepals ca. 1 mm. long, pubescent; petals usually lacking; silicles mostly ca. 3.5 mm. long, glabrous to hairy, reticulate, mostly ovate in outline, with winged apices less than 1 mm. long, usually rounded or obtuse, sometimes acute, not divergent; style lacking; seeds ca. 2 mm. long, with incumbent cotyledons.—More or less alkaline places, below 2500 ft.; Alkali Sink, V. Grassland; San Joaquin V., Livermore and Salinas valleys to San Diego Co.; to Wash., Utah. March–May.

Var. **acùtidens** Gray. [*L. a.* Howell. *L. oxycarpum* var. *a.* Jeps.] Racemes often loosely fld.; silicles mostly over 4 mm. long, with acuminate divergent winged apices over 1 mm. long.—With the sp., but less common; from San Diego to Sacramento V. and Lassen and Siskiyou cos.; to Ore. April–May.

13. **L. oxycárpum** T. & G. Annual, slender, erect to diffuse, 0.5–2 dm. high, subglabrous to pubescent; lvs. 2–6 cm. long, the basal often with 2–4 pairs of linear lobes, the cauline usually linear, entire; racemes rather lax, comprising ca. half the length of the branches; pedicels 2.5–5 mm. long, slightly flattened, but slender; sepals ca. 0.5 mm. long; petals 0.5 mm. long to lacking; silicles ovate, glabrous, finely reticulate, 2.5–3.5 mm. long, the apex widened by the winged widely divergent obtusish lobes less than 1 mm. long; seeds ca. 1 mm. long, with incumbent cotyledons.—Saline flats and alkaline valley floors, below 1500 ft.; V. Grassland and edge of Coastal Salt Marsh; largely about San Francisco Bay, but extending from Yolo and Sonoma cos. to San Benito Co. March–May.

14. **L. látipes** Hook. [*L. Brownii* Heller.] Low spreading pubescent annual, branched from base, the stems 0.3–1(–2) dm. long; lvs. 5–10(–14) cm. long, at least the basal pinnatifid into 3–10 pairs of linear entire or dissected divisions, the upper entire, almost linear; racemes mostly 2–6 cm. long, very dense, almost capitate; pedicels 2–3 mm. long, very wide and flat; sepals pubescent, ca. 1.3 mm. long; petals 2–4 mm. long, greenish; silicles coriaceous, glabrous to pubescent, oblong-ovate, 5.5–7 mm. long, with acute winged apices ca. 2 mm. long and very narrow sinus; seeds ca. 1.6 mm. long, with incumbent cotyledons.—Alkaline flats and beds of winter pools, below 2000 ft.; largely V. Grassland; San Diego to Humboldt cos.; Santa Cruz Id. March–May.

15. **L. flàvum** Torr. [*L. f.* var. *apterum* Henr. & Thell. *Sprengeria minuscula* Greene. *S. f.* Greene.] Prostrate or decumbent glabrous yellowish green annual, the branches 0.3–3 dm. long; basal lvs. spatulate to lanceolate, 2–5 cm. long, irregularly lobed or pinnatifid, the cauline more cuneate, somewhat smaller, entire to pinnatifid; racemes subcapitate to looser; pedicels 2–3 mm. long, terete or nearly so; sepals oblong, yellow-green, ca. 1–1.3 mm. long; petals sulphur-yellow, ca. 2 mm. long; silicles oval, 2–3 mm. long, glabrous, reticulate with distinct divergent apices; style 1.5–2 mm. long; seeds ca. 1 mm. long, with incumbent cotyledons.—Common in washes and semialkaline flats, below 4300 ft.; Creosote Bush Scrub, Joshua Tree Wd.; deserts from Inyo Co. to Imperial Co.; to Nev., L. Calif. March–May.

Var. **felipénse** C. L. Hitchc. Silicles 3–4.5 mm. long, suborbicular, with small winged apices; style 1–1.5 mm. long.—Creosote Bush Scrub; Borrego V., e. San Diego Co.

16. **L. Fremóntii** Wats. Rounded suffrutescent perennial with many branching stems 2–5 dm. high, glabrous and glaucous; lvs. linear, acute, 2–5(–10) cm. long, entire or pinnatifid into few salient elongate lobes; infl. much branched, somewhat leafy; pedicels 5–8 mm. long, slender; sepals 1.5–2 mm. long, glabrous; petals white, ca. 3 mm. long; silicles broadly ovate to obovate, 4–7 mm. long, faintly nerved, with wide winged margins; styles 0.4–0.8 mm. long; seeds almost 2 mm. long, with incumbent cotyledons.—Common in rocky and sandy places, below 5000 ft.; Creosote Bush Scrub, Joshua Tree Wd.; Inyo Co. to n. Riverside Co.; to Utah, Ariz. March–May.

17. **L. montànum** Nutt. ssp. **canéscens** (Thell.) C. L. Hitchc. [*L. scopulorum* f. *c.* Thell.] Biennial or perennial, sometimes in fl. the first year, 2–3 dm. tall, 1- to few-stemmed, sparsely to densely villous with hairs 5–6 times as long as thick; basal lvs. pinnately divided into entire segms. 1–2.5 mm. wide; cauline lvs. few, reduced, entire to lobed; racemes mostly 2–4 cm. long and many-fld.; pedicels terete, slender, 5–8 mm. long; sepals whitish, ca. 1 mm. long; petals white, ca. 2 mm. long; silicles ovate, ca. 2.5 mm. long, glabrous, with very narrow winged margin above and minute apical notch; styles 0.3–0.5 mm. long; seeds ca. 1.2 mm. long, with incumbent cotyledons.—Dry some-

what bushy flats, 4500–6500 ft.; mostly Sagebrush Scrub; Mono Co. to Siskiyou and Modoc cos.; to Ore., Utah. May–July.

Ssp. **cinèreum** (C. L. Hitchc.) C. L. Hitchc. [*L. m.* var. *canescens* f. *cinereum* C. L. Hitchc. *L. alyssoides* Calif. auth., not Gray.] Densely cinereous with hairs ca. twice as long as thick; silicles 3–3.5 mm. long.—Dry alkaline places, 2600–5000 ft.; Creosote Bush Scrub to Pinyon-Juniper Wd.; New York Mts. and Mesquite V. near Kingston, e. Mojave Desert; nw. Ariz. April–May.

18. **L. Jarédii** Bdg. Simple to diffusely branched annual, 1–6 dm. tall, from glabrous below to pubescent; lvs. lanceolate, 3–10 cm. long, mostly entire, some with few teeth; racemes lax, 5–20 cm. long; pedicels terete, ca. 1 cm. long; sepals yellow, pilose, ca. 2.5 mm. long; petals sulphur-yellow, ca. 2.5 mm. long; silicles ovate, 3–4 mm. long, not emarginate; styles 0.5–1 mm. long.—Heavy soil, washes and slopes; V. Grassland; inner S. Coast Ranges, w. Fresno Co. to se. San Luis Obispo Co. April–May.

19. **L. Thúrberi** Woot. Annual, 1–6 dm. tall, branched, canescent-pubescent, with longer and shorter hairs; basal lvs. 3–6 cm. long, petioled, pinnatifid with 3–8 pairs of segms. which are usually lobed to parted, the cauline reduced, pinnatifid to entire; racemes many-fld.; sepals 1–1.5 mm. long; petals white, 2–3 mm. long; silicles glabrous, round-ovate, 2–3 mm. long, narrowly wing-margined near apex, with shallow notch; style 0.3–0.5 mm. long.—Reported from Marsh Hot Springs, Sonoma Co., and Barstow, San Bernardino Co.; New Mex. and Ariz.

9. Corónopus Trev. WART-CRESS

Strong-smelling diffuse or prostrate annuals or biennials, pubescent with simple hairs. Lvs. pinnately parted. Fls. minute, greenish-white, the capitate clusters elongating in fr. into short racemes. Sepals oval, spreading. Stamens 2 or 4. Silicles flattened contrary to the narrow partition, the 2 valves strongly wrinkled or tuberculate, 1-seeded, indehiscent. Styles not evident. Seeds with narrow incumbent cotyledons. Ca. 6 spp. of wide distribution. (Greek, *korone*, crown, and *pous*, foot, from the deeply cleft lvs.)

1. **C. dídymus** (L.) Sm. [*Lepidium d.* L. *Senebiera d.* Pers. *Carara d.* Britton.] Stems 1.5–2 dm. long, leafy, somewhat hairy; lvs. 1–2 cm. long, with narrow divisions; pedicels 2–3 mm. long; fls. less than 1 mm. long; silicles notched, 1 mm. long, 2 mm. wide, rough-wrinkled; $2n = 32$ (Manton, 1932).—Occasional weed through much of the state; to Atlantic Coast, natur. from Eu. March–July.

2. **C. procúmbens** Gilib. Lvs. less divided; silicles tuberculate, not notched; $n = 16$ (Jaretzky, 1929).—Reported from San Francisco; native of Eu.

10. Ionopsídium Rchb. DIAMOND FLOWER

Small annuals, tufted. Lvs. suborbicular or cordate, long-petioled, entire, basal. Fls. axillary, solitary on scapelike pedicels, violet or sometimes white. Sepals oblong, obtuse, concave above. Petals round-oblong, clawed. Stamens 6, tetradynamous. Silicles globular-oblong, the septum at right angles to the valves, lightly crenulate-winged on margin. Styles very short. Seeds several, somewhat truncate-ovoid, granular, with incumbent cotyledons. (Greek, *ion*, violet, and *opsis*, appearance, *i.e.* violetlike.)

1. **I. acaùle** (Desf.) Rchb. [*Cochlearia a.* Desf.] Almost stemless, 7–10 cm. high; lvs. ca. 10–12 mm. wide; petals ca. 1 cm. long; silicles 4–5 mm. long; $n = 12$ (Chiarugi, 1928).—Adv. at Ferndale, Humboldt Co.; native of Portugal. Cult. in gardens.

11. Thláspi L. PENNY-CRESS

Low erect annual or perennial glabrous herbs. Lvs. undivided, the cauline sagittate and clasping. Fls. small, white or purplish, in terminal racemes. Sepals short, oval, obtuse. Petals oblanceolate to obovate. Stamens 6; anthers short, oval. Silicles rounded or obovate, or obcordate, flattened contrary to the narrow partition, the midrib of the valves extending into a wing. Seeds 2–8 per locule; cotyledons accumbent. Ca. 60 spp. of wide distribution. (Greek, *thlaein*, to crush, from the flattened silicle.)

(Payson, E. B. The genus Thlaspi in North America. Univ. Wyo. Pub. Sci. 1: 145–163, 1926.)
Plants annual; silicles rounded, deeply notched; seeds with concentric ridges . . . . 1. *T. arvense*
Plants perennial; silicles cuneate-obovate, scarcely or not notched; seeds smooth . . . . 2. *T. glaucum*

1. **T. arvénse** L. Erect, 2–5 dm. high; lowest lvs. petioled, narrowly obovate, 3–5 cm. long, the cauline oblong, sessile, clasping, somewhat toothed, 2–4.5 cm. long; pedicels spreading, slender, commonly 6–8 mm. long; sepals white-margined, 1.5–2 mm. long; petals white, 3–3.5 mm. long; silicles suborbicular to round-oblong, 1–1.8 cm. long, broadly winged, deeply emarginate; style almost lacking; seeds compressed, oblong, blackish, concentrically ridged, 2–2.3 mm. long; $2n = 14$ (Manton, 1932).—Occasional adv. in waste places, as in Modoc Co., San Gabriel Mts., Pasadena, San Diego, Tule Lake, etc.; natur. from Eu. May–Aug.

2. **T. glaùcum** A. Nels. var. **hespérium** Pays. [*T. alpestre* auth. *T. califórnicum* Wats.] Perennial, loosely cespitose; stems rather slender, 1–2.5 dm. high; lower lvs. oblanceolate to almost obovate, petioled, 2–4 cm. long, entire to toothed; cauline lvs. few, round-ovate to oblong, scarcely half as long as internodes, clasping; racemes short-pedunculate, to ca. 1 dm. long; pedicels slender, 4–9 mm. long; sepals ca. 2 mm. long, white-margined; petals white, 4–5 mm. long; silicles usually obovate or obcordate, 4–7 mm. long, cuneate at base, truncate to obtusish at apex, narrowly winged; styles slender, 1.5–2 mm. long; seeds dark brown, compressed, ca. 2 mm. long, faintly reticulate.—Dry rocky slopes, below 6500 ft.; Mixed Evergreen F., Yellow Pine F.; Humboldt, Siskiyou, Trinity, and Modoc cos.; to Wash., Ida., Nev. May–July.

12. Sisýmbrium L.

Ours annuals or biennials, erect, mostly branched. Lvs. dentate to pinnatifid or finely dissected. Fls. yellow or white, small, in terminal racemes. Pubescence when present of simple hairs. Sepals spreading, oblong to linear. Stamens 6. Siliques cylindric and prismatic, or long-subulate, the valves 1–3-nerved, dehiscent. Stigma 2-lobed. Seeds oblong, marginless, in 1 row in each cell, with incumbent cotyledons. A genus of some size as here recognized; natives of the temp. parts of the world; ours all introd. from Eu. (Greek name for some crucifer.)

Siliques closely appressed, 1–1.5 cm. long; lvs. pinnatifid . . . . 1. *S. officinale*
Siliques not appressed, 4–10 cm. long.
    Fruiting pedicels ca. as thick as the siliques; petals 6–9 mm. long.
        Upper lvs. pinnatifid with linear divisions; siliques ca. 1 mm. wide . . . . 2. *S altissimum*
        Upper lvs. entire to hastate; siliques ca. 2 mm. wide . . . . 3. *S. orientale*
    Fruiting pedicels more slender than the siliques; petals 3–4 mm. long . . . . 4. *S. Irio*

1. **S. officinàle** (L.) Scop. [*Erysimum o.* L.] HEDGE-MUSTARD. Annual, stiffly erect with few widely divaricate branches, ± hirsute especially near base, 2–10 dm. tall; lvs. of basal rosette lyrate-pinnatifid, hirsute, 5–10 cm. long, the upper hastate with long narrow sinuate-dentate terminal lobe; racemes long and narrow; pedicels erect in fr., ca. 2 mm. long; sepals ca. 2 mm. long; petals yellowish, 3 mm. long; siliques closely appressed, 10–15 mm. long, acuminate; seeds dark brown, ovoid, ca. 1 mm. long; $n = 7$ (Wulff, 1937).—Common weed of waste places and waysides, mostly below 5000 ft.; natur. from Eu. April–July.

2. **S. altíssimum** L. [*Norta a.* Britton] TUMBLE-MUSTARD. Erect annual, branched above, 5–10 dm. high, glabrous or nearly so; lower lvs. 10–15 cm. long, runcinate-pinnatifid with lanceolate lateral lobes and a large subdeltoid terminal one, petioled; upper lvs. reduced, pinnatifid into linear divisions; pedicels spreading-ascending, 5–8 mm. long; sepals 4–5 mm. long; petals yellowish-white, 6–8 mm. long; siliques spreading, rigid, narrow-cylindric, 5–10 cm. long, scarcely 1 mm. thick; seeds 0.6–0.8 mm. long; $2n = 14$ (Manton, 1932).—Common weed in waste places, below 7500 ft., throughout the state, especially on the Mojave Desert; native of Eu. May–July.

3. **S. orientàle** L. Annual or biennial, branched, 2–6 dm. high, ± hirsute-pubescent; lvs. pinnate or the upper pinnatifid, petioled, 3–12 cm. long, with terminal lobe hastate, lance-linear to ovate, the lateral lobes paired; racemes long, lax in fr.; pedicels stout, 3–10 mm. long, ascending; sepals 4 mm. long; petals yellow, 7–8 mm. long; siliques ascending-spreading, 4–9 cm. long, 1–1.5 mm. thick; seeds scarcely 1 mm. long.—Roadside and sidewalk weed, San Francisco, Monterey, Ventura, Dry Morongo Wash, San Diego; natur. from Eu. May.

4. **S. ìrio** L. [*Norta I.* Britton.] LONDON-ROCKET. Erect annual, 2–8 dm. tall, glabrous, branched above; lower lvs. 10–15 cm. long, runcinate-pinnatifid, with lanceolate lateral

lobes and a large subdeltoid terminal one; upper lvs. reduced; racemes long, many-fld.; pedicels filiform, ascending, 6–10 mm. long; sepals 2–2.5 mm. long; petals yellow, 3–4 mm. long; siliques ascending, 3–4.5 cm. long, less than 1 mm. thick; seeds light brown, shining, oblong-ovoid, 0.6–0.8 mm. long.—Common weed in orchards, waste places, etc., cismontane s. Calif.; occasional on desert, reported also from Stanislaus and Glenn cos.; natur. from Eu. Jan.–April.

13. Arabidópsis Heynh.

Annual or perennial herbs with slender erect stems and some forked hairs. Basal lvs. petioled, in rosettes, the cauline short-petioled to sessile or clasping, simple, entire to toothed. Fls. small, white or purplish, sometimes yellow, in terminal racemes. Siliques terete, with septum-midrib so broad and thin as to be wholly obscure, the valves rounded, dehiscent. Seeds in 1 or 2 ranks in each cell, numerous, ovoid, with incumbent cotyledons. Ca. 12 spp. of the N. Hemis. (Greek, *Arabis,* and *opsis,* aspect, because of resemblance to *Arabis.*)

1. **A. Thaliàna** (L.) Heynh. [*Arabis T. L. Sisymbrium T.* J. Gay.] Simple or branched annual, hairy at base, 0.5–4 dm. high; basal lvs. oblanceolate, or oblong, entire to obscurely toothed, petioled, 1–3.5 cm. long; upper lvs. remote, small, sessile; racemes lax in fr.; pedicels slender, spreading, 5–12 mm. long; sepals ca. 1–1.5 mm. long; petals white, 3–4 mm. long; siliques very slender, 1–1.5 cm. long; stigma sessile; seeds barely 0.5 mm. long; $n = 5$ (Jaretzky, 1928).—Reported as weed from Trinity, Alameda, and Lake cos.; natur. from Eu. April–May.

14. Halimolòbos Tausch

Biennial or perennial herbs with terete mostly erect pubescent or glabrous stems, the pubescence mostly branched. Lvs. simple, entire to deeply lobed, the basal often caducous, the cauline usually differentiated. Racemes leafless. Sepals oblong, erect, pubescent. Petals white to yellow, spatulate, with narrow claw. Stamens 6; anthers small. Siliques terete, on slender pedicels, not beaked. Stigma capitate. Seeds many, mostly biseriate, ellipsoid, wingless, crowded, with mostly incumbent cotyledons. Ca. 14 spp. of N. and S. Am. (Greek, *alimos,* of the sea, and *lobos,* pod, name used because of the resemblance to *Alyssum halimifolium.*)

(Rollins, R. C. Generic revisions in the Cruciferae: Halimilobos, Contr. Dudley Herb. 3: 241–265, 1943.)

1. **H. diffùsa** (Gray) O. E. Schulz var. **Jàegeri** (Munz) Roll. [*Sisymbrium d.* var. *J.* Munz.] Diffusely branched perennial, suffrutescent, cinereous-tomentose; stems leafy, 3–5 dm. tall; lower lvs. 5–8 cm. long, cinereous, sharply and deeply sinuate-toothed or -lobed, on short, winged petioles; upper lvs. gradually reduced, sessile; racemes numerous, 5–10 cm. long in fr. with spreading pedicels 4–7 mm. long; sepals 2.5–3.5 mm. long; petals white, 3.5–4 mm. long; siliques widely spreading, pubescent, 1.5–2 cm. long, less than 1 mm. thick, somewhat torulose, with slender beak 1.5–2 mm. long; seeds ca. 0.6 mm. long.—Dry rocky places, 5000–8000 ft.; Sagebrush Scrub, Pinyon-Juniper Wd.; White Mts. and Alabama Hills, Inyo Co. to New York, Providence, and Clark mts., San Bernardino Co.; w. Nev. May–Sept.

15. Descuraìnia Webb & Berthel. TANSY-MUSTARD

Annual or biennial herbs, erect, branched especially above. Lvs. ovate to obovate or oblanceolate in outline, 1–2–3-pinnate, ultimately finely or coarsely dissected, the basal in a rosette withering early. Pubescence stellate or forked. Fls. racemose, yellow or whitish, small. Siliques linear-cylindric to subclavate, somewhat torulose, the valves opening from below upward, 1-nerved. Style short or obsolete; stigma entire. Seeds in 1 or 2 series in each locule, elliptic, yellowish to brown. Ca. 20 spp. of temp. Eurasia and Am. (Named for F. *Descourain,* 1658–1740, French botanist.)

(Detling, L. R. E. A revision of the N. Am. spp. of Descurainia. Am. Midl. Nat. 22: 481–520, 1939.) Upper lvs. 2–3-pinnate, mostly into linear segms.; siliques 10–30 (typically ca. 20) mm. long;

silique-septum with 2–3 longitudinal nerves .... 1. *D. Sophia*
Upper lvs. pinnate, the lfts. often deeply incised; siliques mostly less than 15 mm. long; silique-septum
1-nerved.
  Seeds in 1 row in each locule of silique; style conspicuous.
    Siliques 9–15 mm. long; stems glandular-pubescent, especially above ...... 2. *D. Richardsonii*
    Siliques 3–7 mm. long; stems without gland-tipped hairs .... 3. *D. californica*
  Seeds in 2 rows in each locule; style nearly or quite obsolete.
    Siliques linear, 12–20 mm. long .... 4. *D. obtusa*
    Siliques clavate to oblong-elliptic, mostly less than 12 mm. long .... 5. *D. pinnata*

1. **D. Sòphia** (L.) Webb. [*Sisymbrium S. L. Sophia s.* Britton.] Leafy branched annual, 2.5–6 dm. high, stellate-pubescent; lvs. 2–9 cm. long, 2–3-pinnate with fine linear to oblanceolate segms.; pedicels divaricate-ascending, 7–14 mm. long; sepals mostly equal to or exceeding the greenish-yellow petals, 2–2.5 mm. long; siliques linear, often curved, 1–3 cm. long, ca. 1 mm. thick, loosely ascending; style very short; seeds 10–20 in each cell, oblong-ellipsoid, ca. 0.8 mm. long; $2n = 28$ (Manton, 1932).—Occasional weed in dry waste places, below 8000 ft., from many widely scattered localities in Calif.; to Atlantic Coast; natur. from Eu. May–Aug.

2. **D. Richardsònii** (Sweet) O. E. Schulz ssp. **viscòsa** (Rydb.) Detl. [*Sophia v.* Rydb.] Slender biennials, pubescent with mixed simple and stellate hairs, also glandular-pubescent, 3–12 dm. high, usually branched above; lvs. 2–10 cm. long, the lower pinnate or again pinnatifid, the ultimate segms. rather broad, obtuse, the upper pinnate with simple or toothed segms.; pedicels 6–10 mm. long, divaricate-spreading; sepals 1.5–2.5 mm. long; petals bright yellow, 2–3.5 mm. long; siliques 9–15 mm. long, linear, straight or curved upward, short-beaked; seeds uniseriate, 4–14 per locule, red-brown, oblong-elliptic, 1–1.2 mm. long; $n = 7$ (Baldwin and Campbell, 1940).—Dry benches and slopes, 5000–11,000 ft.; Yellow Pine F. to Subalpine F.; San Bernardino Mts., Sierra Nevada, to Trinity, Siskiyou and Modoc cos.; to Wash., Alta., New Mex. May–Aug.

Ssp. **incìsa** (Engelm.) Detl. [*Sisymbrium i.* Engelm. *Descurainia i.* Britton. *S. i.* var. *Sonnei* Rob. *Sophia S.* Greene.] Subglabrous to moderately pubescent, nonglandular; petals 1.5–2 mm. long; seeds ca. 0.8 mm. long; $n = 21$ (Baldwin and Campbell, 1940).— Dry disturbed and rocky places, 4000–10,000 ft.; Yellow Pine F. to Lodgepole F.; Sierra Nevada from Lassen, Nevada, Tulare and Inyo cos., to mts. of s. Calif.; to L. Calif., Son. and Mont. June–Aug.

3. **D. califórnica** (Gray) O. E. Schulz. [*Smelowskia c.* Gray.] Biennial, 3–8 dm. tall, moderately pubescent, not glandular, openly branched; lvs. 2–6 cm. long, the lower pinnate with 2–4 pairs of lanceolate entire to incised pinnae; sepals 1–1.5 mm. long, yellow or greenish; petals slightly longer, yellow; pedicels spreading, 3–7 mm. long; siliques 3–7 mm. long, narrowed toward both ends, ± erect; style prominent, 0.5–0.7 mm. long; seeds uniseriate, 1–3 per locule, elliptic, brownish, 1–1.5 mm. long.—Dry slopes, 7000–11,000 ft.; Montane Coniferous F., mostly e. slope, Sierra Nevada, Nevada and Mono cos. s.; Pinyon-Juniper Wd., White Mts., Providence Mts., Mojave Desert; to Ore., Wyo., New Mex. May–Aug.

4. **D. obtùsa** (Greene) O. E. Schulz ssp. **adenóphora** (Woot. & Standl.) Detl. [*Sophia a.* Woot. & Standl. *Descurainia a.* O. E. Schulz. *Sisymbrium Cumingianum* auth. in part, not F. & M.] Coarse strict biennials, 5–12 dm. tall, canescent, glandular especially in infl.; lvs. 1–6 cm. long, pinnate with 2–5 pairs of linear to lanceolate obtuse entire to incised pinnae; pedicels divaricate-spreading, becoming 1–2 cm. long; sepals 2–2.5 mm. long; petals whitish to light yellow, 2–3 mm. long; siliques linear, 12–20 mm. long, 1–1.5 mm. wide, straight or subarcuate, very short-beaked; seeds 0.8–1 mm. long, obscurely biseriate, crowded, 24–32 in a locule; $2n = 14, 42$ (Baldwin & Campbell, 1940).—Dry slopes, 3000–7000 ft.; Joshua Tree Wd. to Yellow Pine F.; San Gabriel and San Bernardino mts. to Santa Rosa Mts.; to New Mex., L. Calif. May–June.

5. **D. pinnàta** (Walt.) Britton ssp. **Menzièsii** (DC.) Detl. [*Cardamine M.* DC. *Sisymbrium canescens* var. *californicum* T. & G. *D. M.* and var. *glandulosa* O. E. Schulz.] Pubescent annual, 1–6 dm. tall, simple to short-branched; lower lvs. 3–9 cm. long, bipinnate or again pinnatifid, the ultimate segms. mostly obovate and obtuse; upper lvs. pinnate to bipinnate, the segms. linear to oblanceolate; pedicels wide-spreading, 8–15 mm. long; sepals 1.5–2.5 mm. long; petals yellow, almost as long; siliques clavate, 5–12(–15) mm. long, usually curved, 1.5–2 mm. wide; style minute; seeds biseriate,

6–10 in each locule, ca. 0.8–1 mm. long.—Dry sandy often waste places, below 8000 ft.; many Plant Communities; along the coast from Contra Costa Co. to San Diego, in the San Joaquin V., s. Sierra Nevada, Mojave and Colo. deserts. March–June.

KEY TO SUBSPECIES

Fruiting pedicels at almost right angles to stem; herbage mostly canescent.
  Segms. of upper lvs. narrowly oblong to linear; racemes mostly glandular-pubescent.
    Siliques 4 mm. long or less, 2–4-seeded; plants 1–2.5 dm. tall .... ssp. *paradisa*
    Siliques 5–12 mm. long, 10–30-seeded; plants 1.5–6 dm. tall.
      Petals 2–2.5 mm. long, bright yellow; plants strict .... ssp. *Menziesii*
      Petals 1–2 mm. long, whitish or yellow; plants branched below .... ssp. *halictorum*
  Segms. of upper lvs. ovate to oblanceolate; racemes glabrous; petals scarcely 2 mm. long, yellow;
    seeds 8–12 per locule .... ssp. *glabra*
Fruiting pedicels ± ascending; herbage not canescent.
  Pedicels usually longer than siliques; terminal lft. usually greatly elongate; seeds sometimes uniseriate
                                    ssp. *filipes*
  Pedicels usually not longer than siliques; terminal lft. not greatly elongate; seeds always biseriate
                                    ssp. *intermedia*

Ssp. **paradìsa** (Nels. & Kenn.) Detl. [*Sophia p.* Nels. & Kenn.] Foothills west of Bishop, Inyo Co.; to Ore., Nev.

Ssp. **halictòrum** (Ckll.) Detl. [*Sophia h.* Ckll.] $n = 14$ and 21 (Baldwin & Campbell, 1940).—E. slope of Sierra Nevada and Siskiyou Co. to e. San Diego Co.

Ssp. **glàbra** (Woot. & Standl.) Detl. [*Sophia g.* Woot. & Standl.] $n = 14$ (Baldwin & Campbell, 1940).—Most common on deserts, to Cuyama V., Santa Barbara Co.; to New Mex., Chihuahua.

Ssp. **fílipes** (Gray) Detl. [*Sisymbrium incisum* var. *f.* Gray] $n = 7$ (Baldwin & Campbell, 1940).—Largely in Sagebrush Scrub, Pinyon-Juniper Wd.; Siskiyou and Modoc cos. to Tuolumne and Mono cos.; to Wash., Rocky Mts.

Ssp. **intermèdia** (Rydb.) Detl. [*Sophia i.* Rydb.] $n = 14$ (Baldwin & Campbell, 1940).—Yellow Pine F.; Lassen and Nevada cos.; to B.C. and Rocky Mts.

16. Cakìle Hill. Sea-Rocket

Fleshy branched glabrous annuals. Lvs. deeply crenate to pinnatifid. Fls. purplish or whitish. Silicles short, transversely 2-jointed, fleshy, becoming dry and corky when ripe, sessile, flattened or ridged, the joints 1-celled and 1-seeded or the lower sometimes seedless. Seed erect in the upper, suspended in the lower joint. Cotyledons accumbent. Ca. 4 spp. of sea and lake shores of N. Am., Eu. and Afr. (Old Arabic name.)

Lvs. sinuate-dentate; petals ca. 6 mm. long; pods without hornlike processes at apex of lower joint
                                  1. *C. edentula*
Lvs. pinnatifid; petals 9–10 mm. long; lower joint of pod with 2 triangular protuberances at apex
                                  2. *C. maritima*

1. **C. edéntula** (Bigel.) Hook, ssp. **califórnica** (Heller) Hult. [*C. c.* Heller. *C. e.* var. *c.* Fern.] Branched from base, the branches often decumbent, up to 6 dm. long; lvs. oblanceolate to narrow-obovate, rounded at apex, sinuate-dentate, petioled, 4–8 cm. long; racemes dense; pedicels stout, 3–5 mm. long; sepals 3–4 mm. long; petals tinged purple, 6 mm. long; silicles 12–15 mm. long, the lower joint obovoid, 5–7 mm. long, the upper broadly ovoid, 8–10 mm. long, ribbed, flattened at apex; seeds somewhat ovate, compressed, brown, 5–6 mm. long; $n = 9$ (Kruckeberg, 1948).—Beach sands, San Diego; Channel Ids.; to B.C. May–Sept.

2. **C. marítima** Scop. [*Bunias Cakile* L.] Branching from base, procumbent or decumbent; lvs. 4–8 cm. long, deeply pinnatifid into oblong lobes with rounded apices; pedicels stout, ca. 2 mm. long; sepals 3 mm. long; petals pink to purplish, 8–10 mm. long; silicles ca. 15 mm. long, the upper joint flattened, ca. twice as long as lower which has 2 divergent triangular protuberances at apex; seeds 4–5 mm. long; $n = 9$ (Jaretzky, 1929).—Beach sand, Monterey Co. to Mendocino Co.; natur. from Eu. June–Nov.

17. Isàtis L.

Erect annual or biennial or perennial herbs. Lvs. simple, uncleft. Fls. small, yellow, racemose. Sepals ascending, not gibbous at base. Petals exceeding sepals. Silicles flat,

)endulous, oval to oblong, winged, ribbed on each side, indehiscent, 1-celled, 1-seeded, with sessile stigma. Seed pendulous, marginless, with mostly incumbent cotyledons. Ca. 35 spp. of Old World. (Classical name.)

1. I. tinctòria L. DYERS WOAD. Glabrous, glaucous, 5–9 dm. high, branched from near base; basal lvs. obovate to oblong, petioled, coarsely toothed, 4–10 cm. long; the cauline more narrow, auriculate-clasping; racemes several, forming corymbose clusters; fls. ca. 3 mm. long; silicles firm, becoming dark, 8–15 mm. long, 5–7 mm. wide, on slender pedicels; seeds yellowish, 3–3.5 mm. long; $2n = 28$ (Manton, 1932).—Locally established weed, Humboldt and Siskiyou cos.; native of Eu. April–June.

18. Rapístrum Crantz

Annual to perennial, ± branched. Lvs. mostly lyrate-pinnatifid to bipinnate. Fls. rather small; sepals erect-divaricate, not or scarcely saccate. Petals yellow, rarely white, clawed. Siliques on thickened pedicels, transversely 2-jointed, appressed, the upper joint globose, 8-ribbed, abruptly slender-beaked and much thicker than the lower joint. Seeds ovoid or ellipsoid. Ca. 8 spp., largely Medit. (Greek, *rhapis*, rape, and *astrum*, appearance.)

1. R. rugòsum (L.) All. [*Myagrum r.* L.] Annual or biennial, 2.5–6 dm. high, ± stiff-hairy; lower lvs. 5–15 cm. long, petioled, the upper subentire; sepals 2.5–3.5 mm. long; petals yellow with darker veins, 5–7 mm. long; pedicels thickened; siliques with upper joint conspicuously beaked and 3–4 times as long as lower joint; seeds 1–2 mm. long, smooth, yellow-brown; $2n = 16$ (Manton, 1932).—Reported from San Francisco; native of Eu. Summer.

19. Erùca Mill. GARDEN-ROCKET

Annual or biennial, erect, branched. Lvs. pinnatifid to toothed. Fls. racemose, rather large, ochroleucous to yellowish or purplish, with violet veins. Sepals erect. Silique linear-oblong, thickish, somewhat 4-sided, long-beaked, the valves 3-nerved. Seeds many, in 2 rows in each cell, ellipsoid, slightly compressed; cotyledons conduplicate. Ca. 5 spp. of Medit. region. (Classical Latin name used by Pliny.)

1. E. satìva Hill. [*Brassica E. L. E. E.* Britton.] Rather succulent, glabrous, 3–5 dm. high; the lower lvs. 8–15 cm. long, pinnatifid or pinnately lobed, with oblong-spatulate lobes 1–3 cm. long; the upper smaller, less deeply lobed; sepals 10–12 mm. long; petals 15–18 mm. long; siliques erect-appressed on stout pedicels, 1.5–2.5 cm. long and with valves keeled on back, the beak flat, almost as long as body; seeds 1.5–2 mm. long, $2n = 22$ (Manton, 1932).—Waste places and fields (especially flax and alfalfa), at widely scattered localities in Calif.; to Wash. and Atlantic Coast; natur. from Eu. May–July.

20. Brássica L. MUSTARD

Erect branched annual to perennial herbs. Basal lvs. pinnatifid, those of the stem dentate or subentire. Fls. showy, yellow, in elongated racemes. Lateral sepals ± gibbous at base. Petals with long claw and spreading limb. Siliques elongate, slender or thickish, subterete or 4-sided, with a stout indehiscent flat or conic beak and convex 1–3-nerved valves. Stigma truncate or 2-lobed. Seeds subglobose, marginless, in 1 row in each cell; cotyledons conduplicate. Ca. 100 spp. of Eu., Asia and N. Afr.; many, like cabbage, broccoli, cauliflower, etc., are important food plants. (Latin name for cabbage.)

A. Upper stem-lvs. with clasping base.
 B. Lvs. thin, the basal with scattered hairs; petals 6–8 mm. long; siliques 3–6 cm. long
  1. *B. campestris*
  BB. Lvs. thickish, mostly glabrous; petals 10–25 mm. long; siliques 5–10 cm. long.
   C. Infl. open and 10–25 cm. long at anthesis; petals 12–26 mm. long .... 2. *B. oleracea*
   CC. Infl. crowded, the blooming part 3–5 cm. long; petals 11–14 mm. long .... 3. *B. Napus*
AA. Upper stem-lvs. not clasping.
 B. Beak of silique terete or conic, often seedless.
  C. Pedicels mostly shorter than sepals; siliques appressed against the stem.
   D. Plant annual; petals ca. 8 mm. long .... 4. *B. nigra*
   DD. Plant perennial; petals ca. 5 mm. long .... 5. *B. geniculata*
  CC. Pedicels longer than sepals; siliques divergent.

D. Petals ca. 3 mm. wide; beak of silique 6–10 mm. long, the apex narrower than
the stigma . . . . 6. *B. juncea*
DD. Petals ca. 1.5 mm. wide; beak of silique 10–16 mm. long, the apex as wide as
the stigma . . . . 7. *B. Tournefortii*
BB. Beak of silique flat or conspicuously angled, beak usually 1-seeded.
C. Lvs. petioled, pinnatifid; pedicels 5–15 mm. long; beak ensiform, equal to or longer
than the bristly dehiscent part of the silique . . . . 8. *B. hirta*
CC. Lvs. subsessile, the upper merely toothed; pedicels 3–7 mm. long; beak 4-angled,
2-edged, ca. half as long as the smooth or sparsely hairy body of silique . . . 9. *B. Kaber*

1. **B. campéstris** L. FIELD MUSTARD. Erect annual, 3–12 dm. tall, with slender roots,
glaucous and quite glabrous except for the scattered hairs on the lower lvs.; these
petioled, ± pinnatifid or lobed, 1–2 dm. long; upper lvs. sessile, auriculate-clasping, lance-
oblong, subentire, glabrous; pedicels spreading, 1–2 cm. long; sepals narrow-oblong,
yellowish, 4–5 mm. long; petals yellow, spatulate, 6–8 mm. long; siliques not torulose,
terete, 2–5 cm. long, stout with a stout beak an additional 1–1.5 cm. long; seeds 1.5–2
mm. thick, dark, reticulate; $n = 10$ (Karpechenko, 1924).—Common weed, especially
in orchards and waste places, widely distributed in Calif.; natur. from Eu. Jan.–May, also
in other months. (*B. Rápa* L., the TURNIP, very like this, but with thickened roots, sometimes referred to the same sp.)

2. **B. olerácea** L. CABBAGE. Stout glaucous perennial or biennial, the stem 3–10 dm.
long, often decumbent; lower lvs. thick, fleshy, obovate to oblong, 1.5–3 dm. long;
stem-lvs. narrow, some clasping; fls. in long racemes, whitish-yellow; sepals 6–12 mm.
long; petals 12–26 mm. long; siliques 5–10 cm. long, including the long conical beak;
seeds slightly compressed, 2–4 mm. thick, somewhat angled-striate; $n = 9$ (Karpechenko, 1924).—Reported from headlands, Marin Co., also from San Francisco, etc.;
natur. from Eu. March–June. Often persisting for only a short time.

3. **B. Nàpus** L. RAPE. Much like *B. campestris* but more glabrous; terminal lobe of
basal lvs. very large and obtuse; fls. paler; sepals 6–8 mm. long; petals 9–14 mm. long;
siliques 5–9 cm. long, the beak 1–2(–3) cm. long; seeds 1–1.5 mm. thick, obscurely
purple-brown, minutely reticulate-alveolate; $n = 18$ (Morinaga, 1929).—Occasional
weed; native of Eu. April–June.

4. **B. nìgra** (L.) Koch. [*Sinapis n.* L.] BLACK MUSTARD. Erect annual, branched
above, 5–25 dm. high, sparsely pubescent or subglabrous; lower lvs. 1–2 dm. long,
deeply pinnatifid, with large terminal lobe and few small lateral ones; cauline lvs.
gradually reduced but not clasping, the uppermost pendulous; sepals 3.5–4.5 mm. long;
petals bright yellow, 7–8 mm. long; pedicels 2–3 mm. long, erect; siliques appressed,
1–2 cm. long, the beak subulate, empty, 1–3 mm. long; seeds ca. 1–1.3 mm. thick,
dark red-brown, finely reticulate; $2n = 16$ (Manton, 1932).—Common on dry grassy
slopes, in grain fields, waste places, etc., through much of the state; to Atlantic Coast;
natur. from Eu. April–July.

5. **B. geniculàta** (Desf.) J. Ball. [*Sinapis g.* Desf. *Sinapis incana* L. *B. i.* Meigen, not
Ten. *Hirschfeldia adpressa* Moench. *B. a.* Boiss.] Biennial or perennial, 4–8 dm. tall, ±
canescent-hirsute; basal lvs. lyrate-pinnatifid, 4–10 cm. long, with large terminal lobe,
the upper cauline smaller, dentate to lobed; racemes many, terminal on the branches;
pedicels appressed, 1–3 mm. long at anthesis; sepals ca. 3 mm. long; petals 5–6 mm.
long, light yellow; siliques appressed, 8–12 mm. long, torulose, beak flattened, frequently 3–4 mm. long and 1-seeded; seeds ca. 1 mm. long, ovoid or oblong-ovoid, red-
brown, somewhat alveolate.—Common weed in waste places, along roadsides, etc., in
much of cismontane Calif.; natur. from Eu. May–Oct.

6. **B. júncea** (L.) Coss. [*Sinapis j.* L.] Pale subglabrous annual, 3–12 dm. high;
lower lvs. runcinate-pinnatifid and crenate, petioled, 3–12 cm. long, the upper nearly
sessile, lanceolate or linear, entire or dentate; pedicels slender, divergent, 8–12 mm. long;
sepals 4–5 mm. long; petals yellow, spatulate, 7–8 mm. long; siliques 3–4 cm. long,
ascending, the beak 5–8 mm. long; seeds subglobose, ca. 1.5 mm. in diam., red-brown to
yellowish, weakly reticulate; $n = 18$ (Manton, 1932).—At scattered stations in the
state, where it grows as a weed in grain fields and waste places; natur. from Eu. June–
Sept.

7. **B. Tournefórtii** Gouan. Annual, 1–6 dm. tall, branched at base, ± hirsute below;
basal lvs. lyrate-pinnatifid, short-petioled, the upper reduced, sessile, oblong or linear;
fls. crowded at anthesis; pedicels 3–10 mm. long, ascending; sepals 3 mm. long; petals
pale yellow, 5–7 mm. long; siliques 3.5–6.5 cm. long, ascending, the beak 1–2 cm. long:

seeds brown-purple, finely silvery reticulate, ca. 1 mm. thick; $n = 10$ (Sikka, 1940).—
Roadsides and fields, Imperial Co. to Riverside Co. and in w. San Bernardino Co.; natur.
from N. Afr. Jan.–June.

8. **B. hírta** Moench. [*Sinapis alba* L. *B. a.* Rabenh., not Gilib.] WHITE MUSTARD. Annual, 3–7 dm. high, ± hirsute; lower lvs. broad, lyrately pinnate or pinnatifid, 10–20 cm.
long, petioled, the terminal lobe or lft. large; upper lvs. short-petioled, lanceolate or
oblong; pedicels spreading in fr., 5–15 mm. long; sepals 5–6 mm. long; petals yellow,
8–11 mm. long, 4–5 mm. wide; siliques white-bristly, 2–3 cm. long, (including beak),
the valves prominently 3-nerved, few-seeded, the beak flattened, broad, ca. as long as
the rest of the silique; seeds pale yellow, subglobose, 1.5–2 mm. thick, minutely
alveolate; $2n = 24$ (Manton, 1932).—Cult. as a source of mustard and for greens;
natur. in widely scattered localities; introd. from Eurasia. March–Aug.

9. **B. Káber** (DC.) Wheeler. [*Sinapis K.* DC.] CHARLOCK. Erect annual, ± hispid at
base, 3–10 dm. high; lower lvs. obovate, lyrate-pinnatifid, 5–15 cm. long, petioled, the
upper oblong to lanceolate, toothed; pedicels short, thick, 2–3 mm. long, ascending;
sepals 4–5 mm. long; petals yellow, 8 mm. long; siliques 2–2.5 cm. long, the beak
0.6–1.2 cm. long; seeds globose, 1–1.5 mm. thick, red-brown or darker, minutely
alveolate; $n = 9$ (Yarnell, 1956).—Old World. Represented in the U.S. by:

Var. **pinnatífida** (Stokes) Wheeler. [*Sinapis arvensis* var. *p.* Stokes. *B. a.* (L.) Rabenh.,
not L. *Sinapis a.* L. *B. sinapistrum* Boiss.] Fruiting pedicels 3–7 mm. long; siliques
2.5–4.5 cm. long, 3–4 mm. thick, cylindric, scarcely torulose, nearly or quite glabrous,
the beak 4-angled, 2-edged, 1–1.5 cm. long; $2n = 18$ (Manton, 1932).—Weed, natur.
from Eurasia. March–Oct.

Var. **Schkuhriàna** (Rchb.) Wheeler. [*Sinapis S.* Rchb.] Siliques slender, 1.5–2 mm.
thick, strongly torulose, the beak usually curved.—Natur. from Eurasia. March–Oct.

21. Diplotáxis DC.

Annual to perennial herbs, much like *Brassica* in habit. Lvs. toothed to pinnatifid. Fls.
yellow, white or purplish. Siliques linear, ± flattened parallel with the partition, short-
beaked or beakless. Seeds ovoid, in 2 rows in each locule. Ca. 20 spp. of Medit. region
and cent. Eu. (Greek, *diplous*, double, and, *taxis*, row, because of the biseriate seeds.)

Annual; lvs. mostly basal, oblanceolate .... 1. *D. muralis*
Perennial; lvs. up to the infl. lanceolate .... 2. *D. tenuifolia*

1. **D. muràlis** (L.) DC. [*Sisymbrium m.* L.] SAND-ROCKET. Annual or biennial,
branched at base, glabrous to somewhat hispid; stems slender, 3–5 dm. high, leafy only
at base; lvs. oblanceolate, sinuate-lobed, 3–10 cm. long, petioled; racemes elongate, lax
in fr.; pedicels ascending, 1–2 cm. long; sepals 3–5 mm. long; petals yellow, 5–8 mm.
long; siliques erect, flattish, 2–2.5 cm. long, 2 mm. wide, sessile; seeds brownish, smooth,
ca. 1 mm. long; $2n = 42$ (Maude, 1940).—Occasional weed along Santa Ana R. system; native of Eu. Through the year.

2. **D. tenuifòlia** (L.) DC. [*Sisymbrium t.* L.] WALL-ROCKET. Perennial, subglabrous,
bushy, 3–6 dm. tall, leafy to the infl.; lvs. lanceolate, subentire to pinnatifid, 6–12 cm.
long; racemes loose in fr.; pedicels 1–4 cm. long, ascending; sepals 5–8 mm. long; petals
8–10 mm. long; siliques 2–3 cm. long, suberect, stipitate; seeds brownish, smooth, ca.
1.2 mm. long; $2n = 22$ (Manton, 1932).—Occasional weed, Sacramento V. and s.
Calif.; natur. from Eu. March–June.

22. Ráphanus L. RADISH

Annuals or biennials, erect, branched. Lvs. lyrate. Fls. showy, purple or yellow, fading white. Petals long-clawed. Siliques torulose or cylindric, coriaceous, indehiscent,
several-seeded, continuous and spongy in between the seeds, with no proper partition, tapering above into the long persistent slender style. Seeds globose, with conduplicate cotyledons. Ca. 4 spp. of Eurasia. (Greek, *raphanos*, quick appearing, because
of rapid germination of seeds.)

Siliques not longitudinally grooved, 2–3-seeded; fls. mostly purple or white .... 1. *R. sativus*
Siliques longitudinally grooved, 4–6-seeded; fls. yellow, fading white .... 2. *R. Raphanistrum*

1. **R. satìvus** L. WILD RADISH. Freely branched, 3–12 dm. tall, subglabrous to scattered-hispid; lower lvs. pinnately parted, 1–2 dm. long, with large rounded terminal segm.; upper lvs. toothed; pedicels ascending, 1–2 cm. long; sepals narrow, ca. 9–10 mm. long; petals white with rose or purplish veins, or yellowish or purplish, 15–20 mm. long; siliques 2–3 cm. long, including the conical beak, 2–3-seeded, and 6–8 mm. thick; seeds brownish, faintly reticulate, 2–3 mm. in diam.; $n = 9$ (Karpechenko, 1924).— Common weed in waste places, fields, etc., through much of the state; natur. from Eu. Feb.–July. The Garden Radish is a cult. form.

2. **R. Raphanístrum** L. JOINTED CHARLOCK. Like the last, but fls. yellow, aging white; siliques 4–10-seeded, nearly cylindric when fresh, constricted between the seeds when dry, 4–6 mm. thick, slender-beaked, transversely divided into 1-seeded segms. when ripe; $n = 9$ (Karpechenko, 1924).—Occasional weed in waste places; natur. from Eu. April–June.

23. Barbarèa Scop. WINTER-CRESS

Glabrous biennial or perennial herbs with overwintering basal rosettes and angled stems. Lvs. pinnatifid, the cauline with clasping bases. Fls. racemose, yellow. Sepals erect, the 2 outer slightly saccate at base. Petals spatulate, clawed. Siliques linear, often somewhat 4-angled, with pointed style and somewhat bilobed stigma. Seeds in 1 row in each cell, marginless, flat, oblong, with cotyledons accumbent. Ca. 7 spp. of temp. zones. (Named for St. *Barbara*.)

(Fernald, M. L. The N. Am. spp. of Barbarea. Rhodora 11: 134–141, 1909.)

Beak of the silique slender, 1.5–3 mm. long; uppermost lvs. incised to lobed, but rarely pinnatifid
                                                                        1. *B. vulgaris*
Beak of the silique 0.5–1 mm. long, thickish; uppermost lvs. often lyrate-pinnatifid.
   Basal lvs. with 10–20 lateral lfts.; petals deep yellow, 6–8 mm. long; siliques 4–8 cm. long. Weed.
                                                        2. *B. verna*
   Basal lvs. simple or with 2–4 small lateral lfts.; petals pale, 3–5 mm. long; siliques 2–3.5 cm. long.
   Native . . . . 3. *B. orthoceras*

1. **B. vulgàris** R. Br. [*B. stricta* auth., not Andrz.] Lower lvs. simple or with 1–4 pairs of small lateral lobes, the terminal lobe round to elliptic-oblong, larger; upper lvs. coarsely dentate, angulate or lobed; petals 6–8 mm. long; siliques 1.5–3 cm. long (excluding beak), erect to strongly ascending, in a dense raceme, the beak slender, 1.5–3 mm. long; $2n = 16$ (Manton, 1932.)—A Eurasian sp. occasionally introd. in Calif. below 4000 ft. April–May.

2. **B. vérna** (Mill.) Asch. [*Erysimum v.* Mill.] Lvs. all pinnatifid, the basal with rounded-oval to -oblong terminal lobe and 10–20 smaller lateral lobes; petals 6–8 mm. long; pedicels 3–8 mm. long, thick; siliques 4–8 cm. long, slightly flattened, rigid, ascending, with thick beak 0.5–1 mm. long; seeds gray-brown, pitted-reticulate, 1–1.5 mm. broad, 2–2.5 mm. long; $2n = 16$ (Manton, 1932).—Sparingly introd. as a weed in waste places, as in Marin, Napa, Lake, and Fresno cos.; native of Eu. April–July.

3. **B. orthóceras** Ledeb. [*B. americana* Rydb.] Stems rather stout, strict, 2–4 dm. tall, usually few-branched above; basal lvs. 3–10 cm. long, petioled, elliptic to suborbicular, simple or with 2–4 small lfts. and large terminal one; middle and upper cauline lvs. lyrate-pinnatifid; racemes dense in anthesis, looser in fr.; pedicels thick, ascending, 3–6 mm. long; sepals yellow-green, ca. 3 mm. long; petals pale yellow, 4–6 mm. long; siliques erect, appressed, 2.5–3.5 cm. long, 1.5 mm. wide; seeds brown, minutely rugulose, 0.8–1 mm. long.—Banks of streams, springy places, meadows, etc., mostly 2500–11,000 ft.; Yellow Pine F. to Subalpine F.; much of montane Calif.; to Atlantic Coast, Alaska, Asia. May–Sept. Passing gradually into:

Var. **dolichocárpa** Fern. Siliques spreading or ascending, 2.5–5 cm. long, somewhat remote, tending to be curved.—With the sp., to Mex., B.C., Wyo.

24. Roríppa Scop. YELLOW-CRESS

Aquatic to terrestrial annual to perennial herbs. Lvs. usually glabrous, commonly pinnate to pinnatifid. Fls. yellow, in short racemes. Petals with nectariferous glands. Frs. short siliques or silicles, slender to subglobular, terete or nearly so, the valves strongly convex, nerveless. Seeds usually many, small, turgid, marginless, usually in 2 irregular rows in each locule; cotyledons accumbent. A fairly large genus of temp. regions. (Name from an old Saxon word, *Rorippen*.)

A. Plants perennial with creeping rootstocks; petals longer than sepals.
  B. Plants 3–9 dm. high; sepals 1–2 mm. long; frs. globose. Introd. weed in hayfields
                                                                            1. *R. austriaca*
  BB. Plants 1–3 dm. high; sepals 2–4 mm. long; frs. usually longer than thick. Natives.
    C. Fls. 6–8 mm. long; siliques linear-oblong, 7–14 mm. long; stems glabrous
                                                                        2. *R. sinuata*
    CC. Fls. 3–4 mm. long; siliques ovoid to oblong, 3–5 mm. long; stems pubescent.
      D. Frs. puberulent; raceme elongate; stigma not expanded. Siskiyou Co.
                                                                 3. *R. calycina*
      DD. Frs. glabrous; raceme subumbellate; stigma expanded. About Lake Tahoe
                                                     4. *R. subumbellata*
AA. Plants annual; petals shorter than sepals.
  B. Stems diffusely branched from base; pedicels 2–4 mm. long.
    C. Pods strongly curved; lf.-segms. linear to oblong, usually acute; style to 0.5 mm. long,
       stout . . . . 5. *R. curvisiliqua*
    CC. Pods not curved; lf.-segms. obovate or rounded; style 1–2 mm. long, slender
                                                         6. *R. obtusa*
  BB. Stems erect, branched above; pedicels 3–8 mm. long . . . . 7. *R. islandica*

1. **R. austriaca** (Crantz) Bess. [*Nasturtium a.* Crantz. *Radicula a.* Small.] AUSTRIAN FIELD-CRESS. Perennial, with slender erect stems 3–9 dm. high, finely puberulent; lvs. oblong to oblong-obovate, glabrous, unequally serrate, 3–6 cm. long, narrowed below to a petiolelike auriculate base; racemes 7–12 cm. long, in terminal panicles; pedicels spreading-ascending, 4–10 mm. long in fr.; sepals 1–2 mm. long; petals 3–5 mm. long; silicles globose, 1.5–3 mm. long; seeds light brown, ca. 1 mm. long, finely tuberculate-reticulate; $2n = 16$ (Manton, 1932).—Pernicious weed in hay fields of Modoc Co.; introd. from Eu. Summer.

2. **R. sinuata** (Nutt.) Hitchc. [*Nasturtium s.* Nutt. *Radicula s.* Greene.] Perennial from creeping deep-seated rootstocks; stems several, decumbent, glabrous, 1–3 dm. long; lvs. 4–8 cm. long, oblong to lance-elliptic, regularly sinuate- or pectinate-pinnatifid, with subentire linear-oblong lobes; pedicels slender, 6–10 mm. long; sepals 3–4 mm. long; petals ca. 6–7 mm. long; siliques linear-oblong, sometimes ± curved, 7–14 mm. long; style slender, 2–3 mm. long; seeds ca. 1 mm. long.—At 3000–5000 ft., Little Lake, Inyo Co. and in Modoc Co.; to Wash., Ill., Tex. May–Sept.

3. **R. calycina** (Engelm.) Rydb. var. **columbiae** (Suksd.) Roll. [*Nasturtium sinuatum* var. *columbiae* Suksd. *R. c.* Howell. *Radicula c.* Greene.] Perennial from slender creeping rootstocks, pubescent throughout, the stems branched, 1.5–3 dm. long; lvs. 2–5 cm. long, pinnatifid with oblong, often toothed segms., the lower petioled, the upper mostly sessile; racemes terminal and axillary; pedicels ascending or spreading, 5–9 mm. long; sepals ca. 3 mm. long; petals pale yellow, 4 mm. long; siliques narrow-ovoid, turgid, puberulent, 4–5 mm. long; styles ca. 1.5 mm. long; seeds tawny, cellular-pitted, ca. 0.6 mm. long.—Lava slopes; N. Juniper Wd.; Lava Beds National Monument, Siskiyou Co.; to Columbia R. June–July.

4. **R. subumbellata** Roll. Perennial with slender underground rootstocks; stem decumbent, branched, hirsute, 5–18 cm. long; lvs. short-petioled to sessile, subpinnatifid, pilose to glabrous, 1–3 cm. long, 3–10 mm. wide; infl. subumbellate to somewhat elongate; pedicels erect to divaricate, 3–6 mm. long; sepals 2–3 mm. long; petals somewhat longer; siliques broadly oblong to subglobose, glabrous, 3–5 mm. long; styles 1–1.5 mm. long; seeds tawny, beaded, slightly over 1 mm. long.—Moist places, 6000–8000 ft.; Yellow Pine F.; about Lake Tahoe. June–July.

5. **R. curvisiliqua** (Hook.) Bessey. [*Sisymbrium c.* Hook. *Nasturtium c.* Nutt. *Radicula c.* Greene. *N. lyratum* Nutt. *Rorippa l.* Greene.] Annual or biennial, glabrous, the stems diffusely branched, 1–3 dm. long; lvs. pinnatifid into usually acute entire or toothed lanceolate to oblong lobes; lower lf.-blades 2–8 cm. long, with somewhat shorter petioles; cauline lvs. gradually reduced and subsessile; pedicels mostly 1–2 mm. long; sepals 1–2 mm. long; petals somewhat shorter; siliques linear, terete, usually curved, 6–10 mm. long, 1–1.5 mm. thick; style less than 1 mm. long; seeds brown, ca. 0.5 mm. long, finely cellular-reticulate.—Frequent in wet or damp places, below 10,000 ft.; many Plant Communities; throughout cismontane and montane Calif.; to B.C., Wyo., L. Calif. April–Sept. Exceedingly variable as to lvs., silique, etc.

6. **R. obtùsa** (Nutt.) Britton. [*Nasturtium o.* Nutt. *Radicula o.* Greene. *R. sinuata* vars. *integra* and *truncata* Jeps.] Glabrous or subglabrous annual, the stems diffusely branched at the base, 1–3 dm. long; lvs. pinnatifid with rounded or obovate sinuately toothed divisions, the lower lvs. less divided, with blades 2–5 cm. long, petioled, the upper more divided, 1–3 cm. long, subsessile; pedicels spreading, 2–4 mm. long; fls. 1 mm. long; siliques subglobose to oblong, 3–8 mm. long, 2–3 mm. thick; style ca. 1 mm. long; seeds light brown, ca. 0.5 mm. long, round-oblong.—Wet places, mostly 5000–8000 ft.; Yellow Pine F., Red Fir F.; San Bernardino and San Gabriel mts., Sierra Nevada to Siskiyou Co.; to B.C. and e. states. June–Sept.

7. **R. islándica** (Oeder) Borb. var. **occidentàlis** (Wats.) Butters & Abbe. [*Nasturtium terrestre* var. *o.* Wats. *R. pacifica* Howell. *R. palustris* var. *p.* G. N. Jones. *R. palustris* ssp. *occidentalis* Abrams. *Nasturtium dictyotum* Greene.] Annual or biennial, nearly or quite glabrous, the stems erect, branched above, 3–8 dm. high; lower lvs. lyrate-pinnatifid, 8–14 cm. long, wing-petioled, the upper nearly sessile, dentate or somewhat lobed; racemes long, lax in fr.; pedicels slender, widely divaricate, 5–6 mm. long; fls. 2 mm. long; siliques oblong-linear, straight or ± curved, 6–12 mm. long; style less than 1 mm. long; seeds brown, ca. 0.4 mm. long.—Occasional in wet places, often with roots immersed, below 6000 ft.; various Plant Communities; San Diego Co. n. to Alaska. April–July.

Var. **Fernaldiàna** Butters & Abbe. Middle and upper lvs. subentire, entire to toothed; siliques ellipsoid to subcylindric, 3–9 mm. long, 1–2.5 mm. thick—At 7200 ft., Mono Co. to B.C. and Atlantic Coast.

Var. **híspida** (Desv.) Butters & Abbe. [*Brachylobus h.* Desv. *R. h.* Britton.] Stems often hispid; siliques 2–5.5 mm. long, 2–4 mm. thick, short-ellipsoid to subglobose.—To be expected in Calif.; Alaska to New Mex., Atlantic Coast.

25. Nastúrtium R. Br. WATER-CRESS

Mostly small annual to perennial glabrous plants. Stems tending to be somewhat succulent. Lvs. mostly pinnate or pinnatifid, sometimes entire. Fls. white or yellow, small, in terminal racemes. Nectariferous glands horseshoe-shaped. Siliques linear-cylindric, plump, the valves convex and usually weakly nerved. Seeds usually in 2 rows in each locule, numerous, small, marginless, turgid, with accumbent cotyledons. Perhaps 50 spp. mostly of wet places in temp. zone. (Latin, *nasus tortus*, twisted or wry nose, in reference to pungent qualities.)

Siliques 7–18 mm. long, mostly 2–2.5 mm. wide, with 2 rows of seeds in each cell; pedicels 6–15 mm. long; style not over 1 mm. long .... 1. *N. officinale*
Siliques 15–25 mm. long, 1–1.5 mm. wide, with 1 row of seeds in each cell; pedicels 8–25 mm. long; style 0.5–2 mm. long .... 2. *N. microphyllum*

1. **N officinàle** R. Br. [*Sisymbrium Nasturtium-aquaticum* L. *Radicula N.-a.* Britt. & Rend. *Rorippa N.-a.* Schinz & Thell. *Nasturtium N.-a.* Karst.] Aquatic perennial, with prostrate or ascending stems 1–6 dm. long, rooting freely; lvs. pinnate, glabrous, 1–10 cm. long, with 3–11 ovate (or terminal rounded) lfts. 5–25(–40) mm. long, subentire; pedicels divergent; fls. 3–4 mm. long; siliques ascending, curving; style scarcely evident; seeds ca. 1 mm. long, brown, reticulate; $2n = 32$ (Howard & Manton, 1946).—Common in quiet water, on wet banks, etc., below 8000 ft.; many Plant Communities; natur. from Eu. March–Nov. The cult. Water-Cress.

Var. **silifòlium** (Rchb.) Koch. [*N. s.* Rchb.] Terminal and elongate lateral lfts. almost oblong.—Said to occur in Calif.; native of Old World.

2. **N. microphýllum** Boenn. ex Rchb. [*N. officinale* var. *m.* Thell. *N. uniseriatum* Howard & Manton.] Habit and stature of *N. officinale*, but lfts. tending to be smaller; pedicels longer; siliques 15–25 mm. long, 1–1.5 mm. wide, with seeds uniseriate; style 0.5–2 mm. long; $2n = 64$ (Howard & Manton, 1946).—Jonesville, Butte Co.; occasional to Atlantic Coast; natur. from Eu. June–Aug.

26. Armoràcia Gaertn., Mey. & Scherb.

Glabrous perennial herbs with deep roots or rhizomes. Lvs. undivided or pinnatifid, mostly docklike. Fls. many, white, in naked panicled racemes, on slender pedicels.

Stigma 2-lobed or capitate, broad. Silicle subglobose, obovoid or ellipsoid, with nerveless, convex valves. Seeds in 2 rows in each locule, turgid, wingless, with accumbent cotyledons. Ca. 3 spp. of Eurasia. (Ancient Latin name of Horse-Radish.)

1. **A. rusticàna** Gaertn., Mey. & Scherb. [*A. lapathifòlia* Gilib. *Cochlearia A. L. C. r.* Lam. *Nasturtium A.* Fries. *Radicula A.* Rob. *Rorippa A.* Hitchc.] Horse-Radish. Coarse long-lived perennial with deep branched woody root used for condiments; stem erect, 6–12 dm. high; lower lvs. to 4 dm. long, petioled, oblong to oblong-ovate, crenate; cauline lvs. smaller, lanceolate, the lower often pinnatifid; pedicels spreading; fls. 5–7 mm. long; silicles 4–6 mm. long; style very short; stigma capitate; seeds smooth; $n = 16$ (Manton, 1932).—Occasionally escaped from cult. and maintaining itself, as at Point Reyes; native of Eurasia. April–May.

27. Lunària L. Moonwort

Erect branched annual to perennial herbs, with scattered simple hairs. Lvs. ovate-deltoid to ovate, simple, toothed, petioled. Fls. in terminal racemes, violet or purple to white. Sepals usually purplish at tips, the lateral saccate. Petals clawed. Silicles elliptic or rounded, very flat, up to 2.5 cm. wide, the valves parallel to the septum and falling away at maturity. Stigmas minute. Seeds few, flat, winged, with accumbent cotyledons. Two-3 spp. of Eurasia, grown for ornament. (*Luna,* the moon, suggested by the shining septum visible after the valves fall away.)

1. **L. ánnua** L. Honesty. Annual to biennial; stems 4–9 dm. high; lvs. toothed; sepals 6–9 mm. long; petals 18–20 mm. long; silicles rounded at both ends, erect to nodding, 30–45 mm. long, 20–25 mm. wide; style 5–8 mm. long; seeds 5–8 mm. wide; $2n = 28 + 2B$ (Manton, 1932).—Occasional garden escape in moist shaded woods, Marin and Napa cos.; native of Eu.

28. Cardámine L. Bitter-Cress

Annual or perennial herbs, mostly glabrous, with leafy stems. Lvs. entire, lobed or divided. Fls. white or purple, in racemes or corymbs. Sepals equal at base, erect or ascending. Petals obovate to spatulate. Siliques linear, narrow, flattened, many-seeded, opening elastically from the base, the valves nearly or quite nerveless and veinless. Seeds uniseriate, not margined, with slender funicle and accumbent cotyledons. Over 100 spp. of temp. regions. (Greek, *kardamon,* used for some cress.)

(Detling, L. R. E. The Pacific Coast species of Cardamine. Am. Jour. Bot. 24: 70–76, 1937.)
Lvs. all simple, not divided.
  Stems arising from a slender rootstock; lvs. cordate or reniform, with sinuate margin .. 1. *C. Lyallii*
  Stems arising from a taproot; lvs. ovate or elliptical, tapering at base, often entire .. 5. *C. bellidifolia*
Lvs., at least the cauline, compound.
  Petals 8–14 mm. long; lfts. angularly lobed. Humboldt Co. .... 2. *C. angulata*
  Petals 2–6 mm. long; lfts. not angularly lobed.
    Plants from slender rootstocks; petals 4–6 mm. long.
      Some lower lvs. simple, others 3–7-foliolate; petals ca. 4 mm. long. High montane
                                              3. *C. Breweri*
      Lvs. 7–11-foliolate, but never all 7-foliolate; petals 5–6 mm. long. Below 3000 ft.
                                              4. *C. Gambelii*
    Plants from taproots or clusters of fibrous roots; petals 2–3 mm. long.
      Siliques ca. 0.5 mm. wide and with 24–36 seeds; seeds 0.5 × 0.8 mm., scarcely winged. Nevada Co. n. .... 6. *C. pensylvanica*
      Siliques ca. 1 mm. wide and mostly with 8–20 seeds; seeds 0.8 × 1 mm., winged. Widely distributed in Calif. .... 7. *C. oligosperma*

1. **C. Lỳallii** Wats. [*C. cordifolia* var. *L.* Nels. & Macbr. *C. L.* var. *pilosa* O. E. Schulz.] Perennial from slender rootstock, glabrous or sparsely pilose, erect, 2–5 dm. high; lvs. ca. 5–12, mostly scattered along stem, simple, orbicular to ovate in outline, with reniform to cordate base, 1–7 cm. wide, sinuate to shallowly lobed, petioled; raceme dense at anthesis; pedicels spreading, 8–15 mm. long in fr.; sepals 3 mm. long; petals white, 7–9 mm. long; siliques spreading, 2–3 cm. long, 1–1.5 mm. wide, 8–12-seeded; style 1–1.5 mm. long.—Along streams, 5500–7300 ft.; Yellow Pine F., Red Fir F.; Placer Co. to Siskiyou Co.; to B.C., Ida., Nev. June–July.

2. **C. angulàta** Hook. Perennial from slender running rootstock, glabrous or somewhat pubescent, suberect, 2–8 dm. high; lvs. ca. 4–7, scattered along the stem, 3–5-foliolate,

the lower long-petioled, the upper less so; lfts. ovate to broadly lanceolate, usually cuneate at base, 2–8 cm. long, angularly 3–5-lobed or -toothed, the lateral subsessile; raceme dense at anthesis; fruiting pedicels spreading, 10–18 mm. long; sepals 2–3 mm. long; petals white, 8–14 mm. long; siliques spreading, 1.5–3 cm. long, 1.5–2 mm. wide; style scarcely 1 mm. long.—Moist shaded places, below 1000 ft.; Redwood F., Mixed Evergreen F.; Humboldt Co.; to B.C. May–June.

3. **C. Brèweri** Wats. [*C. modocensis* Greene.] Perennial from creeping rootstocks; stems often procumbent at base and rooting at nodes, usually glabrous, 2–6 dm. tall; lvs. mostly cauline, simple or 3–5-foliolate, the simple ones usually near the base, terminal lfts. ovate, sometimes with cordate base, 1–4 cm. long, lobed to toothed, lateral lfts. smaller, ovate to lanceolate; raceme lax; fruiting pedicels 7–15 mm. long; sepals 2 mm. long; petals white, ca. 5 mm. long; siliques ascending or erect, 1.5–2.5 cm. long, 1–1.5 mm. wide; style less than 1 mm. long; seeds 10–20, round-oblong, ca. 1 mm. long.—Along streams, 4000–10,200 ft.; Montane Coniferous F.; San Bernardino Mts., Sierra Nevada to mts. of Humboldt Co.; to B.C., Wyo. May–July.

Var. **orbiculáris** (Greene) Detl. [*C. o.* Greene.] Rootstock often short and vertical; lfts. orbicular to ovate; sepals 1–2 mm. long; petals 3–4 mm. long.—Redwood F.; Humboldt Co.; w. of Cascade Range to B.C. May.

4. **C. Gámbelii** Wats. [*Nasturtium G.* O. E. Schulz.] Perennial from horizontal rootstock; stems erect or procumbent and rooting at base, glabrous to somewhat villous, 5–9 dm. long; lvs. numerous, scattered along stem, 7–13-foliolate, lfts. sessile, oblanceolate to obovate, or suborbicular on lowest lvs., 5–20 mm. long, entire to few-toothed; racemes densely many-fld.; fruiting pedicels spreading to reflexed in age, 10–15 mm. long; sepals 3–4 mm. long; petals white, 6–7 mm. long; siliques erect, somewhat curved, 1.5–2.5 cm. long, ca. 1 mm. wide; style 1.5–2 mm. long; seeds 20–36.—Occasional in swampy places, mostly below 4000 ft.; Chaparral, Coastal Sage Scrub; Santa Barbara Co. to San Diego Co.; L. Calif. April–June.

5. **C. bellidifòlia** L. var. **pachyphýlla** Cov. & Leib. Glabrous perennial from a taproot and branched caudex, 3–10 cm. high; lvs. fleshy, long-petioled, simple or sometimes 3-lobed, ovate or obovate, 4–12 mm. long; infl. subumbellate, ca. 10-fld.; fruiting pedicels mostly 5–10 mm. long; petals white, 3–5 mm. long; siliques erect, 1.2–3 cm. long, 1–1.5 mm. wide; style stout, 1–2 mm. long; seeds brown, oblong, scarcely winged, 1.3–1.6 mm. long.—At 7000–8000 ft.; Alpine Fell-fields; Lassen, Shasta and Siskiyou cos.; to Ore. June–July.

6. **C. pensylvánica** Muhl. ex Willd. Perennial from a slender taproot or cluster of fibrous roots; stems glabrous or minutely puberulent below, erect or somewhat procumbent, 2–10 dm. long; lvs. 5–9-foliolate, 2–10 cm. long; lfts. of lower lvs. orbicular to ovate or obovate, 5–15 mm. long, with terminal lft. largest and often sinuately 3-lobed; lfts. of upper lvs. oblanceolate to linear, 6–25 mm. long; raceme densely many-fld.; fruiting pedicels slender, ascending, 8–10 mm. long; petals white, 2–5 mm. long; siliques erect, 2–2.5 cm. long, less than 1 mm. wide; style not over 0.5 mm. long; seeds 24–36, scarcely winged; $n = 32$ (Smith, 1938).—Infrequent in moist places, 3000–6000 ft.; Yellow Pine F., Red Fir F.; Amador and Nevada cos. to B.C. and Atlantic Coast. May–June.

7. **C. oligospérma** Nutt. Annual or biennial, from a slender taproot; stems glabrous to hispidulous, 1–3 dm. high; lvs. thin, frequently in a basal rosette and cauline, 5–11-foliolate, 2–9 cm. long; lfts. of rosette round to ovate, crenately 3–5-lobed, 3–9 mm. long, those of upper lvs. usually more elongate, oblanceolate to narrow-ovate, 4–15 mm. long; raceme 2–10-fld.; fruiting pedicels ascending to suberect, 3–7 mm. long; sepals ca. 1 mm. long; petals white, ca. 2 mm. long; siliques erect, 1.2–2 cm. long, 1–1.5 mm. wide, 8–28-seeded; style less than 1 mm. long; seeds winged.—Moist open woods or canyons, occasional in open places, below 3000 ft.; Chaparral, Oak Woodlands, Foothill Wd., Yellow Pine F., etc.; Los Angeles Co. n. through Coast Ranges, Sierra Nevada from Madera Co. n.; to B.C. March–July.

29. Dentària L. Toothwort

Perennial from a thickened fleshy horizontal rhizome. Stems erect, mostly simple, leafless below. Lvs. rhizomal, arising independently of base of flowering stem and on upper stem. Infl. racemose. Corolla white to rose or purple, campanulate, much exceeding calyx. Siliques linear, ± flattened, the valves not nerved. Style persistent. Seeds in 1 row in each locule, wingless, with thick accumbent cotyledons. Ca. 10 spp. of damp wds. of the N. Hemis. (Latin, *dens*, tooth, perhaps referring to toothed rhizomes of some spp.)

(Detling, L. R. E. The genus Dentaria in the Pacific States. Am. Jour. Bot. 23: 570–576, 1936.)
Rhizomal lvs. generally simple.
  Siliques 2–4 mm. wide; rhizomal lvs. coarsely toothed at apex. Mts. bordering the Great Cent. V. from Tulare and Lake cos. n. . . . . 1. *D. pachystigma*
  Siliques 1–2 mm. wide; rhizomal lvs. with whole margin sinuate. Solano Co. to Del Norte Co.
                                                    2. *D. californica* vars. *sinuata* and *cardiophylla*
Rhizomal lvs. typically compound.
  Corolla deep purple, 10–15 mm. long; rhizome 7–10 mm. thick, orange-yellow. Del Norte Co.
                                                       3. *D. gemmata*
  Corolla white to rose, 6–14 mm. long; rhizome more slender.
    Rhizomes not more than 3 mm. thick; rhizomal lvs. palmately 3–5-foliolate. Placer Co. to Siskiyou Co. . . . . 4. *D. tenella*
    Rhizomes 5–6 mm. thick; rhizomal lvs. mostly pinnately 3–7-foliolate. Mostly Coast Ranges
                                                  2. *D. californica*

1. **D. pachystígma** Wats. [*D. californica* var. *p.* Wats. *D. corymbosa* Jeps. and var. *grata* Jeps. *D. p.* var. *corymbosa* Abrams.] Rhizome ovoid, ca. 3 mm. thick; rhizomal lvs. fleshy, simple, ovate, cordate at base, the blades 4–5 cm. long, coarsely 5–9-toothed in upper half; stems glabrous, erect, 1.5–3 dm. tall; cauline lvs. 2–3, usually thickish, simple or with very small lateral lfts., ovate, 1.5–4.5 cm. long, undulate to irregularly toothed; racemes short, dense; fruiting pedicels 5–25 mm. long; petals pink, 7–11 mm. long; siliques 3–5 cm. long, 3–4 mm. broad; styles 2–10 mm. long.—Rocky slopes, 5000–9400 ft.; Montane Coniferous F.; Sierra Nevada from Tulare Co. to Shasta Co. and in Coast Ranges s. to Lake Co. May–June.

Var. **dissectifòlia** Detl. Cauline lvs. pinnately 3–5-foliolate.—Serpentine, at ca. 2300 ft., Magalia, Butte Co. and Mendocino Co. March–April.

2. **D. califórnica** Nutt. [*Cardamine c.* Greene. *C. paucisecta* Benth. in part. *C. c.* vars. *Robinsoniana, fecunda,* and *brevistyla* O. E. Schulz. *D. integrifolia* var. *c.* Jeps.] Glabrous perennial from deep-seated ovoid rhizomes 4–8 mm. thick; rhizomal lvs. mostly 3-foliolate, the lfts. broadly ovate, often cordate, 2–5 cm. broad, sinuate to dentate; stems slender, erect, 1–4 dm. high; cauline lvs. 2–3(–5), 3–5-foliolate or -lobed, the lfts. lanceolate to ovate, toothed or sinuate to entire; racemes many-fld.; fruiting pedicels ascending, commonly 1–2.5 cm. long; petals pale rose to white, mostly 9–14 mm. long; siliques 2–5 cm. long, 1–2 mm. broad; style stout, 2–6 mm. long; $2n = 32$ (Manton, 1932).—Shady banks and slopes, mostly below 2500 ft., sometimes much higher; many Plant Communities; San Diego Co. mostly through the Coast Ranges to sw. Ore., occasional in Sierra Nevada; L. Calif. Feb.–May.

KEY TO VARIETIES

Rhizomal lvs. typically compound.
  The rhizomal lvs. 3–5-foliolate, pinnate, sometimes simple.
    Foliage thin; cauline lfts. mostly ovate, variously toothed. Plants of shady woods . . *D. californica*
    Foliage fleshy; cauline lfts. mostly oblanceolate and entire. Plants of open fields . var. *integrifolia*
  The rhizomal lvs. 5–7-foliolate, commonly bipinnate . . . . var. *cuneata*
Rhizomal lvs. typically simple.
  Margin of rhizomal lvs. deeply sinuate; cauline lvs. mostly compound . . . . var. *sinuata*
  Margin of rhizomal lvs. crisped, undulate; cauline lvs. mostly simple . . . . var. *cardiophylla*

Var. **integrifòlia** (Nutt.) Detl. [*D. i.* Nutt. *Cardamine i.* Greene.] Rhizomal lvs. simple or 3-foliolate; petals 8–14 mm. long.—Open fields, mostly in heavy wet soils; Coastal Prairie, Mixed Evergreen F., N. Oak Wd.; w. of main Coast Ranges, Monterey Co. to Humboldt Co. Feb.–April.

Var. **cuneàta** (Greene) Detl. [*Cardamine c.* Greene. *D. c.* Greene. *C. californica* ssp *c.* O. E. Schulz.] Rhizomal lvs. 5–7-foliolate, commonly bipinnate; cauline lvs. 5–9-foliolate or -cleft, the lfts. linears or lanceolate, mostly entire.—Open woods; Foothill Wd.; inner S. Coast Ranges, San Benito and Monterey cos. Feb.–April.

Var. **sinuàta** (Greene) Detl. [*Cardamine s.* Greene. *D. s.* Greene. *D. integrifolia* var. *Tracyi* Jeps. *C. c.* prol. *i.* var. *s.* O. E. Schulz.] Rhizomal lvs. largely simple, orbicular to oval, usually with cordate base, 3–10 cm. broad, deeply sinuate; petals often deep rose.—Open forest floor; Redwood F., Mixed Evergreen F.; Sonoma Co. to s. Ore. March–May.

Var. **cardiophýlla** (Greene) Detl. [*Cardamine c.* Greene. *Dentaria c.* Rob. *D. integrifolia* var. *c.* Jeps.] Rhizomal lvs. simple, round, cordate, 5–10 cm. wide, undulate; cauline lvs. mostly simple.—Moist places; Redwood F., Mixed Evergreen F.; Vaca Mts., Solano Co. to s. Humboldt Co.

3. **D. gemmàta** (Greene) Howell. [*Cardamine g.* Greene. *C. californica* var. *g.* O. E. Schulz.] Rhizome ovoid, orange-yellow, 5–10 mm. thick; basal lvs. 3–5-foliolate, the lfts. ovate, 7–35 mm. long, sinuate or coarsely serrate, thickish, subsessile to petioluled; fl.-stem stoutish, erect, 1–2.5 dm. high, simple; stem-lvs. ca. 2, pinnately 3–7-foliolate, with sessile often confluent lanceolate entire to toothed lfts.; raceme short, dense; petals deep purple, 10–15 mm. long; siliques 3–5 cm. long, on pedicels 1.5–2 cm. long; style ca. 5 mm. long.—Wet places; Yellow Pine F., Mixed Evergreen F.; Del Norte and Siskiyou cos.; sw. Ore. April–May.

4. **D. tenélla** Pursh var. **palmàta** Detl. Rhizome slender, 1–4 mm. thick; basal lvs. palmate, 5-(3-)foliolate, the lfts. sessile or nearly so, oblong with rounded apex, mucronate, 1–5 cm. long, entire; stems slender, 1–1.5 dm. high, simple; stem-lvs. petiolate, palmate, 3–5-foliolate, the lfts. sessile, lance-oblong, 1–3.5 cm. long, mostly entire; petals pink to rose, 6–10 mm. long; siliques 2–5 cm. long; style slender, 3–6 mm. long.—Moist woods, 4000–7000 ft.; Yellow Pine F., Red Fir F.; Humboldt, Siskiyou and Trinity cos. to Placer Co. April–June.

30. Idahòa Nels. & Macbr.

Low scapose glabrous annuals with slender scapes. Lvs. basal, lyrate. Fls. solitary, small, white. Sepals broad, erect. Silicles suborbicular, flattened parallel to the broad partition. Seeds in 2 rows in each locule, broadly winged, reticulate. One sp. (Named for the state, *Idaho*.)

1. **I. scapígera** (Hook.) Nels & Macbr. [*Platyspermum s.* Hook.] Lvs. rosulate, ovate, petioled, usually lyrate, sometimes entire, 2–3 cm. long; scapes 2–15 cm. high, glabrous; fls. ca. 2 mm. long; silicles 6–10 mm. in diam., 8–12-seeded; seeds brown, 4–5 mm. wide including the broad wing.—Moist places, 2000–6000 ft.; mostly Yellow Pine F., Red Fir F.; Mt. Hamilton Range, Mt. St. Helena, Sierra Nevada from Sierra Co. n. to Siskiyou and Modoc cos.; to Wash., Ida., Nev. Feb.–April.

31. Tropidocárpum Hook.

Slender branched annuals, erect or more generally diffusely spreading, pubescent with simple and forked hairs. Lvs. pinnatifid, well distributed on stems. Fls. small, in loose leafy racemes. Sepals ovate-oblong, spreading. Petals yellow, spatulate-obovate. Stamens tetradynamous. Style slender; stigma obscurely lobed. Silique elongate, completely or partially 2-celled or 1-celled, flattened contrary to the narrow partition; valves 2 or 4, opening from above. Seeds in 2 or 4 rows, flattened, not winged, with incumbent cotyledons. Spp. 2, Calif. (Greek, *tropis*, keel, and *karpos*, fr., referring to the keeled siliques.)

Plants mostly with spreading branches; siliques 2-valved and 2-celled. Widespread . . . . . 1. *T. gracile*
Plants mostly erect; siliques 4-valved and 1-celled. Mt. Diablo and region . . . . . . 2. *T. capparideum*

1. **T. grácile** Hook. [*T. scabriusculum* Hook. *T. g.* var. *s.* Greene. *T. macrocarpum* Hook. & Harv. ex Greene.] Plant from simple and erect to (more usually) several-stemmed and decumbent or prostrate; stems 1–3 dm. long, pubescent throughout, the hairs sometimes branched; lvs. 1–5 cm. long, largely basal, pinnatifid into linear or oblong segms., these entire to cleft; upper lvs. reduced; pedicels axillary, 6–15 mm. long; sepals ca. 3 mm. long; petals ca. 4 mm. long; siliques linear, strongly obcompressed throughout, with narrow partition most of the length, glabrous to more commonly strigulose or pubescent, 2–5 cm. long, tardily dehiscent; style to ca. 1 mm. long; seeds in 2 rows, brown, narrow-oblong, 1.2–1.4 mm. long, cellular-reticulate.—Common on dry grassy slopes and in open places, below 3500 ft.; V. Grassland, Foot-

hill Wd., Joshua Tree Wd.; about Great Cent. V., N. and S. Coast Ranges, cismontane s. Calif., w. Mojave Desert. March–May.

Var. **dùbium** (A. Davids.) Jeps. [*T. d.* A. Davids.] Siliques 1-celled and not so strongly obcompressed below, twisted, then 2-celled and obcompressed in upper portion.—W. side of San Joaquin V. to s. Calif. and w. Mojave Desert.

2. **T. capparídeum** Greene. Stems tending to be erect, pilose, 2–2.5 dm. high; lvs. as in *T. gracile;* siliques linear-oblong, 1.5–2 cm. long, turgid, 4-valved, 1-celled with 4 parietal placentae, 6-nerved; style 1–2 mm. long; seeds in 4 rows.—Alkaline low hills, below 500 ft.; V. Grassland; in region about foot of Mt. Diablo (Contra Costa, Alameda, and w. San Joaquin cos.). March–April.

32. Lyrocárpa Hook. & Harv.

Herbaceous annuals or perennials, densely pubescent with dendritic hairs. Stems 2–8 dm. high, branched. Lvs. petioled, repand to pinnatifid. Infl. loosely racemose. Calyx cylindrical, pubescent, the outer sepals subsaccate, longer than inner. Petals much longer than sepals, the blade often twisted. Stamens tetradynamous; 2 discoid nectar-glands at base of each stamen. Silicles obcordate to panduriform, pubescent, flattened contrary to the narrow partition; stigma large, bifid; style short or obsolete. Seeds 3–10 per locule, orbicular, brown, wingless, with accumbent cotyledons. Two spp. of sw. N. Am. (Greek, *lyra,* a lyre, and *karpos,* fr.)

(Rollins, R. C. A revision of Lyrocarpa. Contr. Dudley Herb. 3: 169–173, 1941.)

1. **L. Còulteri** Hook. & Harv. var. **Pálmeri** (Wats.) Roll. [*L. P.* Wats.] Stems several, 3–5 dm. high, with flexuous leafy branches, suffrutescent; lvs. mostly canescent, lyrately pinnatifid, 3–5 cm. long, short-petioled, the segms. linear to oblong; fruiting pedicels 2–6 mm. long; sepals ca. 1 cm. long; petals tawny, 18–20 mm. long; silicles obcordate, 10–15 mm. long, nearly as wide, ± curled; seeds 3–5 in each locule, round, flat, ca. 3 mm. wide.—Among rocks and in canyons, below 2000 ft.; Creosote Bush Scrub; w. edge of Colo. Desert, Borrego region to L. Calif. Dec.–April.

33. Dithýrea Harv. SPECTACLE-POD

Annual or perennial, stellate-pubescent. Stems leafy, erect or decumbent. Lvs. sinuate-dentate to nearly entire. Fls. racemose. Sepals erect, stellate-tomentose, connivent above. Petals white to purplish or yellowish, broadly spatulate, slender-clawed. Stamens 6; anthers linear, sagittate. Silicles indehiscent or tardily dehiscent, strongly obcompressed and didymous, *i.e.,* with 2 round locules side by side. Style almost none; stigma large. Seeds 1 in each cell, with accumbent cotyledons. Two or 3 spp. of sw. N. Am. (Greek, *dis,* two, and *thureos,* shield, referring to the twin fr.)

1. **D. califórnica** Harv. [*Biscutella c.* Brew. & Wats.] Annual, with several decumbent stems from base, 1–3 dm. long; lvs. 2.5–7 cm. long, thickish, the lower obovate to oblong-ovate, coarsely sinuate-dentate, petioled, the upper oblong-ovate, subsessile; racemes dense at anthesis; pedicels ca. 2 mm. long; sepals 6–7 mm. long; petals white, 10–12 mm. long; silicles 8–10 mm. broad, notched above and below, with thickened tomentose margin; seeds flat, oblong, almost 3 mm. long.—Common in sandy places, below 4000 ft.; Creosote Bush Scrub; both deserts from Inyo Co. to Nev., Ariz., L. Calif. March–May.

Var. **marítima** (A. Davids.) A. Davids. ex Rob. in Gray. [*Biscutella c.* var. *m.* A. Davids. *D. m.* A. Davids.] Lower lvs. suborbicular, fleshy, subentire.—Coastal Strand; Los Angeles Co. to San Luis Obispo Co.

34. Physària Gray. DOUBLE BLADDER-POD

Low · silvery-stellate cespitose perennials; stems simple. Basal lvs. usually many, petioled, oblanceolate to obovate or almost round, entire to dentate; cauline lvs. few, entire to dentate. Fls. racemose, yellow. Sepals linear-oblong, pubescent, often cucullate at apex. Petals usually spatulate. Stamens 6. Silicles didymous, pubescent, dehiscent, inflated in ours, with apical sinus. Style slender, persistent. Seeds brown, wingless, several in each locule, with accumbent cotyledons. Ca. 14 spp. of sw. N. Am. (Greek, *phusa*, bellows, because of the inflated pod.)

(Rollins, R. C. The cruciferous genus Physaria. Rhodora 41: 392–415, 1939.)

1. **P. Chàmbersii** Roll. Stems 5–15 cm. long; radical lvs. round to obovate, entire or dentate, petioled, 3–6 cm. long; cauline lvs. entire, spatulate, 1–2 cm. long; sepals 6–8 mm. long; petals 10–12 mm. long; silicles 1–1.5 cm. long; style 6–8 mm. long; seeds round, flat, brown, 2–3 mm. broad.—Dry limestone slopes, 5000–8000 ft.; Pinyon-Juniper Wd.; Clark Mt., e. San Bernardino Co., Grapevine Mts., Inyo Co.; to Utah, Nev. May.

35. Lesquerélla Wats. BLADDER-POD

Low annual to perennial herbs, densely stellate-pubescent. Lvs. simple. Fls. racemose, yellow in ours. Sepals oblong to elliptical. Petals longer than sepals, obovate to spatulate. Stamens 6; anthers sagittate. Silicles generally inflated, subglobose to obovoid or ellipsoid; valves nerveless, dehiscent; locules 2–15-seeded; style slender; stigma entire or nearly so; seeds flattened, marginless or narrowly winged, with accumbent cotyledons. Ca. 40 spp., mostly N. Am. (L. *Lesquereux*, 1805–1889, American botanist.)

(Payson, E. B. Monograph of the genus Lesquerella. Ann. Mo. Bot. Gard. 8: 103–236, 1921.)
Plants annual, without basal rosette; style shorter than fr. . . . . 1. *L. Palmeri*
Plants perennial, with basal rosette; style equal to or longer than fr.
  Siliques subglobose, not flattened at apex and margins. White Mts. to San Bernardino Mts. and Nev.
                                                   2. *L. Kingii*
  Silicles ovoid, flattened at apex and margins. Placer Co. n. . . . . 3. *L. occidentalis*

1. **L. Pálmeri** Wats. [*L. Gordonii* var. *sessilis* Wats.] Annual, with slender ascending stems 1–3 dm. long; basal lvs. entire to lyrate with few lobes, oblanceolate, 2–5 cm. long, petioled; cauline lvs. linear to oblanceolate, subsessile, entire; fruiting pedicels ascending to recurved, sigmoid, 1–1.5 cm. long; sepals ca. 3 mm. long; petals broadly spatulate, ca. 5–7 mm. long; silicles subglobose, 3–4 mm. thick, sparsely stellate-pubescent; seeds round-oblong, flat, not winged, almost 2 mm. long.—Occasional in sandy places, below 3500 ft.; Creosote Bush Scrub; e. Mojave and ne. Colo. deserts; to Utah, Ariz. March–May.

2. **L. Kíngii** Wats. [*Vesicaria K.* Wats.] Silvery-stellate perennial; stems decumbent, 5–15 cm. long; basal lvs. ovate to suborbicular, entire, obtuse, 2–5 cm. long, petioled; cauline lvs. oblanceolate, 0.5–1.5 cm. long; pedicels spreading to recurved, sigmoid, 8–10 mm. long; sepals ca. 5 mm. long; petals narrowly spatulate, 6–7 mm. long; silicles subglobose, 3–5 mm. thick, stellate-pubescent; style 3–4 mm. long; seeds 2–4, not winged, brown, flattened with thickened edge, ca. 2 mm. long.—Dry rocky slopes, 5000–9000 ft.; mostly Pinyon-Juniper Wd.; desert mts. (White, Panamint, Providence, New York, Clark); to Nev. March–June.

Var. **cordifórmis** (Roll.) Maguire & Holmgren. [*Physaria c.* Roll.] Lvs. often narrow-spatulate; silicles 2–4 mm. long.—At 10,500–12,000 ft.; Bristle-cone Pine F.; White Mts., Mono Co.; Nev. July.

Ssp. **bernardìna** (Munz) Munz. [*L. b.* Munz.] Petals 9–10 mm. long; styles 6–9 mm. long.—Dry flats, at ca. 6600 ft.; Yellow Pine F.; Bear V., San Bernardino Mts. May–June.

3. **L. occidentàlis** Wats. [*Vesicaria o.* Wats.] Silvery-stellate perennial, the stems ascending, 0.5–2 dm. long; basal lvs. round to elliptic, entire or more often repand, petioled, 2–7 cm. long; cauline lvs. oblanceolate, entire, 1–1.5 cm. long; pedicels sigmoid, 8–15 mm. long; sepals ca. 4 mm. long; petals narrowly spatulate, 9–10 mm. long; silicles obovoid, stellate, the apex and margin flattened, 4–6 mm. long; style 4–5 mm. long; seeds usually 2 per locule, not margined, brown, 2 mm. long.—Dry slopes, 5500–8000 ft.; Red Fir F., Lodgepole F.; Siskiyou Co.; Ore. May–July. Plants with flowering stems prostrate and the basal portion of the lf.-blade entire or once shallowly toothed are ssp. *diversifòlia* (Greene) Maguire & Holmgren. [*L. d.* Greene].—Lake Co. and Placer Co. to n. Calif.; to Ore., Nev.

36. Phoenicaùlis Nutt.

Perennial herbs with simple or branched caudex, basal lvs. and leafy stems bearing racemose clusters of purple or rose-colored or yellow fls. Sepals oblong, erect, the lateral gibbous at base. Petals with broad blades and long claws. Anthers sagittate at base. Siliques flattened parallel with the partition, the valves 1-nerved. Stigma subentire, capitate. Seeds biseriate, convex or turgid, marginless with accumbent cotyledons. Two spp. of w. N. Am. (Greek, *phaino*, to show, and *kaulos*, stem.)

Siliques linear-lanceolate, divaricately spreading; racemes 10–15 cm. long ...... 1. *P. cheiranthoides*
Siliques ovate-lanceolate, suberect; racemes 2–3 cm. long .... 2. *P. eurycarpa*

1. **P. cheiranthoìdes** Nutt. in T. & G. [*Parrya c.* Jeps. *Hesperis Menziesii* Hook. as to description, but not as to type. *Phoenicaulis M.* auth. *Parrya M.* auth. *P. M.* var. *lanuginosa* Wats. *Phoenicaulis M.* ssp. *l.* Abrams.] Caudex thick, covered with remains of dead lvs.; scapose stems 6–20 cm. long, almost glabrous; lvs. spatulate or oblanceolate, entire, acute or obtuse, 3–10 cm. long, petioled, stellate-tomentose; fls. many; pedicels spreading, 10–20 mm. long in fr.; sepals ± glabrous, ca. 4 mm. long; petals 8–10 mm. long; siliques horizontal, 2–4 cm. long, glabrous, attenuate to the short slender style; seeds 2–4, brown, oblong, somewhat flattened, ca. 3 mm. long.—Dry granitic slopes and benches, 3800–10,700 ft.; Sagebrush Scrub, Yellow Pine F. to Subalpine F.; e. slope of Sierra Nevada from Inyo Co. to Modoc Co., Siskiyou Co. to Mendocino Co.; to Wash., Ida., Nev. May–July.

Ssp. **glàbra** (Jeps.) Abrams. [*Parrya Menziesii* var. *g.* Jeps. *P. cheiranthoides* var. *g.* Jeps.] Herbage glabrous; petals 6–7 mm. long.—Sandy alkaline areas, 4600 ft.; Shadscale Scrub; Modoc Co. May–June.

2. **P. eurycárpa** (Gray) Abrams. [*Draba e.* Gray. *Anelsonia e.* Macbr. & Pays. *Parrya e.* Jeps.] Stems 2–5 cm. high from very long well developed root systems; lvs. oblanceolate, 1–1.5 cm. long, stellate-pubescent; fls. few; pedicels 5–9 mm. long, ascending; petals yellow, siliques oblong-ovate, 2–2.5 cm. long, ca. 8 mm. wide, glabrous; seeds ca. 8–10, silvery, with crisped scales.—Slides and flats, 10,500–14,000 ft.; Alpine Fellfields; Sierra Nevada, Tuolumne Co. to Tulare and Inyo cos., White Mts.; Ida. July–Aug.

37. Hutchínsia R. Br.

Low annuals or winter annuals, ± pubescent with forked hairs. Lvs. entire to pinnately lobed. Fls. racemose, small, white. Stamens 6. Silicles elongate-ovate to elliptic or lanceolate, entire at apex, compressed contrary to the narrow partition, the valves wingless, 1-nerved. Seeds 2–many in each locule, with incumbent cotyledons. Ca. 8 spp. of N. Am. and Eurasia. (Named for Ellen *Hutchins*, Irish botanist, 1785–1815).

1. **H. procúmbens** (L.) Desv. [*Lepidium p.* L. *Capsella p.* Fries. *C. elliptica* C. A. Mey. *C. p.* var. *Davidsonii* Munz. *H. californica* and *desertorum* A. Davids.] Slender annuals, the stems branching from base, erect to procumbent, glabrous to pubescent, 5–18 cm. high; lower lvs. entire to pinnately lobed, oblanceolate, short-petioled, 1–2 cm. long; cauline lvs. scattered, linear or oblanceolate; pedicels slender, spreading, becoming 6–12 mm. long; fls. ca. 1 mm. long; silicles 2–4 mm. long, obtuse, elliptical or oval; seeds light brown, smooth, oblong, 0.5–0.6 mm. long; $2n = 12$ (Manton, 1932).—Scattered, moist alkaline places, up to 8600 ft.; many Plant Communities; San Diego Co. n., from desert edge to coast, White Mts.; to B.C. and Atlantic Coast; Old World. March–July.

38. Capsélla Medic. SHEPHERD'S-PURSE

Annuals or winter annuals, with forked pubescence. Basal lvs. tufted. Fls. small, white ór pink, racemose. Silicle obcordate-triangular, flattened contrary to the narrow partition, the valves boat-shaped, keeled. Style almost none. Seeds several, marginless, with incumbent cotyledons. Ca. 4 spp. of Eurasia. (Latin, *capsella*, a little box.)

1. **C. Búrsa-pastoris** (L.) Medic. [*Thlaspi B.-p.* L. and *Bursa B.-p.* Britton.] Erect, 2–5 dm. tall, branched or almost simple, ± hirsute at base; basal lvs. in a rosette,

runcinate-pinnatifid, petioled, 3–8 cm. long; cauline lvs. lanceolate, sessile with auricled base; pedicels slender, spreading or ascending, 10–15 mm. long; fls. white, ca. 2 mm. long; silicles obcordate, flattened, ca. 6 mm. wide, with straight or slightly convex sides; seeds narrow-oblong, brownish, smooth, ca. 1 mm. long; $n = 8$ (Shull, 1937).— A variable sp. growing as a common weed below 7000 ft., almost throughout Calif.; to Alaska and Atlantic Coast, natur. from Eu. Throughout the year.

39. Camelìna Crantz. FALSE-FLAX

Erect annual caulescent herbs. Lvs. entire to pinnatifid, sagittate-clasping. Fls. small, yellowish, in terminal racemes. Sepals equal. Stamens 6. Style slender. Silicle obovoid or pyriform, slightly flattened parallel with the broad septum, margined, the valves convex, 1-nerved. Seeds several to many in each locule, biseriate, oblong, marginless, with incumbent cotyledons. Ca. 5 spp. of Eurasia. (Greek, *camai*, dwarf, and *linon*, flax.)

1. **C. satìva** (L.) Crantz. [*Myagrum s.* L.] Erect annual, 3–8 dm. high, usually branched; stems glabrous or with minute appressed stellate trichomes; lower lvs. ob-lanceolate, petioled, 5–8 cm. long, toothed or entire, acutish; cauline lvs. largely clasping, entire, smaller; racemes with many fls.; pedicels 1–2.5 cm. long; petals yellowish, 4–5 mm. long; silicles mostly 7–9 mm. long, 6–7 mm. broad, with rather hard walls, 3–4 times as long as style; seeds yellow-brown, 3-angled, 1–2 mm. long, finely tubercled; $2n = 40$ (Manton, 1932).—Occasional weed of fields and waysides, through much of the state; natur. from Eu. May–July.

2. **C. microcárpa** Andrz. Stems with simple and branching rather elongate hairs; silicles ca. 5 mm. long, 4–5 mm. wide, and twice as long as style; $2n = 40$ (Manton, 1932).—Reported from San Gabriel Mts., Los Angeles Co.; more common in Wash., Ore. and e. states; native of Eu.

40. Dràba L.

Annual or perennial herbs, mostly low, tufted, and often with stellate or branched hairs. Stems scapose or leafy. Lvs. simple. Fls. racemose, perfect, white or yellow. Sepals equal at base. Petals entire to bifid. Silicles elliptical, oblong, or rarely linear, usually flat, with nerveless dehiscent valves compressed parallel with the partition. Stigma nearly entire. Seeds biseriate, numerous, usually marginless, with accumbent cotyledons. Ca. 250 spp. of the N. Hemis. (Greek, *drabe*, acrid, applied by Dioscorides to some cress.)

(Hitchcock, C. L. A revision of the Drabas of w. N. Am. Univ. Wash. Pub. Biol. 11: 7–132, 1941.)
A. Plants annual; style scarcely if at all evident.
  B. Lvs. all in a basal rosette; petals bifid .... 1. *D. verna*
  BB. Lvs. not all in a basal rosette; petals not bifid, but entire to emarginate.
    C. Fls. white, dimorphous, the smaller cleistogamous; pedicels usually shorter than silicles.
      D. Pedicels mostly pubescent; lvs. usually dentate .... 2. *D. cuneifolia*
      DD. Pedicels always glabrous; lvs. usually entire .... 3. *D. reptans*
    CC. Fls. yellow, not dimorphous; pedicels usually equaling or longer than silicles.
      D. Silicles acute, mostly 8–12 mm. long. Modoc Co. to Tulare Co. .. 4. *D. stenoloba*
      DD. Silicles rounded or obtuse, mostly 4–8 mm. long. Siskiyou Co. .. 5. *D. nemorosa*
AA. Plants perennial; style usually well developed.
  B. Silicles leathery, inflated, 1–2-seeded; lvs. leathery; fls. white or faintly yellowish.
    C. Silicles 4–6 mm. long, the valves not costate; lvs. 5–12 mm. long, the midrib prominent. San Bernardino Mts. and Sierra Nevada .... 6. *D. Douglasii*
    CC. Silicles 2–3 mm. long, the valves costate; lvs. 2–4 mm. long, the midrib obscure. Masonic Mt., Mono Co. .... 7. *D. quadricostata*
  BB. Silicles membranous, seldom inflated, or plants otherwise not like *D. Douglasii*.
    C. Lvs. glabrous or with mostly unforked hairs only; lvs. all in a basal rosette.
      D. Seeds winged; lvs. obovate; silicles 5–10 mm. long. Siskiyou Co.
        8. *D. pterosperma*
      DD. Seeds wingless. Sierra Nevada.
        E. Styles not over 0.2 mm. long; lvs. not imbricated; petals 2–3 mm. long. Rare .... 17. *D. crassifolia*
        EE. Styles 0.5–1 mm. long; lvs. imbricated; petals mostly 3–6 mm. long.
          F. Lvs. linear to linear-lanceolate, the midrib prominent on lower side; silicles 2–7 mm. long; seeds nearly 2 mm. long ..... 9. *D. densifolia*
          FF. Lvs. lanceolate to obovate, the midribs not prominent; silicles 4–11 mm. long; seeds 1–1.5 mm. long .... 10. *D. Lemmonii*

CC. Lvs. pubescent, at least some of the hairs forked or stellate.
  D. Flowering stems with at least 1 lf.; seeds not winged.
    E. Fls. white, 2–3 mm. long; styles less than 0.3 mm. long.
      F. Flowering stems glabrous except sometimes at base; silicles oblong-ovate, mostly glabrous. Cirque Peak, Tulare Co. ... 11. *D. fladnizensis*
      FF. Flowering stems mostly pubescent; silicles linear- to oblong-lanceolate, pubescent.
        G. Basal lvs. 5–15 mm. long; stems 3–15 cm. high; frs. 4–9 mm. long, twisted, with stellate hairs. Sierra Nevada .. 12. *D. Breweri*
        GG. Basal lvs. 10–30 mm. long; stems 10–30 cm. high; frs. 8–14 mm. long, plane, with simple or forked hairs. Dana Plateau, Mono Co. 13. *D. praealta*
    EE. Fls. yellow.
      F. Petals 2–3 mm. long; styles not over 0.2 mm. long .. 17. *D. crassifolia*
      FF. Petals 3–8 mm. long; styles 1 mm. long or more.
        G. Mature silicles over 12 mm. long, usually over 4 mm. broad; style 1–1.5 mm. long. Sierra Co. and Mt. Lassen ...... 14. *D. aureola*
        GG. Mature silicles shorter, or if as long, less than 4 mm. broad; style mostly 1.5–3.5 mm. long.
          H. Silicles 2–3 mm. wide; hairs of lvs. simple, forked and stellate. S. Calif. .... 15. *D. corrugata*
          HH. Silicles 3–5 mm. wide; hairs of lvs. all stellate. N. Calif. 16. *D. Howellii*
DD. Flowering stems leafless, or if with 1–2 lvs. near base, the seeds are winged.
  E. Lower side of lvs. with pectinately branched sessile appressed hairs, the axis of many of them paralleling the axis of the lvs. Sierra Nevada 18. *D. oligosperma*
  EE. Lower side of lvs. lacking hairs of above type.
    F. Lvs. less than 2 mm. wide.
      G. The lvs. with many unbranched as well as branched cilia on their margins; flowering stems 1–5 cm. tall. Lake Tahoe region 19. *D. Paysonii*
      GG. The lvs. with no unbranched cilia on margins.
        H. Flowering stems 1–3 cm. tall; pedicels 1–4 mm. long. Inyo Co. .... 20. *D. sierrae*
        HH. Flowering stems 5–8 cm. tall; pedicels 5–10 mm. long. Tulare Co. to Lake Tahoe .... 21. *D. cruciata*
    FF. Lvs. 2–7 mm. wide.
      G. Silicles lanceolate, 1.3–3.5 mm. wide; seeds not winged, 1–1.4 mm. long .... 21. *D. cruciata*
      GG. Silicles ovate, 3–6 mm. wide; seeds winged, ca. 2 mm. long 22. *D. asterophora*

1. **D. vérna** L. WHITLOW-GRASS. Annual, with lvs. all basal, spatulate to oblanceolate, 1–2.5 cm. long, entire or denticulate, short pilose-hirsute with branched hairs; scapes slender, 5–20 cm. high, glabrous or pubescent below; racemes 3–30-fld.; pedicels ascending, 15–25 mm. long; sepals ca. 1.5 mm. long; petals white, ca. 2.5 mm. long, bifid; silicles elliptic to elliptic-oblanceolate, 5–10 mm. long, 1.5–4 mm. broad, glabrous; styles ca. 0.1 mm. long; seeds 30–60, brown, 0.4–0.7 mm. long; n = 7, 15, 32 (Manton, 1932).—Sterile often gravelly soil, below 6000 ft.; mostly Mixed Evergreen F. and Yellow Pine F.; Mariposa and Napa cos. n.; widely natur. in U.S. from Eu. Feb.–May.

Var. **aestivàlis** Lejeune. Silicles elliptic-obovate to obovate, 3–5 mm. long and 3–4 mm. broad.—With the sp.

2. **D. cuneifòlia** Nutt. ex T. & G. Annual, simple or branched from base, 5–25 cm. tall; at least part of pubescence on lower stems not branched; lvs. ovate to obovate or oblanceolate, 1–5 cm. long, mostly with some teeth at least in upper half; infl. stellate-pubescent, 10–70-fld., usually less than half the height of the plant; pedicels 2–5 mm. long; sepals with branched hairs and 1.5–2.5 mm. long; petals white, 3.5–5 mm. long; silicles glabrous or hispidulous with simple hairs, linear to oblong-obovate, 5–15 mm. long, ca. 2–2.5 mm. broad; style nearly lacking; seeds 40–200, ca. 0.7 mm. long.—Moist sandy soil, 5000–6000 ft.; Pinyon-Juniper Wd.; New York Mts., e. Mojave Desert; to Tex. May.

Var. **integrifòlia** Wats. [*D. i.* Greene. *D. sonorae* Greene. *D. s.* var. *i.* O. E. Schulz. *D. c.* var. *s.* Parish. *D. c.* var. *brevifolia* Wats. ex Munz.] Pubescence practically all branched; racemes half as long as whole stem; silicles with evident style, stellate or glabrous.—Rather common in shaded places, below 5000 ft.; Chaparral, Coastal Sage Scrub, Creosote Bush Scrub, Pinyon-Juniper Wd.; w. Fresno Co., Los Angeles Co. to L. Calif., Inyo Co. to Imperial Co.; to Utah and n. Mex. Feb.–May.

3. **D. réptans** (Lam.) Fern. [*Arabis r.* Lam.] Annual, much like *D. cuneifolia* but with lvs. usually entire; stems with few stalked stellate hairs below and glabrous above; pedicels always glabrous; silicles usually nearly erect, linear, seldom as much as 2 mm. broad. The forma *micrántha* (Nutt.) C. L. Hitchc. [*D. m.* Nutt.] with upper surface of lvs. having simple hairs and with silicles hispidulous, has been reported from Emigrant Gap, Placer Co. Var. *stellífera* (O. E. Schulz) C. L. Hitchc. [*D. caroliniana* var. *micrantha* f. *s.* O. E. Schulz. *D. cuneifolia* var. *californica* Jeps.] with at least some branched hairs on upper surface and silicles with stiff appressed hairs, has been found on Crooked Creek, White Mts., Inyo Co., and in Surprise Canyon, Panamint Mts.

4. **D. stenolòba** Ledeb. var. **nàna** (O. E. Schulz) C. L. Hitchc. [*D. nitida* var. *n.* O. E. Schulz.] Winter annuals; lvs. mostly basal, obovate to oblanceolate, 1–4 cm. long, 4–8 mm. broad, usually denticulate, hispidulous with simple or branched hairs; stems simple or branched, 5–30 cm. tall, glabrous below or with simple or forked hairs, glabrous above, with 1–8 ovate lvs.; racemes 10–30-fld.; lower pedicels ca. as long as silicles; sepals pilose, 1–2 mm. long; petals yellowish, spatulate, 2–4 mm. long; silicles usually glabrous, sometimes with rather stiff hairs, linear to linear-oblong, acute, 8–22 mm. long, 1.5–2.3 mm. broad; styles quite lacking; seeds 16–40, brownish, 0.8–1 mm. long.—Mostly rather damp, often shaded places, 7000–12,000 ft.; Lodgepole F., Subalpine F., Alpine Fell-Fields; Sierra Nevada from Tulare Co. n. to Modoc Co., White Mts.; to Canada and Rocky Mts. June–Aug.

Var. **ramòsa** C. L. Hitchc. Stems branched from each lf.-axil; lower pedicels longer than silicles; silicles evenly and finely short-pubescent.—Region of Lake Tahoe.

5. **D. nemoròsa** L. Winter annual, simple or branched, 5–25 cm. tall; lvs. ovate-lanceolate to obovate-spatulate, mostly on lower third of plant, 1–3 cm. long, hirtellous with branched and simple hairs; cauline lvs. denticulate to dentate; stems chiefly glabrous above; fls. 10–40; pedicels 1–5 times as long as silicles; sepals ca. 1.5 mm. long; petals pale yellow, 4 mm. long, usually emarginate; silicles elliptic to oblong, mostly 5–10 mm. long, 2–3 mm. broad, rounded to obtuse, glabrous to hispidulous with short simple hairs; style 0; seeds 25–50, ca. 0.7 mm. long.—Collected near Yreka, Siskiyou Co.; to B.C., Colo. April–June.

6. **D. Douglásii** Gray var. **Cróckeri** (Lemmon) C. L. Hitchc. [*D. C.* Lemmon.] Tufted perennial; lvs. basal, thick and leathery, oblanceolate with prominent midrib, 5–12 mm. long, 1–2 mm. wide, ciliate with stiff simple or forked hairs and often with some hairs on surfaces; scapes 1–3 cm. tall, pubescent with slender straight unbranched fairly stiff hairs; racemes 2–10-fld.; pedicels ca. as long as silicles; sepals 2–2.5 mm. long, subglabrous; petals white, 4–5 mm. long; silicles ovoid, with leathery walls, little flattened, 4–7 mm. long, with short spreading, simple hairs; styles 0.5–1.5 mm. long; seeds 1–2, ca. 2 mm. long.—Occasional on dry rocky slopes, 5000–8000 ft.; Yellow Pine F., Lodgepole F.; Bear V., San Bernardino Mts., Sierra Nevada, Sierra and Plumas cos., Round V., Mendocino Co.; w. Nev. May–June. Plants from Warner Mts., Modoc Co. with strigulose silicles are apparently *D. Douglasii* proper, which ranges to Wash.

7. **D. quadricostàta** Roll. Cespitose perennial, forming dense mats 2–30 cm. broad; caudex much branched; lvs. oblong, obtuse, pubescent above and below with simple and branched hairs, obtuse, 2–4 mm. long, densely clothing the lower stems; scapes slender, 2–5 cm. long, hirsute with simple and forked hairs; racemes 5–10-fld.; lower fls. subtended by leaflike bracts; pedicels 2–3 mm. long; sepals pubescent, 2–3 mm. long; petals faintly yellowish, 3–5 mm. long; silicles ovoid, 4-ribbed, 3–4 mm. long, ca. 2 mm. broad, pubescent with short mostly simple hairs; seeds 1–2, plump, ca. 1.5 mm. long.—Dry decomposed granite and rock-crevices, 8100–8600 ft.; Sagebrush Scrub; region of Masonic Mt., Mono Co. July.

8. **D. pterósperma** Pays. Tufted perennial with many prostrate matted stems; lvs. oblong-spatulate to obovate-oblanceolate, 3–6 mm. long, 1.3–2 mm. broad, ciliate with slender, tangled, simple and branched hairs and with similar hairs beneath; scapes several, without lvs., 3–10 cm. tall, soft-pubescent with slender branched hairs, 5–10-fld.; pedicels 3–6 mm. long; sepals 3–4 mm. long, hirsute with branched hairs; petals yellow, ca. 6 mm. long; silicles ovate to lance-ovate, 5–10 mm. long, 3–5 mm. wide, with forked or stellate hairs; style 1.5–3.5 mm. long; seeds 4–12, the body ca. 1.7 mm. long, surrounded by a thin wing ca. 0.25 mm. wide.—Marble and limestone crevices and gravel, 5000–7000 ft.; largely Red Fir F.; Marble Mts., Siskiyou Co. June–July.

9. **D. densifòlia** Nutt. ex T. & G. [*D. glacialis* var. *pectinata* Wats. *D. oligosperma*

var. *p.* Jeps. *D. Nelsonii* Macbr. & Pays. *D. caeruleomontana* Pays. & St. John. *D. sphaerula* Macbr. & Pays.] Cespitose perennial; lvs. linear to linear-oblanceolate, 2–9 mm. long, 0.5–3 mm. broad, usually with midrib prominent beneath, ciliate with stiff straight cilia, otherwise glabrous or with few coarse stellate hairs beneath, usually glabrous above; scapes leafless, 1–12 cm. tall, glabrous or with few simple or branched hairs near base or hirsute throughout; racemes 3–15-fld.; pedicels 3–5 mm. long; sepals 2–3 mm. long, glabrous or nearly so; petals yellow, 2–6 mm. long; silicles ovate to somewhat elliptic, 2–7 mm. long, 2–3.5 mm. broad, rather coarse-pubescent with simple and stellate hairs; styles 0.5–1 mm. long; seeds 2–12, ca. 2 mm. long.—Dry rocky places, 8500–13,000 ft.; Subalpine F., Alpine Fell-fields; Sierra Nevada from Fresno Co. n. to Modoc Co.; to Wash., Rocky Mts. July–Aug.

10. **D. Lemmònii** Wats. [*D. longisquamosa* O. E. Schulz.] Cespitose perennial, spreading by decumbent stems; lvs. basal, tufted, thickish, obovate to oblanceolate, 5–20 mm. long, 2–7 mm. wide, hirsute on both surfaces and ciliate with stiff simple and once-forked hairs, also rather stiffly hirtellous; scapes leafless, 2–12 cm. tall, hirsute with simple or branched hairs; pedicels hirsute, equaling silicles; sepals ca. 2.5 mm. long; petals yellow, 4–6 mm. long; silicles oval to lance-ovate, mostly hirsute, 5–10 mm. long, 2.5–5 mm. wide, usually curved or contorted; styles 0.5–1 mm. long; seeds 4–16, ca. 1–1.5 mm. long.—Gravelly and stony slopes, crevices, talus, etc., 8500–13,000 ft.; Subalpine F., Alpine Fell-Fields; Sierra Nevada from Tulare to Tuolumne and Mono cos. July–Aug.

Var. **incrassàta** Roll. Scapes, pedicels and silicles glabrous; lvs. hairy on margins only and with simple hairs.—Sweetwater Mts., Mono Co.

11. **D. fladnizénsis** Wulf. [*D. Pattersonii* O. E. Schulz.] Low perennials from simple or multicipital caudex; lvs. nearly all basal, mostly oblanceolate, 5–10 mm. long, 1–2 mm. broad, ciliate with long simple hairs and sometimes moderately pubescent with forked ones, the midribs quite prominent; stems 2–7 cm. high, glabrous or pubescent near base with simple or forked hairs, leafless or with 1–2 small denticulate or entire lvs.; fls. 3–12; pedicels 1–4 mm. long; sepals 1–2 mm. long, usually glabrous; petals white, 2–3 mm. long; silicles oblong-ovate, 3–6 mm. long, 1.5–2 mm. broad, mostly glabrous; styles scarcely evident; seeds 10–20, ca. 0.8 mm. long; $2n = 16$ (Löve & Löve, 1948).—Rare, dry scree and talus, 11,000–12,000 ft.; Alpine Fell-fields; Sierra Nevada, Tulare Co.; B.C., Colo., Utah; Eurasia. Aug.

12. **D. Brèweri** Wats. [*D. B.* var. *sublaxa* Jeps.] Somewhat cespitose perennial, usually from branched caudex; lvs. mostly basal, tufted, linear-oblanceolate to obovate-oblanceolate, 8–12 mm. long, 2–4 mm. wide, grayish with short branched hairs, usually with a few long simple cilia; cauline lvs. 1–3 or 4, smaller, remotely denticulate; stems 3–12 cm. high, mostly with branched hairs; racemes 10–25-fld.; pedicels 1–5 mm. long; sepals ca. 1.5 mm. long; petals white, ca. 3 mm. long; silicles suberect, usually contorted, lance-oblong, 4–9 mm. long, 2–3 mm. wide, stellate-pubescent; styles ca. 0.25 mm. long; seeds 10–30, ca. 0.7 mm. long.—Rocky slopes, 8500–13,000 ft.; Subalpine F., Alpine Fell-fields; Sierra Nevada from Tulare Co. n. to Mt. Shasta. July–Aug.

13. **D. praeálta** Greene. Slender-stemmed biennial or perennial, with simple or branched caudex; lvs. mostly basal, forming compact rosettes, usually oblanceolate, 1–3 cm. long, 2–6 mm wide, entire or few-toothed, pubescent with branched stalked hairs; stems 1–several, 1–3 dm. tall, pubescent; racemes 5–30-fld.; pedicels not exceeding silicle; sepals 1.5–2 mm. long, soft-pilose to stellate; petals white, 2.5–3.5 mm. long; silicles lance-linear to lanceolate, 8–14 mm. long, 2–2.5 mm. wide, soft-pubescent with simple or forked hairs; styles very short; seeds 20–70, ca. 0.8 mm. long.—Among willows, etc., 9600 ft. near Lake Sabrina, Inyo Co., and at 11,200 ft., Dana Plateau, Mono Co.; to Wash. and Rocky Mts. July–Aug.

14. **D. aurèola** Wats. Biennial or perennial; lvs. numerous, in dense basal cushions and on lower stems, oblanceolate, 10–25 mm. long, 2–4 mm. wide, grayish with coarse branched hairs; stems simple or branched, 6–15 cm. tall, densely hirsute with coarse simple and branched hairs; racemes densely 20–80-fld.; pedicels 5–8 mm. long; sepals ca. 3 mm. long, stellate; petals yellow, ca. 5 mm. long; silicles oblong or oblong-ovate, 11–15 mm. long, 3–6 mm. wide, flat, coarsely pubescent with branched hairs; styles 1–1.5 mm. long; seeds 20–30, ca. 1.5 mm. long.—Volcanic soil and cinders, 9000–11,000 ft.; Alpine Fell-fields; Mt. Lassen; to Ore., Wash. July–Aug.

15. **D. corrugàta** Wats. Perennial, tufted, with simple or branched crown; lvs. mostly

in cushion-like rosettes, oblanceolate, 15–30 mm. long, 3–10 mm. wide, entire, grayish with dense coarse stiff simple and branched or stellate hairs; stems 1–several, mostly branched, leafy, hirsute, 5–20 cm. high; racemes usually many-fld.; pedicels pubescent, 2–10 mm. long; sepals ca. 2 mm. long, with branched hairs; petals pale yellow, fading white, 3–5 mm. long, emarginate; silicles flat or contorted, often purplish, with minute branched hairs, linear-elliptic or somewhat wider, 7–15 mm. long, 2–3 mm. wide; styles 2–3.5 mm. long; seeds 20–30, ca. 1.2–1.5 mm. long.—Shaded slopes and rocky places, 7000–11,480 ft.; Red Fir F., Subalpine F.; San Bernardino Mts. July–Aug. Forma *vestìta* (A. Davids.) C. L. Hitchc. [*D. vestita* A. Davids.] has the silicles grayish with looser branched hairs; mostly San Gabriel Mts., but also on Mt. San Gorgonio.

Var. **saxòsa** (A. Davids.) M. & J. [*D. saxosa* A. Davids.] Stems mostly simple, leafless; silicles with some branched hairs.—Dry slopes, above 8000 ft.; Red Fir F., Subalpine F.; Santa Rosa and San Jacinto mts. July–Aug.

16. **D. Howéllii** Wats. Cespitose perennial, with rather loosely branched caudex; lvs. obovate, 5–20 mm. long, 2–7 mm. wide, entire or remotely denticulate, densely hispidulous with 4-rayed hairs and with marginal stalked cruciform hairs; fruiting stems 4–12 cm. high, mostly simple, leafless or with 1–3 lvs., soft-pubescent; racemes 3–30-fld.; pedicels usually longer than silicles; sepals 2–3 mm. long; petals yellow, 5–8 mm. long; silicles ovate-elliptic, 5–10 mm. long, 3–5 mm. broad, soft-stellate or glabrous; styles 1.2–2 mm. long; seeds 10–20, 1.2–1.6 mm. long.—Crevices of rocks, 7000–8300 ft.; Subalpine F.; Salmon-Trinity Alps, Marble Mts., Siskiyou Mts. of nw. Calif.; s. Ore. July–Aug.

Var. **carnosùla** (O. E. Schulz) C. L. Hitchc. [*D. c.* O. E. Schulz.] Plant glabrous except for marginal stellate hairs.—At 6800 ft.; Red Fir F.; Mt. Shasta, Devil's Canyon Mts., Trinity Co.

17. **D. crassifòlia** Grah. Biennial or perennial, with simple or branched crowns; lvs. nearly all basal, linear-spatulate to narrowly oblanceolate, 10–25 mm. long, 2–4 mm. broad, ciliate, the upper surfaces usually with few to several simple appressed hairs, or with few forked ones, the midribs not greatly thickened; stems 1–several, 2–20 cm. tall, simple, usually sparsely hairy near base, glabrous above, leafless or with 1 lf. below; racemes 3–20-fld.; pedicels 2–10 mm. long; sepals ca. 1 mm. long; petals yellow, white on fading, 2–3 mm. long; silicles narrowly elliptic to somewhat lanceolate, 5–12 mm. long, 2–3 mm. wide, glabrous; styles almost lacking; seeds 10–60, ca. 1 mm. long; $2n = 40$ (Heilborn, 1941).—Dry rocky places, 10,700–11,700 ft.; Subalpine F., Alpine Fell-fields; Sierra Nevada (Bullfrog Lake, Sky Blue Lake, and Rock Creek, Tulare Co.); Wash. and Ore. to Rocky Mts. July–Aug.

Var. **nevadénsis** C. L. Hitchc. Stems pubescent throughout and with 1–2 lvs.; silicles pubescent with short simple hairs.—At 11,500 ft., Coyote Ridge, nw. Inyo Co.; Nev.

18. **D. oligospérma** Hook. Cespitose matted perennial; lvs. imbricate, linear-lanceolate to -oblanceolate, 3–11 mm. long, 1–1.7 mm. wide, at least the lower surface with appressed pectinately branched hairs, their axis largely parallel with the midrib, margins of lvs. somewhat ciliate with ± pectinately branched hairs; scapes 1–10 cm. high, sometimes with pectinate hairs at base, otherwise glabrous; racemes 3–15-fld.; pedicels 3–10 mm. long; sepals 2–2.5 mm. long; petals yellow, 3–4.5 mm. long; silicles elliptic to ovate, 3–7 mm. long, 2–4 mm. wide, flat, with short stiff simple or branched retrorsely appressed hairs; styles 0.1–1 mm. long; seeds 2–10, 1.5–1.8 mm. long.—Open rocky places, 11,000–13,000 ft.; Alpine Fell-fields; Sierra Nevada, Eldorado Co. to Inyo Co.; to B.C., Rocky Mts. July–Aug.

Var. **subséssilis** (Wats.) O. E. Schulz. [*D. s.* Wats.] A dwarf plant; petals little longer than sepals; styles 0.1–0.4 mm. long; silicles hispidulous with hairs tending to be branched.—At 11,400–14,200 ft.; Alpine Fell-fields; e. slope of Sierra Nevada, Mono and Inyo cos. July.

19. **D. Paysònii** Macbr. var. **Treleàsii** (O. E. Schulz) C. L. Hitchc. [*D. barbata* var. *T.* O. E. Schulz. *D. densifolia* auth., not Nutt.] Matted perennial; lvs. densely imbricate, curved, linear to linear-oblanceolate, 4–14 mm. long, ca. 1 mm. wide, with rather prominent midrib, ciliate with simple and curved or branched hairs ca. 1 mm. long, the lower surface with long tangled branched hairs; scapes leafless, 1–5 cm. tall; racemes 3–10-fld.; pedicels 2–6 mm. long; sepals 1.5–2.5 mm. long; petals light yellow, 2–4 mm. long; silicles ± ovate, 3–5 mm. long, 2.5–3.5 mm. wide, somewhat

inflated, hispidulous; styles 0.5–0.7 mm. long; seeds 4–10, ca. 1.3–1.8 mm. long.—
Rocky ridges, 9000 ft. and higher; Subalpine F., Alpine Fell-fields; Eldorado, Placer,
Nevada, and Mono cos.; to Wash., Mont. July–Aug.

20. **D. siérrae** C. W. Sharsm. Pulvinate cespitose perennial, with persistent lvs.
densely imbricate, 2–6 mm. long, ca. 1 mm. wide, grayish with fine short branched
hairs and ciliate with pectinately branched ones; scapes leafless, 2–8-fld., 1–3 cm.
tall, densely stellate-pubescent; pedicels 1–4 mm. long; sepals stellate-pubescent, ca.
3 mm. long; petals yellow, 4–5 mm. long; silicles ovate, 3–8 mm. long, 2–4 mm. wide,
finely stellate-pubescent, ± contorted; styles 0.4–0.9 mm. long; seeds 3–6, ca. 1 mm.
long.—Dry stony places, 11,000–12,500 ft.; Alpine Fell-fields; e. slope of Sierra Nevada,
Inyo Co. July.

21. **D. cruciàta** Pays. [*D. nivalis* var. *californica* Jeps.] Cespitose perennial, with
many slender prostrate stems; lvs. oblanceolate to almost obovate, 5–12 mm. long, 1–3
mm. wide, entire or denticulate, hispidulous with stiffish forked hairs and with stalked
stellae on margins; scapes slender, 5–8 cm. tall, stellate below, glabrous above; racemes
5–20-fld.; pedicels ca. as long as silicles; sepals ca. 2 mm. long, glabrous; petals yellow,
5–6 mm. long; silicles lanceolate, 6–10 mm. long, 2–3.5 mm. wide, glabrous or slightly
hispidulous; styles 0.3–0.6 mm. long; seeds 8–16, 1–1.4 mm. long.—At ca. 9000–10,000
ft.; Subalpine F.; Mineral King, Tulare Co., and near Lake Tahoe. July–Aug.

Var. **integrifòlia** Hitchc. & Sharsm. Lvs. entire, usually with long simple or forked
cilia as well as stellae; silicles 8–14 mm. long; styles 1–2 mm. long.—Stony places,
10,000–12,400 ft.; Alpine Fell-fields; University Peak, Fresno Co., and Mt. Whitney
region, Inyo Co. July–Aug.

22. **D. asteróphora** Pays. Diffuse cespitose perennial, the vegetative branches prostrate;
lvs. in basal rosettes and on short sterile branches, obovate, 5–12 mm. long, 2–7 mm.
wide, finely pubescent with stellate or cruciform stalked hairs; scapes many, 3–8 cm.
tall, often stellate below, glabrous above; racemes 10–25-fld.; pedicels shorter than
to longer than silicles; sepals 2–2.5 mm. long, glabrous or slightly pilose; petals yellow,
4–5 mm. long; silicles ovate to somewhat elliptic, flat, 5–10 mm. long, mostly glabrous;
styles 0.5–0.8 mm. long; seeds 4–10, the body ca. 2 mm. long, with a wing an additional 0.5 mm. wide.—Rock-crevices and talus, 8000–10,200 ft.; Lodgepole F., Subalpine F.; Sierra Nevada, Eldorado and Tuolumne cos. July–Aug.

Var. **macrocárpa** C. L. Hitchc. Sepals ca. 3 mm. long; petals ca. 6 mm. long; silicles
broadly lanceolate, 10–15 mm. long; styles 1–2 mm. long.—Cup Lake, Sierra Nevada.
July.

41. Athýsanus Greene

Diffuse slender-stemmed annual, branched from near base, pubescent with forked
spreading hairs. Lvs. all near base, simple, few-toothed, ovate-oblong. Fls. minute, in
loose unilateral elongate racemes. Pedicels slender, recurved. Sepals equal. Petals linear
or 0. Stamens 6, almost equal. Ovary 1-celled, 2-ovuled. Silicles orbicular, indehiscent,
1-seeded, uncinate-hispid. Seeds round, glabrous, flat, light brown. One sp. of w. U.S.
(Greek, *a*, without, and *thusanos*, fringe, the fr. wingless.)

1. **A. pusíllus** (Hook.) Greene. [*Thysanocarpus p.* Hook. A. *p.* var. *glabrior* Wats.]
Plants 1–3 dm. high; lvs. few, 5–20 mm. long; fls. ca. 1.5 mm. long; silicles 1.5–2 mm.
long.—Common but inconspicuous, in grassy and brushy places, mostly below 5000
ft.; many Plant Communities; almost throughout cismontane Calif. to desert edge; n.
to B.C., Ida. March–June.

42. Heterodràba Greene

Much like *Athysanus* in habit. Lvs. cuneate-obovate to oblanceolate, entire to few-
toothed. Racemes lax, many-fld.; fls. minute, white. Pedicels recurved in fr. Silicles
round-oval, compressed, 2-celled, finally dehiscent, twisted, pubescent with somewhat
stellate hairs. Seeds 6–12, hispidulous, flat, oval-oblong. One sp. (Greek, *heter*, different,
and *Draba*, the name of a genus.)

1. **H. unilateràlis** (Jones) Greene. [*Draba u.* Jones. *Athysanus u.* Jeps.] Stems 1–4
dm. long, ascending to trailing; lvs. 1–2.5 cm. long; fls. 2 mm. long; silicles round-oval,

3–5 mm. long, hispidulous but not uncinate.—Grassy slopes and flats, below 1800 ft.; V. Grassland, Foothill Wd.; L. Calif., Tehachapi Mts. in and about Great Cent. V. to Ore. March–April.

43. Thysanocárpus Hook. Lace-Pod. Fringe-Pod

Erect slender simple or branched annuals. Fls. minute, white to purplish, in slender racemes. Sepals ovate, spreading. Petals cuneate to spatulate. Stamens mostly 6, subequal. Ovary 1-celled, compressed, 1-ovuled. Silicle indehiscent, plano-convex or biconvex, strongly compressed, rounded in outline, with the wing ± crenate or even perforate and with ± highly developed radiating nerves. Ca. 4 or 5 spp. of w. N. Am. (Greek, *thusanos*, fringe, and *karpos*, fr.)

Fruiting pedicels straight or recurved at tip only; silicles 8–10 mm. wide, the wing with narrow almost filiform rays. About the Sacramento V. .... 1. *T. radians*
Fruiting pedicels ± recurved their whole length; silicles 3–7 mm. wide, the wings with broad or no rays.
  Cauline lvs. auriculate, the basal lvs. rosulate, usually hirsute .... 2. *T. curvipes*
  Cauline lvs. not auriculate, the basal not rosulate, usually glabrous .... 3. *T. laciniatus*

1. **T. ràdians** Benth. [*T. r.* var. *montanus* Jeps.] Glabrous, 1.5–4.5 dm. high, with few ascending rather simple branches; basal lvs. runcinate-pinnatifid, 1.5–5 cm. long; cauline lvs. oblong-ovate to lance-ovate, auriculate, subentire; racemes elongate, with scattered fls.; pedicels ascending, the tips only recurved, ca. 1–1.5 cm. long; sepals ca. 2 mm. long; petals slightly longer; silicles orbicular, 8–10 mm. wide, glabrous to tomentose on the body and with the broad wing entire, white-membranous, with slender dark radiating nerves.—Moist grassy places, below 3000 ft.; V. Grassland, Foothill Wd.; Sacramento V. and bordering foothills w. to Stanford Univ., Santa Rosa, etc.; to s. Ore. March–May.

2. **T. cúrvipes** Hook. [*T. pulchellus* F. & M. *T. c.* var. *p.* Greene. *T. c.* var. *involutus* Greene. *T. c.* var. *cognatus* Jeps. *T. foliosus* Heller.] Erect, 2–5 dm. tall, ± pubescent or hirsute at base, branched above; basal lvs. in rosette, sinuate-dentate to subentire, oblong in outline, 2–5(–10) cm. long; cauline lvs. lanceolate, sagittate, clasping, the upper entire; pedicels slender, recurved, 3–7 mm. long; fls. ca. 1 mm. long; silicles round-obovate, often plano-convex, glabrous to pubescent, mostly 3–4 mm. wide including the wing, the latter entire or crenate, sometimes perforate, with broad rays; style ca. 0.5 mm. long; $2n = 28$ (Manton, 1932).—Common on grassy or brushy slopes, mostly below 5000 ft.; V. Grassland, Coastal Sage Scrub, Chaparral, Foothill Wd., etc.; most of cismontane Calif.; to B.C., Mont. March–May. Exceedingly variable.

KEY TO VARIETIES

Wing of silicle rayed. Plants mostly pubescent below. Cismontane.
  Style 1.2–2 mm. long, persistent. Pine belt of w. base of Sierra Nevada .... var. *longistylis*
  Style mostly less than 1 mm. long, often deciduous in age.
    Silicles 3–4 mm. wide, the wing usually not perforate .... *T. curvipes*
    Silicles 5–6 mm. wide, the wing often perforate .... var. *elegans*
Wing of silicles mostly plane, without rays. Plants glabrous and glaucous. Deserts ..... var. *eradiatus*

Var. **eradiàtus** Jeps. [*T. laciniatus* var. *eremicola* Jeps.] Plant glabrous and glaucous; wing of silicles membranous, without rays; silicles pubescent, ca. 4–5 mm. wide.— Washes and canyons, below 5000 ft.; Creosote Bush Scrub, Joshua Tree Wd., Pinyon-Juniper Wd.; deserts of Calif.; to Utah, Ariz. April–May.

Var. **longístylis** Jeps. Styles 1.3–2 mm. long; silicles 3–5 mm. wide, the wings rayed. —Dry openings, 3000–5000 ft.; mostly Yellow Pine F.; w. base of Sierra Nevada from Tuolumne Co. to Kern Co. May.

Var. **elegáns** (F. & M.) Rob. in Gray. [*T. e.* F. & M. *T. hirtellus* Greene.] Silicles 5–7 mm. wide, the wings often perforated; style ca. 1 mm. long.—Grassy slopes, below 5000 ft.; especially in V. Grassland and Foothill Wd., less common in Yellow Pine F., Chaparral, etc.; cismontane Calif. March–May.

3. **T. laciniàtus** Nutt. ex T. & G. [*T. affinis* Greene. *T. l.* var. *a.* Munz. *T. crenatus* Nutt. *T. l.* var. *c.* Brew. *T. emarginatus* Greene. *T. l.* var. *e.* Jeps. *T. curvipes* var. *e.* Jeps.] Stems slender, simple or branched, usually glabrous and glaucous, 1–4 dm. high;

lvs. not rosulate at base, 1–5 cm. long, the lower narrow, subentire or pinnatifid into narrow segms., the upper narrowly linear, narrowed at base and not or scarcely auriculate; pedicels recurved in fr., capillary, 2–5 mm. long; silicles usually glabrous, elliptical to orbicular, ca. 3–4 mm. wide, the wing continuous, subentire, lacking well defined rays.—Open places such as grassy slopes, below 3500 ft.; V. Grassland, Coastal Sage Scrub, Chaparral, etc.; cismontane s. Calif. and from the w. edge of the desert s. of Santa Barbara and Inyo cos. to L. Calif. March–May.

KEY TO VARIETIES

Pedicels recurved in fr.; silicles mostly 3–5 mm. wide.
  Silicles essentially flat.
    Wing lacking well-defined rays.
      Silicles usually glabrous. Mostly cismontane .... *T. laciniatus*
      Silicles with minute clavate hairs. Desert .... var. *Hitchcockii*
    Wing with well-defined rays.
      Silicles 3–4 mm. wide. Widely distributed .... var. *crenatus*
      Silicles 4–4.5 mm. wide. Insular .... var. *ramosus*
  Silicles cup-shaped. Santa Cruz Id. .... var. *conchuliferus*
Pedicels spreading, almost straight; silicles 2.5 mm. wide. San Diego Co. .... var. *rigidus*

Var. **crenàtus** (Nutt.) Brew. [*T. c.* Brew.] Wing rayed and notched or perforate between the rays.—With the sp. and in S. Coast Ranges to Contra Costa Co.

Var. **ramòsus** (Greene) Munz. [*T. r.* Greene.] Silicles glabrous, larger, 4–4.5 mm. wide, the wings strongly rayed, often perforate.—Santa Cruz and Santa Rosa ids.

Var. **rígidus** Munz. Plants compact, stiff, 3–12 cm. high, purplish, with pinnatifid lvs.; pedicels spreading, almost straight, not recurved; silicles glabrous, 2.5 mm. wide, subcrenate.—At 4500–5500 ft.; Yellow Pine F.; Laguna Mts., San Diego Co.; n. L. Calif. May.

Var. **Hitchcóckii** Munz [probably *T. desertorum* Heller. *T. l.* ssp. *d.* Abrams.] Much branched from base, 1–1.5(–2) dm. high, glabrous and glaucous; silicles 2.5–3 mm. wide, scabrellous with minute clavate hairs, the wing subentire and scarcely rayed.—Mostly from 2500–7500 ft.; Joshua Tree Wd., Pinyon-Juniper Wd.; w. Mojave Desert. April–May.

Var. **conchulíferus** (Greene) Jeps. [*T. c.* Greene.] Plant 4–10 cm. high; fls. pink; silicles boat-shaped, glabrous, the wings lavender, perforate or parted into spatulate lobes.—Santa Cruz Id. March–April.

44. Smelòwskia C. A. Mey.

Low cespitose perennials, with suffruticose caudex covered by lf.-bases. Lvs. pinnatifid, petiolate, pliable, whitish-tomentose. Fls. racemose, small, white. Sepals subequal, somewhat spreading. Petals exserted, obovate with rounded blade. Silicles lance-oblong, ovoid, or almost linear, slightly flattened parallel to septum or subterete, somewhat obcompressed, the valves strongly keeled. Style very short. Seeds 2–10, marginless, with short stout funicle and incumbent cotyledons. Ca. 6 spp. of n. Asia and w. N. Am. (*T. Smielowski*, 19th-century Russian botanist.)

(Rollins, R. C. Smelowskia and Polyctenium. Rhodora 40: 294–305, 1938.)

1. **S. ovàlis** Jones var. **congésta** Roll. Stems 5–15 cm. long, densely pubescent with long simple hairs; basal lvs. petioled, 2–6 cm. long, the segms. obovate, with whitish chiefly simple hairs; cauline lvs. few, smaller; infl. corymbose; pedicels 4–8 mm. long; sepals oblong, scarious-margined, otherwise hairy, 3–3.5 mm. long; petals white or pinkish, 4–5 mm. long; silicle ovate, truncate at base, tapering above, 4–6 mm. long, 4 mm. broad, glabrous; style ca. 1 mm. long; seeds 2–6, oblong, but pointed at distal end, ca. 2 mm. long.—Dry volcanic slopes, at ca. 10,000 ft.; Alpine Fell-fields, Mt. Lassen, Shasta Co. July–Aug.

45. Polyctènium Greene

Low cespitose perennials, the caudices free of lf.-bases. Lvs. nonpetiolate, highly dissected, wiry, with linear segms., sparsely pubescent. Fls. racemose, small, white. Petals cuneate, truncate at apex. Siliques linear, compressed contrary to the septum or subterete. Style evident. Seeds 12–28, marginless, with weak rather long funicle and incumbent cotyledons. One sp. (Greek, *polys,* many, and *kteis, ktenos,* comb, because of pectinate hairs.)

(Rollins, R. C. *See* Smelowskia.)

1. **P. Fremóntii** (Wats.) Greene. [*Smelowskia F.* Wats. *Braya pectinata* Greene.] Stems few to several, usually simple, 5–15 cm. long, pubescent with simple or branched hairs; basal lvs. sessile, 1–2 cm. long, pinnatifid into pungent linear segms., stiff-pubescent with simple to dendritic hairs; cauline lvs. smaller; pedicels 4–6 mm. long; sepals oblong, glabrous or slightly pubescent, 2–3 mm. long; petals 5–6 mm. long; siliques glabrous, 6–13 mm. long, 1–1.5 mm. wide; style less than 1 mm. long; seeds 1 mm. long.—Open flats and dried swales, 4000–5400 ft.; Sagebrush Scrub; Plumas Co. to Modoc and Siskiyou cos.; to Ore., Ida. May–June.

46. Arabis L. Rock-Cress

Biennial or perennial herbs, glabrous or pubescent with simple forked or stellate hairs. Stems leafy, simple or branched. Basal lvs. petioled, entire, dentate or sometimes dissected, the cauline sessile or sometimes petioled, often auricled. Infl. racemose, bractless, elongating. Sepals erect, oblong to subovate, uniform, or the outer pair sometimes saccate. Petals spatulate to oblong, white to purplish, rarely yellowish. Stamens 6, tetradynamous; anthers oblong. Siliques linear, straight or curved, erect to reflexed, sessile or rarely stipitate, flattened parallel to partition, 2-valved, the valves mostly 1-nerved. Style prominent to none. Stigma mostly entire. Seeds many, pendulous, orbicular to almost oblong, flat or plump, winged or not, uniseriate to biseriate, with accumbent cotyledons. Large genus of Am. and Eurasia. (Named for *Arabia.*)

(Rollins, R. C. Monographic study of Arabis in w. N. Am. Rhodora 43: 289–325, 348–411, 425–481, 1941 [Contr. Gray Herb. 138].)
A.   Siliques 3–8 mm. wide; seed-wing 1–3 mm. wide, or, if slightly less, then cauline lvs. petioled; seeds (including wings) 2.5–5 mm. long.
   B.   Siliques and pedicels reflexed.
      C.   Lvs. and stems green, glabrous or pubescent only below; siliques attenuate at apex; seeds uniseriate. Fresno Co. n. .... 33. *A. suffrutescens*
      CC. Lvs. and stems hoary; siliques obtuse at apex; seeds biseriate. Inyo Co. s.
           29. *A. glaucovalvula*
   BB. Siliques and pedicels erect or ascending.
      C.   Lower stem-lvs. with winged petioles; basal lvs. oblanceolate to broadly spatulate, 1–3 cm. wide; petals scarcely longer than sepals .... 28. *A. repanda*
      CC. All stem-lvs. sessile; basal lvs. linear to oblanceolate, less than 0.8 cm. wide; petals definitely exceeding sepals.
         D.   Lvs. and lower stems hoary with a minute pubescence; pedicels pubescent. Mts. bordering Mojave Desert .... 29. *A. dispar*
         DD. Lvs. and stems green, not hoary; pedicels glabrous. Sierra Nevada to San Jacinto Mts.
            E.   Basal lvs. oblanceolate to spatulate, 3–8 mm. wide, glabrous or finely pubescent, deciduous, not forming successive rosettes on crown
               34. *A. platysperma*
            EE. Basal lvs. linear, ca. 2 mm. wide, densely hirsute with coarse hairs, persistent and forming successive hemispherical rosettes on the elongate crown
               35. *A. pygmaea*
AA. Siliques usually less than 3 mm. wide; seed-wing less than 1 mm. wide or seeds wingless; seeds (including wings) less than 2 mm. long.
   B.   Basal lvs. obovate to broadly oblanceolate, obtuse and rounded at apex, often forming a flat rosette at base of stems; outer sepals saccate, except in *A. glabra.*
      C.   Seeds biseriate; siliques semi-terete; fls. cream-yellow or rarely lilac; stem-lvs. ample, ± ovate, usually glaucous .... 1. *A. glabra*
      CC. Seeds uniseriate; siliques flat; fls. white to purple; stem-lvs. small except sometimes in *A. hirsuta,* obovate to oblong, rarely glaucous.
         D.   Plants nearly or quite glabrous.
            E.   Basal lvs. 3–8 cm. long, broadly oblanceolate; pedicels 10–15 mm. long; petals white to pinkish. Sierra Nevada .... 7. *A. Davidsonii*
            EE. Basal lvs. 1–2 cm. long, spatulate; pedicels 8–10 mm. long; petals rose-purple. Mendocino Co. .... 6. *A. McDonaldiana*
         DD. Plants pubescent at least below.
            E.   Petals white, 5–9 mm. long; pedicels glabrous .... 2. *A. hirsuta*
            EE. Petals purplish, 12–18 mm. long; pedicels pubescent.

   F. Plants 0.5–2 dm. high, rather coarse-pubescent with branched appressed hairs. Sonoma Co. to San Mateo Co. . . . . 3. *A. blepharophylla*
  FF. Plants 2.5–4.5 dm. high. Siskiyou Co.
   G. Lower stems fine-pubescent with small appressed stellate hairs; basal lvs. not ciliate . . . . 4. *A. modesta*
   GG. Lower stems hirsute with coarse hairs; basal lvs. ciliate
                      5. *A. oregana*
BB. Basal lvs. linear to linear-oblanceolate, acute to rarely obtuse (if broader, then minutely pubescent or with reflexed siliques, or both); outer sepals not saccate.
 C. Plant hoary with a very minute pubescence. Mostly of the desert or bordering mts.
  D. Siliques reflexed; pedicels reflexed to pendulous.
   E. Seeds biseriate; cauline lvs. linear, entire, not crowded . . . . . 27. *A. pulchra*
   EE. Seeds uniseriate; cauline lvs. oblong to broadly lanceolate, often subpinnatifid, crowded.
    F. Siliques blunt at apex; style 0; petals 7–10 mm. long . 22. *A. puberula*
    FF. Siliques acuminate; style ca. 1 mm. long; petals 10–14 mm. long
                     23. *A. subpinnatifida*
  DD. Siliques erect or ascending to widely spreading; pedicels erect to perpendicular to axis.
   E. Styles 1–8 mm. long; basal lvs. linear.
    F. The styles 4–8 mm. long; seeds with narrow wings; siliques 1–2 cm. long. San Bernardino Mts. . . . . 32. *A. Parishii*
    FF. The styles 1–3 mm. long; seeds broadly winged; siliques 3–5 mm. long. San Jacinto Mts. . . . . 31. *A. Johnstonii*
   EE. Styles less than 1 mm. long; basal lvs. narrowly oblanceolate.
    F. Seeds biseriate, essentially wingless, plump, ca. 1 mm. broad
                   25. *A. Shockleyi*
    FF. Seeds uniseriate, winged, flat, 1.5–2.5 mm. broad.
     G. Seed-wing over 0.5 mm. wide; siliques 2.5–3.5 mm. wide, divaricately ascending . . . . 30. *A. dispar*
     GG. Seed-wing less than 0.5 mm. wide; siliques ca. 2 mm. wide, spreading at right angles . . . . 26. *A. inyoensis*
 CC. Plant greenish, not hoary.
  D. Mature fruiting pedicels erect to ascending; siliques erect, ascending, sometimes arcuate.
   E. Lower fruiting pedicels 2–4 cm. long, glabrous; siliques somewhat curved. Santa Cruz Id. . . . . 16. *A. Hoffmannii*
   EE. Lower fruiting pedicels less than 2 cm. long, pubescent or glabrous; siliques straight or arcuate.
    F. Lower lvs. densely pubescent with dendritic hairs, gray, often felty; stems usually many, from a well-branched caudex.
     G. Petals 4–6 mm. long.
      H. Basal lvs. felty with minute dendritic hairs; cauline lvs. narrow-ovate to oblong, crowded at base.
       I. Cauline lvs. auricled; pedicels 2–5 mm. long; style 0; Lassen Peak to Tulare Co. . . . . 12. *A. Lemmonii*
       II. Cauline lvs. not auricled; pedicels 5–10 mm. long; style ca. 1 mm. long. Masonic Peak, Mono Co.
                 18. *A. Fernaldiana*
      HH. Basal lvs. acute, not felty; cauline lvs. narrow-lanceolate, remote; pedicels 5–15 mm. long. Masonic Peak, Mono Co.
                 19. *A. microphylla*
     GG. Petals mostly 7–14 mm. long.
      H. Plants 0.6–2 dm. high; basal lvs. broadly spatulate, pubescent with 3-forked hairs, 1–3 cm. long . . . . 17. *A. Breweri*
      HH. Plants 2–9 dm. high.
       I. Basal lvs. 3–10 cm. long; siliques 6–12 cm. long
                 15. *A. sparsiflora*
       II. Basal lvs. 2–3 cm. long; siliques 4–6 cm. long
                 26. *A. inyoensis*
    FF. Lower lvs. pubescent to glabrous, greenish; stems 1–several.
     G. Siliques less than 1.5 mm. wide; seeds orbicular, ca. 1 mm. broad; stems sparsely hirsute at base . . . . 19. *A. microphylla*
     GG. Siliques 1.5–3.5 mm. wide; seeds orbicular to oblong, 1–2 mm. broad; stems glabrous to strigose.
      H. Seeds biseriate, oblong, winged on one side and the distal end; siliques and pedicels erect; fls. usually white
                 9. *A. Drummondii*
      HH. Seeds mostly uniseriate, orbicular, winged all around; siliques and pedicels divaricately ascending to erect; fls. pink to purple.
       I. Plants less than 3 dm. high; stems several from a branched caudex; siliques erect to slightly divergent
                 8. *A. Lyallii*
       II. Plants 3–9 dm. high; stems usually single from a simple caudex; siliques ± widely spreading.

J. Siliques 1.5–2.5 mm. wide, with straight margin;
seeds 1–1.5 mm. wide . . . . 10. *A. divaricarpa*
JJ. Siliques 2.5–3.5 mm. wide, with undulate margin;
seeds 2–2.5 mm. broad . . . . 11. *A. rigidissima*
DD. Mature fruiting pedicels diverging at right angles to strictly reflexed; siliques
straight to arcuate, diverging at right angles to stem or even deflexed.
E. Basal lvs. always ciliate with large hairs, oblanceolate; siliques reflexed,
appressed to rachis. Siskiyou Co. to San Bernardino Co. . . 13. *A. rectissima*
EE. Basal lvs. not ciliate.
F. Seeds biseriate; cauline lvs. linear; petals 8–20 mm. long
27. *A. pulchra*
FF. Seeds uniseriate; cauline lvs. lanceolate to ovate; petals less than 12
mm. long.
G. Basal lvs. linear, minutely pubescent; siliques pendulous on slender
widespread pedicels. E. Mono. Co. . . . . 24. *A. cobrensis*
GG. Basal lvs. spatulate to oblanceolate.
H. Pedicels 2–4(–6) mm. long; siliques spreading at right angles
to stem; cauline lvs. mostly ovate and glabrous
12. *A. Lemmonii*
HH. Pedicels 6–20 mm. long; siliques spreading to reflexed; cauline
lvs. oblong to lanceolate, usually pubescent.
I. Mature fruiting pedicels descending to strictly deflexed,
straight, not widely spreading; siliques mostly straight,
pendulous to appressed . . . . 21. *A. Holboellii*
II. Mature fruiting pedicels spreading at right angles to stem,
straight or arched downward; siliques straight and spreading at right angles or arcuate.
J. Plants 1–2(–3) dm. high, cespitose; stems many,
slender; cauline lvs. few, remote, small
19. *A. microphylla*
JJ. Plants 3–9 dm. high, rarely cespitose; stems 1–few,
rather stout; cauline lvs. usually many, crowded and
overlapping below.
K. Basal lvs. entire, finely pubescent; stems mostly
strigose below.
L. Pedicels and siliques at right angles to
rachis; siliques straight . . 26. *A. inyoensis*
LL. Pedicels and siliques curved downward;
siliques usually curved . . . 20. *A. lignifera*
KK. Basal lvs., at least in part, dentate, coarsely
pubescent; stems mostly hirsute below.
L. Outer basal lvs. broadly oblanceolate, obtuse; pedicels slender, 10–20 mm. long,
glabrous; petals 6–9 mm. long
14. *A. perennans*
LL. Outer basal lvs. narrowly oblanceolate, acute;
pedicels stout, 5–12 mm. long, pubescent;
petals 8–12 mm. long . . 15. *A. sparsiflora*

1. **A. glàbra** (L.) Bernh. [*Turritis g.* L. *A. perfoliata* Lam.] Tower-Mustard. Biennial or occasionally perennial; stems 1–few, simple or rarely branched above, usually
hirsute below with simple or forked spreading hairs, glabrous and glaucous above, 4–12
dm. high; lower lvs. oblanceolate to oblong, petioled, usually ± dentate or even repand,
with coarse forked or dendritic hairs, 6–15 cm. long, 1–3 cm. wide; cauline lvs. lanceolate
to ovate, sessile, auricled, mostly entire and glabrous; racemes long, strict; pedicels
suberect, commonly 1–1.5 cm. long; sepals oblong, yellowish, 3–5 mm. long; petals
narrow, 5–7 mm. long, yellowish-white or rarely purplish; siliques erect, semi-terete,
glabrous, 4–10 cm. long, ca. 1 mm. wide, valves nerved at least in lower half; style
short, stout; stigma expanded; seeds oblong to almost round, wingless or narrowly
winged, mostly biseriate, ca. 1 mm. long.—Shaded canyons and mts., mostly below 7000
ft.; many Plant Communities; throughout cismontane Calif.; to B.C. and Atlantic Coast.
March–July.
Var. **furcatipìlis** M. Hopk. Stems pubescent below with appressed several-branched
hairs.—Occasional with the sp., as in Santa Lucia Mts.
2. **A. hirsùta** (L.) Scop. var. **glabràta** T. & G. [*A. pycnocarpa* var. *g.* M. Hopk.]
Biennial or perennial, the stems erect, 1–few, simple or branched above, hirsute below,
glabrous above, 2–7 dm. high; basal lvs. obovate to oblanceolate, mostly entire, obtuse,
petioled, 3–7 cm. long, 1–2.5 cm. wide, hirsute to subglabrous; cauline lvs. oblong to
obovate, entire to rarely few-toothed, sessile, auricled; racemes rather long in fr.;

pedicels erect to ascending, 5–15 mm. long; sepals mostly glabrous, 3–4.5 mm. long; petals white, spatulate, 5–9 mm. long; siliques erect to slightly divaricate, glabrous, flat, 3–6 cm. long, 1 mm. wide; style 0.5–1 mm. long; stigma subentire; seeds brown to dark, rounded to rectangular, winged at distal end to wingless, uniseriate, 1–1.5 mm. long.—Moist places in mts., 4000–8200 ft.; Yellow Pine F., Red Fir F., Lodgepole F.; San Bernardino Mts., Sierra Nevada to Modoc Co.; to B.C., Wyo., Utah. May–July.

Var. **Eschschóltziàna** (Andrz.) Roll. [*A. E.* Andrz. in Ledeb.] Cauline lvs. usually dentate; siliques 1.5–2 mm. wide; stigma bifid.—At 3950 ft., Rail Creek, Siskiyou Co.; to Alaska. June.

3. **A. blepharophýlla** H. & A. [*Erysimum* b. Kuntze.] Perennial; stems simple, 1–few, coarse-pubescent with branching hairs, 0.5–2 dm. high; lower lvs. in rosettes, obovate to oblanceolate, petioled, obtuse, entire or dentate, coarse-pubescent on surface and margin with forked or dendritic hairs, or on margin only, 2–8 cm. long, 0.5–2 cm. wide; cauline lvs. few, oblong to ovate, sessile but not auriculate; pedicels erect, stout, pubescent, 5–10 mm. long; sepals oblong, purplish, pubescent, 6–8 mm. long; petals rose-purple, broadly spatulate, 12–18 mm. long; siliques erect, glabrous, nerved to middle or above, 2–4 cm. long, 2–2.5 mm. wide; style 1–2 mm. long; seeds orbicular, dark brown, uniseriate, narrowly winged, 1.5–2 mm. long.—Rocky places, below 1000 ft.; N. Coastal Scrub, Mixed Evergreen F.; Sonoma Co. to Santa Cruz Co. Feb.–April.

4. **A. modésta** Roll. Perennial with a simple or branched base, 1–few stems 2.5–4.5 dm. high, simple or branched above, pubescent throughout with small appressed stellate hairs; lower lvs. petioled, obovate, obtuse, repand to entire, stellate-pubescent, 2–6 cm. long, 0.8–1.6 cm. wide; cauline lvs. reduced, sessile but not clasping; pedicels becoming 6–12 mm. long, divaricate; sepals oblong, pubescent, 5–7 mm. long; petals spatulate, purple to pinkish-purple, 12–15 mm. long; siliques glabrous, ascending, 3–4.5 cm. long, ca. 1.5 mm. wide; style 1–2 mm. long.—Rocky walls and bluffs, ca. 1500 ft.; Yellow Pine F., Mixed Evergreen F.; Seiad V., Klamath R., Siskiyou Co.; sw. Ore. April–May.

5. **A. oregàna** Roll. [*A. purpurascens* Howell, not Presl. *A. furcata* var. *p.* Wats.] Much like *A. modesta*, but the lower stems and lvs. coarsely pubescent with large forked and smaller dendritic hairs; siliques erect, straight, glabrous, nerved nearly to tip, 4–5 cm. long, ca. 2 mm. wide; seeds dark brown, oblong, narrowly winged on sides.— What seems to be an unusually glabrous form of this grows at ca. 2000 ft.; Mixed Evergreen F., Yellow Pine F.; 10 mi. sw. of Scott Bar, Scott R., Siskiyou Co.; the normal form of the sp. in Jackson Co., Ore. May.

6. **A. McDonaldiàna** Eastw. [*A. blepharophylla* var. *M.* Jeps.] Perennial with branched caudex and several stems, these simple, glabrous, 5–20 cm. high; lower lvs. in rosettes, spatulate, 1.2 cm. long, 4–7 mm. wide, repand to toothed, glabrous or with some teeth bristle-tipped; cauline lvs. small, oblong, remote, sessile; pedicels ascending, glabrous, 8–10 mm. long; sepals oblong, dark purple, glabrous, 5–6 mm. long; petals rose-purple, 9–11 mm. long; siliques divaricate, 3–4 cm. long.—Serpentine soil, at 4000 ft.; Yellow Pine F.; Red Mt. near Bell Spring, Mendocino Co. May–June.

7. **A. Davidsònii** Greene. [*A. Lyallii* Wats. var. *D.* Smiley. *A. Brucae* Jones. *A. cognata* Jeps.] Perennial with heavy caudex; stems 1–several, slender, glabrous, simple, 5–15 cm. high; lower lvs. cuneate-oblanceolate to spatulate, obtuse, entire or few-toothed, glabrous, 3–8 cm. long, petioled; cauline lvs. ± oblong, entire, few, glabrous, sessile, not auricled; pedicels divaricate, glabrous, 8–15 mm. long; sepals oblong, glabrous, 4–5 mm. long; petals spatulate, white to pinkish, 8–10 mm. long; siliques glabrous, divaricate, straight to somewhat curved, nerved to middle or beyond, 3–5 cm. long, 1.5–2 mm. wide; style short, barely evident; seeds orbicular, uniseriate, narrowly winged all around, 1.5 mm. long.—Rocky places, 5000–11,500 ft.; Lodgepole F., Subalpine F., Alpine Fell-fields; Sierra Nevada from Plumas Co. to Tulare and Inyo cos.; e. Ore. July–Aug.

8. **A. Lyállii** Wats. [*A. Drummondii* var. *alpina* Wats. *A. D.* var. *L.* Jeps.] Cespitose perennial with rather slender usually branched caudex; stems few or more, glabrous, 4–25 cm. high; lower lvs. oblanceolate, entire, glabrous, or with small dendritic hairs, petioled, 1–3 cm. long, 3–6 mm. wide; cauline lvs. few, lanceolate, remote, sessile, not or slightly auricled; pedicels suberect, 4–10 mm. long; sepals oblong, glabrous, ca. 4 mm. long; petals rose to purplish, 6–10 mm. long; siliques erect or nearly so, 1-nerved to middle, glabrous, 3–5 cm. long, 2–3 mm. wide; stigma nearly or quite sessile; seeds orbicular,

winged, 1–2 mm. broad, mostly uniseriate.—Rocky places, 8000–12,000 ft.; Alpine Fell-fields; Sierra Nevada from Tulare and Inyo cos. to Siskiyou and Modoc cos.; to B.C. and Rocky Mts. July.

Var. **nubígena** (Macbr. & Pays.) Roll. [*A. n.* Macbr. & Pays. *A. microphylla* var. *n.* Roll. *A. paupercula* Greene.] Lower lvs. 1–2.5 mm. wide; cauline lvs. narrowly linear; petals 5–7 mm. long, pink.—Dry stony slopes, 10,000 to 12,000 ft.; Alpine Fell-fields; Alpine, Mono, and Tuolumne cos. to Tulare Co. July–Aug.

9. **A. Drummóndii** Gray. [*Turritis stricta* Grah., not *A. s.* Host. *Streptanthus angusti-folius* Nutt., not *A. a.* Lam.] Biennial or perennial with simple caudex and 1–few stems, simple or branched above, glabrous to sparingly strigulose, 3–9 dm. high; lower lvs. oblanceolate, entire to dentate, petioled, usually acute, 2–7 cm. long, glabrous or with malpighiaceous hairs; cauline lvs. oblong to oblong-lanceolate, acute, sessile, auricled, usually clasping, crowded toward base; pedicels erect, glabrous, 1–2 cm. long; sepals narrowly oblong, glabrous, 3–5 mm. long; petals white to pinkish, 7–10 mm. long; siliques erect, usually rather crowded, 4–10 cm. long, 1.5–3 mm. wide, the valves 1-nerved to middle or above; styles short or 0; seeds biseriate, oblong, winged on distal end and at least one side, ca. 1 mm. wide, 1.5–2 mm. long.—Dryish or dampish benches and slopes, 5500–10,500 ft.; Red Fir F., Lodgepole F., Subalpine F.; Tulare and Inyo cos. n.; to B.C. and Atlantic Coast. June–July.

10. **A. divaricárpa** A. Nels. [*Turritis brachycarpa* T. & G. *A. Drummondii* var. *b.* Gray. *A. b.* Britton, not Rupr. *A. nemophila* Greene.] Biennial or rarely perennial from simple or branched caudex; stems 1–few, sometimes branched, glabrous above, glabrous to sparsely pubescent below with malpighiaceous hairs, 3–9 dm. high; lower lvs. oblanceo-late to spatulate, usually acute, dentate to subentire, loosely pubescent with 3–several-rayed hairs, petioled, 2–6 cm. long, 4–8 mm. wide; cauline lvs. lanceolate to narrow-oblong, glabrous or sparsely pubescent, auriculate; pedicels glabrous, 6–12 mm. long, divaricate to somewhat descending; sepals 3–5 mm. long, with scarious margin; petals spatulate, pink to purplish, 6–10 mm. long; siliques mostly straight, ± divaricate, mostly nerved, glabrous, 2–8 cm. long, 1.5–2.5 mm. wide; style short or 0; seeds suborbicular to broadly oblong, ca. 1 mm. wide, narrowly winged, mostly uniseriate.—Dry slopes under pines, 7000–11,000 ft.; Red Fir F., Lodgepole F., Subalpine F.; Tulare Co. to Siskiyou and Trinity cos.; to Alaska and Quebec. July–Aug.

Var. **interpósita** (Greene) Roll. [*A. i.* Greene.] Stems usually simple, pubescent throughout; pedicels sparsely pubescent; siliques nearly or quite nerveless.—Trinity Summit and Marble Mts., Humboldt and Siskiyou cos. June–July.

11. **A. rigidíssima** Roll. Perennial from a suffruticose caudex; stems 1–several, some-what pubescent below or glabrous throughout, simple or branched below, 2–4 dm. high; lower lvs. spatulate, entire, obtuse, short-petioled, sparsely pubescent with dendritic or forked hairs, or glabrous, 1.5–3 cm. long, 4–8 mm. wide; cauline lvs. ovate to oblong, sessile, auricled, entire; pedicels glabrous, divaricate, 5–10 mm. long; sepals glabrous, oblong, 4–5 mm. long; petals spatulate, pink, 7–9 mm. long; siliques straight, divaricate, glabrous, nerved to middle or above, 5–7 cm. long, 2.5–3.5 mm. wide; style less than 1 mm. long; seeds suborbicular, 2–2.5 mm. wide, narrowly winged, uniseriate.—Open rocky places, 6000–7000 ft.; Red Fir F.; Trinity Mts., Trinity and Humboldt cos. July–Aug.

12. **A. Lemmònii** Wats. [*A. canescens* var. *latifolia* Wats. *A. polyclada* Greene.] Perennial with branched caudex; stems several, slender, pubescent throughout or usually glabrous above, 6–20 cm. high; basal lvs. spatulate, entire to few-toothed, pubescent with minute dendritic hairs, petioled, 1–2 cm. long; cauline lvs. sessile, oblong-lanceolate, auricled; pedicels mostly glabrous, 2–5 mm. long; sepals oblong, subglabrous, 2–3 mm. long; petals pink to purple, spatulate, 4–6 mm. long; siliques horizontal or slightly de-scending, in a somewhat secund raceme, straight to slightly curved, nerved to middle, 2–4 cm. long, 2–2.5 mm. wide; style very short or 0; seeds orbicular, narrowly winged, uniseriate, slightly over 1 mm. wide.—Dry stony or rocky places, 8000–11,900 ft.; Lodgepole F. to Alpine Fell-fields; Tulare Co. to Lassen Peak and Modoc Co.; to B.C., Mont., Colo. June–Aug.

Var. **depauperàta** (Nels. & Kenn.) Roll. [*A. d.* Nels. & Kenn.] Basal lvs. narrowly oblanceolate; pedicels and siliques divaricately ascending, the raceme not secund.—

Dry granitic gravel and stony places, 10,500–14,000 ft.; Subalpine F., Alpine Fell-fields; Tulare Co. to Placer Co.; Nev. July–Aug.

13. **A. rectíssima** Greene. Biennial, with 1–several stems simple to branched above, glabrous to sparsely hirsute with simple hairs, 2–8 dm. high; basal lvs. spatulate to oblanceolate, petioled, entire, 1–3 cm. long, 4–10 mm. wide, hirsute with simple and forked coarse hairs; cauline lvs. crowded below, oblong to sublanceolate, ± auricled; pedicels glabrous, strictly reflexed, 4–12 mm. long; sepals oblong, sparsely hirsute at apex, 2–3 mm. long; petals white or pinkish, 4–6 mm. long; siliques many, straight, strictly reflexed, appressed to stem, glabrous, 1-nerved near base, 5–8 cm. long, 1.5–2.5 mm. wide; style almost none; seeds orbicular, ca. 1.5 mm. wide, uniseriate, winged all around.—Dry slopes and benches, 4000–9000 ft.; Yellow Pine F., Red Fir F.; Siskiyou Co. to Tulare Co., San Bernardino Co.; Ore., Nev. June–July.

14. **A. perénnans** Wats. [*A. arcuata* var. *p.* Jones.] Perennial with simple or branched woody caudex; stems few to many, simple or branched above, 1.5–6 dm. high, pubescent below with dendritic hairs; lower lvs. oblanceolate to wider, petioled, dentate to entire, 2–6 cm. long, 4–20 mm. wide, densely pubescent with dendritic hairs; cauline lvs. lanceolate, auricled, entire or sparsely toothed; pedicels spreading and arched down in fr., glabrous, 1–2 cm. long; sepals pubescent, ca. 4 mm. long; petals purple to pinkish, 6–9 mm. long; siliques wide-spreading to pendulous, glabrous, curved inward, usually nerveless, 4–6 cm. long, 1.2–2 mm. wide; style 0; seeds orbicular, 1–1.5 mm. wide, winged, uniseriate.—Dry stony slopes, 1000–7000 ft.; Creosote Bush Scrub, Joshua Tree Wd., Pinyon-Juniper Wd., Yellow Pine F.; San Gabriel and San Bernardino mts. to Panamint Mts., Laguna Mts. (e. San Diego Co.), and across the deserts to Colo., New Mex., L. Calif. April–July.

15. **A. sparsiflòra** Nutt. in T. & G. Perennial, the caudex often branched; stems rather slender, simple or branched above, hirsute below, glabrate above, 3–8 dm. high; lower lvs. long-petioled, entire, linear-oblanceolate, pubescent with dendritic hairs, 3–10 cm. long, 3–8 mm. wide; cauline lvs. linear-oblong, obtuse, sagittate-auriculate; pedicels ascending, loosely pilose to subglabrous, 5–15 mm. long; sepals 4–6 mm. long; petals pink to purple, 8–14 mm. long; siliques divaricately ascending, ± arcuate, glabrous, nerved below the middle, obtuse, 6–12 cm. long, 1.5–2 mm. wide; stigma subsessile; seeds orbicular, uniseriate, narrowly winged, 1 mm. wide.—Dry slopes, ca. 4500–9000 ft.; Sagebrush Scrub, Yellow Pine F., N. Juniper Wd.; Mono and Lassen cos. to Modoc Co.; to Ore., Ida., Utah. May–July.

KEY TO VARIETIES

Pedicels with spreading hairs or glabrous; lower stems hirsute or rarely glabrous.
  Upper lvs. and stems glabrous to very sparsely hirsute.
    Basal lvs. entire, linear-oblanceolate; pedicels divaricately ascending; stems usually branched
      above. Ne. Calif. . . . . *A. sparsiflora*
    Basal lvs. dentate, oblanceolate to wider; pedicels horizontal to somewhat ascending; stems usually
      simple. Mostly nw. Calif. . . . . var. *subvillosa*
  Upper lvs. and stems hirsute; basal lvs. linear-lanceolate, coarsely pubescent; pedicels spreading.
    Sierra and Santa Clara cos. to San Diego Co. . . . . var. *arcuata*
Pedicels with closely appressed hairs; fruiting pedicels usually widely pendulous; stems densely pubes-
  cent with appressed hairs. Los Angeles Co. to L. Calif. . . . . var. *californica*

Var. **subvillòsa** (Wats.) Roll. [*A. arcuata* var. *s.* Wats. *A. polytricha* Greene. *A. campyloloba* Greene.] Stems hirsute below with large simple or branched hairs, glabrous above; basal lvs. usually dentate, acute, harshly pubescent; pedicels spreading horizontally, hirsute; siliques arcuate.—Dry slopes, 1500–3500 ft.; Mixed Evergreen F., Yellow Pine F.; Humboldt and Siskiyou cos.; to Wash., Wyo., Mont. May–June.

Var. **arcuàta** (Nutt.) Roll. [*Streptanthus a.* Nutt. *A. a.* Gray. *A. Holboellii* var. *a.* Jeps. *A. a.* var. *rubicundula* Jeps.?] Stems pubescent throughout, the lower part hirsute with large simple or branched hairs; basal lvs. linear-oblanceolate, acute, coarsely pubescent; siliques strongly arcuate.—Dry slopes, 2500–6000 ft.; Chaparral, Foothill Wd., Yellow Pine F.; Shasta and Santa Clara cos. to Kern and Ventura cos., rare farther s. to San Diego Co. March–May.

Var. **califórnica** Roll. [*A. maxima* in large part of Munz.] Stems coarse, pubescent throughout with fine dendritic hairs; basal lvs. large, coarsely toothed, densely pubes-

cent; pedicels pubescent with appressed hairs.—Common on dry stony places, mostly
below 5000 ft.; Chaparral, Coastal Sage Scrub; Los Angeles Co. to L. Calif. Feb.–May.

16. **A. Hoffmánnii** (Munz) Roll. [*A. maxima* var. *H.* Munz.] Perennial with scaly
caudex; stems 1–few, branched above, subglabrous, 5–7 dm. high; basal lvs. many,
linear-lanceolate, sinuate-dentate, obtuse, dendritic-pubescent on lower surface, 5–10 cm.
long, 6–10 mm. wide; cauline lvs. sessile, crowded, linear-oblong; pedicels divaricately
spreading, glabrous, 1–4 cm. long; sepals 4–5 mm. long; petals white, 8–10 mm. long,
narrow; siliques divaricate, straight or slightly arcuate, nerveless, glabrous, 6–10 cm.
long, 2–3.5 mm. wide; style almost obsolete; seeds orbicular, biseriate, narrowly winged,
ca. 1 mm. wide.—Rocky cliffs; Chaparral; Santa Cruz Id. Feb.–April.

17. **A. Brèweri** Wats. [*A. epilobioides* Greene. *A. B.* var. *figularis* Jeps.] Cespitose
perennial from a woody branched caudex; stems several, simple, densely hirsute below
with mostly simple hairs, often glabrous above, 6–20 cm. high; basal lvs. broadly spatulate, mostly entire, obtuse, short-petioled, pubescent with 3-forked hairs, 1–3 cm. long,
4–6 mm. wide; cauline lvs. sessile, auriculate, usually less than 2 cm. long; pedicels
5–9 mm. long, mostly pubescent; sepals pubescent, 4–5 mm. long; petals red-purple to
pink, 6–9 mm. long; siliques divaricate, arcuate or almost straight, 3–7 cm. long, ca.
2 mm. wide, 1-nerved on lower third; style 0; seeds orbicular, uniseriate, 1 mm. or
wider, narrowly winged.—Dry rocky slopes and summits, 1500–7400 ft.; Foothill Wd.,
Yellow Pine F., Red Fir F.; Siskiyou Co. to Monterey Co., Sutter and Yuba cos.; s. Ore.
March–July.

Var. **Aústinae** (Greene) Roll. [*A. A.* Greene.] Cauline lvs. 2–4 cm. long; pedicels
10–15 mm. long; petals 10–13 mm. long.—Little Chico Creek, Butte Co. Feb.–March.

Var. **pecuniària** Roll. Pedicels 3–4 mm. long; siliques 2–3 cm. long.—Dry rocky places,
9100–10,500 ft.; Subalpine F.; San Bernardino Mts. July–Aug.

18. **A. Fernaldiàna** Roll. var. **stylòsa** (Wats.) Roll. [*A. canescens* var. *s.* Wats.]
Cespitose perennial, 1–3 dm. high, with fine and sometimes large dendritic hairs; basal
lvs. many, entire, spatulate to oblanceolate, with dense dendritic pubescence, canescent,
1–2 cm. long, 2–3 mm. wide, the petioles usually ciliate; cauline lvs. sessile, auriculate,
oblong to lanceolate; pedicels pubescent, divaricate, 5–10 mm. long; sepals 3–6 mm.
long, pubescent; petals 5–7 mm. long, pink; siliques erect, acuminate, glabrous, nerveless, 4–6 cm. long, 1.5–2 mm. wide; style ca. 1 mm. long; seeds uniseriate, oblong,
nearly or quite wingless, ca. 1 mm. long.—Rock-crevices, 8000–9100 ft.; Pinyon-Juniper
Wd., Sagebrush Scrub; Masonic Peak, Mono Co.; Nev. June–July.

19. **A. microphýlla** Nutt. Perennial from a subterranean branching caudex, few-fld.,
the stems slender, mostly simple, 1–2 dm. high, glabrous above, usually somewhat hirsute
below with spreading simple or forked hairs; lower lvs. linear to narrowly oblanceolate,
entire, acute, densely dendritic-pubescent, 0.5–2 cm. long; cauline lvs. few, linear-
lanceolate, auriculate; pedicels slender, divaricate to ascending, 5–15 mm. long; sepals
2–3 mm. long; petals rose to purplish, 4–6 mm. long; siliques few on each stem, straight,
erect, almost nerveless, 2–6 cm. long, 1–1.5 mm. wide; style almost absent; seeds
orbicular, uniseriate, ca. 1 mm. wide, narrow-winged.—Rock-crevices, ca. 8000–9000 ft.;
Pinyon-Juniper Wd.; Masonic Peak, Mono Co. July.

20. **A. lignífera** A. Nels. Perennial; stems 1–few, erect, from simple or branched
caudex, minutely appressed-stellate below, glabrous above, 2–5 dm. high; basal lvs.
linear-oblanceolate, entire, 2–5 cm. long, 3–7 mm. wide, petiolate, densely dendritic-
pubescent; cauline lvs. oblong, entire, auricled, 1–3 cm. long; infl. loosely racemose;
sepals oblong, pubescent, 3–4 mm. long; petals pink to purplish, ± spatulate, 5–8 mm.
long; fruiting pedicels arched downward, subglabrous, 5–12 mm. long; siliques laxly
pendulous, glabrous or nearly so, 3–6 cm. long, 1.5–2 mm. wide, without or with short
style; seeds roundish, 1–1.2 mm. broad.—At ca. 7000 ft., summit at n. end of Owens V.;
to Ida., Wyo., Ariz. May.

21. **A. Holboèllii** Hornem. var. **retrofrácta** (Grah.) Rydb. [*A. r.* Grah. *A. secunda*
Howell. *A. H.* var. *s.* Jeps.] Biennial or perennial from simple or branched caudex;
stems 1–several, erect, simple or branched above, 2–8 dm. high, densely pubescent with
fine appressed dendritic hairs to glabrous above; basal lvs. ± pannose, usually
entire, oblanceolate to spatulate, 1–5 cm. long, 3–6 mm. wide; cauline lvs.
revolute-margined, auriculate, the upper finely pubescent; pedicels pubescent, strongly
reflexed, usually geniculate, 6–12 mm. long; sepals 2–4 mm. long; petals pinkish to whit-

ish, 6–10 mm. long; siliques glabrous or sometimes finely pubescent, strongly reflexed, usually appressed to the stem, straight or nearly so, 3.5–8 cm. long, 1–1.5 mm. wide; style almost obsolete; seeds mostly uniseriate, orbicular, ca. 1 mm. wide, narrowly winged all around.—Dry stony places, 1800–5000 ft. in N. Coast Ranges and 6000–10,500 in Sierra Nevada; Mixed Evergreen F., Pinyon-Juniper Wd., Yellow Pine F. to Subalpine F.; Humboldt and Siskiyou cos. to Modoc Co. and s. to San Bernardino Mts.; to Alaska, Colo. May–July.

Var. **pinetòrum** (Tides.) Roll. [*A. p.* Tides.] Stems hirsutulous below, glabrous above, 3–9 dm. high; basal lvs. densely pubescent with coarse dendritic hairs; pedicels arched downward, rarely geniculate, usually glabrous; siliques slightly curved inward, sometimes straight, pendulous, 4–7 cm. long, 1.5–2 mm. wide.—With preceding, Siskiyou Co. to San Diego Co.; to B.C., Nebr. May–July.

Var. **pendulocárpa** (A. Nels.) Roll. [*A. p.* A. Nels.] Stems 1–2 dm. high; cauline lvs. not auricled; basal lvs. 1.5–3 mm. wide; siliques pendulous, 2.5–4 cm. long.—Rocky places, 5000–10,500 ft.; Siskiyou and Modoc cos., Mono Co.; to B.C., Mont., Colo. May–July.

22. **A. pubérula** Nutt. [*A. Beckwithii* Wats. in part. *Erysimum p.* Kuntze. *A. subpinnatifida* var. *B.* Jeps. *A. sabulosa* vars. *frigida* and *colorata* Jones.] Biennial or perennial from a simple caudex; stems 1–few, mostly simple, hoary with dense dendritic pubescence, 1.5–5 dm. high; basal lvs. oblanceolate, entire or few-toothed, acute, hoary, petioled, 1–2.5 cm. long, 3–6 mm. wide; cauline lvs. crowded, densely pubescent, lanceolate to oblong, entire or irregularly toothed, with small or no auricles; pedicels curved downward in fr., pubescent, 4–8 mm. long; sepals pubescent, 4–6 mm. long; petals mostly rose to purple, 7–11 mm. long; siliques many, pendulous or even strictly reflexed, straight, mostly obtuse, pubescent, 3–6 cm. long, 2–3 mm. wide; stigma sessile; seeds orbicular, uniseriate, scarcely 2 mm. wide, narrowly winged all around.—Dry stony places, 4000–10,300 ft.; Sagebrush Scrub to Lodgepole F.; Inyo Co. to Modoc and Siskiyou cos.; to Wash., Wyo. June–July.

23. **A. subpinnatífida** Wats. Perennial from simple or branched caudex; stems 1–few, simple or branched above, 1.5–4 dm. high, densely pubescent with fine dendritic hairs or glabrous above; basal lvs. linear to narrowly oblanceolate, dentate to almost incised, hoary-pubescent, 1–3 cm. long, 2–4 mm. wide, somewhat larger on sterile shoots; cauline lvs. lanceolate to lance-linear, revolute, nearly or quite sessile, usually toothed or subpinnatifid, hoary, 1–3 cm. long; pedicels erect to divaricate in fl., arched downward in fr. and 6–12 mm. long; sepals pubescent, 5–7 mm. long; petals purple to lavender, 10–14 mm. long; siliques pendent, straight to somewhat curved, subglabrous, 5–7 cm. long, 2–3.5 mm. wide, acuminate at apex; style ca. 1 mm. long; seeds uniseriate, suborbicular, winged on sides or all around, 1.5–2.5 mm. wide.—Dry stony places, 2500–7000 ft.; Yellow Pine F., Red Fir F.; Humboldt and Glenn cos. to Siskiyou Co.; s. Ore. May–June.

24. **A. cobrénsis** Jones. [*A. canescens* Nutt., not Brocchi.] Perennial from branched caudex; stems slender, several, mostly simple, soft-pubescent with minute dendritic hairs below, glabrate above, 2–5 dm. high; basal lvs. many, linear, entire, densely pubescent, hoary, 2–5 cm. long, 1–3 mm. wide; cauline lvs. few, linear, entire, sessile, slightly auricled; fruiting pedicels recurved, 3–5 mm. long; sepals sparsely pubescent, 2–3 mm. long; petals white, ca. 4 mm. long; siliques almost straight, pendulous or widely descending, glabrous, obtuse, 3–5 cm. long, ca. 2 mm. wide; style 0 or very short; seeds suborbicular, uniseriate, ca. 2 mm. long, plainly winged.—Dry slope, 8500–9200 ft.; Sagebrush Scrub; near Masonic, Mono Co.; to Ore., Wyo. June–July.

25. **A. Shóckleyi** Munz. Perennial from thick simple caudex; stems 1–few, simple, hoary, 1–3 dm. tall, densely pubescent with minute dendritic hairs; basal lvs. crowded, spatulate, entire, hoary, 1–2 cm. long, 4–6 mm. wide; cauline lvs. broadly lanceolate, auricled, 1–1.5 cm. long; pedicels ascending, densely pubescent, 8–10 mm. long; sepals pubescent, 5–7 mm. long; petals pink, 8–11 mm. long; siliques divaricate, crowded, straight to slightly curved, subglabrous, 5–8 cm. long, ca. 2 mm. wide; stigma subsessile; seeds oblong, biseriate, essentially wingless, ca. 1 mm. wide.—Rare, dry rocky places; Pinyon-Juniper Wd.; n. slope San Bernardino Mts.; w. Nev. May–June.

26. **A. inyoénsis** Roll. Perennial from branched caudex; stems several, erect, densely pubescent below with fine dendritic hairs, glabrate above, 2–5 dm. high; basal lvs.

many, narrowly oblanceolate to spatulate, entire, acute, densely pubescent, 2–3 cm. long, 2–5 mm. wide; cauline lvs. sessile, oblong, auricled; fruiting pedicels usually horizontal, sometimes ascending, 6–12 mm. long, subglabrous; sepals pubescent, ca. 4 mm. long; petals pink to purplish, 7–9 mm. long; siliques wide-spreading, sometimes ± descending or ascending, nerved below, glabrous, 4–6 cm. long, ca. 2 mm. wide; seeds orbicular, uniseriate, ca. 1.5–2 mm. wide, narrowly winged.—Dry stony or rocky places, 5000–11,000 ft.; Pinyon-Juniper Wd. to Subalpine F.; mostly on e. slope of Sierra Nevada from Mono Co. to Inyo Co., Tulare Co., Argus and Panamint mts. May–July.

27. **A. púlchra** Jones. Perennial, rather woody at the simple or branched caudex; stems 1–several, simple or branched, densely pubescent with minute appressed dendritic hairs or glabrous above, 2–6 dm. high; basal lvs. linear, entire or slightly toothed, obtuse, densely pubescent, 4–8 cm. long, 3–6 mm. wide; cauline lvs. linear, sessile, not auricled; pedicels ascending at anthesis, geniculately reflexed in fr., densely pubescent, 8–20 mm. long; sepals pubescent, 5–8 mm. long; petals purple, 10–20 mm. long; siliques strictly appressed to the stem, densely pubescent, straight, 4–7 cm. long, 2.5–3.5 mm. wide; stigma subsessile; seeds suborbicular, biseriate, 1.5–2 mm. wide, rather prominently winged.— Rocky slopes, 2000–7000 ft.; Creosote Bush Scrub, Joshua Tree Wd., Pinyon-Juniper Wd.; deserts from Inyo Co. to L. Calif., Nev. April–May.

Var. **grácilis** Jones. [*A. trichopoda* Greene, not Turcz. *A. pulchra* var. *glabrescens* Wiggins. *A. p.* var. *viridis* Jeps.] Pubescence coarser and less dense; upper stems and pedicels glabrous; pedicels arched downward, not geniculately reflexed; siliques pendulous, glabrous.—With the sp.

Var. **munciénsis** Jones. Pedicels gently spreading downward, densely pubescent; petals purple, less than 10 mm. long.—Darwin, Inyo Co.; Nev., Utah. April.

28. **A. repánda** Wats. Biennial or possibly perennial, with forked or dendritic hairs; stems ascending, mostly 1 or 2, densely pubescent below, sparsely so or glabrous above, 2–8 dm. tall; basal lvs. obovate to broadly oblanceolate, petioled, repand-toothed, 3–8 cm. long, 1–3 cm. wide; cauline lvs. few, oblanceolate; pedicels stout, straight, erect to ascending, usually pubescent, 3–8 mm. long; sepals pubescent, 4–5 mm. long; petals white to pinkish, linear, 4–6 mm. long; siliques divaricately ascending, straight or often falcate, pubescent to glabrous, 4–10 cm. long, 2–4 mm. wide; style slender to 1 mm. long; seeds roundish, 2–4 mm. wide, broadly winged.—Dry slopes under pines, 4600–9000 ft.; Yellow Pine F., Red Fir F., Lodgepole F.; Sierra Nevada from Nevada Co. to Tulare Co., Glenn Co., Mt. Pinos to San Jacinto Mts.; Nev. June–Aug.

Var. **Greènei** Jeps. [*A. inamoena* Greene 1911, not 1908.] Stems 1–3 dm. high; lvs. subentire; pedicels pubescent, 5–10 mm. long; siliques glabrous, 4–5 cm. long, 2–2.5 mm. wide; style ca. 1 mm. long; seeds ca. 2 mm. wide.—Dry slopes, 8800–11,600 ft.; Lodgepole F., Subalpine F.; e. side of Sierra Nevada, Mono and Inyo cos. July–Aug.

29. **A. glaucoválvula** Jones. Perennial from a woody branched caudex; stems 1–several, mostly simple, hoary throughout, 1.5–4 dm. high; basal lvs. narrowly oblanceolate, entire, obtuse, densely pubescent with dendritic hairs, hoary, 2–5 cm. long, 2–5 mm. wide; cauline lvs. lanceolate or lance-linear, sessile, not auricled; pedicels stout, strongly recurved, pubescent, 5–10 mm. long; sepals pubescent, 4–5 mm. long; petals whitish to pink, 6–8 mm. long; siliques reflexed, oblong, obtuse, glabrous, glaucous, 2–4 cm. long, 5–7 mm. wide; style less than 1 mm. long; seeds biseriate, orbicular, 5–6 mm. wide, with broad wing.—Dry stony places, 2500–5300 ft.; Creosote Bush Scrub, Joshua Tree Wd.; desert from Bishop Creek, Inyo Co. to Little San Bernardino, New York, and Eagle mts. (Riverside Co.). March–May.

30. **A. díspar** Jones. [*A. salubris* Jones. *A. juniperina* Jones. *A. nardina* Greene.] Perennial from a cespitose base; stems several, simple or branched above, densely pubescent below, less so above, 1–2.5 dm. high; basal lvs. many, entire, erect, spatulate to narrowly oblanceolate, hoary, with dense fine dendritic pubescence, 1.5–2.5 cm. long, 2–4 mm. wide; cauline lvs. sessile, broadly linear, hoary; pedicels suberect to divaricate, pubescent, 1–2 cm. long; sepals pubescent, ca. 4 mm. long; petals purplish, obovate, 5–6 mm. long; siliques divaricate to ascending, glabrous, 5–7 cm. long, 2.5–3.5 mm. wide, with midnerve in lower half; stigma subsessile; seeds suborbicular, imperfectly uniseriate, ca. 2 mm. wide, broadly winged.—Rare, gravelly slopes, 4000–8000 ft.; Creosote Bush Scrub, Joshua Tree Wd., Pinyon-Juniper Wd.; Panamint Mts., Argus Mts., and near Bishop, Inyo Co., to San Bernardino and Little San Bernardino mts. April–May.

31. **A. Johnstònii** Munz. Perennial, densely pubescent with fine dendritic hairs; stems 1–few, ascending, simple, densely pubescent, 1–2 dm. high; basal lvs. narrowly spatulate to oblanceolate, entire, hoary with fine dense pubescence, 1–2 cm. long, 1.5–3.5 mm. wide; cauline lvs. few, sessile, not auricled; pedicels ascending, pubescent, 6–10 mm. long; sepals pubescent, 4.5–6 mm. long; petals purple, spatulate, 8–10 mm. long; siliques suberect, glabrous, 3–5 cm. long, 2–3 mm. wide, acuminate at apex; style slender, 1–2 mm. long; seeds suborbicular, uniseriate, ca. 1.5 mm. wide, winged.—Dry rocky knolls, 4500–5000 ft.; lower edge of Yellow Pine F.; s. end of San Jacinto Mts., Riverside Co. May–June.

32. **A. Paríshii** Wats. Tufted perennial from branched caudex; stems simple, slender, densely pubescent below with dendritic trichomes, less so above, 3–12 cm. high; basal lvs. many, linear-oblanceolate, entire, short-petioled, hoary with fine dendritic hairs, 5–15 mm. long, 1–2 mm. wide; cauline lvs. few, sessile, not auricled, linear; pedicels erect to slightly divergent, pubescent, 3–7 mm. long; sepals green or purplish, 3–4 mm. long; petals purple or lavender with white base, spatulate, 8–12 mm. long; siliques ascending, glabrous, 1–2 cm. long, 2–3 mm. wide, acuminate; styles filiform, 4–8 mm. long; seeds elliptical to suborbicular, imperfectly uniseriate, 1–1.5 mm. wide, narrowly winged.— Dry stony slopes, 6500–9800 ft.; Yellow Pine F., Red Fir F.; Bear V. and Sugarloaf Peak, San Bernardino Mts. April–May.

33. **A. suffrutéscens** Wats. [*A. duriuscula* Greene.] Suffruticose perennial, with several stems, simple or branched above, glabrous, 2–5 dm. high; basal lvs. linear to almost spatulate, acute to obtuse, glabrous, 1–4 cm. long, 2–6 mm. wide; cauline lvs. few, sessile, auriculate; pedicels slender, glabrous, reflexed, 4–10 mm. long; sepals glabrous, ca. 4 mm. long; petals rose to purplish, 6–8 mm. long; siliques pendulous to strictly reflexed, glabrous, acuminate, 4–7 cm. long, 3–6 mm. wide; style none to almost 1 mm. long; seeds orbicular, imperfectly uniseriate, 2–3.5 mm. wide, the wings ca. 1 mm. wide.— Dry, often stony places, 5500–9000 ft.; Sagebrush Scrub, Yellow Pine F., Red Fir F.; Yolla Bolly Mts. and Siskiyou Co., along the e. Sierra Nevada to Fresno Co.; to Wash., Ida. June–July.

Var. **perstylòsa** Roll. Plants subglabrous; cauline lvs. not auriculate; styles 2–3.5 mm. long.—Serpentine slope, Middle Fork of Feather R., Plumas Co. June.

34. **A. platyspérma** Gray. [*Erysimum p.* Kuntze. *A. inamoena* Greene, 1908, not 1911. *A. oligantha* Greene.] Perennial, the lower stems, basal lvs. and sepals with dendritic pubescence; stems erect to subdecumbent, simple or branched, 1–4 dm. high; basal lvs. many, oblanceolate or narrower, entire, 2–5 cm. long, 3–8 mm. wide; cauline lvs. few, remote, oblong to linear-lanceolate, sessile, not auriculate; pedicels divaricately ascending, straight, 5–15 mm. long; sepals 3–4 mm. long; petals pink to white, spatulate, 4–6 mm. long; siliques erect to divaricately ascending, straight, flat, acuminate, 3–7 cm. long, 3–5 mm. wide; style 0 to almost 1 mm. long; seeds orbicular, uniseriate, 3–4 mm. wide, winged.—Dry stony flats and slopes, 5500–11,200 ft.; Red Fir F. to Subalpine F.; Siskiyou and Glenn cos., through the Sierra Nevada to Tulare Co., San Gabriel Mts. to San Jacinto Mts.; Ore., Nev. June–July.

Var. **Howéllii** (Wats.) Jeps. [*A. H.* Wats. *A. platyloba, conferta,* and *Covillei* Greene. *A. platysperma* var. *imparata* Jeps. *A. inamoena* var. *acutata* Jeps.] Glabrous except sometimes with few hairs on petioles of basal lvs.; stems 0.5–3 dm. high; cauline lvs. sometimes auriculate; sepals glabrous, 3–5 mm. long; petals 5–7 mm. long, obtuse.—Dry granitic gravel, benches, and slopes, 10,000–12,000 ft.; Subalpine F., Alpine Fell-fields; Tulare Co. to Mt. Shasta; lower in high mts. of Trinity and Siskiyou cos.; Ore., Nev. July–Aug.

35. **A. pygmaèa** Roll. [*A. inamoena* auth., not Greene.] Perennial, the caudex usually with series of hemispherical clusters of dead lvs.; stems several, slender, erect to somewhat decumbent, simple, pubescent with forked hairs or glabrate above, 5–10 cm. high; basal lvs. tufted, entire, linear, 1–2 cm. long, 1–2 mm. wide, hispid with forked hairs, or the marginal simple; cauline lvs. few, sessile, not auriculate, linear; pedicels ascending, sparingly pubescent, 5–8 mm. long; sepals pubescent, ca. 2 mm. long; petals white, spatulate, ca. 3.5 mm. long; siliques erect, straight, subacuminate, glabrous, 2–4 cm. long, 2.5–3.5 mm. wide; stigma sessile; seeds orbicular, 2.5–3.5 mm. wide, broadly winged.—Dry flats of volcanic sand or gravel, 8500–11,000 ft.; Subalpine F.; Tulare Co. from Rock Creek to Templeton Meadows. June–July.

47. Sibàra Greene

Annual or biennial. Stems single or several from base, divaricately branched, glabrous or sparsely pubescent below with simple or branched hairs. Lvs. pectinate to runcinate-pinnatifid, the upper cauline sometimes almost entire, glaucous. Fls. small, in lax racemes. Sepals narrowly ovate to almost oblong, nonsaccate or the outer pair slightly saccate. Petals white to pink or purplish, spatulate to almost oblong. Nectar-glands small, subtending or surrounding single stamens, otherwise absent or obsolete. Siliques linear, flattened parallel to septum or subterete, the valves nerveless to nerved below. Seeds oblong to suborbicular, winged or not, uniseriate, with accumbent or incumbent cotyledons. Ca. 11 spp. of N. Am. (Anagram of *Arabis*.)

(Rollins, R. C. Generic revisions in the Cruciferae: Sibara. Contr. Gray Herb. 165: 133–143, 1947.)
Lvs. runcinate-pinnatifid, the lobes oblong to obovate; siliques divaricate, 1.5–2 mm. wide; seeds
winged. Cismontane .... 1. *S. virginica*
Lvs. pectinate; siliques ascending to reflexed, less than 1.5 mm. wide; seeds wingless. Santa Cruz Id.
and deserts.
    Mature siliques divaricately ascending, mostly 2–5 cm. long; pedicels glabrous.
        Pedicels 6–12 mm. long; basal lvs. caducous; styles not expanded toward apex .... 2. *S. filifolia*
        Pedicels 2–3 mm. long; basal lvs. persistent; styles expanded toward apex .... 3. *S. rosulata*
    Mature siliques pendulous to reflexed, less than 2 cm. long; pedicels somewhat pubescent ......
                                                        4. *S. deserti*

    1. **S. virgínica** (L.) Roll. [*Cardamine v.* L. *Arabis v.* Poir.] Annual or biennial, hirsute below, 1–4 dm. high; lvs. lyrate-pinnatifid, 3–8 cm. long, 7–12 mm. wide; fruiting pedicels glabrous, 3–7 mm. long; sepals 1–2 mm. long; petals white to faintly pink, 1.5–3 mm. long; siliques 2–2.5 cm. long, almost straight or slightly curved, erect to divaricate, glabrous; style 0.3–0.5 mm. long; seeds orbicular, uniseriate, ca. 1.5 mm. long, narrowly winged.—Rare about drying pools; V. Grassland, Coastal Sage Scrub; San Diego, Inglewood, Gardena, Tracy, Stockton; e. U.S. April–May.
    2. **S. filifòlia** (Greene) Greene. [*Cardamine f.* Greene. *Arabis f.* Greene.] Slender glabrous glaucous annual, 1.5–3 dm. high; basal lvs. early caducous; cauline pinnate with narrowly linear segms., petiolate, 2–4 cm. long, the segms. 5–10 mm. long; pedicels slender, divaricate, 6–12 mm. long; sepals glabrous; petals spatulate, pink to purplish, 3–5 mm. long; siliques slender, flattened, divaricate, 2.5–4 cm. long, less than 1 mm. wide; style ca. 1 mm. long; seeds oblong, wingless, ca. 0.8 mm. long.—Shady n. slopes; Chaparral; Santa Cruz Id. April.
    3. **S. rosulàta** Roll. Annual with few divaricately branched stems 1–3 dm. high, glabrous to sparsely pubescent; basal lvs. rosulate, deeply pinnately lobed, petiolate, 3–5 cm. long, the lobes 4–8 mm. long, 1–2 mm. wide; cauline lvs. somewhat lobed or the upper entire, linear; fruiting pedicels divaricately ascending, glabrous, 2–3 mm. long; sepals glabrous to sparsely pubescent, 1.5–2 mm. long; petals white, narrowly spatulate, 2.5–3 mm. long; siliques divaricately ascending, glabrous, flattened, 1.5–3 cm. long, 1–1.5 mm. wide; seeds wingless, oblong, ca. 1 mm. long.—Sandy and rocky places, ca. 3000 ft.; Creosote Bush Scrub; about Death V. March–April.
    4. **S. déserti** (Jones) Roll. [*Thelypodium d.* Jones. *Arabis d.* Abrams.] Annual, 1-stemmed, 1–3 dm. high, sparsely pubescent with minute branched hairs; basal lvs. caducous; cauline lvs. petiolate, 2–4 cm. long, sparsely pubescent, the lower pinnate with segms. 4–8 mm. long, the upper entire; pedicels widely spreading to descending, sparsely pubescent, 3–4 mm. long; sepals pubescent, ca. 2 mm. long; petals white, spatulate, 2–3 mm. long; siliques flattened, sparsely pubescent, slightly descending to loosely reflexed, 1–1.5 cm. long; style stout, 1–1.5 mm. long; seeds oblong, wingless, ca. 1 mm. long; $n = 13$ (Rollins, 1947).—Rare, 2000–4000 ft.; Creosote Bush Scrub; Death V.; Nev. March–April.

48. Erýsimum L. WALLFLOWER

Annual, biennial or perennial, leafy-stemmed herbs, stout, with appressed 2–3-forked hairs. Lvs. narrow, entire to dentate. Fls. in ours yellow to orange, in terminal racemes. Sepals erect, narrow, the 2 outer usually saccate at base. Petals clawed, with obovate spreading blades. Silique linear, 4-sided or subterete or compressed, the valves keeled by a strong midrib. Stigma broadly lobed. Seeds uniseriate, oblong, marginless or ± winged, with incumbent or accumbent cotyledons. Ca. 90 spp. of wide distribution in temp. zone. (Greek, *eryomai*, help or save, because of supposed medicinal value of some spp.)

(Rossbach, George B. New taxa and new combinations in the genus Erysimum in North America. Aliso 4: 115–124, 1958. Our key and treatment were adapted from materials kindly prepared by Dr. Rossbach.)

A. Plants annual; petals 3–8 mm. long, less than 2 mm. wide.
    B. Pedicels slender, ⅓–½ as long as siliques; siliques ± ascending, 1–3.5 cm. long; axis of mature raceme straight .... 1. *E. cheiranthoides*
    BB. Pedicels nearly as thick and less than ⅛ as long as siliques; siliques divaricate, 4–8.5 cm. long; axis of mature raceme geniculate .... 2. *E. repandum*
AA. Plants biennial or perennial; petals 12–32 mm. long, 3–15 mm. wide.
    B. Lvs. nearly filiform, 0.3–1.7 mm. broad; stems suffused with ± metallic purplish hue, normally simple; caudex ± elongate, herbaceous to subligneous, single or divided. Santa Cruz Mts. .... 3. *E. teretifolium*
    BB. Lvs., or at least some of them, 2 mm. or more broad.
        C. Plants not suffrutescent; caudex not notably elongate or long-branched above ground, erect, or at least not sprawling or widely spreading, mostly without or with only very short sterile branches; lowest lvs. mostly not marcescent.
            D. Seeds with small scarious distal appendage or wingless; siliques strongly compressed to tetragonal. Not coastal.
                E. Stems low, mostly 2–30 cm. to base of raceme; upper foliar hairs 2-parted or dominantly so; petals yellow.
                    F. Lower lvs. spatulate to bluntly oblanceolate, not acute, mostly thin, green; foliar hairs delicate, sparse; siliques strongly compressed, unequally keeled, 2–3 mm. broad, sparsely and finely pubescent; style 1.8–5 mm. long. Mostly from above 7000 ft. .... 3. *E. perenne*
                    FF. Lower lvs. linear-oblanceolate, acutish, thickish, usually cinereous; foliar hairs mostly strigose, crowded; siliques ± tetragonal, subequally keeled, ca. 1.5 mm. broad, mostly abundantly pubescent; style 0.5–2 mm. long. Mostly from below 7000 ft. .... 4. *E. argillosum*
                EE. Stems taller, commonly 20–70(–130) cm. to base of raceme; upper foliar hairs mostly 3-parted; petals orange, yellow or otherwise.
                    F. Stems robust, commonly 4–12 mm. thick at base, mostly much-branched above; lvs. and siliques variable. Widespread .... 5. *E. capitatum*
                    FF. Stems slender, mostly simple, less than 4 mm. thick near base; lvs. narrowly elongate, acute; siliques slender, 6–13 cm. long, 1.3–1.8 mm. wide; fresh petals orange to orange-yellow. Sandy mesas, n. Santa Barbara Co. .... 11. *E. suffrutescens* var. *lompocense*
            DD. Seeds winged about distal end and ± along one side; siliques, at least when dry, very strongly compressed, usually keeled. Coastal, if native.
                E. Siliques stiffly divaricate, usually upcurved; pedicels divaricate.
                    F. Lvs. oblong-spatulate or at least blunt, the lowest lvs. broadest, 4–13(–30) mm. broad; stems low, commonly 3–7 cm. to base of raceme. Coastal Strand; Mendocino and Humboldt cos.
                                    6. *E. Menziesii*
                    FF. Lvs. ± oblanceolate, acute, the lowest lvs. narrowest, 1.5–3(–6) mm. broad; stems commonly 15–50 cm. tall to base of raceme. Coastal Strand; Monterey Co., Santa Rosa Id., San Diego Co.
                                    7. *E. ammophilum*
                EE. Siliques stiffly ascending, straight to slightly upcurved; pedicels variable.
                  F. Stigma not deeply divided, merely bilobed; lvs. not soon deciduous, usually regularly sinuate-dentate; foliar hairs usually mostly 3-parted; petals rich yellow to creamy-white.
                    G. Lvs. shortly oblanceolate, abruptly contracted to apex; siliques distally blunt or at least tapering abruptly to style; plants fleshy. Mostly coastal bluffs, etc., Marin Co. n. .... 8. *E. concinnum*
                    GG. Lvs. linear-oblanceolate, elongate, gradually narrowed at ends; siliques usually tapering gradually to style; plants mostly not fleshy. Serpentine or sandy soil, San Mateo Co., Mt. Tamalpais, Bodega Bay .... 9. *E. franciscanum*
                FF. Stigma deeply divided with long arching lobes; lvs. soon progressively deciduous along aging stout stem, mostly subentire; foliar hairs 2-parted; petal-color various. Garden escape ... . 10. *E. Cheiri*
        CC. Plants suffrutescent; caudex elongate, long-branched above ground, sprawling or widely spreading, with elongate sterile branches; lower lvs. mostly marcescent.
            D. Seeds extensively winged about distal end and ± along one side, strongly compressed, oval; dry silique very strongly compressed.
                E. Stigma deeply divided, with long arching lobes; lvs. always soon and progressively deciduous along aging stem, entire or sparingly serrulate-denticulate; foliar hairs 2-parted. Garden escape .... 10. *E. Cheiri*
                EE. Stigma slightly bilobed; lvs. not soon deciduous below, regularly sinuate-dentate; foliar hairs commonly 3-parted. Coastal natives .. 9. *E. franciscanum*

DD. Seeds distally winged or wingless, not strongly compressed, convex, variously shaped; siliques variable.
E. Plants fleshy, notably suffrutescent and sprawling-ascending, much branched, with long vegetative stems; lvs. notably marcescent below; siliques coarse.
F. Branched base usually spreading-upcurved; lvs. commonly 2–3 mm. broad; siliques compressed parallel to septum or squarish in cross section. Coastal Strand; Los Angeles to San Luis Obispo cos.
11. E. suffrutescens
FF. Branched base sprawling; siliques plump, squarish in cross section to compressed perpendicular to septum.
G. Lvs. commonly 3–12 mm. broad; foliar hairs often dominantly 3-parted. Coastal dunes, Point Arguello to Point Purisima, Morro Rock . . . . 11. E. suffrutescens var. grandifolium
GG. Lvs. 1.5–3 mm. broad; foliar hairs crowded, 2-parted. Insular
12. E. insulare
EE. Plants not at all fleshy, moderately suffrutescent, not sprawling, with few shortish sterile stems; lvs. only moderately marcescent below; siliques slender, compressed parallel to septum. Sandy mesas in n. Santa Barbara Co.
11. E. suffrutescens var. lompocense

1. **E. cheiranthoìdes** L. [*Cheiranthus c.* Heller.] WORMSEED. Annual, 1–10 dm. high, simple or branched, minutely roughened; lvs. lanceolate to oblanceolate, 2.5–8 cm. long, 0.6–1.5 cm. wide, entire or somewhat dentate, finely pubescent with trifid hairs; pedicels subfiliform, spreading to divergent, 0.6–1 cm. long; petals yellow, mostly 3–5 mm. long; siliques subterete, 1–3.5 cm. long, 1–1.5 mm. wide, subglabrous; seeds brown, oblong, ca. 1 mm. long; $2n = 16$ (Manton, 1932).—Collected long ago along a railroad in Placer Co.; waste places to Alaska and Atlantic Coast; natur. from Eu. June–Aug.

2. **E. repándum** L. TREACLE-MUSTARD. Annual, simple or usually divergently branched, 1–5 dm. high, with scattered bifid hairs; lvs. lanceolate to linear-lanceolate, repand-denticulate, 3–6 cm. long, 0.3–0.7 cm. wide; pedicels thick, 2–4 mm. long; petals pale sulphur-yellow, 6–8 mm. long; siliques appearing continuous with the pedicels, 4-sided, 4–8.5 cm. long, 1.5–2 mm. thick, glabrous or remotely strigose; seeds not winged, plump, oblong, ca. 1 mm. long; $2n = 14–16$ (Manton, 1932).—A weed with tumbling habit, troublesome in alfalfa fields in Modoc and Lassen cos.; to Wash. and Atlantic Coast; natur. from Eu. May–June.

3. **E. perénne** (Wats. ex Cov.) Abrams. [*E. asperum* var. *p.* Wats. ex Cov. *Cheiranthus p.* Greene. *E. nevadense* Heller.] Short-lived perennial, the root-crown clothed with the remains of the old lvs.; plant mostly rather sparingly pubescent with appressed 2-forked hairs, the stems mostly single, simple, 1–3 dm. high; lower lvs. oblanceolate to spatulate, 3–5 cm. long, mostly 3.5–8 mm. wide, ± runcinate-denticulate, short-petioled; upper lvs. narrower and shorter; fruiting pedicels ascending, ca. 1 cm. long; sepals 7–9 mm. long; petals yellow, 15–17 mm. long, 5–7 mm. wide; siliques ascending or somewhat spreading, 6–8 cm. long, ca. 2 mm. wide, flattened, sparsely and finely pubescent, the beak slender, 2–5(–8) mm. long; seeds not or scarcely winged at one end, ca. 2–3 mm. long.—Dry slopes and knolls, mostly 7000–12,000 ft.; Lodgepole F. to Alpine Fell-fields; Sierra Nevada from Tulare Co. n., to Mt. Shasta, Scott Mt. (Trinity Co.), Yolla Bolly Mts. June–Aug.

4. **E. argillòsum** (Greene) Rydb. [*Cheiranthus a.* Greene.] Biennial, mostly simple, 1–3 dm. high; lvs. narrow, 1–2 mm. wide, cinereous with closely appressed pubescence, very crowded on lower stems; petals pale yellow, 16–18 mm. long; siliques 5–7 cm. long, 2 mm. wide; style or beak mostly 0.5–2 mm. long; seeds mostly not over 2 mm. long.— Disintegrated travertine about hot springs, 5000–6750 ft.; Pinyon-Juniper Wd., Sagebrush Scrub; Mono and Inyo cos., to Nev., Ida. May–June.

5. **E. capitàtum** (Dougl.) Greene. [*Cheiranthus c.* Dougl. in Hook. *E. elatum* Nutt. *C. e.* Greene. *E. californicum* Greene. *E. moniliforme* Eastw. *E. asperum* Calif. auth.] Biennial, erect, relatively simple, coarse-stemmed, 2–8 dm. high, strigose; basal lvs. lanceolate, the lower 4–15 cm. long, 0.4–1 cm. wide, usually dentate or denticulate, acute to subacute, some or all of the upper foliar hairs 3-parted (2-parted hairs frequently also present, or exclusively so in southern plants); pedicels stout in fr., 4–6 mm. long; sepals 8–12 mm. long; petals orange, yellow, brick-red, orange-brown, or locally purplish-maroon, 1.5–2 cm. long; siliques 5–10 cm. long, 1.5–2 mm. broad, 4-angled, fairly protrusively keeled on flatter surfaces; style thick, 1–2 mm. long; seeds oblong-elliptic, distally winged, ca. 1.5 mm. long—Frequent in dry stony places below 8000 ft.;

many Plant Communities; largely away from the ocean, through most of cismontane and montane Calif.; to B.C., Ida. March–July. Exceedingly variable.

Var. **angustàtum** (Greene) G. Rossb. [*Cheiranthus a.* Greene.] Caudex elongate; lvs. lance-linear, gradually tapering, acute; siliques 1.5–2 mm. broad.—Coastal Strand; dunes along San Joaquin R. just e. of Antioch, Contra Costa Co.

Var. **Bealiànum** (Jeps.) G. Rossb. [*E. asperum* var. *B.* Jeps.] Stems coarse; lvs. oblance-olate, rather blunt, cinereous, minutely denticulate; siliques mostly 2–2.5(–3) mm. broad.—Dry plains, 2000–4000 ft.; largely Joshua Tree Wd.; Little San Bernardino Mts. through w. San Bernardino Co. to Los Angeles and Kern cos.

6. **E. Menzièsii** (Hook.) Wettst. [*Hesperis M.* Hook. *E. grandiflorum* Nutt. in T. & G. *Cheiranthus g.* Heller.] Biennial or short-lived perennial from a long taproot and with a mostly simple caudex 1–6 cm. long; flowering stems usually several and 0.5–1.5 dm. long; basal lvs. green, oblong-spatulate or at least blunt, 3–9 cm. long, 0.4–1.5 cm. wide, sparingly strigose, entire or obscurely few-toothed; cauline lvs. spatulate, reduced; pedicels thickened in fr., 3–7 mm. long; sepals 8–11 mm. long; petals bright yellow, 15–20 mm. long, 6–8 mm. wide; siliques stiffly divaricate, flattened, 4–8 cm. long, 2.5–3.5 mm. wide; style stout, barely 1 mm. long; stigma-lobes prominent, divergent; seeds narrowly winged, suborbicular, ca. 2 mm. long.—Local on dunes; Coastal Strand; Point Pinos (Monterey Co.), and from Ft. Bragg to n. of Humboldt Bay. March–May, but somewhat through many other months.

7. **E. ammóphilum** Heller. Near to *E. Menziesii*, but with lvs. linear-oblanceolate to oblanceolate, acute, narrowest toward base of stem where narrowly elongate, 1.5–3(–6) mm. wide; stems almost always taller, 0.3–8 (commonly 1.5–5) dm. to base of raceme; fls. bright golden yellow; pedicels ca. 6–10 mm. long.—Coastal Strand; Monterey Bay, and in an atypical form, San Diego Co., Santa Rosa Id. Feb.–May.

8. **E. concínnum** Eastw. Near to *E. Menziesii;* lvs. shortly oblanceolate, abruptly contracted to apex; plants fleshy; pedicels mostly 1–2 cm. long; fls. mostly pale yellow; siliques stiffly ascending, straight to slightly incurved, ca. 6–11 cm. long, distally blunt or at least tapering abruptly to the very short style.—Coastal bluffs and headlands, rarely on dunes; mostly N. Coastal Scrub; Point Reyes (Marin Co.) to Curry Co., Ore. March–May.

9. **E. franciscànum** G. Rossb. Biennial or short-lived perennial, simple or branched, 0.5–4 dm. high; lvs. not fleshy, the lower linear-oblanceolate, sharply and often deeply sinuate-dentate, gradually tapered to petiole, 4–9(–17) cm. long, 2–8(–14) mm. broad, sparsely pubescent; petals yellow to cream, 14–29 mm. long; pedicels ± ascending, 5–14(–24) mm. long; siliques crowded, usually ascending, 4–10(–13.5) cm. long, ca. 2–3.5 mm. broad, commonly tinged purple; style ca. 1–2 mm. long; seeds oblong to oval, 2–3.5 mm. long.—Open or wooded or brushy places in rocky to sandy soil, often on disintegrated serpentine; Chaparral, N. Coastal Scrub, etc.; near the coast, San Mateo and San Francisco cos., near Mt. Tamalpais and Bodega Bay; sw. Ore. March–June.

Var. **crassifòlium** G. Rossb. Suffrutescent, usually basally spreading, with elongate caudex; lvs. fleshy, crowded below, where narrower and ± marcescent; petals rich yellow; siliques fleshy, tapering more abruptly to the 1 mm. style.—Coastal bluffs and headlands of San Mateo Co. and upper sandhills near coast, n. Santa Cruz Co. March–April.

10. **E. Cheìri** (L.) Crantz. [*Cheiranthus C.* L.] WALLFLOWER. Erect perennial, 3–7 dm. high, strigose, grayish; lvs. lanceolate to narrow-lanceolate, mostly entire, acute, 4–8 cm. long, usually crowded beneath the fls. and at ends of sterile shoots, with 2-parted hairs; fls. 2–2.5 cm. long, yellow, orange, brown-orange or purplish; siliques erect, 5–6 cm. long, rather thick, angled, the style short, bearing a bicornate stigma with long arching lobes.—Occasional garden escape; native of s. Eu.

11. **E. suffrutéscens** (Abrams) G. Rossb. [*Cheiranthus s.* Abrams. *E. concinnum* ssp. *s.* Abrams.] Suffrutescent, the branched base usually spreading-upcurved and with long vegetative stems; plants succulent; lvs. narrowly linear-lanceolate, 3–7 cm. long, 2–3(–5) mm. wide, notably marcescent below; petals bright yellow, 15–20 mm. long; siliques coarse, squarish in cross section or compressed parallel to septum, divergent-spreading, 5–9 cm. long, 1.5–2 mm. wide; seeds ca. 1.5–2 mm. long, compressed, slightly winged distally.—Mostly Coastal Strand; Los Angeles Co. and s. San Luis Obispo Co. Jan.–May.

Var. **grandifòlium** G. Rossb. More sprawling and massive, more succulent, with more vegetative branches; lvs. often broader, more sparsely pubescent with more 3-parted hairs; siliques often compressed perpendicularly to the septum or nearly square, the seeds angular, barely compressed, often wingless, smaller.—Coastal Strand; between Points Arguello and Purisima, Santa Barbara Co., also bluffs of Morro Rock, San Luis Obispo Co. Feb.–June.

Var. **lompocénse** G. Rossb. Plants not fleshy, moderately suffrutescent, not sprawling, only 1- or few-times branched, with few rather short sterile stems; lvs. only moderately marcescent below; siliques slender, compressed parallel to septum, ca. 8–10(–13) cm. long, 1.3–1.8 mm. broad.—Sandy mesas; Coastal Sage Scrub, Chaparral; Lompoc, Nipomo, Guadalupe, Santa Barbara Co. Feb.–May.

12. **E. insulàre** Greene. [*Cheiranthus i.* Greene.] Plants succulent, strongly suffrutescent, with a much branched sprawling base, tufted, densely leafy, ca. 2–3 dm. high, cinereous with minute appressed 2-forked hairs; lvs. crowded, lanceolate, entire, firm, 2–3(–5) mm. wide; petals yellow, ca. 15 mm. long; siliques rather strict, quadrangular, 2–6 cm. long, 2.5–3.5 mm. broad; style stout, ca. 1 mm. long; seeds not winged, turgid, ca. 1.5 mm. long.—Coastal Strand; San Miguel and Santa Rosa ids. March–May.

13. **E. teretifòlium** Eastw. [*E. filifolium* Eastw., not F. Muell.] Biennial or winter annual, from a slender taproot; stems simple or branched from base, 3–9 dm. high, with scattered 2-branched hairs; lvs. filiform, 3–10 cm. long, 0.3–1.7 mm. wide, entire or with scattered minute teeth; pedicels divaricate, 5–10 mm. long; petals orange, later yellow, ca. 12–22 mm. long; siliques ascending-divaricate, 8–11 cm. long, mostly 4-sided, ca. 1–1.5 mm. broad, with largely 3-parted hairs; style short; seeds ca. 1.5 mm. long, slightly winged distally.—Old inland marine sands, Ben Lomond and Glenwood region, Santa Cruz Co. April–June.

49. Alýssum L.

Low branching herbs with stellate pubescence, often cespitose or suffrutescent. Lvs. small, simple, entire to toothed. Fls. small, usually yellow, in lengthening terminal racemes. Petals entire or retuse. Fils. toothed or appendaged by a single nectary gland at each side of base of short stamens. Fr. a silicle, orbicular, flattened parallel to septum, the valves nerveless, somewhat convex, the margin flattened. Seeds wingless, 1–2 in each locule. A large genus, mainly of Medit. region. (Greek name of plant supposed to cure hydrophobia, from *a*, without, and *lussa*, rabies.)

1. **A. alyssoìdes** L. Hoary annual, with arched-ascending stems 1–3 dm. long; lvs. linear-spatulate, 1–2 cm. long; petals pale yellow, cuneate, 2–3 mm. long; silicles on spreading pedicels, orbicular, 3 mm. wide, margined, notched at apex; style minute; seeds 4; $2n = 32$ (Manton, 1932).—Natur. in Siskiyou Co.; native of Eu. April–July.

50. Lobulària Desv. Sweet Alyssum

Low perennials with narrow entire lvs. and appressed 2-pointed hairs. Petals white, entire. Silicles as in *Alyssum*. Ca. 5 spp. of Medit. region. (Latin, *lobulus*, small lobe, possibly referring to 2-lobed hairs.)

1. **L. maritima** (L.) Desv. [*Clypeola m.* L. *Alyssum m.* Lam. *Koniga m.* R. Br.] Much-branched decumbent strigose perennial, the stems 0.5–2.5 dm. long; lvs. linear to linear-lanceolate, 1–5 cm. long; fls. sweet-smelling, 3–4 mm. wⁱde; pedicels 5–8 mm. long; silicles ca. 2.5 mm. long; seeds 2, wingless, brown, round-oblong, ca. 1.3 mm. long; $n = 12$ (Jaretzky, 1928).—Common escape from gardens, in waste places, along roads, etc., at low elevs.; native of Eu. Flowering much of year.

51. Matthìola R. Br. Stock

Biennial or perennial herbs with a close stellate tomentum, erect, branched at least above. Lvs. oblong to linear. Fls. purple to white, sweet-scented, racemose. Sepals erect, the lateral saccate at base. Petals with long claws and broad rounded blades. Siliques elongate, subterete, torulose; the valves dehiscent, keeled, 1-nerved. Stigmas

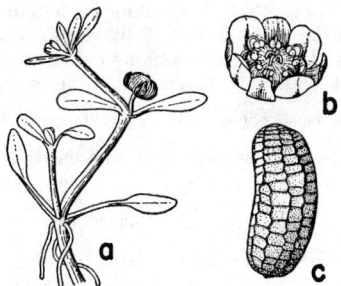

Fig. 32. ELATINACEAE. *Elatine californica:* a, habit, × 3; b, fl., × 15, sepals 4, petals 4, stamens 8, pistil with 4 styles; c, seed, × 40, pitted.

thickened or horned at back. Seeds uniseriate, wing-margined, suborbicular. Ca. 50 spp. of w. Asia and Medit. region. (Named for P. A. *Matthioli*, 1500–1577, Italian botanist and physician.)

1. **M. incàna** (L.) R. Br. [*Cheiranthus i.* L.] Plants 3–5 dm. tall; lvs. entire to sinuately dentate, 5–15 cm. long; sepals ca. 1 cm. long; petals ca. 2 cm. long; siliques 8–16 cm. long, 4–5 mm. thick; seeds almost 2 mm. wide; $n = 7$ (Allen, 1924).— Sandy places and bluffs along the seacoast, San Diego Co. to Monterey Co.; natur. from Eu. March–May. Garden escape.

52. Choríspora DC.

Annual branched glandular or pilose herbs. Lvs. entire to pinnatifid. Fls. yellow or purple, in loose, elongate racemes. Calyx narrow-cylindric, only the tips of the sepals spreading. Petals with long erect claws and narrow spreading blades. Anthers sagittate. Stigma minute. Silique elongate, slender, curving, the lower part torulose, with transverse partitions between the seeds, the upper portion (ca. half the whole pod) seedless, beaklike, slender, subulate. Seeds pendulous, margined or marginless. Cotyledons accumbent. Asiatic herbs. (Latin, *choris*, asunder, and *spora*, a seed, because of the portions of the silique breaking away between the constrictions and enclosing the seed.)

1. **C. tenélla** (Pall.) DC. [*Raphanus t.* Pall.] Plant 1–3 dm. high, glandular-puberulent, branched; lower lvs. runcinate, petioled, oblanceolate, 3–7 cm. long, the middle and upper lanceolate to oblong, undulate-dentate; sepals 4–6 mm. long; petals purple, 10–12 mm. long; siliques 3–4 cm. long; $n = 7$ (Jaretzky, 1929).—Grainfield weed, 1 mi. nw. of Gazelle, Siskiyou Co.; adv. in Wash., Ore., Ida., Nev., Ariz.; native of Asia.

53. Conríngia Link. HARE'S-EAR

Erect glabrous annual. Lvs. sessile, clasping, elliptic, the lower sometimes narrowed at base. Fls. in elongate, terminal racemes. Sepals and petals narrow, the latter light yellow. Siliques long, slender, 4-angled, somewhat rigid, the dehiscent valves 1–3-nerved. Style short; stigma entire or 2-lobed. Seeds uniseriate, oblong, marginless. Ca. 7 spp. of Eurasia. (Named for H. *Conring*, 1606–1681, professor at Helmstadt.)

1. **C. orientàlis** (L.) Dumort. [*Brassica o.* L. *C. perfoliata* Link.] Simple or slightly branched, 3–5 dm. tall; lvs. elliptic, 3–8 cm. long, deeply cordate-clasping; petals ca. 8 mm. long; siliques ascending, 6–10 cm. long, ca. 2 mm. thick; beak ca. 1.5 mm. long; seeds ovate, dark brown, 2–2.5 mm. long; $n = 7$ (Jaretzky, 1929).—Occasional weed in waste places, as w. of Yuma, Upland, Independence; native of Eurasia. April–June.

34. Elatinàceae. WATERWORT FAMILY (Fig. 32)

Low herbs or shrubs. Lvs. opposite or whorled, simple, with paired membranaceous stipules. Fls. small, regular, axillary, perfect, solitary or cymose. Sepals 3–5, free, imbricate, persistent. Petals 3–5, persistent, imbricate. Stamens as many or twice as many, free, hypogynous; anthers 2-celled, opening by longitudinal slits. Ovary superior, 3–5-loculed; placentation axile; styles free, 3–5, introrsely stigmatose, or stigmas sessile. Ovules many. Fr. a septicidal caps.; seeds oblong-cylindric, straight or curved, without endosperm; cotyledons thick and short. Cosmopolitan. Two genera.

Plants glabrous; fls. 2–4-merous; sepals obtuse, without midrib; caps. globose or depressed. Aquatic or semiaquatic . . . . 1. *Elatine*
Plants glandular-pubescent; fls. 5-merous; sepals pointed, with thickened midrib; caps. ovoid. Terrestrial
2. *Bergia*

1. Elatìne L. WATERWORT

Dwarf glabrous annuals or subperennials, often rooting at the nodes. Sepals 2–4; petals 2–4; stamens mostly 2–4; styles or sessile capitate stigmas 2–4. Caps. membranaceous, 2–4-celled, several–many-seeded, the partitions left attached to the axis, or evanescent. Ca. 10 spp. of temp. regions. (Classical name of some low creeping plant.)

(Fassett, N. C. Elatine and other aquatics. Rhodora 41: 367–377, 1939. Mason, H. L. New species of E. in Calif. Madroño 13: 239–240, 1956.)
Caps. with 2–3 carpels; seeds straight or slightly curved, almost alike at both ends.
Seeds with pits in rows of 16–35.
  Fls. with pedicels up to 2.5 mm. long; sepals 3, equal. Ricefield weed . . . . 2. *E. ambigua*
  Fls. sessile; sepals 2–3, when 3, then 1 is reduced.
    Pits of seeds much broader than long, the transverse ridges more prominent than the longitudinal.
      Seeds 6–10 in each locule; lvs. all linear-spatulate . . . . 1. *E. gracilis*
      Seeds 15–40 in each locule; lvs. linear-spatulate to round-obovate . . . . 3. *E. chilensis*
    Pits of seeds nearly as long as broad, the longitudinal and transverse ridges almost equally prominent . . . . 4. *E. rubella*
Seeds with pits in rows of 9–15.
  Lvs. linear to narrow-oblong. S. Calif. . . . . 5. *E. brachysperma*
  Lvs. obovate to elliptic-oblong. Cent. & n. Calif. . . . . 6. *E. obovata*
Caps. with 4 carpels, pedicelled; seeds very strongly curved, differently rounded at the 2 ends
7. *E. californica*

1. **E. grácilis** Mason. [*E. triandra*, Calif. refs.] Plants slender, erect, 2–4 cm. high; lvs. ca. equal to internodes, narrowed to petiolelike base with attenuate lacerate stipules; fl. 1 to a node, sessile; sepals 2, a 3d sepal reduced or wanting; petals 3, thin-membranous, suborbicular; stamens 3–1(–0), alternate with carpels; seeds nearly or quite straight, 7–8 to a locule, the areoles in 9–10 rows of 20–30 each, the horizontal ridges more conspicuous than the longitudinal.—Shallow water or mud-banks, below 9500 ft.; many Plant Communities; cismontane Calif. from Riverside Co. to Modoc Co.; to Canada, Atlantic Coast. April–Sept.

2. **E. ambígua** Wight. Much like no. 1., taller; lvs. ovate; pedicels at first drooping, later erect; sepals 3, linear, obtuse; petals pale rose, acutish.—Reported as weed in rice fields; native of s. Asia. Summer.

3. **E. chilénsis** Gay. Plant to 1 dm. long, aquatic or, if on wet mud, creeping and rooting; lvs. obovate to broadly spatulate, 3–4 mm. long, 1–3 mm. broad, with entire stipules; fls. 1–2 to a node, sessile; sepals 2 or with a third reduced; petals round; stamens 3, alternate with carpels; seeds slightly curved, 24–33 in a locule, erect, with 25–35 short broad pits and appearing transversely rugose.—Ponds and muddy shores; Madeline Plains (Lassen Co.) and Sierra V. (Plumas Co.); S. Am.

4. **E. rubélla** Rydb. [*E. triandra* auth., not Schkuhr.] Prostrate to erect and 1.5 dm. high, often reddish; lvs. opposite, lanceolate to linear-spatulate, 2–15 mm. long; fls. sessile, 1–2 at a node, 3-merous; sepals 2–3, often unequal; stamens 3, alternate with carpels; seeds 12–30 in a locule, cylindrical or slightly curved, longitudinally ribbed and with rows of 16–25 hexagonal pits.—Ponds, vernal pools, ditches, etc., through much of Calif.; w. Am.

5. **E. brachyspérma** Gray. [*E. triandra* var. *b.* Fassett.] Plants densely matted; lvs. linear to narrow-oblong, 1.5–4 mm. long (or longer in submerged plants); fls. 2–3-merous; caps. depressed; seeds short-oblong, 0.4–0.6 mm. long, with 6–8 rows of 9–15 pits in each irregular row.—Many Plant Communities; mostly in s. Calif.; to Ore., Ill., Tex. April–Sept.

6. **E. obovàta** (Fassett) Mason. [*E. triandra* var. *o.* Fassett.] Near to *E. brachysperma*.

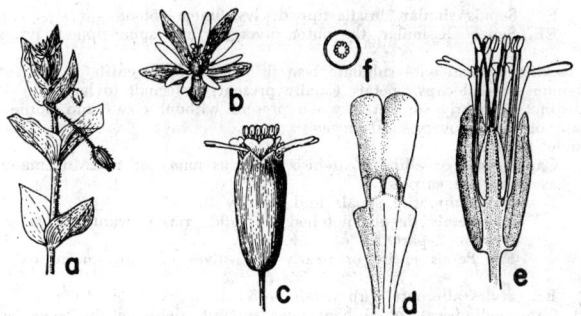

Fig. 33. CARYOPHYLLACEAE. *Stellaria media: a,* habit, × ¾; *b,* fl., × 2½, sepals 5, petals 5, cleft, the stamens 5, styles 3. *Silene Douglasii: c,* fl., × 1, calyx tubular; *d,* petal, × 3, blade bilobed, with 2 appendages below and expanded basal claw; *e,* fl. in section, × ½, carpophore bearing ovary, stamens, petals; *f,* cross section of ovary with free cent. placenta.

the lvs. obovate, 2–5 mm. long; seeds 0.35–0.6 mm. long, with 6–8 rows of 9–15 pits.—Uncommon, cent. Calif.; cent. Mex. *E. heterándra* Mason seems near this, but seems to have 9–10 rows of seed-pits. It is described from Sierra and Lake cos.

7. **E. califórnica** Gray. Matted, 2–14 cm. across; lvs. oblanceolate or wider, rounded or obtuse at apex, 2–4 mm. long; fls. pedicelled, 4-merous; seeds V- to U-shaped, rounded at one end, truncate and subapiculate at other, with ca. 25 pits in each row.—Water borders and mud flats, below 5000 ft.; San Diego Co. to Modoc and Lake cos. March–Aug.

2. Bérgia L.

Diffuse or ascending plants, glandular-pubescent. Fls. pedicelled, solitary or fascicled, 5-merous. Sepals cuspidate, with thickened midrib and scarious margin. Petals oblong. Stamens 5 or 10. Caps. firm. Ca. 20 spp., largely trop. (Named for P. J. *Bergius,* 1723–1817, Swedish botanist.)

1. **B. texàna** (Hook.) Seub. [*Merimea t.* Hook.] Annual, diffusely branched, 5–40 cm. high; lvs. ovate to obovate, 1–2 cm. long, glandular-serrate; sepals 3–4 mm. long; petals white; seeds slightly curved, ca. 0.5 mm. long.—Occasional on mud flats; largely V. Grassland; Elsinore, Murietta Hot Springs, Los Angeles, Sacramento and San Joaquin valleys, Modoc Co.; to Wash., Miss. V. July–Aug.

35. Caryophyllàceae. PINK FAMILY (Fig. 33)

Annual or perennial herbs. Lvs. mostly opposite, entire, simple, often connected at base by a transverse line, with or without stipules, these if present often scarious. Fls. symmetrical, 4–5-merous, with or without petals, solitary or in cymes, mostly perfect. Sepals free or united into a tube. Petals as many as sepals, often small or none. Stamens up to 10, free from one another; anthers 2-celled, dehiscing longitudinally. Ovary superior, sessile or stipitate, 1-loculed (rarely 3–5-loculed) with free cent. placentation; styles 2–5, rarely united into 1. Fr. a dry caps., usually opening by valves or apical teeth, or sometimes a utricle. Seeds many or 1, with endosperm and a ± curved peripheral or excentric embryo. Ca. 1500 spp. in 75 genera, most abundant in temp. and cooler regions; many grown for their fls.

A. Fr. a 1-seeded indehiscent utricle; petals 0; fls. very small, greenish or whitish. (*Illecebraceae,* of some auth.)
    B. Stipules present, scarious.
        C. Sepals united below into a short tube.
            D. Annual, prostrate or spreading; styles 2-cleft .... 17. *Achyronychia*
            DD. Perennial, erect; styles 3-cleft .... 18. *Scopulophila*
        CC. Sepals quite or almost distinct.
            D. Annual; lvs. elliptic-oblanceolate; stipules minute .... 20. *Herniaria*
            DD. Perennial; lvs. linear-oblong or subulate; stipules conspicuous.

E.   Sepals similar, bristle-tipped; lvs. linear-oblong . . . . 19. *Paronychia*
EE. Sepals dissimilar, the outer divergent and spine-tipped; lvs. subulate
22. *Cardionema*
BB. Stipules lacking; annual with subulate lvs.; fls. clustered, greenish . . . . 21. *Scleranthus*
AA. Fr. a several–many-seeded caps.; petals usually present; fls. small to large.
B.   Sepals distinct or nearly so; petals when present without claws and borne on a basal disk
or at base of sessile ovary. (*Alsineae*)
C.   Stipules wanting.
D.   Caps. ovoid or ellipsoid, dehiscent by as many or twice as many valves or teeth
as there are carpels.
E.   Styles opposite sepals and usually 3.
F.   Petals deeply notched or bifid, rarely wanting; valves of caps. bifid
or 2-parted . . . . 1. *Stellaria*
FF. Petals entire or nearly so; valves of caps. entire or bifid or 2-parted
4. *Arenaria*
EE. Styles alternate with sepals, 4–5 . . . . 3. *Sagina*
DD. Caps. cylindrical, often bent near summit, dehiscent by twice as many teeth as
there are carpels . . . . 2. *Cerastium*
CC. Stipules present, scarious.
D.   Styles 3–5, distinct; petals usually present.
E.   Styles and caps.-valves 5; lvs. appearing whorled . . . . 5. *Spergula*
EE. Styles and caps.-valves 3; lvs. opposite . . . . 6. *Spergularia*
DD. Style 1, 3-cleft or -toothed; petals minute or 0.
E.   Lvs. flat, oblong or obovate; stipules scarious . . . . 7. *Polycarpon*
EE. Lvs. subulate; stipules setaceous . . . . 8. *Loeflingia*
BB. Sepals united into a tubular or cuplike calyx; petals clawed and borne on the carpophore
(stipe of ovary). (*Sileneae*)
C.   Styles 3–5; caps. 3-, 5-, 6-, or 10-valved or -toothed.
D.   Calyx-lobes leaflike, much longer than the tube and alternating with the 5 styles
9. *Agrostemma*
DD. Calyx-lobes not leaflike, opposite the styles when 5.
E.   Styles 3, rarely 4; caps. dehiscent by 6, rarely 3, 4, or 8 apical teeth
10. *Silene*
EE. Styles 5, rarely 4, the caps. opening by as many or twice as many teeth. Rare
11. *Lychnis*
CC. Styles 2; caps. mostly 4-valved.
D.   Calyx without involucrelike bracts at its base.
E.   Stamens 10; calyx more than 2 mm. thick.
F.   Calyx ovoid, 5-ribbed, wing-angled; petals not appendaged
12. *Vaccaria*
FF. Calyx tubular, 20-nerved, not wing-angled; petals appendaged at base
of blades . . . . 13. *Saponaria*
EE. Stamens 5; calyx narrowly cylindrical, ca. 1 mm. thick . . . . 16. *Velezia*
DD. Calyx subtended by 1–3 pairs of bracts.
E.   Calyx 30–40-nerved . . . . 14. *Dianthus*
EE. Calyx 5-ribbed or 15-nerved . . . . 15. *Tunica*

1. **Stellària** L. Chickweed. Starwort

Low diffuse annuals or perennials. Fls. white, cymose or solitary, terminal or seemingly
lateral. Sepals 4–5. Petals 4–5, deeply 2-cleft, rarely 0. Stamens 8, 10, or fewer. Styles
3, rarely 4, opposite as many sepals. Caps. ovoid to globose, 1-celled, dehiscent by
twice as many valves as there are styles. Seeds several to many, smooth or rough. (Latin,
stella, a star, because of star-shaped fls.)

A. Plants annual.
B.   Internodes with longitudinal line of hair; lvs. ovate . . . . 1. *S. media*
BB. Internodes lacking lines of hair; upper lvs. lance-linear . . . . 2. *S. nitens*
AA. Plants perennial.
B.   The plants glandular-puberulent.
C.   Lvs. lance-ovate, 1.5–4 cm. long; petals cleft almost to base . . . . 3. *S. littoralis*
CC. Lvs. narrow-lanceolate, 5–10 cm. long; petals not so deeply cleft . . . . . . 4. *S. Jamesiana*
BB. The plants not glandular.
C.   Petals equal to or longer than sepals.
D.   Fls. many, the fruiting pedicels divergent; caps. pale . . . . 5. *S. graminea*
DD. Fls. solitary to few, the fruiting pedicels suberect; caps. dark . . . . . 6. *S. longipes*
CC. Petals much shorter than sepals or none.
D.   Margins of lvs. ± serrulate-ciliate especially toward base; lvs. lanceolate.
E.   Mature sepals 2–3 mm. long; lvs. ca. 1–2 cm. long, ciliate especially near
base; fruiting pedicels ascending . . . . 7. *S. calycantha*
EE. Mature sepals 4–5 mm. long; lvs. 1.5–4 cm. long, serrulate but not ciliate;
fruiting pedicels reflexed . . . . 8. *S. sitchana*
DD. Margins of lvs. smooth, glabrous; lvs. ovate to oblong.

E. Cymes, when well developed, with small scarious bracts; lf.-margins plane
9. *S. umbellata*
EE. Cymes with foliaceous bracts.
F. Lf.-margins crisped; sepals acute, with scarious margins . . . 10. *S. crispa*
FF. Lf.-margins plane; sepals obtuse, the margins not scarious
11. *S. obtusa*

1. **S. média** (L.) Vill. [*Alsine m.* L.] COMMON CHICKWEED. Annual with weak procumbent stems 1–4 dm. long and with a line of hair running down each internode; lvs. ovate, acute, 1–3 cm. long, short-petioled or the upper sessile; cymes leafy; sepals pubescent, ovate, 4–5 mm. long; petals somewhat shorter, 2-parted or wanting; caps. ovoid, slightly exceeding calyx; seeds minute, roughened; $2n = 28$ (Pal, 1952), 40 (Negodi, 1935), 42, 44 (Peterson, 1936).—Common weed in shaded places, about orchards, etc., through most of cismontane Calif.; native of Eurasia. Feb.–Sept.

2. **S. nìtens** Nutt. [*Alsine n.* Greene.] Erect annual forked with filiform stems 1–2 dm. high, glabrous or slightly pubescent at base; lvs. near base linear to lance-linear, or basal even wider, acute, 5–10 mm. long; pedicels erect, mostly 5–20 mm. long; sepals very acute, scarious-margined, subulate-lanceolate, 3–4 mm. long; petals half as long or wanting; caps. oblong, almost equaling calyx; seeds brown, angled, ca. 0.5 mm. long, somewhat roughened.—Common but inconspicuous in grassy places, below 5500 ft.; V. Grassland to Yellow Pine F.; most of cismontane Calif.; to B.C., Rocky Mts. March–June.

3. **S. littoràlis** Torr. [*Alsine l.* Greene.] Glandular-pubescent perennial, the forking stems decumbent, 2–5 dm. long; lvs. numerous, ovate, acute, sessile, 2–4 cm. long; cymes terminal, compound, leafy-bracted; pedicels 5–20 mm. long; sepals 5–6 mm. long, lanceolate, acute, with broad scarious margins; petals ca. as long, deeply cleft; caps. oblong, slightly exceeding sepals; seeds brown, ca. 1 mm. long, faintly reticulate. —Moist or wet places on dunes; Coastal Strand, N. Coastal Scrub; San Francisco to Humboldt Co. March–July.

4. **S. Jamesiàna** Torr. [*Alsine J.* Heller. *A. glutinosa* Heller.] Glandular-pubescent perennial, sometimes glabrate below, diffusely branched; stems 1–3.5 dm. long, erect or ascending, from slender rootstocks which may have thickened tuberlike enlargements; lvs. sessile, lanceolate to lance-ovate, 4–10 cm. long, widely spreading; cymes loose, terminal and axillary; pedicels 5–20 mm. long; bracts green; sepals 3–5 mm. long; petals broadly notched, 6–10 mm. long; caps. broadly ovoid, shorter than the calyx; seeds dark brown, ca. 2 mm. long, muriculate.—About meadows and damp places, 4000–8500 ft.; Yellow Pine F., Red Fir F.; San Bernardino Mts., Frazier Mt., Tehachapi Mts., Sierra Nevada to Modoc Co., N. Coast Ranges from Tehama and Mendocino cos. n.; to Wash., Rocky Mts. May–July.

5. **S. gramínea** L. [*Alsine g.* Britton.] Glabrous, weak, ascending from creeping rootstocks; stems 1.5–3 dm. long, 4-angled; lvs. sessile, lanceolate, 2–2.5 cm. long, acute, ciliolate at base; cymes diffuse, terminal; bracts small, scarious; pedicels slender, spreading; sepals lanceolate, 3-nerved, 3.5–5 mm. long; petals ca. as long, 2-cleft; caps. pale brown, oblong-ovoid, exceeding sepals; seeds finely roughened.—Reported as a weed in lawns, Claremont, La Verne, San Gabriel, also from Lodi; native of Eu. May–July.

6. **S. lóngipes** Goldie. [*Alsine l.* Cov.] More or less tufted perennial from creeping rootstocks, glabrous, 1–2.5 dm. high, the stems erect or ascending, rather simple; lvs. lance-linear, 1–2.5 cm. long, acute, rigid, erect or ascending, mostly green; fls. solitary or few, terminal on slender suberect pedicels; bracts scarious, small; sepals lance-oblong, 3–5 mm. long, acute, with scarious margins; petals cleft, ca. as long as sepals; caps. narrow-ovoid, dark, exceeding calyx; seeds nearly smooth, ca. 0.8 mm. long, rounded; $n = 26$ (Bøcher & Larsen, 1950).—Frequent in moist places, 4500–10,500 ft.; Yellow Pine F. to Subalpine F.; San Bernardino Mts., Sierra Nevada, Coast Ranges from Yolla Bolly Mts. n.; to Alaska and Atlantic Coast. May–Aug. Occasional plants are met that are glaucous and are the var. *laèta* of Jeps. They seem to agree with *S. monantha* var. *altocaulis* Hult., but in Calif. are not separable by Hultén's characters.

7. **S. calycántha** (Ledeb.) Bong. ssp. **intèrior** Hult. [*S. borealis* auth., in part.] Perennial from slender rootstocks, the stems strongly branched, scabrous, 1–3 dm. long; lvs. lanceolate, somewhat serrulate, ca. 1 cm. long; fls. in terminal cymes with leafy bracts, the lower fls. in upper lf.-axils; pedicels ascending; sepals ca. 2 mm. long·

petals minute or none; caps. ovoid, obtuse, pale brown, exceeding calyx; seeds light brown, faintly reticulate.—Rare, moist places, 4500–6000 ft.; Montane Coniferous F.; Tuolumne Co. to Modoc Co.; to Alaska, Atlantic Coast, Eurasia. June–Aug.

Var. **simcòei** (Howell) Fern. [*Alsine s.* Howell. *S. borealis* var. *s.* Fern.] Branches densely pilose.—To be looked for in n. Calif.; to Wash., Mont.

8. **S. sitchàna** Steud. var. **Bongardiàna** (Fern.) Hult. [*S. longifolia* Bong., not Muhl. *S. borealis* var. *B.* Fern.] Perennial, rather coarse, the stems 4-angled, rather simple, 2–4 dm. long, somewhat scabrous; lvs. lanceolate, 2.5–4 cm. long, the upper not much reduced, scarcely ciliated on margins, but somewhat serrulate under a lens; fls. few, terminal or axillary, the cymes leafy-bracted; pedicels reflexed in fr.; sepals 3.5–4.5 mm. long when mature, sharply acute; petals minute or none; caps. oblong-ovoid, exceeding calyx; seeds brown, somewhat reticulate, ca. 0.6 mm. long.—Wet places, 0–6000 ft.; N. Coastal Scrub, Yellow Pine F.; Marin Co. and Mariposa Co. n. to Siskiyou Co.; to Alaska. June–July.

9. **S. umbellàta** Turcz. [*Alsine baicalensis* Cov. *S. gonomischa* Boiv.] Perennial with slender leafy underground rootstocks, the stems glabrous, slender, weak, branched, 1–2 dm. long; lvs. ovate to oblong, acute, 1–2 cm. long, glabrous except for the ciliate margins, thin; fls. in upper axils and in terminal umbellate cymes with small scarious bracts; pedicels filiform, recurved at tip; sepals 1.5–2 mm. long, acute, with scarious margins; petals minute or none; caps. oblong-ovoid, ca. twice as long as calyx; seeds light brown, obscurely reticulate, ca. 0.5 mm. long.—Infrequent in damp shaded places, 6000–11,500 ft.; Lodgepole F., Subalpine F.; Glenn Co. and Sierra Nevada from Tulare Co. n.; to Ore., Rocky Mts.; Siberia. July–Aug.

10. **S. crispa** Cham. & Schlecht. [*Alsine c.* Holz.] Perennial with slender leafy rootstocks; stems glabrous, weak, ± simple, 1–4 dm. long; lvs. ovate, sessile or nearly so, thin, short-acuminate, 0.8–2 cm. long, usually crisped on margins; pedicels axillary, 6–20 mm. long, sometimes deflexed in fr.; sepals 3–4 mm. long, lanceolate, acute, scarious-margined; petals shorter than sepals or wanting; caps. ovoid, pale, exceeding calyx; seeds brown, ca. 0.5 mm. long, rounded, slightly reticulate.—Moist banks and meadows; N. Coastal Scrub?, Redwood F., Mixed Evergreen F., Marin Co. n.; below 11,000 ft., Red Fir F., Lodgepole F., San Bernardino Mts., San Jacinto Mts., Sierra Nevada n.; to Alaska. May–Aug.

11. **S. obtùsa** Engelm. [*Alsine o.* Rose.] Glabrous perennial with numerous decumbent or prostrate angled stems 0.5–1.5 dm. long; lvs. ovate, thin, acute, 8–10 mm. long; fls. solitary, axillary; sepals ovate, obtuse, scarcely scarious on margins; petals minute or none; caps. ovoid, obtuse, exceeding sepals.—At 6300 ft.; Red Fir F.; Plaskett Meadows, Glenn Co.; Ore. to Alaska and Rocky Mts. July.

2. Cerástium L. Mouse-Ear Chickweed

Pubescent herbs, annual or perennial. Fls. in terminal dichotomous cymes, white. Bracts green or scarious. Sepals 5, rarely 4. Petals as many, 2-lobed or -cleft, or wanting. Stamens 10 or 5. Styles mostly 5, opposite the sepals. Caps. slender, elongate, usually exceeding the calyx, often curved, dehiscent at apex by usually 10 teeth. Seeds many, rough. Ca. 50 spp., widely dispersed in cool and temp. regions. (Greek, *cerastes*, horned, referring to shape of caps.)

Perennials, with prostrate or creeping basal branches or offshoots.
  Bracts of infl. broadly scarious-margined or only the lower wholly green. Common.
    Petals 2–3 times as long as sepals; basal branches and offshoots becoming dry and withered and
      with many axillary fascicles, their lvs. linear to oblong, not hirsute .... 1. *C. arvense*
    Petals ca. as long as sepals; basal branches or offshoots green, with few or no axillary fascicles,
      their lvs. oblong, hirsute .... 2. *C. vulgatum*
  Bracts of infl. green or only the upper with narrow scarious margins. Rare in high mts.
                                                                 4. *C. beeringianum*
Annual, without basal persistent sterile offshoots; petals lacking or ca. as long as sepals
                                                                3. *C. viscosum*

1. **C. arvénse** L. [*C. patulum* Greene. *C. Sonnei* Greene. *C. a.* var. *maximum* Hollick & Britt.] Perennial from running rootstocks, the flowering stems tufted, erect or ascending, pubescent, usually glandular, 1–3 dm. high, with basal tough branches and withered firm lvs. and axillary fascicles; main lvs. linear-subulate to -oblong, 1.5–3.5 cm. long,

reduced up the stem; fls. loosely cymose, not numerous, with scarious-margined bracts; sepals broadly lanceolate, 5–7 mm. long; petals 2–3 times as long as sepals, broadly lobed; caps. not much longer than sepals; seeds red-brown, tuberculate, almost 1 mm. in diam.; $2n = 36$, 38, 72 (Brett, 1955).—Moist rocky or grassy banks and cliffs; N. Coastal Scrub, Coastal Prairie; along the coast, Monterey Co. n.; also occasional in wet places, 5000–8000 ft.; Red Fir F., Lodgepole F.; Sierra Nevada, Siskiyou Co.; to Alaska, Atlantic Coast, Eurasia. Feb.–Aug.

2. **C. vulgàtum** L. Short-lived perennial, matted and with basal leafy offshoots; flowering stems glandular-pubescent, 1–4 dm. high, ascending or decumbent; lvs. oblong or the lower oblong-spatulate, 1–2.5 cm. long; bracts similar but smaller, the fls. in loose cymes; pedicels 2–4 times as long as calyx; sepals 4–6 mm. long, lance-ovate, acute, scarious-margined; petals 4–8 mm. long, 2-lobed; caps. ca. twice as long as calyx, curved; seeds ca. 0.6 mm. in diam., reddish-brown, tubercled.—Mostly in lawns in Calif., occasional in wet meadows in mts.; natur. from Eurasia. March–Aug.

3. **C. viscòsum** L. [*C. glomeratum* Thuill.] Erect annual, viscid, simple to freely branched, 1–3 dm. high; lvs. elliptic to narrow-obovate, obtusish, hairy, 1–2.5 cm. long; bracts small, green; infl. a glomerate cyme, becoming lax in age; pedicels scarcely or not longer than calyx; sepals ovate-lanceolate, sharply acute, with scarious margins, 3.5–4.5 mm. long; petals 2-cleft, ca. as long as or slightly shorter than sepals; caps. slender, slightly curved, 5–9 mm. long; seeds pale brown, muriculate, ca. 0.3 mm. long.—Common in waste places, along roads, in pastures, etc., at low elevs. through most of cismontane Calif.; natur. from Eu. Feb.–May. The apetalous form [forma *apétalum* (Dumort.) Mert. & Koch] has been reported from San Diego.

4. **C. beeringiànum** Cham. & Schlecht. [*C. alpinum* var. *b.* Regel.] Matted perennial, the stems spreading, glandular-pubescent, 4–10 cm. long; lvs. mostly 2–7 pairs, oblong, obtusish, pilose, 7–20 mm. long; infl. leafy-bracted, simple to dichotomous, 1–few-fld.; pedicels mostly ascending, 5–20 mm. long; sepals lance-oblong to almost ovate, obtuse, 4–8 mm. long in fr., with conspicuous hyaline margins; petals 6–8 mm. long, bluntly 2-lobed; caps. ca. 8 mm. long; seeds bluntly papillose, 0.6–1 mm. in diam.—Occasional, near snowbanks, 9500–12,200 ft.; Subalpine F., Alpine Fell-fields; Tuolumne, Mono and Alpine cos.; Alaska to Nfld. and Ariz. July–Aug.

3. Sagìna L. PEARLWORT

Low annual or perennial herbs, tufted or matted. Lvs. filiform or subulate, scarious-connate at base. Fls. small, whitish, terminal on stems or branches. Sepals 4 or 5. Petals 4 or 5 or 0, undivided. Stamens as many as or twice as many as sepals. Styles as many as sepals and alternate with them. Caps. many-seeded, 4–5-valved to base, the valves opposite the sepals. Seeds many, smooth or resinous-dotted. Ca. 10 spp. of cool or temp. regions. (Latin, *sagina,* fattening, once applied to Spergula, used in Eu. for forage.)

Plants annual, without sterile basal rosettes; pedicels straight; stems filiform.
    Lf.-bases ciliolate; fls. apetalous, usually 4-parted . . . . 1. *S. apetala*
    Lf.-bases not ciliate; fls. with petals, 5-parted . . . . 2. *S. occidentalis*
Plants perennial, with sterile basal rosettes; pedicels often curved at summit; stems ± fleshy.
    Sepals mostly 4, spreading in fr.; caps. 2–3 mm. long . . . . 3. *S. procumbens*
    Sepals mostly 5, appressed in fr.
        The sepals 1.5–2 mm. long; caps. ca. 3 mm. long. Montane . . . . 4. *S. saginoides*
        The sepals 3 mm. long; caps. ca. 4 mm. long. Seacoast . . . . 5. *S. crassicaulis*

1. **S. apétala** Ard. var. **barbàta** Fenzl. [*Alsinella ciliata* Greene. *S. c.* Heller.] Minutely glandular-pubescent annual 2–5 cm. high, with 1–few filiform stems; lvs. linear subulate, 3–7 mm. long, long-ciliolate near base; pedicels capillary, 3–12 mm. long; sepals 4, lance-ovate, 1.5 mm. long, obtusish; petals usually lacking; caps. ovoid, ca. 2 mm. long; seeds minute.—Infrequent inconspicuous weed on beaten soil along roads, etc., cismontane Calif.; natur. from Eu. April–June.

2. **S. occidentàlis** Wats. [*Alsinella o.* Greene.] Minute inconspicuous annual, glabrous save for the hispidulous-glandular calyx and pedicels, 2–8(–12) cm. high; lvs. filiform or the upper subulate, 6–10 mm. long; pedicels 6–12(–18) mm. long; sepals 5, ovate, blunt, 1.5 mm. long; petals 5, almost as long; caps. 2–2.5 mm. long; seeds light brown,

almost smooth, ca. 0.3 mm. long.—Uncommon or perhaps overlooked, wooded and brushy or grassy places, below 8000 ft.; many Plant Communities; San Diego Co. n.; to B.C. and Ida. March–June.

3. **S. procúmbens** L. Matted perennial or perhaps annual, the stems prostrate and rooting at the nodes, glabrous, 3–8 cm. long; rosette-lvs. bristle-tipped, 0.5–2 cm. long; cauline lvs. subulate, 2–6 mm. long; pedicels recurved at summit after flowering, sometimes straight again later; sepals 5, oval, obtuse, ca. 2 mm. long; petals a little shorter; caps. ovoid, 2–3 mm. long, the valves divergent; seeds finely reticulate but not roughened, ca. 0.3 mm. long; $2n = 22$ (Rohweder, 1939).—Moist shaded banks, near the beach; N. Coastal Scrub, Redwood F.; Point Reyes, Marin Co. n.; to B.C., Atlantic Coast; native of Eurasia. May–Sept.

4. **S. saginoìdes** (L.) Karst. var. **hespéria** Fern. [*S. Linnaei* auth., not Presl.] Low matted perennial, or tufted, glabrous, the stems 2–9 cm. long, numerous; lvs. thickish, linear, 5–10 mm. long; pedicels 5–12 mm. long, often curved at tip; sepals oblong-ovate, obtuse, 1.5–2 mm. long; petals ca. 1 mm. long; caps. 2.5–3 mm. long; seeds barely 0.3 mm. long, papillose.—Frequent on moist banks, 4000–12,000 ft.; Yellow Pine F. to Alpine Fell-fields; mts. from San Diego Co. through Sierra Nevada to Modoc Co., N. Coast Ranges from Glenn Co. n.; to B.C. and Rocky Mts. May–Sept.

5. **S. crassicaùlis** Wats. [*Alsinella c.* Greene.] Glabrous perennial with fleshy branched stems 4–12 cm. long; basal lvs. in permanent rosette, linear, 15–30 mm. long; upper shorter, scarious-connate; pedicels 1–4 cm. long, straight or curved at apex; sepals 5, broadly oval, 3 mm. long; petals almost as long; caps. ovoid, 4 mm. long; seeds brown, ca. 0.5 mm. long.—Moist places back of dunes and on ocean bluffs; N. Coastal Scrub, Closed-cone Pine F.; Monterey Co. n.; to Alaska. June–Dec.

4. Arenària L. SANDWORT

Low branched annual to perennial herbs, commonly tufted or matted. Lvs. sessile, usually exstipulate, subulate to ovate. Fls. small, white, occasionally rose or cream, terminal, cymose or capitate, rarely solitary and axillary. Sepals 5. Petals 5, entire or somewhat emarginate, sometimes ·wanting. Stamens 10. Styles usually 3, opposite as many sepals. Caps. globose to oblong, splitting into as many or twice as many valves as there are styles. Seeds few to many, reniform to globose. Ca. 150 spp., the genus almost world-wide. (Latin, *arena*, sand, in which many spp. grow.)

(Williams, F. N. A revision of the genus Arenaria. Journ. Linn. Soc., Bot., 33: 326–437, 1898. Maguire, B. Arenaria in Am. n. of Mex. A conspectus. Amer. Midl. Nat. 46: 493–511, 1951.)
A. Plants annual.
   B. Petals shorter than sepals.
      C. Lvs. ovate, 3–5-nerved; stems 10–20 cm. long .... 10. A. *serpyllifolia*
      CC. Lvs. narrowly lanceolate, 1-nerved; stems 2–5 cm. long .... 4. A. *pusilla*
   BB. Petals longer than sepals, or at least as long.
      C. Lvs. somewhat lanceolate, 2–5 mm. long, obtuse .... 3. A. *californica*
      CC. Lvs. filiform, 7–30 mm. long.
         D. Petals broadly obovate; plants greenish; lvs. well distributed; seeds smooth,
            broadly margined. Widely distributed .... 1. A. *Douglasii*
         DD. Petals oblong; plants purplish when mature; lvs. mainly basal; seeds minutely
            tuberculate-crested. Del Norte Co. .... 2. A. *Howellii*
AA. Plants perennial.
   B. Lvs. linear-lanceolate to lance-oblong, 2–7 mm. wide, not pungent.
      C. Petals 2–4 mm. long; fls. in terminal cymes; stems puberulent, scarcely angled.
         D. Stems from running rootstocks; lvs. mostly 2–8 cm. long; fls. 1–few; seeds ap-
            pendaged with a pale spongy strophiole at the hilum .... 18. A. *macrophylla*
         DD. Stems from a branching root-crown; lvs. 0.5–2 cm. long; fls. numerous; seeds not
            so appendaged .... 11. A. ʹ*confusa*
      CC. Petals 5–6 mm. long; fls. solitary, axillary; stems glabrous, angled .... 5. A. *paludicola*
   BB. Lvs. filiform or subulate, ca. 1–1.5 mm. wide, often pungent.
      C. Sepals distinctly 3-nerved.
         D. The sepals glabrous; lvs. 1–2 cm. long .... 15. A. *congesta*
         DD. The sepals glandular-pubescent; lvs. mostly less than 1 cm. long.
            E. Sepals obtuse, ca. 4 mm. long; fls. 1–3 .... 8. A. *obtusiloba*
            EE. Sepals sharp-pointed; fls. usually 3–many.
               F. Calyx ca. 3–4 mm. long; petals 2–3 mm. long .... 7. A. *rubella*
               FF. Calyx 4.5–6.5 mm. long; petals ca. 5–6 mm. long .... 6. A. *Nuttallii*
      CC. Sepals 1-nerved or, if with lateral nerves, these indistinct.
         D. The entire plant glandular-pubescent; lvs. 0.3–1 cm. long.

E.   Caps.-valves entire; lvs. 3-nerved .... **6. A. Nuttallii**
EE. Caps.-valves 2-toothed; lvs. 1-nerved .... **17. A. Kingii**
DD. The plant glabrous or at most slightly glandular-puberulent in infl.
E.   Lvs. 0.3–0.8 cm. long; fls. often solitary. High alpine .... **9. A. Rossii**
EE. Lvs. mostly 1 cm. or more long; fls. few—many.
F.   Lvs. 0.4–2 cm. long.
G.   Pedicels 5–20 mm. long, the cymes open.
H.   Plants glaucous; lvs. rigid, straight. N. Calif.
14. A. aculeata
HH. Plants not glaucous.
I.   Lvs. ± recurved. S. Calif. .... **12. A. ursina**
II. Lvs. rigid, straight. Cent. and n. Calif. .... **17. A. Kingii**
GG. Pedicels usually shorter, the cymes condensed .... **15. A. congesta**
FF. Lvs. 2–6 cm. long.
G.   Plants less than 2 dm. high; sepals 3–4 mm. long.
H.   Fls. densely clustered .... **15. A. congesta**
HH. Fls. in open cymes .... **13. A. pumicola**
GG. Plants 2–3 dm. high; sepals 4–6.5 mm. long .. **16. A. macradenia**

1. **A. Douglásii** Fenzl. ex T. & G. [*Alsinopsis D.* Heller. *Minuartia D.* Mattf.] Delicate erect annual with slender freely branched stems 0.5–2(–3) dm. high, subglabrous to somewhat glandular-pubescent, loosely cymose above; lvs. filiform, 0.5–1.5(–3) cm. long; pedicels filiform, 6–25(–40) mm. long; sepals ovate, 1–3-nerved, scarious-margined, 2–3 mm. long; petals white, conspicuous, obtuse, obovate, 3–6 mm. long; anthers 0.5–0.7 mm. long; caps. subglobose, slightly exceeding sepals; seeds brown, smooth, reniform, broadly margined, 1–1.5 mm. broad.—Locally common in dry sterile places, below 7000 ft.; many Plant Communities; cismontane Calif. to desert edge; to Ore., L. Calif. April–June.
Var. **emarginàta** H. K. Sharsm. Petals mostly emarginate, scarcely longer than fruiting calyx; anthers 0.2–0.3 mm. long.—Serpentine or shale talus, Red Mts., Mt. Hamilton Range. April–May.
2. **A. Howéllii** Wats. [*Alsinopsis H.* Heller. *Minuartia H.* Mattf.] Slender-stemmed annual, branched from base, 0.5–3 dm. high, ± glandular-puberulent; lvs. few, mostly basal, narrowly linear, 7–15 mm. long; infl. diffuse, with filiform branches; sepals ovate, 2–2.5 mm. long, nerveless, acute; petals oblong, ca. 4 mm. long; caps. ca. as long as calyx; seeds dark, somewhat flattened, ca. 1 mm. long, tubercled on margin.—Dry open places on serpentine, 1800–3200 ft.; Yellow Pine F., Mixed Evergreen F.; Del Norte Co.; sw. Ore. April–June.
3. **A. califórnica** (Gray) Brew. [*A. brevifolia* var. *c.* Gray. *Alsinopsis c.* Heller. *Minuartia c.* Mattf.] Glabrous erect simple or branched annual 3–10 cm. high; lvs. lanceolate, obtuse, somewhat fleshy, 2–5 mm. long; fls. loosely cymose; pedicels 6–15 mm. long; sepals oblong-ovate, 2–3 mm. long, obtusish, 1–3-nerved, scarious-margined; petals oblong-ovate, 3–4 mm. long; caps. ovoid, slightly surpassing calyx; seeds ca. 0.4 mm. long, minutely roughened.—Grassy slopes and disintegrated rock, below 2500 (4000) ft.; V. Grassland, Foothill Wd., Coastal Prairie, N. Coastal Scrub, Yellow Pine F.; Los Angeles Co. to s. Ore., Tehachapi Mts. to Shasta Co. March–June.
4. **A. pusílla** Wats. [*Alsinopsis p.* Rydb. *Minuartia p.* Mattf.] Slender-stemmed glabrous annual, simple or few-branched, 2–5 cm. high; lvs. lanceolate, 2–5 mm. long, the cauline 1–3 pairs; pedicels capillary, 2–10 mm. long; sepals narrow-lanceolate, acute, 2–3.5 mm. long; petals shorter or none; caps. shorter than calyx; seeds smooth, minute.—Occasional, dry woods and open slopes, below 6800 ft.; Yellow Pine F., Red Fir F.; Humboldt, Siskiyou and Tuolumne cos. n.; to Wash., Ida. April–July.
Var. **diffùsa** Maguire. Cauline lvs. 3–6 pairs; sepals obtusish, 1.8–2.5 mm. long; petals longer.—Dry places; largely in Chaparral; Sutter Co. to Monterey Co., Laguna Mts. in San Diego Co.
5. **A. paludícola** Rob. [*Alsine palustris* Kell. *Arenaria p.* Wats., not Gay. *Alsinopsis p.* Heller.] Glabrous flaccid perennial with several subsimple procumbent stems rooting at lower nodes, angled, leafy throughout, 3–7 dm. long; lvs. rather uniform, flat, 1-nerved, lance-linear, 1.5–4 cm. long, acute, somewhat connate; fls. solitary, axillary; pedicels 1.5–4 cm. long; sepals lance-ovate, 3–4 mm. long, nerveless; petals oblong-obovate, 5–6 mm. long; caps. oblong, ca. as long as calyx.—Occasional, in swamps; Freshwater Marsh; San Bernardino, Los Angeles, Santa Barbara, Santa Cruz Co., San Francisco, etc.; to Wash. May–Aug.

6. **A. Nuttállii** Pax ssp. **grácilis** (Gray) Maguire. [*A. pungens* var. *g.* Gray. *A. N.* var. *g.* Rob.] Perennial with many loosely matted prostrate branching stems 5–15 cm. long, densely leafy, glandular-puberulent; lvs. narrowly subulate, pungent, overlapping, strict or slightly recurved, 5–8 mm. long; secondary lvs. in numerous fascicles; fls. rather few, in open cymes; pedicels 4–10 mm. long; sepals lance-subulate, pungently attenuate, 4–5 mm. long, 1-nerved; petals shorter; caps. ovoid, shorter than calyx; seeds brown, 1–1.2 mm. broad, transversely low-papillate.—Dry granitic gravel, mostly at 10,000–12,000 ft.; Subalpine F., Alpine Fell-fields; Sierra Nevada from Alpine and Tuolumne cos. to Tulare Co., San Bernardino Mts. July–Aug.

Ssp. **frágilis** Maguire & Holmgren. Lvs. arcuate or squarrose, 8–10 mm. long; sepals 3-nerved, 5.5–6.5 mm. long; petals shorter than sepals.—Rocky places, 5500–8200 ft.; Sagebrush Scrub, Pinyon-Juniper Wd.; Mono Co. to Modoc Co.; w. Nev., e. Ore. May–July.

Ssp. **gregària** (Heller) Maguire. [*A. g.* Heller. *A. N.* var. *g.* Jeps.] Lvs. ascending, strict, 3–10 mm. long; sepals 1-nerved, or indistinctly 3-nerved, 3.5–4.5 mm. long; petals exceeding sepals.—Gravelly ridges and slopes, 3500–8000 ft.; Yellow Pine F. to Subalpine F.; Coast Ranges from Lake Co. to Siskiyou Co. June–Aug.

A. Ròsei Maguire & Barneby, like *A. Nuttallii,* with 1-nerved sepals 3–4.5 mm. long and petals 4.5–6 mm. long, has recently been described from serpentine at Peanut, Trinity Co.

7. **A. rubélla** (Wahl.) Sm. [*Alsine r.* Wahl. *A. propinqua* Richards.] Perennial, tufted, densely glandular-puberulent, the stems 2–5(–8) cm. high; lvs. crowded at base of stems, linear-subulate, 3-nerved, 3–10 mm. long, pungent; fls. 2–few, cymose; sepals lanceolate, 2.5–3 mm. long, 3-nerved; petals somewhat shorter; caps. narrow-ovoid, ca. as long as calyx; seeds red-brown, ca. 0.5 mm. long, rugose.—Dry rocky and gravelly places; Alpine Fell-fields; 8000 ft., Siskiyou Co. and 10,000–12,000 ft., San Bernardino Mts. and s. Sierra Nevada; to Alaska and Atlantic Coast; Eurasia. July–Aug.

8. **A. obtusilòba** (Rydb.) Fern. [*Alsinopsis o.* Rydb. *Minuartia o.* House. *A. obtusa* Torr., not All. *A. biflora* Wats., not L.] Perennial from a woody caudex, cespitose, the stems decumbent, slender, 1–6 cm. long, leafy below, glandular-pubescent; lvs. subulate, rigid, 4–8 mm. long, 1-nerved, obtuse, short-ciliate; fls. 1–2(–3); sepals lance-oblong, 3-nerved, glandular-pubescent, 4 mm. long, obtuse; petals spatulate, 6–7 mm. long; caps. exceeding calyx; seeds red-brown, ca. 0.7 mm. long, smooth.—Dry rocky places, 10,500–12,500 ft.; Alpine Fell-fields; Sierra Nevada from Fresno and Inyo cos. to Alaska, Rocky Mts. July–Aug.

9. **A. Róssii** R. Br. in Richards. [*Alsinopsis R.* Rydb. *Minuartia R.* Graebn.] Glabrous perennial, densely tufted, 1–5 cm. high; lvs. linear, fleshy, 5–7 mm. long; fls. solitary to 2 or 3; pedicels 5–10 mm. long; sepals 2–2.5 mm. long, acute, weakly 3-nerved, hyaline-margined; petals scarcely as long; caps. almost equaling calyx; seeds brown, ca. 0.5 mm. long, minutely reticulate.—At 12,000–12,500 ft.; Alpine Fell-fields; Mt. Langley and Cirque Peak, Tulare Co., Mono Mesa, Inyo Co.; to Ore., Wash., Rocky Mts., Canada. Aug.

10. **A. serpyllifòlia** L. Puberulent annual, simple to branched, 5–20 cm. high; lvs. ovate, acuminate, ciliate, scabrous, 3–5-nerved, 3–7 mm. long; cymes open, leafy-bracted; pedicels 3–10 mm. long, capillary; sepals lance-ovate, hispidulous, 3–3.5 mm. long; petals oblong, ca. 2 mm. long; caps. ovoid, exceeding calyx; seeds globose-reniform, rugose, 0.6 mm. long; $2n = 20$, 40 (Woess, 1941).—Weed in lawns and moist places, mostly below 5000 ft.; cismontane Calif. especially toward the n.; natur. from Eu. April–June.

11. **A. confùsa** Rydb. [*A. saxosa* auth., not Gray.] Retrorsely puberulent perennial with branched root-crown and slender spreading stems 1–3 dm. long, leafy in lower half; lvs. lance-oblong, 1–2 cm. long, 2–3(–5) mm. wide; cymes paniculate; pedicels subfiliform, 5–15 mm. long; sepals ovate, carinate, sharply acute, 3 mm. long; petals ca. as long; caps. ovoid, 3–4 mm. long; seeds black, shining, almost 1 mm. long.—Moist or damp places near meadows, 6300–8200 ft.; Red Fir F., Lodgepole F.; San Bernardino Mts.; San Pedro Martir Mts. in L. Calif., s. Rocky Mts. July–Aug.

12. **A. ursìna** Rob. [*A. capillaris* var. *u.* Rob.] Perennial with cespitose caudex, the stems many, erect or ascending, 6–15 cm. high, glandular-puberulent above; lvs. subulate, straight or recurved, glaucous, ciliate, 4–12 mm. long, rigid; cymes few-fld.; pedicels

mostly 4–12 mm. long; sepals ovate, obtusish, 3–4 mm. long, nerveless, with broad scarious margin; petals 4–5 mm. long; caps. 3–5 mm. long; seeds brown, faintly reticulate, ca. 1.2 mm. long.—Dry slopes, 6000–7000 ft.; Yellow Pine F.; Bear V., San Bernardino Mts. June–July.

13. **A. pumícola** Cov. & Leib. var. **califórnica** Maguire. Cespitose perennial, woody below; stems 1.5–3 dm. high, glabrous below, glandular above with several pairs of lvs.; basal lvs. 1.5–2 cm. long, strictly ascending, not stiffly pungent; pedicels 10–25 mm. long; sepals ovate, obtusish, 3–4 mm. long, 1-nerved, scarious-margined; petals 5–6 mm. long; caps. 4.5–5.5 mm. long; seeds flat, winged, ca. 2 mm. long.—Occasional, 6000–9000 ft.; Lodgepole F.; cent. Sierra Nevada. July–Aug.

14. **A. aculeàta** Wats. [*A. congesta* var. *a.* Jones.] Perennial from a woody caudex, much-branched and matted at base, the erect stems glabrous, or glandular-pubescent toward apex, 1–1.8 dm. high; lvs. many, mostly basal, glaucous, stiffly pungent, 1–2(–2.4) cm. long; cyme open; pedicels 5–20 mm. long; sepals ovate, obtusish to acute, 4–5 mm. long, broadly scarious-margined; petals 6–7 mm. long; caps. slightly exceeding calyx to twice as long, the valves 2-toothed; seeds light brown, flat, broadly winged, 2.5–3 mm. long, with pattern of fine radiating reticulations.—Rocky slopes, 4700–8000 ft.; Sagebrush Scrub, N. Juniper Wd.; Nevada, Mono, Plumas, and Modoc cos.; to Ore., Mont., Utah. May–July.

15. **A. congésta** Nutt. ex T. & G. Perennial from a short branched woody caudex, with lvs. bunched at base and stems 1–2.5 dm. high, glabrous; basal lvs. subfiliform, glabrous, 3–6 cm. long, suberect, not pungent; cauline lvs. few, somewhat reduced; infl. capitate, many-fld.; sepals ovate, obtuse, ca. 4 mm. long, with broad scarious margin; petals ca. 6 mm. long; caps. somewhat exserted; seeds brownish, ca. 2 mm. long, winged, flat, reticulately and radiately marked.—Dry rocky slopes, 4400–6500 ft.; Sagebrush Scrub, Yellow Pine F.; Plumas, Lassen, and Siskiyou cos.; to Wash., Mont., Colo. June–July.

KEY TO VARIETIES

Sepals obtuse.
  Infl. capitate.
    Lvs. 3–6 cm. long, subfiliform. Plumas, Lassen & Siskiyou cos. n. . . . . *A. congesta*
    Lvs. 2–3 cm. long, 1–2 mm. wide, almost fleshy. Glenn Co. to Modoc Co. . . . . var. *crassula*
  Infl. subumbellate. Tulare Co. to Siskiyou Co. . . . . var. *suffrutescens*
Sepals acute.
  Fls. with short pedicels. Ne. Calif.
    Infl. subcapitate; sepals 5.5–6.5 mm. long. Lassen and Modoc cos. . . . . var. *simulans*
    Infl. subumbellate; sepals 4–5 mm. long. Modoc, Placer, and Inyo cos. . . . . var. *subcongesta*
  Fls. capitate. E. Mojave Desert . . . . var. *charlestonensis*

Var. **cràssula** Maguire. Lvs. 1.5–3 cm. long, 1–2 mm. wide, thickish; infl. subcapitate; sepals obtuse, 3.5–4 mm. long.—Dry slopes, 4000–7000 ft.; Yellow Pine F. to Lodgepole F.; Siskiyou Co. to Modoc and Glenn cos.; s. Ore. July–Aug.

Var. **suffrutéscens** (Gray) Rob. [*Brewerina s.* Gray. *A. s.* Heller.] Infl. somewhat umbellate; sepals obtuse, 3–4 mm. long.—Dry slopes, mostly 6000–10,500 ft.; Red Fir F. to Subalpine F.; Tulare Co. to Placer Co. in a pubescent form, and Plumas Co. to Siskiyou Co. in a glabrous form; s. Ore. June–Aug.

Var. **símulans** Maguire. Stems 1–1.5 dm. high; infl. subcapitate; sepals acute, 5.5–6.5 mm. long.—Dry slopes, 4500–5500 ft.; Sagebrush Scrub, Yellow Pine F.; Modoc and Lassen cos.; adjacent Nev. May–June.

Var. **subcongésta** (Wats.) Wats. [*A. Fendleri* var. *s.* Wats.] Infl. slightly umbellate; sepals 4–5 mm. long, acute.—Dry slopes, 7000–8500 ft.; Pinyon-Juniper Wd., Yellow Pine F.; Modoc and Lassen cos., Panamint Mts., Inyo Co.; to Nev., Utah. June–July.

Var. **charlestonénsis** Maguire. Stems 4–10 cm. high; lvs. 1–1.5 cm. long, spreading-recurved, largely basal; fls. few, capitate; sepals 4.5–5.5 mm. long, acute, pungent.—Sandy ridge, 7300 ft.; Pinyon-Juniper Wd.; New York Mts., e. Mojave Desert; s. Nev. June.

16. **A. macradènia** Wats. [*A. congesta* var. *m.* Jones.] Perennial from a branched woody caudex, the stems woody at base, ascending, mostly 2–4 dm. tall and with 5–8 pairs of lvs., usually glabrous; basal lvs. subulate, rather stout, ascending, straight, 2–5 cm. long, pungent, rigid, scabrous-ciliolate; cauline lvs. ascending, 0.8–1.2 cm.

long; infl. an open cyme; pedicels 1–3 cm. long; sepals ovate, broadly acute, 5.5–6.5 mm. long, hyaline-margined; petals conspicuously longer; caps. oblong, entire to emarginate, exceeding calyx; seeds dark brown, flat, ca. 1.5 mm. long, papillose.— Dry rocky slopes, below 6500 ft.; Creosote Bush Scrub, Joshua Tree Wd., Pinyon-Juniper Wd.; deserts from Inyo and Los Angeles cos. to San Bernardino and Riverside cos.; to Utah, Ariz. May–June.

KEY TO VARIETIES

Sepals 5.5–6.5 mm. long; stems woody at base.
  Petals conspicuously exceeding sepals; cauline lvs. 5–12 pairs.
    Cauline lvs. ascending, 0.8–1.2 mm. broad .... A. *macradenia*
    Cauline lvs. strongly arcuate, 1.2–2 mm. broad.
      Infl. and sepals glabrous .... var. *arcuifolia*
      Infl. and sepals glandular .... var. *Kuschei*
  Petals scarcely or not longer than sepals; cauline lvs. fewer than 5 pairs .... var. *Parishiorum*
Sepals 4–5 mm. long; stems scarcely or not woody at base .... ssp. *Ferrisiae*

Var. **arcuifòlia** Maguire. Cauline lvs. 6–12 pairs, strongly arcuate, 1.2–2 mm. broad; infl. glabrous; sepals 5.5–6 mm. long; petals exserted.—Dry slopes, 2000–5000 ft.; largely Joshua Tree Wd.; Kernville, Kern Co. to n. base San Gabriel Mts., Los Angeles Co. May–July.

Var. **Kùschei** (Eastw.) Maguire. [*A. K.* Eastw.] Plants 1.5 dm. high; lvs. 1–3 cm. long; infl. glandular, congested, the pedicels 2–5 mm. long; sepals densely glandular, 6–7 mm. long; petals somewhat longer.—Known from a single collection, Forest Camp, Mojave Desert, July 12, 1929.

Var. **Parishiòrum** Rob. Cauline lvs. mostly fewer than 5 pairs; sepals 5.5–6.5 mm. long; petals shorter than or ca. equal to sepals.—Our most common form, on dry slopes, 3000–6900 ft.; mostly in Joshua Tree Wd. and Pinyon-Juniper Wd.; Inyo Co.; Nev., Ariz. April–June.

Ssp. **Ferrísiae** Abrams. Stems scarcely woody at base; infl. glabrous; sepals 3.5–5 mm. long; petals somewhat to much longer.—Dry stony slopes, 5000–10,000 ft.; Sagebrush Scrub, Pinyon-Juniper Wd., Yellow Pine F.; e. slopes of Sierra Nevada (Inyo and Kern cos.), apparently in San· Jacinto Mts., Riverside Co.; to Nev., Utah. May–July.

17. **A. Kíngii** (Wats.) Jones var. **glabréscens** (Wats.) Maguire. [*A. Fendleri* var. *g.* Wats. *A. capillaris* Calif. refs., not Poir.] Perennial with woody caudex; stems slender, 1–2 dm. high, glandular-pubescent, somewhat leafy; lvs. largely basal, acerose, strict or ascending, 1–2 cm. long; cymes open, few–several-fld.; pedicels mostly 1–2 cm. long, almost filiform, glandular-pubescent; sepals narrow-ovate, subacuminate, glandular-pubescent, very broadly hyaline-margined, 3.5–4.5 mm. long; petals oblong-obovate, entire or emarginate, exceeding sepals; caps. 4.5–6 mm. long; seeds 1.5–2 mm. long, brown, almost smooth, not winged.—Frequent on dry rocky slopes, 6000–11,000 ft.; Red Fir F., Lodgepole F., Subalpine F.; Sierra Nevada from Nevada Co. to Inyo Co., White Mts. and Inyo Mts.; Nev., Ore. June–Aug.

Ssp. **compácta** (Cov.) Maguire. [*A. c.* Cov.] Stems subscapose, usually with 1 pair of lvs. and 2–6 cm. high; lvs. 3–6 mm. long; fls. solitary; sepals 2.5–3.5 mm. long.— Dry rocky places, 10,000–12,500 ft.; Alpine Fell-fields; Mt. Whitney region, Inconsolable Range, White Mts. July–Aug.

18. **A. macrophýlla** Hook. [*Moehringia m.* Torr.] Slender-stemmed perennial from running rootstocks, ascending to suberect, puberulent, mostly 0.5–1.5 dm. high; lvs. lanceolate to oblanceolate, bright green, acute at both ends, 1.5–6 cm. long, mostly 4–8 mm. wide; fls. 1–5, in short cymes; pedicels capillary, mostly 5–20 mm. long; sepals ovate, acute to acuminate, ca. 3–4 mm. long; petals exceeding sepals in ♂ fls., shorter in ♀; caps. ovoid, shorter than calyx; seeds 1.3–1.6 mm. long, usually with a pale spongy strophiole at the hilum.—Occasional on shaded slopes, 1500–6500 ft.; Yellow Pine F., Red Fir F., Mixed Evergreen F.; Cuyamaca Mts., San Diego Co., Coast Ranges from Mt. Hamilton n., n. Sierra Nevada; to B.C., Atlantic Coast. April–June.

5. **Spérgula** L. SPURREY

Annuals, with narrowly linear apparently whorled lvs. (because of crowding in axils). Stipules small, scarious. Fls. in terminal cymose panicles. Sepals 5. Petals 5, white, entire. Stamens 5 or 10. Styles 5. Caps. 5-valved, the valves opposite the sepals. Seeds compressed, acutely margined. Two or 3 spp. of Old World. (Latin, *spargere*, to scatter, because of sowing seeds to produce forage.)

1. **S. arvénsis** L. Erect, branched at base, 1–4 dm. high, glabrous or glandular-pubescent; lvs. linear, 1–3 cm. long; pedicels slender, often reflexed in fr.; sepals 4–5 mm. long; petals as long or slightly longer; caps. ovoid, slightly exceeding calyx; seeds dark, round, plump, ca. 1 mm. in diam., minutely papillose, with very narrow wing-margin; $2n = 18$ (Rohweder, 1939).—Commonly natur. in fields and vacant lots through much of cismontane Calif., especially toward the coast; native of Eu. Most months.

6. Spergulària J. & C. Presl. SAND-SPURREY

Low annual to perennial herbs, usually of saline areas. Lvs. narrowly linear or subfiliform, often fleshy, semiterete, with scarious stipules. Fls. whitish to pink, in terminal racemose cymes. Sepals 5. Petals 5, entire. Stamens 2–10. Styles 3. Caps.-valves 3, rarely 5. Seeds compressed to reniform-globose, smooth or roughened, often wing-margined. Ca. 40 spp., widely distributed. (Name derived from *Spergula*.)

(Rossbach, R. B. Spergularia in N. and S. Am. Rhodora 42: 57–83, 105–143; 158–193; 203–213, 1940.)

Lvs. densely fascicled; stipules lance-acuminate; plants perennial (sometimes annual in *S. rubra*).
Sepals mostly 5–10 mm. long; seeds 0.7–0.9 mm. long; styles 0.6–3 mm. long .... 1. *S. macrotheca*
Sepals mostly 3–5 mm. long; seeds 0.4–0.6 mm. long; styles 0.2–0.8 mm. long.
    Stem below the infl. mostly glabrous; caps. equaling calyx; seeds not winged, reticulate
                                                                     3. *S. rubra*
    Stems below the infl. mostly glandular-pubescent; caps. longer than calyx; seeds usually winged,
        papillate .... 8. *S. villosa*
Lvs. not densely fascicled; stipules mostly deltoid, not long-acuminate; plants annual.
    Seeds black, 0.6–0.8 mm. long, rounded, often iridescent .... 2. *S. atrosperma*
    Seeds brown.
      Stamens 6–10.
        Seeds smooth, usually winged, 0.8–1 mm. long; caps. usually 5.5–7 mm. long .... 6. *S. media*
        Seeds papillose, not winged, 0.4–0.6 mm. long; caps. 2.8–5.4 mm. long ...... 4. *S. Bocconii*
      Stamens 2–5.
        Sepals, 1.6–5 mm. long; seeds 0.6–1.4 mm. long; cymes not much compounded, leafy.
          Seeds 0.9–1.4 mm. long, often winged; styles 0.3–0.4 mm. long; caps. 1.5–2 times as long as
            calyx .... 5. *S. canadensis*
          Seeds 0.5–0.8 mm. long, not winged; styles 0.4–0.6 mm. long; caps. 1–1.5 times as long as
            calyx .... 7. *S. marina*
        Sepals 0.8–1.6 mm. long; seeds ca. 0.4 mm. long; cymes much compounded, bracteate
                                                          9. *S. platensis*

1. **S. macrothèca** (Hornem.) Heynh. [*Arenaria m.* Hornem. *Tissa pallida, Talinum* and *valida* Greene. *T. m.* var. *scariosa* Britton. *S. m.* var. *s.* Rob. *S. m.* var. *T.* Jeps.] Perennial from a heavy branched caudex and stout fleshy root; stems 1–many, prostrate to ascending, stout, 1–4 dm. long, glabrous or glandular-pubescent, branched; lvs. linear, mostly fascicled, fleshy, 1–3.5 cm. long; stipules conspicuous, triangular-acuminate, 5–10 mm. long; infl. glandular-pubescent, lax or dense; sepals broadly lanceolate, 5–10 mm. long; petals mostly pink, ovate, 4–7 mm. long; stamens 10; styles 0.6–1.2 mm. long; caps. usually 6–8 mm. long; seeds dark, red-brown, dull, mostly smoothish, 0.6–1 mm. long, usually with a narrow wing.—Near salt marshes along the coast, occasionally inland, and on sea bluffs; Coastal Scrub, Coastal Strand; L. Calif. to B.C. Most of year.

Var. **leucántha** (Greene) Rob. [*Tissa l.* Greene. *T. l.* var. *glabra* A. Davids. *S. m.* var. *g.* Munz.] Sepals usually 5–6 mm. long; petals usually white; styles 1.2–1.8 mm. long.—Alkaline places; Alkali Sink, Freshwater Marsh; interior cismontane valleys and w. Mojave Desert, from San Diego Co. to Colusa Co. April–June.

Var. **longistỳla** R. P. Rossb. Sepals lance-ovate, 6–7 mm. long; petals white; styles 2–3 mm. long; seeds sculptured.—Moist alkaline places; Freshwater Marsh; inner Coast Ranges from Alameda Co. to Napa Co. March–Oct.

2. **S. atrospérma** R. P. Rossb. [*S. diandra* of Calif. refs., not Guss.] Few-stemmed annual, 5–18 cm. high, glabrous or glandular-villous; lvs. fleshy, not fascicled, 1–2.5 cm. long; stipules broadly triangular, 2–2.8 mm. long; sepals ovate-lanceolate, 3–4 mm. long; petals ovate, white to pink, 2–2.6 mm. long; stamens 4–8; styles 0.5–0.8 mm. long; caps. 3–5 mm. long; seeds black, finely areolate, 0.6–0.8 mm. long, iridescent, not winged.—Alkaline places; Alkali Sink; w. Riverside Co., Tulare Co. to Sierra Co.; w. Nev. April–May.

3. **S. rùbra** (L.) J. & C. Presl. [*Arenaria r.* L. *Tissa r.* Britton. *S. r.* var. *perennans*

(Kindb.) Rob.] Short-lived perennial or annual, much branched at crown, ± matted; stems 0.6–3 dm. long, slender, glabrous below, glandular-pubescent toward infl.; lvs. mostly 6–12 mm. long, fascicled; stipules triangular-acuminate, 3–5 mm. long; cymes many-fld., leafy; sepals lanceolate, 4–5 mm. long; petals pink, ovate, 2.5–3.8 mm. long; stamens 6–10; styles 0.6–0.8 mm. long; caps. 3.5–5 mm. long; seeds dark brown, rounded, minutely papillate, 0.4–0.6 mm. long; $2n = 36$ (Löve & Löve, 1942).—Waste places, roadsides, etc. below 7000 ft., cismontane Calif., to B.C., Atlantic Coast; natur. from Eu. April–Sept.

4. **S. Boccònii** (Scheele) Foucaud. [*Alsine B.* Scheele. *Tissa luteola* Greene.] Annual 0.5–3 dm. high, glabrous below, glandular-pubescent in infl.; lvs. not or but slightly fascicled, 1–2 cm. long; stipules deltoid, 2–4 mm. long, scarcely acuminate; bracts foliaceous; sepals ovate, glandular-pubescent, 2.5–5.4 mm. long; petals white to pink, shorter than calyx; stamens 6–10; styles 0.4–0.6 mm. long; caps. 3–5 mm. long; seeds light brown, plump with broad swollen rim, minutely papillate, not winged, 0.4–0.6 mm. long. —Along paths, beach dunes and alkaline places, cismontane Calif.; natur. from Eu. April–Sept.

5. **S. canadénsis** (Pers.) G. Don var. **occidentàlis** R. P. Rossb. Erect annual, glabrous to sparsely glandular-villous, 0.5–2.5 dm. high; lvs. fleshy, 1–4 cm. long; stipules triangular, 1.6–3 mm. long; infl. open, leafy; sepals ovate, 3–4.5 mm. long; petals white or pink, ovate, 1.6–2.6 mm. long; stamens 2–5; styles 0.3–0.6 mm. long; caps. 4.4–6.4 mm. long; seeds brown, dull, usually smooth on sides, 0.9–1.4 mm. long, winged or not, sometimes glandular-papillose at summit.—Salt marshes, Humboldt Co.; to B.C. June–Aug.

6. **S. mèdia** (L.) Presl. [*Arenaria m.* L.] Annual or short-lived perennial, the stems 0.5–4 dm. long, erect or prostrate, usually glabrous below the infl.; lvs. slightly fascicled, 1–5 cm. long; stipules deltoid, 3–6 mm. long; sepals narrow-ovate, glabrous or glandular-pubescent, 3–6 mm. long; petals white, ovate, 2.5–4.5 mm. long; stamens 9–10; styles 0.5–1 mm. long; caps. 4.5–8 mm. long; seeds dark brown, smooth or nearly so, 0.8–1 mm. long, usually winged.—Low ground bordering salt marshes or low alkaline pastures, Marin Co.; introd. from Eu. May–July.

7. **S. marìna** (L.) Griseb. [*Arenaria rubra* var. *m.* L. *Tissa m.* Britton. *S. salina* J. & C. Presl. *Tissa s.* vars. *Sanfordii* and *.sordida* Greene. *S. s.* var. *sordida* Jeps.] More or less diffuse annual; stems 0.5–3 dm. long, fleshy, ± glandular-pubescent; lvs. scarcely fascicled, linear, 2–4 cm. long; stipules deltoid, 2–4 mm. long; infl. lax, usually glandular-pubescent; sepals ovate, blunt-tipped, 2.5–5 mm. long; petals white to pink, ovate, 2–4 mm. long; stamens 2–5; styles 0.4–0.6 mm. long; caps. 3.6–6 mm. long; seeds brown, smooth or glandular-papillose, 0.6–0.8 mm. long, usually not winged; $2n = 36$ (Löve & Löve, 1942).—Common along seashore and alkaline places of interior and occasional on deserts; Coastal Strand, Alkali Sink, near Freshwater Marsh, etc.; to Wash., Atlantic Coast; Eurasia. March–Sept.

Var. **ténuis** (Greene) R. P. Rossb. [*Lepigonum t.* Greene. *S. t.* Rob. *S. salina* var. *t.* Jeps. *S. t.* var. *involucrata* Rob.] Stems usually more branched; infl. more crowded; sepals 1.6–3.5 mm. long; petals 1.5–2 mm. long; caps. 3–4.4 mm. long.—Low alkaline places; Alkali Sink and near Freshwater Marsh; Kern Co. to Colusa Co. April–May.

8. **S. villòsa** (Pers.) Camb. [*Spergula v.* Pers. *Tissa Clevelandii* Greene. *Spergularia C.* Rob.] Perennial from a rather heavy taproot; stems usually viscid-glandular, 1–3 dm. long; lvs. fascicled, narrow, 1–4 cm. long; stipules broadly lanceolate, acuminate, 3–8 mm. long; cyme many-fld., lax, glandular-pubescent; sepals linear-lanceolate, 3–5 mm. long; petals white, ovate, 2.6–5 mm. long; stamens 7–10; styles 0.4–0.6 mm. long; caps. mostly 5–6 mm. long; seeds dark brown, pyriform, black-papillate to smooth, winged or sometimes not, 0.4–0.6 mm. long.—Generally in sandy soil and near the coast, near Coastal Salt Marsh and Coastal Strand; occasional in interior valleys, Coastal Scrub; L. Calif. to Ore.; introd. from S. Am. April–July.

9. **S. platénsis** (Camb.) Fenzl. [*Balardia p.* Camb. in St. Hil. *Lepigonum gracile* Wats. *Tissa g.* Britton.] Glabrous annual, the stems slender, 0.5–3 dm. long; lvs. linear-filiform, 1–3 cm. long; stipules deltoid, subacuminate, 1.5–3.5 mm. long; infl. much compounded, glabrous; sepals broadly lanceolate, 1–1.6 mm. long; petals white, narrowly ovate, 0.6–1 mm. long; stamens 5; styles 0.3–0.4 mm. long; caps. 1.5–2.5 mm. long; seeds brown, 0.35–0.4 mm. long, not winged, tuberculate.—Drying mud flats of vernal pools; Chaparral, Coastal Sage Scrub; San Diego Co. to Riverside and Los Angeles cos.; to Tex.; S. Am. April–June.

7. Polycárpon L.

Low much-branched annuals. Cauline lvs. numerous, flat; stipules and bracts scarious. Fls. small, cymose, numerous. Sepals 5, ± keeled and with scarious margins. Petals 5, hyaline, entire or emarginate, smaller than sepals. Stamens 3–5. Style 1, short, 3-cleft. Caps. 3-valved. Seeds ovoid, several. Ca. 6 spp. of wide distribution. (Greek, *polus*, many, and *karpos*, fr., because of the many capsules.)

Sepals almost 2 mm. long; lvs. obovate to oblong, often appearing to be in 4's .... 1. *P. tetraphyllum*
Sepals ca. 1 mm. long; lvs. spatulate, opposite .... 2. *P. depressum*

1. **P. tetraphýllum** (L.) L. [*Mollugo t.* L.] Glabrous, diffuse or prostrate, the stems 4–10 cm. long; lvs. oblong to obovate, 5–10 mm. long, short-petioled; cymes leafless, many-fld.; stipules and bracts lance-acuminate; sepals obovate or oblong, strongly keeled, almost 2 mm. long; petals thin, oblanceolate; caps. ovoid, almost as long as calyx; seeds somewhat angled, almost transparent, light brown, ca. 0.5–0.8 mm. long.—Occasional in beaten soil of paths and along roads, in waste places, etc., cismontane Calif.; natur. from Eu. May–July.

2. **P. depréssum** Nutt. Prostrate, with many slender stems 1–5 cm. long, much-branched, glabrous; lvs. spatulate, 3–6 mm. long, with slender petiole; sepals 1 mm. long, inconspicuously keeled; petals linear; caps. spherical.—Occasional in sandy open places, below 2500 ft.; Coastal Sage Scrub, Chaparral; Monterey Co. to L. Calif. April–June.

8. Loeflíngia L.

Low spreading rigid glandular-pubescent annuals, branched from base. Lvs. inconspicuous, subulate; stipules setaceous. Fls. small, sessile, axillary, solitary or fascicled. Sepals 5, carinate, narrow, with rigid setaceous straight or recurved tips. Petals 3–5 and small, or none. Stamens 3–5. Style 1 or none; stigmas 3. Caps. 3-valved. Seeds several, oblong. Ca. 5 spp. of Eu., Asia, w. N. Am. (Peter *Loefling*, Swedish naturalist of 18th century.)

Outer sepals recurved and with setaceous tooth on each side .... 1. *L. squarrosa*
Outer sepals straight, entire .... 2. *L. pusilla*

1. **L. squarròsa** Nutt. Stems 3–14 cm. long, glandular-pubescent; lvs. cuspidate, 4–6 mm. long; calyx ca. 4 mm. long, the sepals squarrose, rigid, toothed on each side; petals very minute; caps. shorter than calyx; seeds many, semitranslucent, ca. 0.4 mm. long.— Locally frequent in sandy places, below 2000 ft.; Coastal Sage Scrub, Chaparral, V. Grassland; especially in interior valleys from San Diego to Sacramento V.; L. Calif., Ariz. April–May.

2. **L. pusílla** Curran. Much like the preceding, but the sepals not rigid or squarrose, entire; petals none; caps. equaling calyx.—A local sp. from ca. 4000 ft., Tehachapi Mts. April–May.

9. Agróstemma L. Corn-Cockle

Tall silky annual or biennial. Lvs. linear. Fls. few, solitary, purplish-red. Calyx ovoid, with 10 strong ribs, the elongated teeth (in ours 2–3 cm. long) exceeding the 5 large unappendaged petals. Stamens 10. Styles 5. Caps. coriaceous, dehiscing by 5 teeth. Two spp., Eurasian. (Greek, *agros*, field, and *stemma*, crown.)

1. **A. Githàgo** L. Plants 3–10 dm. high; lvs. 5–8 cm. long; calyx-tube 10–15 mm. long, the lobes 2–3 times as long; $2n = 24$ (Rohweder, 1939), 48 (Favarger, 1946).—Occasional in grainfields, particularly in the interior valleys, sometimes nearer the coast; native of Eurasia. May–June.

10. Silène L. Catchfly. Campion

Annual or perennial herbs, usually somewhat viscid. Lvs. opposite, exstipulate. Fls. solitary or more often cymose. Calyx cylindrical to ovoid or campanulate, 5-toothed, 10–many-nerved. Petals 5, clawed, usually with a scalelike appendage at base of the blade which is usually cleft or toothed. Stamens 10. Styles 3, rarely 4. Ovary on a well developed stipe which with connate fil.-bases and petal-bases forms a carpophore. Caps. 1- or incompletely 2–4-locular, opening apically by 3 or 6 teeth. Seeds reniform to globose, striately muricate to tuberculate. Perhaps 250 spp., chiefly distributed in temp. and cold regions of N. Hemis. (Name said to have come from *Silenus*, foster-father of Bacchus, supposedly covered with foam, many spp. having a viscid secretion.)

(Hitchcock, C. L., & B. Maguire. A revision of the N. Am. species of Silene. Univ. Wash. Pub. Biol. 13: 1–73, 1947.)
A. Annuals, mostly introd. weedy plants.
  B. Calyx with 20–30 well developed ribs.
    C. The calyx 8–12 mm. long; petals lacking appendages . . . . 1. *S. multinervia*
    CC. The calyx 20–30 mm. long; petals with well developed appendages . . . . 2. *S. conoidea*
  BB. Calyx usually with 10 ribs, if more, the ribs indistinct.
    C. Plant subglabrous or minutely puberulent, the upper internodes usually with sticky bands
                                                         3. *S. antirrhina*
    CC. Plant densely pubescent.
      D. Petals not deeply bilobed; appendages linear . . . . 4. *S. gallica*
      DD. Petals deeply bilobed; appendages usually as wide as long.
        E. Calyx not strongly inflated, 10–15 mm. long, the nerves stiffly hirsute
                                               5. *S. dichotoma*
        EE. Calyx strongly inflated in fr. becoming 25–50 mm. long, the nerves not stiffly hirsute . . . . 6. *S. noctiflora*
AA. Perennials; native (except *S. Cucubalus*).
  B. Corolla bright red.
    C. Stems usually less than 2.5 dm. long; lvs. oval or elliptic to obovate; appendages usually not more than 1 mm. long; caps. ovoid, included. Ore. to Los Angeles Co.
                                                      7. *S. californica*
    CC. Stems usually 5–15 dm. long; lvs. largely lanceolate; appendages 1–1.5 mm. long; caps. oblong, exserted. Mostly s. Calif. . . . . 8. *S. laciniata*
  BB. Corolla white or pink to purplish.
    C. Petal-blades bilobed half their length; appendages lacking; carpophore 2–3 mm. long, glabrous; calyx much inflated in fr. and ± papery. Introd. weed . . . . . . 9. *S. Cucubalus*
    CC. Petals, appendages and calyces not as above.
      D. Blades of petals essentially entire, scarcely exserted; stamens included; appendages lacking. Cent. Sierra Nevada . . . . 10. *S. invisa*
      DD. Blades of petals variously lobed, incised or lacerate.
        E. Petals mostly 2.5–3.5 cm. long; calyx 1.5–3 cm. long.
          F. Corolla white to pink or purplish, well exserted. Nw. Calif.
                                           23. *S. Hookeri*
          FF. Corolla yellowish, scarcely exserted. S. Calif. . . . . 25. *S. Parishii*
        EE. Petals and often the calyx shorter.
          F. Appendages of petals usually 4, the claws with more than 3 veins at point of attachment of appendages. N. Calif.
            G. Calyx campanulate, almost as broad as long, green, the lobes ⅓ to ½ as long as tube; petal-blades 3.5–7 mm. long
                                             11. *S. campanulata*
            GG. Calyx tubular, much longer than broad, scarious between the green nerves, the lobes not more than ¼ as long as tube; petal-blades ca. 3 mm. long . . . . 14. *S. oregana*
         FF. Appendages of petals usually 2 or 0, if 4, the claws with but 3 veins at point of attachment of appendages.
            G. Calyx cleft nearly half its length.
               H. Claws woolly on back; appendages lacking; petals 4-lobed. Tulare Co. . . . . 12. *S. aperta*
               HH. Claws not woolly; appendages present; petals bilobed. N. of Alpine Co. . . . . 24. *S. nuda*
            GG. Calyx not so deeply cleft; claws and appendages not both as above.
               H. Corollas mostly less than 10 mm. long; calyx 5–8 mm. long; lower lvs. 2–8 mm. long . . . . 26. *S. Menziesii*
               HH. Corollas usually more than 12 mm. long; calyx 8–18 mm. long; lower lvs. longer.
                 I. Fls. nodding at anthesis; stamens and style long-exserted.
                   J. Petals 2-cleft, the claws glabrous on outer surface; lvs. 2.5–7.5 cm. long . . . . 16. *S. Bridgesii*
                   JJ. Petals 4-cleft, the claws hairy on outer surface; lvs. 1.5–2.5 cm. long . . . . 17. *S. Lemmonii*
               II. Fls. erect or nearly so, sometimes deflexed in *S. Grayi;* stamens little exserted or included.
                   J. Basal lvs. usually less than 2 cm. long and 1–4 mm. wide; plants tufted, usually not more than 2 dm. high.
                     K. Blades of petals equally 4-lobed and 4–6 mm. long; calyx 13–16 mm. long . . . 15. *S. montana*
                     KK. Blades of petals 2-lobed or with 4 unequal lobes.

L. Calyx 8–10 mm. long; blades of petals 3–5 mm. long; basal lvs. 2–5 mm. broad. N. of Sierra Co. .... 19. *S. Grayi*
LL. Calyx 10–14 mm. long; blades of petals 2.5–3.5 mm. long; basal lvs. 1–2 mm. wide. Sierra Co. .... 20. *S. Sargentii*
JJ. Basal lvs. mostly more than 3 cm. long, often more than 4 mm. wide; plants mostly more than 3 dm. tall.
K. Blades of petals with 4 subequal linear lobes.
L. Calyx 13–16 mm. long; petal-blades 4–6 mm. long .... 15. *S. montana*
LL. Calyx 15–35 cm. long; petal-blades 7–18 mm. long .... 18. *S. occidentalis*
KK. Blades of petals bilobed, the lobes in turn sometimes unequally lobed.
L. Lvs. fleshy, oblanceolate; calyx mostly 8–10 mm. long; petal-blades mostly 3–4.5 mm. long .... 19. *S. Grayi*
LL. Lvs. not fleshy and oblanceolate; calyx mostly 10–18 mm. long.
M. Basal lvs. mostly 10–40 cm. long, 1–3 cm. broad, the blades and petioles subequal; cauline lvs. 1–2 pairs; carpophores 1–2 mm. long, subglabrous; appendages linear, 1–1.5 mm. long; corolla rose. Mts. bordering deserts 24. *S. nuda*
MM. Basal lvs. mostly 2–15 cm. long, but if longer and as wide as 1–3 cm., then plants otherwise unlike the above; cauline lvs. usually more numerous. (Go to N.)
N. Stems usually 1–4, from a heavy crown; basal lvs. 15–30 mm. wide. N. coastal. 13. *S. Scouleri*
NN. Stems usually more numerous; basal lvs. 2–12 mm. wide.
O. Calyx not constricted below the ovary, without glands; appendages 1.5–3 mm. long. Trinity and Alpine cos. n. .... 21. *S. Douglasii*
OO. Calyx usually markedly constricted below the ovary, usually glandular; appendages 0.6–2 mm. long. Lake and Mono cos. s. .... 22. *S. verecunda*

1. **S. multinérvia** Wats. Erect annual, simple or branched, 2–6 dm. high, villous and rather viscid; lvs. lance-linear, or the lowest oblanceolate, 2–6(–10) cm. long, 5–12(–25) mm. wide, acute or obtusish; infl. ± open, the bracts mostly 8–15 mm. long, the pedicels 1–4 cm. long, densely glandular; calyx ovoid, 8–12 mm. long, little inflated, ca. 25-nerved, the nerves prominent and stipitate-glandular, the lobes 2–3 mm. long, acuminate; petals white to pinkish, ca. as long as calyx, the blades 2-cleft, 1–3 mm. long, without appendages; carpophore ca. 1 mm. long, glabrous; caps. ovoid, 7–9 mm. long; seeds many, 0.6–0.9 mm. long, scarcely angled, with 3 dorsal rows of raised papillae.—Mostly in disturbed places, such as burns, below 3000 ft.; Chaparral; Sonoma Co. to San Diego Co. April–May.

2. **S. conoìdea** L. Erect annual 5–8 dm. high, retrorsely puberulent, glandular toward top; lvs. lanceolate to oblanceolate, 5–12 cm. long, 8–12 mm. broad, acute; pedicels 1–3 cm. long, glandular and puberulent; bracts 1–2 cm. long; calyx becoming conic-ovoid, 20–30 mm. long, with ca. 30 nerves, the lobes attenuate, 8–12 mm. long; petals white to reddish, conspicuously exceeding calyx, the blades 8–12 mm. long retuse with 2 bilobed appendages 2–4 mm. long; carpophore glabrous, ca. 2 mm. long; caps. 15–20 mm. long; seeds many, 1.2–1.5 mm. long, gray-brown, laterally reniform, with 5 dorsal rows of flattish tubercles.—Occasional weed, as at Blythe; native of Eurasia.

3. **S. antirrhìna** L. Mostly erect annual, simple or usually branched, 2–8 dm. high, mostly retrorse-pubescent below and glabrous above, sometimes puberulent above or all glabrous, usually with glutinous bands on some of the internodes; basal lvs. oblanceolate to spatulate, the cauline oblanceolate to linear, 3–6 cm. long, 2–12 mm. wide, mostly acute; pedicels slender, 1–4 cm. long; bracts subulate, up to 1 cm. long; calyx 10-nerved, glabrous, somewhat contracted at orifice, 4–8 mm. long, the triangular teeth ca. 1 mm. long; petals white to pink, ca. as long as or slightly exceeding calyx, the blades 2-lobed, appendages reduced; caps. 4–8 mm. long; seeds gray-black, 0.5–0.8 mm. long, not angled, with 3–4 dorsal rows of papillae.—Common weed in open sandy places, burns, etc., mostly below 5000 ft., cismontane and desert Calif., to B.C. and Atlantic Coast.

April–Aug. Variable; plants with mostly branched stems, broad basal lvs., linear cauline lvs., prominently exserted petals, have been called var. *confinis* Fern.

4. **S. gállica** L. [*S. anglica* L.] Usually erect annual, simple to branched, 1–4 dm. high, hirsute and strigulose, glandular-pubescent above; basal lvs. oblanceolate or spatulate, obtuse, mucronate, the cauline somewhat narrower, 1–3.5 cm. long; infl. leafy-bracted, 1-sided, the pedicels 2–5 mm. long; calyx 10-nerved, 6–9 mm. long, inflated in age, glandular-pubescent and hirsute, constricted at orifice; petals whitish to pinkish, slightly longer than calyx, the blades elliptic, entire or toothed, slightly twisted, the linear appendages 1 mm. long; carpophore 1 mm. long, pubescent; caps. 6–8 mm. long; seeds slate-gray, almost 1 mm. long, finely rugulose.—Common weed on vacant lots, in fields and waste places; below Yellow Pine F.; cismontane Calif. to B.C., less common e.; from Eu. Feb.–June.

5. **S. dichótoma** Ehrh. Simple or branched annual 3–8 dm. tall, strikingly hirsute; lvs. lanceolate to oblanceolate, 3–8 cm. long, 5–35 mm. wide; pedicels 1–3 mm. long, usually recurved; calyx narrow-tubular, 10–15 mm. long, 10-nerved; petals white to reddish, the blades 5–9 mm. long, bilobed, with short truncate appendages; carpophore 2–4 mm. long, glabrous; seeds 1–1.3 mm. long, dark gray-chocolate, finely rugose.—Once reported at Berkeley; from Eu.

6. **S. noctiflòra** L. Coarse annual 2–6 dm. high, coarse-hirsute; lvs. lance-ovate to elliptic-oblanceolate, 5–10 cm. long, 2–4 cm. broad; pedicels 3–12 mm. long; calyx 10-nerved, becoming 25–40 mm. long in fr.; petals white to pinkish, the blades 7–10 mm. long, bilobed, the appendages 0.5–1.5 mm. long; carpophore 1–3 mm. long; seeds 0.8–1 mm. long, red-brown, papillose-rugose.—Occasional on Pacific Coast, as at Shasta Springs, Wawona, Alhambra; from Eu.

7. **S. califórnica** Durand. [*S. laciniata* var. *c.* Gray. *S. c.* var. *subcordata* Rob.] Perennial from a stout taproot, the stems several, leafy, suberect to decumbent, puberulent and somewhat glandular, 1.5–4 dm. long; lvs. ovate to oblanceolate, 3–8 cm. long, 1–3 cm. broad; fls. few to many; pedicels mostly 2–3 cm. long; calyx broadly tubular, 1.5–2.5 cm. long, distended or ruptured in fr., the lobes 2.5–5 mm. long; petals crimson, much exserted, 2–3 cm. long, the blades 1–1.5 cm. long, deeply 4-lobed, with appendages 1–2 mm. long; carpophore 2–3 mm. long, puberulent or glabrous; caps. ovoid; seeds red-brown, broadly rounded, ca. 2 mm. broad, conspicuously papillate; *n* = 24, 48 (Kruckeberg, 1954).—Open brushy or wooded places, below 5000 ft.; Chaparral, Foothill Wd., Mixed Evergreen F., N. Coastal Scrub, Yellow Pine F., etc.; cismontane Calif. from Los Angeles and Kern cos. n. and from w. slope of Sierra Nevada to the coast; Ore. March–Aug.

8. **S. laciniàta** Cav. ssp. **màjor** Hitchc. & Maguire. [*S. simulans* Greene.] Perennial from a deep fleshy taproot, the stems weak, 2–7 dm. long, retrorsely puberulent and glandular-pubescent; lvs. narrowly linear to lanceolate or lance-elliptic, mostly 5–10 cm. long, 2–14 mm. wide; infl. open, mostly 5–9-fld.; pedicels 2–4 cm. long, glandular-pubescent; calyx 15–20 mm. long, tubular-cylindrical, the lobes lanceolate or wider; petals scarlet, conspicuous, the blades 8–15 mm. long, deeply cleft into 4 linear or lanceolate lobes, the appendages 2, broad, toothed, 1–1.5 mm. long; carpophore 2–4 mm. long, glabrous; caps. oblong-ovoid, 14–18 mm. long; seeds reddish brown, 1.2–1.5 mm. broad, with prominent marginal papillae; *n* = 48 (Kruckeberg, 1954).—Frequent on grassy or brushy somewhat shaded slopes, below 5000 ft.; Coastal Strand, Coastal Sage Scrub, Chaparral; Santa Cruz Co. to L. Calif. May–July. Hitchcock and Maguire have proposed 2 vars. in this ssp.: *angustifòlia* for plants from near the coast, with lvs. linear or linear-lanceolate and calyces 18–20 mm. long; and *latifòlia* from the interior and with lvs. lanceolate and calyces 16–18 mm. long. Typical *S. laciniata*, with longer more pubescent calyx, occurs in the Mex. Cordillera.

9. **S. Cucùbalus** Wibel. [*S. inflata* (Salisb.) Sm. *S. latifolia* (Mill.) Britt. & Rend.] Robust perennial from heavy creeping rootstocks; stems 2–8 dm. tall, usually glabrous and glaucous, often subdecumbent; lvs. ovate-lanceolate to oblanceolate, 3–8 cm. long, 1–3 cm. broad; pedicels slender, 1–3 cm. long; calyx becoming much inflated, 1–2 cm. long, pale green to purplish, delicate, the nerves reticulate; petals white, the blades 3.5–6 mm. long, deeply cleft, with greatly reduced appendages; carpophore glabrous, 2–3 mm. long; caps. ovoid-globose; seeds brown, 1–1.5 mm. long, papillose; $2n = 24$

(Blackburn, 1928).—Occasional weed, as at Anaheim, San Francisco, Santa Rosa, Sisson; from Eurasia. Summer, fall.

10. **S. invìsa** Hitchc. & Maguire. Perennial from a slender taproot; stems few, 1–4 dm. tall, subsimple, puberulent, somewhat glandular above; lvs. mostly basal, oblanceolate to spatulate, 2–4 cm. long, 2–6 mm. broad, the cauline 1–2 pairs; fls. few, the pedicels 0.5–2.5 cm. long; calyx narrowly tubular, ca. 9 mm. long, somewhat larger in fr., 10-nerved, finely glandular-pubescent; petals pale lavender, equal to or slightly exceeding calyx, not appendaged, entire or retuse, scarcely differentiated into blade and claw; carpophore less than 1 mm. long; seeds ca. 1 mm. long, with tessellate margins.—At ca. 7000–8600 ft.; Lodgepole F., Subalpine F.; Sierra Nevada, Nevada and Alpine cos. July.

11. **S. campanulàta** Wats. [*S. c.* var. *angustifolia* F. N. Williams.] Perennial with large woody taproot, many-branched crown and numerous erect slender stems 0.5–1.5 dm. high, scabrous-pubescent, glandular at least in infl.; lvs. linear to lanceolate, 2–6 cm. long, 2–13 mm. wide, the cauline 2–5 pairs; pedicels 5–10 mm. long; calyx campanulate, 10-nerved, 6–8 mm. long at anthesis, almost twice that in fr., the lobes triangular-ovate, ciliate; petals cream to greenish or pinkish, the blades 4–7 mm. long, cleft into 2 or more lobes; appendages usually 2, 1.5–2 mm. long, erose or divided; stamens exserted; carpophore 1–2 mm. long, pubescent; seeds brown, ca. 2.2 mm. long, prominently papillose.—Known only from Red Mountain, Mendocino Co.

Ssp. **glandulòsa** Hitchc. & Maguire. [*S. c.* var. *orbiculata* Rob.] Plants mostly 1.5–3.5 dm. high, at least infl. and calyx glandular; lvs. lanceolate to broadly ovate, usually over 10 mm. broad; $n = 24$ (Kruckeberg, 1954).— Forest floor, 2000–6000 ft.; Red Fir F., Yellow Pine F., Foothill Wd.; Lake and Glenn cos. to Shasta, Siskiyou, and Humboldt cos.; s. Ore. May–Aug.

Ssp. **Greènei** (Wats.) Hitchc. & Maguire. [*S. c.* var. *G.* Wats. *S. c.* var. *latifolia* F. N. Williams, *S. c.* var. *petrophila* Jeps.] Like ssp. *glandulosa,* but glandless throughout.—At similar elevs., Siskiyou Mts. to Del Norte, Humboldt, Trinity, and Shasta cos.

12. **S. apérta** Greene. [Spelled *S. aptera* in Jeps.] Perennial from a several-branched caudex and taproot, the stems several, retrorsely pubescent, scarcely if at all glandular, 1.5–3.5 dm. tall, slender, quite simple; lvs. linear to narrow-oblanceolate, 3–10 cm. long, 1–3.5 mm. wide; pedicels 5–30 mm. long; calyx tubular-campanulate, 10-nerved, 8–10 mm. long, the lobes lanceolate, ca. as long as tube, glandless, ciliate and pubescent; petals whitish, usually lanate on claws, the blades 1.5–3 mm. long, shallowly 4-lobed, without appendages; carpophore 1–2 mm. long, hairy; seeds light brown, papillose, ca. 0.7 mm. long.—Dry benches under pines, 8000–9000 ft.; Red Fir F., Lodgepole F.; Sierra Nevada, Tulare Co. July–Aug.

13. **S. Scoùleri** Hook. ssp. **grándis** (Eastw.) Hitchc. & Maguire. [*S. g.* Eastw. *S. pacifica* Eastw. *S. g.* var. *p.* Jeps.] Perennial with 1–few stems from a heavy crown, ± puberulent, often almost velvety, 1.5–7 dm. tall; lvs. ovate or obovate to elliptic, oblanceolate, 2.5–15 cm. long, 1–3 cm. broad, passing upward into lanceolate leafy bracts; pedicels mostly 5–10 mm. long; calyx 10-nerved, oblong-campanulate, 10–14 mm. long, glandular-pubescent, the lobes 2–4 mm. long; petals greenish-white to rose, the blades 4–7 mm. long, suboval, deeply bilobed, the appendages 1–2.5 mm. long, erose; carpophore 3–6 mm. long, pubescent; seeds reniform, 1–1.5 mm. broad, gray-brown, with elongate papillae; $n = 24$ (Kruckeberg, 1954).—Bluffs near the coast; N. Coastal Scrub; San Mateo Co. to B.C. May–Aug. Typical *S. Scouleri,* with narrower lvs. and less ciliate petal-claws, occurs in Ore., Wash., Ida.

14. **S. oregàna** Wats. Few-stemmed perennial 3–5 dm. high, moderately puberulent; lvs. oblanceolate, the cauline 4–6 pairs, 6–8 cm. long, 6–15 mm. wide; infl. glandular-pubescent, few-fld.; calyx narrow-campanulate, 10–14 mm. long, the lobes 2–3 mm. long; petals whitish-pink, the blades 3–5 mm. long, deeply 4-lobed, the appendages usually 4, linear, acute, 1–1.5 mm. long; carpophore 3–4 mm. long, thinly pilose; seeds brown, irregularly reniform, 1.2–1.8 mm. broad, with low tubercles; $n = 24$ (Kruckeberg, 1954).—Open places, 5000–8000 ft.; Sagebrush Scrub to Subalpine F.; Modoc Co., to e. Wash., Wyo., Nev. July–Sept.

15. **S. montàna** Wats. Perennial from long woody taproot; stems many, 1.5–4.5 dm. tall, finely retrorsely pubescent, glandular only above; lvs. oblanceolate to lance-linear,

2–6 cm. long, 2–6 mm. broad, rather strict, the cauline 2–4 pairs; infl. mostly several-fld., with slender glandular pedicels 1–2 cm. long; calyx tubular, usually contricted below ovary, 13–16 mm. long, glandular-pubescent, the ribs prominent, greenish or purplish, the lobes ca. 3 mm. long; petals white with some pink or purple, the blades 4–6 mm. long, cleft into mostly 4 linear subequal lobes, the appendages 1–2.5 mm. long, lacerate; carpophore 2–5 mm. long, finely pubescent; caps. exceeding calyx; seeds 1.5–1.8 mm. long, ash-brown, with low radially elongate tubercles.—Among rocks and on dry slopes, 4500–9200 ft.; Sagebrush Scrub, Pinyon-Juniper Wd., to Red Fir F.; Mono Co. to Lassen, Tehama, Mendocino, and Siskiyou cos.; Ore., Nev. June–Aug.

Var. **siérrae** Hitchc. & Maguire. Stems often glandular below; lvs. thinner, less strict, 6–15 mm. broad.—From 5000–9000 ft.; Yellow Pine F., Red Fir F.; mostly w. slope of Sierra Nevada, from Tulare Co. to Sierra Co. June–Aug.

Ssp. **bernardìna** (Wats.) Hitchc. & Maguire. [*S. b.* Wats. *S. Shockleyi* Wats.] Plant glandular to base or nearly so; lvs. linear-lanceolate to linear-oblanceolate, 1.5–2(–3) mm. broad.—Dry slopes, 5000–11,000 ft.; Pinyon-Juniper Wd. to Alpine Fell-fields; mostly e. slope of Sierra Nevada in Mono and Inyo cos. June–Aug.

16. **S. Bridgèsii** Rohrb. [*S. Engelmannii* Rohrb. *S. incompta* Gray.] Several-stemmed perennial, puberulent below, glandular-pubescent above; basal lvs. usually oblanceolate, the cauline elliptic to oblong-lanceolate, 3–8 cm. long, 5–15 mm. broad, acute or acuminate; infl. long, open, the pedicels slender, 5–10 mm. long; calyx tubular to campanulate in fr., 8–15 mm. long, viscid-puberulent, the lobes 2–3 mm. long; petals dirty white, the blades 7–12 mm. long, 2-cleft, the appendages 1–2.5 mm. long, entire to erose; carpophore 2–3 mm. long, puberulent; seeds 1.2–1.5 mm. long, brown, papillose; *n* = 24 (Kruckeberg, 1954).—Dry rather open slopes, 2600–8000 ft.; Yellow Pine F., Red Fir F.; Tulare Co. to Shasta Co. June–July.

17. **S. Lemmònii** Wats. [*S. Palmeri* Wats. *S. longistylis* Engelm.] Perennial from a multicipital caudex, the stems slender, 1.5–4.5 dm. long, ± retrorsely fine-pubescent, glandular toward top; lvs. lance-elliptic to oblanceolate, 2–3 cm. long, 5–10 mm. wide, the cauline mostly 2 pairs; infl. open; pedicels slender, 1–2 cm. long; calyx narrow-campanulate, 8–10 mm. long, glandular-pubescent, with prominent green ribs, the lobes 1–1.5 mm. long; petals yellowish-white to pinkish, the blades 4.5–8 mm. long, 2-lobed, the appendages 2, 0.5–1 mm. long; carpophore 2–2.5 mm. long; seeds 1.2–1.4 mm. broad, reddish, papillose.—Open woods, 3500–8000 ft.; Yellow Pine F., Red Fir F.; Cuyamaca Mts. to Sierra Nevada, Santa Ynez Mts. to Siskiyou Mts.; s. Ore. June–Aug.

18. **S. occidentàlis** Wats. Few-stemmed perennial 3–6 dm. high, subpilose and somewhat glandular-pubescent near base, more glandular upward; lower lvs. broadly oblanceolate to spatulate, 1–2.5 cm. long, 5–8(–10) mm. broad, the cauline mostly 3 pairs; infl. rather open, 9–25-fld., the pedicels glandular-pubescent to -villous, rather short; calyx tubular, 15–22 mm. long, densely glandular-pubescent and with some longer hairs, the lobes 2–4 mm. long, broadly obovate; petals flesh-colored to deep rose, the blades 7–12 mm. long, almost equally 4-lobed, the appendages 2–4 mm. long, linear; carpophore 4–7 mm. long, pubescent; caps. oblong-ovate; seeds gray-brown, 1.2–1.5 mm. wide, angled, with low radially elongate tubercles.—Dry open places, 4000–7000 ft.; Yellow Pine F. to Lodgepole F.; Sierra Nevada from Placer and Alpine cos. to Shasta and Modoc cos. June–July.

Ssp. **longistipitàta** Hitchc. & Maguire. Calyx 30–40 mm. long; blade of petal 15–20 mm. long; carpophore 13–18 mm. long.—Chaparral, Yellow Pine F.; Butte Co. July.

19. **S. Gràyi** Wats. [*S. deflexa* Eastw.] Retrorse-puberulent perennial from a multi-cipital caudex, 1–2.5 dm. high, glandular in the infl.; lvs. tufted, the basal oblanceolate or spatulate, 2–4 cm. long, 2–5 mm. wide, the cauline 2–4 pairs; pedicels slender, with ± deflexed fls.; calyx tubular-campanulate to subcylindric, 8–10 mm. long, glandular-puberulent, the lobes ca. 2 mm. long; petals pinkish to rose or purplish, the blades 3–5 mm. long, 2- or almost 4-lobed, the appendages 2, oblong-ovate, 1–1.5 mm. long; carpophore 2–3 mm. long, pilose-woolly; seeds almost smooth, ca. 1.5 mm. long.—Open rocky slopes or talus, 5000–9200 ft.; Red Fir F. to Alpine Fell-fields; Mt. Lassen to mts. of Siskiyou and Trinity cos.; s. Ore. July–Aug.

20. **S. Sargéntii** Wats. [*S. Watsonii* Rob. *S. lacustris* Eastw. *Lychnis californica* Wats.] Perennial from a multicipital caudex, mostly 1–1.5 dm. high, retrorse-puberulent below, glandular above or even lower; basal lvs. tufted, linear-oblanceolate, 1.5–2.5 cm. long,

1.5 mm. broad, puberulent, often glandular, the cauline usually 1–2 pairs; fls. 1–few; calyx tubular to somewhat campanulate, with prominent somewhat purplish ribs, the lobes 2–3 mm. long; petals whitish to rose-purple, the blades 2.5–3.5 mm. long, bilobed and each lobe usually with a small lateral tooth, the appendages 2, oval, 1–1.5 mm. long; seeds light brown, ca. 1.5 mm. long, mostly smooth; $n = 24$ (Kruckeberg, 1954).— About rocks and talus, 6500–12,000 ft.; Subalpine F., Alpine Fell-fields; Sierra Nevada from Tulare Co. to Sierra Co.; w. Nev. July–Aug.

21. **S. Douglásii** Hook. Few–many-stemmed perennial 1–4 dm. high, finely strigulose, or puberulent; lvs. numerous in lower part, mostly oblanceolate, sometimes narrowly so, 2–5 cm. long, 2–7 mm. broad, the cauline 1–8 pairs; infl. rather few-fld., strigulose, not at all glandular, the pedicels 5–30 mm. long; calyx tubular, quite inflated in fr., 12–15 mm. long, the lobes ca. 2 mm. long; petals creamy white or with some color, the blades 4–6 mm. long, bilobed, the appendages linear or oblong, 1 mm. long; carpophore 3–4 mm. long, puberulent; seeds laterally flattened, brown, ca. 1.2–1.5 mm. long, rugose-tessellate on sides and with scales or papillae on margins; $n = 24$ (Kruckeberg, 1954).—Dry flats or slopes, 5000–9500 ft.; Sagebrush Scrub to Lodgepole F.; Alpine Co. to Modoc and Humboldt cos.; to B.C., Mont. June–Aug.

Var. **monántha** (Wats.) Rob. [*S. m.* Wats.] Calyces subglabrous or sparsely puberulent or granular, the hairs very minute; lvs. often linear.—With the sp., cent. Sierra Nevada, Trinity Co.; Mt. Hood, Ore.

22. **S. verecúnda** Wats. [*S. Behrii* (Rohrb.) F. N. Williams.] Several-stemmed perennial 1–3 dm. high, densely retrorse-pubescent, with gland-tipped hairs also in the infl.; basal lvs. linear-oblanceolate to obovate, 2–6 cm. long, 2–10 mm. broad, the cauline lanceolate to oblanceolate; infl. elongate; calyx tubular-campanulate, constricted below the ovary, often purplish, densely glandular-pubescent and almost villous, 10–12 mm. long, the lobes 2–5 mm. long; petals pink to rose, the blades 4–6 mm. long, bilobed, the appendages 2, oblong, 1.2 mm. long; carpophore 2–5 mm. long, puberulent to almost villous; seeds dark brown, ca. 1.5 mm. long, with concentric rows of papillae.—Sandy hills and dunes; Coastal Strand, N. Coastal Scrub; San Francisco to Santa Cruz Co. March–June.

Ssp. **platyòta** (Wats.) Hitchc. & Maguire. [*S. p.* Wats. *S. verecunda* var. *p.* Jeps. *S. luisana* Wats. *S. occidentalis* var. *nancta* Jeps.] Lvs. 5–9 cm. long, 2–6 mm. wide; calyx more sparsely pubescent, quite glandular, usually greenish; petals white to greenish or rose, the appendages 1–2 mm. long; $n = 24$ (Kruckeberg, 1954).—Dry benches and slopes, 5000–11,000 ft.; Yellow Pine F. to Subalpine F.; mts. of s. Calif. to Sierra Nevada of Fresno and Tulare cos., somewhat lower in Lake Co., Mt. Diablo and in Coast Ranges from San Benito Co. s.; L. Calif. June–Aug. Eglandular plants have been designated var. *eglandulòsa* Hitchc. & Maguire and occur in San Bernardino and other s. mts.

Ssp. **Andersònii** (Clokey) Hitchc. & Maguire. [*S. A.* Clokey.] Lower stems scabrous-puberulent; infl. and calyx short-glandular-pubescent; corolla greenish-white, the appendages usually less than 1 mm. long.—Open slopes, 5600–9000 ft.; Sagebrush Scrub, Pinyon-Juniper Wd.; desert slopes of mts. from Mono and Inyo cos. to San Bernardino Co.; Nev., Utah. June–July.

23. **S. Hóokeri** Nutt. ex T. & G. Perennial with many stems, their bases slender, prostrate, usually buried, often rooting at nodes, the upright part 0.5–1.5 dm. long, grayish with subappressed hairs, not at all glandular; lvs. obovate to oblanceolate or spatulate, somewhat grayish, 4–7 cm. long, 8–14 mm. broad; infl. 1–9-fld.; pedicels 1–4 cm. long; calyx tubular, 13–20 mm. long, expanding in fr., the lobes 4–7 mm. long; petals white to pink or violet, the blades flabelliform, 10–12 mm. long, the middle 2 lobes larger than the lateral 2; appendages 2, linear, 2–3.5 mm. long; carpophore 2–5 mm. long, subglabrous; seeds ca. 2 mm. long, purplish-black, uniformly papillose; $n = 36$ (Kruckeberg, 1954).—Dry rocky ground, talus, serpentine, etc., below 4000 ft.; Mixed Evergreen F., Yellow Pine F.; Trinity, n. Humboldt, Siskiyou, and Del Norte cos.; Ore. May–June.

Ssp. **Bolánderi** (Gray) Abrams. [*S. B.* Gray.] Fls. longer, the petal-blades 20–25 mm. long, with linear lobes all of ca. the same width.—Rocky knolls and slopes, often on serpentine, below 5000 ft.; N. Oak Wd., Douglas-Fir F., Yellow Pine F.; Humboldt, Mendocino, and Trinity cos. May–June.

24. **S. nùda** (Wats.) Hitchc. & Maguire ssp. **insectívora** (Henders.) Hitchc. & Maguire. [*S. i.* Henders.] Stout perennial with weak and longer gland-tipped hairs as well

as shorter nonglandular ones; stems several, simple, 1.5–5 dm. high; lvs. essentially basal, or with some cauline, lanceolate to lance-elliptic, 4–20 cm. long, 1–3 cm. wide; infl. open; pedicels stout, 0.5–2 cm. long; calyx tubular, ca. 15 mm. long, not much enlarged in fr., the lobes ⅓–⅖ as long as tube; petals pink, the blades 5–8 mm. long, bilobed, the appendages linear, entire or bifid, 1–1.5 mm. long; carpophore glabrous or nearly so, 1–2 mm. long; seeds brown, papillose, 1.3–1.5 mm. long; *n* = 24 (Kruckeberg, 1954).— Subsaline habitats or not, 4000–6000 ft.; Sagebrush Scrub, Yellow Pine F.; e. slope of Sierra Nevada, Mono Co. to e. Ore. May–June. Typical *S. nuda*, with more glandular infl., occurs in Nev.

25. **S. Paríshii** Wats. Many-stemmed perennial with woody base; stems 2–4 dm. tall, glandular-pubescent, or eglandular below, the hairs mostly 1–3-celled; lvs. subelliptic, lanceolate or oblanceolate, mostly 3–6 cm. long, 4–12 mm. wide, the cauline 5–9 pairs and mostly less than 8 mm. wide; infl. 5–15-fld., rather compact; calyx tubular, constricted at base, 25–30 mm. long, very glandular, the lobes ½–⅓ the length of the tube; petals whitish-yellow, the blades mostly 7–8 mm. long, deeply cleft into several linear lobes, the appendages oblong, ca. 2 mm. long, cleft into several lobes; carpophore glabrous, ca. 3 mm. long; seeds brown, ca. 1.8 mm. long, the margins tessellate; *n* = 24 (Kruckeberg, 1954).—Dry gravelly slopes under trees, or rocky places, 6000–11,000 ft.; Yellow Pine F. to Subalpine F.; San Bernardino Mts. July–Aug.

Var. **víscida** Hitchc. & Maguire. Glandular-pubescent throughout, the hairs several-celled; larger cauline lvs. mostly over 10 mm. wide.—At 7000–10,700 ft., San Jacinto and Santa Rosa mts. July–Aug.

Var. **latifòlia** Hitchc. & Maguire. Plants mostly less than 2 dm. high, strigose and eglandular below; lvs. oblong-elliptic to obovate-oblanceolate, 2–3.5 cm. long, 8–15 mm. broad.—At 6800–9000 ft., San Gabriel Mts. July–Aug.

26. **S. Menzièsii** Hook. [*Anotites M.* Greene. *A. diffusa* Greene.] Perennial from slender rootstocks; stems 0.5–2 dm. long, usually decumbent, nonglandular below, with 3–7-celled hairs having walls with peglike thickenings on outer surface, glandular-pubescent above; lvs. many, 2–6 cm. long, 3–18 mm. wide, lanceolate to elliptic or oblanceolate; infl. leafy; calyx tubular-campanulate, 5–8 mm. long, the lobes ca. 2 mm. long; petals white, the blades 1.5–3 mm. long, cleft or lobed, the appendages 2, entire, very short; carpophore ca. 1.5 mm. long, glabrous; seeds ca. 0.8 mm. long, almost smooth, the sides obscurely reticulate; *n* = 12, 24 (Kruckeberg, 1954).—Moist woods and near small streams, 3000–9000 ft.; Yellow Pine F. to Lodgepole F.; Modoc Co.; to Canada and Rocky Mts. June–Aug. Our more common form, var. *viscòsa* (Greene) Hitchc. & Maguire [*Anotites v.* Greene, *A. costata* Greene] with lower internodes having gland-tipped thin-walled hairs, not sculptured on outer surface, ranges from Siskiyou and Lassen cos. n. with the sp.

Ssp. **Dórrii** (Kell.) Hitchc. & Maguire. [*S. D.* Kell. *Anotites D.* Greene.] Lower stems pubescent with shorter hairs, 1–3-celled, gland-tipped.—From 6000–10,800 ft.; mostly Red Fir F. to Subalpine F.; cent. Sierra Nevada, White Mts., San Bernardino Mts.; Nev. June–July.

11. Lýchnis L. CAMPION

Much like *Silene*, but with styles 5, rarely 4; caps. dehiscing by twice as many valves as styles. Ca. 35 spp. of n. temp. and arctic regions. (Ancient Greek name for a red-fld. sp., from *lychnos*, a flame.)

1. **L. Coronària** (L.) Desr. [*Agrostemma C.* L.] MULLEIN-PINK. White woolly perennial, erect, 4–8 dm. high; lvs. oblong or oval, 3–7 cm. long; fls. few, on stiff ascending peduncles; calyx ovoid-ellipsoid, the tube 10–12 mm. long, the teeth subulate, twisted, 5–6 mm. long; petals red-purple, 2–3 cm. long, the limb rounded and with 2 acute basal scales; caps. sessile; seeds black, ca. 1 mm. long, papillose; $2n = 24$ (Rohweder, 1939).— Occasional as escape from gardens; native of Eu. June–Aug.

2. **L. álba** Mill. [*Melandrium a.* Garcke.] Glandular-pubescent; fls. white to pink.— Occasional, cult. fields at low elevs.; n. Calif.; introd. from s. Eu.

12. Vaccària Medic. Cow-Herb. Cockle

Glabrous annuals with erect forked stems. Lvs. clasping, ovate to lanceolate. Cymes open, terminal, with slender pedicels. Calyx tubular at anthesis, inflated, ovoid and 5-angled in fr. Petals red or pink, surpassing calyx, without appendages. Stamens 10. Styles 2. Caps. 4-toothed. Seeds rounded, laterally attached. Ca. 3 spp. of Eurasia. (Latin, *vacca*, cow, because of use as fodder.)

     1. **V. segetàlis** (Neck.) Garcke ex Asch. [*Saponaria s.* Neck. *S. Vaccaria* L. *V. V.* Britton. *V. pyramidata* Medic. *V. vulgaris* Host.] Plants 3–10 dm. high; lvs. 3–7 cm. long, connate at base; calyx 10–12 mm. long, the angles green, terminal lobes 2–3 mm. long; petals reddish, crenulate, exserted 5–10 mm.; seeds blackish, 2 mm. broad and long, angled; $2n = 24$ (Löve, 1942).—Largely a weed of grainfields, below 5000 ft., widely scattered in Calif.; adv. from Eu. May–Aug.

13. Saponària L. Bouncing Bet

Annuals or perennials, erect or diffuse. Stems leafy. Lvs. mostly broad. Calyx-tube ovoid to cylindric, obscurely 15–25-nerved. Petals 5, long-clawed, entire or emarginate, crowned with an appendage or scale at base of blade. Stamens 10. Styles 2. Caps. 4-toothed. Seeds flat, reniform. Ca. 20 spp. of Eurasia and n. Afr. (Latin, *sapo*, soap, some spp. having saponaceous juice.)

     1. **S. officinàlis** L. Stout perennial 4–7 dm. tall, forming quite large patches; lvs. oval-lanceolate, 5–7 cm. long; fls. pink, commonly double; calyx 15–22 mm. long, the tube cylindric, the lobes lanceolate, 2 mm. long; petals 3.5–4 cm. long; seeds black, 1.6–1.8 mm. in diam., regularly pitted; $2n = 28$ (Favarger, 1946).—Persisting from old gardens and sparingly natur., especially in cool damp places; native of Eu. June–Sept.

14. Diànthus L. Pink

Annual or perennial herbs. Stems rather stiff. Lvs. usually narrow. Fls. solitary or more commonly in rather crowded terminal cymes. Calyx cylindrical, nerved or striate, 5-lobed, or -toothed, subtended by 2 or more imbricate bractlets. Petals long-clawed, toothed. Stamens 10. Styles 2. Caps. 4-valved at apex. Seeds compressed, laterally attached. Ca. 200 spp. of Old World, many ornamental, as the Garden Pinks and Carnations. (Greek *Dios*, Jupiter, and *anthos*, fl., the fl. of Jove.)

     1. **D. barbàtus** L. Sweet William. Tufted perennial with simple glabrous stems 3–7 dm. high; lvs. lanceolate; cyme corymbiform; the fls. crowded; calyx cylindric, 15–18 mm. long; petals reddish to white, with dark markings; $2n = 30$ (Favarger, 1946).—A garden plant sometimes escaping in n. Calif.; introd. from Eu. Summer.

15. Tùnica Scop.

Annual or perennial herbs. Fls. in small clusters surrounded by an invol. of 2–6 scale-like bracts which conceal the sessile calyces within. Calyx tubular or campanulate. Petals without appendages. Stamens 10. Styles 2. Seeds elliptic, concave. Ca. 20 spp., largely Medit.

     1. **T. prolífera** (L.) Scop. [*Dianthus p.* L.] Slender glabrous annual 3–5 dm. high, simple or branched from below; lvs. linear to oblanceolate, mostly toward base of the plant, 2–2.5 cm. long; peduncles long, subscapose, with terminal invol. of 2–3 pairs of broad scarious bracts; calyx 10–13 mm. long; petals red, entire or crenulate, scarcely exserted; seeds 1.5 mm. broad, minutely papillose.—Occasional as a weed, reported from Butte, Nevada, and Yuba cos. May.

16. Velèzia L.

Diffuse annuals. Lvs. subulate. Fls. subsessile, axillary and solitary or crowded at ends of branches. Calyx tubular, 5-toothed, 5–15-ribbed. Petals long-clawed, the blades small, entire or 4-toothed, with small ciliate crests. Stamens 5. Styles 2. Caps. linear, 4-valved at apex. Seeds round or ovoid. Four Medit. spp. (Cristobal *Velez*, friend of the botanist Loefling.)

1. **V. rígida** L. Branched from base, the branches 1–4 dm. long, glandular-puberulent; lvs. 10–15 mm. long, ciliate near base; calyx 12–14 mm. long; petals purple-tipped; $2n =$ 28 (Favarger, 1946).—Established on dry hills, Humboldt, Stanislaus, Eldorado, and Tuolumne cos.; natur. from Eu. May–June.

17. Achyronýchia T. & G.

Low glabrous annuals. Lvs. spatulate, those in each pair unequal; stipules hyaline. Fls. in dense axillary cymose clusters. Calyx 5-lobed, scarious. Petals 0. Stamens 10–15, only 1–5 fertile. Style 2-cleft. Fr. a utricle, included in calyx. Seeds minute, black, round-reniform. Two spp. of sw. U.S. and Mex. (Greek, *achuron*, chaff, and *onyx, onychos*, fingernail, referring to the silvery chaffy calyx.)

1. **A. Coòperi** T. & G. Stems several, prostrate, 3–15 cm. long; lvs. spatulate, 5–15 mm. long; fls. 2 mm. long, white, green at base; calyx-lobes oval, white-scarious above the greenish base; seeds shining.—Common on sandy flats and in washes, below 3000 ft.; Creosote Bush Scrub; Colo. and e. Mojave deserts; Ariz., L. Calif. Jan.–May.

18. Scopulóphila Jones

Low perennial, arising from a dense woody root-crown. Stems glabrous, erect, few-branched. Lvs. oblance-linear; stipules scarious, lacerate. Fls. 1–3, sessile in axillary clusters. Sepals 5, united at very base, hyaline with cent. elongate green spot. Petals 0. Stamens 10, the 5 fertile ones alternating with 5 longer lanceolate staminodia. Style 3-cleft. Seed 1. Monotypic. (Greek, *skopla*, a high place, and *philos*, fond of.)

1. **S. Rixfórdii** (Bdg.) M. & J. [*Achyronychia R.* Bdg. *S. nitrophiloides* Jones. *Eremolithia R.* Jeps.] Root-crown woolly; plant pallid; stems several, 6–18 cm. high; lvs. 5–15 mm. long; calyx-tube scarcely 0.5 mm. long, the lobes ca. 1.5 mm. long; staminodia ca. the same.—Dry rocky places, probably only on limestone, 3600–7500 ft.; Creosote Bush Scrub, Joshua Tree Wd.; Owens V., Death V. region; sw. Nev. April–July.

19. Paronýchia Mill.

Annual or perennial herbs, low, tufted. Lvs. usually linear, with scarious stipules. Fls. small, clustered in dense or loose cymes, or in forks of stem. Calyx 5-parted, the lobes awn-tipped. Petals 0 or obsolete. Stamens 2–5. Styles 2-cleft or stigmas 2 and sessile. Fr. a utricle, enclosed in or barely exceeding calyx. Seed 1. Ca. 40 spp. of temp. and warmer parts of Old and New World. (Greek, name for *whitlow* or *felon*, a disease of the nails, some plants with whitish scaly parts once considered a cure.)

(Core, E. L. The N. Am. spp. of Paronychia. Am. Midl. Nat. 26: 369–397, 1941.)

1. **P. franciscàna** Eastw. Perennial from woody taproot; stems glabrous, wiry, prostrate, forming mats 4 dm. across; lower stems with knotted leafless nodes and ragged persistent stipules; lvs. linear-oblong, 5–8 mm. long, 1–2 mm. wide, bristle-tipped, short-hairy; stipules silvery-scarious; fls. few, in the axils; sepals 1–2 mm. long, pubescent at top, spine-tipped, brown or purplish to green; stamens minute; styles 2; utricle papillate. —Shallow rocky soil on open hills; N. Coastal Scrub, Coastal Prairie; San Francisco Co. to Sonoma Co.; supposedly introd. from Chile. April–June.

20. Herniària L.

Annual or perennial herbs, with much branched prostrate or spreading stems. Lvs. opposite to alternate, small, entire; stipules minute, scarious. Fls. minute, sessile in crowded axillary clusters. Calyx 4–5-parted, the segms. unequal. Petals 0. Stamens 2–5. Style minute, 2-cleft. Fr. a utricle, membranaceous, included. Seed 1. Ca. 20 spp. of Eurasia and N. Afr. (Latin, *hernia*, rupture, one of the spp. being a supposed cure.)

1. **H. cinèrea** DC. [*Paronychia pusilla* Greene.] Forming mats 5–20 cm. across, annual, leafy, hispidulous throughout; lvs. oblong-oblanceolate, 4–10 mm. long; fls. in all

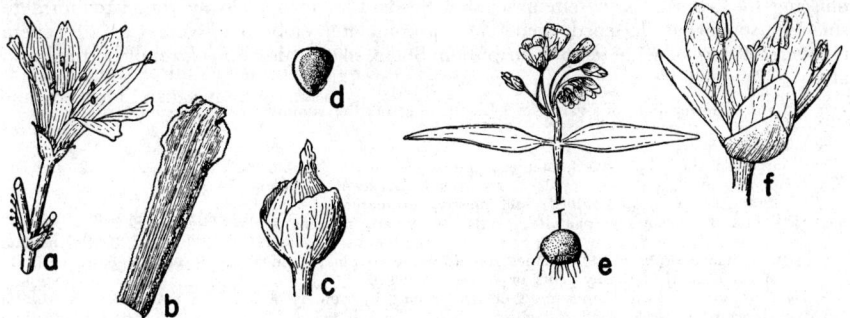

Fig. 34. PORTULACACEAE. *Lewisia Cotyledon* var. *Howellii: a*, fl., × 1, sepals 2, gland-dentate, petals many; *b*, basal fimbriate-toothed lf. *L. nevadensis: c*, old fl., × 1½, sepals broad, petals withered; *d*, seed, × 5. *Claytonia lanceolata: e*, habit, × ½; *f*, fl. × 2½, petals 5, stamens 5, stigmas 3.

the axils, 1 mm. long, hispidulous.—Sparingly natur. in s. Calif., more frequent below 5000 ft. in foothills bordering Cent. V., to Siskiyou and Humboldt cos.; s. Eu. April–Sept.

21. Scleránthus L. KNAWEL.

Low annual or perennial herbs, glabrous or pubescent, the stems rigid, forking. Lvs. connate at base, subulate, pungent. Stipules none. Fls. small, greenish, clustered in axils or terminal and cymose. Calyx 4–5-toothed or -lobed. Petals 0. Stamens 1–10. Styles 2, distinct. Utricle included; seed 1, lenticular. Ca. 10 spp., of Old World. (Greek, *scleros*, hard, and *anthos*, fl., from the hardened calyx-tube.)

1. **S. ánnuus** L. Much-branched, spreading, somewhat pubescent, 2–10 cm. high; lvs. 5–8 mm. long; fls. sessile in the forks, 3–4 mm. long; calyx-lobes scarcely margined; $2n = 22$ (Rohweder, 1939, Ehrenberg, 1945).—Sparingly natur. in dry places, orchards, roadsides, etc., below 4000 ft., San Diego Co. to Shasta and Siskiyou cos.; native of Eu. March–June.

22. Cardionèma DC.

Low tufted perennial herbs, with numerous short branched ± woolly stems, the upper internodes short and usually covered with persistent papery hyaline commonly bifid stipules. Lvs. subulate, pungent, with fascicles of secondary lvs. and of fls. in the main axils. Calyx 5-parted, the segms. unequal, hooded, woolly about base, with erect lower portion and with terminal portion (especially of the 3 outer) spreading and ending in a spine. Petals minute, scalelike. Stamens 3–5, inserted at base of calyx-segms. Style short, bifid. Fr. a utricle, inclosed in a rigid persistent calyx. Several spp. of w. N. and S. Am. (Greek, *cardia*, heart, and *nema*, thread, because of the obcordate stamens.)

1. **C. ramosíssimum** (Weinm.) Nels. & Macbr. [*Loeflingia r.* Weinm. *Pentacaena r.* H. & A.] Stems 5–20 cm. long, prostrate; lvs. 8–13 mm. long; stipules 4–6 mm. long; calyx 4–6 mm. long, pubescent below the spines.—Sandy places; Coastal Strand, N. Coastal Scrub, Coastal Sage Scrub; near the coast, Wash. to L. Calif., w. Mex., Chile. April–Aug.

36. Portulacàceae. PURSLANE FAMILY (Fig. 34)

Annual to perennial herbs, ± succulent. Lvs. entire, alternate or opposite, or mostly basal. Fls. perfect, nearly or quite regular. Sepals 2, rarely more. Petals 3–16, commonly 5, sometimes 0, generally hypogynous, ephemeral. Stamens few–many, sometimes only 1, opposite the petals if of the same number and usually adnate to their bases. Ovary commonly superior, 1-locular with few to many ovules on free cent. or basal placenta; styles 2–8, united below or distinct, stigmatic along the inside. Caps. mostly 1-locular, dehiscing by 2–3 valves, or circumscissile. Seeds 1 to many, mostly round-reniform to orbicular, somewhat flattened, sometimes strophiolate; embryo curved. Ca. 20 genera and over 200 spp. of wide temp. distribution. Some, like garden *Portulaca*, of horticultural value.

Calyx adnate to lower part of ovary, its lobes coming off the summit of the circumscissile caps
                                                1. *Portulaca*
Calyx and ovary free.
  Caps. circumscissile; sepals 2–8; stamens 5–many .... 2. *Lewisia*
  Caps. opening by 2–3 valves; sepals 2; stamens 1–5, except in *Calandrinia*.
    Style-branches 3; caps. 3-valved; infl. mostly not markedly secund.
      Fls. in leafy racemes or panicles; petals mostly red; stamens 5–many; seeds 18–25
                                               3. *Calandrinia*
      Fls. in naked or bracteate racemes; petals white to pink; stamens 1–5; seeds 1–6.
        Stems from thick fleshy roots or corms; ovules 6 .... 4. *Claytonia*
        Stems from mostly fibrous roots or reproducing by runners or bulblets; ovules 3 . 5. *Montia*
    Style 1; caps. 2-valved; infl. secund .... 6. *Calyptridium*

1. Portuláca L.

Fleshy herbs, with mostly scattered alternate or partly opposite lvs. Fls. terminal, sessile, mostly opening only in sunshine. Sepals 2, united below, the tube cohering with the ovary. Petals 5, rarely 4 or 6–many, inserted on the calyx. Stamens 7–20. Style deeply 3–8-parted. Caps. 1-locular, globular, many-seeded, circumscissile near the middle. Seeds many, round-reniform. Ca. 100 spp. of warmer regions. (Old Latin name, of uncertain application.)

Petals yellow; lvs. flat; plant nearly or quite glabrous. Introd. weed .... 1. *P. oleracea*
Petals pink or purplish; lvs. subterete; plant conspicuously hairy in axils. Desert native . 2. *P. mundula*

1. **P. olerácea** L. PURSLANE. Prostrate annual, the stems 1–2 dm. long; lvs. obovate or cuneate, 5–25 mm. long; fls. sessile, 3–6 mm. broad; sepals keeled, acute, 3–4 mm. long; petals pale yellow; style deeply 5–6-parted; caps. 4–8 mm. long; seeds black, finely rugulose-pitted, ca. 0.6 mm. long; $2n = 54$ (Steiner, 1944).—Widely scattered weed in gardens, waste places, etc.; natur. from Eu. May–Sept.

2. **P. múndula** Jtn. Annual, branched from base, the stems ascending, 5–10 cm. long, copiously long-hairy in the axils; lvs. linear-subulate, 6–12 mm. long; petals pink or purplish, 3–4 mm. long; caps. 4–5 mm. long; seeds granular, ca. 0.3 mm. long—Sandy wash, at 3700 ft.; Joshua Tree Wd.; V. Wells, e. Mojave Desert; Ariz. to Mo. and Mex. Sept.

2. Lewísia Pursh

Fleshy perennials from thick fleshy root and short caudex, or from a globose corm. Lvs. largely in a basal rosette, entire, with wide mostly hyaline petiole-base; cauline lvs. few, often bracteate. Infl. bracteate, sometimes disjointing in age. Sepals 2–6(–8), persistent. Petals 4–18, often unequal. Stamens 5–many. Styles 3–8, united at base. Caps. circumscissile near base, globose or ovoid. Seeds 6–many, dark, shining, smooth or finely tuberculate. Ca. 18 spp. of w. N. Am. (Named for Capt. Meriwether *Lewis,* of the Lewis and Clark expedition of 1806/07.)

(Rydberg, P. A. N. Am. Fl. 21: 321–328, 1932.)
Plant without basal lvs., but with 2–5 linear cauline lvs., small, from globose tuber .. 1. *L. triphylla*
Plant with several to many basal lvs.
  Stems 1–3 dm. high, much surpassing the basal rosette (except sometimes in *oppositifolia*).
    Stem-lvs. similar to basal, but smaller; infl. 2–6-fld.; sepals glandless .... 6. *L. oppositifolia*
    Stem-lvs. reduced to bracts; infl. mostly many-fld.; sepals gland-toothed.
      Petals 5–8 mm. long; pedicels slender, mostly 5–15 mm. long.
        Lvs. subterete; petals 5–6.5 mm. long .... 2. *L. Leana*
        Lvs. distinctly flattened, oblanceolate to spatulate; petals 7–8 mm. long.
          Lf.-blades entire, acute. Fresno and Mariposa cos. .... 3. *L. Congdonii*
          Lf.-blades dentate, obtuse to retuse. Plumas and Nevada cos. .... 4. *L. Cantelowii*
      Petals 12–20 mm. long; pedicels stout, mostly 3–5 mm. long .... 5. *L. Cotyledon*
  Stems less than 1 dm. high, scarcely if at all surpassing the basal lvs.
    Bracts remote from sepals and not resembling them.
      Sepals herbaceous or hyaline, not petaloid; infl. not disjointing in fr.
        The sepals glandular-toothed and conspicuously veined in age .... 7. *L. pygmaea*
        The sepals entire or with a few nonglandular teeth, not so conspicuously veined.
          Petals white, 9–18 mm. long; lvs. mostly 3–8 cm. long .... 8. *L. nevadensis*
          Petals pink, 4–5 mm. long; lvs. 2.5–3.5 cm. long .... 9. *L. sierrae*

Sepals petaloid, becoming scarious in age; infl. readily disjointing in fr.
 The sepals 2, only 7–8 mm. long; bracts 2–3, ovate .... 12. *L. disepala*
 The sepals several, 10–25 mm. long; bracts 5–8, subulate .... 13. *L. rediviva*
Bracts just below sepals and like them.
 Sepals entire, 6–8 mm. long. S. Calif. .... 10. *L. brachycalyx*
 Sepals gland-dentate on margin, 8–15 mm. long. Sierra Nevada .... 11. *L. Kelloggii*

1. **L. triphýlla** (Wats.) Rob. [*Claytonia t.* Wats. *Oreobroma t.* Howell. *Erocallis t.* Rydb.] Perennial from a globose corm 3–6 mm. in diam.; stems slender, 1–few, 3–10 cm. high, but ca. half underground; basal lvs. none at anthesis; cauline lvs. 2–5, whorled, linear, 2.5–4.5 cm. long; infl. subumbellate, bracteate, mostly 3–15-fld.; pedicels slender, 5–10 mm. long; sepals entire, oval, 3–4 mm. long; petals 5–8, white or pink, 4–5 mm. long; caps. ovoid, 3–4 mm. long; seeds black, shining, 10–25, ca. 1 mm. long. —Damp gravelly places, 4900–10,700 ft.; Red Fir F. to Subalpine F.; from Tulare Co. through the Sierra Nevada and from Glenn Co. to Siskiyou Co.; to Wash., Rocky Mts. June–Aug.

2. **L. Leàna** (Porter) Rob. in Gray. [*Calandrinia L.* Porter. *Oreobroma L.* Howell.] Perennial from a thick fleshy caudex and branched roots; stems 1–4, scapiform, 1–2 dm. high; basal lvs. many, linear, subterete, glaucous, 2–6 cm. long; stem-lvs. reduced to laciniate gland-toothed bracts in lower part of panicle; infl. paniculate, many-fld.; pedicels slender, 3–15 mm. long; sepals rounded, ca. 2 mm. long, lacerate-dentate, with dark glands on the teeth; petals 6–8, red or white with red veins, 5–6.5 mm. long; caps. ovoid, 4–5 mm. high; seeds mostly 3, dark brown, shining, 2–2.5 mm. long.— Cliffs and rocks; Subalpine F.; 9000–10,000 ft., Fresno Co. and 6000–7500 ft., Yolla Bolly Mts. through Trinity and Siskiyou cos.; adjacent Ore. July–Aug.

3. **L. Congdònii** (Rydb.) J. T. Howell. [*Oreobroma C.* Rydb. *L. columbiana* ssp. *C.* Ferris.] Thick-rooted perennial with short caudex; lvs. many, mostly basal, oblanceolate, acute, the blades flat, entire, 5–10 cm. long, on petioles ca. as long; stem-lvs. gradually reduced upward into glandular-dentate bracts; stems scapiform, 2–4 dm. high, ending in widely branched rather few-fld. panicles; pedicels slender, ca. 1 cm. long; sepals roundish, 2 mm. long; petals rose, ca. 8 mm. long; caps. ovoid, 4 mm. long; seeds black, shining, 2 mm. long.—Rocky places, 6000–9000 ft.; Red Fir F.; Fresno and Mariposa cos. May–June.

4. **L. Cantelòwii** J. T. Howell. Perennial with subglobose caudex; basal lvs. many, plane, oblanceolate, scarcely petioled, 2–4.5 cm. long, sharply dentate, obtuse or sub-truncate at apex to retuse; stems scapiform, 1.5–4 dm. high, the stem-lvs. reduced to lanceolate denticulate bracts, the upper gland-margined; panicles open, many-fld.; pedicels slender, 5–10 mm. long; sepals broadly elliptic, 3 mm. long, gland-dentate; petals light pink with deep pink veins, ca. 7 mm. long.—Wet cliffs, ca. 3000 ft.; Yellow Pine F.; Plumas and Nevada cos. May–June.

5. **L. Cotylèdon** (Wats.) Rob. in Gray. [*Calandrinia C.* Wats. *Oreobroma C.* Howell. *L. Finchae* Purdy.] From a short thick caudex and taproot; stems several, rather thick, 1–3 dm. high, scapiform; basal lvs. many, fleshy, broadly spatulate to oblanceolate, petioled, 4–10 cm. long, entire to undulate, but quite plane at margin; stem lvs. 2–4-toothed, bracteate; infl. short-paniculate, many-fld.; pedicels stout, mostly 3–5 mm. long; sepals round, ca. 4 mm. long, with gland-tipped teeth; petals 8–10, white with red tinge or stripe, obovate to spatulate, 12–14 mm. long; caps. ovoid, 5–6 mm. high; seeds black, shining, ca. 1.5 mm. long.—Rocky places, 4000–7500 ft.; Yellow Pine F., Red Fir F.; Siskiyou and Trinity cos.; s. Ore. June–July.

Var. **Howéllii** (Wats.) Jeps. [*Calandrinia H.* Wats. *Oreobroma H.* Rob.] Lf.-margins strongly crisped, sometimes fimbriate-toothed; petals ca. 15 mm. long.—At 500–1000 ft.; N. Oak Wd.; Humboldt Co. to s. Ore. April–June.

Var. **Héckneri** (Mort.) Munz. [*Oreobroma H.* Mort. *L. H.* J. T. Howell.] Lvs. plane, strongly toothed; petals 16–20 mm. long.—At 4000–7000 ft.; Yellow Pine F. to Sub-alpine F.; Trinity Co. May–July.

6. **L. oppositifòlia** (Wats.) Rob. in Gray. [*Calandrinia o.* Wats. *Oreobroma o.* Howell.] Caudex short and thick, the root often branched; stems 1–4, slender, 0.5–1.5 dm. long; basal lvs. few, linear-spatulate, 5–10 cm. long, petioled; stem-lvs. 1–2 pairs near base of plant, similar but smaller; infl. subumbellate, 2–6-fld.; pedicels 2–4 cm. long; sepals roundish, 4–7(–10) mm. long, fimbriate-dentate, glandless; petals white or

pinkish, oblanceolate, 10–15 mm. long; caps. 5–6 mm. long; seeds smooth, shining.
—Moist rocky places, 1200–4000 ft.; Yellow Pine F.; Del Norte Co. and sw. Ore.
April–May.

7. **L. pygmaèa** (Gray) Rob. in Gray. [*Talinum p.* Gray. *Calandrinia Grayi* Britton.
Oreobroma p. Howell. *O. G.* Rydb.] Root fleshy, fusiform; caudex short; stems several
to many, 2–6 cm. long, partly underground; lvs. many or fewer, basal, linear to
narrowly oblanceolate, 3–8 cm. long; bracts hyaline, lanceolate, 6–10 mm. long, opposite,
and near middle of stem; fls. 1–3 per stem; sepals rounded, 4–5 mm. long, glandular-
dentate (glands light in color, not stipitate); petals white to pink, 7–10 mm. long; caps.
4–6 mm. long; seeds 15–20, dark, shining, 1–1.2 mm. long.—Damp gravel, 9000–12,250
ft.; mostly Alpine Fell-fields; San Bernardino Mts., Sierra Nevada from Tulare Co. n.;
to Wash., Mont., New Mex. July–Sept.

Ssp. **glandulòsa** (Rydb.) Ferris. [*Oreobroma g.* Rydb.] Marginal glands of sepals
stipitate, dark.—Mostly 11,000–12,750 ft., Tuolumne Co. to Tulare Co.

Ssp. **longipétala** (Piper) Ferris. [*Oreobroma l.* Piper.] Petals rose, 12–18 mm. long.—
Sierra Nevada, w. of Truckee.

8. **L. nevadénsis** (Gray) Rob. in Gray. [*Calandrinia n.* Gray. *Oreobroma n.* Howell.
L. bernardina A. Davids. *L. pygmaea* var. *n.* Fosb.] Habit and foliage much like in *L.
pygmaea;* sepals broadly ovate, acute, 5–10 mm. long, the margin quite entire or if
slightly dentate, then not gland-toothed; petals 9–15(–18) mm. long, white; seeds ca.
1.3 mm. long.—Wet banks and meadows, 4500–12,000 ft.; Yellow Pine F. to Subalpine F.; San Bernardino Mts., Mt. Pinos, Sierra Nevada to Modoc Co., N. Coast Ranges
from Glenn Co. n.; to Wash., Rocky Mts. May–July.

9. **L. siérrae** Ferris. Small, tufted, with fusiform root and short caudex; stems 1–2.5
cm. long, 1–3-fld.; lvs. basal, several, linear, 2.5–3.5 cm. long; bracts 2, ovate, opposite;
sepals rounded, obscurely repand-dentate, 2.5–3 mm. long; petals pink, 4–4.8 mm.
long; caps. 2.5–3 mm. high; seeds shining, faintly rugulose, 0.5 mm. broad.—Moist
flats and banks, 9100–13,500 ft.; Subalpine F., Alpine Fell-fields; Tuolumne Co. to
Tulare Co. July–Aug.

10. **L. brachycàlyx** Engelm. ex Gray. [*Oreobroma b.* Howell. *L. brachycarpa* Wats.]
Caudex short, thick; stems many, 2–6 cm. long; lvs. many, fleshy, 3–6 cm. long,
oblanceolate; fls. usually solitary, sessile in pair of calyxlike bracts; sepals ovate, acute,
entire, 6–8 mm. long; petals white, 12–18(–25) mm. long; caps. 8–9 mm. long; seeds
many, black, shining, 1.5 mm. long.—Wet meadows, 4500–7500 ft.; Yellow Pine F.;
San Bernardino and Cuyamaca mts.; to Utah, Ariz., New Mex. May–June.

11. **L. Kellóggii** K. Bdg. [*Oreobroma K.* Rydb. *L. yosemitana* Jeps.] Caudex short;
root stout; stems 1–5 cm. long, 1-fld., usually shorter than lvs.; basal lvs. many,
spatulate, 3–6 cm. long; fls. sessile between pair of sepallike bracts; sepals lance-ovate,
acute, 8–10 mm. long, gland-dentate on margin; petals creamy-white, 8–15 mm. long;
caps. ovoid, ca. 8 mm. long; seeds black, tuberculate, 2 mm. long.—Sandy places on
ridges, 4500–7700 ft.; Yellow Pine F., Red Fir F.; Plumas Co. to Mariposa Co. June–
July.

12. **L. disèpala** Rydb. [*L. rediviva* var. *yosemitana* K. Bdg.] Caudex short; root
branched, fleshy; stems 2–3 cm. long; lvs. basal, fleshy, clavate, 8–20 mm. long; bracts
2–3, ovate, scarious, 2–3 mm. long; fls. solitary, the pedicels 1–2 mm. long, readily
disjointing above the bracts; sepals 2, obovate, 7–8 mm. long; petals pinkish, 13–18 mm.
long; caps. ellipsoid; seeds black, round-reniform, 1.2–1.5 mm. long.—Rocky places, ca.
6500–8500 ft.; Red Fir F.; summits bordering Yosemite V. May–June.

13. **L. redivìva** Pursh. [*L. alba* Kell.] Bitterroot. Caudex short; taproot fleshy, much
branched; stems many, 1–3 cm. long, 1-fld.; lvs. basal, linear, 2–5 cm. long, obtuse,
fleshy; bracts 5–8, whorled, subulate, scarious, at base of pedicel; pedicel readily disjointing in age, mostly 1–1.5 cm. long; sepals several, entire, petaloid, rose to white,
oblong-oval, rounded at apex, mostly 12–25 mm. long; petals rose or white, many, 2–2.5
cm. long; caps. ellipsoid, 5–6 mm. long; seeds dark, shining, round-reniform, 2–2.5 mm.
long.—Loose gravelly slopes and rocky places, mostly 2500–6000 ft.; Yellow Pine F.,
Sage Brush Scrub, Foothill Wd., Mixed Evergreen F.; Santa Lucia Mts. to Siskiyou
Co., Mono Co. to Modoc Co.; to B.C., Rocky Mts. March–June. Intergrading with:

Var. **mìnor** (Rydb.) Munz. [*L. m.* Rydb.] Lvs. clavate to narrow-oblanceolate; pedicels
mostly 3–8 mm. long; sepals mostly ca. 1 cm. long; petals ca. 1.5 cm. long.—From

6500–9000 ft.; Pinyon-Juniper Wd., Yellow Pine F.; San Bernardino Mts. to Mt. Pinos, Panamint Mts., White Mts.; w. Nev. May–June.

3. Calandrínia HBK.

Annual or perennial herbs. Lvs. alternate, entire. Fls. red, rarely white, ephemeral, in rather leafy racemes or panicles. Sepals 2, persistent. Petals 3–7, usually 5. Stamens 3–14, shorter than the petals, seldom of the same no. as the petals. Style-branches 3. Caps. 3-valved from apex. Seeds somewhat flattened, rounded, many, very dark, usually minutely pitted but shining. Ca. 75 spp. of W. Am. and Australia. (Named for J. L. *Calandrini*, Swiss botanist of the 18th century.)

Fls. ± racemose; seeds with a strophiole; petals usually red. Cismontane.
  Plants green; seeds black and shining.
    Caps. scarcely exceeding calyx; lvs. well distributed; pedicels ascending to erect ..... 1. *C. ciliata*
    Caps. ca. twice as long as the calyx; lvs. mostly basal; pedicels often reflexed in fr. ... 2. *C. Breweri*
  Plants glaucous; seeds dull and gray .... 3. *C. maritima*
Fls. in panicle of umbels; seeds naked at hilum; petals white. Desert .... 4. *C. ambigua*

1. **C. ciliàta** (R. & P.) DC. var. **Menzièsii** (Hook.) Macbr. [*Talinum M.* Hook. *C. caulescens* var. *M.* Gray. *C. elegans* Spach. *C. heterophylla, filifolia,* and *muricata* Rydb.] RED MAIDS. Annual, simple or more usually with several spreading stems from base, these 1–4 dm. long, subglabrous; lvs. well distributed, petioled, narrowly oblanceolate to linear, 2–8 cm. long, somewhat fleshy; fls. in leafy racemes, the pedicels suberect, 0.4–2 cm. long; sepals ovate, short-acuminate, glabrous or hispidulous on margin and midrib, 3–8 mm. long; petals 5, rose-red, 4–10(–14) mm. long; caps. ovoid, pointed, 4–7 mm. long; seeds many, black, shining, minutely tuberculate, 0.8–1 mm. in diam.; $2n = 24$ (Sugiura, 1936).—Common, especially in open grassy places and cult. fields, below 6000 ft.; mostly V. Grassland, Foothill Wd., to a lesser extent in other Communities; cismontane Calif., occasional at desert edge; to B.C., L. Calif., Ariz., Son. Mostly Feb.–May. Exceedingly variable.

2. **C. Brèweri** Wats. [*C. Menziesii* var. *macrocarpa* Gray.] Annual, glabrous, the stems several from base, 1–4 dm. long, prostrate or nearly so; lvs. lance-ovate or spatulate, the basal 2–8 cm. long, petioled, the cauline reduced and nearly or quite sessile; fls. in elongate leafy-bracted racemes, not numerous; pedicels 7–20 mm. long, generally reflexed in fr.; sepals glabrous or sometimes ciliate, 5–6 mm. in fr., scarious-margined toward base; petals rose-red, 4–5 mm. long; caps. 10–12 mm. long, well-exserted; seeds many, black, shining, minutely tuberculate, ca. 1 mm. long.—Mostly on burns and in disturbed places, gravelly slopes, below 3500 ft.; Chaparral; Sonoma and Mariposa cos. to L. Calif. March–June.

3. **C. marítima** Nutt. [*Claytonia m.* Kuntze.] Glaucous annuals, branched at base, the stems spreading or ascending, 7–25 cm. long; lvs. mostly at base, spatulate-obovate, 2–6 cm. long, petioled, fleshy; fls. in a lax naked terminal panicle of cymes; pedicels 5–15 mm. long; sepals round-ovate, dark-veined, short-acute, 4–5 mm. long; petals red, ca. 4 mm. long; caps. ovoid, 5–6 mm. long; seeds dull, grayish, minutely roughened, ca. 0.6 mm. long.—Sandy places, sea bluffs; Coastal Sage Scrub; Santa Barbara Co. to L. Calif. March–May.

4. **C. ambígua** (Wats.) Howell. [*Claytonia a.* Wats. *Calandrinia sesuvioides* Gray.] Very fleshy annual with several stems from base, erect or spreading, 3–15 cm. long; lvs. linear-spatulate, 2–5 cm. long, well distributed, obtuse, semiterete, channeled beneath; fls. in rather compact lateral and terminal umbellate clusters; pedicels 2–7 mm. long; sepals ovate, white-scarious at margin, 3–5 mm. long, obtuse; petals white, 3–5, obovate, ca. as long as or shorter than sepals; caps. equal to calyx; seeds shining, black, not strophiolate, 0.6–0.8 mm. long.—Somewhat alkaline washes and slopes, below 2000 ft.; Creosote Bush Scrub; Colo. Desert and drainage of Mojave and Amargosa rivers; Ariz. Jan.–April.

4. Claytònia L. SPRING BEAUTY

Glabrous perennial herbs with deep-seated corms or fleshy roots. Basal lvs. 1 to several; cauline lvs. 2, opposite, rarely 3. Infl. terminal, racemose, the lowest pedicel subtended by a small bract; bractlets 0. Sepals 2, herbaceous, persistent. Petals mostly 5, rose to white or yellowish. Stamens 5, adnate to the petals. Ovules 6; styles 3, united to near apex. Caps. ovoid, 3-valved, the valve-margins involute in age. Seeds 2–6, dark, shining, lenticular, round-reniform. Ca. 15 spp. of N. Am. and the Arctic. (Named for John *Clayton*, 18th-century Am. botanist.)

(Rydberg, P. A.   N. Am. Fl. 21: 296–302, 1932.)
Plant with a globose corm; basal lvs. 1 or 0.
  Cauline lvs. distinctly petioled, obtuse or rounded . . . . 1. *C. umbellata*
  Cauline lvs. sessile or nearly so, acute or acuminate . . . . 2. *C. lanceolata*
Plant with rootstocks or fleshy taproot; basal lvs. several to many.
  The plant with a fleshy taproot and numerous basal lvs. . . . . 3. *C. bellidifolia*
  The plant with fleshy rootstocks and rather few basal lvs. . . . . 4. *C. nevadensis*

1. **C. umbellàta** Wats. [*C. obovata* Rydb.] Corm globose, 1–2.5 cm. in diam.; stems largely subterranean, 5–10 cm. long; basal lf. thick, ovate to obovate, the blade 1–1.5 cm. long, the petiole 6–8 cm. long; cauline lvs. 3-nerved, the blade 1.5–2.5 cm. long, with a petiole shorter than to ca. as long as blade; fls. 3–5, subumbellate; pedicels 5–15 mm. long; sepals 4–5 mm. long, obtuse; petals white to rose, oval to obovate, 6–7 mm. long; seeds dark, shining.—Exposed slopes, 5000–11,400 ft.; Red Fir F., Lodgepole F.; Mono, El Dorado and Tehama cos. to s. Ore., w. Nev. June–Aug.

2. **C. lanceolàta** Pursh. Corm globose, 1–2 cm. in diam.; stems 1–several, 6–15 cm. long; basal lvs. 1–2 (0), petioled, oblanceolate, 5–8 cm. long, acute; cauline pair sessile or nearly so, narrowly to broadly lanceolate to ovate, 3–7 cm. long, mostly 6–20 mm. wide; fls. few to 15, the pedicels 1–2.5 cm. long, recurved in fr.; sepals ovate, 3.5–5 mm. long, entire to somewhat repand; petals usually pink, 8-12 mm. long, retuse or emarginate at apex; caps. ovoid, 4 mm. long; seeds shining, black, smooth, ca. 2 mm. long.—Moist woods and along streams, 4500–8400 ft.; Red Fir F., Lodgepole F.; Sierra Nevada from Eldorado Co. to Modoc and Humboldt cos.; to B.C., Rocky Mts. May–July.

Var. **sessilifòlia** (Torr.) A. Nels. [*C. caroliniana* var. *s.* Torr. *C. s.* Henshaw.] Cauline lvs. narrow, 2–4 mm. wide, long-acuminate, mostly surpassing the infl.; petals 6–8 mm. long.—At same elevs., Tuolumne Co. to s. Ore. May–July.

Var. **Peirsònii** M. & J. Cauline lvs. short-petiolate, widest below the middle, 7–12 mm. wide; fls. subumbellate.—Dry ridges; Lodgepole F.; San Gabriel Mts., w. San Bernardino Co. May–June.

3. **C. bellidifòlia** Rydb. Perennial with large stout taproot and thick caudex; basal lvs. many, spatulate to broadly oblanceolate, 3–8 cm. long, petioled, rounded or acutish at apex, faintly pinnately nerved; cauline lvs. near infl. subopposite, narrow, bractlike; fls. 2–6; pedicels 1–1.5 cm. long; sepals ovate, 5–7 mm. long; petals white, spatulate, 6–10 mm. long; caps. ovoid, 4–5 mm. high; seeds usually 4, minutely tubercled, 2 mm. long.—Rare, on talus and loose rock or gravel, 9000–11,000 ft.; Subalpine F., Alpine Fell-fields; Fresno Co. to Modoc Co.; Ore. to Wyo. July–Aug.

4. **C. nevadénsis** Wats. [*Montia n.* Jeps. *M. alpina* Eastw. *C. chenopodina* Greene.] Perennial with fleshy slender tangled rootstocks; stems 4–10 cm. long; basal lvs. 5–10 cm. long, fleshy, obovate to suborbicular, the blades abruptly narrowed to rather long petioles; cauline lvs. ovate, sessile, 8–15 mm. long; fls. 2–6; pedicels 1–2(–4) cm. long; sepals ovate, acute, 5–6 mm. long; petals white with some pink, spatulate, 5–8 mm. long; caps. ovoid, 3–3.5 mm. long; seeds 4–6, shining, black, ca. 2 mm. long.—Gravelly wet places, 8000–12,000 ft.; Alpine Fell-fields; Tuolumne, Alpine, and Mono cos. July–Aug.

5. Móntia L.

Annuals or perennials, glabrous, often glaucous, with fibrous roots or reproducing by runners or bulblets. Lvs. somewhat fleshy, basal, alternate or opposite. Infl. racemose or paniculate; pedicels enlarged at base of fl. and often recurved in fr. Sepals 2, persistent, often somewhat unequal. Petals 2–5(–6), pink or white, often unequal, free or somewhat connate. Stamens as many, adhering to base of petals. Style-branches 3. Ovules 3 (4). Caps. globose to ovoid, 3-valved from apex. Seeds 1–3, smooth or tuberculate or foveolate, often shining. Ca. 40 spp., of both hemis. (Named for Giuseppe *Monti*, 1682–1760, Italian botanist.)

(Rydberg, P. A. N. Am. Fl. 21: 302–316, 1932.)
A. Cauline lvs. alternate.
  B. Plant perennial; petals 7–10 mm. long .... 1. *M. parvifolia*
  BB. Plant annual; petals 2–5 mm. long or wanting.
    C. Lvs. ovate to deltoid; infl. paniculate, leafy .... 2. *M. diffusa*
    CC. Lvs. linear to linear-spatulate; infl. racemose to subumbellate, not leafy.
      D. Plants 5–20 cm. high; lvs. 20–35 mm. long; sepals 4 mm. long ... 4. *M. linearis*
      DD. Plants 2–5 cm. high; lvs. 7–20 mm. long; sepals 1.5–2 mm. long.
        E. Lvs. linear; plant sparingly branched, the infl. terminal, exceeding lvs. Shasta
          and Siskiyou cos. .... 3. *M. dichotoma*
        EE. Lvs. linear-spatulate; plant diffusely branched, the infl. axillary, shorter
          than lvs. Humboldt Co. n. .... 5. *M. Howellii*
AA. Cauline lvs. opposite.
  B. Stem-lvs. 2 to several pairs.
    C. Annual, delicate; petals united to above the base, 1–3 mm. long.
      D. Seeds dull, muricate, with mostly acute tubercles.
        E. Lower lvs. spatulate to oblanceolate, 8–20 mm. long; sepals ca. 1.5 mm.
          long; seeds strongly turgid, 1–1.4 mm. in diam. .... 6. *M. verna*
        EE. Lower lvs. almost linear, mostly 5–10 mm. long; sepals ca. 1 mm. long;
          seeds flattish, 0.6–0.9 mm. in diam. .... 7. *M. Hallii*
      DD. Seeds shining, smoothish, with low flat tubercles. Rare, mts. of Tulare and San
        Bernardino cos. .... 8. *M. Funstonii*
    CC. Perennial, rather fleshy; petals scarcely united, 5–8 mm. long .... 9. *M. Chamissoi*
  BB. Stem-lvs. 1 pair.
    C. Infl. bracteate with many bracts 4–10 mm. long; plant perennial; petals 7–10 mm. long
                      15. *M. sibirica*
    CC. Infl. bractless or with 1 basal bract.
      D. Plants perennial; petals 9–12 mm. long .... 10. *M. cordifolia*
      DD. Plants annual; petals 2–7 mm. long.
        E. Cauline lvs. not at all united; seeds finely pitted .... 11. *M. saxosa*
        EE. Cauline lvs. united at least on one side; seeds granular, with low tubercles.
          F. Stem-lvs. usually united on both sides, forming a flattish rounded some-
            what angled disk .... 12. *M. perfoliata*
          FF. Stem-lvs. united on part of 1 side, not forming a flattish disk.
            G. Petals 2.5–4 mm. long; infl. mostly 3–6-fld. .... 13. *M. spathulata*
            GG. Petals 5–7 mm. long; infl. mostly 8–15-fld. .. 14. *M. gypsophiloides*

1. **M. parvifòlia** (Moç. in DC.) Greene. [*Claytonia p.* Moç. in DC. *Naiocrene p.* Rydb. *C. filicaulis* Dougl. ex Hook. *M. obtusata* Heller.] Perennial with fleshy rootstock, short caudexlike stems with rosettes of fleshy lvs., and slender ascending or erect stems 1.5–3 dm. long; caudices also producing slender stolons; basal lvs. obovate to oblanceolate, 1–3 cm. long, petioled; cauline lvs. reduced upward and bractlike above, often bearing bulblets in the axils; fls. 1–10, racemose; pedicels commonly 5–15 mm. long; sepals rounded, unequal, 2–2.5 mm. long; petals white with pink or lavender veins, or often pink, emarginate, 7–10 mm. long; caps. ca. 3 mm. long; seeds 1–2, black, dull, minutely pitted, over 1 mm. long.—Moist, often rocky places, below 8500 ft.; Redwood F., Mixed Evergreen F., Douglas-Fir F., Yellow Pine F., Red Fir F.; Sierra Nevada from Tuolumne Co. n. and Coast Ranges from Monterey Co. n.; to Alaska, Mont. May–July.

2. **M. diffùsa** (Nutt.) Greene. [*Claytonia d.* Nutt. in T. & G. *Limnalsine d.* Rydb.] Annual, diffusely branched from base, 5–15 cm. high; basal and cauline lvs. alike, ovate to deltoid, 2–5 cm. long, the blades decurrent on the equally long petiole; infl. paniculate, terminal, leafy; pedicels 1–2 cm. long; sepals obovate, rounded, entire, ca. 2 mm. long; petals white or pinkish, 3–4 mm. long, emarginate; caps. obovoid, slightly exceeding calyx; seeds 1–3, black, broadly ellipsoid, finely reticulately grooved, ca. 1.5 mm. long.—In woods, below 2500 ft.; Redwood F., Douglas-Fir F.; Marin Co. to Humboldt Co. and Wash. May–July.

3. **M. dichótoma** (Nutt.) Howell. [*Claytonia d.* Nutt. in T. & G. *Montiastrum d.* Rydb.] Annual, sparingly branched, 2–5 cm. high; lvs. alternate, linear, 7–15 mm. long, sheathing at base; racemes with ovate scarious bracts, secund, with 6–12 rather crowded nodding fls.; pedicels reflexed, 2–5 mm. long; sepals rounded-ovate, ca. 2 mm. long; petals scarcely if at all longer, unequal; caps. ovoid, 2 mm. long; seeds 3, black, shining, faintly reticulate, ca. 1 mm. long.—About vernal pools and on slopes, below 5000 ft.; Yellow Pine F.; Shasta and Siskiyou cos.; to Wash., Ida. April–June.

4. **M. lineáris** (Dougl.) Greene. [*Claytonia l.* Dougl. ex Hook. *Montiastrum l.* Rydb.] Erect branched annual 5–20 cm. high; lvs. alternate, linear, 2–3.5 cm. long, the petioles with enlarged scarious bases; racemes terminal, secund, lax, mostly 2–7-fld.; pedicels 6–15 mm. long, ± recurved; sepals reniform-orbicular, ca. 4 mm. long, white-

margined; petals unequal, obovate, white, ca. 5 mm. long; caps. ovoid, 4 mm. long; seeds 3, lenticular, microscopically muricate especially on margin, 1.5–2 mm. in diam. —Moist banks, meadows, etc., below 7500 ft.; Mixed Evergreen F., Yellow Pine F., Red Fir F.; Cuyamaca Mts., San Bernardino Mts., Sierra Nevada to Modoc Co., Coast Ranges from Contra Costa Co. n.; to B.C., Mont. April–June.

5. **M. Howéllii** Wats. [*Claytonia H.* Piper. *Montiastrum H.* Rydb.] Diffusely branched annual with slender stems 2–5 cm. high; lvs. alternate, linear-spatulate, 1–2 cm. long, petioled; racemes axillary, few-fld., subumbellate; pedicels recurved, 2–3 mm. long; sepals round-ovate, 1.5–1.8 mm. long; petals 2 or 0, minute, white; caps. obovoid, 1 mm. high; seeds 2–3, black, smooth, shining, ca. 0.8 mm. long.—Wet shaded places near the coast; Redwood F.; Humboldt Co. to B.C. March–May.

6. **M. vérna** Neck. [*M. minor* Gmel. *M. fontana,* many refs., probably not L.] Annual with stems slender, spreading, branched, 3–10(–20) cm. long, often rooting at nodes; lvs. opposite, the lower spatulate to oblanceolate, 8–20 mm. long, petioled, the upper oblong to oblanceolate, sessile; fls. axillary and terminal, nodding, solitary or in small terminal clusters; pedicels 3–15 mm. long; sepals reniform, ca. 1.5 mm. long; petals largely 5, ca. twice as long as sepals; caps. round-obovoid, ca. as long as calyx; seeds black, suborbicular, strongly turgid, ca. 1.1–1.4 mm. in diam., muricate and dull with mostly acute tubercles; $2n = 18$ (Hagerup, 1941).—Floating in sluggish shallow water or ponds, later in season on mud banks, mostly below 5000 ft.; many Plant Communities; San Benito and Placer cos. n.; to Wash.; Eu. March–June.

7. **M. Hállii** (Gray) Greene. [*Claytonia H.* Gray. *M. fontana* var. *tenerrima* (Gray) Fern. & Wieg. *M. dipetala* Suksd. *M. stenophylla* Rydb.] Much like *M. verna* and with it included in former manuals under *M. fontana,* but more delicate; lvs. almost linear, to oblanceolate, mostly 5–10(–14) mm. long; sepals ca. 1 mm. long; petals 2–5; seeds sharply muriculate, 0.6–0.9 mm. long.—Rainpools, etc., below 6000 ft., L. Calif. to B.C. March–June.

8. **M. Funstònii** Rydb. Much like *M. Hallii,* but seeds with low flat tubercles so as to be more shining.—Wet places, 8500–11,000 ft., Tulare Co. and Inyo Co., and at 6750 ft., Baldwin Lake, San Bernardino Co. July–Aug.

9. **M. Chamíssoi** (Ledeb.) Dur. & Jacks. [*Claytonia C.* Eschs. in Ledeb. in Spreng. *C. Chamissonis* Eschs. *Crunocallis Chamissonis* Rydb.] Perennial with creeping or floating stems, slender bulblet-bearing runners, and ascending to erect branches or tips commonly 5–15 cm. long; lvs. opposite, oblong-spatulate, 1.5–4(–5) cm. long; fls. in axillary or subterminal 3–8-fld. racemes; pedicels slender, recurved in fr., 1–2.5 cm. long; sepals unequal, rounded, ca. 2 mm. long; petals pink or white, 5–8 mm. long; caps. 1–1.5 mm. long; seeds 1–3, black, rounded, ca. 1.5 mm. in diam., densely muriculate.—Wet places in meadows, along streams, etc., 4000–11,000 ft.; Yellow Pine F. to Alpine Fell-fields; San Diego Co. to Sierra Nevada and Modoc Co., Lake Co. n.; to Alaska, Minn., New Mex. June–Aug.

10. **M. cordifòlia** (Wats.) Pax & K. Hoffm. [*Claytonia c.* Wats. *Limnia c.* Rydb. *C. asarifolia* Gray, not Bong. *M. a.* Howell.] Perennial with a horizontal rootstock; stems 1–few, 1–3 dm. high; basal lvs. round-reniform to -ovate, the blades 2–5 cm. long, at least as wide, on longer petioles; stem-lvs. 2, opposite, sessile, 1.5–3 cm. long; racemes naked, simple, 4–9-fld., becoming 1–2 dm. long in fr.; pedicels 1–2 cm. long, spreading-recurved in age; sepals suborbicular, ca. 4 mm. long; petals white, obovate, 9–12 mm. long; caps. 4 mm. high; seeds black, roundish, shining, 2 mm. long.—Along streams and in wet places, 3500–7000 ft.; Douglas-Fir F., Yellow Pine F., Red Fir F.; Humboldt Co. and w. Siskiyou Co.; to B.C., Mont., Utah. June–Aug.

11. **M. saxòsa** Bdg. ex Rob. in Gray. [*Claytonia s.* Bdg. *Limnia s.* Heller.] Dense succulent annual with several stems 1–2.5 cm. high; basal lvs. broadly spatulate, oblanceolate or obovate, subsessile, 6–18 mm. long; cauline lvs. 2, opposite, sessile, not united, ovate, 8–12 mm. long; fls. subumbellate, 2–5; pedicels 4–8 mm. long; sepals roundish, 2–3 mm. long; petals pink, ovate, emarginate, 5–6 mm. long; caps. globose, 2.5 mm. high; seeds 2–3, black, shining, 1.3–2 mm. long, oblong, finely pitted.—Rocky places, 3000–7000 ft.; Yellow Pine F., Red Fir F.; Lake Co. to Humboldt and Siskiyou cos. March–June.

12. **M. perfoliàta** (Donn) Howell. [*Claytonia p.* Donn. *Limnia p.* Haw. *L. carnosa* (Greene) Heller. *L. platyphylla* Rydb.] MINER's-LETTUCE. Glabrous green annual,

branched from base, 1–3 dm. high; basal lvs. rhombic-ovate to elliptic-obovate, long-petioled, 5–20 cm. long; cauline lvs. 2, opposite, connate into an oblique suborbicular disk 1–8 cm. broad; racemes ± elongate, sessile or peduncled, the fls. usually ± whorled; pedicels commonly 2–8(–10) mm. long, often recurved in fr.; sepals rounded, ca. 3–5 mm. long; petals white, clawed, obovate, 4–6 mm. long; seeds black, shining, rounded, minutely punctate, 1–2 mm. in diam.—Common in ± shaded and vernally moist places, below 5000 ft.; Coastal Sage Scrub, Chaparral, Foothill Wd., Mixed Evergreen F., etc.; L. Calif. to B.C., inland to desert edge. Feb.–May. Exceedingly variable and in need of study. Running into innumerable forms, of which the following are most notable:

KEY TO VARIETIES

A. Basal lvs. largely ovate to deltoid.
  B. Stems suberect, 1–3 dm. high; racemes elongate, often peduncled; sepals 3–5 mm. long.
    Cismontane, mostly below Yellow Pine F. . . . . *M. perfoliata*
  BB. Stems spreading, mostly less than 1 dm. long; racemes short, compact, sessile; sepals ca. 2 mm.
    long. Yellow Pine F. . . . . var. *depressa*
AA. Basal lvs. largely linear to oblanceolate or spatulate.
  B. Plants green.
    C. Stems mostly 1–3 dm. high. Cismontane and Sierran.
      D. Infl. mostly elongate, open; sepals ca. 2 mm. long . . . . f. *parviflora*
      DD. Infl. mostly short, compact; sepals ca. 3 mm. long . . . . f. *angustifolia*
    CC. Stems mostly not over 1 dm. high. Deserts . . . . Var. *utahensis*
  BB. Plants glaucous, less than 1 dm. high.
    C. Petals 2.5–3 mm. long. W. base of Sierra Nevada and from Lake Co. n. . . . . f. *glauca*
    CC. Petals 4–5 mm. long. Mts. about San Francisco Bay . . . . var. *nubigena*

Forma **parviflòra** (Dougl.) J. T. Howell. [*Claytonia p.* Dougl. *M. p.* Howell. *M. perfoliata* var. *parviflora* Jeps.] Earlier basal lvs. linear, the later oblanceolate; infl. tending to be more elongate and open; sepals ca. 2 mm. long.—With the sp.

Forma **angustifòlia** (Greene) J. T. Howell. [*Claytonia perfoliata* var. *a.* Greene.] Basal lvs. linear to oblanceolate; infl. subsessile, short; sepals 3 mm. long.—Chaparral, Foothill Wd.; San Luis Obispo Co. to Sonoma and Placer cos. March–June.

Forma **gláuca** (Nutt.) J. T. Howell. [*Claytonia parviflora* var. *g.* Nutt. *Montia perfoliata* ssp. *g.* Ferris.] Tufted, glaucous, mostly 3–8 cm. high; basal lvs. fleshy, linear-spatulate; infl. short, compact; petals 2.5–3 mm. long.—Rocky places, below 3500 ft.; Yellow Pine F.; from Lake Co. and w. base of Sierra Nevada n.; to B.C. March–June.

Var. **depréssa** (Gray) Jeps. [*Claytonia parviflora* var. *d.* Gray. *M. humifusa* Howell.] Stems spreading, mostly 5–10 cm. long; plant ± reddish; basal lvs. rhombic-ovate to subdeltoid; infl. short, sessile; sepals 2 mm. long.—Rather dry shaded slopes, largely 4000–7500 ft.; Yellow Pine F., Red Fir F.; mts. of s. Calif. through Sierra Nevada and N. Coast Ranges to Ore., B.C., Mont., Utah. April–June.

Var. **utahénsis** (Rydb.) Munz [*Limnia u.* Rydb. *Claytonia perfoliata* var. *u.* Poelln.] Basal lvs. linear to spatulate, otherwise much as in var. *depressa.*—Semishade, 3500–7500 ft.; largely in Pinyon-Juniper Wd., Joshua Tree Wd.; mts. of Mojave Desert from Inyo Co. to Ariz., Utah. April–June.

Var. **nubígena** (Greene) Jeps. [*Claytonia n.* Greene. *Limnia n.* Heller.] Glaucous, 4–8 cm. high; basal lvs. linear to narrow-oblanceolate; petals pinkish, 4–6 mm. long.—Dry rocky slopes, 1000–4000 ft.; Chaparral, Mixed Evergreen F.; mts. about San Francisco Bay, Santa Lucia Mts. March–April.

13. **M. spathulàta** (Dougl.) Howell. [*Claytonia s.* Dougl. ex Hook. *Limnia s.* Heller.] Densely tufted glaucous annual, the stems 2–6 cm. high; basal lvs. numerous, linear to linear-spatulate, 3–9 cm. long, mostly surpassing the stems; cauline lvs. 2, opposite, linear to lance-ovate, usually joined on one side, 1–2 cm. long; racemes mostly 3–6-fld., not over 2 cm. long; sepals ovate, acute, ca. 1.5 mm. long; petals 5, white or pinkish, 2.5–3 mm. long; caps. ca. 2 mm. high; seeds 2–3, black, shining, minutely tuberculate, 0.6–1 mm. long.—Rather open dry slopes, below 2500 ft.; Foothill Wd., Chaparral, V. Grassland, Mixed Evergreen F., etc.; San Luis Obispo Co. through the Coast Ranges to B.C. Feb.–April. Exceedingly variable and intergrading freely with vars.

KEY TO VARIETIES

A. Cauline lvs. 1–2 cm. long, not exceeding infl.
  B. Petals 2.5–3 mm. long; basal lvs. mostly surpassing stems.

C. Basal lvs. 3–9 cm. long. Widely distributed .... *M. spathulata*
CC. Basal lvs. 1–2 cm. long. Mt. Tamalpais .... Var. *rosulata*
BB. Petals ca. 4 mm. long; basal lvs. shorter than stems .... Var. *exigua*
AA. Cauline lvs. 2–6 cm. long, surpassing the infl.
B. Plant glaucous; basal lvs. linear or nearly so .... Var. *tenuifolia*
BB. Plant green; basal lvs. oblanceolate .... Var. *viridis*

Var. **rosulàta** (Eastw.) J. T. Howell. [*M. r.* Eastw. *Limnia r.* Heller.] Very compact; lvs. 1–2 cm. long, the cauline lance-ovate; seeds almost 2 mm. long.—Serpentine, Rock Spring, Mt. Tamalpais. April.

Var. **exígua** (T. & G.) Rob. [*Claytonia e.* T. & G. *M. e.* Jeps. *M. gypsophiloides* var. *e.* J. T. Howell.] Plants 5–15 cm. high, the stems much surpassing the basal lvs.; stem-lvs. mostly 1–2 cm. long, 1–2 mm. wide; petals ca. 4 mm. long; seeds 1.5–2 mm. long.— Dry rocky or gravelly places, often on serpentine, below 5500 ft.; Foothill Wd., Chaparral, etc.; mts. of San Diego Co. through Coast Ranges to Wash., Nev. April–July.

Var. **tenuifòlia** (T. & G.) Munz. [*Claytonia t.* T. & G. *M. t.* Howell. *C. spathulata* var *t.* Gray. *Limnia t.* Rydb.] Glaucous, 5–10 cm. high; basal lvs. linear or nearly so; cauline lvs. linear to oblance-linear, 2–4 cm. long, usually exceeding infl.; petals 3–4 mm. long; seeds ca. 1–1.2 mm. long.—Dry somewhat shaded places, below 6500 ft.; Yellow Pine F., Chaparral, Foothill Wd.; mts. of San Diego Co. to Siskiyou and Tuolumne cos. April–June.

Var. **víridis** A. Davids. [*M. exigua* var. *v.* Jeps. *Limnia v.* Rydb.] Green, not glaucous, 5–15 cm. high; basal lvs. narrow-oblanceolate; stem-lvs. lanceolate, 2–5 cm. long, surpassing infl.; petals ca. 3 mm. long.—Shaded somewhat damp places, 4000–7200 ft.; largely Yellow Pine F.; San Diego Co. to Ventura Co. May–July.

14. **M. gypsophiloìdes** (F. & M.) Howell. [*Claytonia g.* F. & M. *Limnia g.* Heller. *L. diaboli* Rydb.] Glaucous few–many-stemmed annual 5–20 cm. high; basal lvs. numerous, linear, 3–8 cm. long; stem-lvs. 2, lanceolate to ovate, 1–2.5 cm. long, partially united on one side; fls. usually in a lax raceme, ca. 8–15; sepals rounded, 2–2.5 mm. long; petals 4, pinkish or white, 5–7 mm. long; caps. ca. 2.5 mm. high; seeds 2–3, blackish, shining, minutely punctate, ca. 0.8–1 mm. long.—Moist, loose often serpentine soil, below 4000 ft.; Foothill Wd., N. Oak Wd., Chaparral; San Luis Obispo Co. to Mendocino Co. March–May. Doubtfully specifically distinct from *M. spathulata.*

15. **M. sibírica** (L.) Howell. [*Claytonia s.* L. *Limnia s.* Haw. *C. asarifolia* Bong. *C. alsinoides* Sims. *L. bracteosa* Rydb. *C. bulbifera* Gray. *M. bulbifera* Howell.] Rather succulent perennial from a slender rootstock, the stems few to several, 1.5–4.5 dm. high, glabrous; basal lvs. several, 8–20 cm. long, long-petioled, the blades 1.5–5 cm. long, broadly to narrowly ovate, mostly acute at apex; cauline lvs. 2, distinct, sessile, 2–5 cm. long, ovate; racemes open, 1–few, bracteate, 5–25 cm. long; pedicels slender, 1–3 cm. long, divergent in fr.; sepals round-ovate, 4–6 mm. long; petals pink or white with pink lines, 7–10 mm. long; caps. not exceeding calyx; seeds 1–3, black, shining, finely tuberculate, 2–2.5 mm. long.—Moist shaded places, below 5000 ft.; Redwood F., N. Coastal Scrub, Douglas-Fir F., Red Fir F., Mixed Evergreen F.; Santa Cruz Co. to Del Norte and Siskiyou cos.; to Alaska, Mont.; Siberia. March–Aug.

Var. **heterophýlla** (Nutt.) Rob. in Gray. [*C. unalaschkensis* var. *h.* Nutt. ex T. & G. pro synon. *C. alsinoides* var. *h.* T. & G. *M. h.* Jeps. *Limnia h.* Rydb.] Stems 1–2; basal lvs. few; cauline lvs. ± petiolate, lance-ovate to oblanceolate.—Moist places, 5000–7300 ft.; largely Red Fir F.; Sierra Nevada from Fresno Co. to Siskiyou Co.; to Wash. May–July.

6. **Calyptrídium** Nutt. in T. & G.

Annual or perennial herbs with alternate or basal spatulate lvs. Fls. small, perfect, in scorpioid spikes or groups of spikes. Sepals 2, rounded or round-reniform, scarious or scarious-margined. Petals 2 or 4. Stamens 1–3. Style simple, with 2 stigmas. Caps. membranous, 2-valved, few–many-seeded. Ca. 6 spp. of w. N. Am. (Greek, *kaluptra*, a cap or covering, because of the way the petals close over the caps. in age.)

(Rydberg, P. A. N. Am. Fl. 21: 316–320, 1932. Howell, J. T. Leafl. W. Bot. 3: 262–266, 1943.)
Style short; petals in age folding as a cap over the caps.; caps. elongate, linear-oblong, or ovoid, 6–many-seeded. (*Eucalyptridium*)

Caps. linear-oblong to elliptic, the valves widest near middle and with many nerves near base; fls. sessile or subsessile, deciduous.

    Caps. 2–4 times as long as sepals; sepals 1–2 mm. long ....1. *C. monandrum*
    Caps. less than 2 times as long as sepals; sepals 2–3.5 mm. long ....2. *C. Parryi*
Caps. broadly to narrowly ovoid, the valves widest near base, with 5–6 main nerves; lowest fls. distinctly pedicelled.

    Fls. persistent; sepals ovate, acutish .... 3. *C. pygmaeum*
    Fls. shed early; sepals round or round-reniform in fr.
        Sepals round-reniform, scarious throughout; petals 4 .... 4. *C. quadripetalum*
        Sepals round, herbaceous with scarious margins; petals 2 .... 5. *C. roseum*
Style long-filiform; petals in age twisting about the style; caps. round or nearly so. (Spraguea)
    Caps. 2–10-seeded; sepals 5–8 mm. long .... 6. *C. umbellatum*
    Caps. 1–2-seeded; sepals 2–2.5 mm. long .... 7. *C. monospermum*

1. **C. monándrum** Nutt. in T. & G. Annual, branched at base; stems frequently flat on ground, 5–18 cm. long; lvs. mostly in a basal rosette, linear-spatulate, 2–6 cm. long; cauline lvs. scattered, somewhat reduced; fls. subsessile in short spikes in a terminal panicle, the branchlets scorpioid when young, secund in age; sepals reniform-ovate or deltoid, white-margined, 1–2 mm. long; petals commonly 3, white, ovate, ca. 1 mm. long; caps. compressed, linear-oblong, striate, 4–8 mm. long; seeds 5–10, black, shining, ca. 0.5 mm. long.—Common in sandy and open places, usually below 6000 ft., Creosote Bush Scrub, Joshua Tree Wd.; deserts from Mono Co. s.; also, especially on burns, etc., in Chaparral, lower Yellow Pine F., Coastal Sage Scrub, Foothill Wd., etc.; S. Coast Ranges from San Benito and Monterey cos. to L. Calif.; Nev. to Son. March–June.

2. **C. Párryi** Gray. Habit of *C. monandrum;* lvs. 1–3 cm. long; sepals 2–3.5 mm. long; petals largely 4, shorter than sepals; caps. oblong, ca. 1.5–2 times the length of calyx; seeds dull, muriculate throughout, 0.6–0.7 mm. long.—Disturbed flats and benches, 5000–11,000 ft.; Montane Coniferous F.; Mt. Whitney Road, Inyo Co., Mt. Pinos to San Jacinto Mts. June–July.

Var. **nevadénse** J. T. Howell. Fls. more easily deciduous; abaxial sepal in fr. reniform, with wide scarious margin; seeds shining, almost smooth.—Dry slopes, 7000–10,000 ft.; Pinyon-Juniper Wd.; Panamint Mts.; w. Nev. June–July.

Var. **Hésseae** Thomas. Fls. deciduous; abaxial fruiting sepal ovate, with narrow or no scarious margin; seeds shining, tuberculate on margin.—Dry slopes, 2300–3500 ft.; Chaparral, etc.; Santa Cruz Mts. June–July.

3. **C. pygmaèum** Parish ex Rydb. Diffuse annual; stems 1–2.5 cm. high; lvs. mostly basal, spatulate, 5–10 mm. long; fls. few, in 1-sided racemes; bracts ovate, scarious, early deciduous; pedicels 1–3 mm. long; sepals ovate, acutish, ca. 2 mm. long, scarcely scarious-margined; petals 4, ca. 3 mm. long; caps. ovate, pointed, ca. 4 mm. long; seeds black, smooth, ca. 0.4 mm. long.—Rare, dry to moist sandy or gravelly places, 7500–11,500 ft.; Lodgepole F. to Subalpine F.; San Bernardino Mts., se. Sierra Nevada to n. Inyo Co. June–July.

4. **C. quadripétalum** Wats. [*C. tetrapetalum* Greene.] Annual, branched from base, the stems erect to ascending, 3–12 cm. long; lvs. basal and cauline, spatulate-oblong, 2–6 cm. long, petiolate; racemes axillary and terminal, scorpioid, dense, 2–4 cm. long; lower pedicels 1–2.5 mm. long, upper lacking; sepals round-reniform, white or pink, scarious, 3–4 mm. long; petals 4, ovate, ca. 2 mm. long; caps. oblong-ovate, ca. as long as calyx; seeds black, shining, round, muriculate, 0.5 mm. in diam.—Open places, below 2700 ft.; Chaparral; inner Coast Ranges of Lake, Napa, Sonoma, and Glenn cos. April–June.

5. **C. ròseum** Wats. Several-stemmed annual, depressed or spreading, the stems 2–10 cm. long; lvs. few, oblong-spatulate, basal and scattered on branches, 1–4 cm. long; fls. in panicles of short scorpioid clusters; lowest fls. short-pedicelled; sepals round, scarious-margined, 2.5–4 mm. long; petals 2, scarcely 1 mm. long; caps. ovate-oblong, not exceeding lower sepal; seeds black, shining, muriculate, scarcely 0.5 mm. long.—Moist often somewhat alkaline places, 5000–10,500 ft.; Sagebrush Scrub to Lodgepole F.; e. slope of Sierra Nevada (Lassen, Sierra, Mono, and Inyo cos.); to e. Ore., Ida., w. Nev. June–Aug.

6. **C. umbellàtum** (Torr.) Greene. [*Spraguea u.* Torr. *S. paniculata* Kell. *S. u.* var. *montana* Jones. *S. m.* Heller. *C. nudum* Greene. *S. eximia* and *pulchella* Eastw. *S.*

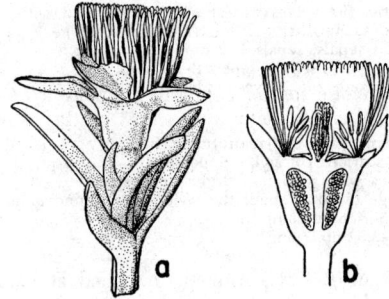

Fig. 35. AIZOACEAE. *Mesembryanthemum: a,* fl., × ½, ovary inferior, sepals 5, fleshy, petals many, linear; *b,* section of fl. with styles and stamens.

caespitosa, irregularis, and *Hallii* Rydb. S. *pulcherrima* Heller. C. *pulchellum* Hoov.] Pussy Paws. Annual to perennial; stems several, spreading to suberect, 5–25 cm. long, ± scapelike; lvs. largely basal, in dense rosette, spatulate, 2–7 cm. long; cauline lvs. almost none or several, reduced; infl. umbellate-cymose, rather lax to subcapitate, consisting of scorpioid spikes; fls. deciduous, imbricate-crowded, very short-pedicellate; sepals pink or white, orbicular-reniform, mostly 5–8 mm. long, with broad scarious margins; petals 4, oblong or ovate, 3–6 mm. long; caps. ovoid, 3–4 mm. long; seeds compressed, round-reniform, smooth, shining, 0.6–1 mm. long.—Rather common locally, in loose sandy or gravelly places, 2500–11,000 ft.; mostly Yellow Pine F. to Subalpine F.; San Jacinto Mts. n. through Sierra Nevada, Santa Cruz Mts. through Coast Ranges to B.C., Rocky Mts., L. Calif. May–Aug. Very variable as to height, foliage, infl., fls.

Var. **caudiciferum** (Gray) Jeps. [*Spraguea u.* var. *c.* Gray.] Caudex much branched, heavy, woody, clothed with dead lvs.; lvs. 5–10 mm. long; stems 5–12 mm. long; infl. capitate, scarcely 1 cm. thick.—Dry gravelly places, 11,000–13,000 ft.; Alpine Fell-fields; Sierra Nevada, also volcanic cinders at ca. 7000 ft.; Yellow Pine F.; Mono Craters; to Wash., Wyo. July–Aug.

7. **C. monospérmum** Greene. [*Spraguea m.* Rydb.] Diminutive annual 2–3 cm. high; basal lvs. spatulate, 1–2 cm. long; cymes scorpioid; sepals 2–2.5 mm. long, round-reniform; petals spatulate-oblong; caps. orbicular, 1–2-seeded.—At ca. 10,000 ft.; Lodgepole F.; Big Cottonwood Meadows, Inyo Co. Probably only a depauperate form of *C. umbellatum.*

37. Aizoàceae. Carpet-Weed Family (Fig. 35)

Herbs or low shrubs, erect or prostrate, often fleshy. Lvs. alternate or opposite or whorled, with or without stipules. Fls. usually perfect, regular. Fl.-tube free or adnate to the ovary. Sepals 4–8, imbricate. Petals 0 to many, linear. Stamens perigynous, many in several series, few in one series, or 1; anthers small, 2-celled, opening lengthwise. Ovary superior or inferior, 3–several-loculed; styles 3–many. Fr. a caps. or drupaceous or nutlike, often clasped by the persistent calyx. Seeds usually many; embryo annular; endosperm copious to scant. Ca. 20 genera and over 1000 spp., mainly trop. and of S. Hemis.

Petals none; calyx sometimes colored within and petaloid.
  Lvs. opposite or whorled; fl.-tube free from ovary.
    The lvs. whorled; fl.-tube nearly or quite wanting; caps. loculicidal.
      Plants glabrous; sepals 1–2 mm. long .... 1. *Mollugo*
      Plants pubescent; sepals 4–7 mm. long .... 2. *Glinus*
    The lvs. opposite; fl.-tube evident; caps. circumscissile.
      Stipules present, scarious; stamens 5–10.
        Stems less than 1 dm. long; stipules laciniate; lvs. 3–10 mm. long .... 3. *Cypselea*
        Stems 2–6 dm. long; stipules not laciniate; lvs. 10–30 mm. long .... 4. *Trianthema*
      Stipules absent; stamens many .... 5. *Sesuvium*
  Lvs. alternate; fl.-tube united to ovary; lvs. triangular-ovate .... 6. *Tetragonia*
Petals present, numerous; ovary inferior .... 7. *Mesembryanthemum*

1. Mollùgo L. INDIAN-CHICKWEED

Low glabrous annuals, much branched. Lvs. whorled or alternate. Fls. axillary on slender pedicels. Sepals 5, white inside. Petals 0. Stamens hypogynous, 5 and alternate with the sepals, or 3 and alternate with the locules of the ovary. Stigmas 3. Caps. 3-locular, 3-valved, loculicidal, the partitions breaking away from the many-seeded axis. Seeds without a strophiole. Ca. 12 spp., mostly trop. (Old name for Galium *Mollugo*, transferred to this genus, perhaps because of whorled lvs.)

Lvs. spatulate; plant prostrate .... 1. *M. verticillata*
Lvs. linear; plant erect or ascending .... 2. *M. Cerviana*

1. **M. verticillàta** L. Prostrate, forming mats 1–3(–5) dm. across; lvs. spatulate, 5–6 in a whorl, unequal, 5–20 mm. long, short-petioled; pedicels filiform, several at a node; sepals oblong, 2 mm. long; caps. slightly longer than sepals, ovoid; seeds reniform, shining or slightly granular, brown, ca. 0.6 mm. long; $2n = 64$ (Sugiura, 1936).—Waste places and fields, becoming widely natur. in Calif.; from trop. Am. May–Nov.

2. **M. Cerviàna** (L.) Ser. [*Pharnaceum C.* L.] Stems very slender, 4–10(–15) cm. high; lvs. glaucous, linear, 5–10 in a whorl, the basal linear-spatulate, 3–15 mm. long; fls. whorled; pedicels filiform; sepals 1.5 mm. long; caps. subglobose; seeds 0.4–0.5 mm. long; $2n = 18$ (Sugiura, 1936).—Sandy places, Thomas V. (San Jacinto Mts.), Twentynine Palms; natur. from Old World. Sept.–March.

2. Glìnus L.

Annuals, with general habit of *Mollugo*, pubescent or glabrous. Lvs. entire, falsely whorled. Fls. densely clustered, on short peduncles from upper nodes. Fl-tube campanulate. Sepals 5. Petals 0. Stamens 3–5. Caps. 3–5-valved, loculicidal. Seeds strophiolate, numerous, minute, the slender funicle large and coiled about the seed. Ca. 10 spp. of warmer regions. (Greek, *glinos*, sweet juice, of uncertain application.)

1. **G. lotoìdes** L. Stellate-pubescent, the branches prostrate or ascending, 1–3 dm. long; lvs. 1–3 cm. long, obovate, petiolate; fls. pedicellate or subsessile; sepals 4–7 mm. long, broadly oblong; caps. ellipsoid, 4 mm. long; seeds dark, tuberculate, 0.6 mm. long.—Dried soil about ponds, etc., below 3500 ft., Santa Clara and San Joaquin cos. to Lake and Butte cos.; native of Eu. June–Nov.

3. Cypselèa Turp.

Small prostrate herbs, with slender branched stems. Lvs. opposite, unequal, with scarious laciniate stipules. Fl.-tube short, campanulate. Sepals 4–5, unequal. Petals 0. Stamens 1–3, alternate with sepals. Ovary superior, 1-loculed; styles 2. Caps. circumscissile at base. Seeds many, minute, smooth. One or 2 spp. of W.I. (Greek, *kupsele*, beehive, because of shape of caps.)

1. **C. humifùsa** Turp. Much-branched annual forming mats 3–7 cm. wide; lvs. glabrous, elliptical, the larger member of the pair 8–10 mm. long, petioled, the smaller ca. half as long and with a fascicle of small lvs.; calyx ca. 1.5 mm. long; caps. subglobose; seeds brown, subreniform, ca. 0.3 mm. long.—Desiccated mud banks, lower San Joaquin R. and Santa Cruz and Marin cos.; from W.I. July–Oct.

4. Triánthema L. HORSE-PURSLANE

Prostrate herbs or subshrubs, branched from base. Lvs. opposite, those of each pair unequal; stipules present. Fls. solitary, axillary. Fl.-tube short. Sepals 5. Petals 0. Stamens 5–10, perigynous. Ovary truncate, 1-2-locular; styles 1–2. Caps. short-cylindric or turbinate, 1–5-seeded, tardily circumscissile, the upper part thick and tough, usually with 2 rounded marginal crests. Seeds reniform, roughened. Ca. 15 spp., mostly trop. (Greek, *treis*, three, and *anthemon*, fl.)

1. **T. Portulacástrum** L. Diffusely branched herb, annual, glabrous except for the rows of soft hairs, the branches rather succulent, 2–6 dm. long; lvs. obovate, 1–2 cm. long, smaller on axillary branchlets; petioles short, dilated at base into bidentate stipular

expansions; fls. small, purplish within, solitary in axils; sepals 2.5 mm. long; caps. cylindrical, 4 mm. long; seeds black, rough, ca. 2 mm. in diam.—Occasional in saline spots; Alkali Sink; Imperial V., Desert Center, Yermo; to Tex., trop. Am. June–Nov.

5. Sesùvium L. SEA-PURSLANE

Usually prostrate or decumbent fleshy herbs or subshrubs. Lvs. opposite, exstipulate. Fls. solitary, axillary. Fl.-tube turbinate. Sepals 5, persistent, purplish or rose-pink inside. Petals 0. Stamens 5–many, inserted on the floral tube. Styles 3–5, separate. Caps. 3–5-locular, circumscissile at middle. Seeds minute, smooth, several to many in each locule. Ca. 5 spp., largely maritime. (Name unexplained.)

1. **S. verrucòsum** Raf. [*S. sessile* auth., not Pers.] Perennial, glabrous, freely branched; stems 1–5 dm. long or more, papillose; lvs. broadly spatulate, 1–4 cm. long; fls. almost sessile, 8–10 mm. long; sepals scarious-margined, short-horned near apex; caps. conic, ca. 5 mm. high; seeds black, shining, 1 mm. long.—Low, ± saline places; V. Grassland, Coastal Sage Scrub, Alkali Sink, etc.; cismontane Calif. from Sacramento V. s., deserts; to L. Calif., Tex., Mex. April–Nov.

6. Tetragònia L. SEA-SPINACH

Annual or perennial herbs or subshrubs. Lvs. alternate, often fleshy. Fls. axillary, solitary or few, sessile or not. Fl.-tube adnate to ovary, fleshy. Sepals 3–5. Petals 0. Stamens 1–many, perigynous. Ovary finally inferior, 3–9-loculed; styles 3–9. Ovule 1 in each locule. Fr. nutlike, indehiscent, in ours 2–5-horned. Seeds subreniform. Ca. 60 spp. of S. Hemis., especially S. Afr. (Greek, *tetra*, four, and *gonu*, knee or angle, referring to fr.)

1. **T. expánsa** Murr. NEW-ZEALAND-SPINACH. Succulent annual with many spreading or procumbent branches 3–several dm. long; herbage with small crystalline vesicles; lvs. triangular-ovate, entire or undulate, 2–5 cm. long, abruptly contracted at base to a cuneately winged petiole; fls. subsessile, solitary in axils, yellow-green; sepals spreading, 1.5–2.5 mm. long; fr. 8–10 mm. long, 2–5-horned; seeds brownish, ca. 2 mm. long; $2n = 32$ (Sugiura, 1936).—Natur. along the beaches and near salt marshes, along the coast to Ore.; native of se. Asia, Australasia. April–Sept.

7. Mesembryánthemum L. ICE PLANT

Mostly low succulent herbs without stipules. Lvs. opposite or alternate. Fls. axillary and terminal, often showy. Fl.-tube adnate to the ovary. Sepals 5, unequal. Petals many, linear, sometimes in several series, inserted together with the numerous and indefinite stamens upon the tube. Fils. slender. Ovary 5–12-loculed; styles as many as locules; ovules many. Fr. a fleshy caps., dehiscing when moist by stellate valves at the flattened summit. Seeds many, minute, somewhat compressed. Several hundred spp., largely African. (Greek, *mesembria*, midday, and *anthemon*, fl.)

A. Lvs. largely alternate.
  B. Annuals; herbage with shining colorless conspicuous papillae; fls. white to reddish.
    C. Lvs. linear, semiterete .... 1. *M. nodiflorum*
    CC. Lvs. ovate to broadly spatulate, flat .... 2. *M. crystallinum*
  BB. Perennial; herbage smooth; fls. bright yellow .... 3. *M. elongatum*
AA. Lvs. opposite; perennials; herbage usually smoothish.
  B. Lf.-blades flat, expanded; fls. red, 1 cm. wide .... 4. *M. cordifolium*
  BB. Lf.-blades 3-sided, or semiterete, linear.
    C. Stigmas and ovary-locules 8–20; lvs. 3-angled.
      D. Lvs. straight, not serrate, 3–5 cm. long; fls. rose-magenta, 3–5 cm. broad
                                       5. *M. chilense*
      DD. Lvs. curved, serrate on lower angle, becoming 7–10 cm. long; fls. yellow or purple, 8–10 cm. broad .... 6. *M. edule*
    CC. Stigmas and ovary-locules 4–6; lvs. mostly semi-terete.
      D. Fls. orange-red ....7. *M. speciosum*
      DD. Fls. purplish or rose to almost white.
        E. Lvs. 2.5–3.5 cm. long; fls. 3–4 cm. in diam. .... 8. *M. crassifolium*
        EE. Lvs. 1–1.5 cm. long; fls. 2–2.5 cm. in diam. .... 9. *M. floribundum*

1. **M. nodiflòrum** L. [*Cryophytum n.* L. Bolus.] Annual, freely branched from base, the branches 5–20 cm. long, suberect to procumbent, covered with fine colorless vesicles;

lvs. linear, subterete, 1–2 cm. long, 1–2 mm. thick; fls. solitary in axils, subsessile or on short thick pedicels; fl.-tube turbinate, 4–5 mm. long; sepals 4–5 mm. long; petals white, ca. 4 mm. long; ovary 5-loculed; seeds brown, finely papillate, ca. 1 mm. long.—Bluffs and low ground along the shore; Coastal Strand, Coastal Sage Scrub; Santa Barbara Co. to L. Calif.; native of Afr. and probably early introd. here. April–Nov.

2. **M. crystallinum** L. [*Cryophytum c.* N. E. Br.] Very succulent annual, prostrate, much-branched, the branches 2–6 dm. long, with large vesicles; lvs. ovate or spatulate, 2–10 cm. long, narrowed to a short amplexicaul base or somewhat petioled; fl.-tube campanulate, 8–12 mm. long; petals white to reddish, 6–8 mm. long; caps. 5-loculed; seeds brown, round-papillate, ca. 0.8 mm. long; $2n = 18$ (Sugiura, 1936).—More or less saline places; Coastal Strand, Coastal Sage Scrub; along the coast, Monterey Co. to L. Calif. Apparently natur. from Afr. March–Oct.

3. **M. elongàtum** Haw. [*Conicosia e.* Schwant. *M. pugioniforme* auth., not L.] Perennial, simple or branched, the stems 2–3 dm. long; lvs. largely alternate, linear, becoming 10–15 cm. long, semi-cylindric, channelled or semiterete; peduncles 1-fld., 8–12 cm. long; fl.-tube 1–1.5 cm. long, open-bowl-shaped; sepals ca. 2 cm. long, subequal; petals bright yellow ca. 2.5 cm. long.—Natur. on dunes, San Francisco, Pacific Grove, Pismo Beach; native of S. Afr. May–Oct.

4. **M. cordifòlium** L. f. [*Aptenia c.* N. E. Br.] Perennial, prostrate, the stems 3–6 or more dm. long; lvs. petioled, flat, ovate, 1–3 cm. long, acute; peduncles 8–15 mm. long; fl.-tube obconic, 6–7 mm. long; sepals unequal, the 2 larger flat, the others subulate; petals purple, ca. 5 mm. long; caps. 1.3–1.5 mm. high; seeds 1 mm. or more long; $2n = 18$ (Snoad, 1951).—Occasional escape from cult., as at Catalina Id., Ventura, La Jolla, San Diego; native of S. Afr. April–May.

5. **M. chilénse** Mol. [*Carpobrotus c.* N. E. Br. *M. aequilaterale* auth., not Haw.] Sea-Fig. Perennial with stems 1 m. or so long, forming extensive mats; lvs. opposite, 3-sided, 3–5 cm. long; fls. terminal, sessile or short-peduncled, 3–5 cm. broad; fl.-tube turbinate, 2–3 cm. long; sepals unequal, the larger foliaceous; petals magenta; ovary 8–10-loculed; seeds obovoid, somewhat compressed.—Sand dunes and bluffs along the coast; Coastal Strand, Coastal Sage Scrub, etc.; Ore. to L. Calif., Chile. April–Sept.

6. **M. édule** L. [*Carpobrotus e.* L. Bolus.] Hottentot-Fig. Like *M. chilense*, but with lvs. somewhat curved, serrate on lower angle, 6–10 cm. long; fls. 8–10 cm. in diam., yellow, drying pink; fr. becoming yellow, edible.—Much planted along highways and banks to control erosion; natur. on dunes along the coast; from S. Afr. April–Oct.

7. **M. speciòsum** Haw. [*Drosanthemum s.* Schwant.] Perennial, suffruticose; stems to 6 dm. long; lvs. semiterete, opposite, 2–4 cm. long; fls. solitary, terminal, peduncled, orange-red, ca. 4 cm. across.—Plants seemingly this sp. natur. along bluffs at Doheny Beach, Orange Co.; native of S. Afr. Oct.

8. **M. crassifòlium** L. [*Disphyma c.* Schwant.] Prostrate perennial with terete stems 3 or more dm. long; lvs. opposite, erect, bright green, 2.5–3.5 cm. long, semicylindric at base, triquetrous at apex; peduncles solitary, subcompressed, 2.5–3.5 cm. long, thickened upwards; fls. ca. 3–4 cm. wide; calyx subturbinate, the 5 segms. unequal; petals 2-seriate, purplish-red to pale, ca. twice as long as calyx-segms.; $2n = 36$ (Snoad, 1951).—Escape at Ventura and Ojai; native of S. Afr. Aug.–Oct.

9. **M. floribúndum** Haw. [*Drosanthemum f.* Schwant.] Perennial with slender subdecumbent branched stems 1–2 dm. long; lvs. opposite, cylindrical, somewhat curved, 1–1.5 cm. long, obtuse, glittering from minute papillae; peduncles solitary, 2–3 cm. long; fls. 2–2.5 cm. in diam.; calyx subturbinate, 2–2.5 cm. in diam., papillose, the 5 segms. linear, obtuse, equal; petals 1–2-seriate, rose, 2–3 times as long as calyx.—Established near beach, Santa Barbara and Ventura cos.; native of S. Afr. Summer.

38. Cactàceae. Cactus Family (Fig. 36)

Perennial, herbaceous or woody succulent plants, with columnar globose terete or flattened stems, often jointed. Ours leafless, except for small subulate, early caducous lvs. in *Opuntia*. Branches, spines, fls. and other parts developed from special structures called areoles which are situated in the lf.-axils when lvs. are present. Fls. usually perfect, solitary and sessile. Perianth with or without a tube and consisting of numerous outer segms. or sepals that commonly intergrade with the inner parts or petals, all

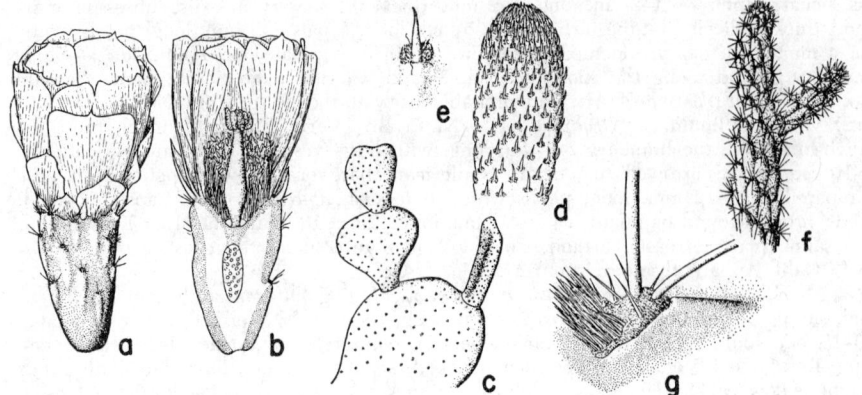

Fig. 36. CACTACEAE. *Opuntia occidentalis: a,* fl., × ½, inferior ovary, sepals grading into petals; *b,* fl. in section. *O. basilaris: c,* habit of a Platyopuntia; *d,* young joint with ephemeral fleshy lvs.; *e,* lf., × 2. *O. Parryi,* of Cylindropuntia group: *f,* habit with cylindrical joints; *g,* details of areole with heavier spines and finer glochids.

imbricated in several rows. Stamens many, the fils. inserted on the perianth-throat. Style 1; stigmas 2 to many. Ovary inferior, 1-celled; placentae 3 or more, parietal, many-ovuled. Fr. a berry or dry, often spiny or glochidiate, usually many-seeded. An Am. family with perhaps 1500 spp. and varying greatly as to generic treatment; found largely in dryer trop. and subtrop. regions; many with very showy fls.

(Baxter, E. M. Californica cactus, 1–93, 1935. Britton, N.L., and J. N. Rose. The Cactaceae. Carnegie Inst. Wash. Publ. 248, 1919–1923.)

Stems jointed, flattened or cylindrical; areoles with numerous minute barbed bristles (glochids) as well as spines; lvs. small, subulate, shed early . . . . 1. *Opuntia*
Stems not jointed, leafless; areoles without glochids.
  Spines borne in bundles on definite straight ribs; fls. borne on or contiguous with the spiniferous areole.
    Fls. nocturnal, borne on the spiniferous areoles; stems many times longer than thick; plant branched above base . . . . 2. *Cereus*
    Fls. diurnal, not borne on the spiniferous areoles; stems often not greatly elongate; plant simple or cespitose.
      The fls. lateral, not at stem-apex; frs. spiny . . . . 3. *Echinocereus*
      The fls. almost terminal; frs. not spiny . . . . 4. *Echinocactus*
  Spines borne on tubercles arranged in spiral rows; fls. borne between, not on or near the spiniferous areoles . . . . 5. *Mammillaria*

1. Opúntia Mill. PRICKLY-PEAR. CHOLLA

Somewhat woody plants with short-jointed stems; joints cylindrical or flattened, often tubercled, but never ribbed. Lvs. small, fleshy, subulate, soon shed. Areoles at axils of lvs. and bearing many short easily detached bristles or glochids and usually longer and stouter spines. Fls. on joints of previous year and on same areoles with the spines. Perianth-tube bearing scales resembling the lvs., short, cup-shaped. Outer perianth-segms. or sepals thick, green or partly colored and grading into the longer colored petals. Stamens shorter than petals. Ovary spiny or spineless; stigma-lobes short. Fr. fleshy or dry; seeds bony, pale, ± discoid or angled. Probably ca. 300 spp., from s. Canada to Straits of Magellan. (Old Latin name used by Pliny, formerly belonging to some other plant.)

The genus very difficult taxonomically, the spp. variable with different environment and with much natural hybridization. Only some of the plants found can be identified definitely by any key. The "prickly-pears" are those with flat joints and if they have edible frs. they are "tunas." Those with cylindrical joints are commonly called "cane-cactus" or "cholla." Many spp. are of ornamental value; some are grown as hedges and for the edible fr.

°A. Joints terete, cylindrical or clavate, tuberculate (Subgenus Cylindropuntia)
  B. Spines acicular, smooth, covered with loose papery sheaths.
    C. Spines solitary (or wanting), acicular; branches slender, 6–10 mm. thick, the old ones
      with solid woody axis; tubercles flat .... 1. *O. ramosissima*
    CC. Spines more than 1; ultimate branches stouter, the woody axis a reticulate cylinder;
      tubercles elevated.
      D. Ultimate joints disarticulating readily; fr. fleshy when first mature.
        E. Tubercles almost as wide as long. Widespread on the deserts .. 2. *O. Bigelovii*
        EE. Tubercles scarcely half as wide as long.
          F. Frs. not proliferous; petals ± yellowish-green.
            G. Spines reddish-brown; plants mostly 1–2 m. tall, with a crown
              of short horizontal branches. W. Colo. Desert .. 3. *O. × Fosbergii*
            GG. Spines yellowish; plants mostly 2–4 m. tall, with well-distributed
              subpendulous branches. Chocolate Mts. .... 4. *O. × Munzii*
          FF. Frs. persistent, proliferous; petals rose-red to red-purple. Near the coast
            5. *O. prolifera*
      DD. Ultimate joints not disarticulating very readily; fr. dry on maturity.
        E. Tubercles elongated, 2–3 times as long as wide.
          F. Frs. strongly tuberculate, the spines 10–12 per areole, 1–1.5 cm. long.
            Deserts .... 6. *O. acanthocarpa*
          FF. Frs. scarcely tuberculate, the spines fewer, less than 1 cm. long. Coastal
            drainage .... 7. *O. Parryi*
        EE. Tubercles short, less than twice as long as wide.
          F. Principal spines 3–10, with silvery to golden sheaths and 20–30 mm.
            long. Deserts .... 8. *O. echinocarpa*
          FF. Principal spines 7–20, with brownish sheaths and 6–18 mm. long. About
            San Diego .... 9. *O. serpentina*
  BB. Spines ± flattened, rugose or papillate, without sheaths.
    C. Tubercles distinct; joints clavate, mostly 3–7 cm. long. Mostly at elevs. above 2500 ft.
      10. *O. Parishii*
    CC. Tubercles tending to be confluent in rows; joints subcylindric, 10–20 cm. long. Near
      Colo. R. .... 11. *O. Wrightiana*
AA. Joints, at least some of them, flat or compressed, not tuberculate (Subgenus Platyopuntia)
  P Fr. dry; plants small or low; areoles 0.5–1.5 cm. apart.
    C. Fls. orchid to rose, rarely white; plants mostly spineless (except in ssp. *Treleasei*)
      12. *O. basilaris*
    CC. Fls. yellow to apricot-pink; plants spiny.
      D. Joints rather firmly attached, mostly 5–20 cm. long, flat. From Mono Co. s.
      13. *O. erinacea*
      DD. Joints easily detached, 2–4 cm. long, rather turgid. Siskiyou Co. ... 14. *O. fragilis*
  BB. Fr. fleshy; plants fairly large; areoles mostly 1.5–5 cm. apart.
    C. Plants mostly treelike with definite trunk, 1–5 m. tall.
      D. Spines yellow, many. Native desert plants .... 15. *O. chlorotica*
      DD. Spines white to brown, few. Introd. spp.
        E. Fr. yellowish or reddish, umbilicus flat; spines brownish .. 16. *O. megacantha*
        EE. Fr. purple, umbilicus depressed; spines white .... 17. *O. Ficus-indica*
    CC. Plants shrublike, much branched from base, mostly less than 1.5 m. tall.
      D. Stems subhorizontal, the joints 1.5–3 dm. long. E. Mojave Desert
      18. *O. mojavensis*
      DD. Stems ascending, or if subhorizontal, the joints mostly less than 2 dm. long. W.
      desert and coast .... 19. *O. occidentalis*

1. **O. ramosíssima** Engelm. [*Cylindropuntia r.* F. M. Knuth. *O. tessellata* Engelm.]
PENCIL CACTUS. Freely branched erect shrub 3–15(–20) dm. tall, bushy; branches
cylindrical, grayish, 6–10 mm. thick, with quite solid woody core, spiny to almost
spineless, covered with low diamond-shaped or obovate plates (tubercles), each of these
with areole at the apical notch; spines solitary, 2–5 cm. long, yellow-sheathed, each with
a deflexed bristlelike spine ca. 6 mm. long at the base; glochids minute; fls. yellow-
green, tinged red, perianth ca. 1 cm. long; frs. at end of short branches, dry, subcylindric, 2–2.5 cm. long, covered with tawny spines ca. 1 cm. long; seeds light tan,
discoidal, 4–5 mm. wide.—Dry washes, slopes and mesas, below 4000 ft.; Creosote
Bush Scrub, Joshua Tree Wd.; Calif. deserts to Nev., Son. April–May. Variable as to
stature, color (stems sometimes purplish) and spininess.

2. **O. Bigelòvii** Engelm. [*Cylindropuntia B.* F. M. Knuth.] JUMPING CHOLLA. BALL
CHOLLA. Erect, usually with a single trunk 6–15(–20) dm. tall, with short lateral
branches forming a close group above, old basal part of plant black; younger branches
cylindrical, 5–25 cm. long, 3.5–4.5 cm. thick, densely set with straw-colored spines;
branches readily detached; tubercles pale green, somewhat 4-sided, ca. 1 cm. long,

° Characters given for joints and tubercles are most applicable in the normal condition, not when
highly turgid at the end of the rainy season.

almost as wide; spines 6–8, 1.5–2.5 cm. long, barbed near apex, with persistent sheaths; petals yellow to pale green or with some lavender, 3–4 cm. long, several at end of year-old branches; frs. green, obovoid, 2–2.5 cm. long, spiny or almost spineless, with many glochids, with light brown seeds ca. 3–4 mm. long, or sterile and propagating vegetatively after falling to the ground.—Locally often very common on fans, benches and lower slopes, mostly below 3000 ft.; Creosote Bush Scrub; Colo. Desert and s. Mojave Desert; s. Nev., Ariz., L. Calif. April.

3. **O.** × **Fosbérgii** C. B. Wolf. [*O. Bigelovii* var. *Hoffmannii* Fosb.] Ascending to erect, 1–2.5 m. tall, the dead lower branches deciduous, the trunk thus bare of branches except at summit, there with a crown of mostly horizontal short branches; tubercles ca. 1 cm. long, half as wide; spines reddish-brown, 1.5–2 cm. long; fr. as in *O. Bigelovii* but red-spined, sterile.—On alluvial fans and open flats; Creosote Bush Scrub; from near Vallecito to head of Mason V., e. San Diego Co. April. Supposed hybrid between *O. Bigelovii* and *O. echinocarpa*.

4. **O.** × **Múnzii** C. B. Wolf. Plant erect, arborescent, 2–4 m. tall, almost as broad, the main trunk 10–15 cm. thick; lower branches rather bare; ultimate joints near ends of branches, somewhat pendulous, 6–25 cm. long, 3–5 cm. thick, readily detached; tubercles strongly raised, 10–16 mm. long, 5–6 mm. wide; areoles with short tan glochids and 10–12(–15) yellowish acicular subequal spines 1–2 cm. long; fls. few; petals 1.5–2 cm. long, yellowish-green, tinged red; frs. mostly sterile, subglobose, 2.5–3.5 cm. long, the tubercles with short glochids and 5–8 slender spines 5–10 mm. long, deciduous, occasionally bearing pale subglobose seeds ca. 4 mm. thick.—Dry gravelly places; Creosote Bush Scrub; along Beal's Well Wash, Chocolate Mts., Imperial Co. May. Apparently a hybrid between *O. Bigelovii* and *O. acanthocarpa*.

5. **O. prolífera** Engelm. Plant erect, bushy with from one to few trunks, 1–2 m. tall and of greater width, with many spreading branches; older stems to ca. 1 dm. thick, the terminal joints 3–15 cm. long, 2–3(–4) cm. thick, fleshy, easily detached, the tubercles 1–2 cm. long, ca. half as broad, with 4–12 rusty-yellow spines 1–2 cm. long and many long glochids; fls. light rose-red to red-purple, the inner petals ca. 1.5 cm. long; fr. subglobose, not tuberculate, fleshy, proliferous, usually sterile, 2–3 cm. long, spineless or nearly so, the glochids soon wearing off.—Forming vegetatively propagated thickets on arid slopes, below 600 ft.; Coastal Sage Scrub; near and along the coast, Ventura Co. to L. Calif.; Santa Rosa, Anacapa, Santa Catalina, San Clemente ids. April–June.

6. **O. acanthocárpa** Engelm. & Bigel. BUCKHORN CHOLLA. Spreading to erect, typically 1–2 m. tall and wider, or even 3 m. tall and openly branched, or sometimes ca. 3 dm. high and sprawling; main trunk short, branching open; branches cylindrical, the joints light green, to 3 dm. long, 2–3 cm. thick; tubercles 2–3 cm. long, ca. 6–8 mm. wide, laterally somewhat compressed; spines 10–12, stout, 2–3.5 cm. long, straw-colored; petals red to yellow or greenish-yellow, 2.5–3.5 cm. long; frs. obovoid, dry, shriveled, not proliferous, spiny, ca. 20 mm. long; seeds gray-brown, 5–6 mm. long.—Dry mesas and slopes, below 4500 ft.; Creosote Bush Scrub, Joshua Tree Wd.; Mojave and Colo. deserts e. of Twentynine Palms and Imperial V.; to Utah, Son. May–June.

Ssp. **Gánderi** C. B. Wolf. Joints 3–5 cm. thick, greener; spines (except the cent. one) more delicate, 15–20, 5–20 mm. long; inner petals ca. 2 cm. long.—Dry slopes, below 3500 ft.; Creosote Bush Scrub; w. edge of Colo. Desert (Riverside and San Diego cos.).

7. **O. Párryi** Engelm. [*O. bernardina* Engelm.] VALLEY CHOLLA. Plants few-stemmed, erect, 6–24 dm. tall; branches ascending, cylindric, the branching open; joints to 3 dm. or more long, 1–3 cm. thick, detachable; tubercles prominent, 2–2.5 cm. long, compressed, 4–6 mm. wide, the areole at the upper end white-woolly, with long yellow glochids and 6–10(–25) unequal yellow to brown slender spines 1–2.5 cm. long; fls. clustered at ends of branches, the outer perianth-segms. tinged red, the inner yellow or green, 2.5–3 cm. long; frs. globular to broadly obovoid, 1.5–2.5 cm. long, the areoles with glochids and the uppermost sometimes 1- to few-spined; sterile frs. drying early, the fertile green; seeds whitish, irregularly thick-discoid, ca. 5 mm. wide.—Common on dry gravelly fans and slopes, below 6000 ft.; Coastal Sage Scrub, Chaparral, Yellow Pine F.; Santa Barbara Co. (Cuyama V.) to w. edge of Colo. Desert in San Diego Co. May–June.

8. **O. echinocárpa** Engelm. & Bigel. [*O. e.* var. *robustior* Engelm. *O. deserta* Griffiths.]

SILVER CHOLLA. GOLDEN CHOLLA. Erect, intricately branched, 6–12(–15) dm. tall, with a short woody trunk and dense crown of cylindrical branches; joints mostly 1–2 dm. long, and 2–3 cm. thick, detachable, the tubercles conspicuous, 8–10 mm. long, 5–6 mm. wide and high, with ca. 3–10 silvery to golden spines 2–3 cm. long; glochids minute; fls. clustered at ends of branches; petals greenish-yellow, the outer sometimes streaked with red, the inner ca. 2.5 cm. long; frs. dry, many-spined, often not maturing, 1.5–2 cm. long; seeds white, thick-discoid, ca. 3 mm. wide.—Dry washes and mesas, below 6000 ft.; Creosote Bush Scrub, Joshua Tree Wd., Pinyon-Juniper Wd.; deserts from Mono Co. to L. Calif., Utah, Ariz. April–May.

Var. Párkeri Engelm. ex Coult. [*O. P.* Engelm. in syn.] Plants more sparingly branched; joints 3–4 cm. thick.—At 500–2500 ft. Creosote Bush Scrub; w. edge of Colo. Desert (Mt. Springs, Borrego V.).

9. **O. serpentìna** Engelm. [*Cereus? californicus* Nutt. ex T. & G. *O. c.* Cov., not Engelm.] Prostrate or ascending, 3–4 dm. high; branches to ca. 1 m. long, cylindrical, 1.5–2.5 cm. thick; young joints easily detached; tubercles 1–2 cm. long, 4–6 mm. wide, prominent, light green with 7–20 acicular brown spines 6–18 mm. long and with yellow-brown sheaths; fls. crowded at tips of joints; petals yellow-green with red outer sepals; frs. dry, broadly obovoid, spiny, 1–1.5 cm. long, often sterile; seeds whitish, irregularly thick-discoid, ca. 5 mm. long.—Dry slopes; Chaparral, Coastal Sage Scrub; canyons about San Diego; L. Calif. April–May.

10. **O. Parìshii** Orcutt. [*O. clavata* of Calif. refs. *O. Stanlyi* var. *P.* L. Benson.] DEVIL'S CACTUS. Stems low, creeping, forming dense patches 1–1.5 dm. high, 1–3 m. across; terminal joints clavate, ascending, 3–7(–12) cm. long, 2.5–4 cm. thick, the tubercles 1–1.5 cm. long, high, almost concealed by the dense armature of stout flattened rugulose spines, the areoles bearing a broad cent. spine up to 4 cm. long, then a ring of rounded or angled spines almost as long, then an outer set of slender spines 1–2 cm. long, surrounded by stiff glochids; young spines reddish, later whitish-gray; fls. yellow, or with a reddish tinge, the inner petals ca. 2.5 cm. long; frs. clavate, 4–5 cm. long, half as thick, yellowish-green, strongly tubercled, spineless but covered with yellowish bristles; seeds light brown, irregularly discoid, ca. 4 mm. broad.—Local on dry flats, 3000–5000 ft.; Creosote Bush Scrub, Joshua Tree Wd.; Little San Bernardino Mts. to Clark Mt.; w. Nev. May–June.

O. pulchélla Engelm., differing from *O. Parishii* by its more slender joints, more terete spines and purplish petals, has been collected in the White Mts. in Nev. just across the state line and may occur in Calif.

11. **O. Wrightiàna** (Baxter) Peeb. [*Grusonia W.* Baxter. *O. Stanlyi* var. *W.* L. Benson.] Low plants with erect or ascending branches 2–3 dm. high and forming clumps 1 m. or more across; joints 1–2 dm. long, 3–4 cm. thick; tubercles 12–20 mm. long, 4–5 mm. wide, arranged in rows and simulating ribs on the stems; spines flattened, ca. 10–20, pinkish to gray or straw-color, 15–35 mm. long, rough-papillate, the sheath wanting or vestigial; glochids large; petals yellow, few; frs. yellow, subcylindric, 3.5–5 cm. long, ca. 1.2 cm. thick, with short light spines and white glochids; seeds yellowish, ca. 4 mm. wide.—Valley floors; Creosote Bush Scrub; e. Imperial Co.; to Ariz., Son.

12. **O. basilàris** Engelm. & Bigel. BEAVER TAIL. Stems low and spreading, 1–3 dm. long, branched at base of joints; joints flattened, obovate, canescent to glabrous, somewhat papillate, 7–15(–20) cm. long, 5–12 cm. wide, glaucous, often purplish, frequently transversely wrinkled; areoles small, close together, spineless, filled with many short brown glochids; fls. orchid to rose, rarely white, clustered at upper edge of joints; petals to ca. 4 cm. long; frs. brownish to gray and dry when ripe, spineless, obovoid, to ca. 3 cm. long; seeds whitish, ± angled, 6–8 mm. wide; $2n = 22$ (Takagi, 1938).—Frequent, dry benches and fans, below 6000 ft.; Creosote Bush Scrub, Joshua Tree Wd.; Mojave and Colo. deserts; to Utah, Ariz., Son. March–June.

KEY TO VARIETIES

Areoles mostly bearing spines; branching largely from base of joints. Se. of Bakersfield, Kern Co.
var. *Treleasei*
Areoles mostly lacking spines.
  Joints ¼ as thick as wide. E. base Sierra Nevada ....var. *whitneyana*
  Joints mostly not so thick (except *brachyclada*).

The joints branched above, narrow-obovate. Interior valleys of coastal drainage ..... var. *ramosa*
The joints basally branched.
  Main joints 3–6 cm. long, from subterete to flat. Desert slopes, San Gabriel and San Bernardino
  mts. .... var. *brachyclada*
  Main joints 7–20 cm. long, flat, mostly broadly obovate. Widespread on deserts ... *O. basilaris*

Var. **brachyclada** (Griffiths) Munz. [*O. b.* Griffiths.] Joints small, reddish, thick, 3–6 cm. long, 2.5–5 cm. broad, branched near base, varying from subterete to distinctly flat; frs. 1–2 cm. long, subglobose.—Dry slopes, 4000–7500 ft.; Joshua Tree Wd.; desert slopes of San Gabriel and San Bernardino mts., Providence Mts. May–June.

Var. **ramòsa** Parish. [*O. basilaris* var. *albiflora* Walton, probably. *O. intricata* & *humistrata* Griffiths. *O. b.* var. *h.* Marshall & Bock. *O. brachyclada* ssp. *h.* Wiggins & Wolf.] Joints branched above, narrow-obovate, 20–30 cm. long, 10–15 cm. wide, glabrous to pubescent; areoles spineless, 1–1.5 cm. apart.—Washes and slopes, below 6000 ft.; Chaparral, S. Oak Wd., Coastal Sage Scrub; coastal drainage, San Gorgonio Pass, Riverside Co. w. to Claremont, Los Angeles Co., also in Greenhorn Range, Kern Co. and Vulcan Mt., San Diego Co. May–June.

Var. **whitneyàna** (Baxter) Marshall & Bock. [*O. w.* Baxter. *O. w.* var. *albiflora* Baxter. *O. basilaris* var. *a.* Marshall & Bock, not Walton.] Joints 4–15 cm. long, 3–10 cm. wide, 0.5–2.5 cm. thick, branched above or below; areoles spineless; fls. white to red.— Among rocks, 4300–9000 ft.; Creosote Bush Scrub, Pinyon-Juniper Wd.; Alabama Hills and e. slope of Sierra Nevada, Mono and Inyo cos. May–June.

Var. **Trelèasei** (Coult.) Toumey. [*O. T.* Coult. *O. T.* var. *kernii* Griffiths & Hare.] Branching largely from base, the joints narrowly to broadly obovate, 5–15 cm. long; areoles many, commonly with 1–3 spines 5–18 mm. long.—Common in dry open places, 500–1500 ft.; V. Grassland; se. of Bakersfield, Kern Co. May.

13. **O. erinàcea** Engelm. & Bigel. OLD MAN. PRICKLY-PEAR. Forming small low clumps with erect or ascending branches; joints ± oblong to obovate, flat, 6–15(–20) cm. long, 4–7 cm. wide, green but thickly beset with mostly grayish spines; areoles ca. 8–10 mm. apart, practically all spine-bearing; spines 4–7(–9), white to pale gray, or brownish especially when young, slender but ± stiff, 0.5–5(–6) cm. long, spreading-reflexed; glochids brownish, 1.5–3(–6) mm. long; fls. yellow, sometimes reddish in age, the inner petals 2–2.5 cm. long; frs. dry, short-spiny, subcylindric, 2–3 cm. long; seed 5–6 mm. wide.—Dry gravelly and rocky slopes, 2500–8000 ft.; Creosote Bush Scrub, Pinyon-Juniper Wd.; Mojave Desert from San Bernardino Mts. to Mono Co., Santa Rosa Mts. below 4000 ft.; to Utah, Ariz. May–June. Exceedingly variable.

KEY TO VARIETIES

A. All or almost all areoles bearing spines; plants usually grayish because of the spines; spines
   mostly 4–14 to an areole.
   B. Spines stiff and rigid, mostly 2–4(–5) cm. long; petals 2–2.5 cm. long .... *O. erinacea*
   BB. Spines, at least some of them, threadlike and flexible, 3–10(–15) cm. long; petals 3–3.6 cm.
       long .... var. *ursina*
AA. Areoles of lower half or ⅔ of joints mostly lacking spines; plants not so thickly beset with spines;
    spines of upper areoles 1–4 .... var. *xanthostemma*

Var. **ursìna** (A. Weber) Parish. [*O. u.* A. Weber.] GRIZZLY BEAR CACTUS. Like the sp. but spines 6–14 in number and 3–15 cm. long, the longer ones flexible and threadlike, strongly reflexed, almost matted; petals yellow, 3–3.5 cm. long.—Dry slopes, 3000–4500 ft.; Creosote Bush Scrub, Joshua Tree Wd., Pinyon-Juniper Wd.; Mojave Desert from Ord. Mts. to Ariz. May.

Var. **xanthóstemma** (K. Schum.) L. Benson. [*O. x.* K. Schum. *O. rhodantha* K. Schum. *O. e.* var. *r.* L. Benson. *O. e.* var. *paucispina* Dunkle.] Joints broadly obovate to, elliptic or oblong, 6–15 cm. long, 3.5–8 cm. wide; areoles of lower half of joints spineless or nearly so; spines of upper areoles 1–4, acicular to slightly flattened, white to brownish, mostly 2–7(–8) cm. long, reflexed or spreading; fls. reddish to salmon-yellow or yellow, the petals 2–3 cm. long; fr. subglobose to obovoid, 2–3 cm. long.—Dry slopes, 4000–11,000 ft.; Chaparral, Pinyon-Juniper Wd., Yellow Pine F.; Santa Rosa Mts., e. slope of Sierra Nevada and White Mts., Inyo and Mono cos.; to Nebr., Ariz. June–July.

14. **O. frágilis** (Nutt.) Haw. [*Cactus f.* Nutt.] PIGMY TUNA. Forming mats 3–12 dm. across and 1.5–2 dm. high; joints easily detached, subglobular to flattened or obovoid, dark green, 2–4 cm. long; areoles closely set, circular, woolly, with 1–5 spines, these

5–25 mm. long, whitish to brownish, slender; fls. pale yellow, the petals ca. 1.5–2 cm. long; frs. dry, spiny, 1.5–2 cm. long, the umbilicus very little depressed; seeds 5–7 mm. wide.—Dry places, 2900–6000 ft.; N. Juniper Wd.; Siskiyou Co.; to B.C., Wis., Tex. May–June.

15. **O. chlorótica** Engelm. & Bigel. [*O. curvospina* Griffiths.] PANCAKE-PEAR. Erect, treelike, 1–2.5 m. tall and almost as broad, with stout trunk and ascending branches; joints circular to broadly obovate, yellow-green, 1–2 dm. long, flat; areoles 1–2 cm. apart, prominent, each with many yellow glochids ca. 5–6 mm. long and 3–6 slender unequal yellow mostly reflexed spines 2–3.5 cm. long; fls. yellow, the inner petals 2–2.5 cm. long; frs. subglobose, tinged purple, fleshy, not spiny, ca. 3.5–5 cm. long; seeds thickish, 2.5–3 mm. long.—Usually on dry rocky walls, 3000–5500 ft.; Creosote Bush Scrub, Joshua Tree Wd., Pinyon-Juniper Wd.; e. Mojave Desert w. to Little San Bernardino Mts., along w. edge of Colo. Desert; to L. Calif., Nev., New Mex., Son. May–June.

16. **O. megacántha** Salm-Dyck. TUNA. Treelike plants 3–5 m. tall; joints obovate to oblong, 4–6 or more dm. long, thick; areoles 4–5 cm. apart, spineless or with few brown spines, the glochids soon deciduous; fls. yellow to orange, the petals 4–5 cm. long; fr. yellowish or reddish, ovoid, 7–12 cm. long, fleshy; seeds brownish.—Formerly much cult. in Calif. for hedges and the edible frs.; sometimes still persisting; native trop. Am. May–June.

17. **O. Fìcus-índica** (L.) Mill. [*Cactus F.-i.* L.] INDIAN-FIG. Much like *O. megacantha* in habit, the spines white, if present, somewhat flattened; fls. yellow to reddish; fr. purple, 5–9 cm. long, edible.—Cult., rarely natur.; native to trop. Am. May–June.

18. **O. mojavénsis** Engelm. & Bigel. [*O. phaeacantha* var. *m.* Fosb.] Plants open, the main stems subhorizontal, to ca. 1.5 m. long; joints erect, suborbicular to broadly elliptic, 1.5–3 dm. long, the areoles remote; spines red-brown at base, white to yellow or yellow-brown distally, principal one porrect, 3–6 cm. long, twisted near the middle, often quite angled and 1 mm. broad, the secondary spreading and shorter; fls. pale yellow to orange, the inner petals 3.5–4.5 cm. long; frs. spineless, obovoid, red-purple, 3.5–7 cm. long; seeds 3–4 mm. broad.—Dry washes and slopes, 4000–5000 ft.; Joshua Tree Wd., Pinyon-Juniper Wd.; New York, Providence, and Clark mts., e. Mojave Desert. May–June.

19. **O. occidentàlis** Engelm. & Bigel. [*O. Lindheimeri* var. *o.* Coult.] PRICKLY-PEAR. TUNA. Erect or spreading, forming great masses to ca. 1 or 1.5 m. high and 2–3 m. broad; joints narrow-obovate, 1.5–3 dm. long, glaucous, sometimes brownish in age, with remote areoles; spines slender, straight, somewhat flattened, dark brown at base, usually white toward tips, 1–3(–5) cm. long, the longest porrect, the other 2 principal ones almost as long, spreading downward; fls. yellow, the petals 4–5 cm. long; frs. narrow-pyriform to -obovoid, red-purple, 5–7 cm. long; seeds round, ca. 4–6 mm. broad.—Dry fans, washes, etc., below 2000 ft.; Coastal Sage Scrub, lower Chaparral; coastal drainage from San Luis Obispo Co. to L. Calif. May–June. Exceedingly variable and apparently freely hybridizing with the vars. treated below.

KEY TO VARIETIES

A. Spines white to brown.
  B. Fls. yellow; plants mostly highly spinose; joints mostly 1.5–3 dm. long.
    C. Plants ± erect, 5–15 dm. high; spines mostly more than 1.
      D. Principal spines usually not twisted, 2–3(–5) cm. long; frs. 5–7 cm. long. San Luis Obispo Co. to L. Calif. .... *O. occidentalis*
      DD. Principal spines often twisted, mostly 3–6 cm. long.
        E. Spines white (or with brownish base); fr. 6–12 cm. long. San Bernardino Mts. to e. San Diego Co. .... var. *megacarpa*
        EE. Spines red-brown; fr. ca. 5 cm. long. San Fernando to San Bernardino var. *Covillei*
    CC. Plants prostrate, scarcely 5 dm. high; joints 1–2 dm. long; spines frequently solitary, twisted, 4–8 cm. long. Desert slopes, San Gabriel Mts. to e. San Diego Co. .. var. *Piercei*
  BB. Fls. salmon to reddish-apricot; plants weakly spinose, the spines less than 2 cm. long; joints mostly less than 2 dm. long. S. slope San Gabriel Mts. to Santa Ana Mts. and Santa Rosa Mts. var. *Vaseyi*
AA. Spines yellow, 1–2.5 cm. long; fls. yellow. Coastal and insular ....var. *littoralis*

Var. megacárpa (Griffiths) Munz. [*O. m.* Griffiths. *O. Engelmannii* var. *m.* Fosb.] Plants to 5 dm. tall; joints 2–3 dm. long; spines heavier, more apt to be white throughout, the cent. one ± twisted at center; fr. 6–12 cm. long; seeds 6–8 mm. in diam.—Dry slopes at edge of desert, 3000–7000 ft.; Creosote Bush Scrub, Joshua Tree Wd., Pinyon-Juniper Wd.; mts. w. end of Mojave Desert to e. San Diego Co. May–June.

Var. **Covíllei** (Britt. & Rose) Parish in Jeps. [*O. C.* Britt. & Rose, *O. phaeacantha* var. *C.* Fosb.] Plants erect, bushy, to 1 m. or more tall; joints obovate to suborbicular, light green, 1–2 dm. long; spines red-brown, up to 6 cm. long, the principal one porrect, usually twisted in middle, the secondary shorter and somewhat spreading; fls. yellow, the inner petals 3–4 cm. long; fr. slender, red, ca. 5 cm. long.—Dry fans and washes at low elevs.; Côastal Sage Scrub; interior valleys, San Bernardino to San Fernando. May–June.

Var. **Pièrcei** (Fosb.) Munz. [*O. phaeacantha* var. *P.* Fosb. *O. demissa* Griffiths.] Plants prostrate; joints obovate to suborbicular, 1–2 dm. long; spines frequently solitary, 4–8 cm. long, dark red-brown to whitish, porrect on edges of joints, often deflexed on flat surfaces, sometimes twisted; inner petals yellow, 3.5–4 cm. long; fr. red, 4–5 cm. long.—Dry rocky slopes, 3000–7000 ft.; Chaparral, Joshua Tree Wd., Pinyon-Juniper Wd.; desert side of San Gabriel and San Bernardino mts. to e. San Diego Co. May–June.

Var. **Vàseyi** (Coult.) Munz. [*O. mesacantha* var. *V.* Coult. *O. V.* Britt. & Rose. *O magenta* Griffiths. *O. V.* var. *magenta* Parish. *O. rubriflora* A. Davids.] Stems prostrate; joints 1–2(–2.5) dm. long; spines 1–2, sometimes 0, less than 2 cm. long, whitish, deflexed; fls. deep salmon to orange-red; petals ca. 4 cm. long; frs. ellipsoid, 4–6 cm. long.—Washes, fans and dry slopes, below 4500 ft.; Coastal Sage Scrub, Chaparral; s. slope of San Gabriel Mts. to Puente Hills, Santa Ana Mts. and Santa Rosa Mts. May–June.

Var. **littoràlis** (Engelm.) Parish. [*O. Engelmannii* var. *l.* Engelm. *O. Lindheimeri* var. *l.* Coult. *O. l.* Ckll.] Plants large, to 1.5 m. tall; joints elliptic to orbicular, 2–3.5 dm. long; spines yellow, spreading, up to 2.5 cm. long; fls. yellow, the petals 4–5 cm. long; fr. pear-shaped to rounded, 3–5 cm. long.—Dry bluffs and slopes near the coast; Coastal Sage Scrub; Santa Barbara Co. (including ids.) to L. Calif. June.

2. Cèreus Mill.

Plants small to very large. Stems elongate, not jointed, leafless, ± branched above base, with continuous spine-bearing ribs; spines not hooked. Areoles not producing glochids. Fls. funnelform or salverform, borne on the mature spiniferous areoles below the stem apex. Fl.-tube ± elongate above the ovary. Ovary with distinct scales. Fr. a berry. Seeds small, black, numerous, usually tessellate and shining. As here treated, a rather large genus, with many plants of horticultural value. (Latin, candle or torch.)

Plant erect, 4–10 m. tall; ovary and fr. scaly; fls. white .... 1. *C. giganteus*
Plant prostrate with ascending branches 2–5 dm. long; ovary and fr. spiny; fls. yellow ... 2. *C. Emoryi*

1. **C. gigántèus** Engelm. [*Carnegiea g.* Britt. & Rose.] SAHUARO. GIANT CACTUS. Stems simple, or with few erect branches above, 4–10 m. tall, 3 or more dm. thick; ribs 12–24, 5–10 cm. apart and 1–3 cm. high; spines in clusters of 10–25, the cent. up to 5 cm. long; fls. nocturnal, white, in crownlike clusters near the ends of the brs. and 10–12 cm. long; fr. ovoid, fleshy, 6–9 cm. long, naked or sparsely spiny, green without, red within; seeds tessellate, black, shining, *n* = 11 (Stockwell, 1935).—Local in Calif. on gravelly slopes and flats, below 1500 ft.; Creosote Bush Scrub; near Colo. R., as in Whipple Mts. and near Laguna Dam; to Ariz., Son. May–June. Important food plant to Indians in Ariz.

2. **C. Emòryi** Engelm. [*Echinocereus E.* Ruempl. *Bergerocactus E.* Britt. & Rose.] Stems prostrate, 1–2 m. long, 3–3.5 cm. thick, with erect or ascending brs. 2–5 dm. long; ribs 15–25, the areoles closely set and bearing 10–30 straight slender unequal yellow spines 0.5–3(–5) cm. long; fls. greenish-yellow, ca. 3–3.5 cm. long; fr. spiny, globose, 1–1.5 cm. long; seeds obovoid, minutely tuberculate.—Dry bluffs and cliffs; Coastal Sage Scrub; along the coast, Orange Co. to L. Calif., Santa Catalina and San Clemente ids. May–June.

3. Echìnocèreus Engelm.

Stems low, erect or prostrate, sometimes pendent, single or cespitose, globular to cylindric, usually 1-jointed, strongly ribbed. Areoles borne on the ribs; spines of sterile and flowering areoles similar. Fls. usually large, sometimes small, diurnal, solitary at lateral areoles. Perianth campanulate to short-funnelform, scarlet, crimson, or purplish, the tube and ovary spiny, segms. few to many. Stigma-lobes green. Fr. fleshy, spiny, ± colored, thin-skinned. Seeds black, tuberculate. Ca. 60 spp. of arid w. U.S. and Mex.; some cult. for their fls. (*Echinos,* hedgehog and *cereus,* in reference to the spiny fr.)

Fls. scarlet; radial spines almost as long as and much like the centrals .... 1. *E. mojavensis*
Fls. purplish-red; radial spines much shorter than and unlike the centrals.
  Stems many, in mounds or clumps; spines curved .... 2. *E. Munzii*
  Stems 1–few; spines relatively straight .... 3. *E. Engelmannii*

1. **E. mojavénsis** (Engelm. & Bigel.) Ruempl. [*Cereus m.* Engelm. & Bigel. *E. triglochidiatus* var. *m.* L. Benson.] MOUND CACTUS. Cespitose, forming clumps or mounds; stems many, globose to oblong, 5–20 cm. long, pale green; ribs mostly 10–12, indistinct on old parts of stem, strongly tuberculate at the areoles; spines white to grayish, curving, tortuous, somewhat flexible, the centrals 1–3, 2.5–4 cm. long, the radials slightly shorter, ca. 6–10; fls. dull scarlet, 5–7 cm. long; perianth-segms. broad, obtuse; fl.-tube 1.7–2 cm. long; fr. oblong, spiny, 2–3 cm. long.—Rocky slopes, 3000–7000 ft.; Creosote Bush Scrub, Joshua Tree Wd., Pinyon-Juniper Wd.; desert slopes of San Bernardino Mts. to White and Clark mts.; Nev., Ariz. April–June.

2. **E. Múnzii** (Parish) L. Benson. [*Cereus M.* Parish. *E. Engelmanii* var. *M.* Pierce & Fosb.] Stems 8–60, oblong, erect, 1–2 dm. high, in compact clumps; ribs rather low; cent. spines 2–4, subulate, unequal, curved, 2.5–5 cm. long, gray or whitish, the radials 10–12, straightish, 1–2 cm. long; fls. crimson-magenta to pinkish, tubular-campanulate, 4–7 cm. long; fr. rose-red, obovoid, 2.5 cm. long.—Dry stony places, 4500–7000 ft.; Pinyon-Juniper Wd., lower edge Yellow Pine F.; San Bernardino Mts. (e. of Baldwin L.), San Jacinto Mts. (Kenworthy), San Diego Co. (se. of Julian); L. Calif. (se. of Tecate). May–June.

3. **E. Engelmánnii** (Parry) Ruempl. [*Cereus E.* Parry.] HEDGEHOG CACTUS. Stems 1–few, cylindrical, 1–3 dm. high, with 10–13 ribs; cent. spines 2–6, red, yellow, white, brown, or gray, mostly not much curved or twisted, 3–8 cm. long; radial spines 6–12, much shorter; fls. crimson-magenta to paler, 4–8 cm. long, the tube ca. 1.2 cm. long; fr. red, suglobose to ovoid, ca. 3 cm. long; $n = 22$ (Stockwell, 1935).—Common, gravelly slopes and benches, below 7200 ft.; Creosote Bush Scrub, Joshua Tree Wd.; Mojave and Colo. deserts to White Mts.; Utah, Ariz., Son., L. Calif. April–May.

4. Echinocáctus Link & Otto

Stems cylindric, oblong, obovoid or globose, large or small, not jointed, branching only at the base; tubercles usually coalesced to form continuous ribs, spiniferous in ours; leafless. Areoles not producing glochids. Fls. short-tubular, campanulate in ours, each from a well developed special areole just above an immature spine-producing areole at the stem apex; no distinct tube above the ovary. Fr. spineless, scaly and sometimes hairy. Seeds mostly black, small, smooth, pitted or tuberculate. Ca. 250 spp. of N. and S. Am.; some cult. (Greek, *echinos,* hedgehog or spine, and *cactus.*)

Principal spines annulate; fr. fleshy when first mature.
  Axils of ovary-scales naked; stem normally solitary.
    Plants cylindrical, 6–25 dm. high. Deserts .... 1. *E. acanthodes*
    Plants subglobose, 1–3 dm. high. Coast .... 2. *E. viridescens*
  Axils of ovary-scales copiously woolly; stems several .... 3. *E. polycephalus*
Principal spines not annulate; fr. dry, thin-walled.
  Spines subulate; ovary-scales naked in axils .... 4. *E. Johnsonii*
  Spines flattened; ovary-scales tufted in axils .... 5. *E. polyancistrus*

1. **E. acanthòdes** Lem. [*E. cylindraceus* Engelm. *Ferocactus a.* Britt. & Rose.] BARREL CACTUS. At first globular, later cylindric, to 2 m. or more tall, simple and erect, or 1–2-branched near base, stout, 3–4 dm. thick; ribs mostly 20–28, 1–2 cm. high, not clearly tubercled; principal cent. spine red to white, pink, or yellowish, spreading, often twisted, 5–14 cm. long, compressed, often hooked at apex, the accessory centrals 2–3, shorter and not recurved; radial spines 5–7 stout and 3–7 slender, shorter and less curved; fls. yellow, 3–6 cm. long; fr. greenish, 3 cm. long, cylindric-ovoid; seeds irregularly obovoid, cellular-pitted, ca. 2 mm. long.—Rocky slopes and walls, gravelly fans, below 5000 ft.; Creosote Bush Scrub, Joshua Tree Wd.; Mojave Desert except extreme

w. part, Colo. Desert; to Utah, Ariz., L. Calif. April–May. Variable, one poorly defined form being var. *Róstii* (Britt. & Rose) Munz. [*Ferocactus R.* Britt. & Rose.] Stems more slender; ribs 16–22; spine-clusters closely set and almost concealing body of plant; $n = 11$ (Stockwell, 1935).—Jacumba to L. Calif. *E. Lecontei* Engelm. [*Ferocactus a.* var. *L. Lindsay*], with the cent. spines less twisted and hooked and turned more definitely downward, occurs near the Colo. R. It is doubtfully distinct from *E. acanthodes*.

2. **E. viridéscens** T. & G. [*Ferocactus v.* Britt. & Rose.] Simple or rarely branched, globose or nearly so, 2–3(–4) dm. high; ribs 10–21, tuberculately irregular; cent. spines 3–4, stout, brown, flattened, 2–5 cm. long; radial spines 10–20, acicular, unequal, 1–2 cm. long; fls. yellow-green, 3–4 cm. long, each petal with a cent. reddish stripe; fr. greenish-red, ovoid to subglobose, 1.5–2 cm. long; seeds irregularly obovoid, black, 1 mm. long.—Dry hills; Coastal Sage Scrub, V. Grassland; about San Diego; L. Calif. May–June.

3. **E. polycéphalus** Engelm. & Bigel. NIGGER HEADS. Stems several, subglobose to short-cylindric, 2–3.5 dm. thick, in clumps of 10–30; ribs 10–20, scarcely tubercled; cent. spines 3–4, red or gray, spreading or curved, 4–7 cm. long; radial spines ca. 6–8, like the centrals but shorter; fls. yellow, 3–5 cm. long, enveloped in abundant wool; fr. densely woolly, ovoid, 2–4 cm. long; seeds angled, minutely tuberculate, 3–4 mm. long.— Rocky slopes, 2000–5000 ft.; Creosote Bush Scrub; n. Colo. Desert, Mojave Desert from Randsburg, San Bernardino Mts. and Panamint Mts. e.; to Utah, Ariz. March–May.

4. **E. Johnsònii** Parry. [*Ferocactus J.* Britt. & Rose. *Echinomastus J.* Baxter. *Echinocactus J.* var. *ocrocentrus* Coult.] Oblong, simple, sometimes few-branched, 1–2 dm. high, 0.8–1.2 dm. thick; ribs usually 18–20, narrow, somewhat tubercled; spine-clusters close, concealing the surface; spines reddish with some gray, the centrals 4–8 or 9, nearly straight, widely divaricate, tapering, stout, rigid, 2.5–3.5 cm. long, the radials ca. 10–12, much shorter; fls. 5–7 cm. long, deep red to pink; fr. pale green, cylindrical, 2 cm. long; seeds black, pitted, ca. 3 mm. long.—Infrequent, dry rocky slopes and washes, 2000–3000 ft.; Creosote Bush Scrub; n. of Kingston Mts., e. Inyo Co.; to Utah, Ariz. April–May. A form with yellow fls., chocolate-brown at base, from near Searchlight, Nev., is var. *lutéscens* Parish.

5. **E. polyancístrus** Engelm. & Bigel. [*Sclerocactus p.* Britt. & Rose.] Simple, oblong to columnar, 2–3(–4) dm. high, 0.7–1.2 dm. thick; ribs 10–15, somewhat tubercled; spines dense, hiding the surface; cent. spines ca. 8–10, almost perpendicular to the stem, the upper ones flattened, white, 4–12 cm. long, the others reddish, terete, often hooked; radial spines white, acicular, 1–2.5 cm. long; fls. magenta, 4–5 cm. long; ovary-scales few; fr. red-purple, 3–6 cm. long, obovoid; seeds black, pitted, ca. 4 mm. long.—Occasional, gravelly mesas and slopes, 2000–6000 ft.; Creosote Bush Scrub, Joshua Tree Wd.; Mojave Desert from Red Rock Canyon and Inyo Co. e.; Nev. April–June.

5. **Mammillària** Haw. PINCUSHION CACTUS. FISHHOOK CACTUS

Small plants, the stems globose to oblong, 1-jointed, single or few to many. Tubercles, many, teatlike, in crossing spiral rows. Areoles borne on the tubercles, spiniferous, not producing glochids. Fls. campanulate or funnel-shaped, borne in the axils of the tubercles or on the tubercles at the base of a groove, diurnal, not forming a tube above the ovary, usually with ciliate scales below the perianth. Fr. a smooth berry. Seeds black, usually pitted. A large genus of N. and S. Am., many with beautiful fls. (Latin, *mammila*, nipple.)

Fls. cent., borne in the axils of young usually nascent tubercles; none of spines hooked.
  Cent. spines 2–9; tubercles 1.2–2.5 cm. long.
    Fls. straw-colored or tinged with red or purple, 2.5–3.5 cm. long; radial spines 20–25
                                                               1. *M. deserti*
    Fls. rose, 3.5–5.5 cm. long; radial spines 15–20 . . . . 2. *M. arizonica*
  Cent. spines 12–16; tubercles less than 1 cm. long . . . . 3. *M. Alversonii*
Fls. lateral, borne in axils of old and mature tubercles; some spines hooked.
  Seeds with large corky appendage; radial spines 30–60; axils with bristles . . . . . . 4. *M. tetrancistra*
  Seeds without corky appendage; radial spines 11–30.
    Outer perianth-segms. ciliate; axils naked; stigmas cream . . . . 5. *M. microcarpa*
    Outer perianth-segms. not ciliate; axils with bristles; stigmas greenish . . . . 6. *M. dioica*

1. **M. déserti** Engelm. [*Coryphantha d.* Britt. & Rose. *M. vivipara* var. *d.* L. Benson. *M. radiosa* var. *d.* K. Schum. *M. arizonica* Calif. refs.] Stems mostly single, sometimes few from base, globose to oblong, 7–20 cm. high; cent. spines 2–4(–6), whitish with red-brown tips, straight, 1–1.5 cm. long; radial spines 15–25, somewhat shorter, gray, bristlelike, interlocking and ± obscuring the plant-surface; fls. ca. 2.5 cm. long, yellowish or amber, tipped with pink; outer perianth-segms. fimbriate; stigmas yellow or tinged pink; fr. oblong, green; seeds brown, minutely pitted.—Dry stony slopes, 1500–6000 ft.; mostly Joshua Tree Wd.; e. Mojave Desert; Ariz. April–May.

2. **M. arizónica** Engelm. [*Coryphantha a.* Britt. & Rose.] Mostly with solitary heads 6–10 cm. in diam. and scarcely as high; tubercles 2–2.5 cm. long, cylindric; cent. spines 3–6, deep brown above, 1–2.5 cm. long; radial spines 12–20, slenderer, rigid, unequal, 1–2.5 cm. long, white or ashy; fls. 3.5–5.5 cm. long, rose-pink; outer perianth-segms. linear, conspicuously fimbriate, the inner lance-linear; fr. ovoid; seeds light brown, pitted, compressed.—On dry slopes, 4000–6000 ft.; Pinyon-Juniper Wd.; e. Mojave Desert (Cima, New York Mts., Providence Mts., etc.); Nev., Ariz. May–June.

3. **M. Alversònii** (Coult.) Zeissold. [*Cactus radiosus* var. *A.* Coult. *M. r.* var. *A.* K. Schum. *M. arizonica* var. *A.* Engelm. ex Davids. & Mox. *M. vivipara* var. *A.* L. Benson. *Corphyantha A.* Orcutt.] FOXTAIL CACTUS. Stems 1 or few, short-cylindrical, 1–2 dm. high; cent. spines 12–16, straight, stout, divergent, purplish or dark above the white base, unequal, 1–2 cm. long; radial spines ca. 30–35, more slender, as long or longer, nearly concealing the plant-surface; fls. ca. 3 cm. long, magenta with deeper red midvein, the outer segms. ciliate; stigmas white; fr. clavate; seeds minutely tuberculate.—Stony slopes, 2000–5000 ft.; Joshua Tree Wd., Creosote Bush Scrub; Little San Bernardino Mts. to Eagle and Chuckawalla mts. May–June.

4. **M. tetrancístra** Engelm. [*Phellosperma t.* Britt. & Rose. *Neomammillaria t.* Fosb. *M. phellosperma* Engelm.] Body usually simple, oblong, 1–2.5 dm. high; tubercles 4–7 mm. high, the axils with some wool when young and a few long bristles, cent. spines 1–4, acicular, dark, one or more of them hooked, 1–1.5 cm. long; radial spines 30–60, mostly very slender and white, ca. 1 cm. long; fls. 2.5 cm. long, the inner petals white with rose or lavender mid-stripe, outer segms. ciliate; stigma-lobes cream; fr. scarlet, subcylindric, 1.2–2 cm. long; seed black, rugose, 2–3 mm. long, the corky hilum ca. as large as the body.—Occasional on dry slopes, below 2000 ft.; Creosote Bush Scrub; Colo. and Mojave deserts; to Utah, Ariz. April.

5. **M. microcárpa** Engelm. [*Neomammillaria m.* Britt. & Rose. *M. Grahamii* Engelm.] Low, the stems solitary or in clumps, 5–15 cm. high; tubercles cylindrical, 7 mm. high; axils naked; cent. spines 1–3, purplish-brown to light tan, 1–1.8 cm. long, one hooked, the others straight; radial spines 20–30, whitish or with brown tip, 6–12 mm. long; fls. 2–2.5 cm. long, the inner petals pink or lavender with darker mid-stripe, the outer ciliate; stigmas 6–8, light greenish; frs. scarlet and 2–2.5 cm. long and cylindric-clavate or sometimes green, subglobose, and ca. 1 cm. long; seeds black, glossy, ca. 1 mm. long.—Dry gravelly places; Creosote Bush Scrub; near Parker Dam, Colo. R.; to Tex., Chihuahua. April.

6. **M. dioìca** K. Bdg. [*Neomammillaria d.* Britt. & Rose. *M. Goodridgei* of Calif. ref.] Stems frequently branched, especially below, globose to cylindric, 5–25 cm. high; areoles 4–6 mm. high, woolly when young; axils with 5–15 bristles; cent. spines 1–4, brown, acicular, the lowest spine stout, ca. 6–10 mm. long, hooked; radial spines 10–20, white, slender, 3–8 mm. long, concealing the plant surface; fls. incompletely dioecious, ca. 2.5 cm. long, the petals creamy or yellowish with purplish or pinkish midrib, the outer segms. entire; stigma-lobes green; fr. scarlet, 1–2 cm. long, ovoid-cylindric; seeds black, minutely pitted; $n = 22$ (Remski, 1954).—Sandy places, below 500 ft.; Chaparral, Coastal Sage Scrub; near San Diego; L. Calif. Feb.–April.

Var. **incérta** (Parish) Munz. [*M. i.* Parish.] Cent. spines 1–2 cm. long, the radials ca. 1.2 cm. long; fls. 1.5–2 cm. long, the petals whitish with red-brown median stripe; fr. 1–2 cm. long; $n = 22$ (Lenz).—Dry slopes and canyon walls, 500–5000 ft.; Creosote Bush Scrub, Pinyon-Juniper Wd.; w. edge Colo. Desert, Santa Rosa Mts. to L. Calif. March–April.

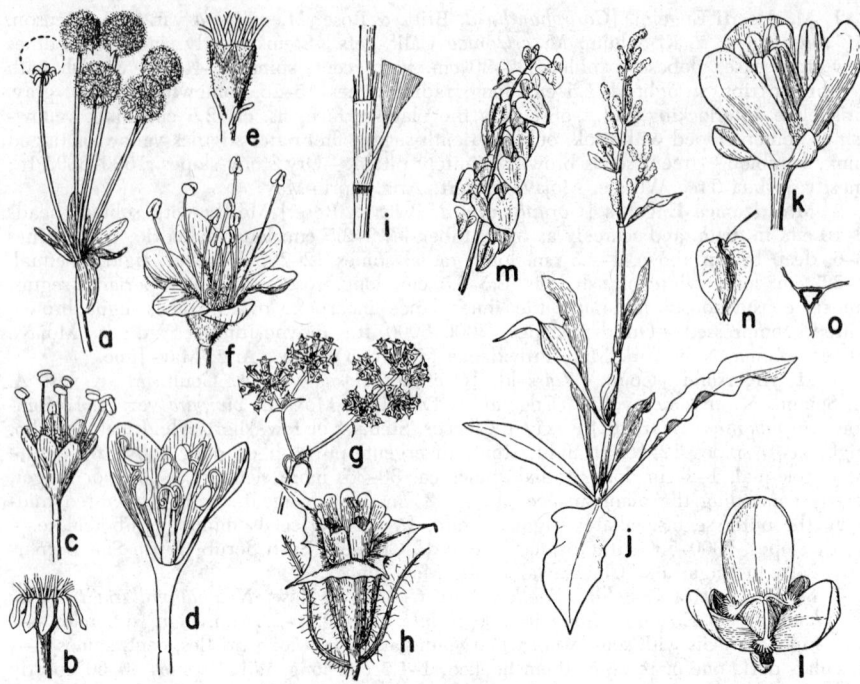

Fig. 37. POLYGONACEAE. *Eriogonum umbellatum: a,* an umbel with heads of fls., each head from an invol., as in *b,* with narrow reflexed lobes; *c,* fl., × 2, with calyx (perianth) 6-lobed and narrowed at base into cylindrical tube; *d,* fl. in section, × 5, with 6 stamens, 3 styles. *E. fasciculatum: e,* invol. with short lobes; *f,* fl., × 5, not drawn into a tube at base. *Chorizanthe: g,* infl., × ½; *h,* fl., × 4, invol. with 3 longer and 3 shorter flaring lobes. *Polygonum punctatum: i,* node to show cylindrical ocrea in lf.-axil and fringed above. *Rumex hymenosepalus: j,* habit; *k,* ♂ fl., × 5, with 3 outer sepals, 3 inner sepals, 6 stamens; *l,* ♀ fl., × 5, 3 styles, ovary, 3 inner sepals (valves), 3 outer sepals; *m,* fruiting branch, the valves enlarged; *n,* ♀ fl. in age, × 1, the 3 valves well developed; *o,* same in cross section.

39. Pólygonàceae. Buckwheat Family (Fig. 37)

Herbs, shrubs, or climbers, rarely trees. Lvs. alternate or rarely opposite, simple, with stipules sometimes dilated into a membranous sheath (ocrea) above the swollen node. Fls. mostly perfect, small, rarely solitary. Calyx ± persistent, 2–6-cleft or -parted, in 1 or 2 series, often petaloid. Petals 0. Stamens usually 6–9, inserted near the base of the calyx. Pistil solitary, superior, mostly 1-celled; styles 2–4, usually free. Ovule 1, basal. Fr. a lenticular or angled ak., sometimes enclosed in the ripening perianth which may become fleshy. Ca. 30–40 genera and 800 spp., widely distributed in cold and temp. regions. A few spp. are used for ornament and food (*Rheum,* rhubarb; *Fagopyrum,* buckwheat, etc.).

A. Lvs. without stipules.
    B. Fls. subtended by bracts or not, but not enclosed in an invol.
        C. Fls. without subtending bracts; stamens 9 .... 1. *Gilmania*
        CC. Fls. with 1 or more subtending bracts; stamens 3–9.
            D. Bracts woolly; fls. in headlike clusters.
                E. Calyx glabrous; stamens 3 .... 2. *Nemacaulis*
                EE. Calyx woolly; stamens 9 .... 3. *Hollisteria*
            DD. Bracts not woolly; fls. not in heads.
                E. Bracts solitary, 2-lobed, enlarged, reticulate and saccate in fr.; lvs. rounded
                        4. *Pterostegia*
                EE. Bracts whorled, hooked at tip; lvs. linear .... 5. *Chorizanthe*
    BB. Fls. enclosed in a tubular to campanulate invol.
        C. Invol. with spine- or bristle-tipped teeth.

D. Invol. commonly 1-fld. and mostly 4–5-toothed, with the tube generally cylindric or prismatic and the teeth often hooked at their tips . . . . 5. *Chorizanthe*
DD. Invol. commonly 2–several-fld. and mostly 4–5-lobed, with the tube turbinate and the lobes ending in straight spines or bristles . . . . 6. *Oxytheca*
CC. Invol. with 3–8 teeth or lobes, these not bristle- or spine-tipped . . . . . . 7. *Eriogonum*
AA. Lvs. with evident stipular sheaths.
B. Calyx 4- or 6-parted or urn-shaped and 6-lobed.
C. Calyx urn-shaped, becoming hard and burlike in fr.; outer lobes spine-tipped . . 8. *Emex*
CC. Calyx 4- or 6-parted, the outer lobes not spine-tipped.
D. Lvs. reniform; sepals 4; stigmas 2 . . . . 9. *Oxyria*
DD. Lvs. not reniform; sepals 6; stigmas 3 . . . . 10. *Rumex*
BB. Calyx 5-parted.
C. Aks. enclosed by the somewhat enlarged fruiting calyx, or exserted only in spp. with narrow lvs. . . . . 11. *Polygonum*
CC. Aks. much exserted from the scarcely enlarged calyx; rare annual with sagittate lvs.
12. *Fagopyrum*

1. Gilmània Cov.

Prostrate annual with divergent branches and yellowish herbage. Basal lvs. in a rosette, petioled, 3-nerved, ± obovate. Cauline lvs. 3 at the nodes, the uppermost sessile. Fls. yellow, pedicelled, fascicled at the nodes, without invols. or bracts. Pedicels jointed at base. Calyx 6-parted. Stamens 9. Styles 3. Ak. ovoid-triangular; cotyledons rounded. One sp. (M. French *Gilman*, 1871–1944, California naturalist.)

1. G. lutèola (Cov.) Cov. [*Phyllogonum l.* Cov. *Eriogonum l.* Jones.] Stems several, 3–15 cm. long, glabrous except for the thinly pilose infl.; lf.-blades 10–15 mm. long, entire, obtuse; pedicels 2–5 mm. long; sepals ca. 1.5 mm. long, linear-oblong; ak. buff-color, slightly longer than calyx, smooth, shining.—Near sea level, barren alkaline slopes; Alkali Sink; Death V. March–April. Local and developing only in wet years.

2. Nemacaùlis Nutt.

Slender-stemmed diffuse but sparingly branched annual. Lvs. mostly radical, spatulate, white-woolly, without stipules. Fls. small, in crowded sessile subglobose heads, perfect, each fl. with a free herbaceous bract. Calyx 6-cleft, enclosing the ak. Stamens 3. Styles 3. Ak. short-ovoid, obscurely 3-angled. One sp. (Greek, *nema*, thread, and *kaulos*, stem.)

1. N. denudàta Nutt. [*N. Nuttallii* Benth.] Stems several from the base, prostrate or usually ascending, reddish, 1–3.5 dm. long, glabrate; lvs. oblance-spatulate, 2–5 cm. long, woolly, ± crisped; cauline lvs. bractlike; bracts of infl. oblong-obovate, 2 mm. long, whorled, glabrous without, woolly within; fls. yellowish or pinkish, 1 mm. long, short-petioled; ak. brown, shining, ovoid, ca. 0.6 mm. long, acute.—Sandy places such as sea beaches, dunes, etc.; Coastal Strand; Los Angeles Co. to L. Calif. April–Sept.

Var. grácilis Goodm. & L. Benson. Wool of infl. very long and copious, almost or quite concealing fls.—Creosote Bush Scrub; w. Colo. Desert. March–May.

3. Hollistèria Wats.

Slender-stemmed diffuse white-woolly annual. Lvs. well distributed, the basal and lower oblanceolate, petioled, acute; the cauline cuspidate, alternate, gradually reduced upward, the uppermost ovate; stipules lanceolate, spine-tipped. Invol. solitary, sessile in axils, consisting of 3 almost distinct acicular bracts, each 2-fld. Calyx turbinate, 6-cleft to below middle, greenish-yellow, densely woolly without, the lobes scarious-margined. Stamens 6–9. Style divided to base. Aks. ovoid, glabrous, shining, dark gray, abruptly contracted to a 3-angled scabrous beak. One sp. (Col. W. W. *Hollister*, California pioneer.)

1. H. lanàta Wats. [*Chorizanthe floccosa* Jones.] Branches prostrate, 0.5–2 dm. long; basal lvs. 2.5–4 cm. long, the uppermost 0.7–1.5 cm. long; calyx ca. 2 mm. long; aks. ca. 1.7 mm. long.—Open slopes and plains, below 2500 ft.; V. Grassland; inner Coast Ranges and San Joaquin V., Merced and Monterey cos. to Kern Co. April–June.

4. Pterostègia F. & M.

Annual with dichotomous diffusely branched slender stems Lvs. opposite, fan-shaped to broadly elliptical, entire or 2-lobed. Bracts foliaceous, often reddish, each subtending a fl., 2-lobed and enlarged in fr., scarious and reticulate, loosely enclosing the ak., 2-gibbous on the back. Calyx reddish, mostly 6-parted into oblong-lanceolate segms. Stamens 3 or 6. Style 3-cleft. Aks. ovoid-triangular, glabrous. One sp. (Greek, *pteron*, a wing, and *stege*, covering, referring to the bract.)

1. **P. drymarioìdes** F. & M. [*P. diphylla* and *microphylla* Nutt.] Stems pubescent, mostly prostrate, 1–4 dm. long; lvs. short-petioled, 0.5–2 cm. long, often broader; bract 1.5–2 mm. long in fr.; calyx ca. 1 mm. long; ak. ca. 1 mm. long; $2n = 28$ (Sugiura, 1936).—Common, especially in shade of shrubs, rocks, etc., below 5000 ft.; many Plant Communities; cismontane Calif. to Ore., L. Calif.; occasional in desert. March–July.

5. Chorizánthe R. Br. ex Benth.

Low annuals (or S. Am. perennial), dichotomously or trichotomously branched, erect to prostrate. Lvs. basal or cauline and alternate, entire, the upper commonly reduced to opposite or whorled bracts. Infl. capitate or cymose. Invols. sessile, cylindric, urn-shaped or funnelform, sometimes lacking, 1-fld. or sometimes 2–3-fld., 3–6-angled, or -ribbed, 3–6-toothed or -cleft, the teeth divaricate, awned or uncinate. Fls. pedicellate or subsessile, included or partly exserted; bractlets lacking. Calyx 6-parted or -cleft (rarely 5), colored. Stamens 9, or 6 or 3. Styles 3. Ak. glabrous, 3-angled. An Am. genus of 50–60 spp. (Greek, *chorizo*, to divide, and *anthos*, fl., because of the divided calyx.)

(Goodman, G. J. A revision of the N. Am. spp. of the genus Chorizanthe. Ann. Mo. Bot. Gard. 21: 1–102, 1934.)

A. Invols. lacking; lvs. linear, the cauline in whorls of 4–5 .... 1. *C. coriacea*
AA. Invols. present.
    B. Bracts entire.
       C. Invols. 6-toothed and -ribbed (See also 5. *C. polygonoides*.)
          D. Teeth of the invol. without a scarious membrane along their margin.
             E. Involucral teeth equal or the outer 3 subequal, the anterior not greatly enlarged.
                F. Calyx-lobes entire or erose, but not fimbriate.
                   G. Outer and inner calyx-lobes alike or nearly so.
                     H. Plants erect; stems easily breaking up into joints when mature. Desert plants .... 2. *C. brevicornu*
                   HH. Plants prostrate to ascending; stems not very fragile.
                     I. Involucral teeth straight. Coast n. of San Francisco
                                    11. *C. cuspidata* var. *villosa*
                     II. Involucral teeth uncinate.
                       J. Calyx scarcely exserted, its lobes not apiculate.
                         K. Plants grayish, almost hoary, rather loosely pubescent. Coast of San Luis Obispo and Santa Barbara cos. .... 12. *C. angustifolia*
                         KK. Plants yellow-green, at least in upper parts, closely strigulose. From Los Angeles Co. s.
                                      29. *C. procumbens*
                    JJ. Calyx definitely exserted, the lobes apiculate. Coast, Santa Cruz Co. to Sonoma Co. .... 11. *C. cuspidata*
                 GG. Outer and inner calyx-lobes definitely unlike, the inner much smaller.
                   H. Involucral teeth straight, not hooked at tip; calyx-lobes erose.
                     I. Plants decumbent; invols. 3 mm. long. S. Calif.
                              28. *C. Parryi* var. *fernandina*
                     II. Plants erect; invols. 5–6 mm. long. Sonoma and Marin cos.
                              16. *C. valida*
                HH. Involucral teeth uncinate.
                   I. Bracts not leaflike, or those of the first node rarely so, but soon deciduous.
                     J. Lf.-blades 1–3 cm. long; invols. mostly in subcapitate clusters; calyx not half-exserted. San Diego to Monterey Co. and inland to San Bernardino and s. Mojave Desert ....· 19. *C. staticoides*
                     JJ. Lf.-blades 0.5–1.5 cm. long; invols. scattered, not crowded; calyx almost half-exserted. San Jacinto Mts. s. in the interior .... 20. *C. leptotheca*

II. Bracts leaflike on main stems and branches.
J. Calyx 5–6 mm. long. Plants from e. of the Coast
  Ranges .... 21. *C. Xanti*
JJ. Calyx 3–4 mm. long. Plants from near the coast.
  K. Foliaceous bracts mostly ovate to suborbicular
                 23. *C. Breweri*
  KK. Foliaceous bracts lanceolate.
    L. Heads of invols. 1–1.5 cm. in diam.; plants
      gray-pubescent, suberect. Coast from Santa
      Monica Mts. to Santa Barbara, also insular
                22. *C. Wheeleri*
    LL. Heads of invols. less than 1 cm. in diam.;
      plants greenish-yellow above, prostrate or
      decumbent. E. Los Angeles Co. to River-
      side Co. .... 28. *C. Parryi*
FF. Calyx-lobes, at least the inner three, fimbriate.
  G. Outer calyx-lobes entire or bilobed, the inner fimbriate
                 25. *C. Palmeri*
  GG. Outer and inner calyx-lobes fimbriate .... 24. *C. fimbriata*
EE. Involucral teeth very unequal, the anterior one usually longer than the
  involucral tube, the others relatively short.
  F. Elongate anterior involucral tooth straight; outer calyx-lobes linear-
    oblong .... 26. *C. uniaristata*
  FF. Elongate anterior involucral tooth uncinate; outer calyx-lobes ovate
              27. *C. Clevelandii*
DD. Teeth of the invol. with a scarious membrane along their margins.
  E. Membranous margins parted or divided at the sinuses, not continuous.
    F. Involucral teeth straight.
      G. Calyx ca. 4 mm. long; plants spreading to decumbent. Ft. Bragg,
        Mendocino Co. .... 14. *C. Howellii*
      GG. Calyx 5–6 mm. long; plants erect. Sonoma and Marin cos.
              16. *C valida*
    FF. Involucral teeth uncinate.
      G. Plants erect; invols. more than 4 mm. long.
        H. Lvs. 3–8 cm. long; involucral membrane pinkish to purplish.
          Coast, Alameda Co. to Monterey Co. .... 15. *C. robusta*
        HH. Lvs. 1–2 cm. long; involucral membrane with white margin.
          Interior, around Cent. V. .... 18. *C. stellulata*
      GG. Plants procumbent to ascending; invols. not exceeding 4 mm. in
        length.
        H. Calyx-lobes erose, the outer obovate to oblong
              10. *C. pungens*
        HH. Calyx-lobes not erose, the outer oblong to lanceolate.
          I. The calyx-lobes tipped with a short cusp.
            11. *C. cuspidata* var. *marginata*
          II. The calyx-lobes not tipped with a short cusp.
            13. *C. diffusa*
  EE. Membranous margins of the involucral teeth continuous through the sinuses.
    F. Calyx-lobes obcordate to bifid; membranous margins continuous but cleft
              18. *C. stellulata*
    FF. Calyx-lobes entire; membranous margins not at all cleft.
      G. Invols. somewhat hirsute, the membrane purplish; lvs. basal,
        oblanceolate .... 17. *C. Douglasii*
      GG. Invols. woolly, the membranes whitish; lvs. basal and on lower
        stems, sublinear .... 9. *C. membranacea*
CC. Invols. not both 6-ribbed and 6-toothed.
  D. Involucral tube angled or ribbed, not transversely corrugated.
    E. Involucral teeth and ribs 4–5, one tooth much exceeding the others; plants
      prostrate, with leafy bracts spinose-tipped. W. Mojave Desert .. 8. *C. spinosa*
    EE. Involucral teeth and ribs 3, no one tooth greatly exceeding others.
      F. The involucral teeth 6, with alternate shorter. Cismontane
              5. *C. polygonoides*
      FF. The involucral teeth 3.
        G. Plants prostrate; involucral teeth uncinate, equal. Point Loma.
          4. *C. Orcuttiana*
        GG. Plants erect; involucral teeth straight, unequal. Deserts
          6. *C. rigida*
  DD. Involucral tube cylindric, corrugated but not ribbed or angled.
    E. Involucral teeth 5, one leaflike and much enlarged. W. Mojave Desert to
      Lassen Co. ....3. *C. Watsonii*
    EE. Involucral teeth 3, equal. Death V. to the s. .... 7. *C. corrugata*
BB. Bracts 3-lobed.
  C. Invols. with 3–6 divaricate spurs at base; bracts small.
    D. Spurs 3, saccate, rather broad; terminal involucral teeth not hooked
              33. *C. Thurberi*
    DD. Spurs 6, spinelike; terminal involucral teeth hooked .... 32. *C. leptoceras*

CC. Invols. not spurred at base.
   D. Bracts conspicuous; involucral teeth unequal.
      E. The bracts unilateral; invols. cylindric, 2–3 in the axils .... 30. *C. californica*
      EE. The bracts orbicular-perfoliate; invols. 4-angled, mostly solitary
                                           31. *C. perfoliata*
   DD. Bracts minute; involucral teeth unequal.
      E. Invols. 4-angled, 4-toothed. Santa Lucia Mts. .... 34. *C. Vortriedei*
      EE. Invols. cylindric, 5-toothed. Inner S. Coast Ranges .... 35. *C. insignis*

1. **C. coriàcea** Goodm. [*C. Lastarriaea* Parry, in part. *C. L.* var. *californica* (H. Gross) Goodm. *Lastarriaea chilensis* auth., not Remy. *L. c.* ssp. *californica* H. Gross.] Brittle slender-stemmed prostrate to ascending plants, the stems diffusely branched, 0.5–2.5 dm. long, pubescent; lvs. linear, basal and whorled, ciliate, 2–2.5 cm. long, scarcely 1 mm. wide, passing above into whorled green bracts with hooked awns, these connate at base, 4–8 mm. long; fls. solitary in the axils, without invols.; calyx coriaceous, tubular, mostly 5-cleft to ca. the middle, 4 mm. long; lobes divergent, long-spinose, hooked, 2 lobes shorter than others; stamens 3; ak. triangular.—In dry sandy places and openings, below 2500 ft.; largely Coastal Sage Scrub, Chaparral; cismontane Calif. from Contra Costa Co. and Monterey Co. to L. Calif. April–June.

2. **C. brevicórnu** Torr. Erect, yellowish, glabrous to strigulose or pubescent, the stems several, ascending, 0.5–2 dm. long, breaking at nodes when dry; lvs. oblanceolate, largely basal, the blades 1.5–3 cm. long, petioles ca. as long; lower bracts foliaceous, lanceolate to oblanceolate, upper reduced, all opposite, apiculate; invols. solitary in axils of branches of cymose infl., narrow-subcylindric, conspicuously ridged, slightly curved, ca. 4 mm. long, with 6 subequal short uncinate teeth; fls. short-pedicelled, 3–4 mm. long, glabrous, the calyx-lobes whitish, similar, scarcely exserted, linear-oblong, ca. 1 mm. long; stamens 3; aks. narrow, ca. 2 mm. long.—Dry gravelly and stony slopes, below 5000 ft.; Creosote Bush Scrub, Joshua Tree Wd.; deserts from Mono Co. to Nev., Utah, L. Calif. March–June.

Ssp. **spathulàta** (Small) Munz. [*C. s.* Small.] Stems and invols. with more red; lvs. broadly spatulate; invols. more broadly cylindric, less conspicuously ridged.—Dry sandy and gravelly places, 5000–7500(–10,000) ft.; Sagebrush Scrub, Pinyon-Juniper Wd.; Panamint Mts. to Mono Co.; Ida., Nev. June–July.

3. **C. Watsònii** T. & G. Erect to ascending, canescent-strigose, the stems branched, 0.4–1.5 dm. long, sometimes reddish; basal lvs. oblanceolate, petioled, 2–3 cm. long, the lower cauline somewhat reduced, the upper bractlike; lower invols. solitary, upper in small clusters, canescent, cylindric, inconspicuously ribbed, ca. 4 mm. long, 5-toothed, the teeth short, recurved to hooked except the anterior which is foliaceous, lanceolate, 3–12 mm. long, terminally hooked; calyx 3–4 mm. long, yellow, scantily pubescent, the lobes oblong, acute, 0.5–1 mm. long, scarcely exserted; stamens 9; ak. narrow, beaked, ca. 3 mm. long.—Dry sandy or gravelly places, 2500–7500 ft.; Joshua Tree Wd., Sagebrush Scrub, Pinyon-Juniper Wd.; w. Mojave Desert to Lassen Co.; to e. Wash., Ida., Utah. April–July.

4. **C. Orcuttiàna** Parry. Branched from base, somewhat pubescent, the stems prostrate, 0.3–1 dm. long; lvs. basal, narrowly oblanceolate, petioled, 1–3 cm. long; lower bracts opposite, foliaceous, the upper reduced, acerose; invols. mostly solitary in axils of the cymes, 3-angled, the tube ca. 2 mm. long, the 3 teeth almost as long, squarrose, hooked at tip; calyx cylindric, 2–2.5 mm. long, with erect lance-linear pubescent lobes; stamens 9; aks. smooth, dark brown, narrow, 3-angled, almost 2 mm. long.—Sandy places; Coastal Sage Scrub; Point Loma and Kearney Mesa, San Diego Co. March–April.

5. **C. polygonoìdes** T. & G. [*Acanthogonum p.* Goodm.] Villous, the stems prostrate, forked, 0.5–1.5 dm. long; basal lvs. oblanceolate to elliptic, petioled, 2–5 cm. long; bracts opposite, the lower like the lvs. but smaller, the upper acicular; invols. 5 mm. long, solitary in axils or in small clusters, the tube obpyramidal, 3-angled, transversely corrugated, the 3 outer teeth 2–2.5 mm. long, hooked, the alternate short and inconspicuous; calyx whitish, scarcely exserted, the lobes oblong, erect; stamens 9; aks. ovoid-triangular, ca. 2 mm. long, gradually narrowed into the rather long beak.—Occasional but widely scattered, dry sandy or gravelly places below 5000 ft.; many Plant Communities; Modoc to Calaveras, Lake, and Marin cos. April–July.

Var. **longispìna** (Goodm.) Munz. [*Acanthogonum p.* var. *l.* Goodm.] Invols. smaller, more glabrate, the spine portion longer.—Chaparral; w. Riverside and San Diego cos.

6. **C. rígida** (Torr.) T. & G. [*Acanthogonum r.* Torr.] Stem erect, usually simple below, often short-branched above, 0.2–1 dm. high, ± woolly; principal lvs. broadly elliptic to obovate, petioled, 2–5 cm. long, woolly beneath, well distributed along stem; secondary lvs. bractlike, lanceolate or narrower, spine-tipped, becoming hard and thornlike; infl. very dense, with clusters of invols. in axils of bracts; involucral tube 2 mm. long, 3-angled, reticulate between angles, with 3 lanceolate spreading unequal straight or spine-tipped teeth 4–15 mm. long; calyx yellowish, scarcely exserted, the lobes oblong, hairy on back; stamens 9; aks. brown, ovoid, prominently beaked, ca. 1.8 mm. long.—Common in dry open stony places, such as desert mosaics, below 3000 ft.; Creosote Bush Scrub; deserts from Inyo Co. to Utah, Ariz., Son., L. Calif. March–May. The old dried plants persist as dense spiny masses for some time.

7. **C. corrugàta** (Torr.) T. & G. [*Acanthogonum c.* Torr.] Erect, tufted, the stems several from base, much-branched above, ± tomentose, 0.5–1.3 dm. high; basal lvs. round-ovate, 1–2 cm. broad, petioled, woolly especially beneath, the lower cauline narrower, spatulate, the upper more bractlike, acicular; invols. solitary in axils, numerous, the upper congested, the tube cylindrical, transversely corrugated, 2–3 mm. long, glabrate, the teeth 3, lanceolate, squarrose, as long as tube or longer; calyx subcylindric, 2–2.5 mm. long, white, the segms. oblong, pubescent, subequal; stamens 6; aks. slightly exserted, subcylindric, ca. 1 mm. long, papillose at apex.—Dry rocky places, below 3000 ft.; Creosote Bush Scrub; deserts from Death V. region to L. Calif., Ariz. Feb.–May.

8. **C. spinòsa** Wats. [*Eriogonella s.* Goodm.] Prostrate, loosely branched, the stems 0.5–2.5 dm. long, puberulent; lvs. basal, broadly oblong, petioled, 2–3 cm. long, woolly beneath, villous above; bracts in whorls of 3, lanceolate, stiff, 5–15 mm. long, spine-tipped, the floral subulate-acicular; invols. in small axillary clusters, canescent, 3–4 mm. long, 4–5-toothed, the teeth straight, unequal, 1 usually much exceeding the others; calyx white, well-exserted, the 3 outer lobes broader than long, the 3 inner narrower and smaller; stamens 9; aks. dark brown, ca. 2 mm. long, subovoid, with a stout, 3-angled, papillose beak.—Occasional, dry sandy and gravelly places, 2500–3500 ft.; Creosote Bush Scrub, Joshua Tree Wd.; Mojave Desert from Rabbit Springs w. to Red Rock Canyon and Mojave. April–July.

9. **C. membranàcea** Benth. [*Eriogonella m.* Goodm.] Erect, mostly simple below, few-branched above, woolly, floccose in age, 1–4.5 dm. high; lvs. basal and alternate on lower stem, linear to narrowly oblanceolate, 1.5–6 cm. long, glabrate above, white-woolly beneath; bracts opposite or whorled, like the lvs. but reduced; invols., except for the very lowest, in dense heads at upper nodes and terminal, woolly, urn-shaped, 4–5 mm. long, 3-angled, the 6 teeth reddish, slender, hooked, subequal, united by a broad pale membrane; calyx woolly, the outer segms. obovate, the inner spatulate, scarcely exserted; stamens 9; aks. brownish, shining, lance-ovoid, somewhat 3-angled, ca. 2 mm. long, tapering into a pointed beak.—Dry rocky slopes, below 4500 ft.; Foothill Wd., Chaparral, V. Grassland; inner Coast Ranges from Mendocino and Siskiyou cos. to Ventura Co., Sierran foothills from Butte and Tuolumne cos. to Kern Co. April–July.

10. **C. púngens** Benth. [*C. Douglasii* var. *albens* Parry.] Stems prostrate or ascending, several from base, grayish-villous, 1–3 dm. long, branched; lvs. basal, petioled, oblanceolate, 3–5 cm. long; bracts opposite, the lower like the lvs., the uppermost acerose; invols. in dense capitate clusters, villous-hirsute, the tube subcylindric, 2–3 mm. long, the teeth 6, divergent, mostly hooked at tip, the alternate 3 larger, all with a broad pale membrane below separated at the sinuses; calyx hairy, partially exserted, 3–3.5 mm. long, the lobes subsimilar, erose, the 3 outer obovate or oblong, the inner slightly narrower and shorter; stamens 9; aks. red-brown, narrow, slightly angled, ca. 2 mm. long, gradually narrowed into a stout beak.—Sandy places near the coast; Coastal Strand, Closed-cone Pine F.; Monterey Peninsula. April–June.

Var. **Hartwégii** (Benth.) Goodm. [*C. Douglasii* var. *H.* Benth.] Involucral membrane usually purplish; outer calyx-lobes more obviously obovate.—Sandy places in hills; N. Coastal Scrub; Santa Cruz Mts. and San Francisco. April–June. An ill-defined variety.

11. **C. cuspidàta** Wats. [*C. pungens* var. *c.* Parry.] Prostrate or decumbent, villous, the stems 1–2.5 dm. long; lvs. basal, petioled, oblanceolate, 2–3.5 cm. long; bracts opposite, the lower like the lvs., the upper acerose; invols. in capitate clusters, campanu-

late to urceolate, the tube ca. 2 mm. long, ± pubescent, the teeth spreading, usually hooked at tip, almost or quite without membranous margin, the 3 alternate larger; calyx partly exserted, subcylindric, 2–2.5 mm. long, hairy, the segms. oblong, entire and tipped with a short cusp, the outer segms. slightly larger than inner; stamens 9; aks. brownish, narrowly ovoid-triangular, ca. 2 mm. long.—Sandy soil and dunes; Coastal Strand, N. Coastal Scrub; Santa Cruz Co. to Sonoma Co. May–July.

Var. villòsa (Eastw.) Munz [*C. v.* Eastw.] Involucral teeth straight and without marginal membrane.—Similar places, coast of Sonoma and Marin cos.

Var. marginàta Goodm. Involucral teeth hooked at tip, with evident membranous margins.—Coastal sandhills, San Francisco to Monterey Co. May–July.

12. **C. angustifòlia** Nutt. [*C. a.* var. *Eastwoodae* Goodm.] Prostrate or decumbent, subappressed-canescent, the stems several from base, 1–4 dm. long; lvs. basal, petioled, oblanceolate, 2–5 cm. long, gray-villous; bracts opposite, similar, the upper reduced; invols. in capitate clusters, the tube cylindric, 3-angled, ca. 2 mm. long, hairy, the teeth 6, slender, spreading, hooked at apex, without membranes, the alternate larger; calyx 2–3 mm. long, scarcely exserted, cylindric, the lobes oblong, erose at the acute to truncate summit; stamens 3 to 9.—Sandy places near the coast; Coastal Strand, Coastal Sage Scrub; San Luis Obispo and Santa Barbara cos. April–June.

13. **C. diffùsa** Benth. [*C. pungens* var. *d.* Parry. *C. p.* var. *nivea* Curran. *C. n.* Jeps. *C. Andersonii* Parry.] Stems 1 to several, decumbent to ascending, thinly pubescent, 0.6–2.5 dm. long; lvs. basal, petioled, oblanceolate, 2–4 cm. long, villous-tomentose beneath, less so above; bracts opposite, the lower leafy, the upper acerose; invols. in small cymose groups, ca. 3 mm. long, the tube with wide-spreading hairs, 3-angled, ca. 2 mm. long, the teeth 6, spreading, hooked, with a white marginal membrane parted at the sinuses, the alternate 3 teeth shorter than others; calyx subcylindric, ca. 2.5 mm. long, partly exserted, the lobes similar, oblong, entire, acutish; stamens 9; aks. reddish brown, narrow, triangular, scarcely beaked, ca. 2 mm. long.—Sandy and gravelly places, at low elevs.; N. Coastal Scrub, Chaparral, Closed-cone Pine F.; San Mateo Co. to Santa Barbara Co. April–July.

14. **C. Howéllii** Goodm. Stems several from base, spreading to decumbent, 1–2 dm. long, conspicuously villous; lvs. basal, petioled, spatulate to broadly obovate, 3–5 cm. long, gray, long-villous beneath, sparsely so above; bracts opposite, like the lvs., but the uppermost reduced; invols. in dense clusters, the tube subcylindric, 3 mm. long, hairy, the teeth 6, spreading, margined with a membrane parted at the sinuses, not hooked at tip, the 3 alternate shorter; calyx subcylindric, ca. 4 mm. long, the lobes oblong, truncate, denticulate at apex, the outer slightly longer than inner; stamens 9; aks. light-colored, narrow, 3-angled, ca. 3 mm. long.—Sand dunes; N. Coastal Scrub; Ft. Bragg, Mendocino Co. May–July.

15. **C. robústa** Parry. [*C. pungens* var. *r.* Jeps.] Erect or ascending, 1.5–5 dm. tall, the stems 1 to few from base, branched, villous-hirsute; lvs. basal or near base, petioled, oblanceolate, 3–8 cm. long, hirsute-villous; bracts opposite or whorled, similar to lvs., the upper reduced; invols. in large dense clusters, the tube subcylindric, 3.5–4 mm. long, hirsute, the teeth 6, margined by a pinkish or purplish membrane which is parted at sinuses, the spines straight to hooked, minute, the 3 outer teeth slightly longer than inner; calyx narrowly obconic, 3–4 mm. long, the lobes oblong-elliptic, rounded-erose, subequal; stamens 9; aks. brownish, narrow, dull, ca. 2 mm. long, gradually attenuate into a somewhat 3-angled beak.—Mostly dry sandy places, below 1000 ft.; N. Coastal Scrub, Coastal Strand; Alameda Co. to Monterey Co. May–Sept.

16. **C. válida** Wats. Erect, simple below, mostly few-branched above, 1–3 dm. tall, villous; lvs. basal or subbasal, broadly oblanceolate, petioled, 3–6 cm. long, villous especially beneath; bracts like the lvs., the uppermost reduced, acerose; invols. in large dense capitate leafy-bracted clusters, the tube 4 mm. long, subcylindric, somewhat strigose to glabrate, the teeth 6, suberect, bordered with inconspicuous membrane, straight-spined, straw-colored; calyx cylindric, 5–6 mm. long, the outer lobes oblong, truncate, erose, almost 2 mm. long, the inner narrower, ca. 1 mm. long; stamens 9.— Sandy places; N. Coastal Scrub; Sonoma and Marin cos. June–Aug.

17. **C. Douglàsii** Benth. [*C. Nortonii* Greene.] Erect, 1–4 dm. tall, villous-tomentose, the stems 1–few; lvs. basal, oblanceolate, 2–8 cm. long, petioled, short-villous; bracts whorled, like the lvs. or the upper acerose; invols. in dense rounded headlike clusters,

red-purple, the tube 3 mm. long, hirsute, the teeth 6, spreading, short-hooked, bordered by a broad purplish membrane continuous through the sinuses; calyx 3–5 mm. long, the outer lobes obovate to oblong, denticulate, ca. 1 mm. long, apiculate, the inner a little shorter, emarginate; stamens 9; aks. narrow, 3-angled, ca. 3.5 mm. long.—Sandy or gravelly slopes, below 5000 ft.; Foothill Wd., Yellow Pine F.; San Benito and Monterey cos. to San Luis Obispo Co. April–July.

18. **C. stellulàta** Benth. Erect, 0.5–2.5 dm. tall, hirsute, simple below the infl.; lvs. basal, sessile or short-petioled, lanceolate to oblanceolate, 1–2 cm. long; bracts opposite or whorled, like the lvs. or the upper acerose; invols. in dense clusters, 4.5–5 mm. long, the tube 6-ribbed, transversely wrinkled between the ribs, ca. 4 mm. long, the teeth 6, short, similar, recurved or hooked, united by a white or purplish membrane which is lobed but extends through sinuses; calyx subcylindric, 4–4.5 mm. long, pale yellow, the outer segms. broadly obovate, obcordate to bilobed, 2–2.5 mm. long, the inner similar but somewhat smaller; stamens 9; aks. slender, grayish, 3-angled, ca. 3 mm. long.—Dry sandy or gravelly places, below 2500 ft.; Foothill Wd.; hills bordering Cent. V. from Shasta Co. to Lake and Tulare cos. May–June.

19. **C. staticoìdes** Benth. [*C. nudicaulis* Nutt. *C. discolor* Nutt. *C. s.* var. *n.* Jeps. *C. s.* forma *bracteata* Goodm.] Turkish Rugging. Stems erect or ascending, 1–2 dm. high, 1–few from base, mostly simple below, branched above, usually reddish purple, ± pubescent; lvs. basal, oblong to oblong-ovate, green above, tomentose beneath, petioled, 2–6 cm. long; bracts subulate, often recurved; invols. solitary in the forks or congested at the ends of the branchlets, cylindric, pubescent to glabrate, 6-ribbed, 3–4 mm. long, 6-toothed, the teeth spreading, not membraned on margins, hooked, 1–2 mm. long, 3 larger than the alternate; calyx mostly rose, sometimes paler, 4–5 mm. long, hairy externally, the lobes narrowly oblong, subentire, obtuse; stamens 9; aks. very narrow, ca. 3 mm. long, not much narrowed toward the somewhat 3-angled apex.—Dry slopes and flats, below 4000 ft.; Coastal Sage Scrub, Chaparral; Monterey Co. to San Bernardino and San Diego cos. April–July. Exceedingly variable and with many segregates proposed, such as:

Var. **brevispìna** Goodm. Involucral teeth less than 1 mm. long.—S. face of San Gabriel Mts. from Pasadena to Claremont and up to 5000 ft.

Var. **latilòba** Goodm. Outer calyx-lobes obovate, truncate.—Desert slopes of San Gabriel Mts.

Var. **elàta** Goodm. Plants 3–6 dm. high, much-branched; lvs. 7–10 cm. long.—Santa Ana Mts. to San Bernardino and San Gabriel Mts. Often on burns in chaparral and may be ecological.

Ssp. **chrysacántha** (Goodm.) Munz [*C. c.* Goodm. *C. c.* var. *compacta* Goodm.] Infl. with few large dense clusters; invols. 5–7 mm. long, the tube 4–5 mm. long; calyx 5–5.5 mm. long, the outer lobes broadly oblong-ovate, ca. 2 mm. long, the inner oblong, shorter; aks. 3 mm. long.—Ocean bluffs; Coastal Sage Scrub; Orange Co. April–May.

20. **C. leptothèca** Goodm. Resembling *C. staticoides* in habit, but with more numerous stems, 1–2 dm. long, scantily pubescent; invols. 4–5 mm. long, not crowded, the tube very slender, curly-pubescent, the 6 teeth spreading-uncinate, the 3 inner smaller than the outer; calyx well exserted, 4.5–5 mm. long, the outer segms. linear, obtuse, the inner smaller; stamens 9; aks. very slender, 2.5 mm. long, scarcely narrowed into the stout beak.—Dry slopes, at 3000–6000 ft.; Chaparral; San Jacinto Mts. to e. San Diego Co.; adjacent L. Calif. May–July.

21. **C. Xánti** Wats. Erect plants, simple or branched from base, 0.5–2.5 dm. high, with gray appressed pubescence; lvs. basal, petioled, oblong-ovate to ovate, 2–5 cm. long; lower bracts leaflike, upper acerose; invols. not in subcapitate clusters but in more open cymes, cylindric, 4.5–6 mm. long, the tube ca. 4 mm. long, canescent, the teeth 6, divergent, hooked, the 3 outer larger than inner; calyx white to pink, subcylindric, 5–6 mm. long, the outer lobes oblong or elliptic, ca. 2 mm. long, the inner somewhat shorter and narrower; stamens 9; aks. very slender, scarcely narrowed into the stout beak, ca. 2.5 mm. long.—Dry slopes and washes, 1000–6000 ft.; Pinyon-Juniper Wd., Foothill Wd., Joshua Tree Wd.; Bakersfield region and Inyo Co. to San Bernardino Mts. April–June.

Var. **leucothèca** Goodm. Invols. white-woolly, 4 mm. long; calyx scarcely evident.—

Below 3500 ft.; Creosote Bush Scrub; edge of Colo. Desert from n. base of Santa Rosa Mts. to foot of San Bernardino Mts. April–May.

22. **C. Wheèleri** Wats. [*C. insularis* R. Hoffm.] Erect to spreading, simple or branched at base, 0.6–2.5 dm. high, gray-pubescent; lvs. basal, 2–3 cm. long, oblong to ovate-spatulate, petioled, tomentose beneath, villous above, lower bracts leaflike, floral bracts acerose; invols. in crowded clusters, cylindric, 2.5–3 mm. long, pubescent to glabrate, the teeth short, widely divergent, uncinate; calyx cylindric, 3–3.5 mm. long, the lobes entire or nearly so, the outer elliptic-oblong to lanceolate, obtuse, 1.3–1.5 mm. long, the inner shorter and narrower; stamens 6; aks. linear, 3-angled upward, ca. 3 mm. long.—Dry slopes; Chaparral; Santa Cruz and Santa Rosa ids., mainland coast from near Santa Barbara to Santa Monica Mts. April–July.

23. **C. Brèweri** Wats. Plants ascending to decumbent, branched from near base, 0.5–1.5 dm. high, appressed-cinereous; lvs. basal, spatulate to ovate, pubescent, petioled, 1.5–4 cm. long; lower bracts leaflike, upper acerose; invols. in small dense clusters, reddish, subcylindric, 3–4 mm. long, pubescent, the teeth short, divergent, uncinate, the inner shorter than the outer; calyx 3–3.5 mm. long, the outer lobes elliptic to obovate-oblong, the inner shorter, broadly ovate; stamens 9.—Dry rocky places, mostly of serpentine; Chaparral, Foothill Wd.; San Luis Obispo to Atascadero. May–June.

24. **C. fimbriàta** Nutt. Stems 1 to several from base, erect or ascending, pubescent, 0.5–3 dm. high; lvs. basal, sparsely tomentose beneath, pubescent above, petioled, obovate-spatulate, 2–5 cm. long; bracts rarely leafy, spreading-acerose; invols. in small dense terminal clusters, cylindric, mostly thinly pubescent, 4–6 mm. long, the teeth widely divergent, uncinate, the 3 inner noticeably smaller than the outer; calyx white or pink, 6–7 mm. long, the segms. with fimbriate margins and linear entire terminal portion; stamens 9; aks. linear, slightly 3-angled, 3 mm. long.—Dry slopes, below 3000 ft.; Chaparral, Coastal Sage Scrub; w. San Diego Co. to e. Orange and w. Riverside cos.; n. L. Calif. April–June.

Var. **laciniàta** (Torr.) Jeps. [*C. l.* Torr.] Calyx 7–9 mm. long, the segms. finely laciniate so that the terminal lobe is scarcely broader than the others.—Below 5000 ft.; Chaparral, Pinyon-Juniper Wd.; s. San Jacinto Mts. to e. San Diego Co. May–June.

25. **C. Pálmeri** Wats. [*C. obovata* Goodm. *C. o.* f. *prostrata* Goodm.] Stems erect or ascending, 1–3 dm. long, single or several from base, mostly strigose; lvs. basal or nearly so, oblanceolate to spatulate, petioled, somewhat hairy beneath, less so above, 2–5 cm. long; lower bracts like the lvs., the upper subulate; invols. in flat-topped cymes or dense clusters, urn-shaped or subcylindric, 4.5–5 mm. long, grayish-pubescent, the 3 inner and 2 of the outer teeth uncinate, strongly divergent, the anterior or 6th noticeably longer and scarcely hooked; calyx 4–5 mm. long, glabrous, the outer lobes roundish, entire, ca. 1 mm. long, the inner smaller, fimbriate; stamens 9; aks. narrowly ovoid, 3-angled, ca. 3.5 mm. long.—Dry rocky places and hillsides, below 2500 ft.; Chaparral, Foothill Wd.; at least sometimes on serpentine, San Benito Co. to Santa Barbara Co. May–Aug.

Var. **bilòba** (Goodm.) Munz. [*C. b.* Goodm.] Outer calyx-segms. ± deeply bilobed.—With the sp.

Var. **ventricòsa** (Goodm.) Munz. [*C. v.* Goodm.] Outer lobes ± erose, ca. 1.5 mm. long.—Near Priest V., e. Monterey Co. and adjacent region.

26. **C. uniaristàta** T. & G. [*C. rectispina* Goodm.] Stems several from base, decumbent, 1–2.5 dm. long, cinereous-tomentose; lvs. basal, oblanceolate, petioled, villous-tomentose, 2–4 cm. long; bracts leaflike to narrower, awn-tipped; invols. in small clusters, narrowly urn-shaped, the tube 2–3 mm. long, pubescent, with 5 short spreading uncinate teeth and 1 much longer, erect or divergent, straight; calyx sparsely pubescent externally along the veins, 3–3.5 mm. long, the outer lobes linear-oblong to -obovate, obscurely erose at summit, the inner linear and shorter; stamens 3–9; aks. slender, grayish to dark, ca. 1.5–2 mm. long.—Dry slopes, often of serpentine, below 4000 ft.; Chaparral, Foothill Wd., V. Grassland, etc.; Coast Ranges from San Benito Co. to Santa Barbara Co. and plains and Sierran foothills in Kern Co. June–July.

27. **C. Clevelándii** Parry. Stems several from base, decumbent, tomentose, 0.5–2.5 dm. long; lvs. basal, oblanceolate, petioled, grayish-pubescent, 1–3 cm. long; lower

bracts leafy, upper acerose; invols. in scattered headlike clusters, the tubes urn-shaped, 3-angled, hoary-pubescent, 3–3.5 mm. long; teeth all uncinate, the anterior much longer than the others; calyx 3.5 mm. long, strigose on veins, the outer lobes ovate, ca. 1 mm. long, minutely erose, the inner shorter and conspicuously erose; stamens 3; aks. grayish, narrow, almost wing-angled above, 2.5 mm. long.—Dry rocky and sandy places, below 5000 ft.; Chaparral, Foothill Wd., N. Oak Wd.; inner Coast Ranges, Mendocino and Lake cos. to Ventura Co. and foothills of Sierra Nevada in Tulare and Kern cos. June–July.

28. **C. Párryi** Wats. Prostrate or decumbent, the stems several from base, 0.4–2(–3) dm. long, repeatedly forked, strigose; lvs. basal, oblong-lanceolate to -oblanceolate, narrowed to the petiole, strigose, 2–7 cm. long; lower bracts leafy but more mucronate, the upper acicular; invols. rather openly distributed in small clusters, urn-shaped, 6-ribbed, the tube ca. 2.5–3 mm. long, appressed-canescent, the teeth long, spreading, uncinate, the 3 outer much longer than the inner; calyx white, 2.5–3 mm. long, the outer lobes oblong-obovate to oblong, obtuse, erosulate, the inner linear-lanceolate; stamens 9; aks. lance-ovoid, grayish, 3-angled upward, ca. 2 mm. long.—Dry sandy places, below 2500 ft.; mostly Coastal Sage Scrub; e. Los Angeles Co. to San Gorgonio Pass, w. Riverside Co. April–June.

Var. **fernandìna** (Wats.) Jeps. [*C. f.* Wats.] Involucral teeth not hooked; calyx lobes subequal.—San Fernando V. to Orange Co. and Del Mar, San Diego Co.

29. **C. procúmbens** Nutt. [*C. uncinata* Nutt.] Stems several, procumbent, few-branched, brittle, 0.5–2.5 dm. long, strigose-tomentulose; lvs. mostly basal, oblong-oblanceolate, petioled, 2–7 cm. long, curly-strigulose; lower bracts like the lvs., upper acicular; invols. in small terminal clusters and axillary, yellowish-green, 2–2.5 mm. long, the tube cylindric, 6-ribbed, pubescent to glabrate, the teeth 6, spreading, with uncinate or curved tips, mostly somewhat unequal; calyx yellow, 1.5–2 mm. long, the lobes mostly oblong, obtuse, entire, scantily strigulose, the outer lobes slightly larger than inner; stamens 9; aks. lance-ovoid, grayish, somewhat 3-angled upward, ca. 1.5 mm. long.—Mostly sandy places, below 2500 ft.; Coastal Sage Scrub, Chaparral; San Fernando and San Bernardino to San Diego and L. Calif. April–June.

Var. **albiflòra** Goodm. Calyx white.—Pala to Fallbrook, San Diego Co.

30. **C. califórnica** (Benth.) Gray. [*Mucronea c.* Benth.] Stem or stems erect or ascending, branching near base, glandular-hirsute, 1–2.5 dm. long; lvs. basal, spatulate to obovate, short-petioled, 1–3 cm. long; bracts deeply 3-lobed, mostly 5–10 mm. long, somewhat broader, sessile, clasping at base, the lobes lance-ovate, spine-tipped; invols. 1–3 at each node, the tube cylindric, 2.5–3 mm. long, obscurely ribbed, the teeth usually 3, unequal, curved outward, ending in a straight spine; calyx white, somewhat exserted, the lobes oblong, obtuse, entire, pubescent; aks. grayish-brown, narrow, somewhat 3-angled and papillose upward.—Occasional, dry sandy places, below 3000 ft.; Coastal Sage Scrub, Chaparral, Coastal Strand; San Luis Obispo Co. to San Bernardino and San Diego. April–July.

Var. **Suksdórfii** Macbr. [*Mucronea c.* var. *S.* Goodm.] Bracts 1–2 cm. long, oblong and more obtuse; invols. urceolate.—Sand dunes; Surf, Playa del Rey, Ballona Harbor.

31. **C. perfolìata** Gray. [*Mucronea p.* Heller.] Stems branching mostly just above base, diffuse, 1.5–3 dm. long, sparingly glandular-puberulent; lvs. basal, spatulate, 2–5 cm. long; bracts perfoliate, orbicular or 3-lobed, spine-tipped, the larger bracts to 2 cm. across; invols. mostly solitary at each node, 4–8 mm. long, the tube 4-angled, corrugated between the ribs, the teeth 4, divergent, spine-tipped; calyx puberulent, white, exserted, the lobes ± fimbriate or erose at summit; stamens 6.—Dry flats and slopes, at low elevs.; Chaparral, Coastal Sage Scrub, Joshua Tree Wd.; inner Coast Ranges, from Mt. Hamilton (Stanislaus Co.) to head of San Joaquin V., Kern Co. and w. Mojave Desert. April–June.

32. **C. leptóceras** (Gray) Wats. [*Centrostegia l.* Gray. *Eriogonella l.* Goodm.] Stems few, slender, prostrate or decumbent, glabrous, forked, 0.5–1.5 dm. long; lvs. basal, oblanceolate, glabrous, 1.5–5 cm. long, short-petioled; bracts 3-lobed, pubescent, 3–6 mm. long, the lobes spine-tipped; invols. 1–3 at an axil, glandular-pubescent, 4–6 mm. long, · the tube cylindric, with 6 terminal long-awned unequal ciliate teeth and a ring of 6 hooked spinelike spurs near base; calyx partly exserted, pubescent, the lobes

spatulate, subequal; stamens 6; aks. dark gray, ovoid, ca. 1.3 mm. long, the beak pyramidal.—Occasional, sandy places; Coastal Sage Scrub; San Fernando V. to San Bernardino V. and Elsinore. April–June.

33. **C. Thúrberi** (Gray) Wats. [*Centrostegia T.* Gray ex Benth. *Chorizanthe T.* var. *cryptantha* Curran.] Erect, branched at or near base, forked above with spreading branches, glandular-hispidulous, 0.5–2 dm. high; lvs. in basal rosette, oblong to broadly spatulate, almost sessile, 1–3 cm. long, subglabrous; bracts 3-lobed, spine-tipped, 2–4 mm. long; invols. solitary, scattered but numerous, 4–6 mm. long, 5-toothed at apex, with erect spiny straight tips, and 3-angled with 3 basal spreading saccate horns near base; calyx included, pubescent, deeply parted into narrow segms.; stamens 6 or 9; aks. lance-ovoid, ca. 1.5 mm. long.—Common, dry sandy places, below 7500 ft.; Creosote Bush Scrub, Pinyon-Juniper Wd.; deserts; and occasional in inner S. Coast Ranges to San Benito Co.; e. to Utah, Ariz. April–June.

34. **C. Vortrièdei** Bdg. [*Centrostegia V.* Goodm.] Stem trichotomously branched at first node, then dichotomous, spreading, 0.6–1.5 dm. long, sparsely glandular; lvs. basal, spatulate, glabrous, 1–3 cm. long; bracts deeply 3-lobed, perfoliate, 2–3 mm. long; invols. solitary at nodes, ca. 3 mm. long, the tube 4-angled, 4-toothed, the teeth deltoid-ovate, cuspidate; fls. 2, pedicelled; calyx yellowish, 5-parted, the segms. with 2 white lobes; stamens 9; aks. black, globose, apiculate.—Dry places; Foothill Wd.; Santa Lucia Mts., Monterey Co. June–Sept.

35. **C. insígnis** Curran. [*Oxytheca i.* Goodm.] Erect and simple below, forked above with divergent branches, 0.5–1 dm. high, glandular-puberulent; lvs. basal, linear-spatulate, glabrous, 0.8–1.5 cm. long; bracts 3-lobed, the lobes oblong to linear-acicular, 3–4 mm. long; invols. solitary at nodes, glandular, the tube prismatic-cylindric, 3–4 mm. long, slightly corrugate, the teeth 5, divergent, spinelike; fls. 4–6, pedicelled; calyx rose, pubescent, exserted, with oblong segms.; stamens 9.—Sandy places; Foothill Wd.; inner Coast Ranges, Monterey and San Luis Obispo cos. June–Sept.

6. Oxythèca Nutt.

Slender-stemmed dichotomously branched annuals with small stipitate glands at the nodes. Lvs. rosulate at the base. Bracts ± connate, often in 3's, foliaceous. Invols. few-fld., ± pedicellate, campanulate to turbinate, 4–5-cleft, each lobe terminated by a bristle or awn, or 7–20-ribbed and each rib ending in a long awn. Fls. pedicelled; calyx 6-parted, glabrous or pubescent. Stamens 9, inserted at base of calyx. Ak. ± ovoid, the cotyledons accumbent, orbicular. Ca. 8–9 spp. of w. N. and S. Am. (Greek, *oxus*, sharp, and *theke*, a case, referring to the spiny invol.)

Invol. not lobed, its tube short, 7–30-ribbed, each rib ending in a long bristle .... 1. *O. Parishii*
Invol. lobed.
    Involucral divisions unequal, 5; plant prostrate; calyx woolly, yellow .... 2. *O. luteola*
    Involucral divisions subequal; plants erect or ascending; fls. not yellow.
        Bracts of upper nodes connate-perfoliate into a conspicuous disk 1–2 cm. broad; invols. sessile or
        nearly so .... 3. *O. perfoliata*
        Bracts not so united; invols. peduncled.
            The invols. 4-lobed.
            Lvs. revolute, linear to linear-oblanceolate, acute, hirsutulous. Mono Co. to Lassen Co.
                                4. *O. dendroidea*
            Lvs. plane, spatulate, rounded at apex, hispidulous. San Bernardino Co. .... 5. *O. Watsonii*
            The invols. 5-lobed.
                Invols. deeply lobed.
                    Calyx-lobes entire, greenish or reddish, very short .... 6. *O. caryophylloides*
                  Calyx-lobes deeply cleft, white, almost as long as tube .... 7. *O. trilobata*
                Invols. very shallowly lobed, reddish, forming a round disk with scarious margins
                                    8. *O. emarginata*

1. **O. Paríshii** Parry. [*Acanthoscyphus P.* Small. *Eriogonum Abramsii* ssp. *A.* S. Stokes.] Stems erect, with ascending branches, 2–6 dm. high, glaucous, glabrous except for the remote internodes; lvs. broadly spatulate, 2–4 cm. long, short-ciliate; bracts 3-lobed, 1–2 mm. long; invols. solitary on slender pedicels 1.5–5 cm. long, broadly turbinate, the basal part scarcely lobed, ca. 2 mm. long and 14–30-ribbed, each rib ending in a long bristle 3–5 mm. long; calyx whitish, ca. 2 mm. long, 6-cleft, the segms. oblong, entire, strigulose on back; aks. brownish, elliptic-ovoid, ca. 2 mm. long,

the base and apex quite contracted.—Dry granitic slopes and flats, 4000–8300 ft.; mostly Yellow Pine F.; Topatopa, San Gabriel and San Bernardino mts. June–Sept.

Var. **Abrámsii** (McGreg.) Munz. [*Acanthoscyphus A.* McGreg. *Eriogonum A.* S. Stokes.] Invol. plainly lobed and with 4–12 bristles, these mostly 2–3 mm. long.—Similar situations, Big Pine Lookout, Santa Barbara Co., Mt. Pinos to San Bernardino Mts. June–Sept.

2. **O. lutèola** Parry. [*Gymnogonium spinescens* Parry, in syn. *Eriogonum s.* S. Stokes.] Prostrate, with several stems from base, 0.3–1 dm. long, yellowish-green, pubescent; lvs. at base and in pairs at lower nodes, the blades green above, woolly beneath, rounded, 2–5 mm. long, on petioles 10–20 mm. long; bracts linear, acerose, 4–5 mm. long; invols. in forks and at the nodes, 5-parted, 3–5 mm. long, the divisions unequal, lance-linear, long-awned; fls. several; calyx 1 mm. long, the tube globose, woolly, the lobes glabrous, yellow; aks. dark brown, ca. 1 mm. long, the stout beak ca. as long as the rounded body.—Alkaline flats, dry lakes, etc.; Alkali Sink, V. Grassland, Creosote Bush Scrub, etc.; San Joaquin V., Madera and Fresno cos. to Kern Co., e. of the Sierra Nevada, Mono Co. to w. Mojave Desert. May–Aug.

3. **O. perfoliàta** T. & G. [*Eriogonum p.* S. Stokes.] Stem or stems erect, then dichotomous or trichotomous with horizontal branches 1–3 dm. long, green then reddish, glandular in lower part of each internode; basal lvs. oblong-oblanceolate, glabrous, 1.5–4 cm. long; bracts of upper nodes mostly 3 and connate-perfoliate into a slightly angled cupulate disk 1–2 cm. broad, with short spinose tips; invols. solitary, narrowly turbinate, 3–4 mm. long, 4-lobed to middle, the lobes ending in spines ca. 3 mm. long; fls. several; calyx whitish, 1.5 mm. long, minutely scaly; aks. brownish, ca. 2 mm. long, granular, ovoid, abruptly narrowed at both ends into a stout stipe and beak.—Sandy or gravelly places, 2400–6000 ft.; Creosote Bush Scrub to Pinyon-Juniper Wd.; Mojave Desert to Lassen Co.; Nev., Ariz. April–July.

4. **O. dendroìdea** Nutt. [*Eriogonum d.* and var. *Hillmanii* S. Stokes.] Stem erect, then dichotomously or trichotomously branched, 1.5–4 dm. high, the ultimate branchlets very fine, somewhat stipitate-glandular; lvs. basal, linear to linear-oblanceolate, 1.5–3 cm. long, acute, short-petioled, thinly hirsute; bracts entire; invols. solitary, narrowly turbinate, 2–4 mm. long (including spines), 4-lobed, the spines 1–2 mm. long; fls. 2–3; calyx pinkish to whitish, 1.5 mm. long, hispidulous.—Dry, sandy and gravelly flats and slopes, 4000–7000 ft.; Sagebrush Scrub, Pinyon-Juniper Wd., Yellow Pine F.; e. slope of Sierra Nevada, White Mts., Inyo Co. to Lassen Co.; to Wash., Wyo., Nev. June–Sept.

5. **O. Watsònii** T. & G. [*Eriogonum cuspidatum* S. Stokes.] Stems erect, dichotomously branched, glaucous, 1–2 dm. high, sparingly stipitate-glandular; lvs. spatulate, 1–2.5 cm. long, hispidulous; bracts lance-ovate, awned; invols. turbinate, 4-lobed, ca. 4 mm. long including the awns; fls. several, pedicelled; calyx white, puberulent, 1–1.5 mm. long.—Joshua Tree Wd.; Cushenberry Springs, sw. Mojave Desert; Nev. May–July.

6. **O. caryophylloìdes** Parry. [*Eriogonum c.* S. Stokes.] Erect, 1- or several-stemmed at base, diffusely branched above, 1–4 dm. high, minutely glandular; lvs. oblong-spatulate or wider, 2–5 cm. long, subglabrous, short-petioled; bracts 3-parted; invols. in forks and terminal, peduncled, deeply 5-parted, glabrous, the divisions oblanceolate to narrow-oblong, 3 mm. long, with awns an additional 1 mm. long; fls. 2–3; calyx greenish, strigose, barely 1 mm. long, the lobes short, entire; aks. triangular-ovoid, red-brown, ca. 1 mm. long.—Occasional, 4000–7000 ft.; Yellow Pine F.; San Gabriel Mts. to San Jacinto Mts. July–Sept.

7. **O. trilobàta** Gray. [*Eriogonum t.* S. Stokes.] Stems or stem erect, trichotomous, with spreading forked branches 1–6 dm. long, sparingly glandular; lvs. basal, spatulate, 1–6 cm. long, somewhat hairy; bracts deeply 3-lobed, the lobes lanceolate, 2–3 mm. long, spine-tipped; invols. mostly on slender peduncles, broadly turbinate, deeply 5-lobed, the lobes lanceolate to subovate, ca. 3 mm. long with terminal bristly awns ca. as long; fls. few; calyx white, with reddish base, the segms. 3-cleft into lanceolate lobes with erosulate sides; aks. brownish, triangular-ovoid, ca. 1.5 mm. long, with rather definite beak.—Fairly frequent, dry slopes, mostly 3000–7000 ft.; Chaparral, Yellow Pine F.; San Gabriel Mts. to Little San Bernardino Mts. and n. L. Calif. July–Sept. Varying considerably especially in width of involucral lobes.

8. **O. emarginàta** Hall. [*Eriogonum e.* S. Stokes.] Stem or stems erect, then dichotomous or trichotomous, with widely spreading branches 5–15 cm. long, reddish, glandular-pubes-

cent; lvs. oblanceolate, pubescent, the blades 1.5–4.5 cm. long on petioles ca. as long or shorter; bracts deeply 3-lobed; invols. mostly peduncled, broadly funnelform, shallowly 5-lobed, reddish, 4–6 mm. long, scarious-margined, the awns ca. 1 mm. long; fls. 3–4; calyx whitish, 6-parted, pubescent, with fimbriate lobes.—Dry slopes, 4000–8000 ft.; Pinyon-Juniper Wd., Yellow Pine F.; San Jacinto Mts. and Santa Rosa Mts., Riverside Co. July–Aug.

7. Eriógonum Michx. WILD BUCKWHEAT

Annual or perennial herbs or shrubs with basal or cauline alternate to whorled entire lvs. without stipules. Fls. perfect or sometimes also imperfect, borne in invols. Invols. campanulate to turbinate or cylindric, 4–8-lobed or -toothed, awnless, few- to many-fld., sessile or peduncled. Pedicels ± exserted, intermixed with setaceous bracts or bractlets and jointed at summit with the base of the perianth. Perianth commonly called "calyx", 6-parted or -cleft, petaloid, with 2 series of 3 segms. each. Stamens 9; fils. filiform, inserted at base of perianth. Ovary 1-celled, 3-angled or -winged; styles 3; stigmas capitate. Aks. mostly 3-angled, sometimes lenticular. A N. Am. genus of perhaps 150 spp., mostly w.; some of importance as bee-plants, others with some horticultural possibility. (Greek, *erion*, wool, and *gonu*, knee or joint, some spp. being hairy at nodes.)

(Stokes, Susan G., The genus Eriogonum, 1–128, 1936.)
A. Calyx stipelike at the attenuated base (see also *saxatile* and *crocatum*); bracts leafy, indefinite in
     number (2–several). (Subgenus Eriogonum)
    B. Invols. with lobes at least half as long as tube and usually reflexed or spreading.
        C. Calyx pubescent externally.
            D. Flowering stems or scapes without bracts and with solitary terminal invols. Inyo
                 Co. to Modoc Co., e. of the Sierra Nevada crest . . . . 1. *E. caespitosum*
            DD. Flowering stems with whorled bracts near the middle or at base of umbel.
                E. Lvs. ± glabrate above; calyx, including the stipe, 7–8 mm. long
                                                                            3. *E. sphaerocephalum*
                EE. Lvs. permanently tomentose on both surfaces; calyx 4–6 mm. long.
                    F. Scape bearing a single invol. Above 4000 ft. . . . . 2. *E. Douglasii*
                    FF. Scape bearing an umbel. Below 2000 ft. . . . . 4. *E. tripodum*
        CC. Calyx glabrous externally.
            D. Flowering stems with a whorl of bracts near their middle and also at base of
                 umbel. Modoc Co. . . . . 6. *E. heracleoides*
            DD. Flowering stems naked save for the bracts at the base of the umbel (which may
                 be simple, so that the bracts appear to be in middle of stem).
                E. Invols. solitary, without subtending bracts. Subalpine, Siskiyou and Trinity cos.
                                                                            5. *E. siskiyouense*
                EE. Invols. in simple or compound umbels.
                    F. Lf.-blades mostly 4–10 cm. long and cordate at base; fls. ochroleucous to
                         pale yellow; aks. 5–6 mm. long. N. Coast Ranges . . . . 9. *E. compositum*
                    FF. Lf.-blades less than 4 cm. long, not cordate at base.
                        G. Scapes erect or nearly so; lf.-blades mostly less than 2 cm. long;
                             fls. frequently bright yellow.
                            H. Lvs. not cordate at base; stems usually not prostrate; aks.
                                 wrinkled only at very base. Throughout Calif.
                                                                            7. *E. umbellatum*
                            HH. Lvs. often cordate at base; stems prostrate; aks. wrinkled
                                 and cellular-papillose on lower half. S. Sierra Nevada
                                                                            8. *E. polypodum*
                        GG. Scapes usually flat on the ground; lf.-blades mostly 2–4 cm. long;
                             fls. white to rose. N. Coast Ranges and n. Sierra Nevada
                                                                            10. *E. Lobbii*
    BB. Invols. with lobes much shorter than tube, toothlike and suberect.
        C. Calyx pubescent externally.
            D. Scapes 15–30 cm. high; calyx cream-colored. Inyo Co. . . . . 11. *E. latens*
            DD. Scapes 4–8 cm. high; calyx white to rose. Shasta and Siskiyou cos.
                                                                            12. *E. pyrolaefolium*
        CC. Calyx glabrous externally.
            D. Bracts in a whorl near middle of flowering stem, which bears single invol.
                E. Calyx yellow; lvs. rounded. Siskiyou Co. . . . . 13. *E. alpinum*
                EE. Calyx white or pink; lvs. oblanceolate to spatulate. Mendocino Co.
                                                                            14. *E. Kelloggii*
            DD. Bracts subtending the umbel or head of several invols.
                E. Fls. white to cream.
                    F. Calyx ochroleucous; styles ca. 2.5 mm. long; lower bracts mostly 1–2
                         cm. long; fils. tomentulose-hairy . . . . 16. *E. ursinum*

FF. Calyx white with red veins; styles ca. 1 mm. long; lower bracts ca.
     5 mm. long; fils. sparsely hairy .... 8. *E. polypodum*
  EE. Fls. yellow; main bracts less than 1 cm. long.
    F. Styles less than 0.5 mm. long. Sierra Nevada to cent. Ore.
                              15. *E. marifolium*
    FF. Styles 3 mm. long. Extreme nw. Calif. .... 17. *E. ternatum*
AA. Calyx not stipelike at base; bracts not leafy, regularly 3 in number.
  B. Invols. usually 3 at a node; stems internally jointed. Death V. (Subgenus Clastomyelon)
                          76. *E. intrafractum*
  BB. Invol. 1 at a node; stems not so jointed.
    C. The invols. bell-shaped to turbinate, not angled, 4- or 5-toothed or -lobed, rarely
      obscurely nerved, the teeth rounded, often with membranous margins, mostly on scat-
      tered peduncles. (Subgenus Ganysma)
      D. Lvs. basal and also on the lower nodes, tomentose except in *E. spergulinum.*
        E. Invols. 4-lobed or 4-toothed.
          F. Plants mostly 1–3 dm. high; lf.-blades 1–3 cm. long.
            G. The lvs. linear, revolute, pilose; invols. 0.5–1 mm. long
                            21. *E. spergulinum*
            GG. The lvs. oblong or rounded to lanceolate, woolly, not revolute;
               invols. 1.5–2.5 mm. long.
              H. Fls. not concealed by cottonlike tomentum.
                I. Outer calyx-segms. broader than inner, incurved
                            18. *E. angulosum*
                II. Outer calyx-segms. essentially like the inner, plane
                            19. *E. gracillimum*
              HH. Fls. concealed by tufts of cottonlike tomentum of inner sur-
               face of invol. .... 20. *E. gossypinum*
          FF. Plants 4–7 dm. high; lf.-blades 2–8 cm. long .... 27. *E. Ordii*
        EE. Invols. 5-lobed or -toothed.
          F. Plants glabrous or thinly tomentose; calyx glabrous .. 38. *E. argillosum*
          FF. Plants densely white-tomentose; calyx papillose .... 39. *E. vestitum*
      DD. Lvs. strictly basal, sometimes pilose, sometimes tomentose.
        E. Invols. 4-lobed or -toothed.
          F. Calyx pubescent with hooked hairs.
            G. Invols. 2-fld.; ak. exserted; calyx ca. 1 mm. long
                            22. *E. hirtiflorum*
            GG. Invols. 4- to 6-fld.; ak. not exserted; calyx ca. 1.5 mm. long
                            23. *E. inerme*
          FF. Calyx puberulent or hispidulous, the hairs not hooked.
            G. The calyx white, 1.5 mm. long, the segms. notched or apiculate
                          24. *E. apiculatum*
            GG. The calyx pink or yellow, 0.5–1 mm. long, the segms. not apiculate.
              H. Lvs. spatulate; fls. pink. Montane .... 25. *E. Parishii*
              HH. Lvs. rounded; fls. yellow. Deserts .... 26. *E. trichopes*
        EE. Invols. 5-lobed or -toothed.
          F. Calyx pubescent or puberulent.
            G. Invols. glabrous on outer surface.
              H. Outer calyx-segms. not saccate-dilated.
                I. Lf.-blades short-hirsute; calyx 2 mm. long, yellow
                            28. *E. inflatum*
                II. Lf.-blades woolly; calyx 1 mm. long, yellow to white
                            33. *E. reniforme*
              HH. Outer calyx-segms. saccate-dilated at each side of the cordate
                base, ca. 1 mm. long, becoming white or rose
                            30. *E. Thomasii*
           GG. Invols. glandular-puberulent.
             H Calyx glandular-puberulent at base only.
              I. Outer calyx-segms. rounded and narrowed abruptly to a
                narrow claw .... 31. *E. Thurberi*
              II. Outer calyx-segms. oval, not clawed ..... 36. *E. nutans*
            HH. Calyx glandular-puberulent on entire outer surface, the outer
              segms. obovate with broad base .... 32. *E. pusillum*
        FF. Calyx glabrous.
          G. Lvs. pilose-hispid, not woolly. E. Inyo Co. .. 29. *E. esmeraldense*
          GG. Lvs. woolly, at least beneath.
            H. Stems glabrous or glandular, not woolly. Deserts.
              I. Peduncles capillary; outer calyx-segms. not cordate at
                base. Local and uncommon.
                J. The peduncles 5–25 mm. long; calyx-segms. oblong-
                  obovate .... 34. *E. cernuum*
                JJ. The peduncles 0.5–6 mm. long; calyx-segms. lance-
                  oblong .... 35. *E. Hoffmannii*
              II. Peduncles stout, sometimes almost lacking; outer calyx-
               segms. usually cordate at base. Common and widespread
                            37. *E. deflexum*
            HH. Stems woolly. W. Fresno Co. .... 46. *E. Eastwoodianum*

CC. The invols. cylindric or cylindric-turbinate or prismatic, often angled, 5–6 nerved,
    mostly sessile, solitary or congested in heads, the teeth short. (Subgenus Oregonium)
  D. Plants annual; lvs. mostly in a basal rosette.
    E. Flowering branches elongate, virgate, bearing invols. at the nodes, the
       lateral ones appressed.
      F. Outer calyx-segms. fan-shaped, their sides incurved below the broad
         truncate apex; branches usually incurved at summit in age; invols. ca.
         1 mm. long. Deserts .... 40. *E. nidularium*
      FF. Outer calyx-segms. not as above.
        G. Calyx densely hairy without. Inner N. Coast Ranges
                                                             51. *E. dasyanthemum*
        GG. Calyx glabrous or minutely puberulent without.
          H. Invols. 1–1.5 mm. long; calyx often glandular-puberulent
                                                             50. *E. Baileyi*
          HH. Invols. 2–5 mm. long; calyx glabrous.
            I. Lvs. oblong-obovate to oblanceolate; stems ± tomentose.
              J. Invols. 2–3 mm. long, turbinate, with prominent
                 teeth; outer calyx-segms. broadly obovate; aks. ca.
                 1 mm. long .... 41. *E. gracile*
              JJ. Invols. 4–5 mm. long, cylindric, with minute teeth;
                  outer calyx-segms. narrowly obovate; aks. ca. 2 mm.
                  long .... 42. *E. virgatum*
            II. Lvs. round or nearly so; stems often quite glabrous.
              J. Invols. 5–7 mm. long. S. Calif. .... 49. *E. molestum*
              JJ. Invols. 2–4 mm. long.
                K. Stems mostly simple below, glabrous except at
                   base; outer calyx-segms. more than twice as long
                   as wide. Mts. of s. Calif.
                                         49. *E. molestum* var. *Davidsonii*
                KK. Stems mostly branched from base, glabrous or
                    tomentose; outer calyx-segms. less than twice as
                    long as wide. Cent. and n. Calif.
                                                     43. *E. vimineum*
  EE. Flowering branches with short branchlets, usually of a single internode; invols.
      in axils and terminal.
    F. Invols. 3–4 mm. long.
      G. Lf.-blades distinctly longer than broad.
        H. Invols. glabrous, 5-ribbed or -toothed; stems glabrous. Marin
           Co. and Oakland Hills .... 47. *E. caninum*
        HH. Invols. tomentose, scarcely 5-ribbed or toothed; stems tomen-
            tose. E. base of Mt. Diablo .... 48. *E. truncatum*
      GG. Lf.-blades round to reniform; invols. with 8 short blunt teeth.
          Monterey and San Benito cos. .... 44. *E. Nortonii*
    FF. Invols. 1.5–2 mm. long; lf.-blades round or nearly so.
      G. Stems floccose-tomentose; calyx glabrous. W. Fresno Co.
                                                     46. *E. Eastwoodianum*
      GG. Stems and calyx glabrous.
        H. Fls. rose to white with red veins. Coast Ranges
                                                     45. *E. Covilleanum*
        HH. Fls. yellowish to cream. Desert. .... 52. *E. mohavense*
DD. Plants perennial to shrubby; lvs. often cauline as well as basal.
  E. Invols. solitary at the nodes, the lateral ones appressed to the branchlets.
    F. Calyx silky-villous, yellowish; plant a shrub 6–15 dm. high. Imperial Co.
                                                     53. *E. deserticola*
    FF. Calyx glabrous.
      G. Calyces 5–7 mm. long; invols. 3–4 mm. long.
        H. Infl. 1–1.5 dm. across, open and lax; calyx narrowed to a
           3-angled base, pinkish to white or yellowish; aks. ca. 2 mm.
           long. Montane. .... 54. *E. saxatile*
        HH. Infl. 0.3–0.8 dm. across, dense; calyx narrowed to a tubular
            base, sulphur yellow; aks. ca. 3 mm. long. N. base of Santa
            Monica Mts. .... 55. *E. crocatum*
      GG. Calyces 2–4 mm. long.
        H. Invols. 6–7 mm. long; loosely branched whitish tomentulose
           herb 6–18 dm. high. S. Coast Ranges .... 60. *E. elongatum*
        HH. Invols. 2–4 mm. long.
          I. Infl. with invols. placed racemosely along the branches;
             invols. tomentose.
            J. Invols. on main ascending branches of infl.
              K. Lvs. round to broadly ovate, white-tomentose on
                 both surfaces, basal; plants scapose from a
                 caudex. E. desert mts. ...... 58. *E. racemosum*
              KK. Lvs. lance-elliptic to oblanceolate, on lower
                  third of plant; plants subshrubs. Montane and
                  cismontane .... 57. *E. Wrightii*
            JJ. Invols. on lateral divaricate branchlets; plants shrubby.

W. edge of deserts . . . . 59. *E. nodosum*
II. Infl. with invols. in cymes or panicles.
  J. The infl. a compact terminal cyme; invols. woolly or
     glabrous; outer calyx-segms. subcordate at base
                                              56. *E. microthecum*
  JJ. The infl. a divaricately branched panicle; invols.
      glabrous.
    K. Outer calyx-segms. round, subcordate at base;
       branches of infl. dichotomous, ascending
                                              61. *E. Heermannii*
    KK. Outer calyx-segms. obovate, narrowed at base;
        branches of infl. mostly horizontal, tiered
                                              62. *E. Plumatella*
EE. Invols. mostly in heads.
  F. Calyx-segms. dissimilar, the outer much broader and often with cordate
     base.
    G. The calyx-segms. ± plane.
      H. Infl. mostly capitate, each stem ending in a single head
                                              63. *E. ovalifolium*
      HH. Infl. cymose-umbellate with divaricate rays.
                                              65. *E. proliferum*
    GG. The calyx-segms. rounded in back, forming a globose fl.
                                              64. *E. Gilmanii*
  FF. Calyx-segms. similar or essentially so.
    G. Plants herbaceous, cespitose, the caudex much branched and
       forming mats or cushions; flowering stems scapelike.
      H. Fls. yellow; invols. 6–8-toothed . . . . . . 66. *E. ochrocephalum*
      HH. Fls. white to rose; invols. 5-toothed . . . . . . 67. *E. Kennedyi*
    GG. Plants shrubby or herbaceous, but not with a matted cespitose base.
      H. Lvs. linear or nearly so, the blades less than 2 cm. long,
         strongly fascicled; shrubs with terminal cymose or subumbel-
         late infl. . . . . 71. *E. fasciculatum*
      HH. Lvs. ± ovate, or if linear 2–3 cm. long.
        I. Plants essentially herbaceous, only the base woody.
          J. Lvs. spreading, oblong-ovate to ovate, obtusish, mostly
             2–6 cm. long . . . . 68. *E. latifolium*
          JJ. Lvs. erect, ± lanceolate, acute.
            K. Flowering stems glabrous, 4–8 dm. long; lf.-
               blades 4–15 cm. long . . . . 69. *E. elatum*
            KK. Flowering stems tomentose, 2–5 dm. long; lf.-
                blades 2–5 cm. long . . . . 70. *E. pendulum*
        II. Plants definitely shrubby.
          J. Heads in dense compound cymes. Insular.
            K. Lvs. linear or narrowly oblong, revolute, 2–3 cm.
               long . . . . 72. *E. arborescens*
            KK. Lvs. oblong-ovate, plane, 3–10 cm. long
                                              73. *E. giganteum*
          JJ. Heads terminal on 2-forked peduncles or scattered
              along the stem. Mainland for most part.
            K. Calyx glabrous without; lf.-blades 5–15 mm. long
                                              74. *E. parvifolium*
            KK. Calyx white-villous; lf.-blades 15–30 mm. long
                                              75. *E. cinereum*

1. **E. caespitòsum** Nutt. Low compact matted perennial from much-branched woody
caudex; lvs. elliptic to oblong-spatulate, densely white-tomentose, 5–9 mm. long, short-
petioled, crowded on the short branches, ± revolute; flowering stems scapelike, bractless,
slender, 3–8 cm. high, somewhat loosely tomentose; invol. solitary, terminal, the tube
turbinate, ca. 3 mm. long, with somewhat longer linear lobes that become reflexed;
calyx yellow, ca. 3 mm. long, later reddish and 4–5 mm. long, pubescent especially
toward the stipelike base, the segms. similar, oblance-oblong; fils. pilose; aks. lanceolate in
outline, somewhat 3-angled, ca. 3 mm. long, often pubescent at apex.—Dry gravelly
slopes and flats, 5000–8600 ft.; Sagebrush Scrub, N. Juniper Wd., Yellow Pine F.;
White Mts., Inyo Co. to Modoc Co.; to Ida., Mont., Colo. May–July.

2. **E. Douglásii** Benth. [*E. caespitosum* var. *D.* Jones. *E. c.* ssp. *D.* S. Stokes.] Rather
loosely matted perennial from much-branched woody caudex, the plants to 3 or 4 dm.
across; lvs. oblanceolate to obovate, tomentose on both surfaces, 1–2 cm. long, the
petioles often making up one third of this; flowering stems loosely tomentose, 4–12 cm.
high, with a whorl of oblanceolate leafy bracts near middle and a single terminal invol.;
involucral tube turbinate, 3 mm. long, with oblong lobes ca. as long; calyx yellow, later
often reddish, 5–6 mm. long, villous-pubescent on midribs and base, segms. narrowly
obovate; fils. pubescent below; aks. lanceolate in outline, 3-angled, hairy in upper half,

ca. 3 mm. long.—Infrequent, dry rocky places, 4500–8000 ft.; Sagebrush Scrub to Yellow Pine F.; e. slope of Sierra Nevada from Nevada Co. to Modoc Co., w. to Siskiyou Co.; to Wash., Nev. May–July.

3. E. sphaerocéphalum Dougl. ex Benth. Caudex much-branched, woody, with decumbent leafy branches 5–12 cm. long; lvs. oblanceolate, in whorls at upper nodes, ± glabrate above, tomentose beneath, 1–3 cm. long including the slender petioles; flowering stems ascending to erect, 5–15 cm. long, with a whorl of leafy oblanceolate bracts at middle, simple or umbellate above, each peduncle bearing 1 invol. with broadly turbinate tube 3–4 mm. long, the lobes ca. as long; calyx cream to yellow, villous-tomentose, 7–8 mm. long, the stipitate base slender and ca. 2 mm. long, the segms. oblong-obovate; aks. lance-ovoid, 3-angled and pubescent above, ca. 3 mm. long.— Dry rocky places, 3000–7500 ft.; Sagebrush Scrub, N. Juniper Wd.; Yellow Pine F.; Lassen and Modoc cos. to Siskiyou Co.; to Wash., Ida., Nev. May–July. (Calif. material is largely a compact bushy type, umbellate and small-leaved, and would fall into var. *brevifólium* S. Stokes ex Jones.)

4. E. trípodum Greene. Caudex woody, loosely branched; lvs in whorls at tips of branches, narrowly oblanceolate, 1.5–2.5 cm. long, white-tomentose above and beneath, revolute; flowering stems slender, 2–3 dm. tall, bearing a middle whorl of foliaceous bracts and usually a 3-rayed umbel, the rays naked or bracted; invol. solitary, tomentose, the spreading or reflexed lobes shorter than the tube; calyx yellow, 4–5 mm. long, villous-tomentose, the stipelike base ca. 2 mm. long; fils. pilose below; aks. narrow-ovoid, strongly angled, pubescent at apex, ca. 2 mm. long.—Gravelly slopes, often on serpentine, below 2000 ft.; Foothill Wd.; inner Coast Ranges and Sierran foothills, Tehama and Lake cos. and Tuolumne and Mariposa cos. June–July.

5. E. siskiyouénse Small. [E. ursinum var. s. S. Stokes.] Low matted or tufted perennial from a woody base, with short compact branches 3–7(–20) cm. long; lvs. oval to spatulate, crowded at ends of branches, 5–8 mm. long, acutish, short-petioled, glabrate above, tomentose beneath; scapes slender, erect, 3–8 cm. high, with a whorl of leafy bracts near middle; invol. usually solitary, without bracts at its base, campanulate, somewhat arachnoid-tomentose, the tube 3.5–4 mm. high, the lobes ca. as long, reflexed; calyx yellow, glabrous, 5 mm. long, short-stipitate, the outer segms. oblong, rounded at apex, the inner somewhat narrower; fils. hairy at base; aks. narrow-ovoid, glabrous, 3-angled.—Rocky ridges, 3000–5000 ft.; Yellow Pine F.; Tuolumne and Nevada cos.; at 7000–9000 ft., in Subalpine F., Siskiyou and Trinity cos. Aug.–Sept.

6. E. heracleoïdes Nutt. Loosely tufted plant from branched woody caudex; lvs. oblanceolate, 2–5 cm. long, densely white-tomentose especially beneath, short-petioled; flowering stems 2–4 dm. high, tomentose, usually with a whorl of leafy bracts near middle and another at base of the umbel; umbel simple or compound, the rays 2–5 cm. long; invol. solitary, turbinate, woolly, the tube 3 mm. long, the lobes 3–4 mm. long, spreading or reflexed; calyx glabrous, 4.5–6 mm. long, including the stipitate base which is tubular and 1.5–2 mm. long, the segms. oblong-ovate; fils. villous below; aks. light brown, narrow, somewhat 3-angled, ca. equally pointed at both ends, pubescent at apex, ca. 2 mm. long.—Dry places, 6000–7500 ft.; Sagebrush Scrub, N. Juniper Wd., Yellow Pine F.; Warner Mts., Modoc Co.; to B.C., Mont., Utah. June–Aug.

7. E. umbellàtum Torr. ssp. polyánthum (Benth.) S. Stokes. [E. p. Benth. E. umbellatum var. p. Jones. E. Torreyanum Gray.] Perennial with a few- to many-branched woody caudex, the branches leafy at tips; lf.-blades spatulate-obovate to elliptic-ovate, mostly 1–2 cm. long, narrowed abruptly to a mostly shorter petiole, green and glabrate on upper surface, white-tomentose beneath; flowering stems scapiform, floccose, 1–3 dm. high, mostly without bracts except the whorl of leafy oblanceolate bracts at base of the infl.; umbel simple, usually with 5–10 rays 2–6 cm. long, these bractless; invol. turbinate, tomentose, the tube 3–4 mm. long, the lobes reflexed, linear-oblong, to ca. as long as tube; calyx bright yellow, glabrous, ca. 5–8 mm. long, the tube cylindric, 1–1.5 mm. long, the lobes spatulate with obtuse apex; fils. pilose at base; aks. narrow-ovoid, light brown, hairy above, sharply 3-angled, ca. 3.5 mm. long.—Rather common on dry slopes and ridges, 2500–10,000 ft.; Sagebrush Scrub to Yellow Pine F. and Subalpine F.; Coast Ranges from Tehama Co. to Humboldt and Siskiyou cos., Sierra Nevada from Tulare Co. n.; to Ore. June–Aug. E. umbellatum is an exceedingly variable sp., the typical form being from the Rocky Mts.; many different segregates have been proposed.

KEY TO SUBSPECIES

A. Primary rays of umbel simple, not branched or bracteate in middle.
  B. Umbel subcapitate or rays few and scarcely more than 1 cm. long.
    C. Calyx yellow, 6–7 mm. long; scapes 2–4 dm. long. From below 5000 ft.
                      (a) ssp. *dumosum*
    CC. Calyx mostly whitish to red, 3–5 mm. long; scapes 0.4–1.2 dm. long. From 8000 ft. or higher.
      D. Lvs. green and glabrate above; fls. mostly whitish or pale yellow with some red.
        Mt. Shasta to Panamint Mts. .... (c) ssp. *Covillei*
      DD. Lvs. permanently densely white-woolly; fls. deep red. San Gabriel Mts.
                       (e) var. *minus*
  BB. Umbel open, the rays mostly 2.5 or more cm. long.
    C. Fls. pale yellow to cream; lvs. tomentose on both surfaces. E. slope of Sierra Nevada to San Bernardino Mts. .... (d) ssp. *aridum*
    CC. Fls. usually deep yellow, often with some red; lvs. ± glabrate above.
      D. Bracts at base of umbel oblanceolate, 0.5–1.5 cm. long; scapes usually 1–3 dm. high, slender. Common, Sierra Nevada to Siskiyou, Trinity, and Humboldt cos.
                    ssp. *polyanthum*
      DD. Bracts almost obovate, 1.5–2 cm. long; scapes stout, 2–4 dm. long. Below 5000 ft., Plumas Co. to Modoc Co. .... (b) var. *modocense*
AA. Primary rays of umbel usually branched, if not, with bracts near middle.
  B. Basal lvs. glabrate above, the blades usually not longer than petioles.
    C. Fls. yellow or with some red; lvs. tomentose beneath, glabrate above. Trinity to Siskiyou and Del Norte cos. .... (f) ssp. *stellatum*
    CC. Fls. ochroleucous; lvs. glabrous or nearly so. Modoc Co. to Placer Co.
                    (i) var. *Torreyanum*
  BB. Basal lvs. quite permanently tomentose on both surfaces, the blades longer than the petioles.
    C. Fls. yellow or with some red. Coast Ranges from Glenn Co. to Riverside Co.
                    (g) ssp. *bahiaeforme*
    CC. Fls. mostly pale or ochroleucous. Deserts .... (h) ssp. *subaridum*

(a) Ssp. **dumòsum** (Greene) S. Stokes. [*E. d.* Greene. *E. umbellatum* var. *serratum* S. Stokes.] Plants 5–10 dm. high; lvs. somewhat tomentose especially beneath, the blades narrow-obovate, 15–30 mm. long, ca. half as wide; scapes slender, 2–4 dm. long, the umbel few- to 1-rayed, with prominent narrow-obovate bracts 1.5–2 cm. long; invol. 4–10 mm. in diam. with rather short teeth, these often erect; fls. yellow, 6–7 mm. long.—At 1200–4500 ft.; Yellow Pine F., Red Fir F.; Nevada Co. to Siskiyou Co. July–Aug.

(b) Var. **modocénse** (Greene) S. Stokes. [*E. m.* Greene.] Plants to ca. 2–3 dm. high; lf.-blades obovate-elliptic, 6–15 mm. long, almost as wide, short-petioled, white-tomentose especially beneath; scapes mostly 1–2.5 dm. long, with large bracts; rays of simple umbel ca. 3–4 cm. long; calyx yellow, ca. 7 mm. long.—At ca. 4000–5000 ft.; Sagebrush Scrub, N. Juniper Wd.; Modoc Co. to Plumas Co. July–Aug.

(c) Ssp. **Covíllei** (Small) Munz. [*E. C.* Small, not *E. ursinum* var. *C. S.* Stokes. *E. umbellatum* var. *polypodum* S. Stokes, not *E. p.* Small. *E. umbellatum* var. *versicolor* S. Stokes.] Depressed, matted; lvs. often almost as broad as long, the blades less than 1 cm. long, glabrate above; scapes very slender, mostly 5–12 cm. long, bearing a capitate or very short- and few-rayed infl.; calyx 3–4 mm. long, whitish or pale yellow, usually with red midribs, or almost entirely red.—Dry gravelly places, 8000–11,000 ft.; Lodgepole F., Subalpine F.; Mt. Shasta, Sierra Nevada, Panamint Mts.; w. Nev. July–Sept.

(d) Ssp. **áridum** (Greene) S. Stokes. [*E. a.* Greene. *E. azaleastrum* Greene. *E. reclinatum* Greene.] Lf.-blades tomentose on both surfaces, 1.5–2 cm. long, scapes 1–2.5 dm. high; rays few, mostly 1–3 cm. long; fls. pale yellow to cream, ca. 8 mm. long.—Occasional, below 10,000 ft.; Pinyon-Juniper Wd., Lodgepole F.; e. slope of Sierra Nevada, White Mts., Mono Co. s. to San Gabriel Mts., San Bernardino Co.; Nev. June–Aug.

(e) Var. **mìnus** Jtn. [*E. m.* Ewan.] Low, matted; lf.-blades round-ovate, 4–10 mm. long, permanently densely white-woolly; scapes 3–12 cm. long; rays of umbel 1–3, 5–20 mm. long; calyx 4–5 mm. long, often deep red.—Dry stony slopes, 8000–10,000 ft.; Lodgepole F., Subalpine F.; San Gabriel Mts. July–Sept.

(f) Ssp. **stellàtum** (Benth.) S. Stokes. [*E. s.* Benth. *E. umbellatum* var. *s.* Jones.] Basal lvs. spatulate-obovate, glabrate above, the petioles often exceeding the blades (these 1.5–2.5 cm. long); umbel-rays bracteate near middle and often branched at that point, forming subcymose infl.; fls. usually bright yellow, sometimes with reddish tinge, mostly 6–8 mm. long.—Dry rocky slopes, 3000–7500 ft.; Yellow Pine F. to Red Fir F.; Trinity to Siskiyou and Del Norte cos.; to e. Wash., Mont. July–Sept. Variable; a form with unusually large fls., to ca. 10 or 12 mm. was described as *E. speciosum* Drew. [*E. u.* var. *s.* S. Stokes].

(g) Ssp. **bahiaefòrme** (T. & G.) Munz. [*E. polyanthum* var. *b.* T. & G. *E. stellatum* var. *b.* Wats. *E. umbellatum* var. *b.* Jeps. *E. trichotomum* Small. *E. Smallianum* Heller. *E. u.* var. *S.* S. Stokes.] Lvs. ± felty, white-tomentose on both surfaces, the petioles scarcely as long as the blades, these 1–1.5 cm. long; infl. compound, or at least with bracts on the primary rays; fls. 5–8 mm. long, mostly yellow, or with some red.—Dry rocky places, mostly above 3000 ft.; Yellow Pine F., Foothill Wd.; Coast Ranges from Lake and Glenn cos. s.; up to 9500 ft. in San Bernardino, San Jacinto, and Santa Rosa mts., occasional in s. Sierra Nevada. July–Aug.

(h) Ssp. **subáridum** (S. Stokes) Munz. [*E. u.* ssp. *stellatum* var. *subaridum* Stokes.] Lvs. very finely and closely tomentose, not felty, the blades mostly 1–1.5 cm. long; infl. compound; fls. mostly very pale, 6–7 mm. long.—Dry rocky slopes, 5000–9000 ft.; Pinyon-Juniper Wd.; White, Argus, and Panamint mts. across the Mojave Desert to s. Nev. July–Aug.

(i) Var. **Torreyànum** (Gray) Jones. [*E. T.* Gray.] Plants nearly or quite glabrous; infl. compound; fls. ochroleucous, 9–10 mm. long.—At ca. 6000 ft.; Placer Co. to Modoc Co.; se. Ore. July–Aug.

8. **E. polypòdum** Small. [*E. umbellatum* var. *p.* S. Stokes, as to name and type only. *E. ursinum* var. *Covillei* S. Stokes, not *E. C.* Small. *E. ursinum* var. *venosum* S. Stokes.] Near to high mountain forms of *E. umbellatum*, but with a prostrate habit; lvs. ovate with rounded or cordate base, revolute, the blades less than 1 cm. long; petioles stout, shorter than blades; scapes erect, 5–15 cm. tall, slender, ending in a capitate or few-rayed umbel; invol. turbinate, 3–4 mm. high; calyx-segms. 5–7, oblong, spreading, ca. 3 mm. long, chalky-white with reddish midveins; aks. with a wrinkled cellular-papillose adherent basal part which is over half their entire length and free from the seed.—At ca. 8000–10,500 ft.; Subalpine F.; s. Sierra Nevada, mostly in Tulare Co. July–Aug.

9. **E. compósitum** Dougl. ex Benth. [*E. c.* var. *citrinum* S. Stokes.] Perennial with ± branched woody caudex; lvs. basal, ovate to lance-ovate, attenuate at base or abruptly contracted, sometimes cordate, usually glabrate and greenish above, white-tomentose beneath, the blades 2–10 cm. long, petioles as long or longer; scapes stout, 2–4 dm. long, subglabrous, ascending to erect; infl. simple to compound, umbellate, subtended by narrow foliaceous bracts; invol. campanulate, mostly pilose-tomentose, with 5 linear finally reflexed lobes; fls. glabrous, ochroleucous to pale yellow, becoming 5–6 mm. long, the stipe 1–1.5 mm. long; calyx-segms. similar, ± oblong-ovate; fils. pilose at base; aks. narrowly triangular-ovoid, pubescent above, light brown, 5–6 mm. long.—Dry rocky walls and slopes, below 7500 ft.; Yellow Pine F., Red Fir F.; Lake Co. to Del Norte and Siskiyou cos.; to Wash., Ida. May–July.

10. **E. Lóbbii** T. & G. [*E. L.* var. *minus* T. & G.?] Caudex stout, woody, few-branched, covered with hairy bases of dead lvs.; lvs. in tufted rosettes, mostly round-ovate, plane, densely tomentose especially beneath, the blades 1–4 cm. long, abruptly narrowed into petioles as long or longer; scapes flat on ground to decumbent, tomentose, 0.5–2 dm. long, with foliaceous bracts subtending the 2- to several-rayed umbel; rays 1–3 cm. long, woolly-hirsute; invol. campanulate, ca. 1 cm. long, the lobes reflexed; calyx white to rose (especially in age), 5–7 mm. long, the base scarcely stipitate, barely 1 mm. long, the segms. oblong-obovate; fils. hairy below; aks. lance-ovoid, glabrous, shining, olive-green, ca. 5 mm. long, 3-angled upward.—Gravelly slopes and ridges, 5500–8000 ft.; Red Fir F. to Alpine Fell-fields; Coast Ranges, Lake Co. to Humboldt and Siskiyou cos. and 5500–12,000 ft., Sierra Nevada, from Inyo and Mariposa cos. to Plumas Co.; w. Nev. June–Aug.

11. **E. làtens** Jeps. Caudex woody, with short branches; lvs. basal, round-ovate to elliptic-obovate, subglabrous to short-pilose, apiculate, the blades 1–3 cm. long, abruptly narrowed to longer petioles; scapes naked, 1.5–3 dm. high; infl. capitate, 2–3.5 cm. in diam., subtended by membranous rose-colored bracts; invols. few, campanulate, sparsely pilose, the lobes oblong-ovate, becoming recurved; calyx not evidently stipitate, cream-colored, strigose near base, 5–6 mm. long; fils. pubescent near base; aks. lance-ovoid, gla-brous, ca. 4 mm. long.—Dry stony slopes and ridges, 6500–11,000 ft.; Pinyon-Juniper

Wd., Red Fir F.; e. slope of Sierra Nevada, White Mts., Inyo and Mono cos.; w. Nev. July–Aug.

12. **E. pyrolaefòlium** Hook. Caudex woody, stout, few-branched; lvs. basal, ovate to rounded, the blades glabrous, 1.5–2 cm. long, abruptly narrowed to the equally long villous petioles; scapes naked, 4–8 cm. high; infl. capitate or umbellate, subtended by 2 lanceolate bracts; rays simple, to ca. 6 mm. long; invol. campanulate, loosely woolly, the teeth short, erect; fls. whitish to rose, loosely villous, 5–6 mm. long, obscurely stipitate; fils. somewhat pubescent at base; aks. angled-ovoid, pubescent above, ca. 5 mm. long.— Dry gravelly and sandy slopes, 8500–10,500 ft.; mostly Alpine Fell-fields; Mt. Lassen, Little Mt. Hoffmann, and Mt. Shasta; n. in a more pubescent form (var. *coryphaèum* T. & G.) to Wash. and Mont. Aug.

13. **E. alpìnum** Engelm. Caudex rather slender, with elongate underground branches; lvs. basal, rounded, densely white-tomentose, the blades 1–3 cm. long; scapes 4–6 cm. high, densely white-tomentose, bracted well above the middle and each bearing a single campanulate invol. with 5–6 or more short erect teeth; calyx yellow, glabrous, 3–5 mm. long, short-stipitate; fils. somewhat pubescent at base; aks. glabrous.—Loose slopes and ridges, ca. 7500–9000 ft.; Subalpine F., Alpine Fell-fields; Mt. Eddy and Scott Mt., Siskiyou Co. Aug.

14. **E. Kellóggii** Gray. Caudex cespitose, branched and forming mats, the stems loosely tomentose, branched, with rosettes of lvs. at tips; lvs. oblanceolate to spatulate, silky-tomentose beneath and often less so above, 4–10 mm. long, short-petioled; scapes slender, 4–7 cm. high, with whorl of leafy bracts near middle; invol. solitary, turbinate, tomentose, with short erect teeth; calyx whitish or pinkish, glabrous, 5–7 mm. long, stipitate at base; fils. hairy below; aks. angled-conical, glabrous except at summit, ca. 5 mm. long.—Dry ridges, ca. 4000 ft.; Yellow Pine F.; Red Mt., n. Mendocino Co. July–Aug.

15. **E. marifòlium** T. & G. [*E. m.* var. *apertum* S. Stokes. *E. cupulatum* S. Stokes.] Caudex loosely much-branched, forming mats with tomentose branchlets with terminal tufts of lvs.; lvs. ovate to oval, white woolly beneath, ± glabrate above, 7–16 mm. long, with petioles ca. as long; scapes slender, sparsely tomentose, 0.5–2(–3) dm. high, with terminal capitate or more frequently open umbels subtended by linear bracts; cent. ray often shorter than lateral, these up to 2 or 3 cm. long; invol. turbinate, woolly, 2–3 mm. long, with short erect teeth; plants polygamo-dioecious; calyx yellowish, often tinged red along midribs, glabrous without, 4–5 mm. long in fertile fls. with stipe ca. 1 mm. long, the ♂ fls. shorter; aks. glabrous except sometimes at very tip, angled-conic with narrowed base, greenish, ca. 4 mm. long; styles ca. 0.4 mm. long; $2n = 32$ (Stokes & Stebbins, 1955).—Dry gravelly slopes, 3500–11,000 ft.; Red Fir F., Subalpine F.; Sierra Nevada mostly from n. of Fresno Co. to Siskiyou Co. and cent. Ore.; w. Nev. July–Aug. Some plants from Mono, Fresno, and Tulare cos. are more compact, with short internodes, lvs. more permanently tomentose above and the tomentum with a tawny cast; they intergrade with:

Var. **incànum** (T. & G.) Jones. [*E. i.* T. & G. *E. rosulatum* Small. *E. ursinum* var. *r.* S. Stokes.] Mats very dense, the tomentum somewhat tawny; lvs. densely tomentose on both surfaces, oblong-ovate, the blades 5–15 mm. long; scapes tomentose, 0.1–2 dm. high; infl. subcapitate or short-rayed; aks. ca. 3 mm. long.—Gravelly slopes and ridges, 7000–12,700 ft.; Red Fir F. to Alpine Fell-fields; Tulare and Inyo to Tuolumne and Mono cos. July–Aug.

16. **E. ursìnum** Wats. [*E. ovatum* Greene.] Caudex woody, branched, forming mats; lvs. tufted at ends of branchlets, ovate, the blades 8–14(–25) mm. long, obtuse to acute, often glabrate above, densely white-tomentose beneath, short-petioled; scapes 2–4 dm. high, villous-tomentulose to subglabrate; umbels compact, subtended by narrow lfy. bracts, the rays often 1–3 cm. long and bearing smaller bracts; invol. woolly-villous, subcampanulate, with short broad teeth; calyx ochroleucous, glabrous externally, 5–6 mm. long, the basal stipe ca. 1 mm.; fils. woolly; aks. subconic, subglabrous, ca. 4 mm. long, narrowed at base; styles. ca. 2.5 mm. long.—Dry open gravelly places, 3500–8000 ft.; Yellow Pine F., Red Fir F.; Sierra Nevada from Placer Co. to Butte and Shasta cos. May–Aug.

Var. **nervulòsum** S. Stokes. Plants rhizomatous; peduncles erect, 4–6 cm. long; infl. congested, subcapitate.—Stony places; Red Fir F.; Snow Mt., Lake Co. Aug.

17. **E. ternàtum** Howell. [*E. ursinum* var. *confine* S. Stokes.] Caudex woody, much-branched, forming mats; lvs. in terminal tufts on the branchlets, obovate to oblong, glabrate to tomentose above, densely tomentose beneath, the blades 10–15 mm. long, the petioles ca. as long; scapes 1–3 dm. high, bracted at summit and also on the rays, these to 1 or 2 cm. long; invol. funnelform, woolly, ca. 6–8 mm. long, the teeth ca. 2 mm. long; calyx sulphur-yellow, glabrous, 3–5 mm. long, stipitate at base; fils. woolly at base; aks. pilose above; styles 3 mm. long.—Rocky places, 2000–6000 ft.; Yellow Pine F., Red Fir F.; Del Norte and Siskiyou cos.; sw. Ore. June–Aug.

Var. **Congdònii** (S. Stokes) J. T. Howell. [*E. ursinum* var. *C.* S. Stokes.] Lvs. revolute, narrowly elliptic or ± oblong; branches of infl. ebracteolate.—At 5000–7000 ft.; Red Fir F.; Trinity and Siskiyou cos. July–Aug.

18. **E. angulòsum** Benth. Annual, erect, with spreading dichotomous ± angled stems 1–3(–9) dm. long, whitish-tomentose to glabrate; basal lvs. oblanceolate to oblance-oblong, the blades 1–3 cm. long, short-petioled, glabrate above, tomentose beneath, revolute and crisped on edges; stem-lvs. well distributed, sessile, 0.5–2 cm. long, lanceolate to oblanceolate; peduncles arising from most axils, slender, 1–2 cm. long, glabrous or sparsely tomentose; invol. open-turbinate, 1.5–2.5 mm. long, ± puberulent, with broad rounded lobes; fls. many, rose, tipped with white, ca. 1.5 mm. long, minutely glandular-puberulent; outer calyx-segms. obovate to elliptical, deeply concave, sometimes with an inflated area near base, inner segms. narrowly spatulate, longer; stamens 1–2 mm. long, usually well exserted; aks. ca. 1 mm. long, grayish, the body subovoid with a sharply angled triangular beak.—Dry open places, mostly below 2500 ft.; V. Grassland, Foothill Wd., Joshua Tree Wd., Pinyon-Juniper Wd.; S. Coast Ranges from Contra Costa Co. to San Diego (rare in s.), San Joaquin V. and foothills of Sierra Nevada to w. Mojave Desert. May–Nov.

Ssp. **maculàtum** (Heller) S. Stokes. [*E. m.* Heller. *E. a.* var. *m.* Jeps. *E. a.* vars. *rectipes* and *flabellatum* Gand.] Pedicels and invols. glandular but hairs not capitate; fls. often yellow with purplish spots, the outer calyx-segms. elliptical to roundish or obovate, with a generally conspicuous inflated area; inner segms. obtuse to acute; stamens 0.5–1 mm. long, included.—Dry often sandy or gravelly places, below 7000 ft.; Creosote Bush Scrub, Joshua Tree Wd., Pinyon-Juniper Wd., Sagebrush Scrub; deserts, e. Wash. to San Diego Co.; to Utah, Ariz., L. Calif. April–Nov.

Ssp. **viridéscens** (Heller) S. Stokes. [*E. v.* Heller. *E. a.* var. *v.* Jeps. *E. bidentatum* Jeps. *E. a.* ssp. *bidentatum* S. Stokes.] Pedicels and invols. with capitate hairs; involucral lobes long; inner calyx-segms. acute to acuminate.—Dry plains and hills about upper San Joaquin V. from inner Monterey Co. and Merced Co. to w. Mojave Desert. May–Oct.

19. **E. gracíllimum** Wats. [*E. angulosum* var. *g.* Jones. *E. a.* ssp. *g.* S. Stokes. *E. variabile* Heller. *E. a.* var. *v.* Parish. *E. a.* var. *victorense* Jones. *E. a.* ssp. *victorense* S. Stokes.] Annual, freely branched at or just above base, 1–4 dm. high, thinly woolly; basal lvs. oblong to oblanceolate, 2–4 cm. long, short-petioled, densely woolly beneath, less so above, with revolute crisped edges; cauline lvs. lance-oblong, well distributed; peduncles filiform, 8–25 mm. long, glabrous; invol. subcampanulate, angled, 2 mm. long, glandular-puberulent, shallowly lobed; fls. rose, tipped with white, 2 mm. long, the outer and inner segms. essentially alike, oblong to elliptical, frequently crenulate; aks. shining black, ca. 1 mm. long, the subglobose body with a triangular angled beak.—Common on sandy plains, bélow 3500 ft.; V. Grassland, Foothill Wd., Joshua Tree Wd., Coastal Sage Scrub; inner S. Coast Ranges from Merced Co. and Monterey Co. to w. Mojave Desert and interior s. Calif. April–Sept.

20. **E. gossýpinum** Curran. Annual, diffusely dichotomous, 0.5–2 dm. high, tomentose; basal lvs. broadly oblanceolate, 1.5–4 cm. long; stem lvs. lanceolate, somewhat smaller; peduncles 2–15 mm. long; invol. turbinate, 3 mm. high, deeply lobed and filled with a dense cottony tomentum; fls. few, concealed; calyx-segms. linear-oblong, similar, 1.5 mm. long.—Uncommon, sandy places, below 3000 ft.; V. Grassland; about head of San Joaquin V., Kings and Kern cos. April–Sept.

21. **E. spergulìnum** Gray. [*Oxytheca s.* Greene.] Erect slender-stemmed annual, forking freely above with widely spreading branches, 1–3 dm. high, the internodes generally with capitate glands; basal lvs. linear, 2–3 cm. long, short-petioled, hispid; lower stem-lvs. linear, whorled, sessile; upper bracteate; peduncles filiform, 5–12 mm. long; invol. solitary, glabrous, calyxlike, 0.5–1 mm. long, 4-lobed; fl. solitary, white, veined with

rose, glabrous or sparsely pubescent, 2.5–3.5 mm. long, the segms. oblong; anthers 0.35–0.5 mm. long, oblong to elliptic; aks. brownish, ca. 1.5 mm. long, lanceolate in outline, narrowed gradually into the angled beak.—Dry gravelly flats and gentle slopes, 5000–10,000 ft.; Montane Coniferous F.; Sierra Nevada from Eldorado Co. to Tulare Co. June–Aug.

Var. **Reddingiànum** (Jones) J. T. Howell. [*Oxytheca R.* Jones.] Internodes generally stipitate-glandular; calyx 1.5–2.5 mm. long; anthers 0.2–0.33 mm. long, roundish.— More common, 4000–11,000 ft.; Montane Coniferous F.; N. Coast Ranges from Lake Co. n., Sierra Nevada to Mt. Pinos, Ventura Co.; to Ore., Ida., Nev. June–Aug.

Var. **praténse** (S. Stokes) J. T. Howell. [*E. p.* S. Stokes.] Internodes usually glandless; calyx ± hirsutulous, ca. 2 mm. long.—Similar places, 8000–11,500 ft.; Sierra Nevada, Inyo and Tulare cos. July–Aug.

22. **E. hirtiflòrum** Gray. ex Wats. [*Oxytheca h.* Greene.] Annual 5–15 cm. high, repeatedly dichotomously branched, glandular-puberulent; lvs. basal and at lower nodes, obovate to spatulate, 1–2.5 cm. long, ciliate, narrowed into winged petioles; invols. sessile in the forks and along the branches, or short-peduncled, narrow, 2-fld., ca. 1 mm. long; calyx reddish, the segms. oblong, ca. 1 mm. long, hirsutulous with hooked hairs; aks. narrow, ca. 1 mm. long, exceeding calyx, with a broad obtuse angled beak.—Occasional, dry gravelly places, below 6000 ft.; Chaparral, Foothill Wd., Yellow Pine F.; N. Fork Tujunga Creek, Los Angeles Co., Council Rock, Ventura Co.; The Pinnacles, San Benito Co.; N. Coast Ranges and Sierra Nevada. June–Aug.

23. **E. inérme** (Wats.) Jeps. [*Oxytheca i.* Wats. *Eriogonum vagans* Wats.] Annual, dichotomously or trichotomously forked just above or at base, then repeatedly dichotomous, 5–30 cm. high, sparingly stipitate-glandular; lvs. basal, spatulate, 1–2 cm. long, sessile, ciliate, otherwise glabrous; bracts 3-lobed; invols. on pedicellike peduncles, 4-cleft nearly to base, subglabrous, 1.5 mm. long; calyx rose, hispid with hooked hairs, the segms. oblong, 1.5 mm. long, the inner retuse and smaller than the outer; aks. scarcely if at all longer than the calyx.—Uncommon, dry barren soils, below 7000 ft.; Foothill Wd., Chaparral, Yellow Pine F.; Coast Ranges from Lake Co. to Kern Co. June–July.

Var. **hispídulum** Goodm. Invol. hispidulous.—From 3000–6000 ft.; Yellow Pine F.; San Bernardino Mts., Sierra Nevada from Tulare Co. to Tuolumne Co. June–Aug.

24. **E. apículàtum** Wats. [*E. a.* var. *subvirgatum* S. Stokes] Erect annual usually simple at base, dichotomously or trichotomously branched above, spreading, 2–9 dm. high, somewhat glandular-pubescent in lower portion of internodes and peduncles, with ultimate very slender branchlets; lvs. strictly basal, obovate to oblanceolate, the blades, 1.5–4 cm. long, pilose, glandular, the petioles ca. as long; bracts 1–2 mm. long; peduncles in forks and scattered along the branchlets, filiform, 2–35 mm. long, often deflexed; invols. 1.5 mm. long, glabrous, 4-lobed half their length; fls. few, white, 1.5–2 mm. long, puberulent, the segms. oblong-obovate, notched to apiculate; aks. ca. 1.5 mm. long, narrow, not beaked.—Dry open places in disintegrated granite, 3600–8000 ft.; Pinyon-Juniper Wd., Yellow Pine F.; San Jacinto, Santa Rosa, Palomar, and Cuyamaca mts. July–Aug.

25. **E. Parishii** Wats. Annual with 1–3 erect stems 1–3 dm. high, diffusely branched into a dense rounded mass of very slender ultimate branchlets, glaucous, glabrous except for short-stipitate glands above the nodes; lvs. basal, spatulate, 2–6 cm. long, hirsute; peduncles capillary but rigid, 4–12 mm. long; invol. solitary, 5-lobed, ca. 0.6 mm. long; fls. 1–2, pinkish, minutely puberulent, ca. 0.6 mm. long, the outer segms. ovate, the inner oblong-spatulate; aks. ca. 1 mm. long, dark, the body subglobose with a stout somewhat angled beak; $2n = 40$ (Stokes & Stebbins, 1955).—Dry gravelly places, 4000–9000 ft.; Pinyon-Juniper Wd., Yellow Pine F., Lodgepole F.; s. Sierra Nevada (Tulare and Inyo cos.), San Gabriel Mts. to n. L. Calif. July–Sept.

26. **E. tríchopes** Torr. [*E. trichopodum* Torr. ex Benth. *E. clavatum* Small. *E. t.* ssp. *c.* S. Stokes. *E. cordatum* Torr.? *E. t.* ssp. *cordatum* S. Stokes?] Annual, sometimes perennial, with 1–several stems from the base, 3–several-forked at the first node, 1–5 dm. high, glabrous, glaucous, sometimes inflated in the lower internodes; lvs. in a basal rosette, round to somewhat round-oblong, ± cordate at base, hirsute, often somewhat crinkled, the blades 1–2 cm. long, on petioles as long to twice as long; peduncles capillary, 8–15 mm. long; invol. turbinate, scarcely 1 mm. long, glabrous, 2-few-fld.,

4-lobed; calyx yellow to green, 1 mm. long at anthesis, longer in fr., white-strigulose, the segms. ovate, alike; aks shining, brown, 1.5–2 mm. long, narrow-ovoid, 3-angled, with stout beak.—Common in washes and on mesas, below 5500 ft.; mostly Creosote Bush Scrub, Joshua Tree Wd.; Mojave and Colo. deserts, inner S. Coast Ranges; to Utah, Son., L. Calif. April–Aug.

27. **E. Órdii** Wats. [*E. tenuissimum* Eastw.] Annual, diffusely paniculate with many capillary ultimate branches, floccose near base, glabrous above, 4–7 dm. high; basal lvs. oblong-obovate to -oblanceolate, the blades 2–8 cm. long on equally long petioles, ± woolly, especially beneath; bracts of lower nodes often in foliaceous whorls, the upper subulate; peduncles capillary, 1–2 cm. long; invol. solitary, turbinate, 1 mm. long, 4-toothed; fls. 1–3, white tinged with pink, densely pubescent, 1.5 mm. long, the calyx-segms. oblong-ovate; aks. shining, olive-green, ca. 1.5 mm. long, ovoid, 3-angled and stout-beaked.—Dry disturbed and barren places, below 3000 ft.; Foothill Wd.; inner Coast Ranges of Monterey and San Benito cos., n. base of Tehachapi Mts.; also Creosote Bush Scrub and Pinyon-Juniper Wd., in n. Los Angeles Co. and at Split Mt., Colo. Desert; w. Ariz. March–June.

28. **E. inflàtum** Torr. & Frém. DESERT TRUMPET. Perennial, but flowering the first year, 2–8 dm. high, glabrous and glaucous throughout or somewhat hirsute at very base, with basal lvs. and 1 to several suberect stems which are simple below, then trichotomous and dichotomous above to form diffuse panicles, the stems conspicuously inflated in upper portion of lower internodes; lvs. oblong-ovate to rounded or even subreniform, usually cordate at base, 1–2.5 cm. long, short-hirsute, green, somewhat crisped, on petioles 2–5 cm. long; peduncles capillary, in forks and racemosely along the branchlets, 5–20 mm. long; invol. glabrous, 1–1.5 mm. long, several-fld., turbinate, 5-lobed, the lobes with stipitate glands; calyx yellow, or with red-brown midribs, conspicuously and densely pubescent, 2–2.5 mm. long, the segms. lance-ovate; aks. brown, ca. 2 mm. long, sharply 3-angled, lance-ovoid, narrowed gradually toward apex.—Common along washes and on mesas, below 6000 ft.; Creosote Bush Scrub, Joshua Tree Wd., Sagebrush Scrub, Pinyon-Juniper Wd.; Mojave and Colo. deserts, to Mono Co.; to Utah, Ariz., L. Calif. March–July, Sept.–Oct.

Var. **deflàtum** Jtn. [*E. glaucum* Small.] Stems not inflated.—Largely on Colo. Desert, also in Death V.; L. Calif.

E. glandulòsum (Nutt.) Nutt. var. **cárneum** J. T. Howell, with capitate-glandular stems and peduncles and pinkish calyces, occurs in extreme e. Inyo and ne. San Bernardino cos.

29. **E. esmeraldénse** Wats. Glabrous annual, 1–few-stemmed, then repeatedly branched, 1–3 dm. high, the ultimate branches very slender; lvs. basal, somewhat round-obovate, 6–15 mm. long, pilose-hispid, on petioles ca. as long; peduncles filiform, 5–15 mm. long; invol. narrow-turbinate, 1 mm. long, 5-lobed; fls. few, glabrous, white to pink, the segms. similar, ± oblong, obtuse or retuse; aks. narrow-ovoid, brown, shining, ca. 1 mm. long, gradually narrowed into a stout beak.—Dry gravelly places, 6000–9800 ft.; Pinyon-Juniper Wd.; mts. of Inyo Co.; w. Nev. July–Sept.

30. **E. Thomàsii** Torr. [*E. minutiflorum* Wats.] Annual, 1–several-stemmed at base, 1–2.5 dm. high, glabrous, glaucous, repeatedly trichotomous; lvs. basal, round to round-reniform, 1–2 cm. wide, often glabrate above, white-woolly beneath, on petioles 2–5 cm. long; peduncles filiform, 5–15 mm. long; invol. glabrous, deeply 5-lobed, barely 1 mm. high; fls. several, at first yellow, later white to rose, ca. 1 mm. long, hispidulous without, the outer segms. ovate, with a saclike dilation on each side of the cordate base, the inner segms. spatulate; aks. ca. 1 mm. long, dark brown, shining, round-ovoid with a 3-angled beak.—Common in dry sandy places, below 5000 ft.; Creosote Bush Scrub, Joshua Tree Wd.; both deserts n. to Inyo Co.; to Utah, Ariz. March–June.

31. **E. Thúrberi** Torr. [*E. cernuum* ssp. *T*. S. Stokes. *E. c.* ssp. *viscosum* S. Stokes. *E. T.* var. *Parishii* Gand.] Annual, simple or several-stemmed from base, diffusely and trichotomously branched, 1–3 dm. high, floccose at least in lower parts; lvs. basal, oblong-ovate, 1–3 cm. long, densely white-woolly beneath, glabrate above, on petioles 1–3 cm. long; peduncles capillary, 5–25 mm. long, ± glandular-puberulent; invol. broadly turbinate, glandular-puberulent, 2 mm. high, 5-lobed to near middle; fls. rose to whitish, 1–1.5 mm. long, glandular-puberulent near base, the outer segms. roundish or

broader, abruptly narrowed to a clawlike base, the inner narrowly lanceolate; aks. almost black, shining, ca. 0.6 mm. long, round-ovoid with short sharp beak.—Sandy places, below 5000 ft.; Coastal Sage Scrub, Chaparral; Los Angeles region to San Diego Co.; Creosote Bush Scrub, Joshua Tree Wd., Colo. Desert and occasional on Mojave Desert; Ariz., L. Calif. April–July.

32. **E. pusíllum** T. & G. [*E. reniforme* ssp. *p.* S. Stokes.] Annual, erect, simple or branched at base, trichotomously branched above, with glabrous glaucous stems 1–3 dm. high; lvs. basal, the blades rounded to oblong-ovate, 1–3 cm. long, somewhat greenish above, white-woolly beneath, on equally long petioles; peduncles very slender, 1–3 cm. long; invol. broadly turbinate, glandular-puberulent, 1.5 mm. high, with 5 broad rounded lobes; fls. several, 1.5 mm. long, glandular-puberulent, yellow, later with reddish midribs, the outer segms. obovate, the inner oblong, both elongating somewhat in fr.; aks. dark, shining, ca. 0.6 mm. long, round-ovoid with short stout beak.—Common on plains and mesas, mostly 2500–6500 ft.; Creosote Bush Scrub, Joshua Tree Wd., Pinyon-Juniper Wd.; deserts from s. Mono Co. to Palm Springs region; Nev., Utah. March–July.

33. **E. renifórme** Torr. & Frém. Annual with 1–several stems, ± floccose below, glabrous above, divergently trichotomously branched, 0.5–2.5 dm. high; lvs. basal, round-reniform or rounded, mostly 1–2 cm. wide, glabrate above, white-woolly beneath, with somewhat crisped margins; peduncles 4–15 mm. long, capillary; invol. glabrous, subcampanulate, almost 2 mm. long, shallowly and broadly 5-lobed; fls. several, pale yellow, ca. 1–1.5 mm. long, glandular-puberulent, the outer segms. elliptic-ovate, the inner narrower; aks. brown, shining, round-lenticular, scarcely beaked, ca. 1 mm. long.—Common in sandy places, below 4500 ft.; Creosote Bush Scrub, Joshua Tree Wd.; Mojave and Colo. deserts, Inyo Co. to Nev., L. Calif. March–June.

34. **E. cérnuum** Nutt. Annual, glabrous, glaucous, diffusely branched from or above base, 1–3 dm. high; lvs. rounded, basal, 1–2 cm. long, densely white-woolly beneath, subglabrate above; petioles 2–4 cm. long; peduncles capillary, usually deflexed, 5–25 mm. long; invol. turbinate, almost 2 mm. high, glabrous, 5-lobed; calyx white, with rose tinge, glabrous, 1–1.5 mm. long, attenuate at base, the segms. oblong-obovate, undulate, emarginate; aks. slender, ca. 1.5 mm. long.—At 7000–10,000 ft.; Panamint Mts., White Mts., e. slope of Sierra Nevada, Inyo Co.; to Ore., Rocky Mts. June–Sept.

35. **E. Hoffmánnii** S. Stokes. [*E. H.* var. *robustius* S. Stokes.] Erect annual, trichotomous above base, then diffusely branched, 1–8 dm. high; stems glabrate or sparsely glandular, or arachnoid in the axils, glaucous; lvs. basal, round-ovate, the blades ± crisped at edge, 1–5 cm. long, glabrate above, white-tomentose beneath, with petioles 1–5 cm. long or longer; peduncles 0.5–6 mm. long, spreading or deflexed; invol. glabrous, turbinate, 1.5–2 mm. long, toothed; calyx-segms. glabrous, pinkish, lance-oblong, ca. 1.5 mm. long; aks. exserted, ca. 1.5 mm. long, brown, 3-angled, slender, gradually attenuate to apex.—Dry slopes and washes, 1500–7500 ft.; Creosote Bush Scrub, Pinyon-Juniper Wd.; Panamint, Black, and Funeral mts., Inyo Co. Aug.-Sept.

36. **E. nùtans** T. & G. [*E. praebens* Gand.] Low erect annuals, simple or branched at base, trichotomous at first node and usually at some upper ones; stems glabrous, glaucous, slender, 1–2 dm. high; lvs. basal, the blades suborbicular, 8–15 mm. wide, white-woolly especially beneath, somewhat glabrate above, the petioles 1–2½ times the blades; peduncles slender, nodding, 5–10 mm. long; invol. campanulate to hemispherical, 2–3.5 mm. broad, red, glandular-puberulent; fls. minutely glandular-puberulent near base, reddish, 2–3 mm. long, the outer segms. oblong or oval; aks. brownish, almost 2 mm. long, the base subglobose, the beak narrow, pubescent, 3-angled.—Dry places, 4300–5500 ft.; Sagebrush Scrub; Lassen Co.; to Nev., Utah. May–July.

37. **E. defléxum** Torr. Erect widely spreading annual, simple or branched at base and branched above, 1–3 dm. high and with intricate sometimes almost horizontal branches in well developed plants, green, glabrous; lvs. all basal, rounded to round-reniform, 1–2.5 cm. long, somewhat wider, densely white-woolly, especially beneath, the petioles longer than blades; peduncles ca. 1–2 mm. long, commonly pendent; invol. solitary, turbinate, green, glabrous, 1.5–2 mm. long, 5-lobed; calyx whitish, sometimes pink, glabrous, 1.5 mm. long or more in fr., the outer segms. ovate, glabrous, cordate at base, the inner shorter; aks. red-brown, ca. 2 mm. long, the subglobose body bearing a prominent somewhat-angled stout beak.—Common in washes and on adjacent slopes,

below 7000 ft.; occasional higher; Creosote Bush Scrub to Pinyon-Juniper Wd.; Mojave and Colo. deserts, Mono Co. to Imperial Co.; w. Nev., Ariz., L. Calif. May–Oct.

Var. **brachypòdum** (T. & G.) Munz. [*E. b.* T. & G.] Stems with stalked glands.— Range of the sp.

Ssp. **Rixfórdii** (S. Stokes) Munz. [*E. R.* S. Stokes.] Stems glabrous, 2–4 dm. tall, branched so as to form a series of layers one above the other; invol. subsessile or nearly so.—Panamint Mts. to White Mts.

Ssp. **Watsònii** (T. & G.) S. Stokes. [*E. Watsonii* T. & G. *E. baratum* Elmer.] Stems glabrous; peduncles 4–10 mm. long; outer calyx-segms. oblong, often roughened, often cordate at base.—Largely 5000–8500 ft.; Pinyon-Juniper Wd. to Red Fir F.; Mt. Pinos, Ventura Co. to e. slope of Sierra Nevada; w. Nev. July–Oct.

38. **E. argillòsum** J. T. Howell. Erect annual 1–3 dm. high, 1–several-stemmed, glabrous to sparingly tomentose; basal lvs. oblong, 1.5–2.5 cm. long, white-woolly beneath, less so above, slightly revolute, the petioles as long as or longer than blades; stem-lvs. at first node somewhat smaller, others still more reduced; peduncles filiform; invol. turbinate, 2.5 mm. high, obscurely 5-lobed, scarious below the sinuses, glabrous; calyx white or rose with dark midrib, 1.5 mm. long, the outer segms. oblong, somewhat broader than the inner; aks. 2–2.5 mm. long.—Clay and serpentine; Foothill Wd.; inner Coast Ranges, Santa Clara Co. to San Benito and Monterey cos. March–June.

39. **E. vestìtum** J. T. Howell. Erect annual 1–4 dm. high, densely white-tomentose, simple below, branched above, leafy; basal lvs. elliptic to elliptic-oblong, 1–3 cm. long, woolly on both surfaces, the petioles ca. as long; lower stem lvs. similar but gradually reduced upward; peduncles 1–6 cm. long, or almost lacking, axillary or racemosely arranged; invol. campanulate-turbinate, 2 mm. high, 5-lobed, with narrow scarious sinuses; calyx white with red midribs, the segms. oblong-ovate, 1.5–2 mm. long, papillose; aks. 2.5 mm. long, with slender papillose beak.—Dry slopes; V. Grassland; inner Coast Ranges, San Benito and w. Fresno cos. May–June.

40. **E. nidulàrium** Cov. [*E. vimineum* ssp. *n.* S. Stokes. *E. n.* var. *luciense* Jones.] Erect annual, repeatedly forked from near base, 5–15(–20) cm. high, floccose-tomentose throughout, forming a dense mass of numerous branches with short internodes and in age the tips often incurved; lvs. basal, rounded, sometimes cordate at base, 1–2 cm. broad, on much longer petioles; invols. sessile, in all the forks and along the branches, cylindrical, few-fld., 1 mm. long; calyx red, white, or yellow, 1.5–2(–3) mm. long, glabrous, the outer segms. obovate, dilated and truncate at apex, the inner somewhat narrower; fils. glabrous; aks. narrow-ovoid, ca. 1.5 mm. long, ending in a longish scaberulous beak.—Common in dry gravelly and rocky places, below 6000 ft.; Creosote Bush Scrub, Joshua Tree Wd., Pinyon-Juniper Wd.; deserts from Mono Co. to San Bernardino Co.; Nev., Ariz. April–Oct.

41. **E. grácile** Benth. [*E. vimineum* ssp. *g.* S. Stokes. *E. acetoselloides* Torr. ex Benth. *E. leucocladon* Benth. *E. agninum* Greene. *E. verticillatum* Nutt.] Erect annual, strictly or rather diffusely branched from base, 2–5 dm. high, thinly tomentose to floccose throughout, the branchlets slender, ascending; lvs. mostly basal, oblanceolate to oblong, the blades 1–3(–4) cm. long, tomentose, especially beneath, on petioles ca. as long; some lower bracts foliaceous, the upper not exceeding invols.; invol. 1.8–2(–3) mm. long, turbinate, subglabrous, the 5 teeth conspicuous, rigid; fls. few, white, pinkish, or yellowish, glabrous, 1.5–2 mm. long, the outer calyx-segms. broadly obovate, rounded at summit, the inner oblong; aks. ovoid, ca. 1 mm. long, with prominent angled beak; $2n = 22$ (Stokes & Stebbins, 1955).—Common in dry cismontane washes, on mesas, etc., below 3500 (5000) ft.; Coastal Sage Scrub, Chaparral, Foothill Wd., S. Oak Woodland, etc.; inner Coast Ranges and Great V. from Vaca Mts. s. through s. Calif.; L. Calif. July–Oct.

Var. **citharaefòrme** (Wats.) Munz. [*E. c.* Wats. *E. vimineum* var. *c.* S. Stokes.] Lvs. practically all basal, crisped on margins and with conspicuously winged petioles.—San Luis Obispo, Santa Barbara, and Ventura cos. May–Aug.

Var. **polygonoìdes** (S. Stokes) Munz. [*E. vimineum* ssp. *p.* S. Stokes.] Stems glabrous, glaucous; lvs. elliptic to ovate-lanceolate, with narrow petioles.—Grain fields and dry slopes; San Luis Obispo to Santa Ynez, Santa Barbara Co. Sept.–Nov.

42. **E. virgàtum** Benth. [*E. vimineum* ssp. *v.* S. Stokes. *E. roseum* Dur. & Hilg.] Erect annual, simple or with few ascending virgate branches, floccose-tomentose throughout, 1–8 dm. high; lvs. at base and lower nodes, the basal oblong-oblanceolate, 1–3 cm.

long, with equally long petioles; invols. rather remote, sessile, cylindric, 4–5 mm. long, 5-toothed, tomentose; fls. yellow to pink or white, glabrous, 2 mm. long, the outer segms. narrowly obovate, the inner oblong; aks. almost 2 mm. long, narrowly ovoid with broad 3-angled scaberulous beak; $2n = 18$ (Stokes & Stebbins, 1955).—Dry, often sandy or gravelly or rocky places, below 5000 ft.; Chaparral, Foothill Wd., Yellow Pine F., N. Oak Wd., etc.; away from the immediate coast, s. Ore. to n. Ventura and Los Angeles cos., especially in the ranges bordering the Great V. June–Oct.

43. **E. vimíneum** Dougl. ex Benth. [*E. luteolum* Greene. *E. vimineum* var. *l.* S. Stokes. *E. pedunculatum* S. Stokes.] Erect annual with 1–several stems from base, usually branched above, 1–5 dm. high, ± floccose-tomentose below, upper branches glabrous and very slender, ± greenish, glaucous; lvs. generally all basal, round to round-ovate, glabrate above, white-woolly beneath, 1–2 cm. long, on longer petioles; invols. usually sessile along the branches, narrow-cylindric, glabrous, 2–3 mm. long, with short blunt teeth; calyx rose to yellowish, glabrous, 2 mm. long, the outer segms. obovate, with rounded apex, the inner narrower; fils. glabrous; aks. almost 2 mm. long, red-brown, ovoid with prominent muriculate angled beak; $2n = 24$ (Stokes & Stebbins, 1955).—Common in dry rocky and sandy places, below 6000 ft.; Chaparral, Foothill Wd., Yellow Pine F., etc.; Coast Ranges and Sierran foothills, Monterey, Santa Clara, and Mariposa cos. n. to Wash., Ida. June–Sept. A form with stems glabrous to base has been called var. *califórnicum* Gand. [*E. v.* var. *multiradiatum* S. Stokes.] and occurs with the sp. in its Calif. range.

44. **E. Nortònii** Greene. [*E. vimineum* ssp. *N.* S. Stokes.] Erect annual, dichotomously or trichotomously forked, 0.5–2 dm. high, with glabrous reddish stems; lvs. at base and lower nodes, round to reniform, deeply emarginate, green above, white-woolly beneath, 5–15 mm. long, on longer petioles; invols. sessile in forks and at tips of short slender branchlets, solitary, broadly turbinate, glabrous, 3–4 mm. long, with ca. 8 short blunt teeth; calyx white to rose, 1.5 mm. long, the segms. similar, obovate, glabrous; aks. glabrous, ovoid, with stout angled beak.—Dry rocky slopes, 1500–4000 ft.; Chaparral; The Pinnacles, inner Coast Ranges, Monterey and San Benito cos. May–June.

45. **E. Covilleànum** Eastw. [*E. vimineum* var. *C.* S. Stokes.] Erect annual, simple below or branched from base, glabrous, the very slender stems 1–3 dm. long, dichotomously or trichotomously branched above in cymose fashion; lvs. basal, suborbicular, rounded at apex, somewhat subcordate at base, subglabrous above, white-tomentose beneath, 0.5–1.5 cm. long, on longer slender petioles; invols. solitary, ca. 2.5 mm. long, sessile at forks and nodes or terminal, narrow-turbinate, glabrous without, 5-veined, the margin subentire and ciliate; calyx 2 mm. long, rose or white with red veins, the segms. elliptic, puberulent toward base without; fils. slightly pubescent at base; aks. ca. 2 mm. long, ovoid with prominent muriculate beak.—Shale and serpentine talus, 1200–2000 ft.; Chaparral, Foothill Wd.; inner Coast Ranges, Alameda Co. to San Benito Co. April–June.

Ssp. **adsúrgens** (S. Stokes) Abrams. [*E. truncatum* var. *a.* S. Stokes ex Jeps.] Invol. ca. 2 mm. long, somewhat toothed, the membrane not extending to apex of the ribs; $2n = 22$ (Stokes & Stebbins, 1955).—Similar situations, San Benito Co. to San Luis Obispo Co.

46. **E. Eastwoodiànum** J. T. Howell. Erect annual, branched from base, 2–5 dm. tall, floccose-tomentose; lvs. basal, suborbicular, 1–3 cm. in diam., woolly beneath, subglabrate above, on petioles 3–8 cm. long; upper branches forming cymose infl.; invols. peduncled in forks, sessile on branchlets or terminal, tomentose, 2 mm. long, turbinate, plainly 5-toothed; calyx glabrous, 2 mm. long, the outer segms. elliptic to oblong-obovate, obtuse, the inner oblong, smaller; fils. basally pubescent; aks. 2 mm. long, brownish.—Diatomaceous shale; Foothill Wd.; mts. of w. Fresno Co. June.

47. **E. canìnum** (Greene) Munz. [*E. vimineum* var. *c.* Greene.] Widely spreading annual; stems mostly glabrous, reddish, several from base, 1.5–3.5 dm. long, repeatedly dichotomous or trichotomous; lvs. basal and on lower nodes, oblong-ovate, greenish above, white-tomentose beneath, rounded at apex, subcuneate at base, 0.5–3 cm. long, on petioles 2–3 times as long; invol. glabrous, narrowly turbinate, 5-ribbed and -toothed, 3–4 mm. long; calyx rose-red, glabrous, 1.5–2.5 mm. long, the segms. obovate, similar, rounded at apex; fils. apparently glabrous; aks. lance-ovoid, reddish, tapered gradually to the angled beak.—Dry rocky slopes on shale or serpentine, 1000–2200 ft.; Coastal Prairie; Marin Co. and Oakland Hills. June–Sept.

346   *Polygonaceae*

48. **E. truncàtum** T. & G. Erect annual, 1–several-stemmed, floccose-tomentose, 1–3 dm. high, dichotomous or trichotomous; lvs. basal and at lower nodes, oblong-oblanceolate to obovate, greenish above, floccose beneath, 2–5 cm. long, attenuate to petioles ca. as long; infl. open, subcymose; invols. subsessile, 1–few in forks and at ends of branchlets, tomentose, 3 mm. long, oblong-turbinate, shallowly and broadly 5-toothed, the sinuses almost filled with membrane; calyx glabrous, light rose-color, ca. 2 mm. long, the outer segms. elliptic-obovate, the inner slightly narrower; fils. pubescent below; aks. dark, ca. 2 mm. long, narrow-ovoid, with broad angled beak.—Dry slopes, 1000–1500 ft.; edge of Chaparral; e. base of Mt. Diablo, Contra Costa Co. April–June.

49. **E. moléstum** Wats. [*E. vimineum* ssp. *m.* S. Stokes. *E. v.* var. *aviculare* S. Stokes.] Erect annual, few-branched or usually simple at base, glabrous and glaucous except on the lvs., 1–3(–5) dm. high; lvs. all basal, white-woolly, rounded or reniform, 1–2 (–4) cm. broad, crisped or undulate, on longer petioles; invols. sessile, remote, few on a branch, cylindric-turbinate, 4–5(–7) mm. long, glabrous, scarious between the ribs; calyx white or tinged pink, glabrous, 1.5–2 mm. long, the outer segms. oblong-obovate, the inner slightly narrower; fils. pubescent at base; aks. brown, narrow-ovoid, shining, ca. 2 mm. long, narrowed slightly into a muriculate stout angled beak.—Occasional, dry places under pines, 4500–7000 ft.; mostly Yellow Pine F.; Ventura Co. to San Diego Co., probably also in Tulare Co. June–Sept.

Var. **Davidsònii** (Greene) Jeps. [*E. D.* Greene.] Invols. more numerous, 3–4 mm. long, more prismatic; fls. pale pink to whitish.—Common in dry places, 4000–8500 ft.; Chaparral, Yellow Pine F., Pinyon-Juniper Wd., occasional at lower elevs. in Joshua Tree Wd.; mts. from San Diego Co. to Ventura Co. and L. Calif. A form from the general range of the variety with rose-colored fls. has been named *E. vimineum* var. *glàbrum* S. Stokes.

50. **E. Baileyi** Wats. [*E. vimineum* ssp. *B.* S. Stokes.] Erect annual diffusely branched from base, 1–3 dm. high, forming a broad round-topped crown, glabrous except at the white-woolly base of stems; lvs. also white-woolly, suborbicular, basal, 5–15(–18) mm. wide, on somewhat longer petioles; invols. sessile at nodes of subvirgate branchlets, tubular-campanulate, 1–1.5 mm. long, ciliate; fls. several, somewhat constricted near middle and flaring above, white to pink, ca. 1.5 mm. long, glandular, the outer segms. oblong or oblong-obovate, the inner narrower; aks. dark brown, ca. 1 mm. long, the body rounded and gradually narrowed into a stout muriculate angled beak.—Dry sandy or gravelly flats and banks, mostly 2500–7500 ft.; Creosote Bush Scrub, Joshua Tree Wd., to Yellow Pine F.; e. end of San Bernardino Mts. across Mojave Desert; to e. Wash., Nev., Ariz. May–Sept.

Var. **brachyánthum** (Cov.) Jeps. [*E. b.* Cov. *E. vimineum* var. *b.* S. Stokes.] Base of plant usually persistently white-woolly; fls. glabrous, yellowish, 1–1.3 mm. long.—Occasional in dry places, 2000–5000 ft.; Creosote Bush Scrub, Joshua Tree Wd.; e. end of San Bernardino Mts. to n. end of Owens V.; w. Nev., n. L. Calif. May–Sept.

Var. **tomentòsum** Wats. Stems floccose-tomentose; fls. glabrous, pinkish.—Yellow Pine F. to Joshua Tree Wd.; Bear V., San Bernardino Mts.; to Nev., Utah, Ariz. Aug.–Sept.

Ssp. **elegáns** (Greene) Munz. [*E. e.* Greene. *E. vimineum* var. *e.* Jeps.] Stems woolly at base, the branches reddish, glabrous, glaucescent; invol. ca. 1.5 mm. long, turbinate; calyx 1.5–2 mm. long, glandular-puberulent without, rose to paler with red veins, the segms. oblong-obovate.—Dry sandy flats and washes; V. Grassland, Foothill Wd.; S. Coast Ranges, Santa Clara Co. to San Luis Obispo Co. May–Aug.

51. **E. dasyánthemum** T. & G. [*E. d.* var. *Jepsonii* Greene.] Erect annual, branched from base or above, 2–6 dm. high, floccose-tomentose, sometimes glabrate; lvs. basal, roundish, white-woolly beneath, glabrate above, 1–2 cm. wide, on petioles ca. as long; lower nodes also with some lvs.; invols. sessile, scattered along the branchlets, 5-ribbed and -toothed, subcylindric, ca. 4 mm. long, tomentose between the ribs; fls. white or rose, 2 mm. long, pubescent without, the segms. oblong-obovate, rounded; aks. 1.5–2 mm. long, with scabrellous angled beak; $2n = 24$ (Stokes & Stebbins, 1955).—Dry slopes; Foothill Wd., Chaparral, V. Grassland; inner N. Coast Range from Lake Co. to Tehama Co. Aug.–Oct.

52. **E. mohavénse** Wats. [*E. delicatulum* Wats.] Erect annual, diffusely and repeatedly trichotomously or dichotomously branched at or above base, 1–3 dm. high,

glabrous and green except at the nodes and lvs., the ultimate branchlets capillary; lvs. all basal, round or broadly oblong, closely white-woolly, 0.6–2 cm. long, on petioles ca. twice as long; invols. sessile in the forks and often terminal on the branchlets, hence in subcymose infl., glabrous except at the throat, turbinate, almost 2 mm. long; fls. yellow, glabrous, the segms. ca. 1 mm. long, the outer oblance-oblong to subelliptic, the inner narrower; aks. dark, ca. 1 mm. long, the stout beak angled and muriculate.—Dry sandy and gravelly places, mostly ca. 2000–4000 ft.; Creosote Bush Scrub, Joshua Tree Wd.; Mojave Desert from e. base of San Bernardino Mts. to Owens V. May–Aug.

Ssp. **ampullàceum** (J. T. Howell) S. Stokes. [*E. a.* J. T. Howell.] Branches of stems almost erect, the angle of branching narrow; invol. turbinate-campanulate, ca. 1.5 mm. long; fls. yellowish to whitish, the segms. ca. 0.5 mm. long, oblong.—At ca. 6500–7000 ft.; Sagebrush Scrub; Mono Co.; adjacent Nev. Aug.–Sept.

53. **E. deserticola** Wats. Erect shrub 6–12(–15) dm. high, much branched, the ultimate branchlets white-tomentose and leafy when young, becoming glabrous, green and leafless in age; lvs. oblong-ovate to round-oblong, 5–15 mm. long, sometimes wider, white-woolly, on petioles 5–12 mm. long; invol. solitary, turbinate, woolly, 1.5 mm. long, with 4 rounded teeth, short-pedicelled; fls. few; calyx yellow with green or reddish midveins, silky-villous, the segms. oblong-obovate, 2–3 mm. long; fils. pubescent below; aks. dark, lance-ovoid, strigose, scabrellous above, 3 mm. long.—Locally common along sandy washes, dunes, etc.; Creosote Bush Scrub; Imperial Co. from Salton Sink to dunes w. of Yuma. Sept.–Dec.

54. **E. saxàtile** Wats. [*E. Bloomeri* Parish. *E. Stokesae* Jones. *E. s.* var. *S.* Jones.] Perennial with few-branched caudex clothed with the crowded closely white-felted lvs.; lvs. basal, rounded or broadly obovate, the blades 1–2.5 cm. long, petioles ca. as long or longer; flowering stems ascending, rather slender, 1–3(–4.5) dm. high, closely tomentose or floccose, forking above with ascending or spreading branches; invols. solitary at nodes, scattered along the branches, tomentulose, 3–4 mm. long, turbinate, many-fld.; calyx white, pinkish, or yellowish, 5–7 mm. long, glabrous, narrowed to a sharply triangular narrow base, the outer segms. oblanceolate, the inner obovate and larger; aks. brownish, glabrous, ca. 2 mm. long, wing-angled, narrowly elliptic in outline, not beaked.—Dry rocky slopes and ridges, mostly 4000–11,000 ft.; Joshua Tree Wd. to Subalpine F.; San Jacinto and Little San Bernardino mts. w. and n. to San Gabriel, Argus and Panamint mts., s. Sierra Nevada and Santa Lucia Mts. May–July.

55. **E. crocàtum** A. Davids. [*E. saxatile* var. *c.* Munz.] Perennial, the caudex loosely branched and clothed with old lvs., densely white-woolly; lf.-blades broadly ovate, 1.5–3.5 cm. long, obtusish, the petioles shorter to ca. as long; flowering stems terminal, 1–2(–3) dm. high, 1–2-forked at right angles, forming a rather dense cyme 3–8 cm. across; invol. broadly campanulate, 3–4 mm. long, white-woolly; fls. many, sulphur-yellow, glabrous, 5–6 mm. long, narrowed into a tubular stipelike base; calyx-segms. oblance-oblong, the inner slightly wider than outer; aks. brownish, lance-ovoid, ca. 3 mm. long, glabrous, somewhat 3-angled; $2n = 40$ (Stokes & Stebbins, 1955).—Rocky slopes, at ca. 500 ft.; Coastal Sage Scrub; Conejo Grade, n. base of Santa Monica Mts., Ventura Co. April–July.

56. **E. microthècum** Nutt. [*E. effusum* var. *rosmarinoides* Benth. in DC.? *E. m.* var. *panamintense* S. Stokes. *E. tenellum* var. *erianthum* Gand. *E. effusum* var. *limbatum* S. Stokes.] Low bushy half-shrubs 1–3 dm. high, whitish-tomentulose or glabrate; lvs. oblong to oblong-lanceolate or elliptic, 8–20 mm. long, glabrate above, white-woolly beneath, often quite revolute, short-petioled; peduncles 3–10 cm. long, with compound cymose infl. mostly 2–5 cm. broad; invol. narrowly turbinate, somewhat angled, 2.5–3 mm. long, short-toothed, tomentose to glabrous, sessile or pedicelled; calyx yellow to white or pink, glabrous, 2–3 mm. long, the outer segms. round-oblong with subcordate base, the inner elliptic; fils. subglabrous; aks. narrow, angled, ca. 1 mm. long.—Widely distributed in dry stony places, mostly 5000–10,000 ft.; Pinyon-Juniper Wd., N. Juniper Wd., Sagebrush Scrub, Yellow Pine F., Red Fir F.; San Bernardino Co. n. along e. slope of Sierra Nevada to Modoc and Siskiyou cos.; to Wash., Rocky Mts. July–Oct. Variable, but with ill-defined segregates, some of which proposed for Calif. are: (1) Ssp. *confertiflòrum* (Benth.) S. Stokes [*E. c.* Benth.], plants tall; lvs. narrow, revolute; cymes compact, flat-topped; Shasta region. (2) Ssp. *aùreum* (Jones) S. Stokes var. *expánsum* S. Stokes, a name proposed for plants from White Mts., with broad, subovate lvs. and

very open cymes. (3) Var. *MacDoùgalii* (Gand.) S. Stokes [*E. M.* Gand.] Plants 3–5 dm. high; lvs. appearing linear because tightly revolute, sharply acute.—Mts. of e. Mojave Desert to n. Ariz., etc.

57. **E. Wrìghtii** Torr. ex Benth. ssp. **trachygònum** (Torr.) S. Stokes. [*E. t.* Torr. ex Benth. *E. W.* var. *t.* Jeps.] Caudex low, branched, woody; lvs. on lower third of plant rather crowded, densely white-tomentose above and beneath, lance-elliptic, 1–2.5 cm. long, acute, short-petioled; petioles expanded at base and woolly; flowering stems several to many, white-tomentose, 1.5–2.5 dm. high, once or twice dichotomous or trichotomous; invols. mostly solitary, scattered along the branches, 2–2.5 mm. long, ± tomentose; calyx whitish or pink, ca. 3 mm. long, glabrous, the outer segms. broadly obovate, the inner less so; fils. somewhat pubescent below; aks. narrow, 3-angled, somewhat scaberulous above, ca. 3 mm. long.—Gravelly and rocky places, mostly below 5000 ft.; Foothill Wd., Pinyon-Juniper Wd., Yellow Pine F.; inner Coast Ranges and w. base of Sierra Nevada, Shasta Co. to n. Los Angeles Co. and through higher desert mts. to n. Ariz., w. Nev. Aug.–Oct. A chromosome count of $2n = 34$ has been reported for *E. Wrightii* (Stokes & Stebbins, 1955).

Ssp. **subscapòsum** (Wats.) S. Stokes. [*E. Wrightii* var. *s.* Wats. *E. junceum* Greene *E. curvatum* Small.] Leafy branches short, forming a dense mat; lvs. crowded, 5–12 mm. long; flowering stems slender, often simple, often glabrate, 1–2 dm. high.—Rocky and gravelly places, 5000–10,500 ft.; Montane Coniferous F.; Sierra Nevada to San Jacinto Mts. July–Oct.

Ssp. **membranàceum** (S. Stokes) S. Stokes. [*E. Wrightii* var. *membranaceum* S. Stokes ex Jeps.] Woody, branched, not densely cespitose; petiole-bases dilated into a glabrate brownish sheath clasping the stem; lvs. strongly revolute.—Dry stony slopes, below 6000 ft.; Chaparral, Joshua Tree Wd.; Pinyon-Juniper Wd.; Little San Bernardino and Santa Monica mts. to San Diego Co.; L. Calif. Aug.–Oct.

58. **E. racemòsum** Nutt. Caudex woody, few-branched, compact; lvs. basal, oblong to oblong-ovate, the blades mostly 2.5–3.5 cm. long, glabrate above and closely white-tomentose beneath, on petioles as long or longer; flowering stems slender, mostly 1.5–3 dm. high, tomentose, trichotomous once or twice, usually not leafy-bracted; invols. solitary and arranged racemosely along the upper branches, tubular-campanulate, 3–4 mm. long, tomentose; calyx white to pinkish, ca. 3 mm. long, the segms. oblong-oblanceolate; fils. pilose at base; aks. lance-ovoid, brownish, ca. 2 mm. long, 3-angled.—At ca. 6000–7000 ft.; Pinyon-Juniper Wd.; White and Cottonwood mts., Inyo Co.; to Colo., New Mex. July–Aug.

The more common form in California is var. **desertòrum** S. Stokes [*E. panamintense* Mort.] Lf.-blades roundish to broadly ovate, rather loosely and permanently gray-tomentose, 1–2(–3) cm. long; lower forks of the repeatedly trichotomous infl. with roundish leafy bracts.—Dry rocky slopes, 5000–9000 ft.; Pinyon-Juniper Wd., Sagebrush Scrub; Inyo-White to New York and Clark mts.; sw. Nev. May–Oct.

A plant with lvs. strictly basal, the blades 1–1.5 cm. long, almost round, densely white-tomentose and bracts not leafy, lance-ovate, mostly 3–6 mm. long, has been described as *E. mensicola* S. Stokes and has been found in the Inyo-White and Panamint mts., at ca. 6000–7500 ft. July–Aug.

59. **E. nodòsum** Small. Shrubby, 3–10 dm. high, permanently white-tomentulose throughout, intricately dichotomously and trichotomously branched, the ultimate branchlets slender, 1–2 mm. thick; lvs. on lower half of plant, lance-elliptic, 8–15 mm. long, acute, short-petioled; invols. racemosely scattered on short lateral divaricate branchlets, turbinate-cylindric, 2–3 mm. long, tomentose; calyx glabrous, 2–3 mm. long, whitish or pink, the outer segms. oblong-obovate, the inner narrower; fils. pilose below; aks. brown, narrow, ca. 2.5 mm. long, somewhat 3-angled, minutely scabrous above.—Dry stony places, 200–3600 ft.; Creosote Bush Scrub, Pinyon-Juniper Wd.; w. edge of Colo. Desert, to Twentynine Palms and e. San Diego Co. Aug.–Nov.

Ssp. **monoénse** S. Stokes. Lf.-blades oblong, 2–3 cm. long; invols. in subcapitate clusters.—At 6000–8500 ft.; Sagebrush Scrub; Mono and Inyo cos.; w. Nev. July–Oct.

60. **E. elongàtum** Benth. Perennial herb, mostly loosely branched at base, whitish-tomentulose throughout, leafy in lower portion, passing into elongate leafless paniculately forked infls. 6–12(–18) dm. high; lvs. lance-oblong to narrowly ovate, crisped-undulate, somewhat glabrate above, white beneath, 3–5 cm. long, cuneate at base, short-petioled;

invols. remotely scattered, oblong-cylindric, 6–7 mm. long, tomentose, truncate, obscurely 5-toothed; calyx white or pinkish, glabrous, 2.5–3 mm. long, the segms. obovate, the inner slightly longer, somewhat pubescent within; fils. glabrous; aks. dark, narrow, 2–2.5 mm. long, glabrous, somewhat 3-angled; $2n = 34$ (Stokes & Stebbins, 1955).— Dry rocky places, below 6000 ft.; Coastal Sage Scrub, Chaparral, Foothill Wd.; Coast Ranges from Monterey and San Benito cos. to n. L. Calif. Aug.–Nov.

61. **E. Heermánnii** Dur. & Hilg. [*E. H.* ssp. *occidentale* S. Stokes.] Woody and branched below, the stems erect, leafy and floccose in lower portion, glabrous and light green above, 3–7(–10) dm. high; lvs. lance-oblong to oblanceolate, 1–2(–2.5) cm. long, green above, floccose beneath, somewhat undulate, short-petioled; infl. a cymose panicle of dichotomously branched rigid branchlets, almost or quite smooth, subterete, the lower internodes 2–4(–6) cm. long; invols. solitary in the forks or terminal, broadly turbinate, glabrous, 2 mm. long, rather deeply lobed; calyx yellowish-white, glabrous, 3–4 mm. long, the outer segms. orbicular, the inner oblong; fils. pilose below; aks., narrow, 3-angled, ca. 2.5 mm. long.—Dry slopes and ridges, 2000–7000 ft.; Pinyon-Juniper Wd., Foothill Wd., Joshua Tree Wd.; borders of San Joaquin V. (San Benito Co., Kern Co.) and w. Mojave Desert (Little San Bernardino Mts. to Inyo Co.); w. Nev. July–Sept.

Var. **argénse** (Jones) Munz. [*E. sulcatum* var. *a.* Jones. *E. Howellii* S. Stokes.] Low, 1–2 dm. high; infl. very compact and intricate, the lower internodes 0.5–1.2 cm. long, the branchlets glabrous and scabrous under a lens; fls. ca. 2 mm. long.—Range of the sp.

Var. **floccósum** Munz. Lower internodes of infl. 2–3 cm. long, the branchlets floccose-tomentose; fls. 2–3 mm. long.—Largely in Pinyon-Juniper Wd.; mts. of e. Mojave Desert, San Bernardino Co. Aug.–Oct.

Ssp. **sulcátum** (Wats.) S. Stokes. [*E. s.* Wats.] Intricately branched rounded bushes 1.5–4 dm. high; internodes short, strongly angled and grooved, but not scabrellous; fls. as in the sp.—At 6300–6800 ft.; Pinyon-Juniper Wd., Kingston Mts., e. Mojave Desert; to Utah, Ariz. Sept.–Oct.

62. **E. Plumatélla** Dur. & Hilg. [*E. Palmeri* Wats.] Rather woody at base, with several erect stems 3–6 dm. high, leafy in lower portion, white-tomentulose almost throughout, forked above; lvs. oblanceolate to oblong-lanceolate, 8–15 mm. long, revolute, acute, hoary-tomentose, with short slender petioles; invols. borne on mostly horizontal branches which spread in tiers to one side of the main axis and form an intricate mass ending in short-noded branchlets, solitary but close together, glabrous, sessile, turbinate-cylindric, 2.5 mm. long; calyx glabrous, white, 2 mm. long, the outer segms. obovate, the inner narrower; fils. pilose below; aks. narrow, brown, slightly angled, scaberulous above.—Dry stony places, below 4500 ft.; Creosote Bush Scrub, Joshua Tree Wd., Shadscale Scrub, Pinyon-Juniper Wd.; Kern R. region to Walker Pass, Mojave Desert; to Utah, Ariz. Aug.–Oct.

Var. **Jaègeri** (M. & J.) S. Stokes ex Munz. [*E. nodosum* var. *J.* M. & J.] Infl. green and glabrous.—Mojave Desert; to w. Ariz.

63. **E. ovalifòlium** Nutt. Cespitose perennial with closely branched woody caudex thickly beset with lvs., densely white-tomentose; lvs. basal, round to obovate, the blades 5–12 mm. long, short-petioled; flowering stems scapose, slender, white-woolly, 1–2 dm. high; infl. capitate, 1.5–2.5 cm. in diam.; invols. several, campanulate, white-woolly; calyx yellow, glabrous, 4–5 mm. long, the outer lobes elliptic, subcordate at base, the inner spatulate, exserted; fils. hairy at base; aks. glabrous, ca. 2–2.5 mm. long, narrowly triangular-ellipsoid; $2n = 40$ (Stokes & Stebbins, 1955).—Dry slopes and flats, 5000–7000 ft.; Sagebrush Scrub, N. Juniper Wd., Pinyon-Juniper Wd.; Mono Co. n. and e.; to B.C., Mont., Wyo. May–July.

KEY TO SUBSPECIES

Fls. yellow, lf.-blades densely white-woolly, 5–12 mm. long. Mono Co. n. . . . . *E. ovalifolium*
Fls. ochroleucous to pinkish.
  Lvs. round-ovate, white-felty with very dense tomentum.
    Fls. 5–7 mm. long; lf.-blades mostly 7–15 mm. long; flowering stems 10–25 cm. high. From
    5000–8000 ft. . . . . ssp. *vineum*
    Fls. 2–3 mm. long; lf.-blades 3–5 mm. long; flowering stems 1–6 cm. high. Mostly above
    10,000 ft. . . . . var. *nivale*
  Lvs. oblong-ovate, loosely gray-tomentose, sometimes brownish, the blades 1–2 cm. long. White Mts.
                                ssp. *eximium*

Ssp. **vinèum** (Small) S. Stokes. [*E. v.* Small. *E. o.* var. *v.* Jeps.] Lf.-blades round-ovate, white, felty-tomentose, 7–15 mm. long; flowering stems 1–2.5 dm. high, the infl. mostly capitate, 2–3.5 cm. in diam.; calyx ochroleucous to pinkish or rose, 5–7 mm. long.—Dry usually rocky places, 5000–11,600 ft.; Joshua Tree Wd., Pinyon-Juniper Wd., Montane Coniferous F.; desert slopes of San Bernardino Mts., Grapevine Mts., White Mts., Sierra Nevada to Siskiyou Co.; Ore., Nev. May–Aug.

Var. **nivàle** (Canby) Jones. [*E. n.* Canby.] Woody, dense, matted; lf.-blades white-felty, 3–5 mm. long; flowering stems 1–6 cm. high, often ending in solitary invols.; fls. white with red veins to rose, 2–3 mm. long.—Dry granitic flats and ridges, 10,000–12,000 ft.; Alpine Fell-fields; Sierra Nevada, Tulare Co. to Madera Co. July–Aug.

Ssp. **exímium** (Tides.) S. Stokes. [*E. e.* Tides.] Caudex loose; petioles 3–4 cm. long; lf.-blades grayish-tomentose, 10–20 mm. long; flowering stems 15–30 cm. high; fls. ochroleucous.—At 5000–7000 ft.; Pinyon-Juniper Wd.; White Mts., Inyo Co.; w. Nev. May–June.

64. **E. Gilmánii** S. Stokes. Low compact perennial from an elongate woody root, the caudex covered with old lvs.; lvs. crowded in a basal rosette, few to ca. 14, suberect, densely white-tomentose, the blades 2–4 mm. long, elliptic, with petioles margined and ca. as long; peduncles solitary, 1–2 cm. high and with 2–3 cymosely arranged invols.; these turbinate, 1.5 mm. long; fls. reddish, glabrous, the outer calyx-segms. in-flated, rounded on back, orbicular in shape, forming a globose calyx 3–4 mm. in diam., the inner narrow, longer, slightly exserted; aks. 2.5 mm. long, acutely 3-angled.—At 6200 ft., Pinyon Mesa, Panamint Mts., Inyo Co. Aug.–Sept.

65. **E. proliferum** T. & G. [*E. ovalifolium* var. *p.* Wats. *E. strictum* ssp. *p.* S. Stokes. *E. Greenei* Gray. *E. niveum* var. *G.* S. Stokes.] Caudex compactly branched, the plant white-tomentose; lvs. basal, the blades ovate, obtuse, slightly bicolored, 1–2 cm. long, on slender longer petioles; flowering stems erect, naked, slender, floccose, 1.5–2.5 dm. high; infl. cymose-umbellate, the rays divaricate, to 4 cm. long; invols. largely solitary in the forks and terminal, oblong-turbinate, 5-toothed, tomentose, 5–6 mm. long; fls. cream to white, 4–5 mm. long, the outer segms. broadly elliptic, with obcordate base, the inner obovate, somewhat exserted; fils. hairy below; aks. glabrous, narrow, 3-angled. —Dry open places, 3500–6000 ft.; Sagebrush Scrub, N. Juniper Wd., Yellow Pine F.; Siskiyou Co. and Plumas Co.; n. to B.C. and Ida. June–Aug.

Ssp. **anserìnum** (Greene) Munz. [*E. a.* Greene. *E. strictum* ssp. *a.* S. Stokes. *E. flavissimum* Gand. *E. ovalifolium* ssp. *f.* S. Stokes.] Invols. 1–3 in a cluster; fls. yellow.— Similar situations, e. Siskiyou Co., Modoc Co.; adjacent Nev. and Ore. June–July.

66. **E. ochrocéphalum** Wats. Caudex cespitose, branched, low, the plant with dense olive-green to whitish tomentum; lvs. crowded, oblanceolate, plane, the blades 1–2 cm. long, with petioles ca. as long; flowering stems scapelike, 5–15 cm. tall, slender, tomen-tose; infl. capitate, 1–2 cm. across; invol. turbinate-campanulate, 3–4 mm. long, with 6–8 teeth; fls. yellow, glabrous, 2–3 mm. long, the segms. obovate, essentially alike; fils. puberulent; aks. lance-ovoid, ca. 1.5 mm. long.—Dry, loose, often volcanic soils, 4000–9900 ft.; Yellow Pine F., Red Fir F.; Tuolumne and Mono cos.; adjacent Nev. June–July.

Ssp. **agnéllum** (Jeps.) S. Stokes. [*E. o.* var. *a.* Jeps.] Lf.-blades 4–9 mm. long; flower-ing stems 1–5 cm. long.—Dry granitic soils, 10,000–12,000 ft.; Alpine Fell-fields; Inyo and Fresno cos. to Modoc Co. July–Aug.

67. **E. Kénnedyi** Porter in Wats. [*E. Purpusii* Bdg.] Caudex branched, woody, form-ing a dense leafy mat; lf.-blades numerous, oblong, white-woolly, 3–6 mm. long, revolute, subsessile; flowering stems scape-like, wiry, subglabrous, 5–15 cm. high; infl. mostly capitate, less than 1 cm. across; invols. few, tomentose to glabrate, turbinate, angled, 3–4 mm. long; calyx glabrous, white with reddish midribs, 2–3 mm. long, the segms. broadly elliptical, somewhat rounded at base; fils. subglabrous; aks. ca. 2 mm. long, papillose-puberulent, lance-ovoid, angled.—Dry stony slopes and ridges, 4000–6500 ft.; Sagebrush Scrub, Pinyon-Juniper Wd.; e. slope of Sierra Nevada from Mono Co. s. to Argus and Coso mts., Inyo Co. May–June.

Ssp. **gracílipes** (Wats.) S. Stokes. [*E. g.* Wats.] Lf.-blades ovate to oblanceolate, 10–20 mm. long; invol. 2–3 mm. long, flaring at throat; fls. 2–3 mm. long.—Dry slopes, 6000–13,000 ft.; Pinyon-Juniper Wd., Bristle-cone Pine F.; White Mts., Mono and Inyo cos.; w. Nev. June–Aug.

Ssp. **austromontànum** (M. & J.) S. Stokes. [*E. K.* var. *a.* M. & J. *E. K.* ssp. *pinorum* S. Stokes.] Lf.-blades oblanceolate, 6–10 mm. long; stems floccose, 5–15 cm. high; calyx-segms. obovate or oblance-obovate, gradually contracted into a cuneate base.—Dry stony benches and slopes, 5000–7000 ft.; Yellow Pine F.; Mt. Pinos and San Bernardino Mts. June–Aug.

Ssp. **alpígenum** (M. & J.) Munz. [*E. K.* var. *a.* M. & J.] Mats very dense and woody; lvs. 2–3 mm. long; scapes 1–2 cm. high; fls. reddish.—Dry granitic gravel, 11,000–11,500 ft.; Alpine Fell-fields; Mt. San Gorgonio (San Bernardino Co.); somewhat lower in San Gabriel Mts. July–Aug. A similar matted form from Olancha Peak, Tulare Co., but with less capitate infl., is *E. Kennedyi* var. *olanchénse* J. T. Howell.

68. **E. latifòlium** Sm. [*E. oblongifolium* Benth.?] Caudex low, woody, densely leafy, often much-branched; lf.-blades persistent, ovate to almost oblong, cordate or rounded at base, obtuse or acute, often crisped on margins, lanate or somewhat glabrate above, densely white-lanate beneath, 2.5–5 cm. long, the petioles shorter or longer, woolly, expanded at base; flowering stems stout, leafless, tomentulose, 2–6 dm. high, simple or 2–4-forked, the forks simple or again forked; infl. of capitate clusters 1.5–3 cm. across, terminal and also sessile in forks; invols. tomentose, shallowly 5-toothed, numerous, ca. 4 mm. long; calyx glabrous or with a few hairs near base, white to rose, 3 mm. long, the segms. obovate, subequal; fils. villous at base; aks. glabrous, brown, lance-ovoid, 3-angled, ca. 4 mm. long; $2n = 40$ (Stokes & Stebbins, 1955).—Cliffs and sandy places along the coast; Coastal Strand and N. Coastal Scrub; San Luis Obispo Co. to Ore. June–Oct. A woolly form with short dilated petioles and small scattered heads, from Mt. Hermon, Santa Cruz Co., has been described as ssp. *decúrrens* Stokes and suggests influence of *E. parvifolium.*

KEY TO SUBSPECIES

A. Invols. and flowering stems ± tomentose.
  B. Heads 1–few, mostly 1.5–3 cm. across. Immediate seacoast .... *E. latifolium*
  BB. Heads usually several to many, smaller. Interior of n. Calif. .... ssp. *sulphureum*
AA. Invols. and flowering stems mostly glabrous or nearly so.
  B. Lvs. scattered along the woody caudex, the blades very strongly undulate-crisped, 3–9 cm. long.
    C. Invol. 5–6 mm. long. Insular .... ssp. *grande*
    CC. Invol. 3–4 mm. long. Mainland, cent. Calif. .... ssp. *auriculatum*
  BB. Lvs. in basal rosettes, the blades plane or slightly crisped, mostly 1–5 cm. long. Mainland.
    C. Invols. 2–6 in a cluster.
      D. Fls. mostly glabrous externally, white to pink, rarely yellow. Cismontane or montane.
        E. Flowering stems branched above. Lower elevs. .... ssp. *nudum*
        EE. Flowering stems scapose, short, ending in a solitary head. Subalpine var. *scapigerum*
      DD. Fls. pubescent externally, yellow to white. Desert edge and borders of San Joaquin and Salinas valleys .... ssp. *saxicola*
    CC. Invols. solitary, rarely in pairs. Mostly of pine belt.
      D. Branches several from base. High Sierra Nevada .... var. *deductum*
      DD. Branches 1–few from base. Mts. of s. Calif. .... ssp. *pauciflorum*

Ssp. **sulphùreum** (Greene) S. Stokes. [*E. s.* Greene. *E. nudum* var. *s.* Jeps. *E. n.* var. *oblongifolium* Wats. *E. Harfordii* Small. *E. capitatum* Heller.] Lvs. basal, the blades largely oblong-spatulate, 2–4 cm. long; flowering stems 5–10 dm. high, white-tomentose, dichotomously branched; invols. mostly 3–6 in a cluster, tomentose, 3–5 mm. long; calyx white or rose, or yellowish, 3–4 mm. long, pubescent without near base.—Dry slopes, mostly below 4000 ft.; N. Oak Wd., Yellow Pine F., Foothill Wd.; Napa Co. to Humboldt and Siskiyou cos., Nevada Co. to Modoc Co.; adjacent Ore., Nev. May–Aug.

Ssp. **gránde** (Greene) S. Stokes. [*E. g.* Greene. *E. nudum* var. *g.* Jeps.] Woody at base, the stems 8–15 dm. long, leafy for 2–3 dm. at base; lf.-blades strongly undulate-crisped, oblong-ovate, 3–10 cm. long, greenish above, closely white-woolly beneath; flowering stems glabrous, glaucous, forking above; invols. 1–3, subglabrous without, 5–6 mm. long; calyx whitish, ca. 3 mm. long, the segms. oblong-obovate, spreading.—Bluffs and cliffs; Coastal Sage Scrub, Chaparral; Santa Cruz, Santa Catalina, Anacapa and San Clemente ids. June–Oct. A form from San Miguel, Santa Rosa, and w. end of Santa Cruz ids., lower, more decumbent, with tendency to subcapitate cymes and with red calyx, is var. **rubéscens** (Greene) Munz. [*E. r.* Greene. *E. grande* var. *r.* Munz. *E. l.* ssp. *r.* S. Stokes.]

Ssp. **auriculàtum** (Benth.) S. Stokes. [*E. a.* Benth. *E. nudum* var. *a.* Tracy. *E. l.* var. *alternans* S. Stokes.] Stems 2–10(–20) dm. high, caudexlike at base with lvs. on lower part; lf.-blades oblong to elliptic, obtuse at apex, truncate or subcordate at base, green above, white beneath, 3–7 cm. long; flowering stems glabrous, glaucous, often fistulous; invols. solitary or in pairs, 3–4 mm. long; fls. in heads ca. 1 cm. across, mostly glabrous and usually cream to pink, sometimes yellowish.—Dry often stony places; Coastal Strand, Chaparral, V. Grassland; Coast Ranges from Sonoma Co. to Monterey Co. July–Oct. Variable; a form with unusually inflated flowering stems, very robust, with fls. yellow to whitish, is var. *indíctum* (Jeps.) S. Stokes [*E. i.* Jeps.] and is from the inner S. Coast Ranges from Merced Co. to Kern Co.; 2*n* = 80 (Stokes & Stebbins, 1955).

Ssp. **nùdum** (Dougl. ex Benth.) S. Stokes. [*E. n.* Dougl. ex Benth. *E. oblanceolatum* and *longulum* Greene. *E. latifolium* var. *parvulum* S. Stokes.] Caudex short, simple or few-branched, the leaf-bearing area not elongate; lf.-blades oblong to oblanceolate or broadly elliptic-ovate, 1–6 cm. long, rounded at apex, glabrate above, white-lanate beneath, ± undulate-crisped, the petioles often much longer; flowering stems commonly 1–few, erect, slender, 3–10 dm. high, glabrous or nearly so, glaucous, usually forking or trichotomous near middle, then branched again; invols. usually in clusters, sub-cylindric, 3–5 mm. long, glabrous or slightly woolly; calyx ca. 2.5 mm. long, mostly white with some pink, sometimes yellow, usually glabrous without; fils. hairy at base; aks. 1.5–3 mm. long; 2*n* = 40, 80 (Stokes & Stebbins, 1955).—Dry usually rocky places, up to ca. 8000 ft.; many Plant Communities; Coast Ranges from about San Francisco Bay n., Sierra Nevada from Tulare Co. n.; to Wash., Nev. June–Nov. Exceedingly variable as here recognized, and intergrading freely with all other sspp. Subalpine plants from the Sierra Nevada, 9000–11,000 ft., tend to have several stems from base, 2–3 dm. high and very slender; lf.-blades 1–2 cm. long; infl. branched and with invols. largely solitary. They are var. *dedúctum* (Greene) S. Stokes. [*E. d.* Greene. *E. n.* var. *d.* Jeps.] The extreme in such reduction, with stems 1–2 dm. long, simple, and ending in single heads of several invols., occurs at ca. 10,000–12,000 ft. in s. Sierra Nevada and is var. *scapígerum* (Eastw.) S. Stokes. [*E. s.* Eastw. *E. n.* var. *s.* Jeps.] July–Aug.

Ssp. **saxícola** (Heller) S. Stokes. [*E. s.* Heller. *E. l.* ssp. *Westoni* S. Stokes. *E. nudum* var. *pubiflorum* Benth. *E. gramineum* S. Stokes?] Flowering stems glabrous, glaucous, 3–6 dm. high, cymose above; invols. clustered, rarely single, subcampanulate; fls. yellow to white, pubescent without.—Dry hot places, below 6000 ft.; largely Foothill Wd., Joshua Tree Wd., Pinyon-Juniper Wd., Yellow Pine F., Santa Ana Mts., w. Mojave Desert from San Gabriel Mts. n. to Modoc Co., Siskiyou and Humboldt cos. and along inner S. Coast Ranges n. to Monterey Co. and Santa Lucia Mts. June–Sept.

Ssp. **pauciflòrum** (Wats.) S. Stokes. [*E. nudum* var. *p.* Wats. *E. n.* var. *perturbum* Jones.] Caudex rather simple with lvs. crowded; lf.-blades oblong-ovate, 1–5 cm. long, green or glabrate above, white-woolly beneath, on long petioles; flowering stems 3–8 dm. high, slender, glabrous, glaucous, forking several times; invols. 1, rarely 2 at a place, rather few on a branch, ca. 5 mm. long; fls. whitish, glabrous, 2 mm. long.—Dry slopes, 5000–9000 ft.; Yellow Pine F., Red Fir F.; Cuyamaca Mts. to Santa Rosa and San Bernardino mts. Aug.–Oct.

69. **E. elàtum** Dougl. ex Benth. Caudex woody, branched or simple; lvs. basal, erect, the blades ovate to lance-ovate, 4–15 cm. long, acutish, green and glabrate above, somewhat tomentose beneath but not hoary, the petioles ca. as long, villous; flowering stems 4–8 dm. long, glabrous, glaucous, repeatedly trichotomous above, somewhat inflated; invols. in terminal clusters of 2 or 4, sometimes solitary in forks, glabrous or somewhat pubescent, turbinate, 5-toothed, ca. 4 mm. long; calyx white or pinkish, 2.5 mm. long, pubescent without, the lobes obovate; fils. glabrous except at very base; aks. brownish, ca. 4 mm. long, subovoid but rather prominently beaked, somewhat 3-angled.—Dry rocky slopes, 4000–9500 ft.; Sagebrush Scrub, N. Juniper Wd., Yellow Pine F.; Mono and Eldorado cos. to Modoc, Siskiyou, and Trinity cos.; to Wash., Ida., Nev. June–Sept.

Var. **villòsum** Jeps. [*E. e.* var. *incurvum* Jeps. *E. e.* ssp. *glabrescens* S. Stokes.] Fl.-stems villous-pubescent.—Panamint Mts. to Siskiyou Co. and w. Nev.

70. **E. péndulum** Wats. Base woody, few-branched, ascending or decumbent, 1–2.5 dm. high below the flowering branches; lvs. crowded near tips of basal branches, lance-oblong, 2–5 cm. long, obtusish, thinly tomentose above, white-felty beneath, short-petioled; flowering stems white-tomentose, 2–5 dm. high, with long branches, leafy-bracted at forks; invol. solitary, sessile or peduncled, turbinate-campanulate, white-tomentose, 5 mm. high, with small teeth; calyx densely villous, ca. 3 mm. long, the segms. narrow-oblong; aks. villous.—Dry slopes, below 3000 ft.?; Mixed Evergreen F.?; Del Norte Co.; adjacent Ore. Aug.–Sept.

71. **E. fasciculàtum** Benth. [*E. f.* var. *maritimum* Parish. *E. f.* var. *oleifolium* Gand.? *E. aspalathoides* Gand.? *E. f.* ssp. *a.* S. Stokes. *E. rosmarinifolium* Nutt.] CALIFORNIA BUCKWHEAT. Low spreading shrubs, the stems ± decumbent, 6–12 dm. long, branched, leafy; branchlets loosely pubescent to subglabrous, ending in leafless peduncles 3–10(–15) cm. long, bearing ± open cymose infl. with many capitate clusters at tips; lvs. numerous, fascicled, oblong-linear to linear-oblanceolate, green and glabrate above, white-woolly beneath, 6–15 mm. long, strongly revolute; invol. prismatic, 3–4 mm. long, glabrous, with 5 short acute teeth; calyx white or pinkish, ca. 3 mm. long, nearly or quite glabrous without, the outer segms. broadly elliptic, the inner obovate; fils. subglabrous; aks. lance-ovoid, light brown, angled, shining, ca. 2 mm. long; $2n = 40$, 80 (Stokes & Stebbins, 1955).—Dry slopes and canyons near immediate coast; Coastal Sage Scrub; Santa Barbara to n. L. Calif. May–Oct.

Ssp. **foliolòsum** (Nutt.) S. Stokes. [*E. rosmarinifolium* var. *f.* Nutt. *E. f.* var. *f.* S. Stokes. *E. f.* var. *obtusiflorum* S. Stokes.] Upper surface of lvs., outer surface of calyx, invol., etc., pubescent; peduncles 1–2 dm. long.—Common on interior cismontane slopes and mesas, below 3000 (4500) ft.; Chaparral, Coastal Sage Scrub; Monterey and San Benito cos. to n. L. Calif. Variable. March–Oct.

Ssp. **polifòlium** (Benth.) S. Stokes. [*E. p.* Benth. *E. f.* var. *p.* T. & G.] Plants commonly 2–4 dm. tall; lvs. densely canescent to hoary above, commonly less revolute; invol. and calyx pubescent; heads solitary or in reduced cymes.—Common on dry slopes, below 7000 ft.; Sagebrush Scrub to Pinyon-Juniper Wd.; both deserts to San Joaquin V. and Inyo Co. and interior cismontane s. Calif.; to Utah, L. Calif. April–Nov.

Ssp. **flavovíride** (M. & J.) S. Stokes. [*E. f.* var. *f.* M. & J.] Low, 2–3 dm. tall; lvs. yellow-green, subglabrous above, strongly revolute; peduncles glabrous; invol. and calyx subglabrous, the latter quite reddish.—Rocky places, below 4000 ft.; Creosote Bush Scrub; Eagle Mts., e. Riverside Co., to Little San Bernardino and Sheephole mts. March–May.

72. **E. arboréscens** Greene. Loosely branched shrub 6–15(–20) dm. tall, the stems to 1 dm. thick with shreddy bark; branchlets tomentose when young, later glabrate, purplish and glaucous; lvs. in crowded terminal tufts, linear to oblong, revolute, 2–3 cm. long, densely white-tomentose beneath, glabrate above; fls. in dense terminal leafy-bracted cymes 5–15 cm. across; invol. tomentose, 3 mm. long, turbinate, with obtuse oval teeth; calyx whitish to pink, 2 mm. long, villous at base; fils. glabrous; aks. lance-ovoid, shining, angled, ca. 2.5 mm. long.—Rocky slopes and canyon walls; Coastal Sage Scrub, Chaparral; Santa Cruz, Santa Rosa, and Anacapa ids. April–Sept.

73. **E. gigantèum** Wats. ST. CATHERINE's LACE. Coarse rounded branching shrub, open, 3–20(–30) dm. high, the cent. trunk to 1 dm. thick, the younger branches tomentose, then glabrate and dark, with lvs. toward tips; lf.-blades leathery, oblong-ovate to ovate, 3–7(–10) cm. long, closely white-tomentose beneath, cinereous and somewhat glabrate above, on stout petioles 1–3 cm. long; peduncles stout, 1–3 dm. long, tomentose, later glabrate, bearing large 2–3-forked cymes often several dm. across and with leafy bracts at forks; invols. crowded, campanulate, 3–4 mm. long, tomentose, with short obtuse teeth, subsessile or on short slender pedicels; calyx white, ca. 2 mm. long, white-hairy, the segms. obovate; fils. hairy; aks. brown, shining, narrow-ovoid, angled above, ca. 2 mm. long.—Dry slopes; Chaparral, Coastal Sage Scrub; Santa Catalina Id. and apparently rare on San Clemente Id. May–Aug.

Var. **compáctum** Dunkle. Lvs. oblong; tomentum on young growth looser; pedicels of invols. very stout.—Santa Barbara Id.

Var. **formòsum** K. Bdg. [*E. f.* Bdg.] Lvs. oblong-lanceolate.—San Clemente Id.

74. **E. parvifòlium** Sm. in Rees. Shrub with loosely branched decumbent or prostrate

stems 3–10 dm. long, thinly floccose, densely leafy to summit; lvs. fascicled, round-ovate to lance-oblong, thickish, revolute, sometimes cordate at base, 5–15 mm. long, on shorter petioles, the blades glabrate and green above, densely white-tomentose beneath; peduncles few, mostly 2–5 cm. long, simple or forked, bearing compact heads 1–2 cm. in diam.; invol. glabrate or somewhat woolly, turbinate-campanulate, 3–4 mm. long; calyx white or tinged rose, glabrous, ca. 3 mm. long, the segms. obovate; fils. pilose below; aks. ovoid-deltoid, wing-angled, shining, brown, 2.5 mm. long; $2n = 40$ (Stokes & Stebbins, 1955).—Common on bluffs and dunes along the coast; Coastal Strand, Coastal Sage Scrub; Monterey Co. to San Diego Co. Mostly summer, but with some fls. throughout year. Two ill-defined forms are: (1) with greenish-yellow rather slender fls., Point Lobos, Monterey Co.; ssp. *lùcidum* J. T. Howell ex S. Stokes; and (2) diffusely branched, with lanceolate lvs. 15–30 mm. long and white fls. in heads scarcely 1 cm. in diam. and broad infl. 1–2 dm. in diam.; from Santa Paula Canyon, Ventura Co., ssp. *Pàynei* C. B. Wolf ex Munz.

75. **E. cinèreum** Benth. Freely branched shrub 6–15 dm. high, tomentulose, leafy below the infl.; lvs. ovate, 1.5–3 cm. long, obtuse, cuneate at base, greenish-cinereous above, white-tomentulose beneath, crisped-undulate, short-petioled; flowering branches elongate, dichotomous, with scattered heads; invol. cylindric-turbinate, tomentulose, 3–4 mm. long, somewhat angled, 5-toothed; calyx densely white-villous, ca. 3 mm. long, the segms. narrow-obovate, whitish to pinkish; fils. subglabrous; aks. brown, deltoid-ovoid, sharply angled, somewhat roughened, ca. 2 mm. long.—Beaches and bluffs near the coast; Coastal Strand, Coastal Sage Scrub; Santa Barbara to San Pedro, Santa Rosa Id. June–Dec.

76. **E. intrafráctum** Cov. & Mort. Perennial, woody at base from taproot; lvs. basal, oblong-ovate, somewhat whitish-pilose, the blades 2.5–7 cm. long, petioles somewhat greater; flowering stems usually solitary, simple below, sometimes branched in infl., rather stout, glabrous, glaucous, transversely jointed and easily fractured, 6–12 dm. high; infl. usually of 2–3 virgate branches 2–4 dm. long and sometimes with shorter secondary branches; invols. usually in whorls of 3 at each node, usually 1 in the axil of each of the 3 bracts, 5-parted into oblong lobes, short-pilose; calyx yellow, tinged with red, pubescent, ca. 2 mm. long, the lobes oblong-oblanceolate, subequal; ak. flask-shaped, brownish, almost 2 mm. long, 3-lobed in lower part, then abruptly narrowed into a triangular beak.—Local and rare, limestone crevices, 2000–5000 ft.; Creosote Bush Scrub; Grapevine and Panamint mts., Inyo Co. May–Aug.

8. Émex Neck.

Glabrous monoecious annuals. Lvs. petioled, alternate, with membranaceous or scarious sheathing stipules. Fls. small, in axillary fascicles or in racemelike terminal infl.; ♂ fls. sessile and below, ♀ on jointed filiform pedicels and above. Staminate fls. with 5–6-parted calyx with narrow segms.; stamens 4–6. Pistillate calyx urn-shaped, with ovoid tube and 6 lobes in 2 series. Ovary 3-angled; styles 3; stigmas fimbriate. Fruiting calyx hard, 3- or 6-angled, burlike, the 3 outer segms. spine-tipped. Aks. 3-angled, inclosed by the spiny perianth. Two spp. of Old World. (Latin *ex,* out of, and *Rumex,* having been transferred from that genus.)

Lvs. mostly truncate or subcordate at base; inner calyx-segms. of ♀ fls. linear-lanceolate
　　　　　　　　　　　　　　　　　　　　　　　　　　　　　　　　　　1. *E. spinosa*
Lvs. mostly cuneate at base; inner calyx-segms. triangular-ovate .... 2. *E. australis*

1. **E. spinòsa** (L.) Campd. [*Rumex s.* L.] Plants decumbent to ascending, the stems 3–8 dm. long; lvs. oblong-ovate to somewhat triangular, 5–12 cm. long, on somewhat shorter to longer petioles; outer segments of fruiting calyx ca. 5–6 mm. long, tipped with divergent spines, the inner erect, 6–7 mm. long; aks. shining, light brown, ca. 4 mm. long; $n = 10$ (Sugiura, 1937).—Weed in orchards, etc., Ventura Co. to San Diego Co.; native of Medit. region. July–Nov.

2. **E. austràlis** Steinh. Lvs. mostly cuneate at base, 2–4 cm. long; inner calyx-segms. broadly triangular-ovate, mucronate.—Reported from San Francisco and Vallejo; native of S. Afr. and Australia. June–Oct.

9. Oxýria Hill. MOUNTAIN SORREL

Low glabrous perennial, with acid juice. Caudex thick, covered with persistent lf.-bases. Stems erect, slender. Lvs. alternate, but mostly basal, long-petioled, round or reniform, with cylindric stipule-sheaths. Fls. perfect, small, whorled, arranged in compact panicles. Sepals 4, the outer larger than the inner. Stamens 6, included, the fils. short-subulate. Ovary 1-celled, 1-ovuled. Styles 2, short; stigmas fimbriate. Aks. lenticular, nearly flat, broadly winged, the body ovate. Two spp. (Greek, *oxus*, sour.)

1. O. dígyna (L.) Hill. [*Rumex d.* L.] Caudex branched; stems scapiform, simple or few-branched, 6–25 cm. high; basal lvs. round-reniform, 1.5–3 cm. wide; petioles 5–12 cm. long; pedicels slender, recurved; sepals red or greenish, 1.5–2 mm. long, the inner erect in fr. and becoming 4–6 mm. long, the outer reflexed; aks. ca. 3 mm. across; $n = 7$ (Jaretzky, 1928).—Rocky places, 8000–13,000 ft.; Subalpine F., Alpine Fell-fields; San Jacinto and San Bernardino mts., Sierra Nevada, White, and Yolla Bolly mts.; to Arctic Am., Atlantic Coast, Eurasia. July–Sept.

10. Rùmex L. DOCK. SORREL

Text contributed by David D. Keck.

Annual or mostly perennial herbs with simple or branched grooved stems. Fls. numerous, mostly greenish, small, usually crowded and commonly whorled in panicled racemes. Lvs. often mostly basal, the cauline alternate, the petioles sheathing at base. Calyx of 6 sepals, the 3 outer herbaceous, sometimes united at base, usually appressed to the margins of the inner; the 3 inner larger, somewhat colored and convergent over the 3-angled nutlet, veiny, called *valves* in fr., often bearing a grainlike tubercle (callosity) on the back. Stamens 6. Styles 3; stigmas tufted. Fls. perfect or imperfect, varying with the spp. Ca. 140 spp., widely distributed. (The ancient Latin name.)

(Rechinger, K. H., Jr. The North American Species of Rumex. Field Mus. Pub. Bot. 17[1]: 1–151, 1937.)

A. Plants dioecious, low, slender, acid to the taste; valves small, without callosities.
    B. Lvs. (at least some of them) hastate at base; valves ca. 1 mm. long.
        C. Valves connate with the nutlet .... 1. *R. angiocarpus*
        CC. Valves not connate with the nutlet .... 2. *R. Acetosella*
    BB. Lvs. not hastate, but gradually narrowed to petioles; valves 3–3.5 mm. long in fr.
        3. *R. paucifolius*
AA. Plants monoecious, usually tall and weedy, not acid to the taste.
    B. Stems ascending to decumbent, or axillary from a creeping rhizome, with axillary shoots or lf.-tufts; without basal lvs. even when young.
        C. Valves 14–18 mm. long, 15–30 mm. wide, without callosities; sheaths 3–4 cm. long, conspicuous, persistent; nutlets 5–7 mm. long .... 4. *R. venosus*
        CC. Valves 2.5–5 mm. long, not much wider, with or without callosities; sheaths smaller, appressed, ± evanescent; nutlets up to 2.5 mm. long.
            D. Valves without callosities .... 5. *R. californicus*
            DD. One or all valves callosity-bearing.
                E. Callosity occupying nearly the whole breadth of the valve (valve-margin on each side of the callosity narrower than the callosity); plants drying brownish.
                    F. Valves relatively large, 4–5 mm. long; lvs. 2–3 times longer than broad
                    6. *R. crassus*
                  FF. Valves much smaller; lvs. narrower.
                      G. Valves 2.3–3 mm. long, only one with a callosity
                      7. *R. salicifolius*
                      GG. Valves 3–4 mm. long, scarcely longer than the callosities, yellowish, one or all with a callosity; nutlets ca. 2.5 mm. long
                      8. *R. transitorius*
                EE. Callosity much narrower than the breadth of the valve (valve-margin at least as broad as the callosity); plants drying yellowish-green.
                  F. Low subaquatic plant with flexuous branches; lvs. usually elliptic or ovate, often papillose; valves 2.1–2.5 mm. long, with small callosities
                      9. *R. lacustris*
                  FF. Ascendent or suberect; lvs. usually lanceolate, always glabrous; valves usually larger .... 10. *R. triangulivalvis*
    BB. Stems erect, without axillary shoots or tufts; with basal lvs. when young.
        C. Valves without callosities; nutlets 3–6 mm. long.
            D. Lvs. not cordate at base, fleshy; valves 8–14 mm. long .... 11. *R. hymenosepalus*
            DD. Lvs. cordate at base.
                  E. Valves more than 7 mm. long .... 12. *R. fenestratus*
                  EE. Valves up to 6 mm. long .... 13. *R. occidentalis*
        CC. At least one valve with a distinct callosity.

D.   Valves entire.
    E.   Lvs. broad, flat, thickish, leathery when dry, the nerves forming almost a
        right angle with the midrib; callosity fusiform, much longer than broad
                                                14. *R. orbiculatus*
    EE. Lvs. narrower, not leathery when dry, the nerves forming an acute angle
        with the midrib; callosity ovate-oblong, at most 1½ times longer than broad.
        F.   Lvs. small, flat, truncate; valves very small, scarcely broader than the
            thick callosities; whorls remote, nearly all foliate .. 15. *R. conglomeratus*
        FF. Lvs. large, somewhat crisped or undulate, often narrowed at base,
            seldom truncate; valves large, much broader than the callosities; only
            the lower whorls foliate and occasionally remote.
            G.   Lvs. rather narrow, broadest at middle, mostly much undulate,
                gradually narrowed to base; petiole somewhat canaliculate on
                upper side; valves (3.5-)5–6 mm. long (lf.-shape, valves and
                number of callosities most variable) .... 16. *R. crispus*
            GG. Lvs. broader, often broadest below the middle, abruptly narrowed
                toward base, truncate or slightly cordate, less undulate; petiole
                flat on upper side; valves larger; callosities smaller in proportion to
                valves; rare .... 17. *R. Kerneri*
DD. Valves denticulate.
    E.   Plants perennial.
        F.   Lvs. small; pedicels short, not longer than fr., articulate at middle
                                        18. *R. pulcher*
        FF. Lvs. large; pedicels long, slender, nearly twice as long as fr., articulate
            toward base.
            G.   Lvs. more than 3 times longer than broad, cuneate at base; valves
                cordate, with short teeth .... 19. *R. stenophyllus*
            GG. Lvs. at most 2½ times as long as broad, cordate at base; valves
                oblong, usually with long fine teeth .... 20. *R. obtusifolius*
    EE. Plants annual.
        F.   Lvs. up to 3 times longer than broad; plants tall.
            G.   Stem fistulous; pedicels short, thickish; valves small, 2.5–3 mm.
                long, with very short often inconspicuous teeth .. 21. *R. violascens*
            GG. Stem not fistulous; pedicels longer, slender; valves larger, 2.5–5 mm.
                long, with long fine teeth .... 22. *R. dentatus*
        FF. Lvs. more than 3 times as long as broad.
            G.   Valves triangular; callosity fusiform, narrowed (length of teeth
                variable) .... 23. *R. fueginus*
            GG. Valves ovate; callosity thickish, rounded .... 24. *R. persicarioides*

1. **R. angiocárpus** Murbeck. [*R. Acetosella* ssp. *a.* Murbeck.] SHEEP SORREL. Perennial with running slender rootstocks; stems tufted, slender, erect or with decumbent base, 1–4 dm. high, glabrous; lvs. lanceolate or linear, the blades 2–6 cm. long, at least the lower with hastate base, the upper usually entire; lower petioles often longer than blades; fls. nodding, in naked terminal panicles, yellowish, but reddish in age; pedicels jointed at summit; calyx ca. 1 mm. long, green, the sepals scarcely enlarged in fr., united with the nutlet into a single body; $n = 7$ (Löve, 1941).—Weed, especially in damp and cooler places, throughout most of cismontane Calif., more common n.; widely natur. from sw. Eu. March–Aug.

2. **R. Acetosélla** L. SHEEP SORREL. Plant very similar to the preceding; valves in fr. free, not connate with the nutlet; $2n = 42$ (Ono, 1930).—Weed, less common than the preceding, in cismontane Calif., n. and e. to Alaska and Greenland; introd. from cent. Eu. and Asia. March–Aug.

3. **R. paucifòlius** Nutt. ex Wats. Perennial from a stout taproot; stems few, erect, 1.5–7 dm. high, glabrous; lvs. mostly basal, plane, glabrous, broadly lanceolate, entire, 4–10 cm. long, narrowed to petioles of equal or greater length; fls. imperfect or perfect, tending toward dioecism, in contracted reddish naked panicles; pedicels somewhat articulate below middle; outer sepals scarcely 1 mm. long, the valves in fr. 3–3.8 mm. long and wide, cordate, finely veined; nutlets smooth, brown, shining, 1.2–1.8 mm. long.—Damp places, 5000–9000 ft.; Yellow Pine F. to Subalpine F.; typical only in Modoc, Lassen, and Fresno cos., n. to B.C., e. to Rocky Mts. July–Sept. Largely replaced in Calif. by:

Ssp. **graciléscens** (Rech. f.) Rech. f. [*R. p.* var. *g.* Rech. f. *R. p.* var. *minusculus* J. T. Howell.] Stems numerous, low, 0.5–2 dm. high; basal lvs. linear-lanceolate to linear; panicle laxer.—At 10,000–12,250 ft.; Alpine Fell-fields; Sierra Nevada from Alpine Co. to Tulare Co.; Nev. Aug.–Sept.

4. **R. venòsus** Pursh. Perennial with rather thick woody rootstock; stems 1.5–4 dm. high, ascending to suberect, glabrous, rather pale, simple or few-branched; sheaths hyaline, dilated upwards; lvs. ovate to lance-oblong, cuneate at base, acutish to sub-

acuminate at apex, 3–10 cm. long, rather short-petioled; infl. a compact panicle, the pedicels jointed near middle; outer sepals lanceolate, ca. 3 mm. long, the valves in fr. 14–18 mm. long, 15–30 mm. wide, deeply cordate, veiny, reddish.—Dry beds of streams, and sandy places; Sagebrush Scrub; Honey Lake V., Lassen Co.; to Wash., Sask., Mo. May–June.

5. **R. califòrnicus** Rech. f. [*R. salicifolius* var. *denticulatus* Torr. *R. s. f. ecallosus* J. T. Howell.] Perennial from stout taproot; stems 2–6 dm. high, ascending or sub-erect, slender but firm, often branching, leafy; lvs. plane, glabrous, lance-linear to lanceolate, to 10 cm. long, to 1.5 cm. wide, acute; petioles as long as the width of the blades; panicle-branches slender, divergent, the lower multiflowered glomerules ± remote, the upper approximate, or all approximate; pedicels jointed near base; outer sepals ovate-lanceolate, ca. 1.8 mm. long, the valves in fr. 3 mm. long, 2.5 mm. wide, triangular, truncate at base, somewhat denticulate, veiny, brown-red.—Widespread in moist places, 300–11,500 ft. (with several ecotypes); many Plant Communities; Sierra Nevada from Plumas Co. to Tulare Co. (mostly above 8000 ft.), Trinity Co. to San Benito Co., San Luis Obispo Co., mts. of s. Calif. to San Diego Co.; L. Calif., Nev., Ariz. May–Sept.

R. utahénsis Rech. f., with thickish mostly low little-branched stems, broader lvs., small very compact panicle and nearly entire valves, occurs at the Calif. border., as at Mt. Rose, Nev., e. to Colo., n. to Ore.

6. **R. cràssus** Rech. f. [*R. salicifolius* var. *c.* J. T. Howell.] Stems procumbent or flexuous-ascending, 2–5 dm. long; lvs. fleshy, drying coriaceous, lance-oblong to lance-ovate, ca. 2.5–3.5 times as long as wide, broadly cuneate at base, acute at apex; panicle-branches short, simple, compact, the multiflowered glomerules contiguous; pedicels jointed near base; outer sepals ca. 2 mm. long, the valves in fr. 4–5 mm. long, 3–4 mm. wide, ovate, rounded at base, the margin minutely and irregularly crenate-denticulate, one valve with a very prominent callosity which covers nearly the whole surface of the valve.—Coastal dunes and rocky ocean bluffs, up to 500 ft.; Coastal Strand, N. Coastal Scrub; Del Norte Co. to Monterey Co., Los Angeles Co.; n. to Wash. May–Sept.

7. **R. salicifòlius** Weinm. In habit and lf.-shape similar to *R. californicus;* stems 3–9 dm. high; lvs. to 13 cm. long, to 1.7 cm. wide; petioles equaling or slightly exceeding the width of the blades; panicle-branches short, simple, ascending, or often elongated; pedicels jointed near base; outer sepals narrowly lanceolate, 1.2–1.5 mm. long, the valves in fr. 2.3–3 mm. long, 1.7–2.1 mm. wide, deltoid, acute, subentire or minutely denticulate, one valve with a large ovate callosity.—Moist places, 100–6500 ft.; many Plant Communities; rare in Siskiyou and Humboldt cos. and Sierra Nevada from Butte Co. to Kern Co., also Mono Co. to Inyo Co., common in Coast Range valleys, Marin Co. to San Diego Co., Channel Ids.; L. Calif., Nev. May–Sept.

8. **R. transitòrius** Rech. f. [*R. salicifolius* f. *t.* J. T. Howell.] In habit similar to *R. californicus,* 2.5–6 dm. high; lvs. lanceolate, 6–12 cm. long, 2–2.5 cm. wide; petioles ca. equaling the width of the blades; panicle-branches simple or the lower branched, curved-spreading; pedicels jointed below the middle; outer sepals narrowly lanceolate, 1.6 mm. long, the valves in fr. 2.5–3 mm. long, 2–2.3 mm. wide, ovate or ovate-lanceolate, acute, entire or subentire, generally all bearing a prominent ovate callosity. —Marshes and moist ground, uncommon, up to 4800 ft. (with several ecotypes); many Plant Communities; Sierra Nevada from Plumas Co. to Nevada Co., near coast from Monterey Co. n.; to Ore., Alaska. May–Sept.

9. **R. lacùstris** Greene. Aquatic (f. *aquatilis* Rech f.) or terrestrial (f. *terrestris* Rech. f.), the aquatic state with erect fistulous stems 5-9 dm. high emerging from the water, in this case the lower cauline lvs. undeveloped, the emersed lvs. somewhat pubescent beneath, or entirely submerged, the lvs. then glabrous, the terrestrial state with decumbent or ascending stems 2–4 dm. high, the oblong-lanceolate ± obtuse lvs. pubescent usually on both sides; panicle reduced, comparatively few-fld., the branches short; pedicels jointed below the middle; outer sepals lanceolate, 1.3 mm. long, the valves in fr. 2.1–2.5 mm. long, narrowly ovate, entire, all subequally calliferous, the callosities fusiform, 1.5–2 mm. long, 0.5–0.6 mm. wide.—Marshes, alkali sinks, ephemeral ponds, 4500–8000 ft.; N. Juniper Wd., Sagebrush Scrub, Yellow Pine F.; Modoc Co. to Mono Co., Tahquitz V., San Jacinto Mts.; s. Ore. July–Sept.

10. **R. triangulivàlvis** (Danser) Rech. f. [*R. salicifolius* ssp. *t.* Danser.] Stouter and

with broader lvs. than *R. californicus;* stems erect, 4–10 dm. high; lower lvs. 12–15 cm. long; panicle ample, the branches subelongate; pedicels jointed near base; outer sepals 1.6–1.8 mm. long, the valves in fr. mostly 3 mm. long and wide, triangular, entire or irregularly crenulate, all subequally calliferous, the callosities narrowly fusiform, 1.8–2.5 mm. long, 0.6–0.9 mm. wide.—Very polymorphic and widespread, 3500–9000 ft.; many Plant Communities; both sides of the Sierra Nevada, Mono Co. to Modoc Co., Bluff Lake, San Bernardino Mts.; to B.C., Man., Quebec, Chihuahua; introd. into Eu. July–Sept.

11. **R. hymenosèpalus** Torr. [*R. Saxei* Kell. *R. salinus* A. Nels. *R. h.* var. *s.* Rech. f.] CANAIGRE. WILD-RHUBARB. Perennial with cluster of tuberous roots; stems smooth, somewhat reddish, 6–12 dm. high, stout; lvs. very fleshy, oblong or oblong-elliptic, 0.6–3 dm. long, on shorter petioles, acute, crisped on margins, the sheaths 1–3 cm. long; panicle compact, 1–3 dm. long, pinkish; pedicels 8–12 mm. long, jointed near middle; outer sepals ca. 2 mm. long, the valves in fr. 8–14 mm. long, cordate-ovate, reticulate; nutlets 4–6 mm. long; $2n = 100$ (Kihara, 1927).—Common in dry sandy places, mostly below 5000 ft.; V. Grassland, Coastal Sage Scrub, Chaparral, Joshua Tree Wd., Creosote Bush Scrub; Kern and San Luis Obispo cos. to n. L. Calif., Wyo., w. Tex. Jan.–May. The dry roots contain as much as 35% tannin.

12. **R. fenestràtus** Greene. Perennial with a stout taproot; stem stout, to 2 m. high; lvs. lanceolate to lance-ovate, somewhat crisped, the lower to 3 or 4 dm. long, half as wide, with long petioles, the upper reduced; panicle rather dense, 3–6 dm. long, rosy in fr.; pedicels 5–15 mm. long, obscurely articulate in lower third; outer sepals narrowly linear-lanceolate, 3 mm. long, the valves in fr. 10 mm. long, 7–9 mm. wide, rotund-cordate, reticulate; nutlets 3.5–4 mm. long.—Coastal often brackish marshes, San Francisco Bay n., and somewhat alkaline marshes, Lassen, Modoc, and Humboldt cos.; to Alaska and Atlantic Coast. July–Sept.

13. **R. occidentàlis** Wats. [*R. procerus* Greene. *R. fenestratus* var. *p.* Rech. f. *R. o.* var. *p.* J. T. Howell.] In habit, herbage and panicle similar to *R. fenestratus,* the stem 5–15 dm. high; outer sepals linear-lanceolate, ca. 2 mm. long, the valves in fr. 4–5 mm. long, 5–6 mm. wide, cordate- or rotund-triangular, the base shallowly cordate or almost truncate, reticulate; nutlets 3 mm. long.—Marshy land, rare in Calif.; Goose L., Mt. Shasta, Fall River L., Bridgeport; to Colo., N. Dak., B.C., Alta. Aug.–Sept.

14. **R. orbiculàtus** Gray. [*R. Britannica* L., nomen confusum.] Stem stout, deeply canaliculate-sulcate, 6–16 dm. high; basal lvs. oblong-lanceolate, to 5 dm. long, to 2 dm. wide; panicle large, the short branches ascending-erect, the whole becoming ± dense in fr.; pedicels 6–10 mm. long, slender, articulate in lower fourth; outer sepals ovate-lanceolate, subobtuse, 2–2.5 mm. long, the valves in fr. 4–6 mm. long, 4.5–7.5 mm. wide, cordate- or reniform-rotund, the base broadly emarginate, the margin subentire, reddish-brown, reticulate, each with a fusiform callosity; nutlets 3.5 mm. long; $2n = 160$ (Jensen, 1936).—One known station in Calif., bog at Taylorsville, Plumas Co.; otherwise Great Plains e. to Nfld. Aug.–Sept.

15. **R. conglomeràtus** Murr. Perennial with taproot; stem smoothish, rather slender, 8–15 dm. high; lower lvs. oblong to lance-oblong, cordate at base, 1–2 dm. long, long-petioled, slightly crisped, obtuse, the upper reduced; panicle leafy, lax, 1–5 dm. long, the branches interrupted-spicate, subsimple, 1–3 dm. long; pedicels ca. 4 mm. long, stoutish, geniculate and jointed near base; valves 2.5–3 mm. long, oblong, obtuse, entire, each with a large smooth callosity; nutlets ca. 2 mm. long; $2n = 18$ (Sugiura, 1936).—Low moist places, cismontane valleys at low elevs.; many Plant Communities; natur. from Eu. April–Oct.

16. **R. críspus** L. CURLY DOCK. Perennial with taproot; stem smooth, rather slender, 5–12 dm. high; lower lvs. lanceolate to oblong-lanceolate, 1–3 dm. long, with long petioles, strongly crisped marginally, acute, the upper reduced; panicle strict, narrow, 1–5 dm. long; pedicels 5–10 mm. long, with swollen joints near base; outer sepals barely 1 mm. long, the valves in fr. 4–6 mm. long, round-ovate, subcordate, entire to minutely erose, usually with 3 equal or unequal callosities, rarely with only 1; nutlets 2 mm. long; $2n = 60$ (Jensen, 1937).—Common weed in low places, through much of N. Am.; native of Eurasia. Most of year.

17. **R. Kérneri** Borb. Perennial; stem strict or subflexuous, often branching above, 6–15 dm. high; lower lvs. oblong-elliptic, 10–22 cm. long, 3–7 cm. wide, with long

petioles; panicle open, the lower glomerules remote, the upper contiguous at maturity; pedicels jointed in lower third; outer sepals 2–2.5 mm. long, the valves in fr. 6–8 mm. long, rotund-cordate, almost entire to minutely and regularly denticulate, the callosities ovate-globose, 2 mm. long, 1.5 mm. wide; nutlets 3 mm. long.—One known Calif. station, Hope Ranch, Santa Barbara Co.; introd. from se. Eu.

18. **R. púlcher** L. FIDDLE DOCK. Dark green perennial, the stem slender, 3–6 dm. high, with spreading branches; lower lvs. oblong, 5–12 cm. long, long-petioled, obtuse, cordate at base, usually contracted above base, the upper short-petioled, gradually reduced; panicle with divergent branches, lax, the clusters remote; valves 4.5–6 mm. long, 2.5–4.5 mm. wide, the margin toward base long-toothed, the apex linguiform, usually all bearing a callosity, but callosities often unequal; nutlets 3–4 mm. long.— Widely spread in Calif. and s. U.S. as a weed in waste places; native of Medit. Basin. May–Sept. Some of our plants are referred to ssp. *divaricàtus* (L.) Murbeck, with lvs. rarely contracted above base; valves ca. as wide as long, short-toothed.

19. **R. stenophýllus** Ledeb. Perennial with taproot; stem 2–6 or more dm. high; lower lvs. lanceolate, acute or acuminate, the petiole equaling the blade, the cauline gradually becoming linear below infl.; panicle with erect or divergent branches, the clusters approximate; outer sepals linear-lanceolate, to 2 mm. long, the valves in fr. 4–5 mm. long, as wide, cordate-triangular, reticulate, the margin sharply short-toothed, all bearing an ovoid-ellipsoidal callosity 2–2.5 mm. long; nutlets to 2 mm. long.—One known Calif. station, moist soil, introd. from Eurasia.

20. **R. obtusifòlius** L. ssp. **agréstis** (Fries) Danser. [*R. o.* (var.) *a.* Fries.] BITTER DOCK. Perennial with stout taproot; stem stout, 6–12 dm. high, simple or few-branched; lower lvs. broadly lance-oblong, cordate at base, flat, 1–3.5 dm. long, long-petioled, the upper with rounded base and smaller; panicle open, with divergent branches, leafy at base; valves in fr. ca. 6 mm. long, triangular-ovate, the margin with few spreading spinose teeth, usually only one with a callosity; nutlets ca. 2 mm. long.—Low moist places, Los Angeles Co., Santa Cruz Co. to Del Norte and Shasta cos.; a common and spreading weed in N. Am.; native of Eu. June–Dec.

21. **R. violáscens** Rech. f. [*R. Berlandieri* auth., not Meissn.] Annual or biennial with taproot; stem stout, 3–7 dm. high; lower lvs. narrow-oblong to spatulate, 6–10 cm. long, rather flat, long-petioled, the upper reduced and lance-linear; panicle narrow, leafy-bracted, 5–12 cm. long, with ascending branches; pedicels ca. as long as fr., jointed below middle; valves in fr. 2.5–3 mm. long, nearly as wide, triangular-ovate, erose or toothed near base, all with an ovate-oblong callosity, these unequal, the largest 1.5–2 mm. long; nutlets 1.7 mm. long.—Alkaline Sink, Creosote Bush Scrub; San Joaquin V., Colo. Desert, Lake Elsinore; to Tex., Mex. March–Aug.

22. **R. dentàtus** L. ssp. **Klotzschiànus** (Meissn.) Rech. f. [*R. K.* Meissn.] Annual or biennial; stem to 7 dm. high; lower lvs. oblong, the base subcordate or truncate, the apex rounded, flat, petiolate, the upper lanceolate, cuneate at base, the uppermost sublinear; pedicels longer than fr., jointed near base; valves in fr. 3–4 mm. long, 2.5 mm. wide, triangular-ovate, bearing teeth on the margin to 2 mm. long, all with a small ovate callosity.—Wet ground, Gustine, Merced Co., and Stockton, the only known U.S. stations; elsewhere usually a rice-field weed; native of s. Asia.

23. **R. fuegínus** Phil. [*R. maritimus* Meissn., not L. *R. m.* L. var. *f.* Dusén. *R. persicarioides* auth., not L. *R. f.* vars. *brachythrix, tanythrix* and *ovato-cordatus* Rech. f.] GOLDEN DOCK. Annual or biennial, scabrid-pubescent, the stem simple or diffusely branched, 1–6 dm. high; lower lvs. lanceolate, with ± crisped margin and cordate to truncate base, 3–15 cm. long, on shorter petioles, the upper narrower; fl.-whorls crowded in leafy compact or interrupted spikes; pedicels jointed toward base; valves in fr. 1.7– 2.5 mm. long, rhombic-oblong, lance-pointed, each with 2–3 awnlike bristles on each side and a prominent oblong callosity ca. 1 mm. long.—Wet often brackish places, below 5000 ft.; many Plant Communities; frequent in s. Calif., occasional in coastal valleys to n. Calif., e. to Shasta Co., rare e. of Sierra Nevada; to B.C., Atlantic Coast, S. Am. May–Sept.

24. **R. persicarióides** L. Annual, similar in habit to *R. fuegínus;* stems one or more from the base, simple or branched, strict or angular-flexuous, 1–5 dm. high; lvs. mostly linear-oblong, obtuse at base and apex, on rather short slender petioles; infl. branched, the fl.-whorls crowded in leafy spikes; valves in fr. 2–2.5 mm. long, ca. 1 mm. wide

(excepting teeth), lance-ovate, each with 1–3 awnlike bristles exceeding the width of the valve on each side and bearing a thick swollen obtuse callosity nearly hiding the face of the valve.—Freshwater or brackish marshes along the coast; Coastal Salt Marsh, Freshwater Marsh; Ventura Co., San Mateo Co. to Marin Co., Humboldt Co.; to Ore., Mass., e. Canada. July–Sept.

11. Polýgonum L. KNOTWEED. SMARTWEED

Annual or perennial herbs, sometimes shrubs, with fibrous roots or thick rootstocks and often swollen joints. Lvs. entire, alternate; stipules usually scarious, sheathing and forming an ocrea, with which the petiole may be jointed. Fls. on jointed pedicels, variously disposed, mostly perfect. Calyx 4–6(mostly 5)-parted, the divisions often petaloid and brightly colored, enlarging, persistent and surrounding the lenticular or 3-angled ak. Stamens 3–9. Styles or stigmas 2 or 3, deciduous. Embryo curved halfway around the albumen and placed in a groove on the outside of it. Ca. 200 spp. of wide geographical range and falling in several natural groups often recognized as genera. (Greek, *poly,* many, and *gonu,* knee or joint, because of the thickened joints of the stem.)

(Stanford, E. E., Several papers in Rhodora 27–29, 1925–1927.)
A. Fls. in axillary fascicles, or in spikes with leafy bracts; ocreae hyaline and finally mostly 2-lobed
      and lacerate.
    B. Lvs. jointed with the ocreae, 1-nerved; fls. mostly 2 or more in the axils, short-pedicelled.
        (Section Avicularia)
        C. Plants perennial, often suffrutescent.
            D. Lvs. linear-lanceolate, revolute; fls. 4–5 mm. long. Coastal Strand
                                                                                            1. *P. Paronychia*
            DD. Lvs. oblong to elliptic or obovate.
                E. Fls. 5–8 mm. long; lvs. revolute. Subalpine .... 2. *P. shastense*
                EE. Fls. 3–4 mm. long; lvs. plane. Coastal Salt Marsh .... 3. *P. Fowleri*
        CC. Plants annual.
            D. Stems terete or nearly so, not sharply angled, often spreading to prostrate. (See
                also *P. minimum*)
                E. Aks. strongly exserted from calyx. Maritime.
                    F. Lvs. of infl. not reduced .... 3. *P. Fowleri*
                    FF. Lvs. of infl. much reduced and bractlike .... 6. *P. patulum*
                EE. Aks. normally included in calyx.
                    F. The aks. dull or scarcely shining; fls. borne in axils of scarcely reduced
                        lvs.; calyx with pinkish to purplish margins .... 4. *P. aviculare*
                    FF. The aks. smooth and shining.
                        G. Upper lvs. bracteate; calyx with pinkish margins
                                                                            5. *P. argyrocoleon*
                        GG. Upper lvs. not reduced to bracts; calyx with yellowish margins
                                                                            7. *P. ramosissimum*
            DD. Stems and branches strongly angled, usually suberect.
                E. Pedicels recurved, at least in fr. .... 8. *P. Douglasii*
                EE. Pedicels erect (except sometimes in *P. spergulariaeforme*).
                    F. Fls. in small axillary clusters, not in spicate infl.
                        G. Lvs. lanceolate, reduced up the stem
                                                                    8. *P. Douglasii* var. *Johnstonii*
                        GG. Lvs. obovate to ovate, scarcely reduced upward .. 9. *P. minimum*
                    FF. Fls. in terminal leafy-bracteate spikes.
                        G. Stamens 5–8.
                            H. Bracts of infl. green.
                                I. Calyx 2.5–3 mm. long; aks. 3–4 mm. long. From above
                                    2500 ft. .... 10. *P. spergulariaeforme*
                                II. Calyx 2 mm. long; aks. ca. 1.5 mm. long. Monterey
                                                                            11. *P. montereyense*
                            HH. Bracts white-margined. Modoc Co. ...... 14. *P. esotericum*
                        GG. Stamens 3.
                            H. Plants mostly 3–8 cm. high; lvs. 5–10 mm. long; bracts
                                mostly 3–5 mm. long .... 12. *P. Kelloggii*
                            HH. Plants mostly 8–19 cm. high; lvs. 10–35 mm. long; bracts
                                mostly 5–10 mm. long .... 13. *P. confertiflorum*
    BB. Lvs. not jointed with the ocreae, 3-nerved; fls. solitary in the axils, sessile. (Section Duravia)
        C. Plant perennial, 2–6 dm. high .... 15. *P. Bolanderi*
        CC. Plants annual, 0.5–2 dm. high.
            D. Ocreae deeply lacerate into bristlelike segms. Widely distributed.
                E. Stems 6–20 cm. long; calyx 2–2.5 mm. long; segms. of ocrea subulate, rather
                    firm .... 16. *P. californicum*
                EE. Stems usually 2–5 cm. long; calyx 1.5 mm. long; segms. of ocrea capillary
                    and weak .... 17. *P. Parryi*

DD. Ocreae 2-lobed, the lobes serrate. Butte Co. .... 18. *P. Bidwelliae*
AA. Fls. fascicled in terminal dense to open spikelike racemes or in panicles without leafy bracts;
ocreae rarely hyaline or lacerate.
  B. Infl. elongate, spiciform, not in ample panicles or lvs. not cordate-sagittate.
    C. Stems simple, from fleshy rootstocks; lvs. basal and long-petioled or cauline and short-
petioled or sessile; infl. a single terminal spike (Section Bistorta) .. 19. *P. bistortoides*
    CC. Stems branched; lvs. all cauline and similar; spikes terminating main stems and
branches.
      D. Ocreae funnelform, oblique and somewhat open on side facing the leaf. (Section
Aconogonum)
        E. Plants 10–20 dm. high; lvs. 3–20 cm. long; fls. in nearly leafless panicles.
          F. Lf.-blades rounded at base. Montane native .. 20. *P. phytolaccaefolium*
          FF. Lf.-blades with 2 basal lobes. Natur. nw. coast .... 23. *P. polystachyum*
        EE. Plants 1–4 dm. high; lvs. 1–4 cm. long; fls. in small axillary clusters.
          F. Foliage glaucous, glabrous to scabrous; aks. oblong-ovoid, 3–4 mm. long
21. *P. Davisiae*
          FF. Foliage green, soft-pilose; aks. obovoid, 4–4.5 mm. long
22. *P. Newberryi*
      DD. Ocreae cylindric, truncate. (Section Persicaria)
        E. The ocreae eciliate or nearly so, at least in maturity.
          F. Perennial plants with creeping rhizomes and stolons; in and about water.
            G. Peduncles glabrous; spikes ovoid or oblong, mostly less than 3 cm.
long .... 24. *P. amphibium*
            GG. Peduncles pubescent; spikes slender-cylindric, mostly 3–10 cm. long
25. *P. coccineum*
          FF. Annuals; not aquatic.
            G. Peduncles with stalked glands; spikes erect, oblong-cylindric; aks.
suborbicular, 2.2–2.5 mm. in diam. .... 26. *P. pensylvanicum*
            GG. Peduncles with or without sessile glands; spikes often arching,
slender-cylindric; aks. subovate, 1.8–2.2 mm. long
27. *P. lapathifolium*
        EE. The ocreae fringed with bristly cilia.
          F. Plants annual.
            G. Calyx punctate with glands; spikes slender, arching.
              H. The calyx greenish or with purple tips; aks. dull
28. *P. Hydropiper*
              HH. The calyx white; aks. shining .... 31. *P. punctatum*
            GG. Calyx not glandular-punctate; spikes dense, erect
29. *P. Persicaria*
          FF. Plants perennial.
            G. Internodes fusiform-thickened; ocreae somewhat ciliate when young
30. *P. fusiforme*
            GG. Internodes cylindric; ocreae bristly-ciliate .. 32. *P. hydropiperoides*
  BB. Infl. of open panicles and the lvs. broad, or the infl. of small axillary clusters and terminal
spike, with the lvs. cordate-sagittate. (Section Tiniaria)
    C. Stems twining; plant annual; lvs. ovate-sagittate .... 33. *P. Convolvulus*
    CC. Stems erect; plants stout perennials; lvs. ovate.
      D. Lvs. rounded at base, abruptly acuminate at apex, 5–15 cm. long
34. *P. cuspidatum*
      DD. Lvs. cordate at base, gradually narrowed at apex, 15–30 cm. long
35. *P. sachalinense*

1. **P. Paronýchia** Cham. & Schlecht. Stems suffrutescent from large woody rootstocks,
much-branched, prostrate or ascending, 2–10 dm. long; older branches covered with
hyaline torn sheaths, the tips leafy; lvs. usually crowded, linear-lanceolate, 6–25 mm.
long, acutish, revolute, glabrous above, the midrib with a 2-winged ciliate keel beneath;
fls. in upper axils, short-pedicelled, crowded; calyx white to rose with green midveins,
4–5 mm. long, the segms. oblong-ovate; stamens 8, the 3 inner dilated at base; aks.
3-angled, ca. 4.5 mm. long, shining, black.—Coastal Strand; Monterey Co. to B.C.
March–Sept.

2. **P. shasténse** Brew. ex Gray. Stems woody, 1–3 dm. long, branched, prostrate or
ascending, from a woody root; lvs. oblong to subobovate, acute or obtuse, revolute in age,
5–15 mm. long, glabrous; stipule-sheaths scarcely lacerated; fls. 2–3 in the axils, rose
with dark midveins, mostly 5–8 mm. long; stamens 8; aks. 3-angled, ca. 3.5 mm. long,
brown, shining.—Rocky or gravelly slopes, 7000–11,000 ft.; Lodgepole F., Subalpine
F.; Sierra Nevada to Siskiyou Co.; Ore., Nev. July–Sept.

3. **P. Fòwleri** Rob. Perennial, suberect, finally much-branched, succulent, pale green
or slightly glaucous, the stems 1–6 dm. long; ocreae flaring at summit, 3–6 mm. long;
lvs. elliptic to oblong-obovate, 6–25 mm. long, broadly rounded above, the upper
bracteal ones crowded; fls. in axillary clusters of 2–3, the sepals oblong, greenish
with roseate margins; aks. oblong-ovoid, 3-angled, ca. 3.5–4 mm. long, olive-brown,

smooth and shining.—Coastal Salt Marsh; San Francisco Bay region; Wash. to Alaska, Atlantic Coast. July–Sept.

4. **P. aviculàre** L. [*P. a.* var. *angustissimum* Meissn.] COMMON KNOTWEED. Annual with prostrate or ascending bluish-green slender stems 1–12 dm. long; lvs. lanceolate to almost oblong, 5–20 mm. long, blue-green, scattered to approximate, not much reduced upward; stipule-sheaths silvery, soon torn; fls. 1–5 in axillary clusters; calyx 2–3 mm. long, greenish with pinkish to purplish margins; stamens 8, rarely 5; aks. dull or slightly shiny, dark brown, 2–2.5 mm. long, somewhat granular or striate-roughened; $2n = 40$, 60 (Löve & Löve, 1942).—Common weed in dooryards, waste places, etc.; natur. from Eurasia. May–Nov. Variable, some forms perhaps worthy of note: Var. *eréctum* Roth. Stems erect.—Occasional as weed. Var. *littoràle* (Link) Koch. [*P. l.* Koch.] Prostrate or nearly so; lvs. thicker and more fleshy, the larger 10–30 mm. long.—Sandy beaches, salt marshes, near the coast and alkaline places in the interior, Calif.; to Wash., Atlantic Coast; Eu.

5. **P. argyrocòleon** Steud. ex Kunze. Pale green annual, erect with ascending striate branches, rather lax; lvs. linear, 1–3 cm. long, narrowed at both ends; ocreae silvery above and lacerate, brown toward base; upper lvs. bracteate, so that infl. seems racemose, although fls. are several at each node; calyx green with roseate margins, 1.5 mm. long; stamens 6; aks. 3-angled, ovoid, smooth and shining, ca. 1.5 mm. long; $2n = 40$ (Heiser & Whitaker, 1948).—Becoming rather widespread in Calif. as a weed about orchards, alfalfa fields, etc.; native of Asia. June–Oct.

6. **P. pátulum** Bieb. Erect annual with much the same habit and stature as *P. argyrocoleon*, and lvs. of infl. bracteate, but the calyx scarcely if at all roseate and the aks. 2.5–3 mm. long, hence well exserted.—About salt marshes in San Francisco Bay region. June–Oct.

7. **P. ramosíssimum** Michx. Rather bushy erect yellowish-green annual 2–10 dm. high; lvs. narrowly lanceolate to linear, 20–50 mm. long, the upper not bracteate; ocreae silvery, deeply cut; fls. several in each upper axil; calyx yellowish or with yellowish margins, ca. 3 mm. long; aks. ovoid, 3-angled, black, smooth, shining, 2.5–3.5 mm. long.—Dry open places, where occasional as weed; to Wash. and Atlantic Coast. July–Sept.

8. **P. Douglásii** Greene. Slender erect annual, rather pale green, loosely few-branched, 1–4(–6) dm. high, subglabrous; lvs. lance-oblong to linear, rather remote, 1–4 cm. long, subsessile; articulation with sheath evident; fls. 1–3 per axil, drooping, reddish; pedicels 2–3 mm. long, deflexed; calyx 3–4 mm. long; stamens 8; aks. 3-angled, oblong, ovoid, smooth, shining, ca. 3.5 mm. long.—Fairly dry areas as at edge of meadows, 4000–9000 ft.; Yellow Pine F. to Lodgepole F.; mts. from San Diego Co. n. through the Sierra Nevada and N. Coast Ranges; to B.C. and Quebec. June–Sept.

Var. **latifòlium** (Engelm.) Greene. [*P. tenue* var. *l.* Engelm. *P. montanum* Greene.] Lvs. oblong to oblanceolate, often rather crowded.—At 5000–10,500 ft.; largely Subalpine F.; Tulare and Trinity cos. n.; to Wash.

Var. **Aústinae** (Greene) Jones. [*P. A.* Greene.] Lvs. ovate to lance-ovate, 5–15 mm. long; calyx green with white margins, 2–2.5 mm. long; aks. ovoid, angled, black, shining, 2.5–3 mm. long.—Sagebrush Scrub to Subalpine F.; Inyo Co. to Modoc Co.; to Ida. June–Aug.

Var. **Johnstònii** Munz. [*P. sawatchense* Small. *P. exile* Eastw.?] Like the sp., but with pedicels erect, ca. 1 mm. long; calyx green, 2–3 mm. long, pale on margins; aks. oblong, 2.5–3 mm. long.—At 5000–10,500 ft.; Red Fir F. to Subalpine F.; San Bernardino Mts., Sierra Nevada especially on e. side, Trinity Co.; to Wash., Rocky Mts. July–Sept.

9. **P. mínimum** Wats. [*P. Torreyi* Wats.] Somewhat scurfy annual 5–15 cm. high, with 1–several simple or branched stems, leafy throughout; lvs. ovate to obovate, sub-sessile, 5–15 mm. long, obtuse or acutish; fls. greenish white, 2–3 in axils of most lvs.; pedicels ca. 2 mm. long; calyx ca. 2 mm. long, with oblong segms.; stamens 5–8; aks. 3-angled, oblong-ovoid, 2–2.5 mm. long, black, shining.—Damp meadows and banks, 5000–11,200 ft.; mostly Subalpine F., Lodgepole F.; Sierra Nevada from Tulare Co. n., N. Coast Ranges from Tehama Co. to Siskiyou Co.; to B.C., Rocky Mts. July–Sept.

10. **P. spergulariaefórme** Meissn. [*P. coarctatum* Dougl. *P. Howellii* Greene?] Rather slender erect glabrous wiry annual, strict or diffuse, somewhat scurfy, 1–4 dm. high,

lvs. linear to narrowly lanceolate, 0.5–3 cm. long, acute, sessile, often revolute, plainly jointed with the ocreae; upper lvs. subbracteate; fls. 2–4 in the axils, the upper nodes rather crowded; pedicels ca. 1.5 mm. long, at length nodding; calyx white to rose, with dark midribs, 2.5–3 mm. long; stamens 8, sometimes fewer; aks. oblong, 3-angled, 3–4 mm. long, dark, smooth and shining or somewhat granular.—Dry hills, often on serpentine, 2500–6000 ft.; Yellow Pine F., Chaparral, Coastal Prairie; N. Coast Ranges from Lake Co. n.; Sierra Nevada from Nevada Co. n.; to B.C., Rocky Mts. June–Sept.

11. **P. montereyénse** Brenckle. Pale green annual, branched at base, 4–10 cm. high; lvs. linear-lanceolate, 5–15 mm. long, jointed with ocreae; upper lvs. subbracteate, green; fls. 1–several in axils of leafy spike; pedicels very short; calyx petaloid when open, 2 mm. long, pale green on back; stamens 5–6; aks. flattened, slender, 3-angled toward apex, smooth and shining or minutely punctate, dark, 1.6 mm. long.—Dry hard clay, Monterey. July–Aug.

12. **P. Kellóggii** Greene. [*P. Watsonii* auth., not Small.] Glabrous often tufted annuals 3–8 cm. high, with very short internodes; lvs. linear or lance-linear, 0.5–1 cm. long, spreading, not imbricate, acute, sessile; bracts green like small lvs., mostly 3–5 mm. long; infl. spicate, 0.5–3 cm. long; pedicels scarcely 1 mm. long; calyx green, almost 2 mm. long, with white margins; stamens 3; aks. ovoid, angled, brownish, 1.5 mm. long, dull, relatively smooth.—Usually in damp, silty or gravelly places, 4500–11,500 ft.; Sagebrush Scrub, Lodgepole F., Subalpine F., Alpine Fell-fields; San Jacinto Mts., San Bernardino Mts., through the Sierra Nevada and nw. to Mendocino and Lake cos.; to Wash., Colo. June–Sept.

13. **P. confertiflòrum** Nutt. ex Piper. Much like *P. Kelloggii*, but taller, mostly 8–19 cm. high; lvs. narrowly lance-linear, 1–3.5 cm. long; floral bracts mostly 5–10 mm. long, white-margined; stamens 3; aks. striate-granular, black, dull.—Low places, 3400–5000 ft.; N. Juniper Wd.; Shasta Co. to Wash. June–July.

14. **P. esotéricum** Wheeler. Glabrous suberect annual 6–14 cm. high; lvs. linear, 8–10 mm. long, acute; bracts leafy, 2–5 mm. long, white-margined; the spicate racemes 2–7 cm. long; calyx 2–3 mm. long; stamens 8; aks. dark brown, lance-ovoid, ca. 2 mm. long, angled, shining.—Seasonally dry adobe flats and pond basins, 4500–5000 ft.; Sagebrush Scrub, N. Juniper Wd.; Modoc and Lassen cos.; adjacent Ore. July–Aug.

15. **P. Bolánderi** Brew. ex Gray. Perennial from a woody root, with many slender wiry erect stems 2–6 dm. high; lvs. linear to subulate, 3–15 mm. long, cuspidate, minutely punctate; ocreae conspicuously lacerate; fls. 1–2 in the axils, sessile, white to rose with greenish midveins; stamens 8; aks. oblong-ovoid, 3-angled, 2–5 mm. long, dark brown, smooth, shining.—Dry gravelly and rocky places, below 5000 ft.; N. Oak Wd., Chaparral? Yellow Pine F.; Napa Co. to Humboldt Co., Butte Co. June–Nov.

16. **P. califórnicum** Meissn. [*P. Greenei* Wats. *Duravia c.* Greene. *D. G.* Greene.] Glabrous annual with wiry slender stems 6–20 cm. long, diffusely branched; lvs. subfiliform to narrowly linear, 1–3 cm. long, subulate-tipped, 3-nerved; ocreae conspicuously lacerate; fls. solitary in the axils, the early ones scattered, the later more crowded; calyx 2–2.5 mm. long, white with pink midveins; stamens 8; aks. angled, narrowly oblong-ovoid, ca. 2 mm. long, dark brown, smooth, shining.—Dry sandy and gravelly flats and bars and clay slopes, below 4500 ft.; Foothill Wd., N. Oak Wd., Yellow Pine F.; N. Coast Ranges, Sierra Nevada from Mariposa Co. n.; to Wash. May–Oct.

17. **P. Párryi** Greene. Compact tufted annual 2–5 cm. high, sometimes more lax and taller, leafy; lvs. narrowly linear, 0.5–2 cm. long, cuspidate, 3-nerved; ocreae much lacerate and concealing fls.; spikes dense, the fls. solitary at each axil, sessile; calyx 1.5 mm. long; stamens 8; aks. oblong-ovoid, 1.5 mm. long, chestnut-brown, angled, smooth, shining.—Uncommon, dry sandy places, 2000–6000 ft.; mostly Yellow Pine F.; Cuyamaca Mts., Sierra Nevada, N. Coast Ranges from Lake Co. n.; to Wash. June.

18. **P. Bidwélliae** Wats. Annual with rather few divergent branches 3–12 cm. long; lvs. narrowly linear, 0.5–2 cm. long, 3-nerved, subulate-tipped; ocreae oblong-ovate, 2-parted, somewhat lacerate in age; spikes terminal, bracted, 1–3 cm. long, with fls. solitary at axils; calyx pink, 2 mm. long, concealed by the silvery scarious ocreae; stamens 8; aks. angled, oblong, dark brown, 1.7 mm. long, smooth or minutely granular.—Volcanic outcrops, at ca. 1000 ft.; Foothill Wd.; a few miles e. of Chico, Butte Co. and in Tehama Co. May–June.

19. **P. bistortoìdes** Pursh. [*P. bernardinum* and *cephalophorum* Greene. *Bistorta leptophylla* Greene.] Perennial with a thick horizontal rootstock and several erect slender simple glabrous stems 2–7 dm. high; lvs. mostly near base, the lower oblong to oblanceolate, 1–2.5 dm. long, on petioles ca. as long, 2–5 cm. wide, with broad midrib; upper lvs. sessile, lanceolate, reduced; ocreae narrowly cylindric, 3–7 cm. long with oblique summit; spikes terminal, thick-cylindric, 1–6 cm. long, 1–1.5 cm. thick; calyx pink or white, 4–5 mm. long; stamens exserted; aks. pale brown, shining, angled, rhombic-obovoid, 3.5–4 mm. long.—Wet meadows and along streams, mostly 5000–10,000 ft.; Yellow Pine F. to Subalpine F.; San Jacinto and San Bernardino mts., Sierra Nevada, N. Coast Ranges; to Alaska and Atlantic Coast. Also Coastal Marshes as far s. as Marin Co. June–Aug.

20. **P. phytolaccaefòlium** Meissn. [*Aconogonum p.* Small ex Rydb. *P. alpinum* auth., not All.] Subglabrous stout perennial 1–2 m. high, bushy; lvs. lanceolate to lance-ovate, 3–15 cm. long, 1–4 cm. wide, ± acuminate at apex, short-petioled, rather fleshy; ocreae 1–3 cm. long, brown, deciduous; infl. of loose terminal almost leafless panicles; pedicels slender, 2–3 mm. long; calyx white or greenish-white, ca. 3 mm. long; stamens 8, included; aks. angled, ovoid, ca. 4 mm. long, smooth, shining.—Moist often rocky places, 5000–9000 ft.; Red Fir F. to Subalpine F.; Sierra Nevada from Yosemite n.; N. Coast Ranges from Yolla Bolly Mts. n.; to Alaska, Ida., Nev. June–Sept.

21. **P. Davísiae** Brew. ex Gray. Perennial with stout much-branched taproot and several–many decumbent to ascending simple to branched stems 1–4 dm. long, glaucous, glabrous or nearly so; lvs. ovate to oblong-ovate, acutish, sessile or nearly so, scabrous-pubescent, 1–4 cm. long; fls. in short, 3–4-fld. terminal and axillary clusters; calyx whitish to purplish-green, 2.5–3 mm. long; stamens 8; aks. angled, oblong-ovoid, 3–4 mm. long, light brown, smooth, shining.—Talus and rocky places, 5000–9000 ft.; Red Fir F. to Subalpine F.; Sierra Nevada from Alpine Co. n., Coast Ranges from Lake Co. n.; s. Ore. June–Sept.

22. **P. Newbérryi** Small. Near to and much like *P. Davisiae*, but soft-pilose, greener, the lower lvs. more definitely petioled; calyx with more rose; aks. 4–4.5 mm. long, obovoid.—At ca. 5000–7900 ft., Siskiyou and Modoc cos.; to Wash. July–Aug.

23. **P. polystáchyum** Wall. Stout branched perennial 1–1.5 m. high, the stems glabrous; lvs. well distributed, lance-ovate, 1–2 dm. long, long-acuminate, pubescent beneath along the veins, ciliate; ocreae 4–5 cm. long, subglabrous; infl. a terminal panicle of racemes and 1–2 dm. long; calyx white, 3–4 mm. long.—Vacant lots and marshes, Eureka, Ft. Bragg; native of Asia. Sept.–Oct.

24. **P. amphíbium** L. var. **stipuláceum** Coleman. [*P. a.* var. *natans* Michx. *P. n.* Eat. *P. Hartwrightii* Gray. *Persicaria H., purpurata* and *insignis* Greene.] WATER SMARTWEED. Perennial with slender rhizomes; stems elongate, leafy, glabrous, not branched; lvs. lanceolate to lance-oblong, or even broader in the floating form, 5–10 cm. long, petioled, obtusish, with rounded or subcordate base; peduncles glabrous; fls. in thick terminal short-cylindric to -ovoid spikes, rose; calyx 4–5 mm. long; stamens 5; aks. lenticular, suborbicular, 2.5–3 mm. long, black, shining or slightly granular.—Ponds and lakes, below 10,000 ft.; many Plant Communities; San Diego Co. through cismontane and montane Calif.; to Alaska and Atlantic Coast. *P. amphibium* is Eurasian; true *stipuláceum* is a strand form on mud banks, while the floating form is sometimes called f. *flùitans* (Eat.) Fern. and a densely villous plant is f. *hirtuòsum* (Farw.) Fern. July–Sept.

25. **P. coccíneum** Muhl. [*P. Muhlenbergii* Wats. *P. emersum* Britton. *Persicaria franciscana, alismaefolia, Covillei,* and *hesperia* Greene.] Plants like *P. amphibium*, but rather coarse, glabrous to strigose; lvs. 0.5–2 dm. long, lance-ovate to lanceolate, somewhat acuminate at apex, with rounded or subcordate base, short-petioled; spikes narrow-cylindric, 3–10(–15) cm. long, on pubescent sometimes glandular peduncles; fls. rose, ca. 3 mm. long.—Occasional, in and about ponds through much of the state; to B.C. and Atlantic Coast. The typical plant is terrestrial and subglabrous to finely strigose; f. *nàtans* (Wieg.) Stanf. is aquatic with floating lvs.; and var. *pratíncola* (Greene) Stanf. is terrestrial and densely canescent-pubescent. June–Oct.

26. **P. pensylvánicum** L. [*Persicaria p.* Small.] PINKWEED. Ascending to erect annual to ca. 1 m. high, the upper stems and peduncles stipitate-glandular; lvs. lanceolate, 0.4–2 dm. long, acuminate, ciliate, otherwise mostly glabrous; ocreae cylindric, without cilia;

fls. in panicles of racemes, these erect, compact, not interrupted, oblong-cylindric; bracts of infl. subglabrous; calyx reddish to purplish, 3–4 mm. long; stamens included; aks. lenticular, suborbicular, 2.2–2.5 mm. across, somewhat shining, flat on one face, concave on other.—Occasionally introd. from the e. U.S., as in marshes of Sonoma Co. July–Sept.

P. mexicànum Small, with interrupted racemes and ciliate bracts in the infl., reported from Kern and Merced cos.; to Mex.

27. **P. lapathifòlium** L. WILLOW WEED. Erect or ascending annual, the stems simple or branched, usually swollen at nodes, 5–15 dm. high; lvs. lanceolate, acuminate at tip and narrowed at base, 5–20 cm. long, often sparsely scabrous with appressed hairs on midrib and near edge; ocreae cylindric, mostly eciliate; peduncles often with subsessile glands; spikes slender-cylindric, 1–6 cm. long, often nodding; calyx white to pink to purplish, ca. 2 mm. long; stamens included; aks. subovate, flattened with one side concave, 1.8–2.2 mm. long; $2n = 22$ (Löve & Löve, 1948).—Common in moist places at low elevs.; many Plant Communities; throughout temp. N. Am., Eurasia. June–Oct.

Var. **salicifòlium** Sibth. [Var. *incanum* (Willd.) Koch.] Lvs. white-pubescent beneath; spikes erect.—Occasional, especially in cent. Calif.; Eu.

Var. **prostràtum** Wimmer. Prostrate or depressed, with trailing branches and subrhombic lvs.—Said to occur in Calif.; from Eu.

28. **P. Hydropìper** L. [*P. H.* var. *projectum* Stanf.] Ascending or erect annual 2–6 dm. high, glabrous, simple or branched; lvs. lanceolate, 3–9 cm. long, acute to acuminate, basally narrowed to short petiole; ocreae cylindric, truncate, with cilia 1–2 mm. long; spikes nodding, slender, paniculate; calyx greenish or with red tips, glandular-punctate, 2.5–4 mm. long; aks. dull, minutely striate, 2–2.5 mm. long, lenticular to 3-angled; $2n = 20$ (Jaretzky, 1928).—Occasional, moist places, probably natur.; to Wash. and Atlantic Coast; Eu. July–Oct.

29. **P. Persicària** L. LADY'S THUMB. Almost glabrous annual, erect or ascending, 2–8 dm. high; lvs. lanceolate, acuminate, subsessile, 3–10 cm. long; ocreae thin, appressed-villous or subglabrous, fringed with short bristles; spikes 1–several, with glabrous peduncles, densely fld., 1.5–2.5 cm. long; calyx pink or somewhat purplish, 2–3(–4) mm. long; aks. lenticular or 3-angled, black, smooth, shining, 2–3 mm. long; $2n = 44$ (Jaretzky, 1928).—Common in moist waste places, below 5000 ft.; to Wash., Atlantic Coast; native of Eu. June–Nov.

30. **P. fusifórme** Greene. Perennial, the stems dark red, ascending, 6–10 dm. high, the internodes fusiform-thickened above the nodes; lvs. lance-linear, 7–13 cm. long, almost sessile, glabrous or strigulose along the midribs; ocreae strigulose, 1–2 cm. long, when young somewhat ciliate; racemes dense, narrow-cylindric, mostly 3–4 cm. long, not more than 6 mm. thick; calyx red in bud, white at anthesis, 2.5 mm. long; stamens 4–5; aks. trigonous, black, shining, ca. 2 mm. long.—Moist places, Colo. R. V.; Ariz. July–Oct.

31. **P. punctàtum** Ell. [*P. acre* HBK., not Lam.] WATER SMARTWEED. Commonly perennial with slender branching rootstocks; stems erect or ascending, 3–10 dm. high, simple or branched, subglabrous; lvs. lanceolate to lance-elliptic, 5–10 cm. long, punctate, acutish to acuminate, subglabrous, short-petioled; ocreae scarious, subglabrous, ciliate; racemes in a naked or leafy panicle, linear-cylindric, 1.5–5 cm. long, suberect, loosely fld. at base; calyx greenish, conspicuously glandular-punctate, 2–3 mm. long; aks. trigonous or lenticular, shining, 2.5–3.5 mm. long.—Common in moist low places, at low elevs.; many Plant Communities; through Calif.; to Wash., Atlantic Coast, S. Am. July–Oct. Variable and in need of study for Calif., many of our plants being referable to the var. *ellípticum* Fassett, with lvs. subelliptic, 2–4 times as long as wide; calyx to 3 mm. long; aks. all trigonous.

32. **P. hydropiperoìdes** Michx. var. **asperifòlium** Stanf. Perennial, the stems glabrous, erect or ascending, 3–10 dm. high; lvs. lanceolate, 5–15 cm. long, somewhat scabrous-strigulose beneath, attenuate at both ends, scabrous-ciliate; ocreae cylindric, truncate, strigose, bristly ciliate; racemes slender, somewhat interrupted, 2–7 cm. long, in terminal panicles, erect; calyx white to rose, 2–3 mm. long; stamens 8, included; aks. sharply 3-angled, shining, brown to almost black, 2.5–3 mm. long.—Moist places, mostly at low elevs.; many Plant Communities; Cent. V. and s. Calif. June–Oct.

33. **P. Convólvulus** L. BLACK BINDWEED. Slightly roughish annual with twining stems

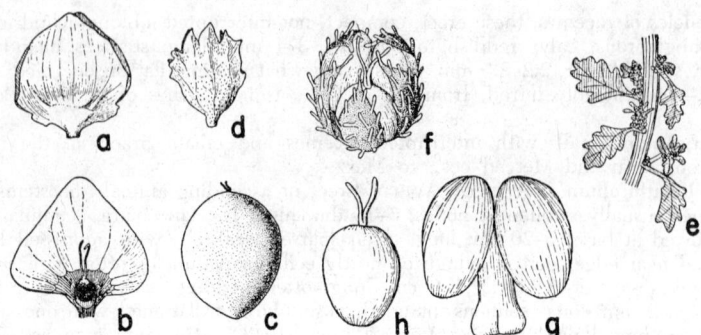

Fig. 38. CHENOPODIACEAE. *Atriplex confertifolia: a*, fr., × 2, pair of appressed bracts, 1 of which is removed in *b* to show pistil; *c*, older fr. *A. expansa: d*, bracts enclosing fr. dentate. *Chenopodium: e*, habit with fl.-clusters; *f*, fl., × 25, 5 sepals enclosing the pistil and single stamen; *g*, stamen; *h*, pistil both × 50.

2–10 dm. long; lvs. ovate-sagittate, 2–5 cm. long, on shorter petioles; ocreae short, with entire margins; fls. few in axillary clusters or in subracemose infl. at ends of branches; calyx green, 3.5–4 mm. long in fr.; aks. dull, minutely roughened, black, 3-angled, 3.5–4 mm. long; $2n = 20, 40$ (Löve & Löve, 1942).—Occasional weed, natur. from Eu. May–Sept.

34. **P. cuspidàtum** Sieb. & Zucc. JAPANESE KNOTWEED. Stout perennial 1–2 m. high, from large rhizomes; stems little-branched, glabrous; lvs. broadly ovate, 7–15 cm. long, abruptly acuminate at apex, rounded at base, on petioles ca. 2.5 cm. long; ocreae short, deciduous; fls. greenish-white, in axillary panicled drooping racemes; aks. 3-sided, shining, ca. 4 mm. long.—Sometimes escaping from cult.; native of Japan.

35. **P. sachalinénse** F. Schmidt ex Maxim. GIANT KNOTWEED. Larger than *P. cuspidatum;* lvs. 10–30 cm. long, cordate at base, acutish at apex; fls. greenish or with some red, in axillary panicles; aks. 3-angled, shining, brown, 2 mm. long.—Sometimes cult.; escaped along Klamath R.; native of Japan.

12. Fagopỳrum Mill. BUCKWHEAT

Annuals with alternate hastate or cordate lvs. and oblique entire ocreae. Fls. perfect, several in corymbiform cymes, the slender pedicels subtended by an ocreola. Calyx petaloid, equally 5-parted. Stamens 8. Styles 3; stigmas capitate. Aks. 3-sided. (Latin, *fagus*, the beech, and Greek, *pyros*, wheat, the seed like the beechnut.)

1. **F. sagittàtum** Gilib. [*F. esculentum* Moench.] Glabrous except at nodes, erect, 2–8 dm. tall; lf.-blades hastate, 3–6 cm. long; calyx white, with honey-bearing yellow glands between the stamens, aks. smooth, shining, 5 mm. long; $n = 8$ (Jaretzky, 1927).—Buckwheat of cult.; reported as escape at La Verne, Los Angeles Co.; native of Eurasia.

40. Chenopodiàceae. GOOSEFOOT FAMILY (Fig. 38)

Herbs or shrubs, often succulent or scurfy, often weedy and frequently of saline or subsaline places. Lvs. simple, without stipules, mostly alternate, sometimes reduced to scales. Fls. perfect, polygamous, or imperfect, small, greenish, usually in small cymose glomerules which may be variously arranged. Calyx free, imbricated in bud, persistent, mostly inclosing the fr., sometimes wanting in ♀ fls., green or membranous, of 5 or fewer sepals. Corolla none. Stamens as many as sepals or fewer, opposite them. Ovary 1-loculed, usually becoming a 1-seeded thin-walled utricle, rarely an ak. Styles 2–3. Seed with or without endosperm; embryo coiled into a ring or conduplicate or spiral. Ca. 100 genera and 1400 spp., world-wide; many are weeds; some, as beet and spinach, grown as vegetables.

A. Sepals strongly imbricate, scarcely united, strongly chartaceous; lvs. opposite, united at base
1. *Nitrophila*
AA. Sepals slightly or not imbricate, herbaceous when young; lvs. mostly alternate.
B. Lvs. foliaceous, flattened, not particularly fleshy or scaly.
C. Fls. perfect, sometimes also ♀, all with calyx and not inclosed in a pair of bracts.
D. Calyx not transversely winged in fr.
E. The calyx with 3–5 segms.; stamens 1–5.
F. Stamen 1; fls. axillary, solitary or clustered .... 2. *Aphanisma*
FF. Stamens 4–5; fls. in clusters.
G. Calyx indurate at base in age, with ovary partly sunk .... 3. *Beta*
GG. Calyx not indurate, with ovary superior.
H. Calyx-lobes not with hooked spines; plants glabrous, mealy, or glandular-pubescent .... 4. *Chenopodium*
HH. Calyx-lobes with stout hooked spines; plants pilose
11. *Bassia*
EE. The calyx of 1 segm.; stamen 1 .... 6. *Monolepis*
DD. Calyx transversely winged in fr. .... 5. *Cycloloma*
CC. Fls. imperfect, the ♀ enclosed in 2 accrescent bractlets.
D. Frs. not hairy; lvs. plane, not revolute.
E. Bracts compressed, the margins never wholly united; lvs. ± farinose
7. *Atriplex*
EE. Bractlets obcompressed, wholly united into a sac; lvs. usually glabrate
8. *Grayia*
DD. Frs. conspicuously hairy; lvs. linear, revolute .... 9. *Eurotia*
BB. Lvs. not or scarcely flattened but fleshy and sublinear, or scaly and spiny.
C. The lvs. scalelike; stems and branches fleshy.
D. Branches opposite; fl.-clusters opposite .... 13. *Salicornia*
DD. Branches alternate; fl.-clusters alternate .... 12. *Allenrolfea*
CC. The lvs. not scalelike, but fleshy and sublinear, or spiny-tipped.
D. Plants with perfect fls., or perfect and ♀.
E. Lvs. not tipped with spine or bristlelike hair.
F. Fruiting calyx transversely winged .... 10. *Kochia*
FF. Fruiting calyx not transversely winged .... 15. *Suaeda*
EE. Lvs. tipped with spine or bristly hair.
F. Lower lvs. mostly 3–5 cm. long; fruiting sepals winged on back.
16. *Salsola*
FF. Lower lvs. mostly 0.6–2 cm. long; fruiting sepals ending in wings
17. *Halogeton*
DD. Plants with ♂ fls. in spikes and ♀ solitary and axillary .... 14. *Sarcobatus*

1. Nitróphila Wats.

Low perennial herb. Lvs. opposite, linear to oblong, fleshy, amplexicaul, entire. Fls. perfect, axillary, 2-bracteolate, small, solitary or in 3's. Calyx chartaceous, 5-parted, the segms. concave, carinate, 1-nerved. Stamens 5, united at base into a short perigynous disk. Style filiform; stigmas 2, subulate. Utricle ovoid, beaked by the persistent style, included by the connivent calyx-segms. Seed vertical, lenticular; embryo annular. Several spp., N. and S. Am. (Greek, *nitron,* carbonate of soda, and *philos,* fond of, i.e., alkali-loving.)

Stems 1–3 dm. high; lvs. linear, 1–2 cm. long ....1. *N. occidentalis*
Stems to 1 dm. high; lvs. ovate, 0.2–0.3 cm. long .... 2. *N. mohavensis*

1. **N. occidentàlis** (Nutt.) Moq. [*Banalia o.* Moq. *Glaux acutifolia* Heller.] Plants glabrous, the stems oppositely much-branched, 1–3 dm. long, decumbent from deep rootstocks; lvs. linear or the lowest oblong, sessile, 1–2 cm. long, mucronate, not much reduced up the stem; bracts like lvs., but shorter, 2–3 times the calyx; fls. quite sessile; calyx-segms. ca. 2 mm. long, broadly oblong, pinkish when fresh, drying straw-colored; fr. brown; seed shining, black, 1 mm. broad.—Moist alkaline places, below 7000 ft.; several Plant Communities; Cent. V. to cismontane and s. Calif. and deserts; to Ore., Nev. May–Oct.

2. **N. mohavénsis** Munz & Roos. Stems 0.5–0.8 dm. high, erect from extensive heavy underground rootstocks; lvs. round-ovate, amplexicaul, concave above, 0.2–0.3 cm. long; calyx-segms. oblong-ovate, rose-colored when fresh; seed shining, black, 1.2 mm. long, 1 mm. broad.—Heavy alkaline mud, 2050 ft.; Alkali Sink; Amargosa Desert, Inyo Co. May–July.

2. Aphanísma Nutt. ex Moq. in DC.

Glabrous succulent annual with slender stems. Lvs. alternate, entire, mostly sessile or subsessile. Fls. perfect, green, solitary or in clusters, axillary, sessile. Calyx 3–5-cleft, the segms. concave, subequal, unchanged in fr. Stamen 1. Style 1; stigmas 3. Fr. a depressed-globose utricle, indurate, finely costate. Seed horizontal, lenticular, rugulose; embryo somewhat annular, surrounding the abundant endosperm. One sp. (Greek, *aphanes,* inconspicuous.)

1. **A. blitoìdes** Nutt. Branched from base, the stems decumbent or ascending, 1–5 dm. long; lvs. ovate, 1–2.5 cm. long, subsessile, or the lower obovate to spatulate, longer and petioled; fr. 1–2 mm. across.—Bluffs; Coastal Strand, Coastal Sage Scrub; along coast, Los Angeles Co. to n. L. Calif., San Clemente Id., Santa Barbara Id. April–May.

3. Bèta L. BEET

Ours glabrous biennial herb with large fleshy roots. Lvs. alternate, the basal rosulate, large, long-petioled, the upper reduced and subsessile. Fls. perfect, in glomerules of 3 or more, in panicled spikes. Calyx 5-parted, the segms. indurate and closed in fr. Stamens 5. Ovary sunk in the succulent base of the calyx; styles 2–3. Fr. ultimately opening by a lid. Seed horizontal, smooth, roundish; embryo annular. Five or 6 spp. of the Old World. (Perhaps Celtic, *bett,* red, because of the red roots.)

1. **B. vulgàris** L. GARDEN BEET. Stems 3–12 dm. tall, paniculately branched above; lower lf.-blades 1–2 dm. long, ovate-oblong; calyx segms. narrow-oblong, ca. 2 mm. long, carinate in fr.; fr. ca. 2.5 mm. long, somewhat wider; $2n = 18$ (Levan, 1942).— Escaping from gardens and sometimes natur. in low damp places; native of Eu. July–Oct.

4. Chenopòdium L. GOOSEFOOT. PIGWEED

Annual or perennial herbs, usually mealy (farinose) or glandular. Lvs. alternate, entire to lobed. Fls. perfect, rarely unisexual, ebracteate, in small clusters or glomerules arranged in panicles of spikes. Calyx 5(rarely 4)-parted or -lobed, the segms. persistent, flat or keeled, ± enveloping the fr. Stamens mostly 5. Styles 2, rarely 3. Fr. a utricle with membranous pericarp free from or adherent to the seed. Seed lenticular, horizontal or vertical, the embryo coiled partly or entirely around the albumen. A large genus, essentially cosmopolitan. (Greek, *chen,* goose, and *pous,* foot, referring to the shape of the lvs.)

(Standley, P. C. N. Am. Fl. 21: 9–21, 1916. Aellen, P., & T. Just. Key and synopsis of the American species of the genus Chenopodium L. Am. Midl. Nat. 30: 47–76, 1943. Wahl, H. A. A preliminary study of the genus Chenopodium *in* N. Am. Bartonia 27: 1–46, 1954.)

A. Plants ± glandular-pubescent or resinous-glandular, especially about the calyx, not mealy or farinose.
    B. Calyx 3–5-toothed, becoming saccate and reticulate; lvs. pinnatifid. . . . . . . 1. *C. multifidum*
    BB. Calyx 5(rarely 4)-parted, not changing in fr. or merely becoming fleshy; lvs. various.
        C. Fls. in glomerules, these in capitate clusters or small spikes.
            D. Glomerules small, headlike, all axillary; seed vertical . . . . 2. *C. pumilio*
            DD. Glomerules in short spikes, the upper in panicles; seed mostly horizontal
                                                   3. *C. ambrosioides*
        CC. Fls. solitary in small cymes, these spreading-recurved and in elongated panicles
                                                    4. *C. Botrys*
AA. Plants mostly mealy or farinose, not glandular-pubescent, but sometimes with nonglandular hairs.
    B. Seeds horizontal, or almost entirely so.
        C. Lvs. shining on upper surface, ± rhombic; only younger parts mealy; plant branched
            from base . . . . 21. *C. murale*
        CC. Lvs. dull on upper surface, various in shape.
            D. Lf.-blades cordate or subcordate at base, bright green, thin, almost glabrous,
                acuminate and with large acuminate teeth . . . . 22. *C. gigantospermum*
            DD. Lf.-blades rounded to truncate or attenuate at base.
                E. Pericarp closely adherent to seed and removable with difficulty; at least
                    lower lvs. conspicuously sinuate-dentate.
                    F. Pericarp and seed-surface essentially smooth.
                        G. Seeds chiefly 1.2–1.5 mm. broad; sepals largely covering fr.
                                                         18. *C. album*

GG. Seeds 0.9–1.2 mm. broad; sepals exposing mature fr.
19. *C. strictum*
FF. Pericarp and seed-surface foveolate-reticulate; seed 1–1.3 mm. broad
20. *C. Berlandieri*
EE. Pericarp free from or easily removed from the seed, or if adherent, the lower
lvs. linear or entire or 3-lobed.
F. Lf.-blades linear to narrow-lanceolate or narrow-oblong, short-petioled,
the blades mostly 1–3-nerved.
G. Lvs. linear, entire, 1-nerved, mostly 2–3 mm. wide
11. *C. leptophyllum*
GG. Lvs. narrow-lanceolate to oblong, the lower 4–18 mm. wide.
H. Pericarp markedly separable; seeds 1–1.2 mm. broad
12. *C. dessicatum*
HH. Pericarp attached, minutely granular-roughened; seeds 1.2–
1.5 mm. wide .... 14. *C. incognitum*
FF. Lf.-blades lance-ovate to ovate or broader, long-petioled, pinnately
veined.
G. Main lf.-blades definitely longer than broad.
H. Plants simple or few-branched; fls. in paniculate spikes
15. *C. atrovirens*
HH. Plants much-branched; fls. in diffuse cymose panicles
16. *C. nevadense*
GG. Main lf.-blades scarcely if at all longer than wide.
H. Plant not ill-scented; calyx-lobes keeled on back; pericarp
free from the shining seed .... 13. *C. Fremontii*
HH. Plant ill-scented; calyx-lobes rounded on back; pericarp ad-
herent to the dull seed .... 17. *C. Vulvaria*
BB. Seeds vertical for the most part.
C. Lvs. densely white farinose beneath, at least when young.
D. The lvs. rhombic to deltoid-rhombic, 1–4 cm. wide; calyx almost completely con-
cealing the utricle .... 5. *C. macrospermum*
DD. The lvs. broadly lanceolate to oblong or subovate, 0.3–1.5 cm. wide; calyx con-
cealing only a small part of the utricle .... 7. *C. glaucum*
CC. Lvs. sparsely or not at all farinose beneath.
D. Plants annual, glabrous; stems up to 6 dm. long.
E. Fls. in small clusters; calyx not both fleshy and bright red in fr.
F. Lf.-blades mostly 3–8 cm. long, sinuate-dentate; fls. in leafy spikes;
calyx fleshy .... 8. *C. rubrum*
FF. Lf.-blades 1–3 cm. long, entire to few-toothed; fls. in axillary glomerules;
calyx not fleshy .... 9. *C. humile*
EE. Fls. in large spicate glomerules; calyx fleshy and bright red in fr.
6. *C. capitatum*
DD. Plants perennial, mealy on upper parts; stems 3–8 dm. long or more
10. *C. californicum*

1. **C. multífidum** L. [*Roubieva m.* Moq.] Glandular annual herb with taproot; strong-
scented; stems branched, prostrate, 2–7 dm. long, villous in younger parts; lvs. oblong
in outline, alternate, pinnatifid, 1–3 cm. long, the lobes mucronulate, narrow; fls. sessile,
1–3 in the axils; calyx urceolate, shallowly 5-dentate, puberulent, ca. 2 mm. long, becoming obovoid, coriaceous and reticulate in age; stamens 5, included; styles 3,
exserted; fr. membranaceous, compressed, gland-dotted, enclosed in perianth.—Uncommon wayside weed, below 4600 ft.; San Francisco Bay region to Cent. V., Sierran
foothills to s. Calif.; natur. from S. Am. June–Nov.

2. **C. pumílio** R. Br. [*C. carinatum* auth., not R. Br.] Annual, branched from base,
glandular-villous, depressed or ascending, the branches 2–4 dm. long; lvs. oblong or
oblong-ovate, the blades 1–2 cm. long, or uppermost reduced, coarsely sinuate-pinnatifid
with obtuse lobes; petioles slender, from shorter than to exceeding blades; fls. in short
axillary clusters; stamen usually 1; calyx 0.6 mm. long, the lobes carinate, only partly
enclosing the fr.; pericarp thin; seed vertical, 0.5 mm. broad.—Occasional weed, mostly
below 5000 ft., through most of cismontane Calif.; natur. from Australia. June–Sept.

3. **C. ambrosioìdes** L. Mexican-Tea. Annual or perennial, erect or ascending, coarse,
strong-scented, the stems 4–10 dm. long, simple or branched, smoothish; lvs. short-
petioled, oblong or lanceolate, subentire to repand-toothed, 2–10 cm. long, gradually
reduced upward; infl. a slender pyramidal panicle of densely-fld. spikes, elongate, leafy
or intermixed with lvs.; calyx ca. 1 mm. long, gland-dotted, enclosing the fr.; pericarp
very thin, deciduous, gland-dotted; seed horizontal or vertical, ca. 0.7 mm. broad;
$2n = 16$, 32, 48? (Kawatani & Ohno, 1950).—Weed in waste, especially damp, places,
particularly in s. Calif., but widely spread on Pacific Coast and to New England; natur.
from trop. Am. June–Dec.

Var. anthelmínticum (L.) Gray. [*C. a.* L.] WORMSEED. Lvs., especially the lower, more strongly toothed, even laciniate-pinnatifid; spikes almost or quite leafless.—With the sp., but more common in the Cent. V.; natur. from trop. Am.

Var. vàgans (Standl.) J. T. Howell. [*C. v.* Standl. *C. chilense* Schrad., not Pers.] Stems and branches loosely white-villous; spikes quite leafless.—With the sp., but most abundant in Sacramento V. and n. Calif.; natur. from S. Am.

4. **C. Bòtrys** L. JERUSALEM-OAK. Erect densely glandular-villous aromatic annual, the stems 2–6 dm. high, viscid; lvs. sinuate-pinnatifid, 1–4 cm. long, oblong to oval, with obtuse angled lobes; petioles short; infl. a virgate elongate panicle of loosely spreading leafless cymes; fls. subsessile, pubescent, ca. 1 mm. long, the calyx-segms. oblong or ovate; pericarp adherent, thin; seed vertical or horizontal, ca. 0.6 mm. broad, subglobose, dull, dark; $2n = 16$? (Kawatani & Ohno, 1950).—Occasional weed, especially in sandy places, below 7000 ft.; to Wash., Atlantic Coast; natur. from Eurasia. June–Oct.

5. **C. macrospérmum** Hook. f. var. **farinòsum** (Wats.) J. T. Howell. [*C. murale* var. *f.* Wats. *C. f.* Standl.] Annual, branched from base, the stems stout, ascending to erect, glabrous, 1–5 dm. long; lvs. rhombic to deltoid-rhombic, 1.5–5 cm. long, obtuse, glabrous above, farinose beneath, sinuate-dentate; petioles shorter than to ca. as long as blades; glomerules in dense spikes, sessile in axils of reduced upper lvs.; calyx barely 1 mm. long, the segms. rounded; seed vertical, ca. 1 mm. long, dark.—Moist places; Coastal Strand; Orange Co. to Humboldt Co. July–Oct.

6. **C. capitàtum** (L.) Asch. [*Blitum C.* L.] STRAWBERRY-BLITE. Glabrous erect simple or branched annual 2–6 dm. high; lower lvs. triangular, 4–10 cm. long, sinuate, somewhat hastate, bright green; petioles slender, usually as long as blades; upper lvs. reduced, narrower; fls. in spherical glomerules, these distinct to confluent and in almost leafless terminal spikes; calyx-lobes fleshy, oblong, reddish; seed ovoid to oblong, erect, 0.8 mm. long.—An infrequent weed in damp places in mts., below 10,000 ft., n. Calif.; from Eurasia. June–Aug. A form with leafy spikes is *C. foliòsum* (Moench) Asch. [*Blitum virgatum* L., not *C. v.* Thunb.]

7. **C. glaùcum** L. ssp. **salìnum** (Standl.) Aellen. [*C. s.* Standl.] Freely branched prostrate to ascending annual, the branches subglabrous, 1–3 dm. long; lvs. many, well distributed, broadly lanceolate to oblong or subovate, the blades 1.5–3 cm. long, acute, sinuately toothed, somewhat hastate, pale green and somewhat farinose above, densely so beneath; petioles short; fls. in small axillary spikes shorter than the lvs., the infl. villous on the branches; calyx green, the lobes obovate, imperfectly enclosing the fr.; pericarp green, free; seed vertical or horizontal, ca. 0.9 mm. broad, dark red-brown, shining, finely tuberculate.—Occasional, alkaline flats and shores as at Elsinore, Baldwin Lake, Alturas, etc.; to Alta., New Mex. July–Oct.

8. **C. rùbrum** L. Erect glabrous annual 2–6 dm. high, simple or branched with stout reddish stems; main lvs. deltoid- or rhombic-ovate, 3–8 cm. long, sinuate-dentate, cuneate at base; petioles ca. as long; fls. in short axillary leafy spikes, crowded; calyx 3–5-lobed, fleshy, not keeled; pericarp green; seed usually vertical, 0.6–0.8 mm. broad, with rounded margin.—Occasional in low rather saline places, widely scattered in the state; to the Atlantic Coast; native of Eu. July–Oct. This should probably be called *C. chenopodioides* (L.) Aellen.

9. **C. hùmile** Hook. [*C. rubrum* var. *h.* Wats.] Low annual, widely branched at base, the branches 3–20 cm. long, àscending, slender, whitish; lvs. almost round to rhombic-ovate, 0.8–3 cm. long, entire or shallowly sinuate, green, glabrous; petioles shorter than blades; fls. green, in dense axillary glomerules, crowded; calyx 3–5-lobed, shorter than fr.; pericarp green; seed vertical, 0.8–1.0 mm. broad, strongly compressed.—Moist alkaline places, mostly above 5000 ft., San Bernardino Mts., n. Calif.; to B.C. and Atlantic Coast. Aug.–Oct.

10. **C. califórnicum** (Wats.) Wats. [*Blitum c.* Wats.] Perennial with stout fleshy root and several decumbent or ascending stems 3–8 dm. long, sparsely farinose on younger parts, stout, not much branched; lvs. deltoid, 3–10 cm. long, truncate or cordate at base, sharply and unequally sinuate-dentate; petioles slender, the lower as long as lvs.; fl.-glomerules small, in long dense terminal spikes; calyx cleft to ca. the middle, green with broad lobes, shorter than fr.; pericarp adherent; seed vertical, compressed-globose, ca. 2 mm. broad.—Common on dryish slopes and plains, below 5000 ft.; many Plant Communities; much of cismontane Calif., to edge of deserts; L. Calif. March–June.

11. **C. leptophýllum** Nutt. [*C. inamoenum* Standl.] Densely farinose annual, erect, simple or branched below, usually much-branched above, with erect branches; lvs. short-petioled, linear, entire, 1-nerved, 1–4 cm. long, usually 2–3 mm. wide, roundish at apex; fls. in dense glomerules, these in interrupted or dense paniculate spikes; calyx completely enclosing the fr., the lobes keeled; pericarp free; seed horizontal, dark, smooth, shining, 1 mm. broad.—Dryish alkaline places, 5000–8000 ft.; Yellow Pine F., N. Juniper Wd., Pinyon-Juniper Wd., Sagebrush Scrub; San Bernardino Mts. and e. of crest of Sierra Nevada; to Ore., Wyo., Tex. July–Sept.

12. **C. desiccatum** A. Nels. [*C. pratericola* ssp. *d.* Aellen.] Much like *C. leptophyllum*, low, diffusely branched from base, 1–3 dm. high; lvs. oblong or oval, entire, 0.8–2 cm. long, greenish above, farinose beneath; calyx-lobes enclosing the fr., obtuse, white-margined.—Dry places, 4000–10,000 ft.; largely Yellow Pine F. to Lodgepole F., Pinyon-Juniper Wd.; San Bernardino and Panamint mts., e. of Sierra Nevada; to Ida., S. Dak., New Mex. July–Sept.

Var. **leptophylloìdes** (J. Murr) H. A. Wahl. [*C. petiolare* var. *l.* J. Murr. *C. pratericola* Rydb.] Erect, 2–8 dm. high; lvs. narrow-lanceolate to oblong, entire or slightly 3-lobed or hastate, the lower 3-nerved, 4–18 mm. wide.—Occasional, mostly in dry places, below 9000 ft.; largely Yellow Pine F., Red Fir F., or lower cismontane valleys; San Jacinto Mts., Sierra Nevada; to Sask., e. U.S. June–Sept.

13. **C. Fremóntii** Wats. Erect annual 3–10 dm. high, slender, usually much-branched throughout with ascending branches, sparsely farinose or glabrous; lf.-blades broadly triangular-hastate, 1–3 cm. long, almost equally wide, bright green above, whitish beneath, obtuse; petioles slender, ½ or more the length of blades; fl.-glomerules small, in rather slender paniculate spikes; calyx ca. 1 mm. broad, deeply cleft, rather sparsely farinose, completely enclosing the fr.; pericarp free; seed horizontal, 1 mm. broad, dark, smoothish, with obtuse margin.—Rather frequent in dry places, mostly at 5000–8500 ft.; Pinyon-Juniper Wd., Yellow Pine F.; Santa Rosa Mts. (Riverside Co.) and Cuyamaca Mts. to Mt. Pinos and Clark Mts. (e. San Bernardino Co.), n. to Mono Co. and B.C., then to N. Dak., and Mex. June–Oct.

Var. **incànum** Wats. [*C. i.* Heller.] Plant 1–3 dm. tall, diffusely branched from base.— At 2500–5000 ft.; Joshua Tree Wd., Pinyon-Juniper Wd.; Mojave Desert to Rocky Mts. April–Aug.

14. **C. incognitum** H. A. Wahl. Plants erect, 3–12 dm. tall, branched from base; lvs. thin, ovate to deltoid-ovate, 1.5–3.5 cm. long, entire or with small basal lobes, farinose beneath; fls. crowded in terminal and axillary spikes; sepals 5, farinose, narrowly keeled, ca. half covering the fr. at maturity; pericarp finely rugulose, attached to the seed; seed black, flattened, 1.2–1.5 mm. in diam., with rounded margin.—Dry gravelly slopes and flats, 6000–8000 ft.; Yellow Pine F., Red Fir F.; San Gabriel Mts. through Sierra Nevada to Siskiyou and Tehama cos.; to Ore., Wyo., New Mex. July–Aug.

15. **C. atròvirens** Rydb. Erect annual 1–5 dm. high, slender, almost simple or few-branched, green, subglabrous; lvs. ovate to triangular-oblong, mostly entire, 1.5–3 cm. long, obtuse; petioles slender, somewhat shorter; fl.-glomerules farinose, in ± paniculate spikes; calyx-lobes sharply carinate, obovate, enclosing the fr.; pericarp free; seed horizontal, 1 mm. broad, dark, shining.—Occasional, dry places, 4000–11,000 ft.; Pinyon-Juniper Wd. to Red Fir F.; Cushenberry Springs, w. Mojave Desert, e. slope of Sierra Nevada, to Siskiyou Mts. and Glenn Co.; e. Ore. to Rocky Mts. July–Sept.

16. **C. nevadénse** Standl. Erect much-branched annual 2–3 dm. high, obscurely farinose; lvs. rhombic-ovate to ovate-oblong, 0.8–2 cm. long, obtuse or rounded, sub-entire; petioles slender, usually shorter than blades; fl.-glomerules very small, in diffuse cymose panicles; calyx farinose, enclosing the fr., the lobes acutish, slightly keeled, subovate; pericarp adherent; seed ca. 0.5 mm. broad, dark brown.—Collected in Owens V. in 1914, by *F. W. Peirson*; Deep Springs V., *Raven 7049*; w. Nev. June–Aug.

17. **C. Vulvària** L. Low ill-scented annual, branched near base, farinose, the branches 1–4 dm. long; lvs. rhombic- to rounded-ovate, 1–3 cm. long, almost as wide, rounded to acutish at apex; petioles mostly shorter; fl.-glomerules usually in dense paniculate spikes, the infl. usually leafy, little branched; calyx farinose, the lobes rounded on back, enclosing the fr.; pericarp adherent; seed horizontal, depressed-globose, dull, black, 1 mm. broad, with rounded margin.—Occasional in waste places, as in Alameda and Siskiyou cos.; to Atlantic Coast; native of Eu. June–Oct.

18. **C. álbum** L. Pigweed. Lamb's-Quarters. Erect annual, pale green, red-veined, 2–20 dm. high, branching, farinose; lvs. glaucous, farinose beneath, rhombic-ovate, or the upper lanceolate, ± sinuate-dentate or -serrate, 1–5 cm. long; petioles slender, ca. half as long; fl.-glomerules thick, in rather dense heavy spikes in upper axils, forming panicles; calyx farinose, enclosing the fr., carinate; pericarp adherent; seed horizontal, black, nearly smooth, shining, 1.3 mm. broad; $2n = 36$ (Cooper, 1935), 54 (Kjellmark, 1934).—Common weed in waste and fallow places, below 6000 ft., widely distributed over N. Am.; natur. from Eu. June–Oct. Very variable and with many named sspp. and vars.

19. **C. strictum** Roth var. **glaucophýllum** (Aellen) H. A. Wahl. [*C. g.* Aellen.] Near *C. album*, the lower lvs. low-serrate, the median oblong, entire; fls. in strict axillary spikes or terminal loose panicles; sepals exposing fr. at maturity; seeds 0.9–1.2 mm. broad, smooth.—Occasional in low places, cismontane Calif.; to Atlantic Coast. Aug.–Oct.

20. **C. Berlandièri** Moq. var. **sinuàtum** (J. Murr) H. A. Wahl. [*C. petiolare* var. *s.* J. Murr.] Differing from *C. album* by its stronger unpleasant odor, slender leafy spikes, dentate but not lobed lvs., more sharply keeled calyx-lobes and mostly puncticulate seeds 1–1.3 mm. in diam.—Reported as occurring in Calif.; sw. U.S., Mex.

21. **C. muràle** L. Rather stout annual, glabrous or sparsely mealy, ill-scented, branched from base, the branches ascending, 2–5 dm. long; lvs. dark green, rhombic-ovate, 2–6 cm. long, irregularly sinuate-dentate; petioles equal to or shorter than blades; fls. in small glomerules in lax or dense axillary and terminal short panicles; calyx 1.5 mm. broad, deeply cleft, the lobes oblong, obscurely keeled, incompletely enclosing the fr.; pericarp green, adherent; seed horizontal, ca. 1.5 mm. broad, puncticulate, with sharp edge; $2n = 18$ (Winge, 1917).—Common weed about orchards and gardens; widespread in N. Am.; natur. from Eu. Most of year, but especially in spring.

22. **C. gigantospérmum** Aellen. [*C. hybridum* L. var. *g.* Rouleau.] Glabrous erect bright green annual 2–14 dm. high, widely branched, the branches slender; lvs. thin, bright green, broadly to triangular-ovate, acuminate, 3–15 cm. long, with 1–few triangular acuminate lobes; petioles slender, ca. half as long as lf.-blades; fl.-glomerules small, in loose terminal leafless panicles; calyx with thin segms., farinose, rounded on back, imperfectly enclosing the fr.; pericarp thin, adherent to seed; seed horizontal, lenticular, black, shiny, ridged, 1.5–2 mm. broad; $2n = 36$ (Löve, 1954).—Occasional in moist somewhat-shaded places, as at Westgaard Summit (Inyo Co.), Goose Lake (Modoc Co.); transcontinental. July–Oct.

5. Cyclolòma Moq. Winged Pigweed

Erect or spreading branched annual. Lvs. alternate, petioled, oblong in outline, early deciduous, coarsely sinuate-dentate. Plants polygamo-monoecious, the fls. small, bracteate, solitary or in glomerules. Calyx 5-lobed, the lobes inflexed, strongly carinate, the basal tube developing in age a broad membranaceous wing. Stamens 5. Stigmas 3. Fr. depressed-globose with membranaceous pericarp. Seed horizontal, smooth; embryo annular. One sp. (Greek, *cyclos*, a circle, and *loma*, a border, referring to the calyx-wing.)

1. **C. atriplicifòlium** (Spreng.) Coult. Diffusely branched, 1–5 dm. tall, villous-tomentose on younger parts; lvs. 2–6(–8) cm. long; infl. a broad panicle; calyx 3–4 mm. broad, villous, reddish in age; pericarp tomentulose; seed black, 1.5 mm. in diam.—Occasional weed in fields and groves, s. Calif.; to Man., Ind., Tex. May–Sept.

6. Monólepis Schrad.

Low branched annual herbs. Lvs. alternate, sessile or petioled, entire or hastate. Plants polygamo-dioecious, the fls. sessile, ebracteate, densely clustered in axillary glomerules or sometimes solitary. Calyx of 1 persistent sepal, not changed in fr. Stamen 1, or 0. Styles 2, slender. Fr. ovoid, with thin pericarp, which may be slightly adherent to the seed. Seed erect, compressed, the embryo annular. Three spp. (Greek, *monos*, one, and *lepis*, scale, because of the single sepal.)

(Standley, P. C. N. Am. Flora 21: 6–7, 1916.)
Stems not dichotomously branched, ± fleshy.
    Lvs. 10–50 mm. long, frequently hastate; pericarp pitted .... 1. *M. Nuttalliana*
    Lvs. 5–15 mm. long, entire; pericarp papillose .... 2. *M. spathulata*
Stems dichotomously branched; lvs. oblong to obovate, 2–8 mm. long .... 3. *M. pusilla*

1. **M. Nuttalliàna** (Schult.) Greene. [*Blitum N.* Schult. *B. chenopodioides* Nutt., not Lam.] Stems several from base, ascending, stout, succulent, mealy when young, 1–2(–3) dm. long; lvs. triangular-lanceolate or narrower, usually hastately lobed at base, the blades 1–4 cm. long, short-petioled or the lower with long petioles; fl.-clusters dense, sessile, often reddish; sepal spatulate or obovate, acutish, ca. 1 mm. long; pericarp 1 mm. broad, minutely pitted; seed 1 mm. broad, dark, with acute margin.—Rather common, dry or moist often saline places and on burns, mostly below 5000 (9000) ft.; many Plant Communities; cismontane to desert valleys; Channel Ids.; to Alaska, Mo., Tex., Son., S. Am.; Asia. April–Sept. A form with narrow entire lvs. has been collected along the lower Colo. R. and in Ariz.

    2. **M. spathulàta** Gray. Branched from base, the stems 3–15 cm. long, decumbent or ascending; lvs. narrowly spatulate to oblanceolate, 5–15 mm. long, fleshy, entire; fl.-clusters many-fld., sessile; sepal spathulate, obtuse, ca. 0.5 mm. long; pericarp papillose, free from the seed, ca. 0.5 mm. broad; seed brown, shining, ca. 0.4 mm. in diam.— Rare, moist subalkaline places, mostly 5000–8000 ft.; Yellow Pine F., Red Fir F.; San Bernardino Mts., along Sierra Nevada to Ida., e. Ore. June–Sept.

    3. **M. pusílla** Torr. ex Wats. Dichotomously much-branched, erect, with slender branches spreading and 0.5–2 dm. high; lvs. oblong, 4–12 mm. long, entire, short-petioled; fls. 1–5 in sessile clusters; sepal obtuse, spatulate; pericarp tuberculate, adherent to seed; seed dull, ca. 0.5 mm. in diam.—Alkaline places, ca. 5000–7000 ft.; Sagebrush Scrub; Mono and Lassen cos.; to Wash., Rocky Mts. May–Aug.

7. **Àtriplex** L. SALTBUSH

Herbs or shrubs, usually grayish or whitish, scurfy with inflated hairs. Lvs. alternate or opposite. Plants monoecious or dioecious, the fls. small, green, in axillary clusters or glomerules or in panicled spikes. Staminate fls. with 3–5-parted calyx, bractless; stamens 3–5. Pistillate fls. consisting of a naked pistil enclosed between a pair of appressed foliaceous bracts which enlarge in fr. and may be partly united, ± expanded, and variously thickened and appendaged. Styles 2. Utricle with a usually free membranous pericarp; seed flattened, erect or inverted, rarely horizontal; embryo annular, surrounding the scanty endosperm. Over 100 spp., the genus essentially cosmopolitan. (The ancient Latin name.)

(Hall, H. M., & F. E. Clements. Carnegie Inst. of Wash. Pub. 326: 235–346, 1923. Standley, P. C. N. Am. Fl. 21: 33–72, 1916.)
A. Plants herbaceous, slightly if at all woody at base, mostly monoecious.
    B. Lvs. green or greenish on both surfaces, sparsely mealy and sometimes grayish when young.
        C. Fruiting bracts roundish, 8–18 mm. broad; lower lf.-blades mostly 6–18 cm. long
                1. *A. hortensis*
        CC. Fruiting bracts not round, less than 5 mm. wide; lf.-blades mostly smaller.
           D. Fruiting bracts united only near base; ♂ fls. mixed with ♀ or in very short spikes; lvs. mostly hastate.
              E. Bracts hastate to rounded or cuneate at base, the tips close together; radicle of embryo pointing downward .... 2. *A. patula*
              EE. Bracts mostly with rounded earlike lobes near base, the tips far apart; radicle pointing upward .... 5. *A. Phyllostegia*
          DD. Fruiting bracts united to above the middle; ♂ glomerules in elongate terminal spikes or panicles; lvs. not hastate .... 9. *A. Serenana*
    BB. Lvs. gray or whitish with a fine scurf, at least on the lower surface.
        C. Bracts thickened, either fleshy or spongy. Introd. perennials.
           D. Fruiting bracts turbinate to globoid, spongy, dry, fibrous, 6–12 mm. long
                4. *A. Lindleyi*
          DD. Fruiting bracts ovate, strongly nerved, fleshy, 3.5–5 mm. long .. 19. *A. semibaccata*
        CC. Bracts not thickened, neither fleshy nor spongy; mostly native annuals and perennials.
           D. Fruiting bracts broadest below their middle.
              E. Staminate glomerules in long naked terminal spikes; perennial.
                F. Lvs. mostly opposite; plant forming tangled mats. S. Coast
                      22. *A. Watsonii*
              FF. Lvs. alternate; plant erect or decumbent. Ne. Calif. .... 31. *A. Nuttallii*

EE. Staminate glomerules in upper lf.-axils or in spikes not more than 1 cm. long; annual (except *californica*).
    F. Lvs. coarsely toothed; bracts hard and indurated. Introd. .... 3. *A. rosea*
    FF. Lvs. entire; bracts not becoming hard. Native.
        G. Plant prostrate with perennial fusiform root; bracts distinct
                                                              18. *A. californica*
        GG. Plants ascending to erect, annual; bracts united to middle or farther.
            H. Stems simple or with a few mostly simple branches.
                I. Lvs. cordate at base.
                    J. The lvs. 5–15 mm. long; fruiting bracts 4–5 mm. long .... 12. *A. cordulata*
                    JJ. The lvs. 2–7 mm. long; bracts 2–3.5 mm. long
                                                              13. *A. vallicola*
                II. Lvs. rounded at base; bracts 2.5–3.5 mm. long
                                                              14. *A. tularensis*
            HH. Stems intricately branched throughout.
                I. Bracts 2.5–3 mm. long; branches scurfy-villous. Glenn Co. to s. Calif. .... 15. *A. Parishii*
                II. Bracts 1–2 mm. long; branches merely scurfy. Ne. Calif.
                                                              16. *A. pusilla*
DD. Fruiting bracts broadest at or above their middle.
    E. The bracts truncate at summit and cuneate at base .... 6. *A. truncata*
    EE. The bracts not truncate at summit or cuneate at base.
        F. Bracts becoming hard and almost bonelike. Introd. weeds .. 3. *A. rosea*
        FF. Bracts not becoming particularly hard. Natives.
            G. Staminate and ♀ fls. mostly mixed in the clusters.
                H. Bracts irregularly toothed at summit.
                    I. Lvs. narrowly ovate to elliptic, narrowed at base, 3–10 mm. wide; bracts mostly 3–4 mm. long ...... 7. *A. coronata*
                    II. Lvs. cordate- to lance-ovate, broad and often subhastate at base, mostly 10–40 mm. wide; bracts 4–8 mm. long
                                                              8. *A. argentea*
                HH. Bracts evenly toothed all around .... 17. *A. elegans*
            GG. Staminate and ♀ fls. mostly in separate clusters.
                H. Bracts not compressed, elliptic-globose, 5–7 mm. long; seed 2.5–3 mm. long. Along immediate coast .. 20. *A. leucophylla*
                HH. Bracts usually ± compressed, 1–4 mm. long; seed 1.5 or less mm. long.
                    I. Fruiting bracts orbicular, toothed all around
                                                              17. *A. elegans*
                    II. Fruiting bracts obovate to round-cuneate, entire near base.
                        J. Plants annual.
                            K. Lvs. mostly dentate, broader in their lower half; bracts dentate above middle .... 9. *A. Serenana*
                            KK. Lvs. entire, broader in their upper half; bracts mostly entire .... 10. *A. pacifica*
                        JJ. Plants perennial.
                            K. Stems 3–10 dm. long; fruiting bracts 2–3 mm. long .... 11. *A. Coulteri*
                            KK. Stems 0.5–3 dm. long; fruiting bracts 3–4 mm. long .... 21. *A. fruticulosa*
AA. Plants shrubby, definitely woody, mostly dioecious.
    B. Fruiting bracts with conspicuous extra wings or crests arising on middle of face
                                                              32. *A. canescens*
    BB. Fruiting bracts not with extra lateral wings.
        C. Plants not spiny.
            D. Lf.-margin entire, flat or crisped but not toothed.
                E. Bracts 2–4 mm. long. Desert plants.
                    F. Lf.-blades 3–18 mm. long, oblong to spatulate .... 24. *A. polycarpa*
                    FF. Lf.-blades 15–50 mm. long, oblong to ovate-deltoid .. 26. *A. lentiformis*
                EE. Bracts 4–10 mm. long. Cismontane.
                    F. Lvs. oblong-ovate; bracts 4–7 mm. long. San Francisco to Orange Co.
                                                              26. *A. lentiformis Breweri*
                    FF. Lvs. round-spatulate; bracts 9–10 mm. long. Playa del Rey
                                                              27. *A. Nummularia*
            DD. Lf.-margin dentate.
                E. Bracts entire, reticulate-veined; lf.-blades 1–3 cm. long. Deserts
                                                              23. *A. hymenelytra*
                EE. Bracts sinuate-dentate, not reticulate; lf.-blades usually 3–5 cm. long. Coast at Playa del Rey .... 27. *A. Nummularia*
        CC. Plants spiny, the spines consisting of sharp-pointed twigs from which the lvs. and bracts have fallen.
            D. Bracts 2–4 mm. long.
                E. The bracts orbicular; lvs. truncate or cuneate at base, the blades 1.5–4 cm. long.

F. Twigs sharply angled. Mostly from above 2000 ft. ..... 25. *A. Torreyi*
FF. Twigs terete, not sharply angled. Mostly from below 2000 ft.
26. *A. lentiformis*
EE. The bracts with broad summit; lvs. cordate at base, the blades 0.5–1.5 cm.
long .... 28. *A. Parryi*
DD. Bracts 6–15 mm. long.
E. Body of bract small, not contracted beneath the free terminal wings; lvs. all
entire .... 29. *A. confertifolia*
EE. Body of bract large, thick, contracted to a neck beneath the free terminal
wings; lvs. sometimes somewhat hastate at base .... 30. *A. spinifera*

1. **A. horténsis** L. GARDEN ORACHE. Stout erect annual 5–15 dm. high, glabrous, green
to yellowish or reddish; lower lvs. somewhat triangular to ovate and subcordate, 5–18
cm. long, entire or denticulate, opposite; upper alternate, lance-oblong, somewhat
smaller; plants monoecious, the fls. in a panicle of terminal and axillary spikes; ♀ fls.
of 2 kinds, some with 3–5-lobed calyx and no bracts, but most with no calyx and enclosed by 2 bracts; fruiting bracts broadly oval, 8–18 mm. long, united only at base,
rounded to acute at apex; seeds 2–4 mm. broad, black; $2n = 18$ (LaCour, 1931).—
Drained marshland; Coastal Salt Marsh; Bay shore of San Francisco Peninsula to Santa
Clara Co.; Ore. and e. U.S.; native of Asia. July–Sept.
2. **A. pátula** L. Annual, simple to much-branched, glabrous or slightly mealy, the
lower branches widely divergent, 3–10 dm. long; lvs. green, lanceolate to ± rhombic,
petioled, cuneate at base, 3–8 cm. long, the lower sometimes hastate, entire to sinuous-
dentate; infl. interrupted-spiciform, only the lower glomerules with leafy bracts; ♀ fls.
without calyx, all enclosed in pair of bracts; fruiting bracts rhombic-oval, green,
herbaceous, 2–5 mm. long, often slightly hastate, denticulate, mostly smooth or occasionally tuberculate on back, united only at cuneate or slightly rounded base; seed 1–2
mm. broad; $2n = 18$ (Wulff, 1936).—Uncommon; Coastal Salt Marsh; cent. Calif. to
B.C.; Atlantic Coast; Eurasia. July–Nov.

KEY TO SUBSPECIES

Fruiting bracts with narrow toothed margins, the teeth sometimes sparse and small.
Lvs. lanceolate to oblong or linear, not hastate; bracts cuneate or narrowly rounded at base
A. patula
Lvs. partly triangular-hastate or rhomboidal, with basal angles or lobes; bracts truncate or broadly
rounded at base .... ssp. *hastata*
Fruiting bracts with wide entire margins.
Bracts 3 mm. long; lvs. rhombic-ovate, coarsely toothed .... ssp. *spicata*
Bracts 4–12 mm. long; lvs. mostly lance-oblong or almost linear .... ssp. *obtusa*

Ssp. **hastàta** (L.) Hall & Clem. [*A. h.* L.] Principal lvs. broadly triangular- to oval-
hastate, often dentate; bracts truncate or broadly rounded at base.—Common in moist
saline places along the coast and in interior valleys; Coastal Salt Marsh, Alkali Sink; to
B.C., Atlantic Coast; Eurasia. June–Nov.
Ssp. **spicàta** (Wats.) Hall & Clem. [*A. s.* Wats., not Stokes. *A. joaquiniana* A. Nels.]
Erect, 3–10 dm. high; lvs. deltoid to rhombic-ovate, mostly sinuate-dentate; bracts ovate-
oblong to round-deltoid, 3 mm. long, entire.—Saline places; Alkali Sink; Cent. V. and
adjacent Coast Ranges. April–Sept.
Ssp. **obtùsa** (Cham.) Hall & Clem. [*A. angustifolia* var. *o.* Cham. *A. Gmelinii* C. A.
Mey.] Usually erect, 1–5 dm. high; lvs. lance-oblong to oblong, the upper sublinear;
bracts 4–12 mm. long, ovate-rhombic to ovate-oblong, entire or rarely subhastate.—
Along the shore; Coastal Salt Marsh, Coastal Strand; San Francisco Bay to Alaska. Aug.–
Nov.
3. **A. ròsea** L. REDSCALE. Erect annual 1–10 dm. high, the branches arched-ascending,
mealy or glabrate; lvs. mostly alternate, short-petioled to subsessile, ovate or rhombic-
ovate, sinuate-dentate above the cuneate base, 2–5 cm. long, the upper reduced, entire;
fls. in axillary glomerules or interrupted terminal spikes; ♂ calyx 5-cleft; fruiting bracts
rhombic or ovate, 4–6 mm. long, united to ca. the middle, firm, often warty; seed 1.5–2
mm. broad.—Alkaline places, fields and waysides, below 7000 ft.; many Plant Communities; widespread in Calif.; to L. Calif., Wash., Atlantic Coast; natur. from Eurasia.
July–Oct.
4. **A. Líndleyi** Moq. [*A. halimoides* Lindl., not Raf.] Erect or procumbent suffrutescent perennial 1.5–3 dm. high; lvs. alternate, crowded, oblanceolate or the lower

rhombic, acute, entire to repand-denticulate, scurfy, 7–20 mm. long; plants monoecious, the ♂ fls. in short axillary spikes, the ♀ solitary or clustered in lower axils; fruiting bracts spongy, broadly turbinate or hemispheric, flattened at summit, 6–12 mm. long, united except at tip.—Sparingly escaped from cult., as in San Diego Co.; native of Australia.

5. **A. Phyllostègia** (Torr.) Wats. [*Obione P.* Torr. ex Wats. *Endolepis Covillei* Standl. *A. C.* Macbr.] ARROWSCALE. Much-branched erect annual 0.5–4 dm. high, rounded, bushy, sparsely mealy; lvs. mostly alternate, the blades 1–4 cm. long, rhombic-triangular to lanceolate, commonly hastate at base, on petioles 1–3 cm. long; plants monoecious or wholly ♀, the ♂ fls. in small axillary glomerules near the ends of the branches, the ♀ in axillary clusters; fruiting bracts 5–20 mm. long, sessile or stalked, united only near base, lanceolate or lance-oblong, 3-ribbed, often hastate, the tips attenuate, widely separated; seed brown, 1.2 mm. long.—Alkaline places; Creosote Bush Scrub, Alkali Sink; w. Mojave Desert, San Joaquin V., e. of Sierra Nevada to Ore., Utah, Ariz. April–Aug.

6. **A. truncàta** (Torr.) Gray. [*Obione t.* Torr. ex Wats.] WEDGESCALE. Erect or somewhat decumbent annual, the branches 2–5 dm. long, grayish-furfuraceous; lvs. alternate, subsessile or the lower petioled, round-ovate or deltoid-ovate, 1.5–4 cm. long; plants monoecious, the fls. in small axillary glomerules; fruiting bracts almost sessile, united to summit, broadly cuneate, truncate at summit, 2–3 mm. long, the sides smooth or obscurely tuberculate; seeds 1–1.5 mm. long, light brown to amber.—Alkaline places, mostly below ca. 8000 ft.; Creosote Bush Scrub to Lodgepole Forest; w. Mojave Desert along e. slope of Sierra Nevada to Siskiyou Co.; to B.C., Rocky Mts. June–Sept.

7. **A. coronàta** Wats. [*A. verna* Jeps. *A. sordida* Standl.] Bushy erect annual 1–3 dm. high, furfuraceous, but glabrate in age; lvs. grayish, crowded, alternate, sessile, ovate to somewhat deltoid, 5–20 mm. long, the lower short-petioled, rounded or cuneate at base, the upper sessile and ovate to lanceolate; plants monoecious, the fls. in axillary glomerules; fruiting bracts sessile, cuneate-orbicular, 4–5 mm. long, united to above middle, the green margins irregularly dentate or laciniate, the faces scurfy but otherwise smooth or cristate with long appendages; seed 1–1.5 mm. in diam., dark brown, shining.—Alkaline flats.; largely V. Grassland; Sacramento and San Joaquin and adjacent valleys, San Jacinto V. May–Aug.

8. **A. argéntea** Nutt. Erect much-branched bushy scurfy-gray annual 1.5–6 dm. high; lvs. all ± petioled, the lowest opposite, the blades 2–5 cm. long, triangular- to rounded-ovate, often slightly hastate, entire to undulate; plants monoecious, the fls. in axillary glomerules and terminal interrupted spikes, with ♂ and ♀ usually mixed in clusters; fruiting bracts obovate, compressed, 4–8 mm. long, united to near summit, the free green margins subentire to variously laciniate and faces smooth to crested; seed 1.5–2 mm. long, brown.—Moist alkaline places; Sagebrush Scrub, N. Juniper Wd.; Mono Co. to Modoc Co.; to B.C., N. Dak., New Mex. June–Sept.

Ssp. **expánsa** (Wats.) Hall & Clem. [*A. e.* Wats. *A. e.* var. *mohavensis* Jones. *A. trinervata* Jeps.] Upper lvs. sessile or nearly so, the lowest alternate.—Alkaline places; several Plant Communities; coast from San Francisco through cismontane s. Calif. to L. Calif., Sacramento V. s. through San Joaquin V.; New Mex., Texas. July–Nov.

9. **A. Serenàna** A. Nels. [*Obione bracteosa* Dur. & Hilg. *A. b.* Wats.] Erect or decumbent annual, usually with branched stems, often forming tangled mats. 5–20 dm. across with ascending twigs, sparsely scurfy; lvs. many, alternate, lanceolate to oblong or oval, nearly or quite sessile, 2–4 cm. long, sharply dentate to entire; plants monoecious, the ♂ glomerules in terminal spikes or panicles, the ♀ clusters small, axillary; fruiting bracts sessile or subsessile, somewhat compressed, united half way, cuneate-orbicular, 2–2.5 mm. long, smooth or tubercled, toothed above middle; seed brown, 1–1.3 mm. long.—Alkaline valleys, at low elevs.; V. Grassland, Coastal Sage Scrub, etc.; Sacramento V. to L. Calif., w. Nev. May–Oct.

Var. **Davidsònii** (Standl.) Munz. [*A. D.* Standl.] Lvs. 1–2 cm. long; terminal ♂ clusters short, subglobose.—Los Angeles to Balboa and Laguna Beach; Coronado Ids. April–Sept.

10. **A. pacífica** Nels. [*Obione microcarpa* Benth. *A. m. D.* Dietr., not Waldst. & Kit.] Prostrate annual forming tangled masses 3–10 dm. in diam., lightly furfuraceous on young parts; lvs. mostly alternate, numerous, sessile or the lower short-petioled, oblanceo-

late or spatulate-elliptic, 5–18 mm. long, entire, greenish above, more scurfy beneath; plants monoecious, the ♂ glomerules largely in the upper leafless axils, thus short-spicate, the ♀ in lower axils; fruiting bracts subsessile, suborbicular to obovate, 1–1.5 mm. long, united to middle, minutely 3–5-toothed at apex, otherwise entire; seed light brown, 0.8 mm. long.—Largely on sea bluffs; Coastal Sage Scrub; Los Angeles Co. to L. Calif.; Santa Rosa, Santa Catalina, and San Clemente ids. March–Oct.

11. **A. Coùlteri** (Moq.) D. Dietr. [*Obione C.* Moq.] Spreading perennial, slightly woody at base, sometimes flowering in first year, much-branched, the stems 3–10 dm. long, sparsely scurfy; lvs. many, alternate, elliptic to lanceolate or ovate, acute, entire, 7–20 mm. long; ♂ glomerules in upper axils and short terminal spikes; ♀ below; fruiting bracts sessile, obovate, 2–3 mm. long, united half way, the free margins sharply dentate, the backs sometimes tubercled; seed brown, 1.3–1.5 mm. long.—Somewhat alkaline low places; V. Grassland, Coastal Sage Scrub; Los Angeles Co. to w. San Bernardino Co. and L. Calif. March–Oct.

12. **A. cordulàta** Jeps. Erect rigid annual herb, simple or branched from base; stems rather coarse, 1–4 dm. high, scurfy but glabrate in age; lvs. sessile, ovate, mostly cordate at base, acute at apex, 5–15 mm. long, thickish and scurfy-tomentose, entire; ♂ and ♀ fls. mixed in small axillary clusters; fruiting bracts sessile or subsessile, round-ovate, 4–5 mm. long, compressed, united to middle, deeply toothed, scarcely if at all tuberculate, thin and soft at margin, hard at center; seed deep red-brown, 1.5–1.8 mm. long.—Hard, trampled, somewhat alkaline soil; V. Grassland; Sacramento and San Joaquin valleys. May–Oct.

13. **A. vallícola** Hoov. White-scurfy annual with slender spreading branches from base 2–20 cm. long; lvs. sessile, entire, the lower lance-ovate, 5–7 mm. long, the upper ovate, 2–4 mm. long, with cordate base; ♂ and ♀ fls. mixed in small axillary clusters; fruiting bracts thick, 2–3.5 mm. long, united almost to apex, with few large irregular teeth at summit and somewhat appendaged on back; seed black, 1.1 mm. long.— Dried rain-pools and flats; V. Grassland; Kern and Fresno cos. June–Aug.

14. **A. tularénsis** Cov. [*A. cordulata* var. *t.* Jeps.] Erect annual with simple or sparsely branched stems 2–8 dm. high, white-scurfy, aging red; lvs. sessile, lanceolate to ovate, 6–20 mm. long, entire, acute or acuminate, the base rounded; ♂ fls. in small axillary dense glomerules; ♀ solitary or in small axillary clusters or mixed with ♂; fruiting bracts sessile, rhombic-ovate, 3–3.5 mm. long, acute or acuminate, united to middle, the thin margins toothed, the faces plane, scurfy; seed dark brown, 1–1.2 mm. long.— Alkaline plains; edge of Alkali Sink; Bakersfield region, Kern Co. June–Oct.

15. **A. Paríshii** Wats. [*A. depressa* Jeps. *A. minuscula* Standl.] Spreading annual, the branches almost horizontal, fragile, white-scurfy, appearing almost pubescent, 5–20 cm. long; lvs. opposite or alternate, numerous, sessile, lanceolate to ovate, 4–10 mm. long, entire, rigid, gray to white, rounded at base; ♂ fls. mostly in upper axils and the ♀ in lower; fruiting bracts sessile, slightly compressed, ovate or rhombic, united half way, 3 mm. long, entire or with few teeth on each side, the faces smooth or tuberculate; seed dark brown or almost black, ca. 1.2 mm. long.—Alkali flats; largely V. Grassland; cismontane s. Calif. to edge of deserts, through Cent. Valley to Glenn Co. June–Oct.

16. **A. pusílla** (Torr.) Wats. [*Obione p.* Torr. ex Wats.] Annual, freely branched from base, sparsely furfuraceous, the branches spreading to erect, slender, 5–20 cm. long; lvs. many, sessile, ovate to subelliptic, acute, entire, 5–12 mm. long, gray to almost green; monoecious, the fls. solitary or in 2's in the axils; fruiting bracts sessile, united to apex, ovate, abruptly acute, compressed, 1–2 mm. long, entire, the faces plane; seed brownish, 0.8 mm. long.—Alkaline soils, 5000–7000 ft.; depressions in Sagebrush Scrub; Mono and Lassen cos. to Ore., Nev. June–Sept.

17. **A. elegáns** (Moq.) D. Dietr. ssp. **fasciculàta** (Wats.) Hall & Clem. [*A. f.* Wats. *A. saltonensis* Parish.] Erect to decumbent annual with many branches 5–20 cm. long, scurfy, slender; lvs. many, mostly alternate, subsessile, elliptic-spatulate to oblong, entire, 5–20 mm. long, densely scurfy; ♂ and ♀ fls. mixed in the same small axillary clusters; fruiting bracts subsessile, strongly compressed, round, united throughout except at the margins, 2–4 mm. wide, minutely toothed to subentire; seed brown, 1–1.4 mm. long.— Rather saline places; Creosote Bush Scrub; deserts from Inyo Co. to Mexican border and w. Ariz. March–July. Plants from e. Mojave Desert with deeply toothed fruiting bracts probably represent *A. elegans* itself.

18. **A. califórnica** Moq. in DC. Prostrate perennial from fusiform taproot, the many much-branched stems 2–5 dm. long, white-scurfy, later glabrate; lvs. many, alternate or the lowest opposite, crowded, sessile, lanceolate to oblanceolate, gray-scurfy, acute, 5–18 mm. long; fls. in mixed axillary clusters or the ♂ in terminal spikes; fruiting bracts ovate, acute, sessile, scarcely united, entire, ca. 3 mm. long; seed dark, ca. 2 mm. long.— Sea bluffs and sandy coast; Coastal Strand, edge of Coastal Salt Marsh, Coastal Sage Scrub, etc.; Marin Co. to L. Calif. April–Nov.

19. **A. semibaccàta** R. Br. AUSTRALIAN SALTBUSH. Prostrate suffrutescent perennial, the stems much-branched, 2–12 dm. long, at first scurfy, then glabrate; lvs. many, alternate, short-petioled, elliptic-oblong, 1–3 cm. long, acute or obtuse, irregularly repand-dentate to subentire; ♂ fls. in small terminal glomerules, the ♀ 1–few in the axils; fruiting bracts fleshy, becoming red, sessile, rhombic, 3.5–5 mm. long, united at base, compressed, entire to denticulate, the faces nerved, otherwise plane; seed black or brown, 1.5–2 mm. long.—Abundant in saline waste places, along roads, marshes, etc.; many Plant Communities; cismontane Calif. from San Luis Obispo Co. to L. Calif. and in San Joaquin and Imperial valleys.; natur. from Australia. April–Dec.

20. **A. leucophýlla** (Moq.) D. Dietr. [*Obione l.* Moq. in DC.] Prostrate perennial, somewhat woody at base, the branches many, coarse, 3–5 dm. long, white-scurfy; lvs. many, sessile, mostly alternate, crowded, round-ovate to -elliptic or oblong, 1–4 cm. long, entire, obtuse or rounded; ♂ fls. in dense terminal spikes, the ♀ in few-fld. axillary clusters; fruiting bracts sessile, almost completely united, broadly ovate, 5–7 mm. long, spongy, not compressed, entire or dentate, usually with wartlike projections on the faces; seed dark red-brown, 2.5–3 mm. long.—Sea beaches; largely Coastal Strand; Humboldt Co. to L. Calif., occasional inland, as at Lake Elsinore, Riverside Co. April–Oct.

21. **A. fruticulòsa** Jeps. Spreading or nearly erect perennial, woody at base, the stems 0.5–3 dm. high, with slender branches; lvs. many, alternate, the lower mostly short-petioled, the upper sessile, linear-lanceolate to -elliptic, 5–20 mm. long, entire, subacute, scurfy; ♂ fls. in dense terminal spikes, the ♀ in small, axillary, clusters; fruiting bracts sessile, scarcely compressed, round-obovate, 3–4 mm. long, united to above middle, acutely toothed from middle upward, the faces muricate or tooth-crested; seed dark brown, ca. 1.5 mm. long.—Alkaline flats, below 2000 ft.; V. Grassland; Sacramento and San Joaquin valleys. April–Nov.

22. **A. Watsònii** A. Nels. [*A. decumbens* Wats., not R. & S.] Prostrate suffrutescent perennial forming tangled mats 1–3 m. across, white-scurfy; lvs. numerous, mostly opposite, sessile, broadly elliptic to ovate, entire, thick, white-scurfy, 8–25 mm. long, acutish; plants dioecious, the ♂ glomerules in naked terminal spikes; ♀ clusters small, axillary; fruiting bracts sessile, ovate to rhombic, 5–8 mm. long, entire to erose, united to above the middle, the sides plane; seed light brown, ca. 1 mm. long.—Coastal bluffs and beaches; Coastal Strand, Coastal Salt Marsh, Coastal Sage Scrub; Santa Barbara Co. to L. Calif.; Channel Ids. March–Oct.

23. **A. hymenélytra** (Torr.) Wats. [*Obione h.* Torr.] DESERT-HOLLY. Low rounded compact shrub, white-scurfy, 2–10 dm. tall; lvs. numerous, persistent, alternate, rounded to rhombic, obtuse at apex, rounded to subcordate at base, 15–35 mm. long, silvery, irregularly and sharply deep-dentate, with petioles somewhat shorter; plants dioecious; ♂ fls. in short dense leafy panicles, the ♀ in short dense spikes; fruiting bracts strongly compressed, roundish, entire, 6–10 mm. long, quite distinct, reticulate-veined; seed brown, ca. 2 mm. long.—Dry alkaline slopes and washes; Creosote Bush Scrub; deserts to Utah, Son., L. Calif. Jan.–April.

24. **A. polycárpa** (Torr.) Wats. [*Obione p.* Torr.] Erect intricately branched shrub 1–2 m. tall, gray-scurfy, with slender divaricate branches; lvs. crowded on young twigs, early deciduous, alternate, usually sessile, oblong to spatulate, 3–20 mm. long, gray-scurfy, entire; plants dioecious, ♂ fls. in dense or interrupted, simple or paniculate naked spikes, ♀ crowded in small sparsely leafy clusters arranged in paniculate spikes; fruiting bracts sessile, somewhat compressed, ± united, cuneate-orbicular, 2–4 mm. long, shallowly to deeply toothed, plane or tuberculate; seed pale brown, 1–1.5 mm. long.—Alkaline soils, below 5000 ft.; Creosote Bush Scrub, Shadscale Scrub, Sagebrush Scrub, Alkali Sink; San Joaquin and adjacent interior valleys, Owens V. to Utah, Son., L. Calif. July–Oct.

25. **A. Tórreyi** (Wats.) Wats. [*Obione T.* Wats. *A. lentiformis* ssp. *T.* Hall. & Clem.]

Erect much-branched gray-scurfy shrub 6–15 dm. high, rather stiff and erect, the twigs acutely angled by prominent striae; stiff and somewhat spiny by loss of lvs. and bracts; lvs. oblong to ovate-hastate, short-petioled, gray-scurfy, 1.5–3 cm. long, entire; plants dioecious; fls. in panicles of dense narrow spikes; fruiting bracts flattish, suborbicular, 2–4 mm. in diam., crenulate, distinct; seed brown, ca. 1.3 mm. in diam.—Occasional in alkaline places, 2000–5500 ft.; Alkali Sink, Shadscale Scrub; Mojave Desert to Owens V.; Utah, Ariz. June–Oct.

26. **A. lentifórmis** (Torr.) Wats. [*Obione l.* Torr. in Sitg.] Erect widely spreading shrub 1–3 m. high, often broader, the twigs terete, gray-scurfy when young, mostly not spinose; lvs. oblong to ovate-deltoid, sessile or short-petioled, 1.5–4 cm. long, mostly 1–2.5 cm. wide, entire, gray-scurfy; plants dioecious, the fl. clusters in profuse terminal panicles; fruiting bracts flattish, round-ovate, 3–4 mm. long, united to about middle, entire to minutely crenulate; seed brown, ca. 1.4 mm. long.—Alkaline places, mostly below 2000 ft.; Alkali Sink; San Joaquin and adjacent interior valleys, Mojave and Colo. deserts; to Utah, Son., L. Calif. Aug.–Oct.

Ssp. **Brèweri** (Wats.) Hall & Clem. [*A. B.* Wats. *A. orbicularis* Wats.] Lf.-blades mostly 3–5 cm. long, 1.5–5 cm. wide; plants monoecious or dioecious; fruiting bracts convex, 4–7 mm. long, entire or undulate.—Saline places; V. Grassland, near Coastal Salt Marsh, Coastal Sage Scrub; San Francisco Bay to Orange Co. and inland to Hollister, Salinas V. and w. Riverside Co. July–Oct.

27. **A. Nummulària** Lindl. [*A. Johnstonii* C. B. Wolf.] Erect shrub 2–3 m. tall, gray-scurfy, with striated twigs; lvs. many, blue-green, 3–6.5 cm. long, ca. as wide, slightly crisped, entire to low-serrate, rounded to obtuse, short-petioled; plants semidioecious; ♂ fls. crowded into short spikes forming large panicles; ♀ in dense compound panicles; fruiting bracts sessile, thick and corky, united halfway, roundish, 5–12 mm. long, subentire to coarsely few-toothed; seeds brown, ca. 2 mm. in diam.—Sandy coastal bluffs, Playa del Rey, Los Angeles Co.; natur. from Australia. Dec.–June.

28. **A. Párryi** Wats. Erect much-branched rounded shrub 2–4 dm. high, white-scurfy, with slender rigid spiny twigs; lvs. crowded, alternate, somewhat petiolate or the upper sessile, round-ovate to almost reniform, entire, 5–18 mm. long; dioecious, the glomerules of fls. in interrupted leafy panicles of spikes; fruiting bracts compressed, thick, partly united, 3–4 mm. long, entire, with broad summit; seed brownish, 1–1.5 mm. in diam.— Alkaline places, mostly below 4000 ft.; Alkali Sink; Mojave Desert to Owens V.; Nev. May–Aug.

29. **A. confertifòlia** (Torr. & Frém.) Wats. [*Obione c.* Torr. & Frém.] Erect rigidly branched spiny rounded shrub 2–10 dm. high, with stout scurfy branchlets; lvs. crowded, deciduous, alternate, round-ovate or -obovate to elliptic, obtuse, subsessile or short-petioled, 1–2 cm. long, entire, gray-scurfy; plants dioecious, the ♂ glomerules in the upper axils, almost spicate, the ♀ 1–few in each of upper lf.-axils; fruiting bracts sessile, convex over the seed, suborbicular, free-margined, entire, 6–12 mm. long; seed red-brown, 1.5–2 mm. in diam.—Common, alkaline flats and slopes, below 7000 ft.; Shadscale Scrub, Sagebrush Scrub, Creosote Bush Scrub; Mojave Desert to e. Ore., N. Dak., Chihuahua. April–July.

30. **A. spinífera** Macbr. Much like *A. confertifolia,* taller than broad, the twigs becoming rigid divergent spines; lvs. deltoid-ovate to elliptic, 1–2 cm. long, short-petioled or subsessile, entire or subhastate, gray-scurfy; fruiting bracts strongly convex below, 7–15 mm. long, oblong to orbicular, entire or dentate, the faces plane or somewhat cristate; seed 2–2.8 mm. in diam.—Alkaline soils, below 2500 ft.; largely Alkali Sink; w. side of San Joaquin V. from Fresno and San Luis Obispo cos. to Kramer and Daggett, w. Mojave Desert. April–June.

31. **A. Nuttállii** Wats. ssp. **falcàta** (Jones) Hall & Clem. [*A. N.* var. *f.* Jones. *A. f.* Standl.] Suffrutescent, much-branched, 2–5 dm. high, the woody portion at base short; stems stout, terete, erect or decumbent; lvs. linear- to oblong-spatulate, sessile or subsessile, entire, firm, 2–4 cm. long, closely scurfy; dioecious, the fls. in axillary glomerules or interrupted paniculate spikes; fruiting bracts lanceolate to lance-ovate, making a subfusiform fr., 5–8 mm. long, united to near apex, entire or somewhat toothed and tubercled or crested on faces; seed brown, 1.5–2 mm. long.—At ca. 4000–5000 ft., alkaline places; Sagebrush Scrub; Lassen Co.; to e. Wash., Utah, Nebr. June–Aug.

32. **A. canéscens** (Pursh) Nutt. [*Calligonum c.* Pursh. *Obione tetraptera* Benth.]

Erect much-branched shrub 4–20 dm. high, grayish-scurfy, with spreading or ascending terete branches; lvs. linear-spatulate to narrowly oblong, sessile, 1.5–5 cm. long, 2–8 mm. wide, revolute; dioecious, the ♂ glomerules in terminal panicles of dense spikes, the ♀ in dense leafy-bracted spikes and panicles; fruiting bracts stalked, the body of the bract hard, ovoid not compressed, the bracts 6–15 mm. long, 4–8(–10) mm. wide, with a second pair of longitudinal wings from middle of each bract, the 4 wings entire to deeply and coarsely dentate, smooth or appendaged; seed brown, 1.5–2.5 mm. long.— Common on dry slopes, flats and washes, below 7000 ft.; Alkali Sink, Creosote Bush Scrub to Pinyon-Juniper Wd.; both deserts; less frequent in subsaline places, Coastal Strand to V. Grassland; cismontane valleys, San Benito and San Luis Obispo cos. to San Diego Co.; to e. Wash., S. Dak., Kans., Tex., Mex. June–Aug.

Var. laciniàta Parish. Poorly marked form with bracts laciniate, 7–8 mm. wide.— Barstow and Newberry (Mojave Desert) to Salton Sea.

Var. macilénta Jeps. Bracts 1.5–3 mm. wide, dentate.—About Holtville, Colo. Desert.

Ssp. lineàris (Wats.) Hall & Clements. [*A. l.* Wats.] Lvs. linear, usually not more than 2 mm. wide; fruiting bracts 4–8 mm. long, irregularly dentate or laciniate, each bract with 2 thin wings 2–3 mm. wide.—Saline places, Alkali Sink; region of Salton Sea through e. Colo. Desert to L. Calif., Son. May–July.

8. Gràyia H. & A.

Stiff much branched shrubs, scurfy-pubescent. Lvs. alternate, entire, sessile, rather fleshy. Dioecious or rarely monoecious. Staminate fls. small, pedicellate, glomerate, bractless; calyx mostly 4-parted; stamens 4–5. Pistillate fls. racemose, with 2 orbicular flattened bracts which are dorsally winged in fr. and united to form a reticulate sac; calyx none; stigmas 2, filiform. Utricle compressed, included in the accrescent bracts; seed free, orbicular, with annular embryo. Two spp. of w. U.S. (Named for Asa *Gray*, 1810–1888, distinguished American botanist.)

1. **G. spinòsa** (Hook.) Moq. [*Chenopodium s.* Hook. *G. polygaloides* H. & A.] Hop-Sage. Erect, 3–10 dm. high, gray-green, often spinose, with mealy younger parts; lvs. oblanceolate to oblong-lanceolate, 1–3(–4) cm. long; ♂ clusters axillary and forming dense terminal spikes; ♀ in dense terminal racemose spikes; fruiting bracts membranous, reddish to whitish, glabrous, 6–12 mm. long, the wings thin, entire; seed flat, brown, round, cellular-reticulate, ca. 2 mm. across.—Common on mesas and flats, mostly 2500–7500 ft.; Creosote Bush Scrub to Pinyon-Juniper Wd.; Mojave Desert and rare in w. and extreme nw. Colo. Desert, to Lassen and Siskiyou cos.; to e. Wash., Wyo., Ariz. March–June.

9. Euròtia Adans.

Low stellate-tomentose shrubs. Lvs. alternate or fascicled, slender, entire. Dioecious or monoecious. Staminate fls. bractless in dense axillary clusters forming spikes; calyx 4-parted; stamens 4. Pistillate fls. without calyx, but with a pair of conduplicate obcompressed partly connate bracts which form a membranous silky-hairy sac that enlarges in fr., becoming 4-angled and beaked above with 2 short horns; styles 2 elongate; ovary ovoid. Fr. a utricle. Seeds vertical, obovoid. Two spp., one Asiatic. (Greek, *euros*, mould, because of hairy covering.)

1. **E. lanàta** (Pursh) Moq. [*Diotis l.* Pursh.] Winter Fat. Erect shrubs 3–8 dm. high, white- or rusty-stellate, often with longer unbranched hairs interspersed; lvs. linear to lanceolate, 1.5–5 cm. long, 2–8 mm. wide, those of the fascicles shorter; fruiting bracts lanceolate, 5–7 mm. long, with dense spreading tufts of long silvery or rusty hairs; ovary hairy; seeds brown, ca. 2 mm. long.—Common on flats and rocky mesas, above 2000 ft.; Creosote Bush Scrub to Pinyon-Juniper Wd.; Mojave Desert, Sunset in w. Kern Co., n. along e. base of Sierra Nevada to Lassen Co.; Wash., Rocky Mts., Tex., Son. March–June. Important as a forage plant. The plants with longer hairs among the stellate, and stems woody only at base, are true *lanàta*, while those with short stellate hairs only, and woodier stems, are sometimes called var. *subspinòsa* (Rydb.) Kearn. & Peeb. Both forms commonly grow together.

10. Kòchia Roth

Annual or perennial herbs or low shrubs. Lvs. alternate or opposite, narrow, often terete. Fls. perfect or ♀, small, sessile, solitary or glomerate in the axils of the lvs. Calyx 5-lobed, the lobes incurved, mostly coriaceous in age and then developing horizontal wings, these membranaceous or scarious, distinct or confluent. Stamens 5, exserted, the fils. compressed. Ovary subsessile, stigmas 2, rarely 3. Utricle depressed-globose. Seed horizontal; embryo annular, surrounding the scanty endosperm. Ca. 35 spp., mostly Old World. (Named for W. D. J. *Koch*, 1771–1849, German botanist.)

Plants perennial, native.
  Stems branched mostly at base, relatively simple above; lvs. subterete . . . . 1. *K. americana*
  Stems paniculately branched above; lvs. flat, 1–3 mm. wide . . . . 2. *K. californica*
Plants annual; introd. weeds . . . . 3. *K. Scoparia*

1. **K. americàna** Wats. GREEN-MOLLY. RED-SAGE. Perennial from a woody branching crown; stems many, erect, 1–3 dm. high, usually villous-tomentose when young, then glabrate; lvs. many, 6–25 mm. long, terete, fleshy, erect or ascending, ca. 1 mm. wide, acutish, sparsely silky to glabrous; fls. solitary or in 2's or 3's, white-tomentose; calyx 2 mm. broad in fr., the wings fan-shaped, 2 mm. long, membranous, striate, crenulate; utricle glabrate; seed brown, flat, roundish in outline, ca. 2 mm. in diam.—Occasional and local on alkaline flats, 4000–6000 ft.; Shadscale Scrub, Joshua Tree Wd.; Barstow and Inyo Co. to se. Ore., Sask., Kans. May–Aug. Plants more densely and permanently pubescent have been designated as var. *vestìta* Wats. [*K. v.* Rydb.]

2. **K. califórnica** Wats. [*K. americana* var. *c.* Jones.] Perennial 1.5–5 dm. high, erect; stems simple below, paniculately branched above, grayish-rusty throughout with dense pubescence; lvs. spreading, linear-oblong, 5–15 mm. long, 1.5–3 mm. wide, flat, sericeous; fls. solitary or in clusters of 2–5, tomentulose; calyx 2 mm. in diam. in fr., with flabellate wings 2 mm. long, scarious, finely nerved; seed 2 mm. in diam.—Alkaline flats; Creosote Bush Scrub, Joshua Tree Wd., V. Grassland, etc.; Mojave Desert (Victorville, Rabbit Springs, Lancaster), San Joaquin V. (Kern Co. to Madera Co.) to se. Inyo Co.; Nev. May–Sept.

3. **K. scopària** (L.) Schrad. [*Salsola s.* L.] SUMMER-CYPRESS. Erect annual 4–15 dm. high, pyramidal to ovoid, densely leafy, pubescent; lvs. lanceolate to lance-linear, often ciliate at base, soft-strigose especially beneath, 1–4 cm. long; fls. in small axillary clusters; midrib of each sepal becoming prominently thickened.—Becoming a weed at scattered stations, below 1000 ft.; Los Angeles, Santa Barbara, and Contra Costa cos.; e. U.S.; introd. from Eurasia. Aug.–Oct.

11. Bássia All.

Annual to suffrutèscent plants. Lvs. alternate, narrow, flat or subterete, entire. Fls. minute, solitary or glomerate in the axils, sessile, perfect and ♀. Calyx globose or depressed, the 5 lobes incurved, armed on the back with a spine, this usually hooked. Stamens mostly 5. Ovary ovoid; style short; stigmas 2–3. Utricle enclosed by the coriaceous calyx; seed free from pericarp, orbicular, horizontal; embryo annular. Several spp. of Old World. (Name for F. *Bassi*, 1710–1774, Italian botanist.)

1. **B. hyssopifòlia** (Pall.) Kuntze. [*Salsola h.* Pall. *Echinopsilon h.* Moq.] Plant a grayish densely pilose annual with stems branched from base, prostrate, 3–5 dm. long and less hairy in maturity; lvs. narrowly linear-lanceolate, flat, 2–4 cm. long; fls. in small axillary glomerules; calyx-lobes broadly ovate, villous, ca. 1 mm. long, each with a spreading stout hooked spine; seed lenticular, ca. 1 mm. across.—Becoming a common weed in rather alkaline places through much of Calif. and in other w. states; from Eurasia. July–Oct.

12. Allenrólfea Kuntze

Much-branched glabrous erect succulent shrub or half-shrub, with alternate articulate green branches. Lvs. short, scalelike. Fls. perfect, sessile, arranged spirally by 3's or 5's in the axils of fleshy peltate bracts and forming cylindrical spikes. Perianth small, angled, of 4 or 5 concave, carinate lobes. Stamens 1 or 2, exserted. Stigmas 2 or rarely 3, short, usually distinct. Utricle ovoid, compressed, with free membranous pericarp. Seed erect, oblong, smooth; embryo partly enclosing the copious endosperm. One sp. (Name for *Allen Rolfe*, English botanist.)

1. **A. occidentàlis** (Wats.) Kuntze. [*Halostachys o.* Wats. *Spirostachys o.* Wats.] IODINE BUSH. Plants 5–12(–20) dm. high, woody below or nearly throughout, somewhat glaucous, the stems jointed; lvs. very short, triangular, some deciduous; spikes many, 6–25 mm. long; calyx closely enveloping the fr.; seed brown, ca. 0.6 mm. long. —Moist alkaline places; Alkali Sink; deserts and San Joaquin V. n. to Byron Springs, Contra Costa Co.; to Ore., Utah, Son., L. Calif. June–Aug.

13. Salicórnia L. GLASSWORT. SAMPHIRE. PICKLEWEED

Annual to suffrutescent herbs with succulent leafless jointed stems, opposite branches and short internodes. Lvs. opposite, reduced and scalelike. Fls. mostly perfect, in cylindric fleshy spikes made up of very short internodes with fls. sunk in groups of 3–7 on opposite sides of the joints. Calyx fleshy, with a truncate or 3–4-toothed margin. Stamens 1 or 2. Styles 2, united at base. Utricle oblong or ovoid, included in spongy calyx. Seed vertical, mostly without endosperm; embryo thick, the cotyledons incumbent upon the radicle. Ca. 10 spp., of saline places; world-wide. (Greek, *sal*, salt, and *cornu*, a horn, being saline plants with hornlike branches.)

(Standley, P. C., *in* N. Am. Fl. 21: 82–85, 1916.)
Plants perennial from creeping rootstocks.
    Seeds glabrous; spikes slender, the upper part sterile .... 1. *S. subterminalis*
    Seeds pubescent; spikes stoutish, fertile to the top .... 2. *S. virginica*
Plants annual.
    Joints of spike thicker than long, the scales acuminate .... 3. *S. Bigelovii*
    Joints of spike longer than thick, the scales barely acute .... 4. *S. europaea*

1. **S. subterminàlis** Parish. [*Arthrocnemum s.* Standl.] Perennial; stems widely spreading or erect and compact, 1.5–3 dm. high, the joints 2–20 mm. long; branchlets many, crowded, 2–3 mm. in diam., the joints 5–15 mm. long; spikes 2–3.5 cm. long, 2–2.5 mm. thick at base, with few to several fl.-bearing scales below and as many slender sterile ones above; fls. subequal; seeds glabrous, brown, ca. 1 mm. long.—Salt marshes and low alkaline places; Coastal Salt Marsh, Alkali Sink, Coastal Sage Scrub, etc.; San Francisco Bay and San Joaquin V. to Mex. April–Sept.

2. **S. virgínica** L. [*S. ambigua* Michx. *S. pacifica* Standl.] Forming extensive colonies, perennial, suffrutescent, decumbent, 2–6 dm. long, rooting along the trailing bases; branches stout, erect or ascending, the joints 6–20 mm. long, 2–4.5 mm. thick; spikes 1.5–5 cm. long, ca. 3 mm. thick, the fls. almost subequal; seed brown, ca. 1 mm. long, pubescent with short slender incurved hairs.—Coastal Salt Marsh and nearby alkaline flats; L. Calif. to B.C., Mex., Atlantic Coast. Aug.–Nov.

S. utahénsis Tides., with spikes 4 mm. thick, has been reported from near Panamint Mts.

3. **S. Bigelòvii** Torr. [*S. mucronata* Bigel., not Lag.] Erect annual, simple-stemmed or with strongly ascending branches mostly from above base, 1–5 dm. high, the joints 1–2.5 cm. long, 2–3 mm. thick, mostly thicker than long; spikes 2–10 cm. long, 4–6 mm. thick; scales of spike triangular-ovate, sharply mucronate; middle fl. half higher than the lateral and reaching nearly to summit of the joint; seed pubescent, 1–1.5 mm. long. —Coastal Salt Marsh; Seal Beach, Los Angeles Co. to San Diego; Atlantic Coast. July–Nov.

4. **S. europaèa** L. [*S. rubra* A. Nels. *S. depressa* Standl.] Low densely bushy annual, branched from base, 1–2 dm. high, with slender joints 1–2.5 mm. thick, longer than thick; spikes 2–6 cm. long, 2–3 mm. thick, with central fl. definitely higher than the lateral ones; seed 1–2 mm. long, pubescent.—Occasional in saline places; Coastal Salt Marsh, Alkali Sink, etc.; San Diego, Point Mugu, near Tehachapi, Modoc Co., Marin Co., Sonoma Co., etc.; across Great Basin to Atlantic Coast; Old World. July–Nov.

14. Sarcobàtus Nees. GREASEWOOD

Much-branched spinescent shrubs. Lvs. alternate or opposite, linear, fleshy, sessile. Monoecious or dioecious; ♂ fls. arranged spirally in terminal spikes, ebracteate, without perianth; stamens 2–3, covered by a peltate scarious scale. Pistillate fls. sessile, 1 or 2, each in axil of a lf.; calyx compressed, ovoid or oblong, adnate to base of the 2 recurved subulate stigmas. Fr. coriaceous, with broad crenulate scarious wing at middle, the lower part turbinate, the upper conic. Seed erect, flat, orbicular; embryo spirally coiled; endosperm none. Only the following known. (Greek, *sarx*, flesh, and *batos*, bramble, possibly because of the fleshy lvs. and spiny stems.)

1. **S. vermiculàtus** (Hook.) Torr. [*Batis v.* Hook. *S. Maximiliani* Nees. *Fremontia vermicularis* Torr.] Rounded erect or spreading shrub 1–2.5 dm. high, the older branches gray, stout, the younger yellowish-white, glabrous or with short white branched hairs; lvs. 0.5–3 cm. long, glabrous or sparsely pubescent, acute to obtuse; staminate spikes 1–3 cm. long, 3–4 mm. thick, the peltate scales rhombic-round, glabrous or pubescent; fruiting calyx-wing 8–12 mm. in diam., the fr.-body glabrate, 4–5 mm. long; seed brown, ca. 2 mm. high.—Alkaline places, 3000–7000 ft.; Alkali Sink to Pinyon-Juniper Wd.; Rabbit Springs (Mojave Desert), Owens Lake (Inyo Co.) to Modoc Co.; to e. Wash., Alta., N. Dak., Tex. May–Aug.

Var. **Bàileyi** (Cov.) Jeps. [*S. B.* Cov.] Lvs. more densely and permanently pubescent, mostly 5–14 mm. long; ♂ spikes 1 cm. or less in length; fr.-body 8–9 mm. long, glabrous; 10–18 mm. in diam.—To be expected in Death V. region; w. Nev.

15. Suaèda Forsk. SEA-BLITE. SEEP-WEED

Annual or perennial herbs or shrubs. Lvs. alternate, entire, terete to spatulate. Fls. small, mostly perfect, solitary or glomerate in the lf.-axils, bracteate. Calyx 5-lobed, fleshy, globose, turbinate or urceolate, the lobes equal and unappendaged, or 1 or more larger and corniculate-appendaged. Stamens 5, with short fils. Ovary subglobose; stigmas usually 2, short, subulate, recurved. Utricle enclosed in the calyx, compressed or depressed. Seed horizontal or vertical; endosperm none or scanty; embryo coiled into a flat spiral. Ca. 50 spp., of wide distribution. (An Arabic name.)

(Standley, P. C. N. Am. Fl. 21: 86–92, 1916.)

Plants annual; calyx-lobes unequal, 1 or more corniculate-appendaged or transversely winged.
  Lvs. broadest at base; calyx-lobes corniculately appendaged .... 1. *S. depressa*
  Lvs. not widened at base; calyx-lobes transversely winged .... 2. *S. occidentalis*
Plants perennial; ± woody at base; calyx-lobes equal, not appendaged.
  Lvs. strongly flattened; branches of infl. very slender .... 3. *S. Torreyana*
  Lvs. terete; branches of infl. stout.
    Fls. 2–3 mm. wide; lvs. of infl. crowded; calyx cleft halfway. Coastal .... 4. *S. californica*
    Fls. 1–1.5 mm. wide; lvs. of infl. not crowded; calyx cleft more than halfway. Alkaline places
    mostly away from coast .... 5. *S. fruticosa*

1. **S. depréssa** (Pursh) Wats. [*Salsola d.* Pursh. *Dondia d.* Britton] Mostly annual, branched from base, glabrous, with stout decumbent branches 2–5 dm. long; lvs. linear, subterete, 1–2.5 cm. long, numerous, becoming shorter in infl.; fls. crowded, barely 2 mm. wide, one or more of calyx-lobes longer than others and corniculate-appendaged; seed horizontal or vertical, black, 1 mm. broad.—Alkaline places, at 6900 ft.; Pinyon-Juniper Wd.; Baldwin Lake, San Bernardino Mts.; to Wash., Minn., Tex. July–Sept.

Var. **erécta** Wats. [*S. minutiflora* Wats. *Dondia m.* Heller.] Erect, the stem simple or branched above, 3–6 dm. high; lvs. 2–5 cm. long.—Alkaline flats and saline places, below 5000 ft.; Coastal Salt Marsh, Coastal Sage Scrub, Sagebrush Scrub, etc.; cismontane s. Calif. to Modoc Co.; Man., Nebr., Tex. July–Oct.

2. **S. occidentàlis** Wats. Annual; branches slender, flexuous; lvs. narrowly linear, not widened at base, mostly spreading, 1–1.5 cm. long; calyx-lobes obtuse, with lobed transverse wings when older.—Alkaline places, Litchfield, Lassen Co.; to Wash., Rocky Mts. July–Sept.

3. **S. Torreyàna** Wats. [*Dondia T.* Standl.] Green glabrous erect suffrutescent perennial with usually slender ascending or spreading branches, sparsely leafy; lvs. linear, 1.5–3 cm. long, strongly flattened, much shortened in infl.; fls. 1–4 in the axils; calyx deeply cleft, green, with obtuse lobes, rounded on back; seeds 1–1.5 mm. broad, minutely tuberculate.—Occasional in alkaline places mostly of the interior, below 5000 ft.; Coastal Sage Scrub, Creosote Bush Scrub, Alkali Sink, etc.; San Diego Co. to San Joaquin V., Inyo Co., Lassen Co., etc.; to Wyo., New Mex. May–Sept.

Var. **ramosíssima** (Standl.) Munz. [*Dondia r.* Standl. *S. r.* Jtn.] Stems and lvs. finely and densely pubescent.—Similar places, Mojave and Colo. deserts, Tulare Co.; Ariz., L. Calif.

4. S. califórnica Wats. [*Dondia c.* Heller.] Glaucous quite glabrous branching suffrutescent perennial, the stems ascending or decumbent, stout, 3–8 dm. long; lvs. many, rather crowded, spreading, subterete, 1.5–3.5 cm. long, not much reduced in infl.; fls. 2–3 mm. wide; calyx equally 5-cleft, not appendaged, glaucous; seed black, shining, 1.5–2 mm. broad.—Coastal Salt Marsh; San Francisco Bay to L. Calif. July–Oct.

Var. taxifòlia (Standl.) Munz. [*Dondia t.* Standl. S. *t.* Standl.] Stems, lvs. and calyces pubescent, sometimes almost tomentosely so; lvs. 1.5–3 cm. long; fls. 2–3 mm. wide.—Coastal Salt Marsh; Santa Barbara Co. to San Diego.

Var. pubéscens Jeps. [*Dondia brevifolia* Standl. S. *taxifolia* ssp. *b.* Abrams.] Lvs. shorter, 3–10 mm. long, densely pubescent; fls. 1–1.5 mm. broad.—Coastal Salt Marsh; San Diego and Orange cos., occasional to Ventura Co.

5. S. fruticòsa (L.) Forsk. [*Chenopodium f.* L. *Dondia f.* Druce. *Suaeda Moquini* Greene.] Glaucous quite glabrous perennial, shrubby at base, 2–8 dm. high, much-branched, erect or ascending; lvs. many, narrow-linear, subterete, spreading, 1–2 cm. long, gradually reduced upward; branches of infl. slender, ascending; calyx deeply cleft, ca. 1 mm. broad; seed mostly horizontal, black, ca. 0.8 mm. broad; $2n = 36$ (Joshi, 1935).—Alkaline and saline places, below 4500 ft.; Coastal Salt Marsh, Coastal Sage Scrub, Creosote Bush Scrub, Sagebrush Scrub, etc.; coastal s. Calif. through interior valleys to Santa Clara Co. and deserts to Lassen Co.; Alta., Mex., W. Indies; Old World. July–Oct.

16. Salsòla L.

Annual or perennial herbs, usually branched. Lvs. usually alternate, sessile or clasping, narrow, often with pungent apex. Fls. perfect, small, solitary or fascicled, axillary, bibracteolate; calyx 5(4)-parted, with lanceolate to oblong segms., concave, usually transversely keeled and often winged in fr. Stamens 5, sometimes fewer. Stigmas 2 (3), subulate. Utricle flattened, included in the calyx. Seed horizontal, orbicular; embryo spirally coiled; endosperm lacking. Ca. 50 spp., widely dispersed. (Latin *salsus*, salty.)

1. S. Kàli L. var. tenuifòlia Tausch. [*S. pestifer* A. Nels.] RUSSIAN-THISTLE. Annual 3–10 dm. high, densely and intricately branched, forming a round bushy clump, usually quite glabrous; lower lvs. terete, fleshy, linear, 3–5 cm. long, pungent-tipped; bracts ovate, short-acuminate, prickly-pointed, indurate in fr.; fruiting calyx 3–6 mm. broad, the wings membranous, conspicuously veined, often reddish; seed black, shining, 1.5–2 mm. broad.—Common as a tumbleweed in cult. fields and waste places; many Plant Communities below Montane Coniferous F.; through most of Calif. and w. N. Am.; to Atlantic Coast; native of Eurasia. July–Oct.

17. Halogèton C. A. Mey.

Annual, glabrous or pubescent, fleshy. Lvs. alternate, terete, bristle-tipped. Polygamous, the fls. small, few, in glomerules in lf.-axils, bibracteate. Sepals 5, in some fls. inconspicuous, in some becoming membranaceous and fanlike. Stamens 5 or 3. Stigmas 2. Utricle ovoid, thin-walled. Seed vertical; embryo spiral. Three spp. of Eurasia. (Greek, *hals*, sea, salty, and *geiton*, a neighbor, from the habitat.)

1. H. glomeràtus (Bieb.) C. A. Mey. in Led. [*Anabasis g.* Bieb.] Branched from base, the stems divergent then ascending, 0.5–5 dm. long; lvs. sessile, 6–20 mm. long, blunt then tipped with conspicuous bristlelike hair; fls. greenish-yellow, exceedingly numerous, the sepals usually conspicuously membranous; seed ± compressed, ca. 1 mm. long.—Alkaline or disturbed places, etc.; Sagebrush Scrub; se. Lassen Co.; Nev., Utah; native of Eurasia. Summer.

41. Amaranthàceae. AMARANTH FAMILY (Fig. 39)

Herbs or shrubs. Lvs. alternate or opposite, exstipulate, simple, entire. Fls. small, perfect or unisexual, inconspicuous, in ours congested in clusters or spikes, with 3 dry scarious persistent often colored bracts. Calyx commonly of 5 sepals, sometimes 1–4, persistent, usually scarious. Petals none. Stamens usually as many as sepals. Ovary

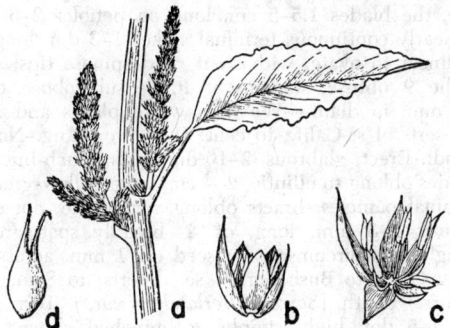

Fig. 39. AMARANTHACEAE. *Amaranthus: a,* habit, × ½, with infls. and lf.; *b,* ♂ fl., × 5, with 5 unequal sepals; *c,* ♂ fl. opened; *d,* pistil.

superior, 1-celled, usually with 2–3 stigmas. Fr. a membranous or fleshy utricle, indehiscent, irregularly dehiscent or circumscissile. Seed 1, with copious endosperm. Often weedy plants, a few grown for ornament, as pot-herbs, or in Latin Am. for their edible seed. Ca. 40 genera and 500 spp., largely of warm countries.

Lvs. alternate; anthers 4-celled; plants nearly or quite glabrous .... 1. *Amaranthus*
Lvs. largely opposite; anthers 2-celled; plants white stellate-woolly or villous.
 Fls. glomerate, with an invol. of upper lvs. Mostly deserts .... 2. *Tidestromia*
 Fls. in axillary headlike spikes, without invol. Mostly near beaches .... 3. *Alternanthera*

1. Amaránthus L. Amaranth

Annual usually coarse herbs, erect to prostrate, usually branched. Lvs. alternate, entire, petioled. Fls. 3-bracted, small, green to purplish, in glomerules in axillary or terminal spiked clusters, commonly with ♂ and ♀ fls. in same cluster. Calyx glabrous. Stamens 5, 2, or 3, separate. Stigmas 2 or 3. Fr. a 1-seeded ovoid utricle, 2–3-beaked at apex and usually longer than calyx. Seed erect, compressed, smooth; embryo coiled into a ring around the endosperm. Ca. 50 spp., widespread except in cold regions. (Greek, *amarantos,* unfading, because of the dry persistent calyx and bracts.)

(Standley, P. C. N. Am. Fl. 21: 101–119, 1917.)
A. Sepals of the ♀ fls. broadened upward, the calyx ± urceolate.
  B. Calyx of ♀ fls. not fimbriate.
   C. Infl. not leafy; bracts rigid and spinose; plants dioecious .... 1. *A. Palmeri*
   CC. Infl. leafy at least below; bracts not spinose; plants monoecious ...... 2. *A. Watsonii*
  BB. Calyx of ♀ fls. fimbriate; plants monoecious .... 3. *A. fimbriatus*
AA. Sepals of ♀ fls. narrowed upward, the calyx not urceolate; plants monoecious.
  B. Axils of many lvs. bearing pair of sharp spines .... 11. *A. spinosus*
  BB. Axils of lvs. without spines.
   C. Utricle indehiscent, fleshy; spikes terminal; prostrate plant with slender stems
    4. *A. deflexus*
   CC. Utricle dehiscent, the top falling away like a lid.
    D. Infl. of terminal or axillary spikes; sepals 5.
     E. Pistillate sepals (at least the longer in each fl.) conspicuously exceeding the utricle, all spreading at maturity, oblong-linear, obtuse or truncate or emarginate and sub-erose, often aristate .... 5. *A. retroflexus*
     EE. Pistillate sepals shorter than to slightly longer than the utricle, erect, lanceolate, tapering gradually to the terminal arista.
      F. Seeds elliptical to broadly ovate in outline, 1.1–1.3 mm. long, brown-black; stamens 3 in some fls., sometimes 4–5; ♀ calyx 3 mm. long
       6. *A. Powellii*
      FF. Seeds circular in outline, not over 1 mm. in diam., jet black; stamens 5; ♀ calyx 2 mm. long .... 7. *A. hybridus*
    DD. Infl. wholly of axillary glomerules.
     E. Sepals 4–5 in both ♂ and ♀ fls.; stems prostrate; seeds 1.5 mm. broad
      8. *A. graecizans*
     EE. Sepals 1–3; seeds less than 1 mm. broad.
      F. The sepals of ♀ fls. 2–3, but only 1 well developed; stems slender, prostrate .... 9. *A. californicus*
      FF. The sepals of ♀ fls. 3, subequal; stems ascending ....10. *A. albus*

1. A. Pálmeri Wats. Stems stout, erect, 2–10(–15) dm. high, pale green, glabrous, or villous about the spikes, simple or with short erect branches; lvs. broadly ovate to rhombic or lanceolate, the blades 1.5–5 cm. long on petioles 2–6 cm. long; dioecious, the fls. in long erect nearly continuous terminal spikes 1–3 dm. long; bracts usually 2–3 times as long as fls., linear to ovate, with stout rigid spinose tips; calyx 2–3 mm. long, the ♂ sepals acute, the ♀ obtuse; stamens 5; utricle subglobose, circumscissile, rugose at summit; seeds 1.3 mm. in diam.—Weed in waste places and in fields, low elevs.; interior valleys and deserts of s. Calif.; to cent. U.S., Mex. Aug.–Nov.

2. **A. Watsònii** Standl. Erect, glabrous, 2–10 dm. high, much-branched; petioles stout, 5–20 mm. long; lf.-blades oblong to elliptic, 2–4 cm. long, yellow-green; fls. in glomerules, in stout spikelike terminal panicles; bracts oblong, acuminate, not exceeding fls.; sepals of ♂ fls. oblong, acute, 2–3 mm. long, of ♀ broadly spatulate, 2–2.5 mm. long; stamens 5; utricle subglobose, circumscissile; seed ca. 1 mm. across, black.—Occasional below 5500 ft.; mostly Creosote Bush Scrub; se. deserts; to Son., L. Calif. Aug.–Sept.

3. **A. fimbriàtus** (Torr.) Benth. [*Sarratia Berlandieri* var. *f.* Torr. *Amblogyna f.* Gray.] Stems slender, erect, 1–5 dm. high, simple or branched above, glabrous or slightly puberulent above; lvs. narrowly lanceolate to linear, bright green, the blades 2–6 cm. long, the petioles 1–2 cm. long; monoecious; fls. in loose clusters, these scattered or in loose spikes; bracts ovate, half as long as calyx; sepals 2.5–3 mm. long, fimbriate, often rose or lavender, obtuse in ♀, acute in ♂ fls.; stamens 3; utricle circumscissile; seeds roundish, 0.8 mm. in diam.—Occasional, dry gravelly places, below 5000 ft.; Creosote Bush Scrub, Sagebrush Scrub, Joshua Tree Wd.; Colo. and e. Mojave deserts; to Utah, Mex. Aug.–Nov.

4. **A. defléxus** L. Stems slender, much-branched, ascending or decumbent, subglabrous or pubescent above; lf.-blades rhombic-ovate to lanceolate, obtuse, 0.5–2 cm. long, on slender petioles ca. as long; monoecious; fls. in lobed dense terminal spikes or short-branched panicles to 1 dm. long; bracts ovate, acute, cuspidate, 2 mm. long; sepals 2–3, oblong, ca. as long as bracts; utricle oblong, indehiscent, much exceeding calyx, 3–5-nerved, with fleshy walls; seeds dark red-brown, oval, 1 mm. long.—Weed in gardens, along streets, etc., at scattered locs.; natur. from Old World. May–Nov.

5. **A. retrofléxus** L. Stout, usually branched, erect, rough-pubescent, 3–15 dm. tall, with erect or ascending branches; lf.-blades ovate or oblong-ovate, 3–10 cm. long, the uppermost narrower; petioles 1–7 cm. long; monoecious; fls. in dense spikes 8–20 mm. thick and crowded into terminal lobulate panicles with crowded lateral spikes 1–5 cm. long; bracts ovate, subulate, ca. twice as long as sepals; sepals 5, linear-oblong, often mucronate, ca. 3 mm. long; stamens 5; utricle rugulose, shorter than sepals; seeds compressed, rounded or obovate, black, ca. 1 mm. long; $2n =$ ca. 32 (Heiser & Whitaker, 1948).—Garden and orchard weed, also in waste places; native of trop. Am. June–Nov.

6. **A. Powéllii** Wats. Much like *A. retroflexus* but with the terminal panicle of few erect stiff spikes, the cent. one much elongate, 1–2.5 dm. long, the lateral 4–12 cm. long; bracts 2–3 times the length of calyx; sepals of ♀ fls. ca. 3 mm. long, acute, tapering gradually at apex; stamens usually 3; seeds 1.1–1.3 mm. long, not as wide.— Waste places at low altitudes, much of cismontane Calif.; to Ore., Wyo., Mex. July– Oct.

7. **A. hýbridus** L. Much like *A. retroflexus,* the stems more slender, often reddish at base, 5–15 dm. high, glabrous or pubescent; lvs. rather dark green, 2–12 cm. long, on somewhat shorter petioles; infl. very many-fld., of few to mostly numerous spikes in panicles, the spikes often crowded, tawny-green, the lateral ± ascending, 2–6 cm. long, 6–12 mm. thick, the terminal usually longer and thicker; bracts almost twice as long as sepals; ♀ fls. 1.5–2 mm. long, acute; utricle from shorter than to ca. as long as calyx; seed round, black, 1 mm. in diam.; $2n = 32$ (Covas & Schnock, 1946).—Waste and cult. ground, at low elevs., through much of cismontane Calif.; semicosmopolitan weed. June–Nov.

8. **A. graecìzans** L. [*A. blitoides* auth., not L. *A. b.* var. *crassior* Jeps.] Stems prostrate, branched from base, 1–5 dm. long, quite glabrous, often purplish; lvs. many, often crowded, the petioles 2–15 mm. long, the blades spatulate to obovate, 8–25 mm. long, pale green; monoecious; fls. in dense axillary clusters usually shorter than petioles; bracts ovate-oblong, ca. as long as sepals, short-acuminate; sepals 4–5, oblong, acute or acuminate, 2–3 mm. long; utricle scarcely rugose, equaling calyx; seeds round, rather dull black, 1.5 mm. broad.—Occasional native weed in waste and cult. places, widely scattered in Calif.; to Wash., Wyo., Tex., Mex. July–Nov.

9. **A. califórnicus** (Moq.) Wats. [*Mengea c.* Moq. *A. carneus* Greene. *A. albomarginatus* Uline & Bray.] Stems prostrate, glabrous, much-branched, 5–20 cm. long, forming mats whitish or often tinged red; lvs. many, crowded, pale green, the petioles 2–10 mm. long, the blades obovate to oblong, obtuse or rounded, 5–15 mm. long; monoecious, the fls. in small axillary clusters; bracts lanceolate, subulate-tipped, ca. 1 mm. long; sepals 2–3 in ♂ fls., usually only 1 well developed in ♀ fls. and barely 1 mm. long; utricle smooth, bursting irregularly, subglobose, often red or purple; seeds round, dark red-brown, ca. 0.7 mm. in diam.—On dried mud, moist flats, etc., often at low elevs., but reaching 8000 ft. at Soda Springs, Nevada Co.; many Plant Communities; native from San Diego Co. through cismontane Calif. to Modoc Co.; to Wash., Alta., Nev. July–Oct.

10. **A. álbus** L. [*A. graecizans* auth., not L.] TUMBLEWEED. Stems erect or ascending, bushy-branched, pale, 2–15 dm. long; lvs. elliptic to spatulate or obovate, obtuse or retuse, 1–6 cm. long, the petioles short; monoecious, the fls. greenish, in small axillary clusters; bracts subulate, green, rigid, 2–4 times as long as sepals; sepals 3, scarcely 1 mm. long; stamens 3; utricle rugose, exceeding calyx; seeds round, dark red-brown, shining, ca. 0.8 mm. in diam.; $2n = 32$ (Heiser & Whitaker, 1948).—Common weed in cult. and waste areas through much of N. Am.; native of trop. Am. June–Oct. Blowing about as a tumbleweed when dry.

11. **A. spinosus** L. Glabrous, bushy, with reddish stems 2–12 dm. long; lvs. lance-ovate to rhombic-ovate, 2–10 cm. long, dull green, with pair of axillary spines; fertile clusters round, axillary; upper clusters sterile, spicate; fls. yellow-green, 1.5 mm. long; utricle thin, bursting irregularly; seeds smooth, black, shining, 0.8–1 mm. in diam.; $2n = 34$ (Takagi, 1933).—Occasional, as in Antelope V., Los Angeles Co.; native of Old World.

2. Tidestròmia Standl.

Annual or perennial herbs or small shrubs, closely grayish stellate-tomentose or villous. Lvs. largely opposite, petioled. Fls. minute, perfect, axillary, mostly glomerate, subtended by leafy involucral bracts and 3 small bractlets. Sepals 5, equal, thin, pubescent. Stamens 5, the fils. united below into a short cup, with 5 short teeth alternating with them. Utricle subglobose, 1-seeded, indehiscent. Three spp. of sw. U.S. and Mex. (I. *Tidestrom*, Am. botanist.)

Plant perennial; staminodia almost half as long as fils. Deserts .... 1. *T. oblongifolia*
Plant annual; staminodia almost lacking. W. San Diego Co. .... 2. *T. lanuginosa*

1. **T. oblongifòlia** (Wats.) Standl. [*Cladothrix o.* Wats. *C. cryptantha* Wats. *T. o.* ssp. *c.* Wiggins.] Perennial from a stout woody taproot, erect to decumbent, much-branched, forming low broad plants 1.5–3 dm. high and ca. twice as wide; lf.-blades oblong to broadly ovate, 1–3 cm. long, obtuse, prominently veined, on somewhat shorter petioles; invol. of 3 broad partly united bracts 3–4 mm. long; calyx 1–2 mm. long; seeds 0.5 mm. long.—Dry sandy places, such as washes, mostly below 2000 ft.; Creosote Bush Scrub; Colo. Desert through e. Mojave Desert to Death V.; Nev., Ariz. April–Dec.

2. **T. lanuginòsa** (Nutt.) Standl. [*Achyranthes l.* Nutt. *Alternanthera l.* Moq.] Prostrate or procumbent annual, much-branched, the branches 1–5 dm. long, stellate-pubescent; lf.-blades round to round-ovate, 0.5–3 cm. long, stellate-pubescent, on slender petioles as long or shorter; glomerules few-fld.; perianth 1–3 mm. long, 3 times the bracts in length; staminodia minute or wanting; seed 0.5 mm. long.—Established on banks of Otay Lake, San Diego Co.; native from Utah to Kans. and Mex. July–Oct.

3. Alternánthera Forsk.

Herbs or subshrubs, erect to prostrate. Lvs. opposite, sessile or petioled. Fls. bracteate and bibracteolate, perfect, in short axillary spikes. Calyx of 5 often unlike sepals. Stamens mostly 5, sometimes 2 or 3; the fils. united into a tube with staminodia alternating with the stamens. Utricle membranaceous, indehiscent; stigma mostly capitate. Seeds inverted, smooth. Ca. 170 spp. of the warmer parts of the New World. (Latin, *alternus*, alternate, referring to alternate stamens and staminodia.)

1. **A. rèpens** (L.) Kuntze. [*Achyranthes r.* L. *Illecebrum Achyranthes* L. *Alternanthera Achyrantha* R. Br.] Perennial with prostrate stems and thick woody vertical root; stems branched, pubescent, 2–5 dm. long; lvs. many, rhombic-ovate, green, those of the pair very unequal, the blades 1–3 cm. long, petioles shorter; heads short-cylindric or ovoid, 5–15 mm. long, sessile, whitish; bracts ovate, ciliate; sepals unequal, ovate, villous on the 3 nerves, 3–5 mm. long.—Occasional as escape, as at Oceanside, Los Angeles, Ventura; native of Mex. Fall.

2. **A. philoxeroìdes** (Mart.) Griseb. ALLIGATOR WEED. Subglabrous aquatic plant; lvs. 3–11 cm. long; heads on long peduncles.—Introd. in Rio Hondo at San Gabriel Blvd., Los Angeles Co.; se. U.S.; S. Am.

42. Phytolaccàceae. POKEWEED FAMILY

Herbs to trees. Lvs. alternate, entire, usually exstipulate. Fls. bisexual or unisexual, mostly in axillary or terminal racemes, regular. Calyx 4–5-parted, persistent. Petals usually 0. Stamens as many as calyx-segms. and alternate with them, or more numerous, hypogynous, distinct or basally united. Ovary usually superior with 1–many distinct or united pistils; styles as many as pistils or none; stigmas linear or filiform. Ovules solitary. Fr. in ours of several locules and forming a berry. Ca. 17 genera and 100 spp., of warmer parts of Am. and of S. Afr.

1. Phytolácca L. POKEWEED. POKEBERRY

Tall stout perennial herbs. Lvs. large, petioled in ours, ovate to oblong-lanceolate. Fls. small. Sepals 5, petaloid. Stamens 5–30. Ovary of 5–12 carpels united in a ring. Fr. a 5–12-locular berry, with a single vertical seed in each locule. Ca. 35 spp., of trop. or subtrop. regions, especially Am. (Greek, *phyton*, plant, and Latin, *lacca*, crimson lake, because of the color in the berries.)

1. **P. americàna** L. [*P. decandra* L.] Strong-smelling, 1–3 m. high; lvs. 1–3 dm. long; fls. bisexual, white, or pinkish, ca. 5–6 mm. across, in peduncled racemes 5–20 cm. long; fr. dark purple, ca. 10–12 mm. across; $2n = 36$ (Suzuka, 1950).—Occasional as weed in cult. areas, San Diego Co. to Siskiyou Co.; native of e. U.S. Aug.–Oct.

43. Nyctáginaceae. FOUR-O'CLOCK FAMILY (Fig. 40)

Largely herbs or shrubs, with ± swollen nodes and fragile stems. Lvs. subentire, petiolate, exstipulate, mostly opposite. Fls. perfect, regular, subtended by bracts which are frequently united into calyxlike invols. Perianth corollalike, usually campanulate or funnelform or salverform, 4–5-lobed or -toothed, constricted above the ovary. Petals none. Stamens hypogynous; fils., filiform; anthers 2-celled. Ovary superior, enclosed by the persistent perianth-tube, 1-celled, 1-ovuled; style short or long; stigma capitate to linear. Fr. an anthocarp, closely invested by the hardened base of the perianth, usually striate, angled or winged and enclosing the free ak. Seed erect, the embryo curved in ours, the cotyledons enclosing the mealy or fleshy endosperm. Ca. 20 genera and 150 spp., of warmer parts of world.

Stigma linear; fr. usually winged; perianth salverform .... 8. *Abronia*
Stigma subglobose; fr. rarely winged; perianth usually funnelform or campanulate.
  Fr. lenticular, the margins inrolled, dentate, the dorsal surface with 2 rows of stipitate glands; invols.
  3-parted .... 5. *Allionia*
  Fr. not lenticular or with dentate margins or glands.
    Invol. of distinct bracts, 1 at base of each fl.
      Perianth 10 or more cm. long, white, tubular, usually solitary .... 3. *Acleisanthes*
      Perianth not over 4 cm. long.
        The perianth not more than 1 cm. long; fr. 5-angled or -ribbed .... 4. *Boerhaavia*
        The perianth 2–4 cm. long.
          Fr. ellipsoid, smooth, with vertical lines; each pedicel partially attached to the leafy sub-
          tending bract; perianth purplish-red .... 1. *Hermidium*
          Fr. with 3–5-hyaline wings; pedicels not attached to subtending bracts; perianth greenish
                                               2. *Selinocarpus*
    Invol. of united bracts, resembling a gamosepalous calyx.
      Fr. 5-ribbed; invol. enlarged and papery in fr. .... 6. *Oxybaphus*
      Fr. smooth or nearly so; invol. but little changed in fr. .... 7. *Mirabilis*

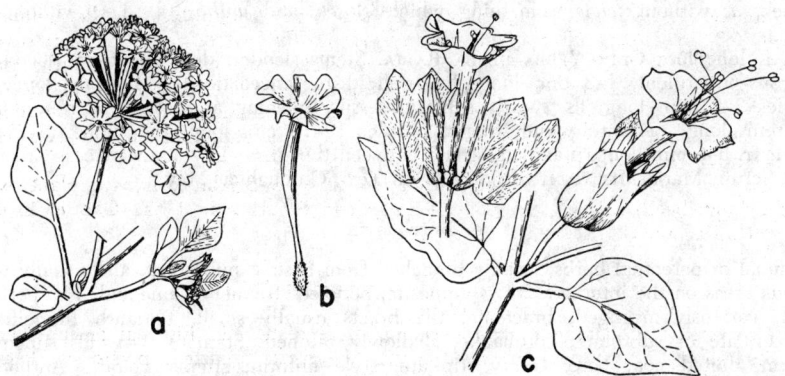

Fig. 40. NYCTAGINACEAE. *Abronia villosa: a,* habit, × ½, with cluster of fls. subtended by separate bracts; *b,* fl., × 1, salverform perianth and inferior pubescent ovary. *Mirabilis Froebelii: c,* habit, × ½, cluster of fls. surrounded by a calyxlike invol. (the one to upper left opened up).

1. Hermídium Wats.

Perennial nearly glabrous herbs with erect dichotomous stems. Lvs. opposite, thick, entire, short-petioled. Fls. in axillary or terminal peduncled headlike clusters, each fl. subtended by a broad foliaceous bract of which the midvein is partially attached to the pedicel of the fl. Perianth campanulate-funnelform; stamens 5–7; fils. unequal, united at base. Ovary globose; style capillary; stigma capitate. Fr. ellipsoid, smooth, usually with 10 vertical lines. Seeds adherent to pericarp; embryo uncinate, the cotyledons orbicular. Monotypic. (Diminutive of *Hermes,* Greek god.)

1. **H. álipes** Wats. Plants 2–4 dm. high, the stems sparsely branched, glaucous, glabrous or obscurely puberulent upward; lf.-blades 2–7 cm. long, round to broadly ovate, on petioles 1–10 mm. long; peduncles 3–12 mm. long; heads 4–6-fld.; bracts oblong to ovate, 1–2 cm. long, green; perianth purplish-red, 2–2.5 cm. long; fr. 6–7 mm. long, almost globose.—Dry slopes and flats, 4000–6500 ft.; Sagebrush Scrub, Pinyon-Juniper Wd.; Panamint Mts. to White Mts.; to Nev., Utah. May–June.

2. Selinocárpus Gray

Perennial herbs with dichotomous branching. Lvs. opposite, petioled, the blades thick. Fls. in clusters, sessile or pedicelled, each subtended by 2–3 narrow bracts, axillary or terminal. Perianth tubular-funnelform. Stamens 3–5. Ovary oblong. Fr. with 3–5 hyaline wings; seed with testa adherent to pericarp; embryo conduplicate. Several spp. of sw. N. Am. (*Selinum,* a genus of Umbelliferae, and *karpos,* fr.)

1. **S. diffùsus** Gray. [*S. d.* ssp. *nevadensis* Standl.] Diffusely branched, 1–3 dm. high, with short appressed white inflated hairs and sparsely glandular; lf.-blades oval to roundish, 1–2.5 cm. long; petioles 3–20 mm. long; fls. solitary or in 2's, often cleistogamous, short-pedicellate, the bracts linear-subulate, 3–6 mm. long; perianth greenish, 3–4 cm. long; fr. 6–7 mm. long, 4–5-winged, truncate, the body puberulent.— Dry often rocky places; Creosote Bush Scrub, Joshua Tree Wd.; San Bernardino Co.– Nev. border. June–Sept.

3. Acleisánthes Gray

Perennial herbs or low shrubs. Lvs. thick, opposite, the blades unequal, entire, petioled. Fls. axillary or terminal, each subtended by 1–3 small narrow bracts, solitary or in 2–3-fld. cymes. Perianth funnelform, with long tube and shallowly 5-lobed limb, constricted above the ovary. Stamens 2–5, unequal, exceeding perianth-tube. Ovary ovoid. or oblong; style exserted; stigma capitate. Fr. narrow-ellipsoid, 5-ribbed or -angled, the ribs smooth, sometimes ending in a gland. Seed with testa adherent to pericarp; cotyledons broad, enclosing the copious endosperm. Several spp. of sw. N. Am.

(Greek, *a,* without, *cleis,* something which closes, and *anthos,* fl., i.e., without an invol.)

1. **A. longiflòra** Gray. YERBA DE LA RABIA. Stems slender, decumbent or ascending, scabrous-puberulent, 1–3 dm. long; lvs. deltoid to lanceolate, 1.5–4.5 cm. long, on petioles 3–8 mm. long; fls. mostly solitary, sessile, opening at night; bracts subulate, 2–3 mm. long; perianth white, tinged purple, 10–16 cm. long, the limb 1.5–2 cm. broad; fr. 5–6 mm. long, 5-angled, striate between the ribs.—Dry stony places; Creosote Bush Scrub; Maria Mts., e. Riverside Co.; to Tex., Chihuahua. May.

4. Boerhaàvia L.

Annual or perennial herbs, usually branched from base, ± pubescent, and usually with viscous areas on the internodes. Lvs. opposite, petioled, frequently unequal. Fls. perfect, small, variously arranged, bracteate, the bracts usually small. Perianth funnelform, campanulate or subrotate, corollalike, shallowly 5-lobed. Stamens 1–5; fils. unequal, filiform, united near base. Ovary stipitate; style filiform; stigma peltate. Anthocarp obovoid or obpyramidal, 3–5–10-nerved, -angled, or -winged. Seed with curved embryo. About 30 spp., in the warm parts of both hemis. (Hermann *Boerhaave,* 1668–1738, botanist of Leiden.)

Fr. 10-nerved; perianth funnelform .... 1. *B. annulata*
Fr. 3–5-nerved or -angled; perianth campanulate.
  Plant perennial; fr. pubescent .... 2. *B. coccinea*
  Plant annual; fr. glabrous.
    Ultimate branches of infl. with fls. in racemose spikes.
      Bracts persistent, as long as the mature fr., which is 4-angled; stamens 3–4 .... 3. *B. Wrightii*
      Bracts deciduous, shorter; fr. 5-angled; stamens 1–2 .... 4. *B. Coulteri*
    Ultimate branches of infl. with fls. in umbels or solitary on slender pedicels.
      Fls. in umbels; fr. ca. 3 times as long as thick .... 5. *B. erecta*
      Fls. mostly solitary, forming cymes; fr. ca. twice as long as thick .... 6. *B. triquetra*

1. **B. annulàta** Cov. [*Anulocaulis a.* Standl.] Coarse perennial with 1 to few sub-erect stems, 3–10 dm. tall, these subglabrous, with reddish viscous semioblique band on each internode; lvs. oblong-ovate to suborbicular, bright green above, paler beneath, reddish-veined, hirsute, the hairs with small dark pulvini at base; blades 2–8 cm. long, rounded to acute at apex, irregularly repand-dentate, often cordate at base; petioles 1–6 cm. long; infl. paniculate, leafless, 3–8 dm. long, the fls. in dense headlike clusters at the ends of the branches; bracts short, persistent, hairy; perianth 6–8 mm. long, pinkish or greenish; stamens 3, exserted; fr. glabrous, 4–5 mm. long, thick-fusiform.—Sandy washes and gravelly slopes, below 3000 ft.; Creosote Bush Scrub; Death V. region of Calif. and Nev. April–May.

2. **B. coccínea** Mill. [*B. caribaea* Jacq. *B. hirsuta* Willd. *B. viscosa* Lag. & Rodr.] Perennial with branching decumbent or prostrate ± glandular-pubescent stems 2–9(–14) dm. long; lvs. round-ovate to oblong-ovate, mostly obtuse, irregularly and shallowly undulate, green above, paler beneath, glabrous to hirsute, even viscid, the blades mostly 1–3 cm. long, the petioles 0.5–2 cm. long; infl. cymose, with many slender glandular-pubescent branches, 1–4 dm. long, the fls. in heads on slender peduncles; perianth purplish-red, 2 mm. long; stamens 1–3, short-exserted; fr. clavate, glandular-pubescent, 5-ribbed, the sulci not wrinkled, broader than the ribs, 2.5–3.5 mm. long.—Dry disturbed places, such as road banks, railroads, burns, washes, etc., mostly below 3000 ft., Creosote Bush Scrub, Colo. Desert; V. Grassland, S. Oak Wd.; San Jacinto V., Santa Ana R., and Tulare Co.; to s. U.S. and trop. Am. April–July.

3. **B. Wrightii** Gray. Erect annual 2–6 dm. high, branched from base, with stems slender, glandular-pubescent; lvs. toward base of plant oblong-ovate to lanceolate, green above, paler and glandular-punctate beneath, undulate, the blades 1–4 cm. long, the petioles 0.5–2 cm. long; infl. glabrous to glandular-pubescent, open and branched, the fls. in loose spikes; bracts 2–3 mm. long, ovate or almost orbicular, often reddish, villous, persistent; perianth pink, 1–2 mm. long; stamens 3–4, included; fr. broadly clavate, 2 mm. long, mostly 4-angled, the angles broad, acute, with broad rugulose sulci.—Sandy and gravelly flats and washes, below 4500 ft.; Creosote Bush Scrub, Joshua Tree Wd.; Mojave and Colo. deserts; to Nev., Tex., Mex. Aug.–Dec.

4. **B. Coùlteri** (Hook. f.) Wats. [*Senkenbergia C.* Hook. f.] Erect or decumbent annual 3–8 dm. high, much-branched, puberulent; lvs. lanceolate to rhombic-ovate,

acutish, entire to sinuate, crisped, green above, paler beneath, glabrous or nearly so, the blades 1.5–5 cm. long, the petioles 0.5–2 cm. long; infl. much branched, with slender branches ending in very slender spicate racemes; bracts ca. 1 mm. long, lanceolate or wider, deciduous, obscurely ciliolate; perianth pink or white, glabrous, 1–1.5 mm. long; stamens 1–2, included; fr. 2.5 mm. long, narrow-obovoid, 5-angulate, the angles broad, smooth, almost touching each other, the sulci linear, scarcely rugose.—Rare, washes and flats, below 4000 ft.; Creosote Bush Scrub; Colo. Desert (Hayfields, Chuckawalla Mts., Valley Wells, San Felipe V.); to Ariz., Mex. Sept.–Nov.

5. **B. erécta** L. var. **intermèdia** (Jones) Kearn. & Peeb. [*B. i.* Jones.] Erect or ascending annual 2–5 dm. high, freely branched, puberulent; lvs. oblong-lanceolate or broader, obtuse to acute, glabrous or nearly so, entire or somewhat sinuate, green above, paler beneath, frequently brown-punctate, the blades 1.5–4 cm. long, the petioles 0.3–2 cm. long; infl. cymose, with small umbels at tips of very slender branches; bracts lanceolate, minute, persistent; perianth pink, 1.5–2 mm. long; stamens 2–3, scarcely or slightly exserted; fr. narrowly obpyramidal, 2–2.5 mm. long, truncate, 5-angled, the sulci narrow, transverse-rugulose.—Infrequent, dry washes and flats, below 4000 ft.; Creosote Bush Scrub; s. San Bernardino Co. to Imperial Co.; to Tex., Mex. Aug.–Oct.

6. **B. triquètra** Wats. Slender-stemmed ascending to spreading annual, branched from base, 1.5–6 dm. high, minutely puberulent; lvs. oblong to narrowly lanceolate, green above, paler beneath, obtuse to slender-pointed, undulate-crisped, ± brown-punctate, the blades 1–3 cm. long, the petioles 0.3–1.5 cm. long; infl. open-cymose, with slender glabrous branches; fls. pink, mostly solitary, short-pedicellate; bracts minute; perianth 1–1.3 mm. long; stamens 2–3, ca. as long as perianth; fr. obpyramidal, truncate, 3–5-angled, 2–2.5 mm. long, more than half as thick, the angles broad, smooth, acute, with open rugulose furrows.—Sandy washes and open gravelly flats, below 5500 ft.; Creosote Bush Scrub, Joshua Tree Wd.; Little San Bernardino Mts. s. and e.; Ariz., L. Calif. Sept.–Dec.

5. Alliònia L.

Annual or perennial herbs, usually glandular-pubescent, with dichotomous prostrate branches. Lvs. petioled, very unequal in the pair, oblong to broadly ovate. Fls. perfect, in axillary peduncled clusters of 3, each subtended by a bract that encloses the fr. Perianth campanulate-rotate, 4–5-lobed. Stamens 4–7, exserted. Stigma capitate. Fr. coriaceous, flattened, the dorsal (seemingly inner) face bearing 2 rows of stipitate glands. Embryo curved, the broad cotyledons enclosing the endosperm. Ca. 3 spp., extending into S. Am. (C. *Allioni*, 1725–1804, Italian botanist.)

1. **A. incarnàta** L. [*Wedelia i.* Kuntze. *Wedeliella i.* Ckll. *Wedelia i.* sspp. *villosa* and *nudata* Standl. *A. i.* var. *n.* Munz.] WINDMILLS. TRAILING-FOUR-O'CLOCK. Perennial or winter annual with slender trailing simple or branched villous to pubescent glandular stems; lf.-blades ovate or broader, usually rounded at base, 2–3 cm. long (or upper reduced), on petioles shorter to as long; peduncles 5–25 mm. long; bracts round-ovate, 5–9 mm. long; perianth rose-magenta, rarely white, opening in morning, 6–15 mm. long; frs. 3–4.5 mm. long, straw-colored, the inner side 3-nerved, the margins usually toothed, incurved.—Dry stony benches and slopes, below 5000 ft.; Creosote Bush Scrub; Mojave and Colo. deserts; to Colo., Chihuahua, S. Am. April–Sept.

6. Oxýbaphus L'Hér. ex Willd.

Perennial herbs, the stems tall and erect to low and decumbent, mostly from a woody root. Lvs. opposite, usually thickish, entire, sessile or petiolate. Fls. perfect, 1–5 in a calyxlike invol. which is gamophyllous, 5-lobed, enlarged and papery in fr., reticulate-veined. Perianth campanulate to short-funnelform, slightly oblique, the limb 5-lobed. Stamens unequal, 3–5, slightly united below. Stigma depressed-capitate. Fr. ± obovoid, constricted at base, 5-angled or -ribbed, mucilaginous when wet. Seed adherent to pericarp; embryo curved, the cotyledons enclosing the abundant endosperm. Ca. 25 spp., of the warmer parts of Am.; 1 Himalayan.

Lf.-blades sessile or subsessile, at least 5 times as long as wide.
   Perianth bright red, 3–4 times as long as invol. E. Mojave Desert . . . . 1. *O. coccineus*
   Perianth pink or purplish-red, ca. twice as long as invol. Cismontane . . . . 2. *O. linearis*

Lf.-blades plainly petioled, usually not more than 3 times as long as wide.
Invols. usually glabrous at anthesis except at very base, 15–20 mm. broad in fr. . . 3. *O. nyctagineus*
Invols. viscid-pilose at anthesis, not over 15 mm. wide in fr.
    Stems much branched, viscid and pilose or villous to base; lvs. scarcely longer than wide
                                                                                    4. *O. pumilus*
    Stems branched only in infl., glabrous below or puberulent in lines; lvs. longer than wide
                                                                                    5. *O. comatus*

1. **O. coccíneus** Torr. [*Mirabilis c.* Benth. & Hook. *Allionia c.* Standl.] Stems 1–many from an elongate woody root, erect or ascending, glaucous, glabrous, 3–6 dm. long; lf.-blades linear or filiform, 2–12 cm. long, 1–5 mm. wide; infl. loosely cymose; peduncles slender; invols. short-pilose, 4–5 mm. long at anthesis, deeply lobed, 5–6 mm. long at maturity; perianth opening at night, deep red, 12–15 mm. long, the tube 3–4 mm. in diam., the limb 11–15 mm. broad; stamens exserted; fr. 5-ribbed, somewhat rugose, 5 mm. long; seed broadly obovoid, 2.5 mm. long.—Dry rocky slopes and along washes, 4000–5300 ft.; Pinyon-Juniper Wd.; Providence, New York, and Ivanpah mts., San Bernardino Co.; to New Mex., Son. May–July.

2. **O. lineàris** (Pursh) Rob. [*Allionia l.* Pursh.] Much like *O. coccineus* in habit, the stems glabrous or puberulent below, more viscid-pubescent above; lvs. linear to lance-linear, 3–10 cm. long, mostly 1–5 mm. wide; invol. ca. 4 mm. long at anthesis, 6–10 mm. long in fr., viscid-villous; perianth pale pink to purple-red, ca. 10 mm. long; fr. ca. 5 mm. long, the angles smooth, the sides transversely rugose.—Occasional escape along Santa Ana R. system (Riverside, Rancho Santa Ana); native, Ariz. to S. Dak., Mo., Mex. July–Oct.

3. **O. nyctagíneus** (Michx.) Sweet. [*Allionia n.* Michx.] Stems many, erect to decumbent, 3–10 dm. high, subglabrous; lf.-blades lance-ovate to subdeltoid, 2–8 cm. long, on petioles 1–3 cm. long; infl. cymose; invol. 5–6 mm. long in anthesis, 10–15 mm. long in fr., puberulent or short-pilose near base, the glabrous lobes ciliate; perianth white or pale pink, ca. 10 mm. long, the limb 12–15 mm. broad; fr. 5 mm. long, short-pilose, the angles broad, the sides rugulose; $2n = 58$ (Bowden, 1945).—Adv., as at Upland, San Bernardino Co.; native from Wis. and Mont. to Mex. June–Aug.

4. **O. pùmilus** (Standl.) Standl. [*Allionia p.* Standl. *A. Brandegei* Standl. *O. B.* Weath.] Stems few to many from a woody root, ascending to procumbent, 1–5 dm. long, short-pilose throughout; lvs. deltoid to ovate-deltoid, 2–6 cm. long, succulent, puberulent or short-pilose, on petioles 1–2 cm. long; infl. axillary or in narrow cymes, viscid-pilose, plainly bracteate; invols. 3–4 mm. long at anthesis, 7–8 mm. long in fr., densely viscid-pilose; perianth 8–10 mm. long, pale pink, pilose; stamens exserted; fr. 5 mm. long, short-pilose, rugose; seed 2.5–3 mm. long, obovoid.—Dry rocky and gravelly places, 4500–7800 ft.; Pinyon-Juniper Wd., Yellow Pine F.; about the Mojave Desert, as San Bernardino Mts., Ivanpah Mts., Clark Mt., Kingston Mts., Inyo Mts.; to Nev., New Mex. June–Aug.

5. **O. comàtus** (Small) Weath. [*Allionia c.* Small.] Stems 1–few, 1–5 dm. long, ascending, simple at least below the infl., glabrous below or with lines of hair; lf.-blades narrow-deltoid, 1.5–4 cm. long, ± puberulent, on petioles 5–15 mm. long; invols. few to many, 3–5 mm. long at anthesis, ca. 8 mm. long in fr., viscid-pilose; perianth purplish-red, 10–12 mm. long, the limb 1–2 cm. broad; stamens long-exserted; fr. obovoid, 3–4 mm. long, the angles broad, tuberculate; seed 2–3 mm. long.—Dry slopes, Pinyon-Juniper Wd.; Ivanpah Mts., e. San Bernardino Co.; Nev. to Tex., Mex. May–June.

7. Mirábilis L. Four-O'Clock

Perennial often suffrutescent herbs with repeatedly dichotomous stems. Lvs. opposite, petiolate to sessile, entire. Fls. 1–several in a 5-lobed calyxlike usually campanulate invol.; invols. axillary, often clustered near ends of branches. Perianth surpassing invol., funnelform to campanulate, white to red or red-purple. Stamens 5(–3), unequal; fils. capillary. Fr. subglobose to elongate, smooth or obscurely angled. Ca. 20 spp., of warmer parts of Am. (Latin, *mirabilis*, wonderful.)

(Standley, P. C. N. Am. Fl. 21: 231–240, 1918.)
Invol. 1-fld., less than 1.5 cm. long; perianth campanulate.
    Perianth 3–5 cm. long; fr. 5-angled. Garden escape . . . . 1. *M. Jalapa*
    Perianth 1–1.5 cm. long; fr. smooth. Native.

Involucral lobes lanceolate, longer than tube; invol. 11–13 mm. long; lf.-blades mostly 3–4 cm. long. W. Colo. Desert .... 2. *M. tenuiloba*
Involucral lobes subovate, not exceeding tube; invol. 5–9 mm. long; lf.-blades mostly less than 2.5 cm. long.
    Fls. white to pale pink; lvs. viscid-villous. Deserts .... 3. *M. Bigelovii*
    Fls. rose to red; lvs. subglabrous to rough-pubescent. Cismontane .... 4. *M. laevis*
Invol. 3–10-fld., 3.5–5 cm. long; perianth funnelform.
    Fr. 5-angled, tubercled; pubescence minute to none. Nw. Calif. .... 5. *M. Greenei*
    Fr. not angled, but with 10 vertical lines; pubescence usually present, glandular. Deserts
                                                    6. *M. Froebelii*

1. **M. Jalàpa** L. Erect, 3–9 dm. high, glabrous or nearly so; lvs. ovate, 5–10 cm. long, truncate or cordate at base, acuminate at apex, the petioles half as long as blades; fls. ca. 2.5 cm. across, of many different colors, opening in late p.m.; $2n = 58$ (Showalter, 1935).—The cult. Four-O'Clock, occasionally reported as garden escape; native of trop. Am. Summer.

2. **M. tenuilòba** Wats. [*Hesperonia t.* Standl.] Woody at base, erect to decumbent, 3–5 dm. high, glandular-pubescent to -villous; lf.-blades ovate to broadly deltoid, 2.5–5 cm. long, rounded or cordate at base, usually acute, glandular-pubescent; petioles 2–8 mm. long; invols. almost sessile, 11–13 mm. long, narrow-campanulate, the lobes slightly unequal, slightly longer than tube, narrow-lanceolate; perianth whitish, pubescent, 12–15 mm. long; fr. 5 mm. long, ovoid, smooth, brown.—Occasional, rocky slopes, below 1500 ft.; Creosote Bush Scrub; w. edge of Colo. Desert; L. Calif. March–May.

3. **M. Bigelòvii** Gray. [*Hesperonia B.* Standl. *M. laevis* var. *glutinosa* Jeps., in part.] Erect or ascending, much-branched, villous-viscid, 3–5 dm. high, the stems slender; lf.-blades ovate to reniform-ovate, 1–3(–4) cm. long, obtuse, viscid-villous; petioles 3–10 mm. long; invols. clustered at ends of branches, short-peduncled, 5–6 mm. long, the lobes ovate, shorter than the tube; perianth white to pale pink or pale lavender, 8–12 mm. long; fr. dark, ovoid, elongate, often mottled, smooth or slightly rugulose, ca. 3 mm. long.—Rocky places, especially in canyons, below 7000 ft.; Creosote Bush Scrub to Pinyon-Juniper Wd.; mostly e. Colo. and e. Mojave deserts to Inyo Co.; Nev., Ariz. March–June; Oct.–Nov.

Var. **áspera** (Greene) Munz. [*M. a.* Greene. *Hesperonia a.* Standl.] Stems viscid-villous; fr. subglobose, with 10 pale vertical lines.—Above 3000 ft., with the sp. and also in w. part of deserts.

Var. **retrórsa** (Heller) Munz. [*M. r.* Heller. *Hesperonia glutinosa* Standl. *M. g.* A. Nels., not Kuntze. *M. limosa* A. Nels.] Stems subglabrous below, scabrous-puberulent above with short retrorse hairs, also somewhat glandular; frs. subglobose, sometimes striate.—Common, mostly below 4000 ft.; Creosote Bush Scrub; both deserts to Mono Co.; to Utah, Ariz. March–June.

4. **M. laèvis** (Benth.) Curran. [*Oxybaphus l.* Benth. *Quamoclidion l.* Rydb. *Hesperonia l.* Standl. *M. californica* Gray.] WISHBONE BUSH. Somewhat woody at base with many decumbent stems, slender, subglabrous or commonly viscid-puberulent, repeatedly forked, 1.5–8 dm. long; petioles 2–15 mm. long; lf.-blades ovate, mostly acute, 1–2.5(–3.5) cm. long, glabrate to glandular-pubescent; invols. clustered near ends of branches, short-peduncled, campanulate, 5–8 mm. long, the lobes subovate, shorter than the tube; perianth rose to purplish-red, 10–14 mm. long; fr. broadly ovoid, 5 mm. long, dark, sometimes mottled or pale-striate, smooth.—Common in dry stony washes and on slopes, below 2000 (3000) ft.; Coastal Sage Scrub, Chaparral, Foothill Wd.; Monterey Co. and w. Merced Co. to San Diego Co. and to the edge of the desert, also on Santa Cruz and Santa Catalina ids. Mostly Dec.–June.

Var. **cedrosénsis** (Standl.) Munz. [*Hesperonia c.* Standl.] Stems scabrous-puberulent with short conical hairs; perianth ca. 12 mm. long; fr. subglobose, dark brown with 10 light lines.—San Clemente Id.; L. Calif. A form from San Clemente Id. with perianth 5–8 mm. long and stems viscid-puberulent has been named var. *cordifòlia* Dunkle.

5. **M. Greènei** Wats. [*Quamoclidion G.* Standl.] Erect or ascending, 3–6 dm. high, sparsely branched, glabrous or with some fine pubescence about the infl.; petioles stout, 4–25 mm. long; lf.-blades round-ovate to ovate-oblong, acute, 4–10 cm. long; invols. 5–8-fld., campanulate, 2.5–4 cm. long, 5-lobed, the lobes ½ to ⅓ the length of the whole; perianth rose-purple, 3.5–5 cm. long; fr. elliptic-oblong, obscurely 5-angled, dark brown, 5 mm. long, rugulose.—Dry slopes and gravelly flats, below 3500 ft.; largely N.

Juniper Wd., Foothill Wd.; inner Coast Ranges from Colusa Co. to Siskiyou Co. May–June.

6. **M. Froebèlii** (Behr) Greene. [*Oxybaphus F.* Behr. *Quamoclidion F.* Standl.] From a thick woody tuberous root, erect or ascending, 3–8 dm. high, much-branched, densely villous-viscid throughout; petioles stout, 3–20 mm. long; lf.-blades broadly ovate, 3–8 cm. long, often subcordate at base, acute to rounded at apex; invols. peduncled, solitary in lower axils, or clustered at tips of branches, campanulate, 2.5–4 cm. long, the lobes ca. ⅓ the whole; perianth rose-purple to deep pinkish, 3.5–4.5 cm. long, funnelform, the limb 2–2.5 cm. broad; fr. elliptic-ovoid, smooth, 8 mm. long, with 10 vertical light-colored lines.—Dry stony places, below 6500 ft.; Creosote Bush Scrub to Pinyon-Juniper Wd.; deserts from L. Calif. n., to interior San Luis Obispo Co., Kern Co. and Mono Co.; Nev. April–Aug.

Var. **glabràta** (Standl.) Jeps. [*Quamoclidion F.* ssp. *g.* Standl.] Herbage glabrous.—Occasional, Providence Mts., New York Mts. (e. Mojave Desert) and w. edge of Colo. Desert; Nev., Ariz., L. Calif.

8. Abrònia Juss. Sand-Verbena

Annual or perennial often viscid branching herbs, usually prostrate or decumbent. Lvs. opposite, unequal, slightly to quite fleshy, petioled, the base often unequal. Fls. perfect, capitate, the heads on axillary or terminal peduncles and subtended by an invol. of 5–8 scarious bracts. Perianth fragrant, often showy, salverform with slender tube, enlarged at throat, the limb with 4–5 emarginate lobes and withering-persistent. Stamens 4–5, unequal, included in the tube and inserted upon it. Stigma fusiform, included, on a filiform style. Fr. an anthocarp, turbinate or biturbinate, deeply lobed or winged. Seed adherent to the pericarp, dark brown, elliptic-oblong or narrower, one cotyledon broad, the other abortive. Perhaps 25 spp., of w. N. Am. (Greek, *abros*, graceful.)

(Standley, P. C. N. Am. Fl. 21: 240–254, 1918.)
Perianth yellow; lvs. as wide as long. Seashore .... 1. *A. latifolia*
Perianth white to rose or purple-red.
  Frs. with 2–5 conspicuous wings.
    Fls. deep dark red; fr. 12–20 mm. broad. Seashore .... 2. *A. maritima*
    Fls. white to rose; fr. often narrower.
      Plants with a perennial branched caudex; lvs. all basal; fls. scapose. Montane .... 7. *A. nana*
      Plants annual or perennial, but not with a branched caudex; lvs. largely on elongate stems; fls. not scapose.
        Wings of fr. translucent, extended around the body (above and below it); fls. mostly 4-merous.
          Perianth 20–25 mm. long; body of fr. villous. Lassen Co. .... 9. *A. Crux-Maltae*
          Perianth 15 mm. long; body of fr. not villous. San Bernardino Co. ...... 10. *A. micrantha*
        Wings of fr. opaque, interrupted above and beneath the body; fls. mostly 5-merous.
          The wings of the fr. usually 2; bracts broadly ovate; plants annual. Deserts
                                                         4. *A. pogonantha*
          The wings of the fr. usually 3–5; bracts commonly lanceolate.
            Stems and perianth densely long-villous; body of fr. rugose-veined, the veins coarse and extending into the wings. Annual. Interior valleys and deserts ..... 5. *A. villosa*
            Stems and perianth glabrous to short-villous; body of fr. not rugose-veined. Mostly perennial. Seashore .... 6. *A. umbellata*
  Frs. not winged, merely angled; annuals.
    Fls. 1–5 in each head; plants perennial, matted .... 3. *A. alpina*
    Fls. 15–30 in a head; plants annual, open in growth .... 8. *A. turbinata*

1. **A. latifòlia** Eschs. Much-branched perennial with stout fleshy roots and prostrate stems 3–10 dm. long, densely glandular-pubescent; lf.-blades round to oval, 1–4(–6) cm. long, thick and succulent, glandular-puberulent to glabrate; petioles stout, 1–6 cm. long; peduncles 2–6 cm. long, viscid; bracts 5–6, ovate to lance-ovate, 6–8 mm. long; perianth 13–18 mm. long, the tube slender, glandular-villous, yellow-green, the limb 5–8 mm. broad, yellow, with shallowly emarginate lobes; fr. 8–15 mm. long, turbinate, with 5 winglike reticulate-veined lobes attenuate upward on the fr.-body; seed elliptic-oblong, brown, 4–5 mm. long.—Coastal Strand; San Miguel Id. and Surf, Santa Barbara Co. to B.C. May–Oct. Hybridizing occasionally with *A. umbellata.*

2. **A. marítima** Nutt. ex Wats. Perennial with fleshy roots and succulent much-branched prostrate glandular-puberulent and somewhat villous stems 2–10 dm. long; lf.-blades oval to oval-oblong, thick, 2–6 cm. long, viscid-puberulent; petioles 1–2 cm. long; peduncles 3–8 cm. long; bracts lanceolate to somewhat wider, 6–8 mm. long;

perianth 11–14 mm. long, dark crimson to red-purple, the limb 3–5 mm. broad; fr. 10–14 mm. long, puberulent above, turbinate, with mostly 5 winglike lobes truncate above, irregular, attenuate below, coarsely reticulate-veined; seed oblong, dark, 5 mm. long.—Coastal Strand; San Luis Obispo Co. to L. Calif. Feb.–Oct.

3. **A. alpìna** Bdg. Perennial, forming compact mats 1–2 dm. across, viscid-puberulent; lf.-blades 4–9 mm. long, round-oval, on slender petioles 1–2 cm. long; peduncles shorter than petioles, slender; bracts lance-ovate, 2–3 mm. long; fls. 1–5, the perianth lavender-pink, 10–15 mm. long, the limb 6–8 mm. broad; fr. 3–4 mm. long, narrowed at both ends, obtusely or acutely 5-angled, reticulate-veined, puberulent.—Rare, sandy meadows, 8000–9000 ft.; Lodgepole F.; Tulare Co. July–Aug.

4. **A. pogonántha** Heimerl. [*A. angulata* Jones.] Much-branched annual with ascending to decumbent stems 1–4 dm. long, villous and viscid-puberulent; lf.-blades oblong-ovate to suborbicular, rounded to obtuse at apex, villous on midveins, 1–4 cm. long; petioles 1–3 cm. long, villous and glandular; peduncles slender, 2–5 cm. long; bracts scarious, oval to rounded, 6–9 mm. long; perianth glandular, white to rose-pink, 12–18 mm. long, the tube very slender, the limb 6–8 mm. broad; fr. 3–6 mm. long, finely reticulate-veined, orbicular-obcordate with 2 (3) thin wings faintly veined; seed obovate, dark brown, 2 mm. long.—Sandy places, mostly 2000–5000(–7500) ft.; Creosote Bush Scrub, Joshua Tree Wd., Pinyon-Juniper Wd.; Mojave Desert, head of San Joaquin V., inner S. Coast Ranges, e. slope of s. Sierra Nevada to Olancha Pass; w. Nev. April–July.

5. **A. villòsa** Wats. Much-branched annual with stout stems 1–5 dm. long, procumbent to ascending, villous, usually viscid; lf.-blades 1–3.5 cm. long, rhombic-ovate to almost round, often very unequal at base, sparingly glandular; petioles 1.5–3 cm. long, viscid; peduncles slender, 2–7 cm. long, viscid-villous; bracts lanceolate to lance-linear, 6–8 mm. long, viscid-viscous, scarious, attenuate; perianth 12–16 mm. long, purplish-rose, viscid-villous, the limb ca. 1 cm. broad; fr. 5–8 mm. long, the body indurate, rugose-veined, so as to appear somewhat pitted, villous above, with 3–4 wings truncate or prolonged slightly above the body; seed narrow-oblong, 2.5 mm. long, dark brown.—Common in open sandy places, mostly below 3000 ft.; Creosote Bush Scrub; Mojave and Colo. deserts; to Nev., Ariz., Son., L. Calif. Feb.–July.

Var. **aurìta** (Abrams) Jeps. [*A. a.* Abrams. *A. pinetorum* Abrams.] Perianth 16–26 mm. long; body of fr. with almost no transverse veins, hence not pitted; wings thin, broad, prolonged above the body; n = ca. 45 (Snow).—Sandy places, below 5000 ft.; Coastal Sage Scrub, Chaparral; from head of Coachella V. to interior cismontane Riverside, Orange, and San Diego cos. March–Aug.

6. **A. umbellàta** Lam. [*Tricratus admirabilis* L'Hér. *A. californica* Raeusch. *A. rotundifolia* Gaertn. f.] Perennial with slender prostrate stems 2–10 dm. long, sparsely to much-branched, succulent, often reddish, viscid-puberulent to glabrous; lf.-blades oval to oval-rhombic, to lance-oblong, irregular in outline, 1.5–6 cm. long; petioles slender, 1–5 cm. long; peduncles slender, 2–12 cm. long, viscid-puberulent; bracts lanceolate, 4–6 mm. long; perianth rose, sometimes whitish, the tube glandular-pubescent, 12–15 mm. long, the limb 8–10 mm. broad, with emarginate lobes; fr.-body rather hard, 7–12 mm. long, glandular-villous above, short-beaked, with 2–5, usually 4, wings faintly net-veined, widened then truncate or narrowed above, not or scarcely prolonged above the body; seed elliptic-oblong, brown, ca. 4 mm. long.—Coastal Strand; Los Angeles Co. to Sonoma Co. Most of year. Variable and with rather poorly defined segregates:

KEY TO SUBSPECIES

A. Perianth ca. 12 mm. long. Mendocino Co. n. .... ssp. *breviflora*
AA. Perianth 15–20 mm. long. Sonoma Co. s.
  B. Wings of fr. not prolonged beyond the body, the whole fr. (including wings) usually longer than broad. Marin Co. to Los Angeles Co. .... *A. umbellata*
  BB. Wings of fr. prolonged beyond body, the whole fr. often broader than long.
    C. The wings of fr. indurate and woody; lf.-blades not very irregular in outline. Insular ssp. *alba*
    CC. The wings of fr. membranous; lf.-blades strongly irregular in outline. Mostly mainland.
      D. Stems puberulent to glabrate. San Luis Obispo Co. to San Diego .. ssp. *variabilis*
      DD. Stems copiously villous. San Diego Co. to L. Calif. .... ssp. *platyphylla*

Ssp. **breviflòra** (Standl.) Munz. [*A. b.* Standl.] Lv.-blades 2–3.5 cm. long; perianth-tube 6–8 mm. long, the limb ca. 5 mm. broad; fr. ca. 8 mm. long, the wings not prolonged above body.—Coastal Strand; Mendocino Co. to Del Norte Co. June–Sept.

Ssp. variábilis (Standl.) Munz. [*A. v.* Standl. *A. alba* var. *v.* Jeps. *A. minor* Standl. *A. umbellata* var. *m.* Munz.] Often annual; stems viscid-puberulent to glabrate; lf.-blades 1.5–4 cm. long, irregular in outline, oval to lance-oblong; bracts lanceolate; perianth 16–20 mm. long; fr. 6–12 mm. long, the body rugose-veined to almost smooth, the wings thin, broad, prolonged beyond the body.—Coastal Strand; San Luis Obispo Co. to San Diego Co.

Ssp. platyphýlla (Standl.) Munz. [*A. p.* Standl. *A. alba* var. *p.* Jeps. *A. u.* var. *p.* Munz. *A. gracilis* ssp. *p.* Ferris.] Stems mostly viscid-villous; lf.-blades roundish to broadly elliptical, only slightly sinuate, 1.5–3.5 cm. long; perianth ca. 20 mm. long; fr. 7–10 mm. long, puberulent, ribbed, the wings very broad, thin and soft, prolonged above the body.—Coastal Strand; San Diego Co. to L. Calif.

Ssp. álba (Eastw.) Munz. [*A. a.* Eastw. *A. u.* var. *a.* Jones. *A. insularis* and *neurophylla* Standl.] Stems viscid-puberulent to glabrate; lf.-blades broad, not strongly sinuate; perianth white to rose, 15–20 mm. long; fr. indurate, ca. 1 cm. long, the wings coriaceous, conspicuously veined, prolonged above the body.—Coastal Strand; Santa Rosa, San Nicolas, San Miguel, and San Clemente ids.

7. **A. nàna** Wats. ssp. **Covíllei** (Heimerl) Munz. [*A. C.* Heimerl. *A. n.* var. *C.* Munz.] Densely cespitose perennial from a thick woody root and branched caudex, the branches 2–6 cm. long; lvs. densely clustered, the blades glaucous, oblong-ovate, 0.5–2 cm. long, obtuse or rounded at apex, minutely puberulent; petioles 1–4 cm. long, puberulent and sometimes with longer hairs; peduncles scapelike, slender, reddish, puberulent, sometimes pubescent, 2–10 cm. long; bracts lanceolate, scarious, 2–3 mm. wide; perianth white or pinkish, 11–15 mm. long, the limb 6–8 mm. wide; fr. turbinate, 7–8 mm. long, obcordate with 5 thin-walled regular winglike lobes; seed dark, narrow, 2.5 mm. long, longitudinally veined.—Dry sandy places, 5000–9400 ft.; Pinyon-Juniper Wd., Yellow Pine F.; San Bernardino Mts., New York Mts. (Mojave Desert), Inyo Mts. to Mono Co.; sw. Nev. June–Aug. The material from the e. Mojave intergrades toward *A. nana.*

8. **A. turbinàta** Torr. [*A. latiuscula* Greene.] Annual, much branched, the stems 1.5–5 dm. long, viscid-puberulent when young, glabrate later; lf.-blades round to oblong-ovate, 1–4 cm. long, rounded to obtuse at apex, irregular at base; petioles slender, 1–5 cm. long; peduncles 3–9 cm. long, slender, viscid-puberulent; bracts lanceolate, 5–8 mm. long, acute; perianth white or pinkish, 15–20 mm. long, the limb 5–6 mm. broad; frs. short-villous, 4–7 mm. long, often beaked, the inner broadly turbinate, deeply lobed, the lobes compressed and winglike, wrinkled; seed dark brown, lance-oblong, 2 mm. long.—Occasional in sandy places, mostly at 4000–8000 ft.; Pinyon-Juniper Wd., Sagebrush Scrub, Creosote Bush Scrub; Mt. Pinos region (Kern Co.) along the e. slope of the Sierra Nevada to Ore., Nev. May–July. A form occurring with the sp., and having ovate obtusish bracts and with frs. wingless or with 2 incurved wings, has been described as *A. exalàta* Standl.

9. **A. Crúx-Máltae** Kell. [*Tripterocalyx C-M.* Standl.] Erect or procumbent much-branched viscid-villous to glabrate annuals, the stems stout, 1–3 dm. long; lf.-blades broadly ovate to elliptic-oblong, 2–6 cm. long; petioles 1–4 cm. long; peduncles 1–6 cm. long; bracts 6–10 mm. long, lanceolate to ovate; perianth rose with green throat, 2–2.4 cm. long; the limb ca. 1 cm. broad; fr. 1–1.5 cm. long, as broad or broader, the body indurate, glandular-villous, transversely veined, the wings 2(3), thin, translucent, coarsely net-veined.—At 4000–5000 ft.; Sagebrush Scrub; Lassen Co.?; adjacent Nev. May–June.

10. **A. micrántha** Torr. [*Tripterocalyx m.* Hook.] Much like *A. Crux-Maltae*, but the stems viscid-puberulent; peduncles 1–2 cm. long; bracts lanceolate to lance-ovate, 6–10 mm. long, attenuate; perianth ca. 1.5 cm. long, viscid-puberulent, the limb greenish-white, 3–4 mm. broad; fr. 1.5–3 cm. long, the body puberulent to subglabrous, the wings 2–3, with finer veins; seed brown, slightly angled, ca. 6–8 mm. long.—Sanddunes; Creosote Bush Scrub; Kelso, Mojave Desert; to N. Dak., Kans., New Mex. April–May.

44. Batidàceae. Batis Family (Fig. 41)

Low maritime bushy plant. Lvs. opposite, entire, fleshy, stipulate, with a small basal loose flange. Dioecious, the fls. crowded in bracteate catkinlike spikes which are

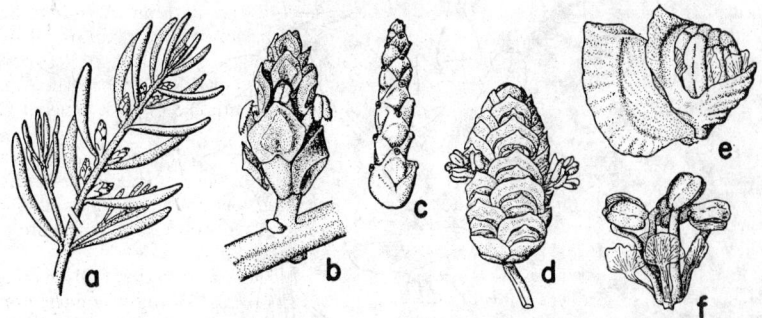

Fig. 41. BATIDACEAE. *Batis maritima: a,* habit, × ½, with catkins of fls.; *b,* ♀ catkin, × 5, showing bracts and capitate stigmas; *c,* same in fr., × 1; *d,* ♂ catkin, × 3, with overlapping bracts and exserted stamens; *e,* young ♂ fl., × 8, with subtending bract; *f,* ♂ fl., × 5, with 4 petaloid staminodia and 4 stamens.

sessile, axillary. Staminate fls. with 2-lobed calyx; stamens 4; petaloid staminodia 4. Pistillate fls. without calyx or corolla, consisting of a 4-loculed ovary with solitary ovule in each locule, and a sessile capitate stigma. Seed without endosperm. A single genus of warmer parts of New World.

1. Bàtis L. Saltwort

Characters of family. One sp. (Greek name of some seashore plant.)

1. **B. marítima** L. [*B. califórnica* Torr.] Stems prostrate or ascending, woody at base, 1–10 dm. long, glabrous; lvs. subterete, linear-oblanceolate, 1–2 cm. long; ♂ spikes sessile, ovoid-cylindric, 5–10 mm. long; bracts rounded, broader than long; calyx shorter than bracts; stamens exserted, longer than the white triangular staminodia; ♀ spikes short-peduncled, ca. 1 cm. long in fr.; pistils coalescent to form a fleshy fr.—Coastal Strand, Coastal Salt Marsh; Los Angeles to San Diego; L. Calif., W.I., Atlantic Coast, S. Am. July–Oct.

45. Styracàceae. Storax Family (Fig. 42)

Shrubs or trees. Lvs. simple, alternate, exstipulate, entire or dentate. Fls. bisexual, regular, in axillary or terminal racemes or fascicles, rarely solitary. Calyx campanulate or tubular, 4–8-lobed, in ours adherent to base of the ovary. Corolla 4–8-lobed, the petals often united only at base. Stamens twice as many as corolla-lobes, sometimes more, in 1 series, the fils. united at base. Ovary partly inferior, 2–5-loculed. Ovules 1–few in each locule. Stigma simple or 2–5-lobed. Fr. a berry or drupe, or dry and dehiscent by 3 valves. Seeds usually 1 in each locule; endosperm copious. Ca. 6 genera and 120 spp., warmer parts of N. and S. Am., e. Asia, Medit. region.

1. Stỳrax L. Storax. Snowdrop Bush

Ours deciduous shrubs, with axillary or leafy-racemose showy fragrant fls. on drooping peduncles. Calyx truncate. Corolla white. Stamens 10–16, monadelphous at base. Fr. globular, nearly dry, 1-loculed, commonly 3-valved. Ca. 100 spp., some grown for the attractive fls. (Ancient Greek name used for the sp. producing the gum storax.)

1. **S. officinàlis** L. var. **califórnica** (Torr.) Rehd. [*S. c.* Torr. *Darlingtonia rediviva* Torr. *S. r.* Wheeler.] Erect, 1–4 m. high, with grayish twigs; lf.-blades round-ovate to obovate, obtuse to subcordate at base, obtuse to rounded at apex, 2–7 cm. long, glabrous above, paler and stellate-cinereous beneath; petioles slender, 3–10 mm. long; fls. few, in terminal clusters on the branchlets; peduncles 6–12 mm. long; calyx unequally obscurely toothed, persistent, truncate; corolla 4–10-lobed, 12–18 mm. long; fr. ca. 12–14 mm. long; seed globose-ovoid, light brown, smooth.—Scattered in dry rocky places, below 3000 ft.; Chaparral, Foothill Wd., Yellow Pine F.; inner Coast Ranges

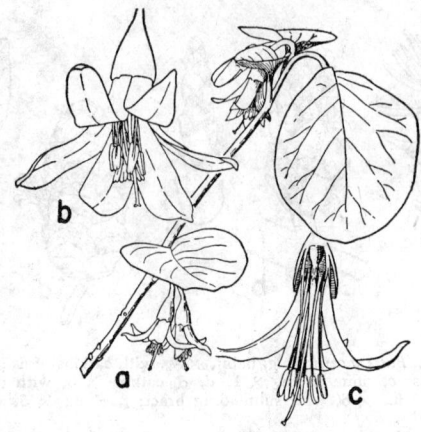

Fig. 42. STYRACACEAE. *Styrax: a,* branch, × ½, with lvs. and fls.; *b,* fl., × 1, ovary partly inferior, calyx quite gamosepalous, ca. 6 long corolla-lobes and twice as many stamens; *c,* fl. in section.

from Shasta Co. to Lake Co. and in foothills of Sierra Nevada from Tulare Co. n. April–May.

Var. **fulvéscens** (Eastw.) M. & J. [*S. californica* var. *f.* Eastw.] Lf.-blades usually pubescent above and stellate-tomentose beneath, the pubescence sometimes tawny; calyx-teeth more prominent.—On slopes and in canyons, below 5000 ft.; Chaparral, S. Oak Wd.; San Luis Obispo Co. to San Bernardino and San Diego cos. April–May.

46. Primuláceae. PRIMROSE FAMILY (Fig. 43)

Scapose or caulescent annual to perennial herbs. Lvs. simple, exstipulate, opposite, whorled or alternate. Fls. perfect, regular, mostly 5-merous. Calyx deeply lobed, often persistent. Corolla deeply parted to lobed, the lobes spreading or reflexed. Stamens of same number as corolla-lobes and inserted opposite them on the corolla-tube. Ovary 1-celled, usually superior, sometimes half-inferior; ovules on a basal or free cent. placenta. Style 1; stigma capitate. Fr. a caps., commonly 2–6-valved, sometimes circumscissile at apex. Seeds few to many, with endosperm. Ca. 25 genera and 600 spp., widely distributed but most common in N. Hemis.

Ovary partly inferior, the lower portion included in and adnate to calyx-tube; plant with leafy stems; fls. small, racemose .... 9. *Samolus*
Ovary superior and free.
  Plants scapose, the lvs. all basal; fls. in bracteate umbels.
    Corolla-lobes spreading or erect, emarginate or obcordate.
      Lobes of corolla 8–10 mm. long .... 1. *Primula*
      Lobes of corolla 1–2 mm. long .... 2. *Androsace*
    Corolla-lobes reflexed, acute to obtuse .... 3. *Dodecatheon*
  Plants with leafy stems; fls. not. in umbels.
    Lvs. opposite or whorled; corolla evident if present.
      Main lvs. in a single whorl at summit of stem .... 5. *Trientalis*
      Main lvs. distributed along the stems.
        Corolla lacking, the calyx petaloid; fls. solitary, axillary, sessile .... 6. *Glaux*
        Corolla present, the calyx not petaloid; fls. not solitary and sessile in axils.
          Perennial; corolla yellow, often with purple dots; caps. valvate .... 4. *Lysimachia*
          Annual; corolla salmon or blue; caps. circumscissile .... 7. *Anagallis*
    Lvs. alternate (except the lowest); corolla shorter than calyx .... 8. *Centunculus*

1. Prímula L. PRIMROSE

Perennial herbs, mostly low, from rhizomes. Lvs. basal; stems scapose. Fls. small to showy, in terminal umbels. Calyx angled, tubular, 5-cleft. Corolla funnelform or salverform, with entire emarginate or 2-cleft imbricate lobes. Stamens 5, included, the fils. very short; anthers oblong, obtuse. Ovary free from calyx, globose or ovoid. Caps.

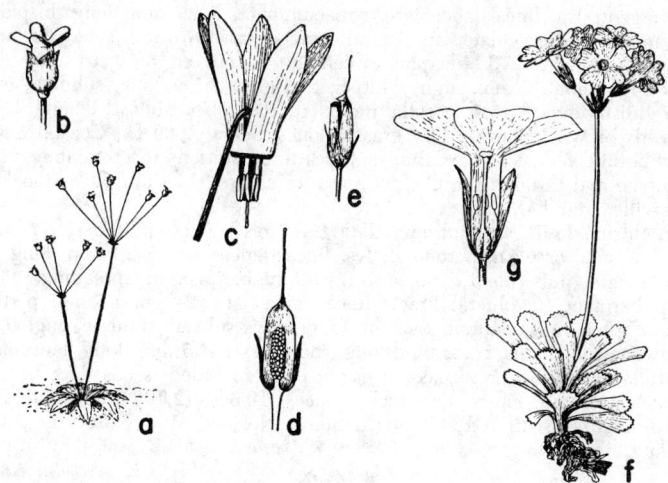

Fig. 43. PRIMULACEAE. *Androsace: a,* habit, × ½; *b,* fl., × 5, with 5-lobed calyx and 5-lobed corolla. *Dodecatheon Hendersonii: c,* fl., × 1, corolla with 5 reflexed lobes, 5 stamens with united fils.; *d,* × 1½, section through ovary with free cent. placenta; *e,* calyx and circumscissile caps. *Primula suffrutescens: f,* habit, × ½; *g,* sect. through fl., × 1½.

many-seeded, 5-valved at summit. Seeds peltate, punctate. Spp. 150–200, mostly of N. Hemis., many prized in cult. (Diminutive of Latin, *primus,* spring, because of early flowering.)

1. **P. suffrutéscens** Gray. Sierra Primrose. Stems woody, branched, creeping; lvs. glabrous, crowded, spatulate, 1.5–3.5 cm. long, the apex rounded and dentate, the base gradually narrowed to broad petioles; scapes slender, 4–10 cm. high, glandular-puberulent above; pedicels 2–several, slender, to ca. 1 cm. long, glandular-puberulent; calyx 6–8 mm. long, the lobes lanceolate, glandular-puberulent; corolla magenta with yellow throat, the tube 8–10 mm. long, the lobes ca. as long; caps. ovoid, scarcely as long as calyx; seeds brown, ca. 1.5 mm. long, including the irregularly winged angles.— Frequent under overhanging rocks and about cliffs, mostly 8000–13,500 ft.; Subalpine F., Alpine Fell-fields; Sierra Nevada and in Siskiyou and Trinity cos. July–Aug.

2. Andrósace L.

Small annual or perennial herbs. Lvs. basal, rosulate. Scapes 1–several, with involucrate umbels of small fls. Calyx 5-cleft, the tube becoming scarious. Corolla salverform, with short tube, constricted throat, and emarginate or obcordate lobes. Stamens 5, included; fils. short; anthers oblong. Ovary globose; style short. Caps. 5-valved. Seeds few to many, ovoid to triquetrous, minutely pitted. Ca. 60 spp., of N. Hemis., especially in the colder parts.

(Robbins, G. Thomas. N. Am. Species of Androsace. Am. Midl. Nat. 32: 137–163, 1944.)
Bracts of the invol. lance-ovate to ovate-obovate. From below 6000 ft.
 Calyx-lobes broadly lanceolate to deltoid; scapes mostly solitary . . . . 1. *A. occidentalis*
 Calyx-lobes broadly subulate-acerose, pungently acute; scapes 1–several . . . . 2. *A. elongata*
Bracts of invol. narrowly triangular to lanceolate, attenuate. Alpine . . . . 3. *A. septentrionalis*

1. **A. occidentàlis** Pursh var. **símplex** (Rydb.) St. John. [*A. s.* Rydb. *A. o.* forma *s.* Robbins.] Annual 3–6 cm. high; scapes mostly solitary; lvs. elliptic-lanceolate, 5–10 mm. long; bracts oblanceolate to oval, 2–5 mm. long; fls. 1–4, the pedicels slender, 5–15 mm. long; calyx-tube ca. 2 mm. long, whitish, the lobes lanceolate, ca. as long, puberulent; corolla white, included; caps. round, included; seeds blackish, minute.—At ca. 5500 ft.; Red Fir F.; Emigrant Gap, Placer Co.; Ariz. through Rocky Mts. Aug.–Sept.

2. **A. elongàta** L. ssp. **acùta** (Greene) Robbins. [*A. a.* Greene. *A. occidentalis* var. *a.* Jeps. *A. asprella* Greene.] Annual, puberulent, 2–8 cm. high, the scapes mostly 1–6,

erect to divergent; lvs. linear-lanceolate, subacuminate, 5–20 mm. long, hispidulous on margins, sometimes denticulate; involucral bracts narrow-ovate, acute to acuminate, 3–5 mm. long; pedicels 2–6, unequal, ascending to curved outward, 1–4 cm. long; calyx-tube obpyramidal, 2 mm. high, whitish, the lobes ca. as long, subulate to acerose, apiculate, reddish toward tips; corolla included, the lobes almost 1 mm. long; caps. included; seeds brown, minute.—Dry grassy slopes, below 4000 ft.; Coastal Sage Scrub, Chaparral, Foothill Wd., V. Grassland, etc.; scattered stations in cismontane s. Calif. to Kern R. Canyon and through the Coast Ranges to s. Ore., L. Calif. Occasional at desert edge, Victorville. March–May.

3. **A. septentrionàlis** L. ssp. **subumbellàta** (A. Nels.) Robbins. [*A. s.* var. *s.* A. Nels. *A. s.* Small.] Annual or weak perennial; lvs. linear-lanceolate, 5–20 mm. long, entire or weakly denticulate, puberulent; scapes 1–3 cm. high, erect or spreading, glabrous to somewhat puberulent; involucral bracts linear-lanceolate, 2–3 mm. long; pedicels 2–7, stout, erect or spreading, unequal, 1–3 cm. long; calyx subcampanulate, angled, the base scarious between the ridges, ca. 2 mm. long, the lobes 1–1.5 mm. long, lanceolate, often reddish; corolla as long as or surpassing calyx; caps. included; seeds dark brown, somewhat angled, ca. 1 mm. long.—Dry rocky places, 10,000–12,700 ft.; Alpine Fell-fields; San Bernardino Mts., scattered stations in Sierra Nevada, White Mts.; to B.C., Rocky Mts. July–Aug.

3. **Dodecátheon** L. SHOOTING STAR

Key contributed by H. J. Thompson

Rather low perennial herbs, glabrous or glandular-puberulent, with short rootstocks and fleshy-fibrous roots. Lvs. basal. Stems scapose, naked, ending in an umbel of few to many fls. Fls. 4–5-merous, nodding, on slender pedicels. Calyx deeply 5-cleft, persistent, the segms. reflexed at anthesis, later erect. Corolla 5-parted, with short tube, dilated throat and reflexed lobes. Stamens opposite the corolla-lobes, exserted, the fils. short, broad, often united; anthers basifixed, mostly erect, approximate. Style filiform, exserted. Caps. partially 5-valved or circumscissile. Seeds many, small, ovoid or angled, punctate. Ca. 14 spp. of N. Am., mostly western. (Greek, *dodeca*, twelve, and *theos*, god, a name given by Pliny to the Primrose as being under the care of the 12 leading gods.)

(Thompson, H. J. The biosystematics of Dodecatheon. Contr. Dudley Herb. 4[5]: 73–154, 1953.)
A. Stigma enlarged, more than twice the diam. of the style; connective rugose, dark maroon to black; fils. 1 mm. long, free or joined by a thin membrane, but not united into a tube.
  B. Anthers 4 (rarely 5 in Siskiyou and Trinity cos.), subulate, obtuse to truncate; plant glandular-pubescent to glabrous; corolla-tube not covering base of anthers.
    C. Lvs. linear to linear-lanceolate, glabrous; infl. sparingly glandular-pubescent to glabrous; anthers 4; caps. valvate. San Jacinto Mts. to Sierra Nevada, n. to Modoc and Humboldt cos. .... 1. *D. alpinum*
    CC. Lvs. linear-oblanceolate to oblanceolate, sparingly to heavily glandular-pubescent; infl. glandular-pubescent; anthers 4, rarely 5; caps. opening by an irregular operculum, occasionally valvate, both types sometimes on same plant. N. Coast Ranges, Sierra Nevada s. to Tulare Co. .... 2. *D. Jeffreyi*
  BB. Anthers 5, lanceolate, acute; entire plant heavily glandular-pubescent; corolla-tube covering base of anthers. High elevs., s. Sierra Nevada, White, Panamint, San Gabriel, San Jacinto mts. 3. *D. redolens*
AA. Stigma not enlarged; connective rugose to smooth, dark or yellow; fils. united into a tube 2–4 mm. long (or short and free in D. *conjugens*).
  B. Connective rugose with transverse folds; caps. operculate.
    C. Rice-grain bulblets produced at flowering time; connective dark maroon to black, never yellow.
      D. Anthers acute; fil.-tube narrow, 1.2–2 mm. wide; plant reddish, glandular-pubescent to glabrous.
        E. Scape 1.2–3.5 dm. high; lvs. spatulate to elliptic; rice-grain bulblets usually less than 10; roots white. Below 7000 ft. .... 4. *D. Hendersonii*
        EE. Scape 0.7–1.4 dm. high; lvs. oblanceolate to spatulate; rice-grain bulblets many; roots reddish. Above 8000 ft. .... 5. *D. subalpinum*
      DD. Anthers obtuse to retuse; fil.-tube broad, 1.5–4 mm. wide; plant entirely green or sparingly red-flecked, glabrous. Foothills of Sierra Nevada .. 6. *D. Hansenii*
    CC. Rice-grain bulblets never produced; connective usually yellow or dark with a yellow spot, sometimes dark maroon to black.
      D. Fils. united into a dark maroon to black rugose tube. Coast Ranges from San Francisco Bay s., Cent. V. from Tehama Co. to Kern Co. ...... 7. *D. Clevelandii*

DD. Fils. free or united into a short yellow smooth tube. Lassen and Modoc cos.
8. *D. conjugens*
BB. Connective smooth, often longitudinally wrinkled upon drying; caps. operculate or valvate.
C. Rice-grain bulblets produced at flowering time; roots reddish; fil.-tube dark maroon to black; caps. operculate. W. slope of Sierra Nevada above 8000 ft. . . 5. *D. subalpinum*
CC. Rice-grain bulblets never produced; roots white; fil.-tube yellow or dark maroon to black; caps. valvate. Modoc and Mono cos. . . . . 9. *D. pulchellum*

1. **D. alpìnum** (Gray) Greene. [*D. Meadia* var. *a.* Gray. *D. Jeffreyi* var. *a.* Gray. *D. a.* f. *nanum* Hall.] Plants glabrous; roots white, without bulblets; lvs. linear to linear-oblanceolate, acute, entire, including the petiole 2–6 cm. long, 0.3–0.9 cm. wide; scape 0.4–1.4 dm. high, umbels 1–3-fld.; pedicels 1–2 cm. long at anthesis; calyx-tube 1.5–2.5 mm. long, the lobes 4–5 mm. long; corolla-tube maroon, yellow above, the lobes 8–11 mm. long, magenta to lavender; fils. free, black, 0.5 mm. long; anthers 5–6 mm. long, linear, with obtuse or blunt apex; caps. 6–8 mm. long, ovoid; seeds ca. 1.5 mm. long, wing-angled; *n* = 22 (Thompson, 1951).—Boggy meadows and wet banks, 8800–12,000 ft.; Lodgepole F. to Alpine Fell-fields; Sierra Nevada from Tulare Co. n., Sweetwater Mts.; to Ore., Nev. July–Aug.

Ssp. **màjus** H. J. Thomps. Lvs. 6–16 cm. long, 0.9–1.6 cm. wide; scape 1.4–3 dm. high, ± glandular above; umbel 4–10-fld.; pedicels 1–3 cm. long; corolla-lobes 9–16 mm. long.—More common; 4000–11,000 ft.; Montane Coniferous F.; San Jacinto Mts., San Bernardino Mts., Inyo and Tulare cos. to Modoc and Humboldt cos.; to Ore., Utah, Ariz. May–Aug.

2. **D. Jéffreyi** Van Houtte. [*D. J.* var. *tetrandrum* Jeps., var. *odoratum* Eastw. and var. *viviparum* Abrams.] Roots white, without bulblets; lvs. including petiole 9–50 cm. long, 1–3.5 cm. wide, oblanceolate, entire or crenate; scape 1.5–6 dm. high, the umbel 3–18-fld.; pedicels 3–7 cm. long at anthesis; fls. 4–5-merous; calyx-tube 2–5 mm. long, the lobes 5–10 mm. long; corolla-tube closely reflexed, with a maroon ring below and yellow above, the lobes 10–25 mm. long, magenta to lavender or white without intermediates; fils. less than 1.5 mm. long, dark, free or partially united; anthers 7–10 mm. long, maroon to yellow on pollen-sacs, mostly dark on connective; caps. 7–12 mm. long; seeds brown, wing-angled, ca. 2 mm. long; *n* = 21, 22 (Thompson, 1953).—Wet places, 2300–10,000 ft.; Montane Coniferous F.; Sierra Nevada from Tulare Co. n., Coast Ranges from Glenn Co. n.; to Alaska, Mont. June–Aug.

Ssp. **pygmaèum** (Hall) H. J. Thomps. [*D. Jeffreyi* f. *p.* Hall. *D. glandulosum* Eastw.] Plants heavily glandular-pubescent; lvs. 4–11 cm. long, 0.5–1.7 cm. wide; scape 0.8–1.6 dm. high; umbels 1–4-fld.; fls. 4-merous.—Dry parts of meadows, 7000–9000 ft.; Montane Coniferous F.; Sierra Nevada from Eldorado Co. to Mariposa Co.

3. **D. rédolens** (Hall) H. J. Thomps. [*D. Jeffreyi* var. *r.* Hall.] Plant heavily glandular-pubescent; roots white, without bulblets; lvs. including petiole 20–40 cm. long, 2.5–6 cm. wide, oblanceolate, obtuse, entire; scape 2.5–6 dm. high; umbel 5–10-fld.; pedicels 3–5 cm. long in fl.; fls. 5-merous; calyx-tube 3–4 mm. long, the lobes 5–8 mm. long; corolla-tube yellow, never maroon, covering the anther-bases, the lobes 15–25 mm. long, magenta to lavender; fils. less than 1 mm. long, free, black; anthers 7–11 mm. long, lanceolate, dark maroon to black; caps. 8–14 mm. long, valvate.—Moist places, 8000–11,500 ft.; Montane Coniferous F.; San Jacinto Mts. to Sierra Nevada of Fresno and Inyo cos., White Mts., Panamint Mts.; to Utah, Nev. July–Aug.

4. **D. Hendersònii** Gray. [*D. integrifolium* var. *latifolium* Hook.] Roots white, with bulblets at flowering time; lvs. including petioles 5–15 cm. long, 2–6 cm. wide, spatulate to elliptic, obtuse to truncate; scape 1.2–4.8 dm. high, glandular to glabrous, the umbel 3–17-fld., the pedicels 2–7 cm. long at anthesis; fls. 5-merous; calyx-tube 2 mm. long, the lobes 3–5 mm. long; corolla-tube maroon, yellow above, the lobes 6–23 mm. long, magenta to deep lavender or white (without intermediates); fil.-tube 1.5–3 mm. long, 1.5–2.5 mm. wide, as broad as anther-whorl, dark; anthers 4–5 mm. long, lanceolate, the pollen-sacs dark red to black; caps. 8–15 mm. long, operculate; seeds brown, pitted, angled, ca. 1.3 mm. long; *n* = 22, 33, 66 (Thompson, 1951).—Mostly in shaded places, below 4000 (6700) ft.; many Plant Communities; San Bernardino Mts., Coast Ranges from San Benito Co. n., Sierra Nevada from Tulare Co. n.; to B.C. Feb.–May.

Ssp. **parvifòlium** (Knuth) H. J. Thomps. [*D. patulum* var. *p.* Knuth.] Glabrous; roots white, with bulblets at flowering time; lvs. 3–7 cm. long, 0.5–2 cm. wide, crisped on edge; scape 0.8–2.1 dm. high, usually reddish above; corolla-lobes white to magenta, with intermediates usually present; fil.-tube 1–2 mm. long, 1–2 mm. wide, narrower than anther-whorl.—V. Grassland, Foothill Wd.; e. side of Sacramento V. in Butte, Yuba, San Joaquin and Amador cos. March–April.

Ssp. **cruciàtum** (Greene) H. J. Thomps. [*D. c.* Greene.] Fls. 4-merous.—N. Oak Woodland, N. Coastal Scrub; Coast Ranges about San Francisco Bay from Sonoma and Napa cos. to Santa Clara Co. March–April.

5. **D. subalpìnum** Eastw. [*D. Hendersonii* var. *yosemitanum* Mason. *D. Cusickii* var. *y.* Jeps.] Roots and bulblets numerous, reddish; plant glabrous, reddish except for lf.-blades; lvs. including petioles 3–7 cm. long, 0.5–1.8 cm. wide, oblanceolate to spatulate, entire; scape 0.7–1.5 dm. high; umbel 1–5-fld.; pedicels 1–1.5 cm. long at anthesis; fls. 5-merous; calyx-tube 2–3 mm. long, the lobes 2.5–4.5 mm. long; corolla-tube dark maroon with yellow band above, the lobes 5–9 mm. long, magenta varying continuously to white; fil.-tube slender, 2–3.5 mm. long; pollen-sacs yellow to purple, with dark maroon to purple connective; caps. 6–13 mm. long, operculate.—Moist shaded places, 7000–10,000(–13,000) ft.; Lodgepole F. to Alpine Fell-fields; Sierra Nevada from Tuolumne Co. to Tulare Co. May–July.

6. **D. Hansènii** (Greene) H. J. Thomps. [*D. Hendersonii* var. *H.* Greene.] Glabrous; lvs. including petiole 4–11 cm. long, 1–2 cm. wide, oblanceolate to spatulate, entire; scape 0.8–2.5 dm. high, the umbels 1–7-fld.; pedicels 2–7 cm. long at anthesis; calyx-tube ca. 2 mm. long, the lobes 3–4 mm. long; corolla-tube dark maroon with yellow band above, the lobes 7–20 mm. long, magenta varying into white; fil.-tube 1.5–3 mm. long, dark maroon to black, rugose; anthers 2.5–4 mm. long, linear, yellow to red, the connective dark; caps. 8–13 mm. long, operculate.—From 1000–6000 ft.; Yellow Pine F. to V. Grassland; foothills of Sierra Nevada from Placer Co. to Kern Co. April–May.

7. **D. Clevelándii** Greene. [*D. C.* var. *splendens* Orcutt.] Glandular-pubescent; roots white, without bulblets; lvs. including petiole 5–11 cm. long, 0.8–3 cm. wide, oblanceolate to spatulate, crisped, dentate, rarely entire; scape 1.8–4 dm. high; umbel 5–16-fld.; pedicels 2–5 cm. long in fl.; calyx-tube 1.5–2.5 mm. long, the lobes 3–5 mm. long; corolla-tube dark maroon with yellow band above, the lobes 10–20 mm. long, magenta varying into white; fil.-tube 2.5–4 mm. long, 3–4 mm. wide, dark maroon to black; anthers 3–5 mm. long, lanceolate, the connective yellow, the anther-sacs yellow each with dark line; caps. 8–13 mm. long, operculate; seeds brown, angled, rough, ca. 1.3 mm. long; $n = 22$ (Thompson, 1951).—Cismontane grassy slopes and flats, below 2000 ft.; V. Grassland, Coastal Sage Scrub; Los Angeles Co. to n. L. Calif. Jan.–April.

KEY TO SUBSPECIES

Connective yellow; fil.-tube without a yellow spot below each anther .... *D. Clevelandii*
Connective maroon to black.
   Yellow spot on fil.-tube lacking .... ssp. *insulare*
   Yellow spot present below each anther on fil.-tube.
      Anther-tip acute to obtuse; pollen sac usually yellow .... ssp. *sanctarum*
      Anther-tip retuse to obtuse; pollen sac dark .... ssp. *patulum*

Ssp. **insulàre** H. J. Thomps. Lvs. 6–18 cm. long, 1–6 cm. wide; scape 1.2–4.5 dm. tall; umbels 5–9-fld.; pedicels 2–10 cm. long; calyx-lobes 4–7 mm. long; corolla-lobes 10–25 mm. long; anther-connective dark, the anther-sacs usually yellow; $n = 22$ (Thompson, 1951).—Below 2500 ft.; Chaparral, Foothill Wd., V. Grassland; s. Monterey Co. to Santa Barbara Co.; Channel Ids.; Guadalupe Id. March–April.

Ssp. **sanctàrum** (Greene) Abrams. [*D. s.* Greene. *D. Meadia* var. *macrocarpum* Gray. *D. patulum* var. *gracile* Greene. *D. laetiflorum* Greene.] Lvs. 4–6 cm. long, 1–2 cm. wide; scape 1.5–3 dm. high; umbels 3–7 fld.; calyx-lobes 3–4 mm. long; corolla-lobes 10–20 mm. long; fil.-tube dark with a light spot at base of each anther; $n = 22$, 33, 44 (Thompson, 1953).—Below 2000 ft.; N. Oak Wd., Foothill Wd.; San Francisco to nw. Los Angeles Co.

Ssp. **pátulum** (Greene) H. J. Thomps. [*D. p.* Greene. *D. p.* var. *bernalinum* Greene.] Lvs. 2–5 cm. long, 1–2 cm. wide; scape 0.4–2 dm. high; umbels 1–6-fld.; pedicels 1–3 cm. long; fls. mostly 5-merous; corolla-lobes 6–15 mm. long, white to rose-purple; fil.-tube dark with light spot at base of each anther; anthers 1.5–4 mm. long, bending away from style after anthesis, the pollen-sacs and connective dark; $n = 22$, 44 (Thompson, 1951).—Moist places, often on serpentine or in subalkaline places; V. Grassland, N. Coastal Scrub; Sacramento and San Joaquin valleys, from Tehama Co. to Kern Co., and in inner Coast Ranges from San Francisco Bay to San Benito Co. Jan.–April.

8. **D. cónjugens** Greene. [*D. glastifolium* Greene.] Glabrous throughout; roots white, without bulblets; lvs. including petiole 3–14 cm. long, 0.8–2 cm. wide, entire; scape 0.8–2.6 dm. high; umbels 1–7-fld.; pedicels 1–4 cm. long at anthesis; fls. 5-merous; calyx-tube 2–4 mm. long, the lobes 4–7 mm. long; corolla-tube maroon, yellow above, the lobes 7–20 mm. long, magenta to white; fils. 0.5–1.5 mm. long, free or united, yellow to dark, smooth; anthers 5–7 mm. long, linear-lanceolate, the pollen-sacs dark maroon to black, rarely yellow, the connective rugose, dark; caps. 8–20 mm. long; seeds brown, wing-angled, ca. 1 mm. long; $n = 22$ (Thompson, 1951).—Moist slopes and meadows, 4500–6000 ft.; Sagebrush Scrub, Yellow Pine F.; Lassen and Modoc cos.; to Wash., Mont., Wyo. May–June.

9. **D. pulchéllum** (Raf.) Merr. [*Exinia p.* Raf. *D. pauciflorum* and *D. radicatum* Greene.] Roots white, without bulblets; lvs. including petiole 4–25 cm. long, 1–6 cm. wide, oblanceolate to ovate, mostly entire; scape 0.6–5 dm. high; umbels 3–25 fld.; pedicels 1–5 cm. long at anthesis; fls. 5-merous; calyx-tube 2–3.5 mm. long, the lobes 2.5–6 mm. long; corolla-tube maroon, yellow above, the lobes 9–20 mm. long, magenta to lavender; fils. 0.5–3.5 mm. long, united or nearly free, yellow, smooth or rugulose; anthers 3–8 mm. long, lanceolate, the pollen-sacs yellow, sometimes red or maroon, the connective mostly dark, smooth; caps. 7–17 mm. long, valvate.—Damp meadows, at 7000 ft.; Sagebrush Scrub; Long V., Mono Co.; to Alaska, Rocky Mts., Wis., Mo., and Durango. April–May.

Ssp. **monánthum** (Greene) H. J. Thomps. [*D. pauciflorum* var. *m.* Greene. *D. radicatum* ssp. *m.* H. J. Thomps.] Fil.-tube dark maroon to black.—Sagebrush Scrub; Modoc Co.; to Ore., Utah.

4. Lysimàchia L. LOOSESTRIFE

Perennial herbs with leafy stems. Lvs. entire, opposite or whorled, gland-dotted. Fls. yellow or orange, sometimes with purple dots, solitary in the axils or in racemes, corymbs, or panicles. Calyx 5–6-parted, imbricate or valvate in bud. Corolla 5–6-parted, rotate, convolute in bud or each lobe convolute or involute about its stamen. Fils. distinct or nearly so, or monadelphous at base; anthers slender to ovoid. Caps. few-many seeded. One hundred or more spp., of temp. regions. (Greek, *lusis*, loose, and *mache*, strife.)

Fls. solitary in axils; stems creeping .... 1. *L. Nummularia*
Fls. in axillary racemes; stems erect .... 2. *L. thyrsiflora*

1. **L. Nummulària** L. MONEYWORT. Glabrous, the stems creeping, 2–5 dm. long, often rooting at nodes; lvs. opposite, round to broadly ovate, 1.5–2.5 cm. long, on petioles 2–4 mm. long; fls. solitary in the axils, usually nodding; pedicels slender, 1.5–2.5 cm. long; calyx-segms. ovate, 5–9 mm. long; corolla yellow, deeply 5-parted, the lobes cordate-ovate, 7–9 mm. long; fils. slightly connate at base, glandular; caps. shorter than sepals; $2n = 36$ (Wulff, 1938).—Natur. in moist meadows, ca. 3500 ft.; Yellow Pine F.; Quincy, Plumas Co.; e. U.S.; native of Eu. June–Aug.

2. **L. thyrsiflòra** L. [*Naumburgia t.* Rchb.] TUFTED LOOSESTRIFE. Erect, from slender rootstocks; stems 2–8 dm. high, simple, quite glabrous; lvs. opposite, lanceolate, acuminate, narrowed at base, 5–12 cm. long, sessile; fls. in short axillary peduncled spike-like racemes; calyx-segms. 1.5–3.5 mm. long, narrow; corolla yellow, often with purple dots, deeply 5–7-parted into linear segms. 3–5 mm. long and with or without a small tooth in each sinus; fils. distinct; caps. globose, slightly exceeding calyx, with black dots. —Wet places, 3000–4500 ft.; N. Juniper Wd., Yellow Pine F.; Plumas, Shasta and Siskiyou cos.; to Alaska, Atlantic Coast; Eurasia. June–Aug.

5. Trientàlis L. STAR-FLOWER

Low glabrous perennials with tuberous rootstocks and simple erect stems. Lower lvs. few, scalelike, minute, the main lvs. in a single whorl at summit, thin, well developed.

Fls. few, terminal on slender axillary pedicels, small, delicate, white or pinkish. Calyx
of 5–7 sepals, persistent. Corolla spreading, flat, parted almost to base. Fils. connate
at base, slender; anthers oblong. Caps. few-seeded, globose. Seeds trigonous to globose.
Ca. 3 spp., of colder parts of N. Hemis. (Latin, meaning one-third foot, referring to
height of plant.)

1. **T. latifòlia** Hook. [*T. europaea* var. *l.* Torr.] Stems erect, 0.5–2 dm. high, slender;
lvs. mostly 4–6, ovate to obovate, abruptly acute at both ends, 4–8 cm. long, 2.5–5
cm. wide, on petioles 1–4 mm. long; pedicels ca. half as long as lvs.; calyx-lobes narrow,
4–6 mm. long; corolla pinkish, 8–15 mm. wide, the segms. abruptly acuminate; caps.
shorter than calyx; seeds globose.—Shaded places, chiefly in woods, below 4500 ft.;
largely Mixed Evergreen F., Redwood F., Yellow Pine F.; San Luis Obispo and Mariposa
cos. n.; to B.C. April–July.

6. Gláux L. Sea-Milkwort

Low fleshy perennial with slender rootstocks. Lvs. opposite, entire, sessile. Fls. minute,
axillary, almost sessile. Calyx petaloid, 5-lobed, campanulate, the lobes ovate or oblong.
Corolla 0. Stamens 5, alternate with calyx-lobes; fils. subulate; anthers cordate. Caps.
5-valved, few-seeded, globose-ovoid. Seeds flattened-ellipsoid. One sp. of saline or
brackish places. N. Hemis. (Greek, *glaukos*, sea-green.)

1. **G. marítima** L. [*G. acutifolia* Heller.] Stems slender, simple or branched, erect,
often tufted, 0.5–2(–3) dm. long; lvs. linear to oblong, obtuse to acutish, 4–14(–20)
mm. long, fleshy; fls. 3–4 mm. long; calyx-lobes ca. twice as long as tube, white to
reddish or lavender; caps. 2.5 mm. high; seeds brown, pitted, ca. 1 mm. long; $2n = 30$
(Wulff, 1937).—Coastal Salt Marsh, Coastal Strand; San Luis Obispo Co. to Humboldt
Co.; to Alaska; occasional inland, as in saline meadows, Mono and Siskiyou cos.; to
Atlantic Coast; Eurasia. May–July.

⅄ 7. Anagállis L. Pimpernel

Low spreading or procumbent annual or perennial herbs, diffusely branched. Lvs.
opposite or whorled, mostly entire, sessile or nearly so. Fls. small, solitary on axillary
pedicels. Calyx 5-parted, persistent. Corolla rotate, deeply 5-parted. Stamens 5, inserted at corolla-base; fils. puberulent, distinct or united at base; anthers oblong. Ovary
round; style filiform. Caps. membranaceous, globose, circumscissile. Seeds many, angled.
Ca. 15 spp., mostly Eurasian. (Greek, *ana*, again, and *agallein*, to delight in, since the
fls. open again when sun strikes them.)

1. **A. arvénsis** L. Glabrous diffusely branched annual, the stems 1–2.5 dm. long;
lf.-blades ovate to oval, sessile, 0.5–2 cm. long; pedicels slender, 1–3 cm. long, recurved in fr.; calyx-lobes lanceolate, 3–5 mm. long; corolla salmon, rarely white, sometimes blue (forma *caerùlea* [Schreb.] Baumg.), rotate, 8–10 mm. broad; caps. 3–4
mm. in diam.; seed 1 mm. long, triangular, dark, finely pitted; $2n = 40$ (Wulff, 1937).
—Common weed at low elevs.; to Atlantic Coast; natur. from Eu. Mostly March–July.

8. Centúnculus L. Chaffweed

Small annuals. Lvs. entire, alternate or lowest opposite, sessile or subsessile. Fls.
minute, solitary in axils. Calyx 4–5-parted, persistent. Corolla rotate, with short urceolate
tube and acute lobes. Stamens 4–5; fils. short, distinct, glabrous. Caps. globose, circumscissile. Seeds many, angled, minute. Spp. 3, of temp. and trop. regions. (Latin
name of some plant, diminutive of *cento*, patchwork.)

1. **C. mínimus** L. Stems ascending, 3–10 cm. long; lvs. spatulate to broadly obovate,
2–5 mm. long; fls. almost sessile, mostly 4-merous; calyx ca. 3 mm. long; corolla shorter,
pink; seeds red-brown, somewhat triangular, minutely pitted, ca. 0.5 mm. long; $2n = 22$
(Hagerup, 1941).—Occasional, vernal pools and moist places, low elevs.; Coastal
Sage Scrub, Foothill Wd., Chaparral, etc.; San Diego Co. to Humboldt Co., w. base of
Sierra Nevada, Santa Rosa Id.; to B.C., Atlantic Coast, Eu., S. Am., etc. April–July.

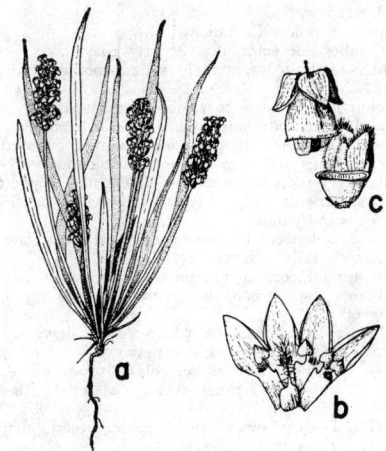

Fig. 44. PLANTAGINACEAE. *Plantago Hookeriana* var. *californica: a*, habit, × ½, with spikes of fls.; *b*, fl., × 4, opened to show 4-lobed corolla, 4 stamens, pistil; *c*, circumscissile caps., × 2½.

9. Sámolus L. Water-Pimpernel. Brookweed

Glabrous perennial herbs. Lvs. alternate, entire. Fls. small, white, in terminal racemes. Calyx-tube adnate to ovary below, the limb 5-cleft. Corolla somewhat campanulate, perigynous, 5-lobed or -cleft. Stamens 5, inserted on corolla-tube opposite the lobes and alternating with 5 staminodia. Caps. globose, 5-valved at summit. Seeds many, minute, triangular. Ca. 10 spp., semicosmopolitan. (Celtic name, supposed to refer to curative properties.)

1. **S. parviflòrus** Raf. [*S. floribundus* HBK.] Stems usually solitary, erect, simple or branched above, 1.5–4 dm. high; basal lvs. often in a rosette, the blades round-obovate to oblong-spatulate, obtuse to rounded at apex, 2–5 cm. long, narrowed gradually into short winged petioles; cauline lvs. similar, shorter; fls. in loose elongate often panicled racemes; pedicels slender, divaricate, 1–2 cm. long; calyx 1–2 mm. long, with short triangular teeth; corolla white, ca. 1.5 mm. broad; caps. ca. 2.5 mm. broad; seeds ca. 0.3 mm. long.—Frequent in moist places, below 4000 ft.; Coastal Sage Scrub, Chaparral; cismontane s. Calif. to Monterey Co., Solano and Contra Costa cos., San Joaquin Co., Santa Cruz Id.; to B.C., Atlantic Coast, Mex., S. Am. June–Aug.

47. Plantáginàceae. Plantain Family (Fig. 44)

Herbs, annual or perennial, mostly with basal longitudinally ribbed lvs. and scapose bracteate spikes, occasionally with branched stems and opposite linear lvs. Fls. regular, 4-merous, sometimes imperfect. Calyx of 4 imbricated persistent sepals, mostly with scarious margins. Corolla salverform or tubular, with 4 erect or spreading scarious persistent veinless lobes. Stamens 4 or 2, alternate with corolla-lobes and inserted on the tube. Anthers 2-loculed. Ovary superior, 2–4-celled, with 1–several ovules in each cell. Style 1, with long hairy stigma. Caps. 2–several-seeded, circumscissile. Seeds mostly with flattened or concave faces, and straight embryo in fleshy endosperm. Ca. 250 spp. and 3 genera, of which the largest and most cosmopolitan is *Plantago*.

1. Plantàgo L. Plantain

Characters as given for family. Seeds with mucilaginous coat, hence of some use as laxative (psyllium). (Latin, from *planta*, footprint.)

(Pilger, R. Das Pflanzenreich, IV, 269: 39–432, 1937.)

A. Lvs. in basal rosettes; infl. scapose.
  B. Corolla-lobes spreading or reflexed; stamens 4.
    C. Tube of corolla pubescent externally. Strictly coastal.
      D. Plants mostly annual; lvs. acutely salient-toothed except in very depauperate in-
          dividuals . . . . 5. *P. Coronopus*
      DD. Plants perennial; lvs. entire to remotely and sparsely denticulate . . 6. *P. maritima*
    CC. Tube of corolla glabrous externally. Generally distributed.
      D. Anterior sepals connate; stamens conspicuously exserted . . . . 11. *P. lanceolata*
      DD. Anterior sepals separate; stamens not conspicuous.
        E. Plants heavy-rooted perennials; bracts not longer than calyx.
          F. Lvs. broadly elliptic to cordate-ovate; caps. circumscissile near middle;
              seeds reticulate . . . . 1. *P. major*
          FF. Lvs. oblanceolate to narrowly elliptic-ovate; caps. circumscissile near
              base; seeds smooth, shining . . . . 4. *P. eriopoda*
        EE. Plants annual, or, if perennial, with lower bracts much exceeding calyx.
          F. Plants dark green; lower bracts 4–10 times as long as sepals. Introd.
              weed . . . . 12. *P. aristata*
          FF. Plants light green to white-woolly; lower bracts 1–3 times the length
              of sepals. Mostly common natives.
            G. Lowest bracts lanceolate to subulate, scarcely membranous on
                sides, 1–3 times as long as sepals; seeds dull, brown
                                                            13. *P. Purshii*
            GG. Lowest bracts ovate with broad membranous margins and not
                exceeding calyx.
              H. Seeds dull, brown; bracts not just like the sepals
                                                            14. *P. Hookeriana*
              HH. Seeds reddish-yellow, shining; bracts almost exactly like the
                  sepals . . . . 15. *P. insularis*
  BB. Corolla-lobes erect in fertile fls. and forming a beak over the caps.
    C. Lvs. linear to subfiliform; stamens 2.
      D. Caps. mostly 6–8-seeded; seeds pitted, ca. 1 mm. long; sepals ovate
                                                            2. *P. heterophylla*
      DD. Caps. mostly 4-seeded; seeds obscurely pitted, ca. 1.5 mm. long; sepals broadly
          obovate . . . . 3. *P. Bigelovii*
    CC. Lvs. linear-oblanceolate to ovate; stamens 4.
      D. Plants annual, rarely biennial; spikes less than 1 dm. long. Introd. weeds.
        E. Bracts lanceolate to narrow-elliptic; lvs. narrow-obovate.
          F. Sepals rounded at apex; seeds pale brown . . . . 7. *P. virginica*
          FF. Sepals long-attenuate; seeds red . . . . 8. *P. rhodosperma*
        EE. Bracts ovate-deltoid; lvs. linear-oblanceolate . . . . 9. *P. truncata*
      DD. Plants perennial; spikes 1–2.5 dm. long. Native . . . . 10. *P. hirtella*
AA. Lvs. in pairs along slender stems; peduncles axillary . . . . 16. *P. indica*

1. **P. màjor** L. Common Plantain. Mostly perennial, acaulescent; lvs. thick, ascending, usually roughish, with minute hairs, broadly elliptic to somewhat cordate-ovate,
the blades 5–15 cm. long, obtuse, with several conspicuous nerves converging at base
and apex; petioles winged, mostly shorter than and rather abruptly expanded into blades;
spikes linear-cylindric, dense, 0.5–4(–7) dm. high, curved-ascending to erect; bracts
broadly ovate, scarious-margined, mostly shorter than calyx; sepals elliptic to almost
round, 1.5–2 mm. long, scarious-margined; corolla-lobes pointed, 0.5 mm. long; caps.
broadly conic, brown or purplish, circumscissile below sepal-tips, mostly 6–10-seeded;
seeds reticulate, brown, papillate, scarcely 1 mm. long; $2n = 12$ (Turesson, 1938).—
Weed of damp waste places; natur. from Eu. April–Sept. Exceedingly variable; two of
principal forms to be expected:
  Var. **Pílgeri** Domin. Lvs. thin, upright, 5–12 cm. long, subglabrous; spikes slender;
caps. conic above, circumscissile near tips of sepals; seeds 6–10.
  Var. **scopulòrum** Fries & Broberg. Villous or pilose; lvs. fleshy, decumbent, undulate
to sinuate-dentate; scapes arched-ascending; caps. rounded at summit; seeds 15–27,
1–1.2 mm. long.—Brackish or saline places, mostly near seacoasts.
  2. **P. heterophýlla** Nutt. [*P. californica* Greene.] Depressed or ascending annual, subglabrous to strigose; lvs. linear, 3–12 cm. long (including blade and petiole), 0.5–4 mm.
wide; scapes few to many, slender, commonly decumbent, 2–10 cm. high, the spikes
proper loosely fld., 1–3 cm. long; bracts somewhat keeled, ovate, broadly scarious-
margined, 1.5–2 mm. long; sepals ovate, with slender green midrib, similar to and mostly
somewhat shorter than bracts; corolla-lobes ca. 0.5 mm. long; caps. ca. twice as long
as calyx; seeds mostly ca. 6–8 per caps., black, pitted, somewhat irregular, 1–1.2 mm.
long.—Occasional on dried mud-flats, etc.; Coastal Sage Scrub, Chaparral, V. Grassland; San Diego Co. to San Joaquin V. and Sacramento V., and Santa Barbara Co.,

Santa Rosa and San Miguel ids.; Ariz. to Atlantic Coast. March–May. (Our form with its few-seeded caps. and large corollas is doubtfully referred to this sp.)

3. **P. Bigelòvii** Gray. Annual 5–12 cm. high, glabrous; lvs. linear to almost filiform, 3–10 cm. long; scapes few to many; spikes loose to dense, 1–5 cm. long; bracts round-ovate, the broad fleshy keel pitted, rugulose when dry, margins hyaline; sepals 1.7–2 mm. long, broadly obovate; corolla-lobes ca. 0.5 mm. long; stamens 2; caps. conic, ca. twice the calyx in length, circumscissile just below sepal-tips; seeds commonly 4, dark, ± obscurely pitted, slender, ca. 1.5 mm. long.—Somewhat saline and alkaline places, as beaches, mud of vernal pools, etc.; Coastal Strand, V. Grassland; San Luis Obispo Co. to Del Norte Co., Kern Co.; to B.C. April–June.

4. **P. eriópoda** Torr. [*P. shastensis* Greene.] Fleshy perennial, usually with yellow wool on the stout root-crown; lvs. thickish, oblanceolate to narrowly elliptic-ovate, the blades 5–12 cm. long, tapered gradually into a broad usually somewhat shorter petiole; scapes ascending, somewhat arcuate at base, 2.5–3.5 dm. high; spikes lax below, denser above, somewhat pilose, 8–12 cm. long; bracts broadly ovate, with broad rounded keel, ca. 2 mm. long; sepals round-elliptic, scarcely keeled, ca. 2 mm. long; corolla-lobes ca. 1 mm. long; caps. conic, with truncate apex, ca. 3 mm. long, dehiscent below the sepal tips; seeds 2–4, red-brown, darker, shining, elliptic, flat, ca. 2 mm. long.—Moist subsaline places; Sagebrush Scrub; interior Humboldt Co., Mono and Siskiyou cos.; to Canada, cent. U.S. July–Aug.

5. **P. Corónopus** L. [*P. Parishii* Macbr.] Mostly annual, coarsely pubescent; lvs. lance-linear in outline, hairy, sharply and acutely salient-toothed or -incised, or sub-entire in depauperate individuals, the blades 2–14 cm. long, gradually attenuate into somewhat shorter winged petioles; scapes 1–4 dm. high, the spikes nodding before anthesis, becoming 2–12 cm. long, dense, with closely appressed fls.; bracts ovate, long-acuminate from a rounded body, green or purplish-red, ciliate, ca. 2 mm. long; sepals ovate, ciliate, green to purplish, ca. 2 mm. long; corolla-tube pubescent; corolla-lobes lanceolate, 1 mm. long; caps. ca. 2 mm. long; seeds commonly 3, subelliptic, glaucous, ca. 1 mm. long.—Sea cliffs, about salt marshes, etc.; Coastal Strand, Coastal Salt Marsh, Closed-cone Pine F., Coastal Sage Scrub; Catalina Id., Monterey Peninsula to Humboldt Co.; native of Eu. April–July.

6. **P. marítima** L. ssp. **juncoìdes** (Lam.) Hult. [*P. j.* Lam. *P. m.* var. *j.* Gray.] Perennial with thickish root; lvs. many, strongly ascending, linear to linear-lanceolate, slightly fleshy, attenuate at tip, 5–15(–20) cm. long including the winged petiole, entire or remotely and sparsely denticulate, glabrous; scapes few, slightly exceeding lvs.; spikes usually dense, 2–10 cm. long; bracts broadly ovate, minutely ciliolate, ca. 1.5 mm. long; sepals broadly oblong, ciliolate, almost 2 mm. long; corolla-tube pubescent; corolla-lobes ca. 1 mm. long; caps. thick-ellipsoid, ca. 3 mm. long, dehiscing below the sepal-tips; seeds oblong to narrowly oval, 2–3, dark brown, ca. 2 mm. long.— Saline places; Coastal Salt Marsh and Coastal Strand; Marin Co. to Alaska. June–Aug.

Var. **califórnica** (Fern.) Pilg. [*P. juncoides* var. *c.* Fern.] Lvs. linear-oblanceolate to subspatulate, obtuse, very fleshy, spreading to depressed, usually 3–6 cm. long; scapes depressed or arched, usually much exceeding lvs.; seeds ca. 1.5 mm. long.—Similar situations, Humboldt Co. to Santa Rosa Id. (Santa Barbara Co.). May–Sept.

7. **P. virgínica** L. Villous annual or biennial; lvs. narrow-obovate, the blades 1–12 cm. long, soft-villous, entire or nearly so, narrowed gradually into somewhat shorter petioles; scapes long-villous, suberect, 5–30 cm. high; spikes dense, 2–12 cm. long; bracts lanceolate to narrow-elliptic, with fleshy green keel, ca. 2 mm. long; sepals elliptic or oblance-elliptic, 2–2.5 mm. long, rounded at apex, the midrib scarcely projected; corollas of fertile fls. closed over maturing caps., the lobes 2–3 mm. long; caps. ovoid, ca. 3 mm. long; seeds 2, pale brown, 1.5–2 mm. long.—Collected near Redding, Shasta Co., Loma Linda, San Bernardino Co.; to Ore., Atlantic Coast. May.

8. **P. rhodospérma** Dcne. Like *P. virginica,* but sepals bristly and attenuate, 2.5–3 mm. long; seeds red, 2.5–3 mm. long.—Occasional weed, as at San Diego, Sweetwater V., Corona del Mar; Mo. and Kans. to Mex. May.

9. **P. truncàta** Cham. ssp. **fírma** (Kunze) Pilg. [*P. f.* Kunze.] Hairy annual 2–6 cm. high; lvs. linear-oblanceolate, 1–6 cm. long, 3–4 mm. wide; scapes 1–few, 1.5–7 cm. high; spikes 1–2.5 cm. long, rather dense; bracts ovate-deltoid, rigid-pilose, ca. 2 mm.

long; sepals narrow-elliptic, rigid-pilose, 2–2.5 mm. long; corolla-lobes ca. 2 mm. long, erect; seeds 2, broadly elliptic, brownish, ca. 1.8 mm. long.—Adventive on open grassy slopes and meadows that are wet in the spring, Sonoma and Marin cos.; native of Chile. April–May.

10. **P. hirtélla** HBK. var. **Galeóttiàna** (Dcne.) Pilg. [*P. G.* Dcne. *P. Durvillei* ssp. *subnuda* (Pilg.) Pilg. *P. s.* Pilg. *P. D. californica* F. & M. *P. D.* var. *angustata* Pilg.] Perennial, ± hirsute-pubescent on scapes, lf.-veins, etc., occasionally almost glabrous; lvs. narrowly elliptic to somewhat oblanceolate, entire to denticulate, the blades 1–2 dm. long, tapered gradually into broad shorter petioles; scapes arcuate-ascending, 1.5–4 dm. high; spikes dense, cylindrical, 1–2.5 dm. long; bracts elliptic-lanceolate, ca. 3 mm. long, ciliate and stiff-pubescent on the sharp keel; sepals unequal, elliptic to broadly lanceolate, 2.5–3 mm. long; corolla-lobes erect, ca. 2 mm. long; stamens 4; seeds 3, oblong-elliptic in outline, flat on 1 face, olive-brown, ca. 1.6 mm. long.— Occasional on moist banks, low elevs.; Coastal Salt Marsh, Coastal Sage Scrub, N. Coastal Scrub, Closed-cone Pine F., etc.; along the coast and inland to San Bernardino, Del Norte Co. to San Diego Co.; to cent. Mex. May–Sept.

11. **P. lanceolàta** L. RIBGRASS. ENGLISH PLANTAIN. BUCKHORN. Usually perennial with strong caudex, somewhat short-villous; lvs. lanceolate to lance-oblong, erect or spreading, the blades 5–20 cm. long, attenuate at apex and gradually narrowed into rather slender somewhat shorter petioles; scapes 2–8 dm. high, arched-ascending, rather slender; spikes dense, ovoid-conic at beginning, cylindric and 2–8 cm. long in fr.; bracts broadly ovate, somewhat pubescent on back, scarious-margined, ca. 2 mm. long; front sepals connate, ca. 3 mm. long; corolla almost rotate; anthers well exserted; caps. oblong-ovoid, dehiscing below middle; seeds 1–2, brown, shining, deeply hollowed on 1 face, ca. 3 mm. long; $2n = 12$ (Nakajima, 1930), 24, 96 (MacCullagh, 1934).— Weed in lawns and moist waste places; introd. from Eu. April–Aug.

12. **P. aristàta** Michx. Annual or perennial, dark green, mostly loosely villous; lvs. linear, 5–15 cm. long, narrowed to margined semiclasping petioles; scapes 1–3 dm. high, slender; spikes 2–8 cm. long; bracts linear, 5–25 mm. long, divergent; sepals spatulate-oblong, ca. 2.5 mm. long; corolla-lobes round-ovate, ca. 2 mm. broad; seeds 2, oblong, finely pitted, brown, 2–2.5 mm. long; $2n = 20$ (Heitz, 1927).—Occasional weed, as at Eureka, in lawns at Loma Linda; cent. U.S. June–July.

13. **P. Púrshii** R. & S. Erect tufted annual, 5–20 cm. tall; lvs. entire, acute, 3–10 cm. long, 1–4 mm. wide, villous, sometimes densely and loosely so; scapes erect or ascending, 2–15 cm. high, mostly villous; spikes cylindrical, dense, 1–8 cm. long; bracts linear-subulate to sublanceolate, scarcely if at all membranous on sides at base, the lowest ca. the length of the calyces; sepals ovate, villous, 2–3 mm. long; corolla-lobes 1–2 mm. long; caps. ellipsoid, circumscissile near middle; seeds dark brown, finely pitted, elliptic, 1.5–2 mm. long.—Near Orleans, Humboldt Co.; to B.C., Rocky Mts., etc. July–Aug.

Var. **pícta** Pilg. [*P. oblonga* Morris. *P. spinulosa* var. *oblonga* Poe. *P. picta* Morris. *P. xerodea* Morris.] Basal bracts 2–3 times as long as calyx.—Poorly defined ssp., dry sandy or rocky places, mostly 2500–7000 ft.; Pinyon-Juniper Wd., Shadscale Scrub, Joshua Tree Wd., Chaparral; mts. of Mojave Desert to w. edge Colo. Desert and interior cismontane s. Calif.; Ariz.; L. Calif. April–June.

14. **P. Hookeriàna** F. & M. var. **califórnica** (Greene) Poe. [*P. patagonica* var. *c.* Greene. *P. erecta* Morris. *P. e.* ssp. *rigidior* Pilg. *P. dura, speciosa, obversa,* and *tetrantha* Morris.] Villous annual; lvs. filiform to linear-lanceolate, entire or with small remote denticulations, 3–12 cm. long; scapes 5–25 cm. tall, erect to arcuate-ascending; spikes capitate to short-cylindric, 0.5–2.5 cm. long, dense; bracts ovate, broad at base, scarious-margined at least half their length, not exceeding calyx; sepals scarious-margined, oblong, 3 mm. long, villous; corolla-lobes spreading, 1–2 mm. long; caps. ellipsoid, ca. 3 mm. high; seeds 2, dull, brown, 2–2.5 mm. long, finely pitted.—Common in dry open places, below 2500 ft.; Coastal Sage Scrub, V. Grassland, Chaparral, etc.; cismontane Calif. to Ore., L. Calif., Santa Barbara Ids. March–May.

15. **P. insulàris** Eastw. [*P. brunnea* Morris. *P. fastigiata* var. *b.* Pilg.] Habit and stature of *P. Hookeriana* var. *californica;* plant villous; bracts usually with brown midribs, ovate to orbicular, 2.5–3 mm. long; sepals elliptic to obovate-elliptic, ca. 2.5 mm. long; seeds reddish-yellow when mature, shining, 2–2.5 mm. long.—Occasional;

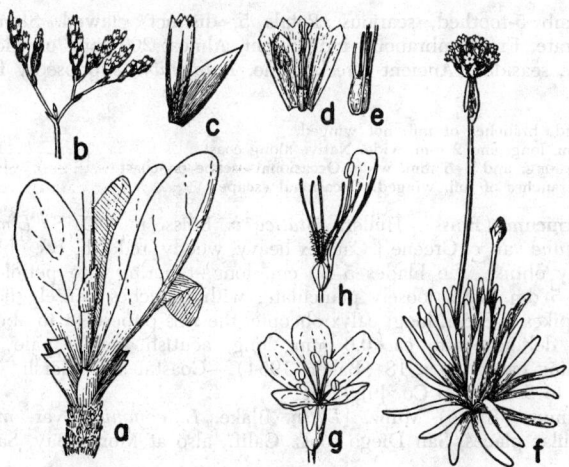

Fig. 45. PLUMBAGINACEAE. *Limonium californicum: a,* base of plant, × ½; *b,* part of infl., × ½; *c,* fl., × 2½, with subtending bract; *d,* corolla opened; *e,* ak. with persistent styles. *Armeria maritima* var. *californica: f,* habit, × ¼; *g,* fl., × 2, with calyx, 5 quite separate petals, 5 stamens; *h,* pistil, × 2, styles 5.

Coastal Strand, Coastal Sage Scrub; along the coast from Santa Barbara Co. to n. L. Calif., Santa Barbara Ids. Feb.–April.

Var. **fastigiàta** (Morris) Jeps. [*P. f.* Morris. *P. minima* A. M. Cunn. *P. scariosa* Morris. *P. insularis* var. *s.* Jeps.] Usually more noticeably white-woolly; bracts usually with green midribs.—Common, dry slopes and flats, below 4500 ft.; Creosote Bush Scrub, Joshua Tree Wd.; Mojave and Colorado deserts; to Utah, Ariz. Jan.–April.

16. **P. índica** L. [*P. arenaria* Waldst. & Kit.] Annual, caulescent, branched, ± pilose, 2–6 dm. high; lvs. opposite, linear, attenuate, 2–4.5 cm. long; peduncles axillary, ascending, 3–7 cm. long; spikes dense, round to ellipsoid, 1–2 cm. long; lower bracts concave, 4–5 mm. long, round-ovate at base with prolonged narrow tip, the upper acute; sepals elliptic to obovate, obtuse, 4 mm. long; corolla-lobes narrow-ovate, 2 mm. long; caps. broadly ellipsoid, 2 mm. long; seeds red-brown, shining, narrow-elliptic, 2.5 mm. long. —Reported from sandy and waste places in many parts of the state, as it is often a constituent of commercial bird-seed, poultry-feed, etc. Native of Old World. July–Nov. Often identified as *P. Psýllium* L., which has the bracts shorter and gradually narrowed.

48. Plumbaginàceae. LEADWORT FAMILY (Fig. 45)

Mostly acaulescent perennial herbs with basal lvs. and scapose panicles, spikes or heads, sometimes with elongate branched stems and some woody. Fls. perfect, regular, 5-merous. Calyx bracted at base, tubular or funnelform, plaited, 5–15-ribbed, often scarious and colored. Corolla usually of nearly distinct petals, convolute or imbricate in bud. Stamens opposite the petals, often adnate to base of the claw. Ovary superior, 1-celled with 1 ovule; styles 5, separate or united. Fr. a utricle or ak., usually inclosed by calyx. Seed 1 with membranous testa; endosperm mealy or lacking; embryo straight. Ca. 10 genera and 300 spp., widely distributed, usually of saline or calcareous places; some grown as ornamentals.

Infl. racemose, corymbose or paniculate; lvs. broad .... 1. *Limonium*
Infl. capitate; lvs. linear .... 2. *Armeria*

1. Limònium Mill. SEA-LAVENDER. MARSH-ROSEMARY

Perennial herbs with broad flat lvs. in a basal tuft. Fls. secund in loose spikes or clusters at ends of much-branched scape. Calyx campanulate or tubular, usually 10-

ribbed, the limb 5-toothed, scarious. Petals 5, distinct, clawed. Stamens 5. Styles mostly 5, separate. Fr. membranous, indehiscent. Almost 200 spp., of wide distribution, often along the seaside. (Ancient Greek name, *Leimonion,* supposedly from *leimon,* a marsh.)

Lvs. not pinnatifid; branches of infl. not winged.
    Fls. ca. 5–6 mm. long, and 2 mm. wide. Native along coast . . . . 1. *L. californicum*
    Fls. ca. 9 mm. long, and 4–5 mm. wide. Occasional escape on coast . . . . 2. *L. Perezii*
Lvs. pinnatifid; branches of infl. winged. Occasional escape . . . . 3. *L. sinuatum*

1. **L. califórnicum** (Boiss.) Heller. [*Statice c.* Boiss. in DC. *S. Limonium* var. *c.* Gray. *L. commune* var. *c.* Greene.] Caudex heavy, woody, reddish; lvs. oblong to oblong-obovate, mostly obtuse, the blades 5–20 cm. long, tapering into petioles ca. as long; scape stout, 2–5 dm. high, loosely paniculate, with branches densely fld. at tips with small secund spikes 1–3 cm. long; calyx obconic, the ribs pubescent to above the middle, lobes whitish, deltoid-ovate, ca. 0.6 mm. long, acutish; petals pale violet, oblong, slightly exceeding calyx; $2n = 18$ (Baker, 1954).—Coastal Salt Marsh, Coastal Strand; Humboldt Co. to San Diego Co. July–Dec.
    Var. **mexicànum** (Blake) Munz. [*L. m.* Blake. *L. commune* var. *m.* Jeps.] Calyx glabrous.—Similar places, San Diego to L. Calif., also at Morro Bay, San Luis Obispo Co. July–Nov.
2. **L. Perèzii** F. T. Hubb. [*Statice P.* Webb.] Woody at base; lvs. rhombic-ovate to deltoid, basal, the blades 8–15 cm. long, subtruncate at base, on petioles exceeding the blades; scapes branched, 4–6 dm. high; calyx purplish-blue, pubescent on ribs; petals pale yellow.—Cult. and occasionally established as at beach at Ventura; native in Canary Ids. March–Sept.
3. **L. sinuàtum** (L.) Mill. [*Statice s.* L.] Rough-hairy perennial or biennial; lvs. basal, lyrate-pinnatifid with rounded lobes and sinuses, the blades 3–12 cm. long, short-petioled; scapes corymbosely panicled, 1.5–4 dm. high, winged, the wings ending in foliose lance-linear appendages 1–4 cm. long; calyx blue to white, funnelform, ca. 1 cm. long; petals yellowish-white.—Beaches and coastal marshes, San Diego and Los Angeles cos., San Luis Obispo Co., as escape; native in Medit. region. June–Oct.

2. Armèria Willd. THRIFT

Tufted acaulescent perennial herbs. Lvs. narrowly linear, basal, persistent. Fls. in a globose head at the end of a naked scape, the heads subtended by 2 or more whorls of usually scarious bracts producing a sort of invol., the 2 outer bracts sheathlike and reflexed. Calyx funnelform, 5-toothed, 10-ribbed, scarious. Petals 5, united somewhat or distinct. Stamens 5. Styles united below. Fr. membranous, rarely dehiscent, 5-pointed apically. Ca. 50 spp., mostly of cooler climates. Several cult. (Name said to be Celtic.)

(Lawrence, G. H. M. The genus Armeria in N. Am. Am. Midl. Nat. 37: 757–779, 1947.)

1. **A. marítima** (Mill.) Willd. var. **califórnica** (Boiss.) G. H. M. Lawr. [*A. andina* var. *c.* Boiss. *A. arctica* ssp. *c.* Abrams. *Statice a.* var. *c.* Blake.] Taproot long, tough; lvs. linear, 4–15 cm. long, 2–2.5 mm. wide, glabrous; scapes 1–several, 0.5–4.5 dm. high; outer sheathing bracts 1–3 cm. long, the inner involucral bracts narrow-deltoid to lance-oblong, acutish, 8–15 mm. long; calyx ca. 6–7 mm. long, with pubescent ribs, the lobes broadly triangular, ca. 1 mm. long, scarious; petals rose-pink, exceeding calyx.—Coastal bluffs and sandy places, below 700 ft.; N. Coastal Scrub, Closed-cone Pine F.; San Luis Obispo Co. to B.C., Santa Rosa Id. April–Aug.

49. Ericàceae. HEATH FAMILY (Figs. 46, 47)

Shrubs or trees, sometimes subshrubs. Lvs. alternate, opposite or whorled, simple, without stipules, persistent or deciduous. Fls. bisexual, regular or slightly irregular, solitary or in axillary or terminal racemes or panicles. Calyx free from or adnate to ovary, 4–5-cleft or -parted, usually persistent. Corolla mostly sympetalous, 4–5-lobed. Stamens as many or twice as many as corolla-lobes, inserted on outer edge of a hypogynous or

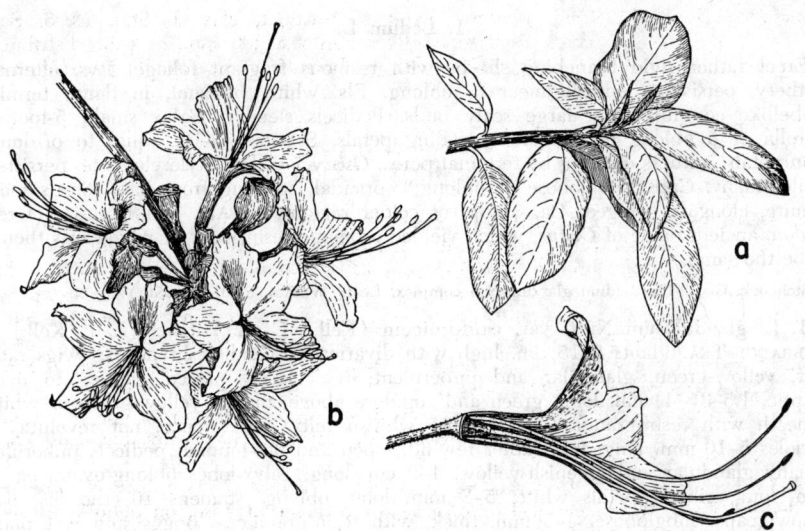

Fig. 46. ERICACEAE. *Rhododendron occidentale:* a, shoot, × ½; b, infl., × ½; c, section of fl., × 1.

epigynous disk; anthers 2-celled, upright, opening by terminal pores or chinks, rarely longitudinally, sometimes awned. Ovary superior or inferior, 2–10-locular, the placentae usually axile. Style and stigma 1. Fr. a caps., berry, or drupe. Ca. 70 genera and 1500 spp., particularly of cooler regions; many (like heathers and rhododendrons and azaleas) of great ornamental value; others (like blueberries and huckleberries and cranberries) valued for their frs.

A. Ovary superior, free from the calyx.
   B. Fr. a dry capsule.
      C. Anthers awnless; caps. septicidal.
         D. Corolla of distinct petals, white; plant evergreen; caps. opening from base upward
                                                  1. *Ledum*
        DD. Corolla of somewhat united petals; caps. opening from apex to base.
           E. Fl.-buds and usually lf.-buds with scaly bracts; caps. longer than thick.
              F. Fls. 5-merous; corolla campanulate to funnelform .... 2. *Rhododendron*
              FF. Fls. 4-merous; corolla globose-campanulate .... 3. *Menziesia*
         EE. Fl.-buds and lf.-buds with coriaceous-foliaceous persistent bracts.
           F. Corolla saucer-shaped, 10-saccate near base, the pouches holding the
              anthers in the bud; fl. mostly 1 cm. or more broad .... 4. *Kalmia*
           FF. Corolla urn-shaped to bell-shaped, without basal pouches; fl. less than
              1 cm. broad .... 5. *Phyllodoce*
      CC. Anthers awned or mucronate; capsule loculicidal.
         D. Corolla campanulate; anthers awned; lvs. scalelike, overlapping, sessile
                                         6. *Cassiope*
        DD. Corolla urn-shaped; anthers mucronate; lvs. broad, petioled ...... 7. *Leucothoe*
   BB. Fr. fleshy.
      C. The proper fr. a caps. enclosed in the fleshy calyx .... 8. *Gaultheria*
      CC. The fr. itself a berry or drupe.
         D. Surface of fr. granular or warty.
           E. Calyx glabrous; plant a tree .... 9. *Arbutus*
           EE. Calyx tomentose; plant a bush .... 10. *Comarostaphylis*
        DD. Surface of fr. smooth, not warty or granular.
           E. Lvs. revolute, not vertical; fils. slender; fr. with a solid 3–5-celled stone
                                      11. *Xylococcus*
           EE. Lvs. plane, mostly vertical; fils. dilated at base; fr. usually with ± separable
              nutlets, rarely with one stone .... 12. *Arctostaphylos*
AA. Ovary inferior, adnate to the fleshy calyx, forming a berry crowned by the calyx-teeth
                                               13. *Vaccinium*

1. Lèdum L.

Erect rather rigid branching shrubs with resinous fragrant foliage. Lvs. alternate, leathery, persistent, entire, linear to oblong. Fls. white, bisexual, in dense terminal umbellike corymbs from large scaly buds. Pedicels slender. Calyx small, 5-toothed. Corolla of 5 oblong to obovate spreading petals. Stamens 5–10, equal to or longer than petals; anthers opening by terminal pores. Ovary 5-celled; style elongate, persistent; ovules many. Caps. subglobose or oblong, septicidal, splitting from base. Seeds many, minute, elongate, winged. Ca. 3 spp. of colder regions, N. Am. and Eurasia. (Greek, *Ledon,* ancient name of *Cistus,* which yielded aromatic resin, odor of which was thought to be the same.)

(Hitchcock, C. L. The Ledum glandulosum complex. Leafl. W. Bot. 8: 1–8, 1956.)

1. **L. glandulòsum** Nutt. var. **califórnicum** (Kell.) C. L. Hitchc. [*L. c.* Kell.] W. LABRADOR-TEA. Plants 5–15 dm. high with divaricate-ascending branches; twigs rather stiff, yellow-green, glandular and puberulent; lvs. closely placed, oblong to ovate-elliptic, 1.5–3(–4) cm. long, green and rugulose above, mostly yellow-green or whitish beneath with resin-granules and minute whitish felty puberulence, not revolute, the petioles 5–10 mm. long; infl. rather few-fld., open and flat-topped; pedicels puberulent, usually glandular and greenish-yellow, 1–2 cm. long; calyx-lobes oblong-ovate, ca. 1.5 mm. long, ciliate; petals white, 5–8 mm. long, oblong; stamens 10, the fils. hairy below; caps. subglobose, 4–5 mm. thick, with resin-granules.—Boggy and wet places, 4000–11,900 ft.; Red Fir F. to Alpine Fell-fields; Sweetwater Mts., Mono Co., Sierra Nevada from Tulare Co. n., to Modoc Co., Trinity Co. June–Aug.

Ssp. **columbiànum** (Piper) C. L. Hitchc. [*L. c.* Piper.] Twigs, pedicels, and under side of lvs. mostly white with very short stiff puberulence and less conspicuous resin-granules than in the sp.; lvs. mostly 3–6 cm. long, usually strongly revolute on margins; infl. with many fls., dense, rounded at top; caps. oblong-ovoid, 5–6 mm. long; $2n = 26$ (Callan, 1941).—Swamps and bogs, below 2000 ft.; Mixed Evergreen F., Closed-cone Pine F., etc.; near the coast from Santa Cruz Co. to Del Norte Co.; to Wash. May–June. Plants from Calif. have been called var. *austràle* C. L. Hitchc. if the lvs. are deep green on upper surface, and ssp. *olivàceum* C. L. Hitchc. if light olive-green.

2. Rhododéndron L. RHODODENDRON. AZALEA

Shrubs or small trees. Lvs. simple, alternate, deciduous or persistent, entire or toothed, rather crowded on branches. Fls. showy, in terminal umbels or corymbs. Calyx small, saucer-shaped, with persistent lobes. Corolla funnelform to subcampanulate, regularly or irregularly 5-lobed. Stamens 5 or 10, the fils. slender, elongate, declined; anthers opening by terminal pores. Ovary 5-celled; style elongate, declined. Caps. septicidally 5-valved. Seeds many, scalelike, wing-margined. Perhaps 200 spp., of wide distribution in N. Hemis., especially in Asia. (Greek, *rhodos,* rose, and *dendron,* tree.)

Lvs. persistent, leathery; fls. rose; stamens 10 .... 1. *R. macrophyllum*
Lvs. deciduous, not leathery; fls. white to cream or with some pink; stamens 5 .... 2. *R. occidentale*

1. **R. macrophýllum** D. Don. [*R. californicum* Hook. *Hymenanthes c.* Copel. f. *H. m.* Copel. f.] CALIFORNIA ROSE-BAY. Evergreen shrub 1–4 m. high with coarse glabrous twigs; lvs. leathery, dark green above, paler and somewhat papillose beneath, oblong to elliptic, entire, 6–20 cm. long; petioles stout, 1–2 cm. long; calyx 5-lobed, the lobes ca. 1 mm. long, considerably broader; corolla broadly campanulate, rose to rose-purple, rarely white, 3–4 cm. long, the obovate lobes undulate; fils. ca. 2 cm. long; caps. cylindric-ovoid, 1.5–2 cm. long, rusty-puberulent and glandular; seeds brown, flat, elongate, ca. 3 mm. long including wing.—Dryish to damp, ± shaded woods, below 4000 ft.; Redwood F., Mixed Evergreen F., Douglas-Fir F., Yellow Pine F.; near the coast from Monterey Co. to Del Norte and Siskiyou cos.; to B.C. April–July.

2. **R. occidentàle** (T. & G.) Gray. [*Azalea o.* T. & G. *A. californica* T. & G., not *Rhododendron c.* Hook.] WESTERN AZALEA. Loosely branched shrub 1–3(–5) m. tall, deciduous, with shredding bark; twigs stiff, divaricate, glabrous to pubescent, sometimes

glandular; lvs. thin, light green, elliptic to obovate, 3–8 cm. long, subglabrous to scattered-pubescent, stiff-ciliate; petioles 4–8 mm. long; infl. terminal, ± glandular-pubescent; calyx-lobes oval to oblong-ovate, ciliate, 2–5 mm. long; corolla funnelform, 3.5–5 cm. long, slightly irregularly lobed, white or with pink tinge, the upper lobe with a yellowish blotch, the lobes lance-oblong; stamens 5, exserted; caps. oblong, pubescent, 1–2 cm. long; seeds tan, flat, narrow, ca. 2.5 mm. long including wing.—Stream banks and moist places, below 7500 ft.; Yellow Pine F., Red Fir F., Mixed Evergreen F., Redwood F.; Cuyamaca Mts. to San Jacinto Mts., Sierra Nevada from Kern Co. and Fresno Co. n. to Siskiyou Co., Coast Ranges from Santa Cruz Co. n.; to Ore. April–Aug. Poisonous to stock. Variable, 2 local types are: var. *sonoménse* (Greene) Rehd. [*R. s.* Greene.] Lvs. 2–2.5 cm. long; fls. rose to whitish with salmon spot in center.—E. side of Napa Range; var. *paludòsum* Jeps. Fls. pink with orange flush.—Near the coast of Humboldt and Del Norte cos.

3. Menzièsia Sm. MOCK AZALEA

Deciduous shrubs with erect or spreading branches. Lvs. alternate, approximate on twigs, thin, short-petioled. Fls. in terminal clusters on nodding pedicels. Calyx small, flattish, entire or shallowly 4-lobed. Corolla cylindric–urn-shaped, 4-lobed. Stamens 8, included; anthers linear, opening by terminal pores. Ovary 4-celled; style filiform; stigma 4-lobed. Caps. ovoid, woody, septicidally 4-valved. Seeds many, slender, pointed or caudate. Ca. 7 spp., N. Am. and Japan. (A. *Menzies,* surgeon and botanist with the Vancouver Expedition.)

1. **M. ferrugínea** Sm. Erect or straggling, 1–4 m. high; twigs glandular-pubescent; lvs. elliptic-oblong to obovate, 3–6 cm. long, thinly rusty-strigose above and mostly on veins beneath, ciliate on the crenate-serrate margin; pedicels glandular-pubescent, 2–2.5 cm. long, becoming erect in fr.; calyx-lobes barely 1 mm. long, ciliate; corolla yellow tinged with red, ca. 7 mm. long; fils. glabrous; caps. oblong-ellipsoid, 6–7 mm. long; seeds light brown, narrow, flat, winged at each end, ca. 2–2.5 mm. long.—Shade of woods, below 1000 ft.; Redwood F.; along coast of Humboldt and Del Norte cos.; to Alaska, Mont. June–July.

4. Kálmia L. AMERICAN-LAUREL

Mostly evergreen smooth branching shrubs. Lvs. alternate, opposite or whorled, entire, usually petioled. Fls. in terminal or lateral corymbs or umbels; pedicels in axils of small thick bracts. Calyx 5-parted. Corolla crateriform, shallowly 5-lobed. Stamens 10, with slender fils., each held back in a little pouch in the corolla and springing up suddenly when the latter expands. Ovary 5-celled; style slender. Caps. subglobose to somewhat ovoid, 5-valved. Seeds many, minute, narrow, winged distally. Ca. 6 spp., N. Am. (P. *Kalm,* 1716–1779, pupil of Linnaeus, who traveled in Am.)

1. **K. polifòlia** Wang. var. **microphýlla** (Hook.) Rehd. [*K. glauca* var. *m.* Hook. *K. m.* Heller.] Low, diffusely branched, 1–2 dm. high, with glabrous to puberulent twigs; lvs. oblong to obovate, opposite, sometimes somewhat revolute, 1–2 cm. long, subsessile, dark green above, paler beneath; corymbs few-fld.; pedicels slender, glabrous, 2–4 cm. long; calyx 5–6 mm. wide; corolla rose-purple, 8–12 mm. wide; caps. depressed-globose, 5–6 mm. across; seeds ca. 0.6 mm. long.—Boggy places and wet meadows, 7000–12,000 ft. in Sierra Nevada, lower in Coast Ranges; mostly Red Fir F. to Alpine Fell-fields; Tulare and Humboldt cos. n.; to Alaska, Rocky Mts. June–Aug.

5. Phyllódoce Salisb. MOUNTAIN-HEATHER

Low evergreen shrubs, much-branched, heathlike. Lvs. linear, needlelike, blunt or subacute, alternate, crowded, revolute. Fls. in umbellike infl., long-pedicelled from axils of persistent herbaceous bracts, nodding or suberect. Calyx-segms. lance-ovate, acute, persistent, usually 5. Corolla urn-shaped to ovoid or open-campanulate, ± lobed. Stamens 7–10; fils. slender; anthers unappendaged, opening by oblique apical pores. Ovary usually 5-celled, globose; style filiform. Caps. ovoid to subglobose, septicidal. Seeds many, minute, narrowly winged. Ca. 8 spp., circumboreal. (Greek name of a sea nymph.)

Stamens exserted; corolla-lobes ca. as long as tube .... 1. *P. Breweri*
Stamens included; corolla-lobes much shorter than tube .... 2. *P. empetriformis*

1. **P. Brèweri** (Gray) Heller. [*Bryanthus B.* Gray] Semi-procumbent shrub, rather lax, 1–3 dm. high; lvs. 6–15 mm. long, somewhat glandular, puberulent at margins, strongly revolute; pedicels 1–1.5 cm. long, glandular-pubescent; calyx-lobes oblong, obtuse, 3–4.5 mm. long, glabrous except for the ciliolate margins; corolla rose-purple to pinkish, open-campanulate, ca. 0.8–1 cm. long; stamens long-exserted; caps. round, 3–3.5 mm. across.—Rocky, sometimes rather moist, places, 6000–12,000 ft.; Subalpine F., Alpine Fell-fields; San Bernardino Mts., Sierra Nevada from Tulare Co. to Mt. Lassen. July–Aug.

2. **P. empetrifórmis** (Sm.) D. Don. [*Menziesia e.* Sm. *Bryanthus e.* Gray.] Cushion-like shrub 1–5 dm. high; lvs. linear to linear-oblong, 6–15 mm. long, minutely glandular-serrulate; pedicels glandular-puberulent, 1–2.5 cm. long; calyx-lobes 2.5 mm. long, glabrous except for the ciliolate margin; corolla campanulate, rose-purple, 7–9 mm. long, the lobes 1.5–2 mm. long; stamens included; caps. 3–4 mm. in diam.—Rocky slopes, 5000–9000 ft.; Subalpine F., Alpine Fell-fields; Trinity and Siskiyou cos.; to Alaska, Rocky Mts. July–Aug.

6. Cassìope D. Don

Small creeping or prostrate alpine shrubs, evergreen, with ascending branches. Lvs. persistent, scalelike or needlelike, closely imbricate in 4 ranks in decussate pairs. Fls. solitary in lf.-axils near ends of branches, nodding on long slender pedicels. Bractlets 4, subtending pedicels. Calyx persistent, 4–5-lobed, the lobes longer than tube. Corolla campanulate, 4–5-lobed, the lobes shorter than tube. Stamens 8 or 10, included; fils. glabrous; anthers opening by terminal pores and with dorsal awnlike appendages. Ovary 4–5-celled; style persistent. Caps. round to ovoid, loculicidally 4–5-valved. Seeds many, minute, winged. Several spp., circumboreal. (*Cassiope*, mother of Andromeda in Greek mythology.)

1. **C. Mertensiàna** (Bong.) G. Don. [*Andromeda M.* Bong. *C. M.* ssp. *californica* and *ciliolata* Piper?.] WHITE-HEATHER. Creeping, with ascending branches, 1–3 dm. high; lvs. ovate-lanceolate, keeled on back, 3–6 mm. long, obtuse; pedicels puberulent, 6–20 mm. long; calyx-lobes ovate, obtuse ca. 2 mm. long; corolla white to pinkish, 5–6 mm. long, with ovate lobes; caps. rounded, ca. 3 mm. long.—Rocky ledges and crevices, ca. 7000–12,000 ft.; Subalpine F., Alpine Fell-fields; Sierra Nevada from Fresno Co. n., Coast Ranges from Trinity Co. to Siskiyou Co.; to Alaska, Mont. July–Aug.

7. Leucóthoe D. Don

Erect shrubs, evergreen or deciduous. Lvs. alternate, entire or toothed, petioled. Fls. white, scaly-bracted, in spiciform axillary or terminal racemes, forming naked panicles. Calyx of 5 almost distinct sepals, imbricate in bud. Corolla urn-shaped or tubular with short lobes. Stamens 10, included; anthers attached near base, opening by terminal pores, naked or with erect awns at apex. Ovary 5-celled; style slender; stigma small. Caps. depressed, ± 5-lobed, 5-valved, with many-seeded placentae on summit of the short columella. Seeds many, minute. Ca. 35 spp., of e. Asia and N. and S. Am. (*Leucothoe*, daughter of King Orchamus of Babylon.)

1. **L. Davísiae** Torr. [*Oreocallis D.* Small.] SIERRA-LAUREL. Evergreen, 5–15 dm. high, with glabrous twigs; lvs. oblong, entire to somewhat serrulate, 3–6 cm. long, on petioles 3–6 mm. long; panicle terminal, 6–15 cm. long, the racemes erect; bractlets scarious, ovate, 2–3 mm. long; pedicels recurved, 3–7 mm. long; calyx-lobes whitish, lance-oblong, ca. 2 mm. long, somewhat fimbriolate; corolla 6–7 mm. long, urn-shaped, the lobes ca. 1 mm. long; caps. depressed-globose, erect, lobed, 4–5 mm. across.—Bogs and springy places, 3200–8500 ft.; Red Fir F., Lodgepole F.; Sierra Nevada from Fresno Co. to Lassen Peak, and in Trinity and Siskiyou cos.; adjacent Ore. June–Aug.

8. Gaulthèria L.

Evergreen shrubs or almost herbaceous. Lvs. coriaceous, alternate. Fls. axillary and solitary to racemose or panicled. Calyx 5-cleft. Corolla 5-toothed or -lobed, cylindric-ovoid to urn-shaped or campanulate. Stamens 8 or 10, included; fils. dilated below; anthers usually 2-awned, with terminal pores. Caps. depressed, many-seeded, 5-loculed, 5-lobed, enclosed when ripe by the calyx or its fleshy base so as to appear a berry. Seeds many, rather small, somewhat angled, subovoid. Ca. 100 spp., mostly of the Andes, fewer in Asia, Australia, N. Am. (Named for J. F. *Gaultier*, 18th-century physician at Quebec.)

Fls. in racemes, the corolla urn-shaped; lvs. 3–10 cm. long .... 1. *G. Shallon*
Fls. solitary, axillary, the corolla campanulate; lvs. 1–3 cm. long.
    Calyx glabrous; lvs. 1–2 cm. long. Sierra Nevada .... 2. *G. humifusa*
    Calyx pubescent; lvs. 1.5–3 cm. long. N. Coast Ranges .... 3. *G. ovatifolia*

1. **G. Shállon** Pursh. SALAL. Spreading subshrub or shrub 4–20 dm. high, with erect or spreading stems and glandular-pubescent branchlets; lvs. ovate to roundish or oblong, mostly 3–10 cm. long, subglabrous except sometimes at edge and near petioles, generally abruptly short-acuminate, with round or subcordate base, finely serrate, on petioles 2–4 mm. long; panicles 3–15 cm. long, glandular-pubescent; bracts subovate, colored, 6–10 mm. long, ± glandular-ciliate; pedicels declined, bibracteolate; calyx pubescent, 6–8 mm. long, the lobes lance-deltoid; corolla white or pink, urn-shaped, 8–10 mm. long, the short lobes recurved; fils. pubescent; fr. dark purple, 7–8 mm. broad; seeds brown, reticulate, ca. 1 mm. long; 2n = 88 (Callan, 1941).—In woods or brushy places, below 2500 ft.; Mixed Evergreen F., Redwood F., Closed-cone Pine F., N. Coastal Scrub; near the coast from Santa Barbara Co. to Del Norte Co.; to B.C. April–July.

2. **G. humifùsa** (Grah.) Rydb. [*Vaccinium h.* Grah. *G. Myrsinites* Hook.] ALPINE WINTERGREEN. Matted scarcely woody plant with creeping rooting stems 1–2 dm. long, glabrous to puberulent; lvs. round to oval, subentire to obscurely serrulate, obtuse or rounded at both ends, mostly 1–2 cm. long, very short-petioled; fls. solitary, axillary; pedicels ca. 1 mm. long, bracteolate; calyx glabrous, ca. 2.5 mm. long; corolla slightly longer, white; fr. 5–7 mm. in diam., red; seeds straw-colored, shining, ca. 0.4 mm. long.—Rare in moist places, ca. 8000–10,500 ft.; Lodgepole F.; Sierra Nevada in Tulare, Fresno, Tuolumne, and Mariposa cos.; to B.C., Rocky Mts. July.

3. **G. ovatifòlia** Gray. Habit of *G. humifusa*, but more pilose, slightly larger; lvs. ovate, 2–3 cm. long, acute at apex, more conspicuously serrulate; calyx pubescent, 2 mm. long; corolla 3.5 mm. long; fr. 4–5 mm. in diam., red.—Wet places, 3000–5500 ft.; Douglas-Fir F., Yellow Pine F.; Humboldt, Siskiyou and Del Norte cos.; to B.C. June–July.

9. Arbùtus L.

Evergreen trees or shrubs with bark of main trunks fissured or smooth and exfoliating. Lvs. alternate, entire or toothed, coriaceous, usually long-petioled. Fls. perfect, in terminal panicles. Calyx 5-lobed, rather persistent. Corolla white or pale, urn-shaped, the 5 lobes spreading or recurved and much shorter than the swollen tube. Stamens 10, included; fils. dilated below, usually pubescent; anthers broad, each with 2 slender awns. Ovary usually 5-celled; stigma obscurely 5-lobed. Fr. hard-stoned, berrylike, globose or depressed-globose, with a rugose or granular surface. Seeds several, obovoid, hard. Ca. 20 spp., of Medit. region, s. Asia, N. and S. Am. (Latin name of *A. Unedo*.)

1. **A. Menzièsii** Pursh. [*A. procera* Dougl.] MADROÑO. MADRONE. Widely branched tree 5–40 m. tall; bark freely exfoliating, leaving a polished reddish or brownish surface, or on very old trunks fissured and dark; lvs. persistent, coriaceous, elliptic to subovate, glabrous except when young, entire or serrulate, dark green above, paler beneath, 5–12 cm. long, on petioles 1–2.5 cm. long; panicles 6–15 cm. long, pubescent or puberulent; bracts ovate, whitish, 4–6 mm. long; calyx-lobes ovate, thin, ciliate, ca. 1 mm. long; corolla white to pink, 6–8 mm. long; style columnar, 5 mm. long; berry red to orange, roundish, 8–10 mm. in diam., rugulose; seeds bony, tightly pressed together, rounded on one margin, almost straight on other, thick, light brown, ca. 2.5 mm. long.—Wooded slopes and canyons, below 5000 ft.; common in Redwood F., Mixed Evergreen F.,

Douglas-Fir F., less so in Foothill Wd., N. Oak Wd., S. Oak Wd.; L. Calif., at scattered localities in s. Calif., and abundant from San Luis Obispo Co. to Del Norte and Siskiyou cos., from Mariposa Co. to Shasta Co.; to B.C. March–May.

10. Comarostáphylis Zucc.

Evergreen erect or spreading shrubs with exfoliating or persistent and shredded bark. Lvs. alternate, coriaceous, entire or toothed, petioled, often revolute. Fls. few to many, in terminal solitary or clustered racemes or panicles, 5(rarely 4)-merous. Calyx persistent, the lobes exceeding the tube, reflexed or spreading at maturity. Corolla urn-shaped, the lobes broad and short, mostly recurved. Stamens 10 (8), included; fils. dilated below, pubescent; anthers broad, awned. Ovary 5(4)-celled, glabrous or pubescent, rounded or ovoid; style columnar; stigma minute. Fr. fleshy, drupelike, warty or papillose, the nutlets united into a round stone. Ca. 20 spp., mostly Mex. (Greek, *komaros*, arbutus, and *staphule*, grape, because of similarity to *Arbutus* and the clustered fr.)

1. **C. diversifòlia** (Parry) Greene. [*Arctostaphylos arguta* var. *d*. Parry. *A. d*. Parry.] SUMMER-HOLLY. Erect, 2–5 m. tall, with shredded bark and canescent-tomentulose twigs; lvs. oblong to elliptic, rounded to acute at apex, serrulate to green and shining above, tomentulose beneath, strongly revolute, 3–8 cm. long, on petioles 2–5 mm. long; racemes mostly solitary and terminal, 3–6 cm. long, tomentose; bracts lance-linear, ca. 3 mm. long; pedicels recurved, 3–5 mm. long; calyx tomentulose, the lobes lanceolate, ca. 2 mm. long; corolla white, puberulent, 5–7 mm. long, the lobes very short; ovary pubescent; drupe globose, 5–6 mm. thick, red, granular-rugose.—Dry slopes, low elevs.; Chaparral; near the coast, San Diego Co.; L. Calif. May–June.

Var. **planifòlia** Jeps. Lvs. not revolute; racemes 5–10 cm. long; bracts oblong-ovate, 3–7 mm. long; pedicels 8–12 mm. long.—Chaparral; Santa Monica Mts., Santa Inez Mts.; Santa Rosa, Santa Cruz, Santa Catalina ids. March–May.

11. Xylocóccus Nutt.

Erect densely branched shrubs with shredding bark. Lvs. alternate or opposite, persistent, entire with revolute margins. Fls. few, in terminal simple or branched panicles. Bracts small, scalelike. Calyx persistent, mostly 5-lobed, the broad lobes reflexed at maturity. Corolla oblong–urn-shaped, mostly 5-lobed. Stamens mostly 10, included; fils. dilated below, pubescent, awned. Ovary mostly 5-celled, pubescent; style elongate; stigma minute. Fr. a dry drupe with smooth surface and a thin pulp; nutlets united into a solid stone. One sp. (Greek, *xylon*, wood, and *kokkos*, berry.)

1. **X. bìcolor** Nutt. [*Arctostaphylos b*. Gray. *A. Clevelandii* Gray.] Two to 3 m. high, with persistent shredding bark and cinereous branchlets; lvs. ovate to oblong, acute at both ends, strongly revolute, dark green and glabrous above, cinereous-tomentose beneath, 3–5 cm. long, short-petioled; panicles recurved, dense, tomentose, 1–2.5 cm. long; bracts ovate, tomentose, ca. 2 mm. long; pedicels 2–4 mm. long; calyx dark red, tomentose, the ovate lobes ca. 1.5 mm. long; corolla white or pink, 8–9 mm. long; fr. round, red, to almost black, 5–8 mm. in diam.; stone solid, smooth; $2n = 26$ (Callan, 1941).—Dry slopes, below 2000 ft.; Chaparral; scattered localities near the coast from Los Angeles Co. to L. Calif. Dec.–Feb.

12. Arctostáphylos. Adans. MANZANITA (Fig. 47)

Woody evergreen plants, varying from low prostrate shrubs to small trees, usually with crooked branches, these usually smooth with thin red to brown bark that exfoliates freely, sometimes rough and shreddy with more persistent fibrous bark. Lvs. simple, alternate, coriaceous, entire to serrulate, sessile or petioled, mostly vertical. Fls. in terminal simple racemes or in panicles, and borne on pedicels with basal bract. Calyx persistent, 4–5-lobed, the lobes broad. Corolla small, urn-shaped, the lobes rounded, recurved, white to pink or rose. Stamens 10 (8), included; anthers with 2 recurved appendages on the back and opening by round terminal pores. Ovary on a hypogynous disk, 4–10-celled, with 1 ovule in each cell. Fr. berrylike, with rather copious granular pulp, or with thin pericarp and dry, enclosing 4–10 nutlets, or these variously fused in

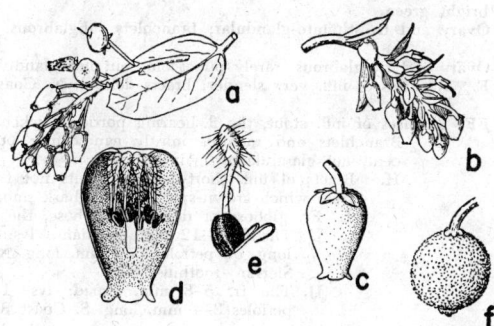

Fig. 47. ERICACEAE. *Arctostaphylos: a*, lf. and infl., × 1, the latter with small bracts; *b*, with foliaceous bracts; *c*, fl., × ½, calyx and urceolate corolla; *d*, section through fl., with 10 stamens; *e*, stamen, × 5, with terminal pores and recurved appendages. *Arbutus Menziesii: f*, rugulose fr.

2's or 3's or even united into a solid stone. A genus of perhaps 50 spp., from Cent. Am. n.; 1 circumpolar. Of some horticultural value. Centering in Calif. Hybridizing freely. Some spp. are characterized by forming at an early stage a basal burl or enlarged root-crown from which sprouting takes place after fire; others lack this. (Greek, *arktos*, bear, and *staphyle,* a bunch of grapes, referring to common name of first known sp.)

(Adams, J. E. A systematic study of the genus Arctostaphylos. Jour. Elisha Mitchell Sci. Soc. 56: 1–62, 1940.)

A. Branches prostrate or procumbent, usually less than 6 dm. high and rooting in contact with the ground.
 B. Lvs., as seen under hand-lens, without stomates on upper surface, but with them evident beneath.
  C. Fls. 4-merous; ovary pubescent .... 1. *A. Nummularia*
  CC. Fls. 5-merous; ovary glabrous.
   D. Lvs. with a rounded or retuse apex, glabrous; fr. red. From Marin Co. n.
                          4. *A. Uva-ursi*
   DD. Lvs. mucronulate and obtusish; fr. brown. Monterey Co.
    E. Branchlets with spreading hairs; lvs. subglabrous, ovate to rounded
                        5. *A. Edmundsii*
    EE. Branchlets finely pubescent; lvs. pubescent, obovate to subspatulate
                        6. *A. pumila*
 BB. Lvs. with stomates on both surfaces.
  C. Ovary with short stiff hairs; young branchlets bristly hairy; fr. splitting open and fall-ing early .... 2. *A. myrtifolia*
  CC. Ovary glabrous; branchlets with short fine hairs; fr. not splitting early.
   D. Hairs of branchlets and infl. not gland-tipped.
    E. Lvs. mostly broadest above the middle; fr. depressed-globose, 5–6 mm. in diam. Pine belt of mts. .... 13. *A. nevadensis*
    EE. Lvs. mostly broadest at or below the middle.
     F. Fr. depressed-globose to round, 4–8 mm. in diam., with separable or irregularly united nutlets.
      G. Terminal infls. simple, few-fld. San Francisco and Monterey cos.
                        7. *A. Hookeri*
      GG. Terminal infls. mostly branched, many-fld.
       H. Fr. 5–6 mm. in diam.; lvs. cuneate at base. Sonoma Co.
                       8. *A. densiflora*
       HH. Fr. 7–8 mm. in diam.; lvs. obtuse at base. Marin Co.
              14. *A. pungens* var. *montana*
     FF. Fr. round-ovoid, 8–12 mm. in diam., with nutlets fused into a solid stone. S. Calif. .... 12. *A. Parryana*
   DD. Hairs of branchlets and infl. gland-tipped.
    E. Plants without a basal burl, hence not stump-sprouting after fire. From Los Angeles Co. s. .... 12. *A. Parryana* var. *pinetorum*
    EE. Plants with a basal burl and stump-sprouting after fire. From Kern Co. n.
                       17. *A. patula*
AA. Branches erect or ascending, taller, not usually rooting.
 B. Bracts of infl. shorter than pedicels, deltoid to almost subulate, even the lower bracts not or scarcely foliaceous.
  C. Pedicels glabrous.
   D. Lvs. pale gray-green; branchlets glabrous or nearly so; fr. 12–14 mm. in diam. W. base of Sierra Nevada .... 16. *A. Mewukka*

DD. Lvs. bright green.
  E. Ovary and fr. stipitate-glandular; branchlets subglabrous. Lake Co.
                                                         10. *A. elegans*
  EE. Ovary and fr. glabrous, rarely pubescent, but not glandular.
    F. Rachises of infl. very slender; bracts deltoid. N. Coast Ranges
                                         9. *A. Stanfordiana*
    FF. Rachises of infl. stout, the fl.-bearing portion thickened.
      G. Branchlets and rachises mostly canescently puberulent or pubescent, not glandular.
        H. Bracts of infl. short-deltoid; plants not forming a basal burl from which crown-sprouts arise; bark smooth.
          I. Fr. globose or depressed-globose, the nutlets ± separable.
            J. The fr. 8–12 mm. in diam.; lvs. mostly 2.5–4.5 cm. long, on petioles 5–9 mm. long. N. Coast Ranges and Sierran foothills . . . . 11. *A. Manzanita*
            JJ. The fr. 5–8 mm. broad; lvs. 1.5–3 cm. long, on petioles 3–4 mm. long. S. Coast Ranges
                                       14. *A. pungens*
          II. Fr. somewhat ovoid, the nutlets united into a solid stone. Tehachapi Mts. to Los Angeles Co. . . . . 12. *A. Parryana*
        HH. Bracts of infl. ± subulate; plants with a basal burl; bark shreddy, rough. San Luis Obispo and Santa Barbara cos.
                                      15. *A. rudis*
      GG. Branchlets and rachises glandular-pubescent.
        H. Fls. 5-merous. Yellow Pine F. or higher.
          I. Plants not forming basal burl, hence not crown-sprouting after fire; fls. white. Mts. of s. Calif.
                            12. *A. Parryana* var. *pinetorum*
          II. Plants forming basal burls; fls. pink. Sierra Nevada and N. Coast Ranges . . . . 17. *A. patula*
        HH. Fls. 4-merous. Below Yellow Pine
                                1. *A. Nummularia* var. *sensitiva*
CC. Pedicels ± glandular-pubescent; ovary mostly glandular or pubescent.
  D. Lvs. bright green, glabrous or nearly so.
    E. Stomates (under a lens) rather equally distributed on both lf.-surfaces; bracts 3–4 mm. long. Tehama Co. . . . . 18. *A. acutifolia*
    EE. Stomates on lower lf.-surface only; bracts 1.5–2.5 mm. long. Insular.
                                      19. *A. insularis*
  DD. Lvs. pale, gray-green, often pubescent.
    E. Bracts of infl. mostly 1–3 mm. long, triangular, persistent.
      F. Fr. pulpy, 6–10 mm. in diam., depressed globose.
        G. Plants without basal burl. Sierra Nevada and N. Coast Ranges.
          H. Lvs. glabrous, smooth; branchlets mostly glabrous; calyx-lobes ciliate . . . . 20. *A. viscida*
          HH. Lvs. glabrous or glandular-hairy, scabrous; branchlets glandular-hairy, calyx-lobes glandular-villous . 21. *A. mariposa*
        GG. Plants with basal burl. Chiefly S. Coast Ranges . . 39. *A. glandulosa*
      FF. Fr. round to round-ovoid, with very little pulp, 12–15 mm. in diam. S. Coast Ranges . . . . 23. *A. glauca*
    EE. Bracts of infl. 5–6 mm. long, lanceolate, deciduous; fr. round-ovoid. San Bernardino Mts. and s. . . . . 22. *A. Pringlei*
BB. Bracts of infl. (at least the lower) as long as or exceeding pedicels, foliaceous.
  C. Stomates, as seen under lens, absent from upper surface of lvs.; young twigs ± bristly.
    D. Lvs. pale green or gray-green; plants without basal burl; bark rough and shreddy. San Luis Obispo Co. . . . . 24. *A. morroensis*
    DD. Lvs. dark or bright green; bark smooth, at least in time, except in *A. tomentosa.*
      E. The lvs. sessile or nearly so, cordate to auriculate at base; plants without basal burl.
        F. Ovary glandular-hairy; fr. depressed-globose; lvs. 3–7 cm. long
                                     36. *A. Andersonii*
        FF. Ovary hairy, but not glandular; fr. globose; lvs. 2–3 cm. long
                                      37. *A. pajaroensis*
      EE. The lvs. petioled, with subcordate to truncate base; plants with basal burl.
        F. Bark smooth, exfoliating; fr. 6–8 mm. in diam.
          G. Branchlets mostly nonglandular, tomentose and ± bristly
                                    41. *A. crustacea*
          GG. Branchlets with long gland-tipped hairs as well as shorter tomentum
                                    43. *A. subcordata*
        FF. Bark shreddy, ± persistent; fr. 8–10 mm. in diam. . . . . 42. *A. tomentosa*
CC. Stomates present on both lf.-surfaces.
  D. Ovary glabrous; lvs. usually pale green.
    E. Branchlets glabrous to minutely puberulent; pedicels glabrous. Base of n. Sierra Nevada . . . . 16. *A. Mewukka*
    EE. Branchlets densely pubescent or tomentose, sometimes also bristly; pedicels glabrous or pubescent.

F. Lvs. usually essentially glabrous, occasionally somewhat pubescent.
    G. The lvs. mostly rounded or truncate at base, petioled. Interior San
          Luis Obispo Co. .... 35. *A. pilosula*
    GG. The lvs. auriculate and subsessile to subcordate and petiolate.
          Coastal and insular. .... 38. *A. pechoensis* var. *viridissima*
FF. Lvs. ± tomentulose.
    G. Branchlets gray-canescent; bracts canescent. Santa Cruz Mts.
                                           28. *A. silvicola*
    GG. Branchlets tomentulose, sometimes also bristly; bracts pubescent
          and viscid. San Luis Obispo and Santa Barbara cos.
                                      38. *A. pechoensis*
DD. Ovary pubescent, often glandular.
  E. Lvs. pale or dull green.
    F. Branchlets not glandular, usually with some sort of pubescence.
      G. Bark shreddy, fibrous; plant low, 6–15 dm. high; fr. 4 mm. in
          diam. Eldorado Co. ....3. *A. nissenana*
      GG. Bark smooth, exfoliating; fr. larger.
        H. Branchlets glabrous to pubescent or tomentose, but not bristly.
          I. Fr. with very little pulp, viscid, globose or slightly
              elongate, 12–15 mm. in diam., with a solid stone; corolla
              8–9 mm. long. S. Coast Ranges. .... 23. *A. glauca*
          II. Fr. with well developed pulp, often depressed-globose,
              6–10 mm. in diam., the nutlets generally separable;
              corolla 6–8 mm. long.
            J. Upper bracts not over 7 mm. long, the lower larger.
              K. Pedicels pubescent, not glandular. Del Norte Co.
                                  25. *A. cinerea*
              KK. Pedicels glandular. From cent. and s. Calif.
                       39. *A. glandulosa* vars.
            JJ. Upper bracts longer, more completely foliaceous.
              K. Lvs. rounded at base; fr. depressed-globose, 7–8
                mm. in diam. Marin Co. n. ... 27. *A. canescens*
              KK. Lvs. truncate to subcordate at base; fr. globose,
                9–10 mm. in diam. Monterey and San Luis Obispo
                cos. .... 29. *A. obispoensis*
        HH. Branchlets bristly with long hairs as well as variously pubes-
          cent.
          I. Lvs. sessile or subsessile, with auriculate base. Alameda
            and Contra Costa cos.
            J. The lvs. with fine white pubescence; bracts tomentose
                           31. *A. auriculata*
            JJ. The lvs. glabrous; bracts subglabrous.
                    36. *A. Andersonii* var. *pallida*
          II. Lvs. petioled, with obtuse or rounded-truncate base.
            J. Plant without basal burl; lvs. with fewer stomates
              above than beneath. .... 33. *A. columbiana*
            JJ. Plant with basal burl; lvs. with ca. as many stomates
              above as beneath. .... 39. *A. glandulosa* vars.
FF. Branchlets with some hairs gland-tipped.
    G. Plants without a basal burl.
      H. Young branchlets not bristly with long hairs; lvs. ± oblong.
        I. Young infl. recurved; lvs. tomentulose. Sonoma Co.
                   27. *A. canescens* var. *sonomensis*
        II. Young infl. erect; lvs. finely pubescent. San Diego Co.
                          32. *A. otayensis*
      HH. Young branchlets bristly with long spreading hairs; lvs. ovate
        to oblong.
        I. Lvs. cordate or auriculate at base. Santa Cruz Mts.
                             30. *A. glutinosa*
        II. Lvs. rounded to obtuse at base. N. Coast Ranges
                         33. *A. columbiana*
    GG. Plants with a basal burl.
      H. Ovary glandular-hairy; stomates ca. as many on upper sur-
        face as beneath. Del Norte Co. to San Diego Co.
                           39. *A. glandulosa*
      HH. Ovary hairy, not glandular; stomates fewer above than be-
        neath. Del Norte Co. .... 40. *A. intricata*
EE. Lvs. bright or dark green.
  F. Branchlets with long spreading hairs as well as shorter pubescence; lvs.
    sessile. San Mateo Co. .... 36. *A. Andersonii* var. *imbricata*
  FF. Branchlets short-hairy only; lvs. petioled.
    G. Pubescence on branchlets not glandular; fr. not viscid. Del Norte
      Co. to Mendocino Co.
      H. Branchlets glabrate to lightly puberulent; lvs. 2–3 cm. long
                              26. *A. parvifolia*

HH. Branchlets tomentose; lvs. 3–6 cm. long
33. A. columbiana var. Tracyi
GG. Pubescence on branchlets glandular; fr. viscid. Marin Co.
34. A. virgata

1. **A. Nummulària** Gray. [*Schizococcus N.* Eastw. *S. N.* var. *latifolius* Eastw.] Fт. BRAGG MANZANITA. Without basal burl; decumbent or prostrate shrub 1.5–5 dm. tall, or erect and up to 1.5 m. tall; branchlets red-brown, puberulent and sometimes pilose-pubescent, densely leafy; lvs. dark green, shining, without stomates on upper surface, elliptic or ovate, sometimes oblong, glabrous, veiny beneath, 8–15 mm. long, obtuse or rounded to short-acute at apex, rounded to subcordate at base, entire or ciliate, sometimes revolute; petioles ca. 2 mm. long; infl. small, terminal, simple or branched, glabrous or puberulent; pedicels ca. 2 mm. long; fls. 4-merous; calyx-lobes ovate, ciliate, ca. 1 mm. long; corolla white, 4–5 mm. long; fr. greenish when ripe, oblong, 4–5 mm. long, pubescent, with thin pericarp and separable into mostly 4, sometimes 2, 1-seeded nutlets.—Acid often moist places near the coast, below 250 ft.; Closed-cone Pine F.; Mendocino and Sonoma cos. March–May.

Var. **sensitìva** (Jeps.) McMinn. [*A. s.* Jeps. *Schizococcus s.* Eastw.] Erect, 1–2 m. high; lvs. suborbicular to broadly elliptic, 12–20 mm. long.—Dry slopes, 1000–2500 ft.; Chaparral; Mt. Tamalpais and Bolinas Ridge (Marin Co.) to Santa Cruz Mts. Jan.–April.

2. **A. myrtifòlia** Parry. [*A. Nummularia* var. *m.* Jeps. *Schizococcus m.* Eastw. *A. Helleri* Eastw.] IONE MANZANITA. Without basal burl; diffusely branched shrub 3–8 dm. high with lateral stems frequently decumbent and rooting, the dark reddish bark of older branches usually with a whitish bloom; branchlets with scaly bark and with rigid and usually some glandular hairs; lvs. light green, shining, minutely papillate, with abundant stomates on both surfaces, elliptic to narrow ovate, mostly acute and mucronate at apex, obtuse to acute at base, 5–15 mm. long, on subhispid petioles 2 mm. long; infl. short, simple or few-branched, puberulent; bracts small, deltoid; pedicels glabrous; calyx-lobes ca. 1 mm. long, ciliate; corolla white or pinkish, 4 mm. long; fr. greenish, globose, stiff-pubescent, 4–5 mm. in diam., with thin pericarp and separable into 3–5 one-seeded nutlets.—Rocky brushy ridges, 300–800 ft.; Chaparral, Foothill Wd.; from e. of Ione, Amador Co. to San Andreas, Calaveras Co.

3. **A. nissenàna** Merriam. [*Schizococcus n.* Eastw. *A. n.* var. *arcana* Jeps.] ELDORADO MANZANITA. Without basal burl; erect shrub 6–15 dm. high with red-brown or grayish rough fibrous bark; branchlets slender, white-villous to subglabrous; lvs. pale green, at first puberulent, later glabrate, with many stomates on both surfaces, elliptic to oblong or broader, 1–2.5 cm. long, acute to obtuse and mucronulate at apex, obtuse to subtruncate at base, on pubescent petioles 2 mm. long; infl. small, dense, short-villous to glabrate; lower bracts foliaceous, the upper reduced, pubescent; pedicels glabrous or nearly so; calyx 5-lobed, reddish, slightly ciliate, ca. 1 mm. long; corolla pale pink, 4–5 mm. long; fr. globular or somewhat angled, thinly hairy, 4 mm. in diam., usually with 5 nutlets and thin pericarp.—Dry ridges, 1500–3500 ft.; Foothill Wd., Chaparral; near Placerville, Eldorado Co. Feb.–March.

4. **A. Ùva-úrsi** (L.) Spreng. var. **coáctilis** Fern. & Macbr. BEARBERRY. SANDBERRY. Without basal burl; prostrate shrub with trailing stems from which arise erect branches 5–15 cm. long; bark dark brown or somewhat reddish; branchlets minutely white-tomentulose; lvs. oval or obovate, coriaceous, shining, somewhat puberulent, without stomates on upper surface, rounded at apex, cuneate below, 1–2.5 cm. long, on petioles 1–2 mm. long; infl. dense, short, puberulent, the bracts green; pedicels puberulent; calyx pinkish, glabrous, ca. 1 mm. long; corolla pinkish to white, 4–5 mm. long; ovary glabrous; fr. bright red, smooth, 6–12 mm. in diam., with separable nutlets.—Sandy and grassy places, below 200 ft.; N. Coastal Scrub, Coastal Strand; San Bruno Mt. (San Mateo Co.), Point Reyes (Marin Co.) along immediate coast to Del Norte Co.; to Alaska, Nfld. March–June. Occasional plants along the n. coast with somewhat hispid stems suggest possible hybridity with *A. columbiana*.

5. **A. Edmúndsii** J. T. Howell. LITTLE SUR MANZANITA. Without basal burl; subprostrate, 1–5 dm. high, 2.5–3.5 m. across, rooting along the stems; branchlets pilose with spreading eglandular hairs; lvs. broadly ovate, elliptic or roundish, 2–3 cm. long, mucronulate and obtusish at apex, truncate to cordate at base, green, subglabrous, pilose-

ciliate, without stomates on upper surface, on petioles 2–4 mm. long; infl. small, compact, simple or nearly so, the rachises pilose, scarcely glandular; lower bracts foliaceous, the upper smaller and scalelike, pilose and somewhat glandular; pedicels 3–4 mm. long, glandular-puberulent; corolla pink, 5 mm. long; ovary glabrous; fr. brown, depressed-globose, ca. 8 mm. in diam., the nutlets separating.—Chaparral; ocean bluffs, mouth of Little Sur R., Monterey Co. Nov.–Dec.

6. **A. pùmila** Nutt. [*Uva-ursi p.* Abrams.] DUNE MANZANITA. Without basal burl; usually decumbent or prostrate, forming mats or mounds 3–8 dm. high, rooting readily; bark rather shreddy; branchlets finely pubescent; lvs. many, dull green and slightly pubescent above, more pubescent beneath, without stomates on upper surface, narrowly obovate to subspatulate, 1.5–2.5 cm. long, obtuse to short-acute or apiculate, on petioles 2–3 mm. long; infl. short, dense, pubescent; bracts lanceolate, acuminate, densely pubescent, 2–3 mm. long; pedicels sparsely pubescent, 3–5 mm. long; calyx white, ca. 1 mm. long, the lobes ciliate; corolla white to pink, ca. 4 mm. long; ovary glabrous; fr. globose, slightly depressed, brown, ca. 5–6 mm. in diam., with 5 separable nutlets.— Sand hills and woods; N. Coastal Scrub, Closed-cone Pine F.; about Monterey Bay. Feb.–April.

7. **A. Hoòkeri** G. Don. [*Andromeda venulosa* DC. *Arctostaphylos acuta* Nutt.] MONTEREY MANZANITA. Without basal burl; low, spreading, forming mounds 4–12 dm. high; bark smooth, red-brown; branchlets tomentose; lvs. ovate to subelliptic, bright green, shining, glabrous, with stomates on both surfaces, short-acute, mucronate, 1.5–2.5 cm. long, on petioles 3–5 mm. long; infl. small, compact, few-fld., tomentose; bracts acuminate, ca. 1–1.5 mm. long; pedicels subglabrous, 3–4 mm. long; calyx white, somewhat ciliate, the lobes ca. 1 mm. long; corolla white to pinkish, 3–4 mm. long; ovary glabrous; fr. bright red, glossy, 4–5 mm. in diam., with separable or irregular nutlets with dorsal ridges.—Sand dunes and woods; Closed-cone Pine F.; near the coast, Monterey Co. Feb.–April.

Ssp. **franciscàna** (Eastw.) Munz. [*A. f.* Eastw. *Uva-ursi f.* Heller.] Branches prostrate, the branchlets minutely puberulent; corolla 5–7 mm. long; fr. 7–8 mm. in diam.—Serpentine outcrops; N. Coastal Scrub; San Francisco Peninsula. Jan.–April.

8. **A. densiflòra** M. S. Baker. SONOMA MANZANITA. Low spreading procumbent shrubs, rooting freely, with blackish branches; branchlets finely pubescent; lvs. elliptic to suboblong, bright green, shining, finely pubescent on margins and veins, obtuse or acute and mucronate, with stomates on both surfaces, 1.5–3 cm. long, on petioles 2–3 mm. long; infl. dense, short, many-fld., puberulent; bracts abruptly acuminate, ca. 2 mm. long; pedicels mostly glabrous, ca. 5 mm. long; calyx whitish, ciliate, ca. 1 mm. long; corolla white or pink, 4–5 mm. long; ovary glabrous; fr. flattened-globose, 5–6 mm. in diam., with separable nutlets.—Banks, along roadside, Vine Hill Schoolhouse, ca. 10 miles w. of Santa Rosa, Sonoma Co. March–April. Apparently forms hybrids with *A. Manzanita* and *A. Stanfordiana*.

9. **A. Stanfordiàna** Parry. [*Uva-ursi* S. Heller.] STANFORD MANZANITA. Without basal burl; erect, much-branched, 1–2 m. high, with rather long slender branches and smoothish red-brown bark; branchlets glabrous or minutely puberulent when young; lvs. bright green, shining, glabrous, with equal numbers of stomates on both surfaces, lance-ovate to oblanceolate, acute at both ends, 3–6 cm. long, mostly erect, on petioles 8–12 mm. long; infl. lax, drooping, with rachises slender, reddish, glabrous or puberulent; bracts subulate, ca. 1.5 mm. long, keeled; pedicels glabrous, slender, 5–7 mm. long; calyx pinkish, ca. 1 mm. long, somewhat ciliolate toward tips; corolla pinkish, 5–6 mm. long; ovary glabrous; fr. bright red, depressed-globose, commonly asymmetrical, ca. 4 mm. high, 5–7 mm. thick, with ± coalescent lightly ridged nutlets.—Common on open ridges and slopes, 1000–4000 ft.; Chaparral; Napa, Lake, Sonoma, and Mendocino cos. Feb.–April. Intermediates with *A. densiflora* and *A. Manzanita* occur.

Ssp. **hispìdula** (Howell) Adams. [*A. h.* Howell.] Branchlets glandular-hispidulous; lvs. dull green; corolla ca. 5 mm. long.—Serpentine areas in open shrubby places; upper Mixed Evergreen F.; near Gasquet, Del Norte Co.; Ore. March–April.

Ssp. **Bàkeri** (Eastw.) Adams. [*A. B.* Eastw.] Branchlets dark, with close viscid puberulence and conspicuous longer hairs; corolla 6–7 mm. long.—Dry serpentine ridges, near Camp Meeker and Occidental, Sonoma Co. Feb.–April.

10. **A. élegans** Jeps. [*A. Manzanita* var. *e.* L. Benson. *Uva-ursi e.* Heller.] KONOCTI

MANZANITA. Without basal burl; erect, with straightish smooth polished branches; branchlets glabrous or nearly so; lvs. bright green, dullish, glabrous, thick, conspicuously veiny, with equal numbers of stomates on both surfaces, broadly ovate to elliptic, acute or obtuse, 3–7 cm. long, on petioles 4–8 mm. long; infl. ample, with thickish-pubescent rachises; bracts deltoid-ovate, spreading, 3–4 mm. long; pedicels glabrous, 5–6 mm. long; calyx whitish, ca. 1.5 mm. long, ciliate; corolla white, 7–8 mm. long; ovary glandular; fr. dark red, stipitate-glandular, 8–10 mm. in diam., with 5–7 nutlets, 2 or 3 consolidated.—Brushy or wooded slopes, 2000–4200 ft.; Chaparral, Foothill Wd.; mts. of Lake Co. March–May.

11. **A. Manzanìta** Parry. [*Uva-ursi M.* Heller. *A. M.* var. *apiculata* Jeps.] PARRY MANZANITA. Without basal burl; erect, 2–4(–7) m. high, with long crooked branches and smooth dark red-brown bark; branchlets mostly densely canescent-pubescent, sometimes puberulent; lvs. bright green, shining or dullish, with equal numbers of stomates on both surfaces, glabrous or lightly puberulent (especially when young), oblong to broad-elliptic, obtusish to acutish, 2.5–4.5 cm. long, on petioles 5–9 mm. long; infl. ample, drooping, with rachises pubescent to almost glabrous and bracts short-deltoid, acute, ca. 2 mm. long; pedicels glabrous, 5–9 mm. long; calyx whitish, ciliate, almost 2 mm. long; corolla white or pink, 7–8 mm. long; ovary glabrous or somewhat pubescent, nonglandular; fr. at first white, later deep red, smooth, slightly depressed-globose, 8–12 mm. in diam., with nutlets irregularly coalescent or separate, ridged on back.—Dry slopes and canyons, 300–4000 ft.; Chaparral, Foothill Wd., N. Oak Wd., Yellow Pine F.; N. Coast Ranges (mostly in interior) from Contra Costa Co. to Humboldt, Trinity and Shasta cos. and foothills of Sierra Nevada from Mariposa Co. n. Feb.–April. Exceedingly variable as to color of lvs., pubescence of branchlets, pedicels, ovaries, etc., with some reason to believe that there is hybridization with other spp.

Ssp. **laevigàta** (Eastw.) Munz. [*A. l.* Eastw.] Low, intricately branched; branchlets puberulent; lvs. lanceolate to oblong, 2–3 cm. long, shining; infl. drooping, with reddish-pubescent rachises; corolla white, 4–7 mm. long; fr. 5–8 mm. in diam., dark brown, glossy.—Rocky places on Mt. Diablo (Contra Costa Co.) and Mt. St. Helena (Lake Co.). Possibly of hybrid origin with *A. Stanfordiana* as one parent. Feb.–March.

12. **A. Parryàna** Lemmon. [*Uva-ursi P.* Abrams.] PARRY MANZANITA. Without enlarged root-crown, diffuse widely spreading shrub 1–2 m. high with lateral branches commonly rooting and decumbent; bark smooth, dark red or red-brown; branchlets canescent to glabrate, not glandular; lvs. ovate to obovatish, bright green, glabrous, somewhat shining, with equal nos. of stomates on both surfaces, acute to rounded at apex, 2–3 cm. long, on petioles 5–10 mm. long; infl. few-fld., simple or few-branched, with puberulent rachises; bracts short deltoid-acuminate, ca. 2 mm. long; pedicels glabrous, 5–9 mm. long; calyx spreading, ca. 2 mm. long; corolla white, 6–7 mm. long; ovary glabrous; fr. dark red, round-ovoid, 8–12 mm. long, with a solid obscurely ridged stone.—Dry stony slopes, 4000–7500 ft.; Chaparral, Yellow Pine F.; Tehachapi Mts., Mt. Pinos region to San Gabriel Mts. March–May.

Var. **pinetòrum** (Roll.) Wies. & Schreib. [*A. p.* Roll.] Branchlets, petioles and infl. glandular-pubescent.—Dry slopes, 5000–9000 ft.; Yellow Pine F., Red Fir F.; San Gabriel Mts. to Santa Rosa Mts. (Riverside Co.); Utah, Colo. May–June.

13. **A. nevadénsis** Gray. [*Uva-ursi n.* Abrams.] PINEMAT MANZANITA. Without basal burl; low sprawling or prostrate shrub with intricately branched stems rooting freely and forming erect branchlets 3–6 dm. high; bark of stems smooth, brownish to deep red; branchlets gray-puberulent, somewhat viscid in age; lvs. bright green, glabrous to somewhat puberulent, with numerous stomates on both surfaces, lanceolate to elliptic or oblanceolate or obovate, obtuse and somewhat mucronulate, 2–2.5(–3.5) cm. long, on puberulent petioles 3–7 mm. long; infl. simple or few-branched, compact, many-fld., erect, puberulent; bracts ca. 3 mm. long, abruptly acuminate, ciliate below; pedicels 2–5 mm. long, pubescent to subglabrous; calyx whitish, ca. 1.5 mm. long; corolla mostly white, 6–7 mm. long; ovary glabrous; fr. depressed-globose, 5–6 mm. in diam., with acid pulp and separable rugose nutlets.—Moist places to dry rocky slopes in woods, mostly 5000–10,000 (sometimes as low as 2000) ft.; Douglas-Fir F., Yellow Pine F. to Subalpine F.; Sierra Nevada from Tulare Co. n., Coast Ranges from Lake Co. n.; to Wash., Nev. May–July.

14. **A. púngens** HBK. [*Uva-ursi p.* Abrams.] MEXICAN MANZANITA. Without basal

burl; erect, branched from base, 2–3 m. high, with smooth red-brown bark; branchlets canescent with a fine dense pubescence; lvs. bright green, glabrous or minutely puberulent, with equal nos. of stomates on both surfaces, elliptic to oblong or obovate or almost lanceolate or oblanceolate, 1.5–3 cm. long, acute to obtuse, on petioles 3–4 mm. long; infl. small, dense, the rachises thickened upward, densely canescent; bracts short-deltoid, ca. 3 mm. long; pedicels glabrous, 5–7 mm. long; calyx-lobes ca. 1.5 mm. long, round-ovate, entire; corolla white, ca. 6 mm. long; ovary glabrous; fr. brownish-red, depressed-globose, 5–8 mm. broad, the nutlets separable or irregularly united, keeled, corrugately roughened; $2n = 26$ (Callan, 1941).—Dry slopes, 3000–7000 ft. in s. Calif. and lower northward; Chaparral, Yellow Pine F., Foothill Wd.; common, San Diego Co. to Los Angeles and San Bernardino cos., occasional in inner S. Coast Ranges to San Benito Co.; to Nev., Tex., Mex. Feb.–March.

Var. **montàna** (Eastw.) Munz. [*A. m.* Eastw.] Low, spreading, bushy to matlike, often rooting where branches touch ground; rachises of infl. not or scarcely enlarged upward.— Serpentine flats and slopes; Chaparral; Marin Co. Feb.–April. Possibly hybrid between *A. canescens* and *A. Hookeri.*

15. **A. rùdis** Jeps. & Wies. SHAGBARK MANZANITA. Shrub 6–15 dm. high with basal burls and reddish-brown rough shreddy bark; branchlets finely gray-puberulent; lvs. elliptic to oblong, occasionally oval, apiculate to obtuse, glabrous or lightly pubescent especially when young, bright green, with equal nos. of stomates on both surfaces, 1.5–3 cm. long, on pubescent petioles 3–5 mm. long; infl. short, dense, finely pubescent; bracts ovate to subulate, 2–3 mm. long; pedicels glabrous, 6–9 mm. long; ovary glabrous; fr. globose, 9–11 mm. thick, brownish-red, the nutlets separable.—Sandy places; Coastal Strand, Coastal Sage Scrub, Chaparral; from Oceano, San Luis Obispo Co. to near Lompoc, Santa Barbara Co. Nov.–Feb.

16. **A. Mewúkka** Merriam. [*A. pastillosa* Jeps. *A. laxiflora* Heller?] INDIAN MANZANITA. Erect, 1–2.5 m. high, from a rounded root-crown or burl and with smooth deep red to purplish bark; branchlets glabrous to minutely puberulent; lvs. pale gray-green, oblong-lanceolate or -elliptic to obovate, with equal nos. of stomates on both surfaces, 2.5–4(–5) cm. long, glabrous, on stout petioles 7–9 mm. long; infl. mostly branched, open, glaucous, glabrous to puberulent, the peduncle and thickened rachises usually dark red; bracts somewhat foliaceous, the lower lanceolate, 8–10 mm. long, the upper deltoid-ovate, acuminate, 3–4 mm. long; pedicels glabrous, 4–5 mm. long; calyx-lobes oval, ciliolate, ca. 1.5 mm. long; corolla white or pinkish, 6–7 mm. long; ovary mostly glabrous; fr. depressed-globose, dark red to red-brown, 12–14 mm. in diam., the nutlets irregularly separable, rarely completely fused.—Dry slopes, mostly 2500–6000 ft.; Chaparral, Yellow Pine F.; base of Sierra Nevada from Butte and Plumas cos. to Tulare Co. March–April.

17. **A. pátula** Greene. [*Uva-ursi p.* Abrams. *A. p.* var. *incarnata* Jeps. *A. pungens* var. *platyphylla* Gray. *A. pl.* Kuntze.] GREENLEAF MANZANITA. Spreading, much-branched, 1–2 m. high, with several stout stems from an enlarged burl or root-crown, and with smooth bright red-brown bark; branchlets resinous-glandular to glandular-puberulent, the glands often yellow; lvs. broadly ovate to almost round, with equal nos. of stomates on both surfaces, bright green, glabrous to minutely glandular near base, obtuse or rounded at apex, 2.5–4 cm. long, on petioles 6–10 mm. long; infl. from compact and corymbose to loosely paniculate, with rachises and peduncle usually glandular-puberulent, the rachises thickened upward; bracts ovate-acuminate to lanceolate-acute, 3–7 mm. long; pedicels glabrous, 5–7 mm. long; calyx-lobes round-ovate, entire, glabrous, ca. 2 mm. long; corolla pink, 5–8 mm. long; ovary glabrous; fr. flattened-globose, 7–10 mm. broad, the nutlets irregularly coalescent, rounded and smoothish on back.—Open forests, 2000–9000 ft.; Yellow Pine F., Red Fir F.; Sierra Nevada from Kern Co. n. and Coast Ranges from Lake Co. n.; to Ore., Nev., Utah. April–June.

18. **A. acutifòlia** Eastw. Stems branched, 5–6 dm. high, smooth, polished; branchlets glandular-villous; lvs. bright green, narrowly to broadly oblong, 2.5–3.5 cm. long, with equal nos. of stomates on both surfaces, sharply acute, glabrous, on glandular-pubescent petioles ca. 1 cm. long; infl. lax, glandular-villous, the rachises thickened upward; bracts triangular-acuminate to lanceolate, 3–4 mm. long, the lowest green, lance-linear, 10–12 mm. long; pedicels sparsely glandular-villous, 5–7 mm. long; calyx-lobes oblong-ovate, ciliate, ca. 2 mm. long, spreading in fr.; corolla white, 5 mm.

long; fr. with stalked or sessile glands, depressed-globose, 7–8 mm. in diam., with nutlets fused, few-ridged dorsally.—Dry places; Yellow Pine F.; Log Springs Ridge, sw. Tehama Co. May.

19. **A. insulàris** Greene. [*Uva-ursi i.* Heller.] ISLAND MANZANITA. Erect, much branched, 1–2.5(–5) m. high, without enlarged root-crown, with bark smooth, dark red-brown and gray-glaucous; branchlets smooth; lvs. bright green, with no stomates on upper surface, shining, glabrous, ovate to elliptical, 2.5–4 cm. long, obtuse to acute, with petioles 3–6 mm. long; infl. an ample spreading panicle, with rachises slender, puberulent; bracts deltoid-acuminate, 1.5–2.5 mm. long, or lower larger and more foliaceous; pedicels glandular-pubescent, 5–12 mm. long; calyx-lobes 1.5 mm. long, oval, glandular-fimbriolate; corolla white, ca. 6 mm. long; fr. depressed-globular, light brown, puberulent to glabrate, 6–8 mm. broad, the nutlets irregularly coalescent, ridged on back.—Rocky hillsides; Chaparral; Santa Cruz and Santa Rosa ids. Jan.–March. More common is

Var. **pubéscens** Eastw. Branchlets and infl. glandular-pubescent to hispid.—Santa Cruz and Santa Rosa ids.

20. **A. víscida** Parry. [*Uva-ursi v.* Heller.] WHITELEAF MANZANITA. Without basal burl; erect, 1–4 m. high, with smooth dark red-brown bark and pale glaucous-green glabrous or sometimes sparsely glandular-pubescent branchlets; lvs. white-glaucous and glabrous, round, ovate or elliptic, with equal nos. of stomates on both surfaces, 2.5–4 cm. long, rounded to acutish and mucronate at apex, on petioles 8–12 mm. long; infl. open, the branches glabrous or glandular-pubescent; bracts deltoid-acuminate, ca. 2 mm. long; pedicels slender, glandular-pubescent, 10–12 mm. long; calyx-lobes ovate, acute, ciliate, pinkish, 1–1.5 mm. long; corolla pink to whitish, 6–7 mm. long; fr. depressed-globose, light brown or red, glabrous or glandular, 6–8 mm. broad, the nutlets usually separable, angled and roughened on back.—Dry slopes, 500–5000 ft.; Chaparral, Foothill Wd., Yellow Pine F.; foothills of Sierra Nevada from Kern Co. n., N. Coast Ranges from Lake and Napa cos. n.; Ore. Feb.–April.

21. **A. maripòsa** Dudl. in Eastw. [*Uva-ursi m.* Abrams.] MARIPOSA MANZANITA. Much like *A. viscida,* but the branchlets and infl. glandular-pubescent to -hairy; lvs. glabrous to glandular-hairy, with equal nos. of stomates on both surfaces, frequently serrulate; calyx-lobes glandular-villous; fr. glandular-viscid.—Dry slopes, mostly 2000–6000 ft.; Yellow Pine F.; w. base of Sierra Nevada from Amador Co. to Kern Co. Feb.–April. A form intergrading toward *A. viscida,* with fr. glabrous or nearly so, from Mariposa and Tuolumne cos. has been named var. **bivìsa** Jeps. [*A. Jepsonii* Eastw.]

22. **A. Prínglei** Parry var. **drupàcea** Parry. [*A. d.* Macbr. *Uva-ursi d.* Abrams.] PINK-BRACTED MANZANITA. Erect shrubs, sweet-scented, 2–4 m. high, without heavy root-crown or burl, with smooth dull red-brown bark; branchlets densely glandular-villous; lvs. oblong-ovate to elliptic, rarely obovate, gray-green, minutely glandular-pubescent or glabrate, with equal nos. of stomates on both surfaces, rounded to acute at apex, 2–4 cm. long, on petioles 3–7 mm. long; infl. ample, almost sessile, usually branched, glandular-pubescent to -hairy; bracts lanceolate, membranaceous, deciduous, pinkish, 5–6 mm. long; pedicels slender, pink, 10–15 mm. long, glandular-villous; calyx-lobes lanceolate, ca. 3 mm. long, glandular-villous, strongly ciliate; corolla rose, 7–8 mm. long; fr. round-ovoid, glandular-villous, red, 6–10 mm. in diam., with solid ribbed stone.—Dry slopes, 4000–7500 ft.; Chaparral, Yellow Pine F.; San Bernardino Mts. to L. Calif. Feb.–April.

23. **A. glaùca** Lindl. [*Uva-ursi g.* Abrams. *A. g.* var. *eremicola* Jeps.] BIGBERRY MANZANITA. Without basal burl; large, erect, shrubby or arborescent, 2–4(–6) m. high, with smooth red-brown bark; branchlets and lvs. glabrous and glaucous, the latter oblong, elliptic or ovate, with equal nos. of stomates on both surfaces, 2–4.5 cm. long, obtuse or acute at apex, short-mucronate, mostly truncate or subcordate, sometimes obtuse at base, on petioles 7–10 mm. long; panicles dense, nodding, broad and short, the rachis glabrous to glandular; bracts 2–3 mm. long, spreading, broadly ovate; pedicels glandular-pubescent; corolla white or with some pink, 8–9 mm. long; ovary glandular; fr. globose, viscid, brownish, 12–15 mm. in diam., with thin leathery pericarp, the nutlets united into a solid smooth apiculate stone.—Common on dry slopes, below 4500 ft.; Chaparral; particularly in the cismontane and coast ranges of s. Calif., less so in inner Coast Ranges to Mt. Diablo, Contra Costa Co.; Joshua Tree Wd., Little San Bernardino Mts.; L. Calif. Dec.–March.

Var. **pubérula** J. T. Howell. Branchlets glandular-puberulent to pubescent.—Santa Clara Co. to Santa Barbara and Ventura cos.

24. **A. morroénsis** Wies. & Schreib. Morro Manzanita. Erect, 1.5–2 m. tall, without burl formation at base, the bark rough and shreddy, red-brown or gray-brown; branchlets strigose, sometimes also with some bristly hairs; lvs. gray-green to yellow-green, without stomates on upper surface, oblong to oblong-lanceolate, 1.5–3 cm. long, truncate to subcordate at base, apiculate to obtuse at apex, glabrous to somewhat pubescent above, tomentose beneath, on petioles 2–4 mm. long; panicles short, mostly pendulous, clustered in upper lvs., finely pubescent; bracts leafy, lanceolate, 7–9 mm. long, finely pubescent, with or without bristles; pedicels glabrous, 3–6 mm. long; calyx-lobes oval, ciliate; corolla white to pinkish, 5–7 mm. long; ovary densely tomentose; fr. slightly depressed-globose, orange-brown, ca. 10 mm. in diam., lightly pubescent, the nutlets separating.—Sandy hills, 100–400 ft.; Chaparral; s. of Morro Bay, San Luis Obispo Co. Jan.–March.

25. **A. cinèrea** Howell. Del Norte Manzanita. With basal burl; semi-erect with decumbent branches, 1–1.5 m. high, the bark light red-brown; branchlets cinereous-pubescent; lvs. dull ashy-green, with equal nos. of stomates on both surfaces, oblong-elliptic to obovate, obtuse to acute at apex, 2.5–3.5 cm. long, on pubescent petioles 2–7 mm. long; panicle spreading, with pubescent rachises and bracts, the latter lanceolate to ovate-acuminate, cinereous, 5–7 mm. long, or the lower leafy and longer; pedicels pubescent, 5–7 mm. long; calyx-lobes oval, somewhat ciliate at the rounded apices; corolla white to pink, 6–7 mm. long; ovary pubescent; fr. flattened-globose, pubescent at least when young, 6–8 mm. in diam., deep red-brown, the nutlets variously separable. —Dry open places, below 2500 ft.; Mixed Evergreen F.; outer Coast Ranges, Del Norte Co.; sw. Ore. March–May.

26. **A. parvifòlia** Howell. Without basal burl, erect, bushy, ca. 1 m. high, the bark dark purple-red; branchlets glabrate to lightly puberulent; lvs. bright green, mostly glabrous except near base, with fewer stomates above than beneath, oblong to elliptic, 2–3 cm. long, on glabrate petioles 5–6 mm. long; infl. a simple compact raceme or few-branched panicle, nodding, lightly puberulent; bracts linear-lanceolate, the lower as long as the sparsely puberulent pedicels; corolla white or pinkish, 5–6 mm. long; ovary ± white-pubescent; fr. glabrate, depressed-globose.—Dry gravelly hillsides, n. Del Norte Co.; sw. Ore. March–April. It has been suggested that plants referred here are of hybrid origin between *A. parvifolia* and *A. nevadensis* and *A. patula*, with which it grows.

27. **A. canéscens** Eastw. Hoary Manzanita. Without basal burl, erect and 1–2 m. high or sprawling and knotty and lower, the bark smooth, dark red-brown; branchlets, lvs. and peduncles densely soft white-pubescent; lvs. pale green and canescent, sometimes glabrate in age, with equal nos. of stomates above and beneath, oblong-ovate to ovate, 3–4 cm. long, acute to obtusish, on petioles 3–8 mm. long; panicles short, densely fld.; bracts foliaceous, ± lanceolate, 9–14 mm. long; pedicels pubescent, sometimes glandular, 5–7(–10) mm. long; calyx-lobes round-oblong, ca. 2 mm. long, apically ciliate; corolla white to pinkish, ca. 8 mm. long; ovary densely white-hairy; fr. depressed-globose, 7–8 mm. in diam., usually pubescent, sometimes glabrate and glaucous, the nutlets separable.—Dry gravelly to rocky slopes, mostly 1000–5000 ft.; Yellow Pine F., Chaparral, Douglas-Fir F.; Santa Cruz and Santa Clara cos., Marin Co. to Del Norte Co.; Ore. Jan.–May.

Var. **sonoménsis** (Eastw.) Adams. [*A. s.* Eastw.] Young growth somewhat glandular as well as densely pubescent; lvs. 2–3 cm. long; fr. viscid-pubescent.—At ca. 500 ft., Rincon Ridge near Santa Rosa, Sonoma Co. Jan.–March.

Var. **candidíssima** (Eastw.) Munz. [*A. c.* Eastw.] Young stems and the lvs. densely white-velvety; lvs. roundish to broadly oblong, 3–4 cm. long, 2–4 cm. wide, truncate to subcordate at base; ovary densely velvety.—At 3500–5000 ft.; Yellow Pine F.; Trinity and Tehama to Lake and Napa cos. May.

28. **A. silvícola** Jeps. & Wies. Silverleaf Manzanita. Without basal burl; erect, silver-gray, 1.5–2.5 m. high, with smooth dark red bark; branchlets densely gray-canescent; lvs. elliptic to oblong-elliptic, 1.5–3.5 cm. long, acutish to obtuse, apiculate at apex, thinly tomentose above, more densely so beneath, occasionally glabrous, with fewer stomates above than beneath, on petioles 3–8 mm. long; infl. short, sessile or short-peduncled, drooping, densely canescent; bracts foliaceous, densely canescent, lanceolate,

5–12 mm. long; pedicels glabrous or nearly so; ovary glabrous or rarely hairy; calyx-lobes ca. 1 mm. long, ciliate; corolla white, 6–7 mm. long; fr. subglobose, light brown, 6–8 mm. in diam., light brown, the nutlets separating.—Old marine deposits, sand hills, below 1800 ft.; Yellow Pine F., Chaparral; Mount Hermon region, Santa Cruz Mts. Feb.–March.

29. **A. obispoénsis** Eastw. Serpentine Manzanita. Without basal burl; erect, 1–3(–6) m. high, with dark purplish-red bark; branchlets gray-tomentulose; lvs. ovate to oblong-ovate, 2–4.5 cm. long, with equal nos. of stomates above and beneath, mostly grayish-tomentulose on both surfaces, truncate to cordate or auriculate at base, sessile or on petioles up to 7 mm. long; infl. usually erect, subsessile, paniculate, on short peduncles; bracts foliaceous, lanceolate, 7–14 mm. long, canescent; pedicels glabrous, 5–6 mm. long; calyx-lobes ca. 2 mm. long, ciliate; corolla white or pinkish, 6–8 mm. long; ovary sparsely hairy; fr. globose, ca. 9–10 mm. in diam., pale orange-brown to red-brown, glabrous or nearly so, the nutlets separable.—Serpentine areas, 500–3000 ft.; Foothill Wd., Closed-cone Pine F.; San Luis Obispo Co. and s. Monterey Co. Feb.–March.

30. **A. glutinòsa** Schreib. Round gray shrub 6–12 dm. high, without basal burl; branchlets densely white-canescent, minutely glandular and usually with long white hairs; lvs. ovate to oblong, 2–4.5 cm. long, canescent, with equal nos. of stomates above and beneath, entire or toothed on basal half, acute to apiculate at apex, cordate or auriculate at base, sessile or on petioles to 4 mm. long; infl. simple, short, canescent; bracts foliaceous, lanceolate, 7–12 mm. long, gray-canescent with white-bristly margins; pedicels glandular-hairy, 5 mm. long; fls. pinkish-white, 6–7 mm. long; ovary densely glandular-hairy; fr. light orange-brown, globose, 9–14 mm. thick, densely stipitate-glandular.—Hillsides of Monterey shale, 1900–2100 ft.; Chaparral; Mill Creek, w. of Bonnie Doon Ridge, Santa Cruz Mts.

31. **A. auriculàta** Eastw. [*A. Andersonii* var. *a.* Jeps. *Uva-ursi a.* Abrams.] Low to erect and arborescent, 2.5–4.5 m. high, without basal burl and with smooth dark red bark; branchlets short-pubescent and bristly-hairy; lvs. ovate to oblong-ovate, with equal nos. of stomates above and beneath, gray-green with a fine canescent pubescence, acute or rounded at apex, entire or serrate toward base, 2–5 cm. long, sessile or sub-sessile, auriculate at base; infl. erect or spreading, small, closely tomentose on branches, bracts, and pedicels; bracts foliaceous, linear or lance-linear, 6–10 mm. long; pedicels 5 mm. long; corolla white or pinkish, 6 mm. long; ovary densely white-hairy; fr. depressed-globose, red-brown or orange-brown, pubescent, 5–7 mm. in diam., the nutlets separable.—Dry slopes of sandstone, 500–2000 ft.; Chaparral; Mt. Diablo region, Contra Costa and Alameda cos. Feb.–March.

32. **A. otayénsis** Wies. & Schreib. Otay Manzanita. Without basal burl; erect, 1–2.5 m. high, with smooth dark red bark; branchlets glandular-pubescent; lvs. elliptic to oblong, 1.5–3.5 cm. long, gray-green to yellow-green, finely pubescent and glandular, with equal nos. of stomates above and beneath, acute or apiculate at apex, rounded or rarely truncate at base, on petioles 4–7 mm. long; infl. of open panicles or simple racemes, glandular-pubescent; bracts leafy, lanceolate, glandular-pubescent, 5–15 mm. long; pedicels glandular-pubescent; corollas white, 5–7 mm. long; fr. globose, 5–8 mm. in diam., pale brown, glabrate or microscopically glandular, rarely sparsely hairy, the nutlets separable or coalesced into a solid nut.—Dry slopes, 1800–5500 ft.; Chaparral; mts. of San Diego Co. Jan.–March.

33. **A. columbiàna** Piper. [*A. setosissima* Eastw.] Hairy Manzanita. Erect, much-branched, often arborescent, 1–3 m. high, without basal burl, the bark smooth, dark red-brown; branchlets densely cinereous-pubescent and with long white bristly hairs some of which may be gland-tipped; lvs. ovate to oblong-ovate or broadly elliptical, with fewer stomates above than beneath, pale gray-green and tomentose on both sides to glabrate above in age, 2.5–6 cm. long, obtuse or rounded-truncate at base, on hairy sometimes glandular petioles 3–6 mm. long; panicles short, dense, tomentulose and somewhat setose; bracts foliaceous, lanceolate, usually hispid-ciliate, 5–15 mm. long; pedicels pubescent to glandular-hairy, 3–4 mm. long; corolla white or with some pink, 6–7 mm. long; ovary tomentose, often glandular; fr. strongly depressed-globose, 7–10 mm. across, bright red, usually lightly pubescent and often viscid, the nutlets irregularly separable.—Dry rocky or clay slopes, below 2500 ft.; Mixed Evergreen F.,

Douglas-Fir F., N. Coastal Scrub; near the coast from Sonoma Co. n.; to B.C. March–May.

Var. **Tràcyi** (Eastw.) Adams. [*A. T.* Eastw.] Branchlets tomentose, neither glandular nor hispid; lvs. brighter green, puberulent when young; bracts glabrous or puberulent, often hispid-ciliate; ovary densely white-hairy, not glandular.—N. Coastal Scrub, edge of Redwood F.; along immediate coast of Humboldt and Mendocino cos. Feb.–March.

34. **A. virgàta** Eastw. in Sarg. [*A. columbiana* var. *v.* McMinn. *A. glandulosa* var. *v.* Jeps.?] BOLINAS MANZANITA. Without basal burl; erect, 2–4.5 m. high, with smooth red-brown bark; branchlets with dark glandular close pubescence; lvs. bright green, thin, with fewer stomates above than beneath, oblong-lanceolate to ovate-lanceolate, acute and mucronate at apex, obtuse at base, 3–5 cm. long, glabrate save for the glandular-hairy midribs and margins; petioles glandular-pubescent, 3–5 mm. long; infl. dense, glandular-hairy; bracts foliaceous, lanceolate, 5–20 mm. long, glandular-hairy on margins particularly; pedicels glandular, 4–7 mm. long; corolla white or pink, ca. 6 mm. long; ovary glandular-hairy; fr. viscid, slightly depressed-globose, the nutlets separable.—Brushy slopes at edge of Closed-cone Pine F. and Redwood F.; Marin Co. Jan.–March.

35. **A. pilosùla** Jeps. & Wies. LA PANZA MANZANITA. Without basal burl; erect, 1–3 m. high, with smooth dark red-brown bark; branchlets pubescent and usually with long or short bristly hairs; lvs. oblong, elliptic to ovate, 1.5–4.5 cm. long, truncate or rounded or rarely subcordate at base, acute to apiculate at apex, with equal nos. of stomates above and beneath, gray-green to yellow-green, glabrous or sparsely pubescent, densely so when young, on bristly petioles 5–7 mm. long; infl. short, dense, mostly drooping, pubescent and bristly; bracts leafy, lanceolate, 5–14 mm. long, pubescent to glabrate and usually also bristly; pedicels glabrous or sparsely villous-hirsute, 2–5 mm. long; corolla white, tinged pink, 4–7 mm. long; ovary usually glabrous; fr. depressed-globose, glabrous, 8–10 mm. broad, orange-brown to red-brown, usually with blue-black vertical stripes, the nutlets separable.—Dry slopes, 200–3600 ft.; Chaparral; La Panza Range, San Luis Obispo Co. Dec.–March.

36. **A. Andersònii** Gray. [*Uva-ursi A.* Abrams. *A. regismontana* Eastw.] HEARTLEAF MANZANITA. Without basal burl; erect, sometimes spreading, 1–4 m. high, with branches often elongated and smooth dark red-brown bark; branchlets densely short-pubescent and bristly, usually glandular; lvs. ovate or oblong-ovate, acute or obtuse at apex, 3–7 cm. long, dark to pale green, without stomates on upper surface, tomentulose to glabrous or sometimes glandular-pubescent beneath, cordate or auriculate at base, usually sessile; infl. large, paniculate, sessile, pubescent or somewhat glandular-villous; bracts mostly leafy, 3–12 mm. long, usually densely glandular-pubescent or -hispid; pedicels glandular-hairy, 5–6 mm. long; corolla white or pink, ca. 6–7 mm. long; ovary densely glandular-hairy; fr. depressed-globose, viscid-pubescent, red-brown, ca. 6–8 mm. broad, the nutlets separable.—Dry sandy and stony ridges, below 2200 ft.; Chaparral, Redwood F., N. Coastal Scrub?; near the coast from Santa Lucia Mts. to San Francisco. Jan.–March.

Var. **imbricàta** (Eastw.) Adams ex McMinn. [*A. i.* Eastw.] Lvs. bright green, with some stomates above, 2–3 cm. long, closely imbricate, glabrous except on midrib toward base; bracts viscid and ciliate; pedicels glandular-hairy; corolla white, 4–5 mm. long; fr. 6 mm. broad.—San Bruno Hills, San Mateo Co.

Var. **pállida** (Eastw.) Adams ex McMinn. [*A. p.* Eastw.] Lvs. pale green, glabrous, with some stomates above, 3–5 cm. long, imbricated; bracts almost glabrous; pedicels glandular-hairy; corolla white, 6–7 mm. long; ovary glandular-hairy; fr. bright red, glandular.—Chaparral; hills of Alameda and Contra Costa cos.

37. **A. pajaroénsis** Adams. [*A. Andersonii* var. *p.* Adams ex McMinn.] PAJARO MANZANITA. Without basal burl; erect, compact, 1–3 m. high, the bark dark red, exfoliating in shreds, slowly becoming smooth; branchlets loosely pubescent and bristly; lvs. ovate-triangular, acute and mucronate-cuspidate at apex, auriculate-clasping at base, 2–3 cm. long, glabrous, green or slightly glaucous, without stomates above, the basal lobes often angular, with 1–3 spinules or sometimes rounded and serrulate; panicles sessile, dense, many-fld., pubescent; bracts oblong, acute, pubescent to glabrate, the lower foliaceous, to ca. 1 cm. long; pedicels pubescent, 5–6 mm. long; corolla white, ca. 5–6 mm. long; ovary densely white-hairy; fr. globular, light red, pubescent, 6–7

mm. broad, the nutlets separable.—Sandy hills, near Prunedale, n. Monterey Co. Dec.–Feb.

38. **A. pechoénsis** Dudl. ex Abrams. [*Uva-ursi p.* Abrams. *A. Andersonii* var. *p.* Jeps.] PECHO MANZANITA. Without basal burl; bushy, 1 m. or so high, sometimes more, with smooth dark red-brown bark; branchlets white-tomentulose, sometimes also bristly; lvs. ovate to ovate-oblong, acute to almost rounded at apex, auriculate and subsessile at base, or subcordate and petiolate, pale green, with fewer stomates above than beneath, tomentulose to glabrate, 1.5–3.5 cm. long; panicle short, few-fld., tomentulose, sometimes also bristly; bracts foliaceous, pubescent and viscid, 8–12 mm. long, broadly lanceolate; pedicels mostly glabrous; corolla white or pinkish, 6–7 mm. long; ovary glabrous, or sparsely hairy near top; fr. depressed-globose, light brown, glabrous, 8–10 mm. in diam., the nutlets irregularly separable.—Dry slopes and ridges, below 2600 ft.; Closed-cone Pine F., Chaparral; w. cent. San Luis Obispo Co. and n. Santa Barbara Co. Jan.–March.

Var. **viridíssima** Eastw. [*A. v.* McMinn. *A. Andersonii* var. *v.* Jeps.] Lvs. dark green, shining, glabrous; bracts broadly lanceolate, almost glabrous save for the bristly ciliation.—Chaparral, Closed-cone Pine F.; Santa Cruz Id., Catalina Id., and Lompoc to Morro Bay.

39. **A. glandulòsa** Eastw. [*A. g.* var. *australis* Adams.] EASTWOOD MANZANITA. With an enlarged woody base or burl; erect, spreading, 1.5–2.5 m. high, the crooked stems smooth, reddish; branchlets coarse, conspicuously glandular-hairy; lvs. ovate to lance-ovate, dull green, equally ± glandular-pubescent and with somewhat similar stomatal numbers on both surfaces, acute to obtuse and mucronate at apex, obtuse to rounded or rarely truncate at base, with glandular-hairy petioles 5–6 mm. long; infl. short-spreading, paniculate, with glandular-hairy rachises and peduncle; bracts leafy, lanceolate, glandular-hairy, 5–15 mm. long; pedicels glandular, ca. 5 mm. long; corolla white, 6–8 mm. long; ovary glandular-hairy; fr. depressed-globose, red-brown, ± viscid, ca. 8 mm. across, the nutlets mostly separable, rugose and ridged on back.—Dry gravelly to rocky slopes and ridges, 1000–6000 ft.; Chaparral, Mixed Evergreen F., Yellow Pine F.; mostly outer Coast Ranges from Ore. to Cuyamaca Mts., San Diego Co. Jan.–April. With many closely related forms, some of which are:

KEY TO VARIETIES

A. Branchlets glandular-hairy.
  B. Lvs. light green, smooth. Widely distributed .... *A. glandulosa*
  BB. Lvs. very pale, scabrous. Santa Barbara Co. .... var. *zacaensis*
AA. Branchlets not glandular-hairy.
  B. Lvs. light or yellow-green or pale.
    C. Infl. and ovary glandular; lvs. yellow-green. Monterey Co. .... var. *Howellii*
    CC. Infl. and ovary non-glandular.
      D. Branchlets not having long white hairs.
        E. Branchlets puberulent. San Diego Co. and Riverside Co. ...... var. *Adamsii*
        EE. Branchlets canescent-tomentulose; lvs. with a light bloom. Marin, Napa, and Sonoma cos. .... var. *Cushingiana*
      DD. Branchlets long white-hairy.
        E. Infl. large, spreading, many-fld. Mt. Hamilton .... var. *Campbellae*
        EE. Infl. small, compact, few-fld. Santa Barbara Co. to Los Angeles Co.
                                                  var. *mollis*
  BB. Lvs. dark green. Coast of San Diego Co. .... var. *crassifolia*

Var. **zacaénsis** (Eastw.) Adams ex McMinn. [*A. z.* Eastw.] Habit and glandular pubescence of the sp.; lvs. pale gray-green, pubescent and somewhat scabrous.—Rocky slopes near Zaca Lake, Santa Barbara Co.

Var. **Howéllii** (Eastw.) Adams ex McMinn. [*A. H.* Eastw.] Branchlets tomentulose-pubescent, not glandular; lvs. pale yellow-green, elliptic-oblong, finely pubescent; infl. glandular, puberulent; bracts linear or upper triangular-subulate, bristly-ciliate; fr. glandular.—Gravelly ridges and canyons, w. cent. Monterey Co. and probably San Luis Obispo Co.

Var. **Cushingiàna** (Eastw.) Adams ex McMinn. [*A. C.* Eastw.] Prostrate to 1 m. high, densely branched; branchlets scarcely if at all glandular; infl. canescent, not glandular; upper bracts deltoid-subulate, finely pubescent; ovary white-hairy, not glandular.—Dry slopes, below 2500 ft.; Chaparral; Marin, Napa, and Sonoma cos.

Var. **Cámpbellae** (Eastw.) Adams ex McMinn. [*A. C.* Eastw.] Branchlets densely puberulent and also with long white hairs; panicle large, spreading, not glandular; upper bracts short-deltoid, the lower foliaceous; ovary white-hairy, not glandular.—Below 4000 ft.; Chaparral; Mt. Hamilton, Santa Clara Co.

Var. **Adámsii** Munz. Branchlets puberulent, without longer hairs; infl. and ovary nonglandular; infl. small, with deltoid bracts.—Dry slopes, 800–6800 ft.; Chaparral, Yellow Pine F.; e. San Diego Co. to w. Riverside Co.

Var. **móllis** Adams. Branchlets dark, puberulent, with scattered long soft hairs; lvs. light green, sparsely puberulent when young, glabrate in age, 2–4 cm. long; infl. small, compact, few-fld., puberulent and with some longer hairs; bracts hirsute-ciliate, the upper deltoid; ovary white-hairy, not glandular.—Rocky and gravelly slopes, 2500–6000 ft.; Chaparral, Yellow Pine F.; mts. from San Luis Obispo Co. to Riverside Co.

Var. **crassifòlia** Jeps. [*A. tomentosa* var. *c.* Jeps.] Six–12 dm. high, spreading; branchlets tomentulose, sometimes with scattered longer hairs, not glandular; lvs. dark green, glabrate above, ± tomentulose beneath, often truncate at base; fr. not glandular. —Sandy mesas and bluffs; Chaparral; coast of San Diego Co.

40. **A. intricàta** Howell. With basal burl; erect, 1–2 m. high, the bark almost black; branchlets tomentose and glandular-hispid; lvs. dull, with fewer stomates above than beneath, lance-oblong to oblong-ovate, 2.5–4 cm. long, somewhat tomentose, mostly obtuse and mucronulate at apex, obtuse at base, on petioles 4–6 mm. long; infl. short, dense, paniculate, tomentose and somewhat bristly; bracts foliaceous, lanceolate, tomentose and bristly-ciliate; pedicels pubescent, 3–5 mm. long; corolla white to rose, 4–5 mm. long; ovary densely hairy, not glandular; fr. depressed-globose, pubescent, 8–10 mm. across.—Rocky places, below 2500 ft.; Mixed Evergreen F.; Del Norte Co. to sw. Ore. March–May.

Var. **oblongifòlia** (Howell) Munz. [*A. o.* Howell.] Branchlets and infl. cinereous-puberulent.—Del Norte Co.; sw. Ore.

41. **A. crustàcea** Eastw. BRITTLELEAF MANZANITA. With basal burl; erect, 1–2 m. high, with several spreading stems and smooth dark purple bark; branchlets tomentose and ± densely setose-bristly, mostly nonglandular; lvs. bright green, without stomates above, sparsely tomentulose, oblong or ovate to lance-ovate, acute or rounded and apiculate at apex, truncatish or subcordate at base, 2–4 cm. long, on hairy petioles 5–8 mm. long; infl. ample, paniculate, tomentose and bristly; bracts foliaceous, lanceolate, 8–15 mm. long, pubescent and often bristly-ciliate, the upper deltoid-acuminate; pedicels mostly pubescent; corolla white or pink, 5–6 mm. long; ovary white-hairy to glabrous, nonglandular; fr. dark red, sparsely pubescent to glabrate, depressed-globose, 6–8 mm. broad, the nutlets variously separable.—Dry slopes and ridges, below 3400 ft.; Chaparral; Contra Costa and San Francisco cos. to Santa Barbara Co., Santa Cruz and Santa Rosa ids. Feb.–April.

Var. **Ròsei** (Eastw.) McMinn. [*A. R.* Eastw.] Branchlets pubescent, not bristly; lvs. 4–6 cm. long.—N. Coastal Scrub; about Lake Merced, San Francisco Co., Calif.

42. **A. tomentòsa** (Pursh) Lindl. [*Arbutus t.* Pursh. *Arctostaphylos cordifolia* Lindl. *A. vestita* Eastw. in Sarg. *Uva-ursi v.* Abrams. *A. glandulosa* var. *v.* Jeps.] SHAGGY-BARKED MANZANITA. With a basal burl; erect, divaricate, 1–2.5 m. high, with shreddy usually persistent bark; branchlets densely white-tomentose; lvs. oblong to broadly elliptic or ovate, truncate or subcordate at base, acute to obtuse, apiculate at apex, 2.5–4.5 cm. long, 1.5–2.5 cm. wide, without stomates above, green and sparsely tomentulose to glabrate above, hoary-tomentose beneath, on tomentulose petioles 2–5 mm. long; panicles congested, tomentose, not glandular; lower bracts leafy, 8–14 mm. long, spreading, tomentulose; pedicels stout, tomentose, 2–4 mm. long; corolla white, 5–6 mm. long; ovary densely tomentose; fr. depressed-globose, brownish, 8–10 mm. across, pubescent, the nutlets irregularly coalescent, wrinkled on back.—Sandy places, below 500 ft.; Closed-cone Pine F., Chaparral; about Monterey Peninsula to n. San Luis Obispo Co. Dec.–March.

Var. **trichoclàda** (DC.) Munz. [*Andromeda bracteosa* and var. *t.* DC. *Arctostaphylos b.* Abrams. *A. tomentosa* ssp. *b.* Adams.] Branchlets glandular-hairy and often also tomentose; lvs., bracts, pedicels also glandular.—With the sp.

Var. **hebeclàda** (DC.) Adams. [*Andromeda bracteosa* var. *h.* DC. *Arctostaphylos b.* var. *h.* Eastw.] Like the sp., but with lvs. 3–5 cm. long and 2–3 cm. wide, glabrate to tomentulose beneath.—Monterey Peninsula.

Var. **tomentòsifórmis** (Adams) Munz. [*A. crustacea* var. *t.* Adams. *A. tomentosa* var.

crinita Adams ex McMinn (without Latin diag.)] Branchlets densely white-tomentose and with white spreading bristly hairs; lvs. densely white-tomentose beneath.—Dry hills and ridges; Closed-cone 'Pine F., Chaparral?; Año Nuevo Point, San Mateo Co. to San Luis Obispo Co.

43. **A. subcordàta** Eastw. SANTA CRUZ ISLAND MANZANITA. With enlarged basal burl; erect, 1–2 m. high, with spreading branches and smooth red-brown bark; branchlets cinereous-tomentose and with long spreading gland-tipped hairs; lvs. dark green, without stomates above, ovate-elliptic to -lanceolate, mostly subcordate to truncate at base, acute to obtuse at apex, subglabrous above, thinly tomentose beneath, 2–5 cm. long, on petioles 2–5 mm. long; panicles dense, subsessile, tomentose and bristly, somewhat glandular; bracts foliaceous, lanceolate, 6–12 mm. long, long-ciliate; pedicels glandular-tomentulose; corolla 4–5 mm. long; ovary white-hairy; fr. depressed-globose, 6–8 mm. across, somewhat pubescent, the angled nutlets readily separable.—Stony ridges; Chaparral; Santa Cruz Id. Feb.–March.

Var. **confertiflòra** (Eastw.) Munz. [*A. c.* Eastw.] Branchlets glandular-hispid, but generally pubescent rather than tomentose; lvs. bright green, less pubescent.—Santa Rosa Id.

13. Vaccínium L. HUCKLEBERRY. BILBERRY. BLUEBERRY

More or less woody plants, from slender and trailing to arborescent. Buds scaly. Lvs. simple, alternate, deciduous or persistent. Fls. perfect, solitary or clustered, frequently bracteate. Calyx-tube adnate to ovary which in fr. becomes a berry or drupe crowned with calyx-teeth. Corolla of 4–5 united petals, often urn-shaped to campanulate, the disk epigynous. Stamens twice as many as corolla-lobes; anthers erect, introrse, their locules partly separated or prolonged into a tubular appendage with apical pore; pollen in tetrads. Ovary mainly inferior, 4–5-celled, many-ovuled. Style 1, filiform; stigma simple. Fr. berrylike, red, blue, or blue-black, sometimes with a bloom. Seeds compressed, with hard coat; embryo straight, imbedded in copious fleshy endosperm. Perhaps 150 spp., quite cosmopolitan. (Ancient name of Bilberry.)

(Camp, W. H. On the structure of populations of the genus Vaccinium. Brittonia 4: 189–247, 1942.)
A.   Plants erect or bushy shrubs; corolla urn-shaped to cylindric, merely toothed.
   B.   Lvs. thick and leathery, persistent; fils. hairy; fls. in racemes . . . . 1. *V. ovatum*
   BB. Lvs. thin, deciduous; fils. glabrous; fls. 1 to 2 or 4, in the axils.
      C.   Calyx distinctly 4–5-lobed or -parted; branchlets terete; fls. sometimes in 2's or 4's.
         D.   Berry globose; lvs. thickish, prominently veined, retuse to obtuse. Humboldt Co. n.
                                                                                       2. *V. uliginosum*
         DD. Berry ellipsoid; lvs. thin, obscurely veined, acute to obtuse. Siskiyou Mts. to Sierra
              Nevada . . . . 3. *V. occidentale*
      CC. Calyx entire or very slightly lobed; fls. solitary.
         D.   Branchlets terete or nearly so; lvs. gradually narrowed toward base.
            E.   Plants 5–10 cm. high, glaucous. Mostly above 7500 ft. . . . . 4. *V. nivictum*
            EE. Plants 20–60 cm. high, not glaucous. Mostly below 7000 ft. . . . 5. *V. arbuscula*
         DD. Branchlets angled; lvs. abruptly narrowed or rounded toward base.
            E.   Lvs. entire or only remotely serrulate and usually rounded at ends
                                                                                       8. *V. parvifolium*
            EE. Lvs. usually distinctly serrulate, at least above the middle, usually acutish
                 at apex.
               F.   The lvs. mostly 2–4 cm. long; plants 1–2 m. high; fr. black
                                                                                       6. *V. membranaceum*
               FF. The lvs. 0.6–1.2 cm. long; plants 0.1–0.4 m. high; fr. red
                                                                                       7. *V. scoparium*
AA. Plants trailing vines; corolla deeply cleft, with spreading lobes . . . . 9. *V. macrocarpon*

1. **V. ovàtum** Pursh. CALIFORNIA HUCKLEBERRY. Stout, erect, much-branched, 1–2.5 m. tall; branchlets pubescent; lvs. persistent, leathery, ovate to lance-oblong, 1.5–4 cm. long, serrate, glabrous and shining above, paler beneath, acute to acutish apically, rounded at base, sometimes somewhat revolute; petioles 2–3 mm. long; fls. white to pink, bell-shaped, in few-fld. axillary racemes; bracts red, deciduous; calyx with 5 deltoid lobes; corolla 5–7 mm. long; anthers without awns; berry broadly ovoid, 6–9 mm. long, black without a bloom, sweet, edible; seeds brown, ca. 1 mm. long, angled, pitted; $2n = 24$ (Darrow et al., 1944).—Dry slopes and canyons, below 2500 ft.; Redwood F., Closed-cone Pine F., Mixed Evergreen F.; common near coast from Del Norte Co. to n. Santa Barbara Co., occasional farther s., as above Santa Barbara, Santa

Rosa and Santa Cruz ids., near Escondido in San Diego Co.; to B.C. March–May. A form with berries glaucous, of better flavor, and more pear-shaped, occurs with the sp. and has been named var. *saporòsum* Jeps.

2. **V. uliginòsum** L. Bog Bilberry. Low, freely branched, 1–6 dm. high; branchlets terete, glabrous; lvs. oval to obovate, 1–2.5 cm. long, entire, firm, green above, paler beneath and reticulate-veined, glabrous, rounded to obtuse at apex, subcuneate at base; fls. in 2's or 4's or solitary, short-pedicelled; calyx 4–5-lobed, the lobes roundish, 1–1.5 mm. long; corolla pink, ovid to almost round, 5–7 mm. long, shallowly 4–5-lobed; stamens 8 or 10; berry dark blue, with a bloom, round, 7–10 mm. in diam., sweet, but of rather poor quality; seeds dark brown, curved-fusiform, ca. 1.6 mm. long, finely pitted; $2n = 24$, 48 (Hagerup, 1933).—Sphagnum bogs, Big Lagoon, Humboldt Co.; to Alaska, circumpolar. July.

3. **V. occidentàle** Gray. Western Blueberry. Low, glabrous, compact, 3–7 dm. high, sometimes decumbent; branchlets terete; lvs. thin, oblanceolate to obovate, 1–2 cm. long, entire, obtusish or acutish, somewhat paler beneath than above, not conspicuously veined; petioles ca. 1 mm. long; fls. mostly solitary, sometimes in 2's or 4's; calyx-lobes 4–5, deltoid, ca. 1 mm. long; corolla white or pinkish, oblong-ovoid, ca. 4 mm. long; stamens 8 or 10; anthers awned; berry blue-black, with a bloom, ellipsoid, ca. 6 mm. long, sweetish, but rather inferior; seeds compressed, straight on one edge, convex on other, ca. 1 mm. long, brown, finely cellular-pitted.—Wet meadows and wet places, 5000–11,000 ft.; largely Lodgepole F., Subalpine F.; Sierra Nevada from Tulare Co. n., to Modoc and Siskiyou cos., w. to Trinity Summit; to B.C. and Rocky Mts. June–July.

4. **V. nivíctum** Camp. [*V. caespitosum* auth., not Michx.] Sierra Bilberry. Depressed, tufted, mostly 5–10 cm. high, from underground rhizomes; branchlets glabrous, terete, glaucous; lvs. obovate, deciduous, glabrous on both surfaces, serrulate (each tooth ending in a gland-tipped hair), glaucous, 1–2(–3) cm. long, rounded to acute at apex, cuneate at base; petioles 1–3 mm. long; fls. solitary in axils, nodding; calyx-lobes obscure; corolla ovoid, pink or white, mostly 5-toothed, 5–6 mm. long; berry blue-black, with a bloom, globose, 5–7 mm. in diam., sweet, of good quality; seeds brown, compressed, with 1 straight and 1 curved margin, ca. 1 mm. long, cellular-pitted.—Wet meadows and near snow banks, 7000–12,000 ft.; mostly Subalpine F., Alpine Fellfields; Sierra Nevada from Tulare Co. n., Mt. Shasta. June–July.

5. **V. arbúscula** (Gray) Merriam. [*V. caespitosum* var. *a.* Gray.] Much like *V. nivictum,* but taller, 2–6 dm. high, scarcely if at all glaucous; branchlets ± reddish, subglabrous to puberulent; lvs. sometimes with gland-tipped hairs on surfaces as well as margins, greener.—Wet places, sea level to ca. 7900 ft.; mostly Coniferous Forests; Sierra Nevada from Sierra Co. n., to Modoc Co. and w. to Humboldt Co.; to B.C., Mont., Utah. May–June.

6. **V. membranàceum** Dougl. [*V. macrophyllum* (Hook.) Piper.] Thinleaf Huckleberry. Branching, erect, 1–2 m. high, glabrous; branchlets slightly angled; lvs. ovate to obovate, 2.5–6 cm. long, thin, membranous, acute or acuminate, finely serrulate, green above and slightly pubescent along the veins, paler beneath and with few hairs on veins; petioles 2–3 mm. long; fls. solitary; pedicels recurved at anthesis, erect in fr.; calyx entire or with undulate margin; corolla yellowish, almost globose, 4–5 mm. in diam.; berry black to dark red, without bloom, globose, 7–10 mm. in diam., of good flavor; seeds ca. 1.5 mm. long, red-brown, almost smooth, compressed.—Shaded slopes and wet places, 4300–7000 ft.; Red Fir F., Subalpine F.; Humboldt, Trinity and Siskiyou cos. to Modoc Co. n.; to B.C., Ida., Mont. June–July.

7. **V. scopàrium** Leib. [*V. myrtillus* auth., not L.] Littleleaf Huckleberry. Tufted, glabrous, with many green angled slender branches, 1–4 dm. high; lvs. ovate to elliptic-oblong, rounded to acute at ends, 5–12 mm. long, paler beneath, serrulate with each tooth ending in a fine hair with gland tip; petioles ca. 1 mm. long; fls. solitary; calyx scarcely lobed; corolla pink, ovoid-urn-shaped, ca. 3 mm. long; berry red, round, 3–5 mm. thick, palatable; seeds light brown, somewhat compressed but plump, ca. 1 mm. long, minutely cellular-pitted.—Shaded rocky woods, 6000–7000 ft.; Lodgepole F., Subalpine F.; Siskiyou Co.; to B.C., Rocky Mts. June–July.

8. **V. parvifòlium** Sm. in Rees. Red Huckleberry. Erect somewhat straggling shrub 1–4(–7) m. high, the branchlets green, sharply angled, glabrous to scaberulous; lvs.

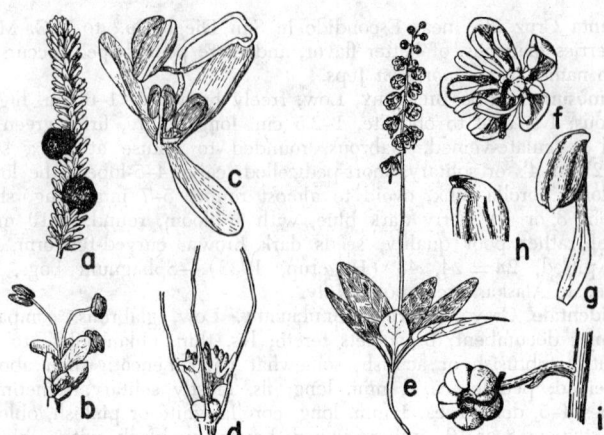

Fig. 48. EMPETRACEAE. *Empetrum: a*, habit, × 1, with berries; *b*, ♂ fl., × 3, with 3 stamens; *c*, section of ♂ fl., × 5; *d*, ♀ fl., × 5, with elaborate style and stigmas.
PYROLACEAE. *Pyrola picta: e*, habit, × ¼; *f*, fl., × 1½, petals 5, stamens 10, 5-lobed ovary and stigma; *g*, stamen, × 5, and *h*, anther-tip, with terminal pores; *i*, 5-lobed caps. with persistent style.

ovate to oblong, usually obtuse at apex, rounded or abruptly narrowed at base, entire, deciduous, almost sessile, 1–3 cm. long; fls. greenish or whitish, in 1's or 2's; calyx ± 5-lobed, the lobes broadly deltoid, to almost 1 mm. long; corolla globular, 4–6 mm. long; berry bright red, rounded, 6–10 mm. in diam., pleasantly flavored; seeds reddish, smooth but slightly angled, 1–1.2 mm. long; $2n = 24$ (Darrow et al., 1944).—Deep woods and moist places, below 5000 ft.; Redwood F., Red Fir F., Mixed Evergreen F.; Coast Ranges from Santa Cruz Mts. to Del Norte Co., through Siskiyou Co. e. and s. in the Sierra Nevada to Fresno Co. May–June.

9. **V. macrocárpon** Ait. [*Oxycoccus m.* Pursh.] CRANBERRY. Stems slender, elongate, much-forked, the flowering branches ascending; lvs. elliptic-oblong, 6–17 mm. long, flat or slightly revolute, green above, paler beneath; pedicels 1–10, from an elongate rachis 1–3 cm. long; pedicels with 2 foliose bracts toward tip; corolla pink to almost rose, deeply 4-parted into segms. 6–10 mm. long; anthers exserted, awnless, ca. twice as long as fils.; berry round, ellipsoid, to pear-shaped, 1–2 cm. thick, red; $2n = 24$ (Darrow et al., 1944).—Reported from a swamp at 3000 ft., 1 mile s. of North Columbia, Nevada Co.; Miss. V. to Atlantic Coast. June–July.

50. Empetràceae. CROWBERRY FAMILY (Fig. 48)

Low evergreen shrubs, heathlike. Stems freely branched, slender. Lvs. rigid, narrow, jointed to short pulvini, revolute. Dioecious, sometimes polygamous; fls. axillary or terminal. Sepals 3. Petals 2–3 or 0. Staminate fls. with 2–3 stamens; fils. filiform; anthers 2-celled, dehiscing by longitudinal slits. Ovary superior, 2–several-celled, with as many branches to styles; ovules 1 in each cell. Fr. a berry, with 2–several 1-seeded nutlets; endosperm abundant. A family of 3 genera and several spp., of colder regions; sometimes placed near Sapindaceae.

1. Émpetrum L. CROWBERRY

Branches many, prostrate or spreading; lvs. crowded. Fls. inconspicuous, solitary in upper axils, usually 3-merous. Staminate fls. with 3 stamens. Pistillate fls. with a 6–9-celled ovary. Berry black or red, with 6–9 nutlets. Two spp. of N. and S. Am., Eurasia. (Greek, *en*, upon, and *petros*, rock.)

1. **E. nìgrum** L. The diffuse procumbent stems 15–35 cm. long, glabrous to glandular-puberulent; lvs. linear-oblong, dark green, 3–7 mm. long; fls. purplish; berry 4–6 mm. in diam; $2n = 26$ (Hagerup, 1927).—Dense beds in rocky places on sea bluffs; N. Coastal Scrub; Del Norte Co.; to Alaska, Atlantic Coast; Eurasia. Spring.

51. Pyrolàceae. WINTERGREEN FAMILY (Fig. 48)

Herbaceous perennials from slender rootstocks and mostly with evergreen simple lvs.; or saprophytes or root-parasites without chlorophyll and varying from white or yellow to brown or red, and with scalelike lvs. Fls. solitary to racemose or corymbose, bracteate, perfect. Sepals 4–5 (2 or 6), free or somewhat united at base. Petals ca. as many, distinct or united. Stamens mostly 8 or 10, hypogynous; fils. distinct or united at base; anthers mostly inverted and with basal pores or opening by slits. Ovary 1–6-celled; style short or elongate; stigma capitate or peltate or short-lobed. Fr. a caps. or fleshy. Seeds mostly numerous. Ca. 18 genera of N. Hemis.

A. Plants with slender subterranean rootstocks, mostly with green lvs.; pollen in tetrads.
  B. Fls. in racemes; valves of caps. with cobwebby margins .... 1. *Pyrola*
  BB. Fls. solitary or in corymbs; valves without threads on margin.
    C. Fls. solitary; fils. glabrous .... 2. *Moneses*
    CC. Fls. in a corymb; fils. hairy .... 3. *Chimaphila*
AA. Plants fleshy-stemmed, from thickened bases, without green lvs.; pollen simple.
  B. Stems striped red and white; not both sepals and petals present .... 4. *Allotropa*
  BB. Stems not striped red and white; both sepals and petals present.
    C. Fls. nodding at anthesis; ovary 4–6-celled, with a cent. column.
      D. Petals distinct; calyx represented by 2–5 bracts .... 5. *Monotropa*
      DD. Petals somewhat united; calyx of 5 regular sepals.
        E. Corolla 7–8 mm. long; anthers horned .... 6. *Pterospora*
        EE. Corolla 12–18 mm. long; anthers not horned .... 7. *Sarcodes*
    CC. Fls. erect at anthesis; ovary 1-celled, without a cent. column.
      D. Petals distinct; sepals mostly 4–5.
        E. Fils. glabrous; anthers elongate; corolla glabrous .... 8. *Pleuricospora*
        EE. Fils. pubescent; anthers as broad as long; corolla pubescent within
                                                  9. *Pityopus*
      DD. Petals united at base; sepals 2 or 4 .... 10. *Hemitomes*

1. Pýrola L. SHINLEAF. WINTERGREEN

Low perennial herbs with slender subterranean rootstocks. Lvs. in a basal cluster (rarely lacking), evergreen, petioled. Fls. in a simple raceme on end of leafless somewhat scaly scape, pedicelled, spreading or nodding, 5-merous. Calyx persistent, 5-parted. Petals 5, concave, ± converging. Stamens 10; fils. naked; anthers extrorse in bud, opening by pair of pores at base. Style elongate; stigma 5-lobed or -rayed. Caps. depressed-globose, 5-lobed, 5-valved from base upward. Seeds numerous, minute, with loose cellular-reticulate coat. Ca. a dozen spp., of N. Hemis. (Diminutive of *Pyrus*, the pear, from resemblance in the lvs.)

(Camp, W. H. Aphyllous forms in Pyrola. Bull. Torr. Club 67: 453–465, 1940.)
Style strongly deflexed at base; stigma subtended by a ring or collar.
  Calyx-lobes much longer than broad; petals red to purplish .... 1. *P. asarifolia*
  Calyx-lobes scarcely longer than broad; petals greenish to whitish, rarely pinkish ...... 2. *P. picta*
Style straight, without a collar or ring below the stigma.
  Raceme spiral; corolla subglobose; style not exserted .... 3. *P. minor*
  Raceme 1-sided; corolla campanulate; style exserted .... 4. *P. secunda*

1. **P. asarifòlia** Michx. var. **bracteàta** (Hook.) Jeps. [*P. b.* Hook. *P. rotundifolia* var. *b.* Gray.] Extensively creeping; basal bracts brown, coriaceous, acuminate; lvs. leathery, basal, elliptic-ovate, 3–8 cm. long, short-acute, shining, distinctly mucronulate-denticulate by the prolongation of the veins at margin, rounded to subcordate at base; petioles ca. as long as blades; scapes slender, 2–4 dm. high; fl.-bracts 8–15 mm. long, reddish, broadly lanceolate, acuminate; sepals lance-deltoid, 4–5 mm. long, reddish; petals ovate to obovate, 6–8 mm. long, rose-purple or dull red; anthers reddish or yellow in age; style declined, 6–8 mm. long, slightly more in fr.; caps. 7–8 mm. broad; seeds yellowish.—Moist woods and bald hills; Coastal Prairie, Redwood F., Yellow Pine F., Douglas-Fir F., etc.; Mendocino and Siskiyou cos. n.; to B.C., Mont. June–July.

Var. **purpùrea** (Bunge) Fern. [*P. rotundifolia* var. *p.* Bunge. *P. r.* var. *incarnata* DC. *P. a.* var. *i.* Fern. *P. uliginosa* Torr.] Lvs. obovate to suborbicular, rounded to subcuneate at base, dull, entire to finely crenulate, not denticulate; petals pink to reddish-purple.—Moist shade of woods, streamsides, etc., 4000–9000 ft.; Yellow Pine F., Red Fir F.; San Bernardino Mts., Sierra Nevada to Modoc Co. and Siskiyou Co.; to Alaska and Atlantic Coast; Asia. July–Sept.

2. **P. pícta** Sm. Rootstock branched; lvs. coriaceous, lance-ovate to broadly elliptic, mottled or veined with white, entire to serrulate, rounded at tip to acute, obtuse to sub-cuneate at base, the blades 2–7 cm. long, the petioles as long or somewhat shorter; scape 1–2 dm. high; bracts lance-deltoid, acuminate, 3–5 mm. long; calyx-lobes ovate, acutish, ca. 2 mm. long; petals oblong-obovate, greenish to cream, with hyaline margin, 7–8 mm. long; anthers yellow; style 5 mm. long; caps. 6–7 mm. broad.—In humus in dry forests, 3000–9500 ft.; Yellow Pine F. to Lodgepole F.; Palomar Mts. to Mt. Shasta and Modoc Co., N. Coast Ranges from Lake Co. to Siskiyou Co.; to B.C., Rocky Mts. June–Aug.

Forma **aphýlla** (Sm.) Camp. [*P. a.* Sm.] Lvs. of flowering stalks scalelike, of sterile stalks sometimes green, 1–2 cm. long; calyx red-purple, ca. 2 mm. long; petals usually greenish or pinkish with white margins, or whitish.—Wooded slopes, below 7500 ft.; Redwood F., Yellow Pine F., Red Fir F.; Cuyamaca and Palomar mts. through Coast Ranges and Sierra Nevada to n. Calif.; Mont. and B.C. June–Aug.

Ssp. **dentàta** (Sm.) Piper. [*P. d.* Sm.] Lvs. obovate to oblanceolate, entire to denticu-late, dull green with little or no whitening along veins; bracts 2–3 mm. long; fls. cream or with some green.—Dry woods, below 5000 ft.; Closed-cone Pine F., Yellow Pine F.; Mendocino Co. to Del Norte Co.; to B.C. June–Aug. In the Siskiyou and Shasta regions intergrading freely with ssp. *íntegra* (Gray) Piper. [*P. dentata* var. *i.* Gray. *P. pallida* Greene.] Lvs. ± glaucous, mostly broadly obovate to roundish, mostly entire, the blades 1–4 cm. long.—Mostly 3000–10,000 ft.; Yellow Pine F. to Subalpine F.; San Jacinto Mts. n.; to Wash. June–Aug.

3. **P. mìnor** L. Rootstock slender; basal bracts oblong, cuspidate; lvs. rounded, slightly crenulate, the blades dull, 1–2(–3) cm. long, on petioles ca. as long; scapes slender, 6–15 cm. high; bracts at base of pedicels lance-oblong, 3–7 mm. long; calyx-lobes triangular, scarcely 2 mm. long; petals white to pink, 3–5 mm. long, with connivent tips; style included; caps. ca. 5 mm. broad; $2n = 46$ (Hagerup, 1928).—Occasional, boggy shaded places, 7000–10,000 ft.; Red Fir F. to Subalpine F.; San Jacinto and San Bernardino mts., Sierra Nevada to Modoc Co.; to Alaska, Atlantic Coast; Eurasia. July–Aug.

4. **P. secúnda** L. Rootstocks long, creeping; caudex woody, branching; lower bracts lanceolate, ciliate; lvs. shining, elliptic to ovate, crenate-serrate, the blades 2–5 cm. long, on petioles usually somewhat shorter; scapes 5–15 cm. high; bracts of the 1-sided infl. ciliate; calyx-lobes ovate, obtuse, ca. 1 mm. long, ciliate-serrulate; petals oblong, erose, yellow-green, 4–5 mm. long; style exserted; caps. 4–5 mm. broad.—Rather dry shaded woods, mostly 3000–10,500 ft.; Douglas-Fir F., Red Fir F. to Subalpine F.; San Jacinto Mts., San Bernardino Mts., Sierra Nevada to Modoc Co., Humboldt and Siskiyou cos.; to Alaska, Atlantic Coast; Eurasia. July–Sept.

2. **Monèses** Salisb.

Glabrous perennial herb from slender rootstocks. Lvs. in 2's or 3's, basally clustered, petioled, persistent. Scape 1–2-bracted, slender, ending in a solitary fl. Sepals 5, per-sistent. Petals 5, orbicular, plane. Stamens 10, the fils. dilated below; anthers con-spicuously 2-horned. Style straight; stigma peltate with 5 radiating lobes. Caps. sub-globose, obtusely 5-angled. Seeds many, minute. One sp., boreal; intermediate between *Chimaphila* and *Pyrola*. (Greek, *monos*, single, and *hesis*, delight, because of the solitary fl.)

1. **M. uniflòra** (L.) Gray. var. **reticulàta** (Nutt.) Blake. [*M. r.* Nutt.] Lvs. thin, ovate, acute, rather sharply serrulate, the blades 1–3 cm. long on petioles almost as long; scapes 4–10 cm. high; sepals elliptic-oblong, 3 mm. long; petals white or pinkish, ovate, obtuse, ca. 1 cm. long; caps. 6–8 mm. across.—Cool moist woods, 500–3500 ft.; Mixed Evergreen F., Redwood F., Douglas-Fir F.; Humboldt Co. to Siskiyou Co.; along the coast to B.C. May–July.

3. **Chimáphila** Pursh. PIPSISSEWA

Low evergreen suffrutescent perennials with running underground rootstocks and branching stems. Lvs. thick, shining, short-petioled. Fls. white to rose-pink, mostly in corymbs or racemes on end of a peduncle. Sepals 5. Petals 5, concave, round. Stamens 10; fils. dilated below, hairy in middle; anther-sacs prolonged into tubes at apex. Ovary 5-lobed; style short, straight; stigma disc-shaped, orbicular. Caps. 5-celled, splitting from apex downward. Seeds numerous, minute. Ca. 7–8 spp., N. Am., Asia. (Greek, *cheima*, winter, and *philein*, to love, referring to one of common names, *wintergreen*.)

Lvs. ovate; peduncle with 1–3 white fls. .... 1. *C. Menziesii*
Lvs. oblanceolate; peduncle with 3–7 pink fls. .... 2. *C. umbellata*

1. **C. Menzièsii** (R. Br. ex D. Don) Spreng. [*Pyrola M.* R. Br. ex D. Don.] Plants 1–1.5 dm. tall, few-branched; lvs. not distinctly whorled, ovate to lance-oblong, 1.5–3.5 cm. long, serrulate, sometimes mottled, dark green above, paler beneath; peduncles mostly 4–5 cm. long, with obovate rather persistent bracts; sepals rounded, erose, ca. 5 mm. long; petals white (to pinkish in age), round, ca. 6 mm. long; caps. 5–6 mm. in diam.—Shaded woods, 2500–6500(–8000) ft.; Yellow Pine F., Red Fir F.; San Diego Co., Sierra Nevada, N. Coast Ranges; to B.C., Ida. June–Aug.

2. **C. umbellàta** (L.) Barton var. **occidentàlis** (Rydb.) Blake. [*C. o.* Rydb.] Stems stoutish, 1.5–3 dm. high; lvs. in whorls of 3–8, oblanceolate, 3–7 cm. long, serrate, mostly yellow-green beneath; peduncles mostly 6–8 cm. long; bracts linear-subulate, early deciduous; sepals ovate, fimbriate, ca. 3 mm. long; petals ciliolate, pink, 5–6 mm. long; caps. 6–7 mm. in diam.—Dry shrubby slopes in forest, 1000–10,000 ft.; mostly Yellow Pine F. to Subalpine F.; San Jacinto and San Bernardino mts., Sierra Nevada, N. Coast Ranges; to Alaska, Mich. June–Aug.

4. Allótropa T. & G. SUGAR STICK

Rather fleshy simple-stemmed saprophytic herb, glabrous. Stems longitudinally striped red and white, densely clothed at base with elongate scalelike lvs., these more scattered upward. Fls. many, in erect spikelike raceme. Sepals 0. Petals 5, triangular-ovate, whitish. Stamens 10, the fils. slender, slightly exserted; anthers short and thick, purple-black, somewhat 2-lobed, each cell opening by a dorsal chink extending from base to middle. Caps. depressed, loculicidal, 5-celled; style short; stigma peltate-capitate. Seeds many, linear, minute. One sp. (Greek, *allos*, different, and *tropos*, turned, the raceme not nodding as in *Monotropa*.)

1. **A. virgàta** T. & G. ex Gray. Erect, 1–5 dm. tall; lvs. ovate-lanceolate to linear-lanceolate, ca. 2–3 cm. long; raceme the upper third to half of the plant; petals 5–6 mm. long, erose-fimbriate, thin; caps. 4–5 mm. in diam.—Occasional in thick humus, 2000–10,000 ft.; Douglas-Fir F., Yellow Pine F. to Subalpine F., sometimes lower in Mixed Evergreen F.; Sierra Nevada from Tulare Co. n., Coast Ranges from Napa and Sonoma cos. n.; to B.C. June–Aug.

5. Monótropa L. INDIAN PIPE. PINESAP

Tawny, white or reddish saprophytic or parasitic herbs, perennial. Stems simple, clustered. Lvs. many, bractlike, 1–several-fld. Fls. nodding at anthesis, erect in fr. Sepals 2–5, lanceolate, bractlike, deciduous. Petals 4–6, spatulate to cuneate, erect, scalelike, somewhat saccate at base, tardily deciduous. Stamens included, 8–12; fils. subulate; anthers becoming 1-loculed, opening by 2 transverse chinks in ours. Caps. ovoid, 8–10-grooved, loculicidal; style short; stigma disklike. Seeds many, minute, with loose coat. Ca. a half-dozen spp., N. Hemis. (Greek, *monos*, one, and *tropos*, turn, as the tip of the flowering stem is turned to one side.)

Fls. solitary at apex of stem; plant practically without odor .... 1. *M. uniflora*
Fls. several in a raceme; plant aromatic-fragrant .... 2. *M. Hypopithys*

1. **M. uniflòra** L. INDIAN PIPE. Waxy-white to reddish, drying black, usually with several stems 1–3 dm. high; lvs. scalelike, 5–12 mm. long; fl. nodding, oblong-campanulate; sepals 2–4, oblong, 14–17 mm. long, ± ciliate near base; petals 5–6, oblong-spatulate, somewhat longer; caps. oblong-ovoid, 10–15 mm. long.—Shaded damp woods, below 2000 ft.; Mixed Evergreen F., Redwood F.; Del Norte and Humboldt cos.; to B.C., Atlantic Coast; Asia. June–July.

2. **M. Hypópithys** L. [*M. lanuginosa* Michx. *Hypopitys l.* Nutt. *M. fimbriata* Gray. *H. f.* Howell.] PINESAP. Tawny or yellowish to somewhat red, pubescent, 1–3 dm. high; lvs. scalelike, oblong-ovate, 6–10 mm. long; raceme drooping when young, later erect, several-fld.; terminal fl. usually 5-merous, the lateral 3–4-merous; sepals 6–10 mm. long, ciliate and sometimes erose; petals cuneate-oblong, ca. 1 cm. long, pubescent and ciliate; fils. and style pubescent; stigma ciliate; caps. 4–5 mm. long.—Humus in deep forests, 1000–7000 ft.; Mixed Evergreen F. to Red Fir F.; n. Lake Co. to Siskiyou Co.; to B.C., Atlantic Coast; Eurasia. July–Aug.

6. Pteróspora Nutt. PINEDROPS

Stout, simple, purple-brown, clammy-pubescent, herbaceous root-parasite. Lvs. scalelike, lanceolate, scattered. Fls. in long terminal raceme, nodding on recurved pedicels. Fls. white to red. Calyx 5-parted. Corolla urn-shaped, persistent, 5-lobed. Stamens 10, included; fils. slender; anthers with 2 dorsal appendages and opening by dorsal slits. Caps. 5-loculed, depressed-globose, loculicidal; style stout, short; stigma capitate-peltate, shallowly 5-lobed. Seeds many, ovoid, tapering to one end, broadly reticulate-winged at apex. One sp. of N. Am. (Greek, *pteros*, wing, and *spora*, seed.)

1. **P. andromedèa** Nutt. Stems 3–10 dm. high; lvs. crowded below, 1.5–3.5 cm. long; sepals lance-linear, 4–5 mm. long; corolla 7–8 mm. long, the lobes rounded; caps. 8–12 mm. in diam.—Humus in forests, 2600–8500 ft.; Yellow Pine F., Red Fir F.; San Jacinto Mts. to Sierra Nevada and Modoc Co., Coast Ranges from Lake Co. n.; to B.C., Atlantic Coast, Mex. June–Aug.

7. Sarcòdes Torr. SNOW PLANT

Red fleshy usually pubescent saprophyte. Stems simple, solitary or in clusters. Lvs. scalelike, lanceolate, more crowded toward base of stem. Fls. many, in a stout spicate raceme, red, subtended by conspicuous red bracts. Sepals 5, glandular-pubescent. Corolla campanulate, 5-lobed, the lobes broad, red, slightly spreading. Stamens 10, included; fils. slender, glabrous; anthers without appendages, opening by terminal pores. Caps. depressed-globose, 5-loculed; style stout; stigma subcapitate, shallowly 5-lobed. Seeds small, pitted-reticulate, subovoid, somewhat angled and with curved beaklike end. One sp. (Greek, *sarx*, flesh, and *oeides*, like.)

1. **S. sanguínea** Torr. Plant 1.5–3(–5) dm. tall; lvs. 2–8 cm. long, ciliate, the lower shorter and broader than the upper; sepals lance-ovate, 10–15 mm. long; corolla slightly longer, the lobes rounded; caps. 1–2 cm. in diam.; seeds ca. 0.7 mm. long, red-brown. —Thick humus of forests, 4000–8000 ft.; Yellow Pine F. to Lodgepole F.; Santa Rosa and San Jacinto mts. to Sierra Nevada and Siskiyou Co., N. Coast Ranges from Trinity Co. n.; Ore., Nev. May–July.

8. Pleuricóspora Gray

White or brownish saprophyte with imbricate scalelike lvs. and simple stem. Fls. at first white, later yellow to dark brown, in terminal spikelike raceme with broad conspicuous bracts. Sepals 4–5, distinct, persistent, lance-ovate, erose-fimbriate. Petals as many, lance-oblong, entire or somewhat fimbriolate. Stamens 8 or 10, with glabrous linear-filiform fils.; anthers dehiscing by a longitudinal slit. Ovary 1-celled, with parietal placentae; style stout; stigma depressed-capitate. Fr. a berry, ovoid. Seeds ellipsoid, chestnut-brown, shallowly pitted. Apparently 1 sp. (Greek, *pleuricos*, at the side, and *spora*, seed, because of the parietal placentation.)

1. **P. fimbriolàta** Gray. [*P. densa* Small.] Rather stout, 1–2 dm. high, glabrous; lvs. ovate or lance-ovate, 8–12 mm. long, entire to erose or fimbriate; sepals lanceolate to lance-ovate, erose-fimbriate, involute, acuminate, 5–10 mm. long; petals elliptic to lance-oblong, entire to fimbriolate, ca. as long as sepals; berry whitish to dark; seeds ca. 0.35 mm. long.—Dry deep humus, 4000–8500 ft.; Yellow Pine F., Red Fir F.; Sierra Nevada from Tulare Co. n., to Siskiyou Co.; at lower elevs., Redwood F.; Santa Cruz and San Mateo cos.; to B.C. June–Aug.

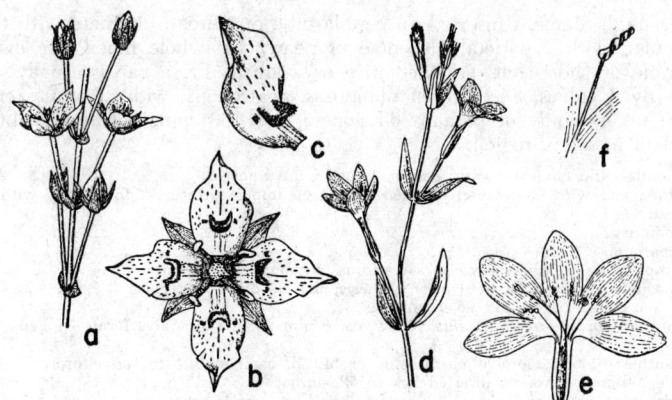

Fig. 49. GENTIANACEAE. *Frasera: a,* × ⅓, portion of panicle; *b,* fl., × 1, 4 sepals, 4 corolla-lobes with U-shaped glands, 4 stamens, pistil; *c,* corolla-lobe with gland. *Centaurium: d,* × ½, upper part of branch; *e,* corolla split lengthwise, × 1; *f,* spirally twisted anther, × 2½.

9. Pityòpus Small

Waxy-white subglabrous saprophytic herb. Stems simple, erect, 1–few in a cluster. Lvs. scalelike, erect, crowded, ovate to lanceolate. Fls. in a terminal dense bracteate spikelike raceme. Sepals 2–4–5, elliptic to narrow-oblong, persistent. Petals 4–5, white, oblong-obovate, pubescent within. Stamens twice as many as petals; fils. slender, pubescent; anthers short, opening from base by longitudinal slits. Ovary ovoid, 1-celled with parietal placentae; style cylindric, pubescent; stigma with heavy collar of hairs beneath the annulus. Fr. somewhat fleshy, ovoid. One sp. (Greek, *pitus,* pine, and *pus,* foot, because of where it lives.)

1. **P. califórnicus** (Eastw.) Copel. f. [*Monotropa c.* Eastw. *Hypopitys c.* Heller. *P. oregona* Small.] Plants 7–20 cm. high; lvs. 1–2 cm. long; sepals ca. 12 mm. long; petals as long or longer.—Rare, deep shade, 1000–5000 ft.; Mixed Evergreen F., Redwood F., Yellow Pine F.; Marin, Humboldt, Del Norte, Mendocino, and Fresno cos.; Ore. May–July.

10. Hemitòmes Gray

Saprophytic fleshy herbs, white to pink and turning brown in age. Stems simple, scaly, often largely subterranean. Lvs. scalelike, imbricated. Fls. in short terminal dense spike or corymbiform head, with broad bracts. Sepals 2 or 4, bractlike, hairy within. Corolla tubular-campanulate, pubescent within, deeply 4–6-lobed. Stamens 8–12; fils. long-hairy; anther-cells with longitudinal slits. Ovary ovoid, 1-celled, with 4–5 two-lobed lateral placentae; style pubescent; stigma depressed-capitate, with retrorse hairs below. Fr. a berry. One sp. (Greek, *hemi,* half, and *tomias,* eunuch, as 1 anther cell may be sterile.)

1. **H. congéstum** Gray. [*Newberrya c.* Torr. *H. pumilum* Greene. *N. subterranea* Eastw. *N. spicata* Gray.] Plant 3–12 cm. high; lvs. ovate, obtuse, erose; corolla 10–14 mm. long; cent. fls. with ovate lobes ca. ⅓ as long as tube, the marginal fls. more deeply lobed; stigma yellow, conspicuous.—Humus in woods, below 2500 ft.; Redwood F.; Monterey Co. to Humboldt Co.; to Wash. May–July.

52. Gentianàceae. GENTIAN FAMILY (Fig. 49)

Herbs with colorless bitter juice. Lvs. opposite or whorled, sometimes alternate, mostly sessile and simple, occasionally petioled, even trifoliolate. Fls. regular, perfect, axillary or in terminal infl. Calyx persistent, free from ovary, 4–12-lobed or toothed. Corolla sympetalous, rotate to funnelform, withering-persistent, the lobes mostly convolute in the bud. Stamens inserted on corolla-tube or -throat, alternate with the lobes. Ovary 1-locular, with 2 parietal placentae or nearly the whole inner face ovuliferous; style 1, simple or short-cleft; stigma entire or 2-lobed. Fr. a caps., usually dehiscent from above by 2 valves. Seeds often numerous, anatropous, with a minute embryo in fleshy albumen. A family of perhaps 65 genera and 600 spp., widely distributed, but most abundant in temp. region.

Lvs. simple, entire, the cauline mostly sessile, opposite or whorled.
  Corolla rotate, and with conspicuous fringed glands on upper surface or, in 1 case, with crownlike
  tubes instead.
    Fls. 5-merous .... 5. *Swertia*
    Fls. 4-merous .... 6. *Frasera*
  Corolla campanulate to funnelform or salverform, without such glands.
    Anthers coiled or spirally twisted after anthesis; fls. mostly red or pink .... 2. *Centaurium*
    Anthers not coiled or twisted after anthesis.
      The anthers cordate-ovate; corolla yellow, ca. 6 mm. long, short-salverform, 4-lobed
                                                               1. *Microcala*
      The anthers oblong; corolla mostly blue or bluish, campanulate to funnelform.
        Style filiform; stamens inserted on corolla-throat .... 3. *Eustoma*
        Style stout, short or lacking; stamens inserted on corolla-tube .... 4. *Gentiana*
Lvs. 3-foliolate, basal or alternate, long-petioled .... 7. *Menyanthes*

1. Microcàla Hoffmsg. & Link

Small annuals. Stems simple or branched, filiform, each ultimate branch terminating in a solitary fl. Lvs. opposite, entire, sessile. Calyx 4-angled and 4-toothed. Corolla short-salverform, 4-lobed, the lobes convolute in the bud. Stamens 4, short, inserted on the corolla-throat; anthers cordate-ovate. Style filiform, deciduous; stigma 2-lobed, the lobes fan-shaped, separating at length. Caps. ovoid, usually covered by the withering persistent corolla. Seeds many, minute, finely reticulate-pitted. Two spp. of Medit. region, etc., and of New World. (Greek, *mikros*, small, and *kalos*, beautiful.)

    1. **M. quadrangulàris** (Lam.) Griseb. [*Gentiana q.* Lam.] Glabrous, 2–7 cm. high; basal lvs. few; cauline lvs. 1–3 pairs, oblong to oval, 4–8 mm. long; calyx 4–5 mm. long, quadrangular, appearing truncate at base and top, the teeth very short, subulate; corolla deep yellow, ca. 6 mm. long, the lobes ca. 2 mm. long, closing in the afternoon; caps. ca. as long as calyx; seeds brownish, round-oblong, ca. 0.3 mm. long.— Grassy places, below 1000 ft.; V. Grassland, N. Oak Wd., Foothill Wd., etc.; foothills of Sierra Nevada from Calaveras Co. to Merced Co. and Coast Ranges from Santa Barbara Co. to Tehama, Shasta, and Humboldt cos.; to Ore.; w. S. Am. March–May.

2. Centaùrium Hill. CENTAURY

Glabrous branching erect annuals. Lvs. opposite, sessile or amplexicaul. Fls. 4–5-merous, mostly pink or rose, in terminal spikes or cymes. Calyx narrow, deeply parted into narrow keeled segms. Corolla salverform or funnelform, with slender tube, the lobes contorted, convolute in the bud. Stamens inserted on corolla-throat, alternate with the lobes; fils. slender; anthers commonly exserted, spirally twisting after shedding pollen. Ovary 1-celled, the parietal placentae sometimes intruded; style filiform, deciduous; stigmas oblong to fan-shaped. Caps. fusiform to oblong-ovoid, 2-valved. Seeds minute, reticulate, oblong to rounded. Perhaps 30 spp., of N. Am., Eurasia and Afr. (From Latin, *Centaurus*, Centaur, who is supposed to have discovered its medicinal properties.)

Fls. sessile or subsessile, the pedicels usually not over 1 mm. long and the cymules densely fld.
  Basal lvs. in a dense tuft; corolla-lobes 5–7 mm. long .... 1. *C. umbellatum*
  Basal lvs. not tufted; corolla-lobes 3.5–5 mm. long.
    Each fl. subtended by a bract with a rudimentary floret in its axil; corolla-tube 6–7 mm. long
                                             2. *C. floribundum*
    Each fl. without subtending rudimentary floret; corolla-tube 10–12 mm. long
                                           3. *C. Muehlenbergii*
Fls. pedicellate, the pedicels 1–many mm. long and the cymules loosely fld.
  Corolla-lobes less than half as long as the tube; anthers 1.5–2.5 mm. long, oblong.
    Pedicels 1–12 mm. long.
      Corolla-lobes narrow, lance-oblong, 3.5 mm. long, 1–1.5 mm. wide. E. of Sierra Nevada
                                          4. *C. curvistamineum*
      Corolla-lobes wider, ovate, 5–5.5 mm. long, 2.5–3 mm. wide. Coastal ...... 5. *C. Davyi*
  Pedicels 10–40 mm. long; corolla-lobes 3–4 mm. long. E. of Sierra Nevada and in s. Calif.
                                              6. *C. exaltatum*

Corolla-lobes more than half as long as tube; anthers 3.5 mm. long, linear.
Stigma-lobes appressed against each other, 0.5 mm. high, narrow, the style not divided below them.
San Mateo Co. to Siskiyou Co. .... 7. *C. trichanthum*
Stigma-lobes divaricate, 1–1.5 mm. high, broad, the style cleft for short way beneath them. Foothills of Sierra Nevada and s. Calif.
Lvs. mostly oblong to ovate, wider in lower half. Largely cismontane .... 8. *C. venustum*
Lvs. mostly oblanceolate to elliptic, tending to be wider in upper half. E. desert
9. *C. calycosum*

1. **C. umbellàtum** Gilib. [*Gentiana Centaurium* L. *Erythraea C.* Pers.] Erect, strictly branched, 2–4 dm. high, the branches ending in dense cymes; basal lvs. densely tufted, 2–4 cm. long, broadly oblong, obtuse; cauline lvs. lance-oblong to lanceolate, acute, gradually reduced upward; fls. many, nearly sessile; calyx-segs. linear-subulate, 4–6 mm. long; corolla rose, the tube 7–10 mm. long, the lobes ovate, obtusish, 5–7 mm. long; anthers 2.5 mm. long; stigmas oval; seeds light brown, rounded, 0.2–0.3 mm. long.—Open fields and waste places, below 500 ft., along immediate coast, Mendocino Co. n.; to Wash.; natur. from Eu. July–Aug.

2. **C. floribúndum** (Benth.) Rob. [*Erythraea f.* Benth.] Mostly branched above base, 1–5 dm. high; basal lvs. obovate to oblong, 1.5–2 cm. long, rounded at tip; cauline lvs. oblong-ovate to -lanceolate, acutish; fls. in rather crowded cymes, those in the forks sessile, the lateral subsessile or very short-pedicelled; calyx-segs. linear, 7–8 mm. long; corolla pink, the tube 6–8 mm. long, the lobes lance-ovate to oblong, 3–4 mm. long; anthers oblong, ca. 1.5 mm. long; stigmas oval; seeds brown, minute, roundish.—Moist places, below 1600 ft.; V. Grassland, Foothill Wd.; inner Coast Ranges from Humboldt Co. to Marin Co. and foothills of Sierra Nevada, from Fresno Co. to Butte Co. May–Aug.

3. **C. Muehlenbérgii** (Griseb.) W. Wight. [*Erythraea M.* Griseb.] Erect, simple or branched from base, 1–3(–4) dm. high; lvs. lance-oblong to ovate, oblong, acutish, 1–2.5 cm. long; infl. loosely paniculate-cymose; fls. sessile or subsessile in the forks, the ultimate fls. in compact 2- to 4-fld. cymules, the subtending bracts lacking abortive florets; calyx-segs. narrowly linear, 8–10 mm. long; corolla-tube 8–12 mm. long, the lobes pink, ca. 3.5 mm. long and 1 mm. wide, acutish; anthers 1–1.5 mm. long; stigma-lobes longer than wide; seeds brown, round-oblong, ca. 0.2 mm. long.—Mostly damp places, below 1500 ft.; Redwood F., N. Oak Wd., Mixed Evergreen F.; Coast Ranges from Monterey Co. to Ore. June–Aug.

4. **C. curvistamíneum** (Wittr.) Abrams. [*Erythraea c.* Wittr. *E. minima* Howell.] Simple or branched, 0.5–1.5 dm. high; lvs. elliptic-oblong, obtusish, 0.7–2 cm. long; fls. borne in forks and at ends of branches on pedicels mostly 3–12 mm. long; calyx-segs. 6–8 mm. long; corolla-tube 8–10 mm. long, the lobes oblong-lanceolate, 3.5–4 mm. long; anthers ca. 1 mm. long; stigma-lobes broader than long; seeds light brown, oblong, ca. 0.4 mm. long.—Moist places, ca. 4000–5000 ft.; Sagebrush Scrub, N. Juniper Wd.; Modoc Co.; to e. Wash., w. Nev. June–Sept.

5. **C. Dàvyi** (Jeps.) Abrams. [*C. exaltatum* var. *D.* Jeps.] Simple or branched from base, 0.5–2 dm. high; lvs. elliptic to narrowly oblong, 0.8–2 cm. long; fls. in forks and at ends of branchlets, on pedicels 4–20 mm. long; calyx-segs. mostly 8–10 mm. long; corolla-tube ca. 8–10 mm. long, the lobes pink, oblong-ovate, often appearing narrow in age, 4–6 mm. long, 2.5–3 mm. wide; anthers 1.5 mm. long; stigmas broad; seeds dark brown, round-oblong, ca. 0.3 mm. long.—Low damp places; Mixed Evergreen F., Closed-cone Pine F., Redwood F.; outer Coast Ranges from San Luis Obispo Co. to Mendocino Co., Santa Cruz Id. May–Aug.

6. **C. exaltàtum** (Griseb.) W. Wight. [*Cicendia e.* Griseb. in Hook. *Erythraea e.* Cov. *E. Douglasii* Gray.] Simple or usually branched, 1–3.5 dm. high; lvs. oblong-elliptic to -lanceolate, 1–3 cm. long, acute; pedicels 1–5 cm. long, in forks and at ends of branchlets; calyx-segs. subulate, 8–10 mm. long; corolla-tube 8–10 mm. long, the lobes pale pink to white, 3–4 mm. long, oblong, obtuse; anthers oblong, ca. 1 mm. long; stigma-lobes fan-shaped; seeds almost round, ca. 0.25 mm. long.—Damp somewhat alkaline places, below 5000 ft.; Coastal Sage Scrub, Chaparral, Creosote Bush Scrub, etc.; cismontane San Diego Co. to Los Angeles Co., w. edge of deserts n. through Inyo and Modoc cos.; to e. Wash., Utah. May–Aug.

7. **C. trichánthum** (Griseb.) Rob. [*Erythraea t.* Griseb.] Simple below or branched, 1–3.5 dm. high; lvs. lanceolate to ovate, 1–3 cm. long; fls. in terminal corymbose open

or crowded cymes, sessile in the forks and sessile or pedicelled on the branches; calyx-segms. 8–14 mm. long; corolla-tube exceeding calyx, the lobes pink, 8–10 mm. long, 3–4 mm. wide, often appearing narrower by involution; anthers 3–3.5 mm. long; stigma-lobes ca. 0.5 mm. long, somewhat narrower, the two closely appressed; seeds 0.5–0.6 mm. long.—Moist often saline places, below 2500 ft.; edge of Coastal Salt Marsh, Mixed Evergreen F., N. Oak Wd., Chaparral; San Mateo Co. to Siskiyou Co. May–Aug.

8. **C. venústum** (Gray) Rob. [*Erythraea v.* Gray.] CANCHALAGUA. Usually simple below, corymbosely branched above, 1–3(–5) dm. high; lvs. ovate to oblong, 1–2.5 cm. long, obtusish; pedicels 2–25 mm. long; calyx-segms. 6–9(–12) mm. long; corolla rose with red spots in the white throat, sometimes albino, the tube ca. 8–12 mm. long, the lobes lanceolate to oblong-ovate, mostly obtuse, 8–15 mm. long, 3–10 mm. wide; anthers 4–6 mm. long; stigmas fan-shaped, broader than long, divaricate; seeds roundish to somewhat oblong, dark, ca. 0.2–0.3 mm. long.—Dry slopes and flats, below 2500 (4000) ft.; Chaparral, Coastal Sage Scrub; cismontane s. Calif. from San Diego Co. to Los Angeles and Ventura cos., to desert edge. May–July. Intergrading with:

Ssp. **Abramsii** Munz. Lvs. broadly lanceolate to ovate, but more sharply pointed; flowers smaller, the corolla-lobes mostly 5–7 mm. long, acutish.—Mostly below 6000 ft.; V. Grassland, Foothill Wd., Yellow Pine F.; base of Sierra Nevada from Kern Co. to Butte and Shasta cos. May–Aug.

9. **C. calycòsum** (Buckl.) Fern. [*Erythraea c.* Buckl.] Much like *C. venustum*, but with more open paniculate branching and often 4–5 dm. high; lvs. tending to be oblanceolate, especially those of lower half of plant; corolla-lobes apiculate.—Damp places, Colo. R. bottom; to Tex., Mex. April–May.

3. Eustòma Salisb. CATCHFLY-GENTIAN

Annual or short-lived perennial herbs, glaucous. Stems erect or ascending, leafy. Lvs. opposite, sessile or clasping. Fls. blue or white to purplish, solitary or paniculate. Calyx deeply cleft, the lobes keeled, long-acuminate. Corolla campanulate-funnelform, 5–6-lobed, the lobes oblong to obovate, convolute in bud and often erose-denticulate. Stamens 5–6, inserted on throat of corolla; anthers oblong, versatile, straight or recurved. Ovary 1-celled; style filiform; stigma 2-lobed, the lobes broad, flattened. Caps. oblong to ellipsoid, 2-valved. Seeds many, small, honeycombed. Spp. 4; from s. U.S., Mex., W. I., n. S. Am. (Greek, *eu*, good, and *stoma*, mouth, the throat of the corolla large.)

1. **E. exaltàtum** (L.) Salisb. [*Gentiana e.* L. *Eustoma silenifolium* Salisb.] Stems solitary to few, erect or ascending, branched above, 4–7 dm. tall; basal lvs. obovate to broadly spatulate, with short broad petiole; cauline lvs. broadly oblong, sessile, sub-cordate-clasping, 4–9 cm. long, the uppermost reduced and bractlike; pedicels stout, 1–10 cm. long; calyx-lobes subulate, 1–1.5 cm. long; corolla blue or deep lavender, sometimes pale, the tube ca. 1 cm. long, the lobes oblong-obovate, 1.5–2 cm. long; style 4–5 mm. long; stigma-lobes ca. 2 mm. long; caps. 8–12 mm. long; seeds brown, round-oblong, ca. 0.5 mm. long.—Occasional along streams, below 1500 ft.; Coastal Sage Scrub, Creosote Bush Scrub; Santa Ana R. of San Bernardino and Orange cos., deserts (Thousand Palms, Palm Springs, San Felipe, etc.); to Fla., W. I., Mex., L. Calif. Most months.

4. Gentiàna L. GENTIAN

Herbs with opposite mostly sessile lvs. Fls. solitary or clustered, axillary or terminal, frequently showy, blue, purple, white, or even yellow, 4–5(–7)-merous. Calyx tubular, lobed. Corolla tubular to funnelform or campanulate, lobed, often with intermediate plaited folds which bear appendages or teeth at the sinuses. Stamens inserted on corolla-tube; anthers versatile, straight or recurved in age. Style short or none; stigmas 2, persistent. Caps. 1-celled, 2-valved, sessile or stipitate. Seeds numerous, minute, winged or wingless, often with a loose cellular coat. Ca. 300 spp., of cool and temp. regions. (Named for King *Gentius* of Illyria, who was supposed to have discovered the medicinal virtues.)

ı

(Gillett, J. M. A revision of the North American species of Gentianella Moench. Ann. Mo. Bot. Gard. 44: 195–269, 1957.)

Plants perennial; calyx-tube with an inner membrane showing between the bases of the lobes.
    Sinus-plaits of corolla entire; at least some fls. without subtending bracts, hence on pedicels 1–several
        cm. long. N. Coast Ranges .... 4. *G. sceptrum*
    Sinus-plaits of corolla toothed or laciniate; fls. sessile in axils of subtending bracts.
        Stems 4–12 cm. high; corolla mostly with dark bands without. Subalpine ...... 1. *G. Newberryi*
        Stems 15–40 cm. high; corolla mostly not dark-banded without.
            Calyx-lobes scaberulous along the margin, the teeth subcolumnar; calyx-tube 4–5 mm. long.
              The calyx-lobes unequal, 1–7 mm. long; floral bracts linear-lanceolate. Del Norte and
              Modoc cos. .... 2. *G. affinis*
              The calyx-lobes subequal, 8–12 mm. long; floral bracts lance-ovate to ovate. Marin Co. to
              Ore. line .... 3. *G. oregana*
            Calyx-lobes merely minutely scaberulous along the margin, the teeth low and broad; calyx-tube
            6–10 mm. long.
              Sinus-plaits of corolla with several bristles in each sinus; corolla-lobes erosulate. Rare,
              Mendocino Co. n. .... 5. *G. setigera*
              Sinus-plaits of corolla with 2 acute to acuminate lobes; corolla-lobes entire. Tulare Co. to
              Siskiyou Co. .... 6. *G. calycosa*
Plants annual or biennial; calyx-tube lacking an inner membrane.
    Corolla 7–20 mm. long.
        The corolla with sinus-plaits; lvs. white-margined; corolla-lobes lacking a basal fimbriate crown.
            Corolla greenish-purple, 7–8 mm. long; caps. ca. 5 mm. long, almost as wide, long-exserted and
            with 2 spreading valves. San Bernardino Mts. .... 7. *G. Fremontii*
            Corolla blue, 12–15 mm. long; caps. 8–10 mm. long, ca. 1.5 mm. wide, usually included and
            the valves not markedly divergent. White Mts. .... 8. *G. prostrata*
        The corolla lacking sinus-plaits; lvs. not white-margined; corolla-lobes with a fimbriate crown
        at base.
            Fls. clustered; plants 5–50 cm. high. Common .... 9. *G. Amarella*
            Fls. solitary; plants 3–8 cm. high. Rare .... 10. *G. tenella*
    Corolla 25–50 mm. long.
        Plants always with a simple stem, with mostly 3–6 pairs of lvs.; calyx-lobes not dark-ribbed; seeds
        subcylindric .... 11. *G. simplex*
        Plants mostly branched at base, the stems with mostly 1–3 pairs of lvs. above base; calyx-lobes
        usually dark-ribbed; seeds oval .... 12. *G. holopetala*

1. **G. Newbérryi** Gray. [*Pneumonanthe* N. Greene. *Dasystephana* N. Arthur. *G. tiogana* Heller. *G. Copelandii* and *eximia* Eastw.] ALPINE GENTIAN. Perennial with a taproot and 1 to few 1-fld. somewhat decumbent stems 4–12 cm. high; lower lvs. broadly spatulate, 2–6 cm. long including the broad petiole, the upper narrower, shorter, subsessile; uppermost lvs. subtending the fl.; calyx-tube turbinate, 8–14 mm. long, the lobes narrowly elliptic-oblong, 6–12 mm. long, acute, with intracalycine membrane entire; corolla broadly funnelform, mostly with dark bands without and commonly white within with greenish spots, sometimes pale blue to purplish, the tube 2–3 cm. long, the lobes narrow-obovate, mucronate, 5–8 mm. long, the plaits 2-cleft with subulate tips; style lacking; caps. ovoid, 10–12 mm. long; seeds light brown, flat, ca. 1.5 mm. long including the all-around wing.—Moist meadows and banks, mostly 7000–12,000 ft.; Subalpine F., Alpine Fell-fields; Sierra Nevada from Tulare Co. n., Siskiyou Co., White Mts.; w. Nev., s. Ore. July–Sept.

2. **G. affinis** Griseb. in Hook. [*Pneumonanthe a.* Greene. *Dasystephana a.* Rydb.] Perennial from a root-crown, with usually several stems, erect or somewhat decumbent, 1.5–4 dm. high; lvs. rather many, the lower lance-ovate, the upper narrower, 2–3 cm. long; fls. several, racemosely arranged in the upper axils, each subtended by 2 linear bracts; calyx-tube 4–5 mm. long, the lobes unequal, somewhat divergent-recurved, 3–7 mm. long, the membrane between the lobes truncate; corolla bluish-purple, 2–3 cm. long, narrow-funnelform, the lobes 3–5 mm. long, somewhat greenish on backs, the plaits in the sinuses divided into usually 2 acuminate teeth; caps. 12–15 mm. long, on a stipe almost as long; seed winged, flat, oval, straw color, ca. 1–1.3 mm. long including wing.—Dry serpentine places, 1500–2500 ft.; Mixed Evergreen F.; near Smith R., Del Norte Co.; to B.C., Mont. July–Aug.

Var. **parvidentàta** Kusnez. Calyx-lobes reduced, two ca. 2 mm. long, the others mere mucronations.—At 4000–6000 ft.; Sagebrush Scrub to Yellow Pine F.; Modoc Co. to Ore.; Ida., etc. July–Sept.

3. **G. oregàna** Engelm. ex Gray. [*Dasystephana o.* Rydb. *G. affinis* var. *ovata* Gray.] OREGON GENTIAN. Perennial with stout taproot; stems erect or somewhat decumbent, simple or few-branched above, 2–4 dm. high; lvs. oblong or ovate, 2–4 cm. long, obtusish to acute; fls. few to several at summit, sessile in axils of 2 lance-oblong to ovate

bracts; calyx-tube 4–5 mm. long, the lobes suberect, lance-oblong to oblong, subequal, 8–12 mm. long, the intracalycine membrane appearing plane and truncate between the calyx-lobes; corolla broadly funnelform, 3–4 cm. long, blue with greenish-white dots within, ± brown-blue without, the lobes round-ovate, 6–10 mm. long, the sinus-plaits mostly with 2 ± toothed lobes; caps. fusiform, 2–2.5 cm. long, on a stipe 12–15 mm. long; seeds winged at one end, light brown, almost 2 mm. long.—Grassy and brushy places, below 3000 ft.; N. Coastal Scrub, Chaparral, Coastal Prairie, Mixed Evergreen F.; Marin Co. to Del Norte and Siskiyou cos.; to B.C., Mont. June–Aug.

4. **G. scéptrum** Griseb. in Hook. [*Pneumonanthe s.* Greene. *G. Menziesii* Griseb. *G. s.* var. *humilis* Engelm.] KING'S GENTIAN. Perennial with short rootstock and thick cord-like roots; stems erect or ascending, solitary or few, mostly simple, 2–7 dm. high; lvs. ovate to lance-ovate, 2–6 cm. long, rather numerous; fls. mostly several, subracemosely arranged near top, some without bracts and on pedicels to 7 cm. long, others closely subtended by ovate bracts; calyx-tube 7–10 mm. long, the lobes lance-oblong, somewhat unequal, 6–12 mm. long, glabrous, the membrane between the lobes truncate; corolla broadly funnelform, 3–4 cm. long, blue, commonly with green dots within, the lobes rounded, 5–6 mm. long, the sinus-plaits entire, subtruncate; caps. fusiform, 1.5–2 cm. long, on a stipe ca. 1 cm. long; seeds winged at tip, light brown, ca. 1.5 mm. long.— Wet meadows and boggy places, below 4600 ft.; N. Coastal Scrub, Closed-cone Pine F., etc. to Yellow Pine F.; near the coast, Sonoma Co. to Del Norte Co.; to B.C. July–Sept.

5. **G. setígera** Gray. [*Pneumonanthe s.* Greene. *G. californica* Kusnez.] MENDOCINO GENTIAN. Perennial with root-crown and thick cordlike roots; stems several, erect or ascending, 2–3 dm. long; lvs. ovate to roundish, 2.5–6 cm. long, obtusish; fls. 1–few, subtended by elliptic to ovate bracts; calyx-tube 8–10 mm. long, the lobes oblong-oblanceolate, 5–8 mm. long; corolla blue, 3–4 cm. long, narrow-campanulate, the lobes 8–12 mm. long, erosulate, apiculate, the plait-membrane with several capillary bristles; caps. ellipsoid, almost 2 cm. long, on a stipe ca. 1.5 mm. long; seeds light brown, winged all way around, ca. 2 mm. long.—Wet places, 4000–6500 ft.; Yellow Pine F., Red Fir F.; Red Mt., Mendocino Co., to w. Siskiyou Co., sw. Ore. July–Sept.

6. **G. calycòsa** Griseb. in Hook. [*Pneumonanthe c.* Greene.] EXPLORER'S GENTIAN. Root-crown stout, bearing several simple erect or ascending stems 1.5–3 dm. high; lvs. ovate to almost round, 2–4 cm. long, obtusish to acute; fls. mostly 1–3, subtended by lance-ovate bracts; calyx-tube 6–9 mm. long, the lobes ovate, with contracted base, unequal, 3–8 mm. long, with scaberulous margin, the membrane subentire between the lobes; corolla deep blue, broadly funnelform-campanulate, 2.5–3.5 cm. long, the lobes round-ovate to almost lance-ovate, acute to rounded at apex, 7–10 mm. long, with yellowish dots, the plait-membrane with acute to acuminate lobes; caps. narrow, ca. 1.5–2 cm. long, on a long stipe; seeds brown, ca. 2 mm. long, the wings narrow and at both ends.—Moist places such as meadows and stream banks, at 4000 ft. in n. Calif. to 10,500 in Sierra Nevada; Red Fir F. to Subalpine F.; Tulare Co. to Siskiyou Co.; to B.C., Mont. July–Sept.

7. **G. Fremóntii** Torr. [*Chondrophylla F.* A. Nels. *G. viridula* Parish. *G. humilis* auth., not Salisb.] MOSS GENTIAN. Annual or biennial, mostly with several simple stems from base, 3–10 cm. high; lvs. oblong, or the lower broader, scarious-margined, 4–6 mm. long, erect; fls. solitary, terminal, subtended by bracts; calyx 6–7 mm. long, narrow-funnelform, the lobes like the upper lvs.; corolla tubular-funnelform, 7–8 mm. long, the lobes greenish-blue with whitish margins, the sinus-plaits rounded, minutely toothed; caps. long-exserted, ca. 5 mm. long, 3–4 mm. wide, with 2 spreading valves; seeds angled-ellipsoid, brown, ca. 1 mm. long.—Boggy meadows, 7800–9100 ft.; Red Fir F.; S. Fork of Santa Ana R., San Bernardino Mts.; Rocky Mts. July–Aug.

8. **G. prostràta** Haenke. [*G. p.* var. *americana* Engelm. *Chondrophylla a.* A. Nels.] PIGMY GENTIAN. Much like *G. Fremontii;* lvs. 2–3 mm. long, ovate to roundish, white-margined; fls. solitary, terminal; calyx 8–10 mm. long; corolla blue, 12–15 mm. long, the lobes lance-ovate, 3–4 mm. long, the sinus-plaits ovate, notched or entire; caps. linear-oblong, 8–10 mm. long, ca. 1.5 mm. wide, usually included in the corolla; seeds ca. 1 mm. long, not winged.—Meadows, at almost 12,000 ft., White Mts.; to Rocky Mts., thence to Alaska. July–Aug.

9. **G. Amarélla** L. [*G. acuta* Michx. *G. A.* var. *a.* Herder. *Amarella californica,*

Copelandii, and *Lembertii* Greene. *Gentianella Amarella* ssp. *acuta* J. Gillett.] FELWORT. Annual, mostly branched, 0.5–5 dm. tall, slender, erect; lvs. lanceolate to oblong or the lower spatulate-obovate, acute to obtuse, 1.5–3.5 cm. long; fls. usually numerous, clustered in axils and terminal, on slender ascending branchlets, the pedicels slender, 5–50 mm. long; calyx-tube 1–4 mm. long, the lobes unequal, 2–10 mm. long, lanceolate; corolla blue to bluish-lavender, tubular-campanulate, 8–20 mm. high, the lobes lanceolate, acute, each with a fimbriate crown at base; caps. sessile, fusiform-cylindric; seeds round-ovoid, smoothish, ca. 0.6 mm. long; $2n = 36$ (D. Löve, 1953).—Moist places, 4500–11,000 ft.; Yellow Pine F. to Subalpine F.; San Bernardino Mts., Sierra Nevada to Siskiyou Co.; to L. Calif., Mex., Alaska, Atlantic Coast; Eurasia. June–Sept. Variable.

10. **G. tenélla** Rottb. [*G. monantha* A. Nels. *Gentianella t.* J. Gillett.] Slender-rooted annual, usually much-branched at base, the stems slender, simple to branched, forming tufts 3–8 cm. high; lvs. oblong to oblance-spatulate, 8–15 mm. long, the cauline somewhat reduced; fls. solitary, terminal on slender elongate pedicels, mostly 4-merous; calyx deeply cleft, 5–9 mm. long, the segms. lance-elliptic; corolla tubular-campanulate, whitish or greenish to bluish, 8–12 mm. long, the lobes lanceolate, acute, with a laciniate crown at base; caps. cylindric; seeds rounded-ovoid, ca. 0.5 mm. long, $n = 5$ (Favarger, 1949).—Uncommon, meadows, 8000–12,200 ft.; Subalpine F., Bristle-cone Pine F., Alpine Fell-fields; Sierra Nevada (Whitney Meadows, Rock Creek Lake Basin), White Mts.; Rocky Mts. to Alaska; Eurasia. July–Aug.

11. **G. símplex** Gray. [*Anthopogon s.* Rydb. *Gentianella s.* J. Gillett.] HIKERS' GENTIAN. Simple, erect, 0.5–2 dm. high; lvs. mostly 3–6 pairs, the lowest clasping, the upper sessile, linear-oblong to lance-oblong, 0.6–2.5 cm. long; stem ending in a single fl. without subtending bracts; calyx 4-lobed to ca. the middle, 1.5–2 cm. high, the lobes lanceolate, acute; corolla blue, 2.5–4 cm. long, the usually 4 lobes ca. as long as tube, rounded, irregularly erose; caps. cylindro-fusiform, ca. 1.5 cm. long, on an equally long stipe; seeds subcylindric, brown, ca. 1 mm. long, narrow-winged at each end.—Meadws, 4000–9500 ft.; Red Fir F. to Subalpine F.; San Bernardino Mts., Sierra Nevada from Tulare Co. n., to Siskiyou and Humboldt cos.; to Ore., Ida., Nev. July–Sept.

12. **G. holopétala** (Gray) Holm. [*G. serrata* var. *h.* Gray. *Gentianella detonsa* ssp. *h.* J. Gillett.] SIERRA GENTIAN. Annual 0.5–4 dm. high, usually branched at base, leafy below and terminating in long naked 1-fld. peduncles; lower lvs. spatulate-obovate, crowded, 1–4 cm. long, those on lower stems linear to somewhat oblong; bracts lacking; calyx mostly 1–2.5 cm. long, with 4 lobes mostly longer than the tube, acuminate, dark-ribbed; corolla blue, funnelform, 3–5 cm. long, the lobes 4, oblong, obtuse, ca. half as long as tube, entire or erose; caps. fusiform, 9–12 mm. long, on a stipe ca. as long; seeds honeycombed, brown, oblong-ovoid, 0.7–1 mm. long.—Wet meadows, 6000–11,000 ft.; Red Fir F. to Subalpine F.; San Bernardino Mts., Sierra Nevada from Tulare Co. to Tuolumne Co.; w. Nev. July–Sept.

5. Swértia L.

Caulescent simple-stemmed perennials. Lvs. opposite or some alternate. Fls. perfect, 5-merous in ours, blue or rarely white, in a thyrsoid infl. Calyx deeply lobed. Corolla rotate, deeply parted, each lobe convolute in the bud and with a pair of nectariferous pits. Stamens inserted on the base of the corolla. Style very short or none; stigma 2-lobed. Ovary ovoid, sessile, 1-celled. Seeds large, flat, commonly margined. Perhaps 50 spp., of N. Am., Eurasia, Afr. (Named for E. *Sweert,* Dutch botanist of 16th century.)

1. **S. perénnis** L. [*S. obtusa* Ledeb. *S. p.* var. *o.* Griseb. *S. Covillei* Greene.] FELWORT. Glabrous perennial with a short rootstock and a single erect simple stem 1–3 dm. high; lvs. mostly basal, obovate to elliptic, the lower 4–12 cm. long with blades ca. equal to the broad petioles, stem-lvs. few, smaller, alternate, or the upper opposite and sessile; panicle or raceme narrow, terminal, elongate, the lowest pedicels mostly 1–4 cm. long; fls. 5(4)-merous; calyx-lobes lanceolate, 4–5 mm. long; corolla-lobes oblong-ovate, greenish-white or with bluish-purple tinge, obtuse, 8–10 mm. long; glands 2 on each lobe, round-elliptic, fringed all around; caps. ellipsoid, flattened, ca. 1 cm. long; seeds compressed, roundish, brown, winged ca. ¾ way round, ca. 1 mm. wide; $n = 12$ (Woycicki, 1937), $2n = 28$ (Favarger, 1952).—Meadows and damp places, 8500–

10,500 ft.; Subalpine F.; Sierra Nevada from Tulare Co. to Mariposa Co.; to Ore., Rocky Mts., Alaska; Eurasia. July–Sept. Variable, the var. *obtùsa* being based on obtuse rather than acute corolla-lobes.

6. Fràsera Walt. GREEN GENTIAN

Biennial or perennial herbs with erect stems from bitter taproots. Lvs. opposite or whorled, entire, thickish. Fls. perfect, in a terminal panicle, 4-merous. Calyx 4-parted, the lobes deeply cleft, subulate, acute. Corolla rotate, parted nearly to base, the lobes convolute in bud, and bearing 1–2 ± fringed glands. Stamens inserted at base of corolla. Ovary 1-celled, gradually attenuate into a filiform style; stigma 2-cleft; placentae parietal, many-ovuled. Caps. ovoid, compressed, 4–20-seeded. Ca. 15 spp., of N. Am. (Named for J. *Fraser*, an English collector.)

(St. John, H. Revision of the genus Swertia of the Americas and the reduction of Frasera. Am. Midl. Nat. 26: 1–29, 1941.)
A.   Glands 2 on each corolla-lobe; lvs. not white-margined .... 1. *F. speciosa*
AA. Glands 1 on each corolla-lobe; lvs. with ± conspicuous white margins.
   B.   Stem-lvs. opposite (except sometimes in *S. puberulenta*).
     C.   Gland lobed at apex or U-shaped; fls. in an open broad panicle.
       D.   Herbage glabrous; plant 6–12 dm. high; corolla-lobes 10–15 mm. long. From
          Los Angeles Co. s. and e. .... 2. *F. Parryi*
       DD. Herbage puberulent; plant 1–3 dm. high; corolla-lobes 6–8 mm. long. Mono and
          Inyo cos. .... 6. *F. puberulenta*
     CC. Gland entire at apex; fls. in a narrow spikelike often interrupted infl.
       D.   Herbage puberulent; caps. 8–12 mm. long including beak. Extreme n. Calif.
                                              7. *F. albicaulis*
       DD. Herbage glabrous.
         E.   Caps. 6–7 mm. long including beak; corolla-lobes 8–10 mm. long; gland
            quadrate. Largely w. end of Mojave Desert .... 5. *F. neglecta*
         EE. Caps. 12–17 mm. long; corolla-lobes 5–8 mm. long; gland oblong. N. Coast
            Ranges and Sierra Nevada .... 7. *F. albicaulis* var. *nitida*
   BB. Stem-lvs. whorled.
     C.   Cauline lvs. lance-linear to narrow-oblanceolate.
       D.   The whorls of 3–4 lvs.; crown at base of corolla lacking. E. Mojave Desert
                                          3. *F. albomarginata*
       DD. The whorls of 5–9 lvs.; crown tubular, deeply 2-lobed. Kern and Tulare cos.
                                          4. *F. tubulosa*
     CC. Cauline lvs. ovate to elliptic. Nw. Calif. .... 8. *F. umpquaensis*

1. **F. speciòsa** Dougl. ex Griseb. in Hook. [*Tessaranthium radiatum* Kell. *Swertia r.* Kuntze.] Stem very stout, from a large taproot, erect, 1–2 m. high; lvs. lance-oblong or the lower oblanceolate to obovate, puberulent, acute, in whorls of 3–7, 1–2.5 dm. long, the lower short-petioled; fls. 4-merous, in a narrow panicle 3–6 dm. long; pedicels stout, 3–8 cm. long; calyx-lobes lance-linear, 1.5–2 cm. long; corolla-lobes ca. as long as calyx, ovate, acute, greenish-white, dotted with purple; glands 2, narrow-oblong, with long fringe all around; scales of crown deeply laciniate; caps. 16–18 mm. long; seeds flat, brown, oblong-elliptic, narrowly winged, ca. 3 mm. long.—Dryish or dampish places, 6800–9800 ft., Yellow Pine F. to Subalpine F., Sierra Nevada from Fresno Co. n., to Modoc Co.; 5000–8000 ft., Coast Ranges from Lake Co. n.; to Wash., Rocky Mts. A form with main lvs. subglabrous has been called *F. macrophylla* Greene and occurs with the usual form. July–Aug.

2. **F. Párryi** Torr. [*Swertia P.* Kuntze.] Stems stout, usually 1, from a heavy taproot, 6–12 dm. high; plant glabrous; lvs. white-margined, the basal oblanceolate, 1–2 dm. long, acute, narrowed at base into a short winged petiole; cauline lvs. lanceolate, sessile, opposite, 4–10 cm. long, the uppermost lance-ovate; panicle broad, 1.5–3 dm. long; pedicels 1–2.5 cm. long, rather slender; calyx-lobes lanceolate, subacuminate,' largely 10–15 mm. long; corolla-lobes ca. as long, greenish-white with black dots, the gland U-shaped, fringed all around; crown-scales wanting; caps. long-conic, 14–16 mm. long; seeds brown, somewhat rhomboid and angled, ca. 3.5–4 mm. long, covered with a spongy pitted layer.—Frequent in rather dry places, 1500–6000 ft.; Chaparral, Coastal Sage Scrub, S. Oak Wd., Yellow Pine F.; cismontane s. Calif. from San Gabriel and San Bernardino mts. to L. Calif., Ariz. April–July.

3. **F. albomarginàta** Wats. [*Swertia a.* Kuntze.] Taproot rather woody; stems 1–few, much-branched, 2–5 dm. high, glabrous; lvs. white-margined, pale, the lower

narrowly oblanceolate, crisped, petioled, 4–8 cm. long; cauline lvs. in whorls of 3–4, lance-linear, largely conduplicate, sessile, 2–6 cm. long; panicle rather broad, diffuse, subcorymbose; pedicels slender, 1–3 cm. long; calyx-lobes lance-linear, acuminate, 5–6 mm. long; corolla-lobes oblong-obovate, acuminate, 8–10 mm. long, greenish-white with dark dots, the gland linear, slightly 2-lobed above, fringed at margin; crown lacking; caps. 10–15 mm. long, flattened-conic, attenuate; seeds brown, subrhomboid, ca. 4 mm. long, spongy-pitted.—Dry rocky and gravelly places, 4500–7000 ft.; Pinyon-Juniper Wd.; New York and Providence mts., e. Mojave Desert; to Colo., Ariz. A form with glandular-puberulent stems is the var. *indùta* (Tides.) Card. [*F. i.* Tides.] Clark Mt., e. Mojave Desert; s. Nev. May–July.

4. **F. tubulòsa** Cov. [*Swertia t.* Jeps.] Stems mostly solitary from a somewhat woody taproot, 2–7 dm. high, mostly glabrous and glaucous, erect, simple; lvs. white-margined, mostly near base of plant, spatulate, 4–8 cm. long, narrow-oblanceolate, with petiole-like base and conduplicate blade; stem-lvs. in whorls of 5–9, reduced, with white margin strongly ruffled; infl. a narrow spicate panicle, dense except for the interrupted base which may extend to near bottom of plant; pedicels 3–20 mm. long; calyx-lobes linear, 6–9 mm. long, the white margin ruffled and widened at base; corolla-lobes white with bluish veins, oblong-obovate, acuminate, 8–10 mm. long, the gland lacking; crown tubular, 2-lobed, laciniate; caps. flattened, elliptic, 12–15 mm. long, including the beaklike tip; seeds 1–2, flattened, brown, irregularly oblong, cellular-muriculate, 6–7 mm. long, 2.5 mm. wide.—Dry granitic and volcanic gravels and slopes, mostly 6000–9000 ft.; Red Fir F., Lodgepole F.; Piute Mts., Kern Co. to Sierra Nevada, Tulare Co. July–Aug.

5. **F. neglécta** Hall. [*Swertia n.* Jeps.] Root-crown branched; stems several, slender, erect, 2–4 dm. high; herbage glabrous, pale; lower lvs. linear to oblance-linear, 6–15 cm. long, inconspicuously white-margined, the cauline opposite, linear, sessile; panicle interrupted, the fls. in dense whorllike clusters; pedicels 5–20 mm. long; calyx-lobes linear-lanceolate, white-margined, 5–7 mm. long; corolla-lobes oblong-obovate, 8–10 mm. long, acute, greenish-white with purple veins, the gland quadrate, fringed; crown much reduced, fimbriate; caps. ovoid, beaked, 6–7 mm. long; seeds ca. 4, rounded on back, with 2 concave surfaces separated by a flat-topped ridge in front, brown, ca. 5 mm. long, cellular-punctate.—Dry slopes, 4500–8000 ft.; largely Yellow Pine F.; desert slopes, San Bernardino, San Gabriel, and San Emigdio mts., Mt. Pinos. May–July.

6. **F. puberulénta** A. Davids. [*Swertia p.* Jeps.] Taproot somewhat woody; stems 1–2, stoutish, 1–3 dm. high, puberulent; lvs. with narrow white margin, conduplicate, oblong-lanceolate, mostly 5–12 cm. long, the basal sometimes oblanceolate, petioled, the cauline opposite, sessile; panicle broad, open, comprising the upper half or more of the plant; pedicels slender, 5–25 mm. long; calyx-lobes lanceolate, 6–8 mm. long; corolla-lobes obovate, short-acuminate, greenish-white with purple dots, ca. as long as calyx; gland oblong, almost covered by the pocket, the opening at top fringed; crown obsolete; caps. round-ovoid, ca. 1 cm. long including the 4 mm. beak.—Dry slopes, 8000–11,000 ft.; largely Red Fir F., Lodgepole F.; e. slope of Sierra Nevada from Inyo and Mono cos. and in White and Inyo mts. June–Aug.

7. **F. albicaùlis** (Griseb. in Hook.) Kuntze. [*Swertia a.* Kuntze. *S. californica, Bethelii, modocensis, shastaensis,* and *sierrae* St. John. *S. modocensis* var. *adglabra* St. John.] Root-crown woody; stems 1–several, erect, 2.5–4.5 dm. high, puberulent; lvs. puberulent, the basal spatulate-oblanceolate, 5–16 cm. long, with narrow white margin, the base narrowed into a winged petiole; cauline lvs. opposite, sessile, mostly 2–3 pairs, the uppermost leafy bracts; panicle narrow, interrupted, 3–16 cm. long, of dense whorllike cymes, these sessile or the lower peduncled; pedicels slender, 4–8 mm. long; calyx-lobes lanceolate, 5–7 mm. long; corolla-lobes lance-ovate, 6–10 mm. long, greenish-white to bluish, acuminate, the gland greenish, oblong to linear or elliptic, fringed all around or not at upper end; scales of crown oblong to obdeltoid, entire to laciniate; caps. ellipsoid-ovate, flattened, beaked, 8–12 mm. long; seeds brown, 4–5 mm. long, flat, cellular-pitted.—Dry to moist places, 4000–6000 ft.; Sagebrush Scrub, N. Juniper Wd., Yellow Pine F.; Lassen, Shasta, Siskiyou, and Modoc cos.; to Wash., Mont. May–July. Exceedingly variable in gland and crown even in fls. on same plant.

Ssp. **nítida** (Benth.) Post. [*F. n.* Benth. *Swertia n.* Jeps. *S. Eastwoodae* and *lassenica* St. John.] Herbage glabrous; pedicels 4–22 mm. long; calyx-lobes linear-lanceolate, 4–8

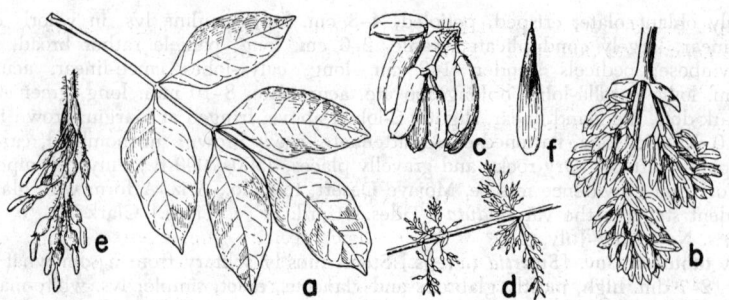

Fig. 50. OLEACEAE. *Fraxinus: a,* lf., × ¼; *b,* ♂ fls., × 1, *c,* × 2½, with abortive calyx and 2-3 stamens; *d* and *e,* clusters of ♀ fls. with calyx and pistil; *f,* samara, × ½.

mm. long; corolla-lobes 5–8 mm. long, elliptic-ovate, acute, whitish, with violet tinge and often violet dots, the gland oblong, greenish, fringed on margin or naked toward tip; crown-scales oblanceolate to oblong or rounded, erose to laciniate, sometimes entire; caps. beaked, flattened, ca. 12–17 mm. long including the 4–6 mm. long beak; seeds 2, pale brown, narrow-oblong, 6–7 mm. long, with a groove along one edge.—Dry slopes, 500–7000 ft.; Yellow Pine F., Foothill Wd., Chaparral; N. Coast Ranges from Lake Co. to Siskiyou Co., Sierra Nevada from Nevada Co. n., to Modoc Co.; s. Ore. May–July.

8. **F. umpquaénsis** Peck & Appleg. Glabrous; stem stout, 6–10 dm. high; basal lvs. oblanceolate to subspatulate, 1.5–2 dm. long, petioled; cauline lvs. in whorls of 3–5, oblong-ovate to elliptic, gradually reduced up the stem; panicle dense, 1–2 dm. long; pedicels erect, 3–10 mm. long; calyx deeply cleft, the segms. linear to oblong, 10–12 mm. long; corolla scarcely as long, pale yellow-green or slightly bluish, each segm. with a roundish impressed glandular pit bordered by a fringed broad membrane.—Horse Ridge, Trinity Co., *Post;* sw. Ore.

7. Menyánthes L. BUCKBEAN

Perennial herbs with creeping rootstocks. Lvs. basal, long-petioled, with 3-foliolate blades. Fls. in racemes or panicles on long lateral scapose peduncles. Calyx 5-parted. Corolla short-funnelform, 5-cleft, the lobes fringed or bearded on inner surface. Stamens 5, the fils. filiform; anthers sagittate. Disk of 5 hypogynous glands. Ovary 1-celled; style persistent; stigma 2-lobed. Caps. ovoid, indehiscent or irregularly rupturing. Seeds few, compressed-globose, shining. One sp. of cooler parts of N. Hemis. (Greek, *men,* month, and *anthos,* fl., referring to length of blooming period.)

1. **M. trifoliàta** L. Rootstock covered by sheathing petiole-bases of old lvs.; plants glabrous; lfts. oblong to obovate, 2–8 cm. long, acute, entire, narrowed to sessile base; petioles 5–20 cm. long; racemes 2–4 dm. high on long naked peduncles; pedicels 0.5–2.5 cm. long; calyx-lobes lance-ovate, 3–5 mm. long; corolla white or pink, the tube slightly exceeding calyx, the lobes ca. 5 mm. long and covered on inner surface with long beard of white hairs; caps. 6–8 mm. long without the filiform style which is ca. as much longer; seeds light brown, shining, ca. 2.5 mm. in diam.—Occasional, bogs and margins of shallow lakes, 3000–10,500 ft.; Yellow Pine F. to Subalpine F.; Sierra Nevada from Tulare Co. n., to Siskiyou Co., then to Alaska; Eurasia. May–Aug.

53. Oleàceae. OLIVE FAMILY (Fig. 50)

Trees, shrubs, or more rarely herbs. Lvs. usually opposite, simple to pinnate, deciduous in ours, exstipulate. Fls. regular, perfect or more commonly unisexual, small, in compact clusters. Calyx free from the ovary, small, usually 4-lobed, sometimes wanting. Corolla sympetalous or polypetalous or none, 2–4-merous when present. Stamens 2, rarely 3 or 4, hypogynous or inserted on the corolla-tube. Ovary superior, 2-locular, with few ovules in each locule; style 1 or the stigma sessile. Fr. a caps., samara, berry, or drupe. Seeds erect or pendulous; endosperm present or lacking; embryo straight. Ca. 25 genera and 300 spp., widely distributed in warmer regions.

Fruit a samara; lvs. largely pinnately compound .... 1. *Fraxinus*
Fruit not a samara; lvs. simple.
    Fruit a drupe; lvs. opposite; large shrubs .... 2. *Forestiera*
    Fruit a capsule; lvs. mostly alternate; small shrubs .... 3. *Menodora*

1. Fráxinus L. Ash

Trees or arborescent shrubs with deciduous usually pinnately compound opposite lvs. Largely dioecious or polygamous, the fls. small, in crowded panicles which appear mostly before the lvs. Calyx 4-cleft or -toothed. Petals in ours 2 or 0. Stamens usually 2, rarely 3 or 4. Fr. a 1-seeded samara, with terminal wing; seed usually 1, oblong. Ca. 40 spp., of N. Hemis. (Latin name of the ash.)

(Miller, Gertrude N. Fraxinus in N. Am. Cornell Univ. Agric. Exp. Sta. Mem. 335, 1955.)
Corolla present, consisting of 2 white petals; fls. bisexual; style obscurely lobed; samara flat, broadly wing-margined to base .... 1. *F. dipetala*
Corolla absent; style conspicuously 2-lobed.
    Lfts. 1 (3); branchlets of the season 4-sided; body of the samara flattened and broadly wing-margined to base .... 2. *F. anomala*
    Lfts. 5–7; branchlets terete; body of samara subterete and narrowly winged along the side.
        Lateral lfts. sessile or subsessile; fr. 3–5 cm. long, the wing usually decurrent downward on each side of the body for ½–¾ the way. Tulare and Santa Clara cos. and n. .... 3. *F. latifolia*
        Lateral lfts. on petiolules 3–12 mm. long; fr. 1–3(–3.5) cm. long, the wing usually decurrent scarcely to middle of the body. Los Angeles and Inyo cos. s. .... 4. *F. velutina*

1. **F. dipétala** H. & A. [*Petlomelia d.* Nieuwl. *F. d.* var. *brachyptera* Gray. *Chionanthus fraxinifolius* Kell.] Flowering Ash. Arborescent shrub or small tree 2–7 m. high, erect, glabrous or the young parts somewhat pubescent; branchlets slender, the youngest somewhat 4-angled; lvs. 4–12 cm. long, the lfts. 3–7(–9), thin, ovate to obovate, serrate, 2–4 cm. long, on short petiolules; panicles 3–12 cm. long, many-fld.; the fls. perfect or polygamous; calyx ca. 1.5 mm. long; petals 2, white, 5 mm. long; samaras 2–3 cm. long, 7–9 mm. wide, the body flattened, winged along the sides, often retuse at apex.—Dry slopes, mostly below 3500 ft.; Chaparral, Foothill Wd.; Coast Ranges from Siskiyou Co. s., Sierra Nevada foothills from Shasta Co. s., cismontane s. Calif. to Orange Co., rare in San Diego Co. April–May.

2. **F. anómala** Torr. ex Wats. Shrub or small tree 2–5 m. high, bushy; branchlets glabrous or somewhat pubescent, 4-sided; lvs. mostly simple, round-ovate, entire or crenulate, 2–5 cm. long, sometimes trifoliolate; petioles slender, ca. as long as blades; fls. perfect or ♀; calyx 1.5 mm. long; petals 0; samaras 1.5–2.5 cm. long, 6–8 mm. wide, the wing decurrent down the sides almost to base.—Dry canyons and gulches, 3000–11,000 ft.; mostly Pinyon-Juniper Wd.; e. Mojave Desert (Panamint, New York, Providence, and Clark mts.); to Colo., Tex. April–May.

3. **F. latifòlia** Benth. [*F. oregona* Nutt. *F. americana* ssp. *o.* Wesmael. *F. o.* var. *l.* Lingelsh. *F. pennsylvanica* spp. *o.* G. N. Miller.] Oregon Ash. Tree 10–25 m. high; branchlets usually stout, ± pubescent; lfts. 5–7, tomentose to glabrate beneath, glabrous or somewhat pubescent above, oblong to oval, the terminal 6–10 cm. long, the lateral smaller, sessile or nearly so; dioecious; calyx ca. 1 mm. long; petals 0; stamens 2; samaras mostly 3–5 cm. long, 5–9 mm. wide, the wing decurrent on the terete body for half or more its length; $2n = 46$ (Sax & Abbe, 1932).—In canyons and near streams, below 5500 ft.; many Plant Communities; w. base of Sierra Nevada from n. Kern Co. to Modoc Co., Coast Ranges from Santa Clara Co. n.; to B.C. March–May.

4. **F. velùtina** Torr. var. **coriàcea** (Wats.) Rehd. [*F. c.* Wats. *F. oregona* var. *glabra* Lingelsh.] Tree 5–10 m. tall; branchlets terete, usually glabrous; lfts. 3–7, lanceolate to ovate or obovate, 2–8 cm. long, thickish, glabrous or pubescent (especially beneath), the lateral commonly on petiolules 3–10 mm. long; dioecious; calyx ca. 1.5 mm. long; petals 0; samaras 1.5–2.5(–3.5) cm. long, 4–7 mm. wide, the wing usually decurrent barely to middle of the body.—Canyons and along streams, below 5000 ft.; Yellow Pine F., Chaparral, S. Oak Wd.; n. Los Angeles Co. to San Diego Co. and on the desert n. to Owens Lake, Inyo Co.; Nev., L. Calif. March–April.

2. Forestiéra POIR.°

Deciduous shrubs with simple opposite lvs. Fls. small, inconspicuous, polygamo-dioecious, appearing before the lvs. Calyx minute, unequally 5–6-cleft, sometimes lacking. Corolla usually wanting, rarely 2–3-petaled. Stamens 2 or 4. Ovary 2-celled, with 2 ovules in each cell; style slender. Fr. a drupe, round to ovoid, usually 1-seeded, glabrous or pubescent. Ca. 15 spp., of N. and S. Am. (Named for M. *Forestier,* French physician.)

1. F. neomexicàna Gray. [*Adelia n.* Kuntze. *A. parvifolia* Cov.] DESERT OLIVE. Glabrous shrub 1.5–3 m. high, with smooth gray bark and spiny branchlets; lvs. spatulate-oblong, obtuse to acuminate, entire or serrulate, 1–5 cm. long, often fascicled; fls. in sessile fascicles; ♂ fls. sessile, ♀ on slender pedicels; drupe blue-black, ellipsoid, 5–7 mm. long; seed longitudinally ridged; $2n = 46$ (Taylor, 1945).—Occasional on dry slopes and ridges, below 6700 ft.; Creosote Bush Scrub, Chaparral, Coastal Sage Scrub, Foothill Wd., etc.; inner Coast Ranges from Contra Costa Co. s. to interior cismontane s. Calif., and on Mojave Desert and bordering ranges, Inyo Co. to Santa Rosa Mts., Riverside Co.; to Colo., Tex. March–April.

3. Menodòra Humb. & Bonpl.

Low subshrubs or suffruticose herbs. Lvs. alternate or the lower opposite, simple, sessile or subsessile. Fls. perfect, solitary and terminal or dichotomously corymbose or paniculate. Calyx deeply 5–15-lobed. Corolla subrotate to salverform, 5–6-lobed. Stamens 2, inserted near base of corolla or in throat; fils. long or short. Ovary superior, bilocular; style slender; stigma depressed-capitate. Ovules 2 or 4 in each locule. Fr. a caps., membranaceous, indehiscent or circumscissile. Seeds 4 (2) in each cell, attached laterally, the outer coat spongy and reticulate; endosperm absent. Ca. 17 spp., of N. and S. Am. and Afr. (Greek, *menos,* force, and *doron,* gift.)

(Steyermark, J. A. Revision of the Genus Menodora. Ann. Mo. Bot. Gard. 19: 87–176, 1932.)
Plants unarmed; corolla-lobes longer than the tube, yellow.
  Lvs. all foliose with well-developed blades; stems 0.5–3 dm. high .... 1. *M. scabra*
  Lvs., at least the upper, reduced to bracts; stems 3–8 dm. high .... 2. *M. scoparia*
Plants spinose; corolla-lobes shorter than the tube, white .... 3. *M. spinescens*

1. M. scàbra Gray. [*M. s.* var. *laevis* Steyerm. *M. l.* Woot. & Standl.] Stems erect, almost herbaceous, 0.5–3 dm. high, ± scabrous; lvs. 5–15 mm. long, 2–5 mm. wide, mucronate, glabrous or sparsely scaberulous, the lower ovate to oblong, the upper lanceolate; calyx glabrous to slightly scabrous, the lobes 7–11, linear, 4–5 mm. long; corolla bright yellow, subrotate, the lobes ovate, 7–8 mm. long, the tube ca. 4 mm. long; caps. 5–7 mm. high, 8–12 mm. broad, thin-walled; seeds 4 in each locule, 4–5 mm. long, ca. 3 mm. broad, flat, obovate, greenish or brownish with a yellowish narrow wing; $2n = 44$ (Taylor, 1945).—Dry rocky places, 3500–5500 ft.; Joshua Tree Wd.; Eagle Mts., New York Mts., Lanfair V., e. Mojave Desert; to Tex., Mex. May.

2. M. scopària Engelm. ex Gray. Paniculately branched erect suffruticose perennial 3–8 dm. high, with many slender mostly glabrous stems; lvs. rather few, oblong-obovate to oblanceolate, the upper remote and rudimentary, the lower 1–2.5 cm. long; fls. few, cymose; calyx-lobes 5–7, subulate, 3–5 mm. long; corolla yellow, subrotate, 10–12 mm. long, the lobes ovate, ca. 7 mm. long; caps. 4–6 mm. long; seeds 4 in each locule, brown, angled on ventral side, rounded on back, 4–5 mm. long, ca. 3 mm. broad; $2n = 22$ (Bowden, 1945).—Dry slopes, 2000–6000 ft.; Joshua Tree Wd., Pinyon-Juniper Wd., Shadscale Scrub; w. edge of Colo. Desert, e. Mojave Desert; to Ariz., Mex. May–July.

3. M. spinéscens Gray. [*M. s.* var. *mohavensis* Steyerm.] Low shrub 1.5–8 dm. high, with irregular divergent branches with short stout spiny puberulent branchlets; lvs. 4–12 mm. long, linear to spatulate-oblong, entire, fleshy, appressed-puberulent, alternate; fls. solitary or clustered, short-pediceled; calyx-lobes 5–7, linear-subulate, ca. 4 mm. long; corolla white, tinged brownish-purple without, 9–15 mm. long, funnelform, the lobes oblong-ovate, spreading, 3.5–4.5 mm. long; caps. 5–7 mm. long, 8–9 mm. broad; seeds 2 in each cell, 5–6 mm. long, 3–4 mm. broad, somewhat rounded, pitted, dark brown.—

° The true Olive (*Olea europaea* L.), with silvery-scaly foliage, is reported as establishing itself in Napa Co.; native of Medit. region.

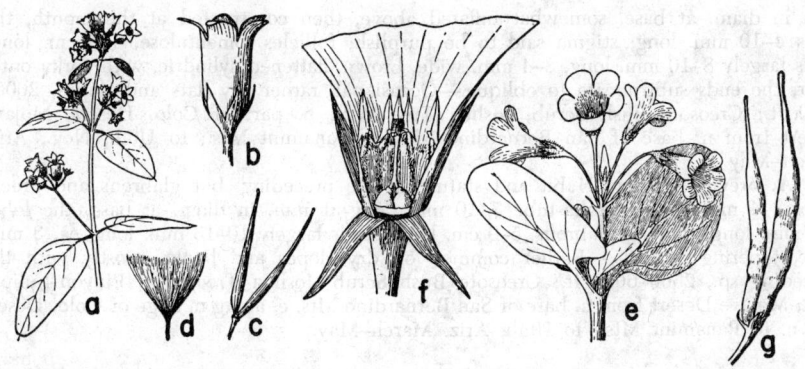

Fig. 51. APOCYNACEAE. *Apocynum: a*, infl., × ½; *b*, fl., × 2; *c.* follicles, × ½; *d*, comose seed, × 1. *Cycladenia: e*, infl. and upper lvs., × ½; *f*, section of fl., × 2, 5-parted calyx, corolla, 5 stamens, pistil with stigma subtended by a globose thickening; *g*, follicles, × ½.

Dry mesas and slopes, mostly 3500–6500 ft.; Shadscale Scrub, Joshua Tree Wd.; e. Mojave Desert to Owens V.; Nev., Ariz. April–May.

54. Apocynàceae. Dogbane Family (Fig. 51)

Perennial herbs, shrubs or vines, some trop. forms arboreal. Juice milky. Lvs. entire, chiefly opposite, sometimes verticillate or alternate, estipulate. Fls. regular, bisexual, solitary and axillary, or cymose to panicled. Calyx free from ovary, persistent, 5-parted, imbricated in bud. Corolla 5-lobed, convolute and often twisted in the bud. Stamens 5, alternate with corolla-lobes, inserted on the tube or throat; anthers often produced at base into a sterile appendage, connivent around the stigma. Carpels 2, distinct, or united and making a 1-celled ovary; styles simple or divided; stigma simple. Fr. of 2 follicles or drupaceous. Seeds often bearing a coma; endosperm fleshy. Ca. 150 genera and 1000 spp.; widely distributed, especially in warmer regions where many are very ornamental (Oleander, Frangipani, etc.).

Lvs. alternate; stamens inserted at summit of corolla-tube; seeds without coma .... 1. *Amsonia*
Lvs. opposite.
   Plant creeping or trailing; stamens inserted at summit of corolla-tube; seeds without coma
                                                         2. *Vinca*
   Plant not trailing; stamens inserted at base of corolla-tube; seeds with coma.
      Corolla 1.5–2 cm. long; style filiform with annular membrane .... 3. *Cycladenia*
      Corolla not over 0.6 cm. long; style short, not appendaged .... 4. *Apocynum*

1. Amsònia Walt.

Perennial herbs; lvs. many, alternate. Fls. somewhat bluish, in terminal compound cymes. Calyx small, 5-parted into narrow acuminate segms. Corolla with narrow funnel-form tube and spreading or reflexed lobes, the tube reflexed-hairy within. Stamens inserted on corolla-throat, included; anthers oblong or ovate. Style filiform; stigma in ours subtended by a globose thickening. Follicles slender, torulose, several-seeded. Seeds cylindric or oblong, without coma. Ca. 17 spp., of N. Am. and Japan. (Dr. Charles *Amson*, an 18th-century resident of Virginia.)

(Woodson, R. Monograph of the genus Amsonia. Ann. Mo. Bot. Gard. 15: 379–434, 1928.)
Plants white-tomentose .... 1. *A. tomentosa*
Plants glabrous and green .... 2. *A. brevifolia*

1. **A. tomentòsa** Torr. & Frém. [*A. brevifolia* var. *t.* Jeps.] Stems 2–4 dm. tall, densely tomentose; lvs. ovate or the upper lance-ovate, white-tomentose, 2–5 cm. long, 1–2 cm. wide, obtuse and sessile or nearly so at base, acute to acuminate at apex; infl. dense, the tomentulose pedicels 0.5–3 mm. long; calyx-lobes linear, 3–9 mm. long, tomentulose; corolla glabrous without, pale lead-blue, the tube 5–12 mm. long, ca. 1.5

mm. in diam. at base, somewhat inflated above, then constructed at the mouth, the lobes 4–10 mm. long; stigma said to be purplish; follicles tomentulose, 4–8 cm. long; seeds largely 8–10 mm. long, 3–4 mm. wide, brown, flattened-cylindric, with corky outer layer, the ends subtruncate to oblique.—Occasional, rather dry flats and slopes, 2000–5500 ft.; Creosote Bush Scrub, Joshua Tree Wd.; n. part of Colo. Desert, Mojave Desert from n. base of San Bernardino Mts. to Panamint Mts.; to Utah, Nev., Ariz. March–May.

2. **A. brevifòlia** Gray. Habit and stature of the preceding, but glabrous and green; calyx 3–4 mm. long; corolla-tube 7–10 mm. long, 1 mm. in diam. at base, the lobes 4–6 mm. long; follicles glabrous, 5–9 cm. long; seeds largely 10–13 mm. long, ca. 3 mm. wide, tapering to ends.—Rather common on dry slopes and banks, mostly with the preceding sp., 2500–6000 ft.; Creosote Bush Scrub, Joshua Tree Wd., Pinyon-Juniper Wd.; Mojave Desert from n. base of San Bernardino Mts. e. along n. edge of Colo. Desert and n. to Panamint Mts.; to Utah, Ariz. March–May.

2. Vínca L. PERIWINKLE

Erect or trailing perennial herbs. Lvs. opposite. Fls. solitary in the axil of every other lf. Calyx 5-parted, the lobes essentially equal. Corolla salverform, equally 5-parted, with tube pubescent within. Stamens 5, included, alternate with the lobes. Ovules several in each carpel; style simple, filiform. Follicles terete, rather slender; seeds many, subcompressed, truncate at each end. Ca. 12 spp., of Old World. (Ancient Latin name.)

1. **V. màjor** L. Plant evergreen; sterile stems trailing, to ca. 1 m. long; fertile stems erect, much shorter; lvs. round-ovate, dark green, with cordate base, acute to obtuse at apex, 2–3 cm. long, on petioles 0.5–2 cm. long; pedicels slender, 3–5 cm. long; calyx-lobes ca. 1 cm. long; corolla violet or blue, 2.5–3 cm. long; follicles cylindric, somewhat torulose, ca. 4–5 cm. long; $2n = 92$ (Rutland, 1941).—Occasional escape about old dwellings and becoming natur. especially in somewhat shaded places; native of Eu. March–July.

3. Cycladènia Benth.

Low perennials from fleshy root; stems 1–several; lvs. opposite, in 2–3 pairs. Fls. 2–5, on axillary peduncles. Calyx 5-lobed. Corolla funnelform, 5-lobed and with small appendages alternate with the lobes. Stamens borne near base of tube; anthers sagittate, connivent. Style filiform, with a conspicuous membranous collar; stigma broad, 5-angled. Ovules many. Follicles slightly fleshy, terete, rather stout; seeds many, comose at apex, narrowly urn-shaped, compressed. One polymorphous sp. (Greek, *kuklos*, ring, and *aden*, a gland, referring to the annular disk.)

1. **C. hùmilis** Benth. Plants 1–2 dm. high, glabrous throughout, glaucous; lvs. rather thick, ovate to roundish, 3–7 cm. long, or the lower smaller, obtuse to rounded at apex, ± cordate at base, on petioles 0.5–3 cm. long; cymes ca. as long as or somewhat exceeding the subtending lvs.; pedicels 7–12 mm. long; bracts narrowly lanceolate, 2–5 mm. long; calyx-lobes linear to broadly lanceolate, 5–8 mm. long, acuminate; corolla rose-purple, ca. 15 mm. long, the lobes broadly ovate, spreading, 5–7 mm. long; follicles 4–6 cm. long, erect; seeds oblong-urn-shaped, red-brown, compressed, 5–6 mm. long, with a somewhat tawny coma 3–4 times as long.—Rocky places, 3500–8500 ft.; Yellow Pine F., Red Fir F.; N. Coast Ranges from Lake Co. to Siskiyou Co. thence s. to Butte and Lassen cos. May–July.

Var. **venústa** (Eastw.) Woodson ex Munz. [*C. v.* Eastw.] Calyx and corolla conspicuously pilose.—At 7000–9900 ft.; Red Fir F.; San Gabriel Mts., s. Calif., somewhat lower in Santa Lucia Mts., Monterey Co., and Sheep Hole Spring, Inyo Co.

Var. **tomentòsa** (Gray) Gray. [*C. t.* Gray.] Plants densely tomentose; calyx hirsute.—With the sp. from Plumas Co. to Siskiyou and Tehama cos.

4. Apócynum L. DOGBANE. INDIAN-HEMP.

Perennial herbs with horizontal rootstocks and upright branching stems with tough fibers. Lvs. chiefly opposite. Fls. small, pale, cymose, on short pedicels. Calyx deeply

5-cleft, the lobes equal. Corolla campanulate to cylindrical or urn-shaped, the tube short, with 5 small sagittate appendages at base opposite the lobes, the limb regular. Stamens inserted on very base of corolla; fils. short and broad; anthers connivent, sagittate. Style none or very short; stigma ovoid. Follicles usually 2, slender, terete; seeds many, truncate, comose. About 7 spp., of N. Am. (Greek, *apo*, from, and *kuon*, dog, ancient name of the dogbane of the Old World.)

(Woodson, R., N. Am. Fl. 29: 188–192, 1938.)

Corolla 4–6 mm. long, 2–3 times the length of the calyx; lvs. drooping or spreading.
  Lvs. drooping; corolla at least 3 times as long as calyx.
    Corolla campanulate, the orifice of tube more than twice the width of base; follicles normally
      pendulous . . . . 1. *A. androsaemifolium*
    Corolla cylindric, the orifice of tube ca. as great as the width of base; follicles suberect
                                                                    2. *A. pumilum*
  Lvs. spreading; corolla ca. twice the length of the calyx . . . . 3. *A. medium*
Corolla 2–3 mm. long; lvs. ascending.
  Bracts of the infl. scarious and aristate; lvs. evidently petiolate; follicles 9–15 cm. long
                                                            4. *A. cannabinum*
  Bracts of the infl. semifoliaceous or blade-bearing; lvs. nearly or quite sessile; follicles 4–10 cm. long
                                                            5. *A. sibiricum*

1. **A. androsaèmifolium** L. Stems erect or ascending, 2–4.5 dm. high, glabrous, diffusely branched; lvs. drooping, ovate to oblong, 2–9 cm. long, subglabrous above, paler and ± tomentulose beneath, subsessile or short-petioled; calyx-lobes ovate to ovate-lanceolate, 1.5–3 mm. long, glabrous or minutely pilosulous; corolla campanulate, 4–7 mm. long, the lobes white, usually with pinkish veins; follicles pendulous at maturity, 4–12 cm. long; seeds almost linear, brown, ca. 1 mm. long, with a pale tawny coma 1–1.5 cm. long; $2n = 16$? (Schürhoff et al., 1937).—Dry flats and slopes, 5000–9500 ft.; Yellow Pine F., Red Fir F.; San Bernardino and San Jacinto mts.; Ariz. to B.C., and Atlantic Coast. June–Aug.

Var. **glàbrum** Macoun. Lvs. glabrous.—Occasional, at ca. same elevs., Sierra Nevada; to B.C.

2. **A. pùmilum** (Gray) Greene. [*A. androsaemifolium* var. *p.* Gray. *A. bicolor* McGreg. *A. cardiophyllum, paniculatum, stenolobum, plumbeum, rotundifolium, Austinae, cercidium,* and *luridum* Greene.] Like *A. androsaemifolium,* but mostly less than 3 dm. high; lvs. more oblong-oval, subglabrous, more obtuse; follicles erect at maturity; coma of seeds white.—More abundant, dry slopes and flats; Yellow Pine F., Red Fir F.; 5000–7000 ft., San Bernardino Mts.; 2000–9500 ft., Sierra Nevada, then to Modoc Co.; 2500–5500 ft., Coast Ranges, Santa Cruz Co. to Del Norte and Siskiyou cos.; to Wash., Mont., Wyo. June–Aug.

Var. **rhomboideum** (Greene) Bég. & Bel. [*A. r.* Greene. *A. tomentellum, pulchellum, arcuatum,* and *diversifolium* Greene.] The plant, or at least the lower surface of the lvs., pubescent.—Less frequent, but with the sp., from Palomar Mts., San Diego Co. through the Sierra Nevada and s. in the Coast Ranges to Napa Co.

3. **A. mèdium** Greene var. **floribúndum** (Greene) Woodson. [*A. f.* Greene. *A. viarum* Heller.] Plants glabrous; stems 3–5 dm. tall; lvs. spreading, lance-ovate, subsessile, 6–10 cm. long; calyx-lobes lanceolate to ovate, 1.5–3 mm. long; corolla broadly urn-shaped to campanulate, 4–5 mm. long, the lobes somewhat spreading and slightly recurved; follicles 7–12 cm. long, reflexed; seeds ca. 1.5 mm. long, the coma slightly tawny, ca. 2 cm. long.—Infrequent in dry to damp places, mostly 3500–8500 ft., sometimes below; Yellow Pine F., Red Fir F. to Foothill Wd.; San Bernardino Mts., Sierra Nevada to Shasta Co., also in Napa and Sonoma cos.; to Wash., Rocky Mts., Tex. June–Aug. Plants with few hairs beneath and ± erosulate calyx-lobes have been called var. *lividum* (Greene) Woodson [*A. l.* Greene]. Those ± densely puberulent throughout are var. *vestìtum* (Greene) Woodson [*A. v.* Greene. *A. incanum* Greene] and have been found in Napa and Glenn cos. and in Ore.

4. **A. cannábinum** L. var. **glabérrimum** A. DC. [*A. Bolanderi* Greene.] Glabrous erect or ascending plants 3–6 dm. tall; lvs. petioled or the lower subsessile, lance-ovate, 4–10 cm. long, mucronate; calyx-lobes scarious, glabrous, 1.5–2 mm. long; corolla greenish to whitish, cylindric to urn-shaped, 2–3 mm. long, almost as wide; follicles pendulous, 12–20 cm. long; seeds ca. 4 mm. long, the whitish coma 2.5–3 cm. long.—Occasional in damp places, below 5000 ft.; many Plant Communities; San Diego Co. through most of Calif. (sometimes in the deserts); to Canada and Atlantic Coast. June–

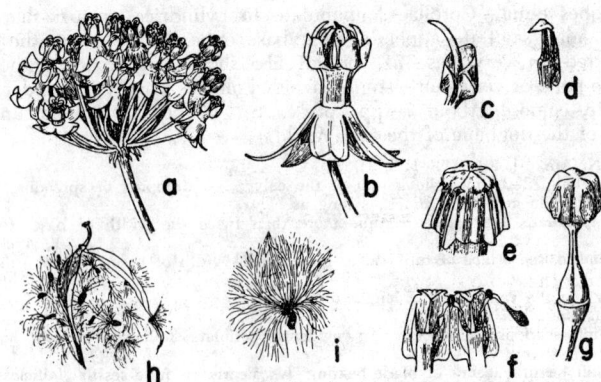

Fig. 52. ASCLEPIADACEAE. *Asclepias fascicularis: a,* umbel, × 1; *b,* fl., × 2½, sepals underneath the spreading corolla-lobes, crown with 5 concave hoods (each with an incurved exserted horn, shown in *c*) and alternating stamens (individual stamen shown in *d* and as group in *e,* × 5, fils. connate into a tube); *f,* stamens from within, showing pollinia and winged anthers; *g,* pistil, × 5, after removal of stamens and horns; *h,* dehiscing follicle, × ¼; *i,* comose seed, × ½.

Aug. Plants ± tomentulose throughout are var. *pubéscens* (R. Br.) A. DC. [*A. palustre* Greene].—Wet places in Butte, Humboldt, and Solano cos.; to Atlantic Coast.

5. **A. sibíricum** Jacq. var. **salígnum** (Greene) Fern. [*A. s.* Greene. *A. hypericifolium* var. *s.* Bég. & Bel. *A. Breweri, densiflorum, thermale,* and *longifolium* Greene.] Glabrous, erect or ascending, 4–7 dm. tall; lvs. mostly opposite, sessile, even subamplexicaul, oblong-ovate, mostly 4–8 cm. long; calyx-lobes 1.5–2 mm. long, scarious; corolla cylindric to urn-shaped, 3–4 mm. long, white or greenish, the lobes erect or slightly spreading; follicles glabrous, pendulous at maturity, 4–9 cm. long; seeds 3.5–4 mm. long, the coma 1.5–2 cm. long.—Moist places, below 7000 ft.; many Plant Communities; through most of the state; to B.C., Minn., Tex. June–Aug.

55. Asclepiadàceae. MILKWEED FAMILY (Fig. 52)

Perennial herbs, vines or shrubs with milky juice. Lvs. opposite, whorled, or rarely alternate, without stipules. Fls. perfect, regular, mostly umbellate, commonly 5-merous. Calyx deeply lobed, the lobes usually imbricate. Corolla 5-lobed to -cleft, the lobes commonly valvate in the bud. A 5-lobed crown is usually present between the corolla and stamens, adnate to either or both. Stamens 5, inserted on the corolla-tube, usually near base; fils. monadelphous or sometimes distinct; anthers united and tipped with a scarious membrane inflexed on the summit of the stylar disk. Pollen grains united into waxlike or granular pollinia. Carpels 2, with distinct superior ovaries and styles, but united above by the peltate discoid stigma. Fr. of 2 follicles; seeds many, compressed, with long coma. Perhaps 200 genera and 2500 spp., widely distributed, but most frequent in warmer regions.

Stems twining.
  Crowns wanting; corolla urn-shaped or campanulate; pollinia strictly pendulous .... 1. *Cynanchum*
  Crowns present; corolla rotate or open-campanulate; pollinia pendulous or horizontal.
    Fls. borne in axillary umbels; pollinia pendulous .... 2. *Sarcostemma*
    Fs. solitary; pollinia horizontal .... 3. *Matelea*
Stems not twining .... 4. *Asclepias*

1. Cynánchum L.

Ours shrubs or suffrutescent plants with slender twining stems. Lvs. opposite, slender or reduced. Fls. small, in axillary umbels or small cymes. Calyx 5-parted, the segms. acute. Corolla campanulate to urn-shaped, 5-lobed. Crown wanting. Pollen-masses solitary in each pollen-sac, pendulous. Follicles long-acuminate, smooth, terete. As here understood, a large genus of Old and New Worlds. (Greek, *cyon,* dog, and *anchein,* to strangle; ancient name of some plant supposed to poison dogs.)

1. **C. utahénse** (Engelm.) Woodson. [*Astephanus u.* Engelm.] Perennial from a branched crown, the stems slender, glabrous, 2–5 dm. tall; lvs. linear, acuminate, 2–3 cm. long, spreading or reflexed; umbels short-peduncled, 3–10-fld., with a few subulate bracts; corolla dull yellow, 2 mm. wide, the lobes ovate, somewhat hooded, puberulent within; anthers unappendaged at apex; follicles long-acuminate, 4–6 cm. long; seeds rough-granulate.—Occasional, dry sandy places, below 3000 ft.; Creosote Bush Scrub; both deserts; to Utah, Ariz. April–June.

2. Sarcostémma R. Br.

Suffrutescent twining or trailing vines. Lvs. usually foliose, opposite, usually with 1 or more glands on the ventral surface of the midrib at base. Fls. 1–30, in cymes or umbels. Calyx deeply 5-lobed. Corolla subrotate to campanulate or salverform, 5-lobed. Stamens 5, the fils. coherent into a column, each fil. bearing an inflated vesicular segm. (corona-vesicle) just below the anther; anthers 2-celled, the membranous dorsal appendage ovate to deltoid; pollinia solitary in each anther-sac, pendulous. Follicle fusiform to clavate. Seeds flattened or unequally biconvex, with a micropylar coma. A rather large genus of the warmer parts of the world. (Greek, *sarx*, flesh, and *stemma*, crown, referring to the fleshy inner corona.)

(Holm, R. W. The Am. species of Sarcostemma. Ann. Mo. Bot. Gard. 37: 477–560, 1950.)

Corolla purplish; plant glabrous or puberulent with subappressed hairs . . . . 1. *S. cynanchoides*
Corolla greenish-yellow; plant canescent-pilose with short spreading hairs . . . . 2. *S. hirtellum*

1. **S. cynanchoìdes** Dcne. ssp. **Hartwégii** (Vail) R. Holm. [*Philibertella H.* Vail. *Funastrum H.* Schlechter. *S. heterophyllum* auth., not Engelm.] Stems 1–2 m. long; lvs. glabrous to puberulent, green, linear to lanceolate, 3–4 cm. long, short-petioled, sometimes auriculately lobed at base; umbels several-fld., on peduncles 1–5 cm. long; pedicels puberulent, 7–14 mm. long; calyx-lobes 2–3 mm. long, narrow-ovate, pilosulose; corolla rotate-subcampanulate, the tube 1 mm. long, the lobes ovate, acute to acuminate, puberulent without, glabrous within, 5–7 mm. long; anthers 1 mm. long, the apical appendage orbicular; follicles slender, fusiform, attenuate, 7–10 cm. long, minutely puberulent; seeds unequally biconvex, 6 mm. long, the coma ca. 3.5 cm. long.—Fairly frequent in dry places, low altitudes; Coastal Sage Scrub, Chaparral, Creosote Bush Scrub; interior cismontane valleys, Los Angeles Co. to San Bernardino and San Diego cos., Colo. Desert and s. Mojave Desert; to Ariz., Utah. April–July.

2. **S. hirtéllum** (Gray) R. Holm. [*S. heterophyllum* var. *h.* Gray. *Philibertella h.* Vail. *Funastrum h.* Schlechter.] Much like the preceding but more canescent, with short spreading hairs; lvs. not at all auriculate; corolla-lobes 4–5 mm. long, glabrous without, subpilose within; anthers 0.8 mm. long, the apical appendage elliptic-ovate; follicles 4–4.5 cm. long, puberulent; seeds somewhat flattened, 8 mm. long, 4 mm. wide, the coma ca. 2.5 cm. long.—Frequent in washes, below 3500 ft.; Creosote Bush Scrub; Colo. Desert and e. Mojave Desert; Ariz., Nev. March–May.

3. Matelèa Aubl.

As here understood a rather large and comprehensive genus of herbs, vines and shrubs mostly with both long eglandular hairs and short bulbose emergences. Ours with petioled opposite usually cordate lvs. Fls. in axillary peduncled cymelike fascicles or umbels. Calyx deeply 5-cleft, glandular within. Corolla rotate to campanulate, 5-parted, the corona consisting of a unit enation of the anther-fil., subtending an additional inner enation. Stamens with connate fils. forming a tube; anthers tipped with a small scarious inflexed membrane in ours. Pollinia solitary in each sac, horizontal. Follicles thick, smooth or tuberculate or angled. Seeds compressed, with coma. Widely distributed. (Name not explained.)

1. **M. parvifòlia** (Torr.) Woodson. [*Gonolobus p.* Torr. *G. hastulatus* Gray. *Vincetoxicum h.* Heller. *G. californicus* Jeps.] Somewhat woody at base, ± twining, 1–4 dm. high, puberulent; lvs. cordate-sagittate, 5–20 mm. long, on petioles ca. half as long, sparsely strigulose; fls. 1, sometimes 2, in the axil, short-pedicelled; calyx-lobes lanceolate, 1–1.5 mm. long; corolla greenish, the lobes oblong, 3–4 mm. long; crown corolla-

like, 5-lobed, borne on the base of the corolla-tube and attached to the stamen-column by 5 thin vertical plates; follicles 5–7 cm. long, tapering, sparsely muricate; seeds brown, flat, ca. 4.5 mm. long, with a coma ca. 2 cm. long.—Rare in dry rocky places, ca. 2000–3000 ft.; Creosote Bush Scrub; Colo. Desert (Corn Springs, Cottonwood Springs, Yaqui [Ironwood] Well), Mojave Desert (Kelso); to Tex., L. Calif. March–May.

4. Asclèpias L. Milkweed

Perennial herbs or shrubs from deep-seated roots. Stems erect or decumbent. Lvs. opposite, alternate, or whorled. Fls. perfect, regular, in axillary or terminal umbels. Calyx small, usually with small glands at base of the 5 lobes. Corolla rotate, deeply 5-lobed, the lobes mostly valvate in bud, reflexed in anthesis, sometimes erect-spreading. Crown usually with a column, the lobes 5, concave and hoodlike, each bearing a hornlike or toothlike projection within, sometimes hornless. Fils. connate into a tube, the anthers winged, the wings broadened below the middle. Pollinia solitary in the sacs, pendulous. Follicles fusiform or narrower, acuminate. Seeds compressed, comose. A rather large genus of the New World. (The Greek name of *Aesculapias* for whom the genus is named.)

(Woodson, R. E., Jr. The N. Am. species of Asclepias. Ann. Mo. Bot. Gard. 41: 1–211, 1954.)
A. Corolla-segms. spreading during anthesis; hoods of crown joined to each other by a lobed disk
1. A. asperula
AA. Corolla-segms. reflexed during anthesis; hoods distinct from each other or lacking.
  B. Stems strongly flattened, prostrate; horns none .... 2. A. Solanoana
  BB. Stems terete (flattened in A. cryptoceras), erect or decumbent; horns often present.
    C. Hoods without horns on inner surface; pedicels deflexed in fr.
      D. Plants white-tomentose; hoods spheroid, open at top and part way down back
3. A. californica
      DD. Plants green, minutely puberulent; hoods oblong-cylindric, open down inner side
4. A. cordifolia
    CC. Hoods with horns on inner surface.
      D. The hoods 2–3 times as long as the stamens and stigma; fil.-column almost lacking; pedicels deflexed in fr.
        E. Lvs. ovate; stems herbaceous.
          F. Plant 4–12 dm. high; hoods open, plane, lanceolate above the broad base
5. A. speciosa
          FF. Plant 1–2 dm. high; hoods with their sides closely appressed
6. A. nyctaginifolia
        EE. Lvs. filiform or lacking; stems woody below .... 7. A. subulata
      DD. The hoods ca. as long as stamens and stigma; fil.-column well developed.
        E. Corolla red-purple; hoods orange. Escape from cult. .... 14. A. curassavica
        EE. Corolla greenish-white to cream, sometimes tinged purple; hoods not orange. Natives.
          F. Lvs. narrow, 1 cm. or less wide, often early deciduous; pedicels erect in fr.
            G. Plant herbaceous, rarely 1 m. high, with persistent green lvs.; stems green .... 8. A. fascicularis
            GG. Plants shrubby, 1–2 m. high, the lvs. soon deciduous; stems white
9. A. albicans
          FF. Lvs. broader, mostly 2–7 cm. wide; pedicels deflexed in fr.
            G. Lateral umbels sessile or subsessile, the terminal peduncled; horns included within the hoods.
              H. Corolla-lobes 10–12 mm. long; column of crown none; horns falcate-subulate .... 10. A. cryptoceras
              HH. Corolla-lobes 6–7 mm. long; column present; horns rather blunt, only slightly incurved .... 11. A. vestita
            GG. Lateral umbels well peduncled.
              H. Lvs. usually 3 or more at a node; horns mostly included within the hoods. Cismontane .... 12. A. eriocarpa
              HH. Lvs. opposite; horns exserted. Desert. .... 13. A. erosa

1. A. aspérula (Dcne.) Woodson. [*Acerates a.* Dcne. in DC. *A. capricornu* Woodson ssp. *occidentalis* Woodson. *Ananthrix decumbens* Nutt. *Asclepiodora d.* Gray.] Stems several, herbaceous, decumbent or ascending, 3–6 dm. long, puberulent; lvs. green, ascending, scabrous-puberulent, lanceolate, 4–12 cm. long, acuminate at apex, narrowed at base into a very short petiole, mostly alternate, sometimes in 3's; umbels many-fld., largely solitary and terminal, peduncled; pedicels, calyx, and outer surface of corolla-lobes puberulent; corolla-lobes widely spreading, greenish-white, ca. 8 mm. long; hoods purplish, 5–6 mm. long, strongly incurved; anther-wings broad, angled above; follicles

7–8 cm. long, erect on recurved pedicels, puberulent; seeds light brown, flat, ca. 6 mm. long, the coma ca. 3 cm. long.—Dry open rocky places, 5000–6500 ft.; Pinyon-Juniper Wd.; New York, Providence, and Clark mts., e. Mojave Desert; to Colo., Texas. May–July and Sept.

2. **A. Solanoàna** Woodson. [*Gomphocarpus purpurascens* Gray, not Rich. *Solanoa p.* Greene. Not *A. p.* L.] Stems 2 or 3 from a woody taproot, 2–3 dm. long, flattened, prostrate, flexuous; herbage canescently puberulent; lvs. elliptic-ovate to broadly cordate-ovate, the lower with base rounded to obtuse, the upper cordate, 2.5–4.5 cm. long, on petioles shorter to almost as long; umbels terminal or also in upper axils, peduncled, many-fld.; pedicels 7–10 mm. long; calyx scarcely 1 mm. long; corolla purple outside, whitish inside, the lobes reflexed, 2.5 mm. long; hoods saccate, pale, 1.5 mm. long, attached from base to top of fil.-column; horns none; anthers 2 mm. long; follicles ca. 5 cm. long, and 1 cm. thick, usually only 1 maturing.—Dry serpentine outcroppings, 2000–5000 ft.; Yellow Pine F.; N. Coast Ranges from Lake Co. to Trinity Co. June.

3. **A. califórnica** Greene. [*Acerates tomentosa* Torr., not *Asclepias t.* Ell. *Gomphocarpus t.* Gray, not Burch. *G. t.* var. *Xanti* Gray. *G. Torreyi* Macbr.] Herbaceous, soft white-woolly, the stems ascending, 1.5–5(–7) dm. long; lvs. mostly opposite, ovate to lance-ovate, 5–15 cm. long, short-petioled, acute to acuminate at tip, obtuse to cordate at base; lateral umbels if present tending to be sessile; fls. 6–12; calyx and outer surface of corolla white-woolly; corolla-lobes purplish, 8–10 mm. long; hoods dark maroon, broadly ovoid, centrally attached to the column, strongly pendulous at base, shorter than the anthers, cleft down the back to below middle; horns lacking; follicles ovoid, hoary, 5–8 cm. long, acuminate; seeds ca. 1 cm. long, the coma ca. 3.5 cm. long.—Frequent on dry slopes, below 7500 ft.; Yellow Pine F., Pinyon-Juniper Wd., Chaparral; n. L. Calif. through cismontane s. Calif. to Kern and Inyo cos. April–July.

Ssp. **Greènei** Woodson. Hoods scarcely pendulous at base.—Dry slopes, below 3500 ft.; Foothill Wd.; Coast Ranges, Contra Costa Co. to Monterey Co.; Sierra Nevada, Mariposa Co. to Fresno Co.

4. **A. cordifolia** (Benth.) Jeps. [*Acerates c.* Benth. *Gomphocarpus c.* Gray. *Asclepias ecornuta* Kell. *Acerates atropurpurea* Kell.] Herbaceous, from a stout woody root, glabrous to somewhat puberulent; stems 3–8 dm. high, often tinged purple in upper parts; lvs. mostly opposite, ovate and mostly acute, cordate-clasping, 5–15 cm. long; umbels loosely many-fld., 1–several at apex and in upper axils, short-peduncled; pedicels filiform, 1.5–3.5 cm. long; calyx-lobes purple, somewhat pubescent, 2.5–3 mm. long; corolla-lobes dark red-purple, 8–9 mm. long, oblong, glabrous; hoods purplish, short-cylindric, open at top and cleft down inner surface, truncate above and the inner angles produced into an ascending toothlike cusp; follicles lanceolate to oblong, straight, long-acuminate, 10–14 cm. long, glabrous; seeds 7–8 mm. long, the coma 2.5–3 cm. long.—Open or wooded slopes, 500–6300 ft.; Chaparral, Foothill Wd., Yellow Pine F., Mixed Evergreen F.; Coast Ranges from Siskiyou Co. to Solano Co., and Sierra Nevada from Kern Co. n., to Modoc Co.; Ore., w. Nev. May–July.

5. **A. speciòsa** Torr. Stout, herbaceous, soft-tomentose or sometimes glabrate, the stem 5–12 dm. high, leafy to summit; lvs. opposite, oval to oblong, short-petioled, acute to obtuse at apex, rounded to cordate at base, 8–15 cm. long; umbels few, peduncled, the lower with fewer fls. than the many-fld. terminal one; pedicels and calyx heavily tomentose, the former 1.5–3.5 cm. long, the latter 4–5 mm. long; corolla-lobes rose-purple, 8–10 mm. long, woolly on the back; hoods pinkish, aging yellowish, much longer than stamens, with wide involute base and then abruptly contracted into a nearly flat lance-shaped part; horns much exserted, incurved; follicles narrow-ovoid, densely woolly, 6–10 cm. long, scattered soft-spiny; seeds ca. 8 mm. long; $2n = 22$ (Moore, 1946).—Dry gravelly and stony places, mostly below 6000 ft.; Yellow Pine F., Mixed Evergreen F.; Coast Ranges from Solano Co. to Siskiyou Co., w. base of Sierra Nevada s. to Fresno Co., e. slope from Inyo Co.; to Nev., Wash., Miss. V. May–July. A plant from Tuolumne Co. described as *A. Giffórdii* Eastw., with pale yellow corollas 8 mm. long combining characters of *A. speciosa* and *A. eriocarpa*, may have been a hybrid.

6. **A. nyctaginifòlia** Gray. [*Podostemma n.* Greene.] Stems few from crown of slender rootstock, herbaceous, decumbent to ascending, 1–2 dm. long, puberulent, slender; lvs. ovate to lance-ovate, light green, puberulent, somewhat crisped, 4–7 cm. long, on petioles 1–2 cm. long, acute at top, obtuse below; umbels subsessile; calyx ca. 2.5 mm.

long, pubescent; corolla thin, greenish-white, 12–14 mm. long; hoods narrow-oblong, 8–10 mm. long; horns winglike, attached above the middle and produced into slender somewhat exserted points; follicles narrow-ovoid, attenuate, puberulent, 5–6 cm. long. —Dry slopes, 4000–5000 ft.; Pinyon-Juniper Wd.; Providence and New York mts., e. Mojave Desert; Ariz. May–June.

7. **A. subulàta** Dcne. in A. DC. Almost shrubby, the greenish-white stems in clusters, erect, rushlike, 1–1.5 m. tall, leafless or with few filiform lvs., 2–5 cm. long; umbels few to several, at and near summit of the branches, rather short-peduncled; pedicels rather stout, ascending, somewhat tomentose, mostly 1.5–2.5 cm. long; calyx tomentose, ca. 3 mm. long; corolla thin, greenish-white, 7–8 mm. long; hoods ca. as long, twice the length of the stamens; horns scarcely exserted, attached near middle; follicles slender, long-acuminate, 6–12 cm. long, smoothish; seeds rather narrow, 6–7 mm. long; coma tawny, ca. 2.5 cm. long.—Occasional in desert washes and sandy places, below 2000 ft.; Creosote Bush Scrub; Colo. Desert, e. Mojave Desert; L. Calif., Ariz. April–Dec.

8. **A. fasciculàris** Dcne. in A. DC. [*A. mexicana* auth., not Cav. *A. macrophylla* var. *comosa* Dur. & Hilg.] Herbaceous, the stems several, erect, 5–9 dm. tall, glabrous or sparsely puberulent; lvs. linear to linear-lanceolate, usually in whorls of 3–6, or lower and upper fewer, 4–12 cm. long, 3–10 mm. wide, short-petioled, commonly folded along midrib; umbels several in upper axils, many-fld.; peduncles 2–5 cm. long; pedicels slender, 6–15 mm. long; calyx pubescent, ca. 2 mm. long; corolla greenish-white, often tinged purple, the lobes oblong, 4–5 mm. long; hoods ca. as long as stamens, broadly ovate; horns slender, exserted, incurved; follicles smooth, narrow, acuminate, 6–9 cm. long; seeds ca. 6 mm. long, the coma 3 or more cm. long.—Frequent as colonies in dry places, mostly below 7000 ft.; many Plant Communities; most of cismontane Calif. away from the immediate coast and uncommon on desert; to Wash., Ida., Utah, L. Calif. June–Sept.

9. **A. albicáns** Wats. Shrubby, 1–2 m. tall, the stems few, white-waxy, slender, erect, rather few-branched; lvs. in 3's, filiform, 1–2 cm. long, early deciduous; umbels somewhat woolly, many-fld., on peduncles ca. 1.5–2 cm. long; pedicels longer; calyx 3–4 mm. long; corolla-lobes greenish-white with some brown or pink, 5–6 mm. long; hoods yellowish, shorter than the anthers; horns incurved, barely exserted; follicles on erect pedicels, tomentulose when young, later glabrate and whitish, smooth, lanceolate in outline, ca. 1 dm. long; seeds narrow, ca. 6 mm. long; coma ca. 1.5 cm. long.— Occasional, dry rocky places, below 2500 ft.; Creosote Bush Scrub; Colo. Desert, e. Mojave Desert (Sheephole Mts., Whipple Mts., etc.); Ariz., L. Calif. March–May.

10. **A. cryptocèras** Wats. ssp. **Davísii** (Woodson) Woodson. [*A. D.* Woodson.] Glabrous, with ± flattened decumbent stems 1.5–3 dm. long; lvs. opposite, roundish to broadly ovate, 3–8 cm. long, short-petioled; umbels 1–3, few-fld., the terminal peduncled, the others sessile; pedicels 2–2.5 cm. long, slender; calyx-lobes ca. 4.5 mm. long; corolla greenish-yellow, the lobes ovate, 10–12 mm. long; hoods subrhombic, truncate to tips, flesh-colored, usually shorter than the anthers, abruptly bi-acuminate; horns included, slender, falcate-subulate; fruiting pedicels recurved; follicles ovoid, 2–5 cm. long, short-acuminate; seeds deep brown, ca. 7 mm. long, somewhat reticulate, the coma whitish, ca. 2 cm. long.—Sandy and gravelly slopes, 5000–8200 ft.; Sagebrush Scrub, Pinyon-Juniper Wd.; Mono Co.; to e. Ore., Wyo., Colo. May–June. *A. cryptoceras* itself, with oblong-ovoid hoods, gradually rounded to the tips, has been reported from Mono Co.

11. **A. vestìta** H. & A. Herbaceous, white-woolly, finally glabrate, the stems simple, ascending, 2–6 dm. long; lvs. opposite, ovate to oblong-lanceolate, acuminate to sharply acute, 4–10 cm. long, on short petioles; umbels 1–4, in upper axils, many-fld.; peduncles short or 0; pedicels slender, 2.5–3 cm. long; calyx ca. 4 mm. long; corolla-lobes slightly purplish to greenish-white, 6–7 mm. long, hoary on back; hoods nearly erect, entire and truncate at summit, white with outer brown stripe, 3 mm. long, slit down inner side; horns ca. as long, blunt, slightly incurved; anther-wings subtended by 2 small teeth; follicles at first hoary, later glabrate, ovoid, ca. 5–6.5 cm. long; seeds ca. 10 mm. long, the coma 1.5 cm. long, tawny.—Dry flats and slopes, below 3000 ft.; V. Grassland, Foothill Wd., Chaparral; borders of San Joaquin V. and adjacent ranges to Calaveras and Stanislaus cos. May–June. A form with lvs. more glabrate, infl. usually solitary and less arachnoid, somewhat shorter pedicels, more purplish corollas and sometimes more

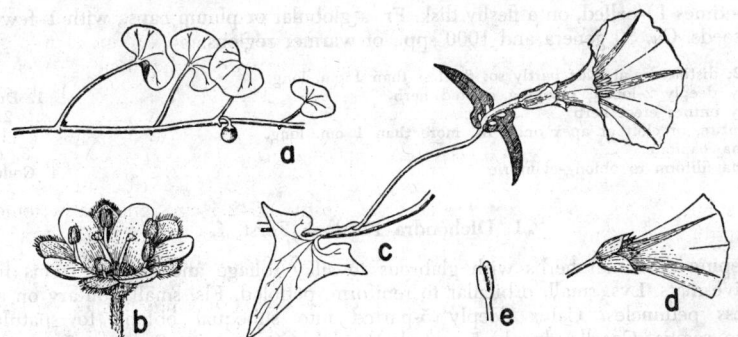

Fig. 53. CONVOLVULACEAE. *Dichondra occidentalis: a,* habit, × ½; *b,* fl., × 5, 5 calyx-segms., 5 corolla-lobes, 5 stamens, 2 styles. *Convolvulus fulcratus: c,* habit, × ½, with shaded hastate bracts resembling lvs. and below the calyx. *C. cyclostegius: d,* fl. to show the sepallike bracts just below the calyx; *e,* stigmas, × 2.

reduced teeth subtending the anther-wings, is found about the w. edge of the Mojave Desert and has been designated as ssp. *Paríshii* (Jeps.) Woodson. [A. v. var. *P.* Jeps.]

12. **A. eriocárpa** Benth. [*A. Fremontii* Torr. ex Gray. *A. Kotolo* Eastw.] Herbaceous, hoary-tomentose throughout, the stems quite simple and erect, 4–9 dm. tall; lvs. usually at least in part in whorls of 3 or 4, some opposite, elongate-oblong, 6–15(–20) cm. long, on very short petioles, truncate to subcordate at base, obtusish at apex; umbels few or several, many-fld., stout-peduncled; pedicels 2–5 cm. long; calyx 2.5–3 mm. long; corolla-lobes cream, oblong-ovate, 4–5 mm. long; hoods shorter than stamens, cream or tinged purplish; horns broad at base, barely exserted, pointed, curved; follicles hoary, oblong, 6–9 cm. long, rather short-acuminate; seeds ca. 8 mm. long, rather narrow, the coma whitish, ca. 3 cm. long.—Frequent in dry barren places, below 7000 ft.; many Plant Communities; from Mendocino and Shasta cos. s. to L. Calif. June–Aug. Poisonous to sheep, sometimes baled in hay and toxic to rabbits and other stock. A form of uncertain value taxonomically, with follicles 3–4 cm. long and from Deep Creek, San Bernardino Mts., was described as var. *microcárpa* M. & J.

13. **A. eròsa** Torr. [*A. e.* var. *obtusa* Gray. *A. obtusata* and *Rothrockii* Greene.] Finely woolly, later glabrate, the stems herbaceous, 5–8 dm. tall; lvs. opposite, oblong-ovate to lance-ovate, 8–15 cm. long, coriaceous, short-petioled, acuminate to acute; umbels peduncled, many-fld.; pedicels slender, 1.5–2 cm. long; calyx woolly, ca. 2.5 mm. long; corolla greenish-white, ± tomentose on outside, 5–6 mm. long; hoods broadly obovoid, little exceeding the stamens, truncate at apex; horns exserted, incurved, attached near base of hood; follicles tomentulose, short-acuminate, 5–8 cm. long; seeds ca. 9 mm. long, pubescent; coma whitish, ca. 2 cm. long.—Dry slopes and washes, below 5000 ft.; Creosote Bush Scrub, Joshua Tree Wd., Pinyon-Juniper Wd., V. Grassland; deserts to Inyo Co., upper San Joaquin V. and adjacent mts.; to Utah, Ariz. May–July.

14. **A. curassávica** L. Glabrous, or puberulent above, 3–8 dm. high, somewhat woody below; lvs. in 2's or 3's, oblong to lance-oblong, 5–15 cm. long, acuminate, short-petioled; umbels axillary and terminal, peduncled; pedicels 1–2 cm. long, erect in fr.; corolla-lobes red-purple, 5–6 mm. long; hoods ovoid, orange, ca. 3 mm. long; horns broad, curved; follicles slightly pubescent, or glabrous, slender, erect, 4–10 cm. long, acuminate.—Occasionally cult. and escapes, as at Mandeville Canyon, Santa Monica Mts.; native in Trop. Am. April–June.

56. Convolvulàceae. MORNING-GLORY FAMILY (Fig. 53)

Annual or perennial herbs, sometimes woody, chiefly twining or trailing. Lvs. alternate. Fls. regular, 4–5-merous, axillary, solitary, or cymose. Pedicels often bracted. Calyx of ± united imbricated sepals, persistent. Corolla hypogynous, ± campanulate to urceolate or cylindrical, mostly convolute or twisted, sometimes imbricate in bud, the limb entire or lobed. Stamens borne on corolla. Pistil of 2 or 4 ± united carpels, the ovary superior,

2(sometimes 1)-celled, on a fleshy disk. Fr. a globular or plump caps., with 1–few rather large seeds. Ca. 50 genera and 1000 spp., of warmer regions.

Styles 2, distinct or at least partly so; fls. less than 1 cm. long.
    Ovary deeply 2-lobed; creeping matted herb .... 1. *Dichondra*
    Ovary entire; erect herb .... 2. *Cressa*
Style entire, or cleft at apex only; fls. more than 1 cm. long.
    Stigma capitate .... 3. *Ipomoea*
    Stigma filiform to oblong-cylindric .... 4. *Convolvulus*

1. Dichóndra Forst. & Forst. f.

Creeping perennial herbs with glabrous to silky foliage and slender stems forming extensive mats. Lvs. small, orbicular to reniform, petioled. Fls. small, solitary on axillary bractless peduncles. Calyx deeply 5-parted into subequal oblong to spatulate or obovate segms. Corolla deeply 5-parted, the lobes spreading. Stamens 5, short, with slender fils. Ovary deeply 2-parted. Capsules 2, utricular, 1–2-seeded. Seeds rather large, round-obovoid. Several spp., of warmer regions; some grown as ground-covers. (Greek, *di*, two, and *chondros*, a grain, from the fr.)

Corolla barely exceeding the calyx; lf.-blades mostly not prominently decurrent on petioles
                                                       1. *D. repens*
Corolla well exserted from calyx; lf.-blades mostly conspicuously decurrent on petioles
                                                      2. *D. occidentalis*

1. **D. rèpens** Forst. & Forst. f. Rooting freely and forming dense mats; lf.-blades mostly 1–2 cm. wide and suborbicular, usually with rather narrow sinus, glabrous or nearly so above, ± strigose beneath and on petiole; petioles much longer than blades; peduncles usually shorter than petioles; calyx-lobes oblong-obovate to spatulate, rounded at apex, silky-pubescent, 1.5–2 mm. long; corolla-lobes ca. as long as calyx, whitish, ovate, acutish; caps.-lobes subglobose, 1.5–2 mm. in diam., pilose; seeds round-obovoid, black-brown, 1.5–2 dm. long, dull.—Open grassy or brushy places, below 1500 ft.; Coastal Prairie, N. Coastal Scrub; near the coast from Humboldt Co. to Monterey Co., reported also from foothills of Eldorado Co. *D. repens* is common in cult. in Calif. as a ground cover and is a widely distributed trop. and subtrop. plant; whether this apparently native cent. Calif. plant is the same is uncertain. March–June.

2. **D. occidentàlis** House. [*D. repens* var. *o.* Jeps.] Much like the preceding but coarser, the lvs. more broadly reniform, 2–5 cm. broad, with broad open sinus and mostly prominently decurrent on petioles, glabrate on both surfaces or scattered-hairy; petioles sparsely strigose, the rhizomes more densely so; peduncles much shorter than petioles; calyx-lobes obovate, 1.5 mm. long, pubescent, rounded at apex; corolla white to purplish, 2–3 mm. long, well-exserted; caps. silky-pubescent, ca. 4 mm. high; seeds rounded, blackish, dull, ca. 2 mm. in diam.—Mostly dry sandy banks in brush or under trees; Coastal Sage Scrub, Chaparral, S. Oak Wd.; along the coast of Los Angeles, Orange, and San Diego cos., on Santa Cruz, Santa Rosa, and Santa Catalina ids.; L. Calif. March–May.

2. Créssa L. Alkali Weed

Low much-branched perennial herbs with erect or diffuse nontwining stems. Lvs. alternate, small, canescent, entire. Fls. solitary in upper axils, small, perfect, 5-merous. Sepals 5. Corolla funnelform, white, persistent, 5-lobed. Stamens 5, exserted. Ovary 2-celled, 4-ovuled; styles 2, distinct; stigmas entire, capitate. Fr. a caps., often 1-seeded by abortion. A small genus of warm temp. and trop. regions around the world. (Greek, *Kressa*, a Cretan woman.)

1. **C. truxillénsis** HBK. var. **vallícola** (Heller) Munz. [*C. v.* Heller.] Low gray tufted woolly-villous plants 1–2 dm. tall; lvs. oblong-ovate, 5–10 mm. long, sessile; pedicels mostly 2–5 mm. long; calyx canescent, 4–5 mm. long, the lobes ca. 2 mm. wide; corolla ca. 6 mm. long, the lobes ca. 2 mm. long, spreading or reflexed, ovate-lanceolate, acute, 1–1.5 mm. wide; ovary and caps. pubescent; $2n = 28$ (Heiser & Whitaker, 1948).— Saline and alkaline places, mostly below 4000 ft.; many Plant Communities from seacoast to the interior of most of cismontane Calif.; to Ore., Tex., Mex. May–Oct.

Var. **mínima** (Heller) Munz. [*C. m.* Heller.] Plant appressed-silky; pedicels 1–4 mm. long; calyx-lobes 2.5–3 mm. wide; corolla-lobes round-ovate, obtuse, 2–2.5 mm. wide. —Alkali flats and swamps; Alkali Sink; Death V. region to Modoc Co.; w. Nev.

3. Ipomoèa L. MORNING-GLORY

Ours twining or trailing herbs similar to *Convolvulus*. Calyx ebracteate (although the pedicels may bear alternate bracts below the calyx). Corolla salverform to campanulate, the limb entire or slightly lobed. Style undivided; stigma capitate or of 2–3 globular lobes. Caps. globular, 2–4-valved. A large genus of warm regions; one sp. (*I. Batatas*) the sweet potato; others grown for the showy fls. (Greek, *ips,* a worm, and *homoios,* like; from the twining habit.)

Sepals acute, ca. 12 mm. long; lvs. entire or nearly so .... 1. *I. purpurea*
Sepals long-acuminate, 20–25 mm. long; lvs. 3-lobed or 3-parted .... 2. *I. hederacea*

1. **I. purpùrea** (L.) Roth. [*Convolvulus p.* L.] COMMON MORNING-GLORY. Annual with twining hairy stems; lvs. broadly cordate-ovate, 7–12 cm. long, entire, short-acuminate, pubescent; petioles shorter; peduncles 1–5-fld., with evident pedicels; sepals lanceolate to oblong, acute, 12–16 mm. long, pubescent; corolla funnelform, purple to blue, pink or white, 5–6 cm. long; seeds glabrous.—Frequent escape from gardens and established in waste places, city-dumps, etc.; native of trop. Am. June–Nov.

2. **I. hederàcea** (L.) Jacq. [*Convolvulus h.* L.] Annual with slender twining ± retrorsely pubescent stem; lvs. cordate-ovate in outline, deeply 3-lobed, 3–8 cm. long, sparsely strigose; peduncles short, axillary, 1–3-fld.; calyx-lobes 12–20 mm. long, hirsute below, with long linear tips; corolla 2–4 cm. long, the limb light blue or purple; seeds glabrous.—Occasional adventive as garden escape; native of se. U.S. July–Nov.

4. Convólvulus L. MORNING-GLORY. BINDWEED

Twining or trailing herbs, usually perennial, sometimes suffrutescent. Lvs. alternate, usually petioled, commonly cordate or sagittate. Peduncles axillary, usually with a pair of bracts just below and covering the calyx or remote from it. Fls. showy, white to purplish or rose or yellowish. Sepals equal or unequal. Corolla funnelform to campanulate, plaited, usually 5-angled or -lobed. Stamens included, inserted on corolla-tube. Ovary 2-celled, rarely 1-celled, 4-ovuled; style filiform; stigmas 2, linear to oblong or ovate. Caps. ± globose; seeds ± ovoid, sometimes somewhat angled, usually 4, glabrous. Ca. 200 spp., of wide distribution; some of horticultural value. (Latin, *convolvere,* to entwine.)

A. Plants annual; corolla ca. 6 mm. long, deeply cleft .... 16. *C. simulans*
AA. Plants perennial; corolla 2–6 cm. long, not cleft.
  B. Calyx enclosed or closely subtended by a pair of large sepallike bracts.
    C. Corolla purple to rose.
      D. Stigmas oblong to oval; lvs. not of 2 distinct types.
        E. Lvs. reniform, obtuse, fleshy, 2–5 cm. broad; prostrate seaside herbs
                        1. *C. Soldanella*
        EE. Lvs. ovate-hastate, thin, acute to acuminate, 6–10 cm. long; climbing swamp
          plant .... 2. *C. sepium*
      DD. Stigmas linear; lvs. of 2 distinct types, the lower cordate-ovate, the upper deeply
        lobed. Introd. weed .... 17. *C. althaeoides*
    CC. Corolla white to cream, sometimes pinkish in age.
      D. Stems mostly over 1 m. long, twining or trailing.
        E. Plant of swamps and marshes, the stems entirely herbaceous ... 2. *C. sepium*
        EE. Plants of dry places, the stems ± woody at base.
          F. Bracts 2–3 cm. long; corolla 5–6 cm. long. Insular . 3. *C. macrostegius*
          FF. Bracts 1–1.5 cm. long; corolla 2–4.5 cm. long. Mostly mainland.
            G. Bracts mostly subcordate at base, membranous and purplish. Near
              the coast .... 4. *C. cyclostegius*
            GG. Bracts mostly rounded at base, firm and greenish. Mostly away from
              the coast .... 5. *C. aridus*
      DD. Stems mostly 1–5 dm. long, erect to prostrate.
        E. Herbage glabrous; corolla 4–5 cm. long. N. Calif. ...... 6. *C. nyctagineus*
        EE. Herbage pubescent to tomentose.
          F. Corolla 4–5 cm. long; plant stemless or nearly so. Cent. Coast Ranges
                      7. *C. subacaulis*

1. **C. Soldanélla** L. [*Calystegia* S. and *reniformis* R. Br.] BEACH MORNING-GLORY. Prostrate perennial with fleshy stems 1–5 dm. long from deep-seated rootstocks; herbage glabrous, rather fleshy; lf.-blades reniform, shining, sometimes somewhat angled, 2–5 cm. broad, scarcely as long, on petioles 4–6 cm. long; bracts subtending the calyx round-oval, 8–14 mm. long, membranous; sepals oval-ovate, mostly 15–20 mm. long in fr.; corolla rose to purplish, short-funnelform, 4–6 cm. long; stigmas ovoid; caps. subglobose, 12–15 mm. long; seeds dull black, round-obovoid, cellular-papillose, ca. 5 mm. in diam.—Common in sand; Coastal Strand; San Diego n.; to Wash., S. Am.; Old World. April–Aug.

2. **C. sèpium** L. HEDGE BINDWEED. Perennial from creeping rootstocks, twining or creeping, the stems glabrous or pubescent, 1–3 m. long; lvs. broadly ovate, longer than broad, pointed, 4–6 cm. long, 3.5–6 cm. wide, with angulate truncate or rounded basal lobes, glabrous or pubescent, on shorter or equal petioles; fls. from many axils, on 4-angled peduncles; paired bracts cordate, ovate, pointed, 1.5–2.5 cm. long; sepals enclosed in bracts; corolla white to pinkish, ca. 5 cm. long.—Occasional in swampy saline places; Coastal Salt Marsh; Marin, Solano, and Contra Costa cos., Orange Co.; e. N. Am.; natur. from Eu. June–Aug.

Var. **rèpens** (L.) Gray. [*C. r.* L.] Plant pubescent to subglabrous; lvs. ca. 5–8 cm. long, 1.5–3.5 cm. wide, the blade proper lance-ovate, sagittate at base with rounded lobes.—Saline swamps; Freshwater Marsh; San Bernardino, Riverside, Chino, Los Angeles, Huntington Beach, etc.; Atlantic Coast; Eu.

Var. **dumetòrum** Pospichal. [*C. Binghamiae* Greene. *C. s.* var. *B.* Jeps.] Lvs. oblong to broadly oval, mostly obtuse, 3–7 cm. long, the basal lobes only slightly divergent; bracts oblong, 8–10 mm. long; sepals 12–16 mm. long; corolla 3–5 cm. long.—Swampy and marshy places; Coastal Salt Marsh; Orange Co. to Santa Barbara Co.; from Eu. April–June.

3. **C. macrostègius** Greene. [*C. occidentalis* var. *m.* Munz. *Volvulus m.* Farw.] Sparsely tomentulose perennial with wiry trailing or twining stems 1–4 m. long; lvs. rather fleshy, deltoid-hastate, broader than long to almost as broad, 4–10 cm. long, acute to acuminate, with somewhat spreading coarsely 2–3-toothed basal lobes; petioles from shorter than to longer than blades; peduncles stout, 1–2 dm. long, mostly 1(sometimes few)-fld.; bracts thin, membranous, pinkish, 2–3 cm. long, round-ovate to suborbicular, not or scarcely mucronate; calyx completely concealed; corolla 5–6 cm. long, white but pinkish in age; stigmas broadly linear; caps. 7–8 mm. long; seeds dull black, round, ca. 3 mm. in diam., minutely reticulate-papillose.—Rocky slopes and canyon-walls; Coastal Sage Scrub, Chaparral; Santa Cruz, Santa Rosa, Santa Barbara, San Nicolas, San Clemente, and Santa Catalina ids.; Guadalupe and San Martin ids. in L. Calif. April–July. Closely approached by some mainland material from Santa Barbara Co. referred to the next sp.

4. **C. cyclostègius** House. [*C. occidentalis* var. *c.* Jeps.] Much like *C. macrostegius* as to habit; lf.-blades triangular-lanceolate to -ovate, 2–5 cm. long, the basal lobes usually toothed or angled; petioles mostly much shorter than blades; peduncles 3–10 cm. long, mostly 1-fld.; bracts near to and like the calyx, membranous, usually with some purple, oval or rounded to ovate, 10–15 mm. long, mostly ± pointed, sometimes rounded, usually subcordate at base and not much longer than calyx; sepals lance-ovate; corolla white, often with purple stripes on outside, sometimes pink in age, 2.5–4.5 cm. long; seeds black, rounded on back, ca. 3 mm. high, reticulate-ridged.—Frequent on dry slopes, below 1000 ft.; Coastal Strand, Coastal Sage Scrub, Chaparral; mostly near the coast from Monterey Co. to Los Angeles Co., Santa Catalina Id. March–Aug.

5. **C. áridus** Greene. Much like the preceding in habit, somewhat cinereous with a dense tomentulose puberulence; lvs. mostly 3–6 cm. long, the middle lobe triangular-lanceolate or narrower, with basal lobes ca. ⅓ as long as middle one and ½ as wide, entire or with 1–2 broad teeth; petioles usually shorter than blade; peduncles 1-fld., 6–10 cm. long; bracts mostly greenish, closely investing calyx, lance-ovate, 12–15 mm. long, mostly acute; sepals like bracts but shorter; corolla cream-white, 3–3.5 cm. long; stigmas linear.—Dry slopes, below 3000 ft.; Coastal Sage Scrub, Chaparral; interior s. Calif. from San Gabriel and San Bernardino mts. to San Jacinto and Santa Ana mts. May–June.

KEY TO SUBSPECIES

Lvs. and stems cinereous with dense tomentulose puberulence .... *C. aridus*
Lvs. and stems glabrous or subglabrous.
  Middle lobe of lvs. narrowly to broadly lanceolate; corolla 3–3.5 cm. long.
    Basal lobes of lvs. less than half as long as middle lobe, not strongly divergent .. ssp. *intermedius*
    Basal lobes of lvs. at least half as long as middle lobe, strongly divergent ...... ssp. *longilobus*
  Middle lobe of lvs. narrowly linear; corolla 2–2.5 cm. long .... ssp. *tenuifolius*

Ssp. **intermèdius** Abrams. Coastal slopes, Topatopa, Santa Monica, and Santa Susanna mts., Ventura Co. to w. Orange Co., Santa Catalina Id. March–Aug.

Ssp. **longilòbus** Abrams. Dry hills about San Diego and toward w. Riverside and e. Orange cos. March–May.

Ssp. **tenuifòlius** Abrams. [*C. occidentalis* vars. *tenuissimus* and *angustissimus* auth., not Gray.] Dry places back from immediate coast, San Diego Co. to L. Calif. Feb.–May.

6. **C. nyctagíneus** Greene. [*C. atriplicifolius* (Hallier f.) House, not Poir.] Low perennial, the stems 3–5 dm. long, or subacaulescent, from fleshy branching rootstock; herbage glabrous; lvs. triangular-hastate, 2–6 cm. long, the basal lobes short or almost as long as the median lobe; petioles from shorter than to longer than blades; peduncles equal to or exceeding lvs.; bracts green, oval or ovate, obtuse, slightly cordate at base, 1–2 cm. long, concealing the calyx; corolla cream-white to pinkish, 4–5 cm. long.— Dry slopes, 2000–3500 ft.; Chaparral?; Lake and Butte cos. w.; to Wash. Rare in Calif. May–July.

7. **C. subacaùlis** (H. & A.) Greene. [*Calystegia s.* H. & A. *Convolvulus californicus* Choisy. *C. s.* var. *dolosus* Jeps.] Perennial, ± pilose, with slender branching rootstocks, stemless or very short-stemmed; lvs. ovate-deltoid, truncate to somewhat cuneate at base, 2–3.5 cm. long, with or without basal divergent lobes; petioles mostly 1.5–3 times as long as blades; peduncles 1-fld., 1–3 cm. long; bracts oblong-ovate to -lanceolate, 12–16 mm. long, equaling but not concealing calyx; corolla white or cream with purple on outside of the folds, 4–5 cm. long; caps. ca. 1 cm. long; seeds black, reticulate, round-ovoid, ca. 4 mm. long.—Dry mostly open hills, below 1500 ft.; N. Coastal Scrub, N. Oak Wd.; Sonoma and Napa cos. to San Luis Obispo Co. April–June.

8. **C. malacophýllus** Greene. [*C. villosus* (Kell.) Gray, not Pers. *C. chartaceus* Jeps.] Rootstocks slender; stems ascending or trailing, mostly 1–3 dm. long; herbage densely grayish-tomentose; lvs. triangular-hastate, 2–4 cm. long and ca. as broad, the basal lobes usually 1–2-toothed; petioles longer than to somewhat shorter than blades; peduncles 1-fld., ca. as long as the petioles; bracts close to and concealing calyx, triangular-ovate, 10–15 mm. long; sepals lance-ovate, mucronate, tomentose without, the outer ca. as long as bracts, the inner shorter; corolla cream-white, 2.5–3.5 cm. long; seeds dark brown, finely tomentose, 5 mm. long.—Steep dry slopes, 3000–7500 ft.; Yellow Pine F., Red Fir F.; Siskiyou and Trinity cos. to the Sierra Nevada and s. to Tulare Co. June–Aug.

Ssp. **collìnus** (Greene) Abrams. [*C. c.* Greene. *C. tridactylosus* Eastw.] Stems tufted, mostly less than 1 dm. long; lvs. mostly reniform-hastate, broader than long; outer sepals pubescent down the middle only.—Rocky or gravelly places, often on serpentine, below 2000 ft.; Chaparral; Coast Ranges from Mendocino Co. to Santa Clara and San Benito cos. April–June.

Ssp. **pedicellàtus** (Jeps.) Abrams. [*C. villosus* var. *p.* Jeps.] Stems 1–3 dm. long; lvs. deltoid-hastate, the basal lobes commonly entire; outer sepals hairy-tomentose.— Dry slopes and ridges, mostly 1500–6000 ft.; Foothill Wd., Chaparral; Mt. Hamilton Range, Alameda Co. to Mt. Pinos and vicinity, Ventura Co. April–July.

9. **C. tomentéllus** Greene. Rootstocks slender, the stems several, ascending to prostrate, 2–4 dm. long; herbage grayish-green, slightly tomentose; lvs. deltoid-hastate, 1.5–3 cm. long, the basal lobes entire to shallowly 2-toothed; petioles from shorter than to longer than petioles; peduncles 1-fld., ca. as long as petioles; bracts near to calyx, lanceolate to lance-ovate, mostly 8–10 mm. long; corolla cream, 3–3.5 cm. long; stigmas linear-oblong.—Dry slopes, 4000–6500 ft.; Foothill Wd., Yellow Pine F.; Greenhorn and Tehachapi mts., Kern Co. May–July.

10. **C. fulcràtus** (Gray) Greene. [*C. luteolus* var. *f.* Gray. *C. auriculaefolius* Eastw.? *C. gracilentus* Greene.] Rootstocks slender; stems slender, trailing, 1–5 dm. long; herbage light green, minutely puberulent to subvillous; lvs. triangular-hastate, 1.5–5 cm. long, the basal lobes half as long as to nearly as long as the lanceolate to lance-ovate cent. lobe; petioles shorter than blades except on lowest lvs.; peduncles slender, 1-fld., mostly exceeding lvs.; bracts well below the calyx, ± hastate like the lvs.; sepals unequal, oblong, blunt to truncate, pubescent, usually mucronate; corolla white to cream, 2.5–3.5 cm. long; stigmas linear; seeds dark, ca. 4 mm. long, minutely reticulate-ridged.—Dry slopes, below 8500 ft.; Chaparral, Foothill Wd., Yellow Pine F.; foothills of the Sierra Nevada from Shasta Co. to Fresno Co., also from San Gabriel Mts. to Santa Rosa Mts. and Laguna Mts. May–Aug.

Var. **Bérryi** (Eastw.) Jeps. [*C. B.* Eastw.] Stems conspicuously villous, 6–8 dm. long; lvs. 2.5–5 cm. long and wide, the lobes squarish to poorly developed; bracts large, 1.5–2 cm. long; sepals 1.5–1.8 cm. long.—Yellow Pine F.; Fresno and Tulare cos.

Var. **deltoìdes** (Greene) Jeps. [*C. d.* Greene.] Lvs. 1.2–1.5 cm. long, deltoid, somewhat broader; herbage densely short-canescent.—At 3000–5000 ft.; Foothill Wd., Yellow Pine F.; Tehachapi Mts., Mt. Pinos.

11. **C. polymórphus** Greene. Perennial from slender rootstocks; stems slender, trailing or erect, 3–6 dm. long, the herbage pale, puberulent; lvs. reniform-hastate to narrow-sagittate, 2–4.5 cm. long, the basal lobes rounded to shallowly 2-toothed; petioles mostly not longer than blades; peduncles 1-fld., equaling or slightly exceeding lvs.; bracts 1–6 mm. below the calyx, green, lanceolate to narrow-elliptic, 4–10 mm. long; sepals unequal, the outer rounded at apex, the inner obscurely mucronate; corolla white, cream within, 2.5–3.5 cm. long; seeds dark, ca. 4 mm. long, minutely reticulate-ridged.—Dry slopes, mostly 1000–6000 ft.; Chaparral, Foothill Wd., Yellow Pine F.; Marin and Nevada cos. n. to Del Norte and Modoc cos.; s. Ore. May–July.

12. **C. occidentàlis** Gray. [*C. luteolus* Gray, not Spreng. *C. o.* ssp. *fruticetorum* Abrams. *C. purpuratus* var. *f.* Heller. *C. illecebrosus* House. *C. f.* Greene.] Tall perennial, mostly climbing over bushes and trees, ± puberulent; lf.-blades triangular-hastate, mostly 1–4 cm. long, blunt, the basal lobes broad, usually 2-toothed, the cent. lobe ± ovate; petioles shorter than blades; peduncles 1–3-fld., much longer than lvs.; bracts 5–15 mm. below the calyx, lanceolate to lance-oblong, 5–10 mm. long; sepals unequal, mostly lance-ovate, 6–12 mm. long, acuminate; corolla white or pinkish, aging purplish, 3–4 cm. long; stigmas linear; seeds dark, ca. 4 mm. long, with small lighter papillae.—Open or bushy slopes and ridges, below 3500 ft.; Chaparral, Foothill Wd., Yellow Pine F.; inner Coast Ranges from Napa Co. to s. Ore. and foothills of Sierra Nevada from Amador Co. n.

KEY TO VARIETIES

Lvs. pubescent beneath . . . . *C. occidentalis*
Lvs. glabrous beneath.
    Lf.-blades with rounded lobes, the lateral scarcely distinct from the cent. . . . . var. *saxicola*

Lf.-blades with sharply pointed lobes, the lateral quite distinct from the cent.
Corolla ochroleucous, ca. 3–3.5 cm. long .... var. *solanensis*
Corolla white to purple-white, 4–4.5 cm. long .... var. *purpuratus*

Var. **solanénsis** (Jeps.) J. T. Howell. [*C. luteolus* var. *s.* Jeps.] Lvs. glabrous, sharp-pointed; fls. ochroleucous, ca. 3 cm. long.—Dry slopes away from immediate coast, Monterey Co. to Humboldt Co.

Var. **purpuràtus** (Greene) J. T. Howell. [*C. luteolus* var. *p.* Greene. *C. p.* Greene.] Glabrous; lf.-lobes acute, the cent. one lanceolate; bracts entire; fls. white to purplish, 4–4.5 cm. long.—Below 1000 ft.; Chaparral; near the coast from Humboldt Co. to Ventura Co.

Var. **saxícola** (Eastw.) J. T. Howell. [*C. s.* Eastw. *C. luteolus* var. *s.* Jeps. *C. purpuratus* var. *s.* Jeps.] Glabrous; lf.-blades 2.5–4 cm. long, the lobes obtuse or rounded; bracts hastate-lobed to entire; corolla white, ca. 4–5 cm. long.—N. Coastal Scrub; Mendocino Co. to Marin Co.

13. **C. Peirsònii** Abrams. Rhizomes branching, slender; stems trailing to somewhat twining, 2–4 dm. long; herbage glabrous and glaucous; lvs. lance-hastate, the cent. lobe 1.5–2 cm. long, the lateral lobes almost to quite as long, entire or with 1–2 shallow teeth; petioles 1–2.5 cm. long; peduncles 1-fld., 1–5 cm. long; bracts narrowly to broadly oval, closely subtending the calyx, 5–8 mm. long, roundish to obtuse at apex, greenish to purplish; sepals oblong-ovate, 10–12 mm. long, rounded at apex and usually recurved-mucronate; corolla white, 3–4 cm. long; stigmas oblong.—Dry slopes, 3000–4500 ft.; Creosote Bush Scrub, Joshua Tree Wd.; n. base of San Gabriel Mts., Los Angeles Co. May–June.

14. **C. lóngipes** Wats. Base woody, branched; stems erect to ascending, branched, glabrous, slender, 3–10 dm. high, sometimes twining; lvs. remote, linear to lance-hastate, 1–3(–5) cm. long, the lower with well-developed spreading basal lobes, the upper gradually reduced to linear bracts; petioles 1–2 cm. long; peduncles slender, 1–2-fld., 5–15(–18) cm. long; bracts lance-linear, near to remote from calyx; sepals broadly oval, rounded and mucronate at apex, unequal, 6–10 mm. long; corolla white to cream, often with lavender veins without, 2.5–3.5 cm. long; stigmas linear-oblong; seeds dark, ca. 4 mm. long, smooth.—Dry slopes, mostly 2000–4000 ft.; Creosote Bush Scrub, Chaparral, Joshua Tree Wd., Foothill Wd.; w. edge of desert from e. San Diego Co. to Inyo Co., head of San Joaquin V. from Kern Co. to San Luis Obispo Co.; w. Nev. May–July.

15. **C. arvénsis** L. BINDWEED. Deep-rooted perennial with prostrate or somewhat twining stems 3–10 dm. long; herbage glabrous to pubescent; lf.-blades oblong-sagittate to ovate, obtuse to rounded at apex, mostly 1.5–3.5 cm. long; petioles slender, shorter than the blades; peduncles mostly 1-fld., shorter than lvs.; bracts well below calyx, mostly subulate to narrowly spatulate; sepals oblong, obtuse, ca. 3 mm. long; corolla white or with some pink, 1.5–2 cm. long; seeds dark, finely punctate, obovoid, 3–4 mm. long; $2n = 50$ (Wolcott, 1937).—A very deep-rooted and troublesome weed in orchards and fields and waste places, below 5000 ft.; to Atlantic Coast; natur. from Eurasia. May–Oct.

16. **C. símulans** L. M. Perry. [*Breweria minima* Gray, not *C. m.* Aubl. *C. pentapetaloides* auth., not L.] Diffuse minutely puberulent to pubescent annual 1–3 dm. high; lvs. oblong to linear-lanceolate, 1.5–4 cm. long, narrowed gradually at base to a shorter somewhat winged petiole; peduncles 1-fld., shorter than lvs.; bracts spatulate to subulate, 4–5 mm. below the calyx, 3–8 mm. long; sepals oblong-ovate, pubescent, 3–4 mm. long, scarious-margined; corolla pinkish, deeply 5-cleft, 6 mm. long; seeds dark, ca. 2 mm. long, minutely papillate.—Occasional, grassy and rocky places, below 1000 ft.; V. Grassland, Coastal Sage Scrub; from Contra Costa Co. along both sides of San Joaquin V. to cismontane s. Calif., Santa Catalina Id.; L. Calif. March–May.

17. **C. althaeoìdes** L. Perennial; stems climbing, 1–3 m. high; herbage ± strigose; lower lvs. cordate-ovate, crenate, the upper ovate in outline, 3–6 cm. long, 3–7-lobed into irregularly lobed or toothed divisions; petioles mostly 1–2.5 cm. long; peduncles 1–2-fld., exceeding lvs.; bracts subulate, 1–1.5 cm. below the calyx; sepals oval, 7–8 mm. long; corolla purple to rose, 2.5–3 cm. long.—Locally established as a weed, Ventura, Los Angeles, and Orange cos.; native of Medit. region. May–Oct.

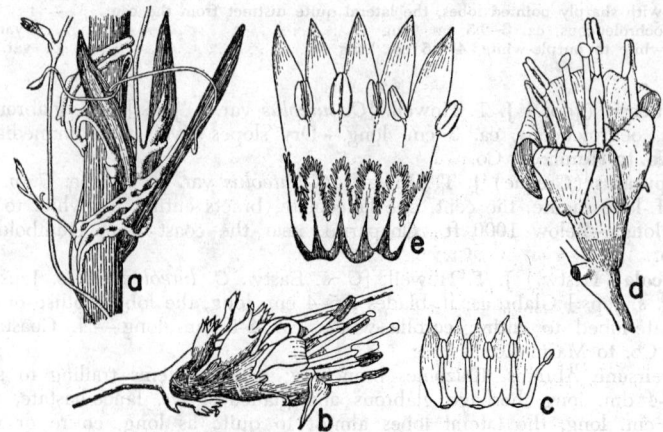

Fig. 54. CUSCUTACEAE. *Cuscuta: a,* slender stems on host, × 2½, showing knoblike haustoria; *b,* fl., × 1, 5-lobed calyx, 5-lobed corolla, 5 stamens, 2 styles; *c,* corolla opened to show stamen insertion. *C. subinclusa: d,* fl., × 5; *e,* corolla opened to show fringed scales in tube.

57. Cuscutàceae. DODDER FAMILY (Fig. 54)

Parasitic plants without chlorophyll; the stems slender, twining, yellow to orange, and fastened to their hosts by knobs or haustoria. Lvs. reduced to minute scales. Fls. small, perfect, mostly waxy-white, in lateral often compact cymose clusters. Calyx 4–5-cleft, colored like the corolla. Corolla 4–5-lobed, campanulate to urn-shaped or cylindrical, the lobes imbricate in bud, the tube usually with small crenulate or appendaged scales alternating with the corolla-lobes. Stamens alternate with the lobes and inserted in the throat or sinuses above the scales; anthers ovoid to ellipsoid, 2-celled. Pistil 1, superior; ovary 2-celled; styles 2, distinct or united below; stigmas capitate or linear. Caps. globose or ovoid, circumscissile or irregularly dehiscent or indehiscent. Seeds 1–4, globose to oblong or ovoid, sometimes angular; endosperm fleshy; cotyledons none, but embryo-apex with 1–4 minute scales.

1. Cuscùta L. DODDER

A single genus in the family. Ca. 100 spp., of wide distribution. (Name supposed to be of Arabic derivation.)

(Yuncker, T. G. The genus Cuscuta. Mem. Torrey Bot. Club 18: 113–331, 1932.)
A. Stigmas capitate; caps. bursting irregularly.
   B. Caps. globose, ± depressed, sometimes thickened about the style-bases.
      C. Ovary and caps. without a thickened stylopodium at apex.
         D. Corolla with scalelike appendages attached to the tube below the stamens.
            E. Fls. 4-merous; caps. capped at apex by the withered corolla
                                                        2. *C. Cephalanthi*
            EE. Fls. 5-merous; caps. having the withered corolla about the base.
              F. Corolla-lobes obtuse, the tips not inflexed .... 1. *C. obtusiflora*
              FF. Corolla-lobes acute, spreading but with the tips inflexed.
                  G. Fls. as broad as or broader than long; corolla subglobose.
                      H. Calyx-lobes broadly overlapping at sinuses to form angles;
                          fls. 1.5–2 mm. long .... 3. *C. pentagona*
                    HH. Calyx-lobes not overlapping at sinuses to form angles; fls.
                        2–3 mm. long .... 4. *C. campestris*
                GG. Fls. longer than broad; corolla campanulate to funnelform
                                              5. *C. suaveolens*
         DD. Corolla lacking scalelike appendages at base of stamens.
            E. Corolla-lobes triangular, ca. as broad as long, shorter than the tube
                                              6. *C. Jepsonii*
            EE. Corolla-lobes oblong or lanceolate, longer than broad.
               F. The corolla-lobes mostly shorter than the tube; calyx-lobes broadly
                  ovate, mostly less than half as long as corolla-tube .. 7. *C. brachycalyx*

FF. The corolla-lobes longer than the tube; calyx-lobes lanceolate to lance-
ovate, longer than half the corolla-tube.
  G. Anthers oval; corolla-lobes spreading, starlike in fr.; fls. sessile
                                                   8. *C. occidentalis*
  GG. Anthers linear-oblong; corolla not star-shaped; fls. pedicelled
                                                  9. *C. californica*
  CC. Ovary and caps. with a thickened stylopodium surrounding base of the styles
                                             10. *C. indecora*
BB. Caps. ovoid, conic, or beaked, commonly longer than thick.
  C. Calyx-lobes acute to acuminate, the edges entire.
    D. Calyx shorter than the corolla-tube; fls. ca. 5 mm. long; calyx-lobes overlapping
                                         11. *C. subinclusa*
    DD. Calyx as long as or longer than corolla-tube; fls. mostly smaller; calyx-lobes not
markedly overlapping.
      E. Corolla-lobes long-attenuate at tips; stems pale; seeds ca. 1 mm. long. On
plants of mt. meadows .... 12. *C. Suksdorfii*
      EE. Corolla-lobes not markedly long-attenuate; stems orange; seeds ca. 1.5 mm.
long. On plants at saline places at low elevs. .... 13. *C. salina*
  CC. Calyx-lobes round, obtuse, denticulate .... 14. *C. denticulata*
AA. Stigmas linear; caps. circumscissile.
  B. Fls. membranous rather than fleshy; calyx-lobes not thickened and turgid at apex; seeds
rounded .... 15. *C. Epithymum*
  BB. Fls. fleshy; calyx-lobes thickened and turgid at apex; seeds oblong, ± angled
                                         16. *C. approximata*

1. **C. obtusiflòra** HBK. var. **glandulòsa** Engelm. [*C. g.* Small.] Stems medium, light yellow; fls. glandular, subsessile, in compact glomerulate clusters ca. 2 mm. long; calyx-lobes broadly ovate, obtuse, ca. as long as corolla-tube; corolla campanulate, the tube short, lobes triangular-ovate, acutish, spreading or reflexed; scales large, ovate, prominently fringed, mostly exserted; stamens shorter than corolla-tube; styles stoutish, subulate, ca. as long as globose ovary; caps. depressed-globose, bursting irregularly; seeds ovoid, ca. 1.5 mm. long, the hilum oblong, diagonal.—Often on *Polygonum;* reported from El Monte, San Bernardino; to Fla., Mex., W.I. May–Oct.

2. **C. Cephalánthi** Engelm. Stems medium, yellow; fls. subsessile or sessile, clustered, ca. 2 mm. long, mostly 4-merous, sometimes glandular; calyx shorter than corolla-tube, deeply divided, with overlapping lobes oblong-ovate, obtuse; corolla cylindric-campanulate, or suburceolate in age, the lobes ovate, obtuse, erect to spreading, much shorter than tube; scales oblong, narrow, fringed; stamens ca. as long as lobes; styles equaling or slightly exceeding the subglobose ovary; caps. globose, often glandular, without stylopodium; seeds ca. 1.6 mm. long, light brown, globose to ovoid, slightly compressed, the hilum oblong, linear, oblique; $2n = 60$ (Fogelberg, 1938).—On many hosts; Shasta and Siskiyou cos.; to Wash. and Atlantic Coast. July–Sept.

3. **C. pentagòna** Engelm. [*C. arvensis* Beyr. ex Hook. in syn.] Stems slender, pale yellow; fls. mostly in small loose clusters, ± glandular, ca. 1.5 mm. long, on pedicels ca. as long as fls.; calyx mostly ca. as long as corolla-tube, the lobes broadly ovate, obtuse, broadly overlapping and calyx appearing angular; corolla-lobes spreading, lanceolate, the tips acute, inflexed, ca. as long as tube; stamens shorter than lobes, the fils. slender, equaling or longer than anthers; scales becoming exserted, ovate, fringed at top; styles slender, not longer than ovary; caps. globose or ± depressed, protruding from withered corolla; seeds pinkish- or grayish-tan, smoothish, depressed-globose, often flattened on 1 surface, the hilum short, oblong, terminal, transverse, ca. 1 mm. long; $2n =$ ca. 56 (Fogelberg, 1938).—On many hosts; *Xanthium, Trifolium,* etc.; reported from Antioch; to e. U.S. July–Oct.

4. **C. campéstris** Yunck. [*C. arvensis* auth. in part. *C. pentagona* var. *calycina* Engelm.] Stems medium, light yellow; fls. 2–3 mm. long, often glandular, mostly on shorter pedicels and in compact globular clusters; calyx ca. equal to corolla-tube, the lobes oval or round, mostly overlapping when young, but not forming angles; corolla-lobes broadly triangular, acute, the tips often inflexed, ca. as long as the campanulate tube; stamens shorter than corolla; scales ovate, fringed, exserted; styles slender, shorter than ovary; caps. depressed-globose; seeds dull pinkish- or grayish-tan, ovoid, ca. 1.5 mm. long, usually flattened on one side, the hilum short, oblong, terminal, transverse; $2n = 56$ (Fogelberg, 1938).—On many hosts, especially of family *Compositae;* cismontane and sometimes transmontane Calif.; to Wash., Atlantic Coast, W. Indies, S Am. July–Nov.

5. **C. suaveólens** Ser. [*C. racemosa* var. *chiliana* Engelm.] Stems slender, straw-colored; fls. 3–4 mm. long, ± glandular, membranous, on mostly shorter pedicels, in racemose clusters; calyx-lobes shorter than corolla-tube, triangular-ovate, acutish, not overlapping, the sinuses ± rounded; corolla campanulate to funnelform, globular in age, the lobes triangular-ovate, erect, with tips inflexed and ca. half or more the length of the tube; stamens shorter than lobes; scales mostly not reaching the stamens, oblong-ovate to triangular, fringed; styles slender, mostly ca. as long as ovary; caps. globose, bursting irregularly; seeds 1.5–2 mm. long, roundish, the hilum oblong, perpendicular. —Mostly on legumes, often a pest in alfalfa fields; introd. from S. Am. Aug.–Oct.

6. **C. Jepsònii** Yunck. Stems pale, slender; fls. 2–2.5 mm. long, 5-parted, on pedicels shorter than fls., in cymose clusters, the infl. fleshy and papillate; calyx-lobes triangular, acute, scarcely half as long as corolla-tube; corolla globular, becoming urceolate, the lobes erect, ± connivent, triangular, acute, not overlapping, less than half the length of the tube; scales represented by ridges; stamens much shorter than corolla-lobes; styles ± subulate, much shorter than ovary; caps. depressed-globose; seeds rounded, compressed.—Uncommon, apparently largely on *Ceanothus;* Yellow Pine F.; N. Coast Ranges, Lake Co. to Trinity Co. July–Aug.

7. **C. brachycàlyx** (Yunck.) Yunck. [*C. californica* var. *b.* Yunck.] Stems medium, pale yellow; fls. ca. 4 mm. long, thin, semitransparent, pedicelled in loose clusters; calyx turbinate, much shorter than corolla, the lobes short, broadly ovate, acute to obtuse; corolla campanulate-globose, saccate between stamen-attachments, the lobes reflexed, oblong, obtuse or abruptly acute, not longer than tube; scales lacking; stamens shorter than lobes; styles long, exserted; caps. globose, thin; seeds ca. 2 mm. long, rounded, but flattened on both surfaces, the hilum short, oblong.—Mostly on *Eriogonum,* sometimes on *Compositae;* V. Grassland to Yellow Pine F.; in and about the Cent. V. June–Sept.

Var. **apodánthera** (Yunck.) Yunck. [*C. californica* var. *a.* Yunck.] Calyx much shorter than the corolla-tube, the lobes triangular-ovate, acute; corolla very thin, transparent, the lobes lanceolate, acute, shorter than the tube; fils. and styles mostly shorter than in the sp.—Often on *Eriogonum;* Yellow Pine F.; Yosemite, etc.

8. **C. occidentàlis** Millsp. [*C. californica* var. *breviflora* Engelm.] Stems medium, yellow; fls. ca. 3 mm. long, often glandular, mostly sessile in small compact clusters; calyx equaling or exceeding corolla-tube, the lobes lance-ovate, acuminate, fleshy and thickened at base; corolla globose, saccate between the stamen attachments, becoming globular about the developing caps.; lobes lanceolate, acuminate, spreading in age; stamens shorter than the lobes; scales lacking; styles longer than ovary; caps. globose; seeds ca. 1.5 mm. long, oval, flattened on 2 surfaces, the hilum short.—On various hosts, like *Grindelia, Solanum,* etc.; many Plant Communities; from the coastal cos. of Calif. to Wash., Colo. June–Aug.

9. **C. califórnica** H. & A. [*C. c.* vars. *graciliflora* and *longiloba* Engelm.] Stems medium, yellow; fls. 3–5 mm. long, on short pedicels, forming loose cymose-paniculate clusters; calyx somewhat shorter than to longer than corolla-tube, the lobes lanceolate to triangular, acuminate to acute, overlapping; corolla cylindric-campanulate, often saccate between the stamen-attachments, the lobes narrow, lanceolate, acute, reflexed, longer than tube (erect and connivent in the bud contrasting with the spreading calyx-lobes); scales lacking; stamens shorter than lobes; styles much longer than ovary; caps. globose, not circumscissile; seeds slightly over 1 mm. long, light brown, rounded but flattened somewhat on 2 sides, the hilum short.—On many herbs and shrubs, sea level–8200 ft.; many Plant Communities; most of cismontane Calif., occasional on deserts; to Wash., L. Calif. May–Aug.

Var. **papillòsa** Yunck. Fls. and pedicels densely papillate.—Uncommon, with the sp. in Calif.

Var. **apiculàta** Engelm. Corolla granulate; ovary ovoid, pointed.—Known from a single collection on the Colo. River.

10. **C. indecòra** Choisy. [*C. i.* var. *neuropetala* (Engelm.) Hitchc. *C. n.* Engelm. *C. decora* Engelm. *C. pulcherrima* Scheele.] Stems rather coarse, yellow; fls. 3–4 mm. long, loosely or compactly clustered on pedicels as long or longer; calyx shorter than the corolla-tube, the lobes triangular-ovate, mostly obtusish, ± granular; corolla broadly campanulate, the lobes erect or somewhat spreading, fleshy and glandular toward the

tips; stamens shorter than the lobes; scales equaling or exceeding corolla-tube, fringed; styles equaling or longer than ovary; caps. globose; seeds tan to brownish or grayish, ca. 1.7 mm. long, roundish or broader than long, somewhat scurfy, the hilum small, transverse or somewhat oblique.—On many hosts, particularly *Ambrosia, Aster, Asclepias, Artemisia, Medicago,* etc., and often in moist places like alfalfa fields; much of cismontane and transmontane Calif.; to Ill., Mex., se. U.S., W. I., S. Am. July–Oct.

11. **C. subinclùsa** Dur. & Hilg. [*C. Ceanothi* Behr.] Stems rather coarse, orange; fls. ca. 5–6 mm. long, sessile or on short pedicels, the clusters few- to several-fld. and scattered or in dense continuous masses; calyx usually less than half as long as corolla-tube, the lobes lanceolate to ovate, acute, overlapping; corolla cylindrical, the lobes slightly overlapping, ovate, erect to spreading, shorter than tube, edges of lobes cellular-crenulate; scales oblong, ca. half as long as tube, short-fringed; styles slender, much longer than ovary; caps. ovoid, pointed, with a collarlike thickening about the intrastylar aperture; seeds ca. 1.8 mm. long, globose, the hilum short, oblong, oblique.—Mostly on shrubs and trees (like *Ceanothus, Rhus, Prunus, Salix, Quercus*), below 5000 ft.; many Plant Communities; cismontane Calif., occasional e. of Sierra Nevada; Ore., L. Calif. June–Oct.

12. **C. Suksdórfii** Yunck. var. **subpedicellàta** Yunck. Stems slender, pale; fls. 2–3 mm. long, 4–5-merous, sessile or subsessile, in few-fld. umbellate clusters; calyx enclosing the corolla, the lobes triangular, acute; corolla-tube campanulate, the lobes triangular-ovate, acuminate, erect; stamens shorter than lobes; scales oblong, shorter than tube, represented by 2 shallowly dentate wings; styles shorter than ovary; caps. globose or depressed-globose, glandular; seeds ca. 1 mm. long, globose, the hilum oblong.—On *Calyptridium, Aster, Trifolium,* etc. at 5000–8000 ft.; Red Fir F., Lodgepole F.; Siskiyou, Alpine, Nevada, Tulare, and San Bernardino cos. July–Sept.

13. **C. salìna** Engelm. [*C. californica* var. *squamigera* Engelm. *C. sq.* Piper. *C. s.* var. *sq.* Yunck. *C. subinclusa* var. *abbreviata* Engelm.] Stems very slender, orange; fls. 2–3 mm. long, 5-merous, on mostly shorter pedicels, in umbellate-cymose clusters; calyx-lobes lance-ovate, acute to acuminate, ca. as long as corolla-tube; corolla-lobes ca. as long as the subcampanulate tube, lance-ovate, sometimes granulate, acute to acuminate, erect or spreading, ± overlapping; scales oblong, narrow, shorter than tube, fringed; fils. not longer than anthers; styles not longer than ovary; caps. broadly ovoid; seeds ca. 1.5 mm. long, round-ovoid, the hilum short, oval, transverse.—On *Chenopodiaceae, Cressa,* etc., in saline places at fairly low elevs., in most of Calif.; to B.C., Utah, Ariz. May–Sept.

Var. **màjor** Yunck. Fls. 3–4.5 mm. long, broadly campanulate, the lobes of the corolla more ovate.—Largely on *Salicornia;* B.C. to San Diego.

Var. **papillàta** Yunck. Fls. ca. 3 mm. long; calyx finely papillate, the lobes acuminate. —Coast of Mendocino Co.

14. **C. denticulàta** Engelm. Stems very slender, pale yellow; fls. ca. 2 mm. long, subsessile, in shortened few-fld. infls., often with lance-ovate acute bracts; calyx-lobes roundish, obtuse, denticulate, overlapping, enclosing the corolla-tube; corolla campanulate (or urn-shaped in fr.), the lobes oval to ovate, spreading, somewhat overlapping, ca. as long as tube; scales denticulate, oblong-ovate; styles shorter than ovary; caps. conic; seeds ca. 1 mm. long, round-ovoid, the hilum small.—On desert shrubs like *Larrea, Hymenoclea, Haplopappus,* etc., mostly below 4000 ft.; Creosote Bush Scrub, Joshua Tree Wd.; Mojave Desert to Mono Co. and n. edge of Colo. Desert; to Utah, Ariz. May–Oct.

15. **C. Epithỳmum** Murr. [*C. Trifolii* Bab.] Stems very slender, often reddish; fls. ca. 3 mm. long, 5-parted, sessile in many-fld. dense clusters; calyx equaling or shorter than corolla-tube, the lobes triangular, acute; corolla-lobes triangular, acute, spreading, shorter than campanulate tube; scales ± spatulate, fringed about upper part; styles ca. twice the length of ovary; caps. globose, circumscissile; seeds ca. 1 mm. long, rough, angled, compressed-ovoid, dull gray or gray-brown, the hilum short, oblong, transverse. —Largely on legumes like clover and alfalfa; rare in Calif.; native of Old World.

16. **C. approximàta** Bab. var. **urceolàta** (Kunze) Yunck. [*C. u.* Kunze. *C. planiflora* auth., not Ten.] Stems medium; fls. 2.5–3.5 mm. long, in compact globular clusters; calyx-lobes mostly broader than long, the tips fleshy and turgid; corolla campanulate, soon becoming globose about the developing fr., the spreading lobes ovate-orbicular; scales

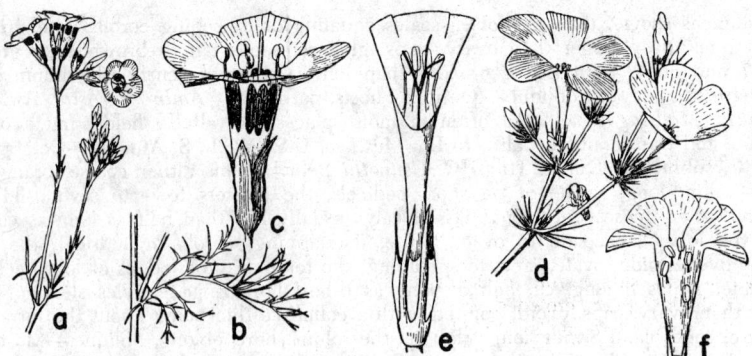

Fig. 55. POLEMONIACEAE. *Gilia tricolor: a,* infl., × ½; *b,* dissected lf., × ½; *c,* fl., × 1, 5-lobed calyx, narrowly campanulate corolla, 5 stamens. *Linanthus: d,* branch, × ½, with opposite palmately parted lvs., salverform corollas; *e,* fl., × 2, calyx with hyaline membranes in sinuses, 3-lobed stigma. *Phlox Douglasii: f,* corolla split lengthwise, × 1½, with unequal insertion of stamens.

mostly entire; anthers and stigmas often red; caps. depressed-globose, circumscissile in a definite line; seeds finely roughened, yellowish to green-brown, 1 mm. long, somewhat angled, ovoid, the hilum short, oblong.—On introd. weeds and alfalfa; in the Cent. V., and Lake, Mendocino, and Modoc cos. July–Oct.

58. Polemoniàceae. PHLOX FAMILY (Fig. 55)

Annual or perennial herbs or shrubs or vines. Lvs. alternate or opposite, simple or dissected or compound. Infl. usually a paniculate glomerate or flat-topped cyme, or the fls. in heads or even solitary. Fls. complete, regular, mostly 5-, rarely 4- or 6-, merous as to perianth and stamens. Calyx herbaceous to variously membranous or chartaceous, accrescent after anthesis or distended or ruptured by the caps., variously cleft, the lobes often with hyaline membrane between. Corolla gamopetalous, campanulate to funnelform or salverform, usually regular. Stamens equally or unequally inserted, the fils. equal or unequal. Ovary mostly 3-loculed; style simple; stigma-lobes usually 3. Fr. a caps., usually regularly dehiscent. Seeds 1 to many. Ca. 18 genera and 315 spp.; most numerous in w. N. Am.

A. Calyx growing with the caps. and not ruptured by it, becoming chartaceous in age.
  B. Calyx wholly green-herbaceous, the sinuses not distended; lvs. pinnately compound; mostly perennial . . . . 1. *Polemonium*
  BB. Calyx ± scarious between the lobes, the sinuses distended or replicate; lvs. entire to bipinnately dissected; mostly annual . . . . 3. *Collomia*
AA. Calyx at length ruptured by the maturing caps., commonly with a membranous pseudotube formed by the coalescence of the membranes bordering the sepals.
  B. Lvs. all involucral or bracteate, connate or perfoliate at base; true foliage-lvs. lacking; small annuals . . . . : . . . 4. *Gymnosteris*
  BB. Lvs. either cauline or basal or both, sometimes also bracteate.
    C. Lvs. usually opposite, at least near base of plant, the upper sometimes alternate (if the lower are alternate, plant usually woody and lvs. prickly).
      D. Stamens unequally inserted; lvs. entire.
        E. Plants perennial; fls. showy . . . . 5. *Phlox*
        EE. Plants annual; fls. inconspicuous . . . . 6. *Microsteris*
      DD. Stamens equally inserted; lvs. mostly palmately cleft.
        E. Plants annual or perennial herbs, usually not very woody below; lvs. not spinose . . . . 13. *Linanthus*
        EE. Plants subshrubby; lf.-lobes mostly ending in ± spinose tips
                                                          12. *Leptodactylon*
    CC. Lvs. mostly all alternate, entire to pinnately dissected.
      D. Calyx-lobes subequal; fls. solitary, cymose, or in heads.
        E. Corolla-lobes essentially alike; lf.-lobes mostly not setose- or spine-tipped.
          F. Upper cauline lvs. usually much reduced (see also *G. capillaris, G. filiformis*); tips of lf.-lobes usually not conspicuously mucronate; bracts mostly subtending groups of fls.; seeds small, rounded, brown
                                                            7. *Gilia*

FF. Upper cauline lvs. quite well developed.
    G.   Tips of lf.-lobes not conspicuously mucronate; bracts mostly sub-
           tending groups of 2–8 fls.; upper lvs. digitately lobed; seeds
           round-ovoid, mostly dark brown to black . . . . 2. *Allophyllum*
    GG. Tips of lf.-lobes bearing horny mucros; bracts subtending individ-
           ual fls.; upper lvs. pinnatifid to entire; seeds long and slender to
           shorter and angled, ± curved, pale brown to straw-colored
                                                  8. *Ipomopsis*
  EE. Corolla-lobes ± unlike; lf.-lobes setose- or spine-tipped . . . . 10. *Langloisia*
DD. Calyx-lobes unequal; fls. in dense bracteate heads.
    E.   Plants cobwebby-pubescent, at least the infl. with a feltlike mass of inter-
       laced hairs; caps. dehiscent from top; lvs. and bracts rarely with rigid
       spinose lobes . . . . 9. *Eriastrum*
    EE. Plants not having a feltlike mat of interlaced hairs; caps. dehiscent mostly
       from below or indehiscent; lvs. and bracts usually with rigid spinose lobes
                                           11. *Navarretia*

1. Polemònium L.

Erect spreading or decumbent annuals, or cespitose or rhizomatous perennials, with simple or branched stems. Lvs. pinnate, the lfts. entire or divided. Fls. in terminal or axillary cymes, congested or lax. Calyx green, ± campanulate, the lobes deltoid to acuminate. Corolla rotate-campanulate to narrow-funnelform, the tube and throat not sharply distinct, the lobes spatulate to roundish, blue or purple to pink, white or yellowish. Stamens usually equally inserted on the tube; fils. and style included to exserted. Caps. ovoid, trilocular, each locule 1–10-seeded. Seeds elongate, oblong or pointed at ends, in some spp. mucilaginous when wet. Ca. 20 spp. of cooler parts of N. Hemis. and of S. Am. (Ancient name, possibly for *Polemon*, early Athenian philosopher.)

(Davidson, J. F. The genus Polemonium. Univ. Calif. Pub. Bot. 23: 209–282, 1950.)
A.   Plants annual; corolla 2–6 mm. long . . . . 7. *P. micranthum*
AA. Plants perennial; corolla mostly longer.
    B.   Lfts. entire, mostly 5–30 mm. long; infl. cymose.
        C.   Stems mostly solitary, erect, 2–9 dm. long; lvs. mostly scattered along the stem.
           D.   Corolla-lobes ca. twice as long as tube, the corolla 13–15 mm. long. High montane
                                                 1. *P. caeruleum*
           DD. Corolla-lobes but slightly longer than tube, the corolla mostly 12–25 mm. long.
              Coastal and inland to mts. of Siskiyou Co. . . . . 2. *P. carneum*
        CC. Stems mostly clustered, cespitose, 1–3 dm. long; lvs. largely near base of plant.
           D.   Calyx-lobes ca. 1.5 times as long as tube; the 3 terminal lfts. confluent. Below tim-
              ber line . . . . 3. *P. californicum*
           DD. Calyx-lobes ca. as long as calyx-tube; terminal lfts. separate. Largely above timber
             line . . . . 4. *P. pulcherrimum*
    BB. Lfts. deeply lobed or divided, mostly 1–5 mm. long; infl. subcapitate.
        C.   Stamens and style ca. as long as corolla; petioles with a broad chartaceous base. Trinity
           and White mts. . . . . 5. *P. chartaceum*
        CC. Stamens and style shorter than corolla; petioles expanded at base but not chartaceous.
           Sierra Nevada . . . . 6. *P. eximium*

1. **P. caerùleum** L. ssp. **amygdalìnum** (Wherry) Munz. [*P. occidentale* Greene. *P. o.* ssp. *a.* Wherry. *P. c.* ssp. *o.* J. F. Davids. *P. Helleri* Brand.] Perennial from a rootstock; stems solitary, erect, glandular-pubescent, 2–9 dm. high; lvs. pinnate, the lfts. mostly 19–27, lanceolate to elliptic, acute or acuminate at tips with obliquely rounded base, 1–3.5 cm. long; infl. cymose, strict; bracts pinnatifid to entire; pedicels mostly shorter than calyx; calyx campanulate, 4–10 mm. long, the lobes ca. as long as tube, glandular-pubescent; corolla blue, rotate-campanulate, ca. 13–15 mm. long, the lobes rounded, ca. twice as long as tube; stamens ca. as long as corolla; fils. pubescent below; style exceeding corolla; caps. ca. 3 mm. long; seeds ca. 2 mm. long, not mucilaginous when wet.—Wet places, 3000–11,000 ft.; Montane Coniferous F.; San Bernardino Mts., Sierra Nevada from Tulare Co. n. to Modoc and Siskiyou cos.; to Alaska and Rocky Mts. June–Aug.

2. **P. cárneum** Gray. Perennial, 4–8 dm. high, branching from base, decumbent in age; stems stoutish, glandular-pubescent; lfts. 13–21, ovate to lance-oblong, 2–4 cm. long, the 3 terminal often confluent; infl. cymose; pedicels 2–10(–20) mm. long; calyx 8–18 mm. long, the lobes lance-ovate, slightly longer than tube, pubescent; corolla purple to pink or flesh-color, rotate-campanulate, 1–2.5 cm. broad, ca. as long, the lobes broad, rounded, slightly longer than tube; stamens ca. as long as corolla, the slender fils.

pubescent below; style longer than corolla; caps. 6–8 mm. long; seeds dark, ca. 2 mm. long; $n = 9$ (Davidson, 1950).—Brushy and grassy places, below 2000 ft. along the coast and up to 6000 ft. farther inland; N. Coastal Scrub, Coastal Prairie, Yellow Pine F.; San Mateo Co. to Del Norte and Siskiyou cos.; to Wash. April–Aug.

3. **P. califórnicum** Eastw. [*P. calycinum, tricolor,* and *Tevisii* Eastw.] Perennial from a slender horizontal rootstock; stems solitary to subcespitose, 1–2(–3) dm. high, glandular-pubescent; lvs. largest at base of plant, reduced upward, the lfts. 11–23, lanceolate to ovate, mostly acute, the upper 3 often confluent, 5–20 mm. long, subglabrous to glandular-pilose; infl. cymose, the bracts entire to pinnatifid; pedicels ca. as long as calyx; calyx narrow-campanulate, 5–8 mm. long, the segms. 1½ to 2 times as long as tube; corolla rotate-campanulate, 8–15 mm. broad, the blue lobes ca. twice as long as the white tube; stamens ca. as long as corolla, the fils. pubescent at base; style exceeding corolla; caps. ca. 3 mm. long; seeds dark, ca. 1.5 mm. long; $n = 18$ (Davidson, 1950).—Moist shaded places, mostly 6000–10,000 ft.; Red Fir F. to Subalpine F.; Sierra Nevada from Fresno Co. n., Trinity and Siskiyou cos.; to Wash., Ida., Mont. June–Aug.

4. **P. pulchérrimum** Hook. [*P. parvifolium* Nutt. *P. Berryi* and *shastense* Eastw.] Perennial from a branched root-crown, the stems suberect, 0.5–3 dm. high; herbage glabrous to minutely glandular-puberulent; lvs. largely basal, the lfts. 11–23, ovate to round, 4–8 mm. long, the terminal discrete; infl. a cyme, with entire to pinnatifid bracts; pedicels slender, often longer than calyx; calyx campanulate, 4–6 mm. long, the segms. ca. as long as tube; corolla rotate-campanulate, 5–8 mm. long, ca. as wide, the blue or purplish lobes ovate, obtuse, subequal to the yellowish tube; stamens ca. as long as corolla, the fils. basally pubescent; style ca. as long as corolla; caps. 3–4 mm. long; seeds dark, ca. 1.5 mm. long; $n = 9$ (Davidson, 1950).—Dry rocky often volcanic slopes, mostly 8000–11,000 ft.; Subalpine F., Alpine Fell-fields; Sierra Nevada from Mono and Mariposa cos. n., to Modoc, Siskiyou, and Trinity cos.; to Alaska, Mont., Wyo. June–Aug.

5. **P. chartàceum** Mason. [*P. confertum* var. *c.* Jeps.] Perennial from a woody caudex; stems several, cespitose, leafy, 1–3 dm. high; herbage glandular-pubescent; petioles with broad sheathing chartaceous base; lfts. many, verticillate, 3–5-lobed, the lobes spatulate, 1–3 mm. long; infl. subcapitate, the pedicels ca. half as long as calyx; calyx narrow-campanulate, 6–8 mm. long, the segms. rounded, obtuse, shorter than the tube; corolla funnelform, 11–13 mm. long, ca. as broad, the lobes rounded, obtuse, spreading, blue, ca. ⅓ as long as the white tube; stamens and style ca. as long as corolla.—Rocky slopes; Subalpine F., Alpine Fell-fields; ca. 6000 ft., Scott Mt. and Mt. Eddy, Siskiyou Co. and 11,000–14,000 ft., White and Sweetwater mts., Mono Co. July–Aug. Doubtfully specifically distinct from the next.

6. **P. exímium** Greene. [*P. confertum* var. *e.* Jeps.] Perennial from a woody caudex, cespitose, the stems 1–3 dm. high; herbage glandular-viscid, with strong musky odor; petioles expanded at base but not chartaceous; lfts. many, verticillate, 3–5-parted, 1–5 mm. long, the lobes spatulate to oblanceolate; infl. subcapitate, the pedicels shorter than calyx; calyx narrow-campanulate, 5–10 mm. long, the segms. slightly shorter than tube, rounded at apex; corolla narrow-funnelform to cylindrical, 12–15 mm. long, the spreading blue limb of rounded lobes not more than half as long as the whitish tube; stamens and style shorter than corolla; caps. ca. 4 mm. long, the seeds brown, ca. 1.5 mm. long.—Dry rocky ridges and slopes, 10,000–14,000 ft.; Alpine Fell-fields, Tuolumne Co. to Tulare and Inyo cos. July–Aug.

7. **P. micránthum** Benth. [*Polemoniella m.* Heller.] Annual, the stems 0.5–2.5 dm. long, solitary to diffusely branched, slender, glabrate to glandular-pilose, leafy; lfts. 7–15, narrow-elliptic to spatulate, 1–5 mm. long, glandular-pubescent; infl. cymose, each fl. on a branch with 1 lf.; pedicels 2–15 mm. long; calyx campanulate, 2–3 mm. long, the segms. mostly 1.5 times as long as the tube; corolla white to bluish, 2–6 mm. long, open-campanulate, the rounded lobes ca. as long as the tube; stamens shorter than corolla; caps. ca. 3 mm. in diam.; seeds dark, ca. 2 mm. long, mucilaginous when moistened; $n = 9$ (Grant, 1958).—Dry open places, 2000–6000 ft.; Sagebrush Scrub, N. Juniper Wd.; Siskiyou, Sierra, and Modoc cos., and Santa Barbara, Ventura, and Kern cos.; to B.C., Utah, Mont.; S. Am. March–May.

2. Allophýllum (Nutt.) A. & V. Grant

Erect annual herbs 1–6 dm. tall with slender or stouter stems, leafy, usually well branched cymosely, sometimes simple, lightly pubescent to villous or glutinous. Lvs. alternate, simple and entire to irregularly toothed, or pinnately to bipinnately lobed, the uppermost digitately 3–5-lobed; segms. linear-lanceolate to oblong, sessile or petiolulate. Fls. in loose to somewhat congested 2–8-fld. cymose glomerules, each glomerule subtended by a leaf; pedicels usually much elongate in fr. Calyx glandular-pubescent, accrescent, the lobes joined by the sinus-membrane to ca. the middle. Corolla funnelform, regular to ± bilabiate, light to dark blue-violet, red-violet, or white. Stamens unequally or equally inserted in upper half of tube, subequal to very unequal in length, included or exserted. Caps. subspherical, 3-loculed, with 1–3 seeds in each locule. Seeds black when ripe, 1.3–2.8 mm. long, plump, rounded on side, or angular on 1 end when 2–3 seeds occur in same locule. Five spp. of sw. N. Am. (Greek, *allos*, other, and *phullon*, lf.)

(Grant, Alva, and V. Grant. The genus Allophyllum [Polemoniaceae]. El Aliso 3: 93–110, 1955.)
A. Segms. of cauline lvs. lanceolate to oblong, 3–15 mm. wide, the terminal segms. 3–8 times longer than wide; fls. pale to medium blue-violet, white, or red-violet to purple with pink lobes.
    B. Corolla regular, white to red-violet to purple with pink lobes; at least some of stamens included; lvs. usually not in a basal tuft.
        C. Corolla-tube 6–16 mm. long, red-violet to purple with pink lobes; infl. usually of 4–8 fls. subtended by a lf. At 1000–5600 ft. .... 1. *A. divaricatum*
        CC. Corolla-tube 5–8 mm. long, it and the limb white; infl. usually of only 2 fls. subtended by a lf. At 4500–9000 ft. .... 2. *A. integrifolium*
    BB. Corolla slightly bilabiate, pale to medium blue-violet; stamens all usually exserted; basal tuft of lvs. usually well developed .... 3. *A. glutinosum*
AA. Segms. of lvs. linear to linear-lanceolate, not over 3 mm. wide, the terminal segms. of upper lvs. 8–14 times longer than broad; fls. dark blue-violet.
    B. Infl. glomerate with 4–8 fls. subtended by a lf.; many lvs. pinnately lobed; infl.-branches usually arising above the middle of a plant. Below 6000 ft. .... 4. *A. gilioides*
    BB. Infl. loose with 2–3 fls. subtended by a lf.; few lvs. pinnately lobed; infl.-branches arising throughout the plant. At 4000–8500 ft. .... 5. *A. violaceum*

1. **A. divaricàtum** (Nutt.) A. & V. Grant. [*Gilia d.* Nutt. *G. d.* var. *volcanica* Brand.] Robust, to 6 dm. high, with stout stems lightly pubescent to villous and viscid and with strong skunklike odor; lower part of plant not densely leafy; lower lvs. entire to pinnately lobed with 1–6 pairs of lobes; infl. congested in early stages, but with pedicels elongating in fr.; calyx in fr. 4–5.5 mm. long in n. Calif. and 6–7 mm. in s. Calif.; corolla regular, 7.5–22 mm. long, the tube red-violet and lobes pink, or tube purple and lobes pink-violet, the tube 2.5–4.5 times as long as lobes; stamens unequally inserted, unequal in length; stigma included or exserted; locules of caps. 1(2–3 in s. Calif.)-seeded; *n* = 8, 9 (Grant and Grant, 1955).—In open ± disturbed places, burns, etc., 1000–6000 ft.; Chaparral, Foothill Wd., etc.; Coast Ranges from Humboldt Co. to Monterey Co., Shasta Co. s. along w. slope of Sierra Nevada; Mt. Pinos, San Gabriel and San Bernardino mts. April–June.

2. **A. integrifòlium** (Brand) A. & V. Grant. [*Gilia gilioides* var. *i.* Brand.] Plant 1–2.5 dm. high, the stems villous, with 1 main leader; lower lvs. few, not forming a basal tuft, oblong to lanceolate, entire or irregularly toothed, the segms. of the cauline lvs. oblong, the terminal one broadest, 5–15 mm. wide and 3–6 times as long; fls. mostly in loose pairs; pedicels slender, unequal, 1–4 mm. in fl. and 6–18 mm. long in fr.; calyx in fr. 5–6.4 mm. long; corolla white or pale blue, regular, 6.5–11 mm. long, the tube 5.5–8 mm. long; stamens unequal, 0.4–2 mm. long, included; caps.-locules mostly 1-seeded; *n* = 8, 9 (Grant & Grant, 1955).—Open sandy or rocky places, 4500–9000 ft.; Yellow Pine F. to Lodgepole F.; San Gabriel and San Bernardino mts. through Sierra Nevada to Shasta Co. June–Aug.

3. **A. glutinòsum** (Benth.) A. & V. Grant. [*Collomia g.* Benth. *Gilia Traskiae* Eastw.] With 1–several rather stout stems from near base 1–6 dm. long, villous and viscid; with strong skunklike odor; lower lvs. forming a basal tuft, pinnately lobed with 3–10 pairs of lobes which are often irregularly toothed or lobed, the segms. 1–2 mm. wide, those of upper lvs. 3–11 mm. wide; infl. loose, 2–8-fld., the pedicels subequal, 5–20 mm. long in fr.; calyx 4.7–7.5 mm. long in fr. or shorter in plants with 1-seeded locules; corolla pale to medium violet-blue, 6–11 mm. long, the lobes 2.5–5 mm. long, spreading widely in

bilabiate manner; style and stamens lying along lower lip and curving upward at tip; stamens ± equally inserted and usually all exserted; stigma usually conspicuously exserted; caps.-locules mostly 2–3-seeded; $n = 9$ (Grant & Grant, 1955).—Shaded brushy slopes and ravines, below 5000 ft.; mostly Chaparral and Coastal Sage Scrub; Coast Ranges from Monterey Co. to L. Calif.; Santa Catalina Id. April–June.

4. **A. giliòides** (Benth.) A. & V. Grant. [*Collomia g.* Benth. *Gilia. g.* Greene. *G. g.* var. *Benthamiana* Brand and var. *ianthina* Jeps. & Hoov.] Main stems 1–several, 1–4 dm. high, usually arising from a leafy base and with many pinnately lobed lvs. up to the infl.; herbage puberulent throughout, ± glandular in the infl.; lower lvs. with 2–5 pairs of linear lobes, the rachis and the lobes 2–3(–4) mm. wide, the lobes 5–15 mm. long; upper lvs. shorter, pinnately to almost digitately lobed with 2 pairs of lobes 2–4 mm. broad; infl. glomerate, each fl.-branch subtended by a lf. and terminated by mostly 4 short-pedicellate fls.; calyx in fr. ca. 5 mm. long; corolla dark blue-violet, 6.5–9.5 mm. long; stamens subequal and subequally inserted, mostly included; stigma included; locules of caps. 1-seeded; $n = 9$ (Grant & Grant, 1955).—Open rather dry slopes and flats, below 6000 ft.; mostly Foothill Wd., Yellow Pine F.; Coast Ranges, s. Sierra Nevada, to mts. of San Diego Co., Panamint Mts.; Ariz. April–June.

5. **A. violàceum** (Heller) A. & V. Grant. [*Gilia v.* Heller.] Plants 0.7–4 dm. high with 1–several slender stems ± dichotomously branched to form a diffuse infl.; herbage puberulent throughout, glandular in infl.; basal and lower stem-lvs. few, 1–5 cm. long, entire or pinnately lobed with 1–2 pairs of lobes 1.5–8 mm. long; lvs. of infl. shorter, digitately lobed; infl. dense, the fls. typically in 2's on a fl.-branch, 1 borne laterally, the other terminally; fruiting calyx 3–5 mm. long; corolla regular, dark blue-violet, 5–8 mm. long; stamens subequally inserted and subequally long, mostly included; stigma included; caps.-locules 1-seeded; $n = 9$ (Grant & Grant, 1955).—Sandy and gravelly places, 4000–8500 ft.; Pinyon-Juniper Wd., Yellow Pine F., Red Fir F.; mts. from San Diego Co. to Santa Barbara Co. and through Sierra Nevada to Plumas Co., Death V. region; w. Nev. May–July.

3. Collòmia Nutt.

Annual or rhizomatous perennial herbs, erect to prostrate. Stems simple or branched. Lvs. alternate, linear to subovate, entire to pinnately dissected. Fls. in cymes or capitately congested or solitary and axillary. Calyx herbaceous or rarely with a chartaceous membranelike area below the sinuses, campanulate to obconic, the sinuses in age distended at base into a projecting fold. Corolla narrow-funnelform to trumpet-shaped, red, purplish, blue, white or salmon-yellow; the lobes spatulate to lanceolate. Stamens equally or unequally inserted on the throat, the fils. equal or unequal in length. Pistil included or exserted. Caps. ellipsoid or obovoid, the locules mostly 1–2-seeded. Seeds oblong, often ± mucilaginous when wetted. Ca. 15 spp., w. N. Am. and s. S. Am. (Greek, *kolla*, glue, because of mucilaginous wetted seeds of most species.)

(Wherry, E. T. Review of the genera Collomia and Gymnosteris. Am. Midl. Nat. 31: 216–231, 1944.)
A.   Plants annual.
    B.   Leaf outline oblong, elliptic or spatulate, the lower lvs. pinnatifid to incised.
        C.   Mature calyx 7–8 mm. long; lvs. usually markedly lobed to incised; fls. pink
                                                                    1. *C. heterophylla*
        CC. Mature calyx 10–12 mm. long; lvs. 3-toothed at summit to entire; fls. blue to purple
                                                                  2. *C. diversifolia*
    BB. Leaf outline linear to lanceolate, mostly entire.
        C.   Fls. paired or solitary, in open cymes; stems simple below, divaricately branched above.
            D.   Stamens unequally inserted; corolla 15–25 mm. long . . . . 3. *C. Tracyi*
            DD. Stamens equally inserted; corolla 8–12 mm. long . . . . 4. *C. tinctoria*
        CC. Fls. in terminal heads, sometimes also in stalked heads in upper lf.-axils; stems simple
            and erect, usually not branched.
            D.   Corolla 5–15 mm. long, the tube little exceeding calyx, the limb pink or white
                                                                     5. *C. linearis*
            DD. Corolla 15–30 mm. long, the tube much longer than calyx, the limb salmon-yellow
                                                               6. *C. grandiflora*
AA. Plants perennial.
    B.   Stems 2–7 cm. high; lvs. 2–3-dissected, hirsute-villous; fls. blue to pink, 12–15 mm. long
                                                                          7. *C. Larsenii*
    BB. Stems 10–60 cm. high; lvs. coarsely serrate or incised, glandular-villous; fls. orange-red,
        25–40 mm. long . . . . 8. *C. Rawsoniana*

1. **C. heterophýlla** Dougl. ex Hook. [*Gilia h.* Dougl. *Navarretia h.* Benth.] Annual, diffuse, branched from base, glandular-villous throughout, the stems slender, 5–20 cm. long; lvs. remote except about fl.-clusters, the lower blades 1–3 cm. long, ovate, pinnate to laciniately cleft with toothed or cleft segms. and on petioles half to as long as blade; upper blades crowded, subsessile, entire or nearly so; fls. sessile, congested; calyx campanulate, 5–8 mm. long, the tube chartaceous, with attenuate-lanceolate lobes; corolla rose to white, narrow-funnelform, 10–14 mm. long, the lobes spreading, 1–2 mm. long; stamens unequally inserted, the fils. unequal, the upper exserted; stigma reaching the lower stamens; caps. ellipsoid, 6–7 mm. long, the locules 2–3-seeded; seeds brown, swollen-oblong, somewhat roughened, ca. 1.3 mm. long; $n = 8$ (Flory, 1937).—Disturbed places, mostly below 4000 ft.; many Plant Communities; Sierra Nevada from Kern Co. n., Coast Ranges from Monterey Co. n.; to B.C., Ida. April–June.

2. **C. diversifòlia** Greene. [*Navarretia d.* Kuntze.] Erect divaricately branched annuals 3–10 cm. high, ± glandular-villous; lvs. entire or 3-toothed or -incised at tips, those of infl. mostly entire; fls. subsessile, congested in terminal clusters; calyx ca. 10–12 mm. long when mature, the sinuses chartaceous, the segms. lance-acuminate to lanceolate; corolla tubular-funnelform, yellow at throat, the limb ± purple; stamens unequally inserted, included; caps. oblong-ellipsoid; seeds 2 in each locule.—Serpentine outcrops, below 2000 ft.; Chaparral, N. Oak Wd.; inner N. Coast Ranges, Napa Co. to Colusa and Mendocino cos. May–June.

3. **C. Tràcyi** Mason. [*C. tinctoria* subvar. *luxuriosa* Brand.] Erect or spreading annual 5–20 cm. high, with forking glandular-puberulent stems; lvs. lanceolate to linear, acutish at both ends, subsessile or short-petioled, 2–6 cm. long, entire, the upper reduced and barely longer than fls.; fls. in clusters of 2–5, terminal or in lf.-axils and branch-forks, the clusters subtended by few leafy bracts; calyx 6–8 mm. long, the lobes lance-attenuate; corolla white to pink, 15–25 mm. long, the very slender tube sometimes purple; stamens unequally inserted, the lower often subsessile, the upper on long fils.; stigma included; caps. obovoid; seeds solitary in the locules, light brown, ca. 2 mm. long.—At 1000–6800 ft., mts. in drainage basin of Van Duzen, Mad, and Klamath rivers of Siskiyou, Humboldt, and Trinity cos. June–July.

4. **C. tinctòria** Kell. [*C. linearis* var. *subulata* Gray. *C. t.* var. *s.* Brand. *Gilia aristella* Gray.] Erect or spreading mostly cymosely forked annuals, glandular-viscid; lvs. linear to lanceolate, entire, mostly acutish at both ends, sessile or the lower short-petioled, the upper surpassing the fls.; fls. 1–3 in axils or in forks, those at ends of branches in subcapitate clusters; calyx campanulate, 5–7 mm. long, the lobes lance-aristate; corolla slender-funnelform, 8–12 mm. long, pink with red-violet tube; stamens almost equally inserted at top of throat with unequal fils., 1 or 2 exserted; stigma barely exserted; caps. obovoid, locules 1-seeded; seeds brown, ca. 2 mm. long.—Dryish open often disturbed places, below 9000 ft.; mostly Red Fir F., Lodgepole F.; Sierra Nevada from Mariposa Co. n. and Coast Ranges from Lake Co. n.; Ore., Nev. June–July.

5. **C. lineàris** Nutt. [*Gilia l.* Gray. *C. l.* var. *humilis* Brand.] Annual, usually simple and erect, 1–6 dm. high, sometimes lower, sometimes branched, puberulent throughout, often glandular above; lvs. lanceolate to linear, entire, sessile, 1.5–6 cm. long, mostly acute; fls. sessile in bracteate heads, terminal on branches and sometimes also in upper axils; bracts leafy; calyx campanulate, 4–7 mm. long, the tube chartaceous in age, the lobes green, lanceolate, attenuate; corolla pink to somewhat purplish, narrow-funnelform, 8–15 mm. long; stamens included, unequally inserted; stigma included; caps. ellipsoid, ca. 5 mm. long; seeds 1 in a locule, brown, ca. 2 mm. long; $n = 8$ (Flory, 1937).—Dry places, mostly 3000–10,500 ft.; Montane Coniferous F.; San Bernardino Mts., Sierra Nevada n. to Modoc Co.; to Alaska, Quebec, New Brunswick, Ariz. May–Aug.

6. **C. grandiflòra** Dougl. ex Lindl. [*Gilia g.* Gray. *C. g.* var. *tenuiflora* Benth.] Erect annual, usually simple and 1–10 dm. high, sometimes branched from base or even above, leafy throughout, glabrous or puberulent or glandular; lvs. lanceolate to linear, entire, sessile, 3–5 cm. long, passing upward into ovate leafy bracts; fls. sessile in dense terminal sometimes axillary heads; calyx obconic, 7–10 mm. long, becoming chartaceous, the lobes lanceolate; corolla narrow-funnelform to almost salverform, yellow-salmon to almost white, 15–30 mm. long; stamens unequally inserted, unequal in length; stigma included; caps. obovoid, ca. 5 mm. long; seeds 1 in each locule, brown, ca. 3 mm. long; $n = 8$ (Flory, 1937).—Dry open and wooded slopes, below 8000 ft.; many Plant

Communities; mts. from San Diego Co. n., through Sierra Nevada to Modoc and Siski-you cos. and s. in Coast Ranges to Lake Co.; to B.C., Rocky Mts. April–July.

7. **C. Larsénii** (Gray) Pays. [*Gilia L.* Gray. *C. debilis* var. *L.* Macbr.] Tufted peren-nial from slender short rootstocks from a deep-seated root; stems simple or branched from base, 3–8 cm. high; herbage pilose-hirsute; lf.-blades 5–15 mm. long, round-ovate in outline, 2 or 3 times pedately dissected into linear or obovate lobes; petioles shorter than or equal to blades; fls. in compact terminal cymes; calyx 6–8 mm. long, the lanceolate lobes longer than tube; corolla narrow-funnelform, 12–15 mm. long, the limb pink to violet, the tube very narrow; stamens equally inserted at base of throat, unequal in length, somewhat exserted; stigma exserted; caps. obovoid, the locules 1-seeded.—Loose volcanic material, ca. 11,400 ft., the type from Alpine Fell-fields, Mt. Lassen; Ore., Wash. July–Sept.

8. **C. Rawsoniàna** Greene. [*Gilia R.* Macbr.] Perennial from slender intricately branched rootstocks; stems slender, 1–4 dm. high, simple or few-branched above, glandular-pubescent, strong-scented; stems leafy, the blades lance-ovate to elliptic, coarsely deeply serrate, mostly 4–8 cm. long, the apex acutish, the cuneate base entire; petioles 3–20 mm. long; fls. crowded at ends of branches, short-pedicelled, subtended by a few leafy bracts; calyx campanulate, 8–12 mm. long, the lobes lanceolate, longer than tube, chartaceous only in age; corolla tubular-funnelform, 2.5–4 cm. long, the limb 1.5–2 cm. broad, orange-red, the throat and tube yellow; stamens equally inserted, but unequal in length and exsertion; stigma exserted; caps. ellipsoid, ca. 4–5 mm. long, the locules 1-seeded; seeds not becoming mucilaginous.—Local along streams, at 3700–7000 ft.; Yellow Pine F.; Madera Co. July–Aug.

4. Gymnósteris Greene

Diminutive annuals with simple slender stems with basal cotyledons, then naked save for the terminal heads of few fls. subtended by an invol. of 4–5 leafy bracts free and green above but united and scarious at base. Calyx urn-shaped, scarious below, the tips green. Corolla salverform or narrow-funnelform, white or yellow, persistent. Stamens sessile, borne in throat of corolla. Caps. many-seeded, dehiscent. Seeds obliquely cubical, with margined or winged angles, mucilaginous when wet. Two spp. (Greek, *gymnos,* naked, and *steros,* foundation, in reference to the leafless stem.)

1. **G. párvula** (Rydb.) Heller. [*Gilia p.* Rydb. *Gymnosteris nudicaulis* var. *p.* Jeps. *G. minuscula* Jeps.] Simple or sometimes with more than 1 stem, 1–5 cm. high; bracts 5–10 mm. long, purple-tinged; calyx 3–4 mm. long; corolla tube ca. 5 mm. long, the limb 1.5–3 mm. broad, pale yellow, sometimes purplish in age.—Gravelly or sandy slopes, 8300–11,800 ft.; largely Bristle-cone Pine F.; White Mts., Mono Co.; to Ore., Ida., Wyo., Colo. May–July.

5. Phlóx L.

Perennial or annual, herbaceous to suffrutescent, erect, diffuse or cespitose. Lvs. largely opposite, entire. Fls. in terminal corymbiform or paniculate cymes or sometimes solitary, often showy, red or purplish or blue to white. Calyx tubular or narrow-campanulate, 5-cleft and ribbed, the lobes acute or acuminate, usually scarious on the margins and with a scarious membrane below the sinuses. Corolla salverform, the tube slender and throat constricted. Stamens short, included, unequally inserted. Ovary 3-celled, ovoid to oblong; style mostly slender. Seeds 1–few in each cell, oblong, usually not emitting spiral threads when wet. Ca. 45 spp., of N. Am. and n. Asia, some of garden use. (Greek, *phlox,* flame, ancient name for Lychnis.)

(Brand, A., *in* Das Pflanzenreich, IV, 250: 55–87, 1907. Wherry, E. T. The genus Phlox. Morris Arb. Mon. III, 1–174, 1955.)

A. Lvs. round-ovate to elliptic, 1–4 cm. broad .... 5. *P. adsurgens*
AA. Lvs. linear to lanceolate, less than 1 cm. broad.
    B. Style 1–3 cm. long; plants usually lax, not cespitose or tufted, not with pungent or acerose lvs.
        C. Lvs. linear, 1–2 mm. wide; calyx-membrane carinate. Modoc Co. .... 1. *P. longifolia*
        CC. Lvs. ± lanceolate, 2–6 mm. wide.
            D. Corolla-tube 3.5–4.5 cm. long; calyx-membrane not markedly carinate below
                sinuses. San Bernardino Mts. .... 2. *P. dolichantha*

DD. Corolla-tube 1–3 cm. long; calyx-membrane mostly carinate below the sinuses. Siskiyou and Modoc cos. to e. San Bernardino Co. . . . . 3. *P. Stansburyi*
BB. Style less than 1 cm. long.
  C. Plants lax, not densely cespitose; corolla-lobes obcordate to deeply 2-lobed; lvs. 4–10 mm. broad . . . . 4. *P. speciosa*
  CC. Plants cespitose or tufted; lvs. crowded, subulate to linear, mostly rigid and acerose.
    D. The plants forming large patches from slender creeping underground rootstocks; stems 2–5 cm. high, tufted. Alpine in s. Sierra Nevada . . . . 6. *P. dispersa*
    DD. The plants cespitose, not conspicuously from creeping rootstocks.
      E. Pubescence glandular.
        F. Lf.-surfaces glandular-hispidulous, the lvs. deeply grooved beneath between the midrib and thickened margins. Mono Co. to San Bernardino Mts.
          G. Lvs. 3–4 mm. long, concave above; styles 2–3 mm. long
                                        7. *P. Covillei*
          GG. Lvs. 4–8 mm. long, flat above; styles 3–6 mm. long
                                        8. *P. caespitosa*
        FF. Lf.-surfaces subglabrous, the lvs. not deeply grooved beneath. Modoc and Siskiyou cos. . . . . 9. *P. Douglasii*
      EE. Pubescence not glandular.
        F. Lvs. mostly 10–15 mm. long, mostly not arched-spreading.
          G. The lvs. acerose but not very stiff or pungent, mostly yellowish-green; calyx-membrane not strongly keeled below the sinuses. San Gabriel Mts. to n. Calif. . . . . 10. *P. diffusa*
          GG. The lvs. stiff, ± pungent, gray-green; calyx-membrane strongly keeled below the sinuses. San Gabriel Mts. to L. Calif.
                                  11. *P. austromontana*
        FF. Lvs. 5–10 mm. long, pungent, ± arched-spreading. Inyo Co. to Modoc Co.
                                  12. *P. Hoodii*

1. **P. longifòlia** Nutt. var. **pubérula** E. Nels. [*P. p.* A. Nels. *P. viridis* E. Nels. *P. l.* ssp. *v.* Wherry. *P. Stansburyi* ssp. *compacta* Brand. *P. l.* ssp. *c.* Wherry.] Suffrutescent perennial, dwarf, compactly branched, mostly 6–10(–15) cm. high, rather densely leafy; herbage densely glandular-pubescent except toward base of plant; lvs. linear, 2–4 cm. long, 1–2 mm. wide, acuminate; cymes few-fld.; pedicels 1–2(–3) cm. long; calyx 8–10 mm. long, the sinus-membrane broad, inflated, replicate, the lobes linear, acuminate, slightly shorter than the tube; corolla pink or whitish, the tube 12–14 mm. long, the limb 15–18 mm. broad, with obovate, subentire or slightly irregular lobes; style filiform, ca. 1 cm. long including the 1 mm. long stigmas; caps. 3–4 mm. long, 1–2-seeded; seeds oblong, ca. 2.5 mm. long.—Reported from dry flats and slopes, ca. 4000–5000 ft.; Sagebrush Scrub; Modoc Co.; to Wash., Ida., Nev. May–June. (Typical *P. longifolia* is a taller more lax longer-leaved plant from much the same region. Various segregates have in recent years been recognized, on the basis of lf.-length and pubescence, as sspp., but are for the most part very trivial and not geographic.)

2. **P. dolichántha** Gray. [*P. bernardina* M. & J. *P. d.* var. *b.* Jeps.] Perennial herb from underground rootstocks, scarcely if at all suffrutescent; stems 1–3 dm. high, branching, subglabrous to somewhat glandular-pubescent in upper parts; internodes mostly from ⅓ to as long as lvs.; lvs. sessile, lance-linear, bright green, 2–4.5 cm. long, 2–6 mm. wide, acute to acuminate; fls. few, in loose cymose clusters, white or rose to pinkish-lavender; pedicels slender, 3–20 mm. long; calyx 10–12 mm. long, glandular-puberulent, the lobes subulate, ca. as long as tube; corolla-tube 35–45 mm. long, the limb 1.5–2 cm. broad, the oblong lobes 8–10 mm. long, subentire; stamens somewhat exserted; style filiform, 2.5–3 cm. long; caps. 5–8 mm. long, 1-seeded.—Open places in forest, 6000–8000 ft.; Yellow Pine F., Red Fir F.; Bear V., San Bernardino Mts. June–July.

3. **P. Stansbúryi** (Torr.) Heller. [*P. speciosa* var.? *S.* Torr. *P. longituba* Heller. *P. superba* Brand.] Suffrutescent, 1–2 dm. high from a branched root-crown, ± glandular-pubescent to -villous throughout, but especially so above; internodes mostly ⅛–½ as long as lvs.; lvs. gray-green, linear-lanceolate, 1.5–3 cm. long, gradually attenuate at apex; fls. rather few, in loose cymose clusters; pedicels 5–25 mm. long; calyx glandular-pubescent, 8–12 mm. long, the lobes subulate, considerably shorter than tube, the sinuses scarious, mostly carinate; corolla rose to whitish, the tube mostly 2–3 cm. long, the lobes oblong-spatulate, entire to slightly emarginate, 5–10 mm. long; stamens included; style almost as long as corolla-tube; caps. 5–7 mm. long, mostly 1-seeded.—Dry gravelly slopes and washes, 5000–9000 ft.; Sagebrush Scrub, Pinyon-Juniper Wd.; e. slope of Sierra Nevada from Mono and Inyo cos. through White and Inyo mts. to Nev., Ariz., New Mex. April–June.

Var. brevifòlia (Gray) E. Nels. [*P. longifolia* f. *b.* Gray. *P. l.* var. *b.* Gray. *P. l.* ssp. *b.* Wherry. *P. Grayi* Woot. & Standl.] Plants mostly compact, 5–10(–15) cm. high; corolla-tube mostly 12–18 mm. long.—With the sp., but more widespread; Modoc and Lassen cos. to e. San Bernardino Co.; thence to Utah, New Mex.

Var. hirsùta (E. Nels.) Jeps. [*P. h.* E. Nels.] Cespitose from stout woody base, the stems 5–15 cm. long; herbage hirsute throughout; lvs. crowded, broadly to narrowly lanceolate, 1–2 cm. long; corolla-tube 12–15 mm. long.—Dry slopes, ca. 3000–4000 ft.; N. Oak Wd.; near Yreka and Etna Mills, Siskiyou Co.

4. **P. speciòsa** Pursh ssp. **occidentàlis** (Durand) Wherry. [*P. divaricata* var. *o.* Durand. *P. o.* Durand ex Torr. *P. s.* var. *o.* Peck.] Suffrutescent, the stems 2–4 dm. high, foliose; herbage puberulent, or lower parts of plant subglabrous, the upper glandular-puberulent; lvs. rather coriaceous, lance-linear, 1–5 cm. long, 4–10 mm. broad, short-acuminate; cymes few-fld., the lower fls. leafy-bracted; pedicels 3–20 mm. long, very slender; calyx mostly 7–10 mm. long, glandular-puberulent, the lobes linear, erect, shorter than the tube; corolla bright pink, the tube 1–1.5 cm. long, the lobes obcordate to deeply 2-lobed, 7–12 mm. long; stamens included; style including stigmas 2–4 mm. long; caps. 6–7 mm. long, mostly with 1 seed ca. 3 mm. long.—Rocky hillsides and wooded slopes, 1500–7000 ft.; Mixed Evergreen F., N. Oak Wd., Yellow Pine F., Red Fir F., Sagebrush Scrub; Sierra Nevada from Fresno Co. n., N. Coast Ranges from Sonoma Co. n.; Ore. April–June.

Var. nítida Suksd. [*P. s.* ssp. *n.* Wherry.] Plant nearly or quite glabrous.—Sonoma, Trinity, Siskiyou, and Plumas cos.; to Wash.

5. **P. adsúrgens** Torr. Perennial from slender underground rootstocks, the stems slender, ascending or creeping, mostly 1–3 dm. long; herbage glabrous, but the infl. glandular-puberulent to -villous; lvs. round- to narrow-ovate, sessile, mostly 1–2.5 cm. long; infl. open, usually few-fld.; pedicels slender, mostly 0.5–2 cm. long; calyx 10–13 mm. long, glandular-pubescent, the lobes subulate, spreading, equaling or longer than tube; corolla bright pink to paler, the tube 12–20 mm. long, the lobes obovate, rounded at tip, sometimes emarginate, 7–12 mm. long; style ca. 14–16 mm. long; caps. ca. 5 mm. long; seeds ca. 3 mm. long; $2n = 14$, ca. 21 (Flory, 1934).—Open wooded slopes, 1500–6000 ft.; Mixed Evergreen F., Yellow Pine F.; Mendocino and Siskiyou cos.; w. Ore. June–Aug.

6. **P. dispérsa** C. W. Sharsm. Perennial from widely creeping slender underground rootstocks, the stems slender, tufted, mostly 2–5 cm. high; herbage green, with short rather harsh gland-tipped hairs; lvs. lance-linear, coriaceous, pungent, 5–10 mm. long, acerose, sessile; fls. few, subsessile; calyx ca. 7 mm. long, the lobes subulate, acerose, longer than the tube, the membrane of the sinuses not carinate; corolla white, the tube ca. 1 cm. long, the lobes obovate, irregular-margined, rounded, 5–6 mm. long; style ca. 2 mm. long.—Dryish flats of loose disintegrated granite, mostly 11,000–12,500 ft.; Alpine Fell-fields; Sierra Nevada of Tulare and Inyo cos. July–Aug.

7. **P. Covíllei** E. Nels. [*P. caespitosa* var. *C.* Brand. *P. Douglasii* var. *C.* Jeps.] Cushion-like from a woody densely cespitose perennial base; branchlets 1–3 cm. long, sub-glabrous; lvs. crowded, linear-oblong, appressed or ascending, thick, mostly 3–5 mm. long, abruptly narrowed to an apiculate tip, the margins thick, riblike, ciliate, the sur-faces glandular-puberulent, the lower surface grooved; fls. mostly solitary, sessile, terminal; calyx 5–6 mm. long, glandular-puberulent, the lobes like the lvs., shorter than or equaling the tube; corolla white or pale pink, the tube 8–10 mm. long, the lobes round-obovate, ca. 4–6 mm. long; style ca. 2 mm. long; caps. ca. 4 mm. long.—Dry slopes and benches, especially on travertine and limestone, 6000–10,000(–12,000) ft.; Pinyon-Juniper Wd., Yellow Pine F. and above; e. slope of Sierra Nevada and White and Inyo mts., mostly in Mono and Inyo cos., San Bernardino Mts.; sw. Nev. June–Aug.

8. **P. caespitòsa** Nutt. ssp. **pulvinàta** Wherry. Much like *P. Covillei,* forming clumps 3–7 cm. high with short divergent branches; lvs. 4–8(–10) mm. long, coarse-ciliate, flatter; calyx 7–8 mm. long; style ca. 4.5–5 mm. long.—Dry stony places, 10,000–13,000 ft.; Subalpine F., Alpine Fell-fields; mostly Inyo and Mono cos., less common in Fresno and Tuolumne cos.; to Ida., Colo. July–Aug.

9. **P. Douglásii** Hook. ssp. **rígida** (Benth.) Wherry. [*P. r.* Benth. *P. caespitosa* var. *r.* Gray.] Compact cespitose perennial from a woody base; stems 3–8 cm. high; herbage glandular-pubescent; lvs. linear-subulate, rigid with prominent midrib, pungent, glaucous-

green, mostly 4–8 mm. long; fls. 1–3 at ends of branches, sessile or nearly so; calyx 7–9 mm. long, the lobes ca. as long as tube, pungent, spreading, glandular-villous, the sinus narrow, not carinate; corolla pink to pale lavender or whitish, the tube ca. 1 cm. long, the lobes narrow-obovate, 5 mm. long; style 4–6 mm. long; caps. rounded or somewhat irregular at apex, ca. 4 mm. long.—Dry rocky and gravelly places, 4500–6000 ft.; Sagebrush Scrub, N. Juniper Wd., Yellow Pine F.; Modoc and Siskiyou cos.; to s. Wash. April–May.

10. **P. diffùsa** Benth. [*P. Douglasii* var. *d.* Gray. *P. Douglasii* var. *modocensis* Jeps.?] Suffrutescent perennial from stout branching base; stems 1–3 dm. long, prostrate or decumbent; herbage subglabrous to thinly tomentose, not glandular; lvs. linear-subulate, yellowish-green, 10–15 mm. long, acerose, not markedly pungent; fls. mostly solitary at ends of short leafy branches, on very short pedicels; calyx 8–10 mm. long, somewhat villous, the subulate lobes ca. as long as tube, the sinus-membrane not folded; corolla pink to lilac or white, the tube 9–13 mm. long, the lobes broadly to narrowly obovate, 6–7 mm. long, rounded at apex; style 3–5 mm. long; caps. 4–5 mm. long; $2n = 28$ (Flory, 1937).—Dry slopes and flats, 3300–11,500 ft.; Douglas-Fir F., Yellow Pine F. to Alpine Fell-fields; San Gabriel Mts. and Sierra Nevada n., Coast Ranges from Glenn Co. n.; Ore. May–Aug.

Ssp. **subcarinàta** Wherry. Lvs. tending to be flatter and 1–1.5 mm. wide; intercostal membrane of calyx somewhat carinate.—Uncommon, dry slopes, 4000–11,000 ft., San Bernardino Mts., Mt. Pinos (Ventura Co.), White Mts. (at 13,500), Sierra Nevada, to Siskiyou and Glenn cos.; to Wash. and w. Nev.

11. **P. austromontàna** Cov. [*P. Douglasii* var. *a.* Jeps. & Mason.] Cespitose perennial from a woody caudex, the stems 5–10(–15) cm. long; herbage gray-green, not glandular, but ± white-pubescent; lvs. mostly 1–1.5 cm. long, acerose to pungent, pubescent above, often glabrous beneath; fls. solitary at the ends of branchlets, short-pedicelled; calyx plicate in sinuses, 6–10 mm. long, the acerose teeth slightly longer than the angled tube; corolla white to pink or lavender, the tube 11–14 mm. long, lobes obovate, rounded, 5–7 mm. long; styles 4–5 mm. long; caps. 4–5 mm. long.—Dry rocky places, 4500–8000 ft.; Yellow Pine F., Pinyon-Juniper Wd.; San Gabriel Mts. to Santa Rosa Mts., Cuyamaca Mts.; to Ariz., Utah, L. Calif. May–July.

12. **P. Hoódii** Richards. ssp. **canéscens** (T. & G.) Wherry. [*P. c.* T. & G. *P. H.* var. *c.* Peck.] Pulvinate perennial 5–20 cm. broad, densely leafy; herbage woolly-villous, but internodes relatively glabrous; lvs. 5–10 mm. long, subulate, often arched-spreading, pungent, gray-green, subglabrous above, ± woolly toward base; fls. usually solitary, terminal, sessile or nearly so; calyx mostly 7–8 mm. long, woolly about base of pungent lobes, these longer than tube; corolla white to lilac, the tube 10–12 mm. long, the lobes obovate, rounded, ca. 6 mm. long; style 3–6 mm. long, $2n = 28$ (Flory, 1934).—Uncommon in dry rocky places, 4500–8000 ft.; Sagebrush Scrub, Pinyon-Juniper Wd., etc.; Modoc Co. to Inyo Co.; to Mont., Utah. May–July.

Ssp. **lanàta** (Piper) Munz. [*P. l.* Piper.] Densely white-woolly throughout; lvs. mostly 4–5 mm. long; calyx 6–7 mm. long.—Occasional, dry sandy and rocky places, 4000–6000 ft.; Sagebrush Scrub, N. Juniper Wd.; Modoc Co.; to se. Ore., sw. Ida., nw. Nev. May–June.

6. Micrósteris Greene

Small branching annuals. Lvs. entire, the lower opposite, spatulate to oblanceolate, the upper alternate, lanceolate. Fls. small, pedicelled, usually in pairs in upper axils, white to purplish-pink. Calyx cylindrical, 5-lobed, the herbaceous lobes acute, subequal, ± united by a hyaline membrane. Corolla salverform, with a slender tube, the lobes ovate to obcordate. Stamens included, unequally inserted in the corolla-tube. Ovary globose, 3-loculed; style slender; stigma 3-lobed. Caps. ovoid, 3-valved, rupturing the calyx in age. Seeds solitary in locules, lenticular, mucilaginous when wet. One polymorphous sp. of N. and S. Am. (Greek, *mikros*, small, and *aster*, star.)

1. **M. grácilis** (Hook.) Greene. [*Gilia gracilis* Hook. Greene. *Phlox g.* Greene. *M. californica, glabella,* and *stricta* Greene *G. g.* var. *villosa* Jeps. & Hoov.] Usually with erect slender stems 1–2 dm. high, generally simple and glabrous to pilose below, branched and glandular-pubescent above; lvs. 1–3 cm. long, short-petioled, the upper

sessile, somewhat reduced; infl. cymose, glandular, usually dense; calyx 5–8(–10) mm. long, the free lobes ca. as long as tube; corolla 8–12 mm. long, well exserted, the tube ± yellowish, the lobes rose to white or lavender, 1–2 mm. long; stamens included; caps. ca. 5 mm. long; seeds brownish, ca. 3 mm. long; $n = 7$ (Kinch, 1956).—Frequent in open grassy places and woods, below 10,000 ft.; many Plant Communities; montane and cismontane Calif.; to B.C., Ida., Mont. March–Aug. Variable, intergrading with

Ssp. **hùmilis** (Greene) V. Grant. [*M. h.* Dougl. ex Greene. *Collomia g.* var. *humilior* Hook. *C. micrantha* Kell.] Diffuse from base, usually broader than high, ± glandular-canescent; corolla largely 5–8 mm. long, barely exserted.—Siskiyou and Modoc cos., mostly along e. slope of Sierra Nevada to desert borders of s. Calif., occasional to inner Coast Ranges as in Lake and Monterey cos.; to e. Wash., Rocky Mts., Ariz.

7. Gília R. & P. GILIA

Key prepared by Alva R. Grant

Annual to rarely perennial herbs. Lvs. alternate, entire or more commonly pinnately lobed or dissected, mostly well developed below, strongly reduced above, often largely in a basal rosette. Fls. solitary in the lf.-axils or in paniculately branched or thyrsoid infls. or crowded into compact heads. Calyx-lobes usually equal, cleft deeply and often margined by membranes which may unite them into a continuous tubular calyx, the membrane often distended or ruptured by the caps. Corolla tubular-funnelform to salverform, mostly regular, blue, lavender, pink, to yellow or white. Stamens equally inserted in the corolla-tube or the throat, often in the sinuses of the corolla-lobes, equal or unequal in length. Caps.-valves remaining joined below, spreading above at dehiscence. Seeds few to many in a locule. A genus of ca. 50 spp., Calif., and s. S. Am. (Felipe *Gil*, Spanish botanist.)

(Grant, V. [or in collaboration with A. Grant]. Genetic and taxonomic studies in Gilia. A series of 10 papers in El Aliso, vols. 2 & 3, 1950–1956.)
A.   Pollen cream or yellow; calyx-lobes broadest at base, narrowing to a point, joined by sinus-membrane at base only, or, in *G. latifolia* and *Ripleyi*, ca. to middle; corolla campanulate, rotate or funnelform. Mainly deserts. (Section Giliastrum)
   B.   Corolla funnelform, pink.
      C.   Plant annual; corolla pink within, buff or white without . . . . 36. *G. latifolia*
      CC. Plant perennial; corolla pink within and without . . . . 37. *G. Ripleyi*
   BB. Corolla campanulate to turbinate, white or cream to yellow.
      C.   Corolla white or cream, the throat yellow; lvs., at least the lower, ovate to lance-ovate.
         D.   Corolla-tube and -throat shorter than the calyx as well as shorter than the corolla-lobes; stamens equal . . . . 35. *G. inyoensis*
         DD. Corolla-tube and -throat longer than calyx and longer than corolla-lobes; stamens unequal . . . . 33. *G. campanulata*
      CC. Corolla golden yellow; lvs. linear-filiform . . . . 34. *G. filiformis*
AA. Pollen blue (sometimes white); calyx-lobes narrow or broad, not wider at base than middle, joined by sinus-membrane to above middle; corolla funnelform or salverform.
   B.   Plants arachnoid woolly on lower stems and lvs. Mainly deserts and mts. of s. Calif. (Section Arachnion)
      C.   Stems glabrous at base (except sometimes in *G. latiflora*), rarely glandular-pubescent at base.
         D.   Corolla-throat dark purple like the tube (sometimes striated with yellow or with a white region around sinuses of lobes), narrow and flaring; cauline lvs. serrate or entire or pinnately lobed with narrow rachis (0.4–1.5 mm. wide), except broad and clasping in *G. tenuiflora* ssp. *arenaria* . . . . 19. *G. tenuiflora*
         DD. Corolla-throat yellow and white or purple in lower part and yellow and white above, usually full and broadly expanded; cauline lvs. shallowly dentate or entire with broad clasping base tapering to a narrow apex.
            E.   Fls. small, 7–12 mm. long with short lobes 1.5–4 mm. long; stamens and styles short, maturing at orifice; pollen white . . . . 27. *G. sinuata*
            EE. Fls. mostly larger, 8.2–35 mm. long, with large lobes 3.5–11 mm. long; stamens unequal, the longest well exserted, but shorter than corolla-lobes; style long, maturing well beyond longest stamens; pollen blue or white.
               F.   Corolla-tube and -throat combined 1½–2 times as long as calyx; corolla 8.2–12 mm. long. Mainly s. of San Gabriel Mts. . . . . . . 25. *G. diegensis*
               FF. Corolla-tube and -throat combined 2½–6 times as long as calyx; corolla 9.5–35 mm. long. Mainly n. of San Gabriel Mts. . . . . . . 26. *G. latiflora*
      CC. Stems cobwebby-pubescent at base (glabrous in *G. ochroleuca* and sometimes in *G. modocensis*).
         D.   Upper corolla-throat violet or white, sometimes yellow; throat lacking dark purple except sometimes streaked with purple in *G. aliquanta;* fls. 4–32 mm. long; calyx

glabrous to glandular-dotted or cobwebby-pubescent; caps. mostly globular to broadly ovoid.

 E. Fls. 4–13.5 mm. long; calyx glabrous to lightly pubescent with small glandular dots or cobwebby hairs.

  F. Calyx glabrous to sparsely cobwebby-pubescent; pedicels divaricate. Diploids.

   G. Corolla-throat broadly expanded and with 2 bands of color (yellow at base, blue-violet above); pedicels and calyces glabrous or pedicels glandular immediately beneath the fls. and then the calyces cobwebby-pubescent .... 15. *G. ochroleuca*

   GG. Corolla-throat narrowly flaring and without a clear banding of color; pedicels glandular beneath the glabrous or sometimes cobwebby calyx .... 18. *G. aliquanta*

  FF. Calyx sparsely to moderately glandular; pedicels strict or at least not strongly divaricate. Tetraploids.

   G. Calyx-lobes acuminate, prolonged ca. 1–1.5 mm. beyond the bordering sinus-membrane; calyx very sparsely pubescent; basal lvs. bipinnately dissected .... 23. *G. ophthalmoides*

   GG. Calyx-lobes acute, terminating less than 0.5 mm. beyond the bordering sinus-membrane; calyx moderately glandular; basal lvs. once pinnate .... 24. *G. transmontana*

 EE. Fls. 10–32 mm. long; calyx moderately pubescent with large glandular dots or cobwebby hairs.

  F. Upper corolla-throat light violet (more intense in *leptantha vivida*); stamens short, barely exserted or to ⅔ as long as corolla-lobes.

   G. Corolla-tube long-exserted from calyx, or if included, then infl. full, divaricately branched, with shortest pedicels at least half as long as longest .... 16. *G. cana*

   GG. Corolla-tube included or but slightly exserted from calyx
                      17. *G. leptantha*

  FF. Upper corolla-throat white or yellow; stamens long, subequal to or exceeding corolla-lobes .... 17. *G. leptantha*

 DD. Upper corolla-throat yellow or purple or partly purple; fls. 4–20 mm. long; calyx moderately to heavily glandular or cobwebby; caps. ± ovoid.

  E. Infl. ± congested or glomerate; calyx moderately to heavily glandular, and at maturity as long as or longer than caps.

   F. Stamens and style long-exserted .... 28. *G. brecciarum*

   FF. Stamens and style maturing at orifice or slightly exserted.

    G. Cauline lvs. deeply lobed, the lateral lobes usually much longer than width of rachis; corolla-veins beneath the sinuses violet. Diploid .... 28. *G. brecciarum*

    GG. Cauline lvs. coarsely dentate, or deeply lobed and then the lateral lobes longer than the width of the rachis; corolla-veins beneath sinuses white, or rarely violet and then cauline lvs. shallowly dentate. Tetraploid .... 29. *G. modocensis*

  EE. Infl. diffuse; calyx moderately glandular, at maturity shorter than to longer than the caps.

   F. Stamens very short, maturing at orifice; corolla-throat purple or if yellow, then purple-spotted or -veined; lf.-parts very narrow, 0.5–1.2 mm. wide; calyx shorter than caps. at maturity .... 22. *G. minor*

   FF. Stamens unequal, the longest exserted; corolla-throat yellow or with some purple; lf.-parts very narrow to broader, 0.5–3 mm. wide; calyx ca. as long as or longer than caps. at maturity.

    G. Plants spreading, with strong flexuous stems; corolla-throat pale yellow or sometimes suffused with purple. Tetraploid
                      20. *G. inconspicua*

    GG. Plants erect, with delicate stems; corolla-throat bright yellow with a pair of dark purple markings just below each lobe. Diploid.
                      21. *G. interior*

BB. Plants not arachnoid-woolly on lower stems or lvs.

 C. Infl. with loose to headlike glomerules of 3-many fls. subtended by a single lf.; upper cauline lvs. well developed and pinnately dissected. Cismontane. (Section Gilia)

  D. Fls. in dense spherical heads; calyx composed of relatively narrow herbaceous bands and narrow to broad hyaline sinuses; stamens as long as or longer than corolla-lobes .... 1. *G. capitata*

  DD. Fls. solitary, in loose cymes, or glomerate to loosely capitate with 12–25 fls. in a head; calyx composed of broad herbaceous bands and narrow hyaline sinuses; stamens shorter than to subequal to the corolla-lobes.

   E. Corolla blue-violet throughout, rarely white, the tube sometimes darker than lobes.

    F. Corolla-throat full and equal to or exceeding tube; lf.-segms. broad. Mainland .... 2. *G. achilleaefolia*

    FF. Corolla-throat narrow and exceeded by tube; lf.-segms. very narrow and linear. Insular .... 5. *G. Nevinii*

   EE. Corolla bi- or tri-colored with yellow tube and purple-spotted or yellow throat.

F.   Corolla campanulate, 5–14 mm. wide when pressed; style exserted, the
      stigmas beyond the anthers; corolla deciduous after anthesis.
  G.   Corolla-throat not purple-spotted, but tube and throat pale yellow,
        the limb white to pale violet; corolla 7–8 mm. long. S. Calif.
                                                              3. *G. angelensis*
  GG. Corolla-throat with 5 pairs of purple spots, the limb blue-violet
        and tube bright yellow or orange; corolla 3–16 mm. long. From
        Tehachapi Mts. n. .... 4. *G. tricolor*
FF. Corolla funnelform, 3–5 mm. wide when pressed; style included, the
      stigmas scarcely beyond the anthers; corolla often ± persistent.
  G.   Stems densely glandular; fruiting calyx 8–11 mm. long; pollen
        white. N. of San Francisco .... 6. *G. millefoliata*
  GG. Stems floccose below, not densely glandular; fruiting calyx 5–6
        mm. long; pollen mostly blue. S. of San Francisco .. 7. *G. clivorum*
CC. Infl. loose, consisting of 1–8 fls. subtended by a single lf.; upper cauline lvs. much
      reduced, or, if well developed, then linear to linear-oblong.
  D.   Plant leafy throughout, the basal rosette absent or very little developed; lvs. linear
        to linear-oblong, usually entire. (Section Saltugilia)
    E.   Fls. 8–18 mm. long; some stamens inserted in middle of corolla-throat; style
          exserted; calyx usually glabrous .... 11. *G. leptalea*
    EE. Fls. 3–6 mm. long; stamens all inserted in corolla-sinuses; style included; calyx
          glandular .... 12. *G. capillaris*
  DD. Plant scapose with reduced lvs. above and basal rosette well developed; lvs. sinuate
        to much dissected or the cauline linear and entire.
    E.   Basal lvs. 1–3-times pinnate with narrow rachis; cauline lvs. pinnately lobed
          except sometimes the uppermost bracts entire. (Section Saltugilia)
      F.   Caps. oblong-ovoid; lvs. finely divided; basal lvs. pubescent with multi-
            cellular hairs; uppermost lvs. usually entire.
        G.   Stamens inserted in corolla-sinuses, shorter than corolla-lobes; fls.
              pink to whitish or pale violet. Monterey Co. to Mojave Desert and
              L. Calif.
          H.   Corolla 10–36 mm. long, usually pink; style exserted
                                                              8. *G. splendens*
          HH. Corolla 5–9 mm. long, whitish or pale violet; style included.
                                                              10. *G. australis*
        GG. Stamens inserted in middle of corolla-throat, exserted beyond
              corolla-lobes; fls. blue-violet. San Diego Co. .... 9. *G. caruifolia*
      FF. Caps. broadly ovoid; lvs. finely to coarsely divided; basal lvs. pubescent
            with multicellular or geniculate hairs; uppermost lvs. toothed or lobed.
        G.   Corolla-tube included in the calyx; lower lvs. with white geniculate
              hairs .... 13. *G. stellata*
        GG. Corolla-tube well exserted; lower lvs. with straight translucent
              hairs .... 14. *G. scopulorum*
    EE. Basal lvs. strap-shaped and sinuate to bipinnatifid; cauline lvs. narrowly linear
          and entire, or the lowermost sometimes like the basal. Deserts. (Section
          Giliandra)
      F.   Corolla 2.5–7.5 mm. long; basal lvs. pinnately toothed or lobed and the
            lobes not exceeding twice the width of the lf. between sinuses; lf.-rachis
            or -blade 2–10 mm. wide.
        G.   Corolla-throat narrow, the tube and throat threadlike; sinus-
              membrane of calyx U-shaped between lobes in fr.
                                                              30. *G. leptomeria*
        GG. Corolla-throat full, the tube and throat not thread-like; sinus-
              membrane of calyx V-shaped between lobes in fr.
                                                              31. *G. micromeria*
      FF. Corolla 8–14 mm. long; basal lvs. bipinnately lobed and the lobes longer
            than twice the rachis-width; lf.-rachis 1–2 mm. wide
                                                              32. *G. hutchinsifolia*

1. **G. capitàta** Sims. [*G. pallida, glandulifera,* and *tenuisecta* Heller. *G. c.* var. *trisperma* Bdg.] Tall slender annuals with relatively shallow roots and glabrous to glandular or ± floccose stems 1–3 mm. in diam. below, 2–8 dm. high; basal and lower cauline lvs. 4–10 cm. long, bipinnately dissected, with narrow rachis, the ultimate pinnules 1–7 mm. long, 0.2–1 mm. wide; upper third of plant corymbosely branched, bearing terminal heads 1.4–4 cm. thick and with 50–100 subsessile fls.; heads glabrous or sparsely floccose, rarely glandular; calyx mostly 3–4 mm. long, the lobes ca. 2 mm. long, joined below by a sinus-membrane, the tips straight; corolla 6–8 mm. long, the linear lobes 0.7–1 mm. wide, 2–4 mm. long, pale to light blue-violet; fils. 2–3.8 mm. long; anthers ca. 0.7 mm. long; style 4–7.5 mm. long; stigmas 0.2–0.4 mm. long; caps. subglobose, 3–4 mm. in diam., not or tardily dehiscent, disarticulating on the stipe above the calyx, 1–6(–10)-seeded; seeds ovoid or ± angled, 1.6–2 mm. long; $n = 9$ (Grant, 1950).—Open well drained slopes, below 6000 ft.; Mixed Evergreen F.

to Yellow Pine F.; Coast Ranges from Marin Co. n.; to B.C., Ida. Natur. locally in Santa Cruz Mts. May–July.

KEY TO SUBSPECIES

A. Heads 50–100-fld.; pedicels 1 mm. long or less. San Joaquin V. and n.
  B. Corolla-lobes linear, 0.7–2 mm. wide; calyx-lobes acute, the tip nearly or quite straight; caps. disarticulating on the stipe above the calyx.
    C. Heads glabrous or sparsely floccose, but not densely floccose at base.
      D. Calyx-sinuses colorless; stigmas 0.2–0.4 mm. long; corolla-lobes 0.7–1.1 mm. wide; caps. mostly 1–6-seeded .... *G. capitata*
      DD. Calyx-sinuses blue-violet; stigmas 0.4–0.9 mm. long; corolla-lobes 1–2 mm. wide; caps. 10–25-seeded .... ssp. *pacifica*
    CC. Heads densely floccose at base.
      D. Stems floccose; stigmas 0.5–0.6 mm. long. Coast Ranges .... ssp. *tomentosa*
      DD. Stems glandular; stigmas 0.6–1 mm. long. Sierra Nevada .... ssp. *mediomontana*
  BB. Corolla-lobes oval, 1.5–3.3 mm. wide; calyx-lobes acuminate with a recurved tip; caps. not disarticulating.
    C. Caps. tardily dehiscent, 6–15-seeded; stigmas 0.6–1 mm. long; anthers 0.8–1 mm. long; corolla-lobes 1.5–2.5 mm. wide; plants shallowly rooted. Sierran foothills
                                               ssp. *pedemontana*
    CC. Caps. freely dehiscent, mostly 10–25-seeded; stigmas 0.9–1.2 mm. long; anthers 1–1.2 mm. long; corolla-lobes mostly 2.0–3.3 mm. wide; plants with taproots.
      D. Corolla deep blue-violet; lvs. bipinnately dissected; plants leafy at base; herbage with skunklike odor. Immediate coast, San Francisco to Sonoma Co.
                                            ssp. *Chamissonis*
      DD. Corolla light blue-violet; lvs. once-pinnately dissected; plants not leafy at base; herbage not with skunklike odor. San Joaquin V. and inner Coast Ranges
                                            ssp. *staminea*
AA. Heads 25–50-fld.; pedicels 1–2 mm. long. Mostly s. of San Joaquin V. .... ssp. *abrotanifolia*

Ssp. **pacífica** V. Grant. Stems glandular, 1–5 mm. in diam. at base and usually branched from below, 2.5–5 dm. high; basal lvs. bipinnately dissected, the ultimate segms. 3–10 mm. long, 0.2–1 mm. wide; heads 1.2–4 cm. in diam.; calyx floccose to glabrous, the lobes acute, 1–1.5 mm. wide at base, 2–2.5 mm. long; corolla 6–8 mm. long, the linear lobes 1–2 mm. wide, 3–4 mm. long; fils. 2.5–4 mm. long; anthers 0.6–1 mm. long; style 5.5–9 mm. long; stigmas 0.4–0.9 mm. long; caps. ovoid, dehiscent, 10–25-seeded; seeds 1.3–2 mm. long; $n = 9$ (Grant, 1950).—Exposed slopes and bluffs, below 1000 (4000) ft.; largely N. Coastal Scrub, Coastal Prairie; Mendocino Co. to Ore. May–Aug.

Ssp. **tomentòsa** (Eastw.) V. Grant. [*G. achilleaefolia* ssp. *Chamissonis* var. *t.* Eastw. ex Brand.] Plants usually from well developed taproot, the stems floccose, 1–10 mm. in diam. below, branching below or at the middle, 1–2.5 or to 20 dm. high; basal lvs. bipinnately dissected, the ultimate segms. 4–7 mm. long, ca. 1 mm. wide; heads 1.5–3 cm. in diam.; calyx densely floccose, the acute lobes 1–1.2 mm. broad at base, ca. 2.5 mm. long, with slightly recurved tip; corolla 7–10 mm. long, the lobes ca. 1 mm. wide, 4 mm. long; fils. 3–4 mm. long; style 6–8 mm. long; stigmas 0.5–0.6 mm. long; caps. ovoid, 3–10-seeded; seeds 1.9–2.2 mm. long; $n = 9$ (Grant, 1950).—N. Coastal Scrub, etc. along the coast of Marin and Sonoma cos., and to 3000 ft. on Mt. Diablo, Contra Costa Co. and Mt. Hamilton, Santa Clara Co. May–July.

Ssp. **mediomontàna** V. Grant. Plants shallowly rooted; stems mostly glandular, 1–2 mm. in diam. below, 1.5–4 dm. high, cymosely branched above; lower lvs. unipinnately or bipinnately dissected, the ultimate segms. 1–4 mm. wide; heads 1.2–2.5 cm. in diam.; calyx densely floccose, the acute lobes 0.8–1.2 mm. wide at base, ca. 2 mm. long; corolla 5–8 mm. long, the linear lobes 1–1.5 mm. wide, 2–4 mm. long; fils. 2.5–4 mm. long; style 6–7 mm. long; stigmas 0.6–1 mm. long; seeds 3–10, ± ovoid, 1.6–2.3 mm. long; $n = 9$ (Grant, 1950).—Open rocky slopes or open woods, 3000–7000 ft.; Yellow Pine F., Red Fir F.; Mariposa Co. to Plumas Co. May–July.

Ssp. **pedemontàna** V. Grant. Stems glandular, mostly 2–10 mm. in diam. below, simple or cymosely branched above, 3–9 dm. high; lower lvs. unipinnately or bipinnately dissected, the ultimate segms. 1–5 mm. wide; heads 1.5–3 cm. in diam.; calyx densely floccose, the lobes acuminate, 1–2 mm. wide at base, 2.5–3 mm. long, with recurved tip; corolla 7–11 mm. long, the oval lobes 1.5–2.5 mm. wide, 3–5 mm. long; fils. 2–3 mm. long; style 6–7 mm. long; seeds 6–15, ca. 1.5 mm. long; $n = 9$ (Grant, 1950).—Rocky open places, below 5000 ft.; Foothill Wd., Yellow Pine F.; Butte Co. to Tulare Co. April–June.

Ssp. **Chamissònis** (Greene) V. Grant. [*G. C.* Greene. *G. capitata* var. *regina* Jeps.] Stems very glandular, 2–5 mm. in diam. below, branching from near base, 1.5–7 dm. long; basal lvs. bipinnately dissected, the segms. 1–4 mm. long, 0.5–2 mm. wide; heads 2.5–3.5 cm. in diam.; calyx densely floccose, the lobes acuminate, 1.5–2.2 wide at base, ca. 2.5 mm. long, with recurved tip; corolla 9–10 mm. long, the oval lobes 2.2–3.2 mm. wide, 3–3.5 mm. long; fils. 2.5–5 mm. long; style 6.5–9.5 mm. long; stigmas 0.9–1.2 mm. long; seeds 10–25, ca. 1–1.7 mm. long; *n* = 9 (Grant, 1950).—Coastal Strand; San Francisco to Bodega Bay. May–July.

Ssp. **stamínea** (Greene) V. Grant. [*G. s.* Greene. *G. capitata* var. *s.* Brand. *G. achilleaefolia* ssp. *s.* Mason & A. Grant.] Stems mostly glandular, 2–10 mm. in diam. at base, simple or cymosely branched above, 1–6 dm. high; lower lvs. unipinnately dissected, the segms. 0.8–3.5 cm. long, 0.3–2 mm. wide; heads 1.5–3 cm. in diam.; calyx densely floccose, the lobes acuminate, 1.5–2.5 mm. wide at base, ca. 2.5 mm. long, with recurved tips; corolla 7–13 mm. long, the oval lobes 2.2–3.3 mm. wide, 3–5 mm. long; fils. 3–5 mm. long; anthers 1–1.2 mm. long; style 10–12 mm. long; seeds 10–25, ca. 1–1.5 mm. long; *n* = 9 (Grant, 1950).—Sandy flats and slopes, below 1000 ft.; V. Grassland, Coastal Strand; San Joaquin V., Coast Ranges from San Benito Co. to Sonoma and Contra Costa cos. March–May.

Ssp. **abrotanifòlia** (Nutt. ex Greene) V. Grant. [*G. a.* Nutt. ex Greene. *G. achilleaefolia* ssp. *a.* Brand.] Commonly branched from base, 2–8 dm. high, floccose or glabrous below and glandular or glabrous above; lower lvs. unipinnately or bipinnately dissected, the ultimate segms. 0.5–2 mm. wide, the axils of lvs. usually floccose; heads several, on naked peduncles 4–25 cm. long, 1.5–3 cm. in diam.; pedicels 1–2 mm. long; calyx 3–5 mm. long, lightly floccose, the acuminate lobes recurved at tip, the midribs greenish to purplish-brown, the wings blue-violet to lavender; corolla 8–12 mm. long, light blue-violet to whitish, the oval lobes 1.5–3 mm. wide; fils. 2–4 mm. long; style 1–3 mm. longer than corolla; seeds 9–18, ovoid, 1–1.5 mm. long; *n* = 9 (Grant, 1952). —Loose sandy or gravelly places, below 6000 ft.; Foothill Wd., Coastal Sage Scrub, Chaparral, Yellow Pine F., etc.; Sierra Nevada from Madera Co. s. through Tehachapi Mts. to mts. of L. Calif., near the coast from Santa Barbara Co. s. April–May.

2. **G. achilleaefòlia** Benth. Erect annuals, mostly ± branched above, 1–7 dm. high, often ± floccose below and glandular above; lower lvs. largely bipinnate, 4–10 cm. long, the ultimate segms. ± falcate, 1–25 mm. long, 0.5–2 mm. wide; lf.-axils floccose; infl. a fairly dense fan-shaped head or heads (dense cymes) on naked peduncles 1–15 cm. long; fls. 8–25, subsessile on pedicels 1–2 mm. long; calyx 4.5–7 mm. long, floccose or with glands at tips of lobes or ± throughout, the lobes acute with hyaline blue-violet wings; corolla funnelform, mostly 10–20 mm. long, blue-violet, the lobes oval; stamens included, the fils. 1.5–3 mm. long; styles 1–2 mm. longer than mature corolla; caps. ovoid, dehiscent, 10–18-seeded; seeds ovoid, angular, 1–2 mm. long; *n* = 9 (Grant, 1954).—Loose soils of open places, below 4000 ft.; Foothill Wd., Mixed Evergreen F., Chaparral, Coastal Strand, etc.; Coast Ranges, n. Santa Barbara Co. to Marin and Contra Costa cos. May–June.

Ssp. **multicáulis** (Benth.) V. & A. Grant. [*G. m.* Benth. *G. peduncularis* Eastw. ex Mlkn. *G. pedunculata* Eastw. and vars. *calycina* and *minima* Eastw.] Infl. loosely cymose, with groups of 2–7 fls. on pedicels 1–60 mm. long; corolla 5–10 mm. long; *n* = 9 (Grant, 1954).—Shaded woods; with the sp.

3. **G. angelénsis** V. Grant. Erect annual 0.7–7 dm. high, usually with spreading branches; stems glabrous or floccose below, sometimes glandular above; lower lvs. unipinnately or bipinnately dissected, 2–5 cm. long, the ultimate segms. ± falcate, 5–15 mm. long, 0.5–1.5 mm. wide, the axils of the lvs. floccose; infl. cymose, the heads usually ca. 5-fld., several, on naked peduncles 1–5 cm. long; pedicels 2–3 mm. long; calyx 3–4 mm. long, mostly floccose, the lobes acute, the cent. portion green, flanked by hyaline margins 0.1–0.3 mm. wide at sinus; corolla campanulate, 7–8 mm. long, the limb blue-violet to white, the tube pale yellow, the lobes ovate, 2–3 mm. wide; fils. ca. 1 mm. long; caps. ovoid, dehiscent, 20–30-seeded; seeds 0.5–1 mm. long; *n* = 9 (Grant, 1952). —Loose, often sandy or gravelly soil, below 6300 ft.; Coastal Sage Scrub, Chaparral, Yellow Pine F., etc.; cismontane s. Calif. to Santa Barbara Co., Mt. Hamilton Range; L. Calif. March–May.

4. **G. tricolor** Benth. Divaricately branched annuals 1–4 dm. high, floccose or glabrous

below, glandular above; lower cauline lvs. unipinnately or bipinnately dissected, 1–4 cm. long, the segms. narrow to subfiliform; fls. solitary or in 2–5-fld. glomerules; pedicels 2–30 mm. long; calyx 4–7 mm. long in fl., 6–8 mm. in fr., floccose, with purple sinuses; corolla 11–16 mm. long, campanulate, the limb 9–14 mm. broad when pressed, the lobes mostly 5–8 mm. wide, pale to deep blue-violet, the tube and lower parts of throat yellow to orange, the throat bearing 5 pairs of purple spots; stamens inserted equally in sinuses; fils. unequal in length; style exserted; stigmas 2–3 mm. long; caps. ovoid, dehiscent, 16–36-seeded; seeds ovoid, 1–1.5 mm. long; $n = 9$ (Grant, 1952).— Open grassy plains and slopes, below 2000 ft.; largely V. Grassland; Coast Ranges from San Benito Co. to Humboldt Co., Cent. V. from Tulare Co. to Tehama Co., Sierran foothills of Plumas and Lassen cos. to Shasta Co. Mostly March–April.

Ssp. **diffùsa** (Congd.) Mason & A. Grant. [*G. d.* Congd. *G. t.* var. *longipedicellata* Greene.] Fls. solitary, on pedicels 1–4 cm. long; calyx 3–4 mm. long in fl., 4–7 mm. in fr., floccose or glandular, the sinus purple or colorless; corolla 8–10 mm. long, the limb 7–9 mm. wide when pressed, the lobes 3–4 mm. wide; $n = 9$ (Grant, 1952).— Open plains and foothills, below 4000 ft.; V. Grassland, Foothill Wd.; Cent. V. and bordering hills from Kern Co. to Placer Co.

5. **G. Nevínii** Gray. [*G. multicaulis* ssp. *N.* Mason & A. Grant.] Erect simple or branched annual, finely glandular-pubescent throughout, leafy, 1–4 dm. high; lvs. 1–7 cm. long, bipinnate or tripinnate into very fine and numerous linear segms.; fls. in capitate clusters; calyx cylindric-campanulate, 5–6 mm. long, lobed ⅓–½ the way down, 7–8 mm. long in fr.; corolla narrowly funnelform, 10–14 mm. long, the tube exceeding the calyx, the lobes 2 mm. long, with blue limb and tube, but throat yellowish and not spotted; caps. oblong, 6.5–8 mm. long; $n = 18$ (Grant, 1958).—Open grassy slopes; Santa Catalina, San Clemente, and Guadalupe ids. March–May.

6. **G. millefoliàta** F. & M. [*G. multicaulis* ssp. *m.* Mason & A. Grant. *G. millefoliata* var. *maritima* Brand.] Annual, ± divaricately branched, usually from base, densely glandular-pubescent, 0.2–2 dm. high; lvs. ± fleshy, 1–5 cm. long, once to twice pinnate into ultimate deltoid-lanceolate lobes up to 4 mm. long; fls. 2–6 in heads on peduncles mostly 2–10 cm. long; calyx 5–7 mm. long in fl., 8–11 mm. in fr., glandular-pubescent; corolla blue-violet, 5–7 mm. long, the throat yellow with 5 prominent dark purple spots, the limb 5–6 mm. wide when pressed, the lobes erect, acute; stamens scarcely exserted; caps. 6–8 mm. long; seeds 25–50, brown, 1–1.5 mm. long; $n = 9$ (Grant, 1954).—Coastal Strand; San Francisco Bay to Del Norte Co.; s. Ore. April–June.

7. **G. clivòrum** (Jeps.) V. Grant. [*G. multicaulis* var. *c.* Jeps.] Erect or divaricately branched annual 0.6–3.5 dm. high, sparsely to densely floccose below, glandular above; lvs. unipinnate or bipinnate, 1–6 cm. long, the ultimate segms. frequently elliptical, sometimes filiform, 0.5–2 mm. wide; fls. 2–5 in glomerules on peduncles 0.5–3 cm. long; pedicels 1–10 mm. long; calyx floccose or glandular, 4–5 mm. long in fl., 5–6 mm. in fr.; corolla funnelform, 6–8 mm. long, the limb 3–5 mm. wide when pressed, the tube yellow, the throat with 5 pairs of purple spots, the limb blue-violet with oval spreading lobes 1–2 mm. wide; stamens included; caps. ovoid, 24–36-seeded; seeds ovoid, 1–1.5 mm. long; $n = 18$ (Grant, 1954).—Open fields and slopes; V. Grassland, Coastal Prairie, Foothill Wd., Mixed Evergreen F., etc.; Coast Ranges from Lake and Solano cos. s. to Ventura Co. and in Riverside Co. March–May.

8. **G. spléndens** Dougl. ex Lindl. [*G. tenuiflora* var. *altissima* Parish. *G. Grinnellii* Brand.] Robust subscapose branched annual 1–6(–12) dm. high; stems glabrous to glandular (especially below); basal lvs. in a rosette, villous, 2–14 cm. long, with narrow rachis, bipinnatifid or tripinnatifid with deeply cut segms.; upper lvs. reduced to linear bracts; infl. cymose, the fls. in pairs on unequal pedicels subtended by a bract and 3–20 mm. long; calyx 3–4 mm. long, with hyaline sinuses and erect acute lobes 1 mm. long; corolla funnelform, the limb usually pink, sometimes pinkish-violet, 12–20 mm. in diam., the acute broadly oval lobes 6–10 mm. long, 4–8 mm. wide; throat white to pale violet, 4–7 mm. long, broad; tube 3–6 mm. long, slender, red to purple; stamens unequal in length, the longest fils. 1–2 mm. long; stigmas 2–3 mm. long; caps. cylindrical, 4–8 mm. long, 35–70-seeded; seeds ellipsoidal, 1 mm. long; $n = 9$ (Grant & Grant, 1954).—Openings in brush or woods, 1000–7000 ft.; Chaparral, Foothill Wd., Yellow Pine F.; Coast Ranges from Monterey Co. to San Jacinto Mts. April–July.

Ssp. **Grántii** (Brand) V. Grant. [*G. collina* var. *G.* Brand.] Corolla-limb intense pink;

tube 10–18 mm. long; longest fils. 2–3 mm. long; *n* = 9 (Grant and Grant, 1954).—
Open slopes and woods, 2700–7000 ft.; mostly Yellow Pine F.; San Gabriel and San
Bernardino mts. May–July.

9. **G. caruifòlia** Abrams. [*G. latiflora* var. *c.* Jeps. *G. tenuiflora* var. *c.* Munz.] Scapose
annual 3–6(–12) dm. high, with erect cent. axis and many branches spreading from
base; base of plant with many large bipinnate or tripinnate lvs. 3–7(–30) cm. long,
with narrow rachis and deeply cut segms.; upper stems with reduced linear bracts;
herbage glabrous except for the glandular upper branches of the cymose infl.; fls. in
pairs on unequal pedicels 1–10 mm. long; calyx 3–4 mm. long; corolla funnelform;
tube 3–5 mm. long, pale blue-violet; throat 3 mm. long, whitish with yellow spots in
upper part; limb 15–20 mm. in diam., pale blue-violet, often with a pair of purple spots
at base of each lobe; stamens unequal, the longer fils. 6–7 mm. long; stigmas 2–3
mm. long; caps. 4 mm. long, 12–20-seeded; seeds 1 mm. long; *n* = 9 (Grant & Grant,
1954).—Dry hills and openings in brush and woods, 4500–7500 ft.; Chaparral, Yellow
Pine F.; mts. of San Diego Co.; n. L. Calif. May–Aug.

10. **G. austràlis** (Mason & A. Grant) V. Grant. [*G. splendens* ssp. *a.* Mason & A.
Grant.] Branched scapose annuals 1–3 dm. high, with basal rosette of villous lvs.;
stems glabrous; infl. glandular, sometimes also sparsely pubescent throughout; lvs. 1-,
2-, 3-pinnate, 2–7 cm. long, the lateral lobes 3–6 on each side, well spaced, narrow,
5–15 mm. long, usually again dissected; infl. cymose; fls. in 2's on unequal pedicels
4–26 mm. long; calyx 2–4 mm. long; corolla funnelform, 5–10 mm. long when pressed;
tube 2–3 mm. long; limb 4–9 mm. in diam., pale violet or whitish, with yellow spots
in throat, the lobes commonly streaked purple on outer surface; stamens subequal, 0.5–
1.5 mm. long; stigmas 1 mm. long, included; caps. ellipsoidal, 4–6 mm. long, 20–30-
seeded; seeds ovoid, 0.5–1 mm. long; *n* = 9 (Grant & Grant, 1954).—Sandy places,
below 4000 ft.; Coastal Sage Scrub and Chaparral, San Bernardino V. to e. San Diego
Co.; Creosote Bush Scrub and Joshua Tree Wd., Little San Bernardino Mts. to Palm
Springs; L. Calif. March–June.

11. **G. leptàlea** (Gray) Greene. [*Collomia l.* Gray.] Erect annuals 0.5–3.5 dm. high,
the stems subsimple to well branched, ± glandular throughout; lvs. linear, entire, 1–5
cm. long, glandular; infl. cymose, the fls. in pairs on pedicels 3–40 mm. long; calyx
slender, 3–4 mm. long, the acuminate lobes 1 mm. long; corolla funnelform; throat 5–7
mm. long, pink or violet, slender, tapering gradually to the tube; lobes oval, pink or
violet, 3–5 mm. long, 2–2.5 mm. wide; stamens unequally inserted, the unequally long
fils. 1–3 mm. long, some anthers slightly exserted; stigma exserted 2–3 mm. beyond
anthers; caps. 3–4 mm. long, 6–12-seeded; seeds angular, 1 mm. long; *n* = 9 (Grant
& Grant, 1954).—Openings in woods, 4500–8000 ft.; Yellow Pine F., Red Fir F.; Tulare
Co. to Lassen and Shasta cos.; Ore. June–Aug.

Ssp. **bìcolor** Mason & A. Grant. Lvs. simple and entire; corolla-throat 4–5 mm. long,
yellow or white, tapering ± abruptly to the tube; *n* = 9 (Grant, 1958).—At 6000–9700
ft.; mostly Red Fir F. to Lodgepole F.; Fresno Co. to Plumas Co. July–Sept.

Ssp. **pinnatisécta** Mason & A. Grant. Lvs. pinnatifid with 1–3 pairs of simple lateral
lobes.—Open places, 1000–7000 ft.; Chaparral, Yellow Pine F.; Coast Ranges from
Humboldt Co. to Lake Co. June–Aug.

12. **G. capillàris** Kell. [*Navarretia c.* Kuntze. *G. sinistra* Jones. *G. subalpina* Greene
ex Brand.] Erect annuals 0.2–3 dm. high, well branched to subsimple, stipitate-glandular
throughout; lvs. simple or rarely with a pair of basal lateral lobes, linear, 0.5–5 mm.
wide, the lower and middle lvs. 1–4 cm. long; infl. cymose, the first pedicels shorter;
calyx slender, 3–4 mm. long, the lobes acuminate, glandular; corolla funnelform, 3–6
mm. long, ca. twice the calyx, pale violet to pink or white, often streaked purple, the
tube sometimes yellow, the throat sometimes with purple spots; fils. 0.5 mm. long;
stigmas included, 0.5–1 mm. long; caps. ovoid, 3–4 mm. long; seeds 6–12, brown,
1–1.6 mm. long; *n* = 9 (Grant, 1958).—Sandy slopes and flats, 2500–7500 ft. or below
in nw. Calif., to 10,500 in Sierra Nevada; mostly Montane Coniferous F.; mts. of s.
Calif. through Sierra Nevada to N. Coast Ranges from Mendocino Co. n.; to Wash.,
Ida. June–Aug.

13. **G. stellàta** Heller. [*G. tenuiflora* var. *Newloniana* Jeps.] Erect annual 1–4 dm.
high; stems stout, paniculately branched, with bent hairs below and stipitate-glandular
above; lower lvs. 2–3-dissected, 2–10 cm. long, the ultimate segms. callous-tipped,

pubescent; upper lvs. reduced; infl. paniculate, the fls. solitary on short stout pedicels; calyx 3–4 mm. long, larger in fr., the lobes joined ⅔ their length by sinus-membrane; corolla funnelform, 6–10 mm. long, pale blue or lavendar to white, with row of purple spots in throat; stamens subequal, shorter than corolla-lobes; stigma 1–1.5 mm. long; caps. broadly ovoid, 5–7 mm. long; seeds many, brown, pitted, irregularly angled; $n = 9$ (Grant, 1958).—Sandy and gravelly places, below 4000 ft.; Creosote Bush Scrub; Colo. & Mojave deserts; to Utah, nw. Mex. March–May.

14. **G. scopulòrum** Jones. [*G. s.* var. *Covillei* Brand.] Glandular-villous annual, erect and simple or with ascending branches 1–3 dm. long, the stems paniculately branched; lower lvs. 3–9 cm. long, viscid, thin, broadly oblanceolate in outline, 2–8 cm. long, 1–2-pinnatifid into ovate coarsely toothed segms.; infl. paniculate, the fls. solitary at ends of stipitate-glandular pedicels; calyx 3–4 mm. long, enlarging in fr., the lobes joined below by a sinus-membrane; corolla funnelform, rose-lavender, 10–14 mm. long, the tube and throat paler or yellowish; stamens ± equal, shorter than corolla-lobes; caps. broadly ovoid, 4.5–5.5 mm. long; seeds many, brown, irregularly angled; $n = 9$, 18 (Grant, 1958).—Dry washes and slopes, mostly below 4000 ft.; Creosote Bush Scrub; Chuckawalla Mts. to Death V. region, Owens V.; to sw. Utah. April–May.

15. **G. ochroleuca** Jones. [*G. inconspicua* var. *o.* Brand.] Delicate annual with a short cent. leader 0.6–1.5 dm. high and many secondary ± spreading branches to 3 dm. long, glaucous and glabrous except for the glandular infl.; lvs. mostly basal, sparsely cobwebby-pubescent, 2–3 cm. long, pinnately lobed with narrow rachis and lobes 1 mm. wide and 3–10 mm. long; bracts entire or with a small pair of lobes; pedicels 3–10 mm. long, in open cymose panicles; calyx glabrous, 2.5–3 mm. long; corolla yellowish, tubular-funnelform, 4–6 mm. long, pale violet in upper throat; stamens maturing at orifice of corolla; caps. globular, 3–4.5 mm. long; locules 1–5-seeded; $n = 9$ (Grant, Beeks & Latimer, 1956).—Sandy slopes and plains, 2500–5000 ft.; Creosote Bush Scrub, Joshua Tree Wd.; w. half of Mojave Desert. April–June.

KEY TO SUBSPECIES

Fls. small, 4–6 mm. long; style maturing at orifice. Mojave Desert .... *G. ochroleuca*
Fls. larger, 8–13 mm. long; style exserted. Mostly cismontane.
    Corolla-tube slightly exserted from calyx and slightly longer than throat; lower stems usually cobwebby-pubescent; pedicels glandular just below fls. San Bernardino Co. to San Diego Co.
                                                 ssp. *exilis*
    Corolla-tube usually included in calyx, shorter than throat; lower stems usually glabrous; pedicels glabrous just below fls. San Luis Obispo Co. to San Bernardino Co. .... ssp. *bizonata*

Ssp. **éxilis** (Gray) A. & V. Grant. [*G. latiflora* var. *e.* Gray. *G. e.* Abrams. *G. Abramsii* and ssp. *integrifolia* Mason & A. Grant. *G. lineata* A. Davids.] Plants 1.5–3 dm. high; stems leafy below the middle; lower lvs. 2–6 cm. long; pedicels mostly glandular throughout, 4–27 mm. long; corolla 8–10.5 mm. long, shallow-throated; stigma at anther-level or 1–2 mm. beyond; $n = 9$ (Grant, Beeks & Latimer, 1956).—Sandy or gravelly places, 800–5000(–8000) ft.; Coastal Sage Scrub, Chaparral, Yellow Pine F.; largely interior cismontane Calif. from San Bernardino Co. s.

Ssp. **bizonàta** A. & V. Grant. Erect, 1.5–5 dm. high; basal lvs. pinnate to bipinnate, 2.5–6 cm. long, the primary lobes 1–15 mm. long; pedicels 5–30 mm. long; corolla 8–13.5 mm. long, the throat a little longer than tube, the lobes light pink; style well exserted; $n = 9$ (Grant, Beeks & Latimer, 1956).—Sandy flats, 2800–6700 ft.; Pinyon-Juniper Wd., Foothill Wd., Yellow Pine F., etc.; mts. from San Luis Obispo Co. to Kern and w. San Bernardino cos.

16. **G. càna** (Jones) Heller. [*G. latiflora* var. *c.* Jones. *G. tenuiflora* var. *c.* Jeps. *G. collina* var. *coronata* Brand.] Erect annuals with 1 to several rather stout stems 1–3 dm. high; lvs. in a basal rosette, densely matted with cobwebby pubescence, bipinnately lobed or toothed, the rachis 0.5–3 mm. wide, the lobes broader; uppermost lvs. reduced to leafy bracts; infl. somewhat crowded, the pedicels glandular-pubescent, 1–13 mm. long; calyx glandular-pubescent, 3.5–5.5 mm. long; corolla 14–26 mm. long, the tube and throat together 10–19 mm. long, the tube purple, the throat narrow, yellow below, violet above, the lobes pinkish-violet, 4–6 mm. long; stamens slightly unequal, slightly to well exserted; caps. 3–4.5 mm. long; $n = 9$ (Grant, Beeks & Latimer, 1956).—Dry gravelly places, 6000–10,000 ft.; largely Pinyon-Juniper Wd. to Lodgepole F.; e. slope of Sierra Nevada in Inyo and Mono cos. June–Aug.

KEY TO SUBSPECIES

Infl. somewhat crowded, with short internodes; shortest pedicels usually much less than half as long as longest; infl. strict, the branches not divaricate.
  Caps. 3–4.5 mm. long; stamens unequal, the longest well exserted. E. slope of Sierra Nevada
                                                                                        G. cana
  Caps. 5–9 mm. long; stamens equal or unequal, not or scarcely exserted.
    Purple color in corolla-tube extending close to orifice; corolla-form narrow. W. edge of Mojave
    Desert .... ssp. *speciosa*
    Purple color of corolla-tube extending only into lower throat; corolla-form broader. San Bernar-
    dino Mts. .... ssp. *bernardina*
Infl. loose, with long upper internodes; shortest pedicels mostly at least half as long as longest; infl.
with divaricate branches.
  Corolla-throat narrow, flaring gradually from tube; corolla-lobes broad, wider than orifice. Rare,
  Mojave Desert .... ssp. *speciformis*
  Corolla-throat full, abruptly expanded; corolla-lobes narrow, not as wide as orifice. Common on
  Mojave Desert .... ssp. *triceps*

Ssp. **speciòsa** (Jeps.) A. & V. Grant. [*G. tenuiflora* var. *s.* Jeps. *G. latiflora* ssp. *s.* Mason & A. Grant.] Densely cobwebby-pubescent below; corolla 20–32 mm. long, with narrow orifice, the lobes broadly oval, 4–8 mm. wide; $n = 9$ (Grant, Beeks & Latimer, 1956).—Canyons and fans, 2500–4000 ft.; Joshua Tree Wd., Creosote Bush Scrub; w. edge of Mojave Desert from Red Rock Canyon to Nine-Mile Canyon. April–May.

Ssp. **bernardìna** A. & V. Grant. Corolla 17–23 mm. long; $n = 9$ (Grant, Beeks & Latimer, 1956).—Sandy flats and washes, 2700–4800 ft.; Joshua Tree Wd.; n. base of San Bernardino Mts. April–May.

Ssp. **specifórmis** A. & V. Grant. Basal lvs. moderately cobwebby-pubescent; corolla 15–29 mm. long, the stout tube 2–4 times the calyx, gradually widened into the slender throat; lobes 3–8 mm. wide.—Sandy places, 2800–3800 ft.; largely Joshua Tree Wd.; Mojave Desert from Ord and Avawatz mts. to Amargosa and Black mts. April–May.

Ssp. **trìceps** (Brand) A. & V. Grant. [*G. tenuiflora* var. *t.* Brand. *G. latiflora* ssp. *t.* Mason & A. Grant.] Basal lvs. moderately cobwebby-pubescent; corolla 8–23 mm. long, the slender tube expanding abruptly into the throat; lobes 2–6 mm. wide; $n = 9$ (Grant, Beeks & Latimer, 1956).—Dry slopes and washes, 2800–5200 ft.; Creosote Bush Scrub, Joshua Tree Wd., Pinyon-Juniper Wd.; Barstow and Kelso to Panamint and White mts.; Nev. April–May.

17. **G. leptántha** Parish. [*G. arenaria* ssp. *l.* Brand. *G. latiflora* ssp. *l.* Mason & A. Grant.] Erect plants 1.5–4.5 dm. high, cobwebby-pubescent below, glandular above; basal lvs. 2.5–6.5 cm. long, bipinnately to almost tripinnately lobed; infl. diffuse, with 2–3 fls. above a bract; pedicels very unequal, 1–27 mm. long; calyx lightly gland-dotted, 3.5–4 mm. long at anthesis; corolla 12–23 mm. long, the tube very slender, expanding into a narrow throat, the lobes 1.4–2.8 mm. wide, 4–6 mm. long, bright pink, the tube and throat ± yellow; stamens unequal, the longest well exserted; caps. 3–4 mm. long; $n = 9$ (Grant, Beeks & Latimer, 1956).—Sandy and gravelly places, 5000–7700 ft.; largely Yellow Pine F.; upper Santa Ana R. system, San Bernardino Mts. June–Aug.

KEY TO SUBSPECIES

Corolla-tube 2–4 times as long as calyx. Pine belt.
  Corolla-lobes 1.4–2.8 mm. wide, ca. 2–2.5 times longer. San Bernardino Mts. .... *G. leptantha*
  Corolla-lobes 2.5–5 mm. wide, 1–2 times longer. S. Sierra Nevada .... ssp. *Purpusii*
Corolla-tube 1–1.5 times as long as calyx.
  Stamens shorter than the corolla-lobes, the fils. not more than 2.5 mm. long. San Gabriel Mts. to
  Little San Bernardino Mts.
    Corollas 13.5–17 mm. long; plants erect with tall cent. stem. Below 6000 ft. .... ssp. *transversa*
    Corollas 7.5–14 mm. long; plants low, spreading, with many stems from base. Above 6000 ft.
                                                                                  ssp. *vivida*
  Stamens equally or exceeding corolla-lobes, the fils. at least 4 mm. long.
    Corolla-tube violet, exserted from calyx; calyx glandular. Mt. Pinos .... ssp. *pinetorum*
    Corolla-tube yellow, included in calyx; calyx not glandular. Lassen Co. to Alpine Co.
                                                                                  ssp. *salticola*

Ssp. **Purpùsii** (Mlkn.) A. & V. Grant. [*G. tenuiflora* var. *P.* Mlkn. *G. collina* Eastw.] Cent. stem to 6 dm. high; infl. diffuse, glandular-pubescent; basal lvs. 2–6 cm. long; pedicels very unequal; corolla 13–29 mm. long, widening abruptly into the short throat; lobes 2.5–4.5 mm. wide, pinkish-violet; longest stamens long-exserted; $n = 9$ (Grant,

Beeks & Latimer, 1956).—Open stream beds and slopes, 2500–8500 ft.; Foothill Wd. to Lodgepole F.; s. Sierra Nevada. May–July.

Ssp. **transvérsa** A. & V. Grant. Stems densely glandular just above basal lvs., less so above; corolla 13–17 mm. long, the tube violet, the lobes pale pink-violet; $n = 9$ (Grant, Beeks & Latimer, 1956).—Sandy places, 3000–6000 ft.; Joshua Tree Wd., Pinyon-Juniper Wd.; desert slopes from San Gabriel Mts. to Little San Bernardino Mts. April–July.

Ssp. **vívida** A. & V. Grant. Many-stemmed, 1–2 dm. high; calyx 2.8–3.5 mm. long at anthesis, glabrous to ± glandular or cobwebby; corolla 9–14 mm. long, the tube 3–5 mm. long, violet, the lobes deep pinkish-violet, 4–5 mm. long; $n = 9$ (Grant, Beeks & Latimer, 1956).—Sandy or gravelly places, 6000–8200 ft.; Yellow Pine F., Red Fir F.; San Gabriel and Tehachapi mts. May–July.

Ssp. **pinetòrum** A. & V. Grant. Plants low, glandular in infl.; calyx 2.8–4.2 mm. long at anthesis; corolla 11–13 mm. long, the tube 4.2–6.4 mm. long, pale violet, the lobes pale violet, 3–4 mm. long; $n = 9$ (Grant, Beeks & Latimer, 1956).—Between pines, 5100–9000 ft.; Yellow Pine F.; Mt. Pinos, Ventura Co. May–June.

Ssp. **saltícola** (Eastw.) A. & V. Grant. [*G. s.* Eastw. *G. alpina* Eastw., not Brand.] Small, branched, the basal lvs. grayish-matted; calyx 2.2–3.8 mm. long, glabrous to cobwebby; corolla 6–9 mm. long, the tube yellowish, the lobes violet.—Sandy plains and slopes, 5000–8800 ft.; Sagebrush Scrub to Lodgepole F.; Lassen and Alpine cos.; w. Nev. May–June.

18. **G. aliquánta** A. & V. Grant. Small erect plant with a main stem and basal rosette of lvs., later secondary stems 1–1.5 dm. long, glabrous to cobwebby or glandular below, lightly glandular above; basal lvs. 1–3 cm. long, with opposite linear lobes 2–3 mm. long; middle cauline lvs. somewhat reduced; infl. loose; pedicels unequal, 1–11 mm. long; calyx glabrous to sparsely cobwebby, 3.5–5 mm. long at anthesis; corolla 6–12 mm. long, 2–3 times the calyx, bright violet, with some yellow in throat, the lobes ca. as long as tube and throat combined; stamens long exserted; $n = 9$ (Grant, Beeks & Latimer, 1956).—Rocky and gravelly slopes, 2500–4100 ft.; Joshua Tree Wd., Pinyon-Juniper Wd.; desert base of San Bernardino and San Gabriel mts. to se. base of Sierra Nevada in n. Kern Co. April–May.

Ssp. **brevilòba** A. & V. Grant. Corolla-lobes 1.3–2.5 mm. long, the tube included to well exserted from calyx; stamens scarcely exserted; $n = 9$ (Grant, 1958).—Rocky and gravelly slopes, 2800–6200 ft.; Joshua Tree Wd., Pinyon-Juniper Wd.; White Mts., Inyo Co., Providence Mts., e. San Bernardino Co.; to Utah. March–May.

19. **G. tenuiflòra** Benth. [*G. arenaria* ssp. *leptantha* var. *Aliciae* Brand.] Erect but somewhat spreading, 1.5–4 dm. high, glabrous below, glandular-pubescent in infl.; basal lvs. 2–6 cm. long, somewhat strap-shaped, with narrow rachis and 4–6 pairs of short lobes, or bipinnately lobed; cauline lvs. with shorter lobes; infl. diffuse; pedicels very unequal; calyx 3.3–4.3 mm. long at anthesis, lightly cobwebby; corolla 12–16 mm. long, the tube slender, purple, widening into a moderately narrow purple throat; lobes 2.5–4 mm. wide, 4–5 mm. long, pinkish-violet; stamens and style well exserted; caps. 3.5–5 mm. long; $n = 9$ (Grant, Beeks & Latimer, 1956).—Sandy washes and canyons, 1500–2100 ft.; largely Foothill Wd., Pinyon-Juniper Wd.; Santa Cruz Mts. to La Panza Range, largely bordering Salinas V. April–June.

KEY TO SUBSPECIES

Corolla-lobes 2–4 mm. wide; throat narrow, the orifice 2.2–3.8 mm. wide; infl. diffuse; stems several, slender.
  Longest stamens well exserted; style long, the stigmas above the stamens; caps. 3.5–5 mm. long.
    Inner Coast Ranges . . . . *G. tenuiflora*
  Longest stamens slightly exserted; style short, the stigmas among the stamens; caps. 5–6 mm. long.
    Monterey Peninsula . . . . ssp. *arenaria*
Corolla-lobes 4–5.8 mm. broad; throat full, the orifice 3.7–5 mm. wide; infl. somewhat congested; stems stout, usually with a cent. leader.
  Corolla-tube and throat together 7–11 mm. long; 2–4 fls. borne above a bract; style exceeding stamens. Upper Salinas R. V. . . . . ssp. *amplifaucalis*
  Corolla-tube and throat together 13–14 mm. long; 1–2 fls. borne above a bract; style equaling stamens. Santa Rosa Id. . . . . ssp. *Hoffmannii*

Ssp. **arenària** (Benth.) A. & V. Grant. [*G. a.* Benth. *G. t.* var. *a.* Jeps.] Low, 0.6–1.7 dm. tall, with erect cent. stem and several others; lvs. in basal cluster; plant densely glandular-pubescent throughout, sometimes cobwebby below.—Coastal Strand; Monterey Bay. April–May.

Ssp. **amplifaucàlis** A. & V. Grant. Erect, 1.6–2.5 dm. tall, glabrous below, ± glandular-pubescent in infl.; calyx 4–5.8 mm. long; corolla 13–17 mm. long.—V. Grassland, Foothill Wd.; upper Salinas R. to Cholame V. March–April.

Ssp. **Hoffmánnii** (Eastw.) A. & V. Grant. [*G. H.* Eastw.] Six–12 cm. tall; rather densely glandular-pubescent; calyx 4.6–5.7 mm. long; corolla 18–20 mm. long.—Sandy soil; Coastal Sage Scrub; Santa Rosa Id. April.

20. **G. inconspícua** (Sm.) Sweet. [*Ipomopsis i.* Sm. *Cantua parviflora* Pursh.] One-half–2 dm. high, with several main stems from among basal and lower cauline lvs., cobwebby-pubescent below, lightly glandular-pubescent above; lower lvs. deeply pinnately lobed into 3–5 or more linear lobes; infl. rather lax; calyx 3–4.5 mm. long at anthesis; corolla 5–8 mm. long, the tube purple, the throat yellow; lobes 1.6–2 mm. long, light pinkish-violet, drying blue; longest stamens well exserted; caps. 4–5.5 mm. long; $n = 18$ (Grant, 1958).—Sandy flats, ca. 6000 ft.; Sagebrush Scrub, Yellow Pine F., etc.; Mono Co. to e. Wash., Colo., Ariz. April–June.

21. **G. intérior** (Mason & A. Grant) A. Grant. [*G. tenuiflora* ssp. *i.* Mason & A. Grant. *G. inconspicua* ssp. *austro-occidentalis* A. & V. Grant.] Annual 8–22(–30) cm. high, with cent. erect stem often shorter than the secondary stems, cobwebby-pubescent below, ± glandular above; basal and lower lvs. 1- to 2-pinnately lobed with narrow rachis 0.5–1.8 mm. wide, rarely wider, and closely spaced lobes 0.5–5 mm. long; pedicels slender to filiform, unequal, the terminal one 1–5 mm. long, the lateral 6–21 mm.; calyx 2–4.3 mm. long, cobwebby or glandular; corolla 6–10 mm. long; tube slender, included or slightly exserted, purple-veined, flaring into a short throat, the throat yellow, with purple markings in upper part, the lobes pinkish violet; stamens unequal, the longest exserted; caps. broadly ovoid; $n = 9$ (Grant, Beeks & Latimer, 1956).—Sandy places, 2000–6000 ft.; V. Grassland, Joshua Tree Wd., Yellow Pine F., etc.; Cuyama and Kern valleys, e. of Sierra Nevada; w. cent. Nev. Variable; plants of Kern R. drainage, mostly with cobwebby calyces, are true *interior;* those of S. Coast Ranges, with mostly glandular calyces and rather light pubescence on lower stem and lvs., have been named *austro-occidentalis;* and those from Walker R., Mono Co., with mostly glandular calyces and rather dense pubescence on lower stems and lvs. are unnamed.

22. **G. mìnor** A. & V. Grant. Mostly 1–2 dm. high, with 3–8 stems from a cluster of basal lvs., cobwebby below, lightly glandular above; basal lvs. 1.5–4 cm. long, pinnately lobed with 4–6 pairs or lobes 1.5–4 mm. long, rarely bipinnatifid; cauline lvs. reduced upward; infl. strict; pedicels unequal; calyx 2–3 mm. long in fl.; corolla 4–7.5 mm. long, the upper tube and throat dark violet or with some yellow, the lobes pale violet; stamens barely exserted; caps. 5–7 mm. long; $n = 9$ (Grant, Beeks & Latimer, 1956).—Sandy flats and washes, 1000–3500 ft.; V. Grassland, Pinyon-Juniper Wd., Creosote Bush Scrub, etc.; inner Coast Ranges from San Benito Co. s. to w. Mojave Desert. March–April.

23. **G. ophthalmoídes** Brand. [*G. o.* ssp. *Clokeyi* auth., not *G. C.* Mason.] Much-branched annual with many stems from near base, 1.5–3 dm. high, cobwebby-pubescent below, glandular-pubescent above; basal lvs. in ± dense rosette, bipinnately lobed with primary lobes 2–3 times as long as width of rachis; infl. diffuse; pedicels very unequal; calyx 2.5–4.5 mm. long at anthesis, glabrous or sparsely glandular; corolla 7–12 mm. long, the tube well exserted from calyx, filiform, light violet, the lower throat ± yellow, pale violet above, the lobes pinkish-violet or pink, 1.4–2.8 mm. long; caps. 3.5–6.5 mm. long; $n = 18$ (Grant, Beeks & Latimer, 1956).—Sandy places, 3800–8500 ft.; Joshua Tree Wd., Pinyon-Juniper Wd.; desert mts. from e. San Bernardino Co. to Mono Co.; Nev., Utah. May–July.

24. **G. transmontàna** (Mason & A. Grant) A. & V. Grant. [*G. ochroleuca* ssp. *t.* Mason & A. Grant.] Erect to ascending, 1–3 dm. high, 1–several-stemmed, cobwebby-pubescent near base, leafy below middle; lower lvs. cobwebby, pinnately lobed with loosely spaced linear lobes 3–9 times as long as rachis-width; infl. strictly branched; calyx moderately gland-dotted; corolla 4.8–7 mm. long with tube and often part of throat included in calyx, the tube violet or white, the throat yellow and white, the lobes pale violet or white; stamens subequal, slightly exserted; caps. 4.5–6 mm. long; $n = 18$ (Grant, Beeks & Latimer, 1956).—Sandy or gravelly places, 2000–6500 ft.; Creosote

Bush Scrub, Shadscale Scrub; Cuyama V. (Santa Barbara Co.) through the Mojave Desert to Nev., Ariz. March–May.

25. **G. diegénsis** (Munz) A. & V. Grant. [*G. inconspicua* var. *d.* Munz.] Erect, 1–4 dm. high, with stout cent. leader; lvs. in dense basal cluster, glabrous to cobwebby beneath, 1–7 cm. long, strap-shaped, sinuately toothed to pinnately lobed, the rachis 1–6 mm. wide; cauline lvs. much shorter; infl. glandular-pubescent, congested or loose, with unequal pedicels; calyx glandular-pubescent, 3–5 mm. long in fl.; corolla 8–12 mm. long, the tube largely included, purple, the throat yellow above; lobes light violet, 2.5–4.4 mm. long; longest stamens slightly exserted; caps. 4–6.7 mm. long; $n = 9$ (Grant, Beeks & Latimer, 1956).—Sandy openings, 1800–7200 ft.; Coastal Sage Scrub, Chaparral to Yellow Pine F.; interior s. Calif. from Los Angeles Co. to n. L. Calif. April–June.

26. **G. latiflòra** (Gray) Gray. [*G. tenuiflora* var. *l.* Gray.] Erect plant 1–3 dm. high, with 1–many stems, glabrous and glaucous at base, ± glandular upward; basal lvs. 2–7 cm. long, strap-shaped, sinuately toothed to pinnately lobed with serrate divisions, lightly cobwebby; cauline lvs. much smaller; infl. loose, with unequal gland-dotted pedicels; calyx 4–5 mm. long at anthesis; corolla 15–22 mm. long, with a slender purple tube 1–1.5 mm. wide, slightly exserted; throat full, abruptly expanded, with orifice 6–9 mm. wide; lobes white at base, pale violet at tips, 6–8 mm. long; stamens unequal, the longest ± exserted; caps. 6–9 mm. long; $n = 9$ (Grant, Beeks & Latimer, 1956).—Sandy washes and flats, 2500–3600 ft.; Creosote Bush Scrub, Joshua Tree Wd.; sw. Mojave Desert. April–May.

KEY TO SUBSPECIES

Stems glabrous and glaucous at base; corolla 10–23 mm. long.
  Corolla-throat yellow and white or more than half yellow and white.
    Calyx 2.3–3.3 mm. long; corolla 10–15 mm. long. S. Coast Ranges .... ssp. *cuyamensis*
    Calyx 4–5 mm. long; corolla 15–20 mm. long. Mojave Desert .... *G. latiflora*
  Corolla-throat purple in its lower half or three-fourths .... ssp. *Davyi*
Stems cobwebby-pubescent near base.
  Corolla-tube and throat together 14–20 mm. long; erect plants. Nw. Mojave Desert.
    Corolla 2½–3½ times as long as the lobes, the form stout and full .... ssp. *excellens*
    Corolla 3–6 times longer than the lobes, the form slender .... ssp. *elongata*
  Corolla-tube and throat together 7.5–8.5 mm. long; low spreading plants. Coso Mts. .... ssp. *cosana*

Ssp. **cuyaménsis** A. & V. Grant. Basal lvs. moderately cobwebby, 1.5–3.5 cm. long; calyx 2.3–3.3 mm. long at anthesis; corolla pale, 10–15 mm. long, the tube and throat 2½–3 times as long as calyx; $n = 9$ (Grant, Beeks & Latimer, 1956).—Sandy places, 2000–5000 ft.; V. Grassland, Pinyon-Juniper Wd.; Cuyama V. to Kern and Ventura cos.

Ssp. **Dàvyi** (Mlkn.) A. & V. Grant. [*G. D.* Mlkn. *G. tenuiflora* var. *D.* Mason ex Jeps.] Basal lvs. 2–7 cm. long; calyx 4–6.6 mm. long; corolla richly colored, 18–24 mm. long, the tube and throat combined 2–3 times the calyx; $n = 9$ (Grant, Beeks & Latimer, 1956).—Open fields and flats, 2500–4000 ft.; Joshua Tree Wd., Creosote Bush Scrub, V. Grassland, etc.; w. Mojave Desert to Cuyama V. and s. San Luis Obispo Co. March–May.

Ssp. **excéllens** (Brand) A. & V. Grant. [*G. tenuiflora* ssp. *l.* var. *e.* Brand.] Basal lvs. bipinnately lobed; calyx 4.7–7 mm. long; corolla 21–30 mm. long, the tube and throat combined 2–4 times the calyx; $n = 9$ (Grant, Beeks & Latimer, 1956).—Open sandy soil, 2500–3000 ft.; Creosote Bush Scrub, Joshua Tree Wd.; w. Mojave Desert from near Mohave to Kramer Hills and beyond El Paso Mts. April.

Ssp. **elongàta** A. & V. Grant. Basal lvs. pinnately to bipinnately lobed; calyx 4–6 mm. long; corolla 21–34 mm. long, the combined tube and throat 3–6 times as long as calyx.—Sandy places, 2300–3300 ft.; mostly Creosote Bush Scrub; Mojave Desert from n. of Barstow to Rand and El Paso mts. April–May.

Ssp. **cosàna** A. & V. Grant. Basal lvs. 2–3.5 cm. long; calyx 3–3.5 mm. long; corolla 12–13 mm. long, tube twice as long as calyx.—Coso Mts., Inyo Co. June.

27. **G. sinuàta** Dougl. ex Benth. [*G. inconspicua* var. *s.* Gray. *G. tenuiflora* var. *s.* Jeps.] Stiffly erect plant with 1 main stem or several, from basal rosette, glabrous and glaucous near base, moderately glandular-pubescent above, 1–3 dm. high; basal lvs. cobwebby-pubescent, strap-shaped, the rachis 2–6 mm. wide, the lobes toothed or short-lobed; cauline lvs. much reduced; infl. ± loose; pedicels very unequal in fr.; calyx 2.5–3.6 mm. long in fl.; corolla 7–11 mm. long, the tube well exserted; lobes 1.5–3 mm. long, pale

violet to pinkish or whitish; stamens very short; caps. 4–7 mm. long; $n = 18$ (Grant, Beeks & Latimer, 1956).—Sandy desert plains, below 6000 ft.; Creosote Bush Scrub, Joshua Tree Wd., Sagebrush Scrub, etc.; Mojave Desert to Plumas Co.; to Wash., Colo., Ariz. April–June.

28. **G. brecciàrum** Jones. [*G. inconspicua* ssp. *sinuata* var. *deserti* Brand.] Erect or decumbent, 1–3 dm. high, leafy, cobwebby-pubescent below, glandular above; basal rosette semi-erect, loose, the lower lvs. bipinnately to 'tripinnately lobed, the rachis 1.5–3 mm. wide, the lobes ca. as wide; calyx glandular, 2.5–4 mm. long at anthesis, the broad herbaceous lobes connected by narrower sinus-membranes; corolla 7–11 mm. long; throat and orifice narrow to broad, the veins beneath sinuses violet; lobes 2–3.5 mm. long, violet to white; stamens and style short, not or slightly exserted; caps. 4–6.8 mm. long, broadly oval; $n = 9$ (Grant, Beeks & Latimer, 1956).—Sandy slopes, below 7500 ft.; Creosote Bush Scrub to Pinyon-Juniper Wd., Sagebrush Scrub, Yellow Pine F., etc.; Cuyama V. to w. Mojave Desert, and e. of Sierra Nevada to Modoc Co.; Ore., Nev. April–June.

Ssp. **negléecta** A. & V. Grant. Corolla 9–20 mm. long, corolla-form full and stout, the tube not or scarcely exserted; stamens and style long-exserted.—Sandy places, 2400–7000 ft.; Creosote Bush Scrub to Pinyon-Juniper Wd.; w. borders of Mojave Desert in Kern and Inyo cos.

Ssp. **argusàna** A. & V. Grant. Corolla 9–20 mm. long, the form long and slender, the tube well exserted; stamens and style long-exserted.—At 2100–5800 ft.; Creosote Bush Scrub, Joshua Tree Wd., etc.; Mojave Desert from Red Rock Canyon and El Paso Mts. to Darwin and Avawatz Mts. March–May.

29. **G. modocénsis** Eastw. [*G. tetrabreccia* A. & V. Grant.] Small to large often spreading plant branching from near base, the ± decurrent secondary branches 1.5–4.5 dm. long, ± cobwebby-pubescent near base, ± glandular above; basal rosette usually well developed, the lvs. 2–7 cm. long, pinnately lobed with rachis 1–5 mm. wide; cauline lvs. ± clasping at base, dentate to deeply lobed; bracts entire or with a pair of short teeth; infl. ± congested in fl., looser in fr., with 1–5 fls. borne above a bract; pedicels very unequal in fr.; calyx glandular equaling or exceeding mature caps.; corolla 6–11 mm. long, 2–2½ times the calyx, the tube and usually lower throat purple, the middle and upper throat with lemon-yellow spots and white rarely violet vein-areas between; lobes violet or white; stamens short, unequal; style barely exserted; caps. 4.4–6.5 mm. long; $n = 18$ (Grant, Beeks & Latimer, 1956).—Sandy places, 1000–7000 ft.; Yellow Pine F. to Sagebrush Scrub, etc.; Mt. Pinos region and Mojave Desert along e. side of Sierra Nevada to Ore., Ida. April–June. Included here are 3 types: (1) *modocénsis* proper, lightly cobwebby; lvs. coarsely dentate; corolla 8–11 mm. long, from Plumas and Lassen cos. n. (2) An unnamed form, lower stems glabrous to moderately cobwebby; lvs. shallowly to coarsely dentate; corolla 6–9 mm. long; San Bernardino Mts. and Mojave Desert to Tulare Co. (3) *tetrabréccia*, densely matted with cobwebby pubescence below; cauline lvs. deeply lobed; corolla 6.5–7.8 mm. long; Mt. Pinos region.

30. **G. leptomèria** Gray. [*G. l.* var. *myriacantha* Jones. *Aliciella Triodon* var. *humíllima* Brand. *G. inconspicua* var. *dentiflora* A. Davids.] Erect annual 0.5–2 dm. high, 1–several-stemmed, ± glandular-puberulent throughout; basal lvs. in rosette, numerous, broadly strap-shaped, coarsely dentate to pinnatifid with broad rachis, mostly 2–5 cm. long; infl. corymbose-paniculate; fls. pedicelled; calyx 1.6–2.7 mm. long at anthesis, becoming ca. equal to caps.; corolla tubular, 4.7–6.5 mm. long, the tube well exserted from calyx, 2.5–3 mm. long, throat narrow and lobes frequently tridentate, 1–1.5 mm. long, usually each with a diffused purple streak; stamens shorter than corolla-lobes; caps. 3–4 mm. long, many-seeded.—Open sandy places, 2600–6750 ft.; Creosote Bush Scrub to Pinyon-Juniper Wd.; Mojave Desert and Mono Co.; to e. Wash., Wyo., Colo., New Mex. April–June.

31. **G. micromèria** Gray. [*G. leptomeria* ssp. *m.* Mason & A. Grant.] Much like *G. leptomeria* but frequently larger, 1–3 dm. high; basal lvs. narrowly or broadly strap-shaped, coarsely dentate to pinnatifid with narrow or broad archis; calyx 1.4–2.8 mm. long at anthesis, becoming much exceeded by caps.; corolla tubular, 2.5–7.5 mm. long, the tube included or slightly exserted; lobes usually oval, acute, rarely tridentate, 1.5–3 mm. long; caps. 3–5.5 mm. long; $n = 8, 17, 25$ (Grant, 1958).—Sandy places, 2100–9000 ft.; Creosote Bush Scrub to Pinyon-Juniper Wd.; Mojave Desert to Ore., Ida., Colo.

Variable; small-fld. form most common in Mojave Desert; larger-fld. form in n. April–June.

32. **G. hutchinsifòlia** Rydb. [*G. leptomeria* ssp. *rubella* (Brand) Mason & A. Grant.] Near to *G. leptomeria,* the basal lvs. more deeply cut and mostly bipinnatifid; middle cauline lvs. well developed, 0.5–3.5 cm. long, narrowly linear and entire or with a few linear lobes; corolla 8–14 mm. long, the tube 3.5–7 mm. long, violet, lower throat yellow, upper white, the lobes white with pale violet streaks on outside; $n = 9$ (Grant, 1958).—Sandy places, largely 3000–5000 ft.; Creosote Bush Scrub, Joshua Tree Wd., etc.; Red Rock Canyon (w. Mojave Desert), Death V. region, Fish Lake V. (e. of White Mts.); to sw. Utah, ne. Ariz. April–May.

33. **G. campanulàta** Gray. [*Navarretia c.* Kuntze.] Spreading annual 0.2–1.5 dm. high, sparsely glandular-puberulent, with very slender branches; lvs. lanceolate to linear, 0.6–1.5 cm. long, entire or nearly so; fls. solitary on slender pedicels, each opposite a lf.; calyx deeply cleft, 2–4 mm. long, the linear-lanceolate lobes united below by a hyaline membrane; corolla funnelform to narrow-campanulate, 5–10 mm. long, the tube to ca. 1 mm. long, the throat yellow, 5–7 mm., the lobes 2–4 mm. long, white; stamens unequal; caps. ovoid, 3–4 mm. long; $n = 9$ (Grant, 1958).—Sandy places, 5000–7000 ft.; Sagebrush Scrub, Shadscale Scrub; Inyo and Mono cos.; w. Nev. May.

34. **G. filifòrmis** Parry ex Gray. [*Tintinabulum f.* Rydb.] Erect annual, mostly simple below, branched above, glabrous, the stems very slender; lvs. mostly alternate, simple, entire, linear, 1–5 cm. long; fls. few, each solitary and opposite a lf. on a slender pedicel 6–12 mm. long; calyx deeply cut, ca. 3 mm. long; corolla campanulate, yellow, 5–6 mm. long, the tube obsolete, the lobes 3–4 mm. long; stamens extending beyond the throat; pistil exserted; caps. ca. as long as calyx; $n = 9$ (Grant, 1958).—Sandy or gravelly places, below 6000 ft.; Creosote Bush Scrub to Pinyon-Juniper Wd.; Mojave Desert; to Utah, Ariz. March–May.

35. **G. inyoénsis** Jtn. [*G. campanulata* var. *breviuscula* Jeps.] Annual 0.3–1 dm. high, glabrous to ± stipitate-glandular, the slender stems spreading; lvs. oblanceolate, entire or ± coarsely toothed, reduced up the stem; fls. solitary in axils of or opposite the bracts; pedicels slender, stipitate-glandular; calyx 1.5–2 mm. long at anthesis, the lobes linear-lanceolate; corolla white with yellow throat, broadly funnelform, 3–6 mm. long, the ovate lobes 2–2.5 mm. long; stamens ca. 2 mm. long; caps. ellipsoid, many-seeded.—Dry sandy and gravelly places, 6500–8500 ft.; Pinyon-Juniper Wd. to Lodgepole F.; e. slope of Sierra Nevada from Rock Creek, n. Inyo Co. to e. Fork of Kern R., Tulare Co. April–July.

36. **G. latifòlia** Wats. [*Navarretia l.* Kuntze.] Erect annual with rank odor, simple or branched, 1–3 dm. high, glandular-pilose throughout; lvs. mostly on lower part, sometimes quite basal, ovate to roundish, 2–8 cm. long, coarsely serrate to laciniate with mucronate teeth, sessile or the lower lvs. short-petioled; fls. many in corymblike panicles, sessile or with slender pedicels 1–2 cm. long; calyx campanulate, 6–8 mm. long, the lobes subulate, 3–4 mm. long; corolla narrow-funnelform, 6–11 mm. long, bright red-pink within, pale or buff outside, the lobes 2–3 mm. long; stamens unequal, barely exserted; pistil included; caps. 5–7 mm. long; seeds many, deep red-brown; $n = 18$ (Grant, 1958).—Common on desert mosaics and in washes, below 2000 ft.; Creosote Bush Scrub; both deserts n. to Inyo Co.; to Utah. March–May.

37. **G. Rípleyi** Barneby. [*G. Gilmanii* Jeps.] Suffrutescent perennial branching from base, glutinous with capitate hairs; lvs. hollylike, 2–6 cm. long, 1–4 cm. wide, mostly simple with a broad obovate blade and long aristate teeth; petioles 0.5–2 cm. long; infl. cymose-paniculate with linear to 3-lobed bracts and divaricate pedicels; calyx 4–5 mm. long, the lobes aristiform; corolla vespertine, 5–7 mm. long, the lobes elliptical, 2–3 mm. long, pink, the tube white, the throat narrow; stamens unequal, subequally inserted in lower tube; fils. stout, unequal; style included, barely lobed; caps. ellipsoid, 4–5 mm. long, included; seeds deep red-brown, small, many.—Limestone crevices, 3000–4500 ft.; Creosote Bush Scrub; Panamint Mts. to sw. Nev. May–June.

8. Ipomópsis Michx.

Annual to perennial herbs, often with a basal rosette of lvs. and with leafy stems; one sp. a shrub. Herbage frequently villous, sometimes with stipitate glands, sometimes glabrous. Lvs. pinnatifid, the lateral segms. well developed, sometimes reduced, the segm.-tips with horny mucros. Infl. cymose, varying from loosely racemose to tightly congested, bracteate, each fl. usually subtended by a bract. Calyx of equal herbaceous mucronate lobes joined by broad hyaline sinuses, villous, glandular or glabrous. Corolla salverform or tubular, the veins branching in the tube and limb, usually not anastomosing, violet, red, white, or varying shades of yellow or pink. Stamens inserted in the tube or sinuses of corolla, often unequal in length and point of insertion, sometimes declined, the anthers included or exserted; pollen blue, white or yellow. Caps. tardily or freely dehiscent, the locules 1–2–several-seeded. Seeds long and slender, often bent, to spheroidal, rounded or angular, white or brown, smooth or corrugated. Ca. 23 spp. from Rocky Mt. region to Pacific Coast, Fla.; 1 sp. in s. S. Am. (Greek, *ipo*, to strike, and *opsis*, appearance.)

(Grant, V. A synopsis of Ipomopsis. El Aliso 3: 351–362, 1956.)
Plants perennial or biennial.
  Corolla 2–5 cm. long, red or pink to yellowish; infl. not subcapitate.
    Plant with a ± woody base, the stems much branched; corolla-lobes unequal and unequally cleft; mature plants lacking a basal rosette of lvs. S. San Diego Co. . . . . 1. *I. tenuifolia*
    Plant scarcely woody at base, the stems largely simple; corolla-lobes subequal, with lanceolate lobes; plants usually with basal rosettes. San Bernardino Co. n. . . . . 2. *I. aggregata*
  Corolla 0.6–1 cm. long, white; infl. subcapitate . . . . 3. *I. congesta*
Plants annual.
  Lvs. pinnatifid; fls. all in terminal clusters . . . . 4. *I. polycladon*
  Lvs. entire or irregularly toothed; lower fls. solitary in lf.-axils . . . . 5. *I. depressa*

1. **I. tenuifòlia** (Gray) V. Grant. [*Loeselia t.* Gray. *Gilia t.* Gray. *G. truncata* A. Davids.] Suffruticose perennial 1–4 dm. high, much-branched from base, subglabrous below, minutely glandular above, leafy throughout; lvs. narrowly linear, entire, or some pinnately dissected into few linear acerose lobes; fls. solitary in axils of the upper lvs., forming subcorymbose clusters; calyx narrow-campanulate, ca. 6 mm. long, membranous between ribs, the lobes ca. ¼ the tube; corolla scarlet, 1.5–2.5 cm. long, tubular-funnelform, the limb ± irregular, the lobes 5–6 mm. long, oblong, retuse with middle triangular tooth; stamens and style long-exserted; *n* = 7 (Grant, 1956).—Dry rocky slopes, 1500–3500 ft.; Creosote Bush Scrub, Pinyon-Juniper Wd., Chaparral; Mt. Springs Grade to Jacumba and Campo, se. San Diego Co.; n. L. Calif. March–May.

2. **I. aggregàta** (Pursh) V. Grant. [*Cantua a.* Pursh. *Gilia a.* Spreng. *Callisteris a.* Greene.] Biennial, mostly 3–8 dm. high, the stems erect, simple or branched from the base, glandular-puberulent to pilose; lvs. pinnately dissected, mostly 3–5 cm. long, the lobes 1–2 cm. long, with ultimate acutish segms.; midrib often villous; infl. a ± thyrsoid elongate panicle, the fls. short-pedicelled; calyx subcampanulate, 6–9 mm. long, the sepals united less than halfway, subulate, moderately awned, the base ± broad; corolla tubular-funnelform, bright red with yellow mottling, rarely yellow, 2–3.5 cm. long, the lobes lanceolate, rotately spreading, often becoming reflexed, ca. ⅓ as long as tube; stamens equally or unequally inserted in or below the corolla-sinuses and exserted from the throat; caps. ovoid, ca. as long as calyx; seeds several in each locule; *n* = 7 (Grant, 1956).—Open places like sandy flats, rocky ridges, etc., 3500–10,300 ft.; Montane Coniferous F.; Sierra Nevada from Inyo and Tulare cos. n., Coast Ranges from Trinity Co. n.; to B.C., Ida., etc. June–Sept.

KEY TO SUBSPECIES

Lf.-segms. acutish; plants strict; sepal-tips usually plainly awned.
  Corolla-tube stout, with marked upward expansion; fls. mostly deep red with yellow mottling.
    Sepal-blade subulate from a ± broad base, the sepals united for less than half the total length of calyx; stamens long-exserted . . . . I. *aggregata*
    Sepal-blade deltoid to oblong, ± abruptly contracted to a subulate tip, the sepals united for ½–⅗ the total calyx-length; stamens included . . . . ssp. *arizonica*
  Corolla-tube slender with but slight upward expansion; fls. pink, white or yellow. Sonora Pass and Bishop Creek (Sierra Nevada), Warner Mts. . . . . ssp. *attenuata*
Lf.-segms. obtusish; plants lax; sepal-tips barely awned. Sierra Nevada from Yosemite region s.
                                                               ssp. *Bridgesii*

Ssp. **arizónica** (Greene) V. & A. Grant. [*Callisteris a.* Greene. *Gilia aggregata* var. *a.* Fosb.] Plant biennial to perennial, 1–3 dm. high, often lanate-pubescent; lvs. mostly 1–2 cm. long; calyx united ½–⅗ its total length of 4–6 mm.; corolla red, 2–2.5 cm. long; stamens mostly included.—Dry washes and rocky places, mostly 4500–10,500 ft.; Pinyon-Juniper Wd., Bristle-cone Pine F.; Providence, New York, Clark, Panamint, and Inyo mts.; to Utah, Ariz.

Ssp. **attenuàta** (Gray) V. & A. Grant. [*Gilia aggregata* var. *a.* Gray.] Plants mostly 2–4 dm. high, ± lanate-pubescent; lvs. 2–5 cm. long; calyx 6–9 mm. long, the sepals united ⅓–½ their length, lance-subulate in free part, rather well awned; corolla pale pink to white or with yellowish tinge, 2.5–3.5 cm. long, the tube slender, with very slight upward expansion.—Dry rocky slopes, ca. 9000–10,000 ft.; Subalpine F., etc.; Sonora Pass and Masonic Mts., Mono Co., Warner Mts., Modoc Co.; to Wash., Colo.

Ssp. **Bridgèsii** (Gray) V. & A. Grant. [*Gilia aggregata* var. *B.* Gray. *G. B.* Wherry.] Plants largely perennial, with a woody branched caudex, of lax habit, the several stems from the base slender, 2–4 dm. long, curved-ascending; lvs. 2–3 cm. long; calyx 5–7 mm. long, the sepals narrow-deltoid, barely awned, united ½–⅗ their length; corolla from deep magenta to slightly salmon-tinged magenta, 2.5–3.5 cm. long; stamens included or barely exserted.—Open places, 6500–9000 ft.; Red Fir F., Lodgepole F.; Sierra Nevada from Mariposa and Mono cos. to Tulare Co.

3. **I. congésta** (Hook.) V. Grant. [*Gilia c.* Hook. *G. c.* var. *crebrifolia* Gray.] Erect perennial with several stems to 1–2.5 dm. long, with arachnoid-floccose pubescence and a persistent basal tuft of lvs.; lvs. pinnately or bipinnately lobed from a broad petiole, or laciniate, 1–4 cm. long, gradually reduced up the stems and reduced to subpalmate leafy bracts in infl., sometimes glabrate on upper surfaces; fls. in single or clustered heads; calyx cylindric, the lobes united below by a membrane, densely arachnoid, ca. 4.5 mm. long; corolla white, salverform, 4–6 mm. long, the yellow tube 3–4 mm. long; stamens exserted; caps. obovoid; seeds 1–2 in a locule; $n = 7$ (Grant, 1956).—Dry cinder slopes, etc., 4000–7000 ft.; Sagebrush Scrub to Yellow Pine F.; Mono Co. to Modoc and Siskiyou cos.; to Ore., Ida., S. Dak., Utah. June–July.

Ssp. **montàna** (Nels. & Kenn.) V. Grant. [*Gilia m.* Nels. & Kenn. *G. c.* var. *m.* Const. & Roll.] Densely matted plants forming cushionlike rosettes; basal lvs. palmately divided into short broad segms.—Mostly at 7000–12,000 ft.; Subalpine F., Bristle-cone Pine F., Alpine Fell-fields, etc.; White Mts., Sierra Nevada to e. Ore. July–Sept.

4. **I. polyclàdon** (Torr.) V. Grant. [*Gilia p.* Torr.] Annual with several naked spreading branches from base, these 0.5–1.5 dm. long, puberulent throughout, ± glandular-villous above; lvs. few, 1–2 cm. long, mostly basal and crowded about the subcapitate fl.-clusters, pinnatifid into short mucronulate segms.; calyx 3–4 mm. long; corolla white, tubular, 4–6 mm. long, the lobes ca. 1 mm. long; stamens inserted in sinuses of corolla; caps. ellipsoid; seeds 2 in each locule; $n = 7$ (Grant, 1956).—Sandy, gravelly or rocky slopes, below 5200 ft.; Creosote Bush Scrub, Pinyon-Juniper Wd., etc.; Mojave Desert to Ore., Colo., Tex. April–June.

5. **I. depréssa** (Jones) V. Grant. [*Gilia d.* Jones.] Divaricately branched annual 2–10 cm. high, glandular-pubescent throughout, even canescent; lvs. well distributed, linear to elliptic-lanceolate, 1–2 cm. long, short-petioled, the upper entire or with 2 teeth; fls. crowded in leafy subcapitate racemes; calyx scarious, 5–6 mm. long, with subulate teeth; corolla white, tubular-funnelform, sometimes ± irregular, ca. 5 mm. long; caps. subglobose; seeds 4–5 in a locule; $n = 7$ (Grant, 1956).—Occasional in dry disturbed places, 3500–5000 ft.; Creosote Bush Scrub, Joshua Tree Wd., Sagebrush Scrub; w. Mojave Desert from Rabbit Springs to Owens V.; to Utah, Nev. April–June.

9. Eriástrum Woot. & Standl.

Annual or perennial, herbaceous or suffrutescent, ± erect to spreading, simple to branched, puberulent to arachnoid or woolly. Lvs. alternate, entire and linear or pinnately divided into linear divisions. Fls. mostly sessile in bracteate heads, rarely solitary and pedicelled. Calyx deeply cleft into subulate pungent lobes, the sinuses ± filled with hyaline membrane. Corolla blue to white or yellow, even pink, funnelform to subsalverform. Stamens inserted at base of corolla-throat or just below sinuses, exserted to included. Anthers often sagittate, sometimes cordate or elliptic. Caps. ellipsoid or obovoid, often splitting into valves. Seeds 1–several in a locule, usually mucilaginous when wetted. A genus of w. N. Am., with 14 spp. (Greek, *erion*, wool, and *astrum*, star, the plants woolly and with starlike fls.)

(Mason, H. L. The Genus Eriastrum, etc. Madroño 8: 65–91, 1945. Craig, T. A revision of the subgenus Hugelia of the genus Gilia [Polemoniaceae]. Bull. Torrey Bot. Club 61: 385–396, 411–428, 1934.)

A. Plants perennial, with woody base; corolla 15–30 mm. long, the tube 2–3 times as long as calyx
                                                                  1. *E. densifolium*
AA. Plants annual, not at all woody; corolla 7–18 mm. long, the tube rarely much longer than calyx.
  B. Stamens inserted in the sinuses of the corolla; anthers 2–2.5 mm. long. W. Mojave Desert and about San Joaquin V. . . . . 5. *E. pluriflorum*
  BB. Stamens inserted well below sinuses.
    C. Corolla 15–16 mm. long, irregular, the lobes ca. 6 mm. long; stamens very unequal; bracts 5–9-lobed. Deserts . . . . 6. *E. eremicum*
    CC. Corolla smaller, or if as long, regular and with stamens subequal and bracts 3-lobed.
      D. Corolla-lobes 6–8 mm. long; corolla dark blue, almost regular. Monterey coast
                                                       2. *E. virgatum*
      DD. Corolla-lobes 2–5 mm. long.
        E. Lobes of corolla ca. as long as tube. Mostly s. Calif. . . . . 3. *E. sapphirinum*
        EE. Lobes of corolla conspicuously shorter than tube.
          F. Fils. of stamens long-exserted.
            G. Corolla yellow. Santa Lucia Mts. . . . . 4. *E. luteum*
            GG. Corolla bluish. Santa Barbara Co. to L. Calif. . . . 8. *E. filifolium*
         FF. Fils. of stamens not long-exserted, the anthers sometimes showing.
          G. Anthers barely exserted.
            H. Corolla 9–12 mm. long, the lobes ca. 4 mm. long. E. of the Sierra Nevada . . . . 10. *E. Wilcoxii*
            HH. Corolla 7–8 mm. long, the lobes 2–3 mm. long.
              I. Stems spreading from base, thinly pilose, later glabrate. Mojave and Colo. deserts . . . . 7. *E. diffusum*
              II. Stems erect, mostly floccose. Little San Bernardino Mts. to e. Ore. . . . . 9. *E. sparsiflorum*
         GG. Anthers included.
           H. Corolla blue with yellow tube and throat; lvs. pinnate. Coast Ranges from Lake Co. to San Benito Co. . . 11. *E. Abramsii*
           HH. Corolla pale to white; lvs. 1–3-parted.
             I. Corolla 8–10 mm. long, exceeding longest calyx-lobe.
               J. Branching racemose; corolla-throat 1 mm. long; stamens 0.7 mm. long. Trinity Co. . . . . . . 14. *E. Tracyi*
               JJ. Branching virgate-corymbose; corolla-throat 2 mm. long; stamens 1.5 mm. long. Lake Co.
                                            12. *E. Brandegeae*
             II. Corolla 5–7 mm. long, subequal to the longest calyx-lobe. Fresno Co. to Kern Co. . . . . 13. *E. Hooveri*

1. E. densifòlium (Benth.) Mason. [*Huegelia d.* Benth. *Gilia d.* Benth. *Navarretia d.* Kuntze. *Welwitschia d.* Tides.] Erect perennial, much-branched from the woody base, 2–3 dm. high, glabrous to subglabrous; lvs. well distributed along the stems, irregularly pinnatifid into linear ± pungent segms. or entire, 2–5 cm. long; fls. often 25 or more, sessile in terminal bracteate arachnoid heads, the heads usually in groups; bracts 3–5-lobed, the terminal lobe much the longest, acerose; calyx 6–7 mm. long, cleft, the lobes subequal, pungent, united below by a white-lanate membrane; corolla salverform, 22–25 mm. long, the tube white to yellow, the lobes blue, ca. 8 mm. long; stamens inserted in throat just below the sinuses, the anthers white, sagittate, exserted; stigmas exserted; caps. ellipsoid, 3–4 mm. long, the locules several-seeded; seeds cellular-reticulate, ca. 2 mm. long; *n* = 7 (Grant, 1958).—Sandy places; Coastal Strand, Coastal Sage Scrub; Monterey Co. to Santa Barbara Co. May–July.

KEY TO SUBSPECIES

Corolla 22–32 mm. long.
  The corolla 22–25 mm. long, the lobes almost half as long as tube; plants subglabrous. Coastal, Monterey Co. to Santa Barbara Co. . . . . *E. densifolium*
  The corolla 25–32 mm. long, the lobes ca. ½ as long as tube; plants lanate. Santa Ana R. drainage, s. Calif. . . . . ssp. *sanctorum*
Corolla 14–19 mm. long.
  Lf.-lobes often ca. as long as width of the rachis; lvs. commonly recurved; fls. pale blue or lavender. Mojave Desert . . . . ssp. *mohavense*
  Lf.-lobes 2–10 times as long as width of the rachis; lvs. mostly ascending; fls. deep blue. Montane and cismontane.

Bracts mostly 1–5-lobed; many fl.-clusters axillary, mostly not more than 10(15)-fld.; lvs. lobed to entire. Below 4000 ft. . . . . ssp. *elongatum*
Bracts commonly 5–9-lobed; fl.-clusters mostly terminal, commonly 15–20-fld.; lvs. 2–15-lobed. From 4000–8000 ft. . . . . ssp. *austromontanum*

Ssp. **sanctòrum** (Mlkn.) Mason. [*Gilia d.* var. *s.* Mlkn. *Huegelia d.* var. *s.* Jeps.] Entire plant lanate even in age, 2.5–7.5 dm. high; fls. blue, 25–32 mm. long.—Below 1500 ft.; Coastal Sage Scrub; along Santa Ana R. June–Aug.

Ssp. **mohavénse** (Craig) Mason. [*Gilia densifolia* var. *m.* Craig. *Huegelia d.* var. *m.* Jeps.] Plant 1–3 dm. high, canescent-lanate; lvs. with broad rachis and short teeth; corolla pale blue to lavender, ca. 15 mm. long.—Dry slopes, 2500–8500 ft.; Joshua Tree Wd., Pinyon-Juniper Wd.; Mojave Desert from Kelso and Little San Bernardino Mts. to Kern and Inyo cos. June–Oct.

Ssp. **elongàtum** (Benth.) Mason. [*Huegelia e.* Benth. *Gilia e.* Steud. *G. d.* var. *e.* Gray.] Plants 2–9 dm. high, ± canescent-lanate even in maturity; lvs. ascending; fl.-clusters rarely as much as 15-fld.; corolla deep blue to blue-violet, 15–16 mm. long.—Dry slopes and washes; Chaparral, Coastal Sage Scrub; mostly away from the coast, from San Luis Obispo Co. to L. Calif., extending inland to edge of desert. June–Sept.

Ssp. **austromontànum** (Craig) Mason. [*Gilia d.* var. *a.* Craig. *Huegelia d.* var. *a.* Jeps.] Plants 1–4 dm. high, glabrate; fl.-clusters often as much as 20-fld.; corolla deep blue, 15–19 mm. long.—Dry slopes, 4000–8000 ft.; Yellow Pine F., Red Fir F.; L. Calif. to Santa Barbara and Inyo cos. May–Sept.

2. **E. virgàtum** (Benth.) Mason. [*Huegelia v.* Benth. *Gilia v.* Steud.] Annual, erect, white-floccose or glabrate in age, usually simple-stemmed, sometimes branched from base, erect, 1–4 dm. high; lvs. linear, 1–4 cm. long, entire or with 2–4 lateral linear lobes; infl. subcapitate, densely arachnoid; bracts tripartite, 1.5–3 cm. long; calyx 7–9 mm. long, the segms. linear, unequal, the sinuses with a hyaline membrane running up on the lobes; corolla blue, regular, 1.5–2 cm. long, sometimes with yellow throat and tube, the tube ca. 8 times the throat; stamens at base of throat, exserted; style exserted; $n = 7$ (Flory, 1937).—Sandy places; Coastal Strand; region of Monterey. May–July.

3. **E. sapphirìnum** (Eastw.) Mason. [*Gilia s.* Eastw. *Huegelia virgata* var. *s.* Jeps.] Annual, erect, 1–3 dm. high, loosely paniculately branched, sparsely leaved, commonly viscid-glandular, sometimes lightly floccose in infl.; lvs. linear, commonly entire, 1–3 cm. long, the upper sometimes with 2 short lateral lobes; fls. sessile in few-fld. cymes, the bracts broadly ovate, 3-lobed, often hyaline-margined; calyx ca. 8 mm. long, with broad hyaline membrane between the lobes; corolla funnelform, 8–10 mm. long, sapphire-blue with yellow tube and throat, the tube 1–2 times the throat; stamens inserted at base of throat, 7–8 mm. long; stigma exserted; caps. ca. 4 mm. long, the locules several-seeded; seeds ca. 1 mm. long, concavo-convex, wing-margined; $n = 7$ (Grant, 1958).—Dry places, mostly 4000–8000 ft.; Chaparral, Yellow Pine F.; San Bernardino Mts. to L. Calif. June–Aug.

KEY TO SUBSPECIES

Flower-heads 1–3(–4)-fld.; corolla usually less than 12 mm. long, the lobes as long as the tube.
  Calyx rarely at all lanate, frequently glandular; bracts 1–3-lobed; corolla usually 10–12 mm. long. Montane.
    Fls. sessile, mostly more than one . . . . E. *sapphirinum*
    Fls. pedicelled, solitary . . . . ssp. *gymnocephalum*
  Calyx always lanate; bracts 2–5-lobed; corolla usually less than 10 mm. long. Deserts
                                                             ssp. *ambiguum*
Flower-heads 3–10-fld.; corolla ca. 14 mm. long. Cismontane . . . . ssp. *dasyanthum*

Ssp. **gymnocéphalum** (Brand) Mason. [*Navarretia virgata* ssp. *g.* Brand. *N. v.* var. *oligantha* Brand.] Fls. solitary and pedicelled, rarely in pairs.—Chaparral and Yellow Pine F.; s. San Diego Co.; L. Calif. May–Sept.

Ssp. **ambíguum** (Jones) Mason. [*Gilia floccosa* var. *a.* Jones. *G. virgata* var. *a.* Craig.] Bracts broad, 3–5-lobed; corolla pale to deep blue, 8–10 mm. long.—Dry places, 2500–7000 ft.; Joshua Tree Wd., Pinyon-Juniper Wd.; w. Mojave Desert and bordering mts.; Santa Ana Mts. May–Sept.

Ssp. **dasyánthum** (Brand) Mason. [*Navarretia virgata* var. *d.* Brand. *Gilia v.* var. *d.* Craig.] Heads dense, mostly 5–10-fld., white-woolly; corolla blue, 14–15 mm. long.—

Common in dry places, below 2500 ft.; Coastal Sage Scrub, Chaparral; cismontane L. Calif. to Ventura Co., rare to San Benito Co. May–Sept.

4. **E. lùteum** (Benth.) Mason. [*Huegelia l.* Benth., not *Gilia l.* Steud. *G. lutescens* Steud.] Erect annual 0.5–2 dm. high, few to many branched, arachnoid to glabrate; lvs. linear, simple, 1–2.5 cm. long, entire or with 2 lateral linear lobes; heads congested, 2–8-fld., white-woolly; bracts pinnately cleft; calyx deeply cleft; corolla nearly or quite regular, yellow, 8–10 mm. long, the lobes almost as long as tube; stamens inserted at base of throat below sinuses, exserted; seeds mostly solitary in the locules.—Dry slopes, below 3000 ft.; Mixed Evergreen F., Foothill Wd.; Santa Lucia Mts., Monterey and San Luis Obispo cos. May–June.

5. **E. pluriflòrum** (Heller) Mason. [*Gilia p.* Heller. *G. virgata* var. *floribunda* Gray, not *G. f.* Gray. *G. Brauntonii* Jeps. & Mason. *Huegelia B.* Jeps. *H. p.* Ewan.] Annual, erect, 1–4 dm. high, simple to branched, floccose to glabrate; lvs. sessile, 1.5–5 cm. long, pinnately parted into 3–9 filiform lobes; infl. densely arachnoid, dense, 1.5–4 cm. broad, 8–50-fld.; calyx 8–11 mm. long, cleft into linear unequal lobes, densely arachnoid; corolla funnelform, nearly or quite regular, 1–2 cm. long, deep blue-violet or with some yellow in throat and tube, the lobes ⅔ as long as tube; stamens inserted in sinuses, 4–5 mm. long; stigma exserted; caps. 4 mm. long, the locules 2–3-seeded.—Dry plains and slopes, below 6000 ft.; V. Grassland, Foothill Wd., Chaparral, Yellow Pine F.; hills about the San Joaquin V., Cuyama V., Santa Barbara Co. May–July.

Ssp. **Sherman-Hoỳtae** (Craig) Mason. [*Gilia S.* Craig.] Erect, 6–12 cm. high; infl. 2–8-fld.; corolla pale blue to lavender, the lobes not more than ½ as long as tube; stamens 3–4 mm. long.—Sandy flats, ca. 2500 ft.; Joshua Tree Wd.; Antelope V. region, w. Mojave Desert. May.

6. **E. erèmicum** (Jeps.) Mason. [*Huegelia e.* Jeps. *Gilia e.* Craig. *Navarretia densifolia* var. *jacumbana* Brand.] Annual, erect or spreading, 0.3–2.5 dm. high, floccose to glabrate, usually much-branched from base; lvs. 1–4 cm. long, pinnatifid into 5–9 linear lobes; fl.-clusters lanate, 2–10-fld.; bracts recurved at tips; calyx-segms. linear, subequal; corolla ca. 15–16 mm. long, ± bilabiate, violet, 1.5–2.5 times as long as calyx, the corolla-lobes ca. 6 mm. long; stamens inserted at base of throat, unequal, the longest exserted; stigma exserted; caps. 4–6 mm. long, the locules several-seeded; seeds ca. 2 mm. long.—Common in sandy places, below 4000 (6000) ft.; Creosote Bush Scrub, Joshua Tree Wd., Pinyon-Juniper Wd.; Colo. and Mojave deserts, Inyo Co. to Imperial Co.; Nev. April–June.

7. **E. diffùsum** (Gray) Mason [*Gilia filifolia* var. *d.* Gray. *Huegelia d.* Jeps.] Annual, diffusely branched, 0.3–1.5 dm. high, thinly pilose, then glabrate; lvs. simple, linear to pinnatifid into 3–5 linear lobes; fl.-clusters 3–20-fld., woolly; bracts 3–7-lobed, ± arched; calyx 5–7 mm. long, the lobes unequal; corolla pale blue to white, 7–8 mm. long, almost regular, the lobes ca. half as long as tube; stamens 1–2 mm. long, inserted in middle of throat; caps. 3–4 mm. long, the locules 2–3-seeded.—Dry sandy places, below 6000 ft.; Creosote Bush Scrub, Joshua Tree Wd., Pinyon-Juniper Wd.; deserts from Inyo Co. s.; to L. Calif., Tex. March–May.

Ssp. **Harwoòdii** (Craig) Mason. [*Gilia filifolia* var. *H.* Craig.] Plant erect, canescentlanate; corolla pale, 7.5 mm. long, with apiculate lobes.—Sand hills; Creosote Bush Scrub; deserts near Kelso and Blythe and e. of Twentynine Palms. March–May.

8. **E. filifòlium** (Nutt.) Wòot. & Standl. [*Gilia f.* Nutt. *Huegelia f.* Jeps.] Annual, pilose to subglabrous, 0.4–4 dm. high, simple and virgate or branched; lvs. linear, filiform, entire to pinnately 3–5-lobed; heads 3–15-fld., lanate; bracts 1–2 cm. long, lanate; calyx 5–7 mm. long; corolla narrow-funnelform, ± regular, 8–9 mm. long, blue, the lanceolate lobes 2–2.5 mm. long, the tube yellow; stamens 2.5 mm. long, attached at base of throat, exserted; caps. 4–5 mm. long, cylindric; seeds several in a locule, red-brown, ca. 1 mm. long, wing-angled.—Occasional, dry slopes, below 5000 ft.; Coastal Sage Scrub, Chaparral; cismontane, Santa Barbara Co. to L. Calif. May–June.

9. **E. sparsiflòrum** (Eastw.) Mason. [*Gilia s.* Eastw. *G. filifolia* var. *s.* Macbr.] Annual, erect, 1–4 dm. high, mostly floccose, usually branched from below; lvs. linear or with a pair of short lobes at base; bracts 3–5-lobed to simple; fl.-clusters 2–5-fld., arachnoid; calyx 5 mm. long, deeply cleft into equal lobes; corolla 7–8 mm. long, pale blue to whitish or pinkish, the lobes ca. 3 mm. long; stamens 2–2.5 mm. long, inserted at base of throat, scarcely evident externally; caps. 3–5 mm. long; seeds light brown, ca.

1 mm. long, slightly wing-angled.—Dry slopes, below 8000 ft.; Joshua Tree Wd., Sagebrush Scrub, Pinyon-Juniper Wd., Yellow Pine F.; Little San Bernardino Mts. to Panamint Mts., Tehachapi Mts. to Sierra Nevada of Fresno Co. and along e. slope to Oregon, Nev. June–July.

10. **E. Wilcóxii** (A. Nels.) Mason. [*Gilia W. A. Nels.*] Branched annual, floccose, 1–2 dm. high, the lowest branches longest, making a flat-topped crown; lvs. with 5 (3) linear lobes; bracts 3–5-lobed; heads 3–5-fld., often several heads crowded together; calyx unequally lobed; corolla ± funnelform, 9–11 mm. long, blue to pale blue, the lobes ca. 4 mm. long; stamens inserted at base of throat, slightly exserted; caps. ca. 5 mm. long, ellipsoid; seeds several to a locule, ca. 1 mm. long, ± wing-angled.—Dry places, below 9000 ft.; Sagebrush Scrub, Pinyon-Juniper Wd.; Panamint Mts., along e. face of Sierra Nevada and White Mts. to Wash., Ida., Utah. June–Aug.

11. **E. Abrámsii** (Elmer) Mason. [*Navarretia A.* Elmer. *Huegelia A.* Jeps. & Bail.] Annual, erect or spreading, 0.3–1.5 dm. high, simple or branched, ± floccose; lvs. 1.5–3 cm. long, pinnately divided into filiform segms.; heads densely bracteate, arachnoid; calyx unequally with linear subequal lobes, arachnoid about the middle; corolla 6–8 mm. long, ± regular, blue with yellow tube and throat, the lobes spreading; stamens equally inserted on throat, ± equal in length, included; caps. 4–5 mm. long; seeds solitary in locules, linear-oblong, somewhat winged at one end.—Dry slopes, below 3500 ft.; Chaparral; Coast Ranges, San Benito Co. to Lake Co. June–July.

12. **E. Brandègeae** Mason. Annual 0.5–3 dm. high, with erect branches, ± floccose; lvs. 3-parted into filiform divisions; heads densely arachnoid; bracts 3–5-lobed, longer than heads; calyx 8–10 mm. long, arachnoid, with unequal linear-acerose lobes; corolla subsalverform, ca. 10 mm. long, white to pale blue, the lobes 3 mm. long; stamens inserted at base of throat, 1–2 mm. long, ± equal, included; caps. 4 mm. long; locules mostly 1-seeded.—Volcanic material; Chaparral, Foothill Wd.; mts. of Lake Co. May–Aug.

13. **E. Hoòveri** (Jeps.) Mason. [*Huegelia H.* Jeps.] Annual, erect, 1–2 dm. high, with virgate branching, glabrate; lvs. linear, entire to 3-cleft with 2 linear lateral lobes; outer bracts 3–5-lobed; heads 1–3-fld., somewhat lanate; calyx unequally cleft, 5–6 mm. long, the tips connivent over the caps.; corolla pale blue to white or yellow, 6–7 mm. long, the narrow lobes 2 mm. long; stamens inserted at base of throat, ± equal, included; caps. oblong-ellipsoid, 3 mm. long, each locule with 2–4 seeds.—Rolling plains, below 500 ft.; V. Grassland; San Joaquin V. from Fresno Co. to Kern Co. April–June.

14. **E. Tràcyi** Mason. Erect annual 1–2 dm. high, simple or racemosely branched, lightly floccose; upper lvs. 3-cleft with filiform segms.; heads congested, arachnoid; bracts 3–5-cleft; calyx 6–8 mm. long, arachnoid; corolla 8–9 mm. long, light blue to white, the lobes 2–3 mm. long; stamens inserted at base of throat, included; caps. 5 mm. long, the locules 1–2-seeded.—At ca. 2600 ft.; Hayfork V., Trinity Co. June–July.

10. Langloìsia Greene

Diffuse low rigid annuals. Lvs. alternate, cuneate to linear, pinnately toothed, the lower divisions reduced to slender bristles, the upper bristle-tipped. Fls. in few-fld. terminal bracteate heads, the bracts leafy, with bristly teeth. Calyx lobed, the lobes equal, spinescent-tipped, the tube scarious between the lobes and splitting to base. Corolla tubular-funnelform, almost regular to 2-lipped (with 3 lobes in upper lip). Stamens 5, inserted in corolla-throat. Caps. 3-sided. Seeds 2–9 in each caps., slightly angled, mucilaginous when wet. A small genus of our deserts. (Named for Father Langlois of Louisiana.)

Corolla distinctly 2-lipped; lvs. pinnatifid with ligulate or spatulate rachis and single marginal bristles.
  Calyx 5–6 mm. long; corolla-lobes rounded or truncate at apex, almost as long as tube
    1. *L. Matthewsii*
  Calyx 3–4 mm. long; corolla-lobes pointed at apex, ⅓ to ½ as long as tube . . . . 2. *L. Schottii*
Corolla almost regular; lvs. abruptly dilated toward apex, the marginal bristles mostly in pairs.
  Calyx 8–9 mm. long; corolla-lobes purple-dotted, almost as long as tube . . . . 3. *L. punctata*
  Calyx 6 mm. long; corolla-lobes not dotted, ca. ⅓ as long as tube . . . . 4. *L. setosissima*

1. **L. Matthèwsii** (Gray) Greene. [*Loeselia M.* Gray. *Gilia M.* Gray. *Navarretia M.* Cov.] Branched from base or above, tufted, 3–15 cm. high, somewhat broader; stems

whitish, thinly tomentulose; lower lvs. broadly linear, 1.5–4 cm. long, pinnately toothed, the narrow lobelike teeth bristle-tipped; bracts somewhat shorter and wider; calyx 5–6 mm. long; corolla 2-lipped, whitish to pink, the upper lobes with an elaborate red and white pattern, the tube 8–12 mm. long, the lobes nearly as long, oblong, sometimes shallowly toothed, rounded or retuse at tips; stamens and style well exserted; caps. ca. 3 mm. long; seeds irregular, plump, ± wing-angled, ca. 1.2–1.5 mm. long.—Common in great masses, sandy and gravelly areas, below 5000 ft.; Creosote Bush Scrub, Joshua Tree Wd.; deserts from Inyo Co. to Imperial Co.; to Nev., Son. April–June.

2. **L. Schóttii** (Torr.) Greene. [*Navarretia S.* Torr. *Gilia S.* Gray. *G. setosissima* var. *exigua* Gray. *L. flaviflora* A. Davids.] Tufted, 2–10 cm. high, scattered villous-tomentulose; lvs. linear to linear-oblanceolate, pectinately pinnatifid, 1–3 cm. long, the teeth mostly bristle-tipped; bracts cuneate-oblanceolate, 1–1.5 cm. long, 3–5-toothed; calyx 3–4 mm. long; corolla pale lavender to white, yellowish, or pinkish, 8–12 mm. long, the upper lip 3-lobed with purple spots, the lower lip 2-lobed, ½–⅓ as long as tube, usually spotted near base; stamens well exserted; caps. 3–4 mm. long, 2–6-seeded; seeds plump, ± angled, 0.6–1.2 mm. long; $n = 7$ (Grant, 1958).—Dry sandy or gravelly places, below 5000 ft.; Creosote Bush Scrub, Joshua Tree Wd.; inner Coast Ranges, Fresno Co. to Kern and Santa Barbara cos., Inyo Co. to Imperial Co.; to Utah, Ariz., Son., L. Calif. March–June.

3. **L. punctàta** (Cov.) Goodd. [*Gilia setosissima* var. *p.* Cov. *L. lanata* Brand.] Tufted or simple, 3–15 cm. high, or flat-topped and 15–20 cm. wide, thinly tomentulose to glabrate; lower lvs. linear, 2–3.5 cm. long, finely bristle-toothed, the upper wider, with deltoid 3–5-toothed apex; calyx 8–9 mm. long; corolla almost regular, 15–20 mm. long, lilac, the lobes entire, almost as long as tube, obtuse, purple-dotted; stamens exserted; caps. narrow-oblong, 7–8 mm. long; seeds angled, plump, ca. 1 mm. long.—Dry gravelly places, below 5000 (8400) ft.; Creosote Bush Scrub, Joshua Tree Wd., Pinyon-Juniper Wd.; Inyo Co. (White Mts.) s. through Mojave Desert to San Bernardino Mts.; Nev., Ariz. April–June.

4. **L. setosíssima** (T. & G.) Greene. [*Navarretia s.* T. & G. *Gilia s.* Gray. *L. s.* var. *campyloclados* Brand.] Tufted, 2–7 cm. high, or with prostrate branches to 1 dm. long, tomentulose to glabrate; lvs. 1–2 cm. long, cuneate, with 3 large apical teeth and one pair of lateral, or the basal lvs. linear-subulate, not apically dilated; bracts like upper lvs.; calyx ca. 6 mm. long; corolla almost regular, light violet, not spotted, 12–16 mm. long, the lobes oval-oblong, ca. ⅓ as long as tube; stamens shorter than corolla-lobes; caps. ca. 6 mm. long; seeds angled, ca. 1 mm. long; $n = 7$ (Grant, 1958).—Dry sandy places, below 3500 ft.; Creosote Bush Scrub; Mojave and Colo. deserts; to Ore., Ida., Utah, Son., L. Calif. April–June.

11. Navarrètia R. & P.

Annuals, mostly with rigid stems, erect and simple to divaricately branched from base or along the stems, or in some cases branched just below the terminal heads, the branches then spreading to prostrate. Lvs. alternate, entire to once or twice pinnate, sometimes palmately lobed, the upper bracteate, acerose or spine-tipped; lobes of lvs. or bracts sometimes proliferating on the backs with extra segms. from the rachis. Fls. 5- or 4-merous, sessile or nearly so; in spiny densely bracted heads, the bracts usually with broad coriaceous rachis. Calyx cleft to base into ± unequal entire to toothed segms. which are usually acerose and united by a scarious membrane in their lower ¼ to ⅔. Corolla funnelform to salverform, white or yellow to pink, blue or purple, the lobes with 1 or 3 main veins, which may be simple or branched. Stamens equally or unequally inserted in throat or in sinuses of corolla-lobes, exserted or included; fils. glabrous; anthers mostly oval. Stigma entire to 2–3-lobed, included or exserted. Caps. ovoid to obovoid, 1–3-loculed, 3–8-valved, with irregular or regular dehiscence, membranous to chartaceous, usually circumscissile near base or above and then splitting into valves from the line of circumscission or sometimes from the top. Seeds mostly brown, ovoid or angled, often minutely pitted, 1–many in each cell. Ca. 30 spp., mostly w. N. Am., one in Chile and Argentina. (Named for Dr. F. *Navarrete*, a Spanish physician.)

(Brand, A., *in* Das Pflanzenreich, IV, 250: 151–168, 1907. Mason, H. L., *in* Abrams, Ill. Fl. Pacific States, 3: 440–452, 1951. Crampton, B. Morph. and ecol. considerations in the classification of Navarretia. Madroño 12: 225–238, 1954.)

A. Main stems finely retrorse-pubescent with crisped white hairs; mature caps. thin-membranous (the walls disintegrating on dehiscence, not splitting into distinct valves) in the first 10 spp.
  B. Stigma entire or minutely 2-lobed; lobes of the foliaceous bracts soft-herbaceous when fresh, the rachis little expanded at base; each corolla-lobe with 1 main vein.
    C. Vein of corolla-lobes branched; corolla 7–9.5 mm. long, white. San Benito and Eldorado cos. n. .... 1. *N. leucocephala*
    CC. Vein of corolla-lobes not branched; corolla 4–7 mm. long (longer in *N. prostrata*).
      D. Stamens inserted on the corolla-throat. Placer Co. to Modoc and Siskiyou cos.
        2. *N. minima*
      DD. Stamens inserted in or just below the sinuses of the corolla-lobes.
        E. Foliaceous bracts 1–2 times as long as the head; corolla 5–6 mm. long. N. Coast Ranges.
          F. Plants prostrate. Lake Co.
            G. Heads 2–15-fld.; hypocotyl spongy, 2–4 times as thick as stem
              3. *N. pauciflora*
            GG. Heads 20–50-fld.; hypocotyl less spongy, not much thicker than stem .... 4. *N. plieantha*
          FF. Plants erect or spreading. Lake Co. to Trinity and Lassen cos.
            5. *N. Bakeri*
        EE. Foliaceous bracts 2–5 times the length of the head; corolla ca. 8 mm. long; cent. head sessile, with prostrate branches from beneath. Monterey Co. to Los Angeles Co. .... 6. *N. prostrata*
  BB. Stigma deeply 2-lobed or 3-lobed; lobes of the foliaceous bracts at base of the head rigidly acerose, the rachis expanded at base.
    C. Stigma 2-cleft.
      D. Each corolla-lobe with 1 main vein.
        E. Corolla exceeding calyx, the lobes 4–6 mm. long; bracts 7–20 mm. long, coarsely shaggy white-pilose. San Diego Co. to B.C., but w. of the Sierran crest .... 7. *N. intertexta*
        EE. Corolla ca. as long as calyx, the lobes ca. 1 mm. long; bracts 5–10 mm. long, pilose. Lake Tahoe to Modoc Co., e. of the Sierran crest ... 8. *N. propinqua*
      DD. Each corolla-lobe with 3 main veins.
        E. Corolla yellow or cream-yellow, with purple spots.
          F. Corolla-lobes 1.5–2 mm. long; seeds 4–5; caps. circumscissile at middle
            12. *N. nigellaeformis*
          FF. Corolla-lobes 3 mm. long; seed 1; caps. circumscissile at base
            13. *N. eriocephala*
        EE. Corolla white to blue or purple; caps. circumscissile at base.
          F. Corolla 5–7 mm. long, usually 4-lobed .... 14. *N. heterandra*
          FF. Corolla 10–14 mm. long, 5-lobed
            G. Rachis of bracts linear; corolla purple with dark spots
              15. *N. Jepsonii*
            GG. Rachis of bracts expanded toward tips; corolla blue with darker veins .... 18. *N. pubescens*
    CC. Stigma 3-cleft; each corolla-lobe with 3 main veins.
      D. Stamens exserted from throat; corolla 10–12 mm. long .... 9. *N. tagetina*
      DD. Stamens included; corolla 5–6 mm. long .... 10. *N. subuligera*
AA. Main stems with spreading hairs to almost glabrous, or if the hairs retrorse, not crisped and appressed against the stems; mature caps. with thicker wall, with a definite scheme of dehiscence.
  B. Seeds 1–2 (if more, the fls. yellow); caps. chartaceous.
    C. Fls. 4-merous.
      D. Corolla yellow, 9–10.5 mm. long; stamens exserted .... 11. *N. cotulaefolia*
      DD. Corolla blue or white, 5–7 mm. long; stamens included ...... 14. *N. heterandra*
    CC. Fls. 5-merous.
      D. Corolla yellow or cream-yellow, with purple spots.
        E. Corolla-lobes 1.5–2 mm. long; seeds 4–5; caps. circumscissile at middle
          12. *N. nigellaeformis*
        EE. Corolla-lobes 3 mm. long; seed 1; caps. circumscissile at base
          13. *N. eriocephala*
      DD. Corolla white to blue or purple.
        E. Rachis of bracts linear; corolla 10 mm. long, purple with dark spots
          15. *N. Jepsonii*
        EE. Rachis of bracts expanded, at least toward tips.
          F. Corolla mostly 8–14 mm. long; lvs. pinnately or bipinnately dissected.
            G. Stems erect, simple or branched.
              H. Bracts constricted in middle, broad above and below; caps. circumscissile at middle. Kern Co., se. of Bakersfield
                16. *N. setiloba*
              HH. Bracts not as above but laciniately linear-lobed from a slightly expanded base; caps. circumscissile at base. Kern and San Luis Obispo cos. n. .... 18. *N. pubescens*
            GG. Stems prostrate or spreading, several from base; bract-rachis broadest above the middle. Monterey Co. to Santa Barbara Co.
              17. *N. mitracarpa*
          FF. Corolla 5–6 mm long; lvs. linear, entire or with few linear lobes. Mariposa Co. to Shasta Co. .... 23. *N. filicaulis*
  BB. Seeds few to many; caps. coriaceous.

C.   Lvs. with filiform rachis, the lateral linear lobes much exceeded by the elongate terminal segm.; branches slender and almost leafless; plants not heavily glandular.
   D.   Branches typically proliferating from beneath terminal heads; bracts palmate.
       E.   Corolla 10 mm. long, longer than calyx; bracts dorsally coarse-villous
                                                                                        19. *N. prolifera*
       EE.   Corolla 4–7 mm. long, shorter than calyx; bracts dorsally subglabrous
                                                                                        20. *N. divaricata*
   DD.   Branches not from beneath terminal heads; bracts pinnate.
       E.   Corolla pale purple; plant sparsely pilose. Tehachapi Mts. to L. Calif.
                                                                                        21. *N. peninsularis*
       EE.   Corolla yellow; plant minutely puberulent. Tulare Co. to Ore.
                                                                                        22. *N. Breweri*
CC.   Lvs. (at least the upper) with broader rachis, the terminal lobe not greatly elongate;
branches stout and leafy; plants heavily glandular.
   D.   Bracts with a broad ovate base.
       E.   Terminal segm. of bract with 3 diverging acerose teeth; caps. few-seeded.
           F.   Corolla-throat narrow; stamens included, unequally inserted
                                                                                        24. *N. atractyloides*
           FF.   Corolla-throat ample; stamens exserted, equally inserted . . 25. *N. hamata*
       EE.   Terminal segm. of bract palmately lacerate into simple lobes; caps. many-
seeded . . . . 26. *N. heterodoxa*
   DD.   Bracts with a linear to lanceolate base.
       E.   Lvs. mostly pinnatifid into opposite equal regularly spaced simple segms.;
corolla mostly 10–17 mm. long; seeds few . . . . 27. *N. viscidula*
       EE.   Lvs. mostly with many unequal irregular ± divided segms.; corolla 6–12 mm.
long; seeds many.
           F.   Corolla 9–12 mm. long, the lobes 2–3 mm. long; plant with skunklike
odor . . . . 28. *N. squarrosa*
           FF.   Corolla 6–7 mm. long, the lobes 1.5 mm. long; plant with sweet odor
                                                                                        29. *N. mellita*

1. **N. leucocéphala** Benth. Erect to sometimes prostrate, 5–15 cm. high, simple or
racemosely branched, glabrous or nearly so below, more densely pubescent above with
white retrorse hairs; lower lvs. linear, entire to pinnate, 1–8 cm. long, glabrous, with
lobes 3–10 mm. long, blunt to cuspidate, the upper lvs. lanceolate to ovate, 1–2-pinnate,
1–5 cm. long, glabrous to ± pubescent near base, the linear lobes acerose to cuspidate;
fls. nearly or quite sessile in heads; bracts 0.5–1.5 cm. long, 1–2-pinnate, ± ciliate,
membranous on margin and between lobes, some lobes usually proliferating from back;
calyx 4–7 mm. long, the segms. entire, acerose, linear, the sinuses filled with truncate
ciliate membrane ca. ⅔ the length; corolla funnelform, 7–9 mm. long, white, the lobes
1.5–2 mm. long, with a single branched main vein; stamens inserted near middle of
throat, exserted, 2–5 mm. long; style exserted, minutely 2-lobed; caps. ovoid, 2.5 mm.
long, indehiscent; locules 4–6-seeded; seeds brown, 1.5 mm. long.—Vernal pools, below
1500 ft.; V. Grassland, Foothill Wd.; inner Coast Ranges from San Benito Co. n., Cent.
V., foothills of Sierra Nevada from Eldorado Co. n., ascending to 5500 ft. in Sagebrush
Scrub, e. Lassen Co.; Ore. April–May.

2. **N. mínima** Nutt. [*Gilia m.* Gray.] Simple or branched from base, 3–10 cm. high,
prostrate to almost erect, the stems slender, mostly whitish, somewhat retrorse-pubescent;
lower lvs. linear, entire or dissected, 1–1.5 cm. long, cuspidate, the upper wider, 1–2.5
cm. long, entire to pinnate with 1–3 pairs of acerose lobes; bracts ± ovate, to 1.5 cm.
long, acerose-pinnate, pubescent inside the membranous-winged base; calyx 4–5 mm.
long, the segms. unequal, acerose, sparsely pubescent below, the intersegmental membrane extending half or more the length, ciliate at summit; corolla white, funnelform,
4–6 mm. long, the lobes 1–2 mm. long, each with 1 unbranched vein; stamens 2.5–3
mm. long, inserted in middle of throat; style exserted, the stigma minutely 2-lobed;
caps. ovoid, indehiscent, 2 mm. long; *n* = 9 (Grant, 1958).—Vernal pools, etc., below
7000 ft.; Sagebrush Scrub, Yellow Pine F., Red Fir F.; Placer Co. to Modoc and
Siskiyou cos.; to Wash., Ida. June–Aug.

3. **N. pauciflòra** Mason. Prostrate, the stems 1–4 cm. long, slender, white with purple
streaks, subglabrous to densely retrorse-pubescent; lvs. 1–2.5 cm. long, linear and
entire or with 1–2 pairs of lateral linear cuspidate lobes, glabrous; bracts several-lobed;
calyx 4–5 mm. long, mostly membranous, the lobes very narrow, the membrane truncate,
ciliate; corolla funnelform, 5–6 mm. long, white to bluish, the lobes 1.5 mm. long,
with a single unbranched vein; stamens inserted in sinuses, somewhat longer than the
lobes; stigma minutely 2-lobed; caps. irregularly dehiscent, the seeds 1–several, minutely

pitted; $2n =$ ca. 18 (Grant, 1958).—Vernal pools in volcanic rubble; Chaparral; 5 miles n. of Lower Lake, Lake Co. June.

4. **N. plieántha** Mason. Prostrate, branched from base, the stems slender, 2–6 cm. long, retrorse-pubescent; lvs. 2–4 cm. long, linear and entire, or broader and pinnately few-lobed, the lobes linear, remote; bracts pinnate, often proliferated, ciliate-membranous below; calyx 4–5 mm. long, somewhat constricted above, quite membranous except for the narrow green lobes, the membrane truncate, ciliate; corolla pale blue, funnelform, 5–6 mm. long, the lobes 2 mm. long, with a single unbranched vein; stamens inserted in sinuses, 2.5 mm. long; stigma entire to 2-cleft, exserted; caps. irregularly dehiscent; seeds ca. 3, red-brown, minutely pitted.—Peaty margin of Bogg's Lake, Mt. Hannah, Lake Co., at 3000 ft.; Yellow Pine F. May–June.

5. **N. Bàkeri** Mason. Erect, 2–15 cm. high, racemosely branched, retrorse-pubescent; lower lvs. linear, entire to pinnatifid, the upper 2–4 cm. long, 1–2-pinnate into linear segms.; bracts pinnatifid with dissected proliferations, mostly glabrous; calyx ca. 5 mm. long, the lobes unequal, slender-aristate, weakly pubescent, the intercostal membrane ciliate; corolla white, 5–7 mm. long, the lobes 1–1.5 mm. long, with a single unbranched vein; stamens inserted in sinuses, 2.5 mm. long; stigma minutely 2-lobed, exserted; caps. ca. 2 mm. long, irregularly circumscissile; seeds few, red-brown.—Vernal pools, below 5500 ft.; N. Oak Wd., Yellow Pine F.; inner Coast Ranges from Sonoma and Lake cos. to Trinity Co., Lassen Co. May–July.

6. **N. prostràta** (Gray) Greene. [*Gilia p.* Gray.] Cent. head nearly or quite sessile, usually with prostrate branches from beneath, these 4–10 cm. long, retrorse-pubescent; lvs. 3–7 cm. long, glabrous, pinnate to bipinnate, the rachis and lobes linear; bracts leafy, pinnate, 1–4 cm. long, glabrous or somewhat pubescent, membranous-margined, ciliate; calyx 4–5 mm. long, the lobes entire to trifid, pubescent, the sinus-membrane ciliate; corolla broadly funnelform, ca. 8 mm. long, white to violet, the lobes 1.5–2 mm. long; stamens inserted in sinuses, 2.5 mm. long; stigma minutely 2-lobed; caps. indehiscent, 2-celled, several-seeded.—Vernal pools and moist places, below 1500 ft.; mostly Coastal Sage Scrub; cismontane Los Angeles Co. to Monterey Co. April–May.

7. **N. intertéxta** (Benth.) Hook. [*Aegochloa i.* Benth. *Gilia i.* Steud.] Simple and erect or branched from base, the stems brown, 5–20 cm. high, retrorse-pubescent below the heads; lvs. 1–5 cm. long, 1–2-pinnate, the rachis linear, glabrous or pubescent on upper surface; bracts foliaceous, pinnate, 1–2 cm. long, ± villous on upper surface, broadly membranous-margined; calyx 8–10 mm. long, the entire or toothed lobes narrow, the sinus-membrane conspicuously pubescent; corolla pale blue to white, slender-funnelform, 5–9 mm. long, the lobes 4–6 mm. long, 1-veined; stamens inserted at base of throat, unequally exserted; stigma deeply 2-lobed; caps. 1-celled, several-seeded; seeds irregularly angled.—Vernal pools and moist places, below 6000 ft.; many Plant Communities; cismontane and montane Calif. w. of the Sierran crest from San Diego Co. n.; to B.C. May–July.

8. **N. propínqua** Suksd. [*N. intertexta* var. *p.* Brand. *N. i.* var. *alpina* Brand.] Spreading, the stems 2–10 cm. long, brown, retrorse-pubescent; lvs. 1–2-pinnate; bracts 5–10 mm. long, with broad coriaceous base, the margin and upper surface pilose; calyx 7–9 mm. long, pilose, the segms. unequal, the 2 short ones often sharply toothed near apex; corolla white, slender-funnelform, ca. as long as calyx, the lobes 0.5–1 mm. long; stamens inserted at base of throat, unequally exserted; stigma deeply 2-lobed; caps. indehiscent, 1-celled, several-seeded; the seeds irregular.—Moist places, 3000–6700 ft.; Yellow Pine F., Red Fir F.; e. of the Sierran crest from Lake Tahoe to Modoc and Siskiyou cos.; to Wash., N. Dak., Ariz. June–Aug.

9. **N. tagetìna** Greene. [*N. pubescens* var. *t.* Jeps. *N. erecta* Heller.] Erect, 5–25 cm. high, simple or racemosely branched, ± retrorse-pubescent, especially up the stems; lower lvs. 1–2-pinnate, 2–5 cm. long, the lobes cuspidate, narrow; upper lvs. 2–3-pinnate, passing into the bracts, the rachis rather broad, puberulent on upper surface; bracts rigid, cup-shaped at base, the lobes proliferating on under surface, densely pubescent; calyx 8–10 mm. long, the lobes unequal, acerose, 3–5-toothed, pubescent on outer surface; corolla pale blue, funnelform, ca. 10 mm. long, the lobes 2–2.5 mm. long, 3-veined; stamens inserted in lower throat, 2–2.5 mm. long; stigma 3-lobed, included; caps. not regularly dehiscent, 3 mm. long; seeds 1–several in each locule, pitted, oblong, almost 2 mm. long.—Moist and grassy places, below 2000 ft.; Yellow Pine F., Foothill

Wd., Mixed Evergreen F., etc.; borders of Sacramento V. from Napa and Amador cos.
n., outer Coast Ranges from Sonoma Co. n.; Cuyamaca Mts., San Diego Co. at 4300
ft.; to s. Wash. April–June.

10. **N. subulígera** Greene. Erect, simple or branched, 3–12 cm. high, puberulent to
retrorse-pubescent; lvs. subglabrous, 1–3 cm. long, pinnatifid into filiform lobes, the
upper lvs. rigid; bracts ovate, coriaceous, spiny-pinnatifid, some proliferating, the margin ciliate; calyx with 2 simple acerose lobes and 3 reduced, the lobes hairy within at
summit of membrane; corolla white, narrow-funnelform, 5–6 mm. long, the lobes 1
mm. long, 3-veined; stamens included; style included, 3-lobed; caps. not regularly
dehiscent, membranous; seeds several, agglutinated into a mass.—Rocky plains and
slopes, below 6500 ft.; Foothill Wd., Yellow Pine F.; Mt. St. Helena and borders of
Sacramento V., San Rafael Mts., Santa Barbara Co.; s. Ore. May–Aug.

11. **N. cotulaefòlia** (Benth.) H. & A. [*Aegochloa c.* Benth. *N. Bowmanae* Eastw.,
not Jeps.] Erect, simple or divaricately branched above base, the stems reddish, ±
pubescent and glandular, 5–25 cm. tall; lower lvs. 2–3 cm. long, 1–2-pinnate into linear
puberulent lobes; upper lvs. with broader rachis; bracts glandular-villous, broad,
coriaceous, with 2–7 pairs of acerose lobes; calyx 4-lobed, 6–7 mm. long, with 2 lobes
shorter; corolla funnelform, yellowish, 4-lobed, 9–10 mm. long, the lobes 1–1.5 mm.
long, 3-veined; stamens 4, inserted in or below sinuses, 3 mm. long; stigma 2-lobed;
caps. circumscissile at base, 1-loculed, 2 mm. long, 1–2-seeded; seeds obovoid, light
brown, almost 2 mm. long.—Moist soil, heavy adobe, etc., below 1500 ft.; Chaparral,
Foothill Wd., V. Grassland; inner Coast Ranges, San Benito Co. to Mendocino Co.
May–June.

12. **N. nigellaefórmis** Greene. [*N. ocellata* Eastw. *N. n.* var. *radians* J. T. Howell.]
Simple or branched from base, the stems erect or spreading, 3–20 cm. long, white or
reddish, retrorse-puberulent; lvs. 1–3 cm. long, with slender rachis, bipinnate into
crowded filiform acerose lobes; bracts 5–20 mm. long, spinosely 1–2-pinnatifid, glandular-
puberulent and coarsely white-pubescent; calyx 5-merous, 8–12 mm. long, coarsely
pubescent, the lobes entire or pinnately toothed, the white sinus-membrane half or
more their length; corolla yellow with purple spot at base of each lobe, funnelform, 10–
14 mm. long, the lobes 1.5–2 mm. long, 3-veined; stamens inserted in lower throat,
1–5 mm. long, usually exserted; stigma 2-lobed; caps. circumscissile near middle, 4–
5-seeded, 2–3 mm. long; seeds irregularly angled, brown, ca. 1.5 mm. long.—Dried
vernal pools, below 1000 ft.; V. Grassland, Foothill Wd.; Great Cent. V. and bordering
foothills, San Luis Obispo Co. to Contra Costa Co., Tulare Co. to Butte Co. April–
May.

13. **N. eriocéphala** Mason. Erect, 5–25 cm. high, the stems tan or red-brown, simple
or racemosely branched, with white retrorse short hairs; lvs. 1–5 cm. long, puberulent,
bipinnate, usually with broad rachis and linear lobes; bracts stiff, bipinnate, with linear
acerose lobes, the rachis expanded toward base and densely white-villous; calyx 5-
merous, unequally cleft, the lobes 3-toothed to entire, coarsely white-villous above, 6–8
mm. long; corolla funnelform, 8–12 mm. long, cream-yellow, often spotted purple, the
lobes 3 mm. long, often tipped blue; stamens inserted on throat, 1–3 mm. long, unequal;
stigma 2-lobed, included; caps. chartaceous, circumscissile at base, obovoid, 1-celled,
1-seeded; seed brown, smooth or somewhat furrowed, 2 mm. long.—Dry, open flats,
below 1000 ft.; V. Grassland, Foothill Wd.; Sierran foothills from Calaveras Co. to
Placer and Sacramento cos. May–June.

14. **N. heterándra** Mason. Near to *N. eriocephala*, but the fls. mostly 4-merous;
corolla 5–7 mm. long, white or blue, the lobes 1 mm. long; stamens unequally inserted.
—Dried vernal pools; V. Grassland, Foothill Wd., N. Oak Wd.; edge of Sacramento
V., Shasta, Tehama, Yuba, Colusa, and Lake cos. May–June.

15. **N. Jepsònii** V. Bailey ex Jeps. Near to *N. eriocephala*, the lvs. with linear rachis;
fls. 5-merous; calyx-lobes entire or the 3 longer ones toothed; corolla purple with darker
spot at base of each lobe, funnelform, 10 mm. long, the lobes 2.5 mm. long; stamens
8 mm. long, inserted at base of throat; style long-exserted; caps. circumscissile at base,
1-seeded.—Drying flats, below 2500 ft.; Foothill Wd., N. Oak Wd.; inner Coast Ranges
from Glenn Co. to Napa and Lake cos. May–June.

16. **N. setilòba** Cov. Erect, usually racemosely branched, 1–2 dm. high, puberulent;
lvs. bipinnate, 1–3 cm. long, the terminal lobe broad, often purple, with irregularly

and finely laciniate margin; bracts with lanceolate terminal lobe; calyx 7–10 mm. long, puberulent, with coarser hairs in middle, the sinus-membrane extending over halfway, the lobes entire or with small teeth; corolla purple, funnelform, ca. 10 mm. long, the lobes 2.5–3 mm. long; stamens inserted in lower throat, 5 mm. long, well exserted; stigma 2-lobed; caps. chartaceous, circumscissile at middle, 1-seeded; seed ovoid, 4-angled, almost 3 mm. long.—Slopes and flats, 5000–7000 ft.; Yellow Pine F., Foothill Wd.; Piute Mts., Kern Co. April–June.

17. **N. mitracárpa** Greene. Erect to spreading, 2–15 cm. high, the stems slender, brown, pubescent, often glandular; lvs. 1–2-pinnate, 1–3 cm. long, with acicular spinescent lobes and flattened rachis having marginal teeth above the middle; bracts 5–15 mm. long, more rigid, less dissected; infl. 1.5–2.5 cm. broad; calyx 5–9 mm. long, unequally lobed, the segms. entire to 3–5-lobed, sinus-membrane narrow; corolla funnelform, 8–11 mm. long, the lobes purple to pink, 2–3 mm. long, 3-veined; stamens inserted at base of throat, 5–8 mm. long; stigma exserted, 2-lobed; caps. obovoid, chartaceous, circumscissile at base, 1-seeded; seed 4-angled, ca. 2 mm. long.—Dry gravelly and clayey slopes, below 3500 ft.; Foothill Wd., Chaparral; S. Coast Ranges from Santa Barbara Co. to Monterey Co. and inland to Fresno and Tulare cos. April–June.

Ssp. **Jarédii** (Eastw.) Mason. [*N. J.* Eastw.] Stem stout; infl. 2–3.5 cm. broad.—Wet clays, Paso Robles, San Luis Obispo Co.

18. **N. pubéscens** (Benth.) H. & A. [*Aegochloa p.* Benth. *Gilia p.* Steud.] Erect, simple or branched mostly above, retrorse-canescent and somewhat glandular; lvs. 1.5–5 cm. long, 1–2-pinnate into linear divisions, the rachis sometimes expanded toward tip especially in upper lvs.; bracts with prominent rachis and pungent lobes; calyx ca. 1 cm. long, the segms. unequal, the longer often toothed, the sinus-membrane densely pilose; corolla funnelform, blue with darker veins, 10–14 mm. long, the lobes 3-veined, 2.5 mm. long; stamens inserted at base of throat, unequal, ± exserted; stigma 2-cleft; caps. chartaceous, circumscissile at base, 1–2-seeded; seeds brown, ca. 2 mm. long.—Dried depressions and dry slopes, below 3500 ft.; Foothill Wd., Chaparral, N. Oak Wd., etc.; from Humboldt and Butte cos. to San Luis Obispo and Kern cos. May–July.

19. **N. prolífera** Greene. Erect, 5–18 cm. high, divaricately branched, the stems brown, glabrous to puberulent, almost destitute of lvs.; lvs. 2–4 cm. long, entire to pinnate with few pairs of short remote lobes; bracts to ca. 1 cm. long, with broad coriaceous rachis, palmately 5–8-cleft into acerose lobes, ± coarse-villous; calyx 5–7 mm. long, the lobes unequal, acerose, hyaline in lower half, pubescent about the middle; corolla funnelform, 10 mm. long, the lobes 3-veined, 2.5 mm. long, blue or purple; stamens inserted in upper half of throat, 2–4 mm. long, unequal; stigma exserted, 3-lobed; caps. coriaceous, dehiscent from base up, several-seeded, 3-celled, 2 mm. long; seeds ovoid, shallowly pitted, scarcely 1 mm. long.—Dry slopes, 2000–4000 ft.; Chaparral; foothills of Sierra Nevada, Amador and Tulare cos. May–June.

Ssp. **lùtea** (Brand) Mason. [*N. p.* var. *l.* Brand.] Corolla bright yellow.—At 2700–4500 ft.; Yellow Pine F.; Eldorado Co.

20. **N. divaricàta** (Torr.) Greene. [*Gilia d.* Torr. ex Gray.] Erect or spreading, 3–15 cm. high, the stems simple or divaricately branched from below the terminal head, glabrous to puberulent; lvs. 1–2 cm. long, simple or with few pairs of linear lateral lobes; bracts palmately lobed into 3–5 acerose divisions, pilose in the sinuses; calyx 4–7 mm. long, the lobes unequal, acerose, simple, white-pilose and with sinus-membrane on lower half; corolla short-funnelform, 3–4 mm. long, the lobes pink to purple or white, scarcely 1 mm. long, 3-veined; stamens inserted on throat, included, the anthers at least as long as fils.; stigma included, 3-lobed; caps. coriaceous, few-seeded, 3-celled, dehiscent from base up; seeds 1–3 per locule, lenticular to angled, scarcely 1 mm. long.—Dry open places and at edge of meadows, 1000–5500 ft. in Coast Ranges, 3000–8500 in Sierra Nevada; Yellow Pine F., Red Fir F.; from Napa, Lake and Tulare cos. n., also in e. Santa Barbara Co.; to Ida., Wash. June–Aug.

Ssp. vivídior (Jeps. & Bail.) Mason. [*N. d.* var. *v.* Jeps. & Bail.] Inner bracts and calyces glandular-puberulent almost to tips.—In most of the range of the sp.

21. **N. peninsulàris** Greene. [*Gilia p.* Munz. *N. MacGregorii* Brand. *N. divaricata* var. *p.* Jeps.] Near to *N. divaricata*, but not branching from below the terminal heads, sparsely glandular-pubescent; lvs. with a rachis ca. 1 mm. wide; bracts pinnate; calyx 4–5 mm.

long, the segms. unequal; corolla 6 mm. long, pale purplish, the lobes ca. 2 mm. long; stamens unequal, the fils. longer than anthers; caps. many-seeded, dehiscent from top down, chartaceous; seeds obscurely pitted, ca. 1 mm. long.—Rare, damp disturbed places, 5000–7500 ft.; Yellow Pine F.; Tehachapi Mts. to n. L. Calif. June–Aug.

22. **N. Brèweri** (Gray) Greene. [*Gilia B.* Gray.] Erect, 2–12 cm. high, the slender brown stems puberulent throughout; lvs. filiform, 1–3 cm. long, entire or with 1–4 short slender lobes; bracts 4–10 mm. long, pinnately 3–5-cleft, acerose, the divisions tripartite; calyx 6–10 mm. long, the segms. unequal, glandular-puberulent, united by the sinus-membrane; corolla yellow, funnelform, 6–7 mm. long, the lobes ca. 1 mm. long; stamens ca. 2.5 mm. long, inserted on throat; stigma 3-lobed, included; caps. coriaceous, few-seeded, dehiscent from base up, 2–3 mm. long; seeds brown, ca. 1.5 mm. long.—Damp to semidry flats and valleys, 4000–11,000 ft.; Pinyon-Juniper Wd. to Subalpine F.; Inyo and Tulare cos. n., mostly near or on e. slope of Sierra Nevada, to Modoc Co.; to Wash., Ida., Utah, Ariz. June–Aug.

23. **N. filicaùlis** (Torr.) Greene. [*Gilia f.* Torr. ex Gray. *N. dubia* Brand.] Erect, 1–several-stemmed from base, sparsely puberulent to glandular, ± racemosely branched; lvs. filiform, 1–3 cm. long, entire or with 1–4 short lobes near base; bracts 4–10 mm. long, palmately 3–5-cleft, the middle lobe acuminate, acerose; calyx 3–5 mm. long, glandular-puberulent, the lobes acerose, unequal; corolla funnelform, violet, 5–6 mm. long, the lobes 1–1.5 mm. long; stamens inserted at base of throat, 4–6 mm. long, exserted; stigma capitate to minutely 2-lobed; caps. 2 mm. long, incompletely 2-valved, thin-walled, 1–2-seeded; seeds brown, ovoid, irregularly angled, ca. 1 mm. long.—Dry plains and slopes, below 2700 ft.; V. Grassland, Foothill Wd., Yellow Pine F.; foothills of Sierra Nevada from Shasta Co. to Mariposa Co. June–July.

24. **N. atractyloìdes** (Benth.) H. & A. [*Aegochloa a.* Benth. *Gilia a.* Steud. *N. hirsutissima* Brand.] Erect, usually freely branched, glandular-hirsute, 5–20 cm. high; lvs. well distributed, sessile, mostly 2–5 cm. long, pinnately dissected from a broad rachis, with pungent lobes or teeth and usually 3-terminal lobes; bracts ovate, coriaceous, with divergent spines; calyx 5–7 mm. long, ± membranous in lower part, the lobes entire or toothed; corolla narrow-funnelform, ca. 1 cm. long, the lobes 2.5 mm. long, white or blue or yellow; stamens included, 1–1.5 mm. long, unequally inserted; stigma included; caps. 3–4 mm. long, ovoid, coriaceous, few-seeded, dehiscent from base up; seeds red-brown, pitted, irregularly angled, ca. 1 mm. long; $n = 9$ (Grant, 1958).— Mostly in dry places, below 1500 ft.; many Plant Communities; Coast Ranges from Humboldt Co. to L. Calif. May–July.

25. **N. hamàta** Greene. [*N. macrantha* Brand. *N. atractyloides* var. *h.* Jeps. *Gilia h.* Munz.] Close to *N. atractyloides,* but tending to have a skunklike odor; corolla with an ample throat and lobes usually purple, to 4.5 mm. long; stamens exserted, equally inserted but unequal in length.—Dry often rocky slopes, mostly below 3000 ft.; Chaparral, Coastal Sage Scrub; San Luis Obispo Co. to L. Calif., Santa Catalina and San Clemente ids. April–June.

Ssp. **foliàcea** (Greene) Mason. [*N. f.* Greene. *N. atractyloides* vars. *f.* and *flavida* Jeps.] Lvs. softer, more herbaceous; fls. white.—Occasional with the sp.

Ssp. **leptántha** (Greene) Mason. [*N. l.* Greene.] Corolla-tube long exserted.—San Diego region to L. Calif.

26. **N. heterodóxa** (Greene) Greene. [*Gilia h.* Greene. *G. viscidula* var. *h.* Gray. *G. parvula* Greene. *N. p.* Greene.] Erect, 5–25 cm. high, divaricately branched, puberulent, glandular, the stems brownish; lvs. 1–4 cm. long, slender, pinnate with many short acerose lobes and narrow rachis; bracts ovate to lanceolate, glandular-puberulent, coriaceous, with basal cuspidate teeth; calyx 5–8 mm. long, the segms. entire, acerose, with basal sinus-membrane; corolla funnelform, ca. 8 mm. long, blue, the lobes 1.5 mm. long; stamens inserted at base of throat, ca. 6 mm. long, exserted; stigma exserted; caps. ovoid, 2–3 mm. long, 3-celled, many-seeded, dehiscing from top down; seeds brown, plump, angled, ca. 0.8 mm. long.—Dry open rocky places, on sandstone, shale and clay, below 2800 ft.; Chaparral, Mixed Evergreen F.; Coast Ranges from Sonoma and Napa cos. to Santa Clara Co. June–July.

Ssp. **rosulàta** (Brand) Mason. [*N. r.* Brand. *N. fallax* Brand.] Stamens included, unequally inserted.—On serpentine, Marin Co.

27. **N. viscídula** Benth. [*Gilia v.* Gray.] Erect, glandular-pubescent, 3–15 cm. high,

the stems white or purplish; lvs. narrow, remotely pinnatifid, 1–4 cm. long, the rachis and lobes very narrow; bracts 7–15 mm. long, little dilated, palmately parted into toothed or entire lobes; calyx 5–9 mm. long, glandular-pubescent, with sinus-membrane on lower half; corolla funnelform, 12–16 mm. long, blue to purple, the lobes 4–6 mm. long; stamens inserted in lower throat, 3–8 mm. long, mostly exserted; stigma 3-lobed, exserted; caps. 3–3.5 mm. long, ovoid, 3-celled, few-seeded, coriaceous, dehiscent from base up; seeds irregularly angled, brown, almost 3 mm. long.—Dry clayey to rocky places, below 2000 ft.; Mixed Evergreen F., N. Oak Wd.; Humboldt Co. to San Mateo Co. June–July.

Ssp. **purpùrea** (Greene) Mason. [*N. p.* Greene.] Corolla 9–12 mm. long, the lobes 2.5–4 mm. long.—Below 4000 ft.; Foothill Wd.; foothills of Sierra Nevada, Tulare Co. to Sutter Co. June–July.

28. **N. squarròsa** (Eschs.) H. & A. [*Hoitzia s.* Eschs. *Gilia pungens* Dougl. *N. pterosperma* Eastw. *N. s.* var. *agrestis* Brand.] Erect, branched at or above base, glandular-pubescent, 0.5–5 dm. high, with skunklike odor; lvs. pinnately to bipinnately unequally cleft, the segms. rigid, lanceolate, often crowded; bracts 1–1.5 cm. long, pinnately to palmately cleft; calyx 8–12 mm. long, glandular-pubescent, with sinus-membrane in lower half; corolla broadly funnelform, 10–12 mm. long, purple to blue, the lobes 2–3 mm. long; stamens inserted in lower half of tube, 1–4 mm. long, included; stigma 3-lobed, included; caps. ovoid, 3–4 mm. long, 3-celled, dehiscent from top down, many-seeded; seeds dark brown, irregular, ca. 0.5 mm. long.—Dry flats and fields, below 2500 ft.; Mixed Evergreen F., N. Oak Wd., Foothill Wd., etc.; Coast Ranges from Monterey Co. n.; Sierran foothills from Amador Co. n.; to B.C. June–Aug.

29. **N. mellìta** Greene. [*N. Eastwoodiae* Brand.] Near to *N. squarrosa*, the odor more pleasant; corolla 6–7 mm. long, with lobes ca. 1.5 mm. long, pale blue; stamens ca. 1.5 mm. long; seeds wing-angled, ca. 0.8 mm. long.—Dry slopes and flats, below 3500 ft.; Mixed Evergreen F., N. Oak Wd., Yellow Pine F., Chaparral; Coast Ranges, Humboldt Co. to San Luis Obispo Co. May–July.

12. Leptodáctylon H. & A.

Subshrubs or perennial herbs, straggly or compact, leafy, often tufted. Lvs. opposite or alternate, palmately or subpinnately parted into linear pungent lobes, usually glandular and with axillary fascicles. Fls. mostly in terminal congested cymes or glomerules, rarely solitary in axils, ± subsessile. Calyx-lobes entire, pungent, the sinuses ca. ⅔ filled with membrane. Corolla narrow-funnelform or salverform, usually showy, white or cream to rose or lilac. Stamens included, inserted in tube or throat; fils. ca. as long as anthers. Style included. Caps. ± cylindric; locules several- to many-seeded. A small genus of w. N. Am. (Greek, *leptus,* narrow, and *dactylon,* finger, because of lf.-lobing.)

(Wherry, E. T. Two Linanthoid genera. Am. Midl. Nat. 34: 381–387, 1945.)

Lvs. opposite, 3-cleft; plants less than 1 dm. high. San Jacinto Mts. . . . . 3. *L. Jaegeri*
Lvs. mostly alternate; plants mostly 3–10 dm. high. More widely distributed.
    Plants mostly 1–3 dm. high; corolla tubular-funnelform, mostly pale or whitish; corolla-lobes 7–10
    mm. long . . . . 1. *L. pungens*
    Plants mostly 3–10 dm. high; corolla salverform, rose to lilac; corolla-lobes 10–15 mm. long
                                             2. *L. californicum*

1. **L. púngens** (Torr.) Rydb. ssp. **pulchriflòrum** (Brand) Mason. [*Gilia p.* ssp. *p.* Brand. *L. lilacinum* Greene. *G. l.* Brand.] Low branching shrub 1–4 dm. high, with several to many stems, quite glandular-villous throughout; lvs. many, scarcely equal to internodes, mostly alternate, palmately cleft into 3–7 lobes, the middle lobe 8–15 mm. long, ca. 1.5 times as long as the next outer, pungent or rigidly acerose; fls. sessile or nearly so, tending to open in evening, in upper axils and in terminal or subterminal glomerules; calyx 8–10 mm. long, deeply cleft into unequal linear acerose lobes, sinuses ca. ⅔ filled with hyaline membrane that runs halfway up the sides of the lobes; corolla narrow-funnelform, 15–25 mm. long, white with some pinkish or purplish tint, the tube very slender, long-exserted, the lobes narrow-obovate, 7–10 mm. long; stamens glabrous, inserted on throat, included, the fils. scarcely longer than anthers; pistil ca. ¼ as long as corolla-tube; caps. cylindric, the locules many-seeded; seeds brown, oblong, ca. 1.2 mm. long.—Dry rocky and sandy places, mostly 5000–12,000 ft.; mostly Montane

Coniferous F.; Lake Co., Sierra Nevada to s. Ore. and San Bernardino Mts.; w. Nev. May–Aug. Intergrading freely with other sspp. along the desert edge.

KEY TO SUBSPECIES

Middle lf.-segm. less than twice as long as next outer; fls. mostly with some pink or purple.
    Membrane of calyx-sinuses running halfway up the lobes. Sierra Nevada to San Bernardino Mts.
                                                               ssp. *pulchriflorum*
    Membrane of calyx-sinuses not running up the lobes. N. Calif. . . . . ssp. *Hookeri*
Middle lf.-segm. 2–4 times as long as next outer; fls. mostly cream. Desert mts. . . . . ssp. *Hallii*

Ssp. **Hállii** (Parish) Mason. [*Gilia H.* Parish. *G. pungens* var. *H.* Mlkn. *G. p.* ssp. *H.* Brand. *L. H.* Heller. *G. tenuiloba* Parish. *G. p.* var. *t.* Mlkn. *L. p.* vars. *t.* and *subflavidum* Jeps.] Stems more grayish-pubescent, less viscid; middle lf.-segm. 2–4 times as long as next outer; corolla-lobes mostly cream, sometimes with pinkish tinge, narrowobovate to oblanceolate.—Dry slopes, 4000–9000 ft.; Pinyon-Juniper Wd., sometimes Yellow Pine F.; mts. bordering and in the deserts, Inyo Co. to San Diego Co.; L. Calif., Nev. April–July.

Ssp. **Hoókeri** (Dougl.) Wherry. [*Phlox H.* Dougl. ex Hook. *Gilia H.* Benth. *G. p.* var. *H.* Gray. *L. p.* var. *shastense* Jeps.] Plant conspicuously glandular; membrane of calyx-sinuses truncate and not running up the calyx-lobes.—At 4000–7000 ft.; Sagebrush Scrub, Yellow Pine F., N. Juniper Wd.; Trinity and Siskiyou cos. to Modoc Co.; to Wash., Ida. June–July.

2. **L. califórnicum** H. & A. [*Gilia c.* Benth.] PRICKLY-PHLOX. Erect widely branched shrub 3–10 dm. high, the stems tomentose not glandular, thickly set with fascicles of prickly lvs. in axils of current and old main lvs.; the latter palmately 5–9-lobed into acerose segms. 3–12 mm. long, unequal, linear, acerose, glabrous or subglabrous; fls. sessile, in terminal congested clusters with ± villous leafy bracts; calyx ca. 1 cm. long, deeply cleft into subequal linear acerose lobes, the sinuses ca. half or ⅔ filled with hyaline membrane; corolla salverform, bright rose or rose-lavender, sometimes almost white, 2–2.5 cm. long, the continuous tube and throat 10–15 mm. long, the lobes 10–15 mm. long, elliptic-obovate to almost round; stamens inserted near middle of tube, the fils. ca. 1 mm. long; stigma included; caps. elongate, the locules severalseeded; $n = 9$ (Flory, 1937).—Dry slopes and banks, below 5000 ft.; Chaparral, Foothill Wd.; San Luis Obispo Co. to w. San Gabriel Mts. and Santa Ana Mts. March–June.

Ssp. **glandulòsum** (Eastw.) Mason. [*Gilia c.* var. *g.* Eastw. *L. c.* var. *g.* Abrams.] Hairs of upper stems and infl. gland-tipped.—Below 3000 ft.; Chaparral; San Gabriel Mts. to San Bernardino and Santa Ana mts.

3. **L. Jaègeri** (Munz) Wherry. [*Gilia J.* Munz. *L. pungens* var. *J.* McMinn.] Low, cespitose, with woody base, the branches many, 0.2–1 dm. high, glandular-pubescent, densely leafy; lvs. mostly opposite, 1–1.5 cm. long, mostly 3-cleft into flat linear prickly lobes, the terminal 6–15 mm. long, the lateral 4–10 mm. long; fls. crowded, solitary in upper axils; calyx 8–9 mm. long, narrow-cylindric, membranous between the 5–6 unequal spiny lobes; corolla whitish, funnelform, 25–30 mm. long, the tube exserted, the 5–6 broadly oblanceolate lobes 7–9 mm. long; stamens 5–6, inserted at base of throat; fils. 2–3 mm. long; caps. 4-loculed.—Dry rocky places, ca. 8800 ft.; Red Fir F.; San Jacinto Mts. July–Aug.

13. Linánthus Benth.

Annuals or perennials, with simple or divaricately or dichotomously branched stems. Lvs. opposite or alternate, palmately or somewhat pinnately parted into linear segms., rarely simple. Infl. from open cymose to subcapitate at ends of the branches, sometimes also with solitary fls. in upper axils. Fls. sessile or nearly so, or on long pedicels. Calyx deeply cleft, with or without a pseudotube of hyaline membrane in the sinuses, the free margins of the lobes with or without such membrane. Corolla campanulate to funnelform or salverform. Stamens equally inserted in the throat or more rarely in the tube, the fils. equal or sometimes unequal, included or exserted. Pistil from short and included to long and exserted; stigma 3–4-lobed. Caps. ellipsoid to cylindrical; the locules adhering at base at dehiscence, 1–several-seeded. Seeds unaffected by water or producing spiracles or mucilage when wetted. As here constituted, a genus of ca. 40 spp., mostly of w. N. Am. and Chile. (Greek, *linon*, flax, and *anthos*, flower.)

(Brand, A., *in* Das Pflanzenreich, IV, 250: 124–147, 1907. Mason, H. L., *in* Abrams: Illus. Fl. Pacific States 3: 413–433, 1951.)

A. Plants annual.
B. Calyx with a conspicuous hyaline membrane in the sinuses, either forming a pseudotube or present on the margins of the lobes.
C. Fls. on pedicels at least 5 mm. long.
D. Corolla barely exceeding calyx, glabrous within; fils. glabrous.
E. Lf.-segms. 5–15 mm. long; calyx 1.5–2 mm. long; corolla-lobes ca. as long as tube plus throat; stamens included. Fresno and Lake cos. n.

1. *L. Harknessii*

EE. Lf.-segms. 2–6 mm. long; calyx 3–5 mm. long; corolla-lobes shorter than tube plus throat; stamens exserted. San Diego Co. to Lake and Butte cos.

2. *L. pygmaeus*

DD. Corolla mostly 2–5 times as long as calyx, with a hairy ring within or fils. hairy at base.
E. Fils. usually pubescent at base, inserted on lower part of corolla-throat.
F. Corolla 2–4 mm. long; hairs on throat at base of stamens

3. *L. septentrionalis*

FF. Corolla 4–15 mm. long; hairs on base of fils.
G. Lf.-segms. 10–30 mm. long; corolla mostly 10–30 mm. long

4. *L. liniflorus*

GG. Lf.-segms. 3–7 mm. long; corolla 4–6 mm. long .... 5. *L. filipes*
EE. Fils. glabrous, inserted in the middle or upper parts of corolla-throat.
F. Corolla yellow, sometimes whitish. S. Calif. .... 7. *L. aureus*
FF. Corolla pink or lilac, sometimes whitish. San Benito Co. n.
G. Stamens inserted in or just below sinuses of the corolla-lobes.
H. Calyx-sinuses half filled with membrane, the lobes puberulent; corolla 6–10 mm. long; fils. 1–2 times the anthers. Sierra Nevada and from Contra Costa Co. n. in Coast Ranges.

8. *L. Bakeri*

HH. Calyx-sinuses nearly filled with membrane, the lobes pilose-hispid; corolla mostly 10–12 mm. long; fils. 3–4 times the anthers. Santa Clara Co. to San Benito Co. ... 9. *L. ambiguus*
GG. Stamens inserted considerably below sinuses of the corolla-lobes.
H. Corolla 10–15 mm. long, the tube 1.5–2 times the calyx; stamens inserted near base of throat. Lake Co. to Mendocino Co. .... 10. *L. Rattanii*
HH. Corolla 6–10 mm. long, the tube included in calyx; stamens inserted on upper half of corolla-throat. Contra Costa Co. to Mendocino Co. .... 11. *L. Bolanderi*
CC. Fls. sessile or on pedicels less than 5 mm. long, i.e. with subtending lvs. or bracts near.
D. Calyx-lobes free to base, membrane-margined but not with an obvious hyaline membrane below the sinuses; corolla campanulate. Small desert plants.
E. Lvs. 3–5-cleft or entire, 6–10 mm. long; corolla 6–8 mm. long with 2 dark lines below base of each lobe. Inyo Co. to Utah, Ariz. ...... 19. *L. demissus*
EE. Lvs. entire, 2–5 mm. long; corolla 4–5 mm. long with vermilion spot on each lobe. Little San Bernardino Mts. .... 20. *L. maculatus*
DD. Calyx-lobes united by their bordering membranes to form a pseudotube which may be quite short; corolla funnelform to salverform.
E. Corolla bearing ± reniform arches in throat below each lobe, blue, white or yellow; internodes usually short, concealed by lvs. Tufted desert plants

16. *L. Parryae*

EE. Corolla without reniform arches below the lobes.
F. Fls. yellow or cream.
G. Calyx 4–5 mm. long; corolla 5–8 mm. long.
H. Stamens inserted just below the corolla-sinuses; plant pubescent throughout. Cismontane and desert s. Calif.

6. *L. Lemmonii*

HH. Stamens inserted at base of throat; plant glabrous or minutely puberulent. Rare, Mojave Desert .... 15. *L. arenicola*
GG. Calyx 6–8 mm. long.
H. Lvs. entire; pedicels and calyces with stalked glands. Deserts

14. *L. Jonesii*

HH. Lvs. 5–9-parted; pedicels and calyces without stalked glands. Mostly cismontane .... 30. *L. androsaceus*
FF. Fls. white to pink or lilac.
G. The corolla-tube proper (below the throat) included in calyx.
H. Calyx 3–4 mm. long; lvs. 3-cleft, 2–4 mm. long. Se. San Diego Co. .... 17. *L. bellus*
HH. Calyx 6–14 mm. long.
I. Calyx-tube almost wholly membranous, the teeth villous.
J. Calyx ca. 10 mm. long; corolla-tube 1–2 mm. long, the lobes with 2 dark linear basal spots. San Gabriel Mts. .... 21. *L. concinnus*
JJ. Calyx ca. 6 mm. long; corolla-tube 4–5 mm. long, the lobes with a single elongate basal spot. San Bernardino Mts. .... 22. *L. Killipii*

II. Calyx-tube membranous only below sinuses.
   J. Fls. white to cream or with some purplish tinge, mostly vespertine, the corolla-lobes entire.
      K. Calyx glabrous.
         L. Lvs. usually 3–7-parted; each stamen with a basal hairy pad; corolla 1.5–3 cm. long
                            12. *L. dichotomus*
        LL. Lvs. mostly entire, rarely 2–3-cleft; stamens without basal pads; corolla 1–1.5 cm. long
                            13. *L. Bigelovii*
      KK. Calyx not glabrous.
         L. Calyx with stalked glands; corolla 9–10 mm. long .... 14. *L. Jonesii*
        LL. Calyx white-hairy; corolla 15–30 mm. long 24. *L. grandiflorus*
   JJ. Fls. pink to lilac, rarely white, diurnal, the corolla-lobes dentate.
      K. Lvs. entire, 10–20 mm. long; calyx 10–16 mm. long .... 17. *L. dianthiflorus*
      KK. Lvs. mostly 3-lobed, 2–4 mm. long; calyx 3–4 mm. long .... 18. *L. bellus*
GG. The corolla-tube proper well exserted from calyx.
   H. Corolla-tube stout, the corolla funnelform.
      I. Calyx-membrane only at base of sinuses and then running up on lobes; calyx 6–10 mm. long. San Diego Co.
                            23. *L. Orcuttii*
      II. Calyx-membrane filling the sinuses ca. ⅔ their length; calyx 10–14 mm. long. Santa Barbara and Merced cos. n.
                            24. *L. grandiflorus*
  HH. Corolla-tube slender and subfiliform, usually 2–4 times as long as calyx, the corolla salverform.
      I. Bracts of the infl. pilose to short-hispid but not conspicuously coarsely ciliate (sometimes so in *L. nudatus* with calyx only 5 mm. long).
         J. Calyx 5 mm. long; corolla mostly 8–18 mm. long.
            K. Corolla-lobes rounded; lf.-lobes narrowed upward. Mostly below 7000 ft. ... 26. *L. nudatus*
            KK. Corolla-lobes truncate or emarginate; lf.-lobes mostly widened upward. From 8500–11,000 ft.
                            27. *L. oblanceolatus*
        JJ. Calyx mostly 7–8 mm. long; corolla mostly 18–35 mm. long.
            K. Corolla-tube glabrous; sinus-membrane of calyx extending more than half way; seeds rugose
                          25. *L. breviculus*
            KK. Corolla-tube ± puberulent; sinus-membrane of calyx very short; seeds tuberculate
                          30. *L. androsaceus*
      II. Bracts of the infl. conspicuously and coarsely ciliate; calyx 7–10 mm. long.
         J. Corolla 12–25 mm. long, the lobes 2–4 mm. long
                          28. *L. ciliatus*
         JJ. Corolla 25–30 mm. long, the lobes 5–8 mm. long
                          29. *L. montanus*
BB. Calyx not membranous in sinuses or on margins of lobes, or very inconspicuously so.
   C. Corolla-lobes 5–8 mm. long.
      D. Corolla 20–35 mm. long; locules of caps. 3–6-seeded .... 30. *L. androsaceus*
      DD. Corolla 10–15 mm. long; locules 1–3-seeded .... 31. *L. serrulatus*
   CC. Corolla-lobes 3–5 mm. long.
      D. Fls. pink to white, the throat and tube yellow; middle lobe of lf. usually broadened upward; stigma-lobes 0.5–1.5 mm. long .. .. 32. *L. bicolor*
      DD. Fls. yellow; middle lf.-lobe linear, acicular; stigma-lobes 2–4 mm. long
                          33. *L. acicularis*
AA. Plants suffrutescent perennials, 1–4 dm. high; lvs. 3–9-cleft; calyx not membranous below sinuses
                          34. *L. Nuttallii*

1. **L. Harknéssii** (Curran) Greene. [*Gilia H.* Curran. *G. pharnaceoides* var. *H.* Jones.] Stems slender, erect, 5–15 cm. high, branching above base, subglabrous to puberulent; lvs. 3–5-parted into linear lobes 5–15 mm. long, subglabrous to pubescent; pedicels filiform, 5–20 mm. long, subtended by simple or parted bracts and in dichotomous paniculate cymes; calyx turbinate, 1.5–2 mm. long, deeply cleft into lance-linear segms., the sinus-membranes hyaline and about half the length of the lobes; corolla white to pale blue, short-funnelform, not much exceeding calyx, the lobes ca. as long as tube and throat; stamens included, glabrous; caps.-locules 1-seeded; seeds

straw-colored, dull, almost 2 mm. long.—Open sandy and gravelly places, 3000–8500(–10,400) ft.; Yellow Pine F. to Subalpine F.; Coast Ranges from Lake Co. n., Sierra Nevada from Fresno Co. n.; to w. Ore. and Wash. June–Aug.

Ssp. **condensàtus** Mason. Plants very compact and densely branched; stamens subsessile.—At 6000 ft., Plaskett Meadows, Glenn Co.

2. **L. pygmaèus** (Brand) J. T. Howell. [*Gilia p.* Brand. *Gilia pusilla* auth., not Benth.] Simple or with several very slender branches 2–10 cm. long, erect or diffuse, minutely hispid-puberulent; lvs. remote, 3–5-cleft into linear segms. 2–6 mm. long; fls. in open leafy dichotomous cymes, the capillary pedicels 4–15 mm. long; calyx narrow-cylindric, 3–5 mm. long, the sinuses between the linear lobes over half filled with membrane; corolla narrow-funnelform, 3–5 mm. long, white to pale blue, the lobes definitely shorter than tube plus throat; stamens inserted in throat, glabrous, ca. as long as corolla-lobes; locules of caps. several-seeded.—Occasional on dry slopes, below 5000 ft.; Yellow Pine F., Chaparral, etc.; mts. of interior San Diego and w. Riverside cos., Tehachapi Mts., Coast Ranges n. to Lake and Yolo cos., Butte Co.; L. Calif. April–June.

3. **L. septentrionàlis** Mason. [*L. Harknessii* var. *s.* Jeps. & Bail.] Erect annual 0.5–3 dm. high; lvs. 5–7 cleft into linear segms. 0.5–2 cm. long; pedicels filiform, 1-fld.; calyx ca. 1–2 mm. long, the sinuses with a hyaline membrane extending on the lobes; corolla 2–4 mm. long, ca. 1.5 times the calyx, with a hairy ring at or above the insertion of stamens; stamens 2–3 times as long as throat, exserted, the fils. glabrous or slightly hairy at base; caps. cylindric, the locules 2–4-seeded.—Gravelly places, 6000–8000 ft.; Sagebrush Scrub, Yellow Pine F.; Mono Co. to Modoc Co.; to Rocky Mts. May–July.

4. **L. liniflòrus** (Benth.) Greene. [*Gilia l.* Benth. *Dactylophyllum l.* Heller.] Annual, erect, 1–5 dm. high, the stem branched above base, the branches largely in pairs at the nodes, glabrous to puberulent; lvs. 3–9-cleft into linear segms. 1–3 cm. long; pedicels slender, in cymose panicles and 1–2.5 cm. long; calyx 3–4 mm. long, turbinate, the lobes linear, pilose-ciliate to subglabrous, the sinus-membrane ca. ⅔ as long as lobes; corolla short-funnelform, 1–3 cm. long, white or with some pink or lilac, the throat 3–4 times as long as tube, the lobes obovate, longer than tube and throat; fils. pubescent at base, inserted at bottom of throat; seeds 1–2 in each locule.—Occasional on dry slopes, often of serpentine; N. Coastal Scrub, Chaparral, N. Oak Wd.; outer Coast Ranges from Contra Costa Co. to San Luis Obispo Co. April–July.

Ssp. **pharnaceoìdes** (Benth.) Mason. [*Gilia p.* Benth. *L. p.* Greene. *L. liniflorus* var. *vallicola* Jeps.] Branching ± dichotomous; corolla-throat 1–2 times as long as tube.—Common, below 5000 ft.; many Plant Communities; cismontane and montane s. Calif., w. Mojave Desert, ranges about the Great Cent. V. and n.; to e. Wash. and Ida. Not very distinct from the sp.

5. **L. fílipes** (Benth.) Greene. [*Gilia f.* Benth. *G. pusilla* var. *californica* Gray.] Annual, erect and branched above or diffuse from base, 5–20 cm. high, the stems filiform, cymosely branched, puberulent, sometimes villous below; lvs. palmately 5-parted into subulate segms. 3–7 mm. long; pedicels filiform, 4–12 mm. long; calyx narrow-turbinate, 2–3 mm. long, hispidulous, deeply cleft into linear lobes, the sinuses ca. ⅔ filled with membrane; corolla white to rose or lilac, 4–6 mm. long, glabrous within; fils. hairy at base, inserted low; stigma exserted; seeds several in each locule.—Slopes and flats, below 4000 ft.; V. Grassland, Foothill Wd., Yellow Pine F.; Humboldt Co. to Shasta and Solano cos. and foothills of Sierra Nevada s. to Kern Co. April–July.

6. **L. Lemmònii** (Gray) Greene. [*Gilia L.* Gray.] Annual, with 1–several very slender stems from base, 5–15 cm. long, short-pubescent throughout, glandular above; lvs. remote, 3–5-cleft into linear lobes 2–5 mm. long; fls. sessile or subsessile in subcapitate terminal and subterminal clusters; calyx 4–5 mm. long, deeply cleft into linear-hispid puberulent lobes, the sinuses with a white membrane ca. halfway; corolla short-funnelform, 5–8 mm. long, cream to yellowish, the lobes 2–3 mm. long; the tube with an inner hairy ring at top; stamens inserted just below the corolla-sinuses, the fils. glabrous; caps.-locules many-seeded.—Sandy open places, below 5500 ft.; Chaparral, Coastal Sage Scrub, Yellow Pine F.; interior cismontane s. Calif. from San Bernardino Co. to n. L. Calif., reaching w. edge of Colo. Desert. April–June.

7. **L. aùreus** (Nutt.) Greene. [*Gilia a.* Nutt. *Leptosiphon a.* Benth. *Dactylophyllum a.* Heller. *G. a.* f. *laeta* and f. *pallescens* Brand.] Annual, subglabrous or puberulent or

± glandular, 1–several-stemmed from base, the stems very slender, simple or few-branched, 5–16 cm. long, erect or ascending; lvs. remote, 3–7-cleft into linear-oblong lobes 3–6 mm. long; cyme open, the pedicels filiform, 5–15 mm. long; calyx narrow-campanulate, 4–6 mm. long, deeply cleft into lance-ovate lobes, the hyaline membrane between the sinuses extending ca. ⅔ their length; corolla funnelform, 6–13 mm. long, deep to pale yellow, with orange to brownish-purple throat; tube included in calyx, shorter than throat and with hairy ring at summit; stamens glabrous, inserted on upper half of throat; stigma exserted; caps.-locules several-seeded; $n = 9$ (Flory, 1937).—Locally common over large areas, in sandy places, below 6000 ft.; Creosote Bush Scrub, Joshua Tree Wd.; Mojave and Colo. deserts, occasional interior cismontane valleys of s. Calif., from Inyo and Ventura cos. to L. Calif., Nev., New Mex. March–June.

Var. decòrus (Gray) Jeps. [*Gilia a.* var. *d.* Gray. *L. a.* ssp. *d.* Mason. *G. a.* f. *d.* Brand.] Corolla whitish, often with a dark throat and 7–15 mm. long.—General range of sp., but often in separate areas of some extent.

8. **L. Bàkeri** Mason. Slender erect annual 5–20 cm. high, glandular-puberulent in upper parts and near the nodes, cymosely branched; lvs. remote, 3–7-parted into linear lobes mostly 2–5 mm. long; pedicels filiform, 5–25 mm. long; calyx narrow, 4–5 mm. long, the lobes linear, puberulent towards tips, the sinuses half filled with membrane; corolla white or pink to lilac, slender-funnelform, 6–10 mm. long, the tube 1–4 times the throat, usually somewhat exserted, usually with a hairy band within, the lobes 2–3 mm. long; stamens inserted just below sinuses of corolla-lobes, the glabrous fils. 1–2 times the length of the anthers; locules of caps. several-seeded.—Dry slopes, mostly below 3000 ft.; Yellow Pine F., Foothill Wd.; Sierra Nevada and Coast Ranges from Fresno and Contra Costa cos. n.; to Wash. March–June.

9. **L. ambíguus** (Rattan) Greene. [*Gilia a.* Rattan. *G. Bolanderi* var. *a.* Brand. *Dactylophyllum a.* Heller. *L. Rattanii* var. *a.* Jeps.] Much like *L. Bakeri* in habit; calyx 3–6 mm. long, cylindrical, the sinuses nearly filled with the hyaline membrane, the lobes pilose-hispid; corolla 9–12 mm. long, the tube not much longer than the purple throat, the pink to blue lobes with a yellow basal band and 4–6 mm. long; stamens inserted just below the sinuses, the glabrous fils. 3–4 times as long as anthers.—Largely on serpentine soil, below 2800 ft.; V. Grassland, Foothill Wd., Chaparral?; San Mateo Co., inner Coast Ranges from Santa Clara Co. to San Benito Co. April–June.

10. **L. Rattánii** (Gray) Greene. [*Gilia R.* Gray.] Habit of *L. Bakeri*; calyx sub-cylindric, the sinuses nearly filled with the membrane; corolla slender-funnelform to subsalverform, 10–15 mm. long, lilac or pink to white, the tube 1.5–2 times as long as calyx and 3–4 times the length of throat; stamens inserted at base of throat, long-exserted, the glabrous fils. 5–6 times the length of anthers; caps. narrow, 4–5 mm. long; seeds solitary in the locule, ca. 1 mm. long, rounded-rugulose.—Dry banks, 3000–5000 ft.; Yellow Pine F., N. Oak Wd.; inner Coast Ranges, Lake Co. to Tehama and e. Mendocino cos. May–July.

11. **L. Bolánderi** (Gray) Greene. [*Gilia B.* Gray.] Much like *L. Bakeri;* calyx 5–6 mm. long, the sinuses almost filled with the hyaline membrane; corolla funnelform, 6–10 mm. long, white or pale pink or lilac, the tube included; stamens inserted on upper half of throat, the glabrous fils. 6–8 times the length of the anthers.—Mostly below 4000 ft.; Chaparral, N. Oak Woodland; inner Coast Ranges, Mendocino Co. to Contra Costa Co.; s. Ore. April–June.

12. **L. dichótomus** Benth. [*Gilia d.* Benth. *G. d.* var. *uniflora* Brand.] EVENING SNOW. Erect annual, simple or commonly dichotomously branched, 5–20 cm. high, the stems slender, usually glabrous and ± glaucous; lvs. opposite, usually 3–7-parted into linear-filiform lobes 1–2.2 cm. long; fls. on short pedicels in the forks or terminal, vespertine; calyx glabrous, glaucous, cylindric, 8–14 mm. long, the green linear subulate divaricate tips 3–5 mm. long above the hyaline membrane which is subtruncate in the sinuses; corolla funnelform, 1.5–3 cm. long, white or with some purple in the throat and usually on the lobes in closing, the tube included in calyx, the lobes convolute in bud, broadly obovate; stamens included, inserted in lower tube and each with a basal dilated hairy pad; caps. several-seeded; seeds pale, ± angled-cubic, ca. 0.7 mm. long, with a loose minutely alveolate coat, $n = 9$ (Flory, 1937).—Dry often sandy or gravelly places, mostly below 5000 ft.; Creosote Bush Scrub, Joshua Tree Wd. on Mojave Desert and

w. edge of Colo. Desert; V. Grassland and Foothill Wd. in S. Coast Ranges and foothills of Sierra Nevada; Ariz., Nev. April–June. A form from near Victorville, San Bernardino Co., with undivided lvs. has been named *Gilia dichótoma* var. *íntegra* Jones.

Ssp. **meridiànus** (Eastw.) Mason. [*L. d.* var. *m.* Eastw.] Fls. diurnal, strictly white. —Foothill Wd.; serpentine outcrops, Coast Ranges from Lake, Napa, Tehama, Butte cos., etc.

13. **L. Bigelòvii** (Gray) Greene. [*Gilia B.* Gray. *G. dichotoma* var. *parviflora* Torr.] Habit of *L. dichotomus;* lvs. mostly entire, rarely 2–3-cleft, 1–3 cm. long; calyx 8–12 mm. long; corolla 1–1.5 cm. long, the lobes half as long as tube; stamens 2–3 mm. long, on upper half of tube, glabrous, included; seeds red-brown, angular-cylindric, ca. 1 mm. long, with a bladdery transparent ± closely fitting testa which makes whitish angles.—Occasional, dry slopes and valleys, mostly below 5000 ft.; Creosote Bush Scrub, Joshua Tree Wd., w. Colo. Desert and Mojave Desert to Inyo Co.; V. Grassland and Foothill Wd., e. Monterey and w. Stanislaus cos. to Ventura Co.; to Utah, Tex. March– May.

14. **L. Jònesii** (Gray) Greene. [*Gilia J.* Gray. *G. Bigelovii* var. *J.* Brand. *L. B.* var. *J.* Jeps. & Mason.] Much like *L. Bigelovii*, 3–15 cm. high; lvs. entire, filiform, 1–2 cm. long; pedicels and calyces with stalked glands; calyx 7–8 mm. long, the membrane filling ca. ¾ of sinus; corolla tubular-funnelform, ca. 9–10 mm. long, whitish or yellowish, the obovate lobes ca. 4 mm. long; stamens glabrous, inserted in tube, ca. 2 mm. long; seeds reniform or nearly so, red-brown, with rounded broad irregularities, but not angled or with loose testa.—Uncommon, sandy slopes and washes, below 2500 ft.; Creosote Bush Scrub; from Death V. region through e. Mojave and Colo. deserts to Ariz., L. Calif. March–May.

15. **L. arenícola** (Jones) Jeps. & Bail. [*Gilia a.* Jones. *L. mohavensis* Mason.] Small erect annual 1–8 cm. high, compactly branched, glabrous to minutely puberulent; lvs. mostly 3–5-cleft above their base, pilose above, glabrous beneath, 3–12 mm. long; fls. vespertine, solitary and sessile in forks of cymes or at tips of branches; calyx 4–5 mm. long, the sinuses ca. ⅔ filled with membrane; corolla yellow, sometimes with purple in throat, 5–7 mm. long; stamens inserted in base of throat, glabrous; seeds short-reniform, brownish, ca. 0.5 mm. long, with bordered whitish angles.—Rare gypsophilous plant, at 2500–4000 ft.; Joshua Tree Wd.; Mojave Desert (Barstow, Kelso, Searles Lake, near Trona, Ubehebe Crater) and Nipton, Nev. March–April.

16. **L. Párryae** (Gray) Greene. [*Gilia P.* Gray. *G. Kennedyi* Porter. *Dactylophyllum P.* Heller.] Tufted annual 1–6(–9) cm. high, mostly compactly branched, glandular-villous; lvs. crowded, 3–7-parted, the linear lobes 5–15 mm. long; fls. crowded, sessile to subsessile, in leafy cymes; calyx deeply cleft, 6–8 mm. long, the segms. scarious-margined, the membranes united at base only; corolla funnelform, white, less frequently blue-purple or even cream, 10–15 mm. long, the tube included in calyx, the lobes 6–12 mm. long, entire or erose, the throat bearing below each lobe ± reniform arches; stamens glabrous, inserted near base of throat; seeds red-brown, obovoid to oblong, plump, ⅓–½ mm. long, with a whitish somewhat angled outer membrane.—Common over large areas, sandy flats and plains, mostly below 3500, sometimes to 6000 ft.; Creosote Bush Scrub, Joshua Tree Wd.; w. Mojave Desert to Inyo and Mono cos., arid inner Coast Ranges, e. Monterey Co. to Greenhorn Range, Kern Co. March–May.

17. **L. dianthiflòrus** (Benth.) Greene. [*Fenzlia d.* Benth. *Gilia dianthoides* Endl. *Fenzlia speciosa* and *F. concinna* Nutt.] GROUND-PINK. Annual, usually with several very slender at first spreading then erect stems 5–12 cm. long, puberulent; lvs. entire, filiform, mostly opposite, 1–2 cm. long, shorter than internodes; fls. solitary or in few-fld. leafy cymes, subsessile or with short pedicels; calyx 10–16 mm. long, cleft deeply into linear lobes, with hyaline membrane filling lower half of sinus and for short distance on lobes; corolla short-funnelform, pink to lilac or white, 10–25 mm. long, with dark basal spots and yellow throat, the lobes dentate, longer than tube and throat; stamens inserted at base of throat; fils. puberulent at base; caps. short-oblong; seeds many, round-oblong, ca. 0.5 mm. long, red-brown, plump, with whitish angled membrane, $n = 9$ (Flory, 1937).—Common, open sandy places, below 4000 ft.; Coastal Sage Scrub, V. Grassland, Chaparral; Santa Barbara to San Diego and to w. edge of Colo. Desert; L. Calif. Feb.– April.

SSp. **farinòsus** (Brand) Mason. [*Gilia dianthoides* var. *f.* Brand.] Lvs. oblong-linear,

5–10 mm. long; calyx-lobes spatulate, farinose-pubescent.—Not common, 1500–4500 ft.; San Bernardino, Hemet Valley (San Jacinto Mts.) to Oak Grove (San Diego Co.).

18. **L. béllus** (Gray) Greene. [*Gilia b.* Gray. *L. Peirsonii* Mason.] With general aspect of *L. dianthiflorus;* stems capillary, subglabrous; lvs. mostly tripartite, 2–4 mm. long; calyx 3–4 mm. long, sessile, the lobes with purple basal band and margined to tip; corolla 10–15 mm. long, lilac to pink, with yellow throat with purple spots; glabrous stamens inserted near base of throat; caps. almost as long as calyx; seeds ca. 0.3 mm. long, light brown, rather sharply white-angled.—Dry slopes and flats, ca. 3000–4000 ft.; Chaparral; se. San Diego Co. (Jacumba, Tecate, etc.); adjacent L. Calif. April–May.

19. **L. demíssus** (Gray) Greene. [*Gilia d.* Gray. *G. Dactylophyllum* Torr., nomen confusum.] Annual, diffusely branched from base, 2–10 cm. high, sparingly glandular-puberulent; lvs. 3–5-cleft or entire, the lobes acicular, 6–10 mm. long; fls. sessile or subsessile, in leafy terminal clusters; calyx ca. 3 mm. long, deeply cleft into lanceolate scarious-margined lobes, the membranes sometimes united below; corolla campanulate, white, 6–8 mm. long, the tube very short, the throat ca. 4 times the tube, the lobes oblong, with 2 dark streaks below base; stamens inserted at base of throat, glabrous, included; caps. ellipsoid, ca. half as long as calyx; seeds several in each locule, yellowish-brown, ca. 0.6 mm. long, slightly irregular in shape but not tubercled.—Occasional on dry plains and in washes, mostly below 4000 ft.; Creosote Bush Scrub, Joshua Tree Wd.; Mojave Desert from Victorville to Inyo Co.; e. to Utah, Ariz. March–May.

20. **L. maculàtus** (Parish) Mlkn. [*Gilia m.* Parish.] Minute annual 1–3 cm. high, with few spreading branches from base and ± pilose; lvs. alternate, oblong, entire, cuspidate, thick, 2–5 mm. long, puberulent above, glabrous beneath; fls. crowded in leafy clusters at ends of slender branches; calyx 2–3 mm. long, parted almost to base, the segms. unequal, scarious-margined, ciliate, oblong-ovate; corolla campanulate, 4–5 mm. long, white with vermilion spot on each spreading oblong-obovate lobe; stamens included; caps. round-ovoid, ca. 1.5 mm. long; seeds brownish, ca. 0.6 mm. long, unequally low-tubercled.—Rare, sandy places, 500–4000 ft.; Creosote Bush Scrub, Joshua Tree Wd.; Little San Bernardino Mts. April–May.

21. **L. concínnus** Mlkn. [*Gilia c.* Munz. *Gilia modesta* Hall, not Phil.] Annual, tufted and 1–2 cm. high or loosely branched and 5–12 cm. high, glandular-puberulent, especially in infl.; lvs. opposite or alternate, cleft into 3–5 lobes, 8–15 mm. long, with some weak curly white hairs; fls. subsessile in 3–7-fld. crowded cymes; calyx ca. 1 cm. long, deeply cleft into linear lobes, the broad hyaline membrane extending over halfway; corolla white, funnelform, 10–15 mm. long, the tube 1–2 mm., the narrow throat 6–8 mm. long, the lobes entire or denticulate, with 2 linear dark basal spots; stamens inserted near base of throat, pubescent; caps. included; seeds several, red-brown, ellipsoid.—Dry rocky slopes, 5000–8500 ft.; Yellow Pine F., Red Fir F.; San Gabriel Mts., s. Calif. May–July.

22. **L. Killípii** Mason. With aspect of *L. concinnus;* lvs. 3–10 mm. long, 5–7-cleft; fls. sessile; calyx ca. 6 mm. long, the membrane extending onto the lobes; corolla 10–15 mm. long, the tube 4–5 mm. long, each lobe with 1 elongate spot near base; fils. glabrous.—Dry slopes, 5000–7000 ft.; Pinyon-Juniper Wd.; Cactus Flat to Baldwin Lake, San Bernardino Mts. May–July.

23. **L. Orcúttii** (Parry & Gray) Jeps. ssp. **pacíficus** (Mlkn.) Mason. [*L. p.* Mlkn.] Several-stemmed, suberect, puberulent, 5–10 cm. high; lvs. remote, 3–7-parted into linear lobes 5–12 mm. long, sparsely hairy; fls. solitary or few in small heads; calyx 6–10 mm. long, deeply cleft, the linear segms. membrane-margined and with sinus membrane below; corolla funnelform, 15–25 mm. long, pink to blue, the stout tube 5–15 mm. long, the short throat yellow, the lobes 5–8 mm. long and each with a purple basal reniform spot; stamens inserted at base of throat; caps. 5–8 mm. long, 6–12-seeded.—Uncommon, 4000–5000 ft.; Chaparral; Palomar Mt., San Diego Co. May–June.

24. **L. grandiflòrus** (Benth.) Greene. [*Leptosiphon g.* Benth. *L. densiflorus* Benth. *Gilia g.* Steud.] Annual, erect, simple or branched from base, 1–5 dm. high, subglabrous to puberulent, sometimes villous below nodes; lvs. remote or almost as long as internodes, 5–11-cleft into linear lobes 1–3 cm. long; fls. few to several in dense heads; calyx 10–14 mm. long, deeply cleft into linear segms., densely white hairy, the sinuses with mem-

brane ca. ⅔ their length; corolla white to pale lilac, stout-funnelform, 15–30 mm. long, the tube 1–2 times the calyx, with an inner hairy ring; stamens inserted at middle of throat, glabrous; caps. ellipsoid; seeds 1–5 in each locule, elongate, ± angled and irregular, pale brown, ca. 1.5–2 mm. long, with a whitish papery outer membrane; $n = 9$ (Flory, 1937, for *densiflorus*).—Open woods and sandy places, below 3500 ft.; Coastal Strand, N. Coastal Scrub, Closed-cone Pine F., Foothill Wd.; Sonoma Co. to Santa Barbara Co. and w. Merced Co. April–July.

25. **L. breviculus** (Gray) Greene. [*Gilia b.* Gray. *L. androsaceus* var. *b.* Mlkn. *Gilia royalis* Brand. *L. b.* ssp. *r.* Mason.] Annual, erect, simple or branched from base, 1–2.5 dm. high, the stems slender, ± cymosely branched above, puberulent or subglabrous; lvs. 3–5-parted, 3–10 mm. long, minutely pilose; fls. sessile in small bracteate compact clusters; calyx 7–8 mm. long, glandular, stiff-pilose, deeply cleft, the sinus-membrane extending more than halfway; corolla salverform, 15–25 mm. long, the tube glabrous, slender, 2–4 times the calyx, the lobes 4–6 mm. long, white to pink or bluish, the throat often purple; stamens 2 mm. long, glabrous, inserted on the throat; anthers at the orifice; caps.-locules several-seeded; seeds light brown, angled, ± rugose.—Dry open slopes, below 7000 ft.; Yellow Pine F., Pinyon-Juniper Wd., Joshua Tree Wd.; San Gabriel and San Bernardino mts. to Liebre and Ord mts. and adjacent Mojave Desert. May–Aug.

26. **L. nudatus** Greene. [*Gilia n.* Brand. *L. Nashianus* Jeps.] Like *L. breviculus;* lvs. 5–11-cleft into linear lobes 3–12 mm. long, the upper lvs. and bracts densely hirsute-ciliate, ± glandular, the divisions joined by a hyaline membrane in their lower half or third; fls. sessile in compact heads; calyx ca. 5 mm. long; corolla 1–2 cm. long, the tube pubescent; stamens exserted; caps.-locules 1-seeded.—Open slopes, 2000–7000 ft.; Jeffrey Pine F., Foothill Wd., Chaparral; Tehachapi Mts., Greenhorn Range to Sierra Nevada of Tulare Co. May–July.

27. **L. oblanceolatus** (Brand) Eastw. ex Jeps. [*Gilia o.* Brand. *G. o.* var. *Culbertsonii* Brand. *G. tularensis* Brand. *L. t.* Mason. *G. t.* ssp. *C.* Mason.] Annual, erect, simple or with few short branches, 2–12 cm. high, the stems slender, puberulent, sometimes cymosely branched; lvs. 3–5-cleft into linear-oblanceolate lobes 5–15 mm. long; fls. sessile in bracteate heads or 1 or 2 in upper axils; calyx ca. 5 mm. long, deeply cleft into lance-linear lobes, the sinus-membrane extending ca. halfway; corolla white, salverform, 8–12 mm. long; tube very slender; throat yellow, ± pubescent without, ca. half as long as tube, the lobes ca. 2 mm. long, oblong; stamens 2–4 mm. long, glabrous, inserted near middle of throat; caps. subgloboid; seeds solitary in the locules.—Mountain flats near edges of meadows, 8400–11,000 ft.; Subalpine F.; Sierra Nevada of Fresno, Tulare and Inyo cos. July–Aug.

28. **L. ciliatus** (Benth.) Greene. [*Gilia c.* Benth. *Leptosiphon c.* Jeps.] Annual, erect, simple or branched at base, rather stiff, 1–3 dm. high, rather stiff-pubescent; lvs. rather remote, 5–11-cleft into linear hispid ciliate lobes 5–20 mm. long; fls. sessile in leafy-bracteate heads; bracts hispid-ciliate; calyx 7–10 mm. long, deeply cleft into ciliate hispid acerose segms., the hyaline membrane of the sinuses extending ca. halfway; corolla rose to white, salverform, 12–25 mm. long, the tube slender, long-exserted, pubescent without, the throat yellow, short, the lobes obovate, 2–4 mm. long; stamens glabrous, inserted ca. midway on throat, slightly exserted; caps. oblong, ca. 6 mm. long, few-seeded; seeds brownish, with narrow whitish wings on the angles, ± rugose, ca. 2 mm. long.—Dry open places, mostly below 8000 ft.; many Plant Communities; Coast Ranges from w. Siskiyou Co. to San Diego Co., inland to Sierra Nevada, San Bernardino Mts., etc.; Ore., Nev. April–July. A depauperate form 2–5 cm. high, from ca. 7000–9600 ft. in the Sierra Nevada, has been called var. *neglectus* (Greene) Jeps. [*L. n.* Greene.]

29. **L. montanus** (Greene) Greene. [*L. ciliatus* var. *m.* Greene. *Gilia m.* Parish. *G. ciliata* var. *m.* Munz.] MUSTANG-CLOVER. Annual, erect, the stems stout, simple or few-branched at base, 1–6 dm. high, pubescent; lvs. remote, 5–11-cleft into linear ± hispid-ciliate lobes 2–3 cm. long; fls. sessile in bracteate heads, the bracts coarsely bristly-ciliate; calyx ca. 1 cm. long, deeply divided into linear acerose hispid-ciliate lobes, the sinuses ca. half membranous; corolla salverform to funnelform, 25–30 mm. long, lilac-pink or white with a purple spot on each lobe, the tube long-exserted, pubes-

cent, the throat short, yellow, the lobes 5–8 mm. long; stamens glabrous, inserted near middle of the throat; caps. ca. 5 mm. long, few-seeded; seeds brownish, ca. 2 mm. long, angled.—Dry gravelly places, 1000–5000 ft.; Foothill Wd., Yellow Pine F.; Sierra Nevada from Nevada Co. to Greenhorn Range, Kern Co. May–Aug.

30. **L. androsàceus** (Benth.) Greene. [*Leptosiphon a.* Benth. *Gilia a.* Steud.] Annual, mostly erect, the stems simple or with few branches from base, mostly pubescent, 0.5–3 dm. high; lvs. remote, 5–9-parted into oblanceolate or narrower segms., 1–3 cm. long, mostly ± stiff-pubescent or -ciliate; fls. sessile in terminal dense bracteate heads; calyx 6–8 mm. long, deeply cleft into subulate lobes, these mostly glabrous on backs, coarsely stiff-ciliate, ± subulate, the sinuses with a short narrow membrane; corolla salverform, rose or lilac to white or yellowish, the tube ca. 1 mm. broad, 1.5–3(–4) times the calyx, often dark, ± puberulent, the throat short, frequently yellow, the lobes 5–8 mm. long, obovate, often subapiculate; stamens glabrous, inserted near middle of the throat, the anthers barely exserted; stigmas scarcely if at all exserted; caps. ellipsoid, 4–5 mm. long; seeds several, ca. 1.3 mm. long, somewhat wing-angled and tuberculate; $n = 9$ (Flory, 1937).—Grassy slopes and open places, below 3500 ft.; N. Oak Wd., Chaparral, Coastal Prairie, etc.; Monterey Co. to Humboldt, Trinity, and Shasta cos. April–June.

KEY TO SUBSPECIES

Calyx-lobes mostly glabrous on the back, coarsely ciliate; corolla-tube rather coarse, ca. 1 mm. thick, mostly 1.5–3 times as long as calyx; corolla-limb 12–15 mm. broad; stems mostly simple or with few erect branches from base. Grassland and woods from Monterey Co. to Humboldt and Shasta cos.
L. androsaceus
Calyx-lobes pubescent on backs as well as edges; corolla-tube mostly subfiliform.
  Stems erect, simple or few-branched from base.
    Styles scarcely if at all exserted.
      Corolla-limb 10–15 mm. broad, always white. Foothills of Sierra Nevada .... ssp. *laetus*
      Corolla-limb 6–10 mm. broad, lilac to pink or yellow, rarely white. Coast Ranges from Santa Barbara Co. to Mendocino Co. .... ssp. *luteus*
    Styles conspicuously long-exserted; fls. yellow to pink or lilac. Monterey Co. to L. Calif.
ssp. *micranthus*
  Stems decumbent to ascending, mostly several-branched.
    Corolla-limb mostly 12–15 mm. broad. Immediate coast from Monterey Co. to Sonoma Co.
ssp. *croceus*
    Corolla-limb mostly 6–10(–12) mm. broad. Interior Monterey Co. to L. Calif. .... ssp. *luteolus*

Ssp. **laètus** Mason. Largely below 1500 ft.; V. Grassland, Foothill Wd.; Butte Co. to Fresno Co. April–May.

Ssp. **lùteus** (Benth.) Mason. [*Leptosiphon l.* Benth. *Gilia l.* Steud. *G. micrantha* var. *aurea* Benth.] Open places in woods; N. Oak Wd., Foothill Wd., Chaparral; Coast Ranges, Mendocino and Lake cos. to Santa Barbara Co. April–May.

Ssp. **micránthus** (Steud.) Mason. [*Gilia m.* Steud. *G. lutea* ssp. *m.* Brand. *Leptosiphon parviflorus* Benth. *G. l.* var. *longistylis* Munz.] $n = 9$ (Flory, 1937).—Dry places, below 4000 ft.; Chaparral, Foothill Wd., etc.; Monterey Co. through the S. Coast Ranges to San Diego Co.; L. Calif. April–May.

Ssp. **cròceus** (Mlkn.) Mason. [*L. parviflorus* var. *c.* Mlkn. *L. c.* Eastw. *Gilia longituba* Benth. *L. l.* Heller. *Leptosiphon parviflorus* var. *rosaceus* Hook. *G. lutea* var. *rosea* Regel. *Linanthus rosaceus* Greene.] Coastal bluffs and hills; Closed-cone Pine F., N. Coastal Scrub; Monterey Co. to Sonoma Co. May–June.

Ssp. **lutèolus** (Greene) Mason. [*L. l.* Greene. *L. parviflorus* var. *l.* Mlkn. *L. Plaskettii* Eastw. *L. androsaceus* ssp. *P.* Mason. *L. graciosus* Mlkn. *G. g.* Brand. *G. tassajarae* Brand.] Dry slopes, below 5000 ft.; Yellow Pine F., Chaparral, Foothill Wd., Pinyon-Juniper Wd.; inner Coast Ranges from Monterey Co. to San Diego Co. and to w. edge of Mojave Desert. April–May.

31. **L. serrulàtus** Greene. [*Gilia androsacea* ssp. *s.* Brand. *L. mariposanus* Mlkn. *G. m.* Brand.] Annual, erect, simple or few-branched from base, the stems slender, 5–18 cm. high, puberulent; lvs. remote, 5–7-cleft into linear ± hispid lobes 4–10 mm. long; fls. sessile, mostly in terminal bracteate heads; calyx 5–7 mm. long, deeply cleft into lance-linear lobes, sparsely hispid, almost without any hyaline membrane; corolla 1–1.5 cm. long, salverform, with purple puberulent tube 7–8 mm. long and yellow throat 4–6 mm. long and white narrow-obovate lobes 5–7 mm. long; stamens ca. 2 mm. long, glabrous, inserted on throat; stigma barely exserted; caps. 4–5 mm. long, the locules 1–3-seeded;

seeds 1.5–2 mm. long, tuberculate.—Dry slopes, 1000–4000 ft.; Yellow Pine F., Foothill Wd.; foothills of Sierra Nevada from Madera Co. to Tehachapi Mts., Kern Co. April–May.

32. **L. bìcolor** (Nutt.) Greene. [*Leptosiphon b.* Nutt. *Gilia exigua* Brand. *G. tenella* Benth., not Nutt. *L. b.* ssp. *minimus* Mason. *L. Eastwoodae* Heller.] Annual, simple or few-stemmed from base, erect or ascending, 3–15 cm. high, puberulent; lvs. remote, 3–7-cleft into linear segms. 3–10 mm. long, hispid-ciliate, the middle segm. broader than others; fls. sessile in leafy-bracteate heads; calyx hispid, 6–8 mm. long, deeply cleft into subulate lobes, not or scarcely scarious in the sinuses; corolla salverform, 15–30 mm. long, bicolored, the lobes red to pink or white, 2–3 mm. long, round-obovate, the throat yellow, 1–2 mm. long, the tube yellow, stout, 12–22 mm. long, puberulent; stamens inserted in throat, exserted; stigma-lobes 0.5–1.5 mm. long; caps. 3–4 mm. long, the locules 2–4-seeded; seeds round-oblong, ca. 1 mm. long, tuberculate.— Grassy and open wooded places, mostly below 5000 ft.; V. Grassland, Foothill Wd., Chaparral; Coast Ranges from Humboldt Co. to San Luis Obispo Co., Sierra Nevada s. to Kern Co., Santa Catalina and San Clemente ids.; n. to Wash. March–June.

33. **L. acícularis** Greene. [*Leptosiphon a.* Jeps.] Habit of *L. bicolor;* lf.-lobes alike; calyx-sinuses with inconspicuous basal hyaline membrane; corolla golden-yellow, 1–2 cm. long, the tube filiform; throat ca. 1 mm. long; lobes 3–4 mm. long; stigma-lobes 2–4 mm. long; seeds scarcely 1 mm. long.—Dry grassy places, below 2000 ft.; Coastal Prairie, N. Oak Wd.; e. Humboldt Co. to Alameda Co. April–June.

34. **L. Nuttàllii** (Gray) Greene ex Mlkn. [*Gilia N.* Gray. *Leptodactylon N.* Rydb. *G. N.* var. *montana* Brand. *Siphonella N.* Heller. *Linanthastrum N.* Ewan.] Bushy perennial from a woody base, the erect stems thickly branched, mostly 1–2 dm. high, ± villous-hispid; lvs. opposite, 5–9-cleft into linear-oblanceolate lobes 1–1.5 cm. long and with fascicled smaller lvs. in axils; fls. sessile or subsessile in upper axils, forming subcapitate infl.; calyx narrow-campanulate, 8–9 mm. long, the tube ca. 2 mm. long, the lobes lance-subulate, puberulent, with scant hyaline membrane in the sinuses; corolla funnelform to subsalverform, 12–15 mm. long, the usually yellowish tube ca. 8 mm. long, pubescent, the throat short, the lobes white or nearly so, 4–5 mm. long, oblanceolate; stamens inserted at base of throat, glabrous, barely exserted; caps. ca. 5 mm. long, oblong, with 1–4 seeds in a locule; seeds yellowish-brown, oblong, ca. 2 mm. long, shining but shallowly rugose-reticulate.—Dry rocky or brushy places, 4000–12,000 ft.; Sagebrush Scrub through various Forest Communities to Subalpine F.; San Bernardino Mts., mostly on e. slope of Sierra Nevada and in White and Inyo mts. to Modoc and Siskiyou cos., then in the Coast Ranges to Humboldt and Trinity cos.; to Wash., Rocky Mts. May–Aug.

Ssp. **floribúndus** (Gray) Munz [*Gilia f.* Gray. *L. f.* Greene. *Leptodactylon f.* Tides. *L. Nuttallii* var. *f.* Jeps. *Linanthastrum N.* ssp. *f.* Ewan. *Siphonella f.* Jeps. *Linanthus saxiphilus* A. Davids.] Stems 1–4 dm. high; lvs. 3–5-cleft into almost filiform lobes 8–20 mm. long, often without axillary fascicles; fls. subsessile or pedicelled in more open cymose clusters.—Dry slopes and benches, below 7000 ft.; Chaparral, Yellow Pine F., S. Oak Wd.; San Jacinto, Santa Rosa, and Santa Ana mts. to L. Calif., Colo., New Mex., Chihuahua. May–July. A form from the s. slope of the Santa Rosa Mts. with lvs. mainly entire, has been called *Siphonella floribunda* var. *Hállii* Jeps. and *L. floribundus* ssp. *Hállii* Mason. A compact almost matted form with fls. ca. 8–10 mm. long, from the White Mts. of Calif. and Nev. and the Toiyabe Mts., Nev., and the Sierra San Pedro Martir of L. Calif. is *Linanthastrum Melíngii* (Wiggins) Wherry. [*Leptodactylon M.* Wiggins.]

59. Hydrophyllàceae. Waterleaf Family (Fig. 56)

Herbs or shrubs with opposite or alternate lvs. Fls. usually 5-merous, cymose or solitary. Calyx deeply lobed, the divisions alike or unlike, with or without basal auricles between the lobes. Corolla gamopetalous, rotate to campanulate or tubular, usually with a pair of scalelike appendages at base of each fil. Stamens hypogynous, inserted near base of the corolla, equal or unequal, exserted or included. Pistil 1, consisting of 2 carpels; style subentire to deeply bifid; stigmas 2, mostly capitate. Fr. a caps., loculicidal with

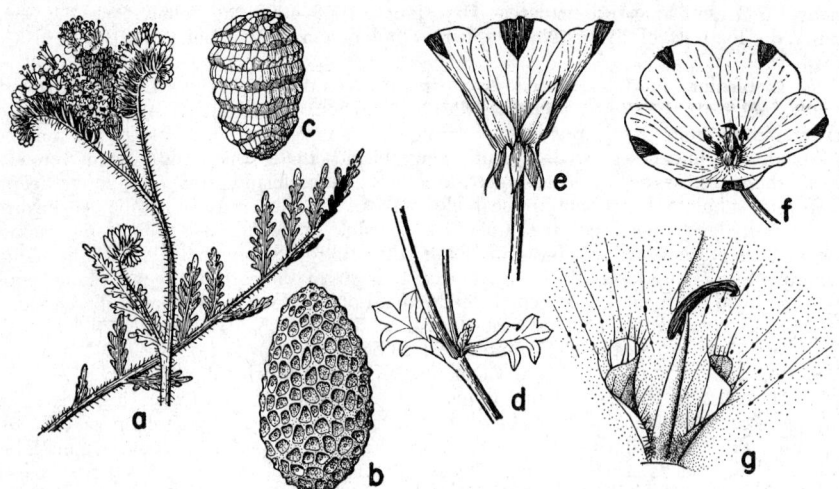

Fig. 56. HYDROPHYLLACEAE. *Phacelia tanacetifolia: a,* dissected lf., $\times$ ½, and scorpioid cymes; *b,* seed, $\times$ 10, pitted. *P. Fremontii: c,* seed, $\times$ 17, corrugated. *Nemophila maculata: d,* lvs., $\times$ ½; *e,* fl., $\times$ 1, with sepallike appendages or auricles between the calyx-lobes; *f,* fl., from inside; *g,* appendages at base of fil., $\times$ 5.

2 valves, or both loculicidal and septicidal and dehiscent by 4 valves, or irregularly dehiscent, 1-loculed or partially 2-loculed by intrusion of the parietal placentae. Ovules few to many. Seeds with endosperm; cotyledons entire. Ca. 25 genera and 300 spp., largely w. Am.

A. Styles ± united.
    B. Perennial herbs.
        C. Calyx-lobes similar.
            D. Lvs. entire or toothed, but not deeply divided or lobed.
                E. Plants caulescent; fls. in cymes or clustered, sometimes solitary and axillary.
                    F. Herbs with a bulblike or tuberous base; lvs. reniform .. 14. *Romanzoffia*
                    FF. Herbs not bulblike at base; lvs. longer than wide.
                        G. Lvs. all opposite; stamens unequal and unequally inserted
                                        7. *Draperia*
                      GG. Lvs mostly alternate.
                          H. Stamens unequally inserted, unequal in length .... 9. *Nama*
                          HH. Stamens equally inserted, subequal in length .... 5. *Phacelia*
                EE. Plants acaulescent; fls. solitary in lf.-axils of basal rosette
                                12. *Hesperochiron*
            D. Lvs. deeply divided or lobed.
                E. Ovary 1-celled, nearly filled by the broad placentae; lvs. largely basal; cymes
                    subcapitate, congested .... 1. *Hydrophyllum*
                EE. Ovary ± falsely 2-celled by intrusion of the narrow placentae; lvs. largely
                    cauline; cymes mostly not congested .... 5. *Phacelia*
        CC. Calyx-lobes of 2 kinds, the 3 outer cordate, enlarged and veiny in fr., the 2 inner nar-
            rower .... 13. *Tricardia*
    BB. Annual herbs.
        C. Herbage viscid and scented; ovules borne on both faces of the placentae; calyx lacking
            auricles between the lobes .... 3. *Eucrypta*
        CC. Herbage not viscid or scented; ovules borne only on the front of the placentae.
            D. Ovary 1-celled, with broad placentae; infl. not usually a clearly marked cyme;
                calyx usually with ± evident auricles between the lobes.
                E. Stems recurved-prickly; caps. prickly or bristly; seeds without a cucullus
                              2. *Pholistoma*
                EE. Stems usually not recurved-prickly; caps. pubescent; seeds with a ± terminal
                    cucullus .... 4. *Nemophila*
            DD. Ovary falsely 2-celled by intrusion of narrow placentae; infl. usually an evident
            coiled cyme.
                E. Stamens equally inserted, and if unequal in length, the seeds corrugate.
                    F. Corolla mostly deciduous, blue to purplish or white, if persistent the
                      fls. erect .... 5. *Phacelia*
                    FF. Corolla persistent, yellowish or reddish, pendulous .... 6. *Emmenanthe*

EE. Stamens unequally inserted and unequal in length; seeds ovoid, reticulate,
often pitted .... 9. *Nama*
AA. Styles separate to base.
  B. Corolla constricted at insertion of stamens, the fls. coherent laterally by their dilated bases;
low annual .... 8. *Lemmonia*
  BB. Corolla not constricted at point of stamen-insertion, the fls. distinct.
    C. Low annuals; seeds pitted or alveolate .... 9. *Nama*
    CC. Perennial or shrubby.
      D. Herbaceous, the fls. in an elongate thyrsoid panicle of cymes; seeds longitudinally
striate .... 10. *Turricula*
      DD. Shrubby, the fls. not paniculate; seeds transversely corrugate .... 11. *Eriodictyon*

1. Hydrophýllum L. WATERLEAF

Ours perennial herbs with horizontal rootstocks and fleshy-fibrous or tuberous roots.
Lvs. alternate or mainly basal, pinnate to pinnatifid, the cauline lobed or divided. Fls.
greenish to white to violet-blue, several to many, ± pedicellate, in terminal open or
subcapitate cymes. Calyx divided almost to base, the lobes subequal, sinuses in ours
naked. Corolla deciduous, campanulate, divided to middle or farther. Stamens exserted,
equal, the fls. hairy at middle, each with a pair of linear ciliate appendages with one
edge free. Style exserted, shallowly bifid. Caps. membranaceous, unilocular, loculicidal,
with a pair of ovules on front of each of the 2 parietal placentae. Seeds 1–3, subglobose,
brown, reticulate. Eight spp., of N. Am. (Greek, *hudor*, water, and *phullon*, lf.)

(Constance, L. The Genus Hydrophyllum L. Am. Midl. Nat. 27: 710–731, 1942.)
Rhizome conspicuous, with fleshy-fibrous roots; plants caulescent; anthers 1–2 mm. long; lfts. usually
toothed on lower edge.
  Lvs. roundish to ovate, the lfts. usually 5; pedicels 3–12 mm. long .... 1. *H. tenuipes*
  Lvs. oblong to oblong-oval, the lfts. usually 7–19; pedicels 2–6 mm. long.
    Lfts. acuminate, usually 8–12 in no., the teeth acuminate. Trinity, Humboldt, and Siskiyou cos. n.
2. *H. Fendleri*
    Lfts. abruptly acute to obtuse, usually 3–6, the teeth obtuse to acute. Kern Co. n.
3. *H. occidentale*
Rhizome very short, with fascicled fingerlike roots; plants almost acaulescent; anthers 0.6–1 mm. long;
lfts. not toothed on lower edge .... 4. *H. capitatum*

1. **H. tenùipès** Heller. [*H. t.* var. *viride* Jeps.] Rhizome elongate; stems 2–5 dm. tall,
retrorse-hispid; lvs. suborbicular in outline, 1–2 dm. in diam., green above, paler beneath, pinnately divided into 5 ovate to lance-ovate principal divisions, the lowest pair ±
distinct, the upper 3 somewhat confluent, all serrate-incised, sparsely strigose; lower
petioles much exceeding blades, the upper equal to or shorter than blades; cymes open,
with pedicels 5–12 mm. long, coarse-pubescent; calyx-lobes narrow, 4–7 mm. long,
hispid-ciliate; corolla in ours mostly greenish to whitish, 5–7 mm. long, the oblong
lobes 3–4 mm. long; caps. 3–5 mm. in diam.; seeds solitary, yellowish to red-brown,
ca. 3.5 mm. thick; *n* = 9 (Cave and Constance, 1942).—Moist shaded places, below
5000 ft.; Redwood F., Mixed Evergreen F., Douglas-Fir F.; Mendocino Co. to Del
Norte Co.; to B.C. April–June.

2. **H. Féndleri** (Gray) Heller var. **álbifrons** (Heller) Macbr. [*H. a.* Heller.] Rootstock elongate; stems 2–5 dm. high, hirsutulous; lvs. often exceeding stems, the blades
5–15 cm. long, paler beneath than above, soft-pubescent, oblong to oval in outline,
pinnatifid usually into 7–9 lanceolate to ovate divisions, the lower ± distinct, the upper
confluent, all coarsely serrate to incised with lance-ovate lobes; infl. hirsutulous, open, the
pedicels 2–6 mm. long; calyx-lobes 3–5 mm. long, strigulose, weakly ciliate; corolla
white to violet, 7–10 mm. long, the lobes 4–5 mm. long; caps. ca. 4 mm. in diam.; seeds
1–3, light brown, 2.5–3 mm. thick; *n* = 9 (Cave & Constance, 1942).—Damp shady
places, mostly 5000–7000 ft.; Yellow Pine F., Red Fir F.; Tehama, Trinity, Siskiyou,
and Humboldt cos.; to B.C., Ida. May–July.

3. **H. occidentàle** (Wats.) Gray. [*H. macrophyllum* var. *o.* Wats.] Rhizome elongate;
stems 1–6 dm. high, short-pubescent to ± retrorse-hispid; lvs. oblong in outline, 5–16 cm.
long, on shorter to equally long petioles, strigulose, paler beneath, pinnatifid into 7–15
principal oblong to ovate divisions with ovate lobes; cymes globose, the pedicels 2–5
mm. long; calyx-lobes strigulose, hispid-ciliate, narrow, 3–4 mm. long; corolla white
to violet, 7–10 mm. long, the lobes 4–6 mm. long, oblong; caps. ca. 4 mm. in diam.; seeds
1–2, brown, ca. 3 mm. thick; *n* = 9 (Cave & Constance, 1942).—Dryish or moist ±
shaded slopes, 2500–8000 ft.; Yellow Pine F., Red Fir F., N. Oak Wd.; Tehachapi Mts.,

Sierra Nevada from Tulare Co. n., Coast Ranges from Monterey Co. n.; to Ore., Ida., Utah, Ariz. May–July.

4. **H. capitàtum** Dougl. var. **alpìnum** Wats. [*H. alpestre* Nels. & Kenn.] Rhizome very short; stems very short, spreading-pubescent; lf.-blades subovate in outline, pubescent, not conspicuously paler beneath, 4–7 cm. long, pinnately parted into 5–7 lanceolate to obovate principal divisions, the larger cleft into entire oblong lobes; petioles longer; peduncles conspicuously shorter than lvs., 1–5 cm. long; pedicels 5–20 mm. long; calyx-lobes narrow, 3–4 mm. long, densely soft-hairy; corolla purple or lavender to white, 5–9 mm. long, the lobes 3–4 mm. long, oblong-obovate; caps. ca. 4 mm. in diam.; seeds usually 2, light brown, 2–3 mm. in diam.; *n* = 9 (Cave & Constance, 1942).—Moist places, 3000–7000 ft.; Sagebrush Scrub, Yellow Pine F.; Placer Co. to Modoc and Siskiyou cos.; to Ore., Ida., Utah. May–June.

2. Pholístoma Lilja ex Lindbl.

Annual, succulent, prostrate or reclining, with brittle stems often having prickly angles. Lower lvs. opposite, the upper alternate, pinnate, the petioles often winged and clasping. Fls. solitary, axillary and also several in open terminal cymes. Calyx lobed nearly to base, the sinuses naked or with a sepaloid auricle. Corolla subrotate, deciduous, white to violet or blue, lobed to about the middle. Stamens included, inserted on the corolla; appendages broad and triangular to minute or reduced to glands. Style cleft less than halfway. Caps. globose, 1-loculed, exceeding the enveloping or spreading calyx, loculicidal. Ovules 2–several on front of each of 2 large placentae. Seeds mostly 1–6, light brown, globose, reticulate or pitted. Three spp. (Greek, *pholis*, scale, and *stoma*, mouth.)

(Constance, L., Bull. Torr. Bot. Club 66: 341–352, 1939.)

Calyx with auricles between the lobes, enveloping the caps.; stems with retrorse prickles on the angles.
   Petioles broadly winged and auriculate-clasping; corolla 1–3 cm. broad . . . . 1. *P. auritum*
   Petioles narrowly winged, scarcely auriculate-clasping; corolla 0.6–1 cm. broad . . 2. *P. racemosum*
Calyx without auricles, stellate-rotate under the caps.; stems slightly scabrous, without retrorse prickles
                                                                     3. *P. membranaceum*

1. **P. aurìtum** (Lindl.) Lilja. [*Nemophila a.* Lindl. *Ellisia a.* Jeps.] Straggling, with coarse loosely branched stems 3–10(–12) dm. long, with retrorse prickles as well as pubescent; lower lvs. oblong to lance-ovate, 6–15 cm. long, with 7–13 divisions, these oblong or lanceolate, ± retrorse, the petioles broadly winged and auriculate-clasping; fls. cymose and terminal or solitary and axillary; pedicels 2–3 cm. long; calyx 5–10 mm. long, the lobes lanceolate or lance-ovate, 4–7 mm. long, the auricles 1.5–3 mm. long; corolla lavender to purple with darker markings, 1.5–3 cm. broad, the lobes oval to obovate, the tube pale; appendages purple, triangular, 2–3 mm. long, often fimbriate on free edge; style ca. 5 mm. long; caps. 5–10 mm. in diam., enclosed by calyx; seeds 1–4, globose, pitted-reticulate, 2–3 mm. in diam.; *n* = 9 (Cave & Constance, 1942).— Shaded slopes and deep canyons, below 4500 ft.; Coastal Sage Scrub, Chaparral, Oak Wd., Foothill Wd.; Coast Ranges from Lake Co. s., foothills of Sierra Nevada from Calaveras Co. s.; to L. Calif. March–May.

2. **P. racemòsum** (Nutt.) Const. [*Nemophila r.* Nutt. ex Gray. *Ellisia r.* Jeps.] Straggling, loosely branched, retrorse-scabrous and pubescent, the stems 2–6 dm. long; lvs. thin, sparsely pubescent, 3–10 cm. long, rather regularly pinnatifid into 5–9 ovate entire or round-toothed lobes and narrowed into petioles scarcely winged or auriculate; fls. solitary and cymose; calyx 5–7 mm. long, the lobes ± lanceolate, 2–3 mm. long, the auricles 1–2 mm. long; corolla 0.6–1 cm. broad, white or sometimes bluish, the lobes obovate; appendages small, narrowly triangular; style 1.5–2 mm. long; caps. 5–8 mm. thick, enclosed by calyx; seeds usually 4–6, angular-subglobose, pitted, 1–2 mm. in diam.; *n* = 9 (Cave & Constance, 1947).—Shaded places, below 1000 ft.; Coastal Sage Scrub, Chaparral, S. Oak Wd.; Channel Ids., coastal San Diego Co.; n. L. Calif. March–May.

3. **P. membranàceum** (Benth.) Const. [*Ellisia m.* Benth. *E. m.* var. *hastifolia* Brand.] Stems weak, procumbent, 2–5 dm. long, glaucous and glabrous except for a few scattered stiff hairs; lvs. 2–6 cm. long (or lower more), the divisions 5–11, linear-oblong to ovate, ± hispidulous, the petiole narrowly winged but not auriculate; fls. mostly cymose; calyx-lobes oval, 2–3 mm. long, the auricles lacking; corolla 0.4–1 cm. broad, white,

often with narrow purple spot on each oval lobe; appendages triangular, minute; style 1.5–2 mm. long; caps. 2–4 mm. in diam., projecting above the stellate-spreading calyx; seeds 1–2, reticulate, globose, 2–3 mm. in diam.; $n = 9$ (Cave & Constance, 1942).— Occasional in shady places, below 3500 ft.; Foothill Wd., S. Oak Wd., Chaparral; inner Coast Ranges from Contra Costa Co. s., San Joaquin V., foothills of Sierra Nevada from Stanislaus Co. s., to San Diego Co.; in shade of rocks, etc., below 5000 ft.; Creosote Bush Scrub, Joshua Tree Wd.; deserts from Inyo Co. to Imperial Co.; Nev., L. Calif. March–May.

3. Eucrýpta Nutt.

Annual, branched, erect or diffuse, viscid, scented, hispid. Lower lvs. opposite, the others alternate, pinnately divided, petioled or the upper sessile or clasping. Fls. several in open, terminal or axillary cymes, with filiform pedicels. Calyx divided halfway or more, the sinuses naked. Corolla white or yellowish to blue, campanulate, deciduous, shallowly lobed, longer than calyx. Stamens included, equally inserted on corolla, the appendages minute or none and with a V-shaped transverse fold between each pair of fils. near the throat. Style shortly bifid. Mature caps. 1-loculed, round to ovoid, surpassed by the enveloping or stellate-spreading calyx. Seeds usually 5–15, dimorphic (the inner lens-shaped, smooth, the outer terete, corrugated) or all alike and corrugated; brown or black. A genus of two spp. (Greek, *eu*, true or well, and *crypta*, secret, referring to the extra, hidden seeds.)

(Constance, L., Lloydia 1: 143–152, 1939.)

Lvs. 2–3-pinnatifid, mostly opposite; mature calyx stellate-rotate beneath the caps.
                                        1. *E. chrysanthemifolia*
Lvs. 1-pinnate, alternate, the lobes mostly entire or few-toothed; mature calyx erect and enclosing caps.
                                        2. *E. micrantha*

1. **E. chrysanthemifòlia** (Benth.) Greene. [*Ellisia c.* Benth. *Eucrypta paniculata* and *foliosa* Nutt.] Erect, branched, 2–5 dm. tall, somewhat hirsute and glandular, with characteristic rather pleasant odor; lvs. 3–10 cm. long, broadly ovate to oblong in outline, pinnatifid, the 9–13 lance-oblong lobes again 1–2-pinnatifid; lvs. short-petiolate or subsessile, auriculate at base; fls. loosely clustered, the fruiting pedicels mostly recurved; calyx pilose and finely pubescent, minutely glandular, the lobes obtuse, 1–2 mm. long; corolla yellowish-white, open-campanulate, 6–8 mm. broad, the lobes roundish; style 1–1.5 mm. long; caps. 2–4 mm. in diam., hirsute, the inner seeds 1–1.5 mm. long, the outer 0.8–1 mm. long, both dark brown; $n = 10$ (Cave & Constance, 1944).—Common on burns and in partly shaded places, below 3000 ft.; Coastal Sage Scrub, Chaparral, Oak Wd.; Marin Co. to San Diego Co., Channel Ids.; L. Calif. March–June.

Var. **bipinnatífida** (Torr.) Const. [*Phacelia micrantha* var.? *b.* Torr. *Ellisia Torreyi* Gray. *Eucrypta T.* Heller.] Weaker and more diffuse, glandular-pubescent only in infl., hairy below; lvs. 2–4.5 cm. long, 7–9-lobed, the lobes shallowly pinnatifid; fls. mostly 4–8; corolla 2–3 mm. broad, not exceeding calyx.—Occasional, in shelter of rocks, below 5000 ft.; Creosote Bush Scrub, Joshua Tree Wd.; deserts from Inyo Co. to Nev., Ariz., L. Calif. March–May.

2. **E. micrántha** (Torr.) Heller. [*Phacelia m.* Torr. *Ellisia m.* Brand.] Erect but rather weak and diffuse, the stems slender, often stipitate-glandular, 1–2.5 dm. tall; lower lvs. oblong to oval, 1.5–3 cm. long, pinnately parted into 7–9 oblong to spatulate entire or few-toothed divisions; upper lvs. auriculate-clasping; fls. 4–12 on each branch of infl., the fruiting pedicels mostly ascending; calyx-lobes 1.5–2 mm. long, spatulate or oblong; corolla 2–4 mm. broad, white to bluish, the lobes oblanceolate; style 1–1.5 mm. long; caps. 2–3 mm. in diam.; seeds 7–15, dark, homomorphic, oblong to ± curved, corrugate to tubercled, ca. 1 mm. long; $n = 6$ (Cave & Constance, 1950).—Usually in partly shaded places, below 7000 ft.; Creosote Bush Scrub to Pinyon-Juniper Wd.; deserts from Inyo Co. to Imperial Co.; to Nev., Tex., L. Calif. March–June.

4. Nemóphila Nutt. ex Barton

Annual, usually branched and diffuse, sometimes prostrate, often weak, hispid to glabrous. Lvs. mostly opposite, variously toothed, lobed or pinnately divided, petioled. Fls. pedicelled, solitary in upper axils or opposite the lvs. Calyx deeply divided, the sinuses usually armed with sepallike spreading or reflexed auricles. Corolla white to blue, sometimes spotted, deciduous, rotate to campanulate, usually with a pair of appendages within at base of each fil. Style ± deeply bifid. Caps. uniloculed, round to ovoid, loculicidal. Ovules 2–several on each of the 2 large parietal placentae. Seeds 1–20, ovoid, smooth to corrugate-tubercled, regularly or irregularly pitted or not, with a ± evident papillaelike group of colorless cells (cucullus). Ca. 13 spp. of N. Am., some used as garden annuals. (Greek, *nemos*, grove, and *phileo*, to love.)

(Constance, L., Univ. Calif. Pub. Bot. 19: 341–398, 1941.)

Stems minutely recurved-prickly; corolla often shorter than calyx; seeds globose, solitary. Modoc Co.
                                                            1. *N. breviflora*
Stems not recurved-prickly; corolla as long as or longer than calyx; seed ± ovoid, more than 1.
  Corolla 1–4 cm. broad; lvs. all opposite.
    Fls. white, with each corolla-lobe bearing a purple blotch; seeds smooth or pitted
                                                      2. *N. maculata*
    Fls. white to blue, but the corolla-lobes not purple-blotched; seeds corrugate-tuberculate
                                                      3. *N. Menziesii*
  Corolla mostly 0.1–1 cm. broad.
    Lvs. all opposite.
      Auricles ⅓ as long as sepals in fr.; corolla ± shallowly campanulate.
        Lvs. oblong or oval, deeply divided, weakly cuneate to truncate at base . . 4. *N. pedunculata*
        Lvs. spatulate, shallowly lobed, strongly cuneate at base . . . . 5. *N. spatulata*
      Auricles minute or obsolete.
        Lvs. mostly divided almost to midrib; corolla rotate . . . . 6. *N. pulchella*
        Lvs. not so deeply divided; corolla bowl-shaped . . . . 8. *N. parviflora*
    Lvs. opposite on lower stems, alternate above.
      Corolla prominently exserted from calyx.
        Auricles minute or obsolete; seeds 1–4, smooth or slightly rough.
          Corolla rotate, blue to violet with white center or white . . . . 6. *N. pulchella*
          Corolla bowl-shaped to campanulate, mostly white . . . .\. 7. *N. heterophylla*
        Auricles evident; seeds 4–10, corrugate-tuberculate . . . . 3. *N. Menziesii*
      Corolla scarcely exserted from calyx.
        Stamens exceeding corolla-tube. E. Monterey Co. to Kern Co. . . . . 6. *N. pulchella Fremontii*
        Stamens shorter than corolla-tube. From Monterey and Greenhorn Mts. n. . . 8. *N. parviflora*

1. **N. breviflòra** Gray. [*Viticella b.* Macbr.] Stems weak, 3–20 cm. long, angled and with minute recurved prickles; lvs. largely alternate, broadly ovate in outline, sparsely hirsute, 1–3 cm. long, pinnately divided into 3–7 lance-oblong falcate acute divisions, these mostly entire, glaucous beneath; petioles ca. as long; pedicels short; calyx-lobes narrow-lanceolate, 3 mm. long, the reflexed auricles half as long; corolla narrow-campanulate, 1.5–3 mm. broad, whitish with purple veins; fils. shorter than tube, the appendages linear or cuneate or reduced to hairy lines; style to 1 mm. long; caps. 3–5 mm. in diam.; seed 1, globose, brick-red, 2–4 mm. in diam., regularly pitted in rows; cucullus persistent, reduced; $n = 9$ (Cave & Constance, 1942).—Moist places, 4000–7000 ft.; Sagebrush Scrub to Red Fir F.; Modoc Co.; to B.C., Rocky Mts. May–July.

2. **N. maculàta** Benth. ex Lindl. [*Viticella m.* Macbr.] FIVESPOT. Stems several, decumbent, loosely hairy or glabrate, 1–3 dm. long; lvs. opposite, oblong to oval in outline, 1–3 cm. long, pinnately and deeply 5–9-lobed (or the upper entire to 3-lobed), hirsute, the petioles equally long; pedicels stout, 2–5 cm. long; calyx-lobes lanceolate to almost ovate, 5–8 mm. long, the reflexed auricles 1–4 mm. long; corolla bowl-shaped to sub-rotate, 1.5–4.5 cm. broad, white and veined or dotted, the obovate lobes longer than tube and each with a large purple blotch; fils. slightly longer than corolla-tube; appendages linear to oblong, ciliate on free edge; style 3–6 mm. long; caps. slightly exceeding calyx; seeds 5–12, ovoid, smooth or shallowly pitted, ca. 2 mm. long, olive; cucullus deciduous, usually papillaeform; $n = 9$ (Cave & Constance, 1950).—More or less moist slopes and flats, below 7500 ft.; V. Grassland, Foothill Wd., Yellow Pine F., Red Fir F.; w. base and slopes of the Sierra Nevada from Plumas Co. to Kern Co. April–July.

3. **N. Menzièsii** H. & A. [*N. insignis* Dougl. ex Benth. *N. M.* ssp. *i.* Brand. *N. Brandegei* and *Evermannii* Eastw.] BABY BLUE-EYES. Diffuse, ± succulent, the stems obscurely winged or angled, pubescent, 1–3 dm. long, often growing up among other plants; lvs. opposite, oval to oblong in outline, mostly 2–5 cm. long, pinnately divided into usually 9–11 rounded to oblong divisions, these mostly again toothed, sparingly appressed-hispid; petioles shorter to as long; pedicels slender, longer than lvs.; calyx-lobes lanceolate, 4–6 mm. long, the auricles narrow, 1.5–2.5 mm. long; corolla semirotate to bowl-shaped,

1.5–4 cm. broad, typically bright blue with a light center, the lobes obovate to oblong, longer than the tube; fils. ca. as long as tube, the appendages broad or narrow, partly free or adherent along one edge or reduced to hairy lines; style 3–5 mm. long; caps. round to ovoid, 5–12 mm. in diam.; seeds usually 10–20, oblong to ovoid, dark, corrugate-tuberculate, ca. 2 mm. long, the cucullus deciduous, often papillaeform; $n = 9$ (Cave & Constance, 1942).—Moist flats and slopes, below 2500 (5000) ft.; Foothill Wd., Oak Wd., V. Grassland, Coastal Sage Scrub, Chaparral, Yellow Pine F., etc.; inner N. Coast Ranges from Tehama Co. s., foothills of Sierra Nevada from Butte Co. s., to San Diego Co. Feb.–June. A very variable complex; worthy of mention are: (1) var. *intermèdia* (Bioletti) Brand. [*N. i.* Bioletti. *N. insignis* var. *i.* Jeps. *N. liniflora* F. & M. *N. modesta* Kell.] Corolla pale blue, striped with conspicuous blue or purplish veins.—Outer Coast Ranges from Mendocino Co. s. to Santa Clara and San Mateo cos.; (2) var. *venòsa* (Jeps.) Brand. [*N. v.* Jeps.] Corolla with a large purple blotch on each lobe and in throat, the upper part of each lobe pale blue and conspicuously purple-veined.—S. Napa Range. The following are more distinct.

KEY TO SUBSPECIES

Upper lvs. like the lower and all divided into 5–13 lobes, these usually in turn toothed.
    Corollas with blue periphery or conspicuously dark-veined, often black-spotted at center; stems ±
      pubescent .... *N. Menziesii*
    Corollas white with conspicuous dark dots nearly to margin; stems subglabrous ...... ssp. *atomaria*
Upper lvs. not like the lower, subentire or shallowly few-lobed, the lobes entire .... ssp. *integrifolia*

Ssp. **atomària** (F. & M.) Brand. [*N. a.* F. & M. *N. M.* var. *a.* Chandl. *N. Johnsonii* and *macrocarpa* Eastw.] Plant succulent, subglabrous; corolla 1.5–3 cm. broad, white with black dots radiating from center toward margin; appendages usually reduced to hairy lines; $n = 9$ (Cave & Constance, 1947).—Moist places, below 2500 ft.; Redwood F., Mixed Evergreen F., N. Oak Wd.; near the coast from Santa Clara Co. to Del Norte and Siskiyou cos.; Ore. March–June.

Ssp. **integrifòlia** (Parish) Munz. [*N. M.* var. *i.* Parish. *N. i.* Abrams.] Lower lvs. with ca. 5–7 mostly entire lobes; upper lvs. entire or 3-lobed, rhomboid, spatulate or oblong; corolla subrotate, mostly pale blue, 7–10 mm. long, mostly with narrow scales with very long hairs; $n = 9$ (Cave & Constance, 1947).—Shaded damp slopes, 2500–6000 ft.; Chaparral, Yellow Pine F., S. Oak Wd.; e. San Gabriel and San Bernardino mts. to e. San Diego Co. April–June. Running into and poorly separated from the following tendencies: var. *incàna* Brand. [*N. M.* var. *minima* Brand.] Plant quite densely villous; upper lvs. less different from lower; corolla 6–10 mm. long, pale blue, the appendages linear, with short hairs.—At 500–5000 ft.; San Bernardino Co. to Ventura Co.; var. *rotàta* (Eastw.) Chandl. [*N. r.* Eastw.] Lvs. commonly 3–5-lobed; corolla light blue, 3–6 mm. long, the appendages triangular, attached only at base, with hairs longer than scales.— Chaparral, Coastal Sage Scrub; Santa Ana Mts. to San Diego; var. *annulàta* Chandl. Fls. deep blue, 3–6(–10) mm. long; appendages oblong, attached along one side, with very short hairs.—Between 3000–5000 ft.; Creosote Bush Scrub, Joshua Tree Wd.; Mojave Desert from Victorville region to Providence Mts. April–May.

4. **N. pedunculàta** Dougl. ex Benth. [*Viticella p.* Macbr. *N. sepulta* Parish. *N. humifusa* Kell. *N. alata, exigua, insularis* and *nana* Eastw.] Stems weak, low, often prostrate, sparingly pubescent, 1–3 dm. long; lvs. all opposite, oblong to oval in outline, 1–3 cm. long, deeply pinnately divided into 5–9 short oblong to obovate divisions, these appressed-hispid, entire to 1–2-toothed; petioles to as long as blades, expanded upward into the cuneate lf.-base; pedicels short, axillary; calyx-lobes narrowly to broadly lanceolate, 1–3 mm. long, the reflexed auricles ca. as long; corolla campanulate, 3–6 mm. broad, white to pale blue, mostly dark-spotted or -veined, each lobe with purple blotch; appendages linear or reduced to hairy lines; style barely 1 mm. long; caps. exceeding calyx; seeds mostly 2–8, ovoid, olive, smooth to slightly corrugate, 1–4 mm. long, the deciduous cucullus usually papillaeform; $n = 9$ (Cave & Constance, 1942).—Moist open or shaded places, below 7500 ft.; many Plant Communities; much of cismontane Calif., Modoc and Lassen cos.; to B.C., Nev., L. Calif. April–Aug.

5. **N. spatulàta** Cov. [*Viticella s.* Macbr. *N. Congdonii, humilis,* and *pratensis* Eastw.] Stems few, weak, 1–2 dm. long, ± hispidulous; lvs. all opposite, thin, 1–3 cm. long, cuneate and passing into the equally long petioles, 3–5-lobed or -toothed, appressed-

hairy; pedicels rather short; calyx-lobes 2.5–5 mm. long, the reflexed auricles 1–2 mm. long; corolla white or bluish, shallowly bowl-shaped, often dotted, sometimes with purple blotches on the lobes; fils. shorter than tube; appendages broad to narrow or lineate; caps. longer than calyx; seeds mostly 5–6, light brown, ovoid, shallowly pitted, 3 mm. long, the cucullus deciduous, mostly papillaeform; *n* = 9 (Cave & Constance, 1947).—In shaded damp places, 4000–10,500 ft.; Yellow Pine F. to Lodgepole F.; mts. from Riverside Co. to Plumas and e. Tehama cos.; w. Nev. May–July.

6. **N. pulchélla** Eastw. [*Viticella p.* Macbr.] Stems ± hirsute, slender, 1–4 dm. long; lvs. all opposite or sometimes the upper alternate, oblong to ovate in outline, 2–4 cm. long, on petioles ca. as long, pinnately divided into 5–7 remote divisions, these rounded, usually toothed, uppermost lvs. entire to shallowly lobed; pedicels slender; calyx-lobes lance-oblong, 2–4 mm. long, the auricles scarcely if at all evident; corolla 5–12 mm. broad, rotate, deep blue to violet, with white center, the rounded lobes longer than tube; fils. exceeding tube; appendages linear to obsolete; style 2–3 mm. long, evidently exserted; caps. rounded; seeds 2–4, brown to olive, ± smooth, ca. 1.8 mm. long, the cucullus deciduous, rather shallow; *n* = 9 (Cave & Constance, 1942).—Partial shade, moist places, below 5500 ft.; Yellow Pine F. to Foothill Wd.; w. base of Sierra Nevada, Madera Co. to Kern Co. April–June.

Var. **grácilis** (Eastw.) Const. [*N. g.* Eastw.] Most lvs. alternate, shallowly lobed; corolla white, scarcely exceeding calyx; style 1–1.5 mm. long; seed usually 1.—Light shade, largely Foothill Wd.; w. base of Sierra Nevada from Mariposa Co. to Kern Co.

Var. **Fremóntii** (Elmer) Const. [*N. F.* Elmer.] Lvs. all opposite; corolla white, rotate, ca. as long as calyx; style 0.5–1 mm. long; seeds usually 2–4.—Largely Foothill Wd.; Kern R. region, Tehachapi Mts., inner Coast Ranges to Monterey and Stanislaus cos.

7. **N. heterophýlla** F. & M. [*Viticella h.* Macbr. *N. decumbens, divaricata, diversifolia, exilis, fallax, flaccida, glauca, hispida, inaequalis, nemorensis,* and *tenera* Eastw.] Stems slender, hirsute or hirsutulous, ± erect, 1–3 dm. long; lower lvs. opposite, ovate to oblong in outline, 1.5–2.5 cm. long, on petioles ca. as long, 5–7 pinnate into rounded rather remote divisions, these entire or 1–3-toothed; upper lvs. alternate entire to 3–5-lobed; pedicels slender, mostly longer than lvs.; calyx-lobes lance-ovate, 2.5–3.5 mm. long, the spreading or reflexed auricles 0.5 mm. long; corolla basin-shaped, white or bluish, 5–10 mm. broad, with obovate lobes; appendages triangular to lineate; style 2.5–3.5 mm. long; caps. rounded, 3–5 mm. in diam.; seeds 2–4, yellow-brown, ovoid, ± smooth, 1–2 mm. long, the cucullus deciduous, often papillaeform; *n* = 9 (Cave & Constance, 1942). —Light shade, slopes and canyons, below 5000 ft.; Chaparral, Foothill Wd., N. Oak Wd., Yellow Pine F.; inner Coast Ranges from San Benito Co. n., Sierra Nevada from Madera Co. n.; sw. Ore. March–July.

8. **N. parviflòra** Dougl. ex Benth. [*Viticella p.* Macbr. *N. macrophylla, Kelloggii, micrantha* and *Plaskettii* Eastw.] Stems weak, hispid to subglabrous, 1–6 dm. long; lower lvs. opposite, roundish to ovate, 1–4 cm. long, usually 5-pinnate into entire or lobed divisions, ± appressed-hispid, on petioles ca. as long; upper lvs. alternate, subsessile; pedicels often not longer than lvs.; calyx-lobes ± lanceolate, 1–3 mm. long, the reflexed auricles ca. 1 mm. long; corolla campanulate, white to bluish, 2–4 mm. broad; appendages linear; fils. shorter than tube; style ca. 1 mm. long; caps. round, 3–5 mm. in diam.; seeds mostly 2–4, yellow to reddish, shallowly pitted, ovoid, ca. 2 mm. long, the cucullus deciduous, ± papillaeform; *n* = 9 (Cave & Constance, 1942).—Moist shade, below 6000 ft.; Redwood F., Mixed Evergreen F., Yellow Pine F., Foothill Wd.; Coast Ranges from Monterey Co. n.; to Wash. April–June.

Var. **Austínae** (Eastw.) Brand. [*N. A.* Eastw.] Thinly hirsute; lvs. mostly opposite, the lower roundish, shallowly lobed; auricles less than 0.5 mm. long; corolla barely exceeding calyx, 2–3 mm. broad; *n* = 9 (Cave & Constance, 1942).—Between 3500–7000 ft.; Red Fir F., Yellow Pine F.; Sierra Co. to Modoc, Siskiyou, and Trinity cos.; to Wash., Ida., Utah. May–June.

Var. **quercifòlia** (Eastw.) Chandl. [*N. quercifolia* Eastw.] Densely villous; lvs. all opposite, roundish; corolla 3–5 mm. broad; *n* = 9 (Cave & Constance, 1942).—Dry shade, 2500–6500 ft.; Foothill Wd., Yellow Pine F.; Kern Co. to Madera Co. May–June.

5. Phacèlia Juss.

Herbs, varying from annual to perennial, usually ± pubescent and often glandular. Lvs. mostly alternate, ranging from entire or dentate to pinnate or bipinnate. Fls. few to many, in dense to lax (especially in fr.) simple or branched cymes. Calyx lobed almost to base. Corolla rotate to campanulate or tubular, ± lobed, purple or blue to white or yellow, deciduous or withering-persistent. Corolla-scales attached in pairs to the corolla-tube at base of each fil. and ± adnate to tube, sometimes absent. Stamens equally inserted at base of corolla-tube, ± equal, included or exserted. Style bifid at apex to divided almost to base. Caps. unilocular or nearly bilocular, round to ovoid or oblong, with 2–many ovules on the 2 linear placentae. Seeds round to oblong or ovoid, reticulate, pitted or transversely corrugate, sometimes excavated on each side of a ridge. Perhaps 200 spp., of New World, especially w. N. Am.; some of horticultural value. (Greek, *phakelos,* a cluster, because of crowded fls.)

A. Plants perennial or biennial.
  B. Ovules several to many on each placenta.
    C. Lf-blades round or nearly so, 1–2 cm. long; stems white-woolly below. Deserts
          7. *P. perityloides*
    CC. Lf.-blades longer than broad, mostly more than 2 cm. long; stems not white-woolly.
      D. Corolla-lobes revolute; stamens well exserted.
        E. Stems 1–3 dm. high; fls. in subcapitate cymes . . . . 1. *P. hydrophylloides*
        EE. Stems 5–15 dm. high; fls. in a panicle of cymes . . . . 2. *P. procera*
      DD. Corolla-lobes plane.
        E. Stems simple; infl. narrow, thyrsoid; corolla withering-persistent, 5–6 mm. long; stamens exserted. Modoc Co. . . . . 3. *P. sericea*
        EE. Stems branched; infl. a corymbose panicle; corolla deciduous, 10–12 mm. long; stamens barely exserted. Nw. coast . . . . 4. *P. Bolanderi*
  BB. Ovules 2 to each placenta.
    C. Lvs. pinnate into toothed or pinnatifid lobes; calyx-lobes mostly spatulate.
      D. Calyx 5–6 mm. long; corolla 6–8 mm. long. Widespread . . . . . 5. *P. ramosissima*
      DD. Calyx 2–3 mm. long; corolla 3–5 mm. long. San Nicolas Id. . . . . . . 6. *P. cinerea*
    CC. Lvs. entire to pinnate into entire lobes or lfts.; calyx-lobes mostly not widened upward.
      D. The lvs. without conspicuous parallel veins; corolla subrotate; stamens not conspicuously exserted; fls. few, the cymes simple, scarcely scorpioid. Scott Mts., Trinity Co. . . . . 35. *P. Dalesiana*
      DD. The lvs. with conspicuous parallel lateral veins; corolla ± campanulate; stamens mostly well exserted; fls. many, in crowded often branched scorpioid cymes.
        E. Infl. mostly elongate, narrow, dense, consisting of many short lateral cymes along a cent. axis; stems mostly simple.
          F. Lower lvs. mostly with 2–several pairs of lateral lfts.
            G. Plants greenish, biennial; calyx-lobes linear to linear-oblong. N. Calif. . . . . 29. *P. heterophylla*
            GG. Plants grayish, perennial; calyx-lobes oblanceolate to ovate. S. Calif. 25. *P. imbricata* ssp. *bernardina*
          FF. Lower lvs. mostly entire, or with 1 pair of basal pinnae; plants perennial.
            G. Lvs. silvery, mostly entire; hairs mostly appressed. N. montane 30. *P. hastata*
            GG. Lvs. greenish, drying brownish, often with 1 (2) pairs of lateral lfts.; hairs spreading, stinging. Coastal . . . . 31. *P. nemoralis*
        EE. Infl. usually shorter, more open, widely branched; stems often branched except in reduced and alpine forms.
          F. Lvs. mostly entire, silvery to gray-green.
            G. Stems appressed-hirsute, the lf.-blades roundish to oval; calyx-lobes oval to oblong. N. Coast . . . . 28. *P. argentea*
            GG. Stems somewhat hispid; lf.-blades lanceolate; calyx-lobes linear. High montane.
              H. Stems mostly less than 2 dm. tall . . . . 34. *P. frigida*
              HH. Stems mostly over 2 dm. tall . . . . 32. *P. mutabilis*
          FF. Lvs. mostly lobed, green to gray but not silvery.
            G. Calyx-lobes broadly lanceolate to ovate, overlapping in fr. Los Angeles Co. to n. Calif. . . . . 25. *P. imbricata*
            GG. Calyx-lobes linear-lanceolate to linear, not overlapping in fr.
              H. Plants not conspicuously cespitose; stems not very glandular.
                I. Lvs. grayish, the cauline mostly pinnatifid.
                  J. Corolla blue to purplish or lavender, 5–6 mm. long. Santa Clara Co. to Del Norte Co. . . 26. *P. californica*
                  JJ. Corolla white, 7–9 mm. long. Tehama and Kern cos. n. 27. *P. egena*
                II. Lvs. greenish, the cauline mostly entire. Montane, from Kern and Lake cos. n. . . . . 32. *P. mutabilis*

                    HH. Plants conspicuously cespitose from much branched caudices
                         and making patches; stems glandular .... 33. *P. corymbosa*

AA. Plants annual.
  B. Seeds not transversely corrugated, or if so, the ovules only 2 to a placenta.
    C. Lvs. pinnately toothed to compound, their divisions further toothed or divided.
      D. Seeds 10–40, less than 1 mm. long.
        E. Lower lvs. ovate, scattered; stems 3–10 dm. tall; fls. sessile. Insular
                                                7. *P. Lyonii*
        EE. Lower lvs. oblong, rosulate; stems 1–4 dm. long; fls. plainly pedicelled.
           Widely distributed .... 53. *P. Douglasii*
      DD. Seeds 1–8, 1.5–3 mm. long.
        E. Calyx-lobes pinnatifid into 3–5 segms. San Clemente Id. .. 8. *P. floribunda*
        EE. Calyx-lobes entire or at the most crenate.
          F. Lvs. not deeply divided, the blades ovate to rounded, shallowly lobed;
            stems prickly-hispid.
            G. Stamens 5–10 mm. long, exserted; corolla broadly campanulate,
               5–7 mm. long. Coastal .... 15. *P. malvifolia*
            GG. Stamens 2–3 mm. long, included; corolla narrow-campanulate, 4–5
               mm. long. Mostly away from immediate coast .... 16. *P. Rattanii*
        FF. Lvs. mostly deeply divided nearly or quite to midrib; stems mostly less
          hispid.
          G. Calyx much enlarged and ± chartaceous and veiny in fr., the lobes
            ovate to lance-oblong.
            H. Corolla broadly campanulate, 8–10 mm. long; stamens 9–13
               mm. long. Rather general in cismontane Calif. .. 17. *P. ciliata*
            HH. Corolla more tubular, 3–4 mm. long; stamens 1.5–2 mm.
               long. Lassen, Shasta, and Modoc cos. .... 18. *P. thermalis*
          GG. Calyx not much enlarged in fr., or if so, not coriaceous and veiny,
            the lobes then linear to spatulate.
            H. Calyx-lobes dimorphic, 2 ovate and entire to crenate, the
             other 3 lanceolate to narrow-spatulate. Sierran foothills.
                                         10. *P. platyloba*
            HH. Calyx-lobes not dimorphic, always entire, sometimes somewhat
             unequal.
             I. Seeds terete or angled, usually pitted, but not excavated
              on either side of a salient ridge; plants not markedly
              viscid or ill-scented.
              J. Corolla lavender or white; calyx-lobes linear or nearly
               so, up to 1 cm. long in fr., often conspicuously clawed
               and loosely enveloping the caps.
               K. Corolla open-campanulate, 8–12 mm. long, much
                 exceeding calyx.
                 L. Fls. many, crowded; caps. with pustulate
                   bristles; stems mostly not strongly purple.
                   Mostly cismontane ..... 12. *P. cicutaria*
                 LL. Fls. few, remote; caps. hirsutulous; stems
                   mostly purple. Mostly deserts
                          14. *P. vallis-mortae*
               KK. Corolla tubular-campanulate, 4–7 mm. long, ca.
                 as long as calyx. Deserts .... 13. *P. cryptantha*
             JJ. Corolla blue or bluish; calyx-lobes linear to obovate,
              shorter than 1 cm. and closely investing the caps.
              K. Calyx-lobes mostly oblanceolate to obovate, 4–5
                mm. long; corolla promptly deciduous; caps. hairy
                in whole upper half; stamens mostly not long-
                exserted .... 9. *P. distans*
              KK. Calyx-lobes linear to lance-linear, 6–8 mm. long;
                corolla tardily deciduous; caps. hairy only at tip;
                stamens long-exserted ...... 11. *P. tanacetifolia*
          II. Seeds flattened, excavated on each side of a salient ridge;
            plants very viscid, ill-scented.
            J. Stamens and style plainly exserted.
              K. Pedicels shorter than calyx; lobes of lf.-blades ±
               oblong.
                L. Corolla 6–10 mm. long, open-campanulate.
                 M. Corolla mostly purple to violet; seeds
                   thick, transversely corrugated. Wide-
                   spread on deserts . .. 19. *P. crenulata*
                MM. Corolla white; seeds thin, not corru-
                 gated. Saline V., Inyo Co.
                             21. *P. amabilis*
               LL. Corolla 4–5 mm. long, tubular-campanulate
                         20. *P. minutiflora*
              KK. Pedicels at least as long as calyx; lobes of lf.-
               blades almost round .... 22. *P. pedicellata*
            JJ. Stamens and style included.

K. Stems greenish; calyx-lobes oblanceolate, 3–4 mm. long; corolla 6 mm. long   23. *P. Anelsonii*

KK. Stems purplish; calyx-lobes lanceolate, 2–3 mm. long; corolla 3–4 mm. long .... 24. *P. coerulea*

CC. Lvs. entire to pinnate, but then with entire divisions.

D. Corolla-scales present, attached to tube at base of each fil.

E. Corolla rotate to campanulate.

F. Infl. lax, scarcely scorpioid.

G. Lvs. ± glabrous, glaucous, mostly opposite ...... 38. *P. racemosa*

GG. Lvs. pubescent, only the lower opposite.

H. Plants glandular-puberulent throughout. Trinity and Siskiyou cos.

I. Corolla 3–5 mm. long; stamens 3–4 mm. long
36. *P. Pringlei*

II. Corolla 2–3 mm. long; stamens 1.7–2 mm. long
37. *P. Leonis*

HH. Plants soft hairy, not very glandular. Sierra Nevada.

I. Corolla 2–4 mm. long; stamens usually included
39. *P. Eisenii*

II. Corolla 4–6 mm. long; stamens usually exserted
40. *P. orogenes*

FF. Infl. usually dense, always scorpioid.

G. Corolla tardily deciduous; style glandular-pubescent.

H. Lvs. elliptic to oval. Nevada and Placer cos.
43. *P. marcescens*

HH. Lvs. linear-oblong.

I. Calyx-lobes unequal in fr.; caps. subglobose. Eldorado Co. to Fresno Co. .... 44. *P. Quickii*

II. Calyx-lobes subequal; caps. ovoid. Siskiyou Co.
45. *P. Greenei*

GG. Corolla promptly deciduous; style not glandular.

H. Plants glandular.

I. Calyx-lobes obovate to narrow-spatulate; lvs. oblong to ovate.

J. Plant hirsute; corolla nearly white, 5–6 mm. long. San Luis Obispo and Monterey cos. . 55. *P. grisea*

JJ. Plant hirsutulous; corolla lavender to violet, 6–7 mm. long. Kern Co. to Modoc Co. ...... 56. *P. Purpusii*

II. Calyx-lobes linear to linear-oblanceolate.

J. Lvs. linear to oblanceolate, mostly entire; corolla 5–8 mm. long. Tulare Co. to Los Angeles Co.
46. *P. mohavensis*

JJ. Lvs. broader, the lower 1–6-pinnate; corolla 3–4 mm. long. Sw. Inyo Co. .... 48. *P. novenmillensis*

HH. Plants not glandular except sometimes in infl.

I. Pedicels elongating noticeably in fr., spreading or recurving; seeds not over 1 mm. long.

J. Lvs. mostly entire, or if pinnate, plants tending to be montane and with the corolla 10–15 mm. long; seeds 0.5–1 mm. long .... 52. *P. curvipes*

JJ. Lvs. mostly pinnate to pinnatifid, mostly below the Yellow Pine belt and often with a smaller corolla; seeds ca. 1 mm. long .... 53. *P. Douglasii*

II. Pedicels not noticeably elongated in fr., erect or ascending.

J. Corolla 6–15 mm. long.

K. Lvs. linear to lanceolate; cymes in thyrsoid panicles; caps. terete. Siskiyou Co. to Modoc and Plumas cos. .... 49. *P. linearis*

KK. Lvs. elliptic to narrow-ovate; cymes simple or few-branched; caps. flattened. Mendocino Co. to Santa Barbara Co., w. base of Sierra Nevada
54. *P. divaricata*

JJ. Corolla 3–6 mm. long.

K. Flowers white to lavender or light blue.

L. Calyx-lobes 2–3 mm. long in fl., 4–5 mm. in fr.

M. Style 2–3 mm. long; stamens 4–4.5 mm. long, glabrous. Contra Costa Co. to w. Fresno Co.   41. *P. Breweri*

MM. Style 1–2 mm. long; stamens 2–4 mm. long, puberulent. Montane, Kern Co. to Riverside Co. 47. *P. austromontana*

LL. Calyx-lobes 4–5 mm. long in fl., 10–12 mm. in fr. Mt. Diablo to Mt. Hamilton
51. *P. phacelioides*

KK. Fls. violet or purple.

                    L.   Calyx-lobes linear to linear-oblong; corolla
violet; style 4–8 mm. long, subglabrous.
Kern Co. to Modoc Co. e. of Sierra Nevada
42. *P. humilis*

         LL. Calyx-lobes obovate to narrow-spatulate;
corolla purple; style 2.5 mm. long, pubescent.
W. base of Sierra Nevada . . 50. *P. vallicola*

EE. Corolla tubular to tubular-campanulate.

    F.   Style parted to the middle or farther.

        G.   Plants glabrous or nearly so and glaucous below the infl.; all but
uppermost lvs. opposite . . . . 38. *P. racemosa*

       GG. Plants pubescent throughout, not glaucous; most cauline lvs. alter-
nate.

           H.   Plants prickly-hispid; lvs. flaccid, shallowly lobed, the lobes
crenate to dentate. From San Luis Obispo Co. n.
16. *P. Rattanii*

          HH. Plants hirsutulous to hirsute, usually glandular; lvs. firm,
entire or with few basal lobes.

             I.   Caps. terete; corolla purple; stamens papillate. W. base of
Sierra Nevada . . . . 50. *P. vallicola*

            II. Caps. flattened; corolla white to pale lavender; stamens
hairy. Mt. Diablo to Mt. Hamilton . . 51. *P. phacelioides*

    FF. Style bifid at apex only or not more than ⅓ its length.

        G.   Corolla 6–14 mm. long.

           H.   Stamens glabrous; seeds 1–1.5 mm. long. Cismontane
58. *P. suaveolens*

          HH. Stamens ± pubescent; seeds 0.5–0.7 mm. long. Deserts.

             I.   Cymes not pedunculate. Alkaline flats, near Kingston, e.
Mojave Desert . . . . 59. *P. pulchella*

            II. Cymes pedunculate. Rocky places, Death V. region
61. *P. mustelina*

       GG. Corolla 3–6 mm. long.

           H.   Petioles ca. as long as or longer than blades; fls. in lax cymes.

             I.   Corolla tardily deciduous, 3–4 mm. long; seeds roundish
66. *P. saxicola*

            II. Corolla early deciduous, 4.5–5 mm. long; seeds oblong
to ovoid.

               J.   Lf.-blades roundish.

                 K.   Lf.-blades plainly toothed, 0.5–2 cm. wide; stems
hirsutulous and glandular; stamens glabrous. Inyo
Co. to Riverside Co. . . . . . . 60. *P. rotundifolia*

                KK. Lf.-blades less toothed, 1.5–3.5 cm. wide; stems
scarcely if at all hirsutulous; stamens hairy at
base. Westgard Pass to Mono Co.
62. *P. Peirsoniana*

            JJ. Lf.-blades oblong-oval to ovate; stamens glabrous.
Inyo Mts. to Clark Mts. . . . . 63. *P. Barnebyana*

          HH. Petioles shorter than lf.-blades; fls. in compact cymes.

             I.   Cymes pedunculate; seeds ca. 25 and 1–1.5 mm. long
65. *P. Parishii*

            II. Cymes subsessile; seeds 50–100, about 0.5 mm. long
64. *P. Lemmonii*

DD. Corolla-scales lacking, the filament-bases sometimes dilated or winged.

    E.   Fils. with a dilation at base; style parted to the middle.

        F.   Corolla open-campanulate, the tube ca. as long as the limb.

           G.   Fls. purple to white.

             H.   Corolla purple to violet, with paler center, 10–20 mm. long;
fil.-dilations pubescent. Mostly below 3000 ft. . 67. *P. Parryi*

            HH. Corolla white, sometimes bluish, 7–12 mm. long; fil.-dilations
glabrous. From 3000–8000 ft. . . . . 69. *P. longipes*

          GG. Fls. deep blue. W. Mojave Desert . . . . 68. *P. Nashiana*

      FF. Corolla tubular-campanulate or campanulate-funnelform, the tube ca.
twice as long as the limb.

        G.   The corolla purple, slightly constricted at throat. Cismontane
70. *P. minor*

        GG. The corolla deep blue, not constricted at throat. Deserts
71. *P. campanularia*

    EE. Fils. without a basal dilation; style parted to below the middle.

        F.   Cymes lax in fr., the pedicels becoming 1–2 cm. long; corolla 8–18 mm.
long, blue with purplish or whitish center; fils. pubescent . 72. *P. viscida*

        FF. Cymes dense, the pedicels less than 1 cm. long; corolla 12–30 mm.
long, the center colored like periphery; fils. glabrous . 73. *P. grandiflora*

BB. Seeds transversely corrugated; ovules many.

  C.   Lvs. entire to crenate-dentate, not deeply lobed. Deserts.

    D.   Corolla 5–7 mm. long; style 2–3 mm. long.

      E.   Plant with black-headed glands throughout; racemes subpedunculate
74. *P. pachyphylla*

EE. Plants grayish pubescent, without black-headed glands; racemes ± concealed
by foliage .... 76. *P. neglecta*
DD. Corolla ca. 10 mm. long; style 5–6 mm. long .... 75. *P. calthifolia*
CC. Lvs. pinnate to bipinnatifid.
D. Corolla early deciduous, the limb not yellow.
E. The corolla 2.5–6.5 mm. long (the larger fls. from ne. Calif.), not longer
than calyx.
F. Plants not with conspicuous dark stalked glands throughout. Kern and
Inyo cos. s.
G. Calyx-lobes linear to narrow-oblanceolate; corolla white with yel-
lowish tube .... 77. *P. Ivesiana*
GG. Calyx-lobes spatulate; corolla lavender with yellowish tube
79. *P. affinis*
FF. Plants with conspicuous dark stalked glands throughout; corolla lavender
with yellowish tube. Mono Co. to Modoc and Siskiyou cos.
78. *P. glandulifera*
EE. The corolla mostly 7–15 mm. long, at least twice as long as calyx.
F. Lvs. deeply pinnately lobed to bipinnatifid; seeds 12–20.
G. The lvs. bipinnatifid; calyx-lobes linear to narrow-oblanceolate;
stamens usually pubescent near base. San Bernardino Co. to Modoc
Co. .... 80. *P. bicolor*
GG. The lvs. pinnate or pinnatifid; calyx-lobes spatulate; fils. glabrous.
H. Corolla bright blue to deep lavender. Deserts
81. *P. Fremontii*
HH. Corolla white to pink. Cismontane ...... 82. *P. brachyloba*
FF. Lvs. shallowly pinnately lobed; seeds 5–8. Inyo and Mono cos.
83. *P. gymnoclada*
DD. Corolla withering-persistent, the limb ± yellow; corolla enclosing the maturing
caps.
E. The corolla 4–7 mm. long.
F. Corolla-tube pubescent within; fils. pubescent; style 2–4 mm. long.
Lassen and Modoc cos. .... 84. *P. adenophora*
FF. Corolla-tube glabrous within; fils. glabrous; style ca. 1 mm. long. Lassen
and Modoc cos. .... 85. *P. inundata*
EE. The corolla 1.5–3.5 mm. long.
F. Corolla pale yellow, 3–3.5 mm. long; stamens 1.5–2 mm. long; fls. 5–
merous. Inyo Co. .... 86. *P. inyoensis*
FF. Corolla whitish, 1.5–2 mm. long; stamens 1 mm. long; fls. mostly 4–
merous. Mono Co to Modoc Co. .... 87. *P. tetramera*

1. **P. hydrophylloìdes** Torr. ex Gray. Perennial with a thick woody taproot and few
to several spreading to decumbent stems 1–3 dm. high and ± branched, with both coarse
and finer hairs, as well as some gland-tipped ones in upper parts; lvs. equably distributed,
oblong-ovate in outline, softly strigose, the blades 1.5–6 cm. long, coarsely toothed or
lobed or the basal pinnatifid, shorter to longer than petioles; fls. in subcapitate cymes 1–3
cm. long, on peduncles 1–7 cm. long; pedicels 1–2 mm. long; calyx-lobes oblong-linear,
3–5 mm. long (to 7 mm. in fr.), hirsute on margins; corolla violet-blue to whitish,
broadly campanulate, 5–6 mm. long, the ovate lobes spreading, pointed because of the
revolute margins; scales adnate, oblong to oval, united over base of fil.; stamens and
style well exserted; style 8–10 mm. long, cleft to ca. middle; caps. ovoid, 5–6 mm. long;
seeds 3–8, dark brown, ca. 2 mm. long, oblong-ovoid, pitted-reticulate; $n = 11$ (Cave &
Constance, 1947).—Occasional, dry open woods, 5000–9800 ft.; Yellow Pine F., Red
Fir F., Lodgepole F.; Sierra Nevada from Tulare Co. n.; Ore., Nev. June–Aug.

2. **P. procèra** Gray. Perennial from heavy root-crown; stems stout, simple, erect,
5–15(–20) dm. high, equably leafy, finely pubescent; lvs. alternate, ovate, the lower
blades 5–12 cm. long, incised to pinnatifid into few lanceolate mostly entire lobes, finely
pubescent on both surfaces, paler beneath; lower petioles ca. as long; cauline lvs.
gradually reduced upward; infl. a compact terminal panicle of cymes, glandular-
pubescent, 5–12(–16) cm. long; fls. subsessile to short-pedicelled; calyx-lobes linear,
3–5 mm. long at anthesis, 7 mm. in fr., finely pubescent; corolla brownish to greenish-
white, open-campanulate, 5–6 mm. long, the lobes ovate, obtuse, revolute; scales oblong,
adnate; stamens exserted; style ca. 10 mm. long, cleft halfway; caps. ovoid, glandular,
6–7 mm. long; seeds 12–16, dark, 3-angled, ca. 2 mm. long, reticulate-pitted; $n = 11$
(Cave & Constance, 1942).—Stony places, about meadows, etc., 4000–7000 ft.; Yellow
Pine F., Red Fir F.; Coast Ranges from Glenn, Trinity, and Tehama cos. to Del Norte
and Siskiyou cos., s. in the Sierra Nevada to Placer Co.; to Wash., Ida. June–Aug.

3. **P. serícea** (Grah.) Gray. [*Eutoca s.* Grah.] Perennial with woody root-crown;
stems equably leafy, 1–4 dm. high; herbage appressed-silky throughout, often silvery;
lf.-blades oblong to ovate in outline, 2–6 cm. long, pinnately incised to pinnatifid, the

lobes lance-oblong, entire to incisely toothed; lower petioles to ca. as long as blades, upper lvs. reduced and subsessile; fls. many in a crowded narrow panicle of short cymes, 3–12 cm. long, subsessile; calyx-lobes linear-oblong, 3–4 mm. long; corolla bluish-purple to whitish, open-campanulate, 5–6 mm. long; scales adnate, oblong, free from fils.; stamens well exserted, 10–15 mm. long, hairy below; style as long, purplish, deeply cleft; caps. ovoid, 4–5 mm. long, hairy; seeds 8–19, black, ovoid-angled, ca. 2 mm. long, pitted-reticulate; $n = 11$ (Cave & Constance, 1942).—Talus slopes, 7000–8500 ft.; Red Fir F.; Modoc Co.; to B.C., Rocky Mts. June–July.

4. **P. Bolánderi** Gray. Perennial from a root-crown; stems few, decumbent or ascending, 3–6(–10) dm. long, mostly hirsute and pubescent, glandular-pubescent especially above, usually branched; lvs. alternate, oblong to broadly ovate, the blades 5–10 cm. long, coarsely toothed to lobed, sometimes with basal pair of lobes; lower petioles ca. as long; upper cauline lvs. reduced; infl. a corymbose panicle of few racemes, hirsute and glandular-pubescent; pedicels 1–3 mm. long; calyx-lobes oblong to almost linear, acute to obtuse, 6–7 mm. long (8–9 in fr.), hispid and glandular; corolla lilac to pale blue, rotate-campanulate, 10–12 mm. long; scales adnate, narrow; stamens very slightly exserted, the fils. pilose; style ca. 1 cm. long, deeply cleft; caps. ovoid, 6–8 mm. long, pubescent; seeds many, dark, angled, 1–1.5 mm. long, foveolate.—Dry to moist places, below 1000 ft.; Redwood F., Douglas-Fir F., etc.; immediate coast, Sonoma Co. to Ore., inland to Siskiyou Co. May–July.

5. **P. ramosíssima** Dougl. ex Lehm. [*P. decumbens* Greene.] Perennial with rather woody root-crown, with few to several branching stems 5–10 dm. long, rather coarse, puberulent, glandular-pubescent and with some longer rather fine hairs; lvs. oblong to broadly ovate in outline, 4–10 cm. long, pinnate, the lobes oval to oblong, toothed to pinnatifid, variously pubescent; petioles mostly quite short; infl. of few rather dense short cymes; fls. subsessile; calyx-lobes mostly spatulate, setose and glandular, 5–6 mm. long, subacute; corolla dirty-white to bluish, campanulate, 6–8 mm. long, deciduous; scales ovate, adnate only at fil.-base; stamens exserted, glabrous; style glabrous, exserted, deeply cleft; caps. 3–4 mm. long, ovoid, pubescent; seeds usually 2 or 4, dark, deeply pitted, somewhat oblong, 2–3 mm. long; $n = 11$ (Cave & Constance, 1942).—Rocky places, mostly below 5000 (7000) ft.; many Plant Communities; Santa Cruz Mts. and Fresno Co. n. on both sides of Sacramento V.; to Wash., Ida. May–July. Exceedingly variable, the following being among the most marked tendencies:

KEY TO VARIETIES

Main stems with some of the hairs gland-tipped.
  Stems finely pubescent and with some longer hairs, but these not setose. Fresno Co. n.
                                                     P. ramosissima
  Stems finely pubescent and with longer setalike hairs with basal pustules. S. Calif. away from the
    immediate coast .... var. *suffrutescens*
Main stems lacking gland-tipped hairs.
  Stems practically glabrous. E. of Sierra Nevada .... var. *eremophila*
  Stems not glabrous.
    The stems puberulent and with some longer fine hairs.
      Plants with slender stems mostly 2–4 dm. long. Sierra Nevada, mostly above 7000 ft.
                                                  var. *valida*
      Plants with coarse stems mostly considerably longer. Seacoast, from Point Conception to Marin
      Co. .... var. *montereyensis*
    The stems ± puberulent and setose-hispid with basal pustules. Seacoast s. of Point Conception
                                        var. *austrolitoralis*

Var. **suffrutéscens** Parry. [*P. s.* Parry. *P. bifurca* and *polystachya* Greene.] Stems 5–10 or more dm. long, ± suffrutescent, glandular-pubescent and hispid; lvs. 5–15 cm. long, pinnate with oblong, crenate to serrate divisions up to 2.5 cm. wide; calyx-lobes oblanceolate, 6–7 mm. long; $n = 10$ (Cave & Constance, 1947).—Frequent in canyons, below 8000 ft.; mostly Chaparral and Yellow Pine F.; interior s. Calif. from Tehachapi Mts. to inner Coast Ranges of Santa Clara Co. and to L. Calif., Ariz. May–Aug.

Var. **eremóphila** (Greene) Macbr. [*P. e.* Greene.] Stems practically glabrous except in the infl., mostly 3–10 dm. long.—Sagebrush Scrub, etc.; Mono Co. to Modoc and Lassen cos.; w. Nev. June–July.

Var. **válida** Peck. [*P. subsinuata* Greene?] Stems mostly 2–4 dm. long, slender, puberulent, with occasional longer fine hairs.—Dry slopes, 7000–9000 ft.; Yellow Pine F., Red Fir F.; Sierra Nevada from Tulare Co. n.; s. Ore. July–Aug.

Var. **montereyénsis** Munz. Stems coarse, to 1 m. or more long, puberulent and with some fine longer spreading hairs, glandular only in infl.—Mostly Coastal Strand; near Surf, Santa Barbara Co. to Marin Co., Santa Rosa Id., Santa Cruz Id. April–July.

Var. **austrolitorális** Munz. Like the last, but the long hairs coarse and with pustulate bases.—Coastal Strand; Santa Barbara Co. to L. Calif. May–Aug.

6. **P. cinèrea** Eastw. ex Macbr. Perennial ca. 5 dm. high, near to *P. ramosissima* but with lvs. more finely divided; calyx-lobes 2–3 mm. long; corolla blue, 3–5 mm. long; seeds 1.5 mm. long.—San Nicolas Id. March–April.

7. **P. Lyònii** Gray. Annual, densely glandular and heavy-scented, sparsely hispid and strigulose throughout; stems simple or branched, 3–10 dm. high; lvs. ovate, alternate, 5–10 cm. long, short-petioled, pinnate, then pinnatifid into crenate-dentate divisions; fls. many, in a short dense panicle of cymes; fls. subsessile; calyx-lobes spatulate, glandular and hirsute, ca. 5 mm. long in fl., 8 mm. in fr.; corolla pale blue, campanulate, 5–7 mm. long, deciduous; scales broadly ovate, attached to fil. at base; stamens included, 4–5 mm. long; style ca. 3 mm. long, deeply cleft; caps. oblong, 5–7 mm. long, pubescent; seeds many, flattened, oval, 0.5–0.7 mm. long, dark, pitted.—Rocky places; Chaparral, Coastal Sage Scrub; Santa Catalina and San Clemente ids. April–June.

8. **P. floribúnda** Greene. [*P. phyllomanica* var. *interrupta* Gray.] Annual 3–5 dm. high, puberulent and hirsute throughout, somewhat glandular above; lvs. ovate in outline, 5–12 cm. long, short-petioled, pinnate, then pinnatifid into crenate-dentate lobes; fls. many, subsessile, on cymes crowded into dense panicles; calyx-lobes short-hirsute, obovate in outline, 4–5 mm. long, pinnatifid into 3 or 5 oblong lobes, or 1 or 2 entire, narrow-spatulate; corolla pale blue, campanulate, 5–6 mm. long, deciduous; scales adnate, oblong-lanceolate; stamens 5–6 mm. long, slightly exserted; style included, deeply cleft; caps. ovoid, 2–3 mm. long, pilose; seeds 1–4, dark brown, oblong, ca. 1.5 mm. long, pitted.—Sheltered places; Coastal Sage Scrub; San Clemente, Santa Barbara, and Guadalupe ids. March–May.

9. **P. dístans** Benth. [*P. scabrella, Arthurii, leptostachys,* and *commixta* Greene. *P. ammophila* Greene ex Brand. *P. d.* var. *australis* Brand.] WILD-HELIOTROPE. Annual 2–8 dm. high, erect and simple or branched and decumbent, finely pubescent and scatteringly stiff-hairy, glandular mostly in upper parts; lvs. ovate to ovate-oblong in outline, 2–10 cm. long, short-petioled, 1–2-pinnate into oblong to lanceolate toothed to pinnatifid divisions; upper lvs. reduced, subsessile; cymes few, simple to somewhat branched; fls. many, subsessile; calyx-lobes mostly unequal, oblanceolate to almost linear, 4–5 mm. long, glandular-pubescent and hirsute; corolla blue or bluish, broadly campanulate, 6–8 mm. long, deciduous; scales half-ovate, free at tips; stamens included to short-exserted, the fils. glabrous to somewhat pubescent; style 7–12 mm. long, very deeply cleft, glabrous or short-pubescent below; caps. globose, 2–3 mm. long, somewhat hairy; seeds 2–4, brown, 2–2.5 mm. long, curved on back, triangular or half-circular in cross section, coarsely pitted; $n = 11$ (Cave & Constance, 1942).—Common, fields and slopes, mostly below 4000 ft. (7000 ft.); many Plant Communities; deserts n. to Mono Co., cismontane s. Calif., in Coast Ranges n. to Tehama and Mendocino cos., and in San Joaquin V. to San Joaquin Co.; Santa Barbara Ids.; to Nev., Ariz., Son., L. Calif. March–June. Variable as to division of lvs., width of calyx-segms., length of stamens, etc.

10. **P. platylòba** Gray. Resembling *P. distans* but more slender-stemmed, glandular-puberulent throughout; fls. fewer; calyx-lobes dimorphic, with 3 lanceolate to narrowly spatulate, 2–3 mm. long, entire, and with 2 broadly ovate, 2–3 mm. long and wide, entire or crenate to lobed, petiolulate; corolla 4–5 mm. long; caps. cylindrical, 2–2.5 mm. long; seed usually 1, dark brown, 2–2.5 mm. long, finely pitted; $n = 11$ (Cave & Constance, 1950).—Partly shaded brushy or rocky places, 1000–3500 ft.; Foothill Wd., Chaparral; foothills of Sierra Nevada from Mariposa Co. to Fresno Co. April–May.

11. **P. tanacetifòlia** Benth. [*P. t.* var. *cinerea* Brand, var. *pseudo-distans* Brand, subvar. *tenuisecta* Brand f. *staminea* Brand.] Resembling *P. distans,* glandular-puberulent and hirsute throughout, often stouter; fls. many, short-pedicelled in prominent compact cymes; calyx-lobes mostly linear, 6–8 mm. long, densely pubescent and hispid; corolla blue, broadly campanulate, 6–9 mm. long, quite persistent; scales usually completely adnate; stamens 1.5–2 times as long as corolla, glabrous; styles glabrous, deeply cleft; caps. ovoid, 3–4 mm. long, pubescent at apex; seeds usually 2, grayish-brown, convex on outer surface, flattened on inner, 2–3 mm. long, deeply and coarsely pitted in trans-

verse rows; n = 11 (Cave & Constance, 1942).—Open flats and slopes, below 4000 ft. (6000 in desert); many Plant Communities; Cent. V. and bordering ranges from Lake and Butte cos. s. to L. Calif., Mojave Desert; Santa Cruz Id.; Nev., Ariz. March–May.

12. **P. cicutària** Greene. [*P. hispida* var. *c.* Macbr. *P. heterosepala* Greene.] Annual 2–6 dm. high, setose-hispid and pubescent throughout and somewhat glandular-pubescent; stems erect or ascending or weaker in shade; lvs. ovate to oblong-ovate in outline, 4–15 cm. long, pinnate into oblong or lanceolate toothed to pinnatifid divisions; petioles mostly shorter than blades; fls. many, short-pedicelled, in dense few-branched cymes which become lax and up to 20 cm. long in fr.; calyx-lobes linear to narrow-spatulate, densely ± yellowish and bristly-hispid and glandular, 6–8 mm. long in fl., to 10 or 12 mm. in fr.; corolla dirty yellowish-white, broadly campanulate, 8–12 mm. long, deciduous; scales adnate to corolla on one side and to fil. on other, with free oval part; stamens 8–12 mm. long, glabrous; style ca. as long, cleft to middle; caps. round-ovoid, 3–4 mm. long, sparsely hispid with dark spot at base of each hair; seeds 2–4, ca. 3 mm. long, dark brown, reticulate-pitted; n = 11 (Cave & Constance, 1944).—Dry rocky slopes, below 4000 ft.; Foothill Wd., V. Grassland; foothills of Sierra Nevada from Butte Co. s. to Tehachapi Mts. March–May.

Var. **híspida** (Gray) J. T. Howell. [*P. ramosissima* var. *h.* Gray. *P. h.* Gray. *P. eximia* Eastw.] Plants more widely branched; sepal-lobes grayish-hispid; corolla lavender; scales more adnate; seeds usually 4, ca. 1.5 mm. long.—Common in dry rocky canyons, etc., below 4000 ft.; Chaparral, Coastal Sage Scrub, S. Oak Wd., Coast Ranges from San Luis Obispo Co. s. through cismontane s. Calif. to edge of desert, Santa Barbara Ids.; L. Calif. March–June. A stout grayish-shaggy form with dense cymes has been named var. **Húbbyi** (Macbr.) J. T. Howell. [*P. hispida* var. *H.* Macbr. *P. tanacetifolia* var. *H.* Jeps. & Hoov.] It occurs from Santa Barbara Co. to Los Angeles Co.

13. **P. cryptántha** Greene. [*P. hispida* var. *brachyantha* Cov. *P. eremica* Jeps.] Resembling *P. cicutaria* but mostly lower; calyx-lobes linear-oblanceolate, 4–7 mm. long (8–10 mm. in fr.); corolla tubular-campanulate, 4–7 mm. long; scales elongate, narrow, ending in a sharp point; stamens included; style 3–4 mm. long, pubescent below; caps. globose, 2 mm. long, hispid; seeds four, 1.5–2.5 mm. long, pitted; n = 11 (Cave & Constance, 1944).—Dry rocky slopes, below 5700 ft.; Shadscale Scrub, Joshua Tree Wd., Creosote Bush Scrub; w. Colo. Desert, Mojave Desert n. to Mono Co.; Nev., Ariz. March–May. A form occurring throughout the range and with calyx-lobes plainly broader above (1.5–2 mm.) and corolla-scales without a sharp tip has been named var. *derivàta* J. Voss.

14. **P. vállis-mórtae** J. Voss. Resembling *P. cicutaria*, but stems slender, purplish, tending to be zigzag and leaning on other plants for support, the stiffer hairs rather slender, somewhat retrorse, finer hairs gland-tipped or not; lvs. rather few; calyx-lobes narrow-oblanceolate, 4–6 mm. long (to 1 cm. in fr.); corolla lavender, open-campanulate, 8–10 mm. long, deciduous; scales 1.5–2 mm. long, truncate to slightly pointed at tips, the transverse portion inconspicuous; stamens 4–8 mm. long; styles 6–10 mm. long; caps. round-ovoid, 3–3.5 mm. long, hirsutulous; seeds usually 4, brown, ca. 3 mm. long, evenly pitted; n = 11 (Cave & Constance, 1950).—Common, dry gravelly and rocky places, 2000–7500(–8000) ft.; Creosote Bush Scrub, Joshua Tree Wd., Pinyon-Juniper Wd.; Mojave Desert from Twentynine Palms to Death V. and near Bishop; Nev., Ariz. April–May.

Var. **helióphila** (Macbr.) J. Voss. [*P. hispida* var. *heliophila* Macbr.] Corolla pale with darker veins, 10–12 mm. long; stamens exserted; n = 11 (Cave & Constance, 1942).—V. Grassland; w. edge of San Joaquin V. from Merced Co. to Kern Co. April–May.

15. **P. malvifòlia** Cham. STINGING PHACELIA. Annual, often coarse, erect, subsimple to freely branched, 2–10 dm. high, bristly hispid (the hairs with basal pustules) and finely pubescent throughout, glandular especially above; lf.-blades ovate to rounded or deltoid, 2–10(–12) cm. long, shallowly lobed, the lobes serrate to subcrenate; petioles equal to or shorter than blade; fls. sessile or nearly so, in dense short cymes; calyx-lobes unequal, spatulate, 4–6 mm. long, hispid; corolla dull white, broadly campanulate, 5–7 mm. long, deciduous; scales semiovate, the apex free; stamens 5–10 mm. long; style 8–12 mm. long, pubescent; caps. globose, 2–3 mm. in diam., pubescent; seeds 2–6, dark, pitted, 2–3.5 mm. long; n = 11 (Cave & Constance, 1942).—Mostly sandy and gravelly places, below 3500 ft.; Redwood F., Mixed Evergreen F., Closed-cone Pine F., N. Coastal

Scrub; near the coast from San Luis Obispo Co. n.; to Ore. April–July. Plants from the s. part of the range have seeds ca. 1.5 mm. long and more finely pitted and have been called var. *loasifòlia* (Benth.) Brand. [*Eutoca l.* Benth.]

16. **P. Rattánii** Gray. With the general aspect of *P. malvifolia*, but stems weaker, less hispid; calyx-lobes 3–4 mm. long, spatulate, but with 1 or 2 much wider than others; corolla narrow-campanulate, 4–5 mm. long; scales lanceolate, wholly adnate; stamens 2–3 mm. long; seeds usually 2 or 4, brown, ca. 1.5 mm. long, reticulate-pitted; *n* = 11 (Cave & Constance, 1947).—Brushy places, below 4000 ft.; Yellow Pine F., Redwood F., N. Oak Wd.; mostly away from the coast, Coast Ranges from San Luis Obispo Co. to Ore., s. to Shasta Co.; sw. Ida. May–July.

17. **P. ciliàta** Benth. [*P. acanthominthoides* Elmer.] Annual, branched from base, 1–5 dm. high, glandular-puberulent throughout, also hispidulous; lvs. ovate to oblong in outline, 3–10 cm. long, pinnate or pinnatifid into toothed or incised divisions; petiole rather short except in lower lvs.; fls. subsessile, crowded in few cymes; calyx-lobes broadly lanceolate, stiff-ciliate, 4–6 mm. long in fl., broader, venulose and subcoriaceous, semitransparent and to 1 cm. long in fr.; corolla blue with paler center, broadly campanulate, 8–10 mm. long, deciduous; scales suborbicular, partly free; stamens 9–13 mm. long; style 6–8 mm. long, pubescent below; caps. round-ovoid, 4–5 mm. long, hispidulous; seeds mostly 4, dark brown, 2.5–3 mm. long, deeply pitted; *n* = 11 (Cave & Constance, 1942).—Grassy slopes, gravelly places, cultivated fields, mostly below 5000 ft.; V. Grassland, Foothill Wd., N. Oak Wd., Coastal Sage Scrub; Coast Ranges and adjacent valleys, from Glenn Co. to Kern Co., uncommon s. to L. Calif. March–May. Plants from near Merced with narrow opaque fruiting calyx-lobes have been named var. *opàca* J. T. Howell.

18. **P. thermàlis** Greene. [*P. ciliata* var. *t.* Jeps.] With general aspect of *P. ciliata* but more densely pubescent; calyx slightly smaller; corolla more tubular, 3–4 mm. long; scales partly free; stamens 1.5–2 mm. long; style 2 mm. long; caps. slightly smaller, hispidulous and finely pubescent; seeds darker; *n* = 11 (Cave and Constance, 1950).—Dry disturbed places, 3500–5000 ft.; Sagebrush Scrub, N. Juniper Wd.; Lassen, Shasta, and Modoc cos.; to Ore., Ida. May–June.

19. **P. crenulàta** Torr. Annual, stems simple or usually few-branched, 1–4(–6) dm. high, hispid-pubescent but green and very glandular and strongly scented; lf.-blades mostly oblong in outline, 3–12 cm. long, the lower petioled, pinnately divided or undulately lobed, the uppermost subsessile, merely coarsely crenate-dentate; cymes dense, in flattish terminal panicles; fls. many, short-pedicelled; calyx-lobes oblong, 4–5 mm. long, obtuse, not much longer in fr., densely glandular and hispid; corolla deep violet or bluish-purple, with white throat, open-campanulate, 6–10 mm. long; scales rectangular to lunate, acute, wholly adnate; stamens exserted, 10–14 mm. long, glabrous; styles exserted, 12–15 mm. long; caps. round-ovoid, ca. as long as calyx, sparsely hispidulous; seeds mostly 2, oblong-ovoid, brown or darker, excavated on ventral side and with cent. crenulate ridge, rounded on back, shallowly pitted, ca. 3 mm. long.—Common in gravelly washes and open places, mostly below 5000 (9000) ft.; Creosote Bush Scrub, Joshua Tree Wd., Pinyon-Juniper Wd.; Mojave Desert from Barstow region toward Twentynine Palms and to Inyo Co.; cent. Nev. and Utah. March–May.

Var. **ambígua** (Jones) Macbr. [*P. a.* Jones.] Stems often purplish, grayish hirsute, less glandular; infl. flat-topped; *n* = 11 (Cave & Constance, 1944).—Mojave Desert east of Twentynine Palms, Colo. Desert; s. Nev. March–May.

Var. **funèrea** J. Voss ex Munz. Stems purplish, mostly simple, slender, glandular-pubescent, mostly 4–8 dm. high; cymes scattered along upper stem, not in flat-topped cluster.—At 4000–9500 ft.; mostly Joshua Tree Wd., Pinyon-Juniper Wd.; Death V. region to White Mts.; adjacent Nev. April–June.

20. **P. minutiflòra** J. Voss. Resembling *P. crenulata*, branched from base, 1–3 dm. high, glandular-pubescent, puberulent, and short-hirsute; lf.-lobes especially the terminal almost round, often not much divided; calyx-lobes unequal, 3–4 mm. long, 1 or 2 wider than others; corolla lavender to violet-purple, tubular-campanulate, 4–5 mm. long; stamens 3–4 mm. long, exserted; caps. ca. 3 mm. in diam.; scales-pubescent; seeds ca. 4, dark, with ventral crenulate ridge, ca. 2 mm. long, and shallowly reticulate-pitted.—Uncommon, sandy and rocky places, below 1500 ft.; Creosote Bush Scrub; Whipple Mts., se. San Bernardino Co. through Colo. Desert; to Ariz., Son. March–April.

21. **P. amábilis** Const. Like *P. crenulata,* the corolla white, 7–8 mm. long; stamens 9–15 mm. long; scales with a broad free portion; seeds 3–4 mm. long, pale, with a thin broad margin.—At 1800 ft., Saline V., Inyo Co. April–May.

22. **P. pedicellàta** Gray. Robust openly branched annual 2–5 dm. high, densely glandular-pubescent and short-hirsute throughout; lf.-blades broadly ovate in outline, 4–12 cm. long, pinnate with 3–7 distinct rounded lfts., these lobed to serrate, often with crinkled margins; petioles shorter than blades; upper lvs. less deeply divided, smaller, but still petioled; fls. many, in short dense paniculately arranged cymes; pedicels filiform, 3–6 mm. long; calyx-lobes linear-oblanceolate, 3–4 mm. long (–5 mm. in fr.), glandular and finely hirsute; corolla pinkish to bluish or paler, open-campanulate, 5–7 mm. long, deciduous; scales short-rounded, often basally auriculate; stamens exserted, 6–8 mm. long, glabrous; style 6–8 mm. long; caps. ovoid, 3–3.5 mm. long, pubescent; seeds 4, brown, 2.5–3 mm. long, excavated and with median ridge on ventral side, shallowly pitted on curved back.—Desert washes and canyons, below 4500 ft.; Creosote Bush Scrub, Joshua Tree Wd.; Panamint Mts. to e. Mojave Desert, Colo. Desert; w. Nev., Ariz., L. Calif. March–May.

23. **P. Anelsònii** Macbr. Viscid-pubescent annual 2–4 dm. high, with simple erect green stems; lf.-blades oblong to oblanceolate, 2–8 cm. long, pinnately lobed or divided into crenate lobes; petioles to almost as long as blades; fls. many, short-pedicelled, in dense elongate cluster of cymes; calyx-lobes oblanceolate, 3–4 mm. long, 1–2 being wider than others; corolla blue or violet, funnelform-campanulate, 6 mm. long, deciduous; scales lunate, narrowed upward; stamens included; style included; caps. ovoid, glandular, puberulent at apex, 2.5–3 mm. long; seeds 4, ca. 3 mm. long, with median ridge on ventral surface, dark brown, favose-pitted.—Rare in dry places, 4000–5500 ft.; Pinyon-Juniper Wd., Joshua Tree Wd.; e. Inyo and San Bernardino cos.; to Nev., Utah. April–May.

24. **P. coerùlea** Greene. Near to *P. Anelsonii,* but stems purplish, more slender; fls. subsessile; calyx-lobes lanceolate, 2–3 mm. long; corolla blue to white, broadly campanulate, 3–4 mm. long; stamens included, 2–3 mm. long; style as long; caps. roundish, 2–3 mm. in diam., somewhat pubescent; seeds 4, dark to light brown, 2–3 mm. long, with salient ventral ridge, favose-pitted dorsally; $n = 11$ (Cave & Constance, 1950).—Dry rocky places, at 5500 ft.; Pinyon-Juniper Wd.; Clark Mt., e. Mojave Desert; to Texas, Chihuahua. April–May.

25. **P. imbricàta** Greene. [*P. circinata* var. *calycosa* Gray. *P. californica* var. *calycosa* Dundas. *P. stimulans* Eastw. *P. i.* vars. *condensata, caudata* and subvar. *Hansenii* Brand.] Perennial, from a branched woody caudex, mostly 2–4(–6) dm. high, densely puberulent and somewhat glandular-puberulent, and hispid; stems simple; lvs. mostly basal, ± ovate in outline, the lower blades 5–12 cm. long, strigose, green or somewhat grayish, prominently veined, pinnate to pinnatifid into 5–9 acute or acuminate lfts.; petioles hispid, to almost as long as blade; upper lvs. reduced, even simple; fls. subsessile or short-pedicelled, many, in dense cymes in open panicles; calyx-lobes broadly lanceolate to ovate, hispid-ciliate, 3–4 mm. long in fl., twice that and overlapping in fr. and ± unequal; corolla white, campanulate to somewhat tubular, 4–7 mm. long, the lobes tending to be involute and to enfold the stamens after anthesis; scales oblong, adnate; stamens 9–13 mm. long, well exserted, the fils. pubescent; style 9–14 mm. long, pubescent near base; caps. narrow-ovoid, 3–4 mm. long, hispid; seeds usually 1, oblong-ovoid, dark brown, ca. 2 mm. long, coarsely pitted in vertical rows; $n = 11$ (Cave & Constance, 1942). —Dry rocky places, below 3000 ft. (to 6500 in s. Calif.); mostly Chaparral, Foothill Wd.; San Gabriel Mts., Los Angeles Co. to Mt. Pinos, then to Coast Ranges of Humboldt Co. and along foothills of Sierra Nevada north to Shasta Co. April–June. Plants from Ventura Co. s. are unusually gray.

KEY TO SUBSPECIES

Stems several, slender.
    Calyx-lobes broadly lanceolate to obovate; terminal lft. mostly more than 1 cm. wide. San Gabriel
    Mts. to n. Calif. . . . . *P. imbricata*
    Calyx-lobes linear to lance-linear; terminal lft. less than 1 cm. wide. Los Angeles Co. to San Diego
    Co. . . . . ssp. *patula*
Stems mostly solitary, coarse. Santa Barbara Co. to San Bernardino Co. . . . . ssp. *bernardina*

Ssp. **pátula** (Brand) Heckard. [*P. magellanica* f. *patula* & f. *Ballii* Brand. *P. californica* var. *p.* Jeps.] Stems slender, ascending to decumbent, 1–3 dm. long; lvs. whitish, 1–5-lobed; calyx-lobes linear to lance-linear, becoming 4–7 mm. long; corolla white, 5–6 mm. long; *n* = 11 (Cave & Constance, 1950).—Dry places, 4500–7500 ft.; mostly Yellow Pine F.; Mt. Pinos to San Jacinto and Santa Rosa mts., Cuyamaca Mts. June–July.

Ssp. **bernardina** (Greene) Heckard. [*P. virgata* var. *b.* Greene. *P. californica* var. *b.* Jeps.] Stems stout, mostly solitary, 2–6 dm. tall, hispid; lvs. 3–9(–13)-pinnate, pale; infl. to 4 dm. long in fr.; calyx-lobes oblanceolate to ovate, to ca. 7 mm. in fr.; corolla white to yellowish, 5–7 mm. long.—Dry washes and slopes, below 7000 ft.; Coastal Sage Scrub, Chaparral, Yellow Pine F.; Santa Barbara Co. to L. Calif. April–July.

26. **P. califórnica** Cham. Perennial, from a branched caudex, mostly 2–8 dm. high, puberulent, somewhat glandular and somewhat hirsute; lvs. greenish, entire to pinnately 5(7)-lobed, the lobes mostly ovate, abruptly acute; fls. many, in dense cymes, subsessile or short-pedicelled; calyx-lobes linear-oblong to -lanceolate, 3–5 mm. long, hirsute-ciliate, unequal, somewhat longer but not imbricated in fr.; corolla mostly lavender, campanulate, 5–6 mm. long; scales oblong, adnate; stamens 7–10 mm. long, the fils. pubescent; style 8–10 mm. long, pubescent near base; caps. narrow, 3–4 mm. long, hispid; seeds 1 or 2, dark brown, ca. 2 mm. long, shallowly pitted in vertical rows; *n* = 11 (Cave & Constance, 1942).—Rocky bluffs and canyons, below 1500 ft.; N. Coastal Scrub, Chaparral, etc.; near coast from Del Norte Co. to Santa Clara Co. April–July.

27. **P. egèna** (Greene ex Brand) Const. [*P. magellanica* ssp. *barbata* f. *e.* Brand. *P. californica* var. *e.* Dundas.] Stems several, ascending to erect, rather slender, 2–6 dm. high; lvs. usually 3–7-pinnatifid into acute divisions; calyx-lobes oblanceolate to lance-linear, 4–6 mm. long in fl., 8–10 mm. long in fr. and venulose; corolla white, 7–9 mm. long.—Dry places, mostly below 5000 ft.; Yellow Pine F., Foothill Wd., Chaparral; w. base of Sierra Nevada from Kern Co. n. to Siskiyou and Humboldt cos. and thence s. to Tehama and Glenn cos., Inyo Co. May–June.

28. **P. argéntea** Nels. & Macbr. [*P. heterophylla* var. *rotundata* Dundas.] Perennial, the stout stems prostrate to ascending, 1–3 dm. long, ± appressed-hirsute and puberulent; lvs. silvery, roundish to oval, 2–5 cm. long, entire or with 2 basal broad lobes, the veins parallel, deeply impressed, silky-strigose; racemes in terminal crowded cluster; calyx-lobes oval to oblong, 3–4 mm. long, hirsute; corolla yellowish-white, 5–6 mm. long; stamens 6–8 mm. long, with pubescent fils.; style 6–8 mm. long, pubescent below; caps. ovoid, hirsute, ca. 3 mm. long; seeds 1–2, 1.5–2 mm. long; *n* = 22 (Cave & Constance, 1942).—Coastal Strand; Del Norte Co. to sw. Ore. June–Aug.

29. **P. heterophýlla** Pursh. [*P. virgata* Greene. *P. v.* var. *ampliata* Greene. *P. californica* f. *vinctens* Macbr.] Biennial or possibly perennial; stems mostly solitary, simple, coarse, erect, 3–12 dm. high, strigulose and hirsute; lower lf.-blades lanceolate to ovate in outline, 5–10 cm. long, grayish-green with appressed longer hairs and puberulence, mostly 5–7 pinnate into prominently veined acuminate lfts.; petioles from less than to as long as blades; upper lvs. reduced, simple and passing into leafy bracts in the infl.; infl. dense, elongate (1–5 dm.), of many short cymes; fls. subsessile; calyx-lobes linear to linear-oblong, 4–7 mm. long, not much changed in fr., hirsute with dense yellowish-green shining hairs; corolla yellowish or greenish-white, campanulate, 4–5 mm. long; scales adnate; stamens 8–10 mm. long, the fils. pubescent; style 8–10 mm. long, puberulent near base; caps. ovoid, 2.5–3 mm. long, hispid; seeds 1–2, ovoid, 1.5–2 mm. long, brown, pitted; *n* = 11, 22 (Cave & Constance, 1942).—Dry rocky places, below 7000 ft.; Red Fir F., Yellow Pine F., N. Juniper Wd.; from Lake, Mendocino, and Trinity cos. n. and to Modoc and Mono cos.; to Wash., Rocky Mts. May–July.

30. **P. hastàta** Dougl. ex Lehm. [*P. leucophylla* Torr. ex Frém.] Perennial with a branched caudex, grayish with ± appressed hairs, the stems ascending, simple or few-branched, 2–5 dm. high; lf.-blades ± silvery, lanceolate to narrow-ovate, mostly entire, 3–6 cm. long, prominently veined, the petiole often ca. as long as blade; cauline lvs. few, reduced upward; infl. ± pedunculate, of rather many cymes in rather dense panicle; fls. subsessile; calyx-lobes linear-oblong or sublanceolate, 3–7 mm. long, canescent and hirsute-ciliate; corolla white to lavender, campanulate, 4–7 mm. long; stamens 6–10 mm. long, the fils. pubescent; style 7–10 mm. long, pubescent near base; caps. ovoid, 3 mm.

long, gray-pubescent; seeds 1 or 2, brown, ovoid, 1.5–2 mm. long, regularly pitted; $n = 11$, 22 (Cave & Constance, 1942).—Dry rocky places, 2500–10,500 ft.; Sagebrush Scrub, N. Juniper Wd., Yellow Pine F., Red Fir F.; Siskiyou, Plumas, Lassen, and Modoc cos. and Yolla Bolly Mts.; to B.C. and Rocky Mts. May–July.

Var. **compácta** (Greene) Macbr. [*P. c.* Greene.] Cespitose, densely white-hirsute, 1–2 dm. high; lvs. oblong to oval, entire or with pair of basal pinnae.—At ca. 7000–10,000 ft.; Yellow Pine F.; Mono and e. Shasta cos.; adjacent Nev. to Ore. July–Aug.

31. **P. nemorális** Greene. [*P. Biolettii* Greene.] Biennial or weakly perennial, the stems 1 to few, 5–15 dm. high, from a taproot, coarsely stinging-hispid and with finer pubescence; lower lf.-blades oblong to ovate, 4–10 cm. long, simple or with 1–2 pairs of lateral ovate lfts., bright green, drying brownish; petioles often longer than blade; upper cauline lvs. gradually reduced; infl. ± elongate, of several to many cymes, ± leafy; fls. many; calyx-lobes green and pubescent on back, hispid along margin, 4–5 mm. long, lanceolate to oblanceolate; corolla whitish, ca. 4–5 mm. long; stamens 7–9 mm. long, the fils. pubescent; style 6–9 mm. long, pubescent below; caps. round-ovoid, densely short-hirsute; seeds 2, brown, ovoid, ca. 2 mm. long, regularly pitted; $n = 22$ (Cave & Constance, 1942).—Damp slopes, below 2500 ft., sometimes higher; N. Oak Wd., Mixed Evergreen F., Redwood F.; Coast Ranges from San Benito Co. n.; to B.C. April–July.

32. **P. mutábilis** Greene. [*P. magellanica* var. *griseophylla* (Brand) Jeps. *P. nemoralis* var. *pseudohispida* Brand. *P. pinnata* var. *pseudohispida* Dundas. *P. californica* var. *rubacea* Jeps. *P. c.* var. *jacintensis* Dundas.] Biennial or weakly perennial, finely pubescent and weakly hirsute, green or grayish; stems usually several, slender, simple, 1–4.5 dm. high; lower lvs. ± tufted, lanceolate to ovate, rather thin, 2–8 cm. long, mostly with 1–2 pairs of lateral lfts.; petiole usually at least as long as blade; fls. many in mostly few cymes, open-paniculate; calyx-lobes linear, 3–4 mm. long, pubescent and hirsute, 6–10 mm. in fr.; corolla lavender to whitish, 4–6 mm. long, tardily deciduous; stamens 6–8 mm. long, with pubescent fils.; style 6–8 mm. long, pubescent; caps. ovoid, 2–3 mm. long, stiff-hairy; seeds 1–4, somewhat ovoid, brown, rather regularly pitted; $n = 22$ (Cave & Constance, 1942).—Rocky rather dry or damper places, 4000–8000(–10,000) ft.; Yellow Pine F. to Subalpine F.; San Jacinto Mts., Sierra Nevada from Kern Co. n., Coast Ranges from Lake Co. n.; to Wash. June–Aug.

33. **P. corymbòsa** Jeps. [*P. magellanica* var. *c.* Jeps. *P. dasyphylla* var. *ophitidis* Macbr.] Cespitose perennial from a much-branched caudex, pubescent, glandular-pubescent and stiff-hirsute throughout; stems rather many, 2–4 dm. high; basal lf.-blades greenish, with brownish veins, 2–6 cm. long, entire or with 1–2 pairs of lateral lfts.; petioles ca. as long; fls. many, in rather dense corymbose cymes; calyx-lobes linear to lance-linear, hispid, 4–5 mm. long, 6–10 mm. in fr.; corolla white, 5–7 mm. long; stamens ca. 10 mm. long, the fils. mostly pubescent; style ca. as long; caps. narrow-ovoid, bristly, ca. 4 mm. long; seeds 1–2, brown, 1.5–2 mm. long, rather regularly pitted.—Dry open places on serpentine, below 7000 ft.; mostly Yellow Pine F. and Red Fir F.; Sonoma, Mendocino, and Trinity cos. to Del Norte and Siskiyou cos.; sw. Ore. May–Aug.

34. **P. frígida** Greene. [*P. dasyphylla* Greene. *P. heterophylla* var. *pygmaea* Jeps.] Perennial from branched heavy caudex, gray-pubescent and hispid throughout; stems several, slender, decumbent to ascending, 0.5–2.5 dm. high; lower lf.-blades grayish, densely rosulate, mostly lanceolate, entire, 2–4 cm. long, prominently veined; petioles ca. as long; cauline lvs. few, reduced; fls. many, in dense short cymes in rather compact mostly elongate panicles; calyx-lobes linear, 3–4 mm. long, often purplish, hispid-ciliate, 6–8 mm. long in fr.; corolla lavender to white, 4–6 mm. long; stamens 6–8 mm. long, the fils. glabrous to somewhat pubescent; style ca. 8 mm. long; caps. slender-ovoid, ca. 4 mm. long, sharply pointed, short-bristly; seeds 1–2, dark brown, narrow-ovoid, almost 2 mm. long, regularly pitted; $n = 22$ (Cave & Constance, 1947).—Gravelly to rocky places, mostly 7000–13,000 ft.; Sagebrush Scrub, Lodgepole F. to Alpine Fell-fields; Sierra Nevada from Kern Co. n., White Mts.; Mt. Lassen, Mt. Shasta; w. Ore. July–Sept.

35. **P. Dalesiàna** J. T. Howell. Perennial from a stout root-crown, the stems few, slender, decumbent, 5–15 cm. long, densely glandular-pubescent; basal lf.-blades oval to oblong, 1–3 cm. long, entire, on slender petioles ca. as long; cauline lvs. few, somewhat reduced; fls. few, in lax simple cymes, the pedicels becoming 1–2 cm. long; calyx-lobes oblanceolate, 3 mm. long, unequal, tangled-pubescent, to 6 mm. in fr.; corolla

white with purple markings in throat, almost rotate, 6–9 mm. long, deciduous; scales almost round, adnate, 2 mm. long; stamens 8 mm. long, the fils. glabrous; style 7 mm. long, pubescent below; caps. rounded, pubescent, 4 mm. thick; seeds 2–4, ovoid, 2.5–3 mm. long, reticulate; $n = 8$ (Cave & Constance, 1950).—Meadows, ca. 5000–6000 ft.; Red Fir F.; Scott Mts., Trinity Co. May–June.

36. **P. Prínglei** Gray. Annual; stems erect, slender, 5–15 cm. high, glandular-puberulent; lower lf.-blades linear to oblanceolate, entire, 1–3 cm. long, short-petioled; fls. few, on short pedicels, in lax simple or branched cymes; calyx-lobes linear to oblanceolate, 1–2 mm. long, glandular-pubescent and hispidulous; corolla lavender, rotate-campanulate, 3–5 mm. long, deciduous; scales oblong, adnate; stamens 3–4 mm. long, the fils. somewhat pubescent; style 3–4 mm. long, pubescent; caps. rounded, ca. 3 mm. in diam., hispidulous; seeds 2–8, oblong-ovoid, ca. 1.6 mm. long, coarsely pitted; $n = 11$ (Cave & Constance, 1947).—Moist places, 4000–8000 ft.; Yellow Pine F., Red Fir F.; Scott Mts., Trinity, and Siskiyou cos. June–Aug.

37. **P. Leònis** J. T. Howell. Like *P. Pringlei*, but calyx-segms. more coriaceous, broader; corolla 2–3 mm. long; stamens 1.7–2 mm. long; style 2 mm. long, glabrous; caps. broadly elliptic, 3–3.5 mm. long; seeds ca. 1.75 mm. long.—At 5500–6500 ft.; Red Fir F.; Trinity and Siskiyou cos. June–July.

38. **P. racemòsa** (Kell.) Bdg. [*Nama r.* Kell. *P. namatoides* Gray.] Annual, slender, erect, simple below, commonly few-branched above, subglabrous below, glandular-pubescent above, 5–20 cm. high; lower lvs. opposite, lance-oblong, 1–3 cm. long, short-petioled, entire, the cauline few; racemes loose, in narrow cymes; pedicels short; calyx-lobes linear to narrow-oblanceolate, 1–2 mm. long, ± unequal, more than twice as long in fr.; corolla bluish, narrow-campanulate, 2–4 mm. long, deciduous; scales unequal, mostly adnate; stamens 1–2 mm. long, glabrous; style 1–1.5 mm. long, pubescent; caps. round, hirsutulous, 2–3 mm. in diam.; seeds mostly 4, brown, regularly pitted, 1–1.5 mm. long; $n = 7$ (Cave & Constance, 1947).—Dry gravelly or stony places, 5000–9000 ft.; Yellow Pine F., Red Fir F.; Sierra Nevada from Fresno Co. to Lassen and Tehama cos. June–Aug.

39. **P. Eisènii** Bdg. [*P. minima* Macbr.] Annual, softly hirsutulous, sometimes somewhat glandular, the stems 3–15 cm. high, ascending, usually branched; lower lvs. usually opposite, oval to linear-oblong, mostly entire, 1–2 cm. long, on petioles ca. as long, the cauline often alternate; cymes short, open, with filiform pedicels to ca. 1 cm. long; calyx-lobes narrow-oblanceolate, 1.5–2.5 mm. long, hirsutulous, 3–5 mm. and ± unequal in fr.; corolla lavender or paler, open-campanulate, 2–3 mm. long; scales narrow, adnate, united at base; stamens 3–5 mm. long, glabrous; style 1–2 mm. long, glabrous or hirtellous below; caps. round-ovoid, 2–3 mm. in diam., hirtellous; seeds 2–4, suboblong, brown, 1.5 mm. long, coarsely pitted; $n = 9$ (Cave & Constance, 1947).—Mostly gravelly places, 4000–11,000 ft.; Yellow Pine F. to Subalpine F.; Sierra Nevada from Eldorado Co. to Inyo and Tulare cos. June–Aug.

Var. **Brandegeàna** J. T. Howell. Lower lvs. usually lobed; corolla 3.5–4 mm. long; style 2–4 mm. long.—Tulare Co.

40. **P. orògenes** Brand. [*P. Pringlei* var. *o.* Jeps.] Annual, branched or simple, 2–10 cm. high, soft-hirsutulous and somewhat glandular; lower lvs. opposite, lance-linear, 1–3 cm. long, entire, petioled; fls. few, in open cymes, with slender pedicels to 7 or 8 mm. long; calyx-lobes linear, unequal, 2–3 mm. long at anthesis, 6–8 mm. in fr.; corolla violet, open-campanulate, 4–6 mm. long; scales oblong; stamens 3–5 mm. long, glabrous; style 3–4 mm. long, hairy; caps. rounded, 3 mm. in diam., hirsutulous; seeds 3–6, angled-ovoid, brown, pitted, 1.5–2 mm. long.—Meadows, ca. 8500–10,300 ft.; Subalpine F.; near Mineral King, Tulare Co. July–Aug.

41. **P. Brèweri** Gray. Annual, the stems usually branched, prostrate to ascending, 1–3(–4) dm. long, puberulent and hirsute throughout; lower lvs. alternate, lanceolate to ovate, mostly pinnate or pinnatifid, 1–4 cm. long, slender-petioled, the cauline reduced upward, becoming entire, short-petioled; cymes few at ends of branches, with many fairly crowded short-pedicelled fls.; calyx-lobes linear-oblong, 2–3 mm. long, subequal, hirsute and puberulent, 4–5 mm. in fr.; corolla light blue to paler, open-campanulate, 4–6 mm. long, deciduous; scales narrow, adnate; stamens 4–4.5 mm. long, glabrous; style 2–3 mm. long, pubescent; caps. ovoid, acuminate, 2–3 mm. long, compressed, hirsutulous; seeds 1–2, narrow-ovoid, brown, finely pitted, ca. 2 mm. long;

$n = 11$ (Cave & Constance, 1942).—Dry rocky places, below 4200 ft.; Foothill Wd., Chaparral; inner Coast Ranges from Contra Costa Co. to San Benito and Fresno cos. April–May.

42. **P. hùmilis** T. & G. [*P. irritans* and *violacea* Brand.] Annual, simple and erect or branched and wide-spreading, hirsute and hirsutulous, scarcely glandular, 5–20 cm. high; lowest lvs. opposite, others alternate, linear-oblong to ovate, 1–4(–10) cm. long, entire, short-petioled or the upper subsessile; racemes 1–3, the fls. many, short-pedicelled; calyx-lobes linear to linear-oblong, 3–5 mm. long, white-hirsute; corolla violet, open-campanulate, 4–6 mm. long, deciduous; stamens 4–6 mm. long, the fils. pubescent; style 4–8 mm. long, subglabrous; caps. ovoid, 2–3.5 mm. long, apiculate, hirsutulous; seeds 1–2, lance-ovoid, brown, 1.5–2.5 mm. long, finely pitted; $n = 11$ (Cave & Constance, 1942).—Flats and borders of meadows, 5000–9400 ft.; Sagebrush Scrub, N. Juniper Wd., Yellow Pine F., Red Fir F.; Sierra Nevada mostly along the e. side, from Inyo Co. to Modoc and Shasta cos.; w. Nev. to e. Wash. May–July.

Var. **Dúdleyi** J. T. Howell. Stamens 6–8 mm. long; calyx 8–12 mm. long in fr.; seeds 2.5–3 mm. long.—Tehachapi Mts., Kern Co.

43. **P. marcéscens** Eastw. ex Macbr. Annual, hirsutulous and glandular-pubescent, simple or branched, 5–20 cm. high, ± mephitic in odor; lvs. elliptic to oval, 1–3.5 cm. long, entire or lobed, the petioles of lower lvs. as long as blades; racemes dense, the pedicels 1 mm. long; calyx-segms. 2 mm. long, linear to oblanceolate, subequal, 3–3.5 mm. in fr.; corolla violet, open-campanulate, 4 mm. long, marcescent; scales narrow, attached on one edge; stamens ca. 6 mm. long, glabrous; style 6 mm. long, hairy; caps. ovoid, 3 mm. long, hirsutulous; seeds 2, lance-ovoid, sharply acute, brown, ca. 1.5 mm. long, coarsely pitted.—Dry gravelly places, 4000–7000 ft.; Red Fir F., Lodgepole F.; Sierra Nevada, in Nevada and Placer cos. May–July.

44. **P. Quíckii** J. T. Howell. [*P. Dociana* Jeps. & Hoov.] Much like *P. marcescens*, but the lvs. linear-oblong; calyx-lobes 2–4 mm. long, unequal and 4–7 mm. in fr.; corolla blue to lavender, drying whitish; stamens 3–6 mm. long; caps. subglobose; seeds plumper, $n = 8$ (Cave & Constance, 1947).—Open sandy places, 4000–7200 ft.; Yellow Pine F., Red Fir F.; w. slope of Sierra Nevada from Eldorado Co. to Fresno Co. May–June.

45. **P. Greènei** J. T. Howell. Near the preceding, but the calyx-lobes subequal, 2–3 mm. long at anthesis, 3.5–5 mm. in fr.; corolla deep violet to purple, subrotate, 5–6 mm. long, deciduous; stamens ca. 5 mm. long; caps. ovoid, 3 mm. long, the beak prominent; seeds almost 2 mm. long; $n = 10$ (Cave & Constance, 1950).—Rare on serpentine, ca. 2500–3000 ft.; Yellow Pine F.; cent. Siskiyou Co. June.

46. **P. mohavénsis** Gray. [*P. m.* var. *exilis* Gray.] Annual, hirsute, pubescent and somewhat glandular throughout, 5–25 cm. high, simple or branched; lvs. linear to oblanceolate, 1–3 cm. long, mostly entire, short-petioled; fls. many, in 1 to few dense cymes; pedicels very short; calyx-lobes linear to linear-oblanceolate, 3–5 mm. long, unequal, glandular-hirsutulous, 5–15 mm. in fr.; corolla lavender, open-campanulate, 5–8 mm. long, deciduous; scales lanceolate; stamens 5–8 mm. long, the fils. puberulent; style 5–8 mm. long, pubescent; caps. ovoid, 3–5 mm. long, hirsutulous, glandular; seeds 4–8, ovoid, brown, 1.5 mm. long, pitted; $n = 9$ (Cave & Constance, 1942).—Dry slopes, 4000–7300 ft.; Yellow Pine F.; Sierra Nevada of Kern and Tulare cos., San Bernardino and San Gabriel mts. June–July.

47. **P. austromontàna** J. T. Howell. [*P. humilis* var. *lobata* A. Davids. *P. l.* Jeps.] Much like *P. mohavensis*, the lvs. often pinnately lobed; pedicels to 2–3 or more mm. in fr.; calyx-lobes 4–5 mm. in fr.; corolla 3–5 mm. long; stamens 2–4 mm. long; style 1–2 mm. long, hairy near base; seeds 2–4.—Dry loose slopes, 6000–9000 ft.; Yellow Pine F., Red Fir F.; s. Sierra Nevada, Panamint Mts., Mt. Pinos, San Gabriel, San Bernardino, and San Jacinto mts.; sw. Utah. May–July.

48. **P. novenmillénsis** Munz. Several-branched from base, 5–10 cm. high, setulose and finely pubescent with some hairs gland-tipped; basal lvs. 2–7 cm. long, mostly with 1–5 ± decurrent lanceolate to lance-elliptic lateral pinnae 0.8–1.2 cm. long and with a terminal broadly oblanceolate lft. 1.5–2.5 cm. long, greenish but setulose on veins underneath; upper lvs. reduced, entire; cymes compact, ca. 2 cm. long in fr., 8–14-fld.; pedicels 2–5 mm. long; calyx cleft to base, the lobes subequal, linear, 2–3 mm. long in fl., 9–10 mm. in fr.; corolla lavender, broadly tubular-campanulate, 3–4 mm. long, the

rounded lobes ca. 1.3 mm. across; scales ca. 1.1 mm. long, lanceolate, adnate to tube on one edge, connivent with adjacent scale near base; stamens glabrous, scarcely exserted; style glabrous, ca. 5 mm. long above the ovary; caps. sharply ovoid, 2.5–3 mm. long, pubescent and finely hirsute; seeds narrow-ovoid, yellow-brown, 1.6–2 mm. long, ± angled, finely pitted.—Dry disturbed bank, 6500 ft.; Pinyon-Juniper Wd.; Nine-Mile Canyon, e. slope of Sierra Nevada, s. Inyo Co. May.

49. **P. lineàris** (Pursh) Holz. [*Hydrophyllum l.* Pursh. *P. Menziesii* Torr.] Annual, strigulose and somewhat hirsute, not glandular; stems stiff, erect, simple or bushy-branched, 1–6 dm. high, dark, leafy; lvs. sessile or nearly so, linear to lanceolate, entire or sometimes pinnately lobed, 2–8 cm. long, the lobes linear; fls. many, short-pedicellate, in leafy thyrsoid-paniculate cymes; calyx-lobes linear to oblanceolate, 4–6 mm. long, bristly-hirsute; corolla purplish or violet to almost white, open-campanulate, 6–10 mm. long, deciduous; scales linear-oblong, with tips free; stamens 5–6 mm. long, the fils. sparsely pubescent; style 5–8 mm. long, hirsutulous; caps. lance-ovoid, 5–7 mm. long, hirsutulous; seeds 6–15, oblong, dark, pitted, ca. 1.5 mm. long; *n* = 11 (Cave & Constance, 1944).—Dry gravelly slopes and flats, mostly 3000–6000 ft.; Sagebrush Scrub, N. Juniper Wd., Yellow Pine F.; Trinity and Siskiyou cos. to Modoc and Plumas cos.; to B.C. and Rocky Mts. May–June.

50. **P. vallícola** Congd. ex Brand. [*P. curvipes* var. *yosemitana* Brand.] Annual, simple or branched, 1–3 dm. high, finely pubescent, glandular and somewhat hirsute; lvs. ovate to lance-ovate, entire, 1–3 cm. long, on somewhat shorter petioles, the upper lvs. somewhat smaller; cymes many-fld., with very short pedicels; calyx-lobes obovate to narrow-spatulate, 3–5 mm. long, ± unequal, 8–10 mm. in fr.; corolla purple, narrow-campanulate, 4–5 mm. long, deciduous; scales oblong; stamens 2–3 mm. long, the fils. papillate; style ca. 2.5 mm. long, pubescent; caps. ovoid, 4–6 mm. long, hirsutulous; seeds ca. 10, oblong, angled, brown, deeply pitted, almost 2 mm. long; *n* = 11 (Cave & Constance, 1947).—Rocky places, 1800–7000 ft.; Foothill Wd., Chaparral, Yellow Pine F., Red Fir F.; w. base of Sierra Nevada in Tuolumne and Mariposa cos. May–June.

51. **P. phacelioìdes** (Benth.) Brand. [*Eutoca p.* Benth. *P. circinatiformis* Gray.] Annual, simple or branched, erect or ascending, pubescent and hispid, glandular especially above, 5–20 cm. high; lower lvs. oval to lanceolate, entire, 2–5 cm. long, on sometimes longer petioles; upper lvs. smaller; cymes many-fld., short-pedicellate; calyx-lobes linear-spatulate, 4–5 mm. long, hirsute, somewhat unequal especially in width, 10–12 mm. long in fr.; corolla white to pale lavender with violet median line on each lobe, narrow-funnelform, 4–6 mm. long, deciduous; scales ovate, denticulate; stamens 2–2.5 mm. long, the fils. hairy; style 2 mm. long, hirsutulous; caps. ovoid, flattened, 4 mm. long, hirsutulous; seeds 6–15, ovoid, dark brown, 1.5 mm. long, roughly pitted; *n* = 11 (Cave & Constance, 1947).—Uncommon, in rocky places, 2000–3500 ft.; Chaparral, Foothill Wd.; inner Coast Ranges from Mt. Diablo to Mt. Hamilton. April–May.

52. **P. cúrvipes** Torr. ex Wats. Annual, branched, diffuse or ascending, soft-pubescent, somewhat glandular above, 4–15 cm. high; lvs. largely basal, oblong to oblong-obovate, mostly entire, 1–3 cm. long, narrowed into a petiole ca. as long; upper lvs. few, reduced; cymes lax, few-fld., with pedicels 2–7 mm. long, curved, spreading or deflexed; calyx-lobes linear-spatulate, 3–6 mm. long, hirsute, 7–10 mm. long in fr., somewhat unequal; corolla bluish to violet with white throat, open-campanulate, 5–8 mm. long, deciduous; scales oblong, adnate; stamens 2–6 mm. long, sparsely hairy; style pubescent; caps. ovoid, flattened, 4–5 mm. long, appressed-hirsute; seeds 6–16, oblong, dark, irregularly pitted, ca. 1 mm. long.—Occasional, dry slopes, 3500–8000 ft.; Chaparral, Yellow Pine F., Sagebrush Scrub, Foothill Wd., etc.; Greenhorn Mts., Kern Co., e. slope of Sierra Nevada from Inyo Co. and White Mts. from Mono Co. s. to mts. of San Diego Co.; to Nev., Utah. April–June. Intergrading with:

Var. **macrántha** (Parish) Munz. [*P. Davidsonii* Gray. *P. D.* var. *m.* Parish. *P. curvipes* var. *D.* Brand. *P. Aldersonii* and *nemophiloides* Greene. *P. pratensis* Heller.] Lower lvs. commonly pinnate; corolla 9–15 mm. long; *n* = 10 (Cave & Constance, 1944).—Chaparral, Foothill Wd., Yellow Pine F.; mts. from San Diego Co. to s. Sierra Nevada.

53. **P. Douglàsii** (Benth.) Torr. [*Eutoca D.* Torr.] Annual, mostly several-stemmed from base, the branches prostrate or ascending, simple or branched, 0.5–4 dm. long, softly hirsutulous to stiffer haired, glandular in infl.; lvs. mostly basal, oblong in outline, pinnate to pinnatifid into unequal oblong or subovate lobes, the blades 3–8 cm. long,

the petioles ca. as long; cauline lvs. few, reduced; cymes lax, few-fld., mostly solitary at
ends of branches; pedicels longer than fls., often recurved; calyx-lobes spatulate to
oblanceolate, 4–7 mm. long, hirsute; corolla light blue to purplish, open-campanulate,
10–12 mm. long; scales lanceolate, adnate; stamens 3–7 mm. long, with fils. hairy or not;
style 2–7 mm. long, hairy; caps. ovoid, pointed, 5–7 mm. long, flattened, hairy; seeds
mostly 10–20, ovoid, brown, pitted, to 1 mm. long; $n = 11$ (Cave & Constance, 1950).—
Sandy places, below 5000 ft.; many Plant Communities; Riverside Co. to San Francisco
Bay, Cent. V. and bordering foothills. March–May.

Var. **petróphila** Jeps. A poorly marked variant, with lvs. entire or with 1 or 2 pairs of
coarse teeth on the sides.—Occasional from Kern Co. to Mt. Hamilton. March–April.

Var. **cryptántha** Brand. [*P. stellaris* Brand. *P. Palmeri* Vasey & Rose, not Wats.]
Calyx 2.5–3 mm. long in fl., 5–6 mm. in fr.; corolla 4–5 mm. long; style 1–2 mm. long.—
Sandy places; Coastal Sage Scrub; Downey, Los Angeles Co. to San Diego Co.; L. Calif.
March–April.

54. **P. divaricàta** (Benth.) Gray. [*Eutoca d.* Benth. *E. Wrangeliana* F. & M.] Annual,
erect and simple or branched, 0.5–3 dm. high, puberulent and somewhat hirsute, rarely
glandular; basal lvs. mostly elliptic to narrow-ovate, 1.5–5 cm. long, entire or rarely
few-lobed, on petioles to 6 cm. long; cauline lvs. scattered, reduced upward; cymes many-
fld., pedunculate, the pedicels shorter than calyx even in fr.; calyx-lobes lanceolate to
obovate, 5–7 mm. long, hirsute-ciliate, somewhat unequal, 10–12 mm. in fr.; corolla
lavender to violet, pubescent, rotate-campanulate, mostly ca. 10–12(–18) mm. long;
scales ovate, adnate; stamens 6–9 mm. long, the fils. glandular, sometimes hairy; style
6–10 mm. long, hairy; caps. ovoid, flattened, 6–10 mm. long; seeds 8–15, ovoid, dark
brown, pitted, 1.5 mm. long; $n = 10$ (Cave & Constance, 1942).—Open flats and slopes,
below 2000 (4000) ft.; V. Grassland, Mixed Evergreen F., Foothill Wd., Chaparral;
coastal flats to inner Coast Ranges, Mendocino Co. to San Benito and Monterey cos.
March–May.

Var. **insulàris** (Munz) Munz. [*P. i.* Munz. *P. curvipes* var. *i.* Jeps.] Lowest pedicels
longer than fruiting calyx; fls. rather few, in open cymes; calyx-lobes spatulate; corolla
rotate-campanulate; caps. obovoid; seeds 1–1.3 mm. long.—Sand dunes; Coastal Strand;
Santa Rosa and San Miguel ids. March–April.

Var. **continéntis** (J. T. Howell) Munz. [*P. insularis* var. *c.* Howell. *P. divaricata* var.
Wrangeliana Jeps. as to plants, not type.] Corolla open-campanulate; seeds 1.3–2 mm.
long; $n = 10$ (Cave & Constance, 1947).—Sand dunes; Coastal Strand; Mendocino Co.
to n. Marin Co. March–May.

Var. **Congdònii** (Greene) Munz. [*P. C.* Greene. *P. humilis* var. *C.* Macbr.] Lowest
pedicels usually not longer than fruiting calyx; stamens 3–5 mm. long; fils. not long-
hairy; style 4–5 mm. long; seeds ca. 6; $n = 10$ (Cave & Constance, 1947).—Dry slopes,
2000–5000 ft.; Chaparral, Foothill Wd., Yellow Pine F.; Tehachapi Mts. to w. base of
Sierra Nevada in Mariposa Co. April–May.

55. **P. grísea** Gray. Annual, erect, branched, 2–6 dm. high, glandular-hirsute and
hirsutulous; lvs. mostly basal, lanceolate to broadly ovate, 1–3 cm. long, entire or dentate-
lobed, on petioles to 3 cm. long; cymes densely fld.; pedicels 1–2 mm. long; calyx-lobes
narrowly spatulate to obovate, 3–4 mm. long, unequal and 6–8 mm. in fr.; corolla white
to pale lavender, open-campanulate, 5–6 mm. long; scales oblong, adnate; stamens 7–8
mm. long, with papillate fils.; style 5–9 mm. long, hairy; caps. ovoid, acute, 4–5 mm.
long, bristly hairy; seeds 5–10, ovoid, brown, ca. 1.5 mm. long, pitted.—Gravelly slopes,
below 3500 ft.; Chaparral; Santa Lucia Mts., San Luis Obispo and Monterey cos. May–
June.

56. **P. Purpùsii** Bdg. [*P. humilis* var. *calycosa* Gray.] Annual 1–4 dm. high, simple or
branched, finely hirsutulous and glandular-villous; lower lvs. oblong to ovate, 2–5 cm.
long, entire or with 1–3 salient teeth or lobes on each side, the petioles 1–4 cm. long;
upper lvs. few, reduced; fls. many, subsessile, in dense simple or few-branched cymes;
calyx-lobes very unequal, narrow-spatulate to obovate, 3–3.5 mm. long in fl., 4–7 mm.
in fr.; corolla lavender to violet, open-campanulate, 6–7 mm. long, the tube with thin
translucent areas behind the stamens; scales oblong, the free margins of adjacent pairs
connivent; stamens 7–8 mm. long, with papillate fils.; style 6–9 mm. long, sparsely hairy;
caps. ovoid, acute, 4–5 mm. long, glandular-hirsutulous; seeds 3–6, oblong, dark brown,
coarsely pitted, 1.5–2 mm. long; $n = 9$ (Cave & Constance, 1947).—Occasional, gravelly

or sandy places, 3000–7000 ft.; Yellow Pine F., Red Fir F.; Kern Co. along w. base of Sierra Nevada to Modoc Co. May–July.

57. **P. perityloìdes** Cov. Perennial, with branched woody root-crown, often with many diffusely branched stems, white woolly below, glandular-pubescent above, 1–4 dm. long; lvs. well distributed, round to ovate in outline, 1–2 cm. across, coarsely dentate to shallowly lobed, on longer slender petioles; cymes lax, few-fld., with pedicels ca. 1 cm. long; calyx-lobes oblong-spatulate, 5–6 mm. long, 1–2 mm. wide, hirsutulous and glandular; corolla white with yellowish or purplish throat, campanulate-funnelform, 10–12 mm. long, deciduous; scales linear; stamens 3–6 mm. long, glabrous; style 4–5 mm. long, hairy below, barely divided at apex; caps. oblong-ovoid, 3–4 mm. long, pubescent; seeds numerous, oblong-angled, brown, ca. 0.5 mm. long, shallowly pitted.— Dry crevices in limestone cliffs, 3000–7500 ft.; Creosote Bush Scrub and higher; Titus Canyon and Panamint Mts. to White Mts., Inyo Co. April–June.

Var. **Jàegeri** Munz. [*P. geraniifolia* Brand.] Calyx-lobes to 1 mm. wide; corolla 12–15 mm. long, the tube white or pale yellow.—Limestone crevices, 6000–7700 ft.; Pinyon-Juniper Wd.; Clark Mt., e. San Bernardino Co.; s. Nev. May–July.

58. **P. suaveòlens** Greene. Annual, hirsutulous and glandular throughout, mephitic, erect or spreading, 0.5–4 dm. high, simple or branched; lvs. well distributed, oblong to ovate or elliptic, 1–6 cm. long, serrate to shallowly lobed, on petioles to ca. 3 cm. long; cymes sessile or short-peduncled, many-fld., the pedicels 1–2 mm. long; calyx-lobes oblanceolate, 4–5 mm. long, glandular-hirsutulous, 6–8 mm. in fr.; corolla lavender to purple with yellow tube, tubular-campanulate, 7–11 mm. long, deciduous; scales partly adnate to fils.; stamens 3–5 mm. long, glabrous; style 3–4 mm. long, glandular-hirsutulous; caps. oblong-ovoid, 3–5 mm. long, hirsute; seeds 10–16, ovoid, dark, 1–1.5 mm. long, coarse-pitted; $n = 12$ (Cave & Constance, 1950).—Disturbed places like burns, 1000–4000 ft.; Chaparral, Closed-cone Pine F.; Coast Ranges from Lake Co. to Santa Clara Co.; Sierran foothills in Amador Co. May–Aug.

Var. **Kéckii** (M. & J.) J. T. Howell. [*P. K. M. & J.*] Corollas mostly 10–14 mm. long; ovules 8–10.—Dry slopes, 4000–5000 ft.; Chaparral; Santa Ana Mts. May–June.

59. **P. pulchélla** Gray var. **Gooddíngii** (Brand) J. T. Howell. [*P. G.* Brand.] Widely and openly branched annual 5–20 cm. high, finely glandular-puberulent and hirsutulous; lvs. oblong-ovate to roundish, 0.5–2 cm. long, crenate or dentate; lower petioles ca. as long; fls. many, short-pedicellate in lax cymes; calyx-lobes oblanceolate, 4–5 mm. long, glandular and hirsutulous, 6–9 mm. in fr.; corolla violet or mauve, with yellow tube, funnelform-campanulate, 8–12 mm. long, deciduous; stamens 3–5 mm. long, the fils. sparsely pubescent at base; style 4–5 mm. long, pubescent below; caps. oblong, 4–5 mm. long, hirsutulous; seeds many, dark brown, subovoid, 0.5–0.7 mm. long, coarsely pitted.— Rather alkaline flats, at 2600 ft.; Creosote Bush Scrub; Mesquite V. n. of Kingston, e. Inyo Co.; s. Nev., nw. Ariz. April–June.

60. **P. rotundifòlia** Torr. ex Wats. Annual, glandular-hirsutulous and hirsute through-out, 0.5–3 dm. high, freely branched, the stems slender and fragile; lvs. roundish, mostly 0.5–2 cm. wide, coarsely toothed, on somewhat longer slender petioles; cymes few-fld., with pedicels 1–5 mm. long; calyx-lobes linear-oblanceolate, 2–4 mm. long, 5–6 mm. in fr.; corolla white to pinkish or pale violet, pale yellow below, tubular, ca. 5 mm. long, deciduous; scales lanceolate or wider, attached to base of fils.; stamens 2–3 mm. long, glabrous; style 1.5–2 mm. long, sparsely hirsutulous; caps. oblong, 4 mm. long, puberu-lent; seeds many, round-oblong, brownish, less than 0.5 mm. long, pitted.—Rocky places, crevices, etc., 300–6000 ft.; Creosote Bush Scrub, Joshua Tree Wd., Pinyon-Juniper Wd.; deserts from Inyo Co. to Riverside Co.; to Utah, Ariz. April–June.

61. **P. mustelìna** Cov. Like *P. rotundifolia* but coarser; lf.-blades 0.5–3.5 cm. long; corolla violet, 6–10 mm. long; fils. sparsely short-hairy; style 3–5 mm. long; seeds 0.5–0.7 mm. long.—Rocky places, often in limestone, 3000–6000 ft.; Creosote Bush Scrub; mts. about Death V. March–June.

62. **P. Peirsonìana** J. T. Howell. Like *P. rotundifolia*, the lvs. less prominently toothed, roundish, 1.5–3.5 cm. wide; stems densely glandular-pubescent but scarcely if at all hirsutulous; calyx-lobes more oblong, 3–4 mm. long in fl., 7–8 mm. in fr.; corolla dull purple to white, 5 mm. long; stamens sparsely hairy at base; seeds 1–1.5 mm. long.— Desert canyons, 4500–8000 ft.; Pinyon-Juniper Wd.; Mono Co. to Westgard Pass, Inyo Co.; adj. Nev. May–Aug.

63. **P. Barnebyàna** J. T. Howell. Like *P. rotundifolia;* stems very slender, hirsutulous and glandular, mostly 3–9 cm. high; lvs. oblong-oval to ovate, 0.5–1.5 cm. long; fls. few, on pedicels 4–10 mm. long; calyx-lobes 2–2.5 mm. long in fl., 3–4 mm. in fr.; corolla pale lavender, 4.5–5 mm. long; stamens glabrous; style glabrous; seeds ca. 1 mm. long.— Rare, 5000–8000 ft.; Pinyon-Juniper Wd.; Inyo Mts. to Clark Mt., e. San Bernardino Co.; w. Nev. May–June.

64. **P. Lemmònii** Gray. [*P. heterosperma* Parish. *E. polysperma* Brand.] Annual, glandular-puberulent, erect, usually branched from base, 7–20 cm. high, the stems ± reddish; lvs. well distributed, oblong-oval to broadly ovate, 1–2.5 cm. long, subentire to repand or dentately lobed; petioles mostly less than 1 cm. long; cymes many-fld., with short pedicels; calyx-lobes oblong-oblanceolate to oblong, very slightly unequal, 2.5–4 mm. long, glandular-hirsutulous, 5–7 mm. in fr.; corolla whitish to pale purple, narrow-campanulate, 4–6 mm. long, deciduous; scales free from fils.; stamens 2–3 mm. long, glabrous; style 2–2.5 mm. long, glandular-pubescent at base; caps. oblong to ovoid, 3–4 mm. long, glandular-hirsutulous; seeds almost round to ovoid, numerous, 0.5 mm. long, dark brown, coarsely pitted; *n* = 22 (Cave & Constance, 1944), 24 (Cave & Constance, 1947).—Damp places, 1400–7000 ft.; Creosote Bush Scrub to Pinyon-Juniper Wd.; deserts from Inyo Co. to Riverside Co.; Nev., Ariz. March–June.

65. **P. Paríshii** Gray. Annual diffusely branched from base, the stems decumbent then ascending, 5–15 cm. long, glandular-puberulent; lvs. largely basal, oblong to obovate, 1–3 cm. long, entire to shallowly dentate, the lower on petioles 1–2 cm. long, the upper subsessile; cymes many-fld., with short pedicels; calyx-lobes very unequal, 4 of them linear-oblong, ca. 4 mm. long, 1.5–2.5 mm. wide, the 5th spatulate-obovate, 2.5–4 mm. wide, all enlarged in fr.; corolla lavender with the base of the tube yellowish, tubular-campanulate, 5–6 mm. long; scales semioval or variable; stamens 2.5–3.5 mm. long, the fils. sparsely hairy at base; style 1.5–2 mm. long; caps. oblong-ovoid, ca. 4 mm. long, hirsutulous; seeds ca. 25, oblong-ovoid, dark, 1–1.3 mm. long, finely pitted.—Uncommon, rather alkaline places, 2000–6000 ft.; Creosote Bush Scrub, Joshua Tree Wd.; Mojave Desert from e. of Victorville; s. Nev. April–June.

66. **P. saxícola** Gray. Annual, branched, slender-stemmed, hirsute and glandular-pubescent, 5–15 cm. high; lvs. oblance-spatulate to ovate, 0.5–0.7 cm. long, entire, on petioles ca. as long; cymes few-fld., with short pedicels; calyx-lobes linear to narrow-oblanceolate, 3–4 mm. long, subequal, becoming 5–7 mm. in fr.; corolla bluish with a white tube, narrow-campanulate, 3–4 mm. long, not readily deciduous; scales linear; stamens 1.3–2 mm. long, mostly glabrous; style 1.5 mm. long; caps. round-oblong, 2.5 mm. long, hispid; seeds many, roundish, black, ca. 0.3 mm. long, shallowly pitted or smooth.—Rare, in limestone areas, 3000–7000 ft.; Creosote Bush Scrub, Joshua Tree Wd.; Mono and Inyo cos.; Nev., Ariz. April–Aug.

67. **P. Párryi** Torr. [*P. P.* var. *celata* Jeps. & Hoov.] Annual, 1–4(–6) dm. high, hirsute and glandular-pubescent throughout, simple or few-branched; lvs. ovate to somewhat oblong, 1.5–5 cm. long, irregularly singly or doubly dentate, the lower petioles longer than blades; cauline lvs. somewhat reduced upward, subsessile; cymes open, many-fld., the pedicels becoming 1–2 cm. long; calyx-lobes linear-spatulate, 5–8 mm. long, enlarging somewhat in fr.; corolla dark purple to violet, with paler center, open-campanulate to almost rotate, 1–2 cm. long, deciduous; stamens 1–2 cm. long, with ± quadrangular basal wings on pubescent fils.; style 1–2 cm. long, pubescent, cleft in upper third; caps. oblong-ovoid, 6–10 mm. long, acuminate, glandular-hirsutulous; seeds many, angled-ovoid, brown, pitted, ca. 0.7 mm. long; *n* = 11 (Cave & Constance, 1947).—Dry slopes and disturbed places like burns, below 2500 (4000) ft.; Chaparral, Coastal Sage Scrub, Creosote Bush Scrub; Coast Ranges from Monterey Co. to L. Calif., w. edge of Colo. Desert. March–May.

68. **P. Nashiàna** Jeps. Plant 1–2 dm. high, glandular-pubescent and with some longer nonglandular hairs; lvs. crowded in lower part of stem, roundish with cordate base or oval with obtuse base, crenate, thickish, mostly 1–2 cm. long, on petioles 2–3 times as long, or upper somewhat scattered, smaller, on shorter petioles; stems branched above with a few cymes elongating in fr.; pedicels to ca. 1 cm. long in fr.; calyx-lobes linear, ca. 5 mm. long in fl., 8 mm. in fr.; corolla deep bright blue with 5 pale basal spots, rotate-campanulate, 1.5–2.2 cm. broad, the lobes broad, rounded; stamens ca. as long as or shorter than corolla; fils. pilose, dilated and with a spreading subulate tooth on each side

near base; caps. many-seeded, 8–10 mm. long; seeds yellowish, reticulate-favose, irregular, ca. 1 mm. long.—Fine loose sand or gravel on steep slopes, below 6500 ft.; Pinyon-Juniper Wd., Joshua Tree Wd.; w. Mojave Desert from Red Rock Canyon, Kern Co. to Nine-Mile Canyon, e. slope of s. Sierra Nevada, Inyo Co. May–June.

69. **P. lóngipes** Torr. ex Gray. Annual, freely branched from base, 1–4 dm. high, hirsute and glandular-pubescent throughout; lvs. ovate to rounded, coarsely and obtusely crenate, the blades 1.5–4 cm. long; petioles longer; cymes lax, several-fld., with pedicels 5–10 mm. long; calyx-lobes linear-oblong, 3–6 mm. long, somewhat more in fr.; corolla white, sometimes bluish, rotate-campanulate, 7–12 mm. long; stamens 10–15 mm. long, the fils. pubescent, with basal ovate or deltoid dilations; style 8–15 mm. long, pubescent, divided halfway; caps. oblong-ovoid, 5–6 mm. long, acuminate, glandular-hispidulous, with longer hairs near apex; seeds 8–15, oblong-ovoid, brown, pitted, 1–1.5 mm. long; $n = 11$ (Cave & Constance, 1944).—Dry loose slopes, mostly 3000–7000 ft.; Chaparral, Yellow Pine F.; mts. from Santa Barbara Co. to San Gabriel Mts. April–July.

70. **P. mìnor** (Harv.) Thell. [*Whitlavia m.* Harv. *P. Whitlavia* Gray. *P. W.* var. *Jonesii* Brand.] WILD CANTERBURY-BELL. Annual, erect, simple or branched from base, 2–6 dm. high, hirsute and glandular-pubescent throughout; lvs. well distributed, the lower ovate to somewhat oblong, coarsely serrate, 2–7 cm. long, on somewhat longer petioles, the upper gradually reduced; cymes many-fld., lax, with pedicels becoming 1–1.5 cm. long; calyx-lobes linear-oblong, 6–8 mm. long; corolla purple, rarely white, tubular-campanulate, slightly constricted at throat, 1.5–4 cm. long; stamens somewhat exserted, 2–4.5 cm. long, the fils. with an oblong usually hairy basal dilation; style 1.5–3.5 cm. long; caps. oblong-ovoid, 8–12 mm. long, acuminate, glandular-hirsutulous and almost hispid; seeds ca. 100, dark brown, angled-ovoid, ca. 1 mm. long, coarsely pitted; $n = 11$ (Cave & Constance, 1942).—Common in dry disturbed places like burns, below 5000 ft.; Chaparral, Coastal Sage Scrub; cismontane s. Calif. from Santa Monica Mts. to edge of desert and to L. Calif. March–June.

71. **P. campanulària** Gray. [*P. minor* var. *c.* Jeps. *P. m.* var. *celata* Jeps. & Hoov.] Much like *P. minor*, the corolla deep blue, rarely white, not constricted at throat, 1.5–3 cm. long; dilation at base of stamens usually glabrous; style 2–3.5 cm. long, cleft ⅓–½ its length; seeds 1–1.5 mm. long; $n = 11$ (Cave & Constance, 1947).—Dry sandy and gravelly places, below 4000 ft.; Creosote Bush Scrub, w. Colo. Desert from Whitewater R. to Collins V. Feb.–April.

Ssp. **vasifórmis** Gillett. Corolla mostly 25–40 mm. long; style 3–4.5 cm. long, cleft ¼–½ its length.—Deserts from Victorville region and Morongo V. to Providence Mts. and Cottonwood Springs. March–May.

72. **P. víscida** (Benth.) Torr. [*Eutoca v.* Benth. *E. albiflora* Nutt.] Annual 1–7 dm. high, erect, simple or few-branched, hirsute and glandular throughout; lvs. oblong-ovate to rounded, 4–9 cm. long, doubly serrate or irregularly doubly dentate, with somewhat shorter petioles; cymes terminating the branches, many-fld., dense in fl., lax in fr., with pedicels becoming 1–2 cm. long; calyx-lobes linear-spatulate, 5–6 mm. long, somewhat enlarged in fr.; corolla blue with purplish or whitish center, rotate-campanulate, 8–18 mm. long, deciduous; stamens included, the fils. slender, pilose; style 5–12 mm. long; caps. oblong-ovoid, 8–12 mm. long, acuminate, hirsute and glandular; seeds many, angled-ovoid, ca. 0.6 mm. long, brown, finely pitted; $n = 11$ (Cave & Constance, 1950).—Open sandy and disturbed places, below 2000 ft.; largely Coastal Sage Scrub, Chaparral; near the coast from Monterey Co. to San Diego Co.; Channel Ids. March–June.

73. **P. grandiflòra** (Benth.) Gray. [*Eutoca g.* Benth. *E. speciosa* Nutt.] Like *P. viscida*, but coarser, 5–10 dm. high; lf.-blades 5–15 cm. long; cymes very dense, with pedicels remaining less than 1 cm. long; corolla violet to bluish, 12–30 mm. long; seeds 0.8–1 mm. long; $n = 11$ (Cave & Constance, 1944).—Dry slopes and disturbed places, especially burns, below 2500 ft.; Chaparral, Coastal Sage Scrub; along coast from Santa Barbara Co. to n. L. Calif., inland to Claremont. April–June.

74. **P. pachyphýlla** Gray. Fleshy annual, simple or few-branched, erect, 1–1.5 dm. high, hirsutulous and with stalked black-headed glands throughout; stems and lvs. brittle; lvs. roundish, thick, entire to crenulate, 2–2.5 cm. in diam., largely near base of plant, with petioles longer than blades; upper cauline lvs. reduced; cymes dense, short, subpaniculate, the pedicels to 1 or 2 mm. long; calyx-lobes oblong, 2–3 mm. long, densely glandular-hirsutulous; corolla purplish or violet, funnelform-campanulate, 5–7 mm. long,

deciduous; scales linear, 1–2 mm. long; stamens 2.5–4 mm. long, the fils. pubescent; style 2–3 mm. long; caps. 5–7 mm. long, almost globose, glandular-puberulent; seeds over 100, brown, 1–1.2 mm. long, transversely corrugated.—Occasional on alkaline flats, mostly below 3000 ft.; Creosote Bush Scrub, Joshua Tree Wd.; deserts, Kern Co. to Imperial Co.; L. Calif. April–May.

75. **P. calthifòlia** Brand. Much like *P. pachyphylla*, but often 1.5–3 dm. high; corolla purple, broadly campanulate, ca. 1 cm. long; stamens 5–6 mm. long; style 5–6 mm. long; seeds ca. 50, ovoid, ca. 1 mm. long; *n* = 11 (Cave & Constance, 1950).—Below 3000 ft.; Creosote Bush Scrub; Inyo and n. San Bernardino cos.; w. Nev. March–May.

76. **P. neglécta** Jones. Like *P. pachyphylla* but more grayish-pubescent and lacking black-headed glands; racemes ± concealed by the foliage; corolla creamy-white, 5 mm. long; stamens 3 mm. long; style 2 mm. long; caps. tending to be deflexed, 4 mm. long; seeds ca. 100, less than 1 mm. long.—Heavy alkaline soil, below 3000 ft.; Creosote Bush Scrub; deserts from San Bernardino Co. to Imperial Co.; Nev., Ariz. March–May.

77. **P. Ivesiàna** Torr. Annual with several to many erect or spreading stems 0.5–2.5 dm. long, ± hirsutulous and glandular-puberulent; lvs. scattered, the lower oblong to oblanceolate, 1–3.5 cm. long, usually divided to midrib into narrow-oblong segms., these entire or toothed; petioles ca. as long; upper lvs. reduced; cymes sessile, the pedicels becoming 2–6 mm. long; calyx-lobes linear to narrow-oblanceolate, unequal, 3–5 mm. long, somewhat more in fr., hirsutulous on margins; corolla white with yellowish tube, tubular-funnelform, 2.5–4 mm. long; scales minute or 0; stamens included, 1–2 mm. long, glabrous; style ca. 1 mm. long, subglabrous; caps. ± oblong, ca. 4 mm. long, hirsutulous; seeds 10–15, narrow-oblong, brown, with 8–12 transverse corrugations.—Occasional, dry sandy places, ca. 2500 ft.; Creosote Bush Scrub; near Kelso, San Bernardino Co.; to Utah, Colo. March–June.

Var. **pediculoìdes** J. T. Howell. Lvs. divided or lobed into deltoid segms.; seeds with 5–7 transverse corrugations.—Occasional, sandy places, below 4000 ft.; Creosote Bush Scrub, Joshua Tree Wd.; Colo. Desert through Mojave Desert to Inyo Co.; Ariz., Nev. March–May.

78. **P. glandulìfera** Piper [*P. Ivesiana* var. *g.* Nels. & Macbr.] Like *P. Ivesiana* but with conspicuous dark stalked glands throughout; lvs. usually bipinnatifid, the main divisions deeply parted or lobed and often alternating with shorter intermediate entire lobes; calyx 4–7 mm. long, becoming 6–10(–15) mm. in fr.; corolla 5–6.5 mm. long, the limb lavender, the tube yellowish; scales united to base of fils.; seeds with 8–10 corrugations, gray-brown, sharp and ribbed; *n* = 13 (Cave & Constance, 1950).—Dry sandy places, ca. 5000–7500 ft.; Sagebrush Scrub, Pinyon-Juniper Wd., Yellow Pine F.; Mono Co. to Modoc and Siskiyou cos.; to e. Wash., Wyo. May–June.

79. **P. affìnis** Gray. Annual with 1–several ± erect stems, 0.5–3 dm. high, hirsutulous and cinereous, with retrorse hairs below, more glandular-puberulent upward; lvs. narrow-oblong, 2–6 cm. long, pinnatifid to pinnate into oblong or ovate segms., these entire to lobed, not noticeably glandular; petioles shorter; cymes simple or few-branched, few- to many-fld.; fls. subsessile or on pedicels 1–2 mm. long; calyx-lobes spatulate, 4–5 mm. long, accrescent and 6–10 mm. in fr., hirsutulous and capitate-glandular; corolla pale lavender or whitish with pale yellow tube, narrow-campanulate, 3–5 mm. long, deciduous; scales inconspicuous or obsolete; stamens 2–2.5 mm. long, included, glabrous; style ca. 2 mm. long, mostly glabrous; caps. oblong or elliptic, 5–5.5 mm. long, obtuse, hirsutulous near apex; seeds plump, ± ovoid, obtuse, ca. 1 mm. long, buff, with 6–8 transverse thick rounded corrugations; *n* = 11 (Cave & Constance, 1947).—Occasional in sandy and gravelly places, below 4000 (6000) ft.; Creosote Bush Scrub, Joshua Tree Wd., Pinyon-Juniper Wd.; deserts from Ventura and Kern cos. to San Bernardino and e. San Diego cos.; to L. Calif., New Mex., Utah. March–June.

Var. **pàtens** J. T. Howell. Nonglandular hairs at base of stems spreading, not retrorse; basal lvs. glandular; corolla ca. 5 mm. long; caps. 4 mm. long.—Dry slopes, 7000–10,000 ft.; Pinyon-Juniper Wd., Bristle-cone Pine F.; Death V. region; w. Nev. June–July.

80. **P. bìcolor** Torr. ex Wats. Annual; stems erect or spreading, 0.5–4 dm. long, commonly branched, hirsutulous and capitate-glandular, leafy; lower lvs. few, oblong, 2–4 cm. long, bipinnatifid into linear divisions; petioles shorter; cauline lvs. gradually reduced upward; cymes few- to many-fld., not elevated above leafy part of pl.; pedicels 1–4 mm. long; calyx-lobes linear to narrow-oblanceolate, 3–6 mm. long, somewhat more

in fr.; corolla ± purplish with yellow tube, funnelform to campanulate, 9–16 mm. long; scales adnate; stamens included, 5–8 mm. long, the fils. usually pubescent below; style included, 4–5 mm. long, pubescent; caps. oblong-ovoid, ca. 4 mm. long, obtuse, hirsutulous at apex; seeds brown, ovoid, ca. 1.5 mm. long, with 8–10 sharp or rounded corrugations; $n = 13$ (Cave & Constance, 1950).—Sandy places, mostly 3000–10,000 ft.; Joshua Tree Wd., Pinyon-Juniper Wd., Sagebrush Scrub; deserts from w. San Bernardino Co. to Modoc Co.; Ore., Nev. May–Aug.

81. **P. Fremóntii** Torr. [*P. Brannani* Kell. *P. Hallii* Brand.] Annual, the stems few to many, ± spreading, 0.5–3 dm. long, retrorsely hirsutulous below, glandular above; lvs. mostly near base, 2–5 cm. long, pinnately divided into oblong to roundish coarse entire or few-toothed divisions; petioles often longer; cauline lvs. scattered, reduced upward; cymes many-fld., dense, well projected above the lvs.; pedicels 1–4 mm. long; calyxlobes spatulate, 4–6 mm. long, sparsely hirsutulous, densely glandular, slightly enlarged in fr.; corolla bright blue to deep lavender, occasionally paler, with yellow tube, funnelform to subcampanulate, 8–15 mm. long, deciduous; scales linear-lanceolate, adnate; stamens included, 3–8 mm. long, glabrous; style included, 3–5 mm. long, pubescent; caps. oblong-ovoid, 5–6 mm. long, hirsutulous especially toward apex; seeds ca. 12, ovoid, brown, 1–1.5 mm. long, with 6–8(–10) coarse corrugations; $n = 13$ (Cave & Constance, 1942).—Sandy or clayey slopes and flats, below 7000 ft.; Creosote Bush Scrub, Joshua Tree Wd., Pinyon-Juniper Wd.; Mojave Desert from Inyo Co. to Riverside Co. and Colo. Desert at n. base of Santa Rosa Mts.; V. Grassland, along w. edge of San Joaquin V. to San Joaquin Co.; to Utah, Ariz. March–May.

82. **P. brachylòba** (Benth.) Gray. [*Eutoca b.* Benth. *P. Cooperae* and *Orcuttiana* Gray. *P. leucantha* Lemmon.] Annual 1–6 dm. high, simple or branched from base and above, hirsutulous and capitate-glandular (especially above); lvs. oblanceolate to narrowelliptical, 3–7 cm. long, mostly near base of plant, pinnately lobed or pinnatifid into entire or few-toothed obtuse segms.; petioles shorter; upper cauline lvs. scattered, reduced; cymes paniculately clustered, forming upper half of plant, many-fld.; pedicels 1–4 mm. long; calyx-lobes linear-oblanceolate, 4–5 mm. long, glandular and hirsutulous, not much enlarged in fr.; corolla white to pink with yellow tube, broadly funnelform to campanulate, 7–10 mm. long, deciduous; scales usually obsolete; stamens 4–5 mm. long, glabrous; style 3–4 mm. long, pubescent; caps. oblong, 4–5 mm. long, hirsutulous above; seeds ca. 20, plump, ovoid, brownish, ca. 0.5 mm. long, pitted and with 5–7 transverse thick rounded corrugations; $n = 12$ (Cave & Constance, 1944).—Frequent in dry sandy or gravelly ± disturbed places, such as burns, below 7000 ft.; Chaparral, Coastal Sage Scrub; cismontane, from Monterey Co. to L. Calif. May–June.

83. **P. gymnoclàda** Torr. ex Wats. Annual, the stems branched from base, spreading, 5–20 cm. long, hirsutulous, somewhat glandular; lower lvs. oblong to oval, 1.5–2.5 cm. long, shallowly pinnately lobed to almost entire, on petioles as long or longer; cauline lvs. few, subtending the infl.; cymes few-fld., the pedicels 1–5(–9) mm. long; calyx-lobes linear to linear-spatulate, 3–5 mm. long, hirsutulous, sometimes glandular, 5–8 mm. in fr.; corolla blue to lavender with yellow tube, campanulate-funnelform, 6–10 mm. long; scales inconspicuous; stamens included, 3–6 mm. long, the fils. pubescent; style 2.5–4 mm. long, pubescent; caps. oblong-ovoid, 3–4 mm. long, pilose; seeds 5–8, oblong-ovoid, 1–1.7 mm. long, with 7–9 rounded corrugations; $n = 13$ (Cave & Constance, 1950).— Uncommon, dry places, 5000–7000 ft.; Pinyon-Juniper Wd.; White Mts., Inyo and Mono cos.; Nev., Ore. May–June.

84. **P. adenóphora** J. T. Howell. [*Emmenanthe glandulifera* Torr. *Miltitzia g.* Heller, not *P. g.* Piper. *M. g.* var. *californica* Brand.] Annual, the stems few to several, spreading, 1–3 dm. long, pubescent and glandular; basal lvs. oblong to subovate, 1.5–3 cm. long, pinnately lobed or divided into oblong entire or toothed segms.; petioles shorter or ca. as long; cauline lvs. scattered, reduced upward; cymes mostly exceeding leafy part of plant; pedicels 1–3 mm. long; calyx-lobes oblong-oblanceolate, 3–4 mm. long, somewhat more in fr.; corolla yellow or tinged with lavender, tubular to campanulate, 4–7 mm. long, pubescent without and within the tube; scales 1–2 mm. long, attached at base to fils.; stamens 3–5 mm. long, the fils. pubescent; style 2–4 mm. long, pubescent; caps. oblong, 4–5 mm. long, puberulent; seeds ca. 8, oblong, dark, 1–1.5 mm. long, with 8–12 rounded transverse corrugations; $n = 12$ (Cave & Constance, 1950).—Alkaline plains and slopes, 4400–7600 ft.; Sagebrush Scrub; Mono, Lassen, and Modoc cos.; adjacent Ore. and Nev.

May–June. Some plants from near Bodie, Mono Co., approach *P. scopulina* (A. Nels.) J. T. Howell in their smaller fls.

85. **P. inundàta** J. T. Howell. [*Emmenanthe parviflora* Gray. *Miltitzia p.* Brand, not *P. p.* Pursh.] Annual, diffuse, the stems 1–4 dm. long, hirsutulous and glandular; basal lvs. in a rosette, oblong, 1–3 cm. long, subentire to pinnate, the divisions mostly entire, oblong or somewhat deltoid; petioles 0.5–2 cm. long; cauline lvs. rather numerous; cymes mostly exceeded by the lvs., many, densely fld.; pedicels 1–2 mm. long in fr.; calyx-lobes narrow-oblong, 4 mm. long; corolla yellow, 4–5 mm. long, pubescent without, glabrous within; scales narrow, 1.2 mm. long or obsolete; stamens 2–3 mm. long, glabrous; style ca. 1 mm. long, pubescent; caps. oblong, 5–7 mm. long, sparsely pubescent; seeds 10–15, ovoid, brown, ca. 1.6 mm. long, finely transversely striate; *n* = 12 (Cave & Constance, 1950).—Subalkaline flats, inundated in early season, 4500–6000 ft.; Sagebrush Scrub, Yellow Pine F.; Lassen and Modoc cos.; adjacent Ore., Nev. May–July.

86. **P. inyoénsis** (Macbr.) J. T. Howell. [*Miltitzia i.* Macbr.] Annual, erect or spreading, 1- to several-stemmed, 3–10 cm. high, hirsutulous and finely capitate-glandular; lvs. few, scattered, elliptic to oblong or obovate, entire to coarsely few-lobed, 1–1.5 cm. long, on petioles ca. as long, or less in upper lvs.; cymes few-fld., loose, the pedicels becoming 2–5 mm. long; calyx-lobes linear-spatulate to linear, 2–3 mm. long, somewhat more in fr.; corolla pale yellow, 3–3.5 mm. long, glabrous within, pubescent without; scales obsolete or minute; stamens 1.5–2 mm. long, glabrous; style 1 mm. long; caps. rounded, 3–4 mm. long, sparsely pubescent; seeds ca. 6, brown, oblong, nearly 1 mm. long, with 6–7 transverse corrugations.—Near dry bogs, 4000–9500 ft.; Lone Pine to Bishop, Inyo Co. and e. Mono Co. May–Aug.

87. **P. tetramèra** J. T. Howell. [*Emmenanthe pusilla* Gray. *Miltitzia p.* Brand, not *P. p.* Buckl.] Annual with several spreading or ascending stems 2–10 cm. long, glandular-pubescent and near the base hirsutulous; lvs. oblong, 1–2 cm. long, entire or few-toothed, on petioles ca. as long; cymes few-fld., lax, the pedicels becoming 3–7 mm. long; fls. largely 4-merous; calyx-lobes linear-spatulate, 3–5 mm. long; corolla whitish, 1.5–2 mm. long; stamens 1 mm. long, glabrous; style ca. 0.3 mm. long; caps. round to oblong, ca. 3 mm. long; seeds ca. 10, brown, ovoid, 1 mm. long, with 7–9 ± evident transverse corrugations.—Alkaline flats and washes, 4500–7000 ft.; Sagebrush Scrub; Mono Co. to Modoc Co.; Nev., Ore. May–June.

6. Emmenánthe Benth.

Annuals, much like *Phacelia*, glandular-viscid, rather agreeably scented, erect. Fls. soon pendulous, the corolla cream-colored, persistent, without inner appendages. Stamens included. Style included, somewhat 2-cleft, deciduous. Caps. unilocular, partly divided by the intrusion of the placentae, loculicidal. Ovules many. Seeds oval, flat, pitted-reticulate. One sp. (Greek, *emmeno*, to abide, and *anthos*, fl., because of the persistent corolla.)

1. **E. penduliflòra** Benth. WHISPERING BELLS. Erect, usually much-branched, villous-pubescent and minutely viscid-glandular, 1–5 dm. tall; lvs. linear-oblong, 3–10 cm. long, the lower short-petioled, the upper sessile, pinnatifid into many oblong entire or toothed lobes; cymes in terminal paniclelike infl.; pedicels recurved, filiform, ca. 1 cm. long; calyx-lobes lance-ovate, 6–10 mm. long; corolla yellowish, tubular-campanulate, 8–12 mm. long, the lobes rounded; style ca. 2 mm. long; caps. ca. 1 cm. long; seeds ca. 15, light brown, ca. 2 mm. long; *n* = 18 (Cave & Constance, 1942).—Common in dry places, particularly after burns or disturbance, below 5000 (6000) ft.; Chaparral, Coastal Sage Scrub, Creosote Bush Scrub, etc.; Coast Ranges from Tehama Co. s., foothills of Sierra Nevada from Nevada Co. s., to L. Calif. and across the deserts from Inyo Co. to Utah, Ariz. April–July.

Var. **ròsea** Brand. [*E. r.* Const.] Herbage brownish; corolla pink, drying white.—Inner S. Coast Ranges from Santa Clara Co. to Ventura Co.

7. Drapèria Torr.

Low diffuse perennial herb with slender stems from slender horizontal woody branches of the root-crown. Lvs. opposite, entire. Fls. in naked terminal subsessile branched cymes.

Calyx deeply divided. Corolla deciduous, tubular-funnelform, without appendages. Stamens included, unequal. Style included, apically 2-lobed. Caps. bilocular, globose. Ovules 2 in each locule, pendulous. Seeds 1–4, ovoid, angular, alveolate. One sp. (Named for J. W. *Draper,* American historian.)

1. **D. systỳla** (Gray) Torr. [*Nama s.* Gray. *D. s.* var. *minor* Brand.] Stems few or more, 1–4 dm. long, hirsute; lvs. ovate, 2.5–5 cm. long, soft-hairy, sessile or petioled; pedicels 1–3 mm. long, the cymes compact; calyx-lobes linear, 4–7 mm. long; corolla pale violet, 10–14 mm. long, pubescent without, the roundish lobes 2–3 mm. long; caps. 2–3 mm. in diam.; seeds dark brown, ca. 2 mm. long; *n* = 9 (Cave & Constance, 1947). —Dry slopes in woods, 2400–8000 ft.; Yellow Pine F.; Red Fir F.; Siskiyou and Trinity cos. through the Sierra Nevada to Tulare Co. May–Aug.

8. Lemmònia Gray

Small depressed annual, dichotomously branched. Lvs. alternate, entire, clustered in a basal rosette and near tips of branches. Fls. sessile, solitary in upper axils and forks and in terminal capitate cymes. Calyx deeply divided into linear lobes. Corolla deciduous, campanulate, without appendages, shorter than calyx. Stamens unequal, included, the fils. coherent laterally at base. Style deeply divided. Caps. falsely bilocular by intrusion of placentae, membranaceous, loculicidal. Placentae narrow, each with 2–3 ovules. Seeds ca. 4, oblong-ovoid, irregularly corrugated. One sp. (J. G. *Lemmon,* early Calif. botanist.)

1. **L. califórnica** Gray. Branches slender, pubescent, 2–10 cm. long; lvs. oblanceolate, 5–15 mm. long; calyx 3 mm. long; corolla white, 2.5 mm. long; seeds dark, ca. 1 mm. long; *n* = 7 (Cave & Constance, 1950).—Dry sandy places, mostly 3000–8200 ft.; many Plant Communities; inner Coast Ranges from Lake Co. to Kern Co., thence in mts. along w. edge of Mojave Desert se. to San Bernardino Mts.; Nev. April–June.

9. Nàma L.

Low branching annuals to somewhat woody perennials. Lvs. alternate, entire to sinuate-dentate, well distributed on the stems. Fls. in reduced terminal nonscorpioid cymes and axillary or in angles of branches, subsessile. Calyx deeply divided. Corolla deciduous, funnelform to ± tubular, longer than calyx. Stamens included, unequal or unequally inserted, appendaged at base or appendages obsolete. Style included, ± bifid. Caps. falsely bilocular by intrusion of the narrow to broad placentae, loculicidal and sometimes septicidal. Ovules many. Seeds usually many, ovoid, usually reticulate, sometimes shallowly pitted. Perhaps 45 spp., of sw. U.S., Mex., S. Am., and 1 in Hawaii. (Greek, *nama,* a spring.)

(Hitchcock, C. L., Am. Jour. Bot. 20: 415–430, 518–534, 1932–1933.)

Plants perennial, with ± woody base; fls. many, in terminal cymes.
  Lvs. coarsely dentate. Inyo and Tulare cos. to San Bernardino Co. . . . . 1. *N. Rothrockii*
  Lvs. entire. Siskiyou and Trinity cos. to Nevada Co. . . . . 2. *N. Lobbii*
Plants annual; fls. solitary or few, in reduced cymes.
  Calyx divided ca. ¾ its length, the tubular base adnate to the somewhat inferior ovary; calyx-lobes
  indurate and recurved in fr. Muddy places, s. Calif. . . . . 3. *N. stenocarpum*
  Calyx divided nearly to base, the ovary superior; calyx-lobes mostly erect and not indurate.
    Styles shallowly 2-lobed at very apex.
      Corolla 3–5 mm. long, 1–3 mm. broad . . . . 4. *N. densum*
      Corolla 10–17 mm. long, 7–12 mm. broad . . . . 5. *N. aretioides*
    Styles divided almost to base.
      Corolla campanulate, 8–15 mm. long, with expanded limb.
        Stems quite erect; adnate portion of fils. not winged; seeds many, dark brown
                                                         6. *N. hispidum*
        Stems prostrate; adnate portion of fils. with narrow margin; seeds 15–25, yellowish-brown
                                                         7. *N. demissum*
      Corolla tubular or nearly salverform, 3–5 mm. long.
        Upper lvs. obovate, abruptly narrowed into a petiole; calyx densely hirsute . . 8. *N. pusillum*
        Upper lvs. oblanceolate, gradually narrowed into petiole; calyx sparsely pubescent
                                                       9. *N. depressum*

1. **N. Rothróckii** Gray. Plants perennial from slender running underground rootstocks; stems 1.5–3 dm. tall, hispid and glandular, simple or few-branched; lvs. lanceolate to oblong, 2–5 cm. long, short-petioled, coarsely sinuate-dentate, hispid and glandular, veiny

beneath; fls. many, in subcapitate terminal cymes; calyx-lobes linear, 10–15 mm. long, hirsute; corolla purplish-lavender, funnelform, 10–15 mm. long, 6–9 mm. broad; stamens unequal; style 8–10 mm. long, divided to base; caps. ovoid, 4–5 mm. long, loculicidal, membranous; seeds ca. 15, ovoid, angular, dark brown, minutely reticulate, ca. 1.5 mm. long.—Dry sandy flats and benches, 7000–10,000 ft.; Pinyon-Juniper Wd. to Subalpine F.; Sierra Nevada from Mono and Fresno cos. to Inyo and Tulare cos., n. slope of San Bernardino Mts.; w. Nev. July–Aug.

2. **N. Lóbbii** Gray. Suffruticose tomentose leafy perennial 0.5–3 dm. tall; lvs. linear-oblong to broadly oblanceolate, subsessile, 1–6 cm. long, more tomentose beneath than above, those of some shoots broader and plane, of others narrow and revolute; fls. many, in densely leafy reduced cymes, subsessile; calyx-lobes linear, 3–7 mm. long; corolla purple, broadly funnelform, ca. 10 mm. long, the round lobes 2–3 mm.; stamens unequal; style ca. 3 mm. long, deeply divided; caps. round-ovoid, ca. 3 mm. long, cartilaginous, loculicidal and septicidal; seeds 10–12, ovoid, angular, dark brown, minutely papillate-rugose, ca. 1.5 mm. long; n = 14 (Cave & Constance, 1947).—Dry rocky or sandy slopes and ridges, 4000–7000 ft.; Yellow Pine F., Red Fir F.; Eldorado Co. to Siskiyou and Trinity cos.; w. Nev. June–Aug.

3. **N. stenocárpum** Gray. [*Conanthus s.* Heller. *N. humifusum* Brand.] Annual, sometimes probably of longer duration; stems many from base, decumbent, 1–3 dm. long, hirsute throughout; lvs. oblanceolate, 1–2.5 cm. long, the lower petioled, the upper sessile and clasping; fls. in terminal leafy clusters; pedicels 1–3 mm. long; calyx-lobes 4–5 mm. long, subspatulate, erect to recurved, indurate in fr.; corolla funnelform, 6–7 mm. long, pale violet; style 1.5–2 mm. long; caps. linear-oblong, half inferior, loculicidal; seeds many, irregularly ovoid, straw-colored, ca. 0.3 mm. long, finely alveolate; n = 7 (Cave & Constance, 1950).—Occasional, muddy shores and banks, below 1000 ft.; Los Angeles Co. to San Diego Co., across Colo. Desert; to Texas, n. Mex., L. Calif. March–May.

4. **N. dénsum** Lemmon. [*Conanthus d.* Heller.] Prostrate hirsute annual, few-branched from base, the branches densely leafy at tips, 2–8 cm. long; lvs. narrow-spatulate, 8–15 mm. long, gray-hirsute; fls. solitary, sessile; calyx-lobes linear-lanceolate, 2.5–3.5 mm. long; corolla lavender, tubular, 3–4 mm. long; style to 1 mm. long; caps. ovoid, 2–3 mm. long; seeds dark brown, ca. 0.6 mm. long, shallowly pitted and minutely transversely corrugated; n = 7 (Cave & Constance, 1947).—Dry sandy or loose soil, 3000–11,700 ft.; Sagebrush Scrub, Pinyon-Juniper Wd. to Lodgepole F.; e. slope of Sierra Nevada from Inyo Co. n. to Siskiyou and Modoc cos., White Mts.; Ore. Nev. May–July.

Var. **parviflòrum** (Greenm.) C. L. Hitchc. [*Conanthus p.* Greenm. *N. p.* Const.] Branches rather uniformly leafy, hirsute-hispid, 3–15 cm. long; lvs. 1–4 cm. long; corolla 4–5 mm. long; styles 1–1.5 mm. long; n = 7 (Cave & Constance, 1950).—Sagebrush Scrub, Modoc Co.; to Wash., Utah.

5. **N. aretioìdes** (H. & A.) Brand. [*Eutoca a.* H. & A.] With general aspect of *N. densum;* calyx-lobes 4–7 mm. long; corolla purple to rose-red, tubular-funnelform, 7–15 mm. long; styles 2–6 mm. long; caps. 10–35-seeded; seeds ovoid, 0.5–1.5 mm. long, more deeply corrugate; n = 7 (Cave & Constance, 1950).—Sandy to rocky places, 3500–6500 ft.; Sagebrush Scrub, Pinyon-Juniper Wd.; Argus Mts., Inyo Co.; to Wash., Ida., Nev. May–June.

Var. **multiflòrum** (Heller) Jeps. [*Conanthus m.* Heller.] More densely leafy; fls. larger, the corolla 12–16 mm. long, more expanded.—A poorly marked variant from Inyo Co. to Modoc Co.; w. Nev.

6. **N. híspidum** Gray var. **spathulàtum** (Torr.) C. L. Hitchc. [*N. biflorum* var. *s.* Torr.] Hirsute annual, branched from base, branches 5–30 cm. long, short-hispid throughout; lvs. narrow-oblanceolate, 1–2 cm. long, plane or somewhat revolute, subsessile or short-petioled; fls. in terminal crowded rather leafy cymes; calyx-lobes linear-spatulate, 4–6 mm. long; corolla bright purple-red, narrow-campanulate, 10–12 mm. long; style 2–5 mm. long; caps. linear-oblong, 5–6 mm. long; seeds many, ovoid, yellow-brown, reticulate, ca. 0.4 mm. long.—Occasional, flats and washes; Creosote Bush Scrub; s. Mojave Desert, Colo. Desert; to Tex., Mex. March–May.

Var. **revolùtum** Jeps. Lvs. grayish, soft-hirsute as well as hispid, strongly revolute.—Colo. Desert; to Ariz., Son., L. Calif.

7. **N. demíssum** Gray. [*Conanthus d.* Heller.] Prostrate annual, few-branched, the

stems slender, strigose and villous-hirsute, 3–15 cm. long, with long internodes; lvs. largely in terminal compact clusters, narrow-spatulate to obovate, 1–2 cm. long, 2–5 mm. wide, green; fls. solitary in axils of branches and numerous at leafy tips; calyx-lobes lance-linear, gray-hirsute, 5–8 mm. long; corolla purplish-red, tubular-campanulate, 9–12 mm. long; styles 3–5 mm. long; caps. ca. 4 mm. long, 8–18-seeded; seeds brown, 0.5 mm. long, shallowly pitted and minutely reticulate; $n = 7$ (Cave & Constance, 1950).— Occasional on dry flats and slopes, 3000–5500 ft.; Creosote Bush Scrub to Pinyon-Juniper Wd.; e. Mojave Desert (Eagle Mts., Kelso, Lavic, etc.); to Utah, Ariz. April–May.

Var. **déserti** Brand. Pubescence and aspect of the sp., but lvs. 1–4 cm. long and mostly 1–2 mm. wide; $n = 7$ (Cave & Constance, 1947).—Common on flats and in washes, below 4000 ft.; Creosote Bush Scrub; deserts from Inyo Co. to Imperial Co.; Ariz. March–May.

Var. **Covíllei** Brand. Plant gray-villous throughout; lvs. rhombic-obovate, 1–2 cm. long, 3–6 mm. wide.—Below 1500 ft.; Creosote Bush Scrub; Death V. Region.

8. **N. pusíllum** Lemmon ex Gray. [*Conanthus p.* Lemmon ex Heller.] Diffuse to matted annual, short-hirsute throughout, the branches 2–10 cm. long; lvs. ovate to spatulate-ovate, 5–8 mm. long, short-petioled; fls. nearly or quite sessile in upper forks; calyx-lobes linear, 3–4 mm. long; corolla whitish, cylindrical, 4–5 mm. long; style 1–1.5 mm. long; caps. ovoid, 2.5–3 mm. long, 20–40-seeded; seeds ovoid, angular, reticulate, dark brown, ca. 0.4 mm. long.—Occasional in open and sandy places, below 4000 ft.; Creosote Bush Scrub; Death V. region to s. Mojave Desert. March–May.

9. **N. depréssum** Lemmon ex Gray. [*Conanthus d.* Lemmon ex Heller.] With general habit and size of *N. pusillum,* softly appressed-pubescent throughout; lvs. oblanceolate, 0.5–1 cm. long, short-petioled; calyx-lobes 3–4 mm. long; corolla 4 mm. long, pubescent without; seeds rhomboid-ovoid, with ca. 20 irregular pits.—Sandy places, below 3000 ft.; Creosote Bush Scrub; Mojave and Colo. deserts from Inyo Co. to San Bernardino and Imperial cos.; w. Nev. April–May.

10. Turrícula Macbr.

Very glandular stout erect ill-scented suffruticose perennial, branched from base. Lvs. alternate, entire or toothed, sessile, lanceolate, numerous. Fls. many in terminal thyrsoid panicle of subsessile scorpioid cymes. Calyx deeply divided. Corolla purple, deciduous, funnelform, shallowly lobed, exceeding calyx. Stamens unequal, included; appendages obsolete. Style deeply divided. Caps. membranaceous, falsely bilocular, 4-valved at maturity. Ovules 6–8 in each locule. Seeds 6–10, black, oblong-ovoid, longitudinally finely ridged and minutely transversely reticulate. One sp. Causes severe dermatitis to many persons. (Latin, little tower.)

1. **T. Párryi** (Gray) Macbr. [*Nama P.* Gray. *Eriodictyon P.* Greene.] Coarse, 1–2.5 m. tall, viscid-villous; lvs. crowded, 5–12 cm. long; calyx-lobes 3–4 mm. long; corolla pubescent, 13–18 mm. long; fils. adnate; style 4 mm. long; caps. ovoid, ca. 3 mm. long; seeds 1–1.5 mm. long; $n = 13$ (Cave & Constance, 1947).—Occasional in dry places, particularly after fire, 1000–8000 ft.; Chaparral, Yellow Pine F.; Sierra Nevada in Fresno and Kern cos., inner Coast Ranges in San Luis Obispo Co., Panamint Mts., Tehachapi Mts. to L. Calif. June–Aug.

11. Eriodíctyon Benth. Yerba Santa

Aromatic shrubs, evergreen, with shredding bark and open weedy growth from woody running underground rootstocks. Lvs. alternate, somewhat coriaceous, toothed or entire, sessile to petioled, mostly crowded toward ends of branches. Fls. many, in terminal branched scorpioid open to subcapitate cymes. Calyx deeply divided. Corolla white to purple, deciduous, funnelform to campanulate, without appendages. Stamens included, equally inserted, the fils. often adnate. Style divided to base. Caps. cartilaginous, falsely bilocular by intrusion and union of narrow placentae, dehiscing by 4 valves. Seeds usually 2–6, ovoid, angled or flattened, finely longitudinally ridged. Ca. 8 spp., of sw. U.S. and Mex. (Greek, *erion*, wool, and *diktuon*, net, referring to under-surface of lvs.)

Upper surface of lvs. and young stems glabrous or subglabrous, glutinous.
  Lvs. usually entire or subentire, subsessile, mostly 2–10 mm. wide.

Calyx 2–4 mm. long; cyme open; corolla sparsely pubescent, 4–7 mm. long. E. Mojave Desert
                                        1. *E. angustifolium*
Calyx 6–8 mm. long; cyme subcapitate; corolla villous, 7–11 mm. long. Coastal Santa Barbara Co.
                                          2. *E. capitatum*
Lvs. usually toothed, petioled, 5–50 mm. wide.
  Calyx sparsely pubescent or ciliate; corolla sparsely pubescent, 9–15 mm. long. Kern Co. n.
                                          3. *E. californicum*
  Calyx densely pubescent to hirsute; corolla densely pubescent, mostly 5–10 mm. long. Los Angeles
  Co. s. . . . . 4. *E. trichocalyx*
Upper surface of lvs. and young stems pubescent to tomentose (except in a form of *E. crassifolium*
of Santa Barbara and Ventura cos.).
  Calyx and corolla glandular; corolla sparsely pubescent, constricted at throat.
    Corolla 3–4 mm. long; calyx densely white-hairy throughout . . . . 5. *E. tomentosum*
    Corolla mostly 6–9 mm. long; calyx dark-colored, thinly hairy on upper part of lobes, merely
    ciliate on lower half . . . . 6. *E. Traskiae*
  Calyx and corolla not glandular; corolla densely pubescent, not constricted at throat
                                        7. *E. crassifolium*

1. **E. angustifòlium** Nutt. To ca. 1 m. tall, the branches glabrous and glutinous, rather densely leafy toward ends; lvs. linear to lance-linear, 4–10 cm. long, glabrous and glutinous above, sparsely pubescent to tomentose beneath, reticulate, short-petioled, entire to crenulate, ± revolute; cymes glutinous and sparsely pubescent, lax in fr.; calyx-lobes 3–4 mm. long, sparsely hirsutulous; corolla narrow-campanulate, white, 5–6 mm. long, pubescent; caps. sparsely pubescent in upper half, ca. 2 mm. long; seeds almost 1 mm. long, finely transversely reticulate; $n = 14$ (Cave & Constance, 1950).—Dry slopes, 5000–5500 ft.; Pinyon-Juniper Wd.; New York Mts., e. Mojave Desert; to Utah, Ariz., L. Calif. June–July.

2. **E. capitàtum** Eastw. To almost 2 m. tall, the branches glabrous and glutinous; lvs. linear, glabrous and glutinous above, white-tomentose beneath, 4–9 cm. long, 2–4 mm. broad, revolute, tapering into a subpetiolate base; cymes glabrous, capitate; calyx-lobes linear, 6–8 mm. long, villous; corolla lavender, tubular-funnelform, 8–12 mm. long; style 3–4 mm. long; caps. densely pubescent on whole surface, 3 mm. long; seeds ca. 1 mm. long, finely transversely reticulate; $n = 14$ (Cave & Constance, 1950).—Brushy slopes, below 1000 ft.; Closed-cone Pine F.; Santa Barbara Co. n. of Lompoc. May–June.

3. **E. califórnicum** (H. & A.) Torr. [*Wigandia c.* H. & A. *E. glutinosum* Benth.] Plants 5–22 dm. high, the branches glabrous to sparsely pubescent, glutinous; lvs. lanceolate to oblong, 5–15 cm. long, entire to serrate, glabrous and glutinous above, veiny and tomentulose beneath, tapering into short petioles; panicles 0.5–2 dm. long, glabrate, lax in fr.; calyx-lobes lance-linear, sparsely pubescent or ciliate, ca. 2 mm. long; corolla lavender to white, tubular-funnelform, 9–15 mm. long, sparsely pubescent on outer upper part, the round lobes 2–3 mm. long; style 4–5 mm. long; caps. 2–3 mm. long, subglabrous; seeds 2–8, nearly black, 1–1.5 mm. long; $n = 14$ (Cave & Constance, 1947).—Dry rocky slopes and ridges, below 5500 ft.; Chaparral, Foothill Wd., Yellow Pine F., N. Oak Wd., Mixed Evergreen F., Redwood F.; Coast Ranges from San Benito and Marin cos. n., Sierra Nevada from Kern Co. n.; to Ore. May–July. Older lvs. may become blackened from a sooty fungus. The plant has been used, as has the next sp., by the Indians and early settlers as a remedy for colds, grippe, asthma, etc.

4. **E. trichocàlyx** Heller. [*E. angustifolium* var. *pubens* Gray.] Much like *E. californicum;* lvs. evidently reticulate beneath; the cymes hirsutulous; calyx-lobes linear to lanceolate, 3–4 mm. long, mostly densely hirsute, not glandular; corolla pale purplish to white, 5–8 mm. long, the round lobes ca. 2 mm. long; style 3–4 mm. long; caps. 2–3 mm. long, densely hispidulous; seeds dark brown, 1–1.5 mm. long; $n = 14$ (Cave & Constance, 1947).—Dry rocky slopes and fans, below 8000 ft.; Chaparral, Yellow Pine F., Pinyon-Juniper Wd., Joshua Tree Wd.; Ventura Co. near and through the San Gabriel and San Bernardino mts. May–Aug.

Var. **lanàtum** (Brand) Jeps. [*E. californicum* var. *l.* Brand. *E. l.* Abrams.] Twigs pubescent; lvs. white-tomentose, obscurely reticulate beneath; calyx-lobes 2–3 mm. long, densely hirsute and somewhat glutinous; corolla-lobes ovate.—Dry slopes and ridges, 1000–6000 ft.; Chaparral, Yellow Pine F., Pinyon-Juniper Wd.; w. edge of Colo. Desert from Santa Rosa Mts. to L. Calif. April–June.

5. **E. tomentòsum** Benth. [*E. niveum* Eastw. *E. crassifolium* var. *n.* Jeps.] Shrub 1–2 m. high, covered throughout with a thick white felt (sometimes olive-green); lvs. oval to oblanceolate, 4–6 cm. long, entire to dentate, plane, short-petioled; cymes remaining

rather compact; calyx-lobes linear, 2–3 mm. long, densely short-hirsute and glandular; corolla white to lavender, tubular-urn-shaped, 3–4 mm. long, the ovate lobes ca. 0.5 mm. long; style ca. 2 mm. long; caps. hirsute, 2 mm. long, 10–12 seeded; seeds 1–1.5 mm. long.—Dry slopes and ridges, below 4000 ft.; Chaparral, Foothill Wd.; inner Coast Ranges from Monterey and San Benito cos. to San Luis Obispo Co. June–July.

6. **E. Tráskiae** Eastw. [*E. crassifolium* var. *T.* Brand.] Much like *E. tomentosum* in its white woolliness, but less tomentose on stems; calyx-lobes 4–5 mm. long, glandular-hirsute; corolla 4–7 mm. long, the lobes ca. 1 mm. long; caps. hairy on upper half, 4-seeded.—Dry hills; Chaparral; near the coast, San Luis Obispo Co. to Ventura Co., Santa Catalina Id. May–June.

7. **E. crassifòlium** Benth. Shrub 1–3 m. tall, leafy above; twigs and both leaf-surfaces and calyx hoary-tomentose; lvs. lance-ovate to oval, 5–15 cm. long, crenate to coarsely dentate, plane, short-petioled; cymes tomentose, lax in fr.; calyx-lobes lance-linear, 3–5 mm. long; corolla lavender, broadly funnelform, 10–15(–17) mm. long, the round lobes 2–3 mm. long; style 4–5 mm. long; caps. 2–3 mm. long, hirsute; seeds ca. 8–12, dark brown, ca. 1 mm. long; $n = 14$ (Cave & Constance, 1947).—Dry gravelly and rocky places, below 6000 ft.; Chaparral, Pinyon-Juniper Wd., etc.; cismontane s. Calif., mostly from Santa Monica and San Gabriel mts. to w. edge of Colo. Desert and to San Diego. April–June. Variable, the following poorly defined: (1) var. *denudàtum* Abrams. Lvs. greenish and glabrate above; corolla 8–10 mm. long.—Ventura Co. (Fillmore, Ojai, etc.) to Santa Ynez Mts.; (2) var. *nigréscens* Brand. Lvs. dull green and short-tomentose above; corolla 6–7 mm. long.—Ventura and Los Angeles cos. in mts. bordering w. end of Mojave Desert (Liebre Mts., Acton, etc.).

12. **Hesperochìron** Wats.

Acaulescent perennial herbs from a short vertical thick root. Lvs. in a rosette, entire, ovate to spatulate, petioled. Fls. solitary in the lf.-axils, on long slender erect or spreading pedicels. Calyx 5-parted, the lobes ± unequal. Corolla white to bluish, funnelform to rotate, deciduous, exceeding the calyx. Stamens included, often unequal, inserted on corolla-tube, the fils. dilated at base. Style included, shortly 2-cleft at apex. Caps. unilocular, ± ovoid. Seeds many, dark brown, ovoid, angular, somewhat pitted. Two spp. (Greek, *hesperos*, western, and *Chiron*, a centaur skilled in medicine.)

Corolla rotate to shallowly bowl-shaped, densely long-hairy within .... 1. *H. pumilus*
Corolla funnelform, subglabrous or short-hairy within .... 2. *H. californicus*

1. **H. pùmilus** (Griseb.) Porter. [*Villarsia p.* Griseb. ex Hook. *Capnorea campanulata* Greene.] Plants glabrous to pubescent; lvs. few (mostly 2–10), narrow-oblong to oblanceolate or slightly broader, 2.5–5 cm. long, on narrow petioles 0.5–2.5 cm. long; pedicels few, 2–5(–8) cm. long; calyx-lobes linear-oblong to lance-ovate, 4–9 mm. long, nearly or quite glabrous except for the ciliate margin; corolla rotate or nearly so, 6–15 mm. long, 1–3 cm. broad, densely long-hairy within, the roundish lobes 4–10 mm. long; caps. 5–9 mm. long; seeds 1–1.5 mm. long.—Moist sometimes subalkaline flats and meadows, 1200–9000 ft.; Sagebrush Scrub, Yellow Pine F., Red Fir F., N. Oak Wd.; Coast Ranges from Lake Co. n., Sierra Nevada from Kern Co. n.; to Wash., Ida., Nev., Ariz. April–July.

2. **H. califórnicus** (Benth.) Wats. [*Ourisia c.* Benth. *H. latifolius* Kell. *Capnorea leporina* Greene.] Plants somewhat pubescent or subglabrous save for the ciliate lvs. and calyx-lobes; lvs. many, narrow-oblong to oval, 1–5 cm. long, on mostly somewhat shorter petioles; pedicels usually many, 2–8 cm. long, spreading; calyx-lobes oblong to subovate, 4–7 mm. long, ± glabrous on backs, ciliate; corolla funnelform to narrow-campanulate, 1–2.5 cm. long, 1–2 cm. broad, subglabrous or sparsely short-hairy within, the lobes ± oblong, 3–6 mm. long; caps. 5–10 mm. long; seeds ca. 2 mm. long; $n = 8$ (Cave & Constance, 1950).—Moist often subsaline places, 4000–9000 ft.; Yellow Pine F. to Lodgepole F.; San Bernardino Mts., Mt. Pinos, mostly on w. slope of Sierra Nevada to Siskiyou Co. May–July. Variable and intergrading completely with var. *Watsoniànus* (Greene) Brand. [*Capnorea W.* Greene.] More pubescent, the lf.-surfaces and backs of calyx-lobes pubescent.—Sierra Nevada, largely along e. slope or where its influence is felt, Tulare and Mono cos. n.; to Ore., Mont., Utah.

13. Tricárdia Torr. ex Wats.

Perennial caulescent herbs, branched from heavy base. Lvs. alternate, largely in a basal rosette, petioled, entire. Fls. purplish, rather few in loose short racemelike cymes, pedicelled. Calyx parted into 5 unequal lobes, the 3 outer large and cordate, becoming scarious and reticulate-veiny in fr., the 2 inner linear. Corolla broadly campanulate, deciduous, with 10 narrow appendages near base of the stamens. Stamens included, equally inserted but unequal in length. Style included, 2-cleft. Caps. uniloculed, the walls scarious in maturity. Seeds 4–8, dark brown, oblong, minutely alveolate. Monotypic. (Greek, *tri*, three, and *cardia*, heart, because of calyx.)

1. **T. Watsònii** Torr. ex Wats. Stems few, 1–3 dm. tall, finely hairy; lvs. oblong- to lance-elliptic, 3–5 cm. long, the lower petioled; cymes usually simple; calyx 5–6 mm. long in fl., 15–25 mm. in fr.; corolla ca. 4 mm. long, 6–8 mm. broad, with a purplish throat and white limb with purple veining, the roundish lobes ca. 2 mm. long; caps. ca. 8 mm. long; seeds 3–4 mm. long.—Occasional on dry slopes, below 7500 ft.; Creosote Bush Scrub, Joshua Tree Wd., Pinyon-Juniper Wd.; about Palm Springs, Riverside Co. across Mojave Desert to e. slope of Sierra Nevada and White Mts.; to Utah, Ariz. April–June.

14. Romanzóffia Cham. ex Nees

Low perennial herbs from a bulbous or tuberous base. Stems somewhat scapelike with few reduced lvs. Lvs. mostly basal, long-petioled, round-reniform, crenately lobed or toothed. Fls. in loose racemelike cymes, pedicelled. Calyx deeply divided. Corolla white, deciduous, broadly funnelform. Stamens included, adnate to corolla-tube, subequal. Style filiform, included, the stigma scarcely if at all lobed. Caps. partially 2-locular, ovoid to oblong. Seeds many, ovoid, angled, brown, pitted. Ca. 4 spp., e. Asia, w. N. Am. (Count Nikolai von *Romanzoff*, promoter of the Russian expedition to Calif. in 1816.)

Plants without tubers, but with dilated petiole bases which overlap to form a bulblike base
                                                                                                       1. *R. sitchensis*
Plants with tomentose basal tubers.
  Herbage glabrous, the infl. conspicuously surpassing the lvs. .... 2. *R. Suksdorfii*
  Herbage villous, the infl. little longer than lvs. .... 3. *R. Tracyi*

1. **R. sitchénsis** Bong. Simple or few-branched, 3–20 cm. tall, without tubers; stems slightly villous to subglabrous, the lf.-sheaths dilated, often somewhat ciliate; petioles 2–6 cm. long, conspicuously dilated at base; lf.-blades round-reniform, 1–2.5 cm. in diam., crenately few-toothed or -lobed; cymes few, 5–20 cm. high; pedicels slender, 1–3 cm. long; calyx-lobes oblong, 2–3 mm. long; corolla 6–8 mm. long, the oval lobes 2–3 mm. long; caps. oblong-ovoid, 4–6 mm. long; seeds 1–1.5 mm. long; n = 11 (Cave & Constance, 1950).—Moist clefts in rocks, 6000–7000 ft.; Subalpine F., Alpine Fell-fields; Trinity and Siskiyou cos.; to Alaska, Mont. July–Sept.

2. **R. Suksdórfii** Greene. [*R. californica, mendocina,* and *spergulina* Greene.] Underground tubers ovoid, tomentose, clustered, ca. 1 cm. in diam.; stems slender, subglabrous, 10–30 cm. long; basal lvs. 1–4 cm. in diam., on petioles 3–10 cm. long; cauline lvs. much reduced; infl. branched, open, elongate, often with reduced tubers in axils; pedicels 1–3 cm. long; calyx-lobes lance-linear, 2–3 mm. long; corolla 5–12 mm. long, with a yellow band below the throat; caps. oblong, ca. 1 cm. long; seeds ca. 2 mm. long; n = 11 (Cave & Constance, 1942).—Moist ± rocky places, below 2500 ft.; Redwood F., Douglas-Fir F., Yellow Pine F., Mixed Evergreen F.; near the coast from Santa Cruz Co. to Del Norte and Siskiyou cos.; to Wash. March–May.

3. **R. Tràcyi** Jeps. Tubers ovoid, tomentose; plant villous, particularly on stems and petioles; stems 5–10 cm. high; petioles 1–8 cm. long; lf.-blades subglabrous, reniform to obovate, 1.5–3.5 cm. in diam., crenate; cymes few; pedicels 2–6 mm. long; calyx-lobes lanceolate, 3–5 mm. long; corolla with pale yellow band in throat, 7–8 mm. long, the oval lobes 2–3 mm. long; caps. ovoid, ca. 5 mm. long; seeds ca. 1.5 mm. long; n = 11 (Cave & Constance, 1947).—Moist places on ocean bluffs; N. Coastal Scrub; Humboldt Co.; to Wash. March–April.

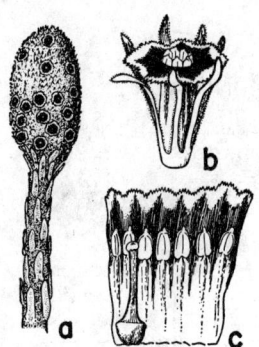

Fig. 57. LENNOACEAE. *Pholisma arenarium: a,* stem and infl., × ½; *b,* fl., × 3, with linear calyx-lobes, corolla with flaring limb; *c,* corolla opened up to show stamen insertion and the pistil.

60. Lennoàceae (Fig. 57)

Fleshy herbs, parasitic on roots and without chlorophyll, ± brown when dry. Lvs. reduced to bractlike scales. Fls. perfect, in spikes or heads. Calyx deeply parted into almost distinct lobes. Corolla tubular with ± flaring 5–8-lobed limb. Stamens as many as the corolla-lobes and inserted on the throat; anthers 2-celled, dehiscent by longitudinal slits. Pistil of 6–14 completely united carpels; style 1, simple; stigma crenulate or obscurely lobed, ± peltate. Fr. a caps., concealed in persistent perianth, finally breaking up into 12–28 nutlets. Seeds with endosperm and a rounded rather undifferentiated embryo. Three genera and 5–6 spp., w. N. Am.

Fls. in a dense terminal spike or somewhat paniculate; sepals glandular-puberulent . . . . . 1. *Pholisma*
Fls. on a dilated concave receptacle; calyx plumose . . . . 2. *Ammobroma*

1. Pholísma Nutt. ex Hook.

Spike simple or compactly branched. Calyx-lobes 5–7, linear. Corolla narrow-funnelform, the limb expanded, obscurely lobed, purplish with white border, undulate-plicate. Ovary subglobose, 6–10-celled, each cell divided into two by a false partition. Stigma peltate, crenately 6–10-lobed. One sp. (Greek, *pholis*, scale, because of the scale-like lvs.)

1. **P. arenàrium** Nutt. ex Hook. Part above ground 1–2 dm. tall, the stem whitish (drying brown), fleshy, with brownish bractlike lvs. 8–14 mm. long; spikes oblong to branched, 2–8 cm. across; calyx shorter than corolla, glandular-puberulent; corolla 3–4 mm. broad; $n = 18$ (Carlquist, 1953).—Occasional in sandy places, on roots of shrubs like *Eriodictyon, Haplopappus, Chrysothamnus, Hymenoclea, Franseria,* etc., below 5000 ft.; Coastal Strand, Creosote Bush Scrub, Joshua Tree Wd.; coast from San Luis Obispo Co. to San Diego, Mojave and Colo. deserts to L. Calif. April–July, Oct. The coastal form tends to differ from the desert plants by more branched infl. and the style longer than stamens; it has been named *P. paniculatum* Templeton.

2. Ammobròma Torr. ex Gray. Sand Food

Stem simple, mostly buried in sand, fleshy, scaly, expanded at summit into a hollow saucer-shaped receptacle which is thickly covered with the fls. Sepals 6–10, filiform, plumose. Corolla regular, purple, tubular, 6–9-lobed, the lobes erect, plicate. Stamens 6–9. Carpels 6–10, divided by false partitions, the ovary 12–20-celled; stigma sub-capitate, crenate on margin. Fr. globose, with fleshy exocarp and hard endocarp; nutlets 12–20. One sp. (Greek, *ammos,* sand, and *broma,* food.)

1. **A. sonòrae** Torr. ex Gray. Stems to 1 m. or so long, 1–2 cm. thick; lvs. oblong-linear, 1–3 cm. long, the lower glabrous, brown-purple, those below the head woolly-

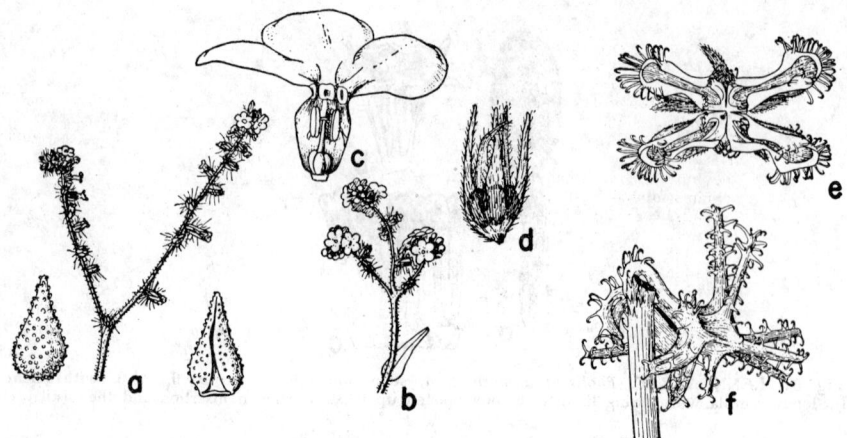

Fig. 58. BORAGINACEAE. *Cryptantha: a* and *b,* cymose infls., × ½, and dorsal and ventral views of nutlets, × 10, with slit on latter view; *c,* fl. in section, × 5, corolla with intruded appendages in throat, stamens, pistil with 4-lobed ovary. *Coldenia: d,* calyx, × 5, with developing nutlets within. *Pectocarya penicillata: e,* nutlets, × 5, divergent and margined. *Harpagonella Palmeri: f,* nutlets, × 4.

villous; disk of infl. 3–10 cm. across, densely covered with fls. and the color of light sand because of the calyx-hairs; calyx-lobes filiform; corolla purple, ca. 8 mm. long, tubular, slightly longer than calyx.—Sand hills, below 1000 ft., parasitic on roots of *Coldenia,* etc.; Creosote Bush Scrub; Colo. Desert e. of Imperial V.; to Ariz., Son., L. Calif. April–May. Once an important food for the local Indians.

61. Boraginàceae. Borage Family (Fig. 58)

Herbs, shrubs or trees; ours small, usually rough-hairy. Lvs. simple, alternate, or sometimes opposite or whorled, mostly entire. Fls. perfect, mostly regular, axillary or in 1-sided scorpioid cymes or racemes. Bracts usually between, to one side, or opposite the fls. Calyx usually 5-parted or -lobed, usually slightly irregular. Corolla 5-lobed, sometimes crested or appendaged in the throat. Stamens 5, alternate with the corolla-lobes, inserted mostly in the corolla-tube. Ovary superior, bicarpellate, 2- or 4-ovulate, becoming tough or even bony, globose and entire or divided, in ours breaking up to form uniovulate nutlets. Nutlets 1–4, smooth to variously roughened and even winged or appendaged. Style lobed or entire, seated in the pericarp at the apex of the fr. or borne between the nutlets directly upon the receptacle or upon an upward prolongation of the receptacle called the gynobase. Endosperm usually absent. Almost 2000 spp., of world-wide distribution, but particularly in the w. U.S.

Style deeply cleft, the branches each with a capitate stigma........................ 1. *Coldenia*
Style entire, terminated by a single simple or obscurely lobed stigma.
    The style borne on the summit of the fr. on the pericarp, falling away with the nutlets; stigma
    discoid, usually surmounted with a short sterile appendage .... 2. *Heliotropium*
    The style borne between the lobes of the fr. (nutlets) and attached directly and independently
    to the gynobase or receptacle; stigma capitate, lacking a sterile apical appendage.
        Mature calyx very irregular, burlike, two of the lobes united and enclosing the fr., becoming
        horned with 7–9 long glochidiate appendages; ovules 2 .... 3. *Harpagonella*
        Mature calyx not conspicuously irregular, not armed with barbed hornlike appendages; ovules
        usually 4.
            Nutlets widely spreading in fr., armed with barbed or hooked prickles.
                The nutlets flat, armed on the margins with hooked bristles; fls. white; plant annual.
                                                      4. *Pectocarya*
                The nutlets subglobose, armed all over with barbed prickles; fls. mostly blue; plants perennial
                                                  5. *Cynoglossum*
        Nutlets erect, often not armed with prickles.
            Attachment of nutlet surrounded by an annular rim, strongly convex and leaving a pit upon
            the low receptacle.
                Corolla tubular-campanulate, the lobes erect, short; plant perennial ..... 6. *Symphytum*

1. Coldènia L.

Low herbs or subshrubs, mostly depressed and canescent to hispid. Lvs. small, alternate, broad, usually with impressed veins. Fls. sessile, usually at forks of stems, but also axillary or terminal, clustered or solitary. Calyx 5-lobed. Corolla white, blue, or pink; the tube short, naked or scaly within; throat open; limb spreading. Stamens 5, included, attached in the tube. Style 2-cleft or 2-parted. Ovary glabrous or pubescent, 2-celled or sometimes 4-celled by the septumlike placentae, entire and subglobose or lobed, breaking up at maturity into uniovulate nutlets, 1–3 of these frequently abortive. Ca. 20 spp., mostly of W. Hemis. (Cadwallader *Colden*, lieutenant governor of N.Y. and correspondent of Linnaeus.)

Fr. globose until completely mature, not lobed, bearing the style on its rounded summit, at last breaking into quarters to form the angular nutlets; lvs. obscurely veined; fls. borne singly in lf.-axils or at
forks of stems .... 1. *C. canescens*
Fr. early divided into distinct nutlets and bearing the style between the apices; lvs. with evidently impressed veins; fls. in clusters at forks.
   Plant annual; corolla pink or white; sepals with short pungent hairs; style shorter than calyx
                                         2. *C. Nuttallii*
   Plants perennial; corolla blue or bluish; sepals villous; style longer than calyx.
     Lvs. with ca. 6 pairs of deeply impressed veins; calyx long-villous within .... 3. *C. plicata*
     Lvs. with only 2–3 pairs of shallowly impressed veins; calyx glabrous or short-pubescent within
                                          4. *C. Palmeri*

1. **C. canéscens** DC. [*C. c.* var. *subnuda* Jtn.] Low, spreading, often matted, the older stems woody and gnarled, 5–25 cm. long, white-tomentose throughout, with intermixed pallid bristles; lf.-blades 6–10 mm. long, ovate to lance-oblong, with shorter petioles; calyx 4–6 mm. long; corolla white, 6–7 mm. long; style shortly exserted from calyx; fr. ca. 2 mm. in diam., glabrous or sparsely hairy near summit.—Rocky slopes and ridges, below 4000 ft.; Creosote Bush Scrub, Joshua Tree Wd.; e. Mojave and Colo. deserts; to Nev., Tex., Mex. March–May.

Var. **pulchélla** Jtn. Corolla 9–12 mm. long, 5–8 mm. across, blue or lavender.—E. Colo. Desert (Chocolate Mts., Chuckwalla Mts., Ogilby); adjacent Ariz.

2. **C. Nuttállii** Hook. Prostrate cinereous annual, finely pubescent and short-hispid, somewhat glutinous; stems slender, 5–15 cm. long; lf.-blades ovate to roundish, 4–8 mm. long, with 2–3 pairs of veins, frequently surpassed by the petioles; fls. in compact clusters in forks and at ends of branches; calyx 4–5 mm. long, finely hispidulous with scattered bristles; corolla pink or almost white, 3–4 mm. long, 2–2.5 mm. broad; nutlets oblong-ovoid, almost 1 mm. long, smooth, shining.—Sandy places in washes and on slopes or subalkaline flats, below 7000 ft.; Creosote Bush Scrub, Joshua Tree Wd., Sagebrush Scrub; Mojave Desert to Kern R., along e. slope of Sierra Nevada to Lassen Co.; to Wash., Utah, Wyo. May–Aug.

3. **C. plicàta** (Torr.) Cov. [*Tiquilia brevifolia* var. *p.* Torr. *C. Palmeri* of many auth.] Matted perennial from a deep woody root, the stems prostrate, 1–4 dm. long, the

branchlets dichotomous, puberulent to subtomentose; lf.-blades obovate to round-ovate, 4–9 mm. long, densely canescent- or almost silvery-strigose, conspicuously plicate with 4–7 pairs of deeply impressed veins; petioles ca. as long; fls. clustered; calyx 2–3 mm. long; corolla 4–6 mm. long, blue or lavender; nutlets ca. 1 mm. long, smooth, shining, ovoid to rounded, usually 1 or more aborted.—Sandy places, below 3000 ft.; Creosote Bush Scrub; Colo. Desert and e. Mojave Desert; Nev., Ariz., L. Calif. April–July.

4. **C. Pálmeri** Gray. [*C. brevicalyx* Wats.] Perennial 1–3 dm. tall, 2–10 dm. broad, slightly glutinous, with suffrutescent trailing stems, thinly hirsutulous, exfoliating in age; lf.-blades obovate to ovate, 4–9 mm. long, usually longer than petioles, sparsely hispid and strigose, irregularly veined with 2–3 pairs of veins; calyx 2–3.5 mm. long; corolla 5–7 mm. long; nutlets ca. 1 mm. in diam., rounded, 1 or more frequently aborted.— Common in sandy places, below 500 ft.; Creosote Bush Scrub; Colo. Desert and along the Colo. R. to above Needles; w. Ariz., L. Calif. April–June.

2. Heliotròpium L. HELIOTROPE

Herbs or shrubs with lvs. usually alternate, petioled, mostly entire. Fls. in spicate or racemose scorpioid cymes or borne along leafy stems, usually between or opposed to the lvs. Calyx 5-toothed or 5-lobed. Corolla white, blue, or yellow, funnelform or with a spreading limb and short tube, rarely subtubular, the throat open. Stamens 5, attached to corolla-tube, included, the anthers obtuse to acuminate. Ovary glabrous or pubescent, with 4 fertile cells or these fewer by abortion, entire or rounded or with 2–4 lobes, at maturity breaking up into uniovulate or biovulate nutlets. Style apical, seated in the pericarp, simple; stigma consisting of a fertile discoid base and a superimposed sterile apical frequently bifid appendage. Ca. 125 spp., of warmer regions. (Greek, *helios*, sun, and *trope*, turning, referring to the summer solstice when the spp. were supposed to come into fl.)

Corolla white or pale. Native plants.
   Perennial, glabrous, succulent; fls. sessile .... 1. *H. curassavicum*
   Annual, rough hairy, not fleshy; fls. short-pedicellate .... 2. *H. convolvulaceum*
Corolla purplish. Garden escape .... 3. *H. amplexicaule*

1. **H. curassávicum** L. var. **oculàtum** (Heller) Jtn. [*H. o.* Heller. *H. spathulatum* ssp. *o.* Ewan.] Perennial with underground rootstocks which send up scattered shoots; stems diffuse, 1–5 dm. long, fleshy, glaucous, glabrous throughout; lvs. succulent, usually glaucous, oblanceolate to spatulate, cuneate at base, 1–4 cm. long, short-petioled; fls. in dense geminate or solitary scorpioid cymes 2–10 cm. long, bractless; calyx 2–3 mm. long; corolla 3–6 mm. broad, white with yellow spots in throat and usually becoming ± purple about the center; fr. glabrous, 1.5–2 mm. broad, subglobose, at last separating into 4 nutlets.—Common in saline or alkaline soils, below 6700 ft.; many Plant Communities; throughout Calif.; to Ariz., Utah, Nev. March–Oct.

2. **H. convolvulàceum** (Nutt.) Gray var. **califórnicum** (Greene) Jtn. [*H. c.* Greene. *Euploca convolvulacea* ssp. *c.* Abrams. *E. albiflora* var. *californica* Jeps. & Hoov.] Loosely branched annual, hispid with spreading and appressed yellowish-white hairs with pustulate bases; stems 5–30 cm. long; lvs. ovate to oblong-ovate, rounded at base, 1.5–3 cm. long, distinctly short-petioled; fls. solitary in lf.-axils, scattered, fragrant; calyx-lobes linear, 4–5 mm. long; corolla white, appressed-hispid without, 8–14 mm. broad; fr. 3–4 mm. broad, silky.—Occasional, in sandy places, below 2000 ft.; Creosote Bush Scrub; e. Mojave Desert, Colo. Desert; to Ariz., Son. April–May.

3. **H. amplexicaúle** Vahl. Heavy-rooted perennial; stems decumbent, 3–6 dm. long, branched, rather soft-hirsute; lvs. deep green, lance-oblong, sinuate-dentate, 2–6 cm. long, subsessile, hirsute on veins beneath; cymes peduncled, 3–5-branched, dense at first, later loose, scorpioid, to ca. 1 dm. long; calyx stiff-pubescent and with shorter gland-tipped hairs; corolla lilac-purple, ca. 5–6 mm. across; fr. shallowly 2-lobed, separating into two 2-locular 2-seeded carpels.—Occasionally established and persisting, as n. of Claremont and about Santa Barbara; native of S. Am. Much of year.

3. Harpagonélla Gray

Small annuals with stems branching from base. Lvs. narrow. Fls. in loose leafy-bracted false racemes. Pedicels twisted and laterally deflexed at maturity. Calyx becoming burlike, with 3 narrow distinct lobes, the 2 others fused, accrescent and indurate to form a galeate structure enclosing the upper part of 1 nutlet and armed dorsally with 5–9 soft hooked spines. Corolla minute, subbracteate. Style entire. Ovary 2-parted. Nutlets 2, dissimilar: the enclosed nutlets flattened on inner surface to form a margined strigose areole, otherwise round and glabrous; the free nutlet angled, completely strigose, bent so as to parallel the enclosed. Monotypic. (Diminutive of Latin *harpago*, grappling hook.)

1. **H. Pálmeri** Gray. Loosely spreading, with strigose disarticulating stems 5–30 cm. long; lvs. strigose, ± linear, 1–3.5 cm. long; pedicels short, stout, recurved in fr.; bracts 2–8 mm. long; calyx-lobes 1–1.5 mm. long in fl., 2–3.5 mm. in fr.; corolla white, ca. 2 mm. long; enclosed nutlet ca. 3 mm. long, minutely muriculate; free nutlet smooth.— Rare and local, on dry slopes and mesas, below 1500 ft.; Chaparral; cismontane s. Calif. from Los Angeles Co. to San Diego; L. Calif. March–April.

4. Pectocárya DC. ex Meissn.

Low mostly spreading annuals, with slender stems and narrow lvs., strigulose-canescent. Fls. in leafy-bracted false racemes which constitute most of the plant. Pedicels decurved in fr. Calyx 5-parted, the lobes finally spreading. Corolla white, small, the lobes short, ascending, the throat with small intruded appendages. Stamens included. Style very short. Stigma capitate. Nutlets 4, linear to obovate, divaricate, commonly paired, usually margined, bearing hooked hairs, radially spreading or recurving. Gynobase low and broadly pyramidal. Ca. 10 spp. of w. N. Am. and w. S. Am. (Greek, *pectos*, combed, and *karua*, nut, because of comblike margins on some nutlets.)

A.   Nutlet-margins lacerate or undulate or uncinate-bristly; nutlet-body linear or oblong.
    B.   Nutlets with margins pectinately lacerate or toothed most of their length, also usually uncinate-bristly about the distal end.
        C.   Nutlet-margins narrow and inconspicuous, the teeth nearly or quite distinct.
            D.   Nutlets straight or somewhat falcate; marginal appendages stout . 1. *P. linearis*
            DD.   Nutlets conspicuously recurved; marginal appendages very elongate
                                                2. *P. recurvata*
        CC.   Nutlet-margins broad and conspicuous, the teeth obviously confluent . . 3. *P. platycarpa*
    BB.   Nutlets with margins entire or undulate along sides, armed only at distal end where densely uncinate-bristly.
        C.   Nutlets all wing-margined .... 4. *P. penicillata*
        CC.   Nutlets heteromorphic, 1 of each divergent pair winged, the other wingless or merely margined .... 5. *P. heterocarpa*
AA.   Nutlet-margins entire, lacking fringelike teeth and uncinate hairs; nutlet-body obovate or rhomboidal.
    B.   Nutlets equally divergent; calyx strigose with hooked hairs toward the tips; stems slender, branched mostly near base, sparsely strigose .... 6. *P. pusilla*
    BB.   Nutlets divergent in pairs; calyx appressed- or spreading-hirsute with straight hairs; stems branched upward, somewhat hispid .... 7. *P. setosa*

1. **P. lineàris** DC. var. **ferócula** Jtn. Stems very slender, prostrate to widely ascending, cinereous-strigose; lvs. filiform-linear to oblance-linear, 1–4 cm. long, strigose, numerous; calyx strigose, 1.5–2 mm. long; corolla 2 mm. long; nutlets homomorphous, the body linear- or somewhat spatulate-oblong, 2–3.5 mm. long, 0.5–1 mm. broad, divaricate or slightly falcate-recurved, the margin generally not conspicuous, divided into short crowded teeth, each of which is abruptly terminated by an uncinate bristle equal to or surpassing the total width of the cartilaginous margin beneath it.—Open grassy mesas and slopes, below 2500 ft.; Coastal Sage Scrub, Chaparral, V. Grassland, Foothill Wd.; San Benito Co. s. through cismontane s. Calif. to L. Calif.; Channel Ids. March–May.

2. **P. recurvàta** Jtn. With the habit of *P. linearis;* calyx surpassed by mature nutlets, these with a linear body, reflexed, distinctly falcate or even scorpioid, 3 mm. long, 0.8 mm. broad; nutlet-margin dissected, reduced to a series of quite distinct short teeth which are crowned by elongate hooked hairs of equal or greater length, the slender subulate marginal appendages thus formed well spaced and surpassing the width of the fr.—Sandy and gravelly slopes, below 4000 ft.; Creosote Bush Scrub, Joshua Tree Wd.; Mojave and Colo. deserts; to Nev., Ariz., Son., L. Calif. March–May.

3. **P. platycárpa** (M. & J.) M. & J. [*P. gracilis* var. *p.* M. & J.] Stems slender, stiffish, prostrate or widely ascending, 8–16 cm. long, cinereous-strigose; lvs. linear or oblance-linear, 1–3.5 cm. long, 0.5–1.2 mm. broad; calyx ca. as long as nutlets; corolla 2 mm.

long; nutlets usually heteromorphous, the body linear-or spatulate-oblong, 2.5–3 mm. long, 0.6–1 mm. broad, at least 3 nutlets with conspicuous wide coarsely toothed pallid margin, of which the irregular teeth are tipped with short hooked bristles that are much shorter than the cartilaginous margin beneath them, the odd nutlet may be narrower, with more dissected wing and more pubescent body.—Dry gravelly benches and slopes, below 4000 ft.; Creosote Bush Scrub, Joshua Tree Wd.; Mojave and Colo. deserts from cent. San Bernardino Co. s.; to Utah, Son., L. Calif. March–May.

4. **P. penicillàta** (H. & A.) A. DC. [*Cynoglossum p.* H. & A.] Stems many, slender, prostrate or widely ascending, cinereous-strigose, 5–20 cm. long; lvs. linear to narrow-spatulate, 1–3 cm. long, 0.5–2 mm. wide; calyx almost as long as nutlets; corolla 2 mm. long; nutlets homomorphous, with body oblong, divaricate, straight, 1.6–2.4 mm. long, 0.5–0.8 mm. wide, with a distinct upturned or incurved margin broadest near base and apex, subentire and armed only at rounded distal end with crowded hooked bristles.— Dry sandy and gravelly places, below 3500 (4500) ft.; many Plant Communities; cismontane Calif. away from the coast and to desert edge; to L. Calif., B.C., Wyo. March–May.

5. **P. heterocárpa** (Jtn.) Jtn. [*P. penicillata* var. *h.* Jtn.] Habit of *P. penicillata;* fruiting nutlets on principal branches heteromorphic, with 2 evidently margined and like those of *P. penicillata* and 2 unmargined and somewhat reflexed; nutlets from base of plant all unmargined and strongly reflexed.—Gravelly or sandy slopes, below 3000 ft.; Creosote Bush Scrub, Joshua Tree Wd., V. Grassland; Mojave and Colo. deserts, inner S. Coast Ranges, Stanislaus Co. to Kern Co., to Utah, Son., L. Calif. Feb.–May.

6. **P. pusílla** (A. DC.) Gray. [*Gruvelia p.* A. DC. *P. chilensis* var. *californica* Torr.] Stems slender, erect or ascending, somewhat flexuous, simple or few-branched, 3–20 cm. long, strigose; cotyledons persistent, conspicuous; lvs. ± linear, 8–20 mm. long, 1–2 mm. wide; calyx definitely longer than nutlets, hispidulous and with terminal hooked hairs; corolla ca. as long as calyx; nutlets homomorphous, 1–4, obovate or subrhomboid, uniformly divergent, 2–2.5 mm. long, ca. 2 mm. wide, marginless or narrow-margined at broad distal end, the back keeled, tuberculate, ± hirsutulous and with hooked hairs along the edge.—Wooded or open slopes, below 3000 ft.; Foothill Wd., N. Oak Wd., Chaparral; Kern and Monterey cos. to Siskiyou Co.; to Wash.; Chile. April–June.

7. **P. setòsa** Gray. [*P. s.* vars. *aptera* and *holoptera* Jtn.] Stems rather stiff and coarse, erect or ascending, dichotomous, scattered-setose and thinly strigose, 5–20 cm. high; lvs. oblong to spatulate or oblanceolate, 5–25 mm. long, 1–3.5 mm. broad, thickish, pustulate, stiff-hairy; calyx surpassing fr.; corolla 2 mm. long; nutlets divergent in pairs, ± heteromorphous, 3 usually with broad entire thin margin, rarely all 4 margined or all marginless; body of nutlets broadly cuneate-obovate, 2 mm. long, the margin and upper face with slender hooked bristles.—Dry gravelly or sandy places, below 5000 (7000) ft.; V. Grassland of inner Coast Ranges, San Joaquin Co. to Ventura Co.; Creosote Bush Scrub, Joshua Tree Wd., Pinyon-Juniper Wd., Sagebrush Scrub; Lassen Co. to Inyo Co. and Mojave and Colo. deserts; to L. Calif., Wash., Ida., Utah, Ariz. April–May.

5. Cynoglóssum L. Hound's Tongue

Perennial or biennial, mostly coarse tall herbs. Basal lvs. long-petioled, the upper sessile or nearly so. Fls. purple, blue or white, in usually bractless terminal panicles of scorpioid racemes. Calyx deeply 5-parted, the lobes often enlarged in fr. Corolla funnelform or salverform, with short tube; throat closed by 5 scales. Stamens included. Ovary deeply 4-lobed, becoming 4 nutlets; style slender. Nutlets equally divergent, horizontal, or obliquely ascending, large, covered with short barbed prickles. Ca. 75 spp.; widely distributed. (Greek, *kuno*, dog, and *glossa*, tongue, because of lf.-texture in some spp.)

Stems with spreading hairs; lower lvs. oblanceolate to oblong-linear, gradually narrowed to petioles
                                                      1. *C. occidentale*
Stems glabrous; lower lvs. ovate, abruptly narrowed to petioles .... 2. *C. grande*

1. **C. occidentàle** Gray. [*C. viride* Eastw.] Perennial from a heavy woody root-crown; stems erect, 2–4 dm. high, hirsute; lower lvs. oblanceolate, hirsute, the blades 5–15 cm. long; petioles winged, 5–10 cm. long; upper lvs. reduced, sessile or cordate-clasping; infl. rather compact, long-peduncled; pedicels ca. 4–8 mm. long; calyx 5–7 mm. long, the lance-linear lobes hirsute; corolla blue, tinged brown or pink, the tube 4–6 mm. long,

the limb ca. as broad; nutlets roundish, ca. 8 mm. long, evenly covered with short glochidi-ate spines.—Dryish openings in woods, 4000–7000 ft.; Yellow Pine F., Red Fir F.; Sierra Nevada from Kern Co. n., to Modoc, Siskiyou, Trinity, and Humboldt cos.; Ore. May–July.

2. **C. gránde** Dougl. ex Lehm. [*C. laeve* Gray. *C. Austinae* Eastw.] Perennial from a heavy root; stems erect, glabrous, 3–9 dm. high; lvs. mostly basal or on lower stem, ovate, glabrous or sparsely short-hirsute, the blades ovate, 8–15 cm. long, abruptly narrowed into petioles often as long; peduncle well developed; panicle lax; pedicels commonly 1–2.5 cm. long; calyx-lobes 5–7 mm. long, narrow-oblong, pubescent; corolla 8–12 mm. long, blue, the tube often purple, the lobes rounded; nutlets 5–6 mm. long, depressed, glochidiate-spinose.—Dry shaded slopes, below 4000 (5000) ft.; N. Oak Wd., Mixed Evergreen F., Yellow Pine F., etc.; Coast Ranges from San Luis Obispo Co. n., Sierra Nevada from Tulare Co. n., to Butte and Siskiyou cos.; to Wash. March–June.

6. Sýmphytum L. COMFREY

Coarse hairy perennial herbs with thick roots and entire lvs. Fls. purple or yellow-blue, in scorpioid racemes which lengthen in age. Calyx deeply 5-cleft. Corolla tubular, shallowly 5-lobed, its throat with 5 lanceolate crests alternating with and equaling the stamens. Stamens included, inserted on corolla-tube. Nutlets 4, wrinkled, erect, attached basally to a flat receptacle and with concave attachment scar. Ca. 15 spp., Old World. (Greek, *symphyton*, grown together, because of supposed healing virtues.)

1. **S. ásperum** Lepechin. [*S. asperrimum* Donn.] Root thick, deep, branched; stems branched, 6–12 dm. high, uncinate-spinulose; lvs. harsh, lance-ovate to somewhat oblong, 5–15 cm. long, all but the upper petioled; calyx 3–5 mm. long; corolla bluish-purple, 1–1.5 cm. long, with erect lobes; nutlets 4 mm. long, constricted above the base, tuberculate.—Natur. about Humboldt Bay; native of Asia. May–July.

7. Lycópsis L. BUGLOSS

Coarse setose annuals with alternate lvs. Racemes leafy-bracted, spikelike, scorpioid. Calyx 5-parted. Corolla salverform with curved tube and slightly unequal limb, the throat closed by stiff hairs. Stamens and style included. Nutlets 4, rough-wrinkled, erect, with concave attachment-scar. Four to 5 spp. of Old World. (Greek, *lycos,* wolf, and *opsis,* appearance.)

1. **L. arvénsis** L. Erect, 3–6 dm. high, rough-bristly; lvs. lanceolate, 3–5 cm. long; calyx 3–4 mm. long; corolla blue or purplish, the tube 3–4 mm. long, the limb 4–5 mm. broad.—Once collected at Upland, San Bernardino Co.; native of Eu.

8. Myosòtis L. FORGET-ME-NOT

Slender rather low annual or perennial herbs. Lvs. alternate, entire. Fls. small, in loose racemes, blue, white or pink. Calyx 5-cleft, the narrow lobes erect, or spreading in fr. Corolla salverform, 5-lobed, with prominent crests in throat. Stamens 5, included. Nutlets 4, small, ovoid, smooth, shining, sharply margined, the attachment-scar flat. Ca. 35 spp., widely distributed in temp. regions. (Greek, *mus,* mouse, and *otos,* ear, because of appearance of lvs. of some spp.)

Hairs of calyx appressed, short, straight; perennials.
  Corolla-limb 5–9 mm. wide; corolla-tube definitely exceeding calyx; style overtopping nutlets, ca.
  as long as calyx-tube . . . . 1. *M. scorpioides*
  Corolla-limb 2.5–5 mm. wide; corolla-tube ca. as long as calyx; style much exceeded by nutlets and
  calyx-tube . . . . 2. *M. laxa*
Hairs of calyx ± spreading and some are hooked at end; annuals or biennials.
  Calyx very unequally cleft, ± 2-lipped; corolla white, 1–2 mm. broad . . . . 3. *M. virginica*
  Calyx equally cleft, not 2-lipped; corolla mostly blue.
    Corolla-limb 5–8 mm. broad; fruiting pedicels longer than calyx . . . . 4. *M. sylvatica*
    Corolla-limb 2–4 mm. broad; fruiting pedicels shorter than calyx.
      Fls. scattered among lvs. to near base of stem, as well as in terminal racemes . 5. *M. micrantha*
      Fls. not scattered among lower lvs., in terminal racemes only . . . . 6. *M. versicolor*

1. **M. scorpioìdes** L. [*M. palustris* Lam.] Perennial, at first erect, later decumbent, appressed-pubescent, the slender stems 1.5–5 dm. long; lvs. oblong to oblanceolate, 3–8

cm. long, the lower with winged petioles; racemes many-fld.; pedicels divergent in fr., longer than calyx; corolla blue with a yellow eye, the limb 5–9 mm. wide; nutlets angled, keeled on inside.—Escape from cult., as in Plumas, Modoc, Eldorado, and Siskiyou cos.; native of Eurasia. May–July.

2. **M. láxa** Lehm. Perennial with slender decumbent stems 1.5–5 dm. long, strigose; lvs. oblong to spatulate, 2–6 cm. long; pedicels divergent and exceeding calyx in fr.; calyx 3–4 mm. long, strigose; corolla pale blue with a yellow eye, the limb 2.5–5 mm. wide; nutlets convex on both surfaces.—Wet places, Lake Talawa, Del Norte Co., and at 3500–4500 ft., Plumas Co.; to B.C. and Atlantic Coast. May–Aug.

3. **M. virgínica** (L.) BSP. [*Lycopsis v.* L.] Annual or biennial, erect, hirsute, 1–3 dm. high; lvs. linear to oblong or oblanceolate, 1–4 cm. long, sessile or the lower short-petioled; racemes bracteate at base or to middle; pedicels suberect, shorter than fruiting calyx; calyx 4–6 mm. long, uncinate-hirsute; corolla white, the limb 1–2 mm. wide; nutlets convex on dorsal side, keeled on ventral.—Adventive in grain fields, Trinity and Humboldt cos.; to B.C., Atlantic Coast. April–June.

4. **M. sylvática** Hoffm. Common Forget-Me-Not. Perennial, the stems 2–5 dm. long, erect to decumbent, ± hirsute; lvs. oblong to lanceolate, 2–6 cm. long, the upper sessile, the lower more spatulate, petioled, all ± strigose; fruiting pedicels ascending to spreading, 6–9 mm. long; calyx 4–5 mm. long, uncinate-pubescent; corolla light blue, the limb 5–8 mm. wide; nutlets 1.5–2 mm. long.—Garden plant, escaping in moist shaded places near the coast from San Mateo Co. n.; native of Eurasia. Feb.–July.

5. **M. micrántha** Pall. in Lehm. Annual with slender stems 1–2 dm. high, ± erect, thinly hairy below, strigose upward; lower lvs. oblanceolate, petioled, 1–2 cm. long; upper lvs. sessile; pedicels in fr. 1–1.5 mm. long; calyx ca. 4 mm. long, uncinate-hirsute; corolla blue, 1.5 mm. wide; style shorter than nutlets; nutlets ca. 1 mm. long.—Reported from coastal nw. Calif.; native of Eu. May–July.

6. **M. versícolor** (Pers.) Sm. [*M. arvensis* var. *v.* Pers.] Annual 1–3 dm. tall, spreading-hirsute below, strigose above; lower lvs. spatulate, 1–2.5 cm. long, petioled; upper oblong to lance-oblong, subsessile; fruiting pedicels 1–2.5 mm. long; calyx 4–4.5 mm. long, with short hooked hairs; corolla yellow becoming blue, ca. 2 mm. broad; style longer than nutlets; nutlets 1–1.4 mm. long.—Moist places, below 4000 ft.; Sonoma and Eldorado cos. n.; to Wash.; native of Eu. April–July.

9. Merténsia Roth. Lungwort

Perennial herbs with erect leafy stems. Lvs. alternate, subglabrous, broad. Fls. blue, purple to white, in terminal panicles. Calyx 5-cleft or -parted, the linear to lanceolate lobes little enlarged in fr. Corolla tubular to campanulate, not appendaged in the throat. Stamens with flattened or filiform fils., included or slightly exserted. Style filiform; stigma entire. Nutlets erect, wrinkled when mature, attached to the convex receptacle by a small scar above their base. Ca. 45 spp., of the N. Hemis. (F. K. *Mertens,* 1764–1831, German botanist.)

(Williams, L. O., Ann. Mo. Bot. Gard. 24: 17–159, 1937.)
Corolla campanulate, not divided into a tube and limb. (Sect. Neuranthia) . . . . 1. *M. bella*
Corolla not campanulate, divided into a tube and limb (Sect. Mertensia).
    Plants with a shallow-seated tuberous root; lower lvs. subsessile or with short broad petioles; corolla-tube 3–4 times the length of rest of corolla. Modoc Co. . . . . 2. *M. longiflora*
    Plants with an elongate stout woody root; lower lvs. long-petioled; corolla-tube 1–2 times the length of rest of corolla.
        Stems mostly 5–15 dm. high; lf.-blades mostly lanceolate to ovate, acute, the cauline with evident lateral veins . . . . 3. *M. ciliata*
        Stems mostly 1–3 dm. high; lf.-blades mostly oblong, obtuse, the cauline without evident lateral veins . . . . 4. *M. oblongifolia*

1. **M. bélla** Piper. [*M. siskiyouensis* Appleg.] Stems solitary, slender, rather sparsely branched to simple, 2–5 dm. high, glabrous to sparsely pilose, arising from a shallow tuberous root; lvs. elliptic to ovate, the upper sometimes opposite, the lower 2–5 cm. long, on petioles ca. as long, glabrous beneath, strigose above; racemes 2–4; pedicels 6–12 mm. long, strigose, slender; calyx strigose, 3–4 mm. long, the lobes acute; corolla bright blue, 5–8 mm. long, the tube ca. 2 mm. long, the limb campanulate, with broadly ovate

lobes 2 mm. long; style ca. half as long as corolla; nutlets ca. 1.5 mm. long.—Moist places, ca. 5000–6000 ft.; Red Fir F.; Siskiyou Mts. of Ore. near state line. May–July.

2. **M. longiflòra** Greene. Stems 1 (2 or 3), erect, simple, slender, 1–2 dm. high, from a shallow tuberous root, glabrous; lvs. mostly cauline, oblong-ovate to spatulate-obovate, the blades glabrous to hairy above, glabrous beneath, 2–6 cm. long, short-petioled to sessile; infl. rather compact; pedicels 1–6 mm. long, subglabrous; calyx-lobes lanceolate, 3–5 mm. long, ciliate; corolla bright blue, the tube 9–14 mm. long, the limb 4–7 mm. long, moderately expanded; fils. ca. as broad as anthers; style ca. as long as corolla; nutlets 3–4 mm. long, rugose.—Dry slopes, 5000–7000 ft.; Yellow Pine F.; Modoc Co.; to B.C., Mont. April–June.

3. **M. ciliàta** (James) G. Don var. **stomatechoìdes** (Kell.) Jeps. [*M. s.* Kell.] Stems erect or ascending, mostly 5–10(–15) dm. high, usually several from the branching rootstock, glabrous; lf.-blades lanceolate to ovate or oblong, 5–12 cm. long, ciliate, often papillate on upper surface, the lower subcordate at base, long-petioled, the upper often attenuate at base, subsessile; infl. lax in age; pedicels 3–14 mm. long; calyx-lobes 2.5–6 mm. long, usually obtuse; corolla-tube 6–8 mm. long, the limb 4–10 mm. long, only slightly expanded; style exceeding corolla; nutlets rugose or mammillate.—Moist places, usually in the shade, 5000–10,200 ft.; Montane Coniferous F.; Sierra Nevada from Tulare Co. to Modoc Co.; Ore., Nev. May–Aug.

4. **M. oblongifòlia** (Nutt.) G. Don. [*Pulmonaria o.* Nutt.] Stems 1–several, erect or ascending, 1–3 dm. tall, glabrous, from the crown of each elongate rootstock; basal lf.-blades oblong or spatulate, 3–8 cm. long, obtusish, strigose above, glabrous beneath, on petioles up to length of blade, the cauline shorter and becoming sessile; infl. congested; pedicels glabrous to strigose, 1–10 mm. long; calyx 3–7 mm. long, the lobes ciliate, sometimes also pubescent; corolla blue, the tube 5–10 mm. long, the limb 4–7 mm. long; styles included; nutlets 3–4 mm. long, rugose.—Moist places, 4500–6500 ft.; Sagebrush Scrub to Yellow Pine F.; Modoc Co.; to Wash., Mont. April–July.

Var. **nevadénsis** (A. Nels.) L. Williams. [*M. n.* A. Nels.] Lvs. essentially glabrous on both surfaces, sometimes somewhat pustulate.—E. slope of Sierra Nevada, Sierra Co. to Modoc Co.; to Wash., Mont., Wyo.

Var. **amoèna** (A. Nels.) L. Williams. [*M. a.* A. Nels.] Lvs. villous-pubescent on both surfaces.—Modoc Co.; to Wash., Mont., Wyo.

10. **Lithospérmum** L. Gromwell. Puccoon

Annual to perennial hairy herbs, the roots usually with red or violet coloring matter. Stems few or several, simple below. Lvs. alternate, mostly rather narrow. Fls. small, yellow, white or blue, in leafy spikes or upper axils. Calyx 5-parted. Corolla funnelform or salverform, with rounded lobes imbricated in bud. Stamens 5, included, inserted in corolla-throat; fils. short; style slender; stigma truncate-capitate or 2-lobed. Nutlets 4 or fewer, erect, attached by their bases to the nearly flat receptacle, the attachment-scar flat. Ca. 50 spp., of all continents except Australia. (Greek, *lithos*, stone, and *sperma*, seed.)

Plants annual; nutlets tubercled, dull . . . . 1. *L. arvense*
Plants perennial; nutlets smooth, shining.
    Upper lvs. crowded, lance-linear, with attenuate apex; fls. greenish-yellow . . . . 2. *L. ruderale*
    Upper lvs. scattered, lance-ovate, acute to obtuse; fls. golden-yellow . . . . 3. *L. californicum*

1. **L. arvénse** L. Strigose annual 2–5 dm. high; lvs. linear to linear-lanceolate, sessile, 2–2.5 cm. long; fls. white, ca. 6 mm. long; nutlets brown, wrinkled and pitted, ca. 2 mm. long; $2n = 28$ (Britton, 1951).—About San Francisco Bay; native of Eu. May–July.

2. **L. ruderàle** Dougl. ex Lehm. Perennial with branched root-crown; stems several, 1.5–4 dm. high, hirsute; lf.-blades linear to narrow-lanceolate, sessile, silky-pubescent, 3–8 cm. long, 2–10 mm. wide; fls. in upper axils of short terminal branchlets, the pedicels stout, 1–3 mm. long; calyx-lobes 7–10 mm. long in fr.; corolla greenish-yellow, 10–12 mm. long, the tube broad, scarcely dilated at throat, the lobes 3 mm. long; nutlets broadly ovoid, 4–5 mm. long, whitish, smooth.—Dry slopes and plains, 4500–6000 ft.; Sagebrush Scrub, N. Juniper Wd.; Placer Co. to Modoc Co.; to B.C., Rocky Mts. May–June.

3. **L. califórnicum** Gray. [*L. ruderale* var. *c.* Jeps.] Stems few to several from the stout root, erect or ascending, 1.5–4 dm. high, hirsute; lower lvs. lance-linear to lanceo-

late, 1–1.5 cm. long, the upper lance-oblong to -ovate, harsh-strigose, 3–6 cm. long; infl. usually subpaniculate, rather congested; calyx ca. 6–9 mm. long; corolla golden-yellow, 12–18 mm. long, the tube slender, evidently dilated into throat, the limb 7–8 mm. broad; nutlets broadly ovoid, white, shining.—Dry slopes and ridges, 1000–5000 ft.; Yellow Pine F., N. Oak Wd.; Coast Ranges from Tehama Co. to Del Norte and Siskiyou cos., and Sierra Nevada from Placer Co. n.; Ore. April–June.

11. Èchium L. Viper's Bugloss

Herbs or shrubs, usually scabrous, hispid, or canescent. Lvs. alternate. Fls. in scorpioid simple or forked cymes, with bracts small or leafy. Calyx with 5 narrow lobes. Corolla tubular to funnelform, with dilated oblique throat, without appendages, the corolla-lobes roundish, unequal. Stamens 5, inserted below the middle of the corolla-tube, unequal and exserted. Style filiform. Nutlets 4, erect, wrinkled. Ca. 35 spp., of Old World. (Greek, *echis*, a viper, the nutlets supposed to resemble a viper's head.)

1. **E. plantagíneum** L. Annual or biennial, erect or diffuse, 4–9 dm. high, villous-hirsute with hairs pustulate at base; lvs. obtuse, strigose, the basal oval and petioled, the upper lance-oblong, subsessile; fls. mostly bracteate; calyx-lobes lanceolate, acuminate, ca. 10 mm. long; corolla blue, 15–20 mm. long, irregular with campanulate throat and oblique limb; $2n = 16$ (Britton, 1951).—Locally adventive as at De Luz, San Diego Co. and Carmel, Monterey Co.; native of S. Eu. May–June.

12. Láppula Moench. Stickseed

Annual or biennial herbs with linear or oblong lvs. Fls. in paniculate leafy-bracted racemes, white or somewhat yellowish, usually on erect pedicels. Calyx 5-parted. Corolla with short tube and rounded ascending lobes, the throat usually appendaged. Stamens short, attached in corolla-tube. Style short, surmounting the columnar gynobase, usually surpassing the nutlets. Stigma subcapitate. Nutlets erect, smooth to roughened, attached along their ventral keel to the elongate gynobase, commonly with 1 or more marginal series of glochidiate appendages. Ca. 14 spp., mostly N. Temp. (Diminutive of Latin, *lappa*, a bur.)

(Johnston, I. M. Contrib. Gray Herb. 70: 47–51, 1924.)

Corolla ca. 3 mm. broad; nutlets with 2 rows of slender marginal spines that are not confluent at base
                                                                            1. *L. echinata*
Corolla ca. 2 mm. broad; nutlets with 1 row of marginal spines that are ± confluent at base
                                                                            2. *L. Redowskii*

1. **L. echinàta** Gilib. Erect annual, simple to freely branched, 1.5–5 dm. high, villous-hirsute with hairs appressed upward; lower lvs. oblanceolate, others linear, sessile, ascending, 2.5–5 cm. long; racemes bracteate; pedicels 1–3 mm. long; calyx-lobes linear, ca. 3 mm. long and spreading in fr.; corolla blue, the tube surpassing the calyx; nutlets 3.5–4 mm. long, muricate-prickly on backs and with 2 rows of slender nonconfluent prickles on margin.—Occasionally adventive (Santa Monica, Upland); native of Eurasia. June–Aug.

2. **L. Redówskii** (Hornem.) Greene. [*Myosotis* R. Hornem. *L. occidentalis* Greene.] Annual, the stems erect, subsimple or with ascending floriferous branches above the middle, 2–8 dm. tall, finely hispid or hispid-villous, somewhat cinereous; lvs. rather numerous, linear to lance-linear or oblong, the upper sessile, 1–2 cm. long, the lower petioled and longer; pedicels 1–2 mm. long; calyx-segms. lance-linear, ca. 3 mm. long; corolla blue to ochroleucous, 2–3.5 mm. long; nutlets 2.5 mm. long, tubercled or muricate and bordered by a single row of barbed prickles 1–2 mm. long and nearly distinct.— Dry open slopes and flats, 5000–8000(–11,150) ft.; mostly Yellow Pine F., Pinyon-Juniper Wd., Sagebrush Scrub; San Bernardino Mts., Panamint Mts., White Mts., e. side of Sierra Nevada to Modoc Co.; to Wash., Dakotas, Tex.; Eurasia, Argentina. April–July.

Var. **desertòrum** (Greene) Jtn. [*L. cupulata* and *L. texana* auth.] Marginal prickles of nutlets confluent, forming a cupulate margin, the body often smoothish.—Dry places, 2000–7500 ft.; Creosote Bush Scrub, Sagebrush Scrub, to Pinyon-Juniper Wd.; e. Siskiyou Co. to San Bernardino Co., across the desert to New Mex., Wyo., Wash.

13. Hackèlia Opiz. STICKSEED

Perennial or biennial herbs with well developed stems and alternate linear to oblong lvs. Infl. naked or bracteate, paniculate; pedicels recurved or deflexed in fr. Calyx 5-parted, spreading or reflexed in fr. Corolla blue, sometimes white or pink, rotate to salverform, with crested throat. Style shorter than nutlets. Nutlets attached below the middle to the broadly pyramidal gynobase by a large ovate or deltoid areola, and armed with flattened barbed prickles over the entire back or mostly along the margin. Ca. 35 spp., mostly N. Temp. (P. *Hackel*, German professor of agriculture.)

A. Dorsal face of the nutlets naked or with few prickles, the margin with a conspicuous border of strongly flattened barbed ones.
    B. Racemes 2–3, terminal, umbellate, not bracteate; stems slender, 1–3 dm. high. Mt. Whitney region .... 1. *H. Sharsmithii*
    BB. Racemes several to many, paniculately arranged, bracteate.
        C. Corolla 5–7 mm. broad.
            D. Plants 1.5–6 dm. high; infl. usually with less than 10 branches, these divergent, open; nutlets usually with a few small prickles on the back.
                E. Stem-lvs. not conspicuously reduced in size up the stems; basal lvs. few; nutlets broadly ovoid, the marginal prickles ca. 10, broadly dilated at base, distinct, often with a shorter one in between; stems 3–6 dm. long, villous-hirsute. Common in mts. from Tulare Co. to Siskiyou and Trinity cos.
                        2. *H. Jessicae*
                EE. Stem-lvs. conspicuously reduced in size up the stems; basal lvs. many; nutlets with marginal prickles somewhat united at base, usually with 1–2 shorter ones between the main ones; stems 1–3 dm. long.
                    F. Corolla blue with a pale eye; stems closely strigose; nutlets narrow-ovoid. Lassen and Modoc cos. .... 3. *H. Cusickii*
                    FF. Corolla whitish or tinged with pale blue; stems with spreading-deflexed hairs; nutlets broadly ovoid. White Mts. .... 4. *H. patens*
            DD. Plants 5–10 dm. high; infl. with many suberect branches; nutlets mostly naked on back. Mono Co. to Modoc Co. .... 5. *H. floribunda*
        CC. Corolla 8–18 mm. broad.
            D. Fls. blue, the corolla-tube not exceeding the calyx.
                E. Herbage short-tomentose, with mostly appressed hairs; corolla 8–10 mm. broad .... 6. *H. amethystina*
                EE. Herbage bristly-hirsute; corolla 12–15 mm. broad .... 7. *H. setosa*
            DD. Fls. white, 12–18 mm. broad, the corolla-tube longer than calyx .. 8. *H. bella*
AA. Dorsal face of the nutlets evenly beset with prickles, these not much dilated and not forming a marginal membrane.
    B. Corolla rotate.
        C. Fls. white, 6–10 mm. broad; prickles stout, up to 2.5 mm. long. Lake and Alpine cos. n.
                        9. *H. californica*
        CC. Fls. pink or bluish; prickles slender, up to 5 mm. long. Tuolumne Co. to Tulare Co.
                        10. *H. mundula*
    BB. Corolla short-salverform, the tube exceeding the calyx.
        C. Fls. 6–8 mm. broad; corolla-appendages papillate-roughened; lvs. thinly hispidulous
                        11. *H. nervosa*
        CC. Fls. 12–20 mm. broad; corolla-appendages hairy; lvs. velvety-pubescent.
            D. Corolla-tube 3–4 mm. long. Tulare Co. to Yosemite .... 12. *H. velutina*
            DD. Corolla-tube 5–6 mm. long. Tuolumne Co. to Placer Co. .... 13. *H. longituba*

1. **H. Sharsmíthii** Jtn. [*Lappula S.* Jeps. & Bail.] Stems several, erect or ascending, slender, 1–3 dm. high, reflexed-strigose, from the stout branched perennial root-crown, of which the branches are conspicuously clothed with old lf.-bases; lvs. bright green, minutely strigulose, elliptic to almost ovate, 1.5–4 cm. long, the lower on petioles as long or longer, the upper sessile or nearly so and reduced; racemes 2–3, mostly 2–5 cm. long; pedicels 1–6 mm. long; calyx 2–2.5 mm. long, strigose; corolla white to bluish with yellow center, ca. 5–6 mm. broad; appendages lunate, ciliate; nutlets ovoid, ca. 2–5 mm. long, the marginal prickles distinct or partly united, the dorsal 1–several, very short.—Local, under overhanging rocks, 10,750–12,000 ft.; Subalpine F., Alpine Fell-fields; region of Mt. Whitney, Inyo and Tulare cos. July–Sept.

2. **H. Jéssicae** (McGreg.) Brand. [*Lappula J.* McGreg. *L. micrantha* Eastw., not Opiz. *Hackelia Eastwoodiae* Jtn. *L. floribunda* var. *J.* Jeps. & Hoov. *L. f.* var. *Geisiana* Jeps.] Stems few, erect or ascending, ± villous-hirsute, 3–6 dm. high, from a heavy root; basal lvs. few, oblanceolate, the blades 7–15 cm. long, villous, with winged petioles ca. as long; upper lvs. sessile, well distributed along the stems, gradually reduced upward; panicle open, of several divergent cymes; pedicels 5–10 mm. long, the rather coarse hairs ± appressed; calyx 2–3 mm. long; corolla pale blue, 4–6 mm. wide,

the appendages rounded, puberulent; nutlets 4–6 mm. long, the marginal prickles ca. 10, broad at base, fairly distinct, with occasional shorter ones between, the dorsal face broadly ovate, becoming ± muriculate and with a few (or 1) short prickles.—Common in moist places, 4500–11,000 ft.; Montane Coniferous F., Sagebrush Scrub; Sierra Nevada from Tulare Co. n., Coast Ranges from Glenn Co. to Humboldt and Siskiyou cos.; to B.C., Nev. July–Aug.

3. **H. Cusíckii** (Piper) Brand. [*Lappula C.* Piper. *L. arida* var. *C.* Nels. & Macbr.] Stems slender, erect, 1–few, 1.5–4 dm. high, from crown of the slender woody taproot; herbage bluish-canescent, densely strigose; basal lvs. many, lanceolate, 2.5–6 cm. long, on petioles up to same length; stem-lvs. few, reduced, lance-linear, sharply acute; panicle few-branched, lax; pedicels 4–6 mm. long; calyx ca. 3 mm. long; corolla blue, 5–10 mm. broad, the appendages short-pilose, emarginate; nutlets narrowly ovate, 4–5 mm. long, the marginal prickles almost as long as the width of the nutlet, 3–5 prickles on each side broad-based and with slender shorter ones between, the dorsal face of nutlet muriculate and with a few short prickles.—Occasional, dry slopes, ca. 5000–6500 ft.; N. Juniper Wd., Sagebrush Scrub; Lassen Co. to Modoc Co.; Ore. May–July.

4. **H. pàtens** (Nutt.) Jtn. [*Rochelia p.* Nutt. *Lappula coerulescens* var. *brevicula* Jeps.] Caudex woody; stems 2–3, ascending-erect, 2.5–4 dm. high, with deflexed-spreading hairs; basal lvs. oblanceolate, roughish-pubescent, 5–10 cm. long, on petioles to ca. as long; stem-lvs. much reduced; cymes few; pedicels 2–6 mm. long; calyx 2–2.5 mm. long; corolla white or tinged pale blue, 6–10 mm. broad, the appendages somewhat broader than long, papillose; nutlets with an ovate face, ca. 3.5 mm. long, the marginal prickles 3–5 on each side, with some shorter between and few or almost none on the dorsal surface.—At 9500–10,000 ft.; Poison Canyon, White Mts., Mono Co.; to Ida., Mont., Utah. July–Aug.

5. **H. floribúnda** (Lehm.) Jtn. [*Echinospermum f.* Lehm. *Lappula f.* Greene.] Stem one, erect from short-lived perennial root, 5–12 dm. high, deflexed-pubescent; lower lvs. oblanceolate, 5–12 cm. long, on petioles ca. as long, with somewhat stiff appressed hairs; cauline lvs. reduced upward, ± oblong, the uppermost sessile; racemes many, rather short, strict, densely fld.; pedicels 5–7 mm. long in fr.; calyx-lobes oblong to lance-ovate, densely strigose, ca. 2.5 mm. long; corolla blue, 6–7 mm. broad, the small appendages obscurely papillose; nutlets ca. 4 mm. long, the face not prickly, but muriculate and hirsutulous, the margin with 4–6 ± confluent flattened prickles on each side.—Occasional, 4000–7000 ft.; Sagebrush Scrub, N. Juniper Wd.; Mono Co. to Modoc Co.; to B.C., Ontario, New Mex. June–Aug.

6. **H. amethystìna** J. T. Howell. Stems stout, erect, 5–7 dm. high, ± canescent with short subappressed or recurved hairs, the root-crown woody; basal lvs. narrow-elliptic, 8–12 cm. long, short-hairy, on petioles ca. as long, the cauline well developed, gradually reduced upward, lanceolate, the uppermost subovate, sessile; infl. short-canescent, with several divergent lax branches; pedicels to 1 cm. in fr.; calyx 2–3 mm. long; corolla blue, 9–10 mm. broad, the tube 2–3 mm. long, the appendages quadrate, conspicuous, papillose; nutlets 5–6 mm. long, the face ovate, tuberculate, minutely granulate, with a few short prickles, the marginal slender prickles 9–13 on each side, 3–5 mm. long.—Wooded slopes, 5000–6500 ft.; Yellow Pine F.; Tehama and Glenn cos. to Mendocino Co. June–July.

7. **H. setòsa** (Piper) Jtn. [*Lappula s.* Piper.] Stems 1–several from the root-crown, suberect, 3–5 dm. high, bristly-hirsute below, ± strigose above; lower lvs. linear-oblanceolate, 5–10 cm. long, on petioles ca. as long; upper lvs. reduced, sessile; infl. few-branched; pedicels 5–10 mm. long; calyx-lobes linear-oblong, 3–4 mm. long; corolla blue, 10–15 mm. broad, the appendages semicircular, pilose; nutlets 4–5 mm. long, the dorsal surface with several short prickles, the margins with long and short alternating prickles united at bases to form a narrow wing.—Open wooded ridges or slopes, 1000–6000 ft.; Yellow Pine F., Red Fir F.; Lake Co. to Del Norte Co., Sierra Co. to Siskiyou Co.; Ore. June–July.

8. **H. bélla** (Macbr.) Jtn. [*Lappula b.* Macbr. *L. Rattanii* Brand.] Stems erect, 5–7 dm. high, from a woody root; herbage strigulose in upper parts, retrorsely so in lower; lower lf.-blades oblanceolate or broader, 10–15 cm. long, on petioles ca. as long, the upper lance-ovate to oblong, reduced and sessile; infl. ample, open; pedicels 10–15 mm.

in fr.; calyx-lobes oblong, 4–5 mm. long; corolla white, the tube ca. 4 mm. long, the limb 12–18 mm. broad; appendages pubescent; nutlets 6–7 mm. long, the dorsal face muriculate, with few short prickles, the margin with 1 row of slender ± distinct obscurely barbed prickles.—Open slopes, 4000–6500 ft.; Red Fir F., Yellow Pine F.; Trinity Co. to Siskiyou and Mendocino cos. June–July.

9. **H. califórnica** (Gray) Jtn. [*Echinospermum c.* Gray. *Lappula c.* Piper. *H. elegans* Brand.] Perennial with woody root-crown; stems suberect, leafy, 4–8 dm. high, densely villous with short spreading or deflexed hairs; basal lf.-blades oblong-oblanceolate, 6–15 cm. long, on petioles to ca. as long, densely to sparsely strigose; upper lvs. lanceolate to narrow-oblong, sessile; panicles large, widely branched in age; pedicels 6–8 mm. in age; calyx-lobes ovate, 2–2.5 mm. long, densely substrigose; corolla white, 6–12 mm. broad, the tube broad, ca. 2.5 mm. long, appendages puberulent, broader than long; nutlets 5 mm. high, the dorsal face dull, muriculate, covered with subequal prickles.—Open mixed woods, 4000–8000 ft.; Yellow Pine F., Red Fir F.; Coast Ranges from Lake Co. to Humboldt Co., Sierra Nevada from Alpine Co. n., to Modoc Co. and Siskiyou Co.; Ore. June–Aug.

10. **H. múndula** (Jeps.) Ferris. [*Lappula californica* var. *m.* Jeps.] Much like *H. californica*, with softer pubescence; pedicels 8–14 mm. in fr.; calyx 3.5–4 mm. long; corolla pink with whitish center, fading blue, 10–18 mm. broad, the tube ca. as long as calyx, the appendages conspicuous, erect-spreading, puberulent; nutlets 6–7 mm. long, the dorsal face shining, smooth between the prickles.—Open rather protected woods, 7000–9500 ft.; Red Fir F., Lodgepole F.; Sierra Nevada from Tulare Co. to Mariposa Co. June–July.

11. **H. nervòsa** (Kell.) Jtn. [*Echinospermum n.* Kell. *Lappula n.* Greene.] Stems 1–few, erect, 2–5 dm. high, glabrous to spreading-pilose; basal lf.-blades oblong-spatulate or narrower, 4–15 cm. long, strigose, on petioles to ca. as long; cauline lvs. shorter, subsessile, the upper lance-ovate; infl. open, of few to several spreading cymes; pedicels slender, to ca. 8 mm. long; calyx-lobes oblong, 2.5–3 mm. long; corolla blue, the tube ca. 4 mm. long, the limb 6–8 mm. broad, appendages ⅙ as long as corolla-lobes; nutlets 5 mm. long, the dorsal face hispidulous and with long barbed prickles.—Rather moist places, 5000–10,000 ft.; Red Fir F., to Subalpine F.; Sierra Nevada from Plumas Co. to Fresno Co. July–Aug.

12. **H. velutìna** (Piper) Jtn. [*Lappula v.* Piper.] Stems 1–few from a woody root, 3–7 dm. high; herbage rather soft-pubescent; basal lvs. oblong-oblanceolate, 5–10 cm. long, on equally long petioles, the cauline lanceolate to somewhat narrow-oblong, gradually reduced upward and sessile; infl. rather compact, of several few-fld. branches spreading in fr.; pedicels 5–10 mm. in fr.; calyx-lobes oblong, canescent, 2.5–3 mm. long; corolla blue to pink, the tube ca. 3.5–5 mm. long, the limb 8–12 mm. broad; appendages spreading, broad, conspicuously exserted; nutlets 5–6 mm. long, the dorsal face shining, densely beset with prickles 2.5–3.5 mm. long and strongly flattened.— Dry wooded slopes, 5000–10,000 ft.; Red Fir F., Lodgepole F.; Sierra Nevada from Tulare Co. to Mariposa Co. June–Aug.

13. **H. longitùba** Jtn. Much like *H. velutina,* the pedicels to 12 mm. in fr.; calyx-lobes 2–3 mm. long; corolla blue, the tube 5–7 mm. long, the limb 10–16 mm. broad, the appendages rather narrow, conspicuously exserted, with revolute tips; nutlets ca. 5 mm. long, muriculate on the dorsal surface and covered with slender blue prickles to 4 mm. long.—Dry flats and slopes, 6000–7500 ft ; Red Fir F.; Sierra Nevada from Tuolumne Co. to Placer Co. June–July.

14. Cryptántha Lehm.

Annual or perennial herbaceous or suffruticose plants, usually setose or hispid. Earliest lvs. opposite, the others alternate. Fls. in bractless or bracteate usually scorpioid spikes or racemes, rarely somewhat cymose-paniculate. Calyx usually 5-parted, persistent or readily deciduous. Corolla white, rarely yellow, usually with a short included tube and spreading limb, the throat with intruded appendages or scales. Ovules 4, rarely 2. Nutlets 1–4, erect, affixed along a ventral slit in the pericarp to a columnar subulate or pyramidal gynobase, rough or smooth, the margin rounded or angled or winged. Ca. 65

spp. of New World, mostly w. N. Am., some in sw. S. Am. (Greek, *cryptos,* hidden, and *anthos,* fl., because of the minute corolla in the spp. first known.)

(Johnston, I. M. The N. Am. species of Cryptantha. Contr. Gray Herb. 74: 1–114, 1925. Payson, E. B. A monograph of the sect. Oreocarya of Cryptantha. Ann. Mo. Bot. Gard. 14: 211–358, 1927.)

A. Plants annual, with slender stems (of longer duration in *C. racemosa*).
    B. Calyx circumscissile at maturity; low diffuse plant; infl. compact, each fl. in axil of leafy bract
                                          1. *C. circumscissa*
    BB. Calyx not circumscissile.
        C. Gynobase subulate, protruding beyond the nutlets, bearing a sessile stigma on its tip; root and base of plant with a purple dye; each fl. in the axil of a leafy bract
                                          2. *C. micrantha*
        CC. Gynobase shorter than the nutlets; style developed; root or herbage usually with very little or no purple dye; fls. all or in part bractless (except *C. maritima*).
            D. Nutlets roughened or (in *C. maritima*) at least one of them so.
                E. Margins of nutlets decidedly winged or knifelike.
                    F. Pedicels usually evident, slender, 1–4 mm. long; lateral angles of nutlets distinctly winged.
                        G. Nutlets homomorphous, broadly winged .... 3. *C. holoptera*
                        GG. Nutlets heteromorphic, narrowly winged .... 4. *C. racemosa*
                  FF. Pedicels obscure or none, less than 1 mm. long.
                    G. Lateral margins of the nutlets usually distinctly winged; nutlets 4; calyx symmetrical.
                      H. Corolla conspicuous, 4–7 mm. broad; nutlets homomorphic
                                            '5. *C. oxygona*
                    HH. Corolla inconspicuous, ca. 1 mm. broad; nutlets heteromorphic, the odd one often wingless .... 6. *C. pterocarya*
                  GG. Lateral margins of the nutlets knifelike or acute.
                    H. Nutlets 1 or 2; odd nutlet axial ....7. *C. utahensis*
                    HH. Nutlets 4; odd nutlet abaxial.
                        I. Nutlets homomorphous, obscurely rugose, the back high-convex .... 8. *C. costata*
                        II. Nutlets heteromorphic, plainly muricate.
                            J. Nutlets 1.3–1.7 mm. long, the margins of the lateral angles knifelike; calyx 2.5–3.5 mm. long in fr.; corolla 1 mm. broad .... 9. *C. inaequata*
                            JJ. Nutlets ca. 1 mm. long, the margins of the lateral angles merely sharp; calyx ca. 3 mm. long in fr.; corolla 1–2.5 mm. broad .... 10. *C. angustifolia*
              EE. Margins of the nutlets rounded or obtuse.
                F. Nutlets decidedly heteromorphous, 1–4, the large nutlet axial and sometimes less roughened.
                    G. Mature calyces conspicuously recurved or deflexed, most hirsute on axial side; nutlet 1, bent; ovules 2 .... 11. *C. recurvata*
                    GG. Mature calyces spreading or erect, most hirsute on the abaxial side; nutlets straight.
                      H. Calyx closely and strictly appressed to the flattened rachis, persistent, gibbous on the axial side due to the basal development of the roughened odd nutlet ...... 12. *C. dumetorum*
                    HH. Calyx ascending or spreading, deciduous, not at all gibbous; odd nutlet somewhat smooth.
                      I. Nutlets 4, triangular-ovate, 0.7–0.9 mm. long; mature calyx subglobose, minute, the lobes scarcely longer than the nutlets .... 13. *C. micromeres*
                      II. Nutlets 1–2, oblong-lanceolate, 1–2 mm. long; mature calyx oblong, medium-sized, the lobes surpassing the nutlets .... 14. *C. maritima*
        FF. Nutlets homomorphic.
            G. Nutlets lanceolate.
                H. Nutlets only 1 or rarely 2 in a normal fr.; style not more than half as long as nutlet.
                    I. Corolla less than 1 mm. broad. Deserts . 15. *C. decipiens*
                    II. Corolla 2–3.5 mm. broad. Cismontane.
                        J. Stems strigose .... 16. *C. corollata*
                        JJ. Stems spreading-hispid .... 17. *C. Rattanii*
                HH. Nutlets 4 in normal fr.; style often more than half as long as nutlets.
                    I. Corolla conspicuous, 2–8 mm. broad; plant hirsute, mostly cismontane .... 18. *C. intermedia*
                    II. Corolla inconspicuous, 1–2 mm. broad; desert plants.
                      J. Fruiting calyx 6–12 mm. long.
                          K. Plant bristly with spreading hairs
                                        19. *C. barbigera*
                        KK. Plant strigose with appressed hairs
                                      20. *C. nevadensis*

JJ. Fruiting calyx 3–4 mm. long .. 10. *C. angustifolia*
GG. Nutlets ovate to triangular-ovate.
  H. Nutlets 4; corolla often inconspicuous.
    I. Dorsal side of nutlets obtuse and with at least a faint median ridge; style surpassing the nutlets
            21. *C. muricata*
    II. Dorsal side of nutlets flat or low-convex, without median ridge; style not surpassing nutlets.
      J. Corolla conspicuous, 4–7 mm. broad. N. Calif.
          22. *C. Hendersonii*
      JJ. Corolla inconspicuous, 0.5–2 mm. broad.
        K. Fls. solitary or in glomerules in lf.-axils, forming a panicle or thyrsus. Cent. Calif.   23. *C. Hooveri*
        KK. Fls. in definite spikes.
          L. Nutlets minute, less than 1.5 mm. long; spikes with numerous bracts. Insular.
            24. *C. Traskae*
          LL. Nutlets larger, 2–2.5 mm. long; spikes naked or bracted at base only. Largely montane.
            M. Plant closely strigose, pallid, usually 2–3 dm. tall; spikes commonly geminate or ternate ..... 25. *C. simulans*
            MM. Plant spreading-hispid, usually 1–1.5 dm. tall; spikes usually solitary, rarely geminate.
              N. Nutlets with low rounded tubercles on the back, nearly smooth toward the base . 26. *C. ambigua*
              NN. Nutlets with elongate papillae or spicules on the back
                27. *C. echinella*
  HH. Nutlets usually solitary; corolla 2–4 mm. broad.
    I. Pedicels evident, 2–3 mm. long; calyx villous with long white hairs. Head of Sacramento V. .... 28. *C. crinita*
    II. Pedicels inconspicuous, less than 1 mm. long.
      J. Nutlets horizontal, bent; calyx distinctly bristly on the midribs. Inner N. Coast Ranges .. 29. *C. excavata*
      JJ. Nutlets erect, straight; calyx with hairs on midribs longer but not bristly. N. Coast Ranges
          30. *C. Milobakeri*
DD. Nutlets smooth and shining, not roughened.
  E. Calyx armed with pale encrusted spreading hooked or curved hairs.
    F. Nutlets with an open areole at base of groove; style ½–⅔ as long as nutlets .... 31. *C. rostellata*
    FF. Nutlets with closed groove; style less than half as high as nutlets.
      G. Nutlets nearly terete, rostrate; hairs on calyx usually encrusted and pale .... 32. *C. flaccida*
      GG. Nutlets decidedly compressed, acute; hairs on calyx smoothish and less pale .... 33. *C. sparsiflora*
  EE. Calyx armed with straight ± tawny unencrusted hairs.
    F. Nutlets with excentric groove, one side of nutlet on lower surface appearing as if somewhat deformed.
      G. Nutlets 4; style at least ⅔ the length of the nutlets; plants erect; fls. in spikes 2–8 cm. long .... 34. *C. affinis*
      GG. Nutlets solitary; style less than half the length of the nutlets; plants spreading; fls. in axillary glomerules .. 35. *C. glomeriflora*
    FF. Nutlets with a centrally placed groove.
      G. Dorsal side of nutlets low-convex or flat; nutlets 1–4, homomorphic.
        H. Nutlets ovoid to lanceolate, their lateral angles obtuse or rounded.
          I. Corolla inconspicuous, ca. 1 mm. broad.
            J. Fruiting calyx 3–7 mm. long. Sierra Nevada and N. Coast Ranges .... 36. *C. Torreyana*
            JJ. Fruiting calyx 2–3 mm. long. Deserts and s. Calif. coast .... 14. *C. maritima*
          II. Corolla showier, 2–6 mm. broad.
            J. Plants stiff, low, widely branched; spikes mostly solitary; calyx 5–7 mm. long .... 37. *C. mariposae*
            JJ. Plants slender, not stiff, the stems erect with slender ascending branches; spikes geminate; calyx ca. 2 mm. long .... 38. *C. incana*
        HH. Nutlets oblong-ovate to lanceolate, their lateral angles sharp especially toward apex.
          I. Corolla 4–7 mm. broad. Inyo Co. s. .. 39. *C. mohavensis*
          II. Corolla ca. 1 mm. broad. Inyo Co. n. .. 40. *C. Watsonii*

GG. Dorsal side of nutlets rounded-convex, the lateral angles rounded or obtuse.
    H. Calyx broadly conic at base, densely appressed-hispid-villous, mostly lacking conspicuous spreading bristles. Deserts
       41. *C. gracilis*
    HH. Calyx rounded at base, with both appressed and spreading bristles. Cismontane.
        I. Hairs on upper part of calyx-lobes conspicuously retrorse. Inner Coast Ranges from Kern Co. to Colusa Co.
           42. *C. nei iaclada*
        II. Hairs on upper part of calyx-lobes spreading or as ending.
            J. Style less than half as long as nutlets; fruiting calyx 1.5–2 mm. long. Widely distributed.
               43. *C. microstachys*
            JJ. Style almost as long as nutlets; fruiting calyx longer.
                K. Spikes bracteate, dense; nutlets mostly 4. Along the coast from Surf to Ore. ... 44. *C. leiocarpa*
                KK. Spikes naked, or the lowest fls. sometimes bracteate.
                    L. Nutlets circular in cross section, solitary; fruiting calyx 2–2.5 mm. long
                       45. *C. hispidula*
                    LL. Nutlets not circular in cross section; fruiting calyx mostly longer.
                        M. Calyx-bristles spreading or reflexed.
                           N. Fruiting calyx 2–4 mm. long, moderately white-bristly
                              46. *C. Clevelandii*
                           NN. Fruiting calyx 6–10 mm. long, conspicuously tawny-bristly
                              47. *C. Ganderi*
                        MM. Calyx-bristles ascending
                            30. *C. Milobakeri*

AA. Plants perennial or biennial, coarse.
    B. Nutlets smooth and shining on the back.
       C. Corolla pale yellow, 12–14 mm. long; stems erect, 1.5–5 dm. tall .. 48. *C. confertiflora*
       CC. Corolla white, 3–4 mm. long; stems decumbent to prostrate, 0.5–1.5 dm. long
           49. *C. Jamesii*
    BB. Nutlets rough or wrinkled on back.
        C. Corolla-tube distinctly longer than calyx; nutlet-scar open and with elevated margin
           50. *C. flavoculata*
        CC. Corolla-tube not exceeding calyx; nutlet-scar various.
            D. The corolla-tube ca. 8 mm. long; calyx 13–15 mm. in fr. Sierra Nevada of Alpine and Tuolumne cos. .... 51. *C. crymophila*
            DD. The corolla-tube 3–5 mm. long; calyx 5–13 mm. in fr.
                E. Inner surface of nutlets smooth or nearly so, the scar narrowly subulate but open at base; calyx 5–7 mm. long in fr.
                   F. Nutlets with conspicuous transverse furrows on upper surface. Siskiyou Co. .... 52. *C. subretusa*
                   FF. Nutlets with distinct tubercles on upper surface. Sierra Nevada
                       53. *C. nubigena*
                EE. Inner surface of nutlets conspicuously rugose or tubercled, the scar open and broadened toward base.
                    F. Limb of corolla 8–10 mm. broad, the tube 4–5 mm. long; margin around nutlet-scar not at all elevated. Sierra Nevada .. 54. *C. humilis*
                    FF. Limb of corolla 3–8 mm. broad, the tube 3–4 mm. long; margin around nutlet-scar somewhat elevated.
                        G. Nutlets 2.5–3 mm. long; calyx-lobes 5–7 mm. in fr. Inyo-White Mts. and Sierra Nevada of Inyo Co.
                           H. Stems less than 15 cm. long, not retrorse-pubescent.
                              I. Stems 0.5–1.5 dm. high, tomentose and appressed-setose; lvs. tomentose and appressed-setose; nutlets muriculate
                                 55. *C. nana*
                              II. Stems 0.1–0.3 dm. high, strigose and appressed-setose; nutlets irregularly ridged .... 59. *C. Roosiorum*
                           HH. Stems 1.5–3 dm. high, conspicuously hirsute and retrorse-pubescent; lvs. retrorse-hirsutulous and sparsely bristly; nutlets tuberculate .... 56. *C. Hoffmannii*
                        GG. Nutlets 4–4.5 mm. long; calyx-lobes 8–12 mm. in fr. Mojave Desert.
                            H. Stems 2–3, from a biennial root; infl. broad; calyx-lobes 10–12 mm. in fr.; nutlets prominently ridged dorsally
                               57. *C. virginensis*
                            HH. Stems several to many, from a perennial root-crown; infl. narrow; calyx-lobes 8–10 mm. in fr.; nutlets not prominently ridged dorsally .... 58. *C. tumulosa*

1. **C. circumscíssa** (H. & A.) Jtn. [*Lithospermum?* c. H. & A. *Greeneocharis c.* Rydb.] Low annual, forming a dense hemispheric mass 2–10 cm. high, appressed-hispidulous to strigose, ± branched above; lower lvs. narrow-oblongish, the upper linear, to 1.5 cm. long; fls. in axils of foliaceous bracts; fruiting calyx 2.5–4 mm. long, united to middle, circumscissile just below middle; corolla white, 1–2 mm. broad; nutlets 4, homomorphous or nearly so, smooth or obscurely muriculate, triangular-ovoid to oblong-lanceolate, 1.2–1.7 mm. long, with angled margins; style ca. as long as nutlets.—Sandy or gravelly places, below 9500 ft.; Creosote Bush Scrub to Lodgepole F.; Mojave Desert and adjacent mt. ranges, e. slope of Sierra Nevada, n. to B.C., mts. along w. edge of Colo. Desert to L. Calif.; to Utah, Colo., Ariz.; S. Am. April–Aug.

Var. **rosulàta** J. T. Howell. Stems strigose, 0.5–2 cm. high; lvs. oblong-oblanceolate; styles shorter than nutlets.—At 10,000–12,000 ft.; Sierra Nevada, Inyo and Tulare cos. July–Aug.

Var. **híspida** (Macbr.) Jtn. [*Greeneocharis c.* var. *h.* Macbr.] Stems spreading-hispid-pubescent.—At 2500–12,000 ft.; Mojave Desert to Nevada Co.; w. Nev. May–July.

2. **C. micrántha** (Torr.) Jtn. [*Eritrichium m.* Torr. *Eremocarya m.* Greene.] Small slender strigose ascendingly branched dichotomous annual 5–15 cm. tall; lvs. oblong-oblanceolate, 3–7 mm. long; spikes many, solitary or geminate, dense, 1–4 cm. long; fruiting calyx ovate-oblong, ca. 2 mm. long; corolla 0.5–1.2 mm. broad, white; nutlets 4, homomorphous or nearly so, 1–1.3 mm. long, smooth or tubercled, the groove narrow, scarcely broadened at base.—Dry sandy places, below 7500 ft.; Creosote Bush Scrub to Yellow Pine F., Coastal Sage Scrub; deserts and inner cismontane valleys of s. Calif.; to L. Calif., Ore., Utah. March–June.

Var. **lépida** (Gray) Jtn. [*Eritrichium m.* var. *l.* Gray. *Eremocarya m.* var. *l.* Macbr.] Plants slightly coarser, taller; corollas 1–3.5 mm. broad.—Montane slopes and valleys, 2000–7000 ft.; Mono Co. to Mt. Pinos, etc., and through mts. of s. Calif. to L. Calif.

3. **C. holóptera** (Gray) Macbr. [*Eritrichium h.* Gray.] Erect hirsute and strigose annuals, 1–5 dm. high, with rather numerous short ascending branches; lvs. linear-lanceolate, the upper sessile, the lower petioled, 2.5–5 cm. long, conspicuously pustulate and hispid; raceme solitary or geminate, sparsely bracted, usually 4–5 cm. long; calyx oblong-ovate, 2.5–3.5 mm. long in fr., hirsute; corolla 1–1.5 mm. broad; nutlets 4, broadly ovoid or triangular-ovoid, 1.5–2.5 mm. long, dark with pale tuberculations, the margins ± winged; style clearly longer than nutlets.—Uncommon in gravelly or rocky places, below 2000 ft.; Creosote Bush Scrub; Colo. and extreme e. Mojave deserts; Ariz. March–April.

4. **C. racemòsa** (Wats.) Greene. [*Eritrichium r.* Wats. *C. r.* var. *lignosa* Jtn. *C. suffruticosa* Piper.] Woody perennial, but flowering the first year, mostly bushy, 2–8 dm. tall, the ultimate branches slender, strigose; lvs. subulate to linear-oblanceolate, 0.8–3.5 cm. long; racemes slender, lax, 2–4 cm. long; pedicels slender, often recurved, 1–4 mm. long; calyx-lobes hirsute and strigose, 2–3 mm. long; corolla white, 1 mm. broad; nutlets 4, triangular-ovate, with pallid tuberculations and knifelike margin; odd nutlet 1–2 mm. long, the others 0.8–1.5 mm. long; style much surpassing nutlets.—Rocky places, below 4500 ft.; Creosote Bush Scrub, Joshua Tree Wd.; Mojave and Colo. deserts, Inyo Co. to Imperial Co.; Nev., Ariz. March–May.

5. **C. oxygòna** (Gray) Greene. [*Eritrichium o.* Gray.] Erect sparsely branched annual 1–4 dm. tall, the stems 1–several, strigose or appressed villous-hispid; lvs. linear to lance-linear, 1–4 cm. long, stiff-pubescent; spikes in 2's or 3's, usually short, dense, bractless; fruiting calyx ovoid, 2.5–4 mm. long, the lanceolate lobes connivent, silky-strigose with ± bristly midribs; corolla white, 4–6 mm. broad; nutlets 4, oblong-ovoid, 2–2.5 mm. long, tubercled or muricate on back, narrowly winged on margin; style clearly surpassing nutlets.—Dry slopes and benches, below 5000 ft.; Joshua Tree Wd., Pinyon-Juniper Wd., Chaparral, V. Grassland; inner S. Coast Ranges from w. Merced and Fresno cos. to Kern Co., w. Mojave Desert to Santa Rosa Mts., Riverside Co.; w. Nev. March–May.

6. **C. ptérocàrya** (Torr.) Greene. [*Eritrichium p.* Torr.] Erect ascendingly branched herb 1–5 dm. tall, finely strigose or short-hirsute; lvs. linear or upper somewhat wider and reduced; spikes in 2's or 3's or occasionally solitary, 2–6(–10) cm. long, usually bractless; mature calyx notably accrescent, ± ovoid, 3–5 mm. long, the connivent lobes ± hispid; corolla white, to 1 (2) mm. broad; nutlets 4, heteromorphous, the lance-oblong or lanceolate body 2–2.8 mm. long, verrucose or muricate, the axial one unmargined, the

other 3 with broad crenulate or lobulate wing-margins; style longer than nutlets.—Sandy and gravelly places, below 6000 ft.; Creosote Bush Scrub, Joshua Tree Wd., Pinyon-Juniper Wd.; deserts to Lassen Co.; to Wash., Utah, Son., L. Calif. March–June.

Var. **cyclóptera** (Greene) Macbr. [*Krynitzkia c.* Greene.] Nutlets all wing-margined.—Below 8000 ft.; deserts to Inyo Co.; to Colo., Tex. March–June.

Var. **Purpùsii** Jeps. One nutlet unmargined, the others with margins narrow, knife-like.—Between 4000–7000 ft.; Argus Mts., Darwin, San Bernardino Mts.

7. **C. utahénsis** (Gray) Greene. [*Krynitzkia u.* Gray.] Erectly branched slender annual 1–3 dm. high, strigose or ± appressed-hirsute; lvs. mostly linear, 3–5 cm. long, appressed short-hispid and pustulate especially beneath; spikes mostly in 2's, bractless; calyx ovoid or oblong-ovoid, 2–3(–4) mm. long, the midrib only rarely setose; corolla 2–3 mm. broad; nutlets broadly lanceolate, usually 1–2 maturing, granulate to muricate-papillate, 1.7–2.5 mm. long, the back low or flat, the margin sharp, knifelike.—Sandy and gravelly places, below 6500 ft.; Creosote Bush Scrub, Joshua Tree Wd.; deserts from Santa Rosa Mts., Riverside Co. to Inyo Co.; to Nev., Utah, Ariz. March–May.

8. **C. costàta** Bdg. [*C. seorsa* Macbr. *C. saxorum* Jeps.?] Coarse stiff relatively few-branched annual 1–2 dm. tall, usually densely villous-strigose and ± hirsute; lvs. linear to lance-linear, 1–3 cm. long; spikes rigid, in 1's or 2's, sparsely bracted, 2–5 cm. long; mature calyx ovate-oblong, 4–6 mm. long, the lance-linear lobes with hirsute midribs, strigose margins; corolla ca. 2 mm. broad; nutlets 4, triangular, barely 2 mm. long, strongly convex on back, the margin sharp, the groove shallow, closed above, dilated toward base; style much surpassing nutlets.—Uncommon, sandy and gravelly places, below 1500 ft.; Creosote Bush Scrub; Colo. Desert, Needles; w. Ariz. Feb.–May.

9. **C. inaequàta** Jtn. Ascendingly branched annual 1–3 dm. tall, strigose and bristly; lvs. linear, 1.5–4 cm. long, hispid or strigose, pustulate; spikes in 1's or 2's, rather dense, minutely sparsely bracted; calyx in fr. 2.5–3.5 mm. long, ovoid-oblong, hirsute; corolla 1 mm. broad; nutlets 4, triangular-ovoid, with pale tuberculations, heteromorphous, the odd one 1.5–2 mm. long, the consimilar 3 slightly shorter; style much surpassing nutlets.—Gravelly and rocky places, below 4000 ft.; Creosote Bush Scrub, Joshua Tree Wd.; Death V. region to e. San Bernardino Co.; w. Nev. March–May.

10. **C. angustifòlia** (Torr.) Greene. [*Eritrichium a.* Torr.] Annual, diffusely branched from base, cinereous with appressed and spreading stiffer hairs; stems 0.5–3 dm. long; lvs. linear, 1.5–3.5 cm. long, hispid and pustulate; spikes in 2's, ca. 3–5 cm. long, mostly bractless, fairly dense; calyx in fr. ovoid-oblong, 3–4 mm. long, hirsute; corolla 1–2 mm. broad; nutlets usually 4, oblong-ovoid, usually brown with pale muriculations, the odd nutlet slightly larger, the others ca. 1 mm. long; style usually surpassing all.—Common, sandy and gravelly places, below 3500 ft.; Creosote Bush Scrub; deserts from Inyo Co. s.; to Utah, Tex., Son., L. Calif. March–May.

11. **C. recurvàta** Cov. Ascendingly branched erect rather slender annual 1–3 dm. high, strigose; the lower lvs. oblanceolate, 1.5–2 cm. long, the cauline remote, linear to lanceolate, 0.5–1 cm. long, strigose and pustulate; spikes bractless, lax, in 1's or 2's; fruiting calyx slender, bent and recurved, 3 mm. long, the thickened midribs hispid; corolla ca. 1.5 mm. broad; nutlets 2, one oblong-lanceolate, curved, brownish, granulate-muriculate, axial, 1.5–2 mm. long, the groove somewhat oblique, narrow or closed, the other abortive; style shorter than perfect nutlet.—Uncommon, in sandy places, 2500–6500 ft.; Creosote Bush Scrub, Joshua Tree Wd.; Owens V., Death V. region to White Mts., Kelso; to Ore., Colo. April–June.

12. **C. dumetòrum** (Greene ex Gray) Greene. [*Krynitzkia d.* Greene ex Gray. *C. intermedia* var. *d.* Jeps.] Laxly branched strigose bright green annual, at first erect, later commonly elongated and sprawling or scrambling up through bushes and 1–4 dm. high; lvs. remote, linear or wider, obtuse, 1–3 cm. long, pustulate, strigose or short-hispid; spikes in 1's or 2's, usually remotely fld., 5–10 cm. long, mostly bractless except sometimes at base; calyx in fr. ovoid, 2–3 mm. long, oblique at base, the 3 outer lobes reflexed-hispid; corolla ca. 1 mm. broad; nutlets 4, heteromorphous, granulate-muricate, the odd one axial, broadly lanceolate, 2–3 mm. long, the others 1.5–2 mm. long, lanceolate, earlier deciduous; style not longer than nutlets.—Sandy and gravelly places, below 4500 ft.; Creosote Bush Scrub; n. Colo. Desert, in scattered localities through Mojave Desert to Inyo Co.; w. Nev. April–May.

13. **C. micrómeres** (Gray) Greene. [*Eritrichium m.* Gray.] Slender dull green usually

erect annual 1–5 dm. tall, short-hirsute throughout; lvs. linear or ± oblong, 1–4 cm. long, rather numerous, short-hispid; spikes commonly ternate, slender, bractless, 2–8 cm. long; calyx in fr. 1–2 mm. long, early deciduous, the lobes connivent, slender-hispid, the hairs on the midribs frequently hooked; corolla 0.5 mm. broad; nutlets 4, triangular-ovoid, the margins subangulate, odd nutlet ca. 1 mm. long, the others 0.7 mm. long and papillate, the groove open, gradually dilated into a small open areola.— Frequent in dry open places, such as burns, below 2000 ft.; Chaparral; scattered places in cismontane s. Calif. to L. Calif. and to Marin and Amador cos., Santa Cruz, Santa Rosa, and Santa Catalina ids. March–June.

14. **C. marítima** (Greene) Greene. [*Krynitzkia m.* Greene.] A stiffish ascending loosely branched annual 1–3.5 dm. high, rather sparsely strigose or sparsely hispid, brown or reddish; lvs. linear to lance-linear, hispid and pustulate, 1–3.5 cm. long; spikes in 1's or 2's, 1–6 cm. long, completely or partially bracted; fruiting calyx asymmetrical, 2–3 mm. long, the lobes connivent, firm, with 3 hispid on midribs and ± villous; corolla 0.5–1 mm. broad; nutlets 1–2, heteromorphous, the axial one smooth, shining, lance-oblong, 1–2 mm. long, the abaxial grayish, minutely tubercled, slightly shorter, often abortive; style ca. as high as smaller nutlet.—Sandy and gravelly places, below 4000 ft.; Creosote Bush Scrub, Joshua Tree Wd., Coastal Sage Scrub; deserts from Inyo Co. to Imperial Co., coastal mesas, Santa Barbara Ids., San Diego; L. Calif., Ariz. March–May.

Var. **pilòsa** Jtn. Calyx conspicuously villous with long white hairs. With the sp.

15. **C. decípiens** (Jones) Heller. [*Krynitzkia d.* Jones.] Slender loosely branched annual 1–4 dm. tall, strigulose and often short-hispid; lvs. few, linear, obtuse, 1–3 cm. long; spikes mostly in 2's, slender, 4–10 cm. long; calyx in fr. 3–4 mm. long, the lobes linear, connivent, with spreading tips; nutlets 1, sometimes 2, lance-ovoid, 1.5–2.4 mm. long, muriculate-granulate to tuberculate, usually brownish, convex on back, rounded on sides, the open or closed groove dilated basally into an areola; style well surpassed by nutlet.—Sandy or gravelly places, below 4000 ft.; Creosote Bush Scrub, Joshua Tree Wd.; deserts from Inyo Co. to Nev., Ariz. March–May.

16. **C. corollàta** (Jtn.) Jtn. [*C. decipiens* var. *c.*] With much the same habit as *C. decipiens;* appressed-hispidulous throughout; calyx in fr. ascending, ca. 3 mm. long, white-strigose, the outer lobes also spreading-bristly, the tips erect; corolla 2–3.5 mm. broad; nutlet 1, lance-ovoid, acuminate, 2 mm. long, brownish, low convex on back with a low broad ridge toward base, granulate and papillate-muriculate on both sides, the groove closed throughout and raised into a narrow keel, basally dilated into a small areola.—Dry slopes and ridges, below 4500 ft.; Foothill Wd., V. Grassland, etc.; inner Coast Ranges from San Benito and Monterey cos. to Ventura Co. March–June.

17. **C. Rattánii** Greene. [*C. corollata* ssp. *R.* Abrams.] With the habit of *C. corollata,* but stems hirsute-hispid with spreading hairs as well as strigose; lvs. oblong-linear, 1.5–6 cm. long, pustulate; calyx in fr. appressed, ca. 3 mm. long; corolla 2–5 mm. broad; nutlets usually 2–3, brownish, granulate-muriculate, lance-ovoid, almost 2 mm. long, rather sharp-edged, the ventral groove closed, ending basally in a small areola.—Open places, below 3000 ft.; Foothill Wd., V. Grassland, etc.; watershed of Salinas and Carmel rivers, Monterey and w. Merced cos. May–July.

18. **C. intermèdia** (Gray) Greene. [*Eritrichium i.* Gray. *C. i.* var. *Johnstonii* Macbr. *C. barbigera* var. *Fergusonae* Macbr. *C. Hansenii* Brand. *C. H.* var. *pulchella* Brand.] Erectly branched commonly stiff annual 1.5–5 dm. tall, frequently somewhat strigose, usually very hirsute; lvs. lanceolate to linear, hispid or strigose, 1.5–5 cm. long, ± pustulate; spikes bractless, in 2's to 5's, 5–15 cm. long; calyx in fr. oblong-ovoid, ca. 4–6 mm. long, the lance-linear lobes connivent with ± spreading tips, the thick midribs pungently hirsute, the margins short hispid or villous; corolla 3–7 mm. broad; nutlets mostly 4, lance-ovoid, 1.5–2.3 mm. long, tubercled or verrucose, the back convex, the groove narrow or closed, with small basal areola; style usually reaching ca. to tip of nutlets.— The common sp. of cismontane areas, mostly below 3000 ft., but reaching 6000 ft.; Chaparral, Coastal Sage Scrub, Pinyon-Juniper Wd., Foothill Wd., etc.; cismontane s. Calif. to desert borders, n. to w. Siskiyou Co. in mts. bordering the Cent. V.; L. Calif. March–July.

19. **C. barbígera** (Gray) Greene. [*Eritrichium b.* Gray.] Erectly branched annual 1–4 dm. tall, very bristly, sparsely if at all strigose except in the infl.; lvs. linear to broadly

oblong, obtuse, ± pilose and hirsute, pustulate; spikes mostly in 2's, bractless, 3–15 cm. long; fruiting calyx oblong-ovoid to -lanceolate, 5–10 mm. long, the lobes linear-lanceolate to lanceolate, connivent with recurved tips, the margins usually conspicuously long-white-villous, midribs thickened and hirsute; corolla 1–2 mm. broad; nutlets 1–4, lance-ovoid, 1.5–2.5 mm. long, strongly verrucose, the back convex, the edges obtusish, the groove open or closed, with a triangular areola below; style reaching to ca. the tip of the nutlets.—Sandy or gravelly places, mostly below 3000 ft., rarely to over 7000; Creosote Bush Scrub, to Pinyon-Juniper Wd.; deserts to n. Inyo Co.; to L. Calif., Son., New Mex., Utah. Feb.–May.

20. **C. nevadénsis** Nels. & Kenn. [*Krynitzkia barbigera* var. *inops* Bdg. *C. b.* var. *i.* Macbr. *C. arenicola* Heller.] Slender-stemmed annuals 1–5 dm. tall, the 1–several stems erect or flexuous, short-strigose; lvs. linear to linear-oblanceolate, 1–3.5 cm. long; spikes slender, in 2's or 3's, sometimes bracteate, somewhat congested or elongate up to 15 cm. long; fruiting calyx lanceolate, 8–12 mm. long, the linear lobes connivent above, with recurved tips, the margin somewhat villous, the midribs hirsute; corolla 1–2 mm. broad; nutlets 4, lanceolate, long-acuminate, ca. 2.5 mm. long, verrucose on back, muricate at tip, the groove open or closed, dilated below into small areola.—Sandy or gravelly places, below 6000 ft.; Creosote Bush Scrub to Pinyon-Juniper Wd.; deserts from Inyo Co. to L. Calif., Ariz., Utah. March–May.

Var. **rígida** Jtn. [*C. intermedia* var. *r.* Brand.] Erect, not flexuous, less slender-stemmed; calyx 5–10 mm. long; nutlets oblong-ovoid, merely acute, ca. 2 mm. long, verrucose but not noticeably muricate.—At 3000–6000 ft.; Creosote Bush Scrub, Joshua Tree Wd., V. Grassland, Foothill Wd.; San Joaquin V. and borders, Santa Lucia Mts., w. Mojave Desert; Ariz.

21. **C. muricàta** (H. & A.) Nels. & Macbr. [*Myosotis m.* H. & A. *Eritrichium? muriculatum* A. DC. *Cryptantha muriculata* Greene. *C. horridula* Greene.] Erect hirsute annual 1–5 dm. tall, loosely branched, usually with several well developed ascending laterals; lvs. linear, cinereous, short-hirsute, 1.5–3 cm. long; spikes in 2's to 5's, bractless, 2–15 cm. long, terminating the well developed sparsely leafy pedunculate stem or branches; calyx in fr. ovoid, 2–4 mm. long, the lanceolate lobes connivent, tawny-hirsute; corolla 2–6 mm. broad; nutlets 4, glossy or dull, verrucose or tuberculate, grayish, the lateral angles usually acute, the ventral groove narrow or closed, broadly forked or opened into basal areola; style surpassing nutlets.—Gravelly or rocky open slopes, washes, etc., below 8000 ft.; many Plant Communities; Coast Ranges from Contra Costa Co. s., Sierra Nevada of Kern Co., Tehachapi Mts., to Orange Co. April–June.

Var. **denticulàta** (Greene) Jtn. [*Krynitzkia d.* Greene.] Loosely and sparsely branched; corolla 1–2 mm. broad.—Dry slopes, 4000–8000 ft.; Yellow Pine F.; cent. Sierra Nevada s. to San Bernardino Mts.; w. Nev.

Var. **Clòkeyi** (Jtn.) Jeps. [*C. C.* Jtn.] Calyx 6–9 mm. long; nutlets whitish.—Mojave Desert, n. of Barstow.

Var. **Jònesii** (Gray) Jtn. [*Krynitzkia J.* Gray. *C. J.* Greene. *C. vitrea* Eastw.] Stems commonly solitary and erect, or several and fastigiate, 2–10 dm. tall, bearing from tip to below middle many short floriferous branchlets forming an elongate leafy paniculate infl.; corolla 1–2.5 mm. broad.—Below 5500 ft.; many Communities; Coast Ranges from Glenn Co. s., Sierra Nevada from Nevada Co. s., to San Diego Co.; L. Calif., Nev., Ariz.

22. **C. Hendersònii** (A. Nels.) Piper. [*Allocarya H.* A. Nels. *C. intermedia* var. *H.* Jeps. & Hoov. *C. trifurca* Eastw.] Erect annual with few ascending branches, hirsute, 2–5 dm. high; lvs. oblanceolate to linear, appressed-hirsute; spikes mostly in 3's, ebracteate except at very base, 5–10(–20) cm. long; calyx in fr. ± ovoid, 3–6 mm. long, the lobes linear to lance-linear, connivent with spreading tips, appressed-hirsute on margins, tawny-hirsute on midribs; corolla 4–6 mm. broad; nutlets 4, sometimes fewer, ovoid, pale, 2–3 mm. long, muriculate, low-convex on back, rounded on margin, the groove closed or narrow, broadly forked at base; style not reaching quite to tip of nutlets.—Sandy or rocky places, 1500–4500 ft.; Yellow Pine F., N. Oak Wd., Foothill Wd.; Del Norte Co. to Lake, Modoc and Lassen cos.; to B.C., Ida. May–July.

23. **C. Hoòveri** Jtn. Annual 1–1.5 dm. high, the stems solitary or branched at base, sometimes the outer branches decumbent, strigose; lvs. narrow-spatulate, 1–2.5 cm. long or upper linear, revolute, hispidulous; fls. solitary or in glomerules in lf.-axils,

forming a narrow elongate rather dense thyrsus; calyx in fr. 4–5 mm. long, the linear lobes ascending-villous on margins, midribs with some yellowish bristles; corolla inconspicuous; nutlets 4, triangular-ovoid, ca. 1.3 mm. long, acute, the back convex, papillate, the groove very narrow above, abruptly dilated into basal deltoid areola.— Coarse sand; V. Grassland; San Joaquin V. from Contra Costa Co. to Madera Co. April–May.

24. **C. Tráskae** Jtn. [*C. Torreyana* var. *T.* Jeps.] Loosely branched slender strigose annual 1–2 dm. high; lvs. linear, 0.5–2 cm. long, strigose or hispidulous; spikes in 1's or 2's, dense, 1–5 cm. long, the lower fls. bracteate; calyx in fr. ovoid, 2–3 mm. long, hirsute; corolla 1.5 mm. broad; nutlets 4, ovoid, finely granulate, the back convex, ± tubercled toward tip, the margin obtuse, the groove closed, slightly dilated to a minute basal areola; style not exceeding nutlets.—Rocky and gravelly places; Coastal Sage Scrub; San Nicolas and San Clemente ids. April–June.

25. **C. símulans** Greene. [*C. Stuebelii* Brand. *C. ambigua* some auth.] Erect slender-stemmed annual 1–4 dm. tall, with very few strictly ascending branches, whitish-strigose; lvs. linear to oblance-linear, 1–3 cm. long, strigose; spikes in 1's to 3's, bractless, mostly loosely fld., elongate; calyx in fr. 3–7 mm. long, the lance-linear lobes connivent above with spreading green tips, the midribs with reflexed or spreading-arcuate bristles, the margins villous-hirsute with ascending hairs; corolla 1–2 mm. broad; nutlets 4, broadly ovoid, 2–2.5 mm. long, tesellate-papillate, the angles rounded, the back low-convex, the groove mostly closed throughout, broadly forked below; style slightly surpassed by nutlets.—Dry slopes in woods, 2500–7500 ft.; Yellow Pine F., Red Fir F.; mts. from San Diego Co. to Modoc, Trinity, and Siskiyou cos.; to Wash., Ida. May–July.

26. **C. ambígua** (Gray) Greene. [*Eritrichium muriculatum* var. *a.* Gray. *C. polycarpa* Greene.] Usually loosely branched from base, erect, 1–2.5 dm. high, strigose and hirsute; lvs. linear to lance-linear, 2–4 cm. long, hirsute, ± pustulate; spikes often solitary, naked or bracted at base, 5–15 cm. long; calyx in fr. 4–7 mm. long, the narrow lobes ± connivent, with tawny-hispid midribs and strigose-hirsute margins; corolla 1–2 mm. broad; nutlets 4, broadly ovoid, 1.6–2 mm. long, granulate and coarsely tuberculate, the low back convex with obtuse sides, the groove mostly closed, broadly forked at base; style not surpassing nutlets.—Gravelly places, 4500–6000 ft.; Sagebrush Scrub to Red Fir F.; Nevada Co., Sierra Nevada to Modoc and Siskiyou cos.; to Wash., Mont., Utah. June–July.

27. **C. echinélla** Greene. [*C. ambigua* var. *e.* Jeps. & Hoov.] Erect, sparsely branched, mostly 0.5–3 dm. high, short-hirsute; lvs. ± oblanceolate; spikes solitary or geminate, often leafy-bracted below, 1–5 cm. long; calyx in fr. oblong-ovoid, spreading, 5–6 mm. long, the lobes lance-linear, connivent above, with ± recurved tips, the midribs pale tawny-hispid, the margins appressed-hirsutulous; corolla 1–1.6 mm. broad; nutlets 4, broadly ovoid, grayish, 2 mm. long, papillate-echinate on the rounded back, the groove closed or nearly so, widely forked at base; style not quite reaching to nutlet-tips.—Dry places, 2500–9000 ft.; Yellow Pine F., Pinyon-Juniper Wd., Red Fir F.; San Bernardino Mts. to Sierra Nevada of Nevada Co., White Mts., Panamint Mts.; w. Nev. June–Aug.

28. **C. crinìta** Greene. Annual, rather strictly branched from base, 1.5–3 dm. high, strigose and hirsute; lvs. linear to oblance-linear, 1.5–3 cm. long; spikes mostly in 2's, white-villous, becoming 4–6 cm. long; pedicels 1–2 mm. long; calyx in fr. 5–6 mm. long, shaggy, white-hirsute, the lobes linear, obtuse; corolla 3–4 mm. broad; nutlets solitary, ovoid, 2.5 mm. long, the back low, rounded with a few scattered tubercles and microscopically papillate.—Sandy and gravelly creek bottoms, below 1000 ft.; Foothill Wd.?; Stillwater and Millville plains, head of Sacramento V., Shasta Co. April–May.

29. **C. excavàta** Bdg. Annual, branched above the base, 1–2 dm. high, appressed-hirsute; lvs. linear, 1–2 cm. long, ± pustulate; spikes in 2's or 3's, 3–10 cm. long; calyx in fr. 2–2.5 mm. long, ca. as thick, villous with ascending hairs and in lower part hispid with subulate bristles; corolla 2–3 mm. broad; nutlet 1, almost horizontal, ovoid-acuminate, muriculate-papillose, and with scattered tubercles, keeled dorsally and above on ventral side, the groove opening basally into a large triangular deep scar.—Sandy and gravelly places, below 1000 ft.; Foothill Wd.; inner Coast Ranges from Yolo Co. to Colusa and Lake cos. April–May.

30. **C. Milobàkeri** Jtn. [*C. Torreyana* var. *scrutata* Jeps.] Erect annual 2–4 dm. tall, with rather strictly ascending branches, strigose; lvs. linear-oblong to -lanceolate, 1–3

cm. long, strigose; spikes mostly in 2's, 5–15 cm. long; calyx in fr. 3–5 mm. long, densely hirsute-pilose with ascending hairs; corolla 2–4 mm. broad; nutlets mostly 1, lance-ovoid, 2–2.5 mm. long, smooth or inconspicuously tessellate, low convex on back, with rounded margins, the groove closed to the forked base; style not reaching to tip of nutlet.—Rocky to gravelly places, below 4000 ft.; Chaparral, Foothill Wd.; Lake Co. to Del Norte, w. Siskiyou and Plumas cos.; s. Ore. May–July.

31. **C. rostellàta** (Greene) Greene. [*Krynitzkia r.* Greene.] Stiffly erect branched strigulose annual 1–2 dm. high; lvs. abundant along main stem, ascending, linear to oblanceolate, 1–1.5 cm. long; spikes in 1's or 2's, stiff, ebracteate, 2–4 cm. long; calyx in fr. oblong-ovoid, 3–4 mm. long, spreading or ascending, the lobes having the midribs armed with stout encrusted hooked or curved hairs, the margins sparsely ciliate or strigose; corolla to 1 mm. broad; nutlets solitary, smooth, compressed, lance-ovoid, 2–3 mm. long, dorsally convex, rounded on sides, the ventral groove closed above, forked at base into a distinct areola; style considerably shorter than nutlet tip.—Dry hills, below 3000 ft.; Foothill Wd., V. Grassland; borders of Sacramento V. from Butte and Colusa cos. n. to Siskiyou Co.; to Wash. April–May.

Var. **spithamèa** (Jtn.) Jeps. [*C. s.* Jtn.] Reflexed bristles wanting on lower part of calyx.—Foothill Wd.; Sierran foothills, Mariposa and Tuolumne cos.

32. **C. fláccida** (Dougl.) Greene. [*Myosotis f.* Dougl. ex Lehm. *C. f.* var. *maior* Brand.] Erect wiry annual 1–5 dm. tall, branched above or below, strigose with encrusted hairs; lvs. linear to oblance-linear, 2–5 cm. long, strict or ascending, strigose; spikes in 1's to 5's, stiffish, ebracteate, 4–12 cm. long; calyx in fr. 2–4 mm. long, asymmetrical, usually appressed to rachis, the lobes connivent with spreading tips, the midribs with spreading or reflexed coarse hooked or curved bristles, the margins ciliate or strigose; corolla 1–4 mm. broad; nutlet mostly 1, lance-ovoid, abaxial, subterete, smooth, the groove closed above, dilated below into a small areola; style not more than ½ the length of the nutlet.—Dry slopes and flats, below 6000 ft.; many Plant Communities; most of cismontane Calif. except the floor of the Cent. V.; to Wash., Ida. April–June.

33. **C. sparsiflòra** (Greene) Greene. [*Krynitzkia s.* Greene.] Annual with very slender stems loosely and widely branched, 1–3 dm. high, sparsely strigose; lvs. narrowly linear, 1–3 cm. long, strigose; spikes in 1's or 2's, nearly bractless, slender, 2–6 cm. long; calyx in fr. ovoid to oblong-ovoid, ascending, 2–3 mm. long, the lance-linear lobes somewhat connivent, armed on midribs with short hooked hairs, the margins sparsely ciliate; corolla less than 1 mm. broad; nutlet 1, ovoid, compressed, smooth, 2 mm. long, low convex on back, angled on margins, the groove closed and broadly forked at base; style not more than ½ the height of the nutlet.—Rocky slopes, below 4000 ft.; Chaparral, Foothill Wd.; hills bordering San Joaquin V. from w. Stanislaus and San Benito cos. and Mariposa Co. to Kern Co. April–May.

34. **C. áffinis** (Gray) Greene. [*Krynitzkia a.* Gray. *C. geminata* Greene.] Erect sparsely branched annual 1–4 dm. high, pubescent with short upwardly curved hairs and scattered longer stiff ones; lvs. oblanceolate, obtusish, mostly 2–3.5 cm. long, sparsely short-hispid, pustulate; spikes in 1's or 2's, slender, 2–8(–10) cm. long, becoming loosely fld., with some leafy bracts below; calyx in fr. 2.5–4 mm. long, ca. as thick, laterally compressed, the lobes lanceolate, weakly hirsute; corolla ca. 1.5 mm. broad; nutlets 4, ovoid or oblong-ovoid, 1.8–2.5 mm. long, smooth, grayish, frequently mottled, low-convex on back, rounded on sides, the ventral groove closed, simple or short-forked at base; style usually surpassed by nutlets.—Dry slopes, 3000–9500 ft.; Yellow Pine F., Red Fir F., Lodgepole F.; Cuyamaca and San Bernardino mts., Sierra Nevada, to Siskiyou Co., Coast Ranges from Lake Co. n.; to Wash., Wyo., Nev. June–Aug.

35. **C. glomeriflòra** Greene. Erect annual, simple or few-branched, 0.3–1 dm. high, strigose; lvs. linear- to lance-oblong, appressed-hispidulous, 0.5–1.5 cm. long; fls. 2–4 in small glomerules in lf.-axils and at ends of branchlets; calyx in fr. 1–1.5 mm. long, the lance-linear lobes hispidulous and bristly with stout straight setae; corolla minute; nutlet 1, ovoid, smooth, rounded dorsally and somewhat keeled on narrowed apex, the ventral groove a little off center, closed, opening at base into a sunken areola.— Occasional, slopes and meadows, 6000–11,000 ft.; Montane Coniferous F.; Sierra Nevada from Nevada Co. to Tulare and Inyo cos., White and Sweetwater mts., Mono Co. June–Sept.

36. **C. Torreyàna** (Gray) Greene. [*Krynitzkia T.* Gray. *K. T.* var. *calycosa* Gray. *C. T.* var. *calistogae* Jtn.] Erect annual 1–4 dm. high, the stems solitary or several with erect or spreading branches, strigose and somewhat hirsutulous; lvs. oblanceolate or linear, 2–6 cm. long, hirsutulous; spikes in 1's, 2's, or 3's, bractless, mostly becoming loose-fld. and elongate; calyx in fr. ovoid or oblong-ovoid, 3–6 mm. long, the lobes lance-linear, connivent above with tips often spreading, the midrib bristly-hispid, the margins appressed-hirsute; corolla to 1 mm. broad; nutlets usually 4, broadly ovoid, smooth, flattish on back, with obtuse sides, the groove and its short basal fork closed; style from ⅔ to as tall as nutlets.—Open or half-shaded slopes, 1500–7500 ft.; Red Fir F., Yellow Pine F., N. Oak Wd., Chaparral, Mixed Evergreen F.; Coast Ranges from Marin Co. n., Sierra Nevada from Kern Co. n.; to B.C., Wyo., Utah. May–Aug.

Var. **pùmila** (Heller) Jtn. [*C. p.* Heller.] Plants ca. 1 dm. high, conspicuously spreading-hirsute; calyx in fr. 2–3.5 mm. long; style shorter.—Marin Co. to Santa Clara Co.

37. **C. maripòsae** Jtn. Erect annual with ascending branches, mostly 1–2 dm. high, strigose; lvs. oblanceolate or oblong, 0.8–2 cm. long, hirsute, somewhat pustulate; spikes in 1's or 2's, mostly bractless, 1.5–5 cm. long; calyx in fr. 4–5 mm. long, the linear lobes erect, villous on sides with ascending hairs, spreading-bristly on midribs; corolla ca. 2 mm. broad; nutlets mostly 4, ovoid, rostrate from a broad body, low-rounded on back, shining, brown or mottled, somewhat tuberculate, obtuse on lateral angles, the ventral groove narrow, opening basally into a rounded scar.—Serpentine outcrops, below 2000 ft.; Chaparral; foothills of Sierra Nevada from Calaveras Co. to Mariposa Co. April–May.

38. **C. incàna** Greene. [*C. Torreyana* var. *i.* Jeps.] Grayish annual 2–4.5 dm. high, with several ascending branches, hispidulous and strigose; lvs. oblong-oblanceolate, appressed-hispidulous, 1.5–3 cm. long; spikes in 2's, bractless, to 5–6 cm. in fr.; calyx in fr. 3–4 mm. long, hirsutulous-hispidulous with ascending hairs; corolla 3–4 mm. broad; nutlets 2, lance-ovoid, rounded below, abruptly attenuate apically, smooth, shining, grayish, mottled with brown, 1.5–1.9 mm. long, the groove slightly opened downward toward and at the short fork.—Dry places, near 6000 ft.; Yellow Pine F.; Tulare Co. July–Aug.

39. **C. mohavénsis** (Greene) Greene. [*Krynitzkia m.* Greene. *C. fallax* Greene.] Usually erect annuals 1–4 dm. tall, short-hispid to hispid-strigose; lvs. linear to lance-linear, hirsute-hispid, minutely pustulate; spikes in 2's or 3's, usually dense, bractless, 2–6 cm. long; calyx in fr. oblong-ovoid, 3–5 mm. long, the lobes lanceolate, connivent, sparsely hirsute, the margins ± silky; corolla 4–7 mm. broad; nutlets 4, oblong-ovoid, or lance-ovoid, 2–2.5 mm. long, smooth, shining, the back low-convex or flattish, the lateral angles obtuse, the groove closed above, opened at fork into a small triangular areola; style surpassing the nutlets.—Dry places, 2500–9000 ft.; Joshua Tree Wd., Pinyon-Juniper Wd.; extreme w. Mojave Desert near Tehachapi Mts. n. along e. slope of Sierra Nevada to n. Inyo Co. May–July.

40. **C. Watsònii** (Gray) Greene. [*Krynitzkia W.* Gray.] Slender-stemmed annuals, erect, branched, 1–3 dm. high, short-hirsute; lvs. linear to linear-oblanceolate, hirsute; spikes in 1's, rarely 2's, mostly bractless, 1–5 cm. long; calyx in fr. ovoid, 2–3.5 mm. long, the lanceolate lobes with connivent tips, ascending-hirsute and with a few spreading bristles on midribs; corolla ca. 1 mm. broad; nutlets 4, lanceolate, 1.5–2 mm. long, mostly smooth and shining, flattish on back, acute on lateral angles, the ventral groove closed or nearly so, shortly forked at base; style almost or quite as high as the nutlets.—Rare, in dry often rocky places, 5000–10,000 ft.; Pinyon-Juniper Wd., Subalpine F.; e. slopes of Sierra Nevada and in White Mts., Mono and Inyo cos.; to Wash., Mont., Utah. June–Aug.

41. **C. grácilis** Osterh. Erectly branched annual 1–2.5 dm. high, densely short-hispid; lvs. linear to oblance-linear, 1–3 cm. long, short-hispid, pustulate; spikes in 1's or 2's, usually dense and 1–2 cm. long; calyx in fr. ovoid, 2–2.8 mm. long, divaricately spreading, the lobes lanceolate, erect at apex, tawny hirsute-villous, with a few bristles on the midribs; corolla to 1 mm. broad; nutlets 1 (2–3), lanceolate, 1.5–2 mm. long, smooth, shining, the back nearly flat, the sides rounded at least toward apex, the groove usually open to above middle, scarcely forked below; style ca. ¾ as high as nutlets.—Dry slopes and mesas, 3000–7000 ft.; Joshua Tree Wd., Pinyon-Juniper Wd., Creosote Bush Scrub;

scattered localities, Little San Bernardino Mts. to Mono Co., Providence and Clark mts., etc.; to Ida., Colo. April–June.

42. **C. nemáclada** Greene. Annual, erect, simple or branched from the base, the stems slender, sparsely strigose and ± short-hirsute; lvs. narrowly linear, strigose and loosely hispid, 1–3 cm. long, minutely pustulate; spikes in 1's or 2's, bractless, very slender, becoming loose and 3–7 cm. long; calyx 2–4 mm. long in fr., oblong-ovoid, ascending, the linear lobes connivent above with spreading tips, the midribs hispid below, retrorsely setulose toward tips, the margins sparsely strigose; corolla less than 1 mm. broad; nutlets 1–4, lanceolate to lance-ovoid, 1.7–2 mm. long, smooth, convex on back, whitish, rounded on angles, the ventral groove closed, not or shortly forked at base; style ca. ¾ the height of the nutlets.—Barren banks and slopes, below 3000 ft.; Pinyon-Juniper Wd., Foothill Wd., Chaparral; Coast Ranges from Colusa Co. to San Luis Obispo Co., Tehachapi Mts. April–May.

43. **C. microstàchys** (Greene ex Gray) Greene. [*Krynitzkia m.* Greene ex Gray.] Erect annual, slender-branched from base or above, usually very hirsute, 1–5 dm. tall; lvs. linear to lance-linear, 1–4 cm. long, obtuse, hirsute-hispidulous, sometimes strigose; spikes commonly in 2's or 3's, slender and loosely fld. in age, 3–8 cm. long; calyx ovoid to somewhat oblong in fr., 1–2 mm. long, usually strict, hirsute, the lobes lance-linear; corolla 0.5–1 mm. broad; nutlet 1, lanceolate or attenuate-ovoid, 1.5 mm. long, smooth, rounded on back and sides, the groove closed, simple or forked at very base; style ca. half as tall as nutlets.—Dry openings, below 4000 ft.; Chaparral, Foothill Wd., etc.; inner and middle Coast Ranges from Glenn Co. s., Sierra Nevada in Kern Co., cismontane s. Calif. to San Diego Co. April–June.

44. **C. leiocárpa** (F. & M.) Greene. [*Echinospermum l.* F. & M. *C. l.* var. *eremocaryoides* Brand.] Erect annual commonly branched from base, sometimes ± decumbent in age, 1–3 dm. long, hirsute-hispid and strigose; lvs. linear or oblanceolate, appressed-pilose-hispid, 1–3.5 cm. long; spikes usually many, leafy-bracted near base, 2–6 cm. long; calyx ± ovoid in fr., 2–3 mm. long, ascending-hirsute and spreading-bristly; corolla 1–2.5 mm. broad; nutlets 4 (sometimes 1), lance-ovoid, 2 mm. long, smooth, shining, mostly mottled brown and gray, rounded on back, the angles obtuse, the ventral groove closed, obscurely or not forked at base; style equal to or higher than nutlets.—Sandy places, below 500 ft.; Coastal Strand, N. Coastal Scrub; coast line from Surf (Santa Barbara Co.) to s. Ore. April–June.

45. **C. hispídula** Greene ex Brand. Erect annual with slender stems widely branched mostly above the base, strigose and spreading-hirsutulous; lvs. linear or the lower wider, 0.6–1.5 cm. long, hispidulous, somewhat pustulate; spikes in 2's or 3's, becoming 5–7 cm. long; calyx 2–2.5 mm. long in fr., ascending, the linear lobes connivent with the tips spreading, hispid on midribs, appressed-hispidulous on margins; corolla 2–2.5 mm. broad; nutlet 1, smooth, lanceolate to ovoid, with slender beak, well rounded on back and sides, the ventral groove open, short-forked at base; style shorter than nutlet.—Serpentine outcrops, below 3000 ft.; Chaparral, Yellow Pine F.; inner Coast Ranges, Lake and Napa cos. April–June.

46. **C. Clevelándii** Greene. [*C. Abramsii* Jtn. *C. Brandegei* Jtn.] Annual, stems erect to decumbent, 1–5 dm. long, strigose or hirsute, usually branched; lvs. usually many at base, scattered upward, lance-linear to linear, 1–5 cm. long, thinly appressed-hispidulous with some setae on margins; spikes in 1's or 2's, slender, bractless, 4–10 cm. long; calyx 2–5 mm. long in fr., ovoid to somewhat oblong, the lobes linear, connivent above with spreading tips, the 3 outer hispid on thick midribs, all densely appressed-hirsutulous; corolla 1–2 mm. broad; nutlets 1–4, oblong-ovoid to broadly lanceolate, 1.5–2 mm. long, convex on back, smooth, shining, the groove closed, broadly forked at base, rarely with a small areola; style ca. ⅔ as high as nutlets.—Slopes and rocky places, below 2500 ft.; Coastal Sage Scrub, Coastal Strand, Chaparral; coastal s. Calif. from Ventura Co. s., Santa Barbara Ids.; L. Calif. April–June.

Var. **floròsa** Jtn. [*C. hispidissima* Greene. *C. C.* var. *h.* Jtn.] Usually stouter, erect, ascendingly branched; infl. more pedunculate; corolla 2–5 mm. broad.—Chaparral, Coastal Sage Scrub, etc.; Point Reyes, Marin Co., Santa Cruz Co. to San Diego Co.

Var. **dissíta** (Jtn.) Jeps. & Hoov. [*C. dissita* Jtn.] Corolla 4–6 mm. broad; calyx 5–6 mm. long in fr.; style slightly exceeding nutlets.—Serpentine outcrops, near Lakeport, Lake Co.

47. **C. Gánderi** Jtn. Bushy-branched annual 1–4 dm. high, hirsute-hispid; lvs. linear, thinly hispidulous, pustulate; spike solitary, bractless, 5–15 cm. long, loosely fld. in age; calyx ascending, 6–10 mm. long in fr., the narrowly linear lobes erect, with yellow bristles on the costate midrib; corolla inconspicuous; nutlets usually 1 (3 abortive), lanceolate, 2–2.5 mm. long, smooth, gray-brown, obscurely mottled, convex on back, rounded on margins, the groove closed above, forked at base into a narrow triangular areola.—Desert washes, below 1200 ft.; Creosote Bush Scrub; Borrego V., w. Colo. Desert. Feb.–May.

48. **C. confértiflòra** (Greene) Pays. [*Oreocarya c.* Greene. *O. lutea* Greene ex Fedde.] Perennial from a branched root-crown and stout woody root; stems erect, 1.5–5 dm. tall, strigose and sparsely setose above, the bases white-hairy; lvs. linear-oblanceolate to oblanceolate, 3–10 cm. long; infl. a terminal cymose head 2.5–5 cm. across and with smaller pedunculate ones from upper lf.-axils; calyx-lobes lance-linear, 10–12 mm. long in fr., loosely strigose and weakly bristly; corolla-tube 9–13 mm. long, the limb 10 mm. broad, pale yellow or cream-colored; nutlets broadly ovoid, 3 mm. long, sharply 3-angled, glossy, smooth, the scar nearly closed.—Dry rocky places in limestone, 4000–9000 ft.; Sagebrush Scrub, Pinyon-Juniper Wd., to Lodgepole F.; White and Inyo mts. and e. slope of Sierra Nevada from Mono Co. s. (also in Tulare Co.), to San Bernardino Mts. and mts. of Mojave Desert; to Utah, Ariz. May–July.

49. **C. Jàmesii** (Torr.) Pays. var. **abortìva** (Greene) Pays. [*Oreocarya a.* Greene. *O. suffruticosa* var. *a.* Macbr.] Stems prostrate or laxly ascending, from a woody perennial root, leafy, pallid, 0.5–1.8 dm. long, tomentulose, sparsely hispidulous; lvs. oblanceolate, 4–10 cm. long, silky-strigose; infl. open, not much elevated above lvs.; bracts conspicuous; calyx-lobes lance-ovate, becoming 4–5 mm. long, tomentulose, sparsely short-setose; corolla-tube 3–4 mm. long, the limb white, ca. 5 mm. broad; nutlets 1–4, lance-ovoid, 2–2.5 mm. long, smooth, glossy, the back strongly rounded from base toward apex, the venter decidedly triangular in cross section, with the narrow groove along the crest of its median inner angle.—Dry gravelly soil in open woods, 6000–10,200 ft.; Sagebrush Scrub, Yellow Pine F., to Subalpine F.; Bear V., San Bernardino Mts., White Mts.; w. Nev. May–Aug.

50. **C. flavóculàta** (Nels.) Pays. [*Oreocarya f.* Nels.] Perennial with cespitose woody caudex; stems 1–many, strigose and hispid, 1–3 dm. long; lvs. oblanceolate, 3–8 cm. long, appressed-silky-pubescent with scattered bristles; infl. ± cylindrical, with upper cymules congested, the lower more scattered; fl.-bracts lance-linear; calyx-lobes densely bristly with ± yellowish setae, lance-linear, ca. 10 mm. long in fr.; corolla-tube 7–10 mm. long, the white or pale yellow limb 7–10 mm. broad; nutlets ovate to lanceolate, 2.5–3.5 mm. long, obtusish, with acute-angled margins, somewhat glossy but muriculate on back, the scar open, conspicuously margined.—Dry rocky slopes, 5000–9500 ft.; Pinyon-Juniper Wd., Sagebrush Scrub; Panamint Mts., Inyo Co. to Sweetwater Mts., Mono Co.; to Wyo., Colo. May–July.

51. **C. crymóphila** Jtn. [*Oreocarya c.* Jeps. & Hoov.] Perennial, stems several, 1.5–3 dm. high, erect, simple, hirsute and minutely villous below, hispid above; lvs. grayish-villous-tomentose and setose, the lower 5–10 cm. long, spatulate-oblanceolate, the cauline reduced; cymes few-fld., in lf.-axils and glomerate at apex; calyx 13–15 mm. long in fr., sparsely setose; corolla white, the tube 8 mm. long, the limb 5 mm. broad; nutlets 4, ovoid, ca. 5 mm. long, irregularly rugose on the rounded back, the ventral side smooth, the scar narrow, subulate at base.—Rocky places, 9000–9500 ft.; Subalpine F.; Alpine Co. July–Aug.

52. **C. subretùsa** Jtn. [*Oreocarya s.* Abrams.] Perennial, cespitose from a compact caudex, the stems 1–2 dm. high, densely yellow-bristly; lvs. congested below, spatulate, subretuse to obtuse, tomentose; fls. crowded, in a narrow thyrsus, the bracts yellow-hispid; calyx-lobes 5–7 mm. in fr.; corolla white, the tube 3–4 mm. long, the limb 3–6 mm. broad; nutlets lance-oblong, 3–4 mm. long, acute to narrow-winged on angles, convex on back, inconspicuously tubercled or with short low rugae, the scar linear with slightly open base.—Dry gravel, pumice, etc., 7000–8000 ft.; Subalpine F.; Mt. Eddy, Redshale Mt., Siskiyou Co.; Ore., n. Nev. July–Aug.

53. **C. nubígena** (Greene) Pays. [*Oreocarya n.* Greene. *C. Clemensae* Pays.] Stems slender, several, from a densely leafy and branched root-crown, erect, 0.5–1.5 dm. high, retrorsely pubescent and spreading-setose; lvs. linear-oblanceolate or spatulate, 2–3 cm.

long, thinly hirsute-pubescent and setose, ± pustulate; infl. short-spicate, leafy-bracteate; calyx-lobes 6–7 mm. long in fr., slender-setose and with shorter finer hairs; corolla white, the tube ca. 3 mm. long, the limb ca. 4 mm. broad; nutlets lance-linear, 3 mm. long, narrowly wing-margined, tubercled or rugose on back, groove straight, narrow but open, extending almost entire length of nutlet.—Rocky and gravelly places, 8000–12,500 ft.; Subalpine F., Alpine Fell-fields; Sierra Nevada from Mono and Tuolumne cos. to Olancha Peak, Tulare Co. July–Aug.

54. **C. hùmilis** (Greene) Pays. [*Eritrichium glomeratum* var. *h.* Gray. *Oreocarya h.* Greene.] Cespitose perennial with 1 to several stems from a woody caudex, densely leafy, 1–3 dm. high, hirsute and finely strigose; lvs. broadly spatulate to oblanceolate, 2–4 cm. long, short-petioled, silky-tomentose and with some slender bristles; infl. rather narrow, thyrsoid, leafy-bracted below; calyx-lobes lanceolate, 8–12 mm. long in fr.; corolla white, the tube 4–5 mm. long, the limb 8–10 mm. broad; nutlets mostly 4, lance-ovoid, 3.5–4.5 mm. long, with acute margins and glossy tuberculate back or sometimes ± rugose, the ventral surface indistinctly tuberculate, the scar triangular, almost closed to open at base.—Montane ridges, 6000–11,400 ft.; Red Fir F. to Subalpine F.; Sierra Nevada from Mono Co. to Nevada, Alpine and Inyo cos.; Nev. June–Aug.

55. **C. nàna** (Eastw.) Pays. var **ovìna** Pays. Densely cespitose perennial, the stems 1–1.5 dm. high, strigulose and slender-setulose; lvs. crowded near base, spatulate or oblanceolate, scattered on stems, white-tomentose and appressed-setulose, 1–2.5 cm. long; infl. narrow-thyrsoid, yellowish-setulose; calyx-lobes ca. 6 mm. long in fr., densely setulose and tomentose; corolla-tube 3–4 mm. long, the limb white, ca. 3–4 mm. broad; nutlets lance-ovoid to ovoid, acute, densely and uniformly muriculate, the scar slightly open and ± triangular.—Disintegrated travertine, at 6200 ft.; Pinyon-Juniper Wd.; Bridgeport, Mono Co.; to Nev., Utah. May–June.

56. **C. Hoffmánnii** Jtn. [*Oreocarya H.* Abrams.] Root biennial or perennial; stems 1–several, 1.5–3 dm. high, hirsute and retrorsely pubescent; lower lvs. spatulate, 2–3 cm. long, on longer petioles, sparsely bristly and retrorsely pubescent, pustulate; infl. 5–15 cm. long, ± interrupted, with many cymules; calyx-lobes 5–7 mm. long in fr., linear-lanceolate, hirsutulous and bristly; corolla white, the tube 3 mm. long, the limb 5–6 mm. broad; nutlets ovoid, ca. 3 mm. long, tuberculate and slightly rugose, the ventral scar open, extending almost to apex.—Dry rocky places, 7000–10,200 ft.; Pinyon-Juniper Wd., Bristle-cone Pine F.; White and Inyo mts. June–Aug.

57. **C. virginénsis** (Jones) Pays. [*Krynitzkia glomerata* var. *v.* Jones. *Oreocarya v.* Macbr.] Biennial or short-lived perennial, the stems tufted, 1–4 dm. tall, hispidulous and hispid; lvs. oblanceolate to spatulate, 5–10 cm. long, the lower crowded, bristly, somewhat tomentulose, pustulate; infl. thyrsoid, loose, 3–6 cm. thick, fulvescent, hirsute; calyx-lobes 7–12 mm. long in fr., lance-linear, setose; corolla white, the tube 3–4 mm. long, the limb 3–7 mm. broad; nutlets ovoid, short-acuminate, 3–4 mm. long, coarsely tuberculate and irregularly rugose, carinate on back, the scar narrowly triangular.—Dry rocky slopes, 3600–10,200 ft.; Creosote Bush Scrub to Pinyon-Juniper Wd., Bristle-cone Pine F.; Panamint Mts., Inyo Mts. to Barstow, Baker, Kingston Mts., Clark Mt.; to Utah. April–June.

58. **C. tumulòsa** (Pays.) Pays. [*Oreocarya t.* Pays.] A strong perennial, the stems 1–2.5 dm. tall, somewhat villous and slender-setose; lvs. oblanceolate to obovate-spatulate, 3–5 cm. long, appressed-setose, obscurely pustulate; infl. becoming loosely cylindrical, 2–2.5 cm. thick, with yellowish bristles; calyx-lobes 7–10 mm. long in fr., setose-spreading with retrorse bristles; corolla white, the tube 3.5–4 mm. long, the limb 6–7 mm. broad; nutlets ovoid, 4 mm. long, irregularly and coarsely tuberculate, sometimes obscurely rugulose, the scar open.—Dry rocky places, in limestone, 4500–6000 ft.; Pinyon-Juniper Wd.; Providence, New York, and Ivanpah mts., e. San Bernardino Co.; Charleston Mts., Nev. April–June.

59. **C. Roosiòrum** Munz. Densely cespitose perennial; stems 0.1–0.3 dm. long; lvs. clustered, spatulate-oblanceolate, acutish, 0.5–1.2 cm. long, grayish-strigose and appressed-setose, scarcely if at all pustulate; infl. compact, ca. 1 cm. long; calyx strigose and ± setose, the lobes linear, 3 mm. long in fl., 4 mm. in fr.; corolla white, the tube 2.5 mm. long, the limb 5 mm. broad; style slightly surpassing the nutlets; nutlets ca. 2.5 mm. long, lance-ovate, whitish and ± longitudinally rugose on back, with ventral

scar open toward base.—Dry rocky slope, at 10,600 ft.; Bristle-cone Pine F.; Inyo Mts. June–July.

15. Plagiobóthrys F. & M.

Annual or perennial herbs, usually with weak slender appressed hairs, at times bristly but not pungently so. Lower lvs. opposite or rosulate and crowded. Fls. generally in slender racemes or spikes, occasionally glomerate or axillary, frequently bracted. Calyx divided to middle or lower, usually persistent, frequently tawny or brown, often accrescent. Corolla white, the tube short and usually included in calyx. Stamens included. Ovules 4. Nutlets 1–4, erect or incurved, roughened or rarely smooth, the margin round or angled, the back occasionally appendaged, attached medianly or basally to the low convex pyramidal or truncated gynobase through a variously developed (mostly caruncular) scar, the scar decurrent on lower part of keel, or projected on a short basal prolongation of keel, or more commonly located at lower end of keel and ± sunken beneath the level of its crest. Ca. 100 spp., w. N. and S. Am. (Greek, *plagios*, on the side, and *bothrys*, a pit, referring to the elevated caruncular scar of *P. fulvus*, the first known sp.)

(Johnston, I. M. A synopsis and redefinition of the genus Plagiobothrys. Contr. Gray Herb. 68: 57–80, 1923; and The Allocarya section of the genus Plagiobothrys in the western U.S. Contr. Arnold Arb. 3: 1–82, 1932.)

A. Lvs. all alternate.
  B. Caruncle of nutlet elongate, apparently extending along crest of ventral keel; nutlets trigonous. E. of Sierra Nevada. (Section Amsinckiopsis.)
    C. Corolla 4–7 mm. broad; nutlets irregularly rugose . . . . 1. *P. Kingii*
    CC. Corolla 1–2.5 mm. broad; nutlets conspicuously tessellate . . . . 2. *P. Jonesii*
  BB. Caruncle round or nearly so, at or below end of ventral keel.
    C. Lowest lvs. not in a rosette; caruncle weakly developed, borne at tip of a short ventral stipe; nutlets lacking a broad transverse ventral groove. Montane and cismontane s. Calif. . . . . 14. *P. californicus*
    CC. Lowest lvs. mostly in a rosette; caruncle well developed, sessile on nutlet; nutlet with a shallow transverse ventral groove.
      D. Infl. glomerate; caruncle at or above middle of nutlet; basal lvs. not present in maturity of plant. E. of Sierra Nevada (Section Sonnea) . . . . 3. *P. hispidus*
      DD. Infl. elongate, racemose; caruncle cartilaginous, at or below middle of nutlet; basal lvs. evident at maturity of plant.
        E. Calyx circumscissile in fr., less than 4 mm. long; lobes usually connivent over fr.; mature nutlets usually only 1 or 2. (Section Plagiobothrys)
          F. Infl. a long simple bracted raceme; nutlets highly arched in lateral outline, 1–2.5 mm. long; corolla 2–3 mm. broad. W. edge of San Joaquin V. to w. desert . . . . 4. *P. arizonicus*
          FF. Infl. forked, bracted at base only if at all; nutlets low and flattened in lateral outline, 2–3 mm. long; corolla 3–9 mm. broad. Widely distributed in cismontane Calif. . . . . 5. *P. nothofulvus*
        EE. Calyx not circumscissile, or if so, the strongly accrescent calyx over 4 mm. long; calyx-lobes erect or spreading; nutlets usually 4.
          F. Bristles of calyx-lobes hooked. Gabilan and Santa Lucia mts.
            6. *P. uncinatus*
          FF. Bristles of calyx-lobes not hooked.
            G. Nutlets with a conspicuous annular caruncle and 2.3–3.3 mm. long; calyx cleft to near base, rusty-hirsute; corolla-tube longer than calyx.
              H. Racemes bractless; areolae on dorsal surface of nutlet regular and rectangular, the dorsal keel not winged . . . 7. *P. fulvus*
              HH. Racemes bracteate; dorsal side of nutlet not areolate or the areolae irregular and unequal with short curved or interrupted rugae, the dorsal keel narrowly white-winged
                8. *P. infectivus*
            GG. Nutlets with a solid caruncle and less than 2.3 mm. long.
              H. Transverse dorsal crests of nutlets very narrow and sharp, with median keel enclosing polygonal granulate areoles.
                I. Corolla-tube shorter than calyx; plants erect to prostrate, rather coarse-stemmed; lvs. 3–7 mm. broad
                  9. *P. canescens*
                II. Corolla-tube at least as long as calyx; plants very slender, strict or ascending; lvs. 1.5–2.5 mm. broad
                  10. *P. myosotoides*
              HH. Transverse dorsal crests of nutlets very low and broad, separated only by low lineate ridges.

       I. Nutlets ovoid, usually constricted at apex only, the base rounded or rarely weakly constricted, dark; plant dye-stained . . . . 11. *P. Torreyi*

      II. Nutlets decidedly cruciform due to the abrupt constrictions at base and apex, glossy; plant rarely dye-stained.

         J. Calyx 5–7 mm. long and almost as thick; spikes bracteate; nutlets 2–2.7 mm. long . . . . . 12. *P. shastensis*

         JJ. Calyx 3–5 mm. long, ca. half as thick; spikes bractless except at very base; nutlets 1.5–2 mm. long
                                      13. *P. tenellus*

AA. Lvs. opposite at least below.

  B. Nutlets attached to gynobase by a ± well developed stipelike ventral projection (Section Echidiocarya) . . . . 14. *P. californicus*

  BB. Nutlets attached directly to gynobase, without a stipelike projection. (Section Allocarya)

    C. Plants perennial, coarse, densely soft-villous. Sierra Co. to Modoc Co., also Sonoma Co.
                                      15. *P. mollis*

    CC. Plants annual, slender-stemmed.

      D. Plant floriferous to near base and with lower pedicels stout and recurved; calyx-lobes with indurate ribs, eventually contorted and irregularly spreading or recurving; stems prostrate.

        E. Nutlets broadly ovoid, glossy, sparsely if at all tuberculate; scar ⅓–⅕ the length of nutlet, surrounded by a high collar. Amador and Stanislaus cos.
                                    16. *P. scriptus*

        EE. Nutlets lance-ovoid, dull, granulate and tuberculate; scar ⅕ the length of nutlet, not surrounded by a high collar. Tehama Co. to Stanislaus Co.
                                    17. *P. humistratus*

      DD. Plant not floriferous to base or if so, the lower pedicels not stout and recurving.

        E. Scar large, deeply excavated, ¼–½ the length of nutlet; nutlets frequently with prickles.

          F. Spikes mostly in 2's; plant erect, nearly or quite glabrous; nutlets ovoid, 1.5 mm. long; corolla 4–6 mm. broad. Upper Napa V. . 18. *P. strictus*

          FF. Spikes not in 2's; corolla 1–4 mm. broad (larger in *P. glyptocarpus*).

            G. Nutlets broad, ca. ⅔–⅘ as broad as long.

               H. The nutlets 2.5–3 mm. long, not transversely rugose dorsally. San Joaquin Co. to Ore. . . . . 19. *P. Greenei*

               HH. The nutlets 1.5–2 mm. long, transversely rugose dorsally.

                   I. Appendages on nutlets short and stout, covered with minute subulate bristles. Solano Co. . . 20. *P. hystriculus*

                  II. Appendages on nutlets slender and elongate, barbed at apex, or almost lacking. San Joaquin V. to San Diego Co.
                                    21. *P. acanthocarpus*

          GG. Nutlets elongate, ca. half as broad as long.

             H. Dorsal keel of nutlets knifelike, bearing coarse bristles; nutlets glossy on back . . . . 22. *P. Austinae*

               HH. Dorsal keel of nutlets not knifelike; nutlets dull and transversely rugose on back.

                 I. Calyx scarcely accrescent; nutlets ca. 1.5 mm. long
                                    23. *P. distantiflorus*

                 II. Calyx evidently accrescent; nutlets ca. 2 mm. long
                                    24. *P. glyptocarpus*

        EE. Scar small, not much if at all excavated, ⅕ the length of nutlet or less.

          F. Nutlet attachment basal, frequently substipitate; plants ± succulent; calyx-lobes strongly ribbed.

            G. Plants prostrate; calyx-lobes connivent and together directed off toward one side of the fl. . . . . 25. *P. leptocladus*

            GG. Plants erect or ascending; calyx-lobes spreading equally.

               H. Stems fistulous-enlarged; nutlets with well developed lateral keel; calyx with indurated ribs . . . . 26. *P. glaber*

               HH. Stems not fistulous-enlarged; nutlets with weakly developed lateral keel; calyx-ribs weakly indurate . . 27. *P. stipitatus*

         FF. Nutlet attachment lateral to obliquely basal; calyx-ribs only rarely thickened.

            G. Scar linear or nearly so, usually borne on a narrow knifelike attachment. Coastal plants.

               H. Ventral keel in an elongate depression which is groovelike only near base. San Diego Co. to Marin Co.
                                    28. *P. undulatus*

               HH. Ventral keel in a deep longitudinal groove throughout its length, the groove sometimes infolding and concealing the keel.

                 I. Nutlets rough and dull, 1.3–1.9 mm. long; plant trailing or prostrate; corolla 5–10 mm. broad . 29. *P. Chorisianus*

                 II. Nutlets smooth and shining, 2.5–3 mm. long; plant erect; corolla 2–4 mm. broad . . . . 30. *P. lithocarpus*

           GG. Scar broad, the attachment not at all knifelike, or if so the nutlets rather asymmetric.

H. Stems with distinctly spreading hairs. Deserts . 31. *P. Parishii*
HH. Stems strigose or appressed-hispidulous.
   I. Nutlet-scar in an areole broader than long or areole want-
       ing; nutlets asymmetric.
     J. Scar ovate to triangular, the thickened margin usually
        divergent or spreading; nutlets mostly dull. From Tuo-
        lumne and Nevada cos. n. . . . . 32. *P. cognatus*
    JJ. Scar elongate, with thick knifelike erect or inflexed
        margin; nutlets shining. Plumas, Lassen, and Modoc
        cos. . . . . 33. *P. Cusickii*
   II. Scar of nutlet in an areole longer than broad; nutlets
      symmetrical.
     J. Ventral keel not in a groove.
       K. Corolla 3–7 mm. broad; racemes essentially bract-
         less . . . . 34. *P. tener*
      KK. Corolla 1–3.5 mm. broad; racemes ± abundantly
        bracted.
       L. Scar near or at base of nutlet, oblique to the
         ventral keel . . . . 35. *P. bracteatus*
       LL. Scar distinctly lateral, parallel with the
         ventral keel.
        M. Nutlets usually muriculate or minutely
          bristly; scar linear-oblong. Sierra Ne-
          vada and mts. of s. Calif.
                         36. *P. hispidulus*
        MM. Nutlets not muriculate or bristly; scar
          ovate or deltoid. Coast Ranges, s. to
          Los Angeles Co. . 37. *P. trachycarpus*
    JJ. Ventral keel in a conspicuous groove at least below
       the middle.
      K. Nutlets densely tuberculate, the pericarp thick and
        rather bony. San Francisco . . . . . 38. *P. diffusus*
      KK. Nutlets sparsely if at all tuberculate, the pericarp
        thin. Marin Co. to Del Norte Co.
                        39. *P. reticulatus*

1. **P. Kíngii** (Wats.) Gray. [*Eritrichium K.* Wats. *Sonnea K.* Greene.] Erect annual 1–4 dm. high, usually with short floriferous branches from about the base of the cent. axis, hirsute and also usually with fine short pubescence; basal lvs. narrowly oblanceolate, the cauline lance-oblong to linear, 1.5–6 cm. long, hirsute; cymes dense, scorpioid, with scattered bracts, more lax in fr., arranged in an elongate infl.; calyx-lobes becoming 5–6 mm. long, stiffly hirsute; corolla 4–7 mm. broad; nutlets 4, cuneate-ovoid, 2.5–3 mm. long, somewhat incurved at apex, with median-dorsal and lateral keels and irregular transverse rugae forming rather broad papillate areolae; scar elongate, keellike, medial. —Sandy desert valleys, 4000–7000 ft.; Sagebrush Scrub, Pinyon-Juniper Wd.; n. Inyo Co., Mono Co.; w. Nev. May–June.

Var. **Harknéssii** (Greene) Jeps. [*Sonnea H.* Greene. *P. H.* Nels & Macbr.] Infl. glomerate or scarcely elongated; plant 1–2 dm. high; scar of nutlet elongate, extending nearly the whole length of the ventral keel.—Owens Lake, Inyo Co. to Mono Lake; e. Ore., w. Nev.

2. **P. Jònesii** Gray. [*Sonnea J.* Greene.] Annual, erect or ascending; stems 1–several, divergently branched, conspicuously bristly and fine-pubescent, pustulate; lower lvs. oblanceolate to linear, 2–6 cm. long, the cauline mostly lanceolate; racemes scorpioid, leafy-bracted, rather dense, terminal and in upper axils, but even lower lvs. may have 1–few fls. in their axils; calyx-lobes linear-subulate, 6–10 mm. long; corolla 1–2 mm. broad; nutlets 2 or 3, incurved and 4-angled by the dorsal and ventral keels and lateral ridges, 2.5–3.5 mm. long, tuberculate on angles, tessellate between; scar narrow, merging into the keel above and with a diverging lateral ridge extending to either side.— Gravelly and rocky places, below 5700 ft.; Creosote Bush Scrub to Pinyon-Juniper Wd.; Whipple Mts. e. San Bernardino Co. to White Mts., Lone Pine, Townsend Pass, etc.; Ariz. April–May.

3. **P. híspidus** Gray. [*Sonnea h.* Greene.] Rather densely leafy annual; stems 1–few, diffusely branched, 0.5–2 dm. high, hispid and sparsely tomentulose; lvs. hispid, some-what pustulate, lanceolate to oblong, 1.5–4 cm. long, obtuse; fls. in small terminal glomerules and in upper axils; calyx-lobes lanceolate, ca. 2 mm. long; corolla 1 mm. broad; nutlet usually 1, ovoid, 1 mm. long, tubercled, ribbed dorsally and on angles; scar slightly above middle, tapering into the sharp keel above.—Gravelly and sandy

places, 4000–9000 ft.; Yellow Pine F. to Lodgepole F.; e. slope of Sierra Nevada from Mono Co. n., to e. Siskiyou Co.; e. Ore., w. Nev. June–Aug.

4. **P. arizònicus** (Gray) Greene ex Gray. [*Eritrichium canescens* var. *a.* Gray.] Erect or loosely ascending annual, usually ± branched below the middle, 1–4 dm. tall, hirsute-hispid and somewhat villous; lvs. with stiff somewhat appressed hairs, ± pustulate, the lower in a rosette, linear-oblanceolate, 1.5–5 cm. long, the upper scattered on the slender stems, lanceolate, reduced; roots, petioles, etc., with purple dye; spikes elongate, lax in fr., mostly bractless; calyx ca. 3 mm. long, lobed to ca. the middle, usually tawny-hirsute, connivent, circumscissile in age; corolla 2–2.5 mm. broad; nutlets 1–4, ovoid, abruptly acute, the back marked off into rectangular areoles by narrow keels and ridges and transverse rugae, the keels often tuberculate, the areolae smooth or minutely papillate; scar median, in a sunken area at base of keel.—Dry slopes and flats under desert influence, below 4000 (7000) ft.; Foothill Wd., Pinyon-Juniper Wd.; hills along w. edge of San Joaquin V. from Merced Co. to Kern Co., Tehachapi Mts. to Olancha, across the desert to Providence Mts., w. edge of Colo. Desert; to New Mex., Son. March–May.

Var. **catalinénsis** Gray. [*P. canescens* var. *c.* Jeps. *P. c.* Macbr.] Calyx-lobes variable, but tending to be less connivent over the nutlets.—Santa Catalina and San Clemente ids.

5. **P. nothofúlvus** (Gray) Gray. [*Eritrichium n.* Gray.] POPCORN FLOWER. Erect annual 2–5 dm. high, the stems branched mostly above, villous with short curly hairs or finely hispid; lvs. largely in a basal rosette, oblanceolate, 3–10 cm. long, sparsely villous, the cauline few, lance-linear, reduced upward; root, stem, petioles, etc., with copious purple dye; infl. once or twice forked, the spikes slender, mostly bractless, elongating and lax; calyx densely appressed-silky-villous, usually tawny, 2–3 mm. long in fr., the lobes erect, ca. as long as tube, this circumscissile in fr.; corolla 6–8 mm. broad; nutlets 1–4, round-ovoid, abruptly constricted into an acute apex, the back with ± rectangular granulate areas between narrow ridges and keels; scar annular, median at base of the narrow ventral keel.—Common, grassy fields and hillsides, mostly below 2500 ft., occasional to 4700; V. Grassland, Foothill Wd., Coastal Sage Scrub; almost throughout cismontane Calif., occasional at desert edge; to L. Calif., Wash. March–May.

6. **P. uncinàtus** J. T. Howell. Annual 1–3 dm. high, tinged purplish, few–many-stemmed, suberect to divergent, spreading-hirsute; basal lvs. linear-oblong, 1.5–2.5 cm. long, hispidulous, the cauline oblong-ovate to ovate, remote; fls. scattered along the stems and in small terminal clusters; calyx deeply divided, 2–2.5 mm. long in fr., densely uncinate-bristly; corolla 1–1.5 mm. broad; nutlets broadly ovoid, 1–1.3 mm. long, shiny, gray, keeled down the back and on the lateral angles, transversely rugulose, scarcely tuberculate in the areolae, the ventral side narrowly keeled above the middle, the scar roundish, small.—Canyon sides, 1000–2000 ft.; Chaparral, etc.; Santa Lucia Mts., Gabilan Range, Monterey and San Benito cos. May.

7. **P. fúlvus** (H. & Á.) Jtn. var. **campéstris** (Greene) Jtn. [*P. c.* Greene.] Erect annual, simple or branching above, the stems 1 or more, 3–6 dm. high, villous-hirsute and sparsely tomentose, with purple dye; lvs. appressed-hirsute, the basal spatulate-oblanceolate, 2–8 cm. long, the cauline linear-oblong, remote, reduced upward; spicate racemes very loose, bractless, to 3 or 4 dm. long; calyx fulvous when young, the lanceolate lobes ca. 5 mm. long in fr., erect; corolla 3–4 mm. broad; nutlets triangular-ovoid, usually 4, abruptly short-acute, 2.5–3 mm. long, thin-keeled dorsally, usually with transverse rugae; scar annular, bordering a deep excavation.—Open gravelly or sandy places, below 1500 ft.; Foothill Wd., V. Grassland; Coast Ranges from Santa Clara Co. n., foothills of Sierra Nevada from Fresno Co. n.; Ore. March–May.

8. **P. infectìvus** Jtn. Purple-stained annual, the stems 1–several, erect or ascending, villous-hirsute; lvs. linear to somewhat oblong, the lower ± crowded, 2–8 cm. long, villous, the upper scattered, reduced; spikes to 1 or 2 dm. long in fr., lax, leafy-bracteate; calyx deeply cleft, 5–7 mm. long in fr., rusty-hirsute when young; corolla 3–4 mm. broad; nutlets 4, broadly ovoid, 3–3.5 mm. long, abruptly constricted into the short acute apex, the median keel prominent upward, the lateral keels most distinct on the beak, the ventral keel prominent above the scar, this raised and annular; back tuberculate but with few or no rugae.—Mostly heavy soils, below 1000 ft.; inner Coast Ranges, Colusa Co. to San Luis Obispo Co. March–May.

9. **P. canéscens** Benth. [*Eritrichium c.* Gray. *P. microcarpus* Greene.] Stems usually several from base, decumbent or prostrate, rarely erect, 1–6 dm. long, villous or finely

hispid; lvs. linear or the basal linear-oblanceolate, 1.5–5 cm. long, the cauline well developed; roots, stems, etc. with some purple dye; spikes elongate and loosely-fld. in age, leafy-bracteate; calyx 4–6 mm. long in fr., densely villous-hirsute, cleft over halfway; corolla ca. 2–4 mm. broad; nutlets usually 4, round-ovoid, narrowed into a short beak-like apex, strongly incurved, obscurely tuberculate, 1–2 mm. long, usually with prominent transverse rugae forming rectangular papillate intervals; scar medial, annular, slightly raised.—Grassy slopes and flats, below 4500 ft.; Foothill Wd., V. Grassland, Coastal Sage Scrub, etc.; Siskiyou Co., s. along w. base of Sierra Nevada, Cent. V., S. Coast Ranges, w. Mojave Desert, cismontane s. Calif.; Santa Barbara Ids. March–May.

10. **P. myosotoìdes** (Lehm.) Brand. [*Lithospermum m.* Lehm. *P. tinctorius* (R. & P.) Gray.] Erect or procumbent ± hispid annual, the stems 0.5–1.2 dm. long, with purple dye; basal lvs. rosulate, 1.5–2 cm. long, the cauline alternate; cymes axillary, short, bracteate; calyx 2–2.5 mm. long; corolla 2 mm. broad; nutlets strongly keeled dorsally, not shining, dark brown, 1.5 mm. long, with ridges separated by broad intervals that are sometimes papillate.—Rare; Chaparral; ridge between Isabel V. and Arroyo Bayo, Mt. Hamilton Range; Big Sandy V., Black Mt., Fresno Co.; Chile and Peru. April–May.

11. **P. Tórreyi** (Gray) Gray. [*Eritrichium T.* Gray. *P. T.* vars. *diffusus* and *perplexans* Jtn.] Annual with 1 to several slender stems, erect and few-branched or more often diffuse, 0.5–1.5 dm. long, spreading-hirsute; purple dye unusually abundant; lvs. oblong to somewhat linear or the uppermost subovate, sessile, 0.5–2 cm. long; spikes lax in fr., leafy-bracted throughout, lax; calyx 2.5 mm. long in fr., hirsute and somewhat hispid; corolla 1.5–2 mm. broad; nutlets mostly 4, round-ovoid, 1.5–2.2 mm. long, abruptly narrowed at apex, keeled on back especially in upper half, shining, with ca. 7 transverse ridges with narrow intervening sinuses, smooth or with few whitish tubercles; scar small, borne in the transverse groove.—Meadows and flats, 4000–11,000 ft.; Yellow Pine F. to Subalpine F.; Sierra Nevada from Sierra Co. to Kern Co., San Bernardino Mts. June–July. Typical *Torreyi* from the Yosemite region is supposed to be less branched than plants from other areas, which have been called var. *diffusus*.

12. **P. shasténsis** Greene ex Gray. Annual; stems 1–few from the base, simple or branched above, 1–3 dm. high, pilose; lvs. largely in a basal rosette, linear-oblanceolate, 1–3 cm. long, appressed-hirsute above and ± pustulate, subglabrous beneath; cauline lvs. sessile, linear, shorter; spikes often in 2's, bracteate throughout, loosely fld., 1–10 cm. long; calyx cleft to middle, ca. 6–7 mm. long in fr., hirsute, often rusty; corolla 2.5 mm. broad; nutlets quadratish-ovoid, abruptly acute, 2–2.5 mm. long, shining, 3-keeled on back and with broad transverse ridges separated by narrow lineate grooves; scar small, round.—Uncommon, slopes and flats, below 2500 ft.; Foothill Wd.; w. base of Sierra Nevada from Stanislaus Co. to Butte Co., inner Coast Ranges from Merced Co. to Siskiyou Co.; s. Ore. May–June.

13. **P. tenéllus** (Nutt.) Gray. [*Myosotis t.* Nutt. ex Hook. *P. parvulus, asper,* and *colorans* Greene. *P. t.* vars. *p.* and *c.* Jtn. *P. humifusus* Jones.] Annual, with 1–several slender stems, erect, 1–3 dm. tall, soft-villous; lvs. of basal rosette lance-oblong, villous, 1.2–5 cm. long, the cauline few, ± ovate, shorter; racemes loosely-fld., elongate, slender, bracted only near base; calyx ca. 3 mm. long in fr., short-villous, whitish or reddish; corolla 2–3 mm. broad; nutlets thick-cruciform, 1.5–2 mm. long, usually light-colored, sharply ridged dorsally and on edges, commonly tubercled on ridges.—Grassy and gravelly places, below 5000 ft.; Chaparral, Foothill Wd., Yellow Pine F.; much of cismontane Calif.; to L. Calif., Ariz., Utah, B.C., Ida. March–May.

14. **P. califórnicus** (Gray) Greene. [*Echidiocarya c.* Gray. *Allocaryastrum c.* Brand. *P. Cooperi* Gray. *P. allocaryoides* Brand.] Light green annual with rather slender spreading or prostrate stems 1–4 dm. long, spreading-hirsute; lvs. often numerous below, oblanceo-late or spatulate, 1–4 cm. long, thinly hirsute or appressed canescent, the upper cauline reduced, linear; racemes simple, elongate, leafy-bracted at least in lower half; calyx deeply lobed, 3 mm. long in fr., hirsute and sparingly hispid; corolla 4–7 mm. broad; nutlets usually 4, ovoid, 1.5–2 mm. long, the ridges and keels usually prominent, the ridges loosely reticulate or almost parallel, sometimes broken up into a series of tubercles; scar elevated on a prominent cylindrical stipe near the ak.-base.—Grassy places, below 1500 ft.; V. Grassland, Coastal Sage Scrub; cismontane s. Calif. from Los Angeles and San Bernardino cos. to L. Calif.; Santa Rosa Id. March–May.

KEY TO VARIETIES

Corolla 4–7 mm. broad; pubescence fine, appressed, somewhat silky .... *P. californicus*
Corolla 1–3 mm. broad; pubescence usually spreading.
    Lvs. narrowly linear, 2–2.5 mm. broad; pubescence fine, canescent .... var. *gracilis*
    Lvs. oblanceolate, 3–5 mm. broad; pubescence coarse, fulvescent.
      Racemes dense, hidden among the lvs. .... var. *ursinus*
      Racemes elongated, projected from among the lvs. and evident .... var. *fulvescens*

Var. **grácilis** Jtn. Mature calyx 2–3 mm. long; corolla 1.5–2 mm. broad; nutlets 1–1.5 mm. long.—About La Jolla and San Diego, San Clemente, Santa Catalina, and Santa Cruz ids.

Var. **ursìnus** (Gray) Jtn. [*Echidiocarya u.* Gray. *P. u.* Gray.] Of dense compact habit; stems 2–8 cm. long, stout, much-branched; racemes short; fls. obscured by lvs. and bracts; nutlets 1.7–2 mm. long, usually with few transverse ridges.—Gravelly and sandy places in montane valleys, 4500–6800 ft.; Yellow Pine F.; San Bernardino and San Jacinto mts.; L. Calif.

Var. **fulvéscens** Jtn. Habit of the sp., but herbage rougher, short-hispid; corolla 2–3 mm. broad.—Dry slopes and benches, 2000–6500 ft.; Chaparral, Yellow Pine F.; inner Coast Ranges, San Benito and San Luis Obispo cos., Santa Barbara Co. to San Diego Co., Santa Rosa and Anacapa ids.; L. Calif.

15. **P. móllis** (Gray) Jtn. [*Eritrichium m.* Gray. *Allocarya m.* Greene.] Coarse perennial from a fleshy taproot, densely spreading-pilose throughout; stems many, spreading, branched, 1–3 dm. long, rather densely leafy; lvs. opposite, oblong to linear, 4–8 cm. long; racemes mostly single at ends of branches, 3–10 cm. long in age, often bracted at base; pedicels 1–2 mm. long; calyx 4–5 mm. long in fr., the lance-cuneate lobes almost erect; corolla 5–10 mm. broad; nutlets ovoid, ca. 1.5 mm. long, the back with irregular transverse ridges that merge toward the sides, the medial keel most distinct toward apex, the ventral side with a conspicuous ovate or triangular submedial scar.—Moist alkaline places, 4000–5500 ft.; Sagebrush Scrub; Sierra Co. to Modoc Co.; Ore., Nev. June–July.

Var. **vestìtus** (Greene) Jtn. [*Allocarya v.* Greene. *A. mollis* var. *v.* Jeps.] Plants more rank, usually decumbent; upper lvs. alternate; nutlets ± reticulate with the interspaces somewhat granulate.—Valley flats, Sonoma Co. (near Petaluma). June–July.

16. **P. scríptus** (Greene) Jtn. [*Allocarya s.* Greene.] Annual with prostrate stems 1–2 dm. long, strigose, the lower internodes short, congested; branches wiry, floriferous to near base; lower lvs. linear, upper oblanceolate, subglabrous above, loosely strigose beneath, 0.5–2 cm. long; spikes leafy-bracted; pedicels stout, recurved, to 2 mm. long; calyx accrescent, loosely hispidulous, to almost 8 mm. long; corolla ca. 2 mm. broad; nutlets ovoid, 2 mm. long, glossy and dark with pale ridges and keels that bear slender branched hairs, the back medially keeled, lateral keels also present, with loose reticular transverse ridges, the ventral side loosely rugose, with prominent keel to below the middle; scar large, suprabasal, suboblique, deltoid, excavated, surrounded by a high collar.—Rare, moist banks; V. Grassland; Ione (Amador Co.) and La Grange (Stanislaus Co.) March.

17. **P. humistràtus** (Greene) Jtn. [*Allocarya h.* Greene. *A. limicola* and *sigillata* Piper.] Somewhat succulent annual, the stems several, mostly prostrate, glabrous or nearly so, 1–4 dm. long; lvs. linear, 3–8 cm. long; spikes somewhat 1-sided, the pedicels stout, ± curved and to ca. 2 mm. long; calyx accrescent in fr., sparsely strigose, 6–10 mm. long; corolla 1–2 mm. broad; nutlets lance-ovoid, 2–2.5 mm. long, the back keeled medially most of its length, irregularly transversely ridged with broad tubercular interspaces, also keeled laterally, the ventral side angulate and keeled; scar suprabasal, somewhat oblique, deltoid to ovate, usually concave, surrounded by a distinct ridge.—Hog wallows and low places, below 550 ft.; V. Grassland; rare in Livermore, Sacramento, and lower San Joaquin valleys. March–April.

18. **P. stríctus** (Greene) Jtn. [*Allocarya s.* Greene. *A. californica* var. *s.* Jeps.] Slender-stemmed erect annual 1–4 dm. tall, subglabrous; lower lvs. linear, 4–9 cm. long, glabrous or with a few short hairs with pustulate base; spikes slender, solitary or not, naked or bracted at base only; pedicels strict, 1 mm. long; calyx weakly accrescent in fr., 3 mm. long, strigose, usually tawny; corolla 4–6 mm. broad; nutlets ovoid, ca. 1.5 mm. long, keeled on the back, weakly so on the edges, granulate, tubercled, somewhat transversely

rugose with papillate-dentate ridges; ventral side obliquely rugose; scar lateral, excavated, narrow, the apical half often closed by the infolding edges.—Wet alkaline places, 400–500 ft.; Mixed Evergreen F.; near sulphur springs, Calistoga, Napa. Co. March–June.

19. **P. Greènei** (Gray) Jtn. [*Echinospernum G.* Gray. *Allocarya G.* Greene. *Echinoglochin G.* Brand. *Allocarya Echinoglochin* Greene.] Erect to spreading annual, strigose, the stems 1–4 dm. long; lower lvs. linear to spatulate-linear, 1–5 cm. long, the upper oblanceolate to linear-oblong; racemes lax in age, bracted in lower part; pedicels slender, 1–2 mm. long; calyx accrescent in fr., loose, broad, 5–8 mm. long, ± tawny toward apex; corolla 2–5 mm. broad; nutlets broadly ovoid, dull, depressed, 2.5–3 mm. long, contracted toward apex, the back strongly medially keeled, usually also with weak lateral keels, finely tuberculate, armed along the keels and between with many subulate apically glochidiate appendages 0.4–0.8 mm. long, without transverse ridges; scar large, deep, ovate or deltoid, broadly flanged.—Grassy clay hills or dry mud-flats, below 2800 ft.; Foothill Wd., V. Grassland; Sierran foothills from Stanislaus Co. to Shasta Co., Cent. V. from San Joaquin Co. n., Coast Ranges from Sonoma Co. to Siskiyou Co.; Ore. March–May.

20. **P. hystrículus** (Piper) Jtn. [*Allocarya h.* Piper. *A. Greenei* var. *h.* Jeps. *Echinoglochin h.* Brand.] Plant like *P. acanthocarpus*, the nutlets ca. 2 mm. long, armed with stout short appendages ± joined by weak transverse ridges and densely covered from base to apex with minute spreading conic-subulate bristles.—Rare, plains and hills; V. Grassland; Solano Co. April–May.

21. **P. acanthocárpus** (Piper) Jtn. [*Allocarya a.* Piper. *Echinoglochin a.* Brand. *A. anaglyptica, Eastwoodae, echinacea, microcarpa, oligochaeta, papillata,* and *spiculifera* Piper.] Strigose erect to spreading annual; the stems 1–4 dm. long; lower lvs. linear or spatulate-linear, 2–6 cm. long, upper linear to oblanceolate; racemes bracted, becoming loose and elongate; pedicels slender, 1–2 mm. long; calyx broad and loose in fr., 3–6 mm. long, ± tawny; corolla 1–2.5 mm. broad; nutlets ovoid, 1.5–2.5 mm. long, the back strongly medially keeled, with lateral keels ± developed and with transverse ridges so as to be reticulate-rugose, tubercled in the interspaces, ± armed with stiff subulate apically glochidiate appendages; ventral side with large deeply excavated deltoid or ovate scar that is broadly flanged.—Moist clay or adobe flats or beds of winter pools, below 2000 ft.; Chaparral, V. Grassland; mesas near San Diego, San Joaquin V. and lower Sacramento V. and adjacent hills; L. Calif. March–May.

22. **P. Aùstinae** (Greene) Jtn. [*Allocarya A.* Greene. *Echinoglochin A.* Brand. *A. cristata* Brand. *A. A.* var. *c.* Jeps. *A. A.* var. *nuda* Hoov.] Plant like *A. Greenei,* but nutlets glossy, 2.5–3 mm. long, the body angular, ovoid, abruptly contracted into an elongate laterally much-compressed beak; dorsal keel knifelike, the lateral keels obtuse; area between keels irregularly tuberculate to ± smooth, the dorsal keels armed with subulate-columnar appendages bearing a plumose array of coarse spreading falcate hairs near their tips; scar triangular, excavated, ⅓–½ the length of the nutlet.—Usually in clay depressions, below 1500 ft.; V. Grassland; e. side of Cent. V. from Shasta Co. to Stanislaus Co. March–May.

23. **P. distantiflòrus** (Piper) Jtn. [*Allocarya d.* Piper. *Glyptocaryopsis d.* Brand.] Subprostrate to ascending strigose annual with slender stems 1.5–3 dm. long; lvs. linear-oblanceolate, 1–2.5 cm. long; spikes 5–12 cm. long, partially bracteate; calyx not much enlarged in fr., 2.5–3 mm. long; corolla 1–1.5 mm. broad; nutlets elongate, angular, ca. 1.5 mm. long, dorsally keeled to below the middle, the edges with dentate lateral keels, transversely wrinkled with tubercles and papillae; ventral side with prominent crowded ridges; scar ⅓–½ of the length of the nutlet, excavated, elongate, lance-ovate to sublinear.—Moist places, below 1200 ft.; V. Grassland; Sierran foothills from Tuolumne Co. to Madera Co. April–May.

24. **P. glyptocárpus** (Piper) Jtn. [*Allocarya g.* Piper. *Glyptocaryopsis g.* Brand.] Erect annual, strigose, slender-stemmed, 1–5 dm. high; lower lvs. linear to spatulate, 4–8 cm. long, the upper oblong-linear to oblanceolate; spikes simple, bracted below, loose in age; mature calyx accrescent, 3–5 mm. long; corolla 5–9 mm. broad; nutlets elongate, angular, ovoid to oblong-ovoid, 2 mm. long, the back with a strongly dentate medial keel and lateral keels, transversely ridged, tuberculate and granular; ventral side with prominent crowded ridges; scar deep, ⅓–½ the length of the nutlet, often somewhat

covered by the ingrown pericarp toward the summit.—Moist places, below 2000 ft.; V. Grassland, Foothill Wd.; Lake and Butte cos. to Ore. April–May.

Var. **modéstus** Jtn. Corolla 2–3 mm. broad.—Yellow Pine F.; Cedar Crest, near Grass V., Nevada Co.

25. **P. leptoclàdus** (Greene) Jtn. [*Allocarya l.* Greene. *A. californica* var. *subglochidiata* (Gray) Jeps. *A. divergens, charaxata,* and *versicolor* Piper.] Somewhat succulent annual, the stems prostrate, strigose, 1–3 dm. long; lvs. subglabrous above, sparsely strigose and pustulate beneath, the lower linear or somewhat spatulate, 3–10 cm. long, the upper oblance-linear; racemes unilateral, rather lax in fr.; mature calyx usually accrescent, sparsely strigose, the lobes with ± indurated ribs, lanceolate to somewhat spatulate, 4–8 mm. long; corolla 1–2 mm. broad; nutlets narrowly to broadly lanceolate, 1.5–2.5 mm. long, keeled dorsally only above the middle, tuberculate, somewhat rugose, smooth, granulate to penicillate-hairy; ventral side angulate, keeled the entire length; scar basal, not surrounded by a ridge, frequently with a downwardly directed dorsal flange.—In heavy usually alkaline soils that have been wet, below 2000 ft.; V. Grassland, Coastal Sage Scrub; coastal valleys of s. Calif., Cent. V.; to Ore., Ida., Utah, L. Calif. March– May.

26. **P. glàber** (Gray) Jtn. [*Lithospermum g.* Gray. *Allocarya g.* Macbr. *A. salina* Jeps.] Yellow-green coarse erect annual, somewhat strigose, 1–1.5 dm. tall, succulent; lower lvs. linear, the upper oblanceolate, glabrous above, sparsely strigose and pustulate beneath; racemes somewhat fistulous, rather densely fld., ± unilateral, leafy-bracted below; mature calyx 8–10 mm. long, united into a thick cylindrical appressed base 2–3 mm. long, the lobes firm with indurated ribs; corolla ca. 3 mm. broad; nutlets lanceolate, 2 mm. long, dorsally medially keeled above the middle, with rather definite lateral keels, tubercled, weakly transversely ridged, the ventral side keeled its entire length; scar basal, substipitate.—Coastal Salt Marsh; s. shore of San Francisco Bay and alkaline flats in Santa Clara V. and near Hollister (San Benito Co.). April–May.

27. **P. stipitàtus** (Greene) Jtn. [*Allocarya s.* Greene. *Lappula s.* Druce.] Erect yellow-green annual 1–5 dm. tall, sparsely strigulose, somewhat succulent; lvs. glabrous above, strigose and somewhat pustulate beneath, the lower linear, 2–11 cm. long, the upper linear to oblanceolate; racemes stiffish, wiry, somewhat unilateral, bracted below, rather loosely fld.; mature calyx accrescent, ± strigose, 5–8 mm. long, the lobes somewhat indurate on midribs; corolla 5–12 mm. broad; nutlets lanceolate to lance-ovate, 1.5–2.5 mm. long, usually distinctly compressed below the middle, thickened and somewhat beaked above, keeled dorsally only near tip and about the margin, obliquely or transversely rugose mostly above middle, tuberculate below; ventral side keeled to base, tuberculate or rugose; scar small, basal, sessile to substipitate.—Low ± alkaline places, like hog wallows, vernal pools, etc., below 1500 ft.; Foothill Wd., V. Grassland; Sacramento V. w. to Sonoma and Napa cos., Hollister (San Benito Co.); to Ore. March–May.

Var. **micránthus** (Piper), Jtn. [*Allocarya s.* ssp. *m.* Piper.] Corollas ca. 2.5 mm. broad.— Below 5000 ft.; Yellow Pine F. to V. Grassland; Lassen, Shasta and Humboldt cos. to San Diego Co. April–July.

28. **P. undulàtus** (Piper) Jtn. [*Allocarya u.* Piper. *A. Chorisiana* var. *u.* Jeps. *A. inornata* and *corrugata* Piper.] Slender-stemmed sparsely strigulose annual, at first erect, later often sprawling, the branches 1–3 dm. long; lvs. usually sparsely strigose or appressed-hispidulous beneath, subglabrous above, 2–6 cm. long, the lower linear, the upper wider; spikes loose, slender, with some leafy bracts below; calyx somewhat accrescent in age, appressed-villous-hispidulous, usually tawny toward tips, ca. 2 mm. long; corolla 1.5–2 mm. broad; nutlets ovoid to narrower, 1–1.6 mm. long, medially dorsally keeled toward apex, transversely rugose with low rounded undulating ridges, keeled laterally ca. ⅗ the length then replaced by the linear scar which is slightly elevated and lying in an elongate depression.—Mud flats of vernal pools or moist stream beds, below 1200 ft.; several Plant Communities; Coast Ranges from San Diego Co. to Mendocino Co., possibly also along e. side of Cent. V. March–June.

29. **P. Chorisiànus** (Cham.) Jtn. [*Myosotis C.* Cham. *Allocarya C.* Greene. *Eritrichium connatifolium* Kell.] Sparsely strigose annual, trailing or erect, usually branched above the base; stems 1–4 dm. long; lower lvs. 3–7 cm. long, conspicuously connate in pairs, sparsely ciliate, the upper lance-oblong, shorter; racemes simple, elongate, bracted, loose in age, slender, 2–15(–25) mm. long; calyx ca. 4 mm. long in fr., usually tawny,

the lanceolate lobes strigose; corolla 6–10 mm. broad; nutlets ovoid, 1.3–1.9 mm. long, keeled dorsally only near apex, loosely tubercled, with scattered irregular transverse ridges, granulate; ventral side with a suprabasal knifelike attachment bearing the linear scar, this and the keel in a longitudinal groove.—Grassy and moist places; N. Coastal Scrub, Chaparral; San Francisco to Santa Cruz Mts. April–June.

Var. **Hickmánii** (Greene) Jtn. [*Allocarya H.* Greene. *A. myriantha* Greene. *A. Jonesii* Brand.] Prostrate, branched from base, without prominent basal sheathing lf.-bases; pedicels mostly shorter than calyx.—Near the coast; Closed-cone Pine F.; Santa Cruz Mts. to San Luis Obispo Co. April–June.

30. **P. lithocáryus** (Greene ex Gray) Jtn. [*Krynitzkia l.* Greene ex Gray. *Allocarya l.* Greene.] Erect annual 1–3 dm. tall, strigulose; lvs. linear, 3–6 cm. long, rather obtuse, the upper oblanceolate, smaller; racemes elongate, lax, bracted, with slender pedicels 1–4 mm. long; calyx ca. 4 mm. long in fr., the subulate lobes usually with tawny tips; corolla 2–4 mm. broad; nutlets 4 or fewer, ovoid, somewhat triquetrous, 2.5–3 mm. long, minutely granular and finely reticulate-rugulose, dorsally distinctly angled; ventral side flat, grooved its entire length with the ventral keel and knifelike attachment-scar in the groove.—Rare, moist places, 1000–1500 ft.; valleys in Lake and Mendocino cos. bordering the Mayacamas Range. April–May.

31. **P. Paríshii** Jtn. [*Eritrichium Cooperi* Gray, not *P. C.* Gray. *Allocarya C.* Greene.] Diffusely branched annual, the stems prostrate, branched, 0.5–3 dm. long, with short stout spreading hairs; lvs. linear or the upper oblong, somewhat hispidulous and pustulate beneath, 1–5 cm. long; racemes slender, elongate, few-bracted in lower part; mature calyx hispidulous, tending to be deciduous, 2–3 mm. long, the lobes ligulate to lanceolate; corolla 3–5 mm. broad; nutlets 1–1.8 mm. long, ovoid or lance-ovoid, ± heteromorphic with the axial nutlet somewhat larger, plumper, with a triangular-ovate scar; other nutlets with sublinear scar; back medially keeled only near apex, transversely ridged, breaking up into tuberculations toward base; scar in a groove.—Occasional in wet places, 2500–4500 ft.; largely Joshua Tree Wd., w. Mojave Desert from Rabbit Springs, Camp Cady, Lovejoy Springs, etc., to Owens V. and near Chalfont, Mono Co. April–June.

32. **P. cognàtus** (Greene) Jtn. [*Allocarya c.* Greene. *A. microcalyx* Brand. *A. filicaulis* Brand.] Erect to spreading annual, the strigose stems 0.5–2.5 dm. long; lvs. strigose and pustulate beneath, less so above, the lower linear, 2–7 cm. long, the upper linear or subspatulate; racemes solitary, slender, usually loosely fld., bracted at least in lower part; calyx in fr. ± appressed-hispidulous, the linear to linear-lanceolate lobes 2–4 mm. long; corolla 1–2 mm. broad; nutlets broadly ovoid to oblong-ovoid, somewhat asymmetric, 1.3–2.2 mm. long, obtuse to rounded at base, the back keeled near apex and upper part, tubercled and with irregular ridges, usually finely granulate on general surface; ventral side with keel ± folded over near base, the scar expanded, triangular to ovate, convex to concave, with spreading margin and surrounded by ridges forming a ± evident broad transversely elongate angular aerola.—Damp places, 1000–7000 ft.; many Plant Communities up to Lodgepole F.; Sierra Nevada from Tuolumne and Lake cos. n.; to Wash., Rocky Mts., Ariz. June–Aug.

33. **P. Cusíckii** (Greene) Jtn. [*Allocarya C.* Greene. *A. ambigens* Piper.] Prostrate or ascending annual, the stems ± strigulose, 0.5–2 dm. long; lvs. pustulate and strigose beneath, less so above, often somewhat ciliolate, the lower linear, 3–10 cm. long, the upper linear to lanceolate; racemes solitary, usually slender, loose or dense, bracted at least in lower part; calyx appressed-hispidulous, weakly accrescent, the linear or lance-linear lobes 1.5–4 mm. long; corolla 1–1.5 mm. broad; nutlets rather asymmetric, lanceolate to oblong-ovate, 1–2 mm. long, glossy, not granulate, usually abruptly rounded at base, the back keeled near apex, irregularly obliquely ridged, tuberculate in lower part; ventral side keeled to well below middle, the lower part oblique and with a small deep scar especially on the 3 homomorphic nutlets, the axial one with a broad flat scar.—Montane flats and meadows, 4000–6100 ft.; Sagebrush Scrub, N. Juniper Wd., Yellow Pine F.; e. slope of Sierra Nevada from Inyo Co. n.; to Wash., Nev. May–Aug.

34. **P. téner** (Greene) Jtn. [*Allocarya t.* Greene. *A. gracilis, laxa, pratensis, scalpocarpa,* and *vallata* Piper. *A. hispidula* var. *t.* Jeps.] Slender-stemmed annual, mostly erect, 1–3 dm. tall, sparsely strigulose or appressed-hispidulous; lvs. sparsely appressed-hispidulous and pustulate at least beneath, the lower usually linear, 2–6 cm. long, the upper shorter; racemes slender, usually loosely-fld., mostly without bracts except at

very base; mature calyx usually sparsely strigose, somewhat accrescent, the narrow lobes 2–3.5 mm. long; corolla 3–7 mm. broad; nutlets lance-ovoid, 1.5–2.2 mm. long, slightly incurved, the back roundish, somewhat keeled toward tip, granulate and tuberculate, commonly with some narrow obliquely transverse papillate-dentate ridges or these broken up into papillae, the surface generally muricate or with short microscopic hairs, the ventral side diagonally rugulose, keeled above the scar which is lance-ovate and extending almost halfway up the nutlet.—Wet places, below 5500 ft.; Chaparral, Foothill Wd., Yellow Pine F.; inner Coast Ranges from Napa Co. to Shasta Co., and to Modoc Co. April–July. Plants from Lake Co., ± succulent, subglabrous, with erect connivent calyx-lobes and corollas 3 mm. broad, have been designated as var. *subglàber* Jtn. and from Plumas Co., with nutlets attached by an oblique base, as var. *fállax* Jtn.

35. **P. bracteàtus** (Howell) Jtn. [*Allocarya b.* Howell. *A. commixta* and *Piperi* Brand. *A. conjuncta* Piper. *A. debilis* Greene ex Brand. *A. Cusickii* vars. *d.* and *vallicola* Jeps.] Sparsely strigose annual, the slender stems with long branches, usually ascending, 1–4 dm. long; lower lvs. 3–10 cm. long, linear, the upper linear to oblong; racemes slender, bracted toward base; mature calyx with linear to lanceolate lobes 2–4 mm. long; corolla 1–3 mm. broad; nutlets 1.2–2 mm. long, the back ± keeled above the middle, granulate, with irregular ± oblique transverse wrinkles or ridges, the lower somewhat broken into tubercles; ventral sides keeled and angulate to beyond the middle, the lower part forming a ridge surrounding the small oblique or subbasal narrowly ovate to elliptical scar.—Moist places or dry beds of pools and ditches, below 5000 ft.; many Plant Communities; much of cismontane Calif.; to Ore., L. Calif. April–June. A form with minute apically barbed trichomes on the nutlets, from Butte Co., has been named var. *aculeolàtus* (Piper) Jtn. [*Allocarya a.* Piper.]

36. **P. hispídulus** (Greene) Jtn. [*Allocarya h.* Jtn. *A. penicillata* Greene. *A. cryocarpa* and *cervina* Piper. *A. nigra* Brand.] Grayish-green annual, usually branched at base, prostrate or loosely ascending, the strigose stems 0.5–4 dm. long; lvs. appressed-hispidulous, ± pustulate, especially beneath, the basal linear, 1–5 cm. long, the upper linear to oblanceolate; racemes usually elongate, loosely fld., single, bracted in lower part; mature calyx strigose, the narrow lobes 2–3 mm. long, ascending; corolla 1–2 mm. broad; nutlets ovoid to lance-ovoid, 1.5–2 mm. long, usually abruptly rounded at base, the back keeled above the middle, obliquely transverse-ridged, the ridges often broken, anastomosing, frequently dentate with papillae, or bearing branched hairs, the intervening spaces tuberculate and granular; ventral side keeled and angulate to below the middle, the lowermost part with a ridge encircling the linear-oblong scar.—Moist meadows or dryish flats, 4000–11,000 ft.; Montane Coniferous F.; mts. of San Diego Co., San Bernardino Mts., Sierra Nevada, N. Coast Ranges from Yolla Bolly Mts. n., to Modoc and Siskiyou cos.; to Wash., Wyo., Nev. June–Sept.

37. **P. trachycárpus** (Gray) Jtn. [*Krynitzkia t.* Gray. *Allocarya t.* Greene. *A. interrasilis* Piper.] Laxly ascending or prostrate annual, strigose or appressed-hispidulous, branched at base, the stems 0.5–4 dm. long; lower lvs. linear, obtuse, 5–10 cm. long, the upper oblanceolate; racemes lax, leafy-bracted throughout; mature calyx strigose, the narrow lobes 3–5 mm. long, usually with tawny hairs near tip; corolla 1–2 mm. broad; nutlets ovoid, rather angulate, ca. 2 mm. long, the back keeled to the middle or beyond as well as laterally and with transverse ridges, the interspaces tuberculate; ventral side obtusely angled with a strong keel; scar suprabasal, broad, ovate or deltoid, concave, closely surrounded by a strong ridge.—Flats and dried pools, below 3200 ft.; Coastal Sage Scrub, V. Grassland, Chaparral, Foothill Wd., etc.; coastal Los Angeles Co. through Coast Ranges to Contra Costa and San Joaquin cos. April–June.

38. **P. diffùsus** (Greene) Jtn. [*Allocarya d.* Greene.] Like *P. trachycarpus* in habit; the nutlets broadly ovoid, 1.3–1.5 mm. long; back conspicuously convex, keeled to middle or farther, the ridges rather irregular, ± reticulately joined, the small interspaces granulate, tuberculate; ventral side with the strong ridge around the scar prolonged forward along the keel, hence the keel in a groove to the middle of the nutlet; scar lateral with the expanded margin strongly upturned, hence narrow, deeply convex.—N. Coastal Scrub; San Francisco. April–June.

39. **P. reticulàtus** (Piper) Jtn. [*Allocarya r.* Piper. *A. minuta* var. *r.* Jeps. *A. minuta* Piper not *P. m.* Jtn. *A. areolata* Piper.] More or less strigose grayish or yellowish-green annual, usually much-branched at base, erect to decumbent, the stems 1–4 dm. long;

lvs. appressed-hispidulous, the lower linear to spatulate-linear, 3–8 cm. long, the upper more oblong; racemes slender, single, elongate, lax, leafy-bracted in lower part; calyx in fr. 2–4 mm. long; corolla 1.5–3.5 mm. broad; nutlets ovoid, plump, with thin pericarp, 0.7–1.7 mm. long, finely and sparsely granulate; back convex, keeled toward apex only, the ridges low, rounded, irregularly anastomosed; ventral side strongly keeled; scar small, narrow-elliptic to ovate, concave; ridge surrounding scar prolonged forward to form ridges parallel to keel and a broad open trough to ca. middle of nutlet.—Fields and moist places, below 1000 ft.; Redwood F.; coast from Mendocino Co. to Ore. May–July.

Var. **Rossianòrum** Jtn. [*Myosotis californica* F. & M. *Allocarya c.* Greene. *A. scalpta* Piper.] Nutlets more granulate, ± oblong-ovoid, the back transverse-rugose, more abundantly tuberculate; ventral keel in a narrower groove.—Near the coast from Mendocino Co. to Marin Co., also in Santa Barbara Co. May–June.

16. Amsínckia Lehm. FIDDLENECK

Annual herbs, usually pungent-bristly, with erect or spreading branched stems. Lvs. alternate. Fls. in usually naked or sparsely bracted scorpioid spikes. Calyx 5-parted or with 1 to several lobes ± united, persistent, often tawny or even brown-hairy. Corolla yellow to orange, tubular to salverform, the tube frequently exserted, the limb narrow to broad, occasionally obscurely irregular, the throat unappendaged. Stamens inserted in throat or tube. Style filiform; stigma capitate, 2-lobed. Nutlets crustaceous, 1–4, smooth or rough, attached by a caruncular scar placed submedially at lower end of ventral keel or attached along lower part of groove or slit in pericarp on ventral angle of nutlet. Cotyledons each 2-parted by a dorsi-ventral longitudinal cleft. Gynobase pyramidal, ca. half as long as nutlets. Perhaps 20 spp., w. N. Am. and s. S. Am. (W. *Amsinck*, early nineteenth-century patron of botanic garden in Hamburg.)

(Suksdorf, W., Werdenda 1: 47–113, 1931. Johnston, I. M., Journ. Arnold Arb. 16: 197–204, 1935. Ray, P. M., and H. F. Chisaki, Am. Jour. Bot. 44: 529–554, 1957.)

Nutlets smooth and shining.
  Lvs. coarsely and densely pustulate above, sparsely so beneath, bristly-ciliate; scar of nutlet poorly developed, sublineate, nearly basal. Mojave Desert to Monterey Co. . . . . 1. *A. vernicosa*
  Lvs. finely and densely pustulate, hirsute above and beneath; scar of nutlets conspicuous, lance-ovate, nearly median. Contra Costa and Alameda cos. . . . . 2. *A. grandiflora*
Nutlets roughened and dull.
  Back of nutlets tessellate, i.e. like a diminutive cobblestone pavement; corolla-tube 20-nerved below stamens; calyx-lobes unequal in width and at least 2 of them fused for some length.
    Corolla 6–14 mm. broad; calyx densely rusty-pubescent and bristly.
      Fls. with ratio of length of style to that of stamens various; pollen with 3 equatorial apertures. Monterey Co. to Ventura Co. . . . . 3. *A. Douglasiana*
      Fls. with ratio of length of style to length of stamens fairly constant; pollen with 4 equatorial apertures. Colusa Co. to Los Angeles Co. . . . . 4. *A. gloriosa*
    Corolla 2–4 mm. broad; calyx thinly white-hirsute. Deserts and San Joaquin V. and inner Coast Ranges n. to Contra Costa Co. . . . . 5. *A. tessellata*
  Back of nutlets not tessellate, the roughenings not crowded to give a pavementlike appearance; corolla-tube 10-nerved below stamens; calyx-lobes various.
    Calyx with the 2 axial lobes united at least to middle (except in the early fls.); lvs. mostly erose-dentate; nutlets dark, 1.5–2 mm. long . . . . 6. *A. spectabilis*
    Calyx with lobes all distinct; lf.-margins usually entire; nutlets usually gray or brown, mostly 2.5–3.5 mm. long.
      Corolla-throat constricted and nearly closed by intruding hairy saccate processes; stamens inserted evenly below the constriction . . . . 7. *A. lycopsoides*
      Corolla-throat open and glabrous; stamens ± irregularly inserted in throat.
        Corolla yellow, pale, 5–7 mm. long, 2–3 mm. broad, the tube almost or quite included in calyx; style 2.5–3 mm. long; plant usually cinereous, the stems finely strigulose as well as spreading-hirsute . . . . 8. *A. Menziesii*
        Corolla orange-yellow, 7–20 mm. long, the tube distinctly exserted; style 5–18 mm. long; stems hispid, but scarcely if at all finely strigose.
          Fls. homostylic, i.e. styles of one length with reference to stamens, nearly radially symmetrical.
            Nutlets 2.5–3.5 mm. long, the back depressed, the scar ovate; calyx becoming 5–11 mm. long . . . . 9. *A. intermedia*
            Nutlets 1.5–2 mm. long, the back rounded, the scar narrow; calyx becoming 4–5 mm. long . . . . 6. *A. spectabilis* var. *microcarpa*
          Fls. heterostylic, i.e. of more than 1 length with reference to stamens, ± asymmetric
                                      10. *A. lunaris*

1. **A. vernicòsa** H. & A. [*A. carnosa* Jones. *A. glauca* Suksd.] Erect, sparsely setose and pustulate, or nearly glabrous, 3–5 dm. tall; lvs. somewhat glaucous and fleshy, abun-

dantly pustulate above, sparsely so beneath, middle and upper lvs. lanceolate to lance-ovate, acuminate, 4–8 cm. long, with somewhat clasping base; spikes 3–12 cm. long; mature calyx setose on back, pale ciliate on margins, 9–18 mm. long, sometimes 2 or more of the lobes partly united; corolla 10–12 mm. long, golden-yellow, 3–6 mm. broad; nutlets gray, smooth, shining, 4–6 mm. long, plane on back and lateral surfaces, the scar lineate; $2n = 14$ (Ray, 1954).—Dry plains and slopes, below 4500 ft.; Joshua Tree Wd., Chaparral, Foothill Wd., V. Grassland; w. Mojave Desert, Greenhorn Mts. (Kern Co.), w. side of San Joaquin V. and adjacent Coast Ranges to Monterey and Fresno cos. March–May.

Var. furcàta (Suksd.) Hoov. in Jeps. [A. f. Suksd.] Corolla orange, 12–18 mm. long, 8–14 mm. broad; $n = 7$ (Ray & Chisaki, 1957.)—W. Fresno Co. to se. San Luis Obispo Co.

2. A. grandiflòra Kleeb. ex Gray. [A. vernicosa var. g. Gray.] Erect, 3–6 dm. high, sparingly hispid below, thinly pilose above and the stiffer hairs weak or represented by their pustulate bases only; lvs. pustulate on both surfaces, but the bristles often not developed from some pustules; spikes becoming 1–1.5 dm. long; calyx with rust-colored bristles that may completely conceal the appressed hairs beneath, the lobes fused into 3 or 4, to ca. 12 mm. long in fr.; corolla orange, 14–18 mm. long, 8–10 mm. broad; nutlets ovate-lanceolate, smooth, shining, plane on back and sides, the scar broadly lanceolate, a little below the middle; $2n = 12$ (Ray, 1954).—Open grassy slopes, below 1200 ft.; V. Grassland; inner Coast Range and adjacent valley, Contra Costa and Alameda cos. April–May.

3. A. Douglasiàna A. DC. [A. Lemmonii Macbr. A. macrantha Suksd.] Erect, usually rather slender, 3–5 dm. high, simple up to infl., or branched below, thin bristly below, more abundantly so and bristles more spreading above; lvs. lance-linear to lanceolate, appressed-hirsute and subcinereous, weakly pustulate; spikes becoming 1–1.5 dm. long; fls. heterostylic; calyx rusty-pubescent, 1 or 2 pairs of lobes united, 6–12 mm. long in fr.; corolla orange, 12–16 mm. long, 6–10 mm. broad; pollen tricolporate; nutlets broadly ovoid, 4 mm. long, flattish and tessellate on the back, sometimes denticulate on lateral angles; scar ovate, median; $n = 6$ (Ray & Chisaki, 1957).—Dry open slopes and valleys, below 2000 ft.; V. Grassland, Foothill Wd.; S. Coast Ranges away from immediate coast, Monterey and San Benito cos. to n. Santa Barbara Co. and w. Kern Co. March–May.

4. A. gloriòsa Eastw. ex Suksd. [A. elegans, pulchra, cuneata (curvata), Munzii, and ampla Suksd.] Resembling A. Douglasiana; fls. homostylic; pollen tetracolporate; corolla 10–12 mm. long; $n = 12$ (Ray & Chisaki, 1957).—Inner Coast Ranges, Colusa Co. to Los Angeles Co.

5. A. tessellàta Gray. [A. collina Greene. A. pustulata Heller. A. conica, etc. Suksd.] Stems 2–6 dm. tall, hispid, pustulate; lvs. linear to lanceolate, or the upper ovate to subcordate, mostly pustulate especially above, with many spreading or appressed hairs; spikes becoming 5–12 cm. long; calyx-lobes fused into 3 or 4, hispid, white-hirsute on the margins, 8–12 mm. long in age; corolla orange, 8–12 mm. long, 2.5–5 mm. broad; nutlets ovoid, 3–3.5 mm. long, the back low, usually pale and crustose, angled on the margin, tessellate, often transversely rugose; $n = 12$ (Kamb in 1952).—Common in dry mostly sandy or gravelly places, below 6000 ft.; Creosote Bush Scrub, Joshua Tree Wd., Sagebrush Scrub, V. Grassland, Foothill Wd.; inner Coast Ranges from Contra Costa Co. to Santa Barbara Co., San Joaquin V., deserts, cismontane s. Calif. in interior as River-side and San Diego cos., e. of Sierra Nevada; to e. Wash., Nev., Ariz., L. Calif. March–June.

6. A. spectàbilis F. & M. [A. maritima Eastw. A. nigricans Brand. A. acuminata, cana, celsa, constricta, etc. Suksd.] Loosely branched, usually becoming spreading or prostrate, the stems 1–6 dm. long, sparsely hispid, pustulate; lvs. lanceolate, acuminate, usually bright green, sparsely but pungently hispid, pustulate beneath; spikes becoming 8–10 cm. long; calyx 3–6 mm. long in fr., the 5 lobes linear-lanceolate (2 or 3 usually partly united), hispid and pilose, usually fulvous; corolla orange, 6–12 mm. long, 4–6 mm. broad; nutlets ovoid, usually somewhat incurved, 1.5–2.0 mm. long, dark but pale-tuberculate, usually also with narrow oblique ridges and scattered papillae; scar sub-median, narrow; $n = 5$ (Kamb in 1952).—Sandy places and edges of marshes; Coastal Strand, Coastal Salt Marsh; along the coast from San Diego Co. to Tillamook Bay, Ore., L. Calif. March–June.

Var. nicòlai (Jeps.) Jtn. ex Munz. [A. intermedia var. n. Jeps. A. st.-nicolai Eastw.]

Spikes bracted throughout.—San Nicolas, San Miguel, and San Clemente ids.

Var. **microcárpa** (Greene) Jeps. & Hoov. [*A. m.* Greene. *Benthamia m.* Druce. *A. ochroleuca* and *dentata* Suksd.] Calyx-lobes all distinct to base or slightly united; corolla 13–16 mm. long; nutlets 1–2 mm. long, muriculate but not usually with dorsal rídges.—Sandy places; near coast, San Luis Obispo and nw. Santa Barbara cos.

7. **A. lycopsoìdes** Lehm. [*A. arenaria, glomerata, simplex* Suksd. *A. Howellii* Brand.] Three–10 dm. high, erect to procumbent, bristly-hirsute with few or no finer hairs; lvs. 3–10 cm. long, linear-oblanceolate below to lanceolate or wider above, with spreading and appressed stiff hairs; spikes to ca. 1 dm. long, often bracteate below; calyx 6–10 mm. long in fr., bristly-hirsute and long-ciliate; corolla deep yellow, usually clearly exserted, 7–10 mm. long, salverform, constricted at throat, almost closed by hairy saccate intrusions; nutlets triangular-ovoid, 2.5–3 mm. long, muricate but not rugose or slightly so; *n* = 15 (Kamb, 1952).—Occasional, moist places at low elevs.; Sagebrush Scrub, Foothill Wd., etc.; San Luis Obispo and San Benito cos. n.; to Wash. April–June.

8. **A. Menzièsii** (Lehm.) Nels. & Macbr. [*Echium M.* Lehm. *A parviflora* Heller, not Bernh. *A. Helleri* Brand. *A. Copelandii, Eatonii, hirticaulis, retrorsa,* and *sparsiflora* Suksd.] Erect with few ascending branches above, mostly simple below, 1–6 dm. tall, bristly hirsute and often also fine-strigose; lvs. linear or upper lanceolate, hirsute on both sides with appressed or ascending hairs, the blades 3–12 cm. long, the lower long-petioled; racemes strict or ascending, bractless, 5–15 cm. long; calyx 5–11 mm. long in fr., the lobes distinct, white- to tawny-hispid; corolla pale yellow, 5–7 mm. long, the tube scarcely exserted, the limb ca. 2 mm. broad; nutlets 2.5–3.5 mm. long, triangular-ovoid to ovoid, usually incurved, 1–2 frequently aborted, tuberculate, usually with a narrow dorsal keel and often with some broken oblique ridges, the surface minutely granulate or spiculiferous; scar rather large, expanded over the pericarp; *n* = 8, 13, 17 (Ray & Chisaki, 1957).—Occasional, in dry grassy places, below 5000 ft.; many Plant Communities; cismontane valleys mostly away from the immediate coast, Santa Catalina Id., San Diego Co. n.; to Wash., Ida., Utah. April–June.

9. **A. intermèdia** F. & M. [*A. campestris* and *obovallata* Greene. *A. attenuata* Eastw. *A. i.* var. *attenuata* Jeps. & Hoov. *A. californica* Suksd. *A. i.* var. *c.* Jeps. & Hoov. *A. intactilis* and *valens* Macbr. *A. Hanseni, irritans, longituba, Parishii,* and *sanctae-barbarae* Brand. *A. abbreviata, Abramsii, ammophila, amplexicaulis, angustifolia,* etc. Suksd.] Erect, slender and simple to widely branched, 2–8 dm. tall, sparsely bristly, otherwise subglabrous except for pubescence toward infl.; lvs. linear to lanceolate, 2–15 cm. long, the lower petioled, thinly hirsute on both sides with spreading often pustulate hairs; racemes leafy-bracteate at base, 5–20 cm. long in age; fls. homostylic, nearly radially symmetrical; calyx 5–10 mm. long in fr., the lobes separate, rusty-hispid on backs, white-hirsute on edges; corolla orange-yellow, 8–10 mm. long, the tube not much exserted, the limb 2–6 mm. broad; nutlets 2–3 mm. long, ovoid or angular-ovoid, tuberculate, usually medially keeled and with some oblique ridges, usually granulate; scar usually small; *n* = 15, 17, 19 (Ray & Chisaki, 1957).—Common in grassy or open places, below 5000 ft.; many Plant Communities; throughout cismontane Calif., occasional on deserts; to Wash., Idah., Ariz., L. Calif. March–June. Exceedingly variable, especially in lf.-shape, pubescence, and in size, shape, and roughening of nutlets, Suksdorf recognizing over 100 segregates.

Var. **echinàta** (Gray) Wiggins. [*A. e.* Gray.] Nutlets ovoid, ca. 2 mm. long, the keel drawn up into a fragile knifelike edge, the surface rough-papillate.—Sandy washes; Creosote Bush Scrub; e. Mojave Desert.

Var. **Eastwoòdae** (Macbr.) Jeps. & Hoov. [*A. E.* Macbr. *A. Johnstonii, Jonesii, jucunda, splendens, stenophylla,* and *verna* Suksd.] Corolla deep orange, 14–18 mm. long; nutlet rather large; *n* = 12 (Kamb, 1952).—Cismontane s. Calif., Great V. and bordering hills.

10. **A. lunàris** Macbr. [*A. anomala, cinerea, disjuncta, longifolia, papillata* and *yosemitensis* Suksd.] Differing from *A. intermedia* in its heterostylic ± asymmetric bilaterally marked fls. with bent corolla-tube; *n* = 4 (Ray & Chisaki, 1957).—Coast Ranges of cent. Calif.

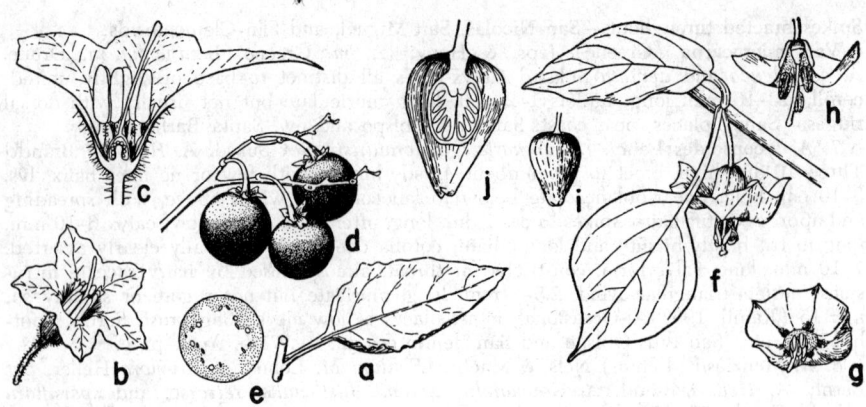

Fig. 59. SOLANACEAE. *Solanum Xanti: a,* lf., × ½; *b,* fl., × 1, rotate corolla; *c,* section of fl.; *d,* berries; *e,* section of 2-loculed fr. with fleshy cent. part. *Physalis: f,* habit, × ½; *g,* fl., × ½; *h,* fl. in section; *i,* young fr. with enlarging calyx, × ½; *j.* section of calyx, × 1, with young berry within.

62. Solanàceae. NIGHTSHADE FAMILY (Fig. 59)

Herbs or shrubs with alternate lvs. Fls. perfect, regular or nearly so, terminal or axillary, solitary, umbellate, cymose or paniculate. Calyx 5-lobed or -toothed. Corolla 5-lobed, rotate to tubular, the lobes valvate or mostly plicate in the bud. Stamens 5, alternate with the corolla-lobes and inserted on the tube. Ovary entire, superior, mostly 2-celled; style 1; stigma terminal. Fr. a berry or caps. Seeds many. A family of over 3000 spp. in almost 100 genera, widely distributed. Many plants like tomato, potato, pepper, petunia, tobacco, etc., of economic importance.

A.   Shrubs with spiny branches; fr. fleshy or dry and bony; corolla funnelform .... 1. *Lycium*
AA. Shrubs without spines or more commonly herbs.
  B.   Corolla rotate, or nearly so.
    C.   Anthers connivent, longer than the fils.; calyx remaining small.
      D.   Anthers dehiscing by terminal pores; plants prickly if with yellow fls.
                                                          6. *Solanum*
      DD. Anthers dehiscing lengthwise; plants not prickly, but with yellow fls.
                                                           7. *Lycopersicon*
    CC. Anthers not connivent, mostly shorter than the fils.
      D.   Calyx herbaceous, not inflated, closely investing the berry but open above; corolla
         with tomentose pads alternating with the stamens. .... 3. *Chamaesaracha*
      DD. Calyx large and bladdery in fr., nearly closed at top; corolla lacking tomentose
         pads .... 4. *Physalis*
  BB. Corolla not rotate.
    C.   Fr. a berry; corolla urceolate. Introd. weedy perennial .... 5. *Salpichroa*
    CC. Fr. a caps.; corolla tubular to funnelform.
      D.   Corolla 5–6 mm. long; plant a low annual, mostly less than 1 dm. high, prostrate.
        E.   Fls. in axillary umbels, yellowish or brownish with purple tinge; seeds
          flattened .... 2. *Oryctes*
        EE. Fls. solitary in the axils, purplish-red; seeds not flattened ...... 10. *Petunia*
      DD. Corolla 15–200 mm. long; plants mostly larger, more erect.
        E.   Caps. prickly, 25–50 mm. long; fls. solitary, axillary .... 8. *Datura*
        EE. Caps. not prickly, 5–13 mm. long; fls. in terminal racemes or panicles
                                                  9. *Nicotiana*

1. Lýcium L. BOX-THORN

Armed shrubs, erect or spreading, sometimes scrambling over supports. Lvs. often fasciculate, entire to minutely dentate, glabrous and glaucous to glandular or pubescent. Fls. solitary or in 2's–4's in the axils. Calyx campanulate to tubular, commonly ruptured by fr. Corolla whitish to purplish or greenish-purple, regular, tubular to funnelform, 4–6-lobed. Anthers affixed near their middle. Fr. 2-celled, from dry and bony to fleshy and juicy, 2–many-seeded. Embryo coiled. Ca. 100 spp., from ± arid regions in all continents. (*Lycia*, ancient country in Asia Minor.)

(Hitchcock, C. L. A monographic study of the genus Lycium of the W. Hemis. Ann. Mo. Bot. Gard. 19: 179–374, 1932.)
A.   Fr. 2-seeded; lvs. fleshy-turgid, subterete in cross section; corolla-tube ca. as long as calyx
                                                         1. *L. californicum*
AA. Fr. several-seeded; lvs. ± flattened; corolla-tube longer than calyx.
    B. Calyx-lobes ⅔ as long as calyx-tube, or at least 2 mm. long.
        C. Fr. hard, greenish or purplish.
            D.   Plant glaucous; fr. not transversely grooved near middle . . . . 2. *L. pallidum*
            DD. Plant not glaucous, densely pubescent; fr. with 2 transverse grooves above the middle . . . . 3. *L. Cooperi*
        CC. Fr. soft, fleshy, red.
            D.   Calyx and rest of plant densely pubescent; fls. mostly borne singly . 4. *L. Parishii*
            DD. Calyx and rest of plant glabrous or sparsely pubescent; fls. mostly in 2's or 3's.
                E.   Fils. glabrous, attached near summit of corolla. San Nicholas Id.
                                                  5. *L. verrucosum*
                EE. Fils. pubescent at base, attached near middle of corolla-tube. Widely distributed . . . . 6. *L. brevipes*
    BB. Calyx-lobes less than ⅔ as long as the tube, or less than 2 mm. long.
        C. Scrambling shrubs, sometimes adventive near habitations of cismontane Calif.
            D.   Corolla-tube longer than limb, much narrower below the middle; lvs. mostly lanceolate to oblanceolate . . . . 7. *L. halimifolium*
            DD. Corolla-tube shorter than limb, rather wide; lvs. rhombic-ovate to lance-ovate
                                                    8. *L. chinense*
        CC. Stiff erect native shrubs of the deserts.
            D.   Corolla-lobes ⅓ as long to as long as tube . . . . 6. *L. brevipes*
            DD. Corolla-lobes less than ⅓ as long as tube.
                E.   Plant densely pubescent.
                    F.   Calyx-tube 3–6 mm. long, the lobes shorter. From alkaline soil
                                            9. *L. Fremontii*
                    FF. Calyx-tube not more than 3 mm. long, the lobes usually as long. Not from alkaline soil . . . . 4. *L. Parishii*
                EE. Plants not densely pubescent.
                    F.   Corolla-lobes lanate-ciliate; some lvs. usually over 5 mm. broad
                                          10. *L. Torreyi*
                    FF. Corolla-lobes glabrous to ciliolate; no lvs. over 5 mm. broad
                                       11. *L. Andersonii*

1. **L. califórnicum** Nutt. Dense intricately branched shrub 1–2 m. tall, the branches with very spinose tips; lvs. glabrous, fleshy, subterete, 3–10 mm. long, subsessile; fls. 1–2 in the axils, on pedicels 1–5 mm. long; calyx campanulate, ca. 2.5 mm. long, 2–4(–5)-lobed, the lobes minute, broadly triangular; corolla white with a purplish tinge, the tube 2–3 mm. long, the lobes ca. as long, rotate to slightly reflexed; fils. pubescent below; fr. ovoid, reddish, firm, 2-seeded, 3–6 mm. long.—Dry bluffs and slopes, near the coast; Coastal Sage Scrub; Los Angeles Co. to L. Calif.; Channel Ids. March–July.

2. **L. pállidum** Miers var. **oligospérmum** C. L. Hitchc. Glabrous somewhat glaucous intricately branched very thorny shrub 1–2 m. tall; lvs. oblong-oblanceolate, 1–5 cm. long; fls. pendent on pedicels 8–12 mm. long; calyx campanulate, 5–8 mm. long, the ovate lobes ca. as long as tube; corolla white to lavender, narrow-campanulate, 15–20 mm. long, the limb 1.4–1.8 cm. broad; stamens exserted; fr. depressed-globose, somewhat fleshy but hard on outside, greenish-purple to -white, 8–10 mm. in diam.—Dry rocky hills and mesas, below 2500 ft.; Creosote Bush Scrub; Mojave Desert from near Barstow to Panamint and Death valleys. March–May.

3. **L. Coòperi** Gray. [*L. C.* var. *pubiflorum* Gray.] Densely leafy compact thorny shrub 1–1.5(–2) m. high, glandular-puberulent; lvs. oblanceolate to spatulate, 1–3 cm. long, 0.4–1 cm. broad; fls. pendent, the pedicels 1–3 in an axil, ca. as long as calyx; calyx bowl-shaped, 8–15 mm. long, the lobes half as long as tube; corolla greenish-white, funnelform, 8–12 mm. long, the limb ca. as broad; stamens ca. as long as corolla-tube; fr. ovoid, dry, greenish, constricted above, 6–10 mm. long, several-seeded.—Dry mesas and slopes, below 5000 ft.; Creosote Bush Scrub to Pinyon-Juniper Wd.; Mojave and Colo. deserts, upper San Joaquin V. and borders in Kern Co.; to Utah, Ariz. March–May.

4. **L. Paríshii** Gray. Erect, intricately branched, spiny, 1–3 m. tall, pubescent, slightly glandular; lvs. 0.5–3 cm. long, oblanceolate to subelliptic, the longer short-petioled; fls.

usually solitary; calyx campanulate, the tube 1.5–2.5 mm. long, the oblong-oval lobes from half as long as to longer than tube; corolla purplish, tubular-funnelform, 8–12 mm. long, the tube 6–10 mm., the rounded lobes ± spreading; stamens ca. as long as corolla-lobes; fr. ovoid, red, several-seeded, 4–6 mm. long—Rare, dry places, below 2000 ft.; Coastal Sage Scrub?, Creosote Bush Scrub; San Bernardino V., w. Colo. Desert (Vallecitos, Mountain Palm Springs); Ariz., Son. March–April.

5. **L. verrucòsum** Eastw. Much like *L. brevipes*, but calyx-lobes shorter than tube, pubescent; corolla 8–10 mm. long, the lobes ca. ¼ as long as tube; fils. adnate to bases of sinuses between corolla-lobes, glabrous; fr. ovoid, reddish.—San Nicolas Id. April.

6. **L. brévipes** Benth. [*L. Richii* Gray.] Thorny, 1–3 m. tall, subglabrous to pubescent, divaricately branched; lvs. oblanceolate to spatulate, 0.5–3 cm. long, 3–10 mm. broad, short-petioled; fls. few to many, on pedicels 1–10 mm. long; calyx campanulate, the tube 2–6 mm. long, the unequal lobes linear and as long as tube to triangular and much shorter; corolla lavender to whitish, funnelform, the tube 6–10 mm. long, the lobes 3–5 mm. long, glabrous or sparsely ciliolate; stamens slightly exserted; fr. ovoid, bright red, 4–9 mm. in diam., many-seeded.—Washes and hillsides, below 1500 ft.; Creosote Bush Scrub, edge of Alkali Sink; w. Colo. Desert and San Clemente Id.; L. Calif., Son. March–April.

Var. **Hássei** (Greene) C. L. Hitchc. [*L. H.* Greene.] Calyx-lobes equal, ± spatulate, 1–3 times as long as tube.—Coastal Sage Scrub; coastal bluffs; Santa Catalina Id., San Clemente Id., mainland at Santa Barbara, San Diego.

7. **L. halimifòlium** Mill. MATRIMONY VINE. Upright or spreading, with arching scrambling branches to 5 m. high; lvs. ovate to spatulate, 2–6 cm. long, glabrous, short-petioled; calyx mostly cupulate, the lobes triangular, obtusish, from ca. half as long as to as long as tube; corolla rotate-campanulate, dull lilac-purple, 8–10 mm. broad; fr. ovoid, salmon to red, 10–15 mm. in diam.—Occasional escape from gardens; native of Eurasia. July–Oct.

8. **L. chinénse** Mill. Rambling, sometimes unarmed, 1–3 m. tall; lvs. rhombic-ovate, 2–4 cm. long; calyx cupulate, the tube 2–3 mm. long, the lobes ⅓ as long; corolla purple, ca. 1 cm. long, the lobes slightly longer than the wide tube; fr. ovoid, scarlet, 1.5–2.5 cm. long.—Adventive in bottom lands of lower Sacramento R.; native of Asia. July–Oct.

9. **L. Fremóntii** Gray. Densely glandular-pubescent, much-branched, erect, 1–3 m. tall, spinose; lvs. oblanceolate-spatulate, 1–2.5 cm. long, 3–10 mm. broad; pedicels slender, 4–25 mm. long; calyx tubular, the tube 4–6 mm. long, the lobes mostly triangular, 1–2 mm. long; corolla lavender to pale violet, tubular to narrow-funnelform, glabrous, 10–15 mm. long, the spreading lobes 1.5–3.5 mm. long; stamens mostly shorter than corolla-lobes; fr. ovoid, red, juicy, 6–8 mm. long, many-seeded.—Rather alkaline places, below 1500 ft.; Creosote Bush Scrub; Colo. Desert from near Mecca s.; to Ariz., Son., L. Calif. March–April.

10. **L. Tórreyi** Gray. Spreading, intricately branched, subglabrous, 1–3 m. tall, heavy-spined; lvs. broadly spatulate, 1–5 cm. long, 3–10 mm. broad; fls. mostly in small fascicles, the pedicels 5–20 mm. long; calyx cupulate to short-cylindric, 2.5–4 mm. long, the lobes 0.5–2 mm. long, ciliolate; corolla lavender-purple, tubular-clavate, 10–15 mm. long, the lanceolate to ovate lobes lanate-ciliate, 3–4 mm. long; stamens ca. equal to corolla-lobes; fr. red, juicy, many-seeded, ovoid, 7–10 mm. long.—Along washes and benchlands, below 2000 ft.; Creosote Bush Scrub; Mojave R., Colo. R., near streams of w. edge of Colo. Desert; to Utah, Tex., Mex. March–May.

11. **L. Andersònii** Gray. Subglabrous or sparsely pubescent rounded shrub, much-branched, 1–2(–3) m. high, the spines needlelike; lvs. spatulate, subterete, ± pear-shaped, 0.3–1.5 cm. long, 1–3 mm. broad; fls. mostly 1–2 in an axil, on pedicels 3–9 mm. long; calyx cup-shaped, 1.5–3 mm. long, glabrous; corolla whitish-lavender, tubular-funnelform, the tube 10–16 mm. long, the lobes 1.5–2.5 mm. long, entire or ciliolate; stamens exserted 2–3 mm.; fr. ellipsoid to ovoid, red, fleshy, 4–8 mm. long.—Dry stony hills and mesas, below 6000 ft.; Creosote Bush Scrub to Pinyon-Juniper Wd., Sagebrush Scrub, Chaparral, Coastal Sage Scrub; common, Mojave and Colo. deserts n. to Mono Co.; occasional, cismontane s. Calif.; to Utah, New Mex., Mex. March–May.

Var. **deserticola** (C. L. Hitchc.) Jeps. [*L. A.* f. *d.* C. L. Hitchc.] Lvs. 2–3.5 cm. long, flattish.—Greenhorn Mts., Kern Co., Mojave and Colo. deserts.

2. Oryctes Wats.

Low erect annual with leafy stems, sparsely scurfy. Lvs. linear to ovate. Fls. pedicelled, in few-fld. axillary umbels. Calyx 5-parted into narrow lobes. Corolla yellow to brownish or tinged purple, tubular, with 5 short teeth. Stamens 5, inserted at base of corolla-tube, unequal, slightly exserted. Style as long as longer stamens. Fr. a caps., membranous, globose, 2-valved. Seeds round, flat, hyaline-margined, muriculate. One sp. (Greek, *oructes*, a digger, applied since the plant came from the land of the Digger Indians.)

1. **O. nevadénsis** Wats. Stems few, 5–20 cm. high; lvs. 1–3 cm. long, narrowed into petioles 0.5–1 cm. long; calyx 2–3 mm. long, accrescent; corolla 6 mm. long; caps. 6–7 mm. in diam.; seeds tawny, the body 2 mm. wide, the membrane 0.5 mm. wide.— Sandy places, 4000–5000 ft.; near Alkali Sink; 4 mi. se. of Aberdeen, Inyo Co.; w. Nev., Ida. May.

3. Chamaesaràcha Gray

Low perennial herbs with leafy decumbent to prostrate branched stems. Lvs. entire to pinnatifid, with margined petioles. Fls. 1–5 in axillary fascicles, pedicelled. Calyx campanulate, 5 toothed or -lobed, slightly accrescent in fr. Corolla rotate, white to cream, with some purple tinge, with pubescent cushionlike appendages in the throat. Stamens inserted near base of corolla. Stigma obscurely 2-lobed. Berries globose, on recurved pedicels. Seeds flattened, reniform, finely rugose-favose. Ca. 9 spp., w. N. and S. Am. (Greek, *chamae*, low, and *Saracha*, a Solanaceous genus of S. Am.)

Corolla 10–15 mm. broad; plant scurfy with white stellate hairs; lvs. lanceolate, partly pinnatifid. E. Mojave Desert .... 1. *C. Coronopus*
Corolla 15–25 mm. broad; plant scaberulous, the hairs not stellate; lvs. ovate, entire. Mono Co. to Siskiyou Co. .... 2. *C. nana*

1. **C. Corónopus** (Dunal) Gray. [*Solanum C.* Dunal.] Diffusely branched, 0.5–1.5 dm. high, from slender underground rootstocks, green but scurfy-puberulent; lvs. many, lanceolate or linear, 2–6 cm. long, with cuneate-attenuate base, entire to laciniate-pinnatifid, short-petioled; pedicels slender, 1–18 cm. long; calyx copiously stellate-puberulent, ca. 5 mm. long in fr., the lobes narrow-triangular, shorter than tube; corolla greenish-white, the large appendages nearly contiguous, almost filling the throat; berry almost hidden in calyx.—Dry clay soil, 4000–5000 ft.; Pinyon-Juniper Wd.; Barnwell, New York Mts., e. San Bernardino Co.; to Utah, Kans., Mex. May–July.

2. **C. nàna** (Gray) Gray. [*Saracha n.* Gray.] Stems 1–few, erect, 0.5–2.5 dm. high, from tough slender rootstocks, scaberulous-pubescent; lvs. ovate, 1.5–5 cm. long, entire, abruptly narrowed into the equally long, narrowly winged petiole; pedicels stout in fr.; calyx 6–10 mm. long in fr., the lobes 2–3 mm. long; corolla white with 5 basal green spots; berry dull white to yellowish, 1–1.2 cm. in diam.; seeds 1.5–2 mm. in diam.— Sandy flats, 5000–9000 ft.; Yellow Pine F., Red Fir F.; Sierra Nevada from Mono and Sierra cos. n., to Modoc and Siskiyou cos.; Ore., Nev. May–July.

4. Phýsalis L. GROUND-CHERRY

Annual or perennial herbs. Lvs. entire to sinuate-dentate. Fls. solitary in axils, less often in clusters of 2–5. Pedicels slender. Calyx campanulate to tubular-campanulate, 5-toothed, in fr. enlarged and bladdery-inflated, membranous, the teeth mostly connivent. Corolla open-campanulate, yellowish or whitish or purplish, obscurely 5-lobed, plicate. Stamens 5, inserted near base of corolla-tube. Style slender; stigma faintly 2-lobed. Fr. a berry, completely and loosely enclosed in the calyx. Seeds many, flattened, finely pitted. Perhaps 100 spp., mostly of New World, some from Eurasia and Australia. (Greek, *physalis*, a bladder, because of inflated calyx.)

(Rydberg, P. A. The N. Am. spp. of *Physalis* and related genera. Mem. Torr. Bot. Club 4: 297–374, 1896.)

A. Lvs. pubescent beneath with at least some stellate or forked hairs; plants perennial.
  B. Corolla 1.5–2 cm. in diam.; pubescence dense, mostly stellate. Introd. weeds.
    C. Lvs. elliptic to oblanceolate, subentire to repand .... 1. *P. viscosa*
    CC. Lvs. broadly ovate, sinuate-dentate .... 3. *P. mollis*

BB. Corolla 0.8–1 cm. in diam.; pubescence largely of forked or simple hairs. Native in e. desert
                                    2. *P. Fendleri*
AA. Lvs. lacking stellate or forked hairs beneath.
 B. Plants perennial; anthers yellow, rarely purplish. Desert plants.
  C. Pedicels shorter than fls.; foliage somewhat canescent; calyx-lobes more than ⅔ the
   length of tube at anthesis .... 4. *P. hederaefolia*
  CC. Pedicels longer than fls.; foliage green; calyx-lobes not more than half as long as tube
   at anthesis .... 5. *P. crassifolia*
 BB. Plants annual; anthers purple, green or blue, rarely yellow (as in *P. Greenei*).
  C. Plant distinctly pubescent.
   D. Calyx-lobes ± lanceolate, at anthesis often as long as the tube; pedicels short, not
    exceeding the fruiting calyx.
    E. Lvs. subentire, acute; stems sharply angled; fruiting calyx membranaceous
                            6. *P. pubescens*
    EE. Lvs. sinuate-dentate, obtuse; stems obtusely angled; fruiting calyx chartaceous.
     F. Upper stems with fine short pubescence; fruiting calyx deeply sunken at
      base; lvs. scarcely oblique at base .... 7. *P. neomexicana*
     FF. Upper stems villous; fruiting calyx shallowly sunken at base; lvs. strongly
      oblique at base .... 8. *P. pruinosa*
   DD. Calyx-lobes deltoid, shorter than the tube at anthesis; pedicels longer, exceeding the
    fruiting calyx .... 9. *P. Greenei*
  CC. Plants subglabrous.
   D. Pedicels exceeding fruiting calyx; corolla without dark center.
    E. Corolla whitish, with yellow center and 10–20 mm. broad; lvs. sinuate-
     dentate; fruiting calyx-lobes acuminate .... 10. *P. Wrightii*
    EE. Corolla yellow, 3–8 mm. broad; lvs. shallowly sinuate; fruiting calyx-lobes
     acute .... 11. *P. lanceifolia*
   DD. Pedicels shorter than fruiting calyx; corolla 10–15 mm. broad, yellow, with dark
    center .... 12. *P. ixocarpa*

1. **P. viscòsa** L. Perennial with prostrate or spreading stems 4–8 dm. long; pubescence of upper parts cinereous, of minute branched or stellate hairs; lvs. ovate to elliptic, 3–10 cm. long, entire, short-cuneate at base, petioled; pedicels 1–2 cm. long; lobes of calyx at anthesis triangular, shorter than tube; calyx ovoid in fr., 2–3 cm. long; corolla greenish-yellow with darker center; $2n = 24$ (Menzel, 1951).—Reported as weed from Ventura Co.; native, S. Am. to se. U.S.

2. **P. Féndleri** Gray. Perennial from a deep fleshy rootstock, the branches ascending, 2–4 dm. high; herbage with some divided and stellate hairs; lf.-blades broadly ovate, somewhat cordate, sinuate-dentate, 2–4 cm. long, with petioles ca. as long; fruiting pedicels ca. 8–20 mm. long; calyx campanulate, 5–7 mm. long at anthesis, the lobes narrow-triangular, ca. as long as tube; corolla yellow with brown center, campanulate-rotate, 8–10 mm. in diam.; fruiting calyx 2–3 cm. long, obscurely angled; berry yellow. —Dry rocky and gravelly places, 3200–5500 ft.; Creosote Bush Scrub to Pinyon-Juniper Wd.; Clark, Providence, and New York mts., e. San Bernardino Co.; to Colo., New Mex. May–July.

3. **P. móllis** Nutt. Like *P. Fendleri*, but more densely, finely, and uniformly stellate-pubescent; lvs. not cordate, grayish with close stellate tomentum, coarsely sinuate-dentate, the blades 1–3 cm. long, almost as wide; calyx-lobes much shorter than tube at anthesis; corolla ca. 1.5 cm. in diam., yellow with purplish center.—Adventive in barley stubble, near Yorba Linda, Orange Co.; native, Okla. and Ark. to Mex. July–Sept.

4. **P. hederaefòlia** Gray. [*P. Palmeri* Gray.] Perennial from underground rootstocks, mostly erect, cinereous-puberulent, 1–2.5 dm. high; lf.-blades ovate-deltoid to ± cordate, 1.5–4 cm. long, coarsely sinuate-dentate, on petioles ca. as long; pedicels 8–12 mm. long at anthesis; calyx tubular-campanulate, 6–8 mm. long at anthesis, the lobes slightly shorter than the tube; corolla campanulate-rotate, 12–14 mm. broad, yellow; fruiting calyx ovoid, obtusely 10-angled, 2–3 cm. long; berry yellow.—Dry gravelly or rocky places, 3400–6000 ft.; Joshua Tree Wd., Pinyon-Juniper Wd.; Jacumba (e. San Diego Co.), New York, Providence, Granite, and Clark mts. (e. San Bernardino Co.); to Tex., Mex. May–July.

5. **P. crassifòlia** Benth. [*P. c.* var. *cardiophylla* (Torr.) Gray.] Diffusely and intricately branched perennial, viscid-puberulent, 2–5 dm. tall; lf.-blades ovate, deltoid or cordate, 1–3 cm. long, entire or shallowly sinuate, on petioles of same length; pedicels slender, 1–2 cm. long; calyx campanulate, 3–5 mm. long at anthesis, the lobes short, broadly deltoid, 1–1.5 mm. long; corolla pale tawny yellow, 10–15 mm. broad; calyx in fr. ovoid, 1.5–2.5 cm. long, obscurely angled; berry greenish.—Sandy and rocky places,

below 4000 ft.; Creosote Bush Scrub; Colo. Desert, e. Mojave Desert; to Utah, Tex., L. Calif. March–May.

6. **P. pubéscens** L. Diffusely branched, angled, villous-pubescent annual, somewhat glandular, 1.5–4 dm. high; lf.-blades thin, 1.5–6 cm. long, broadly ovate, ± asymmetrical at base, subentire, acutish, on petioles ca. as long; pedicels filiform, ca. as long as petioles; calyx at anthesis tubular-campanulate, the lanceolate lobes ca. as long as the tube; corolla rotate-campanulate, 6–10 mm. broad, yellow with dark center; calyx in fr. ca. 2 cm. long; *n* = 12 (Vilmorin & Simonet, 1927).—An escape in sandy cult. fields; San Diego, Visalia, Clear Lake, along Colo. R.; e. U.S.; trop. Am. June–Dec.

7. **P. neoméxicàna** Rydb. Near to *P. pubescens,* the lf.-blades round-ovate, obtuse, coarsely sinuate-dentate, 2–7 cm. long; pedicels ca. 8 mm. long in fr.; corolla 5–6 mm. broad; calyx in fr. 2–3 cm. long.—Adventive at Elsinore, San Gabriel; native of Ariz., Colo., New Mex. July–Sept.

8. **P. pruinosa** L. Stout annual 2–6 dm. tall, villous; lvs. 2–10 cm. long, asymmetrical and cordate at base; pedicels recurved in fr.; calyx at anthesis 4–7 mm. long, the narrow triangular-ovate lobes almost as long as tube; corolla 4–8 mm. in diam., yellow with purplish center; fruiting calyx firm, 2.5–3.5 cm. long.—Moist places, Inyo and w. Riverside cos.; Wash., cent. U.S. Aug.–Oct.

9. **P. Greènei** Vasey & Rose. [*P. pedunculata* Greene, not Mart. & Gal.] Erect-spreading annual 1–4 dm. tall, with puberulent slender somewhat flexuous branches; lf.-blades ovate, acute, sinuate-dentate, 1.5–3 cm. long, puberulent, with equally long petioles; pedicels 2–3 cm. long, slender; calyx in fl. broadly tubular-campanulate, 5–6 mm. long, the lobes deltoid, shorter than the tube; corolla greenish-yellow, 12–15 mm. broad; calyx in fr. 1.5–2 cm. long, hispidulous on the angles.—Uncommon; Coastal Sage Scrub; coastal Orange and San Diego cos.; L. Calif. March–June.

10. **P. Wrìghtii** Gray. Erect or ascending annual, branched, 2–10 dm. high, glabrous or sparsely strigose above; lf.-blades lanceolate, deeply sinuate-toothed, cuneate at base, 2–7 cm. long, on somewhat shorter petioles; pedicels 3–4 cm. long in fr.; calyx at anthesis campanulate, 3–5 mm. long, the narrow-deltoid lobes ca. as long as the tube; corolla whitish with yellow center, rotate, 10–16 mm. broad; calyx in fr. 2–3 cm. long, ovoid, with acuminate lobes.—Orchards, roadsides, and waste places as a weed; cismontane s. Calif. from Los Angeles Co. s.; Colo. Desert; to Tex., Mex. July–Oct.

11. **P. lanceifòlia** Nees. Like *P. Wrightii,* but lvs. entire to slightly toothed; corolla more yellow, 4–6 mm. broad; calyx-lobes in fr. broadly deltoid.—Occasional weed in cult. fields, etc.; to Tex., cent. Mex. June–Sept.

12. **P. ixocárpa** Brot. TOMATILLO. Much-branched erect to spreading annual 3–9 dm. high, glabrous or the younger parts puberulent; lf.-blades ovate, cuneate or cordate at base, entire to sinuate-dentate, 1.5–6 cm. long, on equally long petioles; pedicels 4–8 mm. long; calyx in fl. 3.5–4.5 mm. long, the deltoid lobes shorter than the tube; corolla yellow with dark center, 8–15 mm. broad; calyx in fr. ovoid, 1.5–2 cm. long; berry purple; $2n = 24$ (Menzel, 1951).—Orchard weed and escape from gardens, from Marin and Solano cos. s.; to Atlantic Coast and Mex. June–Sept.

5. Salpichròa Miers

Herbs or shrubs with entire lvs. Fls. perfect, white or yellow, solitary, axillary. Calyx 5-parted. Corolla tubular or urceolate, 5-lobed. Stamens 5, inserted ca. the middle of the corolla-tube; fils. slender; anthers converging about the style, oblong. Ovary 2-celled, many-ovuled; style filiform; stigma entire to bilobed. Fr. an oblong or ovoid berry. Seeds round, compressed, with strongly curved embryo and little endosperm. Ca. 20 spp., of S. Am. (Greek, *salpe,* trumpet, and *chroa,* color or complexion because of form and color of fls.)

1. **S. rhomboìdea** (Gill. & Hook.) Miers. [*Atropa r.* Gill. & Hook.] LILY-OF-THE-VALLEY VINE. Heavy-rooted perennial with running rootstocks; stems branched, spreading or trailing, 5–15 dm. long, sparsely hirsutulous; lf.-blades ovate to broadly elliptic, 1–3 cm. long, on somewhat shorter petioles; pedicels 4–6 mm. long; calyx-lobes lance-linear, 2–3 mm. long; corolla white, urceolate, 6–7 mm. long, the spreading round-ovate lobes 1–1.5 mm. long; anthers slightly exserted; berry oblong, yellowish, 10–12 mm. long; seeds ca. 2 mm. broad; $n = 12$ (Vilmorin & Simonet, 1928).—Established as a difficult

weed in orchards and fields here and there through much of cismontane Calif.; native of S. Am. July–Oct.

6. Solànum L. NIGHTSHADE

Herbs or shrubs, glabrous to pubescent or tomentose, often glandular, sometimes climbing, armed or unarmed. Lvs. simple and entire to lobed or parted. Fls. commonly in umbels or cymes, white or yellow to blue or purple. Calyx 5-cleft or -toothed, rotate to campanulate. Corolla rotate, 5-angled or -lobed, plaited in bud. Stamens 5, inserted on corolla-tube; fils. short; anthers connivent around style, dehiscent by a terminal pore or short introrse slit. Ovary 2-celled; stigma small, capitate or ± bilobed. Fr. a subglobose berry with several to many seeds, these ± flattened, with annular embryo. Over 1000 spp., of all continents, but especially in trop. Am. (Latin, *solamen,* quieting, because of narcotic properties of some spp.)

A. Plants not prickly.
  B. Corolla deeply 5-cleft, whitish except in S. *Dulcamara;* peduncles mostly longer than pedicels.
    C. Lvs. entire or shallowly toothed.
      D. Stems conspicuously and persistently villous or hirsute; ripe berries greenish, yellow or reddish . . . . 1. S. *sarrachoides*
      DD. Stems sparsely pubescent or glabrous, at least in maturity; ripe berries black.
        E. Larger infls. usually bifurcate, the peduncle deflexed at maturity; corolla 6–11 mm. long; fils. unequal in length . . . . 2. S. *furcatum*
        EE. Larger infls. rarely bifurcate, the peduncle erect at maturity; fils. subequal.
          F. Corolla 6–11 mm. long; anthers 2.6–4 mm. long. Native half-shrub
                                                               3. S. *Douglasii*
          FF. Corolla 3–7 mm. long; anthers 1.2–2.6 mm. long. Introd. weeds.
            G.⁻ Infl. umbelliform; calyx-lobes distinct, reflexed at maturity; surface of berry glossy; anthers 1–1.2 mm. long. Frequent
                                                   4. S. *nodiflorum*
            GG. Infl. subracemose; calyx-lobes partly fused, not reflexed at maturity; surface of berry dull; anthers 1.8–2.4 mm. long. Rare
                                                  5. S. *nigrum*
    CC. Lvs., at least in part, pinnatifid or hastate.
      D. Annual with pinnatifid or deeply lobed lvs.; corolla 6–10 mm. broad, white
                                              6. S. *triflorum*
      DD. Perennial vine, with 3-lobed or hastate lvs.; corolla 12–16 mm. broad, blue
                                              7. S. *Dulcamara*
  BB. Corolla angulately shallowly lobed, often purple; peduncles mostly shorter than pedicels.
    C. Lvs., at least in part, pinnatifid; berries ± yellow. Introd. tall shrub . . 8. S. *aviculare*
    CC. Lvs. usually not pinnatifid, sometimes with a pair of basal lobes; berries mostly not yellow. Native herbs or half-shrubs.
      D. The lvs. mostly hastate with a pair of basal linear lobes. Se. San Diego Co.
                                             9. S. *tenuilobatum*
      DD. The lvs. seldom lobed.
        E. Pubescence of upper stems with at least some forked or branched hairs
                                        10. S. *umbelliferum*
        EE. Pubescence of stems lacking or of simple hairs.
          F. Stems distinctly pubescent.
            G. Lvs. densely tawny-viscid-villous, oblong-ovate; berry 10–25 mm. in diam., purple or yellow. Insular . . . . 11. S. *Wallacei*
            GG. Lvs. glabrous to pubescent, not both tawny and viscid-villous, lanceolate to ovate; berry mostly 6–8 mm. in diam., greenish. Widely distributed . . . . 12. S. *Xantii*
          FF. Stems glabrous or nearly so.
            G. Lvs. acute or tapering at base. From Lake and Lassen cos. n.
                                          13. S. *Parishii*
            GG. Lvs. obtuse or subcordate at base. Santa Barbara Co.
                                   12. S. *Xantii* var. *Hoffmannii*
AA. Plants prickly with stiff spines.
  B. Plants annual; lvs. pinnatifid or bipinnatifid; berry partly or wholly invested by spiny calyx.
    C. Corolla bluish or white; anthers equal; fruiting pedicels spreading; lvs. pinnatifid with acute lobes . . . . 14. S. *sisymbriifolium*
    CC. Corolla yellow; anthers unequal; fruiting pedicels erect; lvs. bipinnatifid with rounded lobes . . . . 15. S. *rostratum*
  BB. Plants perennial or shrubby; lvs. entire, sinuate or lobed, but not pinnatifid; berry not enclosed by the spiny or unarmed calyx.
    C. Lvs. entire to repand-dentate, permanently gray-canescent on both surfaces.
      D. The lvs. linear-lanceolate to narrowly oblong, gradually tapering at base; calyx-lobes linear-subulate . . . . 16. S. *elaeagnifolium*
      DD. The lvs. ovate, broadly rounded to cordate at base; calyx-lobes broadly ovate
                                            17. S. *Torreyi*
    CC. Lvs. lobed or deeply dentate, not permanently gray-canescent at least on upper surface.

D.   Berry less than 1 cm. in diam., orange; corolla light purplish-blue, ca. 1.5 cm.
     in diam.; lvs. tomentose, but becoming greenish in age ...... 18. *S. lanceolatum*
DD. Berry 1.5–4 cm. in diam.; corolla whitish with some purple, 2–3.5 cm. in diam.
     E.   Fr. 3–4 cm. in diam., yellowish; lvs. with a whitish marginal band of to-
          mentum above .... 19. *S. marginatum*
     EE. Fr. 1.5–2.5 cm. in diam., orange-yellow; lvs. not so banded
                                                          20. *S. carolinense*

1. **S. sarrachoìdes** Sendt. ex Mart. [*S. villosum* auth., not L.] Much-branched annual with ascending to decumbent stems 1–5 dm. long; herbage short viscid-villous; lf.-blades ovate, 2.5–6 cm. long, gradually to abruptly narrowed at base, acute to obtuse at apex, subentire to sinuate-toothed, the petioles mostly 1–1.5 cm. long; peduncles 5–10(–20) mm. long; pedicels somewhat shorter; calyx 2–2.5 mm. long at anthesis, accrescent in fr.; corolla white, 3–5 mm. in diam., the lanceolate lobes villous outside; anthers ca. 2 mm. long; berry 6–7 mm. in diam., round, ± yellowish; seeds light buff to yellowish, ca. 2 mm. long, with concentric lines; $n = 12$ (Stebbins & Paddock, 1949).—Occasional weed in fields and waste places; native of Brazil. May–Oct.

2. **S. furcàtum** Dunal. Perennial, mostly reclining on other plants, the stems 1 m. or so long, subglabrous except on the narrowly winged angles; lf.-blades ovate, 3–7 cm. long, subentire to sinuate-toothed, slightly decurrent on the petioles which are ca. half as long; peduncles slender, 1.5–4 cm. long, usually dichotomous near apex; pedicels 6–10 mm. long; calyx 2.5–3 mm. long, with narrow-deltoid lobes; corolla 10–18 mm. in diam., white or faintly lavender; anthers 3–3.5 mm. long; berry dark, 5–6 mm. in diam., with many large stone-cell concretions; seeds 12–25, pale yellow, reticulate-pitted, ca. 1.8 mm. long; $n = 36$ (Stebbins & Paddock, 1949).—Fields and waste places near the coast, San Mateo Co. n.; to Ore.; native of S. Am. May–Oct.

3. **S. Douglàsii** Dunal in DC. Perennial, ± woody, 1–2 m. tall; herbage puberulent to subglabrous, the stem-angles rough-pubescent, lf.-blades ovate, 2–10 cm. long, sinuate-dentate, subacuminate at apex, cuneate to subtruncate at base; petioles 1–2.5 cm. long; peduncles 1–3 cm. long; pedicels 0.5–1.2 cm.; calyx 2–3 mm. long at anthesis, with lance-oblong lobes; corolla whitish with greenish basal spots, 1–2 cm. broad, the lobes lance-oblong; anthers ca. 3–4 mm. long; berry black, 6–9 mm. in diam.; seeds pale yellow, ca. 1.5 mm. long, minutely reticulate-pitted; $n = 12$ (Stebbins & Paddock, 1949).—Frequently on partly shaded slopes, in canyons, etc., mostly below 3500 ft.; Coastal Sage Scrub, Chaparral, Coastal Strand, Closed-cone Pine F.; San Mateo Co. to cismontane s. Calif., Channel Ids.; to Mex., L. Calif. Most of year.

4. **S. nodiflòrum** Jacq. [*S. nigrum* of auth., not L.] Annual or perennial, straggling; stems rounded or angled, 3–6 dm. long, glabrous to ± scabrous on the angles; lvs. entire or sinuate-dentate, ovate to elliptic, 4–10 cm. long, with cuneate to truncate base; petioles mostly ⅓–⅔ as long; peduncles slender, mostly 2–3 cm. long; pedicels 5–12 mm. long at anthesis; calyx-lobes quite distinct to base, 1–2.5 mm. long; corolla white or faint purple, 4–6 mm. in diam.; anthers 1.2–1.4 mm. long; berry 5–6 mm. in diam., shining, black, usually without stone-cell concretions; seeds many, pale cream, ca. 1.5 mm. in diam.; $n = 12$ (Stebbins & Paddock, 1949).—Common weed in damp fields and waste places, throughout cismontane Calif.; native of Old World. April–Nov.

5. **S. nìgrum** L. Annual 3–8 dm. high, branched, glabrous or nearly so; lvs. ovate, 3–7 cm. long, entire to somewhat serrate, with cuneate base; petioles mostly 1–3 cm. long; peduncles 1–2.5 cm. long; pedicels 3–8 mm. long at anthesis; calyx 1.5–2 mm. long at anthesis, the deltoid lobes ca. 0.5 mm. long; corolla white, 5–6 mm. in diam.; anthers ca. 2 mm. long; berry dull, dark, 6–9 mm. in diam., without concretions; seeds many, ca. 2 mm. in diam., minutely reticulate; $n = 36$ (Stebbins & Paddock, 1949).—Sparingly established in waste places and about fields, cismontane Calif.; native of Eu. March–Oct.

6. **S. triflòrum** Nutt. Annual, ± pubescent, branched, the stems 1–4 dm. long, decumbent; lvs. oblong, 2.5–4 cm. long, deeply pinnatifid with rounded sinuses, on petioles 1–1.5 cm. long; peduncles 1–3-fld., 5–15 mm. long; pedicels 2–5 mm. long; calyx 2.5–3 mm. long, with lance-ovate lobes; corolla white, 7–9 mm. broad; anthers 3 mm. long; berry 8–12 mm. in diam., green; seeds many, pale, 2.5–3 mm. in diam.—Dry places, 5000–7000 ft.; Sagebrush Scrub, N. Juniper Wd., etc.; Siskiyou and Modoc cos. to Mono Co., Clark Mt. (e. San Bernardino Co.); adventive in citrus grove near Claremont; to B.C., Minn., Okla. June–Sept.

7. **S. Dulcamàra** L. BITTERSWEET. Woody climber to 3 m. high, sparsely puberulent to subglabrous; lvs. ovate, acuminate, with or without auricled or lobed bases, 5–12 cm. long, the petioles 1–4 cm. long; peduncles several-fld., with compound cymes; pedicels ca. 1 cm. long; calyx 3–4 mm. long; corolla blue to purple, 12–16 mm. in diam., deeply 5-cleft; anthers ca. 5 mm. long, connivent; berry red, ovoid or ellipsoid, 8–12 mm. long; seeds roundish, ca. 2 mm. in diam., minutely tessellate; $n = 12$ (Vilmorin & Simonet, 1928).—Occasional escape from gardens in moist places; native of Old World. June–Sept.

8. **S. aviculàre** Forst. f. Glabrous unarmed shrub, 1–3 m. tall; lvs. lanceolate and entire to broadly ovate and pinnatifid with 1–4 ascending or divaricate lanceolate lobes, the blades 8–20 cm. long, the petioles 1–2 cm. long, ± decurrent on stems as raised lines; cymes few-fld., the peduncles 1–3 cm. long; calyx 5–6 mm. long; corolla purplish, shallowly lobed, 3–3.5 cm. broad; berry round to ovoid, 2–2.5 cm. long, ± yellowish; seeds many, ca. 2 mm. long, reticulate with rounded ridges.—Sparingly established near the coast from San Francisco n.; native of New Zealand. June–Oct.

9. **S. tenuilobàtum** Parish. Suffrutescent, glabrescent below, hirsutulous above, 3–10 dm. high; lvs. linear to narrow-oblong, 2–3 cm. long, all but the upper hastate with a pair of linear basal lobes 2–5 mm. long; petioles 3–5 mm. long; fls. few; peduncles ca. 1 cm. long; pedicels 1–1.5 cm. long; calyx 3–5 mm. wide, broadly campanulate, the lobes ca. half as long as tube, deltoid-oval; corolla blue-purple, 15–18 mm. broad; anthers ca. 4 mm. long; berry ca. 6–7 mm. in diam.—Dry open places, 1000–2700 ft.; Chaparral; s. San Diego Co.; L. Calif. March–April.

10. **S. umbellíferum** Eschs. Rounded to spreading subshrub 6–10 dm. high and of greater spread, greenish, angled or ridged, finely pubescent throughout with simple, forked or few-branched nonglandular hairs; lvs. elliptic-ovate, entire or shallowly pinnatifid toward the base, 1.5–4.5 cm. long, tapering toward a petiole 4–10 mm. long; peduncles 5–15 mm. long; pedicels slender, 1–2 cm. long; calyx broadly campanulate, 4–5 mm. long, with short deltoid lobes; corolla mostly blue, 1.5–2 cm. broad, with 2 greenish glands at base below each lobe; anthers ca. 4 mm. long; fils. glabrous; berry round, 10–15 mm. in diam., whitish except for the greenish base; seeds lenticular, ca. 2 mm. in diam., finely reticulate.—Dry brush-covered slopes and in canyons, below 2500 ft.; Chaparral, Mixed Evergreen F., N. Coastal Scrub; Coast Ranges from Mendocino Co. to Santa Barbara Co. Some fls. most of the year.

Var. **incànum** Torr. [*S. californicum* Dunal. *S. umbelliferum* var. *c.* Parish.] Lf.-blades smaller, more narrow-oblong, 1–2 cm. long, ± tomentose; stems white-tomentose with several-branched hairs.—Dry valleys and slopes, below 4500 ft.; Chaparral, Foothill Wd., V. Grassland; interior valleys from Contra Costa Co. to Los Angeles Co.

Var. **glabréscens** Torr. Branches sparingly pubescent, the hairs nonglandular, short, partly simple and partly few-branched.—Dry slopes, below 5000 ft.; Chaparral, Coastal Sage Scrub; cismontane s. Calif. from Santa Barbara Co. to San Diego Co. and Ariz.

11. **S. Wallàcei** (Gray) Parish. [*S. Xantii* var. *W.* Gray.] Suffrutescent, erect-spreading, 1–2 m. tall, densely tawny-villous, with many hairs gland-tipped; lvs. many, thickish, oblong-ovate, 3–14 cm. long, acute, with rounded or subcordate base; petioles stout, 1–2 cm. long; peduncles stout, 1–3 cm. long, with cymes of several fls.; pedicels 1–2 cm. long, slender; calyx campanulate, 5–7 mm. long, the deltoid lobes shorter than the tube; corolla purplish-blue, 2–4 cm. broad; anthers ca. 5 mm. long; berry round, dark purple, 1.5–2.5 cm. in diam.; seeds many, red-brown, lenticular, ca. 2 mm. in diam., coarsely lenticular.—Dry rocky slopes and in canyons; Chaparral; Santa Catalina and Guadalupe ids. March–Aug.

Var. **Clòkeyi** (Munz) McMinn. [*S. C.* Munz. *S. arborescens* Clokey, not Moench.] Foliage less viscid and tawny; corolla 1.5–2 cm. broad; fr. yellow, 1–1.5 cm. in diam.—Santa Cruz and Santa Rosa ids. March–July.

12. **S. Xántii** Gray. Suffrutescent, 4–9 dm. tall, short-villous with white mostly nonglandular hairs, especially on the stems and veins; lvs. ovate, subentire, sometimes lobed at base, 2–4(–7) cm. long, cuneate to subcordate at base, acutish to obtuse at apex, on petioles 5–18 mm. long; fls. mostly 6–10 in lateral subumbellate cymes; calyx 5–6 mm. long, the broadly deltoid lobes mostly shorter than the tube; corolla deep violet to dark lavender, 1.5–2.5 cm. in diam.; anthers 3.5–4 mm. long; berry greenish, round, 6–8 mm. in diam.; seeds broadly ovate-lenticular, ca. 1.7 mm. long, pale brown, minutely reticu-

late.—Dry places about brush or woods, below 4000 ft.; Chaparral, Foothill Wd.; cismontane s. Calif. from San Luis Obispo Co. to w. Kern Co. and through Santa Barbara and Ventura cos.; Santa Rosa and Santa Cruz ids. Feb.–June.

KEY TO VARIETIES

Stems hairy.
 Hairs on stems mostly non-glandular, the stems short-villous.
  Plant suffrutescent, 4–9 dm. tall. Below Montane Coniferous F. .... S. *Xantii*
  Plant herbaceous, 1–3 dm. tall. Montane Coniferous F. .... var. *montanum*
 Hairs on stems largely gland-tipped, the stems minutely pubescent .... var. *intermedium*
Stems glabrous .... var. *Hoffmannii*

Var. **montànum** Munz. Strictly herbaceous, the stems 1–4 dm. long, often in prostrate mats, densely grayish-pubescent; berry 9–10 mm. in diam.—Dry places, 5000–9000 ft.; Montane Coniferous F.; San Bernardino Mts. to Sierra Nevada of Nevada Co. May–Sept.

Var. **intermèdium** Parish. [*S. Wallacei* var. *viride* Parish.] Subshrubs; stems and lvs. short-pubescent with gland-tipped hairs, the lvs. often ± cordate at base.—Dry slopes and canyons, below 3500 (5000) ft.; Chaparral, Foothill Wd., Oak Wd.; most of cismontane Calif. except open Cent. V. Feb.–July. A form from San Luis Obispo Co., with crisped lvs. has been called S. *obispoénse* Eastw.

Var. **Hoffmánnii** Munz. Glabrous; lvs. oblong-ovate, 3–6 cm. long, truncate to subcordate at base.—Chaparral; Gaviota Pass (Santa Barbara Co.), Warner's Ranch, etc. (San Diego Co.). March–May. Very near to S. *Parishii,* but with different lf.-bases and larger fls.

(Plants heretofore referred to S. X. var. *glabrescens* are to be sought under S. *Parishii* if without branched hairs, under S. *umbelliferum* var. *glabrescens,* if with some branched hairs.)

13. **S. Paríshii** Heller. [*S. Xantii* var. *glabrescens* Parish, in part. S. X. var. *Spencerae* Macbr.?] Erect or ascending, suffrutescent, 5–10 dm. high, the branches slender, slightly angled, striate, glabrous or nearly so except for a few short coarse hairs toward tips of stems and on veins and margins of lvs.; lvs. elliptic to lance-oblong-ovate, 2–5 cm. long, cuneate at base, acute to rounded at apex, on petioles to ca. 1 cm. long; peduncles short, few-fld.; calyx 4–5 mm. long, the deltoid lobes to ca. as long as tube; corolla lavender, 15–18 mm. broad; anthers ca. 4 mm. long; berries round, 7–9 mm. in diam.; seeds reticulate-granular, ca. 2 mm. long.—Dry grassy and brushy slopes, below 6000 ft.; Chaparral, Foothill Wd., Yellow Pine F.; from Lake, Yolo, and Lassen cos. n. to Ore.; and in Riverside and San Diego cos. S. material shows some influence of S. *Xantii* var. *intermedium* in the wider lvs. April–July.

14. **S. sisymbriifòlium** Lam. Prickly annual, the unequal flattened yellow prickles abundant on stems, lvs., pedicels and calyx; plant robust, 5–15 dm. tall, viscid-villous throughout; lvs. 1–2 dm. long, deeply pinnatifid with deeply cut or sinuate lobes; fls. cymose; calyx 7–9 mm. long; corolla ca. 3 cm. broad, white to light blue; anthers alike, ca. 1 cm. long; berry red, 1–2 cm. in diam.; seeds orange-yellow, ca. 2 mm. long, finely reticulate-pitted.—Occasional weed; introd. from S. Am. Feb.–Oct.

15. **S. rostràtum** Dunal. BUFFALO-BUR. Annual 4–8 dm. tall, the prickles yellowish, subulate; herbage also stellate-pubescent; lvs. 1–2-pinnatifid, 5–10 cm. long; calyx nearly hidden by prickles; corolla yellow, 2–2.5 cm. broad; anthers dimorphic, some 9–10, others ca. 6 mm. long; berry ca. 1 cm. in diam.; seeds black, finely and deeply pitted, 3–3.5 mm. long.—Occasional weed; native in cent. U.S. May–Sept.

16. **S. elaeágnifòlium** Cav. SILVERLEAF-NETTLE. Short-lived perennial with dense fine stellate silvery canescence throughout, 3–8 dm. high; stems unarmed or with prickles; lvs. lance-oblong to lance-linear, 3–9 cm. long, ± cuneate at base, on much shorter petioles; fls. cymose; pedicels, peduncles and calyces usually with straight yellow spines 3–4 mm. long; calyx-lobes narrow-lanceolate, to 1 cm. long; corolla violet or blue, 2–3 cm. in diam.; anthers 7–9 mm. long; berry round, 10–14 mm. in diam., yellow or brownish; seeds lenticular, biconvex, dark brown, ca. 3 mm. in diam.; $n = 12$ (Heiser & Whitaker, 1948).—Quite widely introd. as weed in s. Calif. and Cent. V.; native of cent. U.S. to Mex. May–Sept.

17. **S. Tórreyi** Gray. Less cinereous than S. *elaeagnifolium;* lvs. ovate with broadly

rounded or cordate bases and sinuate-lobed margins; calyx-lobes broadly ovate, abruptly acuminate; corolla 2–2.5 cm. broad, violet; berry yellow, 2–3 cm. in diam.—Reported from Roseville, Placer Co.; native of s. cent. U.S.

18. **S. lanceolàtum** Cav. Subshrubby, 1–2 dm. high, covered with white tomentum, the lvs. ± glabrescent above in age; main stems spiny; lower lvs. bluntly and irregularly lobed below their middle, 1–2 dm. long, the upper part oblong-ovate, obtuse; upper lvs. entire, lanceolate, acuminate; fls. in long-peduncled cymose panicles in upper lf.-axils; corolla light purplish-blue, ca. 15 mm. in diam.; berries orange, ca. 6–8 mm. in diam. —Reported from Los Angeles; native of trop. Am.

19. **S. marginàtum** L. f. Shrub 1–1.5 m. tall, armed with scattered stout yellow spines 5–10 mm. long; stems and lower surfaces of lvs. densely white tomentose with stellate hairs; lvs. broadly ovate, 1–2 dm. long, coarsely sinuate-lobate, glabrescent above except for the white tomentum near the edges; petioles 1–2 cm. long; peduncles few-fld., stout, 1–2 cm. long; calyx 8–12 mm. long, tomentose, the lobes narrow-deltoid; corolla 2.5–3.5 cm. in diam., white with cent. purple star; berry yellow, 3–4 cm. in diam.; seeds lens-shaped, 3 mm. in diam., papillose-granular; $2n = 24$ (Vilmorin & Simonet, 1928).—Natur. along the coast from San Francisco to Monterey; native of Afr. May–Aug.

20. **S. carolinénse** L. HORSE-NETTLE. Prickly perennial from running underground rootstock; stems erect, 3–10 dm. high, armed with slender yellowish spines 2–5 mm. long and stellate-pubescent; lvs. ovate to oblong, repand to pinnatifid, scabrous, 4–12 cm. long; petioles 0.5–2 cm. long; fls. cymose-racemose; pedicels 5–15 mm. long; calyx-lobes lance-acuminate, 6–7 mm. long; corolla violet, 2–3 cm. broad; berry yellow, 1–1.5 cm. thick; seeds yellow, ca. 2.5 mm. in diam., reticulate-pitted.—Sparingly established as a weed; native of e. U.S. May–Aug.

7. Lycopérsicon Mill. TOMATO

Near to *Solanum*, but always unarmed; lvs. always pinnate or pinnatifid; fls. yellow; anthers projected into narrow or sharp sterile tips and dehiscing from top to bottom; fr. a red or yellow pulpy berry with 2–few cells, that multiply under domestication. Ca. 6 spp., of S. Am. (Greek, *lykos*, wolf, and *persicon*, peach, perhaps because of supposed poisonous properties.)

1. **L. esculéntum** Mill. [*Solanum Lycopersicum* L.] Annual or short-lived perennial, hairy-pubescent and ± strong-smelling and glandular, 1–2 m. tall, much-branched; lvs. 1–3 dm. long, odd-pinnate with small lfts. interspersed between the 5–9 main lfts.; fls. nodding, 2–2.5 cm. broad; fr. of various colors and sizes.—The cultivated tomato escapes from cult. and persists for some time in waste places. Summer–fall.

8. Datùra L. THORN-APPLE. JIMSON-WEED

Ours erect or spreading coarse rank-smelling herbs. Lvs. alternate, short-petioled, entire or sinuate-dentate or lobed. Fls. solitary, large, erect in ours, on short peduncles in the forks of branching stems, whitish-purple to violet, opening in evening at least in some spp. Calyx tubular, 5-toothed, circumscissile near the base in ours, with the lower part persistent as a collar below the caps. Corolla funnelform, convolute-plicate in bud. Stamens included; fils. filiform. Ovary 2- or falsely 4-celled. Caps. 2–4-valved from the top, sometimes splitting irregularly, prickly or spiny. Seeds large. Ca. 25 spp. in warmer parts of all continents. Plants ± poisonous, and with narcotic properties; some spp. used by the Calif. Indians in certain tribal rites. (The Hindoo name, *dhatura*.)

(Safford, W. E. Synopsis of the genus Datura. Jour. Wash. Acad. Sci. 11: 173–189, 1921.)

Corolla 15–20 cm. long; calyx tubular, not prismatic; caps. nodding .... 1. *D. meteloides*
Corolla 6–12 cm. long; caxy prismatic.
  Caps. nodding, pubescent as well as prickly; stems cinereous. Desert native .... 2. *D. discolor*
  Caps. erect, glabrous to sparsely pubescent; stems glabrous to sparsely puberulent, not cinereous.
    Introd. weeds.
      Caps.-spines broad at base, 8–22 mm. long; teeth of calyx 3–4 mm. long, subequal ... 3. *D. ferox*
      Caps.-spines not much broadened at base, 3–10 mm. long; teeth of calyx 5–10 mm. long, unequal
                                    4. *D. Stramonium*

1. **D. meteloìdes** A. DC. Perennial, erect, widely branched, 5–10(–15) dm. tall, minutely grayish-pubescent; lvs. unequally ovate, 4–12 cm. long, irregularly repand to

subentire, on somewhat shorter petioles; calyx 7–10 cm. long, the basal persistent part usually rotate, sometimes reflexed; corolla white, suffused with violet, 15–18 cm. long, 10–20 cm. broad, with 5 subulate teeth ca. 1–2 cm. long; anthers white, ca. 1.5 cm. long; caps. nodding, 2.5–3 cm. long, densely prickly and puberulent, the spines 5–12 mm. long; seeds subreniform, buff to light brown, ca. 5 mm. long, flattened, smooth, with cordlike margin; $n = 12$ (Buchholz et al., 1935).—Sandy or gravelly dry open places, below 4000 ft.; Coastal Sage Scrub, V. Grassland, Joshua Tree Wd., Creosote Bush Scrub, etc.; lower Sacramento V., Salinas V., S. Coast Ranges, etc. to cismontane s. Calif.; deserts from Inyo Co. s.; to Tex., Mex., n. S. Am. April–Oct. Possibly introd. from Mex.

2. **D. díscolor** Bernh. Annual, erect, 2–5 dm. tall, cinereous-pubescent; lvs. broadly ovate, sinuate-dentate, 5–14 cm. long, on somewhat shorter petioles; calyx prismatic, 4–6 cm. long, the persistent basal part mostly rotate; corolla white with purplish tinge in throat, 10–14 cm. long, the limb 4–8 cm. broad, the teeth 4–7 mm. long; anthers white, ca. 8 mm. long; caps. nodding, globose, 3–4 cm. thick, glandular-puberulent, the spines 1–2 cm. long; seeds black, rugulose, finely pitted; $n = 12$ (Buchholz et al., 1935).—Low places, washes, etc., below 1500 ft.; Creosote Bush Scrub, Alkali Sink?; Colo. Desert; Ariz. to Mex., W.I. April–Oct.

3. **D. fèrox** L. Coarse branched herb 3–5 dm. high, sparsely pubescent to glabrate; lf.-blades broadly ovate, 1–2 dm. long, coarsely sinuate-dentate, on shorter petioles; fls. erect; pedicels 1–2 cm. long; calyx 3–4 cm. long, with rounded subequal triangular teeth; corolla 6–8 cm. long, 2.5–3.5 cm. broad, with teeth 3–5 mm. long; caps. ovoid, 4–5 cm. long, heavily spined and sparsely puberulent, the spines 1–3 cm. long and up to 1 cm. broad at base; seeds reniform, ca. 5 mm. long, light brown, rugulose and pitted; $n = 12$ (Buchholz et al., 1935).—Adventive at scattered stations in Cent. V.; native of Asia. Early summer.

4. **D. Stramònium** L. Annual, erect, few-branched, 5–12 dm. tall, sparsely puberulent to glabrate; lvs. elliptic to ovate, sinuately and laciniately lobed, 5–20 cm. long, on petioles ca. half as long; calyx 3.5–4.5 cm. long, the teeth 5–10 mm. long, unequal; corolla white, 6–8 cm. long, 3–5 cm. broad, the teeth 5–8 mm. long; caps. erect, ovoid, 4–5 cm. long, puberulent to glabrate and often with spines 3–10 mm. long; seeds dark, pitted and rugulose; $n = 12$ (Blakeslee, 1928).—Locally adventive, especially in waste places; native of trop. Am. Summer.

Var. **Tátula** (L.) Torr. [*D. T.* L.] Stems purplish; corolla violet; caps. more prickly. —Occasionally adventive in much of Calif.; native of trop. Am. Summer.

9. Nicotiàna L. Tobacco

Narcotic-poisonous heavy-scented annual or perennial herbs or shrubs, usually ± viscid-pubescent. Lvs. large, alternate, entire or repand. Fls. in terminal panicles or racemes. Calyx tubular-campanulate or ovoid, 5-cleft, persistent. Corolla funnelform, salverform, or nearly tubular, the limb usually shallowly 5-lobed and spreading. Fils. filiform. Ovary 2-celled, rarely 4. Caps. 2- or 4-valved at summit. Seeds many, small, ovoid to reniform, minutely reticulate-punctate. Ca. 60 spp., mostly of N. & S. Am.; some important in commerce and for ornamentals. (J. *Nicot*, French ambassador to Portugal, who introduced tobacco into France ca. 1560.)

(Goodspeed, T. H. The genus Nicotiana. Chron. Bot. 16: 1–536, 1954.)
A.   Shrubs; fls. yellow .... 1. *N. glauca*
AA. Herbs; fls. whitish or cream, sometimes tinged violet.
  B.   Lvs. auriculate-clasping at base; fls. open throughout day. Desert perennial or biennial
                                                             2. *N. trigonophylla*
  BB. Lvs. not auriculate-clasping; fls. closing in sun. Plants annual.
    C.   Cauline lvs. sessile or nearly so, except near base of stem.
      D.   Corolla 1.5–2 cm. long, 0.8–1 cm. broad; fils. inserted nearly equally near base of corolla-throat; calyx-lobes very unequal. Cismontane and desert
                                                        3. *N. Clevelandii*
      DD. Corolla 4–7 cm. long, 2–5 cm. broad; fils. inserted ± unequally high in throat; calyx-lobes unequal. Cismontane .... 4. *N. Bigelovii*
    'CC. Cauline lvs. petioled.
      D.   Calyx with 5 dark stripes, the lobes linear-lanceolate, unequal, mostly longer than the tube; calyx-tube not obviously pock-marked; plant densely viscid-glandular
                                                  5. *N. acuminata*
      DD. Calyx not striped, the lobes deltoid, subequal, not longer than the tube; calyx-tube obviously pock-marked; plant not densely viscid-glandular .... 6. *N. attenuata*

1. **N. glaùca** Grah. TREE TOBACCO. Erect glabrous glaucous shrub or small tree 2–8 m. tall, with loose branching and open panicles; lvs. ovate, entire to repand, 3–8(–15) cm. long, on long petioles; fls. greenish-yellow, loosely paniculate; calyx ca. 1 cm. long, unequally 5-toothed; corolla tubular, 3–4 cm. long, somewhat contracted at throat, minutely pubescent, with narrow limb; caps. ovoid, 10–12 mm. long, 4-valved above; seeds brownish, ca. 0.6 mm. long; $n = 12$ (Goodspeed, 1923).—Common, natur. in waste places, below 3000 ft.; native of S. Am. Spring and summer.

2. **N. trigonophýlla** Dunal in A. DC. Viscid-pubescent, 2–8 dm. tall, erect, simple or few-branched; lowest lvs. petiolate, others oblong-ovate to -lanceolate, sessile and auricled, 2–8 cm. long; infl. loosely paniculate-racemose; pedicels 5–10 mm. long; calyx campanulate, 6–12 mm. long with lance-subulate lobes ca. as long as tube; corolla greenish-white, 18–22 mm. long, constricted at orifice, the limb 8–10 mm. broad; caps. 2-valved, 8–10 mm. long; seeds dark brown, ca. 0.6 mm. long; $n = 12$ (Clausen, 1928). —Mostly about rocks, below 4000 ft.; Creosote Bush Scrub, Joshua Tree Wd.; Mojave and Colo. deserts to Mono Co.; to Tex., Mex. Mostly March–June.

3. **N. Clevelándii** Gray. Viscid-pubescent annual 2–6 dm. tall, mostly branched; lvs. ovate or the upper lanceolate, sessile or short-petioled, 3–6(–8) cm. long; calyx 9–12 mm. long, the lobes linear, distinctly unequal, the longest much exceeding the tube; corolla whitish, tinged with violet, 1.5–2 cm. long, the limb 8–10 mm. broad; caps. 4-valved, 5–8 mm. long; seeds red-brown, 0.5–0.6 mm. long; $n = 24$ (Clausen, 1928). —Occasional in sandy places, below 1500 ft.; Coastal Sage Scrub, Coastal Strand; near the coast from Santa Barbara Co. to n. L. Calif.; Colo. Desert (Shavers Well, Corn Springs, etc.); Santa Cruz Id.; Ariz. March–June.

4. **N. Bigelòvii** (Torr.) Wats. [*N. plumbaginifolia* var. *B*. Torr.] Annual, glandular-pubescent, ill-smelling, with ascending branches, mostly 4–12 dm. high; lvs. sessile or the lower petioled, ovate-oblong to lanceolate, 5–20 cm. long; fls. mostly racemose; calyx 1.5–2 cm. long, the linear-lanceolate teeth ± unequal; corolla white, tinged with green, 4–7 cm. long, the limb 3–5 cm. wide; caps. ovoid, ca. 1.5 cm. long; seeds red-brown, ca. 0.7–0.8 mm. long; $n = 24$ (Goodspeed, 1923).—Dryish plains, mesas, valleys, below 3500 ft.; many Plant Communities; cismontane Calif. s. to perhaps n. Los Angeles Co.; s. Ore. May–Oct. Intergrading with:

Var. **Wallàcei** Gray. Corolla-limb 2–3 cm. broad.—Chaparral, Coastal Sage Scrub; Santa Barbara Co. to San Diego Co., especially in sandy interior valleys.

5. **N. acuminàta** Hook. var. **multiflòra** (Phil.) Reiche. [*N. m.* Phil.] Erect, simple to spreading-branched, glandular-pubescent, 5–12 dm. high; lvs. lance-ovate to lanceolate, 1–4 cm. long, on petioles ca. as long; panicles lax; calyx 8–12 mm. long, the lobes lance-linear to linear; corolla 3–4 cm. long, 1–2 cm. broad, greenish-white; caps. ovoid, ca. 1 cm. long; seeds dark brown, ca. 0.8 mm. long.—Natur. in flats and along creeks, below 5000 ft.; cent. and n. cismontane Calif.; native of S. Am. Summer.

6. **N. attenuàta** Torr. Erect, simple or branched, glandular-pubescent to glabrate, 3–19 dm. high; lvs. 5–15 cm. long, ovate to lance-ovate, mostly petioled; infl. racemose or the racemes in a panicle; calyx ovoid-campanulate, 6–8 mm. long, the teeth deltoid; corolla white, 2.5–3 cm. long, ca. 1 cm. broad; caps. ca. 8–12 mm. long; seeds brown, ca. 0.6–0.7 mm. long; $n = 12$ (Clausen, 1928).—Disturbed places, below 8500 (10,000) ft.; many Plant Communities; N. Coast Ranges of Humboldt, Trinity, and Siskiyou cos., San Benito Co., Sierra Nevada s. to Inyo, Fresno and Kern cos., cismontane interior s. Calif., occasional on deserts; to Wash., Ida., Utah, Tex. May–Oct.

10. Petùnia Juss.

Viscid annual or perennial herbs. Lvs. entire, the upper often subopposite. Fls. solitary, axillary. Calyx 5-parted. Corolla funnelform. Stamens 5, unequal, four being didynamous, the fifth shortest. Hypogynous disk fleshy. Caps. ovoid, with 2 undivided valves. Seeds minute, spherical. Perhaps 40 spp., largely S. Am.; some of horticultural value. (*Petun*, an Indian name of tobacco.)

Corolla 4–6 mm. long .... 1. *P. parviflora*
Corolla 25–35 mm. long .... 2. *P. violacea*

1. **P. parviflòra** Juss. Prostrate diffusely branched annual, glandular-puberulent; the stems 1–4 dm. long; lvs. oblong-linear to -spatulate, 0.4–1.2 cm. long, rather fleshy; calyx

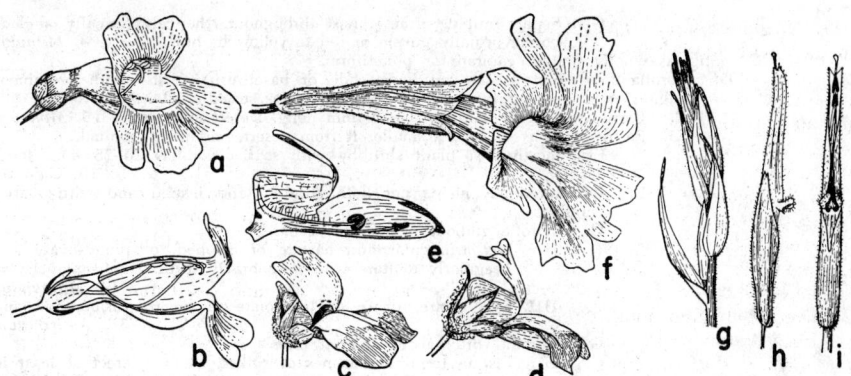

Fig. 60. SCROPHULARIACEAE. *Penstemon spectabilis: a,* fl., × 1, with bilabiate corolla and basal calyx; *b,* section with pistil, fertile stamens and sterile stamen and corolla in position. *Collinsia hetero- phylla: c* and *d,* fl., × 1, calyx, corolla with basal sack, with the middle lobe of lower lip folded into a keel; *e,* fl. in section, × 1½. *Mimulus: f,* fl., × 1, calyx angled, corolla bilabiate. *Castilleja stenantha: g,* bracteate infl., × ½; *h* and *i,* lateral and front views of fl., × 1, with calyx, tubular corolla (upper lip the galea, lower lip a short protuberance).

subsessile, 3–4 mm. long, with linear lobes; corolla 4–6 mm. long, purplish with whitish tube, the short lobes somewhat unequal; caps. 2–3 mm. long; seeds ca. 0.6 mm. long, pale brown, favose-reticulate; $n = 9$ (Ferguson & Coolidge, 1932).—Sandy arroyos and dried beds of pools, below 4000 ft.; many Plant Communities; Sacramento V., cent. Coast Ranges, cismontane s. Calif.; to Ariz., Tex., Fla., trop. Am. April–Aug.

2. **P. violàcea** Lindl. Viscid, the stems 1–1.5 dm. long, slender, branching; lvs. ovate; fls. rose-red or violet, 2.5–3.5 cm. long; $n = 7$ (Ferguson & Coolidge, 1932).—Cult. and sometimes escaping; native of Argentina.

63. Scrophulariàceae. FIGWORT FAMILY (Fig. 60)

Herbs, occasionally shrubs or trees, with rounded stems. Lvs. simple, alternate to rarely whorled, mostly entire, sometimes parted or pinnatifid, estipulate. Fls. perfect, racemose or paniculate. Calyx usually 2-lipped, sometimes nearly regular, with 4–5 lobes, sometimes these grown into a leaflike structure split on one side. Corolla 4–5-lobed, usually bilabiate, sometimes almost regular. Stamens 5, sometimes 4 and didyn-amous, or only 2. Carpels 2, united into a superior 2-loculed ovary with styles usually united. Fr. a 2-celled, mostly 2-valved caps. Seeds few to many, wingless or winged, with fleshy endosperm. A family of ca. 200 genera and 3000 spp., widely distributed.

A. Fertile stamens 5; lvs. alternate; corolla nearly regular, rotate. Tall introd. weedy plants
                                                                                      8. *Verbascum*
AA. Fertile stamens 4 or 2; lvs. opposite or alternate; corolla usually ± labiate.
    B. Plants acaulescent; corolla nearly rotate, 1.5 mm. long; anther-cells wholly confluent
                                                                                      7. *Limosella*
    BB. Plants mostly caulescent; corolla mostly not rotate, larger; anther-cells distinct.
        C. Stigmas distinct, flattened or platelike.
            D. Corolla campanulate, not distinctly 2-lipped.
                E. The corolla 9–10 mm. long, white; lvs. cuneate-obovate .... 6. *Bacopa*
                EE. The corolla 40–50 mm. long, mostly purple; lvs. oblong-lanceolate
                                                                                      20. *Digitalis*
            DD. Corolla nearly 2-lipped.
                E. Connective of stamens wider than the parallel anther-cells; corolla whitish
                    with yellow tube .... 5. *Gratiola*
                EE. Connective of stamens not so expanded, the anther-cells not parallel; corolla
                    purplish, violet, red, to yellow or buff.
                    F. Anther-bearing stamens 2.
                        G. Upper lvs. not much reduced, ovate; herbage glabrous.
                                                                                      1. *Lindernia*
                        GG. Upper lvs. much reduced, scalelike; herbage sparsely stipitate-
                            glandular .... 2. *Dopatrium*
                    FF. Anther-bearing stamens 4.
                        G. Sepals nearly or quite distinct; fls. blue-violet .... 4. *Stemodia*

GG. Sepals united, often almost throughout, the calyx usually angled;
    corolla usually purple or red to yellow or buff ...... 3. *Mimulus*
CC. Stigmas wholly united, capitate or punctiform.
   D. Corolla spurred or saccate on lower side of base, usually also with prominent
     palate; caps. opening by pores or irregularly; lvs. usually alternate.
     E. Fertile stamens 2; corollas 15–30 mm. long. Desert annuals .. 15. *Mohavea*
     EE. Fertile stamens 4; corolla smaller if from desert, and plants annual.
       F. Lvs. mostly 3's; plant shrubby with scarlet corollas 20–25 mm. long.
         Insular .... 13. *Galvezia*
       FF. Lvs. alternate; plants not both shrubby and insular and with scarlet
         corollas.
         G. Corolla gibbous or saccate at base.
           H. Lvs. triangular and hastate or 5-lobed, or round-ovate and
             irregularly dentate with long bristly teeth. Desert plants
             14. *Maurandya*
           HH. Lvs. entire, ovate or lanceolate to linear. Cismontane and
             deserts .... 16. *Antirrhinum*
         GG. Corolla with narrow spur at base.
           H. Fls. in terminal racemes or spikes; stems ± erect at least in
             upper part; lvs. linear to lanceolate or ovate ... 17. *Linaria*
           HH. Fls. in axils of broad lvs.; stems trailing or twining.
             I. Lvs. glabrous, palmately veined and lobed; caps. opening
               by 2 pores; fls. violet .... 18. *Cymbalaria*
             II. Lvs. pubescent, pinnately veined, rounded to hastate; caps.
               ± circumscissile; fls. white or yellowish-white
               19. *Kickxia*
DD. Corolla not spurred or saccate on lower side at base, 2-lipped to nearly regular,
   without prominent palate; caps. valvate.
   E. Corolla with upper lip flattened or widely arched, not forming a galea.
     F. Upper lip of corolla 2-lobed.
       G. Plants annual; corolla gibbous on upper side of base.
         H. Lower lip of corolla appearing 2-lobed, the middle lobe keel-
           shaped and concealed by the lateral lobes; stamens included
           in keel .... 11. *Collinsia*
         HH. Lower lip of corolla 3-lobed; fils. exserted ..... 12. *Tonella*
       GG. Plants perennial; corolla not gibbous on upper side.
         H. Corolla ± brownish, mostly less than 1 cm. long, inflated, with
           4 erect lobes and 1 reflexed; sterile stamen reduced to a scale
           or wanting .... 10. *Scrophularia*
         HH. Corolla mostly blue, lavender, rose, white, rarely
           brownish, 10–40 mm. long, tubular or funnelform, ± bilabiate;
           sterile stamen usually ca. as long as the fertile ones
           9. *Penstemon*
     FF. Upper lip of corolla appearing 1-lobed by fusion of 2.
       G. Plants herbaceous. Mostly natives.
         H. Stems elongating and leafy .... 21. *Veronica*
         HH. Stems scapose, the lvs. basal .... 23. *Synthyris*
       GG. Plants shrubby; escapes from cult. .... 22. *Hebe*
   EE. Corolla with upper lip narrowly arched, forming a galea which encloses the
     anthers.
     F. Anther-cells equal in size and position; seed-coat not loose and reticulate.
       G. Lvs. opposite; caps. symmetrical; plants annual or biennial.
         H. Corolla white and pink; calyx-lobes unequal, obtuse; fls.
           crowded in dense infl. with spreading bracts .. 24. *Bellardia*
         HH. Corolla yellow; calyx-lobes alike, lanceolate; fls. scattered,
           with ascending bracts .... 25. *Parentucellia*
       GG. Lvs. alternate or basal; caps. asymmetrical, usually decurved; plants
         perennial .... 26. *Pedicularis*
     FF. Anther-cells unequally placed, the upper attached by its middle, the
      lower normally attached or abortive; seed-coat usually loose and reticu-
      late.
       G. Calyx-tube surrounding the lower portion of the corolla, its 4 (or
        2 by fusion) lobes placed laterally.
         H. Lower lip of corolla ca. as long as and larger than galea;
           plants annual .... 27. *Orthocarpus*
         HH. Lower lip of corolla definitely shorter and smaller than galea;
           plants mostly perennial .... 28. *Castilleja*
       GG. Calyx-tube surrounding only base of corolla or mostly entirely to
        the dorsal side and consisting of a bractlike structure, entire or
        minutely bidentate at apex .... 29. *Cordylanthus*

1. Lindérnia All. FALSE PIMPERNEL

Small smooth annual herbs, ± diffuse. Lvs. opposite, subentire to denticulate. Fls.
pedicelled, mostly in axils of foliage-lvs., or upper racemed. Sepals 5, distinct. Corolla

2-lipped, blue-violet, the upper lip short, erect, 2-lobed, the lower larger, spreading, 3-cleft. Fertile stamens 2, included, posterior; anterior pair sterile, 2-lobed. Stigma 2-lobed. Caps. septicidal, ovoid or ellipsoid, many-seeded. Seeds smooth or finely lined transversely. Perhaps 50 spp., mostly of warmer parts of Old World. (Named for F. B. von *Lindern,* 1682–1755, an early botanist.)

Lower pedicels shorter than their subtending lvs.; lower lvs. narrowed at base .... 1. *L. dubia*
Lower pedicels exceeding their subtending lvs.; lvs. all widest near base .... 2. *L. anagallidea*

1. **L. dùbia** (L.) Penn. [*Gratiola d.* L. *Ilysanthes d.* Barnh. *L. d.* ssp. *major* Penn.] Glabrous, 0.5–3 dm. tall, simple to much-branched, depressed to ascending; lvs. lance-elliptic to ovate, 1–3 cm. long, the upper rounded at base; pedicels 5–12 mm. long; calyx-lobes linear, ca. as long as caps.; corolla 8–10 mm. long; caps. 4 mm. long; seeds ca. 0.4 mm. long.—Margins of streams, 500–2000 ft.; Mixed Evergreen F.; Humboldt and Siskiyou cos.; to Wash., Atlantic Coast, Mex. July–Sept.

2. **L. anagallídea** (Michx.) Penn. [*Gratiola a.* Michx.] Glabrous, slender-stemmed, diffuse, 0.5–2 dm. tall; lvs. ovate, 0.5–1.2 cm. long; pedicels 10–25 mm. long; corolla 7–9 mm. long; caps. 4–5 mm. long; seeds ca. 0.3 mm. long.—Damp places, below 5000 ft.; Freshwater Marsh, etc.; Cent. V. from Tulare Co. n.; to Wash., Atlantic Coast, Mex. July–Sept.

2. Dopàtrium Buch.-Ham.

Small herbs with scalelike lvs. on upper parts of stems, the lower lvs. opposite, few, fleshy. Fls. solitary in the lf.-axils, yellow or violet. Calyx short, 5-cleft. Corolla with long thin tube and 2-lipped flat limb; upper lip small, erect, slightly 2-lobed; lower lip larger, 3-lobed (the middle lobe much larger than the lateral ones). Fertile stamens 2; staminodia 2. Caps. loculicidal. Ca. 7 spp.; Old World.

1. **D. juncèum** (Roxb.) Buch.-Ham. in Benth. [*Gratiola j.* Roxb.] Rather fleshy small annual, sparingly gland-dotted; lvs. few, oblong, obtuse, 2–5 cm. long, the upper remote, ca. 0.5 mm. long; pedicels capillary, 1–2.5 cm. long, spreading in fr.; calyx-tube ca. as long as the obtuse lobes; corolla pale blue, the tube slightly exceeding the calyx; style 1, short; caps. globose, ca. 4 mm. in diam.; seeds reticulate.—Weed in ricefields; Butte Co.; native of Asia and Australia.

3. Mímulus L. Monkey-Flower

Annual or perennial herbs or low shrubs, often ± viscid or glandular-pubescent or pilose to slimy-viscid. Lvs. opposite, entire to dentate or even laciniate. Fls. axillary to foliage lvs. or bracts, often in open racemes. Bracteoles none. Sepals 5, united into a tubular or campanulate calyx, of which the tube is usually inflated and plicate-angled, the lobes or teeth equal to unequal. Corolla 2-lipped to subregular, purple or red or almost violet to yellow or buff, with a pair of bearded or naked ridges running down the lower side of the throat and forming a palate which may almost close the throat. Stamens 4, didynamous, usually all antheriferous. Stigmas 2, distinct and lamelliform, or adhering to form a funnellike structure. Caps. cylindrical, loculicidal, the septum unruptured or splitting. Seeds many, small, oblong to oval or fusiform, mostly yellowish, reticulate to almost smooth, wingless. Ca. 150 spp., mostly of w. N. Am., but also in S. Am., Asia, Australia, etc. (Diminutive of Latin *mimus,* a comic actor, because of the grinning corolla.)

(Grant, A. L. A monograph of the genus Mimulus. Ann. Mo. Bot. Gard. 11: 99–388, 1924. McMinn, H. E. Studies in the genus Diplacus. Madroño 11: 33–128, 1951.)
A.   Plants perennial.
  B.   Pedicels longer than the calyces; plants not at all woody. Wet places.
    C.   Fls. red or magenta, 4–5 cm. long; caps. 13–18 mm. long.
      D.   Corolla rose to pink or magenta, the lobes nearly equally spreading; stamens included .... 1. *M. Lewisii*
      DD. Corolla scarlet, the upper lobes arched, the lower deflexed-spreading; stamens exserted .... 2. *M. cardinalis*
    CC. Fls. yellow, mostly 1–4 cm. long; caps. 6–9 mm. long.
      D.   Mature calyx little or not much inflated, the lobes practically straight; corolla-throat open, the limb not conspicuously 2-lipped.

    E.   Corolla 3–4 cm. long, the throat widely campanulate. Humboldt Co. and n.
                                             4. *M. dentatus*
   EE. Corolla 1–3 cm. long, the throat narrow-campanulate. Widely distributed.
      F.   Plants ± slimy; lvs. pinnately veined; stems mostly 1–3 dm. long
                                          3. *M. moschatus*
     FF. Plants not slimy; lvs. palmately veined; stems mostly less than 1 dm. long
                                         5. *M. primuloides*
 DD. Mature calyx strongly inflated, the lower lobes curved against the others; corolla-throat ± closed by the palate formed by the large ventral ridges, the limb strongly 2-lipped.
    E.   Stems mostly with definite racemes; pedicels usually shorter than corollas; rootstocks rarely fleshy or yellow .... 34. *M. guttatus*
   EE. Stems mostly 1- to 3-, rarely 5-fld.; pedicels usually longer than corollas; rootstocks fleshy, yellow .... 37. *M. Tilingii*
BB. Pedicels shorter than calyces; plants somewhat woody at base to definitely shrubby. Dry places.
  C.   Plants woody at base only; lvs. finely glandular-pubescent on both surfaces; calyx usually swollen at base; caps. 9–12 mm. long. Santa Ana Mts. and e. San Diego Co.
                                      71. *M. Clevelandii*
  CC. Plants shrubby; lvs. not glandular-pubescent on upper surface; calyx not swollen at base; caps. 12–25 mm. long.
    D.   Corollas orange, copper, yellow, buff, cream, or almost white, never true red.
      E.   Calyx, lower lf.-surfaces, and upper branchlets glabrous or microscopically glandular-puberulent.
        F.   Lvs. slightly to distinctly impressed-veiny above; corolla-tube not usually exserted from calyx .... 73. *M. aurantiacus*
        FF. Lvs. not impressed-veiny above; corolla-tube usually exserted from calyx.
           G.   Corollas light lemon-yellow, 2–2.5 cm. broad, the lobes scarcely notched. Se. San Diego Co. .... 74. *M. aridus*
          GG. Corollas warm buff to orange-yellow, 3.5–5 cm. broad, the lobes distinctly notched. Santa Lucia Mts. and w. Sierra Nevada
                                       75. *M. bifidus*
      EE. Calyx, lower lf.-surfaces, and upper branchlets obviously pubescent to almost woolly .... 72. *M. longiflorus*
    DD. Corollas red or reddish.
      E.   Calyx glandular-hairy; pedicels mostly less than 10 mm. long
                               72. *M. longiflorus var. rutilus*
      EE. Calyx glabrous or nearly so; pedicels mostly 10–25 mm. long.
        F.   Lvs mostly less than 3–4 times as long as wide; corolla not over 1.8 cm. broad, the lobes subequal, only slightly if at all notched. Insular.
                                   76. *M. Flemingii*
        FF. Lvs. mostly 3–4 times as long as wide; corolla 1.8 or more cm. broad, the lobes unequal and notched. Mainland mostly ..... 77. *M. puniceus*
AA. Plants annual.
  B.   Pedicels conspicuous, normally longer than the calyx; corolla mostly deciduous, dropping before shrivelling and leaving the styles exposed (± persistent in *M. Breweri*).
    C.   Mature calyx strongly inflated, sagittally compressed, the lowermost lobes tending to curve up against the upper; corolla yellow.
      D.   Bracts of infl. connate, forming round disks; plants glaucous. Tehama and Butte cos. .... 35. *M. glaucescens*
      DD. Bracts of infl. not connate; plants not glaucous.
        E.   Bracts of infl. narrowed to petioled or sessile bases, but these not clasping or rounded.
          F.   Corolla 1.5–2 cm. long; lvs. denticulate; calyx 10–12 mm. long. Lake Co.
                                      32. *M. nudatus*
          FF. Corolla 0.7–1.5 cm. long; lvs. mostly pinnatifid-lobed; calyx 5–15 mm. long. Tuolumne Co. to Tulare Co. .... 33. *M. laciniatus*
        EE. Bracts of infl. rounded and sessile with clasping bases.
          F.   Uppermost calyx-lobe rarely more than twice the length of the others, the lower teeth in maturity only partly closing the orifice
                                      34. *M. guttatus*
          FF. Uppermost calyx-lobe almost 3 times as long as others, the lower teeth in maturity folding up so as almost to close the orifice .. 36. *M. nasutus*
  CC. Mature calyx not strongly inflated or if somewhat so, cylindric, the lobes permanently straight or nearly so.
    D.   Fls. predominantly yellow, or sometimes partly white, sometimes with reddish spots.
      E.   Calyx not ridged or low-winged along the sepal-midribs; caps. glandular-puberulent .... 70. *M. pilosus*
      EE. Calyx ridged or low-winged along the sepal midribs; caps. glabrous.
        F.   Calyx-lobes 1.5–3 mm. long; corolla mostly 15–25 mm. long.
          G.   Corolla bicolored, the upper lip usually white, the lower yellow; calyx-teeth without cilia .... 9. *M. bicolor*
          GG. Corolla not bicolored, the upper lip yellow; calyx-teeth ciliate.
            H.   Corolla 15–20 mm. long.
              I.   Calyx-lobes subulate, 2–3 mm. long; caps. 4–5 mm. long. Yellow Pine F. .... 15. *M. floribundus*
                                          var. *subulatus*

II. Calyx-lobes acute, 2 mm. long; caps. 6–7 mm. long. Foothill Wd. . . . . 14. *M. Dudleyi*

HH. Corolla 10–12 mm. long . . . . 20. *M. latidens*

FF. Calyx-lobes 0.5–1 mm. long; corolla 4–15 (–20) mm. long.

  G. Anthers pubescent; corolla 13–20 mm. long. Tuolumne Co. to Kern Co. . . . . 12. *M. discolor*

  GG. Anthers glabrous.

    H. Calyx-lobes triangular-acute to -acuminate.

      I. Corolla less than twice as long as calyx; style ½–⅔ as long as caps. Lassen and Modoc cos. . . 19. *M. breviflorus*

      II. Corolla at least twice as long as calyx; style over ⅔ the length of the caps.

        J. Lower corolla-lip slightly longer than but not much deflexed from upper; lf.-blades longer than petioles.

          K. Stem and lvs. strongly pubescent, the cauline lvs. rounded to cordate at base; plants mostly 1–3 dm. tall . . . . 15. *M. floribundus*

          KK. Stem and lvs. glandular-puberulent or finely glandular-pubescent, the cauline lvs. narrowed at base; plants 0.5–1.5 dm. tall.

            L. Lvs. acute or acutish, dentate or denticulate; stem finely glandular-pilose; corolla 13–22 mm. long.

              M. Fls. 13–17 mm. long; calyx-lobes acute. Mariposa Co. to Fresno Co. 16. *M. arenarius*

              MM. Fls. 20–22 mm. long; calyx-lobes acuminate. Calaveras Co. 38. *M. Whipplei*

            LL. Lvs. obtuse, denticulate to entire; stems glandular-puberulent; fls. 7–11 mm. long. Mariposa Co. to Siskiyou Co. . 17. *M. Pulsiferae*

        JJ. Lower lip of corolla longer than and deflexed from upper lip; lf.-blades not longer than the petioles. Humboldt and Siskiyou cos. . . . . 18. *M. alsinoides*

    HH. Calyx-lobes rounded, either bluntly so or mucronate, 0.5–1 mm. long.

      I. Corollas 5–9 mm. long.

        J. Calyx-lobes ciliate; plants rather open in growth; pedicels 7–20 mm. long. Mts. along desert edge to White Mts. . . . . 23. *M. rubellus*

        JJ. Calyx-lobes not ciliate; plants compact, bushy; pedicels 2–7 mm. long. San Jacinto Mts. to Siskiyou and Modoc cos. 24. *M. Suksdorfii*

      II. Corollas 10–17 mm. long.

        J. Stigmas not fringed; corolla-tube little or not surpassing calyx. Below 7000 ft., Inyo and Mono cos. 22. *M. montioides*

        JJ. Stigmas fringed; corolla-tube well exserted. From 7000–11,000 ft., Tulare and Inyo cos. 27. *M. barbatus*

DD. Fls. predominantly purple, red, or magenta.

  E. Calyx 2 mm. long; corolla 2–2.5 mm. long; anthers glabrous. San Bernardino Mts. . . . . 31. *M. exiguus*

  EE. Calyx 4–12 mm. long; corolla 6–20 mm. long.

    F. Calyx-lobes mostly 1–2 mm. long, acute to subulate-tipped (shorter in *M. Palmeri*).

      G. Anthers pubescent; lvs. sessile, or nearly so by narrow base; corolla 15–20 mm. long.

        H. Calyx-ribs corky when mature, the lobes 1–2 mm. long, equal, triangular-acute. Tuolumne and Mariposa cos. 10. *M. Biolettii*

        HH. Calyx-ribs not corky.

          I. Calyx-teeth unequal, not ciliate, 1–2 mm. long. Mariposa Co. . . . . 11. *M. filicaulis*

          II. Calyx-teeth equal, ciliate, 0.5–1 mm. long. Mariposa Co. to Mojave Desert . . . . 13. *M. Palmeri*

      GG. Anthers glabrous; lvs. sessile by rounded base; corolla 10–12 mm. long.

        H. Plant minutely glandular-puberulent; lvs. 1–2 cm. long; calyx inflated in fr. Cismontane, Butte and Lake cos. to San Diego Co. . . . . 20. *M. latidens*

        HH. Plant densely glandular-villous; lvs. 1.5–4 cm. long; calyx not much inflated in fr. Mojave Desert . . . . 21. *M. Parishii*

  FF. Calyx-lobes mostly ca. 0.5 mm. long, broader than long, obtuse to rounded.

    G. Anthers pubescent; plants essentially glabrous.

   H. Corolla 8–10 mm. long. Amador Co. to Tuolumne Co.

                   6. *M. inconspicuus*

  HH. Corolla 13–16 mm. long.

    I. Pedicels shorter than the lvs.; calyx glandular-puberulent.
Mariposa Co. to Tulare Co. . . . . 7. *M. Grayi*

    II. Pedicels longer than the lvs.; calyx glabrous. Fresno and
Tulare cos. . . . . 8. *M. acutidens*

 GG. Anthers glabrous: plants glandular-puberulent or -pubescent
throughout.

   H. Stigmas not fringed, basal tube of corolla scarcely or not exceeding calyx; corollas 6–10 mm. long.

    I. Pedicels 5–20 mm. long, less than twice the length of the
subtending bracts; anthers and stigmas included; lvs. entire, 3-veined from narrow base.

      J. Plants glandular-puberulent; calyx-teeth ciliate,
rounded. Mts. bordering deserts, n. to White Mts.

                  23. *M. rubellus*

      JJ. Plants glandular-pubescent; calyx-teeth not ciliate, del-
toid. San Jacinto Mts. n. through Sierra Nevada, N.
Coast Ranges . . . . 25. *M. Breweri*

    II. Pedicels 15–27 mm. long, 3–5 times the length of the
bracts; anthers and stigmas exposed; lvs. crenate-dentate,
scarcely veined, round-clasping at base. Tehachapi Mts. to
Lake Co. . . . . 26. *M. androsaceus*

  HH. Stigmas ciliate-fringed; basal tube of corolla exceeding calyx;
corollas 10–15 mm. long.

    I. Corolla-lobes distinctly unequal.

      J. Style glabrous.

        K. Corolla-lobes narrow, emarginate; pedicels 1–2
times as long as the bracts. Tulare and Inyo cos.

                 27. *M. barbatus*

        KK. Corolla-lobes broad, not emarginate; pedicels
more than twice as long as bracts. Mariposa Co.

                 28. *M. gracilipes*

      JJ. Style pubescent. San Bernardino Mts.

                 29. *M. purpureus*

    II. Corolla-lobes subequal. San Jacinto and Santa Ana mts.
to L. Calif. . . . . 30. *M. diffusus*

BB. Pedicels normally shorter than calyces; corolla mostly shrivelling and persistent on the developing caps.

 C. Fls. purple to rose or magenta.

  D. Caps symmetrical, membranous, soon dehiscent, not gibbous or oblique at base.

   E. Corolla salverform, the cylindrical tube expanding directly into the rotate
lobes which are purple with pale margins. Mojave Desert . 58. *M. mohavensis*

   EE. Corolla tubular-campanulate, with definite throat wider than the tube below.

    F. Stigmas and usually the longer stamens exserted from the corolla-throat.
(See also 56. *M. Bolanderi*.)

      G. Lvs. elliptic, acuminate or cuspidate; corolla-throat campanulately
inflated, expanding distally. Modoc Co. . . . . 38. *M. Cusickii*

      GG. Lvs. narrower, acute to obtuse; corolla-throat narrower.

        H. Calyx-lobes obscurely glandular-pubescent; plants obscurely
glandular, not strong-scented.

          I. Corolla 13–20 mm. long; mature calyx 7–8 mm. long, dis-
tinctly inflated. Tehama and Lassen cos. . . 40. *M. nanus*

          II. Corolla 9–11 mm. long; mature calyx 4–5 mm. long,
scarcely inflated. Nevada Co. to Siskiyou Co.

                 41. *M. Jepsonii*

        HH. Calyx-lobes strongly glandular-pubescent; plants evidently
glandular-pubescent, strong-scented.

          I. Calyx-lobes subequal, the uppermost less than twice as
long as lowermost; mature calyx little or not at all inflated.

            J. Calyx-teeth acute, ca. ¼ as long as the tube; caps.
usually well exserted.

              K. Calyx 5–7 mm. long; lvs. mostly short-petioled.
Plumas Co. to Sierra Co. . . . . 42. *M. mephiticus*

              KK. Calyx 7–9 mm. long; lvs. mostly sessile. E. slope
of Sierra Nevada (Lassen Co. to Inyo Co.) and
White Mts. . . . . 43. *M. densus*

            JJ. Calyx-teeth subulate, at least ⅓ as long as tube; caps.
scarcely exserted. Subalpine in Sierra Nevada, White
Mts. . . . . 44. *M. coccineus*

          II. Calyx-lobes strongly unequal, the uppermost ca. twice as
long as lowermost; mature calyx distinctly inflated. San
Gabriel Mts. . . . . 45. *M. Johnstonii*

    FF. Stigmas and anthers included in corolla-throat.

G. Corolla 6–10 mm. long; calyx 3–8 mm. long.
  H. Mature calyx distinctly inflated, 6–8 mm. long; stigmas unequal. Lake and Colusa cos. to Marin Co. .. 54. *M. Rattanii*
  HH. Mature calyx scarcely inflated, 3–4 mm. long; stigmas equal. Sierra Nevada .... 47. *M. leptaleus*
GG. Corolla 10–40 mm. long; calyx in maturity 5–25 mm. long.
  H. Corolla-lobes subequal; stigmas subequal.
    I. Mature calyx little or not inflated, nearly plane, the ridges pubescent and narrower than the intervening scarious portions. High elevs. of Fresno and Tulare cos.
      46. *M. Whitneyi*
    II. Mature calyx distinctly inflated, the ridges raised, glandular-pubescent and wider than the intervening pale tissue.
      J. Inner face of corolla-lobes glabrous. Cent. Calif. to cismontane s. Calif.
        K. Corolla 20–25 mm. long.
          L. Anthers glabrous; corolla-lobes abruptly widely spreading. San Benito Co. to L. Calif.
            48. *M. Fremontii*
          LL. Anthers ciliate; corolla-lobes less abruptly spreading. Sierra Nevada .. 50. *M. viscidus*
        KK. Corolla 13–20 mm. long.
          L. Corolla-tube at least 1.7 times as long as calyx; calyx-teeth subequal. From Napa and Fresno cos. n. .... 51. *M. Layneae*
          LL. Corolla-tube less than 1.7 times as long as calyx; calyx-teeth unequal. Santa Lucia Mts.
            49. *M. subsecundus*
      JJ. Inner face of corolla-lobes pilose. Deserts and San Gabriel Mts.
        K. Lowest pair of calyx-lobes more than half as long as uppermost lobe; corolla 12–30 mm. long; bracts acuminate to cuspidate. Deserts
          52. *M. Bigelovii*
        KK. Lowest pair of calyx-lobes not more than half as long as uppermost lobe; corolla 13–20 mm. long; bracts obtuse to acute. San Gabriel Mts.
          45. *M. Johnstonii*
  HH. Corolla-lobes unequal; stigma-lobes unequal.
    I. Corolla 10–13 mm. long; lower stigma-lobe rounded. Lake Co. ... . 55. *M. brachiatus*
    II. Corolla 15–40 mm. long. Widely distributed.
      J. Corolla-tube included in calyx; plicae of calyx thick, dark green, contrasted with intervening pale surface; lower lobe of stigma lanceolate .... 56. *M. Bolanderi*
      JJ. Corolla-tube exserted; plicae of calyx thin, not strongly contrasted with pale intervening surface; lower stigma-lobe rounded .... 53. *M. Torreyi*
DD. Caps. asymmetrical, cartilaginous, tardily or not dehiscent, gibbous or oblique at base.
  E. Fls. cleistogamous, not opening or barely so; subacaulescent
    61. *M. Douglasii*
  EE. Fls. opening normally.
    F. Corolla 6–12 mm. long.
      G. Fls. 6–7 mm. long; stem ca. 1 cm. long. Modoc Co.
        59. *M. pygmaeus*
      GG. Fls. 9–12 mm. long; stems 10–20 cm. long. Tulare and Kern cos.
        60. *M. pictus*
    FF. Corolla 13–50 mm. long.
      G. Lower lip of corolla less than ⅓ as long as upper lip.
        H. Mature calyx 10–12 mm. long; corolla purple; plants 1–6 cm. high.
          I. Corolla 3–4 cm. long, except sometimes in cleistogamous-fld. plants; calyx densely ciliate with glandless hairs. Mainland .... 61. *M. Douglasii*
          II. Corolla 1.3–1.5 cm. long; calyx glandular-pubescent. Santa Cruz Id. . . 69. *M. Brandegei*
        HH. Mature calyx 17–20 mm. long; corolla purple with white upper lip; plants 10–14 cm. high. Catalina Id.
          67. *M. Traskiae*
      GG. Lower lip of corolla more than ⅓ as long as the upper.
        H. The lower lip of corolla at least as long as the upper; stigmas subequal.
          I. Anthers glabrous; corolla-tube ca. as long as calyx; calyx-lobes acute to attenate; pedicels decurved. Death V.
            62. *M. rupicola*

II. Anthers pubescent; corolla-tube much longer than calyx; calyx-lobes obtusish; pedicels straight.
  J. Corolla-tube not more than twice as long as calyx; mature calyx 12–14 mm. long; caps. 6–7 mm. long. Cent. V. and Coast Ranges from Kern Co. n.
                                                         63. *M. tricolor*
  JJ. Corolla-tube more than twice the length of the calyx; mature calyx 7–10 mm. long; caps. 3–4 mm. long. W. base of Sierra Nevada.
    K. Lower lip of corolla yellow; corolla-tube pilose without; calyx-lobes very unequal
                                       64. *M. pulchellus*
    KK. Lower lip of corolla purple; corolla-tube glabrous without; calyx-lobes subequal . 65. *M. angustatus*
 HH. The lower lip of corolla shorter than the upper; stigmas unequal, the lower longer.
  I. Corolla 3–4.5 cm. long; mature calyx 12–15 mm. long; caps. 8–10 mm. long .... 66. *M. Kelloggii*
  II. Corolla 1.5–2.5 cm. long; mature calyx 9–12 mm. long; caps. 5–7 mm. long .... 68. *M. Congdonii*
CC. Fls. yellow.
 D. Corolla 20–50 mm. long; plants mostly 2–8 dm. high. Santa Barbara Co. to L. Calif. .... 57. *M. brevipes*
 DD. Corolla smaller; plants less than 2 dm. tall.
  E. The corolla 6–7 mm. long; plants acaulescent. Plumas and Modoc cos.
                                       59. *M. pygmaeus*
  EE. The corolla 12–20 mm. long; plants caulescent.
    F. Stigmas and usually at least one pair of anthers exserted; plants mostly 4–12 cm. tall.
     G. Calyx-lobes 1–2 mm. long, subequal.
      H. Mature calyx distinctly inflated; plants obscurely glandular, not strong-scented. Tehama and Lassen cos. n.
                                40. *M. nanus*
      HH. Mature calyx not noticeably inflated; plants evidently glandular, strong-scented. Plumas Co. to Tulare Co.
                              42. *M. mephiticus*
     GG. Calyx-lobes 2.5–3.5 mm. long, distinctly unequal. Lassen Co. to Inyo Co. .... 43. *M. densus*
    FF. Stigmas and anthers included in the corolla throat; plants mostly 2–4 cm. tall. Fresno and Tulare cos. .... 46. *M. Whitneyi*

1. **M. Lewísii** Pursh. Perennial from a running rootstock, the stems erect, mostly simple, 3–8 dm. tall, the plant ± viscid-pubescent throughout; lvs. oblong-elliptic, 2–7 cm. long, 3–5-veined from base, sinuate-denticulate, sessile with round-clasping base or the lower short-petioled; pedicels 30–60(–90) mm. long, from upper axils; calyx tubular, 15–25 mm. long, sharply angled, the teeth subequal, triangular-subulate, 4–6 mm. long; corolla 3–5 cm. long, rose to pink with darker lines down the throat, often also blotched maroon and with 2 yellow hairy ridges, the lobes rounded, ± emarginate; anthers ciliate; stigmas fimbriolate; caps. oblong, 13–14 mm. long, dehiscing through the apex of septum; seeds narrow-oblong, apiculate, longitudinally wrinkled, ca. 0.5 mm. long.—Stream banks, 4000–10,000 ft.; Montane Coniferous F.; Sierra Nevada from Tulare Co. n., to Modoc and Siskiyou cos.; to B.C., Alaska, Mont., Colo. June–Sept.

2. **M. cardinàlis** Dougl. ex Benth. [*Erythranthe c.* Spach. *M. c.* vars. *griseus* and *rigens* Greene.] Freely branched viscid-villous perennial, erect or decumbent, the stems 2.5–8 dm. long, from a running rootstock; lvs. obovate to oblong, 2–8 cm. long, sessile, serrate, longitudinally 3–5-veined, at least the upper with broad clasping bases; pedicels 50–80 mm. long; calyx tubular or oblong-prismatic, 20–30 mm. long, angulate-winged, the teeth subequal, 4–5 mm. long, ovate, acute to acuminate; corolla strongly bilabiate, 4–5 cm. long, scarlet, sometimes yellowish, the throat narrow, yellowish, with yellow hairy ridges, the upper lip arched-ascending, the lower decurved-reflexed; anthers ciliate, arched in the upper lip; stigmas fimbriolate; caps. oblong, acuminate, 16–18 mm. long, dehiscing through apex of the septum; seeds narrow-oblong, apiculate, longitudinally wrinkled, ca. 0.5–0.6 mm. long; $n = 8$ (Brozek, 1932).—Frequent on stream banks, seeps, etc., below 8000 ft.; many Plant Communities; through most of cismontane and montane Calif. and sparingly e. of the Sierra Nevada; to Ore., Nev., Ariz. April–Oct.

3. **M. moschàtus** Dougl. ex Lindl. Musk Flower. Perennial from slender creeping rootstocks, glandular-pilose to -villose, ± slimy; stems 0.5–3 dm. long, creeping or

decumbent, subsimple to diffusely branched; lvs. ovate to somewhat oblong, pinnately veined, 1–5 cm. long, subentire to denticulate, truncate or rounded to short petioles or the upper sessile; pedicels slender, 1–5 cm. long; calyx campanulate, 9–12 mm. long, ± plicate-angled, the teeth lanceolate, 2–3 mm. long, often slightly unequal; corolla yellow, 18–26 mm. long, funnelform, the cylindrical tube exserted, ventrally 2-ridged and pilose, the lobes rounded, subequal, spreading; anthers subglabrous to pubescent; caps. 6–8 mm. long, not dehiscing through the septum; seeds subglobose, ca. 0.3 mm. long, brown, tuberculate.—Wet places, below 7500 ft.; Red Fir F., Yellow Pine F., Closed-cone Pine F., N. Coastal Scrub, etc.; mts. from San Diego Co. to Modoc Co., Coast Ranges and along the coast from Santa Cruz Co. to Del Norte and Siskiyou cos.; to B.C., Rocky Mts. and Atlantic Coast. June–Aug. Variable; plants with suberect habit, ovate lvs., and fls. 22–26 mm. long, have been called var. *longiflòrus* Gray [*M. macranthus* Penn.]; and those more diffusely branched, with lvs. somewhat oblong, the upper sessile, and corolla 20–25 mm. long, var. *sessilifòlius* Gray [*M. inodorus* Greene]. Both of these forms occur throughout the range.

Var. **moniliórmis** (Greene) Munz. [*M. m.* Greene. *M. Leibergii* Grant.] Plants glabrous to finely pubescent; lvs. all petioled; corolla 22–28 mm. long.—Moist places, 4000–8300 ft.; Montane Coniferous F.; Sierra Nevada from Eldorado and Alpine cos. n.; mts. of Siskiyou and Humboldt cos. June–Aug.

4. **M. dentàtus** Nutt. ex Benth. in A. DC. Perennial from running slender rootstocks; stems 2–4 dm. tall, pilose or glabrescent, erect or ascending, simple; lvs. ovate, coarsely dentate, pinnately veined, 2–7 cm. long, acute, short-petioled, with rounded or sub-cuneate base; pedicels 2–3 cm. long; calyx campanulate, 1–1.5 cm. long, sparsely villous along the acutely angled ribs, the lobes deltoid-acuminate, 4–6 mm. long, ciliate, sub-equal; corolla yellow, 3–4 cm. long, the throat campanulate, almost as broad as long, spotted brown on lower side; lobes rounded, 6–10 mm. long, spreading, the upper somewhat shorter than lower; anthers villous; caps. 8 mm. long, the septum not splitting; seeds favose-pitted.—Wet places, below 1200 ft.; Redwood F., N. Coastal Scrub; near the coast from Humboldt Co. n.; to Wash. May–Aug.

5. **M. primuloìdes** Benth. Rosulate or short-stemmed perennial, rhizomatous, often stoloniferous, the stems villose to subglabrous, 0.1–0.5 dm. long; lvs. obovate to oblong, nearly or quite glabrous, longitudinally 3-veined, dentate to entire, 1–4 cm. long, some-what cuneate at the subsessile base, usually in several pairs and closely set; pedicels erect, slender, 3–12 cm. long; calyx tubular, mostly 6–8 mm. long, the lobes 1–2 mm. long; corolla yellow, 15–20 mm. long, funnelform with narrow-campanulate throat, the palate deeper yellow, densely hairy, the throat with red-brown spots below the lower lip, lobes spreading, notched; caps. 6–7 mm. long, dehiscing through the apex; seeds reticulate, brown, ± oblong, 0.4–0.5 mm. long.—Common in meadows and on wet grassy banks, 4000–8000 ft.; Yellow Pine F. to Lodgepole F.; Sierra Nevada from Tulare Co. n., Coast Ranges from Trinity Co. n.; to Wash., Rocky Mts. June–Aug.

Var. **pilosèllus** (Greene) Smiley. [*M. p.* Greene. *M. nevadensis* Gand.] Perhaps smaller, the lvs. with long soft white hairs above, which catch dew and glisten in the sun; calyx mostly ca. 5 mm. long; corolla ca. 8–15 mm. long.—At elevs. below 11,000 ft.; Montane Coniferous F.; more common than the typical form in Calif. extending s. to Glenn and Riverside cos.; to Ore.

Ssp. **linearifòlius** (Grant) Munz. [*M. p.* var. *l.* Grant. *M. l.* Penn.] Stems 0.6–1.2 dm. tall; lvs. linear-oblanceolate, 2–5 cm. long, subglabrous; pedicels 6–8 cm. long; calyx 8–11 mm. long; corolla 18–24 mm. long.—Moist places, 2600–7500 ft.; Montane Coniferous F.; Trinity Co. (Scott Mt., Trinity R., Salmon and Mary Blaine mts.), Siski-you Co. (Mt. Eddy, Rail Creek, Shackleford Creek). July–Aug.

6. **M. inconspícuus** Gray. Glabrous annual 0.5–1.5 dm. high, with stems erect to decumbent; lvs. oval to ovate, 0.8–2 cm. long, crisped, the first petioled and in a basal rosette, the latter sessile, in few pairs on the stems, ca. 1 cm. long; pedicels 10–12 mm. long, ± erect; calyx 7–9 mm. long, angled, the lobes 0.5 mm. long, slightly ciliate; corolla 0.8–1 cm. long, rose-pink to purplish, the throat cylindric, the low ventral ridges pubescent, the limb somewhat bilabiate, with spreading lobes; anthers villous; caps. 6–7 mm. long, dehiscing through septum; seeds ovoid, apiculate at one end.—Moist places, 1500–2500 ft.; Yellow Pine F., Foothill Wd.; foothills, Amador Co. to Tuolumne Co. May–June.

7. **M. Gràyi** Grant. Glabrous annual, erect, 0.5–2 dm. tall, simple or branched from base, with long internodes; lvs. few, broadly ovate, 0.7–1.8 cm. long, 3–5-nerved, denticulate, subsessile; pedicels slender, spreading, 5–10 mm. long; calyx 8–10 mm. long, puberulent, ridge-angled, the lobes 0.5 mm. long, apiculate; corolla 13–16 mm. long, rose to rose-purple, with white and yellow cylindric throat, the limb slightly bilabiate, with emarginate lobes; anthers ciliate; caps. 6 mm. long, dehiscing through septum-apex; seeds oblong-ovoid, favose-reticulate, light brown, 0.4–0.5 mm. long.—Moist places, 1800–9500 ft.; Montane Coniferous F.; base of Sierra Nevada from Mariposa Co. to Tulare Co. May–July.

8. **M. acùtidens** Greene. [*M. inconspicuus* var. *a.* Gray.] Glabrous annual, the stems erect to decumbent, slender, 0.5–2 dm. tall; lvs. in few pairs, broadly ovate, 3–5-nerved, denticulate, sessile, 0.8–2 cm. long; pedicels slender, 10–23 mm. long; calyx 7–9 mm. long, plicate-angled, the lobes ca. 0.6 mm. long, ciliolate, triangular; corolla rose to rose-purple, bilabiate, 1.3–1.5 cm. long, the throat cylindric, with 2 ventral ridges yellow, puberulent, and limb with spreading notched lobes; anthers ciliate; caps. 6 mm. long, dehiscing slightly through septum-apex; seeds oblong, yellow-brown, ca. 0.4 mm. long, reticulate.—Moist places, 1000–4000 ft.; Foothill Wd., Yellow Pine F.; base of Sierra Nevada in Fresno and Tulare cos. April–July.

9. **M. bìcolor** Hartw. ex Benth. [*M. Prattenii* Durand.] Glandular-pubescent annual, erect or ascending, 1–3 dm. tall, simple or branched from base; lvs. oblanceolate, to obovate, 1–3 cm. long, obtuse, denticulate to dentate, sessile or semipetiolate; pedicels 10–25 mm. long; calyx 8–10 mm. long, corky-ribbed in maturity, angled, the triangular-acute teeth 2–3 mm. long, equal; corolla somewhat bilabiate, 1.6–2.5 cm. long, the throat short, broad, funnelform, with red spots on the pubescent ridges; lobes emarginate, the upper upcurved, usually white, the lower spreading, yellow, with red dots; anthers ciliate; caps. 5 mm. long, not dehiscing through the septum-apex; seeds reticulate, oval-oblong, yellowish, ca. 0.3 mm. long.—Moist places, 1600–6000 ft.; Foothill Wd., Yellow Pine F., Chaparral; Sierra Nevada from Kern Co. n., to Shasta Co. and Trinity Co. April–June.

10. **M. Bioléttii** Eastw. Erect glandular-pubescent annual 0.5–1.5 dm. high, simple or branched; lvs. few, oblanceolate, subentire, 1–2.5 cm. long, the lowest subpetiolate; pedicels 10–25 mm. long; calyx oblong-campanulate, 8–10 mm. long, glandular, often dotted red, inflated in fr., corky-ribbed, the lobes 1–2 mm. long, equal; corolla red-purple with dark red blotches down the lobes and 2 broad yellow patches dotted with red below the lower lip, 1.5–2 cm. long, the funnelform throat with clavate hairs on the ridges, the limb with subequal lobes, usually emarginate; anthers pubescent; caps. 6 mm. long, not dehiscing through the septum-apex; seeds oval, favose-reticulate.—Wet gravelly places, 4000–5000 ft.; Yellow Pine F.; w. base of Sierra Nevada, Tuolumne and Mariposa cos. May–July.

11. **M. filicaùlis** Wats. Glandular-puberulent annual, 0.4–0.9 dm. tall; lvs. elliptic to oblong, few, 0.8–0.9 cm. long, entire; pedicels 10–15 mm. long; calyx 5–6 mm. long, narrow-campanulate, the teeth triangular-acute, 1 much smaller than others; corolla ca. 2 cm. long, rose, the yellow tube exserted, the throat broad, with yellow ventral ridges and purple spots, the lobes subequal, suberect, emarginate; anthers finely hairy.—Moist places; Montane Coniferous F.; Sierra Nevada, Mariposa Co. May–June.

12. **M. díscolor** Grant. Glandular-pubescent annual 0.3–1 dm. tall, simple or branched; lvs. sublinear to linear-oblanceolate, sessile, 0.5–2 cm. long; pedicels 8–20 mm. long; calyx campanulate, 6–9 mm. long, weakly angled, the deltoid teeth subequal, 1 mm. long, ciliate; corolla 1.3–2 cm. long, yellow, ± tinged or dotted red or red-purple, the slender tube exserted, the throat funnelform with dense clavate hairs on the ventral ridges, the upper lobes rounded, subentire, the lower longer, entire to emarginate, somewhat spreading; anthers obscurely pubescent; caps. 5–7 mm. long, tardily dehiscent through the septum-apex.—Moist places, 3000–9000 ft.; Montane Coniferous F.; Sierra Nevada from Tuolumne Co. to Kern Co. June–July.

13. **M. Pálmeri** Gray. Glandular-pubescent annual, the stem erect, usually branched, 0.6–1.5 dm. tall; lvs. few, linear to oblanceolate, 1.2–1.7 cm. long, subentire, all but the lowest sessile; pedicels 1–2.5 cm. long; calyx cylindrical, 7–8 mm. long, slightly angled, the lobes to 1 mm. long, acute, ciliolate; corolla red-purple, funnelform, 1.5–2 cm. long, the ventral ridges yellow, pubescent, the lobes subequal, rounded, emarginate;

anthers finely pubescent; caps. ca. 5 mm. long, dehiscing slightly through the septum-apex; seeds oblong, favose-reticulate, yellow-brown, 0.3–0.4 mm. long.—Mostly rather damp places, below 7000 ft.; Yellow Pine F., Chaparral, Joshua Tree Wd.; San Bernardino Mts. to Sierra Nevada of Mariposa Co. and edge of Mojave Desert. April–June.

14. **M. Dúdleyi** Grant. Viscid-villous annual, simple to much-branched, the stems 1–1.5 dm. long; lvs. thin, broadly ovate, sharply dentate, 1–3 cm. long, truncate or cordate at base, on petioles up to length of blades; pedicels 2–3 cm. long in age; calyx 6–8 mm. long, wing-angled, usually tinged red, the teeth ciliolate, 1.5–2 mm. long; corolla yellow, ± bilabiate, 1.5–2.2 cm. long, the throat narrow-campanulate, with ventral ridges clavate-pubescent; lobes spreading, rounded, unequal, sparsely pilose; anthers glabrous; caps. 6–7 mm. long, not dehiscing through septum-apex; seeds oblong, slightly papillate, yellowish, ca. 0.3 mm. long.—About rocks, below 2000 ft.; Foothill Wd.; w. base of Sierra Nevada in Madera, Tulare and Kern cos. March–April.

15. **M. floribúndus** Dougl. ex Lindl. [*M. geniculatus* Greene. *M. membranaceus* A. Nels. *M. multiflorus* and *trisulcatus* Penn.] Viscid-villous somewhat slimy annual, much-branched, erect to decumbent, the stems 1–5 dm. long; lvs. thin, ovate to lance-ovate, scattered, 1.5–4 cm. long, mostly sharply dentate, subpalmately veined, rounded or cordate at base, with petioles to almost as long as blades; pedicels 0.5–2.5 cm. long; calyx plicate-carinate, 5–8 mm. long, or more in fr., the lobes 1–1.5 mm. long; corolla yellow, 0.7–1.5 cm. long, the throat narrow-campanulate, the ventral ridges pubescent with clavate hairs and with reddish spots, the lobes unequal, mostly erect, rounded; anthers glabrous; caps. 5–6 mm. long, not dehiscing through the septum-apex; seeds round-ovoid, ca. 0.3 mm. long, yellowish.—Moist places, mostly below 8000 ft.; many Communities; most of cismontane Calif.; to B.C., S. Dak., Chihuahua. April–Aug.

Var. **subulàtus** Grant. [*M. s.* Penn.] Calyx-lobes 2–3 mm. long, ± spreading; corolla 1.5–2 cm. long.—At 3500–6500 ft.; Yellow Pine F.; Sierra Nevada from Tuolumne Co. to Tulare Co.

16. **M. arenàrius** Grant. Glandular-pubescent annual, erect or ascending, simple or branched, the stems 0.6–1.5 dm. long; lvs. few, elliptical to subovate, rounded at base, 0.7–1.5 cm. long, all but the lowest sessile, entire to ± denticulate, 1–3-nerved from base; pedicels reddish, 1.2–2.2 cm. long; calyx 6–9 mm. long, angled, the acute teeth 1 mm. long, subequal, ciliolate; corolla bilabiate, yellow, 1.3–1.6 cm. long, the cylindric throat with 2 ventral stiffly hairy ridges, and red dots extending on to the lower lip, the lobes rounded, spreading; anthers glabrous; caps. 7 mm. long, not dehiscing through the apex; seeds papillate.—Moist places, 4000–7000 ft.; mostly Red Fir F.; w. base of Sierra Nevada from Mariposa Co. to Tulare Co. May–July.

17. **M. Pulsíferae** Gray. Glandular-puberulent annual, erect or loosely branched, 0.5–1.5 dm. tall; lvs. short-petioled, narrow-oblong, subentire, 0.8–2 cm. long; pedicels ± curved-spreading in fr., 1–2 cm. long; calyx 7–10 mm. long, ridge-angled, the acute lobes 1 mm. long, subequal; corolla yellow, 7–11 mm. long, the narrow-campanulate throat pubescent on the ventral ridges, the limb with unequal spreading rounded lobes and some red mottling or spotting on the lower median; anthers glabrous; caps. ca. 7 mm. long, not dehiscent through the apex; seeds smooth, round-ovoid, minute.—Damp places, 2500–5000 ft.; Yellow Pine F.; Mariposa Co. to Humboldt, Siskiyou, and Trinity cos.; to Wash. April–June.

18. **M. alsinoìdes** Dougl. ex Benth. Glandular-pubescent to subglabrous annual, usually branched, the stems decumbent to erect, 0.5–3 dm. long; lvs. broadly ovate, denticulate, 1–2 cm. long, on equal or longer petioles, rounded or truncate at base; pedicels 1–3 cm. long; calyx 6–8 mm. long, plicate-angled, minutely puberulent, the 3 lower teeth slightly longer and more rounded than the 2 upper; corolla yellow, 1–1.2 cm. long, ± bilabiate, the throat narrow-campanulate, ridged and pubescent on floor and with brownish spots, the lobes erose or emarginate; anthers glabrous; caps. 5–6 mm. long, not dehiscing through the apex; seeds smooth, oblong, minute.—Wet places about rocks, below 2800 ft.; Coastal Prairie, Mixed Evergreen F.; Del Norte, Humboldt, and Siskiyou cos.; to B.C. March–May.

19. **M. breviflòrus** Piper. Glandular-puberulent, slender, often much-branched below, 0.3–1.5 dm. high; lvs. elliptic-lanceolate, denticulate, palmately 3-veined, 0.5–2 cm. long, subsessile or short-petioled; pedicels 5–13 mm. long, filiform; calyx 5–8 mm. long, the lobes 0.5–1 mm. long, acute, minutely ciliate; corolla yellow, 4–5 mm. long, almost

regular, the throat tubular-cylindric, with 2 ventral ridges having short knobbed hairs and faint brown spots; caps. ca. as long as calyx, not dehiscing through septum-apex.— Moist places, 5000–7100 ft.; Yellow Pine F.; Lassen and Modoc cos.; to Wash. and Ida. May–June.

20. **M. látidens** (Gray) Greene. [*M. inconspicuus* var. *l.* Gray.] Minutely glandular-puberulent annual, the stem simple or branched, 1–2.5 dm. high; lvs. broadly ovate, the first forming a rosette of petioled dentate lvs., the cauline remote, sessile, 1–2 cm. long, subentire to denticulate; pedicels 2–3 cm. long in fr.; calyx 7–8 mm. in anthesis, 10–12 mm. in fr., plicate, constricted at the oblique orifice, the ovate lobes acute, 1–2 mm. long; corolla white or yellowish, often flushed with pink and pink-dotted, 10–11 mm. long, the throat cylindric, the lobes short, spreading, subtruncate; anthers glabrous; caps. ca. 7 mm. long, not dehiscing apically; seeds papillate, round-ovoid, minute.—Drying mud flats in heavy soil, below 2500 ft.; V. Grassland, Foothill Wd.; interior valleys, from Butte and Lake cos. to San Diego Co.; L. Calif. April–June.

21. **M. Paríshii** Greene. Stout erect densely glandular-villous annual 1–5 dm. high, simple or branched from base; lvs. ovate to lance-oblong, 1.5–4 cm. long, longitudinally 3-veined, remotely denticulate to dentate, sessile with broad base; pedicels ascending-erect, 1.5–1.8 cm. long; calyx oblong, 8–9 mm. long in fl., 10–12 mm. in fr., angled, the acute lobes 1 mm. long; corolla pinkish to lilac or white, 10–12 mm. long, with subequal erect lobes; anthers glabrous; caps. ca. 8–9 mm. long, dehiscing through the apex; seeds oblong, yellowish, ca. 0.4 mm. long, smooth.—Occasional in wet sandy places, below 7000 ft.; mostly Joshua Tree Wd., Pinyon-Juniper Wd.; desert slopes of San Gabriel and San Bernardino mts. to Little San Bernardino and New York mts. May–Aug.

22. **M. montioìdes** Gray. [*M. rubellus* var. *latiflorus* Wats.] Glandular-puberulent annual, usually densely branched, 0.2–0.6 dm. tall; lvs. linear-oblanceolate, to elliptic, entire, 0.6–1.3 cm. long, 1-veined, sessile or the lower subovate, petioled; pedicels 0.5–2.5 cm. long; calyx 5–7 mm. long, ± reddish, the equal teeth broadly ovate, acute to rounded, ca. 0.6 mm. long; corolla yellow, 1–1.7 cm. long, the throat campanulate, minutely pubescent on the ventral ridges, the spreading lobes rounded, subequal; anthers glabrous; caps. dehiscing through apex, 4–5 mm. long; seeds reticulate, minute.—Moist places, 3000–6000 ft.; Pinyon-Juniper Wd.; Death V. region, Inyo Mts., s. Sierra Nevada; w. Nev. April–July.

23. **M. rubéllus** Gray in Torr. Glandular-puberulent annual, simple or branched from base, ± reddish, 0.2–1.5 dm. tall; lvs. oblong to lanceolate, 3-veined, entire or nearly so, 0.5–2 cm. long, sessile or lowest subpetiolate; pedicels 0.7–2 cm. long; calyx 5–8 mm. long, tubular, the broadly ovate teeth ciliate, to 0.5 mm. long; corolla yellow or reddish, 7–9 mm. long, the throat narrow, puberulent on the ventral ridges, the lobes rounded, little spreading, subequal, ± emarginate; anthers glabrous; caps. 4 mm. long, not dehiscent through apex; seeds yellowish, ± oblong, ca. 0.3 mm. long.—Moist sandy places, 3000–8000 ft.; Joshua Tree Wd., Pinyon-Juniper Wd., along w. edge of Colo. and Mojave deserts, n. to White Mts.; w. Nev. to Colo., New Mex., L. Calif. April–June.

24. **M. Suksdórfii** Gray. Bushy compact glandular-puberulent annual, 0.1–0.6 dm. tall, ± reddish; lvs. oblong to oblanceolate, entire, 0.5–1.2 cm. long, subsessile; pedicels 2–7 mm. long; calyx 4–6 mm. long, cylindrical, the rounded-mucronate lobes to ca. 1 mm. long; corolla yellow, funnelform, 5–6 mm. long, the puberulent ventral ridges faintly brown-spotted, the lobes subequal, emarginate; anthers glabrous; caps. ca. 4 mm. long, not apically dehiscent; seeds oblong, yellowish, ca. 0.3 mm. long.—Moist sandy places, mostly 5000–13,000 ft.; Montane Coniferous F., Subalpine Fell-fields; San Jacinto and San Bernardino mts. through Sierra Nevada, White Mts., Grapevine Mts. (Inyo Co.), to Modoc and Siskiyou cos.; to Wash., Mont., Colo., Ariz. May–Aug.

25. **M. Brèweri** (Greene) Cov. [*Eunanus B.* Greene. *M. rubellus* var. *B.* Jeps.] Glandular-pubescent annual, simple or branched, 0.3–1.8 dm. tall; lvs. narrowly oblong-lanceolate to -oblanceolate, entire or nearly so, 0.5–3 cm. long, narrowed to sessile bases; pedicels becoming 3–10 mm. long; calyx cylindrical, 5–7 mm. long, with 5 glandular-pubescent ridges, the acute-deltoid lobes ca. 1 mm. long; corolla funnelform, pale pink to purplish, 6–10 mm. long, persistent, the narrow throat with finely pubescent yellow ventral ridges, the subequal lobes spreading, ± notched; anthers glabrous; caps. 5–6 mm. long, dehiscing through the septum-apex; seeds oblong, yellowish, ca. 0.3 mm. long.— Damp sandy places, 4000–11,000 ft.; Montane Coniferous F.; San Jacinto and San

Bernardino mts. n. through Sierra Nevada, Coast Ranges from Lake Co. n.; to B.C., Mont., Nev. June–Aug.

26. **M. androsàceus** Curran ex Greene. [*M. Palmeri* var. *a.* Gray.] Minutely glandular-puberulent annual, erect, simple or branched, 0.2–0.8 dm. tall; lvs. few, sessile, oblong-ovate to -lanceolate, entire or distally obscurely toothed, 0.3–0.7 cm. long; pedicels 2–8 mm. long; calyx reddish, 6–7 mm. long, lightly ridged, the lobes almost 1 mm. long; corolla red-purple, ca. 8 mm. long, the narrow throat yellowish or purplish, slightly purple-spotted on the yellowish ridges, the lobes rounded, equal, spreading; anthers glabrous; caps. 4 mm. long, not dehiscing through septum-apex; seeds oblong-ovoid, minute, yellowish, favose-reticulate.—Rare, gravelly and stony places, below 4000 ft.; Chaparral, Foothill Wd.; Lake Co., Santa Clara and Santa Cruz cos., Tehachapi Mts. (Kern Co.) March–June.

27. **M. barbàtus** Greene. [*M. deflexus* Wats.] Finely glandular-pubescent ± clammy annual, simple or branched, 0.2–0.8 dm. tall; lvs. linear to linear-oblong, entire, almost veinless, 0.8–1.5 cm. long, narrowed to the sessile base; pedicels 5–15 mm. long; calyx 5–6 mm. long, angled, the lobes rounded-mucronate, almost 0.5 mm. long; corolla bilabiate, 10–15 mm. long, the upper lip purple, the lower yellow, or both yellow or purple, the narrow tube ca. twice the length of the calyx, the throat short and campanulate, the lower lip spotted purple, hairy, spreading, the upper erect; anthers glabrous; caps. 4 mm. long, not dehiscing through the septum-apex; seeds favose-reticulate, short. —Damp gravelly or sandy flats, mostly 7000–11,000 ft.; Subalpine F. to Lodgepole F.; Sierra Nevada of Tulare and Inyo cos. June–Aug.

28. **M. gracílipes** Rob. Simple or branched glandular-puberulent annual, erect or ascending, 0.6–1.5 dm. tall; lvs. lanceolate to oblong, entire or denticulate, 3-veined, 0.5–1 cm. long; pedicels 15–25 mm. long; calyx 4–6 mm. long, the lobes rounded-ovate, ca. 1 mm. long, ciliate; corolla purple, 1–1.5 cm. long, the tube yellow, the throat campanulate, with 2 low yellow ridges, the lobes rounded, entire, the upper shorter than the lower; anthers glabrous; caps. not dehiscing through the apex; seeds favose-areolate, minute, oblong-ovoid.—Disturbed places like burns, at ca. 2000 ft.; region of Mormon Bar, Mariposa Co. April–May.

29. **M. purpùreus** Grant. Glandular-pubescent annual 0.3–1 dm. tall, simple or with few basal branches; lvs. lance-oblong, obtuse, obscurely 3–5-veined, subentire, 1–1.5 cm. long; pedicels 3–4.5 cm. long; calyx 6–8 mm. long, slightly angled, the lobes round-ovate, mucronulate, not ciliate, 0.5–1 mm. long; corolla red-purple, 12–15 mm. long, the narrow throat obscurely or not ridged, glabrous, the lobes round-emarginate, the upper shorter, suberect, the lower longer, spreading; anthers glabrous, red-purple; caps. ca. 6 mm. long, dehiscing slightly through the apex; seeds oblong, ovoid, minutely reticulate.—Moist sandy places, 6000–7300 ft.; Yellow Pine F.; San Bernardino Mts. May–July.

30. **M. diffùsus** Grant. [*M. Grantianus* Eastw.] Glandular-puberulent diffusely branched annual 0.5–2 dm. tall; lvs. oblong-ovate to -lanceolate, entire to deeply dentate-lobed, 1–2 cm. long, sessile or the lowermost petioled; pedicels 2.5–4.5 cm. long; calyx oblong-campanulate, 5–7 mm. long, weakly angled, the obtuse glabrous teeth 0.5–1 mm. long, not ciliate; corolla 12–15 mm. long, purple to rose-violet, the lobes subequal, emarginate, irregularly marked with yellow and purple, the throat with 2 prominent yellow ridges; anthers glabrous; caps. 6 mm. long, dehiscing through the apex; seeds reticulate, oblong, minute.—Damp sandy or gravelly places, 4000–6000 ft.; Chaparral, Yellow Pine F.; San Jacinto and Santa Ana mts. to n. L. Calif. April–June.

31. **M. exíguus** Gray. Minutely glandular-puberulent, usually diffusely openly branched, 0.3–0.7 dm. tall, ± reddish throughout; lvs. few, linear to spatulate, or elliptic, 0.3–0.6 cm. long, subentire, sessile; pedicels 15–20 mm. long; calyx campanulate, 2 mm. long, the acutish lobes ca. 0.5 mm. long, not ciliate; corolla funnelform, reddish-purple, 2–2.5 mm. long, the minute lobes scarcely opening; anthers glabrous; caps. 3 mm. long, tardily dehiscent apically; seeds minute, yellowish, faintly reticulate.—Rare, moist disturbed places, 6000–7200 ft.; Yellow Pine F.; San Bernardino Mts.; n. L. Calif. June–July.

✓ 32. **M. nudàtus** Curran ex Greene. Subglabrous annual, branched from base, erect or ascending, 1–3 dm. high, ± reddish-purple; lvs. very few, lanceolate to oblong, ± denticulate, 0.5–1.5 cm. long, on petioles as long or longer; pedicels 2.5–3 cm. long;

calyx ± campanulate, 10–12 mm. long, plicate-angled, the lower lobes short and up-curved against the uppermost which is ca. 2 mm. long; corolla 1.5–2 cm. long, yellow, with 2 hairy ridges in throat that almost close the orifice, the lower lobes deflexed, the palate red-dotted, the upper lip shorter, ascending; anthers glabrous; stigmas fimbriate; caps. 6 mm. long, not dehiscing through apex; seeds oblong, brown, pitted in longitudinal rows, minute.—Wet places in gravel and serpentine rocks; Chaparral, Foothill Wd.; e. Lake Co. May–June.

33. **M. laciniàtus** Gray. [*M. Eisenii* Kell.] Subglabrous annual 0.5–3 dm. tall, simple or few-branched; lvs. remote, oblong to oval in outline, mostly deeply pinnatifid-lobed, 0.5–2.5 cm. long, on petioles to as long, or uppermost subsessile; pedicels 1.5–4.5 cm. long; calyx oblong, plicate-angled, 0.5–1.5 cm. long, ± spotted, the lower lobes upcurved against the much longer upper tooth, with some fine hairs beneath the sinuses; corolla yellow, 0.7–1.5 cm. long, usually with a large red-brown spot on lower lip, often with smaller dots below, lower lobes deflexed-spreading; anthers glabrous; stigmas ± fimbriate; caps. oblong, short-stipitate, 6 mm. long, not dehiscent through septum-apex; seeds minute, oblong, brown, finely pitted in longitudinal rows.—Damp sandy places, 3300–8700 ft.; Yellow Pine F., Red Fir F.; Tuolumne Co. to Tulare Co. May–July.

34. **M. guttàtus** Fisch. ex DC. [*M. Langsdorfii* Donn ex Greene. *M. microphyllus* Benth. *M. equinus, clementinus, paniculatus, procerus,* and *petiolaris* Greene. *M. platycalyx* Penn. *M. g.* var. *puberulus* Grant.] Usually perennial, glabrous below the ± glandular-puberulent infl., rooting at the nodes, typically with stolons or creeping root-stocks; stems ± fistulous, stout to weak, 0.5–10 dm. tall, mostly simple; lvs. oval, rounded to denticulate or pinnatifid-dentate at base, 1–8 cm. long, the upper sessile, ± connate, the lower long-petioled; infl. ± racemose; pedicels 2–6 cm. long; calyx campanulate, glabrous to puberulent, often tinged or dotted with red, inflated and 1.5–2.5 cm. long in fr., strongly plicate-angled, the lobes acute and the lower infolding fr. so as partly to close the orifice, the upper tooth mostly less than 3 times the length of the others; corolla yellow, usually spotted red, mostly 1.5–4 cm. long, the throat nearly closed by the hairy ridges, the upper lip with reflexed margins and shorter than lower spreading lip; anthers glabrous; stigmas fimbriolate; caps. stipitate, 7–9 mm. long, not dehiscing through septum-apex; seeds brown, oblong, plump, ca. 0.5 mm. long, longitudinally striate; n = 14 (Campbell, 1950).—Exceedingly common in wet places, below 10,000 ft.; many Plant Communities; most of Calif.; to Alaska, Rocky Mts., Mex. March–Aug. Variable and difficult of analysis.

Ssp. **litoràlis** Penn. [*M. Langsdorfii* var. *grandis* Greene. *M. g.* Heller. *M. guttatus* var. *g.*] Plants usually stout, softly pubescent in infl.; lvs. pubescent; calyx pubescent, 17–30 mm. long; corolla 3–4.5 cm. long, n = 14 (Campbell, 1950).—Wet places; Coastal Strand, N. Coastal Scrub; along the coast from Santa Barbara Co. to Wash. April–Oct. A small form from sand dunes at Monterey has been named ssp. *arenicola* Penn.

Ssp. **arvénsis** (Greene) Munz. [*M. a.* Greene. *M. g.* var. *a.* Grant.] Lvs. with a small tomentose area on upper surface, roundish, 1–3 cm. long, often with several pinnules on the rather long petioles; calyx subtruncate at base and top, 10–15 mm. long, the sinuses glabrous; corolla ca. 2 cm. long.—Moist banks and fields, below 2000 ft.; Chaparral, Mixed Evergreen F.; Coast Ranges from Del Norte Co. to Santa Clara Co. April–June.

Ssp. **micránthus** (Heller) Munz. [*M. m.* Heller. *M. nasutus* var. *m.* Grant.] Lvs. sparsely pubescent; pedicels 1–3 cm. long; calyx 8–15 mm. long, the lower teeth only slightly folded over, the sinuses glabrous; corolla 5–10 mm. long.—Moist places; Chaparral, Yellow Pine F.; Siskiyou Co. to Santa Clara Co. April–June.

35. **M. glaucéscens** Greene. [*M. guttatus* var. *g.* Jeps.] Near to *M. guttatus,* but almost glabrous and glaucous, 3–6 dm. tall; upper lvs. sessile, passing into connate bracts forming round disks 1.5–2 cm. in diam.; pedicels 1–3 cm. long; calyx 1.2–1.8 cm. long in fr., the lowest teeth incurved up against the upper tooth, closing the orifice, the sinuses white-villous; corolla 2–3.5 cm. long; anthers sparsely hairy; caps. 1–1.5 cm. long.—Wet places, below 2500 ft.?; V. Grassland, Foothill Wd.; Tehama and Butte cos. March–May.

36. **M. nasùtus** Greene. [*M. luteus* var. *gracilis* Gray. *M. guttatus* var. *g.* Campb. *M. n.* var. *insignis* Grant. *M. lyratus* Benth. *M. cordatus, geniculatus, glareosus, minus-*

culus, cuspidatus, and *subreniformis* Greene. *M. Bakeri* Gand. *M. Parishii* Gand., not Greene. *M. platycalyx* and *pardalis* Penn.] Subglabrous to puberulent annual, usually branched from base, erect or ascending, 1–8 dm. tall; lf.-blades round-ovate, irregularly dentate to somewhat lobed at base, palmately veined, 0.5–6 cm. long, round or cordate at base, with long broad petioles, or uppermost sessile; infl. racemose; pedicels 0.5–3.5 cm. long, strongly recurved in fr.; calyx 1–2 cm. long, usually dark-dotted, plicate-angled, appressed-puberulent, almost closed by the very acute lower lobes being eventually upcurved against the middle upper which is ca. 3 times as long and projects forward like a long index-finger; corolla yellow, 1–2(–3) cm. long, with 2 rounded ventral hairy ridges forming a palate almost closing the orifice, usually with a red-brown blotch on lower lip and smaller dots down the throat; lower lip spreading, the upper ascending and shorter; anthers glabrous; stigmas fimbriate; caps. 5–10 mm. long, not dehiscing through the septum-apex; seeds oblong, ± apiculate, reticulate in longitudinal rows, ca. 0.4 mm. long; $n = 14$ (Campbell, 1950).—Wet sandy and gravelly places, below 7500 ft.; many Plant Communities; through most of Calif.; to B.C., Rocky Mts., n. Mex. March–Aug. Variable and at many places apparently hybridizing with *M. guttatus.*

37. **M. Tilíngii** Regel. [*M. implexus* and *lucens* Greene.] Perennial with mass of yellowish slender rootstocks, the stems glabrous or nearly so, ± branched, 2–4 dm. high, commonly stoloniferous; lvs. few, light green, often slimy, broadly ovate to subelliptic, usually saliently dentate, 1–3 cm. long, the lower short-petioled; the upper sessile, not usually connate; fls. few; pedicels 2.5–5 cm. long; calyx broadly campanulate, 1.5–2 cm. long in fr., often dotted red, strongly plicate-angled, the lowermost pair of teeth eventually curved up against the median upper which is 3–5 mm. long; corolla yellow, 2.5–3.5 cm. long, the orifice almost closed by the brown-spotted ridges; upper lip ascending-erect, the lower deflexed-spreading; anthers glabrous; stigmas slightly fimbriate; caps. short-stipitate, 7–8 mm. long, not dehiscing through the septum-apex; seeds oblong, brown, somewhat apiculate, ca. 0.35 mm. long, reticulate; $n = 14$ (Vickery, 1955).—Wet banks, 6400–11,000 ft.; Red Fir F. to Alpine Fell-fields; San Jacinto and San Bernardino mts. through Sierra Nevada; to B.C., Rocky Mts. July–Sept. A form with pubescent usually less clammy lvs. has been called var. *corallìnus* (Greene) Grant. [*M. c.* Greene. *M. implicatus* and *minusculus* Greene.]

38. **M. Whípplei** Grant. Glandular-pubescent annual 1–1.5 dm. high, branched from base; lvs. broadly ovate, coarsely dentate, 3–5-nerved, ca. 2 cm. long, with petioles almost as long; pedicels ca. 3–4 cm. long; calyx 7–8 mm. long, slightly inflated in fr., the upper tooth almost twice the others, villous at sinuses, often spotted red; corolla ca. 2 cm. long, yellow, the orifice open, the lobes rounded, spreading; anthers glabrous; caps. stipitate, ca. 4–5 mm. long; seeds oblong, papillate.—Rocky places; Yellow Pine F.; Murphy, Calaveras Co. May.

39. **M. Cusíckii** (Greene) Piper. [*Eunanus C.* Greene. *M. Bigelovii* var. *ovatus* Gray.] Glandular-puberulent or -pubescent, aromatic, erect, 0.5–3 dm. tall; lvs. elliptic to oblanceolate, mostly acuminate or cuspidate, entire, cuneate to sessile at base, 2.5–5 cm. long; pedicels 1–3 mm. long; calyx 10–14 mm. long, narrow-campanulate, oblique, the lobes somewhat unequal, 2–3 mm. long, triangular, acuminate; corolla 2–3 cm. long, purplish-red with yellow tube, the throat with 2 yellow red-dotted ridges, the lobes rounded to truncate; caps. oblong, 11–13 mm. long, upcurved and somewhat exserted.— Dry sandy and rocky slopes, 3000–6000 ft.; Sagebrush Scrub to Yellow Pine F.; Modoc Co.; to Ore., Ida. May–Aug.

40. **M. nànus** H. & A. [*Eunanus n.* Holz. *E. Tolmiei* Benth.] Minutely glandular-pubescent annual 0.2–1.4 dm. tall, mostly densely branched; lvs. elliptic-lanceolate to oblong, entire, obscurely longitudinally veined, sessile or the lower petioled, 1–3 cm. long; pedicels 1–4 mm. long; calyx tubular-campanulate, 7–8 mm. long, the lobes puberulent, subequal, ca. 2 mm. long, triangular-acute; corolla purple, rarely yellow, 1.3–2 cm. long, pubescent and with white or yellow patches on the floor of the throat dotted purple, the lobes spreading, the upper lip longer than lower; anthers pubescent, the longer scarcely exserted; stigmas exserted, glandular-ciliolate; caps. slightly exserted, dehiscing throughout; seeds oblong, plump, yellowish, ca. 0.5 mm. long.—Dry gravelly or sandy places, 4500–7500 ft.; Chaparral, N. Juniper Wd., Yellow Pine F., Red Fir F.; Tehama and Lassen cos. to Wash., Nev.. Rocky Mts. May–July. Modoc Co. plants with yellow corollas have been called *M. Austìnae* (Greene) Grant. [*Eunanus A.* Greene.]

41. **M. Jepsònii** Grant. [*M. microcarpus* Penn.] Minutely glandular-pubescent annual, erect, 0.2–1 dm. tall; lvs. not crowded, oblong- to narrow-oblanceolate, entire, attenuate at base, 1–1.5 cm. long; pedicels 1–2 mm. long; calyx 4–5 mm. long, ridge-angled, the acuminate glandular-puberulent lobes 1.5–2 mm. long; corolla purple, 9–11 mm. long, marcescent, with 2 inner yellow ridges, upper lip erect and longer than lower; anthers glabrous, the upper subexserted; stigmas ciliolate; caps. usually exserted; seeds plump, slightly apiculate and reticulate.—Not common, bare disturbed soil, 4000–9000 ft.; Montane Coniferous F.; Nevada Co. to Modoc and Siskiyou cos.; s. Ore. June–Aug.

42. **M. mephíticus** Greene. [*Eunanus m.* Greene.] Glandular-pubescent annual, simple or branched, with strong skunky odor, erect, 0.2–1.2 dm. tall; lvs. oblanceolate to sublinear, entire, longitudinally veined, 1–2 cm. long, subsessile; pedicels 1–5 mm. long; calyx open-campanulate, 5–7 mm. long, ridge-angled, scarious below the sinuses, the lanceolate lobes ca. 2 mm. long; corolla purplish-red to yellow, 1.2–1.7 cm. long, the tube slender, the throat loosely pilose within, ridged and brown-spotted, the lobes rounded, subequal; anthers pubescent, mostly included; caps. ± exserted, dehiscing throughout; seeds oblong, apiculate, reticulate, ca. 0.5 mm. long.—Common in open sandy places, 5000–9000 ft.; Sagebrush Scrub, Montane Coniferous F.; Sierra Nevada from Plumas Co. to Tulare Co., Masonic Mts., Mono Co. June–Aug.

43. **M. dénsus** Grant. Glandular-pubescent annual, densely branched, 0.5–1.5 dm. tall; lvs. lanceolate to oblanceolate, 1–1.7 cm. long, entire, sessile; pedicels 1–4 mm. long; calyx 7–9 mm. long in fr., slightly angled, the ± unequal lobes lanceolate, 2.5–3.5 mm. long; corolla 1.5–2 cm. long, yellow and streaked with red, to red-purple, pubescent within and shallowly ridged and ± spotted; anthers pubescent, scarcely exserted; stigmas ciliate, subequal, ± exserted; caps. 7–10 mm. long, dehiscing throughout; seeds oblong, ± apiculate, reticulate, 0.4–0.6 mm. long.—Dry gravelly or sandy places, 4000–9000 ft.; Sagebrush Scrub, Pinyon-Juniper Wd., Yellow Pine F., Red Fir F.; e. slope of Sierra Nevada from Lassen Co. to Inyo Co., White Mts.; w. Nev. May–July.

44. **M. coccíneus** Congd. [*M. stamineus* Grant. *M. Wolfii* Eastw.] Strong-scented glandular-pubescent annual 0.1–1.8 dm. tall, simple or branched; lvs. oblanceolate to sublinear, 0.6–1 cm. long, entire, sessile; pedicels 1–3 mm. long; calyx 4–8 mm. long in fr., ridge-angled, the linear unequal teeth 1.5–3 mm. long; corolla funnelform, red-purple, 1.2–2 cm. long, with 2 ventral yellow pubescent ridges and dark spots, the lobes rounded; anthers finely pubescent; stigmas ciliolate, the lower somewhat larger; caps. 7–8 mm. long, dehiscing throughout; seeds oblong, apiculate, reticulate, ca. 0.5 mm. long.—Dry gravelly places, mostly 8000–12,000 ft.; Lodgepole F. to Alpine Fellfields; Sierra Nevada from Tulare and Inyo cos. to Alpine Co., White Mts.; w. Nev. June–Aug.

45. **M. Johnstònii** Grant. Glandular-pubescent annual, openly branched, 1–2.5 dm. tall; lvs. obovate to oblanceolate, entire, 1–2 cm. long, the lower petioled; pedicels 1–3 mm. long; calyx 8–9 mm. long in fr., angled, the acuminate lobes unequal, the upper 2–4 mm. long; corolla rose-purple, 1.3–2 cm. long, greenish-yellow and purple-spotted within, the lobes rounded; anthers pubescent; stigmas ciliolate; caps. ca. as long as longest calyx-lobe, dehiscing throughout; seeds oblong, apiculate, reticulate, ca. 0.6–0.8 mm. long.—Dry gravelly slopes, 4000–7000 ft.; mostly Yellow Pine F.; San Gabriel Mts., occasional at lower elevs., to Kern and Santa Barbara cos., San Diego Co. May–July.

46. **M. Whítneyi** Gray. [*Eunanus bicolor* Gray, not *M. b.* Hartw. *M. nanus* var. *b.* Gray.] Finely glandular-pubescent annual 0.1–0.6 dm. tall; lvs. few, linear to elliptical, 1–2 cm. long, entire, sessile; pedicels 1–2 mm. long; calyx 5–6 mm. long in fr., faintly ridged, the acuminate lobes unequal, 1–2 mm. long; corolla funnelform, 1–1.5 cm. long, pale yellow with maroon blotches and lines, or purple with similar paler areas, hairy within and with low yellow ridges, the lobes rounded; anthers pubescent; stigmas ciliate; caps. 5–6 mm. long, dehiscing throughout; seeds oblong, apiculate.—Rare, gravelly places, 6000–11,000 ft.; Montane Coniferous F.; Sierra Nevada of Fresno and Tulare cos. June–Aug.

47. **M. leptàleus** Gray. [*Eunanus l.* Greene.] Glandular-pubescent annual 0.1–0.9 dm. tall, simple or branched; lvs. obovate to sublinear, entire, 0.5–1.5 cm. long, the lower short-petioled; pedicels 1–2 mm. long; calyx slender, 3–4 mm. long, scarcely ridged, the throat slightly oblique, the teeth lance-attenuate, ca. 1 mm. long; corolla

slender, 6–9 mm. long, red-purple, ventrally yellow, glabrous internally with dark purple lines and spots, sometimes the lobes yellow; anthers glabrous; stigma included, the lobes unequal; caps. ca. 5 mm. long, dehiscing throughout; seeds oblong-ovoid, plump, apiculate, reticulate, ca. 0.4 mm. long.—Open gravelly places, 7000–11,100 ft.; Montane Coniferous F.; Sierra Nevada from Inyo and Tulare cos. to Plumas Co. June–Aug.

48. **M. Fremóntii** (Benth.) Gray. [*Eunanus F.* Benth.] Glandular-pubescent annual, ± reddish throughout, usually freely branched, 0.4–2 dm. high; lvs. oblanceolate to oblong, 1–3 cm. long, entire or nearly so, sessile; pedicels 2–4 mm. long; calyx campanulate, 8–10 mm. long, strongly angled, the throat oblique, the subequal triangular lobes ca. 2 mm. long; corolla broadly funnelform, 2–2.5 cm. long, rose-purple, pubescent externally, glabrous within except on the ventral yellow ridges, the lobes truncate-rounded, ± unequal; anthers included, glabrous; stigmas equal, ciliolate; caps. ca. as long as calyx, dehiscing throughout; seeds oblong, reticulate, apiculate, ca. 0.5–0.7 mm. long.—Dry disturbed places, such as burns, etc., below 7000 ft.; Chaparral, Foothill Wd., Joshua Tree Wd., Yellow Pine F., etc.; edge of deserts through cismontane s. Calif. to inner S. Coast Ranges of San Benito Co.; L. Calif. April–June.

49. **M. subsecúndus** Gray. [*Eunanus s.* Greene.] Glandular-pubescent annual, with nonglandular hairs also, branched, erect, 0.5–2 dm. tall; lvs. sessile, oblong or oblanceolate, mostly entire, 0.8–2.5 cm. long; pedicels 2 mm. long; calyx broadly ovoid, 8–10 mm. long, ± constricted at the throat, green, wing-ridged with pale intervening surface, the lobes unequal, deltoid, 2–3 mm. long; corolla red-purple, externally pubescent, 1.5–1.7 cm. long, internally glabrous, with 2 yellow ridges in the ventricose throat, the lobes subequal, rounded; anthers glabrous; stigmas slightly unequal; caps. 7 mm. long, dehiscing throughout; seeds oblong, tuberculate, apiculate.—Occasional in dry places, ca. 3000 ft.; Yellow Pine F.; Santa Lucia Mts. of Monterey and San Luis Obispo cos. May–July.

50. **M. víscidus** Congd. [*M. subsecundus* var. *v.* Grant. *M. Fremontii* var. *v.* Jeps.] Much like *M. subsecundus,* but more nearly erect, more coarsely and heavily glandular-pubescent; fls. scattered along stems; calyx 9–12 mm. long, with a narrow pale line between the green ridges, the lobes acute to acuminate; corolla 2–2.5 cm. long, internally pubescent on the ventral ridges; anthers ciliate; caps. 8–9 mm. long.—Dry hot places, 2000–6000 ft.; Foothill Wd., Yellow Pine F., Chaparral; Sierran foothills from Calaveras Co. to Tulare Co. May–July.

Ssp. **constríctus** (Grant) Munz. [*M. subsecundus* var. *c.* Grant. *M. c.* Penn.] With glandless hairs between the gland-tipped; fls. crowded near branch-tips; calyx-lobes caudate-attenuate, subulate.—At 3000–6000 ft.; Greenhorn Mts. (Kern Co.) to s. Sierra Nevada (Tulare Co.). May–July.

51. **M. Làyneae** (Greene) Jeps. [*Eunanus L.* Greene.] Densely glandular-pubescent annual, erect, 1–2 dm. tall, simple or branched, with strong odor; lvs. narrow-oblong to linear, entire or nearly so, 1–2.5 cm. long, with tapered almost sessile base; pedicels 1–2 mm. long; calyx 5–6 mm. long, somewhat inflated and strongly ribbed in age, the triangular teeth spreading, 1–2 mm. long, subequal; corolla rose to red-purple, 1.3–2 cm. long, pubescent without, internally with ± pubescent ridges and distally white with purple spots, the lobes rounded and with a purple median line; anthers ciliate; stigmas fimbriolate, ± unequal; caps. 7–9 mm. long, dehiscing throughout; seeds oblong, apiculate, 0.4–0.5 mm. long.—Dry sandy and disturbed places, below 7500 ft.; many Plant Communities; Coast Ranges from Napa Co. n. and Sierra Nevada from Fresno Co. n., to Humboldt and Siskiyou cos. May–Aug.

52. **M. Bigelòvii** (Gray) Gray. [*Eunanus B.* Gray.] Densely glandular-pubescent annual, simple or branched, 0.5–2.5 dm. tall, erect; lvs. scattered, elliptic to obovate, ± acute, subentire, 1–3 cm. long, sessile or the lower short-petioled; fls. mostly clustered near tips of stems; pedicels 1.5–4 mm. long; calyx broadly oblong, 8–11 mm. long, wing-ridged, space between the glandular ridges pale; lobes attenuate, unequal, 3–4 mm. long; corolla red-purple, 2–3 cm. long, funnelform, externally finely pubescent, internally pubescent on the rounded lobes of the subrotate limb, the throat pale yellow below with purple dots; anthers ± pubescent; stigmas ciliolate, subequal; caps. 9–10 mm. long, dehiscing throughout; seeds oblong, apiculate, ca. 0.5–0.6 mm. long.—Common in dry sandy or gravelly washes and canyons, below 6500 (9000) ft.; Creosote Bush

Scrub, Joshua Tree Wd., Pinyon-Juniper Wd., Sagebrush Scrub, etc.; Mojave and Colo. deserts and bordering mts. n. to Mono Co.; Nev., w. Ariz. March–June.

Var. **cuspidàtus** Grant. Stout, densely viscid-villous; upper lvs. almost round, abruptly cuspidate; pedicels 3–8 mm. long; calyx 8–10 mm. long; corolla 1.2–2 cm. long, the ventral ridges more strongly constricting the orifice.—Dry rocky places, below 7000 ft.; Barstow and Baker, Mojave Desert, to Death V. region and White Mts. March–May, Sept.–Oct.

Var. **panaminténsis** Munz. Lvs. as in the sp.; calyx 6–8 mm. long; corolla 15–18 mm. long; caps. slender, curved, 10–12 mm. long.—Dry rocky places, 8000–10,000 ft.; Panamint Mts. July–Aug.

53. **M. Tórreyi** Gray. [*Eunanus T.* Greene.] Glandular-pubescent annual, erect, simple or branched, 0.5–3.5 dm. tall; lvs. obovate to narrow-elliptic or oblong, 1.5–3.5 cm. long, entire or nearly so, pinnate-veined, sessile; pedicels 2–3 mm. long; calyx 7–8 mm. long, weakly angled, scarious at base and below sinuses between the narrow green ridges, the lobes unequal, obtuse, 1 mm. long; corolla bilabiate, 1.5–2.2 cm. long, rose-red to -purple, externally pubescent, internally slightly pilose ca. the orifice, the throat ventricose, yellow-ridged, pubescent, margined by dark purple, the lobes rounded, the upper lip shorter; anthers glabrous; stigmas rounded, the upper shorter; caps. 7–8 mm. long, dehiscing throughout; seeds plump, oblong, ca. 0.7 mm. long.—Dry disturbed places, 1500–8000 ft.; Foothill Wd. to Lodgepole F.; w. slope of Sierra Nevada, Plumas Co. to Kern and Mono cos. May–Aug.

54. **M. Rattánii** Gray. [*Eunanus R.* Greene.] Pubescent annual with reflexed glandless hairs and longer spreading gland-tipped ones, simple or branched, 0.4–1.5 dm. tall; lvs. few, sessile, oblong to oblanceolate, 0.8–2 cm. long, slightly dentate in upper half; pedicels 1–2 mm. long; calyx 6–8 mm. long, the lobes triangular, mostly obtuse; corolla red-purple, 8–10 mm. long, finely pubescent without, glabrous within, the lobes rounded; anthers glabrous; stigma-lobes unequal, the lower short; caps. 11 mm. long, much exserted, dehiscing throughout; seeds oblong-oval, apiculate.—Largely on Chaparral burns; Lake, Colusa, and Marin cos. May–July.

Ssp. **decurtàtus** (Grant) Penn. [*M. d.* Grant.] Calyx-lobes blunter, at least the lateral rounded.—Gravelly places; Chaparral, Yellow Pine F.; Santa Cruz Co. June–July.

55. **M. brachiàtus** Penn. Much like *M. Rattanii*, but with more spreading branches and all hairs on stems gland-tipped; calyx 8–9 mm. long, the green ridges contrasting with the white intervening surface, the lobes acute-acuminate, 1–2 mm. long; corolla 10–13 mm. long, pubescent internally and externally; caps. 8 mm. long.—Gravelly serpentine areas; Lake Co. May–July.

56. **M. Bolánderi** Gray. [*Eunanus B.* Greene. *M. B.* var. *brachydontus* Grant.] Densely glandular-pubescent annual, erect, simple or branched, viscid, with strong tobaccolike odor, 2–6 dm. tall; lvs. oblong to obovate, often serrulate toward tips, acute, 2–6 cm. long, cuneate at base, sessile; pedicels 2–3 mm. long; calyx sharply angled, 1.5–1.7(–2.5) cm. long in fr., the ridges dark, glandular, with whitish intervening spaces, the lobes unequal, lanceolate, the uppermost 2–4 mm. longer than the lower pair; corolla bilabiate, 2–2.5(–4) cm. long, pink to red-purple, finely pubescent without, glandular-puberulent within where white with purple spots, the lobes rounded; anthers glabrous; stigmas unequal, the upper somewhat shorter than the lanceolate lower; caps. 12–15 mm. long, dehiscing throughout; seeds oblong, reticulate, apiculate, ca. 0.4–0.6 mm. long.—Dry open places such as burns, below 6500 ft.; Chaparral, Foothill Wd., Yellow Pine F., etc.; Coast Ranges from Santa Barbara Co. to Mendocino Co., Sierra Nevada from Calaveras Co. to Fresno Co., Tehachapi Mts. May–July. A form from Fresno Co. to Amador Co., with calyx 20–25 mm. long in fr. and corolla 25–30 mm. long, has been called *M. platylaèmus* Penn.

57. **M. brévipes** Benth. [*Eunanus b.* Greene.] Viscid-pubescent annual, erect, simple or branched, 1–8 dm. tall; basal lvs. obovate, 5–8 cm. long, on slender petioles, the cauline remote, linear to lanceolate, 2–6 cm. long, entire to denticulate, sessile or nearly so, ± acuminate; pedicels 2–6(–10) cm. long; calyx 2–2.5 cm. long, broadly tubular-campanulate, glandular-pubescent, the ridges green in contrast with white intervening spaces, the lobes lanceolate, acuminate, the uppermost 9–12 mm. long, the lower 3–5 mm.; corolla yellow, bilabiate, 2–5 cm. long, glabrous without, pubescent within on the ridges and base of lower lip, the lobes rounded; anthers glabrous; stigmas ciliolate,

the lobes unequal; caps. 9–13 mm. long, dehiscing throughout; seeds oblong, apiculate, 0.5–0.7 mm. long.—Dry exposed places, mostly below 3500 (5100) ft.; Chaparral, Coastal Sage Scrub, Yellow Pine F.; cismontane s. Calif. from Santa Barbara Co. to L. Calif. April–June.

58. **M. mohavénsis** Lemmon. [*Eunanus m.* Greene.] Puberulent, ± glandular, leafy, simple to many-branched annual, reddish-purple, 0.3–0.8(–1.5) dm. tall; lvs. elliptical to oblong or obovate, 1–1.8 cm. long, acute, entire, sessile or the lower petioled; pedicels 2–3 mm. long; calyx 10–12 mm. long in fr., angled, broadly campanulate, the teeth spreading, ciliate, the upper 3–4 mm. long, the lowest 2–2.5 mm.; corolla salverform, 13–15 mm. long, red-purple in the tube and center of limb, the margin pale, ciliate or erose, internally glabrous except ventrally hirsute, the lobes rounded-truncate; anthers glabrous; stigmas rounded, the lower slightly larger; caps. 9–13 mm. long, dehiscing throughout; seeds narrow-oblong, apiculate, ca. 0.5 mm. long.—Uncommon, sandy and gravelly places, 2000–3000 ft.; Creosote Bush Scrub, Joshua Tree Wd.; Mojave Desert in the Barstow–Victorville–Ord Mts. region. April–June.

59. **M. pygmaèus** Grant. [*M. minutissimus* Eastw.] Glandular-hairy almost acaulescent annual; lvs. linear to spatulate, 0.4–0.8 cm. long; pedicels to 1 mm. long; calyx 4–5 mm. long, cylindrical, pubescent, the lobes linear-lanceolate, glandular-pilose, the uppermost 1.5–2 times as long as the lower pair; corolla pink to yellowish or purplish, funnelform, 6–7 mm. long, the lobes rounded, short; anthers glabrous, those of lower pair of stamens lacking or reduced; stigmas unequal, the lower longer; caps. ca. 4 mm. long; seeds oblong.—Egg Lake, Modoc Co. June.

60. **M. píctus** (Curran) Gray. [*Eunanus p.* Curran.] Viscid-pubescent annual, simple or branched, 1–3 dm. high; lvs. elliptic-oblong to obovate, 1–3 cm. long, slightly serrulate, obtuse, sessile or the lowermost petioled; pedicels 2–3 mm. long; calyx cylindrical, accrescent, 11–12 mm. long in fr., glandular-villous; gibbous, the triangular lobes subequal; corolla regular, white with rose-red to red-purple veining, 9–12 mm. long; anthers glabrous; stigmas unequal, the upper shorter and fimbriolate; caps. 10–12 mm. long, dehiscing throughout; seeds oblong, apiculate, reticulate, ca. 0.3 mm. long.—Dry slopes, 1000–4000 ft.; largely Foothill Wd.; Tehachapi Mts., Greenhorn Mts., Kern Co. April–May.

61. **M. Douglásii** (Benth. in DC.) Gray. [*Eunanus D.* Benth. in DC. *M. nanus* var. *subuniflorus* H. & A. *M. s.* Jeps. *M. atropurpureus* Kell.] Subcaulescent or to 0.6 dm. high, annual, with recurved fine glandless hairs and some longer glandless or gland-tipped; lvs. narrow-elliptic, obtuse, crenulate, longitudinally veined, 1.5–3 cm. long, short-petioled; pedicels 1–2 mm. long; calyx 10–12 mm. long, plicate-ridged, the ridges ± pilose, the intervening surface pale and glabrous or pilose, the lobes lanceolate-triangular, ciliate, the uppermost ca. twice as long as lowermost; corolla 3–4 cm. long, purple, subglabrous externally, the tube ca. twice as long as calyx, the throat dark purple, ventrally streaked and with the ridges pale, the upper lip 7–9 mm. long, erect-arched, the lower rudimentary or lacking; anthers glabrous; stigmas unequal; caps. 6 mm. long, slightly decurved, cartilaginous, not or tardily dehiscent; seeds oblong, apiculate.—Open gravelly places moist in the spring, often on serpentine, below 4000 ft.; Chaparral, Foothill Wd., etc.; Coast Ranges from San Benito Co. n., foothills of Sierra Nevada from Tulare Co. n.; to Ore. March–May. Plants with corollas ± cleistogamous, 2–8 mm. long, from San Benito, San Mateo, and Napa cos., are *M. cleistógamus* J. T. Howell.

62. **M. rupícola** Cov. & Grant. Finely pubescent annual with some gland-tipped hairs, branched from base, tufted, 0.3–1.5 dm. tall; lvs. rhombic-ovate to elliptic-oblanceolate, acute at both ends, entire, sessile or short-petioled, 2–7 cm. long, pinnately veined; pedicels 3–4 mm. long; calyx 14–16 mm. long in fr., distally plicate-ridged, the lance-attenuate lobes glandular-pilose, the uppermost 5 mm. long, the lower 3 mm.; corolla rose, 2.5–3 cm. long, puberulent externally, the tube ca. as long as calyx, glandular-puberulent within the throat and on low ventral ridges and yellow with red-brown spots, the lobes rounded-truncate, each with a purple basal spot; anthers glabrous; stigmas equal; caps. 3–4 mm. long, slightly decurved, tardily dehiscent; seeds ovoid, muriculate, apiculate, ca. 1 mm. long.—Crevices in limestone, below 5000 ft.; Creosote Bush Scrub; mts. e. side of Death V., Inyo Co. March–May.

63. **M. trìcolor** Hartw. ex Lindl. [*Eunanus t.* Greene. *E. Coulteri* Harv. & Gray.] Glandular-pubescent annual, branched from base, 0.1–1.2 dm. tall; lvs. oblanceolate,

obtuse, 1.5–2.5 cm. long, glandular-ciliate, entire to obscurely denticulate, subsessile; pedicels 2–3 mm. long; calyx in fr. 12–14 mm. long, the ridges green, pubescent, the intervening surface pale, the lobes oblong-ovate, obtuse, glandular-puberulent, the longest 3 mm. long, the shortest 1 mm.; corolla 3–4 cm. long, the tube slender, yellow, ca. twice as long as calyx, the throat dorsally dark purple, ventrally white with a yellow patch with dark purple spots, the lobes purple and with a darker median spot; anthers pubescent; stigmas subequal; caps. 7–8 mm. long, tardily if at all dehiscent; seeds oblong, apiculate, ca. 0.4 mm. long.—Drying vernal pools and similar places, below 2000 ft.; V. Grassland, Foothill Wd., etc.; Cent. V. from Kern Co. n., valleys in the Coast Ranges from Sonoma Co. n.; to Ore. April–June.

64. **M. pulchéllus** (E. Drew ex Greene) Grant. [*Eunanus p.* E. Drew ex Greene.] Tufted annual, villous in upper parts, the stems 0.1–0.3 dm. tall; lvs. narrow-oblanceolate, 1–2 cm. long, mostly entire, sessile or petioled; pedicels 1–2 mm. long; calyx 7–10 mm. long in fr., pale, glabrescent between the villous ridges, the lobes obtuse, unequal, the uppermost longest; corolla 2–5 cm. long, ± pilose without, the yellowish tube 2.5–4 times as long as calyx, the narrow-campanulate throat purple except for the ventral yellow part with maroon dots, the lobes rounded, the upper shorter and purple, the lateral purple or yellow, the lowermost yellow; anthers pubescent; stigmas ciliate, subequal; caps. 4–5 mm. long, broadly ovoid, the lower thinner part breaking away from the base; seeds somewhat angled, pitted, 0.4–0.5 mm. long.—Moist places, 3000–5000 ft.; Yellow Pine F.; Sierra Nevada of Calaveras, Tuolumne, and Mariposa cos. May–July.

65. **M. angustàtus** (Gray) Gray. [*Eunanus Coulteri* var. *a.* Gray. *E. a.* Greene. *M. Clarkii* Kell. ex Curran.] Tufted annual, villous and puberulent, the stems scarcely 0.1 dm. high; lvs. linear-oblanceolate, entire, sessile, 1–2(–3) cm. long; pedicels 1–2 mm. long; calyx ca. 10 mm. long, the tube narrow, slightly ridged, the lobes ovate-lanceolate, subequal, 3–3.5 mm. long; corolla 3–4.5 cm. long, glabrous without, the yellowish tube 3–6 times as long as calyx, the throat ± campanulate, purple except paler and dark-dotted below, the lobes purple, each with a darker median spot; anthers pubescent; stigmas subequal; caps. ca. 3 mm. long, broadly ovoid, thinner at base, tardily dehiscent; seeds oblong, pitted.—Dried vernal pools, etc., below 4000 ft.?; Yellow Pine F., Chaparral; Sierran foothills from Fresno Co. to Butte Co., Coast Ranges from Napa and Lake cos. to Mendocino Co. April–June.

66. **M. Kellóggii** (Curran ex Greene) Curran ex Gray. [*Eunanus K.* Curran ex Greene.] Pubescent caulescent annual, with some hairs gland-tipped; stems erect, simple or branched, 0.3–3 dm. tall; lvs. oblanceolate to rhombic-ovate, obtuse, entire or nearly so, 1–5 cm. long, sessile or short-petioled; pedicels 3–5 mm. long; calyx 12–15 mm. long, plicately ridged, glandular-pubescent, the intervening surface as wide as the ridges, the lobes 1–2 mm. long, obtuse; corolla 3–4.5 cm. long, rose-purple, glabrous throughout, the tube 2–2.5 times as long as calyx, white below, the throat short, purple above, yellow ventrally with red dots, the lobes rounded, the upper ca. 7 mm. long, the lower half as long; anthers ciliate; stigmas ciliate, the upper short; caps. 8–10 mm. long, slightly arcuate, tardily dehiscent; seeds oblong, apiculate, scurfy-muriculate, ca. 1 mm. long.— Dampish disturbed places, below 3000 ft.; Yellow Pine F., N. Oak Wd., Foothill Wd.; Coast Ranges from Napa and Lake cos. to Humboldt and Trinity cos., foothills of the Sierra Nevada from Kern Co. to Shasta Co. March–June.

67. **M. Tráskiae** Grant. Glandular-pubescent annual 1–1.4 dm. tall; lvs. few, broadly ovate, subglabrous, 3–4 cm. long, the lower short-petioled; pedicels 3–4 mm. long; calyx 17–20 mm. long, glandular-pubescent, the glandular-pubescent ridges green, the intervening space pale, the lobes lanceolate, very unequal, the uppermost 3–3.5 mm. long; corolla red-purple and white, 2–3.5 cm. long, glabrous, the tube ca. as long as calyx, the 2 upper lobes white, 5–6 mm. long, the lower purple, scarcely 1 mm. long; caps. ca. 7–8 mm. long, tardily or not dehiscent.—Santa Catalina Id. March.

68. **M. Congdònii** Rob. [*Eunanus C.* Greene. *E. Douglasii* var. *parviflorus* Greene. *M. modestus* Eastw.] Annual, acaulescent to 0.6–0.8 dm. tall, coarsely pubescent, some of the hairs gland-tipped; lvs. obovate, 1.5–3.5 cm. long, subentire to bluntly dentate, cuneate at base with ± developed ciliate petioles; pedicels 2–4 mm. long; calyx 9–12 mm. long, straight, the tube pale between the green-pubescent ridges, the lobes triangular, obtuse, somewhat unequal, to 1 or 2 mm. long; corolla rose-pink or -purple, 1.5–2.5 cm.

long, sparsely hairy without, the tube slender, 1.5–2 times the length of the calyx, the throat short, the lobes rounded, the upper 3–5 mm. long, the lower 1.5–3 mm. long; anthers glabrous; stigmas unequal; caps. 5–7 mm. long, slender, tardily dehiscent; seeds slender-oblong, apiculate, scurfy-muriculate, 0.3–0.5 mm. long.—Damp grassy and disturbed places, below 3000 ft.; V. Grassland, Foothill Wd.; Coast Ranges from Mendocino Co. to Ventura Co., Sierran foothills, Mariposa Co. to Tulare Co. March–May.

69. **M. Brandègei** Penn. Glandular-pubescent annual 0.2–0.3 dm. high; lvs. narrow-elliptic to spatulate-oval, obtuse, entire, short-petioled; pedicels 1–2 mm. long; calyx 10–12 mm. long, scarcely ridged, the lobes obtuse, ± unequal; corolla purple, 13–15 mm. long, the tube equal to the calyx, the throat funnelform, the lobes rounded, the 2 upper 3–4 mm. long, the lower ca. 1 mm.; anthers glabrous; caps. 7 mm. long, tardily or not dehiscent.—Santa Cruz Id.

70. **M. pilòsus** (Benth.) Wats. [*Herpestis p.* Benth. *Mimetanthe p.* Greene. *Mimulus exilis* Dur. & Hilg.] White-villous annual, erect or ascending, slightly viscid, simple or branched, 0.5–3.5 dm. high, flowering from near the base; lvs. lanceolate to oblong, entire, 1–3 cm. long, sessile; pedicels 10–15(–30) mm. long; calyx 6–7 mm. long, oblique at orifice, short-campanulate, plane, the lobes unequal, lance-ovate, the lowermost shortest and ca. as long as calyx-tube; corolla yellow, obscurely 2-lipped, 7–8 mm. long, the lower lip usually with maroon spots; anthers glabrous; caps. 4–7 mm. long, loculicidal along whole upper side and on lower side toward apex; seeds scurfy-muriculate, oblong-ovoid, apiculate, ca. 0.5 mm. long.—Common in moist sandy and gravelly places, below 8500 ft.; many Plant Communities; cismontane and montane Calif.; to Wash., Utah, Ariz., L. Calif. April–Sept.

71. **M. Clevelándii** Bdg. [*Diplacus C.* Greene.] Erect freely branched suffrutescent perennial 3–9 dm. tall, glandular-villous; lvs. of early season lanceolate to oblong, 2.5–10 cm. long, with broad sessile base, serrate in upper half or entire, yellow-green impressed-veiny and finely hairy above, paler distinctly hairy and somewhat glutinous beneath; lvs. of late season smaller, usually fascicled in main axils; pedicels 3–4 mm. long, borne in upper axils; calyx campanulate, 20–25 mm. long, ridge-angled, the tube narrow, constricted above the ovary, the lobes lanceolate, unequal, 6–9 mm. long; corolla golden-yellow, 3.5–4 cm. long, glandular-pubescent externally, subglabrous within, ventrally 2-ridged and with rows of orange dots on lower side, the upper lobes notched to sub-entire, the lower entire to finely toothed; anthers glabrous; stigmas unequal, ciliate; caps. 10–12 mm. long, splitting at apex along 4 lines; seeds ellipsoid, apiculate, reticulate-roughened, ca. 0.7–1 mm. long; $n = 10$ (Stebbins in McMinn, 1951).—Dry disturbed places, 3000–6000 ft.; Chaparral, Yellow Pine F.; Santa Ana Mts. to s. San Diego Co. May–July.

72. **M. longiflòrus** (Nutt.) Grant. [*Diplacus l.* Nutt. *M. glutinosus* var. *brachypus* Gray. *D. arachnoideus* Greene. *D. speciosus* Davy.] Much-branched shrub, 3–12 dm. high, with upper stems, branches, under surface of lvs., pedicels, and calyces densely pubescent and glandular-hairy; lvs. lanceolate to oblong, 2.5–8 cm. long, light green, subglabrous, glandular and impressed–veiny above, finely toothed to subentire at base, often revolute, sessile; pedicels 3–7 mm. long; calyx 2.5–3.7 cm. long, the tube gradually expanded upward into a slightly wider throat that is contracted toward apex, the lobes linear-lanceolate, the uppermost 7–9 mm. long, the lower ca. half as long; corolla orange-yellow, sometimes deep orange or buff to almost white, 5–6 cm. long, the tube included and expanding gradually into a throat ca. 1.8 cm. long, the lower surface with 2 deep orange bands, the limb with dorsiventral spread of 3–4.5 cm., the upper lobes with 1–2 deeper notches, the lower 3–7-toothed; anthers included, glabrous; stigmas ciliolate, subequal; caps. ca. 15 mm. long, dehiscing along upper suture; seeds fusiform, ca. 1 mm. long, reticulate-roughened.—Common on dry rocky slopes, to ca. 5000 ft.; Chaparral, Coastal Sage Scrub; Pozo Mts., San Luis Obispo Co. through cismontane s. Calif. to cent. San Diego Co., inland to San Jacinto Mts. and Kern R. region; n. L. Calif. March–July. Hybridizing freely with other spp.

Var. **rùtilus** Grant. [*Diplacus longiflorus* var. *r.* McMinn. *D. r.* McMinn.] Lvs. dark green, villous toward base and beneath; corolla deep velvety red.—With the sp. particularly in interior Los Angeles Co., less so in Ventura and Riverside cos.

Ssp. **calycìnus** (Eastw.) Munz. [*Diplacus c.* Eastw. *M. longiflorus* var. *c.* Grant. *D. l.* var. *c.* Jeps.] Lvs. oblong-elliptic to ovate or lanceolate, yellow-green and glabrous above, glandular-pubescent beneath; calyx-tube abruptly expanded into a more inflated throat, densely pubescent to woolly; corolla light lemon-yellow, the tube usually well exserted, the limb with a dorsiventral spread of 2.5–3 cm.; $n = 10$ (Stebbins ex McMinn, 1951). —Rocky places, below 4500 ft.; Foothill Wd., Chaparral, base of Sierra Nevada from Fresno Co. to Kern Co. and in inner S. Coast Ranges of San Luis Obispo Co.; to 7500 ft., Chaparral, Yellow Pine F., Joshua Tree Wd., e. San Gabriel Mts. to Little San Bernardino Mts., San Jacinto Mts. and mts. of ne. San Diego Co. April–July.

73. **M. aurantìacus** Curt. [*Diplacus a.* Jeps. *M. glutinosus* Wendl. *D. latifolius* and *leptanthus* Nutt.] Erect profusely branched shrub 6–12(–20) dm. tall, the stems, branches and lvs. puberulent to subglabrous, but glandular and glutinous; lvs. 2.5–5(–7) cm. long, oblong-lanceolate or -elliptic to sublinear, dark green, impressed–veiny, glandular but subglabrous above, paler, with gland-tipped and nonglandular hairs beneath, serrate to subentire, often revolute; pedicels 5–16 mm. long; calyx tubular, 20–25 mm. long, slightly inflated upward, the lobes lanceolate, the longest ca. 7 mm., the shorter ca. half as long; corolla deep- to yellow-orange, 3.5–4.5 cm. long, the tube included, gradually enlarged into a funnelform throat, the limb with dorsiventral spread of 2–3.5 cm., the 2 upper lobes ± perpendicular to throat, laterally notched, entire to finely toothed, the 3 lower extending forward, entire to notched or toothed; anthers glabrous; stigmas fimbriolate, somewhat unequal; caps. ca. 20 mm. long, opening along upper suture; seeds fusiform, ca. 1 mm. long.—Rocky places, below 1800 (3000) ft.; Foothill Wd., N. Oak Wd., Mixed Evergreen F., N. Coastal Scrub, Redwood F., Closed-cone Pine F.; Coast Ranges from Del Norte Co. to Santa Barbara Co., foothills of Sierra Nevada from Placer Co. to Tuolumne Co.; w. Ore. March–Aug. Apparently hybridizing in its southernmost range with *M. longiflorus*.

Ssp. **austràlis** (McMinn) Munz. [*Diplacus a.* McMinn. *M. longiflorus* var. *linearis* auth., not *M. l.* Benth.] Calyx tubular, scarcely enlarged upward; corolla orange-yellow to light apricot or buff or white; lvs. only slightly impressed–veiny.—Rocky and disturbed places, below 3500 ft.; Chaparral; Santa Ana Mts. through interior San Diego Co. to n. L. Calif. March–July. Hybridizing with *M. puniceus.*

Ssp. **lompocénsis** (McMinn) Munz. [*Diplacus l.* McMinn.] Calyx expanding somewhat above the tube into the throat which is slightly contracted near the summit; corolla light orange-yellow.—Chaparral; Lompoc Mesa and low hills, s. San Luis Obispo Co. to w. Santa Barbara Co. March–July.

74. **M. áridus** (Abrams) Grant. [*Diplacus a.* Abrams.] Glabrous glutinous decumbent subshrub 2–4 dm. tall, 6–12 dm. across; lvs. 2–4(–6) cm. long, elliptic to ± lanceolate, yellow-green, slightly dentate, ± revolute; pedicels 3–6 mm. long; calyx 2.5–3 cm. long, funnelform, glabrous, flared above, the uppermost lobe 8–10 mm. long, the others ca. 4–6 mm.; corolla light lemon- to golden-yellow, 3.5–5 cm. long, the tube exserted, the limb rotate, with a dorsiventral spread of 2–2.5 cm.; anthers glabrous; stigmas unequal, fimbriate; caps. 12–18 mm. long, splitting along upper suture; seeds fusiform, ca. 1 mm. long; $n = 10$ (Stebbins in McMinn, 1951).—Local in dry rocky places, 2500–3500 ft.; Chaparral; se. San Diego Co.; adjacent L. Calif. April–July.

75. **M. bífidus** Penn. [*Diplacus glutinosus* var. *grandiflorus* Lindl. & Paxt. *D. grandiflorus* Groenl., not *M. g.* Howell.] Spreading subshrub 4–7(–10) dm. high, glutinous, minutely glandular-puberulent; lvs. 2.5–6 cm. long, 0.6–2 cm. wide, oblong-elliptic, dark green above, entire to slightly toothed, ± revolute; pedicels 6–8 mm. long; calyx 2.5–3 cm. long, tubular or with somewhat expanded throat, the lobes attenuate-tipped, the uppermost 8–10 mm. long, the lower 5–7 mm.; corolla buff or cream, 5.5–6.5 cm. long, the tube slightly exserted, the throat evident, funnelform, the limb with a dorsiventral spread of 3.5–4 cm., the 2 upper lobes with deep lateral notch, the 3 lower with deep median notch; anthers glabrous; caps. 15–17 mm. long, opening along upper suture; seeds fusiform, ca. 1 mm. long.—Rocky places, below 5000 ft.; Foothill Wd., Yellow Pine F.; w. base of Sierra Nevada, Plumas and Butte cos. to Placer Co. April–July.

Ssp. **fasciculàtus** Penn. [*Diplacus f.* McMinn.] Lvs. linear-oblong, 0.3–0.5(–1) cm. wide; corolla 4–6 cm. long; $n = 10$ (Stebbins in McMinn, 1951).—Rocky hills, below 4500 ft.; Chaparral, Foothill Wd.; Santa Lucia Mts. April–July. *M. linearis* Benth. with fls. 2.5–4 cm. long, from Monterey and w. San Luis Obispo cos., seems to be a series of hybrids of *M. aurantiacus* and *M. bifidus* ssp. *fasciculatus.*

76. **M. Flemíngii** Munz. [*Diplacus parviflorus* Greene, not *M. p.* Lindl.] Subglabrous semierect or spreading subshrub 1–6 dm. tall, with greater spread; lvs. 2–4.5 cm. long, obovate to oval-elliptic, dark green above, paler beneath, entire to glandular-denticulate, usually revolute; pedicels 10–20 mm. long; calyx tubular, 2–2.3 cm. long, the lobes linear, the uppermost 6–8 mm. long, the lower 3–4 mm.; corolla brick-red or orange-red, 2.5–4 cm. long, the tube included in calyx, the limb with dorsiventral spread of 2–2.4 cm., the upper lobes subentire, the lower entire or slightly toothed; anthers glabrous; stigmas subequal, fimbriolate; caps. ca. 2 cm. long, opening along upper suture; seeds fusiform, ca. 1 mm. long.—Rocky places; Coastal Sage Scrub, Chaparral; Santa Cruz, Santa Rosa, Anacapa, and San Clemente ids. March–July.

77. **M. puníceus** (Nutt.) Steud. [*Diplacus p.* Nutt. *M. glutinosus* var. *p.* Gray.] Freely branched erect shrub 5–15 dm. tall, glabrous to puberulent, glutinous; lvs. 2.5–7 cm. long, linear-lanceolate to elliptic, dark green and glabrous above, paler and glabrous to puberulent beneath, entire to finely toothed, usually revolute; pedicels 10–30 mm. long; calyx 20–25 mm. long, tubular, usually reddish, the lobes linear, the uppermost 6–8 mm. long, the lower 3–4 mm.; corolla brick-red to orange-red, 3–5 cm. long, the tube included in calyx, the limb with a dorsiventral spread of 2.5–3 cm., the upper lobes with a lateral notch and irregularly toothed, the 3 lower lobes subentire to finely toothed at tip; anthers glabrous; stigmas subequal, ciliolate; caps. 15–20 mm. long, splitting along upper suture; seeds fusiform, ca. 1 mm. long; $n = 10$ (Stebbins in McMinn, 1951).—Dry slopes and mesas, below 2500 ft.; Chaparral, Coastal Sage Scrub; Laguna Beach and Santa Ana Mts. to n. L. Calif., Santa Catalina Id. March–July.

4. Stemòdia L.

Perennial herbs with opposite or alternate serrate clasping lvs. Fls. solitary in the axils of bracts of a loose raceme. Bracteoles 2 just beneath the calyx. Calyx 5-parted into linear-lanceolate segms. Corolla blue-violet, with cylindrical tube, 2-lobed arched upper lip, and 3-lobed deflexed-spreading lower. Stamens 4, didynamous. Anther-cells separate and stipitate. Stigmas distinct, lamelliform. Caps. cylindric-ovoid, septicidal and loculicidal. Seeds many, obscurely reticulate. Ca. 20 spp., mostly of trop. Am. (Name abbreviated from *Stemodiacra*, meaning stamens with 2 tips.)

1. **S. durantifòlia** (L.) Sw. [*Capraria d.* L. *S. arizonica* Penn.] Stems several to many, ascending, 1–5 dm. tall, glandular-pubescent; lvs. lanceolate, subsessile with narrowed then dilated base, 2–4 cm. long, smaller upward; infl. spiciform, glandular, of 6–12 remote fascicles; lower pedicels 2–12 mm. long; calyx 5–6 mm. long; corolla 7–8 mm. long, purplish; caps. 5 mm. long; seeds 0.3 mm. long, brown.—Occasional in wet places, as near Palm Springs (Riverside Co.), San Diego region; to Ariz., trop. Am. Much of year.

5. Gratìola L. HEDGE-HYSSOP

Erect or diffuse herbs. Lvs. opposite, entire to denticulate, sessile. Fls. axillary to foliose bracts in a simple loose raceme. Bracteoles present beneath calyx or lacking. Sepals 5, distinct. Corolla-tube quadrangular; upper lip entire or 2-lobed; lower 3-cleft. Fertile stamens 2 (the upper pair); anther-cells parallel on the flat expanded connective. Stigmas 2, lamelliform. Caps. 4-valved, many-seeded. Seeds wingless. Ca. 20 spp., widely distributed. (Diminutive of *gratia*, grace or favor, from supposed medicinal qualities.)

A. Pedicels bibracteolate beneath calyx; lvs. and sepals obtuse or acutish, the latter 5-6 mm. long; plants finely glandular-pubescent . . . . 1. *G. neglecta*
AA. Pedicels without bracts.
    B. Lvs. and sepals attenuate, the latter 8-23 mm. long; plants mostly glabrous; fls. white
                                                                 2. *G. ebracteata*
    BB. Lvs. and sepals blunt, the latter 4-6 mm. long; fls. yellow and white . . . 3. *G. heterosepala*

1. **G. neglécta** Torr. Annual 1–3 dm. tall; lvs. oblong-lanceolate to subovate, round-clasping at base, 0.5–5 cm. long; pedicels 10–20 mm. long, slender, spreading; bracteoles 2, at least as long as sepals; corolla 9–10 mm. long, whitish with yellow tube, the upper lobes joined almost to apex; caps. 5 mm. long; seeds thick-cylindric, 0.4–0.5 mm.

long.—Wet or muddy places, 5000–6500 ft.; Red Fir F., Yellow Pine F.; Sierra Nevada from Tuolumne Co. n., to Lassen, Modoc and Shasta cos.; to B.C., Atlantic Coast. May–Aug.

2. **G. ebracteàta** Benth. Stems 0.3–2.5 dm. high, mostly glabrous; lvs. lanceolate-attenuate, slightly clasping at base, 0.5–2.5 cm. long; pedicels stout, 10–23 mm. long; corolla 5–7 mm. long, white with yellowish tube, the upper lobes united less than half their length; caps. 4–5 mm. long, subglobose; seeds narrow-cylindric, light brown, regularly pitted, ca. 0.8–1 mm. long.—Muddy places, below 5000 ft.; many Plant Communities; from San Joaquin and Tuolumne cos. n. to Humboldt and Siskiyou cos.; to B.C., Ida. April–Aug.

3. **G. heterosèpala** Mason & Bacig. Near *G. ebracteata* but upper lvs. obtuse or emarginate; sepals unequal, the 2 lower 4–6 mm. long, the 3 upper united part way; upper corolla lobes yellow, joined nearly to tip, the lower free, white.—At 2900 ft., shore of Boggs Lake, Lake Co. April–June.

6. Bacòpa Aubl.

Perennial or possibly annual herbs with opposite lvs. and axillary fls. Bracteoles lacking in ours. Calyx of 5 almost distinct sepals, dissimilar, the upper broadest. Corolla campanulate, white in ours, the upper lip emarginate or 2-lobed, the lower 3-lobed. Stamens 4, all fertile. Caps. thin, roundish, 2-valved, the valves 2-parted. Seeds many, wingless, reticulate. Ca. 60 spp., largely of New World trop. (Thought to be an aboriginal name.)

Calyx 6–7 mm. long, erect in age; pedicels mostly exceeding subtending lf. at anthesis .. 1. *B. Eisenii*
Calyx 2 mm. long, spreading in age; pedicels shorter than subtending lf. at anthesis .. 2. *B. Nobsiana*

1. **B. Eisènii** (Kell.) Penn. [*Ranapalus* E. Kell. *B. rotundifolia* auth., not Wettst.] Stems fleshy, branched, 2–3 dm. long, pubescent distally; lvs. rounded-cuneate, palmately 7–11-veined, 1–2 cm. long; pedicels 10–50 mm. long; sepals 6–7 mm. long, the outermost oval, green, many-veined, ca. 3 times as wide as the lance-oblong hyaline inner; corolla 9–10 mm. long; styles separate at tip; caps. 4–5 mm. long; seeds cylindric, regularly pitted, with bladdery testa.—Pools and muddy places, as in rice fields of San Joaquin V., Lone Pine; w. Nev. June–Sept.

2. **B. Nobsiàna** Mason. Like the preceding, but pedicels and fls. together shorter than subtending lvs.; sepals 2 mm. long, spreading in age and exposing caps.; caps. broadly ovoid, grooved apically along the septum, 4 mm. long.—Rice fields, etc., Cent. V.

7. Limosélla L. MUDWORT

Stoloniferous or tufted glabrous perennial or annual herbs without ascending stems. Lvs. entire, petioled, palmately veined, in tufts. Fls. solitary on long pedicels from the tufts. Bracteoles none. Calyx 5-lobed. Corolla subrotate, white or pinkish to violet-tinged, 5-lobed, the lobes acute. Stamens 4, subequal; anther-cells wholly confluent. Stigmas united and capitate. Caps. septicidal, distally 1-celled, the septum not complete. Seeds many, minute. Ca. 12 spp., widely distributed. (Latin, *limus*, mud, and *sella*, seat, the spp. growing in wet places.)

Lvs. flat; caps. ellipsoid; style not longer than ovary.
    Style 0.2–0.4 mm. long, usually sharply decurved at base; corolla-lobes acute, dull white or pinkish; lvs. 2–8 mm. wide .... 1. *L. aquatica*
    Style 0.5–1 mm. long, straight or arcuately curved; corolla-lobes rounded, white or with violet tinge; lvs. 0.5–2 mm. wide .... 2. *L. acaulis*
Lvs. terete to subulate; caps. globose; style longer than ovary .... 3. *L. subulata*

1. **L. aquática** L. Plants consisting of tufts connected by slender stolons; lvs. elliptic to oblong, 1–1.5 cm. long, rather abruptly narrowed into petioles 2–4 times as long; pedicels usually ca. half as long as petioles; fls. ca. 1.5–2 mm. broad; caps. ca. 3 mm. long; $2n = 40$ (Vachell & Blackburn, 1939).—Occasional, muddy shores of ponds and streams, below 10,500 ft.; many Plant Communities; Kern, San Mateo, Plumas, and Inyo cos. n.; to Alaska, Nfld.; Eurasia. June–Sept.

2. **L. acaùlis** Ses. & Moç. Lf.-blades linear-oblanceolate, 0.6–1.2 cm. long, attenuate

to petioles several times as long; pedicels mostly ca. as long as petioles.—Muddy shores, below 8000 (10,900) ft.; many Plant Communities; Plumas, Placer, and Marin cos. s., most abundant in mts. of s. Calif.; to New Mex., L. Calif., and Mex. May–Oct.

3. **L. subulàta** Ives. Tufted, stoloniferous; lvs. not differentiated into petiole and blade; calyx campanulate; corolla campanulate, with oblong minutely papillate lobes; seeds ridged and reticulate.—Sandy shores of lower San Joaquin R. near Antioch Bridge; Atlantic Coast.

8. Verbáscum L. MULLEIN

Biennial or perennial herbs, erect, simple or virgately branched. Cauline lvs. simple, alternate, sessile, clasping or somewhat decurrent. Fls. in racemes or crowded spikes, ephemeral. Bracteoles 0. Calyx 5-parted. Corolla rotate, slightly irregular, commonly yellow. Stamens 5, the fils. villose-pubescent, alike or the lower pair different from the others. Stigmas united, capitate. Caps. ellipsoid to subglobose, septicidally 2-valved. Seeds many, wingless, pitted or roughened. Ca. 250 spp., of Eurasia. (Corrupted from *Barbascum,* the old Latin name.)

Plants very woolly; lvs. entire; pedicels less than 10 mm. long .... 1. *V. Thapsus*
Plants with green herbage, with simple gland-tipped hairs; lvs. dentate.
    Pedicels 10–15 mm. long; lvs. sinuate-dentate, glabrous .... 2. *V. Blattaria*
    Pedicels 3–5 mm. long; lvs. sinuately denticulate, pubescent .... 3. *V. virgatum*

1. **V. Thápsus** L. COMMON MULLEIN. Stem stout, 5–18 dm. tall; basal lvs. in rosettes, oblong-obovate to obovate-lanceolate, 15–40 cm. long including petioles; cauline lvs. elliptic-lanceolate, gradually reduced up the stem, decurrent; pedicels less than 2 mm. long; calyx 7–9 mm. long; corolla 20–25 mm. broad, yellow; caps. 5 mm. long, stellate-pubescent; seeds brown, irregular-oblong, 0.5–0.7 mm. long, with rows of pits; $2n = 34$, 36 (Håkansson, 1926).—Common weed in waste places, along river bottoms, etc., in n. Calif., in pine belt of Sierra Nevada; occasional in s. Calif. mostly above 4000 ft.; natur. from Eurasia. June–Sept.

2. **V. Blattària** L. MOTH MULLEIN. Stem 4–12 dm. high, glandular-pubescent above; cauline lvs. 2–12 cm. long, elliptic to ovate, doubly serrate-crenate, not decurrent; calyx 5–8 mm. long; corolla yellow or white, 25–30 mm. broad; caps. 6–8 mm. long; seeds brown, ± oblong-ovoid, ca. 0.6–1 mm. long, with rows of pits; $2n = 30$, 32 (Håkansson, 1926).—Occasional weed in waste places, through most of cismontane Calif.; natur. from Eurasia. May–Sept.

3. **V. virgàtum** Stokes ex With. Similar to *V. Blattaria,* but somewhat more pubescent and glandular; lvs. crenate; pedicels shorter than caps.; $2n = 32$ (Håkansson, 1926).—Occasional as weed in s. Calif.; natur. from Eurasia. May–Sept.

9. Pénstemon Mitch. BEARD-TONGUE

Perennial herbs or shrubs. Lvs. opposite, rarely whorled or the upper alternate, the lower usually petioled, the upper sessile. Fls. usually showy, in racemose, cymose, or thyrsoid panicles. Calyx 5-parted. Corolla tubular, the throat often inflated, the limb weakly or strongly 2-lipped, the upper lip 2-lobed, the lower 3-cleft. Fertile stamens 4, paired, the fils. arched, a 5th stamen represented by a long sterile fil. (staminode) often dorsally bearded; anthers 2-celled, the cells often confluent. Caps. septicidal, cartilaginous. Seeds many, with irregularly angled cellular coat. Ca. 230 spp., mostly of w. N. Am., 1 of e. Asia; several of horticultural importance. (Greek, *pente,* five, and *stemon,* stamen.)

(Keck, D. D., *in* Abrams: Ill. Fl. Pac. States 3: 733–770, 1951.)
A. Fils. glabrous at base or, at most, two of them puberulent.
    B. Anthers glabrous or sparsely hairy, not comose; corolla rounded dorsally.
        C. Orifice of corolla closed, concealing the stamens, bearded on all sides within; staminode ca. ⅕ the length of the corolla. Butte Co .... 47. *P. personatus*
        CC. Orifice of corolla ± open, revealing the stamens, bearded only ventrally within or glabrous; staminode at least half as long as corolla.
            D. Anther-sacs opening from the free tips throughout or partially, almost always divaricate after dehiscence.
                E. Corolla scarlet.

F.   Herbage glaucous; anther-sacs dehiscent by a continuous slit extending
     across their contiguous parts; corolla subtubular. Cismontane
                                                    30. *P. centranthifolius*
FF. Herbage not glaucous; anther-sacs opening at their distal ends,
     the slits not confluent.
     G.   Corolla obscurely 2-lipped, the short round lobes scarcely spread-
          ing. Desert mts. and borders below the Yellow Pine belt
                                                    33. *P. Eatonii*
     GG. Corolla strongly 2-lipped, the upper lip galeate, erect, the lower
          with sharply reflexed linear lobes. Yellow Pine belt and above.
                                                    34. *P. labrosus*
EE. Corolla not scarlet, although it may be rose or carmine, to blue, purple, etc.
    F.   Lvs. of lower and middle stems entire or obscurely denticulate.
         G.   Infl. not glandular-puberulent.
              H.   Corolla 18–35 mm. long.
                   I.   Fls. blue-purple, 25–35 mm. long; infl. secund; lvs. not
                        glaucous. Borders of Mojave Desert n. along e. face of
                        Sierra Nevada .... 19. *P. speciosus*
                   II. Fls. red-purple to carmine or paler, 18–27 mm. long;
                        infl. not secund; lvs. ± glaucous.
                        J.   Corolla white to flesh-color with lavender limb; lvs.
                             3–6 mm. wide. N. and e. Mojave Desert.
                                                    22. *P. fruticiformis*
                        JJ. Corolla red to red-purple; lvs. wider.
                             K.   Lvs. lance-ovate, 10–25 mm. wide; fls. red-
                                  purple. Cismontane s. Calif. ... 29. *P. Parishii*
                             KK. Lvs. linear to lance-linear, 5–15 mm. broad; fls.
                                  rose-lavender or carmine. Desert.
                                  L.   Thyrsus racemiform; corolla glandular-pubes-
                                       cent, carmine; anther-sacs peltately ex-
                                       planate. San Bernardino Co.
                                                    31. *P. utahensis*
                                  LL. Thyrsus open, ± decompound; corolla gla-
                                       brous, rose-lavender; anther-sacs boat-shaped.
                                       Owens V. region .... 32. *P. confusus*
              HH. Corolla 7–18 mm. long.
                   I.   Fls. blue-purple.
                        J.   Cauline lvs. lanceolate to lance-oblong, 6–15 mm.
                             wide.
                             K.   Corolla ± declined, mostly 7–10 mm. long; anther-
                                  sacs round .... 1. *P. procerus*
                             KK. Corolla horizontal, 10–15 mm. long; anther-sacs
                                  longer than broad .... 3. *P. oreocharis*
                        JJ. Cauline lvs. linear-lanceolate, 3–6 mm. broad; anther-
                             sacs round .... 2. *P. cinicola*
                   II. Fls. pink to lavender-rose.
                        J.   Cauline lvs. cuneate-oblong to spatulate or obovate,
                             5–18 mm. wide.
                             K.   Plant 8–12 cm. high; thyrsus 2–4 cm. long;
                                  lvs. not white-margined. Trinity Co.
                                                    11. *P. Tracyi*
                             KK. Plant 15–30(–40) cm. high; thyrsus 5–15 cm.
                                  long. Mojave Desert.
                                  L.   Lvs. white-margined, rhombic-obovate. San
                                       Bernardino Co. ..... 12. *P. albomarginatus*
                                  LL. Lvs. not white-margined, lance- or linear-
                                       oblong. Owens V. region ... 32. *P. confusus*
                        JJ. Cauline lvs. linear, 1–3 mm. wide; plant 30–60 cm.
                             high. Deserts .... 15. *P. Thurberi*
         GG. Infl. glandular-puberulent or -pubescent.
              H.   Lvs. glabrous.
                   I.   Cauline lvs. lanceolate to oblong or ovate, 6–25 mm.
                        wide.
                        J.   Fls. deep blue-purple, obscurely bilabiate; the palate
                             prominently brownish-yellow-bearded. Sierra Nevada
                             to Siskiyou and Modoc cos.
                             K.   Staminode included; infl. prominently glandular.
                                  Sierra Nevada, White Mts. ... 4. *P. heterodoxus*
                             KK. Staminode reaching orifice; infl. moderately gland-
                                  ular. Shasta Co. to Modoc and Siskiyou cos.
                                                    5. *P. shastensis*
                        JJ. Fls. deep lavender to blue-violet, definitely bilabiate;
                             palate glabrous or sparingly bearded. Humboldt and
                             Glenn cos. n. .... 8. *P. anguineus*
                   II. Cauline lvs. lance-linear, 2–5 mm. wide; corolla violet
                        with blue limb; Inyo Co. to San Bernardino Co.
                                                    27. *P. incertus*

HH. Lvs. puberulent.
   I. The lvs. ashy-puberulent, mostly not much more than 1 cm. long, crowded on the stems.
      J. Lvs. 3–5 mm. wide; corolla 2-ridged on floor within throat, the upper lip erect, the lower spreading. E. Mojave Desert .... 13. *P. Thompsoniae*
      JJ. Lvs. 1.5–3 mm. wide; corolla not plicate in throat, the 2 lips spreading. San Jacinto Mts. to L. Calif.
           14. *P. californicus*
   II. The lvs. usually not ashy, mostly 2 or more cm. long, not crowded on stems.
      J. Fls. blue to blue-purple, the corolla-throat 2-ridged on its floor.
         K. Limb of corolla narrow, less than twice as wide as throat; calyx 3–5 mm. high. Mono Co.
             6. *P. humilis*
         KK. Limb of corolla wider, at least twice as broad as throat; calyx 2–3 mm. high. Siskiyou Co.
             7. *P. cinereus*
      JJ. Fls. rose-red to red-purple or dull purple, the corolla-throat not or scarcely 2-ridged on its floor.
         K. Staminode included; corolla reddish; cauline lvs. 8–25 mm. wide.
           L. Corolla 14–20 mm. long, ± ventricose; anther-sacs not explanate, 1.3 mm. long. Desert ranges e. of Owens V.
             16. *P. monoensis*
           LL. Corolla 11–14 mm. long, narrowly tubular; anther-sacs explanate, 0.5 mm long. Grapevine and Providence mts. . . 17. *P. calcareus*
         KK. Staminode exserted; corolla dull purple; cauline lvs. 4–8 mm. wide. Lassen Co. and n.
            18. *P. miser*
FF. Lvs. of lower and middle stems definitely toothed, usually coarsely so.
  G. Corolla 2-ridged on floor of throat. N. Calif, and cent. Sierra Nevada.
    H. Fls. 24–30 mm. long; anther-sacs 1.2–1.4 mm. long. Santa Cruz Co., Mendocino Co. to Del Norte Co. . . 9. *P. Rattanii*
    HH. Fls. 10–18 mm. long; anther-sacs 0.7–1.1 mm. long.
      I. Corolla deep lavender to blue-violet. Humboldt and Glenn cos. n. .... 8. *P. anguineus*
      II. Corolla ochroleucous. Glenn Co. and cent. Sierra Nevada to Modoc Co. .... 10. *P. deustus*
  GG. Corolla scarcely if at all ridged ventrally. S. Calif.
    H. The corolla abruptly inflated from a short tube ca. equaling the calyx, strongly 2-lipped, white or tinged pink, lavender or purple; staminode long-bearded, uncinate.
      I. Upper lvs. connate-perfoliate; ovary glandular-puberulent; thyrsus virgate. Mojave Desert .... 20. *P. Palmeri*
      II. Upper lvs. distinct; ovary glabrous; thyrsus lax. Mts. of s. Calif. and edge of San Joaquin V. . . 21. *P. Grinnellii*
    HH. The corolla nearly tubular to inflated from tube twice as long as calyx, not whitish; staminode short-bearded or glabrous, not uncinate.
      I. Anther-sacs peltately explanate and glabrous; thyrsus virgate, secund; corolla pink to rose-purple.
         J. Corolla obviously 2-lipped and strongly inflated (except in ssp. *Austinii*) more than 10 mm. wide (except in ssp.); lvs. glaucous .... 23. *P. floridus*
         JJ. Corolla nearly regular and tubular-funnelform, 4–8 mm. wide; lvs. green or glaucous.
           K. Calyx 4–6 mm. long; corolla moderately ampliate, 5–8 mm. wide.
            L. Lvs. thin, glaucous, dentate; corolla rose-purple, with dark guide-lines; stems to 10 dm. tall. E. deserts.
             24. *P. pseudospectabilis*
           LL. Lvs. thick, green or glaucous, entire or dentate; corolla crimson or red-purple, without guide lines; stems to 7 dm. tall. W. border of Colo. Desert. . 25. *P. Clevelandii*
         KK. Calyx 3–4.5 mm. long; corolla essentially tubular, 4–5 mm. wide. E. Mojave Desert.
            26. *P. Stephensii*
      II. Anther-sacs not explanate, ± scabro-ciliate at suture; thyrsus lax; corolla purplish with blue limb, obviously 2-lipped. Cismontane . . .. 27. *P. spectabilis*

DD. Anther-sacs opening across their continuous apices, the free tips remaining saccate, parallel even after dehiscence, glabrous or somewhat hairy.
  E. Corolla scarlet, tubular; limb relatively long, the upper lip erect, the lower sharply reflexed. Montane, Alpine Co. to San Diego Co. .. 46. *P. Bridgesii*
  EE. Corolla bluish or purplish, usually ampliate; limb relatively short, with both lips ± spreading.
    F. Staminode bearded; infl. glandular-pubescent.
      G. Lvs. glabrous; corolla 13–16 mm. long; anther 1 mm. long. Montane from Lake Tahoe n. .... 35. *P. gracilentus*
      GG. Lvs. canescent; corolla 25–35 mm. long; anthers 1.5 mm. or more long.
        H. Calyx-lobes linear-lanceolate, attenuate; corolla-throat rounded and glabrous within; lvs. evenly distributed. E. base of Sierra Nevada .... 36. *P. papillatus*
        HH. Calyx-lobes broadly ovate, acute; corolla-throat 2-ridged and villous within; lvs. mostly basal. White and Inyo mts. 37. *P. scapoides*
    FF. Staminode glabrous.
      G. Lvs. mostly basal; orifice of corolla bearded ventrally. San Bernardino Mts. to s. Sierra Nevada. .... 38. *P. caesius*
      GG. Lvs. well distributed; orifice of corolla glabrous.
        H. Herbage glabrous to finely pubescent; basal lvs. narrowly oblanceolate to obovate.
          I. Infl. glandular-pubescent; peduncles divergent.
            J. Lvs. not glaucous; corolla blue-purple.
              K. Lvs. linear to oblong or oblanceolate, 2–12 mm. wide, not involute, pubescent to subglabrous, without fascicles developing in the axils. Widely distributed. .... 39. *P. laetus*
              KK. Lvs. filiform, 0.5–1.5 mm. wide, tightly involute, essentially glabrous, often with fascicles developing in the axils. Shasta Co. .... 40. *P. filiformis*
            JJ. Lvs. glaucous, glabrous; corolla usually tricolored. Sierra Co. to Lassen and Shasta cos. 41. *P. neotericus*
          II. Infl. not glandular; peduncles appressed.
            J. Lvs. just below infl. amplexicaul, blue-glaucous, glabrous.
              K. Corolla 18–30 mm. long; anthers 1.7–3.2 mm. long .... 42. *P. azureus*
              KK. Corolla 14–20 mm. long; anthers 1.4–1.8 mm. long .... 43. *P. parvulus*
            JJ. Lvs. just below infl. narrow at base, sessile, rarely glaucous, puberulent or glabrous 44. *P. heterophyllus*
        HH. Herbage densely canescent; basal lvs. oval to roundish. N. Coast Ranges. .... 45. *P. Purpusii*
  BB. Anthers densely comose, the cells peltately explanate; corolla strongly 2-ridged on floor of throat and prominently keeled dorsally.
    C. Plants low, often decumbent, woody at base; lvs. coriaceous; infl. subracemose; corolla glabrous externally.
      D. Stems 1.5–3 dm. high; corolla rose-red to rose-purple. ...... 48. *P. Newberryi*
      DD. Stems less than 1 dm. high.
        E. Corolla rose; lvs. glaucous, ± hirtellous, serrate .... 49. *P. rupicola*
        EE. Corolla blue-violet; lvs. green, glabrous, entire .... 50. *P. Davidsonii*
    CC. Plants 3–8 dm. high, herbaceous quite to base, erect; lvs. thin; infl. openly paniculate; corolla glandular externally .... 51. *P. nemorosus*
AA. Fils. all strongly pubescent at base; corolla strongly bilabiate, the upper lip subgaleate, lower reflexed; plants shrubby.
  B. Corolla whitish, yellowish, or fulvous, not distinctly tubular.
    C. Infl. spicate-racemose; pedicels shorter than calyces; fls. solitary or in 2's. Mono, Inyo, and Tulare cos. and San Jacinto Mts. .... 52. *P. Rothrockii*
    CC. Infl. paniculate or thyrsoid; pedicels longer than calyces; fls. usually in 2's or several.
      D. Staminode glabrous; corolla white, tinged with pink, long-hairy externally. From Los Angeles and Tulare cos. n. .... 53. *P. breviflorus*
      DD. Staminode densely bearded; corolla short-pubescent externally.
        E. Corolla ca. 4 mm. wide, fulvous with yellowish lower lip; stems glaucous; lvs. denticulate. Siskiyou Co. to Solano and Eldorado cos. 54. *P. Lemmonii*
        EE. Corolla ca. 10 mm. wide, yellow; stems not glaucous; lvs. usually entire. S. Calif. .... 55. *P. antirrhinoides*
  BB. Corolla red, distinctly tubular.
    C. Lvs. opposite, narrowly elliptic to narrowly cordate; stems not glaucous.
      D. Staminode bearded only apically; lvs. mostly subcordate; scandent shrub. San Luis Obispo Co. .... 56. *P. cordifolius*
      DD. Staminode bearded throughout; lvs. tapering to base; not scandent. From Monterey and Sutter cos. n. .... 57. *P. corymbosus*
    CC. Lvs. ternate, linear-lanceolate; stems glaucous. S. Calif. .... 58. *P. ternatus*

1. **P. prócerus** Dougl. ex Grah. ssp. **formòsus** (A. Nels.) Keck. [*P. formosus* A. Nels. *P. chionophilus* Greene. *P. cacuminis* Penn. *P. Tolmiei* ssp. *f.* Keck.] Densely cespitose, the stems slender, 0.4–1.5 dm. tall; basal rosette well developed, glabrous throughout; lvs. deep green, rather firm, the basal ± ovate, the blades ca. 1 cm. long, sometimes narrower and folded, the petioles ca. as long, the cauline lvs. somewhat narrower; thyrsus of 1–2 dense clusters; calyx 1.7–2.7 mm. long, the lobes obtuse to cuspidate-tipped; corolla deep blue-purple, 7.5–11 mm. long, tubular, the palate lightly bearded to subglabrous; staminode glabrous to lightly bearded.—Rocky, sometimes meadowy slopes, 6500–11,000 ft.; Red Fir F. to Subalpine F.; Sierra Nevada of Alpine, Mono, and Tuolumne cos., Marble Mts. and Mt. Eddy, Siskiyou Co.; Ore., Nev. July–Aug.

Ssp. **brachyánthus** (Penn.) Keck. [*P. b.* Penn. *P. Tolmiei* ssp. *b.* Keck.] Stems 1.5–3 dm. high; thyrsus of 3–5 (sometimes 1) interrupted clusters; calyx 2–3 mm. long; the lobes with prominently scarious erosulate margin; staminode bearded; $n = 8$ (Keck, 1945).—Rocky or grassy places, 3600–7500 ft.; Montane Coniferous F.; Trinity and Siskiyou cos.; Ore. July–Aug.

2. **P. cinícola** Keck. Stems slender, numerous, forming clumps 1.5–3.5 dm. high, glabrous or minutely puberulent, without basal rosette; lvs. green or grayish, not glaucous, entire, lance-linear, folded and somewhat recurved, 2.5–5.5 cm. long, sessile or the lower short-petioled; thyrsus strict, virgate, of 2–7 clusters, the lower well spaced, and on slender erect peduncles; calyx 1.4–2 mm. high, the lobes oblong-obovate, truncate or mucronate, with broad scarious entire margin; corolla purple with deep blue spreading limb, 7–9 mm. long, the palate obscurely ridged, moderately bearded; anther-sacs 0.35–0.5 mm. long; staminode included, with few short hairs apically; caps. 3–4 mm. long; seeds brown, scarcely 1 mm. long; $n = 8$ (Keck, 1945).—Dry or moist volcanic sands, 4000–5000 ft.; Yellow Pine F.; Lassen Co. to Siskiyou and Modoc cos.; se. Ore. June–Aug.

3. **P. oreócharis** Greene. [*P. interruptus* and *lassenianus* Greene. *P. recurvatus* Heller. *P. tinctus* Penn.] Stems 2–5(–7) dm. high, the herbage bright green, essentially glabrous, the basal rosette well developed; lvs. thin, the basal linear-oblanceolate to elliptic, 2.5–8 cm. long, short-petioled, the cauline lance-oblong or oblong, the upper amplexicaul; thyrsus strict, of 1–6 rather distinct many-fld. clusters; calyx 3–5(–7) mm. high, the lobes oblong, abruptly narrowed to the acuminate tip, with narrow to broad usually entire scarious margin; corolla blue-purple, 10–14 mm. long, the lips equal, the palate prominently bearded; anther-sacs 0.6–0.9 mm. long; staminode reaching orifice, usually densely yellow-bearded apically, rarely glabrous; caps. 5–6 mm. long; $n = 8$ (Keck, 1945).—Dry to wet meadows, 4600–8100 ft.; Sagebrush Scrub to Subalpine F.; from Inyo and Fresno cos. n.; local in Glenn Co.; to Wash., Ida., Nev. May–Aug.

4. **P. heterodóxus** Gray. [*P. alsinoides, depressus,* and *geniculatus* Greene. *P. confertus* var. *g.* Jeps.] Stems slender, mostly 1–2(–2.5) dm. high, the basal rosette developed; lvs. deep green, thin, glabrous, the basal linear-oblanceolate to spatulate, 4–8 mm. wide, 1.5–3 cm. long on equally long petioles, the cauline lvs. oblanceolate to broadly lanceolate; thyrsus strict, of 2–4 rather distinct many-fld. clusters, glandular-puberulent; calyx 3–6 mm. long, the lobes oblong, abruptly narrowed to the acute tip, with broad to narrow, ± entire scarious margin; corolla blue-purple, 10–16 mm. long, the lips equal, the palate prominently bearded; anther-sacs 0.7–1 mm. long; staminode included, bearded with short stiff hairs, rarely glabrous; caps. ca. 5 mm. long; seeds dark, to ca. 1 mm. long; $n = 8$ (Keck, 1945).—Rocky slopes and alpine meadows, mostly 8000–12,000 ft.; Lodgepole F. to Alpine Fell-fields, Bristle-cone Pine F.; Sierra Nevada from Tulare Co. to Plumas Co., White Mts.; Nev. July–Aug.

Ssp. **cephalóphorus** (Greene) Keck. [*P. c.* and *P. glastifolius* Greene.] Stems stouter, mostly 1.5–4 dm. tall; basal lvs. 2.5–6 cm. long, long-petioled, 6–12 mm. wide, the middle cauline up to 6 cm. long.—At 6500–10,000 ft.; largely Lodgepole F.; w. of Kern R., Fresno and Tulare cos. July–Aug.

5. **P. shasténsis** Keck. Stems slender to rather stout, 2–5 dm. high, the basal rosette well developed; lvs. deep green, rather thin, glabrous, the basal elliptic, obtuse, 1–3 cm. long on equally long petioles, the cauline lance-oblong, the lower 5–8 cm. long, short-petioled, the upper reduced; thyrsus of 2–6 dense clusters, sometimes only 1, moderately glandular-pubescent; calyx 2.5–5 mm. high, the lobes lance-oblong, acuminate, not abruptly tipped; corolla blue-purple, 10–13 mm. long, like *heterodoxus;*

anther-sacs ovate, boat-shaped, 0.6 mm. long; staminode reaching orifice, densely golden-bearded apically; $n = 16$ (Keck, 1945).—Occasional in meadowy places, 4200–7300 ft.; Yellow Pine F. to Red Fir F.; Siskiyou and w. Modoc cos. to Shasta Co. June–Aug.

6. **P. hùmilis** Nutt. ex Gray. [*P. puberulus* Jones.] Stems densely cespitose, forming clumps 1–3 dm. high; herbage cinereous-puberulent below, grayish, the basal rosette well developed; lvs. rather firm, entire, those of rosette mostly lanceolate, 1.5–3 cm. long, on equally long petioles, the cauline oblong or the upper linear-lanceolate; thyrsus glandular-pubescent, of 3–6 ± confluent few-fld. clusters; calyx 3–5 mm. high, the lobes ovate to broadly lanceolate, obtuse to short-acuminate; corolla azure-blue to blue-lavender with purplish tube, 12–16 mm. long, subtubular; lower lip longer than upper; anther-sacs ovate to rounded, ± explanate, ca. 0.5 mm. long; staminode reaching orifice, prominently golden-tufted at apex.—Dry slopes; Sagebrush Scrub, Pinyon-Juniper Wd.; e. Mono Co.; to Ore., Colo., Wyo. May–July.

7. **P. cinèreus** Piper. Stems slender, often purplish, forming clumps 1–5 dm. high, herbage cinereous-puberulent below the infl., the basal rosette well developed; lvs. rather firm, the basal lanceolate to lance-ovate, acute or obtuse, mostly 1–2 cm. long, on equally long petioles, the cauline narrower, shorter; thyrsus glandular-pubescent, of 3–9 distinct few-fld. clusters, the cymules rather lax; calyx 2–3 mm. long, the lobes ovate, obtuse to acute; corolla bright blue to blue-purple, 9–13 mm. long, subtubular or gradually ampliate; anther-sacs round to ovate, boat-shaped or ± explanate, ca. 0.5 mm. long; staminode reaching orifice, prominently bearded with short golden hairs; $n = 8$ (Keck, 1945).—Volcanic gravels, 3500–6000 ft.; Sagebrush Scrub, N. Juniper Wd.; Modoc and Siskiyou cos.; Ore., Nev. May–July.

8. **P. anguíneus** Eastw. [*P. Rattanii* var. *minor* Gray. *P. m.* Keck.] Stems entirely glabrous below, 3–8 dm. high; lvs. glabrous, serrate to finely denticulate or subentire, the basal ovate to oval, 3–8 cm. long, on petioles almost as long, the lower cauline oblong, the upper triangular-ovate, cordate-amplexicaul, 2–5 cm. long; thyrsus ± glandular-pubescent, of 3–10 dense clusters or more openly paniculate with lower divergent peduncles up to 10 cm. long; calyx 4–7 mm. high, the lobes entire, lanceolate; corolla blue-violet to deep lavender, with bright purple tube, 13–18 mm. long, 4–6 mm. wide pressed, rather abruptly ampliate, the upper lip short, erect, the lower longer, spreading; anther-sacs broadly ovate, nearly explanate, 0.8–1.1 mm. long; staminode exserted, sparsely bearded or glabrous; caps. ca. 5 mm. long; seeds dark, ca. 1 mm. long; $n = 8$ (Keck, 1945).—Dry or moist places, 2300–7000 ft.; Mixed Evergreen F. to Red Fir F.; Glenn and Humboldt cos. to w. Ore. June–Aug.

9. **P. Rattánii** Gray. Stems stoutish, glabrous below, 3–12 dm. high; lvs. glabrous, undulate-serrate to dentate, the basal lanceolate to oval, 4–20 cm. long, short-petioled, the cauline oblong, sessile, upper cordate-amplexicaul; thyrsus glandular-pubescent, of 2–7 clusters, leafy below, the lower peduncles divergent, 1–4 cm. long, the cymules lax; calyx 7–9 mm. long, the lobes lanceolate, attenuate or acute, exceeding mature caps.; corolla pale lavender to red-purple or violet-purple, 24–30 mm. long, 8–10 mm. wide pressed, shaped like *anguineus;* anther-sacs broadly ovate, nearly explanate, 1.2–1.4 mm. long; staminode well exserted, moderately long-bearded; $n = 8$ (Keck, 1945).—Grassy slopes and woods, below 6000 ft.; Mixed Evergreen F., Redwood F., Yellow Pine F.; Mendocino Co. to sw. Ore. May–Aug.

Ssp. **Kleèi** (Greene) Keck. [*P. K.* Greene.] Calyx-lobes ovate-oblong, obtuse, 6–7 mm. long, shorter than mature caps.—At 1300–3200 ft.; Chaparral and Yellow Pine F.; Santa Cruz Mts. May–June.

10. **P. deùstus** Dougl. ex Lindl. Stems woody and much-branched below, forming clumps 2–6 dm. high, erect, glabrous or glandular-puberulent; lvs. bright green, coarsely dentate-serrate, those of sterile shoots 1–4.5 cm. long, 0.6–2 cm. wide, short-petioled, those of fertile shoots linear-lanceolate to elliptic-ovate, sessile or clasping, all acute to acuminate; thyrsus strict, sparingly glandular; calyx-lobes lanceolate to ovate-attenuate; corolla ochroleucous, prominently marked with purplish guide lines, 10–16 mm. long, subtubular, the upper lip shorter than lower, sparingly glandular externally and internally; anther-sacs orbicular, explanate, widely divaricate, 0.7 mm. long; staminode reaching orifice, usually glabrous, sometimes distally short-bearded; caps. ca. 4 mm. long; $n = 8$ (Keck, 1951).—Dry rocky places, below 8200 ft.; mostly Yellow Pine F., Red Fir F.;

Sierra Nevada from Alpine Co. n. and w. to Siskiyou Co. then s. to Trinity and Glenn cos. in Coast Ranges; to Wash., Wyo. May–July.

Ssp. **sudáns** (Jones) Penn. & Keck. [*P. s.* Jones.] Herbage and corolla prominently glandular-pubescent.—Volcanic soils, 4000–5500 ft.; Sagebrush Scrub, Yellow Pine F.; Lassen Co.

Ssp. **heteránder** (T. & G.) Penn. & Keck. [*P. h.* T. & G.] Very woody, with reduced stems; lvs. mostly narrow and fine-toothed, glabrous, glaucescent; corolla obscurely viscid-puberulent without and glabrous within.—Sagebrush Scrub to Yellow Pine F.; Modoc, e. Siskiyou, Shasta, and Lassen cos.; se. Ore., nw. Nev.

11. **P. Trácyi** Keck. Suffrutescent, 0.8–1.2 dm. high, the herbage light green, glaucescent, glabrous throughout; lvs. coriaceous, cuneate-oblong to oval or round, mostly entire, some finely denticulate, those of the basal rosette numerous, short-petiolate, the cauline usually tapering to a sessile base; thyrsus contracted, dense, 2–4 cm. long, of 2–3 clusters; calyx 2.5–3 mm. long, the lobes ovate, acute, with narrowly hyaline erosulate margin; corolla pink, 11–13 mm. long, tubular, the palate densely villous, the limb small; anther-sacs orbicular, explanate, 0.4 mm. long; staminode included, sparsely bearded distally.—Rock-crevices, at 7000 ft., Devil's Canyon Mts., Trinity Co. July–Aug.

12. **P. albomarginàtus** Jones. Stems 1.5–3 dm. high, many, from an elongate fleshy deeply buried root, the whole plant pallid, glaucescent, glabrous; lvs. entire, spatulate to rhombic-ovate, 1–3 cm. long, on shorter petioles, lvs. and calyx-lobes narrowly bordered with a scarious white ± scabrid margin; thyrsus leafy, 5–12 cm. long; calyx-lobes linear-lanceolate to broadly ovate-oblong, 4–5 mm. long; corolla lavender-pink, whitish ventrally, with purple guide lines within, 13–18 mm. long, up to 5 mm. wide at throat, the 2 palatal ridges bearded with flat yellow hairs; anther-sacs broadly ovate, explanate; staminode reaching orifice, glabrous; caps. ca. 7 mm. long.—Deep sand, at ca. 1800 ft.; Creosote Bush Scrub; near Lavic, s. Mojave Desert; s. Nev., w. Ariz. March–May.

13. **P. Thompsòniae** (Gray) Rydb. [*P. pumilus* var. *T.* Gray.] Stems prostrate or ascending, arising from a woody caudex, forming tufts or mats, 0.2–0.5 dm. high and 1–2.5 dm. across; lvs. entire, oblanceolate to spatulate-oblong, 0.5–1.2 cm. long, mucronate, cinereous-whitened with closely appressed hairs, narrowed to short petioles; thyrsus racemiform, leafy, obscurely viscid; calyx-lobes acuminate to attenuate, with or without a narrow scarious margin toward the base, 4–5 mm. long; corolla blue-violet, 13–18 mm. long, up to 5 mm. wide at throat, subtubular, the palate bearded; anther-sacs ovate-oblong; staminode golden-bearded most its length; caps. ca. 5 mm. long.—Dry slopes of limestone, at ca. 6000 ft.; Pinyon-Juniper Wd.; Clark Mt., e. San Bernardino Co.; to se. Utah, n. Ariz. May–June.

14. **P. califórnicus** (M. & J.) Keck. [*P. linarioides* var. *c.* M. & J.] Stems 0.5–1.5 dm. tall, forming matted tufts, densely leafy below, the herbage cinereous-puberulent with retrorse appressed hairs; lvs. linear-oblanceolate, entire, thickish, 5–9 mm. long, mucronate, short-petioled; thyrsus racemiform, minutely glandular; calyx-lobes ovate, acute to acuminate, 3.5–5 mm. long, the margin scarious; corolla blue with purplish cast, 14–18 mm. long, tubular-funnelform, strongly bilabiate; anther-sacs ovate-oblong; staminode yellow-bearded most of its length or mostly apically; caps. 5–7 mm. long.—Local, stony slopes, 3500–7000 ft.; Chaparral, Yellow Pine F.; Hemet V. (San Jacinto Mts.), Santa Rosa Mts., Aguanga; n. L. Calif. May–June.

15. **P. Thúrberi** Torr. [*P. ambiguus* var. *T.* Gray. *Leiostemon T.* Greene.] Intricately branched subshrub 3–6 dm. high, with numerous slender erect mostly unbranched stems, glabrous throughout; lvs. bright green, equally distributed, entire, ± scabrid on margin, mostly narrowly linear, involute, 1–3 cm. long, ca. 1 mm. wide, but some earlier lvs. to 3 mm. wide; thyrsus racemose, the short divergent peduncles mostly 1-fld.; calyx 2–3 mm. long, the lobes broadly ovate, abruptly acuminate, entire; corolla lavender-rose or bluish, 12–15 mm. long, obliquely salverform, the limb prominent; anther-sacs ovate, explanate, 0.7 mm. long; staminode included, glabrous; caps. 6–7 mm. long.—Local, dry gravelly places, 800–5100 ft.; Creosote Bush Scrub, Joshua Tree Wd., Pinyon-Juniper Wd.; San Felipe V. (e. San Diego Co.), Little San Bernardino Mts., Providence Mts. (Mojave Desert); to New Mex., L. Calif. May–June, sometimes also in fall.

16. **P. monoénsis** Heller. [*P. divergens* Jones.] Stems 1.5–3.5 dm. high, densely

cinereous-puberulent; lvs. entire, the margin often crisped, densely scurfy puberulent, oblong-ovate to lance-oblong, the basal 4–12 cm. long, with petioles 2–4 cm. long, the cauline elliptic to deltoid-ovate, the upper broadly clasping; thyrsus densely glandular-pubescent, of 4–8 dense clusters; calyx 7–8 mm. long (to 12 mm. in fr.), the lobes linear-lanceolate; corolla rose-purple or wine-red, 14–20 mm. long, 4–6 mm. wide pressed, tubular-funnelform, glabrous within or sometimes the palate sparingly pilose; anther-sacs divergent, dehiscent quite to the proximal apices, not explanate, 1.5 mm. long; staminode included, strongly bearded for its outer half with fine short yellow hairs; caps. 6–8 mm. long.—Dry stony places, 3800–6000 ft.; Joshua Tree Wd., Sage-brush Scrub; base of White and Inyo mts., Mono and Inyo cos. April–May.

17. **P. calcàreus** Bdg. [*P. desertorum* Jones.] Stems 0.5–2.5 dm. tall, densely pruinose-puberulent, purplish; lvs. firm, mostly entire, largely basal, ovate to elliptic, 1–3.5 cm. long, on petioles as long, the cauline linear- to oblong-lanceolate, smaller, the upper-most often subcordate-amplexicaul; thyrsus densely glandular-pubescent, of 2–8 congested clusters; calyx 6 mm. long in fl., to 11 mm. in fr., the lobes linear-lanceolate; corolla light rose to rose-purple, 12–16 mm. long, 2.5–4 mm. wide pressed, tubular, the palate sparingly pilose; anther-sacs widely divaricate, round, peltately explanate, 0.5 mm. long; staminode included, strongly bearded for ⅔ its length with coarse golden hairs; caps. 5–7 mm. long.—Dry crevices in limestone, 3500–6000 ft.; Creosote Bush Scrub to Pinyon-Juniper Wd.; Grapevine Mts. (Inyo Co.), Providence Mts. (San Bernardino Co.). April–May.

18. **P. mìser** Gray. Stems 1–2.5 dm. high, cinereous-puberulent; lvs. mostly entire, densely cinereous-puberulent, the basal linear-lanceolate to elliptic, the upper linear to oblong; thyrsus densely glandular-pubescent, compact; calyx 8–12 mm. high, the lobes lanceolate, acuminate; corolla dull purple, with purple guide lines, 13–28 mm. long, 5–10 mm. wide pressed; anther-sacs widely divaricate, broadly ovate, very small, peltately explanate; staminode prominently exserted, hooked, strongly bearded with stiffish orange velvety hairs.—Sandy and gravelly places, 3500–7300 ft.; Sagebrush Scrub, N. Juniper Wd., Yellow Pine F.; Modoc and Lassen cos.; Ore., Nev. May–July.

19. **P. speciòsus** Dougl. ex Lindl. [*P. pilifer* (as *piliferus*) Heller. *P. glaber* and var. *utahensis* of Jeps.] Herbage glabrous to pruinose-puberulent, sometimes glaucescent; stems in erect clumps 2–8 dm. high; lvs. entire, thickish, the basal lanceolate to ob-lanceolate or spatulate, 3–8 cm. long, on petioles almost as long, the cauline linear-lanceolate, gradually reduced upward, sessile; thyrsus elongate of many obscurely interrupted showy clusters, ± secund; calyx 4–6(–8) mm. high, the lobes narrowly ovate to broadly oblong or suborbicular with short tip; corolla bright blue-purple, 25–35 mm. long, 8–10 mm. wide pressed, glabrous, the tube rather long, abruptly flaring into the ample throat, the limb large, strongly 2-lipped; anther-sacs divaricate, sigmoid-curved, 2–2.4 mm. long, opening from the apex for ⅔ the distance to the line of contact, finely toothed on suture; staminode glabrous or rarely bearded; caps. 6–12 mm. long; seeds ca. 2 mm. long; *n* = 8 (Keck, 1951).—Dry plains and slopes, 3500–8000 ft.; Sagebrush Scrub to Red Fir F.; borders of Mojave Desert from San Bernardino Mts. to Tehachapi, e. face of Sierra Nevada, White Mts., etc., to Modoc and Siskiyou cos.; to Wash., Ida., Utah. May–July.

Ssp. **Kennédyi** (A. Nels.) Keck. [*P. K.* A. Nels.] Flowering calyx 8–12 mm. high, the lobes long-tipped.—Mostly 8000–10,400 ft.; Sierra Nevada, White Mts.; Nev. June–Aug.

20. **P. Pálmeri** Gray. Gray-glaucous and glabrous below the infl., the stems 5–12 dm. tall; lvs. irregularly spinose-dentate or the uppermost subentire, the basal oblong-ovate, 2–8 cm. long, on petioles almost as long, the cauline lance-ovate, obtuse or acute, auriculate-clasping, the upper pairs usually connate-perfoliate; thyrsus glandular-pubescent, virgate, secund, 2–6 dm. long; calyx 4–6 mm. long, the lobes broadly ovate; corolla whitish suffused with pink or lilac, with prominent guide lines extending into throat from lower lip, 22–35 mm. long, 10–20 mm. wide pressed, the tube short, expanding abruptly into the inflated throat; anther-sacs longer than broad, peltately explanate; staminode exserted, shaggy-bearded; caps. 10–14 mm. long; seeds black, minutely pitted, 1.5–2 mm. long; *n* = 8 (Keck, 1951).—Dry rocky gullies, 4000–6000 ft.; Joshua Tree Wd., Pinyon-Juniper Wd.; mts. w. of Death V., Kingston, Clark, Providence, and New York mts.; to Utah, Ariz. May–June.

21. **P. Grinnéllii** Eastw. [*P. Palmeri* var. G. M. & J.] Stems decumbent at base, ±
branched, 1–4 dm. high, forming low spreading plants, glabrous; lvs. light green, finely
to coarsely spinulose-dentate, or uppermost entire, like those of *Palmeri* but not connate;
thyrsus more lax and open, glandular-pubescent, 4–8 cm. wide, 1–2 dm. long; calyx
4–6 mm. long, the lobes ovate to lance-ovate; corolla whitish with flesh-pink or lavender
tinge, 20–30 mm. long, 10–16 mm. wide pressed, the guide lines prominent, otherwise
like that of *P. Palmeri; n* = 8 (Keck, 1951).—Dry gravelly mostly granitic slopes, 4500–
9500 ft.; mostly Yellow Pine F., Red Fir F.; San Gabriel Mts. to Santa Rosa Mts.
(Riverside Co.). May–Aug.

Ssp. **scrophularioìdes** (Jones) Munz. [*P. s.* Jones. *P. hians* Jtn.] Plants larger, 3–6
dm. tall, glaucous; lvs. blue-green; corolla with violet to blue-violet limb.—Dry slopes,
1800–6000(–8000) ft.; Foothill Wd., Pinyon-Juniper Wd.; Ventura Co. to Kern Co.
and s. Sierra Nevada of Tulare Co., Coast Ranges from Mt. Hamilton to Santa Lucia
Mts. April–July.

22. **P. fruticiformis** Cov. Glabrous glaucous shrub, much-branched, 3–6 dm. high;
lvs. ± entire, narrowly linear-lanceolate, 2–6 cm. long, 3–6(–8) mm. wide, ± involute
on margin; thyrsus lax, short, few-fld.; calyx 5–7 mm. long, the lobes ovate to roundish,
abruptly acute to short-acuminate; corolla 20–27 mm. long, 10–13 mm. wide pressed,
white or flesh-colored, with pale lavender limb, the guide lines purple, glabrous externally, otherwise like *P. Palmeri; n* = 8 (Keck, 1951).—Dry rocky places, 3600–
7500 ft.; Creosote Bush Scrub to Pinyon-Juniper Wd.; Panamint, Argus, and Inyo mts.,
Inyo Co. May–June.

Ssp. **amargòsae** Keck. Calyx-lobes lance-ovate to broadly ovate; corolla glandular-
puberulent externally, scarcely at all within.—Kingston Mts., e. San Bernardino Co.;
w. Nev.

23. **P. flóridus** Bdg. Stems several, erect, virgate, 6–12 dm. high, the herbage blue-
glaucous, glabrous below the infl.; lvs. irregularly spinulose-dentate or the uppermost
subentire, the basal oblong-ovate, 2–6 cm. long, on fairly long petioles, the lower cauline
lance-ovate, 5–10 cm. long, sessile; thyrsus glandular-pubescent; calyx-lobes ± ovate,
4–7 mm. long; corolla rose-pink, with dark guide lines within, 22–30 mm. long, 12–15
mm. wide pressed, abruptly inflated, the orifice oblique, not villous within; staminode
glabrous; *n* = 8 (Keck, 1951).—At 5500–8500 ft.; Pinyon-Juniper Wd.; in and near
White and Inyo mts.; w. Nev. May–July.

Ssp. **Austínii** (Eastw.) Keck. [*P. A.* Eastw.] Corolla gradually ampliate, the orifice
perpendicular, the throat 6–10 mm. wide pressed.—At 3500–6500 ft.; Inyo and Panamint
ranges. May–June.

24. **P. pseudospectábilis** Jones. Habit of *P. floridus;* lvs. glaucous, prominently serrate,
the basal lance-ovate to broadly ovate, the upper cauline connate-perfoliate, forming
disks up to 15 cm. long and 8 cm. broad; thyrsus sparingly glandular; calyx-lobes mostly
ovate, short-acuminate, 4–6 mm. long; corolla rose-purple, 20–26 mm. long, 6–9 mm.
wide pressed, moderately ampliate, viscid-puberulent at orifice; staminode glabrous;
n = 8 (Keck, 1951).—Desert washes and canyons, below 4000 ft.; Creosote Bush Scrub;
Sheephole Mts. (San Bernardino Co.) to Chuckwalla Mts. (Riverside Co.) and Imperial
Co.; Ariz. March–May.

25. **P. Clevelándii** Gray. Stems few to several, 3–7 dm. high; lvs. glaucescent to deep
green, entire to moderately serrate, the basal ovate, 2–5 cm. long, petioled, the upper
cauline triangular-lanceolate to cordate, distinct; thyrsus narrowly racemose, 1–3 dm.
long, 3–6 cm. broad, ± glandular pubescent; calyx-lobes ovate to roundish, purplish,
obtuse to acuminate, 4–5 mm. long; corolla crimson to red-purple, 17–24 mm. long,
5–8 mm. wide pressed, tubular-funnelform, the proper tube shorter than the gradually
ampliate throat, the limb rotately spreading, glandular-puberulent; anther-sacs glabrous,
explanate; staminode 9–11 mm. long, glabrous or feebly bearded; *n* = 8 (Keck, 1951).
—Dry open slopes, 2500–4500 ft.; Chaparral; e. San Diego Co. to L. Calif. March–
May.

Ssp. **connàtus** (M. & J.) Keck. [*P. C.* var. *c.* M. & J.] Glaucous; upper lvs. connate-
perfoliate; infl. glabrous; corolla-limb not glandular; anthers not explanate, ciliolate-
denticulate; staminode bearded.—Below 5500 ft., canyons bordering Colo. Desert,
Riverside Co.

Ssp. **mohavénsis** Keck. Lvs. bright green, coarsely serrate; upper lvs. not connate-perfoliate; corolla contracted at orifice, narrow in throat; staminode 6–8 mm. long, bearded.—Dry rocky places, 3500–4500 ft.; Creosote Bush Scrub, Joshua Tree Wd.; Little San Bernardino Mts., Sheephole Mts., s. Mojave Desert. March–May.

26. **P. Stephénsii** Bdg. [*P. Clevelandii* var. S. M. & J.] Habit of preceding; lvs. mostly finely and sharply denticulate, the upper pairs connate-perfoliate; thyrsus sparingly glandular; calyx 3–4.5 mm. long, the roundish to broadly ovate lobes abruptly acute; corolla rose to magenta, 17–22 mm. long, without prominent guide lines, 4–6 mm. wide pressed, the throat slightly dilated; anther-sacs glabrous, peltately explanate, at least as broad as long; staminode included, glabrous.—Rocky slopes, 5000–6000 ft.; Shadscale Scrub, Sagebrush Scrub; Kingston and Providence mts., San Bernardino Co. April–June.

27. **P. incértus** Bdg. [*P. fruticiformis* var. *i.* M. & J.] Habit of *P. fruticiformis;* lvs. 2–3 mm. wide; thyrsus lax, moderately glandular; calyx 5–7 mm. high, the lobes lance-ovate to roundish; corolla 25–28 mm. long, 8–12 mm. wide pressed, violet with a reddish cast or purple, the limb deep blue without guide lines, strongly 2-lipped; anther-sacs divaricate but not explanate, minutely denticulate-ciliolate at suture; staminode well included, densely bearded.—Dry gravelly slopes and flats, 3500–5500 ft.; Joshua Tree Wd., Sagebrush Scrub; Lone Pine Creek (Inyo Co.) to Walker Pass region (Kern Co.), then to n. Los Angeles Co. and base of San Bernardino Mts. May–June.

28. **P. spectábilis** Thurb. ex Gray. Stems several from the base, erect, 8–12 dm. high, green or glaucescent, glabrous throughout; lvs. coarsely serrate, the lower broadly oblanceolate to ovate, 2–10 cm. long, 2–5 cm. broad, the upper connate-perfoliate; thyrsus lax, often half as tall as the plant; calyx 4–7 mm. high, the lobes lance-ovate to roundish; corolla lavender-purple with blue lobes, whitish within, 25–33 mm. long, 8–12 mm. wide pressed, the tube rather abruptly expanded into the ample throat, the limb strongly bilabiate; anther-sacs not explanate, twice as long as wide; staminode glabrous toward tip; caps. 10–14 mm. long; seeds black, ca. 1.5 mm. long, angled; $n = 8$ (Keck, 1951).—Dry washes and recently disturbed places, below 6000 ft.; Coastal Sage Scrub, Chaparral; cismontane slopes from e. Los Angeles Co. to n. L. Calif. April–June.

Ssp. **subviscòsus** Keck. Pedicels and calyces glandular-puberulent.—San Gorgonio Pass to Santa Monica Mts. and Liebre Mts. (Los Angeles Co.)

29. **P. × Paríshii** Gray. Habit of *P. centranthifolius;* herbage glaucescent and glabrous; lvs. entire to shallowly serrate, the uppermost clasping but not connate; thyrsus virgate; corolla red-purple, the gradually ampliate throat 6–9 mm. wide.—Occasional in region where *P. centranthifolius* and *P. spectabilis* come together and apparently a hybrid between them.

30. **P. centranthifòlius** Benth. Scarlet Bugler. Glabrous, glaucous; stems few to several, virgate, 3–12 dm. high; lvs. entire, the basal spatulate, 3–7 cm. long on much shorter petioles, the cauline lanceolate, the upper auriculate-clasping; thyrsus virgate, half as tall as the plant, secund; calyx 3–6 mm. long, the lobes ovate to rounded, abruptly acute, the margin broadly scarious, entire to erose; corolla scarlet, tubular, 25–33 mm. long, 4.5–6 mm. wide pressed, the lobes scarcely spreading, glabrous; anther-sacs peltately explanate; staminode glabrous; $n = 8$ (Keck, 1951).—Dry disturbed places, below 6500 ft.; Chaparral, etc.; Coast Ranges, Lake Co. to San Diego Co., commonest in s. Calif.; L. Calif. April–July.

31. **P. utahénsis** Eastw. [*P. Eastwoodiae* Heller.] Much like *P. centranthifolius*, 3–6 dm. high; cauline lvs. lance-oblong, 3–8 cm. long, 0.5–1.5 cm. wide, broadest at clasping base; corolla carmine, 18–24 mm. long, 4–6 mm. wide pressed, the lobes glandular-pubescent without, glandular about orifice; staminode uncinate at apex, glabrous or slightly bearded.—Occasional, in rocky places, 4000–5500 ft.; Shadscale Scrub to Pinyon-Juniper Wd.; New York and Kingston mts., e. Mojave Desert; to s. Utah, n. Ariz. April–May.

32. **P. confùsus** Jones ssp. **pàtens** (Jones) Keck. [*P. c.* var. *p.* Jones.] Like *P. utahensis* but lower, more leafy; thyrsus more open, compound; corolla rose-lavender to purplish, 14–20 mm. long, 5–6.5 mm. wide, slightly ampliate, glabrous; anther-sacs not explanate, scabrid-ciliolate at suture; staminode uncinate, papillose-bearded at apex; $n = 8$ (Keck, 1951).—Dry loose soil, washes and slopes, 6000–7500 ft.; Pinyon-Juniper Wd., Sagebrush Scrub; hills surrounding Owens V., Inyo and Mono cos. May–June.

33. **P. Eatònii** Gray. Stems few to several, glabrous, virgate, 3–10 dm. high; lvs. coriaceous, green or glaucescent, glabrous, the basal oblanceolate, 3–10 cm. long, on petioles almost as long, the cauline lance-oblong with clasping base, 4–10 cm. long, 1–3 cm. wide; thyrsus strict, secund, half the height of the stem, glabrous; calyx 4–6 mm. high, the lobes elliptic to broadly ovate, acute to abruptly short-acuminate, the narrow scarious margin entire; corolla scarlet, 25–30 mm. long, 6–8 mm. wide pressed, subtubular, obscurely 2-lipped, glabrous; anther-sacs parallel or divergent, the slits from the free tips for ½ to ⅔ their length, puberulent, the suture finely toothed; staminode glabrous or ± bearded; $n = 8$ (Keck, 1951).—Dry gravelly slopes, below 8000 ft.; Pinyon-Juniper Wd. to Creosote Bush Scrub; desert slopes from San Bernardino Mts. e. to Clark Mt.; to Utah, Nev. March–July.

Ssp. **undòsus** (Jones) Keck. [*P. E.* var. *u.* Jones. *P. Munzii* Jtn.] Stems and lvs. puberulent; anthers often ± exserted.—With the sp., but more frequent; to Utah, Ariz.

34. **P. labròsus** (Gray) Hook. f. [*P. barbatus* var. *l.* Gray.] Mostly bright green, glabrous throughout, the stems 1–few, erect, virgate, 3–7 dm. high; lvs. coriaceous, the basal linear-oblanceolate, obtuse, short-petioled, the blades 3–7 cm. long, 0.4–0.8 cm. wide, on petioles almost as long, the cauline linear, rapidly reduced up the stem; thyrsus strict, slender, somewhat secund; calyx 4–5.5 mm. long, the lobes ovate, acuminate, the margin scarious, erose to entire; corolla scarlet, 32–40 mm. long, 5–6 mm. wide pressed, tubular, the limb ca. ⅜ the length of the whole, the upper lip erect, the lower more deeply divided, with reflexed linear lobes; anther-sacs opening by slits for ⅔ their length, divergent, glabrous save for minute denticulation on suture; staminode glabrous; $n = 8$ (Keck, 1951).—Fairly dry slopes and benches, 5000–10,000 ft.; Montane Coniferous F.; mts. from Ventura Co. to San Diego Co.; n. L. Calif. July–Aug.

35. **P. graciléntus** Gray. Stems many from compact crown, 2–7 dm. high; herbage bright green to ± glaucescent, glabrous below the ± glandular-pubescent infl.; lvs. entire, thin, mostly basal, the lowest oblanceolate, 1–5 cm. long, with petioles equally long; cauline lvs. reduced, linear-lanceolate; thyrsus compact, with ca. 3–5 nodes; calyx 3.5–5 mm. high, the lobes lanceolate to oblong-ovate; corolla purplish-blue to red-purple, 13–16 mm. long, 4–5 mm. wide pressed, slightly ampliate, the lower lip villous within; anther-sacs dark, ca. 1 mm. long, dehiscent less than 0.5 mm.; staminode yellow-bearded; $n = 8$ (Keck, 1951).—Rather dry places, 4200–8300 ft.; Red Fir F., Lodgepole F.; Sierra Nevada from Lake Tahoe n., to Modoc and Siskiyou cos.; Ore., w. Nev. June–Aug.

36. **P. papillàtus** J. T. Howell. Stems few, erect, 2–4 dm. tall; herbage gray-green, cinereous-puberulent; basal lvs. elliptic to spatulate-orbicular, 1–3 cm. long, on petioles ca. as long, the cauline oblong to oblanceolate, the uppermost lanceolate; thyrsus glandular-pubescent, compact, of 3–6 nodes, the pedicels short; calyx 7–10 mm. long, the lobes lanceolate, attenuate; corolla purplish-blue, 24–30 mm. long, 5–10 mm. wide pressed, moderately ampliate, lower lip the longer, glabrous within; anther-sacs pale, 1.5–1.9 mm. long, minutely toothed along suture, otherwise glabrous, dehiscent less than half their length; staminode yellow-bearded.—Rocky open slopes, 7000–9000 ft.; Pinyon-Juniper Wd., Lodgepole F.; e. slope of Sierra Nevada, Mono and Inyo cos. June–July.

37. **P. scapoìdes** Keck. Stems few from a matted branching caudex, erect, slender, 2–4 dm. high, glabrous, glaucous; lvs. mostly basal, ovate to roundish, often folded, 0.7–1.5 cm. long, on petioles as long, 5–10 mm. wide, the cauline much reduced, few, with long internodes; thyrsus glandular-pubescent, few-fld., lax, the solitary fls. on long pedicels; calyx 3–4.5 mm. high, the lobes oblong to broadly ovate, abruptly acute; corolla 26–34 mm. long, 5–7.5 mm. wide pressed, blue, the throat pale beneath and within, only slightly ampliate, the lower lip somewhat exceeding the upper, the palate 2-ridged, yellow-hairy; anther-sacs pale, 1.5–1.7 mm. long, toothed along the suture, dehiscent less than half their length; staminode yellow-pilose below the tip; $n = 8$ (Keck, 1951).—Dry stony ridges and canyons, 7000–10,000 ft.; Pinyon-Juniper Wd., Bristle-cone Pine F.; White and Inyo mts. June–July.

38. **P. caèsius** Gray. Loosely cespitose from a matty wooded caudex, the herbage ± glaucous, glabrous below, the stems erect, 1.5–4.5 dm. high; lvs. mostly basal, coriaceous, the lower roundish, 1–2 cm. long, on petioles as long, the cauline remote, reduced, the upper linear-oblanceolate; thyrsus glandular-pubescent, lax, rather few-fld.; calyx 4–7

mm. high, the lobes oblong or ovate, obtuse or abruptly acute; corolla purplish-blue, 17–23 mm. long, 4.5–6.5 mm. wide pressed, gradually ampliate, the lips equal, small, glabrous within; anther-sacs pale, 1.3–1.5 mm. long, short-toothed on the suture; staminode glabrous.—Dry rocky slopes, 6700–11,200 ft.; Yellow Pine F. to Subalpine F.; San Bernardino and San Gabriel mts., Sierra Nevada of Tulare Co. June–Aug.

39. **P. laètus** Gray. Woody at base, 2–8 dm. high, the herbage gray- or yellow-green, mostly puberulent or canescent, the stems often purplish; lvs. linear to oblanceolate, the upper lanceolate, 1–6 cm. long, 0.2–1.2 cm. wide, the lower petioled; thyrsus glandular-pubescent, narrow, but somewhat lax; calyx 4–8 mm. high, the lobes lanceolate to narrow-ovate or oblong, acute to acuminate; corolla 20–30 mm. long, 8–12 mm. wide pressed, blue-lavender to blue-violet, tubular-campanulate, the limb bright blue, widely gaping, glabrous within; anther-sacs tinged purple, 2–2.8 mm. long, mostly white-hairy, the suture spinose; staminode glabrous; $n = 8$ (Keck, 1951).—Dry rocky and disturbed slopes, 1200–8500 ft.; Foothill Wd. to Red Fir F.; mts. from Ventura Co. through Tehachapi Mts., w. slopes of Sierra Nevada n. to Yuba Co. May–July.

KEY TO SUBSPECIES

Anthers broadly oval in outline, ca. as broad as long, dehiscent ½–⅗ their length; corolla-throat not constricted at junction with lip, the lips gaping.
 Corolla 20–30 mm. long.
  Calyx 4–8 mm. long; sepals lanceolate or broader, acute or acuminate .... *P. laetus*
  Calyx 8–15 mm. long; sepals linear-lanceolate, attenuate .... ssp. *leptosepalus*
 Corolla 14–20 mm. long .... ssp. *Roezlii*
Anthers narrowly sagittate, ½–⅔ as broad as long, dehiscent ¾–⅘ their length; corolla 20–30 mm. long; throat somewhat constricted at junction with lip, the lips scarcely spreading .... ssp. *sagittatus*

Ssp. **leptosèpalus** (Greene ex Gray) Keck. [*P. laetus* var. *l.* Greene ex Gray.] Lvs. 4–15 mm. wide; calyx 8–15 mm. high, the linear-lanceolate lobes attenuate.—At 2000–5500 ft.; Yellow Pine F., Red Fir F.; Tehama Co. to Placer Co.

Ssp. **Roèzlii** (Regel) Keck. [*P. R.* Regel. *P. gracilentus* var. *ursorum* Jeps.] Lvs. 2–12 mm. wide; calyx 4–7 mm. high; corolla 14–20 mm. long, 5–9 mm. wide pressed.—At 2000–8000 ft.; Sagebrush Scrub to Lodgepole F.; Sierra Nevada from Mono Co. and Lake Tahoe n., to Modoc and Siskiyou cos.; Ore., Nev.

Ssp. **sagittàtus** Keck. [*P. s.* Penn.] Lvs. 2–4(–8) mm. wide; corolla 20–30 mm. long, 6–9 mm. wide pressed, ± incurved dorsally; anthers sagittate, often very slender.—Dry slopes, 2000–6000 ft.; Yellow Pine F. to Red Fir F.; Del Norte and Humboldt cos. to Modoc Co.; Ore.

40. **P. filifórmis** (Keck) Keck. [*P. laetus* ssp. *f.* Keck.] Habit of *P. laetus*, 2–5 dm. high; lvs. filiform, tightly involute, 0.5–1 mm. broad, the lower with fascicles in their axils; thyrsus mostly with 1-fld. spreading peduncles; calyx 3.5–5 mm. long, the lobes lanceolate, acuminate; corolla deep blue with purple tube, 14–18 mm. long, 5–7 mm. wide pressed; anthers 1.5–1.6 mm. long, inverted U-shaped, dehiscent half their length; $n = 8$ (Keck, 1951).—Open dry stony places; Foothill Wd., Yellow Pine F.; Shasta Co. June.

41. **P. neotéricus** Keck. Woody below, with branched caudex, the stems many, erect, 2–6 dm. high; herbage glabrous, blue-glaucous; lvs. crowded near base, more remote above, coriaceous, the lower narrowly oblanceolate to spatulate, 1–4 cm. long, 3–9 mm. wide, short-petioled, the upper lanceolate to lance-ovate, the uppermost amplexicaul; thyrsus strict, glandular-pubescent, usually elongate, the divergent peduncles 1–2-fld.; calyx 4–7 mm. long, the lobes lanceolate to ovate-oblong; corolla blue-purple, 25–35 mm. long, 8–12 mm. wide pressed, tubular-campanulate, gaping, the limb azure, glabrous within; anther-sacs 2.5–3.2 mm. long, hairy at sinus, the sacs dehiscent ½–¾ their length, the suture-margin spinose; staminode glabrous; $n = 32$ (Keck, 1951).—Dry places, 3500–6000 ft.; Yellow Pine F.; Sierra Co. to Lassen and Shasta cos. May–Aug.

42. **P. azùreus** Benth. [*P. glaucifolius* Gray. *P. Jeffreyanus* Hook. *P. a.* var. *J.* Gray. *P. heterophyllus* var. *azureus* Jeps.] Woody below, 2–5 dm. high, the herbage blue-glaucous and glabrous throughout; basal lvs. oblanceolate to obovate, 1–5 cm. long, short-petiolate, 5–18 mm. wide, the cauline lanceolate to ovate; thyrsus strict, sub-secund; calyx 3.5–6 mm. high, the lobes oblong or obovate, abruptly contracted to a mucronate tip; corolla deep blue-purple, 20–30 mm. long, 7–12 mm. wide pressed, tubular-campanulate, gaping, glabrous; anthers cordate, 2.2–3.3 mm. long, ca. as

broad, ± hirsute at sinus, spinose on suture margin; staminode usually glabrous, dilated at tip; $n = 24$ (Keck, 1951).—Dry slopes, 3500–7500 ft.; Chaparral, Yellow Pine F.; Fresno Co. to Butte, Shasta, Glenn and Humboldt cos., and n. to sw. Ore. May–Aug.

Ssp. **angustíssimus** (Gray) Keck. [*P. azureus* var. *a.* Gray.] Herbage paler yellow-green; lower lvs. 2–5 mm., the upper 3–9 mm. wide; calyx-lobes oblong or obovate, with a subulate tip 1–3 mm. long.—At 1200–3000 ft.; Sagebrush Scrub, Foothill Wd., Yellow Pine F.; Sierra Nevada foothills from Butte Co. to Fresno Co., rare in w. Glenn Co.

43. **P. párvulus** (Gray) Krautter. [*P. azureus* var. *p.* Gray. *P. Jeffreyanus* var. *p.* Jeps.] Resembling *P. azureus*, the stems 2–3.5 dm. long; basal lvs. oblanceolate to spatulate, 1–3 cm. long, short-petioled, the cauline lanceolate to narrow-ovate, semi-amplexicaul, 6–10 mm. wide; thyrsus narrow, short; calyx-lobes lanceolate to oblong or oval, often mucronate or attenuate; corolla 14–20 mm. long; anthers 1.4–1.8 mm. long; $n = 16$ (Keck, 1951).—Dry slopes, 2400–9500 ft.; Montane Coniferous F.; Siskiyou Mts. to Scott Mts. (Trinity Co.), high Sierra Nevada of Tulare and Fresno cos.; s. Ore. June–Aug.

44. **P. heterophýllus** Lindl. [*P. leucanthus* Greene.] Woody at base, forming clumps 3–5 dm. high, glabrous throughout or sometimes minutely puberulent below, green or glaucous; lvs. linear, usually fasciculate, 2–4.5 cm. long, 2–4 mm. wide; thyrsus strict, subracemose, glabrous; calyx 4–6 mm. long, the lobes mostly oblanceolate to obovate, abruptly acuminate or subulate, glabrous; corolla rose-violet, with blue or lilac lobes, 25–35 mm. long, 9–12 mm. wide pressed, glabrous, gaping; anther-sacs ca. 2.5 mm. long, sagittate, arcuate, usually hirsute at sinus, spinose on suture margin; staminode moderately dilate, glabrous; $n = 8$, 16 (Keck, 1951).—Dry hillsides, below 5500 ft.; Chaparral, Foothill Wd., Yellow Pine F., etc.; Coast Ranges from Humboldt Co. to San Diego Co. April–July.

Ssp. **Púrdyi** Keck. Plant 2.5–7 dm. high, puberulent throughout; lvs. 2.5–9 cm. long, rarely fasciculate; sepals usually glabrous.—Below 5000 ft.; Coast Ranges from San Benito Co. to Humboldt and Trinity cos., Sierran foothills from Butte Co. to Placer Co. May–June.

Ssp. **austràlis** (M. & J.) Keck. [*P. h.* var. *a.* M. & J.] Puberulent almost throughout; lvs. narrow, fasciculate; calyx usually puberulent.—Below 5000 ft.; Chaparral; San Diego Co. to Monterey Co. May–June.

45. **P. Purpùsii** Bdg. Crown ± woody; stems 1–2 dm. long, spreading and rooting in age, the younger ascending to decumbent; herbage densely canescent; lvs. mostly entire, the lower oval to roundish, ca. 1–1.5 cm. long, with petioles half as long, the cauline oval to lanceolate; thyrsus crowded, short, glandular-pubescent; calyx 5–10 mm. long, the lobes linear-lanceolate to broadly oval, attenuate to obtuse; corolla violet shading to blue, 20–30 mm. long, 5–8 mm. wide pressed, the ample throat constricted at orifice, the lips scarcely spreading, glabrous within; anthers sagittate, arcuate, 2.5–2.8 mm. long, hirsute at sinus, spinose on suture margin; staminode glabrous, apically dilated.—Dry open stony slopes, 5000–7600 ft.; Red Fir F., Yellow Pine F.; peaks of Coast Ranges from Humboldt and Trinity cos. to Lake Co. June–Aug.

46. **P. Bridgèsii** Gray. Woody at base, 3–10 dm. high from a branched caudex; herbage yellow-green, glabrous or puberulent; lower lvs. 2–6 cm. long on somewhat shorter petioles, 2–12 mm. wide, linear-oblanceolate to spatulate, the cauline linear to elliptic, sessile; thyrsus subsecund, glandular-pubescent, mostly rather narrow; calyx 4–8 mm. high, the lobes lanceolate to lance-ovate; corolla scarlet to vermilion, 22–35 mm. long, 4–6 mm. wide pressed, tubular, sparingly glandular without and within, the lower lip sharply reflexed; anthers oblong or ovate, ca. 2 mm. long, the sacs dehiscent ¼–⅓ their length, the suture spinulose-ciliate; staminode glabrous; $n = 8$ (Keck, 1951).—Dry slopes, 5000–10,700 ft.; Montane Coniferous F.; Sierra Nevada from Alpine Co. s., White Mts., to San Diego Co. and L. Calif. across Mojave Desert to Colo., Ariz. June–Aug.

47. **P. personàtus** Keck. Stems few, erect, 3–5 dm. high; lvs. not crowded, entire or obscurely denticulate, glaucescent, glabrate above, puberulent beneath, ovate or ovate-oblong, obtuse, 3–5 cm. long, the lower 1.2–3.5 cm. wide, short-petioled; thyrsus lax, 0.7–2.5 dm. long, glandular-pubescent; calyx 5–6 mm. high, the lobes ovate-lanceolate, abruptly long-acuminate; corolla blue-purple?, personate, 20–25 mm. long, glabrous or sparsely viscid externally, densely bearded on all sides within, the limb short; anther-

sacs divaricate, subexplanate, glabrous, 1.2–1.4 mm. long; staminode scarcely 4 mm. long, densely yellow-bearded at tip.—Dry hillsides, 4000–6000 ft.; Yellow Pine F.; known from 3 colonies in Butte Co. July.

48. **P. Newbérryi** Gray. [*P. Menziesii* var. *Robinsonii* Mast. *P. M.* var. *N.* Gray.] Woody below, matted, the stems decumbent or creeping, 1.5–3 dm. high, glabrous or ± puberulent, green or glaucescent; lvs. coriaceous, elliptic to ovate, obtuse, mostly short-petioled, serrulate, 1–3.5 cm. long, 8–16 mm. wide, much reduced on fl. stems; thyrsus racemose, short, subsecund; calyx 7–12 mm. high, the lanceolate lobes attenuate to acuminate; corolla rose-red, 22–30 mm. long, 5–8 mm. wide pressed, the throat only slightly dilated, the ventral ridges bearded; anthers exserted; staminode ¾ as long as fertile fils., slender, yellow-bearded; $n = 8$ (Keck, 1951).—Rocky and gravelly places, 5000–11,000 ft.; Montane Coniferous F.; Mt. Shasta s. through the higher Sierra Nevada to Tulare Co.; w. Nev. June–Aug.

Ssp. **Bérryi** (Eastw.) Keck. [*P. B.* Eastw.] Corolla 27–33 mm. long, 8–12 mm. wide pressed; anthers included.—At 5000–7500 ft.; Red Fir F., Yellow Pine F.; Humboldt and Glenn cos. to sw. Ore.

Ssp. **sonoménsis** (Greene) Keck. [*P. s.* Greene.] Floral lvs. scarcely reduced in size; corolla dark rose-purple.—At 800–2500? ft.; peaks of Lake, Napa, and Sonoma cos.

49. **P. rupícola** (Piper) Howell. [*P. Newberryi* var. *r.* Piper.] Matted, woody at base, the flowering stems mostly less than 1 dm. high, glabrous or ± canescent, usually glaucous; lvs. elliptic to orbicular, coriaceous, short-petioled, serrate-denticulate, 0.8–2 cm. long, 0.6–1.2 cm. wide; raceme few-fld., glandular-pubescent; calyx 6–10 mm. high, the lobes lanceolate to oblong, acute to acuminate; corolla deep rose, 27–35 mm. long, 8–12 mm. wide pressed, the throat moderately dilated, the ventral ridges sparsely villous; anthers slightly exserted; staminode ½–¾ the length of fertile fils., ± bearded at filiform tip; $n = 8$ (Keck, 1951).—Rocky dry places, 6000–7700 ft.; Red Fir F.; Yolla Bolly Mts. (Tehama Co.) n.; to Wash. June–Aug.

50. **P. Davidsònii** Greene. [*P. Menziesii* ssp. *D.* Piper.] Forming creeping mats, from woody branched caudex, the flowering stems less than 1 dm. high, puberulent; lvs. elliptic to round, thick, ± glandular-punctate, entire, 0.5–1.5 cm. long, 0.4–0.8 cm. wide, short-petioled; raceme few-fld., glandular-pubescent; calyx 7–11 mm. long, the lobes linear- to ovate-lanceolate; corolla purple-violet, 18–35 mm. long, 7–12 mm. wide pressed, the throat moderately dilated, the ventral ridges ± villous; anthers included; staminode barely ½ as long as fertile bearded fils.; $n = 8$ (Keck, 1951).—Rocky places, 9000–12,000 ft.; Subalpine F., Alpine Fell-fields; Sierra Nevada from Tulare Co. n., through Siskiyou and Modoc cos. to Wash., Nev. July–Aug.

51. **P. nemoròsus** (Dougl. ex Lindl.) Trautv. [*Chelone n.* Dougl. ex Lindl.] Stems few, erect, from an unbranched woody caudex, puberulent, 3–8 dm. high; lvs. all cauline, equally spaced, thin, serrate, lanceolate to ovate, very short-petioled, mostly rounded at base, 5–10 cm. long, 1.5–4 cm. wide, glabrous or ± puberulent beneath; thyrsus glandular-pubescent, few-fld. and terminal, or larger, open and leafy; calyx 6–13 mm. long, the lobes lanceolate to ovate; corolla rose-purple to pale maroon, paler ventrally, 25–35 mm. long, 8–11 mm. wide pressed, plicate but glabrous within, strongly bilabiate; fertile fils. retrorsely puberulent above, hirsute below, the staminode ca. ⅔ as long, bearded; $n = 15$ (Keck, 1951).—Moist rocky shaded slopes, 4500–5500 ft.; Douglas-Fir F., Red Fir F.; Siskiyou Mts. n.; to B.C. June–Aug.

52. **P. Rothróckii** Gray. [*P. scabridus* Eastw.] Low rounded bush 3–6 dm. high, with many slender strict stems, puberulent throughout; lvs. sparsely scabridulous, 0.5–1.5 cm. long, 0.2–0.7 cm. wide, subsessile, lance-oblong to ovate, entire or undulate-denticulate; raceme ± glandular, spiciform, strict, the lower fls. geminate, the upper often alternate; calyx 4–6 mm. long, the lobes lanceolate; corolla dull yellow with purplish guide lines, 10–12 mm. long, 3–5 mm. wide pressed, the upper lip erect, the lower reflexed; staminode glabrous.—Dry rocky slopes, 7000–10,300 ft.; Pinyon-Juniper Wd. to Lodgepole F.; Mono, Inyo, and Tulare cos. s. to Panamint Mts.; w. Nev. June–Aug.

Ssp. **jacinténsis** (Abrams) Keck. [*P. j.* Abrams.] Corolla 13–16 mm. long.—Dry slopes, 7000–9500 ft.; San Jacinto Mts., Riverside Co.

53. **P. breviflòrus** Lindl. [*P. carinatus* and *canoso-barbatus* Kell.] Shrub 5–20 dm. high, the stems many, virgate, glabrous, glaucous, rather lax in age; lvs. 1–5(–7) cm.

long, 0.3–1.2 cm. wide, lanceolate, subsessile, entire to serrulate; thyrsus pyramidal, 1–5 dm. long, 4–15 cm. wide, many-fld.; pedicels and calyx glandular-pubescent; calyx 5–10 mm. high, the lobes ovate-lanceolate to ovate; corolla white, flushed with rose and with purplish guide lines, 15–18 mm. long, the upper lip arched, galeate, over half the entire length, the lower lip reflexed, lobed almost to its base, glandular-pubescent externally, ± hirsute apically; staminode glabrous; $n = 8$ (Keck, 1951).—Dry rocky slopes, below 8000 ft.; many Plant Communities; Coast Ranges from Alameda Co. to Los Angeles Co., thence n. into s. Sierra Nevada, occasional to Lake Tahoe region. May–July.

Ssp. **glabrisèpalus** Keck. Calyx glabrous.—Coast Ranges from Mendocino Co. to Napa Co., Sierra Nevada from Shasta Co. to Tulare Co.; w. Nev.

54. **P. Lemmònii** Gray. Open shrub 5–15 dm. high; herbage bright green, glaucous (at least on stems), glabrous up to the pedicels; lvs. 1–6 cm. long, 0.5–2.5 cm. wide, ovate-lanceolate to elliptic, subentire to serrulate; thyrsus viscid-pubescent, narrow and less than 2 dm. long, to more open, longer, more compound; calyx 4–7 mm. high, the lobes lanceolate; corolla yellow with brownish galea and purple guide lines, 10–14 mm. long, the limb large, gaping; staminode densely yellow-bearded, exserted; $n = 8$ (Keck, 1951).—Brushy or wooded slopes, 1000–7000 ft.; Yellow Pine F., Red Fir F., Mixed Evergreen F., N. Oak Wd.; Siskiyou Co. to Humboldt, Solano, and Eldorado cos.; w. Nev. June–Aug.

55. **P. antirrhinoìdes** Benth. [*Lepidostemon penstemonoides* Lem. *P. Lobbii* Hort. ex Lem.] Shrub 1–2.5 m. high with spreading much-branched stems, ± puberulent throughout, only the fls. viscid; lvs. entire, linear- to ovate-elliptic, 1–2 cm. long, 0.2–0.7 cm. wide, firm, crowded; panicle broad, leafy; calyx 3–6 mm. long, the lobes ovate to rounded, obtuse or cuspidate-acute; corolla yellow, tinged with brownish-red, 16–20 mm. long, ca. 8–10 mm. broad at throat, this abruptly much dilated, the upper lip broad, arching, the lower reflexed; staminode densely bearded with long yellow hairs, exserted.—Dry often rocky slopes, below 4500 ft.; Chaparral; interior cismontane s. Calif. from San Bernardino Co. s.; to n. L. Calif. April–May.

Ssp **microphýllus** (Gray) Keck. [*P. m.* Gray. *P. Plummerae* Abrams.] Herbage yellowish gray-green, canescent throughout, the twigs cinereous; calyx 5.5–8(–10) mm. high, canescent, viscid, the lobes lance-oblong, acuminate.—Rocky places, below 5000 ft.; Creosote Bush Scrub to Pinyon-Juniper Wd.; w. edge of Colo. Desert, s. and e. Mojave Desert; Ariz. April–June.

56. **P. cordifòlius** Benth. Loosely branched scandent shrub 1–3 m. high; herbage dark green, glabrous to puberulent, more densely hairy and moderately glandular in infl.; lvs. 2–5 cm. long, 1–3 cm. wide, lance-ovate to cordate, remotely serrulate to sharply dentate, shiny, strongly veined; panicle pyramidal, compact, subsecund, drooping, hence the fls. resupinate and peduncles often reflexed, early; calyx 7–10 mm. long, the lobes lanceolate; corolla dull scarlet, 30–40 mm. long, 5–7 mm. wide pressed, tubular, the upper lip galeate, the lower widely spreading; staminode densely bearded with long yellow-brown hairs, well included; $n = 8$ (Keck, 1951).—Dry slopes and canyons, below 4000 ft.; Chaparral; San Luis Obispo Co. through cismontane s. Calif. to n. L. Calif. May–July.

57. **P. corymbòsus** Benth. [*P. intonsus* Heller. *P. c.* var. *puberulentus* Jeps.] Shrub 3–5 dm. high; herbage dark green, glabrous to canescent, the infl. densely glandular-pubescent; lvs. 1.5–4 cm. long, 0.6–1.7 cm. wide, elliptic, entire to remotely serrate, coriaceous, narrowly revolute on margin; corymb terminal, often many-fld.; calyx 6–10 mm. long, the lobes lance-linear to -ovate; corolla brick-red, 25–35 mm. long, 4–6 mm. wide pressed, tubular, the upper lip galeate, the lower spreading; staminode densely yellow-bearded, well included; $n = 8$ (Keck, 1951).—Rocky slopes and cliffs, below 5000 ft.; Redwood F., Mixed Evergreen F., Yellow Pine F.; Del Norte Co. to Monterey Co. and foothills of Sierra Nevada from Shasta Co. to Sutter Co. June–Oct.

58. **P. ternàtus** Torr. ex Gray. Straggly shrub 5–15 dm. high, the wand-like stems glaucous, glabrous, erect or sometimes scandent; lvs. in whorls of 3, or lower opposite, 2–5 cm. long, 0.2–0.9 cm. wide, lanceolate, remotely serrate-dentate, thickish, often folded along midrib; panicle elongate, many-fld.; calyx 3–5 mm. high, the lobes lance-ovate, acuminate; corolla scarlet, 23–30 mm. long, 4–5 mm. wide pressed, narrowly tubular, glandular-puberulent, the upper lip galeate, the lower spreading; staminode densely yellow-bearded; $n = 8$ (Keck, 1951).—Dry slopes and canyons, below 6000 ft.;

Chaparral, Yellow Pine F.; San Gabriel and San Bernardino mts. to San Diego Co.; n. L. Calif. June–Sept.

Ssp. **septentrionàlis** (M. & J.) Keck. [*P. t.* var. *s.* M. & J.] Pedicels and calyces glandular-pubescent.—Liebre Mts., n. Los Angeles Co. to Tehachapi Mts., Kern Co.

10. Scrophulària L. FIGWORT

Coarse perennial herbs with 4-angled stems. Lvs. opposite, petioled with ± irregularly toothed to divided blades. Fls. small, in loose cymes forming terminal panicles. Calyx 5-parted into rather broad to rounded lobes. Corolla greenish-purple to maroon, the tube globular to ellipsoid, ventricose, the upper lip with its 2 lobes projected forward, the lower with the lateral lobes vertical and the cent. deflexed. Stamens 4, declined, usually included, the anther-cells divergent; sterile stamens (uppermost) scalelike to lacking. Caps. septicidal; stigma entire or emarginate. Seeds many, oblong-ovoid, rugose, marginless, plump. Ca. 100 spp., of N. Hemis. (From *scrofula*, since some spp. have rhizomal knobs that were supposed to cure this disease.)

Infl. villose, the hairs with small gland-tips; sepals acuminate to acute; sterile fil. nearly or quite lacking. Santa Catalina Id. . . . . 2. *S. villosa*
Infl. puberulent or short-pubescent, the hairs with large gland-tips; sepals often rounded; sterile fil. developed.
  Corolla dark maroon, 9–11 mm. long, its upper half blackish; sterile fil. blackish-maroon. Coastal Santa Barbara Co. . . . . 3. *S. atrata*
  Corolla ± brownish, 5–15 mm. long; sterile fil. paler.
    Sterile fil. clavate to obovate, longer than wide, usually brown; main lvs. subcordate to rounded at base. Over most of Calif. . . . . 1. *S. californica*
    Sterile fil. flabelliform, wider than long, greenish-yellow; main lvs. cuneate to subtruncate at base. Extreme n. Calif. . . . . 4. *S. lanceolata*

1. S. califórnica Cham. & Schlecht. [*S. oregana* Penn.] Stems coarse, 10–18 dm. tall, finely pubescent, with some hairs gland-tipped especially in upper parts; lf.-blades triangular-ovate to ovate, acute, rather regularly and mostly simply, sometimes doubly, dentate, 3–10 cm. long, 2–8 cm. wide, truncate to subcordate, with petioles 1.5–5 cm. long; panicle 2–4 dm. long, 5–10 cm. wide, open, with ascending-spreading branches; calyx-lobes oblong to ovate, 3–4 mm. long, acuminate to rounded-obtuse; corolla 8–14 mm. long, red-brown to maroon; sterile uppermost fil. brown to purplish, clavate to obovate; caps. conic-ovoid, 6–8 mm. long; seeds 0.6–0.8 mm. long.—Common in ± damp places, especially brushy thickets, etc.; N. Coastal Scrub, Closed-cone Pine F., Redwood F.; near the coast from Santa Monica Mts. n.; to B.C. Feb.–July. Exceedingly variable.

KEY TO VARIETIES

Calyx-lobes oblong to obovate, longer than wide; corolla 8–15 mm. long, the upper lobes almost as long as tube; lvs. rather regularly dentate. Coastal . . . . *S. californica*
Calyx-lobes almost round; corolla 5–8(–10) mm. long, the upper lobes shorter than the tube; lvs. ± incised, the divisions toothed. Interior.
  Main lf.-blades rather dark green, thin, mostly 6–12 cm. long; caps. mostly 5–7 mm. long; infl. largely 3–6 dm. long, diffuse, open. Low elevs., cismontane . . . . var. *floribunda*
  Main lf.-blades yellow-green, thickish, mostly 3–6 cm. long; infl. mostly 1.5–3 dm. long, narrow, compact. Above 5000 ft., ultramontane . . . . var. *desertorum*

Var. **floribúnda** Greene. [*S. f.* Heller. *S. c.* var. *laciniata* Jeps. *S. multiflora* Penn.] Lf.-blades doubly dentate to ± incised, 5–15 cm. long, on petioles 2–6 cm. long; panicle rather lax; calyx-lobes mostly ca. 3 mm. long; corolla mostly 6–10 mm. long.—Dryish to moist open places, below 5000 (6500) ft.; Chaparral, Coastal Sage Scrub, etc.; cismontane s. Calif. to Tehachapi Mts., borders of Cent. V. to Butte and Lake cos.; n. L. Calif. Mostly March–May.

Var. **desertòrum** Munz. Lf.-blades lighter green, smaller, mostly more regularly toothed; calyx 2–3 mm. long; corolla mostly 5–7 mm. long.—Dryish slopes, 5000–10,000 ft.; Pinyon-Juniper Wd., Lodgepole F.; Tulare, Mono, and Inyo cos. mostly on e. slope of Sierra Nevada, Inyo-White Mts., to Panamint Mts.; w. Nev. Late May–Aug.

2. S. villòsa Penn. [*S. californica* var. *catalina* Jeps.] Lvs. ovate, acute, dentate with sharply acute teeth, the blades 10–15 cm. long, 8–12 cm. wide, on petioles 3–5 cm. long; infl. glandular-villous with conspicuous white hairs; panicle with widely spreading branches; calyx-lobes triangular-ovate, acute to acuminate; corolla 8–9 mm. long,

deep maroon, the upper lobes dark, the lower slightly paler and deflexed; uppermost fil.
a minute awnlike rudiment or lacking.—Rocky canyons; Coastal Sage Scrub, Chaparral;
Santa Catalina and San Clemente ids. April–Aug.

3. **S. atràta** Penn. Infl. glandular-puberulent; lvs. ovate, acute, the teeth ± rounded,
the blades 6–10 cm. long, 5–8 cm. wide, on petioles 2–7 cm. long; infl. elongate, the
branches divaricate; calyx-lobes ovate or lance-ovate, 3 mm. long, rounded, erose-
margined; corolla 9–11 mm. long, dark maroon, the upper half blackish, the tube with
constricted orifice, the lowermost lobe deflexed-spreading; sterile fil. lance-oblong,
blackish-maroon, much narrower than the neck of the upper corolla-lip.—Dry rocky
places, diatomaceous shale; Coastal Sage Scrub, Chaparral; coastal Santa Barbara Co.
April–June.

4. **S. lanceolàta** Pursh. [*S. occidentalis* (Rydb.) Bickn.] Stems 5–20 dm. high, usually
puberulent; main lf.-blades lanceolate to ovate, tapering to subtruncate at base, 7–12
cm. long, sharply simply or doubly dentate, on petioles 1–3 cm. long; infl. narrow, 2–5
dm. long; calyx-lobes rounded-ovate, ca. 3 mm. long; corolla greenish-brown, 7–12 mm.
long, the dorsal lobes broadly oblong; rudimentary stamen greenish-yellow, 1.3–1.8 mm.
long; caps. slender-ovoid, acuminate, 6–9 mm. long.—Rare in Calif., as in Siskiyou,
Modoc, and Lake cos.; to B.C. and Atlantic Coast. May–July.

11. Collínsia Nutt.

Annuals with simple mostly opposite entire to crenulate lvs., the upper sessile or
clasping and passing into ± foliose bracts. Fls. solitary to fascicled, in axils of uppermost lvs. Calyx 5-parted. Corolla 5-lobed, gibbous or saccate at base on upper side,
bilabiate, the tube short, the throat well developed; upper lip 2-lobed, the lower 3-lobed,
the middle lobe keel-shaped and enclosing the 4 declined stamens and style. Lower pair
of fils. inserted higher on the corolla-tube than the upper pair. Caps. dehiscent, the
valves 2-cleft. Seeds flattened, convex dorsally, concave ventrally, smooth to reticulate,
± winged. A genus of ca. 17 spp., all (except 2 e. spp.) found on Pacific Coast. (Zaccheus
Collins, 1764–1831, Philadelphia botanist.)

(Newsom, V. A revision of the genus Collinsia. Bot. Gaz. 87: 260–301, 1929.)
A. Fls. congested in whorls, the pedicels mostly shorter than calyces in the lower whorls (except
    in *C. Greenei*).
    B. Lateral lobes of lower lip of corolla bearded; lvs. pubescent beneath; plants strongly glan-
        dular. Shasta Co. to Sonoma and Tulare cos. .... 6. *C. tinctoria*
    BB. Lateral lower lobes of corolla glabrous; lvs. mostly glabrous; plants not strongly
        glandular.
        C. Upper lip of corolla less than half as long as lower; infl. of 1–2(–3) whorls; corolla
            mostly 15–18 mm. long. Coastal, Mendocino Co. to San Francisco .... 5. *C. corymbosa*
        CC. Upper lip of corolla more than half as long as lower; infl. of 3–8 whorls; corolla
            often shorter.
            D. Upper pair of fils. with distinct basal appendages 1–2 mm. long; upper lip of
                corolla usually distinctly paler than lower. Most of cismontane Calif.
                                                      2. *C. heterophylla*
            DD. Upper pair of fils. without or with very rudimentary basal appendages.
                E. Fils. glabrous, the upper pair bearded at very base; pedicels of lower whorls
                    as long as calyces; upper corolla-lip with distinct lateral and transverse
                    callous-crests projecting into throat. Trinity Co. to Sonoma Co.
                                                   13. *C. Greenei*
                EE. Fils. (at least the upper pair) well bearded; upper corolla-lip with trans-
                    verse but not conspicuous lateral callous-crests at throat; pedicels shorter
                    than calyces.
                    F. Keel sparsely bearded without; corolla bluish-lavender, the upper lip
                        almost as long as lower and not evidently veined. San Bernardino Mts.
                        to L. Calif. .... 1. *C. concolor*
                    FF. Keel glabrous; corolla paler, at least on upper lip.
                        G. Corolla-lobes not strongly veined; stems subglabrous below; calyx-
                            lobes acute. Along coast, San Francisco to Monterey Peninsula
                                                3. *C. franciscana*
                        GG. Corolla-lobes strongly veined; stems canescent-puberulent below;
                          calyx-lobes obtuse. Shasta and Lake cos. to Los Angeles Co.
                                                4. *C. bartsiaefolia*
AA. Fls. pedicelled, solitary or in whorls, the lower pedicels as long as or longer than calyces.
    B. Infl. not glandular-pubescent, or if so, the glands minute and scarcely thicker than the
        supporting hairs; calyx-lobes exceeding caps.; seeds usually 3 or more to a locule.
        C. Calyx-lobes obtuse or obtusish; corolla 7–10 mm. long; upper fils. bearded. Santa
            Monica Mts. to San Bernardino Mts. .... 7. *C. Parryi*

CC. Calyx-lobes acute to attenuate.
   D. Upper fls. glabrous; seeds turgid, not winged; calyx-lobes attenuate.
       E. Corolla 8–18 mm. long, the lateral lower lobes ca. as long as keel. Mendocino
           and Siskiyou cos. n. .... 8. *C. grandiflora*
       EE. Corolla 4–7 mm. long, the lower lobes exceeding the keel. Mts. from Del
           Norte and Modoc cos. to San Diego Co. .... 9. *C. parviflora*
   DD. Upper fls. bearded; seeds ± winged; calyx-lobes acute. From Fresno and San
       Luis Obispo cos. n. .... 10. *C. sparsiflora*
BB. Infl. glandular-pubescent; calyx-lobes scarcely or not longer than caps.; seeds mostly 1–2(–3)
    in a locule.
   C. Fruiting pedicels ascending or ascending-spreading; upper bracts of infl. at least 2 mm.
       long.
       D. Cauline lvs. linear, thickened, gray-green; upper corolla-lip much paler than
           lower. Lake Co. to Del Norte and Siskiyou cos.
           E. Corolla 5–8 mm. long, the upper lip somewhat upcurved .. 14. *C. Rattanii*
           EE. Corolla 8–12 mm. long, the upper lip strongly reflexed .... 15. *C. linearis*
       DD. Cauline lvs. oblong to ovate; upper corolla-lip not much paler than lower.
           E. Corolla 6–9 mm. long.
               F. Calyx membranous, rounded and ca. 3 mm. wide at base in fr., the
                   lobes exceeding the caps.; seeds 2, ca. 3 mm. long; upper lvs narrowed
                   to base. San Diego Co. to Mariposa and Monterey cos. .. 12. *C. Childii*
               FF. Calyx thickened, subtruncate and swollen, ca. 5 mm. wide at base in fr.,
                   the lobes ca. as long as caps.; seeds 6–8, ca. 2 mm. long; upper lvs.
                   rounded to base. San Bernardino Co. to Kern and Inyo cos.
                                                                         11. *C. callosa*
           EE. Corolla 10–12 mm. long. Trinity Co. to Sonoma Co. ...... 13. *C. Greenei*
   CC. Fruiting pedicels deflexed-spreading; upper bracts of infl. obsolete, less than 2 mm.
       long; upper corolla-lip white, the lower blue. Mts. from San Bernardino Co. to
       Trinity, Siskiyou, and Modoc cos. .... 16. *C. Torreyi*

1. **C. cóncolor** Greene. [*C. bicolor* var. *c.* Jeps.] Erect, occasionally diffusely branched, puberulent, sometimes minutely glandular in infl., 1.5–4.5 dm. tall; lf.-blades thin, lance-oblong, obtuse, crenate-serrulate to entire, glabrous, 2–5 cm. long, with rounded to almost sessile bases or the lower on petioles to 2 cm. long; fls. sessile or on pedicels to 4 mm. long, several in a whorl; bracts 5–10 mm. long; calyx-tube villous, 2–3 mm. long, the lobes minutely pubescent, 3–4 mm. long, oblong; corolla declined, 10–14 mm. long, bluish-lavender, the upper lip 6–10 mm. long, the lower slightly larger, the lateral lobes loosely pilose within, the keel sparsely bearded without; fils. ± bearded, at least near base; caps. 4 mm. long; seeds ca. 1.5 mm. long, flattened, irregularly round; $n = 7$ (Garber, 1956).—Shade of bushes, etc., below 5500 ft.; Chaparral, Yellow Pine F.; interior s. Calif. from San Bernardino Co. to n. L. Calif. April–June.

2. **C. heterophýlla** Buist ex Grah. [*C. bicolor* Benth., not Raf. *C. multicolor* Lindl. & Paxt. *C. Hernandezii* Elmer.] CHINESE HOUSES. INNOCENCE. Stem simple or diffusely branched, subglabrous to puberulent or pubescent, sometimes glandular above, green to purplish, 2–5 dm. high; lf.-blades lanceolate to lance-oblong, subentire to serrulate, glabrous, 1–7 cm. long, ± obtuse, the lower short-petioled; fls. 2–7 in sessile or short-pedicelled (to 5 mm.) whorls; bracts 5–20 mm. long; calyx green to red-purple, the tube pubescent to villous, ca. 2 mm. long, the lobes pubescent, lanceolate, 5–6 mm. long, ± acute; corolla mostly 1.5–2 cm. long, pubescent, usually the white to lilac upper lip paler than the violet or rose-purple lower lip and from almost as long to as long, the lower lateral lobes mostly glabrous, keel glabrous; upper fils. bearded half their length and with a linear basal bearded appendage ca. 2 mm. long projecting into the nectar pouch; caps. 5 mm. long; seeds 2 mm. long, flattened, ovate, rugose-reticulate, slightly winged; $n = 7$ (Sugiura, 1940).—Common in shaded places, below 2500 ft.; many Plant Communities; through most of cismontane Calif. from Humboldt and Shasta cos. to n. L. Calif. March–June.

Var. **austromontàna** (Newsom) Munz. [*C. bicolor* var. *a.* Newsom. *C. a.* Penn.] Stems and under side of lvs. pubescent; upper lip of corolla mostly ca. half as long as lower.—Dry slopes, mostly 2000–5000 ft.; Chaparral, Yellow Pine F.; San Gabriel and San Bernardino mts. May–July.

3. **C. franciscàna** Bioletti. [*C. sparsiflora* var. *f.* Jeps.] Diffusely branched, subglabrous below, glandular-pubescent and often viscid above, 2–6 dm. tall; lf.-blades lance-ovate to subdeltoid, ± serrulate, glabrous or nearly so, 2–5 cm. long, sessile or the lower short-petioled; bracts gradually reduced, the uppermost linear; pedicels glandular-pubescent, 0.2–4 cm. long, the upper fls. almost sessile, several in a whorl, the lower 1–3 at a node, pedicelled; calyx puberulent, the tube 3 mm. long, lobes ovate-lanceolate,

5 mm. long; corolla 15–20 mm. long, the upper lip whitish, purple-spotted at base, 6–10 mm. long, the lower violet-blue, 9–13 mm. long, the keel distally purple, glabrous; caps. 5–6 mm. long; seeds irregularly round-oblong, 1.5 mm. long, rugose.—Dry stony and grassy slopes; N. Coastal Scrub, Closed-cone Pine F.; San Francisco to Monterey Peninsula. March–May.

4. **C. bartsiaefòlia** Benth. in DC. [*C. stricta* Greene. *C. b.* var. *s.* Newsom. *C. b.* var. *hirsuta* (Kell.) Penn. *C. h.* Kell.] Plants 1.5–4 dm. tall, simple to diffuse; stems canescent-puberulent to distally glandular-puberulent; lf.-blades oblong, obtuse, crenate-serrate, glabrous, 1.5–4 cm. long, rounded at base, short-petioled to sessile; fls. several in a whorl, sessile or nearly so, the bracts shorter than fls.; calyx-tube villous to glabrous, 2 mm. long, the lobes lance-oblong, obtuse, pubescent to glabrous, 3–6 mm. long; corolla rose-purple to almost white, often dark veined and spotted, 8–20 mm. long, the upper lip whitish-lavender to darker, 3–7 mm. long, the lower lip 7–12 mm. long, with whitish to purple lateral lobes and glabrous purple-dotted keel; fils. not or short-appendaged; caps. 3.5–5 mm. long; seeds flattened, rounded, rugose-reticulate, 1–1.5 mm. long; *n* = 7 (Hiorth, 1933).—Often sandy or granitic soils, mostly below 2000 ft.; Foothill Wd., V. Grassland, Closed-cone Pine F., etc.; Shasta and Lake cos. s. to Kern Co. and from Sierran foothills to near coast. March–May. Plants with fls. 8–9 mm. long have been called var. *stricta*, and with fls. 16–20 mm. long, var. *hirsùta*.

Var. **Davidsònii** (Parish) Newsom. [*C. D.* Parish.] Lvs. oblong-ovate; plants 0.5–2 dm. high, not at all glandular.—At 2000–4000 ft.; Foothill Wd., Joshua Tree Wd.; inner Coast Ranges from San Benito Co. to Greenhorn Mts., Kern Co. and Antelope V., w. Los Angeles Co. April–June.

5. **C. corymbòsa** Herder. Stems simple to usually diffusely branched, erect to sub-decumbent, canescent-puberulent, 1–2 dm. long; lv.-blades oblong to elliptic-ovate, obtuse, crenate, ± pilose especially above, 2–3 cm. long, subsessile or the lower short-petioled; whorls 1–3, several-fld., with crenulate villous bracts; pedicels 1–5 mm. long; calyx puberulent to conspicuously villous, 7–10 mm. long, the lobes ovate to oblong, 3–5 mm. long; corolla pale lavender to yellowish-white, 13–20 mm. long, the upper lip bluish, 3–6 mm. long, the lower whitish, 8–12 mm. long; caps. 5–6 mm. long; seeds flattened, oblong, irregularly winged, ca. 1.5 mm. long; *n* = 7 (Garber, 1956).—Occasional, sandy places; Coastal Strand; Humboldt Co. to San Francisco. April–June.

6. **C. tinctòria** Hartw. ex Benth. [*C. barbata* Bosse. *C. septemnervia* Kell.] Stems simple, erect, 2–6 dm. tall, or occasionally sparsely branched, glabrous below, glandular-pubescent in infl.; lvs. thin, ovate to lance-oblong, entire to serrate-crenate, finely pubescent beneath, obtuse, 3–8 cm. long, the upper clasping, the lower short-petioled; whorls in infl. 2–6, many-fld., with pedicels 0–2 mm. long; calyx 5–8 mm. long, glandular-pubescent, the lobes parted almost to base; corolla yellow to greenish-white with purple dots or lines, 12–17 mm. long, the tube 5 mm. long, the upper lip less than half as long as lower, the latter with lateral lobes bearded within, the keel sparsely bearded without; caps. ca. 4 mm. long; seeds flattened, round-oblong, rugose-reticulate, 2.5–3 mm. long, slightly winged; *n* = 7 (Garber, 1956).—Dry or moist mostly stony places, 2000–6000(–7500) ft.; Foothill Wd. to Red Fir F., Chaparral; Shasta Co. to Sonoma and Tulare cos. May–Aug.

7. **C. Párryi** Gray. [*C. cahonis* Jones.] Stem simple to diffusely branched, minutely puberulent, 1–5 dm. tall; lf.-blades lanceolate, obtusish, entire to inconspicuously serrulate, glabrous, 1.5–4 cm. long, with rounded to subsessile bases; lowermost petioles to 2 cm. long; infl. lax, with foliose bracts subtending 1–2 fls. and pedicels 1–4 cm. long; calyx puberulent, 4–7 mm. long, the broadly lanceolate lobes obtusish, 2–3 mm. long, ± ciliate; corolla 7–10 mm. long, glabrous, the lips violet-blue, the upper somewhat shorter than the lower and purple-dotted near base; caps. 4–5 mm. long; seeds round-oblong, rugulose, tan, ca. 1 mm. long, the margins somewhat recurved.—Disturbed places, such as burns, below 5000 ft.; Chaparral; Santa Monica Mts. to San Bernardino Mts. March–June.

8. **C. grandiflòra** Dougl. ex Lindl. Stems simple or branched, mostly erect, 1–4 dm. tall, canescent-puberulent, ± glandular above; lf.-blades oblong to sublanceolate, obtuse, entire to crenulate, 2.5–4 cm. long, mostly sessile, becoming bracteate and linear in infl.; infl. lax, the fls. 3–7 at a node, on puberulent pedicels 3–20 mm. long; calyx subglabrous, 4–7 mm. long, the lobes lance-attenuate, 3–5 mm. long; corolla 12–18 mm. long,

glabrous, the upper lip white to purplish, the lower longer, violet-blue, the keel almost as long as lateral lobes; caps. 4–5 mm. long; seeds smooth, round-oblong, red-brown, 1.5 mm. long.—Open grassy or rocky places, ca. 1400–4000 ft.; largely Coastal Prairie; from Mendocino and Siskiyou cos. n.; to B.C. April–June.

Var. **pusílla** Gray. [*C. Diehlii* Jones. *C. parviflora* var. *D.* Penn.] Corolla 8–10 mm. long, the keel shorter than the lateral lower lobes.—Similar places, Lake and Plumas cos. n.; to Wash., Ida. (Combines characters of *C. grandiflora* and *C. parviflora*.)

9. **C. parviflòra** Dougl. ex Lindl. Branched, ascending to erect, puberulent, 0.5–4 dm. tall; lf.-blades lance-oblong, obtuse, mostly entire, glabrous, 2–4 cm. long, sessile or the lower short-petioled, the upper passing into bracts; fls. 1–2 in bract-axils; pedicels puberulent to ± glandular, 3–15 mm. long; calyx 5–7 mm. long, glabrous, the lobes acuminate, ca. 2–3 mm. long; corolla 4–7 mm. long, glabrous, the upper lip white to violet-blue at the tips, the lower longer, violet-blue, the lateral lobes longer than keel; caps. 3–4 mm. long; seeds 1.5–2 mm. long, round-oblong, smooth, turgid, brown; $n = 7$ (Garber, 1956).—Moist ± shaded places, 2500–11,150 ft.; Sagebrush Scrub to Subalpine F.; mts. from San Diego Co. to Modoc and Del Norte cos.; to B.C., Ontario, Colo. April–July.

10. **C. sparsiflòra** F. & M. [*C. parviflora* var. *s.* Benth.] Branched at base or simple, 0.5–2.5 dm. tall, glabrous to ± puberulent; lf.-blades glabrous to sparsely puberulent, narrow-oblong, entire to crenulate, 1–3 cm. long, or the lower more deeply toothed, and with petioles to 1 cm. long; infl. lax, the bracts linear each mostly subtending 1 fl. on a pedicel 1–3 cm. long; calyx 6–7 mm. long, glabrous except for the setulose margins, the lobes lanceolate, 3–4 mm. long; corolla 8–12 mm. long, purple, the dorsal side of the tube forming a right angle with the pedicels, the upper lip whitish toward base, purple-dotted, the lower longer, the keel somewhat hairy; caps. 5–6 mm. long; seeds roundish, flattened with recurved margins, rugulose, winged, ca. 2.5–3 mm. in diam.; $n = 7$ (Garber, 1956).—Grainfields and grassy places, below 3000 ft.; Coastal Prairie, V. Grassland, Chaparral; Humboldt and Butte cos. to Contra Costa and Tuolumne cos. March–May.

KEYS TO VARIETIES

Seeds mostly 3–4 mm. wide, flat, circular-winged; corolla 8–20 mm. long, the keel mostly bearded.
   Corolla strongly declined, the dorsal side of the saccate base forming a right angle with the pedicel; calyx ⅓–½ as long as corolla.
      The corolla 8–12 mm. long .... *C. sparsiflora*
      The corolla 12–20 mm. long .... var. *arvensis*
   Corolla not strongly declined, the dorsal side of the saccate base often in a straight line with the pedicel; calyx more than half as long as corolla .... var. *Brucae*
Seeds mostly 1.5–2 mm. wide, turgid, often scarcely winged; corolla 5–8 mm. long, the keel glabrous or slightly hairy .... var. *collina*

Var. **arvénsis** (Greene) Jeps. [*C. a.* Greene.] Corolla 12–20 mm. long, the basal pouch more strongly inflated.—With the sp., but more abundant; Mendocino and Sonoma cos. to Colusa and Napa cos. March–May.

Var. **Brùcae** (Jones) Newsom. [*C. B.* Jones.] Calyx-lobes often long-ciliate; corolla scarcely surpassing calyx, 8–12 mm. long, with basal pouch not much inflated; seeds 3–4 mm. wide, winged, rounded.—Open woods and fields, from below 3000 ft.; Foothill Wd.; Lake and Butte cos. to Wash. March–May.

Var. **collìna** (Jeps.) Newsom. [*C. parviflora* var. *c.* Jeps. *C. solitaria* and *divaricata* Kell. *C. s.* var. *solitaria* Newsom.] Plants mostly divaricately branched; calyx 5–6 mm. long; corolla 5–8 mm. long, with low basal pouch; seeds 1.5–2 mm. wide.—Grassy places and open woods, below 5000 ft.; V. Grassland, Foothill Wd., Yellow Pine F., etc.; Lake and Eldorado cos. to San Luis Obispo and Fresno cos. March–June.

11. **C. callòsa** Parish. Stems relatively stout, simple or usually diffusely branched, glandular-pubescent above, 0.5–2 dm. high; lf.-blades thickened, glabrous, oblong to oblong-lanceolate, obtuse, entire, 1–3 cm. long, sessile or short-petioled; infl. lax, the bracts linear, pedicels 5–15 mm. long; calyx conspicuously broad in fr., 4–6 mm. long, the lobes half as long, lance-ovate; corolla 7–9 mm. long, rose-lavender, the upper lip purple-dotted on the whitish base, ca. as long as lower; caps. 4–6 mm. long; seeds oblong, thickened, rugose-reticulate, ca. 2 mm. long.—Dry places, 3000–7500 ft.; Chaparral, Sagebrush Scrub, Yellow Pine F., Pinyon-Juniper Wd., San Bernardino and

San Gabriel mts. to Tehachapi, Greenhorn, Panamint mts. and e. slope of Sierra Nevada in Inyo Co. April–June.

12. **C. Chìldii** Parry ex Gray. [*C. inconspicua* Congd. *C. breviflora* Suksd.] Erect, simple to branched, puberulent below, glandular above, 1–4 dm. tall; lf.-blades oblong-lanceolate, subentire to serrulate, 1–4 cm. long, short-petioled; infl. lax, the lower bracts foliose, the upper reduced; pedicels 5–25 mm. long; calyx 5–7 mm. long, glandular-puberulent, the lobes lanceolate, 3–4 mm. long; corolla 6–8 mm. long, the lips subequal, pale violet to whitish, the lobes rather narrow; caps. 3–4 mm. long; seeds 3 mm. long, round-oblong, turgid, with ± recurved margins; $n = 7$ (Garber, 1956).—Dry shaded places, 3000–7000 ft.; Yellow Pine F., S. Oak Wd., Foothill Wd.; mts. from San Diego Co. to Mariposa and Monterey cos. April–June.

13. **C. Greènei** Gray. Stems slender, mostly branched, 1–3 dm. tall, puberulent, with gland-tipped hairs above; lf.-blades oblong-lanceolate, entire to crenulate, 1–3 cm. long, sessile or lower petioled; infl. of 1–few rather compact whorls of 1–5 fls. each; pedicels glandular-puberulent, 2–10 mm. long; calyx 5–6 mm. long, puberulent and with gland-tipped hairs, the lobes lanceolate, 2–4 mm. long; corolla 10–12 mm. long, purplish-blue, the upper lip erect-spreading, the lower somewhat longer; caps. 4 mm. long; seeds 2.5 mm. long, oblong, flattened with recurved margins.—Rocky and stony places, often in serpentine, below 7000 ft.; Chaparral to Red Fir F.; Trinity and Humboldt cos. to Sonoma and Lake cos. May–July.

14. **C. Rattánii** Gray. [*C. Torreyi* var. *R.* Jeps.] Stems 1–4 dm. tall, puberulent, and in infl. also with gland-tipped hairs; lf.-blades linear-lanceolate to linear, obtuse, entire, ± revolute, 1–5 cm. long, or lower wider and short-petioled, the upper passing into linear bracts; pedicels 2–4 at a node, 3–10 mm. long; calyx 5–8 mm. long, the lobes 4–6 mm. long, lanceolate; corolla 5–8 mm. long, the upper lip white to lavender-violet, somewhat upcurved, the lower longer, purple-violet, decurved; caps. 3–4 mm. long; seeds 1.5–2 mm. long, oblong, turgid, slightly margined.—Open woods, 500–5000 ft.; Yellow Pine F., Mixed Evergreen F.; Lake Co. to Siskiyou and Del Norte cos.; to Wash. April–June.

15. **C. lineàris** Gray. [*C. Rattanii* var. *l.* Newsom. *C. Torreyi* var. *l.* Jeps.] With habit of *C. Rattanii*, perhaps more widely divaricate; pedicels filiform, 5–10 mm. long, glandular-puberulent; calyx 3–4 mm. long, the lobes oblong-lanceolate; corolla 8–12 mm. long, bluish to blue-violet with whitish reflexed upper lip.—Mixed Evergreen F. to Yellow Pine F.; Humboldt Co. to Siskiyou and Del Norte cos.; s. Ore. April–July.

16. **C. Tòrreyi** Gray. [*C. T.* var. *brevicarinata* Newsom.] Erect, widely branched, 0.5–2 dm. high, glandular-pilose on stems and in infl.; lf.-blades broadly linear, many times as long as wide, 1.5–4 cm. long, sessile or short-petioled; infl. lax, with foliose lower bracts and reduced upper; pedicels 5–10 mm. long; calyx 3–4 mm. long, the linear-obtuse lobes half as long; corolla 7–9 mm. long, with broadly rounded basal pouch, the upper lip pale with yellow base having purple dots, the lower lip longer, deeper blue; caps. 3 mm. long; seeds turgid, round-oblong, ca. 2 mm. long.—Damp or half-dry sandy banks and flats, below 10,000 ft.; Montane Coniferous F.; Sierra Nevada from Tulare Co. n., to Trinity and Siskiyou cos. May–Aug.

Var. **latifòlia** Newsom. Lf.-blades lance-ovate to elliptic, 3–4 times as long as wide.—At 4000–7000 ft.; Glenn Co. to Siskiyou and Modoc cos.; s. Ore. June–July.

Var. **Wrìghtii** (Wats.) Jtn. [*C. W.* Wats. *C. brachysiphon* Eastw. *C. monticola* A. Davids.] Plants 0.5–3 dm. tall; fls. 4–6 mm. long.—Granitic sand, etc., mostly 7000–11,000 ft.; Red Fir F., Subalpine F.; San Bernardino and San Gabriel mts. to Siskiyou and Trinity cos. June–Aug.

12. Tonélla Nutt. ex Gray

Slender branching erect annuals. Lvs. opposite, entire or the lower tripartite. Fls. small, solitary or fascicled, in axils of somewhat reduced upper lvs., resembling those of *Collinsia* but the lower middle lobe not carinate. Fils. thickened upward, pubescent. Caps. globose-ovoid. Seeds large, turgid, wingless, 1–4 in a locule. Two spp., w. N. Am. (Origin of name unknown.)

1. **T. tenélla** (Benth.) Heller. [*Collinisa t.* Benth. in DC. *T. collinsioides* Nutt.] Stems very slender, 1–3 dm. tall, glabrous, or minutely pubescent above nodes; lf.-

blades 1–1.5 cm. long, pubescent above, the lower lvs. ovate to roundish, entire or notched on each side, petioled, the upper with 3 lance-oblong to broader segms.; pedicels in 2's or 3's, filiform, 1–1.5 cm. long; calyx 2 mm. long, the 5 lobes half as long, ciliolate; corolla 2–2.5 mm. long, violet toward tips, the upper lobes shorter than lower; caps. 2–2.5 mm. long; seeds 2, 1.5 cm. long.—Mostly shaded slopes, below 3000 ft.; Chaparral, Foothill Wd., N. Oak Wd., Yellow Pine F., etc.; Santa Clara Co. to Siskiyou and Butte cos.; to Wash. March–May.

13. Galvèzia Domb. ex Juss.

Shrubs or herbs, with opposite or whorled lvs. Fls. in a terminal raceme. Calyx 5-parted. Corolla saccate at base, tubular, red, strongly 2-lipped, with rather prominent palate. Stamens 4, didynamous, the fils. with 2 rows of tack-shaped glands. Caps. globose-ovoid, with irregular subterminal dehiscence. Seeds cylindric, with thin irregular wing-like plates. Ca. 4 spp., from Calif. to Peru. (Jose *Galvez*, a Spanish administrator.)

1. **G. speciòsa** (Nutt.) Gray. [*Gambelia s.* Nutt. *Antirrhinum s.* Gray.] Glabrous or pubescent spreading bright green shrub to ca. 1 m. tall and 2 m. broad; lvs. in 3's, thickish, elliptic-ovate, entire, 2–4.5 cm. long, on petioles ca. 5–8 mm. long; fls. in terminal leafy-bracted rather dense soft-pubescent racemes; pedicels slender, 1–2 cm. long; calyx 7–10 mm. long, the lobes lance-attenuate, divided almost to base; corolla 2–2.5 cm. long, the lips ca. 5 mm. long; fils. dilated; caps. 6–7 mm. long; seeds dark, 1 mm. long.—Rocky canyons; Coastal Sage Scrub; Santa Catalina and San Clemente ids.; Guadalupe Id. Feb.–May.

14. Mauràndya Ort.

Perennial herbs with twining or prostrate stems. Lvs. alternate, petioled, coarsely toothed or hastately lobed. Calyx 5-parted. Corolla gibbous or saccate at base, bilabiate, usually with internal plaits, occasionally almost closed by true palate. Stamens 4, didynamous; fils. with 2 rows of tack-shaped glands. Caps. scarcely oblique, irregularly dehiscent near apex. Seeds oblong, with irregular corky ridges and tubercles. Ca. 10 spp., sw. N. Am. (Dr. *Maurandy*, a teacher of Botany at Carthagena.)

Lvs. hastately lobed, triangular-ovate; stems twining or climbing, to ca. 1 m. long; fls. rose to purple. Providence Mts. . . . . **1. *M. antirrhiniflora***
Lvs. rounded, irregularly dentate; stems ± pendent, ca. 1 dm. long; fls. yellow. Death V.
                                                                                                  2. *M. petrophila*

1. **M. antirrhiniflora** Humb. & Bonpl. ex Willd. [*Antirrhinum maurandioides* Gray. *A. a.* Hitchc. *Asarina a.* Penn.] Stems slender, glabrous, much-branched; lvs. glabrous, triangular, hastate to 5-lobed, 1–2.5 cm. long, on equally long somewhat flexuous petioles; fls. solitary, axillary, on slender pedicels 1–2 cm. long; calyx 10–12 mm. long; corolla 2.5–3 cm. long, the tube pale, the lobes röse to purple, the palate yellowish-white with dark lines, pubescent; stamens 17–19 mm. long; caps. 7–8 mm. long, subglobose, thin-walled; seeds oblong, 1 mm. long, brown, with short corky tuberculate ridges; $n = 12$ (Heitz, 1927).—Limestone, 2500–4000 ft.; Joshua Tree Wd., Shadscale Scrub; Providence Mts., e. Mojave Desert; to Tex., Oaxaca. April–May.

2. **M. petróphila** Cov. & Mort. [*Asarina p.* Penn.] Stems slender, branched, forming dense, ± pendent tufts from a woody caudex; plants soft-hairy; lf.-blades round-ovate, 2–3 cm. long, irregularly dentate with long callose bristly teeth; petioles 1–2 cm. long; pedicels 1–3 mm. long; calyx 8–12 mm. long, the lobes lanceolate, irregularly dentate with slender teeth; corolla pale yellow, 15–30 mm. long, the throat cylindric, open with 2 ventral pilose ridges; the lobes spreading; caps. ca. 9 mm. long; seeds pale, 2.5 mm. long, with lines of spongy tubercles.—Limestone crevices, 3500–5800 ft.; Creosote Bush Scrub; Grapevine Mts., Inyo Co. April–June.

15. Mohàvea Gray

Annual herbs, erect, with lanceolate or slightly broader alternate lvs. Fls. in dense leafy spikes; bracteoles none. Calyx 5-parted. Corolla with a short tube, merely gibbous at base, and with ample subcampanulate bilabiate limb, the lips fan-shaped, the lower with large hairy palate. Fertile stamens 2, connivent, the other 3 abortive. Caps. ovoid, thin, bursting irregularly. Seeds ovate-discoid, flattened, each surrounded by its incurved wing. Two spp. (Name of stream where first collected by Fremont.)

Fls. pale yellow, 2.5–3.5 cm. long; palate purple-dotted; lower lip lobed only to 6 or 8 mm. above the palate .... 1. *M. confertiflora*
Fls. lemon yellow, 1.5–2 cm. long; palate not conspicuously dotted; lower lip lobed to within 2 or 3 mm. of palate .... 2. *M. breviflora*

1. **M. confertiflòra** (Benth. in DC.) Heller. [*Antirrhinum c.* Benth. in DC. *M. viscida* Gray.] Ghost Flower. Simple or few branched, viscid-pubescent, 1–4 dm. tall; lvs. 1–6 cm. long, linear- to ovate-lanceolate, short-petioled; pedicels 5–10 mm. long; calyx 9–12 mm. long, enlarging in fr., the segms. narrow; corolla closed at throat, the tube and throat ca. ⅓ entire length; stamens 9–10 mm. long, somewhat pubescent at base; caps. subglobose, 10–12 mm. long; seeds dark, barely 2 mm. long.—Common in sandy washes and on dry gravelly slopes, below 3000 ft.; Creosote Bush Scrub; Colo. Desert and Mojave Desert w. to Daggett and Ord Mts.; to L. Calif., Ariz., Nev. March–April.

2. **M. breviflòra** Cov. With much same habit, 0.5–2 dm. tall, densely glandular; lvs. ovate-lanceolate, 1–4 cm. long; pedicels 2–5 mm. long; calyx 10–11 mm. long; corolla 15–18 mm. long; stamens glabrous; caps. 8–10 mm. long; seeds 2–2.5 mm. long.—Dry sandy and gravelly places, below 2500 ft.; Creosote Bush Scrub; Death V. Region; w. Nev., nw. Ariz. March–April.

16. Antirrhìnum L. Snapdragon

Erect or diffuse annual or perennial herbs. Lvs. alternate, or the lower opposite or whorled, entire. Fls. axillary to foliage lvs. or in terminal racemes. Calyx 5-parted. Corolla 2-lipped, gibbous or saccate at base; palate usually closing the throat. Fertile stamens 4, didynamous, the fils. often dilated toward apex. Caps. opening by 2 or 3 pores below base of the style, or bursting somewhat irregularly. Perhaps 40 spp., mostly in Medit. region and in sw. U.S. (Greek, *anti*, like, and *rhinon*, nose, because of snout-like fls.)

(Munz, P. A. The Antirrhinoideae-Antirrhineae of the New World. Proc. Calif. Acad. Sci., IV, 15: 323–397, 1926. Rothmaler, W. Taxonomische Monographie der Gattung A. Fedde's Repertorium, Beiheft 136: 1–124, 1956.)
A. Caps. ± oblique, dehiscing by fairly definite terminal or subterminal pores; stems self-supporting or supporting themselves by tortile branchlets.
   B. Seeds appearing cup-shaped because of the broad incurved wing; slender erect annual with narrow lvs.; fls. purple or white, ca. 12 mm. long; calyx-segms. linear. Occasional waif .... 1. *A. Orontium*
   BB. Seeds not cup-shaped.
      C. Corolla 3–5 cm. long; short-lived perennial escaping from gardens ...... 2. *A. majus*
      CC. Corolla mostly not over 2 cm. long; usually annual, native.
         D. Corolla-throat gaping; upper lip pink, the lower whitish. Interior San Luis Obispo and Monterey cos. .... 12. *A. ovatum*
         DD. Corolla-throat closed; colors not as above.
            E. Stems self-supporting, lacking filiform tortile branchlets.
               F. Plants stout; fls. reddish, 16–19 mm. long; corolla-tube merely saccate at base; calyx-hairs if present, short and glandular, not ⅔ as long as calyx-segms.
                  G. Plant glabrous throughout its vegetative parts; lvs. linear
                                        3. *A. virga*
                  GG. Plant densely glandular-hirsute; lvs. lanceolate .. 4. *A. multiflorum*
               FF. Plants slender; fls. 10–12 mm. long, bluish with yellow palate; corolla-tube with spurlike base; calyx-hairs to ⅔ as long as calyx-segms.
                                      5. *A. cornutum*
            EE. Stems in mature plants largely supported by tortile branchlets, or at least possessing them.
               F. Plant simple below, erect, glabrous below the glandular-villous minutely bracted spicate raceme; fls. whitish with the lower lip much enlarged. S. Calif. .... 6. *A. Coulterianum*
               FF. Plants usually branched below and ± pubescent along the stem; infl. lax or fairly dense but not set off sharply by its pubescence and leaflessness from the upper stem.
                  G. Palate and corolla-tube with 2 bands of hairs, the tips of which are conspicuously enlarged; pedicels 5–20 mm. long, exceeding calyx; corolla-tube merely gibbous at base, ca. as long as lower lip. S. Calif. .... 7. *A. Nuttallianum*

GG. Palate and corolla-tube minutely and uniformly puberulent or
glandular-puberulent; pedicels mostly shorter than calyx (except
in *A. Kingii*).
  H. Corollas 16–18 mm. long; dorsal calyx-segm. 10–20 mm. long,
    several-ribbed; coarse herb frequently 7–8 dm. high, densely
    leafy. Glenn and Colusa cos. .... 8. *A. subcordatum*
  HH. Corollas 8–16 mm. long; dorsal calyx-segm. not over 10 mm.
    long or more than 3-ribbed; rather slender herbs, usually less
    than 5 dm. high, not densely leafy.
    I. Fls. light purple, 10–15 mm. long; pedicels 2–5 mm.
      long; stems glandular. San Benito and Mariposa cos. n.
      J. Fls. 12–15 mm. long; dorsal calyx-segm. 8–12 mm.
        long .... 9. *A. vexillo-calyculatum*
      JJ. Fls. 10–12 mm. long; dorsal calyx-segm. 4–7 mm.
        long .... 10. *A. Breweri*
    II. Fls. largely whitish, 7–8 mm. long; pedicels 5–20 mm.
      long; stems subglabrous except at woolly base and
      glandular-pubescent infl. E. of Sierra Nevada
                                 11. *A. Kingii*
AA. Caps. not oblique, dehiscing by irregular bursting; stems twining and supported by the long
capillary pedicels.
  B. Fls. yellow, 11–13 mm. long; stems very slender. Deserts. .... 13. *A. filipes*
  BB. Fls. blue, 13–15 mm. long; stems fairly stout at base. Cismontane ...... 14. *A. Kelloggii*

1. **A. Oróntium** L. [*Misopates O.* Raf.] Annual, often branched from base, 3–5 dm.
tall, glandular above; lvs. linear; bracts foliar; calyx 1–2 cm. long; corolla pink-purple,
10–13 mm. long; caps. ca. 1 cm. long, $n = 8$ (Heitz, 1927).—Occasional in cult. and
reported as escape near Santa Cruz; Eurasian.

2. **A. màjus** L. COMMON SNAPDRAGON. Much-branched, 4–8 dm. high, glandular-
pubescent above; lvs. lanceolate; fls. of many colors; calyx-lobes broadly ovate, 3–5
mm. long; caps. glandular, 10–14 mm. long; $n = 8$ (Propach, 1935).—Common garden
plant, occasional about dumps and waste places; native of Medit. region. Most of year.

3. **A. vírga** Gray. Erect perennial with coarse virgate stems 8–15(–20) dm. high,
occasionally branched above; lvs. rather crowded, linear, sessile, acute, 2–9 cm. long,
3–7 mm. wide, gradually passing into linear-subulate bracts; infl. secund, crowded, 1–7
dm. long; pedicels 3–7 mm. long; calyx herbaceous, oblique-campanulate, 6–7 mm.
long; corolla red-purple, 16–18 mm. long, tubular with broad saccate spur 1.5–2 mm.
long, the upper lobes rounded to ovate, 2 mm. long, the lower lip 6–7 mm. long, with
lobes 2.5 mm. long and with prominent palate; caps. 7–8 mm. long; seeds dark, ovoid,
ca. 1.5 mm. long, with several fimbrillate winglike longitudinal ridges.—Open gravelly
and rocky places, below 3500 ft.; Chaparral, N. Oak Wd.; Lake, Sonoma, and
Mendocino cos. June–July.

4. **A. multiflòrum** Penn. [*A. glandulosum* Lindl., not Lejeune.] Stout widely branched
annual or short-lived perennial, viscid, glandular-hirsute throughout, 6–15 dm. tall;
branchlets spreading, nontortile; lvs. many, entire, lanceolate, sessile, 1–6 cm. long,
gradually passing upward into leafy bracts; infl. subsecund, 0.5–5 dm. long; pedicels
appressed, 5–7 mm. long; calyx oblique, herbaceous, the upper segm. 10–13 mm., the
others 7–9 mm. long; corolla rose-red with white or cream palate, glandular-pubescent
without, 17–19 mm. long, saccate at base, the upper lip reflexed, 6–7 mm. long, the
lower erect, 7–8 mm. long; caps. glandular-pubescent, 8–9 mm. long; seeds brown,
ovoid, ca. 1 mm. long, with many broken fimbriate winglike ridges.—Dry slopes, below
4000 ft.; Chaparral, Closed-cone Pine F.; s. face of San Bernardino Mts. to the coast and
n. to Santa Clara Co., Sierran foothills from Calaveras Co. to Tuolumne Co. May–July.

5. **A. cornùtum** Benth. [*A. c.* var. *venosum* Jeps.] Erect rather slender-stemmed
annual, viscid-villous throughout, simple or few-branched, 1–5 dm. high; lvs. linear-
oblong to oblong-ovate, 1–2.5 cm. long, on slightly winged petioles 4–10 mm. long; fls.
solitary in all but lowest axils, subsessile; calyx glandular-villous, the segms. linear-
oblong to lance-oblong, 4–5 mm. long; corolla lilac lined with darker violet and with
paler palate, 10–12 mm. long, pubescent without, the basal pouch spurlike; fertile fils.
all strongly oblique-dilated and ciliate-pubescent toward tip; style ca. 5 mm. long; caps.
6–7 mm. long; seeds ovoid, ca. 0.6 mm. long, echinate-favose.—Occasional in recently
disturbed places, below 5000 ft.; Yellow Pine F., Foothill Wd., etc.; Eldorado and
Napa cos. to Shasta and Humboldt cos. May–July.

Var. **leptàleum** (Gray) Munz. [*A. l.* Gray.] Fils. glabrous except at genicula, the shorter pair scarcely geniculate toward tip; style ca. 4 mm. long.—Similar places, Sierran foothills from Mariposa Co. to Kern Co.

6. **A. Coulteriànum** Benth. in DC. [*A. Nevinianum* Gray. *A. C.* var. *N.* Jeps.] Erect annual 3–12 dm. high, glabrous except in infl., with rather coarse main stem, simple below and with many simple tortile branchlets above; lvs. lance-ovate to ovate, the lower opposite, with blades 1–3 cm. long and petioles 1–2 cm. long, the main cauline lanceolate, alternate, 2–9 cm. long, subsessile or short-petioled; raceme dense, subsecund, glandular-villous, with linear to lance-linear green bracts; pedicels 2–3 mm. long; calyx glandular-villous, the subequal segms. 3–4 mm. long, the dorsal narrower; corolla white, 10–14 mm. long, pubescent without, the saccate spur broad, 1 mm. long, the palate yellowish; caps. 6–8 mm. long, glandular-pubescent; seeds dark, ovoid, ca. 1 mm. long, with many high almost continuous parallel longitudinal ridges or these ± anastamosed.—Common in dry ± disturbed places like burns, below 5000 ft.; Chaparral, Coastal Sage Scrub, etc.; cismontane s. Calif. from Santa Barbara Co. to n. L. Calif., inland to Joshua Tree Nat. Monument. April–July. Forma **Orcuttiànum** (Gray) Munz. [*A. O.* Gray. *A. C.* ssp. *O.* Penn.] Fls. bluish, 8–9 mm. long.—With the sp., San Diego Co.

7. **A. Nuttalliànum** Benth. in DC. [*A. N.* var. *effusum* Gray. *A. subsessile* Gray. *A. N.* var. *s.* Jeps.] Annual or biennial, simple and erect or commonly diffusely branched, leafy, softly viscid, glandular-pubescent, 1–10 dm. high, with tortile horizontal branchlets; lvs. ovate, 0.5–4 cm. long, usually short-petioled; raceme lax, leafy-bracted; pedicels 5–20 mm. long; calyx 3–5 mm. long, the uppermost segm. exceeding others; corolla violet, 10–12 mm. long, glandular-pubescent without, the basal pouch pale, shallow, the palate large, pale with violet reticulation; stamens dilated, glabrous above; caps. glandular-pubescent, 6–8 mm. long; seeds ca. 0.6 mm. long, subcylindrical, dark, alate- or cristate-costate.—Common in dry, especially disturbed places, below 2500 ft.; Chaparral, Coastal Sage Scrub; cismontane s. Calif. from Santa Barbara Co. to L. Calif.; s. Ariz., Santa Cruz, Santa Catalina, and San Clemente ids. March–July. A small form with cleistogamous fls. from Point Loma, San Diego Co. and ids. off L. Calif. has been called forma *pusíllum* (Bdg.) Munz. [*A. p.* Bdg.]

8. **A. subcordàtum** Gray. Bright green annual 3–8 dm. tall, pilose-hispid below, glandular-pubescent above, 1- to few-stemmed, with filiform tortile branchlets; lvs. ovate, crowded, sessile to subsessile, subcordate, 1–4.5 cm. long, 3–several veined; raceme foliose, secund, with short pedicels; calyx sparsely glandular-villous, the ovate upper segm. 10–20 mm. long, the linear-lanceolate lateral segms. 6–10 mm. long; corolla ochroleucous, 16–18 mm. long, glandular-pubescent without, the basal pouch 1–2 mm. deep, the lips 6–7 mm. long; caps. 6–7 mm. long; seeds ovoid, ca. 1 mm. long, reticulate-favose.—Rare, apparently on serpentine; Chaparral; Glenn and Colusa cos. June–July.

9. **A. vexíllo-calyculàtum** Kell. [*A. vagans* Gray. *A. vagans* var. *Bolanderi* Gray. *A. vagans* var. *rimorum* Jeps. *A. Coulterianum* var. *appendiculatum* Dur. & Hilg. *A. Elmeri* Rothm.] Erect or ascending annual, simple and ± glabrate or stiff-hairy below, diffusely branched and glandular-pubescent above, 3–9 dm. tall; lowest lvs. opposite, others alternate, not crowded, ovate to lanceolate, the blades 1–6 cm. long, 3–5-veined; petioles 2–35 mm. long; lvs. on tortile branches roundish, less than 1 cm. long; lower fls. remote, short-pedicelled, the upper in spicate racemes 0.5–3 dm. long; pedicels 2–5 mm. long; calyx glandular-pubescent, 4–12 mm. long, the segms. unlike, the dorsal 8–12 mm. long, broadly elliptic to narrow-ovate, the others 7–9 mm. long, linear-lanceolate; corolla lavender-violet, 12–15 mm. long, pubescent without, with basal saccate spur 1.5–2 mm. deep, the lips 6–7 mm. long, the palate pubescent; caps. 5–6 mm. long, glandular-pubescent, shorter than calyx-segms.; seeds 1 mm. long, ovoid, the winged ridges fimbriate, anastomosed.—Rocky slopes, often on burns, below 4000 ft.; Chaparral, etc.; near the coast from Sonoma Co. to San Benito Co. May–July.

10. **A. Brèweri** Gray. [*A. vagans* var. *B.* Jeps. *A. vexillo-calyculatum* var. *B.* Munz. *A. B.* var. *ovalifolium* Gray.] Near to *A. vexillo-calyculatum*, but stems generally glandular-pubescent throughout; fls. 10–12 mm. long; calyx-segms. less strongly dissimilar, the dorsal 4–7 mm. long, the others 3–5 mm. long; palate hirsute-villose; caps. not much or no shorter than calyx-segms.—More common, in open rocky places, below 4500 ft.; Chaparral, Mixed Evergreen F., Foothill Wd., Douglas-Fir F., Yellow Pine F.; Sonoma and Mariposa cos. n. to sw. Ore. June–Aug.

11. **A. Kíngii** Wats. Annual 1–4 dm. high, erect, simple or branched at base, glabrous except for tomentum at very base and glandular-puberulent infl.; lvs. lanceolate to linear, 0.5–3 cm. long, sessile to petioled; fls. short-pediceled, beginning at lowest axils, the upper in a secund raceme; calyx 4 mm. long in fl., the upper segm. 5–7 mm. in fr., oblong-lanceolate, the others 3–4 mm., lanceolate; corolla white, with purple veins, 7–8 mm. long, the villous palate not quite closing orifice; caps. 3–4 mm. long, glandular-puberulent; seeds ca. 0.5 mm. long, ovoid, deeply fimbriate-costate to alate-tuberculate. —Dry gravelly places, 5000–8000 ft.; Pinyon-Juniper Wd.; Clark Mts., e. San Bernardino Co. to Mono Co.; to Nev., Utah. May–June.

12. **A. ovàtum** Eastw. [*Howelliella o.* Rothm.] Glandular-villous annual, simple or much branched, 1–4 dm. tall, leafy; lf.-blades oblong to oval, sessile or subsessile, 1–3 cm. long, or lowest on petioles 1–2 cm. long; lvs. of lateral branchlets reduced, round-oval; infl. leafy; pedicels 2–3 mm. long; calyx-segms. dissimilar, the uppermost elliptic-oval, 10–15 mm. long, the others lanceolate, 5–7 mm. long; corolla ca. 2 cm. long, the upper lip rose-pink, the lower whitish, the basal sac ca. 1.5 mm. deep, the palate glabrous and not closing orifice; caps. 8–9 mm. long, glandular-pubescent; seeds 1 mm. long, dark brown, cuneate-obovoid, with an open network of low muricate ridges; $n = 8$ (Lenz, 1950).—Rare, often in subalkaline places, at ca. 2000 ft.; V. Grassland, Foothill Wd., etc.; inner Coast Ranges of s. Monterey and San Luis Obispo cos. May–June.

13. **A. fílipès** Gray. [*Asarina f.* Penn. *Neogaerrhinum f.* Rothm. *Antirrhinum Cooperi* Gray.] Filiform-stemmed, 3–8 dm. high, bright green branched glabrous annual with twining stems and pedicels; lowermost lvs. ovate, 0.5–2 cm. long on petioles almost as long, the upper narrower and longer and passing into lance-linear bracts; pedicels capillary, 3–8 cm. long, solitary, axillary; calyx obscurely glandular-puberulent, the lanceolate segms. subequal, ca. 4 mm. long; corolla bright yellow, 11–13 mm. long, glandular-puberulent without, the tube saccate at base, the lips 5–6 mm. long, the prominent palate hairy and with dark spots; stamens 6–8 mm. long, the lower pair dilated; caps. globose, 3–5 mm. long, glandular-puberulent; seeds scarcely 1 mm. long, tuberculate, with corky winglike outgrowths.—Twining among low shrubs in sandy places, below 5000 ft.; Creosote Bush Scrub, Joshua Tree Wd.; Calif. deserts n. to Inyo Co.; to Utah, Nev. Feb.–May.

14. **A. Kellóggii** Greene. [*A. strictum* (H. & A.) Gray, not Sibth. & Sm. *Maurandya s.* H. & A. *Asarina s.* Penn. *Antirrhinum Hookerianum* Penn.] Annual, glabrous except for slight woolliness at very base, 3–10 dm. high, strict or branched, upper parts usually vinelike and climbing by pedicels; lower lvs. crowded, ovate, obtuse, 0.5–2 cm. long, petioled, the upper narrower, sessile, longer; pedicels 4–8 cm. long, solitary in axils of narrow leafy bracts; calyx-segms. 5–6 mm. long, subequal, lance-linear; corolla 13–15 mm. long, blue, with pale violet-reticulated pubescent palate; caps. globose, glabrous, 6–7 mm. long; seeds scarcely 1 mm. long, tuberculate with many broken winglike plates. —Dry slopes, especially on burns, below 2000 ft.; Chaparral; Marin Co. to L. Calif. March–May.

17. Linària Mill. TOADFLAX

Mostly annual or perennial herbs with opposite or whorled lvs. or upper alternate, entire, dentate or lobed. Fls. in terminal spikes or racemes. Calyx 5-parted. Corolla with long tube, long-spurred at base, the throat often nearly closed by the palate. Fertile stamens 4, the fils. and styles filiform. Caps. dehiscent by 4 or more mostly 3-toothed pores or slits below the summit. Ca. 100 spp., many in cult. (Latin, *linum*, flax, as the lvs. of some spp. are flaxlike.)

A. Fls. yellow.
    B. Plants perennial, the stems 3–8 dm. long.
        C. Lvs. linear; corolla (including the spur) 2–3 cm. long . . . . .1. *L. vulgaris*
        CC. Lvs. ovate to lanceolate; corolla (including the spur) 3.5–4 cm. long . . 2. *L. dalmatica*
    BB. Plants annual, the stems 0.5–2 dm. high; corolla 1.5–2 cm. long . . . . 3. *L. supina*
AA. Fls. not yellow.
    B. Plants perennial; lvs. linear; corolla including spur 12–18 mm. long, violet-purple with dark palate . . . . 4. *L. purpurea*
    BB. Plants annual or biennial; lvs. linear or lance-linear.

C. Spur strongly curved, placed transversely or obliquely; corolla including spur 15–20
    mm. long.
D. Corolla violet with whitish ridges instead of well-formed palate. Native
                                                              5. *L. canadensis*
DD. Corolla violet-purple with well-formed orange palate. Occasional waif
                                                              6. *L. bipartita*
CC. Spur nearly straight, vertical.
D. Corolla 15–18 mm. long including spur, purple with yellow palate reticulate with
    purple veins ....7. *L. pinifolia*
DD. Corolla 35–38 mm. long, violet-purple with small yellowish patch on palate
                                                              8. *L. maroccana*

1. **L. vulgàris** Hill. BUTTER-AND-EGGS. Perennial, with several ascending stems to ca.
1 m. long; lvs. pale, numerous, narrow; raceme dense; corolla bright yellow, 2–3 cm.
long, with rounded orange palate; caps. 9–12 mm. long; seeds flattened, 1.5 mm. long;
n = 6 (East, 1933).—Occasional weed or escape; reported from Humboldt, Mendocino,
Sonoma, Marin, and San Bernardino cos.; widely distributed in N. Am.; native of Eu.
Summer.

2. **L. dalmática** (L.) Mill. [*Antirrhinum d.* L.] Rather coarse perennial to 1 m. high;
lvs. ovate to oblong, clasping, 1–4 cm. long; calyx 5–7 mm. long; corolla yellow, 3.5–4
cm. long; caps. 5–6 mm. long; seeds black, rhomboid, angled, ca. 1 mm. long, pitted;
$2n = 12$ (Matsuura & Soto, 1935).—Occasional waif, San Diego Co., Ventura Co.;
native of e. Medit. region. Summer.

3. **L. supìna** (L.) Desf. [*Antirrhinum s.* L.] Annual 0.5–2 dm. high, glabrous; lvs.
linear, 0.5–2 cm. long; racemes short, crowded; calyx glandular-puberulent; corolla ca.
15 mm. long; caps. ca. 4 mm. long; seeds winged.—Has been reported as a waif; native
of Medit. region. Summer.

4. **L. purpùrea** (L.) Mill. [*Antirrhinum p.* L.] Perennial 3–7 dm. high, glabrous;
lvs. many, linear, 2–5 cm. long; racemes dense; calyx ca. 2.5 mm. long; corolla 15–18
mm. long, violet-purple with dark palate; caps. ca. 3 mm. long; seeds ca. 1 mm. long,
angled, rugose-reticulate.—Occasional escape, as in Ventura Co.; native of s. Eu. May–
June.

5. **L. canadénsis** (L.) Dum.-Cours. [*Antirrhinum c.* L.] Slender annual or biennial
1–6 dm. high, with short trailing basal offshoots; cauline lvs. narrowly linear, sessile,
0.5–2.5 cm. long, those of basal sterile stems ovate to linear; racemes spicate, slender,
the pedicels 2–10 mm. long; calyx ± glandular-puberulent, 2–3 mm. long; corolla blue-
violet, 7–9 mm. long without spur, this 2–6 mm. long; caps. ca. 3 mm. long; seeds
cylindro-conic, truncate, ± angled, smooth, less than 0.5 mm. long.—Occasional near
coast, from Santa Clara Co. to Del Norte Co.; to B.C.; mostly in e. U.S. April–June.

Var. **texàna** (Scheele) Penn. [*L. t.* Scheele.] Corolla without the spur 9–12 mm.
long; spur 5–9 mm. long; seeds minutely tubercled.—Dry slopes and waste places,
often on burns below 2000 ft.; many Plant Communities; cismontane Calif.; to B.C.,
L. Calif., Ga., Va., S. Am. March–May.

6. **L. bipartìta** Willd. Annual ca. 3 dm. high, glabrous, glaucous; lvs. linear, 3–5 cm.
long; calyx 2.5–3 mm. long; corolla violet-purple with orange palate, 18–20 mm. long
including spur; caps. ca. 2 mm. long; seeds minute, rugose.—Occasional garden escape,
as at Claremont, Avalon, Santa Barbara; native of Medit. region. Early summer.

7. **L. pinifòlia** (Poir.) Thell. [*Antirrhinum p.* Poir. *A. reticulatum* Sm. *L. r.* Desf.]
Annual 8–10 dm. tall, glabrous except for the glandular-pubescent infl., the main stems
having lvs. linear, 2–4 cm. long; the basal horizontal branches having elliptic lvs.;
corolla 15–18 mm. long, purple with yellow palate with reticulate purple veining; caps.
ca. 4 mm. long; seeds black, minute, rugulose; *n* = 6 (Heitz, 1927).—Has been reported as garden escape; native of w. Medit. region.

8. **L. maroccàna** Hook. f. Annual to 5 dm. tall, glabrous below, viscid-pubescent
above; lvs. remote, linear, 2–4 cm. long; corolla 35–38 mm. long including spur, violet-
purple with whitish patch on palate; seeds obconic with 4–6 ringlike wings; *n* = 6
(Heitz, 1927).—Occasional escape from cult., as at Santa Barbara; native of N. Afr.

18. Cymbalària Hill

Trailing or twining herbs with long-petioled palmately veined and lobed lvs. Fls.
axillary, scattered, long-pedicelled. Calyx 5-parted. Corolla short-spurred at base on

lower side, the throat closed by the prominent palate. Stamens 4, didynamous, the anthers glabrous and somewhat coherent. Caps. globose, opening by 2 pores, the 2 valves each splitting into 3. Seeds rounded, rugose or crested. Ca. 10 spp., mostly Medit. (*Cymbalum*, a cymbal, from the round lvs.)

1. C. muràlis Gaertn., Mey. & Scherb. [*Antirrhinum C. L. Linaria C.* Mill.] KENIL-WORTH–IVY. Stems filiform, glabrous; lvs. 5–9-lobed, reniform-orbicular, 1–3 cm. wide; corolla violet, 7–9 mm. long excluding the spur, which is 3 mm.; caps. 4 mm. long; seeds black, 0.5 mm. long; $n = 7$ (Heitz, 1927).—Cult. and occasionally escaping in moist cool places near the coast; introd. from Eu. May–Sept.

19. Kíckxia Dumort. FLUELLIN

Diffuse repent perennial herbs, hairy, with short-petioled scattered lvs. Fls. scattered, axillary, white or yellowish-white. Calyx 5-parted. Corolla ventrally spurred at base, 2-lipped, with prominent palate closing throat. Stamens 4, didynamous, the anthers ciliate, ± coherent. Caps. globose, circumscissile by 2 pores. Seeds rounded, brown, alveolate with thin winglike irregular convolutions. Ca. 30 spp., mostly from Medit. region. (J. *Kickx*, Belgian professor in early 19th century.)

Sepals ovate; lvs. rounded or cordate at base; corolla (without the spur) 6–8 mm. long . 1. *K. spuria*
Sepals lanceolate; lvs. hastate-lobed at base; corolla 5 mm. long .... 2. *K. Elatine*

1. K. spùria (L.) Dumort. [*Antirrhinum s. L. Linaria s.* Mill.] Much branched, glandular and soft-hairy, forming mats a meter or so across; lvs. ovate to rounded, 2.5–4 cm. wide, on petioles 2–5 mm. long; pedicels 10–15 mm. long; calyx-lobes accrescent; upper lip of corolla violet, the lower yellow; corolla-spur 5 mm. long; caps. 3–4 mm. long; seeds 1 mm. long; $2n =$ ca. 14 (Heitz, 1927).—Occasionally natur. as in Merced, Mariposa, Napa, San Mateo, and Ventura cos.; introd. from Eu. June–Sept.

2. K. Elatìne (L.) Dumort. [*Antirrhinum E. L. Linaria E.* Mill.] Like the preceding; lvs. ovate, 1–3 cm. long, 1–2.5 cm. wide; petioles 2–5 mm. long; pedicels 10–25 mm. long; corolla 5 mm. long, the spur an additional 4 mm.; caps. 3–4 mm. long; seeds 1–1.2 mm. long; $n = 9$ (Bruun, 1932).—Occasionally natur. in old fields, river bottoms, etc., from Orange Co. to Humboldt Co.; introd. from Eu. June–Sept.

20. Digitàlis L. FOXGLOVE

Erect biennial or perennial herbs, with alternate lvs. and showy racemose fls. Calyx 5-parted. Corolla declined, with a somewhat inflated tube and short scarcely spreading limb, the orifice open. Stamens 4, didynamous, included. Stigmas distinct, lamellate. Caps. ovoid, loculicidal. Seeds many, reticulate, not winged. Ca. 30 spp., of Old World. (*Digitalis*, pertaining to the finger, as fingers of a glove, because of tubular corolla).

1. D. purpùrea L. Stoutish pubescent biennial 5–18 dm. high, distally glandular; lower lvs. oblong-lanceolate, acute, dentate, 1–3 dm. long, somewhat petioled at base; fls. many, from linear-lanceolate bracts; pedicels 1.5–2.5 cm. long; calyx-lobes ovate, becoming 15–18 mm. long; corolla purple to white, spotted on lower paler side, 4–5 cm. long; caps. 12 mm. long; seeds 0.5 mm. long; $2n = 56$ (Buxton & Newton, 1928).—Natur. in ± shaded places near the coast; Santa Barbara n. to B.C.; native of Eu. May–Sept.

21. Verónica L. SPEEDWELL

Annual or perennial, erect or prostrate herbs, with opposite lvs. or the upper bract-like, alternate. Calyx 4- or 5-parted. Corolla subrotate, white to blue, 4-lobed (the upper broad lobe formed by fusion of the normal upper pair). Stamens 2. Stigma entire. Caps. flattened, loculicidal. Seeds flattened, smooth or rarely roughened. Ca. 250 spp., of N. Temp. Zone. (Possibly named for *St. Veronica*.)

(Pennell, F. W. Veronica in N. and S. Am. Rhodora 23: 1–22, 29–41, 1921.)
A.   Main stem ending in a single racemelike infl.
   B.   Plants perennial from underground rootstocks; fls. in upper axils only.
      C.   Caps. at least as long as wide, shallowly or not notched; corolla glabrous within.
         D.   Style longer than caps.; corolla 8–13 mm. wide.

E. Calyx-lobes 5; corolla 8–10 mm. wide; lf.-blades pubescent. Scott Mts., n.
  Calif. . . . . 1. *V. Copelandii*
  EE. Calyx-lobes 4; corolla 10–13 mm. wide; lvs. mostly glabrous. Sierra Nevada
    to Siskiyou Co. . . . . 2. *V. Cusickii*
    DD. Style shorter than caps.; corolla 6–7 mm. wide. Tulare Co. n. . . . . 3. *V. alpina*
    CC. Caps. wider than long, deeply notched; corolla pubescent in the tube. Most of state
      4. *V. serpyllifolia*
  BB. Plants annual; fls. from most axils.
    C. Corolla 2–2.5 mm. wide; pedicels 1–2 mm. long.
      D. Lvs. linear-oblong to spatulate; corolla white . . . . 5. *V. peregrina*
      DD. Lvs. rounded to oval; corolla bright blue . . . . 6. *V. arvensis*
    CC. Corolla 6–11 mm. wide; pedicels 5–25 mm. long.
      D. Lvs. deeply cleft or lobed, with blunt spatulate divisions; pedicels 5–10 mm.
        long . . . . 7. *V. triphyllos*
      DD. Lvs. crenate-serrate or dentate; pedicels 10–25 mm. long . . . . . . 8. *V. persica*
AA. Main stem with lateral racemes below the tip.
  B. Lvs. of main stems definitely short-petioled; corolla 7–10 mm. wide.
    C. Lvs. much longer than wide. Common . . . . 9. *V. americana*
    CC. Lvs. at least as wide as long. Rare . . . . 10. *V. Beccabunga*
  BB. Lvs. of main stems sessile, ± cordate-clasping; corolla 3–7 mm. wide.
    C. Caps. ovoid to roundish, scarcely notched; lvs. lanceolate to broader.
      D. Pedicels in fr. ascending, 6–8 mm. long; calyx mostly exceeding the caps
        11. *V. Anagallis-aquatica*
      DD. Pedicels in fr. ± horizontal, 4–6 mm. long; calyx shorter than caps . 12. *V. comosa*
    CC. Caps. deeply obcordate, much broader than long; lvs. linear to linear-lanceolate
      13. *V. scutellata*

1. **V. Copelándii** Eastw. Low perennial, branched, barely suffrutescent, 0.5–1.2 dm. tall, softly pilose throughout with some hairs gland-tipped; lf.-blades oblong-elliptic, ± acute, entire, sessile, 0.5–1.8 cm. long; fls. in small terminal raceme; pedicels 5–8 mm. long; calyx-lobes 5, elliptic, 2.5 mm. long; corolla 8–10 mm. wide; fls. 4–5 mm. long; style 4–6 mm. long; caps. longer than wide, shallowly notched.—Subalpine F.; Scott Mts., Siskiyou and Trinity cos. Aug.

2. **V. Cusíckii** Gray. Much like *V. Copelandii,* but hairs on stems, pedicels, and calyces shorter; lvs. elliptic-oval, mostly glabrous, obtuse to acutish; calyx-lobes 4, narrow-ovate; corolla 10–13 mm. wide; caps. deeply notched, scarcely longer than wide. —Rare, meadows and moist places, 8500–9200 ft.; Lodgepole F., Subalpine F.; Alpine and Placer cos., Sierra Nevada and Klamath Mts. of Siskiyou Co.; to Wash., Mont. July–Aug.

3. **V. alpìna** L. var. **alterniflòra** Fern. [*V. Wormskjoldii* of Calif. refs.] Perennial with slender caudices and erect stems 1–3 dm. tall, loosely pilose on stems and lvs., glandular-pubescent in infl.; lvs. elliptic or oblong-oval, obtusish, 5–15 mm. long, of 4–7 pairs, the upper often alternate; racemes to 1 dm. long in fr.; pedicels 3–8 mm. long, often alternate; calyx 2.5–4 mm. long; corolla 6–7 mm. wide; caps. 5–7 mm. long, oval, widely retuse; seeds 1 mm. long.—Wet places, mostly 7000–11,500 ft.; Red Fir F. to Subalpine F.; Sierra Nevada from Tulare Co. n., to Modoc and Siskiyou cos.; to B.C., Wyo. June–Aug.

4. **V. serpyllifòlia** L. Much-branched perennial, with creeping base; stems to ca. 1.5 dm. high, closely and finely strigulose, not glandular; lvs. ovate or oblong, obscurely crenate, 0.5–1.5 cm. long, the lower petioled and rounded, the uppermost becoming bracteate; racemes terminal, small, with pedicels 4–5 mm. long; calyx 3–4 mm. long; corolla whitish or pale blue, with darker lines, 6–7 mm. broad; caps. ca. 3 mm. long, 4 mm. broad, obtusely notched; seeds 0.5 mm. long.—Occasional weed as in lawns; native of Eu. April–June.

Var. **humifùsa** (Dickson) Vahl. [*V. h.* Dickson.] Stouter, 2–4 dm. high, upper stems and pedicels with longer often spreading hairs, some being gland-tipped; corolla 5–8 mm. wide, mostly bright blue; caps. mostly 4–5 mm. long.—Common in moist places, mostly 6000–11,000 ft. (lower in n.); Montane Coniferous F.; San Jacinto and San Bernardino mts., Sierra Nevada, N. Coast Ranges from Glenn Co. n.; to Alaska, Atlantic Coast, S. Am., Eurasia. April–Aug.

5. **V. peregrìna** L. ssp. **xalapénsis** (HBK.) Penn. [*V. x.* HBK.] Erect annual, simple to few-stemmed, 1.5–3 dm. tall, glandular-pubescent on stems and caps.; lvs. linear-oblong to spatulate, obtuse, entire to ± dentate, sessile or the lower petioled, 1–2.5 cm. long; raceme lax, leafy-bracted; pedicels 1–2 mm. long; calyx-segms. subequal, 3 mm. long; corolla white, 2–2.5 mm. wide; caps. 3–3.5 mm. long, shallowly emarginate; seeds

0.5 mm. long.—Moist places, from sea level to 10,200 ft.; many Plant Communities; cismontane Calif.; to w. Canada, Mex., S. Am. April–Aug.

6. V. arvénsis L. Erect or ascending annual, the slender stems 1–3 dm. long, pilose, nonglandular; lower lvs. rounded to oval, crenate-dentate, 5–15 mm. long, petioled, the upper ovate to lanceolate, smaller, sessile, subtending the fls.; pedicels 1–2 mm. long; calyx 3.5–4 mm. long; corolla bright blue, 2–2.5 mm. wide; caps. 2–2.5 mm. long, rounded, deeply emarginate, pilose; seeds 1 mm. long; $2n = 14$, 16 (Yamashita, 1937). —Occasional weed in lawns, waste places, etc.; native of Eu. April–July.

7. V. triphýllos L. Erect annual, usually branched at base, 5–12 cm. high, finely glandular-pubescent throughout; lower lvs. roundish, mostly 5-cleft or deeply lobed, sessile, 5–12 mm. long, the upper 5-lobed with narrow blunt spatulate divisions; bracts 3-lobed; pedicels 5–10 mm. long; calyx 4–5 mm. long; corolla deep blue, 6–9 mm. broad; caps. 5–6 mm. long, deeply emarginate; seeds 2 mm. broad, hemispheric, concave on one side.—Gravelly places, at 2700 ft., near Yreka, Siskiyou Co.; natur. from Eu. March–April.

8. V. pérsica Poir. [V. Buxbaumii Ten.] Annual, usually with several procumbent stems 1–4 dm. long, spreading-pubescent; lf.-blades roundish-ovate, crenate-serrate or coarsely dentate, 6–20 mm. long, on shorter petioles; racemes lax, pedicels 10–25 mm. long; calyx 5 mm. long; corolla deep blue, 7–11 mm. wide; caps. 2.5–3 mm. long, widely notched, with strongly divergent lobes; seeds 1.5 mm. long; $2n = 28$ (Beatus, 1936).—Occasional weed in lawns, waste places, etc.; natur. from Eu. Feb.–May.

9. V. americàna (Raf.) Schw. [V. Beccabunga var. a. Raf.] BROOKLIME. Glabrous ± succulent perennial, rhizomatose and creeping at base, the main stems 1–10 dm. long; principal lvs. lance-ovate to lanceolate, acutish, serrate or denticular, 1–9 cm. long, short-petioled, the lowermost ± rounded; racemes lax, few–many-fld., the lower pedicels to ca. 12 mm. long; calyx 3 mm. long; corolla 7–10 mm. wide, violet-blue to lilac; caps. turgid, suborbicular, 3–4 mm. long; seeds 0.5 mm. long; $2n = 36$ (Schlenker, 1936).—Wet places along streams, sea level to 10,500 ft.; many Plant Communities; cismontane and montane Calif.; in s. Calif. also occasional on deserts; to Alaska, Atlantic Coast, Mex., Asia. May–Aug.

10. V. Beccabúnga L. Like V. americana but with rounded lvs.—Introd. near Bridgeport, Mono Co.; native of Eu.

11. V. Anagállis-aquática L. Habit of V. americana; lvs. of flowering stems ± cordateclasping, lance-oblong, acute; racemes mostly many-fld., the fruiting pedicels ascending, 6–8 mm. long; calyx 4–4.5 mm. long; corolla 4–5 mm. broad, pale lavender, with violet lines; caps. ovoid to roundish, scarcely notched, ca. 4 mm. long; seeds 0.5 mm. long; $2n = 36$ (Ehrenberg, 1945).—Occasionally natur. in wet places; native of Eu. May–Sept.

12. V. comòsa Richt. [V. connata auth. V. connata var. glaberrima Penn. V. catenata Penn.] Similar to V. Anagallis-aquatica; lvs. lance-oblong; pedicels 4–6 mm. long, horizontally divergent; calyx shorter than caps.; corolla whitish or pale rose, 3–4 mm. wide; caps. 2.5–3 mm. long, obcordately notched; seeds 0.4 mm. long.—Occasional in ditches and wet places; San Bernardino and Santa Clara cos.; to Canada, Pa., etc. July–Sept.

13. V. scutellàta L. Weak glabrous perennial, rhizomatose at base; stems slender, 1–6 dm. tall; lvs. sessile, linear to· narrow-lanceolate, entire or minutely denticulate, 1.5–8 cm. long; racemes divergent, lax, small-bracted, rather few-fld.; pedicels 6–16 mm. long; calyx 3 mm. long; corolla lilac or blue-lavender, 5–7 mm. wide; caps. 3–4 mm. long, deeply obcordate, broader than long; seeds flat, oblong, ca. 1.3–1.5 mm. long; $2n = 18$ (Hagerup, 1944).—Swales and wet places; many Plant Communities; Coast Ranges from Marin Co. n., Sierra Nevada, at 3500–7000 ft., from Fresno Co. n., to Siskiyou Co., possibly also in San Bernardino Mts. (Peirson); to B.C., New England; Eurasia. May–Aug.

22. Hèbe Comm.

Shrubs or small trees, with persistent leathery opposite lvs. Fls. white to blue or pink, in axillary racemes or small heads. Calyx usually 4-parted. Corolla with a short tube, the limb commonly 4-lobed. Stamens 2, exserted. Caps. flattened, septicidal. Seeds few to many. A large genus of S. Hemis. (Greek, *hebe*, small.)

Corolla deep blue-purple, 8 mm. wide; lvs. 5–10 cm. long .... 1. *H. speciosa*
Corolla lilac, 10 mm. wide; lvs. ca. 3 cm. long .... 2. *H. franciscana*

1. **H. speciòsa** (R. Cunn.) Cockayne & Allan. [*Veronica s.* R. Cunn.] Stout glabrous shrub 6–15 dm. high, with angled spreading branches; lvs. obovate-oblong, 5–10 cm. long, sessile or nearly so; racemes densely fld., not longer than lvs.; calyx-lobes oblong-ovate; corolla reddish- to blue-purple, 8 mm. broad; stamens and style very long; caps. 6 mm. long, broadly ovate; $2n = 40$ (Simonet, 1934).—Sparingly natur. along coast as in San Francisco; native of New Zealand. Summer.

2. **H. × franciscàna** (Eastw.) Souster. [*Veronica f.* Eastw.] Shrub ca. 1 m. high, densely branched; lvs. elliptic, decussate, sessile, green above, paler beneath, ca. 3 cm. long, 2 cm. wide; racemes densely fld., clustered toward ends of the branches; corolla lilac or violet, ca. 1 cm. broad.—Natur. on ocean bluffs, San Francisco; native country unknown. Summer.

23. Synthỳris Benth. in DC.

Perennial herbs from rootstocks, with a basal tuft of rounded petioled lvs. Fls. racemose, blue or violet-blue. Calyx 4-parted. Corolla rotate to subcampanulate, 4-lobed. Stamens 2, exserted. Anther-cells parallel. Stigmas united, minutely capitate. Caps. flattened, loculicidal. Seeds many or few, flattened or with incurved margins. Ca. 14 spp., of w. N. Am. (Greek, *sun*, together, and *thuris*, a door, because of the adherence of the base of the caps.-valves to the placentae.)

(Pennell, F. W. A revision of Synthyris and Besseya. Proc. Phila. Acad. Nat. Sci. 85: 77–106, 1933.)
Corolla 6–9 mm. long; lvs. ± pilose. Coastal .... 1. *S. reniformis*
Corolla ca. 4 mm. long; lvs. glabrous. Modoc Co. .... 2. *S. missurica*

1. **S. renifórmis** (Dougl.) Benth. var. **cordàta** Gray. [*S. rotundifolia* var. *c.* Gray.] Stems 0.6–1.4 dm. high; herbage sparingly appressed-pilose; lf.-blades ovate-cordate, shallowly and crenately lobed, paler beneath, 1.5–5.5 cm. long, on longer petioles; infl. scapose, ± pilose, few-fld.; pedicels 7–10 mm. long; calyx-lobes oval, 4–4.5 mm. long; corolla campanulate, pale bluish-lavender, 6–9 mm. long, the lobes shorter than tube; style 5–8 mm. long; caps. 2–4 mm. long, with widely divaricate lobes; seeds with thick incurved margins; $n = 12$ (C. McMillan, 1949).—Moist shaded forests, below 2800 ft.; Mixed Evergreen F., Douglas-Fir F., Redwood F.; Marin Co. to Del Norte Co.; sw. Ore. Feb.–April.

2. **S. missùrica** (Raf.) Penn. [*Veronica m.* Raf. *V. reniformis* Pursh, not Raf.] Stems 1–2 dm. high; basal lf.-blades round-cordate, shallowly doubly dentate-lobed, 2–3 cm. wide, long-petioled; racemes ± puberulent; pedicels 3–6 mm. long; calyx-lobes linear to ± oblong, 3–4 mm. long; corolla subrotate, blue to purplish, the lobes longer than tube; style 4–5 mm. long; caps. ca. 5 mm. long, distally notched; seeds flat.—Grassy places, at 6500–7000 ft.; Yellow Pine F., Red Fir F.; Warner Mts., Modoc Co.; to Wash., Ida. May–June.

24. Bellárdia All.

Erect pubescent annuals, with opposite lvs. and spikelike racemes. Fls. sessile in axils of broad imbricated bracts. Calyx 4-lobed. Corolla 2-lipped, the upper lip galeate, slightly shorter than the lower which is broad, shallowly 3-lobed and with a 2-ridged palate. Stamens 4, didynamous, the fils. broad, the anthers pubescent and mucronate-tipped. Caps. ovoid, mucronate, hirsute, loculicidal. Seeds irregularly ovoid, with white lines. Two spp., of Medit. region.

1. **B. Trixàgo** (L.) All. [*Bartsia T.* L.] Stems 2–5 dm. high, with stiff retrorse hairs; infl. ± glandular; lvs. lanceolate, coarsely dentate-lobed, sessile; bracts ovate-cordate; calyx 8–9 mm. long, the obtuse lobes 1–1.5 mm. long; corolla 16–18 mm. long, the galea pale purple, the lower lip white; caps. 6–7 mm. long; seeds 0.4 mm. long.—

Natur. in old fields, grassy places, etc.; counties about San Francisco Bay, Claremont; native of Medit. region. April–May.

25. Parentucéllia Viv.

Erect annual or biennial hairy herbs with opposite toothed lvs. Fls. in terminal leafy spikes. Calyx campanulate, with 4 lanceolate lobes. Corolla 2-lipped, the upper lip galeate, the lower with a 2-ridged palate and short spreading lobes. Stamens 4, didynamous, the anthers lanate, mucronate-tipped. Caps. cylindric, acute, loculicidal. Seeds ellipsoid-oblong, smooth. Two spp., of Medit. region. (T. *Parentucelli*, founder of the botanic garden in Rome.)

1. **P. viscòsa** (L.) Caruel. [*Bartsia v.* L.] Glandular-pubescent, 3–5 dm. tall; lf.-blades oblong-lanceolate, sessile, 2–4 cm. long; corolla yellow, 16–17 mm. long; caps. ca. 8 mm. long, brown-hairy; seeds 0.3 mm. long.—Roadside weed, near the coast, Sonoma Co. to Ore. April–June.

26. Pediculàris L. Lousewort

Perennial herbs. Lvs. in ours alternate or basal, pinnatifid or pinnate, rarely simple. Fls. in a usually spikelike raceme, white to yellow, red or purple. Calyx 2–5-cleft or -toothed. Corolla strongly 2-lipped, the upper lip galeate, strongly arched; lower lip shorter, with spreading or appressed lobes. Stamens 4, didynamous; anthers glabrous, the equal cells obtuse to subulate at tips. Caps. flattened, glabrous, loculicidal. Seeds several, turgid, often slightly winged. Ca. 500 spp., mostly of N. Temp. Zone. (Latin, *pediculus*, a louse; application not certain.)

A. Calyx-lobes 5 or 4, all distinct at tips; lvs. deeply pinnatifid or pinnate.
   B. Segms. of lf.-blades sharply toothed; stems with a tuft of basal lvs.
      C. Upper lip of corolla blunt and beakless; raceme often little exceeding the lvs.
         D. Corolla pale yellow.
           E. Stem not over 1 dm. tall, much exceeded by the basal lvs.
                                                    1. *P. semibarbata*
           EE. Stem 3–9 dm. tall, longer than basal lvs. . . . . 4. *P. flavida*
         DD. Corolla mostly purple-red.
           E. Lower lip of corolla ca. ⅛ as long as upper; corolla ca. 25 mm. long; stem
             exceeding lvs. Widely distributed . . . . 2. *P. densiflora*
           EE. Lower lip of corolla ca. half as long as upper; corolla ca. 18 mm. long;
             stems shorter than lvs. Santa Cruz Mts. . . . . 3. *P. Dudleyi*
      CC. Upper lip of corolla produced above the galea into a slender curving proboscislike beak;
         raceme much exceeding lvs.
         D. Beak decurved, shorter than body of galea; corolla white or faintly yellow, ca.
           15 mm. long . . . . 5. *P. contorta*
         DD. Beak upcurved, longer than body of the galea; corolla ± purplish, 7–10 mm. long
           (without the beak).
           E. Infl. villous; calyx-teeth ½ to as long as calyx-tube; beak of galea 3–6 mm.
             long . . . . 6. *P. attollens*
           EE. Infl. glabrous; calyx-teeth shorter; beak of galea 6–12 mm long
                                             7. *P. groenlandica*
   BB. Segms. of lf.-blades crenate-serrulate; stems without basal lvs. Siskiyou Mts. . . 8. *P. Howellii*
AA. Calyx-lobes seemingly 2, the laterals on each side united; lvs. serrate or crenate.
   B. Stems with longitudinal·pubescent lines; lf.-margins usually white-callose; corolla purple.
      Mono Co. . . . . 9. *P. crenulata*
   BB. Stems glabrous; lf.-margins not white-callose; corolla whitish or purple-tinged. Placer Co. to
      Trinity and Siskiyou cos. . . . . 10. *P. racemosa*

1. **P. semibarbàta** Gray. Stems few, mostly underground, from the root-crown, not over 1 dm. long; lvs. in a rosette, bipinnatifid, 5–15 cm. long, the pinnules ovate and deeply cut into irregularly toothed segms.; petioles as long as blades; pedicels 4–5 mm. long; calyx ca. 10 mm. long, the 5 lobes linear, entire or ± dentate; corolla 15–20 mm. long, yellowish with purplish tips, white-tomentulose without, the lower lip ca. 2 mm. shorter than the upper; anther-cells sharply acuminate; caps. ca. 9 mm. long, decurved, dehiscent dorsally; seeds ca. 3–4 mm. long, brownish, smooth.—Dry slopes, mostly 5000–11,000 ft.; Yellow Pine F. to Subalpine F.; Cuyamaca Mts. (San Diego Co.) to Sierra Nevada and Shasta and Butte cos., Glenn Co. n.; Ore., Nev. May–July.

2. **P. densiflòra** Benth. ex Hook. Indian Warrior. Stems several, simple, 1–5 dm.

tall, pubescent, exceeding lvs.; lvs. largely in a basal rosette or well distributed on stems, pubescent to glabrate, the blades 5–15 cm. long on shorter petioles, twice-pinnatifid into many laciniate-dentate lobes; spike dense, oblong, the bracts ca. as long as fls., oblong-lanceolate, with sharp salient distal teeth; calyx 8–12 mm. long, deep red, deeply 5-toothed; corolla 2–2.5 cm. long, deep purple-red, cylindric, glabrous, strongly arched; the galea 12–16 mm. long; anther-cells acute; caps. ca. 7 mm. long, dehiscing dorsally and ventrally; seeds 2–4, ca. 4 mm. long.—Dry slopes, below 6000 ft.; Chaparral, Foothill Wd., Yellow Pine F., etc.; San Diego Co. to Del Norte Co.; s. Ore., n. L. Calif. Jan.–June.

Ssp. **aurantìaca** E. Sprague. Galea 12–18 mm. long, orange to yellow; bracts oblanceolate.—Yellow Pine F.; Plumas Co. to Sierra Co.

3. **P. Dúdleyi** Elmer. Stem 1–1.5 dm. tall; plant pubescent or lvs. ± glabrous; lvs. mostly basal, 12–20 cm. long, the pinnules elliptic-oblong to ovate, deeply cut into sharply dentate segms., the blades longer than petioles; bracts oblong-lanceolate, ca. as long as fls., sharply serrate in upper portion; calyx 10–11 mm. long, the 5-lobes lanceolate; corolla 17–18 mm. long, purplish-red, glabrous, the upper lip 10–11 mm. long, the lower 6–7 mm. long and paler than upper; anthers obtusish; caps. ca. 8 mm. long.—Redwood F.; Santa Cruz and San Mateo cos. April–June.

4. **P. flàvida** Penn. Stems subglabrous, 3–9 dm. high; lvs. ± ovate in outline, 5–10 cm. long, the lower segms. quite distinct, lanceolate, deeply dentate to somewhat lobed, the lobes denticulate; racemes rather dense, 5–15 cm. long, short-pedicelled; bracts entire or denticulate above on the narrowed terminal portion; calyx 5-cleft, the lobes lanceolate, the lateral shorter than the tube; corolla 15–18 mm. long, the galea erect, strongly hooded, with apex not beaked or toothed; lower lip much shorter, incurved, toothed.—Damp places, 5000–7000 ft.; Red Fir F.; Siskiyou Mts. to s. Ore. July–Aug.

5. **P. contórta** Benth. ex Hook. Glabrous; stems 3–5 dm. tall; lvs. basal and on lower stem, 3–8 cm. long, the pinnules oblong-lanceolate, cut into rounded or acutely dentate segms.; petioles mostly shorter than blades; bracts narrowly lanceolate, few-lobed; pedicels 3–5 mm. long; calyx 6–8 mm. long, the 5 lobes ovate, attenuate; corolla ca. 15 mm. long, glabrous, white or faintly yellow, with small dark purple spots, the upper lip decurved with a terminal upcurved beak; anther-cells acute; caps. 9–10 mm. long; seeds ca. 2 mm. long.—Edge of meadows and in woods, 6200–7500 ft., Canyon Creek, Trinity Co., Mt. Eddy, Siskiyou Co.; to B.C., Mont. July–Aug.

6. **P. attóllens** Gray. [*Elephantella a.* Heller.] Little Elephant's Head. Glabrous below the white-woolly infl., 1.5–4 dm. high; lvs. basal and on lower stem, the blades 3–9 cm. long, the pinnules narrow, serrate-dentate; petioles shorter than blades; bracts shorter than or ca. as long as fls., few-lobed; pedicels 1–2 mm. long; calyx ca. 5 mm. long, with 5 lance-linear lobes; corolla body ca. 7 mm. long, glabrous, lavender or pink, with purplish markings, the tube recurved through the ventral cleft of calyx, then recurved and narrowed into an upward beak; anthers obtuse; caps. ca. 9 mm. long; seeds ca. 3 mm. long.—Common in meadows and moist places, 5000–12,000 ft.; mostly Montane Coniferous F. above Yellow Pine F.; White Mts., Sierra Nevada, n. to Siskiyou and Modoc cos.; to Ore. June–Sept. A form from Lassen Co. with slightly larger corollas has been called ssp. *protogỳna* Penn.

7. **P. groenlándica** Retz. [*Elephantella g.* Rydb.] Elephant Heads. Glabrous throughout; stems 3–6 dm. tall; lvs. basal and on lower stems, the blades 3–10 cm. long, with pinnules narrow-lanceolate, serrate-dentate; petioles shorter than blades; bracts narrow-lanceolate, shorter than fls., few-lobed; calyx 4–5 mm. long, the 5 lobes subulate; corolla-body 8–10 mm. long, glabrous, red-purple, the slender beak 4–8 mm. long; anther-cells acute; caps. 6–8 mm. long; seeds ca. 3 mm. long.—Occasional, meadows and wet places, 6000–11,200 ft.; Montane Coniferous F. above Yellow Pine F.; Sierra Nevada to Humboldt, Siskiyou, and Modoc cos.; to boreal Am., Atlantic Coast. June–Aug. A form with the sp. and having the beak 8–12 mm. long, is ssp. *surrècta* (Benth.) Piper. [*P. s.* Benth.]

8. **P. Howéllii** Gray. Stems 3–4 dm. tall, minutely pubescent below the villous infl.; lvs. cauline, glabrous except on the pubescent midrib; lfts. few, the lateral oblong-ovate, 1.5–2 cm. long, the terminal larger; bracts almost equal to fls., entire, ciliate; pedicels 1 mm. long; calyx 6–7 mm. long, the 5 lobes ovate, woolly, attenuate; corolla glabrous, yellow, 8–9 mm. long, strongly arched with a short recurved beak; caps. ca. 8 mm.

long; seeds 1.5–2 mm. long.—Dry ridges, at 6000 ft.; Red Fir F.; Siskiyou Mts. June–Aug.

9. **P. crenulàta** Benth. in DC. Stems 2–3.5 dm. high, pubescent below, ± villous above; lvs. all cauline, 3–6 cm. long, short-petioled, linear to narrow-oblanceolate, crenate-dentate with callous white edges; bracts shorter than fls.; pedicels 3–4 mm. long; calyx 10–11 mm. long; corolla 20–22 mm. long, glabrous, purple, the upper lip 11–12 mm. long, ascending, curved and round-hooded; caps. 15–16 mm. long; seeds 1.5 mm. long.—Near streams, ca. 7000 ft.; Convict Creek, Mono Co.; to Wyo., Colo. June–July.

10. **P. racemòsa** Dougl. ex Hook. Stems 3–5 dm. high, glabrous below the finely pubescent infl.; lvs. all cauline, 3–7 cm. long, short-petioled, glabrous, lanceolate, serrate-dentate; bracts equal to or longer than fls.; pedicels 2–4 mm. long; calyx 5–8 mm. long; corolla 10–12 mm. long, glabrous, pink or whitish; upper lip erect, strongly incurved, prolonged into a hooked beak; lower lip deflexed-spreading, 10–12 mm. wide; caps. 10–12 mm. long, attenuate.—Dry slopes, 4000–7000 ft.; Douglas-Fir F., Yellow Pine F., Red Fir F.; Nevada Co. to Humboldt and Siskiyou cos.; to B.C., Rocky Mts. June–Aug.

27. Orthocárpus Nutt.

Erect or diffuse annuals, with sessile alternate narrow entire to pinnate or parted lvs. Infl. spicate, prominently bracted. Calyx tubular-campanulate, 4-cleft, or cleft before and behind with the divisions lobed. Corolla narrow-tubular, strongly 2-lipped, valvate, the upper lip (galea) erect, entire, beaklike, not much longer than the inflated saccate lower lip which is tipped with 3 small teeth. Stamens 4, inserted near summit of tube; anthers 1- or 2-celled, thin, usually explanate and ciliate. Caps. loculicidal. Seeds few to many, with reticulate or alveolate often loose coat. Ca. 25 spp., of w. N. Am. and 1 in the Andes. (Greek, *orthos,* straight, and *karpos,* fr.)

(Keck, D. D. A revision of the genus Orthocarpus. Proc. Calif. Acad. Sci., IV, 16: 517–571, 1927.)
A.   Anthers 2-celled. (Subgenus Orthocarpus)
   B.   Lower lip of corolla ± 3-saccate; seed coat loose-fitting, except in *O. campestris.*
     C.   Bracts green throughout; lower lip deeply 3-saccate; galea equaling or barely exceeding lower lip.
       D.   Bracts entire; lower cell of anthers ¼–⅕ as long as upper cell .. 1. *O. campestris*
       DD. Bracts cleft into lanceolate or linear lobes; lower cell of anthers at least ½ as long as upper cell.
         E.   Galea finely pubescent or puberulent.
           F.   Lower lip of corolla 5–8 mm. wide . . . . 2. *O. lithospermoides*
           FF. Lower lip of corolla 4 mm. wide or less.
             G.   Ventral margins of galea pubescent; sacs of lower lip 3–5 mm. deep
                                                    3. *O. lacerus*
             GG. Ventral margins of galea glabrous; sacs of lower lip ca. 2 mm. deep
                                                    4. *O. hispidus*
         EE. Galea densely white-villous . . . . 5. *O. lasiorhynchus*
     CC. Bracts tipped with purple or yellow; galea exceeding lower lip.
       D.   Corolla widened upward, the lower lip more than 2 mm. deep; spike usually broad and conspicuous.
         E.   Stems pubescent or nearly glabrous; spike showy.
           F.   Galea nearly straight, pubescent.
             G.   ·Lvs. oblong, more than 3 mm. wide, entire or with rounded teeth; stems usually ascending or the plants forming mats. Saline places
                                                6. *O. castillejoides*
             GG. Lvs. lanceolate, less than 3 mm. wide, with linear divisions; stems erect. Grassy fields . . . . 7. *O. densiflorus*
          FF. Galea hooked at tip, densely bearded; stems purple; lf.-divisions filiform
                                                     8. *O. purpurascens*
        EE. Stems villous-pubescent above; spike pale. Foothills of Sierra Nevada
                                                 9. *O. linearilobus*
       DD. Corolla linear, the lower lip not more than 2 mm. deep; spike narrow, pale and rather inconspicuous . . . . 10. *O. attenuatus*
   BB. Lower lip of corolla simple saccate or nearly so; seed coat tight-fitting or ridged.
     C.   Bracts and calyx glandular-pubescent; bracts gradually differing from upper lvs.; style glabrous.
       D.   Corolla yellow; lower lip relatively shallow, 3.5–4 mm. long .. 11. *O. luteus*
       DD. Corolla rose-purple; lower lip relatively deep, 5–7 mm. long .. 12. *O. bracteosus*
     CC. Bracts and calyx not glandular; bracts abruptly differing from upper lvs.; style microscopically pubescent.

<pre>
      D. Galea pubescent, even at tip.
          E. Corolla 20–25 mm. long; galea exceeding lower lip by 3–5 mm.
                                                                13. O. cuspidatus
          EE. Corolla 12–15 mm. long; galea equaling or exceeding lower lip by usually
              less than 2.5 mm. . . . . 14. O. Copelandii
      DD. Galea puberulent, glabrate at very tip.
          E. Corolla 20–30 mm. long; tip of galea inflexed 1 mm., surpassing lower lip
              by 2.5–3 mm. . . . . 15. O. pachystachyus
          EE. Corolla 10–18 mm. long, tip of galea inflexed 0.5 mm., surpassing lower
              lip by ca. 1 mm. . . . . 16. O. imbricatus
 AA. Anthers 1-celled; seed coat tight-fitting or ridged. (Subgenus Triphysaria)
      B. Stamens shorter than galea; each lobe of lower lip less than 1.5 times as deep as long.
          C. Branches divergent from the cent. erect axis; fls. showy, usually conspicuously exserted;
              galea straight or gradually curved; bracts pinnatifid.
              D. Galea purple; herbage pubescent . . . . 17. O. erianthus
              DD. Galea yellowish; herbage glabrous, or puberulent in infl. . . . . 18. O. faucibarbatus
          CC. Branches many, weak, ascending from the base, the cent. stem indistinct; fls. minute,
              inconspicuous; galea sharply curved; bracts often bipinnatifid . . . . 19. O. pusillus
      BB. Stamens exceeding galea; each lobe of lower lip a deep sac, 1.5 times as deep as long
                                                                20. O. floribundus
</pre>

1. O. campéstris Benth. [*O. c.* var. *succulentus* Hoov. *O. columbinus* Jones.] Stem simple or branched, 1–2.5 dm. high, the lower herbage glabrous; lvs. entire, narrowly lance-linear, 1.5–4 cm. long; spike dense; bracts like the lvs. but shorter, glabrous, ciliate, or ± hirsute at base; calyx hirsute, 2-cleft to middle, each half 2-lobed with subulate teeth; corolla bright yellow, sometimes whitish, 1.5–2.5 cm. long; lower lip ample, 4–5 mm. long, abruptly widened from narrow tube, villous within anteriorly, the teeth oblong, 1.5–2 mm. long; galea straight, narrow, slightly longer than lower lip; anthers 2-celled, the lower cell vestigial or reduced; caps. ovoid, 5–7 mm. long; seeds many, irregular, black, narrow, 0.7 mm. long, with thin close-fitting coat.—Winter pools and moist places, below 5000 ft.; V. Grassland, Foothill Wd.; Fresno Co. to Sonoma, Shasta, and Modoc cos.; s. Ore. April–July.

2. O. líthospermoìdes Benth. Cream Sacs. Stem stoutish, erect, simple or strictly branched, 2–7 dm. high, the herbage hirsute and glandular-pubescent, especially above; lvs. 2–8 cm. long, lanceolate, the upper pinnatifid into 3–7 linear lobes; spike heavy; bracts 1–2.5 cm. long, palmately lobed; calyx 8–14 mm. long, subequally 4-lobed; corolla clear yellow, usually with 2 purple spots at base of lower lip, 1.5–2.5 cm. long, puberulent, the lower lip large, abruptly dilated, floccose within; caps. ovoid, 5–6 mm. long; seeds many, in thin loose-fitting reticulate coats, ca. 0.7 mm. long, yellowish.—Open grassy places, below 3000 ft.; Coastal Prairie, N. Oak Wd., Chaparral, Mixed Evergreen F.; Santa Clara Co. to Douglas Co., Ore. April–June.

Var. **bìcolor** (Heller) Jeps. [*O. b.* Heller. *O. rubicundulus* Jeps.] Corolla white, turning pinkish in age.—Grassy places; V. Grassland, Foothill Wd.; Sacramento V. and neighboring foothills, from Lake and Napa cos. to Siskiyou and Butte cos. April–June.

3. O. lácerus Benth. [*O. Brownii* Eastw.] Stem slender, erect, simple or with ascending branches, the herbage finely pubescent, becoming villous and minutely glandular above; lvs. filiform to lance-linear, 1–5 cm. long, the upper pinnately parted into 3–7 filiform lobes; spike rather lax; bracts palmately 3–7-cleft, 1–2 cm. long; calyx subequally 4-lobed, 7–10 mm. long; corolla 1–1.8 cm. long, bright yellow with 2 brown dots at base of lower lip, finely pubescent; galea straight; caps. oblong to elliptical, 5–7 mm. long; seeds many, in a loose-fitting reticulate coat, yellowish, ca. 0.8 mm. long.— Grassy slopes and swales, below 5000 (7000) ft.; Foothill Wd., Yellow Pine F., Sagebrush Scrub, w. base of Sierra Nevada from Fresno Co. n., to Modoc and Siskiyou cos.; s. Ore. May–July.

4. O. híspidus Benth. [*Triphysaria h.* Rydb. *O. tenuis* Heller. *O. falcatus* Eastw.] Stems slender, erect, simple or with few erect branches, 1–4 dm. high, pubescent below, ± glandular and hirsute above; lvs. narrowly linear-lanceolate, 1–4 cm. long, the upper 3–5-cleft; spike slender; bracts ovate, 1–2.5 cm. long, palmately 3–7-cleft; calyx 8–10 mm. long; corolla yellow or white, 1.2–2 cm. long, pubescent, the lower lip small, ca. 2 mm. deep; galea usually straight, narrow; caps. oblong, 5–8 mm. long; seeds many, brown, in a loose reticulate coat, ca. 1 mm. long.—Meadows, 3000–7000 ft.; Yellow Pine F., Red Fir F., Sagebrush Scrub; rare in San Diego and San Bernardino cos., more frequent from Tulare, Glenn, and Mendocino cos. n.; to Alaska, Ida., Nev. May– Aug.

5. **O. lasiorhýnchus** Gray. Stem slender, erect, often with erect branches, 1–3 dm. high, pilose; lvs. lance-linear, at least the upper with 2 small lateral lobes; spike lax; bracts 3–5-parted, 0.6–1.2 cm. long; calyx cleft halfway into 2 lobes each with 2 teeth; corolla yellow with 2 minute blackish dots at base of lower lip, 1.2–2.5 cm. long, the tube slender, gradually inflated into the obovoid lower lip; caps. ellipsoid, 6–9 mm. long; seeds many, in loose-fitting alveolate coats.—Meadows, 4000–7500 ft.; Yellow Pine F.; San Bernardino Mts. to Cuyamaca Mts. June–July.

6. **O. castillejoìdes** Benth. [*O. maculatus* Eastw. *O longispicatus* Elmer. *O. sonomensis* Eastw. *O. c.* var. *insalutatus* Jeps.] Stem simple, erect or with ascending or decumbent branches, 1–3 dm. long, pubescent; lvs. 1–5 cm. long, lance-linear to oblong, obtuse, entire or with 1–3 pairs of lateral lobes; spike congested, rather broad; bracts, 1.4–2.2 cm. long, oblong to ovate in outline, palmately cleft into 3–7 oblong rounded yellow to white or pinkish lobes; calyx 1.2–2 cm. long, subequally cleft ca. half its length into narrowed lobes with colored tips; corolla 1.4–2.5 cm. long, yellow with purple markings, narrow or broader, the lower lip 4–6 mm. long not including the 1–3 mm. long teeth; caps. ellipsoid, 8–12 mm. long; seeds many, yellowish, in loose reticulate coats, ca. 1–1.2 mm. long.—Low saline places; N. Coastal Scrub, edge of Coastal Salt March, etc.; near the coast from Monterey Co. n.; to B.C. May–Aug.

Var. **humboldtiénsis** Keck. Bracts and calyx purple-tipped; corolla purplish, the lower lip tipped with yellow.—Salt marshes about Humboldt Bay.

7. **O. densiflòrus** Benth. [*O. noctuinus* Eastw.] OWL's-CLOVER. Stem erect, slender, often corymbosely branched above, 1–3.5 dm. high, puberulent, yellowish; lvs. linear or linear-lanceolate, 2–8 cm. long, the lower entire, the upper with a pair of lateral lobes; spike dense; bracts usually 3-lobed, 1–2 cm. long, the upper purple-tipped, finely pubescent; calyx 1–2.5 cm. long, pubescent, cleft ⅓–½ its length into 4 subregular linear lobes; corolla purplish, 1–2.5 cm. long, the galea subulate, puberulent, the lower lip longer than deep, often yellowish with 3 prominent purple spots, the erect teeth 1.5–2.5 mm. long, purple; caps. ovoid, 7–10 mm. long; seeds black, ca. 0.5 mm. long in a loose-fitting reticulate coat.—Grassy places, below 2500 ft.; Foothill Wd., Mixed Evergreen F., N. Coastal Scrub; Coast Ranges from Los Angeles Co. to Mendocino Co., occasional in Sierra Nevada foothills. March–May.

Var. **grácilis** (Benth.) Keck. [*O. g.* Benth. *O. Parishii* Gray.] Bracts mostly 8–10 mm. long; corolla more exserted, the lower lip much inflated, as deep as or deeper than long, often white, the teeth minute.—Chaparral, Foothill Wd., Yellow Pine F.; Santa Lucia Mts. to L. Calif.

Var. **obispoénsis** Keck. Spike white; corollas creamy-white with few purple dots; lower lip deeply saccate.—Grassy slopes, coastal San Luis Obispo Co.

8. **O. purpuráscens** Benth. [*O. p.* var. *Palmeri* Gray.] OWL's-CLOVER. Stem erect, ± slender, often corymbosely branched from base upward, 1–4 dm. high, ± reddish, hirsute; lvs. 1–5 cm. long, deeply pinnatifid into several filiform thickened divisions; spikes dense, pale to deep purple; bracts 1–2 cm. long, palmately 5–7-lobed, typically with greenish hirsute base, greenish-purple in middle, and velvety and rose-purple at tips; calyx 1–2 cm. long, with 4 linear equal lobes, with pubescence and color of bracts; corolla 1.2–3 cm. long, crimson or purplish, scarcely exserted, the lower lip mostly purplish often with white or yellow at tip and with purple dots; galea slender, ± hooked at tip, bearded; caps. ovoid, 10–15 mm. long; seeds ovoid to ellipsoid, less than 1 mm. long, smooth, in a loose-fitting alveolate coat.—Open fields, grassy slopes, etc., below 3000 ft.; V. Grassland, Coastal Sage Scrub, Foothill Wd.; Great V. to the coast, from Mendocino Co. s.; to n. L. Calif., Ariz., Son. March–May.

KEY TO VARIETIES

Lower lip of corolla purple, sometimes tipped with yellow.
  The lower lip ± tipped with yellow or white. Mostly cismontane.
    Floral bracts cleft into filiform or linear lobes 1 mm. or less wide. Widely distributed, n. of Los
    Angeles Co. . . . . *O. purpurascens*
    Floral bracts cleft into linear-spatulate lobes 1–2 mm. wide. Immediate coast . . var. *latifolius*
  The lower lip with outer third bright orange-yellow. W. Mojave Desert . . . . . var. *ornatus*
Lower lip of corolla yellow or white, without any purple. From e. San Luis Obispo Co. to Riverside Co.
                                                                  var. *pallidus*

Var. **latifòlius** Wats. [*O. p.* var. *multicaulis* Jeps.] Branches procumbent or ascending, or stem simple; lobes of fl.-bracts usually broad and showy, tipped with lavender, pink, or greenish-white and making the spike look banded; galea rich purple.—Old dunes; Coastal Strand, N. Coastal Scrub; Humboldt Co. to San Luis Obispo Co.

Var. **ornàtus** Jeps. [*O. venustus* Heller. *O. p.* var. *v.* Keck.] Spike deep purple; corolla deep velvet-red, the outer third of lower lip orange-yellow.—Open flats, 2000–3000 ft.; Creosote Bush Scrub, Joshua Tree Wd.; w. Mojave Desert.

Var. **pállidus** Keck. [*O. exsertus* Heller.] Lower lip of corolla white or pale yellow.— Sandy and disturbed places, below 4000 ft.; Coastal Sage Scrub, V. Grassland, Chaparral; s. San Luis Obispo Co. to w. Riverside and n. San Diego cos.

9. **O. linearilòbus** Benth. [*O. mariposanus* Congd.] Stem erect, simple or with erect branches above, 1.5–3.5 dm. high, pubescent, villous above; lvs. linear, the lower entire, the upper with 2–3 pairs of subfiliform lobes; spike dense; bracts 1.2–2 cm. long, the linear lobes purplish or yellowish-tipped; calyx subequally 4-lobed; corolla cream or yellowish, 1.5–2.5 cm. long, pubescent, the galea sometimes rose-lilac, the lower lip yellow with 2 small purple dots at base and 3 larger toward summit; caps. ovoid, 7–10 mm. long; seeds black, shiny, 1 mm. long, loosely coated.—Grassy slopes, below 5000 ft.; V. Grassland, Foothill Wd., Yellow Pine F.; from Kern Co. along foothills of Sierra Nevada to Shasta Co. April–June.

10. **O. attenuàtus** Gray. Erect, simple or few-branched, 1–4 dm. high, canescent; lvs. lance-linear, 2–6 cm. long, the upper with 2 filiform lobes; spike narrow; bracts 1.5–2 cm. long, 3-lobed, white-tipped, sometimes purplish; calyx 2–2.3 cm. long, the subequal lobes white-tipped; corolla 1–2.5 cm. long, whitish or sometimes purplish-tinged, narrow, the shallow lower lip purple-dotted; caps. ellipsoid, 6–10 mm. long; seeds brown, smooth, the loose reticulate coat ca. 1 mm. long.—Grassy places, below 5000 ft.; V. Grassland, Foothill Wd., Chaparral, Yellow Pine F., Mixed Evergreen F., etc.; cis-montane Calif. from San Diego Co. and n. L. Calif. to B.C., Chile. March–May.

11. **O. lùteus** Nutt. [*O. strictus* Benth.] Stem erect, simple or branched above, 1–3.5 dm. high, pilose and glandular-pubescent; lvs. linear-lanceolate, 1.5–4 cm. long, hispidulous and glandular, mostly entire, dark; spike very narrow; bracts dark green, 1–1.5 cm. long, 3-lobed, the cent. lobe ovate or lance-ovate; calyx 6–8 mm. long, the subequal lobes deltoid-lanceolate; corolla golden-yellow, 9–12 mm. long, glandular-pubescent, the lower lip 3.5–4 mm. wide, ca. equal to galea; caps. ellipsoid, 5–7 mm. long; seeds several, yellow-brown, the reticulate wrinkled coats tight-fitting, 1–1.25 mm. long.— Rare in Calif., in moist places; Sagebrush Scrub, N. Juniper Wd., Yellow Pine F.; mostly e. of Sierra Nevada, Fresno and Mono cos. n.; to B.C., Rocky Mts. July–Aug.

12. **O. bracteòsus** Benth. Stem erect, simple or with erect branches above, pubescent, somewhat glandular, 1–4 dm. high; lvs. linear-lanceolate, 2–4 cm. long, the lower entire, the upper with a pair of narrow lateral lobes; spike dense; bracts 1–2 cm. long, green or tinged purple, with 2–4 divaricate lanceolate lobes; calyx green, 6–10 mm. long, soft-hairy; corolla 1.2–2 cm. long, rose-purple, the tube exceeding calyx; galea broad, pubescent, scarcely exceeding lower lip; caps. ellipsoid, 5–7 mm. long; seeds few, 2–3 mm. long, the coat ridged, not alveolate.—Moist meadows, below 5000 ft.; Sagebrush Scrub, N. Juniper Wd.; Plumas Co. to Modoc Co.; to B.C. June–July.

13. **O. cuspidàtus** Greene. [*O. pachystachyus* var. *c.* Jeps.] Stem erect, short-branched above, 1–3.5 dm. high, canescent; lvs. linear-lanceolate, 1–5 cm. long, the lowermost entire or all parted to below the middle with 3 linear-attenuate lobes; spike dense; bracts abruptly different from lvs., 1.2–2 cm. long, broadly ovate-oblong with a basal pair of divaricate lobes, the middle lobe 8–12 mm. wide, strongly veined, purple-tipped; calyx 8–12 mm. long; corolla 2–2.5 cm. long, rose-purple with white lower lip, which is 3–5 mm. shorter than the pubescent arched galea; caps. ellipsoid, 6–8 mm. long; seeds black, few, with a tight-fitting alveolate coat, 2 mm. long.—Open grassy places, 3000–6000 ft.; Yellow Pine F. to Subalpine F.; Siskiyou Mts., Siskiyou Co.; adjacent Ore. June–July.

14. **O. Copelándii** Eastw. Stem erect, simple or corymbosely branched, 1–3.5 dm. high, finely canescent and viscidulous; lvs. entire or the upper with a pair of lateral lobes, acuminate, 1–6 cm. long; spike dense, often subcapitate; bracts abruptly different from lvs., ca. 1 cm. long with a near-basal pair of attenuate lobes, the mid-lobe 5–8 mm.

wide, roseate-tipped, scabrid-puberulent, obtuse to subtruncate; corolla 1.2–1.5 cm. long, rose-purple with white lower lip, the galea slightly curved, pubescent; lower lip 3–4 mm. deep; caps. ellipsoid, 5–7 mm. long; seeds few to several, ca. 2 mm. long, with tight-fitting alveolate coat.—Open places, 4500–7000 ft.; Red Fir F., Subalpine F.; Glenn Co. to Siskiyou Co.; sw. Ore. June–Aug.

Var. **cryptánthus** (Piper) Keck. [*O. c.* Piper.] Spike elongated; corolla 1–1.2 cm. long, nearly hidden by bracts, the lower lip 2 mm. deep.—Rocky slopes, 7000–8500 ft.; Sagebrush Scrub to Red Fir F.; Mono Co. to Modoc Co.; se. Ore., w. Nev.

15. **O. pachystáchyus** Gray. Stout, 1.5–2.5 dm. high, glandular-pubescent; lvs. entire or 3–5-lobed, scabrous, 1.5–3.5 cm. long; spike heavy, compact; bracts differing abruptly from lvs., broad, 1.5–2.8 cm. long, 3–7-lobed, the mid-lobe ovate-oblong, 5–9 mm. wide; calyx 1.5–2 cm. long, parted nearly to base behind; corolla rose-purple, the lower lip glabrate with pubescent margin, ca. 2–3 mm. shorter than the slender subglabrous galea; caps. ellipsoid, 5–7 mm. long; seeds several, flattened, 3 mm. long, with tight-fitting finely alveolate coat.—Rare, Sagebrush Scrub? plains of the Shasta River, Siskiyou Co. at ca. 2500 ft. June.

16. **O. imbricàtus** Torr. ex Wats. Stem erect, puberulent, simple or corymbosely branched, 1–3.5 dm. high; lvs. linear to lance-linear, entire, somewhat scabrid, 2–4 cm. long; spike compact, short; bracts abruptly different from lvs., entire or with 2 small lobes, ovate or broadly oblong, puberulent, ciliate below, purple-tipped, 8–14 mm. long; calyx 4–6 mm. long, deeply cleft before and behind; corolla purple with partly white lower lips, partly concealed by bracts, 1–1.3 cm. long, glandular-puberulent; caps. ovoid, flattened; seeds few to several, 2 mm. long, with tight-fitting reticulate coat.—Dry open places, 5000–7000 ft.; Red Fir F., Subalpine F.; Shasta, Trinity, and Humboldt cos. n.; to Wash. July–Aug.

17. **O. eriánthus** Benth. [*O. Bidwelliae* Gray.] Butter-and-Eggs. Johnny-Tuck. Stem corymbosely branched from below upward, rarely simple, finely canescent and glandular, 0.5–3.5 dm. high; lvs. 1–5 cm. long, narrow-linear, pinnately divided into several filiform divisions, ± purplish; spike lax below, more congested above, the fls. well exserted; bracts 0.5–1.8 cm. long, with 4–10 linear divisions; calyx 5–8 mm. long; corolla 1–2.5 cm. long, the tube long, slender, pubescent, much exserted from calyx, the lower lip abrupt, purplish at base, the sacs inflated, light yellow, 3–4 mm. deep, the teeth purplish, inconspicuous; galea dark purple, subulate; caps. 4–8 mm. long, oblong; seeds many, brown, with loose reticulate coat, scarcely 1 mm. long.—Common in open places like valley floors, below 2000 (4000) ft.; V. Grassland, Foothill Wd.; Mendocino and Shasta cos. to San Luis Obispo and Kern cos., rare to San Diego Co. March–May.

Var. **gratiòsus** Jeps. & Tracy. Lower lip of corolla with middle lobe yellow and lateral lobes white, the purple band around the throat prominent.—Sandhills, Mendocino Co. to s. Ore.

Var. **ròseus** Gray. [*Triphysaria versicolor* F. & M. *O. v.* Greene. *O. e.* var. *inopinus* Jeps. and var. *v.* Jeps.] Lower lip of corolla 4–5 mm. deep, rose-pink or white turning rose.—Open grassy hills; N. Coastal Scrub, Coastal Prairie; San Luis Obispo Co. to Mendocino Co.

Var. **micránthus** (Gray) Jeps. [*O. Bidwelliae* var. *m.* Gray. *O. m.* Greene ex Heller.] Corolla 8–15 mm. long, the lower lip yellow, 1–2 mm. deep.—Grassy plains; V. Grassland; Mariposa Co. to Kern Co., e. Monterey and San Benito cos.

18. **O. faucibarbàtus** Gray. [*O. erianthus* var. *laevis* Gray.] Stem simple or corymbosely branched, straw-colored, glabrous or nearly so except for the puberulent infl.; lvs. 2–8 cm. long, linear-lanceolate, pinnately divided into filiform lobes; spike elongate; bracts gradually differing from lvs.; corolla 1.2–2.2 cm. long, the slender tube much exserted from calyx, the lower lip abruptly inflated, 2–4 mm. deep, pale sulphur-yellow with purple dots on margins; galea yellowish; caps. and seeds like *O. erianthus.*—Open places, below 1500 ft.; Mixed Evergreen F.; away from immediate coast, Marin Co. to Mendocino Co. April–June.

Ssp. **álbidus** Keck. Lower lip of corolla white, fading rose, with greenish yellow spot at base of each tooth.—N. Coastal Scrub, back of Coastal Strand, etc.; near the coast, Monterey Co. to Lane Co., Ore.

19. **O. pusíllus** Benth. Stem slender, with weak ascending branches from base,

0.5–2.5 dm. high, the herbage hispidulous and with purplish tinge; lvs. 1–3 cm. long, linear, the upper pinnatifid with filiform divisions; spike racemose, extending almost to base of stems; bracts 5–12 mm. long, 1–2-pinnatifid with filiform divisions; calyx 5–7 mm. long, subequally 4-lobed; corolla 4–6 mm. long, dull brownish-purple, glabrous or sparingly hispidulous, the tube slender, swelling into the trisaccate lower lip 1 mm. long and wide, the galea broad, uncinate; caps. subglobose, 4–6 mm. long; seeds many, black, with tight-fitting reticulate coat, ca. 1 mm. long.—Grassy fields, below 2000 ft.; Mixed Evergreen F., Foothill Wd.; Coast Ranges from San Luis Obispo Co. n., occasional from Stanislaus Co. to Butte Co.; to B.C. April–May.

20. **O. floribúndus** Benth. Stem openly branched, straw-colored, subglabrous, 1–3 dm. high; lvs. 1–4 cm. long, pinnatifid above with many filiform divisions; spike short, compact; bracts gradually differing from the lvs., scabrid, lobed; calyx 4–6 mm. long; corolla 1–1.4 cm. long, white or cream, the tube slender, glabrous, exceeding calyx, the lower lip abruptly inflated, with 3 divergent spreading oblong sacs 2 mm. deep, the small erect teeth ciliolate; galea hyaline except for the midrib; caps. ovoid, 4–5 mm. long; seeds many, black, with tight-fitting reticulate coat, ca. 0.6 mm. long.—Open places; N. Coastal Scrub; San Mateo Co. to Point Reyes, Marin Co. April–May.

28. **Castillèja** Mutis. PAINT-BRUSH

Herbs or suffrutescent plants, partially root-parasites. Lvs. alternate, sessile, entire or laciniate, passing above into usually more incised and colored conspicuous bracts of the terminal spikelike raceme. Bracteoles none. Calyx tubular, 4-lobed or seemingly 2-lobed by fusion of those on each side. Corolla tubular, ± compressed and greenish, 2-lipped, the upper lip (galea) elongate, entire, enclosing the style and stamens; lower lip shorter, often rudimentary, 3-toothed, 3-carinate or -saccate below the teeth. Stamens 4, didynamous, all with unequal 2-celled anthers. Caps. many-seeded, glabrous, loculicidal, ovoid or cylindric-ovoid. Seeds many, with a loose reticulate coat. Ca. 200 spp., of New World, with 1 going into Asia. (D. *Castilleja*, a Spanish botanist.)

(Pennell, F. W. Castilleja, *in* L. Abrams: Illus. Fl. Pacific States, 3: 819–846, 1951.)
A. Plants annual; lvs. and bracts entire, linear-lanceolate. Wet places.
  B. Corolla 2.5–3.5 cm. long, the galea ca. as long as tube. Cismontane . . . . . 32. *C. stenantha*
  BB. Corolla 1.5–1.8 cm. long, the galea ca. half as long as tube. Owens V. . . 33. *C. exilis*
AA. Plants perennial; lvs. often and upper bracts usually divided.
  B. Calyx-lobes united much farther dorsally than ventrally and their tips usually curved upward; corolla usually curved forward through lower calyx-sinus.
    C. Lvs. yellow-green, fleshy, not crowded; stems 6–15 dm. long; galea greenish. Transmontane . . . . 29. *C. linariaefolia*
    CC. Lvs. grayish-green, not fleshy, ± crowded; stems 2–7 dm. long; galea yellowish. Cismontane.
      D. Calyx-lobes 3–5 mm. long; lvs. ± obtuse. Coast Ranges . . . . . . 30. *C. franciscana*
      DD. Calyx-lobes 5–7 mm. long; lvs. sharply tipped. Sierran foothills
                                      31. *C. subinclusa*
  BB. Calyx-lobes not united much farther dorsally than ventrally, the tips straight.
    C. Lower corolla-lip ca. half to almost as long as galea and with thin often whitish usually not incurved lobes.
      D. Calyx-lobes joined little or no farther laterally than medianly, linear to narrow lanceolate; bracts and calyces yellowish to dull reddish ventrally.
        E. Lower corolla-lip thin and whitish, at least ⅔ as long as galea. From n. of Kern Co.
          F. Herbage woolly with appressed matted hairs.
            G. Galea lanose-pubescent dorsally, the thin margins narrow; lower lip of corolla lanose-pubescent. Siskiyou Co. . . 1. *C. arachnoidea*
            GG. Galea dorsally minutely pubescent, the thin margins wide; lower corolla-lip minutely pubescent. Mt. Lassen to Modoc Co.
                                     2. *C. Payneae*
        FF. Herbage with distinct and spreading hairs.
          G. Galea 4–6 mm. long, its thin margins yellowish-translucent throughout; lower lip of corolla usually exserted from calyx, its lobes 2–2.5 mm. long. From Plumas Co. n. . . 3. *C. psittacina*
          GG. Galea 6–8 mm. long, its thin margins ± purplish; lower lip of corolla only tardily if at all exserted, its lobes often shorter.
            H. Bracts rounded distally; corolla 17–22 mm. long. Mostly below 9000 ft., Mono Co. to Plumas Co. . . . . 4. *C. pilosa*
            HH. Bracts sharp-pointed; corolla 13–16 mm. long. Mostly above 10,000 ft., from Eldorado and Mono cos. to Tulare and Inyo cos. . . . . 5. *C. nana*

EE. Lower corolla-lip green and thickened at apex, ca. half as long as galea.
    San Bernardino Co. .... 6. *C. cinerea*
DD. Calyx-lobes joined much farther laterally than medianly, the free tips ± ovate;
    bracts and calyces ± purplish.
    E. Corolla 18–20 mm. long, the galea dorsally glandular-hairy, the lower lip
        6–7 mm. long, exserted from calyx and with 3 evident pale lobes
                                                 7. *C. Lemmonii*
    EE. Corolla 16–18 mm. long, the galea dorsally puberulent or glabrescent, the
        lower lip ca. 4 mm. long, included in calyx and with rudimentary terminal
        lobes .... 8. *C. Culbertsonii*
CC. Lower corolla-lip less than half as long as galea, usually green or dark and with minute
    and incurved lobes.
    D. Herbage with simple (not branched) hairs, green to grayish-green.
      E. Plant evidently glandular-pubescent below the infl.
        F. Stems mostly less than 2 dm. long; galea 6–10 mm. long, the lower
          lip mostly ca. ⅓ as long.
          G. Calyx-lobes 1–2 mm. long, rounded; corolla 15–16 mm. long,
            the galea ca. 6 mm. long. Siskiyou and Del Norte cos.
                                    9. *C. brevilobata*
          GG. Calyx-lobes 3–4 mm. long, lanceolate; corolla 16–22 mm. long, the
            galea 7–10 mm. long. Sierra Nevada.
            H. Stems quite pilose; calyx-lobes lanceolate, 3–4 mm. long;
              galea not conspicuously exserted .... 10. *C. Breweri*
            HH. Stems scarcely pilose; calyx-lobes linear, 5 mm. long; galea
              conspicuously exserted .... 11. *C. Peirsonii*
        FF. Stems mostly more than 2 dm. long; galea 10–20 mm. long, the lower
          lip mostly less than ⅕ as long.
          G. Plants quite woody at base, bushy with much-branched stems; lvs.
            mostly rounded to blunt at tip; galea usually rather strongly
            pubescent. Near seacoast .... 12. *C. latifolia*
          GG. Plants scarcely or not woody at base; stems mostly not much-
            branched; galea usually puberulent or finely pubescent. Away from
            immediate coast.
            H. Calyx-lobes usually rounded; galea 15–24 mm. long, stout,
              conspicuously exserted. Inner Coast Ranges from Lake Co. to
              San Diego Co. .... 15. *C. Roseana*
            HH. Calyx-lobes sharp-pointed; galea 10–17 mm. long, more
              slender, less conspicuously exserted.
              I. Calyx-lobes 0.5–2 mm. long, almost or quite as wide;
                lvs. 2–5 cm. long. Los Angeles and San Bernardino cos.
                to L. Calif. .... 16. *C. Martinii*
              II. Calyx-lobes 2–4 mm. long, not nearly as wide; lvs 2–3.5
                cm. long. Ventura Co. and Tulare Co. n.
                J. Lvs. lanceolate; lower corolla-lip not exserted. Coast
                  Ranges .... 17. *C. Applegatei*
                JJ. Lvs. almost linear; lower corolla-lip exserted. Sierra
                  foothills .... 18. *C. disticha*
      EE. Plant not glandular-pubescent below the infl.
        F. Calyx-lobes mostly less than 2 mm. long, blunt; lvs. mostly rather blunt,
          oval to broadly lanceolate. Along immediate coast.
          G. Lvs. rounded or oblong, less than 3 times as long as wide; stems
            practically without gland-tipped hairs .... 12. *C. latifolia*
          GG. Lvs. lance-oblong, at least 3 times as long as wide; stems mostly
            with some gland-tipped hairs .... 14. *C. Wightii*
        FF. Calyx-lobes 3–6 mm. long, pointed; lvs. narrower.
          G. Galea 9–10 mm. long; calyx ca. 15 mm. long, it and the bracts
            mostly yellowish-tipped. Tiburon Peninsula, Marin Co.
                                     23. *C. neglecta*
          GG. Galea 12–23 mm. long; calyx mostly longer, it and bracts mostly
            purplish- or reddish-tipped.
            H. Main cauline lvs. entire.
              I. Herbage glabrous or inconspicuously pubescent; lvs.
                mostly 2–5 cm. long, without axillary fascicles; lower
                corolla-lip mostly exserted. Plants of damp places.
                J. Infl. not subcinereous; plant less than 1 m. tall.
                  K. Spikes mostly with much scarlet. Widely dis-
                    tributed .... 19. *C. miniata*
                  KK. Spikes somewhat yellowish. Pitkin Marshes
                              20. *C. uliginosa*
               JJ. Infl. subcinereously pubescent; plant 1 m. tall. Point
                 Reyes .... 21. *C. Leschkeana*
              II. Herbage stiff-pubescent; lvs. 3–9 cm. long, with axillary
              fascicles; lower corolla-lip not exserted. Dry places
                                  22. *C. affinis*
        HH. Main cauline lvs. mostly with 1–2 pairs of lobes.

I. Galea 16–23 mm. long, the lower lip dark green to
brownish, 1.5–2 mm. long. Coast Ranges . . 22. *C. affinis*
II. Galea 12–15 mm. long, the lower lip dark green, 2–3 mm.
long. Deserts . . . . 24. *C. chromosa*
DD. Herbage with branched hairs, ± grayish to white-woolly.
E. Calyx cleft laterally more deeply than medianly, the upper lip 4–7 mm. long,
the lower 6–9 mm. long; corolla 15–20 mm. long; infl. yellow. Desert slopes
and base of San Gabriel Mts. . . . . 28. *C. plagiotoma*
EE. Calyx cleft medianly more deeply than laterally; corolla usually longer; infl.
mostly reddish.
F. Herbage not densely tomentose; calyx-lobes acute.
G. Galea 7–9 mm. long; corolla 18–20 mm. long . . . . . 13. *C. mollis*
GG. Galea 15–20 mm. long; corolla 25–30 mm. long . . 25. *C. pruinosa*
FF. Herbage densely tomentose; calyx-lobes rounded or wholly united
laterally.
G. Galea greenish dorsally with pale thin reddish margins; calyx
equally cleft dorsally and ventrally. Coast of mainland, Santa
Catalina Id. . . . . 26. *C. foliolosa*
GG. Galea yellowish with pale thin margins; calyx more deeply cleft
dorsally than ventrally. San Miguel, Santa Rosa, Santa Cruz,
Anacapa, and San Clemente ids. . . . . 27. *C. hololeuca*

1. **C. arachnoìdea** Greenm. [*Orthocarpus pilosus* var. *a.* Jeps. *C. Eastwoodiana* Penn.]
Stems several, clustered on a perennial root-crown, erect, 1–2.5 dm. long, simple or
branched, arachnoid-lanate; lvs. less hairy, lance-linear in outline, 2–6 cm. long including
petioles, with 1–2 pairs of linear spreading lobes; bracts and calyces yellowish at
tips or dull reddish, the bracts closely appressed, broad, cleft to below middle; calyx
12–14 mm. long, cleft into lance-linear lobes; corolla 10–12 mm. long, the galea
dorsally lanose-pubescent, 3–4 mm. long and with thin pale narrow margins, the lower
lip slightly shorter, with a lanose-pubescent green trisulcate pouch and short pale yellow
lobes; caps. 8–9 mm. long; seeds yellowish, ca. 1 mm. long, the loose reticulate coat
deeply pitted.—Dry rocky slopes, 5000–7300 ft.; Red Fir F. to Alpine Fell-fields; Glenn
Co. to Humboldt and Siskiyou cos.; sw. Ore. July–Aug.

Ssp. **schizótricha** (Greenm.) Penn. [*C. s.* Greenman. *Orthocarpus s.* Jeps.] Stems
0.8–1 dm. long; lvs. entire or with a distal pair of lobes; pubescence browner, more
evidently dendritic.—Decomposed marble at ca. 5000–6000 ft.; Red Fir F.; Marble
Mts., Siskiyou Co.; Ore.

Ssp. **shasténsis** Penn. Bracts and calyces dull red at tips, the bracts ± divergent; lower
lobe of corolla reddish.—Dry slopes, 7500–9000 ft.; Subalpine F., Alpine Fell-fields; Mt.
Shasta, Siskiyou Co.

2. **C. Pàyneae** Eastw. [*C. pumicicola* Penn.] With aspect of *C. arachnoidea;* stems
arachnoid at base; corolla 12–14 mm. long, the galea 4–6 mm. long, minutely pubescent
dorsally, with thin pale or reddish margins ca. half the width of the galea, the lower lip
shorter, minutely pubescent on the purplish trisaccate pouch; $n = 12$ (Gillett in 1954).
—Gravelly pumice or granitic soil, 7000–9000 ft.; Red Fir F. to Alpine Fell-fields; Spanish Peak, Plumas Co., Lassen Peak (Shasta Co.) to Warner Mts. (Modoc Co.); to Cascades of Ore. June–Aug.

3. **C. psittacìna** (Eastw.) Penn. [*Orthocarpus p.* Eastw. *C. pratensis* Heller. *C.
ochracea* Eastw.] Villous-hirsute perennial, the clustered stems 1.5–3.5 dm. tall, simple
or few-branched; lvs. 2–6 cm. long, including petioles, linear or linear-lanceolate in
outline, mostly with 2 pairs of slender divaricate lobes; infl. dense, the bracts with 2
pairs of ascending-spreading lobes, distally yellowish; calyx 12–20 mm. long, yellowish
distally; corolla 16–20 mm. long, the galea 4–6 mm. long, attenuate, pale yellowish-
green, dorsally finely pubescent, with pale thin margins, the lower lip slightly shorter,
with a pubescent trisaccate pouch and white appressed lobes 2–2.5 mm. long; caps.
7–8 mm. long; seeds ca. 1 mm. long, obovoid, the loose coat yellowish.—Dry places,
4000–5700 ft.; Sagebrush Scrub, N. Juniper Wd.; Plumas Co. to e. Siskiyou and Modoc
cos.; e. Ore. May–July.

4. **C. pilòsa** (Wats.) Rydb. [*Orthocarpus p.* Wats.] Stems several from a stout
perennial root-crown, leafy, villous-hirsute, 1.5–3.5 dm. high; lvs. linear and entire, to
subovate in outline and pinnately 3–5-parted into linear lobes; bracts and calyces distally white- or yellow-margined, the former with a pair of lateral linear lobes and the
cent. lobe 2.5–5 mm. wide, blunt; calyx 15–16 mm. long, deeply cleft into linear lobes;

corolla 17–22 mm. long, white or pinkish, the galea 7–8 mm. long, acutish, somewhat recurved at tip, dorsally greenish and finely pubescent, the ventral margins wide, purple at base, white toward apex, the lower lip 1–1.5 mm. shorter, slightly saccate, strongly 3-toothed; caps. ca. 9 mm. long; seeds ca. 1 mm. long.—Mostly open dry rocky places, 5000–9000 ft.; Sagebrush Scrub to Lodgepole Pine F.; Plumas Co. along the e. slope of the Sierra Nevada to Placer Co.; w. Nev. June–Aug.

Ssp. **Jussèlii** (Eastw.) Munz. [*C. J.* Eastw.] Bracts with lobes narrow, the cent. one 1.5–2.5 mm. wide, acutish; plant less hairy.—At 7000–9500 ft.; e. slope of Sierra Nevada, Mono Co.

5. **C. nàna** Eastw. [*C. inconspicua* Nels & Kenn. *C. ambigua* Jones. *Orthocarpus pilosus* var. *monensis* Jeps.] Much like *C. pilosa;* lvs. mostly 3–5-lobed; bracts spreading, they and calyces distally dull yellow to dull purplish-red; calyx 15 mm. long, deeply cleft into linear lobes; corolla 13–16 mm. long, the galea ca. 6 mm. long, dorsally greenish and puberulent, the thin margins wide, dark purple proximally, white distally, the lower lip ca. 0.5–1 mm. shorter than galea; caps. ca. 10 mm. long.—Dry rocky places, mostly 8000–12,000 ft., sometimes lower; Bristle-cone Pine F., Subalpine F., Alpine Fell-fields; Sierra Nevada from Eldorado and Mono cos. to Tulare and Inyo cos.; White and Inyo mts. July–Aug.

6. **C. cinèrea** Gray. [*Orthocarpus c.* Jeps.] Stems several from the perennial root-crown, densely cinereous-pubescent, 0.5–1.5 dm. long; lvs. many, overlapping, lance-linear, cinereous, entire or the upper sometimes with 3 linear lobes; bracts and calyces purple-red to greenish-yellow, the former rounded, with 1–2 pairs of lobes; calyx 15–20 mm. long, equally cleft into narrow lobes; corolla 16–20 mm. long, yellowish, the galea 4–5 mm. long, with narrow purplish thin margins, the lower lip 2–2.5 mm. long, greenish with minute incurved lobes; caps. ca. 6–10 mm. long; seeds 0.8–1.3 mm. long.—Local on dry benches and slopes, 6000–9300 ft.; Yellow Pine F., Red Fir F.; San Bernardino Mts. May–Aug.

7. **C. Lemmònii** Gray. [*C. lassenensis* Eastw.] Glandular-pubescent perennial with many simple stems 1–2 dm. tall; infl. villous; lvs. linear or lance-linear, mostly entire; bracts and calyces purple at ends, the bracts mostly rounded and with 1–2 pairs of lobes; calyx 16–18 mm. long, the lateral clefts only 1–2 mm. deep and the lobes rounded; corolla 18–20 mm. long, the galea 8–9 mm. long, glandular-hairy dorsally, the ventral margins thin, purple, the lower lip exserted from calyx, 6–7 mm. long, subglabrous, green and somewhat inflated proximally, with 3 short but obvious pale purple lobes distally; caps. 7–8 mm. long; seeds dark, ca. 1 mm. long; $n = 12$ (Gillett in 1954 for *lassenensis*).—Moist meadows, 7000–11,000 ft.; Lodgepole F. to Alpine Fell-fields; Sierra Nevada from Fresno and Inyo cos. n., to Mt. Lassen. July–Aug.

8. **C. Culbertsònii** Greene. Much like *C. Lemmonii,* but fewer-stemmed; corolla 16–18 mm. long, the galea dorsally puberulent or glabrescent with purple thin margins, the lower mostly included in calyx, ca. 4 mm. long, the distal lobes rudimentary.—Wet meadows, 8000–11,500 ft.; Lodgepole F. to Alpine Fell-fields; Tulare and Inyo cos. to Madera and Mono cos. July–Aug.

9. **C. brevilobàta** Piper. Stems many, finely glandular-pubescent, 1–2 dm. long, mostly simple; lvs. lanceolate, entire or the upper with a pair of lobes, glandular-pubescent, 1.5–2 cm. long, ± wavy; bracts and calyces red-tipped, the bracts with a pair of lobes; calyx 12–14 mm. long, medianly cleft ca. ⅓ way, laterally cleft 1–2 mm. with rounded lobes; corolla 15–16 mm. long, the galea ca. 6 mm. long, puberulent and green above, the wide thin margins reddish, the lower lip ca. 2 mm. long, green, rudimentary; caps. 6–7 mm. long.—Dryish open stony places, 1000–5000 ft.; Yellow Pine F., Mixed Evergreen F.; Siskiyou and Del Norte cos.; sw. Ore. May–July.

10. **C. Brèweri** Fern. [*C. adenophora* Eastw.] Stems many, 1–2.5 dm. tall, glandular-pubescent and usually with longer nonglandular hairs also; lvs. lanceolate to lance-linear, entire or upper broader and with a pair of lobes, glandular; bracts and calyces distally red, the former 3-lobed; calyx 13–15 mm. long, medianly cleft more than ⅓ its length, laterally cleft 3–4 mm. into acutish lanceolate lobes; corolla 16–22 mm. long, the galea 7–10 mm. long, puberulent and greenish dorsally with wide thin red ventral margins, the lower lip 2–3 mm. long, dark green, rudimentary, quite concealed by calyx; caps. 9–10 mm. long.—Dry stony places and about meadows, 7000–11,000 ft.; Red Fir F. to Alpine Fell-fields; Inyo and Tulare cos. to Mono and Eldorado cos. June–Aug.

Var. **pállida** Eastw. [*C. glandulifera* ssp. *pallida* Penn.] Bracts and calyces yellow-tipped.—With the sp.

11. **C. Peirsònii** Eastw. [*C. Carterae* Beane.] Near to *C. Breweri,* but less pilose; lvs. 3–5-fid, the segms. filiform-linear, the rachis 3–5 mm. broad; calyx-segms. linear, 5 mm. long; corolla 20–28 mm. long, the galea 10–12 mm., conspicuously exserted, as is the stigma.—Dampish rocky banks, 8700–11,000 ft.; Subalpine F.; Fresno and Inyo cos. to Madera and Mono cos. July–Aug.

12. **C. latifòlia** H. & A. [*C. macrocarpa* Benth. *C. l.* ssp. *carmelensis* Eastw.] Suffrutescent perennial with ascending to suberect ± branched stems 3–5 dm. long, rough-pubescent below, villous-hirsute in infl.; lvs. oblong to rounded, entire, less than 3 times as long as wide, blunt, sessile, mostly less than 2 cm. long; bracts and calyces distally red, the former entire and subtruncate or 3-lobed; calyx 18–22 mm. long, yellowish below, cleft medianly ca. halfway, laterally less than 2 mm. into blunt lobes; corolla 24–27 mm. long, the galea 15–20 mm. long, strongly pubescent dorsally, with wide reddish thin margins, the lower lip ca. 2 mm. long, dark green, appressed; caps. 12–15 mm. long; seeds ca. 2 mm. long.—Sandy places; Coastal Strand, Closed-cone Pine F., etc.; along the coast of Monterey and Santa Cruz cos. Feb.–Sept.

Ssp. **mendocinénsis** Eastw. [*C. m.* Penn.] Stems soft-pubescent below the villose-hirsute infl., 4–6 dm. long; bracts 3-lobed; calyx cleft laterally into lobes 2–5 mm. long; corolla 3–3.2 cm. long, the galea 17–18 mm. long, the dark green lower lip ca. 3 mm. long.—Coastal Strand, N. Coastal Scrub; Humboldt and Mendocino cos. May–Aug.

13. **C. móllis** Penn. Suffrutescent, the stems diffusely branched, to 3 or 4 dm. long, with short axillary leafy shoots; lvs. oblong, mostly entire, 3–7 cm. long, distally rounded, soft-tomentose with branched hairs; axis of infl. villous-hirsute and with some gland-tipped hairs; bracts pale green or yellowish, obovate, distally rounded or 3-lobed; calyx 16–18 mm. long, cleft medianly almost half its length and laterally into ovate acute lobes ca. 2 mm. long; corolla 17–18 mm. long, the galea 7–8 mm. long, truncate, strongly pubescent dorsally, with wide pale thin margins, the lower lip 2 mm. long, appressed, green.—Sand dunes; Coastal Strand, Coastal Sage Scrub; Santa Rosa Id., coast of San Luis Obispo Co. April–Aug.

14. **C. Wìghtii** Elmer. [*C. latifolia* var. *W.* Zeile. *C. l.* sspp. *pinnatifida* Eastw. and *insularis* Eastw. *C. inornata* Eastw. *C. anacapensis* Dunkle.] Pubescent or pilose perennial with gland-tipped hairs as well, the stems densely pubescent and villous in infl., decumbent to erect, branched, suffrutescent, 3–8 dm. long; lvs. oblong, entire or upper 3-lobed, these obtusish to rounded; main lvs. crowded, mostly 2–6 cm. long and with short fascicles in their axils; bracts distally yellowish or dull red, with a pair of rounded lobes; calyx 18–20 mm. long, the ± obtuse lobes 2–3 mm. long; corolla 21–25 mm. long, the galea 13–15 mm. long, bluntish, dorsally pubescent, with pale thin margins, the lower lip ca. 2 mm. long, dark green; caps. 8–11 mm. long.—Dry slopes and banks; Coastal Strand, N. Coastal Scrub, Mixed Evergreen F., etc.; near the coast from Mendocino Co. to San Mateo Co., Santa Cruz and Anacapa ids. March–July. A form with brighter red bracts and corollas 2.3–2.8 cm. long has been named ssp. *rubra* Penn. [*C. episcopalis* Penn.]

Ssp. **inflàta** (Penn.) Munz. [*C. i.* Penn.] Herbage not glandular below the infl.; calyx-tube somewhat inflated, pale yellow, cleft laterally little less than medianly; lvs. grayish-pubescent.—Coastal Strand, Marin Co.

Ssp. **litoràlis** (Penn.) Munz. [*C. l.* Penn.] Herbage not glandular below the infl.; calyx-tube not inflated, laterally cleft much less than medianly; lvs. glabrous to somewhat pubescent.—Coastal Strand, N. Coastal Scrub; Humboldt Co. to Wash.

15. **C. Roseàna** Eastw. [*C. gyroloba* Penn.] Stems several from a woody root-crown, villous and with shorter gland-tipped hairs, erect or ascending, 3–6 dm. tall, simple or branched; lvs. lanceolate, mostly entire, plane or ± crisped, 3–5 cm. long; bracts and calyces distally scarlet, the former with a pair of lateral spreading lobes with blunt apices; calyx 15–25 mm. long, medianly cleft ca. halfway, laterally cleft 2–3 mm. with rounded lobes; corolla 2.5–3.5 cm. long, the galea 15–24 mm. long, well exserted, dorsally puberulent, ventrally with thin reddish margins, the lower lip 1.5–2 mm. long; caps. ca. 12–13 mm. long.—Dry slopes, below 3500 ft. in n. part of range, below 8500 ft. in s. part; Chaparral, Foothill Wd., Yellow Pine F., etc.; inner Coast Ranges from Lake Co. to San Diego. April–Aug.

16. **C. Martìnii** Abrams. Several-stemmed from a woody root-crown, the stems with gland-tipped and glandless hairs, branched, 3–7 dm. tall; lvs. lanceolate to broadly lanceolate, ± crisped on margins, 2–5 cm. long, entire or sometimes the upper with a pair of lobes; bracts and calyces distally scarlet, the former 3-lobed; calyx 15–18 mm. long, cleft medially ca. ⅓ way, laterally into ovate lobes ca. 1–2 mm. long; corolla 2–2.5 cm. long, the galea ca. as long as tube, with narrow red thin margins, the lower lip 1–2 mm. long, not exserted; caps. ca. 10–12 mm. long.—Dry slopes, 1000–8000 ft.; Chaparral, Yellow Pine F.; mts. from Los Angeles Co. to L. Calif. May–Sept.

Ssp. **Ewànii** (Eastw.) Munz. [*C. E.* Eastw.] Much like *C. Martinii*, but lvs. lance-linear to almost linear, less crisped, 2–3 cm. long, more stiff-pubescent; calyx 16–17 mm. long; corolla 2.3–2.5 cm. long, the galea 16–17 mm. long, the lower lip 1–2 mm. long, exserted.—Dry rocky slopes, 6500–8500 ft.; Pinyon-Juniper Wd., Yellow Pine F.; San Bernardino Mts., largely in e. end. June–July.

17. **C. Applegàtei** Fern. [*C. pinetorum* Fern. *C. angustifolia* var. *adenophora* Fern. *C. Brooksii* Eastw. *C. trisecta* Greene. *C. fragilis, excelsa, dolichostylis,* and *Hoffmannii* Eastw.] Glandular-pubescent perennial, with simple to branched clustered stems 2–5 dm. tall from a woody base; lvs. lanceolate 2–3.5 cm. long, crisped on margins, entire or 3-lobed, with acutish segms.; bracts and calyces distally scarlet, sometimes with more yellow, the upper bracts 3–5-lobed; calyx 12–22 mm. long, cleft medially ca. halfway, laterally into lanceolate lobes mostly 2–4 mm. long; corolla 2–3 cm. long, the galea 12–15 mm. long, with thin red margins, the lower lip ca. 2 mm. long, greenish, mostly included; caps. ca. 10 mm. long; seeds ca. 1.3 mm. long.—Dry places, mostly 2000–11,000 ft.; Sagebrush Scrub to Subalpine F.; N. Coast Ranges s. to Tehama Co., Siskiyou and Modoc cos. s. through Sierra Nevada to Mt. Pinos (Ventura Co.) and Panamint Mts.; to Ore., Ida., Nev. May–Aug.

18. **C. dísticha** Eastw. [*C. Quibellii* Beane.] Near to *C. Applegatei;* lvs. mostly lance-linear; corolla 2.5–3 cm. long, the lower lip mostly exserted from the calyx.— Dry slopes, 6000–9500 ft.; Red Fir F., Lodgepole F.; Sierra Nevada from Tuolumne Co. to Tulare Co. May–Aug.

19. **C. minìata** Dougl. ex Hook. [*C. montana* Congd. *C. oblongifolia* Gray.] Erect herbaceous few-stemmed glabrous to somewhat pubescent perennial; stems simple or somewhat branched above, 4–8 dm. tall; lvs. lanceolate, mostly 2–5 cm. long, acute, entire or the upper occasionally lobed; infl. villous-pubescent; bracts and calyces distally mostly scarlet, at least the upper bracts with a pair of slender lobes; calyx mostly 2–2.7 cm. long, medially cleft ½–⅔ its length, laterally cleft 3–7 mm. into lance-linear acuminate lobes; corolla mostly 2.5–3.5 cm. long, the galea as long as the tube, with thin red margins; lower lip 1–2 mm. long, green, ± exserted; caps. 10–12 mm. long; $n = 24$, (Gillett, 1954).—Along streams and in wet places, below 11,000 ft.; Montane Coniferous F.; Coast Ranges from Glenn Co. n., Siskiyou and Modoc cos. s. through Sierra Nevada to mts. of San Diego Co.; to B.C., Rocky Mts. May–Sept.

Ssp. **elàta** (Piper) Munz. [*C. e.* Piper.] Stems more slender; bracts and calyces distally dull red-purple or dull red-orange; fls. 1.5–2.7 cm. long; caps. 6–8 mm. long.—Moist places, often on serpentine, below 5200 ft.; Mixed Evergreen F., Yellow Pine F.; Del Norte and Siskiyou cos.; sw. Ore. May–Aug.

20. **C. uliginòsa** Eastw. Perennial with simple stems 3–5 dm. high, villous; lvs. oblong to lance-oblong, ca. 4 cm. long, 8–12 mm. wide, obtuse; upper bracts trifid, with acuminate segms.; lower fls. concealed by bracts, the upper surpassing their bracts, ca. 3 cm. long; calyx ca. 23 mm. long, the segms. shorter than the tube, glandular-villous, bidentate or bisected; galea ca. as long as corolla-tube, dorsally scabrous-hispid; lower lip small, protuberant, the outer segms. acuminate, 1 mm. long, the interior smaller.— Pitkin Marsh, Sonoma Co. June–July.

21. **C. Leschkeàna** J. T. Howell. Perennial ca. 1 m. tall, erect, branched below the infl., sparsely pilose in upper parts; lower lvs. oblong-linear, 0.5–1.5 cm. long; middle lvs. narrowly elliptic-oblong, entire, acute or lobed near tip, 6–7 cm. long, subcinereous; bracts reddish, acute, 2.5–4 cm. long, 2–3 cm. wide, 3–5-fid, the median lobe broadest; calyx 2–2.5 cm. long, ca. equally cleft before and behind, the segms. lobed ca. halfway, densely pilose, reddish; corolla 2.5–3 cm. long, the galea equal to the tube, the lower lip dark green, 1.5 mm. long.—Swampy ground behind the coastal dunes; Point Reyes, Marin Co. June.

22. **C. affìnis** H. & A. [*C. Douglasii* Benth. *C. multisecta* Eastw. *C. californica* Abrams. *C. polytoma* Eastw.] Stems rather few from a ± woody base, often purplish, rather stiff-pubescent with glandless hairs, simple or little branched, 3–5 dm. tall; infl. hirsute; lvs. rather scabrous-pubescent, lanceolate, 3–9 cm. long, entire or with 1–3 pairs of slender lobes, often with axillary fascicles; bracts and calyces distally scarlet, the former with 2–3 pairs of lobes; calyx 18–25 mm. long, cleft medially ca. halfway, laterally into linear-oblong to lance-ovate lobes 3–6 mm. long; corolla 2.5–3.5 cm. long, the galea 16–23 mm. long, finely pubescent dorsally with thin red margins, the lower lip 1.5–2 mm. long, dark green to brownish; caps. ca. 12–13 mm. long.—Dry wooded or brushy slopes; Closed-cone Pine F., Mixed Evergreen F., Foothill Wd., Chaparral, Coastal Sage Scrub, etc.; Coast Ranges from Sonoma and Napa cos. to n. L. Calif. March–May.

Ssp. **insulàris** (Eastw.) Munz. [*C. latifolia* ssp. *i.* Eastw. *C. Douglasii* ssp. *i.* Penn.] Corolla 1.5–2 cm. long.—Closed-cone Pine F., Chaparral; Santa Cruz Id.

23. **C. neglécta** Zeile in Jeps. Suffrutescent perennial, the stems erect, branched, 3–6 dm. tall, rather sparsely pilose; lvs. lanceolate, with 1–2 pairs of spreading narrow lobes; bracts and calyces distally yellowish, the former sometimes tipped red, deeply cleft; calyx 15 mm. long, cleft medially ⅔ its length, laterally into oblong-ovate ciliolate lobes 5–6 mm. long; corolla 18–20 mm. long, yellow to red, the galea 9–10 mm. long, the lower lip 2 mm. long, pale or translucent green; caps. ca. 8–10 mm. long.— Open serpentine slopes; Coastal Prairie; Tiburon Peninsula, Marin Co. April–June.

24. **C. chromòsa** A. Nels. [*C. angustifolia* Calif. refs.] Herbaceous perennial from woody root-crown, the stems 1–4 dm. tall, simple or few-branched, subcinereous with rather stiff glandless hairs; lvs. lanceolate, 1–4 cm. long, entire or with 1–2 pairs of spreading narrow lobes; bracts and calyces distally scarlet, the former with 2 pairs of lobes; calyx 15–20 mm. long, cleft medially ca. ⅓ its length, laterally into ovate or oblong lobes 4–5 mm. long; corolla 2.5–3 cm. long, the galea 12–15 mm. long, sparsely puberulent dorsally, with wide reddish thin margins, the lower lip dark green, 2–3 mm. long, included in calyx; caps. ca. 15–17 mm. long; seeds ca. 2.5 mm. long.—Frequent on dry brushy slopes, 2000–7000(–11,000) ft.; mostly Sagebrush Scrub, Shadscale Scrub, Joshua Tree Wd., Pinyon-Juniper Wd., etc.; deserts from Pinto Mts., Riverside Co., e. of Sierra Nevada to ne. Calif.; e. Ore., Wyo., Colo., New Mex. April–Aug.

25. **C. pruinòsa** Fern. [*C. nevadensis* and *globosa* Eastw. *C. muscipula* and vars. *armeniaca* and *angustifolia* Eastw.] Perennial, ± grayish with mostly branched hairs, the stems simple or branched, 3–7 dm. tall; lvs. linear-lanceolate, 2–6 cm. long, entire or the upper with a pair of lobes; bracts and calyces distally scarlet, the former with a pair of lobes; calyx ca. 13–16 mm. long, yellow between the basal green and distal red portions, cleft medially ca. ⅔ its length, laterally 3–5 mm. into lanceolate lobes; corolla 2.5–3 cm. long, the galea 15–20 mm. long, yellowish and puberulent dorsally, with wide red thin margins, the lower lip dark green, ca. 2 mm. long, included in calyx; caps. 10–15 mm. long.—Dry rocky places, 1400–8000 ft.; Yellow Pine F., Red Fir F.; Del Norte and Siskiyou cos. to Tuolumne, Eldorado, and Modoc cos.; Ore. April–Aug.

Ssp. **gleasònii** (Elmer) Munz. [*C. g.* Elmer. *C. Douglasii* var. *contentiosa* Macbr.] Calyx-lobes ± ovate, 1–3 mm. long.—Rocky places, 5000–7100 ft.; Yellow Pine F.; about Mt. Gleason, San Gabriel Mts. May–June.

26. **C. foliolòsa** H. & A. [*C. Clementis* Eastw.] Suffrutescent bushy perennial, white-woolly throughout, 3–6 dm. tall; lvs. linear or oblong-linear, 1–2.5 cm. long, entire or upper 3-lobed, obtuse, fairly crowded and with fascicles in lower axils; bracts and calyces distally scarlet, the former with 1–2 pairs of lobes, the latter ± yellow in middle; calyx 15–20 mm. long, cleft medially ca. ⅔ its length, laterally barely 0.5 mm. into rounded truncate lobes; corolla 1.8–2.2 cm. long, the galea 7–15 mm. long, minutely pubescent and greenish dorsally, with pale thin reddish margins, the lower lip ca. 2 mm. long, dark green, included in calyx; caps. 10–14 mm. long; seeds ca. 2 mm. long.—Dry open ± rocky places, below 5000 ft.; many Plant Communities; Coast Ranges from Humboldt Co. to n. L. Calif., less common along w. base of Sierra Nevada, Santa Catalina Id. March–June.

27. **C. hololeùca** Greene. Covered with a dense white-woolly felt, shrubby, 3–8 dm. high, much branched; lvs. linear, dense, 1–3 cm. long, entire, obtuse, mostly with fascicles in axils; bracts and calyces distally red, the former with a pair of lobes; calyx 15–18 mm. long, cleft medially ⅔ its length, not at all laterally, the 2 sides rounded at

their tips; corolla 2–2.5 cm. long, the yellowish galea 12–13 mm. long, puberulent dorsally, with pale thin margins, the dark green lower lip 2–3 mm. long, included; caps. ca. 1 cm. long.—Canyon walls and rocky slopes; Chaparral, Coastal Sage Scrub; San Miguel, Santa Rosa, Santa Cruz, and Anacapa ids. March–Aug.

Ssp. grísea (Dunkle) Munz. [_C. g._ Dunkle.] Indumentum grayish; bracts green or brownish-green; calyx ca. 13 mm. long; corolla ca. 1.5 cm. long, the galea ca. 7 mm. long.—Coastal bluffs, San Clemente Id. Feb.–April.

28. **C. plagiótoma** Gray. Stems erect or spreading, growing up through low shrubs, sparsely pubescent, 3–6 dm. tall; infl. more pubescent; lvs. narrowly linear, 1–3 cm. long; the upper lvs. and bracts divided into 3 linear lobes; bracts white-pubescent, green, the middle segms. broad, rounded; calyx 10–15 mm. long, white-woolly, cleft laterally more deeply than medianly, the upper lip 4–7 mm. long, the lower 6–9 mm. with lobes 2–3 mm. long; corolla 1.5–2 cm. long, the galea ca. as long as tube, straight, yellowish, pubescent dorsally, with wide thin pale margins, the lower lip 1.5 mm. long, yellowish, included in calyx; caps. 9–10 mm. long.—Dry flats and ridges, 2500–7500 ft.; Sagebrush Scrub, Joshua Tree Wd.; desert slopes and base of San Gabriel Mts. April–June.

29. **C. linariaefòlia** Benth. [_C. candens_ Dur. & Hilg. _C. Howellii_ and _C. salticola_ Eastw.] Stems several to many, from a ± woody root-crown, 6–8(–15) dm. long, simple or little-branched, subglabrous or finely pubescent especially near base and ± hirsute in infl.; lvs. linear, 1.5–8(–10) cm. long, entire or with a pair of slender lobes; bracts and calyces distally red, the former with 1–2 pairs of slender lobes; calyx 2.5–3.5 cm. long, medianly cleft more deeply ventrally than dorsally, laterally cleft 5–7 mm. into lanceolate lobes which upcurve together; corolla 3–4 cm. long, decurved, the galea ca. as long as the tube, dorsally puberulent, greenish, with thin red margins, the lower lip dark green, ca. 3 mm. long; caps. 10–15 mm. long; seeds 1.5–2 mm. long.—Moist places, 2500–10,000 ft.; Joshua Tree Wd., Pinyon-Juniper Wd., Sagebrush Scrub, etc.; s. edge of Mojave Desert n. to Modoc Co.; to Ore., Mont., New Mex. June–Sept.

30. **C. franciscàna** Penn. [_C. affinis_ Calif. refs., not H. & A.] Grayish-green pubescent perennial, with several herbaceous stems simple or ± branched, 2–7 dm. high; infl. more hairy; lvs. lanceolate to lance-linear, 2–8 cm. long, entire or with a pair of slender lobes, ± obtuse at tips; bracts reddish especially at tips, simple or 3-parted; calyces red, 25–35 mm. long, cleft medianly more deeply ventrally than dorsally, cleft laterally into lanceolate or lance-oblong lobes 3–5 mm. long; corolla 3.5–4.5 cm. long, the galea 2–2.5 cm. long, dorsally pubescent and ± greenish-yellow, with wide thin yellowish or reddish margins, the lower lip ca. 3 mm. long, dark green, usually exserted; caps. 10–15 mm. long.—Dry slopes, below 1000 ft.; Mixed Evergreen F., Coastal Strand, etc.; Coast Ranges from San Mateo Co. to Sonoma Co. March–June.

31. **C. subinclùsa** Greene. Near to _C. franciscana;_ lvs. linear, more sharply tipped; calyx-lobes 5–7 mm. long, lanceolate to lance-linear; galea slender-attenuate, 18–19 mm. long.—Dry slopes, below 4000 ft.; Chaparral, Foothill Wd.; foothills of Sierra Nevada from Butte Co. to Tulare Co. April–June.

32. **C. stenántha** Gray. Erect simple or somewhat branched annual, pubescent with some hairs gland-tipped, mostly slender, 3–15 dm. high; lvs. lanceolate to ascending, lance-linear, entire, 2–8 cm. long; bracts leaflike, the lower green, the upper red-tipped, attenuate-acute; calyx green, 15–25 mm. long, cleft medianly ⅔ its length, laterally 1–3 mm. into narrow lobes; corolla 2.5–3.5 cm. long, the galea 15–20 mm. long, dull reddish-yellow dorsally and ± puberulent, with narrow pale thin margins; lower lip 2–3 mm. long, yellow, well exserted; caps. 10–15 mm. long; seeds ca. 1.5 mm. long.—Frequent in moist places, below 7000 ft.; many Plant Communities; cismontane s. Calif. to the desert edge, less common n. to Lake and Fresno cos. May–Sept.

Ssp. spiràlis (Jeps.) Munz. [_C. s._ Jeps.] Bracts obtusish to rounded; lower lip of corolla purple-red.—Occasional in moist places; Foothill Wd.?; Lake and Napa cos. to Tuolumne Co. June–Oct.

33. **C. éxilis** A. Nels. Near to _C. stenantha,_ but coarser and frequently more and coarsely pubescent; lvs. lanceolate, 4–8 cm. long, glandular and scabrous; calyx 14–18 mm. long, the lanceolate lobes 1–2 mm. long; corolla 16–18 mm. long, scarcely exserted, the galea ca. half as long as tube.—Occasional in wet places, 4000–6500 ft.; Sagebrush Scrub, etc.; near Bishop, Inyo Co., Mono Lake to Wash., Mont., Colo. July–Sept.

C. minor (Gray) Gray. [*C. affinis* var. *m.* Gray.] Differing from *C. exilis* in having the corolla well exserted from calyx, more slender spike and finer pubescence.—Reported from alkaline places e. of the Sierra Nevada; to Rocky Mts.

29. Cordylánthus Nutt. ex Benth. in DC. BIRD's-BEAK

Branched annuals with yellow roots. Lvs. alternate, entire or pinnatifid into narrow divisions. Fls. dull yellow or purple, in spikes and subtended by outer bracts that are seldom colored. Inner flowering bract mostly foliose or modified into calyxlike structure. Calyx forming a single piece, split almost to base ventrally, extending dorsally into a tonguelike structure, entire or bifid apically. Corolla 2-lipped, the upper lip galeate and hooding the stamens, the lower as long or shorter, ± inflated and minutely lobed. Stamens 4 or 2, the anther-cells placed unequally, with the lower often smaller or abortive. Caps. turgid, glabrous, loculicidal. Seeds many, wingless, with loose reticulate testa. Ca. 35 spp., of w. N. Am. (Greek, *cordule,* club, and *anthos,* fl.)

(Ferris, R. S. Taxonomy and distribution of Adenostegia. Bull. Torrey Bot. Club 45: 399–423, 1918. Pennell, F. W., *in* Abrams: Ill. Fl. Pac. States, 3: 846–859, 1951.)

A.   Lvs. oblong to lanceolate, at least the lower entire; fls. in an elongate spike; calyx spathelike, enclosing proximal part of corolla.
  B.   Bracts entire or short-lobed; lvs. acute.
    C.   Corolla shorter than calyx, purple on lower lobes and thin margin of galea; bracts usually with a pair of short distal lobes. Coastal . . . . 1. *C. maritimus*
    CC. Corolla longer than calyx. Transmontane.
      D.   Corolla yellowish on lower lobes and edges of galea; bracts entire. E. of Sierra Nevada and n. . . . . 2. *C. canescens*
      DD. Corolla lavender; bracts lobed. E. Mojave Desert . . . . 3. *C. tecopensis*
  BB. Bracts deeply 3–7-lobed; lvs. obtusish.
    C.   Corolla purplish, the galea dorsally with reflexed hairs. Cent. V. . . . . 4. *C. palmatus*
    CC. Corolla yellowish or cream, the galea dorsally with spreading hairs.
      D.   Calyx entire at the obtuse apex; corolla 15 mm. long, the galea concave-rounded. San Joaquin V. . . . . 5. *C. hispidus*
      DD. Calyx with acute teeth 1 mm. long; corolla 17–18 mm. long, the galea compressed. Marshes, San Francisco Bay . . . . 6. *C. mollis*
AA. Lvs. or their segms. linear to filiform; fls. in heads or scattered; calyx proper narrow, enclosing corolla at base only (the bract below the fl. often confused with calyx).
  B.   Galea longer than and curved up away from lower lip, rather bright purple; fls. solitary, their outer bracts 3-lobed. Dry places, e. Mojave Desert . . . . 7. *C. parviflorus*
  BB. Galea not or scarcely longer than lower lip, mostly pale or dull in color.
    C.   Corolla 8 mm. long; flowering bracts entire, pilose with stout yellowish glands. Lake Co. 8. *C. Pringlei*
    CC. Corolla 10–30 mm. long; flowering bracts with slender hairs.
      D.   Corolla wholly glabrous, less than to about twice as long as wide, inverted with ventral side up; flowering bracts setose; anthers 1-celled. Mts. of s. Calif. 9. *C. Nevinii*
      DD. Corolla ± pubescent, more than twice as long as wide, the dorsal side up; flowering bracts with hairs like those of other lvs. of infl.; anthers 2-celled.
        E.   Flowering (inner) bract with 1–3 pairs of lobes; calyx teeth ca. 1.5 mm. long. E. of Sierra Nevada . . . . 10. *C. Helleri*
        EE. Flowering (inner) bract entire or toothed at apex; calyx entire or with teeth to 0.5 mm. long.
          F.   Outer bracts palmately 3–7-lobed; corolla-throat ventrally pubescent within.
            G.   Outer bracts nearly to quite as long as inner flowering bract and corolla; fls. green or dull purplish. From Mono Co. n.
11. *C. ramosus*
            GG. Outer bracts much shorter than flowering bract and corolla.
              H.   Infl. purplish, contrasted with whitish foliage; main lvs. 2–2.5 cm. long. Panamint Mts. . . . . . . 11. *C. ramosus* ssp. *eremicus*
              HH. Infl. yellowish, only slightly contrasted with foliage; main lvs. 1–1.5 cm. long. San Bernardino Mts. . . . . 12. *C. bernardinus*
          FF. Outer bracts entire or 3-lobed; corolla-throat glabrous within ventrally or nearly so.
            G.   Infl. a dense head of 3–15 fls. and subtended by several 3-lobed outer bracts; no hairs gland-tipped.
              H.   Outer bracts and their lobes not noticeably widened at their tips; spikes not markedly stiff setose-ciliate; plants widely branched.
                I.   Corolla 13–16 mm. long, the lower lip sparsely pubescent without; plants finely pubescent, the bracts ciliate. Sierra Nevada of Fresno and Tulare cos. . . 13. *C. Ferrisianus*
                II. Corolla 17–21 mm. long, the lower lip pubescent ex-

ternally; plants soft-pubescent throughout. Coastal hills, Monterey and Santa Barbara cos. .... 14. *C. littoralis*
HH. Outer bracts and their lobes distinctly widened at the tips; spikes rather harshly setose-pilose; plants with ascending branches.
    I. Throat of corolla longer than wide, scarcely distinguishable from tube; width of bract-divisions more than half the length of the spreading setae; corolla 12–14 mm. long, ± yellowish. Cent. Calif. .... 15. *C. rigidus*
    II. Throat of corolla wider than long, strongly contrasted with tube; width of bract-divisions less than half the length of the spreading setae; corolla white with purple lines. S. Calif. .... 16. *C. filifolius*
GG. Infl. of racemosely arranged clusters of 1–3 fls. and subtended by 1–few simple or 3-lobed outer bracts; hairs often gland-tipped.
    H. Outer bracts entire or angulate-dilated at apex.
        I. Infl. hirsute, the bracts relatively long-ciliate; lvs. usually flat, linear; stem pubescent to hirsute .. 17. *C. pilosus*
        II. Infl. pubescent, the bracts finely ciliate; lvs. mostly involute, subfiliform; stem glabrous or nearly so 18. *C. tenuis*
    HH. Outer bracts 3-lobed, the segms. with enlarged tips.
        I. Plants diffusely spreading, with procumbent slender stems; galea quite pubescent; fls. solitary. Serpentine, Mt. Diablo 19. *C. nidularius*
        II. Plants erect, with ascending branches; galea finely pubescent; fls. 1–3 in a cluster.
            J. Fl.-clusters dark, hirsute.
                K. Lobes of outer bracts linear, flat. Shasta Co. to Tuolumne Co. .... 20. *C. Hansenii*
                KK. Lobes of outer bracts subfiliform, often involute. Tehama Co. to Del Norte Co. 21. *C. viscidus*
            JJ. Fl. clusters yellow-green, finely pubescent. Base of Mt. Shasta .... 22. *C. pallescens*

1. **C. marítimus** Nutt. ex Benth. in DC. [*Adenostegia m.* Greene. *Chloropyron m.* Heller. *Chloropyron palustre* Behr.] Loosely and corymbosely much-branched, often decumbent, the stems 2–4 dm. long; herbage pubescent, some of the hairs gland-tipped; lvs. and bracts glaucous-green, the former 0.5–2.5 cm. long, lance-oblong, the latter ± oblong and often with a pair of teeth near summit; calyx 15–22 mm. long, oblong-lanceolate, the terminal sharp teeth less than 0.5 mm. long; corolla 15–20 mm. long, the galea finely pubescent dorsally, with wide purplish thin margin, the lower lip pilose-pubescent; stamens 4, the upper fils. slender, with small anthers, the lower fils. thicker, longer, with connivent anthers; caps. 7–9 mm. long; seeds brown, curved, 1.5–2 mm. long.—Coastal Salt Marsh; s. Ore. to n. L. Calif. May–Oct.

2. **C. canéscens** Gray. [*C. maritimus* var. *c.* Jeps. *Adenostegia c.* Greene. *C. Parryi* Wats. *Adenostegia P.* Greene. *C. m.* var. *P.* Jeps.] Plants 2–4 dm. tall, corymbosely much-branched, pubescent with fine hairs; lvs. and bracts glaucous-green, the former lanceolate, 1–2.5 cm. long, the latter somewhat wider; calyx 13–16 mm. long, broadly lanceolate, canescent, the teeth ca. 0.5 mm. long; corolla 15–17 mm. long, the minutely pubescent galea pale yellow with wide pale thin margin, the lower lip with distally pubescent pouch and minute glabrous lobes; stamens 4, the upper fils. slender, with reduced anthers, the lower thicker, longer, with connivent anthers; caps. ca. 10 mm. long; seeds round-oblong, ca. 1 mm. long.—Alkaline flats and marshes, 3700–5500 ft.; Sagebrush Scrub, Shadscale Scrub; Inyo Co. to Modoc Co.; to e. Ore., Utah. June–Sept.

3. **C. tecopénsis** Munz & Roos. Grayish, glandular, erect, 3–6 dm. high, diffusely branched; lvs. glaucous, lance-linear to narrowly lance-oblong, 5–15 mm. long; bracts 10–12 mm. long, narrow-ovate, with 1 pair of linear lobes ca. 2 mm. long; calyx 10–13 mm. long, entire; corolla pale lavender, ca. 1 cm. long, the galea puberulent, with glabrous thin margin, the lower lip ± saccate, pubescent, with very short teeth; stamens 4, the upper pair sterile, the lower fertile; caps. 7–8 mm. long; seeds ca. 1 mm. long.—Alkaline meadows, below 2500 ft.; Creosote Bush Scrub; Tecopa Hot Springs, Saratoga Springs, etc., se. Inyo Co.; Nye Co., Nev. Aug.–Oct.

4. **C. palmàtus** (Ferris) Macbr. [*Adenostegia p.* Ferris.] Low, 1–2 dm. high, loosely much-branched, pilose or pubescent, some of the hairs gland-tipped; lvs. pale green, oblong, 1–2 cm. long, mostly incised, the lower sometimes entire; bracts ovate, with

3 pairs of ascending lobes radiating from basal portion; calyx 12–15 mm. long, oblong-lanceolate, entire or bidentate with teeth ca. 1 mm. long; corolla 12–16 mm. long, the finely reflexed-pubescent galea with wide glabrous thin margin, the lower lip finely pubescent on the pouch, glabrous on the minute lobes; stamens 2, the upper pair lacking.—Alkaline overflowed lands; V. Grassland; Colusa and Madera cos. June.

Ssp. **carnulòsus** (Penn.) Munz. [*C. c.* Penn.] Herbage hirsute-pubescent with glandless hairs to glabrescent; lvs. and bracts yellowish-green, the upper lvs. with 1–2 pairs of lobes, the bracts with 2–3 pairs; calyx 14–15 mm. long, lance-oblong, the teeth 1 mm. long; corolla ca. 15 mm. long; caps. ca. 6–7 mm. long.—Alkaline soil; V. Grassland; Fresno Co. July–Aug.

5. **C. híspidus** Penn. [*C. mollis* var. *viridis* Jeps.] Plant 1.5–2 dm. high, diffusely branched, hirsute-hispid with spreading glandless hairs; lvs. and bracts pale green, the former entire, oblong, 0.5–2 cm. long; the latter lance-ovate, with 2–3 pairs of lobes; calyx 13–14 mm. long, obtuse and entire; corolla ca. 15 mm. long, the pubescent galea with wide glabrous thin margin, the lower lip with finely pubescent pouch and minute glabrous lobes; stamens 2 (the lower pair present).—Alkaline places; V. Grassland; Merced to Kern cos. June–July.

6. **C. móllis** Gray. [*Adenostegia m.* Greene. *Chloropyron m.* Heller.] Plants 2–4 dm. high, branched, hirsute-hispid with glandless hairs; lvs. and bracts pale green, the lower lvs. entire, oblong, 0.5–1 cm. long, the upper broader, 1–2 cm. long, with 1–2 pairs of small lobes; bracts usually with 3 pairs of lobes; calyx 15–16 mm. long, lance-oblong, the teeth ca. 1 mm. long; corolla 16–17 mm. long, the galea spreading-pubescent, with wide glabrous thin margin, the lower lip with yellowish-pubescent pouch and rounded glabrous lobes; stamens 2 (the lower pair); caps. ca. 8 mm. long.—Coastal Salt Marsh; n. shore of San Francisco Bay. July–Nov.

7. **C. parviflòrus** (Ferris) Wiggins. [*Adenostegia p.* Ferris.] Plants 2–4 dm. tall, branched, loosely pubescent with some hairs gland-tipped; lvs. linear, mostly 1–1.5 cm. long; outer bracts 3-lobed, the flowering bracts 11–12 mm. long, purplish; calyx 10–14 mm. long, lanceolate, slightly 2-toothed; corolla purplish, 14–18 mm. long, inverted, the galea with thin margin, the lower lip pubescent; stamens 4, with bearded fils.; caps. ca. 7 mm. long.—Dry rocky slopes of limestone, 4000–5700 ft.; Pinyon-Juniper Wd.; New York Mts., e. San Bernardino Co.; to Utah, Ariz. Aug.–Oct.

8. **C. Prínglei** Gray. [*Adenostegia P.* Greene.] Minutely pubescent to glabrescent, 5–9(–16) dm. high, much-branched; lvs. linear, 5–10 mm. long, obtuse, involute; fls. 3–5 in heads, subtended by 3–5-lobed outer bracts; inner or flowering bract lance-oblong, ca. 1 cm. long, with yellow glands; calyx lanceolate, 9–11 mm. long, entire or nearly so; corolla ca. 8 mm. long, the lips densely hairy, the galea with broad wide white margin; stamens 4, the fils. villous; caps. rounded.—Dry hills, below 2500 ft.; Foothill Wd., Chaparral; Lake Co. Aug.–Sept.

9. **C. Nevínii** Gray. [*Adenostegia N.* Greene.] Slender, freely paniculately branched, 2–5 dm. tall, pubescent to ± hirsute with spreading glandless hairs; lower lvs. crowded, 2–2.5 cm. long, mostly with 3 linear divisions, the upper more remote, linear, shorter; fls. in 1–3-fld. heads subtended by several 3-lobed outer bracts; inner bract 13–15 mm. long, oblong-lanceolate, stiff hairy; calyx 10–13 mm. long, ± purplish, linear-lanceolate, slightly 2-toothed; corolla 12–16 mm. long, purplish, glabrous, the galea dark with white lateral margins; stamens 4, the fils. bearded, the anthers 1-celled; caps. 6–7 mm. long.—Dry slopes, 5000–8000 ft.; Yellow Pine F., Red Fir F.; San Gabriel Mts. to Little San Bernardino Mts. and mts. of San Diego Co. July–Sept.

10. **C. Hélleri** (Ferris) Macbr. [*Adenostegia Kingii* var. *involucrata* Kuntze. *A. H.* Ferris.] Plant much-branched, 1–3 dm. tall, soft grayish-pubescent, some of hairs gland-tipped; lvs. 0.5–1.5 cm. long, linear or with pair of linear lobes; bracts usually with 2 pairs of lobes; heads 1–4-fld.; calyx 2–2.2 cm. long, the 2 acute teeth 1–2.5 mm. long; corolla ca. 2 cm. long, the galea purplish, hairy-striate, glabrous on margin and tip, the lower lip with a purple-striped pouch externally hairy; stamens 4; fils. ± bearded; caps. 6–7 mm. long.—Rocky slopes, 5000–8000 ft.; Sagebrush Scrub, Pinyon-Juniper Wd.; Inyo-White Mts. to Mono Co. and w. Nev. July–Sept.

11. **C. ramòsus** Nutt. ex Benth. in DC. [*Adenostegia r.* Greene.] Plants 1.5–3 dm. high, with many slender branches, gray-puberulent with minute recurved-spreading glandless hairs; lvs. filiform, involute, 1–3.5 cm. long, entire or with 1 pair of filiform

lobes; fls. 3–5 in dense clusters; bracts several, 5-lobed; flowering bract 14–15 mm. long, oblong-lanceolate, entire, obtusish; calyx ca. as long, the teeth ca. 0.5 mm. long; corolla 15–17 mm. long, brownish-yellow, the galea pubescent dorsally, glabrous and yellow at apex and margin, the lower lip with a yellowish finely pubescent pouch; stamens 4, the fils. bearded; caps. ca. 8 mm. long.—Dry flats and slopes, 4000–6000 ft.; Sagebrush Scrub, Yellow Pine F.; Lassen Co. to Modoc Co.; e. Ore. to Wyo., Colo. July–Aug.

Ssp. setòsus Penn. Heads 5–10-fld., the bracts ± setose.—Dry slopes, 8000–10,000 ft.; Sagebrush Scrub; e. slope of Sierra Nevada and in Sweetwater Mts., Inyo and Mono cos.; w. Nev. July–Aug.

Ssp. erèmicus (Cov. & Mort.) Munz. [*Adenostegia e.* Cov. & Mort. *C. e.* Munz.] Outer bracts much shorter than flowering bract and corolla; corolla purplish.—Dry rocky places, ca. 7000 ft.; Pinyon-Juniper Wd.; Panamint Mts., Inyo Co. Aug.–Oct.

12. **C. bernardìnus** Munz. Plants 2–4.5 dm. high, diffusely branched, yellow-green, finely and scabrous-pubescent throughout; lvs. many, linear, entire, involute, 0.5–2 cm. long; outer bracts olive green with purple tips, 3–5-parted, 0.5–1.7 cm. long, the lobes linear; fls. mostly 2–5 in heads; fl.-bract ca. 18 mm. long, bulbous-scabrous and glandular-pubescent and ciliate-villous, ± blunt at apex; calyx 15–16 mm. long, finely pubescent, entire or obscurely bidentate; corolla 14–16 mm. long, yellowish, the galea minutely pubescent with white wide margin, the lower lip pubescent; stamens 4, the fils. bearded; caps. ca. 8 mm. long; seeds ± arcuate, ca. 1.5 mm. long.—Alkaline seeps, 3000 ft.; Joshua Tree Wd.; n. base of San Bernardino Mts. Oct.

13. **C. Ferrisiànus** Penn. Plants 3–6 dm. tall, branched, finely pubescent with gland-less recurved hairs; lvs. 1–3 cm. long, linear or with a pair of linear lobes; fls. 3–5 in heads subtended by several 3-lobed dorsally ± glabrous, but ciliate outer bracts; flowering bract 13–15 mm. long, lance-oblong, rounded at tip, ± setose-pilose apically; calyx 14–16 mm. long, lanceolate, entire; corolla 13–16 mm. long, white, the galea minutely pubescent, apically glabrous with white membranous edges, the lower lip sparsely pubescent; stamens 4, the fils. bearded; caps. 9–10 mm. long.—At 4500–7000 ft.; Red Fir F., Yellow Pine F.; Sierra Nevada of Fresno and Tulare cos. July–Sept.

14. **C. littoràlis** (Ferris) Macbr. [*Adenostegia l.* Ferris. *C. rigidus* var. *l.* Jeps.] Diffusely branched, 3–6 dm. high, finely pubescent with minute recurved glandless hairs; lvs. linear, 13–25 mm. long, entire or with a pair of linear lobes; heads 5–10-fld. subtended by several finely pubescent ± ciliate 3-lobed bracts 14–22 mm. long; fl.-bract 18–19 mm. long, yellow-green, oblong, rounded at tip; calyx 17–19 mm. long, lance-oblong, minutely bidentate; corolla 17–20 mm. long, white, the throat with 2 dull purplish lines, the galea finely pubescent, distally and on margin glabrous, ± yellowish; lower lip equal to upper, with pubescent pouch and yellow apex; stamens 4, the fils. bearded; caps. 9–10 mm. long.—Back of Coastal Strand, Closed-cone Pine F.; Monterey Peninsula. July–Sept.

Ssp. platycéphalus (Penn.) Munz. [*C. p.* Penn.] Plants 4–12 dm. high; bracts setose-pilose and finely ciliate.—Chaparral, S. Oak Wd.; hills back of Santa Barbara.

15. **C. rígidus** (Benth.) Jeps. [*Adenostegia r.* Benth. in Lindl. *C. r.* var. *sylvaticus* Jeps.] Plants 3–7 dm. tall, much-branched, not glandular, finely pubescent and sometimes with some longer hairs on stems; lvs. mostly 1–2 cm. long, linear and entire or with a pair of linear lobes, obtuse or ± truncate at tips; heads mostly 5–6-fld., subtended by several ± setose-pilose 3-lobed outer bracts; flowering bract 15–17 mm. long, lance-oblong, ± purplish, acute, ciliate and ± setose-pilose; calyx 12–15 mm. long, lanceolate, subentire; corolla 12–15 mm. long, whitish with some purple, the galea finely pubescent, distally and marginally glabrous, the lower lip moderately inflated, finely pubescent with glabrous tip; stamens 4, the fils. bearded; caps. 8–10 mm. long.—Dry slopes, below 3000 ft.; Chaparral, Foothill Wd., Mixed Evergreen F.; Coast Ranges from Santa Cruz and Santa Clara cos. to San Luis Obispo Co. Aug.–Sept.

Ssp. brevibracteàtus (Gray) Munz. [*C. filifolius* var. *b.* Gray. *Adenostegia rigida* var. *b.* Greene. *C. r.* var. *b.* Macbr. *C. compactus* Penn.] Branches more strict and erect; outer bracts setose-ciliate, otherwise but slightly setose; corolla light yellow.—Dry granitic slopes, 3000–6000 ft.; Foothill Wd., Chaparral, Yellow Pine F.; Liebre and Tehachapi mts. to Mariposa Co. July–Aug.

16. **C. filifòlius** Nutt. ex Benth. in DC. [*Adenostegia f.* Abrams. *C. rigidus* var. *f.* Macbr.] Plants 3–10 dm. tall, much-branched, pubescent with some longer hairs, not

glandular; lvs. 1–3 cm. long, with 3 subfiliform lobes, the segms. mostly involute, obtuse; outer bracts several, with 3 subfiliform lobes widened toward tips, ± setose-pilose; heads 5–15-fld.; flowering bract 15–17 mm. long, lance-oblong, obtusish, pubescent and ± setose-pilose; calyx 14–16 mm. long, lanceolate, minutely denticulate; corolla 13–16 mm. long, white with 2 wide purplish lines, the galea minutely pubescent, greenish-yellow, with purplish margins, the lower lip inflated, the pouch pubescent, apex obscurely lobed; stamens 4, the fils. bearded; caps. ca. 10 mm. long.—Dry slopes and open places, below 6500 ft.; Coastal Sage Scrub, Chaparral, S. Oak Wd.; Los Angeles Co. to n. L. Calif. May–Aug.

17. **C. pilòsus** Gray. [*Adenostegia p.* Greene.] Plants 5–7 mm. tall, many-branched, hirsute and glandular-pubescent; lvs. entire, linear, obtuse, mostly 1–3 cm. long, 1–2 mm. wide; fls. in clusters of 1–3, subtended by 1–2 entire lanceolate subtruncate outer bracts with enlarged usually angulate tips; flowering bract purplish, lance-oblong, obtusish, 18–22 mm. long; calyx 15–19 mm. long, linear-lanceolate, bidentate; corolla 14–16 mm. long, yellowish or purplish, the galea dorsally glabrescent, with puberulent decurved beak, the lower lip slightly pouched, minutely pubescent, with shallow incurved lobes; stamens 4, the fils. bearded; caps. ca. 8 mm. long.—Dry open hillsides; Chaparral, Mixed Evergreen F.; about San Francisco Bay. July–Sept.

Ssp. **diffùsus** (Penn.) Munz. [*C. d.* Penn.] Plant canescent-pubescent with fine mostly glandless hairs, except glandular toward infl.; cauline lvs. often less than 1 mm. wide.— Dry slopes, especially on serpentine, below 4000 ft.; Mixed Evergreen F., N. Oak Wd., Douglas-Fir F.; Coast Ranges from Lake Co. to Del Norte and Siskiyou cos. July– Sept.

Ssp. **Bolánderi** (Gray) Munz. [*C. p.* var. *B.* Gray. *Adenostegia B.* Kuntze. *C. B.* Penn.] Plant pubescent, ± glandular at least above; lvs. linear to narrowly linear; outer bracts scarcely or not enlarged at tips; flowering bract 14–16 mm. long; corolla-throat dark purple.—Dry open places in woods, 2500–7000 ft.; Yellow Pine F. to Red Fir F.; Sierra Nevada, Mariposa Co. n.; to s. Ore. July–Sept.

18. **C. ténuis** Gray. [*Adenostegia t.* Greene. *C. pilosus* var. *t.* Jeps.] Plants 3–6 dm. tall, open-branched, glabrous or minutely pubescent, not glandular; lvs. narrow-linear, entire, 1–3 cm. long, mostly involute, finely pubescent; fls. 1–3 in a cluster, the outer bracts filiform-linear, slightly if at all enlarged upward, pubescent, minutely ciliate; flowering bract 13–14 mm. long, ± purplish, oblong, obtusish; calyx 14–15 mm. long, linear-lanceolate, purplish, entire or bidentate; corolla 12–13 mm. long, ± greenish-yellow, the brown galea purple apically, dorsally puberulent, with yellowish membranous margin, the lower lip wider, yellowish, finely pubescent; stamens 4, the fils. bearded; caps. ca. 6–7 mm. long.—Dry open slopes, 4500–8500 ft.; Sagebrush Scrub, Yellow Pine F. to Lodgepole F.; Sierra Nevada from Fresno Co. n., to Butte and Plumas cos.; w. Nev. July–Sept.

Ssp. **brùnneus** (Jeps.) Munz. [*C. pilosus* var. *b.* Jeps. *C. b.* Penn.] Tips of outer bracts enlarged, often ± angulate-lobed; corolla pale or white with dark purplish streaks.—Often on serpentine; Chaparral, Foothill Wd.; inner Coast Ranges of Napa, Sonoma, and Lake cos. July–Aug.

19. **C. nidulàrius** J. T. Howell. Plant prostrate, glandular-viscid, greenish-purple, the stems 1–4 dm. long, pubescent with recurved-spreading hairs; lvs. narrowly linear, 1–5 cm. long, entire, obtuse; fls. ± solitary, the outer bracts deeply 3-lobed, 1.5 cm. long, the linear segms. spatulate-thickened, with dark margins; inner flowering bract 15–16 mm. long, lance-oblong, obtusish; calyx 16–17 mm. long, linear-lanceolate, acute; corolla 14–15 mm. long, white or tinged lilac, the throat with purple lines, the galea pubescent almost to rounded apex, the margin glabrous, thin, the lower lip widened, pouched, finely pubescent; stamens 4, the fils. bearded; caps. 7 mm. long.—Serpentine slope, at ca. 2000 ft.; Chaparral; near Deer Flat, Mt. Diablo, Contra Costa Co. July–Aug.

20. **C. Hansénii** (Ferris) Macbr. [*Adenostegia H.* Ferris. *C. pilosus* var. *trifidus* Rob. & Greenm.] Plants stout, 5–10 dm. high, much branched, with long spreading hairs, some of which are gland-tipped; lvs. 1–3 cm. long, linear, obtusish; fls. 1–3 in a cluster, the outer bracts 1.5–3 cm. long, 3-parted to entire, with linear to linear-oblanceolate segms.; flowering bract dark purple, 16–19 mm. long, linear-oblong, rounded at tip; calyx 18–20 mm. long, linear-lanceolate, acute, bidentate; corolla 16–18 mm. long, the galea dorsally pubescent, with glabrous thin margin, the lower lip widened, minutely

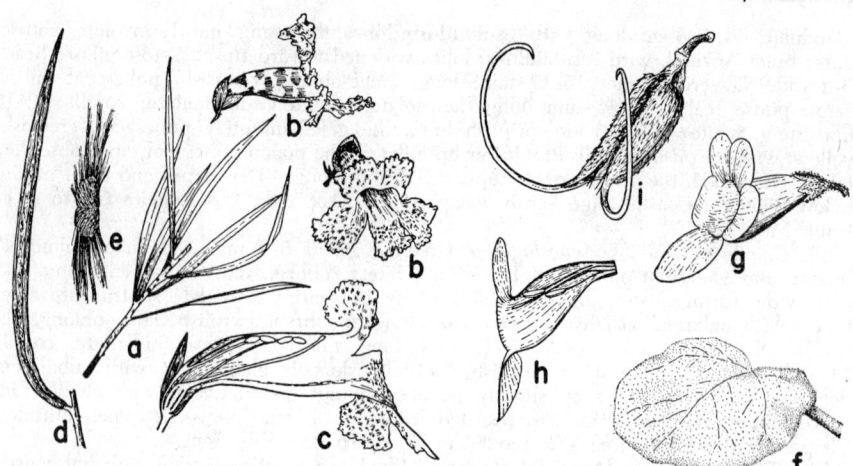

Fig. 61. BIGNONIACEAE. *Chilopsis linearis: a,* lvs., × ½; *b* and *b,* lateral and front views of fl., × ½; *c,* fl. in section, × 1, bilabiate corolla; *d,* caps., × ¼; *e,* seed, × 1, with wing dissected into long hairs.
MARTYNIACEAE. *Proboscidea: f,* lf., × ¼; *g,* bilabiate fl., × ½; *h,* fl. in section; *i,* horned caps., × ¼, dehiscent from apex.

pubescent, shallowly lobed; stamens 4, the fils. bearded; caps. 8–9 mm. long.—Dry gravelly places, 1000–2000 ft.; Foothill Wd.; foothills of Sierra Nevada from Tuolumne Co. n., to Shasta Co. July–Aug.

21. **C. víscidus** (Howell) Penn. [*Adenostegia v.* Howell. *C. tenuis* var. *v.* Macbr.] Plants 1.5–6 dm. tall, openly branched, pubescent, some of hairs gland-tipped; lvs. subfiliform, 1–4 cm. long, obtuse; fls. 1–3 in clusters subtended by 3-lobed outer bracts; inner bract purplish, ca. 14–15 mm. long, linear-oblong, rounded and subentire at apex; calyx ca. as long, linear-lanceolate, acute, slightly bifid, purplish; corolla 13–16 mm. long, dark on throat, greenish-yellow beneath, the galea red-brown at base, greenish-yellow and finely pubescent upward, with pale thin margins; lower lip widened, embracing much of galea, greenish-yellow with maroon lines; stamens 4, the fils. bearded; caps. 6–7 mm. long.—Dry flats and slopes, often of serpentine, 3000–5000 ft.; Yellow Pine F.; Coast Ranges from Tehama and Trinity cos. to Del Norte Co.; sw. Ore. July–Aug.

22. **C. palléscens** Penn. [*C. capillaris* Penn.?] Much like *C. Hansenii;* fl.-clusters finely pubescent, light yellow-green, the bracts ciliate; corolla 12–13 mm. long, garnet-brown on the sides of the throat, the galea white with yellowish apex and white margin, lower lip embracing galea.—W. base of Mt. Shasta, Siskiyou Co., Volta, Merced Co.? July–Aug.

64. Bignòniàceae. BIGNONIA FAMILY (Fig. 61)

Trees or erect to scandent shrubs. Lvs. mostly opposite. Fls. large and showy, ± bilabiate. Stamens 4, in 2 sets. Style 1; stigmas 2. Fr. a long woody 2-valved caps. resembling a silique, the valves pulling away from the broad partition which bears the numerous winged seeds. Ca. 100 genera and 600 spp., widely distributed in warmer regions.

1. Chilópsis G. Don. DESERT-WILLOW

A large shrub, with simple, entire, narrow, usually alternate lvs. Fls. in short terminal racemes. Calyx inflated, deeply 2-lipped, the upper lip 3-, the lower 2-toothed. Corolla funnelform, 5-lobed, the lobes erose. Antheriferous stamens 4, the 5th rudimentary. Caps. terete, linear. Seeds with wings dissected into copious long hairs. One sp. (Greek, *cheilos,* lip, and *opsis,* resemblance.)

(Fosberg, F. R. Varieties of the desert willow, Chilopsis linearis. Madroño 3: 362–366, 1936.)

1. **C. lineàris** (Cav.) Sweet. [*Bignonia l.* Cav. *C. saligna* D. Don.] Two–6 m. high, few- to many-stemmed; twigs very slender, glabrous or ± puberulent; lvs. deciduous, linear or lance-linear, 10–15 cm. long, ± arcuate, attenuate at ends; racemes ± pubescent to almost woolly especially on calyces, to ca. 1 dm. long; calyx ca. 1 cm. long; corolla 3–3.5 cm. long, lavender or pink or whitish, with purplish lines and markings; caps. 1.5–3 dm. long; seeds ca. 8 mm. long, 2–3 mm. wide, oblong, flat, thin, bearing a coma 1–1.5 cm. long at each end; 2n = 40 (Bowden, 1945).—Common along washes and watercourses, below 5000 ft.; Creosote Bush Scrub, Joshua Tree Wd.; Mojave Desert from Cushenbury Springs and Daggett e., Colo. Desert, San Jacinto V., Aguanga, etc.; to Tex., Mex. Our form is referred to var. *arcuàta* Fosb. by its author. May–Sept.

65. Martyniàceae. MARTYNIA FAMILY (Fig. 61)

Viscid-pubescent herbs with simple opposite or alternate lvs. and terminal racemose infl. Calyx 4–5-lobed. Corolla campanulate, obscurely 2-lipped. Stamens 4, didynamous, or 2 with the other pair forming staminodia. Ovary superior, of 2 carpels, but 1-celled with 2 parietal placentae; style 1; stigma with 2 flat lobes. Caps. horned, with fleshy deciduous exocarp and woody endocarp crested on median line above and sometimes below. Seeds few to many, black, nearly oblong, sculptured, somewhat compressed. A small family of the warmer parts of New World.

(Van Eseltine, G. P. A preliminary study of the unicorn plants [Martyniaceae]. N. Y. State Agric. Exp. Sta. Tech. Bull. 149, 1929.)

Calyx composed of 5 free sepals; body of fr. echinate .... 1. *Ibicella*
Calyx ± spathaceous, dentate or lobed above, cleft to base below; body of fr. roughly sculptured
2. *Proboscidea*

1. Ibicélla Van Es.

Lvs. broadly ovate or suborbicular, entire. Infl. dense, terminal. Calyx of 5 sepals, the 3 upper linear-lanceolate to obovate, the 2 lower much broader. Corolla yellow, oblique-campanulate. Fertile stamens 4. Fr. including beak 1.5–2 dm. long, the body 8–10 cm. long, cylindrical-ovoid, echinate. Two spp., native of S. Am. (Diminutive of *ibex*, the wild goat or chamois, because of the curved horns of the fr.)

1. **I. lùtea** (Lindl.) Van Es. [*Martynia l.* Lindl. *Proboscidea l.* Stapf.] Glandular-pubescent, spreading, branched, the stems 3–5 dm. long; lvs. opposite or alternate, the blades rounded, dentate, ± angularly incised or subcordate, 8–14 cm. wide, on petioles 10–15 cm. long; racemes few-fld., dense, scarcely surpassing lvs.; calyx ca. 2 cm. long; corolla deeper colored within, often dotted with red, ca. 3 cm. long; fils. thick, with purple spots; caps. with a slender horn longer than the short-spined body; seed 6–12 mm. long, compressed, rugose; 2n = 32 (Covas & Schnack, 1947).—Occasional weed, in Sacramento and San Joaquin valleys and s.; natur. from S. Am. Late summer.

2. Proboscídea Keller in Schmid. UNICORN-PLANT

Coarse viscid-pubescent annuals. Lvs. petioled, the lower mostly opposite, large, entire to shallowly lobed. Fls. few, large, showy, purplish, pinkish or yellowish. Calyx 5-lobed, split ventrally to base, viscid-pubescent. Corolla limb flaring. Stamens 4, didynamous. Ovary 1-celled; style long, cylindrical; stigmas with 2 flat lobes. Caps. with an ovoid or cymbiform sculptured body and ending in a long incurved dehiscent horn. Seeds many, large. Ca. 9 spp. (Greek, *proboscis*, beak.)

Lvs. averaging 5 cm. or less wide; fls. yellowish or copper-colored .... 1. *P. altheaefolia*
Lvs. averaging 10 cm. or more across; fls. deep reddish-purple to pinkish, or at least with purplish blotches.
  Calyx 1–1.5 cm. long; corolla 2.5–3.5 cm. long .... 2. *P. parviflora*
  Calyx ca. 2 cm. long; corolla 4–5 cm. long .... 3. *P. louisianica*

1. **P. altheaèfolia** (Benth.) Dcne. [*Martynia a.* Benth.] Coarse spreading perennial 3–4 dm. tall; lvs. reniform to suborbicular, usually broader than long, 3–5 cm. wide, on petioles 4–8 cm. long; raceme 3–7-fld. on an axis 5–8 cm. long; pedicels often as long as axis of infl.; calyx ca. 1 cm. long; body of caps. 5–6 cm. long, the horns ca. twice as

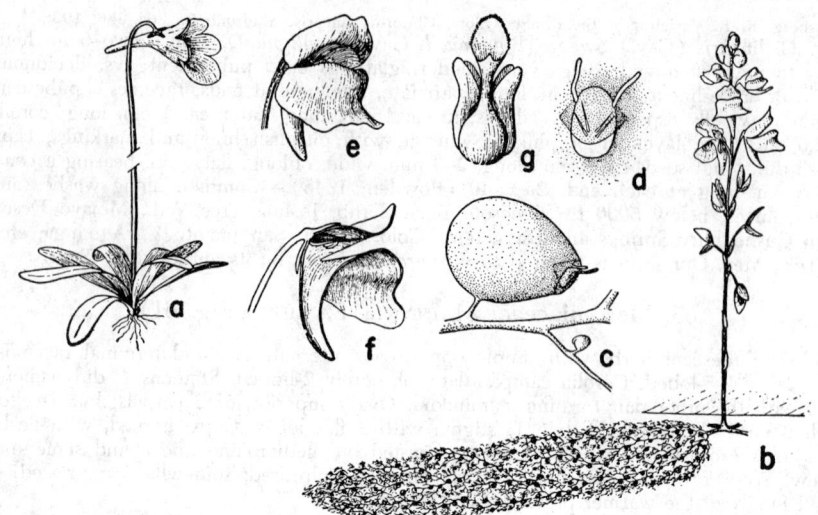

Fig. 62. LENTIBULARIACEAE. *Pinguicula vulgaris: a*, habit, × ½. *Utricularia vulgaris: b*, habit, × ⅜, submerged dissected lf. with small bladders, emergent infl.; *c*, traplike bladder, × 6; *d*, front of bladder, × 10; *e*, fl., × ½, bilabiate, spurred; *f*, fl. in section; *g*, pistil and 2 stamens, × 5.

long, crested on both dorsal and ventral edges.—Occasional, sandy places; Creosote Bush Scrub; Colo. Desert; to Ariz., Mex. Summer.

2. **P. parviflòra** (Woot.) Woot. & Standl. [*Martynia p.* Woot.] Annual, spreading or matted, the stems up to 8 dm. long; lvs. ± opposite, broadly triangular to round-ovate, subentire to obtusely 5–7-lobed, cordate, obtuse, 8–12(–25) cm. wide; corolla red-purple to almost white, often dotted or blotched with red-purple and streaked with yellow; body of fr. 5–7 cm. long, the horns ca. twice as long; seeds black, 6–8 mm. long.—Occasional, as near Whittier, La Verne, Panamint Mts.; Ariz., and s. Nev. to Mex., Tex. Summer.

3. **P. louisiánica** (Mill.) Thell. [*Martynia l.* Mill. *P. Jussieui* auth.] Annual 5–8 dm. high; lvs. suborbicular, opposite or upper ± alternate, entire or sinuate-margined, 1–3 dm. wide, on petioles 1–2 dm. long; fls. dull white or yellowish, mottled or blotched with red-purple; body of caps. 6–10 cm. long, the horns 8–20 cm. long; $n = 15$ (Gaiser et al, 1943).—Occasional as natur. plant, especially in Sacramento V.; native of s. states. Summer.

66. Lentíbulariàceae. BLADDERWORT FAMILY (Fig. 62)

Rather small herbs, many insectivorous, and in water or wet places. Calyx often bilabiate. Corolla 2-lipped, the lower lip in ours with a narrow basal spur. Stamens 2, inserted on the corolla-tube near its base, the anther-sacs of each stamen confluent. Ovary 1-locular with a free cent. placenta. Caps. 2–4-valved or often bursting irregularly. Seeds several, anatropous. Five genera and ca. 300 spp., rather cosmopolitan.

Lvs. dissected or very fine, they or special branches bearing bladders or traps; calyx mostly 2-lobed; fls. racemose to solitary, each subtended by a bract .... 1. *Utricularia*
Lvs. entire in basal rosette; calyx 5-lobed; fls. solitary, bractless .... 2. *Pinguicula*

1. Utriculària L. BLADDERWORT

Stems mostly submerged. Lvs. simple to much dissected and bearing small urn-shaped bladders which possess a kind of valvelike opening and trap insects and minute crustacea. Scapes emergent, with small auricled scales, 1–several-fld. Fls. perfect, racemose, yellow in ours. Calyx with 2 entire lips. Corolla with a projecting palate on lower lip, the upper lip erect. Stamens 2, twisted, flattened. Caps. irregularly dehiscent, many seeded. Some spp. form terminal winter-buds of very crowded lvs. A ± cosmopolitan genus. (Latin, *utriculus*, a little bladder.)

(Rossbach, C. B. Aquatic Utricularias. Rhodora 41: 113–128, 1939.)

Lf.-segms. terete or capillary.
 Lvs. 3 times dichotomous; scape stout, with 6–20 fls.; spur longer than lower lip of corolla; stems at least 1 mm. in diam. . . . . 1. *U. vulgaris*
 Lvs. mostly once dichotomous; scape slender, mostly 1–3-fld.; spur shorter than lower corolla-lip; stems less than 0.5 mm. thick.
  Bladders on lvs. only; corolla-spur shorter than lower lip; body of seed smooth . . . . 2. *U. gibba*
  Bladders on both stems and lvs.; corolla-spur longer than lower lips; body of seed tubercled
   3. *U. fibrosa*

Lf.-segms. flattened.
 Traps borne on the lvs.; margins of terminal lf.-segms. entire . . . . 4. *U. minor*
 Traps borne mostly on separate elongate branches; margins of terminal lf.-segms. serrulate
  5. *U. intermedia*

1. U. vulgáris L. Immersed stems 3–10 dm. long, coarse, few-branched; lvs. mostly tripinnately divided, 1.5–4.5 cm. long, with numerous traps ca. 2–2.5 mm. in diam.; scapes 1–3 dm. long; calyx 3.5–4 mm. long; lower lip of corolla 12–15 mm. long, yellow with brown stripes on the large palate; fruiting pedicels mostly arched-recurving, 1–1.8 cm. long; winter-buds ellipsoid, 1–2 cm. long, the lf.-margins hairy; $2n = 36$–40 (Reese, 1952).—Ponds and quiet water of many Plant Communities in scattered localities from San Bernardino Mts. n.; to Alaska, Atlantic Coast, Eurasia. July–Sept.

2. U. gíbba L. More delicate, the creeping or floating branches filiform, to 2 dm. long; lvs. less than 1 cm. long, mostly with 2 filiform segms. and scattered traps; scapes to 1 dm. high; calyx 2–3 mm. long; corolla 6–10 mm. long; fruiting pedicels ascending, less than 1 cm. long.—Shallow water, below 5200 ft., San Joaquin, Tuolumne, Sonoma, Lake, and Contra Costa cos.; transcontinental and to Cent. Am. July–Sept.

3. U. fibròsa Walt. Differing from *U. gibba* in having traps mostly on stems, seed rough-tuberculate.—Sacramento V. and Mendocino Co.; to Atlantic Coast.

4. U. mìnor L. Slender-stemmed, creeping on bottom; lvs. 2–4-times forked, ca. 5–10 mm. long; scapes filiform, 3–20 cm. high, 2–9-fld.; calyx 2–3 mm. long; corolla 5–8 mm. long, gaping; fruiting pedicels arched-recurving, 3–8 mm. long; winter-buds lax; $2n = 36$–40 (Reese, 1952).—Reported from Plumas Co.; transcontinental; Eurasia.

5. U. intermèdia Hayne. Stems creeping under water, several dm. long; lvs. 4–10 mm. long, mostly 3 times forked; bladders borne on branches separate from the lvs.; scapes to 2.5 dm. high, 1–4-fld.; pedicels ascending; calyx ca. 3 mm. long; corolla ca. 1.5 cm. long, the spur almost as long; winter-buds dense, ± ovoid, hairy.—Shallow water, 4000–7500 ft.; Fresno, Plumas, and Modoc cos. to B.C., Atlantic Coast; Eurasia. July–Aug.

2. Pinguícula L. BUTTERWORT

Acaulescent perennial herbs with a rosette of basal entire lvs. Fls. solitary on bractless scapes. Calyx 5-lobed, the upper lip 3-cleft, the lower 2-cleft. Corolla with an open hairy or spotted palate and spreading lobes. Caps. 2–4-valved. Ca. 45 spp., mostly of temp. zones. (Diminutive of Latin, *pinguis*, fat, because of the lvs.)

1. P. vulgàris L. Lvs. few, elliptic to ovate, ± fleshy, obtuse, 2–5 cm. long, viscid above; scapes 1–3, 0.5–1.5 dm. tall; corolla violet, 1.5–2 cm. long excluding the spur which is ca. 1 cm. long, the lips very unequal; $2n = 64$ (Doulat, 1947).—On mossy bank, at 300 ft., French Flat, Del Norte Co.; to Alaska, New England. April.

67. Orobanchàceae. BROOM-RAPE FAMILY (Fig. 63)

Root-parasitic rather fleshy herbs, without chlorophyll, having alternate scales in place of lvs. Fls. complete, with persistent calyx. Corolla tubular, ± 2-lipped, the upper lip mostly 2-lobed, the lower 3. Stamens 4, in 2 pairs. Ovary 1-celled, ovoid, pointed with a long style. Caps. 2–4-valved, each valve with 1 or 2 placentae. Seeds many, very small:

Fils. not hairy; caps. 2-valved; anther-cells deeply separated from below, mucronate or aristulate at base . . . . 1. *Orobanche*
Fils. hairy at base; caps. 4-valved; anther-cells closely parallel, blunt at base . . . . . . 2. *Boschniakia*

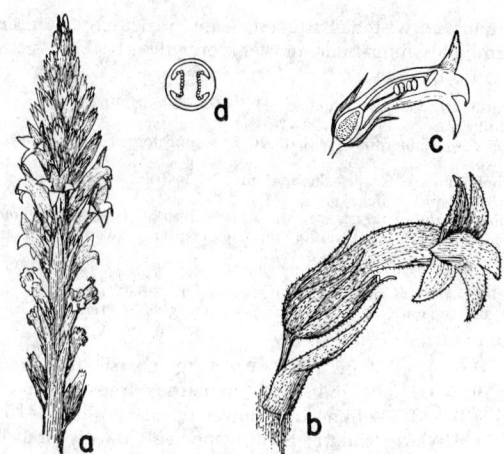

Fig. 63. OROBANCHACEAE. *Orobanche Ludoviciana* var. *latiloba: a,* infl., × ½; *b,* fl., × 1½, calyx 5-lobed, corolla bilabiate; *c,* fl. in section, × 1; *d,* ovary in section, × 2, to show 2 placentae.

1. Orobánche L. Broom-Rape

Usually viscid-pubescent plants with purplish to yellowish fls. Calyx 5-cleft into acute or acuminate lobes. Corolla curved, the upper lip erect or arched, the lower spreading or erect. Stamens mostly included. Stigma 2-lobed or peltate. Placentae 4. Ca. 100 spp.; widely distributed. (Greek, *orobos,* vetch, and *anchone,* choke, because of the parasitic habit.)

(Beck, G. Orobanchaceae. Das Pflanzenreich, IV. 261: 1–348, 1930. Achey, D. A revision of the section Gymnocaulis of the genus Orobanche. Bull. Torr. Bot. Club 60: 441–451, 1939. Munz, P. A. The N. Am. species of Orobanche, section Myzorrhiza. Bull. Torr. Bot. Club 57: 611–624, 1931.)

Fls. on elongate scapelike pedicels, the pedicels without bractlets. (Section Gymnocaulis)
  Pedicels 1–3 times as long as the stem; calyx-lobes longer than tube . . . . 1. *O. uniflora*
  Pedicels many, not exceeding length of stem; calyx-lobes not longer than tube . . 2. *O. fasciculata*
Fls. on shorter pedicels, these subtended by bractlets.
  Calyx 5-dentate or -lobed. Native spp. (Section Myzorrhiza)
    Stems with a thickened tuberlike base and branched above.
      Calyx 1 cm. long; corolla without palatal ridges in throat. Chaparral from Solano and Eldorado cos. s. . . . . 3. *O. bulbosa*
      Calyx 6–8 mm. long; corolla with palatal ridges in throat. From Tehama co. n.
                                                                          4. *O. pinorum*
    Stems not having a thickened tuberlike base.
      Fls., especially the lower, on pedicels up to 2 or 3 cm. long; infl. often corymbose.
        Lower lip of corolla spreading, 7–15 mm. long; anthers woolly . . . . 6. *O. Grayana*
        Lower lip of corolla erect, 4–6(–8) mm. long; anthers subglabrous except in material from e. of the Sierra Nevada . . . . 6. *O. californica*
      Fls. sessile or nearly so; infl. spicate.
        Corolla-lobes pointed, ± triangular.
          Upper corolla-lip 5–8 mm. long; plant yellowish to brown; anthers pubescent
                                                                           7. *O. ludoviciana*
          Upper corolla-lip 3 mm. long; plant dark; anthers glabrous . . . . 8. *O. valida*
        Corolla-lobes rounded; anthers glabrous . . . . 9. *O. multiflora*
  Calyx 4-dentate. Introduced parasite in tomato fields. (Section Trionychon) . . . . . . 10. *O. ramosa*

1. **O. uniflòra** L. var. **purpùrea** (Heller) Achey. [*Thalesia p.* Heller. *O. porphyrantha* G. Beck.] Stems slender, 2–4 mm. thick, 0.5–5 cm. long, largely subterranean with 2–6 crowded bracts; pedicels slender, 3–10 cm. long, 1–3; calyx-lobes narrow-subulate, from a broad base; corolla usually deep purple, 22–30 mm. long, constricted at base of tube and 5–8 mm. wide at throat; anthers mostly pubescent; caps. ovoid, usually capped by the persistent dry corolla; seeds coarsely reticulate, ca. 0.3 mm. long.—Mostly 3000–7000 ft.; Yellow Pine F., Red Fir F.; Santa Cruz Mts. and Sierra Nevada n.; to B.C., Ida. May–Aug.

Var. **minùta** (Suksd.) Achey. [*Aphyllon m.* Suksd.] Pedicels 2–7(–10) cm. long; corolla 15–20 mm. long, strongly curved, usually deep purple, the tube 4–5 mm. wide at throat, the lobes rounded; anthers glabrous.—On *Saxifragaceae,* below 6000 ft.; from San Gabriel Mts., and San Benito Co. and Sierra Nevada n. to B.C., Santa Cruz Id. March–May.

Var. **Sèdi** (Suksd.) Achey. [*Aphyllon* S. Suksd. *O. S.* Fern.] Corolla 15–20 mm. long, usually straw-colored, slightly curved, the tube 3–5 mm. wide at throat, the lobes truncate or oblong.—On *Crassulaceae, Compositae,* mostly below 7000 ft.; San Bernardino and Los Angeles cos. n. through Coast Ranges and Sierra Nevada; to B.C., Rocky Mts. March–July.

2. **O. fasciculàta** Nutt. Plant purplish, the caudex 3–10 mm. thick, 3–12 cm. long with 5–12 bracts and several erect axillary pedicels 3–10 cm. long; calyx 6–8 mm. long, the triangular lobes not longer than the tube; corolla usually purple, 15–22 mm. long, constricted at base, 2–5 mm. wide at throat, the lobes suborbicular; anthers glabrous to pubescent.—Occasional, mostly at 4000–10,650 ft.; on *Artemisia, Eriogonum, Eriodictyon,* etc., through much of Calif.; to B.C., Mich., New Mex. April–July.

Var. **franciscàna** Achey. Plant yellowish; corolla yellowish or tinged with purple, 22–30 mm. long, 4–8 mm. wide at throat; calyx-lobes usually acuminate; anthers usually pubescent.—Many Plant Communities; cismontane Calif. and s. Ore. April–July.

Var. **lùtea** (Parry) Achey. [*Phelipaea l.* Parry.] Plant yellowish; corolla more densely pubescent, 15–20 mm. long; calyx-lobes mostly acute.—Sagebrush Scrub, Pinyon-Juniper Wd.; Mono Co. to e. San Bernardino Co.; to Alta., Nebr., Chihuahua. May–June.

3. **O. bulbòsa** (Gray) G. Beck. [*Aphyllon tuberosum* Gray. *O. tuberosa* Heller, not Vell.] Stout, dark, 1–3 dm. tall, pruinose-puberulent throughout, from thickened tuber-like base; cauline bracts lanceolate to acuminate, closely placed, ca. 1 cm. long; infl. densely pyramidal, thyrsoid-paniculate, the fls. nearly or quite sessile; calyx ca. 1 cm. long, somewhat exceeding the subtending bracts, unequally divided into lanceolate segms. equal to or exceeding the tube; corolla 12–15 mm. long, yellowish to purplish or brownish, the lips erect, 2.5–3.5 mm. long, usually with acute and sublanceolate lobes and without palatal folds in throat; anthers white, subglabrous, somewhat apiculate; stigma discoid; caps. ca. as long as calyx.—On *Adenostoma* and associates, below 5000 ft.; Chaparral; from Solano, Marin, and Eldorado cos. to San Diego Co., Santa Cruz, Santa Rosa, and Santa Catalina ids. Mostly April–July.

4. **O. pinòrum** Geyer ex Hook. [*Aphyllon p.* Gray.] Plant 1–3 dm. high, glandular and cinereous-pubescent, from a tuberous base, simple or few-branched above; cauline scales lanceolate, 5–7 mm. long; pedicels 3–6 mm. long; calyx 6–8 mm. long, the lobes lance-subulate, ca. 3–4 mm. long; corolla 13–19 mm. long, yellowish, glandular-pubescent, the lips erect, 3–4 mm. long, the lobes acute, the upper 1.5 mm. long, the lower 3 mm.; anthers glabrous; stigma disciform; caps. equaling and rupturing calyx.—On conifers, 7500 ft., Red Fir F.; Tehama and e. Humboldt cos. to Wash., Ida.

5. **O. Grayàna** G. Beck. [*O. comosa* Hook., not Wallr.] Usually corymbose, sometimes paniculate, 5–10 cm. tall, glandular-puberulent throughout, usually few-fld.; cauline bracts lanceolate, 5–12 mm. long; pedicels 5–25 mm. long; calyx 12–16 mm. long, with lance-linear lobes; corolla 25–30 mm. long, strongly 2-lipped, pale with darker veins, the lips 10–14 mm. long, the upper reflexed, the lower spreading with palatal folds well developed; anthers woolly-pubescent; stigma peltate; seeds oblong-ovoid, ca. 0.3 mm. long, favose-reticulate.—At 6000–8000 ft.; Red Fir F.; Eldorado Co. n.; to Wash. June–Sept.

KEY TO VARIETIES

Corolla 25–35 mm. long, whitish to rose-purple or purplish-brown.
  Cauline bracts lanceolate to lance-linear, acute to acuminate; corolla whitish to pinkish with darker veins.
    Calyx 12–16 mm. long; plants 5–10 cm. tall. On *Erigeron,* etc. in meadows, from Lake Tahoe n. . . . . *O. Grayana*
    Calyx 15–20 mm. tall; plants 10–30 cm. tall. On *Baccharis, Grindelia,* etc., at low elevs., cent. Calif. . . . . var. *Jepsonii*
  Cauline bracts mostly ovate and obtuse; corolla purplish.
    Plants coarse, 12–20 cm. tall; corolla tube broadish, the lips purplish-brown; calyx 16–20 mm. long. On *Artemisia tridentata,* mts. of s. Calif. to Mono Co. . . . . var. *Feudgei*
    Plants slender, commonly 8–15 cm. tall; corolla tube slender, the lips purplish; calyx 8–15 mm. long. On *Grindelia,* etc. along coast from Monterey to B.C. . . . . var. *Nelsonii*
Corolla 40–45 mm. long, lurid purple. Coastal, San Luis Obispo Co. to Mendocino Co. . var. *violacea*

Var. **Jepsònii** Munz. Cauline bracts 8–12 mm. long; corolla mostly pinkish, 28–35 mm. long, the lips 10–12 mm. long.—Below 3000 ft., Sonoma and Colusa cos. to Santa Cruz and Tulare cos. June–Oct.

Var. **Feùdgei** Munz. Cauline bracts ca. 5 mm. long; infl. capitate-corymbose; corolla commonly purplish-brown, with lower lip yellow at base, the upper 10–12 mm. long and almost as broad, its lobes 2.5–3 mm. long, the lower lip 9–10 mm. long.—At 3000–8000 ft.; Sagebrush Scrub, Pinyon-Juniper Wd., Yellow Pine F.; Mono Co. through mts. of s. Calif. to n. L. Calif. May–July.

Var. **Nelsònii** Munz. Corolla purplish, 25–30 mm. long, the lips 7–10 mm. long.—Coastal Strand; along the coast, Monterey Co. n.; to B.C. July–Oct.

Var. **violàcea** (Eastw.) Munz. [*Aphyllon v.* Eastw.] Calyx 20–25 mm. long; upper corolla-lip 12–18 mm. long, the lobes obtuse, the lower lip 10–15 mm. long.—N. Coastal Scrub; sea bluffs, mostly Mendocino Co. to Marin Co.; occasional s. to San Luis Obispo and Los Angeles cos. May–Aug.

6. **O. califórnica** Cham. & Schlecht. [*Aphyllon c.* Gray.] Mostly spicate to paniculate, 1–3 dm. tall, glandular-pubescent; lower pedicels 5–20 mm. long; cauline bracts remote, ovate to lanceolate, 8–12 mm. long; fls. crowded, many; calyx 1–2 cm. long, the segms. narrow and acuminate; corolla pale with darker veins, 20–25 mm. long, the upper lip ca. 8 mm. long, its lobes 2.5–3 mm. long, the lower lip 6–8 mm. long with acute lobes; anthers subglabrous; caps. and seeds like those of *O. Grayana*.—On *Compositae*, below 5000 ft.; various Plant Communities; Monterey Co. to Shasta and Butte cos., Santa Rosa Id. Aug.–Sept.

KEY TO VARIETIES

Upper corolla-lip divided only part way.
  Corolla pale, yellowish with darker veins; anthers subglabrous.
    Plants 1–3 dm. tall; fls. crowded, numerous. Cent. Calif. *O. californica*
    Plants 0.5–1.5 dm. tall; fls. fewer, more scattered. S. Calif. . . . . var. *Parishii*
  Corolla purplish on upper lip at least; anthers woolly. E. of Sierra Nevada . . . . var. *corymbosa*
Upper corolla-lip divided almost to base, its lobes ca. 8 mm. long, acute; plants corymbosely paniculate, 1–2 dm. tall. Los Angeles Co. . . . . var. *claremontensis*

Var. **Paríshii** Jeps. Spicate or weakly paniculate; cauline bracts ovate, 5–8 mm. long; calyx ca. 10 mm. long; corolla 18–22 mm. long, yellowish with darker veins, the lips 5–7 mm. long; stamens subglabrous.—On *Eriodictyon, Haplopappus*, etc., below 6000 ft.; Ft. Tejon, Kern Co. to e. San Diego Co. May–July.

Var. **corymbòsa** (Rydb.) Munz. [*Myzorrhiza c.* Rydb.] Plant mostly 0.5–1.2 dm. tall, light colored, subcorymbose; cauline bracts lance-ovate, 8–10 mm. long; calyx 12–18 mm. long; corolla 24–28 mm. long, the upper lip purple, ca. 5 mm. long, the lower light or purplish, ca. 4 mm. long; anthers rather hairy.—Mostly on *Artemisia tridentata*, 6000–8500 ft.; Panamint Mts., Inyo Co. to e. Wash., and Rocky Mts. June–Aug.

Var. **claremonténsis** Munz. Corymbosely paniculate, 1–2 dm. tall, many-fld.; bracts on stems few, the lower suborbicular; calyx ca. 1 cm. long with lance-linear segms.; corolla ca. 25 mm. long, light-colored with pale purplish tinge; anthers subglabrous.—On *Quercus agrifolia*, Claremont, Los Angeles Co. July–Feb.

7. **O. ludoviciàna** Nutt. var. **Coòperi** (Gray) G. Beck. [*Aphyllon C.* Gray.] Viscid-pubescent, 1–3 dm. tall, quite simple and spicate; cauline scales obtuse, 5–10 mm. long; pedicels up to 1.5 cm. long; calyx 5–8 mm. long, the segms. ± unequal; corolla 15–28 mm. long, the lobes purplish, lanceolate, acute, the lips 3–6 mm. long, erect; anthers pubescent after dehiscence; caps. equaling or slightly exceeding calyx.—Largely on *Franseria dumosa*, below 4000 ft.; Creosote Bush Scrub; deserts of s. Calif. to Nev., New Mex., Son. Jan.–May. Parasitizing tomato plants, especially in Coachella and Imperial valleys.

Var. **latilòba** Munz. Calyx 10–12 mm. long; corolla-lips 6–9 mm. long, the lobes broad, abruptly narrowed at tip.—On *Hymenoclea* and *Franseria*, deserts, to Utah, New Mex.

8. **O. válida** Jeps. [*O. ludoviciana* var. *v.* Munz.] Whole plant dark, brownish-purple, spicate, 1–2 dm. high, slender, simple or few-branched; cauline bracts lanceolate, ca. 1 cm. long; pedicels short, the longest to 1 cm.; calyx 7–9 mm. long, with lance-acuminate segms.; corolla 12–14 mm. long, brownish-purple, the upper lip 3–4 mm.

long, purple, the 2 lobes acute, the lower lip yellowish, 3–4 mm. long, each lobe with dark purple midvein; anthers glabrous.—On *Eriodictyon*, etc., 5500–7000 ft.; Yellow Pine F.; San Gabriel Mts., Los Angeles Co. June–July.

9. **O. multiflòra** Nutt. var. **arenòsa** (Suksd.) Munz. [*Aphyllon a.* Suksd.] Slender, 0.5–1.5 dm. tall, viscid-pubescent; fls. subsessile; cauline bracts lance-ovate, 5–10 mm. long; calyx 9–12 mm. long, with lance-linear lobes; corolla purplish or yellowish, 15–20 mm. long, the upper lip ca. 5 mm. long, the lower lip ca. 4 mm.; anthers quite glabrous. —On *Artemisia*, etc., 3500–5500 ft.; Joshua Tree Wd., Pinyon-Juniper Wd.; mts. of e. Mojave Desert; to e. Wash., Colo. May–June.

10. **O. ramòsa** L. Stems several, slender, 1–3 dm. tall, simple or few-branched, densely pubescent; scales few, deltoid; spikes rather lax, slender; fls. sessile; calyx 6–7 mm. long, with 4 equal acuminate lobes ca. as long as the broad tube; corolla ca. 12 mm. long, the limb blue, the tube pale yellow; $2n = 24$ (Gardé, 1952).—On roots of *tomatoes, Xanthium, Amaranthus*, etc., between Centerville and Alvarado, Alameda Co.; introd. from Eu. Oct.–Nov.

2. Boschniàkia C. A. Mey.

Stem simple, glabrous, brown, thick, from a cormlike basal thickening and junction with root of host. Lvs. scalelike, mostly imbricated. Fls. nearly or quite sessile, in a dense spike, ± concealed by the subtending bracts. Calyx cup-shaped, 5-toothed, somewhat truncate. Corolla with throat broader than tube, ± curved, bilabiate, the upper lip entire or emarginate, the lower lip 3-lobed. Stamens 4, from almost as long as corolla to somewhat exserted. Stigma generally lobed (2–4) according to no. of carpels. Caps. 2–4-valved. Seeds many, favose. Few spp., widely distributed. (*Boschniaki,* a Russian botanist.)

(Gilkey, H. M. Northwestern American plants. Ore. State Mon. 9: 7–15, 1945.)
Plant 8–12 cm. tall; bracts generally acute, widest at middle or below; stigma 2–3-lobed. On *Gaultheria*, near the coast . . . . 1. *B. Hookeri*
Plant 15–20 cm. tall; bracts mostly obtuse to truncate, widest toward apex; stigma mostly 4-lobed. On *Arctostaphylos, Arbutus* . . . . 2. *B. strobilacea*

1. **B. Hoòkeri** Walp. [*Orobanche tuberosa* Hook., not Vell. *Kopsiopsis t.* G. Beck. *B. t.* Jeps.] Plant yellowish to dark, not more than 3 cm. thick in flowering part; bracts ca. 1 cm. long, ± acute; fl. 1–1.5 cm. long; calyx mostly with 2–3 short abruptly slender teeth; corolla somewhat bent above, the upper lip longer than lower; seeds ca. 2 mm. long, mostly angled.—On *Gaultheria Shallon*; Mixed Evergreen F., Redwood F., N. Coastal Scrub; Marin Co. to B.C. June–July.

2. **B. strobilàcea** Gray. [*Kopsiopsis s.* G. Beck.] Plants 1–2.5 dm. tall, the spike 3.5–6 cm. thick through flowering parts, dark reddish-brown; bracts 1.5–2 cm. long; fl. 1.5 cm. or more long; calyx truncate or 1–3-toothed, the teeth mostly shorter than the tube; corolla noticeably bent at middle, the lower lip as long as upper.—Reported from spp. of *Arctostaphylos* and from *Arbutus*, below 10,000 ft.; many Plant Communities; from San Jacinto Mts. n. through Coast Ranges and Sierra Nevada; to s. Ore. May–July.

68. Acanthàceae. ACANTHUS FAMILY (Fig. 64)

Ours low shrubs with simple opposite lvs. Fls. perfect, in ours bibracteolate at base of 5-parted calyx. Corolla tubular, bilabiate, the upper lip 2-, the lower 3-lobed. Stamens 2 in ours. Ovary 2-celled. Fr. a loculicidal caps. Ca. 200 genera and 2000 spp., mostly of warmer regions. Many are prized ornamentals.

1. Beloperòne Nees

Ours a subshrub with spreading often leafless branches. Fls. axillary, forming short 4-rowed racemes. Corolla dull scarlet. Anther-cells somewhat unequal and oblique. Caps. clavate, the locules 2-seeded. Ca. 30 spp., of trop. Am. (Greek, *belos,* an arrow and *perone,* something pointed.)

1. **B. califórnica** Benth. CHUPAROSA. Branches greenish-canescent, 3–15 dm. long, ±

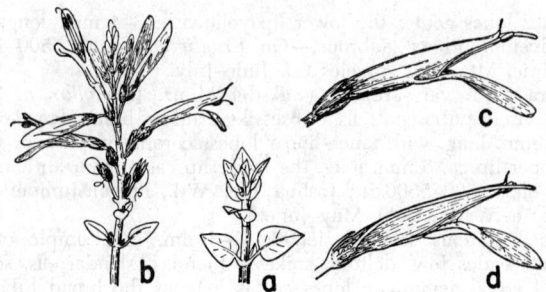

Fig. 64. ACANTHACEAE. *Beloperone californica: a,* young leafy shoot, × ½; *b,* infl., × ½; *c,* fl., × 1, with bilabiate corolla; *d,* fl. in section (stamens 2, but only 1 showing).

arched, slender; lvs. ovate, 1–1.5 cm. long, short-petioled, deciduous; calyx canescent, 4–5 mm. long, the segms. lanceolate; corolla straight, 3–3.5 cm. long, the lips 12 mm. long, oblong, truncate; stamens exserted; caps. 1.5–2 cm. long; seeds yellowish with purplish mottling, roundish and somewhat flattened, ca. 3.5–4 mm. across.—Common along watercourses, below 2500 ft.; Creosote Bush Scrub; w. and n. edges of Colo. Desert; to L. Calif., Ariz., Son. March–June.

69. Verbenàceae. Vervain Family (Fig. 65)

Herbs, shrubs or trees, with opposite or whorled lvs. Fls. perfect, mostly ± irregular, in terminal or axillary spikes, racemes or panicles. Calyx persistent, 2-, 4-, or 5-toothed or -lobed. Corolla gamopetalous, the tube cylindrical, the limb 4–5-lobed. Stamens 4 and didynamous, or rarely 2 or 5, inserted on corolla-tube. Ovary 2–4-celled, ovules usually 1 in a cell; style simple; stigmas 1–2. Fr. dry, forming 2 or 4 bony nutlets or a drupe with 2–4 nutlets. Ca. 80 genera and 800 spp., mostly of warmer regions.

Calyx 5-toothed; fls. in terminal spikes; nutlets 4 .... 1. *Verbena*
Calyx 2–4-toothed; fls. in short, usually axillary spikes or heads; nutlets 2 .... 2. *Lippia*

1. Verbèna L. Verbena. Vervain

Annual or perennial herbs or subshrubs. Lvs. mostly opposite. Fls. bracteate, in terminal corymbose or paniculate spikes. Calyx tubular, 5-ribbed, 5-toothed. Corolla with straight or incurved tube, the limb spreading, 5-lobed, regular or slightly 2-lipped. Stamens 4, in 2 pairs, rarely only 2. Ovary entire or slightly 4-lobed at apex, 4-celled, the cells 1-ovuled. Fr. dry, inclosed in calyx, separating into 4 narrow nutlets. Ca. 100 spp., chiefly in warmer parts of Am. (Ancient Latin name of the common European Vervain.)

(Perry, L. M. Revision of the N. Am. species of Verbena. Ann. Mo. Bot. Gard. 20: 239–362, 1933.)
A. Fls. in ± slender spikes; corolla 3–6 mm. long.
  B. Bracts shorter than fls., inconspicuous.
    C. Spikes short, dense and in terminal cymose clusters with stems almost leafless just beneath.
      D. Lvs. sessile, ± auriculate-clasping .... 1. *V. bonariensis*
      DD. Lvs. short-petioled, not auriculate-clasping .... 2. *V. litoralis*
    CC. Spikes elongate in fr., open at least below, often with stems leafy just below; lvs. petioled.
      D. Spikes panicled, much elongate, slender.
        E. Petioles wingless or obscurely winged, expanded abruptly into the blade.
          F. Corolla 2.5–3 mm. broad; lvs. mostly not or slightly scabrous. Cent. to n. Calif. .... 3. *V. hastata*
          FF. Corolla 2 mm. broad; lvs. decidedly scabrous. S. Calif. .. 4. *V. scabra*
        EE. Petioles definitely winged, expanded gradually into the blade.
          F. Infl. strigose-canescent; nonglandular. S. Calif. .... 5. *V. menthaefolia*
          FF. Infl. densely glandular-pubescent. Cent. and n. Calif. .. 6. *V. officinalis*
      DD. Spikes in 3's at the end of the branches, often congested at anthesis.
        E. The spikes 1–3 dm. long in fr.; lvs. canescent, not scabrous above; inner face of nutlets without whitish papillae .... 7. *V. lasiostachys*

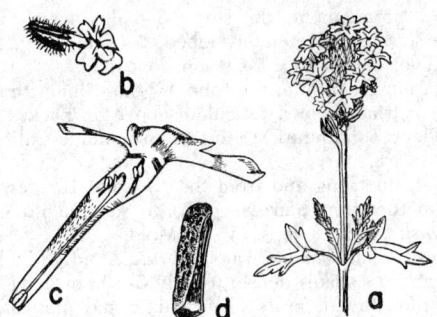

Fig. 65. VERBENACEAE. *Verbena Gooddingii: a,* infl., × ½; *b,* fl., × 1, the corolla-limb spreading, slightly irregular; *c,* fl. in section, × 2½, ovary slightly 4-lobed, stamens 4, 2 showing; *d,* nutlet, × 5, much elongate.

EE. The spikes 0.3–1 dm. long in fr.; lvs. bright green, scabrous above, inner
face of nutlets with whitish papillae .... 8. *V. robusta*
BB. Bracts longer than or equaling fls., conspicuous .... 9. *V. bracteata*
AA. Fls. mostly in headlike clusters; corolla 10–20 mm. long.
B. The ultimate lf.-divisions oblong, flat; corolla ca. 10 mm. long. Native of e. Mojave Desert
10. *V. Gooddingii*
BB. The ultimate lf.-divisions linear, revolute; corolla ca. 15 mm. long. Escape from gardens.
11. *V. tenera*

1. **V. bonariénsis** L. Stems erect, 3–10 dm. high, simple below, nearly square in cross section, subglabrous below or sparsely hirsutulous, scabrous on angles; lvs. lanceolate, sessile, 3–10 cm. long, dentate, strigulose above; spikes dense, purplish toward apex, 2–4 cm. long, in crowded cymes; calyx ca. 3 mm. long, pubescent; corolla lilac, the tube ca. 5 mm. long, the limb ca. 1 mm. broad; nutlets 2 mm. long, muricate; $n = 14$ (Dermen, 1936).—Waste places, Marin, Yuba, Fresno cos.; native of S. Am. June–Oct.

2. **V. litorális** HBK. [*V. Hansenii* Greene.] Like *V. bonariensis;* stem scabrous on angles; lvs. lanceolate to oblong, 2–6 cm. long, short-petioled or sessile; spikes more lax; calyx 2–2.5 mm. long; corolla-tube 2.5–3 mm. long, the limb 2.5–3 mm. broad; nutlets ca. 1.5 mm. long; $n = 14$ (Dermen, 1936).—Waste places, Shasta, Amador, San Joaquin, Stanislaus cos.; native Mex. to S. Am.

3. **V. hastàta** L. Erect perennial, 4–12 dm. high, strigose-hispidulous; lvs. lanceolate to oblong-lanceolate, the lower sometimes hastate, acute, serrate or incised-dentate, 5–12 cm. long, short-petioled; spikes short-peduncled, densely fld., crowded on short branches; calyx 2 mm. long, strigose-hispidulous; corolla blue, pink, or white, the tube ca. 3 mm. long, the limb 2.5–3 mm. wide; nutlets smooth or faintly striate, brownish; $n = 7$ (Dermen, 1936).—Moist waste places, along Sacramento and San Joaquin rivers; to B.C. June–Sept.

4. **V. scàbra** Vahl. Perennial with underground rootstocks, 4–12 dm. high, hispidulous, branched above; lvs. oblong to lance-ovate, serrate, 2–8 cm. long, scabrous, short-petioled; spikes very slender, 6–20 cm. long; calyx 2 mm. long, hispidulous; corolla white or purplish, the tube ca. 2 mm. long, the limb 2 mm. broad; nutlets 1.5 mm. long, striate.—Moist places, Coastal Sage Scrub, V. Grassland; Los Angeles and Orange cos. to San Bernardino and Riverside cos.; to Atlantic Coast, W.I. Sept.–Oct.

5. **V. menthaefòlia** Benth. Annual or short-lived perennial; stems 3–6 dm. tall, ± strigose; lvs. obovate to oblong in outline, 1–5 cm. long, ± canescent-strigose, irregularly pinnatifid or serrate, the petioles to ca. 1 cm. long, winged; spikes becoming 5–20 cm. long, slender; calyx 2.5–3 mm. long; corolla purple, the tube 3.5–4 mm. long, the limb 5–6 mm. broad; nutlets 2–2.5 mm. long, striate, reticulate above.—Dry places; Coastal Sage Scrub; Riverside and San Diego cos.; Ariz., L. Calif. to s. Mex. April–June.

6. **V. officinàlis** L. Much like *V. menthaefolia,* but glandular-pubescent in infl.; calyx 2–2.5 mm. long; corolla ca. 4 mm. broad; $n = 7$ (Dermen, 1936).—Waste places, Amador, Trinity cos.; natur. from Eu. June–Aug.

7. **V. lasiostàchys** Link. [*V. prostrata* R. Br., not Savi.] Perennial, much-branched and

ultimately diffuse and ± procumbent, the stems 3–8 dm. long, villous; lvs. oblong- to broadly-ovate, coarsely serrate to laciniately lobed, 2–6 cm. long, the cuneate base narrowed into a short petiole; spikes 1–3, 5–20 cm. long and lax after anthesis; calyx 4–5 mm. long, hairy; corolla mostly purple, the tube 4–5 mm. long, the limb 3–4 mm. wide; nutlets oblong-trigonous, striate below, reticulate above on backs.—Dry to moist places, below 8000 ft.; many Plant Communities; most of cismontane Calif.; to Ore., n. L. Calif. May–Sept.

A form with calyx 3–4 mm. long and from the N. Coast Ranges, has been named var. *septentrionàlis* Mold.; another from San Diego Co. to Sacramento V., and with calyx 2–3 mm. long, is var. *Abrámsii* (Mold.) Jeps. [*V. A.* Mold.]

8. **V. robústa** Greene. [*V. lasiostachys* var. *scabrida* Mold.] Much like *V. lasiostachys,* but lvs. greener and scabrous; spikes dense, usually 3–10 cm. long in fr.; calyx 3–4 mm. long; corolla-limb 2–3 mm. broad; nutlets muriculate and gray on commissural face.—Moist places; several Plant Communities; Marin and Tuolumne cos. to San Diego Co.; n. L. Calif. May–Oct.

9. **V. bracteàta** Lag. & Rodr. [*V. b.* Michx.] Annual to perennial, diffuse or spreading, the stems 1–5 dm. long, ± hirsute; lvs. oblong to obovate, pinnately lobed or incised, 1–4 cm. long, narrowed at base into a winged petiole 0.5–1.5 cm. long; stems leafy up to the spikes which are 3–6(–10) cm. long in fr.; bracts 5–10 mm. long; calyx 3–4 mm. long; corolla blue, the tube 3–4 mm. long, the limb 2.5–3 mm. broad; nutlets ca. 2 mm. long, reticulate above; $n = 7$ (Dermen, 1936).—Occasional in waste places below 5000 ft.; many Plant Communities; cismontane s. Calif., deserts from Imperial V. n., to San Joaquin V., Salinas V., and in Inyo Co.; to B.C. and across continent. May–Oct.

10. **V. Gooddíngii** Briq. [*V. bipinnatifida* var. *G.* Jeps.] Perennial, with several ascending stems 2–4.5 dm. high, hairy, ± glandular; lvs. rounded in outline, 1–2 cm. long, palmately 3-parted, then pinnately cleft, on petioles 0.5–1.5 cm. long; spikes capitate; bracts lance-linear, ca. 8 mm. long; calyx 7–8 mm. long; corolla purplish, the tube ca. 1 cm. long, the limb 8–10 mm. broad with retuse lobes; nutlets 3–3.5 mm. long, reticulate to near the striate base, retrorsely hispidulous on commissural face.—Dry canyons and slopes, 4000–6500 ft.; Joshua Tree Wd., Pinyon-Juniper Wd.; Providence, New York, and Ivanpah mts. to Clark Mt., e. Mojave Desert; to Utah, Son. April–June.

11. **V. ténera** Spreng. Perennial, cespitose, the stems branched, slender, sparsely hairy, 1.5–3 dm. long; lvs. 1–3 cm. long, 3-parted, then pinnately cut into linear entire acute lobes, short-petioled; spikes pedunculate, elongating in fr.; calyx narrow, strigose, ca. 7 mm. long; corolla rose-violet, the tube ca. 10–12 mm. long, the limb 6–8 mm. broad; nutlets linear-oblong, ca. 3.5 mm. long, reticulate toward apex, whitish-muriculate on commissural face; $n = 15$ (Beale, 1940).—Escape in Ventura and Los Angeles cos.; native of S. Am. April–Oct.

2. Líppia L.

Herbs or shrubs with opposite or verticillate lvs. Fls. in spikes or heads on slender axillary peduncles. Calyx small, 2–4-cleft. Corolla-tube cylindrical, the limb bilabiate. Nutlets 2. Rather a large genus (over 100 spp.), often divided into several. (Dr. A. *Lippi,* French naturalist.)

Low herbs; fls. capitate or nearly so; calyx 2-lobed. (*Phyla*)
  Lvs. broadest below the middle, the blades 2–4 cm. long, serrate from below the middle to
    the apex; calyx-lobes ca. as long as tube .... 1. *L. lanceolata*
  Lvs. broadest above the middle, the blades mostly 0.5–2 cm. long, serrate only above the middle;
    calyx-lobes shorter than tube.
      Lf.-margins with spreading teeth; lvs. apically obtuse, tapering gradually to the ± sessile base
                                                   2. *L. incisa*
      Lf.-margins with teeth pointing forward; lvs. mostly acute at apex, abruptly narrowed to a
        petioled base .... 3. *L. nodiflora*
Erect shrub; fls. in narrow spikes; calyx 4-lobed. (*Aloysia*) .... 4. *L. Wrightii*

1. **L. lanceolàta** Michx. [*Phyla l.* Greene.] Perennial with procumbent stems rooting at base, 2.5–4 dm. long, strigulose-canescent; lvs. opposite, lanceolate to ovate, 3–6 cm. long, acute at apex, cuneately narrowed at base to short petioles; peduncles in upper axils, exceeding lvs.; spikes 1–1.5 cm. long; bracts broadly ovate; calyx 2 mm. long;

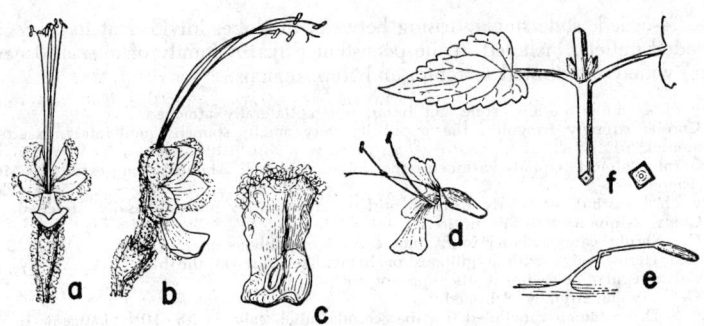

Fig. 66. LABIATAE. *Trichostema lanatum:* a and b, fl. in front and side views, × 1, woolly calyx and irregular corolla with long exserted stamens and style; c, nutlet, × 5. *Salvia sonomensis: d,* fl., × 1, calyx with arched upper lip, corolla bilabiate, fertile stamens 2 with upper anther-sac fertile, connective long and filamentlike, lower anther-sac abortive. *S. spathacea:* e, stamen, × 1, lower anther-sac represented by a linear structure. *Agastache: f,* habit, × ¾, to show opposite lvs. and 4-angled stem.

corolla pale blue to lavender, 2.5–3 mm. long, ± strigose toward summit externally; fr. subglobose, ca. 2 mm. long.—Occasional in low wet places; V. Grassland, Freshwater Marsh, etc.; Cent. V. and valleys of cismontane s. Calif.; to Atlantic Coast. May–Sept.

2. **L. incìsa** (Small) Tides. [*Phyla i.* Small.] Spreading or creeping, rooting at nodes, much-branched, sparingly strigose; lvs. linear-cuneate, 1–3 cm. long; peduncles 3–7 cm. long; heads becoming 1–2 cm. long in age; bracts rhomboidal; calyx almost 2 mm. long; corolla white or bluish, 2.5–3 mm. long; fr. broadly obovoid, 1.5–2 mm. long.— Low wet places, V. Grassland, etc.; San Joaquin V., San Diego and Imperial cos.; New Mex., Tex. April–Oct.

3. **L. nodiflòra** Michx. var. **ròsea** (D. Don) Munz. [*Zappania n.* var. *r.* D. Don. *Phyla n.* var. *r.* Mold. *Lippia filiformis* Schrad.] GARDEN LIPPIA. Plants matted, ± woody near base, cinereous-strigillose; lvs. pale green, narrow-oblanceolate to -obovate, entire to toothed, 1–2 cm. long (including short petiole), ± acutish at apex; peduncles mostly 1.5–3 cm. long; spikes ovoid, mostly 5–8(–10) mm. long, 6 mm. thick; bracts ovate; calyx ca. 1 mm. long; corolla rose to white, 4–5 mm. long.—Used in cult. for a ground-cover and well established in ± moist waste places in cent. and s. Calif.; from S. Am. May–Oct.

Var. **canéscens** (HBK.) Kuntze. [*L. c.* HBK. *Phyla n.* var. *c.* Mold.] Spikes cylindric in age, 4–5 mm. thick; calyx 2 mm. long; corolla 3 mm. long.—Occasional as an introd. in San Joaquin and Imperial valleys; S. Am.

Var. **réptans** (HBK.) Kuntze. [*L. r.* HBK. *Phyla n.* var. *r.* Mold.] Lvs. oval to obovate-cuneate; peduncles 2–6 cm. long; spikes ovoid, 10–15 mm. long, 6 mm. thick.—Sacramento V., Lake Co.; native of S. Am.

4. **L. Wrìghtii** Gray. [*Aloysia W.* Heller.] Aromatic shrub 6–15 dm. high with slender spreading tomentose branchlets; lvs. canescent-tomentose, 5–10 mm. long, ± ovate, crenate; spikes slender, 2–2.5 cm. long; bracts lanceolate; calyx 2.5 mm. long; corolla white, 3 mm. long, 2 mm. wide; nutlets oblong, almost 2 mm. long.—Dry rocky places, 3200–5000 ft.; Joshua Tree Wd., Pinyon-Juniper Wd.; Little San Bernardino Mts., Clark and Providence mts., e. Mojave Desert; to Tex., n. Mex. Spring and fall.

70. Labiàtae. MINT FAMILY (Fig. 66)

Aromatic herbs or shrubs, rarely trees, mostly with 4-angled stems and opposite or whorled simple lvs. Fls. mostly irregular, bilabiate, variously clustered in ± cymose and bracteolate infls. Calyx persistent, 2-lipped or regular, mostly 5-toothed or -lobed. Corolla usually 2-lipped with the upper lip 2-lobed to entire, the lower lip 3-lobed, sometimes almost regular. Stamens on the corolla-tube, mostly 4 and didynamous, or one of the pairs abortive; anthers 2-celled, or sometimes one of the cells abortive. Ovary 4-lobed or -parted, 4-celled; style single, arising between the lobes, divided at its summit. Fr. of four 1-seeded nutlets, included in the persistent calyx. A family of over 150 genera and 3000 spp., widely distributed in temp. and trop. regions.

A. Ovary of 4 united nutlets; style not basal; nutlets laterally attached.
  B. Corolla strongly irregular, the upper lip very small; stamens moderately exserted. Desert annual. . . . . 1. *Teucrium*
  BB. Corolla almost equally 5-lobed, the lobes declined; stamens long-exserted. Montane or cismontane . . . . 2. *Trichostema*
AA. Ovary of 4 distinct or nearly distinct nutlets; style basal; nutlets basally attached.
  B. Calyx 2-lipped, the lips entire.
    C. Shrub; calyx inflated in fr., not crested or gibbous on the back . . . . 3. *Salazaria*
    CC. Herbs; calyx with a gibbous or helmetlike crest on the back . . . . 4. *Scutellaria*
  BB. Calyx regular or 2-lipped, the lips not entire.
    C. Corolla strongly 2-lipped.
      D. Stamens included in the corolla-tube; calyx with 10 spinescent hooked teeth at the tip . . . . 5. *Marrubium*
      DD. Stamens exserted from the corolla or included in throat; calyx-teeth not hooked at tip.
        E. The stamens ascending, not declined and enveloped by the lower lip.
          F. Upper lip of corolla concave.
            G. Fertile stamens 4.
              H. Lvs. 3-parted; corolla white or violet, 2 cm. long 7. *Cedronella*
              HH. Lvs. simple.
                I. Upper pair of stamens longer than lower pair.
                  J. Anther-sacs parallel or nearly so; upper stamens declined; fls. rose-purple. Native . . . . . . 6. *Agastache*
                  JJ. Anther-sacs divergent. Introd.; in waste places.
                    K. Plants erect; corolla whitish, 10–12 mm. long 8. *Nepeta*
                    KK. Plants creeping; corolla purplish-blue, 16–22 mm. long . . . . 9. *Glecoma*
                II. Upper pair of stamens shorter than or equal to lower pair.
                  J. Calyx 2-lipped, closed in fr., the upper lip truncate; fls. in dense spike, purplish; low herb . . 10. *Prunella*
                  JJ. Calyx 5-toothed, not closed in fr., if 2-lobed the upper lip not truncate.
                    K. Plants shrubby; fls. 2.5–4 cm. long.
                      L. Lvs. coarsely serrate, lanceolate; corolla orange, 3.5–4 cm. long . . . . . . 11. *Leonotis*
                      LL. Lvs. entire to crenate, oblong to ovate; corolla yellow, 2.5 cm. long . . 12. *Phlomis*
                    KK. Plants herbaceous; fls. shorter.
                      L. Calyx becoming membranous and with large spreading limb 3 cm. broad; plant annual. 14. *Molucella*
                      LL. Calyx not becoming membranous and spreading.
                        M. Calyx-teeth not spine-tipped; corolla-tube without a hairy ring within; nutlets sharply 3–sided; plants annual. 13. *Lamium*
                        MM. Calyx-teeth spine-tipped; corolla-tube with hairy ring within; nutlets ovoid; plants mostly perennial . . 15. *Stachys*
            GG. Fertile stamens 2.
              .H. Calyx 2-lipped, or in one sp. of *Salvia* entire and oblique at the orifice.
                I. Pollen-sacs in each anther 2, with very short connective between; plants annual with spines on bracts and calyx-teeth . . . . 16. *Acanthomintha*
                II. Pollen-sacs in each anther 1 or 2, at the ends of elongate arms of connective jointed to the fils.; plants shrubs to annuals, usually not spiny on bracts or calyx-teeth . . . . 17. *Salvia*
            HH. Calyx equally 5-cleft; plant annual. Desert . . 18. *Monarda*
        FF. Upper lip of corolla plane.
           G. Fertile stamens 2; low purple-fld. plant of e. Mojave Desert 20. *Hedeoma*
           GG. Fertile stamens 4, or if 2, on cismontane plants.
             H. Fls. solitary or few in the lf.-axils.
                I. Plants quite woody.
                  J. Corolla 2.5–4 cm. long; lvs. ± ovate or deltoid. Cismontane . . . . 19. *Lepechinia*

JJ. Corolla 1–1.3 cm. long.
    K. Lvs. round-ovate; corolla white. S. San Diego
          Co. .... 22. *Satureja*
    KK. Lvs. linear to linear-oblong; corolla pale purplish-
          blue. Mojave Desert .... 23. *Poliomintha*
II. Plants strictly herbaceous.
   J. Corolla with recurved-ascending tube and 10–14 mm.
     long .... 21. *Melissa*
   JJ. Corolla with straight tube, and 6–8 or 30–40 mm.
     long .... 22. *Satureja*
HH. Fls. in dense heads.
   I. Plants annual, with round to spatulate lvs. not more than
     2 cm. long .... 24. *Pogogyne*
   II. Plants perennial, with ovate to lance-ovate lvs. mostly
     2–8 cm. long.
     J. Calyx bearded in the throat; bracts purplish. Introd.
        25. *Origanum*
     JJ. Calyx naked in the throat; bracts green, leafy. Native
        27. *Pycnanthemum*
  EE. The stamens declined and enveloped by the lower lip of the corolla; scurfy-
    tomentose desert shrub .... 30. *Hyptis*
CC. Corolla regular or nearly so, the lobes subequal.
  D. Fls. in terminal heads; stamens 4. Plants of dry places ...... 26. *Monardella*
  DD. Fls. in axillary whorls which may be aggregated in terminal spikes. Plants of
    wet places.
    E. Fertile stamens 2; fls. white .... 28. *Lycopus*
    EE. Fertile stamens 4; fls. usually lavender to purplish .... 29. *Mentha*

1. Teùcrium L. Germander

Herbs or subshrubs with entire to laciniate lvs. Fls. solitary in the axils of ± modified fl. lvs. or crowded into linear bracteate terminal spikes, subsessile or pedicelled. Calyx campanulate, 10-veined, with deltoid or lance-deltoid teeth. Corolla with short or a longer funnelform tube, the lower lip prominent with oblong to obovate middle lobe and smaller lateral lobes, the upper lip small, 2-cleft. Stamens 4, paired. Style exserted with stamens, 2-parted at apex. Nutlets oval, smooth or sculptured, glabrous or pubescent. Ca. 100 spp., world wide. (*Teucer*, a Trojan king.)

(McClintock, E., and C. Epling. A revision of Teucrium in the New World. Brittonia 5: 491–510, 1946.)

Pedicels 4–12 mm. long; annual herb .... 1. *T. cubense*
Pedicels 15–40 mm. long; low shrub .... 2. *T. glandulosum*

1. **T. cubénse** Jacq. ssp. **depréssum** (Small) McClint. & Epl. [*T. d.* Small. *T. c.* var. *densum* Jeps.] Mostly much-branched from base, 1–3 dm. high, glabrous below, pubescent above; median lvs. 1.5–2.5 cm. long, 3–5-lobed nearly to midvein, obovate- cuneate in outline, petioled; fl.-lvs. 3-parted, sessile; calyx 5–6 mm. long, the teeth spreading, ca. 4 mm. long, pubescent; corolla pale blue, 7–8 mm. long; nutlets corky-thickened, mostly pubescent apically, reticulate.—Rare, semialkaline flats; Creosote Bush Scrub; Colo. Desert (Hayfields, Palo Verde V., etc.); to Tex., nw. Mex. March–May.

2. **T. glandulòsum** Kell. Branching, to ca. 1 m. high, glabrous; lvs. oblong-lanceolate, entire, or some of the lower obscurely 3-lobed, 2–5 cm. long; fls. in elongate racemes; calyx campanulate, 7–11 mm. long, gland-dotted, the teeth lanceolate, 4–8 mm. long; corolla white, with violet streaks, pubescent, 15–20 mm. long, the lower lip 10–17 mm. long; nutlets oval, 2.5–3 mm. long, grooved, apically hairy.—Rocky places, at 1450 ft.; Creosote Bush Scrub; Whipple Mts., e. Colo. Desert; L. Calif. April–May.

2. Trichostèma L. Bluecurls

Strong-scented annual or perennial herbs, or subshrubs with entire lvs. Fls. 1–many in axillary cymes or racemose. Calyx equally deeply 5-lobed, to unequally so. Corolla blue or lavender to pink or whitish, 5-lobed, the slender tube often exceeding the calyx, the limb oblique, deeply 5-cleft into oblong, ± declined segms. Stamens 4, usually ascending from near the throat, arched or nearly straight, anther-sacs divergent at maturity. Nutlets joined ca. ⅓ their length, alveolate to rugose-reticulate, the ridges ± prominent. Ca. 16 spp., N. Am. (Greek, *trichos*, hair and *stemon*, stamen.)

(Lewis, H. A revision of the genus Trichostema. Brittonia 5: 276–303, 1945.)
Plants annual, with pale pubescence.
  Stamens 10–20 mm. long.
    Lower petioles 5–15 mm. long; lvs. rather remote, not apiculate .... 5. *T. laxum*
    Lower petioles not more than 4 mm. long; lvs. crowded, apiculate.
      Lvs. lanceolate, at least 3 times as long as wide; corolla-tube 5–10 mm. long
                                                                          7. *T. lanceolatum*
      Lvs. ovate, not more than twice as long as wide; corolla-tube 2.5–5.5 mm. long
                                                                          8. *T. ovatum*
  Stamens 2–7 mm. long.
    Calyx-lobes ca. twice as long as the tube, approximately equal to each other in width.
      Lvs. mostly 2–3.5 times as long as broad, oblong to oval; calyx-teeth slightly constricted
        at base. From Tulare Co. n. .... 1. *T. oblongum*
      Lvs. mostly 3.5–6 times as long as wide, linear-lanceolate; calyx teeth broadest at base.
        Mono Co. and s. Calif. .... 2. *T. austromontanum*
    Calyx-lobes 1–2 times as long as the tube, the uppermost tooth more slender than the other 4.
      Corolla-tube not distinctly longer than calyx.
        Calyx-lobes lanceolate; hairs of stem all appressed downward; lower petioles 7–16 mm.
          long .... 6. *T. micranthum*
        Calyx-lobes triangular-lanceolate; hairs of stem in part spreading; lower petioles to
          5 mm. long .... 3. *T. simulatum*
      Corolla-tube distinctly longer than calyx; lower petioles not more than 5 mm. long. Cent.
        Calif. .... 4. *T. rubisepalum*
Plants low shrubs, with ± purplish wool in infl.
  Corolla-tube 9–14 mm. long; calyx ca. 8 mm. long, with lanceolate lobes; fils. exserted 2.5–3 cm.
                                                                          9. *T. lanatum*
  Corolla-tube 4–7 mm. long; calyx ca. 5 mm. long, with triangular lobes; fils. exserted 1–1.5 cm.
                                                                          10. *T. Parishii*

1. **T. oblóngum** Benth. Soft-villous, 1–3(–5) dm. high, simple to somewhat branched; lvs. oblong to lance-oblong, 2–3 cm. long, mostly obtuse on petioles to 5 mm. long; cymes axillary, few–many-fld., horizontal and spreading, very short-peduncled; calyx 3–5 mm. long in fr., the tube barely 1 mm. long, the teeth lanceolate, subacuminate; corolla 5–6 mm. long, the tube slightly exserted, the lobes villous without, lower lip 2–3.5 mm. long; stamens arched, 2.7–5 mm. long; nutlets 1–2 mm. long, reticulate, hirtellous, with sharp ridges; *n* = 7 (Lewis, 1945).—Dry margins of meadows, etc., below 10,000 ft.; mostly Montane Coniferous F., Mixed Evergreen F.; from Lake and Tulare cos. n.; to Wash., Ida. June–Sept.

2. **T. austromontànum** Lewis. Resembling *T. oblongum*, the lvs. linear- to oblong-lanceolate, 1.5–3.5 cm. long, acute, 2–8 mm. wide; calyx-lobes lance-subulate, the teeth longer than the tube; corolla 3 mm. long, the tube equal to the calyx; stamens 5–6 mm. long; nutlets 1.2–2 mm. long, irregularly reticulate, puberulent; *n* = 14 (Lewis, 1945).— Drying edges of meadows, etc., 3500–7500 ft.; Yellow Pine F., Red Fir F.; mts. from San Diego Co. to Los Angeles Co., Mono Co. e. of Sonora Pass. July–Sept.

Ssp. **compáctum** Lewis. Compact plant, to ca. 1 dm. high; lvs. mostly less than 2 cm. long; nutlets subglabrous.—Hidden Lake, San Jacinto Mts. at 8000 ft.

3. **T. simulàtum** Jeps. Erect, 1–3 dm. high, mostly branched below, ± reddish, hirtel- lous with downwardly curled appressed hairs, glandular-pubescent; lvs. lanceolate to ovate-lanceolate, 2–5 cm. long; cymes axillary, few–many-fld.; calyx 4 mm. long, the lobes deltoid, 1–2 mm. long (to 3.5 mm. in fr.); corolla blue, ca. 4 mm. long, the lower lip 2–3 mm. long; stamens exserted ca. 1.5 mm.; nutlets 2–2.5 mm. long, deeply and irregularly alveolate, short-hirtellous.—Dry open places, 2000–6000 ft.; Yellow Pine F.; from Trinity and Plumas cos. to Siskiyou and Modoc cos.; s. Ore. June–Aug.

4. **T. rubisépalum** Elmer. [*T. laxum* var. *r.* Jeps.] Erect, 1–4 dm. tall, branched below, pilose and glandular-pubescent; lvs. lance-ovate to lanceolate, 2–5 cm. long; cymes axillary, few–several-fld.; calyx with narrow-deltoid lobes 1.3–1.8 mm. long at anthesis; corolla blue, 5–6 mm. long, the tube longer than calyx; stamens 4–6 mm. long; nutlets 1.4 mm. long, irregularly reticulate, sharp-ridged, hirtellous.—Dry places, below 4500 ft., Foothill Wd., Yellow Pine F.; Sierra foothills in Mariposa and Tuolumne cos., inner Coast Range of San Benito Co. June–Aug.

5. **T. láxum** Gray. TURPENTINE WEED. Glandular-pubescent or -puberulent, erect, 1–5 dm. tall, simple or branched below, often reddish on stems, etc.; lvs. lanceolate, 3–7 cm. long, attenuate; fls. many, in axillary cymes; calyx 2.5–4 mm. long at anthesis, the teeth deltoid, 1.7–3 mm. long; corolla-tube 4–8 mm. long, exserted, lower lip 4–7 mm. long; stamens 7–16 mm. long, strongly arched; nutlets irregularly alveolate, mostly hirtellous,

1.4–2.2 mm. long.—Damp gravelly or sandy places, below 6600 ft.; Mixed Evergreen F., Redwood F., Chaparral, Foothill Wd., Yellow Pine F.; Coast Ranges from Napa and Sonoma cos. to sw. Ore. July–Sept.

6. **T. micránthum** Gray. Erect, 1–3 dm. tall, branched below, hirtellous and glandular-pubescent; lvs. lance-linear, 2–2.5 cm. long, 2–5 mm. wide, acute; cymules 2–10-fld.; calyx at anthesis 1.5–2.5 mm. long, the lobes scarcely longer than the tube; corolla 2–3 mm. long; stamens exserted ca. 1.5 mm.; nutlets 1.2–1.7 mm. long, reticulate, the ridges sharp and hirtellous.—Drying margins of wet places, 6500–7500 ft.; Yellow Pine F.; San Bernardino Mts., and n. L. Calif. July–Sept.

7. **T. lanceolàtum** Benth. VINEGAR WEED. Annual, mostly branched from base, 1–6 or more dm. high, mostly bushy, glandular-villous, strong-smelling; lvs. lanceolate to lance-ovate, mostly 2–5 cm. long, 4–14 mm. wide, acute to acuminate, sessile or nearly so; cymes in mostly simple, ± secund racemes; calyx 2.5–4 mm. long at anthesis, the teeth deltoid to lanceolate, the uppermost often narrower; corolla light blue, the tube 5–8 mm. long, the lower lip 4–8 mm. long, strongly deflexed; stamens 13–20 mm. long, arched; nutlets alveolate, usually with several peglike projections, hirtellous, 1.6–3 mm. long; $n = 7$ (Lewis, 1945).—Dry fields and open places, mostly below 3500 ft.; Coastal Sage Scrub, Chaparral, Oak Wd., Foothill Wd., etc.; most of cismontane Calif. to n. L. Calif. and to B.C. Aug.–Oct.

8. **T. ovàtum** Curran. Villous-pubescent and canescent, erect, 1.5–6 dm. high, branched below; lvs. ovate to round-ovate, 1–2 cm. long, crowded, subsessile; cymes axillary, ± hidden by lvs., 1–8-fld.; calyx 2.5–3.5 mm. long at anthesis, the teeth lance-ovate, ± acuminate; corolla light blue, the tube ca. 6 mm. long, the lobes externally villous, the lower lip 3.5–5 mm. long; stamens 11–16 mm. long, arched; nutlets 1.5–2 mm. long, alveolate with podlike projections, hirtellous; $n = 7$ (Lewis, 1945).—Waste places and fields below 1000 ft.; V. Grassland; San Joaquin V. from Kern Co. to Fresno Co. July–Oct.

9. **T. lanàtum** Benth. WOOLLY BLUE-CURLS. ROMERO. Rounded shrub 5–15 dm. tall, many-branched, hirtellous; lvs. lance-linear, sessile, 3.5–7.5 cm. long, revolute, green above, lanate beneath, usually with axillary fascicles; infl. an interrupted spike of dense subsessile cymes, woolly with blue, pink, or whitish hairs 2–3 mm. long and concealing pedicels; calyx 5–8 mm. long at anthesis, the teeth subequal, ovate, acuminate to acute, equal to or longer than tube; corolla floccose without, blue, the tube 9–14 mm. long, the lower lip 7–12 mm. long; stamens arched; nutlets 2–4 mm. long, irregularly reticulate, hirtellous; $n = 10$ (Lewis, 1945).—Dry slopes, mostly below 3500 ft.; Chaparral, Coast Ranges near the coast from Monterey and San Benito cos. to Orange and San Diego cos. May–Aug.

10. **T. Paríshii** Vasey. [*T. lanatum* var. *denudatum* Gray.] Resembling *T. lanatum*, the lvs. pubescent beneath; infl. open, the tomentum short and not concealing the pedicels; calyx 3–5 mm. long at anthesis, the teeth lanceolate to ovate; corolla-tube 4–7 mm. long, the lower lip 5–9 mm. long; nutlets 1.8–3.5 mm. long, irregularly reticulate, pubescent; $n = 10$ (Lewis, 1945).—Dry slopes, 2000–6000 ft.; Chaparral, Yellow Pine F., Joshua Tree Wd.; away from the coast, Acton (Los Angeles Co.) to n. L. Calif. May–Aug.

3. Salazària Torr. BLADDER-SAGE

Low dense and divaricately branched shrub with spinescent branchlets. Lvs. small, entire or rarely toothed, short-petioled. Fls. in loose spicate racemes. Calyx equally 2-lobed, the lips entire, becoming inflated and globular in fr. Corolla purple, bilabiate, the upper lip arched, the lower with recurved sides and small lateral lobes. Anthers ciliate, those of upper pair of stamens 2-celled, of lower 1-celled. Nutlets tubercled. Monotypic. (Don Jose *Salazar,* Mexican commissioner on the Boundary Survey.)

1. **S. mexicàna** Torr. Six to 10 dm. tall, intricately branched, with canescent twigs; lvs. green, subglabrous or minutely hispidulous, oblong or broadly lanceolate, 1–1.5 cm. long, short-petioled; racemes 5–10 cm. long; calyx 8 mm. in fl., 2 cm. and papery in fr.; corolla 15–18 mm. long, pubescent without, the throat pale, lips darker; nutlets olive-brown, peltate on a raised gynobase, tessellate-tubercled.—Common in dry washes and canyons, below 5000 ft.; Creosote Bush Scrub, Joshua Tree Wd.; deserts from Inyo Co. to Riverside Co.; to Utah, Tex., n. Mex. March–June.

4. Scutellària L. SKULLCAP

Herbs with opposite entire or toothed lvs. Fls. 1–3 in the axils or in bracted racemes or spikes. Calyx campanulate, gibbous, 2-lipped, the lips entire, the upper with a crest-like projection on the back. Corolla blue, violet or white, well exserted, bilabiate, dilated above the throat, the upper lip arched, entire or emarginate, the lower spreading or deflexed, with lateral lobes small and middle large. Stamens 4, in 2 pairs; anthers ciliate-pilose, the upper pair 2-, the lower 1-celled. Style unequally 2-cleft at apex. Nutlets 4, subglobose or depressed, on an elongate or short gynobase, papillose or tubercled. Ca. 100 spp., of wide distribution. (Latin, a tray or platter, referring to fruiting calyx.)

(Epling, C. Notes on Scutellariae of w. N. Am. Madroño 5: 49–72, 1939. Epling, C. The American species of Scutellaria. Univ. Calif. Pub. Bot. 20: 1–146, 1942.)

Corolla 5–7 mm. long; fls. in slender, axillary bracteate racemes . . . . 9. *S. lateriflora*
Corolla 12–33 mm. long; fls. solitary in axils of lvs.
  Petioles of all but lowermost lvs. 1–3 mm. long; blades truncate-cordate at base.
    Corolla blue, its lower lip and throat glabrous; pubescence of stems curled downward
                                     1. *S. galericulata*
    Corolla whitish, its lower lip and throat pilose; pubescence of stems spreading
                                       4. *S. Bolanderi*
  Petioles 5–30 mm. long; blades of all but lowest lvs. not truncate-cordate.
    Stems loosely villous with spreading hairs . . . . 2. *S. tuberosa*
    Stems with short ± curved hairs.
      Hairs of stem curled downward and closely appressed; stems less than 1 dm. high
                                        3. *S. nana*
      Hairs of stem curved upward or outward, not appressed; stems more than 1 dm. high.
        Corolla white with yellow tinge, 15–20 mm. long; longer stamens attached 5–6 mm.
        above base of corolla-tube . . . . 5. *S. californica*
        Corolla blue with white markings or blue-purple.
          Corolla 10–15 mm. long, the throat closed by the palate . . . . 6. *S. antirrhinoides*
          Corolla 20–30 mm. long, the throat open.
            Calyx glandular-pubescent; median lvs. averaging ca. ⅕ as wide as long
                                   7. *S. siphocampyloides*
           Calyx puberulent, not glandular; lvs. averaging ca. ⅐ as wide as long
                                     8. *S. Austinae*

1. **S. galericulàta** L. [*S. epilobiifolia* A. Ham.] Perennial with slender stolons, the stems mostly 3–6 dm. tall, simple or branched, mostly pubescent with small curled hairs; lvs. ovate-oblong, 3–5 cm. long, truncate-subcordate at base, crenate-serrate; pedicels to 2.5 mm. long; calyx ca. 4 mm. long; corolla blue, 14–20 mm. long, the galea 9–11 mm.; lower stamens attached 8–10 mm. above base; nutlets light brown, rugose, ca. 1.5 mm. in diam.; $2n =$ ca. 32 (Scheel, 1931).—Swamps and wet places, 4000–7000 ft.; Red Fir F., Lodgepole F.; Eldorado Co. to Shasta Co.; to Alaska, Atlantic Coast; Eurasia. June–Sept.

2. **S. tuberòsa** Benth. [*S. t.* var. *similis* Jeps.] Perennial with tuber-bearing rhizomes; stems 0.5–2 dm. tall, usually branched at base, ± viscid with rather long spreading hairs; lvs. ovate, 1–2 cm. long, mostly coarsely dentate, the lower on petioles 0.5–1.5 cm. long; fls. solitary in axils, the pedicels 2–3 mm. long; calyx pilose, the lower lip 4–6.5 mm. long; corolla 15–20 mm. long, the palate pilose, the tube glabrous within below the middle; lower stamens 8–11 mm. long; nutlets black, irregularly and coarsely papillate.— Borders of brush and open woods, below 5000 ft.; Foothill Wd., Chaparral, Yellow Pine F.; Sierran foothills, Fresno Co. n., Coast Ranges from Marin Co. n.; to sw. Ore. March–July. Intergrading freely with:

Ssp. austràlis Epl. Palate glabrous, the tube hairy within.—Dry slopes, below 2000 ft.; Chaparral, Closed-cone Pine F.; coastal s. Calif. to n. L. Calif. and occasional to Santa Clara and Alameda cos. April–May.

3. **S. nàna** Gray. Rootstocks with slender tuberous offshoots; stems 0.4–0.8(–1.2) dm. high, branching at base, cinereous with short recurved hairs; lvs. oblong-elliptical, entire, 1–1.5 cm. long, obtuse, short-petioled; fls. axillary; calyx 3.5–5 mm. long, cinereous, the lower lip ca. 6 mm. long, the upper 3.5 mm.; corolla 16–20 mm. long, puberulent; nutlets black, strongly rugose, somewhat angled.—Dry slopes, 4000–6500 ft.; Yellow Pine F., Sagebrush Scrub, Lodgepole F.; Plumas Co. to Siskiyou and Modoc cos.; to Ore., Ida., Nev. June–Aug.

4. **S. Bolánderi** Gray. Spreading by slender rhizomes; stems 2–4 dm. high, simple or loosely branched; lvs. thin, spreading, deltoid-ovate, 1–2 cm. long, on petioles 0.5–1 cm.

long, the upper ± sessile; fls. few, solitary in axils; calyx at anthesis 4.5 mm. long, pubescent, ± glandular, the lower lip becoming 5–6 mm. long; corolla white, with spreading ± gland-tipped hairs, tube and galea 16–18 mm. long, the lower lip blotched with violet; lower fils. 9–11.5 mm. long, attached 6–7 mm. above the base of the tube; nutlets smoky, rugose, banded, 1 mm. long.—Moist ± gravelly places, 1000–4000 ft.; Foothill Wd., Yellow Pine F.; base of Sierra Nevada from Tulare Co. to Plumas Co. June–July.

Ssp. austromontàna Epl. Corolla 12–15 mm. long, the lower lip not purple dotted.— Rare, 2000–6000 ft.; Chaparral, Yellow Pine F., etc.; interior s. Calif. (Victorville to San Diego Co.)

5. **S. califórnica** Gray. [*S. antirrhinoides* var. *c.* Gray. *S. Bolanderi* var. *c.* Penl.] Rootstocks slender; stems 1–3 dm. high, puberulent with upwardly incurved hairs and ± glandular; lvs. linear-oblong to oblong-ovate, 15–25 mm. long, on petioles 5–20 mm. long, subcrenate to entire, rounded at apex; fls. in upper axils; pedicels ca. 3 mm. long; calyx puberulent, ca. 4 mm. long; corolla white with ± yellow, 15–20 mm. long; lower stamens attached below middle of tube; nutlets black, rugose, obscurely banded.— Drying, ± gravelly places, below 7000 ft.; Douglas-Fir F., Yellow Pine F., Red Fir F., Mixed Evergreen F.; from Alameda and Tuolumne cos. to Siskiyou Co. June–July.

6. **S. antirrhinoìdes** Benth. [*S. viarum* Heller. *S. sanhedrensis* Heller.] Resembling *S. californica*, the lvs. oblong-ovate to oblong-elliptic, 1–2(–3) cm. long; corolla violet-blue, 12–22 mm. long, the throat closed; lower stamens seated near the middle of the tube; nutlets black, rugose, somewhat banded.—Moist or drying places, below 7000 ft.; Coastal Prairie, Red Fir F., Yellow Pine F., Mixed Evergreen F., etc.; Amador and Sonoma cos. n.; to Ore., Ida., Nev. June–Sept.

7. **S. siphocamyploìdes** Vatke. [*S. angustifolia* var. *canescens* Gray.] Stems 2–4 dm. tall, glandular-pubescent, the hairs but little curved; lf.-blades linear-oblong, 1.5–2.5 cm. long, obtuse, narrowed to a ± sessile base, entire or rarely subserrate; fls. axillary, on pedicels 3–5 mm. long; calyx glandular-pubescent, 5–6 mm. long; corolla deep violet-blue, 25–30 mm. long; lower stamens seated above the middle; nutlets black, rugose, obscurely banded.—Dry places, below 5000 ft.; Mixed Evergreen F., Foothill Wd., Yellow Pine F.; base of Sierra Nevada from Butte Co. to Tulare Co., inner Coast Ranges from Alameda Co. to San Benito Co. May–July.

8. **S. Áustinae** Eastw. [*S. linearifolia* Eastw.] Rhizomes slender; stems 1–3 dm. tall, glabrous to puberulent, with short ascending curved hairs; lvs. oblong-lanceolate to -linear, 15–25 mm. long, rounded apically, entire, subsessile; fls. axillary in upper half of plant; pedicels 3–5 mm. long; calyx puberulent, 4.5–6 mm. long; corolla deep violet-blue, 24–29 mm. long, the tube slender, curved upward; lower stamens seated above middle of tube; nutlets black, rugose, obscurely banded.—Gravelly or rocky places, below 6500 (7800) ft.; Yellow Pine F., Foothill Wd., Chaparral; San Jacinto Mts. and Santa Rosa Mts. n. to and along Sierra Nevada to Shasta and Siskiyou cos., thence s. in Coast Ranges to Lake Co. May–July.

9. **S. lateriflòra** L. Glabrous or puberulent above; stems 2–6 or more dm. tall, slender, branched above; lvs. 3–7 cm. long, deltoid-ovate, acutish, crenate-serrate, on petioles 0.5–2 cm. long; fls. in lateral axillary racemes; calyx 1.5–2.5 mm. long at anthesis, hirtellous and glandular; corolla blue, 5–7 mm. long; nutlets smooth, buff, subcompressed.—Bouldin Id., San Joaquin Co.; Saline V., Inyo Co.; to Atlantic Coast, B.C. May–July.

5. Marrùbium L. HOREHOUND

Perennial, mostly woolly herbs, caulescent, mostly with bitter sap. Lvs. rugose, toothed, petioled. Fls. small, white or purple, in dense whorls. Calyx cylindric, 5–10-nerved, regularly 5–10-toothed, the teeth aristate to acute, spreading or recurved, subequal or the alternate smaller. Corolla bilabiate, the upper lip erect, entire or 2-cleft, the lower spreading, 3-cleft, the middle lobe often emarginate. Stamens 4, included, didynamous, the posterior pair shorter; anthers 2-celled. Nutlets smooth, rounded. Ca. 30 spp. of Old World. (From Hebrew word meaning bitter.)

1. **M. vulgàre** L. Branched from base, erect, white-woolly, 2–6 dm. tall; lvs. roundish, ovate, crenate, 1.5–4 cm. long, canescent above, tomentose beneath, on petioles ca. as long as blades; calyx 4–5 mm. long, the calyx-teeth subulate, recurved; corolla white,

5–6 mm. long; $2n = 34$ (Rutland, 1941).—Common pestiferous weed in waste places and old fields, the dried calyx forming a bur; natur. from Eu. Spring and summer.

6. Agástache Clayt.

Tall perennial herbs. Lvs. ovate, serrate, petioled. Fls. in dense sessile whorls which are mostly in continuous or interrupted cylindrical or tapering spikes. Calyx tubular to turbinate, 15-veined or more, 5-dentate, the teeth equal or the 3 posterior ± united, deltoid to subulate, often thin and colored. Corolla rose, violet, or white, with oblique orifice, the upper lip erect, 2-lobed, the lower spreading, 3-lobed. Stamens 4 the lower pair shorter. Nutlets ovoid, smooth with hispidulous apex. Ca. 20 spp., mostly of N. Am. (Greek, *agan*, much, and *stachus*, ear of grain, because of many spikes.)

(Lint, H., and C. Epling. A revision of Agastache. Am. Midl. Nat. 33: 207–230, 1945.)
Lower lf.-surface densely minutely pubescent, the individual hairs obscure in the general felt; lvs. mainly 1–1.5 cm. broad. Ne. Calif. .... 1. *A. parvifolia*
Lower lf.-surface glabrous to pubescent, the individual hairs evident if present; lvs. mainly 3–4 cm. broad. Widely distributed .... 2. *A. urticifolia*

1. **A. parvifòlia** Eastw. Stems slender, 4–8 dm. high, ± branched, thinly puberulent; lf.-blades deltoid or deltoid-ovate, the median 2–3.5 cm. long, mostly acute, truncate or subcordate at base, coarsely serrate, on petioles 1–2 cm. long; infl. compact, tapering, 5–9 cm. long; calyx rose, minutely pubescent, the tube 5–6 mm., the teeth 4–7 mm. long, lance-linear; corolla rose, tubular, 10–11 mm. long; nutlets 1.5–1.8 mm. long.—Lava rock, 3000–5000 ft.; Foothill Wd., Yellow Pine F.; Lassen, Shasta, Siskiyou, and Modoc cos. June–Aug.

2. **A. urticifòlia** (Benth.) Kuntze. [*Lophanthus u.* Benth. *A. glaucifolia* Heller.] Stems several, branched above, 1–2 m. tall, subglabrous; lf.-blades ovate or deltoid-ovate, the median 3.5–8 cm. long, obtuse or acute, with truncate or subcordate base, mostly coarsely serrate; petioles 1–2.5 cm. long; infl. 4–15 cm. long; calyx green or rose, hirtellous or glabrous, the tube 4–7 mm. long, the teeth deltoid-lanceolate, 2.5–5 mm. long; corolla rose or violet, 10–15 mm. long; nutlets brown, dull, ± striate, 1.5–2 mm. long.—Moist places, below 9000 ft.; many Plant Communities but mostly in Montane Coniferous F.; San Jacinto Mts. n., Sierra Nevada, Coast Ranges from San Luis Obispo Co. n., to Del Norte and Modoc cos.; to B.C. and Rocky Mts. June–Aug. Plants from Coast Ranges are mostly more pubescent than those from the Sierra Nevada.

7. Cedronélla Moench

Tall erect herb with 3-parted lvs. Infl. terminal, spicate, bracteate. Calyx broad-tubular, almost straight, 13–15-nerved, with 5-equal pointed teeth. Corolla with cylindrical exserted gradually widened tube and 2-lipped limb. Stamens 4. Nutlets smooth, ovoid. One sp. (Diminutive of Greek, *kedros*, cedar.)

1. **C. canariénsis** (L.) Willd. [*Dracocephalum c.* L.] HERB OF GILEAD. Sweet-scented, ca. 1 m. high, suffrutescent; lvs. trifoliolate, the lanceolate to oblong lfts. 3–8 cm. long, pubescent beneath; calyx ca. 1 cm. long, pubescent; corolla white or violet, ca. 2 cm. long.—Occasional escape from gardens, as in San Francisco region; native of Canary Ids.

8. Népeta L.

Perennial herbs with toothed lvs. Fls. usually white or blue, in crowded whorls. Calyx tubular, often incurved, 15-nerved, 5-toothed, obscurely 2-lipped, the upper teeth usually longer. Corolla dilated in throat, the upper lip erect, notched or 2-cleft, the lower 3-cleft with middle lobe the largest. Stamens 4, didynamous. Nutlets ovoid, compressed, smooth. Ca. 150 spp., of Eurasia. (Latin name of Catnip.)

1. **N. Catària** L. CATNIP. Erect, 3–8 dm. high, with ascending branches; herbage densely canescent; lvs. ovate to oblong, acute, usually cordate at base, coarsely crenate-serrate, the blades 3–9 cm. long, the petioles 1–4 cm. long; calyx urceolate, very pubescent, ca. 6 mm. long, the teeth subulate; corolla whitish, dotted with purple, 10–12 mm. long; $2n = 36$ (Sugiura, 1940).—Occasional as natur. weed; native of Eu. July–Sept.

9. Glecòma L.

Low mostly creeping herbs with petioled ovate to suborbicular lvs. Fls. blue or blue-purple in verticillate clusters. Calyx oblong-tubular, 15-nerved, oblique at throat, unequally 5-toothed. Corolla-tube exserted, enlarged above, the limb 2-lipped, the middle lower lobe enlarged. Stamens 4, didynamous. Nutlets smooth, ovoid. Ca. 6 spp. of Eurasia. (Greek, *glechon*, old name for pennyroyal.)

1. **G. hederàcea** L. GILL-OVER-THE-GROUND. GROUND-IVY. Stems 1–5 dm. long, retrorsely puberulent; lvs. round-reniform, crenate, 1–2.5 cm. broad, long-petioled; fls. axillary, short-pedicelled; calyx 5–6 mm. long, puberulent; corolla 16–22 mm. long, purplish-blue; $2n = 18$, 24, 36 (various auth., 1941–1947).—Natur. in moist shaded places, mostly in n. Calif.; to Wash., Atlantic States; native of Eu. March–May.

10. Prunélla L. SELFHEAL

Low perennial herbs with nearly simple stems from slender rootstocks. Lvs. petioled. Fls. rather small, in 3's in axils of round bractlike membranaceous fl.-lvs. that are imbricated in a dense spike or head. Calyx tubular-campanulate, usually 10-nerved, deeply 2-lipped, the upper lip truncate or with 3 short teeth, the lower cleft into 2 lanceolate teeth. Corolla slightly contracted at throat, 2-lipped, the upper lip erect, arched, entire, the lower reflexed-spreading, 3-cleft with lateral oblong lobes and rounded middle denticulate lobe. Stamens 4, didynamous. Nutlets smooth, obovoid. Ca. 5 spp., worldwide. (Origin of name doubtful, often written *Brunella*.)

1. **P. vulgàris** L. Stems mostly tufted or loosely ascending, 1–5 dm. high, simple or branched; lvs. in basal tufts and cauline, the main ones ovate to ovate-oblong, rounded at base, mostly obtuse, the blades 1–3 cm. long, on somewhat shorter petioles; calyx purplish, 4–5 mm. long; corolla bluish, violet, or lavender, ca. 8–10 mm. long, $n = 16$ (Hruby, 1932).—Weed in lawns, etc.; native of Eu. Summer.

Ssp. **lanceolàta** (Barton) Hult. [*P. pennsylvanica* var. *l.* Barton. *P. v.* var. *l.* Fern.] Stems and lvs. commonly pilose; lvs. lance-ovate to -oblong, 2.5–5 cm. long, mostly acutish, narrowed to short petioles; bracts ± ciliate, often tinged purple; calyx 5–10 mm. long, often with ciliate teeth; corolla 10–20 mm. long, mostly violet; nutlets brown, shining, ca. 2 mm. long.—Moist woods, about ditches, etc., below 7500 ft.; many Plant Communities, but mostly Montane Coniferous F.; most of cismontane and montane Calif.; to Alaska, Atlantic States. May–Sept.

11. Leonòtis R. Br. LION'S-EAR

Herbs or shrubs with lanceolate to ovate, sessile to petioled lvs. Fls. white to orange, in dense axillary whorls. Calyx 8–10-ribbed, the tube funnel-shaped, arched, teeth often acerose-tipped. Corolla-tube as long as calyx or longer, the upper lip long and concave, hairy outside, the lower lip shorter, deflexed, the 3 lobes subequal. Stamens 4, arched. Ca. 12 spp. of S. Afr. and farther n. (Greek, *leo*, lion, and *otis*, ear.)

1. **L. Leonùrus** (L.) R. Br. [*Phlomis L. L.*] Shrubby, 1–2 m. tall; lvs. lanceolate, 4–6 cm. long, coarsely serrate; calyx ca. 12 mm. long; corolla orange, 3.5–4 cm. long, densely pilose; stamens not exserted.—Occasional escape from cult.; native of S. Afr. July–Sept.

12. Phlòmis L.

Shrubs or perennial herbs, ± woolly; lvs. all similar or the upper bractlike. Fls. sessile, few to many in whorls; calyx regular, truncate or 5-toothed. Corolla densely hairy, erect or curved, lower lip spreading with 3 rounded lobes. Stamens 4, one pair often appendaged at base. Nutlets 4, ovoid, 3-angled. Ca. 70 spp., of Old World. (An old Greek name.)

1. **P. fruticòsa** L. JERUSALEM-SAGE. Shrub 6–15 dm. high, widely branched, yellowish-tomentose; lvs. ovate to oblong, entire to crenate, pale green and stellate-pubescent above, white-tomentose and rugose beneath, 2–10 cm. long, the lower short-petioled,

the floral subsessile; calyx ca. 16–18 mm. long, with short cuspidate teeth; corolla yellow, ca. 25 mm. long, densely stellate-pubescent, the upper lip strongly arched, the lower spreading.—Occasional escape from cult.; native of E. Eu.

13. Làmium L. HENBIT

Low herbs, with lvs. mostly toothed, petiolate. Fls. small, verticillate in axillary and terminal clusters. Calyx tubular-campanulate, usually 5-nerved, 5-toothed, the teeth sharp-pointed, equal or the upper longer. Corolla 2-lipped, the tube somewhat longer than the calyx, upper lip ascending and concave, the lower spreading, 3-lobed, the middle lobe emarginate, contracted at the base. Stamens 4, didynamous. Nutlets smooth or tubercled. Ca. 40 spp., of the Old World. (Greek, *laimos*, throat, because of the gaping corolla.)

Upper lvs. sessile or clasping .... 1. *L. amplexicaule*
Upper lvs. petioled .... 2. *L. purpureum*

1. **L. amplexicáule** L. Mostly annual, sparsely pubescent, the stems branched from the base, ± decumbent, 1–4 dm. long; lvs. broadly ovate to roundish, truncate or cordate at base, coarsely crenate, the lower petioled, the upper not, 2–2.5 cm. wide; fls. few, in axillary and terminal clusters; calyx pubescent, ca. 4–5 mm. long, the teeth erect; corolla purple-red, 12–16 mm. long, the tube very slender, the upper lip pubescent; nutlets ca. 2 mm. long, brownish, mottled; $2n = 18$ (Bernström, 1944).—Natur. in waste places, fields, etc.; native of Eu. March–Aug.

2. **L. purpùreum** L. Like the preceding, but upper lvs. short-petioled, broadly ovate, usually acutish at apex, cordate at base; corolla lilac, 10–16 mm. long, the tube rather stout, the lateral lobes reduced to short teeth; $2n = 18$ (Bernström, 1944).—Occasionally natur. in waste places of n. Calif.; to Wash.; native of Eu. April–Sept.

14. Molucélla L. SHELL-FLOWER

Annual herbs with toothed or cut petioled lvs. Fls. axillary, in whorls of 6–10. Bracts subulate, spinulose. Calyx campanulate, with broadly dilated limb. Corolla white, tipped pink, scarcely exserted from calyx, upper lip erect, lower trilobed with middle lobe broad, notched. Stamens 4, ascending under the hood. Nutlets 4, convex on one side, angular on the other, enlarged upward, truncate. Two spp., of Medit. region. (Diminutive of *Molucca*.)

1. **M. laèvis** L. Glabrous, the stems simple or branched below, 3–5 dm. high; lvs. rounded, coarsely toothed, 2–4 cm. long, long petioled; fls. fragrant, in whorls of 6: bracts shorter than calyx; calyx with large spreading membranaceous border, ± 5-angled, ca. 3 cm. broad in fr. and subtended by several spines; corolla included, white, the upper lip arched.—Occasional escape from gardens in cent. Calif.; native of w. Asia. Spring and summer.

15. Stàchys L. HEDGE-NETTLE

Commonly pubescent or hispid herbs. Fls. few to many, borne in whorls in dense or interrupted terminal spikes, or also in upper axils. Calyx usually ± campanulate, 5–10-nerved, 5-toothed, the teeth subequal, erect or spreading. Corolla with cylindrical tube, little or not dilated at throat; upper lip erect, concave, entire or emarginate; lower longer, spreading, 3-lobed, the middle lobe larger, the lateral often deflexed. Stamens 4, didynamous, ascending under the upper lip. Nutlets ovoid or oblong. Ca. 150 spp., mostly of N. Temp. Zone, some of S. Temp. (Greek, *stachus*, ear of grain, hence a spike.)

(Epling, C. Preliminary revision of the Am. Stachys. Fedde Rep. Sp. Nov. Regni Veg. 80: 1–75, 1934.)
A.   Plants annual. Natur. in waste places .... 1. *S. arvensis*
AA. Plants perennial; natives.
    B.   Corolla-tube 13–24 mm. long; calyx 11–15 mm. long. Near the Coast .. 3. *S. Chamissonis*
    BB. Corolla-tube 6–11 mm. long.
        C.   Upper and lower lvs. sessile, the median on petioles 1–4 mm. long. Ne. Calif.
                        5. *S. palustris*

CC. Upper lvs. usually sessile, the lowermost usually with petioles to 1 cm. long.
  D. Ring of hairs on inner surface of corolla-tube horizontal and not indicated on the
     outside by a constriction (except sometimes slightly so in *S. Emersonii*).
    E. Ring of hairs within the corolla-tube near the base of the tube. Coast Ranges
       from San Francisco s. .... 2. *S. bullata*
    EE. Ring of hairs within the corolla-tube near the middle of the tube. From
        Mendocino Co. n. .... 4. *S. Emersonii*
  DD. Ring of hairs on inner surface of corolla-tube oblique and indicated on the outside
      by a constriction of the tube, this most noticeable on the anterior side.
    E. Upper lip of corolla 2–2.5 mm. long; stamens exserted 1–1.5 mm.
                                                                        6. *S. stricta*
    EE. Upper lip of corolla 3–6 mm. long; stamens exserted 2–3 mm.
      F. Pubescence of very slender soft hairs that become tangled and cob-
         webby in age; corolla whitish .... 10. *S. albens*
      FF. Pubescence of straight usually stiffish hairs.
        G. Lf.-blades oblong, narrowed at base, usually silky-strigose; corolla
           whitish .... 9. *S. ajugoides*
        GG. Lf.-blades oval or ovate to oblong, scarcely narrowed at base,
            never silky.
          H. Fl.-whorls forming a continuous spike; corolla whitish
                                                             8. *S. pycnantha*
          HH. Fl.-whorls forming an interrupted spike; corolla pale rose-
              purple .... 7. *S. rigida*

1. **S. arvénsis** L. Hirsute annual; stems freely branched, 2–6 dm. long, decumbent to ascending; lvs. ovate to oblong-ovate, obtuse, serrate, with subcordate to obtuse base, 1.5–2.5 cm. long, the upper short- the lower long-petioled; fls. in upper axils and in short bracted spikes; calyx 5–6 mm. long, purple, the teeth lanceolate and ca. as long as tube; corolla pale purple, 6–10 mm. long, often scarcely surpassing calyx; $2n = 10$ (Lang, 1940).—Sparingly natur. in Marin, Humboldt and Del Norte cos.; native of Eu. Spring.

2. **S. bullàta** Benth. [*S. californica* Benth. *S. acuminata* Greene.] Stems procumbent at base, slender, simple or branched, 4–8 dm. long, sparsely retrorse-hispid on angles, otherwise ± glandular and pubescent; lvs. ovate to oblong-ovate, 3–18 cm. long, mostly obtuse, crenate-serrate, the base subcordate, on petioles up to 6 cm. long; whorls 6-fld., rather distinct; calyx 6–7.5 mm. long, pilose, the teeth triangular, with apical spines; corolla purple, the tube 7–10 mm. long, upper lip 3.5–5.5 mm., lower 6–10 mm. long; stamens exserted 2–4 mm.; nutlets ca. 1.5 mm. long.—Dryish slopes and canyons, below 2000 (4000) ft.; Chaparral, Mixed Evergreen F., Redwood F., Coastal Sage Scrub, etc.; coastal hills from San Francisco Co. to Orange Co., Santa Cruz and Santa Rosa ids. April–Sept.

3. **S. Chamissònis** Benth. [*S. flaccida* Eastw.] Stems stout, simple or branched, 6–10 dm. high, retrorsely hispid with pustulate bristles on angles, ± pubescent on sides; lvs. mostly narrow-ovate, 6–18 cm. long, acute, coarsely crenate-serrate, cordate at base, on petioles to 5 cm. long; fls. mostly 2–5 in a whorl, with all but lowest pair of subtending lvs. bracteate; calyx 11–15 mm. long, densely villous-pubescent and glandular, the teeth deltoid, short-spined; corolla rose-purple, the tube mostly 18–20 mm. long, mostly transversely annulate, upper lip 7–9, lower 12–13 mm. long; stamens exserted 4–5 mm.; nutlets ca. 2.5–3 mm. long, mottled.—Moist places at low elevs.; Redwood F., N. Coastal Scrub, Mixed Evergreen F.; Humboldt Co. to San Mateo Co., San Luis Obispo Co. June–Oct.

4. **S. Emersònii** Piper. [*S. pubens* (Gray) Heller.] Four–12 dm. high, stems mostly simple, hirsute, the hairs on the angles pustulate at base; lvs. ovate-lanceolate, 6–12 cm. long, acute, coarsely crenate-serrate, ± cordate at base, on petioles 2–4 cm. long; fls. 6–20 in whorls in interrupted spikes, with lower bracts foliose; calyx 5–7 mm. long, the lobes deltoid, short-spinose; corolla red-purple, the tube 8–11 mm. long, lower lip 6–8 mm. long and with white spots.—Moist places; Redwood F., Douglas-Fir F. Mendocino to Del Norte cos.; to B.C. June–Aug.

5. **S. palústris** L. ssp. **pilòsa** (Nutt.) Epl. [*S. pilosa* Nutt.] Slender-stemmed, simple or few-branched, 3–6(–9) dm. high, ± villous-pubescent and glandular-pubescent; lvs. sessile or subsessile, lance-oblong, 4–8 cm. long, obtuse, truncate or subcordate at base; spikes 0.5–1.5 dm. long, interrupted except at top; calyx villous, 6–7 mm. long, with lanceolate cuspidate teeth; corolla pale rose, veined with deeper red, the tube ca. 8–9 mm. long, the lower lip 8 mm. long, villous on back.—Wet places, 4000–5000 ft.; Sagebrush Scrub; Modoc Co.; to Alaska, Great Lakes. June–Aug.

6. **S. strícta** Greene. [*S. ajugoides* var. *s.* Jeps.] Stems simple, 3–8 dm. long, strict or ± decumbent, villous-hirsute and glandular; lvs. oblong-deltoid, 5–15 cm. long, obtusish, crenate-serrate, truncate-cordate at base, on petioles to 5 cm. long; fls. 8–12 in whorls, these in spikes interrupted in age; calyx 5–6 mm. long, glandular-hirsute, the deltoid teeth weakly spinose; corolla white, the tube ca. 6 mm. long, the inner hairy ring well below the middle; upper lip less than 2 mm. long, lower 4–5 mm. long; fils. scarcely exserted.—Wet meadows, etc., below 2000 ft.; Foothill Wd.; Coast Ranges from Lake and Sonoma cos. to Mendocino and Glenn cos., foothills of Sierra Nevada from Merced to Butte cos. June–Sept.

7. **S. rígida** Nutt. ex Benth. [*S. bracteata* Greene. *S. ajugoides* var. *r.* Jeps. & Hoov.] Stems erect or ± decumbent, simple or branched above, mostly 6–10 dm. high, villous-hirsute; lvs. deltoid-oblong to lance-oblong, acute, crenate-serrate, rounded or sub-cordate at base, 5–9 cm. long, the lower on petioles 2.5–4 cm. long; spikes interrupted in age, often 1–2 dm. long; fls. 1–3 in axil of each bract; calyx 5–8 mm. long, the teeth narrow-deltoid, ca. as long as tube, villous-hirsute, mucronulate; corolla rose-purple or veined with purple, 12–16 mm. long, the upper lip 3–5.6 mm., the lower 7–9 mm. long; stamens exserted 2–3 mm.—Mostly low moist places, below 8000 ft.; Yellow Pine F., Red Fir F.; from Plumas and Butte cos. n. and from Humboldt and Trinity cos. n., also mts. of Riverside, Orange, and San Diego cos.; to Wash. July–Aug.

KEY TO SUBSPECIES

Petioles of lower lvs. usually 2.5–4 cm. long.
    Stems mostly 6–10 dm. long; lvs. ± oblong to deltoid-oblong. N. Calif. and extreme s. Calif.
                                                          S. rigida
    Stems mostly less than 6 dm. high; lvs. ± ovate to cordate. Coast Ranges, Humboldt Co. to
    San Diego Co. .... ssp. *quercetorum*
Petioles of lower lvs. usually less than 2.5 cm. long; lvs. oblong to oblong-ovate.
    Both lf.-surfaces subglabrous and green or at most thinly hirute. Sierra Nevada and Coast Ranges
    and Napa and Lake cos. .... ssp. *rivularis*
    Both lf.-surfaces softly lanate, silvery. Del Norte Co. .... ssp. *lanata*

Ssp. **quercetòrum** (Heller) Epl. [*S. q.* Heller. *S. Nuttallii* var. *leptostachya* Benth. *S. gracilenta, viarum* and *ramosa* Heller.] Lf.-blades 4–8 cm. long.—Mostly below 5000 ft.; Mixed Evergreen F., N. Coastal Scrub, Foothill Wd., Coastal Sage Scrub, etc.; Coast Ranges from Humboldt Co. to s. Ore. and to n. L. Calif.

Ssp. **rivulàris** (Heller) Epl. [*S. r.* Heller. *S. Prattenii* Durand. *S. littoralis, ingrata* and *striata* Greene. *S. veronicaefolia* Davy ex Jeps.] Plants 1.5–4 dm. high; lvs. 4–9 cm. long, round-truncate at base.—At 2500–7500 ft.; Yellow Pine F., Red Fir F.; Sierra Nevada from Tuolumne Co. n., Warner Mts., Coast Ranges from Napa and Lake cos. to Trinity Co. to Ore., Nev.

Ssp. **lanàta** Epl. Stems 2.5–3 dm. high; lvs. oblong, 5–7 cm. long, lanate, on petioles 3–10 mm. long.—At 1000–2000 ft.; Mixed Evergreen F.; Del Norte Co.

8. **S. pycnántha** Benth. [*S. cymosa* Heller. *S. ajugoides* var. *c.* Jeps.] Strong-scented, simple or branched, 3–10 dm. high, soft-villous and glandular-puberulent; lvs. ovate to lance-oblong, 5–12 cm. long, crenate-serrate, rounded to subcordate at base, obtusish at apex, the lower on petioles to 4 cm. long; spike subcapitate, 4–5 cm. long; calyx villous, 6–8 mm. long, the teeth lance-deltoid, cuspidate; corolla white with purple veins, the tube 6–7 mm. long, the upper lip 3.5–4 mm., the lower 5–7 mm. long.—Moist places, below 3500 ft.; Foothill Wd., Mixed Evergreen F., Closed-cone Pine F., Yellow Pine F.; Marin and Contra Costa cos. to San Luis Obispo and San Benito cos. and in e. Tehama, Butte and Sierra cos. June–Sept.

9. **S. ajugoìdes** Benth. Stems simple and erect, or ± branched below and decumbent, 1–6 dm. long, villous and glandular; lvs. oblong, 1.5–7 cm. long, rounded at apex, narrowed at base, the lower on petioles to 3 cm. long; fls. in whorls of 6, the spikes bracted, approximate or interrupted, 1–2 dm. long in age; calyx 6–8 mm. long, villous, the lanceolate or ovate-deltoid teeth cuspidate; corolla white, pale rose or with purple veins, 10–15 mm. long, the upper lip 5–6 mm., the lower 6–7.5 mm. long.—Moist places, below 2500 ft.; Mixed Evergreen F., N. Coastal Scrub, Closed-cone Pine F., Coastal Sage Scrub; Sonoma and Glenn cos. to Los Angeles and Orange cos. May–Sept.

10. **S. álbens** Gray. [*S. lanuginosa, malacophylla,* and *velutina* Greene. *S. a.* var. *juliensis* Jeps.] Stems stout, 3–10 dm. high, usually branched, ± densely white-woolly;

lvs. 3–12 cm. long, narrowly to broadly ovate, ± cordate at base, villous-tomentose beneath, crenate-serrate, on petioles to 5 cm. long; spikes 1–2 dm. long, ± interrupted in age; calyx 7 mm. long, woolly, the deltoid-ovate teeth cuspidate; corolla white or pinkish with purple veins, the tube 6–8 mm. long, upper lip 3.5–5.5 mm., lower 6–8 mm. long; fils. woolly.—Moist places, below 8000 ft.; V. Grassland and Coastal Sage Scrub to Lodgepole F.; mostly away from coast, Lake and Tuolumne cos. to Riverside Co., White Mts., Inyo Co. May–Oct.

16. Acanthomíntha Gray. THORNMINT

Glabrous or pubescent, aromatic annuals, mostly branched from base, sometimes simple, with entire to denticulate or serrulate lvs. Fls. in distinct or at length remote whorls, each whorl subtended by a pair of lvs. and several conspicuous membranous broad bracts armed with needlelike spines. Calyx 2-lipped; upper lip with 3 aristate teeth, lower with 2 oblong acute or spine-tipped lobes. Corolla 2-lipped, white or tinged with rose, the palate cream; upper lip entire or retuse, ± hooded; lower lip 3-lobed; tube exserted, with funnelform throat. Stamens inserted high on the throat, the lower pair antheriferous, the upper shorter and the anthers smaller or imperfect. Style slender, 2-lobed, the lower longer. Nutlets ovoid, smooth. Three spp. (Greek, *acantha*, thorn, and *mentha*, mint.)

Bracts and axis of infl. glandular-villous; style hairy; bracts mostly definitely longer than wide; upper lip of corolla ± falcate-incurved. Alameda Co. to s. Monterey Co. . . . . 1. *A. lanceolata*
Bracts and axis of infl. ± puberulent, but not glandular; style glabrous; bracts mostly almost as broad as long; upper lip of corolla erect.
    Anthers woolly or pubescent, all 4 polleniferous. San Mateo Co. to Ventura Co. . . . . 2. *A. obovata*
    Anthers glabrous, 2 fertile, 2 obsolete. San Diego Co. . . . . 3. *A. ilicifolia*

1. **A. lanceoláta** Curran. Stems 1–3 dm. long, retrorsely puberulent, ± glandular-villous above; herbage glandular, ill-scented, the lvs. lance-oblong to broadly ovate, 1–2 cm. long, entire to denticulate, the upper spine-tipped; petioles 0.5–1.5 cm. long; bracts ± oblong-ovate, 9–12 mm. long, with aristate prickles divergent and 10–12 mm. long; calyx ca. 12 mm. long including the spinescent teeth; corolla white, or tipped with pink, 2–2.5 cm. long, glandular-pubescent, the upper lip 8–10 mm. long, 2-lobed at tip, the lower lip spreading, almost as long; all 4 anthers fertile, glabrous.—Dry open slopes of serpentine, below 3500 ft.; Foothill Wd., Chaparral; inner Coast Ranges from Alameda and Santa Clara cos. to se. Monterey Co. and San Benito Co. March–June.

2. **A. obováta** Jeps. Stems 1–2 dm. high, minutely canescent; lvs. lance-oblong to obovate, denticulate or the upper spinose-serrate, blades 8–12 mm. long, the petioles often longer; bracts broadly ovate to almost round, sparingly puberulent, the spines 5–8 mm. long; calyx puberulent, to pubescent, 8–9 mm. long; corolla white or with lavender tinge at tips, 1.5 cm. long, the upper lip entire, 4 mm. long, the lower ca. as long; anthers woolly-pubescent, those of upper stamens smaller.—Dry slopes, 1000–5000 ft.; Foothill Wd., Chaparral; inner Coast Ranges from San Benito Co. to Ventura Co. April–June.

    Ssp. **Duttònii** Abrams. Infl. and calyx subglabrous or microscopically puberulent; anthers short-pubescent.—Dry slopes of serpentine; Mixed Evergreen F.; San Mateo Co. May.

3. **A. ilicifòlia** (Gray) Gray. [*Calamintha i.* Gray.] Stems 0.5–1.5 dm. long, glabrous to ± puberulent; lf.-blades 0.5–1.5 cm. long, ovate, serrate-denticulate, on petioles ca. as long; bracts roundish, 7–9 mm. long, the spines 4–6 mm. more; calyx ca. 5 mm. long; corolla 12 mm. long, white with some rose, upper lip 3–4 mm. long, lower 5–6 mm.; antheriferous stamens 2, glabrous or papillate, the other pair abortive.—Clay depressions on mesas and slopes; Chaparral, Coastal Sage Scrub; sw. San Diego Co.; adjacent L. Calif. April–May.

17. Sálvia L. SAGE

Herbs or shrubs, usually strongly aromatic. Lvs. opposite, sometimes mostly basal. Fls. usually in whorls, these in ± interrupted spikes or panicles or racemes. Calyx tubular to campanulate or ovoid, 2-lipped, the upper lip usually concave or arched, entire to

2-lobed, the lower 2-toothed or -cleft, sometimes teeth and lips suppressed. Corolla strongly 2-lipped, the upper usually erect, concave or arched, entire to 2-lobed, the lower lip spreading or drooping, 3-lobed. Stamens inserted in throat, the upper sterile or rudimentary, the lower pair fertile with anther-cells widely separate on a long filamentlike connective which exceeds the fil. itself and is jointed to it, sometimes the lower arm rudimentary or suppressed and the anther deformed or obsolete. Style 2-cleft at summit; ovary deeply 4-parted. Nutlets smooth. A genus of over 500 spp., widely distributed in temp. and warmer regions, some spp. cult. as ornamentals or for flavoring. (Latin, *salveo*, the verb to save, because of medicinal use.)

(Epling, C. The Californian Salvias. Ann. Mo. Bot. Gard. 25: 95–188, 1938.)

A. Both branches of the stamen-connective evident and bearing fertile anthers; lvs. pinnatifid or spinose-sinuate or -lobed (sometimes entire in *S. funerea*).
  B. Plants herbs with lvs. largely basal.
    C. Fls. yellow. Natur. in n. Calif. .... 1. *S. Aethiopis*
    CC. Fls. blue or lavender.
      D. Plants perennial; corolla blue, 10–15 mm. long; lvs. not spiny. Natur. near San Francisco Bay .... 2. *S. verbenacea*
      DD. Plants native annuals.
        E. Corolla lavender, 25–35 mm. long, the lower lip fimbriate; lvs. spiny, tomentose .... 3. *S. carduacea*
        EE. Corolla blue, 12–16 mm. long, the lower lip not fimbriate; lvs. not spiny or tomentose .... 4. *S. Columbariae*
  BB. Plants shrubby, with lvs. all along the current year's twigs.
    C. Calyx subglobose with subequal teeth; lvs. tomentose with long terminal spine and often lateral ones. Mojave Desert .... 5. *S. funerea*
    CC. Calyx elongate with unequal teeth; lvs. with greenish spinulose teeth. Colo. Desert
                                                                                    6. *S. Greatae*
AA. Lower end of stamen-connective obsolete or forming only a tooth, rarely bearing an anther.
  B. Plants perennial herbs.
    C. Fls. 30 mm. long, purplish-red; lvs. 10–20 cm. long .... 7. *S. spathacea*
    CC. Fls. 15–18 mm. long, bluish; lvs. 2–5 cm. long .... 8. *S. sonomensis*
  BB. Plants shrubby.
    C. Infl. with dense glomerules in a ± interrupted spike, the spikes occasionally branched; corolla-tube usually longer than lower lip; upper lip well developed.
      D. Lf.-blades entire, mostly broader in upper half, not rugose; fl.-bracts membranous, usually colored.
        E. Corolla 13–15 mm. long, the tube pubescent within, but the hair not forming a definite band; upper lip of corolla at least half as long as tube
                                                                                    9. *S. Dorrii*
        EE. Corolla 15–22 mm. long, the tube with a definite inner band of hairs; upper lip of corolla ¼–⅕ as long as tube ....10. *S. pachyphylla*
      DD. Lf.-blades crenulate to crenate, broader in their lower half, rugose; bracts often not colored.
        E. Stamens not or scarcely exceeding the upper corolla-lip, lying close under it; corolla 8–12 mm. long.
          F. Lvs. not revolute; corolla-tube with narrow band of hair within.
            G. Lf.-blades oblong-elliptical, mostly 3–6 cm. long, ashy beneath; corolla mostly pale blue to whitish. Widely distributed
                                                                                    11. *S. mellifera*
            GG. Lf.-blades oblong-obovate, mostly 1–2 cm. long, ashy on both surfaces; corolla clear blue. S. San Diego Co. .... 12. *S. Munzii*
          FF. Lvs. revolute; corolla-tube with inner hair ± scattered. Santa Rosa Id.
                                                                                    13. *S. Brandegei*
        EE. Stamens well exserted from upper corolla-lip and not lying close under it; corolla 15–25 mm. long.
          F. Calyx-teeth not tipped with long spines.
            G. Pubescence of simple hairs; calyx-teeth of lower lip evident.
              H. Lvs. green and rugose on upper surface, not ashy beneath.
                I. Fl.-bracts purplish-green, spinulose-tipped; corolla 20–25 mm. long. Colo. Desert .... 14. *S. eremostachya*
                II. Fl.-bracts pale, mucronulate; corolla 15–16 mm. long. Mojave Desert .... 15. *S. mohavensis*
              HH. Lvs. rugose, but tomentose beneath; corolla ca. 2 cm. long. Cismontane San Diego Co. .... 16. *S. Clevelandii*
            GG. Pubescence of short much-branched hairs; calyx-teeth and lips completely united. Orange Co. to San Luis Obispo and Kern cos.
                                                                                    17. *S. leucophylla*
          FF. Calyx-teeth tipped with long spines; lvs. whitish; corolla white, 2 cm. long. Colo. Desert .... 18. *S. Vaseyi*
    CC. Infl. thyrsoid-paniculate with lax few-fld. glomerules; corolla-tube shorter than lower lip; upper lip very short. Cismontane and to desert edge .... 19. *S. apiana*

1. **S. Aethiòpis** L. MEDITERRANEAN SAGE. Rather bushy stout-branched biennials, 3–6 dm. high, floccose-tomentose; lower lf.-blades ovate to orbicular in outline, 1–3 dm. long, on petioles almost as long, incised and dentate; upper lvs. much reduced; infl. a terminal widely spreading panicle, the branches with distinct whorls of fls.; bracts roundish, clasping, subulate-spinose at apex; calyx white-woolly, 8–10 mm. long, the teeth spine-tipped; corolla 15–20 mm. long, pale yellow, the upper lip strongly arched, as long as tube, the middle lobe of lower lip saccate; $2n = 24$ (Felfoldy, 1947).—Natur. about old towns, at 4000–5000 ft., Lassen, Modoc and Plumas cos.; native of N. Afr. June–Aug.

2. **S. verbenàcea** L. Perennial, 1–5 dm. high, villous; principal lvs. in basal rosette, oblong to elliptical in outline, subglabrous, sinuately lobed, 4–6 cm. long, on equally long petioles; cauline lvs. few, often merely toothed, sessile; fls. in a panicle with whorls of 4–8, separated by internodes 1–3 cm. long; calyx ca. 6 mm. long; corolla blue, 10–15 mm. long, the upper lip almost straight; $2n = 54$ (Benoist, 1937).—Waste places, vacant lots, etc. about Berkeley; native of Eu. June–Sept.

3. **S. carduàcea** Benth. [*S. gossypina* Benth.] THISTLE SAGE. Annual, simple or few-branched at base, the stems erect, scapelike from a rosette of basal lvs., 1–5 dm. high, ± white-woolly, bearing 1–4 capitate whorls of fls.; lvs. oblong in outline, sinuate-pinnatifid, spinulose-toothed, 3–10(–15) cm. long, on shorter winged petioles; fls. 1–4 in a whorl, equaled or surpassed by the foliaceous lanceolate spinescent bracts; calyx woolly, 10–15 mm. long, the lobes spine-tipped; corolla lavender, 2–2.5 cm. long, the upper lip 2-cleft, with laciniate or denticulate segms., the lower with small erose lateral lobes and a large fan-shaped fimbriate middle lobe; fils. short, the connectives 12–14 mm. long; anthers vermilion or brick-red; nutlets tan-gray, ± mottled, dorsally flattened, 2.5–3.5 mm. long; $n = 12$ (Stewart, 1939).—Sandy and gravelly places, below 4500 ft.; Creosote Bush Scrub, Coastal Sage Scrub, V. Grassland, etc.; from interior Contra Costa, San Joaquin, and Stanislaus cos. through the Cent. V. and inner Coast Ranges to Kern Co., then through interior cismontane s. Calif. and w. Mojave Desert to n. L. Calif. March–June.

4. **S. Columbàriae** Benth. [*Pycnosphace C.* Rydb.] CHIA. Annual, simple or branching below, 1–5 dm. tall, ± cinereous with mostly retrorse hairs; lvs. mostly basal, finely pubescent, oblong-ovate in outline, the blades 2–10 cm. long, 1–2-pinnatifid into toothed or incised divisions, the petioles ca. as long; upper lvs. reduced; fls. in capitate glomerules, these 1–3, subtended by rounded, glabrous or hispidulous, colored awn-tipped bracts; calyx 8–10 mm. purplish, arcuate, the middle spinose tooth of the upper lip suppressed; corolla blue, 12–16 mm. long, upper lip small, emarginate, lower with small lateral lobes and larger middle lobe ± 2-lobed; nutlets tan-gray, ± mottled, and dorsally flattened, ca. 2 mm. long; $n = 8$ (Stewart, 1939).—Common in dry open disturbed places, below 4000 ft.; occasional to 7000 ft.; Coastal Sage Scrub, Chaparral, Foothill Wd., Creosote Bush Scrub, etc.; inner Coast Ranges from Mendocino Co. s., Sierran foothills from Calaveras Co. s., throughout s. Calif.; to Utah, Ariz., Son., L. Calif. March–June.

S. × bernardìna Parish ex Greene. [*S. Columbariae* var. *b.* Jeps.] Annual to somewhat woody; lvs. with texture of *S. mellifera*, but pinnately lobed, strongly bicolored; upper lip of calyx with 3 spinose-teeth.—Apparent hybrid between *S. C.* and *S. mellifera* and found only where both spp. occur.

5. **S. funèrea** Jones. [*S. f.* var. *fornacis* Jeps.] Compact densely branched shrub, 5–8(–12) dm. tall, white-woolly, densely leafy; lvs. ovate to lance-ovate, 1.5–2 cm. long, subsessile, acuminate-spinose at apex, mostly entire, sometimes with a lateral pair of spiny teeth, fls. mostly 3 in axils of iliciform lvs. and crowded into leafy spikes 3–8 cm. long; calyx white-woolly, 4.5–6 mm. long, with subequal deltoid teeth; corolla violet, 12–16 mm. long, the lower lip 5–6 mm. long, with large erosulate middle lobe.—Dry washes and rocky places, 1000–3000 ft.; Creosote Bush Scrub; w. slope of Amargosa Range, in Grapevine Mts. and n. part of Panamint Mts. March–May.

6. **S. Greàtae** Bdg. Shrubby, much like *S. funerea*, but with lvs. less crowded, less tomentose, the blades 2–3 cm. long and mostly with more than 1 pair of lateral spinulose teeth; whorls of infl. more remote; calyx 9–11 mm. long; corolla lavender, 14 mm. long, the lower lip 3 mm. long.—Dry washes and fans, below 600 ft.; Creosote

Bush Scrub; Orocopia Mts., Riverside Co. to Chocolate Mts., Imperial Co. March–April.

7. **S. spathàcea** Greene. [*Audibertia grandiflora* Benth., not S. *g.* Etling. *Audibertiella g.* Briq. *Ramona g.* Briq.] PITCHER SAGE. Coarse perennial herb with creeping rhizomes, the annual stem stout, glandular-villous, 3–7(–10) dm. high; lvs. numerous below, oblong-hastate, 8–20 cm. long, obtuse, rugose, green above, ± ashy-tomentose beneath, mostly on petioles 3–8 cm. long; stem lvs. reduced upward, subsessile with truncate base; infl. spicate, consisting of several close or remote whorls, coarse, viscid, 1.5–3 dm. long; bracts ovate to lance-ovate, acuminate, 1.5–4 cm. long, mostly purplish; calyx 1.5–2(–3) cm. long, hispidulous within; corolla purplish-red, 3 to almost 4 cm. long, with exserted stamens; nutlets brown, plump, dull, ± reticulate, seemingly gland-dotted, ca. 5 mm. long; *n* = 13 (Stewart, 1939).—Grassy and shaded slopes; below 2000 ft.; Chaparral, Oak Wd., Foothill Wd., Coastal Strand, etc.; Solano Co. s. through Coast Ranges to Orange Co. March–May.

8. **S. sonoménsis** Greene. [*Audibertia humilis* Benth., not S. *h.* Benth. *Ramona h.* Greene. *Audibertiella h.* Briq.] Herbaceous perennial with creeping leafy somewhat matted stems and erect quite leafless flowering stems 1–4 dm. tall; lf.-blades elliptical-obovate, mostly 3–6 cm. long, rounded at tip, attenuate at base into subequal petioles, crenulate, green and rugulose above, closely whitish tomentulose beneath; infl. a spike of 4–6 remote whorls, hispid, ± glandular; bracts 5–8 mm. long; calyx ca. 1 cm. long, lobes of upper lip wholly united, trimucronate; corolla blue-violet, ca. 15 mm. long, the lower lip 7–8 mm. long, with lateral lobes obsolete; nutlets oblong, brownish, ca. 2.5 mm. long; *n* = 16 (Stewart, 1939).—Dry slopes, below 6500 ft.; Chaparr., N. Oak Wd., Yellow Pine F.; foothills of Sierra Nevada from Shasta Co. to Calaveras Co., Coast Ranges, Siskiyou to Napa cos., Monterey and San Luis Obispo cos., San Diego Co. May–June.

9. **S. Dórrii** (Kell.) Abrams. [*Audibertia D.* Kell. *Audibertiella D.* Briq. *Ramona D.* Briq. *A. incana* var. *pilosa* Gray. S. *p.* Merriam. S. *carnosa* var. *p.* Jeps.] Low broad much-branched shrub 3–8 dm. tall; branches densely scurfy-canescent, punctate-glandular, with short axillary fascicles; lvs. round-obovate to spatulate, ca. 7–15 mm. in diam., rounded to emarginate at apex, ± abruptly narrowed at base to a petiole 5–8 mm. long; infl. hispidulous and ± pilose between the 3–4 glomerules which appear ± contiguous in age, 1.5–2.5 cm. in diam.; bracts purplish or greenish, oblong-elliptic to roundish, pilose on backs, conspicuously ciliate; calyx ca. 6 mm. long, the lower lip deeply 2-toothed, the upper entire; corolla blue, ca. 10–12 mm. long, the upper lip erect, 2–3 mm. long, the lower 3-lobed with middle lobe erose; stamens long-exserted, the upper pair short, sterile; nutlets dark brown, ± mottled, ca. 3 mm. long; *n* = 16 (Stewart, 1939).—Dry flats and slopes, 2500–8800 ft., Joshua Tree Wd., Pinyon-Juniper Wd., Sagebrush Scrub; Mojave Desert (Los Angeles and San Bernardino cos.) to Lassen Co.; w. Nev. May–July.

KEY TO SUBSPECIES

Lf.-blades mostly round to spatulate, abruptly narrowed to petioles, mostly 4–15 mm. long.
  Bracts pilose on outer surface; glomerules mostly 3–4, usually approximate. San Bernardino Co.
  to Lassen Co. . . . . *S. Dorrii*
  Bracts usually ± glabrous except for the ciliate margins; glomerules mostly 2–3.
    Glómerules mostly ca. 1.5 cm. in diam., usually 0.5–1.5 cm. distant on the slender rachis.
    Death V. region . . . . ssp. *Gilmanii*
    Glomerules mostly 1.5–2.5 cm. in diam., generally crowded. E. Mono and Inyo cos.
                                                      subsp. *argentea*
Lf.-blades usually oval or elliptical, rarely obovate, mostly 15–30 mm. long, gradually narrowed at base. Siskiyou and Modoc cos. . . . . ssp. *carnosa*

Ssp. **Gilmánii** (Epl.) Abrams. [S. *carnosa* ssp. *G.* Epl.] Lvs. very scurfy-hoary; lvs. 4–7 mm. in diam., on petioles ca. 2.5 mm. long; bracts mostly opaque, glabrous or hispidulous, usually rose; *n* = 11 (Stewart, 1939).—Dry places, 3000–7000 ft.; mostly Pinyon-Juniper Wd.; Panamint, Argus, and Inyo mts.; w. Nev.

Ssp. **argéntea** (Rydb.) Munz. [*Audibertiella a.* Rydb. S. *carnosa* ssp. *a.* Epl.] Lvs. 8–20 mm. in diam.; bracts mostly glabrous, purple or blue, shining.—5000–10,000 ft.; Sagebrush Scrub, Pinyon-Juniper Wd.; Topaz, Mono Co., Nelson Range and Inyo-White Mts., Inyo Co.; to Ida., Utah, Ariz.

Ssp. **carnòsa** (Dougl. ex Jeps.) Abrams. [*S. c.* Dougl. ex Benth. as synonym. *Audibertia incana* Benth., not *S. incana* Mart. & Gal.] Lvs. larger, narrower; bracts opaque, usually greenish, hispidulous-ciliate.—Mostly at 3000–5000 ft.; Sagebrush Scrub, N. Juniper Wd.; Siskiyou and Modoc cos.; to e. Wash., Ida.

10. **S. pachyphýlla** Epl. ex Munz. [*Audibertia incana* var. *pachystachya* Gray. *A. p.* Parish, not Trautv. *Ramona p.* Heller. *S. carnosa* var. *compacta* Hall. *S. compacta* Munz, not Kuntze.] Resembling *S. Dorrii,* but more sprawling; lf.-blades oblanceolate to obovate, 2–4 cm. long, on petioles 5–15 mm. long; whorls in infl. rather crowded, forming a dense spike to 1.5 dm. long; bracts mostly purplish, 1–2.5 cm. long, obovate to oblong, rounded-truncate at apex, ± ciliate, glabrous to hirtellous on back; calyx 9–12 mm. long; corolla dark- to violet-blue, 15–22 mm. long, the tube cylindrical, the lower lip ca. ⅓ as long as tube and with erose middle lobe; stamens well exserted; nutlets brownish, ca. 3 mm. long.—Dry rocky slopes, 5000–10,000 ft.; Pinyon-Juniper Wd., Yellow Pine F.; San Bernardino Mts. to n. L. Calif.; Panamint, Kingston, Clark, New York mts. July–Sept. The sp. hybridizes easily in the Bot. Garden with *S. Clevelandii,* making plants with stature and odor of the latter and with large beautiful blue fls.

11. **S. mellífera** Greene. [*Audibertia stachyoides* Benth. *Audibertiella s.* Briq. *Ramona s.* Briq., not *S. s.* Kunth.] BLACK SAGE. Openly branched shrub 1–2 m. tall, with herbaceous leafy strigulose twigs, ± glandular; lf.-blades oblong-elliptical, 2–6 cm. long, obtuse, green and somewhat rugulose above, cinereous-tomentulose beneath, crenulate, subsessile or on petioles to 12 mm. long; fls. many in compact glomerules, 2–4 cm. in diam. with ovate greenish cuspidate bracts 5–10 mm. long, intervals between glomerules mostly 2–6 cm. long; calyx 6–8 mm. long, villous, ± glandular, the lower teeth free, 1.5–2 mm. long, the upper connate; corolla pale blue or whitish, sometimes lavender, ca. 12 mm. long, the lower lip almost equal to tube, with notched large middle lobe; stamens exserted; nutlets oblong, dark brown, ca. 2 mm. long; $n = 16$ (Stewart, 1939).—Common on dry slopes and benches, below 2000 ft.; Coastal Sage Scrub, Chaparral; Coast Ranges from Contra Costa and w. Stanislaus cos. to n. L. Calif. April–July. It hybridizes freely with *S. apiana, S. Columbariae* (*S.* × *bernardina*) and *S. leucophylla,* sometimes with *S. carduacea* and *S. Clevelandii.*

12. **S. Múnzii** Epl. [*S. mellifera* var. *Jonesii* Munz.] Like *S. mellifera,* but young stems more slender; lf.-blades oblong-obovate, mostly 1–2 cm. long, ashy on both surfaces, the upper hirtellous, the lower with minute appressed hairs; fl.-glomerules 1–1.5 cm. in diam., with oblong-elliptic bracts; calyx ca. 5 mm. long, hirtellous; corolla clear blue, ca. 10 mm. long; $n = 12$ (Stewart, 1939).—Coastal Sage Scrub; San Miguel Mt., San Diego Co. to n. L. Calif. Feb.–April.

13. **S. Brandègei** Munz. [*Audibertia stachyoides* var. *revoluta* Bdg. *S. mellifera* var. *r.* Munz.] Near to *S. mellifera;* the twigs with branching hairs; lf.-blades linear, to linear-oblong, 2–6 cm. long, 2–5 mm. wide, obtuse, subsessile, revolute, green and rugose above, white-tomentose beneath; glomerules few-fld., 1.5–2 cm. in diam. with ovate bracts with branching hairs; calyx 7–8 mm. long; corolla lavender with broader throat, the tube 7–8 mm. long, the lips 3–4 mm. long; stamens included.—Dry places; Coastal Sage Scrub; Santa Rosa Id. Reported also from Santo Tomás, n. L. Calif. April–May.

14. **S. eremostáchya** Jeps. Intricately branched shrub, 6–8 dm. high, the branchlets ashy with spreading glandular hàirs; lf.-blades lance-oblong, 1.5–3.5 cm. long, 4–10 mm. wide, obtuse, subtruncate at base, crenulate, green and rugose above, hispidulous beneath, on petioles 3–8 mm. long; glomerules 2–3 in interrupted spikes, with thin round-ovate bracts; calyx ca. 11 mm. long, lower lobes free, weakly spinose, upper united; corolla pale blue to rose, 2–2.5 cm. long, the lower lip half as long as tube and with large notched middle lobe; stamens exserted; nutlets yellowish-brown, 3 mm. long; $n = 12$ (Stewart, 1939).—Dry rocky and gravelly places, 1200–4500 ft.; Creosote Bush Scrub; w. edge of Colo. Desert. March–May.

15. **S. mohavénsis** Greene. [*Audibertia capitata* Gray, not *S. c.* Schlecht. *Audibertiella c.* Briq. *Ramona c.* Briq.] Compact many-branched shrub, 2–6(–9) dm. tall, the branchlets hispidulous; lf.-blades lance- to ovate-oblong, 1–2 cm. long, obtuse, crenulate, rugose and glandular above, hispidulous beneath, on petioles 5–10 mm. long; capitate whorls mostly solitary, with outer whitish bracts membranaceous and ovate, inner narrower; calyx 8–10 mm. long, hirtellous, the upper teeth joined; corolla pale blue or

lavender, 15–16 mm. long, the lobes oblong, entire; stamens exserted; nutlets light
brown, flat, 2.5–3 mm. long.—Dry rocky washes and canyons, 1000–5000 ft.; Creosote
Bush Scrub, Joshua Tree Wd.; deserts from Little San Bernardino and Sheephole mts. to
Clark and Turtle mts.; to sw. Nev., nw. Son. April–June.

16. **S. Clevelándii** (Gray) Greene. [*Audibertia C.* Gray. *Audibertiella* and *Ramona C.*
Briq.] A fragrant rounded shrub to 1 m. tall, ashy with retrorse hairs; lf.-blades elliptic-
oblong, 1–3 cm. long, obtuse, crenulate, rugose and grayish-tomentulose, on petioles
3–6 mm. long; glomerules 1–3, separated, many-fld., with ovate bracts 7–8 mm. long;
calyx 8–10 mm. long, only the lower teeth free; corolla dark blue-violet, ca. 2 cm. long,
the tube narrow, well exserted, the middle lobe of the lower lip oblong, 3–4 mm. long,
plane, retuse; stamens exserted; nutlets 1.5–2 mm. long, light yellow with small brown
dots; *n* = 16 (Stewart, 1939).—Dry slopes below 3000 ft.; Chaparral; San Diego Co.;
n. L. Calif. April–July. With very characteristic and wafted odor. A good substitute
for *S. officinalis* in cookery.

17. **S. leucophýlla** Greene. [*Audibertia nivea* Benth., not *S. n.* Thunb. *Audibertiella*
and *Ramona n.* Briq.] Much-branched grayish-white tomentose shrub, 1–1.5 m. high;
lf.-blades lance-oblong, or the lower somewhat wider, obtuse, crenulate, rugose, paler
beneath, 2–6 cm. long, on petioles 3–8 mm. long; whorls capitate, 3–5, crowded or
scattered, with oval to oblong bracts ca. 9–11 mm. long; calyx 10–12 mm. long, the
teeth wholly united into 1 lip; corolla rose-lavender, ca. 2 cm. long, the lips subequal,
the lower middle lobe 4–5 mm. long, oblong; stamens exserted; nutlets brownish-gray,
± mottled, rather flat, 3–3.5 mm. long; *n* = 12 (Stewart, 1939).—Common on dry
barren slopes, mostly below 2000 ft.; Coastal Sage Scrub; Orange Co. to San Luis
Obispo and Kern cos. May–July.

18. **S. Vàseyi** (Porter) Parish. [*Audibertia V.* Porter. *Audibertiella* and *Ramona V.*
Briq.] Whitish rounded shrub to 1.5 m. tall; lf.-blades oblong-ovate, mostly 2–6 cm.
long, crowded at base of wandlike branchlets, obtuse, crenulate, on petioles 5–10 mm.
long; glomerules 4–8(–14), in long, interrupted, long-peduncled spikes, with white bracts
tipped with a stout bristle; calyx white with minute hairs, 8–10 mm. long not counting
the terminal bristles, these 3–7 mm. long; corolla white, ca. 2 cm. long, the tube 11–14
mm., the upper lip roundish, retuse, 3–4 mm. wide, the lower lip 7–12 mm. long, with
erose subreniform middle lobe; stamens exserted; nutlets light brown, shining, ± keeled,
2.5–3 mm. long.—Dry rocky slopes and canyons, below 2500 ft.; Creosote Bush Scrub;
w. edge of Colo. Desert from Morongo V. to Mountain Springs; n. Lower Calif. April–
June.

19. **S. apiàna** Jeps. [*Audibertia polystachya* Benth., not *S. p.* Ort. *Audibertiella* and
Ramona p. Briq. *S. californica* Jeps., not Bdg.] WHITE SAGE. Shrubby below, 1–2(–3)
m. tall, white with minute appressed hairs, the growth of the current year of long erect
branches with lvs. crowded at base and flowering above; lf.-blades lance-oblong, 3–9
cm. long, obtuse, crenulate, on petioles 0.5–2 cm. long; fls. few in lax glomerules in
open thyrsoid panicles 5–15 dm. long; bracts ovate-lanceolate, 5–8 mm. long; calyx
5–7 mm. long, the lower teeth mostly free, ovate, ca. 1.5 mm. long, the upper united;
corolla white, usually with some lavender, 12–22 mm. long, the tube 5–7 mm. long,
upper lip entire or retuse, 1.5–2 mm. long, the lower 8–18 mm. long, abruptly bent at
base, closing the orifice and with rounded erose, cupped middle lobe; stamens exserted;
nutlets light brown, shining, keeled, 2.5–3 mm. long; *n* = 16 (Stewart, 1939).—Common
on dry benches and slopes, mostly below 5000 ft.; Coastal Sage Scrub, Chaparral, Yellow
Pine F.; cismontane Calif. from Santa Barbara Co. to n. L. Calif. April–July. A form
along the desert edge with condensed spicate panicles is var. *compácta* Munz. *S. apiana*
hybridizes with most other spp. with which it comes in contact.

18. **Monárda** L. HORSEMINT

Aromatic herbs with erect stems and dentate or serrate petioled lvs. Fls. in dense
capitate bracteate axillary or terminal clusters. Calyx tubular, 15-ribbed, nearly regularly
5-lobed. Corolla 2-lipped, the throat dilated, upper lip narrow, erect or arched, entire
or notched; lower lip spreading, 3-lobed, the middle lobe much the larger. Anther-
bearing stamens 2, ascending close under the upper lip, the upper pair reduced or

wanting; anthers narrow, 2-celled, versatile, with divergent sacs. Styles 2-cleft at apex. Nutlets ovoid, smooth. Ca. 12 spp., of N. Am. (Dedicated to N. *Monardes,* a 16th-century Spanish botanist.)

1. **M. pectinàta** Nutt. Stout annual, 2–4 dm. high, retrorsely puberulent; lvs. narrowlanceolate to oblanceolate, distantly serrulate, ciliate, punctate, 2–4 cm. long; bracts pale, lanceolate, 10–14 mm. long, mucronate-aristate; calyx puberulent, with aristate teeth; corolla yellowish-white to pink, 16–18 mm. long, puberulent; stamens scarcely equaling the upper lip.—Dry slopes, 4000–5000 ft.; Joshua Tree Wd., Pinyon-Juniper Wd.; New York Mts., e. San Bernardino Co.; to Nebr. and Tex. July–Sept.

19. Lepechìnia Willd. PITCHER SAGE

Shrubby or suffrutescent, aromatic. Fls. showy, solitary in axils of reduced upper lvs. and forming short racemes, or 2–6 in a verticil and forming spikes. Calyx campanulate, subequally 5-toothed, often enlarged in fr. Corolla with a broad tube, piloseannulate at base within, 5-lobed, the lobes broad, rounded and plane, ± erect, the upper bifid, the lateral smaller, entire, the middle lower emarginate. Stamens 4, subequal or didynamous; fils. glabrous; anther-sacs divergent. Nutlets ovoid, black, stony, smooth. Ca. 38 spp., of w. S. Am., Calif., Hawaii. (Named for *Lepechin,* a Russian botanist and traveler.)

(Epling, C. A synopsis of the tribe Lepechinieae. Brittonia 6: 352–364, 1948.)
Calyx-teeth deltoid, mostly shorter than the mature tube.
  Lvs. narrowed toward the base, narrowly ovate; pedicels seldom 1.5 cm. long; nutlets puberulent.
  Ventura Co. to Lake Co., Mariposa Co. to Butte Co. 1. *L. calycina*
  Lvs. strongly cordate, broadly ovate; pedicels 1–2.5 cm. long; nutlets glabrous. Santa Ana Mts.
                                                                                   2. *L. cardiophylla*
Calyx-teeth lanceolate or acicular, equal to or longer than the mature tube; nutlets glabrous.
  Lvs. deltoid or subhastate, less often subcuneate, viscid-villous; calyx-teeth lanceolate, membranous. Los Angeles Co. and nearby ids. . . . . 3. *L. fragrans*
  Lvs. narrowed toward base, glabrate and green; calyx-teeth acicular, rigid. S. San Diego Co.
                                                                                   4. *L. Ganderi*

1. **L. calycìna** (Benth.) Epl. in Munz. [*Sphacele c.* Benth. *S. c.* var. *glabella* Gray. *S. gracilis* Eastw. *S. c.* var. *g.* Jeps.] Erect, 3–12 dm. high, pubescent to somewhat woolly; lf.-blades oblong-ovate, obtuse, veiny, punctate, 4–12 cm. long, the lower on petioles 0.5–2(–4) cm. long, the upper sessile; pedicels mostly 5–15 mm. long; calyx at anthesis campanulate, 10–15 mm. long, the lobes variable, deltoid to lanceolatedeltoid, calyx enlarging, reticulate, membranous, 2.5–3 cm. long in fr.; corolla 2.5–3 cm. long, white or pink with purplish blotches and veins; nutlets broadly ellipsoid, ca. 3.5 mm. long.—Open slopes below 3000 ft.; Chaparral, Foothill Wd.; Coast Ranges from Ventura to Lake cos., Sierran foothills from Mariposa to Butte cos. April–June.

2. **L. cardiophýlla** Epl. Near to *L. calycina,* differing in the broader lvs. with cordate base; pedicels 1.5–3.5 cm. long; corolla 3–3.5 cm. long; nutlets glabrous, 3–4 mm. long. —Dry slopes, 2000–4000 ft.; Chaparral; Santa Ana Mts., Orange Co.

3. **L. fràgrans** (Greene) Epl. [*Sphacele f.* Greene. *S. calycina* var. *Wallacei* Gray. *S. cordifolia* Gand.] Like the preceding, but with lvs. deltoid or subhastate, spreadingvillous-tomentose; calyx-teeth lanceolate, villous, longer than the tube; corolla purplish, 2.5–3.5 cm. long; nutlets glabrous, 3.5 mm. long.—Occasional in canyons, below 3000 ft.; Chaparral; San Gabriel and Santa Monica mts., Santa Cruz, Santa Rosa, and Catalina ids. March–May.

4. **L. Gánderi** Epl. Stems hirtellous; lvs. oblong-ovate, cuneate at base on petioles to 1 cm. long; pedicels 1–3 cm. long; calyx teeth rigid, acicular, 6–11 mm. long, the tube 8–10 mm. long; corolla lilac; nutlets glabrous, 2–3.5 mm. long.—Dry slopes, 2500–3500 ft.; Chaparral; Otay Mt. and San Miguel Mt., San Diego Co. June–July.

20. Hedeòma Pers. MOCK PENNYROYAL

Strongly aromatic herbs; the small entire or toothed lvs. sessile or short-petioled. Fls. in small cymules or solitary in axils of upper lvs. Calyx tubular, 13-ribbed, gibbous at base, pubescent, 5-toothed. Corolla blue or purple, 2-lipped, the upper lip erect, entire or 2-lobed, the lower 3-lobed and spreading. Stamens 4, usually only 1 pair fertile. Style 2-cleft at apex, glabrous. Nutlets ovoid, smooth. Ca. 25 spp., all Am. (Greek, *hedus*, sweet, and *osme*, odor.)

(Epling, C., & W. S. Stewart. A revision of Hedeoma with a review of allied genera. Rep. Spec. Nov. Beih. 115: 1–50, 1939.)

1. **H. nàna** (Torr.) Briq. ssp. **califórnica** W. S. Stewart. [*H. thymoides* of Calif. auth.] Perennial, cinereous-puberulent, 1–2 dm. tall, diffusely branched from base; lvs. ovate, 6–10 mm. long, petioled, the upper gradually reduced into bracts; upper axils with 1–2 short, 1–2-fld. peduncles; calyx tubular, 4–5 mm. long, the teeth subulate-aristate; corolla 7–8 mm. long, light purple, the lower lip with a white blotch and purple lined.—Dry rocky slopes, 2800–6000 ft.; Joshua Tree Wd., Pinyon-Juniper Wd.; Providence, New York, Clark, and Kingston mts. of e. Mojave Desert; to Nev. and Ariz. May–June.

21. Melíssa L. BALM

Perennial herbs, upright, branched, with broad dentate lvs. and rather small white or yellowish fls. in small lax clusters. Bracts few, ovate, leaflike. Calyx tubular-campanulate, declined in fr., 13-nerved, nearly naked in throat, bilabiate, the flat upper lip 3-toothed, the lower 2-parted. Corolla with recurved-ascending tube, enlarged above, the upper lip emarginate, the lower 3-lobed, spreading. Stamens 4, didynamous, ascending under the upper lip. Style 2-cleft, with subulate lobes. Nutlets ovoid, smooth. Ca. 4 spp. of Eurasia. (Greek, *melissa*, a bee.)

1. **M. officinàlis** L. GARDEN or LEMON BALM. Pubescent, lemon-scented, 4–8 dm. high; lvs. ovate, obtuse to subcordate at base, obtuse to acutish at tip, 2–6 cm. long; calyx ca. 6 mm. long; corolla white or with lavender tinge, 10–14 mm. long; $2n = 32$ (Reese, 1952).—Waste places, open woods; natur. in cent. and n. Calif.; from Eu. June–Sept.

22. Sature̊ja L.

Perennial herbs or suffrutescent, with entire or toothed lvs. Fls. large to small, solitary, clustered in the lf.-axils. Calyx narrow-campanulate or tubular, 10–13–15-nerved, 5-toothed. Corolla small and little-exserted to larger and well exserted; upper lip erect, lower spreading. Stamens 4, all perfect. Style glabrous or hairy. Nutlets ovoid, smooth. Perhaps 150 spp., widely distributed. (Ancient Latin name used by Pliny.)

Trailing herb; calyx 4 mm. long .... 1. *S. Douglasii*
Erect or decumbent plants; calyx 8–12 mm. long.
  Lvs. 4–6 cm. long; corolla 3–4 cm. long. Los Angeles Co. and n. .... 2. *S. mimuloides*
  Lvs. 1 cm. long; corolla 10–13 mm. long. Riverside and San Diego cos. .... 3. *S. Chandleri*

1. **S. Douglásii** (Benth.) Briq. [*Thymus D.* Benth. *T. Chamissonis* Benth. *Micromeria D.* Benth. *M. C.* Greene.] YERBA BUENA. Trailing evergreen perennial herb, the stems slender, rooting, 2–6 dm. long; lvs. round-ovate, 1.5–2.5 cm. long, short-petioled, crenate, ± pubescent, obtuse at apex, obtuse to subcordate at base; fls. solitary in axils on pedicels 1–1.5 cm. long; calyx tubular; corolla white or ± purplish, 6–8 mm. long, pubescent; nutlets brown, shining, ca. 1 mm. long.—Shaded woods, below 3000 ft.; N. Coastal Scrub, Closed-cone Pine F., Redwood F., Chaparral, Mixed Evergreen F., etc.; Coast Ranges from Los Angeles Co. n.; to B.C., Santa Catalina Id. April–Sept.

2. **S. mimuloìdes** (Benth.) Briq. [*Calamintha m.* Benth. *Clinopodium m.* Kuntze.] Slender-stemmed perennial herb, villous and glandular, 8–15 dm. high; lvs. ovate, coarsely serrate-dentate, 4–6 cm. long, on petioles ca. half as long, upper narrower, subentire; fls. solitary in axils on pedicels 1.5–2.5 cm. long, or with additional short-pedicelled fls.; calyx tubular, 12–14 mm. long; corolla orange to reddish, 3–4 cm. long, ± pubescent; nutlets apparently seldom maturing.—Creek banks, below 6000 ft.; Chaparral, Douglas-Fir F.; Coast Ranges from Monterey Co. to Los Angeles Co. June–Oct.

3. **S. Chándleri** (Bdg.) Druce. [*Calamintha C.* Bdg.] Shrubby, branched, 2–5 dm. high, upper parts pubescent; lvs. round-ovate, 1–2 cm. long, short-petioled, entire to ±

crenate, densely pubescent beneath, puberulent above; fls. 1–5 in axillary clusters; pedicels 1–2 mm. long; calyx tubular-campanulate, 6–7 mm. long; corolla white, pubescent, 10–13 mm. long; nutlets dark brown, cellular-reticulate, ca. 1.2 mm. long.— Rocky canyons, below 2500 ft.; Chaparral; Santa Ana Mts. near Murietta (Riverside Co.) and San Miguel Mt. and Jamul Mt. (San Diego Co.). March–May.

23. Poliomíntha Gray

Hoary-canescent shrub with linear or linear-oblong lvs. Fls. in small axillary clusters toward the ends of the branches. Calyx 15-veined, the tube shaggy-pilose, the teeth subequal, ± connivent, throat strongly annulate. Corolla pale purplish-blue, ± annulate in the tube with coarse ascending hairs. Fertile stamens 2, ascending against the upper lip. Nutlets oblong, smooth. One sp. (Greek, *polios*, hoary and *mentha*, mint.)

1. **P. incàna** (Torr.) Gray. [*Hedeoma i.* Torr.] Six to 10 dm. high; lvs. 1–2 cm. long, nearly sessile; calyx 6–7 mm. long; corolla ca. 12 mm. long.—Collected at 5400 ft., above Cushenbury Springs, n. base of San Bernardino Mts.; Ariz. to Utah, w. Tex. June–July.

24. Pogógyne Benth.

Small aromatic annuals with suborbicular to spatulate lvs. Fls. congested in cymules in axils of bracts, forming dense terminal spikes or the lower whorls distant. Bracts and calyces glabrate to hirsute-ciliate. Calyx deeply 5-cleft, 15-nerved, the 2 lower teeth longer. Corolla blue or purple, tube exserted, the upper lip entire, lower spreading, 3-lobed. Stamens 4, all antheriferous or the upper 2 sterile, ascending under the upper lip; fils. pubescent. Style ± hairy below the branches. Nutlets ± obovoid, brown, concolorous or mottled. Ca. 5 spp. (Greek, *pogon*, beard, and *gune*, female, because of the bearded style.)

(Howell, J. T. The genus Pogogyne. Proc. Calif. Acad. Sci. [IV], 20: 105–128, 1931.)

All 4 stamens with fertile anthers; style conspicuously bearded below the subequal stigma lobes; corolla 9–20 mm. long, plainly exserted.
    Fl.-bracts and calyx-lobes conspicuously hirsute, bristly-ciliate.
        Infl. 1–3 cm. broad; calyx-tube 3–4 mm. long; fl.-whorls in dense spikes. From Kern and San Luis Obispo cos. n. . . . . 1. *P. Douglasii*
        Infl. less than 1 cm. broad; calyx-tube 2 mm. long; fl.-whorls ± remote. San Diego
                                                      3. *P. Abramsii*
    Fl.-bracts and calyx-lobes subglabrous; infl. 1–1.5 cm. broad. San Diego . . . . . . 2. *P. nudiuscula*
Upper stamens sterile; style sparsely hairy immediately below the unequal stigma lobes; corolla 3–12 mm. long, ± hidden in the heads.
    Stems mostly prostrate or ascending, slender; infl. 8–12 mm. broad; corolla 3–5 mm. long
                                                4. *P. serpylloides*
    Stems erect or nearly so, robust; infl. mostly 12–20 mm. broad; corolla 4–8 mm. long
                                                5. *P. zizyphoroides*

1. **P. Douglásii** Benth. [*P. multiflora* Benth. *P. D.* ssp. *minor* and ssp. *ramosa* J. T. Howell.] Stems simple or freely branched from base, 0.5–4 dm., glabrous or puberulent; lvs. elliptic to narrow-oblong, mostly obtuse, entire or coarsely serrate, glabrous, 1–2 cm. long, narrowed into a short petiole; bracts linear to oblanceolate, mostly apically pungent, bristly ciliate; calyx 8–12 mm. long, the tube 3–4 mm. long, glabrous or with pubescent nerves, the lower calyx-teeth 1.5–2.5 times as long as the tube; corolla 1–2 cm. long, lavender or purple, often mottled with pale yellow on palate of lower lip; anthers and upper part of fils. pubescent; style pubescent for 2–6 mm. below the stigma-lobes; nutlets 1–1.7 mm. long.—Low places like dry beds of winter pools, etc., below 3500 ft.; V. Grassland, Foothill Wd., Yellow Pine F., N. Oak Wd., etc.; from Lake and Butte cos. to San Luis Obispo and Kern cos. May–July.

Ssp. **parviflòra** (Benth.) J. T. Howell. [*P. p.* Benth.] Infl. dense and subcapitate; lower calyx-lobes 1–1.5 times as long as tube; corolla 11–15 mm. long.—Low places; Mixed Evergreen F., N. Oak Wd.; valleys in Coast Ranges; Mendocino, Lake, and Sonoma cos.

2. **P. nudiúscula** Gray. Erect and simple, or more commonly with ascending branches from the base, 1–3 dm. tall, minutely strigulose; lvs. spatulate or narrower, obtuse, subglabrous, 1–2 cm. long, short petioled; fl.-whorls distant or crowded into a short

subcapitate spike; bracts linear-oblanceolate, 10–14 mm. long; calyx ca. 8 mm. long, the lobes linear-subulate, the lower 3–5 mm. long; corolla 11–14 mm. long, lavender, sparsely pubescent; upper stamens pubescent, 2–3 mm. long; lower glabrous, 5–6 mm. long; style pubescent 1–4 mm. below the stigma-lobes; nutlets 1.5 mm. long.—Moist flats; Chaparral, Coastal Sage Scrub; mesas from San Diego and Loma Alta, San Diego Co. May–June.

3. **P. Abrámsii** J. T. Howell. Like *P. nudiuscula*, but fl.-bracts and calyx-lobes conspicuously hirsute and bristly-ciliate; calyx 6 mm. long; corolla 10–12 mm. long, rose-purple; nutlets 1–1.5 mm. long.—Beds of dried pools; Chaparral, Coastal Sage Scrub; mesas from San Diego to Miramar. April–June.

4. **P. serpylloìdes** (Torr.) Gray. [*Hedeoma? s.* Torr. *Hedeomoides s.* Briq. *P. s.* ssp. *intermedia* J. T. Howell.] Stems slender, 0.3–2.5 dm. long, puberulent; lf.-blades oblanceolate to oblong-ovate, 0.2–1.2 cm. long, entire or minutely crenulate, on petioles 1–6 mm. long; fl.-bracts spatulate to linear-oblanceolate, ± ciliate, ca. as long as calyx; infl. mostly with lower whorls ± spaced, upper forming a subcapitate spike; calyx pubescent on nerves, the tube 1–3.5 mm. long, lower lobes 2–4 mm., upper 1.5–3 mm. long; corolla lavender or lilac, 3–5 mm. long; lower stamens 0.5–1.5 mm. long, fertile; style lightly pubescent below the stigma-lobes; nutlets round-ovoid, 1–1.3 mm. long.— Moist grassy or brushy places, below 4000 ft.; Foothill Wd., Yellow Pine F., Chaparral, Mixed Evergreen F.; Coast Ranges from Humboldt to San Luis Obispo cos. and Sierran foothills from Tulare Co. to Marysville Buttes, Sutter Co. April–June.

5. **P. zizyphoroìdes** Benth. [*Hedemoìdes z.* Briq.] Stems several, 0.5–3 dm. tall, ± pubescent; lvs. ovate to oblong, entire, 0.5–1.5 cm. long, on slender petioles ca. as long; bracts spatulate to narrow-oblanceolate, ciliate in their lower part; fls. in dense short spikes; calyx glabrous or rarely bristly on the nerves, the tube 3–5 mm. long, the lower lobes 3–6 mm., the upper 1.5–4 mm. long; corolla lavender, 4–8 mm. long; lower stamens fertile, 1–2 mm. long; style 3–7 mm. long, lightly hairy below the branches; nutlets obovoid, 1.6–2.5 mm. long.—Dry beds of winter pools, below 1000 ft.; V. Grassland, Foothill Wd.; Santa Clara and Mariposa cos. to Shasta and Humboldt cos. and at 5000 ft. in Modoc Co.; s. Ore. March–May in valleys, July–Aug. in Modoc Co.

25. Oríganum L. Wild-Marjoram

Perennial or ± shrubby plants with nearly entire fairly small lvs. Fls. purplish or pink, small, in cylindrical or ellipsoid spikes of dense whorls with colored imbricated bracts. Calyx ovoid or campanulate, ca. 13-nerved, 5-toothed, hairy in throat, ± 2-lipped. Corolla 2-lipped, the tube ca. as long as the calyx, the upper lip suberect, slightly notched, the lower longer, of 3 subequal spreading lobes. Stamens 4, didynamous, exserted, with diverging anther-sacs. Style 2-cleft at summit. Nutlets ovoid or oblong, smooth. Ca. 30 spp., all of Old World. (Greek, *oros*, mountain, and *ganos*, ornament.)

1. **O. vulgàre** L. With ± horizontal rootstocks, villous-pubescent, 3–8 dm. high; lf.-blades round-ovate, 1–3 cm. long, petioled, often with small fascicles in the axils; fl.-clusters 3–5 cm. broad, the bracts ovate, purplish, obtuse, ca. 3 mm. long; calyx 2.5–3 mm. long, the teeth ciliate; corolla 5–6 mm. long, with rounded lobes; $n = 16$ (Scheerer, 1942).—Reported from borders of woods and thickets, Santa Cruz Mts.; w. Ore. and e. U.S.; natur. from Eu. Aug.–Oct.

26. Monardélla Benth.

Annual or perennial herbs, some ± woody at base, of pleasant odor. Lvs. rather small, entire or serrate. Fls. borne in terminal heads subtended by broad involucral bracts, which are frequently colored. Calyx tubular, narrow, ca. equally 5-toothed, 10–15-nerved, the teeth mostly erect, ± triangular. Corolla rose to purplish, lavender, or white, the upper lip erect, 2-lobed, the lower 3-lobed, horizontal or declined; lobes linear-oblong. Stamens 4, all fertile, anther-cells divergent. Style unequally 2-cleft at apex. Nutlets broadly oblong, smooth. Ca. 20 spp., w. N. Am. (Diminutive of *Monarda*, which it resembles.)

(Epling, C. Monograph of the Genus Monardella. Ann. Mo. Bot. Gard. 12: 1–106, 1925.)
A. Fls. in rather loose heads; limb of corolla ⅕–½ the length of the tube; calyx 10–25 mm. long.
    B. Corolla 20–45 mm. long; lvs. pubescent.
        C. Corolla 35–45 mm. long, scarlet or yellowish; calyx 20–25 mm. long . 1. *M. macrantha*

CC. Corolla 25–30 mm. long, pale to pinkish; calyx 12–15 mm. long ...... 2. *M. nana*
BB. Corolla 15–18 mm. long; lvs. subglabrous on both surfaces .... 3. *M. Palmeri*
AA. Fls. in dense heads; limb of corolla ½ to ⅔ the length of the tube; calyx 5–10 mm. long.
  B. Lvs. strongly crisped along the margins. Coastal cent. Calif.
    C. Stems ± tomentose with white curly hairs; bracts 7–8 mm. long ...... 12. *M. crispa*
    CC. Stems sparsely strigulose; bracts ca. 1 cm. long .... 13. *M. undulata*
  BB. Lvs. plane, not strongly crisped.
    C. Plants perennial, mostly woody at base.
      D. Outer bracts reflexed, leaflike in texture and shape .... 5. *M. villosa*
      DD. Outer bracts erect, sheathing, not leaflike.
        E. The lvs. with well developed feltlike tomentum on under surface
                                6. *M. hypoleuca*
        EE. The lvs. lacking a feltlike tomentum beneath.
          F. Lvs. 0.5–1.2 cm. long, round-ovate to lance-ovate, ± toothed; plants
             low cespitose dwarfs, mostly less than 1 dm. high. San Gabriel Mts.
                                4. *M. cinerea*
          FF. Lvs. 1-several cm. long, mostly narrower, entire; plants taller.
            G. Stems subglabrous to pubescent, but not silvery-puberulent.
              H. Bracts firm, not membranaceous or chaffy; corolla-lobes blunt
                                7. *M. viridis*
              HH. Bracts membranaceous; corolla-lobes rounded to a point.
                I. Plant quite glabrous, shining, 1–1.5 dm. high. Del
                   Norte Co. .... 8. *M. purpurea*
                II. Plant ± villous or puberulent, if subglabrous than glaucous
                    and not shining.
                  J. Stems and lvs. glabrous or puberulent. Widely dis-
                     tributed .... 9. *M. odoratissima*
                  JJ. Stems and lvs. villous with spreading hairs. Little
                      San Bernardino Mts. .... 10. *M. Robisonii*
            GG. Stems silvery with a dense minute puberulence. S. Calif. and deserts
                                11. *M. linoides*
    CC. Plants annual.
      D. Bracts not white even on tips or margins; calyx-teeth not white-tipped; corolla
         mostly colored.
        E. The bracts puberulent.
          F. Bracts acute, with green secondary veins forming a prominent network.
             Shasta Co. along the Sierra to Kern Co., San Luis Obispo Co. s.
                                14. *M. lanceolata*
          FF. Bracts abruptly acuminate, with secondary veins lacking or not con-
              spicuous. Alameda Co. to Los Angeles Co. .... 15. *M. Breweri*
        EE. The bracts villous throughout. Near Colton, San Bernardino Co.
                                16. *M. Pringlei*
      DD. Bracts white or with white margins; calyx-teeth white or with white tips; corolla
          mostly white.
        E. Bracts with the lateral veins ± parallel to the midvein.
          F. Calyx-teeth erect, triangular-lanceolate; stamens exserted.
            G. Bracts acute, narrowly white-margined; calyx 13-nerved; con-
               nective distinctly notched on lower margin. Nevada Co. to Kern Co.
                                17. *M. candicans*
            GG. Bracts abruptly acuminate with white tips; calyx 15-nerved; con-
                nective entire on lower margin. Mojave Desert .... 18. *M. exilis*
          FF. Calyx-teeth divergent, subulate; stamens included. Merced and Stanislaus
              cos. .... 19. *M. leucocephala*
        EE. Bracts with lateral veins diverging at almost right angles to the midvein;
            interstices with translucent membrane .... 20. *M. Douglasii*

1. **M. macrántha** Gray. [*Madronella m.* Greene.] Perennial from slender woody root-stocks; stems branched or simple, 1–3(–5) dm. long, pubescent with ± recurved hairs; lf.-blades ovate to elliptic, obtuse, 1–3 cm. long, on petioles half as long, entire or nearly so, subglabrous above, pubescent beneath; heads 10–20-fld., 3–4 cm. broad, the bracts purplish, ciliate-villous, 1–1.5 cm. long; calyx 2–2.5 cm. long, glandular-pubescent, the subulate teeth 4 mm. long; corolla scarlet to yellowish, 3–4.5 cm. long, slender-funnelform, sparsely pubescent, the limb 5–6 mm. long; stamens well exserted, anthers 1.3–1.5 mm. long; nutlets straw color, ca. 3 mm. long.—Dry slopes and ridges, 2500–6000 ft.; Chaparral, Yellow Pine F.; Santa Lucia Mts., Monterey Co., San Gabriel Mts., mts. of San Diego Co., to n. L. Calif. June–Aug.
Var. **Hállii** Abrams. [*M. m.* ssp. *H.* Abrams. *M. m.* var. *longiloba* Abrams.] Herbage densely villous-pubescent throughout; corolla frequently yellowish, the limb 6–10 mm. long.—San Gabriel and San Bernardino mts., to Palomar and Santa Ana mts.
2. **M. nàna** Gray. [*M. macrantha* ssp. *n.* Epl. *M. villosa* var. *leptosiphon* Torr. *M. n.* var. *l.* Abrams. *Madronella nana* Greene.] Stems from creeping woody rootstocks, simple or few-branched, 1–2(–3) dm. long, pubescent; lvs. mostly 0.5–1.5 cm. long, ovate,

almost glabrous above, ± cinereous beneath, with shorter petioles; glomerules 2–3.5 cm. broad, the bracts narrow-ovate, 1.5–2 cm. long, membranaceous, pale with purple tinge; calyx 12–15 mm. long, teeth ca. 3 mm. long; corolla pinkish or paler, 2–3 cm. long, pubescent, the lobes 5–6 mm. long; stamens ca. as long, the anthers 0.7 mm. long; nutlets ca. 2 mm. long.—Dry slopes, 4000–6000 ft.; Chaparral, Yellow Pine F.; Laguna and Cuyamaca mts. and n. L. Calif. June–July.

KEY TO SUBSPECIES

Stems with a short ± retrorse pubescence.
  Lf.-blades subglabrous above. Laguna and Cuyamaca mts. ....*M. nana*
  Lf.-blades cinereous above.
    Corolla-lobes acute, narrower at base than toward the middle, ca. 6–8 mm. long. Yellow Pine F.,
    San Jacinto and Santa Rosa mts. ....ssp. *tenuiflora*
    Corolla-lobes often obtuse, oblong, 4–6 mm. long. Desert slopes, San Jacinto and Santa Rosa mts.
                                            ssp. *arida*
Stems villous with spreading hairs. Palomar Mts. .... ssp. *leptosiphon*

Ssp. **tenuiflòra** (Wats.) Abrams. [*M. t.* Wats. ex Gray.] Lf.-blades ca. 1 cm. long, the petioles half as long; corolla-tube ca. 20 mm. long.—At 4500–8000 ft.; Yellow Pine F.

Ssp. **árida** (Hall) Abrams. [*M. macrantha* var. *a.* Hall.] Lf.-blades 5–10 mm. long, the petioles ca. ¾ as long; corolla-tube mostly 15–18 mm. long.—Canyons at 4000–5000 ft.; Pinyon-Juniper Wd.

Ssp. **leptosiphon** (Torr.) Abrams. [*M. villosa* var. *l.* Torr.] Stems and lvs. villous with spreading hairs; corolla-tube 20–25 mm. long, the limb 1 cm. long.—Chaparral and Yellow Pine F.; Palomar Mts., San Diego Co.

3. **M. Pálmeri** Gray. [*M. linoides* var. *P.* Jeps.] Perennial from a slender rhizomatous stem, 1–1.5 dm. high, purplish; lvs. subcoriaceous, subglabrous, 1–2 cm. long, oblong, subentire, obtuse, sessile or short-petioled; glomerules 2.5–3 cm. broad, purplish, the oblong bracts longer than calyces; calyx 9–11 mm. long, 13-veined, the slender teeth 1.5–2 mm. long; corolla 15–17 mm. long, the slender tube twice the limb, lips subequal; nutlets 2.5 mm. long.—Rare; on serpentine; Foothill Wd.; Santa Lucia Mts., Monterey Co. to San Luis Range, San Luis Obispo Co. June–Aug.

4. **M. cinèrea** Abrams. Dwarf perennial, woody at base, 0.5–1 dm. high, cinereous; lvs. narrowly to broadly ovate, 5–12 mm. long, denticulate, soft subvillous, sessile; heads 1.5–2 cm. broad, purplish, the bracts broadly ovate, ca. 8 mm. long; calyx 7 mm. long, the subulate teeth 2 mm. long; corolla rose-purple, 13–14 mm. long, the tube not exserted, the lobes 4 mm. long; stamens exserted.—Dry slopes, 6000–10,000 ft.; Yellow Pine F., Red Fir F.; e. half of San Gabriel Mts. July–Aug.

5. **M. villòsa** Benth. [*M. globosa* Greene. *M. involucrata* Heller.] COYOTE-MINT. Decumbent perennial from a branching woody base, the stems simple or branched, 1–6 dm. long, villous in upper part; lvs. ovate, entire or commonly unevenly serrate, 1–2.5 cm. long, dark green and thinly pubescent above, paler and ± villous-pubescent beneath, on slender petioles half as long; heads 1.5–2 cm. broad, often closely subtended by 1–2 pairs of ordinary lvs., the bracts ovate to broadly lanceolate, 6–9 mm. long, usually purplish; calyx 7–8 mm. long, glandular-villous; corolla purple to pink or whitish, 12–15 mm. long, pubescent, the linear lobes 4–5 mm. long; stamens not much exserted; nutlets 1.5–2.2 mm. long.—Dry rocky or gravelly places below 3000 ft.; Closed-cone Pine F., N. Coastal Scrub, Mixed Evergreen F., Humboldt Co. to San Luis Obispo Co. June–Aug. Plants from w. San Luis Obispo Co. with lvs. densely white-tomentose beneath, are var. *obispoénsis* Hoov. in Jeps.; and from near Cambria, San Luis Obispo Co. with lvs. almost glabrous, are var. *subglàbra* Hoov. Plants from San Luis Obispo Co. to Marin Co., with upper parts ± woolly-pubescent and lvs. likewise, especially underneath, but with hairs simple, are var. *franciscàna* (Elmer) Epl. [*M. f.* Elmer.]

KEY TO SUBSPECIES

Upper parts of plant and lower surface of lvs. pubescent to villous.
  Lvs. ovate to roundish, mostly green above .... *M. villosa*
  Lvs. lanceolate, grayish to cinereous .... ssp. *subserrata*
Upper parts of plant and lower lf.-surfaces puberulent at most.
  Bracts leaflike, not markedly ciliate .... ssp. *Sheltonii*
  Bracts membranous, purple, ciliate .... ssp. *neglecta*

Ssp. **subserràta** (Greene) Epl. [*M. s.* Greene. *M. tomentosa* Eastw. *M. mollis* Heller. *M. villosa* var. *interior* Jeps.] Stems villous or pubescent above; lvs. lanceolate, the upper canescent; corolla 15–18 mm. long.—Dry ridges; N. Oak Wd., Foothill Wd., Yellow Pine F.?; Coast Ranges from Mendocino Co. to Monterey Co., Sierran foothills from Amador Co. to Tuolumne Co.; s. Ore.

Ssp. **Sheltònii** (Torr.) Epl. [*M. S.* Torr. *M. v.* var. *glabella* Gray. *M. coriacea* Heller.] Stems puberulent or glabrous above; lvs. ovate to lanceolate, puberulent to subglabrous, mostly 2–2.5 cm. long, on petioles 2–5 mm. long; bracts lanceolate, reflexed, short-pubescent; corolla 12–20 mm. long.—Dry places, often on serpentine, below 6000 ft.; Foothill Wd., Douglas-Fir F., Redwood F., etc.; Coast Ranges from Monterey Co. n. and Sierra Nevada from Tulare Co. n.; to s. Ore.

Ssp. **neglécta** (Greene) Epl. [*M. n.* Greene.] Stems purple, puberulent to glabrous; lvs. 1–1.5 cm. long, glabrous at least above; bracts pubescent to glabrous, strongly ciliate.—On serpentine; Chaparral; Marin and Sonoma cos.

6. **M. hypoleùca** Gray. [*M. robusta* Elmer.] Suffrutescent perennial from creeping rootstocks, 2–5 dm. tall, simple or branched, pubescent; lvs. narrowly to broadly rhomboid-lanceolate, 2–4 cm. long, entire, obtuse, green and mostly glabrous above, white beneath with feltlike tomentum, slightly revolute, on petioles 3–10 mm. long; heads 3–4 cm. broad; bracts ovate, tomentose, 8–12 mm. long; calyx 6–7 mm. long, with triangular-ovate teeth; corolla white to pale lavender, 15–16 mm. long, the limb 5–6 mm. long; stamens well exserted.—Occasional on dry slopes, below 4500 ft.; Chaparral; mts. from Santa Barbara to Orange cos. July–Sept.

Ssp. **lanàta** (Abrams) Munz. [*M. l.* Abrams.] Branchlets villous to lanate; lvs. short-pubescent to lanate above, white-tomentose beneath.—Chaparral; e. San Diego Co. June–July.

7. **M. víridis** Jeps. [*M. ledifolia* Greene.] Perennial from creeping rootstocks, suffrutescent; stems branched below, 1–3 dm. high, puberulent; lvs. rhomboid-lanceolate, 1–2 cm. long, subglabrous and green above or ± pubescent toward the base, white-strigose beneath, subentire, revolute, on petioles 2–3 mm. long; heads 1.5–2 cm. broad; bracts membranous to subfoliar, lance-ovate, ca. 7 mm. long, villous on margins; calyx 7–8 mm. long, villous or shaggy-pubescent, with triangular-subulate teeth; corolla 14–16 mm. long, rose-purple, the tube scarcely exserted; stamens ca. as long as corolla.—Dry slopes and ridges, below 1500 ft.; Foothill Wd.; Napa and Lake cos. July–Sept.

Ssp. **saxícola** (Jtn.) Ewan. [*M. s.* Jtn. *M. linoides* var. *s.* Jeps.] Lvs. 1.5–3 cm. long, glabrous above or nearly so; heads 2.5–3 cm. broad; calyx 8–10 mm. long; stamens well exserted.—Dry rocky places, 1700–6000 ft.; Chaparral, Yellow Pine F.; San Gabriel Mts. June–Sept.

8. **M. purpùrea** Howell. Stems branched at base from a woody caudex, purplish, minutely puberulent, 1–1.5 dm. high; lvs. oblong-ovate to -lanceolate, 1.5–3 cm. long, obtuse, glabrous and shining, deep green, short-petioled; heads 1.5–2 cm. broad; bracts ovate-oblong to ovate, glabrous or nearly so, but ciliate, purplish, 1–1.5 cm. long; calyx 8–9 mm. long, purplish, the subulate teeth hirsute; corolla rose-purple, ca. 2 cm. long, with well exserted tube, lobes linear, 6–7 mm. long.—Dry open slopes, 1400–4000 ft.; Mixed Evergreen F., Yellow Pine F.; Del Norte Co.; sw. Ore. July–Sept.

9. **M. odoratíssima** Benth. Branched perennial, ± woody at base, 1.5–3.5 dm. high, thinly pubescent toward tips; lvs. lanceolate, entire, 1.5–3 cm. long, green on both surfaces but ± short-pubescent under a lens, tapering into a subsessile base; heads 1.5–2.5 cm. broad; bracts roundish to ovate, pubescent, mostly obtuse, ciliate, ca. as long as calyx; calyx 6–8 mm. long, woolly-pubescent around the ovate teeth; corolla ca. 15 mm. long, pale purple, exserted, the tube retrorsely puberulent, lobes lanceolate; stamens retrorsely pubescent at base, exserted; nutlets oblong-ovoid, ca. 2 mm. long.—Wash., Ore. etc. No California plants probably belong here, but among the following sspp.:

A. Lvs. appearing nearly or quite glabrous, ± glaucous, mostly over 2 cm. long.
    B. Bracts roundish to ovate, pubescent .... *M. odoratissima*
    BB. Bracts oblong to elliptical, puberulent .... ssp. *glauca*
AA. Lvs. distinctly pubescent.
    B. Bracts usually exceeding calyces, short-acuminate .... ssp. *australis*
    BB. Bracts not longer than calyces, acute to obtuse.
        C. Heads 0.5–1.5 cm. broad, rarely 2 cm. .... ssp. *parvifolia*
        CC. Heads 2–3 cm. broad, rarely less.

D.   Calyces woolly; pubescence ca. the same on both lf.-surfaces . . . . ssp. *pallida*
DD.   Calyces hirsute; pubescence on lower lf.-surface mostly longer and softer than on
upper . . . . ssp. *pinetorum*

Ssp. glaùca (Greene) Epl. [*M. g.* Greene. *M. modocensis, ovata,* and *rubella* Greene. *M. o.* var. *Follettii* Jeps.] Branchlets puberulent, glaucous, often purple; lvs. lance-ovate to oblong, 1.5–4 cm. long, puberulent under a lens, on petioles 1–5 mm. long; bracts purplish; calyx pubescent; corolla reddish purple, 1–2 cm. long.—Dry slopes, mostly 3000–10,500 ft.; Montane Coniferous F., Sagebrush Scrub; Fresno Co. to Siskiyou and Modoc cos.; Ore., Nev. June–Aug.

Ssp. austràlis (Abrams) Epl. [*M. a.* Abrams.] Branches decumbent or ascending, subvillous; lvs. lanceolate to oblong, green or cinereous, 1–2.5 cm. long, on petioles 1–3 mm. long; bracts lanceolate, longer than calyces; calyx ca. 8 mm. long; corolla rose-purple, 1.5 cm. long.—Dry slopes, 4500–9500 ft.; Montane Coniferous F.; San Gabriel Mts. to San Jacinto Mts. June–Aug.

Ssp. parvifòlia (Greene) Epl. [*M. p.* Greene. *M. muriculata* Greene.] Branches slender, puberulent and ± muriculate; lvs. lanceolate to oblong, 1–2 cm. long, on petioles 1–3 mm. long, cinereous; heads 1–2 cm. broad; bracts purplish, seldom longer than calyces; calyx 5–6 mm. long, pubescent; corolla ca. 1 cm. long, rose-lavender.—Dry slopes, 8000–10,500 ft.; largely Subalpine F.; Tulare Co. to Tuolumne and Mono cos.; to s. Rocky Mts. July–Aug.

Ssp. pállida (Heller) Epl. [*M. p.* Heller. *M. o.* var. *ovata* Jeps., in large part.] Branches scurfy-pubescent, cinereous; lvs. lance-oblong, 2–3 cm. long, on petioles 2–8 mm. long, minutely cinereous-puberulent; heads 1.5–2.5 cm. broad; bracts broadly ovate, inconspicuous, short-pubescent, purplish, often decurved; calyx woolly, 5–7 mm. long; corolla pale, often whitish, 1–1.5 cm. long.—Dry slopes below 11,000 ft.; Montane Coniferous F.; White Mountains, Sierra Nevada from Tulare Co. n., to Siskiyou Co. and Yolla Bolly Mts.; to Ore., Nev. July–Aug.

Ssp. pinetòrum (Heller) Epl. [*M. p.* Heller.] Branchlets soft-pubescent; lvs. lanceolate to ovate, 1.5–2.5 cm. long, soft-pubescent, on petioles 2–8 mm. long; heads 1.5–3 cm. broad; bracts ovate, short-pubescent, purplish, not conspicuous; calyx pubescent, 7–8 mm. long; corolla rose, 1–1.5 cm. long.—Dry slopes, 4000–5000 ft.; Yellow Pine F.; Lake and Glenn cos., and Sierra Nevada of Tulare and Fresno cos. July–Sept.

10. **M. Robisònii** Epl. in Munz. Perennial with branching woody caudex, stems 3–5 dm. high, cinereous-hirtellous with short spreading hairs; lf.-blades lanceolate to narrowly ovate-elliptic, cinereous-hirtellous, 6–15 mm. long, with petioles 1–3 mm. long; heads 1–2 cm. broad; bracts membranaceous, pallid, ovate, as long as calyces; calyx 7–8 mm. long, the teeth narrow-deltoid, 1–1.5 mm. long; corolla pale, tube slightly exserted; stamens exserted.—About rocks, 3800–4500 ft.; Joshua Tree Wd.; Little San Bernardino Mts. June.

11. **•M. linoìdes** Gray. [*M. anemonoides* Greene.] Stems several from a woody base, rather dense, erect, 4–6 dm. high, silvery with fine retrorse appressed pubescence; lvs. linear to linear-oblong, entire, thick, minutely silvery-pubescent, 1.5–2 cm. long, subsessile or short-petioled; heads mostly 2–3 cm. broad; bracts broadly ovate, membranous, whitish, puberulent, soft-ciliate, 8–15 mm. long; calyx ca. 8 mm. long, short-pubescent, the teeth with longer hairs; corolla pale rose-lavender, 12–15 mm. long; stamens exserted.—Dry slopes, 3000–9500 ft.; Pinyon-Juniper Wd. to Red Fir F.; desert slope of Sierra Nevada in Mono and Inyo cos., desert slopes, Little San Bernardino and San Jacinto mts. to Laguna Mts., mts. of Mojave Desert; to Nev.? June–Aug.

KEY TO SUBSPECIES

Lvs. silvery-gray.
The lvs. linear to linear-lanceolate, 1.5–2.5 cm. long. Desert slopes . . . . *M. linoides*
The lvs. lance-oblong, 1–1.5 cm. long. Ventura and Kern cos. . . . . ssp. *oblonga*
Lvs. greenish.
The lvs. 2–4 cm. long. W. San Diego Co. . . . . ssp. *viminea*
The lvs. 1–1.6 cm. long, San Gabriel and San Bernardino mts. . . . . ssp. *stricta*

Ssp. oblónga (Greene) Abrams. [*M. o.* Greene.] Bracts ovate, pale purple, 1.5–2 cm. long.—Dry slopes, 3000–7000 ft.; Yellow Pine F., Foothill Wd.; mts. of Ventura and Kern cos.

Ssp. vimínea (Greene) Abrams. [*M. v.* Greene.] Plants less silvery, stems puberulent;

lvs. narrowly linear-lanceolate; bracts lanceolate, greenish-white with tips often rose, to 14 mm. long.—Rocky washes below 1000 ft.; Coastal Sage Scrub, Chaparral; sw. San Diego Co.

Ssp. **stricta** (Parish) Epl. [*M. l.* var. *s.* Parish. *M. epilobioides* Greene. *M. e.* var. *erecta* Abrams.] Lvs. mainly linear-lanceolate, greenish; bracts lanceolate, purplish, acuminate, ca. 1 cm. long.—Dry banks and slopes, 5000–9000 ft.; Yellow Pine F., Red Fir F.; San Gabriel and San Bernardino mts.

12. **M. crispa** Elmer. [*M. undulata* var. *c.* Epl.] Suffrutescent perennial, bushy, the branches ± decumbent, 2–4 dm. long, ± tomentose with white curly hairs; lvs. oblong-spatulate, 1.5–2.5 cm. long, often crisped on margins, short-pubescent, fleshy, with petioles to ca. 8 mm. long and with fascicles in axils; heads ca. 2–3 cm. broad; bracts round-ovate, 7–8 mm. long, membranous and tinged with reddish-purple; calyx 6 mm. long; corolla purplish, 10–12 mm. long; stamens well exserted.—Dunes and back beaches; Coastal Strand; San Luis Obispo and Santa Barbara cos. n. of Point Conception. April–Aug.

13. **M. undulàta** Benth. Annual, with slender taproot, diffusely or loosely branched, usually from near the base, 2–4 dm. high; stems reddish-brown, subglabrous to minutely strigose; lvs. green, oblong-oblanceolate to almost linear, crisped on margins, 2–5 cm. long, short-petioled, subglabrous; heads 2–3 cm. broad; bracts round-ovate, membranous between the green veins, ± purplish toward tips, subglabrous to villous, ca. 1 cm. long; calyx 6–9 mm. long, ± villous, with deltoid-lanceolate teeth; corolla rose-purple, 15–20 mm. long; stamens well exserted; nutlets dull yellow to brownish-mottled, ca. 1.4 mm. long.—Sandy places, below 500 ft.; Coastal Strand, N. Coastal Scrub? Closed-cone Pine F.? near the coast from Marin Co. to Santa Barbara Co. May–July.

Var. **frutéscens** Hoov. Perennial with thick woody taproot, often not flowering until the second year.—Sandy fields, coastal San Luis Obispo Co.

14. **M. lanceolàta** Gray. [*M. sanguinea* Greene.] Erect annual, 2–5 dm. tall, simple or branched, glabrous below, puberulent above; lf.-blades lanceolate to lance-oblong, entire, obtuse, 1.5–5 cm. long, on petioles 5–15 mm. long; heads 1.5–3 cm. broad; bracts lance-ovate, acute, scabrous, membranous but green, often purplish toward tips, 5–15 mm. long, pinnately veined; calyx 6–8 mm. long, pubescent, the teeth ciliate; corolla rose-purple or paler, 12–15 mm. long, the tube puberulent; stamens well exserted; nutlets brownish, ± mottled.—Locally common in dry places, especially disturbed areas, below 8000 ft.; many Plant Communities; Sierra Nevada from Shasta Co. to Kern Co., San Luis Obispo Co. to San Diego Co.; w. Nev. May–Aug. A form from the San Gabriel Mts., with upper part of plant with stalked glands, is var. *glandulífera* Jtn. Another with heads 1–2 cm. broad, from Laguna Mts., San Diego Co. to n. L. Calif., is var. *microcéphala* Gray.

15. **M. Brèweri** Gray. [*M. Elmeri* Abrams.] Erect annual, 1.5–3 dm. tall, branched, puberulent above, usually purplish; lvs. lanceolate, 1.5–4.5 cm. long, puberulent, short-petioled; heads 2–3 cm. in diam.; bracts broadly ovate, 1–1.5 cm. long, abruptly acuminate, with 5–9 subparallel hispidulous veins, purplish above; calyx 6–8 mm. long, ± hirsute, the teeth lanceolate-deltoid; corolla rose, 12–14 mm. long; stamens well exserted.—Sandy flats, below 4500 ft.; largely Foothill Wd.; inner Coast Ranges from Alameda Co. to n. Los Angeles Co. May–Aug.

16. **M. Prínglei** Gray. Cinereous-puberulent annual, branched, 2–4 dm. high; lvs. oblong to lance-ovate, 1.5–3.5 cm. long, ± finely pubescent, short-petioled, entire; heads 2–2.5 cm. broad; bracts ovate, 8–10 mm. long, abruptly acuminate, with 5–7 subparallel finely villous veins; calyx 6–7 mm. long, pubescent with slender hirsute teeth; corolla rose, 11–13 mm. long; stamens exserted.—Sandy places; Coastal Sage Scrub; near Colton, San Bernardino Co. May–June.

17. **M. cándicans** Benth. Erect annual, mostly simple below, branched above, purplish, retrorsely puberulent, 2–4 dm. high; lvs. lanceolate to lance-oblong, puberulent to subglabrous especially above, 2–4 cm. long, short-petioled; heads 1.5–2.5 cm. broad; bracts broadly ovate, acute, 8–10 mm. long, whitish with greenish hirsutulous veins and an evident network of secondary veins; calyx ca. 5 mm. long, villous above, with white scarious teeth; corolla white or with purple dots, 10–11 mm. long, the limb 4–5 mm. long; stamens slightly exserted; nutlets mottled, ca. 1.5 mm. long.—Sandy places, below 2500 ft.; Foothill Wd.; Sierran foothills, from Nevada Co. to Kern Co. May–July.

18. **M. éxilis** (Gray) Greene. [*M. candicans* var. *e.* Gray.] Near to *M. candicans*, but

with more obvious peduncles without green lvs. just below the heads; bracts often purplish, abruptly acuminate, with few or no secondary veins; corolla limb shorter, 2.5–3 mm. long.—At 2000–3500 ft.; Joshua Tree Wd.; w. Mojave Desert. May–June.

19. **M. leucocéphala** Gray. Erect annual, 1–2 dm. high, cinereous-puberulent, dichotomously branched throughout; lvs. lanceolate to ± oblong, 1–1.5 cm. long, pubescent, short-petioled; heads 1.5 cm. broad; bracts ovate to round, pure white even on the sparsely hispidulous veins; calyx 5–6 mm. long, 15-nerved, with white teeth spreading or recurved and ± hirsute; corolla white, 5–6 mm. long; stamens included; nutlets ca. 2 mm. long.—Sandy places; V. Grassland; Merced and Stanislaus cos. June–July. Almost extinct.

20. **M. Douglásii** Benth. Erect annual, simple or divaricately branched, 1–3 dm. high, puberulent; lvs. lanceolate to narrow-oblong, 1–3 cm. long, strigose, short-petioled; heads 1.5–3 cm. broad; bracts lance-ovate, 1–1.5 cm. long, acuminate, pubescent, the veins purple, intervening spaces translucent; calyx 7–9 mm. long, pubescent to hirsute, with rigid subcuspidate teeth; corolla reddish-purple, 11–12 mm. long, at least some of the lobes ending in glands; stamens ± exserted.—Slopes below 3500 ft.; largely Foothill Wd.; inner Coast Ranges from Contra Costa Co. to se. Monterey Co. June–July.

Var. **venòsa** (Torr.) Jeps. [*M. candicans* var. *v.* Torr. *M. D.* var. *Parryi* Jeps.] Stouter; bracts broadly ovate, 15–18 mm. long.—Known from very few collections from plains of e. side of Sacramento V. in Butte and Sutter cos.

27. Pycnánthemum Michx. MOUNTAIN-MINT

Aromatic perennial herbs, mostly with branching erect stems. Whorls remote, many-fld., leafy-bracted. Calyx ovoid to cylindric, with equal teeth and 10–13-nerved. Corolla bilabiate, short, white or purple-dotted, the upper lip entire to emarginate, the lower 3-cleft with obtuse lobes. Stamens 4, didynamous; anther-sacs parallel. Nutlets smooth, pubescent or roughened. Ca. 17 spp., of N. Am., mostly in the e. (Greek, *pychnos*, dense, and *anthemon*, fl.)

1. **P. califórnicum** Torr. [*Koellia c.* Kuntze.] Stems from creeping rootstocks, canescent, simple or few-branched, 6–10 dm. tall; lvs. ovate to lance-ovate, finely pubescent, the upper canescent, serrulate, 3–8 cm. long, sessile or subsessile; heads 1–4, compact; calyx 4–5 mm. long, pubescent, the teeth woolly-villous; corolla 6–7 mm. long, white, the throat hairy, lobes ca. 2 mm. long; nutlets smooth, narrow-oblong, subterete, yellow-brown, ca. 1.3 mm. long.—Moist places in canyons, etc., below 5500 ft.; Chaparral, Yellow Pine F., N. Oak Wd., Foothill Wd., etc.; San Diego Co. to Humboldt, Siskiyou, and Shasta cos. June–Sept.

28. Lỳcopus L. WATER-HOREHOUND

Perennial herbs with slender rootstocks and erect or diffuse stems. Lvs. opposite, punctate, toothed to pinnatifid, usually petioled. Fls. small, white or whitish, in sessile densely capitate glomerules in axils of upper scarcely reduced lvs. Calyx campanulate, 10-ribbed, regular or nearly so, 4–5-lobed. Corolla funnelform or campanulate, nearly regular, the upper lip slightly broader, entire or notched, the lower 3-lobed. Fertile stamens 2, anterior, the posterior pair wanting or rudimentary; anther-sacs parallel. Nutlets smooth, truncate at top, narrowed below. Ca. 15 spp., of N. Temp. regions. (Greek, *lukos*, wolf, and *pous*, foot.)

Calyx-teeth subulate- or cuspidate-tipped, longer than the nutlets.
  Lower and main lf.-blades tapering to petioles or petiolelike bases; lvs. unevenly incised
                                        1. *L. americanus*
  Lower and main lf.-blades sessile, evenly serrate .... 2. *L. lucidus*
Calyx-teeth ± ovate, almost obtuse, not longer than nutlets .... 3. *L. uniflorus*

1. **L. americànus** Muhl. [*L. lacerus* Greene.] Plant with stout nontuberous stolons; stem erect, slender, 2–9 dm. high, acutely 4-angled, glabrous or nearly so; lvs. lanceolate or oblong, acuminate, 3–10 cm. long, irregularly incised or pinnatifid, short-petioled, the upper narrow and merely sinuate; calyx-teeth with long subulate tips; corolla 2 mm. long; nutlets 1–1.5 mm. long, with entire or slightly undulate angles, the dorsal angular face soft, dark, summit entire; $2n = 22$ (Ruttle, 1932).—Wet places, below 2000 ft.;

many Plant Communities; Los Angeles, Riverside, Orange, and San Bernardino cos., Cent. V., Humboldt, Siskiyou, and Butte cos.; to B.C., Atlantic Coast. Aug.–Sept.

2. **L. lùcidus** Turcz. ex Benth. [*L. maritimus* Greene.] Perennial with stolons; stems stout, 4–8 dm. high; herbage mostly puberulent; lvs. lance-oblong to lanceolate, 4–8 cm. long, acute to acuminate, subsessile, coarsely but evenly toothed; calyx-lobes lance-subulate, acuminate, hispidulous on margins; corolla scarcely longer than calyx; nutlets with inner angle granulose to base.—Swampy places, below 1000 ft.; many Plant Communities; from Solano, Inyo, and Shasta cos. n. to Wash.; e. Asia. June–Oct.

3. **L. uniflòrus** Michx. Rootstocks tuberous; stems erect, 1–5 dm. high, sparingly puberulent; lf.-blades lanceolate to rhomboid-lanceolate, acuminate to acute, serrate, 2–6 cm. long, narrowed to sessile or subsessile base; calyx-teeth oblong-ovate to subtriangular, obtuse; corolla ca. 3 mm. long; nutlets with smooth or shallowly rugose summit, the apical ridge erose to tuberculate.—Boggy places below 100 ft.; coastal Humboldt Co.; to B.C., Atlantic Coast. July–Sept.

29. Méntha L. MINT

Aromatic caulescent perennial herbs from rootstocks. Stems erect or diffusely branched. Lvs. opposite, punctate, toothed, usually petioled. Fls. in dense axillary clusters or in terminal spikes. Calyx campanulate to cylindric, 10-nerved, regular or slightly bilabiate, 5-toothed. Corolla funnelform or campanulate, bilabiate, the tube shorter than the calyx, upper lip entire or emarginate, lower 3-lobed. Stamens 4, equal, included or exserted; fils. glabrous; anther-sacs parallel. Nutlets ovoid, smooth. Ca. 30 spp., of N. Temp. regions. (*Minthe*, a Greek nymph supposed to have been changed into Mint.)

Fl.-whorls distant and in fl.-axils, mostly exceeded by the subtending lvs.
    Lvs. 1–2 cm. long, grayish; calyx-teeth dissimilar, the 2 lower lanceolate-subulate .. 1. *M. Pulegium*
    Lvs. 2–7 cm. long, light green; calyx-teeth quite alike, subequal .... 2. *M. arvensis*
Fl.-whorls crowded, usually in terminal spikes, or some of the lower more distant.
    Plants glabrous or nearly so.
        Lvs. sessile or subsessile; spikes narrow, mostly interrupted .... 3. *M. spicata*
        Lvs. petioled; spikes thick, dense.
            Calyx hirsute; lvs. lanceolate, acute .... 4. *M. piperita*
            Calyx glabrous; lvs. ovate, obtuse .... 5. *M. citrata*
    Plants villous or canescent.
        Lvs. round-ovate, 2.5–5 cm. long, woolly .... 6. *M. rotundifolia*
        Lvs. elliptic to oblong-ovate, 1–2 cm. long, pubescent .... 1. *M. Pulegium*

1. **M. Pulègium** L. PENNYROYAL. Stems erect and simple to ± decumbent, slender, 2–5 dm. long, white-pubescent; lvs. elliptic to oblong-ovate, 1–2 cm. long, canescent; fls.-whorls often many, rather remote, with reduced subtending lvs.; calyx 3–4 mm. long, short-hirsute on nerves and teeth; corolla lavender, ca. twice as long as calyx, the lobes villous.—Low moist places, below 3000 ft.; many Plant Communities; San Diego Co., San Joaquin V., Monterey Co., Santa Clara Co., Sonoma Co., Mendocino Co., Marin Co., Humboldt Co., etc.; to Ore.; natur. from Eu. June–Sept.

2. **M. arvénsis** L. Stems usually branched, 1–8 dm. high, the angles in region of the first-flowering infl. always more pubescent than the sides; stem, petioles and lower surface of lvs. slightly to very pubescent; lvs. lanceolate to oblong or ovate, serrate, petioled, 2–5 cm. long, the upper ovate to elliptic, ± rounded at base; fl.-whorls all axillary; calyx ca. 3 mm. long, pubescent, the teeth triangular-subulate, ca. as long as tube; corolla lilac-pink to purplish, 5–6 mm. long, subglabrous; $2n = 12, 54, 60, 64, 72, 92$ (various auth.).—Moist places, below 7500 ft.; many Plant Communities, much of Calif. in one of several forms that are questionable as to whether native here or not; to Atlantic Coast, Eurasia. July–Oct. Plants with angles and sides of stem ± equally pubescent with hairs spreading, 1–3.5 mm. long, are f. *lanàta* (Piper) S. R. Stewart, but with hairs 0.2–1.5 mm. long and appressed, are f. *pubérula* S. R. Stewart.

Var. **villòsa** (Benth.) S. R. Stewart. [*M. canadensis* L. and var. *villosa* Benth. *M. a.* var. *c. Briq.*] Lvs. in region of infl. lanceolate, with ± cuneate bases; in region of first flowering infl. the angles of stem more pubescent than sides.—More common than the sp.; to Alaska, Atlantic Coast. Plants with angles and sides of stem ± equally pubescent and hairs 1–3 mm. long, spreading, are f. *lanígera* S. R. Stewart, but with hairs 0.2–1.5

mm. long, appressed, are f. *brevipilòsa* S. R. Stewart. Those with stem in region of first flowering infl. glabrous on sides, are f. *glabràta* (Benth.) S. R. Stewart.

3. **M. spicàta** L. SPEARMINT. Nearly or quite glabrous, 3–12 dm. high, often purplish; lvs. oblong- or ovate-lanceolate, ± rounded at base, pubescent to villous, serrate, 2–6 cm. long, subsessile or nearly so; spikes slender, to 4 or 6 cm. long in fr., the bracts lance-subulate, ciliate, 4–8 mm. long; calyx with glabrous base, ca. 1.5 mm. long, the subulate teeth ciliate; corolla pale lavender, glabrous, 2.5–3 mm. long; $2n = 36, 48$ (Löve & Löve, 1942).—Moist fields and marshes, below 5000 ft.; many Plant Communities; natur. from Eu. July–Oct.

4. **M. piperìta** L. PEPPERMINT. Much like *M. spicata*, but with petioled lvs.; spikes 2–12 cm. long; calyx with hirsute teeth; corolla rose-purple to white; $2n = 36, 64$ (Glotov, 1940), 66, 68, 70 (Ruttle, 1931).—Natur. occasionally in wet places at low elevs.; native of Eu. July–Oct.

5. **M. citràta** Ehrh. BERGAMOT MINT. Stems rather weak, glabrous, 3–5 dm. long; lvs. glabrous, ovate to round-ovate, obtuse, 2–4 cm. long, slender petioled; whorls terminal and in upper axils, the mature spikes 2–2.5 cm. long; calyx-teeth subulate, glabrous, shorter than the tube; corolla rose-purple, 6–7 mm. long.—Occasional escape; natur. from Eu. July–Oct.

6. **M. rotundifòlia** (L.) Huds. [*M. spicata* var. *r*. L.] Herbage ± tomentose and viscid; stems 5–8 dm. tall, simple or branched, mostly erect; lvs. round-ovate to broadly elliptical, cordate, subsessile, round-tipped, rugose-reticulated, 2.5–5 cm. long; spikes slender, 5–10 cm. long in maturity; calyx campanulate, puberulent, almost 2 mm. long, the subulate teeth ca. as long as tube; corolla white, puberulent, ca. 4 mm. long; $2n = 18, 24, 54$ (several auth.).—Sparingly natur. in moist places; from Eu. July–Oct.

30. Hýptis Jacq.

Herbs or shrubs with opposite commonly toothed lvs. Fls. bilabiate, usually in dense axillary clusters. Calyx mostly equally 5-toothed. Corolla 2-lipped, declined, the lower lip saccate, abruptly deflexed at the contracted and callous base, the lobes of the upper lip and the lateral ones of the lower similar, flat, equal. Fertile stamens 4, the upper pair shorter; all included in the sac of the middle lower corolla-lobe. Nutlets smooth or slightly roughened. Ca. 350 spp., of the New World, widely distributed in S. Am. (Greek, *huptios*, turned back, referring to the lower lip.)

1. **H. Émoryi** Torr. DESERT-LAVENDER. Erect aromatic shrub, with numerous slender fairly straight branches, 1–3 m. tall, whitish scurfy-tomentose; lvs. ovate, crenulate, 1–2 cm. long, on petioles less than half as long; fls. in axillary short-peduncled cymes which are in a ± paniculate arrangement at ends of branchlets; pedicels 1–4 mm. long; calyx stellate-woolly, 4–6 mm. long, with setaceous teeth; corolla violet, 4–6 mm. long; nutlets light brown, flattish-oblong, ca. 1.5 mm. long, smooth.—Common in washes and canyons, below 3000 ft.; Creosote Bush Scrub; Colo. Desert and s. Mojave Desert; to Ariz., Son., L. Calif. Jan.–May.

71. Crassulàceae. STONECROP FAMILY (Fig. 67)

Herbaceous or somewhat woody, mostly succulent and glabrous. Lvs. in ours entire, estipulate. Fls. in cymes or panicles, rarely solitary, regular, usually perfect. Calyx free from the ovary, mostly persistent, 4–5-parted or -lobed. Petals 4–5, distinct or ± united, usually persistent. Stamens as many or twice as many as petals. Carpels as many as calyx-segms., distinct or united below, usually with a scale at base of each; styles filiform or subulate. Fr. of 1-loculed follicles, dehiscent along ventral suture. Seeds minute, mostly narrow, pointed at both ends. Ca. 25 genera and 900 spp., widely distributed, many of horticultural value.

A. Plants annual; stamens mostly 3–5.
   B. Fls. 1–several in axils, the petals 1–1.5 mm. long .... 1. *Tillaea*
   BB. Fls. in definite terminal cymes, the petals 2–4 mm. long .... 2. *Parvisedum*
AA. Plants perennial; stamens twice as many as petals.
   B. Fls. many, in cymose infl.
      C. Flowering stems lateral, arising from axils of lvs. of basal rosette .. 3. *Dudleya*
      CC. Flowering stems terminal, not arising in axils of basal rosette

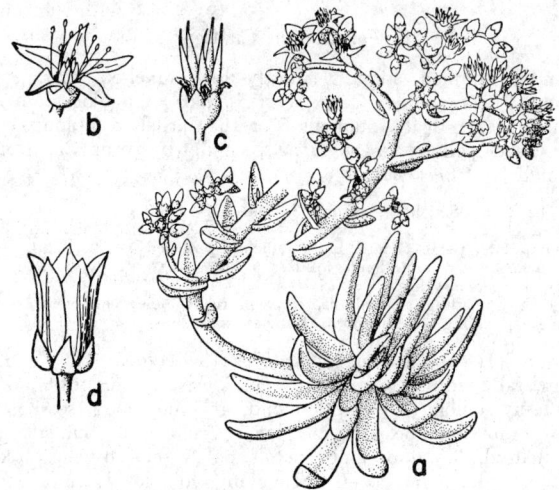

Fig. 67. CRASSULACEAE. *Dudleya virens: a,* habit, × ¼, with fleshy basal lvs. and branched infl.; *b,* fl., × ½, with gamosepalous calyx, spreading petals of subgenus Stylophyllum; *c,* older fl., showing maturing carpels. *Dudleya: d,* erect petals of subgenus Dudleya, × 1.

D.   Corolla with petals connate at base; lvs. 5–10 cm. long. Introd. . . . 6. *Cotyledon*
DD. Corolla with parts free or essentially so; lvs. mostly smaller. Native . . 4. *Sedum*
BB. Fl. solitary; root tuberous . . . . 5. *Congdonia*

1. Tillaèa L. Pigmy-Weed

Diminutive much-branched glabrous annual herbs, with opposite entire minute lvs. Fls. very small, axillary, subsessile to pedicelled. Calyx 3–5-parted (in ours mostly 4-). Petals distinct, or united at very base, 3–5 (mostly 4). Carpels 3–5, distinct, with short subulate styles and 1–12-seeded, becoming follicles. Ca. 45 spp., of wide geographical distribution. (Named for M. A. *Tilli,* 1655–1740, Italian botanist.)

Fls. clustered; carpels 1–2-seeded. Dry places.
  Sepals gradually narrowed to a slender tip, mostly 4 in number. Native . . . . 1. *T. erecta*
  Sepals abruptly narrowed to a long acerose tip, 3 in number. Rare introd. . . . . 2. *T. muscosa*
Fls. solitary; carpels several-seeded. Mud flats . . . . 3. *T. aquatica*

1. **T. erécta** H. & A. [*T. minima* Miers. *T. leptopetala* Benth.] Tufted, erect, branched, 2–6(–8) cm. high, becoming reddish; lvs. fleshy, ovate to oblong, connate at base, 1.5–3 mm. long; fls. in axillary bracted clusters; sepals mostly 4, ovate, 1 mm. long; petals lanceolate, scarcely as long as sepals; seeds mostly 1 in each carpel.—Common, often in great masses in open dry places, burns, etc., below 2500 ft.; many Plant Communities; throughout cismontane Calif.; to Ore., Ariz., L. Calif.; Chile. Feb.–May. A very slender plant with pedicels to 3 mm. long and from w. Colo. Desert has been called var. *erèmica* Jeps.

2. **T. muscòsa** L. Like the preceding, but with 3-parted fls., the sepals with a long aristate apex; carpels usually 2-seeded.—Reported from a clay flat at Ione, Amador Co.; native of Eu. March–April.

3. **T. aquática** L. [*Tillaeastrum a. Britton. T. Bolanderi* Greene.] Stems spreading or decumbent, 1–6 cm. long; lvs. oblong-linear, entire, 4–6 mm. long, connate at base; fls. solitary, axillary, sessile or short-peduncled; fl. parts mostly 4; sepals 1 mm. long, ovate; petals longer than sepals; carpels several-seeded.—On dry mud flats, below 2000 ft.; many Plant Communities; widely scattered localities, cismontane Calif.; to Wash., e. U.S., Mex.; Eurasia. Occasional at higher elevs., such as Hidden Lake, San Jacinto Mts., at 8000 ft. March–July. Apparently it is the matured plants with peduncles longer than lvs., that have been called var. *Drummóndii* Jeps. [*T. D.* T. & G.]

2. Parvisèdum Clausen

Diminutive slender-stemmed annuals, mostly few-branched. Lvs. small, ovoid to oblong-ovoid, fleshy. Fls. in cymes. Calyx with 5 small triangular teeth. Petals 5, connate at base, linear to lance-ovate. Stamens 5 or 10. Carpels 5, oblong, erect or spreading; styles slender. Seeds solitary in the carpels, slender-fusiform. Ca. 3 or 4 spp., Calif. (Latin, *parvus*, small, and *Sedum*.)

(Sharsmith, H. K. The genus Sedella. Madroño 3: 240–248, 1936.)

Stamens 5; style short, erect.
    Carpels glandular-papillate; petals 2 mm. long; follicles appressed .... 1. *P. pentandrum*
    Carpels glabrous; petals 3–3.5 mm. long; follicles ± spreading .... 2. *P. leiocarpum*
Stamens 10; style longer or recurved.
    Petals 3–4 mm. long, spreading at anthesis, erect in fr.; follicles connivent .... 3. *P. pumilum*
    Petals 2 mm. long, spreading at all times; follicles spreading .... 4. *P. Congdonii*

1. **P. pentándrum** (H. K. Sharsm.) Clausen. [*Sedella p.* H. K. Sharsm.] Erect, glabrous, 3–13 cm. tall, reddish-green to green; lowest lvs. opposite, others alternate, early deciduous, fleshy, oblong- to elliptic-ovoid, 4–7 mm. long, sessile; infl. bracteate, usually with 2–5 virgate branches 2–3 cm. long; bracts leafy but small; fls. crowded; sepals 5, fleshy, deltoid, 0.5 mm. long; petals pale greenish-yellow, sometimes with reddish median line; styles erect, 0.3–0.4 mm. long; follicles 1.5 mm. long, red to yellowish, 1.2–1.5 mm. long; seeds light brown, erect, 0.8 mm. long, microscopically striate; *n* = 9 (Baldwin, 1940).—Rocky places, often on serpentine, 800–2500 ft.; openings in Foothill Wd., Chaparral; inner Coast Ranges, Lake Co. to San Benito Co. April–May.

2. **P. leiocárpum** (H. K. Sharsm.) Clausen. [*Sedella l.* H. K. Sharsm.] Glabrous, erect or spreading, 3–5 cm. high; cauline lvs. sessile, fleshy, oblong-ovoid, 4–5 mm. long, caducous; fls. crowded in a bracteate falsely racemose cyme, secund in 2 rows, subsessile; sepals triangular-acute, 0.5 mm. long; petals pale yellow dorsally diffused with red, lanceolate, acuminate; style erect, 0.3–0.4 mm. long; follicles reddish, 2–3 mm. long; seeds light brown, 1–1.3 mm. long.—Dry rocky places and vernal pools; Chaparral; region of Lower Lake, Lake Co. April–May.

3. **P. pùmilum** (Benth.) Clausen. [*Sedum p.* Benth. *Sedella p.* Britt. & Rose.] Glabrous, succulent, 3–17 cm. tall, with several to many branches; lvs. sessile, entire, obtuse, 4–7 mm. long, caducous; fls. crowded, secund on virgate branches of a bracteate cyme; calyx with minute triangular teeth; petals lanceolate, straw-colored; styles erect, 1 mm. long; follicles 2–2.5 mm. long; *n* = 9 (Baldwin, 1940).—Rocky places and beds of vernal pools, below 4000 ft.; Foothill Wd., Yellow Pine F.; Napa Range and Sierran foothills from Sutter Co. to Merced Co. March–May.

4. **P. Congdònii** (Eastw.) Clausen. [*Sedum C.* Eastw. *Sedella pumila* var. *Congdonii* Jeps. *S. C.* Britt. & Rose.] Three–9 cm. high, diffusely branched; lvs. 3–5 mm. long; fls. scattered on the tortuous lax branches of the bracteate cyme; petals bright yellow with reddish median line; style recurved, 0.5 mm. long; follicles 1.2–1.5 mm. long; *n* = 9 (Baldwin, 1940).—Rocky places, below 5000 ft.; Foothill Wd., Yellow Pine F., V. Grassland; Sierran foothills, El Dorado Co. to Tulare Co. March–May.

3. Dúdleya Britt. & Rose. LIVE-FOREVER

Perennial herbs with simple or branched rootstocks or small globose to oblong corms; lvs. principally in basal rosettes, fleshy, flattened to subterete, ± ovate to linear. Flowering stems axillary, with cauline lvs. much reduced and ± like fleshy bracts. Fls. in terminal paniculate or cymose clusters, the branches of which are here termed "cincinni." Calyx deeply 5-lobed into erect, lance-linear to ovate segms. Corolla white to yellow or red, cylindric or campanulate, the petals united near the base, erect or spreading from near the middle or near the tips. Stamens 10, borne on the corolla tube. Carpels 5, ± united below, erect or divergent. Seeds many, minute, narrow, pointed. Perhaps 40 spp. of sw. N. Am., often very variable and puzzling, apparently freely hybridizing. (Named for W. R. *Dudley*, early professor of botany at Stanford Univ.)

(Moran, R. V. A revision of Dudleya, subgenus Stylophyllum. Desert Plant Life 14: 190–193; 15: 9–14, 24–28, 40–45, 55–60, 1943. Uhl and Moran. The cytotaxonomy of Dudleya and Hasseanthus.

Am. Journ. Bot. 40: 492–501, 1953. Moran, R. V., A revision of Dudleya, 1–295, 1951. [Unpublished thesis at Univ. of Calif., Berkeley.] Innovations in Dudleya, Madroño: 14: 106–108, 1957.)

A. Primary stem a ± elongating epigaeous caudex; rosette lvs. mostly persistent.
  B. Corolla tubular, the petals erect with the tips only slightly spreading (Dudleya).
    C. Petals united for ⅓ or more of their length.
      D. Fls. spreading to pendent at anthesis, erect in fr. by a sharp bending of the pedicel; corolla red, the lobes ca. as long as the tube; plant densely white-pulverulent.
        E. Caudex 4–9 cm. thick; rosette-lvs. 8–25 cm. long; pedicels 5–30 mm. long. Coastal .... 11. *D. pulverulenta*
        EE. Caudex 1–4 cm. thick; rosette-lvs. 5–15 cm. long; pedicels 5–15 mm. long. Deserts .... 12. *D. arizonica*
      DD. Fls. erect or ± spreading, the pedicels not sharply bent.
        E. Rosettes large, 1–5 dm. in diam., usually of 30–80 or more lvs.; rosette-lvs. mostly 6–25 cm. long, 1.6–8 cm. wide. Insular ...... 6. *D. candelabrum*
        EE. Rosettes smaller, less than 1 dm. in diam., mostly of 10–25 lvs.; rosette-lvs. mostly 2–8 cm. long, 0.5–3 cm. wide. San Bernardino Mts. to L. Calif.
        8. *D. Abramsii*
    CC. Petals united for less than ⅓ of their length.
      D. Plant branching by stolons; predehiscent carpels ascending; petals erect, yellow. San Joaquin Hills, Orange Co. .... 1. *D. stolonifera*
      DD. Plant with dichotomous branching or none; predehiscent carpels mostly erect.
        E. Rosettes large, 1–5 dm. in diam., usually solitary, mostly with 20–45 lvs.; petals white to pale yellow or pink. Insular ...... 6. *D. candelabrum*
        EE. Rosettes smaller, solitary to many, mostly less than 1.5 dm. in diam., or if larger, with not more than 25 rosette lvs.
          F. Plants leafless in summer; rosette-lvs. 5–12(–15), 1.5–4 cm. long, 0.3–1.2 cm. wide; infl. mostly of 1–2 simple cincinni. Santa Monica Mts.
            G. Petals bright yellow, sharply acute, 10–14 mm. long; rosette-lvs. 5–12 mm. wide; pedicels 5–12 mm. long . 2. *D. cymosa marcescens*
            GG. Petals pale yellow, acute, 8–12 mm. long; rosette-lvs. 3–6 mm. wide; pedicels 1–3 mm. long .... 3. *D. parva*
          FF. Plants evergreen; rosette-lvs. mostly 10–25 or more, if fewer, then more than 12 mm. wide.
            G. Caudex elongate, mostly 1–6 dm. long, often much branched; rosette-lvs. usually rather thick. Maritime.
              H. Rosette-lvs. oblong-ovate, 2–6 cm. long, 1–2.5 cm. wide; petals pale yellow, the exposed margin of each petal commonly separated from each adjacent petal. Los Angeles Co. n. on sea bluffs .... 4. *D. farinosa*
              HH. Rosette-lvs. longer or narrower; petals pale yellow to bright yellow or red, the exposed margin of each petal usually appressed against the adjacent petal.
                I. Petals bright yellow to red, 8–16 mm. long, erect; rosette-lvs. thick to thin. Monterey Co. to Los Angeles Co.
                5. *D. caespitosa*
                II. Petals pale yellow, 8–12 mm. long, sometimes curved outwardly at apex; rosette-lvs. thick. Insular
                7. *D. Greenei*
             GG. Caudex short, erect, rarely over 0.5 dm. long, simple or cespitosely few-branched; rosette-lvs. relatively thin. From near the coast to inland.
              H. Pedicels 0.5–6 mm. long.
                I. Petals thin, often erose, 8–13 mm. long, 2–3 mm. wide, pale yellow and with red lines, connate for 1.5–4.5 mm.; rosette-lvs. 1.5–6(–11) cm. long, 0.5–2 cm. wide. San Luis Obispo Co. and San Bernardino Mts. to L. Calif.
                8. *D. Abramsii*
                II. Petals thick, entire, 10–16 mm. long, 3.5–5 mm. wide, pale or bright yellow to red, connate for 1–2 mm.; rosette-lvs. 5–20(–30) cm. long, 1–4 cm. wide. Santa Barbara Co. to L. Calif. .... 9. *D. lanceolata*
             HH. Pedicels 5–20 mm. long.
              I. Rosette-lvs. oblanceolate to spatulate, acuminate to cuspidate, and sometimes tapering from base and acute; petals thin, sharply acute; fl.-stems 0.5–2.5(–4) dm. tall; infl. rather dense, the cincinni mostly 1–5 cm. long. N. and cent. Calif. to Santa Monica, San Gabriel, and San Bernardino mts. .... 2. *D. cymosa*
              II. Rosette-lvs. oblong-lanceolate, acute to subacuminate; petals thick to thin, acute; cincinni often elongate. S. Calif.
                J. Rosette-lvs. 5–30 cm. long, 1–4 cm. wide; fl.-stems 1.5–7.5 dm. tall; petals yellow to red, connate for 1–2 mm. Santa Barbara Co. to L. Calif.
                9. *D. lanceolata*

JJ. Rosette-lvs. 3–15 cm. long, 0.5–2.5 cm. wide; fl.-stems
0.5–4 dm. tall; petals yellow, connate for 1–4 mm.
10. *D. saxosa*
  BB. Corolla with the segms. widely spreading from near the middle. (Stylophyllum).
    C. Lvs. flattened.
      D. Petals bright yellow; lvs. 4–6 times as broad as thick. Santa Barbara Ids.
13. *D. Traskae*
      DD. Petals white or tinged with red; lvs. often narrower.
        E. Herbage viscid. Mainland .... 14. *D. viscida*
        EE. Herbage not viscid, either green or glaucous. Mostly insular .. 15. *D. virens*
    CC. Lvs. terete or nearly so, except at very base.
      D. Branches of infl. usually 2 or 3, simple; caudex mostly less than 1 cm. thick;
mature carpels ascending, slightly gibbous. S. San Diego Co. .. 16. *D. attenuata*
      DD. Branches of infl. several, once or twice bifurcate; caudex mostly more than 1 cm.
thick; mature carpels abruptly divergent, strongly gibbous.
        E. Herbage mealy; styles 2.5–3 mm. long; pedicels 2–4 mm. long. San
Gabriel Mts. .... 17. *D. densiflora*
        EE. Herbage slightly glaucous but not mealy; styles 1.5–2 mm. long; pedicels
0–2 mm. long. W. Riverside and Orange cos. s. .... 18. *D. edulis*
AA. Primary stems hypogaeous, usually cormlike; rosette-lvs. vernal. (Hasseanthus)
  B. Fls. white, with a sweet odor.
    C. Petals erect to ascending, 7–14 mm. long; lf.-bases 4–12 mm. wide. Santa Cruz Id.
19. *D. nesiotica*
    CC. Petals wide-spreading, 5–10 mm. long; lf.-bases 1–4 mm. wide. Mainland and Santa
Rosa Id. .... 20. *D. Blochmanae*
  BB. Fls. yellow, odorless.
    C. Rosette-lvs. oblanceolate, strongly narrowed below, 1–7 cm. long; petals connate for
0.5–1 mm. San Diego region .... 21. *D. variegata*
    CC. Rosette-lvs. linear-lanceolate, not narrowed at base, 4–15 cm. long; petals connate for
1–2 mm. Los Angeles, Riverside, and Orange cos. .... 22. *D. multicaulis*

1. **D. stolonífera** Moran. Caudex branching by slender stolons that become 1 dm.
long; rosette-lvs. green, with purple tinge, not glaucous, oblong-obovate, 3–7 cm. long,
1.5–3 cm. wide; fl.-stems 0.8–2 dm. tall; cauline lvs. cordate-ovate, acute, 0.5–1.3 cm.
long; infl. with cincinni 1–6 cm. long and 3–9-fld.; pedicels erect, 3–8 mm. long; calyx
6–7 mm. wide, 3–4 mm. high, truncate at base; petals yellow, 10–11 mm. long, 3–3.5
mm. wide, connate for 1–2 mm.; carpels 5–7 mm. high, separating before dehiscence;
$n = 17$ (Moran, 1949).—Cliffs in Coastal Sage Scrub; Aliso and Laguna canyons, near
Laguna Beach, Orange Co. May–July.

2. **D. cymòsa** (Lem.) Britt. & Rose. [*Echeveria c.* Lem. *Cotyledon nevadensis* Wats.
C. Purpusii K. Schum.? *C. Plattiana* Jeps. *C. angustiflora* Fedde. *Dudleya a.* Rose. *D.
Sheldonii* Rose.] Plants ± glaucous; rosette-lvs. mostly oblong-oblanceolate, acute to
acuminate, 5–10 cm. long, 1.5–4 cm. wide, in a very dense rosette; fl.-stems ± red, 1–2
dm. high; cauline lvs. lanceolate to oblong-lanceolate, 1.5–3.5 cm. long, rounded at
base, acute; infl. dense, of several short cincinni; pedicels rather slender, 5–12 mm.
long; calyx-lobes ± deltoid-ovate, 3–4 mm. long; petals mostly bright yellow to red,
10–12 mm. long, lanceolate; carpels 4–5 mm. long; $n = 17$ (Uhl & Morgan, 1953).—
Rocky cliffs, etc., below 3500 ft. in N. Oak Wd., Foothill Wd., Chaparral, etc. in Coast
Ranges from Humboldt Co. to Santa Clara Co., and up to 9000 ft., Foothill Wd.,
Yellow Pine F., etc., in Sierra Nevada from Butte to Kern cos. April–June.

KEY TO SUBSPECIES

A. Rosette-lvs. evergreen, 2–17 cm. long, 1–6 cm. wide; caudex mostly 1 cm. or more thick;
fl.-stems 0.5–4.5 dm. tall.
  B. Fl.-stems mostly 0.5–2.5 dm. tall; rosette-lvs. green or glaucous, 2–12 cm. long; petals pale
yellow to red.
    C. Fl.-stems commonly 1–2.5 dm. tall; rosette-lvs. mostly 4–12 cm. long. Cent. and n. Calif.
      D. Petals mostly bright yellow to red; rosette-lvs. oblong-oblanceolate, acute to
acuminate, rarely spatulate and cuspidate, 1–5 cm. wide. Coast Ranges from Santa
Clara Co. n., Sierra Nevada .... *D. cymosa*
      DD. Petals pale yellow; rosette-lvs. oblong-oblanceolate to triangular-oblong, acute
to acuminate, 0.5–2 cm. wide. Inner S. Coast Ranges ...... ssp. *Setchellii*
    CC. Fl.-stems mostly less than 1.5 dm. tall; rosette-lvs. mostly less than 5 cm. long; petals
bright yellow to red. Santa Lucia Mts. to s. Calif.
      D. Rosette-lvs. green or glaucous, usually 10–25, mostly short-acuminate to
cuspidate; petals yellow to red. Santa Lucia Mts. to San Bernardino Mts.
ssp. *minor*
      DD. Rosette-lvs. green, 6–10, acute to acuminate; petals yellow. Santa Monica Mts.
ssp. *ovatifolia*

BB. Fl.-stems 1.5–4.5 dm. tall; rosette-lvs. glaucous, 4–17 cm. long; petals red. Amador and
    Calaveras cos. . . . . ssp. *gigantea*
AA. Rosette-lvs. withering in summer, 1.5–4 cm. long, 0.5–1.2 cm. wide; caudex 0.2-0.7 cm. thick;
    fl.-stems 4–10 cm. tall. Santa Monica Mts. . . . . ssp. *marcescens*

Ssp. Setchéllii (Jeps.) Moran. [*Cotyledon laxa* var. S. Jeps. *C. caespitosa* var. *paniculata* Jeps. *Dudleya* S. Britt. & Rose. *D. p.* Britt. & Rose. *D. humilis* Rose. *Echeveria
Jepsonii* Nels. & Macbr. *E. diaboli* Berger.] Caudex 1–2 cm. thick; rosette-lvs. oblong-
oblanceolate to -triangular, 3–10 cm. long, 0.5–2 cm. wide; fl.-stems 0.5–2.5 dm. tall;
cincinni 1–5 cm. long, 4–10-fld.; petals pale yellow, 1.5–2.5 mm. wide; $n = 17$ (Uhl &
Moran, 1953).—Dry rocky outcrops, below 5000 ft.; Chaparral, Foothill Wd.; Contra
Costa Co. to the Pinnacles, San Benito Co. May–June.

Ssp. minor (Rose) Moran. [*D. m.* Rose. *D. nevadensis* ssp. *m.* Abrams. *D. pumila*
Rose. *D. bernardina* Britton. *D. Goldmanii* Rose.] Caudex 1–2(–3.5) cm. thick; plant
glaucous; rosette lvs. rhomboid-oblanceolate to spatulate, 2–6(–10) cm. long, mostly
1–3 cm. wide, abruptly acuminate; fl.-stems slender, 0.5–1.5 dm. high; cauline lvs.
ovate, cordate, 0.4–1.2 cm. long; cincinni mostly 1–3 cm. long, 3–6-fld.; pedicels slender,
5–14 mm. long; calyx 2.5–4 mm. wide, often truncate at base, 3–4 mm. long; petals
orange to salmon-red to bright yellow, 10–12 mm. long, connate for ca. 2 mm., lanceolate; $n = 17$ (Uhl & Morgan, 1953).—Dry rocky places, 2000–8500 ft.; Foothill Wd.,
Chaparral, Yellow Pine F., etc.; mts. from Monterey Co. to San Bernardino Co. April–
July.

Ssp. ovatifòlia (Britton in Britt. & Rose) Moran. [*D. o.* Britton.] Caudex 1–1.5 cm.
thick; rosette-lvs. green, glabrous, ovate to elliptic, 6–10, shining, acute, 2–5 cm. long,
1.5–2.5 cm. wide; fl.-stems 0.4–1.5 dm. high; cauline lvs. 0.5–1 cm. long, cordate;
cincinni 1–3 cm. long, 3–5-fld.; calyx 2.5–3 mm. high; petals bright yellow, ca. 10 mm.
long, lanceolate, 2–2.5 mm. wide; $n = 17$ (Uhl & Moran, 1953).—Rocky places; Coastal
Sage Scrub, Chaparral; Santa Monica Mts., Los Angeles Co. March–May.

Ssp. gigántea (Rose) Moran. [*D. g.* Rose. *Echeveria amadorana* Berger. *E. lanceolata*
var. *incerta* Jeps.] Rootstock 1–3 cm. thick; rosette-lvs. very glaucous, oblong-oblanceolate, 4–17 cm. long; fl.-stems stoutish, 1.5–4.5 dm. high; cincinni 4–17 cm. long, 5–20-fld.;
pedicels rather slender, 5–10 mm. long; petals red, 9–10 mm. long, erect, acute, connate
for ca. 1.5 mm.; $n = 17$ (Uhl & Moran, 1953).—Rocky banks, 800–1500 ft.; Foothill
Wd.?; Amador and Calaveras cos. May–July.

Ssp. marcéscens Moran. Caudex 0.2–0.7 cm. thick; rosette-lvs. withering in summer,
1.5–3.5 cm. long, 0.5–1.2 cm. wide, 1–2 mm. thick; fl.-stems 0.4–1 dm. tall; infl. of 1–2
simple cincinni each 3–5 fld.; pedicels 5–12 mm. long; petals bright yellow, often marked
with red, 10–14 mm. long, 2.5–3.5 mm. wide; $n = 17$ (Uhl & Moran, 1953 as affin. *D.
ovatifolia*).—Shaded rocky slopes, at 1100 ft.; Chaparral; Little Sycamore Canyon, Santa
Monica Mts., Ventura Co. May–June.

3. D. párva Rose & Davids. Caudex 3–5 cm. long, branched, purplish; rosette-lvs.
linear to oblanceolate, 1.5–4 cm. long, 0.3–0.6 cm. wide, acute, slightly glaucous, rather
early fugacious; fl.-stems 0.5–1.5 dm. high; cauline lvs. well distributed, lanceolate to
triangular-ovate, acute, 0.5–1.5 cm. long; infl. of 1–2 cincinni 3–4(–8) cm. long, each
5–10-fld.; pedicels stout, 1–3 mm. long; calyx 3–5 mm. broad, ± rounded below, glaucous,
3–5 mm. high, the lobes triangular-ovate; petals pale yellow, sometimes with red flecks,
8–12 mm. long, connate for 1–2 mm., elliptic-oblong; $n = 17$ (Moran, 1948).—Bare
rocky slopes, 1000 ft.; Chaparral, Coastal Sage Scrub; Conejo Grade and Arroyo Santa
Rosa, s. Ventura Co. May–June.

4. D. farinòsa (Lindl.) Britt. & Rose. [*Echeveria f.* Lindl. *D. septentrionalis, D.
Eastwoodiae,* and *D. compacta* Rose. *Sedum Cotyledon* of Jeps., not Jacq. *C. Palmeri*
Wats. *C. lingula* Wats.?] Caudex stout, 1–3 cm. thick, usually with several rosettes, each
with 15–30 lvs.; rosette-lvs. densely white-mealy to green, ovate-oblong, acute, flat on
upper surface, lightly rounded beneath, 2.5–6 cm. long, 1–2.5 cm. wide; fl.-stems stout,
mostly white-mealy, 1–3.5 dm. high; cauline lvs. many, triangular-ovate, auriculate, 1–2.5
cm. long, concave; cincinni 1–3.5 cm. long, 3–11-fld.; pedicels stout, 1–5 mm. long;
calyx mostly 5–6 mm. wide, rounded at base, 5–8 mm. high, the lobes deltoid-ovate;
petals lemon-yellow, 10–14 mm. long, oblong, acute; carpels 5–8 mm. long; $n = 17, 68,$
85, 119 (Uhl & Moran, 1953).—Sea bluffs; Coastal Sage Scrub, N. Coastal Scrub; Los
Angeles Co. to s. Ore. Mostly May–Sept.

5. **D. caespitòsa** (Haw.) Britt. & Rose. [*Cotyledon c.* Haw. *Sedum Cotyledon* Jacq. *C. califórnica* Baker. *D. Helleri* Rose. *Echeveria laxa* Lindl.] Much like *D. farinosa;* rosette-lvs. ± shining, only the inner glaucous, 5–20 cm. long, 1–5 cm. wide; cincinni becoming 3–11 cm. long, 3–14-fld.; calyx 4–8 mm. broad, 4–6 mm. high, subtruncate to tapering at base; petals bright yellow to red, 8–16 mm. long; $n = 51$ or 68 (Uhl & Moran, 1953).—Sea bluffs, Monterey Co. to Los Angeles Co.; Anacapa Id. April–July.

6. **D. candelàbrum** Rose. [*Echeveria c.* Berger.] Caudex simple, 2–8 cm. thick, to 2 dm. long; rosette-lvs. green, oblong-lanceolate to obovate, 10–15 cm. long, 3–7 cm. broad; fl.-stems 1.5–3.5 dm. tall; infl. flat-topped, 0.5–2.5 dm. broad, mostly with ca. 3 cincinni, each with 5–25 fls.; calyx 6–9 mm. long; petals 8–12 mm. long; $n = 17$ (Uhl & Moran, 1953).—Rocky places, Santa Cruz and Santa Rosa ids. April–July.

7. **D. Greènei** Rose. [*Echeveria G.* Berger. *D. Hoffmannii, D. regalis,* and *D. echeverioides* Johansen.] Cespitose with branched caudex 2–5 cm. thick and several rosettes; rosette-lvs. densely white-mealy to green, oblong-oblanceolate to -obovate, 3–11 cm. long, 1.5–3 cm. wide, acute, with reddish tinge in age; fl.-stems 1.5–4 dm. tall, stout; cauline lvs. scattered, lance-ovate, cordate, mostly 1–2 cm. long; cincinni several, each 2–15-fld., the infl. ca. 1 dm. broad; pedicels mostly 1–4 mm. long; calyx-lobes triangular-lanceolate, 2–5 mm. long; petals pale yellow, 10–12 mm. long, connate for 2 mm.; $n = 34$ or 51 (Uhl & Moran, 1953).—Rocky bluffs; Chaparral, Coastal Sage Scrub; Santa Cruz, Santa Rosa, San Miguel, and Santa Catalina ids. May–July.

8. **D. Abrámsii** Rose. [*Echeveria A.* Berger. *D. tenuis* Rose.] Caudex short, 1–1.5 cm. thick; rosette-lvs. gray, ± glaucous, lanceolate, 2–6 cm. long, acuminate; fl.-stems slender, 0.5–1.6 dm. tall; cauline lvs. mostly in upper parts, ovate, acute, ± cordate, the lower 4–15 mm., the upper 2–5 mm. long; cincinni 2–3, 2–12 cm. long, rather few-fld.; pedicels 2–5 mm. long; calyx 3–5 mm. long, greenish, the lobes lance-ovate; petals pale yellow with reddish stripes, 7–12 mm. long, connate for 2–4 mm., acute; $n = 17$ (Uhl & Moran, 1953).—Dry rocky places, 2500–9000 ft.; Chaparral to Yellow Pine F.; desert edge from San Bernardino Mts. to n. L. Calif. April–June.

Ssp. **murìna** (Eastw.) Moran. [*D. m.* Eastw.] Caudex 1–3 cm. thick; lower cauline lvs. 10–30 mm. long; petals united for 1.5–3 mm.; $n = 17$ (Uhl & Moran, 1953).—Serpentine; Chaparral, Foothill Wd.; San Luis Obispo region, May–June.

9. **D. lanceolàta** (Nutt.) Britt. & Rose. [*Echeveria l.* Nutt. ex T. & G. *E. monicae* Berger. *D. congesta* and *D. robusta* Britton. *D. elongata, D. Brauntonii, D. Parishii, D. lurida,* and *D. Hallii* Rose.] Caudex short, simple; rosette-lvs. pale green, ± glaucous, lanceolate, long-acuminate, 5–20(–30) cm. long, 1–3 cm. wide; fl.-stems 2–6 dm. tall, fairly stout, ± reddish; cauline lvs. lanceolate, cordate or sagittate at base, 0.5–3 cm. long; cincinni several, 5–12 cm. long, many-fld.; pedicels stout, 3–12 mm. long; calyx-lobes lance-ovate, 3–5 mm. long; petals orange or pale green with red tinge, 10–15 mm. long; $n = 34$ (Uhl & Moran, 1953).—Common on dry stony slopes and banks below 3500 ft.; Chaparral, Coastal Sage Scrub, etc.; Santa Barbara and Kern cos. to n. L. Calif. May–July.

10. **D. saxòsa** (Jones) Britt. & Rose. [*Cotyledon s.* Jones.] Pale green or ± glaucous; caudex 1–1.5 cm. thick; rosette-lvs. 10–25, narrow-lanceolate, semiterete, 3–10 cm. long, ca. 1–1.5 cm. wide; fl.-stems ± reddish, 0.5–2 dm. tall; cauline lvs. ovate-lanceolate, 0.5–2 cm. long, slightly clasping; infl. 0.5–1.5 dm. wide, the cincinni ascending, ± reddish; pedicels 1–2 cm. long; calyx-lobes red, lance-ovate, ca. 5 mm. long; petals yellow, ± reddish in age, 10–12 mm. long; $n = 68$, 85 (Uhl & Moran, 1953).—Dry stony slopes, 3000–7000 ft.; Creosote Bush Scrub to Pinyon-Juniper Wd.; Panamint Mts., Inyo Co. May–June.

Ssp. **aloìdes** (Rose) Moran. [*D. a.* Rose. *D. grandiflora* and *D. delicata* Rose. *Echeveria lanceolata* var. *a.* Munz and var. *composta* Jeps.] Scarcely to quite glaucous; fl.-stem 1–3.5 dm. high; fls. yellow; $n = 17$ (Uhl & Moran, 1953).—Dry rocky places, 800–5500 ft.; Creosote Bush Scrub, Chaparral, Pinyon-Juniper Wd., etc.; desert mts. of San Bernardino Co., desert slopes of San Jacinto and Laguna mts. April–June.

11. **D. pulverulénta** (Nutt.) Britt. & Rose. [*Echeveria p.* Nutt. in T. & G. *E. argentea* Lem.] Plant covered with a white mealy powder throughout; caudex 4–9 cm. thick and up to 4 dm. long; rosette-lvs. 30–80, obovate-spatulate, 8–25 cm. long, 4–10 cm. wide, spreading; fl.-stems stout, 4–8 dm. high; cauline lvs. many, broadly ovate, cordate-clasping, acute, 1–2(–4) cm. long; cincinni 2–several, ascending, 1–4 dm. long, 10–30-fld.;

pedicels slender, 5–30 mm. long, spreading; calyx-segms. lanceolate, acute, red to glaucous, ca. 4–8 mm. long; petals deep red, 12–18 mm. long, connate nearly to the middle; carpels ± distinct, erect, ca. 7–8 mm. long; seeds brown, slender, ca. 0.5 mm. long; $n = 17$ (Uhl & Moran, 1953).—Rocky cliffs and canyons, mostly below 3000 ft.; Coastal Sage Scrub, Chaparral, etc.; near the coast from San Luis Obispo Co. to L. Calif. May–July.

12. **D. arizónica** Rose [*Echeveria lagunensis* Munz. *D. l.* Walth.] Caudex 1–4 cm. thick; lower lvs. obovate to spatulate, 5–15 cm. long, 2–5 cm. wide; fl.-stems 2–6 dm. high; cauline lvs. 1–3 cm. long, clasping, broadly ovate; pedicels rather slender, 5–15 mm. long; calyx 5–7 mm. long, the lobes lanceolate; petals 12–14 mm. long, brick-red; $n = 17$ (Uhl & Moran, 1953).—Dry slopes, 2000–4000 ft.; Creosote Bush Scrub, Joshua Tree Wd.; desert slopes of Laguna and Cuyamaca mts. and of mts. in e. Mojave Desert (Kingston, Old Dad, Old Woman mts.); to Nev., Ariz., Son., L. Calif. May–July.

13. **D. Tráskae** (Rose) Moran. [*Stylophyllum T.* Rose. *Cotyledon T.* Fedde. *Echeveria T.* Berger.] Caudex short, branched to form clumps with 20–100 heads; rosette lvs. 25–35, strap-shaped to ± oblanceolate, subacuminate, 4–15 cm. long, 1–4 cm. broad, glaucous at least when young; flowering stems 2–3 dm. long; cauline lvs. deltoid-ovate, acute, horizontal, mostly 1–3 cm. long; infl. flat-topped; pedicels stout, 1–3 mm. long; calyx-lobes deltoid to deltoid-ovate, acute, 2.5–4 mm. long; petals connate 1–1.5 mm., narrow-ovate, 8–10 mm. long, bright yellow, later ± red veined, curving outward in upper half; carpels ± spreading in age, 7–8 mm. long; $n = 34$ (Uhl & Moran, 1953).—Santa Barbara Id. April–May.

14. **D. víscida** (Wats.) Moran. [*Cotyledon v.* Wats. *Stylophyllum v.* Britt. & Rose. *Echeveria v.* Berger.] Caudex short; rosette lvs. 15–35, linear-deltoid, acute, 6–10 cm. long, dark green, viscid; flowering stems 2–3 dm. long; cauline lvs. deltoid-lanceolate, subacute, 1–3 cm. long; infl. elongate, ± cylindrical or subcorymbose; pedicels mostly 0–3 mm. long; calyx-lobes deltoid-ovate, to oblong-ovate, acute, 1.5–4 mm. long; petals elliptic-oblong, acute, 6–9 mm. long, white strongly marked with red, spreading from near middle; carpels slender, 7–9 mm. long, not widely spreading; seeds caudate, 1.2–1.4 mm. long; $n = 17$ (Uhl & Moran, 1953).—Dry rocky places, below 1200 ft.; Coastal Sage Scrub; from near San Juan Capistrano, Orange Co., to near Oceanside, San Diego Co. May–June.

15. **D. vìrens** (Rose) Moran. [*Stylophyllum v.* Rose. *Cotyledon v.* Fedde. *Echeveria v.* Berger. *S. albidum, insulare,* and *Hassei* Rose.] Caudex short and thick to elongate and procumbent, up to 3 dm. long; rosette lvs. 25–35, strap-shaped, tapering from base or from near middle, acute, 4–25 cm. long, mostly green; flowering stems 1–5 dm. long, 2–8 mm. thick; cauline lvs. 8–25, deltoid-lanceolate, 1–5 cm. long; infl. 0.5–1 dm. wide, elongate and cylindrical to corymbose and flat-topped; pedicels 1–4 mm. long; calyx-lobes deltoid-ovate, acute to ± obtuse, 1.5–3 mm. long; petals elliptic-oblong, acute, white, 7–10 mm. long, connate for 1–2 mm.; carpels ± spreading when mature; seeds 1–1.3 mm. long; $n = 17$ (Uhl & Moran, 1953).—Rocks and cliffs, below 1500 ft.; Coastal Sage Scrub, Chaparral; Point San Vincente and Santa Catalina and San Clemente ids. April–June. More glaucous plants with slender caudex and $n = 34$, from Catalina, San Clemente, and Guadalupe ids., are *S. Hassei* Rose.

16. **D. attenuàta** (Wats.) Moran ssp. **Orcúttii** (Rose) Moran. [*Stylophyllum O.* Rose. *S. Parishii* Britton. *Echeveria palensis* Berger.] Caudex often elongate and branched, 0.5–1 cm. thick; rosette lvs. 10–15, linear-oblanceolate, acute, 3–10 cm. long, glaucous when young; flowering stems 1–2.5 dm. long; cauline lvs. turgid, deltoid-ovate to -lanceolate, acute, 0.5–2.5 cm. long; infl. usually with 1–3 simple branches; pedicels 1–3 mm. long; calyx-lobes deltoid-ovate to oblong-ovate, acute, 1–2.5 mm. long; petals white flushed with rose, 6–8 mm. long, elliptic, acute, connate 1–2 mm.; carpels 4–7 mm. long, ± spreading in age; seeds red-brown, 0.8–1 mm. long; $n = 17$ or 34 (Uhl & Moran, 1953). —Dry gravelly or rocky places, below 1000 ft.; Coastal Sage Scrub, Chaparral; s. San Diego Co. from Pala s.; adjacent L. Calif. May–July.

17. **D. densiflòra** (Rose) Moran. [*Cotyledon nudicaulis* Abrams, not Lam. *Stylophyllum densiflorum* Rose.] Caudex branched, 1–2 cm. thick; rosette lvs. 20–25, linear or ± enlarged upward, acute, 5–10 cm. long, persistently glaucous; flowering stems 1.5–3 dm. long; cauline lvs. turgid, acute, 1–4 cm. long; infl. dense, ± rounded, 3–several-branched; pedicels 2–5 mm. long; calyx-lobes deltoid-ovate, acutish, 1.5–2.5 mm. long;

petals white or tinged pink, narrow-ovate, acute, 5–10 mm. long; carpels abruptly divergent, 5–10 mm. long; seeds 1 mm. long; $n = 17$ (Uhl & Moran, 1953).—Rocky cliffs, between 800 and 2000 ft.; Chaparral; s. base of San Gabriel Mts. in canyons near San Gabriel Canyon. June–July.

18. **D. édulis** (Nutt.) Moran. [*Sedum e.* Nutt. *Stylophyllum e.* Britt. & Rose.] Caudex usually short, 1–3 cm. thick; rosette lvs. 10–20, linear, subacuminate, 5–15 cm. long, slightly glaucous; flowering stem 1.5–5 dm. long; cauline lvs. turgid, deltoid-lanceolate, acute, 1–5 cm. long; infl. elongate, rather open; pedicels to 2 mm. long; calyx-lobes oblong, acute, 2.5–4.5 mm. long; petals oblong-lanceolate, acutish, cream-white, 7–10 mm. long, connate for 1–1.5 mm.; carpels 6–8 mm. long, abruptly divergent; seeds 0.8–1 mm. long; $n = 17$ (Uhl & Moran, 1953).—Rocky hillsides, below 3500 ft.; Chaparral, Coastal Sage Scrub; cismontane s. Calif. from Riverside and Orange cos. to n. L. Calif. May–June.

19. **D. nesiótica** (Moran) Moran. [*Hasseanthus n.* Moran.] Corms subglobose to irregular, 1–2(–3) cm. long; rosette lvs. 8–16, oblanceolate to spatulate, 2.5–5 cm. long, 0.5–2 cm. wide; flowering stems 3–10 cm. tall; cauline lvs. triangular-lanceolate to -ovate, turgid, 1–2.5 cm. long; infl. usually of 2 simple branches, each with 3–8 fls.; pedicels 1–2 mm. long; calyx 5–6 mm. broad, 4–6 mm. high, the lobes triangular-ovate, 3–4 mm. long; petals white above, greenish-yellow on the keel, elliptic, 7–14 mm. long, 3.5–5.5 mm. wide, carpels ascending; $n = 34$ (Uhl & Moran, 1953).—Sea bluffs, Santa Cruz Id. March–April.

20. **D. Blóchmanae** (Eastw.) Moran. [*Sedum B.* Eastw. *Hasseanthus B.* Rose. *H. Kessleri* A. Davids. *Sedum Gertrudianum* Eastw.] Corms globose to fusiform, mostly not more than twice as long as thick; rosette lvs. 5–8, linear-oblanceolate to linear-spatulate, yellow-green, 2–7 cm. long; fl. stems 4–15 cm. long; cauline lvs. deltoid-lanceolate to broadly ovate, 0.5–2.5 cm. long, not more than half as wide; infl. a cyme of 2–several branches; fls. subsessile; calyx 1.5–4 mm. long, the lobes ovate to ± oblong-ovate; petals white, marked with red or purple, 5–9 mm. long, lanceolate; follicles spreading, red, 4–6 mm. long; $n = 17$, 34, or 51 (Uhl & Moran, 1953).—Dry stony places, below 1500 ft., often on serpentine; Coastal Sage Scrub; near the coast from San Luis Obispo Co. to n. L. Calif. May–June.

Ssp. **insulàris** (Moran) Moran. [*Hasseanthus B.* ssp. *i.* Moran.] Rosette lvs. 15–30; flowering stems 3–4 cm. long, more glaucous than in the sp.; $n = 17$ (Uhl & Moran, 1953).—Santa Rosa Id. March–April.

Ssp. **brevifòlia** (Moran) Moran. [*Hasseanthus B.* ssp. *b.* Moran.] Corms oblong, 4–8 times as long as thick; rosette lvs. 0.7–1.5 cm. long, subglobular; flowering stems 2–11 cm. high; cauline lvs. 0.2–1 cm. long, almost as wide; $n = 17$ (Uhl & Moran, 1953).— Dry sandstone bluffs; Chaparral; Del Mar to La Jolla, San Diego Co. April.

21. **D. variegàta** (Wats.) Moran. [*Sedum v.* Wats. *Hasseanthus v.* Rose.] Corms ovoid or rounded to oblong, 1–3 cm. long; rosette lvs. 5–8, oblanceolate to spatulate, blue-green, 1.5–5 cm. long; flowering stems 5–29 cm. long; cauline lvs. lanceolate to narrow-deltoid, 0.5–2 cm. long, 0.3–0.8 cm. wide; infl. open, cymose, 2–3-branched; calyx-lobes deltoid-ovate to ovate-oblong, erect, 2–4 mm. long; petals yellow, with some red, elliptic-lanceolate, 5–7 mm. long; follicles spreading, red; $n = 17$ (Uhl & Moran, 1953).— Dry stony places, below 500 ft.; Coastal Sage Scrub, Chaparral; sw. San Diego Co., n. L. Calif. May–June.

21. **D. multicaùlis** (Rose) Moran. [*Hasseanthus m.* Rose. *H. elongatus* Rose. *Sedum oblongorhizum* Berger. *S. sanctae-monicae* Berger.] Corms oblong, 1.3–5 cm. long, 0.3–1.8 cm. thick; rosette lvs. 6–15, linear, subterete, 3–15 cm. long; flowering stems 4–35 cm. long; cauline lvs. like the basal, but shorter; infl. cymose, 2–3-branched; fls. sessile or short-pedicelled; calyx-lobes deltoid-ovate to linear-oblong, 2–4 mm. long; petals yellow, elliptic-lanceolate, 5–9 mm. long, often flecked with red; follicles spreading, 5–8 mm. long; $n = 17$ (Uhl & Moran, 1953).—Dry stony places, below 2000 ft.; Coastal Sage Scrub, Chaparral; Los Angeles Co. to w. San Bernardino, Riverside, and Orange cos. May–June.

4. **Sèdum** L. STONECROP. ORPIN

Herbs or subshrubs, usually glabrous; lvs. mostly alternate, often small and imbricated, sometimes in terminal or basal rosettes. Fls. in terminal cymes, often with secund branches. Calyx 4–5-parted. Petals 4–5, distinct or partly united. Stamens mostly twice as many, sometimes of same number as petals. Pistils mostly 4–5, ± distinct, mostly many-ovuled. Seeds minute, slender, pointed at both ends, mostly brownish. Ca. 350 spp., N. Temp. Zone and mts. of trop. (Latin name, to assuage, because of the healing properties of Houseleek, to which *Sedum* was applied by some writers.)

(Fröderstrom, H. The genus Sedum L. Med. Göteborgs Bot. Trädgård 5 [Appendix], 1935. Praeger, L. Sedum as found in cultivation. Jour. Royal Hort. Soc. 46: 1–310, 1921. Clausen, R. T., & C. H. Uhl. Taxonomy and cytology of subgenus Gormania of Sedum. Madroño 7: 161–180, 1944.)

A.   Fls. usually unisexual, 4-merous; lvs scattered, not forming rosettes . . . . 7. *S. Rosea*
AA. Fls. usually bisexual, 5-merous; lvs. usually forming rosettes, sometimes also scattered along stems.
    B.   Petals long-acuminate, 10–12 mm. long; cauline lvs. like the basal . . . . 6. *S. oreganum*
    BB. Petals obtuse to acute, 5–10 mm. long.
        C.   Rosette lvs. larger and quite different from cauline lvs.
        D.   Petals separate to their bases, erect for ca. 1/10 their length, then widely spreading; infl. a 3-parted cyme.
            E.   Rosettes loose, not densely compressed; rosette lvs. minutely crenulate and papillose. Widely distributed . . . . 1. *S. spathulifolium*
            EE. Rosettes densely compressed; rosette lvs. prominently papillate on the margins. Klamath Mts., n. Calif. . . . . 2. *S. Purdyi*
        DD. Petals united for at least 1/4 their length, divergent above; infl. a paniculate cyme.
            E.   Fls. yellow, sometimes fading to whitish in age; infl. a paniculate cyme; rosette lvs. 1–3.5 cm. long. Sierra Nevada and N. Coast Ranges
                                                    3. *S. obtusatum*
            EE. Fls. white or cream to pink; infl. a dense paniculate or corymbose cyme; rosette lvs. 1–4.5 cm. long.
                F.   Fls. white or creamy white; sepals ovate, 2–3 mm. long. Siskiyou Co.
                                                4. *S. oregonense*
                FF. Fls. pink or pinkish, rarely white; sepals lanceolate to ovate, 2–5 mm. long. From Lake to Del Norte and Siskiyou cos. . . . . . . 5. *S. laxum*
        CC. Rosette lvs. not much different from cauline lvs.
        D.   Fls. yellow; lvs. broadest usually at or near the base. Cent. and n. Calif.
            E.   Lvs not becoming scarious before falling.
                F.   Lvs. alternate, not very turgid; carpels suberect with divergent tips
                                                  8. *S. lanceolatum*
                FF. Lvs. opposite, very turgid; carpels widely spreading . . 11. *S. divergens*
            EE. Lvs. becoming conspicuously scarious before falling . . . . 9. *S. stenopetalum*
        DD. Fls. white; lvs. narrowed toward base. San Bernardino Mts. . . . . 10. *S. niveum*

1. **S. spathulifòlium** Hook. [*S. californicum* Britton.] Perennial with slender rootstocks and sterile stems 1–8 cm. long and 1.5–3 mm. thick; lvs. in prominent rosettes, usually spatulate, blunt or ± emarginate, crenulate on margins, glaucous, 0.5–3 cm. long; sterile shoots 1.5–2 mm. thick, usually naked except for the terminal rosette; flowering stems erect or decumbent, 0.5–3 dm. high; cauline lvs. spatulate to elliptic-oblong, 0.6–2 cm. long; infl. a simple to compound 3-parted cyme, 12–51-fld.; sepals lanceolate to lance-ovate, 2–4 mm. long; petals yellow, rarely orange or white, lanceolate, 5–8 mm. long; follicles yellow-green, erect or divergent, 4–7 mm. long; styles 1.2–3 mm. long; $n = 15$ (Hollingshead, 1942).—Rocky places, below 7500 ft.; Red Fir F., Yellow Pine F., Mixed Evergreen F., Douglas-Fir F.; Coast Ranges from Santa Cruz Co. n., Sierra Nevada from Eldorado Co. n.; to B.C. May–July.

Ssp. anómalum (Britton) Clausen & Uhl. [*Gormania a.* Britton. *S. yosemitense* Britton.] Lvs. green, minutely crenulate; sterile stems 1–1.5 mm. thick; $n = 15$ (Clausen & Uhl, 1944).—Rocky places, 2500–7000 ft.; Yellow Pine F., Chaparral; San Bernardino and San Gabriel mts. to Eldorado and Monterey cos. June–July. Plants of larger size, of unknown derivation are the var. *màjus* Praeger.

Ssp. pruinòsum (Britton) Clausen & Uhl. [*S. p.* Britton.] Rosette lvs. very pruinose; stems of offsets frequently falling throughout, as well as with terminal rosette; $n = 15$ (Clausen & Uhl, 1944).—Sea bluffs; N. Coastal Scrub; Humboldt Co. to B.C. May–July.

2. **S. Púrdyi** Jeps. Perennial with creeping or procumbent stems 1–4 mm. thick, 4–7 cm. long; rosette lvs. flat, spatulate, rounded or truncate at apex, papillose on margins, to 1.8 cm. long, yellow-green; flowering stems to 1.5 dm. high; cauline lvs. oblong-spatulate, spreading, 0.3–1.5 cm. long; infl. a 3-parted cyme, 10–39-fld.; pedicels to 2 mm. long; calyx-lobes linear-lanceolate, green, 2–2.5 mm. long; petals lanceolate, bright yellow to white, 5–7 mm. long; follicles erect or divergent, 4–7 mm. long; $n = 15$ (Clausen & Uhl, 1944).—Rocky ledges and slopes, 500–5000 ft.; Mixed Evergreen F. to Yellow Pine F.; Klamath Mts., s. Siskiyou, Trinity, and Shasta cos. April–May.

3. **S. obtusàtum** Gray. [*Gormania o.* Britton. *G. Hallii* Britton. *G. Burnhamii* Britton. *S. rubroglaucum* Praeger.] Perennial with rootstocks to 5 mm. thick; rosette lvs. fleshy,

spatulate, glaucous, ± rounded apically, 0.5–2.5 cm. long, usually with some red; flowering stems erect, 3–16 cm. high, often reddish; cauline lvs. oblong-spatulate, spurred; infl. a paniculate cyme, 2–10 cm. long, 2–5 cm. broad; calyx-lobes ovate-lanceolate, obtuse, green, 2–4 mm. long; petals lemon-yellow, fading to buff or pink, erect below, ± spreading above, 6–9 mm. long, united for 2–3 mm.; follicles erect, 6–7 mm. long, red in maturity; $n = 15$ (Hollingshead, 1942).—Rocky ridges and slopes, 5000–13,000 ft.; mostly Lodgepole F., Subalpine F., Alpine Fell-fields; Sierra Nevada from Tulare to Plumas cos.; w. Nev. June–July.

Ssp. **boreàle** Clausen. Basal lvs. larger, 1–3.5 cm. long, ± retuse at apex; stems paler red; $n = 15$ (Hollingshead, 1942).—Rocky places, 3500–7500 ft.; Yellow Pine F. to Subalpine F.; mts. of Trinity, Humboldt, Siskiyou, and Del Norte cos.; s. Ore. May–July.

4. **S. oregonénse** (Wats.) Peck. [*Cotyledon o.* Wats. *Gormania Watsonii* Britton. *S. W.* Tides.] Perennial with branched spreading rootstocks to 7 mm. thick; rosette lvs. fleshy, spatulate, rounded, truncate to ± retuse at apex, glaucous, yellow-green often pinkish, 1–2.8 cm. long; flowering shoots 6–27 cm. high, ± reddish; cauline lvs. oblong-spatulate to rounded, spurred, 0.5–1.2 cm. long; infl. paniculate, 2–9 cm. long; calyx-lobes ovate, erect, 2–3 mm. long; petals oblong, whitish, 7–8 mm. long; follicles erect, 6–7 mm. long; $n = 45$ (Hollingshead, 1942).—Rocky places, 6000–8200 ft.; Subalpine F., Alpine Fell-fields; Siskiyou Co.; adjacent Ore. July.

5. **S. láxum** (Britton) Berger. [*Gormania laxa* Britton. *Cotyledon Brittoniana* Fedde.] Stout, with rootstocks to 15 mm. thick; rosette lvs. stiff, flat, narrow-spatulate, truncate or retuse, dark green to glaucous, 2–4 cm. long; flowering stems stout, to 2–4 dm. high; cauline lvs. oblong-spatulate; infl. ± paniculate, 5–17 cm. long, ca. 10 cm. across; calyx-lobes ovate-lanceolate; petals deep pink, oblong-lanceolate, acute, spreading above, 8–10 mm. long; carpels erect, 7–8 mm. long; $n = 15$ (Hollingshead, 1942).—Dry rocky slopes, ca. 2000 ft., Josephine Co., Ore. June. California plants belong to the following sspp.:

A. Infl. congested; rosettes closely crowded, forming a dense mat with rather thin lvs.; stems of
    sterile shoots green to black, 1.5–2 cm. long .... ssp. *retusum*
AA. Infl. lax; rosettes less crowded, forming a loose mat with thick leathery lvs.
    B. Rosette lvs. 2–3 cm. wide, triangular, obcordate; petals pale pink to white .. ssp. *latifolium*
    BB. Rosette lvs. 0.3–2 cm. wide, spatulate to oblong-oblanceolate; petals pink
        C. Cauline lvs. oblong-spatulate, longer than broad .... ssp. *perplexum*
        CC. Cauline lvs. cordate to subcordate, ca. as broad as long .... ssp. *Heckneri*

Ssp. **retùsum** (Rose) Clausen. [*Gormania r.* Rose. *G. Eastwoodiae* Britton. *Cotyledon mendocinoana* Fedde.] Flowering stems 1–1.5 dm. long; calyx-lobes ca. 3 mm. long; petals 6–7 mm. long; $n = 15$ (Hollingshead, 1942).—Exposed places, 1500–5000 ft., Mixed Evergreen F., Yellow Pine F.; Glenn, Lake, and Mendocino cos.

Ssp. **latifòlium** Clausen. Rosette lvs. not or slightly glaucous, 1.5–4 cm. long; flowering stems 1.5–3 dm. long; calyx-lobes lanceolate, 3–4 mm. long; petals 7–10 mm. long; $n = 15$ (Hollingshead, 1942).—Rocky places, ca. 1500 ft.; Mixed Evergreen F.; near Gasquet, etc., Del Norte Co. July.

Ssp. **perpléxum** Clausen. Rosette lvs. glaucous, 0.5–2.5 cm. long; flowering stems 0.7–1.5 dm. long; calyx-lobes 4–5 mm. long; petals 8–10 mm. long; $n = 15$ (Hollingshead, 1942).—Rocky places, below 6500 ft.; Yellow Pine F., Red Fir F.; Trinity to Del Norte and Siskiyou cos.; sw. Ore. June–July.

Ssp. **Héckneri** (Peck) Clausen. [*S. H.* Peck.] Glaucous; rosette lvs. 1–2.5 cm. long; flowering stems 1–2 dm. high; calyx-lobes 2–4 mm. long; petals 8–10 mm. long.—Dry rocky places, 1500–6000 ft.; Yellow Pine F., Red Fir F.; Siskiyou and Humboldt cos.; s. Ore.

6. **S. oregànum** Nutt. [*Gormania o.* Britton.] Perennial, creeping, the stems bare with many ascending branches; sterile stems 2.5–7 cm. long, leafy; lvs. alternate or opposite, green, shining, often reddish, flat, spatulate, 1–2 cm. long; flowering stems 0.8–1.5 dm. high; cauline lvs. like basal; infl. a ± congested cyme, flat, ca. 4 cm. across, 2–3-forked; calyx-lobes ovate-lanceolate, acute, green, 4 mm. long; petals yellow, mostly tinged red, lanceolate-attenuate, long-acuminate, 10–12 mm. long, united ca. ¼ their length; follicles suberect, 6–8 mm. long; $n = 12$ (Clausen & Uhl, 1944).—Rocky places at high elevs.; said to be in n. Calif.; ranging to Alaska.

7. **S. Ròsea** (L.) Scop. ssp. **integrifòlium** (Raf.) Hult. [*Rhodiola integrifolia* Raf.]

Fleshy perennial from a short scaly rootstock; stems several, 0.7–1.5 dm. high; lvs. equally distributed up the stems, flat, comparatively thin, sessile, obovate, 1–1.5 cm. long, acute, entire or dentate above the middle; infl. a terminal, ± congested cyme; fls. dioecious or polygamous, usually 4-merous; calyx-lobes lanceolate, 1.5–2 mm. long; petals dark purple, 3 mm. long, spreading slightly; carpels erect, dark purple, 3–5 mm. long, tipped with a divergent or recurved beak.–Moist rocky places, 7500–12,500 ft.; Subalpine F., Alpine Fell-fields; Sierra Nevada from Tulare Co. to Eldorado Co., White Mts.; w. Nev. to Colo., Alaska, e. Siberia. May–July.

8. **S. lanceolàtum** Torr. [*S. stenopetalum* auth., not Pursh.] Perennial with slender branching rootstocks, tufted; lvs. crowded, alternate, sessile, linear, subterete, 0.5–1.5 cm. long, glaucous or dull green; flowering stems 0.7–2 dm. long; fls. in an elongate racemelike cyme with short branches; calyx-lobes lanceolate, fleshy, acuminate, 4 mm. long; petals yellow, narrow-lanceolate, 6–7 mm. long; follicles 4 mm. long, slender, suberect with divergent tips.—Rocky places, 6000–12,000 ft.; Subalpine F. to Red Fir F.; Tulare and Inyo cos., Modoc and Siskiyou cos.; to B.C., Rocky Mts. June–Aug.

9. **S. stenopétalum** Pursh. [*S. Douglasii* Hook.] Perennial from slender rootstocks; stems 0.5–2 dm. high; lvs. narrow-lanceolate, with scarious-sheathing dilated base, ± flattened, 1–3 cm. long, infl. racemose, sparsely branched with scattered fls.; calyx-lobes triangular-oblong, acute or acuminate, 2–2.5 mm. long; petals yellow, narrow-lanceolate, spreading, 6–10 mm. long; follicles widely spreading, 5–6 mm. long.—Rocky places, 4500–5500 ft.; Sagebrush Scrub, Red Fir F., Yellow Pine F.; Lassen and Modoc cos.; to B.C., Ida. June–Aug.

Ssp. **radiàtum** (Wats.) Clausen. [*S. r.* Wats.] Lvs. elliptic-oblong, 0.3–1 cm. long, calyx-lobes 3–4 mm. long.—Rocky places, below 8000 ft.; mostly Yellow Pine F., Mixed Evergreen F., N. Oak Wd.; Coast Ranges from Monterey Co. n., Sierra Nevada from Tulare and Fresno cos.; s. Ore. June–Aug.

10. **S. níveum** A. Davids. Prostrate glabrous branching perennial with fleshy rhizomatous stems 1–3 dm. long and short ascending branches; lvs. many, green, alternate, oblong-obovate, 5–6 mm. long, fleshy; fls. solitary or in small cymes; calyx-lobes lanceolate, 2.5–3 mm. long; petals white, with pinkish median stripe, lanceolate, 6–7 mm. long; follicles erect.—Rocky shaded ledges, 7000–9500 ft.; Red Fir F., Lodgepole F.; San Bernardino Mts.; Santa Rosa Mts. June–July.

11. **S. divérgens** Wats. Perennial with slender branching rootstocks and erect or ascending flowering stems 5–12 cm. high; lvs. opposite, obovate to spatulate, 5–8 mm. long, rounded or obtuse at apex, glabrous; cyme 3–5 cm. broad; fls. on short stout pedicels; calyx-lobes triangular; petals yellow, linear-lanceolate, spreading, 5–6 mm. long; follicles widely divergent.—Rocky alpine slopes; reported from extreme n. Calif.; to B.C. July–Aug.

5. Congdònia Jeps.

Minute herb with slender tuber-bearing rootstock. Stem scapoid, naked or with 1–2 inconspicuous lvs. above and bearing a single terminal erect fl. Lvs. ovate, rather thin, closely imbricate in basal rosettes. Petals white, united at base. Carpels erect, several-ovuled. Seeds rubescent. One sp. (Named for J. W. *Congdon,* onetime resident of Mariposa Co. and collector of Sierran flora.)

1. **C. pinetòrum** (Bdg.) Jeps. [*Sedum p.* Bdg.] Stems slender, 1–4 cm. high; lvs. sessile, 2–3 mm. long; petals 3–4 mm. long.—E. slope of Sierra Nevada, near Mammoth, Mono Co.

6. Cotylèdon L.

Plants ± woody at base. Lvs. sessile, opposite or alternate. Fls. erect or drooping, yellow, red, or greenish, in terminal cymes. Calyx 5-parted. Corolla tubular, cylindrical, gamopetalous, usually much longer than calyx. Stamens 10, mostly included. Carpels 5, free, with narrow scale at base of each. Fr. of several-seeded follicles. Ca. 30 spp., largely of S. Afr. (Greek, *cotule,* a cavity, from the cuplike lvs. of some spp.)

1. **C. orbiculàta** L. Stout, branched, 6–12 dm. high; lvs. opposite, oblong to round-obovate or spatulate, 5–10 cm. long, flat, glaucous, powdery, often with red or purple

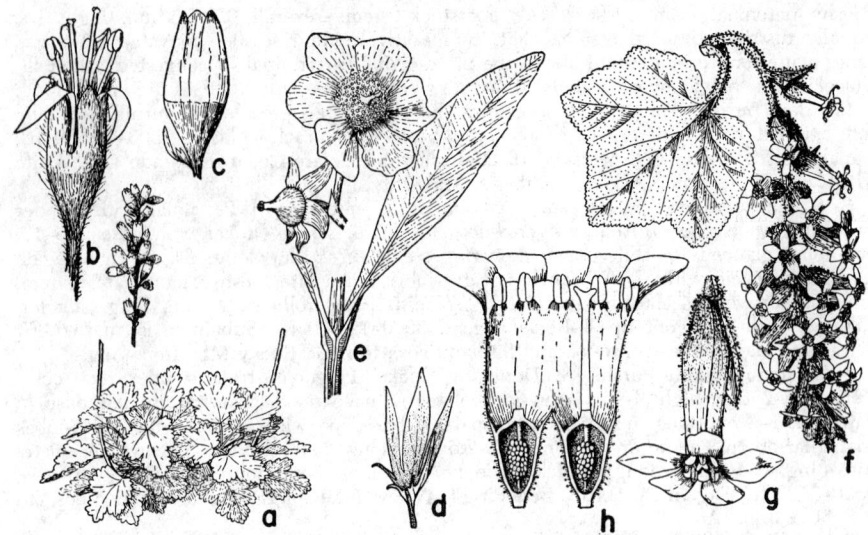

Fig. 68. SAXIFRAGACEAE. *Heuchera rubescens: a,* habit, × ¼, with basal lvs. and spicate infl.;
b, fl., × 4, half-inferior ovary, 5 sepals, 5 petals, 5 stamens. *H. ovalifolia: c,* fl., × 2½, petals and
stamens included. *Lithophragma bulbifera: d,* deeply parted petal. *Carpenteria californica: e,* × ½,
lf., fl. and pistil. *Ribes sanguineum: f,* lf. and raceme, × ½; *g,* fl., × 2, ovary inferior, floral tube,
spreading sepals, erect petals alternating with stamens; *h,* opened fl., × 2½, with parietal placentae.

margins; fls. red, drooping, the cyme on a peduncle ca. 5–6 dm. long; corolla-tube 2–2.5
cm. long, 4–5 times the calyx-length; $n = 9$ (Uhl, 1948).—Natur. on bluffs at Newport
Beach, Orange Co.; native of S. Afr.

72. Saxifragàceae. Saxifrage Family (Fig. 68)

Herbs or shrubs with opposite or alternate lvs., usually without stipules. Fls. perfect,
perigynous, solitary or in racemes, cymes or paniculate clusters. Fl.-tube free or partly
united to the ovary, usually persistent. Sepals 5 or 4. Petals 5 or 4, sometimes more.
Stamens usually 5 or 10, sometimes fewer or more. Carpels mostly fewer than sepals,
± separate or combined into 1 pistil, with ovary superior to inferior. Fr. a caps., follicle
or berry. Seed with endosperm. Ca. 700 spp., mostly of N. Temp. and Arctic regions,
a few Andean.

A. Plants herbaceous, not woody.
    B. Lvs. peltate, 1–4 dm. across; free part of floral tube flattish . . . . 6. *Peltiphyllum*
    BB. Lvs. not peltate, mostly much smaller.
        C. Fls. appearing before the lvs.; petals spatulate; stem from cormlike root . . 5. *Jepsonia*
        CC. Fls. normally appearing with or after the lvs.
            D. Petals absent; stamens 4 or 8; fls. in axils of upper lvs. . . . . . . 9. *Chrysosplenium*
            DD. Petals present; stamens mostly 5 or 10; fls. mostly in definite infls.
                E. Fertile stamens 10.
                    F. Styles 3; petals clawed, usually laciniate or toothed; rootstocks slender,
                        tuberous . . . . 8. *Lithophragma*
                    FF. Styles 2; petals clawed or sessile, entire or toothed; rootstocks various.
                        G. Ovary 2-celled, with axile placentae; petals entire . . 7. *Saxifraga*
                        GG. Ovary 1-celled with parietal placentae.
                            H. Petals entire, almost filiform; caps. unequally 2-valved
                                            10. *Tiarella*
                        HH. Petals laciniate-pinnatifid; caps. equally 2-valved
                                                12. *Tellima*
        EE. Fertile stamens 5 or 3.
            F. Fertile stamens alternating with clusters of gland-tipped sterile staminodia;
                fls. solitary, scapose . . . . 1. *Parnassia*
            FF. Fertile stamens not alternating with sterile staminodia.
                  G. Stamens 3; petals filiform, entire . . . . 11. *Tolmiea*

GG. Stamens 5.
  H. Plants with short bulbiferous rootstocks.
    I. Petals subulate, greenish, with purplish tips and edges;
       ovary free from the fl.-tube . . . . 2. *Bolandra*
    II. Petals elliptic-obovate, white; ovary partly adnate to
       fl.-tube . . . . 3. *Suksdorfia*
  HH. Plants from ± scaly rhizomes.
    I. Petals pinnately or ternately cleft . . . . 13. *Mitella*
    II. Petals entire.
      J. Ovary 2-celled with placentae axile . . 4. *Boykinia*
      JJ. Ovary 1-celled with placentae parietal . 14. *Heuchera*
AA. Plants definitely shrubby.
  B. Lvs. opposite; fr. a caps., partly inferior.
    C. Caps. depressed-globose, beakless; styles distinct, deciduous; petals whitish, 3 mm.
       long. Coast Ranges . . . . 19. *Whipplea*
    CC. Caps. beaked by the persistent styles.
      D. Petals white, 7–30 mm. long; stamens 20 to many.
        E. Plants with deciduous lvs.; petals 7–16 mm. long . . . . . . 15. *Philadelphus*
        EE. Plants evergreen with persistent lvs.; petals 20–30 mm. long
                                                              16. *Carpenteria*
      DD. Petals pink or white, 3–7 mm. long; stamens 10.
        E. Lvs. crenate-serrate, 1–2.5 cm. long; petals 6–7 mm. long . . . . 17. *Jamesia*
        EE. Lvs. entire, 0.6–1.4 cm. long; petals 3–4 mm. long . . . . 18. *Fendlerella*
  BB. Lvs. alternate; fr. a berry, wholly inferior . . . . 20. *Ribes*

1. Parnássia L. GRASS-OF-PARNASSUS

Glabrous perennial scapose herbs, with short rootstocks. Lvs. entire, in basal tuft, or sometimes also with a small sessile lf. on scape. Fls. solitary, terminal, white or pale yellow. Fl.-tube short, poorly developed. Calyx deeply 5-lobed. Petals 5, conspicuously veined, each bearing at its base a group of gland-tipped sterile fils. (staminodia). Stamens 5, alternating with the sepals. Ovary superior to half inferior, 1-celled, with 3–4 projecting parietal placentae. Caps. 3–4-valved; seeds many, winged; endosperm none. Ca. 25 spp., of N. Temp. and Subarctic regions. (Named for Mt. *Parnassus*).

Petals entire, 10–18 mm. long . . . . 1. *P. palustris*
Petals fimbriate on the sides of basal half, mostly 10–12 mm. long.
  Staminodia with 3–5 short lobes, ca. 1 mm. long; lvs. cordate to reniform . . . . 2. *P. fimbriata*
  Staminodia with 7–12 filiform fils. ca. 2–3 mm. long; lvs. oval . . . . 3. *P. cirrata*

1. **P. palústris** L. var. **califórnica** Gray. [*P. c.* Greene.] Basal lvs. ovate, 2.5–4 cm. long, with 5–7 principal veins, cuneate at base, petioles 2–10 cm. long; scape 2.5–5 dm. high, with an ovate bract 5–10 mm. long, sessile above the middle or sometimes wanting; sepals lance-ovate, 4–6 mm. long; petals oblong-ovate to roundish, 3–7-veined; sterile fils. 15–24, capillary, gland-tipped, 3–4 mm. long, united in lower half; stamens ca. 7–8 mm. long.—Wet meadows, below 7000 ft. in Coast Ranges, below 11,000 in Sierra Nevada; Foothill Wd., Yellow Pine F. to Subalpine F., Mixed Evergreen F.; San Bernardino Mts., Sierra Nevada to Mt. Shasta, Coast Ranges from San Benito Co. n.; to sw. Ore. July–Oct.

2. **P. fimbriàta** Konig. Lf. blades 2–4 cm. broad, petioles 5–15 cm. long; scape 2–3.5 dm. high, with a cordate bract near its middle 5–15 mm. long; sepals elliptic to oval, 5–6 mm. long; petals obovate, clawed; staminodia united into lobed scales; fils. 4–5 mm. long; caps. 8–10 mm. long.—Boggy places, 6600–9020 ft.; Red Fir F. to Subalpine F.; Nevada and Placer cos. to Modoc Co.; to Alaska, Rocky Mts. July–Sept.

3. **P. cirràta** Piper. Basal lvs. 1–2 cm. long, on petioles 2–6 cm. long; scape 2–4 dm. high with lanceolate to ovate bract 5–10 mm. long; sepals lanceolate to lance-oblong, 5–7 mm. long; petals oblong-obovate, ca. 1 cm. long; staminodia with filiform gland-tipped fils.; fertile fils. 3–5 mm. long.—Wet places, 2500–8000 ft.; Red Fir F., Yellow Pine F., San Bernardino and San Gabriel mts., upper Sacramento V. Aug.–Sept.

2. Bolándra Gray

Perennial herbs with bulbiferous rootstocks. Stems leafy. Lvs. thin, palmately veined. Fls. loosely paniculate, purplish. Fl.-tube deeply campanulate, free from ovary. Sepals 5, long-attenuate. Petals 5, lanceolate-attenuate. Stamens 5, alternate with petals; fils. subulate to filiform. Carpels 2, united part way; ovary completely 2-loculed; stigmas small, capitate. Seeds many, pendulous. Two spp. (Named for H. N. *Bolander*, early Calif. botanist.)

1. **B. califórnica** Gray. Stems slender, 1.5–2 dm. high, glandular-puberulent above; lower lvs. petioled, the blades round-cordate, glabrous, 2–4 cm. wide, 5–7-lobed into rounded-ovate, ± deeply toothed lobes; petioles 2–10 cm. long; upper lvs. reduced; fl.-tube ca. 5 mm. long; sepals 3–4 mm. long; petals subulate, greenish, with purplish tips and edges, ca. 5 mm. long; carpels 7–8 mm. long.—Wet rocks, 3300–8000 ft., Red Fir F., Yellow Pine F.; Sierra Nevada from Mariposa to Eldorado cos. June–July.

3. Suksdórfia Gray

Leafy-stemmed perennials from short bulbiferous rootstocks. Basal lvs. ± reniform in outline, crenate to ternately divided; cauline lvs. stipulate. Infl. a loose panicle. Fl.-tube ± campanulate, adnate to lower part of ovary. Sepals 5. Petals 5, spatulate to obovate or oval, narrowed to a claw, entire or lobed. Stamens 5, opposite the sepals, with anthers subsessile or on definite fils. Ovary 2-loculed, half to almost wholly inferior; styles short; stigmas truncate or capitate. Caps. dehiscent between styles. Two spp. of Pacific Nw. (Named for W. *Suksdorf*, long a resident botanist in Wash.)

1. **S. ranunculifòlia** (Hook.) Engler. [*Saxifraga r.* Hook. *Hemieva r.* Raf.] Stems 1–3 dm. high, slender, glandular-puberulent, simple; lower lvs. ternately divided to base, the lfts. 1–2 cm. long, cuneate and 3–4-lobed, petioles 2–10 cm. long; upper lvs. smaller; infl. a short compact cyme; fl.-tube 1–1.5 mm. long; sepals triangular-ovate, tinged purplish, 1.5–2 mm. long; petals white, 4–6 mm. long, elliptic-obovate, entire.—Wet rocks, 5000–6000 ft.; Red Fir F.; Plumas Co. to Siskiyou Co.; to B.C., n. Rocky Mts. June–Aug.

4. Boykínia Nutt.

Perennial herbs with scaly rootstocks. Stems simple with several alternate lvs. and a basal tuft of lvs. Lvs. reniform, toothed to lobed, usually stipulate. Infl. a leafy bracted paniculate or corymbose cyme. Fl.-tube campanulate, adnate to lower half of ovary. Sepals 5, lanceolate to lance-ovate. Petals white, 5, spatulate to obovate, usually clawed and early deciduous. Stamens 5, opposite the sepals; fils. short. Ovary and caps. 2-loculed; styles 2, forming 2 divergent beaks. Seeds many, ovoid, minutely punctate. Ca. 8 spp., N. Am. and e. Asia. (Dr. *Boykin*, a resident in Georgia in early 19th century.)

Petals well exserted; stipules evident; lvs. cleft or incised.
    Upper lvs. with large green leafy stipules; petals 5–7 mm. long .... 1. *B. major*
    Upper lvs. with brownish stipules often reduced to bristles; petals 3–4 mm. long ...... 2. *B. elata*
Petals scarcely exceeding calyx; stipules not evident; petals 2 mm. long .... 3. *B. rotundifolia*

1. **B. màjor** Gray. [*Therofron m.* Kuntze.] Stems stout, 3–9 dm. high, ± glandular-pubescent with brown hairs; lower lvs. on petioles 1–2 dm. long, the blades reniform or round-cordate in outline, 1–2 dm. broad, 5–7-cleft, then again cleft and coarsely toothed with gland-tipped teeth; middle cauline lvs. smaller; infl. dense, many-fld., densely glandular-puberulent; fl.-tube 2–3 mm. long, glandular-puberulent; sepals triangular-lanceolate, acute, 3 mm. long; petals broadly oval to obovate; seeds black, ca. 0.5 mm. long.—Moist rocky places and banks, below 7500 ft.; Mixed Evergreen F., Yellow Pine F., Red Fir F.; from Trinity and Madera cos. n. to Wash., Mont. June–Sept.

2. **B. elàta** (Nutt.) Greene. [*Saxifraga e.* Nutt. in T. & G. *Therofron e.* Greene.] Stems slender, erect, 2–6 dm. high, with minute brown gland-tipped hairs; lower lvs. thin, 2–8 cm. wide, reniform to ovate-cordate, somewhat incised into 5–7 lobes with bristle-pointed teeth, petioles 5–15 cm. long; upper lvs. reduced; infl. a many-fld. panicle, densely glandular-puberulent; fl.-tube 2–3 mm. long; sepals lanceolate-acuminate, glabrous, 1.5–2 mm. long; petals ± oblanceolate; $n = 7$ (Hamel, 1953).—Shaded springy places below 5000 ft.; N. Coastal Scrub; Mixed Evergreen F., Redwood F., Chaparral, Yellow Pine F.; Coast Ranges from Santa Monica Mts. n. to Del Norte Co., Sierra Nevada from Amador Co. n.; to Wash. June–July.

3. **B. rotundifòlia** Parry. [*Therophon r.* Wheelock.] Rather stout, 3–8 dm. high, densely glandular-villous; lower petioles 8–16 cm. long, upper shorter; lf.-blades round-cordate, mostly 5–12 cm. wide, shallowly round-lobed with crenate-dentate teeth; panicle

2–15 cm. long; fl.-tube glandular-pubescent, striate, 3–4 mm. long; sepals 4–5 mm. long, lance-ovate; petals obovate-spatulate, with definite claw; $n = 13$ (Hamel, 1953). —Wet places in canyons, below 5500 ft., Chaparral; Elsinore and San Jacinto mts. to s. face of San Bernardino and San Gabriel mts., Cuyama V. in Santa Barbara Co. June–July.

5. Jepsònia Small

Perennial herbs with cormlike rootstocks, few basal mostly vernal lvs., and slender autumnal scapes. Lvs. round-cordate, petioled, shallowly lobed. Fls. in terminal cymes. Fl.-tube ± campanulate, scarcely adnate to the ovary, the upper part free, veiny, mostly purplish-striate. Sepals 5. Petals 5, white, spatulate, clawed, withering-persistent. Stamens 10, shorter than the sepals; fils. filiform. Carpels 2, united to ca. the middle. Follicles veiny, beaked. Seeds small, 4-ridged. Apparently 1 polymorphous sp. (W. L. *Jepson*, longtime professor of Botany at the Univ. of Calif. and authority on flora of the state.)

1. **J. Párryi** (Torr.) Small. [*Saxifraga P.* Torr.] Caudex clothed with the firm dark persistent lf.-bases; lvs. few, 2–6 cm. wide, not so long, pubescent, shallowly several-lobed and toothed, the teeth apiculate, basal sinus open to closed; petioles hirsute to subglabrous, mostly 2–6 cm. long; scapes glandular-puberulent, 1–3 dm. high; fl.-tube and sepals mostly 5–6 mm. long; petals ca. 3.5–6 mm. long, often purple-veined; follicles 5–7 mm. long.—Moist shaded banks, below 2000 ft.; Chaparral, Coastal Sage Scrub, Foothill Wd.; mostly Oct.–Dec. Variable and with perhaps 3 ± geographical forms: (1) plants with sepals tending to be shorter than the fl.-tube and the latter truncate at the base, from sw. San Bernardino Co. to San Diego region and n. L. Calif., are *Párryi*; (2) those from Sierran foothills of Eldorado Co. to Mariposa Co. like the above but with fl.-tube more acute at base and less evidently dark-striped, have been called *J. heterándra* Eastw. [*J. P.* var. *h.* Jeps.]; and (3) those from the Channel Ids., with sepals not shorter than fl.-tube and petals ca. 3–3.5 mm. long, have been named *J. malvaèfolia* (Greene) Small. [*Saxifraga m.* Greene. *J. Neonuttalliana* Millsp.]

6. Peltiphýllum Engl.

Coarse acaulescent herb, perennial, with thick fleshy horizontal rhizome. Lvs. basal, orbicular-peltate, cupped at center, long-petioled. Scapes appearing before the lvs., naked or with 1–2 reduced broadly stipular lvs. Infl. of simple or paniculately compound cymes. Fls. small; fl.-tube flattish, shorter than the sepals; these 5, reflexed in age. Petals 5, white to pink, oblong-obovate to roundish. Stamens 10; fils. subulate. Carpels 2, almost distinct, forming spreading superior follicles. One sp. (Greek, *pelti*, peltate, and *phullon*, lf.)

1. **P. peltàtum** (Torr.) Engl. [*Saxifraga p.* Torr. ex Benth. *Leptarrhena inundata* Behr.] Stout plant, 3–10 dm. tall; lvs. 3–6 dm. broad, 9–15-lobed, mostly glabrous; petioles 3–10 dm. long, glandular-scabrous; scapes brownish-pubescent; sepals 3–4 mm. long, obtuse; petals broad, 5–7 mm. long; follicles 8–11 mm. long, reddish in age; $2n = 34$ (Hamel, 1948).—Banks of mountain streams, below 6000 ft.; Mixed Evergreen F. to Yellow Pine F.; Sierra Nevada from Tulare Co. n., to Siskiyou and Humboldt cos.; s. Ore. April–July.

7. Saxífraga L. SAXIFRAGE

Herbs, largely perennial. Lvs. alternate or opposite or basal, entire to pinnatifid. Fls. perfect, solitary to cymose-paniculate. Fl.-tube ± developed, adnate to at least the base of the ovary. Sepals 5. Petals 5, deciduous, entire. Stamens 10, the fils. subulate to subpetaloid. Ovary from almost superior to partly inferior, 2-loculed, dehiscent between the beaks, sometimes of 2 almost separate follicles. Seeds many, small, smooth to roughish. Ca. 250 spp., mostly of cooler parts of N. Temp. Zone. (Latin, *saxum*, rock, and *frango*, to break; many spp. rooting in rock-crevices.)

(Engler, A., and E. Irmscher. Saxifraga *in* Das Pflanzenreich 4 [117]: 1–709, 1916.)
A.   Lf.-blades roundish to broader than long.
B.   The lvs. 0.6–1.4 cm. broad, mostly rather deeply lobed; plants mostly less than 1 dm. tall
                                                                                        1. *S. debilis*

BB. The lvs. 2–8 cm. broad, shallowly toothed or lobed; plants 1–3 dm. tall.
    C. Petals unequal, 2 of them 3 to 4 times longer than the other 3; plant stoloniferous
                                          16. *S. sarmentosa*
   CC. Petals subequal; plants not stoloniferous.
      D. Lvs. without stipules, the blades simply and evenly dentate; rhizome slender,
          horizontal; petals 2.5–3.5 mm. long .... 3. *S. punctata*
     DD. Lvs. with stipular dilations, the lf.-blades doubly toothed; rhizomes erect, corm-
          like; petals 3–5 mm. long .... 4. *S. Mertensiana*
AA. Lf.-blades definitely longer than broad.
    B. Petals spatulate; petioles jointed with the lf.-blades .... 2. *S. fragarioides*
   BB. Petals ovate to roundish; petioles not jointed with the blades.
      C. Lower stems with ± horizontal perennial branches densely covered with linear semi-
          terete lvs.; plants woody at base .... 5. *S. Tolmiei*
     CC. Lower stems not producing such branches; lvs. almost all basal; plants not woody
         at base.
        D. Petals dissimilar, 3 broader than the other 2; infl. often with bulblets.
           E. The petals 3.5–5 mm. long; lvs. sharply toothed; stoutish perennial
                                          6. *S. ferruginea*
          EE. The petals 2–3 mm. long; lvs. mostly entire; slender annual
                                          7. *S. bryophora*
       DD. Petals alike; infl. lacking bulblets.
          E. Petals 1–2 mm. long, not much longer than sepals.
            F. Scapes glabrous or nearly so except at base; infl. mostly capitate; petals
               with basal claw .... 8. *S. aprica*
            FF. Scapes glandular-pubescent; infl. mostly a narrow panicle; petals scarcely
               clawed at base .... 9. *S. nidifica*
         EE. Petals 2.5–5 mm. long, obviously exceeding sepals.
            F. Lf.-blades nearly or quite glabrous on upper surface.
               G. The lf.-blades less than twice as long as wide, rather coarsely
                  toothed.
                  H. Sepals reflexed at least in age; petals with 2 greenish to
                     yellow spots below the middle, 2.5–3.5 mm. long
                                     13. *S. Marshallii*
                  HH. Sepals not reflexed; petals lacking such spots, ca. 4 mm. long
                                       15. *S. fallax*
              GG. Lf.-blades usually more than twice as long as wide, mostly entire
                 to shallowly sinuate-dentate.
                  H. Lvs. gradually narrowed to the base, often scarcely petioled,
                     usually 7–15 cm. long; scapes mostly 4–8 dm. tall
                                     10. *S. oregana*
                  HH. Lvs. definitely petioled, often smaller; scapes 0.5–4 dm. tall.
                    I. Scapes 2–4 dm. tall, densely glandular-puberulent; lvs.
                     fleshy, not cuneate at base ....11. *S. fragosa*
                  II. Scapes 0.5–1.5 dm. tall, glabrous except at base; lvs.
                     rather thin, cuneate at base .... 12. *S. Howellii*
            FF. Lf.-blades usually quite evidently hairy or pilose on upper surface, petals
               3.5–5 mm. long; scapes 1.5–3 dm. tall.
                G. Lvs. mostly more than twice as long as wide; sepals reflexed after
                  anthesis. Below 3000 ft. .... 14. *S. californica*
               GG. Lvs. less than twice as long as wide; sepals erect after anthesis.
                  From above 5000 ft. .... 15. *S. fallax*

1. **S. débilis** Engelm. [*S. cernua* var. *d.* Engler. *S. rivularis* Calif. refs., not L.]
Glabrous to glandular-puberulent perennial from a small rootstock, tufted, 0.3–1 dm.
tall, slender-stemmed, leafy; lower lvs. thin, reniform in outline, mostly 3–5-lobed, on
slender petioles 1–3 cm. long; upper lvs. smaller, more bractlike; stems 1–few-branched,
each branch with 1 terminal fl.; fl.-tube ± campanulate, 3–4 mm. deep; sepals oblong-
ovate, 1.5–2 mm. long; petals white, oblong-spatulate, 3–6 mm. long; carpels 6–7 mm.
long, united except for the divergent styles.—Damp places in shade of overhanging
rocks, 11,000–12,000 ft.; Alpine Fell-fields; Sierra Nevada of Inyo, Tulare, Madera,
and Tuolumne cos.; B.C. to Rocky Mts. and n. Ariz. July–Aug.

2. **S. fragarioìdes** Greene. [*Saxifragopsis f.* Small.] Perennial with woody root-crown
with stout horizontal branches; lvs. in basal tufts, subglabrous, cuneate-obovate, uni-
foliately compound, 1.5–4 cm. long, dentate at summit; petioles ca. as long as blades;
flowering stems glandular-pubescent, scapelike, 1–2.5 dm. high, with reduced bractlike
lvs., bearing narrow terminal panicles; sepals lance-ovate, reflexed in age, 1.5–2 mm.
long; petals white, obovate, 2–3 mm. long, reflexed in age; fr. ca. 4–5 mm. long, the
carpels united at least to the middle.—Rock-crevices, 5000–7300 ft.; Red Fir F. to
Subalpine F.; mts. of Humboldt and Siskiyou cos. to s. Ore. July–Aug.

3. **S. punctàta** L. ssp. **argùta** (D. Don) Hult. [*S. a.* D. Don.] Perennial from a

horizontal rhizome; lvs. basal, orbicular to orbicular-reniform, 1.5–7 cm. wide, glabrous, coarsely dentate, paler beneath, on petioles 2–20 cm. long; scapes 2–4 dm. tall, slender, glandular-pubescent upward; infl. open, with spreading rather few-fld. cymules; fl.-tube scarcely evident; sepals lance-oblong, 1.5–2 mm. long, reflexed, usually purplish; petals white with 2 yellow dots, rounded, 2.5–5 mm. long, with distinct claws; fils. broadened upward; follicles 6–8 mm. long, ± purplish, with short spreading to recurved tips.— Moist stream banks, 6500–11,200 ft.; Red Fir F. to Subalpine Fell-fields; San Bernardino Mts., Sierra Nevada n., Yolla Bolly Mts. (Tehama Co.) n.; to Wash., Rocky Mts. July–Aug.

4. **S. Mertensiàna** Bong. [*Heterisia Eastwoodiae* Small.] Rootstock scaly, bulblike; lvs. largely basal, orbicular, commonly cordate at base, 2–7 cm. wide, glabrous to ± pubescent, on petioles 3–12 cm. long; fl.-stems slender, 1–3 dm. high, finely glandular-pubescent, with 1–few reduced lvs. in lower part; infl. paniculate, open; fl.-tube scarcely evident; sepals oblong, 1.5–2.5 mm. long, reflexed, mostly glabrous; petals white to pink, 3–5 mm. long, ovate to oblong, short-clawed; fils. broad; carpels half united or more, 5–7 mm. long.—Moist rocky places, below 2000 ft., N. Coastal Scrub, Mixed Evergreen F., Douglas-Fir F. etc., Coast Ranges from Sonoma to Del Norte cos.; at 4000–5000 ft., Yellow Pine F., Sierra Nevada of Mariposa and Nevada cos.; to Alaska, Ida. Feb.–July.

5. **S. Tòlmiei** T. & G. [*S. ledifolia* Greene.] Densely tufted perennial, diffusely branched, the stems ± prostrate, short, densely clothed with oblong-linear closely imbricated glabrous thickish lvs. 0.8–1.5 cm. long; flowering stems slender, 0.3–1.2 dm. high, finely glandular-pubescent, often reddish, few-branched at summit; fl.-tube poorly developed; sepals glabrous, ovate, 2 mm. long; petals white, elliptic-spatulate, 4–5 mm. long; fils. dilated; caps. 8–12 mm. long.—Moist rocky places, 8500–11,800 ft.; Lodgepole F. to Alpine Fell-fields; Sierra Nevada from Tulare Co. n., Coast Ranges from Tehama Co. n.; to Alaska. July–Aug.

6. **S. ferrugínea** Grah. [*S. Bongardii* Presl.] Basal tuft of lvs. from summit of very slender rootstock, their blades spatulate to ± oblanceolate, 1.5–4 cm. long, sharply toothed, ± ciliate and stiff-pubescent, narrowed to broad petioles; flowering stem scape-like, slender, finely glandular-puberulent, 1–3 dm. high, openly and paniculately branched, bulbiferous and floriferous; fl.-tube poorly developed; sepals reflexed, oblong-ovate, 1–2 mm. long; petals white with 2 yellow basal spots, unequal, the 3 upper lanceolate, the 2 lower elliptic-spatulate, 3.5–5 mm. long; fils. narrow upward; caps. ca. 5 mm. long.—Wet banks, 5500–7500 ft.; Red Fir F., Subalpine F.; Yolla Bolly Mts., Siskiyou Mts., Trinity Mts., Marble Mts., Salmon Mts.; to Mont. and Alaska. July–Aug.

7. **S. bryóphora** Gray. [*Spatularia b.* Small.] Apparently wintering over as bulblets or seeds, the plants themselves annual, glandular-pubescent, with lvs. in basal tuft, oblong-oblanceolate to linear-oblong, sessile or nearly so, 0.5–2.5 cm. long, mostly entire; flowering stems slender, 0.5–2 dm. high, finely glandular-pubescent, openly branched above; pedicels filiform, all but the terminal soon deflexed, bulbiferous; fl.-tube obsolete; sepals mostly reflexed, ovate, 1–2 mm. long; petals white with 2 yellow spots, ± unequal, elliptic, 2–3 mm. long; caps. 3–4 mm. long.—Moist gravelly places, 7000–11,200 ft.; Red Fir F. to Alpine Fell-fields; Sierra Nevada from Tulare Co. to Plumas Co. July–Aug.

8. **S. apríca** Greene. [*S. nivalis* of Calif. refs., not L. *S. umbellulata* Greene, not Hook. f. & Thomas.] Plants usually purplish; lvs. in a basal tuft from a perennial simple caudex, ovate to oblong or spatulate, glabrous, 1–4 cm. long, on shorter petioles, entire to dentate; scapes solitary, subglabrous, except for minute glandular pubescence toward base, 0.3–1 dm. high, simple; infl. subcapitate; fl.-tube ca. 1 mm. long; sepals ± ovate, 1–1.5 mm. long; petals white, obovate, 1.5–2 mm. long, with clawlike base; stamens filiform; caps. purple, 2.5–4 mm. long, with divergent tips.—Moist gravelly and stony places, 5500–12,000 ft.; Subalpine F., Alpine Fell-fields; Sierra Nevada from Tulare Co. n., to Siskiyou, Trinity, and Tehama cos.; s. Ore., w. Nev. May–Aug.

9. **S. nidífica** Greene. [*S. montana* (Small) Fedde.] Caudex short, bulbiferous; lvs. in basal tuft, ovate, 1–4 cm. long, on shorter petioles, entire to repand-denticulate, glabrous; flowering stems slender, glandular-pubescent, 1–3 dm. high, mostly few-branched above; fl.-tube ca. 1 mm. long; sepals ovate to oblong, ca. 1.5 mm. long; petals white, roundish-obovate, ca. 2 mm. long; fils. somewhat dilated toward base; caps. ca. 2–2.5 mm. long, the tips ± divergent.—Moist places, 2500–11,000 ft.; Montane

Coniferous F.; Sierra Nevada from Tulare and Inyo cos. n.; Coast Ranges from Lake Co. n.; to Ore. June–Aug.

10. **S. oregàna** Howell. [*S. sierrae* (Cov.) Small. *S. integrifolia* Calif. refs., not Hook.] Rootstock stout, creeping; lvs. in basal tuft, oblanceolate-spatulate to elliptic, mostly glabrous, 3–12 cm. long, narrowed to the broadly margined petiole or scarcely petiolate, ± glandular-toothed; flowering stems stoutish, 3–8 dm. tall, glandular-pubescent, ± branched above; cymules clustered; fl.-tube ca. 1 mm. long; sepals reflexed, broadly ovate, ca. 2 mm. long; petals white, obovate, 2–4 mm. long, sessile; fils. broad; caps. 3.5–5 mm. long, often purplish, with spreading beaks.—Wet meadows and boggy places, 3500–11,000 ft.; Montane Coniferous F.; Tulare Co. to Siskiyou, Trinity, and Modoc cos.; to Wash., Ida., Nev. May–Aug.

11. **S. fragòsa** Suksd. Much like *S. oregana,* the lvs. more definitely petiolate, deltoid-ovate, 3–5 cm. long, on petioles as long or much longer; flowering stems 2–4 dm. high, more slender; sepals 1.5 mm. long; petals 2.5–3.5 mm. long.—Wet rocky places, 3000–8000 ft.; Chaparral, Yellow Pine F., Red Fir F., Mixed Evergreen F.; from Lake and Sonoma cos. n. to Siskiyou and Modoc cos.; to Wash. April–July.

12. **S. Howéllii** Greene. Caudex perennial; lvs. in basal rosette, few, 1–2 cm. long, coarsely toothed, oblong-cuneate, rusty arachnoid beneath, with petioles ca. 1–2 cm. long; flowering stems slender, ± rusty-pubescent especially toward the base; cymules few-fld.; sepals ± oblong, ca. 1.5 mm. long; petals oblong, 2.5–3 mm. long; fils. slender; caps. 2.5–3 mm. long.—Moist places, below 2000 ft.; Mixed Evergreen F., N. Oak Wd.; Del Norte Co., Siskiyou Co. to sw. Ore. March–May.

13. **S. Marshállii** Greene. Caudex short; lvs. basal, oblong to ovate-oblong, thickish, 2–4 cm. long, rather coarsely dentate, reddish-pubescent beneath, on petioles ca. as long; flowering stems 1–3 dm. high, somewhat glandular-puberulent, branched above; cymules lax; fl.-tube almost lacking; sepals reflexed, oblong-ovate, 2–2.5 mm. long; petals 2.5–3.5 mm. long, oblong to oval; fils. broadened upward; caps. 2.5 mm. long with spreading tips.—Moist places, 300–7000 ft.; Mixed Evergreen F. to Subalpine F.; n. Mendocino to Del Norte and Siskiyou cos.; to Wash., Ida. May–June.

14. **S. califórnica** Greene. [*S. virginiensis* var. *c.* Jeps. *S. napensis* Small.] Caudex erect, short; lvs. basal, ovate to oblong-elliptic, denticulate to serrate, 1–5 cm. long, on somewhat shorter petioles, green to purplish; flowering stems 1–3 dm. tall, ± glandular-pubescent, loosely branched above; fl.-tube broad, shallow; sepals ovate, 1.5–2 mm. long, soon reflexed, often purplish; petals obovate to broadly elliptic, 3.5–5 mm. long; fils. ± strap-shaped; caps. 2.5–3.5 mm. long, purplish, the carpels but slightly connate. —Common on shaded often grassy banks, mostly below 2500 (4900) ft.; Coastal Sage Scrub, Chaparral, Foothill Wd., Closed-cone Pine F., etc.; cismontane Calif. from Santa Ana Mts. and San Bernardino Co. n. through Coast Ranges and Sierran foothills to sw. Ore., Santa Cruz Id. Feb.–June.

15. **S. fállax** Greene. [*S. gracillima* Johnson.] Caudex short; lvs. basal, glabrous to ± pubescent above, oblong-oval, rather coarsely sinuate-toothed, 1–4 cm. long, on petioles as long or longer, rounded apically; flowering stems slender, 1–3 dm. high, glandular-pubescent, loosely branched above; fl.-tube broad, 1–1.5 mm. long; sepals ovate, 1–1.5 mm. long, usually not reflexed; petals oblong, ca. 3–4 mm. long, clawed.—Less frequent, rocky places, 3000–7500 ft.; Red Fir F., Yellow Pine F., Foothill Wd.; Sierra Nevada from Fresno Co. to Plumas Co.; to s. Ore. March–May.

16. **S. sarmentòsa** L. STRAWBERRY-GERANIUM. Perennial, with runners like a strawberry's; loosely hairy; lvs. round-cordate, 3–10 cm. broad, veined white above, reddish beneath; fls. white in bracted panicles; longer petals ca. 12 mm., others ca. 3–4 mm. long, often spotted with purple or yellow; $n = 18$ (Skovsted, 1934).—Garden plant from Asia; reported natur. in Sonoma Co.

8. Lithophrágma Nutt. WOODLAND-STAR

Perennial herbs with slender bulblet-bearing rootstocks and simple slender scapose flowering shoots. Lvs. largely basal, petioled, roundish to reniform in outline, mostly 3–5-lobed or parted; petioles with stipulelike dilated bases. Fls. in simple racemes. Floral tube hemispheric or campanulate to obconic; sepals 5, rather short. Petals 5, white to pinkish, clawed, entire to toothed or cleft, usually ± unequal. Stamens 10, included, the fils. short. Ovary 1-celled with 3 many-seeded parietal placentae; styles 3, unequal. Fr. a caps.; seeds ovoid, horizontal. Ca. 12 spp., of w. N. Am. (Greek, *lithos*, rock, and *phragma*, fence or hedge.)

(Small, J. K., and P. A. Rydberg *in* N. Am. Fl. 22: 84–89, 1905.)
A. Basal lvs. not lobed to near their bases.
  B. Petals entire or shallowly toothed.
    C. Fl.-tube with a ± rounded, not definitely acute base; pedicels 1–3 mm. long; stem-lvs. alternate.
      D. Base of petal blade not involute or toothed; fl.-tube almost truncate at base
                                                      1. *L. heterophylla*
      DD. Base of petal blade somewhat involute, minutely toothed or laciniate; fl.-tube obtuse at base . . . . 2. *L. scabrella*
    CC. Fl.-tube with an acute base; pedicels 5–10 mm. long; stem-lvs. mostly opposite
                                                3. *L. Cymbalaria*
  BB. Petals deeply parted.
    C. Pedicels 1–2 mm. long; petals 4–7 mm. long; fl.-tube almost truncate at base
                                                1. *L. heterophylla*
    CC. Pedicels 2–8 mm. long; petals 6–10 mm. long; fl.-tube obconic at base . . 4. *L. affinis*
AA. Basal lvs. lobed almost to their bases.
  B. Petals mostly 3–5 mm. long; fl.-tube with a ± rounded base.
    C. Fls. 3–8; fl.-tube scarcely striate.
      D. Stem-lvs. not bearing axillary bulblets; stems 2–3.5 dm. tall . . . . 5. *L. breviloba*
      DD. Stem-lvs. often bearing axillary bulblets which replace the fls.; stems 1–2 dm. tall
                                             6. *L. bulbifera*
    CC. Fls. 8–20; stems 2–5 dm. tall; fl.-tube ± striate . . . . 7. *L. rupicola*
  BB. Petals mostly 6–10 mm. long; fl.-tube with a ± acute base.
    C. Base of fl.-tube acutish but not obconic. S. Calif. . . . . 8. *L. tripartita*
    CC. Base of fl.-tube obconic. From Kern and San Benito cos. n.
      D. Fl.-tube 3 times as long as broad; petals with 3 broadly oblong lobes
                                            9. *L. trifoliata*
      DD. Fl.-tube twice as long as broad; petals with 3–7 linear-oblong lobes
                                            10. *L. parviflora*

1. **L. heterophýlla** (H. & A.) T. & G. [*Tellima h.* H. & A. *L. Bolanderi* Gray. *L. triloba* Rydb.] Stems glandular-pubescent or ± hirsutulose, 2.5–5 dm. high; basal lvs. round-reniform, 1.5–4 cm. wide, crenately shallowly lobed; cauline lvs. alternate, mostly more deeply 3-cleft; fls. mostly 3–9; pedicels 1–2 mm. long; fl.-tube campanulate, rounded to truncate at base, 2–3 mm. long; sepals triangular-ovate, ca. 2 mm. long; petals white, oblong to cuneate, 4–10 mm. long, entire to rather deeply lobed.—Shaded slopes, below 6500 ft.; many Plant Communities; San Diego Co. to Ore., most common in Coast Ranges of cent. Calif. March–June.

2. **L. scabrélla** (Greene) Greene. [*Tellima s.* Greene. *L. laciniata* Eastw.] Stems ± hispidulose and glandular-scabrous, 2–5 dm. high; basal lvs. 1–3 cm. wide, ± crenate-lobed; stem-lvs. more dissected; fls. 3–15; fl.-tube with an obtuse base; sepals triangular-ovate, 1.5–2 mm. long; petals white, 5–7 mm. long, with long slender claws and oblong blades often bearing minute basal lateral teeth.—Wooded slopes, 2000–8800 ft.; Foothill Wd. to Red Fir F.; n. Los Angeles Co., Tehachapi Mts., Kern Co. to w. base of Sierra Nevada to Tehama Co. May–July.

Var. **Peirsònii** Jeps. Many lvs. bearing bulblets in their axils.—At ca. 3500–5000 ft.; Foothill Wd.; Liebre Mts., n. Los Angeles Co. May–June.

3. **L. Cymbalària** T. & G. [*Tellima C.* Steud.] Stems finely glandular-puberulent, 2–3.5 dm. tall; basal lvs. round-reniform, 3–5-lobed, 1–2.5 cm. wide; stem-lvs. 2, opposite; fls. 3–8; fl.-tube rounded-turbinate; sepals broadly deltoid, ca. 2 mm. long; petals white, 6–8 mm. long, entire, oblong-spatulate with narrow claws.—Dry slopes, below 3000 ft.; Chaparral, Foothill Wd., etc.; Palomar Mts. (San Diego Co.), Ventura and Santa Barbara cos. to Stanislaus Co., Channel Ids. March–May.

4. **L. affinis** Gray. [*Tellima a.* Gray. *L. intermedia* Rydb. *L. catalinae* Rydb. *L. trifida* Eastw.] Stems 2–6 dm. tall, glandular-hirsutulous; basal lvs. roundish, 1–4 cm. wide, shallowly to ± deeply 3-lobed, the divisions crenate to incised; petioles 3–15 cm. long, usually villous; cauline lvs. 1–3, alternate; fls. 3–12; fl.-tube obconic below, densely pubescent with ± gland-tipped hairs; sepals ovate, acutish; petals white, 7–10 mm. long, usually rather equally 3-lobed at apex.—Common on moist grassy banks, mostly below 3500 ft.; many Plant Communities; mostly cismontane Calif.; to s. Ore. March–May.

5. **L. brevilòba** Rydb. Stem glandular-pubescent; basal lvs. with blades 1–2.5 cm.

wide, deeply ternately lobed into broadly cuneate divisions with 3–4 rounded-oblong lobes; stem lvs. few, alternate, deeply divided; fls. 3–8; pedicels 2–5 mm. long; fl.-tube rounded at base, densely glandular-puberulent, 2–3 mm. long, the sepals triangular, acute, half as long; petals pink, 3–5 mm. long, palmately 3–5-parted into narrow segms. —Open slopes and woods, 2000–7000 ft.; Yellow Pine F., Red Fir F.; Sierra Co. to Siskiyou and Modoc cos. April–July.

6. **L. bulbífera** Rydb. [*L. glabra* var. *b.* Jeps. *Tellima b.* Fedde.] Very close to *L. breviloba;* stipules more fimbriate; cauline lvs. and many of fl.-bracts bearing axillary bulblets.—Moist places, 4500–11,000 ft.; Sagebrush Scrub to Subalpine F.; San Bernardino and San Gabriel mts., Sierra Nevada from Tulare Co. n., Coast Ranges from Trinity Co. n.; to Wash., Rocky Mts. April–July.

7. **L. rupícola** Greene. Near to the two preceding, but stems coarser, hispidulous-scabrous, basal lvs. often bulbiferous in their axils; stipules of stem lvs. conspicuous, widened above, fimbriolate; racemes elongate, 8–20-fld.; fl.-tube hemispheric-campanulate, crisped-hispidulous, conspicuously striate, 3–4 mm. long; sepals short-deltoid, scarcely 1 mm. long; petals white, ca. 3 mm. long, with 3–5 spatulate-oblong divisions.—Moist places, 4000–8000 ft.; Sagebrush Scrub, N. Juniper Wd., Yellow Pine F., Red Fir F.; Lassen and Modoc cos.; s. Ore. May–July.

8. **L. tripartìta** (Greene) Greene. [*Tellima t.* Greene.] Stems slender, 1.5–2.5 dm. high, hispidulous, glandular above; basal lvs. grayish-hirsute, 1–2.5 cm. wide, ternately cleft to near their base into incised divisions; stem lvs. few, alternate; fls. 3–7; pedicels 2–5 mm. long; fl.-tube acute at base, ± hispidulous, 2.5–4 mm. long; sepals ca. 1 mm. long deltoid; petals 6–8 mm. long, white, mostly 3-cleft.—Dry slopes, 5000–7500 ft.; Yellow Pine F.; Palomar Mts. (San Diego Co.) to Greenhorn Mts. (Kern Co.) May–June.

9. **L. trifoliàta** Eastw. Stems scabrous-hispid below, glandular above, 2–3 dm. high; basal lvs. 1–2 cm. wide, hirsute, with 3–5 cuneate divisions; stem lvs. mostly 2, alternate; fls. 5–8; pedicels 1–3 mm. long; fl.-tube elongate-obconic, hispidulous, 4–5 mm. long, the sepals ca. 1.5 mm. long, elongate-triangular, sharply acute; petals 8–10 mm. long, with 3 ovate or oblong lobes.—Dry slopes; Foothill Wd., Yellow Pine F.; from Yuba and Tehama cos. to Modoc Co. March–April.

10. **L. parviflòra** (Hook.) Nutt. [*Tellima p.* Hook. *L. austromontana* Heller?] Stems 2–3.5 dm. high, glandular-pubescent and ± scabrous; basal lvs. 1–3 cm. wide, somewhat white-hairy, with 3–5 divisions cuneate and in turn cleft; stem lvs. 2–3, alternate; fls. 3–9; pedicels 2–5 mm. long; fl.-tube elongate-obconic, hispidulous, 3–4 mm. long, with horizontal yellow band at summit; sepals deltoid, ca. 1–1.5 mm. long; petals white or pinkish, 5–10 mm. long, deeply 3–5-cleft into linear-oblong lobes.—Open slopes, 2000–5700 ft.; Foothill Wd., Yellow Pine F., Jeffrey Pine F.; Tehachapi Mts. to San Benito Co. and along Sierra Nevada to Modoc Co.; to B.C., Rocky Mts. March–June.

9. **Chrysosplènium** L. Golden-Saxifrage

Low dichotomously branched herbs with perennial rootstocks. Stems leafy. Lvs. opposite or alternate, petioled, crenate, without stipules. Fls. small, axillary or terminal, greenish, solitary or in small corymbs. Fl.-tube campanulate to saucer-shaped, adnate to lower part of ovary. Sepals mostly 4. Petals 0. Stamens mostly 4–8, inserted on margin of disk lining fl.-tube; fils. short. Ovary 1-celled, flattish, 2-lobed; styles 2, short, recurved. Caps. membranous, ± 2-lobed, 2-valved, few- to many-seeded. Seeds muricate or pilose. Ca. 15 spp., of N. Temp. Zone and s. S. Am. (Greek, *chrysos,* gold, and *splen,* spleen, from reputed medical quality.)

1. **C. glechomaefòlium** Nutt. in T. & G. Glabrous, the stems ± ascending, 1–2 dm. long; lf.-blades roundish, 6–15 mm. broad, abruptly cuneate at base, crenate, the petioles slender, 8–18 mm. long; fls. solitary, axillary; fl.-tube 2–3 mm. broad, the sepals rounded, entire; stamens 8, ca. as long as fl.-tube; caps. exserted in age.—Wet mucky places, below 500 ft.; Redwood F.; Mendocino Co. to Del Norte Co.; to B.C. April–June.

10. Tiarélla L.

Perennial herbs with scaly rootstocks and mostly basal lvs. which are long-petioled, lobed to trifoliate. Stipules small, adnate to petioles. Flowering shoots leafy, slender, lateral, with a terminal raceme or panicle. Fl.-tube small, adnate to base of ovary only. Sepals 5. Petals 5, clawed, linear to oblong or elliptic. Stamens 10, long and slender. Ovary 1-celled, with 2 horns tapering into long filiform styles. Caps. membranous, with 2 unequal valves. Seeds rather few. Ca. 5 spp., N. Am. and Asia. (Diminutive of Greek, *tiara*, a high cap, referring to form of caps.)

1. **T. unifoliàta** Hook. [*Heuchera californica* Kell. *T. u.* var. *procera* Gray. *T. c.* Rydb.] SUGAR-SCOOP. Glabrous to ± white-hirsute, glandular above, 1–5 dm. high; basal lvs. broadly cordate in outline, 3–8 cm. wide, acutely 3–5-lobed with ovate divisions, doubly crenate to coarsely toothed, the petioles 5–18 cm. long; cauline lvs. 1–4, similar but reduced; panicle narrow, 1–2 dm. long; sepals whitish to pinkish, ovate, hispidulous, ca. 1.5 mm. long; petals linear-subulate, white, ca. 3 mm. long; stamens well exserted; carpels oblong, the larger 9–12 mm., the shorter 4–6 mm. long; seeds oblong-obovoid, brown, smooth, shining, 1.5 mm. long.—Shaded places below 2000 ft.; Redwood F., Mixed Evergreen F., Closed-cone Pine F.; from Santa Cruz Mts. along the Coast to Del Norte Co.; to Alaska and Mont. May–July.

11. Tolmièa T. & G.

Perennial herbs with scaly rootstocks and chiefly basal lvs., these cordate, long-petioled, with membranous stipules. Flowering stems with a few lvs. and terminal bracteate, many-fld. racemes. Fl.-tube cylindric-funnelform, free from ovary, 9-veined, unequally 4-cleft. Sepals 5, unequal, 3 central larger than 2 lateral. Petals 4, rarely 5, filiform, subulate, elongate, recurved, persistent. Stamens 3 or 2, with broad fils., scarcely exserted, opposite the 3 larger sepals. Ovary 1-celled, with 2 long subequal apical beaks; placentae 2, parietal, many-seeded. Seeds minute, round, muricate-hispid. Monotypic. (Named for Dr. W. F. *Tolmie*, Hudson Bay Co. physician at Ft. Vancouver in 1832.)

1. **T. Menzièsii** (Pursh) T. & G. [*Tiarella M.* Pursh. *Heuchera M.* Hook.] Basal lvs. obscurely lobed, cuspidate-toothed, hairy, 3–10 cm. wide, on petioles 5–20 cm. long; cauline lvs. similar but smaller; stems slender, 3–8 dm. long; fl.-tube greenish, and with some purple, ca. 5–6 mm. long; 3 upper sepals ca. 3 mm. long, 2 lateral shorter; petals brown, 5–7 mm. long; caps. projecting from slit on lower side of fl.-tube, 12–14 mm. long; *n* = 14 (Skovsted, 1934).—Moist cool places, below 6000 ft.; Red Fir F., Mixed Evergreen F., Redwood F., Douglas-Fir F.; Glenn Co. to Del Norte and Siskiyou cos.; to Alaska. May–June.

12. Tellìma R. Br.

Perennial herbs with horizontal rootstocks and numerous basal lvs. as well as some cauline. Flowering stems simple, with a long terminal, many-fld. raceme. Fl.-tube large, urn-shaped to inflated-campanulate, 10-nerved, adnate to ovary for almost half its length. Sepals 5, ovate, erect. Petals 5, sessile by a broad base, laciniate-pinnatifid, reflexed. Stamens 10, on short fils., included. Ovary 1-celled, with 2 parietal many-seeded placentae, narrowly 2-beaked. Caps. opening longitudinally between the beaks. Seeds oblong, tubercled. A monotypic genus. (Name an anagram of *Mitella*.)

1. **T. grandiflòra** (Pursh) Dougl. [*Mitella g.* Pursh.] FRINGE-CUPS. Flowering branches stout, 4–8 dm. high, spreading-hirsute; basal lvs. roundish to round-reniform in outline, ± hirsute, shallowly 3–7-lobed, cordate at base, serrate to crenate, 5–10 cm. broad, on petioles 5–20 cm. long; cauline few, smaller; raceme glandular-puberulent, few–many-fld.; pedicels 1–3 mm. long; fl.-tube glandular-puberulent, striate, 3–5 mm. long; sepals 1.5–2 mm. long; petals at first whitish, later red, 4–6 mm. long, the segms. filiform; caps. ovoid, open at apex, the styles divergent, hardened; *n* = 7 (Skovsted, 1934).—Moist woods and rocky places, below 5000 ft.; Redwood F., Mixed Evergreen F., Yellow Pine F.; Coast Ranges from San Luis Obispo Co. n. and Sierra Nevada from Placer Co. n.; to Alaska. April–June.

13. Mitélla L. Miterwort. Bishop's-Cap

Low perennials with rootstocks. Lvs. round-cordate, alternate, slender-petioled, mostly basal. Flowering stems slender, often scapelike with terminal spikelike raceme of small fls. Fl.-tube adherent to base of ovary. Sepals 5. Petals 5, slender, pinnately or ternately cleft. Stamens 5 or 10, included. Styles 2, short. Caps. short, 2-beaked, 1-celled, with 2 parietal or basal several-seeded placentae, 2-valved at summit. Seeds minute, smooth, shining, ± obovoid. Ca. 12 spp., of temp. and colder regions of N. Am. and Asia. (Diminutive of Latin, *mitra*, cap, because of the young fr.)

Petals greenish, pinnately cleft or pinnatifid; fl.-tube disk-shaped or saucer-shaped.
    Ovary more than half superior, not flattened above; flowering stems leafy .... 1. *M. caulescens*
    Ovary almost entirely inferior, with flat disklike top; flowering stems leafless.
      Stamens opposite the petals .... 2. *M. pentandra*
      Stamens alternate with the petals.
        Lf.-blades nearly glabrous, reniform .... 3. *M. Breweri*
        Lf.-blades distinctly hairy, oval with cordate base .... 4. *M. ovalis*
Petals whitish, palmately 3-cleft at apex; fl.-tube broadly turbinate.
    Lvs. all basal, obscurely lobed .... 5. *M. trifida*
    Lvs. basal and usually 1 cauline, commonly 3-lobed .... 6. *M. diversifolia*

1. **M. cauléscens** Nutt. in T. & G. [*Mitellastra c.* Howell.] Flowering stems 1–3.5 dm. high, slender, 1–3-leaved, glandular-puberulent; basal lvs. roundish to broadly cordate, shallowly 5-lobed, irregularly crenulate, 1–4 cm. wide, ± scabrous, the petioles 5–10 cm. long, reflexed-hirsute; stem lvs. similar, smaller, with green stipules; fls. yellow-green, the fl.-tube saucer-shaped, ca. 3 mm. in diam.; sepals triangular, 1–1.5 mm. long; petals with 5–9 filiform lobes; fils. purplish, connivent; caps. globose, styles distinct; $n = 7$ (Hamel, 1953).—Moist, ± shaded places, 2000–5000 ft.; Mixed Evergreen F., Yellow Pine F., Douglas-Fir F.; Humboldt Co. to B.C., Mont. May–June.

2. **M. pentándra** Hook. [*Mitellopsis p.* Walp. *Pectiantia p.* Rydb.] Subglabrous, the scapes slender, 1–3 dm. high, leafless; basal lvs. glandular-puberulent above, cordate to round-reniform, 2–6 cm. wide, obscurely round-lobed, crenate to ± serrate, scattered pubescent beneath, on petioles 2–10 cm. long and with straight hairs; fls. 8–20; fl.-tube ca. 3 mm. wide; sepals broad-deltoid, mostly recurved; petals with ca. 5–7 filiform lobes; caps. 4 mm. broad; seeds with rather pointed ends; $n = 7$ (Hamel, 1953).— Uncommon, shaded rocky or wooded places, 5000–8000 ft.; Red Fir F. to Lodgepole F.; Tulare and Inyo cos. n. to Siskiyou and Trinity cos.; to Alaska and Rocky Mts. May–July.

3. **M. Brèweri** Gray. [*Pectiantia B.* Rydb.] Much like *M. pentandra*, petioles ± tawny-hirsute with curled hairs; stamens opposite the sepals; seeds with somewhat rounded ends; $n = 7$ (Hamel, 1953).—Damp shaded slopes, 6000–11,500 ft.; Red Fir F. to Subalpine Fell-fields; Sierra Nevada from Tulare Co. n.; to B.C., Mont. June–Aug.

4. **M. ovàlis** Greene. [*Pectiantia o.* Rydb.] Near to the preceding, but more hirsute with short white, ± retrorse hairs; petals with 3–5-divisions; stamens alternate with petals.—Moist places; Mixed Evergreen F., Redwood F.; Marin Co., Mendocino Co. to B.C. April–May.

5. **M. trífida** Grah. [*Ozomelis t.* Rydb. *M. anomala* Piper?] Rootstock rather long; lvs. basal, cordate to round-reniform, crenate, obscurely lobed, glabrous to short-hispid, 2–6 cm. long, on petioles 3–10 cm. long, retrorse-hirsute; flowering stems 2–4 dm. high, glandular-scabrous, slender, 5–25-fld. toward summit; fl. ca. 3 mm. long, puberulent, the sepals whitish; petals whitish, ca. 2 mm. long, narrow, mostly 3-cleft near apex.—Partly shaded slopes in woods, 3000–6000 ft.; Mixed Evergreen F., to Red Fir F.; from Trinity and Plumas cos. n.; to B.C., Alberta. May–July.

6. **M. diversifòlia** Greene. [*Ozomelis d.* Rydb.] Like the preceding, but the lvs. mostly angularly 3-lobed in upper half, otherwise subentire; flowering stems usually with 1 lf. just above their base.—Wet places, 2500–6000 ft.; Yellow Pine F., Red Fir F.; Trinity and Siskiyou cos.; to Wash. June–July.

14. Heùchera L. Alum-Root

Perennial herbs mostly with stout caudices or rhizomes. Lvs. mostly basal, long-petioled, roundish-cordate, lobed and notched or toothed; stipules united with petiole at base. Flowering stems lateral, scapose or ± leafy, with a terminal paniculate infl. Fl.-tube half epigynous, urceolate or cylindric to turbinate or saucer-shaped. Sepals 5, often unequal. Petals 5, entire, small, spatulate to linear or obovate, sometimes 0. Stamens 5, opposite the sepals; fils. mostly filiform. Ovary half inferior, 1-celled with 2 parietal placentae; styles filamentous to short and cylindrical. Caps. opening between the 2 ± divergent beaks. Seeds small, ovoid, mostly echinate with minute spines. Ca. 50 spp., of N. Am. (Named for J. H. von *Heucher*, 1677–1747, German medical botanist.)

(Rosendahl, Butters, and Lakela. A monograph on the genus Heuchera. Minn. Studies in Plant Science 2: 1–180, 1936.)

A.   Mature styles long and slender, the closed distal portion not less than 1.5 mm. long; fl.-tube urceolate or cylindric to deeply campanulate.
B.   Free portion of the fl.-tube shorter than the part adnate to the ovary; fls. mostly regular.
C.   Fls. turbinate; styles long, slender, much exserted.
D.   Fls. 1–3 mm. long; lvs. 2–7 cm. wide; plants almost or quite acaulescent. Mainland .... 1. *H. micrantha*
DD.   Fls. 3.5–4.5 mm. long; lvs. 6–18 cm. wide; plants caulescent. Insular.
    2. *H. maxima*
CC.   Fls. rounded at the base, soon hemispheric; styles short, stout, only slightly exserted.
D.   Lvs. deeply cordate, deeply lobed and pentagonal in outline; petioles 7–20 cm. long, densely villous. Humboldt to San Luis Obispo cos. ...... 3. *H. pilosissima*
DD.   Lvs. shallow-cordate to truncate, moderately lobed and rounded in outline; petioles 2–4.5 cm. long, hirsute. Trinity and Siskiyou cos. .... 4. *H. Merriami*
BB.   Free portion of fl.-tube equal to or longer than adnate portion; fls. ± irregular.
C.   Stamens long exserted, surpassing the sepals and ca. equal to petals; free portion of fl.-tube less than 1 mm. long; petals very narrow, filamentlike; styles long-exserted.
D.   Fls. almost regular; petals equal, up to 0.3 mm. wide.
E.   Inferior part of ovary narrow obconic, distinctly longer than broad. San Diego Co. .... 7. *H. leptomeria*
EE.   Inferior part of ovary turbinate to broadly obconic, usually broader than long.
F.   Fls. open-campanulate, 3–5 mm. long, 2–3 mm. wide; sepals villous. Plumas Co. to San Bernardino Co. .... 6. *H. rubescens*
FF.   Fls. hemispheric, 4 mm. long and wide; sepals sparsely hirsute. Del Norte and Siskiyou cos. .... 8. *H. Pringlei*
DD.   Fls. irregular; petals unequal, to 0.5 mm. wide. San Bernardino Mts.
E.   Fls. campanulate; sepals crisp-hairy .... 9. *H. Parishii*
EE.   Fls. narrow-campanulate, soon urceolate; sepals glandular-hirtellous
    10. *H. alpestris*
CC.   Stamens mostly included, shorter than petals; free portion of fl.-tube more than 1 mm. long on its shorter side; petals distinctly wider than fils.; styles scarcely exserted.
D.   Fls. glandular-puberulent with a few longer hairs at sepal tips; petioles not hirsute. San Gabriel Mts. .... 11. *H. Abramsii*
DD.   Fls. villous-hirsute; petioles ± hirsute.
E.   Stamens reaching about to the apex of the sepals; free portion of fl.-tube over 1 mm. long on its shorter side.
F.   All sepals with oblong or ovate free parts.
G.   Fl.-tube cylindric, soon urceolate; stamens subequal. San Gabriel Mts. .... 12. *H. elegans*
GG.   Fl.-tube widest above; stamens unequal. Tulare and Kern cos.
    13. *H. caespitosa*
FF.   Sepals on long side of fl. with only very short semicircular tips free. San Jacinto Mts. .... 14. *H. hirsutissima*
EE.   Stamens distinctly shorter than sepals; free portion of fl.-tube ca. 0.7 mm. long. Laguna Mts., San Diego Co. .... 5. *H. brevistaminea*
AA.   Mature styles (true styles, not hollow carpel beaks) cylindrical, not more than 1 mm. long; fl.-tube open-campanulate or saucer-shaped.
B.   Fls. 6–8 mm. long; petals almost wanting. Siskiyou and Modoc cos. .... 15. *H. ovalifolia*
BB.   Fls. 2–2.5 mm. long; petals evident. Mono Co. .... 16. *H. Duranii*

1. **H. micrántha** Dougl. ex Lindl. var. **pacífica** Rosend., Butt. and Lak. With well developed caudex; basal lvs. 5–7-lobed (the lobes with broad, ± cuspidate teeth), shallowly cordate at base, slightly longer than broad, coarsely and sparsely strigose above, more densely so along veins beneath, mostly 3–8 cm. long, on soft-hairy petioles 5–18 cm. long; flowering stems stoutish, 3–7 dm. high, 0–4-leaved, the paniculate infl. lax, puberulent with somewhat stalked glands and mostly linear bracts; fls. 1–3 mm. long at anthesis, the fl.-tube obconic at base, glandular-puberulent, or with some longer hairs, greenish or with some purple; sepals almost 1 mm. long; petals whitish, ca. 2 mm. long, oblanceolate with narrow claw; stamens exserted; caps. ca. 4–5 mm. long, ovoid; seeds black, ca. 0.5 mm. long, ± short-spiny; $n = 7$ (Skovsted, 1934).—Moist

banks of humus and rocks, below 2000 ft.; Closed-cone Pine F., Mixed Evergreen F., Redwood F.; near the coast from San Luis Obispo Co. to Del Norte Co. and s. Ore. May–July. A form with the same distribution and having ± hirsute infl.-axis and peduncles, is var. *Hartwégii* (Wats.) Rosend. [*H. pilosissima* var. *H.* Wats. ex Wheelock.] At higher elevs. in the N. Coast Ranges, intergrading with:

Var. **erubéscens** (A. Br. & Bouché) Rosend. [*H. e.* A. Br. & Bouché.] Lvs. glabrous above; pedicels minutely glandular-puberulent, the glands scarcely if at all stalked.— At 2500–7000 ft.; Yellow Pine F., Red Fir F.; Sierra Nevada from Tulare Co. n., to Siskiyou Co.

2. **H. máxima** Greene. Much like *H. micrantha*, but with heavier longer caudex; flowering branches stouter, often 3–4-lvd., mostly hirsute; basal lvs. round-cordate, 6–18 cm. in diam., with deep basal sinus, 7–9-lobed and with aristate teeth, upper surface subglabrous, lower strigose-villous especially along the veins; petioles 8–20 cm. long, villous-hirsute; panicle narrowly cylindrical, 2–3 dm. long, glandular-puberulent and with some longer hairs; fls. 3.5–4.5 mm. long, the fl.-tube short-campanulate; petals whitish, ca. 2 mm. long; caps. 5–8 mm. long.—Canyon walls and cliffs; Chaparral, Coastal Sage Scrub; Santa Cruz, Santa Rosa, and Anacapa ids. Feb.–April.

3. **H. pilosíssima** F. & M. [*H. hispida* H. & A., not Pursh. *H. hirtiflora* T. & G. *H. hemisphaerica* Rydb. *H. pilosella* Rydb.] Rhizome elongate; flowering stems 2–6 dm. high, brown-hairy with some hairs gland-tipped; basal lvs. round-cordate, 3–7 cm. wide, shallowly lobed, the lobes crenate-toothed with ultimate teeth ± aristate, both surfaces hairy; petioles 8–12 cm. long, with rust-brown hairs; cauline lvs. few, much reduced; infl. rather narrow and compact, glandular and hirsute; fl.-tube globular, villous, 3–4 mm. long with the sepals; petals pinkish-white, oblong-elliptic, clawed, 1.5–2 mm. long; caps. 5–7 mm. long; seeds black, echinate, ca. 0.6 mm. long; $n = 7$ (Schoennagel, 1931).—Wooded slopes, below 1000 ft.; Closed-cone Pine F., Redwood F.; immediate coast from Humboldt Co. to San Luis Obispo Co. April–June. Plants at the s. end of the range tend to have the infl. more lax than more n. ones and are the var. *hemisphaèrica* (Rydb.) Rosend. [*H. h.* Rydb.]

4. **H. Mérriami** Eastw. Caudex branched; basal lvs. roundish, cordate or truncate, crenately 5–7-lobed, the sinuses deep, blades 2–5 cm. in diam., hispid to glabrate above, hirsute beneath, on petioles 2–4.5 cm. long; flowering branches slender, 1–2 dm. high, scapose, villous-hirsute to subglabrous; panicle narrow, compact, glandular-hirtellous; fls. 2–3 mm. long, the fl.-tube subhemispheric, glandular-puberulent and ± hairy; sepals obtuse; petals spatulate, ca. 1.5 mm. long; caps. 3–4 mm. long, rounded; seeds echinate, ca. 0.6 mm. long.—At 6000–7500 ft.; Red Fir F., Subalpine F.; Salmon Mts., of Trinity and Siskiyou cos. July–Aug. Possibly of hybrid origin between *H. pilosissima* and *H. rubescens*.

5. **H. brevistamínea** Wiggins. Caudex short, woody; flowering stems 1–3 dm. high, scapose, glandular-puberulent and with some longer hairs; basal lvs. roundish, 1.5–3 cm. wide, shallowly 5-lobed, the lobes shallowly toothed, white-ciliate; petioles white-hirsute, 2–4 cm. long; infl. narrow, rather lax; fls. 4–5 mm. long, the fl.-tube campanulate, reddish-purple, glandular, 1.7 mm. long; sepals oblong with rounded tips; petals ca. 3.5 mm. long, oblance-spatulate, conspicuously veiny; caps. ca. 5 mm. long; seeds ca. 0.7 mm. long.—Dry rocky places, 5000–6000 ft.; Yellow Pine F.; Laguna Mts., San Diego Co. May–July.

6. **H. rubéscens** Torr. var. **alpícola** Jeps. [*H. r.* var. *pachypoda* (Greene) Rosend., Butt. & Lak. *H. p.* Greene.] Caudex thick, mostly multicipital, cespitose; lvs. basal, coriaceous, 1–2.4 cm. in diam., roundish, subcordate to rounded at base, ± deeply 3–5-lobed, rather sharply dentate, ciliate, the surfaces glandular-puberulent to glabrate in age; petioles slender, 1–5 cm. long, glandular-puberulent and short-hirsute; flowering stems 1–3 dm. high, scapose, ± glandular-puberulent; panicles narrow, 3–11 cm. long, ± secund; lower peduncles 0.5–1.2 cm. long; fls. 4–5.5 mm. long, only slightly irregular; the fl.-tube turbinate, oblique; sepals oblong; petals 3–4 mm. long, oblanceolate, clawed, 1–3-nerved; stamens slightly exceeding petals; styles strongly papillose; caps. 4.5–5.5 mm. long; seeds 0.8 mm. long.—Dry rocky places, 6000–11,900 ft.; Pinyon-Juniper Wd. to Alpine Fell-fields; s. Sierra Nevada (Mono, Inyo, Tulare, Fresno, Mariposa cos.), Inyo-White Mts., Panamint Mts., New York Mts., Clark Mt.; to Nev. May–July.

Var. **Rydbergiàna** Rosend., Butt. & Lak. [*H. r.* var. *nevadensis* Jeps. in part.] Fls.

3.5–3.7 mm. long, sparingly hirtellous; panicle 6–15 cm. long, scarcely secund, narrow, lower peduncles ca. 1 cm. long; pedicels and internodes 1–2 mm. long; petals 3–4 mm. long, 0.35 mm. wide; styles papillose only at their base.—At ca. 6000–11,000 ft.; Red Fir F. to Subalpine F.; cent. Sierra Nevada especially in Yosemite Region. July–Aug.

Var. **glandulòsa** Kell. [*H. lithophila* Heller.] Near to the preceding var., but panicle more lax, lower peduncles 1–3 cm. long, upper pedicels and internodes 3–5 mm. long; styles papillose mostly at their base.—At ca. 6000–10,000 ft.; Red Fir F. to Subalpine F.; largely from Placer to Plumas cos. July–Aug.

7. **H. leptomèria** Greene var. **peninsulàris** Rosend., Butt. & Lak. Near to *H. rubescens* but with lower part of fl.-tube narrow-obconical, distinctly longer than broad, glandular-puberulent, sparingly hirsute; petals 3–5 mm. long, mostly 1-nerved; stamens ca. as long as petals; seeds 0.5 mm. long.—At ca. 5000–6000 ft., Hot Springs Mts., Cuyamaca Mts. of San Diego Co.; L. Calif. May–June.

8. **H. Prínglei** Rydb. Near to *H. rubescens*, but fls. hemispherical, ca. 4 mm. long and wide, scarcely at all zygomorphic, glandular-puberulent with some bristly hairs on sepals; petals 2–3 mm. long, narrow-oblanceolate; styles papillose.—Rock-crevices, 6000–7000 ft.; Subalpine F.; Del Norte and Siskiyou cos. A doubtful sp. Aug.

9. **H. Paríshii** Rydb. [*H. rubescens* var. *P.* Jeps. *H. hirsuta* Rydb.] Much like *H. rubescens* in habit; lvs. 1.5–4 cm. wide; flowering stems 1–3 dm. high, glandular-puberulent and ± hirsute, the panicle narrow, 5–10 cm. long; fls. quite zygomorphic, 4–5 mm. long; fl.-tube with a turbinate base, glandular-puberulent; sepals oblong; petals oblanceolate, 2–3 mm. long; stamens exserted; styles glabrous except at very base; seeds ca. 0.9 mm. long.—Rocky places, 5000–8900 ft.; Yellow Pine F., Red Fir F.; San Bernardino Mts. July–Aug.

10. **H. alpéstris** Rosend., Butt. & Lak. Near to the preceding, lvs. 0.5–2 cm. wide; flowering stems 0.5–2 dm. long; fls. irregular, narrow-campanulate, soon urceolate, ca. 4 mm. long; fl.-tube glandular-puberulent; sepals with some longer hairs near the tips; petals oblanceolate, 2.5–3 mm. long; stamens well exserted; seeds ca. 0.6 mm. long.—At 8500–11,400 ft.; Red Fir F. to Subalpine F.; San Bernardino Mts. July–Aug.

11. **H. Abrámsii** Rydb. [*H. rubescens* var. *A.* M. G. Stewart.] Lvs. 0.5–1.8 cm. wide; deeply lobed; close to the preceding in general appearance, paucity of long pubescence on fls. and very oblique floral tube, but fls. 4.5–6 mm. long; fl.-tube subcylindric; petals ca. 2.5 mm. long, oblanceolate; stamens included; seeds 0.6 mm. long.—At 9000–10,000 ft., Red Fir F.; San Antonio Mts. (e. part of San Gabriel Mts.) July–Aug.

12. **H. elegáns** Abrams. [*H. rubescens* var. *e.* Jeps.] Lvs. 1.5–3.5 cm. wide, shallowly lobed; petioles more hirsute; fls. 4–7 mm. long, narrow campanulate; fl.-tube round-turbinate at base, it and sepals hirsute; petals ca. 5 mm. long; stamens scarcely exserted; seeds 0.7 mm. long.—At 4000 to 8500 ft.; Yellow Pine F. to Red Fir F.; w. San Gabriel Mts. May–June.

13. **H. caespitòsa** Eastw. [*H. rubescens* var. *c.* M. G. Stewart.] Like *H. elegans*, but the fls. ca. 5 mm. long; fl.-tube finely glandular-puberulent, sepals sparsely villous; petals unequal, 2.5–3.5 mm. long.—An uncertain entity, at ca. 8000 ft., in Kern and Tulare cos.

14. **H. hirsutíssima** Rosend., Butt. & Lak. Near to *H. elegans*, but sepals of long side of fl.-tube united almost to top; anthers ± exserted; fl.-tube very oblique.—At 7000–10,800 ft.; Red Fir and Subalpine F.; San Jacinto and Santa Rosa mts. July–Aug.

15. **H. ovalifòlia** Nutt. Cespitose; lvs. all basal, ovate to almost round, 1–3 cm. wide, usually truncate at base, crenate-dentate to 3–5-lobed, glandular-pubescent; petioles 2–7 cm. long, glandular-puberulent and hirtellous; flowering stems 1–3.5 dm. high, glandular-puberulent sometimes also hirtellous; panicle compact, spicate; fls. densely glandular-puberulent and hirsute, 6–8 mm. long, campanulate with broadly turbinate base; sepals 3–4 mm. long; petals almost wanting; stamens included; caps. 8–10 mm. long; seeds ca. 0.7 mm. long.—At 4000–8000 ft.; N. Juniper Wd. to Red Fir F.; e. Siskiyou and Modoc cos. to Wash. and Nev. June–Aug.

16. **H. Duránii** Bacig. From a heavy caudex; lvs. basal, reniform to almost round, 1–2 cm. wide, shallowly crenately 5–9-lobed, glandular-puberulent above and beneath with some short hairs along the veins beneath; petioles 2–5 cm. long; flowering branches 1.5–3.5 dm. high, slender, glandular-puberulent; panicle narrow, 4–8 cm. long, glandular-puberulent; fls. yellowish with some pink, 2–2.5 mm. long, the fl.-tube broadly turbinate, minutely glandular-puberulent; sepals scarcely 1 mm. long; petals ca. 1.5 mm. long;

stamens incurved, not exceeding sepals; caps. 4–5 mm. long, the beaks scarcely exserted.—Rocky hillsides, 8900–11,700 ft.; Bristle-cone Pine F., Subalpine F.; White Mts., Sweetwater Mts., Mono Co.; w. Nev. July–Aug.

15. Philadélphus L. Mock-Orange. Syringa

Deciduous shrubs with opposite entire or toothed, mostly petioled lvs. Fls. bisexual, regular, rather large, white, solitary or in few-fld. cymes in terminal clusters. Sepals 4–5, persistent. Petals 4–5, distinct. Stamens usually many; fils. subulate, free or united below; anthers short. Ovary almost completely inferior, mostly 4-celled; styles short or elongate, ± united. Fruit a many-seeded caps., loculicidal. Seeds with a membranous testa. Ca. 40 spp., of N. Temp. regions; some of considerable horticultural value. (Named for Ptolemy *Philadelphus* of ancient Egypt.)

(Hitchcock, C. L. The xerophyllous species of Philadelphus in sw. N. Am. Madroño 7: 35–56. 1943. Hu, S-y. A monograph of the genus Philadelphus. Journ. Arnold Arb. 35: 275–333, 1954; 36: 52–109, 325–368, 1955; 37: 15–90, 1956.)

Lvs. 0.5–2 cm. long; fl.-clusters mostly 1–3-fld. ....1. *P. microphyllus*
Lvs. mostly 2.5–8 cm. long; fl.-clusters 4–40-fld. ....2. *P. Lewisii*

1. **P. microphýllus** Gray ssp. **stramíneus** (Rydb.) C. L. Hitchc. [*P. s.* Rydb. *P. serpyllifolius* Calif. auth., not Gray.] Much-branched rounded shrub 1–2 m. tall; young branches densely strigose; lf.-blades lance-ovate, 1–2.5 cm. long, strigose above and beneath, entire, on petioles 1–3 mm. long; fl.-tube 2–3 mm. long at anthesis, 3–5 mm. in fr.; ± strigose; sepals 3–5 mm. long, lanate on inner surface, with stiff straight hairs without; petals 7–10 mm. long; stamens 30–50; styles 1–2 mm. long, free above or united to tips.—Dry rocky places, 5000–9000 ft.; Pinyon-Juniper Wd.; White Mts. (Mono Co.) to Clark and New York mts. (e. San Bernardino Co.); to Nev., L. Calif. May–July.

Ssp. **pùmilus** (Rydb.) C. L. Hitchc. [*P. p.* Rydb.] Lvs. mostly 8–12 mm. long, hirsute on upper surface.—At 7000–8500 ft.; Yellow Pine F., Red Fir F.; San Jacinto and Santa Rosa mts. July.

2. **P. Lewísii** Pursh ssp. **califórnicus** (Benth.) Munz. [*P. L.* var. *c.* Gray. *P. c.* Benth. *P. Fremontii* Rydb. *P. L.* var. *Helleri* (Rydb.) Hu. *P. L.* var. *ellipticus* Hu.] Loosely branched, 1–3 m. high; young branches light brown, glabrous, older gray; lf.-blades ovate, 2.6–8 cm. long, entire to minutely denticulate, ± glabrous above, strigose on veins beneath; petioles mostly 3–8 mm. long; fl.-tube glabrous, 2.5–3.5 mm. long; sepals 4–5 mm. long, woolly along the margins; petals 11–16 mm. long, elliptic-oblong to obovate; fils. many; styles united to near the stigmas; caps. 5–6 mm. long; seeds brown, narrow-fusiform, ca. 1 mm. long; $2n = 26$ (Janaki Ammal, 1951).—Rocky slopes and canyons, 1000–4500 ft.; Foothill Wd., Yellow Pine F.; Sierra Nevada from Tulare Co. n., to Humboldt, Siskiyou, and Trinity cos. May–July.

Ssp. **Gordoniànus** (Lindl.) Munz. [*P. G.* Lindl. *P. L.* var. *G.* Jeps. *P. L.* var. *oblongifolius* Hu. *P. oreganus* Nutt. ex Hu.] Lvs. pubescent all over beneath, usually dentate; petals 15–20 mm. long; $n = 13$ (Bangham, 1931).—Below 4000 ft.; Mixed Evergreen F., Yellow Pine F.; Coast Ranges from Lake and Mendocino cos. to Del Norte and Siskiyou cos.; to B.C. May–July.

16. Carpentèria Torr.

Erect shrub with persistent coriaceous opposite lvs. on ± angled branches. Lvs. simple, lance-oblong, bicolored. Fls. bisexual, large, regular, solitary in upper axils or in few-fld. cymes in upper axils and terminal. Fl.-tube broad, shallow, adnate to base of ovary. Sepals 5–7, persistent. Petals white, 5–7, broad, obovate to round-obovate. Stamens many, with filiform fils. Ovary 5–7-celled; style short with 5–7-lobed stigma. Caps. leathery, ± conical, abruptly beaked with the persistent style, loculicidal. Seeds many, minute, brownish, attenuate at both ends, with membranous testa. One sp. (Named for Professor William *Carpenter* of La.)

1. **C. califórnica** Torr. Tree-Anemone. Erect and 1–2(–5) m. high or becoming ± sprawling and 4 m. long, with pale shredding bark; twigs glabrous or nearly so; lvs. 4–9 cm. long, glabrous above, gray beneath with straight appressed hairs and tomen-

tulose; petioles to ca. 1 cm. long; sepals 10–12 mm. long, strigose without, with a lanate band on inner margin; petals 2–3 cm. long; caps. ca. 10–12 mm. long; seeds ca. 1 mm. long.—Dry granite ridges and slopes, 1500–4000 ft.; Foothill Wd., Yellow Pine F.; between San Joaquin and King rivers, Fresno Co. May–July.

17. Jàmesia T. & G.

Low deciduous shrubs with opposite toothed lvs. Fls. in terminal cymose clusters. Fl.-tube hemispheric to turbinate, adnate to lower part of ovary. Sepals 5, lanceolate to subovate. Petals 5, spreading, oblong-obovate, pubescent within. Stamens 10, the alternate shorter; fils. dilated. Ovary imperfectly 3–5-celled; styles 3–5, distinct; stigmas terminal. Caps. conic, beaked by the persistent styles. Seeds many, striate-reticulate. One sp. (Named for Dr. E. *James,* a member of the Long Expedition to the Rocky Mts. in 1820.)

1. **J. americàna** T. & G. var. **califórnica** (Small) Jeps. [*Edwinia c.* Small.] Low, 3–10 dm. high, with shreddy bark, pubescent on young growth; lvs. oblong to roundish, 1–2.5 cm. long, coarsely crenate-serrate, green and pubescent above, gray-strigose beneath, prominently veined; petioles mostly 2–6 mm. long; pedicels, fl.-tube and calyx strigose; sepals lance-ovate, 2–2.5 mm. long; petals rose-pink, narrowly oblong-ovate, 6–7 mm. long; styles 6–7 mm. long in fr.; caps. ca. 4 mm. long; seeds many, brown, attenuate at both ends, reticulate, minute; $n = 16$ (Hamel, 1953).—About rocks, 7800–12,000 ft.; Pinyon-Juniper Wd., mostly Subalpine F. to Alpine Fell-fields; s. Sierra Nevada in Fresno, Tulare, Mono, and Inyo cos., Panamint and Inyo mts.; sw. Nev. and Utah. July–Aug.

18. Fendlerélla Heller.

Low shrub with shreddy bark. Lvs. opposite, entire, small. Fls. bisexual, small, in small dense terminal cymes. Fl.-tube turbinate-campanulate, adnate to lower half of ovary. Sepals 5, lanceolate. Petals 5, white. Stamens 10, alternately longer and shorter; fils. dilated toward base. Ovary 3-celled; styles 3, distinct. Caps. narrow, 3-valved, septicidal. Seeds solitary in each cell. Three spp., sw. N. Am. (Diminutive of *Fendlera* a related genus named for A. *Fendler,* early Texan collector.)

1. **F. utahénsis** (Wats.) Heller. [*Whipplea u.* Wats.] Much-branched, 3–6 dm. high, the young growth strigose; lvs. oblong-elliptic to -spatulate, subsessile, 6–14 mm. long, strigose; sepals ca. 1.5 mm. long; petals 3–4 mm. long; caps. ca. 4 mm. long.—Dry limestone slopes, at ca. 5500–6500 ft.; Pinyon-Juniper Wd.; Last Chance Mts. (Inyo Co.), Clark Mt. (e. San Bernardino Co.); to Utah, Tex. June–Aug.

19. Whípplea Torr.

Trailing slightly woody plant with weak slender branches. Lvs. opposite, deciduous, ovate, slightly toothed. Fls. small, in small crowded terminal cymose clusters. Fl.-tube broadly turbinate. Sepals 5–6, thin, erect. Petals 5–6, white, spreading. Stamens 10 or 12, the alternate shorter; fils. flattened. Ovary globose, 4–5-celled; styles 4–5, distinct, deciduous; stigmas introrse. Ovule 1 in each cell. Caps. globose, with 1 seed in each carpel. One sp. (Named for Lt. A. W. *Whipple,* commander of a Pacific RR. Expedition to Los Angeles in 1853/54.)

1. **W. modésta** Torr. YERBA DE SELVA. Stems 3–18 dm. long, pubescent, rooting freely; lvs. 1–3 cm. long, shallowly few-toothed, subsessile, strigose; peduncles 1–6 cm. long; pedicels 2–10 mm. long; sepals lance-ovate, ca. 2 mm. long; petals oblong-ovate, 3 mm. long; caps. 2–2.5 mm. in diam.—Shaded slopes, below 4500 ft.; Redwood F., Mixed Evergreen F., Yellow Pine F., etc.; Coast Ranges from Monterey Co. to Ore. April–June.

20. Rìbes L. CURRANT. GOOSEBERRY

Shrubs, either unarmed or with nodal spines and sometimes internodal bristles. Lvs. simple, alternate, usually palmately lobed, with stipules adnate to the petiole or lacking. Fls. usually in racemes, sometimes solitary, on 1–2-lvd. axillary shoots; pedicels subtended by bracts and usually bearing 2 bractlets. Fl.-tube adnate to the globose ovary and ± produced above it. Sepals 5, rarely 4; petals as many and usually shorter. Stamens alternate with the petals. Ovary inferior, 1-celled; style 2-lobed or -divided. Fr. a 1-celled pulpy berry, several-seeded; endosperm fleshy; embryo minute, terete. Ca. 120 spp., of the N. Temp. Zone and of the Andes. The gooseberries are sometimes recognized as a separate genus *Grossularia*. (*Ribes*, an ancient Arabic name.)

(Berger, A. A taxonomic review of currants and gooseberries. Tech. Bull. 109 of New York State Agric. Exp. Sta., 1924.)
A. Nodal spines lacking; berry spineless; fls. usually several to many in a raceme.
    B. Lvs. not lobed, persistent, leathery; fl.-tube subrotate beyond the ovary. Insular
                                                                        1. *R. viburnifolium*
    BB. Lvs. lobed, mostly deciduous; fl.-tube beyond the ovary subrotate to campanulate or cylindrical.
        C. Free portion of fl.-tube saucer-shaped or obsolete.
            D. Ovary with sessile glands; stems erect; free portion of fl.-tube saucer-shaped.
                E. Sepals green; berry with whitish bloom; fl.-bracts broader above their middle .... 2. *R. bracteosum*
                EE. Sepals white; berry without bloom; fl.-bracts broader below their middle
                                                                  3. *R. petiolare*
            DD. Ovary with stalked glands; stems ± decumbent; free portion of fl.-tube obsolete
                                                         4. *R. laxiflorum*
        CC. Free portion of fl.-tube campanulate to cylindrical.
            D. Fls. yellow, sometimes tinged with red; lvs. glabrous or nearly so, almost alike on both surfaces .... 8. *R. aureum*
            DD. Fls. not yellow; lvs. ± pubescent or glandular, the 2 surfaces unlike.
                E. The fls. white or greenish-white.
                    F. Free portion of fl.-tube 2.5–3 times as long as broad; berry red.
                        G. Bracts of raceme cuneate-obovate, with dentate truncate summit; styles usually pubescent. Siskiyou to Riverside cos.
                                                            5. *R. cereum*
                      GG. Bracts of raceme narrow-obovate, acute, entire or with lateral teeth; style usually glabrous. E. slope of Sierra Nevada.
                                                        6. *R. inebrians*
                  FF. Free portion of fl.-tube 1–2 times as long as broad; berry blue or black.
                      G. Sepals 4 mm. long; fl.-tube (including ovary) 9–10 mm. long. Tulare Co. to Modoc Co. .... 7. *R. viscosissimum*
                      GG. Sepals 1.5 mm. long; fl.-tube (including ovary) 4–5 mm. long. Santa Barbara Co. s. .... 12. *R. indecorum*
              EE. The fls. pink or red or purplish.
                  F. Anthers with a conspicuous cup-shaped gland at the apex
                                                         7. *R. viscosissimum*
                  FF. Anthers lacking such gland.
                      G. Sepals erect; free part of fl.-tube short, bowl-shaped
                                                    9. *R. nevadense*
                      GG. Sepals spreading or recurved; free part of fl.-tube cylindric to urceolate.
                          H. Styles glabrous; sepals 4–5 mm. long, deep rose. Del Norte Co. to San Luis Obispo Co. .... 10. *R. sanguineum*
                        HH. Styles pubescent; sepals 1.5–4 mm. long.
                            I. Fls. bright rose; lvs. rugose; sepals 3–4 mm. long. Tehama Co. to Riverside Co. .... 11. *R. malvaceum*
                            II. Fls. rose-purple; lvs. scarcely rugose; sepals 1.5 mm. long. S. San Diego Co. .... 13. *R. canthariforme*
AA. Nodal spines present; berry often spiny; fls. mostly 1–few.
    B. Free part of fl.-tube inconspicuous, saucer-shaped; pedicels jointed beneath the ovary.
        C. Lvs. nearly or quite glabrous; berry purplish-black. Humboldt and Siskiyou cos.
                                                              14. *R. lacustre*
        CC. Lvs. glandular-pubescent; berry red. Modoc Co. to Riverside Co. .. 15. *R. montigenum*
    BB. Free part of fl.-tube evident, campanulate to cylindric; pedicels not jointed beneath the ovary.
        C. Sepals 4, red, erect, ca. 1 cm. long; petals as long; stamens 2–3 times as long. Coastal canyons, Santa Clara Co. to L. Calif. .... 31. *R. speciosum*
        CC. Sepals 5, mostly not erect; petals shorter than sepals.
            D. Ovary and berry without spines or prickles.
                E. The ovary and berry glabrous.
                    F. Sepals greenish or purplish, 5–8 mm. long .... 16. *R. divaricatum*
                    FF. Sepals yellowish, 2–3 mm. long.
                      G. Free part of fl.-tube ca. 4 mm. long; berry red
                                                    17. *R. lasianthum*
                    GG. Free part of fl.-tube ca. 2.5 mm. long; berry black
                                                    18. *R. quercetorum*
                EE. The ovary and berry soft-pubescent or glandular ...... 19. *R. velutinum*
            DD. Ovary and berry with spines or bristles.

E. Anthers blunt at apex, almost as broad as long.
  F. Young twigs usually without internodal bristles. N. Coast Ranges.
    G. Lvs. not glandular beneath; sepals 4–5 mm. long
                                                   20. *R. binominatum*
    GG. Lvs. ± glandular beneath; sepals 10–15 mm. long.
      H. Free part of fl.-tube distinctly longer than broad; berry with
           gland-tipped bristles .... 23. *R. Lobbii*
      HH. Free part of fl.-tube as broad as long; berry with glandless
           spines .... 24. *R. Marshallii*
  FF. Young twigs usually with gland-tipped internodal spines or bristles.
    G. Stamens not longer than petals. S. Sierra Nevada
                                    21. *R. tularense*
    GG. Stamens about 2–3 times as long as the petals. Santa Lucia Mts.
                                  22. *R. sericeum*
EE. Anthers apiculate, much longer than broad.
  F. Free part of fl.-tube ca. as broad as long.
    G. Young twigs usually bristly on the internodes; lvs. glandular beneath.
      H. Fls. purplish or reddish; ovary with some hairs or bristles
           that are not gland-tipped .... 25. *R. Menziesii*
      HH. Fls. greenish-white; ovary with all bristles gland-tipped
                                26. *R. Victoris*
    GG. Young twigs not bristly on internodes; lvs. not glandular
                               27. *R. californicum*
  FF. Free part of fl.-tube much longer than broad; young twigs usually not markedly bristly on internodes.
    G. Ovary and berry mostly with all or nearly all of the bristles gland-tipped; lvs. glandular beneath .... 29. *R. amarum*
    GG. Ovary and berry with eglandular spines; lvs. not glandular beneath.
      H. Sepals dull purplish-red; berry 14–16 mm. in diam. Mainland
                                28. *R. Roezlii*
      HH. Sepals pinkish; berry 7–8 mm. in diam. Santa Cruz Id.
                                30. *R. Thacherianum*

1. **R. viburnifòlium** Gray. Straggling, evergreen, to ca. 1 m. high, the branches tending to spread ± horizontally, young growth resinous-glandular; lvs. coriaceous, oval or ± obovate, dark green and glabrous above, pale and resinous-dotted beneath, not lobed, rounded-obtuse, 1.5–3.5 cm. long, on petioles 3–10 mm. long; racemes lax, pubescent, few-fld.; pedicels 2–5 mm. long; fl.-tube turbinate, pubescent; sepals spreading, purplish-brown, 2.5 mm. long; petals minute, greenish; berry red, ca. 6 mm. in diam., glabrous; $n = 8$ (Hamel, 1953).—Among shrubs and in canyons; Chaparral; Santa Catalina Id.; All Saints Bay, L. Calif. Feb.–April.

2. **R. bracteòsum** Dougl. ex Hook. STINK CURRANT. Erect shrub 1–4 m. tall, loosely and sparingly pubescent throughout; lvs. 4–20 cm. wide, thin, deeply 5-lobed with doubly serrate divisions, these ± ovate, acute; petioles 2–10 cm. long; racemes erect, conspicuously bracteate and to 2 dm. in fr.; pedicels mostly 3–4 mm. long; fl.-tube flaring, resin-dotted; sepals green, spreading, 3–4 mm. long; petals small; berry black with whitish bloom, resin-dotted, 8–10 mm. long; $2n = 16$ (Zielinski, 1953).—Canyon bottoms and along streams; N. Coastal Coniferous F., Redwood F., Mixed Evergreen F.; along coast from Mendocino Co. to Del Norte Co.; to Alaska. Feb.–June.

3. **R. petiolàre** Dougl. W. BLACK CURRANT. Erect or spreading, 1–2 m. high, slightly strigulose on young growth; lvs. thin, round-cordate, 3–8 cm. wide, 3–5-lobed, resin-dotted beneath, the lobes ovate-deltoid, obtusish; petioles slender, 3–8 cm. long; racemes erect, 5–12 cm. long, not conspicuously bracteate; pedicels 2–4 mm. long; fl.-tube resin-dotted; sepals white, ovate, 4–5 mm. long; petals white, smaller; berry black, without bloom, roundish, 9–10 mm. in diam.—Moist places, 3000–5000 ft.; Yellow Pine F., etc.; n. Siskiyou and Modoc cos.; to Wash., Mont., Utah. May–July.

4. **R. laxiflòrum** Pursh. Spreading or decumbent, the branches 1–2.5 m. long, young growth retrorse-strigulose and ± glandular; lvs. roundish in outline, deeply 5–7-lobed, 5–10 cm. broad, glabrous and darker green above, sparingly pubescent and paler beneath, the lobes rather bluntly, doubly serrate; petioles slender, 3–8 cm. long; racemes erect-spreading, lax, 6–12-fld.; pedicels 6–12 mm. long, strigulose and with gland-tipped bristly hairs; sepals purplish, ca. 3 mm. long; petals half as long, red; berry with bristly gland-tipped hairs, black or purple, roundish, ca. 4 mm. in diam.—Clambering over logs and stumps, moist places; Redwood F.; Humboldt Co. to B.C. March–May.

5. **R. cèreum** Dougl. [*R. balsamiferum* Kell. *R. viscidulum* Berger.] SQUAW CURRANT. Compact, erect, much-branched, 1–12 dm. high, fragrant, with glandular-pubescent young growth; lvs. round-reniform, 1–4 cm. wide, puberulent and glandular or upper surface subglabrous and ± shining, 3–5-lobed, obtuse, crenulate; petioles 5–15 mm. long; racemes mostly 3–7-fld., glandular-puberulent; the cuneate-obovate bracts toothed on the subtruncate apex; fl.-tube tubular (including ovary), 6–8 mm. long, greenish-white to pink; sepals round-ovate, 1–1.5 mm. long; petals minute, rounded; berry red, slightly glandular-hairy, ca. 6 mm. in diam.; *n* = 8 (Tischler, 1927).—Dry rocky places, 5000–12,600 ft.; Pinyon-Juniper Wd. to Alpine Fell-fields; mts. from Santa Rosa and San Jacinto ranges n. through the Sierra Nevada to Siskiyou and Modoc cos.; Mono, Inyo, Panamint, Clark mts.; to B.C., Mont., Utah, Ariz. June–July. Variable; the s. shining-lvd., 5–10-fld. plants constitute *viscidulum*.

6. **R. inèbrians** Lindl. Much like *R. cereum* in habit and pubescence; bracts of raceme narrow obovate, 5–7 mm. long, acute, entire or sometimes with 1–2 lateral teeth; styles usually glabrous; berry bright red, 6–8 mm. in diam., mostly glandular.— Dry slopes, e. side of Sierra Nevada as Olancha Peak, at 12,135 ft.; to Ida., Nebr., New Mex. June–July.

7. **R. viscosíssimum** Pursh. STICKY CURRANT. Erect shrub, 1–1.5 m. tall, with fragrant glandular foliage, the young growth with long gland-tipped stiff hairs and nonglandular minute ones; twigs yellowish brown, older gray; lvs. roundish, deeply cordate, 3–8 cm. wide, 3(5)-lobed with rounded, crenate-dentate divisions; petioles shorter than blade; racemes 3–10 cm. long at anthesis, 4–15-fld., pubescent and glandular-hairy; pedicels to ca. 1 cm. long, bracts ca. as long; fl.-tube broadly cylindrical, greenish or pinkish, ca. 1 cm. long (including ovary); sepals greenish or pinkish, spreading to reflexed, ca. 4 mm. long; petals white, ca. half as long; berry elliptic-globular, black without bloom, ± glandular-bristly, ca. 10–12 mm. long, disagreeably musky in flavor.—Shaded woods and rocky places, 5000–9500 ft.; Sagebrush Scrub, to Red Fir F.; Sierra Nevada from Tulare Co. to Eldorado Co., e. Modoc Co.; to B.C., Mont., Colo. June–July. Intergrading freely with:

Var. **Hállii** Jancz. [*R. H.* Jancz.] Calyx purplish; ovary glabrous; berry glabrous, with a bloom.—From Mariposa Co. to Humboldt, Siskiyou, and Modoc cos.; to Ore., Nev.

8. **R. aúreum** Pursh. GOLDEN CURRANT. Erect, 1–2.0 m. tall, almost glabrous; lvs. rather firm, cuneate to subcordate at base, obovate to round-reniform in outline, 1.5–5 cm. wide, mostly 3-lobed, the lobes rounded, entire to toothed, light green; petioles to ca. as long as blades; racemes 5–15-fld., 3–7 cm. long; bracts 5–12 mm. long, ± glandular-ciliate; pedicels shorter; fls. with spicy odor; fl.-tube 6–10 mm. long (including ovary), slender, yellow; sepals spreading, 5–8 mm. long, erect after anthesis; petals orange in age, oblong, erose, ca. 2–3 mm. long; berry round, mostly red or black, 6–8 mm. in diam.; *n* = 8 (Darlington, 1929).—Moist banks, 2500–7800 ft.; Sagebrush Scrub to Lodgepole F.; Fresno and Inyo cos., then e. of Sierra Nevada to Modoc Co. and Siskiyou Co.; to Wash., S. Dak., New Mex. Occasional also in bottom lands, San Joaquin V. April–May.

Var. **gracíllimum** (Cov. & Britt.) Jeps. [*R. g.* Cov. & Britt.] Fls. odorless, soon deep red; fl.-tube more slender; sepals 3–4 mm. long; berry orange to yellowish.—Brushy often alluvial places, below 2500 ft.; Foothill Wd., S. Oak Wd.; Coast Ranges from Alameda to w. Riverside cos. Feb.–April.

9. **R. nevadénse** Kell. [*R. ascendens* Eastw. and var. *Jasperae* Eastw. *R. Hittellianum* Eastw. *R. glaucescens* Eastw. *R. Grantii* Heller.] SIERRA CURRANT. Slender-stemmed open deciduous shrub 1–2 m. tall, glabrous to puberulent on young growth; lvs. roundish in outline, rather thin, 3–7 cm. wide, 3–5-lobed, the lobes obtuse, bluntly toothed, resinous-dotted, glabrous above, ± pubescent and paler beneath; petioles 1–4 cm. long; racemes 8–12-fld., spreading or drooping; bracts thin, oblanceolate, glandular-ciliate, 4–8 mm. long; pedicels 3–5 mm. long; fls. rose to rather deep red, the entire tube ca. 5 mm. long, the broad free part above the ovary ca. 2 mm. long; sepals erect, pink to reddish, 4–5 mm. long; petals white, shorter; berry round, blue-black, with a bloom, ± glandular, ca. 8 mm. in diam.—Moist places and along streams, 3000–10,000 ft.; Yellow Pine F. to Subalpine F.; mts. from Palomar Range (San Diego Co.) n., Sierra Nevada to Modoc, Siskiyou, Del Norte, Humboldt, and Trinity cos.; to s. Ore. May–July. Variable: a finely puberulent form from N. Coast Ranges, is var. *glaucéscens* (Eastw.)

Berger. [*R. g.* Eastw.]; another with glabrous lvs., from San Jacinto Mts., is var. *Jàegeri* Berger.

10. **R. sanguíneum** Pursh. [*R. Scuphamii* Eastw.] RED FLOWERING CURRANT. Erect, deciduous, 1–3(–4) m. high, the stems slender, young growth puberulent and often with stalked glands; old bark brownish to gray; lvs. round-reniform in outline, dark green and puberulent above, whitish-pubescent to tomentulose beneath, 2–7 cm. wide, 3–5-lobed, the lobes roundish, serrulate; petioles to ca. as long as blade, puberulent; racemes 10–20-fld., mostly exceeding lvs., finely pubescent and glandular; bracts oblanceolate, 6–8 mm. long, red or purplish; pedicels shorter; fls. deep rose; ovary with short curled hairs among the gland-tipped ones; fl.-tube narrow, 6–7 mm. long; sepals oblong, 4–5 mm. long; petals red or paler, half as long; berry round-oblong, black, with a bloom, glandular; *n* = 8 (Darlington, 1929).—Moist shaded places, 2000–6000 ft.; Yellow Pine F., Red Fir F.; n. Lake Co., Humboldt to Del Norte and Siskiyou cos.; to B.C. April–June.

Var. **glutinòsum** (Benth.) Loud. [*R. g.* Benth. *R. albidum* Paxt. *R. sanctae-luciae* Jancz.] Racemes more drooping; fls. deep to pale pink, rarely whitish; lvs. more sparsely pubescent and greenish beneath.—Open places and among brush or trees, below 2000 ft.; Chaparral, Mixed Evergreen F., Foothill Wd., Closed-cone Pine F., etc.; Coast Ranges from Del Norte and Humboldt cos. to Santa Barbara and San Luis Obispo cos. March–April. Variable: a form about San Francisco Bay with black berries, is the var. *melanocárpum* (Greene) Jeps. [*R. g.* var. *m.* Greene.] and another with whitish or pale fls., is var. *dedúctum* (Greene) Jeps. [*R. d.* Greene.]

11. **R. malvàceum** Sm. [*R. Watkinsii* Eastw.] CHAPARRAL CURRANT. Erect, deciduous, 1–2 m. high, the young growth tomentose and with gland-tipped bristly hairs; lvs. mostly roundish in outline, rather thick, rugose, 2–6 cm. wide, dull green and rough above with stalked glands, glandular and gray-pubescent beneath, 3- to 5-lobed, the lobes obtuse and doubly toothed; petioles 1–5 cm. long; racemes drooping, exceeding lvs., 10–25-fld., pubescent and glandular-hairy; bracts oblanceolate or wider, 6–9 mm. long; fls. bright rose; the fl.-tube 5–8 mm. long; sepals spreading, obovate, 3–4 mm. long; petals erect, roundish, shorter; berry purple-black, with a bloom, subglobose, ca. 6 mm. in diam., ± hairy and glandular.—Dry wooded or open hills, below 2500 ft.; Chaparral, Foothill Wd., Closed-cone Pine F.; Eldorado Co.; inner Coast Ranges from Tehama Co. to Contra Costa Co., outer Coast Ranges from Marin Co. to Los Angeles Co., Santa Cruz Id. Oct.–March.

Var. **viridifòlium** Abrams. [*R. v.* Heller. *R. purpurascens* Heller.] Lvs. greener beneath and ± scabrous with a coarser pubescence, more glandular; fl.-tube 8–12 mm. long. —Dry gullies and canyons, below 5000 ft.; Chaparral; San Jacinto Mts. to Santa Ana and Santa Monica mts.

Var. **clementìnum** Dunkle. Lvs. thin, white-woolly beneath, acutely serrate; fls. light rose.—San Clemente Id.

12. **R. indécorum** Eastw. [*R. malvaceum* var. *i.* Jancz.] WHITE-FLOWERED CURRANT. Erect open shrub, fragrant, largely deciduous, 1.5–2.5 m. tall, new growth densely pubescent and glandular; lvs. 3–5-lobed, 1–4 cm. wide, thickish, dark green, stipitate-glandular and ± rugose above, whitish-tomentose beneath, the lobes rounded, crenulate, obtuse; petioles 5–25 mm. long; racemes 2–5 cm. long, rather densely fld., glandular-pubescent; pedicels 1–2 mm. long; fl.-tube cylindrical-urceolate, whitish, 4–5 mm. long; sepals white, obtuse, recurved, 1.5 mm. long; petals ca. 1 mm. long; berry viscid-pubescent and with some stipitate glands, globose, 6–7 mm. in diam.—Interior washes and canyons, usually below 2000 ft.; Chaparral, Coastal Sage Scrub; Santa Barbara Co. to n. L. Calif. Nov.–March.

13. **R. cantharifórme** Wiggins. Much like *R. indecorum*, with finely pubescent and stipitate-glandular young growth; lvs. 4–6 cm. wide, bright green, cordate at base, 3-lobed, villous and slightly rugulose above, more densely villous beneath; petioles 2.5–3.5 cm. long; racemes many-fld., 3–6 cm. long, the fls. rose-purple; fl-tube villous-pubescent and with some gland-tipped hairs, broadly urceolate, ca. 2 mm. long, much broader; sepals spreading, 1.5 mm. long; petals less than 1 mm.; berry round-ovoid, 5–6 mm. in diam., dark, sparsely white-villous and with some stalked glands.—Chaparral in neighborhood of Moreno Dam, San Diego Co. Feb.–April.

14. **R. lacústre** (Pers.) Poir. [*R. oxycanthoides* var. *l.* Pers.] SWAMP CURRANT. Prostrate to ascending, the stems to 1 m. long, usually with 3 nodal spines and bristly

internodes; lf.-blades 3–5 cm. wide, round-cordate in outline, 3–5-lobed, the lobes deeply cut, acutish, nearly or quite glabrous, paler beneath; petioles slender, 2–5 cm. long, ± glandular-bristly; racemes soon loosely spreading or hanging; pedicels slender, to ca. 1 cm. long; fl.-tube saucer-shaped above the ovary, ± purplish-green, glandular-bristly; sepals short, broad, 1.5 mm. in diam.; petals smaller, more purplish; stamens ca. 1 mm. long; berry black-purple, ca. 4 mm. in diam., with weak gland-tipped hairs; $n = 8$ (Tischler, 1927).—Wet places like meadows, mostly 5000–6000 ft.; Red Fir F.; Humboldt and Siskiyou cos. to Alaska and Atlantic Coast. June–July.

15. **R. montígenum** McClat. [*R. lacustre* var. *molle* Gray. *R. nubigenum* McClat., not Phil. *R. lentum* Cov. & Rose.] Straggling many-branched shrub 3–6 dm. tall, the twigs mostly bristly-prickly, glandular-pubescent; nodal spines 3–5; lvs. 5–25 mm. wide, 5-cleft almost to the base, the lobes incised-serrate, glandular-pubescent on both sides; petioles glandular-pubescent, ca. as long as the blades; racemes few-fld.; pedicels 2–5 mm. long; fl.-tube saucer-shaped above the ovary, glandular-bristly; sepals 3–4 mm. long; petals purplish; berries red, ca. 5 mm. thick, glandular-bristly.—Dry rocky places, 7000–12,500 ft.; Red Fir F. to Alpine Fell-fields; from San Jacinto Mts. to Sierra Nevada, then to Modoc Co.; to interior B.C. and Rocky Mts. June–July.

16. **R. divaricátum** Dougl. [*Grossularia d.* Cov. & Britt. *R. villosum* Nutt.] Widely branched, deciduous, 1–3 m. high, with long straggling and often reflexed branches; spines 1–3 at a node, sometimes 0; young growth pubescent and often also bristly; lvs. thin, roundish, 2–5 cm. wide, 3–5-lobed, the lobes coarsely toothed, subglabrous to finely pubescent above, paler and ± pubescent beneath; petioles 1–3 cm. long; racemes drooping, 2–6-fld.; pedicels 5–12 mm. long, glabrous, subtended by small, roundish, villous-pubescent bracts; fl.-tube greenish or purplish, 3–4 mm. long, ± glabrous, broad at summit; sepals greenish or purplish, oblong, ca. 5 mm. long, spreading-reflexed; petals white, ca. 2 mm. long; berry round, smooth, dark, 6–10 mm. in diam.; $n = 8$ (Tischler, 1927).—Shaded canyons and stream banks, below 2000 ft.; Foothill Wd., Mixed Evergreen F., Redwood F.; Coast Ranges from Santa Barbara Co. to Humboldt and Shasta cos.; to B.C. March–May.

KEY TO VARIETIES

Stamens longer than sepals; sepals 2–4 times as long as the free part of the fl.-tube. Coast Ranges from Santa Barbara Co. n. . . . . *R. divaricatum*
Stamens not longer than sepals; sepals 1–2 times as long as free part of fl.-tube.
Fls. green except for the white petals.
    Lvs. glabrous on both surfaces or nearly so; sepals usually glabrous. Siskiyou to Tulare and Inyo cos. . . . . var. *inerme*
    Lvs. villous; sepals mostly sparingly villous. Humboldt to Shasta cos. . . . . . . var. *klamathense*
Fls. purplish-red, the petals red. S. Calif. . . . . var. *Parishii*

Var. **inérme** (Rydb.) McMinn. [*R. i.* Rydb. *Grossularia i.* Cov. & Britt.] Stems glabrous with short nodal spines and mostly no internodal bristles; lvs. 3–5-lobed, mostly glabrous; free portion of fl.-tube ca. as long as sepals; berry wine-colored, smooth.—Moist often shaded places, 3500–10,900 ft.; Montane Coniferous F.; Sierra Nevada from Inyo and Tulare cos. to Modoc Co. and w. to Siskiyou Co.; to B.C., Mont., New Mex. May–June.

Var. **klamathénse** (Cov.) McMinn. [*Grossularia k.* Cov. *R. inerme* var. *k.* Jeps.] Nodal spines mostly 1, often 0; lvs. 3–5-lobed, villous on both surfaces; calyx usually hairy.—Shaded canyons, etc.; Yellow Pine F., etc.; Humboldt, Siskiyou, and Shasta cos.; s. Ore. April–May.

Var. **Paríshii** (Heller) Jeps. [*R. P.* Heller. *Grossularia P.* Cov. & Britt.] Lvs. densely pubescent beneath, 2–5 cm. broad, 3–5-lobed; racemes 2–5-fld., pubescent; ovary glabrous; free part of fl.-tube 4 mm. long, purple-red, pubescent; sepals 6–8 mm. long.— Willow thickets, swamps; Coastal Sage Scrub; San Bernardino, El Monte, Pasadena, San Gabriel R. March–April.

17. **R. lasiánthum** Greene. [*Grossularia l.* Cov. & Britt. *R. leptanthum* var. *l.* Jeps.] Low spreading shrub to 1 m. high; spines 1–3 at each node; young shoots puberulent; lvs. roundish, 1–2 cm. wide, mostly glandular-puberulent on both surfaces, cleft into 3–5 blunt, toothed lobes; petioles pubescent, 3–6 mm. long; fls. lemon-yellow, 2–4 in a raceme; bracts broad, pubescent; fl.-tube 5–6 mm. long, cylindrical and pubescent in free portion; sepals early reflexed, then erect, ca. 2 mm. long; petals shorter; berry round, red, smooth, ca. 6–7 mm. in diam.—Rocky places, 7000–10,000 ft.; Red Fir F.

to Subalpine F.; Sierra Nevada from Nevada Co. to Tulare Co.; apparently also in San Gabriel Mts. at Big Pines near Jackson Lake. June–July.

18. **R. quercetòrum** Greene. [*R. leptanthum* var. *q.* Jancz. *Grossularia q.* Cov. & Britt. *R. Congdonii* Heller.] Rounded shrub 6–15 dm. tall, with arcuate-spreading branches, the young twigs puberulent, without bristles; nodal spines usually 1; lvs. roundish, glandular-puberulent, light green, 1–2 cm. wide, 3–5-cleft into toothed lobes; petioles glandular-puberulent, ca. as long as blades; peduncles 2–3-fld.; fl.-tube 6–7 mm. long, the ovary subglabrous, the free portion tubular, puberulent, yellow; sepals yellow, ca. 3 mm. long; petals white, ca. 1 mm. long; berry glabrous, black, ca. 7–8 mm. in diam.—Dry slopes, below 4000 ft.; Foothill Wd., S. Oak Wd., etc.; foothills of the Sierra Nevada from Tuolumne Co. to Kern Co. and scatteringly in inner Coast Ranges from Alameda Co. s., to w. edge of Colo. Desert and n. L. Calif. March–May.

19. **R. velutìnum** Greene. [*Grossularia v.* Cov. & Britt. *R. leptanthum* var. *brachyanthum* Gray.] Stout, rigidly branched, 0.6–2 m. tall, the recurved branches soft-pubescent, without bristles or glands; modal spines 1–3, stout; lvs. roundish, 1–2 cm. broad, softpubescent, deeply 5-cleft, the lobes often 3-cleft; fls. 1–4, yellowish to whitish; ovary soft-pubescent; fl.-tube cylindric; sepals ca. 3 mm. long; petals 2–2.5 mm. long; berry dark, 4–6 mm. in diam., velvety-pubescent, but not usually glandular.—Dry slopes, 2500–8500 ft.; Sagebrush Scrub, Pinyon-Juniper Wd., Yellow Pine F.; from Trinity and Siskiyou cos. to Modoc Co. and s. on e. side of Sierra Nevada to Inyo Co., White Mts.; to Ore., Utah, Ariz. May–June.

Var. **glandulíferum** (Heller) Jeps. [*R. g.* Heller. *R. Stanfordii* Elmer.] Lvs. glandularpuberulent; ovary and berry stipitate-glandular and ± bristly.—Similar places, Siskiyou Co. to San Gabriel Mts. (Los Angeles Co.) and New York and Kingston mts. (e. San Bernardino Co.) May–June.

20. **R. binominàtum** Heller. [*Grossularia b.* Cov. & Britt. *R. ambiguum* Wats., not Maxim. *R. montanum* Howell, not Phil.] Low, deciduous, with trailing branches to ca. 1 m. long, pubescent, without internodal bristles; nodal spines mostly 3, to 2 cm. long; lvs. thin, roundish, cordate, 2–5 cm. wide, deeply 3(5)-lobed, pubescent above and beneath, not usually glandular, somewhat paler beneath; petioles ca. as long as blades; fls. 1–3, greenish-white on hairy peduncles; fl.-tube ca. 4 mm. long, the ovary bristly and glandular, the free tube broad, pubescent; sepals 4–5 mm. long, long-pubescent; petals 2–4 mm. long; stamens exceeding petals; berry round, ca. 1 cm. in diam., with long nonglandular stiff spines and shorter gland-tipped bristles.—Trailing on forest floor, 3500–6500 ft.; Yellow Pine F., Red Fir F.; Glenn and Trinity to Humboldt and Siskiyou cos.; s. Ore. May–June.

21. **R. tularénse** (Cov.) Fedde. [*Grossularia t.* Cov.] Very near to *R. binominatum;* young twigs with weak gland-tipped bristles and villous-pubescent; lvs. pubescent and glandular; stamens not longer than petals.—At ca. 5000–6000 ft.; Yellow Pine F., Red Fir F.; Tulare Co. May.

22. **R. seríceum** Eastw. [*Grossularia s.* Cov. *R. s.* var. *viridescens* Eastw.] Spreading, deciduous, 1–2 m. high, pubescent and thickly beset with gland-tipped bristles; nodal spines 3, stout; lvs. thin, round-ovate, 1.5–4 cm. wide, pubescent- and scattered-setose, glandular, 3–5-lobed; petioles glandular-pubescent, ca. as long as blades; fls. 1–3 on glandular-pubescent peduncles; fl.-tube ca. 7–8 mm. long, the ovary densely and heavily glandular-bristly and villous, the free tubal portion cylindric-campanulate, greenish-red, pubescent; sepals red to greenish, 6–8 mm. long, short-villous; petals white, involute; berry round, 1–2 cm. in diam., purple, bristly.—Along streams, below 1000 ft.; Redwood F.; coastal slope of Santa Lucia Mts., Monterey and San Luis Obispo cos. Feb.–April.

23. **R. Lóbbii** Gray. [*Grossularia L.* Cov. & Britt.] GUMMY GOOSEBERRY. Deciduous, 1–2 m. high, without internodal spines or bristles, the young twigs glandular-pubescent; nodal spines mostly 3; lvs. thin, roundish, mostly 1.5–5 cm. wide, ± glabrous above, glandular-pubescent beneath, 3–5-lobed, crenately toothed; petioles glandular-pubescent, ca. as long as blades; fls. 1–3 on short glandular peduncles; ovary glandular-puberulent; free part of fl.-tube 3–4 mm. long, narrow-campanulate, purple-red, finely pubescent; sepals purple-red, reflexed, 10–12 mm. long; petals pale yellow, involute; berries oblong, densely gland-bristly, 12–15 mm. long.—Dryish slopes, below 6500 ft.; Red Fir F., N. Coastal Coniferous F.; mts. from n. Lake Co. to Humboldt and Shasta cos.; to B.C. May–July.

24. **R. Márshalli** Greene. [*Grossularia M.* Cov. & Britt.] Much like *R. Lobbii*, lvs. more glabrous beneath; petioles more glabrous; fls. solitary; ovary with weak glandless bristles and densely pubescent; free part of fl.-tube as wide as long, densely pubescent; sepals 12–15 mm. long, purplish; berry covered with glandless-spines.—Forested slopes, 5000–7000 ft.; Red Fir F., Subalpine F.; Humboldt and Siskiyou cos.; s. Ore. June–July.

25. **R. Menzièsii** Pursh. [*Grossularia M.* Cov. & Britt. *R. ferox* Sm.] CANYON GOOSE-BERRY. Loosely branched shrub 1–2 m. tall, the young twigs densely bristly and pubescent; nodal spines mostly 3; lvs. rather firm, ± rounded in outline, cordate or truncate at base, 1.5–4 cm. wide, subglabrous to roughish with glandular hairs above, velvety-pubescent and with gland-tipped hairs beneath, 3–5-lobed; petioles glandular-hairy and pubescent, 1–2.5 cm. long; fls. purplish, 1–3; pedicels glandular, 3–6 mm. long; bracts shorter; ovary glandular-bristly and pubescent; free part of fl.-tube 2–3 mm. long, ca. as broad, ± glandular-hairy; sepals reflexed, oblong, 7–11 mm. long; petals whitish, ca. half as long as fils.; stamens almost as long as sepals, anthers apiculate; berry globular, ca. 1 cm. in diam., densely clothed with gland-tipped and nonglandular bristles.—Canyons and flats, below 1000 ft.; Redwood F., Mixed Evergreen F.; outer Coast Ranges from San Luis Obispo Co. to Del Norte Co. and s. Ore. March–April.

KEY TO VARIETIES

Lvs. definitely glandular beneath and sometimes above.
  Ovary with both gland-tipped and nonglandular bristles.
    Petals shorter than filaments.
      Plant not strongly aromatic; fr. ca. 1 cm. in diam. Coastal .... R. *Menziesii*
      Plant strongly aromatic; fr. 1.2–2 cm. in diam. Sierra Nevada .... var. *ixoderme*
    Petals as long as fils. Santa Lucia Mts. .... var. *Hystrix*
  Ovary densely covered with gland-tipped bristles only; petals half as long as fils. Sonoma Co. to
  San Mateo Co. .... var. *leptosomum*
Lvs. scarcely at all glandular beneath; ovary with soft glandless hairs and a few gland-tipped or
glandless bristles. Santa Clara and Santa Cruz cos. .... var. *senile*

Var. **ixodérme** Quick. Lvs. with gland-tipped and nonglandular hairs above and beneath, strongly aromatic; berry 15–25 mm. in diam., viscid and aromatic, densely glandular-bristly.—Chaparral, below 3500 ft.; foothills of Sierra Nevada, Fresno and Tulare cos. April.

Var. **Hýstrix** (Eastw.) Jeps. [*R. H.* Eastw.] Young shoots densely bristly; lvs. less rugose than in the sp., thinner, less pubescent beneath; petals as long as the broad fils.—Wooded canyons; Redwood F., Closed-cone Pine F.; coastal slopes of Santa Lucia Mts. from Point Lobos to Point Gorda. Feb.–March.

Var. **leptósmum** (Cov.) Jeps. [*Grossularia l.* Cov. *R. subvestitum* H. & A.] Like the sp., but with fewer prickles on the stems; ovary densely and evenly covered with gland-tipped bristles.—Wooded canyons; Redwood F., Mixed Evergreen F.; Sonoma to San Mateo and Alameda cos. Feb.–March. Plants from summit of Mt. Diablo with more numerous, more slender shorter spines, have been named var. *hystrículum* Jeps. A less hairy form from Berkeley and Oakland is var. *fáustum* Jeps.

Var. **sénile** (Cov.) Jeps. [*Grossularia s.* Cov.] Lvs. beneath and ovaries densely soft-hairy without gland-tipped hairs to any extent.—Wooded slopes, Mixed Evergreen F.; Santa Cruz Mts. April–May.

26. **R. Victòris** Greene. [*R: Menziesii* var. *V.* Jancz. *Grossularia V.* Cov. & Britt.] Deciduous, 1–2 m. tall, young twigs puberulent, ± viscid, densely to very sparsely bristly; lvs. 1.5–5 cm. wide, glandular-hairy and ± finely pubescent, incisely 3–5-lobed; petioles glandular-hairy, to ca. as long as blades; fls. 1–2, greenish-white; ovary thickly beset with gland-tipped hair; free part of fl.-tube ca. 3 mm. long, ca. as broad, densely glandular; sepals 6–10 mm. long, recurved, sometimes purplish; petals ca. 3 mm. long, white; stamens 6–10 mm. long; berry rounded, yellow, densely glandular-bristly and with some nonglandular bristles.—Wooded slopes in shaded canyons, etc.; Mixed Evergreen F., Redwood F.; Marin, Sonoma, Napa, and Solano cos. March–April. Plants with unusually long stamens and lvs. not or scarcely glandular, are the var. *minus* Jancz. [var. *Greeneia-num* (Heller) Jeps. *R. G.* Heller.]

27. **R. califórnicum** H. & A. [*Grossularia c.* Cov. & Britt. *R. occidentale* H. & A.] Compact, 6–14 dm. high, intricately branched, the young twigs usually without bristles, glabrous; nodal spines stoutish, usually 3; lvs. roundish, thin, subglabrous, 1–3 cm. wide,

3–5-cleft, crenate, nonglandular; petioles 0.6–2 cm. long, pubescent; fls. mostly 1, greenish, dull white, or purplish; ovary bristly, the shorter bristles sometimes gland-tipped; free part of fl.-tube ca. 1.5 mm. long, cylindric; sepals 5, reflexed, usually pubescent toward apex, ca. 6–7 mm. long; petals white, ca. 3 mm. long; stamens ca. as long as sepals, anthers apiculate; berry round, reddish, ca. 9–10 mm. in diam., bristly—Open slopes and rocky canyons, below 2500 ft.; Redwood F., Chaparral, Mixed Evergreen F., Foothill Wd.; Coast Ranges from Mendocino Co. to Monterey Co. Feb.–March. Exceedingly variable; sometimes intergrading toward forms of *R. Menzièsii* in the bristly twigs. Plants of Monterey Co. whose berries have very few spines are *R. oligacánthum* Eastw.

Var. **hespèrium** (McClat.) Jeps. [*R. h.* McClat. *Grossularia h.* Cov. & Britt.] Lvs. finely pubescent; sepals reddish, mostly strigose over entire outer surface; petals typically almost as long as fils.—Canyons below 2000 ft.; Chaparral, etc.; Santa Ana and San Gabriel mts. to Santa Barbara Co. Jan.–March. Variable.

28. **R. Roèzlii** Regel. [*Grossularia R.* Cov. & Britt. *R. Wilsonianum* Greene. *R. aridum* Greene.] SIERRA GOOSEBERRY. Stout, 5–12 dm. high, with spreading branches; young growth pubescent but not bristly; nodal spines 1–3, straight; lvs. roundish, 1.2–2.5 cm. wide, dark green and puberulent above, paler, short-pubescent beneath, 3–5-cleft into toothed lobes; petioles 6–18 mm. long, pubescent; fls. 1–2, the peduncles short-pubescent and ± glandular; ovary white-hairy and bristly, some bristles gland-tipped; free portion of fl.-tube ca. 6 mm. long, purplish, pubescent; sepals dull purplish-red, 7–10 mm. long, pubescent, lanceolate; petals whitish, 3–5 mm. long; fils. ca. as long as petals, anthers apiculate; berry purple or lighter, 14–16 mm. in diam., pubescent and with long pubescent spines.—Dry open slopes, mostly 3500–8500 ft.; Yellow Pine F., Red Fir F.; inner Coast Ranges, Trinity and Tehama cos. to Modoc Co. and s. through the Sierra Nevada to mts. of San Diego Co. May–June.

Var. **cruéntum** (Greene) Rehd. [*R. c.* Greene.] Lvs., ovary and sepals mostly glabrous; ovary spiny with glabrous spines.—Open wooded slopes, below 5000 ft.; Mixed Evergreen F., N. Oak Wd., to Red Fir F.; Coast Ranges from Napa and Sonoma cos. to Del Norte and Siskiyou cos.; Ore. April–June.

Var. **amíctum** (Greene) Jeps. [*R. a.* Greene.] Lvs. on under surface, bracts and sepals hoary-pubescent.—Mixed Evergreen F.; upper parts of S. Fork of Eel River and surrounding mts.

29. **R. amàrum** McClat. [*Grossularia a.* Cov. & Britt. *R. mariposanum* Congd.] BITTER GOOSEBERRY. Erect, deciduous, 1–2 m. high, pubescent and glandular on young twigs, but not bristly; nodal spines brown, to 1 cm. long; lvs. roundish, with cordate base, 2–3 cm. wide, pubescent and ± glandular-puberulent above and beneath, 3–5-lobed, crenate; petioles glandular-pubescent, to ca. as long as blades; fls. 1–3, purplish, on glandular-pubescent peduncles; bracts broadly ovate; ovary densely glandular-bristly, free part of fl.-tube 5–6 mm. long, ca. half as wide, glandular-pubescent; sepals pubescent, 7–8 mm. long; petals pinkish white, almost as long as fils.; stamens ca. 1 cm. long, anthers apiculate; berry rounded, 1.5–2 cm. in diam., with short gland-tipped bristles 1–2 mm. long.—Wooded canyons, below 5000 (6500) ft.; Chaparral, Foothill Wd., Yellow Pine F.; Coast Ranges from Monterey Co. s., Sierra Nevada from Eldorado Co. s., to San Diego Co. March–April.

Var. **Hoffmánnii** Munz. Berry distinctly pubescent, its spines not crowded, unequal, 1–3.5 mm. long.—Canyons, Santa Barbara Region.

30. **R. Thacheriànum** (Jeps.) Munz. [*R. Menziesii* var. *T.* Jeps.] Shrub 1–2.5 m. high, the young growth densely white- and soft-pubescent, not glandular, sometimes somewhat bristly; nodal spines 0–3; lvs. mostly 2–3 cm. wide, greenish and subglabrous above, paler and soft-pubescent beneath, rather shallowly 5-lobed, the lobes coarsely crenate-dentate; petioles 1–2 cm. long, copiously white-pilose; peduncles white-pilose, drooping, 2–4 cm. long, bracted, 1–2-fld.; ovary pilose; free fl.-tube subcylindric, 4–5 mm. long, white-pilose; sepals 9–10 mm. long, narrow, pinkish, not reflexed, softpilose; petals ca. ⅗ as long; stamens shorter than the sepals, anthers apiculate; berry dark, ca. 7 mm. in diam., pilose-hairy and with nonglandular spines 1–1.5 mm. long.—Ravines and stream bottoms, Santa Cruz Id. March–April.

31. **R. speciòsum** Pursh. [*Grossularia s.* Cov. & Britt. *R. stamineum* Sm. *R. fuchsioides* Moç. & Ses.] FUCHSIA-FLOWERED GOOSEBERRY. Evergreen, 1–2 m. tall, with long simple

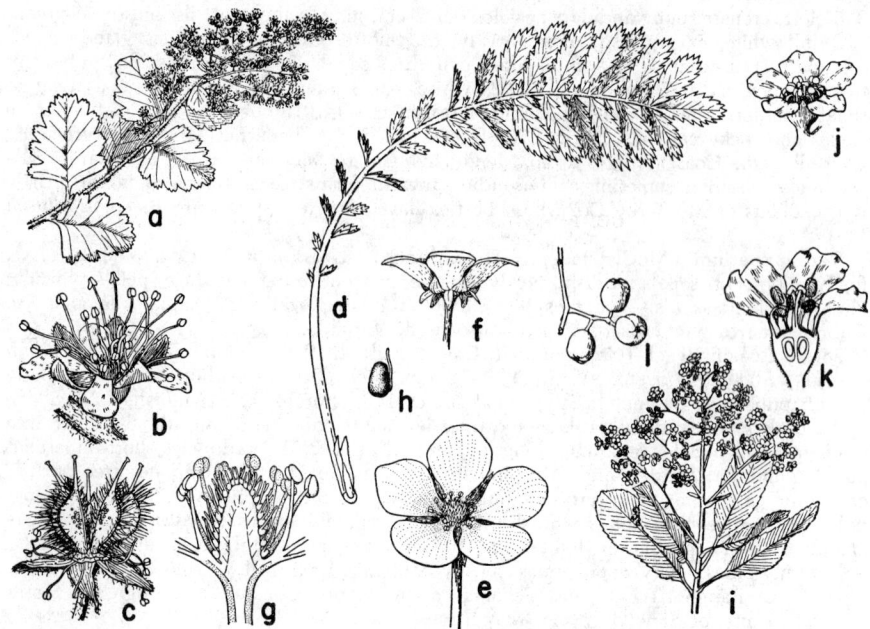

Fig. 69. ROSACEAE. *Holodiscus discolor: a,* habit, × ¼; *b,* fl., × 5, pistils 5, stamens ca. 20, petals 5, sepals 5; *c,* in later stage, the 5 carpels maturing into indehiscent frs. *Potentilla Egedii* var. *grandis: d,* lf., × ¼; *e,* fl., × 1; *f,* side view of fl. with bractlets alternating with sepals; *g,* section through fl., × 2½, with cent. conical receptacle, numerous pistils and stamens; *h,* ak., × 2½, with basal filiform style. *Heteromeles arbutifolia: i,* infl. and lvs., × ¼; *j,* fl., × 2; *k,* fl. in section, × 2½. with inferior ovary that will become a pome, with persistent sepals, petals, stamens, 2 styles; *l,* pomes, × ½.

spreading bristly branches; nodal spines 3, stout, 1–2 cm. long; lvs. roundish, 1–3.5 cm. across, shining and dark green above, paler beneath, quite glabrous to somewhat glandular-pilose, somewhat 3-lobed; petioles shorter than blades; peduncles drooping, 1–4-fld.; ovary glandular-bristly; free fl.-tube red, broadly campanulate, 2–3 mm. long, glandular-hairy; sepals 4, red, ligulate, glandular, erect, ca. 1 cm. long; petals ca. as long; stamens 2–3 times as long as sepals; berry glandular-bristly, ovoid, ca. 1 cm. long.— Common in shaded canyons, below 1500 ft.; Coastal Sage Scrub, Chaparral; near the coast from Santa Clara Co. to San Diego and n. L. Calif. Jan.–May.

73. Rosàceae. ROSE FAMILY (Fig. 69)

Herbs, shrubs, and small ·trees, spiny or unarmed, evergreen or deciduous. Lvs. usually alternate and with stipules. Fls. mostly bisexual and regular, solitary or clustered. Sepals and petals at edge of a fl.-tube (hence perigynous) which is lined or rimmed with a glandular disc. Sepals 5 (4), often with alternating bractlets; petals 5 (4 or 0). Stamens mostly in whorls or cycles of 5, sometimes numerous and indefinite. Pistils 1–many, simple, distinct and free from fl.-tube or united into a 2–5-celled ovary which may be ± inferior. Fr. a follicle, ak., drupe, pome, or cluster of drupelets. Seeds usually without endosperm. A family of over 100 genera and over 3000 spp.; many of great horticultural value.

A. Ovary or ovaries superior; fr. not a pome.
  B. Fr. of 1–5 dehiscent follicles.
    C. Lvs. opposite, 1–2 dm. long; fls. bisexual, in large corymbose panicles; sepals deciduous. Insular . . . . 1. *Lyonothamnus*
    CC. Lvs. alternate, usually smaller; fls. not as above, or if so, unisexual; sepals persistent. Mainland.

D. Plants woody, with bisexual fls.
  E. Foliage stellate-pubescent or -tomentose; petals 3–5 mm. long.
    F. Lvs. simple, generally 3–7 lobed; carpels inflated in maturity
                                                                     2. *Physocarpus*
    FF. Lvs. twice pinnate; carpels coriaceous at maturity . . 7. *Chamaebatiaria*
  EE. Foliage not stellate-pubescent; petals mostly 1.5–2 mm. long.
    F. Lvs. dissected into linear divisions; stems decumbent. N. Calif.
                                                                     5. *Luetkea*
    FF. Lvs. not dissected, but entire or toothed.
      G. Fr. 1-seeded, tardily dehiscent; stamens scarcely exserted; fls.
          mostly whitish . . . . 8. *Holodiscus*
      GG. Fr. several-seeded, soon dehiscent; stamens well exserted; fls. rose
          or white.
        H. Lvs. deciduous; follicles opening on ventral side; erect
            shrubs . . . . 3. *Spiraea*
        HH. Lvs. persistent; follicles dehiscent on both sutures; prostrate
                                                                     4. *Petrophytum*
  DD. Plants tall perennial herbs with large, 2–3-pinnate lvs. and unisexual fls. N. Calif.
                                                                     6. *Aruncus*
BB. Fr. of indehiscent aks., or of drupelets, or of large drupes.
  C. The fr. of dry aks., which may be imbedded in the surface of a fleshy receptacle
      (strawberry) or enclosed in a fleshy cup-shaped structure (rose).
    D. Lvs. simple, sometimes 3-lobed; woody.
      E. Petals present.
        F. Fls. solitary; lvs. cuneate, 3-lobed above . . . . 24. *Purshia*
        FF. Fls. paniculate; lvs. not lobed.
          G. The lvs. sublinear; ovule and seed 1 . . . . 20. *Adenostoma*
          GG. The lvs. obovate to ovate; ovules 2; seed mostly 1
                                                                     8. *Holodiscus*
      EE. Petals absent.
        F. Fl.-tube salver-shaped, the limb deciduous, the tubular base persistent;
            ak. with long feathery tail . . . . 26. *Cercocarpus*
        FF. Fl.-tube campanulate; ak. not long-tailed . . . . 27. *Coleogyne*
    DD. Lvs. compound or pinnately dissected.
      E. Plants shrubby.
        F. Pistil usually 1.
          G. Lvs. 3-lobed; fls. solitary . . . . 24. *Purshia*
          GG. Lvs. pinnately dissected into minute segms.; fls. cymose-paniculate
                                                                     25. *Chamaebatia*
        FF. Pistils 4-many.
          G. Carpels not enclosed in a fleshy fl.-tube; plants not prickly.
            H. Lvs. pinnate; fls. yellow . . . . 12. *Potentilla*
            HH. Lvs. mostly 5-lobed; fls. white or cream.
              I. Sepals alternating with bractlets; petals 13–14 mm. long
                                                                     22. *Fallugia*
              II. Sepals present; bractlets absent; petals 6–8 mm. long
                                                                     23. *Cowania*
          GG. Carpels enclosed in the fl.-tube which is fleshy in fr.; stems mostly
              prickly . . . . 29. *Rosa*
      EE. Plants herbaceous.
        F. Styles jointed with the ovary and deciduous.
          G. Bractlets lacking between the sepals; fls. without a stalked recep-
              tacle; fl.-tube funnelform . . . . 11. *Purpusia*
          GG. Bractlets present, alternating with sepals; fls. with a stalked re-
              ceptacle; fl.-tube not funnelform.
            H. Stamens 5.
              I. Lvs. pinnate, with more than 3 lfts. . . . . 10. *Ivesia*
              II. Lvs. with 3 lfts. . . . . 15. *Sibbaldia*
            HH. Stamens 10, 15, 20 or many.
              I. Fils. dilated, 10 in number; fl.-tube usually deep
                                                                     9. *Horkelia*
              II. Fils. filiform or narrow, mostly 20 or more; fl.-tube
                  shallow or none.
                J. Receptacle not enlarged in fr.
                  K. Uppermost lfts. confluent; petals usually clawed;
                      carpels usually few; stamens inserted some dis-
                      tance above the receptacle and without an an-
                      nular thickening at the base of the fils.
                                                                     10. *Ivesia*
                  KK. Uppermost lfts. not confluent; petals usually ses-
                      sile; carpels mostly many; stamens inserted near
                      the base of the receptacle on a ± evident
                      annular thickening . . . . 12. *Potentilla*
                JJ. Receptacle enlarged in fr.
                  K. Fls. yellow; receptacle spongy . . 13. *Duchesnea*
                  KK. Fls. white; receptacle fleshy . . . . . . 14. *Fragaria*

FF. Styles not jointed with the ovary, persistent.
  G. Pistils many; petals 4–10 mm. long .... 21. *Geum*
  GG. Pistil 1; fls. smaller.
    H. Fl.-tube not prickly; petals O.
      I. Lvs. palmately lobed; plants not over 1 dm. high
                              16. *Alchemilla*
      II. Lvs. pinnate; plants 1–6 dm. high .... 17. *Sanguisorba*
    HH. Fl.-tube armed with prickles.
      I. Prickles on fl.-tube retrorsely barbed; petals O; sepals
        mostly 4 .... 18. *Acaena*
      II. Prickles hooked; petals yellow; sepals 5 .. 19. *Agrimonia*
CC. The fr. of drupelets or drupes.
  D. Carpels becoming drupelets which are ± coherent and form a fleshy "berry";
    plants mostly ± woody, often prickly; lvs. mostly pinnate .... 28. *Rubus*
  DD. Carpels becoming larger solitary drupes; trees or shrubs with simple lvs.
    E. Pistil 1; fls. bisexual; lvs. mostly serrate .... 30. *Prunus*
    EE. Pistils usually 5; fls. unisexual or bisexual; lvs. entire ...... 31. *Osmaronia*
AA. Ovary inferior; fr. a pome.
  B. Lvs. compound; fls. in a compound terminal corymb .... 32. *Sorbus*
  BB. Lvs. simple.
    C. Plant with deciduous lvs.
      D. Fls. in terminal corymbs or corymblike clusters; branches ± thorny.
        E. Styles united at base; fr. with papery or leathery center ...... 33. *Malus*
        EE. Styles entirely distinct; fr. with bony cent. nutlets .... 36. *Crataegus*
      DD. Fls. not in corymbs; branches not thorny.
        E. Petals spreading; styles 2; fls. 1–3 in a sessile umbel ...... 35. *Peraphyllum*
        EE. Petals erect; styles 2–5; fls. racemose .... 34. *Amelanchier*
    CC. Plant with persistent leathery lvs.; fls. in large corymbose panicles .. 37. *Heteromeles*

1. Lyonthámnus Gray. CATALINA IRONWOOD

Evergreen tree with bark exfoliating in narrow strips. Lvs. opposite, petioled, thick-ish, simple and entire to pinnately compound. Stipules deciduous. Fls. many, perfect, in large terminal corymbose compound panicles, with short pedicels. Fl.-tube campanu-late, free from ovary, subtended by 1–3 bractlets. Sepals 5, persistent. Petals 5, clawless. Stamens ca. 15, inserted on a woolly disk lining the fl.-tube. Pistils 2, distinct, 1-loculed. Style stout, with subcapitate stigma. Ovules 4 in each ovary. Fr. a pair of small woody follicles, each usually 4-seeded. Seeds oblong, brown, flat. One sp. Used in Calif. as an ornamental. (Named for W. S. *Lyon*, early resident of Los Angeles, and *thamnos*, shrub.)

1. **L. floribúndus** Gray. Slender, 5–15 m. tall, with red-brown to grayish bark and narrow crown; young growth and twigs pubescent; lvs. lance-oblong, 10–16 cm. long, bicolored, glossy green above, ± pubescent beneath, entire to crenate-serrate, or ± lobed toward base, on petioles 1–2 cm. long; infl. dense, many-fld., 1–2 dm. broad; petals white, rounded, 4–5 mm. long; follicles glandular-pubescent, 3–4 mm. high; seeds ca. 2 mm. long.—Dry slopes; Chaparral; Santa Catalina Id. May–June.

Var. **asplenifòlius** (Greene) Bdg. [*L. a.* Greene.] Lvs. broadly ovate in outline, pin-nate into 2–7 lfts., then pinnatifid into numerous oblique lobes.—Intergrading freely with the sp.; Santa Catalina, San Clemente, Santa Rosa, and Santa Cruz ids.

2. Physocárpus Maxim. NINEBARK

Deciduous shrubs with exfoliating bark. Lvs. simple, alternate, petioled, palmately lobed. Fls. in terminal corymbs. Fl.-tube campanulate, stellate-pubescent. Sepals 5, persistent. Petals white, 5, rounded, spreading. Stamens 20–40, inserted on a disk in the throat of the fl.-tube. Pistils 1–5, ± united at base; styles filiform; stigmas capitate. Follicles inflated, 2–4-seeded, ± inflated. Seeds ovoid, shining, bony, with copious endosperm. Ca. 10 spp., N. Am. and Asia. (Greek, *phusa*, bladder, and *karpos*, fr.)

Pistil 1; alternate stamens longer and their fils. more dilated toward base .... 1. *P. alternans*
Pistils 3–5; stamens similar .... 2. *P. capitatus*

1. **P. altérnans** (Jones) J. T. Howell. [*Neillia monogyna* var. *a.* Jones. *Opulaster a.* Heller.] Densely branched, 0.5–1.5 m. high, with tawny or grayish-white bark, the young twigs stellate-pubescent; lvs. 0.5–1.8 cm. broad, roundish, cordate, 3–7-lobed, the lobes doubly crenate, ± stellate-pubescent on both surfaces; petioles 5–10 mm.

long; fls. 3–12; floral tube stellate without, glabrous within, 3–4 mm. wide; sepals stellate, ovate, 3 mm. long; petals 3–4 mm. long; stamens ca. 20; follicle ca. 5 mm. long, stellate-pubescent.—Dry rocky slopes, 6000–10,000 ft.; Pinyon-Juniper Wd.; White Mts., Inyo Co.; to Utah. June–July. Plants from the Panamint Mts. with lvs. more densely stellate above, have been called ssp. *panaminténsis* J. T. Howell; some from the White Mts. with an inner ring of hair at summit of fl.-tube, ssp. *annulàtus* J. T. Howell.

2. **P. capitàtus** (Pursh) Kuntze. [*Spiraea c.* Pursh. *Neillia opulifolia* var. *mollis* Brew. & Wats.] Erect or spreading, 1–2.5 m. high; pubescent on young growth; lvs. round-ovate, 3–5-lobed, the lobes serrate to incised, glabrous or nearly so above, glabrous to stellate-pubescent beneath, 3–7 cm. long; petioles 1–2 cm. long; fls. many; pedicels and fl.-tube densely stellate; sepals 2.5–3 mm. long, ovate, stellate-pubescent; petals ca. as long; follicles ± stellate to glabrous, 5–6 mm. long; seeds straw-colored, 1.5–2 mm. long; *n* = 9 (Sax, 1931).—Moist banks and north slopes, below 4500 ft.; many Plant Communities from Redwood F. to Red Fir F.; Sierra Nevada from Tulare Co. n.; Coast Ranges from Santa Barbara Co. n.; to B.C., Mont., Utah. May–July.

3. Spiraèa L. SPIRAEA

Shrubs with simple deciduous lvs. usually serrate and without stipules. Fls. perfect, in a corymbose to paniculate infl. Fl.-tube turbinate to campanulate. Sepals 5. Petals 5, white or rose. Stamens 15–many, inserted in 1 or more series under the disk-margin. Pistils 3–8, commonly 5; ovules 2–several. Follicles not inflated, opening along ventral side. Seeds tapering at both ends, with little or no endosperm. Ca. 70 spp., of N. Temp. Zone. (Greek, *speira*, a band or wreath.)

Fls. in elongate panicles; sepals soon reflexed .... 1. *S. Douglasii*
Fls. in flat-topped corymbs; sepals not reflexed .... 2. *S. densiflora*

1. **S. Doúglasii** Hook. Erect shrub, 1–2 m. high; young growth ± close-tomentose; lvs. elliptical to oblong or oblong-oval, green and glabrous above, white-tomentose beneath, serrate above the middle, 3–9 cm. long, on petioles 0.3–1 cm. long; panicle 1–1.5 dm. long, congested; fl.-tube tomentose, to ca. 1 mm. long; sepals to ca. 1 mm. long; petals rose, ca. 1.5 mm. long; follicles ca. 3 mm. long, glabrous, the beak an additional 1.5 mm.; seeds light brown, linear, cellular-reticulate, ca. 2 mm. long; *n* = 18 (Sax, 1936).—Low damp places, below 6000 ft.; Redwood F. to Red Fir F.; from Humboldt, Trinity, Butte, and Plumas cos. n.; to B.C. June–Sept.

2. **S. densiflòra** Nutt. ex T. & G. [*S. Helleri* Rydb.] Glabrous or with lvs. ± ciliate; 2–9 dm. high; lvs. oval to elliptical, 1.5–3 cm. long, crenate or serrate in upper half, on petioles to ca. 3 mm. long; infl. 2–4 cm. broad; fl.-tube glabrous, 2–2.5 mm. broad; sepals ca. 1 mm. long; petals 1.5 mm. long, rose; follicles glabrous, shining, ca. 2.5 mm. long, the beak an additional 1.5 mm.—Moist rocky places, 5000–11,000 ft.; Montane Coniferous F.; Sierra Nevada from Tulare Co. n., to Siskiyou, Trinity, and Humboldt cos. where it descends to ca. 2000 ft.; to B.C. July–Aug. Many Calif. plants have finely pubescent young growth and are the ssp. *spléndens* (Baumann) Abrams. [*S. s.* Baumann.]

4. Petrophỳtum Rydb. ROCK-SPIRAEA

Woody cespitose plants forming dense mats on rocks. Lvs. persistent, crowded, oblanceolate to spatulate, entire. Fls. perfect, in spicate racemes. Sepals 5. Petals 5, white. Stamens ca. 20. Pistils 3–5; ovary pubescent; style filiform. Follicles dehiscent on both sutures. Ca. 4 spp. of w. N. Am. (Greek, *petra*, a rock, and *phyton*, plant.)

1. **P. caespitòsum** (Nutt.) Rydb. [*Spiraea c.* Nutt. *Eriogynia c.* Wats. *Luetkea c.* Kuntze.] Depressed undershrub with prostrate branches forming dense mats 3–8 dm. wide, densely silky-pubescent and clothed with rosulate tufts of spatulate lvs. 5–12 mm. long and 1-nerved; peduncles bracted, 3–10 cm. high, bearing a spike 2–4 cm. long and usually simple; sepals lance-ovate, acute, 1.5 mm. long, canescent; petals 1.5 mm. long, mostly obtuse; follicles 2 mm. long, 2–3-ovuled, 1–2-seeded; seeds brown, smooth, linear-obovoid, ca. 1.5 mm. long.—On limestone ledges and rocks, 5000–9000 ft.; largely Pinyon-Juniper Wd.; Providence, Clark, Panamint, and Inyo mts.; to Rocky Mts. May–Sept.

Ssp. **acuminàtum** (Rydb.) Munz. [*P. a.* Rydb.] Lvs. ± 3-nerved, oblanceolate, 10–18 mm. long, sparingly pubescent; sepals lance-ovate, long-acuminate, sparingly pubescent; petals ca. 2 mm. long, acutish.—Limestone cliffs, 4000–7600 ft.; Yellow Pine F., Red Fir F.; base of Sierra Nevada, Fresno and Tulare cos. Aug.–Sept.

5. Luétkea Bong.

Cespitose plants with somewhat woody decumbent or creeping stems. Lvs. 2–3 times ternately dissected. Fls. perfect, racemose. Fl.-tube hemispheric. Sepals and petals 5. Stamens ca. 20, the fils. united at base. Pistils mostly 5, distinct. Follicles coriaceous, dehiscent on both edges. Seeds more than 1, narrow, smooth, pointed at both ends. Monotypic. (Named for Count F. P. *Luetke*, a Russian sea captain.)

1. **L. pectinàta** (Pursh) Kuntze. [*Saxifraga p.* Pursh.] Flowering stems 5–15 cm. high, subglabrous; lvs. alternate, 0.5–1.5 cm. long, with ultimate linear pointed divisions; raceme narrow, 1–5 cm. long, with entire to ternate bracts; sepals ovate, ca. 2 mm. long; petals white, round-obovate, ca. 3 mm. long; stamens and styles included; follicles smooth, ca. 4 mm. long.—Damp rocky slopes, 6500–9000 ft.; Red Fir F., Subalpine F.; Yolla Bolly Mts. (Trinity Co.) to Salmon Mts., Siskiyou Mts., and Mt. Shasta (Siskiyou Co.); to Alaska and Canadian Rocky Mts. July–Aug.

6. Arúncus (L.) Adans. GOAT'S-BEARD

Tall perennial, dioecious herbs with thick rootstocks. Lvs. large, 2–3-pinnate, without stipules. Fls. white, small, in long slender spikes in large panicles. Sepals 5. Petals 5. Stamens 15–30, exserted. Follicles usually 3, distinct, reflexed in fr., dehiscent along ventral suture and near apex of dorsal. Seeds linear-lanceolate, acute at both ends; endosperm none. Two spp., of N. Temp. zone. (Latin, *aruncus*, meaning goat's beard.)

1. **A. vulgàris** Raf. [*A. sylvester* Kost. *Spiraea Aruncus* L. *A. acuminatus* Rydb.] Stems erect, 1–2 m. high, glabrous; lower lvs. 3-compound, first ternate, then pinnate or again ternate, the lfts. thin, lance-ovate to ovate, acuminate, irregularly serrate, 3–12 cm. long, ± hairy above and beneath; petioles well developed; panicles terminal, 1–4 dm. long; racemes 0.3–1.5 dm. long, slender, continuous; petals white, ca. 1 mm. long, elliptic to obovate, those of ♀ fls. slightly smaller; follicles 3 mm. long, shining; seeds ca. 2 mm. long, cellular-punctate.—Along streams and in moist woods, below 5000 ft.; Mixed Evergreen F., Yellow Pine F., Douglas-Fir F., Red Fir F.; Mendocino and Trinity cos. to Del Norte, Siskiyou, and Shasta cos.; to Alaska, Eurasia. May–July.

7. Chamaebatiària (Porter) Maxim. FERN BUSH. DESERT SWEET

Low shrubs, aromatic, ± stellate-pubescent. Lvs. twice-pinnate, with entire stipules. Fls. white, fairly large, in panicles. Fl.-tube turbinate. Sepals 5. Petals 5. Stamens many, inserted on the margin of the disk, included. Pistils 5, ± united below; ovules pendulous. Follicles coriaceous, few-seeded, dehiscent at apex and down ventral suture. Seeds terete, with endosperm. Monotypic. (Resembling *Chamaebatia*.)

1. **C. millefòlium** (Torr.) Maxim. [*Spiraea m.* Torr.] Stout, densely branched, 0.6–2 m. tall, the young growth and herbage ± glandular and stellate-pubescent; lvs. 2–4 cm. long, oblong in outline, with 15–20 pairs of primary pinnae, these divided into 10–17 pairs less than 1 mm. long; panicle 3–10 cm. long, heavily glandular; sepals lanceolate, acute, 3–5 mm. long; petals roundish, 5 mm. long; follicles 5 mm. long; seeds yellowish, pointed at both ends, ca. 2.5 mm. long.—Dry rocky slopes, 3400–10,200 ft.; Sagebrush Scrub, Pinyon-Juniper Wd., N. Juniper Wd., to Bristle-cone Pine F.; Panamint, White, and Inyo mts., e. slope of Sierra Nevada to Modoc and Siskiyou cos.; to Ore., Wyo., Ariz. June–Aug.

8. Holodíscus Maxim.

Small to arborescent shrubs, often spreading. Lvs. alternate, simple, toothed, without stipules. Infl. terminal, racemose or paniculate, villous. Fls. whitish to pinkish, small, perfect. Sepals 5, 3-nerved, erect. Petals 5, rounded or short-clawed. Fl.-tube saucer-shaped, free from ovaries. Stamens usually 20, inserted on disk, with 3 opposite each petal, 1 opposite each sepal. Pistils 5, distinct, villous; styles terminal; ovules 2, pendulous. Frs. indehiscent, 1-seeded, laterally flattened, villous. Seeds broadly oblong, with thin endosperm. Ca. 8 spp., of w. N. Am. (Greek, *holo,* whole, and *diskos,* a disk, the disk not lobed.)

(Ley, Arlene. A taxonomic revision of the genus Holodiscus. Bull. Torrey Bot. Club 70: 275–288, 1943.)
Lvs. toothed along sides to below the middle, elliptic to orbicular in shape, not obovate.
  The lvs. elliptic to elliptic-ovate or ovate, longer than broad, with 3–6 teeth on each side, usually deeply toothed, glabrous or with a scattered pubescence above. Coastal

  1. *H. discolor*
  The lvs. broadly ovate to roundish, scarcely longer than broad, with 3–4 teeth on each side, not deeply toothed, pubescent to villous above. Montane .... 2. *H. Boursieri*
Lvs. toothed at top, rarely to the middle, obovate to spatulate. Montane .... 3. *H. microphyllus*

1. **H. díscolor** (Pursh) Maxim. [*Spiraea d.* Pursh. *Sericotheca d.* Rydb. *Spiraea ariaefolia* Sm.] CREAM BUSH. OCEAN SPRAY. Spreading, 1.5–6 m. tall; older bark dark red to brownish or gray, exfoliating in age; twigs straw-colored, pubescent or villous; lvs. broadly ovate with a truncate base to ovate-elliptic with cuneate base, 5–9 cm. long, villous to tomentose beneath, deeply 3–7-toothed on each side, each tooth again divided; petioles 0.7–2 cm. long; infl. dense, very compound, villous, 0.7–2 dm. long and broad; pedicels 1.5–3 mm. long with linear bracts 1 mm. long; sepals ± ovate, 1.5–2 mm. long; petals oval, 2 mm. long, with several hairs at outer base; pistils villous, to 1.5 mm. long; styles to 1 mm. long; aks. with straight upper and convex lower edges.—Woods and rocky places, below 4500 ft.; Redwood F., Mixed Evergreen F., Chaparral; Los Angeles Co. to Del Norte Co., Santa Catalina Id., Santa Cruz Id.; to B.C. and Mont. May–Aug. Variable.

KEY TO VARIETIES

Lf.-blades at least 4.5 cm. long, 3–7-toothed on each edge, and with short pubescence beneath
  H. discolor
Lf.-blades less than 4.5 cm. long, 2–4-toothed on either edge, often with many long hairs beneath.
  Teeth of lvs. compound; lvs. grayish beneath; petioles 5–8 mm. long. S. Calif. to Ore.
  var. *franciscanus*
  Teeth of lvs. usually simple; lvs. whitish beneath; petioles 6–12 mm. long. Del Norte Co.
  var. *delnortensis*

Var. **franciscànus** (Rydb.) Jeps. [*Sericotheca f.* Rydb.] Lf.-blades usually thick, broadly ovate with truncate base and rounded apex, villous beneath, mostly 3–4.2 cm. long; infl. 7–12(–18) cm. long, 6–12 cm. broad.—Below 4000 ft.; Chaparral, Douglas-Fir F., Mixed Evergreen F., Redwood F.; Orange Co. north to s. Ore. May–July.
Var. **delnorténsis** Ley. Lf.-blades ovate with cuneate base, gray-pubescent above, white-tomentose beneath, simply dentate, 1.5–3.5 cm. long; infl. 6–14 cm. long, 3–7 cm. broad.—Below 5500 ft.; Redwood F. to Red Fir F.; Del Norte Co., possibly to Trinity Co.; s. Ore. June–July.

2. **H. Boursiéri** (Carr.) Rehd. in Bailey. [*Spiraea B.* Carr. *Sericotheca B.* Rydb. *H. saxicola* Heller. *S. obovata* Rydb.] To 1 m. high, compact; young twigs often angled, pubescent to almost villous; lf.-blades broadly obovate to orbicular, the base sometimes cuneate, apex broad, rounded, many-toothed, the teeth broad, rounded, shallow, blades green or gray-green and finely to villous-pubescent above, grayish or whitish and villous to villous-tomentose beneath, often glandular, 1–3 cm. long; petioles 2–3 mm. long; infl. villous, 2.5–8 cm. long and wide; pedicels 1–2 mm. long, 3-bracted; sepals 1.5–2 mm. long; petals oval, 2 mm. long; stamens longer than sepals; carpels 1.5 mm. long, villous; styles 1 mm. long; aks. straight-edged above, very convex below.—Dry rocky slopes, 4000–9600 ft.; Montane Coniferous F.; mts. from s. Calif. through the Sierra Nevada to Trinity, Tehama, and Lake cos.; w. Nev. June–Aug.
3. **H. microphýllus** Rydb. [*Sericotheca m.* Rydb. *H. discolor* var. *m.* Jeps.] Spreading, bushy, 0.2–2 m. high; young twigs reddish-tan to light tan, from glabrescent with gland droplets to villous-pubescent; lvs. obovate to spatulate with cuneate base, rounded apically, many-toothed, the teeth usually small, broad, rounded and only above the middle of the blade which is finely to densely pubescent above, villous beneath, 0.5–2 cm. long, on very short petioles (1–2 mm.); infl. villous, 2.5–3.5 cm. long, 1–3 cm.

wide; sepals narrow, 1–1.5 mm. long; petals oval to elliptic, 1.5–2 mm. long; stamens
longer than sepals; carpels villous, 1 mm. long, styles 1 mm. long; aks. straight on upper
edge, convex on lower.—Dry rocky places, 5500–11,000 ft.; Pinyon-Juniper Wd., Montane Coniferous F.; San Bernardino and San Gabriel mts. and mts. of e. Mojave Desert,
Sierra Nevada and White Mts. to Plumas and Mono cos.; to Colo., Ariz. June–Aug.

KEY TO VARIETIES

Lvs. pubescent to villous beneath, with hairs masking gland-droplets if these occur, often pubescent
above.
    Lf.-blades villous beneath, finely to densely pubescent above. San Bernardino Mts. n.
                                                                        H. microphyllus
    Lf.-blades densely white-silky beneath, villous above. San Jacinto Mts. . . . . var. *sericeus*
Lvs. glabrescent to glabrous beneath with evident gland-droplets between veins, glabrescent or
scattered-pubescent above . . . . var. *glabrescens*

Var. **sericeus** Ley. Lf.-blades villous above, white-silky beneath, 0.5–2 cm. long;
infl. 3–6 cm. long, 1–4 cm. wide.—At ca. 7500 ft., San Jacinto Mts.; s. Nev., n. L. Calif.

Var. **glabréscens** (Greenm.) Ley. [*Spiraea discolor* var. *g.* Greenm. *H. g.* Heller.
Sericotheca g. Rydb. *H. discolor* var. *g.* Jeps.] Lvs. subglabrous above, glabrescent beneath with conspicuous gland-droplets, the blades 0.5–2 cm. long; infl. 3–10 cm. long,
1–7 cm. wide.—At 4500–8500 ft.; Montane Coniferous F.; Trinity and Butte to Siskiyou
and Modoc cos.; Ore. June–July.

9. Horkèlia Cham. & Schlecht.

Perennial herbs with a thick woody caudex or rootstock. Lvs. pinnate with several pairs
of toothed, cleft or divided lfts., the uppermost confluent. Fls. white, rarely cream or
pink, cymose, ± crowded. Fl.-tube deeply cup-shaped to hemispheric or saucer-shaped.
Bractlets, sepals and petals 5. Petals round to spatulate, often emarginate or short-
clawed. Stamens 10; fils. usually dilated. Carpels many, sometimes few, on a dry or
fleshy conical or hemispherical, mostly hairy receptacle; styles subterminal, glandular-
thickened at base, deciduous. Aks. mostly brown, smooth, ovoid, sometimes rugulose.
Ca. 17 spp., w. N. Am. (Named for J. *Horkel*, a German physiologist.)

(Keck, D. D. Revision of Horkelia and Ivesia. Lloydia 1: 75–142, 1938.)
A. Pistils more than 50; bractlets ovate, sepaloid; petals white; antisepalous anthers (except in
    H. cuneata) 1 mm. or more long.
  B. Fl.-tube cylindrically cup-shaped, 3–5.5 mm. deep; bractlets equaling or exceeding sepals,
    often toothed; stamens clearly in 2 rows.
    C. The fl.-tube glabrous within; sepals greenish within; antisepalous anthers 1–1.4 mm.
      long.
      D. Lfts. 3–5 pairs, serrate; fils. opposite sepals lanceolate to deltoid. Sonoma Co.
        to Monterey Co. . . . . 1. *H. frondosa*
      DD. Lfts. 5–10 pairs, much incised; fils. all essentially linear. Humboldt to Contra
        Costa cos., and Amador to Fresno cos., San Bernardino Co. . . . . . . 2. *H. elata*
    CC. The fl.-tube hairy within; sepals purple-flecked within; antisepalous anthers 1.3–1.8
      mm. long . . . . 3. *H. californica*
  BB. Fl.-tube deeply saucer-shaped, 1.5–2 mm. deep; bractlets shorter than the sepals, entire;
    stamens obscurely in 2 rows.
    C. Lfts. 5–10 pairs, the uppermost confluent; petals oblanceolate to narrowly obovate. San
      Francisco Bay to San Diego Co. . . . . 4. *H. cuneata*
    CC. Lfts. 1–3 pairs, terminal lft. petiolate; petals orbicular. San Diego Co.
                                                  5. *H. truncata*
AA. Pistils less than 30 (sometimes more in *Wilderae, Bolanderi*, and *marinensis*); bractlets lanceolate
  to narrowly linear, much smaller than the sepals; anthers rarely up to 1 mm. long.
  B. Pedicels recurving in fr.; cyme diffuse; fl.-tube glabrous within; petals white.
    C. Pistils 17–52; fl.-tube saucer-shaped; fl. 10–14 mm. across. Eldorado Co. to Calaveras Co.
                                                 6. *H. Parryi*
    CC. Pistils 3–4; fl.-tube cupulate; fl. 5–6 mm. across. San Bernardino Mts.
                                             7. *H. Wilderae*
  BB. Pedicels never recurving, erect.
    C. Lfts. 5–20 pairs, serrate to divided.
      D. Lower stipules not dissected; style glandular at base.
        E. Fl.-tube pubescent within (except sometimes in *Bolanderi*); the fils. oppo-
          site the sepals mostly more than 1.5 mm. long; petals entire. Coast Ranges
          (except *hispidula*).
          F. Lfts. serrate to shallowly lobed, 7–12 pairs.
            G. Herbage densely shaggy-villous; fils. linear-dilated. Maritime.
                                     8. *H. marinensis*

GG. Herbage hoary to moderately villous; fils. lanceolate-dilated. Mon-
tane . . . . 9. *H. Bolanderi*
  FF. Lfts. deeply palmatifid.
    G. Cymes compact, many-fld.; lfts. villous, 10–20 pairs, 5–10 mm.
long. Mendocino Co. to San Luis Obispo Co. . . . . 10. *H. tenuiloba*
    GG. Cymes open, few-fld.; lfts. short-hirsute, 6–12 pairs, 3.5–5 mm.
long. White Mts. . . . . 11. *H. hispidula*
  EE. Fl.-tube glabrous within; fils. less than 1.5 mm. long; petals often emarginate.
Tulare to Siskiyou and Modoc cos. . . . . 12. *H. fusca*
  DD. Lower stipules 2–4 times divided into filiform segms.; style not glandular at base.
    E. Fl.-tube turbinate, 2–2.5 mm. wide; petals white or pinkish; antisepalous fils.
1–1.7 mm. long; lfts. simply cleft or bisected; cyme diffuse . 13. *H. sericata*
    EE. Fl.-tube shallowly cupulate, 3–3.5 mm. wide; petals cream; antisepalous fils.
1.6–2.2 mm. long; lfts. mostly compoundly dissected; cyme usually compact.
14. *H. daucifolia*
  CC. Lfts. 2–5 pairs, short-toothed at apex only or entire; petals white . . 15. *H. tridentata*

1. **H. frondòsa** (Greene) Rydb. [*Potentilla f.* Greene. *P. californica* var. *f.* Jeps.]
Stems stout, erect or decumbent, leafy, 3–8 dm. long; herbage ± pilose, aromatic, glandular especially upward; lower lvs. 1–3.5 dm. long; lfts. 3–6 cm. long, ovate to oblong,
serrate to almost cleft; cyme forked, the fls. in small glomerules; fl.-tube 4–8 mm. wide,
3–4 mm. deep; bractlets ovate, usually 3-toothed; sepals 4–5 mm. long, equalled by
petals; antisepalous fils. 1.5–2 mm. long, lanceolate to deltoid, antipetalous linear; pistils
80–220.—Dry hills below 1000 ft.; Closed-cone Pine F., Chaparral, Mixed Evergreen
F.; Sonoma Co. to Monterey Co. May–Sept.

2. **H. elàta** (Greene) Rydb. [*Potentilla e.* Greene. *H. glandulosa* Eastw. *P. e.* var.
dissita Crum.] Leafy, 3–8 dm. high, pilose, glandular, aromatic; lower lvs. 0.7–2.5 dm.
long; lfts. 5–10 pairs, cuneate-oblanceolate to broadly flabelliform, 1–3-cleft and incised into laciniate teeth; upper lvs. smaller; cyme forked; fl.-tube 4–6 mm. wide,
3–3.5 mm. deep; bractlets mostly entire, lanceolate to ovate; sepals 3–5 mm. long, ca. as
long as petals; fils. mostly linear, 1.5–2 mm. long; pistils 50–100; $2n = 28$ (Gustafsson,
ms.).—Moist often shaded places, below 5000 ft.; Mixed Evergreen F., Yellow Pine
F.; Coast Ranges from Humboldt Co. to Contra Costa Co., Sierra Nevada from Amador
to Fresno cos., San Bernardino Mts. May–Aug.

3. **H. califórnica** Cham. & Schlecht. [*Sibbaldia c.* Spreng. *Potentilla c.* Greene. *P. c.*
var. *carmeliana* Jeps. *H. grandis* H. & A.] Stems stout, erect or spreading, leafy, 2–10 dm.
long; herbage pilose and glandular, pleasantly aromatic; lvs. mostly 1–2 dm. long, often
long-petioled; lfts. 5–8(–10) pairs, 1–4 cm. long, roundish to obovate, doubly toothed
or incised with slender teeth; cyme forked, with terminal glomerules and solitary fls. in
axils; fl.-tube cup-shaped, 6–8 mm. wide, 4–5.5 mm. deep, rust to purplish, ± pilose
within; bractlets lance-oblong, mostly 3-toothed; sepals 4.5–6 mm. long, equaled by
petals; stamens clearly 2-rowed, the antisepalous fils. ± dilated, 2–3 mm. long, with
anthers 1.3–1.8 mm. long; pistils 80–200; $n = 28$ (Gustafsson, ms.).—Common on
grassy slopes near coast; N. Coastal Scrub, Coastal Prairie; Humboldt Co. to Santa Cruz
Co., San Luis Obispo Co. May–Aug.

4. **H. cuneàta** Lindl. [*P. c.* Baill., not Wall. *P. Lindleyi* Greene. *P. L.* var. *lepida*
Crum. *H. californica* var. *cuneata* Gray. *H. platycalyx* Rydb.] Stems erect, 2–5 dm.
high; herbage glandular-villous to glabrate; lower lvs. 1–2 dm. long; lfts. 5–10 pairs,
the uppermost confluent, 1–3 cm. long, cuneate-oblong to broadly obovate or oval, sharply
serrate above base; cymes mostly congested; fl.-tube deeply saucer-shaped, 4.5–7 mm.
wide, 1.5–2 mm. deep, green, densely pilose within; bractlets entire, lanceolate to ovate,
shorter than sepals; sepals 4.5–6.5 mm. long, green or with purple flecks, equalled by
narrowly oblanceolate petals; stamens obscurely 2-rowed, the fils. usually all dilated, the
antisepalous deltoid, 2–2.7 mm. long, with anthers 0.7–1.1 mm. long; pistils 40–80;
$n = 14$ (Gustafsson, ms.).—Open sandy fields and in woods; Coastal Strand, Closed-
cone Pine F., Foothill Wd.; San Francisco to San Diego Co. most common from Santa
Cruz to Santa Barbara cos. April–Sept. Fls. often red in plants from San Luis Obispo
Co.

Fl.-tube densely pilose within; fls. often glomerate.
  Herbage glandular-villous to glabrate . . . . *H. cuneata*
  Herbage densely silky, obscurely glandular . . . . ssp. *sericea*
Fl.-tube obscurely pilose to glabrous within; fls. in open cymes; herbage moderately glandular-
pubescent . . . . ssp. *puberula*

Ssp. **serícea** (Gray) Keck. [*H. californica* var. *s.* Gray. *H. Kelloggii* Greene. *Potentilla K.* Greene. *H. s.* Rydb.] $n = 14$ (Gustafsson, ms.).—Sandy and gravelly places; N. Coastal Scrub, Closed-cone Pine F.; along coast from Sonoma Co. to Santa Barbara Co.

Ssp. **pubérula** (Greene) Keck. [*Potentilla p.* Greene. *H. p.* Rydb. *P. Lindleyi* var. *p.* Jeps.] Sandy and gravelly places away from the coast; Chaparral, Coastal Sage Scrub; San Luis Obispo Co. to San Diego Co.

5. **H. truncàta** Rydb. [*Potentilla t.* M. & J.] Erect, sparsely leafy, 2–5 dm. high, glandular-pubescent throughout; lower lvs. 0.5–1.2 dm. long, with 1–3 pairs of lateral lfts. 1–3 cm. long and a ± petiolulate larger terminal lft., all ± oblong, rounded-truncate and dentate at apex; cyme forked, fls. mostly solitary; fl.-tube saucer-shaped, 5 mm. wide, 1.5 mm. deep, greenish, glabrous within; bractlets ovate, entire, setose-ciliate, shorter than sepals; sepals 4–5.5 mm. long, somewhat shorter than the roundish petals; stamens essentially 1-rowed, fils. dilated, the antisepalous deltoid, 1.5 mm. long, with anthers 1 mm. long; pistils 50–80.—Dry slopes, 2000–4000 ft.; Chaparral, S. Oak Wd.; San Diego Co.; n. L. Calif. May–June.

6. **H. Párryi** Greene. [*Potentilla P.* Greene. *H. platypetala* Rydb.] Rootstocks horizontal; stems slender, leafy, 1–2.5 dm. long; herbage green, glandular-pubescent throughout; basal lvs. 0.4–1 dm. long; lfts. 4–6 pairs, 5–12 mm. long, cuneate-oblong to broadly obovate, deeply incised; cymes diffuse; fl.-tube saucer-shaped, 2.5–3.5 mm. wide, 0.6–1 mm. deep, glabrous within; bractlets linear-lanceolate, ⅓ shorter than the lance-ovate sepals; petals broadly cuneate-oblong, truncatish at apex, 4–6 mm. long, exceeding sepals; stamens essentially 1-rowed, the fils. dilated, ovate-lanceolate, the antisepalous larger, 2–3 mm. long, with anthers 0.6–1 mm. long; pistils 17–52, gray, finely reticulate; $2n = 28$ (Gustafsson, ms.).—Dry hills, below 1000 ft.; openings in Chaparral; Sierran foothills from Eldorado Co. to Calaveras Co. April–June.

7. **H. Wílderae** Parish. [*Potentilla W.* M. & J. *P. Parryi* var. *W.* Jeps.] Caudex from a deep taproot, stems widely spreading, slender, leafy, diffusely branched, 1–3 dm. long; herbage pale green, finely glandular-pubescent; basal lvs. 0.5–1 dm. long; lfts. 4–7 pairs, 5–10 mm. long, broadly obovate, deeply flabellate-dissected into linear-oblong lobes; cyme diffuse; fl.-tube cupulate, 1.5–3 mm. wide, 1–1.5 mm. deep, glabrous within; bractlets narrow-oblong, 0.7–1 mm. long; sepals triangular-ovate, 1.7–2.5 mm. long, ± purplish, stipitate-glandular and somewhat pilose; petals broadly oblanceolate, longer than sepals; stamens biseriate, the antisepalous broadly deltoid, 0.7–0.9 mm. long, the anthers ca. 0.3 mm. long; pistils 3–4, 1.8 mm. long, buff, ± rugulose.—Dry benches in forest, 6000–7500 ft.; Yellow Pine F.; around Barton Flats, San Bernardino Mts. May–Aug.

8. **H. marinénsis** (Elmer) Crum in Keck. [*H. Bolanderi* var. *m.* Elmer. *Potentilla Kelloggii* var. *m.* Jeps. *P. m.* J. T. Howell.] Rootstocks woody, horizontal and forming cushionlike mats 1–5 m. across; flowering stems sparsely leafy, 1–2.5 dm. high; herbage pale yellow-green, shaggy-villous and finely glandular, rankly aromatic; basal lvs. 0.5–1.5 dm. long; lfts. 7–10(–12) pairs, 8–15 mm. long, cuneate-obovate to flabelliform, palmately lobed or incised; cymes congested; fl.-tube shallowly cupulate, 4–5 mm. wide, 1.2–1.5 mm. deep, green, pilose within; bractlets linear-lanceolate, entire, shorter than sepals; sepals 4–6 mm. long; petals somewhat longer, narrowly cuneate-oblong; stamens obscurely 2-rowed, the fils. linear-dilated, the antisepalous 2–2.8 mm. long, with anthers 0.6–1 mm. long; pistils 24–30; $2n = 56$ (Gustafsson, ms.).—Coastal Strand, etc.; Mendocino Co. to Marin Co. May–Sept.

9. **H. Bolánderi** Gray. [*Potentilla B.* Greene.] Cespitose clumps from woody rootstocks; stems few, 1–3 dm. high, sparingly leafy; herbage white-canescent and strigulose; basal lvs. 4–8 cm. long; lfts. 8–12 pairs, 0.4–0.8 cm. long, the uppermost confluent, cuneate to flabelliform, toothed at apex or palmately lobed; cymes subcapitate to more open; fl.-tube cupulate, 3.3–5 mm. wide, 1.3–2 mm. deep; bractlets lance-ovate, ¾ as long as sepals; sepals triangular-lanceolate, 3–4(–5) mm. long, slightly exceeded by petals; stamens obscurely 2-rowed, fils. lanceolate-dilated, the antisepalous 1.3–2 mm. long, their anthers 0.5–0.7 mm. long.—Moist places like meadows and stream banks, 2500–3500 ft.; Yellow Pine F.; Colusa and Lake cos. June–Aug.

Pistils 10–27; inner anthers 0.5–0.7 mm. long; lvs. white-canescent and strigulose; stems slender, 1–3 dm. long; fl.-tube pubescent within. Lake and Colusa cos. . . . . *H. Bolanderi*
Pistils mostly 35–55; inner anthers 0.8–1 mm. long; stems mostly coarser and taller.

Fl.-tube pubescent within; lvs. hoary to moderately short-pilose, ± strigose, obscurely glandular. Santa Lucia Mts., Mt. Pinos, and San Bernardino Mts. . . . . ssp. *Parryi*
Fl.-tube glabrous within; lvs. greenish, usually less hairy and more obviously glandular. San Jacinto Mts. to L. Calif. . . . . ssp. *Clevelandii*

Ssp. **Párryi** (Wats.) Keck. [*H. Bolanderi* var. *P.* Wats. *Potentilla B.* var. *P.* M. & J. *H. P.* Rydb. *H. Rydbergii* Elmer. *H. bernardina* Rydb.] $2n = 28$ (Gustafsson, ms.).— Stems 1.5–7 dm. long; basal lvs. 4–15(–20) cm. long; lfts. 0.5–1.5 cm. long.—Mats in moist places, 3800–9000 ft.; Yellow Pine F.; San Bernardino Mts. to Mt. Pinos, also in Santa Lucia Mt. regions. June–Aug.

Ssp. **Clevelándii** (Greene) Keck. [*Potentilla C.* Greene. *H. C.* Rydb. *P. Bolanderi* var. *C.* M. & J.] Stems 1.5–5 dm. long; rosette-lvs. 0.6–1.8 dm. long; lfts. 0.4–1.5 cm. long.—Dampish places, 4000–6500 ft.; Yellow Pine F.; San Jacinto Mts. to n. L. Calif. June–Aug.

10. **H. tenuilòba** (Torr.) Gray. [*H. fusca* var. *t.* Torr. *Potentilla t.* Greene, not Jordan. *P. Micheneri* Greene. *P. stenoloba* Greene.] Caudex with slender branches bearing leafy stems 1–3 dm. high; herbage pale green, ± reddish, villous to glabrate, viscid with stipitate and punctate glands; basal lvs. 0.6–1.5 dm. long; lfts. 10–16(–20) pairs, crowded, 0.5–1 cm. long, flabelliform, palmately deeply cleft into 3–8 linear lobes; cyme compact; fl.-tube cupulate, 2.5–3.5 mm. wide, 1–1.2 mm. deep, pilose within; bractlets linear to oblong-lanceolate, much smaller than sepals; sepals broadly lanceolate, 3.5–4.5 mm. long, ca. or almost as long as oblanceolate emarginate petals; stamens obscurely 2-rowed, fils. broadly linear to lanceolate, the antisepalous 1.5–2 mm. long, their anthers 0.6–0.7 mm. long; pistils 9–26; $2n = 28$ (Gustafsson, ms.).—Sandy or silty meadows, below 1800 ft.; Chaparral, Mixed Evergreen F.; outer Coast Ranges from Sonoma, Marin, and San Luis Obispo cos. April–July.

11. **H. hispídula** Rydb. [*Potentilla h.* Jeps.] Caudex simple or few-branched; stems slender, erect or ascending, glandular-puberulent, ± villous, scarcely leafy, 1–2.5 dm. high; basal lvs. 0.5–1 dm. long, pale green, densely hispidulous, setose-ciliate, finely and obscurely glandular; lfts. 6–12 pairs, crowded, 3.5–5 mm. long, flabelliform, unequally cleft into 5–7 spatulate to obovate segms; cyme few-fld.; fl.-tube hemispheric, 3.5–4 mm. wide, 2 mm. deep, ± pilose within; bractlets linear, 2–3 mm. long; sepals triangular-lanceolate, acuminate, 3.5–4.2 mm. long; petals obovate, emarginate, somewhat longer; stamens 1-rowed, the fils. lance-dilated, the antisepalous 1.7–2 mm. long; pistils 12–18, the dark brown aks. plump, lightly rugulose.—Brushy and rocky slopes, 9000–11,000 ft.; largely Bristle-cone Pine F.; White Mts., Inyo Co. June–Aug.

12. **H. fúsca** Lindl. ssp. **capitàta** (Lindl.) Keck. [*H. c.* Lindl.] Caudex short, stout; stems erect or ascending, sparingly leafy, 1–5 dm. high; herbage bright green, glandular, glabrate; basal lvs. erect, 2–3.5 dm. long; lfts. 5–10 pairs, oval to roundish, toothed above the base, 0.5–1.5 cm. long, the upper ± confluent; cyme capitate, purplish; fl.-tube ± cupulate, 2–4 mm. wide, 1.2–2.2 mm. deep, glabrous within; bractlets linear, somewhat shorter than the sepals; sepals acuminate, lanceolate, 2.5–4 mm. long; petals obcordate; stamens obscurely biseriate, fils. lanceolate, to 1.6 mm. long, anthers ca. 0.5 mm. long; pistils 15–25, the aks. smooth, brown, less than 2 mm. long; $2n = 28$ (Gustafsson, ms.). —Moist places, below 7000 ft.; Yellow Pine F., Red Fir F.; Modoc Co.; to Ore., Ida. May–July.

KEY TO SUBSPECIES

Petals 4–6.5 mm. long, usually pink, with red veins; bracts subtending the cymes prominent and often exceeding them; herbage bright green, glabrate; lfts. mostly broad and merely toothed. Modoc Co. . . . . ssp. *capitata*
Petals 2.5–4 mm. long, usually white, lightly roseate-veined; bracts subtending the cymes inconspicuous.
  Lfts. variously dentate or lobed to short-incised.
    Lvs. densely pubescent, grayish, slightly glandular. Modoc and Siskiyou cos. to Alpine Co.
        ssp. *pseudocapitata*
    Lvs. sparingly pubescent, dark green, more glandular. Tulare Co. to Shasta and Siskiyou cos.
        ssp. *parviflora*
  Lfts. parted or divided into linear divisions. Volcanic ash, Lassen and Shasta cos. to Plumas Co.
        ssp. *tenella*

Ssp. **pseudocapitàta** (Rydb.) Keck. [*H. p.* Rydb. *H. Brownii* Rydb. *Potentilla Douglasii* var. *p.* Jeps.] Stems slender to stoutish, pilose below or glandular throughout;

young lvs. gray-villous, scarcely glandular; lfts. narrow-cuneate to broadly obovate, bluntly or sharply toothed or lobed, principally apically; cymes dense; sepals broadly lanceolate, much longer than bractlets; petals spatulate, sometimes emarginate; fils. broadly lanceolate to deltoid, to 1.2 mm. long.—Woods, 3500–8000 ft.; Yellow Pine F., Red Fir F.; e. flank of n. Sierra Nevada; Ore., Nev. July–Aug.

Ssp. **parviflòra** (Nutt.) Keck. [*H. p.* Nutt. *Potentilla Andersonii* Greene.] Stems slender; herbage glandular-pubescent, dark green; lfts. cuneate-oblong to flabelliform, toothed to shallowly incised or lobed; cyme dense or somewhat branched; sepals longer than bractlets; petals spatulate, sometimes emarginate; fils. very broad, mostly less than 1 mm. long; $n = 14$ (Gustafsson, ms.).—Open places, 2500–10,500 ft.; Yellow Pine F. to Subalpine F.; w. slope of Sierra Nevada from Tulare Co. n., to Del Norte and Siskiyou cos.; to Wash., Wyo. July–Aug.

Ssp. **tenélla** (Wats.) Keck. [*H. fusca* var. *t.* Wats. *H. t.* Rydb.] Stems ± leafy; herbage softly pilose to glabrate, glandular throughout; lfts. short and broad, palmatifid to palmatisect; cyme dense, mostly capitate; $2n = 28$ (Gustafsson, ms.).—Dry volcanic ash, 4000–6000 ft.; Yellow Pine F., Red Fir F.; n. Sierra Nevada. July–Aug.

13. **H. sericàta** Wats. [*Potentilla s.* Greene. *P. laxiflora* E. Drew. *H. l.* Rydb.] Caudex short, simple or branched; stems few, slender, strict, ± purplish, 1.5–4 dm. high; lvs. mostly basal, densely silky-villous to glabrate, ± glandular, the basal 0.4–1.5 dm. long, the stipules 2–4-times dichotomously divided into filiform segms; lfts. 10–20 pairs, crowded, 0.5–1 cm. long, often divided to base into 2 or 3 narrowly elliptic lobes or merely cleft; cymes loose, fl.-tube turbinate, 2–2.5 mm. wide, 1–1.5 mm. deep, glabrous within; bractlets linear-lanceolate, shorter than the broadly lanceolate acuminate sepals which are 2.5–4.5 mm. long; petals white or pinkish, obcordate, longer than sepals; fils. lanceolate-dilated, the antisepalous larger, 1–1.7 mm. long, their anthers 0.5–0.7 mm. long; pistils 3–6; aks. plump, brown, smooth, 2–2.7 mm. long; $n = 14$ (Gustafsson, ms.). —Dry serpentine and red clay, 1500–3600 ft.; Mixed Evergreen F., Yellow Pine F.; Humboldt Co. to sw. Ore. May–July.

14. **H. daucifòlia** (Greene) Rydb. [*Potentilla d.* Greene. *P. d.* var. *indicta* Jeps. *P. congesta* var. *lobata* Lemmon. *H. pulchra* Rydb.] Caudex short, mostly simple; stems suberect, somewhat leafy, often purplish, pilose below, often merely glandular above, 1–3.5 dm. long; basal lvs. villous to silky, sometimes glandular, 0.8–1.5 dm. long, long-petioled, their stipules divided into filiform segms.; lfts. 6–10 pairs, crowded, 1–2 cm. long, mostly compoundly dissected into linear segms. 1–2 mm. wide; cymes corymbosely congested; fl.-tube shallowly cupulate, 3–3.5 mm. wide, 1–1.5 mm. deep, glabrous within; bractlets linear, much smaller than the sepals; sepals lanceolate, 4–5 mm. long; petals cream, broadly spatulate to round, somewhat longer than sepals; fils. dilated, the antisepalous larger, 1.6–2.2 mm. long, their anthers 0.7–0.8 mm. long; pistils 6–9, aks. dark, pyriform, glossy, ± rugulose, 2–2.5 mm. long.—Dry woods, 2700–4000 ft.; largely Yellow Pine F.; Tehama and Trinity cos. to Siskiyou Co.; s. Ore. April–July.

Ssp. **làtior** Keck. Lfts. deeply bifid, the segms. oblong, 3 mm. wide.—At ca. 5300 ft.; Scott Mt., Siskiyou Co. July.

15. **H. tridentàta** Torr. [*Ivesia t.* Gray. *H. Tilingii* Regel. *Potentilla T.* Greene.] Caudex short, cespitose; stems decumbent or ascending, sparingly leafy, usually purplish, 1.5–3 dm. high; herbage silky or cobwebby-pilose, upper surface of lvs. often glabrous; basal lvs. 0.3–0.8 dm. long, with inconspicuous stipules; lfts. 2–7 pairs, not crowded, 0.8–2 cm. long, spatulate to broadly oval, usually 3-toothed at apex; cyme compact; fl.-tube glabrous within, 2–3.5 mm. wide; bractlets linear, smaller than sepals; sepals deltoid, 2–3 mm. long; petals white, sometimes pinkish, narrow, ca. as long as sepals; fils. subulate to filiform; $n = 14$ (Gustafsson, ms.).—Woods at 2000–6500 ft.; Yellow Pine F., Red Fir F.; Tulare Co. through Sierra Nevada to Siskiyou Co.; sw. Ore. May–July.

Ssp. **flavéscens** (Rydb.) Keck. [*H. f.* Rydb.] Fl.-tube pilose within, 2.5–4.5 mm. wide.—Woods, 2500–5000 ft.; Yellow Pine F.; Coast Ranges from Lake Co. to Siskiyou Co. and in Plumas, Sierra, and Nevada cos. May–July.

10. Ivèsia T. & G.

Perennials with mostly heavy caudex. Lvs. pinnate, mostly basal; lfts. mostly parted or divided into narrow lobes, usually small, many and imbricated. Fls. yellow or white or purple, cymose, usually crowded. Fl.-tube turbinate or campanulate to saucer-shaped or disciform. Bractlets, sepals and petals usually 5. Petals linear to spatulate or obovate, often clawed. Stamens 5 (antisepalous) or 20 (1 antipetalous to 3 antisepalous), rarely 10 or 15; fils. filiform, rarely subulate, usually inserted on rim of fl.-tube; anthers with lateral sutures. Carpels few (1–15, rarely more), on a dry usually low hirsute receptacle; styles subterminal; aks. smooth or carunculate. Twenty-two spp., w. Am. (named for Lt. E. *Ives*, leader of a Pacific Ry. Survey.)

(Keck, D. D. Revision of Horkelia and Ivesia. Lloydia 1: 75–142, 1938.)

A. Inner row of stamens inserted at margin of receptacle; stamens 20; cespitose dwarfs with open few-fld. cymes; anthers 0.3 mm. long; aks. caruncled.
    B. Petals white, obovate; lfts. 2-lobed or entire. San Jacinto Mts. .... 1. *I. callida*
    BB. Petals yellow, linear; lfts. 2–5-lobed. Clark Mt. .... 2. *I. Jaegeri*
AA. Inner row of stamens inserted well away from the margin of the receptacle.
    B. Stamens 5 or 10.
        C. The stamens 10; pistils 12–30; petals broad. Sierra Nevada ...... 3. *I. pygmaea*
        CC. The stamens 5; pistils mostly 1–10; petals usually narrow.
            D. Lvs. densely silvery-silky, vermiform. Sierra Nevada .... 8. *I. Muirii*
            DD. Lvs. green, not silvery-silky.
                E. Lfts. 12–40 pairs.
                    F. Pistils 8–18, the receptacle short-hairy; petals obovate to rounded; fl.-tube shallowly cupulate. Sierra Nevada .... 4. *I. lycopodioides*
                    FF. Pistils 1–8; receptacle prominently white-hirsute; petals linear or spatulate; fl.-tube campanulate. Tuolumne Co. to Modoc and Siskiyou cos. 5. *I. Gordonii*
                EE. Lfts. 5–10 pairs.
                    F. Fl.-tube disciform, thickened; fils. 1 mm. long; anthers less than 1 mm. long; cyme open. Placer Co. to Inyo Co. .... 6. *I. Shockleyi*
                    FF. Fl.-tube hemispheric, not thickened; fils. 2–3 mm. long; anthers more than 1 mm. long; cyme dense. Plumas and Sierra cos. ... 7. *I. Webberi*
    BB. Stamens usually 15 or 20.
        C. Stamens 20; styles glandular at base.
            D. Pistils 25–50; fils. subulate-dilated; herbage obviously glandular. Tulare to Mono and Inyo cos. ....9. *I. purpurascens*
            DD. Pistils 2–9; herbage obscurely, if at all glandular.
                E. Fils. subulate-dilated; stems subscapose; herbage densely silvery-silky; lvs. vermiform. San Bernardino Mts. .... 10. *I. argyrocoma*
                EE. Fils. filiform; stems leafy; lvs. not vermiform.
                    F. Fl.-tube campanulate; sepals 3.5–5.5 mm. long; bractlets thin; herbage densely villous or tomentose.
                        G. The fl.-tube glabrous within, 2–2.5 mm. deep; pistils 4–7; lfts. 20–35 pairs, 4–15 mm. long. Plumas Co. to Nevada Co. 11. *I. sericoleuca*
                        GG. The fl.-tube pilose within; 1.5 mm. deep; pistils 2–3; lfts. 25–50 pairs, 2–5 mm. long. Siskiyou and Trinity cos. .. 12. *I. Pickeringii*
                    FF. Fl.-tube saucer-shaped; sepals 2.5–3 mm. long; bractlets thickened; herbage glabrous to canescent. Mono Co. .... 13. *I. Kingii*
        CC. Stamens 15 (sometimes 10 to 20); styles glandular or not.
            D. Pistils 3 or more; anthers broadly oblong, yellow, ± explanate; fls. glomerate; aks. less than 1.5 mm. long.
                E. Fls. regularly 5-merous; petals white; fl.-tube turbinate, 1.7–2.2 mm. deep. Mariposa Co. to Fresno Co. .... 14. *I. unguiculata*
                EE. Fls. usually 4-merous; petals yellow; fl.-tube hemispheric, 1.5 mm. deep. Tulare Co. .... 15. *I. campestris*
            DD. Pistil 1; anthers broadly obcordate, dehiscent by narrow subapical slits, purple-black; fls. scattered; aks. 2 mm. long. Eldorado Co. to Riverside Co. 16. *I. santolinoides*

1. **I. cállida** (Hall) Rydb. [*Potentilla c.* Hall. *P. Shockleyi* var. *c.* Jeps.] Stems 0.2–0.5 dm. long, spreading; herbage hirsute and finely glandular throughout; lvs. 0.2–0.3 dm. long; lfts. 6–8 pairs, 2–3 mm. long; cyme open, 1–6-fld.; fl.-tube ca. 2 mm. wide; bractlets lance-linear; sepals lanceolate, 2.5–3.5 mm. long; petals white, equaling sepals; pistils 4–6.—Rock crevices at ca. 8000 ft.; Red Fir F.; Tahquitz Peak, San Jacinto Mts., Riverside Co. July–Aug.

2. **I. Jàegeri** M. & J. [*Potentilla J.* Wheeler.] Stems subdecumbent, 0.5–1.2 dm. long, subscapose; herbage puberulent and finely glandular; lvs. many, thin, 0.3–0.8 dm. long;

lfts. 4–8 pairs not crowded, 3–6 mm. long; cyme open, few-fld., with filiform pedicels; fl.-tube disciform, ca. 2.5 mm. wide; bractlets ovate, minute; sepals deltoid-lanceolate, 2–3 mm. long; petals yellow, linear, 1.5–2 mm. long; fils. narrowly filiform, 1.5 mm. long; pistils 3–9; aks. ca. 2 mm. long.—Limestone crevices, at ca. 7500 ft.; Pinyon-Juniper Wd.; Clark Mt., e. San Bernardino Co.; sw. Nev. June–July.

3. **I. pygmaèa** Gray. [*I. Gordonii* var. *p.* Wats. *Potentilla decipiens* Greene. *P. nubigena* Greene. *Horkelia chaetophora* Rydb. *P. Gordonii* vars. *p.* and *c.* Jeps.] Stems scapose, 0.4–1.2 dm. long; herbage ± puberulent and villous, densely glandular; lvs. 0.2–0.8(–1.2) dm. long; lfts. 10–20 pairs, crowded, 1–5 mm. long, divided mostly to base into 3–6 linear-oblong or oblanceolate bristle-tipped segms.; cyme rather dense; fl.-tube shallowly cupulate, ca. 3 mm. wide, 0.5 mm. deep; bractlets oblong or elliptic, half as long as deltoid-lanceolate sepals; petals yellow, broadly spatulate to broadly obovate, 2.5–3.5 mm. long; fils. filiform; pistils 12–30, with thickened glandular styles; aks. green-brown, smooth, carunculate, 1.3–1.5 mm. long.—Rocky slopes, 9500–13,000 ft.; mostly Alpine Fell-fields; Sierra Nevada of Fresno, Inyo and Tulare cos. July–Aug.

4. **I. lycopodioìdes** Gray. [*I. Gordonii* var. *l.* Wats. *Potentilla G.* var. *l.* Jeps. *Horkelia l.* Rydb.] Stems to ca. 1 dm. long, from a heavy woody caudex; herbage usually glabrate; lvs. vermiform, 0.2–0.6 dm. long; lfts. thick, 1–2 mm. long, with cuneate-orbicular segms.; cyme ± capitate; fl.-tube shallowly cupulate, 2–2.5 mm. wide, 0.5–1 mm. deep; bractlets linear-oblong to oval, shorter than sepals; sepals deltoid-lanceolate, ca. as long as petals; petals yellow, oblong-lanceolate, ca. 2 mm. long; fils. filiform; pistils 8–18, the styles thickened, 1 mm. long; aks. greenish-brown, smooth, plump, ca. 1.5 mm. long; $2n = 28$ (Gustafsson, ms.).—Moist gravelly places, 10,000–12,500 ft.; Alpine Fell-fields; Sierra Nevada from Fresno Co. to Eldorado Co.; w. Nev. July–Aug.

Lfts. tightly imbricated, 1–2 mm. long, the lvs. 2–6 cm. long; fls. 5–6 mm. across.
  Herbage glabrate to sparingly puberulent, glutinous .... *I. .lycopodioides*
  Herbage densely puberulent, the lfts. somewhat villous-ciliate, less glutinous .. ssp. *scandularis*
Lfts. not imbricated, 2–6 mm. long, the lvs. 4–15 cm. long; fls. ca. 10 mm. across .. ssp. *megalopetala*

Ssp. **scandulàris** (Rydb.) Keck. [*Horkelia s.* Rydb. *I. s.* Rydb. *Potentilla Gordonii* var. *s.* Jeps.] Meadows at 11,000–13,000 ft.; Bristle-cone Pine F.; White Mts. of Mono and Inyo cos. July–Aug.

Ssp. **megalopétala** (Rydb.) Keck. [*Horkelia Gordonii* var. *m.* Rydb. *I. m.* Rydb. *Potentilla G.* var. *m.* Jeps.] Stems 1–2.5 dm. long; herbage glabrous to sparsely hairy, glandular; petals ca. 3.5 mm. long; fils. 2 mm. long.—Meadows, 7500–11,000 ft.; Lodgepole F. to Subalpine F.; Sierra Nevada from Tuolumne Co. to Tulare Co. July–Aug.

5. **I. Gordònii** (Hook.) T. & G. [*Horkelia G.* Hook. *Potentilla G.* Greene. *P. G.* var. *ursinorum* Jeps.] Stems subscapose, 0.5–2 dm. long, from a thick woody caudex; herbage ± puberulent and viscid-glandular to subglabrous; lvs. 0.3–1.8 dm. long; lfts. 10–25 pairs, 2–8 mm. long, ± crowded, divided to base into 2–5 cuneate to oblong segms. or forked; cyme capitate, many-fld.; fl.-tube campanulate, 2.5–3 mm. long, 2–3 mm. deep, yellowish; bractlets linear, ca. half as long as the deltoid-lanceolate erect sepals, 2.5–4 mm. long; petals yellow, spatulate to oblanceolate, usually as long as bractlets; fils. filiform to linear-subulate, 1.5–2.5 mm. long; pistils 1–6, the styles filiform, 3–4 mm. long; aks. gray-brown, smooth, ovoid, 1.7–2 mm. long.—Dry rocky places, 7500–13,000 ft.; Subalpine F. to Alpine Fell-fields; Yolla Bolly Mts. (Trinity Co.) to Siskiyou and Modoc cos., in the Sierra Nevada s. to Tuolumne and Mono cos.; to Wash., Mont., Colo. July–Aug.

6. **I. Shóckleyi** Wats. [*Horkelia S.* Rydb. *Potentilla S.* Jeps.] Stems 0.3–1 dm. long; herbage pallid, densely glandular-puberulent, ± hispid; lvs. 0.2–0.7 dm. long; lfts. 7–10 pairs, crowded, 1.5–3 mm. long, divided to base into 2–5 oblanceolate to rounded segms.; cyme open, few-fld., with filiform pedicels; fl.-tube disciform, yellow within, ca. 2.5 mm. across; bractlets lance-oblong to ovate, very short; sepals 1.5–3.5 mm. long; petals pale yellow, oblanceolate to oval, shorter than sepals; fils. linear-subulate, ca. 1 mm. long; pistils usually 3; aks. brown, 2–2.5 mm. long.—Gravelly and rocky places, 9000–13,000 ft.; Subalpine F. and Bristle-cone Pine F., to Alpine Fell-fields; Sierra Nevada and White Mts. from Placer Co. to Inyo Co. July–Aug.

7. **I. Wébberi** Gray. [*Potentilla W.* Greene. *Horkelia W.* Rydb.] Stems 0.8–1.5 dm.

long, wiry, with 1 pair of lvs. at middle, villous to glabrate; lvs. canescent to short-pilose, ± glandular; basal lvs. 0.4–1 dm. long; lfts. crowded, 5–10 pairs, 3–8 mm. long, parted into 2–5 linear lobes; cyme capitate or subcapitate; fl.-tube hemispheric, 2.5–3.5 mm. broad, ca. 1.5 mm. deep; bractlets linear; sepals longer, broadly lanceolate, 3.5–4.5 mm. long; petals narrow, 1–1.5 mm. long; fils. filiform, 2–3 mm. long; pistils 3–8; aks. smooth, plump, 2.2–5 mm. long.—Open patches of volcanic ash, 5000–6000 ft.; Sagebrush Scrub; e. Lassen Co. to Sierra Co.; w. Nev. May–July.

8. **I. Mùirii** Gray. [*Potentilla M.* Greene. *Horkelia M.* Rydb. *H. Chandleri* Rydb. *I. C.* Rydb.] Stems slender, 1–2-bracteate, purplish, 0.5–1.5 dm. long, herbage densely silvery-silky; lvs. basal, many, 0.2–0.6(–1) dm. long, terete, vermiform; lfts. 25–40 pairs, densely imbricated, ca. 1 mm. long, divided into 3–5 segms.; fls. many, in congested cymes; fl.-tube shallowly cupulate, 2 mm. broad, 1 mm. deep, glabrous within; bractlets subulate; sepals deltoid, 1.5–2 mm. long; petals yellow, linear, shorter; fils. filiform, short; pistils 1–4; aks. smooth, dull, compressed, 1.5–2 mm. long.—Gravelly slopes, 9500–12,000 ft.; Alpine Fell-fields; Sierra Nevada from Tuolumne and Mono cos. to Fresno Co. July–Aug.

9. **I. purpuráscens** (Wats.) Keck. [*Horkelia p.* Wats. *Potentilla p.* Greene. *P. p.* var. *pinetorum* Cov. *H. pinetorum* Rydb.] Stems erect, 2–6 dm. high; herbage pubescent, ± glandular; lvs. mostly basal, 0.7–2 dm. long; lfts. 15–28 pairs, crowded, divided almost to base into 2–8 oblanceolate rounded segms.; cyme few-fld.; fl.-tube campanulate, often purplish, 3–5 mm. wide, 2.5–3.5 mm. deep; bractlets elliptic-lanceolate; sepals longer, lanceolate, ca. 4 mm. long, acuminate; petals white, oblong-oblanceolate, ca. 5 mm. long; fils. subulate-dilated to linear, 2–3 mm. long; pistils 25–50; aks. dark brown, gray-spotted, reniform, 1.3 mm. long.—Meadow borders and creek banks, 6000–9000 ft.; Red Fir F. to Lodgepole F.; Sierra Nevada of Tulare Co. July–Aug.

Ssp. **Cóngdònis** (Rydb.) Keck. [*Horkelia C.* Rydb. *H. myriophylla* Greene. *Potentilla p.* var. *C.* Jeps.] Strongly glandular; lfts. 2–5 mm. long, usually imbricated; cyme narrow, often subcapitate, the stouter peduncles stiffly erect.—At 6500–9000 ft.; n. Sequoia Park (Tulare Co.) to Mono and Inyo cos. July–Aug.

10. **I. argyrocòma** (Rydb.) Rydb. [*Horkelia a.* Rydb. *Potentilla a.* M. & J.] Stems slender, 1–2 dm. long; herbage densely silvery-silky; basal lvs. many, vermicular, 0.3–1 dm. long; lfts. 20–35 pairs, tightly imbricate, 1–3 mm. long, divided to base into 2–3 oval lobes; cyme often paniculately branched, sometimes subcapitate; fl.-tube turbinate to campanulate, 2.5–3.5 mm. wide, 1.5–2.5 mm. deep, glabrous within; bractlets linear, half as long as sepals; sepals lanceolate, 2.5–3.5 mm. long; petals white, obovate, longer; fils. subulate-dilated, to 1.3 mm. long; pistils 4–8; aks. smooth, shining, ca. 2.2 mm. long.—Dry meadows, 6000–9600 ft.; Yellow Pine F., Red Fir F.; Bear V., San Bernardino Mts. June–Aug.

11. **I. sericoleùca** (Rydb.) Rydb. [*Horkelia s.* Rydb.] Stems slender, 1.5–4 dm. long, leafy; herbage ± densely white-silky to tomentose, not glandular; basal lvs. 1–2 dm. long; lfts. 20–35 pairs, crowded, 0.4–1(–1.5) cm. long, divided to base into 2–4 narrowly oblong to elliptic, acute to rounded, entire lobes; cyme flat-topped, dense. many-fld.; fl.-tube turbinate to campanulate, 2.5–3 mm. wide, 2–2.5 mm. deep, glabrous within; bractlets narrowly lanceolate, half as long as sepals; sepals lanceolate to sub-deltoid, acuminate, 3.5–5.5 mm. long; petals white or yellow, broadly spatulate, ca. as long as sepals; fils. filiform, to 2.5 mm. long; pistils 4–7; aks. brown, smooth, pyriform, ca. 2.5 mm. long; $n = 14$ (Gustafsson, ms.).—Dry ± alkaline flats and slopes, 4500–6000 ft.; mostly Sagebrush Scrub; e. base of Sierra Nevada from Plumas Co. to Placer Co. June–Aug.

12. **I. Pickeríngii** Torr. ex Gray. [*Potentilla P.* Greene. *Horkelia P.* Rydb.] Stems leafy, 2.5–4.5 dm. long; herbage densely and finely villous, ± viscid; basal lvs. 0.8–1.8 dm. long; lfts. 25–50 pairs, crowded, 2–5 mm. long, parted or divided to base into 2–5 elliptic-oblong or obovate entire lobes; cyme corymbosely branched, open; fl.-tube campanulate, 2.5–3.0 mm. wide, 1.5 mm. deep, pilose within; bractlets lanceolate, half as long as sepals; sepals lanceolate, acuminate, 4–5 mm. long; petals white, broadly spatulate, slightly longer; fils. filiform, 2 mm. long; pistils 2–3; aks. dark brown, smooth, ellipsoid, 2.2–2.9 mm. long; $2n = 28$ (Gustafsson, ms.).—Rare; Yellow Pine F.; Siskiyou and Trinity cos. July–Aug.

13. **I. Kíngii** Wats. [*Potentilla K.* Greene. *Horkelia K.* Rydb.] Stems leafy, 1.5–3.5

dm. long, divaricately branched toward summit; herbage glabrous to canescent, glaucous, not glandular; basal lvs. 0.5–1.5 dm. long; lfts. 12–30 pairs, close, 3–6 mm. long, usually ternately divided into elliptical or oblanceolate lobes; cyme dichotomous, open; fl.-tube saucer-shaped, 2.2–3 mm. wide; bractlets lance-ovate, less than half the sepals; sepals broadly lanceolate, 2.5–3 mm. long; petals white, obovate to broadly spatulate, somewhat longer; fils. filiform; pistils 2–9; aks. pyriform, light brown, smooth, ca. 2 mm. long.—Edges of alkaline meadows and flats, 5000–6500 ft.; Sagebrush Scrub; Mono Co. to Utah. June–July.

14. **I. unguiculàta** Gray. [*Potentilla u.* Hook. f. *P. ciliata* Greene. *Horkelia u.* Rydb.] Stems leafy, 2–3.5 dm. long; herbage silky-villous but green, obscurely glandular; basal lvs. 0.8–1.4 dm. long; lfts. 15–25 pairs, crowded, 3–6 mm. long, divided to base into 2–5 linear acute lobes; cyme branched, the fls. in dense clusters; fl.-tube turbinate, purplish, 2.5–3 mm. wide, 1.7–2.2 mm. deep; bractlets linear to oblong, to almost as long as sepals; sepals broadly lanceolate, 2.5–3.5 mm. long; petals white, oblanceolate to cuneate, ca. as long; fils. filiform, to 1 mm. long; pistils 3–9; aks. gray-brown, smooth, ovoid-oblique, 1.2 mm. long.—Open slopes, 5000–8000 ft.; Red Fir F. to Lodgepole F.; Sierra Nevada, from Mariposa Co. to Fresno Co. June–Aug.

15. **I. campéstris** (Jones) Rydb. [*Potentilla utahensis* var. *c.* Jones. *Horkelia c.* Rydb. *Potentilla c.* Jeps. *H. mollis* Eastw. *I. m.* Rydb.] Stems decumbent, slender, leafy, 1.5–3.5 dm. long; herbage silky-villous, greenish, minutely glandular; basal lvs. 0.5–1.2 dm. long, lfts. 15–30 pairs, crowded, 3–10 mm. long, divided to base into 2–5 linear lobes; cyme dense; fls. usually 4-merous; fl.-tube hemispheric, green, ca. 3 mm. wide, 1.5 mm. deep; bractlets linear to oblong, shorter than sepals; sepals lance-deltoid, 2.2–3.5 mm. long; petals light yellow, oblanceolate to spatulate-obovate, somewhat longer; stamens 15 in 4-merous fls., to 20 in 5-merous; fils. filiform, to 1 mm. long; pistils 4–20; aks. gray-brown, smooth, pyriform, 1–1.4 mm. long.—Meadows, 6500–11,000 ft.; Red Fir F. to Subalpine F.; Sierra Nevada of Tulare Co. July–Aug.

16. **I. santolinoìdes** Gray. [*Potentilla s.* Greene. *Stellariopsis s.* Rydb.] Stems suberect, slender, scarcely leafy, 1–4 dm. high, diffusely branched above, subglabrous except at base and axils; lvs. mostly crowded in a basal rosette, densely silvery-silky, vermiform, terete, 0.3–1 dm. long; lfts. very numerous, tightly imbricate, minute, divided to base into 4–5 oval segms. 1–1.5 mm. long; cyme many-fld., diffuse; fl.-tube broadly funnelform to disciform, ca. 2.5 mm. wide; bractlets roundish to oblong, minute; sepals lance-ovate, less than 2 mm. long; petals white, cuneate-obovate to roundish, 2–2.5 mm. long; fils. filiform, to 1.8 mm. long; pistils 1; aks. quadrate-orbicular, globose to ± compressed, smooth, pale, 2 mm. long; *n* = 14 (Gustafsson, ms.).—Dry gravelly slopes and ridges, 5000–12,000 ft.; Yellow Pine F. to Alpine Fell-fields; Sierra Nevada from Eldorado Co. s. to San Jacinto Mts. June–Aug.

11. Purpùsia Bdg.

Cespitose glandular perennials, with short caudex and thick roots. Lvs. mostly basal, odd-pinnate. Lfts. 5–11, roundish to elliptical or oblanceolate, crenate to palmately lobed or divided, glandular-pilose. Infl. cymose, few-fld., leafy-bracted. Fl.-tube funnelform. Bractlets mostly lacking. Sepals 5. Petals 5, white to yellowish. Stamens 5, opposite the sepals; fils. filiform. Pistils 6–13, on a stalked receptacle; styles subterminal. Monotypic. (Named for J. A. *Purpus*, 1860–1932, botanical collector.)

1. **P. saxòsa** Bdg. [*P. arizonica* Eastw. *P. Osterhoutii* A. Nels. *Potentilla O.* J. T. Howell.] Stems 0.5–2 dm. long; lfts. 0.5–1.5 cm. long; pedicels 1–2 cm. long; fl.-tube glabrous within, glandular and pilose without, ca. 2 mm. deep; sepals 2.5–3 mm. long, lance-ovate, acuminate; petals ± oblanceolate, acuminate, 3–4 mm. long; receptacle 1.5–2 mm. long.—Limestone rocks and crevices, 4000–10,000 ft.; Pinyon-Juniper Wd. to Bristle-cone Pine F.; Inyo Mts., Grapevine Mts., and Funeral Mts., Inyo Co.; to Nev., n. Ariz. May–Aug.

12. Potentílla L. CINQUEFOIL. FIVE-FINGER

Perennial, sometimes annual, herbs, rarely shrubs. Lvs. pinnately or digitately compound. Fls. perfect, solitary or cymose. Fl.-tube persistent, flat to hemispherical, 5-

bracteolate. Sepals 5. Petals 5. Stamens few to many; fils. not flattened. Pistils many, inserted on a hemispherical or conical receptacle; style terminal, lateral or basal, deciduous. Fr. aks. Perhaps 250 spp., of N. Hemis. (Diminutive of *potens*, powerful, because of supposed medicinal powers.)

(Rydberg, P. A., *in* N. Am. Fl. 22 [4]: 293–355, 1908. Clausen, Keck, and Hiesey, Carnegie Inst. of Wash. 520: 26–195, 1940.)
A.   Plants shrubs with woody branches .... 22. *P. fruticosa*
AA. Plants herbaceous.
   B.   Fls. solitary, axillary, long-pedicelled; stem slender, prostrate, often rooting at nodes.
      C.   Sepals and petals 4; styles terminal or subterminal .... 1. *P. anglica*
      CC. Sepals and petals 5; styles lateral.
         D.   Stems and petioles pubescent; basal lvs. spreading; aks. with deep dorsal groove. Montane .... 23. *P. anserina*
         DD. Stems and petioles subglabrous; basal lvs. erect; aks. without groove. Coastal
                                        24. *P. Egedei*
   BB. Fls. cymose, short-pedicelled; stems not stoloniferous.
      C.   Styles terminal or nearly so.
         D.   The styles fusiform and glandular.
            E.   Basal lvs. with 3 lfts.
               F.   Stems soft-pubescent; stamens usually 10; petals much shorter than sepals.
                  G.   The stems diffusely branched from base, 1–3 dm. long, decumbent; lfts. deeply serrate .... 2. *P. rivalis* var. *millegrana*
                  GG. The stems mostly simple at base, often 3–5 dm. long, erect; lfts. coarsely crenate .... 3. *P. biennis*
               FF. Stems stiffly hirsute below; stamens 15–20; petals ca. as long as sepals
                                    4. *P. norvegica*
            EE. Basal lvs. with 5–11 or more lfts.
               F.   Plants annual or biennial; lfts. not deeply cleft; petals shorter than sepals
                                    2. *P. rivalis*
               FF. Plants mostly perennial; petals not shorter than sepals; lfts. deeply cleft.
                  G.   Cymes many-fld., diffuse; lfts. 0.4–0.6 cm. long. Modoc Co.
                                    5. *P. Newberryi*
                  GG. Cymes few-fld., dense; lfts. 1.5 cm. long. Mono and Inyo cos.
                      H.   Lfts. greenish above, tomentose beneath, the segms. revolute; lvs. pinnate .... 6. *P. pennsylvanica*
                      HH. Lfts. grayish-silky on both sides, not revolute; lvs. subpalmate.
                                     • 7. *P. pseudosericea*
         DD. The styles filiform, not glandular.
            E.   Basal lvs. definitely pinnate into 5 or more lfts.
               F.   Petals acutish or acuminate at apex. Plants from desert regions.
                  G.   Stamens 15–35; anthers ca. 0.2 mm. long .... 14. *P. saxosa*
                  GG. Stamens 5–9; anthers ca. 0.7–1 mm. long .... 15. *P. patellifera*
               FF. Petals emarginate or rounded at apex.
                  G.   Lfts. coarsely toothed but not deeply divided. Coastal Los Angeles Co. .... 8. *P. multijuga*
                  GG. Lfts. deeply toothed or divided.
                      H.   Lvs. white with a dense tomentum, the lfts. divided almost to base. Montane from Tulare to Modoc and Siskiyou cos.
                                    10. *P. Breweri*
                      HH. Lvs. green and subglabrous or strigose, not white-tomentose.
                        I.   Lfts. pinnately toothed or incised; petals 6–10 mm. long. Montane .... 9. *P. Drummondii*
                        II. Lfts. palmately toothed or incised; petals 5–7 mm. long.
                          J.   The lfts. 9–13; bractlets ca. half as long as sepals. Sonoma Co. to Monterey Co. .... 11. *P. Hickmanii*
                          JJ. The lfts. 13–25; bractlets ca. as long as sepals. Nevada Co. to Modoc and Siskiyou cos. .... 12. *P. millefolia*
          EE. Basal lvs. palmate or if pinnate, 3-foliolate.
            F.   Lfts. 3, bright green, subglabrous, fan-shaped. High montane
                                  13. *P. flabellifolia*
            FF. Lfts. 5–9.
               G.   Stems ± prostrate; lvs. silky-villous; petals 4–5 mm. long. Mts. from Tulare Co. to Riverside Co. .... 20. *P. Wheeleri*
                GG. Stems erect or ascending.
                  H.   Stamens ca. 30; aks. ± reticulate. Occasional escape
                                    21. *P. recta*
                  HH. Stamens mostly 20; aks. smooth. Natives.
                      I.   Anthers round or oval, ca. 0.5 mm. long; lvs. sometimes subpinnate, the lfts. sharply few-toothed apically, cuneate basally. Montane, from Inyo and Tulare cos. n.
                                  16. *P. diversifolia*
                      II. Anthers ovate to lance-cordate, mostly 1 mm. long.

J. Lfts. divided ⅔ or more of the way to the midrib and
with linear segms., always digitate.

K. The lfts. strongly bicolored, densely tomentose beneath, dark green above, if sericeous only along
the veins, the teeth spreading, not pectinately
crowded. E. slope of Sierra Nevada
18. *P. flabelliformis*

KK. The lfts. scarcely bicolored, densely silky and
beneath often ± tomentose, gray-green, pectinately
toothed. San Bernardino Mts. to Mono Co.
19. *P. pectinisecta*

JJ. Lfts. merely crenate or serrate, or if toothed halfway
to the midrib the segms. broader and less crowded;
pubescence various. Widely distributed
17. *P. gracilis*

CC. Styles lateral or nearly basal.
D. Fls. purplish; styles filiform .... 25. *P. palustris*
DD. Fls. yellow to whitish; styles fusiform .... 26. *P. glandulosa*

1. **P. ánglica** Laicharding. [*P. procumbens* Sibth.] Perennial with a stout caudex;
stems slender, soon trailing, 2–7 dm. long, mostly forked, ± pilose; lower lvs. long
petioled, palmate with 3–5 lfts., 0.5–2 cm. long, obovate-cuneiform, coarsely dentate;
upper lvs. shorter, petioled, mostly ternate; stipules of lower lvs. with lance-ovate, acute
auricles; fls. solitary on filiform pedicels 3–15 cm. long, mostly 4-merous, 12–18 mm. in
diam.; sepals acute, lance-ovate; bractlets lanceolate or oblong, equaling sepals; petals
yellow, obcordate, 6–10 mm. long; receptacle pilose; stamens 15–20; pistils 20–50,
rugose; $n = 14$ (Wulff, 1939).—Natur. from Eu. April–July.

2. **P. rivális** Nutt. RIVER CINQUEFOIL. Annual or biennial, rather slender, 1–5 dm.
tall, softly pubescent, simple below, branched above; lower lvs. pinnate with 2–3 pairs
and a terminal lft. which may be 3-parted; lfts. cuneate-obovate to -oblong, 2–5 cm.
long, crenate to incised-serrate; infl. paniculate-cymose, leafy; fl.-tube hirsute, ca. 5
mm. broad; bractlets oblong, ca. 3 mm. long; sepals acute, ovate, 3–4 mm. long; petals
yellow, cuneate, much shorter than sepals, stamens 5–20; pistils many; aks. smooth.—
Bottom lands, below 500 ft.; Freshwater Marsh, V. Grassland; lower Sacramento and
San Joaquin valleys, Colo. R.; to B.C., Miss. V., Mex. Much of year.
Var. **millegràna** (Engelm.) Wats. [*P. m.* Engelm., not Dougl. *P. leucocarpa* Rydb.]
Stems diffusely branched from base; lvs. all 3-foliolate.—Moist disturbed places, below
6000 ft.; many Plant Communities; at scattered stations throughout Calif.; to Wash.,
Miss. V., New Mex. April–Oct.

3. **P. biénnis** Greene. Stems 1–several, mostly erect, with erect branches, pubescent
and glandular, 2–5 dm. tall; lvs. 3-foliolate, ± pubescent; lfts. roundish to broadly
obovate, 2–4 cm. long, coarsely crenate; petioles 1–7 cm. long; cymes often appearing
racemose; pedicels 4–8 mm. long; fl.-tube glandular-pubescent; bractlets ± oblong,
shorter than sepals; sepals ovate-acute, 2.5–3 mm. long; petals yellow, ca. 2 mm. long;
stamens ca. 10; aks. whitish, smooth, numerous, 0.5–0.6 mm. long.—Moist mostly sandy
places, 4500–9800 ft.; Sagebrush Scrub, Montane Coniferous F.; e. slope of Sierra
Nevada from Modoc Co. to Inyo Co., mts. about w. part of Mojave Desert; to B.C.,
Rocky Mts., L. Calif. May–Aug.

4. **P. norvègica** L. ssp. **monspeliénsis** (L.) Asch. & Graebn. [*P. m.* L.] Annual to
short-lived perennial, stems erect or ascending, stout, branched above, 2–7 dm. tall,
hirsute with stiff mostly spreading hairs; lvs. 3-foliolate; lfts. oblanceolate to obovate,
coarsely serrate, ± hirsute, 3–5 cm. long; cyme leafy; fl.-tube hirsute, 7–8 mm. broad;
bractlets acutish, ca. as long as sepals; sepals 4–5 mm. long; petals obovate, mostly
slightly shorter; stamens 15–20; aks. rugulose, almost 1 mm. long; $n = 28$ (Löve,
1954).—Moist places, 4500–7500 ft.; Montane Coniferous F.; scattered stations in
Calif.; to Alaska, Canada, Atlantic Coast, Mex.; Eurasia. June–Sept.

5. **P. Newbérryi** Gray. Perennial or biennial, the slender stems 1–4 dm. long, ±
silky-villous; basal lvs. pinnate, silky-villous, the lfts. 3–10 pairs, divided to near the
base into 3–5 oblong-spatulate segms. 3–5 mm. long; stem-lvs. reduced; cyme diffuse,
with slender recurved pedicels to ca. 1 cm. long; fl.-tube villous, ca. 5 mm. wide;
bractlets much like sepals which are lance-ovate, ca. 4 mm. long; petals white, obcordate,
ca. 5 mm. long; stamens 20; receptacle bristly; pistils many, ± veiny, almost 1 mm.

long.—Alluvial places near lakes and streams, ca. 4500–6500 ft.; Sagebrush Scrub, N. Juniper Wd.; Mono Co., Modoc Co.; e. Ore. May–July.

6. **P. pennsylvánica** L. var. **strigòsa** Pursh. Perennial with short cespitose caudex; stems several, ascending, 1–4 dm. high, densely puberulent and with longer hairs; basal lvs. pinnate; petioles 3–7 cm. long; lfts. 7–11, oblanceolate or obovate, 1–5 cm. long, silky on both sides, but greenish above and whitish-tomentose beneath, deeply cleft into lanceolate or linear, revolute divisions; cymes dense; fl.-tube villous, 6–7 mm. broad in fr.; bractlets lanceolate, 4–5 mm. long; sepals ovate, slightly longer, ribbed; petals roundish, equal to sepals, yellow; stamens 20; pistils many.—Meadows, etc., 9000–12,000 ft.; Subalpine F., Bristle-cone Pine F.; White Mts., Sierra Nevada of Inyo Co.; to Alaska, e. Asia, Hudson Bay. June–Aug.

7. **P. pseudosericea** Rydb. Perennial, from short cespitose caudex, the stems to 1.5 dm. long, few-lvd., grayish-silky; lvs. palmate, grayish-silky, the lfts. 5–9, 1–2 cm. long, obovate, divided to near the midrib into linear segms.; cymes few-fld.; fl.-tube grayish-silky, ca. 5 mm. wide in fr.; bractlets oblong, shorter than the ovate sepals, these 4–5 mm. long; petals obovate, light yellow, 4–5 mm. long; stamens 20; pistils many.— Dry rocky places, 10,500–13,000 ft.; Subalpine F., Bristle-cone Pine F., Alpine Fell-fields; Sierra Nevada from Tuolumne, Mono, and Inyo cos. n., White Mts.; to Wyo., Colo. July–Aug.

8. **P. multijùga** Lehm. Perennial with a taproot and almost no caudex; stems few, erect, 3–7 dm. high, slightly silky-strigose, somewhat leafy; lvs. pinnate, the lower numerous, 1–3 dm. long, the petioles 0.6–1.2 dm. long; lfts. 11–27, sparsely strigose to glabrate, 1–4 cm. long, cuneate-obovate, with few coarse teeth above middle; fls. in strict cymes; fl.-tube 4–6 mm. wide, sparsely pubescent; sepals oblong-ovate, acute, 5–6 mm. long, ⅓ longer than the bractlets; petals yellow, broadly obcordate, ca. 7 mm. long; stamens ca. 20; pistils many.—Originally in brackish meadows; Coastal Sage Scrub; near Ballona, Los Angeles Co. June–July. Probably extinct.

9. **P. Drummóndii** Lehm. Perennial, caudex short; stems erect, 3–6 dm. high, slightly hairy, few-lvd., branched above; stipules ca. 2 cm. long, lance-ovate, acuminate, subentire; basal lvs. ovate-oblong in outline, pinnate; lfts. 2–5 pairs, moderately spaced, 2–6 cm. long, cuneate-obovate, dark green, almost glabrate to moderately strigose, deeply and sharply serrate; stem-lvs. 1–5-foliolate; fls. long-pedicelled; fl.-tube hirsute, veined in fr., 7–9 mm. broad; bractlets lanceolate, shorter than the ovate-lanceolate, acuminate, 6–7 mm. long sepals; petals 6–10 mm. long, obcordate; stamens ca. 20; pistils many, with filiform styles; $2n =$ ca. 92–108 (Clausen, Keck and Hiesey, 1940).—Often in wet places, 6000–11,500 ft.; Lodgepole F. to Subalpine F.; Sierra Nevada from Tulare Co. n., to Modoc, Siskiyou, and Humboldt cos.; to B.C. July–Aug.

Ssp. **Brùceae** (Rydb.) Keck. [*P. B.* Rydb.] Lf.-blades round-ovate in outline; lfts. gray-green, densely villous or pilose, sometimes ± tomentose when young, 2–4 pairs, closely approximate, 1.5–3.5 cm. long, irregular in size, broadly obovate, the basal pair usually divided to the base; $2n =$ ca. 64–98 (Clausen, Keck and Hiesey, 1940).—At 7000–13,000 ft., Alpine Fell-fields, Bristle-cone Pine F., etc.; Sierra Nevada, White, and Sweetwater mts. from Tulare and Inyo cos. n.; to Lake Co., Ore. July–Aug.

10. **P. Brèweri** Wats. [*P. B.* var. *expansa* Wats. *P. B.* var. *viridis* Jeps. *P. plattensis* var. *leucophylla* Greene.] Much like *P. Drummondii* but lf.-blades lance-oblong in outline, white with a dense tomentum at least when young; lfts. 4–6 pairs, usually crowded, 1–2.5 cm. long, quite regular in size, broadly flabelliform, several pairs often divided to base; $2n =$ ca. 72–102 (Clausen, Keck and Hiesey, 1940).—At elevs. of 4500–12,000 ft.; Red Fir F. to Alpine Fell-fields; from Tulare and Inyo cos. to Modoc and Siskiyou cos.; to Ore. June–Aug.

11. **P. Hickmánii** Eastw. Taproot woody; stems slender, decumbent, 1–3 dm. long, sparingly strigose; basal lvs. pinnate, sparingly strigose; lfts. 4–6 pairs, 0.6–1.5 cm. long, palmately cleft to ca. middle into 3–6 lanceolate divisions; cyme open, few-fld.; pedicels slender, 4–25 mm. long, recurved in fr.; fl.-tube 5 mm. broad; bractlets oblong, obtuse, ca. half as long as sepals, these lanceolate, acute, ca. 5 mm. long; petals yellow, obcordate, 6 mm. long; stamens ca. 20; pistils many, with filiform styles; aks. smooth, plump, yellow-green, ca. 2 mm. long.—Rare, in ± marshy places, at scattered stations; Freshwater Marsh; Sonoma Co. to Monterey Co. April–Aug.

12. **P. millefòlia** Rydb. [*P. plattensis* Greene, not Nutt. *P. p.* var. *m.* Jeps. *P. m.* vars. *algida* and *densa* Jeps.] Caudex short; stems several, 1–1.5 dm. long, slender, spreading to prostrate; herbage long-strigose; basal lvs. pinnate, oblong in outline, somewhat shorter than to as long as stems; lfts. 6–12 pairs, divided nearly to base into linear divisions; cauline lvs. few, smaller; cyme open, few-fld.; pedicels slender, often reflexed in fr.; fl.-tube shallow, ca. 4 mm. wide; bractlets and sepals lanceolate, ca. 5 mm. long, bractlets slightly the smaller; petals yellowish, obcordate, longer than sepals; stamens ca. 20; aks. smooth, 25–40.—Damp grassy places, 2500–6000 ft.; Sagebrush Scrub to Yellow Pine F.; Sierra Nevada from Nevada Co. n., to Modoc and Siskiyou cos.; Ore. June–July.

Var. **klamathénsis** (Rydb.) Jeps. [*P. k. Rydb.*] Plant loosely hairy; sepals linear-lanceolate, 6–7 mm. long.—On ± alkaline moist flats, 4000–5000 ft.; Sagebrush Scrub; Goose Lake and Klamath valleys; ne. Calif. and adjacent Ore. June–July.

13. **P. flabellifòlia** Hook. ex T. & G. [*P. gelida* Wats., not C. A. Mey.] Rootstock slender, branched, scaly; stems slender, 1–3 dm. high; herbage finely pubescent to subglabrous; lvs. few, mostly basal, on petioles 2–12 cm. long, the lfts. 3, subsessile, bright green, cuneate-flabelliform, thin, 1–3.5 cm. long, incised-serrate; cymes few-fld.; fl.-tube 5–9 mm. wide, puberulent to glabrous; bractlets elliptic-lanceolate, ca. as long as the ovate, acute sepals which are 5–6 mm. long; petals yellow, obovate, cuneate, deeply emarginate, 8–10 mm. long; stamens ca. 20; pistils many; aks. red-brown, smooth, plump, ca. 1.2 mm. long.—Moist places, 5800–12,000 ft.; Red Fir F. to Alpine Fell-fields; Sierra Nevada from Tulare Co. n., to Shasta, Trinity and Siskiyou cos.; to B.C. June–Sept.

Var. **Gràyi** (Wats.) Jeps. [*P. G.* Wats. *P. Clarkiana* Kell.] Lfts. roundish, glaucous-green, coarsely 5–7-toothed, the terminal lft. long-petiolulate.—At 8000–10,000 ft.; Lodgepole F. to Subalpine F.; Sierra Nevada from Mariposa Co. to Tulare Co. June–Aug.

14. **P. saxòsa** Lemmon ex Greene. [*Horkelia s.* Rydb. *P. rosulata* Rydb. *P. acuminata* Hall.] Low cespitose perennial, usually with thick woody root and caudex; stems slender, glandular-pubescent, 0.3–2.5 dm. long, leafy; basal lvs. pinnate, villous and somewhat glandular; lfts. 5–15, flabelliform, 0.5–1.5 cm. long, strongly toothed to dissected into rounded or oblong-acutish segms.; cymes few-fld.; pedicels spreading, slender, 0.8–1.5 cm. long; fl.-tube 2–4 mm. broad; bractlets oblong, erect, 1.2–2 mm. long; sepals ovate to deltoid-ovate, acute, spreading, 2–3 mm. long; petals light yellow, mostly 1–2 mm. wide, 2–3 mm. long; stamens 28–40, the anthers 0.15–0.2 mm. long; pistils 10 or more.—Rock crevices, dry places, 3000–6000 ft.; Joshua Tree Wd., Pinyon-Juniper Wd.; Inyo Mts., Kingston Mts., Cottonwood Springs, Little San Bernardino and e. San Bernardino mts., along w. edge of Colo. Desert to L. Calif. April–June.

Ssp. **siérrae** Munz. Plant viscid with dense glandular pubescence; stamens 15–25.—Rock-crevices, 8000–9500 ft.; Jeffrey Pine F.; e. slope of Sierra Nevada, White Mts., Inyo Co. July–Aug.

15. **P. patellífera** J. T. Howell. Much like *P. saxosa*, villous, slightly glandular, the stems 5–20 cm. long; fl.-tube 1.5–2 mm. in diam.; bractlets 0 to ca. 0.5 mm. long; petals 0.5–1 mm. wide, 2–3 mm. long; stamens 5–9, the anthers 5–9, 0.8–1 mm. long.—Crevices of rocks, 5500–6500 ft.; Pinyon-Juniper Wd.; Kingston Mts., ne. San Bernardino Co. June and Oct.

16. **P. diversifòlia** Lehm. [*P. dissecta* Nutt., not Pursh. *P. glaucophylla* Lehm. *P. diversifolia* var. *g.* Lehm.] Perennial with a cespitose caudex; stems erect, slender, 1–3-lvd., glabrous or strigose; basal lvs. digitate or crowded-pinnate; lfts. 5–7, obovate to cuneate-oblanceolate, ± deeply toothed with deltoid-lanceolate teeth, strigose-villous to glabrate; cymes many- to few-fld.; fl.-tube shallow, 5–9 mm. in diam. in fr.; bractlets lanceolate, acute, 3–5 mm. long; sepals ovate to lanceolate, acuminate, 5–6 mm. long; petals yellow, obcordate, 6–10 mm. long; stamens ca. 20; pistils many; aks. ca. 1.5 mm. long; $2n = 82$–ca. 101 (Clausen et al, 1940).—Moist rocky places, 8000–11,600 ft.; Lodgepole F. to Alpine Fell-fields; Sierra Nevada from Inyo and Tulare cos. n.; to B.C., S. Dak., Colo., Ariz. July–Aug.

17. **P. grácilis** Dougl. ex Hook. Cinquefoil. Perennial with a short caudex; stems slender, 4–7 dm. high, erect or ascending, ± silky-villous; basal lvs. digitate; lfts. 5–7, oblanceolate, 3–6 cm. long, green but ± silky above, finely white-tomentose beneath, divided ca. halfway into coarse lanceolate teeth; cymes many-fld.; fl.-tube long-silky,

8–10 mm. in diam. in fr.; bractlets lanceolate, slightly shorter than sepals; these ovate, acuminate, 6–7 mm. long; petals yellow, obcordate, ca. 8–11 mm. long; stamens ca. 20; pistils many; aks. smooth, 1–1.3 mm. long.—Mostly moist places, 1500–3000 ft.; Coastal Prairie, N. Oak Wd., Mixed Evergreen F., etc.; Mendocino Co. to Del Norte Co.; to Alaska. May–July.

Ssp. **Nuttállii** (Lehm.) Keck. [*P. N.* Lehm. *P. fastigiata* Nutt. *P. g.* var. *f.* Wats. *P. Blaschkeana* Turcz. *P. etomentosa, Hallii, angustata, amadorensis, Parishii, lasia,* and *Hassei* Rydb. *P. glomerata* A. Nels.] Lvs. less prominently discolored, pilose, hirsute to tomentulose to glabrate, variously cut; $2n = 52$–109 (Clausen et al, 1940).—Moist places like meadows and along streams, 2500–9000 ft.; Montane Coniferous F.; Coast Ranges from Lake Co. n. to Siskiyou Co., thence s. through Sierra Nevada to mts. of San Diego Co.; to Alaska, Alta., S. Dak. June–Aug.

18. **P. flabellifórmis** Lehm. [*P. gracilis* var. *f.* Nutt. *P. f.* var. *inyoensis* Jeps.] Perennial from a short caudex; stems slender, 4–6 dm. high, silky-strigose, few-lvd.; basal lvs. digitate, lfts. ca. 6–7, strongly discolored, dark green above and glabrous or silky-villous along the veins, densely tomentose beneath, 3–5 cm. long, divided nearly to the midrib into linear, spreading lobes; cymes open; fl.-tube silky villous, ca. 8 mm. in diam. in fr.; bracts linear-lanceolate, ca. 3 mm. long; sepals triangular-lanceolate, acuminate, 4–5 mm. long; petals obcordate, ca. 5–6 mm. long; stamens 20; pistils many; aks. smooth, ca. 1 mm. long; $2n = $ ca. 60–65 (Clausen et al., 1940).—Moist places, 4000–7000 ft.; Sagebrush Scrub to Red Fir F.; e. slope of Sierra Nevada from Inyo Co. to Modoc Co.; to B.C., Saskatchewan, Wyo. June–July.

19. **P. pectinisécta** Rydb. [*P. Elmeri* Rydb. *P. comosa* Rydb. *P. gracilis* var. *Elmeri* Jeps.] Much like *P. flabelliformis,* mostly 3–4 dm. tall; lfts. 5–9, appressed-silky on both sides, sometimes ± tomentose beneath, obovate, the linear segms. evenly and pectinately arranged; sepals 5–6 mm. long; petals 7–8 mm. long; $n = 21$ (Clausen et al., 1940).—Moist places, 4000–7600 ft.; Pinyon-Juniper Wd., Yellow Pine F.; mts. bordering w. end of Mojave Desert (San Bernardino and San Gabriel mts., Mt. Pinos, Tehachapi Mts.) to White Mts. and e. slope of Sierra Nevada, n. to Mono Co.; to s. Ida., sw. Mont., Colo. May–July.

20. **P. Wheèleri** Wats. [*P. W.* var. *viscidula* Rydb. and var. *paupercula* Jeps.] Perennial with spreading to prostrate stems 0.5–2 dm. long, silky-villous; basal lvs. many, digitate, 1–8 cm. long; lfts. 5, silky-villous on both surfaces, cuneate to obovate, 1–2.5 cm. long, with few large broad terminal teeth; stem lvs. few, reduced; cyme lax, few-fld.; fl.-tube saucer-shaped, silky-villous, 5–7 mm. broad in fr.; bractlets oblong, 1.5–2 mm. long; sepals ovate, 2–3 mm. long; petals yellow, obcordate, 4–5 mm. long; stamens 20; pistils ca. 20; aks. pale, ± striate.—Edge of meadows, 6500–11,500 ft.; Montane Coniferous F.; Sierra Nevada of Tulare Co., San Bernardino Mts. June–Aug.

Var. **rimícola** M. & J. Lvs. not conspicuously silky, rather green and more glandular; branches of cyme and pedicels very slender.—Rock crevices, 7500–9000 ft.; Red Fir F.; San Jacinto Mts., San Pedro Martir Mts. July–Aug.

21. **P. récta** L. Stem erect, very leafy, 2–7 dm. high, loosely hirsute; basal lvs. digitately 5–7-foliolate; lfts. oblanceolate, 3–14 cm. long, ± hirsute above and beneath, paler beneath, with narrowly deltoid teeth; cymes many-fld. with strongly ascending branches; bractlets lance-linear, 6–7 mm. long; sepals lanceolate, acuminate, ca. as long; petals pale yellow, ca. 7–8 mm. long; stamens mostly 30; pistils many, with short thick styles; $2n = 28$ (Popoff, 1935), 42 (Shimotomai, 1930).—Once adventive in Santa Clara Co.; e. U.S.; native of Eu. May–July.

22. **P. fruticòsa** L. [*Dasiophora f.* Rydb. *Pentaphylloides f.* Schwarz. *Potentilla floribunda* Pursh. *Pentaphylloides f.* (Pursh) A. Löve.] Bush Cinquefoil. Much-branched shrub, 2–12 dm. high; branchlets densely lfy., silky when young; lvs. pinnate, to ca. 2 cm. long; lfts. 3–7, linear-oblong, entire, acute, 0.5–2 cm. long, green but silky above, somewhat paler and silky beneath, revolute; fls. 1 or few in small cymes; fl.-tube ca. 5–6 mm. wide; bractlets narrow-elliptic, ca. 5–7 mm. long; sepals ovate, acuminate, 5–6 mm. long; petals yellow, round, 5–15 mm. long; stamens ca. 25; pistils many, styles lateral; aks. pubescent, $n = 7$ for Am. plants (Sax, 1931; Bowden, 1957).—Moist places, 6500–12,000 ft.; Lodgepole F. to Alpine Fell-fields; White Mts., Sierra Nevada from Tulare Co. n., to Siskiyou and Modoc cos.; Alaska, Labrador, New Jersey, New Mex.; Eurasia. June–Aug. Our material possibly separable from European.

23. **P. anserìna** L. [*P. argentina* Huds. *Argentina anserina* Rydb.] SILVERWEED. Low stoloniferous perennial, with rosettes of horizontal pinnate lvs. 1–2 dm. long; lfts. ca. 9–31, with smaller ones interspersed, 1–4 cm. long, oblong to lance-oblong, deeply and sharply serrate, green and subglabrous above, white silky-tomentose beneath; petioles 1–5 cm. long, silky-villous; lvs. on stolons much reduced; peduncles axillary, solitary, 1-fld.; fl.-tube saucer-shaped; bractlets lanceolate to broader, simple to divided, mostly slightly exceeding sepals; these broadly ovate, 3–5 mm. long; petals yellow, oval, 7–10 mm. long; stamens 20–25; aks. many, corky, thick, deeply dorsally grooved; $n = 14$, 21 (Erlandsson, 1942).—Moist ± alkaline, places, 4000–8000 ft.; Montane Coniferous F.; mts. from San Bernardino Mts. to Modoc and Siskiyou cos. mostly along the e. side of the Sierra Nevada; to Alaska, Nfld., New York, New Mex.; Eurasia. May–Oct. Many of our plants are white-silky on both surfaces of the lvs. and belong to the var. *sericea* Hayne.

24. **P. Egèdei** Wormsk. var. **grándis** (Rydb.) J. T. Howell. [*P. anserina* var. *g.* T. & G. *P. pacifica* Howell. *Argentina occidentalis* Rydb.] Habit of *P. anserina*, but with stolons, petioles, peduncles subglabrous; lvs. suberect, 2–5 dm. long; lfts. 7–31, oblong to obovate, 3–6 cm. long, serrate, green above, white-tomentose to glabrate beneath, the pubescence opaque and dull, not silky or shiny; bractlets lanceolate, 6–8 mm. long; sepals ovate, 5–6 mm. long; petals 10–12 mm. long; aks. less plump, not corky or grooved.—Coastal Strand and Salt Marsh; Los Angeles Co. to Alaska; Asia. April–Aug.

25. **P. palústris** (L.) Scop. [*Comarum p.* L.] Perennial with creeping rootstock; stems stout, ascending, 1–6 dm. high, glabrous below, ± pilose and glandular above; lvs. pinnate, 5–7-foliolate; lfts. oblong-lanceolate to -oblanceolate, sharply serrate, 2–6 cm. long, green and glabrous above, pale and glabrous to ± silky beneath; cymes few-fld.; fl.-tube 7–8 mm. in fl., 10–12 mm. wide in fr.; bractlets narrow-lanceolate, ca. half the length of the sepals; these 10–15 mm. long, purplish especially within, ovate, acuminate; petals wine-purple, ca. half as long as sepals; stamens 20–25; pistils many.— Swamps and bogs, below 8000 ft.; many Plant Communities; Sierra Nevada from Eldorado Co. n., to Modoc, Mendocino, and Del Norte cos.; to Alaska and Atlantic Coast, Eurasia. May–Aug. The form with lvs. strigose beneath has been called the var. *villòsa* Lehm.

26. **P. glandulòsa** Lindl. [*P. Wrangelliana* Fisch. & Avé-Lall. *Drymocallis g.* and *W.* Rydb.] Perennial from a woody caudex; stems erect, 3–8 dm. high, leafy, glandular- or viscid-villous, branching above, often reddish; basal lvs. pinnate, sparsely long-pubescent, glandular, dark green above, lighter beneath; lfts. 5–9, obovate, 1–4 cm. long, serrate; stem-lvs. reduced; cyme open, many-fld.; fl.-tube glandular-hirsute, 4–8 mm. broad; bractlets linear, 4–6 mm. long; sepals lance-ovate, 6–8 mm. long (more in fr.), acuminate, glandular-pubescent; petals pale yellow to creamy-white, ca. as long as sepals; stamens ca. 25; pistils many; aks. brownish, veiny, ca. 1 mm. long; $n = 7$ (Clausen et al, 1937).—Dryish to moist open places, mostly at lów elevs., but up to ca. 8000 ft.; many Plant Communities; along the coast from n. L. Calif. to B.C., Sierran foothills; to Ida. May–July. *Variable.*

KEY TO SUBSPECIES

A. Petals much longer than sepals.
  B. Lfts. ± densely beset with stalked glands. E. side of Sierra Nevada, Siskiyou co.
        ssp. *pseudorupestris*
  BB. Lfts. pilose but not glandular.
    C. Herbage glabrate to moderately pilose; cyme open; stolons elongated; aks. ca. 0.8 mm. long . . . . ssp. *nevadensis*
    CC. Herbage strongly hirsute; cyme congested; stolons very short; aks. ca. 1 mm. long
        ssp. *ashlandica*
AA. Petals slightly if at all longer than sepals.
  B. Lfts. average more than 15 mm. long, principally on anterior half of rachis; stems mostly more than 1 dm. long; sepals mostly 7 or more mm. long.
    C. Petals equaling or slightly exceeding the sepals, broad.
      D. Branches erect, not leafy above, scarcely anthocyanous or glandular; sepals broadly lanceolate, to 10 mm. long . . . . ssp. *Hansenii*
      DD. Branches divaricate, leafy-bracted above, anthocyanous, prominently glandular; sepals ovate, to 12 mm. long . . . . *P. glandulosa*
    CC. Petals shorter than the sepals, narrow.
      D. Petals cream white, erect or ascending at anthesis; branching erect; cyme condensed . . . . ssp. *globosa*

DD. Petals deep yellow, reflexed or spreading at anthesis; branching divaricate; cyme
open . . . . ssp. *reflexa*
BB. Lfts. up to 6 mm. long, extending most of the length of the rachis; herbage ± silky pubescent;
stems only 0.8–1.2 dm. long, from slender stolons; sepals 5 mm. long . . . . ssp. *Ewanii*

Ssp. **pseudorupéstris** (Rydb.) Keck. [*P. p.* Rydb. *Drymocallis p.* Rydb.] Plants 1.5–2.5 dm. high; petals ca. 8–9 mm. long.—Open rocky places, 9500–12,400 ft.; Subalpine F., Alpine Fell-fields; e. slope of Sierra Nevada, Inyo and Eldorado cos.; Siskiyou Co.; to B.C., Mont., Wyo. July–Aug.

Ssp. **nevadénsis** (Wats.) Keck. [*P. g.* var. *n.* Wats. *P. g.* var. *lactea* Greene. *P. l.* Greene. *P. g.* var. *monticola* Rydb. *P. g.* var. *Austinae* Jeps. *Drymocallis cuneifolia* Rydb. *D. gracilis* Rydb. *P. pumila* Fedde. *P. Peirsonii* Munz.] Stems slender, 2–4 dm. high.— Moist places, mostly 5000–11,000 ft.; Montane Coniferous F.; mts. from San Jacinto Mts. n. through Sierra Nevada, Coast Ranges from Lake Co. n.; to Wash. June–Aug.

Ssp. **ashlándica** (Greene) Keck. [*P. a.* Greene. *Drymocallis a.* Rydb.] Stems slender, 2–3 dm. high.—Uncommon, 4000–6500 ft.; Montane Coniferous F.; Siskiyou Mts. to Ore. June–July.

Ssp. **Hansènii** (Greene) Keck. [*P. H.* Greene. *Drymocallis H.* Rydb.] Stems 5–8 dm. high, villous; lfts. 9–11, 1–4 cm. long, coarsely serrate; sepals 5–7 mm. long.—Meadows, 4000–6000 ft.; Red Fir F.; w. slope of Sierra Nevada from Amador Co. to Mariposa Co. May–July.

Ssp. **globòsa** Keck. Stems 2–4 dm. high, robust, leafy; lvs. densely villous or canescent, scarcely glandular; infl. leafy-bracted; sepals erect; aks. dark brown.—Woods, 4300–7000 ft.; Yellow Pine F., Red Fir F.; South Fork Mt., Humboldt Co. to Mt. Ashland, Ore. June–July.

Ssp. **refléxa** (Greene) Keck. [*P. g.* var. *r.* Greene. *P. r.* Greene. *Drymocallis r.* Rydb. *D. viscida* Parish. *D. laxiflora* Rydb.] Stems 3–6 dm. high, villous, slightly glandular; lfts. 7, densely pubescent, 1–3 cm. long; fls. few; bractlets lanceolate, shorter than ovate, mucronate sepals, these 5–6 mm. long; petals ca. 4 mm. long.—Mostly dryish slopes, to ca. 7000 ft.; many Plant Communities; N. Coast Ranges, Sierra Nevada, to L. Calif. and Ore. May–July.

Ssp. **Éwanii** Keck. Stems 0.8–1.2 dm. high, not leafy; basal lvs. viscidulous, hirsute, 2.5–5 cm. long; lfts. 7–11, round-flabelliform, deeply incised-serrate, 4–7 mm. long; fls. few; petals cuneate-obovate, slightly exceeding sepals, which are ca. 4–5 mm. long.— Seeps at 6500–7500 ft.; Yellow Pine F.; Mt. Islip, San Gabriel Mts., Los Angeles Co. June.

13. Duchésnea Sm. MOCK-STRAWBERRY

Strawberrylike plants with trailing rooting stems and trifoliolate lvs. Fls. axillary, yellow, on slender pedicels. Sepals, bractlets and petals 5, the bractlets toothed and exceeding the sepals. Receptacle dry, enlarged in fr. but not pulpy. Stamens and pistils many. Aks. superficial. Two spp., of s. Asia. (Named for A. N. *Duchesne,* monographer of Fragaria.)

1. **D. índica** (Andr.) Focke. [*Fragaria i.* Andr.] Plants ± silky pubescent; lfts. 2–4 cm. long, rhombic-ovate; fls. 15–20 mm. broad; fr. red; $2n = 84$ (Ichijima, 1926).—Used to some extent as a ground cover and escaping; native of India. May–Aug.

14. Fragària L. STRAWBERRY

Perennial herbs with scaly rootstock and runners which root at the nodes. Lvs. and fls. in basal tufts; lfts. 3; stipules membranous. Fls. white to pinkish, borne in cymes on a naked stalk. Fl.-tube almost flat. Bractlets and sepals 5. Petals 5, round or elliptical. Stamens 20, in 3 series, sometimes abortive; fils. short. Pistils many, borne on an elevated conical receptacle which enlarges and becomes fleshy in fr. Styles lateral. Aks. small, turgid, borne on the surface of the pulpy fr. Perhaps 30 spp., of N. Temp. Zone and Andes. (Latin, *fragum,* fragrant.)

(Rydberg, P. A., *in* N. Am. Fl. 22: 356–365, 1908.)
Lvs. thick, coriaceous; fls. 20–35 mm. broad.
The lvs. strongly reticulate with ridged veinlets, crenate with the terminal tooth much smaller

than lateral; petioles and peduncles ± strigose, at least when young. Immediate seacoast
1. *F. chiloensis*
The lvs. less coriaceous and less reticulate, more sharply toothed, with terminal tooth seldom much
smaller than lateral; petioles and peduncles with spreading hairs. N. Coast Ranges .... 2. *F. crinita*
Lvs. thin, scarcely reticulate; fls. mostly 10–20 mm. broad.
The lvs. densely silky beneath, with subsessile lfts.; aks. superficial or in shallow pits
3. *F. californica*
The lvs. slightly silky beneath to glabrate, with slightly petioluled lfts.; aks. in deep pits in receptacle
4. *F. platypetala*

1. **F. chiloénsis** (L.) Duchn. [*F. vesca* var. *c.* L. *F. chilensis* Mol. *F. c.* var. *Scouleri*
Wats.] BEACH STRAWBERRY. Rootstocks stout, thick; lvs. many, with stipules 1–2 cm.
long; petioles stout, 0.5–2 dm. long, densely silky or with hairs spreading in age; lfts.
glabrous and shiny above in age, densely silky and tomentulose beneath, the terminal
2–5 cm. long, broadly obovate, petiolulate, the lateral smaller, usually subsessile, oblique;
plants largely dioecious; peduncles shorter or longer than lvs.; fl.-tube and calyx silky;
petals broadly obovate, 8–12 mm. long, ca. 1.5 times the sepals; fr. 1.5–2 cm. in diam.;
aks. superficial; $2n = 56$ (Staudt, 1953).—Coastal Strand, N. Coastal Scrub; San Luis
Obispo Co. to Alaska; S. Am., Hawaii. Used as a ground cover, the Chilean form of the
sp. one of the parents of domestic strawberries. March–Aug.
2. **F. crinìta** Rydb. Much like *F. chiloensis,* but with lvs. thinner, less reticulate, more
sharply toothed; peduncles and petioles more villous-hirsute.—Occasional, up to 2000
ft.; Mixed Evergreen F.; N. Coast Ranges; Del Norte Co. to Wash. April–June. Intermediate between *F. chiloensis* and *F. californica.*
3. **F. califórnica** Cham. & Schlecht. Rootstock short, not thick; lvs. rather few, with
brownish stipules to ca. 1 cm. long; petioles slender, 0.3–1.3 dm. long, sparingly villous;
terminal lfts. 2–5 cm. long, rounded-obovate, obtuse, coarsely serrate, subsessile, glabrate
above, silky beneath; lateral similar, very oblique, shorter; peduncles usually several,
slender, few-fld., villous; fl.-tube and calyx silky; petals obovate, 5–8 mm. long, not
much longer than sepals; fr. to 1 or 1.5 cm. thick; $n = 7$ (Ichijima, 1930).—Shaded,
fairly damp places, below 7000 ft.; Closed-cone Pine F., Chaparral, Mixed Evergreen F.,
etc., to Yellow Pine F.; Coast Ranges from Santa Barbara Co. to Del Norte Co., Siskiyou
Co. through Sierra Nevada to Tulare Co., mts. of s. Calif.; L. Calif. March–June. A form
with thicker lvs. and growing in exposed areas in region of San Francisco Bay has been
called var. *franciscàna* Rydb.
4. **F. platypétala** Rydb. [*F. virginiana* var. *illinoensis* Wats. *F. v.* var. *p.* Hall. *F.
truncata* Rydb.] Rootstock thick, woody; lfts. glabrous and glaucous above, silky beneath, rather firm, broadly cuneate or obovate, 2–8 cm. long, coarsely crenate or serrate
above the middle, nearly always petiolulate, the lateral not strongly oblique; petioles
0.2–2 dm. long, silky-villous; scape ca. 0.5–1 dm. high, several-fld., often leafy-bracteate;
sepals and bractlets lanceolate to elliptic; petals 6–10 mm. long, longer than sepals;
fr. 1–1.5 cm. in diam.; $n = 28$ (Yarnell, 1931).—Damp banks and woods, 4000–10,500
ft.; Montane Coniferous F.; Sierra Nevada from Tulare Co. n.; to Modoc and Siskiyou
cos.; to B.C. May–July. Plants of higher elevs. with small lfts., ± truncate at apex, have
been called var. *sibbaldifòlia* (Rydb.) Jeps. [*F. s.* Rydb.]

15. Sibbáldia L.

Low tufted perennial herbs with short cespitose rootstocks. Lvs. ternate. Fls. cymose,
on scapelike peduncles. Fl.-tube small, saucer-shaped or cup-shaped. Bractlets, sepals
and petals 5. Petals yellow. Stamens 5; fls. short, filiform. Pistils 5–20; styles lateral.
Ca. 5 spp., of n. regions. (Named for R. *Sibbald,* a Scotch botanist.)
1. **S. procúmbens** L. [*Potentilla Sibbaldii* Haller f.] Rootstocks creeping or cespitose;
petioles slender, strigose, mostly 1–4 cm. long; lfts. of basal lvs. obovate to oblanceolate,
1–2 cm. long, cuneate at base, 2–5-toothed at apex, sparsely pubescent; stem-lvs.
smaller; flowering stems to ca. 8 cm. high; cymes dense; fl.-tube ca. 3–4 mm. in diam.;
bractlets narrow, ca. 1–1.5 mm. long; sepals ± ovate, 2.5–3 mm. long; petals narrow,
ca. 1.5 mm. long; $n = 7$ (Böcher, 1938).—Dry stony places, 6000–12,000 ft.; Subalpine F. to Alpine Fell-fields; San Bernardino Mts., Sierra Nevada, n. to Modoc and
Siskiyou cos.; to Alaska, New England, Greenland, Eurasia. June–Aug.

16. Alchemílla L.

Annual to perennial, herbaceous. Lvs. alternate, palmately lobed, with sheathing stipules. Fls. small, cymose or in ours in small axillary clusters. Fl.-tube campanulate to urn-shaped. Bractlets 4–5 or sometimes obsolete. Sepals 4–5. Petals 0. Stamens 1–4. Pistils 1–8, free from fl.-tube; styles nearly basal, persistent. Widely distributed genus of perhaps 50 spp. (Valued in *alchemy*.)

1. **A. occidentàlis** Nutt. [*A. cuneifolia* Nutt. *Aphanes macrosepala* Rydb. *A. arvensis* of Calif. refs.] Small slender-branched annual, 0.3–1 dm. high, floriferous throughout; lvs. petioled, cuneate-flabelliform, 5–8 mm. long, ± short-hirsute, deeply parted into 3–5-cleft divisions; fl.-tube urn-shaped, ca. 1 mm. long; bractlets ± developed; sepals ovate, ca. 1–1.5 mm. long; stamen 1; pistils 1–2; aks. glabrous.—Open grassy or wooded places, below 2000 ft.; V. Grassland, Foothill Wd., Mixed Evergreen F., etc.; much of cismontane Calif.; to Wash., L. Calif. March–June.

17. Sanguisórba L. BURNET

Chiefly perennial herbs with unequally pinnate lvs. and stipules adherent to the petiole. Lfts. toothed or divided. Fls. small, often polygamous or dioecious, crowded in a dense head or spike at the end of the long naked peduncle, each bracteate and bibracteolate. Fl.-tube urn-shaped, angled, usually winged. Sepals 4. Petals 0. Stamens 4–12 or more. Pistils 1–3, with terminal slender styles tipped by brushlike or tufted stigma. Ak. mostly solitary, enclosed in the 4-angled dry thickish fl.-tube. N. Temp. Zone. (Name for *sanguis,* blood, and *sorbere,* to absorb, some spp. supposed to be styptic.)

Plants annual or biennial; lfts. incisely pinnatifid .... 1. *S. occidentalis*
Plants perennial; lfts. toothed.
   Lateral lfts. almost round, 6–12 mm. long; fruiting heads subglobose or short-oblong, 0.6–1 cm. long. Natur. plant .... 2. *S. minor*
   Lateral lfts. oblong-ovate, 10–40 mm. long; fruiting heads rounded-elliptic to oblong-cylindric, 1.5–2.5 cm. long. Native, n. Calif. .... 3. *S. microcephala*

1. **S. occidentàlis** Nutt. [*Poteridium o.* Rydb. *S. annua* of Calif. refs.] Glabrous, the stems branching, leafy, 1–4 dm. tall; stipules of cauline lvs. foliaceous, pectinately divided; lower lvs. with 11–15 lfts., these obovate in outline, 0.7–2 cm. long, pectinately pinnatifid into linear segms.; spikes roundish to oblong, 0.5–2.5 cm. long; fls. perfect; bracts and bractlets ovate, concave, green on midrib, with broad scarious margins; sepals oval, white-margined, obtuse, ca. 2 mm. long; stamens 2, opposite inner sepals; fils. filiform, short; fruiting fl.-tube lance-ovoid, 4-angled with narrow thick wings and reticulate faces.—Dry or semidamp grassy places, 2500–6500 ft.; Montane Coniferous F.; Cuyamaca Mts., Nevada Co. to Modoc and Siskiyou cos.; to B.C., Mont. May–July.

2. **S. mìnor** Scop. [*Poterium Sanguisorba* L.] Glabrous or sparsely hairy perennial, branched, 2–5 dm. high, leafy; stipules of cauline lvs. lunate, coarsely toothed; lfts. 7–21; lower fls. ♂, upper perfect or ♀; bracts and bractlets ovate, green, ciliate; sepals oval, purple-tinged, 3.5–4 mm. long; stamens and stigmas purple-tinged; pistils 2; fruiting fl.-tube ovoid, 4 mm. long, 4-angled, with thick ridges and alveolate-favose faces; $2n = 28$ (Lindenbein, 1937).—Occasional as natur. plant escaped from cult.; native of Eurasia. May–July.

3. **S. microcéphala** Presl. [*S. officinalis* of Calif. refs.] Glabrous perennial with a rootstock; stems 3–6 dm. high; lvs. odd-pinnate, the lower with 9–13 lfts.; upper stipules rounded, foliaceous, toothed; lfts. oblong-ovate, 1–4 cm. long, cordate at base, serrate; spike ellipsoid to oblong, 1.5–2.5 cm. long; sepals dark purple, 2–2.5 mm. long; fils. 3–7 mm. long; fruiting fl.-tube narrowly 4-winged.—Swamps, below 5000 ft.; Red Fir F.; Mendocino to Del Norte cos.; to Alaska. July–Aug.

18. Acaèna L.

Perennial herbs with ± woody caudices. Lvs. odd-pinnate into pinnatifid lfts. Stipules adnate to petioles. Fls. green, in spikes or racemes. Fl.-tube ellipsoid, contracted at throat, armed with retrorsely barbed prickles. Sepals 3–5, mostly 4. Petals 0. Stamens

3–5, inserted in mouth of fl.-tube. Pistil usually 1; style terminal; stigma feathery. Aks. enclosed in the indurated fl.-tube. Perhaps 40 spp. of S. Hemis. and n. to Mex., Calif., Hawaii. (Greek, *akaina*, a thorn, because of the spines on the fl.-tube).

1. **A. califórnica** Bitter. [*A. pinnatífida* var. *c.* Jeps. *A. trífida* of Calif. refs.] Stems mostly simple, 1–6 dm. high, sparingly villous; lvs. crowded near base, glabrous or ± ciliate above, silky beneath, lfts. 11–17, cleft into 3–7 segms., 6–8 mm. long; fl-tube hairy, ca. 4 mm. long in fr.; spike lax at base, crowded above; sepals linear-oblong, 2–3 mm. long; stamens exserted, dark purple.—Sandy and rocky places, below 1500 ft.; N. Coastal Scrub, Coastal Strand, etc.; Sonoma Co. to Santa Barbara Co. April–June.

19. Agrimònia L. AGRIMONY

Perennial herbs with rootstocks. Lvs. pinnate, with crenate-serrate lfts. Fls. small, spicate-racemose. Fl.-tube turbinate or hemispherical, constricted at the throat where beset with hooked bristles, indurate in fr. and enclosing the 2 aks. Sepals 5, connivent after flowering. Petals 5, small, yellow. Stamens 5–15. Styles terminal. Perhaps 12–15 spp., of N. Hemis. (Possibly from Greek, *argema*, an eye disease, because of supposed medicinal value.)

1. **A. gryposèpala** Wallr. [*A. Eupatoria* of Calif. refs., not L.] Stems 3–15 dm. high, hirsute and glandular-puberulent; lvs. remote, extending to base of infl.; larger lfts. 5–9, mostly lanceolate or oblanceolate, 4–12 cm. long, resinous-glandular beneath, sparingly hirsute on veins; racemes 2–4 dm. long; pedicels 2–10 mm. long; fruiting fl.-tube 4–5 mm. long, ca. as broad, the hooked bristles spreading from a thin horizontal flange, the short outer bristles often reflexed; petals obovate, ca. 3 mm. long.—Borders of woods, 2500–5500 ft.; mostly Yellow Pine F.; Palomar and Cuyamaca mts., San Bernardino Mts., mts. from Lake to Medocino cos., Plumas and Shasta cos.; e. U.S. and Mex. July–Aug.

20. Adenóstoma H. & A.

Unarmed evergreen shrubs with ± resinous herbage. Lvs. small, entire, alternate or fascicled, linear, rigid, numerous. Fls. small, white, crowded, in terminal panicled racemes. Fl.-tube obconical, 10-striate. Sepals 5. Petals 5, roundish. Stamens 10–15, inserted 2–3 together, alternating with petals. Pistil 1; style lateral. Ovary 1-celled, 1–2-ovuled. Aks. enclosed by the indurated fl.-tube. Two spp. (Greek, *aden*, gland, and *stoma*, mouth, because of glands at mouth of fl.-tube.)

Lvs. fascicled; bracts lance-linear, not scarious; fls. sessile .... 1. *A. fasciculatum*
Lvs. scattered; bracts broadly lanceolate, scarious-margined; fls. pedicelled ...... 2. *A. sparsifolium*

1. **A. fasciculàtum** H. & A. [*A. f.* var. *densifolium* Eastw.] CHAMISE. GREASEWOOD. Diffuse shrub, 0.5–3.5 m. high, with well developed basal burl; bark reddish, subglabrous on the twigs, becoming shreddy with age; stipules small, acute; lvs. linear to narrow-clavate, glabrous, acute, short-petioled, 4–10 mm. long, often resinous; panicles 4–12 cm. long; bracts ca. 1 mm. long; fl.-tube green, almost 2 mm. long; sepals barely 1 mm. long; petals ca. 1.5 mm.; ovary obliquely truncate.—Common dominant on dry slopes and ridges, below 5000 ft.; Chaparral; Coast Ranges from Mendocino Co. to L. Calif., foothills of Sierra Nevada. May–June.

Var. **obtusifòlium** Wats. [*A. brevifolium* Nutt.] Twigs pubescent; lvs. obtuse, 4–6 mm. long.—Dry mesas, sw. San Diego Co. and adjacent L. Calif.

2. **A. sparsifòlium** Torr. RIBBON BUSH. RED SHANK. Erect, arborescent, 2–6 m. high, the trunks red-brown and freely exfoliating; twigs green, resinous-glandular; lvs. alternate, filiform, 6–15 mm. long, resinous-glandular; fls. in open showy panicles less than 1 dm. long; fl.-tube ca. 1.5 mm. long; sepals ca. 2 mm. long; petals elliptic, ca. 2 mm. long; ovary truncate.—Dry slopes and mesas, below 6000 ft.; Chaparral; mostly away from the coast, local in San Luis Obispo and n. Santa Barbara cos., Santa Monica Mts., common from San Gorgonio Pass to L. Calif. July–Aug.

21. Gèum L. Avens

Perennial herbs with rootstocks. Lvs. pinnate, stipulate. Stipules adnate to the sheathing petioles. Fls. rather large, solitary or cymose, yellow, white to purple. Fl.-tube persistent, turbinate or hemispheric, usually 5-bracteolate. Sepals 5. Petals 5, round to cuneate. Stamens many; fils. filiform. Pistils many, on a short clavate receptacle; ovules solitary. Styles filiform, elongate in fr. Fr. an ak. tipped with the elongate style. Perhaps 50 spp., of temp. and cooler regions. (The ancient Latin name.)

(Bolle, F. Eine Übersicht über die Gattung Geum L., etc. Repert. Sp. nov. regni veg., Beiheft 72, 1933. Gajewski, W. A cytogenetic study on the genus Geum. Polskie Tow. Bot.; Mon. Bot. 4: 1–416, 1957.)

Ultimate divisions of larger basal lvs. at least 1 cm. broad; style conspicuously jointed and kinked above the middle, the upper portion mostly hairy and readily deciduous ...... 1. *G. macrophyllum*
Ultimate divisions of larger basal lvs. much narrower; style straight, jointed or not, the upper portion glabrous, ± persistent.
    Bracts well exceeding sepals; lfts, cuneate, toothed at apex .... 2. *G. ciliatum*
    Bracts mostly shorter than sepals; lfts. dissected at least halfway into narrow divisions
                                                       3. *G. canescens*

1. **G. macrophýllum** Willd. Stems stout, erect, bristly-pubescent, 3–10 dm. high; stipules broad, foliaceous; basal lvs. lyrate-pinnate, 1–4 dm. long, including petiole, with large rounded terminal lft. ± 3-cleft and with 2–6 smaller principal lateral lfts. and lesser ones interspersed; middle cauline lvs. sessile or short-petioled, mostly with ca. 3 rhombic or cuneate lfts. of which the terminal is 3-cleft to ca. the middle and with dentate lobes; fls. few, the cyme open; bractlets small, linear; sepals 3–5 mm. long; petals yellow, 4–8 mm. long; receptacle oblong, short-pubescent; aks. puberulent below, bristly above, the persistent part of the style hooked, glandular-puberulent, ca. 6–8 mm. long; *n* = 21 (Raynor, 1952).—Moist places like meadows, 3500–10,500 ft.; Montane Coniferous F.; San Bernardino Mts. and mts. of Lake Co. n. to Modoc and Del Norte cos.; to Alaska, e. Asia, and Labrador. May–Aug. Plants with the terminal lft. of the basal lvs. more deeply divided, the cauline lvs. with narrower more deeply cut lfts., and the infl. more glandular, occur and have been called *G. oregonénse* Rydb. Their status and the applicability of the name are uncertain.

2. **G. ciliàtum** Pursh. [*G. triflorum* var. *c.* Fassett. *Sieversia c.* G. Don. *Erythrocoma c.* Greene.] Erect, 2–5 dm. high, soft-hairy and finely glandular; basal lvs. tufted, 1–2 dm. long including the petioles; main lfts. 9–19, obovate in outline, ± crowded, cleft into linear or cuneate toothed segms.; stem-lvs. few, reduced; cymes 1–3-fld.; bractlets linear, 8–20 mm. long; sepals ± purplish-red, lance-ovate, acuminate, 8–10 mm. long; petals yellow or with purplish tinge, ca. as long; styles in fr. with a long plumose part ca. 2–3 cm. long and an upper glabrous finally deciduous part ca. 3–4 mm. long; *n* = 21 (Gajewski, 1957).—Dry to dampish, rocky slopes or flats, between 4000 and 8000 ft.; Sagebrush Scrub, to Red Fir F.; Siskiyou and Modoc cos. to Lassen Co.; to B.C. and Rocky Mts. May–July.

3. **G. canéscens** (Greene) Munz. [*Erythrocoma c.* Greene. *Sieversia c.* Rydb. *G. ciliatum* var. *triflorum* of Jeps., not *G. t.* Pursh.] Much like *G. ciliatum*, finely pilose throughout and glandular; lfts. mostly 2–3-lobed scarcely halfway to their base, the lobes with 2–3 broadly oblong-ovate teeth; bractlets 5–7 mm. long; sepals 8–10 mm. long; petals elliptic, slightly exceeding sepals; styles not or indistinctly jointed, ca. as long as in *G. ciliatum*.—Dryish slopes, 8500–11,000 ft.; Subalpine F.; Sweetwater Mts., e. slope of Sierra Nevada from Alpine Co. to Nevada Co.; w. Nev. June–Aug.

22. Fallùgia Endl. Apache Plume

Low deciduous shrubs with flaky bark. Lvs. pinnately dissected into linear divisions with revolute margins. Fls. terminal, showy, peduncled, solitary or few. Fl.-tube hemispheric, persistent, villous within. Sepals 5, with alternate linear bractlets. Petals 5, white, rounded, spreading. Stamens many, in 3 series. Pistils many, villous, on a conic receptacle; style terminal; ovules solitary, basal. Aks. oblong, villous, tipped by the plumose style. One sp. (Named for V. *Falugi*, Italian abbot.)

1. **F. paradóxa** (D. Don) Endl. [*Sieversia p.* D. Don.] Much branched, 3–15 dm.

tall, the young growth and lvs. ± rusty-lepidote and pubescent; lfts. mostly 5–10 mm. long, pinnatifid with 3–5-linear, obtuse segms.; fl.-tube 3–4 mm. long; sepals 5–7 mm. long, apiculate; petals 13–14 mm. long; fruiting styles 2.5–3 cm. long; $n = 9$ (Baldwin, 1951).—Dry rocky slopes, 4000–5500 ft.; Joshua Tree Wd., Pinyon-Juniper Wd.; mts. of e. Mojave Desert; to Nev., Tex. Mex. May–June.

23. Cowània D. Don

Shrubs or small trees with alternate pinnatifid coriaceous, gland-dotted lvs. Fls. solitary, terminal on short branches. Fl.-tube funnelform, persistent. Sepals 5, imbricate. Petals 5, obovate, spreading. Stamens many, in 2 series. Pistils 4–12, villous-hirsute, the style terminal, plumose, persistent and elongate in fr.; ovules solitary. Aks. striate, villous, nearly included. Ca. 5–6 spp., of sw. N. Am. (Named for J. *Cowan*, British amateur botanist.)

1. **C. mexicàna** var. **Stánsburiàna** (Torr.) Jeps. [*C. S.* Torr.] Freely branched shrub 0.3–3 m. tall with dark shreddy bark and red-brown glandular twigs; lvs. obovate in outline, 6–15 mm. long, glandular-punctate and green above, white-tomentose beneath, pinnately 3–5-divided into narrow revolute segms.; pedicels 3–8 mm. long, glandular-pubescent; fl.-tube ca. 5 mm. long; sepals oblong-ovate, 4–6 mm. long; petals cream, 6–8 mm. long, broadly obovate; pistils 5–10; styles 3–5 cm. long in fr.; $n = 9$ (Baldwin, 1951).—Dry slopes and canyons, 4000–8000 ft.; Joshua Tree Wd., Pinyon-Juniper Wd.; mts. of e. Mojave Desert from White Mts. to Providence Mts.; to Colo., New Mex., Mex. April–July. Occasional plants occur with some fls. ♂ and with perfect fls. having only 2–3 pistils that produce hairy (not plumose) very short fruiting styles. These have been called var. *dùbia* Bdg. and are reported from Providence Mts.

24. Púrshia DC. ex Poir. ANTELOPE BUSH

Shrubs or small trees. Lvs. alternate, crowded, apparently fascicled, deeply 3-cleft with revolute margins. Fls. solitary at ends of short branches. Fl.-tube turbinate to funnelform, persistent. Sepals 5. Petals 5, cream to yellow. Stamens ca. 25, in one series. Pistil mostly 1, short-styled, 1-ovuled. Fr. an ak., tipped with the rather short persistent style. Two spp. (Named for F. *Pursh*, author of an early flora of N. Am.)

Lvs. pubescent above, not punctate with sunken glands. From Tulare and Inyo cos. n.
                                                                                      1. *P. tridentata*
Lvs. glabrous above, conspicuously punctate with sunken glands. Mono Co. s. . . . . . . 2. *P. glandulosa*

1. **P. tridentàta** (Pursh) DC. [*Tigarea t.* Pursh.] Grayish shrub 1–3 m. tall with gray or brown bark and ± glandular and tomentose young twigs; lvs. cuneate, 0.5–3 cm. long, white-tomentose beneath, the lobes oblong-linear; pedicels short; fl.-tube funnelform, ca. 3 mm. long, white-tomentose; sepals oblong, ca. 3 mm. long; petals cream-yellow, spatulate-obovate, 7–8 mm. long; aks. fusiform, ca. 1.5 cm. long including the style; seeds black, pointed-ovoid, 5–6 mm. long.—Dry slopes, 3000–10,000(–11,000) ft.; Sagebrush Scrub to Subalpine F.; White Mts., Sierra Nevada (especially on e. slope) from Tulare and Inyo cos. n. to Modoc, Siskiyou, and Trinity cos., thence to Lake Co.; to B.C., Mont., New Mex. May–July.

2. **P. glandulòsa** Curran. [*P. tridentata* var. *g.* Jones.] Greenish shrub 1–2(–5) m. high, the glabrous twigs prominently glandular; lvs. 0.5–1 cm. long, slightly tomentose beneath; fl.-tube ca. 3–4 mm. long, tomentulose; sepals 3 mm. long; petals 6–8 mm. long, spatulate; ak. canescent, almost 2 cm. long including style; seed black, lanceolate, ca. 6 mm. long.—Dry slopes, 2800–9000 ft.; Chaparral, Joshua Tree Wd., Pinyon-Juniper Wd.; w. edge of Colo. Desert, Cajon Pass, Mojave Desert to Mono Co.; to Nev., Ariz., L. Calif. April–June.

25. Chamaebàtia Benth.

Glandular-pubescent shrubs, evergreen, heavy-scented. Lvs. twice- or thrice-pinnate with numerous minute segms. Fls. white, cymose-paniculate. Fl.-tube persistent, turbinate-campanulate. Sepals 5. Petals 5. Stamens many, in several series. Pistils solitary;

style terminal, villous at base; ovule 1. Aks. coriaceous, obovoid. Two spp. (Greek, *chamae*, low, and *batos*, bramble.)

Lf.-outline obovate, less than twice as long as wide; ovary hairy. Tulare to Shasta cos.
1. *C. foliolosa*
Lf.-outline elliptical, almost 3 times as long as wide; ovary glabrous. San Diego Co .. 2. *C. australis*

1. **C. foliolòsa** Benth. Mountain Misery. Shrub 2–6 dm. high, with many leafy branches and glandular-pubescent young twigs soon exfoliating to leave a smooth bluish to brown bark; lvs. 2–10 cm. long, viscid, mostly thrice pinnate, with ultimate divisions crowded, elliptical, tipped with a stalked gland, very minute; fl.-tube ca. 4–5 mm. long, glandular-hispid; sepals ca. 4 mm. long, lanceolate; petals white, obovate, 6–8 mm. long; ak. obovoid, brown, 5–6 mm. long.—Open forests, 2000–7000 ft.; Yellow Pine F., Red Fir F.; Shasta to Tulare cos. May–July.

2. **C. austràlis** (Bdg.) Abrams. [*C. foliolosa* var. *a*. Bdg.] Shrub 6–20 dm. high, the twigs less pubescent; lvs. 3.5–8 cm. long, twice or weakly thrice-pinnate, the oval ultimate segms. tipped with sessile glands; fl.-tube mostly 3 mm. long; sepals ca. 3 mm. long; petals 4–5 mm. long.—Dry slopes, below 2200 ft.; Chaparral; s. San Diego Co.; n. L. Calif. Nov.–May.

26. Cercocárpus HBK. Mountain-Mahogany

Evergreen shrubs or low trees with alternate simple ± coriaceous, straight-veined lvs. borne on spurlike branchlets. Fls. solitary or fasciculate, small, axillary or terminal. Fl.-tube with a lower persistent subcylindric portion and an upper deciduous bowl-shaped part. Sepals 5, broadly triangular to almost subulate. Petals 0. Stamens 10–45, inserted in 2 or 3 rows. Pistil 1; style terminal; ovule 1. Fr. a cylindric-fusiform ak. with terminal elongate silky-plumose style. Seed cylindric. Ca. 8–10 spp., of w. and sw. N. Am. (Greek, *kerkos*, tail, and *karpos*, fr.)

(Martin, F. L. A revision of Cercocarpus. Brittonia 7: 91–111, 1950.)
Lvs. ± toothed, not strongly inrolled or resinous; anthers hairy.
    Fls. 6–8 mm. broad; lvs. whitish, ± hairy beneath. Widely distributed. ...... 1. *C. betuloides*
    Fls. 2–4 mm. broad; lvs. greenish-yellow and glabrous beneath. S. San Diego Co.
2. *C. minutiflorus*
Lvs. entire, strongly revolute, resinous; anthers glabrous.
    The lvs. mostly 1.2–3 cm. long, elliptic, the margins only slightly revolute .... 3. *C. ledifolius*
    The lvs. mostly less than 1 cm. long, linear, the margins revolute almost to the midrib
4. *C. intricatus*

1. **C. betuloìdes** Nutt. ex T. & G. [*C. betulaefolius* Nutt. ex Hook. *C. parvifolius* var. *glaber* Wats. *C. montanus* var. *g*. F. L. Martin. *C. b*. var. *minor* C. K. Schneid. *C. rotundifolius* Rydb. *C. Douglasii* Rydb.] Erect open shrub to small tree, 2–7 m. high, with stiff erect or graceful spreading terminal branches; bark smooth gray; twigs sub-glabrous; lf.-blades obovate to oval or broadly elliptical, mostly 1–2.5 cm. long, cuneate and entire below the middle, serrate above, dark green and glabrous on upper surface, paler and somewhat pubescent beneath with evident feather veining; petioles 3–6 mm. long; fls. mostly in clusters of 2–3, with short pedicels; fl.-tube silky-tomentose, 5–6 mm. broad at summit, later glabrescent, brownish and split part way down one side, 8–10 mm. long; sepals broad-triangular; styles 4–9 cm. long in fr.; $2n = 18$ (Morley, 1949).—Common on dry slopes and in washes, below 6000 ft.; Chaparral, N. Oak Wd.; cis-montane Calif. to sw. Ore., n. L. Calif. March–May. Variable as to pubescence, lf.-size, etc.; plants from s. Calif. with 5–15 fls. in a cluster and often with lvs. to 5 cm. long, have been designated var. *multiflòrus* Jeps.

KEY TO VARIETIES

Most mature lvs. less than 3 cm. long; lateral veins 3–6 on each side of the midrib
C. betuloides
Most mature lvs. more than 3 cm. long; lateral veins 6–10 on each side of midrib.
    Lvs. flexible, rather thin, mostly not woolly beneath; fl.-tube not densely white-woolly.
        The lf.-blades broadly ovate to obovate with short broad triangular apiculate teeth. Insular
var. *Blancheae*
        The lf.-blades broadly obovate with coarse ovate teeth. N. Calif. .... var. *macrourus*
    Lvs. very coriaceous, woolly beneath; fl.-tube densely white-woolly. Santa Catalina Id.
var. *Traskiae*

Var. **Bláncheae** (C. K. Schneid.) Little. [*C. betulaefolius* var. *B.* C. K. Schneid. *C. montanus* var. *B.* F. L. Martin. *C. alnifolius* Rydb.] Petioles 8–10 mm. long; lf.-blades 4–6 cm. long, dark green above, the veins not deeply impressed, pale and subglabrous or strigose beneath; fl.-tube strigose; tail of mature fr. 5–7 cm. long.—Chaparral; Santa Cruz, Santa Rosa, and Santa Catalina ids. March–April.

Var. **macroùrus** (Rydb.) Jeps. [*C. m.* Rydb. *C. montanus* var. *m.* F. L. Martin.] Petioles 6–9 mm. long; lf.-blades 3–6 cm. long, tail of mature fr. 7–9 cm. long.—Dry rocky slopes, largely at 4000–6000 ft.; Yellow Pine F.; Trinity, Siskiyou and Modoc cos.; s. Ore. June.

Var. **Tráskiae** (Eastw.) Dunkle. [*C. T.* Eastw.] Lvs. 3–6 cm. long, coriaceous, the upper surface with impressed veins.—Salte Verde Canyon, Santa Catalina Id., March.

2. **C. minutiflòrus** Abrams. [*C. montanus* var. *m.* F. L. Martin.] Shrub 2–5 m. high, glabrous throughout; lvs. obovate to almost round, 1–2 cm. long, cuneate at base, serrate above, light green on upper surface, yellow-green beneath, with 3–5 lateral veins on each side of midrib; petioles 2–5 mm. long; fl.-tube 5–8 mm. long at anthesis, 11 mm. in fr., glabrous or sparsely silky; sepals subulate-triangular; tail of mature fr. 3–7 cm. long. —Dry slopes below 3000 ft.; Chaparral; s. San Diego Co., n. L. Calif. March–May.

3. **C. ledifòlius** Nutt. Shrub or tree 2–9 m. high, with red-brown furrowed bark; twigs canescent when young; lvs. lance-elliptic, lanceolate to oblanceolate, 1–3 cm. long, 0.3–1 cm. wide, acute, short-petioled, thick-coriaceous, tomentulose to ± glabrous above, resinous and tomentulose and tan or light green beneath, revolute; fls. 1–3, sessile, 4–5 mm. wide; fl.-tube 4–6 mm. long at anthesis, 6–10 mm. in fr.; sepals woolly, triangular; stamens 18–25; tail of fr. 4–7 cm. long.—Dry rocky slopes, 4000–10,500 ft.; Sagebrush Scrub, Pinyon-Juniper Wd., to Subalpine F.; from Santa Rosa and San Jacinto mts. through mts. of w. Mojave Desert n. to Modoc and Siskiyou cos.; to e. Wash., Mont., Colo., Ariz., L. Calif. April–May.

Var. **intercèdens** C. K. Schneid. Two–5 m. tall; lvs. 0.2–0.5 cm. wide, linear to narrowly lanceolate, strongly revolute, white to tan beneath; fl.-tube 3–4 mm. long at anthesis, 3–7 mm. in fr.—Occasional, with the sp.

4. **C. intricàtus** Wats. [*C. ledifolius* var. *i.* Jones.] Intricately branched, 1–3 m. tall, the young growth pubescent; lvs. 0.3–1(–1.5) cm. long, 0.1–0.2 cm. wide, linear, rolled to midrib, gray beneath, with petioles to 1 mm. long; fl.-tube 3–5 mm. long at anthesis and in fr.; fls. 1–2 mm. wide; stamens 10–15; ak.-tails 1–2 cm. long.—Dry slopes, 4000–9000 ft.; mostly Pinyon-Juniper Wd.; reported from s. Sierra Nevada and the desert from White Mts. to Providence and Clark mts.; to Utah, Ariz. May.

27. Coleógyne Torr. BLACKBUSH

Intricately branched shrub with opposite spinescent branches. Lvs. in opposite fascicles, linear-oblanceolate, entire, deciduous. Fls. solitary, terminating short branchlets, subtended by trifid bracts. Fl.-tube coriaceous, short, persistent. Sepals 4, persistent. Petals 0. Disk at mouth of fl.-tube tubular, separating stamens from pistils. Stamens 30–40; fils. filiform. Pistil 1; ovary 1-celled; ovule 1; style lateral, conspicuously villous at base, exserted, bent and twisted, persistent. Ak. glabrous. One sp. (Greek, *koleos*, sheath, and *gune*, ovary.)

1. **C. ramosíssima** Torr. Three–20 dm. tall, ashy gray, with divergent branches; lvs. 5–15 mm. long, strigose, thickish; stipules persistent after lf.-fall; sepals 7–8 mm. long, oblong-lanceolate, strigose, the inner with scarious margins; sheath between stamens and pistil 4–5 mm. long; ak. brown, ca. 3–4 mm. long.—Dry slopes, below 5000 ft.; Creosote Bush Scrub, to Pinyon-Juniper Wd.; w. Colo. Desert, Mojave Desert; to Colo., Ariz. April–June.

28. Rùbus L. BLACKBERRY, RASPBERRY, etc.

Shrubs to trailing vines, prickly or unarmed. Stems in their first year shooting up and sterile (primocanes), in second year usually flowering and with different foliage (floricanes). Stipules adnate to petioles. Lvs. alternate, simple to pinnately compound or subpalmate, petioled. Fls. ± perfect, racemose or paniculate, sometimes few or solitary. Fl.-tube persistent, rotate to campanulate. Bractlets 0. Sepals mostly 5. Petals mostly 5.

Stamens many. Carpels many, crowded on an elevated receptacle, becoming drupelets which coalesce and form an aggregate fr. Ovules 2, 1 abortive; style terminal, slender. Ssp. estimated between 200 and 700, of n. temp. regions and the Andes. Many of great horticultural value. (Latin name for bramble, related to *ruber*, red.)

(Bailey, L. H. Rubus in N. Am. Gentes Herb. 5: 1–932, 1941–1945.)
A. Plants ± herbaceous, barely 1 dm. high, not prickly, stoloniferous; stipules broad, free or nearly
    so .... 1. *R. lasiococcus*
AA. Plants definitely woody, taller, often prickly or bristly; stipules narrow, adnate to petioles.
    B. Lvs. simple, palmately lobed; stems unarmed, erect, with peeling or flaky bark; styles
        club-shaped .... 12. *R. parviflorus*
    BB. Lvs. mostly 3–5-foliolate; stems mostly prickly; styles filiform.
        C. Drupelets adhering to the fleshy receptacle and falling with it or separately, the fr.
           oblong. (*Blackberries* and *dewberries*)
           D. Fls. borne in small racemiform or cymiform clusters; prickles slender, ± setose.
              Native plants.
              E. Primocanes pruinose or with a "bloom"; fls. mostly functionally unisexual.
                F. Plants mostly glandless.
                  G. Lvs. bright- or green-looking, thin, lightly pubescent to glabrous
                      beneath; primocanes without hairs between the prickles; middle
                      lvs. of flowering shoots mostly 3-lobed; lfts. mostly long-pointed,
                      acutely toothed .... 2. *R. vitifolius*
                  GG. Lvs. dull or gray-green, ± felted-tomentose beneath at least when
                      young; primocanes usually hairy between the prickles; middle lvs.
                      of flowering shoots usually 3-foliolate; lfts. dull- to sharp-pointed
                      3. *R. ursinus*
                FF. Plants with well developed pinhead glands on pedicels and calyx
                    4. *R. macropetalus*
               EE. Primocanes not pruinose; fls. bisexual. Escape from cult. .... 5. *R. almus*
           DD. Fls. borne in large terminal panicles; prickles stout, broad-based. Escapes from
              cult.
              E. Lfts. deeply cut or dissected, not whitish-tomentose or canescent beneath
                  6. *R. laciniatus*
               EE. Lfts. not deeply cut, whitish-tomentose or gray-canescent beneath.
                  F. Canes and infl. quite unarmed; lf.-margins finely serrate
                    7. *R. ulmifolius*
                  FF. Canes and infl. armed; lf.-margins coarsely, unequally serrate or toothed
                    8. *R. procerus*
        CC. Drupelets forming a hollow cone which separates from the dry receptacle as a single
          aggregate fr. (*Raspberries*)
          D. Fls. white or nearly so; petals mostly not more than 1 cm. long; stems mostly
            strongly glaucous.
            E. Sepals deflexed at anthesis; pedicels and sepals glandless or nearly so
                9. *R. leucodermis*
             EE. Sepals not deflexed; pedicels and sepals with stipitate glands
                10. *R. glaucifolius*
        DD. Fls. red-purple, the petals 1.5–2 cm. long; stems not glaucous
           11. *R. spectabilis*

1. **R. lasiocóccus** Gray. [*Comarobatia l.* Greene.] Semiherbaceous unarmed perennial producing short runners, somewhat pubescent, rooting at nodes, fl.-shoots to 1 dm. long; lvs. roundish-reniform, 2–6 cm. broad, sparsely pubescent, serrate-dentate, obtusely 3-lobed to -foliolate, on slender petioles; fls. 1–2 on slender pedicels 1–5 cm. long; sepals ovate, acuminate, 6–7 mm. long; petals white, ca. as long; fr. red, of few juicy tomentose drupelets.—Dry places, 5000–6000 ft.; Red Fir F., Yellow Pine F.; Humboldt and Siskiyou cos.; to B.C. June–Aug.

2. **R. vitifòlius** Cham. & Schlecht [*R. ursinus* var. *v.* Focke. *R. lemurum* W. S. Brown.] CALIFORNIA BLACKBERRY. Green mound-builder or trailer or partial climber, almost glabrous to thinly pubescent, the long running stems tip-rooting, with straightish bristle-prickles; primocane stems soon glabrous between the setae and lvs. 3-foliolate to -lobate, thinly hairy to ± pubescent, but glàbrescent; lfts. ovate, the terminal to 9 or 10 cm. across, the lateral narrower, subsessile, all pointed, sharp-serrate to semilobulate; petioles shorter than lateral lfts.; flowering shoots to 5 dm. long, leafy, their main lvs. sharply acute, notched and sharply serrate, all but the lowermost 3-lobed or the upper not lobed, at first thinly pubescent beneath, ± prickly on ribs beneath; fl.-clusters to 1.5 dm. long; fls. mostly unisexual; sepals ovate-pointed, to acuminate, pubescent without, white-tomentose within, ca. 1 cm. long in ♂ fls., 7 mm. in ♀; petals narrow, 12–13 mm. long in ♂, broader, 7–8 mm. in ♀; fr. black, ± oblong, subglabrous, to ca. 1 cm. long.— Woods and somewhat damp places, below 4000 ft.; Mixed Evergreen F., to Yellow Pine

F., Coastal Strand, etc.; near the coast from Mendocino to San Luis Obispo cos. March–July. A form, var. *titànus* (Bailey) Bailey, larger in stature and lfts. and with berries to 6 cm. long, apparently sometimes escapes from gardens.

Var. Eastwoodiànus (Rydb.) Munz. [*R. E.* Rydb. *R. ursinus* var. *E.* J. T. Howell.] Lfts. with broad or rounded teeth and blunter apices; plants almost glabrous; a local form in azalea thickets on Mt. Tamalpais.

3. **R. ursìnus** Cham. & Schlecht. [*R. vitifolius* ssp. *u.* Abrams. *R. u.* var. *medusae* S. W. Brown and *pentaphyllus* S. W. Brown.] CALIFORNIA BLACKBERRY. Grayish or ± tomentose mound-builder with running or semiscandent stems; primocanes pruinose, pubescent to nearly glabrous, with straightish bristlelike prickles; lvs. of primocanes mostly 3-foliolate, on bristly glandless petioles, the lfts. triangular-ovate to ovate-pointed, the terminal one stalked, 5–12 cm. long, almost as broad, dentate to shallowly lobed; lvs. of flowering stems variable, the lower and middle mostly 3-foliolate, upper 3-lobed, smaller, mostly more blunt than on primocanes; fls. 2–15, at or near summit of lateral leafy shoots, perfect or imperfect, on prickly mostly glandless pedicels; petals of ♂ fls. narrow, to 15 mm. long, of ♀ smaller; sepals tomentose, pointed, bristly; fr. oblong or conical, black, ± pilose, to 2 cm. long; $n = 21$, 28, 42 (Gustafsson, 1943).—Waste places, fields, canyons, etc., below 3000 ft.; many Plant Communities, through most of cismontane Calif.; to Ore., L. Calif. March–July. The Loganberry, Youngberry, and Boysenberry are developments from this sp.

Var. sirbènus (Bailey) J. T. Howell. [*R. s.* Bailey.] The lfts. or lobes usually ca. twice as long as broad, tapering or acuminate, felted-pubescent beneath; pedicels and calyces sometimes weakly glandular.—Occasional with the sp.

4. **R. macropétalus** Dougl. ex Hook. [*R. Helleri* Rydb.] Primocanes to 6–7 m. long, tip-rooting, loose-hairy and with straightish prickles; lvs. on primocanes 3–5-foliolate, sparsely hairy above, hairy or pubescent beneath, terminal lfts. usually stalked, up to 12 cm. long, 3–5 cm. broad, lateral narrow-ovate, long-pointed, coarsely doubly sharp-toothed; lfts on fl. shoots smaller, acute to obtuse; fls. several to many, in open clusters, on slender acicular glandular pedicels, imperfect to perfect, with petals 5–7 mm. long in ♀ fls., longer in ♂; fr. ovoid, oblong or longer, from 15–25 mm. long, black, not pilose.—Fields, banks, disturbed areas, below 4500 ft.; Yellow Pine F., Mixed Evergreen F., etc.; Lake and Butte cos. to B.C. March–July. Some forms in cult.

5. **R. álmus** (Bailey) Bailey. [*R. flagellaris* var. *a.* Bailey.] GARDENA DEWBERRY. Prostrate, dense, the primocanes to 2 m. long, glabrous except near tips; prickles scattered, 3–4 mm. long; lfts. 5–7, soft-pubescent beneath, 7–8 cm. long, elliptic-ovate, narrowly cut-toothed; fls. few on each fl. shoot, on stout pilose pedicels; sepals heavily pubescent, prominently pointed; corolla 2–3 cm. broad; fr. oblong, to 3 cm. long, juicy.—Occasional escape; native of Tex. May–June.

6. **R. laciniàtus** Willd. CUT-LEAF BLACKBERRY. Much-branched, diffuse, glandless shrub to 3 m. high, stoutly armed with wide-based curved hooked prickles 5–6 mm. long; primocane lvs. 5-foliolate, the lfts. 3–8 cm. long, ovate, cut and parted to midrib into laciniate or lobed subleaflets, glabrous above, ± soft-pubescent beneath; armed with hooks on midribs, petiolules and petioles; floricane lvs. usually 3-foliolate, or simple in fl.-cluster which is terminal on floricane, with prickly pedicels; fls. mostly pinkish to rose and ca. 2 cm. across; sepals ± foliaceous, reflexed, with very narrow apices; fr. large, rounded, ca. 1.5 cm. thick, of few succulent drupelets.—Garden escape, especially in n. Calif.; European garden plant. May–July.

7. **R. ulmifòlius** Schott var. **inérmis** (Willd.) Focke. [*R. i.* Willd.] A vigorous plant scrambling over bushes and fences, without prickles throughout, glandless, rooting at tips; canes ± canescent-tomentose, the primocane lvs. 5-foliolate, subglabrous above, cano-tomentose beneath, usually minutely and closely serrate; lfts. ovate to ± obovate, 7–10 cm. long, 5–6 cm. broad, short-acuminate; lvs. of floricanes largely 3-foliolate; infl. a narrow long terminal cluster; corolla ca. 2 cm. across, the petals erose, broad; sepals soon reflexed, acuminate; fr. conic, ca. 1.5 cm. long, with many palatable drupelets.—Occasional garden escape as in Sonoma, Marin, Santa Barbara, and San Diego cos.; European garden plant.

8. **R. prócerus** P. J. Muell. HIMALAYA-BERRY. Robust, sprawling, ± evergreen, glandless shrub, to 3 m. tall; primocanes pilose-pubescent, glabrescent, angled, furrowed, with broad-based straight or curved prickles 6–10 mm. long; primocane lvs. 5-foliolate,

glabrous above when mature, cano-pubescent or -tomentose beneath, with hooked prickles on petioles; lfts. broad, the terminal roundish to broad-oblong, 10–12 cm. long, abruptly narrowed at apex, unequally and coarsely serrate-dentate; floricane lvs. 3–5-foliolate, smaller; infl. large, terminal, branched, prickly and cano-tomentose; fls. white or rose, 2–2.5 cm. across; sepals broad, cano-tomentose, soon reflexed, ca. 7–8 mm. long; petals broad, often pinkish, ca. 1 cm. long; fr. roundish, shiny black, to 2 cm. long, with large succulent drupelets; $2n = 28$ (Crane, 1936).—Becoming widely natur. especially in n. Calif.; to Wash.; native of Eu.

9. **R. leucodérmis** Dougl. ex T. & G. [*Melanobatus l.* Greene.] WESTERN RASPBERRY. Stems to ca. 2 m. long, arched and branched, rooting at tips, with heavy whitish bloom at least when young, the prickles many, ± straight, 4–6 mm. long, ± broad-based; primocane lvs. 3–5-foliolate, white-canescent to -tomentose beneath, green, subglabrous above, thin in texture, lfts. ovate to almost lanceolate, the terminal 7–9 cm. long, 4–5 cm. broad, irregularly doubly sharp-serrate, long stalked; fls. mostly 3–10, in rather compact cluster on lateral shoots, 7–10 mm. across, scarcely if at all glandular; sepals deflexed, acuminate, exceeding petals; latter narrow; fr. firm, dark purple to blackish, or yellow-red, depressed-globose, to ca. 1.5 cm. in diam., canescent; $n = 7$ (Darrow and Longley, 1933).—Slopes and canyons, below 7000 ft.; mostly Montane Coniferous F., Mixed Evergreen F.; Sierra Nevada from Tulare Co. n., Coast Ranges from Santa Cruz Co., n., to Siskiyou and Del Norte cos.; B.C., Mont., Utah. April–July.

Var. **bernardínus** (Greene) Jeps. [*Melanobatus b.* Greene. *Rubus b.* Rydb.] Pedicels and fl.-tube and sepals ± stipitate-glandular.—Dry flats and slopes, 4700–7500 ft.; Yellow Pine F.; San Bernardino and San Gabriel mts., Mt. Pinos, Palomar Mts. June–July.

Var. **trinitàtis** Berger. [*R. t.* Bailey.] A little-known plant with lvs. of florocanes simple, small, rounded, ca. 2 cm. long; sepals shorter than in the sp.—Douglas City, Trinity Co.

10. **R. glaucifòlius** Kell. [*Melanobatus g.* Greene. *R. leucodermis* var. *g.* Jeps.] Much like *R. leucodermis*, but with erect flowering shoots from decumbent or prostrate runner-like main canes; prickles few, 1–2 mm. long; lvs. green and finely pubescent to glabrous above, closely white-tomentose beneath, lfts. 3, the terminal 4–8 cm. long, 3–7 cm. wide, oval to ovate or oblong, coarsely and unequally dentate; fls. mostly few, in umbel-like clusters overtopped by lvs., with some glands on pedicels, etc.; sepals lanceolate, pubescent, not reflexed, 6–8 mm. long, attenuate; petals white, ca. as long; fr. hemispheric to conic, red or purplish, with few drupelets, pubescent, less than 1 cm. high.—Dry slopes, 3000–6000 ft.; Yellow Pine F., Red Fir F.; mts. from Tulare and Lake cos. n.; to Ore. June–July.

Var. **Gánderi** (Bailey) Munz. [*R. G.* Bailey.] Plants wholly prostrate; lvs. more acute, more sharply and irregularly dentate; pedicels apparently not glanduliferous.—Shaded woods, 4750–5500 ft.; Yellow Pine F.; Cuyamaca Mts., San Diego Co. June.

11. **R. spectábilis** Pursh. [*Parmena s.* Greene.] SALMON BERRY. Branching glandless leafy shrub 2–4 m. high, with young twigs glabrous to pilose, the older with yellowish shredding bark; primocanes sometimes with small straight prickles, plant otherwise unarmed; lvs. 3-foliolate, deciduous; lfts. thin, green, somewhat pubescent above and beneath, doubly serrate, ovate, the terminal 4–10 cm. long; floricane lfts. 3-foliolate to simple, smaller; fls. 1 or 2–4; sepals ovate, ca. 1 cm. long, pubescent; petals red-purple, broadly elliptical, 1.5–2 cm. long; fr. round to ovoid or conic, red to salmon-color or yellow, 1.5–2 cm. long, with many glabrous drupelets; $n = 7$ (Darrow and Longley, 1933).—Moist spots in and about woods, below 1000 ft.; Redwood F., Mixed Evergreen F.; Mendocino to Del Norte cos.; to Alaska. March–June.

Var. **franciscànus** (Rydb.) J. T. Howell. [*R. f.* Rydb. *R. s.* var. *Menziesii* (Hook.) Wats., not *R. M.* Hook.?] Lvs. more densely pubescent beneath.—Canyons and thickets; Redwood F., Mixed Evergreen F.; Santa Cruz Mts. to Sonoma Co.

12. **R. parviflòrus** Nutt. [*R. nutkanus* Moç. *Rubacer p.* Rydb.] THIMBLEBERRY. Deciduous, mostly 1–2 m. high, without prickles, the bark shreddy in age; lvs. palmately 5-lobed, unequally serrate, 1–1.5 dm. broad, subglabrous to puberulent, ± cordate at base; petioles and peduncles hirsute-glandular; fls. few, in terminal corymbs, white to pink, 2–5 cm. broad; sepals pubescent, evidently glandular, 10–15 mm. long, terminated by a long taillike or leafy appendage; petals elliptic, 1.5–2 cm. long; fr. scarlet, hemispheric, 1–1.6 cm. broad; $n = 7$ (Darrow & Longley, 1933).—Open woods and in canyons, below 8000 ft.; Red Fir F., Yellow Pine F., Redwood F.; from mts. of San

Diego Co. n., through Sierra Nevada to Siskiyou and Humboldt cos.; to Alaska. March–
Aug. Exceedingly variable as to pubescence and glands. Most Calif. plants, according to
one classification, would fall into var. *hypomálacus* Fern. (with glands of pedicels and
peduncles very unequal, the longest up to 1–2 mm. long) or var. *bifàrius* Fern. (with
glands of pedicels and peduncles subequal, rarely more than 0.5 mm. long). A more
clearly marked form is:

Var. velutìnus (H. & A.) Greene. [*R. v.* H. & A.] Petioles and young branches copiously
long-pilose to villous; sepals densely long-villous with pale hairs hiding the glands; lvs.
soft-pubescent above and whitish-velutinous beneath.—Mixed Evergreen F., Closed-cone
Pine F., Redwood F.; near the coast from Santa Barbara Co. to Mendocino Co.

29. Ròsa L. Rose

Erect sprawling or climbing shrubs, usually prickly. Lvs. alternate, deciduous or persistent, mostly odd-pinnate, with stipules adnate to petiole. Fls. solitary or in corymbs or
panicles, rather large, ours mostly rose-pink. Fl.-tube fleshy, cup-shaped to urceolate.
Sepals 5. Petals 5, rounded, spreading. Stamens many, inserted on the disk at the edge
of the fl.-tube. Pistils few to many, free and distinct but included in the fl.-tube. Fr. a
fleshy hip (ripened fl.-tube) containing the hairy aks. More than 100 spp., mostly from
n. temp. regions. (Ancient Latin name.)

(Cole, D. A revision of the Rosa californica complex. Am. Midl. Nat. 55: 211–224, 1956.)
A.   Sepals pectinate-glandular on margins to pinnatifid, especially on the attenuate tips.
   B.   Lfts. resinous-pubescent or glandular beneath; rounded to acute at apex, usually doubly
      serrate.
      C.   Prickles strongly curved or hooked; sepals spreading or reflexed in fr., at length
         deciduous. Introd. sp. .... 1. *R. Eglanteria*
      CC. Prickles straight; sepals erect and persistent in fr. Native sp. .... 2. *R. nutkana*
   BB. Lfts. subglabrous beneath, acuminate, finely and sharply serrate .... 3. *R. canina*
AA. Sepals mostly entire; lvs. glabrous to pubescent beneath, sometimes glandular-hairy.
   B.   Fl.-tube covered with gland-tipped bristles; plants small, 1–3 dm. high .. 4. *R. spithamea*
   BB. Fl.-tube lacking gland-tipped bristles; plants mostly larger.
      C.   Sepals glandular-hispid on back; lvs. simply serrate without gland-tipped teeth
                                                             5. *R. pisocarpa*
      CC. Sepals not glandular-hispid on back.
         D.   The sepals and styles deciduous in fr.; pistils few; pedicels 1–3 cm. long, ±
            reflexed in fr.; fr. 4–8 mm. thick at maturity .... 9. *R. gymnocarpa*
         DD. The sepals and styles persistent in fr.; pedicels usually less than 2 cm. long, not
            reflexed in fr.; fr. more than 7 mm. thick at maturity.
            E.   Stems armed with stout flattened recurved prickles; pedicels villous; fl.-tube
               often externally pilose when young; sepals ± pubescent on backs
                                            6. *R. californica*
            EE. Stems armed with straight or ascending weak, slender prickles; pedicels not
               villous; fl.-tube glabrous; sepals not pubescent.
               F.   Stipules, petioles and rachises copiously glandular; lfts. doubly serrate
                  with gland-tipped teeth .... 7. *R. pinetorum*
               FF. Stipules petioles and rachises not conspicuously glandular; lfts. simply
                  serrate, without gland-tipped teeth .... 8. *R. Woodsii*

1. **R. Eglantèria** L. [*R. rubiginosa* L.] Sweet-Brier. Eglantine. Coarse shrub with
stems to 3 m. high, armed with stout flat recurved prickles 7–13 mm. long and often
more slender straighter ones between; lfts. 7–9, glandular-scurfy on both surfaces,
resinous-aromatic, usually doubly serrate, roundish to elliptic, 1–3 cm. long; stipules
glandular-ciliate; fls. 1–4, pink to whitish, 3–5 cm. broad; pedicels short, glandular-
hispid; sepals lanceolate, usually laciniately lobed, with stipitate glands; fr. ovoid, scarlet
or orange, 12–18 mm. long; $2n = 35$ (Täckholm, 1922).—Natur. in pastures and waste
places, as in Marin, Trinity, Humboldt, and Nevada cos.; European. May–July.

2. **R. nutkàna** Presl. Nootka Rose. Stems stout, 1–2 m. tall, mostly armed with
stout straightish prickles to ca. 8 mm. long or sometimes unarmed; lfts. 5–9, elliptical
to broadly oval, 1.5–5 cm. long, doubly serrate with glandular teeth, glabrous above,
paler ¹ ˉneath; stipules ± glandular-dentate, 1–2 cm. long; fls. mostly solitary, rose-pink,
4–6 cm. wide; pedicels glabrous to glandular-bristly; fl.-tube glabrous; sepals lanceolate,
2–3 cm. long, glabrous or rarely glandular on back; petals broadly obcordate, 2.5–3.5
cm. long; fr. globose, 15–18 mm. in diam.; $n = 21$ (E. Erlanson, 1932).—Damp flats
and slopes, below 1500 ft.; Redwood F., Mixed Evergreen F.; Mendocino and Trinity
cos. n.; to Alaska, n. Rocky Mts. May–July.

Var. **muriculàta** (Greene) G. N. Jones [*R. m.* Greene.] Stipules glandular-muriculate; sepals glandular-hispid on back; fr. 12–15 mm. in diam.—Occasional, with the sp.

Var. **setòsa** G. N. Jones [*R. n.* var. *hispida* auth., not Fern.] Fl.-tube setose or glandular-prickly.—With the sp.

3. **R. canìna** L. Dog Rose. Strongly growing variable sp.; stems to 4 m., with scattered uniform strongly hooked prickles, otherwise glabrous; lfts. 5–7, oval, elliptic or ovate, nearly or quite glabrous, 2–4 cm. long, usually acuminate, sharply simply and finely serrate; stipules adnate for much of their length; fls. solitary to 2 or 5, white to pinkish, 2.5–5 cm. across; sepals reflexed, pinnatifid, glabrous on back, promptly deciduous from the ellipsoid scarlet fr.—Natur. about old settlements, as in Nevada Co.; Eurasia. May–June.

4. **R. spithamèa** Wats. [*R. Lesterae* Eastw.] Ground Rose. Plants low from creeping rootstocks, with straight terete infrastipular and sometimes internodal prickles; lfts. 3–7, oval to roundish, 1–3 cm. long, almost glabrous and green above, paler and glandular-pruinose beneath, doubly serrate with gland-tipped teeth; stipules glandular on margins; fls. 1–few; pedicels, sepals and fl.-tube usually glandular-hispid; petals obcordate, 12–20 mm. long; fr. subglobose, 7–8 mm. broad.—Open woods, below 5000 ft.; Mixed Evergreen F., Douglas-Fir F., Yellow Pine F.; Coast Ranges from Lake and Mendocino cos. to Humboldt and Trinity cos., Sierra Nevada from Tulare to Yuba cos. June–Aug.

Var. **sonoménsis** (Greene) Jeps. [*R. s.* Greene. *R. granulata* Greene.] Lfts. 0.5–1.5 cm. long, glabrous above and beneath, ± glaucous.—Dry slopes, below 4000 ft.; Chaparral, Mixed Evergreen F.; San Luis Obispo Co. to Mendocino Co. May–Aug.

5. **R. pisocárpa** Gray. [*R. Copelandii* Greene. *R. Pringlei* Rydb. *R. Eastwoodiae* Rydb.] Cluster Rose. Stems slender, 1–2 m. tall, with few rather slender straight or ascending prickles to ca. 7 mm. long, sometimes unarmed; lfts. 5–7, oval to elliptic-oblong, glabrous and green above, paler and puberulent beneath, finely serrate, 1–4 cm. long; stipules short-pubescent, ± glandular-dentate; fls. corymbose; pedicels glabrous; fl.-tube smooth; sepals 13–15 mm. long, mostly glandular-hispid on back, caudate-attenuate; petals 12–16 mm. long; fr. globose, 7–10 mm. in diam., glabrous; $n = 7$ (E. Erlanson, 1932).—Shaded slopes, below 5000 ft.; Yellow Pine F.; Lake Co. to Del Norte, Siskiyou, and Shasta cos.; to B.C. June–Aug. A glabrous form with lfts. sometimes puberulent on veins beneath has been described as var. *rivàlis* (Eastw.) Jeps. [*R. r.* Eastw.]

6. **R. califórnica** Cham. & Schlecht. [*R. myriantha* Carr. *R. Aldersonii* Greene. *R. Breweri* Greene. *R. Greenei* Rydb. *R. brachycarpa* Rydb. *R. sanctae-crucis* Rydb. *R. Davyi* Rydb. *R. pilifera* Rydb. *R. Johnstonii* Rydb.] Erect, branched, 1–3 m. tall, armed with stout flattened usually recurved prickles; petiole and rachis pubescent, prickly, nonglandular to glandular; lfts. 5–7, oval, 1–3.5 cm. long, simply or doubly serrate, puberulent above, pubescent and often glandular beneath; stipules narrow, usually pubescent, often glandular-denticulate on margins; fls. in corymbs; pedicels glabrous to villous, sometimes glandular; fl.-tube glabrous or pilose when young; sepals lanceolate, caudate-attenuate, glabrous or pubescent on backs, sometimes glandular; petals 1–2.5 cm. long; fr. globose to ovoid, 8–16 mm. long, 10–15 mm. thick, usually with distinct neck; $n = 14$ (E. Erlanson, 1932).—Fairly moist places, canyons, near streams, etc., below 6000 ft.; many Plant Communities; cismontane Calif. to L. Calif., s. Ore. May–Aug. Exceedingly variable; some authors separate the series with doubly toothed lfts. as *R. Aldersònii*.

7. **R. pinetòrum** Heller. [*R. Bolanderi* Greene. *R. calvaria* Greene. *R. Dudleyi* Rydb. *R. bidenticulata* Rydb. *R. corymbiflora* Rydb.] Erect, 0.5–1 m. high, armed with slender straight weak prickles, the infrastipular stouter; lfts. usually 5, elliptic to broadly oval, rounded at ends, 1–3 cm. long, puberulent and glandular beneath, doubly serrate with gland-tipped teeth; fls. mostly 1, sometimes few; stipules pubescent, often glandular, glandular-ciliate on margins; pedicels glandular-hispid; fl.-tube glabrous; sepals mostly glandular on backs, with foliaceous tips; petals 10–14 mm. long; fr. globose, 6–12 mm. in diam.—Open woods, 2000–6500 ft.; Yellow Pine F., Red Fir F.; Coast Ranges from Monterey Co. to Trinity and Humboldt cos., Sierra Nevada from Tulare Co. to Shasta Co. May–July.

8. **R. Woódsii** Lindl. var. **ultramontàna** (Wats.) Jeps. [*R. californica* var. *u.* Wats. *R. u.* Heller.] Erect, 1–3 m. tall, stout, armed with straight slender prickles, or floral

branches ± unarmed; lfts. 5–7, oval, 1–4 cm. long, coarsely serrate, glabrous above, puberulent and ± pruinose beneath; stipules pubescent below and on margins, glandular-ciliate or -dentate; fls. in corymbs; pedicels glabrous; fl.-tube glabrous; sepals glabrous on back, caudate-attenuate, rarely glandular; petals 1.5–2 cm. long; fr. ellipsoid to globose, 7–10 mm. in diam. at maturity.—Dampish places, 3500–11,000 ft.; Montane Coniferous F.; e. slope of Sierra Nevada from near Bishop (Inyo Co.) n., to Modoc and Shasta cos.; to B.C., Mont., Nev. June–Aug. Odor unusually pleasant.

Var. **gratíssima** (Greene) Cole. [*R. g.* Greene. *R. pisocarpa* var. *g.* Jeps.] Stems more densely armed, even floral branches usually heavily spiny; lfts. 1–2 cm. long; fls. 1–few; petals 1.5 cm. long.—Dry stony slopes, 5000–9000 ft.; Pinyon-Juniper Wd. to Red Fir F.; Sierra Nevada from Fresno and Mono cos. s., to Panamint and San Bernardino mts. April–Aug.

Var. **glabràta** (Parish) Cole. [*R. californica* var. *g.* Parish. *R. mohavensis* Parish. *R. Woodsii* var. *m.* Jeps.] Stems to ca. 1 m. high, well armed; lvs. glabrous; sepals glabrous.—Moist places about springs, 3000–4000 ft.; Joshua Tree Wd.; n. base of San Bernardino Mts. May–July.

9. **R. gymnocárpa** Nutt. in T. & G. [*R. glaucodermis* Greene. *R. prionota* Greene. *R. piscatoria* Greene.] WOOD ROSE. Slender-stemmed, mostly to 1 (3) m. high, armed with slender straight prickles, sometimes with stouter infrastipular prickles; petiole and rachis glandular-hispid, sometimes unarmed; lfts. 5–7, oval to roundish, 1–4 cm. long, doubly serrate, with gland-tipped teeth, glabrous on both surfaces; stipules narrow, glandular-ciliate, usually dentate; fls. usually solitary; fl.-tube smooth; sepals ovate, acuminate, glabrous on back, to ca. 1 cm. long; petals 8–12 mm. long; fr. ellipsoid to globose, glabrous, 5–10 mm. long, red, without sepals at maturity; $n = 7$ (E. Erlanson, 1932).—Shaded woods, below 6000 ft.; many Plant Communities; Palomar Mts. (San Diego Co.), Sierra Nevada from Fresno Co. n., to Modoc Co., and Coast Ranges from Monterey Co. to Humboldt and Siskiyou cos.; to B.C., Mont. May–July.

Var. **pubéscens** Wats. [*R. Bridgesii* Crép. *R. crenulata* Greene. *R. oligocarpa* Rydb.] Lvs. puberulent beneath.—Occasional, with the sp.

30. Prùnus L. STONE-FRUITS

Trees and shrubs with simple, deciduous or persistent, mostly serrate lvs. often bearing glands on the petiole and base of blade. Stipules small, caducous. Fls. umbellate, corymbose or racemose, appearing before or with the lvs. Fl.-tube hemispheric or cup-shaped. Sepals and petals 5. Stamens many. Pistil 1; style terminal; ovules 2, pendulous. Fr. a drupe developed from a fleshy pericarp and a bony endocarp enclosing usually a single seed. Perhaps 150 spp., of temp. climates, mostly in N. Hemis. Many valuable in horticulture. (*Prunus*, ancient Latin name of plum.)

A. Plants evergreen, with ± coriaceous or leathery lvs.; fls. racemose.
   B. Lvs. with crenate-dentate margins; racemes exceeding lvs. Garden escape
                                                        12. *P. lusitanica*
   BB. Lvs. not crenate-dentate; racemes usually not exceeding lvs.
      C. The lvs. coarsely spinose-toothed, crisped, 2–5 cm. long; fr. red. Native.
                                                          13. *P. ilicifolia*
      CC. The lvs. not as above, mostly longer; fr. dark.
         D. Base of lf. acute, the blade ca. ⅓ as wide as long; racemes ca. 2–3 cm. long. Introd. tree . . . . **11. *P. caroliniana***
         DD. Base of lf. rounded, the blade ca. ½ as wide as long; racemes ca. 5–12 cm. long. Insular . . . . **14. *P. Lyonii***
AA. Plants with deciduous lvs.
   B. The plants shrubby, many of the branchlets spinose. Natives.
      C. Lvs. broadly ovate to rounded, evidently serrulate, not fascicled.
         D. Lf.-blades 2–5 cm. long, the petioles 0.6–1.8 cm. long; drupe mostly glabrous. Kern Co. to Modoc Co. **2. *P. subcordata***
         DD. Lf.-blades 1–2 cm. long, the petioles 0.3–0.4 cm. long; drupe puberulent. Riverside and e. San Diego cos. . . . . **3. *P. Fremontii***
      CC. Lvs. ± oblanceolate, obscurely, if at all, serrulate, mostly fascicled.
         D. Petals 5–6 mm. long, pink; pedicels 6–8 mm. long. Inyo Co. to Modoc Co.
                                                  4. *P. Andersonii*
         DD. Petals 2–3 mm. long, white; pedicels ca. 1 mm. long. Inyo Co. to Imperial Co.
                                                    5. *P. fasciculata*
   BB. The plants shrubby to arboreous, not having spinose branchlets.
      C. Fls. 1–10, not in well developed racemes.

D. The fls. 1–2, subsessile; sepals not reflexed. Escapes from cult.
  E. Lvs. lanceolate, acuminate; fr. pubescent at maturity.
    F. Flesh of fr. hard, inedible, splitting to stone at maturity; lvs. mostly
      to 1 dm. long. Almond .... 6. *P. Amygdalus*
    FF. Flesh of fr. soft, usually not splitting to stone; lvs. mostly longer. Peach.
      7. *P. persica*
  EE. Lvs. ovate, abruptly short-pointed; fr. glabrous at maturity. Apricot.
    15. *P. armeniaca*
DD. The fls. several, on slender pedicels; sepals reflexed at anthesis.
  E. Fls. 2–2.5 cm. across; fr. more than 1 cm. in diam. Sweet cherries; escapes
    8. *P. avium*
  EE. Fls. 1–1.5 cm. across; fr. less than 1 cm. in diam.
    F. Corymbs short, subumbellate; drupe red, bitter; lvs. much longer than
      wide. Common native .... 1. *P. emarginata*
    FF. Corymbs elongate; drupe black; lvs. almost as wide as long. Introd.
      9. *P. Mahaleb*
CC. Fls. 12 to many, in elongate racemes. Native Choke Cherry ...... 10. *P. virginiana*

1. **P. emarginàta** (Dougl.) Walp. [*Cerasus e.* Dougl. *C. glandulosa* Kell. *C. californica, arida, prunifolia, rhamnoides, Kelloggiana, obliqua* and *parvifolia* Greene.] Bitter Cherry. Erect deciduous shrub or small tree, 1–6 m. high, with glabrous red shining twigs, older bark smooth; lvs. oblong-obovate to -elliptic, acutish, to obtuse, finely serrulate, subglabrous or slightly pubescent beneath, 2–5 cm. long, short-petioled; corymbs short, 3–10-fld.; fl.-tube campanulate, glabrous, ca. 3 mm. long; sepals 1.5–2 mm. long, oblong; petals 5–7 mm. long, obovate; drupe red, bitter, 6–8 mm. in diam.; stone ellipsoid, somewhat pointed at ends.—Rocky ridges to dampish slopes and canyons, below 9000 ft.; Yellow Pine F., Red Fir F., Chaparral, etc.; mts. of Calif. from San Diego Co. n.; to B.C., Ida. April–May.

2. **P. subcordàta** Benth. Sierra Plum. Deciduous shrub or small tree, 1–3(–6) m. tall, with stiff crooked branches and short thornlike branchlets, young growth glabrous to puberulent; lvs. ovate, elliptic to roundish, obtuse to rounded at tip, obtuse to subcordate at base, serrulate, 2–5 cm. long, the petioles 6–18 mm. long; fls. 2–4, white, pink in age; fl.-tube campanulate, 4–5 mm. high; sepals ca. as long; petals obovate, 4–6 mm. long; fr. broadly ellipsoid, glabrous, 1.5–2 cm. long, red-purple, edible, on pedicels ca. 10–12 mm. long; stone ± flattened, keeled on edges, smooth on faces except for 2–3 low ridges.—Dry rocky or moist slopes, below 6000 ft.; Yellow Pine F.; Coast Ranges from Santa Cruz Mts. to Del Norte and Siskiyou cos., Sierra Nevada from Tulare and Kern cos., n. to Modoc Co.; Ore. March–May. Variable, the following names having been proposed: var. *Kellóggii* Lemmon, for plants with sweet yellow fr. and from Sierra Co. to Mt. Shasta; var. *oregàna* (Greene) W. Wight, [*P. o.* Greene], fr. dark red, ± pubescent and from Modoc Co. to Ore.; and var. *rubicúnda* Jeps., drupe subglobose, bright red, 2–3 cm. long, with bitter pulp and from Modoc Co.

3. **P. Fremóntii** Wats. [*Amygdalus F.* Abrams. *P. F.* var. *pilulata* Jeps. *P. eriogyna* S. C. Mason.] Desert Apricot. Rigidly branched deciduous shrub or small tree, 1.5–4 m. tall, the twigs often spine-tipped, glabrous, red-brown; lvs. roundish to broadly ovate, serrate, 1–2 cm. long, on petioles 3–4 mm. long; fls. 1–few; pedicels slender, 8–12 mm. long; fl.-tube campanulate, ca. 2.5 mm. long; sepals ciliate, ca. 1.5 mm. long; petals white, 4–6 mm. long; fr. elliptic-ovoid, yellowish, puberulent, 8–14 mm. long, dry; stone with a thick ridge on ventral side.—Canyons, below 4000 ft.; Creosote Bush Scrub to Pinyon-Juniper Wd.; w. edge of Colo. Desert from Palm Springs region s.; to L. Calif. Feb.–March.

4. **P. Andersònii** Gray. [*Amygdalus A.* Greene.] Desert Peach. Diffusely branched spreading deciduous shrub 1–2 m. tall with short stiff spinescent branches; lvs. fascicled, oblanceolate, 1–2 cm. long, obscurely serrulate, glabrous, 1–2 cm. long, on very short petioles; fls. usually 1; pedicels 6–8 mm. long; fl.-tube 2.5 mm. high; sepals ca. 3 mm. long; petals broadly obovate, ± rose, 5–6 mm. long; fr. roundish, ca. 12 mm. long, brownish tomentulose, with thin dryish pulp and ± roughened stone.—Dry slopes and mesas, 3500–7500 ft.; Sagebrush Scrub, Yellow Pine F.; from Kern and Inyo cos. along e. slope of Sierra Nevada to Modoc Co., Inyo Mts.; w. Nev. March–April.

5. **P. fasciculàta** (Torr.) Gray. [*Emplectocladus f.* Torr. *Lycium Spencerae* Macbr.] Desert Almond. Divaricately much branched deciduous shrub 1–2(–3) m. tall, with short stiff, ± thornlike twigs, minutely pubescent when young; lvs. fascicled on short stubby branchlets, oblance-spatulate, mostly entire, 6–15 mm. long, pale green, minutely

pubescent; fls. imperfect or perfect, 1–few, subsessile; fl.-tube ca. 2 mm. long; petals oblanceolate, 2–3 mm. long; fr. ovoid, dry, 8–12 mm. long, pubescent; stone smooth, narrowly ridged on ventral side.—Dry slopes and washes, 2500–6500 ft.; Creosote Bush Scrub, Joshua Tree Wd., Pinyon-Juniper Wd.; Mojave and Colo. deserts; to Ariz., Utah. March–May.

Var. **punctàta** Jeps. Young branches more pubescent; lvs. glandular-punctate, not pubescent.—Sandy flats, below 1000 ft.; Coastal Sage Scrub, Chaparral; n. Santa Barbara and s. San Luis Obispo cos. March–April.

6. **P. Amýgdalus** Batsch. [*Amygdalus communis* L. *P. c.* Arcang. not Huds.] ALMOND. Tree like *P. persica;* lvs. 7–10 cm. long, acuminate, glabrous, closely crenate-serrate; fls. 2.5–4 cm. across, pink or whitish, solitary, sessile; fr. oblong-ovoid, compressed, pubescent, 4–5 cm. long, the flesh hard and splitting to reveal the shallowly pitted stone.—Occasional escape from cult.; Eurasia. Feb.–March.

7. **P. pérsica** Batsch. PEACH. Small deciduous tree with glabrous twigs; lvs. lanceolate, 5–20 cm. long, acuminate, coarsely crenate-serrate, glabrous; fls. solitary, sessile, in advance of lvs., mostly pink, 1–5 cm. across; fr. 3–8 cm. in diam., pubescent, yellow or red, with white or yellow flesh and deeply pitted stone; $2n = 16$ (Darlington, 1930). —Occasional escape from cult.; Asia. April–May.

8. **P. àvium** L. SWEET CHERRY. Deciduous tree; lvs. oblong-ovate to -obovate, 10–15 cm. long, taper-pointed, doubly glandular-serrate; fls. white, ca. 2.5 cm. across, appearing with first lvs. from clusters of buds on lateral spurs; fl.-tube constricted at top; drupe globular, yellow to red, with sweet or ± bitter flesh; $2n = 16, 24, 32$ (Darlington, 1928).—Occasional escape from cult.; Eurasia.

9. **P. Mahaléb** L. Small tree with slender green twigs; lvs. light green, broadly ovate to roundish, 3.5–6 cm. long, abruptly short-pointed, closely crenate-serrate, rather short-petioled; fls. several in racemose corymbs, white, fragrant, ca. 1–1.5 cm. broad; fr. ovoid, black, with little flesh, ca. 6 mm. long, on much longer pedicels; $2n = 16$ (Darlington, 1928).—Cult. especially as budding stock, sometimes escaping; Eu. April.

10. **P. virginiàna** L. var. **demíssa** (Nutt.) Sarg. [*Cerasus d.* Nutt. *P. d.* Walp.] WESTERN CHOKE CHERRY. Erect deciduous shrub or small tree, 1–5 m. high, with smooth gray-brown bark, the young twigs pubescent; lvs. oblong-ovate to obovate, 3–8 cm. long, finely serrate, ± pubescent beneath, abruptly pointed at apex, ± subcordate at base; petioles ca. 1 cm. long; racemes many-fld., 5–10 cm. long, at ends of short lfy. branches; sepals obtuse, barely 1 mm. long, often much broader, glandular-pectinate; petals white, rounded, 5–6 mm. broad; fr. round, 5–6 mm. thick, dark red, bitter but edible, especially late in the season; stone smooth, globose.—Dampish places in woods and on brushy slopes and flats, below 8200 ft.; Chaparral, Yellow Pine F., Foothill Wd.; mostly in mts. from San Diego Co. n., through Coast Ranges and Sierra Nevada; to Wash., Ida. May–June.

Var. **melanocárpa** (A. Nels.) Sarg. [*Cerasus demissa* var. *m.* A. Nels.] Twigs and lvs. glabrous; fr. black.—At 4000–5000 ft.; Siskiyou, Modoc, and Lassen cos.; to Rocky Mts.

11. **P. caroliniàna** Ait. CHERRY-LAUREL. Evergreen tree, 6–12 m. tall; lvs. lance-oblong, acuminate, 5–10 cm. long, entire, slightly revolute, glossy; racemes dense, 2–3 cm. long; fls. ca. 3–4 mm. across; fr. short-ovoid, pointed, black, shining, dryish, 8–10 mm. long.—Occasional escape from cult. Native se. U.S.

12. **P. lusitánica** L. PORTUGAL-LAUREL. Evergreen large shrub or small tree, to 12 m. tall; lvs. thick, leathery, lance-ovate to oblong-ovate, 5–10 cm. long, acuminate, crenate-dentate, glossy above; racemes 10–20 cm. long; fls. ca. 8 mm. across; fr. conic, pointed, dark purple, not fleshy, ca. 8 mm. long; $2n = 64$ (Almeida, 1947).—Occasional escape from cult.; native of sw. Eu.

13. **P. ilicifòlia** (Nutt.) Walp. [*Cerasus i.* Nutt. ex H. & A.] HOLLY-LEAVED CHERRY. ISLAY. Dense evergreen shrub or small tree, 1–8 m. tall, glabrous, the twigs soon gray or reddish-brown; lvs. coriaceous, ovate to roundish, 2–5 cm. long, coarsely spinose-toothed, crisped, the petioles ca. 8–12 mm. long; fls. few to many, racemose; racemes mostly 3–6 cm. long; sepals ± deltoid, to ca. 1 mm. long; petals white, round-oblong, 2–3 mm. long; fr. red, rarely yellow, ovoid-ellipsoid, 12–15 mm. long, with thin sweetish pulp; stone smooth, apiculate.—Common on dry slopes and fans, below 5000 ft.;

Chaparral, Foothill Wd.; Coast Ranges from Napa Co. s.; to L. Calif., Santa Catalina and San Clemente ids. April–May.

14. **P. Lyònii** (Eastw.) Sarg. [*Cerasus L.* Eastw. *P. integrifolia* Sarg. not Walp. *P. ilicifolia* var. *occidentalis* (Nutt.) Bdg.] CATALINA CHERRY. Like *P. ilicifolia*, but more arboreous, to 15 m. high; lvs. darker green, more narrowly ovate, mostly entire, plane, 4–10 cm. long; racemes many-fld., 5–12 cm. long; fr. almost black, 12–24 mm. long.— Chaparral; canyons of Santa Catalina, San Clemente, Santa Cruz, and Santa Rosa ids. March–May. Occasionally escaping from cult. on mainland. Hybridizes easily with *P. ilicifolia*.

15. **P. armeniaca** L. APRICOT. Small tree with reddish bark and glabrous twigs; lvs. ovate, sometimes subcordate, 5–8 cm. long, abruptly short-pointed, closely serrate, pubescent beneath on veins; fls. ± pinkish, 2–2.5 cm. across, in advance of foliage; fr. smooth at maturity, pubescent earlier, somewhat flattened, yellow with some red, the stone flat, ridged; $2n = 16$ (Darlington, 1928).—Occasional escape; native of China.

31. Osmarònia Greene. Oso BERRY

Shrubs with simple entire deciduous lvs. and small stipules. Fls. perfect or imperfect, white, fragrant, in nodding racemes at ends of lfy. branchlets. Fl.-tube turbinate-campanulate, deciduous. Sepals and petals 5. Staminate fls. with spreading petals and 15 stamens in 3 series. Pistillate fls. with smaller erect petals, abortive stamens, and 5 simple, free, distinct, glabrous pistils. Fr. of 1–5 drupes, with thin pulp and bony endocarp. One sp. (Greek, *osme*, fragrant, and *Aronia*, a Rosaceous genus.)

1. **O. cerasifórmis** (T. & G.) Greene. [*Nuttallia c.* T. & G. *O. laurina, padiformis, obtusa, demissa,* and *bracteosa* Greene.] Shrubby or treelike, 1–5 m. tall, with rather straight slender stems, smooth gray or red-brown bark; lvs. oblong to oblanceolate, 5–10 cm. long, paler and ± pubescent beneath, thin, short-petioled; raceme 3–10 cm. long; fl.-tube ca. 5 mm. high; sepals 3 mm. long; petals of ♂ fls. obovate, 5–6 mm. long, of ♀ fls. narrower, shorter; drupe ca. 1 cm. long, black, glaucous, the pulp bitter; $n = 8$ (Moffett, 1931).—Canyons, below 5600 ft.; Mixed Evergreen F., Chaparral, Yellow Pine F., Redwood F. etc.; Coast Ranges from Santa Barbara Co. to Del Norte Co., Sierra Nevada from Tulare Co. to Shasta Co.; to B.C. Feb.–April.

32. Sórbus L. MOUNTAIN-ASH

Deciduous shrubs or trees, with alternate stipulate pinnate to simple lvs. Fls. white, in a compound terminal corymb. Sepals and petals 5. Stamens 15–20. Pistils 1–5, each carpel 2-ovuled, the ovaries either partially connate and half inferior or totally connate and inferior; styles usually ± separate. Fr. a small pome, of 2–5 locules, each 1–2-seeded. Ca. 80 spp., of N. Hemis. (The ancient Latin name.)

(Jones, G. N. A synopsis of the N. Am. spp. of Sorbus. Jour. Arnold Arbor. 20: 1–43, 1939.)
A. Winter-buds densely whitish-villous; rachises, pedicels and calyces usually copiously whitish pubescent at flowering time. Occasionally natur. . . . . 1. *S. Aucuparia*
AA. Winter-buds glabrous or pilose, with whitish or rusty trichomes. Natives.
  B. Lfts. 11–13, ± lanceolate; stipules caducous; infl. flat-topped, 80–200 fld. . . 2. *A. scopulina*
  BB. Lfts. 7–11, ± oblong; stipules ± persistent; infl. somewhat convex, 30–60 fld.
    C. Plant 2–5 m. tall; lfts. mostly 5–7 cm. long, 2–3 cm. wide; young twigs pubescent; petals 5–6 mm. long. Modoc to Siskiyou and Trinity cos. . . . . 3. *S. cascadensis*
    CC. Plant 1–2 m. tall; lfts. mostly 2–4 cm. long, 1–2 cm. wide; young twigs glabrous; petals 3–4 mm. long. Siskiyou to Tulare cos. . . . . 4. *S. californica*

1. **S. Aucupària** L. [*Pyrus A.* Gaertn.] Round-headed tree to 15 m. tall, with smooth close bark; lfts. 9–15, oblong-lanceolate, blunt or short-pointed, sharp-serrate from near the base, sessile, 3–5 cm. long; infl. 10–18 cm. broad, 75–100-fld.; fls. 8–9 mm. in diam.; petals orbicular, 4 mm. long; fr. scarlet, 9–11 mm. in diam.; seeds oval, flattened, light brown, 4 mm. long, 2.5 mm. broad; $2n = 34$ (Moffett, 1931).—Occasional escape from cult., as in San Bernardino Mts. (Fredalba); native of Eu.

2. **S. scopulina** Greene. [*S. sitchensis* var. *densa* Jeps.] Shrubby, 1–4 m. tall, with thick reddish bark; winter-buds glossy, glutinous, brown, glabrous or sparsely pilose; lfts. 11–13, the lateral lanceolate, 3–6 cm. long, 1.2–2 cm. wide, cuneate at base,

sharply acute to acuminate at apex, finely and sharply, singly or doubly serrate almost to base, glabrous; peduncles and pedicels sparingly pilose; fls. ca. 1 cm. in diam.; sepals pilose, 1.5 mm. long; petals oval, 5–6 mm. long; styles 2–2.5 mm. long; fr. orange to scarlet, globose, 8–10 mm. in diam.; seeds oblong, light brown, 3.5–4 mm. long, 1.5–2 mm. wide.—Occasional, canyons and wooded slopes, 4000–9000 ft.; Montane Coniferous F.; Sierra Nevada from Tulare Co., to Siskiyou and Modoc cos.; to B.C., Rocky Mts. June.

3. **S. cascadénsis** G. N. Jones. Shrub 2–5 m. tall, with smooth gray bark and pubescent young twigs and winter-buds; lfts. 9–11, oblong-oval, abruptly acute, glabrous, dark and glossy above, coarsely and sharply serrate from top to below middle; peduncles and pedicels sparsely pilose; fls. 8–10 mm. across; sepals 1.5 mm. long; petals round, 5–6 mm. in diam.; styles 2 mm. long; fr. scarlet, globose, 8–10 mm. in diam.; seeds flattened, ovoid, dark brown, 4 mm. long, 2 mm. wide.—Moist places, 4000–6000 ft.; Yellow Pine F.; Modoc to Trinity and Siskiyou cos.; to B.C. Doubtfully distinct from the next.

4. **S. califórnica** Greene. [*Pyrus sitchensis* var. *c.* Smiley.] Many-stemmed shrub 1–2 m. tall with mostly glabrous twigs and glutinous winter-buds; lfts. 7–9(–11), oblong-oval, glabrous; petioles, peduncles, pedicels, and rachises tending to be glabrous; sepals 2 mm. long; petals round, 3–4 mm. long; fr. scarlet, ellipsoid or pyriform, 7–10 mm. long; seeds oval, slightly flattened, light brown, 4 mm. long, 2 mm. wide.—Moist places, 5000–10,900 ft.; Montane Coniferous F.; Santa Cruz Mts., Tulare Co. to Modoc, Glenn, and Humboldt cos.; w. Nev. May–June.

33. Màlus Mill. APPLE

Deciduous trees or shrubs with alternate toothed or lobed lvs. Fls. white or pink, in simple terminal clusters. Fl.-tube urn-shaped, open in anthesis and not closed about the styles which are ± connate basally. Sepals 5. Petals 5. Styles 2–5 (mostly 5); ovules 2 in each cell; carpels papery or leathery. Fr. a pome, usually hollowed at base, ± depressed-globose, lacking stone- or grit-cells. Ca. 25 spp. of N. Temp. Zone. (Classical name of apple.)

Some of lvs. on elongated shoots lobed; fr. ca. 1.5 cm. long, oblong, purplish. Native along n. coast
                                                                                                          1. *M. fusca*
None of lvs. on elongated shoots lobed; fr. normally much larger, ± reddish. Occasional escape from cult .... 2. *M. sylvestris*

1. **M. fúsca** (Raf.) C. K. Schneid. [*Pyrus f.* Raf. *P. diversifolia* Bong. *P. rivularis* Dougl.] OREGON CRAB APPLE. Small tree or large shrub, 5–10 m. tall, ± thorny; lf.-blades 2–8 cm. long, ovate to lance-ovate, acute to acuminate, serrate, sometimes 3-lobed, green and subglabrous above, paler, pubescent and eventually rusty beneath, on petioles to 4 cm. long; fls. several in flat-topped clusters; pedicels 2–3 cm. long in fr., they and fl.-tube tomentose; petals 7–11 mm. long, white, roundish; frs. oblong, 12–14 mm. long, purple-black in age.—Moist open woods, below 1000 ft.; Redwood F., Mixed Evergreen F.; near the coast from Sonoma and Napa cos. to Del Norte Co.; to Alaska. April–June.

2. **M. sylvéstris** Mill. [*Pyrus Malus* L.] COMMON APPLE. Familiar round-headed tree, with heavily pubescent young growth; lvs. oval to elliptic or ovate, thick, veiny, abruptly short-petioled, rather bluntly serrate, becoming ± glabrous above, pubescent beneath, 5–10 cm. long; infl. woolly, the pedicels stout, 1–2.5 cm. long; fls. white or pinkish, 3–5 cm. across; fr. 3 cm. or more in diam.; $2n = 34$, 51 (Darlington & Moffett, 1930).— Occasional escape and maintaining itself; native of Eurasia. April–May.

34. Amelánchier Medic. SERVICE-BERRY

Deciduous shrubs to small trees, with unarmed branches and slender terete branchlets. Lvs. simple, alternate, pinnately veined, entire to serrate. Fls. perfect, regular, mostly in racemes, terminating short leafy branchlets of the season. Fl.-tube campanulate or urceolate, ± adnate to carpels, becoming globose or ellipsoid in fr. Sepals 5, persistent, rather narrow. Petals 5, white, oblanceolate to narrowly oval. Stamens 10–20, short. Carpels 2–5, ± united to form an inferior, 2–5-loculed ovary, each locule divided by a false partition, the cells 1-seeded. Fr. a pome, with firm not bony carpel walls and fleshy edible outer tissue. Seeds small, smooth, dark brown. Perhaps 20 spp. of N. Temp. Zone. (The Savoy name of the Medlar.)

(Jones, G. N. Amer. Spp. of Amelanchier. Univ. Ill. Press, pp. 1–126, 1946.)
Lvs. glabrous at maturity, sometimes slightly pubescent beneath.
  Petals 12–15 mm. long; styles 5; ovary-top tomentose. Mendocino to Del Norte cos. . . 1. *A. florida*
  Petals 6–12 mm. long; ovary-top glabrous.
    Styles 5; petals 8–12 mm. long. Lake Tahoe region .... 2. *A. pumila*
    Styles 2–3; petals 6–8 mm. long. Mojave Desert .... 4. *A. utahensis Covillei*
Lvs. permanently puberulent to tomentulose, especially beneath.
  Petals oval to obovate, 8–11 mm. long, 3–4.5 mm. wide; styles free nearly to the base; lvs. with
    7–9 pairs of lateral veins. Siskiyou Mts. to Cuyamaca Mts. .... 3. *A. pallida*
  Petals mostly 6–8 mm. long, oblanceolate, 1.5–2.5 mm. wide; styles united below; lvs. with 9–13
    pairs of lateral veins. Panamint Mts. to San Bernardino Mts., Laguna Mts., and Providence Mts.
                                                               4. *A. utahensis*

1. **A. flórida** Lindl. Slender, 1–5 m. tall or even higher, with erect branches; bark brownish becoming gray; young twigs red-brown, at first tomentose, later glabrous; lvs. thin, oval to oblong or roundish, tomentulose when young, especially beneath, glabrous when mature, pale beneath, the blades 3–4 cm. long, 2–3 cm. wide, mostly rounded to subtruncate at apex, with 8–12 pairs of lateral veins, usually coarsely toothed above the middle; petioles 1–2.5 cm. long; fls. 2–3 cm. in diam., fragrant, the erect racemes 4–8 cm. long, 5–15-fld.; sepals deltoid-lanceolate, acute or acuminate, 2–2.5 mm. long; petals oblanceolate, obtuse; stamens ca. 20, shorter than calyx; styles five, 2–2.5 mm. long, united to ca. middle; fr. globose, 10–13 mm. in diam., purplish-black when ripe; seeds ca. 5 mm. long, 2 mm. wide.—Moist woods and open places; Redwood F., Mixed Evergreen F.; n. Mendocino Co. to Del Norte Co.; to Alaska. March–May.

2. **A. pùmila** Nutt. [*A. glabra* Greene.] Shrubs 1–3 m. tall, glabrous throughout; bark red-brown, becoming gray; lvs. oval to roundish, 1.5–3 cm. long, truncate to retuse at apex, with 7–9 pairs of lateral veins, coarsely serrate down to the middle; petioles shorter than blades; racemes 4–8-fld., 2–4 cm. long; sepals lanceolate, 3 mm. long; petals oblong-spatulate, 8–12 mm. long, 3–4 mm. wide; stamens 12–15; styles mostly 5, united at base; mature frs. depressed-globose, dark purple, 8–9 mm. in diam.; seeds brown, 4–5 mm. long.—Damp woods, 6000–8000 ft.; Red Fir F., Lodgepole F.; Nevada to Eldorado cos.; to Mont., Colo. May–June.

3. **A. pállida** Greene. [*A. subintegra* Greene. *A. gracilis* Heller. *A. siskiyouensis* C. K. Schneid. *A. recurvata* Abrams. *A. alnifolia* var. *cuyamacensis* Munz. *A. alnifolia* of most Calif. refs., not Nutt.] Shrubs 1–3(–6) m. tall, with rigid erect to spreading branches; bark glabrous, red-brown to gray; lvs. oval to elliptical or rounded or broadly obovate, rather thick, tomentulose or puberulent on both surfaces, the lower usually paler than upper, the blades 2–4 cm. long, 1.5–2.5 cm. wide, acute to roundish at apex, with 7–9 pairs of obscurish veins, entire to toothed to or below the middle; petioles pubescent, 6–12 mm. long; racemes corymbose, 2–4 cm. long, pubescent, 4–6-fld.; sepals lanceolate, acuminate, 2–3 mm. long, villosulous on both sides; petals oval or obovate, 8–11 mm. long, 3–4.5 mm. wide; stamens ca. 15; styles mostly 3–4; fr. subglobose, purplish-black, 4–6 mm. in diam., the upper part of the fl.-tube constricted on the young fr.— Dry gravelly and rocky slopes and flats, below 11,000 ft.; many Plant Communities, but especially in Montane Coniferous F.; Coast Ranges and Sierra Nevada s. to Kern and Ventura cos. and in San Gabriel, Cuyamaca, and Palomar mts.; to Ore., w. Nev. April– June. Exceedingly variable and in need of study.

4. **A. utahénsis** Koehne. [*A. alnifolia* var. *u.* M. E. Jones. *A. venulosa* Greene. *A. alnifolia* var. *v.* Jeps.] Shrubby, much-branched, 1–5 m. high, with ashy-gray bark and rigid twigs, white-pubescent on young growth; lvs. roundish to oval, or elliptical, rounded to acute at apex, grayish-green, tomentulose or cinereous on both sides, the blades 1–3 cm. long, 0.5–2 cm. wide, with 11–13 pairs of lateral veins, coarsely crenate-serrate usually to near the base; petioles 5–15 mm. long; racemes sublanate, 3–6-fld., 2–3 cm. long; sepals linear to lance-linear, tomentulose, 3 mm. long; petals linear-oblanceolate to cuneate, 6–8 mm. long; stamens mostly 10–15; styles mostly 3–4; fr. 6–10 mm. in diam., purplish-black to brownish, juicy to dry; seeds brown, flattened-ellipsoid, 5–6 mm. long, 3 mm. wide.—Dry slopes, 5000–7000 ft.; Yellow Pine F., Pinyon-Juniper Wd.; mts. of San Diego Co. to San Bernardino and San Gabriel mts., Kingston Mts. to Sweetwater Mts.; to Mont., New Mex. April–May.

Ssp. **Covíllei** (Standl.) Clokey. [*A. C.* Standl.] Plant nearly or quite glabrous.—Pinyon-Juniper Wd.; Panamint Mts. to Clark, Old Dad mts., s. Nev. April.

35. Peraphýllum Nutt. SQUAW-APPLE

Low shrub with grayish bark and simple alternate lvs. crowded at ends of spurlike branchlets or scattered along new growth, oblanceolate, entire or serrulate. Fls. appearing with lvs., 1–3, perfect. Fl.-tube subglobose, adnate to ovary. Sepals 5, persistent, reflexed. Petals 5, orbicular. Stamens ca. 20, distinct. Pistil 1, with 2 carpels, but ovary 4-celled by false partitions; styles 2. Fr. a pome, globose, fleshy, bitter, the carpels cartilaginous. Monotypic. (Greek, *pera*, excessively, and *phullon*, leafy, i.e. very leafy.)

1. **P. ramosíssimum** Nutt. Intricately branched, 1–2 m. high; lvs. 2–4 cm. long, acute, strigose; sepals ca. 3 mm. long, woolly on inner surface, sometimes also on outer; petals pale pink, 7–8 mm. long; fr. yellowish, 8–10 mm. thick, bitter.—Dry washes and slopes, 4000–8000 ft.; Sagebrush Scrub, Pinyon-Juniper Wd., N. Juniper Wd., Yellow Pine F.; White Mts., Inyo Co. to Modoc and Shasta cos.; to Ore., Colo. April–May.

36. Crataègus L. HAWTHORN

Shrubs or small trees, usually armed with sharp thorns. Lvs. deciduous, alternate, simple to lobed. Fls. in terminal corymbs; fl.-tube urn-shaped to cup-shaped, adnate to ovary. Sepals 5, reflexed after anthesis. Petals 5, white, sometimes pink, rounded. Stamens 5–25, in 1–3 series; fils. filiform. Ovary inferior, 1–5-celled; styles 1–5, separate; ovules usually 1 in each cell. Fr. a drupelike pome, yellow or red or purplish-black, with 1–5 bony 1-seeded nutlets. A large genus, of N. Temp. Zone. (Greek, *kratos*, strength, referring to wood.)

1. **C. Douglásii** Lindl. [*C. rivularis* Nutt. *C. Gaylusaccia* Heller.] Much-branched, 2–10 m. tall, the thorns stout, 1–2.5 cm. long; twigs reddish; lvs. ovate to obovate, obtuse or acutish, ± doubly serrate above the cuneate entire base, often lobed, glabrous beneath, subglabrous beneath, pubescent above at least on midrib, 2–7 cm. long, rather thick in texture, on petioles to ca. 1 cm. long; fls. rather many, the pedicels glabrous or nearly so; sepals triangular, ± villous, entire to somewhat gland-toothed; petals 4–5 mm. long; fr. purple-black, 10–12 mm. long, smooth; nutlets slightly ridged on the outer face and pitted on the inner; $3n = 51$ (Longley, 1924).—Near streams and meadows, 2500–5500 ft.; Foothill Wd., Mixed Evergreen F., N. Oak Wd., Yellow Pine F., Red Fir F.; mts. from Marin Co. n. to Siskiyou and Modoc cos.; to B.C., Mich. May–June. Plants from Shasta and Modoc cos. are more pubescent and have longer more toothed sepals and approach *C. columbiàna* Howell.

37. Heterómeles M. Roem. TOYON. CHRISTMAS-BERRY

Evergreen arboreous shrub, unarmed, with simple coriaceous toothed lvs. Fls. small, white, many, in large terminal corymbose panicles. Fl.-tube turbinate, partly adnate to ovary. Sepals 5, persistent. Petals 5, spreading, rounded, concave. Stamens 10, in pairs opposite the sepals; fils. dilated at base and ± connate. Ovary 2–3-celled, with 2 ovules in each cell; styles 2–3, distinct. Fr. a berrylike pome, red or yellow, ovoid, the thickened persistent sepals incurved over the carpels. One sp. (Greek, *heter*, different, and *malus*, apple, unlike neighboring genera.)

1. **H. arbutifòlia** M. Roem. [*Crataegus a.* Ait., not Lam. *Photinia a.* Lindl. *P. salicifolia* Presl. *H. s.* Abrams.] Plant 2–10 m. high, freely branched, with gray bark and tomentulose young branchlets; lvs. elliptical to oblong or lance-oblong, 5–10 cm. long, acute at both ends, rather sharply toothed, dark green above, lighter beneath, glabrous or sparsely tomentulose; petioles 1–2 cm. long; fl.-tube ca. 3 mm. high; sepals triangular, 1–1.5 mm. long; petals ca. 4 mm. long; fr. 5–6 mm. long, quite persistent through the winter months; seeds oblong-ovoid, flattened, brown, 2.5–3 mm. long, ca. 1.6–1.8 mm. broad.—Common on semidry brushy slopes and in canyons, below 4000 ft.; Chaparral, Foothill Wd., etc.; mts. of s. Calif. to Humboldt Co. and in foothills of Sierra Nevada from Shasta Co. to Tulare Co.; n. L. Calif. June–July. Occasional plants growing with the sp. and bearing yellow frs., have been called *Photinia a.* var. *cerìna* Jeps.

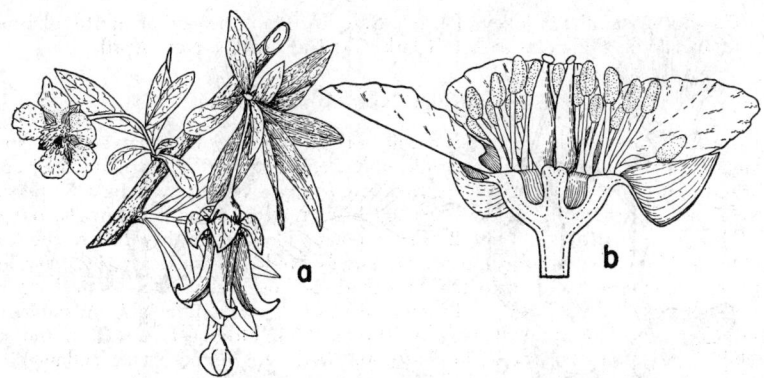

Fig. 70. CROSSOSOMATACEAE. *Crossosoma californicum: a,* habit, × ½, with fl. and young follicles; *b,* section of fl., × 2½, turbinate fl.-tube, outer sepals, larger petals, many stamens, separate pistils (maturing into follicles).

Var. **macrocárpa** (Munz) Munz. [*Photinia a.* var. *m.* Munz.] Fr. red, 8–10 mm. long. —Santa Catalina and San Clemente ids. The most desirable form for cult., with larger infl. and larger frs. less readily eaten by birds.

74. Crossosomatàceae. Crossosoma Family (Fig. 70)

Glabrous shrubs, deciduous or with dry lvs. in dry season. Lvs. alternate, entire, simple. Fls. solitary, perfect, regular, borne at ends of naked or bracted peduncles. Fl.-tube turbinate, lined with a thin glandular disk. Sepals 5, persistent. Petals 5, white, deciduous. Stamens 15–50, inserted in several series on the disk lining the fl.-tube; fils. ± dilated at base; anthers basifixed. Pistils 2–9, unicarpellate, distinct, stipitate, with capitate stigmas. Fr. follicular. Seeds several, with a conspicuous fringed aril; endosperm present. Three to four spp. of sw. N. Am.

1. Crossosòma Nutt.

Characters of the family. (Greek, *krossoi,* fringe and *soma,* body, because of the aril.)

Lvs. 2.5–9 cm. long, scattered; petals rounded, scarcely clawed. Insular ...... 1. *C. californicum*
Lvs. 0.5–1.5 cm. long, fascicled; petals spatulate to oblong, distinctly clawed. Deserts
                                         2. *C. Bigelovii*

1. **C. califórnicum** Nutt. Shrub 1–2 m. tall, or arborescent and to 5 m. tall, with stout gray-brown branches; lvs. oblong, rarely obovate, mucronate, subsessile, pale green; sepals ca. 1 cm. long, round-ovate, scarious on margins; petals 1.2–1.5 cm. long, rounded, white; follicles 2–7(–9), cylindric, 1.5–2 cm. long, ± recurved; seeds ca. 20 or more, black, shining, round, somewhat flattened, ca. 2.5 mm. in diam., with yellowish fringed aril.—Dry rocky slopes and canyons; Chaparral, Coastal Sage Scrub; Santa Catalina and San Clemente ids.; Guadalupe Id. Feb.–May.

2. **C. Bigelòvii** Wats. Spreading, 1–2 m. tall, much-branched with rigid spinescent branchlets; lvs. elliptic to oblong-ovate, subsessile, apiculate, gray-green; sepals 4–5 mm. long, rounded; petals oblong, white to rose, 9–12 mm. long; follicles 1–3, 8–10 mm. long; seeds 2–5, ca. 2 mm. in diam.—Dry rocky canyons, below 3000 ft.; Creosote Bush Scrub; w. Colo. Desert to s. Mojave Desert (Ord Mts., Warren's Well, Sheephole Mts.); to L. Calif., Ariz., Son. Feb.–April.

75. Leguminòsae. Pea Family

A very large family of herbs, shrubs or small trees. Lvs. alternate, usually with stipules, mostly compound, the ultimate lfts. 1–many and usually entire. Fls. mostly perfect, regular or more generally irregular and papilionaceous. Calyx 5-toothed or -cleft. Petals

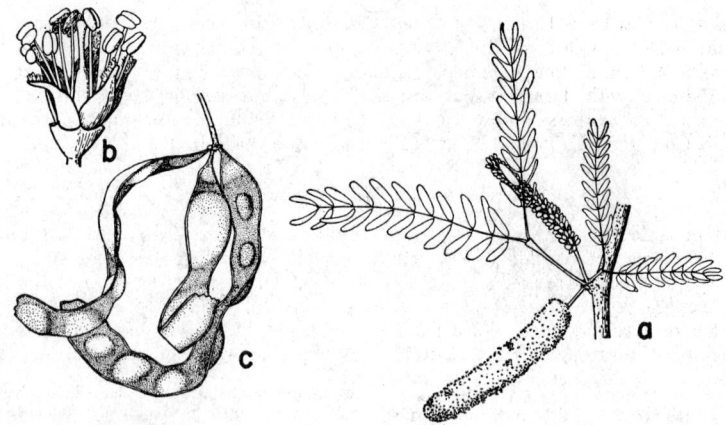

Fig. 71. LEGUMINOSAE-MIMUSOIDEAE. *Prosopis juliflora* var. *Torreyana: a,* habit, $\times$ ½, with lvs. with 1 pair of pinnae and many pinnules, and with spikes of many sessile fls.; *b,* fl., $\times$ 5, calyx campanulate, 5-toothed, petals 5, alike, stamens 10, free. *Acacia Greggii: c,* legumes, $\times$ ½.

usually 5, distinct or partly united, hypogynous, rarely 1 or none (if papilionaceous, the upper petal the "banner" or "standard," the lateral petals the "wings," and the 2 lower petals joined to form the "keel"). Stamens few to many, often 10, distinct or more often the fils. variously united. Pistil 1, simple, free, superior, becoming a legume in fr. (a dehiscent 2-valved pod) or a loment (with several indehiscent segms.) or sometimes entire and indehiscent. Seeds mostly without endosperm. With 450–500 genera and many thousand spp., many of great economic importance for food, forage, dyes, wood, as ornamental plants, etc.

KEY TO SUBFAMILIES

Corolla regular or nearly so, valvate in bud; lvs. bipinnate, the lfts. mostly small; fls. small, in many-fld. heads or spikes; petals inconspicuous; stamens conspicuous .. 1. *Mimusoideae*   p. 796
Corolla ± irregular, imbricate in the bud, the petals unlike in size or shape or both; fls. often larger and petals conspicuous; stamens mostly included.
    Upper petal internal in bud, enveloped by the lateral ones, therefore the corolla not
    papilionaceous .... 2. *Caesalpinioideae*   p. 799
    Upper petal external in bud, the corolla papilionaceous .... 3. *Papilionoideae*   p. 801

I. Mimusoìdeae. Mimosa Subfamily (Fig. 71)

KEY TO GENERA

Stamens united below into a tube.
    Pod evidently dehiscent; fls. rose to reddish-purple. Desert native .... 1. *Calliandra*
    Pod indehiscent; fls. mostly yellowish. Escape from cult. .... 3. *Albizia*
Stamens distinct or nearly so.
    The stamens more than 10; anthers not gland-tipped .... 2. *Acacia*
    The stamens 10; anthers gland-tipped .... 4. *Prosopis*

1. Calliándra Benth.

Herbs, shrubs or trees, with slender branches. Lvs. bipinnate, with small lfts. Fls. in dense pedunculate heads arranged in axillary or terminal racemes. Calyx 5-toothed or -lobed. Petals united to ca. middle into a funnelform or campanulate corolla. Stamens many, long-exserted; fils. united basally into a tube; anthers small. Style filiform; ovules many. Legumes linear, flat, straight, elastically dehiscent from apex, the valves coriaceous with raised margins. Seeds roundish to obovate, flattened. Perhaps 150 spp., of trop. and subtrop. Am., Afr., India. (Greek, *kallos,* beautiful, and *andra,* stamen.)

1. **C. erióphylla** Benth. [*Anneslia e.* Britton.] Fairy Duster. Mock Mesquite. Densely branched ± spreading unarmed shrub ca. 2–3.5 dm. high, with gray pubescent twigs;

pinnae 2–4 pairs with 5–10 pairs of oblong lfts. 3–5 mm. long, strigose beneath, obtuse or acutish; heads few-fld., sparingly strigose, rose to reddish-purple; calyx 1–1.5 mm. long; corolla 4–6 mm. long; stamens reddish, 1.5–2 cm. long; legume 5–7 cm. long, silvery-pubescent with dark red margins; seeds gray, narrow-obovate, 6–7 mm. long.— Sandy washes and gullies, below 1000 ft.; Creosote Bush Scrub; Imperial Co. and e. San Diego Co.; to Tex., Puebla, L. Calif. Mostly Feb.–March.

2. Acàcia Mill. ACACIA

Armed or unarmed shrubs or trees with bipinnate lvs. and small lfts. or with lvs. reduced to simple flat phyllodia. Fls. minute, usually yellow, in heads or spikes. Calyx usually 4–5-toothed, or sepals sometimes distinct. Petals usually 4 or 5, united or distinct, sometimes 0. Stamens many, exserted; fils. filiform, distinct. Legume linear to oval, flat or ± turgid, often constricted between seeds. An immense genus, chiefly in subtrop. regions, but especially in Australia and Afr. Many of horticultural value. (Greek, *akakie*, from *ake*, a point, because of the prickles.)

A. Plants with spiny or thorny branches.
    B. Spines mostly 3–4 mm. long, recurved; lvs. with 2–3 pairs of pinnae, the pinnules 1.5–3
        mm. wide. Deserts . . . . 1. *A. Greggii*
    BB. Spines mostly 8–30 mm. long, straight; lvs. with 2–8 pairs of pinnae, the pinnules scarcely
        1 mm. wide. Cismontane . . . . 2. *A. Farnesiana*
AA. Plants not armed.
    B. Lvs. bipinnate, the pinnules exceedingly numerous . . . . 3. *A. decurrens*
    BB. Lvs. simple and reduced to phyllodes, except sometimes in juvenile foliage as in young
        plants or on sprouts from stumps, pruned places, etc.
        C. Phyllodes 1-nerved; fls. yellow . . . . 4. *A. retinodes*
        CC. Phyllodes mostly 2–5 nerved.
            D. Fls. in heads, cream-color; phyllodia somewhat sickle-shaped . . 5. *A. melanoxylon*
            DD. Fls. in spikes, yellow; phyllodia curved alike on both edges . . . . 6. *A. longifolia*

1. **A. Gréggii** Gray. [*Senegalia G.* Britt. & Rose.] CATCLAW. Spreading straggling deciduous shrub, 1–2 m. high, or ± arborescent and taller (to ca. 5 m.); branches armed with short, stout, curved spines, the young growth ± pubescent; lvs. 2–5 cm. long, each pinna divided into 4–6 pairs of oblong to oblong-ovate pinnules 2–8 mm. long; fls. yellow, in cylindrical spikes 1–4 cm. long; legumes compressed, ± constricted between the seeds, 2–12 cm. long, 1.5–1.8 cm. broad; seeds dark brown, roundish, 7–9 mm. long, biconvex.—Common in washes and canyons, below 6000 ft.; Creosote Bush Scrub to Pinyon-Juniper Wd.; Colo. and s. Mojave deserts; to Tex., Son., L. Calif. Occasional in cismontane s. Calif. April–June.

2. **A. Farnesiàna** (L.) Willd. [*Mimosa F.* L.] Arborescent shrub or small tree, deciduous; lfts. mostly 10–25 pairs on the pinnae, glabrous; fls. deep yellow, very fragrant, in heads on pubescent peduncles; pod cylindric, 4–8 cm. long, scarcely dehiscent, filled with a pith which separates the seeds; $n = 27$ (Ghimpu, 1930).—Dry slopes; Chaparral; near San Diego (San Dieguito V., East San Diego, Otay); probably introd. in mission times from trop. Am. Jan.–March. Fls. very fragrant; widely cult. in warm countries as a source of perfume.

3. **A. decúrrens** Willd. GREEN WATTLE. A tree to 15 m. tall, with angled unarmed branches; lvs. bipinnate, with 8–15 pairs of pinnae, each with 30–40 or more pairs of linear lfts. 4–7 mm. long; fls. yellow, in heads in axillary often compound racemes; legumes 5–10 cm. long, 6–8 mm. wide, ± contracted between the seeds; $n = 13$ (Ghimpu, 1930).—It and var. *móllis* Lindl. [*A. mollissima* Willd.], BLACK WATTLE (with lvs. dark green and shining above and fls. pale yellow), as well as var. *dealbàta* F. Muell., SILVER WATTLE (with silvery-gray to light green lvs. and deep yellow fls.), are all widely cult. in Calif. and occasionally establish themselves in the wild; native of Australia. Feb.–March.

4. **A. retinòdes** Schlecht. Small tree, 5–8 m. tall or ± shrubby, with spreading branches drooping at ends; phyllodia blue or yellow-green, 8–15 cm. long, 0.6–1.8 cm. wide, with a marginal gland 6–12 mm. above the base; fls. yellow, in globose heads on lateral racemes; legumes 7–10 cm. long, ca. 0.6 cm. wide; $2n = 26$ (Tijo, 1948).—Cult. Australian plant, occasionally natur. Fls. much of year.

5. **A. melanóxylon** R. Br. BLACKWOOD ACACIA. Tree, 7–20 m. tall, with spreading thick head; lvs. dull dark green, lanceolate to oblanceolate, 5–10 cm. long, 10–20 mm.

wide, 3–5-veined, with marginal gland near very base; fls. creamy-white, in heads on short racemes; legumes 7–12 cm. long, ca. 1 cm. wide, twisted or bent; seeds encircled by the long red funicle; $2n = 26$ (Atchison, 1948).—Commonly cult. tree from Australia; occasional escape. May–June.

6. **A. longifòlia** Willd. GOLDEN WATTLE. Tree, 5–8 m. high, with light green foliage; phyllodia oblong-lanceolate, 5–15 cm. long, 0.6–1.4 cm. wide, 2–4-veined, the 2 margins convex, the gland near the base; fls. bright yellow, in spikes ca. 2–6 cm. long; legume terete until ripe, 4–12 cm. long; seeds black, shining, the funicle fitting like a cap over one end; $n = 13$ (Ghimpu, 1930).—Australian native, widely cult.; occasional escape. May.

3. Albízia Durazz.

Unarmed trees or shrubs, with alternate usually deciduous bipinnate lvs. and numerous small lfts. Fls. yellow, white or pink, mostly 5-merous, in globose heads or cylindric spikes which are axillary and paniculately arranged at the ends of the branches. Calyx toothed or lobed. Corolla-segms. connate to above the middle. Stamens many, connate at base, exserted. Legume large, strap-shaped, nonseptate, without pulp. Ca. 50 spp. of subtrop. and trop. regions, mostly of Old World. (From *Albizzi*, Italian naturalist.)

1. **A. distàchya** (Vent.) Macbr. [*A. lophantha* Benth.] Shrub or small tree 2–6 m. high; lvs. with 14–24 pinnae, the pinnules 40–60, linear-oblong, obtuse, ca. 7–8 mm. long; spikes mostly 2, yellowish, ca. 4–5 cm. long; fls. distinctly pedicelled; legume ca. 7–8 cm. long, 1–1.2 cm. wide.—An Australian plant in cult. in Calif. and occasionally reported as natur. Early spring.

4. Prosòpis L.

Deciduous shrubs and trees, armed with paired supra-axillary spines. Lvs. bipinnate with 1–2 pairs of pinnae and many small entire pinnules. Fls. small, greenish to yellow, regular, sessile in axillary spikes. Calyx campanulate, 5-toothed. Petals 5, united basally or free. Stamens 10, free and exserted; anthers tipped with a deciduous gland. Pods indehiscent, coriaceous, with numerous seeds separated by spongy partitions. Ca. 30–35 spp., of warm temp. and trop. regions. The pods of our species much eaten by cattle. (Greek, *prosopis*, ancient name for Burdock.)

(Burkart, A. Materiales para una monografía del genero Prosopis. Darwiniana 4: 57–128, 1940. Benson, L. The mesquites and screw-beans of the U.S. Am. Jour. Bot. 28: 748–754, 1941.)

Fr. straight or curved, not spirally coiled; lvs. bright green with 14–30 pinnules . . . . 1. *P. juliflora*
Fr. tightly spirally coiled.
    Fls. in cylindrical spikes; lvs. canescent, with mostly 10–16 pinnules. Native . . . . 2. *P. pubescens*
    Fls. in round heads; lvs. glabrescent, with mostly 12–32 lfts. Escape about Bard
                                                                      3. *P. strombulifera*

1. **P. juliflòra** (Sw.) DC. var **Torreyàna** L. Benson. [*P. chilensis* of Calif. refs., not Stuntz. *P. j.* var. *glandulosa* of Calif. refs., not (Torr.) Ckll. *P. odorata* Torr. & Frém.] MESQUITE. Low tree or large shrub with several trunks and crooked arched branches, 3–7 m. high; lfts. bright green, glabrous or nearly so, linear to oblong, 7–17 pairs, acute, 1.5–2.3 cm. long, mostly 2–2.5 mm. wide; spikes slender, 4–6 cm. long; petals 2.5–3.5 mm. long; legumes glabrous, flat, 1–several in a cluster, 5–15 cm. long, constricted between the seeds; seeds brown or red-brown, 6–7 mm. long, 4–5 mm. broad.—Common in washes and low places, below 3000 (5000) ft.; Creosote Bush Scrub, edge of Alkali Sink; Colo. and Mojave deserts, upper San Joaquin V., Cuyama V., interior cismontane San Bernardino Co. to San Diego region; to Gulf of Mex., Mex. April–June.

Var. **velutìna** (Woot.) Sarg. [*P. v.* Woot.] Pinnules oblong, 2–4 mm. wide, more pubescent.—San Diego R. above Grantland School and Atwood, Orange Co.; perhaps introd.; Ariz. to Tex. and nw. Mex.

2. **P. pubéscens** Benth. [*Strombocarpa p.* Gray.] SCREW-BEAN. TORNILLO. Shrub or small tree, to 10 m. high with pale slender twigs and stout spines ca. 1 cm. long; lvs. canescently pubescent, 2–6 cm. long, the lfts. oblong, 3–8 mm. long; fls. yellow, in cylindrical spikes 4–7 cm. long; legumes many, sessile, coiled with many turns into a tight springlike cylinder 2–4 cm. long; seeds tan, subovoid, 3 mm. long.—Washes and

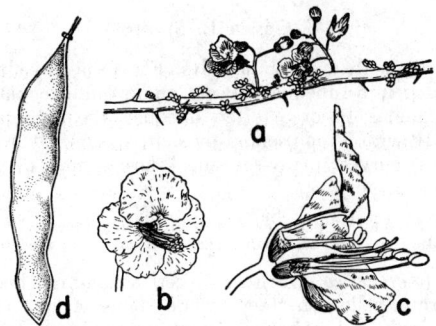

Fig. 72. LEGUMINOSAE-CAESALPINIOIDEAE. *Cercidium floridum:* a, habit, × ½; b, slightly irregular fl., × 1, the upper petal long-clawed; c, section through fl., × 2½, stamens 10, distinct; d, legume × ½.

canyons, below 2500 ft.; Creosote Bush Scrub; San Joaquin V. n. to w. Fresno Co.; Colo. and Mojave deserts; to Utah, Tex., Chihuahua, L. Calif. May–July.

3. **P. strombulífera** (Lam.) Benth. [*Mimosa s.* Lam.] Near to *P. pubescens,* but glabrescent, the fls. in globose heads; legumes twisted, 4–7 cm. long; $2n = 28$ (Schnack & Covas, 1947).—Native of Argentina and grown at Experiment Station at Bard, Imperial Co., from which it is reported as escaped.

II. Caesalpinioideae. Senna Subfamily (Fig. 72)

Lvs. simple; fls. rose-purple .... 5. *Cercis*
Lvs. compound; fls. ± yellowish.
   The lvs. once-pinnate; low subshrubs or herbs .... 6. *Cassia*
   The lvs. twice-pinnate.
      Plants low subshrubs or herbs without spines or thorns .... 9. *Hoffmannseggia*
      Plants large shrubs or trees with spines or thorns.
         Rachis of the pinnae flattened, more than 8 cm. long; lfts. alternate; fls. in racemes up to
         20 cm. long .... 7. *Parkinsonia*
         Rachis of the pinnae terete not more than 4 cm. long; lfts. opposite; fls. in short racemes
         or corymbs .... 8. *Cercidium*

5. Cércis L. Redbud. Judas Tree

Large shrubs to small trees with simple broad lvs. Fls. red-purple in short lateral fascicles, appearing before the lvs. in the spring. Calyx broadly campanulate, 5-toothed. Corolla irregular; petals 5, the standard enclosed by the wings in the bud; keel larger than wings. Stamens 10, distinct. Ovary short-stipitate, many-ovuled. Legumes oblong or linear-oblong, flat, the upper suture margined. Seeds compressed, obovate to rounded. Ca. a half-dozen spp. of N. Am. and Eurasia. (The ancient Greek name.)

(Hopkins, M. Cercis in N. Am. Rhodora 44: 193–211, 1942.)

1. **C. occidentàlis** Torr. ex Gray. [*Siliquastrum o.* Greene. *C. nephrophylla* Greene. *C. latissima* Greene.] Rounded or spreading shrub or small tree, 2–5 m. high, with clustered erect stems and glabrous twigs; lvs. round to reniform, glabrous, glossy, cordate at base with open to almost closed sinus, palmately 7–9-veined, entire, ± coriaceous, 3–9 cm. broad, retuse to emarginate or rounded at apex; petioles 1–3 cm. long; fls. magenta-pink to reddish-purple, occasionally white, 8–12 mm. long; pedicels 7–11 mm. long; pods many, oblong, flat, 4–9 cm. long, 2–2.5 cm. wide, attenuate or abruptly acute, pendent, often ± reddish-purple; seeds round, 3–4 mm. in diam.; $n = 7$ (Atchison, 1949).—Dry slopes and canyons in foothills, below 3500 (4500) ft.; Foothill Wd., Chaparral; inner Coast Ranges from Humboldt Co. to Solano Co., Sierran foothills from Shasta to Tulare cos., desert slopes of Laguna and Cuyamaca mts.; to Utah and Ariz. Feb.–April. Variable in lf.-texture and time of flowering.

6. Cássia L. Senna

Herbs, shrubs, or trees with even-pinnate lvs. Fls. usually yellow, in racemes. Calyx deeply toothed or divided into subequal lobes. Corolla quite regular; petals 5, spreading, imbricated, clawed. Stamens 10 or 5, often unequal, the anthers opening by 2 pores at the summit, in ours 10 with 3 represented by short sterile fils. Ovary sessile or stipitate; pods flat or terete, often curved, many-seeded. A large genus of warm temp. and trop. regions. (Ancient Greek name.)

Herbage nearly glabrous; lfts. in 1–4 pairs, each 4–8 mm. long; racemes terminal .... 1. *C. armata*
Herbage densely white-pubescent; lfts. in 3 pairs, each 1–3 cm. long; racemes axillary . 2. *C. Covesii*

1. **C. armàta** Wats. [*Xerocassia a.* Britt. & Rose.] Much-branched rounded shrub 5–15 dm. high, with numerous yellowish-green striate stems leafless much of the year; lf.-rachis 5–15 cm. long with 1–4 pairs of distant oblong-ovate lfts.; racemes 5–15 cm. long; fls. numerous, bright yellow; calyx ca. 6 mm. long; petals 8–10 mm. long, veiny; legume lance-cylindric, yellowish, slightly compressed, spongy, 2–4 cm. long, 5–7 mm. wide and thick, sparsely strigose to glabrescent; seeds grayish with a thin membrane that breaks away and reveals a brown surface, irregularly obovoid, 7–9 mm. long.— Common in sandy washes and open places, below 3700 ft.; Creosote Bush Scrub; Colo. and Mojave deserts; Ariz., Nev. April–May.

2. **C. Còvesii** Gray. [*Earlocassia C.* Britt. & Rose.] Low suffrutescent plant 3–5 dm. high, densely and finely whitish-pubescent throughout; lvs. 2–7 cm. long with 3 pairs of oblong-obovate lfts. 1–3 cm. long; racemes few-fld., calyx 6–7 mm. long; petals 1 cm. long, yellow, veiny; legume brownish, compressed, 2–5 cm. long, 6–8 mm. wide; seeds angled-wrinkled, irregular, grayish-brown, 3–3.5 mm. long.—Not common, dry washes below 2000 ft.; Creosote Bush Scrub; Colo. Desert; Ariz., L. Calif. April–June.

7. Parkinsònia L. Palo Verde

Low trees with green branches and twigs; spines present in pairs at each node and a spine terminating the primary lf.-rachis. Lvs. bipinnate; primary lfts. in 1–3 pairs, crowded, both the rachis and petiole almost obsolete; secondary lfts. many. Fls. in long racemes. Sepals reflexed. Banner yellow, spotted with red, turning red in age. Stamens 10, all functional; anthers versatile, splitting lengthwise. Legume bulging, strongly constricted between seeds. Two spp.; 1 Am., 1 Afr. (J. *Parkinson*, 1567–1650, English herbalist.)

1. **P. aculeàta** L. Horse-Bean. Mexican Palo Verde. To 10 m. high; pinnae with flattened rachis to ca. 2 dm. long; lfts. scattered, 4–10 mm. long; petals 10–15 mm. long; legumes 5–10 cm. long, short-stiped, sharp-pointed; $n = 14$ (Pantulu, 1942).— Common in cult., especially in desert and sometimes natur.; native from Ariz. to Fla., W.I., Mex. to S. Am. Summer.

8. Cercídium Tulasne. Palo Verde

Shrubs or small trees with green bark and ± spinose twigs. Lvs. bipinnate, the pinnae with terete rachis and clearly borne below the spines. Infl. a short axillary corymb. Calyx-lobes short, valvate. Petals 5, yellow, clawed. Stamens 10, distinct, hairy at base of fils. Legume linear to oblong, flattened or cylindric, several-seeded. An Am. genus of ca. 10 spp. (Greek, *kerkidion*, a weaver's shuttle, referring to fr.)

Pinnae with 1–3 pairs of lfts., glabrous; pods flattish .... 1. *C. floridum*
Pinnae with 4–8 pairs of lfts., pubescent; pods cylindric .... 2. *C. microphyllum*

1. **C. flóridum** Benth. [*Parkinsonia f.* Wats. *P. Torreyana* Wats. *C. T.* Sarg.] Tree up to 10 m. high with short trunk and smooth blue-green bark, leafless most of year; branchlets slender, glaucous, frequently spiny; lvs. few, scattered, 1–2 cm. long, with 1 pair of pinnae; fls. on slender pedicels; calyx ca. 6–7 mm. long; petals yellow, 8–10 mm. long; pod glabrous, 1–8-seeded, 4–8 cm. long, often somewhat constricted between seeds; seeds oblong-obovoid, strongly compressed, olive and brown, 8–10 mm. long, 6–7 mm. broad, smooth.—Washes and low sandy places, below 1200 ft.; Creosote Bush

Scrub; Colo. Desert; Ariz., Son., L. Calif. March–May. Occasionally hybridizing with the next sp.

2. **C. microphýllum** (Torr.) Rose & Jtn. [*Parkinsonia m.* Torr.] Spiny shrub or tree to 6 m. high with yellowish-green bark; lvs. pubescent, the pinnae with 4–8 pairs of minute lfts.; petals pale yellow, 5–6 mm. long; pod 4–8 cm. long, almost 1 cm. wide, turgid, puberulous, 1–4-seeded; seeds brown, oblong-obovoid, scarcely compressed, 7–8 mm. long.—Gravelly slopes and washes, below 1200 ft.; Creosote Bush Scrub; Whipple Mts., e. San Bernardino Co.; Ariz., Son., L. Calif. April–May.

9. Hoffmannséggia Cav.

Herbs or small shrubs. Lvs. bipinnate, glandular-dotted, with small stipules and lfts. Fls. and fr. in glandular racemes. Calyx 5-parted. Petals 5, subequal, yellow. Stamens 10, distinct; fils. often glandular below. Legume flat, several-seeded, linear to ovate, 2-valved. Ca. 20 spp., Am. and Afr. (Count of *Hoffmannsegg*, coauthor of a flora of Portugal.)

Herb; lvs. with several pairs of pinnae; fls. orange .... 1. *H. densiflora*
Shrub; lvs. with 1 pair of lateral pinnae and the terminal pinna; fls. yellow ...... 2. *H. microphylla*

1. **H. densiflòra** Benth. ex Gray. [*H. stricta* Benth. *H. Falcaria* of Calif. refs., probably not Cav.] Stems puberulent, 1–4 dm. tall, several, scattered, from underground root-stocks; lvs. several, often near the base, 6–12 cm. long, with 5–11 pinnae, each with 6–10 pairs of crowded oblong pubescent pinnules 2–6 mm. long; peduncle 3–7 cm. long with equally long raceme, stipitate-glandular and pubescent; calyx 6–8 mm. long; corolla ca. 1 cm. long, the petals glandular on claws and lower margins; stamens puberulent, the alternate glandular; legume compressed, 2–3 cm. long, ± glandular.—Heavy alkaline soil, below 2600 ft.; Creosote Bush Scrub; Mojave and Colo. deserts; cismontane Calif., probably as an adventive, especially along railroads; to Tex., Mex. April–June.

2. **H. microphýlla** Torr. Broad rounded subshrub usually 6–10 dm. tall and with many rushlike stems, sometimes to 2 m. high, pubescent; lvs. scattered, 2–3 cm. long, with 3 pinnae, the terminal with 10–14 pinnules, the 2 lateral each with 8–12; pinnules oblong, 3–5 mm. long; racemes many, slender, terminal and lateral, 8–20 or more cm. long, the fls. 6–7 mm. long; stamens exserted, the fils. glandular-stipitate; legumes lunate, 1.5–2.5 cm. long, with many stipitate glands; seeds brown, shining, somewhat compressed, obovoid, ca. 3 mm. long.—Common about canyons and washes, below 4000 ft.; Creosote Bush Scrub; Colo. Desert; Ariz., Son., L. Calif. March–May.

III. Papilionoideae. BEAN SUBFAMILY (Fig. 73)

KEY TO GENERA

A. Stamens distinct to very base or nearly so.
  B. Fls. yellow; plant an herb .... 10. *Thermopsis*
  BB. Fls. purplish; plant a low shrub .... 11. *Pickeringia*
AA. Stamens all united or 9 or 5 of them united.
  B. The lvs. simple; shrubs, mostly spiny.
    C. Corolla yellow, 15–25 mm. long.
      D. Calyx deeply 2-lipped; branches spinescent; lvs. acicular .... 13. *Ulex*
      DD. Calyx cleft above, hence 1-lipped; branches not spinescent; lvs. ± oblong
            15. *Spartium*
    CC. Corolla blue to red-purple, 8–10 mm. long.
      D. Herbage ± gland-dotted; stamens monadelphous; fls. in terminal spikes or
         racemes .... 22. *Dalea*
      DD. Herbage not gland-dotted; stamens diadelphous; fls. in axillary racemes . 32. *Alhagi*
  BB. The lvs. 3–many-foliolate.
    C. Lvs. palmate or pinnately 1–3-foliolate.
      D. Anthers strongly differentiated, some small and versatile, others larger and
         basifixed; stamens monadelphous and 10 in number; pods dehiscent.
        E. Lfts. 4–many. Native herbs or shrubs .... 12. *Lupinus* √
        EE. Lfts. 3 or fewer. Introd. shrubs .... 14. *Cytisus*
      DD. Anthers not much differentiated, or if they are, the plants with prickly pods;
         stamens 9 and 1, or 5 and monadelphous; pods largely indehiscent.
        E. Plants gland-dotted; lfts. 3–5 .... 20. *Psoralea*

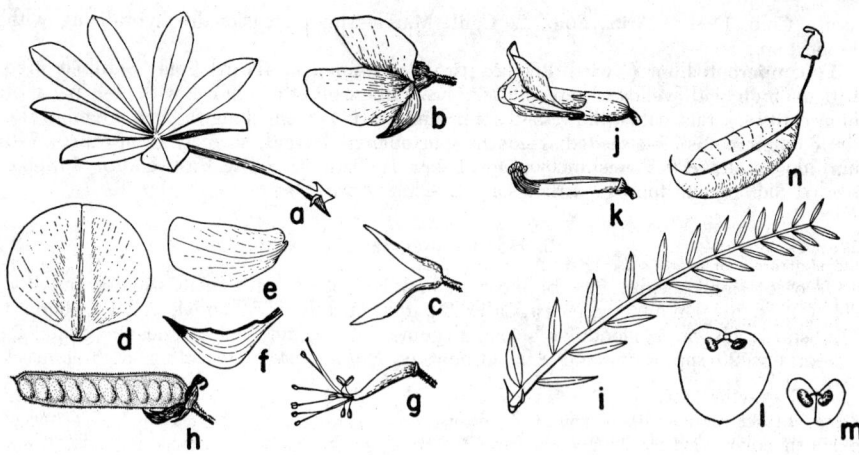

Fig. 73. LEGUMINOSAE-PAPILIONOIDEAE. *Lupinus longifolius:* a, lf., × ½, palmately compound; b, papilionaceous fl. from side, × 1; c, calyx bilabiate; d, banner; e, wing; f, keel of 2 united petals; g, monadelphous stamens; h, legume, × ½. *Astragalus Menziesii:* i, pinnately compound lf.; j, fl., × 1½, with gamosepalous calyx, papilionaceous corolla; k, diadelphous stamens (9 + 1); l, cross section of inflated legume; n, stipitate legume. *Astragalus Bolanderi:* m, legume in cross section to show apparent 2-loculed condition.

10. Thermópsis R. Br. FALSE-LUPINE

Perennial herbs, stout, with sheathing scales at base and mostly erect clustered stems. Lvs. palmately 3-foliolate, petioled, with free foliaceous stipules. Fls. yellow in ours, in a terminal raceme, with persistent bracts. Calyx campanulate, the teeth equal or the 2 upper united. Banner suborbicular, shorter than to ca. as long as the oblong wings; keel almost straight. Stamens 10, distinct. Legume narrow, flat, 2-valved, few- to many-seeded. Perhaps 20 spp., N. Am., Asia. (Greek, *thermos*, lupine and *opsis*, likeness.)

(Larisey, Mary M. A revision of the N. Am. species of the genus Thermopsis. Ann. Mo. Bot. Gard. 27: 245–258, 1940.)

Stems glabrous or nearly so; calyx-teeth broadly triangular .... 1. *T. gracilis*
Stems silky to villous; calyx teeth ± lanceolate.
   Pubescent silky, the hairs straight and appressed; calyx-teeth lance-subulate, ca. as long as the tube; stipules mostly lanceolate. Modoc and Shasta cos. .... 2. *T. argentata*
   Pubescence subtomentose, the hairs ± curled and divergent; calyx-teeth lanceolate, shorter than the tube; stipules ovate to lance-ovate. Coast Ranges .... 3. *T. macrophylla*

1. **T. grácilis** Howell. [*T. venosa* Eastw. *T. g.* var. *v.* Jeps. *T. montana* var. *v.* Jeps.] Stems rather slender, 4–8 dm. high, glabrous or subglabrous, ± glaucous over a dark purplish color, slightly branched above; stipules lanceolate to cordate-ovate, 1–2.5 cm. long; petioles 2–3 cm. long; lfts. oblanceolate to obovate, 2–5 cm. long, glabrous above, sometimes somewhat pubescent beneath; raceme rather loosely fld., 1–1.5 dm. long; bracts broadly ovate, 0.6–1.2 cm. long; pedicels 4–10 mm. long; calyx silky-villous, 6–7 mm. long; standard 10–14 mm. long, wings and keel 16–18 mm. long; pods villous, divaricate, 4–7 cm. long.—Dry to moist ± stony flats, 1000–5000 ft.; Mixed Evergreen F., Yellow Pine F.; Del Norte Co. to Trinity and Shasta cos.; to B.C., Mont. April–June.

2. **T. argentàta** Greene. [*T. gracilis* var. *a.* Jeps. *T. macrophylla* var. *a.* Jeps.] Stems rather slender, 3–6 dm. high, somewhat branched above, the whole plant silvery-canescent with a dense short silky pubescence; stipules narrowly to broadly lanceolate, 1–3 cm. long; lfts. cuneate-obovate, mostly 2–4 cm. long; racemes 6–12 cm. long; bracts 4–10 mm. long; calyx 5–7 mm. long; standard 10–12 mm. long, wings and keel 15–18 mm. long; pods erect to divaricate, 4–5 cm. long, silky-pubescent; seeds smooth, light brown, ca. 4 mm. long.—Damp places, 3000–5000 ft.; Yellow Pine F.; e. Shasta and sw. Modoc cos. May–June.

3. **T. macrophýlla** H. & A. [*T. californica* Wats. *T. robusta* Howell.] Stems stout, 3–8 dm. high, the whole plant ± short villous-tomentose with somewhat spreading hairs; stipules ovate-cordate, 2–3.5 cm. long; petioles 1.5–2.5 cm. long; lfts. obovate-cuneate to oblanceolate, 4–7 cm. long, 3–4 cm. broad; racemes 1.5–2.5 dm. long, compact to lax; bracts 0.5–1 cm. long; pedicels 4–6 mm. long; calyx appressed-pubescent, 5–7 mm. long; fls. 17–19 mm. long, the standard almost as long as wings and keel; pods densely appressed-pubescent, erect, 6–8 cm. long, few-seeded.—Open places, below 4500 ft.; Mixed Evergreen F., Foothill Wd., etc.; Coast Ranges from Ventura Co. to Del Norte Co.; s. Ore. April–June. Much of the material from s. of San Francisco is somewhat more silky than the other and has been called var. *velùtina* (Greene) Larisey [*T. v.* Greene.]

Var. **semòta** Jeps. Herbage densely velvety, almost shaggy.—Grassy slopes and benches, 4000–5500 ft.; Yellow Pine F.; Laguna Mts., Cuyamaca Mts., Pine Hills etc., San Diego Co. May–June.

11. Pickeríngia Nutt. CHAPARRAL-PEA

Spiny evergreen shrub with stiff branches. Lvs. small, palmately 1–3-foliolate, almost sessile, without stipules. Fls. large, purple, subsessile, solitary. Calyx campanulate, the border with 5 low broad teeth. Corolla with equal petals; standard roundish with reflexed sides, wings and keel-petals oblong, the latter distinct and straight. Stamens distinct. Legume linear, flat, stipitate, straight, several-seeded. Monotypic.

1. **P. montàna** Nutt. [*Xylothermia m.* Greene.] Densely and irregularly branched, 0.5–2 m. tall, the branchlets spinescent, greenish, pubescent to subglabrous; lfts. obovate to oblanceolate, entire, firm, subglabrous, 4–12 mm. long; calyx ca. 6 mm. long; petals 16–19 mm. long; legume 3–5 cm. long, somewhat constricted between the seeds; seeds

black, oblong, ca. 3.5 mm. long.—Dry slopes and ridges, below 5000 ft.; Chaparral, Mixed Evergreen F.; Coast Ranges from Santa Monica Mts. to Mendocino Co.; foothills of Sierra Nevada from Butte Co. to Nevada Co.; Santa Cruz Id. May–Aug. Rarely fruiting, but often spreading by underground stems especially after fire.

Ssp. **tomentòsa** (Abrams) Abrams. [*Xylothermia m.* ssp. *t.* Abrams.] Young twigs and lvs. canescent.—San Bernardino Mts. to mts. of e. San Diego Co.

12. Lupìnus L. LUPINE

Annual or perennial herbs or shrubs. Lvs. alternate, palmately compound, rarely unifoliolate. Stipules narrow. Fls. perfect, in terminal racemes. Calyx bilabiate, the lips entire or toothed, or the upper sometimes bifid, often with interstitial bracteoles. Corolla usually bluish, sometimes reddish, whitish, or yellow; banner commonly with reflexed sides, the back glabrous or ± pubescent; wings mostly glabrous; keel curved to almost straight, glabrous or ciliate on upper edges, sometimes on lower. Stamens 10, monadelphous, the anthers alternately of 2 forms, elongate and short. Pods flat, 2–12-ovuled. Seeds with a sunken hilum, which is thus often surrounded by a thickened ring. (Latin, from *lupus*, a wolf, because of an old idea that lupines rob the soil.)

(Smith, C. Piper. Treatment *in* Abrams: Ill. Fl. Pacific States 2: 483–519, 1944. Dunn, David B. Taxonomy of Lupinus, Group Micranthi etc. El Aliso 3: 135–171, 1955.)
A. Plants annual.
   B. Fls. definitely whorled or mostly so.
      C. Keel ciliate on upper margins near the claws.
         D. The keel ciliate on lower edges also, near the claws.
            E. Plants 0.5–1.5 dm. tall; peduncles 2–4 cm. long.
               F. Fls. ascending to suberect at anthesis, 10–11 mm. long. San Bernardino Co. n. . . . . 1. *L. horizontalis*
               FF. Fls. ± spreading, 6–7 mm. long. Riverside Co. s.
                          2. *L. concinnus* var. *brevior*
            EE. Plants 2–10 dm. tall; fls. spreading in anthesis, 12–14 mm. long; peduncles mostly 3–15 cm. long.
               F. Fls. yellowish; pods ca. 1.5 cm. long . . . . 5. *L. luteolus*
               FF. Fls. deep purple-blue, sometimes pink or white; pods 4–5 cm. long
                          14. *L. succulentus*
         DD. The keel not ciliate on lower edges.
            E. Banner 8–9 mm. long, 2–6 mm. wide; fls. suberect at anthesis . . 2. *L. ruber*
            EE. Banner 12–16 mm. long, 8–9 mm. wide; fls. spreading at anthesis.
               F. Fls. ± erect after anthesis, not secund, purple-red to rose-pink
                          3. *L. subvexus*
               FF. Fls. spreading or secund after anthesis, white or yellowish, sometimes violet or rose . . . . 4. *L. densiflorus*
     CC. Keel ciliate on upper margins toward the apex or not ciliate at all.
         D. Pedicels slender, 4–10 mm. long; fls. 8–15 mm. long; banner mostly broader than long.
            E. Plants pubescent or villous with hairs less than 2 mm. long; lvs. cauline as well as basal.
               F. Fls. 10–15 mm. long.
                 G. Keel entire, not toothed; spreading hairs generally many; pods 5.5–6.5 mm. wide . . . . 22. *L. nanus*
                 GG. Keel with a distinct tooth on the upper margin near the middle; spreading hairs generally few; pods 7.5–8.5 mm. wide
                          25. *L. affinis*
               FF. Fls. 6–10 mm. long.
                 G. Reflexed portion of banner not shorter than the appressed portion, the tip reflexed more than 3 mm. from the tip of the wings. Inner S. Coast Ranges   22. *L. nanus*
                 GG. Reflexed portion of banner shorter than the appressed portion, the tip reflexed 2–3 mm. from the wing-tips. Mostly Sierran
                          23. *L. vallicola*
            EE. Plants villous with hairs 2–4 mm. long; pods 6–7 mm. wide; lvs. mostly basal; seeds 6–10. Sierran foothills . . . . 24. *L. spectabilis*
         DD. Pedicels 1–3 mm. long; fls. 4–8 (–12) mm. long.
            E. Keel ciliate on upper edges toward apex.
               F. Banner roundish to obovate, its reflexed portion generally as long as the straight part; keel slender with long narrow ± upturned beak
                          27. *L. bicolor*
               FF. Banner cuneate or spatulate, its reflexed portion much shorter than the straight part; keel short, broad, almost straight.
                 G. Plants mostly 1–4 dm. tall; lfts. linear to oblanceolate, 1.5–4 cm. long; fls. mostly whorled . . . . 28. *L. micranthus*

GG. Plants to ca. 1 dm. tall; lfts. spatulate, 0.4–1 cm. long; fls. mostly
scattered .... 29. *L. Congdonii*
  EE. Keel essentially glabrous.
    F. Pods 6–9 mm. wide; seeds 4–5 mm. long; banner 7–8.5 mm. long
26. *L. pachylobus*
    FF. Pods 3–5 mm. wide; seeds 2–3 mm. long; banner 5.2–6.5 mm. long
27. *L. bicolor* ssp. *Pipersmithii*
BB. Fls. not at all whorled.
  C. Keel not ciliate.
    D. Cotyledons sessile and perfoliate; ovules usually 2 except in *L. odoratus.*
      E. Peduncles 1-fld.; plants 1–2 cm. high; lfts. ca. 5 mm. long . 7. *L. uncialis*
      EE. Peduncles several- to many-fld.; plants taller; lfts. longer.
        F. Fls. mostly 5–8 mm. long; ovules 2.
          G. Pods constricted in middle; peduncles ca. 1 cm. long, the racemes
not surpassing the foliage. Inyo Co. to Modoc Co. .. 11. *L. pusillus*
          GG. Pods scarcely or not constricted; peduncles 3–10 cm. long, the
racemes surpassing the lvs. Inyo Co. s.
            H. Racemes usually subcapitate, 1–2.5 cm. long; pods villous on
sides; banner longer than wide .... 6. *L. brevicaulis*
            HH. Racemes mostly 3–8 cm. long.
              I. Pods loosely villous on sides; banner suborbicular
9. *L. flavoculatus*
              II. Pods scaly on sides; banner longer than wide
8. *L. Shockleyi*
        FF. Fls. ca. 10 mm. long; ovules 2–6 .... 10. *L. odoratus*
    DD. Cotyledons petioled; ovules 2–4 .... 12. *L. concinnus*
  CC. Keel ciliate at least on lower margin near the base.
    D. The plant covered with long stinging hairs; fls. reddish-violet, 13–15 mm. long;
lfts. 12–25 mm. broad .... 15. *L. hirsutissimus*
    DD. The plant not covered with stinging hairs; lfts. 2–12 mm. broad.
      E. Banner bright yellow, the wings rose to purple; keel whitish; racemes
shorter than the peduncles .... 13. *L. Stiversii*
      EE. Banner and other petals not so colored, or if so, with short peduncles.
        F. Keel stout with short blunt acumen; plant subglabrous to sparsely
strigulose .... 16. *L. truncatus*
        FF. Keel with slender acute acumen; plant usually ± villous.
          G. Mature pods ascending; petals blue, lilac or purple.
            H. Lfts. linear-filiform, 2–5 cm. long; bracts much exceeding the
fl.-buds. Cent. Calif. .... 17. *L. Benthamii*
            HH. Lfts. linear to oblanceolate, 1–3 cm. long; bracts scarcely ex-
ceeding the fl.-buds. S. Calif.
              I. Fls. blue to lilac, 10–17 mm. long; lfts. mostly 1.5–3 mm.
wide. Cismontane and deserts.
                J. Pedicels mostly 3–5 mm. long; fls. 10–12 mm. long;
pods 1–2 cm. long. Widely distributed
18. *L. sparsiflorus*
                JJ. Pedicels 6–12 mm. long; fls. 14–16 mm. long; pods
5–6 cm. long. San Clemente Id. .. 19. *L. Moranii*
              II. Fls. pale purplish-pink, 8–10 mm. long; lfts. mostly
4–12 mm. wide. Deserts .... 20. *L. arizonicus*
          GG. Mature pods deflexed; petals orange or golden to white or pinkish.
Sierran foothills. .... 21. *L. citrinus*
AA. Plants perennial.
  B. The plants shrubby or with definite woody stems of some length above the roots.
    C. The banner ± pubescent on the back; plants mostly appressed-silky; keel narrow with
long slender acumen.
      D. Keel not ciliate on the edges or with only an occasional hair; petioles 2–3.5 cm.
long. Immediate coastal dunes and beaches .... 60. *L. Chamissonis*
      DD. Keel definitely ciliate along upper edge toward acumen; petioles 2–8 cm. long.
Away from the immediate coast.
        E. The keel narrowed toward the base. Mostly cent. and n. Calif.
          F. Plants appressed-silky; fls. 10–13 mm. long, or if more, plant strigose
62. *L. albifrons*
          FF. Plants white-woolly-villous; fls. 14–16 mm. long .. 55. *L. Abramsii*
        EE. The keel not narrowed toward base. Mostly s. Calif. .... 63. *L. excubitus*
    CC. The banner glabrous on the back; plants appressed-silky to tomentose or spreading-
villous; keel mostly not markedly slender at the acumen.
      D. Stems erect or nearly so.
        E. Herbage densely tomentose and villous; plants 3–5 dm. high; petioles 6–10
cm. long. San Luis Obispo Co. .... 59. *L. ludovicianus*
        EE. Herbage pubescent to silky or villous, not tomentose; plants mostly 5–25 dm.
high; petioles various.
          F. Herbage often ± silvery; racemes mostly 1–2 dm. long above the
peduncles.
            G. Petioles 2–6 cm. long; petals mostly yellow, sometimes lavender or
blue. Along immediate coast from Ventura Co. n.
61. *L. arboreus*

GG. Petioles 4–10 cm. long; petals blue to violet or orchid. Plants of
 the s. interior .... 63. *L. excubitus*
 FF. Herbage greenish, even though sometimes appressed-pubescent; racemes
  mostly 2–4 dm. long above the peduncles.
  G. Lvs. strigose above; petioles 4–7 cm. long. Ventura Co. to L. Calif.
                    64. *L. longifolius*
  GG. Lvs. glabrous above; petioles 3–5 cm. long. Coast from Mendocino
    Co. n. .... 77. *L. rivularis*
DD. Stems mostly decumbent or prostrate. Immediate coast.
 E. Roots not bright yellow; petioles 4–10 cm. long; pedicels 4–12 mm. long;
  seeds 7–9. Humboldt Co. to San Luis Obispo Co. .... 65. *L. variicolor*
 EE. Roots bright yellow; petioles mostly 2–5 cm. long; pedicels 3–6 mm.
  long; seeds 9–14. Medocino Co. to Del Norte Co. .... 66. *L. littoralis*
BB. The plants strictly herbaceous, i.e. stems not woody above the root-crown or caudex.
 C. Lvs. glabrous on upper surface or essentially so.
  D. Keel naked on upper edges.
   E. Stems 5–15 dm. tall, stout, fistulose; corolla 12–14 mm. long. Plants of wet
    places .... 78. *L. polyphyllus*
   EE. Stems 4–7 dm. tall, more slender; corolla 10 mm. long .. 79. *L. Tracyi*
  DD. Keel ciliate on upper edges (sometimes very sparsely so as in *L. onustus*)
   E. Plants 2–3 dm. high.
    F. Plants appressed-silky; lvs. mostly near the base, with petioles 8–13 cm.
     long; lfts. mostly 2–4 cm. long. Dry slopes, Plumas to Del Norte cos.
                      49. *L. onustus*
    FF. Plants spreading-villous; lvs. well distributed, with petioles 5–10 cm.
     long; lfts. 1–3 cm. long. Modoc Co. .... 80. *L. saxosus*
   EE. Plants mostly taller, and lvs. more evenly distributed.
    F. Lfts. 1.5–4.5 cm. long, 2–5 mm. wide; stems 3–6 dm. tall; banner ±
     pubescent on back. Dry places.
     G. Cauline petioles 2–5 cm. long; calyx short-spurred; racemes 7–12
      cm. long; fls. bluish. Modoc to Siskiyou and Butte cos.
                  45. *L. argenteus*
     GG. Cauline petioles 1–2 cm. long; calyx not spurred; racemes 15–20
      cm. long; fls. yellow. Shasta Co. . 71. *L. Andersonii* var. *Christinae*
    FF. Lfts. 3–10 cm. long, 5–30 mm. wide; stems mostly 4–20 dm. tall;
     banner glabrous. Mostly from moist places.
     G. Plants mostly subglabrous or with appressed hairs. Widely dis-
      t·ibuted   ... 76. *L. latifolius*
     GG. Plants ± villous, especially about the nodes.
      H. Fls. 8–10 mm. long. Modoc Co.
              76. *L. latifolius* var. *barbatus*
      HH. Fls. 12–16 mm. long. Coast, Mendocino Co. to Del Norte Co.
                77. *L. rivularis*
CC. Lvs. ± hairy above.
 D. Banner glabrous on back.
  E. Lvs. mostly in dense basal clusters, the cauline few if any; plants cespitose.
   F. Stems prostrate or nearly so; racemes subcapitate, rarely more than twice
    as long as thick.
    G. Pubescence appressed, the longest hairs few, to 1 mm. long;
     racemes 1–2(–4) cm. long .... 30. *L. Lyallii*
    GG. Pubescence with some longer hairs spreading and 1.5–2.5 mm.
     long, among the appressed ones; racemes 2–6 cm. long
                 31. *L. Lobbii*
   FF. Stems erect or ascending; racemes ± elongate, usually more than twice
    as long as thick.
    G. Racemes not exceeding lvs., but mostly down among them; fls. pale,
     7–8 mm. long. White Mts. .... 33. *L. caespitosus*
    GG. Racemes extending well above the lvs.
     H. The racemes 2–4 cm. long, of 2–3 whorls of fls.; fls. 10–12
      long, purplish-blue. Siskiyou Co. ...... 53. *L. lapidicola*
     HH. The racemes mostly longer, with more whorls.
      I. Fls. 8–10 mm. long.
       J. Plants to ca. 12 cm. high; lfts. almost half as wide as
        long. Volcanic ash, Mono Co. .... 40. *L. Duranii*
       JJ. Plants mostly taller; lfts. definitely narrower. Granitic
        soils, Mariposa Co. to Shasta Co., and N. Coast Ranges
                32. *L. sellulus*
      II. Fls. 10–14 mm. long.
       J. Stems several from a stout root; bracts 8–9 mm. long,
        persistent; banner elliptic-obovate. Dampish places
                34. *L. confertus*
       JJ. Stems from a branched woody caudex; bracts 4–5 mm.
        long, deciduous; banner round. Dry slopes
                58. *L. Grayi*
  EE. Lvs. well distributed along the stems.
   F. Keel ciliate on upper edges at least toward tip.

G. Bracts of infl. quite persistent, usually linear, mostly rather conspicuous, especially before anthesis.
  H. Lfts. 3–8 cm. long; bracts 7–15 mm. long. S. Sierra Nevada.
    I. Fls. 10–12 mm. long; plants strigose, ± hollow-stemmed;
bracts 7–9 mm. long . . . . 36. *L. pratensis*
    II. Fls. 8–9 mm. long; plants spreading-villous, not hollow-
stemmed; bracts 8–15 mm. long . . . . . . 38. *L. Covillei*
  HH. Lfts. mostly less than 3 cm. long.
    I. Fls. 11–14 mm. long; bracts 8–10 mm. long.
      J. Stems erect; plant white-silky; racemes 5–30 cm.
long. Below 8500 ft. . . . . 34. *L. confertus*
      JJ. Stems decumbent; plant rather loosely villous; racemes
5–10 cm. long. Near timber line . . 54. *L. montigenus*
    II. Fls. 7–10 mm. long; bracts 3–5 mm. long. S. Sierra
Nevada.
      J. Lfts. mostly 1.2–3 cm. long; fls. ca. 10 mm. long
35. *L. Culbertsonii*
      JJ. Lfts. mostly 0.6–1.2 cm. long; fls. 7–8 mm. long
37. *L. hypolasius*
GG. Bracts of infl. early deciduous, mostly rather inconspicuous.
  H. Lvs. densely woolly with ± curly and interwoven hairs, the
lfts. mostly 7–10 mm. wide.
    I. Plants mostly 3–5 dm. high; petioles 6–12 cm. long;
peduncles 8–15 cm. long.
      J. Fls. 10–12 mm. long; lfts. cuneate or spatulate.
W. San Luis Obispo Co. . . . . 59. *L. ludovicianus*
      JJ. Fls. 12–14 mm. long; lfts. oblanceolate. Plumas and
Butte cos. to Kern Co. . . . . 56. *L. Grayi*
    II. Plants mostly 5–12 dm. tall; petioles 10–20 cm. long;
peduncles 10–30 cm. long; fls. 10–18 mm. long. Inyo Co.
82. *L. magnificus*
  HH. Lvs. not woolly, but silky-strigose to strigulose.
    I. Fls. 5–6 mm. long; petioles 2–3.5 cm. long. Sierra
Nevada from Madera and Mono cos. to Plumas Co.
42. *L. meionanthus*
    II. Fls. 10–18 mm. long; petioles mostly longer.
      J. Lfts. 4–8 cm. long, 2–4 mm. wide; bracts 8–10 mm.
long. Yosemite National Park region
68. *L. gracilentus*
      JJ. Lfts. mostly 1.5–3.5 cm. long; bracts mostly shorter.
        K. Plants ± sprawling with decumbent or prostrate
stems. Seacoast.
          L. Roots not bright yellow; most petioles ca.
twice as long as their lfts. . 65. *L. variicolor*
          LL. Roots bright yellow; most petioles less than
twice as long as their lfts.
            M. Lfts. mostly 2–3.5 cm. long; petioles
3–5 cm. long; stems usually with some
spreading hairs. Mendocino Co. n.
66. *L. littoralis*
            MM. Lfts. mostly 1.3–2 cm. long; petioles
mostly 1–3 cm. long; stems strigose.
Marin and Monterey cos.
67. *L. Tidestromii*
        KK. Plants with ascending to erect stems. High
montane.
          L. Fls. 10–12 mm. long; plant long-villous. Ne.
Mono Co. . . . . 39. *L. nevadensis*
          LL. Fls. 13–18 mm. long; plant silky-strigose.
            M. Lvs. silvery, crowded near base of
plant; petioles 4–10 cm. long; fls. 14–
18 mm. long. Kern Co. to L. Calif.
63. *L. excubitus* var. *austromontanus*
            MM. Lvs. greenish, silky, not silvery, well
distributed; petioles to 15 cm. long;
fls. 13–15 mm. long. E. Inyo Co.
81. *L. Holmgrenanus*
FF. Keel glabrous on upper edges or practically so.
  G. Plants less than 2 dm. high, matted, with ± prostrate stems from
a woody base; fls. 4–9 mm. long. Montane . . . . . . 41. *L. Breweri*
  GG. Plants normally taller, erect or ascending; fls. mostly larger.
    H. Petioles 1–2.5 dm. long; lfts. ca. 12, from 1–1.3 dm. long;
bracts ca. 10 mm. long. N. Calif.
78. *L. polyphyllus* var. *pallidipes*
    HH. Petioles less than 1 dm. long; lfts. fewer, less than 1 dm.
long; bracts 2–7 mm. long.

I. Banner narrow, acute at apex; wings narrow, the keel
    much exposed. Sierra Nevada from Fresno Co. n., Coast
    Ranges from Trinity Co. n. . . . . 69. *L. albicaulis*
II. Banner ovate to round, mostly obtuse or rounded at apex;
    wings wider, largely covering the keel.
    J. Stipules green and leaflike, lanceolate to oval. Sierra
       Nevada from Eldorado Co. to Fresno Co.
                                      72. *L. fulcratus*
    JJ. Stipules linear to subulate, not leaflike.
       K. Fls. yellow.
         L. The fls. 12–15 mm. long; banner round.
            Siskiyou and Trinity cos. . . 70. *L. croceus*
         LL. The fls. 10–12 mm. long, the banner obovate.
            Del Norte Co. to Santa Clara Co., San
            Diego Co. . . . . 74. *L. adsurgens*
       KK. Fls. blue or violet to yellowish-white.
         L. Lfts. thinly strigose above, greenish.
           M. Fls. 10–12 mm. long. San Bernardino
              Mts. to Sierra Nevada and N. Coast
              Ranges . . . . 71. *L. Andersonii*
           MM. Fls. 13–16 mm. long. San Gabriel Mts.
              to San Jacinto Mts.
                75. *L. formosus* var. *hyacinthinus*
         LL. Lfts. ± silky-strigose above.
           M. The lfts. silvery-silky; fls. 10–14 mm.
              long. Mt. Pinos and San Gabriel Mts.
                    73. *L. elatus*
           MM. The lfts. more loosely silky, less shiny.
             N. Fls. 10–14 mm. long, the banner
               obovate; racemes mostly 4–10 cm.
               long. Coast Ranges from Santa
               Clara n., San Diego Co., Sierra
               Nevada . . . . . . 74. *L. adsurgens*
            NN. Fls. 12–18 mm. long, the ban-
              ner roundish; racemes mostly 10–
              25 cm. long. Humboldt and Butte
              cos. to L. Calif. . 75. *L. formosus*
DD. Banner pubescent on the back, the pubescence sometimes almost concealed by
    the upper lip of the calyx.
    E. Calyx evidently spurred at the base just above the pedicel.
       F. Wing-petals with a patch of short dense pubescence on outer surface
         near upper distal corner . . . . 44. *L. arbustus*
       FF. Wing-petals not pubescent or with a few scattered hairs.
         G. Pubescence appressed, the plants silky-strigose; fls. 10–13 mm.
           long. Inyo to Modoc and Shasta cos. . . . . 46. *L. caudatus*
         GG. Pubescence spreading, or if appressed, the fls. smaller.
           H. Fls. ca. 12 mm. long. E. base of Sierra Nevada
                        48. *L. inyoensis*
          HH. Fls. 8–10 mm. long. Inyo-White Mts. and e. Sierra Nevada
                      47. *L. Palmeri*
    EE. Calyx not spurred, sometimes ± gibbous at base.
       F. Keel naked or scarcely ciliate on upper edges.
         G. Lfts. essentially green, although strigose. Alpine, Placer, and
           Nevada cos. . . . . 71. *L. Andersonii* var. *apertus*
         GG. Lfts. silvery-silky. Mt. Pinos and San Gabriel Mts. . . 73. *L. elatus*
       FF. Keel ciliate along at least the upper margins.
         G. Fls. crowded in very dense racemes. Plants of wet places. Inyo Co.
                   36. *L. pratensis* var. *eriostachyus*
         GG. Fls. in open racemes. Plants of dryish places.
           H. Plants matted, less than 2 dm. tall, with silvery-silky foliage;
            fls. 6–9 mm. long . . . . 41. *L. Breweri*
          HH. Plants not matted, taller; fls. mostly larger.
            I. Lfts. 10–30 mm. wide (cf. also *L. leucophyllus*).
              J. Fls. yellow, 10–12 mm. long; pedicels 1–1.5 mm.
                long. W. Mojave Desert . . . . 50. *L. Peirsonii*
              JJ. Fls. bluish, 14–16 mm. long; pedicels 3–5 mm. long.
                K. Stems and petioles with appressed hairs; fls. pur-
                   plish-blue. Lake and Sonoma cos.
                        56. *L. sericatus*
                KK. Stems and petioles with spreading hairs; fls. light
                   blue. Santa Lucia Mts. . . . . . . 57. *L. cervinus*
          II. Lfts. 3–10 mm. wide, sometimes wider in *L. leucophyllus*.
           J. Plants woolly-villous with ± interwoven hairs.
              K. Plants 5–9 dm. high, if lower, the lfts. 3–5 mm
              wide.
                L. Fls. 10–12 mm. long; lfts. 3–6 cm. long;

pedicels 1–3 mm. long. Colusa and Trinity cos. to Modoc and Siskiyou cos.

51. *L. leucophyllus*

LL. Fls. 15–16 mm. long; lfts. 1–3 cm. long; pedicels 4–7 mm. long. Santa Lucia Mts.

55. *L. Abramsii*

KK. Plants 2–5 dm. high.

 L. Fls. 12–14 mm. long; lfts. 5–7 mm. wide. Sierra Nevada . . . . 58. *L. Grayi*

 LL. Fls. 10–12 mm. long; lfts. 7–10 mm. wide. Santa Lucia Mts. . . . . 59. *L. ludovicianus*

JJ. Plants not at all woolly, but strigose.

 K. The plants ca. 1 dm. high, silvery-silky; fls. 10–12 mm. long. Siskiyou Co. . . . . 53. *L. lapidicola*

 KK. The plants 2–5 dm. tall.

  L. Fls. 14–18 mm. long. Mts. of s. Calif.

   63. *L. excubitus* var. *austromontanus*

  LL. Fls. smaller.

   M. Pedicels appressed-pubescent.

    N. Racemes 10–20 cm. long, lax; banner roundish; calyx gibbous at base. Inyo Co. . . 43. *L. alpestris*

    NN. Racemes 3–7 cm. long, dense; banner broader than long; calyx not gibbous at base. Plumas and Trinity cos. n.

     52. *L. obtusilobus*

   MM. Pedicels spreading-pubescent.

    N. Lfts. 2.5–3.5 cm. long; calyx gibbous at base. Mono Co.

     54. *L. montigenus*

    NN. Lfts. 1–2.5 cm. long; fls. 10–14 mm. long; calyx not gibbous. Coast Ranges

     62. *L. albifrons* var. *collinus*

1. **L. horizontàlis** Heller. [*L. microcarpus* var. *h.* Jeps.] Diffuse short-villous annual, 0.5–1.5 dm. high, the lower branches subprostrate; petioles 4–6 cm. long; lfts. 7–9, oblanceolate, 1–2 cm. long, 5–6 mm. wide, glabrous above, villous, rounded-obtuse at apex; peduncles 3–4 cm. long; racemes 5–8 cm. long, equaling or slightly longer than lvs.; bracts ca. 3 mm. long, persistent, acuminate; pedicels 1–1.5 mm. long; calyx villous, 6–8 mm. long, upper lip cleft, lower bidentate; petals pale violet-blue, the banner almost plane, ca. 7 mm. wide, 10–11 mm. long, wings 4–5 mm. wide, keel straight, wings and keel ciliate on upper and lower edges near claws; pods ovate, villous, 12–15 mm. long, 2-ovuled; seeds 3 mm. long, with small dark spots.—Dry slopes, below 2000 ft.; V. Grassland; inner S. Coast Ranges, e. San Luis Obispo and w. Kern cos. April–May.

Var. **platypétalus** C. P. Sm. Branches all ascending; fls. 13–15 mm. long, whitish to pale violet.—Open sandy and gravelly places, below 3500 ft.; Creosote Bush Scrub, Joshua Tree Wd.; w. Mojave Desert to Barstow region and to Greenhorn Mts. in Kern Co. March–May.

2. **L. rùber** Heller. [*L. microcarpus* var. *r.* C. P. Sm.] Villous annual, 1–3(–5) dm. tall, branched from base; lvs. near the base, the petioles 4–8 cm. long; lfts. 5–8, oblanceolate to spatulate, 1–2 cm. long, glabrous above, villous beneath; peduncles 2–7 cm. long, the racemes 3–6 cm. long, of 2–5 whorls, not much surpassing the foliage; bracts persistent, ca. 8 mm. long; pedicels ca. 1 mm. long; upper calyx-lip barely 2 mm. long, cleft with rounded lobes or slender teeth, the lower lip 8 mm. long, bidentate with slender teeth; petals dull red to pink, the banner almost plane, lance-ovate, 2–4 mm. wide, 8–9 mm. long, wings narrow, scarcely ciliate, the keel straight, ciliate only above near the claws; pods ovate, 12–17 mm. long, 10 mm. wide, ± villous, 2-ovuled; seeds pale, ca. 3 mm. long, ± rugose.—Uncommon, dry open places, below 5000 ft.; V. Grassland, Creosote Bush Scrub; inner S. Coast Ranges from San Benito Co. to L. Calif., inland as far as New York Mts., San Bernardino Co. March–June.

3. **L. subvéxus** C. P. Sm. [*L. microcarpus* of many Calif. refs., not Sims. *L. s.* vars. *insularis* and *nigrescens* C. P. Sm.] Loosely villous annual, 1.5–4 dm. tall, simple or branched; lvs. basal or on lower and middle stem; lfts. mostly 5–9, oblanceolate, 1.5–3.5 cm. long, abruptly apiculate, glabrous above, villous beneath; petioles 5–15 cm. long; peduncles 5–12 cm. long; racemes 5–15 cm. long, with 3–9 whorls; bracts persistent,

soon reflexed; pedicels 2 mm. long; calyx 9–11 mm. long, ± inflated at the base; petals dark violet-purple, lilac or rose-pink, the banner rounded at apex, 12–14 mm. long, 7–8 mm. wide; wings usually nonciliate, keel straight, ciliate above near the claws; pods 2-ovuled, ca. 1.5 cm. long; seeds dark brown, rough, 4–5 mm. long.—Dry fields and slopes below 2500 ft.; V. Grassland, Foothill Wd.; Coast Ranges from Lake and Tehama cos. to Ventura and Kern cos., foothills of Sierra Nevada in Kern and Tulare cos., Santa Cruz Id. April–June. Variable; plants with the calyx 9–11 mm. long, the banner angled at apex and 5–7 mm. wide, occurring to ca. 5000 ft., Sagebrush Scrub, e. Siskiyou to Modoc and Lassen cos., s. Ore., are var. *transmontànus* C. P. Sm. Those with lfts. green, 2–4 cm. long; calyx 7–8 mm. long, banner 10–11 mm. long, 6 mm. wide, Foothill Wd., Valley Grassland, inner Coast Ranges from Santa Clara Co. to Ventura Co., are the var. *phoeníceus* C. P. Sm. Those with the small fls. of the last named, but with lfts. 1–2 cm. long, and ± whitish, woolly-villous, are var. *albilanàtus* C. P. Sm.; they occur in e. San Luis Obispo and Monterey cos. These varieties are poorly marked and not definitely geographical.

4. L. densiflòrus Benth. [*L. d.* vars. *latilabrus, Tracyi* and *stenopetalus* C. P. Sm.] Appressed-pubescent or nearly so, annual, 2–4 dm. tall, simple or branched especially above, stems often hollow; petioles 4–10 cm. long; lfts. oblanceolate, 7–9, glabrous above, villous beneath; peduncles 0.5–2 dm. long, with racemes as long or longer; whorls 5–12, ± separate; calyx ca. 8 mm. long, green, with inconspicuous pubescence perhaps 0.5 mm. long, the lower lip barely 3 mm. wide, acute with 2 slender teeth; petals white, tinted or veined with rose or violet, the banner 14 mm. long, 8 mm. wide, keel slightly shorter, rather slender, ciliate on upper edges near the claws, as are often the wings; pods ovate-oblong, ca. 1.5 cm. long, long-villous; seeds dark to pale, 4–5 mm. long; $2n = 48$ (Tuschnjakowa, 1935).—Grassy and open fields and hillsides, below 2000 ft.; Mixed Evergreen F., Foothill Wd., etc.; Coast Ranges from Santa Clara Co. to Humboldt and Butte cos. April–June. Most variable, some 24 varieties having been recognized by C. P. Smith (Bull. Torrey Bot. Club 45: 167–202, 1918). The more definite variations are ca. as follows:

Pubescence of stems and peduncles very short, ± appressed, barely 0.5 mm. long.
  Calyx 6.5–8 mm. long; petals not yellow.
    Petals white, tinted or veined with violet or rose; lfts. not blackened on drying. Santa Clara Co. to Humboldt Co. . . . . *L. densiflorus*
    Petals light blue, the banner with a white center; lfts. blackening on drying. Mt. Pinos, Ventura Co. . . . . var. *glareosus*
  Calyx 9–12 mm. long; petals often yellow. Lake Co. to Santa Barbara Co. . . . . var. *aureus*
Pubescence of stems and peduncles longer, spreading or retrorse, 1–1.5 mm. long.
  Calyx rather inconspicuously pubescent, the hairs barely 1 mm. long. Sacramento and Placer cos. along w. base of Sierra Nevada to Tehachapi and Mt. Pinos regions, then in the interior to San Diego Co. . . . . var. *lacteus*
  Calyx bushy-villous near the base, the hairs 1.5–4 mm. long, drying tawny.
    Lower calyx-lip ± bent and inflated near the base; banner mostly rounded at apex. N. Los Angeles Co. to Butte and Lake cos. . . . . var. *palustris*
    Lower calyx-lip straight, not inflated at base; banner acute at apex. San Diego region
                                                  var. *austrocollium*

Var. glareòsus (Elmer) C. P. Sm. [*L. g.* Elmer.] Fls. 13–14 mm. long.—Dry slopes, 4000–6000 ft., Pinyon-Juniper Wd., Sagebrush Scrub; Mt. Pinos, Ventura Co.

Var. aùreus (Kell.) Munz. [*L. Menziesii* var. *a.* Kell. *L. M.* J. G. Agardh. *L. d.* var. *M.* C. P. Sm. *L. d.* var. *perfistulosus* C. P. Sm.] Fls. 14–17 mm. long, yellowish or edged with red-purple.—More common than the sp., on dry banks and slopes below 2500 ft.; Foothill Wd., V. Grassland; Mendocino and Colusa cos. to n. Santa Barbara Co.

Var. lácteus (Kell.) C. P. Sm. [*L. l.* Kell. *L. d.* vars. *vastiticola, sublanatus, McGregori, altus, versabilis, latidens, Dudleyi, persecundus* C. P. Sm. *L. arenicola* Heller.] Fls. 11–16 mm. long, white or rose or purple.—Dry open banks and slopes, below 6000 ft.; V. Grassland, Foothill Wd., Yellow Pine F., S. Oak Wd., etc.; away from the coast from Placer Co. to San Diego Co., reaching the edge of the deserts.

Var. palústris (Kell.) C. P. Sm. [*L. p.* Kell. *L. d.* vars. *stanfordianus, curvicarinus, trichocalyx,* C. P. Sm. and *crinitus* Eastw. ex C. P. Sm.] Fls. 12–17 mm. long, white or cream to dark purple.—V. Grassland, Foothill Wd., Mixed Evergreen F., Chaparral, etc.; from Butte and Lake cos. to n. Los Angeles Co.

Var. **austrocóllium** C. P. Sm. Fls. ca. 13 mm. long, white to pinkish.—Dry open places below 2500 ft.; V. Grassland, Chaparral, Coastal Sage Scrub; s. Orange and Riverside cos. to w. San Diego Co.

5. **L. lutèolus** Kell. [*L. Bridgesii* Gray. *L. l.* var. *albiflorus* Eastw. *L. Milo-Bakeri* C. P. Sm.] Much like *L. densiflorus*, but not fistulose, the stems 3–8 dm. tall, widely branched above, strigose; petioles 2–5 cm. long; lfts. mostly 7–9, cuneate-oblanceolate, 2–3 cm. long, strigose on both surfaces, or glabrous above; peduncles mostly 5–15 cm. long; whorls crowded, few to many in racemes 5–20 cm. long; fls. light to pale yellow, sometimes pale lilac at first, 12–14 mm. long; calyx subsaccate, strigose, ca. 10–11 mm. long; banner ovate; wings ciliate above and sometimes below near the base; keel ciliate on upper and lower margins of base; legume hirsute, ca. 1.5 cm. long; seeds dark brown or mottled gray, ca. 4 mm. long, ± tuberculate.—Dry slopes and flats, below 6000 ft.; Yellow Pine F., N. Oak Wd., Pinyon-Juniper Wd., Foothill Wd., etc.; Ventura Co. through Coast Ranges to Humboldt and Siskiyou cos., Shasta Co.; s. Ore. May–Aug.

6. **L. brevicaùlis** Wats. Densely villous annual, 0.3–1 dm. tall, the main stem scarcely 1 cm. long; petioles 3–6(–8) cm. long; lfts. 5–8, spatulate, 0.5–1.5 cm. long, glabrous above, villous beneath; peduncles 3–8(–10) cm. long, spreading to almost prostrate, with subcapitate racemes 1–2.5 cm. long; fls. 6–8 mm. long, the pedicels 1–2 mm.; bracts persistent, often purplish, 2–3 mm. long; calyx villous, the upper lip truncate to bifid, 1–2 mm. long, the lower 4–6 mm. long, entire to tridentate; petals bright blue or yellowish-white on lower half, the banner ca. 6 mm. long, rounded or angled at apex, keel straight nonciliate; pods ovate, ca. 1 cm. long; seeds light brown, smooth, ca. 2 mm. long.—Sandy places, 4000–7700 ft.; Pinyon-Juniper Wd., Sagebrush Scrub, Joshua Tree Wd.; Mojave Desert from e. San Bernardino Co. to White Mts.; to Ore., Colo., Chihuahua. May–June.

7. **L. unciàlis** Wats. Acaulescent tufted plant 1.5–2 cm. high, villous; petioles ca. 1 cm. long; lfts. commonly 5, spatulate, ca. 0.5 cm. long, villous on both sides; peduncles 1-fld., ca. as long as lvs.; bracts oval; calyx ca. 2 mm. long; petals 4–5 mm. long, ochroleucous, the keel tipped with purple; pod ca. 6 mm. long.—Dry slopes, 5000–7000 ft.; Pinyon-Juniper Wd., Sagebrush Scrub; in extreme w. Nev. and to be sought in Modoc, Mono, and Inyo cos. May–June.

8. **L. Shóckleyi** Wats. Acaulescent or with stems to ca. 1 dm., densely pubescent with hairs to ca. 1 mm. long; petioles 4–12 cm. long; lfts. 7–10, subappressed-silky beneath, glabrous above except near margins, ± spatulate, 1–2 cm. long; peduncles 3–10 cm. long; racemes 3–6 cm. long, equal to or slightly longer than lvs.; fls. scattered, 5–6 mm. long, on slender pedicels to ca. 4 mm. long; bracts narrow, 2–3 mm. long; calyx densely white-pubescent, the upper lip 3 mm. long, bifid, the lower 3–4 mm., minutely 3-toothed or bifid; petals blue, purple or pink, the banner 5–6 mm. long, 4 mm. wide, ± pointed; keel nonciliate; pods oblong-ovate, ca. 1.5 cm. long, ciliate, with scaly sides; seeds pale, rough, ca. 3 mm. long.—Dry sandy places, below 4000 ft.; Creosote Bush Scrub; Colo. and Mojave deserts; Ariz., Nev. April–May.

9. **L. flavoculàtus** Heller. [*L. rubens* var. *f.* C. P. Sm. *L. odoratus* var. *f.* Jeps.] Almost acaulescent annual, 0.5–1.5 dm. high, villous; petioles mostly 2–5 cm. long; lfts. mostly 7–9, broadly oblanceolate, 0.5–2 cm. long, glabrous above, sparsely villous beneath; peduncles wide-spreading, 3–10 cm. long; racemes dense, 2–8 cm. long; pedicels 1–4 mm. long; bracts subulate, ca. 3 mm. long; calyx sparsely villous, the upper lip 2–3 mm. long, bidentate, the lower 5–6 mm., bidentate; petals 7–8 mm. long, deep rich violet-blue, the banner with a squarish yellow spot, keel glabrous; pods ovate, ca. 8–11 mm. long, villous; seeds ± pinkish, dull, ± wrinkled, ca. 2 mm. across.—Dry open places, 2600–7000 ft.; Creosote Bush Scrub to Pinyon-Juniper Wd.; Mojave Desert from e. San Bernardino Co. to White Wts.; w. Nev. April–June.

10. **L. odoràtus** Heller. Glabrous subacaulescent ± succulent annual, 1.5–3 dm. high; lvs. basal, the petioles commonly 5–10 cm. long; lfts. cuneate to spatulate, 5–7, glabrous above, glabrous or somewhat strigose beneath; peduncles 8–15 cm. long, the racemes 5–10 cm. long, fairly dense, well surpassing foliage; pedicels glabrous, 4–6 mm. long; bracts narrow, 2–3 mm. long, villous; calyx glabrous, the lips entire, the upper 2 mm., the lower 4 mm. long; fls. fragrant, mostly deep violet-purple with yellow spot on banner, the banner ca. 10 mm. long and wide, keel somewhat curved, glabrous; pods yel-

lowish, 17–20 mm. long, villous on margins; seeds gray-pink, wrinkled, 2–3 mm. across.—Locally common on dry open flats, 2000–4000 ft.; Creosote Bush Scrub, Joshua Tree Wd.; Mojave Desert from Barstow region w. and n. to Lone Pine; w. Nev. April–May.

Var. **piloséllus** C. P. Sm. Plants villous; seeds 4.—Nipton, San Bernardino Co.

11. **L. pusíllus** Pursh var. **intermontànus** (Heller) C. P. Sm. [*L. i.* Heller.] Loosely villous, 5–12 cm. high, with spreading branches; lvs. rather crowded, the petioles 3–7 cm. long; lfts. usually 5, oblong-oblanceolate, 1–3.5 cm. long, glabrous above, appressed-hairy beneath; peduncles barely 1 cm. long; racemes 3–5 cm. long, mostly shorter than foliage; pedicels ± villous, ca. 2–3 mm. long; calyx ± villous, the lower lip 3–4 mm. long, entire to 3-toothed; fls. 7–9 mm. long, bright violet-blue, fading white or pinkish, the banner 7–8 mm. long, 5–6 mm. wide, keel nonciliate; pods ca. 1.5 cm. long, constricted in middle, villous; seeds pale yellow, rugose, ca. 4 mm. across.—Dry places, 4000–5000 ft.; Sagebrush Scrub; Deep Springs V., Inyo Co., e. Modoc Co.; to e. Wash., Wyo., Colo., Ariz. May–June.

12. **L. concínnus** J. G. Agardh. [*L. nipomensis* Eastw.] Densely villous usually much branched annual, 0.5–2 dm. tall, with slender stems to ca. 1.5 mm. thick; lvs. well distributed, the petioles commonly 2–6 cm. long; lfts. 5–9, narrow-oblanceolate, 1–2 cm. long, mostly 1.5–3 mm. wide, soft-hairy; peduncles 2–8 cm. long, the racemes 1–3(–7) cm. long, villous; pedicels 1–2 mm. long, suberect soon after anthesis; fls. 7–9 mm. long, the upper calyx-lip cleft, the lower 3-toothed; petals mostly lilac, edged with red-purple, the banner 7–9 mm. long, 4–5 mm. wide, with yellow center; keel and wings not ciliate; pods 1–1.5 cm. long, 2–4-ovuled, hairy; seeds mostly dull-spotted on a pale ground, subquadrate, 2–3 mm. long.—Dry open and disturbed, often gravelly, places below 5000 ft.; many Plant Communities; cismontane Calif. from Monterey and Fresno cos. to San Diego Co. March–May.

KEY TO VARIETIES

Plants villous with ± spreading hair; fls. mostly lilac with purple-red borders, keel naked.
  The plants densely shaggy-villous, ± grayish or brownish with hair.
    Lfts. mostly 1.5–3 mm. wide; stems largely 1–1.5 mm. thick. Cismontane ...... *L. concinnus*
    Lfts. mostly 4–7 mm. wide; stems largely 1.5–2 mm. thick.
      Fls. 9–11 mm. long. Largely from Yellow Pine F. .... var. *optatus*
      Fls. mostly 6–8 mm. long. Deserts .... var. *Orcuttii*
  The plants sparsely villous, rather bright green.
    Keel not ciliate. Cismontane .... var. *Agardhianus*
    Keel ciliate on both edges of claws. W. edge of Colo. Desert .... var. *brevior*
Plants strigose with closely appressed hairs; fls. pale or yellowish; keel ciliate. Deserts
                                                 var. *pallidus*

Var. **optàtus** C. P. Sm. Dry largely granitic slopes, mostly 4000–7000 ft.—Mts. from San Diego Co. to Santa Lucia Mts. May–July.

Var. **Orcúttii** (Wats.) C. P. Sm. [*L. O.* Wats. *L. micensis* Jones.] Rather coarse plants, densely shaggy; lfts. broad; fls. 6–8 mm. long.—Common in dry sandy and gravelly places, below 5000 ft.; Creosote Bush Scrub, Joshua Tree Wd.; mostly on deserts from Inyo to Imperial cos.; to Nev., Ariz., New Mex. March–May.

Var. **Agardhiànus** (Heller) C. P. Sm. [*L. gracilis* J. G. Agardh. *L. A.* Heller.] Like the typical form of the sp., but much more sparsely hairy and greener.—With the sp.

Var. **pállidus** (Bdg.) C. P. Sm. [*L. p.* Bdg. *L. desertorum* Heller. *L. c.* var. *d.* C. P. Sm.] Mostly rather coarse plants with appressed pubescence; fls. 6–7 mm. long, pale blue to whitish or yellowish.—Occasional in sandy and gravelly places; Creosote Bush Scrub; both deserts; to Baja Calif. March–April.

Var. **brévior** (Jeps.) D. Dunn. [*L. sparsiflorus* var. *b.* Jeps.] Lfts. truncatish, 5–12 mm. long; fls. ca. 6 mm. long.—Creosote Bush Scrub; Palm Springs region to foot of Mt. Springs Grade. March–April.

13. **L. Stìversii** Kell. Minutely pubescent annual, 1–4.5 dm. high, freely branched; lvs. scattered, the petioles 3–8 cm. long; lfts. 6–8, cuneate to obovate, 1–4 cm. long, strigose on both sides; peduncles 3–8 cm. long; racemes 1–3 cm. long; pedicels ca. 2 mm. long; calyx pubescent, the upper lips cleft, the lower entire, 5–8 mm. long; banner 13–15 mm. long, bright yellow, the wings rose-pink or purple, ca. 15–16 mm. long, keel whitish, ciliate above and below at base; pods ca. 2–2.5 cm. long, glabrous; seeds several, flat, angled, ca. 2.5 mm. long, dark-spotted on a pale ground.—Sandy or gravelly places,

1600–4600 ft.; Yellow Pine F., Foothill Wd.; Sierra Nevada from Butte Co. to Kern Co., and locally in the Santa Lucia, San Gabriel and San Bernardino mts. April–July.

14. **L. succuléntus** Dougl. ex Koch. [*L. s.* var. *Layneae* C. P. Sm. *L. affinis* of many refs., not J. G. Agardh.] Stout usually succulent or fistulous annual, mostly ± strigose, simple or branched, 2–6(–10) dm. tall; lvs. well distributed, the petioles 6–12 cm. long; lfts. 7–9, cuneate to cuneate-obovate, 2–7 cm. long, rather dark green, glabrous above, ± strigulose beneath; peduncles 2–10 cm. long; racemes 6–30 cm. long, with pedicels 4–6 mm. long, in whorls or groups; bracts pubescent, subulate, 5–6 mm. long, soon deciduous; calyx substrigose, ca. 7–8 mm. long, the upper lip deeply bifid, the lower entire to 3-toothed; fls. mostly deep purple-blue, sometimes albino, the banner ca. 12–14 mm. long, usually with yellow center; wings slightly ciliate at base on upper edge, keel ± curved, ciliate near base on both edges; pods 4–5 cm. long, ca. 8 mm. wide, loosely pubescent, dark, several-seeded; seeds oblong, 4–5 mm. long, marbled with dark brown.—Usually in heavy soil on grassy flats and slopes, below 2000 ft.; many Plant Communities; Coast Ranges from Mendocino Co. to n. L. Calif., inland to Butte and Shasta cos. and w. side of San Joaquin V. Feb.–May.

15. **L. hirsutíssimus** Benth. Robust annual, with stiff yellowish nettlelike hairs 3–5 mm. long, ± fistulous, 2–8(–10) dm. high, usually few-branched from base; lvs. well distributed, the petioles mostly 5–18 cm. long; lfts. 5–8, broadly cuneate-obovate, 2–5 cm. long, with nettlelike hairs; peduncles 5–8 cm. long; racemes 10–25 cm. long; fls. red-violet to majenta, scattered, 13–15 mm. long, the pedicels 3–4 mm. long; calyx 8–10 mm. long, the upper lip cleft, the lower entire to 3-toothed; banner suborbicular, tending to have a yellow blotch, the keel densely ciliate on lower edges near the claws; pods 2.5–3.5 cm. long, hispid-bristly, several-ovuled; seeds 3–4 mm. long, pale, ± marbled with brown.—Locally abundant in open wooded or brushy places, below 4000 ft.; Coastal Sage Scrub, Chaparral; Coast Ranges from San Mateo Co. to L. Calif. March–May.

16. **L. truncàtus** Nutt. ex H. & A. [*L. t.* var. *Burlewi* C. P. Sm.] Subglabrous to sparsely strigulose annual, branched, 3–7 dm. tall, rather deep green; petioles mostly 5–10 cm. long, ± flattened; lfts. 5–7, linear, truncate to emarginate or toothed, 1–4 cm. long; peduncles 3–10 cm. long; racemes 5–15 cm. long; pedicels 2–3 mm. long; bracts subulate, 3–4 mm. long, rather persistent; fls. not crowded, 10–12 mm. long, mostly purplish-blue, redder in age; calyx 5–6 mm. long, the upper lip cleft, the lower entire to 3-toothed; banner ca. 10 mm. long, 9 mm. wide, keel ± ciliate on lower edge toward claws, more so on upper; pods ca. 3 cm. long, 6 mm. wide, villous, 6–7-ovuled; seeds ca. 3 mm. long, rhomboid, plump, ± plainly mottled with brown on a pale flesh-color background.—Open grassy places, burns, open woods, etc., below 3000 ft.; V. Grassland, Coastal Sage Scrub, Chaparral, etc.; Monterey Co. to n. L. Calif. March–May.

17. **L. Benthàmii** Heller. [*L. leptophyllus* Benth., not Cham. & Schlecht.] Erect villous annual, simple or branched, 3–6 dm. tall; petioles slender, 6–12 cm. long; lfts. 7–10, linear, 2–5 cm. long, villous; peduncles 5–10 cm. long, racemes 10–20 cm. long; bracts linear, longer than the buds; pedicels 3–6 mm. long; fls. 10–15 mm. long, light to deep blue; calyx mostly 4–5 mm. long, the upper lip cleft, the lower 3-toothed; banner suborbicular, ca. 11–14 mm. long, with a yellow spot; keel curved, ciliate on lower edges near claws, not on upper; pods ascending, 2–3.5 cm. long, ca. 6 mm. wide, 3–9-seeded; seeds 2 mm. long, tawny, marbled with brown.—Open slopes and hills, below 3800 ft.; V. Grassland, Foothill Wd.; hills bordering interior valleys from Sacramento and Monterey cos. to Kern and n. Los Angeles cos. March–May. More robust plants with hollow stems and fls. 14–15 mm. long, have been called var. *opimus* C. P. Sm.

18. **L. sparsiflòrus** Benth. [*L. subhirsutus* A. Davids.] Slender-stemmed annual, strigose and spreading-villous, mostly 2–4 dm. high, erect, branched; petioles 3–8 cm. long; lfts. 5–9, linear to oblanceolate, 1–3 cm. long, mostly 1.5–3 mm. wide and acutish at apex; peduncles 2–8 cm. long; racemes 8–20 cm. long; fls. not crowded, light blue to lilac, mostly 10–12 mm. long; pedicels mostly 3–5 mm. long; bracts lance-linear, 3–4 mm. long, early deciduous; calyx ca. 5 mm. long, the upper lip 2-parted, the lower apiculate to 3-toothed; banner suborbicular, with a yellow spot, keel curved, usually ciliate on lower edge near claws and often on upper; pods 1–2 cm. long, 5–7-ovuled, strigose; seeds 2–3 mm. long, angled, dotted or marbled on a pale ground.—Open fields and slopes, below 4000 ft.; Coastal Sage Scrub, Chaparral, Creosote Bush Scrub;

Ventura Co. to n. L. Calif. and inland across the deserts to s. Nev., Ariz. March–May. Variable; plants with subtruncate lfts. have been named var. *inopinàtus* C. P. Sm.

19. L. **Moránii** Dunkle. [*L. aliclementinus* C. P. Sm.] Annual, bushy, 3–9 dm. high, spreading-villous throughout; lvs. well distributed, the stipules 1–2 cm. long, linear; petioles 2–7 cm. long; lfts. 7–10, linear to narrow-oblanceolate, acutish, 1–4.5 cm. long; peduncles 5–8 cm. long; racemes 5–15 cm. long; fls. not in definite whorls, rather few, 14–16 mm. long; pedicels 6–12 mm. long; bracts ca. 1 mm. long; upper calyx bifid, 6–7 mm. long, lower tridentate; petals blue, the banner rounded, glabrous, ca. 12 mm. long; wings ca. 12 mm. long, 7–8 mm. wide, rounded at apex; keel arcuate, ciliate on upper edge; pods 5–6 cm. long, densely villous.—Banks and canyon-slopes; Coastal Sage Scrub; San Clemente Id. Feb.–April.

20. L. **arizònicus** (Wats.) Wats. [*L. concinnus* var. *a.* Wats. *L. sparsiflorus* var. *a.* C. P. Sm. *L. s.* var. *barbatulus* Thornb.] Branched fleshy annual, ± strigulose and with some spreading hair, ± fistulose, 3–6 dm. tall, light green; petioles 2–8 cm. long; lfts. 5–10, broadly oblanceolate, 1–3.5 cm. long, 4–12 mm. wide, thickish, rounded and mucronulate at apex, glabrous or nearly so above, ± villous beneath; peduncles 1–5 cm. long; racemes 5–30 cm. long, with fls. scattered, 8–10 mm. long, pale purplish-pink, often drying violet; pedicels commonly 2–3 mm. long; bracts linear, 6–8 mm. long; calyx 4–5 mm. long; the upper lip deeply cleft; banner suborbicular, keel ciliate near base on lower or both edges; pods and seeds as in *L. sparsiflorus.*—Common in sandy washes and open places, below 2000 ft.; Creosote Bush Scrub; deserts n. to e. Inyo Co.; Nev., Ariz., Son. March–May.

21. L. **citrìnus** Kell. Annual, diffusely branched, 1–2 dm. high, white-villous; lvs. ± crowded, the lfts. 6–8, oblanceolate, obtuse, 1–2.5 cm. long, shorter than the petioles; racemes 10–15 cm. long, the pedicels ca. 3 mm. long, deflexed after anthesis, pubescent; the 2 calyx-lips subequal, the upper 2-cleft, the lower 3-toothed; petals orange or golden, 8–9 mm. long, the banner suborbicular, emarginate, keel ciliate on lower edge near the claws; pods 12–14 mm. long, 3–4 mm. wide, deflexed, glabrate; seeds 2–4, rhomboidal, ca. 2 mm. long, dark-spotted on pale ground.—Rocky hills, 4000–5300 ft.; Yellow Pine F.; Sierran foothills, Mariposa, and Fresno cos. April–June.

Var. **defléxus** (Congd.) Jeps. [*L. d.* Congd.] Plants 3–4 dm. high, less hairy; fls. white or pinkish, ca. 10 mm. long.—Foothills, Mariposa Co. March–May.

22. L. **nànus** Dougl. in Benth. Erect annual, simple or branched at base, ± villous and minutely pubescent or strigulose, 1–5 dm. tall; lvs. mostly well distributed; petioles 4–8 cm. long; lfts. 5–7, linear to spatulate, mostly acute, 1.5–3 cm. long; peduncles 5–8(–14) cm. long, the racemes 6–24 cm. long, with well separated whorls; bracts linear-lanceolate, deciduous, 5–10 mm. long; pedicels 3–5 mm. long; upper calyx-lip bifid, 3–5 mm. long, lower mostly 2–3-toothed, 3.5–5.5 mm. long; petals rich blue except for the white or yellowish spot on banner; banner suborbicular, 8–11.5 mm. long, much reflexed, the wings mostly concealing the slender keel which is not much curved, ciliate on upper edges of slender apex; pods strigose, 2–3.5 cm. long, 4–5.5 mm. wide, usually 4–8-ovuled; seeds 2.6–3.7 mm. long, usually well marked with brown or dark gray mottling; $2n = 48$ (Tuschnjakowa, 1935).—Local, grassy hills and fields and brushy slopes, below 3000 ft.; largely Foothill Wd.; Coastal Ranges from Santa Cruz Co. to Santa Barbara Co. April–May.

Largest lfts. 5–15 mm. wide; pods 6–8.5 mm. wide; seeds 3.5–4 mm. long. Widespread
ssp. latifolius

Largest lfts. 5–7.5 mm. wide; pods 3.5–5.5 mm. wide.
  Banner 8–11.5 mm. long; pods 4–5.5 mm. wide. Coast Ranges, Santa Cruz to Santa Barbara cos.
L. nanus
  Banner 9–15 mm. long; pods 3.5–4.5 mm. wide. S. end of Cent. V. . . . . ssp. *Menkerae*

Ssp. **latifòlius** (Benth.) D. Dunn. [*L. n.* var. *l.* Benth. ex Torr. *L. n.* of Calif. refs.] Lfts. 1.2–3.3 cm. long, 2–7.5 mm. wide, oblanceolate; racemes 5–18 cm. long; pedicels 3.5–9 mm. long; lower calyx lip 4–7 mm. long, the upper 3–5.7 mm.; banner orbicular, 10–15 mm. long; pods 3–4 cm. long, 5–6.5 mm. wide; seeds 3.5–4 mm. long, tan, unevenly stippled.—Grassy and rocky places, below 2000 ft.; V. Grassland, N. Coastal Scrub, Coastal Prairie, Coastal Sage Scrub, etc.; Coast Ranges from Mendocino Co. to Los Angeles Co. and Sierran foothills. April–May.

Ssp. **Ménkerae** (C. P. Sm.) D. Dunn. [*L. n.* var. *M.* C. P. Sm.] Fls. pale lilac, turning bluer on drying; seeds pale flesh-colored, obscurely marked, 2–3 mm. long.—V. Grassland, below 2000 ft.; head of San Joaquin V. in Kern Co. to Coalinga, Fresno Co. March–April.

23. **L. vallícola** Heller. [*L. nanus* var. *v.* C. P. Sm. *L. persistens* Heller, not Rose. *L. Blaisdellii* Eastw.] With habit and pubescence of *L. nanus,* but more frequently simple, 1.5–3.5 dm. high; lfts. largely 6–8, linear, 1.5–2.5 cm. long, 2–3 mm. wide; peduncles 3–7 cm. long; racemes 3–10 cm. long, with fairly well separated whorls; pedicels slender, 4–5 mm. long; upper calyx-lip ca. 2 mm. long, the lower ca. 4 mm. long; fls. 6–10 mm. long, bright blue; banner usually wider than long, with cent. pale spot and the apex not much reflexed from upper margins of wings, keel usually strongly curved, ciliate above toward apex; pods 1.5–2.5 cm. long, 4–5 mm. wide, ± silky; seeds 3–6, pale, flesh-colored, scarcely mottled, ca. 2.5 mm. long.—Grassy banks, below 3500 ft.; Foothill Wd., V. Grassland, Yellow Pine F.; foothills of Sierra Nevada from Kern Co. to Butte and Shasta cos. Mostly March–June.

Ssp. **aprìcus** (Greene) D. Dunn. [*L. a.* Greene. *L. nanus* var. *a.* C. P. Sm.] Banner slightly longer than wide, its apex more than 2 mm. from the upper distal united corners of wing petals; seeds gray, plainly mottled with brown.—Open grassy fields and valleys, below 3000 ft.; V. Grassland, Foothill Wd.; Sierran foothills from Kern Co. n., Coast Ranges from Monterey Co. to Siskiyou Co.; to B.C. March–May.

24. **L. spectábilis** Hoov. [*L. nanus* var. *perlasius* C. P. Sm.] Erect annual, 2–6 dm. tall, densely villous with hairs up to 2 and 3 mm. long; stems stout, simple, 1–few from base; petioles 2–5 cm. long; lfts. 8–12, oblanceolate, 1–4 cm. long; peduncles 2–8 cm. long; racemes 10–25 cm. long; fls. not definitely whorled but ± approximate at intervals; bracts dark, linear, 4–8 mm. long, early deciduous; pedicels 5–10 mm. long; upper calyx-lip ca. 4, lower ca. 6 mm. long; petals bright blue or blue and purple, 12–17 mm. long, the banner and wings very broad, keel strongly arcuate, ciliate on upper edge toward apex; pods 3–4.5 cm. long, 6 mm. wide, villous; seeds 6–10, 4 mm. long, with a prominent dark lateral line.—Uncommon on rocky slopes of serpentine, 800–2000 ft.; Foothill Wd.; foothills of Sierra Nevada in Mariposa and Tuolumne cos. April–May.

25. **L. affinis** J. G. Agardh. [*L. carnosulus* Greene. *L. nanus* var. *c.* C. P. Sm. *L. affinis* var. *c.* Jeps.] Erect annual, 3–6 dm. high, ± succulent, strigose and pubescent or short-villous, mostly several-stemmed from base; petioles 4–10 cm. long; lfts. mostly 6–8, oblanceolate, 1.5–4 cm. long; peduncles 5–12 cm. long; racemes 6–22 cm. long, with rather remote whorls; bracts ca. 5–7 mm. long, early deciduous; pedicels 4–6 mm. long; upper calyx-lip 4–5, lower 5–6 mm. long; petals 9–12 mm. long, deep blue, the banner suborbicular, with cent. light spot, keel not much curved, toothed near middle of upper edge, ciliate on upper edge toward tip; pods 30–50 mm. long, 7–9 mm. wide, villous-hirsute; seeds 5–8, subquadrate, 4–5 mm. long, rather dark brown, mottled.—Banks and grassy slopes, below 2500 ft.; Coastal Prairie, N. Coastal Scrub, Mixed Evergreen F., etc.; Coast Ranges from Santa Cruz Co. to w. Ore., inland to Butte Co. March–May.

26. **L. pachýlobus** Greene. [*L. micranthus* var. *p.* Jeps.] Annual, branched from base, 1–3 dm. tall, villous; petioles slender, 3–7 cm. long; lfts. 6–8, oblanceolate, 1.5–2.5 cm. long, hairy on both sides; peduncles 4–8 cm. long; racemes shorter, lax, with 2–4 whorls; pedicels 1–3 mm. long; bracts 2–3 mm. long, early deciduous; upper calyx-lip 2–4 mm. long, lower 3–5 mm.; petals blue, the banner with a white center and suborbicular, 6–8 mm. wide; wings 6–8 mm. long; keel ± arcuate, nonciliate; pods 2.5–3 cm. long, 6–9 mm. wide, shaggy-villous; seeds 4–6, brown, 4–5 mm. long, ± marked with darker lateral and marginal lines.—Grassy slopes, below 2000 ft.; Foothill Wd., V. Grassland; Shasta Co. to Santa Clara and Fresno cos. March–April.

27. **L. bìcolor** Lindl. [*L. micranthus* var. *b.* Wats. *L. hirsutulus* Greene.] Annual, the stems 1–several from base, erect, 1–4 dm. tall, villous throughout; petioles 2–7 cm. long; lfts. 5–7, oblanceolate to cuneate, 1–3 cm. long; peduncles 3–7 cm. long; racemes 1–7 cm. long, mostly with 1–3 whorls; pedicels 1.5–3 mm. long; bracts subulate, deciduous, 4–6 mm. long; upper calyx-lip 2–4 mm. long, the lower 3-toothed to entire, 4–6 mm. long; petals blue, the banner with cent. white purple-dotted spot and 6–9 mm. long by 6–8 mm. wide, oblong, truncate; keel slender, acute, ciliate along upper edges of slender beak; pods strigose, 1.5–2 cm. long, 4–5.4 mm. wide; seeds 5–8, pale to

pinkish or grayish, 2–3 mm. long, plain to mottled.—Occasional, sandy places, below 3000 ft.; Coastal Strand, N. Coastal Scrub, Mixed Evergreen F., etc.; Humboldt Co. to B.C. March–June.

KEY TO SUBSPECIES

A. Keel not or scarcely ciliate; fls. 5–8 mm. long. Widespread .... ssp. *Pipersmithii*
AA. Keel definitely ciliate on upper edge toward tip.
  B. Lower calyx-lip deeply cleft, the teeth 1–5 mm. long.
    C. Banner linear-oblong, 5–6 mm. long; plants mostly prostrate. Coastal sands, San Francisco to Santa Barbara Co. .... var. *trifidus*
    CC. Banner narrowly to broadly obovate, 6–9 mm. long; plants erect to decumbent. Humboldt Co. to San Diego Co. .... ssp. *umbellatus*
  BB. Lower calyx-lip entire or with teeth not more than 1 mm. long.
    C. Wings shaped like the keel, with a long tapering acumen .... var. *rostratus*
    CC. Wings normal, not shaped like the keel.
      D. Banner oval, lemon-shaped when flattened, 3.6–6 mm. long, mucronate; keel with a short blunt acumen .... ssp. *microphyllus*
      DD. Banner obovate or oblong, rounded to truncate or mucronate; keel with a slender acumen.
        E. The banner obovate, rounded to truncate at tip.
          F. Banner 5.7–8.6 mm. long, mostly rounded at apex; whorls mostly 1–2. Coast Ranges and insular .... ssp. *umbellatus*
          FF. Banner 4.5–7.5 mm. long, truncate to emarginate at apex; whorls 3–9. Mt. valleys of s. Calif. .... ssp. *marginatus*
        EE. The banner oblong, truncate at apex, sometimes mucronate.
          F. Banner 3–7.5 mm. long; racemes usually many-fld.; whorls usually distant. Interior of Coast Ranges .... ssp. *tridentatus*
          FF. Banner 6–9 mm. long; racemes few-fld.; whorls indistinct. From Humboldt Co. n. .... *L. bicolor*

Ssp. **Pipersmíthii** (Heller) D. Dunn. [*L. P.* Heller. *L. b.* var. *P.* C. P. Sm.] Racemes 1–7 cm. long; fls. 5–8 mm. long, the keel nonciliate or with an occasional hair.—At low elevs. in much of cismontane Calif., but especially about the San Joaquin V.

Ssp. **microphýllus** (Wats.) D. Dunn. [*L. micranthus* var. *m.* Wats. *L. b.* var. *m.* C. P. Sm.] Racemes to 5 cm. long; whorls 1–9; fls. 4–7 mm. long, whorled or ± alternate, blue with white spot on banner; pods strigose, 1.5–2 cm. long.—Common in open sandy and gravelly places, below 5000 ft.; Coastal Sage Scrub, V. Grassland, Chaparral, lower Yellow Pine F.; cismontane s. Calif. to edge of deserts and on ids.; less frequent, Foothill Wd., V. Grassland, Yellow Pine F., etc., centr. Calif. to Ore. March–May.

Ssp. **tridentàtus** (Eastw. ex. C. P. Sm.) D. Dunn. [*L. b.* var. *t.* Eastw. ex C. P. Sm. *L. b.* var *tetraspermus* C. P. Sm.] Racemes to 11 cm. long; banner oblong, truncate, 3–7.5 mm. long.—Many Plant Communities; in most of cismontane Calif.

Var. **rostràtus** (Eastw.) Jeps. [*L. r.* Eastw.] Vegetatively like ssp. *tridentatus;* banner ca. 4.5 mm. long; upper calyx-lip deeply split and the lobes largely united to the lower lip; wings slender and like the keel.—V. Grassland; San Luis Obispo Co.

Ssp. **marginàtus** D. Dunn. Racemes compact, distinctly verticillate; banner obovate, 4.5–7.5 mm. long.—V. Grassland, Foothill Wd., Joshua Tree Wd., Yellow Pine F.; head of San Joaquin V. (Kern Co.) and along w. edge of deserts to n. L. Calif.

Ssp. **umbellàtus** (Greene) D. Dunn. [*L. u.* Greene. *L. b.* var. *u.* C. P. Sm. *L. sabulosus* Heller.] Verticils 1–4, distinct; fls. 5.5–8.5 mm. long; banner obovate.—Many Plant Communities; Coast Ranges from Humboldt Co. to San Diego Co.

Var. **trífidus** (Torr. ex Wats.) C. P. Sm. [*L. t.* Torr. ex Wats.] Fls. 5–8 mm. long, the lower calyx-lip deeply trifid; banner elliptic, acute.—Sandy and gravelly places; Coastal Strand, Closed-cone Pine F.; San Francisco to Santa Barbara Co.

28. **L. micránthus** Dougl. in Lindl. [*L. polycarpus* Greene.] With habit and stature of *L. bicolor,* strigose and ± villous; lfts. 5–7, linear to oblanceolate, 1.5–4 cm. long, glabrous or sparsely hairy above, strigose beneath; peduncles 3–10 cm. long, racemes 1–8 cm. long, ± whorled; pedicels 1–2 mm. long; upper calyx-lip bifid, ca. 3 mm. long, lower 3-toothed, 3–4 mm. long; petals deep blue, the banner with a white spot having purple dots, cuneate-obovate to spatulate, 6–8 mm. long, ± emarginate; wings 5–6 mm. long, keel almost straight on upper edge, ciliate beyond the middle, the tip blunt; pods 2–3 cm. long, appressed-hairy; seeds 6–7, oblong, ca. 3 mm. long, gray or brown, usually mottled; $2n = 48$ (Tuschnjakowa, 1935).—Rather heavy soils, below 5000 ft., plains and valleys; many Plant Communities from Yellow Pine F. to Foothill Wd., Mixed

Evergreen F., etc.; Calif. from coast to Modoc and Lassen cos., s. to San Diego Co., but rare in s. Calif.; to B.C. March–June.

29. **L. Congdònii** (C. P. Sm.) D. Dunn. [*L. micranthus* var. *C. C. P. Sm.*] Stems to ca. 1 dm. long, spreading, ± stiff-villous; lvs. few, the petioles 1–4 cm. long; lfts. 6–9, spatulate, 4–10 mm. long, 1–3 mm. wide, ± glabrous above, strigose beneath; peduncles 2.5–3 cm. long, racemes 1–2 cm. long; fls. 6–7 mm. long, few, scattered, the pedicels 1–2 mm. long; upper calyx-lip bifid, ca. 3 mm. long, the lower ca. 4 mm. long; banner obovate, ca. 6.5 mm. long, blue, ± emarginate; the keel straight, ciliate above distally; pods 1.5–2.3 cm. long, ca. 6 mm. wide, strigose; seeds 3–6, mottled, ca. 3 mm. long.— Rare, in slightly depressed places, at ca. 3000 ft.; Yellow Pine F.; Sierran foothills of Mariposa and Nevada cos.; s. Ore. April–May.

30. **L. Lyállii** Gray. [*L. lepidus* ssp. *L.* Detl. *L. oreocharis* Eastw. *L. alpinus* Heller.] Perennial with heavy woody cespitose base, subappressed-silky, the stems semiprostrate, mostly 5–12 cm. long; lvs. crowded near base, the petioles slender, 3–5 cm. long, slender; lfts. 5–6, oblanceolate, 4–12 mm. long, appressed-silky above and beneath, acute; peduncles 5–10 cm. long, slender; racemes ± capitate, 1–3 cm. long, rarely more than twice as long as wide; pedicels 1–2 mm. long; bracts lance-linear, 3–5 mm. long, early deciduous; upper calyx-lip cleft or parted, 2–3 mm. long, lower entire to 3-toothed, slightly longer; petals 8–12 mm. long, the banner elliptical-obovate, dark blue or with large pale center, the wings blue, keel straight, mostly ciliate on upper edges of slender distal half, sometimes nonciliate; pods silky, 1–1.5 cm. long; seeds 3–4, pale, quadrate, ca. 2 mm. long; $n = 24$ (Phillips, 1957).—Dry ridges and summits, 8000–11,000 ft.; Subalpine F., Alpine Fell-fields; s. Sierra Nevada to Wash. July–Sept. Southward, the fls. become paler and there is intergradation with:

Var. **danàus** (Gray) Wats. [*L. d.* Gray.] Pubescence more loose, somewhat spreading; lfts. mostly 4–8 mm. long; fls. 6–8 mm. long, mostly pale lilac to whitish.—At 9500–13,000 ft.; mostly Alpine Fell-fields; Sierra Nevada n. to Warner Mts.; w. Nev.

31. **L. Lóbbii** Gray ex Greene. [*L. aridus* var. *L.* Gray ex Wats. *L. Lyallii* var. *L.* C. P. Sm. *L. L.* var. *villosus* Jeps. *L. pinetorum* Heller not Jones. *L. washoensis* Heller.] Cespitose perennial from a stout taproot, ± loosely villous or spreading-villous, the stems many, procumbent, 0.5–2 dm. long; petioles 2–5 cm. long; lfts. 5–7, oblanceolate, 10–13 mm. long, mostly 4–5 mm. wide, silvery beneath, greenish above; peduncles 3–5 cm. long; racemes cylindrical, mostly 2.5–6 cm. long, dense; pedicels ca. 2 mm. long; bracts 3–5 mm. long, subulate, deciduous; calyx silky-villous, the upper lip ca. 2 mm. long, bifid, the lower ca. 3 mm., 3-toothed; petals violet-blue, the banner elliptic-obovate, ca. 7 mm. long, 5 mm. wide, with large whitish spot; keel rather strongly curved, ciliate above in distal half; pods white-silky, 11–15 mm. long; seeds 3, smooth, grayish.—Dry ridges and plateaus, 6500–10,000 ft.; Subalpine F., Alpine Fell-fields; cent. Sierra Nevada (Mono, Mariposa cos.); w. Nev. June–Aug.

32. **L. séllulus** Kell. [*L. Torreyi* auth., not Gray.] Cespitose perennial from a slender taproot, the stems erect to decumbent, 1–2 dm. long, rather poorly developed, with few subbasal lvs., the hairs appressed; petioles mostly 3–8 cm. long; lfts. 6–8, oblanceolate, 3–6 mm. wide with rather long appressed hairs, but still greenish; peduncles commonly 1–3 cm. longer than the lvs., racemes cylindrical, dense, 4–12(–15) cm. long; bracts subulate, 3–4 mm. long, semipersistent; pedicels ca. 1 mm. long; calyx silky, the upper lip ca. 3 mm., the lower ca. 4 mm. long; fls. violet to blue, 8–9 mm. long, the banner elliptic-obovate, with a large pale spot, wings slightly longer, keel woolly-bearded along upper edge especially toward the purple bluntish apex; pods silky, ca. 1 cm. long; seeds 2–4, smooth, pinkish to whitish, ca. 3.5 mm. long.—Well-drained, often moist granitic soil, 4000–9000 ft.; Yellow Pine F. to Lodgepole F.; Sierra Nevada from Mariposa Co. n., to Shasta Co.; w. Nev. June–Aug.

Var. **ártulus** (Jeps.) Eastw. [*L. lepidus* var. *a.* Jeps.] Racemes narrow; fls. 6–7 mm. long; keel more densely ciliate.—At ca. 5000 ft.; Yellow Pine F.; Modoc Co. May–June.

Ssp. **ursìnus** (Eastw.) Munz. [*L. u.* Eastw. *L. rubro-soli* Eastw.] With a much grayer shaggier silky-villous indument; petioles 4–9 cm. long, slender; lfts. 5–7, oblanceolate to ± oblong, 1–2.5 cm. long, 2–6 mm. wide, acute; peduncles ca. as long as or slightly surpassing foliage; racemes 3–10(–14) cm. long, dense; pedicels ca. 1 mm. long, bracts 3–4 mm. long, persistent; upper calyx-lip 2 mm. long, lower ca. 3 mm. long; petals violet, 8–9 mm. long, the banner elliptic-obovate; seeds ca. 3 mm. long, mottled.—Mostly dry rocky places, 2200–7100 ft.; Sagebrush Scrub, Yellow Pine F., Red Fir F.; Glenn, Trinity, Mendocino, and Siskiyou cos.; s. Ore. June–Aug.

33. **L. caespitòsus** Nutt. ex. T. & G. [*L. lepidus* ssp. *c.* Detl.] Subacaulescent, sub-appressed-silky, 5–12 cm. tall; petioles 5–10 cm. long, silky, ± purplish; lfts. 5–7, oblanceolate, 1–2.5 cm. long, green and sparsely silky above, grayish-silky beneath; peduncles ca. 1 cm. long, the densely-fld. racemes 4–8 cm. long, much exceeded by the lvs.; pedicels ca. 1 mm. long; bracts ± deciduous; upper calyx-lip deeply cleft, ca. 3 mm. long, lower tridentate, ca. 4 mm. long; fls. 7–8 mm. long, pale blue or lilac to whitish, the banner almost twice as long as wide, wings narrow, keel short-ciliate on upper edges, almost straight; pods ca. 1.5 cm. long; seeds 3–4, flesh-color, faintly mottled, ca. 2.5 mm. long.—Rocky slopes and flats, 10,000–11,000 ft.; Sagebrush Scrub; White Mts., Inyo Co.; e. Ore. to Mont., Colo. June–July.

34. **L. confértus** Kell. [*L. lepidus* var. *c.* C. P. Sm. *L. Torreyi* Gray. *L. aridus* var. *T.* C. P. Sm. *L. c.* var. *ramosus* (Jeps.) Eastw. *L. l.* var. *r.* Jeps.] Stout, densely white-silky perennial, with several stems from a stout root, 2–3.5 dm. high, densely leafy below, less so above; petioles 3–9 cm. long, slender; lfts. mostly 7, elliptical-oblanceolate, 1.5–4 cm. long, 5–8 mm. wide, acutish, loosely grayish-silky; peduncles 2–7 cm. long, sur-passing foliage; racemes dense, 5–30 cm. long; pedicels 1–2 mm. long; bracts persistent, subulate, 8–9 mm. long; upper calyx-lip ca. 5 mm., lower 7 mm. long; fls. 10–14 mm. long, violet-purple, the banner elliptic-obovate, with pale cent. spot, keel slightly curved, short-acuminate, woolly-ciliate on upper edges toward the distal part; pods 10–18 mm. long, 2–5-seeded, white-silky-villous; seeds pale, ± mottled, 2.5–3 mm. long.—Meadows and dampish places, 3000–8500 ft.; Yellow Pine F., Red Fir F., Sage-brush Scrub; Plumas Co. to Mt. Pinos and San Bernardino Mts.; w. Nev. June–Aug.

35. **L. Culbertsònii** Greene. [*L. lepidus* var. *C.* C. P. Sm.] A ± cespitose perennial 1.5–3.5 dm. high, with rather slender stems, strigulose and sparsely villous, the plant quite greenish; lvs. fairly well distributed, the petioles slender, 3–9 cm. long; lfts. 5–7, oblanceolate, 1–3 cm. long, mostly 3–5 mm. wide, cuspidately acute; peduncles mostly 3–5 cm. long; racemes 5–10 cm. long, fairly dense; pedicels 1–2 mm. long; fls. ca. 1 cm. long, suberect after anthesis; bracts subulate, 4–5 mm. long, rather persistent; upper calyx-lip ca. 3 mm., lower ca. 4 mm. long; petals deep blue, the banner roundish, with cent. light spot, keel curved, ciliate on upper edges toward purple apex.—Rocky slopes, 7700–12,000 ft.; Lodgepole F., Subalpine F.; Tulare and Fresno cos. July–Aug.

36. **L. praténsis** Heller. [*L. sellulus* var. *elatus* Eastw.] Plants clumped, 3–7 dm. high, strigose especially upward, the stems ± hollow, suberect, rather leafy; petioles slender, 2–8 cm. long; lfts. 5–9, linear to linear-oblanceolate, 3–8 cm. long, 3–8 mm. wide, acute; peduncles 5–13 cm. long; racemes densely fld., 8–16 cm. long; pedicels 2–3 mm. long, bracts persistent, lance-linear, 7–9 mm. long; upper calyx-lip ca. 4, the lower ca. 5 mm. long; petals violet to blue or purple, the banner oval or ovate, ca. 10 mm. long, glabrous, wings slightly longer, keel straight, ciliate on upper edges; pods 18–20 mm. long, ca. 5 mm. wide, loosely pubescent; seeds ca. 5, quadratish, ca. 4 mm. long.—Moist places, 4000–10,500 ft.; Sagebrush Scrub to Subalpine F.; s. Sierra Nevada of Tulare, Fresno, Mono and Inyo cos. May–Aug.

Var. **eriostàchyus** C. P. Sm. Banner pubescent on back.—Big Pine Creek, Inyo Co.

37. **L. hypolàsius** Greene. [*L. brunneo-maculatus* Eastw. *L. danaus* var. *bicolor* Eastw.] A ± cespitose or more loose perennial, the stems decumbent, commonly branched, 0.5–3 dm. long, herbage strigose and villous; petioles slender, 2–6 cm. long; lfts. 5–7, unequal, cuneate to oblanceolate, 6–15 mm. long, 2–4 mm. wide, acute; peduncles 3–5 cm. long; racemes lax, 2–10 cm. long; pedicels 2–3 mm. long; bracts subulate, 3–4 mm. long, persistent; upper calyx-lip ca. 2, the lower ca. 3 mm. long; petals purple or blue, the banner suborbicular, ca. 7 mm. long, with cent. light spot, the keel nearly straight, with ciliated upper edges; pods 1.5 cm. long, villous; seeds 3–4, obscurely mottled, 3 mm. long.—Sandy and rocky, ± damp places, 9600–11,000 ft.; Subalpine F.; Mono, Fresno, and Tulare cos. July–Aug.

38. **L. Covíllei** Greene. [*L. dasyphyllus* Greene.] Loosely villous perennial, with erect simple leafy stems 2–8 dm. high; petioles 1–6 cm. long; lfts. 7–9, almost linear, attenuate, loosely silky-villous; peduncles 2–6 cm. long; racemes 10–20 cm. long, with scattered or semiverticillate fls.; pedicels 2–3 mm. long; bracts linear, villous, persistent, 8–15 mm. long; upper calyx-lip ca. 6 mm., the lower ca. 7 mm. long; petals blue, the

banner suborbicular, glabrous, ca. 8–9 mm. long, with large cent. yellowish or pale spot, wings and keel longer, the latter arcuate, sparsely ciliate above near apex; pods densely hairy, ca. 2.5–3 cm. long, 4–6-ovuled.—Rocky places, 8500–10,000 ft.; Subalpine F.; Sierra Nevada, Fresno, and Tulare cos. to Tuolumne Co. July–Sept.

39. **L. nevadénsis** Heller. Cespitose perennial, with erect or ascending stems, simple to branched, long-villous with shorter pubescence beneath, 2–4 dm. high; lvs. well distributed, the basal petioles to ca. 14 cm., the upper to ca. 3 cm. long; lfts. 7–10, elliptic-oblanceolate, 2.5–3.5 cm. long, shaggy-villous, acute; peduncles 3–6 cm. long; racemes rather lax, whorled, 10–18 cm. long; pedicels 5–8 mm. long; bracts subulate, 5 mm. long, deciduous; upper calyx-lip cleft, 3–3.5 mm. the lower 3-toothed, 4–4.5 mm. long; fls. 10–12 mm. long, blue-violet, the banner broader than long with a whitish cent. spot and purple dots; keel mostly ciliate on upper margins, strongly curved, with attenuate purple apex; pods 3–4 cm. long, hairy; seeds 3–4, flesh-color.—At ca. 4000–6000 ft.; Sagebrush Scrub; ne. Mono Co., Grapevine Mts., Inyo Co.; w. cent. Nev. April–June.

40. **L. Duránii** Eastw. Cespitose, few–several-stemmed, 5–12 cm. high, densely silky-villous throughout and ± shaggy, from a branched root-crown and deep roots; lvs. in basal clusters, the petioles 2–5 cm. long; lfts. 5–8, oblanceolate, 0.5–1.7 cm. long, 0.3–0.9 cm. wide; stipules 6–11 mm. long, adnate ⅓–½ their length; peduncles scapose, mostly slightly surpassing foliage; racemes dense, 2–6 cm. long, many-fld.; bracts lance-linear, 4–7 mm. long, deciduous; pedicels 4–5 mm. long; calyx mostly ca. 6–7 mm. long, upper lip deeply bidentate; fls. 8–11 mm. long, lilac or violet with large cream-white spot on banner; banner glabrous or ± pubescent on back, obovate, wings oblong, keel subglabrous to ciliate on upper edge; pods 1–2 cm. long; seeds 3–5, white.—In dry volcanic sand or gravel, 6500–8500 ft.; Yellow Pine F., Red Fir F.; Mono Co. (Lundy Lake to June Lake.) May–Aug.

41. **L. Brèweri** Gray. Prostrate or decumbent branched matted perennial with woody base and silvery-silky foliage and stems; lvs. crowded, the petioles 1–5 cm. long, the lfts. 7–10, oblanceolate to spatulate, 0.5–2 cm. long; peduncles 1–3 cm. long; racemes ca. 3–5 cm. long, mostly densely fld.; pedicels 1–3 mm. long, bracts deciduous; upper calyx-lip cleft, lower entire to 3-toothed, ca. 3 mm. long; fls. 6–9 mm. long, violet, the banner roundish to obovate with white or yellowish center, mostly glabrous on back; keel straightish on upper edges and glabrous or with few cilia; pods 12–16 mm. long, silky; seeds 3–4, flesh-colored with brownish markings and 3–4 mm. long.—Dry stony slopes and benches, 4000–11,000 ft.; Montane Coniferous F.; Mt. Pinos to Sierra Nevada and Siskiyou Mts.; s. Ore., w. Nev. June–Aug. A form with subcapitate infl. and fls. 5–7 mm. long, has been named var. *párvulus* C. P. Sm.

KEY TO VARIETIES

Keel glabrous or with few cilia; banner mostly ± glabrous on back.
    Fls. mostly 6–9 mm. long; lfts. mostly more than 7 mm. long .... *L. Breweri*
    Fls. mostly 4–6 mm. long; lfts. mostly 3–5 mm. long .... var. *bryoides*
Keel mostly strongly ciliate on upper edges; banner mostly pubescent on back .... var. *grandiflorus*

Var. **bryoìdes** C. P. Sm. in Jeps. [*L. tegeticulatus* Eastw.] Infl. ca. 2–3 cm. high, subcapitate; petioles ca. 1 cm. long.—Dry flats and ridges, 8000–12,000 ft.; Lodgepole F. to Alpine Fell-fields; Ventura Co. to Sierra Nevada of Tulare and Inyo cos., White Mts. July–Aug.

Var. **grandiflòrus** C. P. Sm. [*L. B.* var. *Clokeyanus* C. P. Sm. *L. Campbellae* Eastw. *L. Campbellae* var. *bernardinus* Eastw. *L. monensis* Eastw.] Peduncles 3–8 cm. long; racemes 3–10 cm. long; fls. 9–11 mm. long.—At 6500–10,000 ft.; Montane Coniferous F.; San Bernardino Mts. to Sierra Nevada of Tuolumne Co. June–Aug.

42. **L. meionánthus** Gray. Plants 2–5(–9) dm. high, erect or ascending from a woody branching root-crown, densely subappressed-silky; lvs. well distributed, the petioles 2–3.5 cm. long; lfts. 6–9, oblanceolate, 1.2–2.5 cm. long, silky on both sides; peduncles 1.5–3 cm. long; racemes 5–14 cm. long, rather densely fld.; pedicels 2–3 mm. long; bracts ca. 2 mm. long, deciduous; fls. ca. 6 mm. long; calyx-lips subequal, obscurely 2–3-toothed; petals dull blue or lilac, the banner glabrous, roundish, with yellow center, keel ciliate on upper margins, with short obtuse tip; pods 1.5–2.5 cm. long, hairy; seeds 3–4, pale, ca. 5 mm. long.—Dry open places, 5000–9800 ft.; Sagebrush Scrub, Red Fir F., Lodge-

pole F.; Sierra Nevada from Madera and Mono cos. to Plumas and Modoc cos.; w. Nev. July–Aug.

43. **L. alpéstris** A. Nels. [*L. Munzii* Eastw. *L. Funstonanus* C. P. Sm.] Rather finely strigose, ± canescent perennial, with many erect leafy branched stems 5–7 dm. high; lowermost petioles 8–10 cm. long, those of lvs. present at anthesis mostly 2–3 cm. long; lfts. 5–8, unequal, linear-elliptic, sharply acute, 2–4 cm. long, 4–6 mm. wide, finely strigose on both surfaces but greener above; peduncles ca. 2.5–3.5 cm. long; bracts lanceolate, ca. 2.5–3 mm. long, deciduous; pedicels slender, 4–5 mm. long; upper calyx-lip ca. 4 mm. long, cleft, the lower ca. 4.5 mm. long, often gibbous at base, subentire; corolla 9–12 mm. long, light blue, the banner suborbicular, slightly angled apically, pubescent on back, yellowish toward base, the keel somewhat curved and ciliate on upper edge toward slender apex.—Dry rocky places, 6500–10,800 ft.; Pinyon-Juniper Wd., Bristle-cone Pine F.; Panamint Mts., Inyo Co.; to Mont., Colo., Ariz. June–July.

44. **L. arbústus** Dougl. in Lindl. ssp. **silvícola** (Heller) D. Dunn. [*L. laxiflorus* var. *s.* (Heller) C. P. Sm. *L. s.* Heller. *L. elegantulus* Eastw. *L. lassenensis* Eastw.] Perennial with several erect stems 3–6 dm. high, strigose especially in upper parts; lvs. scattered, the lower petioles 6–9 cm. long, slender, the upper shorter; lfts. 7–9, oblanceolate, 2–4 cm. long, 3–8(–10) mm. wide, strigose on both sides but quite green; peduncles 2–4 cm. long; racemes 6–12 cm. long, rather open-verticillate; bracts lanceolate, 3–4 mm. long, deciduous; pedicels 2–3 mm. long; upper calyx-lip 2-toothed, 3 mm. long, lower 3-toothed, ca. 4 mm. long, spur ca. 1 mm. long; fls. 8–9 mm. long, mostly blue or violet, the banner pubescent on the back, wings pubescent without toward apex, the keel moderately curved, ciliate at short pointed apex; pods 2–2.5 cm. long; seeds 3–5, obscurely marked, ca. 4.5 mm. long.—Dry open woods, 5600–7500 ft.; Montane Coniferous F.; Sierra Nevada from Mariposa Co. n., to Modoc Co.; s. Ore. May–July.

Ssp. **arbústus** var. **montànus** (Howell) D. Dunn. [*L. laxiflorus* var. *m.* Howell. *L. laxiflorus* var. *cognatus* C. P. Sm. in Jeps.] Fls. 10–13 mm. long, the upper calyx-lip largely hidden by the broad banner.—Widely scattered from n. slope of San Gabriel Mts. to Modoc Co.; to Ore. Doubtfully distinct from ssp. *silvicola.*

Ssp. **calcaràtus** (Kell.) D. Dunn. [*L. c.* Kell. *L. Noldekae* Eastw. *L. laxiflorus* var. *c.* C. P. Sm.] Calyx-spur 1–3 mm. long; petals often pale yellow to whitish, sometimes blue.—Dry slopes and meadows, 5000–9600 ft.; Sagebrush Scrub to Lodgepole F.; e. slope of Sierra Nevada from Inyo Co. to Modoc Co.; to Ore., Ida., Utah.

45. **L. argénteus** Pursh. [*L. corymbosus* Heller. *L. laxiflorus* var. *c.* Jeps.] Short-strigose perennial, erect, branched above, 3–6 dm. tall; lvs. well distributed, the lower petioles to 1 dm. long and dying before anthesis, but upper lvs. mostly 2–5 cm. long; lfts. 5–9, linear-oblanceolate, 1.5–4.5 cm. long, 2–5 mm. wide, subglabrous above, strigose beneath, acute; peduncles 2–3 cm. long; racemes 7–12 cm. long, ± lax; bracts lance-ovate, ca. 2 mm. long, deciduous; pedicels 3–6 mm. long; upper calyx-lip ca. 4 mm. long, notched, the lower entire, ca. 5 mm. long, spur short, blunt; fls. blue or lilac, ± subverticillate, 10–13 mm. long, the banner suborbicular, ± pubescent on back, wings glabrous, keel mostly ciliate and strongly curved on upper edges; pods strigose-silky, 2–2.5 cm. long; seeds 2–4, flesh-colored, mostly unmottled, ca. 4 mm. long; $n = 24$ (Phillips in 1957).—Dry flats and slopes, 4000–5000 ft.; Sagebrush Scrub, N. Juniper Wd., Yellow Pine F.; Siskiyou to Butte and Modoc cos.; to Mont., New Mex. June–Oct.

Var. **tenéllus** (Dougl. ex G. Don) D. Dunn. [*L. t.* Dougl. ex G. Don.] Lfts. strigose to sericeus above; stems often reddish, strigose; fls. 8–10 mm. long.—Dry rocky slopes, 9400–10,500 ft.; Bristle-cone Pine F.; Inyo and White mts.; to Ore., Colo., New Mex. July–Aug.

46. **L. caudàtus** Kell. [*L. Rosei* Eastw.] Strigose, ± silky perennial, the stems 2–6 dm. tall, simple or branched; lower lvs. ± persistent until anthesis, with petioles to 1 dm. long, upper 2–5 cm. long; lfts. 5–9, oblanceolate, 2–5 cm. long, silky on both sides, often silvery, acutish; peduncles 2–5 cm. long; racemes 5–15 cm. long; fls. scattered or subverticillate, 9–12 mm. long; bracts subulate, 3.5–5 mm. long, deciduous; pedicels 2–4 mm. long; upper calyx-lip bidentate, ca. 5 mm. long, the lower entire, ca. 7 mm. long, the spur ca. 1 mm. long; petals violet-blue to deep blue or almost white, banner silky on back, wings subglabrous, keel ciliate on upper somewhat curved edges; pods 2.5–3 cm. long, silky, seeds 5–6, red-brown to flesh-color, ca. 4 mm. long; $n = 24$ (Phillips in 1957).—Dry open slopes, 4000–9600 ft.; Sagebrush Scrub, N. Juniper Wd., Pinyon-

Juniper Wd., Yellow Pine F.; n. Inyo Co. to Modoc and Shasta cos.; e. Ore., Ida., Utah. May–Sept.

47. **L. Pálmeri** Wats. [*L. candidissimus* Eastw. *L. fontis-Batchelderi* C. P. Sm. *L. portae-westgardiae* C. P. Sm. *L. Jaegeranus* C. P. Sm. *L. inyoensis* var. *eriocalyx* C. P. Sm. *L. Keckianus* C. P. Sm.] Plants 3–6 dm. tall, the several stems ± hoary with retrorsely spreading hairs 1–2 mm. long; petioles 4–10 cm. long; lfts. 6–9, elliptic-oblanceolate, 2.5–5 cm. long, densely sericeous; peduncles 4–7 cm. long; racemes 10–20 cm. long, lax, subverticillate; pedicels 2–7 mm. long; fls. blue, 8–10 mm. long, the banner suborbicular, reflexed above the mid-point; keel ciliate above toward the tip.—Dry stony places, 6500–7500 ft.; Pinyon-Juniper Wd.; s. White Mts. to Grapevine Mts., Inyo Co.; to Nev., Ariz., New Mex. May–June.

48. **L. inyoénsis** Heller. [*L. Padre-Crowleyi* C. P. Sm.] Perennial, 5–6 dm. high, spreading-pubescent, leafy; lower petioles 8–13 cm. long, upper 2–5 cm.; lfts. 7–9, oblanceolate, 2–3.5 cm. long, densely villous-pubescent on both sides with subappressed hairs, acute; peduncles 2–4 cm. long; racemes 8–15 cm. long; fls. scattered to verticillate, ca. 12 mm. long; bracts ca. 5–6 mm. long, deciduous; calyx-spur ca. 1 mm. long, the upper lip 6–7 mm. long, 2-toothed, the lower 7–8 mm., entire; petals blue to violet, the banner silky on back, wings glabrous, keel sharply curved, ciliate on outer half of upper edges, with short-acuminate apex; pods villous, 1.5–2.5 cm. long; seeds 4–5, compressed, mottled, ca. 4 mm. long.—Dryish sandy places, 7000–9000 ft.; Pinyon-Juniper Wd., Sagebrush Scrub, etc., e. slope of Sierra Nevada, Inyo Co. May–July.

49. **L. onústus** Wats. [*L. pinetorum* Jones. *L. violaceus* Heller and vars. *shastensis* and *delnortensis* Eastw. *L. sulphureus* ssp. *d.* Phillips. *L. Thompsonianus* C. P. Sm.] Perennial from rather slender underground rootstocks, the stems slender, decumbent at base, appressed-silky, 2–3 dm. high; lvs. few, near the base; petioles mostly 8–13 cm. long; lfts. 5–9, oblanceolate, 2–3(–4) cm. long, 5–10 mm. wide, acute to obtuse, glabrous above, appressed-silky beneath; peduncles 5–8 cm. long, racemes 5–15 cm. long; fls. rather few, scattered, 8–11 mm. long; bracts subulate, 3–4 mm. long, deciduous; pedicels 3–5 mm. long; calyx scarcely gibbous, the upper lip notched, ca. 2 mm. long, the lower entire, 3.5–4 mm. long; petals deep violet-blue, the banner and wings glabrous, keel almost straight and ciliate on upper edges; pods 3–4.5 cm. long, ca. 1 cm. wide; seeds 5–6, brown, 6–7 mm. long.—Occasional, dry slopes, 2900–5500 ft.; largely Yellow Pine F.; Trinity and Del Norte cos. to Lassen and Plumas cos.; s. Ore. April–Sept.

50. **L. Peirsònii** Mason. Herbaceous silvery-silky perennial, 3–6 dm. tall, the stems many, ascending; lvs. subbasal; petioles 6–15 cm. long; lfts. 5–8, oblong-oblanceolate, obtuse and apiculate, 3–7 cm. long, 10–15 mm. wide, densely silky on both sides; peduncles 1–2.5 dm. long; racemes 1–1.5 dm. long; fls. subverticillate, 10–12 mm. long, the lower whorls distinct; bracts lanceolate, 6–7 mm. long, deciduous; pedicels 1–1.5 mm. long; calyx-lips subequal, 4–5 mm. long, entire or obscurely toothed; petals yellow, the banner pubescent on back, with a brownish spot above center, keel ciliate on upper straightish edges; pods silky, 3 or more cm. long, 3–5-seeded.—Rare, loose gravelly and rocky slopes, 4000–4500 ft.; Pinyon-Juniper Wd., Joshua Tree Wd.; desert slopes of San Gabriel and Tehachapi mts. April–May.

51. **L. leucophýllus** Dougl. ex Lindl. Grayish woolly-villous perennial, the stout stems erect, simple or branched, 5–9 dm. tall; lvs. many, well distributed; petioles 6–15 cm. long; lfts. 7–9, oblanceolate, 3–6.5 cm. long, acute, silky-velvety; peduncles 3–8 cm. long, racemes dense, 8–30 cm. long; fls. crowded, many, subverticillate, 10–12 mm. long, white, pinkish, bluish or purple; bracts linear-lanceolate, 3–4 mm. long, ± persistent; pedicels 1–3 mm. long; calyx often gibbous, calyx-lips 3–4 mm. long, the upper bi-dentate, the lower subentire; banner pubescent on back, keel stout, ciliate on upper edges; pods 2–2.5 cm. long, silky-woolly; seeds 3–6, red-brown to grayish, obscurely mottled; $n = 24$, 48 (Phillips, 1957).—Dry places, 1500–5300 ft.; mostly Sagebrush Scrub, Yellow Pine F.; Colusa and Trinity cos. to Shasta, Lassen, Modoc, and Siskiyou cos.; to Wash., Mont. May–Aug. Variable, plants with fls. 8–10 mm. long and stems ± villous, have been called var. *Bélliae* C. P. Sm.

Var. **canéscens** (Howell) C. P. Sm. [*L. c.* Howell.] Stems with short appressed hairs; fls. 8–10 mm. long.—Dry places, 4000–5000 ft.; Sagebrush Scrub, Yellow Pine F.; Plumas, Lassen, and Modoc cos.; to e. Wash.

52. **L. obtusilóbus** Heller. [*L. ornatus* var. *o.* C. P. Sm.] Perennial with a woody

branching root-crown, the stems subdecumbent to ascending, 1.5–3 dm. long, herbage appressed silvery-silky; petioles 2–5 cm. long; lfts. oblong-oblanceolate, 2–4 cm. long, 4–6 mm. wide, acute to obtuse; peduncles 1–5 cm. long; racemes dense, 3–7 cm. long; fls. 11–13 mm. long, blue or lilac, ± verticillate; bracts subulate, 3–4 mm. long, caducous; pedicels 3–5 mm. long; calyx silky, the upper lip bidentate, 3.5–4 mm. long, the lower slightly longer, often 3-toothed; banner broader than long, pubescent on back, with a yellow center, keel ciliate on upper edges, moderately curved, with an acuminate apex; pods silky, 2.5–4 cm. long; seeds 4–5, brown, mottled, ca. 4 mm. long.—Gravelly summits, 5500–10,000 ft.; Red Fir F. to Subalpine F.; Yolla Bolly Mts. to Humboldt and Siskiyou cos., then to Plumas and Sierra cos. June–Sept.

53. **L. lapidícola** Heller. Silvery-silky, from branched woody root-crown, to ca. 1 dm. high; lvs. subbasal, the petioles 2–4.5 cm. long; lfts. 6–8, linear-oblanceolate to subspatulate, 1–2 cm. long, silvery on both sides, acute; peduncles 5–10 cm. long, slender; racemes of 2–3 whorls, 2–4 cm. long; bracts subulate, ca. 5 mm. long, deciduous; pedicels slender, 2–4 mm. long; calyx silvery, the upper lip ca. 4, the lower ca. 5 mm. long; fls. 10–12 mm. long, purplish-blue, the banner mostly pubescent on back, with a dull yellow center, keel mostly ciliate on upper edges, the short acumen upturned.— Dry granitic gravel, 4900–8000 ft.; Yellow Pine F. to Subalpine F.; Mt. Eddy and Siskiyou Mts., Siskiyou Co. July.

54. **L. montígenus** Heller. [*L. Olive-Nortonae* and *L. Olive-Brownae* C. P. Sm.] Erect or decumbent, with simple or branched stems 3–4 dm. tall, appressed-pubescent and ± villous; lvs. well distributed, the basal petioles 10–15 cm. long, upper 2–5 cm.; lfts. 7–9, oblanceolate, acute, 2.5–4 cm. long, 4–8 mm. wide, silky on both sides; peduncles 3–4 cm. long; racemes 5–10 cm. long, ± lax; bracts rather persistent, 4–6 mm. long; pedicels 4–5 mm. long; calyx gibbous at base, the upper lip 2-toothed, the lower 3-toothed; petals 9–12 mm. long, blue to violet, the banner longer than broad, sometimes pubescent on back, keel straightish to curved, ciliate on upper edges; pods 1.5–2 cm. long, silky. —Reported from Gaspipe Springs and Sweetwater Mts., 10,300 ft., Mono Co., Rock Creek Lake Basin, Inyo Co., 10,000 ft.; w. Nev. July–Aug.

55. **L. Abrámsii** C. P. Sm. White-woolly-villous perennial, with branched decumbent stems 2–6 dm. tall; petioles 3–5 cm. long; lfts. 8–9, oblanceolate, 2–3 cm. long, 3–7 mm. wide, acute to obtuse; peduncles 6–10 cm. long; racemes 15–25 cm. long, with verticillate fls. 14–16 mm. long; bracts deciduous, 10 mm. long; pedicels 4–7 mm. long; calyx woolly, upper lip bifid, lower 3-toothed; petals broad, blue, banner ± pubescent on back, 15–16 mm. wide, keel slightly curved, ciliate on upper edges; pods 3.5–4 cm. long, silky-strigose.—Open woods, 2000–5000 ft.; Mixed Evergreen F., Yellow Pine F.; Santa Lucia Mts. May–June.

56. **L. sericàtus** Kell. Minutely but densely silky-canescent perennial from a branched root-crown, the stems 1.5–3(–5) dm. long, mostly unbranched, erect to decumbent; lvs. rather scattered; petioles 5–15 cm. long; lfts. 6–7, spatulate-obovate, rounded or retuse at apex, 3–4 cm. long, 1–2 cm. wide; peduncles 8–12 cm. long; racemes 10–30 cm. long, rather dense, the fls. scattered or subverticillate; bracts subulate, 5–7 mm. long, early deciduous; pedicels 4–5 mm. long; calyx strigose, upper lip 2-toothed, lower entire or minutely 3-toothed, 6–7 mm. long; petals purplish-blue, 14–16 mm. long, the banner roundish, ± pubescent on back, keel curved, ciliate on upper edges and on lower free edges near claws; pods pubescent, 2–2.5 cm. long; seeds 5–7, light brown, ± mottled. —Open wooded slopes, 2000–4000 ft.; Yellow Pine F., Mixed Evergreen F.; Napa and Mayacamas ranges, Lake and Sonoma cos. April–June.

57. **L. cervìnus** Kell. [*L. latissimus* Greene.] Plants densely velvety-pubescent or subsilky, erect, 1.5–3 dm. high; lvs. subbasal, the petioles 10–20 cm. long; lfts. 5–8, lance-obovate or wider, 2–6 cm. long, 1–3 cm. wide, silky on both sides; peduncles 15–20 cm. long; racemes 12–20 cm. long, the scattered or subverticillate fls. rather crowded, 14–16 mm. long; bracts deciduous; pedicels 3–5 mm. long, spreading-pubescent; calyx-lips subequal, the upper 2-, the lower 3-toothed; petals light blue, broad, the banner broader than long, ± pubescent on back, keel curved, ciliate on upper edges and on lower near claws; pods silky, ca. 3 cm. long; seeds 4–7, ± mottled and with dark lateral line, rhombic-oblong, 4 mm. long.—Dry places, 1000–4500 ft.; Yellow Pine F.; Santa Lucia Mts. May–June.

58. **L. Gràyi** (Wats.) Wats. [*L. Andersonii* var. *G.* Wats. *L. Louise-Bucariae* C. P.

Sm. *L. Ione-Grisetae* C. P. Sm.] Stems several from a branched woody root-crown, ascending to decumbent, 2–3.5 dm. long, densely grayish-tomentose and ± villous; lvs. largely subbasal, the petioles 5–12 cm. long; lfts. 5–11, oblanceolate, 2.5–3.5 cm. long, 5–7 mm. wide, acute, tomentose; peduncles 5–15 cm. long; racemes 10–15 cm. long, with subverticillate fls. 12–14 mm. long; bracts 4–5 mm. long, early deciduous; pedicels 2–4 mm. long; upper calyx-lip broad, ± deeply 2-toothed, 5–6 mm. long, the lower entire to 3-toothed, slightly longer; petals deep purplish-blue to pale, the banner roundish, glabrous or pubescent on back, with a yellow center, the keel not strongly curved, densely ciliate on upper edges and often on lower edges near base; pods strigose, 2.5–3.5 cm. long; seeds 4–6, obscurely mottled and with dark lateral line.—Dry slopes, 2000–7800 ft.; Yellow Pine F., Red Fir F.; Sierra Nevada from Plumas and Butte cos. to Kern Co. May–July. *L. falsoformis* and *L. falsograyi* C. P. Sm. may be hybrids with *L. Andersonii* Wats.

58a. **L. sublanàtus** Eastw. Resembling *L. Grayi* in pubescence; lfts. ca. 2 cm. long; fls. 1 cm. long.—Mono Co. July.

59. **L. ludoviciànus** Greene. Densely tomentose and villous, branching from a woody base, 3–5 dm. high; petioles 6–10 cm. long; lfts. 4–8, cuneate or spatulate, 1.5–3 cm. long, 7–10 mm. wide, acute to obtuse; peduncles 8–15 cm. long; racemes 10–20 cm. long with separate whorls of fls. 10–12 mm. long; bracts deciduous; pedicels ca. 3 mm. long; upper calyx-lip deeply cleft, the lower 3-toothed; petals purplish, banner glabrous or ± pubescent on back, keel somewhat curved, ciliate on upper edges; pods densely strigose, 2–2.5 cm. long; seeds 4–6, grayish-yellow, ± marked, ca. 4 mm. long.—Dry places, below 1500 ft.; Chaparral?; sw. San Luis Obispo Co. June–July.

60. **L. Chamissònis** Eschs. [*L. C.* var. *longebracteatus* Wats.] Erect branching shrub, 5–20 dm. tall, minutely silky-strigose, with many short leafy branches; petioles 2–3.5 cm. long; lfts. 6–9, oblanceolate, appressed-silky, 1–2.5 cm. long, 4–6 mm. wide; peduncles 2–5 cm. long; racemes 5–15 cm. long, the fls. ± verticillate, 12–16 mm. long; bracts lance-linear, 7–10 mm. long, early deciduous; pedicels 6–8 mm. long, spreading-pubescent; upper calyx-lip cleft, 6–7 mm. long, lower entire, 7–8 mm. long; petals blue or lavender, the banner pubescent on back, broad, with a yellow spot, keel curved, nonciliate or nearly so, with slender upturned acumen; pods 2.5–3.5 cm. long, strigose; seeds 4–7, mottled with brown, 4–5 mm. long.—Sandy beaches and dunes; largely Coastal Strand; Marin Co. to Los Angeles Co. March–July.

61. **L. arbòreus** Sims. [*L. propinquus* Greene. *L. rivularis*, as used by Eastw.] From suffrutescent to shrubby, sometimes low, mostly 1–2(–2.5) m. tall, with numerous short branches; herbage puberulent or pubescent to ± silky; petioles 2–6 cm. long; lfts. 5–12, oblanceolate, 2–6 cm. long, 5–10 mm. wide, strigose above and beneath or only beneath; peduncles 4–10 cm. long; racemes 10–30 cm. long, mostly lax, the fls. scattered or sub-verticillate, 14–17 mm. long; bracts lanceolate, to ca. 1 cm. long, deciduous; upper calyx-lip notched, lower entire, 5–7 mm. long; petals broad, mostly yellow, sometimes lilac to blue or mixed, the banner roundish, glabrous, keel curved, ciliate on upper edges; pods brown, strigose, 4–7 cm. long; seeds 8–12, oblong, dark brown, ± mottled, 4–5 mm. long, with pair of pale spots enclosing micropyle; $2n = 40$ (Savchenko, 1935). —Sandy places, below 100 ft.; Coastal Strand, Coastal Sage Scrub, N. Coastal Scrub, Closed-cone Pine F.; near the coast from Ventura Co. to Del Norte Co. and n. March–June.

Var. **exímius** (Davy) C. P. Sm. [*L. e.* Davy.] Somewhat villous; racemes 5–10 cm. long; banner mostly yellow, wings blue.—N. Coastal Scrub; Montara Mt., San Mateo Co. April–July.

62. **L. álbifrons** Benth. [*L. fragrans* Heller.] Rounded leafy shrub, appressed-silky, much-branched, 6–15 dm. high; petioles 2–4 cm. long; lfts. 7–10, oblanceolate to spatulate, 1–3 cm. long, 4–10 mm. wide, silvery-silky on both sides, acute or rounded apically; peduncles 5–13 cm. long; racemes 8–30 cm. long, not dense, the fls. mostly whorled, 10–14 mm. long; bracts to ca. 5 mm. long, deciduous; pedicels 4–8 mm. long, with spreading hairs; upper calyx-lip cleft, lower entire, ca. 5–6 mm. long; petals blue to red-purple or lavender, the banner ± pubescent on back, with a light center; keel narrowed toward base, ciliate above toward apex; pods 3–5 cm. long, villous-strigose; seeds 5–9, ca. 4 mm. long, mottled or spotted, usually with a marginal line.—Sandy to rocky places, below 5000 ft.; many Plant Communities; Coast Ranges from Humboldt

Co. to Ventura Co.; Sierra Nevada foothills from Tulare Co. n., to Shasta Co. March–June.

KEY TO VARIETIES

Stems not woody, mostly 2–4 dm. tall; petioles mostly 4–8 cm. long. Coast Ranges, San Francisco
Bay region .... var. *collinus*
Stems woody, mostly 6–30 dm. tall; petioles various.
   Bracts mostly 10–15 mm. long, exceeding flower buds. Near coast .... var. *Douglasii*
   Bracts mostly 4–8 mm. long, not or scarcely exceeding buds.
      Fls. mostly 10–13 mm. long; pubescence of pedicels ± spreading. Widely distributed
                                                  L. albifrons
      Fls. mostly 14–16 mm. long; pubescence of pedicels sometimes appressed. Mostly from interior.
         Plants 1–3 dm. high; petioles 3–4 times as long as lfts. Siskiyou Co. .... var. *flumineus*
         Plants mostly 10–25 dm. high; petioles 1–2 times as long as lfts. Butte Co. to San Diego Co.
                                                var. *eminens*

Var. **collìnus** Greene. [*L. c.* Heller. *L. Isabelianus* Eastw.] Plants matted; keel often nonciliate.—Rocky places below 4500 ft.; N. Coastal Scrub, Chaparral, etc. hills in San Francisco Bay region, n. to Lake Co., s. to San Luis Obispo Co.

Var. **Douglàsii** (J. G. Agardh) C. P. Sm. [*L. D.* J. G. Agardh. *L. fallax* Greene. *L. D.* var. *fallax* J. T. Howell.] Lfts. mostly 3–6 cm. long.—Open slopes; Chaparral, Coastal Scrub, etc.; Marin Co., Monterey Co. to n. Santa Barbara Co.; Santa Cruz Id.

Var. **flumíneus** C. P. Sm. Lfts. mostly 1–2 cm. long.—Dry hills, 3400–4500 ft.; Yellow Pine F.; Siskiyou and Mendocino cos.; s. Ore.

Var. **éminens** (Greene) C. P. Sm. [*L. e.* Greene. *L. tricolor* Greene. *L. jucundus* Greene. *L. Brittonii* Abrams. *L. acutilobus* Heller.] Doubtfully distinct from the sp., but tending to be taller, with larger often less distinctly verticillate fls.—Mostly below 3500 ft.; Foothill Wd., Chaparral, Coastal Sage Scrub, etc.; ranges bordering Cent. V. from Siskiyou Co. to Kern Co., s. to San Diego Co., n. L. Calif.

63. **L. excùbitus** Jones. Close to and not certainly distinct from the *L. albifrons* complex; shrubby, mostly 5–15 dm. tall, ± branched, densely silky-appressed-pubescent with short hairs; petioles 4–10 cm. long; lfts. 7–9, oblong-oblanceolate to spatulate, 2–4 cm. long, densely silvery above and beneath; peduncles 4–15 cm. long; racemes 10–25 cm. long; fls. 10–12 mm. long, in separate verticils, 2–5 cm. apart; bracts deciduous, lanceolate, acuminate, 6–7 mm. long; pedicels 4–6 mm. long, strigose; upper calyx-lip notched, lower entire; petals blue to violet or orchid, the banner suborbicular, glabrous or somewhat pubescent on back, with center yellowish; wings and keel broad, the latter curved, ciliate on upper edges; pods 3–5 cm. long, silky; seeds 6–8, pale, ± mottled, with yellow-brown lateral lines and ca. 4 mm. long.—Gravelly and rocky places, washes and slopes, 4000–8700 ft.; Creosote Bush Scrub, Pinyon-Juniper Wd.; mts. of Inyo Co. s. and e. along desert margin to Little San Bernardino Mts. April–June.

KEY TO VARIETIES

Fls. 10–13 mm. long; desert plants.
   Pubescence adpressed; fl.-whorls mostly 2–5 cm. apart. Inyo Co. to San Bernardino Co.
                                                L. excubitus
   Pubescence tomentose; fl.-whorls mostly 1–2 cm. apart. Sw. Colo. Desert .... var. *medius*
Fls. 14–18 mm. long; montane and cismontane.
   Stems woody below.
      Plants 5–15 dm. high; foliage ± greenish, scarcely silvery. Below pine belt ...... var. *Hallii*
      Plants 1–2 dm. high; foliage silvery. Pine belt, San Gabriel Mts. .... var. *Johnstonii*
   Stems scarcely if at all woody, but quite herbaceous. Mostly in pine belt .... var. *austromontanus*

Var. **Hállii** (Abrams) C. P. Sm. [*L. H.* Abrams. *L. Paynei* A. Davids.] Plant coarser and greener, the lvs. less silky, whorls in infl. usually not more than 2.5 cm. apart; fls. 14–18 mm. long.—Gravelly and sandy washes, etc., below 4000 ft.; Chaparral, Coastal Sage Scrub; Ventura and San Bernardino cos. to n. L. Calif. April–June.

Var. **austromontànus** (Heller) C. P. Sm. [*L. a.* Heller.] Stems mostly herbaceous, 2–5 dm. high; lvs. crowded near base, silvery; racemes mostly 5–10 cm. long, congested; fls. 14–18 mm. long.—Dry slopes and rocky places, ca. 4000–8500 ft.; upper Chaparral, Yellow Pine F.; mts. from Tehachapi Mts. to n. L. Calif. May–July.

Var. **Johnstònii** C. P. Sm. in Jeps. Basal stems woody, branched, to ca. 1–2 dm.

high; racemes 6–12 cm. long, congested.—Dry slopes under pines, 5500–6600 ft.; Yellow Pine F.; San Gabriel Mts. May–July.

Var. **mèdius** (Jeps.) Munz. [*L. albifrons* var. *m.* Jeps. *L. Grayi* var. *m.* C. P. Sm.] Plants 3–7 dm. high, woody at base, finely white-tomentulose; fls. 11–13 mm. long, bluish-violet.—Washes; Creosote Bush Scrub, Pinyon-Juniper Wd.; Mt. Springs Grade, e. San Diego and w. Imperial cos. March–April.

64. **L. longifòlius** (Wats.) Abrams. [*L. Chamissonis* var. *l.* Wats. *L. mollisifolius* A. Davids.] Erect, shrubby below, appressed-pubescent, greenish, 1–1.5 m. tall; petioles 4–7 cm. long; lfts. 6–9, elliptic- or oblong-oblanceolate, obtuse, 2.5–6 cm. long, 5–15 mm. wide, subsilky on both sides; peduncles 6–12 cm. long; racemes 20–40 cm. long, the fls. scattered or subverticillate, 14–18 mm. long; bracts deciduous, 4–6 mm. long; pedicels 5–10 mm. long, with spreading pubescence; upper calyx-lip cleft or 2-toothed, low or entire or minutely 2-toothed; petals blue to violet, the glabrous banner suborbicular, with cent. yellowish spot, wings broad, keel curved, ciliate along upper edges; pods brownish, pubescent, 4–6 cm. long; seeds 6–8, gray, ca. 6 mm. long, with some mottling and brown lateral lines.—Coastal bluffs and canyons; Chaparral, Coastal Sage Scrub, S. Oak Wd.; Ventura Co. to San Diego Co., inland to San Dimas and Santa Ana Mts., etc.; n. L. Calif. Mostly April–June. *L. mollisifòlius* was proposed for plants from Sierra Madre, La Cañada, etc., with more upright growth, ± fistulose stems, and spring flowering, while coastal plants have some fls. most of year.

65. **L. variícolor** Steud. [*L. versicolor* Lindl., not Sweet. *L. franciscanus* Greene. *L. Micheneri* Greene.] Stems slender, usually branched, decumbent or prostrate, 2–8 dm. long, leafy, from a woody base; herbage ± long-pubescent or villous; petioles 4–10 cm. long; lfts. 7–9, oblanceolate, subglabrous to strigose above, ± silky beneath, 2–3.5 cm. long; peduncles 4–12 cm. long; racemes 6–15 cm. long, few-whorled; bracts subulate, 4–5 mm. long, deciduous; pedicels 4–12 mm. long, ± spreading-pubescent; fls. 11–16 mm. long, yellow, whitish, pink, purple, or blue, often with several colors in 1 fl.; upper calyx-lip entire or notched, lower entire; banner glabrous, broad, often paler than the broad wings; keel curved, ciliate on upper edges; pods 3–4 cm. long, brown, loosely hairy to strigose; seeds 7–9, ± mottled, 3–4 mm. long.—Grassy fields, slopes and sand dunes, below 1000 ft.; Coastal Strand, N. Coastal Scrub, Coastal Prairie, etc.; near the coast from Humboldt Co. to San Luis Obispo Co. April–July.

66. **L. littoràlis** Dougl. in Lindl. Near *L. variicolor*, but more pronouncedly spreading-villous especially about the nodes; roots bright yellow; petioles mostly 3–5 cm. long; lfts. 5–9, strigose on both sides, 2–3.5 cm. long; fls. 10–13 mm. long, the petals blue or lilac, fading brown; pods 3–3.5 cm. long; seeds 9–14, linear-oblong, mottled, ca. 3 mm. long; $n = 24$ (Phillips, 1957).—Coastal Strand and N. Coastal Scrub; Mendocino Co. to B.C. May–Aug.

67. **L. Tidestròmii** Greene. Near to *L. littoralis*, with appressed-silky slender stems 1–3 dm. long from a branched root-crown, roots yellow; petioles 1–3 cm. long; lfts. 3–5, oblanceolate, 1.3–2 cm. long, silvery-silky on both sides; peduncles 4–8 cm. long; racemes 2–10 cm. long, the bracts lance-ovate, 5 mm. long, early deciduous; pedicels 3–5 mm. long; fls. 11–13 mm. long, whorled, blue; calyx ca. 8 mm. long, the upper lip deeply lobed, lower entire or notched; banner glabrous, roundish, the light center violet in age; keel curved, ciliate on upper edges; pods 2–2.5 cm. long, yellowish; seeds 5–8, ± mottled with black, ca. 3 mm. long.—Coastal Strand; Monterey Peninsula. May–June.

Var. **Làyneae** (Eastw.) Munz. [*L. L.* Eastw.] Pubescence more shaggy; peduncles 1–4 cm. long; seeds not mottled.—Coastal Strand; Point Reyes, Marin Co. May–June.

68. **L. graciléntus** Greene. Stems slender, suberect, 2–4 dm. tall, leafy, subglabrous to ± strigose; petioles very slender, 3–6(–12) cm. long; lfts. 5–8, linear, 4–8 cm. long, 2–4 mm. wide, acuminate, strigose but green; peduncles 6–12 cm. long; racemes 10–20 cm. long; bracts filiform, 8–10 mm. long, deciduous; pedicels 2–4 mm. long; fls. 10–13 mm. long, in a few distinct whorls; upper calyx-lip bidentate, lower entire to bidentate; petals blue, the banner glabrous with yellowish center, wings broad, glabrous, keel ciliate above in outer half; pods 2–3 cm. long, copiously villous; ovules 7–8.—At 8000–10,500 ft.; Subalpine F.; Yosemite National Park to Rock Creek Lake Basin, nw. Inyo Co. July–Aug.

69. **L. albicaùlis** Dougl. ex Hook. [*L. Wolfianus* C. P. Sm. *L. purpurascens* Heller. *L. ochroleucus* Eastw. *L. sylvestris* E. Drew.] Herbaceous from a heavy root-crown, the stems mostly slender, erect, branched above, 4–9 dm. high, leafy, thinly strigose; stipules linear or subulate; petioles 2–4 cm. long; lfts. 5–9, oblanceolate, thinly strigose above and beneath, 2–5 cm. long, obtuse and mucronate; peduncles 3–8 cm. long; racemes 10–30 cm. long, spreading-pubescent, lax; bracts lanceolate, ca. 6–7 mm. long, early deciduous; pedicels 4–6 mm. long; fls. 12–16 mm. long, whitish to purple, fading brown; calyx-lips subequal, notched or the lower entire; banner ovate, acute, glabrous, wings narrow, keel slender, much-curved, practically nonciliate; pods 3–4 cm. long, silky-villous; seeds 5–7, compressed, ca. 4 mm. long, mottled with gray; $n = 24$ (Phillips in 1957).—Dry slopes and openings, 2000–8500 ft.; Yellow Pine F., Red Fir F.; Sierra Nevada from Fresno Co. n., Coast Ranges from Trinity Co. to Siskiyou Co.; to w. Wash. May–Aug. A possible hybrid with *L. formosus* is *L. Whiltonae* Eastw.

Var. **shasténsis** (Heller) C. P. Sm. [*L. s.* Heller.] Fls. 8–11 mm. long.—With the sp., from Kern and Trinity cos. to s. Ore.

70. **L. cròceus** Eastw. Erect herb 4–6 dm. tall, the stems slender, ± strigose, leafy; petioles mostly 3–8 cm. long; lfts. 5–8, oblanceolate, obtuse and mucronate at apex, 4–6 cm. long, subsilky to glabrate; peduncles 4–6 cm. long; racemes 6–10 cm. long, with several ± definite whorls; bracts linear, 3–4 mm. long, deciduous; pedicels ± strigose, 3–4 mm. long; fls. yellow, 12–15 mm. long; lower calyx-lip ca. 4 mm. long, entire, the upper shorter, bidentate; banner glabrous, round, wings broad, covering the curved nonciliate keel; pods 2–3.5 cm. long, hirsute; seeds 4–5, ca. 6–7 mm. long, clay-color with darker dots.—Dry places, 5000–8000 ft.; Yellow Pine F., Red Fir F., Subalpine F.; Siskiyou and Trinity cos. May–July.

Var. **piloséllus** (Eastw.) Munz. [*L. p.* Eastw.] Pubescence spreading; calyx-lips subequal.—At ca. 2700–5000 ft.; Yellow Pine F.; Trinity Co. to Siskiyou Co.

71. **L. Andersònii** Wats. [*L. Louise-Grisetae* C. P. Sm. *L. Leland-Smithii* C. P. Sm. *L. rimae* Eastw.] Near to *L. albicaulis* in stature and pubescence; stipules linear; racemes 6–18 cm. long, rather lax; fls. 10–12 mm. long; petals blue, purplish, or yellowish-white, banner rounded, glabrous, wings covering most of the curved nonciliate keel; pods strigose, 3–4 cm. long; seeds 4–6, compressed, 4–5 mm. long, yellowish to brown, ± obscurely mottled.—Dry slopes under pines, 4000–8500 ft.; Yellow Pine F. to Lodgepole F.; San Bernardino Mts., Mt. Pinos, Sierra Nevada to Humboldt and Siskiyou cos.; s. Ore., w. Nev. June–Sept.

Var. **apértus** (Heller) C. P. Sm. [*L. a.* Heller. *L. angustiflorus* Eastw?] Banner ± pubescent on back near apex.—5500–7900 ft., Red Fir F.; Alpine, Placer, and Nevada cos.

Var. **Christìnae** (Heller) Munz. [*L. C.* Heller.] Stems and lvs. subglabrous or slightly strigose; petioles 1–2 cm. long; lfts. 2–4 cm. long; corolla yellow, 10 mm. long.—Open gravelly ridge, 6000–7000 ft.; se. Shasta Co.

72. **L. fulcràtus** Greene. [*L. albicaulis* var. *f.* Jeps. *L. Andersonii* var. *f.* C. P. Sm. *L. fraxinetorum* Greene. *L. Beaneanus* C. P. Sm.] Much like *L. Andersonii*, but with spreading pubescence; stipules green and leaflike, lanceolate to oval; banner ± pointed apically.—Dry places among pines, 6000–9500 ft.; Red Fir F., Lodgepole F.; Nevada and Eldorado cos. to Tulare and Fresno cos. June–Aug. Possible hybrids with *L. Andersonii* are: *L. lingulae, L. Ione-Walkerae* and *L. cymba-egressus* C. P. Smith.

73. **L. elàtus** Jtn. [*L. albicaulis* var. *e.* Jeps. *L. formosus* var. *e.* C. P. Sm.] Much like *L. Andersonii;* erect, 5–9 dm. tall, branched above, densely short-silvery-silky; petioles mostly 2–4 cm. long; lfts. 2–6 cm. long, silvery-silky above, duller beneath; fls. 10–14 mm. long, lavender to blue, the banner suborbicular, sometimes pubescent near middle of back.—Dry slopes among pines, 6000–8700 ft.; Yellow Pine F., Red Fir F.; Mt. Pinos and San Gabriel Mts. June–Aug.

74. **L. adsúrgens** E. Drew. [*L. Pendletonii* E. Drew. *L. debilis* Eastw. *L. arvenseplaskettii* C. P. Sm.] Subappressed to appressed-pubescent, ± silky, 2–6 dm. tall, erect or nearly so; petioles 1–4(–7) cm. long; lfts. 5–8, oblanceolate, 2–3.5 cm. long; peduncles 2–6 cm. long; racemes 4–10 cm. long; bracts ca. 5 mm. long, deciduous; pedicels 2–4 mm. long, spreading-pubescent; upper calyx-lip notched or entire, lower entire; fls. 10–12 mm. long, pale yellow or blue to lilac, the banner obovate, glabrous, keel strongly curved, nonciliate; pods strigose, 2.5–3 cm. long; seeds 4–6, ca. 5 mm. long, brown, faintly

mottled.—Dry slopes to ca. 5000 ft.; Chaparral, Yellow Pine F., Douglas-Fir F.; San Diego Co. and Coast Ranges from Santa Clara Co. to s. Ore. May–July.

Var. **lilacìnus** Heller ex C. P. Sm. in Jeps. [*L. antoninus* Eastw. *L. alcis-montis* C. P. Sm. *L. Aliciae* C. P. Sm.] Pubescence spreading; fls. 12–14 mm. long, wings much exceeding keel.—Yellow Pine F.; Lake, Mendocino, Glenn cos.

Var. **undulàtus** C. P. Sm. in Jeps. [*L. Dalesae* Eastw. *L. formosus Clemensae* C. P. Sm. *L. salticola* Eastw. *L. mariposanus* Eastw. *L. klamathensis* Eastw. *L. finitus* Eastw.] Pubescence spreading; wing petals scarcely exceeding keel.—Dry slopes, 3000–10,000 ft.; Yellow Pine F., Foothill Wd.; Kern and Inyo cos. to Shasta and Siskiyou cos.

75. **L. formòsus** Greene. [*L. proximus* Heller. *L. pasadenensis* Eastw. *L. lutosus* Heller. *L. marinensis* Eastw.] Stems several, decumbent or ascending, 3–8 dm. long, 3–4 mm. thick, appressed-silky-pubescent, but somewhat greenish, leafy; petioles 3–7 cm. long; lfts. 7–9, silky on both sides, oblanceolate, 3–7 cm. long, mostly 5–15 mm. wide; peduncles 1–4 cm. long; racemes 10–25 cm. long, mostly rather dense; bracts deciduous, lance-linear, 5–7 mm. long; pedicels spreading-pubescent, 3–4 mm. long; fls. mostly whorled, 12–14 mm. long, violet or blue to lilac or white, the banner roundish, glabrous, wings covering keel which is slender, curved, nonciliate; pods silky-hairy, ca. 3–3.5 cm. long; seeds 5–7, mottled with gray, 3–4 mm. long.—Dry open fields and sandy places, mostly below 2500 ft., but up to 9000 ft.; many Plant Communities; Humboldt, Siskiyou, and Butte cos. to n. L. Calif. April–Oct. Variable; common throughout the range are plants with pubescence spreading, the var. *Bridgèsii* (Wats.) Greene. [*L. albicaulis* var. *B.* Wats. *L. B.* Heller. *L. Greenei* Heller.] An apparently aberrant plant from near Olema, Marin Co., with hairy petals was named *L. punto-reyesensis* C. P. Sm.

Var. **robústus** C. P. Sm. [*L. sonomensis* Heller. *L. albopilosus* Heller. *L. Brandegei* Eastw. *L. caeruleus* Heller. *L. navicularis* Heller.] Stems spreading-pubescent, stout, 5–7 mm. thick; fls. 16–18 mm. long; banner oblong.—V. Grassland; Tehama Co. to Kern Co.

Var. **hyacinthìnus** (Greene) C. P. Sm. [*L. h.* Greene. *L. albicaulis* var. *h.* Jeps.] Pubescence thin, plant decidedly green, erect, 5–10 dm. tall; lfts. mostly sharp-pointed, 3–8 cm. long; fls. 13–16 mm. long, the banner rounded, blue or purplish with yellow center.—Dry slopes under pines, 7000–10,000 ft.; Yellow Pine F. to Subalpine F.; San Jacinto Mts., less frequent in Santa Rosa Mts. and e. San Gabriel Mts. June–Aug.

76. **L. latifòlius** J. G. Agardh. [*L. perennis* ssp. *l.* Phillips. *L. cytisoides* J. G. Agardh. *L. rivularis,* as used by Jeps. *L. Edwin-Livingstoni* C. P. Sm.] Herbaceous perennial, erect, 3–12 dm. tall, subglabrous to minutely strigose, leafy; basal lvs. often dry by anthesis, the largest at mid-stem; petioles 5–20 cm. long; lfts. mostly 7–9, sometimes 5–12, broadly oblanceolate, mostly acute, sometimes obtuse, 4–10 cm. long, 1–3 cm. wide; peduncles 8–20 cm. long; racemes 15–45 cm. long, rather lax, the fls. whorled or scattered, 10–14 mm. long; bracts deciduous, linear-subulate, 8–12 mm. long; pedicels spreading-pubescent, 6–12 mm. long; upper calyx-lip notched, lower entire; petals blue or purplish or with some pink, fading brown, the banner suborbicular, glabrous, 9–10 mm. wide; wings truncate or incurved on lower edge, ± exposing the curved keel which has a slender acumen and is ciliate on upper margins from middle toward claws; pods brown, ca. 3 cm. long, hairy; seeds 7–10, mottled with dark brown, ca. 4 mm. long; $n = 24, 48$ (Phillips, 1957).—Common in open woods and thickets, below 7000 ft.; many Plant Communities, from coast to Red Fir F.; Santa Monica Mts., n. Los Angeles Co. to Del Norte and Siskiyou cos., less frequent in Sierra Nevada; to Wash. April–July.

KEY TO VARIETIES

Fls. 8–14 mm. long.
  The fls. 8–10 mm. long.
    Stems ± villous, especially about the nodes. Modoc Co. .... var. *barbatus*
    Stems strigose. Siskiyou Co. ....—..... var. *viridifolius*
  The fls. 10–14 mm. long; stems strigulose to subglabrous.
    Keel somewhat exposed, the wing petals incurved on lower free edges. Mostly Coast Ranges
                                        L. latifolius
    Keel almost covered, the wing petals outcurved on lower free edges. Sierra Nevada
                                          var. *columbianus*

Fls. 14–18 mm. long.
Plants densely villous with spreading hairs. San Mateo Co. .... var. *Dudleyi*
Plants strigose to subglabrous. S. Calif. .... var. *Parishii*

Var. **columbiànus** (Heller) C. P. Sm. [*L. c.* Heller. *L. c.* var. *simplex* C. P. Sm. *L. confusus* Heller, not Rose. *L. longipes* Greene. *L. agninus* Gand. *L. Pennellianus* Heller. *L. Wyethii* var. *Hansenii* C. P. Sm.] Plants 4–20 dm. tall; keel practically covered.— On ± moist slopes and along streams, 3000–11,000 ft.; Montane Coniferous F.; Sierra Nevada; to Wash. May–Sept.

Var. **Dúdleyi** C. P. Sm. Stems decumbent, spreading-villous; lfts. broadly oblong-oblanceolate, 2.5–3.5 cm. long; fls. 13–16 mm. long.—Chaparral at ca. 400 ft.; Montara Mts., San Mateo Co. April–May.

Var. **Paríshii** C. P. Sm. Stems stout, mostly 0.5–2 m. tall, ± fistulous; racemes quite dense, 1.5–3 dm. long; fls. 14–18 mm. long, rose to lavender.—Moist places, 1000–7500(–10,000) ft.; Coastal Sage Scrub, Chaparral, Yellow Pine F., Red Fir F.; mts. from San Diego Co. to Los Angeles Co. to Kern and Tulare cos. May–Aug.

Var. **barbàtus** (Henders.) Munz. [*L. ligulatus* var. *b.* Henders. *L. rivularis* var. *b.* Jeps. *L. b.* Heller. *L. latifolius* var. *ligulatus* C. P. Sm.] Stems stout, ± fistulous, glabrous to ± villous; stipules and bracts villous, long, conspicuous; fls. mostly pale.—Wet places, 5000–7000 ft.; Coniferous F.; Lassen and Modoc cos. to se. Ore. June–July.

Var. **virídifolius** (Heller) C. P. Sm. [*L. v.* Heller. *L. rivularis* var. *v.* Jeps. *L. caudiciferus* Eastw.?] Branched above; stipules short, inconspicuous; bracts not exceeding buds; pedicels strigose.—3000–5000 ft.; Yellow Pine F., Mixed Evergreen F.; Shasta, Siskiyou, and Humboldt cos.; to sw. Ore. June–Aug.

77. **L. rivulàris** Dougl. ex Lindl. [*L. lignipes* Heller.] Resembling *L. latifolius*, but ± villous; petioles 3–5 cm. long; lfts. 3–4 cm. long, strigulose especially beneath; fls. 12–16 mm. long, the keel ciliate along whole upper edge or ± in the middle; pods 4–5 cm. long, subappressed-hairy; seeds 8–12, mottled, usually with a diagonal line on each side, and 3–4 mm. long.—Wet sandy or gravelly places; Coastal Strand, N. Coastal Scrub; Mendocino Co. to Del Norte Co.; to B.C. May–Aug.

78. **L. polyphýllus** Lindl. [*L. grandifolius* Lindl. in J. G. Agardh. *L. p.* var. *g.* C. P. Sm. *L. macrophyllus* Benth. in Sweet. *L. magnus* Greene.] Erect, mostly unbranched, 5–15 dm. high, stout, fistulous, with minute spreading or appressed pubescence, few-lvd.; petioles 1.5–3 dm. long, stout; lfts. 10–17, oblanceolate, 7–15 cm. long, 1.5–3 cm. wide, mostly glabrous above, sparingly hirsute beneath; peduncles 3–8 cm. long; racemes 15–60 cm. long, rather dense; fls. subverticillate, 12–14 mm. long; bracts early deciduous, linear, to ca. 1 cm. long; pedicels 5–15 mm. long, semistrigose to spreading-pubescent; calyx-lips entire; petals blue, purple, or reddish, glabrous, the banner ± oblong-obovate, keel curved, naked, with long pointed acumen; pods brown, loosely hairy, 2.5–4 cm. long; seeds 5–9, variously spotted, ca. 4 mm. long; $2n = 48$ (Cooper, 1936).—Moist places, below 7500 ft.; Coastal Strand to Red Fir F.; coast and Coast Ranges from Santa Cruz Co. to Del Norte and Siskiyou cos.; to B.C. May–Aug. Plants from Sonoma Co. s. are supposed to have more compact racemes and more spreading pubescence on pedicels and would be var. *grandifòlius*.

Var. **pallídipes** (Heller) C. P. Sm. [*L. p.* Heller.] Lvs. hirsute-pubescent on both surfaces.—Redding, Shasta Co.; to B.C. May–June.

Ssp. **supérbus** (Heller) Munz. [*L. s.* Heller. *L. elongatus* Heller. *L. s.* var. *e.* C. P. Sm. *L. piperitus* A. Davids. *L. p.* var. *sparsipilosus* Eastw. *L. perglaber* Eastw. *L. alilatissimus?* Carolus-Bucarii, lacus-huntingtoni, meli-campestris, prato-lacunosum, sabuli, s.* var. *subpersistens* and *L. viridicalyx* C. P. Sm.] Lfts. mostly 5–9, mostly 4–7 cm. long, 0.5–1.5 cm. wide, sometimes larger, strigose beneath; fls. 11–14 mm. long, in ± elongate lax racemes; keel minutely papillose on upper margin, but not ciliate.—Wet places, 4000–8500 ft.; Yellow Pine F. to Lodgepole F.; Sierra Nevada from Tulare Co. to Siskiyou and Modoc cos.; w. Nev. May–July.

Ssp. **bernardìnus** (Abrams) Munz. [*L. superbus* var. *b.* Abrams ex C. P. Sm. in Jeps. *L. b.* Abrams ex Eastw.] Lfts. glabrous or nearly so on both surfaces; fls. 9–11 mm. long. —Wet places, 6500–8500 ft.; Red Fir F.; San Bernardino and San Jacinto mts. June–Aug.

79. **L. Tràcyi** Eastw. Glabrous glaucous perennial from a woody base, 4–7 dm. tall; lower petioles to ca. 1 cm. long, fused with the stipules their entire length and lfts.

dwarfed; principal cauline lvs. with 6–7 obovate lfts. 1–3.8 cm. long, 1–1.3 cm. wide, obtuse, mucronate, the petioles ca. as long; racemes 4–15 cm. long, short-peduncled; bracts lanceolate, ciliate, deciduous, surpassing the buds; pedicels 5–6 mm. long; calyx gibbous at base, ± villous, 6–8 mm. long, the upper lip bifid, the lower 3-toothed with filiform segms. 2 mm. long; corolla whitish, ca. 10 mm. long, the reflexed banner shorter than the wings; keel glabrous, covered by the wings; pods white-villous.—Near snow banks; Montane Coniferous F.; Klamath Mts., ne. Humboldt Co. July.

80. **L. saxòsus** Howell. Perennial with a compact root-crown and erect stems 1–3 dm. high, rather stiff-villous; lvs. mostly basal; petioles 5–10 cm. long; lfts. 8–12, oblanceolate, 1–3 cm. long, stiff-villous beneath, glabrous above; peduncles 2–4 cm. long; racemes 5–12 cm. long, dense; bracts deciduous; pedicels 4–6 mm. long, spreading-pubescent; upper calyx-lip notched to deeply lobed, the lower 3-toothed; petals blue, 14–18 mm. long, the banner almost as wide as long, with yellow center; keel ciliate on upper edges; pods villous, 2–2.5 cm. long; seeds 4–5; $n = 48$ (Phillips, 1957).—Gravelly or rocky places, 3500–6600 ft.; Sagebrush Scrub, N. Juniper Wd.; Lassen and Modoc cos.; to e. Wash. May–June.

81. **L. Holmgrenànus** C. P. Sm. Perennial, 4–7 dm. high, ± silky-pilose, erect; lower petioles 8–15 cm. long, suberect, slender, strigose; the blades with 4–7 lfts., oblanceolate, 2–4 cm. long, silky-strigose, greener above than beneath, sharply acute; stipules 0.5–2 cm. long; upper lvs. somewhat reduced; peduncles ± fistulose, 6–10 cm. long, 3–4 mm. thick; racemes rather densely many-fld., 1–2 dm. long; bracts linear, 8–10 mm. long; pedicels 8–10 mm. long, spreading-ascending; fls. 13–15 mm. long, bluish-purple on drying; calyx slightly gibbous at base, appressed-villous, the upper lip bidentate, the lower ca. 6 mm. long; banner roundish, 10 mm. long, glabrous, with a yellowish center; wings ca. 9 mm. long, 6 mm. wide; keel ± ciliate on upper edge below the acumen, arcuate, narrow.—Dry slopes, 6000–7500 ft.; Pinyon-Juniper Wd.; Last Chance Range, e. Inyo Co.; w. Nev. May–June.

82. **L. magníficus** Jones. Perennial, erect, 5–12 dm. tall, tomentulose or tomentose and with longer spreading hairs; lvs. near the base; petioles 1–2 dm. long; lfts. 5–9, elliptic-oblanceolate, 2–4.5 cm. long, 6–10 mm. wide; peduncles 1–3 dm. long; racemes 2–4 dm. long; bracts deciduous; pedicels stout, 3–4 mm. long; fls. 16–18 mm. long, ± whorled; upper calyx-lip bifid, lower entire; petals pinkish-purple, banner glabrous, roundish, the cent. yellow spot becoming dark purple; keel yellow, curved, ciliate above on the short upturned acumen; pods 3–7 cm. long, hairy; seeds 5–8.—Dry gravelly banks, 5500–7500 ft.; Pinyon-Juniper Wd., Panamint Mts., Inyo Co. May–June.

Var. **glarecola** Jones. [*L. Kerrii* Eastw.] Fls. 10–13 mm. long.—Coso Mts., Sierran foothills, Inyo Co.

Var. **hespérius** (Heller) C. P. Sm. [*L. h.* Heller.] Racemes ca. 1 dm. long; fls. 10–13 mm. long; keel straight.—Near McGee's Meadows, w. of Bishop, Inyo Co.

13. Ùlex L. Furze. Gorse

Densely branched shrubs with stiff spinescent branches. Lvs. simple, stiff, spinose. Fls. yellow, showy, solitary or racemose. Calyx yellow, deeply 2-lipped, the upper lip 2-toothed, the lower 3-toothed. Banner ovate, wings and keel oblong, obtuse. Stamens monadelphous. Pods short-oblong, several-seeded. Ca. 20 spp., of w. Eu. and N. Afr. (Old Latin name of some similar plant.)

1. **U. europaèus** L. Shrubs 1–2 m. tall, ± pubescent; lvs. acicular, 5–15 mm. long; calyx 10–15 mm. long, pubescent; corolla 15–18 mm. long; pods villous, dark, 15–18 mm. long; seeds dark, smooth, shining, ca. 2 mm. long, with a yellow basal strophiole; $n = 48$ (Tschechow, 1931).—Natur. especially along the sandy coast, sparingly in s. Calif., more abundantly in cent. and n. Calif.; to B.C.; native of Eu. Feb.–July.

14. Cýtisus L. Broom

Evergreen or deciduous shrubs, mostly unarmed, sometimes nearly leafless. Lvs. 3-foliolate, sometimes unifoliolate, the lfts. entire. Fls. terminal, solitary or racemose, yellow, white or purple; calyx 2-lipped with short teeth; banner ovate to rounded; wings oblong or ovate; keel straight or curved. Stamens monadelphous. Pods compressed, several-seeded, with a callous appendage or strophiole near the base. Ca. 80 spp., Eurasia and N. Afr. (Greek, *kutisus,* a kind of clover.)

Fls. white; banner hairy on the back .... 1. *C. proliferus*
Fls. yellow; banner mostly glabrous.
    Stems sharply angled, leafless or nearly so; pods hairy along margins only ...... 2. *C. scoparius*
    Stems obtusely angled or ridged, leafy; pods hairy all over.
        Racemes nearly capitate, 3–9-fld., at the end of short lateral branchlets; fls. ca. 10 mm. long
                                                                        3. *C. monspessulanus*
        Racemes elongate, 6–many-fld., secund, terminal and lateral; fls. 12–14 mm. long.
            Petioles 8 mm. or less long; lfts. 6–12 mm. long .... 4. *C. canariensis*
            Petioles 8 or more mm. long; lfts. 8–25 mm. long ....5. *C. maderensis*

1. **C. proliferus** L. Shrub to 4 m. tall, with long slender pubescent branches; petioles 6–12 mm. long; lfts. oblanceolate, 2.4–4 cm. long, oblanceolate, green and sparsely pubescent above, silky-pubescent beneath; fls. axillary, 3–8 on tomentose pedicels; calyx tomentose, 8–9 mm. long; petals ca. 1.5 cm. long; pod tomentose-villous, 3.5–5 cm. long; *n* = 24 (Duarte de Castro, 1949).—Occasional escape, as in Marin Co.; native of Canary Ids. Jan.–March.

2. **C. scopàrius** (L.) Link. [*Spartium s.* L.] SCOTCH BROOM. Shrub 1–2 m. high, deciduous, with angular naked or almost naked branches; petioles 2–10 mm. long; lfts. 1–3, obovate to oblanceolate, 6–12 mm. long, strigose; fls. usually solitary in axils; calyx glabrous, flaring, ca. 7 mm. long; petals ca. 2 cm. long; pods 3.5–5 cm. long, brownish-black, villous on margins; *n* = 24 (Duarte de Castro, 1949).—Becoming extensively natur., from Santa Cruz Co. n.; to Wash.; native of Eu. April–June.

3. **C. monspessulànus** L. FRENCH BROOM. Shrub 1–3 m. tall, with villous branchlets; petioles 3–5 mm. long; lfts. 3, ± obovate, 1–2 cm. long, subglabrous above, pubescent beneath; racemes subcapitate, terminating short lateral branches; calyx pubescent, 4–5 mm. long; petals bright yellow, 10 mm. long; pods densely villous, 2–2.5 cm. long; *n* = ± 23 (Duarte de Castro, 1949).—Natur., often perniciously, near the coast from Ventura Co. n.; to Wash.; native of Canary Ids. March–May.

4. **C. canariénsis** (L.) Kuntze. [*Genista c.* L.] Evergreen, much-branched, to 2 m. tall, strigose and slightly villous; petioles short; lfts. 3, cuneate, obovate, pubescent on both sides; racemes densely many-fld., terminal; calyx appressed-pubescent, 4–5 mm. long; petals bright yellow, ca. 12–14 mm. long; pods 12–20 mm. long, pubescent; *n* = 23 (Duarte de Castro, 1949).—Reported as occasionally natur., as at Catalina Id.; native of Canary Ids. March–April.

5. **C. maderénsis** Masf. To 6 m. tall, finely pubescent; lfts. silky-pubescent on both sides or smooth above, oblong-obovate to lanceolate; racemes 6–12-fld.; calyx silky, 5–6 mm. long; petals ca. 13–14 mm. long, the keel silky; pods ca. 2.5 cm. long, hairy; $2n = 48$ (Santos, 1945).—Natur., as at Carmel Highlands, Monterey Co.; native of Madeira. Feb.–March.

15. Spártium L. SPANISH BROOM

Mostly a rather tall virgately branched shrub with long slender, leafless or few-lvd., green, rushlike branchlets. Lvs. alternate, simple, entire, small. Fls. yellow, in loose terminal racemes. Calyx split above, hence 1-lipped, with 5 minute teeth. Banner and keel longer than the wings, keel pubescent along its lower edge. Stamens monadelphous. Pod linear, compressed, many-seeded; seeds with basal strophiole. Monotypic. Medit. Region. (Greek, *sparton,* broom.)

1. **S. juncèum** L. To ca. 3 m. high; lvs. oblance-oblong or narrower, 1–3 cm. long, ± strigose, short-petioled; fls. 2–2.5 cm. long, fragrant; pod 5–10 cm. long, ± strigose; *n* = 24–26 (Tschechow, 1931).—Natur. in dry and waste places at widely scattered localities. April–June.

16. Medicàgo L. MEDICK

Annual or perennial herbs. Lvs. pinnately 3-foliolate, the lfts. mostly dentate; stipules adnate. Fls. in small heads, racemes or umbels. Calyx-teeth subequal, short. Banner obovate to oblong; wings oblong; keel obtuse. Stamens diadelphous, the upper 1 free. Style subulate. Pods small, 1–several-seeded, curved to spirally coiled, indehiscent, reticulate or spiny. Seeds small, ± beanlike. Ca. 50 spp., of Eurasia and Afr.; many important for hay and forage. (Greek, *medice*, name of alfalfa, since it came to Greece from Medea.)

Plant perennial; fls. blue .... 1. *M. sativa*
Plants mostly annual; fls. yellow.
    Pods subreniform, 1-seeded; fls. many, in dense elongate spikelike racemes .... 2. *M. lupulina*
    Pods spirally coiled; fls. few, not in spikes.
        The pods ca. 10–12 mm. in diam., tightly coiled, flat, without spines; stipules divided almost
        to the base. Rare .... 3. *M. orbicularis*
        The pods (excluding spines) smaller, loosely coiled, ± barrel-shaped; stipules usually less deeply
        cut.
            Pods without prickles .... 4. *M. hispida* var. *confinis*
            Pods with well developed marginal prickles.
                Lfts. at least as broad as long, with a dark cent. blotch; prickles curved from the base
                        5. *M. arabica*
                Lfts. narrower than long, without dark cent. blotch; prickles straight, divergent, sometimes
                hooked at tip.
                    Stipules pectinate, with linear lateral segms.; plants subglabrous or nearly so.
                        Lfts. 8–20 mm. long; fls. 4–5 mm. long. Common .... 4. *M. hispida*
                        Lfts. 2–4 mm. long; fls. 2 mm. long. Rare .... 7. *M. praecox*
                  Stipules lanceolate, short-dentate at base; plant pilose .... 6. *M. minima*

1. **M. satìva** L. ALFALFA. Erect or ascending smooth perennial from an elongate taproot, much-branched, 4–9 dm. high; petioles 5–20 mm. long; stipules 5–8 mm. long, entire to toothed, lanceolate; lfts. oblanceolate to obovate, 10–25 mm. long, sharply serrate at the ± truncate tip; racemes short, dense, subcapitate, on peduncles ca. 2–3 cm. long; bracts subulate, 2–3 mm. long; calyx 5–6 mm. long, sparsely villous; corolla 8–12 mm. long, blue-violet to purple; pods unarmed, coiled loosely 2–3 times, pubescent; $n = 16$ (Fryer, 1930); 32, 64 (Tomé, 1947).—Commonly cult. and frequently established in waste places, along roadsides, etc.; native of Old World. April–Oct.

2. **M. lupulìna** L. BLACK MEDICK. Annual or sometimes perennial, many-branched from base, prostrate to decumbent, the stems 2–6 dm. long, pubescent; petioles 2–15 mm. long; lfts. obovate to roundish, 6–15 mm. long, denticulate above; stipules lance-ovate, few-toothed; peduncles slender, 1–2.5 cm. long, with short terminal few-fld. spikes; fls. yellow, 1.5–2 mm. long, the calyx villous; fr. subreniform, smooth, black when ripe, 1-seeded, 1.5–3 mm. long, unarmed.—Natur. in waste places; native of Eu. April–July.

Var. **cupanìana** (Guss.) Boiss. [*M. c.* Guss.] Perennial, forming dense mats that root at the nodes; fls. larger.—A bad weed in lawns; introd. from Eurasia.

3. **M. orbiculàris** All. Sparsely villous-pubescent and glandular annual, branched from base; petioles 1–2 cm. long; lfts. obovate, 8–12 mm. long, sharply toothed above; peduncles slender, recurved, 1–2-fld., ca. 1 cm. long; fls. ca. 4 mm. long; pods with 5–6 closely coiled spirals, reticulate; $2n = 16$ (Fryer, 1930).—Reported as introd.; native of Eu.

4. **M. híspida** Gaertn. [*M. polymorpha* var. *nigra* L.] BUR-CLOVER. Subglabrous annual, branched from base, the stems procumbent, 1–4 dm. long; petioles 1–4 cm. long; lfts. obovate or obcordate, 8–20 mm. long, sharply denticulate; stipules deeply divided with long acicular teeth; peduncles slender, 5–25 mm. long, 2–5-fld.; fls. 4–5 mm. long, the calyx sparsely villous; pods coiled 2–3 times, 4–6 mm. in diam., glabrous, with 2–3 rows of spines arising from a raised ridge and without any furrow between the rows, the spines usually hooked; $n = 7$ (Fryer, 1930).—Common in grassy places of most of cismontane Calif. and valued for pasturage; native of s. Eu. March–June.

Var. **confinis** (Koch) Burnat. [*M. apiculata* auth., not Willd.] Pods without prickles or with only short knobs.—With the sp., but less common.

5. **M. aràbica** (L.) Huds. SPOTTED MEDICK. Similar to *M. hispida*, but petioles 4–9 cm. long; lfts. broadly obcordate, 1–3 cm. broad, usually with a dark blotch on upper surface; peduncles 2–5-fld., ca. 1–2 cm. long; fls. 3.5–4 mm. long; pods 5–6 mm. broad, with 4–6 spirals, the edges grooved between the rows of prickles; $2n = 16$ (Ghimpu, 1928).—Occasional in grassy places; natur. from Eurasia. April–June.

6. **M. mínima** (L.) Desr. Resembling *M. hispida*, but softly pubescent; stipules short-dentate; lfts. 5–15 mm. long, pubescent on both sides; peduncles short; fls. 3–4 mm. long; pods subglobose, tightly coiled, pubescent between the prickles; $n = 8$ (Ghimpu, 1929).—Occasional, as in Ventura and Riverside cos.; native of Eurasia. April–May.

7. **M. praècox** DC. Stems slender, 1–3 dm. long, subprostrate, weakly pubescent; lfts. cuneate-obovate, 2–4 mm. long; petioles 5–6 mm. long; fls. 1–2, on peduncles shorter than

the lvs.; calyx ca. 1.6 mm. long; corolla ca. 2 mm. long; pods subglabrous, the faces strongly veined, the spines ca. 2 mm. long, hooked at the tips.—Natur. in Butte and Tehama cos.; native of Medit. region. April–May.

17. Melilòtus Mill. SWEET-CLOVER

Annual or biennial herbs, with pinnately trifoliate lvs. Fls. small, white or yellow, in spikelike racemes. Calyx subequally 5-toothed. Banner obovate; wings oblong; keel obtuse. Stamens diadelphous. Pod ovoid, coriaceous, wrinkled, longer than the calyx, scarcely dehiscent, 1–2-seeded. Ca. 20 spp., of Eurasia and Afr. (Greek, *meli*, honey, and *lotos*, some leguminous plant.)

Fls. white .... 1. *M. albus*
Fls. yellow.
    The fls. 2–3 mm. long, on pedicels less than 1 mm. long .... 2. *M. indicus*
    The fls. 5–7 mm. long, on pedicels 1.5–2 mm. long .... 3. *M. officinalis*

1. **M. álbus** Desr. Erect, 1–2 m. tall; petioles mostly 5–15 mm. long; stipules subulate, 5–7 mm. long; lfts. mostly lanceolate to oblanceolate, truncate, 1–2 cm. long, serrate; peduncles commonly 3–5 cm. long; racemes 5–10 cm. long; pedicels 1–2 mm. long; fls. 4–6 mm. long; pods ovoid, glabrous, ca. 3 mm. long; $n = 8$, 12 (Atwood, 1936).—Abundantly natur. in waste places, especially in damp places in s. Calif.; to B.C. and Atlantic Coast; native of Eurasia. May–Sept.

2. **M. índicus** (L.) All. Stems erect, 2–8 dm. high, glabrous or ± appressed-pubescent; petioles commonly 0.5–2 cm. long; stipules lance-subulate, 4–6 mm. long; lfts. cuneate-oblanceolate to -obovate, 1–3 cm. long, obtuse or truncate, denticulate; racemes 2–10 cm. long including the peduncles; pods ovoid, reticulate, glabrous, 1.5–2 cm. long; $n = 8$ (Tschechow, 1932).—Common in waste places at low elevs., most of Calif.; native of Eurasia. April–Oct.

3. **M. officinàlis** (L.) Lam. Like *M. indicus*, but taller; fls. larger; pods 2–3 mm. long; $n = 8$ (Tschechow, 1932).—Less commonly natur. in Calif.; native of Eurasia. May–Aug.

18. Trifòlium L. CLOVER

Herbs with mostly palmately trifoliolate lvs. and adnate stipules. Lfts. sometimes 4–7. Heads white, yellow, pink to rose or purple, capitate to short-spicate or sub-umbellate. Calyx 5-toothed, the lobes equal or unequal, entire to bifid or trifid. Petals usually persistent. Stamens diadelphous. Pods globose to elongate, 1–2(–8)-seeded, included within the persistent calyx, dehiscent or not. Ca. 300 spp., most abundant in N. Temp., but also in S. Am. and Afr. (Latin, *tres*, three, and *folium*, lf.)

(McDermott, L. F. Illus. key to the N. Am. spp. of Trifolium, 1–325, 1910.)
A.   Heads without an invol. at base of fls.
    B.   Plants annual.
        C.   Fls. pedicellate, reflexed in age; calyx mostly glabrous (except in *T. bifidum*).
           D.   Fls. yellow; calyx 5-nerved. Introd. spp.
             E.   Stipules lance-oblong; terminal lft. subsessile; fls. 5–6 mm. long
                  1. *T. agrarium*
             EE. Stipules ovate; terminal lft. petiolulate; fls. 2.5–4.5 mm. long.
               F.   Heads 8–15 mm. long; banner dilated, not folded over the pod, conspicuously veined .... 2. *T. procumbens*
               FF. Heads 5–10 mm. long; banner not dilated, folded over the pod, not conspicuously veined .... 3. *T. dubium*
          DD. Fls. purple; calyx 10-nerved. Native spp.
             E.   Plants ± villous on peduncles and calyx .... 4. *T. bifidum*
             EE. Plants nearly or quite glabrous.
               F.   Calyx not ciliolate on the lobes.
                 G.   Stipules lance-ovate, 8–10 mm. long; lfts. obovate, the serrations not setaceous. Widely distributed .... 5. *T. gracilentum*
                 GG. Stipules lance-linear, 15–20 mm. long; lfts. almost linear, serrulate with setaceous teeth. Insular .... 6. *T. Palmeri*
               FF. Calyx ciliolate on the lobes with short flat appendages 7. *T. ciliolatum*
        CC. Fls. sessile, not reflexed in age; calyx densely villous.
           D.   Heads narrow, ± cylindrical; fls. white to red. Introd. spp.
             E.   Calyx 5–6 mm. long; corolla shorter; lfts. linear to oblanceolate
                  25. *T. arvense*

EE. Calyx ca. 10 mm. long; corolla longer; lfts. obovate to obcordate
                26. *T. incarnatum*
DD. Heads globose to ± ovoid; fls. purple.
 E. Calyx 20-nerved; stipules with a long bristle-tip; heads 1.5–2.5 cm. in diam.
   Introd. sp. . . . . 27. *T. hirtum*
 EE. Calyx 10-nerved; stipules usually not bristle-tipped; heads often smaller.
   Native spp.
   F. Heads sessile, usually in pairs, subtended by the upper lvs. and their
     stipules . . . . 28. *T. Macraei*
   FF. Heads peduncled, solitary.
    G. Corolla exceeding or ca. as long as calyx.
     H. Calyx 10–12 mm. long; corolla 12–15 mm. long
                    29. *T. amoenum*
     HH. Calyx 5–10 mm. long; corolla 6–10 mm. long.
      I. Corolla well exserted from the calyx, 8–10 mm. long
                30. *T. dichotomum*
      II. Corolla ca. as long as calyx, 6–7 mm. long
               31. *T. albopurpureum*
    GG. Corolla much shorter than the calyx, quite obscured by it
                32. *T. olivaceum*
BB. Plants perennial.
 C. Peduncles axillary; fls. on pedicels mostly 4–5 mm. long, recurved in fr.
  D. Plant villous-pubescent throughout; peduncles recurved at top in age. Native
                     8. *T. Breweri*
  DD. Plant glabrous or slightly pubescent; peduncles not recurved at apex. Introd.
   E. Stems erect or ascending, not rooting at nodes; stipules 10–25 mm. long,
     with long hairlike tip; corolla pink . . . . 9. *T. hybridum*
   EE. Stems creeping, rooting at nodes; stipules 4–10 mm. long, lance-ovate;
     corolla mostly white . . . . 10. *T. repens*
 CC. Peduncles terminal or subterminal; fls. sessile or on shorter pedicels.
  D. Calyx glabrous.
   E. Rachis not prolonged above the heads.
    F. Calyx 7–8 mm. long; corolla red, 12–15 mm. long. Nevada Co. to Modoc
     Co. . . . . 11. *T. Beckwithii*
    FF. Calyx 3–4 mm. long.
     G. Corolla white, 9–12 mm. long; stipules 20–30 mm. long. Humboldt,
      Del Norte, and Siskiyou cos. . . . . 12. *T. Howellii*
     GG. Corolla lavender, 8–9 mm. long; stipules ca. 8 mm. long. Mariposa
      Co. to Fresno Co. . . . . 13. *T. Bolanderi*
   EE. Rachis prolonged above the heads; corolla rose to purplish, 9–12 mm. long.
     Tuolumne Co. to Siskiyou and Modoc cos. . . . . 14. *T. productum*
  DD. Calyx villous or hairy.
   E. Lfts. mostly 5–9.
    F. Corolla purple to pink; pedicels to ca. 1 mm. long.
     G. Fls. 20–30 mm. long; stipules 8–19 mm. long. Sierra Co. n.
                  18. *T. macrocephalum*
     GG. Fls. 8–15 mm. long; stipules 5–10 mm. long.
      H. Corolla purple, 12–15 mm. long; calyx 9–10 mm. long;
       peduncles not exceeding lvs. Sierra Co. n. . . 19. *T. Andersonii*
      HH. Corolla pink, 8–10 mm. long; calyx 7–8 mm. long; peduncles
       well exceeding lvs. Mono Co. . . . . 20. *T. monoense*
    FF. Corolla yellow to reddish, 10 mm. long; pedicels 1–2 mm. long.
     G. Lfts. obovate, coarsely few-toothed; peduncles surpassing lvs.;
      calyx curved at base. Sierra Co. to Mt. Lassen . . 21. *T. Lemmonii*
     GG. Lfts. elliptic, serrulate; peduncles shorter than lvs.; calyx straight.
      Lassen Co. n. . . . . 22. *T. gymnocarpon*
   EE. Lfts. 3.
    F. Heads with long peduncles. Native.
     G. Peduncles ± recurved at apex; calyx with widely spreading hairs;
      petals yellowish, 12–14 mm. long. Trinity to Del Norte and Shasta
      cos. . . . . 15. *T. eriocephalum*
     GG. Peduncles erect; calyx with appressed or ascending hairs; corolla
      not yellow.
      H. Fls. pedicellate, reflexed in age. Trinity to Humboldt and
       Shasta cos. . . . . 6. *T. oreganum*
      HH. Fls. sessile or nearly so, erect. Del Norte Co. to Riverside Co.
                   17. *T. longipes*
    FF. Heads sessile, subtended by uppermost lvs. Introd.
     G. Lfts. oval to obovate; heads ± ovoid; calyx not spiny-tipped
                   23. *T. pratense*
     GG. Lfts. linear to lance-linear; heads elongate; calyx-teeth spiny
      tipped . . . . 24. *T. angustifolium*
AA. Heads with an invol. at the base of the fls., this reduced to a mere ring in *T. depauperatum*
 B. Corolla not inflated in age; involucral bracts united.
  C. Invol. campanulate to bowl-shaped.
   D. Lobes of the invol. entire; heads pubescent, small . . . . 37. *T. microcephalum*

DD. Lobes of the invol. toothed.
  E. Calyx-teeth 1–3 times trichotomously forked, glabrous .. 33. *T. cyathiferum*
  EE. Calyx-teeth simple, or the upper one forked, all hairy or ciliate.
    F. The calyx-teeth plumose, the upper one forked; corolla purple.
      G. Corolla 5–6 mm. long, shorter than the calyx .. 34. *T. barbigerum*
      GG. Corolla 10–12 mm. long, ca. twice as long as calyx
                                                                    35. *T. Grayi*
    FF. The calyx-teeth ciliate, all simple; corolla white to pink
                                                                    36. *T. microdon*
CC. Invol. flat, rotate.
  D. Plants perennial.
    E. Corolla cream-color with purple-tipped keel; fls. 1–8 in a head
                                                                    38. *T. monanthum*
    EE. Corolla white to light purple; fls. many .... 39. *T. Wormskioldii*
  DD. Plants annual.
    E. The plants viscid-pubescent and clammy; corolla 12–14 mm. long, pale with
        dark cent. spot .... 46. *T. obtusiflorum*
    EE. The plants glabrous.
      F. Calyx conspicuously pilose; corolla pale purple, 7–8 mm. long. Monterey
          region .... 42. *T. trichocalyx*
      FF. Calyx glabrous.
        G. Calyx-teeth dilated and 3-toothed to simple; fls. purple with paler
            tips.
          H. Corolla 12–15 mm. long; plants erect to ± decumbent. Widely
              distributed ....   45. *T. tridentatum*
          HH. Corolla 8–10 mm. long; plants subprostrate. Monterey
              Peninsula .... 42. *T. polyodon*
        GG. Calyx-teeth not dilated, but subulate and mostly entire.
          H. Corolla 8–10 mm. long, the keel with a long-apiculate beak.
              Humboldt Co. to Monterey .... 40. *T. appendiculatum*
          HH. Corolla 5–8 mm. long, the keel not so beaked.
            I. The petals not much longer than the 10-nerved calyx
                                                                    41. *T. oliganthum*
            II. The petals considerably longer than the 20-nerved calyx
                                                                    44. *T. variegatum*
BB. Corolla conspicuously inflated in age; involucral bracts distinct.
  C. The fls. 12–25 mm. long; involucral divisions 6–18 mm. long .... 47. *T. fucatum*
  CC. The fls. 5–8 mm. long; involucral bracts to ca. 4 mm. long.
    D. Involucral lobes evident .... 48. *T. amplectens*
    DD. Involucral lobes lacking, the invol. consisting of a mere ring
                                                                    49. *T. depauperatum*

1. **T. agràrium** L. Hop Clover. Erect annual, 2–5 dm. high, somewhat strigose; petioles 1–1.5 cm. long, adnate to the lance-oblong stipules; lfts. subsessile, oblanceolate or wider, 1–2 cm. long; peduncles mostly surpassing lvs.; heads short-cylindric, 1–2 cm. long, not involucrate; fls. 5–6 mm. long, yellow; calyx glabrous, 2-lipped; petals striate-sulcate in age; style and pod subequal; seeds ovoid.—Collected at Huntington Lake, Fresno Co.; Ida., Mont., etc. to Atlantic Coast. Native Eu. July–Aug.

2. **T. procúmbens** L. Hop Clover. Similar to *T. agrarium*, but the stipules ovate, broadly rounded at base; terminal lft. petiolulate; heads 8–15 mm. long, globose to short-cylindric; fls. mostly 20–30, and 3.5–4.5 mm. long; style much shorter than the pod.—Reported from lawns, Santa Barbara; sparingly natur. from Humboldt Co. n.; to B.C., Atlantic Coast; native of Eu. May–Sept.

3. **T. dùbium** Sibth. Much like *T. procumbens;* heads 5–10 mm. long; fls. 5–15, ca. 2.5–3.5 mm. long; banner much less conspicuously veined; $2n = 14$, 16, 28 (several auth.).—Waste places, fields, lawns, etc., below 2500 ft.; Santa Barbara, San Francisco Bay region to B.C. and Atlantic Coast; natur. from Eu. May–July.

4. **T. bìfidum** Gray. Pale green annual, with erect very slender stems 1.5–4 dm. high, usually with some long slender hairs on upper parts; petioles mostly 2–6 cm. long; stipules lance-ovate, entire, 8–15 mm. long; lfts. linear-cuneate, 0.6–2 cm. long, remotely sharply serrulate to subentire on sides, the lower and middle lvs. with lfts. deeply bifid at apex, the 2 lobes ± denticulate; peduncles slender, 3–10 cm. long; fls. 6–15; pedicels recurved, 1–3 mm. long; calyx-teeth subulate-setaceous, almost equaling the corolla; petals pale pink to purplish, 4–5 mm. long, the banner strongly veined; pod included, 1-seeded.—Grassy places, below 2000 ft.; Mixed Evergreen F., Coastal Prairie, etc.; Coast Ranges from Santa Clara Co. n. to Ore. April–June.

Var. **decípiens** Greene. [*T. Greenei* House.] Lvs. oblanceolate to obcordate, rounded to retuse at apex.—Grassy places, San Diego, Santa Barbara Co. n. and up to Yellow Pine F. in Sierra Nevada from Madera Co. n.; to Wash.

5. **T. graciléntum** T. & G. [*T. denudatum* Nutt. *T. exile* Greene. *T. g.* var. *reductum* Parish.] Practically glabrous annual, the stems slender, erect to procumbent, 1–4 dm. long; petioles commonly 2–7 cm. long; stipules lance-ovate, entire, 8–10 mm. long; lfts. obovate, 0.6–1.5 cm. long, serrulate, emarginate at apex; peduncles slender, mostly 2–6 cm. long; heads 6–10 mm. long; pedicels reflexed in age, 1–3 mm. long; calyx-teeth subulate-lanceolate, entire, shorter than the corolla; petals pink to reddish-purple, 5–6 mm. long, not conspicuously veined; pods 1–2-seeded.—Common in open and grassy places, below 5000 ft.; V. Grassland, Chaparral, Foothill Wd., etc.; most of cismontane Calif. and occasional along w. edge of deserts; to B.C., L. Calif. April–June.

Var. **inconspícuum** Fern. Depauperate, mostly less than 1.5 dm. high; fls. mostly pinkish, scarcely exceeding calyx.—Poorly marked, mostly toward s. part of range of the sp.

6. **T. Pálmeri** Wats. [*T. gracilentum* var. *P.* McDer.] Glabrous branched annual, the stems 1–3 dm. long; petioles mostly 2–5 cm. long; stipules 15–20 mm. long, lance-linear, with long setaceous tips; lfts. lance-linear, 1–3 cm. long, acute, serrulate with setaceous teeth; heads 10–15 mm. broad; calyx almost as long as purple corolla, the latter rose-purple, 6–7 mm. long; pools 2-seeded.—Grassy places, Santa Catalina and San Clemente ids.; Guadalupe Id. March–May.

7. **T. ciliolàtum** Benth. [*T. ciliatum* Nutt., not Clarke. *T. c.* var. *discolor* Loja.] Glabrous pale green annual with erect ± fistulous stems 2–5 dm. high; petioles to ca. 10 cm. long; stipules lanceolate-acuminate, 15–30 mm. long; lfts. oblong to ± obovate, 1–3 cm. long, obtuse, entire to serrulate; peduncles 5–15 cm. long; heads ovoid, 10–20 mm. long; pedicels reflexed in age, ca. 1 mm. long; calyx shorter than the corolla, the teeth lance-acuminate, ciliate with flat short appendages; corolla pinkish-purple, 6–7 mm. long, the banner inflated at the base; pods 1–2-seeded.—Frequent on open and grassy slopes, below 5000 ft.; many Plant Communities; most of cismontane Calif.; to Wash. March–June.

8. **T. Bréweri** Wats. Glaucous pubescent perennial, with slender stems 1–3 dm. long; petioles slender, mostly 1–3 cm. long; stipules lanceolate or lance-ovate, 6–8 mm. long; lfts. obovate, 0.4–1.5 cm. long, denticulate, obtuse to emarginate at apex; peduncles slender, 2–4 cm. long, recurved at top in age; heads few-fld.; pedicels 2–3 mm. long; calyx pubescent, shorter than the corolla, the teeth subulate-lanceolate; corolla rose to cream-white, 5–6 mm. long; pods pubescent, 1-seeded.—Wooded slopes, below 6500 ft.; Mixed Evergreen F., N. Coastal Coniferous F., Yellow Pine F.; Trinity Co. to Del Norte and Siskiyou cos., Sierra Nevada from Madera Co. n.; to sw. Ore. May–Aug.

9. **T. hýbridum** L. ALSIKE CLOVER. Subglabrous perennial, with stems 3–6 dm. long, ± stout and succulent; petioles commonly 2–8 cm. long; stipules lance-ovate, 1–2.5 cm. long, with long hairlike tips; lfts. obovate to ovate, 1–3 cm. long, sharp-serrulate; peduncles to ca. 1 dm. long; heads many-fld., globose, 1.5–2 cm. in diam.; pedicels 4–8 mm. long; calyx 3–4 mm. long, pubescent in the sinuses between the subulate teeth; corolla pink, 7–8 mm. long; pods 3–4-seeded; $2n = 16$ (Kawakami, 1930).—Natur. in lawns, meadows and damp places, below 6000 ft.; Santa Barbara and Coast Ranges, Sierra Nevada from cent. Calif. n.; to Wash., Atlantic Coast; native of Eu. May–Oct.

10. **T. rèpens** L. WHITE CLOVER. Glabrous perennial with creeping stems 1–3 dm. long and rooting at nodes; stipules lance-ovate, 4–10 mm. long; lfts. 1–2 cm. long, obovate with cuneate base, serrulate, ± emarginate at apex; peduncles 5–30 cm. high; heads globose, 1.5–2.5 cm. broad; pedicels 2–4 mm. long, reflexed in age; calyx 4–6 mm. long, the teeth subulate; corolla white to pale pinkish, 7–10 mm. long; $n = 16$ (Atwood & Hill, 1940).—Planted in lawns and meadows; natur. in wet places especially in the mts. and in n. parts of the state; native of Eu. April–Dec.

11. **T. Beckwíthii** Brew. ex Wats. [*T. altissimum* T. & G., not Lois.] Glabrous perennial with stems stout, ascending, 0.5–3 dm. long; petioles 2–10 or more cm. long; stipules lanceolate to ovate, acute, 1–2.5 cm. long; lfts. lanceolate to oblong, 0.5–4 cm. long, serrate, obtuse to acute; peduncles 3–15 cm. long; heads roundish, dense, many-fld., 1.5–3 cm. across; pedicels 1–1.5 mm. long, reflexed in age; calyx 7–8 mm. long, the subulate teeth ca. as long as tube; corolla red, 12–15 mm. long; ovules 2–6.—Valleys and meadows, 4000–7000 ft.; Yellow Pine F., Red Fir F.; Nevada Co. to Modoc Co.; to Ore., Ida., Nev. May–Aug.

12. **T. Howéllii** Wats. Glabrous perennial with stout subsimple stems 5–7 dm. high; petioles 2–10 cm. long; stipules green, ovate, 2–3 cm. long, entire; lfts. elliptic-ovate

to elliptic, 3–8 cm. long, serrulate; peduncles 5–15 cm. long; heads many-fld., elongate-ovoid, 1.5–2.5 cm. long; pedicels reflexed, ca. 1 mm. long; calyx ca. 4 mm. long, the subulate teeth ca. as long as tube; corolla white, 9–12 mm. long; pods stipitate, mostly 1-seeded.—Moist places, 4000–5000 ft.; Yellow Pine F., Red Fir F.; Humboldt, Del Norte, and Siskiyou cos.; sw. Ore. July–Aug.

13. **T. Bolánderi** Gray. Glabrous perennial with many decumbent or ascending slender stems 1–2 dm. long; petioles mostly basal, 1–7 cm. long; stipules ovate, acute, entire, ca. 8 mm. long; lfts. obovate, ± oblong, 0.6–1.5 cm. long, somewhat serrulate; peduncles slender, 7–15 cm. long; heads ovoid, 10–15 mm. long, the pedicels recurved, ca. 1 mm. long; calyx 3–4 mm. long, the lanceolate teeth somewhat shorter than the tube; corolla lavender, narrow, ca. 8–9 mm. long; ovules 2.—Wet meadows, ca. 7000 ft.; Red Fir F.; Sierra Nevada from Mariposa Co. to Fresno Co. June–July.

14. **T. prodúctum** Greene. [*T. Kingii* var. *p.* Jeps.] Glabrous perennial with slender ascending stems 1–4 dm. high; petioles largely basal and to ca. 12 cm. long, the few cauline shorter; stipules entire, broadly lanceolate, 6–12 mm. long; lfts. lanceolate, to ± oblong or ovate, 1–4 cm. long, acuminate, spinulose-serrate; peduncles slender, 5–12 cm. long; heads many-fld., short-ovoid, mostly 1–1.5 cm. long, with rachis produced above the fls.; pedicels reflexed in age, scarcely 1 mm. long; calyx 3–4 mm. long, the subulate teeth scarcely equal to the tube; corolla rose to purplish, 9–12 mm. long; pods stipitate, 1–2-seeded.—Moist ± wooded places, 3800–7500 ft.; Yellow Pine F., Red Fir F., Lodgepole F.; Tuolumne Co. to Modoc and Siskiyou cos.; Ore. June–Aug.

15. **T. eriocéphalum** Nutt. [*T. e.* var. *Butleri* Jeps. *T. scorpioides* Blasd.] Perennial, subglabrous to densely villous, the stems erect or ± decumbent, 1–5 dm. long; petioles to ca. 1 dm. long; stipules linear, 2–3.5 cm. long, long-acuminate; lfts. lance-oblong, 2–4 cm. long, finely serrulate, obtuse; peduncles 5–15 cm. long, ± recurved at top; heads round to ovoid, many-fld., 1.5–2.5 cm. long; pedicels reflexed in age, barely 1 mm. long; fls. yellowish, 12–14 mm. long; calyx 7–8 mm. long, the teeth ca. 3 times as long as the tube, filiform, plumose with long spreading hairs; pods long-hairy toward apex, 1–2-seeded.—Grassy and open places, below 6000 ft.; Redwood F., Mixed Evergreen F., Yellow Pine F., Coastal Prairie, etc.; Trinity to Del Norte and Shasta cos.; to Wash. May–July. Many California plants are more glabrous than more n. ones.

16. **T. oregànum** Howell. Glabrous to ± villous-pubescent perennial with several stems 1–2 dm. long; stipules lance-ovate, 8–18 mm. long, acuminate; lfts. lance-linear to obovate, 0.8–2 cm. long, serrulate, mucronate; peduncles 3–12 cm. long; heads 1.2–2 cm. broad; calyx 8–10 mm. long, the teeth subulate, ca. twice as long as the tube, villous-pubescent with ± appressed hairs; corolla pink to purple, 10–15 mm. long.—Moist places in open woods, 4000–6000 ft.; Yellow Pine F., Red Fir F.; Trinity and Humboldt cos. to Shasta Co. and Wash. June–Aug. A ± uncertain entity.

17. **T. lóngipes** Nutt. [*T. Elmeri* Greene. *T. l.* var. *E.* McDer. *T. Hansenii* Greene. *T. l.* var. *H.* Jeps. *T. l.* var. *nevadense* Jeps. *T. shastense* House. *T. l.* var. *s.* Jeps.] Perennial from branching creeping rootstocks from a cent. taproot; glabrous or sparsely strigose, the stems decumbent to erect, 0.5–4 dm. long; stipules green, lanceolate, entire, 0.5–2 cm. long; lfts. linear-lanceolate to oblanceolate or obovate, 0.5–3(–6) cm. long, mostly glabrous above and strigose beneath, obtuse to acute, serrulate; peduncles mostly 3–10 cm. long, ± villous above; heads mostly ovoid, dense, 1–2 cm. long; fls. erect, very short-pedicelled; calyx mostly 7–9 mm. long, the teeth subulate, much longer than the tube, ± villous-pubescent; corolla whitish or tinged with purple, (8–)10–12 mm. long; pods villous at apex, 2–4-seeded.—Moist places, below 9000 ft.; mostly Montane Coniferous F.; mts. from Mendocino and Tehama cos. n. and from Tulare Co. n.; to Wash., Ida., Nev. June–Sept. Variable and with poorly defined segregates: Robust, large-lvd. plants from Trinity Co. etc., are the var. *Elmeri;* dwarf, slender-stemmed plants with small heads, from Amador Co., the var. *Hansenii;* those with long linear acuminate lfts. and elongate heads, from Nevada Co., the var. *nevadense;* and those with lfts. 4–5 cm. long, banner and wings long-acuminate, from Del Norte and Siskiyou cos., the var. *shastense.*

Var. **atrorùbens** (Greene) Jeps. [*T. Rusbyi* var. *a.* Greene.] Calyx-teeth densely white-villous; fls. from pale to dark purple.—Wet meadows, 6000–8000 ft.; Yellow Pine F., Red Fir F.; San Bernardino and San Jacinto mts.

18. **T macrocéphalum** (Pursh) Poir. [*Lupinaster m.* Pursh.] Perennial with underground rootstocks and sparsely villous stems 1–3 dm. long forming loose mats; petioles mostly 2–6 cm. long; stipules oblong-ovate, green, 8–18 mm. long; lfts. 5–9, cuneate-

oblong to obovate, obtuse, mucronate, 0.8–2 cm. long, serrulate; peduncles mostly 1–3 cm. long; heads short-ovoid, many-fld., 2.5–4 cm. broad; pedicels to ca. 1 mm. long; calyx 12–20 mm. long, the teeth subulate, villous-plumose; corolla purplish, mostly 2–3 cm. long; pods stipitate, 1-seeded, glabrous.—Dry slopes and valleys, etc., 2000–5500 ft.; Sagebrush Scrub, N. Juniper Wd., Yellow Pine F.; Siskiyou to Modoc and Sierra cos.; to B.C., Ida., w. Nev. April–May.

19. **T. Andersònii** Gray. Cespitose perennial with a woody caudex, 5–10 cm. high, densely silky-villous throughout; stipules lanceolate, acuminate, entire, to ca. 1 cm. long; lfts. 3–5, oblanceolate to obovate, 0.8–2 cm. long, entire, ± mucronate; peduncles to ca. 7 cm. long; heads subglobose, 2–4 cm. broad, subtended by a scarious vestige of an invol.; calyx ca. 1 cm. long, the subulate teeth densely plumose; corolla purple, ca. 1.2–1.5 cm. long; pods tomentose, 1–2-seeded.—Dry slopes and valleys, 3500–8000 ft.; Sagebrush Scrub to Yellow Pine F.; Sierra Co. to Modoc and Shasta cos.; to Ore., Nev. May–June.

20. **T. monoénse** Greene. With habit and stature of *T. Andersonii*, but grayish villous; stipules ca. 5–7 mm. long; lfts. 4–6, narrowly obovate, abruptly acute or acuminate, 6–10 mm. long, entire, strigose; peduncles slender, 3–7 cm. long; heads subglobose, 1.2–1.8 cm. broad; calyx densely plumose, 7–8 mm. long, the teeth filiform; corolla pink, 8–10 mm. long.—Dry rocky flats and slopes, 7000–13,000 ft.; Sagebrush Scrub to Bristle-cone Pine F.; White Mts. to near Bridgeport, Mono Co. May–Aug.

21. **T. Lemmònii** Wats. A ± strigose perennial from a deep taproot and with slender stems 1.5–2 dm. high; stipules ovate, coarsely few-toothed, to ca. 1 cm. long; lfts. 4–6, obovate, obtuse, mucronate, coarsely toothed, 6–12 mm. long; peduncles 5–12 cm. long, far surpassing lvs.; heads 1.5–2 cm. broad, many-fld., the rachis swollen; pedicels reflexed in age, 1–2 mm. long; calyx curved at base, ca. 4 mm. long, pubescent, the subulate teeth slightly longer than the tube; corolla bright yellow, 1 cm. long; pods somewhat pubescent.—Slopes and valleys, ca. 5000–7000 ft.; Yellow Pine F., Sagebrush Scrub; Sierra Co. to Lassen Peak; w. Nev. June–July.

22. **T. gymnocárpon** Nutt. var. **Plúmmerae** (Wats.) J. S. Martin. [*T. P.* Wats.] Much like *T. Lemmonii*; lfts. usually elliptic, serrulate, strigose on both surfaces; peduncles shorter than lvs.; heads few-fld., the fls. in 1–4 whorls, the rachis not swollen; calyx straight; corolla yellowish-white to reddish.—Plains and slopes, 5000–6000 ft.; Sagebrush Scrub; Lassen Co.; to Ore., Mont., Utah. May–June.

23. **T. praténse** L. RED CLOVER. Perennial, several-stemmed, 2–6 dm. high, pubescent; upper stipules ovate, membranous, conspicuously veined, subulate-tipped; lfts. oval to obovate, obtuse, pubescent, entire or crenulate, 2–5 cm. long, often with a pale blotch; heads subtended by 1–2 lvs., ovoid to obovoid, 2–3 cm. broad; fls. many, sessile; calyx 5–8 mm. long, villous, the subulate teeth slightly longer than tube; corolla red or pink, 10–20 mm. long; pod 2-seeded; $n = 7$ (Levan, 1942).—A variable sp., widely cult. for hay and pasturage; natur. in many places in Calif. especially in the cooler and moister parts, native of Eurasia. April–Oct.

24. **T. angustifòlium** L. Erect annual, 1–4 dm. high, strigose; stipules ca. 2 cm. long, with long subulate free tips; lfts. linear to lance-linear, 2–5 cm. long, entire; heads many-fld., short-peduncled, elongate, 2–6 cm. long; fls. sessile, rose; fruiting calyx ca. 1 cm. long, the teeth plumose, divaricate, spiny-tipped; corolla ca. 1 cm. long, the upper part deciduous; $2n = 14$ (Karpechenko, 1925).—Natur. in Larkin V., Santa Cruz Co. and Santa Lucia Mts., Monterey Co.; native of Medit. region. Late spring.

25. **T. arvénse** L. RABBIT'S-FOOT CLOVER. Silky-pubescent annual, erect, branched, 2–4 dm. high; stipules narrow, subulate-tipped; lfts. linear to oblanceolate, obtuse to emarginate, denticulate above, 1–2.5 cm. long; heads short-peduncled, ovoid-cylindric, 1–4 cm. long, densely many-fld.; fls. sessile, white or pinkish; calyx 5–6 mm. long, the setiform teeth plumose, much longer than the tube; corolla shorter than calyx; $n = 7$ (Karpechenko, 1925).—Possible weed in Calif.; native of Eu.

26. **T. incarnàtum** L. Rather coarse villous-pubescent annual, erect or ascending, 2–9 dm. high; lower petioles long; stipules broadly ovate-oblong; lfts. obovate to obcordate, 1.5–2.5 cm. long; heads on stout peduncles, oblong, 3–6 cm. long; fls. sessile; calyx ca. 1 cm. long, villous, the teeth subulate, ca. as long as tube; corolla crimson, mostly exceeding calyx; $2n = 14$ (Karpechenko, 1925).—Sparingly introd. as in Mendocino and Butte cos.; native of Eu. May–Aug.

27. **T. hírtum** All. Densely long-villous annual, 1–4 dm. high; stipules narrow, abruptly

narrowed to a long bristle-tip; lfts. narrowly cuneate-obovate, 0.8–2 cm. long; heads globose, sessile, subtended by paired lvs. or their stipules, 1.5–2.5 cm. in diam.; calyx 20-nerved, densely hirsute with ascending hairs, the setaceous lobes 4–6 mm. long, exceeding the tube; corolla purple, slender, longer than the calyx.—Reported as a weed in Butte Co.; natur. from Eu. April–May.

28. **T. Macràei** H. & A. [*T. catalinae* Wats. *T. bicephalum* Elmer. *T. mercedense* Kenn.] Decumbent or spreading, soft-pubescent mostly strigose annual with several stems 1–3 dm. long; stipules ovate, often acuminate, green; lfts. oblanceolate to obovate, 5–15 mm. long, obtuse to emarginate, serrulate; heads sessile, often in pairs, in axils of uppermost sessile lvs.; calyx 6–7 mm. long, the subulate teeth densely plumose, linear-subulate; corolla purple, ca. as long as calyx.—Open grassy places; various Plant Communities; near the coast, Santa Barbara Co. to Del Norte Co., Santa Catalina, Santa Rosa and Santa Cruz ids.; Chile. March–May. Much of the insular material tends to have more spreading hairs than that of the mainland.

29. **T. amoènum** Greene. Pubescent coarse annual, erect, 1–6 dm. high; petioles mostly long; stipules large, ovate, hairy, with an elongate subulate tip; lfts. broadly cuneate-obovate, 2–3 cm. long, often obscurely denticulate; peduncles 3–10 cm. long; heads globose-ovoid, densely fld., 2–3 cm. long; calyx ca. 10–12 mm. long, the plumose teeth longer than the tube; corolla purple, tipped with white, 12–15 mm. long.—Rather rare in low rich fields, swales, etc.; Marin to Solano cos. April–June.

30. **T. dichótomum** H. & A. [*T. Macraei* var. *d.* Brew. *T. d.* var. *turbinatum* Jeps. *T. petrophilum* Heller. *T. californicum* Jeps.] Sparsely to densely pubescent annual with erect or ascending stems 1.5–4 dm. high; petioles slender, rather long; stipules ovate, abruptly acuminate, commonly ca. 5 mm. long; lfts. elliptical to cuneate-obovate, 0.6–2 cm. long, obtuse to emarginate, ± serrulate; peduncles 5–15 cm. long, ± recurved at apex; heads cylindric-ovoid, 1–2.5 cm. long; calyx 5–7 mm. long, the subulate-filiform teeth plumose, ca. twice as long as tube; corolla purple, tipped with cream-white, 8–10 mm. long; pods 1-seeded.—Dry rocky or sandy fields and slopes, below 3500 ft.; Coastal Prairie, Mixed Evergreen F., etc.; Coast Ranges from Santa Clara Co. to Humboldt Co. March–May.

31. **T. albopurpùreum** T. & G. [*T. Macraei* var. *a.* Greene. *T. neolagopus* Loja. *T. a.* var. *n.* McDer. *T. columbinum* var. *argillorum* Jeps. *T. Helleri* Kenn. *T. Traskae* Kenn. *T. pseudoalbopurpureum* Kenn. *T. insularum* Kenn.] Villous-pubescent decumbent to ascending annual, with slender stems 1–4 dm. long; stipules entire, ovate-lanceolate; lfts. obovate to cuneate-oblong, obtuse, denticulate above the middle, 0.6–1.8 cm. long; peduncles 5–15 cm. long, slender; heads ovoid, 8–15 mm. long; fls. sessile; calyx-teeth plumose-villose, subulate; corolla 6–7 mm. long, purple, scarcely as long as calyx; pods ± hairy at apex, 1-seeded; $2n = 16$ (Wexelsen, 1928).—Open grassy valleys and slopes, below 4000 ft.; V. Grassland, Foothill Wd., Coastal Prairie, etc.; cismontane Calif.; to Wash., B.C. March–June. Plants with rigidly erect stems and corollas ca. as long as calyx have been called var. *neolagopus*.

32. **T. olivàceum** Greene. Glaucous villous-pubescent annual with erect or ascending stems 2–3 dm. long; stipules entire, lance-ovate, abruptly aristate; lfts. cuneate-obovate, serrulate, 1–3 cm. long; peduncles stoutish, mostly 2–7 cm. long; heads ovoid, olive-green, 1.4–2 cm. long; calyx ca. 8–10 mm. long, villous-plumose with ascending hairs on the subulate teeth; corolla purple with paler tips, ca. half as long as calyx; pods glabrous, 1-seeded.—Open grassy places, below 1200 ft.; mostly V. Grassland, Foothill Wd.; interior valleys bordering N. Coast Ranges and n. Sierran foothills. April–May. Var. *columbìnum* (Greene) Jeps. [*T. c.* Greene.] Heads dove-colored; ovaries pubescent.—With the sp. Var. *griseum* Jeps. Heads gray; corollas less completely concealed.—San Carlos Range of S. Coast Ranges.

33. **T. cyathíferum** Lindl. Glabrous annual with erect or decumbent stems 1–3 dm. long; stipules lanceolate to ovate, laciniately toothed; lfts. obovate to elliptic-oblong, acute or obtuse, ± spinulose-denticulate, 1–2 cm. long; peduncles slender, mostly 2–9 cm. long; invol. 8–20 mm. broad, bowl-shaped, membranous, nerved and unequally toothed; calyx-teeth 1–3 times trichotomously forked, equaling the pink to paler corolla which is 7–10 mm. long; pods 2-seeded.—Moist places, below 8000 ft.; Red Fir F., Yellow Pine F., Mixed Evergreen F., etc.; Mt. Pinos, Sierra Nevada from Tulare Co. n., Coast Ranges from Lake Co. n.; to B.C., Ida. May–Aug.

34. **T. barbígerum** Torr. Pubescent to subglabrous annual, with several rather stout procumbent stems 1–3 dm. long; stipules broadly ovate, ca. 10 mm. long, scarious, laciniate; lfts. obovate to deltoid, setate-serrulate, 0.4–1.2 cm. long; peduncles slender, 4–9 cm. long; invol. 10–14 mm. wide, shallowly bowl-shaped when young, flatter later, short-lobed and with many setaceous teeth; calyx-teeth awned, plumose, the upper one once or twice forked, slightly exceeding the dark purple corolla which is ca. 5–6 mm. long; pods 2-seeded.—Open low moist places; Coastal Prairie, Mixed Evergreen F., Closed-cone Pine F., etc.; near the coast from Del Norte Co. to Santa Barbara Co.; San Miguel and Santa Rosa ids. April–June.

35. **T. Gràyi** Loja. [*T. barbigerum* var. *Andrewsii* Gray. *T. A.* Heller. *T. lilacinum* Greene. *T. b.* var. *l.* Jeps.] A. ± pilose annual with ascending stem 2–4 dm. long; stipules ca. 10–20 mm. long, laciniate; lfts. broadly oblanceolate, 0.8–2 cm. long; peduncles 8–20 cm. long; invol. 12–20 mm. wide, shallowly bowl-shaped, the lobes laciniate; calyx 5–8 mm. long, the teeth plumose and the upper 3-parted; corolla 10–12 mm. long, purple with white tip.—Wet meadows, below 2000 ft.; Mixed Evergreen F., Redwood F., etc.; Mendocino to Monterey cos. April–June.

36. **T. mìcrodon** H. & A. Rather slender-branched ± erect subglabrous annual, 1–5 dm. high; stipules lance-ovate, abruptly acuminate, entire or somewhat toothed; lfts. obcordate to oblanceolate, 0.5–2 cm. long, ± setate-serrulate; peduncles slender, 1–8 cm. long; invols. 5–10 mm. wide, deeply campanulate, scarious and subglabrous below, ± pubescent on the green several-toothed lobes; calyx 4–5 mm. long, the short triangular teeth abruptly apiculate, ciliate and scarious-margined; corolla 5–6 mm. long, white to pale rose; pod 1-seeded.—Open valleys and slopes, below 2500 ft.; Coastal Prairie, V. Grassland, Mixed Evergreen F., Closed-cone Pine F., etc.; Del Norte Co. to San Luis Obispo Co.; to B.C., Chile. March–June.

Var. **pilòsum** Eastw. Two–several cm. high, densely woolly-pubescent on peduncles and invols.—San Nicolas and Santa Cruz ids. March–May.

37. **T. microcéphalum** Pursh. Mostly slender-stemmed annual, ± villous, the stems procumbent to ascending, 2–4 dm. long; stipules ovate, acuminate, ± toothed; lfts. obcordate to oblanceolate, retuse, serrate, 0.5–1.5 cm. long; peduncles very slender, commonly 3–7 cm. long; invols. ca. 5–8 mm. broad, deeply campanulate, the lobes 7–10, lanceolate, entire, with scarious weblike margins; calyx ca. 4 mm. long, pubescent, the teeth subulate; corolla rose to white, ca. 6 mm. long; pods 1–2-seeded; $2n = 16$ (Wexelsen, 1928).—Open grassy places, below 8500 ft.; many Plant Communities; cismontane Calif.; to B.C., L. Calif., Nev. April–Aug.

38. **T. monánthum** Gray. [*T. m.* f. *spatiosum* McDer.] Glabrous to sparingly villose perennial from a taproot and with slender decumbent to suberect stems 1–10(–20) cm. long; stipules lanceolate, mostly subentire; lfts. obcordate to oblanceolate, retuse to truncate or rounded at the apex, 0.4–1.2 cm. long, 0.2–0.5 cm. wide, sparingly toothed; infl. 1–2(–4)-fld.; peduncles straight, the fls. erect; invol. small, 2–4-lobed, the lobes 1–5 mm. long; calyx 2–4 mm. long, the teeth not longer than the tube; corolla creamcolored with purple-tipped keel, 8–12 mm. long; pod 1–3-seeded.—Wet places, 5000–11,500 ft.; Montane Coniferous F.; San Gabriel Mts. (Los Angeles Co.) to the Sierra Nevada of Plumas Co.; Nev. June–Aug.

KEY TO VARIETIES

Upper lvs. with lfts. mostly rounded or retuse at tip. Mostly in the Sierra Nevada.
    Stems mostly less than 1 dm. long, subglabrous; involucral lobes mostly 2–5 mm. long
                                                      T. monanthum
    Stems mostly 1–3 dm. long, noticeably villose; involucral lobes mostly 0.5–2 mm. long
                                                      var. *parvum*
Upper lvs. with lfts. mostly acute at tip.
    Plants glabrous to sparingly villous; infl. mostly erect on straight peduncles. Mts. of s. Calif.
                                                   var. *Grantianum*
    Plants moderately to strongly villous; infl. mostly at right angles to the bent peduncles. Sierra Nevada
                                                 var. *Eastwoodianum*

Var. **párvum** (Kell.) McDer. [*T. pauciflorum* var. *p.* Kell. *T. p.* Heller. *T. multicaule* Jones.] Mostly moderately villous; lfts. 0.4–2 cm. long, 0.2–0.7 cm. wide; infl. mostly 4–8-fld., the peduncle usually bent below the infl.; calyx-teeth usually equal to or longer than the tube.—At 5000–9000 ft.; Montane Coniferous F.; Nevada Co. to Fresno Co.

Var. **Grantiànum** (Heller) Parish. [*T. G.* Heller. *T. monanthum* var. *tenerum* auth., not *T. t.* Eastw. *T. simulans* House.] Glabrous to sparingly villous; stems 0.3–3 dm. long; infl. mostly 3–6-fld., usually on straight peduncles; involucral lobes mostly 3–5 mm. long.—At 5000–9500 ft.; Montane Coniferous F.; San Gabriel, San Bernardino, and San Jacinto mts.

Var. **Eastwoodiànum** J. S. Martin. [*T. tenerum* Eastw.] Mostly villous, the stems 1–3.5 dm. long; lfts. conspicuously setate-serrate, 0.7–1.8 cm. long, 0.2–0.5 cm. wide; infl. mostly 3–4-fld., with the peduncles strongly bent below the invol.; involucral lobes 3–4 mm. long; calyx-teeth usually slightly longer than the tube.—At 5000–10,000 ft., Montane Coniferous F.; Tuolumne Co. to Tulare Co.

39. **T. Wormskióldii** Lehm. [*T. involucratum* Ort., not Lam. *T. fimbriatum* Lindl. *T. heterodon* T. & G. *T. calocephalum* Nutt. *T. W.* var. *Kennedianum* (McDer.) Jeps.] Glabrous perennial with creeping rootstocks and branched decumbent stems 1–3 dm. long; stipules lanceolate, laciniately toothed; petioles commonly 2–7 cm. long; lfts. oblanceolate to wider, 1–3 cm. long, acutish and mucronulate to obtuse, setulose-serrulate; peduncles mostly 2–6 cm. long; invol. mostly 12–15 mm. broad, flattish, mostly ± lobed and then toothed; calyx 10-nerved, ca. 8–9 mm. long, the teeth subulate, ± longer than the tube; corolla ca. 12 mm. long, the banner broad, white to light purple, the wings and keel dark; pods 2–6-seeded; $2n =$ ca. 48 (Wexelsen, 1928).—In wet places below 10,000 ft.; many Plant Communities; cismontane and montane Calif.; to B.C. and Rocky Mts. May–Oct. Variable: plants from the immediate coast tend to have the involucral lobes pronounced, then toothed, and are the sp.; coastal plants with reduced invols. are var. *Kennedianum;* plants of the interior and mts. tend to have the involucral lobes less pronounced or scarcely evident, but sometimes the whole invol. rather uniformly toothed and have been called *T. spinulosum* Dougl.

40. **T. appendiculàtum** Loja. [*T. splendens* Heller. *T. rostratum* Greene.] Glabrous annual with erect or decumbent stems 1–4 dm. long; stipules round-ovate, laciniate, ± spreading; lfts. obovate, obtuse or retuse, 0.6–2 cm. long; peduncles mostly 2–5 cm. long; invol. ca. 1 cm. broad, 7–9-lobed, the lobes with 3–5 lance-subulate teeth; calyx ca. 6 mm. long, the teeth subulate, longer than tube; corolla 8–10 mm. long, purple with cream tip, the keel with a long-apiculate peak; pods 2-seeded.—Occasional in low wet fields and grassy places; Coastal Prairie, N. Coastal Scrub, Mixed Evergreen F., etc.; near the coast from Humboldt Co. to Monterey. April–June.

41. **T. oligánthum** Steud. [*T. pauciflorum* Nutt., mss. *T. filipes* Greene. *T. triflorum* Greene. *T. hexanthum* Greene.] Mostly glabrous, slender-stemmed annual 1–4 dm. high, erect, sparsely leafy; stipules narrow, few-toothed; lfts. linear to cuneate-oblong, entire to setate-serrate, 0.6–2 cm. long, acute; peduncles slender, 2–10 cm. long; invol. flat, small, divided almost to base into entire lance-subulate teeth; calyx 10-nerved, 4–5 mm. long, the teeth broadly subulate, scarcely as long as tube; corolla not much longer, lavender with white tips and purple keel; pod 2–3-seeded.—Open woods, below 3500 ft.; Yellow Pine F., Douglas-Fir F., Mixed Evergreen F., etc.; Coast Ranges from San Luis Obispo Co. n., Sierra Nevada from Fresno Co. n.; to B.C. March–June.

42. **T. trichocàlyx** Heller. [*T. oliganthum* var. *t.* McDer.] Stems branched, prostrate or decumbent, ± pilose above; stipules green, ± ovate, laciniately toothed; lfts. obovate-cuneate, 0.4–1.2 cm. long, truncate or retuse at apex, spinulose-denticulate; invol. cut into simple lance-subulate lobes; calyx ca. 7 mm. long, conspicuously pilose, the lance-subulate teeth longer than the tube; corolla pale purple, 7–8 mm. long.—Sandy places; Closed-cone Pine F.; Monterey Peninsula. April–June.

43. **T. polyòdon** Greene. Glabrous annual with subprostrate stems 2–4 dm. long; stipules round-ovate, with laciniate margins; lfts. obovate, serrulate, 0.8–2 cm. long; invol. 10–13 mm. across, shallowly lobed, each lobe several-toothed; calyx 10-nerved, 6–7 mm. long, each tooth 3–more-toothed, glabrous; corolla 8–10 mm. long, pale-tipped; pods 2-seeded.—Closed-cone Pine F.; Monterey Peninsula. May–June.

44. **T. variegàtum** Nutt. [*T. melananthum* H. & A. *T. v.* var. *m.* Greene. *T. v.* var. *major* Loja. *T. Morleyanum* Greene. *T. pauciflorum* Nutt. *T. v.* var. *p.* McDer. *T. phaeocephalum* Greene. *T. geminiflorum* Greene. *T. trilobatum* Jeps. *T. calophyllum* Greene.] Glabrous annual with usually several stems, decumbent or ascending, 1–6 dm. long; stipules ovate, laciniately toothed; lfts. obovate to ± oblanceolate or oblong, 0.5–1.5 cm. long, setose-serrulate; peduncles slender, 1–8 cm. long; fls. 1–many; invol. with 4–12

lobes, these 3–7-toothed; calyx 5–20-nerved, the teeth subulate-setaceous, simple or 1 bifid; corolla purple with white tip, 5–8 mm. long; pods 1–2-seeded; $2n = 16$ (Wexelsen, 1928).—Moist places, below 8000 ft.; many Plant Communities; common throughout cismontane Calif. April–July. Exceedingly variable: plants with 1–few-fld. heads and invols. 1–4-lobed, largely montane, are var. *pauciflorum;* those with lfts. 1.5–2.5 cm. long; fls. many, corolla dark purple, 8–9 mm. long, are var. *major* (*melananthum*) of the Coast Ranges; while similar plants with yellow-tipped or lilac-tipped corollas and from Sierran foothills, are var. *trilobatum* (Jeps.) Jeps.

45. **T. tridentàtum** Lindl. [*T. trimorphum* Greene. *T. segetum* Greene.] Glabrous annual, with erect to ± decumbent stems 1–4 dm. long; lower stipules lanceolate, acuminate, entire, the upper round-ovate, laciniate; lfts. linear to lance-oblong, 1.5–3.5 cm. long, ± setate-denticulate; peduncles 5–9 cm. long; invol. flat, unevenly laciniate but not lobed, commonly 10–15 mm. broad; calyx-teeth shorter than the 10-nerved tube, dilated below, often with 2 small lateral teeth or shoulders; corolla 12–15 mm. long, red-purple, the banner pale toward tip, the wings dark; pods mostly 2-seeded.—Common in mostly grassy places below 5000 ft.; many Plant Communities; cismontane Calif. to B.C. March–June. Variable and with many poorly defined tendencies, the most outstanding being:

Var. **aciculàre** (Nutt.) McDer. [*T. a.* Nutt. *T. Watsonii* Loja. *T. scabrellum* Greene. *T. polyphyllum* Nutt.] Calyx-tube 20–25-nerved, the teeth mostly entire.—Commonest form in Cent. V. and s. Calif.

46. **T. obtusiflòrum** Hook. [*T. tridentatum* var. *o.* Wats. *T. roscidum* Greene. *T. majus* Greene.] Soft-pubescent annual, clammy, the stems ± fistulous, 3–5 dm. long, erect to decumbent-ascending; stipules rather broad, conspicuously setulose; lfts. lance-linear to narrow-obovate, 2–3 cm. long, setulose-serrate; peduncles 3–8 cm. long; invol. flat, 12–16 mm. wide, deeply laciniate with subulate divisions; calyx-tube mostly 20-nerved, the entire teeth dilated near their base; corolla ca. 12–14 mm. long, pale with cent. dark spot; pods 2-seeded; $2n = 16$ (Wexelsen, 1928).—Moist places, below 5000 ft.; many Plant Communities; cismontane Calif. to sw. Ore. April–July.

47. **T. fucàtum** Lindl. [*T. physopetalum* F. & M. *T. flavulum* Greene.] Glabrous annual with stout fistulous diffuse stems 1–8 dm. long; stipules ovate, 1.5–2 cm. long, entire; lfts. rhombic-ovate to obovate, 0.6–3 cm. long, denticulate to almost setulose; peduncles 3–15 cm. long, stout; invol. with 5–9 entire setulose-tipped scarious lobes; calyx scarious, 4–5 mm. long, the subulate teeth simple, shorter than to longer than the tube; corolla cream, becoming pink and much inflated in age, then 20–25 mm. long, the keel dark purple; pods stipitate, 3–8-seeded; $2n = 16$ (Wexelsen, 1928).—Moist places, ± brackish, below 3000 ft.; many Plant Communities; cismontane Calif. to Ore. April–June.

Var. **viréscens** (Greene) Jeps. [*T. v.* Greene.] Lfts. sharply serrulate; well developed calyx-teeth only 3, much exceeding the tube; corolla 12–15 mm. long, yellow with green or purple tinge.—Moist valleys and dry hills; several Plant Communities; N. Coast Ranges.

Var. **Gámbelii** (Nutt.) Jeps. [*T. G.* Nutt.] Calyx 8–10 mm. long, the teeth 5, much elongate, bristlelike, 1 or more being 2–3-cleft; corolla 15–20 mm. long, white with yellow and purple tinge.—Grassy slopes, Los Angeles Co. to Contra Costa Co.; Santa Catalina and Santa Rosa ids.

48. **T. ampléctens** T. & G. [*T. depauperatum* var. *a.* McDer.] Light green annual, glabrous, with slender decumbent stems 1–2.5 dm. long; stipules broad, entire, with a subulate point; lfts. obovate to oblanceolate, emarginate to truncate, serrulate, 0.6–2 cm. long; peduncles mostly to ca. 4–5 cm. long; invol. deeply lobed, the lobes 3–4 mm. long, broad, scarious-margined, entire or 1–2-toothed; calyx ca. 4–5 mm. long, the lower teeth setose, longer than the tube; corolla ca. 6 mm. long, white to reddish or purple, much inflated in age; pods 4–6-seeded.—Grassy places, below 2500 ft.; V. Grassland, Foothill Wd., Coastal Sage Scrub, etc.; Coast Ranges (chiefly in cent. Calif.) and Central V., Lake Co. to San Diego Co. April–June.

Var. **truncàtum** (Greene) Jeps. [*T. franciscanum* var. *t.* Greene. *T. f.* Greene. *T. amplectens* var. *stenophyllum* (Nutt.) Jeps. *T. s.* Nutt. *T. diversifolium* Nutt. *T. anodon* Greene. *T. brachyodon* Greene.] Involucral bracts oblong, ca. as long as calyx, without or with very narrow hyaline margins.—With the sp. and much more common.

Var. **hydrophìlum** (Greene) Jeps. [*T. h.* Greene.] Involucral bracts oblong, ca. half as long as calyx.—Saline and alkaline places, San Luis Obispo to Santa Clara and Colusa còs.

49. **T. depauperàtum** Desv. [*T. laciniatum* var. *angustatum* Greene.] Glabrous annual with ascending stems to ca. 1.2 dm. long; stipules entire, subulate-tipped; lfts. cuneate-oblong, denticulate, 0.6–2 cm. long; peduncles slender, 3–6 cm. long; invol. reduced to a small ring; calyx ca. 2–2.5 mm. long, the teeth unequal, triangular-subulate; corolla whitish to purple, 7–8 mm. long in age; pods 1–2-seeded.—Moist ± alkaline places, below 2500 ft.; V. Grassland, Mixed Evergreen F., etc.; Coast Ranges and Sacramento V. from Humboldt and Shasta cos. to Alameda and San Joaquin cos.; Ore., Chile. March–May. A poorly marked form is var. *laciniàtum* (Greene) Jeps. [*T. l.* Greene.] Lfts. deeply laciniately toothed.—With the sp.

19. Lòtus L. Bird's Foot Trefoil

Annual or perennial herbs or suffrutescent. Lvs. alternate, pinnately 1- to several-foliolate. Stipules foliaceous, scarious, or glandlike. Fls. axillary, solitary to umbellate, sessile or on short to long, mostly leafy-bracteate peduncles. Calyx-teeth 5, subequal. Corolla white or yellow, often with some red or purple. Stamens diadelphous. Pod flattened or terete, straight to strongly arcuate, 1–many-seeded, dehiscent or indehiscent. Perhaps 150 spp., of all continents, but mostly of N. Hemis. (Ancient Greek name.)

(Ottley, A. M. A revision of the California spp. of Lotus. Univ. Calif. Pub. Bot. 10: 189–305, 1923, and The Amer. Loti with special consideration of a proposed new section, Simpeteria. Brittonia 5: 81–123, 1944.)
A. Stipules expanded, not glandlike.
  B. The stipules green, resembling and almost as large as lfts.
    C. Lfts. 9–19; bract 5–7-foliolate. Native .... 2. *L. stipularis*
    CC. Lfts. 3; bract 1–3-foliolate. Introd. ssp.
      D. Plants perennial; calyx-teeth ca. as long as the tube; corolla 8–12 mm. long.
        E. Umbel mostly 3–6-fld.; calyx 3–4 mm. long; keel abruptly narrowed to the beak .... 34. *L. corniculatus*
        EE. Umbel mostly 8–12-fld.; calyx 5–6 mm. long; keel gradually narrowed to the beak .... 35. *L. uliginosus*
      DD. Plants annual; calyx-teeth mostly exceeding the tube; corolla ca. 4 mm. long
                                    36. *L. angustissimus*
  BB. The stipules membranaceous or hyaline, not like the lfts.
    C. Pods 3–5 mm. wide; bract mostly distant from the umbel, usually 3–5-pinnate.
      D. Plant densely short-villous or almost tomentose; calyx-teeth 2–3 mm. long; banner reddish, wings white .... 1. *L. neo-incanus*
      DD. Plant glabrous or nearly so; calyx-teeth 1–1.5 mm. long.
        E. Corolla white and red, or pinkish; calyx ca. 3.5 mm. long .. 3. *L. aboriginum*
        EE. Corolla greenish-yellow with some purplish-red; calyx 4–4.5 mm. long
                                 4. *L. crassifolius*
    CC. Pods less than 3 mm. wide; bracts near base of umbel and 1–3-foliolate or lacking.
      D. Herbage glabrous or nearly so.
        E. Corolla 12–14 mm. long, the claws long, well exserted.
          F. Bract mostly lacking; banner and keel yellow, the wings white; calyx ca. 7 mm. long .... 5. *L. pinnatus*
          FF. Bracts 3–5-foliolate; banner yellow, wings and keel rose to purple; calyx ca. 5 mm. long .... 6. *L. formosissimus*
        EE. Corolla 7–8 mm. long, the claws short, mostly not exserted in the calyx.
          F. Plants with a heavy branched root-crown; bract 1–3-foliolate. Dry places, Trinity and Humboldt cos.   7. *L. yollabolliensis*
          FF. Plants with slender rootstocks; bract mostly lacking. Meadows, Tulare Co.
                                    9. *L. cupreus*
      DD. Herbage pubescent or strigose; plants with slender rootstocks; claws of petals scarcely exserted .... 8. *L. oblongifolius*
AA. Stipules reduced to dotlike often dark or reddish glands.
  B. Pods dehiscent, straight or nearly so, abruptly short-beaked.
    C. Plants perennial.
      D. Lf.-rachis 2–5 cm. long; corolla 1.5–2.4 cm. long .... 10. *L. grandiflorus*
      DD. Lf.-rachis not more than 1 cm. long.
        E. Plants erect, 3–9 dm. tall; corolla 1.3–2 cm. long; pods 3–4 cm. long; lvs. sparsely strigose .... 11. *L. rigidus*
        EE. Plants prostrate to ascending, the stems 1–3 dm. long; corolla 8–12 mm. long; lvs. silvery-silky to appressed-canescent .... 12. *L. argyraeus*
    CC. Plants annual.
      D. Fls. yellow, often aging red.

  E. Wings conspicuously longer than the keel; styles hairy at the base of the
    stigma; seeds not smooth.
    F. Lfts. linear-oblong to elliptic, not noticeably succulent; herbage
      strigulose. Cismontane .... 13. *L. strigosus*
    FF. Lfts. mostly ± obovate, quite succulent; herbage canescently tomentulose.
      Deserts .... 14. *L. tomentellus*
  EE. Wings ca. as long as or shorter than keel; styles glabrous; seeds smooth.
    F. Fls. 1–5 on a peduncle; pods 1.5–2 mm. wide .. 15. *L. salsuginosus*
    FF. Fls. solitary, subsessile; pods ca. 3 mm. wide.
      G. Plant subglabrous to strigose; calyx-teeth ca. as long as the tube;
        pods 10–15 mm. long .... 17. *L. subpinnatus*
      GG. Plant densely villous-pubescent; calyx-teeth ca. twice as long as the
        tube; pods 5–10 mm. long .... 18. *L. humistratus*
 DD. Fls. cream-white or red or pink.
  E. Fls. subsessile, the keel not prolonged into an incurved beak.
    F. Plants erect, 2–5 dm. high, subglabrous to strigose; calyx-teeth ca. 1.5
      mm. long. From Butte and Lake cos. n. .... 16. *L. denticulatus*
    FF. Plants decumbent, the stems to 1 dm. long; plants pilose; calyx-teeth
      4–5 mm. long. Mt. Hamilton Range .... 19. *L. rubriflorus*
  EE. Fls. peduncled; keel prolonged into an incurved beak.
    F. Calyx-teeth shorter than the tube; corolla pinkish or pale salmon,
      tinged or turning red; pods not deflexed; lfts. mostly 3–10 mm. long
                            20. *L. micranthus*
    FF. Calyx-teeth longer than the tube; corolla whitish, tinged with rose;
      pods deflexed; lfts. mostly 10–15 mm. long .... 21. *L. Purshianus*
BB. Pods indehiscent, often strongly curved, tapering to an elongate beak.
 C. Plants mostly annual.
  D. Umbels pedunculate; bract usually present, unifoliolate; corolla 5–7 mm. long; beak
    of pod glabrous .... 28. *L. Nuttallianus*
  DD. Umbels subsessile; bract lacking; corolla 3–4.5 mm. long; beak of pod strigose
                             29. *L. hamatus*
CC. Plants perennial.
 D. Corolla mostly 3–9 mm. long.
  E. Pubescence spreading, ± woolly-villous .... 22. *L. Heermannii*
  EE. Pubescence appressed or almost lacking.
    F. Fls. on peduncles mostly more than 5 mm. long.
      G. Plants ascending to suberect, ± strigose to almost glabrous.
        H. Lfts. 5–15 mm. long; fls. mostly more than 1 or 2 on a
          peduncle. Plants from near the coast.
          I. Calyx-teeth short-triangular, scarcely 1 mm. long; pod
            strigose .... 23. *L. junceus*
          II. Calyx-teeth subulate, ca. 2 mm. long; pods sublabrous
                          30. *L. Benthamii*
        HH. Lfts. 2–5 mm. long; fls. 1–2 on a peduncle. Desert edge
                         33. *L. Haydonii*
      GG. Plants prostrate, densely strigose to silvery.
        H. The plant ± canescent throughout, but scarcely silvery; calyx-
          teeth ca. half as long as the tube. Mts. of San Diego Co. and
          from Inyo and Kern cos. to Plumas Co. Rare in Riverside and
          San Bernardino cos. .... 24. *L. nevadensis*
        HH. The plant more silvery.
          I. Calyx-teeth almost as long as the tube; banner reflexed or
            at right angles to the wings. Mt. Pinos to Santa Rosa Mts.
                         26. *L. Davidsonii*
          II. Calyx-teeth ca. half as long as tube; banner ascending,
            at an angle acute to the wings. San Gabriel and San
            Bernardino mts. .... 27. *L. argophyllus* var. *decorus*
    FF. Fls. sessile or subsessile, in lf.-axils.
      G. Plant plainly grayish or whitish with appressed pubescence.
        H. Body of the pod scarcely exserted beyond the calyx, arcuate,
          the beak ca. as long as the body; herbage densely silvery-
         tomentose .... 27. *L. argophyllus*
        HH. Body of the pod well exceeding the calyx, somewhat curved,
          the beak shorter than the body; herbage with closely ap-
         pressed pubescence .... 32. *L. procumbens*
      GG. Plant green, subglabrous except at growing tips; body of pod well
        exserted .... 31. *L. scoparius*
 DD. Corolla 10–12 mm. long.
  E. Umbels on well developed peduncles; pods well exserted from calyx. Insular.
    F. Plant decumbent to ascending, herbaceous; lfts. 3–7; fls. 12–20
                  27. *L. argophyllus* var. *ornithopus*
    FF. Plants erect, woody; lfts. 3; fls. 7–12 .. 31. *L. scoparius* var. *dendroideus*
  EE. Umbels sessile or nearly so; plants prostrate or decumbent.
    F. Calyx-teeth ca. as long as the tube; lvs. loosely villous-tomentose. From
      Lassen and Mendocino cos. n. .... 25. *L. Douglasii*

FF. Calyx-teeth ca. half as long as the tube; lvs. more closely pubescent or
    if not, plants insular.
  G. Body of the pod scarcely exserted from the calyx, ± arcuate.
    H. Herbage with short closely appressed pubescence. Santa Lucia
       Mts. and w base of Sierra Nevada.
                              27. *L. argophyllus* var. *Fremontii*
    HH. Herbage silvery silky, with ± spreading hairs. Insular.
                              27. *L. a.* var. *niveus*
  GG. Body of the pod well exserted from the calyx; plant silky-strigose.
    San Miguel Id. . . . . 31. *L. scoparius* var. *Veatchii*

1. **L. neo-incànus** Munz. [*Hosackia incana* Torr. *L. i.* Greene, not Dougl.] Stems
several from the root-crown, ± erect, 1–3 dm. tall; herbage densely short-villous, almost
silky-tomentose; stipules ovate, scarious; lfts. 5–15, usually 7 or 9, elliptic to obovate,
7–15 mm. long; peduncles 1–3 cm. long, slightly shorter than lvs.; fls. few to many; bract
typically stipulate and of 5 lfts.; calyx-tube cylindric, the linear-subulate teeth hairy,
less than ½ as long as tube; petals 12–15 mm. long, the banner reddish, long-clawed;
wings white, longer than the obtuse keel; pods reddish or dark yellowish-brown,
glabrous, reticulate, 1.5–4 cm. long, 0.3–0.7 cm. wide; seeds several, ovoid, smooth,
mottled, ca. 3 mm. long.—Open woods, 2500–5000 ft.; Yellow Pine F.; w. Sierra Nevada
from Placer Co. to Butte Co. May–June.

2. **L. stipulàris** (Benth.) Greene. [*Hosackia s.* Benth. *H. macrophylla* Kell. probably
L. purpurascens Eastw.] Much like *L. neo-incanus,* but 2–5 dm. high, loosely villous or
the lfts. almost glabrous; stipules herbaceous, lanceolate with auriculate base; lfts. 9–19,
oblong-elliptic to oval, 6–20 mm. long; umbels 4–10-fld.; fls. 10–12 mm. long, the calyx-
tube 4 mm. long, teeth subulate or short-triangular, 1–2 mm. long; banner and keel
red-purple with white tips, wings white; pods 2–3 cm. long, 2–3 mm. wide.—Dry
usually wooded slopes, below 4000 ft.; Mixed Evergreen F., Chaparral, etc.; Coast
Ranges from Monterey Co. n., Sierra Nevada from Tulare Co. n. April–June. A fragrant
form with resinous glands among the hairs is *L. balsamiferus* (Kell.) Greene. [*Hosackia
b.* Kell.]

3. **L. aboríginum** Jeps. (spelled aboriginus) [*L. stipularis* var. *subglaber* Ottley.
Hosackia rosea Eastw.] Much like *L. stipularis,* but almost glabrous; stipules triangular-
lanceolate, membranous; peduncles equaling or exceeding lvs.; bract sometimes much
reduced; calyx purple, 3.5–4 mm. long, the short triangular teeth ciliolate; corolla 8–9
mm. long, pink; pods 1.5–3.5 cm. long, 3–4 mm. wide.—Wood borders, below 2500 ft.;
Mixed Evergreen F., Closed-cone Pine F., etc.; Sonoma Co. to Del Norte Co.; to Wash.
May–July.

4. **L. crassifòlius** (Benth.) Greene. [*Hosackia c.* Benth. *H. stolonifera* Lindl. *H.
platycarpa* Nutt.] Perennial, glabrous or puberulent, the stems stout, 4–12 dm. high,
erect; stipules triangular-lanceolate, membranous; lfts. 7–15, oval, rhombic or obovate,
1–3 cm. long; peduncles 3–8 cm. long, shorter than lvs.; bract usually present, remote
from umbel, 1–5-foliolate; fls. 8–15; pedicels 2–4 mm. long; calyx-teeth short, subulate-
triangular, the tube ca. 3–4 mm. long; corolla 9–12 mm. long, greenish-yellow with some
purplish-red; pods 3.5–6.5 cm. long, 4–5 mm. wide; seeds 7–12, dark, 2–2.5 mm. long.—
Dry banks and flats, 2000–8000 ft.; Yellow Pine F., Red Fir F., Lodgepole F.; mts. of
s. Calif., Sierra Nevada, Coast Ranges from San Luis Obispo Co. n.; to Wash. May–Aug.

5. **L. pinnàtus** Hook. [*Hosackia bicolor* Dougl. ex Benth.] Perennial, nearly or quite
glabrous, rather stout, 2–4 dm. tall; stipules scarious, narrowly ovate; lfts. 5–9, oval
to obovate, 1–2.5 cm. long; peduncles 5–10 cm. long, exceeding the lvs.; bract mostly
absent; fls. 3–7; calyx-tube ca. 5 mm. long, the lower teeth subulate, 2–3 mm. long, the
upper shorter; corolla 12–14 mm. long, banner and keel yellow, the wings white; pods
4–6 cm. long, ca. 2 mm. wide; seeds oblong, several.—Moist places, below 6000 ft.;
Foothill Wd., Yellow Pine F., Red Fir F.; Sierra Nevada from Calaveras Co. n. and
Coast Ranges from Lake Co. n.; to Wash., Ida. May–July.

6. **L. formosíssimus** Greene. [*Hosackia gracilis* Benth., not *L. g.* Salisb.] Much like
L. pinnatus, more nearly prostrate, glabrous; peduncles equaling or exceeding lvs.;
bract subtending the umbel and 3–5-foliolate; calyx-tube ca. 3 mm. long, the teeth
subulate, almost equal, 2 mm. long; corolla 12–14 mm. long, banner yellow, wings and
keel rose to purple; pod 2.5–3.5 cm. long, ca. 2 mm. wide; seeds oblong, 1–1.5 mm.

long.—Moist places, below 1500 ft.; Mixed Evergreen F., N. Coastal Scrub, Closed-cone Pine F., etc.; near the coast from Monterey Co. n.; to Wash. March–July.

7. **L. yollabolliénsis** Munz. Perennial with heavy almost woody root-crown; stems slender, branched, decumbent, 5–10 cm. long; herbage glabrous; stipules membranaceous, ovate, 1–2 mm. long; lfts. 3–5, oblanceolate, petiolulate, acute-mucronulate, 3–10 mm. long; peduncles 2–3 cm. long; bract subtending umbel, 1–3-foliolulate; umbels 1–3-fld.; pedicels to ca. 1 mm. long; calyx-tube campanulate, 1.5–2 mm. long, the teeth subequal, triangular to triangular-subulate, 0.5–1 mm. long, ± finely ciliolate; corolla 7–9 mm. long, yellow, the claws short, mostly included in the calyx; pods brown, 1.8–2.4 cm. long, 1.5–2 mm. wide; seeds narrow-oblong, 2–2.5 mm. long, finely mottled green and black.—Dry barren open slopes, 5500–7000 ft.; Red Fir F.; N. Yolla Bolly Mts. (Trinity Co.) to S. Fork Mt. (Humboldt Co.). June–Aug.

8. **L. oblongifòlius** (Benth.) Greene. [*Hosackia o.* Benth. *H. lathyroides* Dur. & Hilg. *H. o.* var. *angustifolia* Wats. *L. Torreyi* var. *seorsus* Macbr.] Perennial with slender rootstocks, erect or ascending, 2–5 dm. high, ± strigose throughout; stipules membranaceous, lanceolate; lfts. mostly 7–11, lance-linear to elliptical, mostly acute at both ends, 0.5–2 cm. long; peduncles mostly exceeding lvs., 5–12 cm. long; bract 1–3-foliolate, closely subtending the 1–5-fld. umbel; calyx-tube 2–3 mm. long, the teeth narrowly subulate, almost as long; corolla 9–14 mm. long, banner yellow, ± veined with purple, as may be the keel; pods 2.5–4 cm. long, 1.5–2 mm. wide; seeds narrow-oblong, mottled, 1.5–2 mm. long.—Wet places, below 8500 ft.; Chaparral, Yellow Pine F., Red Fir F.; Monterey Co. to cismontane s. Calif. and foothills of e. and w. sides of Sierra Nevada to Panamint Mts. May–Sept.

Var. **nevadénsis** (Gray) Munz. [*Hosackia Torreyi* var. *n.* Gray. *H. T.* Gray. *L. o.* var. *Torreyi* Ottley.] Pubescence looser, ± divergent; lfts. broader, often obtuse, those of the uppermost lvs. ± acute.—Wet places; mostly Montane Coniferous F., also Mixed Evergreen F., etc.; Sierra Nevada to Modoc Co., N. Coast Ranges from Lake Co. n.; to Ore.

9. **L. cùpreus** Greene. [*Hosackia c.* Smiley. *L. oblongifolius* var. *c.* Ottley.] Essentially glabrous perennial, with slender rootstocks and subprostrate slender stems 1–2 dm. long; lfts. 3–5, oblanceolate to obovate-cuneate, 5–15 mm. long; peduncles 1–2 cm. long; bract mostly 1-foliolate; fls. 1–3, yellow, turning copper-color in age; calyx-tube ca. 1.5 mm. long, the teeth subulate, 1 mm. long; petals 7–8 mm. long, short-clawed.—Wet meadows, 8000–9000 ft.; Lodgepole F.; Sierra Nevada of Tulare Co. July–Aug.

10. **L. grandiflòrus** (Benth.) Greene. [*Hosackia g.* Benth. *Anisolotus g.* Heller. *H. macrantha* Greene. *H. ochroleuca* Nutt. in T. & G.] Perennial from ± woody base, 2–6 dm. high, strigose throughout with upwardly appressed incurved hairs; stipules gland-like; lfts. mostly 7 or 9, obovate to elliptical, obtuse, mucronate, 0.7–2 cm. long; peduncles exceeding lvs., 4–8 cm. long; umbels 2–several-fld.; bract 1-, sometimes 3-foliolate, 1–2 cm. long; calyx-tube ca. 5 mm. long, the teeth subulate, almost as long; corolla yellow, aging red, 1.5–2.4 cm. long, the claws not exceeding calyx; pods 3–4 cm. long, ca. 2 mm. wide; seeds many, broadly ± ovoid, dark brown, ca. 2 mm. long.—Dry slopes, below 6000 ft.; Yellow Pine F., Foothill Wd., Chaparral, etc.; cismontane Calif. from Shasta and Mendocino cos. s. April–July. Plants with more spreading short hairs have been called var. *mutábilis* Ottley. [*L. leucophaeus* Greene. *L. confinis* Greene. *Hosackia grandiflora* var. *anthylloides* Gray, which would be the oldest varietal name.]

11. **L. rígidus** (Benth.) Greene. [*Hosackia r.* Benth. *Anisolotus r.* Rydb. *L. argensis* Cov.] Perennial, usually several-stemmed, erect, woody at base, coarse, broomlike, sparsely strigose, 3–9 dm. tall; lvs. remote, 1–2 cm. long, with 3–5 narrowly oblong lfts.; peduncles 5–12 cm. long, 2–3-fld.; bract 0 or 1-foliolate; calyx-tube 4–5 mm. long, strigose, the teeth subulate, nearly as long; corolla 1.3–2 cm. long, yellow, with some red or purple in age; pods subglabrous, 3–4 cm. long, ca. 3 mm. wide; seeds subglobose, granulose, light brown with some darker mottling, ca. 1.5 mm. in diam.—Common on dry slopes and in washes, below 5000 ft.; Creosote Bush Scrub to Pinyon-Juniper Wd.; deserts from Inyo Co. s.; to Utah, Ariz., L. Calif. March–May.

12. **L. argyraèus** (Greene) Greene. [*Hosackia a.* Greene. *Anisolotus a.* Heller.] Prostrate perennial, forming silvery mats, the stems slender, branched, from a woody base, silky pubescent, 1–3 dm. long; lfts. 3–5, cuneate-oblanceolate to -obovate, mostly 4–10 mm.

long, silvery, mostly obtuse, on a very short rachis; peduncles exceeding lvs., ebracteate, 1–3-fld.; calyx-tube 2.5–3 mm. long, the teeth subulate, almost as long; corolla 8–10 mm. long, yellow, aging red; pods 1–1.5 cm. long, 2–3 mm. wide, strigose; seeds 2–several, roundish-oblong, smooth, olive-brown with faint darker mottling, 1.5–1.8 mm. long.— Dry slopes and benches, 3500–8200 ft.; mostly Yellow Pine F.; San Bernardino, San Jacinto, Santa Rosa mts.; n. L. Calif. May–Aug.

Ssp. multicáulis (Ottley) Munz. [*L. Wrightii* var. *m.* Ottley. *Hosackia W.* subsp. *m.* Abrams.] Stems prostrate to ascending, 1–2 dm. long; herbage appressed-canescent, not silvery; lfts. linear-oblong to cuneate-obovate, 5–12 mm. long; fls. 1–2; calyx-tube 3–3.5 mm. long; corolla 8–12 mm. long.—Dry slopes and flats, 4000–7200 ft.; mostly Pinyon-Juniper Wd.; New York, Providence mts. (e. San Bernardino Co.)

13. **L. strigòsus** (Nutt. in T. & G.) Greene. [*Hosackia s.* Nutt. *H. nudiflora* Nutt. *Anisolotus s.* Heller.] Slender-stemmed decumbent to ascending annual, the several branches mostly 0.5–3 dm. long; herbage strigose; lvs. 1–2.5 cm. long, with flattened rachis, the lfts. 6–10, linear-oblong to elliptic, acute, 5–12 mm. long; peduncles mostly exceeding lvs., the lower 1-fld., the upper 2–3-fld.; bract 0 or 1–3-foliolate; calyx-tube ca. 2 mm. long, the teeth subulate, almost as long; corolla mostly 6–10 mm. long, yellow, turning rose-red or somewhat reddish on back; pods 2–3 cm. long, 2–3 mm. wide, strigose; seeds quadrate, ± irregularly rugose, brown, ca. 1 mm. long.—Common in dry disturbed places, below 5000 ft.; many Plant Communities; cismontane Calif. from Marin, Sutter, and Tuolumne cos. s., e. to edge of desert; to L. Calif. March–June. Plants with fls. 5–7 mm. long apparently are *L. rubéllus* (Nutt.) Greene. [*Hosackia r.* Nutt.] and occur with larger fld. ones. A strand form of the sp. makes dense prostrate mats.

Var. **hirtéllus** (Greene) Ottley. [*L. h.* Greene.] Pubescence spreading; seeds granulose, mostly not rugulose.—With the sp., but mostly in the interior.

14. **L. tomentéllus** Greene. [*Hosackia t.* Abrams.] Prostrate; pubescence canescently tomentulose; lvs. to ca. 2 cm. long, with 4–8 thick cuneate-oblanceolate, obovate or oblong lfts. 3–10 mm. long; peduncles mostly shorter than subtending lvs.; umbels 1–2-fld.; bract 0 to 1–3-foliolate; calyx-tube 2–2.5 mm. long, the teeth lanceolate, somewhat shorter; corolla 5–7 mm. long, yellow, mostly lacking red on back, but aging red; pod 1–2 cm. long, 2 mm. wide; seeds globose to ovoid, rarely cubical, finely granulose, brownish or greenish, ± mottled.—Sandy places, below 4500 ft.; mostly Creosote Bush Scrub; deserts from Inyo Co. s.; to Nev., Ariz., Son., L. Calif. Feb.–May.

15. **L. salsuginòsus** Greene. [*Hosackia maritima* Nutt., not *L. m.* *L. Anisolotus m.* Heller.] Decumbent or prostrate annual, the stems 1–3 dm. long; herbage subglabrous to ± strigose; lvs. slightly succulent, to 4 cm. long; lfts. 5–8, obovate to suborbicular, 4–14 mm. long; peduncles 1–4 cm. long, ca. equal to lvs., 1–5-fld.; bract broadly ovate, unifoliolate; calyx-tube ca. 3 mm. long, the teeth broadly subulate, slightly shorter; corolla 6–9 mm. long, yellow, wings with petal-blades narrowed abruptly to short claws; pod 1.5–3 cm. long, 1.5–2 mm. wide, glabrous; seeds oblong-ovoid, smooth, brown, 1–1.5 mm. long.—Dry slopes and fields, below 3500 ft.; Chaparral, Coastal Sage Scrub, Foothill Wd., etc.; Santa Clara Co. to San Diego Co.; L. Calif.; Channel Ids. March–June.

Var. **brevivexíllus** Ottley. [*L. humilis* Greene.] Corolla 3–5 mm. long, the keel exposed by the shorter banner and wings; pod more constricted between the seeds.— Sandy washes and fans, below 6000 ft.; Creosote Bush Scrub; deserts from Panamint Mts. s. to L. Calif., Ariz.

16. **L. denticulàtus** (E. Drew) Greene. [*Hosackia d.* E. Drew. *Anisolotus d.* Heller.] Erect annual, branched from base, 2–5 dm. high, subglabrous to strigose; herbage pale green, the lvs. often scantily villous, 3–4, obliquely obovate, 5–30 mm. long, ± denticulate; fls. solitary, subsessile, axillary; calyx pilose, the tube ca. 1 mm. long, the teeth subulate, somewhat longer; corolla 6–8 mm. long, whitish-cream, with back of banner tinged purple, keel attenuately beaked; pods 1–2 cm. long, 3–4 mm. wide, strigose; seeds asymmetric, notched at hilum, smooth, 2–4 mm. by 1–2 mm.—Grassy slopes and gravelly places, below 4000 ft.; Mixed Evergreen F., Foothill Wd.; Butte Co. to Modoc Co. and Lake to Humboldt and Siskiyou cos.; to B.C. May–July.

17. **L. subpinnàtus** Lag. [*Hosackia s.* T. & G. *L. Wrangelianus* F. & M.] Annual with decumbent to ascending stems 1–3 dm. long, subglabrous to strigose; lfts. 3–5, obovate, entire, 5–15 mm. long; fls. subsessile, solitary in axils; calyx-tube ca. 2 mm. long, the

teeth ca. as long; corolla yellow, tinged red-purple in age, 5–7 mm. long; keel with attenuate beak; pod 10–15 mm. long, ca. 3 mm. wide, compressed, sparsely strigose; seeds ca. 1.5–2 mm. long, notched at hilum, smooth.—Dry grassy slopes, below 2500 ft.; many Plant Communities; most of cismontane Calif., Chile. March–June.

18. **L. humistràtus** Greene. [*Hosackia brachycarpa* Benth., not *L. b.* Hochstetter & Steud.] Much like *L. subpinnatus*, but densely villous throughout; calyx-lobes ca. twice as long as calyx-tube; pods 5–10 mm. long, densely villous.—Common below 6000 ft.; many Plant Communities; cismontane Calif. and occasional on deserts; to L. Calif., New Mex. March–June.

19. **L. rubriflòrus** H. K. Sharsm. Decumbent, slender annual, 0.3–1 dm. high; herbage pilose throughout; lvs. 1–1.6 cm. long, with 4 lanceolate acute lfts. 2–11 mm. long; fls. solitary, subsessile; calyx-tube 1.5–2 mm. long, the teeth linear, acuminate, 4–5 mm. long; corolla 6–7 mm. long, bright pinkish-red, with short claws; pods pilose, 8–9 mm. long, 2.5 mm. wide, straw-colored; seeds 2–4, ± lens-shaped, 2 mm. long and wide, notched at hilum, olive green to brownish, ± mottled.—Rolling hills, at 1600 ft.; Mt. Hamilton Range, Stanislaus Co. April–May.

20. **L. micránthus** Benth. [*Hosackia parviflora* Benth., not Desf. *H. microphylla* Nutt.] Diffuse slender-stemmed annuals 1–3 dm. high, glabrous or sparsely strigose; lvs. 1–1.5 cm. long; lfts. 3–5, oblong to oblanceolate or elliptical, mostly 3–10 mm. long; peduncles shorter than to longer than lvs., 1-fld.; bract of 1–3 lfts.; calyx-tube turbinate-campanulate, ca. 1 mm. long, the subulate teeth shorter; corolla 4–5 mm. long, pinkish or pale salmon, tinged or turning red; pods 1.5–2 cm. long, ca. 2 mm. wide, constricted between seeds, glabrous; seeds suborbicular to short-oblong, less than 2 mm. long.—Open plains and slopes, below 5000 ft.; many Plant Communities; cismontane Calif.; to Wash. March–May.

21. **L. Purshiànus** (Benth.) Clem. & Clem. [*Hosackia americana* Benth. *L. sericeus* Pursh, not Moench. *Trigonella a.* Nutt. *L. a.* Bisch., not Vell. *Acmispon gracilis* Heller. *A. sparsiflorus* Heller. *A. aestivalis* Heller.] Erect or ascending much-branched annual, 1.5–8 dm. tall, glabrate to villous; lvs. 1–2.5 cm. long, mostly 3-foliolate lfts. lance-oblong to elliptical, 10–15 mm. long, the terminal petiolulate; peduncles 10–15 mm. long, 1-fld.; bract 1-foliolate; calyx-tube ca. 1.5 mm. long, the subulate teeth longer; corolla whitish, tinged with rose, 4–7 mm. long; pods 1.5–2.5 cm. long, 2–2.5 mm. wide, glabrous, deflexed; seeds oblong, smooth, olive with darker mottling to dark brown, ca. 3 mm. long; $2n = 14$ (Larsen, 1956).—Common in dry fields and disturbed places, below 7000 ft.; many Plant Communities; cismontane Calif. and sometimes on the desert; to B.C., Dak., Mex. May–Oct.

Var. **glàber** (Nutt.) Munz. [*Hosackia elata* var. *g.* Nutt. *Acmispon g.* Heller. *L. americanus* var. *minutiflorus* Ottley.] More or less decument, pilose to nearly glabrous; corolla 4–5 mm. long.—Mostly in Yellow Pine F.

22. **L. Heermánnii** (Dur. & Hilg.) Greene. [*Hosackia H.* Dur. & Hilg. *Syrmatium H.* Greene. *L. eriophorus* var. *H.* Ottley. *H. decumbens* var. *glabriuscula* H. & A.] Perennial, with several to many prostrate stems 3–10 dm. long, much-branched and forming mats; herbage villous especially when young, later ± glabrate; lvs. 1–3 cm. long; lfts. 4–6, oblanceolate to obovate, 5–12 mm. long; umbels 4–10-fld., on peduncles up to 5 mm. long; bract unifoliolate; calyx ca. 3 mm. long, villous, the setaceous teeth ca. as long as tube; corolla 3–5 mm. long, yellow, aging red; pods usually 1-seeded, villous, arcuate, with a long incurved beak.—Moist banks and canyons, below 6000 ft.; Coastal Sage Scrub, Chaparral, V. Grassland, Foothill Wd., etc.; mostly away from the coast, Santa Cruz Co. to San Diego Co. and to edge of desert; to L. Calif. March–Oct. Plants from near the coast and from Sonoma Co. to San Diego Co., with more permanent villous covering and fls. 5–7 mm. long, pods usually 2-seeded, are the var. *eriòphorus* (Greene) Ottley [*Hosackia tomentosa* H. & A. *L. t.* Greene, not Desr. *L. eriophorus* Greene.]

23. **L. juncèus** (Benth.) Greene. [*Hosackia j.* Benth. *Syrmatium j.* Greene. *L. Bioletti* Greene. *L. j.* var. *B.* Ottley.] Slender-stemmed perennial, much-branched, suffrutescent, strigulose but glabrate, suberect to decumbent, the stems 2–5 dm. long; lfts. 3–5, rounded-oblanceolate, 5–10 mm. long; peduncles slender, 2–20 mm. long, 1–5-fld.; bract mostly lacking; calyx-tube ca. 3 mm. long, the teeth short-triangular to triangular-subulate, scarcely 1 mm. long; corolla 6–7 mm. long, yellow, tinged red, wing shorter than keel and banner; pods ± strigose, 1–2-seeded, the strongly arcuate beak ca. as long

as the body; seeds elongate, olive, smooth, ca. 2.5 mm. long.—Dry hills, below 1500 ft.; Closed-cone Pine F., Mixed Evergreen F., etc.; near coast from Mendocino Co. to San Luis Obispo Co. April–July. Plants north of San Francisco Bay tend to be more wiry, less erect and more pubescent and have been called var. *Bioléttii*.

24. **L. nevadénsis** Greene. [*Hosackia decumbens* var. *n.* Wats. *Syrmatium n.* Greene. *Hosackia n.* Parish.] Prostrate perennial with many slender wiry branches forming mats 3–8 dm. across; herbage villous-strigose, glabrate in age; lfts. 3–5, obovate, obtusish, 7–15 mm. long; lower peduncles 1–2.5 cm. long, upper shorter; umbels 1–several-fld.; bract 1–3-foliolate; calyx-tube ca. 2 mm. long, the teeth subulate, ca. 1 mm. long; corolla yellow or tinged red, 5–7 mm. long, wings at least as long as keel; pods strongly arcuate, ca. 6 mm. long, the slender curved beak longer than the body; seed 1, slender, ca. 2.5 mm. long.—Dry sandy and gravelly slopes and benches, 3500–8500 ft.; Yellow Pine F. to Lodgepole F.; Sierra Nevada from Plumas Co. s., Panamint, Argus mts., etc., to San Diego Co.; w. Nev., L. Calif. May–Aug.

25. **L. Doúglasii** Greene. [*Hosackia decumbens* Benth., not *L. d.* Poir. *L. nevadensis* vars. *Douglasii* and *congestus* Ottley. *L. Leonis* Eastw.] Much like *L. nevadensis*, more loosely villous-tomentose; fls. 10–12; calyx-teeth ca. as long as tube; corolla 10–12 mm. long; pod ca. 10 mm. long, the seeds 2.—Dry slopes, 1500–4400 ft.; Mixed Evergreen F., Foothill Wd., Yellow Pine F.; Lassen Co. to Siskiyou, Lake, and Mendocino cos.; to Wash., Ida. May–Aug.

26. **L. Davidsònii** Greene. [*L. sulphureus* Greene, not Boiss. *Syrmatium D.* Heller.] Much like *L. nevadensis*, but more silvery when young; fls. 4–8; bract unifoliolate; calyx 3 mm. long, the subulate teeth almost as long as tube; corolla sulphur yellow, aging red, 5–8 mm. long, the banner obovate, its blade at a right angle to the wings or reflexed; body of pod ca. as long as calyx, 1-seeded, with long curved beak; seed narrow, elongate, ca. 3 mm. long.—Common on dry slopes, 4500–9000 ft.; Yellow Pine F., Red Fir F.; Santa Rosa Mts. to Mt. Pinos, largely replacing *L. nevadensis* in this area. May–July.

27. **L. argophýllus** (Gray) Greene. [*Hosackia a.* Gray. *Syrmatium a.* Greene. *Hosackia argentea* Kell.] Prostrate or decumbent perennial, many-branched, densely silvery-tomentose, the stems 2–10 dm. long; lfts. 3–5, broadly oblanceolate to obovate, mostly 4–12 mm. long, obtusish; umbels sessile or nearly so, 3–8-fld.; bract unifoliolate; calyx ca. 4 mm. long, the subulate teeth half as long as the tube; corolla yellow, 8–9 mm. long, the banner brown or purple in age, broadly oblong, ascending; pods 1-seeded, the body short, arcuate, silky, scarcely exceeding the calyx, the beak ca. as long; seed elongate, scarcely 2 mm. long.—Dry hills and slopes, mostly below 5000 ft.; Foothill Wd., Coastal Sage Scrub, Chaparral, Pinyon-Juniper Wd., etc.; Coast Ranges from Monterey Co. s., Sierran foothills from Fresno Co. s., to desert edge and L. Calif. April–July.

KEY TO VARIETIES

Fls. 7–9 mm. long; banner-claw shorter than the blade.
    Foliage rather loosely silky-canescent; calyx-teeth ca. half as long as tube. Below Pine Belt
                                                                     L. argophyllus
    Foliage more closely silvery-strigose; calyx-teeth almost as long as tube. Largely from Pine Belt
                                                                 var. *decorus*
Fls. 10–12 mm. long; banner-claw scarcely shorter than the blade.
    Calyx-teeth ca. 1.5–2.5 mm. long; corolla well exceeding calyx. Insular . . . . var. *ornithopus*
    Calyx-teeth ca. 3.5 mm. long; corolla scarcely exceeding calyx.
        Pubescence silvery-silky, the hairs ± spreading. Insular . . . . var. *niveus*
        Pubescence closely appressed. Mainland . . . . var. *Fremontii*

Var. **decòrus** (Jtn.) Ottley. [*Hosackia a.* var. *d.* Jtn.] Glistening, satiny-canescent, at least near the tips; peduncles to ca. 1 cm. long.—Dry slopes, 4000–7000 ft.; largely Yellow Pine F.; San Gabriel and San Bernardino mts.

Var. **ornithòpus** (Greene) Ottley. [*Hosackia o.* Greene. *L. o.* Greene. *L. a.* vars. *adsurgens, argenteus* and *Hancockii* Dunkle. *H. venusta* Eastw.] Stems decumbent to ascending; lfts. acute; peduncles 0.5–4 cm. long; fls. 12–20, yellow; pods conspicuously exserted, 2-seeded.—Dry slopes; Coastal Sage Scrub, Chaparral; Santa Catalina, San Miguel, San Clemente, and Santa Barbara ids.; Guadalupe Id. March–June.

Var. **níveus** (Greene) Ottley. [*Syrmatium n.* Greene. *Hosackia n.* Wats.] Pubescence dense, silky; umbels nearly or quite sessile; body of pod little or not exserted.—Santa Cruz Id.

Var. **Fremóntii** (Gray) Ottley. [*Hosackia a.* var. *F.* Gray.] Lfts. rather broad, acute; umbels densely fld., subsessile near the ends of the branches.—Below 4000 ft.; Foothill Wd., Yellow Pine F., etc.; Sierra Nevada from Placer Co. to Mariposa Co.; Santa Lucia Mts.

28. **L. Nuttalliànus** Greene. [*Hosackia prostrata* Nutt., not *L. p.* L. *Syrmatium p.* Greene.] Prostrate annual, or of longer duration, the branches slender, 3–8 dm. long, strigose when young; herbage thinly hirsutulous, later glabrate; lfts. 3–5(–7), oblong-obovate, 4–12 mm. long, acute or obtuse; peduncles slender, 0.5–2.5 cm. long; usually with a unifoliolate bract; fls. 1–9; calyx 2–2.5 mm. long, the teeth triangular, 0.5 mm. long; corolla yellow, 5–7 mm. long, often red on banner and wing-tips; pods much longer than calyx, slender, arcuate, constricted between the 2 seeds, the slender glabrous beak shorter than the body; seeds narrow-oblong, ca. 2.5 mm. long.—Sandy soil, below 100 ft.; Coastal Sage Scrub, Coastal Strand; s. San Diego Co.; n. L. Calif. March–June.

29. **L. hamàtus** Greene. [*Hosackia micrantha* Nutt., not *L. m.* Benth. *Syrmatium m.* Greene.] Similar to *L. Nuttallianus;* umbels sessile or nearly so, bractless; fls. 3–4.5 mm. long; calyx-teeth subulate, shorter than tube; banner shorter than the keel; pods strigose, even on the beak.—Dry slopes, below 5000 ft.; Coastal Sage Scrub, Chaparral; from Los Angeles and San Bernardino cos. to n. L. Calif. March–June.

30. **L. Benthàmii** Greene. [*Hosackia cystoides* Benth., not *L. c.* L. *Syrmatium c.* Greene.] Suffrutescent perennial, decumbent to suberect, glabrous except on young foliage; stems 3–6 dm. long; lfts. oblong or cuneate-oblanceolate, 3 to 5, obtusish, 5–10 mm. long; peduncles 1–2.5 cm. long; bract sometimes 0, mostly of 1–2 lfts.; fls. mostly 3–7; calyx ca. 6 mm. long, the ± recurved teeth ca. half as long as the tube; corolla 7–9 mm. long, reddish with darker veins or yellow tinged with red or purple; pods 2-seeded, glabrous, the body exceeding calyx, tapering to a slender beak; seeds subcylindric, olive with brown mottling, 1.5–2 mm. long.—Sandy and rocky places; Coastal Strand, N. Coastal Scrub; near the coast from Sonoma Co. to Santa Barbara Co. March–July.

31. **L. scopàrius** (Nutt. in T. & G.) Ottley. [*Hosackia s.* Nutt. *H. s.* var. *diffusa* Gray. *Syrmatium glabrum* Vog. *L. glaber* Greene, not Mill. *H. crassifolia* Nutt., not Benth.] Suffruticose, bushy, mostly suberect, 4–12 dm. high, with rather virgate green branches, subglabrous except at growing ends; lfts. 3 (4–5), oblong to oblanceolate, 4–10 or more mm. long; umbels sessile in axils, 1–5-fld.; calyx commonly ca. 3 mm. long, the subulate teeth ⅓–½ as long as the tube; corolla 7–10 mm. long, yellow or tinged with red, the claws slightly exserted, wings as long as keel; pods often ± curved, the body well exceeding calyx, glabrous, tapering to the subulate beak; seeds 2, narrow, ca. 1.5–2 mm. long, dark brown.—Dry slopes and fans, especially after burns, below 5000 ft.; Chaparral, Coastal Sage Scrub, Coastal Strand, etc.; cismontane Calif. from Humboldt and Plumas cos. s.; to L. Calif. March–Aug.

Var. **Vèatchii** (Greene) Ottley. [*Hosackia V.* Greene. *Syrmatium patens* Greene.] Decumbent, silky-strigose; fls. 10–12 mm. long.—San Miguel Id.; L. Calif.

Var. **dendroìdeus** (Greene) Ottley. [*Syrmatium d.* Greene. *S. Traskiae* Eastw. *L. s.* var. *T.* Ottley.] Erect, 1–2 m. tall, woody; lfts. 8–15 mm. long; umbels pedunculate; fls. 10–12 mm. long; pods 2–3-seeded.—Coastal Sage Scrub, Chaparral; Santa Rosa, Santa Cruz, Santa Catalina, and San Clemente ids. Feb.–Aug.

Var. **brevialàtus** Ottley. Like the sp., but the wings shorter than the keel.—Chaparral, Joshua Tree Wd., etc.; interior valleys and w. edge of desert from San Diego Co. to Kern Co.

32. **L. procúmbens** (Greene) Greene. [*Hosackia p.* Greene. *H. sericea* Benth., not *L. s.* Moench. *L. leucophyllus* Greene.] Silky-canescent, suffrutescent, much-branched, the stems procumbent to ascending, 3–8 dm. long; lfts. 3, narrowly elliptical to oblanceolate, 5–12 mm. long, acute; umbels sessile or nearly so, scattered, mostly 1–3-fld.; calyx-tube ca. 3 mm. long, the triangular-subulate teeth ca. 1 mm. long; corolla 6–8 mm. long, yellow, with exserted claws, wings exceeding keel; pods reflexed, exceeding calyx, straight or curved, strigose; seeds 2 or more, round-oblong, ca. 1.5 mm. long.— Dry slopes, 3000–7500 ft.; Creosote Bush Scrub to Pinyon-Juniper Wd.; desert slopes from Little San Bernardino Mts. to Kern Co., inner Coast Ranges n. to San Benito Co. April–June.

Var. **Jepsònii** (Ottley) Ottley. [*Hosackia sericea* subsp. *J.* Abrams.] Fls. 1–2 in axils, 9–12 mm. long; calyx-teeth ca. 2 mm. long.—At 6000–8500 ft.; s. Sierra Nevada.

33. **L. Haydònii** (Orcutt) Greene. [*Hosackia H.* Orcutt. *L. Spencerae* Macbr.] Bushy erect perennial, 2–4 dm. high, with many slender wiry branches, sparsely strigose; lfts. 3, elliptic, 2–5 mm. long, the lvs. rather remote; fls. 1–2, on very short axillary peduncles without bracts; calyx strigose, 2.5 mm. long, the broadly linear teeth shorter than the tube; corolla yellow, 4–5 mm. long, the wings shorter than the keel; pods curved, ca. 5 mm. long, sparsely strigose, 1-seeded; seed terete, slightly curved, greenish, narrow, ca. 2.5 mm. long.—Dry rocky places, 2000–4000 ft.; Pinyon-Juniper Wd. to Creosote Bush Scrub; sw. Colo. Desert and adjacent L. Calif. March–June.

34. **L. corniculàtus** L. Perennial, with ascending to procumbent slender stems 1–5 dm. long, glabrous or ± strigose; stipules foliar, almost as large as lfts.; lfts. 3, obovate to oblong, 5–15 mm. long; peduncles ca. 1–1.3 cm. long; bract 1–3-foliolate; fls. mostly 3–6; calyx 3–4 mm. long, the teeth ca. as long as the tube; corolla 8–12 mm. long, yellow or the banner reddish; pods linear, 2–2.5 cm. long, terete, straight, glabrous, 2–3.5 mm. thick; seeds round-ovoid, 1–1.3 mm. long, brownish, often with darker mottling; $2n = 24$ (Milovidov, 1941).—Occasional in lawns, on roadsides, etc., as a waif; introd. from Eu. June–Sept.

35. **L. uliginòsus** Schk. [*L. corniculatus* var. *major* Ser.] Perennial with root-bearing stolons; stems ascending, 3–9 dm. high, branched, glabrous or ± pubescent; stipules foliar, like and as large as the 3 lfts.; these oblong-ovate, ± long-ciliate; 1–2 cm. long, peduncles 5–15 cm. long; bract 1–3-foliolate; fls. mostly 8–12, crowded; calyx 5–6 mm. long, the linear teeth ca. as long as tube; corolla yellow, often with some red, ca. 12 mm. long; pods 2–3 cm. long, terete, ca. 2 mm. thick; seeds almost 1 mm. long, brown to dark green, almost globose; $2n = 24$ (Milovidov, 1941), 12 (Dawson, 1941). —Occasionally established, as in Marin and Del Norte cos.; native of Eu. June–Sept.

36. **L. angustíssimus** L. Prostrate villous branched annual, the stems 1–3 dm. long; stipules foliar and equal to lfts.; lfts. 3, lance-ovate, 5–12 mm. long; peduncles 0.4–1.2 cm. long, bearing 1–2 fls.; bract 3-foliolate; calyx ca. 4 mm. long, the teeth subulate, exceeding the tube; corolla ca. 4 mm. long, yellow; pods linear, 1.5–2 cm. long, ca. 1 mm. thick; seeds round, light brown, ca. 0.6 mm. in diam.—Established in Sonoma Co. near the coast; from Old World. June–Aug.

20. Psoràlea L.

Herbs or shrubs with heavy-scented foliage punctate with dots or glands. Lvs. alternate, 3–5-foliolate; stipules large. Fls. purple or whitish, mainly in pedunculate racemes or spikes. Calyx 5-cleft into subequal lobes or the lower longer, the upper sometimes united. Banner ovate to orbicular, clawed; wings oblong or falcate; keel obtuse, incurved. Stamens monadelphous or diadelphous. Ovary sessile or short-stipitate, 1-ovuled. Pod ovoid, indehiscent or circumscissile or bursting irregularly, seldom exceeding the calyx. Ca. 125 spp., widely distributed, especially in warmer regions. (Greek, *psoraleos*, scurfy or rough, from the glandular dots.)

(Rydberg, P. A., *in* N. Am. Flora 24 [1]: 1–24, 1919.)

Lvs. 4–5-foliolate, subpalmate; plants almost acaulescent, from deep-seated fusiform roots.
  Calyx-lobes almost alike; pedicels ca. 4 mm. long; seeds smooth . . . . 1. *P. californica*
  Calyx-lobes unlike, the lower much larger than others; pedicels not more than 2 mm. long; seeds reticulate . . . . 2. *P. castorea*
Lvs. 3-foliolate; plants from rootstocks.
  The lvs. palmately 3-foliolate; lfts. mostly oblanceolate to linear. Lassen Co. . . 3. *P. lanceolata*
  The lvs. pinnately 3-foliolate; lfts. lance-ovate to orbicular.
    Stems prostrate; petioles and peduncles erect, much elongated . . . . 4. *P. orbicularis*
    Stems erect.
      Corolla ochroleucous or whitish with purple-tipped keel; calyx-teeth subequal.
        Lfts. 1–1.5 times as long as wide, broadly ovate; calyx inflated in fr., the lobes 1–2 mm. long . . . . 5. *P. physodes*
        Lfts. 2–3 times as long as wide, lance-ovate; calyx not inflated in fr., the lobes 3–5 mm. long . . . . 6. *P. rigida*
      Corolla purple; lower calyx-lobe much longer than others.
        The corolla ca. 1 cm. long; stipules lanceolate-subulate . . . . 7. *P. macrostachya*
        The corolla ca. 15 mm. long; stipules lance-ovate . . . . 8. *P. strobilina*

1. **P. califórnica** Wats. [*Pediomelum c.* Rydb. *Psoralea monticola* Greene.] Crown branched, the stems to ca. 1 or 2 dm. long; herbage silvery-villous with appressed hairs; lower lvs. with petioles 4–8 cm. long, the blades mostly 5–6-foliolate; lfts. cuneate-

obovate, 1–3 cm. long; upper lvs. reduced; racemes in upper axils, on peduncles 1–4 cm. long; racemes dense, 1.5–3.5 cm. long; calyx densely white-villous, the campanulate tube ca. 4 mm. long, the lobes 5–6 mm. long, the lower somewhat broader than the other 4 lance-linear ones; corolla ca. 12 mm. long, the banner whitish, wings and keel purple; pod oblong-ovoid, 5–6 mm. long, the beak slightly more; seeds dark brown, smooth, ± mottled, ca. 4 mm. long.—Dry slopes, 1500–5500 ft.; Chaparral, Foothill Wd., Pinyon-Juniper Wd., etc.; scattered stations in inner Coast Ranges from Glenn Co. to Ventura Co., Kern Co., San Bernardino Mts., San Jacinto Mts.; L. Calif. May–June.

2. **P. castòrea** Wats. [*Pediomelum c.* Rydb.] Resembling *P. californica*, but main stem more erect, densely leafy, to ca. 5 cm. high; herbage more closely strigose; petioles stout, 5–12 cm. long; lfts. broadly cuneate-obovate, 2–4 cm. long; pedicels stout; calyx-tube ca. 2 mm. long, the lobes 8–10 mm. long, the lowest one spatulate to obovate, 1–2 mm. longer than the ± subulate other 4; corolla bluish, 1 cm. long; body of pod ca. 8 mm. long, the beak 14–15 mm.; seed light brown, flattened, transversely wrinkled, ca. 7 mm. long.—Sandy flats and washes, 1500–3000 ft.; Creosote Bush Scrub, Joshua Tree Wd.; Mojave Desert from Victorville to Yermo; to Utah, Ariz. April–May.

3. **P. lanceolàta** Pursh ssp. **scàbra** (Nutt.) Piper. [*P. s.* Nutt. in T. & G. *P. Purshii* Vail. *Psoralidium P.* Rydb.] From a creeping rootstock, the stem suberect, branched, 1–5 dm. high, strigose and gland-punctate; stipules lance-subulate; petioles 1.5–2.5 cm. long; lfts. oblanceolate, conspicuously gland-dotted, 1.5–3.5 cm. long; peduncles 2–4 cm. long; racemes 1–2 cm. long; calyx 3 mm. long, hairy, the triangular teeth shorter than tube; corolla white, 5 mm. long; pod globose, densely villous, ca. 5 mm. long.—Dry sandy places, at ca. 4000–4500 ft.; Sagebrush Scrub; Amedee, Doyle, Lassen Co.; to Wash., Wyo., Ariz. May–July.

4. **P. orbiculàris** Lindl. [*Hoita o.* Rydb.] Stems prostrate, creeping at nodes; petioles 10–50 cm. long; lfts. round-obovate, 3–8 cm. long, glabrous to short-pubescent; peduncles 20–70 cm. high; racemes dense, 5–30 cm. long; bracts lanceolate, deciduous, 1–2 cm. long; calyx densely villous, the tube 4–5 mm. long, the lobes linear-lanceolate, the lowest one ca. 15 mm. long, the other ca. 10 mm.; corolla reddish-purple, ca. 15 mm. long, banner often with a white spot on each side; pod ca. 8 mm. long, hirsute, the beak small, straight; seed dark, ellipsoid, ca. 5 mm. long, smooth.—Moist places, below 5000 ft.; many Plant Communities; much of cismontane Calif. May–July.

5. **P. physòdes** Dougl. [*Hoita p.* Rydb.] CALIFORNIA-TEA. From a creeping rootstock, the stems erect, 3–7 dm. high, glabrous or ± black-hairy, prominently grooved; stipules lanceolate, 4–6 mm. long; petioles 2–5 cm. long; lfts. ovate, 2–6 cm. long, mucronate, sparsely puberulent; peduncles 3–10 cm. long; racemes dense, 1.5–2.5 cm. long, black-villous; calyx black- and white-hairy, the tube 4 mm. long in fl., 6–8 mm. in fr.; lobes deltoid, shorter than tube, acuminate; corolla 10–12 mm. long; pod ca. 6 mm. long, flat-ovoid, hairy and punctate.—Open places in brush or woods, below 7500 ft.; many Plant Communities; Coast Ranges from Orange and San Bernardino cos. to Humboldt Co.; to B.C. April–June.

6. **P. rígida** Parish. [*Hoita r.* Rydb.] From a rootstock, erect, 3–6 dm. high, sparsely puberulent; stipules 1 cm. long, lance-subulate; petioles 2–4 cm. long; lfts. 3–10 cm. long; peduncles 3–7 cm. long; racemes dense, 2–3 cm. long; calyx short-pubescent, the tube 4–5 mm. long, the lobes 4–5 mm. long, the lowest slightly exceeding others; corolla 10–12 mm. long; pod 8–10 mm. long, strigose.—Dry slopes, 1000–7000 ft.; Chaparral, Yellow Pine F.; Laguna Mts. to San Bernardino Mts. June–July.

7. **P. macrostàchya** DC. [*Hoita m.* Rydb. *H. rhomboidea, H. longiloba, H. Hallii* and *H. villosa* Rydb. *P. Douglasii* Greene. *P. D.* var. *Hansenii* Jeps. *P. Hallii* Jeps. and var. *media* Jeps.] Stems 0.5–3 m. tall, subglabrous to finely pubescent; stipules subulate, 3–5 mm. long; petioles 3–10 cm. long; lfts. 2–8 cm. long, lance-ovate to ovate-rhombic, subglabrous to cinereous-pubescent, obscurely gland-dotted; peduncles 4–10 cm. long; spikes 5–12 cm. long, silky villous with white or blackish hairs; bracts ovate to rhombic, sharply and abruptly pointed, deciduous, 5–8 mm. long; calyx villous, unequally lobed, 6–8 mm. long, the lowest lobe equaling or surpassing corolla; latter purple, 8–10 mm. long, the banner suborbicular; pod 6–8 mm. long, pubescent; $n = 10$ (Kreuter, 1930). —Moist places, below 5000 ft.; many Plant Communities; most of cismontane Calif.; n. L. Calif. May–Aug. Variable; typical plants are those with pubescent lvs., long racemes, lowest calyx-lobe ca. as long as corolla, infl. white-hairy. If racemes are short,

it is var. *rhombifòlia* Torr. If lowest calyx-lobe exceeds corolla, we have var. *longilòba* (Rydb.) Macbr. If bracts are broadly ovate, longer than fls. and infl. is black-hairy, with long-villous bracts, we have *P. Douglásii,* but if pubescence is sparse and oil-glands in infl. conspicuous, *P. Hállii.*

8. **P. strobilìna** H. & A. [*Hoita s.* Rydb.] From a rootstock, the stems erect, 0.5–1 mm. high, densely stipitate-glandular, with long dark interspersed hairs; stipules ovate, cuspidate; petioles 3–7 cm. long; lfts. 3–6 cm. long, broadly ovate, pubescent and densely glandular-punctate, obtuse; peduncles 4–6 cm. long; racemes dense, 3–6 cm. long; bracts lance-oblong, 1–2 cm. long; calyx densely villous, the tube 4–5 mm. long, the lance-linear acuminate lobes unequal, the lowest 15 mm., the others ca. 10 mm. long; petals dark purple, 15 mm. long, whitish toward the base; pods ovoid, black-reticulate, short-hairy, ca. 9 mm. long.—Brushy and wooded slopes, below 2000 ft.; Mixed Evergreen F., Chaparral, etc.; near the coast from Contra Costa Co. to Santa Clara and Santa Cruz cos. May–June.

21. Amórpha L. FALSE INDIGO

Deciduous shrubs with gland-dotted and heavy-scented foliage. Lvs. odd-pinnate, with setaceous stipules and entire petiolulate lfts. Fls. many, small, ± purplish, in long narrow terminal spikes. Calyx turbinate or obconic, 5-lobed, slightly oblique, persistent. Petals wanting except the banner, this erect, clawed. Stamens united at very base only. Style slender, bearded. Pod 1–2-seeded, exceeding the calyx, indehiscent. Seeds oblong or ± curved, rounded and broadened at apex. Ca. 15 spp., of N. Am. (Greek, *amorphos,* deformed, because of the corolla.)

(Palmer, E. J. A conspectus of the genus Amorpha. Journ. Arnold Arb. 12: 157–197, 1931.)
Branchlets and lf.-rachises with pricklelike glands; calyx-teeth ½ to ¾ as long as the tube
                                               1. *A. califórnica*
Branchlets and lf.-rachises lacking pricklelike glands; calyx-teeth low-triangular, very short
                                               2. *A. fruticosa*

1. **A. califórnica** Nutt. Shrub 1.5–3 m. high, the herbage pubescent; lvs. 1–2 dm. long, on short petioles; lfts. 11–27, oblong-elliptical, rounded or retuse at apex, mucronulate, 1–3 cm. long; spikes 5–20 cm. long, the axis pilose; calyx pubescent, 4–5 mm. long, 10-nerved, the teeth triangular-lanceolate; banner red-purple, obovate-cuneate, ca. 5 mm. long; pod 6–8 mm. long, curved on back, turgid, puberulent, gland-dotted; seeds olive-green to brownish, smooth, turgid, 3–4 mm. long; *n* = 10 (Kreuter, 1930). —Dry wooded or brushy slopes, below 7500 ft.; Yellow Pine F., Chaparral, etc.; Santa Rosa Mts. and Santa Ana Mts. to Santa Lucia Mts. May–July.

Var. **napénsis** Jeps. [*A. hispidula* Greene. *A. c.* var. *h.* Palmer.] Plants less pubescent or almost glabrous; calyx-lobes shorter.—Mixed Evergreen F., N. Oak Wd., etc.; Shasta and Placer cos. to Marin and Monterey cos.

2. **A. fruticòsa** L. var. **occidentàlis** (Abrams) Kearn. and Peeb. [*A. o.* Abrams.] Minutely pubescent, 1–2.5 m. high; lvs. 1–2 dm. long; lfts. ovate to oblong, 2–5 cm. long, rounded and mucronate at apex; racemes 1 to few together, 1–2 dm. long, the axis ± pubescent; calyx canescent, especially at the margin, glandular, 2.5–3 mm. long, the teeth triangular, ca. ¼ the length of the tube; banner ca. 5 mm. long; pod 6 mm. long, glabrous, gland-dotted toward apex.—Stream banks and canyons below 5000 ft.; Coastal Sage Scrub, Chaparral; San Bernardino Co. to San Diego Co.; Ariz., New Mex., Chihuahua. May–July. Plants with subglabrous lvs. and truncate to retuse lfts. 1–2 cm. long, have been called *A. occidentàlis* var. *emarginàta* Palmer.

22. Dàlea Juss. INDIGO BUSH

Herbs, shrubs or small trees, usually gland-dotted. Lvs. odd-pinnate, rarely entire, with small subulate stipules. Fls. purple, white or yellow, in terminal spikes or racemes. Calyx persistent, ± equally 5-toothed. Petals clawed. Banner usually cordate or auriculate, mostly exceeded by the wings and keel. Stamens monadelphous. Pod ovoid, compressed, sessile or short-stipitate, generally 2(rarely 4–6)-ovuled. Seeds 1–2, reniform. Over 100 spp., of N. and S. Am. (T. *Dale,* an early English botanist.)

(Rydberg, P. A., *in* N. Am. Fl. 24: 41–65, 1919. Wiggins, I. L. Contr. Dudley Herb. 3: 41–64, 1940.)
A. Plants herbaceous; petals, except the banner, adnate to the stamens for ca. half their length.
    B. Fls. in a dense spike; calyx-lobes longer than the tube.
        C. Fls. 4–5 mm. long; calyx 3–4 mm. long; pubescence of whitish hairs, rarely turning
            rusty on drying .... 1. *D. mollis*
        CC. Fls. 6–8 mm. long; calyx ca. as long as corolla; pubescence of infl. often turning rusty
            brown on drying .... 2. *D. mollissima*
    BB. Fls. in loose spikelike racemes; calyx-lobes not longer than the tube .... 3. *D. Parryi*
AA. Plants shrubs or small trees; claws of petals free or adnate only to very base of stamen-tube.
    B. Lvs. simple, sometimes wanting.
        C. Herbage silvery-silky, the lfts. often lacking except on young growth .. 4. *D. spinosa*
        CC. Herbage green, subglabrous, the lfts. mostly present during vegetative period
                                        5. *D. Schottii*
    BB. Lvs. pinnate, present during vegetative periods.
        C. Fls. sessile in dense capitate spikes; corolla 5–6 mm. long.
            D. Lfts. 2–4 mm. long; glands many, conspicuous, flat .... 6. *D. polyadenia*
            DD. Lfts. 5–20 mm. long; glands rather few, small, prickle-shaped .. 7. *D. Emoryi*
        CC. Fls. pedicelled, in loose spikes; corolla 8–12 mm. long.
            D. Plant densely villous-tomentose; fls. 10–12 mm. long ...... 8. *D. arborescens*
            DD. Plant strigose to subglabrous; fls. ca. 8 mm. long.
                E. Lfts. not decurrent on rachis or confluent .... 9. *D. Fremontii*
                EE. Lfts. decurrent on rachis or the upper ones confluent .... 10. *D. californica*

1. **D. móllis** Benth. [*Parosela m.* Heller.] Short-lived perennial, blooming the first year, herbaceous, with taproot; stems many, 1–3 dm. long, branched, decumbent, soft-villous, dotted with small flat brown glands; lfts. 9–13, cuneate-oblong to obcordate, 3–8 mm. long, gland-dotted along margins; spike dense, 1.5–3.5 cm. long; calyx densely villous, 3–4 mm. long, the filiform teeth ca. as long as tube; corolla slightly longer, white with pinkish tinge; wing-petals usually widest near base of blade, narrowed toward a rounded or acute tip below which is a gland-bearing notch; pods obovoid, 3 mm. long, hirsute.—Open sandy places, below 3000 ft.; Creosote Bush Scrub; deserts s. and e. of Twentynine Palms; to L. Calif., Son., Ariz. March–June.

2. **D. mollíssima** (Rydb.) Munz. [*Parosela m.* Rydb. *D. mollis* var. *m.* Munz. *D. neomexicana* ssp. *m.* Wiggins.] With habit and pubescence of the former; calyx 6–8 mm. long, the filiform plumose teeth longer than the tube; corolla included, the wing-petals rounded, without a gland near tip.—Creosote Bush Scrub; Colo. Desert n. to Inyo Co. and s. Nev., w. Ariz.

3. **D. Párryi** T. & G. [*Parosela P.* Heller.] Diffuse, slender-stemmed, suffruticose perennial 3–5 dm. high, the stems purplish, ashy-strigose, gland-dotted; lvs. pinnate, the lfts. round to elliptical, 13–35, 2–4 mm. long, ± strigose, also glandular beneath; spikes loose, 3–5 cm. long; calyx canescent, 3 mm. long, the ovate teeth equal to tube; corolla blue-purple, with some white, ca. 6–8 mm. long, the keel longer than banner and wings; pod ca. 2 mm. long.—Dry gravelly or rocky places below 2500 ft.; Creosote Bush Scrub; se. Mojave Desert, Colo. Desert; Nev., Ariz., Son., L. Calif. Feb.–June.

4. **D. spinòsa** Gray. [*Asagraea s.* Baill. *Parosela s.* Heller. *Psorodendron s.* Rydb.] SMOKE TREE. Intricately branched shrub or small tree, spinose, 1–6 m. high, nearly leafless, ashy-gray throughout with close canescence, dotted with glands; lvs. early deciduous, simple, few, cuneate-oblong, 1–2 cm. long; spikes numerous, spinescent 2–3 cm. long; calyx 3–4 mm. long, gland-dotted, the ovate teeth scarcely equal to tube; corolla bright blue-purple, 8–10 mm. long, the keel not longer than other petals; pods obliquely ovoid, ca. 6 mm. long, with amber-colored glands.—Locally common in sandy washes, below 1500 ft.; Creosote Bush Scrub; s. Mojave Desert from Daggett e., Colo. Desert; Ariz., Son., L. Calif. June–July.

5. **D. Schóttii** Torr. [*Parosela S.* Heller. *Psorodendron S.* Rydb.] Intricately branched shrub, spinescent, 1–3 m. high, the young growth bright green, subglabrous to minutely strigose, nearly glandless; lvs. usually simple, linear, 1–3 cm. long, 0.5–2 mm. wide, subglabrous, gland-dotted; racemes 5–10 cm. long, lax; calyx 4–5 mm. long, ± strigose toward apex, the triangular teeth hardly 1 mm. long; corolla blue, 8–10 mm. long; pod gland-dotted, almost 1 cm. long.—Washes and benches, below 1000 ft.; Creosote Bush Scrub; Coachella V. to Ariz., L. Calif. March–May.

Var. **pubérula** (Parish) Munz. [*Parosela S.* var. *p.* Parish.] Young branches, young lvs. and calyces canescent-puberulent.—W. edge of Colo. Desert.

6. **D. polyadènia** Torr. ex Wats. [*Parosela p.* Heller. *Psorothamnus p.* Rydb.] Intricately branched shrub 3–20 dm. high, with canescent spine-tipped branches having

amber glands; lvs. 1–2 cm. long, the pinnae 5–11, obovate, 2–4 mm. long; calyx 4 mm. long, canescent and gland-dotted, the teeth subulate, scarcely equal to tube; corolla rose to purplish, 5–6 mm. long; pods 3–5 mm. long.—Dry slopes and mesas, 2500–6000 ft.; Creosote Bush Scrub, Joshua Tree Wd.; Mojave Desert from se. of Twentynine Palms to Inyo and Mono cos.; Nev. May–Sept. Plants from Owens V., with the calyx-tube glabrous are the var. *subnùda* Wats.

7. **D. Emòryi** Gray. [*Parosela E.* Heller. *Psorothamnus E.* Rydb.] Densely branched shrub 3–15 dm. high, white with feltlike tomentum and sprinkled with orange glands; lvs. 1–8 cm. long, the lfts. mostly 5–7, linear to oblong or obovate, 5–15 mm. long; calyx rusty-villous, 5–7 mm. long, the teeth ca. as long as tube; corolla rose to purplish, 5–6 mm. long; pod obliquely obovoid, villous and glandular above.—Dry open places, below 1000 ft.; Creosote Bush Scrub; Colo. Desert; to L. Calif., Son., Ariz. March–May.

8. **D. arboréscens** Torr. ex Gray. [*Parosela a.* Heller. *P. neglecta* Parish. *Psorodendron a.* Rydb.] Spinescent much-branched shrub 1–1.5 m. high, hoary-tomentose; lvs. 1.5–2.5 cm. long, the pinnae mostly 5–7, obovate, 5–10 mm. long, obscurely glandular; racemes 3–5 cm. long; calyx-villous-tomentose, ca. 7 mm. long, the lance-subulate teeth ca. as long as tube; corolla deep violet-blue, 10–12 mm. long; pod villous, conspicuously gland-dotted, the body ca. 1 cm., the beak 6 mm. long.—Low hills and mesas, 2000–2600 ft.; Creosote Bush Scrub; Barstow region, Mojave Desert. April–May.

9. **D. Fremóntii** Torr. [*Parosela F.* Vail. *Psorodendron F.* Rydb.] Intricately branched shrub 5–20 dm. tall, younger growth strigose-canescent; lvs. 2–4 cm. long, closely strigose, the lfts. mostly 3–5, narrowly oblong, ca. 6–8 mm. long, gland-dotted; racemes ca. 7–12 cm. long, rather lax; calyx ca. 6 mm. long, with small glands, the lobes ca. 2 mm. long, the upper 2 triangular, ca. twice as wide as the lowest one; corolla dark purple-blue, the petals subequal; pods ovoid, with brown glands, the body 8 or more mm. long, equaled by the beak.—Occasional, dry places, 2500–4500 ft.; Creosote Bush Scrub, Basin Sagebrush; Owens V., Inyo Co.; to Nev., Utah. April–May.

KEY TO VARIETIES

Pubescence closely appressed, the lvs. and young growth quite canescent.
   Lfts. oblong, mostly ca. 6 mm. long. Owens V.......................... *D. Fremontii*
   Lfts. linear to lanceolate, 6–20 mm. long, Death V. region to Colo. Desert .... var. *minutifolia*
Pubescence ± spreading, the lvs. and the young growth greener. Victorville to Owens V.
                                                   var. *Saundersii*

Var. **minutifòlia** (Parish) L. Benson. [*Parosela Johnsonii* var. *m.* Parish. *D. J.* Wats. *P. J.* Vail. *P. F.* var. *J.* Jeps.] Lfts. 1–3 mm. broad; pods ca. 8 mm. long.—Common, dry slopes and plains, below 5000 ft.; Creosote Bush Scrub, Joshua Tree Wd.; Mojave Desert from Death V. region to near Indio and Whipple Mts., Riverside Co.; to Utah, Ariz.

Var. **Saundérsii** (Parish) Munz. [*D. S.* Parish.] Lfts. 2.5–4 mm. broad, lanceolate to oblong or elliptic-lanceolate, 6–13 mm. long; pubescence rather sparse, short, not closely appressed; body of pod 8–12 mm. long.—Dry plains and slopes, 2500–6000 ft.; Creosote Bush Scrub, Joshua Tree Wd., Sagebrush Scrub; Victorville region w. and n. to White Mts.

10. **D. califórnica** Wats. [*Parosela c.* Vail. *P. c.* var. *simplifolia* Parish. *D. Fremontii* var. *c.* McMinn. *D. F.* var. *s.* L. Benson.] Much like var. *minutifolia* above; lfts. 1–7, ± decurrent and the upper confluent, densely silky-pubescent; calyx-lobes not markedly different in width; pods subglabrous, conspicuously glandular.—Dry places below 3500 ft.; Creosote Bush Scrub; Colo. Desert from Morongo V. to Shavers Well, etc., Twentynine Palms region, San Jacinto V.; April–May.

23. Petalóstemon Michx. PRAIRIE-CLOVER

Herbs, mostly perennial, ± glandular-dotted. Lvs. odd-pinnate, the lfts. entire. Fls. perfect, in terminal spikes; bracts deciduous. Calyx campanulate, 10-nerved, the lobes triangular to lanceolate. Corolla indistinctly papilionaceous, pink to purple or yellowish; banner free, the wings and keel alternating with the free portion of the anthers and borne on the top of filamentous sheath. Stamens 5, monadelphous below. Style filiform. Pod membranous, subglobose to obovoid. Ca. 50 spp., of N. Am. (Greek, words for *petal* and *stamen*, because of the peculiar union of these parts.)

1. **P. Séarlsiae** Gray. [*Kuhnistera* S. Kuntze.] Caudex branched, woody, the year's growth 3–5 dm. high, glabrous, glandular-dotted; lvs. 3–5 cm. long, with 3–7 oblong to oblanceolate lfts. 10–15 mm. long; peduncles 5–12 cm. long, the spikes ca. 1 cm. thick, 1–4 cm. long; calyx 4 mm. long, villous, the lobes lanceolate, almost as long as tube; corolla rose, ca. 4 mm. long, the petals distinctly narrow-clawed; pod 4 mm. long, pubescent; seed 1, brown, kidney-shaped, ca. 1.5 mm. long.—Dry gravelly and stony banks, 5000–7000 ft.; Pinyon-Juniper Wd.; Death V. region, Inyo Mts., New York and Providence mts.; to Utah, n. Ariz. May–June.

24. Robínia L. Locust

Trees or shrubs, often with spiny stipules. Lvs. odd-pinnate, the lfts. ovate or oblong and with stipulelike appendage at base. Fls. showy, in drooping axillary racemes. Calyx with 5 short teeth, the 2 upper ± united. Banner large and rounded, not much exceeding wings and keel. Stamens 10, diadelphous. Pod linear, flat, several-seeded, tardily 2-valved. Ca. 8 spp. of e. N. Am. (Named for Jean and Vespasian *Robin*, 16th century, who first cultivated the Locust Tree in Europe.)

1. **R. Pseùdo-Acàcia** L. Tree with rough bark and subglabrous twigs and lvs.; lfts. 9–19, petiolulate, 2–5 cm. long; pedicels 6–12 mm. long; fls. fragrant, whitish, 15–20 mm. long; pods 5–10 cm. long, glabrous; $n = 10$ (Kreuter, 1930).—Natur., especially in n. Calif. as an escape from cult.; easily spreading by underground parts. May–June.

25. Ólneya Gray. Desert-Ironwood

Spinose tree with thin scaly bark and pairs of spines below the lvs. Lvs. pinnate or odd-pinnate, densely canescent, the lfts. 8–24, oblong-cuneate to wider; stipules obsolete. Fls. few, in axillary racemes, appearing before the new growth of lvs. Calyx campanulate, 5-lobed, the 2 upper quite united. Petals short-clawed, the banner roundish, emarginate. Stamens 10, diadelphous. Ovary several-ovuled; style bearded above. Pod thick, torulose, puberulent, glandular-hispid. Seeds broadly ellipsoid. One sp. (Named in honor of S. T. *Olney*, New England botanist of 19th century.)

1. **O. Tesòta** Gray. Broad-crowned grayish tree 5–8 m. high; lvs. 3–10 cm. long; lfts. 5–20 mm. long; racemes 3–5 cm. long; bracts minute, deciduous; calyx 4–5 mm. long, the lobes ovate, shorter than tube; corolla pale rose-purple, 1 cm. long; pod 4–6 cm. long; seeds black, 8–9 mm. long.—Desert washes, below 2000 ft.; Creosote Bush Scrub; Colo. Desert; to Ariz., Son., L. Calif. April–May.

26. Sesbània Adans.

Herbs or shrubs with abruptly pinnate lvs. and numerous lfts. Stipules small, scarious, caducous. Fls. solitary or in axillary racemes, commonly yellow. Calyx campanulate, the lobes shorter than the tube. Banner with a rounded reflexed blade; wings oblong; keel lunate, blunt. Stamens diadelphous. Pod slender, linear, terete, short-stipitate, many-seeded, partitioned between the seeds. Seeds many, narrowly oblong. Ca. 15 spp. of warm and trop. regions. (*Sesban*, Arabic name of 1 sp.)

1. **S. exaltàta** (Raf.) Cory. [*Darwinia e.* Raf. S. *macrocarpa* Muhl.] Colorado-River-Hemp. Glabrous annual 4–30 dm. high, the stems striate; lfts. 20–80, linear-oblong, 1–2 cm. long, pale green; fls. few; pedicels 5–10 mm. long, slender; calyx 4–5 mm. long, the lobes triangular-subulate; corolla ca. 15 mm. long, yellowish, the banner purple-dotted; pod 10–20 cm. long, 3–4 mm. wide; seeds 4 mm. long; $2n = 12$ (Atchison, 1949).—Frequent in overflow lands, along ditches, etc.; Alkali Sink; Imperial Co. and e. Riverside Co.; to Atlantic Coast and Cent. Am. Reported from LaVerne, Los Angeles Co. April–Oct.

27. Astrágalus L. Milkvetch. Rattleweed. Locoweed

The treatment of *Astragalus* and *Oxytropis* by R. C. Barneby

Ours annual or perennial, caulescent or acaulescent herbs, the stems arising from a taproot or oblique rhizome, often persistent at base as an indurated caudex. Pubescence simple and basifixed, or dolabriform, composed of hairs attached above the base, with an ascending and shorter descending arm. Stipules petiolar-cauline or connate opposite the petiole. Leaves imparipinnate, rarely 3-foliolate. Infl. axillary, racemose or spicate. Calyx 5-toothed, naked or bracteolate. Corolla papilionaceous, the keel, except in one annual, muticous. Stamens diadelphous, 9 and 1. Style glabrous. Pod sessile, stipitate or jointed to a stipelike gynophore, highly variable in outline, texture, and compression, the valves often infolded dorsally to form a double-walled, complete or partial partition or false septum. A genus of nearly 2000 herbs and subshrubs, chiefly of Medit. and Continental climates in the N. Hemis., most abundant in Asia, about 400 in N. Am., 100 S. Am. and 1 S. Afr. Many Astragali are notorious stock-poisons, which fact the vernacular name Locoweed refers. *A. lentiginosus* is in this category, and all are under suspicion until proved otherwise. A few provide excellent forage, and the toxic species are rarely eaten, except by already sick and addicted animals, where other pasture is available. (Greek, *astragalos*, ankle-bone or [pl.] dice, early applied to some leguminous plant, and perhaps analogous to Rattleweed, from the rattling of the seeds in an inflated pod).

A. Annual plants with small fls., the keel not more than 8 mm. long; racemes capitate, sub-umbellate or, if racemose, then not more than 10-fld., nor over 3 cm. long in fr.
   B. Pod less than 5 mm. long, ovoid or suborbicular in outline, transversely wrinkled, 2-locular, mostly 2-seeded.
      C. Fls. spreading or nodding at anthesis, at length racemose; pod deflexed, broadly ovate or suborbicular in dorsal view, obcompressed, the valves spreading from an open dorsal groove and incurved at base toward the concavely arched ventral suture
                                                                          1. *A. Gambelianus*
      CC. Fls. ascending in dense ovoid-oblong heads; pod erect, ovoid, not obcompressed, the ventral suture convex, the body divided by a deep narrow dorsal groove into 2 saclike lobes .... 2. *A. didymocarpus*
   BB. Pod larger, either strongly turgid or elongate, several-seeded.
      C. Calyx subsessile, the fls. and pods borne in dense globose or oblong heads on commonly divaricate peduncles; pod ovoid-acuminate, bladdery-inflated, ascending-pilose or villous-hirsute .... 81. *A. Hornii*
      CC. Calyx pedicelled, the fls. racemose or subumbellate.
         D. Keel-petals equaling or surpassing the wings; herbage densely silvery-strigulose throughout. Mojave Desert.
            E. Pod 13–18 mm. long, 3–3.5 mm. in diam., compressed-triquetrous, falcate, the valves papery .... 10. *A. albens*
            EE. Pod 15–30 mm. long, 5–8.5 mm. in diam., subterete, becoming variously compressed in age, the valves leathery or woody ...... 11. *A. mohavensis*
        DD. Keel-petals surpassed by the wings or, if exserted, then the herbage green and glabrate, or distribution cismontane.
            E. Pod obliquely ovoid or subglobose, inflated or turgid, without septum, the beak deltoid, shorter than the body. Desert spp.
               F. Vesture of the herbage villous-hirsutulous; valves of the pod leathery or stiffly papery, hirsutulous .... 75. *A. sabulonum*
               FF. Vesture appressed, strigulose or silky (shortly incurved-ascending in *A. Gilmani* of the Panamint Mts.); valves of the pod membranous or thinly papery, strigulose or appressed-silky.
                  G. Fls. and pods ascending, the latter densely silky-strigulose, 3–6(–7)-ovulate. Colo. Desert .... 76. *A. aridus*
                  GG. Fls. and pods horizontal or declined, the latter merely strigulose, 7–28-ovulate.
                     H. Petals (in Calif.) whitish.
                       I. Pod very strongly oblique, half-ovate or lunately elliptic in profile. Mono Co. .... 78. *A. Geyeri*
                       II. Pod subsymmetrically ovoid-ellipsoid. E. Mojave Desert
                                                   81. *A. Wootoni*
                HH. Petals pinkish-purple, drying violet.
                    I. Hairs of the herbage straight and appressed. Deserts, 1500–6400 ft.
                       J. Lfts. 11–17; banner 5–7.5 mm. long; pod 7–14-ovulate, the valves nearly always purple-speckled
                                     77. *A. insularis* var. *Harwoodii*
                       JJ. Lfts. 7–13; banner 8–10.5 mm. long; pod 19–24-ovulate, the valves green turning straw color
                                           79. *A. nutans*
                    II. Hairs of the herbage incurved-ascending. Panamint Mts., 6500–10,000 ft. .... 80. *A. Gilmani*
         EE. Pod linear or narrowly oblong or, if the body ovoid, then the beak subulate and nearly as long; septum (except sometimes in *A. Nuttallianus*) well developed.
            F. Stipules herbaceous, white-strigulose; calyx nearly always white-strigulose. Desert ssp.

G. Keel-petals obliquely acuminate or cuspidate at apex; pod narrowly crescentic, evenly incurved, resupinate on contorted pedicels; lfts. all obtuse or retuse .... 8. *A. acutirostris*

GG. Keel petals abruptly rounded at apex; pod more strongly arched near base than distally, spreading or declined on curved pedicels; lfts. of upper or all lvs. acute. Vars. of ...... 9. *A. Nuttallianus*

FF. Stipules with membranous margins, sparsely black-ciliate; calyx nearly always dark-hairy. Cismontane ssp.

G. Keel-petals much shorter than the wings; pod variable in shape, sessile.

H. Pod fertile in the lower half, this ovoid-oblong, silky-strigulose, abruptly passing into a narrow subulate-acuminate beak nearly as long .... 3. *A. Breweri*

HH. Pod fertile along the middle ⅔ or nearly the whole length, linear-oblong to narrowly lunate, sparsely strigulose or glabrate.

I. Pod 6–20 mm. long, rounded at base. Great V. and coast s. from Monterey.

J. Racemes subcapitate, not over 8 mm. long and if over 5 mm. long the fls. 7 or more; pod not mottled, never resupinate. Lower Sacramento V. southward. 6. *A. tener*

JJ. Racemes loosely 2–5(–7)-fld., the axis 7–20 mm. long in fr.; pod brightly mottled or suffused with purple, often resupinate on contorted pedicels. Upper Sacramento V. and foothills to the e. 7. *A. pauperculus*

II. Pod (15–)18–60 cm. long, if less than 21 mm. long cuneately tapering at base. N. Coast Ranges and w. foothills of Sacramento V. .... 4. *A. Rattani*

GG. Keel-petals longer and much wider than the wings; pod linear, attenuate at both ends and elevated on a slender gynophore 1.5–2.5 mm. long. Napa and Sonoma cos. ...... 5. *A. Clarianus*

AA. Perennial plants, or, if short-lived, or flowering the first season, or monocarpic, then the fls. larger, with keel 9 mm. long or more, or the raceme loosely many-fld.

B. Lfts. of at least some developed lvs. (especially those subtending the peduncles) more than 23 in number. Robust plants of the cent. Sierra foothills, Great V., Coast Ranges, cismontane s. Calif., the coast and ids.

C. Calyx subsessile, the fls. and bladdery-inflated pods ascending in dense globose or oblong heads. Plants of alkaline flats and lake shores .... 82. *A. Hornii*

CC. Calyx pedicelled, the fls. racemose, or if subsessile then both fls. and pods deflexed.

D. Stipules, at least at the lowermost 1–3 (often leafless) nodes, ± united opposite the petiole; pod papery, inflated.

E. Stipules herbaceous, tomentulose externally; pod sessile. Insular plants 58. *A. miguelensis*

EE. Stipules scarious or early becoming so, thinly pubescent or glabrous. Mainland or, if insular, then the pod stipitate or jointed to a stipelike gynophore.

F. Ovary and pod stipitate or borne on a stipelike gynophore.

G. Calyx-tube cylindric or deeply campanulate, 5–7 mm. long; either the stipe 1.5–4 cm. or the pubescent gynophore 3–11 mm. long. S. Coast Ranges and Great V.

H. Herbage and calyx subappressed-silky; body of the pod bladdery-inflated, not laterally compressed, continuous with a stipe 1.5–4 cm. long .... 62. *A. asymmetricus*

HH. Herbage and calyx villous or villosulous; body of the pod strongly turgid but somewhat laterally compressed, the sutures both prominent, jointed to a gynophore 3–11 mm. long 63. *A. oxyphysus*

GG. Calyx-tube campanulate, 3.5–5.5 mm. long; stipe 5–10 mm. or glabrous gynophore 2.5–6 mm. long. Coastal or insular (inland only locally in San Benito Co.).

H. Pod jointed to the gynophore .... 64. *A. curtipes*

HH. Pod continuous with the stipe. A rare insular form of 65. *A. leucopsis*

FF. Ovary and pod sessile. Plants of the immediate coast.

G. Fls. large, the banner more than 1 cm. long; pod 2.5–5 cm. long. Dunes and ocean bluffs .... 59. *A. Nuttallii*

GG. Fls. smaller, the banner 8–10 mm. long; pod not over 1 cm. long. Salt marshes . ... 61. *A. pycnostachyus*

DD. Stipules all free from each other opposite the petiole, or so early ruptured by the growing stem as to appear so.

E. Fls. very small and numerous, 20–95 to the raceme, reflexed, the banner 5–6 mm. long; racemes very narrow and elongate, the axis 1–3 dm. long 51. *A. Clevelandii*

EE. Fls. fewer or, if nearly as numerous then the banner at least 9 mm. long; racemes shorter.

F. Petals not very strongly incurved; blades of the keel equaling or
    shorter than their claws or, if slightly longer, then the banner at least
    12 mm. long.
    G. Lfts. linear-oblong, often more densely pubescent above than be-
        neath; pod arcuate-erect on a stout stipe, 2-locular, the glabrous
        valves woody .... 21. *A. pachypus* var. *Jaegeri*
    GG. Lfts. broader or, if linear-oblong then equally pubescent on both
        sides or more densely so beneath than above; pod either papery,
        or sessile, or 1-locular, or deflexed.
        H. Fls. small, the keel 8.5 mm. long or less; fls. and pods re-
            trorsely imbricated in dense racemes; pod sessile or sub-
            sessile, 2-locular.
            I. Stems white-tomentose; petals purplish; pod 6.5–9 mm.
                long. Los Angeles and Orange cos. .. 55. *A. Brauntonii*
            II. Stems straw color, glabrescent or thinly villous; petals
                whitish; pod 11–15 mm. long. Humboldt Co.
                                                        54. *A. agnicidus*
        HH. Fls. larger, the keel over 8.5 mm. long (if slightly less,
            the pod 1-locular).
            I. Ovary and pod exactly sessile; pod greatly inflated.
                J. Calyx-tube campanulate, gibbous dorsally at base (sub-
                    symmetric at base in *A. Deanei* of sw. San Diego
                    Co.); neither of the Great V. nor inner Coast Range
                    to the w.; petals whitish, never truly yellow.
                    K. Pubescence villosulous, usually copious, but if
                        sparse and the herbage green then the stems
                        erect (but this only on San Francisco Peninsula);
                        stipules scarious, connate in vernation, but the
                        sheath often ruptured; Santa Barbara Co. n.
                        along the immediate coast .... 59. *A. Nuttallii*
                    KK. Pubescence appressed, sparse or nearly lacking,
                        the whole plant green; stipules herbaceous, free
                        from the first; stems diffuse n. of Los Angeles;
                        San Luis Obispo Co. s. away from the imme-
                        diate coast.
                        L. Peduncles 5.5–14 cm. long; racemes dense,
                            3.5–9 cm. long in fr.; petals not strongly
                            graduated, the wings as long to 1 mm.
                            shorter than the banner; San Luis Obispo
                            Co. s. .... 60. *A. pomonensis*
                        LL. Peduncles 12–20 cm. long; racemes loose,
                            9–16 cm. long in fr.; petals strongly grad-
                            uated, the wings about 2 mm. shorter than
                            the banner; local in sw. San Diego Co.
                                                    72. *A. Deanei*
                JJ. Calyx-tube narrowly campanulate to cylindric, not gib-
                    bous at base; pod 2-locular, mottled; petals either
                    purple or pale yellow; valley-floor and Coast Ranges
                    to the w.; vars. *idriensis* and *nigricalycis* of
                                                    83. *A. lentiginosus*
            II. Ovary and pod stipitate, the stipe at least 1 mm. long
                (at least 3 mm. long except in *A. Congdoni* of the Sierran
                foothills).
                J. Pod 2-locular, triquetrously compressed, the valves
                    firmly papery or leathery.
                    K. Stipe of the pod 4–8.5 mm. long, the body 8–16
                        mm. long, 3–5.5 mm. in diam.; whole plant
                        softly villous-tomentulose. Insular
                                                    56. *A. Traskiae*
                    KK. Stipe of the pod 1–2.5 mm. long, the body
                        2–3.5 cm. long, 2.5–3 mm. in diam.; plant vil-
                        losulous, but the herbage greenish. Sierran foot-
                        hills .... 52. *A. Congdoni*
                JJ. Pod 1-locular, with papery or papery-membranous
                    valves, either bladdery-inflated, or compressed and
                    2-sided.
                    K. Calyx-tube subcylindric, 5–7 mm. long. Inner S.
                        Coast Ranges and Great V.
                            See 62. *A. asymmetricus* and 63. *A. oxyphysus*
                    KK. Calyx-tube campanulate 3–4 mm. long or, if up
                        to 5 mm. long then plants of the immediate
                        coast from Santa Barbara Co. s. or insular.
                        L. Body of the pod 1–2 cm. in diam., strongly
                            oblique and convex dorsally, nearly always
                            puberulent; pubescence usually composed of
                            incurved or twisted hairs, appearing minutely

woolly .... 65. *A. leucopsis*
LL. Body of the pod 4–10 mm. in diam., less or
little oblique, commonly glabrous; pubescence
mostly of straight subappressed hairs.
M. Pod turgid or decidedly inflated, the sutures not prominent
66. *A. trichopodus*
MM. Pod laterally compressed, 2-sided, 2-
carinate by the acute sutures
67. *A. Antiselli*
FF. Petals strongly incurved and short-clawed, none of them over 12.5 mm.
long, the blades of the keel longer than their claws; pod greatly inflated.
G. Stems 6–10 dm. long, mostly erect; herbage nearly glabrous, the
lfts. thick, the larger ones pinnate-veined; pod stiffly papery or
leathery, nearly always glabrous, ascending, persistent until dehiscence on thickened pedicels. Interior San Diego Co.
73. *A. oocarpus*
GG. Stems rarely over 5 dm. long, diffuse; herbage ± pubescent, especially beneath, the lfts. thin-textured, not veined; pod spreading,
thin-papery, strigulose or villosulous, falling from the receptacle
and dehiscent on the ground.
H. Herbage strigulose; pod strigulose. Widespread s. from Yolo
Co. .... 69. *A. Douglasii*
HH. Herbage and pod villosulous. Local in S. Coast Ranges,
Salinas V. to Kern Co. .... 70. *A. macrodon*
BB. Lfts. not more than 23 in any one lf. or, if occasionally more, then plants of montane
elevs. in the Sierra Nevada, of the Klamath Basin, the deserts, or transmontane.
C. Petals marcescent, the banner and wings villosulous dorsally; densely tufted, silky-
villosulous plants, with subcapitate racemes and scarcely exserted, 2-locular pods. Local
in high n. Sierra Nevada .... 32. *A. Austinae*
CC. Petals glabrous, usually promptly deciduous.
D. Stipules, at least at the lowest 1–3 nodes, connate into a sheath opposite the
petiole.
E. Hairs of the herbage dolabriform; stout plants with creeping rootstocks; fls.
and pods in dense racemes, the former nodding, the latter erect, oblong,
2-locular .... 28. *A. canadensis* var. *brevidens*
EE. Hairs basifixed.
F. Fls. very small, the keel 3.5–5.5 mm. long.
G. Lfts. 3–9, continuous with the rachis, the stiff midrib running out
as a callous mucro or spinule at apex; plants matted or pulvinate;
pod lenticular, sessile, 1-locular; vars. of .... 93. *A. Kentrophyta*
GG. Lfts. 7 or more, jointed to the rachis, not mucronate; stems diffuse;
pod either partially or fully 2-locular, or bladdery-inflated.
H. Lfts. 13–23; racemes 1.5–4.5 cm. long; pod shortly stipitate,
triquetrous, 2-locular. Mono Co. .. 46. *A. Johannis-Howellii*
HH. Lfts. 7–13; racemes not over 1.5 cm. long; pod sessile.
I. Pod lenticular, neither turgid nor inflated, laterally compressed, fully 2-locular, about 8 mm. long. Plumas Co.
45. *A. lentiformis*
II. Pod obliquely ovoid, either greatly inflated or obcompressed, 1-locular or with narrow partial septum.
J. Stems 1–3 dm. long; herbage either long-villous or
villosulous with incurved hairs; pod papery-membranous, bladdery-inflated, not mottled, 3–9-ovulate.
Ne. Calif., below 6000 ft. .... 50. *A. Pulsiferae*
JJ. Stems 1.5–10 cm. long; herbage strigulose; pod papery,
turgid but scarcely inflated, not bladdery, mottled, 16–
20-ovulate. Alpine in s. Sierra Nevada
49. *A. Ravenii*
FF. Fls. larger, the keel 6–14 mm. long.
G. Herbage villosulous or tomentulose, the hairs spreading or some
of them so, straight, incumbent or contorted.
H. Pod sessile, the body hirsute or villosulous, partly or fully
2-locular.
I. Calyx-teeth linear- or subulate-setaceous, 2.5–6.5 mm.
long, over half as long as the tube; racemes oblong,
10–30-fld.; pod oblong-falcate, compressed, shaggy-villous.
J. Root-crown superficial; stems tomentose at base; keel
6.5–9 mm. long. Mono Co. n. .. 30. *A. Andersonii*
JJ. Root-crown buried, the stems glabrous and subterranean at base; keel 10–12 mm. long. Inyo Co.
31. *A. sepultipes*
II. Calyx-teeth subulate, 1.5–2 mm. long, not more than
half as long as the tube; racemes subcapitate, 6–12-fld.;
pod obliquely ovoid, obcompressed, shortly hirsutulous
48. *A. monoensis*

HH. Pod stipitate, the stipe at least 3 mm. long, either glabrous or, if pubescent, 1-locular.
    I. Pod erect, incurved toward the axis of the raceme, 2-locular in the lower half; rare villosulous form of
        22. *A. Bolanderi*
    II. Pod pendulous, 1-locular.
        J. Body of the pod bladdery-inflated, balloon-shaped, the valves papery-membranous; var. *lenophyllus* and var. *confusus* of .... 92. *A. Whitneyi*
        JJ. Body of the pod linear-oblong or ellipsoid, not inflated, the valves stiffly papery or leathery.
            K. Lfts. oblong-oblanceolate; petals ochroleucous; pod nearly straight, linear-oblong, laterally flattened, 3.5–5 mm. in diam. Klamath Basin
                88. *A. californicus*
            KK. Lfts. obovate-cuneate; petals dull yellow; pod lunately oblong-ellipsoid, transversely dilated, the sutures both arched; Sierra to Lassen Co.
                87. *A. Gibbsii*
GG. Herbage appressed-pubescent, the hairs usually straight, rarely wavy, sometimes nearly lacking.
    H. Fls. erect, subsessile, in ovoid heads; bracts oblong mostly obtuse; pod erect, ovoid, triquetrous, 2-locular, densely white-pilose .... 29. *A. agrestis*
    HH. Fls. spreading or declined, pedicelled, racemose; bracts acute; pod otherwise.
        I. Calyx large, the tube and teeth together 6–9 mm. long; pod of leathery texture.
            J. Lfts. with a silky sheen from the close coat of fine closely set hairs; calyx-tube subcylindric. Sierra and Plumas cos. .... 24. *A. Webberi*
            JJ. Lfts. strigulose, the pubescence dull, the hairs spaced. Plumas Co. s.
                K. Pod erect and strongly incurved toward the axis of the raceme, the body turgid, truncate-rounded at base above the stipe, 2-locular in the lower half. Sierra Nevada .... 22. *A. Bolanderi*
                KK. Pod declined, the body solid, cuneate or acuminate above the stipe, 1-locular. San Bernardino and San Antonio mts.
                23. *A. bicristatus*
        II. Calyx smaller, the tube and teeth together less than 6 mm. long; pod of thinly papery or membranous texture.
            J. Pod balloon-shaped, bladdery-inflated, 1–2.5 cm. in diam., not compressed .... 92. *A. Whitneyi*
            JJ. Pod linear-oblong, not inflated, laterally compressed and 2-sided.
                K. Lfts. 9–19, all jointed to the rachis; petals ochroleucous, the keel abruptly incurved to a broad obliquely deltoid apex; stems erect
                90. *A. filipes*
                KK. Lfts. 5–11, the terminal one, at least of the upper leaves, confluent with the rachis; petals dull reddish-pink, the keel attenuate into a narrow acutish apex; stems diffuse .... 91. *A. inversus*
DD. Stipules all petiolar-cauline or petiolar, the lowest sometimes amplexicaul but none connate into a sheath opposite the petiole.
    E. Pubescence of the herbage, and of the stems when developed, hirsute, villous, villosulous or tomentulose, the hairs curly or incumbent or, when straight, spreading at a wide angle, or some of them so.
        F. Fls. small, the banner not more than 8 mm. or the keel more than 6.5 mm. long.
            G. Densely tufted or matted plants, the internodes short or sub-obsolete; herbage cottony-tomentose; pod shaggy-hirsute; vars. of
                42. *A. Purshii*
             GG. Stems developed; pubescence not cottony.
                H. Perennials, softly long-villous. Sierra Co. n. . 50. *A. Pulsiferae*
                HH. Winter-annuals or biennials, sometimes appearing perennial late in the season. Mono Co. s.
                    I. Petals bright purple; pod papery, greatly inflated.
                        J. Pod 1-locular; pubescence of incurved hairs. Panamint Mts. .... 80. *A. Gilmani*
                        JJ. Pod 2-locular; pubescence of spreading hairs. White Mts.; a form of var. *Fremontii* of 83. *A. lentiginosus*
                    II. Petals ochroleucous tinged with dull lilac; pod merely

turgid, leathery, hirsutulous. Colo. Desert; perennating
form of .... 75. *A. sabulonum*

FF. Fls. larger, the banner at least 10 mm., the keel at least 8 mm. long.

    G. Stems none or very short, the plants tufted or matted; pod sessile,
1-locular (except in cottony-tomentose *A. leucolobus*).

        H. Petals scarlet, about equal in length, 3.5–4 cm. long; pod
densely shaggy-hirsute .... 40. *A. coccineus*

        HH. Petals purple or whitish, less than 3 cm. long.

           I. Pod strigulose or thinly hirsutulous, neither tomentulose
nor shaggy-hirsute; lvs. not softly cottony-tomentose.

             J. Pod crescentic, acuminate and laterally compressed at
both ends, 2.5–5 cm. long, arched through at least half
a circle .... 37. *A. Tidestromii*

             JJ. Pod oblong-ovoid or -ellipsoid, laterally compressed in
beak only, usually less than 2.5 cm. long, straight or
little incurved .... 38. *A. tephrodes* var. *brachylobus*

          II. Pod densely shaggy-hirsute or tomentulose, the vesture
usually hiding the valves or, if thinner, the lvs. cottony-
tomentose.

             J. Lfts. silvery-pilose with appressed hairs, only the
petioles ascending-hirsute at anthesis, but the herbage
becoming hirsute late in the season; strictly acicules-
cent, the petioles persistent as a thatch on the thick
root-crown .... 39. *A. Newberryi*

             JJ. Lfts. silky- or cottony-tomentose with fine entangled
hairs; either acaulescent or shortly caulescent, but the
petioles weakly if at all persistent on the slender root-
crown or divisions of the caudex.

                K. Wing-petals equaling or shorter than the keel,
the latter 21–26 mm. long; pod 2.5–4 cm. long.
Death V. .... 41. *A. funereus*

                KK. Wing-petals surpassing the keel; pod not over
2.5 cm. long.

                  L. Pod densely villous-hirsute, resembling a floc
of silky or cottony wool, the vesture con-
cealing the valves and at least 2 mm. thick;
petals usually strongly graduated, the keel
(except sometimes in var. *longilobus*) much
shorter than the banner; pod obliquely ovoid,
rigid .... 42. *A. Purshii*

                  LL. Pod shortly villous-tomentulose, the vesture
composed of fine entangled tomentum not
more than 1 mm. thick, scarcely or not concealing the valves; petals little graduated;
pod either lance-oblong or of papery texture.

                    M. Banner 16–18 mm. long, racemes sur-
passing the lvs.; pod lance-oblong
in outline, deeply and narrowly grooved
dorsally, the valves rigid, inflexed as a
narrow partition. San Bernardino and
Santa Rosa mts. ... 43. *A. leucolobus*

                  MM. Banner 11–13 mm. long; racemes
shorter than the lvs.; pod obliquely
ovoid, shallowly depressed dorsally in
the lower half, the valves thin, not inflexed. S. Sierra Nevada
44. *A. subvestitus*

GG. Stems developed or, if short, then the pubescence not cottony-
tomentose.

    H. Petals white, yellow or ochroleucous; keel, except in insular
A. Nevinii, more than 1 cm. long; pod stipitate, the stipe at
least 4 mm. long.

        I. Herbage densely and softly white-tomentulose. Insular
57. *A. Nevinii*

        II. Herbage at most villosulous with incurved hairs, not
tomentulose. Mainland.

           J. Calyx-tube campanulate, 4–5.5 mm. long, not gibbous
at base; fls. spreading; ovary and pod glabrous, the
latter 2-locular, its body obliquely oblong, about twice
longer than broad. Upper Sacramento and Trinity
valleys to the Klamath Basin
27. *A. accidens* var. *Hendersonii*

           JJ. Calyx-tube deeply campanulate or cylindric, the tube
7–10 mm. long, gibbous at base; fls. declined; ovary
and pod pubescent, the latter 1-locular, its body of

narrower outline, falcately incurved. Transmontane.
K. Stems densely villosulous with horizontal hairs;
    calyx-teeth 2–3.5 mm. long; petals dull yellow;
    pod-valves laterally distended, 5–8 mm. wide,
    villosulous .... 87. *A. Gibbsii*
KK. Stems strigulose with subappressed or incurved
    hairs; calyx-teeth rarely over 2 mm. long; petals
    whitish or cream-colored; pod more strongly
    flattened laterally, the sides flat or low convex,
    3–4 mm. wide, thinly strigulose
                                    89. *A. curvicarpus*
HH. Petals purple or purple-tipped, if white the keel less than
    1 cm. long; pod sessile or subsessile, the stipe not over 2 mm.
    long.
    I. Calyx ebracteolate; lowest internodes developed, the
       stipules not imbricated at base of the stem; pod inflated,
       papery-membranous or if not inflated, then densely silky-
       strigulose; vars. of .... 83. *A. lentiginosus*
    II. Calyx bracteolate at base; lowest internodes inhibited, the
        large scarious stipules amplexicaul and imbricated at base
        of the stem; pod not inflated, firm, villous or hirsute.
        J. Pod erect or ascending at a narrow angle, obong, nearly
           straight, 12–23 mm. long, not mottled, densely and
           softly long-villous .. 34. *A. Minthorniae* var. *villosus*
        JJ. Pod either deflexed or at least 3 cm. long and strongly
            incurved, in any case usually mottled and coarsely
            shaggy-hirsute.
            K. Pod narrowly oblong, abruptly acute at both ends,
               1.5–3 cm. long, moderately falcate, deflexed;
               stems arising together from the superficial crown
               of the taproot .... 35. *A. malacus*
            KK. Pod crescentic or arched through nearly a circle,
                cuneate at base and long-acuminate distally, 3–6
                cm. long, commonly incurved-ascending, some-
                times declined; stems arising singly and few
                together from buds on slender deeply buried
                cordlike rhizomes .... 36. *A. Layneae*
EE. Pubescence of the stems and herbage strictly appressed or subappressed, the
hairs usually straight, rarely wavy, sometimes almost lacking.
    F. Calyx-tube cylindric, 6–13.5 mm. long, pubescent even though
       sparsely so.
       G. Acaulescent plants, the stipules imbricated. E. Mojave Desert.
          H. Lfts. 3–13, densely silky on both faces; pod densely and
             stiffly cottony-tomentose .... 39. *A. Newberryi*
          HH. Lfts. 11–many, sparingly strigulose or glabrescent above; pod
              strigulose or obscurely villosulous
                                       38. *A. tephrodes* var. *brachylobus*
       GG. Caulescent plants, with developed internodes, the stipules not
           imbricated.
           H. Lfts. linear-oblong or narrowly elliptic, acute or obtuse; calyx
              not gibbous above the pedicel, commonly bracteolate, be-
              coming membranous in fr.; fls. purple.
              I. Pod erect, persistent, broadly oblong, subterete, 1.7–3 cm.
                 long, abruptly cuspidate-beaked, glabrous; lfts. mostly
                 acute .... 25. *A. Serenoi*
              II. Pod declined or deflexed, readily deciduous, narrowly
                  oblong, strongly obcompressed in the lower ⅔, acuminate
                  distally into a laterally compressed beak, 2–5.5 cm.
                  long, strigulose; lfts. obtuse .... 26. *A. Casei*
           HH. Lfts. obovate-cuneate, obtuse or retuse.
               I. Petals purple; bracteoles usually present; pod erect or
                  ascending, ovoid or oblong-ovoid, 1-locular; herbage
                  malodorous. Mojave and Colo. deserts.
                  J. Body of the pod oblong-ellipsoid or fusiform, 6–10
                     mm. in diam., the glabrous valves firmly papery.
                     Mojave Desert .... 86. *A. Preussii*
                  JJ. Body of the pod ovoid, more strongly inflated, 10–14
                      mm. in diam., the thinner-textured valves strigulose.
                      Colo. Desert .... 85. *A. Crotalariae*
               II. Petals white; bracteoles absent; herbage nearly scent-
                   less; pod 2-locular. Ne. Calif.; var. *platyphyllidius* of
                                                   83. *A. lentiginosus*
    FF. Calyx-tube campanulate or cylindro-campanulate, less than 6 mm. long or,
        if up to 6.5 mm. long, glabrous.
        G. Acaulescent or nearly so, with no internodes longer than the im-
           bricated stipules, the peduncles scapiform.

H. Pubescence dolabriform; wing-petals deeply 2-lobed at apex; pod oblong, strongly compressed, 2-locular . 12. *A. calycosus*
HH. Pubescence basifixed; wings entire or obscurely emarginate.
  I. Banner at least 12 mm. long, far exceeding the keel; lfts. 5–7, distantly disposed along a filiform rachis. Death V. region . . . . 15. *A. panamintensis*
  II. Banner not over 1 cm. long, about equaling the keel; lfts. either more numerous or crowded. Mono Co. n.
    J. Keel abruptly truncate; pod spreading from reclining peduncles, greatly inflated. Subalpine in Mono Co. 14. *A. platytropis*
    JJ. Keel obliquely acuminate; pod erect from erect or stiffly divaricate peduncles, linear-oblong, not inflated. Lassen Co. n. . . . . 13. *A. obscurus*
GG. Caulescent, with at least one evident internode.
  H. Calyx glabrous; lvs. glabrous or nearly so; pod pendulous, stipitate, bladdery-inflated . . . . 74. *A. oophorus*
  HH. Calyx pubescent, even though sparsely so.
    I. Terminal lft. continuous with the rachis; densely silky plants with bladdery pods. S. Colo. Desert 68. *A. magdalenae* var. *Peirsonii*
    II. Terminal lft. jointed to the rachis.
      J. Racemes narrowly 20–90-fld., the minute whitish fls. reflexed, the keel ca. 4 mm. long; pod reflexed, 4.5–7 mm. long, 2-locular, glabrous . . 51. *A. Clevelandii*
      JJ. Racemes not more than 30-fld. or, if slightly more, the fls. either purple, or larger, and the pod not as above.
        K. Peduncles paired in most of the axils; fls. minute, the banner not more than 6 mm. long; pod lenticular, 2-locular, 4–7 mm. long 47. *A. Lemmonii*
        KK. Peduncles solitary in the axils; fls. nearly always larger; pod not as above. (Go to L.)
L. Pod greatly inflated, the body ovoid to subglobose, the valves membranous or papery.
  M. Pod 2-locular.
    N. Calyx bracteolate; pod stipitate; herbage essentially glabrous; petals purple; var. *sufflatus* of . . . . 33. *A. cimae*
    NN. Calyx ebracteolate; pod sessile; herbage nearly always pubescent, if glabrate then the petals whitish; vars. of . . . . 83. *A. lentiginosus*
  MM. Pod 1-locular.
    N. Racemes at most 10-fld. E. Mojave Desert. See 79. *A. nutans* and 80. *A. Gilmani.*
    NN. Racemes 10-many-fld. Cismontane to w. Colo. Desert.
      O. Pod 2.5–6 cm. long, rarely less; petals ochroleucous. Cismontane (just reaching the edge of the sw. Mojave Desert) . . . . 69. *A. Douglasii*
      OO. Pod 1–2 cm. long, rarely slightly more; petals purple (sometimes ochroleucous veined with purple in var. *Johnstonii*). Mts. about the head of Coachella V. s. on the desert slope . . . . 71. *A. Vaseyi*
LL. Pod at most turgid, either of much thicker texture or of much narrower outline, usually both.
  M. Ovary and pod pubescent; pod (except in *A. inyoensis*) sessile.
    N. Racemes openly 15- or more-fld., 6–25 cm. long in fr.; pod erect, subterete, densely silky-strigulose; petals purple. Colo. and e. Mojave deserts; var. *borreganus* of 83. *A. lentiginosus*
    NN. Racemes 1–15(–20)-fld., 1–5, rarely 7, cm. long in fr.; pod spreading or pendulous or, if erect, grooved dorsally, merely strigulose, and distribution not as above.
      O. Racemes 1–4-fld.; stems and caudex-branches very slender, beset with wiry persistent petioles . . . . 15. *A. panamintensis*
      OO. Racemes at least 5-fld., if some occasionally less then the petioles deciduous.
        P. Lfts. 5–11, densely silvery on both faces; petals bright purple. Colo. and Mojave deserts. See 10. *A. albens* and 11. *A. mohavensis*
        PP. Lfts. 13 or more, glabrescent above or, if either less numerous, or cinereous on both faces, the petals whitish, or distribution transmontane n. from Mono Co.
          Q. Pod erect; keel-petals acuminate distally into a narrow triangular apex 13. *A. obscurus*
          QQ. Pod spreading or pendulous; keel rounded or deltoid at apex.
            R. Pod either stipitate or, if sessile, pendulous; wing-petals bidentate or obscurely emarginate at apex. Rare sp. of n. Mojave Desert.
              S. Calyx-tube deeply campanulate, 4.5–5.5 mm. long; pod sessile, narrowly oblong, slightly arched downward 16. *A. mensanus*
              SS. Calyx-tube shallowly campanulate, 2.5–3.5 mm. long; pod stipitate, the body oblong-ellipsoid, strongly incurved 17. *A. inyoensis*
            RR. Pod sessile and horizontally spreading; wings rounded at apex. Either cismontane or transmontane n. from Mono Co.
              S. Pod lance-acuminate in outline, abruptly hooked near the

middle and bent through half a circle or more; septum none
or very narrow .... 84. *A. iodanthus*
SS. Pod with ovoid or subglobose body and ± incurved deltoid
beak; septum broad and complete; vars. of
83. *A. lentiginosus*
MM. Ovary and pod glabrous; pod stipitate or elevated on a gynophore, the stipe or
gynophore sometimes concealed by the calyx and not more than 1 mm. long.
N. Stems glabrous; lfts. glabrous or merely ciliate.
O. Stipe short, 1–2.5 mm. long, the body of the pod falcate, compressed-triquetrous, ca. 3 mm. broad; lvs. thin; fls. white. Humboldt Co. n.
53. *A. umbraticus*
OO. Stipe 6–8 mm. long, the body obcompressed, strongly incurved, about 1 cm.
broad; lvs. thick; fls. purple. Mojave Desert .... 33. *A. cimae*
NN. Stems pubescent; lfts. strigulose at least on one surface.
O. Stems weak, flexuous, diffuse; fls. small, the banner not over 1 cm. long; lfts.
equally pubescent on both faces or less so above. Mojave Desert.
P. Pod pendulous on a stipe 3–5 mm. long, laterally compressed and 2-sided,
2-carinate by the prominent sutures .. .. 18. *A. Jaegerianus*
PP. Pod arcuate-erect on a stipelike gynophore 1–1.5 mm. long, the body
sharply triquetrous .... 19. *A. bernardinus*
OO. Stems stouter, not diffuse; fls. larger, the banner 13–22 mm. long; lfts. nearly
always more densely pubescent above than beneath.
P. Lfts. oval-ovate, often retuse, 3–12 mm. long; pod borne on a stipelike
gynophore 1–2.5 mm. long, sharply triquetrous, the valves papery, the
sutures filiform .... 20. *A. tricarinatus*
PP. Lfts. linear-oblong, 8–25 mm. long; pod long-stipitate, the body laterally
compressed, 2-carinate by the prominent cordlike sutures
21. *A. pachypus*

1. **A. Gambeliànus** Sheld. [*A. nigrescens* Nutt., not Pall. *A. Elmeri* Greene, a rare form, from Marin and Amador cos., with banner up to 4–6.5 mm. long. *A. Gambelianus* var. *Elmeri* (Greene) J. T. Howell.] Annual from a slender taproot, thinly pilosulous with ascending hairs, green or cinereous; stems 1–several, erect and ascending, 5–30 cm. long; stipules free, membranous with green midrib, 1.5–3 mm. long; lvs. 1–4 cm. long, with 7–13 narrowly cuneate-oblanceolate retuse lfts. 2–9 mm. long, often glabrous above; peduncles 1.5–6 cm. long; racemes 4–15-fld., becoming oblong and 5–20 mm. long in fr., the fls. at length recurved; calyx usually dark-pilosulous, the tube 1–1.5 mm., the teeth 0.5–0.9 mm. long; petals whitish tinged with violet, the banner mostly 2.5–3, rarely up to 6.5 mm. long; pod deflexed, sessile, broadly ovate in dorsiventral view, obcompressed, 3–4 mm. long, the lateral angles incurved toward the ventral suture, sulcate dorsally, the valves cross-wrinkled, hirsutulous or appressed-pilosulous, inflexed as a narrow but complete septum; 2n = 22 (L. E. James, 1951).—Grassy hillsides, sometimes in shade of oaks, 50–2900, reportedly up to 4000 ft.; Foothill Wd., S. Oak Wd., Coastal Sage Scrub; Coast Ranges from sw. Ore. to n. Lower Calif. and ids.; Great V. (rare); Sierran foothills from Shasta s. to Fresno Co. March–June.

2. **A. didymocárpus** H. & A. [*A. catalinensis* Nutt. *Hesperastragalus compactus* Heller.] Similar to the last, the stems erect, ascending or prostrate, up to 3 dm. long; lfts. 7–17, commonly glabrous above; fls. subsessile in globose or ovoid heads, not deflexed; calyx-tube 1.5–2.5 mm. long, the teeth as long or shorter, 0.8–1.4 mm. long; banner 3–6 mm. long; pod erect, sessile, invested by the calyx, ovoid to subglobose, 2.5–4 mm. long and nearly as broad, keeled ventrally and divided dorsally by a narrow groove into two cross-wrinkled saclike 1-seeded lobes, the valves thinly hispidulous; n = 12 (L. E. James, 1951).—Grassy hillsides and valley floors, 30–3100 ft.; Coastal Sage Scrub, V. Grassland; San Joaquin V. s. from Contra Costa Co.; cismontane s. Calif. and ids.; also on the sw. Mojave Desert and occasional between 3800 and 4400 ft. in Creosote Bush Scrub and Joshua Tree Wd. in the e. Mojave Desert; to s. Nev. and n. Lower Calif. March–May.

Var. **dispérmus** (Gray) Jeps. [*A. d.* Gray. *A. d.* var. *albus* L. E. James.] Stems prostrate, radiating; herbage gray or silvery-pubescent; calyx-teeth 1.5–2.5 mm. long, longer than the tube; keel 3.5–4.5 mm. long; n = 13 (L. E. James, 1951).—Sandy flats, 0–3000 ft.; Creosote Bush Scrub, sw. Mojave and Colo. deserts; also rarely in the e. Mojave Desert (Kingston Peak, about 4000 ft.); Ariz., L. Calif. Feb.–May.

Var. **daleoìdes** Barneby. Like the typical form, but calyx-tube narrowly campanulate, 2.5–3 mm., the teeth nearly 2 mm. long.—V. Grassland; Upper Salinas and Cholame valleys, San Luis Obispo Co. Poorly known, perhaps a mere forma. April–May.

Var. **Milesiànus** (Rydb.) Jeps. [*Hesperastragalus M.* Rydb.] Like the typical form, but fls. larger, the keel 5.5–7 mm. long; pod glabrous or glabrescent.—Along the coast; Coastal Sage Scrub, ca. 300 ft.; San Luis Obispo and Santa Barbara cos. April–June.

Var. **obispénsis** (Rydb.) Jeps. [*Hesperastragalus o.* Rydb.] Usually canescent like var. *dispermus*, but fls. larger, the keel 5–7 mm. long, attenuate into a narrowly triangular apex.—Grassy and sandy hillsides, 1400–3100 ft., S. Oak Wd. and (rarely) Joshua Tree Wd.; interior s. Calif. in a narrow strip mostly w. of the deserts, Antelope V., Los Angeles Co., s. to n. L. Calif. April–June.

3. **A. Brèweri** Gray. Annual, with a single erect stem or several ascending stems 2–30 cm. long, green but sparingly strigulose, the lfts. rarely cinereous beneath; stipules membranous-margined, often black-ciliate, 1.5–4.5 mm. long; lvs. 2–6 cm. long, with 7–13 obovate-cuneate to oblong retuse lfts. 3–10 mm. long; peduncles erect, 2.5–9 cm. long; racemes capitately 3–10-fld., not elongating; calyx dark-strigulose, the tube 2.5–3.5 mm., the teeth 1–2 mm. long; petals lilac, fading paler, the keel purple-tipped, the banner 8–11.5 mm. long; pod ascending, sessile, the body ovoid or fusiform, 6–10 mm. long, grooved dorsally, silvery-strigulose, abruptly passing into a linear-subulate, declined, glabrescent beak 4–10 mm. long, the firm valves inflexed as a complete septum. —Meadows and grassy hillsides, often on serpentine, 300–2400 ft., N. Oak Wd., Foothill Wd.; inner N. Coast Ranges, s. Mendocino and Lake cos. to Mt. Tamalpais, Marin Co. April–May.

4. **A. Rattáni** Gray. Habit of the preceding; fls. pink-purple, rarely white, the banner 10–12.5 mm. long, the keel 6.5–8 mm. long, far surpassed by the wings; pod erect or ascending, sessile, narrowly linear and attenuate at both ends, 2–6 cm. long, 1.5–3 mm. in diam., obscurely triquetrous, straight or gently incurved, the firm strigulose valves inflexed as a complete septum.—Open grasslands, sometimes on sandbars along streams, 100–2500 ft., N. Oak Wd., Foothill Wd.; N. Coast Ranges, cent. Humboldt to cent. Mendocino and nw. Lake cos. April–July.

Var. **Jepsoniànus** Barneby. Fls. smaller, commonly only purple-tipped, rarely pink-purple or white, the banner 7–9.5 mm., the keel 4–5 mm. long; pod averaging shorter, 1.5–3 cm. long.—Meadows and grassy hillsides, commonly on serpentine, 1100–1900 ft., Foothill Wd.; inner N. Coast Ranges, s. Lake, Napa, w. Colusa and w. Tehama cos. April–June.

5. **A. Clariànus** Jeps. [*A. Rattani* var. *C.* Jeps.] Quite like *A. Rattani*, but dwarf, 5–15(–20) cm. high; lfts. 5–9; fls. bicolored, the wings whitish, the banner and keel purple in the upper third; keel prominent, 7.5–9 mm. long, 2.5–3.5 mm. broad, surpassing the narrow wings; pod horizontal, linear and attenuate at both ends, disjointing from the stipelike gynophore 1.5–2.5 mm. long, gently incurved, 2–3 cm. long, 2–3 mm. in diam., strigulose, 2-locular.—Grassy hillsides, 300–550 ft., Foothill Wd.; Napa Range near St. Helena and Santa Rosa. Rare and local. April–May.

6. **A. téner** Gray. [*A. Hypoglottis* var. *strigosus* Kell. *Hamosa Kelloggiana* Rydb. *A. t.* var. *rattanoides* Jones.] Habit of the three last but variable, the stems 3–30 cm. long; lfts. 7–17, all cuneate-obovate retuse or truncate or those of the upper, rarely of all leaves, narrowly oblong to linear and then often acute, 5–16 mm. long; racemes subcapitately 3–12-fld.; calyx-tube 2–3 mm., the teeth 1.3–2.3 mm. long; petals purple, striate, the banner 9–12 mm., the keel 5–6.5 mm. long; pod spreading, sessile, narrowly oblong, straight or lunately incurved, 1–1.5 cm. long, 2–3.5 mm. in diam., flattened and grooved dorsally, the green at length stramineous firmly papery strigulose or glabrous valves inflexed as a complete septum; ovules 10–14.—Grassy alkaline flats, moist in spring, below 500 ft., V. Grassland; cent. Great V. and Delta region, from Solano Co. to San Francisco Bay and s. to the lower San Benito and Salinas valleys. March–June.

Var. **Tìti** (Eastw.) Barneby. [*A. T.* Eastw.] Lfts. always broad; fls. smaller, the banner 5–6 mm. long; pod 6–14 mm. long, 5–11-ovulate, but only 6–9 mm. long and 5–6-ovulate in cent. Calif.—Depressions in dunes and sandy flats near the coast, below 50 ft.; Coastal Strand; Monterey Bay, Los Angeles Bay (probably extinct), and near San Diego, rare and local. April–May.

7. **A. paupérculus** Greene. [*A. tener* var. *Bruceae* Jones. *A. B.* Abrams.] Habit of *A. tener*, but stems short, 1–8 cm. long, the internodes commonly short and the early peduncles often appearing subradical; lfts. 7–11, obovate, retuse, 2–8 mm. long; racemes loosely 2–5-fld., the axis somewhat elongating, 7–20 mm. long in fr.; petals purple, the

banner 5.5–10.5 mm., the keel 4.5–6 mm. long; pod ascending or declined, often resupinate on twisted pedicels, sessile, linear-oblong, gently incurved, 1.2–2 cm. long, 2.5–3.5 mm. in diam., triquetrously compressed, the papery valves greenish and mottled when young becoming purplish-brown, strigulose or rarely glabrous, fully 2-locular.— Stony flats and shallow depressions moist in spring, associated with volcanic rocks, 450–2050 ft., V. Grassland, Foothill Wd.; e. foothills of upper Sacramento V., s. Shasta to Butte cos. March–May.

8. **A. acutiróstris** Wats. [*A. streptopus* Greene. *A. Nuttallianus* var. *a.* (Wats.) Jeps.] Slender annual, strigulose or pilosulous with incurved or contorted hairs, cinereous or greenish; stems solitary and erect or several ascending, 2–25 cm. long; stipules small, deltoid, free; lvs. 1.5–4 cm. long, with 9–15 obovate to cuneate retuse lfts. 2–8 mm. long; peduncles erect, 1.5–6 cm. long; racemes 3–6-fld., lax and 1–4 cm. long in fr.; calyx strigulose, the tube 1.5–2 mm., the teeth 1–1.5 mm. long; petals purplish, the banner 5–7 mm., the keel 4.5–6 mm. long, the blades incurved to a narrowly triangular or cuspidate apex; pod resupinate on contorted pedicels, erect or pendulous, linear-oblong, gently incurved, 1.2–2.2 cm. long, 2.5–3 mm. in diam., compressed-triquetrous, grooved dorsally, fully 2-locular, the papery valves green or purplish, strigulose.—Sandy or gravelly slopes and washes, 2000–5200 ft., Joshua Tree Wd., Creosote Bush Scrub; cent. and w. Mojave Desert, from lower Owens V. to Death V. and the Little San Bernardino Mts.; w. edge of Colo. Desert in San Diego and w. Imperial cos.; s. Nev. April–May.

9. **A. Nuttalliànus** DC. var. **imperféctus** (Rydb.) Barneby. [*A. N.* var. *trichocarpus* and var. *canescens* of Calif. refs., not T. & G.] Like the preceding, strigulose with sub-appressed hairs, silvery or greenish, the stems up to 3 dm. long; lfts. mostly oblong-elliptic, acute or obtuse, rarely retuse-emarginate in the lower lvs.; calyx 3.2–4.5 mm. long, the tube 2–2.8 mm., the teeth 1–1.7 mm. long; banner 4–6.5 mm., keel 4–5.5 mm. long; pod horizontally spreading on arched pedicels, sessile, linear-oblong, slightly to strongly incurved (especially in the lower half), 12–21 mm. long, 2–3 mm. in diam., triquetrous with obtuse lateral angles, grooved or, when fully ripe, flattened dorsally, the papery valves glabrous or strigulose, either fully 2-locular or the septum reduced to a vestige.—Sandy or stony mesas and canyon-washes, 900–5200 ft., Creosote Bush Scrub, Joshua Tree Wd.; Death V. region s. to e. and cent. Mojave Desert; to n.-cent. and sw. Utah and w. Ariz. March–June.

Var. **cedrosénsis** Jones. Stems filiform; lfts. of lower leaves cuneate-obcordate; calyx 2.5–3 mm. long, the tube 1.4–1.7 mm., the teeth 1–1.6 mm. long; pod 1.6–2 mm. in diam., the septum evident but often incomplete.—Desert flats and boulder-strewn washes, 50–900 ft.; Creosote Bush Scrub; w. edge of Colo. Desert, w. Imperial and adjacent San Diego cos.; nw. Son., L. Calif.

10. **A. álbens** Greene. Short-lived perennial flowering the first season, silvery-white throughout with appressed hairs; stems slender, decumbent, 3–30 cm. long; stipules free, deltoid, small; lvs. 1.5–4 cm. long, with 5–9 obovate obtuse-truncate lfts. 2–11 mm. long; peduncles about equaling the lvs.; racemes loosely 5–14-fld., 2.5–4.5 cm. long in fr.; calyx-tube campanulate, ca. 2.5 mm. long, the teeth 1.5–2 mm. long; petals pink-purple, the banner 7–9.5 mm. long, the prominent broad keel nearly as long or slightly longer, exserted from the wings; pod spreading, sessile, oblanceolate in profile, lunately incurved, 13–18 mm. long, 3–3.5 mm. in diam., triquetrous, grooved dorsally, the papery purplish valves strigulose, inflexed as a complete septum.—Gravelly flats and washes, 4000–5800 ft.; typically in Pinyon-Juniper Wd., sometimes washed down to the edge of Creosote Bush Scrub; ne. slope of San Bernardino Mts. in and near Cushenbury Canyon, fairly common locally. March–May.

11. **A. mohavénsis** Wats. Winter-annual or short-lived perennial, like *A. albens* but coarser; lvs. 2–10 cm. long, with 5–11 mostly obovate to elliptic lfts. 3–18 mm. long; racemes 3–16-fld., 1–7 cm. long in fr.; calyx-tube 2.5–4.5 mm., the teeth 2–3 mm. long; banner 9–12.5 mm. long; wings narrower than the broad obtuse keel but of equal length, 7.5–10 mm. long; pod spreading or deflexed, sessile, oblong-obovate in outline, abruptly cuspidate-beaked, 1.5–3 cm. long, 5–8.5 mm. in diam., subterete and fleshy when young, the valves shrinking in age and leaving both sutures prominent, straight or a little incurved and the dorsal suture then depressed, the valves leathery, reticulate, strigulose or hirsutulous, fully 2-locular.—Rocky slopes in canyons or on cliff-ledges,

preferring limestone, 2500–7500 ft. and in Death V. occasionally as low as 1200 ft.; Joshua Tree Wd., Creosote Bush Scrub; Mojave Desert from Pinto Basin, Riverside Co., n. to Darwin Mesa, Death V., Providence Mts. and s. Nev. April–June.

Var. **hemigȳrus** (Clokey) Barneby. [*A. h.* Clokey.] Pod relatively longer and narrower, oblanceolate in outline, 3.5–5.5 mm. in diam., definitely or strongly falcate, the dorsal suture depressed and sulcate at maturity; fls. slightly smaller, the banner 7–9 mm., the keel 6–8 mm. long.—Similar habitats, 4100–5200 ft., rare and local, near Darwin, Inyo Co.; Charleston Mts., Nev. Our form somewhat transitional to the sp.

12. **A. calycòsus** Torr. Low acaulescent tufted perennial, silvery-strigulose with dolabriform hairs; lvs. 2–7 cm. long, with 3–7 oblanceolate-obovate lfts. 5–19 mm. long; racemes shortly exserted, loosely 2–6-fld.; calyx-tube campanulate, 4–6.5 mm. long; petals whitish or pink-tipped, fading bluish, the banner 10–17 mm. long, the wings deeply cleft at apex; pod sessile, oblong, slightly incurved or straight, 1–1.5 cm. long, 3–4 mm. in diam., laterally compressed, narrowly grooved dorsally, fully 2-locular, the valves papery, strigulose-canescent.—Stony knolls and clay draws, mostly on limestone, in Calif. 6000–10,000 ft.; Pinyon-Juniper Wd.; Panamint, Inyo, and White mts.; to cent. Ida., s. Wyo., Utah, n. Ariz. May–June.

13. **A. obscùrus** Wats. Low perennial with a woody caudex, strigulose-canescent or greenish, exceptionally villosulous; stems almost none, the peduncles then subscapose, or developing 2–4 internodes and up to 15 cm. long, slender and wiry; stipules deltoid, the lower scarious and amplexicaul but free; lvs. 3–10 cm. long, with 7–15 oval and obtuse or linear-elliptic and acute lfts. 2–14 mm. long; peduncles 5–15 cm. long; racemes 3–14-fld., the fls. ascending, the axis becoming 2–8 cm. long in fr.; calyx-tube campanulate, 2.5–4 mm., the teeth 0.5–1.5 mm. long; petals straw color often tinged with lurid purple, the banner and keel 7–10 mm. long, the wings slightly shorter, the keel incurved into a narrowly triangular porrect apex; pod stiffly erect, sessile, linear-oblong, 13–25 mm. long, ca. 3 mm. in diam., obscurely triquetrous, deeply grooved dorsally, straight or nearly so, the stiffly papery valves strigulose, inflexed as a nearly complete septum.—Stony flats, dry plains and hillsides, 2650–6500 ft., mostly on basaltic substrata; Sagebrush Scrub; Lassen and Modoc cos., w. to the foot of Mt. Shasta; to n. and w. Nev., e. Ore. and s. Ida. May–July.

14. **A. platytròpis** Gray. Diminutive tufted perennial, silvery-strigulose throughout, the divisions of the cespitose caudex either very short or sometimes elongating on shifting rock-slides, commonly beset with a thatch of persistent petioles; stems almost none, rarely 1 cm. long, clothed in imbricated stipules; lvs. 1–9 cm. long, with 5–15 obovate to elliptic mostly obtuse lfts. 2–11 mm. long; peduncles scapiform, 2–6.5 cm. long, reclinate in fr.; racemes subcapitately 2–9-flowered; calyx-tube campanulate, 2–3 mm., the teeth 0.5–2 mm. long; petals ochroleucous tinged with leaden purple, of nearly equal length, 7–9.5 mm. long, the broad keel abruptly incurved to a rectangular blunt apex; pod spreading, sessile, bladdery-inflated, ovoid to broadly ellipsoid with short conic beak, 1.5–3.5 cm. long, the papery valves strigulose, mottled, inflexed as a partition produced across the cavity to meet a wide flange issuing from the ventral suture. —Bare ridges above timber line, in pumice gravel, in Calif. 9300–11,750 ft.; Alpine Fell-fields; Sweetwater Mts., Mono Co. and "mountain near Sonora Pass," perhaps in the same range; to s. Nev., sw. Mont., there descending into Sagebrush Scrub down to 5550 ft. July–Aug.

15. **A. panaminténsis** Sheld. [*A. atratus* var. *p.* (Sheld.) Jeps.] Low perennial, silvery-strigulose throughout, the slender divisions of the branched and matted caudex beset with wiry persistent incurved petioles, young plants sometimes subacaulescent; stems of the year 1–5 cm. long, the internodes all short; stipules small, free; lvs. 2–12 cm. long, with filiform rachis and 5–9 distant linear-elliptic acute involute readily deciduous lfts. 2–12 mm. long; peduncles filiform, 1–6 cm. long; racemes loosely 1–4-fld.; calyx-tube campanulate, 3–4.5 mm., the teeth 1.5–3 mm. long; petals pink-purple, the banner 9.5–14 mm. long; pod spreading, sessile, oblong in outline, slightly arched downward, 8–18 mm. long, 3–4.5 mm. in diam., triquetrous with blunt lateral angles, grooved dorsally, semibilocular, the papery purple-speckled valves strigulose.—Forming entangled bird's-nestlike tufts on ledges and in fissures of canyon walls, mostly on limestone, 4000–7100 ft.; Pinyon-Juniper Wd.; Panamint, Last Chance and Inyo mts. April–June.

16. **A. atràtus** Wats. var. **mensànus** Jones. [*A. m.* Abrams ("mensarus").] Slender

perennial, greenish or thinly strigulose-cinereous, with slender ascending stems 10–30 cm. long, from a taproot; stipules small, free, the lowest scarious; lvs. 5–13.5 cm. long, with 9–15 oblanceolate obtuse or retuse lfts. 3–16 mm. long, glabrous above; peduncles 4.5–13 cm. long; racemes loosely 7–16-fld., the fls. early nodding and secund; calyx purplish, the deeply campanulate tube 4.5–5.5 mm., the teeth 0.8–2.5 mm. long; petals lilac-purple, the banner 10–13.5 mm. long, the wings slightly longer, toothed at apex, the keel 8–10 mm. long; pod pendulous, sessile, narrowly oblong, 16–22 mm. long, 3.5–4.5 mm. in diam., slightly arched downward, strongly compressed, the sutures both prominent, semibilocular, the stiffly papery valves strigulose.—Rare and local, dry hillsides and mesas, often sheltering under and entangled in sagebrush, 5400–6050 ft.; Pinyon-Juniper Wd., Sagebrush Scrub; Darwin Mesa and Nelson Range, Inyo Co. April–June.

17. **A. inyoènsis** Sheld. Perennial from a taproot, except for the upper lf.-surface strigulose-cinereous throughout; stems diffuse or prostrate from a short caudex, 1.5–6 dm. long, zigzag upward; stipules deltoid, small, free; lvs. 1.5–5 cm. long, often shorter than the internodes, with 11–21 crowded obovate obtuse or retuse lfts. 3–10 mm. long; peduncles divaricate, 2.5–7 cm. long; racemes loosely 6–15-fld., becoming 2.5–7 cm. long in fr.; calyx-tube campanulate, 2.5–3.5 mm., the teeth 1–2.5 mm. long; petals dull pink-purple, the banner 8.5–11 mm. long, the broad keel scarcely shorter, equaling or exserted from the obscurely emarginate wings; pod stipitate, the horizontally spreading stipe 2–5 mm. long, the body incurved to erect, obliquely oblong-ellipsoid, cuneate at base and obcompressed, passing upward into a laterally compressed, deltoid, long-cuspidate beak, 12–16 mm. long, 3.5–5 mm. broad, grooved dorsally, semibilocular, the leathery purplish valves strigulose.—Open gravelly flats and hillsides, (5000–)5800–7800 ft., Pinyon-Juniper Wd.; s. end of White Mts., Inyo Mts. and Darwin Mesa, Inyo Co. May–July.

18. **A. Jaegeriànus** Munz. Perennial, strigulose-cinereous or greenish; stems arising from a buried root-crown, flexuous, weak, reclining or scrambling through bushes, 3–7 dm. long, branched; stipules small, free; lvs. 2–6 cm. long, with 9–15 linear-oblong mostly obtuse lfts. 3–20 mm. long; peduncles divaricate, 2.5–8 cm. long; racemes loosely 5–15-fld., becoming irregularly secund and up to 7 cm. long; calyx-tube campanulate, 2.5–4 mm., the teeth 1.5–2 mm. long; petals ochroleucous, veined or suffused with lavender, the banner 7–10 mm. long, the keel slightly shorter, acutish; pod pendulous, stipitate, the stipe 3–5 mm. long, the body narrowly oblong, 16–23 mm. long, 3.5–5 mm. in diam., straight, laterally compressed, 2-carinate by the sutures, fully 2-locular, the fleshy glabrous ± mottled valves becoming brown and leathery.—Low granite hills, 3000–3800 ft.; Joshua Tree Wd.; known only from Coolgardie Mesa and about Goldstone, cent. Mojave Desert. April–June.

19. **A. bernardìnus** Jones. Slender perennial from a branched caudex, strigulose but greenish, the stems sometimes nearly glabrous but the lfts. often canescent above; stems wiry, flexuous, branched, 1–5 dm. long; stipules small, free; lvs. 4–12 cm. long, with 7–17 oblong to narrowly elliptic obtuse or retuse lfts. 5–25 mm. long; peduncles 3–10 cm. long; racemes loosely 10–25-fld., the fls. ascending, the axis becoming 4–15 cm. long; calyx-tube campanulate, 2.5–4 mm., the teeth 1–3 mm. long; petals sordid lilac, drying bluish, the banner 7–10 mm. long, the keel 7–9.5 mm. long, incurved to a narrow acutish apex; pod arcuate-erect, sessile on and readily deciduous from a stipe-like gynophore 1–1.5 mm. long, narrowly oblanceolate in profile, sharply triquetrous, 2.5–3.1 cm. long, 4–4.5 mm. in diam., fully 2-locular, the papery valves glabrous and glaucescent when fresh.—Stony washes and dry mesas, often scrambling up through low bushes, 3000–6700(–7500) ft., on granite or limestone; Joshua Tree and Pinyon-Juniper Wd.; s. and se. Mojave Desert, from e. slope of San Bernardino Mts. to Pinto Basin, Riverside Co., New York and Ivanpah mts.; (reportedly) s. Nev. April–June.

20. **A. tricarinàtus** Gray. Perennial, forming loose bushy plants, greenish and thinly strigulose except for the silvery upper surface of the lfts.; stems flexuous, 1–2.5 dm. long, greatly surpassed by the infls.; stipules free, reflexed; lvs. 7–20 cm. long, with 17–27 distant elliptic or obovate mostly retuse lfts. 3–12 mm. long, deciduous from the stiff persistent petioles; peduncles stout, erect, 9–20 cm. long; racemes loosely 6–15-fld., 6–18 cm. long in fr.; calyx-tube broadly campanulate, 4–5 mm., the teeth 2–3 mm. long; petals whitish, drying ochroleucous, the banner 13–16 mm. long; pod like

that of the preceding, the gynophore 1–2.5 mm. long, the body 2.5–4.2 cm. long, 3.5–5.5 mm. in diam., turning straw color, not glaucescent.—Gravelly washes and canyons, 1500–4000 ft., Creosote Bush Scrub, Joshua Tree Wd.; foothills and desert mts. about the head of Coachella V., from Whitewater and Morongo Pass to the Orocopia Mts. Feb.–May.

21. **A. páchypus** Greene. Perennial with a woody caudex, cinereously strigulose throughout, the stems and usually the upper face of the lfts. whitened with a dense coat of short hairs; stems erect and ascending, forming bushy plants, rarely hanging down from steep banks, 2–4.5 dm. long, flexuous distally; stipules small, free, ultimately deciduous and leaving a prominent scar; lvs. 6–15 cm. long, with 11–21 linear-oblong retuse lfts. 8–25 mm. long; peduncles stiff, erect, 7–20 cm. long; racemes loosely 8–20(–28)-flowered, becoming 3–13 cm. long in fr.; calyx-tube campanulate 4.3–5.2 mm., the teeth 2.5–4.3 mm. long; petals white, the banner sometimes pink-tinged, drying yellowish, the banner 17–22 mm., the keel 12.5–15 mm. long; pod stipitate, the stout spreading stipe 4–8 mm. long, the body arcuate-erect, oblong, (12–)15–25 mm. long, (4–)5–7.5 mm. in diam., cuspidate-beaked, at first fleshy and subterete, when ripe laterally compressed and 2-carinate by the thick prominent cordlike sutures, fully 2-locular, the valves woody, transversely wrinkled, glabrous; $2n = 22$ (Head, 1957).—Dry hillsides, in stiff alkaline clay soils, 1650–6300 ft., V. Grassland, Creosote Bush Scrub; S. Coast Ranges, in scattered stations, from Griswold Hills and New Idria, San Benito Co., to San Rafael Mts., Ventura Co., e. around the head of San Joaquin V. to Tehachapi, and to the desert-edge in Kern and Los Angeles cos. March–July.

Var. **Jàegeri** Munz & McBurn. Lfts. 17–27; fls. slightly smaller, the banner 15–17 mm., the keel 10.5–12.5 mm. long; petals clear lemon-yellow.—Dry ridges and gullied clay hillsides, 1700–2500 ft.; interior Riverside Co. (Banning to Temecula and Aguanga).

22. **A. Bolánderi** Gray [*A. supervacaneus* Greene] Perennial, thinly strigulose-villosulous, the stems commonly nearly glabrous, exceptionally villosulous throughout with crisped hairs; stems ascending from a buried root-crown, 1.5–4.5 dm. long; stipules dimorphic, the lower connate into a scarious sheath, the upper free or joined by a line; lvs. 4–15 cm. long, with 17–25 oval to narrowly oblong-elliptic mostly acute lfts. 5–23 mm. long; peduncles 3–8 cm. long; racemes loosely 5–18-fld., little elongating, 1–3(–5) cm. long in fr.; calyx-tube campanulate, 5–6.5 mm., the teeth 2–5.5 mm. long; petals whitish, rarely suffused with lavender, the banner 13–17.5 mm. long; pod stipitate, the erect or spreading stipe 4–12 mm. long, the body obliquely ovoid, inflated, strongly incurved toward the raceme-axis, slightly obcompressed and openly grooved along both sutures, 13–32 mm. long, 6–12 mm. in diam., fully 2-locular below the deltoid-acuminate beak, the valves stiffly papery or leathery, glabrous.—Dry stony meadows, sandy flats and openings in Red Fir-Lodgepole F., sometimes on lake shores, 5200–10,000, reportedly 10,800 ft.; Sierra Nevada from Kern-Kaweah Divide in cent. Tulare Co. n. to s. Plumas Co., and extending e. of the crest around Lake Tahoe to Mt. Rose, Nev. June–August.

23. **A. bicristàtus** Gray. Perennial, strigulose nearly throughout, cinereous or greenish, the lfts. sometimes silvery above; stems weakly ascending, 2–4.5 dm. long, naked at base; stipules dimorphic, the lowest 1–3 pairs connate into a short scarious sheath, the rest herbaceous, small, free; lvs. 3–13.5 cm. long, with 13–21 linear to narrowly oblong, rarely oblong-oblanceolate, acutish to obtuse lfts. 4–25 mm. long; peduncles 5–12 cm. long; racemes loosely 5–15 fld., 1.5–6(–9) cm. long in fr.; calyx-tube campanulate, 6–7.5 mm., the teeth 2–3.5 mm. long; petals greenish-white or lilac-tinged, the banner 15.5–19.5 mm. long; pod spreading or declined, stipitate, the stout stipe 6–12 mm. long, the body obliquely oblong-ellipsoid, cuneate or acuminate at either end, lunately or hamately incurved, 2–4 cm. long, 5–9 mm. in diam., subterete and fleshy when fresh becoming roughly quadrangular when ripe, with transversely dilated obtuse lateral angles and prominent or almost winglike sutures, the valves stiffly leathery or almost woody, reticulate, glabrous, not inflexed.—Rocky or sandy soil of dry coniferous wd., 5800–8000(–9000) ft., Yellow Pine F.; San Bernardino and San Antonio mts. May–August.

24. .**A. Wébberi** Gray. Perennial from a taproot and knotty root-crown, the herbage satiny-strigulose with a coat of fine short hairs, greenish or silvery in age; stems many, diffuse and ascending, 2.5–5 dm. long, slender and leafless at base; stipules dimorphic,

the lowest scarious, connate into a short sheath, the rest herbaceous, 2–7 mm. long, free, reflexed; lvs. 4–15 cm. long, with 13–25 elliptic-oblanceolate obtuse or retuse lfts. 1–3.5 cm. long; peduncles 6–15 cm. long; racemes loosely 7–14-fld., 2.5–10 cm. long in fr.; calyx-tube subcylindric, 6.5–7.5 mm. long, becoming membranous and circumscissile, the teeth 2.5–4.5 mm. long; petals whitish, the banner 15.5–19 mm. long; pod ascending, humistrate, sessile, lunately oblong-ellipsoid, 2.5–3.5 cm. long, 8–12 mm. in diam., somewhat obcompressed, the sutures thick and prominent, the fleshy glabrous valves becoming woody, brownish or rarely mottled, not or but vestigially inflexed.— Uncommon, well drained brushy slopes and thickets, 2700–3550, reportedly 5000 ft.; Yellow Pine F.; Sierra Nevada, in Sierra and upper Feather River valleys, Sierra and Plumas cos. May–July.

25. **A. Serènoi** (Kuntze) Sheld. [*A. nudus* Wats., not Clos. *A. Watsonianus* Speg., not Sheld.] Sparsely leafy bushy-branched perennial, thinly strigulose, but the upper face of the lfts. often canescent; stems erect, flexuous, junceous, 1.5–4.5 dm. long; stipules free, reflexed; lvs. 2–15 cm. long, with 5–11 remote linear acute lfts. 0.5–4 cm. long, these sometimes much reduced or wanting; peduncles stout, 0.5–2.5 dm. long; racemes loosely 5–25-fld., up to 2 dm. long in fr.; calyx-tube cylindric, 9–13.5 mm. long, becoming membranous and circumscissile, the teeth 2.5–4 mm. long; petals purple with white wing-tips, the banner 17–26 mm. long; pod erect, sessile, broadly and plumply oblong, 17–30 mm. long, 7–10.5 mm. in diam., abruptly cuspidate, nearly straight, terete and fleshy, becoming slightly obcompressed with prominent thick sutures, the ultimately woody glabrous valves inflexed as a partial septum; $2n = 22$ (Trelease & Beath, 1949, as *A. canonis* Jones).—Bare gravelly hillsides, 5000–7400 ft.; Pinyon-Juniper Wd., Sagebrush Scrub; e. Inyo and Mono cos. (Grapevine, Cottonwood, Inyo, and White mts.); w. Nev. May–July.

26. **A. Càsei** Gray. Wiry sparsely leafy perennial, strigulose nearly throughout, the lfts. often canescent above; stems from a buried root-crown, 1.5–4 dm. long, erect or assurgent, flexuous; stipules small, free, the lowest scarious, amplexicaul but not connate; lvs. 3–10 cm. long, with 7–15 distant elliptic to linear-oblong obtuse lfts. 5–25 mm. long; peduncles erect, 3–9 cm. long; racemes loosely 6–26-fld., 3–15 cm. long in fr.; calyx-tube cylindric, 6–7.5 mm., the teeth 1–2 mm. long; petals purple with white wing-tips, the banner 12–18 mm. long; pod deflexed, sessile, narrowly oblong- or lance-ellipsoid, 2–5.5 cm. long, 5–10 mm. in diam., acuminate distally into a laterally flattened cuspidate beak, obcompressed in the lower ⅔, gently sigmoidally arched or incurved, the ventral suture cordlike, the leathery valves reticulate, mottled, strigulose, not inflexed; $2n = 22$ (Trelease & Beath, 1949).—Hillsides and valley-floors, in sandy soil, 4000–7200 ft.; Sagebrush Scrub, Pinyon-Juniper Wd.; Panamint Mts., n. through Owens V. to Mono Lake; w. Nev., reported from se. Wash. April–June.

27. **A. áccidens** Wats. var. **Hendersòni** (Wats.) Jones. [*A. H.* Wats., not Baker. *A. pacificus* Sheld. *A. Watsoni* Sheld. *A. cymatodes* Greene. *A. pruniformis* Jones.] Perennial from a forking caudex, thinly villosulous with incurved or ascending hairs, usually greenish; stems many, decumbent, 3–4.5 dm. long, forming low clumps; stipules free, acuminate, 3–6 mm. long; lvs. 5–12 cm. long, with 15–27 oblong-oblanceolate, rarely obovate-cuneate, retuse or truncate thin-textured lfts. 6–22 mm. long, glabrous or sparsely pubescent above; peduncles 7–15 cm. long, reclining under the pods' weight; racemes loosely 7–15-fld., 3–7.5 cm. long in fr.; calyx villosulous, the campanulate tube 4–5.5 mm., the teeth 2–4 mm. long; petals whitish, the banner 14–19.5 mm. long; pod pendulous (or when humistrate variably oriented), stipitate, the stout stipe 6–12 mm. long, the body obliquely oblong-obovoid, obtuse at both ends but cuspidate-beaked, 16–25 mm. long, 8–12 mm. in diam., at first fleshy and subterete, becoming somewhat compressed, with cordlike ventral and sinuate dorsal sutures, the glabrous ultimately woody valves rugulose-reticulate, inflexed as a complete septum.—Dry open woods and thickets, 900–4000 ft.; N. Oak and Foothill Wd.; around upper Sacramento V. (Butte and Tehama cos. to upper Trinity R.) n. through Siskiyou Co.; to sw. Ore. Late April–July.

28. **A. canadénsis** L. var. **brévidens** (Gand.) Barneby. [*A. Mortoni* Calif. refs., not Nutt.] Coarse perennial, spreading by running rootstocks, strigulose throughout with dolabriform hairs or glabrate; stems solitary or in clumps, mostly erect, 1.5–5.5 dm. long; stipules all connate into a scarious sheath 3–10 mm. long; lvs. 5–15 cm. long,

with 13–25 mostly oblong obtuse lfts. 5–30 mm. long; peduncles erect, 5–15 cm. long; racemes densely many-fld., the fls. retrorsely imbricated in oblong heads 4–8 cm. long; calyx-tube deeply campanulate, saccate at base, 5–8.5 mm. long, the upper teeth broadly deltoid, 1 mm. long or less; petals greenish-white or sordid, the banner 12.5–17 mm. long; pod erect, sessile, oblong-ellipsoid, 1–1.5 cm. long, 3–4 mm. in diam., abruptly contracted into a cuspidate beak 1.5–3 mm. long, shallowly grooved dorsally, carinate ventrally by the thick suture, the stiffly leathery or woody strigulose valves inflexed as a complete septum.—Stiff alkaline soil moist in spring, especially along ditches or about ephemeral pools, seeps, or depressions, ca. 4000–7000 ft. in Calif.; Sagebrush Scrub; Mono V., e. Placer to e. Siskiyou Co.; to B.C., S.D., n. Colo. and sw. Utah. May–August.

29. **A. agréstis** Dougl. [*A. goniatus* Nutt.] Low perennial, except for the pod and calyx thinly appressed-pilosulous; stems arising from the divisions of a deeply buried caudex, 3–20 cm. long; stipules 3–10 mm. long, connate into a loose sheath, the lower scarious, the upper with herbaceous blades; lvs. 2–10 cm. long, with 9–21 narrowly oblong to ovate mostly retuse lfts. 4–15 mm. long; peduncles 2–10 cm. long, bearing an ovoid-oblong head of 6–12 erect, conspicuously bracteate fls.; calyx villous-hirsute, the subcylindric tube 4–6 mm., the teeth 2–5 mm. long; petals purple or whitish, the banner 18–21 mm. long; pod erect, minutely stipitate, obliquely ovoid-oblong, abruptly cuspidate-beaked, 8–10 mm. long, 3–4 mm. in diam., triquetrously compressed with blunt lateral angles, grooved dorsally, fully 2-locular, the valves chartaceous, densely white-pilose.—Stiff clay soil moist in spring, ca. 5500 ft. in Calif.; Sagebrush Scrub; Lassen Co. (Madeline Plains); to Yukon, Man., New Mex., where found in many habitats. May–July.

30. **A. Andersònii** Gray. Perennial from a taproot and shortly forking aerial caudex, gray villous-hirsute throughout; stems erect and ascending in tufts, 8–20 cm. long; stipules dimorphic, the lower connate into a scarious sheath, the upper herbaceous, free or nearly so; lvs. 3–10 cm. long, with 11–21 oblanceolate to obovate, mostly acutish lfts. 5–13 mm. long; peduncles equaling the lvs.; racemes rather densely 12–26-fld., 3–8 cm. long in fr., the fls. at length declined; calyx-tube campanulate, 3.5–4.5 mm., the subulate-setaceous or filiform teeth 2.5–4.5 mm. long; petals ochroleucous or tinged with lavender, the banner 9.5–14.5 mm. long; pod sessile, loosely reflexed, lance-oblong in profile, incurved, 10–18 mm. long, 3–4.5 mm. broad, strongly compressed, narrowly grooved dorsally, fully 2-locular, the papery valves straw color, villosulous; $2n = 24$ (Head, 1957).—Sandy flats, hillsides and valley floors, 4300–7200 ft.; Sagebrush Scrub; e. slope of Sierra Nevada in Mono Co., thence n. through w. Nev. to Lassen Co. May–June.

31. **A. sepúltipes** (Barneby) Barneby. [*A. Andersonii* var. *s.* Barneby.] Perennial from a buried root-crown, strigulose-villosulous nearly throughout, the herbage silvery-silky, the infl. villous; stems ascending, subterranean for a space of 1–several cm., 1.5–3.5 dm. long; stipules dimorphic, the lower amplexicaul and connate into a scarous sheath, the upper ones nearly free; lvs. 2–8 cm. long, with 11–17 obovate-cuneate or oblanceolate truncate-retuse lfts. 5–13 mm. long; peduncles erect, 5–11 cm. long; racemes 10–30-fld.; calyx-tube campanulate, 5–6.5 mm., the linear-caudate teeth 3.5–6.5 mm. long; petals pale pinkish-lilac, the banner 13–17.5 mm. long; pod like that of *A. Andersonii* but a little broader.—Dry benches and canyon banks, in granite sand, 6000–6400 ft.; Pinyon Juniper Wd., Sagebrush Scrub; foothills of Sierra Nevada from Independence Creek s. to Lone Pine, Inyo Co. May–June.

32. **A. Aústinae** Gray. Related to the two preceding, but dwarf, tufted, matted or subpulvinate, the stems 0–11 cm. long; stipules all scarious and connate; lvs. 1–4 cm. long, with 7–13 crowded elliptic mostly acutish lfts. 1.5–9 mm. long; peduncles 1–4 cm. long; racemes subcapitately 4–10-fld.; petals whitish or lilac-tinged, the banner 8.5–11.5 mm. long, like the wings villosulous dorsally; pod scarcely exserted, oblong-ovoid, 6–7 mm. long, 3–3.5 mm. in diam., fully 2-locular, the valves tomentulose.—Dry exposed crests and ridges, 8800–10,500 ft.; Subalpine F.; peaks of the n. Sierra Nevada (Tinker's Knob, Placer Co.; Mts. Stanford and Lola, Nevada Co.; Mt. Tallac, Eldorado Co.); Mt. Rose, Nev. June–August.

33. **A. cìmae** Jones. Perennial from a stout taproot, except for the calyx nearly glabrous, the herbage glaucescent; stems diffuse, 6–25 cm. long; stipules membranous,

free, the lowest ovate obtuse, up to 1 cm. long; lvs. 5–11 cm. long, with 11–21 obovate to suborbicular retuse lfts. 5–20 mm. long; peduncles 3–8.5 cm. long; racemes loosely 10–25-fld., 4–12 cm. long in fr.; calyx-tube deeply campanulate, 4.5–5.5 mm., the teeth 1.5–2.5 mm. long; petals purple with white wing-tips, the banner 12–15 mm. long; pod stipitate, the thick stipe 6–8 mm. long, horizontally spreading, abruptly expanded into the incurved-erect, oblong-ellipsoid obcompressed body 1.5–2.5 cm. long, 8–12 mm. in diam., the green purple-dotted fleshy glabrous valves becoming woody and straw color, inflexed as a complete septum.—Very local, stiff calcareous soils of mesas or stony hillsides, 4700–6000 ft.; Sagebrush Scrub, Joshua Tree Wd.; e. Mojave Desert (New York Mts. and Mid Hills). April–May.

Var. **sufflàtus** Barneby. Pod larger, greatly inflated, the stipe 5–12 mm. long, the body 3–3.7 cm. long, 13–21 mm. in diam., the valves papery.—Rocky hillsides, 5000–6000 ft.; Pinyon-Juniper Wd.; e. slope of Inyo Mts. about the head of Saline V., Inyo Co. May.

34. **A. Minthórniae** (Rydb.) Jeps. var. **villòsus** Barneby [*A. M.*, Calif. refs.] Perennial, usually robust, villous throughout with spreading or curly hairs, the stems canescent at least at base; stems erect or ascending 4–35 cm. long; stipules free, the lowest conspicuous, several-nerved, imbricated, 8–12 mm. long, the upper shorter; lvs. 5–18 cm. long, with 9–17 ovate-elliptic subtruncate to acute lfts. 5–26 mm. long; peduncles erect, 5–16 cm. long; racemes loosely 5–35-fld., 2.5–25 cm. long in fr.; calyx-tube subcylindric, 4.5–6.5 mm., the teeth 1.5–2.5 mm. long; petals pink-purple with white wing-tips or livid, the banner 12–17 mm. long; pod erect or ascending, subsessile, narrowly oblong, straight or slightly incurved, 12–23 mm. long, 4–6 mm. in diam., subterete becoming a little laterally compressed, the leathery valves softly long-villous with pale or dark hairs, inflexed as a complete septum.—Open stony hillsides and along draws in canyons of desert mts., chiefly on limestone, 4400–7800 ft., Pinyon-Juniper Wd.; e. Mojave Desert (New York, Clark, Kingston, and Inyo mts.); e. slope of San Bernardino Mts. (Cushenbury canyon); sw. Nev. Late March–June.

35. **A. málacus** Gray. Perennial from a taproot, hirsute throughout with coarse spreading hairs, the lfts. sometimes glabrous above; stems stout, erect or assurgent, 3–23 cm. long; stipules free, membranous, conspicuous, ovate to acuminate, 6–15 mm. long; lvs. 4–15 cm. long, with 9–21 obovate-cuneate, obtuse or truncate lfts. 5–15 mm. long; peduncles stout, 4–10 cm. long; racemes many-fld., rather dense, becoming 3–15 cm. long in fr., the fls. declined; calyx-tube cylindric, 6–8 mm., the teeth 2–4 mm. long; petals reddish-violet, the banner 16–18 mm. long; pod reflexed, shortly stipitate, the stipe 1–2 mm. long, the body narrowly oblong, distally acuminate, falcately curved, 1.5–3 cm. long, 4–6 mm. in diam., compressed-triquetrous, narrowly grooved dorsally, fully 2-locular, the papery mottled valves shaggy-hirsute.—Plains and stony hills, in sand or clay on volcanic substrata, 4000–7500 ft.; Pinyon-Juniper Wd., Sagebrush Scrub; Lassen and Modoc cos., s. through w. Nev. to near Mono Lake, Mono Co.; to e. Ore. and sw. Ida. May–June.

36. **A. Làyneae** Greene. [*A. malacus* var. L. Jones.] Perennial from slender deeply seated horizontally creeping rhizomes beset with pubescent buds and adventitious roots, villous-hirsute with ascending or tangled hairs, greenish or canescent; stems arising singly or few together, above ground stout and angular, 2–16 cm. long; stipules free, acuminate, 4–10 mm. long; lvs. 6–16 cm. long, with 13–21 obovate obtuse or emarginate lfts. 5–23 mm. long; peduncles stout, often reclinate in fr., 5–14 cm. long; racemes 15–45-fld., loose and 5–20(–30) cm. long in fr.; calyx-tube deeply campanulate, black-hirsute, 5–7 mm., the teeth 1–2 mm. long; petals ochroleucous tipped with dull purple, the banner 13–18 mm. long; pod arcuate-ascending or rarely reflexed, sessile, narrowly sickle-shaped, cuneate at base, long-acuminate distally, 3–6.5 cm. long, 3.5–8 mm. in diam., incurved through ⅓ to nearly a complete circle, triquetrous with blunt angles, grooved dorsally, the leathery mottled valves hirsute, inflexed as a narrow but commonly complete septum.—Sandy flats and mesas, often in large colonies, 1500–5100 ft.; Creosote Bush Scrub; well distributed over the Mojave Desert, from Twentynine Palms to Owens V. and Death V.; to sw. Nev., ne. Ariz. March–May.

37. **A. Tidestròmii** (Rydb.) Clokey. [*A. Marcus-Jonesii* Munz, in part.] Subacaulescent tufted perennial, villous-tomentulose with curly hairs; stems rarely up to 7 cm. long; stipules free, 3–7 mm. long, often imbricated; lvs. 4–15 cm. long, with 7–19 suborbicular to obovate obtuse lfts. 4–13 mm. long; peduncles 3–12 cm. long, reclinate

in fr.; racemes loosely 5–13-fld., 2.5–6 cm. long in fr.; calyx-tube subcylindric, 5–8 mm., the teeth 1–2 mm. long; petals sordid, the keel dull purple-tipped, the ochroleucous or lurid-purple banner 13–16 mm. long; pod spreading, sessile, oblong-ellipsoid incurved through half to more than a complete circle, laterally compressed at base and in the acuminate beak, elsewhere obcompressed, 2.5–5 cm. long, 8–15 mm. in diam., 2-carinate by the prominent sutures, 1-locular, the fleshy strigulose valves becoming stiffly leathery and rugulose on the angles.—Sandy or gravelly washes and desert fans, mostly on limestone, 1000–3000 ft.; Creosote Bush Scrub; e. Mojave Desert (New York and Clark mts.), e. slope of San Bernardino Mts. (Cushenbury Canyon); s. Nev. April–May.

38. **A. tephròdes** Gray var. **brachylòbus** (Gray) Barneby. [*A. remulcus* Jones.] Habit of the preceding, but less densely pubescent, the hairs sometimes straight; calyx commonly strigulose; pod oblong, straight or nearly so, truncate-obtuse at base, obcompressed except for the deltoid laterally compressed beak, 2–2.5 cm. long, the leathery valves strigulose.—Widespread and variable sp. of Ariz. and adjoining states, usually of Pinyon-Juniper Wd., Yellow Pine Forest, once collected on the Colo. R. near Needles, perhaps a waif. April–May.

39. **A. Newbérryi** Gray. [*A. N.* var. *castoreus* Jones.] Low acaulescent tufted perennial, appressed silky, except the pod, or with some hirsute hairs on the petioles, which persist as a thatch on the heavy root-crown; stipules free, imbricated, 5–9 mm. long; lvs. 2–13 cm. long, with 3–13 ovate-elliptic obtuse lfts. 4–15 cm. long; peduncles scapose, shorter than the lvs.; racemes shortly 3–8-fld.; calyx-tube cylindric, 10–11 mm., the teeth 2–5 mm. long; petals pink-purple, sometimes pale, the banner 22–27 mm. long; pod spreading, sessile, ovoid and incurved into a stiff deltoid laterally compressed beak, obcompressed below, 1.5–2.5 cm. long, 8–10 mm. in diam., 1-locular, the leathery valves concealed by a dense hirsute-tomentose vesture about 2 mm. thick.—Dry stony hills, mostly above 4000 ft.; Sagebrush Scrub, Pinyon-Juniper Wd.; e. Mojave Desert (New York and Providence mts.) to Death V., the Inyo and White mts.; to e. Ore., s. Ida., n. Ariz. and nw. New Mex. April–June.

40. **A. coccíneus** Bdg. [*A. grandiforus* Wats., not L. *A. Purshii* var. *c.* Parry.] Habitally similar to the last, but the dense pubescence more tangled and tomentose; lfts. 7–15; calyx-tube reddish, 12–16 mm. long; petals narrow, nearly erect, scarlet drying crimson, of equal length, 3.5–4 cm. long; pod ovoid-acuminate or fusiform, 2.5–4 cm. long, 9–12 mm. in diam., gently incurved, somewhat obcompressed, 1-locular, silky-villous and tomentose with hairs 2–3 mm. long.—Gravelly ridges and canyons of desert-bordering mts., 2100–7000 ft., occasionally lower on outwash fans; mostly Pinyon-Juniper Wd.; Owens V. to Death V., s. along the edge of the deserts to n. Lower Calif.; sw. Ariz. March–May.

42. **A. Púrshii** Dougl. Low acaulescent or shortly caulescent tufted or matted perennials, similar to the last 3 in habit; pubescence arachnoid-villous or tomentose with fine tortuous hairs; stipules membranous, often attenuate; racemes subcapitately 2–10-fld.; pod obliquely ovoid to ovoid-acuminate, variably incurved into a stiff deltoid, laterally compressed beak, the fleshy valves becoming leathery, concealed by the dense shaggy vesture of silky or cottony usually entangled hairs 2–5 mm. long, not or little inflexed; $2n = 22$ (Head, 1957).

a. Fls. large, the calyx-tube and teeth together 10–15 mm., the keel 14–19 mm. long or, if smaller, the lfts. less than 9 in most lvs.
  b. Petals ochroleucous, only the keel purple-tipped .... a. var. *Purshii*
  b. Petals pink or bright purple.
    c. Lfts. mostly 7–11; peduncles shorter than the lvs. .... b. var. *tinctus*
    c. Lfts. mostly 11–17; peduncles equaling or longer than the lvs. .... c. var. *longilobus*
a. Fls. smaller, the calyx-tube and teeth together 6–9 mm., the keel 8–12 mm. long.
  b. Pod moderately arched, 24–32-ovulate, the ventral suture nearly straight below, the beak incurved .... d. var. *lectulus*
  b. Pod strongly incurved throughout, usually through over half a circle, 14–20-ovulate
                                              e. var. *lagopinus*

a. Var. **Púrshii.** [*A. P.* var. *incurvus* (Rydb.) Jeps. *A. i.* Abrams, not Desf.] Dry plains and hillsides, 3500–7000 ft.; Sagebrush Scrub, Yellow Pine F.; ne. Calif. (Sierra to Trinity, Siskiyou, and Modoc cos.); to s. B.C., s. Alta., n. Colo. April–June.

b. Var. **tínctus** Jones. [*A. candelarius* Sheld. *Xylophacos subvillosus* Rydb. *A. Purshii* var. *gavisus* Jeps.] Dry hillsides, mostly 3000–8000 ft., but descending w. down the

Klamath R. as low as 1500 ft.; Sagebrush Scrub, Pinyon-Juniper Wd., Yellow Pine F.; ne. Calif. (Nevada to Siskiyou and Modoc cos.), White Mts., Mono Co.; w. Nev., n. to Wash. Intergrades with var. *lectulus.* May–June.

c. Var. **longílobus** Jones. [*A. leucolobus* ssp. *consectus* (Sheld.) Abrams. *A. inflexus* var. *flocculatus* and var. *ordensis* Jeps. *A. Purshii* var. *o.* Jeps.] Two races, perhaps distinct, one of the S. Coast Ranges, 4000–8000 ft. (San Carlos Range, San Benito Co., s. to Mt. Pinos and the Tehachapi Mts.), with petals of nearly equal length and pale in color, the other of the Mojave Desert, 4000–7000 ft. (Ord, Argus, and Panamint mts. and Darwin Mesa), with strongly graduated bright purple petals. April–June.

d. Var. **léctulus** Jones. [*Xylophacos argentinus* Rydb. *A. Jonesii* Abrams.] Dry rocky ridges above timber line, openings in pinewoods, descending from Alpine Fell-fields into Pinyon-Juniper Wd. and Yellow Pine F., rarely into Sagebrush Scrub, 6000–11,000 ft.; e. slope and crest of Sierra Nevada from Alpine to Inyo cos.; Sweetwater Mts., Mono Co.; San Bernardino Mts.; N. Coast Ranges, Mendocino to Siskiyou cos.; Mt. Shasta; Warner Mts., Modoc Co.; to w. Nev. and e. Ore. In ne. Calif. passing into var. *tinctus.* April–August.

e. Var. **lagopìnus** (Rydb.) Barneby. [*A. l.* Peck.]—Plains, often in volcanic sand, 4000–4500 ft. in Calif.; Sagebrush Scrub; n. Modoc Co. to e. Ore. May–June.

43. **A. leucólobus** Jones. [*A. Purshii* var. *l.* Jeps.] Closely resembling the cismontane form of *A. Purshii* var. *longilobus,* the petals poorly graduated, the banner 16–18 mm. long; pod lance-oblong in outline, 13–22 mm. long, about 7 mm. in diam., obcompressed, grooved dorsally, the leathery valves densely but quite shortly tomentulose with matted hairs about 1 mm. long, inflexed as a narrow but complete septum.—Openings in sandy wds. and stony lake-shores in the mts. overlooking the deserts, 6000–7800 ft.; Yellow Pine F., Sagebrush Scrub; San Antonio, San Bernardino, and Santa Rosa mts. May–July.

44. **A. subvestìtus** (Jeps.) Barneby. [*A. leucolobus* var. *s.* Jeps.] Like the two preceding in habit and vesture; fls. small, the calyx-tube 5.5–6.5 mm., the teeth 1.5–3 mm. long; pod obliquely ovoid, obtuse at base, strongly obcompressed in the lower half, abruptly incurved into a deltoid beak, 8–15 mm. long, 4–6.5 mm. in diam., 1-locular, the thinner stiffly papery valves thinly villous-tomentulose.—Sandy meadows and Sagebrush flats, s. Sierra Nevada (S. Fork Kern River, 8000–8500 ft., Tulare Co.; Piute Mts., about 5000 ft., Kern Co.). June–July, perhaps earlier s.

45. **A. lentifórmis** Gray. Perennial from a woody taproot, finely villosulous throughout with curly hairs; stems many, diffuse, ca. 1 dm. long; stipules dimorphic, the lowest connate, the upper free; lvs. 1.5–3 cm. long, with 11–13 oblanceolate obtuse lfts. 3–8 mm. long, glabrescent along the midrib above; peduncles 5–18 mm. long; racemes loosely 5–8-fld., less than 1 cm. long; calyx-tube campanulate, 2 mm. long, ca. equaling the teeth; petals whitish, the banner about 7 mm. long, pod pendulous, sessile, oblong-ellipsoid, scarcely beaked, 8–9 mm. long, 3 mm. broad, strongly compressed, carinate by the ventral suture, narrowly grooved dorsally, fully 2-locular, the valves papery, hirsutulous.—Dry hills, ca. 5000 ft.; Sagebrush Scrub; se. Plumas Co. (Red Clover V.; Portola), rare. May–June.

46. **A. Johánnis-Howéllii** Barneby. Slender perennial with a branched caudex, loosely strigulose, greenish; stems diffuse, wiry, branched below, 5–20 cm. long; stipules scarious, 1.5–4 mm. long, the lowest connate into a sheath, the upper free through half their length; lvs. 4–6 cm. long, with 13–23 oblanceolate obtuse lfts. 1.5–6 mm. long, glabrous above; peduncles 1–2.5 cm. long; racemes loosely 6–12-fld., 1.5–4 cm. long in fr.; calyx-tube campanulate 1.5–2 mm. long, the teeth about as long; petals ochroleucous, the banner 5–5.5 mm. long; pods pendulous, stipitate, the stipe 0.5–2 mm. long, the body lunately ellipsoid, 7–10.5 mm. long, 2.5–3 mm. in diam., triquetrous with blunt lateral angles, widely grooved dorsally, fully 2-locular, the papery, translucent valves straw color, strigulose.—Sandy ground, ca. 7000 ft.; Sagebrush Scrub; Upper Owens R., Mono Co. (Long V.). June–July.

47. **A. Lémmoni** Gray. Slender prostrate perennial from a woody taproot, green but thinly strigulose; stems radiating, 1–4 dm. long, floriferous from near the base; stipules free, 2–5 mm. long; lvs. 1.5–4.5 cm. long, with 9–15 elliptic mostly acute lfts. 2–11 mm. long, the rachis subfiliform; peduncles much shorter than the leaves, often paired in the axils; racemes loosely 5–13-fld., 1 cm. long or less in fr.; calyx-tube campanulate, ca. 2 mm., the teeth ca. 1.5 mm. long; petals whitish or lilac-tinged, the banner 5–6 mm.

long; pod spreading, sessile, ovoid-oblong, 4–7 mm. long, 1.5–2.5 mm. in diam., compressed-triquetrous, deeply grooved dorsally, fully 2-locular, the papery valves strigulose.—Moist grassy or sedgy stream- and lake-shores, 4200–7000 ft.; Sagebrush Scrub; e. base and lower slopes of the Sierra-Cascade axis from Mono Co. (Hilton Creek) and Sierra Co., n. to e. Ore., nw. Nev. Late May–August.

48. **A. monoénsis** Barneby. Perennial from a short buried caudex, silky-cinereous with contorted hairs; stems prostrate or weakly ascending, naked at base, 1–2 dm. long; stipules dimorphic, the lowest connate into a loose scarious sheath, the upper herbaceous, nearly free; lvs. 1–3 cm. long, with 9–15 crowded obovate obtuse folded lfts. 2–8 mm. long; peduncles 1–4.5 cm. long; racemes subcapitately 6–12-fld., ca. 1 cm. long in fr.; calyx-tube campanulate, 3–4.5 mm., the teeth 1.5–2 mm. long; petals whitish suffused with lilac, the banner 10–13 mm. long; pod spreading, sessile, obliquely lance-ovoid, strongly incurved, 1.5–2 cm. long, 6–9 mm. in diam., somewhat obcompressed except in the acuminate beak, the firmly papery shortly villous valves inflexed as a partial septum.—Gravelly or sandy flats, sometimes sheltering under and scrambling through low sage, 7500–7900 ft., Pinyon-Juniper Wd.; e. slope of Sierra Nevada in Mono Co. (Crestview; Mammoth), very local. June–August.

49. **A. Ravènii** Barneby. Very slender delicate perennial from a taproot and buried root-crown, silvery-strigulose, the lfts. a trifle more densely and loosely pubescent above than beneath; stems 1.5–10 cm. long, subterranean for 1–6 cm., prostrate above ground, simple; stipules small, connate; lvs. 0.5–2.5 cm. long, with 7–13 broadly oblong-obovate or suborbicular retuse lfts. 1–3.5 mm. long; peduncles subfiliform, 1–5.5 cm. long, mostly longer than the lf.; racemes loosely 2–6-fld., the axis less than 1 cm. long; calyx-tube campanulate 2.5–3.3 mm., the teeth 0.6–1.2 mm. long; petals whitish, the banner 6–8.5 mm., the keel 4.5–5.5 mm. long; pod ascending, humistrate, sessile, obliquely ovoid, turgid, gently incurved, obtuse at base, contracted at apex into a laterally flattened deltoid beak, otherwise obcompressed, carinate by the prominent but depressed ventral suture, plano-convex dorsally, the papery mottled valves thinly strigulose, inflexed as a narrow partial septum.—Open stony slopes, on metamorphic substrata, about 11,250 ft.; Alpine Fell-fields; crest of the Sierra Nevada in e. Fresno Co. (n. of Sawmill Pass). July–August.

50. **A. Pulsíferae** Gray. Slender perennial, from a deeply buried rarely superficial root-crown, softly villous with long fine widely spreading hairs up to 1–1.6 mm. long; stems diffuse, naked below, branched above, 1–3 dm. long; stipules small, the lowest amplexicaul and either free or connate into a short sheath; lvs. 2–4.5 cm. long, with 7–13 oblanceolate to cuneate-obovate, mostly retuse lfts. 2–12 mm. long; peduncles subfiliform, 4–30 mm. long; racemes shortly but loosely 3–12-fld., 4–15 mm. long in fr.; calyx-tube campanulate 1.5–2.5 mm., the subulate-setaceous teeth mostly 2–3.5 mm. long; petals whitish, purple-veined, the banner 6–8.5 mm. long; pod horizontal or declined, sessile, bladdery-inflated, asymmetrically ovoid, obtuse or cuneate at base, abruptly deltoid-beaked, 0.8–2 cm. long, 6–11 mm. in diam., the ventral suture straight to strongly concave, the dorsal gibbous-convex, the thinly papery valves at length straw color, long-villous, not inflexed.—Sandy valley floors and hillsides, on basaltic substrata, 4300–5500 ft.; Sagebrush Scrub; Sierran foothills and adjoining plains, ne. Sierra, e. Plumas Co. and Lassen Co.; adjacent Nev. June–August.

Var. **Suksdórfii** (Howell) Barneby. [A. S. Howell.] Herbage shortly villosulous, the hairs incurved-ascending or subappressed and not over 0.5–0.75 mm. long; calyx-teeth mostly shorter, 1.4–2.1 mm. long.—Stony plains and openings in pine forests, on basalt, 4300–5000 ft.; Sagebrush Scrub, Yellow Pine F.; nw. Plumas to n. Lassen and probably e. Shasta cos.; also near Mt. Adams, e. Wash. June–August.

51. **A. Clevelándii** Greene. Tall perennial from a taproot, appearing glabrous, but the leaves thinly strigulose beneath; stems erect, 3–10 dm. long; stipules free; lvs. 4–14 cm. long, with 13–25 narrowly elliptic or oblanceolate obtuse or acutish lfts. 5–23 mm. long; peduncles from the upper axils, 5–15 cm. long; racemes very narrow, elongate, 20–95-fld., 1–3 dm. long in fr., the fls. early reflexed; calyx-tube campanulate ca. 2 mm., the teeth ca. 2 mm. long; petals whitish, the banner 5–6 mm. long; pod reflexed, sessile, obliquely ovoid, somewhat incurved, acute, 4.5–7 mm. long, 1.5–2 mm. in diam., triquetrous with blunt lateral angles, fully 2-locular, the thin-leathery valves green turning straw color, glabrous.—Gravel-bars and sandy banks of streams, chiefly on

serpentine; Foothill Wd.; local, 650–3000 ft., in n. Napa and e. Lake cos.; also at 4450 ft. in San Carlos Range, San Benito Co. June–Sept.

52. **A. Cóngdoni** Wats. Perennial, loosely villous-pilosulous throughout, the stems canescent, the herbage cinereous or greenish; stems diffuse or ascending from a short caudex, 1.5–7.5 dm. long; stipules free, 2–8 mm. long; lvs. 4–12 cm. long, with 17–35 obovate-elliptic to suborbicular, truncate or shallowly retuse lfts. 4–15 mm. long; peduncles ascending, 7–18 cm. long; racemes loosely 12–35-fld., the fls. declined, the axis 5–20 cm. long in fr.; calyx-tube campanulate 4–5 mm., the teeth 1.5–2.5 mm. long; petals whitish, the banner 10.5–15.5 mm. long; pod spreading or declined, shortly stipitate, the stipe 1–2.5 mm. long, the body linear in outline, 2–3.5 cm. long, 2.5–3 mm. in diam., falcate to nearly straight, triquetrously compressed with blunt lateral angles, narrowly grooved dorsally, 2-locular or nearly so, the firmly papery valves strigulose-villosulous; $2n = 26$, 52 (Head, 1957).—Dry rocky and brushy hillsides, on metamorphic bedrock, 550–2000 ft., Foothill Wd.; Sierran foothills, from Mokelumne s. to Tule R., Amador to Tulare cos. March–May.

53. **A. umbràticus** Sheld. [*A. sylvaticus* Wats., not Willd.] Slender perennial, glabrous except for a few scattered hairs on calyx and lf.-margins, pale or dark green; stems ascending, stramineous, 2.5–5 dm. long; stipules free, the lowest amplexicaul, the upper acuminate, 3.5–9 mm. long; lvs. 4–12 cm. long, with 13–23 oblong or ovate retuse thin-textured lfts. 6–18 mm. long; peduncles 5–12 cm. long; racemes loosely 10–25-fld., little elongating, 2–5 cm. long in fr., the fls. declined; calyx-tube campanulate 3–4 mm., the teeth 2–3 mm. long; petals white, the banner 10–14 mm. long; pod horizontal or declined, stipitate, the stipe 1–2 mm. long, the body narrowly oblong in profile, gently incurved, abruptly subulate-beaked, 13–24 mm. long, 2.5–3.5 mm. in diam., strongly compressed, narrowly grooved dorsally, fully 2-locular, the stiffly papery glabrous valves green turning straw color.—Glades and thickets, 600–4000 ft., N. Oak Wd.; n. Humboldt Co. (Bald Mt.; Redwood Creek); w. Ore. May–July.

54. **A. agnícidus** Barneby. Robust perennial, green but thinly villous upward with long loose weak hairs; stems ascending, hollow, 4–9 dm. long, straw color, glabrous below; stipules acuminate, 4–15 mm. long, free; lvs. 5–16 cm. long, with 19–27 thin ovate to lanceolate obtuse or acutish lfts. 5–22 mm. long, glabrous above, ciliate; peduncles 5–13 cm. long; racemes densely 15–40-fld., little elongating, the fls. declined; calyx-tube campanulate, 3–4.3 mm., the teeth 3.3–5 mm. long; petals white, the banner 9–11 mm. long; pod declined, subsessile, lunately lanceolate in profile, 11–15 mm. long, 3–3.5 mm. in diam., compressed-triquetrous, narrowly grooved dorsally, fully 2-locular, the papery valves pilosulous.—Disturbed wds., ca. 2500 ft.; near Bear Buttes, s. of Miranda, Humboldt Co. June–August.

55. **A. Brauntònii** Parish. Stout perennial with an at length indurated woody trunk, villous-tomentulose throughout, the stems canescent, the herbage sometimes greenish; stems erect, 7–15 dm. long, hollow; stipules membranous, free, 3–10 mm. long; lvs. 4–14 cm. long, with 25–33 ovate or oblong lfts. 3–20 mm. long; peduncles from the upper axils, 2.5–8 cm. long; racemes densely 25–60-fld., narrowly oblong, 3.5–14 cm. long, the subsessile fls. reflexed and retrorsely imbricated; calyx-tube campanulate, turgid in age, 3.5–4 mm. long, the filiform teeth 2.5–5 mm. long; petals dull lilac, the banner 9–12 mm. long; pod deflexed, sessile, lunately oblong in profile, abruptly narrowed into a slender cuspidate beak, 6.5–9 mm. long, 3–4 mm. in diam., grooved dorsally, the stiffly papery valves villous-tomentulose, inflexed through the lower ⅔ as a complete septum.—Brushy hillsides, sometimes abundant on firebreaks, 50–1500 ft.; Chaparral; low hills bordering the plain of Los Angeles: Santa Monica Mts., San Gabriel Mts. (Monrovia), Santa Ana Mts. (Sierra Peak). Feb.–June.

56. **A. Tráskiae** Eastw. Diffuse perennial, white villous-tomentose throughout; stems leafy, with short sterile spurs below, 1.5–2 dm. long; stipules free; lvs. 5–10 cm. long, with 21–29 ovate-elliptic obtuse or retuse lfts. 5–15 mm. long; peduncles 4–14 cm. long; racemes loosely 12–30-fld., 2.5–8 cm. long in fr.; calyx-tube deeply campanulate 5–6 mm., the teeth 2.5–3.5 mm. long; petals ochroleucous, the banner 14–17.5 mm. long; pod pendulous, stipitate, the stipe 4–8.5 mm. long, the body obliquely oval-oblong in outline, gently incurved, abruptly acuminate into a narrow beak, 8–16 mm. long, 3–5.5 mm. in diam., triquetrously compressed with blunt lateral angles, narrowly grooved dorsally, fully 2-locular, the leathery valves shortly tomentulose.—Dunes and

sandy bluffs and gulches, 100–650 ft.; Coastal Strand, Coastal Sage Scrub; San Nicolas and Santa Barbara ids. March–July.

57. **A. Nevínii** Gray. Closely related to the last, but lfts. fewer, 11–21; fls. smaller, the banner 10–12 mm. long; pod glabrous.—Dunes and sandy bluffs; Coastal Strand; San Clemente Id. March–July.

58. **A. miguelénsis** Greene. [*A. vestitus* var. *m.* Jeps.] Perennial, densely white-tomentulose throughout; stems decumbent and assurgent, 1.5–3 dm. long; stipules deltoid, 2–5.5 mm. long, shortly connate, tomentulose externally; lvs. 4–12 cm. long, with 21–27 oblong-obovate, obtuse or retuse lfts. 6–22 mm. long; peduncles 5–13 cm. long; racemes loosely 10–30-fld., 2–9 cm. long in fr.; calyx-tube campanulate 4–5 mm., the teeth 2–3.5 mm. long; petals yellowish, the banner 12.5–16 mm. long; pod spreading, sessile, bladdery-inflated, ovoid with deltoid beak, 1.5–3.5 cm. long, 13–20 mm. in diam., the papery valves finely tomentulose, not or very obscurely inflexed.—Dunes, sandy bluffs and talus beneath cliffs, below 500 ft.; Coastal Strand; San Miguel, Santa Rosa, Santa Cruz, Anacapa, and San Clemente ids. Spring, and occasional throughout the year.

59. **A. Nuttálli** (T. & G.) J. T. Howell. [*A. Menziesii* Gray. *A. vestitus* var. *M.* Jones. *Phaca N.* T. & G.] Robust perennial, becoming indurated and woody below, with many ascending branched stems forming wide bushy plants 3–8 dm. in diam., becoming prostrate and matted on wind-swept bluffs; pubescence of the herbage villous or villosulous, either dense and canescent or nearly lacking, the lfts. either glabrous or pubescent above; stipules scarious, 3–14 mm. long, strongly connate in vernation but sometimes ruptured by the growing stems and appearing free; lvs. 4–15 cm. long, with 23–43 cuneate-obovate to oblong retuse lfts. 3–25 mm. long; peduncles 6–15 cm. long; racemes densely many-fld., the fls. nodding, the axis 3–10 cm. long in fr.; calyx-tube campanulate 4–6 mm., the teeth 1–3 mm. long; petals greenish-white, the keel tipped and the banner sometimes suffused with dull lilac, the banner 11–14.5 mm. long, scarcely surpassing the wings; pod horizontal or declined, sessile, obliquely ovoid with broad short deltoid beak, bladdery-inflated, 2.5–6 cm. long, 1.5–2.5 cm. in diam., 1-locular, the ventral suture nearly straight, the dorsal strongly convex, the papery valves straw color, finely strigulose-villosulous, sometimes glabrate in age; ovules 28–36.—Dunes and bluffs along the ocean, 10–300 ft.; Coastal Strand; Monterey Bay s. to Point Conception. Jan.–Oct.

Var. **virgàtus** (Gray) Barneby. [*A. Crotalariae* var. *v.* Gray. *A. Menziesii* ssp. *v.* Abrams. *A. franciscanus* Sheld. *A. C.* var. *f.* Jones.] Stems often stouter, but sometimes prostrate, 4–10 dm. long; stipules often appearing free from the first (except on some slender lateral spurs or branchlets), but the lowest connate in vernation; pubescence sparse; racemes commonly looser, 40–90(–125)-fld., 6–19 cm. long in fr.; pod 2.5–4 cm. long, 1.5–2.5 mm. in diam., glabrous or villosulous; ovules 14–21.—Brushy slopes and sandy fields both along and back from the ocean, below 400 ft.; Coastal Strand; Marin to San Mateo cos.; reported from Mendocino Co.; Angel I., San Francisco Bay; inland to the Bay Shore on San Francisco Peninsula and formerly on the e. shore at Alameda. April–June, occasional later.

60. **A. pomonénsis** Jones. Coarse procumbent perennial, the stout glabrous stems 2.5–8 dm. long, the herbage green, thinly strigulose beneath; stipules subherbaceous, the lowest amplexicaul but free, the upper decurrent, reflexed; lvs. 8–20 cm. long, with 25–41 oblong-oblanceolate truncate or retuse lfts. 6–37 mm. long; peduncles stout, 5.5–14 cm. long; racemes densely 10–45-fld., becoming interrupted below and 3.5–9 cm. long in fr., the fls. declined; calyx-tube campanulate, gibbous dorsally, 3.5–5 mm., the teeth 1–3 mm. long; petals greenish-white or ochroleucous, the banner 11–15 mm. long, the wings ca. as long; pods densely racemose, spreading or ascending, obliquely ovoid, bladdery-inflated, 2.5–4.5(–5) cm. long, 1–2 cm. in diam., the ventral suture nearly straight, the dorsal strongly convex, the stiffly papery valves minutely strigulose, sometimes glabrate in age, not inflexed.—Sandy valleys, fallow fields, grassy hillsides, usually in low ground, sometimes a weed in orchards, 10–650 ft.; Coastal Sage Scrub, Foothill Wd., etc.; near the coast from s. San Luis Obispo to Ventura cos. and interior cismontane s. Calif. from Los Angeles to Banning; to n. L. Calif. March–May, occasional later in irrigated land.

61. **A. pycnostàchyus** Gray. Stout perennial, the herbage canescently pilose-tomentulose

with short twisted hairs; stems several, erect, purplish, hollow, glabrescent, 4–9 dm. long; stipules membranous, glabrescent, dimorphic, the lowest and those of young shoots or spurs connate into a sheath, the sheath often ruptured by the growing stem, the upper free or nearly so; lvs. 3–15 cm. long, with 25–41 narrowly oblong obtuse crowded lfts. 5–30 mm. long; peduncles erect from the upper axils, 3–10 cm. long; racemes many-fld., dense, the declined fls. forming oblong-cylindric heads 2–9 cm. long; calyx-tube campanulate 4–5 mm., the teeth 1.5–3 mm. long; petals ochroleucous, the banner 7–10 mm. long; pods retrorsely imbricated, sessile, ovoid and somewhat inflated, beaked by the persistent style, 6–9 mm. long, 3.5–6 mm. in diam., a little compressed, both sutures acutish, the papery glabrous valves not inflexed; ovules 2–5, commonly 4.—Moist depressions behind dunes or barrier-beaches and banks of creeks opening into the ocean, 5–30 ft.; Coastal Salt Marsh; occasional from Cape Mendocino, Humboldt Co., to San Mateo Co. June–Oct.

Var. **lanosíssimus** (Rydb.) Munz & McBurn. [*Phaca l.* Rydb.] Herbage densely white silky-tomentose; peduncles shorter, 2–4 cm. long; calyx-teeth 1.5 mm. long or less; pod 8–11 mm. long, thinly strigulose, 8–12-ovulate.—Coastal marshes, now very rare or extinct, Ventura and Los Angeles cos.

62. **A. asymmétricus** Sheld. [*A. leucophyllus* T. & G., not Willd. *A. leucopsis* var. *l.* Jones.] Stout perennial, canescent throughout with a coat of short appressed straight or incurved hairs; stems erect and ascending from an indurated base, hollow, 5–12 dm. long; stipules dimorphic, the lowest 5–14 mm. long, connate into a scarious sheath early ruptured by the growing stem, the upper shorter, free, deciduous; lvs. 6–21 cm. long, with 19–35 linear to narrowly oblong obtuse or acute lfts. 5–25 mm. long; peduncles strict, 6–24 cm. long; racemes loosely 15–45-fld., 5–17 cm. long in fr., the fls. nodding; calyx-tube deeply campanulate, gibbous at base, 5–7 mm., the teeth 1.5–4 mm. long; petals ochroleucous, little graduated, the banner 13–17.5 mm. long; pod pendulous on a filiform arched stipe 1.5–4 cm. long, the bladdery-inflated body obliquely ellipsoid to semiovoid, 2.5–4 cm. long, the papery valves strigulose, not or only vestigially inflexed.—Grassy hillsides, fields and rolling plains, often in somewhat alkaline soils, 200–2500 ft.; Foothill Wd., V. Grassland; Lower Great V. (Solano to Madera cos.) and inner S. Coast Ranges from Contra Costa to San Luis Obispo cos. April–July, occasionally Sept.–Oct.

63. **A. oxýphysus** Gray. Similar to the last, but the pubescence looser, villous or villosulous; lfts. 17–27, oblanceolate to elliptic, rarely glabrescent above; calyx villosulous, the deeply campanulate tube not strongly gibbous at base, 6–8.5 mm., the teeth 1.5–3.5 mm. long; banner 14–19 mm. long; pod pendulous, raised on a filiform hirsutulous stipelike gynophore 3–11 mm. long, the body obliquely or rhombically elliptic in outline, acuminate below, deltoid-beaked, 2.5–4.5 cm. long, 8–15.5 mm. in diam., laterally compressed but strongly turgid, the sutures both acute, the papery lustrous valves strigulose or rarely glabrate, not inflexed.—Common and gregarious; plains and low hills, 250–2200 ft.; V. Grassland, ascending rarely s. into Foothill Wd. up to 2750 ft.; w. side of San Joaquin V. and inner S. Coast Ranges from Mt. Hamilton Range, Stanislaus Co., s. to the upper Cuyama V., Santa Barbara Co., and the foothills of the Tehachapi Mts., Kern Co. Feb.–June.

64. **A. cúrtipes** Gray. [*A. leucopsis* var. *c.* (Gray) Jones. *A. l.* var. *brachypus* Greene. *A. l.* var. *c.* Sheld.] Perennial, loosely strigulose or subtomentulose, the stems commonly canescent, sometimes nearly glabrous, the herbage cinereous or green; stems erect and ascending from an indurated base, 2.5–4 dm. long, with leafy spurs in the lower axils; stipules mostly connate into a scarious sheath 2–12 mm. long, the upper often ruptured or free from the first; lvs. 5–16 cm. long, with 25–39 oblong to narrowly obovate obtuse lfts. 4–25 mm. long; peduncles 5.5–25 cm. long; racemes rather closely 15–35-fld., 2–11 cm. long in fr., the fls. declined; calyx-tube campanulate 4–5 mm., the teeth 1.5–3.5 mm. long; petals ochroleucous, the banner 13–16 mm. long; pod ascending, loosely spreading or declined, sessile but elevated on a stipelike gynophore 2.5–6 mm. long, the body obliquely ovoid with deltoid flattened beak, bladdery-inflated, 2.5–7 cm. long, the papery valves strigulose, not inflexed.—Dry rocky and grassy hills, canyon walls, sometimes on serpentine, rarely on dunes; Coastal Sage Scrub, Coastal Strand; perhaps three races: the typical, with canescent stems and pods 2.5–4 cm. long, near the coast, below 500 ft., from Cambria, San Luis Obispo Co., s. to Santa Maria, Santa

Barbara Co.; an insular phase, with nearly glabrous stems and green herbage, otherwise typical, on Santa Rosa and San Miguel ids. and the mainland near Point Arguello, Santa Barbara Co.; the third, with pods 4–7 cm. long, at ca. 1500 ft. in the mts. of San Benito Co. Jan.–June, sometimes again in fall.

65. **A. leucópsis** (Torr.) T. & G. [*Phaca canescens* Nutt., not H. & A. *A. l.* var. *lonchus* Jones.] Bushy-branched perennial, like the last in habit, strigulose-villosulous with mixed longer straight and shorter incurved or twisted hairs, the herbage greenish-cinereous or canescent, the lfts. sometimes glabrous above; stems 2–6 dm. long; stipules all free from the first, rarely (on the Channel Ids.) the lowest or nearly all connate into a sheath; lvs. 5–15 cm. long, with 21–39 ovate-oblong to oblanceolate lfts. 5–23 mm. long; peduncles 6–20(–30) cm. long; racemes 12–36-fld., 3.5–11 cm. long in fr., the fls. nodding; calyx-tube campanulate, oblique or dorsally gibbous at base, 3.5–5.5 mm., the teeth 1.5–3.5 mm. long; petals greenish-white, exceptionally lavender-tinged, the banner 11.5–19 mm. long; pod pendulous, stipitate, the stipe 5–15 mm. long, the obliquely ovoid bladdery-inflated body 2–4.5 cm. long, 1–2 cm. in diam., with nearly straight ventral and strongly convex dorsal sutures, the papery valves strigulose or rarely glabrate, not inflexed.—Sandy bluffs and low hills along the immediate coast, sometimes on shingle-banks behind barrier-beaches, s. extending up to 6 miles inland along seaward-running canyons, below 250 ft.; Coastal Sage Scrub, Coastal Strand; Point Mugu, Ventura Co. s. to n. Lower Calif.; Santa Cruz, Anacapa, and Santa Catalina ids., up to 1000 ft. Feb.–June, occasional in fall and winter.

66. **A. trichópodus** (Nutt.) Gray. [*Phaca t.* Nutt. *A. capillipes* Jones, not Fisch. *A. t.* var. *c.* Munz & McBurn.] Similar to the last; banner 11.5–15.5 mm. long; pod pendulous, stipitate, the filiform, purplish, strigulose stipe 6.5–17 mm. long, the body turgidly ellipsoid, symmetric or somewhat oblique, 1.5–3.5 cm. long, 5–10 mm. in diam., the ventral suture either straight or equally convex with the dorsal, the thinly papery valves straw color tinged with purple, lustrous, glabrous or rarely strigulose.—Shale or sandstone outcrops on brushy hills and ocean bluffs; mostly Coastal Sage Scrub; in three restricted areas of s. Calif.: immediate coastline of se. Santa Barbara and n. Ventura cos.; s. end Catalina Id.; inland on the Puente and Chino hills (Orange, Los Angeles, and San Bernardino cos.) up to 1100 ft. March–June, occasional later.

67. **A. Antisélli** Gray. [*A. trichopodus* var. *A.* Jeps. *A. A.* var. *phoxus* Jones. *A. Hasseanus* Sheld. *A. gaviotus* Elmer. *A. trichopodus* var. *g.* Jeps.] Close to the last; stems up to 10 dm. long, often forming wide bushy plants; racemes (10–)20–50-fld.; stipe of the pod 6–15 mm. long, the body linear-elliptic or oblanceolate in profile, 1.5–3.5 cm. long, 4–9 mm. in diam., laterally compressed and 2-sided, not inflated, carinate by the sutures, the valves glabrous or rarely puberulent.—Dry rocky or brushy hillsides, washes and canyons, mostly away from the immediate coast, 100–3000, rarely 4000 ft., sometimes abundant on firebreaks and road-cuttings; Chaparral, Foothill Wd.; common in the foothills back from the ocean around the plain of Los Angeles and San Fernando V. nw. through the Liebre, Topatopa, Santa Ynez, and San Rafael mts. to the Cuyama V. and n., becoming rarer, to the head of Salinas V., Los Angeles to cent. San Luis Obispo Co.; also extending feebly to the desert-edge in Antelope V. and coming out to coastal bluffs at 100–400 ft. in Coastal Sage Scrub between Santa Barbara and Point Conception. Feb.–June, sometimes in fall and winter.

68. **A. magdalènae** Greene var. **Peirsònii** (Munz & McBurn.) Barneby. [*A. P.* Munz & McBurn. *A. Crotalariae* var. *piscinus* of Jeps., not *A. Douglasii* var. *p.* Jones.] Stout perennial of short duration, flowering as a winter-annual, at length woody below, silky-canescent or cinereous throughout with subappressed hairs; stems erect, 2–7 dm. long, with short leafy spurs in the lower axils; stipules free, early deciduous; lvs. 5–15 cm. long, with broad flattened rachis and 8–12 irregularly inserted oblanceolate folded lateral lfts. 2–14 mm. long, the terminal one decurrent and continuous with the rachis; peduncles 7–10 cm. long; racemes 10–17-fld., becoming 4–6 cm. long in fr.; calyx-tube campanulate 3.5–4 mm., the teeth 1.5–5 mm. long; petals dull purple, the banner 10–14 mm. long; pod spreading, sessile, broadly ellipsoid with short deltoid beak, bladdery-inflated, 2–3.5 cm. long, 1.5–2 cm. in diam., the papery strigulose valves not inflexed.—Mobile dunes, below 800 ft.; Creosote Bush Scrub; s. Colo. Desert (Borrego V.; Algodones Sandhills).

69. **A. Douglásii** (T. & G.) Gray. [*Phaca D.* T. & G. *A. tejonensis* Jones. *A. Douglasii*

var. *megalophysus* (Rydb.) Munz & McBurn. *Phaca vallicola* Rydb. (an intergrade to var. *Parishii*).] Perennial with several procumbent stems radiating from the taproot, forming low clumps or loose mats 4–10 dm. in diam., the herbage green and glabrescent to densely silvery-strigulose; stipules free, the lowest papery and amplexicaul, the upper herbaceous, often deflexed; lvs. 5–15 cm. long, with 11–25 ovate or oblanceolate lfts. 6–25 mm. long, acute, obtuse or retuse; peduncles incurved-ascending, 2.5–9 cm. long; racemes loosely 10–35-fld., 2.5–12(–17) cm. long in fr., the fls. spreading; calyx-tube campanulate 3–4 mm., the subulate teeth 1.5–3 mm. long; petals ochroleucous, the suborbicular, strongly reflexed banner 8–13 mm. long, the keel narrowed at apex; pod spreading, sessile, obliquely ovoid-ellipsoid, 2.5–6 cm. long, the papery valves strigulose or glabrate in age, not inflexed.—Dry fields, grassy and brushy hillsides and rolling plains, from the valley floor upward into the pine-belt, 180–6800 ft.; V. Grassland, Foothill Wd., Yellow Pine F., rarely Joshua Tree Wd.; cent. Great V., Yolo to Merced cos., s. in the inner Coast Ranges from San Benito Co. to the head of San Joaquin V., the San Gabriel and San Bernardino mts., extending to the desert-edge near Cajon and Walker passes. April–July, occasional later.

Var. **Paríshii** (Gray) Jones. [*A. P.* Gray. *Phaca pseudoocarpa* Rydb.] Calyx ± white-silky, the teeth either deltoid and 1.2 mm. long or less, or subulate from a broad triangular base and up to 2 mm. long, the broad sinuses outlined with a band of short silvery hairs; banner 8–12 mm. long; pod 2.5–4(–5) cm. long.—Brushy hillsides and flats in the mountains; S. Oak Wd., Yellow Pine F.; 4000–7700 ft.; San Bernardino Mts. (intergrading with var. *Douglasii*); San Jacinto Mts., s. to n. L. Calif. May–Aug., rarely Oct.

Var. **perstríctus** (Rydb.) Munz & McBurn. [*A. Parishii* ssp. *p.* Abrams.] Calyx and flower of var. *Parishii*, but the habit of *A. oocarpus*, perhaps a self-perpetuating hybrid; stems stout, erect, 4–10 dm. long; lfts. 11–19, of thick texture; peduncles erect; pod semiovoid, 4–6 cm. long, the ventral suture nearly straight, the valves strigulose.—Stony hillsides and sandy flats, 3000–4000 ft.; S. Oak Wd.; local in se. San Diego Co. (Campo, Jacumba, Mountain Springs). May–June.

70. **A. mácrodon** (H. & A.) Gray. [*Phaca m.* H. & A. *A. holosericeus* Jones.] Closely resembling typical *A. Douglasii*, but the whole plant loosely villosulous-canescent with ascending curly hairs; calyx-teeth 2.5–4.5 mm. long; petals yellowish, the banner red-veined; pod 2–4 cm. long.—Shale or sandstone outcrops and barren eroded slopes, 750–3250 ft.; V. Grassland, Foothill Wd.; upper San Benito R., Salinas V. and enclosing foothills from s. San Benito and e. Monterey cos. s. to Atascadero, San Luis Obispo Co., thence se. to Temblor Range, Kern Co., uncommon; reported from the Great V. ("Fresno," about 300 ft.). May–June.

71. **A. Vàseyi** Wats. [*A. metanus* Jones. *A. V.* var. *m.* Munz & McBurn.] Short-lived perennial, flowering the first season, silvery-strigulose throughout, or the lfts. sometimes glabrescent beneath; stems radiating, prostrate and ascending, forming loose mats 2–10 dm. across; stipules free; lvs. 3–15 cm. long, with 11–21 ovate-oblong to narrowly elliptic acute, obtuse or rarely retuse lfts. 5–25 mm. long; peduncles ascending, 4–13 cm. long; racemes narrow, 13–40-fld., 4–21 cm. long in fr.; calyx either white- or black-strigulose, the campanulate tube 2.5–3.5 mm., the teeth 1–2.5 mm. long; petals magenta-purple with pale or white wing-tips, the banner 7–10 mm. long; pod spreading or ascending, sessile, obliquely ovoid-ellipsoid, 7–27 mm. long, 4–9 mm. in diam., turgid or moderately inflated in the lower half, there either a trifle laterally or (when large) dorsiventrally compressed, passing upward into a laterally flattened deltoid beak as long to half as long as the body, the papery yellowish or purple-speckled valves strigulose and often canescent, not inflexed.—Rocky hillsides and washes, (600–)1000–4000 ft.; Pinyon-Juniper Wd., occasional in Creosote Bush Scrub; desert slope of the mts. from near Palm Springs, Riverside Co., s. to L. Calif. March–May. Passes n. imperceptibly into:

Var. **Johnstònii** Munz & McBurn. Commonly greener, the stems sometimes nearly glabrous; petals either purple or ochroleucous and purple-veined; pod 16–32 mm. long, 8–15 mm. in diam.—Gravelly washes and canyons 3100–4000 ft., Joshua Tree Wd., Creosote Bush Scrub; mts. around the head of Coachella V. (Morongo Pass to Little San Bernardino and Eagle mts.). April–May.

72. **A. Dèanei** (Rydb.) Barneby. [*Phaca D.* Rydb. *A. Vaseyi* var. *D.* (Rydb.) Jeps.]

Resembling *A. oocarpus* in growth-habit and *A. Vaseyi* var. *Johnstonii* in the pod; nearly glabrous, the stems 3–6 dm. long; stipules free; lvs. 9–18 cm. long, with 19–29 ovate to lanceolate lfts. 1–2 cm. long; peduncles elongate, mostly 12–20 cm. long; racemes loosely 18–25-fld., 7–16 cm. long in fr.; calyx-tube campanulate 3.5–4.5 mm., the teeth 2–3 mm. long; petals whitish, the banner 12.5–15 mm. long; pod loosely ascending, sessile, obliquely ovoid-ellipsoid, bladdery-inflated, 2–3 cm. long, 1-locular, the papery valves strigulose.—Dry hillsides, sometimes on burns, 800–1000 ft.; Chaparral, S. Oak Wd.; upper Otay and Sweetwater valleys, sw. San Diego Co., very local. March–May.

73. **A. oocárpus** Gray. [*A. Crotalariae* of authors, not *Phaca C.* Benth.] Stout perennial from a woody base, except for the calyx nearly glabrous; stems erect or in age flexuous and supported by bushes, 6–13 dm. long, hollow below; stipules free, decurrent, early reflexed and finally deciduous; lvs. 6–17 cm. long, with 19–35 thick obovate-oblong truncate or retuse lfts. 6–33 mm. long, the larger ones pinnate-veined; peduncles stout, 1.5–7 cm. long; racemes loosely 15–60-fld., 4–19 cm. long in fr.; calyx-tube campanulate 3.5–4.5 mm., the teeth deltoid 0.8–1.8 mm. long, their margins and sinuses often strigulose-canescent; petals ochroleucous, little graduated, the banner 10.5–12.5 mm. long, strongly recurved; pod erect, sessile, persistent until dehiscence, ovoid but abruptly contracted at base into a thick obconic neck, shortly deltoid-beaked, 1.5–2.5 cm. long, 1–1.6 cm. in diam., inflated, 1-locular, the stiffly leathery valves glabrous or thinly strigulose.—Dry brushy slopes, (1500–)2700–5000 ft., S. Oak Wd.; interior San Diego Co., on the cismontane slope (Palomar Mt. s. to Descanso). June–August.

74. **A. oóphorus** Wats. Perennial, nearly glabrous throughout, the fresh herbage subglaucescent; stems ascending from a short caudex, 5–25 cm. long; stipules free, reflexed; lvs. 5–15 cm. long, with 9–23 ovate obtuse, rarely elliptic acute lfts. 4–20 mm. long; peduncles incurved, 4–10 cm. long; racemes loosely 5–12-fld., 2–7 cm. long in fr.; calyx glabrous, the campanulate tube 4.5–6.5 mm., the teeth 2–4 mm. long; petals purple with white wing-tips, the banner 16–23 mm. long; pod pendulous, sessile on a slender stipelike gynophore 4–8 mm. long, the body broadly ellipsoid, bladdery-inflated, 3–5.5 cm. long, 1-locular, the papery, brightly mottled valves glabrous.—Bare places in canyons and on barren hillsides, 5500–10,200 ft.; Sagebrush Scrub, Pinyon-Juniper Wd.; Panamint, Inyo, and White mts.; upper Owens and Walker River valleys; to Nev. and se. Ore.

75. **A. sabulónum** Gray [*Phaca arenicola* Rydb.] Winter annual, rarely becoming indurated below in late summer and appearing perennial, silky hirsute or villosulous throughout, the lfts. sometimes glabrescent above; stems usually several, decumbent and ascending, 2–25 cm. long; stipules small, free; lvs. 1.5–6.5 cm. long, with 9–15 oblong to obovate-cuneate obtuse or retuse lfts. 4–13 mm. long; peduncles 1–3.5 cm. long; racemes loosely 2–6-fld., becoming 5–25 mm. long, the fls. at length declined; calyx-tube campanulate 1.5–2.5 mm., the subulate-setaceous teeth 2–3.5 mm. long; petals purplish, the banner 5–7 mm. long; pod horizontal or declined, sessile, obliquely oblong-ovoid, incurved into a short deltoid beak, 9–15 mm. long, 5–8 mm. in diam., turgid and a little obcompressed below, openly grooved dorsally, 1-locular, the thinly leathery valves hirsutulous.—Sandy valleys, *minus* 350–500 ft.; Creosote Bush Scrub; Colo. Desert; to n. Son.; e., up to 6500 ft., to s. Utah and nw. New Mex. Feb.–May.

76. **A. áridus** Gray [*A. albatus* Sheld.] Like the last, but densely silvery-strigulose throughout, even to the pod; racemes 4–9-fld., the fls. ascending, the axis 1.5–5.5 cm. long in fr.; calyx-teeth 1–1.5 mm. long; petals whitish tinged with pink, the banner 3.5–6.5 mm. long; pod ascending, narrowly lunate-ellipsoid, cuneately acute at both ends, 10–17 mm. long, 4.5–7 mm. in diam., not grooved, the valves papery.—Dunes and sandy flats, *minus* 200–1200 ft., Creosote Bush Scrub; Colo. Desert (Salton Sink to Borrego and Imperial valleys, and along Colo. R. from Palo Verde V. to Yuma); n. Lower Calif., sw. Ariz.

77. **A. insuláris** Kell. var. **Harwóodii** Munz & McBurn. Slender annual, simple or branched from the root-crown, cinereous with short appressed hairs; stems incurved-ascending, 4–35 cm. long; stipules small, free; lvs. 3–11.5 cm. long, with 11–19 narrowly oblong to oblanceolate often folded lfts. 3–18 mm. long; peduncles 1.5–7 cm. long; racemes loosely 3–9-fld., the fls. at length declined, becoming 1–3.5 cm. long in fr.; calyx-tube obconic-campanulate 2–2.7 mm., the teeth 1.3–2.2 mm. long; petals reddish-purple drying violet, the striate banner 5.5–7.5 mm. long; pod declined, sessile, obliquely

semiovoid with nearly straight ventral and gibbous-convex dorsal sutures, 14–23 mm. long, 6–12(–15) mm. in diam., greatly inflated below and passing into a porrect deltoid compressed beak, 1-locular, the papery purple-speckled valves strigulose.— Dunes and sandy valleys, 300–1200 ft.; Creosote Bush Scrub; cent. and sw. Colo. Desert (Pinto Basin, Desert Center and vicinity, Riverside Co., along Carrizo Creek, San Diego Co.); to sw. Ariz. and n. Son.

78. **A. Gèyeri** Gray. Like the last, but the lfts. 3–11, linear-oblong, often retuse and glabrous above; peduncles and racemes together shorter than the lvs., the earliest borne close to the ground in the lower axils; petals usually whitish; pod more strongly incurved, lunate in profile, the pale green valves turning straw color.—Sandy valley-floors and dunes; Sagebrush Scrub; se. Mono Co. (Oasis, about 5000 ft.); to se. Ore., w. Wyo. and Utah.

79. **A. nùtans** Jones. [*A. deserticola* Jeps. *A. chuckwallae* Abrams.] Habit of the two last, but short-lived perennial flowering the first season, silvery-strigulose or at length greenish; stems 6–15 cm. long; stipules small, free; lvs. 3–8 cm. long, with 7–13 oblanceolate-elliptic mostly obtuse lfts. 5–15 mm. long; peduncles 2–5 cm. long; racemes loosely 6–10-fld., 1–3.5 cm. long in fr.; calyx-tube campanulate 2.5–3 mm., the teeth 1.2–2 mm. long; petals pink-purple drying violet, the banner 8–10.5 mm. long; pod spreading, either sessile on a minute boss within the calyx or elevated on a stipelike gynophore up to 2 mm. long, ovoid with conical beak, bladdery-inflated, 1-locular, the papery-membranous valves green turning straw color, strigulose.—Uncommon, foot-hills and washes of desert mts., 1500–6400 ft.; Creosote Bush Scrub, Joshua Tree Wd., Pinyon-Juniper Wd.; e. Mojave and n. Colo. deserts (Clark, New York, Providence, Old Dad, and Chuckawalla mts.). March–June, occasional in fall (Oct.).

80. **A. Gílmani** Tides. [*A. triflorus* var. *morans* Crum. *A. Wootoni* var. *m.* Barneby.] Like the last, but more loosely strigulose-villosulous; stems mostly 8–25 cm. long; peduncles 1.5–3.5 cm. long; racemes 3–9-fld., 1.5–4 cm. long in fr.; calyx-tube 2.5–3.5 mm., the teeth 1–2 mm. long; banner 6–8 mm. long; pod sessile, 1.5–2.5 cm. long, 8–15 mm. in diam., green or minutely purple-dotted, strigulose with incumbent hairs.— Canyons and rocky hillsides, 6500–10,000 ft., Pinyon-Juniper Wd., Sagebrush Scrub; Panamint Mts., Inyo Co. June–July, occasional later.

81. **A. Woòtoni** Sheld. Like the 2 last, but thinly strigulose, the 11–19 oblanceolate to linear-oblong lfts. glabrous above; peduncles 1.5–5 cm. long, shorter than the lvs.; racemes loosely 4–10-fld., 1–5 cm. long in fr.; calyx-tube ca. 2 mm., the teeth 2–3 mm. long; petals (in Calif.) whitish, the banner 5–7 mm. long; pod spreading, sessile, ovoid-ellipsoid, symmetric or little oblique, obscurely deltoid-beaked, 1.5–3 cm. long, the papery valves yellowish-green, strigulose.—Sandy flats and plains; Creosote Bush Scrub; e. Mojave Desert (Goffs, ca. 2600 ft.); to Tex. and n. Mex. April.

82. **A. Hórnii** Gray. [*A. H.* var. *tularensis* (Rydb.) Jeps.] Annual but of long duration, flowering through summer and the root becoming indurated, the stems nearly glabrous, the leaves thinly villosulous or loosely strigulose on one or both faces; stems at first erect becoming diffuse or procumbent, 2–12 dm. long, branched upward or at first simple; stipules herbaceous, free, reflexed; lvs. 2.5–13 cm. long, with 15–33 narrowly elliptic to broadly ovate, acute obtuse or emarginate lfts. 5–23 mm. long; peduncles either strongly divaricate or ascending, 2–15 cm. long; racemes densely 10–35-fld., sub-globose to oblong, 0.7–5 cm. long, scarcely elongating, the fls. subsessile; calyx silky-villosulous, the campanulate tube 3–4.5 mm., the teeth 2–2.5 mm. long; petals whitish, the banner 8–10 mm. long; pod spreading, sessile, ovoid, acuminate into a triangular compressed beak, bladdery-inflated, 9–18 mm. long, 5–9 mm. in diam., the papery valves villous-hirsute to loosely ascending-pilose, not inflexed.—Plains, in alkaline soil, sometimes about desiccating pools or on lake-shores; Alkali Sink; San Joaquin V., 200–500 ft., Kern and adjacent Kings and Tulare cos.; w. edge of Mojave Desert, 2700–3700 ft., near Lone Pine and in Antelope V.; formerly at 1100 ft. in San Bernardino V.; nw. Nev. June–Oct.

83. **A. lentiginòsus** Dougl. Perennials and a few winter-annuals, erect or prostrate, glabrous, villosulous, strigulose or white-silky; fls. loosely or densely racemose, variable in size and color; calyx-tube deeply campanulate to cylindric; pod spreading, sessile, lance-ellipsoid to subglobose, mostly strongly inflated, abruptly deltoid-beaked, the papery-membranous to leathery valves inflexed as a complete septum; $2n = 22$ (Head,

1957).—A complex group of reticulately interrelated forms, most of them intergrading freely at the edge of their ranges with vicariant relatives.

KEY TO VARIETIES

a. Fls. small, the keel not more than 8 mm. long or, if longer, surpassing the wings and the banner whitish.
  b. Beak of the pod nearly linear, tubular, resembling a persistent style. Upper Kern R., Tulare Co.
                                                      var. *kernensis*
  b. Beak of the pod deltoid or triangular, laterally flattened.
    c. Petals whitish or pinkish; herbage green and nearly glabrous; ne. Calif., n. of Lake Tahoe.
      d. Pod of firm texture, leathery, strigulose, obliquely lance-ovoid, 11–17 mm. long, not greatly inflated, the ventral suture often prominent .... var. *carinatus*
      d. Pod thinner, papery or membranous, the valves usually diaphanous, the body subglobose, bladdery, usually glabrous.
        e. Fruiting racemes dense, cylindric or globose, ca. half as long as the lf.; stems prostrate, 25–40 cm. long, commonly branched .... var. *floribundus*
        e. Fruiting racemes looser, ca. equaling the lf.; stems ascending, 5–30 cm. long
                                                  var. *salinus*
    c. Petals purple or, if whitish, then the herbage copiously pubescent. S. of Lake Tahoe. Pod papery-membranous.
      d. Beak of the pod decurved; stems elongate, 3–10 dm. long; racemes cylindric, dense in fr.; lfts. canescent above .... var. *albifolius*
      d. Beak of the pod incurved or rarely erect.
        e. Racemes loose, 5 cm. or more in fr.; mostly below 5000 ft. .... var. *Fremontii*
        e. Racemes compact, not over 4 cm. long in fr.; above 5000 ft.
          f. Calyx-teeth at least 1.5 mm. long. Sierra Nevada and White Mts.
            g. Lvs. mostly 5–10 cm. long, the lfts. remote, strigulose beneath .... var. *semotus*
            g. Lvs. mostly 3–5 cm. long, the lfts. crowded, pubescent beneath with curved or villous hairs .... var. *ineptus*
          f. Calyx-teeth 1 mm. long or less. San Bernardino and San Antonio mts.
            g. Lfts. glabrate above, ciliate. San Bernardino Mts. .... var. *sierrae*
            g. Lfts. silvery on both faces. San Antonio Mts. .... var. *antonius*
a. Fls. larger, the keel 8.5 mm. long or more, exserted from the wings only in purple-fld. var. *variabilis*
  b. Pod strongly inflated, the body ovoid or subglobose.
    c. Petals white; plants nearly glabrous. Ne. Calif. Pod leathery .... var. *platyphyllidius*
    c. Petals purple or pale yellow. S. and cismontane Calif.
      d. Petals pale yellow; herbage villosulous. Coarse plants of San Joaquin V.
                                         var. *nigricalycis*
      d. Petals purple; pubescence appressed or incumbent.
        e. Racemes compact, not over 4 cm. long in fr. S. Coast Ranges e. to Tehachapi Mts.
                                      var. *idriensis*
        e. Racemes loose, open, mostly much more than 4 cm. long in fr. Deserts, extending at low elevs. w. to the head of San Joaquin V.
          f. Calyx-teeth 1–1.4 mm. long; pubescence variable, but very rarely densely white-silky; pod usually not more than strigulose; w. and s. Mojave Desert to upper San Joaquin V. .... var. *variabilis*
          f. Calyx-teeth 1.4–3 mm. long; whole plant, including the pod, white-silky; either n. Colo. or n. Mojave Desert.
            g. Winter-annual or short-lived perennial; longest hairs of the herbage 0.5–1.2 mm. long. Coachella V., Riverside Co. .... var. *Coulteri*
            g. Strong perennial, indurated at base; longest hairs 1.2–2 mm. long. Eureka V., Inyo Co. .... var. *micans*
  b. Pod not strongly inflated, lance-acuminate in profile, silky-strigulose; racemes elongate, loose; herbage silky. Deserts .... var. *borreganus*

Var. **carinàtus** Jones. [*A. lentiginosus* var. *cuspidocarpus* of Jeps., not Jones.]—Plains and hillsides, on basalt, 3500–4800 ft.; Sagebrush Scrub; Lassen to Siskiyou and Modoc cos.; Ore. May–June.

Var. **floribúndus** Gray.—Sandy valleys at the foot of the Sierra Nevada, 3800–5200 ft.; Sagebrush Scrub; e. Placer, Sierra and Lassen cos.; in n. Mono Co., at 6450 ft., passing into var. *ineptus;* w. Nev., se. Ore. May–June.

Var. **salinus** (Howell) Barneby. [*A. s.* Howell.]—Plains and hillsides, often but not exclusively in alkaline soils, 3500–4500 ft.; Sagebrush Scrub; Lassen to Siskiyou cos.; to Mont. and Utah. Intergrades with var. *floribundus.* May–June.

Var. **albifòlius** Jones [*A. a.* Abrams.]—Clay flats and seeps, prostrate or trailing over low shrubs, 2100–4700 ft., Shadscale Scrub, Alkali Sink; along e. base of Sierra Nevada from Owens V. to Antelope V., Inyo to n. Los Angeles cos. Sharply marked. April–July.

Var. **semòtus** Jeps.—Sandy flats and ridges, 7000–11,000 ft.; mostly in Sagebrush Scrub; Inyo and White mts., to adjacent Nev. Intergrades with var. *ineptus.* June–August.

Var. **inéptus** (Gray) Jones. [*A. i.* Gray.]—Sandy flats and exposed gravelly crests and talus-slopes, 7000–11,400(–12,000) ft., Subalpine Forest, Basin Sagebrush Scrub; e. slope and crest of Sierra Nevada, Alpine to Inyo Co. June–August.

Var. **siérrae** Jones [*A. s.* Tides.]—Stony meadows and openings in pinewoods, 6000–7000 ft.; Sagebrush Scrub, Yellow Pine F.; e. end San Bernardino Mts. April–July.

Var. **antònius** Barneby.—Open slopes in Yellow Pine F.; 5000–8500 ft., Mt. San Antonio, Los Angeles Co.

Var. **kernénsis** (Jeps.) Barneby. [*A. k.* Jeps.]—Dry sandy soil of mountain meadows, often among sagebrush, 8000–8500 ft.; upper Kern River, Tulare Co. (Monache Mdws.; Templeton Mdws.; Volcano Creek); Charleston Peak, s. Nev. June–August.

Var. **platyphyllídius** (Rydb.) Peck. [*A. araneosus* of Abrams, not Sheld.]—Plains and low hills, often on basalt, 3500–4500 ft., Sagebrush Scrub; Lassen and Modoc cos.; to Ore., Wyo., nw. Colo. May–June.

Var. **idriénsis** Jones. [*A. i.* Abrams. *A. tehatchapiensis* (Rydb.) Tides. *A. lentiginosus* var. *t.* (Rydb.) and var. *caesariatus* Barneby.]—Grassy and brushy flats or hillsides, sometimes on sandstone outcrops, 2100–7000 ft., Foothill Wd., Yellow Pine F.; mts. around the head of San Joaquin V., from Tehachapi to Mt. Pinos, the San Rafael and Temblor mts., n. in the Coast Ranges to the Santa Lucia Mts., Monterey Co., the San Carlos Range, San Benito Co., and Mt. Hamilton, Santa Clara Co. April–July.

Var. **nigricàlycis** Jones. [*A. n.* Abrams.]—Plains and foothills, 300–2500(–3190) ft.; V. Grassland; San Joaquin V. and adjacent Coast Ranges, sw. Fresno Co. to upper Cuyama V., Santa Barbara Co. and the valley floor in Kern Co., there passing into var. *variabilis.* March–May.

Var. **variábilis** Barneby [*Cystium pardalotum* Rydb.]—Sandy plains, sometimes on dunes, 800–4000 ft.; Creosote Bush Scrub; s. and sw. Mojave Desert, abundant; occasional in V. Grassland on the Bakersfield Plain, Kern Co. April–June.

Var. **Fremóntii** (Gray) Wats. [*A. F.* Gray. *A. eremicus* Sheld. *A. F.* ssp. *e.* Abrams.]—Sandy plains, washes and fans of desert mts., sometimes on alkaline flats; Creosote Bush Scrub, Shadscale Scrub, Joshua Tree Wd., Pinyon-Juniper Wd.; 1500–6500 ft.; e. Mojave Desert to Owens V. and the White Mts., exceptionally ascending the e. face of the Sierra Nevada in Inyo Co. into the Jeffrey Pine Belt up to 9000–9500 ft.; s. Nev. April–July, occasional Sept.–Oct.

Var. **Còulteri** (Benth.) Jones. [*A. C.* Benth.]—Dunes and sandy flats, 20–1200 ft.; Creosote Bush Scrub; Coachella V. February–May.

Var. **mìcans** Barneby.—Dunes, 3050–3100 ft., s. end Eureka Valley, Inyo Co. April–June.

Var. **borregànus** Jones. [*A. Arthu-Schottii* Gray. *A. agninus* Jeps.]—Dunes and sandy valley-floors, 100–800 ft.; Creosote Bush Scrub; e. Mojave and s. Colo. deserts; to sw. Ariz. and n. Son. Feb.–May.

84. **A. iodánthus** Wats. Perennial, thinly strigulose, the lfts. sometimes white-ciliate; stems decumbent, 1–3.5 cm. long; stipules free; lvs. 3–7 cm. long, with 11–19 obovate to elliptic, obtuse or retuse lfts. 5–15 mm. long; peduncles ascending, 2–4.5 cm. long; racemes 7–20-fld., 1–4.5 cm. long in fr.; calyx-tube campanulate to subcylindric, 4–5 mm., the teeth 1.5–3 mm. long; petals ochroleucous with purple keel-tip (in Calif.), the banner (10–)12–15.5 mm. long, the prominent keel (in Calif.) ca. 1 mm. shorter, equaling or surpassing the wings; pod declined, sessile, narrowly lance-acuminate, 2–4 cm. long, 5–8.5 mm. in diam., triquetrously or dorsiventrally compressed with prominent ventral and depressed or openly sulcate dorsal sutures, nearly straight in the lower third, abruptly hooked or arched through half to a complete circle, the leathery strigulose valves commonly inflexed as a narrow partial septum.—Dry plains and hillsides, in sandy often alkaline soils, 4000–7000 ft.; Sagebrush Scrub; Sierra foothills in e. Mono Co.; Sierra and Lassen cos.; to se. Ore. and nw. Utah. April–June.

85. **A. Crotalàriae** (Benth.) Gray. [*Phaca C.* Benth. *A. limatus* Sheld. *A. Preussii* var. *l.* Jeps.] Coarse malodorous annual or short-lived perennial, greenish but finely strigulose or pilosulous, the lfts. commonly glabrous above; stems 1–many, erect, 1.5–6 dm. long; stipules free, deflexed; lvs. 5–14 cm. long, with 7–19 thick obovate-cuneate or oblong retuse lfts. 5–35 mm. long; peduncles erect, 7–17 cm. long; racemes 10–25-fld., 3–9.5 cm. long in fr.; calyx reddish drying dark, the cylindric tube 7.5–12 mm., the teeth 1.5–2.5 mm. long; petals bright purple, drying violet (not rarely all white), the banner 21–28 mm. long; pod ascending, sessile but contracted at base into a stout stipelike

neck about 1 mm. long, the body ovoid-ellipsoid, inflated, 2–3 cm. long, 10–14 mm. in diam., the stiffly papery valves greenish-yellow, not or obscurely inflexed; $2n = 24$ (Trelease & Beath, 1949).—Sandy flats and desert fans, 220 to 800 ft.; Creosote Bush Scrub; Colo. Desert (Salton Sink to Carrizo V., e. to Yuma); sw. Ariz., n. Lower Calif. Jan.–April. This and the next sp. occur only where selenium is present in the soil and heavy concentrations of the element are commonly found in the green tissues and seeds, the plants therefore highly toxic.

86. **A. Preússii** Gray. Resembling and related to the last, but of longer duration and nearly glabrous; racemes 4–16-fld.; calyx-tube 6–10 mm. long; petals pink- or lilac-purple with pale claws, the banner 15–21 mm. long; pod incurved-ascending or sub-erect, stipitate, the stout stipe 2–5.5 mm. long, the body oblong-ellipsoid or fusiform, cuspidate-beaked, 1.5–4 cm. long, 6–10 mm. in diam., subterete, the stiffly papery valves glabrous; $2n = 24$ (Trelease & Beath, 1949).—Barren clay flats; Shadscale Scrub, Alkali Sink; extreme e. Mojave Desert (Kingston; Mesquite V., 2500–2600 ft.); to n. Ariz. and Utah. April–May.

Var. **laxiflòrus** Gray. [*Phaca Davidsonii* Rydb. *A. Crotalariae* var. *D.* Munz & McBurn.] Pod sessile or nearly so.—Alkaline flats, Antelope V., sw. Mojave Desert (about Lancaster, 2400 ft.); s. Nev. and immediately adjoining Utah and Ariz.

87. **A. Gíbbsii** Kell. [*Homalobus Plummerae* Rydb.] Stout perennial from a buried root-crown, villosulous with incurved and spreading hairs, canescent or greenish; stems decumbent, 1.5–3.5 dm. long; stipules dimorphic, the lowest connate into a short scarious sheath, the upper free, herbaceous, 3–5 mm. long; lvs. 3–9.5 cm. long, with 11–19 obovate-cuneate retuse lfts. 4–20 mm. long; peduncles 3–10 cm. long; racemes at first densely 10–30-fld., becoming 2.5–10 cm. long in fr., the fls. nodding; calyx-tube broadly and deeply campanulate, gibbous dorsally at base, 7.5–10 mm., the teeth 2–3.5 mm. long; petals dull yellow, the banner 14–18 mm. long, scarcely or not as long as the wings; pod pendulous, stipitate, the stipe 7–20 mm. long, the body lunately oblong-ellipsoid incurved through $\frac{1}{4}$–$\frac{1}{2}$ a circle, 2–3 cm. long, 5–8 mm. in diam., laterally compressed with rounded lateral faces and prominent cordlike sutures, the valves fleshy becoming leathery, villosulous with incurved hairs, not inflexed.—Sandy valleys, silty meadows and thinly wooded hillsides, 4150–6000 ft.; Sagebrush Scrub, Yellow Pine F.; e. foothills of the Sierra Nevada from Lassen to Nevada cos.; extreme w. Nev. and reëntering Calif. in Alpine and Mono cos. May–June.

88. **A. califórnicus** (Gray) Greene. [*A. collinus* var. *c.* Gray.] Perennial, from a buried root-crown, greenish but thinly villous-villosulous with fine contorted hairs, the lfts. often glabrous above; stems ascending or diffuse, 2–5 dm. long; stipules dimorphic, the lowest connate into a short scarious sheath, the upper free, 2–6.5 mm. long; lvs. 3–8.5 cm. long, with 13–21 oblong-oblanceolate obtuse or emarginate lfts. 6–20 mm. long; peduncles 5–14 cm. long; racemes 10–25-fld., 3–10.5 cm. long in fr., the fls. early declined and secund; calyx-tube deeply campanulate, oblique at base, 5–7 mm., the teeth 1–3 mm. long; petals ochroleucous, the banner 12.5–17.5 mm. long; pod pendulous, stipitate, the stipe 8–14 mm. long, the body linear-oblong, straight or nearly so, 3–4 cm. long, 3.5–5 mm. in diam., laterally compressed, 2-carinate by the sutures, 1-locular, the stiffly papery valves mottled, strigulose or rarely glabrous.—Dry rocky hillsides and benches in canyons, 1500–4550 ft.; Sagebrush Scrub, N. Oak Wd.; upper Sacramento V. and w. edge of Klamath Basin, Shasta and Siskiyou cos.; sw. Ore. May–July.

89. **A. curvicárpus** (Heller.) Macbr. [*A. Gíbbsii* var. *falciformis* Gray. *A. Whitedii* auth., not Piper.] Perennial from a buried root-crown, pubescent with short incurved hairs or villosulous, cinereous or early greenish; stems ascending, 1.5–4 dm. long; stipules all free, the upper deflexed; lvs. 2.5–9 cm. long, with 7–19 obovate-cuneate retuse lfts. 5–20 mm. long; peduncles 4–15 cm. long; racemes 10–25-fld., the fls. pendulous, the axis 2–12 cm. long in fr.; calyx-tube deeply campanulate or broadly cylindric, gibbous dorsally at base, 7–9.5 mm., the teeth 0.5–2.5 mm. long; petals white or ochroleucous, the banner 16.5–21 mm. long; pod pendulous, stipitate, the stipe 1–2 cm. long, the body linear-oblong, strongly falcate, hamate, or coiled into a complete ring, 2–3.5 cm. long, 3–4 mm. in diam., strongly compressed with prominent cordlike sutures, 1-locular, the leathery or stiffly papery valves puberulent or glabrate.—Valleys, plains and low ridges, usually in sandy soil, sometimes on dunes, 3100–9200 ft.; Sagebrush Scrub; Sierra to Siskiyou and Modoc cos.; Mono Co.; to Nev., e. Ore. and s. Ida. May–June.

90. **A. filipes** Torr. [*A. stenophyllus* auth., not T. & G. *A. MacGregorii* (Rydb.) Tides.

A. filipes var. *residuus* Jeps. *A. stenophyllus* var. *r.* Barneby.] Perennial from a buried root-crown or caudex, thinly strigulose or glabrate, the herbage green or rarely cinereous, the lfts. glabrous or rarely pubescent above; stems usually many, commonly erect, leafless at base, 3–7 dm. long; stipules dimorphic, the lowest connate into a scarious sheath, the upper herbaceous, free or nearly so; lvs. 3–10 cm. long, with 9–19 linear to linear-oblong lfts. 5–25 mm. long; peduncles strict, 8–25 cm. long; racemes loosely 7–25-fld., 4–10 cm. long in fr., the fls. nodding; calyx-tube campanulate 3–4 mm., the teeth usually deltoid and less than 1 mm. long; pod pendulous, stipitate, the strigulose stipe 8–15 mm. long, the body linear-oblong, straight or nearly so, 1.7–2.5 cm. long, 4–5 mm. in diam., strongly compressed, 2-carinate by the sutures, 1-locular, the papery lustrous valves glabrous or rarely puberulent.—Dry hillsides, plains and valley-floors, usually on volcanic substrata, 3400–5000 ft.; Sagebrush Scrub; Lassen to Siskiyou and Modoc cos.; to Nev., s. Ida. and s. B.C.; also disjunctly in the mts. of interior s. Calif., mostly above 4000 ft. (Mt. Pinos; n. slope San Bernardino Mts.; w. slope San Jacinto and Santa Rosa mts.); n. Lower Calif. April–June.

91. **A. invérsus** Jones. [*A. filipes* var. *i.* Jeps.] Resembling the last, but the stems diffuse, junceous and sparsely leafy; lfts. 5–11, distant, linear-filiform, those of the upper lvs. much reduced, the terminal one often confluent with the rachis; petals reddish-pink, the keel and wing-tips yellowish, the banner 9.5–12 mm. long, the keel attenuate into a narrowly triangular apex; stipe of the pod 6–16 mm., the body 2–3.5 cm. long, 3–5 mm. in diam., the mottled valves either strigulose or glabrous.—Plains and sparsely wooded hills, chiefly on basalt, 4000–6000 ft.; Sagebrush Scrub, Yellow Pine F.; Lassen to e. Siskiyou, ne. Shasta and Modoc cos. May–Aug.

92. **A. Whítneyi** Gray. [*A. Hookerianus* var. *W.* Jones. *A. W.* ssp. *pinosus* (Elmer) Abrams.] Perennial from a buried caudex or root-crown, green and nearly glabrous to densely silvery-strigulose; stems decumbent or ascending, dwarfed at high altitudes, 0.3–2.5 dm. long; stipules dimorphic, the lowest connate into a subtruncate scarious sheath 2–4 mm. long, the upper free or nearly so; lvs. 3–8.5 cm. long, with 9–19 linear-elliptic to oblong-oblanceolate obtuse or acute lfts. 2–15 mm. long; peduncles 2–10 cm. long; racemes 4–16-fld., 1–4 cm. long in fr.; calyx-tube campanulate 3.5–4.5 mm., the teeth 0.5–1.5 mm. long; petals purple with white wing-tips, early fading, sometimes whitish, the banner 8.5–14 mm. long; pod pendulous, stipitate, the stipe 2–4.5 mm. long, the bladdery-inflated body balloon-shaped, broadest just below the obtuse almost beakless apex, scarcely or not oblique, 1.5–3(–4) cm. long, the papery-membranous translucent valves brightly mottled, glabrous, not inflexed.—Dry gravelly crests and mountain slopes, 6800–12,000 ft.; Sagebrush Scrub, Subalpine F., Alpine Fell-fields; summits and w. slope of Sierra Nevada, Alpine to Inyo cos.; White Mts.; summit of Mt. Pinos; w. and n. Nev. May–Sept.

Var. **siskiyouénsis** (Rydb.) Barneby. [*A. Hookerianus* var. *s.* Peck. *A. Whitneyi* ssp. *s.* Abrams.] Racemes looser, 3–7 cm. long in fr.; petals ochroleucous, perhaps sometimes pink-tinged; pod longer, the body 2.5–6 cm. long.—Open rocky slopes and crests, commonly on serpentine, 3500–8500 ft.; Yellow Pine F., N. Oak Wd.; inner N. Coast Ranges, Trinity to Siskiyou cos.; sw. Ore. Along Pit River in Shasta Co. and near Susanville, Lassen Co., passing into var. *confusus.* May–Aug.

Var. **lenóphyllus** (Rydb.) Barneby. [*A. l.* Tides. *A. Whitneyi* var. *Sonneanus* of Jeps., not *A. S.* Greene.] Leaves short, with 11–17 crowded lfts. densely villous-hirsutulous with spreading hairs; racemes short, ca. 1 cm. long in fr.; body of the pod 2–4 cm. long, glabrous.—Easterly summits of the Sierra Nevada, 7000–9000 ft., Placer and Nevada cos. July–Aug.

Var. **confúsus** Barneby. Pubescence of the herbage gray- or silvery-silky, loose, incurved or spreading; fls. larger, the calyx-tube 5–6 mm., the teeth 1.5–2.5 mm., the whitish or pinkish banner 13.5–17 mm. long; body of the pod 2.5–6 cm. long, the valves strigulose.—Dry hillsides, 4400–8000 ft., N. Juniper Wd., Sagebrush Scrub; Warner Mts. and vicinity, Modoc Co.; to se. Ore., sw. Ida., ne. Nev. June–July.

93. **A. Kentrophýta** Gray var. **impléxus** (Canby) Barneby. [*A. tegetarius* Wats. *Kentrophyta t.* Rydb.] Prostrate loosely or densely matted perennial, greenish or canescently strigulose-villosulous; stems closely and repeatedly branching, 5–25 cm. long; stipules all scarious and connate; lvs. less than 2 cm. long, with 5–9 crowded linear-elliptic lfts. continuous with the rachis, the stiff midrib produced into a glabrous mucro or spinule up to 1 mm. long; racemes 2–3-fld., subsessile in the axils or very

shortly pedunculate; petals (in Calif.) pink-purple with pale or white wing-tips, the banner 4–8 mm. long; pod sessile, ellipsoid, compressed, 4–8 mm. long, 1-locular, 1–3-seeded.—Stony knolls and gravelly talus-slopes, 9000–11,000 ft.; Sagebrush Scrub, Subalpine F.; White Mts., Mono Co.; to Mont., Colo. and n. New Mex. June–September.

Var. danàus Barneby. [A. tegetarius var. d. Barneby. Kentrophyta montana of Jeps., not Nutt.] Pulvinate or densely matted; lvs. 3–5-foliolate, more rigid, the spinule 1–1.5 mm. long; petals whitish with purple keel-tip.—Alpine summits, commonly on metamorphic bedrock, 11,250–11,500, perhaps 12,000 ft.; Alpine Fell-fields; Sweetwater Mts., Mono Co.; Sierra Nevada in Mono Co. from Mt. Warren s. to Mono Pass and near Sawmill Pass, e. Fresno Co. July–Sept.

28. Oxýtropis DC. Locoweed. Crazyweed. Oxytrope

Perennial caulescent or acaulescent herbs with the aspect of Astragalus. Lfts. asymmetrical at base. Keel-petals abruptly narrowed into a cuspidate or mucronate apex. Pod more deeply grooved ventrally than dorsally, the valves not infolded, but the seminiferous suture often intruded, sometimes forming a false partition. A genus of perhaps some 300 spp., of montane, continental and arctic climates in the N. Hemis., most numerous in cent. Asia, the 22 N. Am. chiefly of the Rocky Mts. and high latitudes. Several are notoriously toxic to cattle, sheep and horses, and are a problem of some magnitude in the mountain and plains states. (Greek, oxus, sharp and tropis, keel, in reference to the beaked lower petals.)

(Barneby, R. C. A revision of N. Am. species of Oxytropis. Proc. Calif. Acad. Sci., Series IV, 27: 177–312, 1952.)

A.   Stipules adnate to the petiole only at base, or nearly free; pod pendulous, stipitate; plants often
     with one or more developed internodes; var. sericea of .... 1. O. deflexa
AA. Stipules adnate through nearly half their length to the dilated petiole-bases; pod spreading or
     erect, sessile.
  B.   Plant green, the herbage glandular and resinous, the calyx-teeth and pod beset with glan-
       dular warts; bracts glabrous dorsally .... 2. O. viscida
  BB. Plant silvery, the herbage silky-pilose, eglandular; bracts pilose dorsally.
    C.   Pod spreading, ovoid, bladdery-inflated, the valves papery-membranous
                                                                        3. O. oreophila
    CC. Pod erect, oblong-acuminate or subcylindric, merely turgid, the valves leathery
                                                                        4. O. Parryi

1. O. defléxa (Pall.) DC. var. serícea T. & G. [O. retrorsa and O. r. var. s. Fern. O. d. var. culminis Jeps.] Silky-pilose, canescent or greenish; stems usually with one developed internode, appearing acaulescent in youth; lvs. 5–20 cm. long, with 23–41 ovate lfts. tapering upward along the rachis; peduncles exceeding the lvs.; racemes loosely 10–25-fld., becoming secund; petals whitish or lilac, the banner 5–9 mm. long; stipe of the pod 1–2 mm. long, the body narrowly oblong, 10–15 mm. long, 3–4 mm. wide, deeply grooved ventrally.—Moist meadows and turfy banks, in Calif. at ca. 9000 ft.; White Mts. (Cottonwood Creek); to Alaska, Man., n. New Mex. June–Aug.

2. O. víscida Nutt. Cespitose, the branches of the stout forking caudex beset with imbricated marcescent stipules; lvs. 6–10 cm. long, with 25–39 green, sparsely pilose lfts. less than 1 cm. long; racemes ca. 10-fld., oblong, slightly exserted on erect or arched scapiform peduncles; petals whitish or red-purple, the banner ca. 12 mm. long; pod oblong-ellipsoid with acuminate beak, ca. 15 mm. long.—Bare crests and talus-slopes, 11,500–12,200 ft.; Alpine Fell-fields; Sierra Nevada in Inyo Co.; White Mts.; to Alaska, Quebec, Colo. July–Aug.

3. O. oreóphila Gray. Densely cespitose or pulvinate, forming cushions up to 15 cm. broad, the herbage silky-pilose; lvs. 1–3 cm. long, with 7–15 lfts. 3–8 mm. long; racemes shortly 2–8-fld., immersed in the foliage or slightly exserted on sigmoid peduncles; petals pink, the banner ca. 1 cm. long; pod ovoid to subglobose with a short conical-compressed beak, bladdery inflated, 8–13 mm. long, the papery valves villous-hirsute.—Barren stony slopes, 11,000–11,450 ft., about the summit of Mt. San Gorgonio, San Bernardino Mts.; to Utah, n. Ariz. July.

4. O. Párryi Gray. Like the last, but merely cespitose, the leaves clustered in a few rosettes and sometimes longer; racemes mostly 2-fld., subcapitate, usually well exserted; pod erect, oblong-acuminate, 13–20 mm. long, scarcely more than turgid, the valves firm, shortly pilosulous.—Dry knolls and rocky ridges, near timber line and above, 11,000–12,000 ft.; Subalpine F., Alpine Fell-fields; e. slope of Sierra Nevada in Inyo

and Mono cos. (Inconsolable Range; Sweetwater Mts.) and White Mts.; to Ida., n. New Mex. June–July.

29. Sphaerophỳsa DC.

Perennial herbs with odd-pinnate lvs. Fls. in axillary racemes, red to bluish. Calyx-tube campanulate, 5-toothed. Banner orbicular; wings falcate-oblong; keel incurved, obtuse. Stamens diadelphous. Ovules many. Style longitudinally bearded above, incurved; stigma terminal, capitate. Pod stipitate, membranaceous, inflated, indehiscent. Three spp., of Russia and Asia. (Greek, *sphaira*, a sphere, and *physa*, bladder, because of the inflated pods.)

1. **S. salsùla** (Pall.) DC. [*Phaca s.* Pall.] Suffrutescent, with a tough running rootstock; stems erect, adpressed-canescent, leafy, 3–5 dm. high; lvs. 4–9 cm. long, with 13–21 oval to obcordate lfts.; racemes 5–8 cm. long, 4–8-fld.; fls. ca. 15 mm. long, purplish-red; pods ovoid, 2.5–3.5 cm. long.—Collected near Bakersfield, becoming more common in w. irrigated lands and difficult to eradicate; native from the Caspian to China.

30. Glycyrrhìza L. WILD LICORICE

Perennial herbs with thick sweet rootstocks and glandular odd-pinnate lvs. Fls. in axillary spikes or racemes, whitish to blue. Calyx 5-cleft, the 2 upper teeth somewhat shorter and ± united. Banner short-clawed, the blade oblong to ovate; wings oblong; keel acute to obtuse. Stamens mostly diadelphous, the alternate anthers shorter. Pods sessile, covered with glands or prickles, indehiscent, several-seeded. Ca. 15 spp., of N. Temp. Zone, S. Am., Australia. (Greek, *glukus*, sweet, and *rhiza*, root.)

Pods burlike with hooked prickles; fls. yellow-white. Native .... 1. *G. lepidota*
Pods glabrous; fls. bluish. Occasional escape .... 2. *G. glabra*

1. **G. lepidòta** Pursh. Stems erect, 3–10 dm. high, ± viscid-puberulent, sometimes minute-scaly; lfts. 11–19, oblong to ovate-lanceolate, 2–3 cm. long, mucronate; stipules linear-subulate; spikes broadly oblong, shorter than the lvs.; calyx very glandular, ca. 6 mm. long, the lance-acuminate teeth longer than the tube; corolla yellowish white, 8–12 mm. long; pod oblong, 12–15 mm. long, burlike with hooked prickles; seeds brown, flat, ± broad-reniform, ca. 2.5 mm. long; $2n = 16$ (Heiser & Whitaker, 1948).—Occasional as patches in low ground and moist waste places, below 7500 ft.; many Plant Communities; cismontane Calif., Owens V., etc.; to B.C., Ont., Mex. May–July. Our plants are more heavily glandular-villous than those in the e. part of the range and are sometimes called the var. *glutinòsa* (Nutt.) Wats.

2. **G. glàbra** L. CULTIVATED LICORICE. Plant subglabrous, rootstocks sweet, woody, immensely developed; lfts. ovate, 2.5–5 cm. long; fls. pale blue; pods red-brown, 1.2–2.5 cm. long, glabrous; $2n = 16$ (Tschechow, 1930).—Occasional escape and well established, as near Pomona and Goose Lake (Kern Co.); native of Eurasia. Spring.

31. Ornithòpus L. BIRD'S FOOT

Rather weak annual with odd-pinnate lvs. and no or small stipules. Fls. rose or yellow, small, 1–8 in umbels or on axillary peduncles. Calyx tubular, with 5 short equal teeth. Petals short-clawed, the keel almost straight, shorter than banner and wings. Stamens diadelphous. Pod linear, curved, beaked, veiny. Ca. 6–7 spp. of Old World and S. Am. (Greek, *ornithos*, bird, and *pous*, foot.)

1. **O. ròseus** Dufour. Stems 1–4 dm. long, pubescent; lfts. 13–31, lance-oblong, fls. 2–5, pale rose, the banner with purple veining, 6–7 mm. long; pods straight, glabrous, narrowed between the seeds.—Established along the Graham Hill road between Felton and Santa Cruz, Santa Cruz Co.; native of sw. Eu. Spring.

32. Alhàgi Desv. CAMEL THORN

Stiff spiny shrubs with small simple entire lvs. Stipules small. Fls. few, in axillary racemes. Calyx campanulate, with short subequal teeth. Corolla red, the banner short-clawed, obovate; wings oblong-falcate; keel incurved, obtuse. Stamens diadelphous. Pod linear, terete, incompletely 2-celled. Spp. 3, Medit. and S. Asia. (The Mauretanian name.)

1. **A. camelòrum** Fisch. Low, forming large impenetrable masses 3–8 dm. high; the

branches striate, greenish, glabrous, with slender spines 1–2.5 cm. long; lvs. 8–15 mm. long, glabrous above, strigose beneath; fls. mostly 4–6, the calyx ca. 2 mm. long, with short broadly triangular teeth; corolla ca. 8–9 mm. long, reddish-purple; pod stipitate, torulose, few-seeded.—Locally a very troublesome weed in low places in San Joaquin V., Colo. Desert, Orange Co.; native of Asia Minor. June–July.

33. Láthyrus L. Pea

Annual or mostly perennial herbs with rootstocks or sometimes taproots; stems erect to twining, angled to winged. Lvs. pinnate, the rachis usually produced into a tendril. Fls. sometimes solitary, mostly in axillary racemes. Calyx obliquely campanulate, the teeth subequal or the 2 upper much the shortest. Corolla showy, the banner usually distinctly clawed; wings with small lunate ridge on inner face which fits into a fold in the keel. Stamens diadelphous. Style curved, flattened, hairy along the inner side. Pods flat or ± terete, 2-valved. Seeds several, ± spherical or compressed. Ca. 100 spp., of N. Hemis. and S. Am. (Ancient Greek name of some leguminous plant.)

(Hitchcock, C. L. A revision of the N. Am. spp. of Lathyrus. Univ. Wash. Pub. Biol. 15: 1–104, 1952.)

A. Lfts. lacking; stipules triangular-ovate, hastate; fls. yellowish; calyx-lobes at least twice as long as tube. Introd. . . . . 1. *L. Aphaca*
AA. Lfts. present; stipules seldom triangular-ovate or hastate; calyx-lobes rarely as much as twice the tube in length.
  B. Lfts. 2. Introd. plants.
    C. Plants annual.
      D. Fls. less than 1.5 cm. long.
        E. Peduncles 1-fld., the rachis extending beyond the pedicel as a prominent bristle; plants glabrous . . . . 2. *L. sphaericus*
        EE. Peduncles usually 2-fld., sometimes 1-, but the rachis not prolonged as a prominent bristle; plants hirsute at least about the nodes . . 3. *L. hirsutus*
      DD. Fls. more than 1.5 cm. long.
        E. Plants glabrous; lfts. linear to linear-lanceolate; fls. deep purple
                                                    4. *L. tingitanus*
        EE. Plants rough-hairy; lfts. oval or oblong; fls. varicolored . . . . 5. *L. odoratus*
    CC. Plants perennial; fls. purple or white, 2–2.5 cm. long . . . . 6. *L. latifolius*
  BB. Lfts. at least 4 on some of the lvs.; native perennial plants.
    C. Tendrils absent or reduced and bristlelike, not prehensile.
      D. Plants densely silky-villous, growing on or near coastal sand dunes
                                                 8. *L. littoralis*
      DD. Plants not densely silky-villous; mostly not on coastal dunes.
        E. Fls. 1–2 on a peduncle, 8–13 mm. long. Open woods of Coastal Ranges
                                               22. *L. Torreyi*
        EE. Fls. more than 2 on a peduncle or usually over 15 mm. long.
          F. The fls. 8–12 mm. long . . . . 16. *L. Lanszwertii*
          FF. The fls. 13–27 mm. long.
             G. Lfts. usually less than 4 times as long as broad; plants with a branched crown from a stout taproot; fls. mostly white or pinkish
                                                21. *L. rigidus*
             GG. Lfts. mostly much more than 4 times as long as wide; plants mostly from rootstocks; fls. mostly blue to purple or red
                                               19. *L. nevadensis*
    CC. Tendrils well developed, prehensile on at least some of the lvs.
      D. Stems winged.
        E. Corolla white, penciled with pink or orchid, aging to yellow or tan, 12–15 mm. long . . . . 18. *L. delnorticus*
        EE. Corolla pink to purple, often over 15 mm. long.
          F. Lfts. mostly 6; calyx 8–12 mm. long; corolla mostly lavender to bluish-purple . . . . 9. *L. palustris*
          FF. Lfts. 10–14; calyx 13–16 mm. long; corolla pink to red . 12. *L. Jepsonii*
      DD. Stems not winged, at most sharply angled.
        E. Stipules as long as or longer than lfts. Plants of sea-beaches . 7. *L. japonicus*
        EE. Stipules definitely shorter than lfts. Plants mostly not on sea beaches.
          F. Corolla deep rich red, 3–4 cm. long, the banner reflexed and lying in nearly a straight line with wings and keel. Se. San Diego Co.
                                              14. *L. splendens*
          FF. Corolla mostly other than deep red, or if so less than 3 cm. long; banner seldom reflexed more than 90°.
             G. Fls. tan to yellowish or orange with purple or greenish penciling, 10–15 mm. long; claw of banner broader than and more than twice as long as the blade . . . . 17. *L. sulphureus*
             GG. Fls. white to blue, if at all yellowish then over 15 mm. long, or the claw of the banner not wider and slightly if at all longer than the blade.

H. Lateral calyx-lobes considerably broadened above their junc-
    ture with the tube, the lowest lobe usually at least as long as
    the tube .... 11. *L. vestitus*
HH. Lateral calyx-lobes usually broadest at their point of juncture
    with tube, the lowest lobe often much shorter than the tube.
    I. Fls. white to flesh-color, often with pink or lavender vein-
        ing and fading to yellow or tan.
        J. Corolla at least 18 mm. long.
            K. Lfts. scattered, usually more than 8; racemes
                5–12-fld.; claw of banner scarcely half as long as
                blade. From Santa Barbara Co. s.
                                              13. *L. laetiflorus*
            KK. Lfts. paired, mostly 4–8; racemes usually 2–6-fld.;
                claw of banner ca. as long as blade. From
                Fresno Co. n. .... 19. *L. nevadensis*
        JJ. Corolla 9–12 mm. long; lfts. mostly 4–6; banner
            broadly cordate. Glenn Co. to Siskiyou Co.
                                              20. *L. Tracyi*
    II. Fls. bluish or red, but often paler and fading to brown.
        J. Lfts. mostly 6; upper calyx-teeth scarcely half as long
            as lower lateral teeth; stems ± winged; stipules usually
            dentate .... 9. *L. palustris*
        JJ. Lfts. usually more than 6; upper calyx-teeth usually
            more than half as long as lower laterals; stems wing-
            less; stipules often entire.
            K. The lfts. 10–16; stipules sagittate-ovate, mostly
                at least half as long as lfts.; plants glabrous except
                the calyx, erect, scarcely at all clambering
                                              10. *L. polyphyllus*
            KK. The lfts. mostly fewer, if more, the stipules smaller
                and narrower, or plants pubescent or otherwise
                not as above.
                L. Fls. over 17 mm. long, if the plants are
                    densely pubescent they are 1–3 dm. tall with
                    lfts. 1–2.5 cm. long.
                    M. Claw of banner scarcely half as long as
                        blade; plants mostly over 5 dm. tall,
                        clambering. S. Calif. . 13. *L. laetiflorus*
                    MM. Claw of banner mostly subequal to
                        blade; plants mostly shorter, not clam-
                        bering. From Fresno Co. n.
                                              19. *L. nevadensis*
                LL. Fls. less than 17 mm. long, or if more, the
                    plants densely pubescent and over 3 dm. tall
                    with lfts. 3–6 cm. long.
                    M. Plants essentially glabrous, the stipules
                        mostly more than half as long as lfts.,
                        the latter usually 8–14
                                              15. *L. pauciflorus*
                    MM. Plants usually pubescent, or stipules
                        linear and less than half as long as lfts.,
                        or latter fewer than 8.
                        N. Fls. pale lavender to pinkish violet
                            or pinkish-orchid, 8–16 mm. long;
                            lfts. 4–12, mostly 3–8 times as
                            long as broad
                                              16. *L. Lanszwertii*
                        NN. Fls. rose, blue or purple; lfts. 4–
                            10, usually 6–8, usually not more
                            than 3 times as long as broad
                                              19. *L. nevadensis*

1. **L. Aphàca** L. Glabrous annual with slender stems, angled but not winged, 2–6
dm. long; stipules 1–4 cm. long, lanceolate to ovate, hastate-sagittate at base; lfts.
lacking, the stiff rachis ending in an unbranched tendril; fls. usually 1, lemon-yellow;
calyx 7–10 mm. long, the narrow lobes ca. twice the tube; banner 10–14 mm. long;
pods 2–4 cm. long, 3–5 mm. broad; $n = 7$ (Senn, 1938).—Occasional escape in n.
Calif.; native of Eurasia. Spring and summer.

2. **L. sphaèricus** Retz. Glabrous annual, the narrowly winged or wingless stems 2–5
dm. long; stipules narrow, ca. ⅓ as long as lfts., the latter 2, narrow, 3–6 cm. long;
upper lvs. with tendrils; fl. 1, brick-red, ca. 1 cm. long; calyx 5–8 mm. long, the narrow
lobes equaling or longer than tube; pods 3–6 cm. long, 3–5 mm. wide, glabrous; $n = 7$
(Senn, 1938).—Reported from Mt. St. Helena, Napa Co.; native of Eurasia.

3. **L. hirsùtus** L. Sparingly hirsute annual, 2–10 dm. tall; stipules narrow, mostly entire, scarcely ¼ as long as the 2 elliptic or narrower, 3–8 cm. long lfts.; tendrils branched; fls. red to bluish, 9–12 mm. long, 1–2(–4), on long peduncles; calyx 5–7 mm. long, the lanceolate or broader teeth ca. as long as tube; banner ca. twice as long as claw; pods 2.5–4 cm. long, 5–8 mm. broad, hirsute with hairs having pustulate bases; n = 7 (Senn, 1938).—Occasional escape as at Smith Flat, Eldorado Co.; native of Eu. May–July.

4. **L. tingitànus** L. TANGIER PEA. Glabrous annual, with narrowly winged stems 8–20 dm. tall; stipules narrowly to broadly lanceolate, entire or toothed, ½–⅔ as long as lfts.; these 4, linear to lance-elliptic, 4–10 cm. long; tendrils pinnate; fls. 1–3, rose-purple, 2.5–3 cm. long; calyx ca. 1 cm. long, the triangular-lanceolate teeth shorter than tube; claws shorter than blades of petals; pods glabrous, 7–10 cm. long, 7–9 mm. wide; n = 7 (Senn, 1938).—Often forming large masses on bluffs etc. along the coast, from Ventura Co. n.; native of Eurasia. May–July.

5. **L. odoràtus** L. COMMON SWEET PEA. Crisp-hairy annuals with winged stems 8–30 dm. tall; stipules linear-lanceolate, 2-lobed, ¼–½ as long as the 2 oval to elliptic lfts.; tendrils pinnate; fls. 2–5, fragrant, varicolored, 2–3 cm. long; calyx 12–15 mm. long; pods 3–6 cm. long, 4–7 mm. broad, rough-hairy; n = 7 (Senn, 1938).—Waste places and banks, especially near the coast, as garden escape; native of Medit. region. April–June.

6. **L. latifòlius** L. EVERLASTING PEA. Glabrous perennial from rootstocks; stems broadly winged, 6–20 dm. tall; stipules lanceolate, 3–5 cm. long, mostly entire; lfts. 2, lance-elliptic to -obovate, 5–10(–15) cm. long; tendrils well developed; fls. 5–15, purplish-red to white, on long peduncles; calyx 8–12 mm. long, the narrow teeth ca. as long as tube; petals short-clawed; pods 6–10 cm. long, 7–10 mm. broad; n = 7 (Senn, 1938).— Garden escape especially in n. Calif.; native of Eu. May–Sept.

7. **L. japónicus** Willd. var. **glàber** (Ser.) Fern. [*Pisum maritimum* var. *glabrum* Ser. ex DC. *L. maritimus* Calif. refs. *L. californicus* Dougl. ex Lindl.] Glabrous or sparsely pubescent perennial, the stems 2–10 dm. long, stout, decumbent, angled; stipules obliquely hastate or sagittate-ovate, equal to or longer than lfts.; lfts. 6–12, oblong to ovate, ± fleshy, 3–6 cm. long; tendrils well developed; fls. 2–8, purple or wings and keel whitish, 2–2.5 cm. long, claw of banner ca. as long as blade; pods 3–7 cm. long, ca. 1 cm. broad, mostly pubescent; n = 7 (Hitchcock, 1952).—Coastal Strand, Del Norte Co. n.; to Alaska, Minn., Mich., Ohio. May–July.

8. **L. littoràlis** (Nutt. ex T. & G.) Endl. [*Astropha l.* Nutt. ex T. & G.] Perennial from widespread rootstocks, densely silky-villous, the stems 1–6 dm. long, prostrate to erect; stipules lanceolate to oblique-ovate, mostly as long as and wider than lfts.; lfts. 4–8, oblance-oblong, 1–2 cm. long; rachis broad, prolonged as a broad leaflike bristle; fls. 2–6(–10), white to pink or purple, 12–18 mm. long; calyx 8–11 mm. long, the teeth subequal, lanceolate, ca. as long as tube; banner 14–18 mm. long, the blade longer than claw; pods villous, ca. 3 cm. long, 1 cm. broad, 1–5-seeded.—Coastal Strand; Monterey Co. to B.C. April–July.

9. **L. palústris** L. Perennial, from rootstocks, glabrous or nearly so; stems 3–10 dm. tall, scandent, mostly winged; stipules not more than half as long as lfts., lanceolate to sagittate-ovate, ± dentate; lfts. mostly 6, linear to ovate, 2–7 cm. long; tendrils well developed; fls. 2–5(–8), pink to bluish-purple, 12–20 mm. long; calyx 8–12 mm. long, the upper teeth broadly triangular-oblong, shorter than the mostly lanceolate, lower laterals; banner 15–22 mm. long, the claw shorter than blade; pods 4–6 cm. long, 4–6 mm. broad, 5–8-seeded, glabrous; n = 7 (Senn, 1938).—Moist or wet places along the coast; Humboldt and Del Norte cos.; to Alaska, e. U.S. June–Aug. Plants with copious short soft pubescence are the var. *pilòsus* (Cham.) Ledeb.; with the sp.

10. **L. polyphýllus** Nutt. ex T. & G. [*L. ecirrhosis* Heller.] Subglabrous perennial from rootstocks; stems 4–10 dm. tall, erect to scandent, not winged, largely leafless below at anthesis; stipules sagittate-ovate, acuminate, erose-dentate, over half as long as lfts.; these 10–16, scattered, 2.5–6 cm. long, lance-ovate to -elliptic; tendrils small; fls. 5–13, in 1-sided racemes, bluish-purple, 16–20 mm. long; calyx 7–13 mm. long, the teeth ciliate, the upper deltoid to lanceolate, the lower laterals longer, narrower; banner-blade ca. as long as claw; wings and keel paler than banner; pods 4–7 cm. long, 4–9 mm. broad, glabrous, many-seeded.—Fairly dryish places below 4000 ft.; N. Coastal Conif-

erous F., Douglas-Fir F., Yellow Pine F., Coastal Prairie; Lake and Mendocino cos. to Del Norte Co.; to Wash. May–July.

11. **L. vestìtus** Nutt. ex T. & G. [*L. quercetorum* Heller.] Mostly whitish-pubescent but sometimes glabrous perennial, seldom over 4 dm. high; lvs. crowded, the stipules ¼–⅔ as long as lfts., usually lanceolate, undulate-dentate; lfts. mostly 10, narrowly oblong-lanceolate, 2–3.5 cm. long; tendrils well developed; fls. 5–20, white or cream with purple to pink lines, to pink or lavender, usually fading yellowish, 15–20 mm. long; calyx 12–15 mm. long, hairy, the upper deltoid-lanceolate teeth less than half as long as lower laterals; banner pink to orchid-lavender or blue-purple, 17–22 mm. long, the claw some shorter than blade; pods 4–6 cm. long, 4–7 mm. broad, pubescent.— Brushy and wooded places, below 4000 ft.; Mixed Evergreen F., etc.; region about San Francisco Bay. April–June.

Ssp. **Bolánderi** (Wats.) C. L. Hitchc. [*L. B.* Wats.] Plants mostly over 4 dm. tall, essentially glabrous, usually scandent; lfts. mostly 2.5–5 cm. long; corollas pale pink to deep orchid or reddish-purple, drying yellowish or orange.—Bushy and wooded places; Redwood F., Mixed Evergreen F., etc.; mostly from San Francisco Bay to sw. Ore.

Ssp. **pubérulus** (White ex Greene) C. L. Hitchc. [*L. p.* White ex Greene. *L. violaceus* Greene, in part. *L. polyphyllus* var. *insecundus* Jeps.] Plants usually over 4 dm. tall, mostly scandent and pubescent; fls. pinkish to bluish-lavender.—Chaparral, Mixed Evergreen F., etc.; Los Angeles Co. to Humboldt Co.

12. **L. Jepsònii** Greene. Glabrous perennial, mostly 10–25 dm. tall, suberect to scandent, with winged stems; stipules lanceolate, ¼–½ the length of the lfts., undulate-dentate; lfts. 10–14, ± scattered, elliptic to lanceolate or ovate-elliptic, 4–7 cm. long; tendrils well developed; fls. 10–20, mostly crimson or rose-purple, ca. 2 cm. long; calyx 13–16 mm. long, subglabrous save for the ciliate teeth, the upper teeth ± deltoid, ⅓–½ as long as lower laterals; banner 18–22 mm. long, the claw almost equal to blade in length; pods 5–9 cm. long, 6–9 mm. wide.—Freshwater Marsh; Suisun and San Pablo bays. May–June.

Ssp. **califórnicus** (Wats.) C. L. Hitchc. [*L. venosus* var. *c.* Wats. *L. Watsoni* White.] Herbage and calyx puberulent to pubescent; banner pale pinkish to lavender or reddish-lavender; $n = 7$ (Hitchcock, 1952).—Along watercourses and on sandy slopes, below 5000 ft.; many Plant Communities; Fresno and San Luis Obispo cos. to Siskiyou and Humboldt cos., coast to Sierra Nevada. April–June.

13. **L. laètiflorus** Greene. Pubescent to subglabrous perennial, scandent, the stems not winged, 1–3 m. high; stipules lanceolate to lance-ovate, usually dentate, ¼ to almost as long as lfts.; these mostly 8–12, linear to ovate, 2–5 cm. long; tendrils well developed; fls. 5–12, whitish to pale flesh-color, 16–22 mm. long; pedicels mostly 2–5 mm. long; calyx 10–14 mm. long, the upper teeth triangular, shorter than the narrower lower laterals; banner obcordate, usually reflexed at less than a right angle to axis of keel, the claw less than half as long as blade; pods 4–8 cm. long, 5–8 mm. wide, many-seeded.— Dry places, below 5000 ft.; Coastal Sage Scrub, Chaparral; Santa Monica Mts. to San Jacinto and Santa Rosa mts. April–June.

Ssp. **bárbarae** (White) C. L. Hitchc. [*L. violaceus* var. *b.* White. *L. strictus* vars. *barbarae* and *Thacherae* Jeps.] Corolla pink to lavender.—Coastal Sage Scrub and Chaparral; near the coast from Santa Barbara Co. to Orange and w. Riverside cos.; Santa Cruz, Santa Rosa, and San Clemente ids.

Ssp. **Aleféldii** (White) Brads. [*L. A.* White. *L. strictus* var. *A.* Jeps. *L. laetiflorus* var. *A.* Jeps. *L. strictus* Nutt., not Grauer.] Pedicels 3–10 mm. long; fls. usually over 2 cm. long, deep red, the banner reflexed at more than a right angle to keel-axis; $n = 7$ (Hitchcock, 1952).—Chaparral, Coastal Sage Scrub; Catalina Id. and w. Riverside and s. Orange cos. to n. L. Calif.

14. **L. spléndens** Kell. PRIDE OF CALIFORNIA. Subglabrous to pubescent perennial, the stems 0.5–3 m. tall, not winged; stipules narrow-lanceolate, ¼–⅓ the length of the lfts., mostly few-toothed; lfts. mostly 6–10, linear to oval, 2–7 cm. long; tendrils well developed; fls. 4–12, deep rich red or crimson, 3–4 cm. long; calyx pubescent, 8–12 mm. long, the upper teeth deltoid-lanceolate, ca. ½ as long as lower laterals; banner reflexed so as to lie in almost the same plane as the keel, obcordate; pods 5–8 cm. long, 5–9 mm. broad, glabrous; $n = 7$ (Hitchcock, 1952).—Dry slopes, below 3500 ft.; Chaparral; near the Mexican border, e. San Diego Co., adjacent L. Calif. April–June.

15. **L. pauciflòrus** Fern. ssp. **Brównii** (Eastw.) Piper. [*L. B.* Eastw. *L. Lanszwertii* var. *B.* Jeps. *L. Schaffneri* Rydb., as to Calif. refs.] Perennial, glabrous except for calyx-teeth; stems angled, erect, 2–6 dm. tall, sturdy; stipules obliquely lance-ovate, mostly dentate, ca. ⅓–⅔ as long as lfts., sometimes narrower; lfts. thickish, subleathery, mostly 8–10, linear to ovate, 1.5–3.5 cm. long; tendrils simple or forked; fls. mostly 4–7, orchid or pinkish-lavender to violet-purple, aging bluish, 13–17 mm. long; calyx strongly 10-nerved, 7–10 mm. long, glabrous or the teeth ciliate, upper 2 teeth triangular to lance-linear, much shorter than the linear to lanceolate lower 3; banner reflexed ca. 90°, orchid to lilac, wings and keel mostly paler; fruits 3–5 cm. long, 3–6 mm. wide, glabrous. —Dry slopes, 4000–6000 ft.; Yellow Pine F.; Tehachapi Mts. through Sierra Nevada to Siskiyou Co. and in Coast Ranges s. to Tehama Co.; Ore. April–June.

16. **L. Lanszwértii** Kell. [*L. Goldsteinae* Eastw.] Mostly soft-pubescent perennial, 2–8 dm. tall, clambering to erect, angled; stipules mostly narrow, ¼–¾ as long as lfts., entire on margins; lfts. 4–10, usually paired, linear-lanceolate to oblong-elliptic, 1.5–3.5 cm. long, rather coriaceous; tendrils usually well developed; fls. 2–8 in no., 13–16 mm. long; calyx 5–8 mm. long, the tube mostly pubescent, the teeth ciliate, lanceolate, shorter than the tube; banner lavender to pinkish-orchid with deeper lines, reflexed at more than right angle; wings lavender to white, keel also pale; pods glabrous, 4–6 cm. long, 3–6 mm. broad; *n* = 7, 14 (Hitchcock, 1952).—Dry slopes, 4000–6500 ft.; Yellow Pine F.; Sierra Nevada from Eldorado Co. n.; to Wash., Ida., Utah. May–July.

Ssp. **áridus** (Piper) Brads. [*L. coriaceus a.* Piper.] Lfts. mostly linear to linear-elliptic, 3–6 cm. long; tendrils bristlelike or short and scarcely prehensile; fls. nearly white, ca. 1 cm. long, the banner with some lavender.—With the sp., n. to Wash., Ida.

17. **L. sulphùreus** Brewer ex Gray. Perennial from taproots and branched crown, glabrous except for the ciliate calyx-teeth; stems 0.5–3 m. tall, angled, mostly clambering; stipules lanceolate to ovate, often longer than lfts., ± dentate; lfts. mostly 8–10, narrowly lanceolate-elliptic to ovate, 2–5 cm. long, the tendrils well developed; fls. 10–20, mostly secund, 10–15 mm. long, tan to yellowish with some purple tinge or orange; calyx glabrous, 9–12 mm. long, teeth ciliate, the upper deltoid, much shorter than lower, narrower 3; banner with very broad claw, keel much recurved; pods 4–7 cm. long, 4–6 mm. broad, glabrous; *n* = 7 (Hitchcock, 1952).—Dry slopes, below 8000 ft.; Foothill Wd., Yellow Pine F., Red Fir F.; Sierra Nevada from Kern Co. n., Coast Ranges from Lake Co. n.; to s. Ore. April–July.

Var. **argillàceus** Jeps. Plants hairy throughout.—Tehama and Shasta cos.

18. **L. delnórticus** C. L. Hitchc. Perennial from rootstocks, 2–8 dm. tall, sprawling to scandent; stipules ¼–½ as long as lfts., lanceolate to ovate, ± undulate-dentate; lfts. 10–16, usually scattered, 2.5–5 cm. long, narrow-elliptic; tendrils usually branched; fls. 8–20, white but aging tan or yellow-brown, 12–15 mm. long; calyx 8–11 mm. long, glabrous except the ciliate teeth, the upper teeth deltoid, ca. ⅓ the length of the lower lance-oblong laterals; banner reflexed at more than 90°, white to orchid, penciled, claw longer and narrower than blade, wings and keel white; pods 3–4 cm. long, 3–5 mm. broad.—Rather moist to dryish places; N. Coniferous F., Mixed Evergreen F., Redwood F.; Del Norte Co. to sw. Ore. June–July.

19. **L. nevadénsis** Wats. [*L. venosus* var. *obovatus* Torr.] Perennial from rootstocks, ± pubescent, the stems erect, 1.5–3.5 dm. high, angled; stipules narrow, less than half as long as lfts.; lfts. 4–8, paired or scattered, linear to lance-ovate, 2–10 cm. long; tendrils lacking or poorly developed; fls. 2–7, mostly pale to dark blue, reddish or purple, 17–27 mm. long; calyx 6–10 mm. long, ± hairy, the teeth subequal, lanceolate to lance-linear, shorter than tube; banner obcordate, strongly reflexed, the claw ca. as wide as blade; wings and keel usually paler than banner; pods 3–7 cm. long, 4–9 mm. broad, glabrous.—Dry slopes, 1500–7000 ft.; mostly Yellow Pine F., Red Fir F.; w. base of Sierra Nevada from Fresno Co. to Butte Co., Coast Ranges from Humboldt Co. n.; to Ore. April–June.

Ssp. **lanceolàtus** (Howell) C. L. Hitchc. [*L. l.* Howell. *L. Nuttallii* subsp. *l.* Piper. *L. Nuttallii* Wats. *L. nevadensis* ssp. *l.* var. *N.* C. L. Hitchc.] Plants mostly more than 3 dm. tall, scandent; lfts. 8–14; tendrils mostly present; fls. 13–20 mm. long, mostly bluish or reddish; *n* = 14 (Hitchcock, 1952).—Sierra Nevada from Madera Co. n. to Siskiyou Co.; to B.C.

20. **L. Tràcyi** Brads. [*L. Bolanderi* var. *T.* Jeps.] Glabrous to pubescent perennial

from rootstocks, the stems ± angled, erect to ± scandent, 1.5–5 dm. tall; stipules lance-olate, less than ½ as long as lfts.; lfts. mostly 4 or 6, linear to oblong or obovate, 1–7 cm. long; tendril from a mere bristle to pinnate, fls. 3–10, white or cream, aging brownish, 9–12 mm. long; calyx glabrous to hairy, 5–8 mm. long, the linear-lanceolate lobes unequal, up to half the tube; banner with very broad claw; pods glabrous, 4–6 cm. long, 4–6 mm. broad; $n = 7$ (Hitchcock, 1952).—Dry slopes, below 3500 ft.; Mixed Ever-green F., Yellow Pine F.; Coast Ranges from Glenn Co. to Siskiyou Co. May–June.

21. **L. rígidus** White. [*L. albus* Wats., not Garcke.] Glabrous, ± glaucous perennial from a root-crown and taproot, 1.5–4 dm. tall, erect, angled; stipules almost as long as lfts., often toothed, the upper lobes broadly lanceolate, the lower much shorter than upper; lfts. mostly 6–10, oblanceolate, 1.5–3 cm. long, mucronulate; tendril bristle-like; fls. 2–5, white or pinkish, 20–27 mm. long; calyx 7–9 mm. long, the ciliate teeth lanceolate, ca. ⅔ as long as tube; banner with broad claw; pods 3–5 cm. long, 6–9 mm. broad, glabrous; $n = 7$ (Hitchcock, 1952).—At 2500–4000 ft., Sagebrush Scrub; Modoc Co. to e. Ore., w. Ida. April–June.

22. **L. Tórreyi** Gray. Finely villous perennial from rootstocks, the stems 1–4 dm. long, erect to decumbent, not scandent, angled; stipules lanceolate, the upper lobe much longer than lower; lfts. 10–16, elliptic to oval, 5–25 mm. long, apiculate; tendril a short unbranched bristle; fls. 1–2, pale lilac to blue, 8–13 mm. long; calyx 6–10 mm. long, the 2 upper teeth much shorter than the 3 lower; banner pale lilac to blue, the claw narrowed somewhat, wings and keel paler; pods ca. 2 cm. long, 4–5 mm. broad, pubes-cent; $n = 7$ (Hitchcock, 1952).—Open woods, below 1500 ft.; Redwood F., Mixed Evergreen F.; Santa Cruz Co. n.; to Wash. May–June.

34. Vícia L. VETCH. TARE

Herbs, mostly vinelike, with pinnate usually tendril-bearing lvs. and evident stipules. Fls. or peduncles axillary, the fls. 1–many, purple to yellowish or white. Calyx 5-cleft or -toothed, the 2 upper teeth often shorter, or the lowest longer. Petals clawed, the wings adhering to the middle of the keel. Stamens ± diadelphous (9 and 1), the orifice of the tube oblique. Style filiform, hairy all around or only on the back at the apex or beneath the stigma. Pod mostly laterally compressed, 2-valved, 2–several-seeded. Seeds globular to somewhat compressed. Ca. 130 spp., of N. Hemis. and temp. S. Am. (The classical Latin name.)

A.   Fls. borne near the ends of evident peduncles.
　　B.   The fls. 2–5 mm. long.
　　　　C.   Pods glabrous; calyx-teeth very dissimilar.
　　　　　　D.   Lfts. 2–6 pairs; pods 3–4 mm. broad, 3–6-seeded.
　　　　　　　　E.   Calyx-teeth ca. as long as the tube; pods 8–12 mm. long ... 1. *V. tetrasperma*
　　　　　　　　EE.   Calyx-teeth ca. ¼ as long as tube; pods 20–30 mm. long .. 2. *V. exigua*
　　　　　　DD.   Lfts. 8–10 pairs; pods 5–7 mm. wide, 2-seeded .... 3. *V. disperma*
　　　　CC.   Pods hirsute-pubescent; calyx-teeth almost alike .... 4. *V. hirsuta*
　　BB.   The fls. 10–18 mm. long.
　　　　C.   Lfts. 4–8 pairs; peduncles 3–9- or 10-fld.
　　　　　　D.   Plants perennial, native; racemes mostly shorter than lvs.
　　　　　　　　E.   Stems subglabrous; pods glabrous; corolla 16–18 mm. long 5. *V. americana*
　　　　　　　　EE.   Stems villous-tomentose; pods pubescent; corolla 12–14 mm. long
　　6. *V. californica*
　　　　　　DD.   Plants annual, introd.; racemes equaling or exceeding lvs.   8. *V. benghalensis*
　　　　CC.   Lfts. mostly 8–12 pairs, sometimes only 6; peduncles with 1-sided dense many-fld. (10 or more) racemes.
　　　　　　D.   Plant perennial; corolla 10–12 mm. long.
　　　　　　　　E.   Stipules entire; pods 2–2.4 cm. long; banner-blade ca. as long as claw. Introd. .... 7. *V. Cracca*
　　　　　　　　EE.   Stipules sharply dentate; pods 3–4 cm. long; banner-blade shorter than claw. Native .... 9. *V. gigantea*
　　　　　　DD.   Plants annual; corolla 14–15 mm. long. Introd.
　　　　　　　　E.   Stems spreading-villous; lfts. to ca. 1 cm. long ...... 10. *V. villosa*
　　　　　　　　EE.   Stems glabrous or ± strigose; lfts. 1–2 cm. long .... 11. *V. dasycarpa*
AA.   Fls. 1–4 in lf.-axils, sessile or subsessile. Introd.
　　B.   Lvs. without tendrils; lfts. 1–3 pairs; stems mostly simple, erect .... 12. *V. Faba*
　　BB.   Lvs. with tendrils; lfts. more than 3 pairs.
　　　　C.   Banner pubescent on back; fls. 2–4, yellow-white, 20–25 mm. long .. 13. *V. pannonica*
　　　　CC.   Banner glabrous on back; fls. mostly 1–2.
　　　　　　D.   Pods long-hairy; fls. 20–25 mm. long, white with some yellow .... 14. *V. lutea*

DD. Pods short-pubescent to glabrous; fls. 10–25 mm. long, violet-purple.
  E. The fls. 1.8–2.5 cm. long, purple with violet wings; pod ± torulose; seeds
    5 mm. broad .... 15. *V. sativa*
  EE. The fls. 1–1.8 cm. long, uniformly purple; pod plane; seeds 3 mm. broad
    16. *V. angustifolia*

1. **V. tetraspérma** (L.) Moench. Annual with very slender branched stems 3–5 dm. long; stipules entire, small; lfts. 2–5 pairs, oblong-linear, mucronulate; tendrils simple or forked; peduncles filiform, exceeding lfts., 1–2-fld.; calyx ca. 1.5 mm. long, upper teeth triangular, lower lanceolate; corolla lilac, ca. 4 mm. long; pods glabrous, 1–1.3 cm. long; seeds 3–5, subglobose, ca. 2 mm. in diam.; $2n = 14$ (Sweschnikova, 1927).—Natur. n. Calif.; native of Eu. May–July.

2. **V. exígua** Nutt. in T. & G. Slender-stemmed annual, 3–6 dm. high, climbing, ± strigose-villous; stipules entire, semisagittate; lfts. 2–6 pairs, linear, 1–2.5 cm. long, rounded to obtuse or emarginate, ± strigose especially beneath; tendrils well developed; peduncles filiform, shorter than the lvs., 1–2-fld.; calyx ca. 2 mm. long, the teeth triangular-subulate; corolla white to purplish, ca. 5 mm. long; pods glabrous, 2–3 cm. long; seeds 4–5, 2–2.5 mm. in diam.—Grassy, brushy or wooded slopes, below 2000 ft.; many Plant Communities; Coast Ranges; to Ore., L. Calif. April–June.

Var. **Hássei** (Wats.) Jeps. [*V. H.* Wats.] Stouter; lfts. linear-oblong to oblong, emarginate; fls. 6–7 mm. long.—With the sp., from Marin and Alameda cos. to L. Calif.

3. **V. dìsperma** DC. Somewhat pilose annual, 2–6 dm. high; lfts. linear-oblong, mucronate, 8–10 pairs; peduncles shorter than lvs., 2–5-fld.; tendrils branched; fls. bluish, 4–5 mm. long; calyx-teeth unequal; pods 12–16 mm. long, oblong, 2-seeded, 5–7 mm. wide; seeds black; $2n = 14$ (Heitz, 1931).—Occasional introd.; native of Medit. region.

4. **V. hirsùta** (L.) S. F. Gray. [*Ervum h.* L.] Slender-stemmed annual, 1.5–6 dm. high, subglabrous; stipules linear, sometimes toothed; lvs. 3–5 cm. long; lfts. mostly 6–8 pairs, linear to linear-oblong, 0.5–1.5 cm. long, glabrous to sparsely pubescent, especially beneath, truncate to emarginate; peduncles slender, ca. half as long as lvs., 2–6-fld.; calyx-teeth somewhat longer than tube; corolla pale blue-purple, 2–4 mm. long; pods hirsute-pubescent, 6–10 mm. long, 2-seeded; $2n = 14$ (Heitz, 1931).—Occasionally natur. in cent. and n. Calif.; native of Eu. May–July.

5. **V. americàna** Muhl. ssp. **oregàna** (Nutt.) Abrams. [*V. o.* Nutt. in T. & G. *V. Copelandii* Eastw.] Trailing or climbing perennial, sparsely pubescent, 6–12 dm. tall; stipules incisely toothed; lfts. 4–8 pairs, ovate- to oblong-elliptic, rounded to obtuse at apex, mucronulate, 1–4 cm. long, ± pubescent; peduncles shorter than lvs., 4–9-fld.; calyx deltoid-lanceolate, somewhat unequal, shorter than the tube; corolla 16–18 mm. long, purplish aging blue; pods glabrous, 3–4 cm. long, 7–9 mm. wide; seeds several, dull black, ca. 4 mm. in diam.—Open places, below 5000 ft.; many Plant Communities; most of cismontane Calif., occasional in Inyo and Mono cos., etc.; to B.C., Ida. April–June. The following rather trivial segregates have been made:

Var. **truncàta** (Nutt.) Brew. [*V. t.* Nutt. in T. & G. *V. pumila* Heller.] Lfts. of upper lvs. truncate at apex and few-denticulate.—With the sp.

Var. **lineàris** Wats. Lfts. narrowly linear.—With the sp.

6. **V. califórnica** Greene. [*V. Durbrowii* Eastw. *V. truncata* var. *villosa* Kell.] Much like *V. americana*, but villous-pubescent throughout, the rather stiff, ± zigzag stems 2–3 dm. long; lfts. 4–6 pairs, elliptic to cuneate-obovate, often 3–5-denticulate at the ± truncate apex; stipules laciniate; fls. 3–6, deep purple, 12–14 mm. long.—Open slopes, mostly 2000–8000 ft.; mostly Montane Coniferous F., Mixed Evergreen F.; cismontane Calif. to s. Ore. April–July. A form from Madera Co. with lfts. strongly serrate above the base has been described as var. *madrénsis* Jeps.

7. **V. Crácca** L. A ± pubescent perennial with slender climbing or trailing stems 5–10 dm. long; stipules entire; lfts. 6–12 pairs, linear, 15–30 mm. long, acute to obtuse; racemes 1-sided, equal to or longer than lvs., densely many-fld.; calyx-teeth unequal, lance-attenuate, pubescent, shorter than tube; corolla bright blue-purple, 10–12 mm. long; pods glabrous 2–2.4 cm. long; seeds few, dull black; $2n = 12, 14, 28$ (Sweschnikova, 1927).—Sparingly natur. in fields, 4000–5000 ft., n. Calif.; native of Eurasia. June–Aug.

8. **V. benghalénsis** L. [*V. atropurpurea* Desf.] Soft-pubescent annual, 3–8 dm. high; stipules toothed; lfts. 5–8 pairs, elliptic to oblong, 1.5–2.5 cm. long; racemes equaling to exceeding lvs., 3–10-fld.; calyx villous, the teeth subulate, unequal, the lower exceeding

the tube; corolla 12–14 mm. long, rose-purple, darker in age; pods 2.5–3.5 cm. long, 8–10 mm. wide, soft-pubescent, 4–6-seeded.—Natur. in waste places, San Francisco Bay region n.; native of Medit. region. May–June.

9. **V. gigantèa** Hook. [*Lathyrus cinctus* Wats.] Stout-stemmed perennial, somewhat pubescent, 6–10 dm. high, turning almost black on drying; stipules broad, to 2 cm. long, sharply-dentate; lfts. mostly 8–12 pairs, lance-oblong to narrowly oblong, obtuse to rounded at apex, 1.5–3.5 cm. long, sparsely strigose; racemes shorter than lvs., 1-sided, densely rather many-fld.; calyx ca. 7 mm. long, the teeth very unequal, shorter than the tube, upper teeth ± deltoid, lower lanceolate; corolla reddish-purple, ca. 12 mm. long; pods oblong, 3–4 cm. long; ca. 1 cm. wide; seeds several, globose, black, 4 mm. in diam.—Moist places, largely Redwood F., Mixed Evergreen F., N. Coastal Scrub; San Luis Obispo Co. to Alaska. March–June.

10. **V. villòsa** Roth. WINTER VETCH. Annual or biennial, spreading-villous, the stems 6–12 dm. long; stipules narrow, entire, to ca. 1 cm. long; lfts. 8–12 pairs, linear to narrow-oblong, mucronate, 1–2.5 cm. long; racemes equaling or exceeding lvs., 1-sided, densely many-fld.; calyx 5–6 mm. long, the unequal teeth linear-acicular, villous, the lower as long as the gibbous tube; corolla violet and white, 14–15 mm. long; pods oblong, 2–3 cm. long, 7–10 mm. wide, glabrous; seeds 4–6, globular; $2n = 14$ (Senn, 1938).—Natur. in waste places, cent. Calif. n., more sparingly so in s. Calif.; native of Eu. April–July.

11. **V. dasycárpa** Ten. Much like *V. villosa*, but glabrous or sparingly appressed-pubescent, lfts. mostly 1–2 cm. long; fls. fewer, 12–17 mm. long; calyx-teeth sub-glabrous, the lower shorter than the tube; corolla mostly purple-violet; pods 2–4 cm. long, 7–10 mm. wide; $2n = 14$ (Schwesnikova, 1927).—Natur. in Marin Co.; native of Medit. region. Spring.

12. **V. Fàba** L. HORSE-BEAN. Annual, erect, glabrous, 4–6 dm. high; stipules half-cordate, mostly entire; lfts. 2–3 pairs, ovate to elliptic, 4–10 cm. long, obtuse to acute; fls. 2–3 cm. long, subsessile, axillary, dull white with a large purple spot on the wings; pods 8–12 cm. long, 1–2 cm. wide, 2–5-seeded, the seeds flat, 1–3 cm. long; $2n = 12$, 14 (several auth.).—Grown widely in Calif. and escaping; native of Eurasia.

13. **V. pannónica** Crantz. HUNGARIAN VETCH. Reclining or climbing annual, the stems 2–5 dm. long, weakly pubescent; stipules small; lfts. 7–9 pairs, linear to oblong, villous, 1–1.5 cm. long; fls. 2–4, short-peduncled, yellowish-white, 15–18 mm. long; pods nodding, 2.5–3 cm. long, 7–9 mm. broad, hairy; $2n = 12$ (Heitz, 1931).—Sparingly natur. as in Sonoma Co.; native of Eu.

14. **V. lùtea** L. Pubescent to suglabrous annual, 2–6 dm. high; stipules small; lfts. 4–8 pairs, lance-linear, 1–2 cm. long, rounded to acutish apically; fls. 1(–3), subsessile, white with some yellow, 2–2.5 cm. long; pods 2.5–3 cm. long, 8–10 mm. wide, hairy; seeds 5–9, ca. 3 mm. in diam.; $2n = 14$ (Heitz, 1941).—Local, established in San Francisco Bay Region; native of S. Eu. and N. Afr. April–June.

15. **V. satìva** L. SPRING VETCH. Pubescent annual or subglabrous, 3–8 dm. high; stipules 5–15 mm. long, ± laciniate-toothed; lfts. 4–8 pairs, lance-oblong to obovate, pubescent when young, truncate to emarginate, 1–3 cm. long; fls. 1–2, subsessile, violet-purple 1.8–2.5 cm. long; calyx 1–1.5 cm. long, the teeth as long as tube; pods 4–8 cm. long, torulose, at first pubescent; seeds compressed-globose, 5 mm. broad; $2n = 12$, 14, (Coutinho, 1940, '45).—Natur. in waste places in much of cismontane Calif.; native to Eu. April–July.

16. **V. angustifòlia** Reichard. COMMON VETCH. Near to *V. sativa*, glabrous or glabrate; lfts. mostly 2–5 pairs, those of lower lvs. oblong, of upper linear, 1.5–3 cm. long; fls. 1–1.8 cm. long; calyx 7–11 mm. long; pods plane, 4–5 cm. long; seeds 3 mm. broad; $2n = 12$ (Schwesnikova, 1927).—Natur. in waste places, some of our plants being the var. **segetàlis** (Thuill.) Koch, with broader lfts. having a truncate to emarginate apex. April–June.

35. Pìsum L. PEA

Annual and perennial herbs with pinnate lvs., the pinnae 1–3 pairs, the rachis ending in a pinnate tendril. Stipules large and foliar. Fls. solitary or in small axillary racemes. Calyx oblique or gibbous at the base, with ± leafy lobes. Corolla white or colored, the

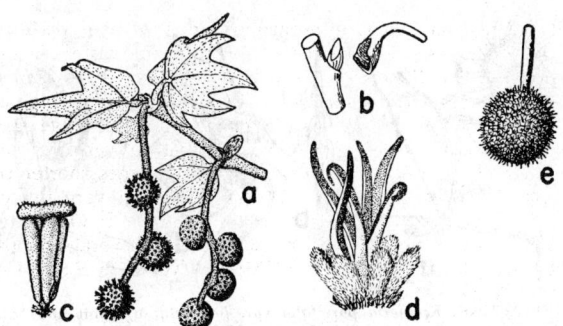

Fig. 74. PLATANACEAE. *Platanus racemosa: a*, young shoot, × ½, with ♂ heads to right, ♀ heads to left; *b*, petiole dilated at base and lifted off winter-bud; *c*, stamen, × 5; *d*, ♀ fl., × 5, pistils elongate, surrounded by bracts variously interpreted as staminodia or petals; *e*, fr., × ½.

wings somewhat adherent to the keel. Stamens 9 and 1. Style bearded down one side. Pod flattened, dehiscent, several-seeded. Ca. 6 spp., of Medit. region and w. Asia; grown for food and fodder. (*Pisum*, the classical name.)

1. **P. sativum** L. GARDEN PEA. Glabrous glaucous annual, 1–2 m. tall, scandent; stipules mostly larger than lfts., the lower part denticulate; lfts. entire, oval to oblong, 2–5 cm. long; peduncles exceeding the stipules; fls. 1–3, white, 1.5 or more cm. long; pod 5–10 cm. long; seeds 2–10, wrinkled or plane, mostly 4–8 mm. in diam.; $n = 7$ (Sansome, 1933).—Occasional waif from cult., especially in waste places. Spring.

Var. **arvénse** (L.) Poir. FIELD PEA. Peduncles short; fls. colored with violet banner and purplish wings; pods smaller.—Occasional escape.

76. Platanàceae. SYCAMORE FAMILY (Fig. 74)

Trees with thin exfoliating bark. Lvs. large, alternate, deciduous, palmately lobed. Stipules thin, sheathing, entire or toothed. Petioles dilated at the base and largely covering the buds. Infl. of spherical unisexual heads, in ours these racemose on long slender peduncles. Fls. imperfect, minute. Calyx of 3–8 minute scalelike sepals. Petals minute, 3–6, in the ♂ fls. cuneiform-sulcate, scarious-pointed and longer than the sepals, in the ♀ fls. acute and longer than the rounded sepals. Stamens as many as the sepals and opposite them. Carpels as many as the sepals, 1-loculed, surrounded by persistent hairs and a few staminodia. Style terminal with the stigmatic surface ventral. Ovule 1 (2). Fr. a dense globose head of aks. with intermingled hairs and staminodia. Seed elongate-oblong, pendulous, with fleshy endosperm.

1. Plátanus L. SYCAMORE. PLANE-TREE

Characters of the family. Nine spp., of the N. Temp.

1. **P. racemòsa** Nutt. [*P. californica* Benth.] Large tree, 10–25 m. tall, with smooth pale bark; young growth rusty-tomentose; lvs. 1.5–2.5 dm. broad, ca. as long, deeply 5-lobed, the lobes subentire, tomentose on both surfaces when young, glabrescent above; petioles 3–8 cm. long; stipules 2–3 cm. long; ♂ heads several, 8–10 mm. in diam.; ♀ heads 3–5, sessile, 2–2.5 cm. in diam. in fr.—Along stream beds and watercourses below 4000 ft.; many Plant Communities; Cent. V. and bordering mts., S. Coast Ranges through cismontane s. Calif. to L. Calif. Feb.–April.

77. Krameriàceae. KRAMERIA FAMILY (Fig. 75)

Shrubs or perennial herbs, mostly pubescent. Lvs. alternate, simple and entire, rarely 3-foliolate. Fls. irregular, axillary or racemose, ours purplish. Peduncles with paired opposite leafy bracts. Sepals 4–5, unequal. Petals 5, the 3 upper long-clawed, the other 2 sessile and smaller. Stamens 4 in ours, free or borne on the united claws of the upper

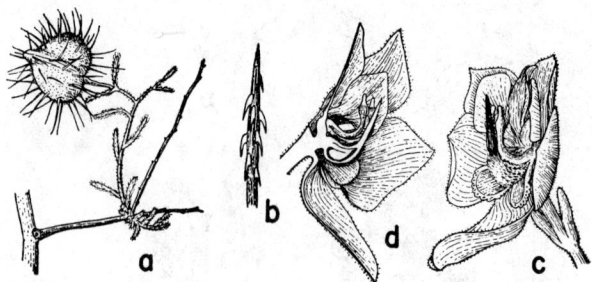

Fig. 75. KRAMERIACEAE. *Krameria parvifolia* var. *imparata: a,* habit and fr., $\times$ 1; *b,* spine of fr. barbed down the sides; *c,* fl., $\times$ 3, with 5 sepals, 5 petals (3 upper and 2 lower), 4 stamens, 1 carpel; *d,* fl. in section.

petals; anthers 2-celled, dehiscent by a pore. Ovary 1-celled; ovules 2; style cylindric. Fr. globose, indehiscent, spiny, 1-seeded. A single genus, with ca. 20 spp.

1. KRAMERIA L.

Characters of the family. (Named for J. G. H. *Kramer,* 18th-century Austrian botanist.)

(Britton, N. L., N. Am. Flora 23 [4]: 195–200, 1930.)

Barbs on spines of frs. in an umbrellalike group at apex; upper petals separate to the base
1. *K. Grayi*
Barbs on spines of frs. scattered along whole upper portion; upper petals partly united at base of claws
2. *K. parvifolia*

1. **K. Gràyi** Rose & Painter. [*K. canescens* Gray, not Willd.] Densely and intricately branched, thorny, spreading, 3–7 dm. high, of wider spread, silky-tomentose on young growth; lvs. linear to lanceolate, 8–18 mm. long; peduncles commonly 15–25 mm. long, with 1 pair of bracts; calyx 7–10 mm. long, red-purple within; upper petals spatulate, 4–5 mm. long, the lower cuneate-obovate, shorter; fr. densely canescent, 7–8 mm. in diam., armed with many slender spines each with a few barbs at very tip.—Dry sandy and rocky places, below 4000 ft.; Creosote Bush Scrub; Colo. Desert and s. Mojave Desert; to Tex., n. Mex. April–May.

2. **K. parvifòlia** Benth. var. **imparàta** Macbr. [*K. i.* Britt.] Like the preceding, but lvs. more narrowly linear, 2–12 mm. long; peduncles sericeous, 10–15 mm. long, with 2–3 pairs of lfy. bracts; sepals sericeous; spines on fr. with weak barbs along upper half or third.—Rather common; cent. Mojave Desert e. and s.; to Nev., Son., L. Calif. March–May.

Var. **glandulòsa** (Rose & Painter) Macbr. [*K. g.* Rose & Painter.] Peduncles and sepals with stalked glands.—At 2000–4000 ft.; Creosote Bush Scrub, Joshua Tree Wd.; Death V. Region, Kingston Mts., New York Mts., Santa Rosa Mts., etc.; to Nev., Tex.

78. Betulàceae. BIRCH FAMILY (Fig. 76)

Deciduous trees and shrubs with alternate simple, usually serrate, petioled lvs. and deciduous stipules. Monoecious, both kinds of fls. in aments and appearing before the lvs. Staminate aments pendulous, elongate, with 1–3 fls. in each bract-axil; calyx membranous, 2–4-parted or 0; stamens 1–10. Pistillate aments erect or drooping, relatively short; calyx 0 or adnate to the 1–2-celled ovary; style 2-cleft. Ovules 1–2 in each cell, pendulous. Fr. a small 1-seeded winged nutlet in conelike clusters or larger and enclosed in an invol. of enlarged bracts. Six genera of N. Hemis.

Lvs. cordate at base; fr. a nut enclosed in a leafy invol.; ♂ fls. solitary in scale-axils .... 1. *Corylus*
Lvs. not cordate at base; fr. conelike; ♂ fls. usually 3 in each axil.
   Pistillate catkins solitary, cylindrical in fr.; fruiting scales thin and 3-lobed, falling away separately
   at maturity . . . . 2. *Betula*
   Pistillate catkins in short clusters, short-ellipsoid in fr.; fruiting scales woody, obscurely toothed,
   falling away as a cone . . . . .3. *Alnus*

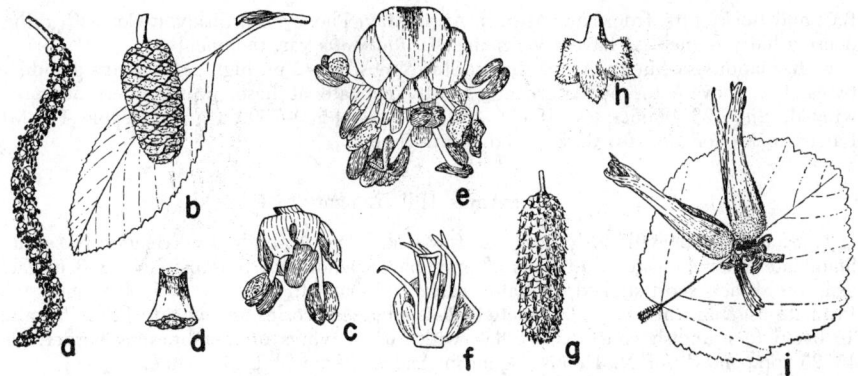

Fig. 76. BETULACEAE. *Alnus: a,* ♂ catkin, × ½, with peltate bracts; *b,* lf. and mature ♀ cone; *c,* ♂ fl., × 5, calyx and stamens; *d,* ♀ fl., × 2½, with connate structure around the pistil. *Betula: e,* ♂ fls., × 5; *f,* ♀ fls., 3 in a cluster, no perianth; *g,* ♀ catkin, × 1; *h,* 3-lobed bract from ♀ catkin. *Corylus: i,* lf., × ½, and tubular invols. (each inclosing a nut).

1. Córylus L. Hazelnut. Filbert

Lvs. thinnish, doubly toothed. Staminate aments single or fascicled from scaly buds in axils of preceding year, the ♀ terminating early leafy shoots; ♂ fls. with 8 half-stamens with 1-locular anthers, each anther-cell and its stalk representing half of a forked fil. Pistillate fls. several from a scaly bud, 2 to each bract, with 2 bractlets. Calyx adherent to ovary; style short; stigmas 2, elongate, bright red. Fr. a smooth nut inclosed in a leafy invol. made of the 2 bractlets enlarged and united. Ca. 15 spp., some in cult. for their nuts. (Greek, *corys,* a helmet, referring to invol.).

1. **C. cornùta** Marsh. var. **califórnica** (A. DC.) Sharp. [*C. rostrata* var. *c.* A. DC. *C. c.* Rose.] Open spreading shrub 2–6 m. high, with smooth bark, the glandular-pubescent branchlets glabrate in age; lvs. rounded to obovate, 4–7 cm. long, ± oblique at cordate base, doubly serrate, sometimes ± 3-lobed, pale beneath, soft-pubescent, ± glandular, glabrate in age; bracts of ♂ catkins pubescent, anthers pubescent at apex; invol. hispid, forming a tube 1.5–2.5 cm. long; nut ovoid, 1.2–1.5 cm. long.—Damp slopes and banks, below 7000 ft.; many Plant Communities; Coast Ranges from Santa Cruz Co. n., Sierra Nevada from Tulare Co. n.; to B.C. Jan.–April. Some plants have less pubescent lvs. and short involucral tubes (to ca. 1 cm.) and have been called *C. rostràta* var. *Tràcyi* Jeps.; they are mostly in nw. Calif.

2. Bétula L. Birch

Bark smooth, aromatic, with long lenticels and often separating in papery sheets. Lvs. variously toothed. Staminate aments solitary or clustered, elongate; fls. 3 in each bract-axil, the 2 lateral subtended by bractlets; calyx 2–4-lobed. Stamens 2, the fils. forked and ending in 2 pollen sacs. Pistillate fls. 3 in axils of bracts of erect catkins; sepals 0. Ovary sessile, style branched. Frs. small compressed nutlets, with winged margins. Cone-scales deciduous with nutlets. Ca. 30 spp., of cooler regions. (Ancient Latin name.)

Lvs. thin, 1.5–7 cm. long, sharply irregularly serrate .... 1. *B. occidentalis*
Lvs. thick, mostly less than 1 cm. long, rounded-crenate .... 2. *B. glandulosa*

1. **B. occidentàlis** Hook. [*B. fontinalis* Sarg.] Water Birch. Tall shrub or tree to 8 m. tall; bark dark bronze, shining, not peeling as layers; twigs rough with large resinous glands; lvs. broadly ovate, sharply serrate except near the entire base, gland-dotted when young, bicolored; fruiting aments 2–3 cm. long, the bractlets glabrous or puberulent, ciliate, with short lateral lobes; nutlet-body not narrower than wings.— Moist places, 2000–8000 ft.; mostly Montane Coniferous F.; Inyo Co. (Panamint Mts., White Mts., s. Sierra Nevada), Tulare Co., thence to Modoc and Humboldt cos.; to

B.C. and Rocky Mts. from New Mex. n. April–May. Plants from Siskiyou Co. with rather densely hairy branchlets have been named *B. fontinàlis* var. *inopìna* Jeps.

2. **B. glandulòsa** Michx. RESIN BIRCH. Shrub mostly 1–2 m. high, with warty glandular twigs; lvs. obovate to round, crenate, mostly cuneate at base, firm; nutlets obscurely winged; $2n = 28$ (Poucques, 1949).—Wet places, 6500–7500 ft.; Lodgepole F.; Mt. Lassen to Modoc Co.; to Alaska, Nfld. April–June.

3. **Álnus** Hill. ALDER

Trees and shrubs with scaly bark and few-scaled long lf.-buds. Lvs. dentate to serrate. Staminate aments clustered at ends of branchlets, pendulous. Staminate fls. 3–6 in each axil, the bract short-stalked, peltate; calyx 3–5-parted; stamens 3–5; bractlets 4–5. Pistillate fls. 2 in axils; calyx 0; bractlets 2–4, adnate to scales and woody in fr., forming an ovoid to roundish cone. Nutlet flat, with lateral wings or membranous border. Ca. 15–25 spp., mostly of N. Hemis., some in Andes. (Ancient Latin name.)

Fls. developed with the expanding lvs. on twigs of the season; peduncles longer than the cones; lvs.
  shining. Largely montane shrub . . . . 1. *A. sinuata*
Fls. developed before the lvs, and on last year's twigs; peduncles shorter than cones; lvs. dull.
  Shrubs 3–5 m. high; lvs. doubly serrate and lobed. High montane . . . . 2. *A. tenuifolia*
  Trees 10–25 m. tall; lvs. not lobed. Mostly below 6000 ft.
    The lvs. rusty pubescent beneath, revolute on edges; nutlets narrowly winged . . 3. *A. oregona*
    The lvs. green beneath, not revolute on edges; nutlets margined . . . . 4. *A. rhombifolia*

1. **A. sinuàta** (Regel) Rydb. [*A. viridis* var. *s.* Regel.] SITKA ALDER. Shrub or small tree 1.5–3 m. high (taller n. of our range), with brown or grayish bark, the twigs shining; lvs. thin, ovate, mostly 6–15 cm. long, acute with ± rounded or cuneate base, yellow-green and shining above, paler beneath with some hair along veins, doubly and unevenly glandular-serrate, the petioles 1–2 cm. long; ♂ catkins 2.5–3 cm. long, 2–4 in a cluster; ♀ 1.2–1.5 cm. long with scales thickened at the truncate apex; nutlet oval, almost as wide as wings.—Wet places, below 7000 ft.; Redwood F. to Red Fir F.; Del Norte, Humboldt, and Siskiyou cos.; to Alaska, Colo., Siberia. May–July.

2. **A. tenuifòlia** Nutt. [*A. incana* var. *virescens* Wats.] MOUNTAIN ALDER. Shrub or small tree 1–3(–7) m. tall, with smooth gray or red-brown bark; lvs. roundish to oblong-ovate, acute at apex, rounded to cordate at base, coarsely toothed then finely serrate, 2.5–6 cm. long, on petioles 1–2 cm. long, pubescent when young, glabrate or ± pubescent in age especially beneath; ♂ catkins 3–4 in a cluster, 2.5–7 cm. long; ♀ cones 0.8–1.2 cm. long, the scales much thickened at their truncate, ± lobed apex; nutlets with thin narrow margin; $2n = 28$ (Gram et al., 1941).—Moist places, 4500–8000 ft. (somewhat lower in N. Coast Ranges); Yellow Pine F. to Red Fir F.; Sierra Nevada from Tulare Co. n., to Modoc and Humboldt cos.; to Alaska, Rocky Mts. April–June.

3. **A. oregòna** Nutt. [*A. rubra* Bong., not Marsh.] RED ALDER. OREGON A. Tree 15–25 m. tall, with thin pale gray or whitish outer and red-brown inner bark; lvs. broadly elliptic-ovate, 6–12 cm. long, acute at apex, rounded to cuneate at base, crenately shallowly lobed or coarsely toothed, then finely glandular-dentate, dark green and glabrous to ± pubescent above, rusty-pubescent beneath, on petioles 1.5–2 cm. long; ♂ catkins 2–4 in a cluster, 10–15 cm. long; ♀ cones 2–2.5 cm. long, with truncate apically thickened scales; nutlet much broader than the wing.—Stream banks and marshy places, below 500 ft.; Mixed Evergreen F., Redwood F., etc.; Santa Cruz Co. along the coast to Del Norte Co.; to Alaska. March–April. In exposed places often reduced to mere shrubs.

4. **A. rhombifòlia** Nutt. WHITE ALDER. Tree 10–35 m. high with whitish or gray-brown bark; lvs. oblong-ovate or -rhombic, apically rounded to acute, basally cuneate, finely or coarsely doubly serrate, dark green and glabrous or ± pubescent above, lighter green and puberulent beneath, 5–11 cm. long, on petioles 1–2 cm. long; ♂ catkins 2–several in a cluster, 3–8 cm. long; ♀ cones 1–1.8 cm. long, the scales somewhat thickened and lobed at apex; nutlets with thin narrow margin.—Along streams, mostly below 5000 ft.; Chaparral, Yellow Pine F., Foothill Wd., etc.; much of cismontane Calif. except immediate strip along n. coast; to B.C., Ida. Jan.–April. Forms with more pubescent lvs. occur; one with cones 0.8–1 cm. long, from the Yellow Pine F. of the San Gabriel, San Bernardino, and San Jacinto mts. has been named var. *bernardìna* M. & J.

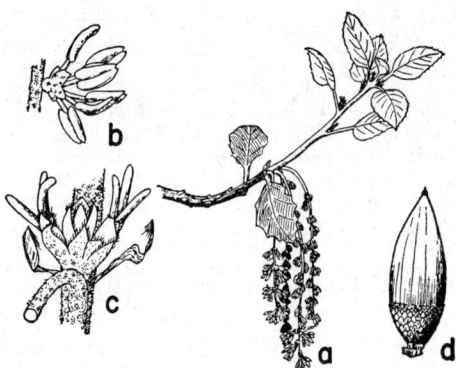

Fig. 77. FAGACEAE. *Quercus agrifolia: a,* shoot, × ½, with pendent ♂ catkins and axillary ♀ fls.; *b,* ♂ fl., × 4, lobed calyx and 6 stamens; *c,* ♀ fls., × 4, each with many-bracted invol. and 3 styles; *d,* acorn, × ½, with cuplike invol.

79. Fagàceae. Beech Family (Fig. 77)

Trees and shrubs with deciduous or persistent alternate petioled lvs. and small mostly deciduous stipules. Plants monoecious; fls. apetalous, the ♂ in catkins or capitate clusters, the ♀ solitary or in small clusters and subtended by an invol. of ± consolidated bracts which become indurated and form a cupule partly or completely enclosing the 1-locular and 1-seeded nut. Calyx of ♂ fls. 4–7-lobed; stamens 4–20. Calyx of ♀ fls. 4–8-lobed, adnate to the 3–7-locular ovary. Ovules 1–2 in each locule, usually only 1 ripening; styles 3. Endosperm none; cotyledons large, fleshy. Five genera and ca. 400 spp., mainly of N. Hemis., *Nothofagus* in the S. Hemis.

Fr. a spiny bur enclosing 1–3 nuts; lvs. persistent, yellowish-gray or rusty-tomentose beneath
                                         1. *Castanopsis*
Fr. an acorn with a cuplike invol.
    Staminate fls. in erect dense catkins with persistent bracts; ♀ fls. at base of ♂ catkins; lvs. with prominent parallel lateral veins . . . . 2. *Lithocarpus*
    Staminate fls. in ± drooping lax catkins with deciduous bracts; ♀ fls. in axillary clusters; lvs. usually not so veined . . . . 3. *Quercus*

1. Castanópsis Spach. Chinquapin

Trees or shrubs, evergreen, the buds with imbricated scales. Lvs. simple. Staminate fls. in groups of 3, arranged in axils of bracts along the elongate catkins. Calyx 5–6-parted; stamens 6–17. Pistillate fls. 1–3 in an invol. usually at base of ♂ catkins or in short separate ones; calyx 6-cleft, with abortive stamens on its lobes. Ovary 3-loculed with 2 ovules in each locule. Fr. maturing in second season, the spiny invol. inclosing 1–3 nuts; these ovoid or rounded, ± angled, mostly 1-seeded. Ca. 30 spp., largely Asiatic. (Greek, *kastanea,* chestnut, and *opsis,* resemblance.)

Lvs. long-pointed, ± folded along the midrib in shrubby forms, 5–15 cm. long; mostly trees with thick rough bark . . . . 1. *C. chrysophylla*
Lvs. mostly obtuse, plane; shrubs with thin smooth bark . . . . 2. *C. sempervirens*

1. **C. chrysophýlla** (Dougl.) A. DC. [*Castanea c.* Dougl.] Giant Chinquapin. Tree 15–45 m. tall, the trunk 1 m. or so thick with heavily furrowed bark and stout spreading branches; lvs. lanceolate to ± oblong, 5–15 cm. long, plane, entire, thick, coriaceous, dark green above, golden-tomentose or later olive-yellow beneath, tapering at both ends, hence mostly abruptly long-pointed at apex; petioles ca. 5–8 mm. long; burs chestnutlike, 4-valved; seeds 8–12 mm. long, hard-shelled, with sweet kernel.— Forested slopes, below 1500 ft.; Redwood F., etc.; near coast, Mendocino to Del Norte cos.; to Wash. Intergrading with:

Var. **minor** (Benth.) A. DC. [*Castanea c.* var. *m.* Benth.] Golden C. Shrub or small tree to ca. 5 (10) m. tall; lvs. commonly folded upward on each side along the midrib.—

Gravelly or rocky ridges and slopes, mostly 1000–6000 ft.; Yellow Pine F., Closed-cone Pine F., Chaparral, etc.; Coast Ranges from Santa Lucia Mts. to Del Norte, Trinity, and Siskiyou cos.; reported also from w. slope of Sierra Nevada in Eldorado Co. June–Sept.

2. **C. sempervìrens** (Kell.) Dudl. [*Castanea s.* Kell.] BUSH CHINQUAPIN. SIERRA C. Low spreading round-topped shrub 0.5–2.5 m. high with smooth brown or gray bark; lvs. oblong or oblance-oblong, mostly obtuse, 3–7.5 cm. long, subentire, yellowish gray-green above, golden or rusty-tomentose beneath, plane, on petioles 10–15 mm. long; ♂ fls. ill-smelling; burs 2–3 cm. thick, like preceding sp.—Dry rocky slopes and ridges, mostly 2500–11,000 ft.; Montane Coniferous F.; San Jacinto Mts. to San Gabriel Mts., Sierra Nevada from Kern Co. n., Coast Ranges from Lake Co. n.; to s. Ore. July–Aug.

2. Lithocárpus Blume. TANBARK-OAK

Evergreen trees or shrubs with alternate lvs. with persistent stipules. Staminate catkins many, dense, elongate, erect, with clusters of 3 fls. in the axils of round-ovate bracts with 2 lateral bractlets; stamens 10. Pistillate fls. at the base of the ♂ ament, solitary in the axils of acute bracts and minute lateral bractlets; calyx 6-lobed; styles 3. Fr. an acorn, the cup with slender spreading scales. Ca. 100 spp., of se. Asia, 1 in w. Am. (Greek, *lithos*, rock, and *karpos*, fr., because of the hard acorn.)

1. **L. densiflòra** (H. & A.) Rehd. [*Quercus d.* H. & A. *Pasania d.* Oerst.] Tree 20–45 m. tall, with narrow conical crown and tomentose young twigs; bark thick, fissured; lvs. oblong, acute, with prominent parallel lateral veins ending in sharp teeth, pale with whitish or yellowish stellate pubescence when young, ± glabrate later, especially on upper surface, 4–12 cm. long, acutish or ± rounded at ends, on petioles 1–2 cm. long; ♂ catkins 5–10 cm. long, ill-smelling; acorn maturing second year, short, thick-cylindric, with rounded base and gradually tapering to apex, mostly 2.5–3.5 cm. long, tomentose when young, glabrate in age; cup shallow, saucer-shaped, 1.5–2.5 cm. in diam., tomentose within, the scales narrow, spreading.—Wooded slopes below 4500 ft.; Redwood F., Mixed Evergreen F.; Coast Ranges from Ventura Co. to Del Norte Co.; s. Ore. June–Oct.

Var. **echinòides** (R. Br.) Abrams. [*Quercus e.* R. Br.] DWARF TANBARK. Shrub to ca. 3 m. high; lvs. entire, oblong, 1.5–6 cm. long, pale beneath, less conspicuously veined; cups with subulate recurving scales.—Dry slopes, 2000–8000 ft.; Yellow Pine F., Red Fir F.; mts. from Del Norte and Humboldt cos. to w. slope of Sierra Nevada in Mariposa Co.; s. Ore. June–Aug.

3. Quércus L. OAK

Deciduous or evergreen trees or shrubs with hard wood, ± contorted branches and fairly slender angled twigs marked by lenticels. Staminate catkins slender, drooping or spreading, one or more from the lower axils of the current season's growth. Staminate fls. solitary in axils of caducous bracts, the calyx usually 6-lobed; stamens 5–12. Pistillate fls. solitary in many-bracted invols., in upper axils; calyx urn-shaped, adnate to the 3-celled inferior ovary; styles 3, short. Fr. an acorn, the nut set in the cuplike invol. Ca. 300 spp., widely distributed in the N. Hemis. and into the mts. of the tropics. (Classical Latin name.)

A. Bark dark, not scaly but smooth, or on old trunks irregularly ruptured; stigmas on slender styles; involucral cups with thin closely imbricated scales; nut tomentose on inner surface. (Black Oaks)
   B. Plant deciduous, the lvs. usually 1 dm. or more long, with large bristle-tipped lobes
                                                              1. *Q. Kelloggii*
   BB. Plant evergreen, the lvs. smaller, coriaceous.
      C. Acorns maturing the second autumn; lvs. plane, glabrous beneath . . . . 2. *Q. Wislizenii*
      CC. Acorns maturing the first autumn; lvs. mostly convex on upper surface, with hair beneath, especially in axils of lvs. . . . . 3. *Q. agrifolia*
AA. Bark light in color, scaly, furrowed only on large trees; stigmas subsessile; involucral cups usually with tuberculate scales. (White Oaks)
   B. Fr. maturing the first season; shell of the nut glabrous on the inner surface.
      C. Deciduous species.
         D. Lvs. not blue-green, distinctly lobed.
            E. Acorn-cups deeply hemispheric; nut tapering at apex.
               F. Branchlets drooping; bark deeply fissured or cuboid-checked; lvs. not prominently reticulate-veined beneath. Mainland . . . . . . 4. *Q. lobata*

FF. Branchlets not drooping; bark scaly; lvs. prominently reticulate-veined
beneath. Insular. . . . . **8.** *Q. MacDonaldii*

EE. Acorn-cups shallow; nut ± oblong with rounded apex. Mostly N. Coast Ranges
5. *Q. Garryana*

DD. Lvs. blue-green, toothed or wavy-margined; cup shallow-cupulate; nut oval, acute.
Slopes bordering interior valleys . . . . **6.** *Q. Douglasii*

CC. Evergreen species, or at least retaining some of the lvs. in winter.

D. Tree; lvs. mostly 3–6 cm. long; cup enclosing nearly half the nut
7. *Q. Engelmannii*

DD. Shrubs, occasionally somewhat arborescent; lvs. mostly 1–2.5 cm. long; cup more
shallow.

E. Lf.-blades shining and subglabrous above.

F. Lvs. not chestnutlike 1.5–2.5 cm. long; acorn-cups thick-walled with
strongly tubercled scales. Most of cismontane Calif. . . . . . **9.** *Q. dumosa*

FF. Lvs. chestnutlike, 5–12 cm. long; acorn-cups thin-walled, the scales
scarcely tubercled. Trinity to Del Norte and Siskiyou cos.
12. *Q. Sadleriana*

EE. Lf.-blades stellate-pubescent above and dull.

F. The lvs. convex above; nuts abruptly pointed. Serpentine, middle Coast
Ranges from San Luis Obispo Co. n. . . . . **10.** *Q. durata*

FF. The lvs. plane; nuts more tapering. Deserts and in inner Coast Ranges
to San Benito Co. . . . . **11.** *Q. turbinella*

BB. Fr. maturing the second season; shell of the nut tomentose on the inner surface.

C. Trees; acorn-cup large, thick-walled, densely tomentose.

D. Lvs. 4–8 cm. long, with prominent parallel lateral veins. Insular
13. *Q. tomentella*

DD. Lvs. mostly 2–5 cm. long, the lateral veins not prominent or parallel. Widely
distributed . . . . **14.** *Q. chrysolepis*

CC. Shrubs; cup smaller, rather thin-walled.

D. Low spreading shrub to ca. 1.5 m. high with slender flexible twigs; lvs. mostly
entire, sometimes dentate; acorn 1–1.5 cm. long. Mts. from Fresno Co. to
Siskiyou, Del Norte, and Mendocino cos. . . . . **15.** *Q. vaccinifolia*

DD. Stiff shrub 2–5 m. high with rigid branchlets; lvs. spinose-toothed; acorn 2–3
cm. long. Mostly desert edge . . . . **16.** *Q. Palmeri*

1. Q. Kellóggii Newb. [*Q. tinctoria* var. *californica* Torr. *Q. sonomensis* Benth.]
CALIFORNIA BLACK OAK. A deciduous tree with broad rounded crown, 10–25 or more
m. high, trunk thick, with dark smooth bark that divides into ridges or checks deeply
in age; young twigs subglabrous; lf.-blades broadly elliptic to obovate in outline, deeply
and mostly sinuately lobed into ca. 3 main divisions on each side, each lobe with 1–4
coarse bristle-tipped teeth, bright green and mostly glabrous above, paler and often finely
stellate-tomentose beneath when young, 7–20 cm. long, on petioles an additional 2.5–5
cm. long; ♂ catkins 3.5–7.5 cm. long; stamens 5–9; ♀ fls. single or several on a
peduncle; acorns maturing in second year, oblong, 2.5–3 cm. long, 1.5–1.8 cm. thick,
pubescent; cups 1.5–2.5 cm. deep, 2–2.8 cm. wide, puberulent within, the thin scales
membranous and ± ragged on the margins; $n = 12$ (Duffield, 1940).—Common in hills
and mts., mostly 1000–8000 ft.; Mixed Evergreen F., Yellow Pine F., N. Oak Wd.; San
Diego Co. n. through Sierra Nevada and Coast Ranges; to Ore. April–May. At higher
elevs. ± shrubby, this form named f. *cibàta* Jeps.

Q. × **mórehus** Kell. ORACLE OAK. Evergreen tree, 4–15 m. high; lf.-blades oblong to
elliptic in outline, 4–12 cm. long, on petioles 0.5–2 cm. long, sinuately rather shallowly
lobed, the lobes pointed forward and spinose-tipped, ± stellate-pubescent beneath; cups
similar to those of *Q. Wislizenii* or deeper, 1.5–2 cm. wide, 1.2–1.5 cm. deep, thin-
scaled; acorns cylindric, ca. 2.5 cm. long, 1.2–1.5 cm. thick, pubescent.—At scattered
stations and usually of 1 to few individuals, below 5000 ft.; usually near *Q. Kelloggii*
and *Q. Wislizenii*, of which spp. it is undoubtedly a hybrid and with which some
retrogression is shown; San Diego Co. to Trinity and Eldorado cos.

2. Q. Wislizènii A. DC. INTERIOR LIVE OAK. Evergreen tree, mostly 10–22 m. high,
with round top and smooth bark 5–6 cm. thick, broadly ridged below in old age; lf.-
blades mostly oblong, varying to elliptic and lanceolate, rather firm, entire to spiny-
toothed, plane, glabrous and shining above and beneath but ± bicolored, 2–4(–7) cm.
long; petioles 0.5–1 cm. long; ♂ catkins ca. 3–6 cm. long; acorns maturing second
year, slender, cylindrical to ± conical, 2–4 cm. long, 0.7–1.3 cm. thick, often with longitudinal dark bands, glabrous; cup turbinate to cup-shaped, 12–20 mm. deep, 12–14
mm. in diam., with thin brownish pubescent and ciliate scales; $n = 12$ (Duffield, 1940).
—Valleys and slopes, below 5000 ft.; mostly Foothill Wd.; lower slopes of Sierra Nevada
and inner Coast Ranges, from Ventura to Shasta and Siskiyou cos. March–May. A form

with acorns ca. 6 mm. thick and 1.5–2 cm. long, from Amador and Tulare cos., has been named f. *extìma* Jeps. (var. *e.* Jeps.)

Var. **frutéscens** Engelm. [*Q. parvula* Greene.] A shrub 1–4 m. high, with very rigid twigs and lvs. 2–4 cm. long.—Chaparral; mts. of s. Calif. to Lake and Shasta cos., uncommon in Sierra Nevada; Santa Cruz Id.

3. **Q. agrifòlia** Neé. [*Q. acroglandis* Kell.] COAST LIVE OAK. ENCINA. Broad-headed evergreen tree 10–25 m. high; trunk smooth or with broad checked ridges in old trees; lf.-blades oblong to oval or elliptic, 2–6 cm. long, harsh, strongly convex above, glabrous or somewhat stellate-pubescent especially in vein-axils on under side; petioles 5–15 mm. long; ♂ catkins ca. 3–6 cm. long; acorns slender, pointed, 2.5–3.5 cm. long, 1–1.4 cm. thick, glabrous, maturing first season; cup turbinate, 8–12 mm. deep, 10–16 mm. in diam., silky within, the scales brownish, thin, puberulent, ciliate; $n = 12$ (Duffield, 1940).—Common in valleys and on not-too-dry slopes, below 3000 ft.; Mixed Evergreen F., Foothill Wd., S. Oak Wd., etc.; Coast Ranges from Sonoma Co. to San Diego Co.; L. Calif. March–April. A shrub form occurring in Chaparral has been named var. *frutéscens* Engelm.

Var. **oxyadènia** (Torr.) J. T. Howell. [*Q. o.* Torr.] Lvs. densely stellate, especially beneath.—At 2000–4600 ft.; interior cismontane Riverside and San Diego cos.; L. Calif.

The sp. hybridizes occasionally with *Q. Kelloggii* and various intermediate forms occur; one of these from Monterey and Santa Clara cos. is a partially evergreen tree with lvs. 5–14 cm. long, shallowly and irregularly lobed; petioles 1–2.5 cm. long; acorns maturing second year; it has been named **Q. × Chàsei** McMinn, Babcock, & Righter. Another occurs in Sonoma Co. and has more coriaceous lvs. deeper green above. A third from San Diego Co. has been named **Q. × Gánderi** C. B. Wolf and is a hybrid between var. *oxyadenia* and *Kelloggii;* its cones mature the first year. For the most part these can all be distinguished from *Q. × morehus* by the tufts of hairs in vein axils on undersurface of lvs.

In many cases smallish trees referred to *Q. agrifolia* seem to show introgression with *Q. dumosa* in their papery bark, small flattish lvs., etc. Yet crossing between the 2 subgenera represented has yet to be established.

4. **Q. lobàta** Neé. [*Q. Hindsii* Benth.] VALLEY OAK. ROBLE. Stately graceful deciduous tree with open head, 12–35 m. high, with trunk to 4 m. thick; bark thick, cuboid-checked; twigs silky at first, glabrescent in second year; lf.-blades oblong to obovate, 5–10 cm. long, deeply divided into 3–5 pairs of obtuse lobes which are mostly coarsely 2–3-toothed at apex, green and glabrescent above, paler and stellate beneath, yellow-veined; petioles stout, 5–12 mm. long, pubescent; ♂ catkins 2.5–7.5 cm. long; acorns maturing the first autumn, long-conical, 3–5 cm. long, 1.2–2 cm. thick, glabrous; cup deeply hemispheric, 12–20 mm. deep, 12–25 mm. in diam., the lower scales conspicuously warty; $n = 12$ (Duffield, 1940).—Rich loam, valleys and slopes, below 2000 ft.; Foothill Wd.; Cent. V. and its borders, inner and middle Coast Ranges to San Fernando V. and San Marino, Los Angeles Co. Said to be the largest Am. oak. March–April. Variable, a number of forms having been named: var. *Wálteri* Jeps. with roundish lvs. having narrow sinuses, nuts very thick.—At ca. 4000 ft., Kaweah River Basin. Var. *turbinàta* Jeps. with unusually large, thick lvs. and turbinate acorns; Little Lake V., Mendocino Co. Var. *argillàra* Jeps. with rather smooth often whitish bark and deeply, narrowly lobed ± persistent lvs.; clay hills, w. Solano Co. And var. *insperàta* Jeps. with narrow lvs., shallow cups and small nuts; foothills of Kaweah R.

5. **Q. Garryàna** Dougl. OREGON OAK. Round-headed deciduous tree 8–20 m. high or more, with stout spreading branches and thin white ± scaly bark; twigs red-pubescent, later glabrate; lf.-blades obovate to oblong in outline, 8–15 cm. long, coriaceous, green and subglabrous above, rusty or pale pubescent beneath, coarsely pinnatifid into oblong-ovate lobes, these entire or 2–3-toothed; petioles 1.5–2.5 cm. long, pubescent; ♂ catkins commonly 3–6 cm. long; acorn ovoid to subglobose, or subcylindric, rounded at tip, maturing the first autumn, 2–3 cm. long, subglabrous; cup shallow, puberulent within, 1.2–1.8 cm. in diam., with tubercled pubescent or tomentose scales; $n = 12$ (Duffield, 1940).—Wooded slopes, mostly 1000–5000 ft.; N. Oak Wd., Mixed Evergreen F.; Santa Cruz Mts., Marin Co. to Siskiyou Co.; to B.C. April–June.

Var. **Brèweri** (Engelm. in Wats.) Jeps. [*Q. B.* Engelm. *Q. lobata* ssp. *fruticosa* Engelm. *Q. Oerstediana* R. Br.] Spreading shrub 1–5 m. high, with smooth gray bark; lf.-blades mostly 3–5 cm. long.—Mt. slopes, 2000–6000 ft.; Yellow Pine F., Red Fir F.; Coast Ranges, Lake to Siskiyou cos.

Var. **semòta** Jeps. Shrub; lf.-blades mostly 5–9 cm. long.—Dry slopes, 2500–5000 ft.; Chaparral, Yellow Pine F.; w. slope of Sierra Nevada from Plumas Co. s., to Liebre Mts., n. Los Angeles Co.

Q. Garryana apparently hybridizes with several other spp.; there have been recorded: *Q.* × *Eplingii* C. H. Mull. (*Douglasii* × *Garryana*), from Lake Co. with lvs. suggesting *Douglasii* in size, shape and texture and cups like *Garryana*; *Q.* × *subconvéxa* Tucker (*durata* × *Garryana*), a shrub or a small tree with rather persistent lvs. 4–8 cm. long, coarsely toothed or shallowly acutely lobed, hemispheric cups, and ellipsoid acorns, from near serpentine in Santa Clara and Marin cos.; and *Q.* × *Howéllii* Tucker (*dumosa* × *Garryana*) with lvs. like the preceding but less convex, and with shorter hairs, from Fish Grade, Mt. Tamalpais, Marin Co.

6. **Q. Doúglasii** H. & A. [*Q. Ransomi* Kell.] BLUE OAK. Deciduous tree with ± rounded crown, 6–20 m. high; bark shallowly checked into small thin scales; twigs ± tomentose when young; lf.-blades oblong, 3–10 cm. long, entire to undulately shallowly few-lobed, blue-green above, minutely stellate-pubescent, paler and pubescent beneath; petioles 3–8 mm. long; acorns oval, acute, 2–3 cm. long, glabrous; cups shallow-cupulate, 10–12 mm. in diam., the scales with small warty processes; *n* = 12 (Duffield, 1940).—Dry rocky slopes, mostly below 3500 ft.; Foothill Wd.; mostly on slopes bordering interior valleys from n. Los Angeles Co. to head of Sacramento V. April–May.

7. **Q. Engelmánnii** Greene. ENGELMANN OAK. Spreading tree with rounded top, 5–18 m. high; bark covered with thin grayish scales; twigs tomentose; lf.-blades gray-green, semipersistent, oblong to obovate, plane, 2–6 cm. long, subentire to sinuate-dentate, obtuse, coriaceous, glabrate to somewhat pubescent; petioles 3–7 mm. long; acorns oblong-cylindric, 1.5–2.5 cm. long, 1.2–1.4 cm. thick, glabrous, ± acute; cup shallow to bowl-shaped, enclosing nearly half the nut, 1–1.5 cm. broad, 0.8–1 cm. deep, puberulent within, the scales light brown, tomentose, the lower tuberculate, the upper ciliate; *n* = 12 (Duffield, 1940).—Dry fans and foothills, below 4000 ft.; S. Oak Wd.; Pasadena region (Los Angeles Co.) inland to San Dimas and s. to e. San Diego Co. but away from the coast; L. Calif. April–May.

Occasionally hybridizing in the Pasadena region with *Q. lobata*, forming large trees with lvs. up to 15 cm. long and ± lobed; also with *Q. dumosa* especially in lower San Gabriel Mts., Santa Ana Mts., and e. San Diego Co. where the hybrids tend to be ± arborescent, with narrow lvs. oblong and entire or often rather regularly and sharply serrate-lobed (*Q. Macdonaldii* var. *elegantula* Greene, *Q. dumosa* var. *elegantula* Jeps., *Q. grandidentata* Ewan.)

8. **Q. MacDónaldii** Greene. [*Q. dumosa* var. *M.* Jeps.] Small deciduous tree, 5–15 m. tall with a compact rounded crown; bark scaly; young twigs densely tomentose; lf.-blades oblong to obovate in outline, 4–7 cm. long, with 2–4 blunt or sharp-pointed, bristle-tipped lobes on each side, glabrous above, pubescent beneath; petioles 3–8 mm. long; acorn oblong-conical, 2–3.5 cm. long, acute, glabrous; cup deeply hemispherical, 1.5–2.5 cm. in diam., with strongly tubercled pubescent scales.—Ravines and canyons; Chaparral, S. Oak Wd.; Santa Cruz, Santa Rosa, Santa Catalina ids.

9. **Q. dumòsa** Nutt. [*Q. acutidens* Torr.] SCRUB OAK. Evergreen, mostly a shrub 1–3 m. high, sometimes arborescent, with stiff pubescent to glabrate and brownish young twigs; lf.-blades oblong to elliptic or roundish, mostly mucronate-dentate to entire or subspinose, coriaceous, 1.5–2.5 cm. long, green and shining above, paler and pubescent beneath; petioles mostly 2–3 mm. long; acorns ovoid, 1–3 cm. long, broad at base, rounded or acute at apex; cups hemispheric to ca. ⅔ spherical, 10–15 mm. in diam., the margins tapering inward, walls thick with the scales mostly strongly tuberculate; *n* = 12 (Duffield, 1940).—Common on dry slopes, mostly below 5000 ft.; Chaparral, Foothill Wd., etc.; n. L. Calif. through cismontane s. Calif. along w. base of Sierra Nevada to Tehama Co. and N. Coast Ranges. March–May. Variable and apparently hybridizing freely with other spp., as *Q. durata*, *Q. turbinella*, *Q. Engelmannii*, *Q. Garryana*, etc.; with much introgression occurring. *Q. dumosa* var. *Kínselae* C. H. Mull. from near Santa Barbara, evidently based on a hybrid between *dumosa* and *lobata*, is ± arborescent, with incised lvs. intermediate in size between the spp. and with mucronately pungent lobes.

10. **Q. duràta** Jeps. [*Q. dumosa* var. *bullata* Engelm. *Q. d.* var. *munita* Greene?] LEATHER OAK. Evergreen shrub 1–3 m. high with stout rigid densely tomentose twigs; lf.-blades oval, oblong or elliptic, 1.5–2.5 cm. long, coriaceous, convex above, dull green and pubescent above until old, pale and tomentose beneath, usually with spinose teeth; petioles 1–3 mm. long; acorns thick, cylindric, abruptly pointed at apex, 1.5–2 cm. long; cups 1.2–1.8 cm. in diam., enclosing ⅓–½ the nut, the scales tuberculate.—On serpentine; Chaparral, Foothill Wd.; middle and inner Coast Ranges from Trinity to San Luis Obispo cos. and Sierra Nevada from Nevada to Eldorado cos. April–May. Supposed to hybridize with *Q. dumosa* and *Q. Garryana*.

11. **Q. turbinélla** Greene. [*Q. dumosa* var. *t.* Jeps.] Near to *Q. dumosa*, the young twigs densely gray-yellow tomentose; lf.-blades dull, gray to gray-green on upper surface, mostly oblong in outline, pointed apically, the margins spinose-dentate with definite short spines to 1 mm. long; fr. usually pedunculate, the acorns yellow-brown to buff, 1.2–2.3 cm. long, cylindric-ovoid to -ellipsoid, tapering abruptly toward apex; cups mostly ± hemispheric, the margins not turned inward.—At 4000–6000 ft.; Pinyon-Juniper Wd.; New York Mts., e. San Bernardino Co.; to s. Nev., w. Tex., n. L. Calif.

Ssp. **califórnica** Tucker. Lf.-blades more irregularly dentate, elliptical to ovate or suborbicular in outline often rounded at apex; fr. mostly sessile; acorns often brown, 2–3 cm. long; cups often turbinate.—Dry slopes, largely from 3000–6500 ft.; Pinyon-Juniper Wd., Joshua Tree Wd., etc.; edge of w. Mojave Desert from Little San Bernardino Mts. w. and n. to inner S. Coast Ranges of San Benito Co. Plants from w. edge of Colo. Desert are mostly nearer *Q. dumosa* than here.

Ssp. *californica* hybridizes freely with *Q. dumosa* along the desert margins and with *Q. Douglasii;* this latter hybrid and segregates form the *Q. Alvordiàna* Eastw. [*Q. dumosa* var. *A.* Jeps.] with ± persistent lvs. of intermediate character, somewhat arborescent habit, etc. It occurs from hills w. of Salinas V., Monterey Co. to Liebre Mts., n. Los Angeles Co.

12. **Q. Sadleriàna** R. Br. Campst. DEER OAK. Evergreen shrub 1–2.5 m. high, the young twigs subglabrous; stipules oblanceolate, 1–1.5 cm. long, silky; lf.-blades elliptic-to oblong-ovate, 5–12 cm. long, coarsely and regularly serrate and prominently pinnate-veined, subglabrous; petioles mostly 1–2 cm. long; acorns ovoid, 1.5–2 cm. long; cup cupulate, thin-walled, 1.5–1.8 cm. in diam., the scales thin, scarcely tubercled, ± tomentose.—Dry slopes and ridges, 3100–7000 ft.; Yellow Pine F., Red Fir F.; mts. from Trinity to Del Norte and Siskiyou cos.; sw. Ore. April–June.

13. **Q. tomentélla** Engelm. ISLAND OAK. Small round-headed evergreen tree, 5–12 m. tall; bark red-brown, scaly; young twigs hoary tomentose, later brown; lf.-blades dark green, oblong to oblong-lanceolate, acutish, often revolute, 4–8 cm. long, glabrate above, ± tomentose beneath, strongly pinnately veined, crenate-dentate; petioles 5–17 mm. long; acorns subglobose, 2–2.5 cm. long, bluntish; cups shallow, 2–3.5 cm. in diam., the ovate acute scales almost hidden in the dense tomentum.—Canyons and ravines; Chaparral, S. Oak Wd.; Santa Cruz, Santa Rosa, Santa Catalina, and San Clemente ids.; Guadalupe Id. April–May.

14. **Q. chrysólepis** Liebm. [*Q. fulvescens* Kell. *Q. crassipocula* Torr.] CANYON OAK. MAUL OAK. An evergreen ± roundish or spreading tree 6–20 m. high, with pale gray rather smooth scaly bark and hoary-tomentose young twigs; lf.-blades coriaceous, grayish- or yellowish-tomentose beneath or later glabrate and glaucous, usually oblong, entire to spinulose-dentate (especially on young and sucker shoots), plane, 2–6 cm. long; petioles mostly 3–8 mm. long; acorns oblong-ovoid, 2.5–3 cm. long, 2–2.5 cm. thick, scanty-tomentose within; cup saucer-shaped, thick-walled, silky within, mostly 2–2.5 cm. in diam., the scales ± hidden by the feltlike tomentum; $n = 12$ (Duffield, 1940).—Common in canyons and on moist slopes, below 6500 ft.; many Plant Communities; cismontane Calif. and occasional about and on the desert at higher elevs.; to Ore., L. Calif., New Mex. April–May. Exceedingly variable, a number of forms having been proposed: f. *grándis* Jeps. (var. *g.* Jeps.) from near Ukiah, Mendocino Co., to 30 m. high; cups 1.2–1.6 cm. wide, less pubescent; f. *péndula* Jeps. (var. *p.* Jeps.) from upper San Benito R. and Amador Co., with pendulous branchlets, lvs. deep shining green above; f. *Hánsenii* Jeps. (var. *H.* Jeps.) from Pine Grove, Amador Co., low tree with slender-cylindric acorn; and var. **nàna** (Jeps.) Jeps. from high exposed chaparral ridges and

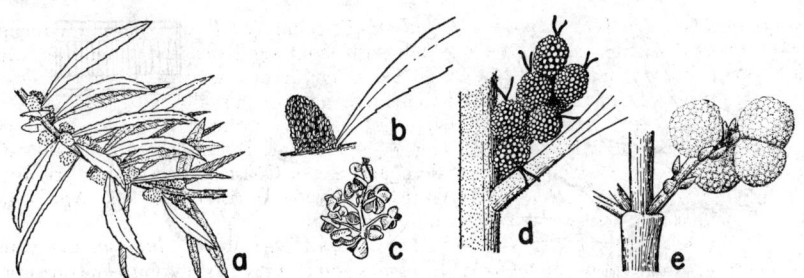

Fig. 78. MYRICACEAE. *Myrica californica:* *a,* ♂ branch, × ¼, lvs. with axillary catkins; *b,* ♂ catkin, × 1; *c,* ♂ fls. with 2-locular anthers; *d,* ♀ catkins, × 2; *e,* ♀ catkins in fr., × 2.

slopes from Trinity and Eldorado cos. to Sierra Nevada, having compact shrubby habit; lvs. 2–4 cm. long; acorn 2–2.5 cm. long. This var. possibly a hybrid with *Q. vaccinifolia.*

15. **Q. vaccinifòlia** Kell. [*Q. chrysolepis* var. *v.* Engelm.] HUCKLEBERRY OAK. Low spreading often prostrate evergreen shrub to ca. 12 dm. high, with slender flexible tufted, glabrous to pubescent branchlets; lf.-blades oblong-ovate, mostly entire, 1.2–3 cm. long, dull gray-green and glabrous above, pale and finely tomentulose to subglabrous beneath, acute to obtuse; petioles mostly 3–6 mm. long; acorns round-ovoid, 1–1.5 cm. long, tomentose on inner surface; cup shallowly bowl-shaped, ca. 1–1.4 cm. in diam., thin-walled, pubescent within, the scales white-tomentose.—Dry ridges and rocky places, 3000–10,000 ft.; Montane Coniferous F.; Sierra Nevada from Fresno Co. n., to Siskiyou, Del Norte, and Mendocino cos.; Ore. May–July.

16. **Q. Pálmeri** Engelm. [*Q. Dunnii* Kell.] PALMER OAK. Stiff evergreen shrub 2–5 m. high with rigid, at first ± pubescent twigs; lf. blades brittle, elliptic to round-ovate, wavy-spinose, crisped, 1–3 cm. long, gray-green above, ± tomentose beneath especially when young; petioles 2–5 mm. long; acorns ovoid, 2–3 cm. long, acute, the shell tomentose within; cup shallow, ca. 12–30 mm. broad, 7–10 mm. deep, the walls thin, silky within, densely tomentose without.—Dry thickets and margin of Chaparral, mostly 3000–5000 ft.; Peachy Canyon Road, San Luis Obispo Co., Deep Creek, n. base San Bernardino Mts., San Jacinto Mts. along w. edge of Colo. Desert to n. L. Calif., Ariz. April–May.

80. Myricàceae. WAX-MYRTLE FAMILY (Fig. 78)

Monoecious or dioecious shrubs or trees with simple alternate deciduous or persistent resinous-dotted often fragrant lvs. without stipules. Both kinds of fls. in short scaly aments. Differing from *Betulaceae* mostly in the 1-locular ovary with a single erect ovule and the drupelike nut. Perianth and invol. none. Two genera and ca. 40 spp., of wide distribution.

1. Myrìca L.

Lvs. entire, dentate or lobed. Staminate aments oblong or thick-cylindrical; stamens 2–20, the fils. somewhat united below; anthers 2-locular. Pistillate aments ovoid or subglobose, the ovary subtended by 2–4 short bractlets. Fr. globose or ovoid. Ca. 30 spp. of temp. and trop. regions. (Greek, *myrike,* an old name of a fragrant shrub.)

Plants dioecious; lvs. deciduous, appearing after the fls., pubescent .... 1. *M. Hartwegii*
Plants monoecious; lvs. persistent, evergreen, glabrous .... 2. *M. californica*

1. **M. Hartwégii** Wats. SIERRA SWEET-BAY. Shrub commonly 1–2 m. high; twigs pubescent and with resin globules, later glabrate; lvs. broadly oblanceolate, 4–8 cm. long, coarsely serrate, lighter green and more persistently pubescent beneath than above, acute or obtuse at apex, attenuate at base to a petiole 5–12 mm. long; ♂ aments 1–2 cm. long, with light brown glabrous bracts and 2–4 stamens in each fl.; ♀ ca. 8 mm. long in fr.; nutlets smooth or with resinous globules, compressed and ca. 2 mm. long,

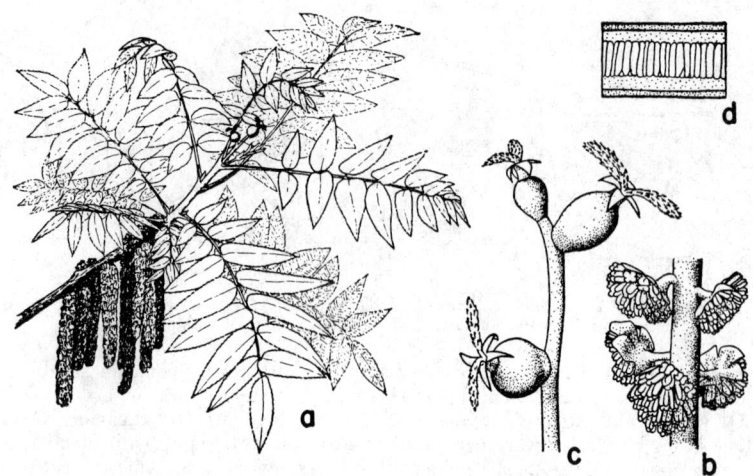

Fig. 79. JUGLANDACEAE. *Juglans californica: a*, habit, × ¼, pinnate lvs., pendent ♂ catkins, terminal ♀ fls.; *b*, ♂ fls., × 2½, many stamens surrounded by lobed calyx; *c*, ♀ fls., × 1, inferior ovary, lobed calyx, 2 stigmas; *d*, longitudinal section of twig with chambered pith, × 1½.

winged by the adnate bractlets.—Stream banks, 1000–5000 ft.; largely Yellow Pine F.; w. base of Sierra Nevada from Yuba to Fresno cos. May–June.

2. **M. califórnica** Cham. & Schlecht. Wax-Myrtle. Shrub 2–4 m. tall or even a small tree to 10 m.; branches slender, ascending; bark gray or light brown; lvs. oblong to oblanceolate, dark green, glossy, persistent, 5–11 cm. long, subentire to remotely serrate, with a petiole 3–10 mm. long; ♂ catkins 1–2 cm. long, borne in lower axils; stamens 7–16; ♀ catkins in upper axils, 8–12 mm. long; ovary ovoid; bractlets minute; fr. brown-purple, 6–8 mm. in diam., covered with a whitish wax.—Canyons and moist slopes, below 500 ft.; Redwood F., Coastal Strand, N. Coastal Scrub, Closed-cone Pine F., etc.; Santa Monica Mts. to Del Norte Co.; to Wash. March–April.

81. Juglandàceae. Walnut Family (Fig. 79)

Aromatic deciduous trees and shrubs. Lvs. alternate, pinnate, without stipules. Monoecious, the fls. opening after the lvs. unfold; ♂ in bracteate aments on twigs of the previous year and consisting of a 3–6-lobed calyx, with 3–many stamens with short distinct fils. and oblong anthers of longitudinal dehiscence. Pistillate fls. 1–several on peduncles at the end of shoots of the season, bracted, then usually with 2 bracteoles and 3–5-lobed calyx. Petals sometimes present. Ovary inferior, 1- or incompletely 2–4-loculed; style short; stigmas 2, plumose. Fr. drupaceous, with a fibrous somewhat fleshy exocarp, indehiscent or 4-valved; endocarp bony, rugose or sculptured, forming a nut. Seed solitary, 2-lobed, with large fleshy, oily cotyledons. Six genera and ca. 40 spp., of N. Temp. Zone and n. Andes.

1. Jùglans L. Walnut

Staminate fls. with 4–40 stamens, in 2 or more series. Pistillate fls. with a 4-lobed calyx and 4 petals. Ca. 10 spp. (Classical name of the Walnut.)

Lfts. oblong-lanceolate, mostly 11–15, acute to acuminate, mostly 2–6 cm. long. Ventura Co. to Orange and San Bernardino cos. . . . . 1. *J. californica*
Lfts. ovate-lanceolate, mostly 15–19, long-pointed, mostly 6–10 cm. long. Lake to Contra Costa cos.
2. *J. Hindsii*

1. **J. califórnica** Wats. Low tree with several trunks, 3–10 or more m. tall; young twigs brownish-tomentose; old bark dark, with broad irregular ridges; lvs. 1.5–2 dm. long, the petioles glandular-pubescent; lfts. finely serrate, cuneate or rounded at base, glabrous

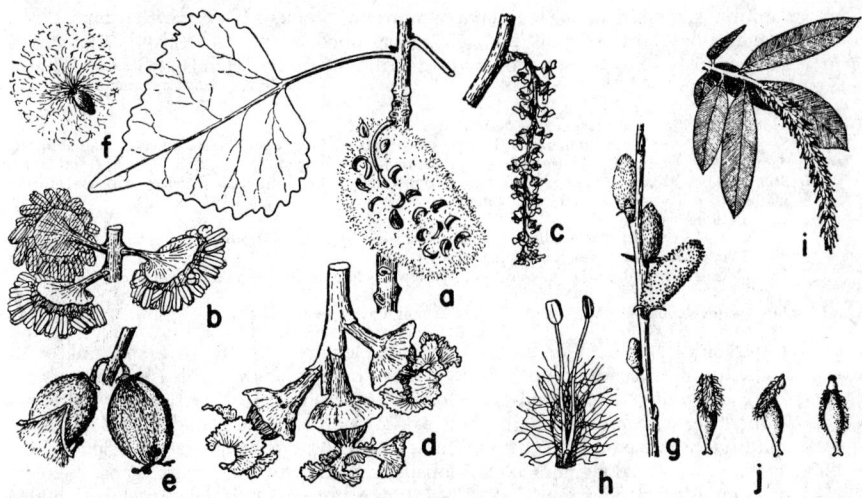

Fig. 80. SALICACEAE. *Populus Fremontii: a*, lf. and mature ♀ catkin, × ½; *b*, ♂ fls., × 3, disk with many stamens; *c*, ♀ catkin, × ½; *d*, ♀ fls., × 2½, disk with pistil with large stigmas; *e*, young capsules, × 1; *f*, seed, × 2, with coma. *Salix: g*, ♂ branch with young catkins; *h*, ♂ fl., × 5, bract with 2 stamens and basal gland; *i*, ♀ branch with catkin; *j*, ♀ fls. in 3 views, pistil with subtending bract and basal gland.

above, with hairy tufts beneath in vein-axils; ♂ aments 5–8 cm. long, brownish-pubescent; ♀ fls. in small spikes, with yellow recurved stigmas; fr. roundish, 2.5–3 cm. in diam., with dark brown pubescent husk; nut globose, thick-shelled, longitudinally shallowly grooved; $2n = 34$? (Babcock, 1915).—Locally common, below 2500 ft.; S. Oak Wd.; Ventura Co. to w. cismontane San Bernardino Co. and Santa Ana Mts., Orange Co. April–May.

2. **J. Hìndsii** (Jeps.) Jeps. [*J. californica* var. *H.* Jeps.] Doubtfully specifically distinct from the former, forming a lofty, mostly 1-trunked tree 15–25 m. high; lvs. mostly 1.8–3 dm. long, the lfts. more lanceolate and long-pointed; fr. 3.5–5 cm. in diam.; nut more faintly grooved.—Mostly about old Indian campsites, Lake and Napa cos. to Contra Costa and Stanislaus cos. April–May. When grown in s. Calif in less favorable conditions it is less definitely a tall tree. Used as a street tree in cent. Calif. and for stock for budding English Walnuts, with which domesticated species both California natives freely hybridize.

82. Salicàceae. WILLOW FAMILY (Fig. 80)

Dioecious deciduous trees or shrubs with simple alternate leaves mostly with stipules. Wood soft, light, mostly pale. Winter buds scaly. Roots often producing new shoots. Fls. early in spring, often before the lvs., in unisexual aments, the entire catkin falling away at once. Perianth none; each fl. subtended by a scalelike bract. Stamens 2–many, sometimes 1 by fusion. Ovary 1-celled, ovoid or globose; stigmas 2–4. Fr. a 2–4-valved caps., with many minute comose seeds.

Scales of the catkins laciniate; fls. surrounded by a cup-shaped often oblique disc; stamens many; buds with numerous scales . . . . 1. *Populus*
Scales of the catkins entire; disc a minute glandlike structure; stamens 1, 2, or more; buds with a single scale . . . . 2. *Salix*

1. Pópulus L. ASPEN. COTTONWOOD. POPLAR

Trees with pale furrowed bark and soft usually whitish wood. Winter-buds resinous. Lvs. alternate, deciduous, stipulate, broad to narrow, petioled. Stipules minute, deciduous. Bracts of the usually pendulous catkins laciniate, narrowed at base. Disk cup-shaped, lobed or entire, symmetrical or lengthened in front. Stamens 6–80, with distinct fils. Ovary sessile; style short; stigmas 2–4, entire or lobed. Caps. 2–4-valved, the valves recurved at maturity. Ca. 30 spp., of wide distribution in N. Hemis. (Classical Latin name.)

A.   Lf.-blades not white-woolly beneath. Native spp.
    B.   Stigma-lobes filiform; stamens 6–12; petioles flattened or compressed toward the summit; caps. oblong-conical, thin-walled; winter buds 3–10 mm. long . . . . 1. _P. tremuloides_
    BB.   Stigma-lobes dilated; stamens 40–80; petioles terete or flattened; caps. ± globose, thick-walled; winter-buds 10–20 mm. long.
        C.   Petioles terete.
            D.   Lvs. dark green above, rusty or silvery beneath. Common . . . . 2. _P. trichocarpa_
            DD.   Lvs. light green on both surfaces. Rare . . . . 3. _P. angustifolia_
        CC.   Petioles laterally compressed; lvs. yellowish-green, alike on both surfaces
                                                           4. _P. Fremontii_
AA.   Lf.-blades conspicuously white-woolly beneath, angulate-lobed. Introd. . . . . 5. _P. alba_

1. **P. tremuloìdes** Michx. Q<small>UAKING</small> A<small>SPEN</small>. Slender tree mostly 3–20 m. tall, with extensive lateral roots sending up sucker shoots; bark smooth, greenish-white; twigs slender, often drooping; lvs. round-ovate or wider, crenulate or serrulate with many teeth, or subentire, with a short sharp point at the apex, broadly rounded or subcordate at base, quite glabrous, pale beneath, the blades 2–4(–6) cm. long; petioles slender, ca. as long; catkins 3–6 cm. long; bracts 3–5-lobed, fringed with long hairs; caps. ± conic, ca. 5–7 mm. long; $2n = 38$ (Smith, 1943).—Stream borders and ± damp slopes, 6000–10,000 ft.; Red Fir F. to Subalpine F.; Fish Creek, San Bernardino Mts., Sierra Nevada, Canyon Creek, Trinity Mts.; to Alaska, Labrador, n. Mex. April–June. W. material is often referred to var. _aúrea_ Daniels.

2. **P. trichocárpa** T. & G. B<small>LACK</small> C<small>OTTONWOOD</small>. Tree 40–60 m. tall with broad open crown, the grayish bark furrowed in age; lvs. ovate, finely serrate, truncate or cordate at the base, acute to subacuminate at apex, dark green above, pale and somewhat glaucous beneath, 3–7 cm. long, on terete petioles 2–5 cm. long; catkins 4–8 cm. long; stamens 40–60; caps. subglobose, 4 mm. thick, pubescent; $2n = 38$ (Smith, 1943).—Along streams, below 9000 ft.; many Plant Communities; San Diego Co. through cismontane Calif.; to Alaska, w. Nev. Feb.–April. At higher elevs. lvs. become lance-ovate and on the upper Santa Ana R. in the San Bernardino Mts. they are lanceolate and less than 2 cm. wide; these are var. _ingràta_ (Jeps.) Parish.

3. **P. angustifòlia** James. Tree 15–20 m. high with rather a narrow head and mostly glabrous slender twigs; lvs. mostly lanceolate, subacuminate at apex, rounded at base, coarsely serrate, 5–8 cm. long, 1–2 cm. wide, yellow-green and glabrous; stamens 12–20; ♀ disk shallow, slightly lobed; caps. abruptly contracted above the middle, short-pointed; $2n = 38$ (Smith, 1943).—At ca. 6000 ft., Division Creek near Independence and Lone Pine Creek, Inyo Co.; Alta., S. Dak. to New Mex., Mex.

4. **P. Fremóntii** Wats. Tree, 12–30 m. tall with broad open crown and whitish roughly cracked bark; twigs stout, quite glabrous; lvs. quite glabrous, deltoid, 4–7 cm. long, 4–9 cm. wide, slightly cordate or abruptly cuneate at the entire base, abruptly sharp-pointed at the apex, coarsely and irregularly serrate-dentate, bright green, lustrous; petioles laterally flattened, 3–6 cm. long; catkins 4–5 cm. long; stamens 60 or more; ovary glabrous; caps. 8–12 mm. long.—Moist places below 6500 ft.; several Plant Communities; cismontane s. Calif.; Mojave Desert, Cent. V. and bordering hills w. to Lake Co.; to Nev., Ariz. March–April. An ill-defined form with pubescent twigs and lvs., from San Bernardino, Orange, and San Diego cos. is var. _pubéscens_ Sarg.

Var. **arizónica** (Sarg.) Jeps. [_P. a._ Calif. auths., not Sarg.] Twigs glabrous, slender, perhaps 1–2 mm. thick; lvs. largely cuneate at base, mostly 2–3 cm. wide; disc 4–6 mm. broad; caps. ca. 6 mm. long.—Reported from Mill Creek, San Bernardino Mts. and Snow Creek, San Jacinto Mts.; Ariz.

Var. **MacDoùgallii** (Rose) Jeps. [_P. M._ Rose.] Twigs coarser, pubescent; lvs. bluish-green, mostly truncate at base and 4–8 cm. wide; disc ca. 3 mm. broad; caps. 10–12 mm. long.—Creosote Bush Scrub; Colo. Desert and n. along Colo. R.; Ariz.

5. **P. álba** L. S<small>ILVER</small> P<small>OPLAR</small>. Tree with mostly grayish-white smooth bark; young twigs and winter-buds and young lvs. white-tomentose; older lvs. 3–5-angulate-lobed with few teeth, white-tomentose beneath; caps. woolly, 3–5 mm. long.—Occasional escape near old dwellings; introd. from Eu.

2. Sàlix L. Willow

Trees or shrubs with rather slender twigs and simple mostly narrow pinnately veined lvs. Winter-buds single-scaled. Fls. unisexual, the ♂ and ♀ borne in separate catkins that appear before, with, or after the lvs. Each fl. subtended by a small scale; sepals and petals none; 1–2 small glands at base. Stamens 1–10, mostly 2–5; pistil 1, forming a caps; style evident or obsolete; stigmas entire or divided. A genus of 300 or more spp., mostly temp. and colder.

A. Lf.-blades entire or nearly so.
  B. Lvs. glabrous beneath, except possibly when very young.
    C. Plants low, creeping, forming woody mats with erect catkins. Alpine.
      D. The lvs. oblanceolate, 2–4 cm. long; catkins appearing with the lvs. 1–6 cm.
        long. Common, Tulare to Lassen cos. .... 20. *S. anglorum*
      DD. The lvs. oblong-obovate to roundish, 0.7–1.2 cm. long; catkins appearing after
        the lvs., to ca. 1 cm. long. Rare, Mono Co. .... 21. *S. nivalis*
    CC. Plants ascending to erect, usually over 1 m. tall.
      D. Lvs. lanceolate to oblanceolate, mostly definitely less than ⅓ as wide as long.
        E. Lf.-blades 1.5–4 cm. wide, mostly 3–4 times longer.
          F. Fils. and caps. glabrous or nearly so.
            G. Scales of catkins tawny; styles less than 0.5 mm. long
                                                       9. *S. lutea*
            GG. Scales of catkins blackish; styles 1–1.5 mm. long
                                  15. *S. commutata* var. *denudata*
          FF. Fils. and caps. hairy; catkin-scales dark .... 23. *S. Lemmonii*
        EE. Lf.-blades 1–2 cm. wide, mostly ca. 5 times as long.
          F. Catkin-scales yellow; stamens 3–9. Trees.
            G. Petioles glandular above .... 1. *S. lasiandra* var. *Abramsii*
            GG. Petioles not glandular .... 3. *S. Gooddingii*
          FF. Catkin-scales dark; stamens 2.
            G. The catkin-scales elliptic-oblong; lvs. lingulate to lanceolate
                                          10. *S. ligulifolia*
            GG. The catkin-scales round-obovate; lvs. oblanceolate
                                          13. *S. lasiolepis*
      DD. Lvs. obovate to elliptic-oval, ½–⅓ as wide as long.
        E. The lvs. 1.5–1.8 cm. wide, 3–5 cm. long; caps. glabrous. N. Coast
                                            14. *S. Tracyi*
        EE. The lvs. 0.8–1.5 cm. wide, 2–3.5 cm. long; caps. silky. Sierra Nevada
                                          22. *S. planifolia*
  BB. Lvs. definitely pubescent or hairy beneath, even when quite mature.
    C. The lvs. linear or lance-linear, less than 1 cm. wide.
      D. Stamen-fils. pubescent near base; catkin-scales yellow.
        E. Stigmas ca. 1 mm. long, on styles ca. 0.5 mm. long; caps. hairy
                                          5. *S. Hindsiana*
        EE. Stigmas ca. 0.5 mm. long, sessile; caps. glabrous or nearly so, at least in
          maturity .... 6. *S. exigua*
      DD. Stamen-fils. glabrous; catkin-scales dark or at least reddish at apex.
        E. Shrubs 2–6 m. tall; petioles 5–15 mm. long; twigs pubescent
                                  13. *S. lasiolepis* var. *Bracelinae*
        EE. Shrubs to ca. 1 m. high; petioles to ca. 3 mm. long; twigs ± glabrous
                                          28. *S. Breweri*
    CC. The lvs. lanceolate to oblanceolate or obovate, mostly more than 1 cm. wide.
      D. The caps. glabrous.
        E. Catkins subsessile; lvs. narrowly to broadly oblanceolate. Mostly below Red
          Fir F. Widely distributed .... 13. *S. lasiolepis*
        EE. Catkins on leafy peduncles; lvs. broadly lanceolate or broadly oblanceolate to
          obovate or elliptic-oval.
            F. Shrub 1–3 m. tall; catkins appearing with the lvs. Red Fir F. in n. Calif.
                                            15. *S. commutata*
            FF. Large shrub or small tree, 4–10 m. tall; catkins appearing before the
              lvs. Near the beaches, extreme nw. Calif. .... 19. *S. Hookeriana*
      DD. The caps. hairy to puberulent.
        E. Catkin-bracts dark or almost black.
          F. Styles 1–1.5 mm. long.
            G. Lvs. ± silky-villous on both surfaces; caps. on pedicels 1–1.5 mm.
              long; fils. hairy below .... 17. *S. orestera*
            GG. Lvs. silky beneath, sparsely pubescent above; caps. subsessile; fils.
              glabrous .... 24. *S. Drummondiana*
          FF. Styles 0.2–0.5 mm. long .... 30. *S. Scouleriana*
        EE. Catkin-bracts brown or yellow or with reddish tips.
          F. Styles evident, mostly 0.5–1 mm. long.
            G. Stamen 1; fil. glabrous. Near the coast.

        H.   Young twigs very slender, somewhat pubescent, soon glabrate;
             lvs. pubescent beneath; caps. short-pedicelled
                                            25. *S. sitchensis*
        HH. Young twigs stout, densely pubescent to tomentose; lvs.
             opaque-tomentose beneath; caps. subsessile .. 26. *S. Coulteri*
      GG. Stamens 2; fils. glabrous or pubescent.
        H.   Shrub 1–3 m. high with shining twigs; lvs. oblanceolate. Red
             Fir F. to Subalpine F. .... 27. *S. Jepsonii*
        HH. Shrub to ca. 1 m. high with gray-tomentose twigs; lvs. obovate.
             Gasquets, Del Norte Co. .... 29. *S. delnortensis*
      FF. Styles obsolete or very short, to ca. 0.3 mm. long .... 31. *S. Geyeriana*
AA. Lf.-blades distinctly serrulate.
  B.   Lvs. glabrous beneath except sometimes when very young.
    C.   Caps. pilose to silky.
      D.   Catkin-scales yellow; tree 6–10 m. tall; lvs. gray-green on both surfaces; stamens
         4–5 .... 3. *S. Gooddingii*
      DD. Catkin-scales dark; shrub 1–5 m. high; lvs. deep green above; stamens 2
                                      23. *S. Lemmonii*
    CC. Caps. essentially glabrous (sometimes with few hairs near summit in *Piperi*).
      D.   Catkin-scales yellow, soon deciduous.
        E.   Lf.-blades 0.6–1.5 cm. wide.
          F.   Caps. 5.5–7 mm. long; stamens 4–5; lvs. gray-green above and be-
             neath .... 3. *S. Gooddingii* var. *variabilis*
          FF. Caps. 4–5 mm. long; stamens 2; lvs. dark green above
                                       7. *S. melanopsis*
        EE. Lf.-blades 1.5–4 cm. wide.
          F.   Stamens 3–9, the fils. hairy near base; twigs mostly red-brown.
            G.   Shrub 1–4 m. high; lvs. not glaucous beneath. At 6000–8500 ft.
                                      2. *S. caudata*
           GG. Tree 5–15 m. high; lvs. glaucous beneath. Below 5000 ft.
                                     4. *S. laevigata*
          FF. Stamens 2, the fils. glabrous; twigs yellowish or brownish .. 9. *S. lutea*
      DD. Catkin-scales dark, ± persistent.
        E.   Styles 0.2–0.7 mm. long.
          F.   Pedicels of caps. 2.5–4 mm. long; twigs dark brown. Red Fir F.
                                   11. *S. Mackenziana*
          FF. Pedicels of caps. 1–2.5 mm. long; young twigs yellowish to reddish.
            G.   Lvs. glaucous beneath.
              H.   Lvs. 5–10 cm. long, ligulate to oblong-lanceolate. Montane
                Coniferous F. .... 10. *S. ligulifolia*
              HH. Lvs. 3–5 cm. long, oblanceolate to obovate. Mixed Evergreen
                F. .... 14. *S. Tracyi*
           GG. Lvs. not glaucous beneath, although pale and veiny, elliptic-
             lanceolate to oblong-oval. Montane Coniferous F.
                                   12. *S. pseudocordata*
        EE. Styles 1–1.5 mm. long.
          F.   Shrub 1–3 m. tall; lvs. ± pubescent on both sides. Red Fir F.
                                    15. *S. commutata*
          FF. Shrub 4–6 m. tall; lvs. glabrous. Coastal Strand ...... 18. *S. Piperi*
  BB. Lvs. rather permanently pubescent beneath.
    C.   Trees or large shrubs mostly 3–15 m. high.
      D.   Stamens 4–6; lvs. 1.5–3 cm. wide, 5–12 cm. long; style and stigmas minute
                                    4. *S. laevigata* var. *araquipa*
      DD. Stamens 2.
        E.   Lvs. 0.2–1.4 cm. wide, 4–8 cm. long; style to ca. 0.5 mm. long; catkin-
          scales yellow.
          F.   Stigma-lobes ca. 1 mm. long; style ca. 0.5 mm. long .. 5. *S. Hindsiana*
          FF. Stigma-lobes very short; style obsolete
                          7. *S. melanopsis* var. *Bolanderiana*
        EE. Lvs. 1–4 cm. wide, 3–12 cm. long; style ca. 1 mm. long; catkin-scales black.
          N. beaches .... 19. *S. Hookeriana*
  CC. Shrubs mostly not over 3 m. tall.
    D.   Catkin-scales yellowish; stigmas subsessile.
      E.   Lvs. linear, 0.2–0.8 cm. wide. S. Calif. .... 6. *S. exigua*
      EE. Lvs. elliptical to oblong, 0.8–2.5 cm. wide. Nw. coast .... 8. *S. Parksiana*
    DD. Catkin scales dark; styles 1–1.5 mm. long.
      E.   Twigs yellow to brownish, glabrate to pilose; stamens glabrous; caps. glabrous.
          Trinity, Siskiyou, and Modoc cos. .... 15. *S. commutata*
      EE. Twigs mostly dark, tomentose; stamens pilose at base; caps. tomentose.
          Tulare Co. to Modoc Co. .... 16. *S. Eastwoodiae*

1. **S. lasiándra** Benth. [*S. speciosa* Nutt., not Host. *S. arguta lasiandra* Anderss.] Tree
6–15 m. tall; bark rough, brown; twigs reddish, shining, glabrous; lvs. lanceolate to
broadly lanceolate, acuminate, closely glandular-serrulate, dark green and shining above,
glaucous beneath, 6–10 cm. long, 1.5–3.5 cm. wide; stipules small, rounded, acute,

glandular; petioles stout, 5–15 mm. long, glandular at upper end; ♂ catkins 2–6 cm. long; ♀ 3–10 cm. long; scales yellow, lanceolate to ovate, usually toothed; stamens 4–5, pubescent at base; caps. light brown to pale straw color, lanceolate, glabrous, 5–7 mm. long; style short; $2n = 76$ (Wilkinson, 1944).—Stream banks, below 8000 ft.; many Plant Communities; most of Calif. except deserts; to Alaska, Ida., n. Mex. March–May.

Var. **lancifòlia** (Anderss.) Bebb. [*S. l.* Anderss.] Young branchlets pilose-pubescent. With the sp.

Var. **Abrámsii** Ball. Twigs glabrous; lvs. 1–1.7 cm. wide, subentire to shallowly serrulate.—An indefinite form, at 4000–8000 ft.; Coniferous F.; San Bernardino Mts., Sierra Nevada; to Ida., Ore.

2. **S. caudàta** (Nutt.) Heller. var. **Bryantiàna** Ball & Bracelin. Erect shrub 1–4 m. high, several-stemmed; twigs and branches brownish-red, glabrous, lustrous; lf.-blades lanceolate, 6–12 cm. long, 1.5–3 cm. wide, dark green and glabrous above, paler but not glaucous beneath, glandular-serrulate, acuminate; petioles 3–6 mm. long, glabrous, glandular in upper part; stipules small or none; catkins appearing with the lvs.; scales yellow, glabrous or nearly so; ♂ catkins 2–4 cm. long; stamens 3–9; fils. hairy below; ♀ 2.5–5 cm. long; caps. glabrous; styles evident.—Along streams and in meadows, 6000–8500 ft.; Red Fir F., Lodgepole F.; San Bernardino Mts., Sierra Nevada to Plumas and Siskiyou cos.; to Alta., S. Dak., New Mex. May. The sp. itself is distinguished by pubescent young growth and occurs at ca. 5000 ft. in Modoc Co.

3. **S. Goóddingii** Ball. [*S. nigra* var. *vallicola* Dudl. *S. n.* var. *venulosa* (Anderss.) Bebb.] Tree 6–10(–20) m. tall with rough dark bark, sometimes shrubby; twigs yellowish, subglabrous to finely pubescent when young; lvs. narrowly lanceolate, acute or acuminate, finely glandular-serrulate, grayish-green on both surfaces, often pubescent when young, 6–10 cm. long, 0.8–1.4 cm. wide; petioles 6–10 mm. long, not glandular; stipules usually ± glandular; catkins 4–8 cm. long, on lateral leafy branchlets 2–4 cm. long; stamens 4–5, pubescent, subtended by yellow bracts; caps. conic-ovoid, 5.5–7 mm. long, pilose; pedicels 2–3 mm. long; styles and stigmas very short.—Stream banks and wet places, mostly below 2000 ft. Creosote Bush Scrub, Foothill Wd., etc.; mostly in drainage of Colo. R., less common in cismontane s. Calif. and Cent. V. to Yolo Co.; to sw. Utah, w. Ariz., Son., L. Calif. March–April. Largely replaced in cismontane Calif. by:

Var. **variábilis** Ball. Ovary and caps. glabrous.—Abundant in wet places; V. Grassland, Foothill Wd., Coastal Sage Scrub, etc.; Cent. V. from Shasta Co. s., cismontane s. Calif.; to New Mex., Tex., Chihuahua.

4. **S. laevigàta** Bebb. Tree 5–15 m. high, with rough bark; twigs glabrous, red-brown to yellowish; lvs. lanceolate to oblong-lanceolate, closely serrulate, light green above, paler and more glaucous beneath, 5–12 cm. long, 1.5–3 cm. wide; petioles 2–10 mm. long, not glandular; stipules small, glandular-toothed; catkins 3–10 cm. long, on short leafy lateral peduncles; stamens 4–6, the fils. pilose below; scales elliptic, yellow, woolly at base; caps. glabrous, ovoid, 3.5–5 mm. long; style and stigmas minute.—Along streams, below 5000 ft.; many Plant Communities; cismontane Calif.; to Utah, Ariz. March–May.

Var. **araquìpa** (Jeps.) Ball. [*S. l.* forma *a.* Jeps.] Young twigs, petioles, etc., pilose.—With the sp., but more common in the s. parts of its range.

√5. **S. Hindsiàna** Benth. [*S. sessilifolia* var. *H.* Anderss.] SANDBAR WILLOW. Erect shrub or small tree, 2–7 m. high, with gray furrowed bark; young twigs gray-tomentose; lvs. linear to lance-linear, tapering at both ends, 4–8 cm. long, 0.3–0.6(–1) cm. wide, mostly entire and gray-silky-villous to subtomentose, sometimes green and glabrate; petioles 1–3 mm. long, not glandular; stipules wanting or small, to arcuate-lanceolate and 8 mm. long on sucker shoots; catkins appearing after the lvs., 2–4 cm. long, on leafy peduncles; stamens 2, with fils. pubescent; scales oblanceolate, villous; caps. subsessile, 5–6 mm. long, villous-tomentose; style ca. 0.5 mm. long; stigmas 1 mm. long.—Common locally along ditches, on sand bars, etc., below 3000 ft.; many Plant Communities; cismontane Calif. to Ore., L. Calif.; less common s. March–May.

Var. **leucodendroìdes** (Rowlee) Ball. [*S. macrostachya* var. *l.* Rowlee. *S. argophylla* Calif. auth. *S. sessilifolia* var. *l.* C. K. Schneid.] Lvs. remotely denticulate, more pointed, mostly narrower; caps. usually less densely pilose; styles 0.5–0.7 mm. long.—Sparingly in Santa Clara and Tulare cos. to Kern Co., more common Ventura to San Diego cos.; to L. Calif.

Var. **Parishiàna** (Rowlee) Ball. [*S. P.* Rowlee. *S. exigua* var. *P.* Jeps.] Lvs. gray to silvery, linear, 2–3.5 mm. wide, ± revolute; style 0.1–0.2 mm. long.—Occasional, up to 5500 ft., San Benito Co. (Pinnacles) to San Diego Co.

6. **S. exígua** Nutt. [*S. longifolia* var. *e.* Bebb. *S. fluviatilis* var. *e.* Sarg.] NARROW-LEAF W. Shrub 2–4 m. tall; lvs. linear, entire or remotely denticulate, canescent on both surfaces, sometimes silvery-tomentose beneath, 5–12 cm. long, 0.2–0.8 cm. wide, subsessile; catkins coming after the lvs. on leafy pedunculate branchlets; scales yellowish, lanceolate, ± hairy; stamens 2; fils. hairy below; caps. ca. 5 mm. long, glabrous or nearly so; stigmas sessile, ca. 0.5 mm. long.—Wet places, below 8000 ft.; Sagebrush Scrub, Creosote Bush Scrub, etc.; deserts and e. of Sierra Nevada, Imperial Co. to Modoc Co.; occasional on cismontane side from Kern Co. s.; to B.C., Tex. March–May.

Var. **stenophýlla** (Rydb.) C. K. Schneid. Lvs. silvery-pubescent, 2–5 mm. wide; caps. silvery-pubescent when young, 5–7 mm. long. With the sp.

7. **S. melanópsis** Nutt. Shrub or small tree to 3 or 5 m. tall; twigs dark; lvs. oblanceolate to elliptic, acute at both ends, 4–7 cm. long, 6–14 mm. wide, rather closely denticulate, dark green and lustrous above, paler to glaucescent, subglabrous beneath, subsessile; stipules lanceolate to semicordate, dentate; aments 3–4 cm. long, slender; scales oblong to obovate, nearly or quite glabrous; stamens 2, fils. hairy below, distinct; caps. 4–5 mm. long, glabrous, subsessile, with obsolete styles and very short stigmas.—Stream banks, below 8000 ft.; many Plant Communities; Sierra Nevada n. to Modoc Co., Coast Ranges from Lake and Sonoma cos. n.; to B.C., Rocky Mts. March–May. Plants from Del Norte and Humboldt cos., with lvs. glabrous beneath, 6–12 mm. wide and caps. pedicelled, have been named var. *gracílipes* Ball; those from Tulare to Butte cos. with similar lvs. and caps., but the lvs. 2–4 mm. wide, are var. *tenérrima* (Henders.) Ball. [*S. longifolia* var. *t.* Henders.]

Var. **Bolanderiàna** (Rowlee) C. K. Schneid. [*S. B.* Rowlee.] Young growth and lvs. pubescent to grayish-pubescent.—With the sp., but extending sparingly s. into San Gabriel and San Bernardino mts.

8. **S. Parksiàna** Ball. Cespitose shrub 1–3 m. tall, with gray bark and brownish ± pilose branchlets; lvs. sessile, elliptical to ± oblong, acute at both ends, 4–12 cm. long, 0.8–2.5 cm. wide, irregularly glandular-denticulate, thinly to densely pilose on both surfaces; catkins on shoots of the season, 4–8 cm. long; scales elliptical-oblong to obovate, yellowish, pilose; stamens 2, basally pilose; caps. 4.5–5.5 mm. long, subglabrous, without styles; stigmas 0.2–0.4 mm. long.—Gravel bars and along streams, below 1000 ft.; Coastal Strand, Redwood F., Mixed Evergreen F.; Humboldt and Del Norte cos. May–June.

9. **S. lùtea** Nutt. [*S. cordata* var. *l.* Bebb.] YELLOW WILLOW. Clustered shrub, 2–5 m. high, with yellowish or brownish glabrous twigs; lvs. narrowly to broadly lanceolate, acute to short-acuminate at apex, ± rounded at base, entire to serrulate, yellowish-green above, paler and ± glaucous beneath, 4–10 cm. long, 1.5–4 cm. wide, on petioles 3–10 mm. long; stipules ovate to lunate, entire to serrulate; catkins appearing before and with the lvs., subsessile, 2–3 or 4 cm. long; scales tawny, oblanceolate, pilose; stamens 2, fils. glabrous, united at base; caps. conic-ovoid, 4–5 mm. long, glabrous, on pedicels 1–2 mm. long; styles less than 0.5 mm. long; stigmas short.—Wet places along streams, etc., 5000–9500 ft.; Red Fir F., Lodgepole F.; Sierra Nevada, from Nevada to Modoc cos.; to Alta., Colo., Ariz. May–June.

Var. **Watsònii** (Bebb) Jeps. [*S. cordata* var. *W.* Bebb.] Twigs shorter and more divaricate, yellowish-white; lvs. 4–6 cm. long.—San Jacinto and San Bernardino mts., White Mts. and near Bridgeport, Mono Co.; Great Basin states.

10. **S. ligulifòlia** (Ball) Ball. [*S. lutea* var. *l.* Ball. *S. l.* var. *nivaria* Jeps.] Clustered shrub 1–5 m. tall, with dark gray bark and dark green foliage; young twigs yellowish, puberulent, the older browner and glabrescent; lvs. ligulate or lanceolate to oblonglanceolate, 5–10 cm. long, 1–2 cm. wide, subentire to glandular-serrulate, dark green above, glaucous beneath, glabrous when mature; petioles 5–10 mm. long; catkins sessile, 2–4 cm. long; scales elliptic-oblong, dark, white-tomentose; stamens 2, the fils. free, glabrous; caps. lanceolate to lance-ovoid, 4.5–5 mm. long, glabrous; pedicels ca. 1 mm. long; styles 0.3–0.5 mm. long; stigmas entire to bifid.—Meadows and along streams, 3000–9500 ft.; Yellow Pine F. to Lodgepole F.; Sierra Nevada from Tulare to Plumas cos.; to Ore., S. Dak., New Mex. April–May.

11. **S. Mackenziàna** (Hook.) Barr. [*S. cordata* var. *M.* Hook.] Shrub or small tree

to 6 m. tall; twigs dark brown, sometimes pubescent, or ± yellowish when young; lvs. lanceolate to ovate-lanceolate, 6–10(–12) cm. long, short-acuminate at apex, rounded to cordate at base, glandular-serrulate, dark green above, glaucous beneath; petioles 5–10 mm. long; stipules reniform to semilunate; catkins 2.5–6 cm. long, appearing with the lvs.; scales oblanceolate, obtusish, pubescent, dark brown; stamens 2, fils. glabrous; caps. glabrous, 4.5–5.5 mm. long; pedicels 2.5–4 mm. long; styles ca. 0.5 mm. long, the stigmas shorter.—Occasional below 7000 ft.; largely Red Fir F.; Sierra Nevada from Tulare Co. n., Siskiyou and Modoc cos.; to Canada, Utah. May.

12. **S. pseudocordàta** Anderss. [*S. pseudomyrsinites,* Calif. auth.] Shrub to 4 m. tall with yellowish to red-brown, glabrous to ± pubescent, shining twigs; lvs. elliptic- or oblong-lanceolate to oblong-oval, acute to short-acuminate at apex, rounded to cordate at base, 3–8 cm. long, 1–3 cm. wide, thickish, glandular-serrulate, green above, paler beneath, veiny, mostly glabrous by maturity; petioles 4–10 mm. long; stipules narrowly ovate, glandular-serrulate; catkins appearing with the lvs., 1.5–4 cm. long, on short bracted peduncles; scales oblong-oblanceolate, 1–1.5 mm. long, dark brown, villous, glabrate on outside; stamens 2, fils. glabrous; caps. 4–6 mm. long, glabrous; pedicels 1–1.5 mm. long; styles 0.3–0.7 mm. long; stigmas short.—Stream banks and moist places, 6000–11,300 ft.; Montane Coniferous F.; Sierra Nevada from Tulare Co. n., to Shasta and Modoc cos.; to Alaska, Ida. New Mex. June–July.

13. **S. lasiólepis** Benth. [*S. Bakeri* von Seem.] ARROYO WILLOW. Erect shrub or small tree, 2–10 m. high, with smooth bark and yellowish to dark brown twigs, usually pubescent; lvs. narrowly to broadly oblanceolate, 6–10 cm. long, 1–2 cm. wide, acute to obtuse, narrowed at base, nearly entire, subrevolute, dark green and glabrous above, pubescent to glabrate and glaucous beneath; petioles 3–12 mm. long; stipules none or ovate to roundish; catkins mostly appearing before the lvs., subsessile, 3–7 cm. long; scales dark, round-obovate, 1 mm. long, densely hairy; stamens 2, fils. glabrous, united below; caps. 4.5–5 mm. long, glabrous; pedicels 0.5–1.2 mm. long; styles 0.5 mm. long; stigmas short.—Common on stream banks and beds, below 7000 ft.; many Plant Communities; throughout cismontane Calif., occasional in the desert; to Wash., Ida., New Mex. Feb.–April.

Var. **Bigelòvii** (Torr.) Bebb. [*S. B.* Torr. *S. franciscana* von Seem.] Lvs. broadly cuneate-oblanceolate to spatulate or obovate, more pubescent beneath; caps. to 6 mm. long; pedicels 1–2 mm. long.—With the sp. in the cent. and n. parts of its range, beginning in San Francisco region.

Var. **Sandbérgii** (Rydb.) Ball. [*S. S.* Rydb.] Densely pubescent throughout; lvs. oblanceolate to obovate-oval, thinly pubescent above, densely so beneath even in maturity.—N. part of range of sp., from Humboldt and Siskiyou cos. n.

Var. **Bracelìnae** Ball. Shrubs 2–6 m. tall, with stoutish pubescent to puberulent twigs; lvs. linear to linear-elliptic or -lanceolate; petioles 5–15 mm. long.—At 2500–7200 ft.; San Diego to Sonoma, Alpine, and Inyo cos.; to Utah, Tex., Chihuahua, L. Calif.

14. **S. Tràcyi** Ball. A slender graceful shrub or small tree 2–6 m. tall, with smooth grayish-green trunks and slender grayish-yellow to brownish glabrous twigs; lvs. oblanceolate, elliptical-oblanceolate, or narrowly obovate, 3–5 cm. long, 1.5–1.8 cm. wide, green above, glaucous beneath, entire or crenulate-serrulate on vegetative shoots; petioles slender, 2–6 mm. long; catkins with or before the lvs., 2–5 cm. long; scales spatulate-obovate, to broadly oblanceolate, 1–1.3 mm. long, dark, ± hairy; stamens 2, fils. glabrous, united ca. ⅓ their length; caps. glabrous, lanceolate, 4.5–6 mm. long, on pedicels 1–3 mm. long; style 0.2–0.5 mm. long; stigmas ca. as long.—Sand and gravel bars of streams, below 500 ft.; Redwood F., Mixed Evergreen F., etc.; coast of Humboldt and Del Norte cos., sw. Ore. April–May.

15. **S. commutàta** Bebb. Shrub 1–3 m. tall with stoutish yellow to brownish glabrate to pilose twigs; lvs. elliptical to broadly oblanceolate or obovate, 6–8 cm. long, 1.5–4 cm. wide, acute to abruptly cuspidate, subentire to minutely glandular-serrulate, green, obscurely veiny, ± pubescent on both sides; petioles stoutish, stipules semicordate to ovate; catkins appearing with the lvs., 2–5 cm. long, on leafy peduncles; scales oblanceolate, obtuse, blackish, 1.5–2 mm. long, villous; stamens 2, fils. glabrous; caps. 5–7 mm. long, glabrous; pedicels 1–2 mm. long; styles 1–1.5 mm. long; stigmas ca. 0.5 mm. long. —At ca. 5000–7400 ft.; Red Fir F.; mts. of Trinity, Siskiyou, and Modoc cos.; to Alaska, Wyo., Utah. May–June.

Var. **denudàta** Bebb. Lvs. glabrate to glabrous when mature. With the sp.

16. **S. Eastwoódiae** Ckll. [*S. californica* auth., not Lesq.] Low shrub 0.5–2 m. high, with twigs mostly dark, tomentose; lvs. elliptical to ± obovate, 4–6 cm. long, 1–2 cm. wide, gray-tomentose on both sides or ± glabrate in age, acute, ± denticulate; petioles ca. 3–6 mm. long; stipules ovate, small, glandular-serrulate; catkins appearing with the lvs., 1–4 cm. long, leafy-pedunculate; scales broadly lanceolate, brownish, tomentose, ca. 1–1.5 mm. long; stamens 2, fils. free, pilose at base; caps. 5–6.5 mm. long, gray-tomentose; pedicels 0.7–1 mm. long; styles 1–1.5 mm. long.—Stream banks, meadows, 7000–10,500 ft.; Red Fir F. to Subalpine F.; Sierra Nevada from Tulare Co. n., to Modoc Co.; Wash. June–July.

17. **S. oréstera** C. K. Schneid. [*S. glauca* var. *o.* Jeps. *S. commutata* var. *rubicunda* Jeps.] Spreading or erect shrub 1–3 m. high with tomentose brown twigs; lvs. oblanceolate to ± elliptic, 4–6.5 cm. long, 1–2 cm. wide, acutish at both ends, ± silky-villous on both sides, green above, glaucous beneath, entire; petioles 3–6 mm. long; stipules none or lanceolate; catkins appearing with the lvs., subsessile, dense, 1–4 cm. long; scales ± oblong, obtuse, dark brown, silky; stamens 2, fils. hairy below, mostly distinct; caps. villous, 7–8.5 mm. long, on pedicels 1–1.5 mm. long; style 1 mm. long; stigmas short, bifid.—Stream banks, meadows and wet places, 8000–12,000 ft.; Lodgepole F. to Alpine Fell-fields; Sierra Nevada from Tulare Co. to Lassen Co.; w. Nev. June–July.

18. **S. Pìperi** Bebb. DUNE W. Large erect shrub 4–6 m. tall, with stout glabrous shining dark branchlets; lvs. broadly elliptical to oblanceolate, 6–12 cm. long, 2–5 cm. wide, obliquely acute to acuminate, dark green and glabrous above, glaucous and glabrous beneath, serrulate; petioles 3–12 mm. long; stipules mostly none, if present reniform to subcordate; catkins with or before the lvs., with short leafy peduncles and brown, long-hairy scales; stamens 2, fils. glabrous, united ca. ⅓ their length; caps. glabrous or thinly pubescent above, 6–7 mm. long; style and pedicel each ca. 1 mm. long.— Wet sandy places at mouth of streams, below 500 ft.; Coastal Strand, Mixed Evergreen F., Redwood F., etc.; Mendocino Co. n.; to Wash. April–May.

19. **S. Hoókeriana** Barr. Erect shrub or small tree 4–10 m. tall, with dark gray rough bark and densely pubescent twigs; lvs. broadly lanceolate to elliptic-oval, acute, dark green and glabrous above, densely white- or rusty-hairy beneath, 3–12 cm. long, 1–4 cm. wide, subentire to crenulate-serrulate; petioles 6–9 mm. long; stipules ± cordate; catkins appearing before the lvs., short-peduncled, dense; scales black, long-villous; stamens 2, fils. glabrous, united below; caps. glabrous, 6–8 mm. long, on pedicels ca. 1 mm. long; style 1 mm. long.—Ocean beach about stream-mouths and lagoons, below 500 ft.; Coastal Strand, Mixed Evergreen F., etc.; Mendocino to Del Norte cos.; to B.C. April–May. Plants with tomentose capsules are var. *tomentòsa* Henry.

20. **S. anglòrum** Cham. var. **antiplásta** C. K. Schneid. [*S. petrophila* var. *caespitosa* (Kenn.) C. K. Schneid. *S. c.* Kenn.] ALPINE W. Creeping shrub from underground stems forming mats with erect branchlets (the younger yellowish, sometimes pubescent, the older brownish to purplish-black, glabrate); lvs. elliptic-oblanceolate to broadly oblanceolate, 2–4 cm. long, 0.7–1.7 cm. wide, acutish, entire, glabrous or hairy when young; petioles 3–10 mm. long; stipules none or minute; catkins peduncled, appearing with the lvs., 1–6 cm. long; scales narrowly obovate, 1.5–2 mm. long, dark brown; stamens 2, fils. glabrous; caps. 5–7 mm. long, subsessile, gray-tomentose; styles 1–1.5 mm. long; stigmas 0.4–0.6 mm. long.—Moist banks and meadows, 8500–12,000 ft.; Subalpine F., Alpine Fell-fields; Sierra Nevada from Inyo and Tulare cos. to Lassen Co.; B.C., Quebec, New Mex. July–Aug.

21. **S. nivàlis** Hook. SNOW W. Creeping shrub from buried stems; lvs. oblong-obovate to almost round, 0.7–1.2 cm. long, 0.4–0.8 cm. wide, shining, with depressed veins above, glaucous and reticulate beneath, entire, somewhat revolute, glabrous at maturity; petioles 2–7 mm. long; stipules none; catkins appearing after the lvs., to ca. 1 cm. long, few-fld.; scales oblong to obovate, 1 mm. long; stamens 2, fils., glabrous; caps. sessile, white-tomentose, 2.5–3.5 mm. long; styles 0.1–0.2 mm. long; stigmas minute.—Uncommon, moist places, 10,000–12,000 ft.; Alpine Fell-fields; Mt. Dana, near Koip Glacier, Tioga Crest, Virginia Lakes, etc., all in Mono Co.; to B.C., Ida., Utah, Colo. July–Aug.

22. **S. plànifòlia** Pursh var. **mònica** (Bebb) C. K. Schneid. [*S. m.* Bebb.] Ascending to erect shrub 0.3–1.5 m. tall, with yellow-green to brown-red glabrous twigs; lvs. narrowly to broadly elliptical-oval or the lower obovate, acute to rounded at base, acute

at apex, 2–3.5 cm. long, 0.8–1.5 cm. wide, subentire, glabrous and bright green above, slightly paler beneath; petioles 2–8 mm. long; stipules none or minute; catkins appearing with the lvs., 1–3 cm. long, subsessile; scales brown to almost black, villous, elliptical to oval; stamens 2, fils. glabrous; caps. 5–7 mm. long, subsessile, subsericeous; styles 0.6–1.4 mm. long; stigmas 0.3–0.6 mm. long.—Moist places, 8000–12,500 ft.; Lodgepole F. to Alpine Fell-fields; Sierra Nevada of Tuolumne, Fresno, Inyo, and Mono cos.; to B.C., Mont., Colo., New Mex. June–Aug.

23. **S. Lemmònii** Bebb. [*S. Austinae* Bebb.] Shrub 1–5 m. high with younger twigs glabrous to thinly pubescent, yellowish and the older brownish-black, somewhat glaucous; lvs. oblanceolate to almost lanceolate, 3–10 cm. long, 0.8–2 cm. wide, acute to acuminate, entire or ± glandular-serrulate, deep green and shining above, ± glaucous beneath; petioles 3–10 mm. long; stipules minute to lanceolate and 8 mm. long; catkins mostly appearing with the lvs., 1–6 cm. long; scales narrowly obovate, dark, 1–1.5 mm. long, somewhat silky; stamens 2, fils. pilose at base; caps. slender, 6–9 mm. long, silky; pedicels 1–2 mm. long; styles 0.6–1.2 mm. long; stigmas ca. 0.5 mm. long.—Moist places, 5000–10,000 ft.; Montane Coniferous F.; San Bernardino Mts., Sierra Nevada from Tulare Co. n., to Modoc and Siskiyou cos.; to Ore., Wyo. May–June.

24. **S. Drummondiàna** Barr. var. **subcoerùlea** (Piper) Ball. [*S. s.* Piper. *S. Covillei* Eastw.] Shrub 1–4 m. tall, with slender, ± subglabrous, dark twigs; lvs. of flowering twigs narrowly oblong-oblanceolate, mostly acute, rather short, those on summer shoots oblong-lanceolate to oval-lanceolate, acuminate, 3–7 cm. long, sparsely pubescent and green above, silky and glaucescent beneath, entire or slightly crenulate; petioles 2–6 mm. long; stipules none or occasional on vigorous shoots; catkins appearing with or before the lvs., 1–4 cm. long, sessile or nearly so; scales oblong to obovate, acute or obtuse, dark, thinly pilose; stamens 2, fils. free, glabrous; caps. silvery-silky, 3.5–5 mm. long, almost sessile; styles 1–1.5 mm. long; stigmas 0.2–0.5 mm. long.—Moist places, 8400–9500 ft.; Lodgepole F., Subalpine F.; Sierra Nevada of Fresno, Inyo, and Tulare cos.; to B.C., Rocky Mts. May–June.

25. **S. sitchénsis** Sanson. Shrubby or arborescent, 2–7 m. tall, twigs slender, yellow-brown and somewhat pubescent when young, darker and ± glabrescent when older; lvs. broadly oblanceolate to mostly cuneate-obovate, 4–10 cm. long, 1.5–3.5 cm. long, 1.5–3.5 cm. wide, acute to obtuse at apex, cuneate at base, thin, entire, ± revolute, dull green above, silky with prominent veins beneath; petioles 5–10 mm. long; stipules mostly none, occasionally lance-ovate and to 5 mm. long; catkins with or before the lvs., 2–8 cm. long on leafy peduncles 1–2 cm. long; scales elliptic-oblanceolate, acutish, 1–1.5 mm. long, brown, ± villous; stamen 1; fil. glabrous; caps. silky, 4–6 mm. long; pedicel 0.3–0.6 mm. long; styles 0.3–0.6 mm. long; stigmas short.—Occasional at low elevs.; Mixed Evergreen F., Redwood F.; Sonoma Co. to Del Norte and Siskiyou cos.; to Alaska, Mont. March–April.

26. **S. Còulteri** Anderss. [*S. sitchensis* var. *C.* Jeps.] Resembling *S. sitchensis,* but with stout brown densely pubescent to tomentose twigs; lvs. oblanceolate to obovate, 4–10 cm. long, 1.5–3 cm. wide, densely white, opaque-tomentose beneath; petioles 3–12 mm. long; stipules reniform; catkins appearing with the lvs.; scales oblanceolate, tawny or fuscous at tip, densely white-woolly; stamen 1; fil. glabrous; caps. 4–4.6 mm. long, silvery-tomentose, subsessile; style ca. 0.5 mm. long.—Stream banks, Redwood F., Mixed Evergreen F., etc.; near the coast from San Luis Obispo Co. n.; to Wash. Feb.–April.

27. **S. Jepsònii** C. K. Schneid. [*S. sitchensis* var. *angustifolia* Bebb. *S. s.* f. *parvifolia* Jeps. *S. s.* var. *Ralphiana* Jeps.] Shrub 1–3 m. high, with chestnut to dark purple shining twigs, ± pubescent when young; lvs. oblanceolate, acute to obtuse at apex, cuneate at base, 3–7 cm. long, 1–2 cm. wide, firm, entire, green but ± silky above, whitish and silky-tomentose beneath; petioles 3–6 mm. long; catkins before or with the lvs., ♂ subsessile, 1–2 cm. long, ♀ longer, short-pedunculate; scales brown, densely pilose, oblong-obovate; stamens 2; fils. glabrous, mostly free; caps. 4.5–5.5 mm. long, densely silky; pedicel 1–1.3 mm. long; style 0.7–1 mm. long; stigmas short.—Damp places, 5500–10,000 ft.; Red Fir F. to Subalpine F.; Sierra Nevada from Tulare Co. n., to Tehama Co., N. Coast Ranges of Trinity, Siskiyou, and Humboldt cos. May–June.

28. **S. Bréweri** Bebb. Low shrub to ca. 1 m. high, with slender dark brown, glabrous or, when young, ± pubescent twigs; lvs. oblong to lance-linear, 2–6 cm. long, entire, dull green and puberulent above, acute, gray tomentose and with reticulate raised veins

beneath; petioles to ca. 3 mm. long; stipules none or small; catkins with or before the lvs., subsessile; scales oblanceolate, yellowish, or reddish at apex, pilose; stamens 2; fils. free, glabrous; caps. sessile, ca. 5 mm. long, short-tomentose; styles ca. 1 mm. long.— Gravel beds and ravines, below 4500 ft.; mostly Foothill Wd.; inner Coast Ranges, San Benito Co. to Lake and Colusa cos. March–April.

29. **S. delnorténsis** C. K. Schneid. Much like *S. Breweri*, the young branches gray-tomentose; lvs. obovate, 1.5–8 cm. long, entire, rounded or abruptly apiculate at apex, tomentose to glabrescent above, densely tomentose and rugose-reticulate beneath; catkins subpedunculate, appearing with the lvs.; anthers ± purple, the fils. ± united; caps. 4–5 mm. long, pilose-tomentose; style ca. 1 mm. long.—At 300–600 ft., Mixed Evergreen F.; Smith River, near Gasquets, Del Norte Co. April–May.

30. **S. Scouleriàna** Barr. [*S. flavescens* Nutt. *S. brachystachys* Benth.] Shrub or small tree 1–10 m. tall, densely to thinly pubescent; twigs stoutish, the younger yellowish, the older brownish to blackish, glabrate; lvs. mostly oblanceolate to obovate, 3–10 cm. long, 1.5–3 cm. wide, mostly abruptly acute and entire, dull green above, glaucous and coarsely veiny beneath, silvery-pubescent to glabrate and glaucous beneath; petioles 5–15 mm. long; stipules subreniform to semiovate; catkins before the lvs., dense, subsessile; scales obovate, 1.5–2.5 mm. long, black, villous; stamens 2; fils. glabrous; caps. 7–9 mm. long, gray-tomentose; styles 0.2–0.5 mm. long; stigmas 0.5–1 mm. long.—Along streams and in moist places, below 10,000 ft.; many Plant Communities; San Jacinto Mts. n. through Sierra Nevada, near the coast from Monterey Co. n.; to Alaska and Rocky Mts. April–June.

Var. **coetàna** Ball. Catkins appearing with the lvs., the ♀ rather lax, both ♂ and ♀ on short leafy-bracted peduncles.—From Humboldt and Modoc cos. n.; to B.C., Mont., Utah.

31. **S. Geyeriàna** Anderss. Shrub to 4 or 5 m. tall with slender leafy olive-green to brown (later darker), mostly glabrous twigs; lvs. narrowly oblanceolate to elliptic-oblong, 2–7 cm. long, 0.6–1.5 cm. wide, acute at both ends, entire, dark green above, ± glaucous beneath, thinly silky on both sides; petioles slender, 4–10 mm. long; stipules none; catkins coming with the lvs., on short leafy peduncles; scales oblong to narrow-oblanceolate, 1–2 mm. long, yellowish with reddish tips; stamens 2; fils. mostly pubescent below; caps. 5–7 mm. long, silky; styles to 0.3 mm. long; stigmas 0.3–0.5 mm. long.—At ca. 10,000 ft.; White Mts., Mono Co.; to B.C., S. Dak., Ariz. May–June.

Var. **argéntea** (Bebb) C. K. Schneid. [*S. macrocarpa* var. *a.* Bebb.] Twigs ± pubescent; lvs. permanently silvery-pilose on both surfaces.—Wet places, 5000–10,500 ft.; Montane Coniferous F.; Sierra Nevada from Tulare Co. n., to Modoc Co.; to Ida., Wyo., Colo. May.

83. Ulmàceae. ELM FAMILY (Fig. 81)

Trees or shrubs with watery juice, terete branchlets, and alternate simple serrate pinnately veined deciduous lvs. with stipules. Perfect or polygamous, the fls. clustered or the ♀ solitary. Calyx 4–9-parted or -lobed, small. Corolla none. Stamens 4–6. Styles 2; ovary 1-celled, with 1 suspended ovule. Fr. a samara, nut, or drupe.

Fr. a dry samara; fls. perfect .... 1. *Ulmus*
Fr. drupaceous; fls. imperfect .... 2. *Celtis*

1. Ulmus L. ELM

Trees or shrubs, the buds with many ovate rounded brown scales imbricated in 2 ranks. Lvs. simply or doubly serrate; stipules linear to obovate. Fls. in fascicles or cymes, appearing in spring before the lvs. Calyx persistent; stamens 5–6, fils. filiform or slightly flattened; fr. surrounded at base by remnants of calyx and winged most of the way around with a terminal notch and persistent style. Ca. 18–20 spp., of N. Temp. and colder regions. (Classical Latin name for elm.)

Lvs. with scattered pubescence, scabrous above; branchlets pubescent until the second year
                                                                                   1. *U. procera*
Lvs. with axillary tufts beneath, smooth above; branchlets subglabrous .... 2. *U. carpinifolia*

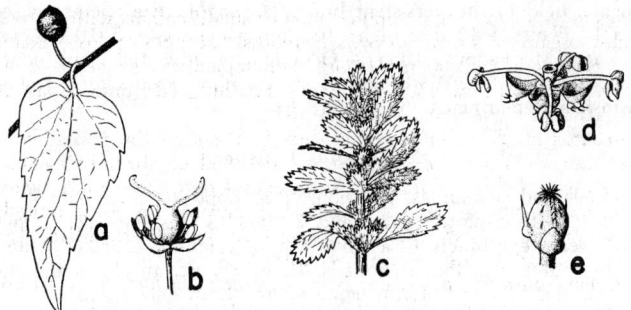

Fig. 81. ULMACEAE. *Celtis Douglasii: a,* lf. and drupe, × ½; *b,* fl., × 2½, lobed calyx, stamens, pistil with 2 styles.
URTICACEAE. *Urtica urens: c,* shoot, × ½, with serrate lvs. and axillary groups of fls.; *d,* ♂ fl., × 5, 4 sepals and 4 stamens; *e,* ♀ fl., × 5, with spiny ovary and tufted stigma.

1. **U. prócera** Salisb. ENGLISH ELM. Tree to 35 m. tall, with oval head; usually suckering abundantly; bark deeply fissured; young twigs pubescent; lvs. 5–7 cm. long, ovate to broadly oval, short-acuminate, very unequal at base, scabrous above, pubescent beneath; samara to 2.5 cm. long; $2n = 28$ (Sax, 1933).—Reported as weed tree in cent. Calif.; native of Eu.

2. **U. carpinifòlia** Gleditsch. SMOOTH-LVD. ELM. Tree to 30 m., suckering; young twigs subglabrous; lvs. 5–8 cm. long, on petioles 6–12 mm. long; samara cuneate at base; $2n = 28$ (Sax, 1933).—Occasional escape; native of Eurasia.

2. Céltis L. HACKBERRY

Deciduous, sometimes evergreen; bark mostly smooth, gray. Lvs. 3-nerved at base, serrate or entire. Fls. greenish, appearing with the lvs. Calyx 5–6-parted, the ♂ fls. in cymose clusters, the ♀ solitary or in few-fld. clusters in upper lf.-axils. Ovary ovoid; fr. a drupe with thick-walled nutlet. Perhaps 70 spp., of N. Hemis. and trop. (Greek name of a tree with sweet fr.)

1. **C. Doúglasii** Planch. Small spreading tree, 2–6 m. tall, with slender glabrous or puberulent twigs; lvs. lance-ovate, acute or acuminate, serrate to almost entire, 3–6 cm. long, on petioles 2–6 mm. long, scabrous above, strongly reticulate-veined beneath; pedicels 1–1.5 cm. long; fr. globose, 6–8 mm. thick, orange-brown.—Occasional about widely scattered damp places, 2800–5000 ft.; Creosote Bush Scrub, Joshua Tree Wd., Pinyon-Juniper Wd., etc.; Laguna Mts. (e. San Diego Co.), Banning, Clark Mt., Providence Mts., etc., Independence (Inyo Co.), Caliente Creek (Kern Co.), etc.; to e. Wash., Utah, Ariz. April–May.

84. Moràceae. MULBERRY FAMILY

Herbs to trees, often with milky juice. Lvs. mostly alternate and simple. Fls. small, inconspicuous, regular, mostly imperfect, usually in heads or spikes, sometimes the ♂ racemose. Perianth single, entire or parted; stamens of same no. as perianth parts and opposite them. Styles or stigmas 2; ovary 1-celled, ovule pendulous. Fr. an ak., or drupe, often imbedded in fleshy perianth or axis.

Plant an erect herb; lvs. digitate, alternate .... 1. *Cannabis*
Plant a twining vine; lvs. lobed, opposite .... 2. *Humulus*

1. Cánnabis L. HEMP. MARIJUANA

Stout erect rough puberulent annual herbs, mostly dioecious. Lvs. 5–11-divided; fls. greenish, axillary, the ♂ panicled, the ♀ spicate. Stamens 5, drooping, sepals free. Pistillate fl. in axil of small enclosing bract, the calyx inconspicuous, adherent to the long ovary. Fr. an ak., held in the persistent bract. (*Cannabis,* the ancient classical name.)

1. **C. satìva** L. Plants 4–12 dm. high; lfts. linear-lanceolate, 5–10 cm. long, coarsely toothed; $2n = 20$ (Medvedeva, 1935).—Occasionally adventive, but now not grown without permit; native of Old World and of use for fiber. Producing a resinuous oil, the source of hashish and marijuana.

2. **Hùmulus** L. HOPS

Twining annuals and perennials, dioecious. Lvs. opposite. Staminate fls. in panicled, tassellike racemes, with 5-parted calyx and 5 stamens. Pistillate fls. 2 together under large imbricate persistent bracts in a spike that becomes a conelike structure or hop, each fl. with an entire calyx that surrounds the ovary. Fr. an ak. in the tight enlarged calyx. Three spp. of N. Temp. (Latin name of uncertain derivation.)

1. **H. Lùpulus** L. Perennial with rough stems; lvs. mostly 3-lobed to ca. the middle, the terminal lobe broadest, margins coarsely serrate, upper lf.-surface rough, lower less so and sparsely glandular; petioles ca. as long as blade; fr.-spikes 3–5 cm. long at maturity with thin light-colored resinous-dotted scales.—Occasional escape from cult.; grown for brewing purposes; native of Eurasia.

85. Urticàceae. NETTLE FAMILY (Fig. 81)

Ours herbaceous; trop. forms may be woody. Stinging hairs often present. Lvs. simple, alternate or opposite, mostly stipulate. Fls. small, greenish, imperfect or perfect, in cymose racemes or fascicles. Corolla none. Calyx 2–5-cleft or of separate sepals. Stamens as many as the calyx-divisions and opposite them. Ovary superior, 1-celled, with a solitary orthotropous ovule. Style and stigma 1. Fr. an ak., with oily endosperm.

Plants with stinging hairs and opposite lvs.
  Pistillate calyx 4-parted, the segms. almost distinct, the inner much the longer and inclosing the ak.
                              1. *Urtica*
  Pistillate calyx saccate, 2--4-toothed at the apex .... 2. *Hesperocnide*
Plants without stinging hairs and with alternate lvs. .... 3. *Parietaria*

1. **Úrtica** L. NETTLE

Annual or perennial herbs with stinging hairs and opposite petioled lvs. with 3–7 nerves, stipulate. Our spp. dioecious or monoecious, the fls. clustered in geminate racemes or heads. Staminate calyx deeply 4-parted and with 4 stamens. Pistillate calyx with unequal sepals, the longer inner ones enclosing the flattened ak. Stigma sessile, tufted. Perhaps 30 spp., of wide distribution. (The ancient Latin name, from urere, *to burn.*)

Plant perennial; ♀ and ♂ fls. in separate clusters.
  Stems whitish-pubescent; lvs. more than twice as long as wide; fl.-clusters mostly exceeding upper lvs.
    Lvs. coarsely velvety-pubescent beneath; stems hispid as well as pubescent. Common
                              1. *U. holosericea*
    Lvs. more sparsely pubescent beneath, not velvety; stems often scarcely if at all hispid. Occasional ....2. *U. Serra*
  Stems glabrous except sometimes for scattered bristles; lvs. less than twice as wide; fl.-clusters mostly shorter than upper lvs.
    Lvs. nearly or quite glabrous beneath, ending in a rather pronounced toothless elongate tip. Mendocino Co. n. ....3. *U. Lyallii*
    Lvs. more heavily pubescent beneath, lacking a prolonged entire tip. Sonoma Co. to San Mateo Co. ....4. *U. californica*
Plant annual; ♀ and ♂ fls. mixed in the same cluster .... 5. *U. urens*

1. **U. holosericea** Nutt. [*U. gracilis* var. *h.* Jeps. *U. h.* vars. *densa* and *Greenei* Jeps.] Perennial, from underground rootstocks, the stems rather stout and simple, 1–2.5 m. tall, bristly and also densely fine-pubescent; lvs. lanceolate to narrow-ovate, 5–12 cm. long, coarsely serrate, densely soft-pubescent and grayish beneath, greener above, attenuate at apex, rounded at base, on petioles 1–4.5 cm. long; stipules narrow-oblong, acute to obtuse, 5–10 mm. long; ♂ clusters rather loose, almost as long as the lvs.; ♀ denser and shorter; fls. ca. 1 mm. long; ak. broadly ovate, tan-colored, smooth.—Low damp places, below 9000 ft.; many Plant Communities; throughout cismontane Calif. to Wash., Ida.; Santa Cruz, Catalina ids. Occasional on desert edge. June–Sept.

2. **U. Sérra** Blume. [*U. Breweri* Wats.] Much like *U. holosericea,* but stems mostly less bristly; lvs. yellow-green and less pubescent beneath; seeds finely tuberculate, ± olive brown.—Wet places, mostly 5000–10,000 ft.; Montane Coniferous F.; occasional Sierra Nevada to Ore., Ida., New Mex., Mex. July–Aug.

3. **U. Lyállii** Wats. Stems 1–2 m. tall, glabrous or sparsely bristly; lvs. ovate, sometimes cordate, thin, 4–15 cm. long, coarsely toothed, with a terminal entire long point, green and almost glabrous to somewhat pubescent beneath; petioles slender, 3–8 cm. long; stipules oblong, rounded at tip, ca. 1 cm. long; ak. smooth, ca. 1 mm. long, broad.—Moist places near the coast; Mixed Evergreen F., etc.; Mendocino Co. to Del Norte Co.; to Alaska, Ida. April–June.

4. **U. califórnica** Greene. [*U. Lyallii* var. *c.* Jeps.] Doubtfully distinct from *U. Lyallii;* differing from it in its more pubescent, more cordate lvs., which are more pubescent beneath and lack the entire elongate tip; seeds ± puncticulate.—Wet brushy thickets along the immediate coast; Mixed Evergreen F., etc.; Sonoma Co. to San Mateo Co. April–June.

5. **U. ùrens** L. DWARF NETTLE. Erect annual, simple or branched from the base, glabrous except for the stinging hairs, 1–5 dm. high; lvs. ovate, glabrous, coarsely laciniate-serrate, the blades 1.5–3 cm. long, petioles 1–2 cm. long; stipules 1 mm. long; fl.-clusters scarcely 1 cm. long; calyx almost 2 mm. long; ak. ca. 2 mm. long, ± yellow; $2n = 24, 26, 52$ (Löve & Löve, 1942).—Garden and orchard weed; natur. from Eu. Jan.–April.

2. Hesperócnide Torr.

Annual herbs similar to *Urtica urens* in appearance. Stipules minute; ♂ fls. in clustered axillary heads, mixed with pistillate. Calyx of ♂ fls. 4-parted, the lobes equal; that of ♀ fls. tubular, 2–4-toothed, with uncinate hairs. Stamens 4. Stigma tufted. Ak. enclosed by the membranous calyx. Two spp., one in Hawaii. (Greek, *hespera,* west, and *knide,* nettle.)

1. **H. tenélla** Torr. Stems slender rather weak, 2–5 dm. long, simple or branched, with stinging hairs; lvs. opposite, ovate, crenate-serrate, the blades 0.5–2.5 cm. long; petioles slender, 0.5–2.5 cm. long; calyces 1–1.5 mm. long in fr.—Occasional on shaded slopes at low elevs.; several Plant Communities; Coast Ranges from Napa Co. to San Diego Co., Sierran foothills, Tulare Co., desert edge. April–June.

3. Parietària L. PELLITORY

Herbs to shrubs, without stinging hairs. Lvs. alternate, entire, 3-nerved, exstipulate. Plants polygamous, the fls. in axillary glomerate clusters, involucrate by small leafy bracts. Staminate calyx 4-parted. Pistillate calyx tubular-ventricose, 4-lobed. Stigma tufted. Ak. ovoid, enclosed in the persistent calyx. Ca. 7 spp., of wide distribution. (Ancient Latin name of the Italian sp.)

Invol. mostly less than twice as long as fls.; lvs. broadly ovate.
    Plant annual; lvs. obtusish at apex. Native .... 1. *P. floridana*
    Plant perennial; lvs. subacuminate. Uncommon introd. .... 2. *P. judaica*
Invol. 2–3 times as long as fls.; lvs. mostly lanceolate .... 3. *P. pensylvanica*

1. **P. floridàna** Nutt. [*P. debilis* Calif. refs., not Forst. f.] Loosely branched annual with slender weak procumbent pilose stems 1–3 dm. long; lvs. ovate to lance-ovate, 0.5–3 cm. long, entire, obtuse, short-petioled; clusters few-fld., the bracts narrow, 3 mm. long; sepals ca. 2 mm. long; ak. 1 mm. long, shining.—Moist shaded slopes about rocks, etc., at low elevs.; several Plant Communities; cismontane and desert s. Calif. n. to Monterey and Inyo cos.; to se. U.S. Feb.–June.

2. **P. judaìca** L. Perennial, 3–5 dm. high, diffuse or ascending; lvs. cuneate at base, lance-ovate to ovate, ± acuminate, blades 1–2.5 cm. long, petioles 0.5–1 cm. long; bracts connate at base, ± decurrent; calyx of perfect fls. strongly accrescent.—Occasional weed in San Francisco, Santa Cruz, and Monterey cos.; native Eurasia.

3. **P. pensylvánica** Muhl. Simple or sparingly branched annual, 1–3 dm. high, minutely downy; lvs. mostly oblong-lanceolate, 2–7 cm. long, long-tapering, on petioles ca. 1 cm. long; involucral bracts 3–4 mm. long.—Occasional, moist shade; Colusa Co. n.; to B.C., Atlantic Coast. May–July.

Fig. 82. LYTHRACEAE. *Lythrum californicum: a,* infl., $\times$ ½, each fl. in axil of bract; *b,* fl., $\times$ 1½, with cylindrical ribbed tube, 6 petals, several stamens; *c,* fl. in section.

86. Lythràceae. LOOSESTRIFE FAMILY (Fig. 82)

Herbaceous or woody, mostly with opposite lvs., no stipules. Fls. perfect, solitary or clustered, regular, in ours minute. Fl.-tube persistent, enclosing but free from the 1–4-loculed ovary, with 4–6 minute sepals and sometimes accessory teeth in the sinuses. Petals when present as many as the primary calyx-teeth, inserted with the 4–17 stamens on the fl.-tube. Anthers versatile, longitudinally dehiscent. Style 1; stigma 2-lobed; ovules many or few, anatropous. Caps. 1–several-celled, variously dehiscent or indehiscent. Seeds without endosperm.

Fl.-tube short, campanulate to globular; lvs. opposite; petals 4.
  Lvs. lance-linear to linear-oblong.
    Fls. mostly 1 in each axil; caps. with 3–4 valves, septicidal .... 1. *Rotala*
    Fls. mostly 3-several in each axil; caps. bursting irregularly .... 2. *Ammannia*
  Lvs. obovate to spatulate; fls. solitary .... 3. *Peplis*
Fl.-tube cylindric; petals usually 6; lvs. mainly alternate .... 4. *Lythrum*

1. Ròtala L. TOOTH-CUP

Low mostly glabrous annuals, with angled stems. Lvs. mostly sessile. Fls. small, 4-merous. Stamens 4; fils. short. Caps. spherical. Seeds many, minute, angled. Ca. 40 spp., largely trop. (Diminutive of *rota,* wheel, from the whorled lvs. of the type sp.)

  1. **R. ramòsior** (L.) Koehne. [*Ammannia r.* L.] Simple or branched from base, mostly 0.5–2 dm. high; lvs. oblong, 1–3 cm. long, narrowed to a short petiole or sessile; fls. minute; subtending bracts ca. half as long as fr.; petals broadly obovate; caps. 3 mm. long; seeds faintly reticulate.—Rare, moist places below 5000 ft.; V. Grassland to Yellow Pine F.; San Joaquin V.; to Wash., Atlantic Coast. June–Aug. Plants from San Joaquin V. with subtending bracts 1–2 times as long as the fr. may be referrable to *R. dentífera* (Gray) Koehne.

2. Ammánnia L.

Subglabrous annuals with angled stems and opposite sessile narrow lvs. which are auricled at base. Fl.-tube 4-angled; sepals 4, with small accessory teeth in the sinuses. Petals 4, deciduous. Stamens 4–8. Ovary 2–4-loculed. Seeds many, angled, minutely pitted. Ca. 20 spp., mostly in warmer regions. (Named for J. *Ammann,* 18th-century German botanist.)

  1. **A. coccínea** Rottb. Ascending or depressed, 1–4 dm. high; lvs. 2–5 cm. long, linear-lanceolate, cordate-auriculate, with acute or acuminate apex; fls. subsessile in lf.-axils; petals purple, 1–2 mm. long, deciduous; caps. 4 mm. long; style persistent.—Occasional in wet places at low elevs.; many Plant Communities; cismontane s. Calif. to Sacramento V., Lake Co., Humboldt Co.; to Wash., Atlantic Coast, Brazil. May–Oct.

  2. **A. auriculàta** Willd. Fls. on pedicels 3–15 mm. long; caps. 2–3 mm. in diam.—Cent. V.; Mex. to Atlantic states and S. Am.

3. Péplis L. Water-Purslane

Small herbs rooting in the mud. Lvs. opposite. Fls. small, greenish, solitary in axils, 5–6-merous, with no appendages between the sepals. Petals 0 or small. Caps. indehiscent or bursting irregularly, globular. (Name used by Pliny.)

1. **P. Pórtula** L. Glabrous annual; lvs. oblong to obovate, 0.5–2 cm. long, entire, narrow at base; fls. minute, sessile; fl.-tube short-campanulate; sepals 0.7–1 mm. long; petals rose to white or none; caps. bursting irregularly; seeds very small, ± pear-shaped, brownish, angled; $2n = 10$ (Hagerup, 1941).—Reported from Summit V., Placer Co., *J. T. Howell;* native of Eu.

4. Lýthrum L. Loosestrife

Slender herbs with angled stems and opposite, alternate or even whorled lvs. Fls. solitary in axils to spicate or subpaniculate. Fl.-tube cylindric, 8–12-ribbed, the sepals 4–6, with as many intervening appendages. Petals 4–6, rarely none. Stamens 4–12. Ovary oblong, 2-loculed; style filiform. Caps. membranous, included, 2-valved or bursting irregularly. Seeds many, minute, flat or angled. Ca. 30 spp., widely distributed. (Greek, *lytron,* a name used by Dioscorides for *L. Salicaria.*)

Fls. sessile or nearly so .... 1. *L. Hyssopifolia*
Fls. short-pedicelled.
    Annual; petals ca. 1 mm. long. Rare introd. .... 2. *L. tribracteatum*
    Perennial; petals 4–6 mm. long. Common native .... 3. *L. californicum*

1. **L. Hyssopifòlia** L. Simple or branched, pale green, glabrous annual or perennial, 1–5 dm. tall; lvs. linear to oblong, 0.6–1.5 cm. long, sessile, obtuse; fls. solitary and sessile, whitish to pale purple, the petals 1.5–2 mm. long; stamens included; fl.-tube ca. 4 mm. long in fr.; seeds obliquely ovoid, scarcely 1 mm. long, almost as wide; $2n = 20$? (Tischler, 1929).—Moist places, below 5000 ft.; many Plant Communities; cismontane Calif.; to Wash., Atlantic Coast; Eu. April–Oct. Perennial, ± stoloniferous plants have been called *L. adsúrgens* Greene.

2. **L. tribrácteàtum** Salzm. ex Ten. Stems prostrate, 0.5–2 dm. long; lvs. oblong to ± linear, 0.5–2 cm. long, obtuse; fl.-tube 4–6 mm. long; petals pale, scarcely 1 mm. long.— Dried rain-pools, near Elmira, Solano Co., *J. T. Howell;* native of Eu. May–June.

3. **L. califórnicum** T. & G. [*L. Sanfordii* Greene.] Erect, somewhat woody at base, 5–18 dm., tall, pale green, glabrous; lvs. linear to linear-oblong, 1–3 cm. long, entire; fl.-tube 5–6 mm. long; petals purple, 4–6 mm. long; seeds linear-lanceolate, ca. 1 mm. long, 0.5 mm. broad.—Moist places, below 6000 ft.; many Plant Communities, cismontane Calif., n. to Marin and Solano cos.; occasional on deserts from Inyo Co. s.; to L. Calif. April–Oct.

87. Onagràceae. Evening-Primrose Family (Fig. 83)

Herbs, shrubs or trees with simple alternate or opposite lvs. Stipules none or ± glandular. Fls. perfect, mostly symmetrical, axillary or in terminal racemes, the parts mostly in 2's or 4's. Fl.-tube adnate to ovary and usually prolonged beyond. Sepals 4, sometimes 2, 5, 6. Petals 4, sometimes 2, 5, 6, inserted at summit of fl.-tube. Stamens as many or twice as many as sepals and petals. Style one, slender; stigma 2–4-lobed or discoid, capitate or elongate. Ovary inferior, mostly 2–4-loculed. Fr. a caps., berry, or nutlike and indehiscent. Pollen grains often connected by cobwebby threads. Seeds 1–many, exalbuminous. Ca. 600 spp., of wide distribution, especially in w. N. Amer.

Sepals persistent.
    Petals 5, 1 cm. or more in length; stamens mostly 10, in 2 series; caps. at length reflexed
                                                               1. *Jussiaea*
    Petals minute or none; stamens 3–6, in 1 series; caps. erect .... 2. *Ludwigia*
Sepals deciduous after anthesis.
    Fls. 4-merous normally, sometimes 3.
        Seeds with tuft of hairs (coma) at one end.
            Fl.-tube 2–3 cm. long, funnelform, with row of 8 scales within at ca. ½ its length; fls. scarlet
                                                                         3. *Zauschneria*

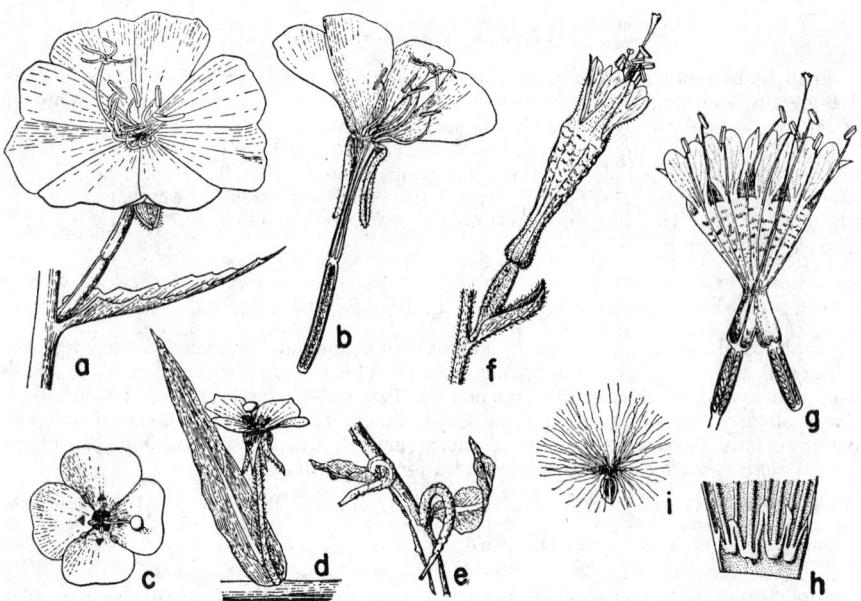

Fig. 83. ONAGRACEAE. *Oenothera deltoides* var. *cognata: a,* fl., × ½, and *b,* section through fl., with reflexed sepals, 4 petals, 8 stamens, inferior ovary, long fl.-tube, pistil with 4 linear stigma-lobes. *O. bistorta* var. *Veitchiana: c* and *d,* fl., ca. ½, elongate inferior ovary, 4 sepals. 4 petals, 8 stamens, spherical stigma; *e,* young caps., × 1, developing from pedicellike ovary. *Zauschneria: f,* fl., × 1, inferior ovary, long fl.-tube surmounted by 4 sepals, 4 petals, 8 stamens, upper part of style; *g,* section through fl.; *h,* detail of inside of narrow part of fl.-tube, × 2½, with lobelike appendages; *i,* seed, × 2½, with coma.

      Fl.-tube less than 1 cm. long or lacking, without internal scales; fls. not scarlet .. 4. *Epilobium*
  Seeds without coma.
    Fr. a dehiscent capsule.
      Ovary 4-loculed, fl.-tube prolonged beyond the ovary.
        Anthers innate, attached near the base, erect; petals not yellow, but ranging from pink
          to lavender or rose, sometimes whitish.
          Sepals erect; petals small or wanting; pollen in tetrads .... 5. *Boisduvalia*
          Sepals reflexed or the tips remaining united and turned to one side at anthesis; pollen
            not in tetrads; fls. mostly showy .... 6. *Clarkia*
        Anthers usually versatile, attached near middle; petals yellow or white, often reddish
          especially in age .... 7. *Oenothera*
      Ovary 2-loculed; fl.-tube not prolonged beyond the ovary; fls. usually minute; stem branches
        capillary .... 8. *Gayophytum*
    Fr. indehiscent, nutlike and hard.
      Plants biennial or perennial; anthers all fertile; fils. usually with a scalelike appendage at
        the base; stigma 4-lobed .... 9. *Gaura*
      Plants annual; anthers opposite the petals sterile; fils. not appendaged at base; stigma dis-
        coid, entire .... 10. *Heterogaura*
  Fls. 2-merous; fr. indehiscent, obovoid, bristly with hooked hairs .... 11. *Circaea*

1. Jussiaèa L.

    Herbs, shrubs or small trees with alternate lvs. Fls. yellow to white, axillary, on short or long pedicels, with 2 bracteoles on pedicel or base of fl.-tube. Fls. regular, 4–5(–6)-merous; sepals acute, persistent; petals caducous; stamens in 2 series, twice as many as petals and inserted with them under the margin of the epigynous usually hairy disc; fils. short; anthers ovate to oblong; pollen mostly in tetrads. Ovary 4–5(–6)-loculed: style simple, ± produced above the disc; stigma subcapitate, slightly lobed. Caps. cylindric, prismatic or obconic, loculicidally and septicidally dehiscent. Seeds pluriseriate and naked with prominent raphe, or uniseriate and surrounded by endocarp. Ca. 40 spp., of warmer regions. (Bernard de *Jussieu,* 1699–1777, botanist.)

(Munz, P. A. A revision of the New World species of Jussiaea. Darwiniana 4: 179–284, 1942.)
Flowering stems usually floating or creeping; lf.-blades 1–4 cm. long, oblong; bracteoles deltoid; caps. mostly 2–3 mm. thick. Widely distributed. Native .... 1. *J. repens*
Flowering stems usually erect; lvs. ± lanceolate, mostly 5–10 cm. long; bracteoles lanceolate; caps. 3–4 mm. thick. Local introd. .... 2. *J. uruguayensis*

1. **J. règpens** L. var. **peploìdes** (HBK.) Griseb. [*J. p.* HBK. *Ludwigia adscendens* var. *p.* Hara. *J. r.* var. *californica* Wats. *J. c.* Jeps.] Mostly glabrous perennial herb with decumbent stems rooting freely at nodes, 3-many dm. long; lvs. oblong to spatulate-oblong, obtuse to acute, subentire, plainly and evenly pinnate-veined, the blades 1–4(–7) cm. long, on petioles 1–2.5 cm. long, or longer especially on floating lvs.; pedicels 1–4 cm. long in fr.; fls. mostly 5-merous, pubescent about the base of the stamens and style; sepals lanceolate, 4–7 mm. long; petals yellow, obovate, pinnately veined, 10–14 mm. long; stamens ca. ⅓ as long; pistil as long as stamens; stigma globose; caps. hard, quite cylindric, ca. 2 cm. long, at length reflexed and the sepals deciduous from the mature fr.; seeds in 1 row in each locule, pendulous, included in the endocarp, oblique-truncate at ends, somewhat triangular in cross section, 1–1.5 mm. long.—Pools and slow streams, below 2000 ft.; several Plant Communities; cismontane s. Calif. and n. through the Great V. and bordering ranges away from the immediate coast; to Ore., S. Am. May–Oct. Occasional plants with unusually long petioles and pedicels approach the var. *glabréscens* Kuntze of the se. U.S.

Var. **montevidénsis** (Spreng.) Munz. [*J. m.* Spreng.] Plants with soft spreading hairs on upper stems and herbage, also viscid; stipules glandular; sepals 6–8 mm. long; petals 8–15 mm. long.—Stanislaus and Tuolumne cos.; apparently introd. from S. Am.

2. **J. uruguayénsis** Camb. in St. Hil. [*J. grandiflora* Michx. *Ludwigia u.* Hara.] Perennial herb from creeping rhizome, rooting freely at nodes; stems slender and floating, or erect to ascending and ± soft-hirsute, often freely branched, or compactly much-branched, creeping and forming mats (depending on whether in deeper water, emergent from shallow water, or growing on sand or mud banks); lvs. on erect flowering branches linear-lanceolate to oblanceolate, short-petioled; fls. on pedicels 1–2 cm. long; bracteoles lanceolate, 5–13 mm. long; sepals 5 (6), lanceolate, 6–13 mm. long; petals yellow, oblong-ovate, 12–20 mm. long; stamens unequal, 3–4 and 2–3 mm. long; caps. sub-cylindric, usually hairy, 1.3–2.5 cm. long, 3–4 mm. thick; seeds enclosed in the hard endocarp, 1.5 mm. long.—Introd. in wet places, Tiburon (Marin Co.), San Diego; native from S. Am. to se. U.S. June–Aug.

2. Ludwígia L.

Perennial herbs of marshes and wet places with opposite or alternate lvs. Fls. solitary and axillary, or in terminal spikes or heads, normally 4-merous, though petals may be lacking. Fl.-tube not prolonged beyond the ovary. Stamens as many as the sepals; fils. rather short. Style short; stigma capitate to ± 4-lobed. Caps. rounded to obpyramidal or elongate and subcylindric, dehiscing terminally or longitudinally. Seeds many, naked with evident raphe and multiseriate in each locule, rarely uniseriate and enclosed in endocarp. Ca. 30 spp., largely e. N. Am. (C. G. *Ludwig*, 1709–1773, German botanist.)

(Munz, P. A. The Am. species of Ludwigia. Bull. Torrey Bot. Club 71: 152–165, 1944.)
Fl.-tube and caps. with 4 evident longitudinal green bands; bracteoles at base of fl.-tube not more than 1 mm. long, sometimes not evident; petals lacking .... 1. *L. palustris*
Fl.-tube and caps. without green bands; braceoles above the base of the fl.-tube, 1–3 mm. long; petals present but easily shed .... 2. *L. natans*

1. **L. palústris** (L.) Ell. var. **americàna** (DC.) Fern & Griscom. [*Isnardia p.* var. *a.* DC.] Glabrous, with floating or creeping stems bearing erect branches 1–3 dm. tall; lvs. lanceolate to elliptic-ovate, subentire, acute, the blades 1–2.5 cm. long, usually at least half as wide, long-petiolate; fls. solitary, axillary, sessile; sepals deltoid, acute, persistent, 1–2 mm. long; petals none; stamens ca. 1 mm. long; stigma 4-lobed; caps. 3–5 mm. long, oblong, 2–3.5 mm. broad, ± 4-angled; seeds yellowish, broadly obovoid, 0.5 mm. long.—Wet places, below 5000 ft., Sierra to Siskiyou cos.; to Canada and Atlantic Coast, Cent. Am. June–Sept.

Var. **pacífica** Fern. & Griscom. Lvs. mostly short-petiolate, the blades mostly 1–3 cm. long, usually not half as wide; sepals acuminate; caps. 2–2.8 mm. thick.—Ponds and

muddy places, below 3000 ft., several Plant Communities; Coast Ranges from San Diego
and Ventura cos. n., Butte, Eldorado, Sonoma, and Humboldt cos.; to B.C. June–Sept.
    2. **L. nàtans** Ell. var. **stipitàta** Fern. & Griscom. Habit of preceding sp.; lf.-blades
1–4.5 cm. long, rhombic-ovate, rather long-petiolate; fls. short-pedicelled; sepals 4,
triangular-acuminate; petals shorter than sepals, quickly shed; pedicel of caps. 2–4 mm.
long; caps. 6–8 mm. long, 3–3.5 mm. broad, light brown, ± 4-angled, without longitudinal green bands; seeds pale yellow or straw-colored, straight on raphal edge, shining, ca. 0.5 mm. long.—Ponds and wet places, below 5000 ft.; several Plant Communities; San Bernardino, Victorville, Lake Arrowhead, Lone Pine Canyon near Cajon
Pass. July–Sept.

3. **Zauschnèria** Presl. CALIFORNIA FUCHSIA

    Erect or decumbent perennials, sometimes woody at base, with shredding epidermis
on lower stems. Lvs. sessile or nearly so, opposite or alternate, ± fascicled. Infl. spicate,
the fls. large, horizontal, fuchsialike. Fl.-tube scarlet, globose at base, narrowed into a
long tube bearing within the narrow part 8 lobelike appendages, 4 erect and 4 deflexed.
Sepals 4, red. Petals 4, red. Stamens 8, the alternate shorter; anthers versatile. Ovary
4-loculed; stigma 4-lobed, peltate to capitate. Caps. imperfectly 4-loculed, many-seeded.
Seeds oblong, narrowed at base, comose at apex. Four spp. (Named for Dr. M. *Zauschner,*
professor of natural history at Univ. of Prague.)

(Clausen, Keck, and Hiesey. The genus Zauschneria. Carnegie Inst. Wash. Pub. 520: 213–259, 1940.)
Plants herbaceous throughout; lvs. broadly lanceolate to ovate, main ones more than 6 mm. wide.
    Broadest lvs. 5–8(–10) mm. wide, subentire, the lower usually white-canescent; plants matted,
    stems mostly less than 2 dm. long. Redwood region, Humboldt and Mendocino cos.
                                                                1. *Z. septentrionalis*
    Broadest lvs. 7–15 mm. wide, not white-canescent.
        Lvs. rather coriaceous, ± sharply denticulate, the lateral veins evident, surface subglabrous to
        pilose, obscurely if at all glandular, broadly ovate. E. Mojave Desert .... 2. *Z. Garrettii*
        Lvs. not coriaceous, the lateral veins obscure, subentire to denticulate, variously pubescent, often
        tomentose, often glandular, broadly lanceolate to ovate. General montane
                                                                3. *Z. californica latifolia*
Plants suffrutescent at base; lvs. linear to lanceolate, mostly less than 6 mm. wide.
    Stems slightly woody; lvs. mostly 3–5 mm. wide. ± pilose, not canescent ...... 3. *Z. californica*
    Stems obviously woody at base; lvs. 2–3.5 mm. wide, densely tomentose-canescent.
        Lvs. linear, 2.5–3.5 mm. wide, moderately fasciculate; fls. 30–40 mm. long
                                                                3. *Z. californica angustifolia*
        Lvs. subfiliform, not more than 2 mm. wide, very densely fasciculate; fls. 25–35 mm. long
                                                                4. *Z. cana*

    1. **Z. septentrionàlis** Keck. Herbaceous perennial, matted, the stems 1–2 dm. long;
lvs. broadly lanceolate to oval, 4–8 mm. wide, 1–2.5 cm. long, entire or sometimes
obscurely denticulate, at least the lower lvs. white-canescent, the upper sometimes
greenish and villous; fls. 2.8–3.2 cm. long; caps. as in *Z. californica; n* = 15 (Clausen et
al., 1940).—Dry sandy or rocky places, below 2500 ft.; Mixed Evergreen F., Redwood
F., Douglas-Fir F.; from sw. Trinity Co. and Laytonville, Mendocino Co. to Hoopa V.,
Humboldt Co. Aug.–Sept.
    2. **Z. Garréttii** A. Nels. [*Z. latifolia* var. *G.* Hilend.] Herbaceous, the stems 1.5–3 dm.
long; lvs. quite coriaceous, subglabrous to pilose, sharply denticulate, broadly ovate,
1–3.5 cm. long, 0.7–1.5 cm. wide; fls. 2.8–3.2 cm. long; *n* = 15 (Clausen et al., 1940).—
Dry rocky places, at 5500 ft.; Pinyon-Juniper Wd.; Kingston Mts., e. Mojave Desert;
to Utah, Wyo. June–July.
    3. **Z. califórnica** Presl. [*Z. mexicana* Presl. *Z. Eastwoodae* Mox. *Z. velutina* Eastw.]
Suffrutescent at base, often much-branched, green to gray-pilose, ± glandular, the stems
3–9 dm. long; lvs. green to grayish, lanceolate to linear-lanceolate or -oblong, 0.5–4 cm.
long, 1.5–6 mm. wide, entire or remotely denticulate, the lower sometimes opposite, the
upper usually alternate; fl.-tube funnelform, globular at base, then narrowed, then
gradually ampliate, 2–3 cm. long; sepals erect, lanceolate, 8–10 mm. long; petals 2-cleft,
8–15 mm. long; caps. sessile to short-pedicelled, linear, 4-angled, 8-nerved, with short
beak, often curved, 1.5–2 cm. long; seeds broad, 1.5 mm. long; *n* = 30 (Clausen et al.,
1940).—Dry, mostly stony or gravelly places, mostly below 3500 ft.; many Plant Communities; from Sonoma and Lake cos. to Baja Calif. Aug.–Oct. Variable; plants from

the Santa Barbara Ids. with long soft hairs, so as to appear shaggy-villous, have been called var. *villòsa* (Greene) Jeps. [Z. *v.* Greene.]

Ssp. **angustifòlia** Keck. Suffrutescent at base, the stems to 7 dm. long; lvs. linear, densely tomentose-canescent; fls. 3–4 cm. long; *n* = 30 (Clausen et al., 1940).—Dry slopes, below 2000 ft.; Coastal Sage Scrub, Chaparral, etc.; Coast Ranges, Monterey Co. to San Diego Co., Catalina Id. Aug.–Oct.

Ssp. **latifòlia** (Hook.) Keck. [Z. *californica* var. *l.* Hook. Z. *l.* Greene. Z. *tomentella* Greene. Z. *glandulosa* Mox. Z. *pulchella* Mox. Z. *canescens* Eastw.] Plants herbaceous, the stems 1–5 dm. long, slender; lvs. mostly opposite, ovate to lance-ovate, tapering to both ends or rounded at base, 7–17 mm. wide; plant green to grayish-green, villous to tomentose, often silky, often ± glandular; *n* = 30 (Clausen et al., 1940).—Dry slopes and ridges, below 10,000 ft., mostly Montane Coniferous F.; mts. from Tulare Co. to Plumas and Lassen cos., Trinity Co. to sw. Ore. Aug.–Sept. Variable, plants with broadly ovate to elliptic lvs. more crowded on stem and stems 1–3 dm. long, herbage more viscid; ranging from Sierra Nevada of Kern Co. to San Jacinto Mts., have been named Z. *latifolia* var. *viscòsa* (Mox.) Jeps. [Z. *v.* Mox., Z. *Hallii* Mox., Z. *orbiculata* Mox., Z. *elegans* Eastw.] They occur mostly above 7000 ft.; those at 3500–6500 ft., 2.5–5 dm. tall, villous, ± glandular, with lvs. lance-ovate to elliptical and from L. Calif. and Palomar Mts. to San Gabriel, Little San Bernardino and Eagle mts., have been called Z. *latifolia* var. *Johnstònii* Hilend. Both these vars. are here treated as synonymous to ssp. *latifolia.*

4. **Z. càna** Greene. [Z. *californica* var. *microphylla* Gray. Z. *m.* Mox.] Suffrutescent at base; stems 3–6 dm. long; herbage tomentose-canescent, entirely gray, mostly not very glandular; lvs. narrowly linear-lanceolate to almost filiform, not over 2 mm. wide, entire or nearly so, much fascicled; fl.-tube 2–3 cm. long; sepals 8–10 mm. long; petals 8–12 mm. long; caps. 1.5–2 cm. long, curved or almost straight, beaked or not; *n* = 15 (Clausen et al., 1940).—Dry slopes, below 2000 ft.; Chaparral, Coastal Sage Scrub, etc.; Monterey Co. to Los Angeles Co., Santa Cruz, Anacapa and Catalina ids. Aug.–Oct.

4. Epilòbium L. Willow-Herb

Mostly herbs, sometimes suffruticose; annual, or usually perennial and wintering over by *turions* (tuberlike buds with swollen scales which persist about the base of the next year's stem) or *rosettes* (at first ± fleshy) or slender *stolons*. Lvs. opposite or alternate, nearly or quite sessile, entire or denticulate. Fls. axillary or in terminal racemes or panicles, perfect. Fl-tube short or not prolonged beyond the ovary. Sepals 4. Petals 4, usually notched, purplish, pink or white, sometimes yellow. Stamens 8, the alternate shorter. Stigma oblong to 4-lobed. Caps. elongate, subcylindric to fusiform or clavate, 4-loculed, loculicidal. Seeds with coma (tuft of silky hairs) at upper end. Over 100 spp., cosmopolitan except in trop. (Greek, *epi-*, upon, and *lobon*, a caps.)

(Trelease, W. A revision of the N. Am. spp. of Epilobium occurring n. of Mex. Mo. Bot. Gard., 2d Ann. Rept., pp. 69–117, pls. 1–48, 1891.)

A. Fl.-tube not prolonged above the ovary; fls. slightly irregular, the petals 1–2 cm. long, entire, spreading. (Subgenus Chamaenerion)
   B. Style pilose at base, exceeding stamens; lvs. 5–20 cm. long, membranous, reticulate-veiny beneath, with lateral veins confluent in marginal loops; racemes many-fld., elongate; seeds 1–1.3 mm. long. Common .... 1. *E. angustifolium*
   BB. Style glabrous, shorter than stamens; lvs. 2–6 cm. long, thick and fleshy, glaucous, not veiny; racemes short, few-fld.; seeds fusiform, 2 mm. long. Rare ...... 2. *E. latifolium*
AA. Fl.-tube prolonged above the ovary; fls. regular, mostly smaller. (Subgenus Epilobium)
   B. Fls. large, the petals usually 14–20 mm. long; stigma evidently lobed.
      C. Lvs. rounded at base, denticulate, 1–2 cm. long, subsessile; fl.-tube 2–4 mm. long. Tulare and Inyo cos. to Modoc, Siskiyou, and Trinity cos. .... 3. *E. obcordatum*
      CC. Lvs. acute at base, mostly entire, 3–4 cm. long, petioled; fl.-tube 1–1.5 mm. long. Del Norte Co. .... 4. *E. rigidum*
   BB. Fls. smaller, the petals 2–12 mm. long, stigma usually oblong.
      C. Plant suffrutescent with several stems from a woody caudex and 1–2 dm. tall, pubescent throughout. Lake and Mendocino cos. .... 5. *E. nivium*
      CC. Plants not suffrutescent.
         D. The plants annual; stems with exfoliating epidermis. Of dry situations.
            E. Stems 3–20 dm. tall, glabrous except in upper parts; lvs. usually alternat with fascicles in axils; fl.-tube 1–8 mm. long .... 6. *E. paniculatum*
            EE. Stems 0.5–3 dm. tall, puberulent throughout; lvs. mostly opposite, without axillary fascicles; fl.-tube scarcely 1 mm. long .... 7. *E. minutum*

DD. The plants perennial (sometimes blooming the first year); epidermis not exfoliating from the stems, mostly of wet situations.
    E. Rootstocks bearing globose or ovoid turions or winter-buds with fleshy over-
lapping scales that persist at base of next year's stem.
        F. Petals 5–10 mm. long; stems simple to divaricately branched.
            G. Fl.-tube narrow; sepals suberect; stems coarse, simple or virgately
few-branched at summit .... 8. *E. glandulosum*
           GG. Fl.-tube ca. as wide as long; sepals more divaricate; stems slender,
spreading-branched above .... 9. *E. exaltatum*
        FF. Petals 2–5 mm. long; stems simple.
            G. Stems glabrous to pubescent but not with decurrent lines of hair
from lf.-bases.
                H. Stems 2–6 dm. tall; lvs. ovate to lanceolate, denticulate; petals
purplish or paler, 3–5 mm. long. From San Jacinto Mts. n.
10. *E. brevistylum*
              HH. Stems mostly 0.5–2 dm. long, very slender; lvs. oblong-linear
to lanceolate, mostly entire; petals mostly white, 2–2.5 mm.
long; Sierra Nevada n. .... 11. *E. Pringleanum*
           GG. Stems with decurrent lines of hair from the lf.-bases; lvs. lance-
linear, erect. From Trinity and Mono cos. n. .. 12. *E. Halleanum*
    EE. Rootstocks not turioniferous.
        F. Plant pallid, glaucous and almost glabrous throughout; rootstocks
branched, scaly, rather tough .... 14. *E. glaberrimum*
        FF. Plant not glaucous but green or canescent with pubescence.
            G. Stems mostly 1–3 dm. tall, simple above, with few pairs of oppo-
site lvs. High montane.
                H. Lvs. sessile, oblong or linear, suberect; stem slender; perennial
with subfiliform stolons .... 13. *E. oregonense*
              HH. Lvs. ± distinctly petioled and spreading.
                    I. Plant densely cespitose, stoloniferous; stems sigmoidally
bent, 1–1.5 dm. tall; petals purplish to rose, 4–6 mm.
long; lvs. 1–2 cm. long.
                      J. Caps. linear, slender, 1 mm. or less thick, 2–4 cm.
long; seeds smooth, 1 mm. long; buds nodding. From
Tulare and Inyo cos. n. ..... 15. *E. anagallidifolium*
                      JJ. Caps. subclavate, stouter, to 2 mm. thick, 2–2.5 cm.
long; seeds papillose, 1.5–2 mm. long; buds erect.
Siskiyou Co. .... 16. *E. clavatum*
                II. Plant not so densely cespitose; stems erect, 1–3 dm. tall;
lvs. 1.5–5 cm. long.
                    J. Petals purplish, 5–10 mm. long; seeds papillose, 1 mm.
long .... 17. *E. Hornemannii*
                    JJ. Petals white or with pink tips; 3–4 mm. long; seeds
smooth, 1 mm. long .... 18. *E. lactiflorum*
        GG. Stems mostly 3–10 dm. tall and freely branched especially above,
if shorter, the upper lvs. alternate and more numerous. Mostly
midmontane to low elevs.
              H. Petals 3–6 mm. long, often pale to whitish; stems green to
light-colored, glandular to canescent-strigose above; many of
upper lvs. alternate.
                  I. Lvs. firm, sessile or with short broad petioles; coma mostly
persistent; papillae of seeds ± rounded. Widespread
19. *E. adenocaulon*
                  II. Lvs. thin, flaccid, pale green, tapered to definite slender
petioles; papillae of seeds ± conical; coma caducous.
E. Calif. .... 20. *E. ciliatum*
             HH. Petals 6–10 mm. long, red-purple; stems reddish, canescent
above; lvs. mostly opposite. Immediate coast
21. *E. Watsonii*

1. **E. angustifòlium** L. [*Chamaenerion a.* Scop. *E. spicatum* Lam.] Fireweed. Perennial from underground rootstocks, the stems mostly simple, few, 6–25 dm. tall, glabrous
below, commonly puberulent above, rather densely leafy; lvs. alternate, lanceolate,
nearly entire, acute to subacuminate, paler beneath, with lateral veins confluent in
submarginal loops, sessile or nearly so, 7–20 cm. long; fls. many in long terminal racemes
with small almost linear bracts; pedicels 5–12 mm. long; fl.-tube not prolonged above the
ovary; sepals lance-linear, 8–12 mm. long, commonly canescent-puberulent, tinged
lavender; petals lilac-purple, rose, rarely white, clawed, obovate, 8–18 mm. long;
stamens 8, in a single series, often unequal, shorter than petals; fils. dilated below; style
hairy at base, exceeding stamens; stigma-lobes slender and elongate; caps. 5–8 cm. long,
canescent; seeds oblong, 1–1.4 mm. long, with long dingy coma; $2n = 36$ (Löve & Löve,
1948).—In disturbed areas such as burns, clearings, etc., in fairly moist places, mostly

below 9000 ft.; Montane Coniferous F., Mixed Evergreen F., Redwood F., etc.; mts. from San Diego Co. n. to Modoc and Siskiyou cos., N. Coast Ranges to the seacoast; to Alaska, Atlantic Coast; Eurasia. July–Sept. Exceedingly variable, a few of the more recognizable forms are: (1) plants from mostly 9000–11,000 ft., 1–4 dm. high, with median cauline lvs. 3–6 cm. long, infl. to 1.3 dm. long, petals 1–1.5 cm. long, have been named var. *intermèdium* (Wormsk.) Fern. [var. *pygmaeum* Jeps.]; (2) larger plants with median lvs. 2–5 cm. wide, and prominently veined, infl. ± leafy-bracted, have been called var. *macrophýllum* (Hausskn.) Fern. White-petaled plants with sepals red or sepals white are to be expected.

2. **E. latifòlium** L. [*Chamaenerion l.* Sweet.] Stems several from a cespitose rootstock, depressed or arched-ascending, 1–5 dm. long, glabrous below, puberulent above; lvs. elliptic-ovate to lanceolate, subopposite, fleshy, glaucous, entire, not veiny, sessile or nearly so, 2–6 cm. long; racemes short, few-fld., leafy-bracted; pedicels 5–10 mm. long; sepals lanceolate, purplish, 13–18 mm. long; petals purple, rose or white, purple-veined, rhombic-obovate, 1.5–2.5 cm. long; style glabrous, shorter than stamens; stigma-lobes oblong; caps. 5–8 cm. long; seeds fusiform, 2 mm. long; $2n = 72$ (Löve & Löve, 1948).— Rare, wet stony places, 7600–11,400 ft.; mostly Lodgepole F. to Subalpine F.; Sierra Nevada of Inyo, Fresno, Tuolumne, and Mono cos.; to Alaska, Atlantic Coast; Eurasia. July–Aug.

3. **E. obcordàtum** Gray. [*E. o.* var. *puberulum* Jeps.] Stems several from a cespitose suffrutescent base, decumbent to matted, mostly 0.5–1.5 dm. long, the branches simple, glabrous below, ± puberulent at summit, leafy; lvs. opposite, crowded, glabrous and glaucous, ovate, obscurely and remotely denticulate, mostly 6–12 mm. long, obtuse, sessile or short-petioled; fls. 1–few, solitary in upper axils; pedicels 2–20 mm. long; petals rose-purple, broadly obcordate, 1–2 cm. long; stamens 8, in 2 series, the shorter half the longer and ⅔ as long as petals; style purplish, glabrous; stigma-lobes short; fl.-tube prolonged beyond ovary 2–4 mm.; sepals lanceolate, purplish, 9–12 mm. long; caps. cylindric-clavate, 2.5–3.5 cm. long; seeds 1.5 mm. long, finely papillose.—Dry ridges and flats, 7000–13,000 ft.; mostly Subalpine F., Alpine Fell-fields; Sierra Nevada from Tulare and Inyo cos. to Modoc Co.; Nev., e. Ore., Ida. July–Sept.

Var. **láxum** (Hausskn.) Dempst. ex Jeps. [*E. o.* f. *l.* Hausskn.] A poorly defined var. with somewhat larger lvs., 10–22 mm. long, more acute.—Mostly at 6000–8000 ft.; Subalpine F. and Red Fir F.; Trinity, Siskiyou, and Placer cos.

4. **E. rígidum** Hausskn. [*E. r.* var. *canescens* Trel.] Much like *E. obcordatum*, but taller (to 4 dm.), canescently pubescent above or throughout; lvs. 3–4 cm. long, mostly entire, cuneate at base, acute at apex; fl.-tube 1–1.5 mm. long; sepals 9–10 mm. long; caps. densely white-glandular, 2–2.5 cm. long; seeds ca. 2 mm. long, appearing smooth.— Sandy and rocky benches below 3100 ft.; Mixed Evergreen F., Yellow Pine F.; Del Norte Co.; sw. Ore. July–Aug.

5. **E. nívium** Bdg. Suffrutescent with several stems from a short branched caudex, 1–2 dm. tall, pubescent throughout; lvs. oblong- to elliptic-lanceolate, thick, not veiny, entire to denticulate, sessile or short-petioled, 8–15 mm. long; fls. few, in upper axils; pedicels 3–5 mm. long; fl.-tube 5–7 mm. long above the ovary, reddish; sepals lanceolate, 3–5 mm. long; petals violet-purple, obcordate, 7–10 mm. long; stamens in 2 sets; pistil equaling petals; stigma-lobes short; caps. subfusiform, stout, 10–12 mm. long; seeds rather few, smooth, with dingy coma.—Dry talus and shaly slopes, 6000–8000 ft.; Red Fir F.; inner N. Coast Ranges of n. Lake and Mendocino cos. Sept.

6. **E. paniculàtum** Nutt. ex T. & G. Erect annual with stem simple and shreddy below, paniculately branched above, 3–20 dm. tall, glabrous except for the tips of the infl. which are slightly glandular-puberulent; lvs. linear-lanceolate to linear, 2–5 cm. long, usually alternate, short-petioled, remotely denticulate with thickened acute tip and teeth, quite early caducous and with fascicles of smaller lvs. in the axils; fls. in lax racemes on filiform branches of the panicle; bracts subulate; pedicels 5–15 mm. long, usually slightly glandular-puberulent; fl.-tube 2–3 mm. long, subglabrous; sepals 2–3 mm. long; petals pink to almost white, 3–6 mm. long, rotate, deeply 2-cleft; stamens ca. ⅓ as long; caps. 2–2.5 cm. long, 4-angled, linear-clavate, beaked, usually slightly glandular-puberulent; seeds obovoid, flattened, 2 mm. long, with tawny coma.—Open, usually dry, disturbed ground, below 7500 ft.; many Plant Communities; most of cis-montane and montane Calif.; to B.C., S. Dak., New Mex. June–Sept. Plants with pedicels

and caps. densely glandular-puberulent have been named f. **adenocládon** Hausskn., $n = 12$ (Lewis et al, 1958); those with pedicels and caps. quite glabrous, f. **subulàtum** Hausskn., $n = 12$ (Lewis et al, 1958).

Var. **Tràcyi** (Rydb.) Munz. [*E. T.* Rydb. *E. p.* f. *T.* St. John.] Pedicels glabrous or nearly so; fl.-tube 1–1.5 mm. long, subglabrous; sepals 1–2 mm. long; petals ca. 2 mm. long, mostly white; caps. glabrous; $n = 12$ (Lewis et al., 1958).—Occasional in n. Calif.; to Wash., Rocky Mts. More common is a form with glandular puberulence on pedicels and caps., s. Calif. to Wash., N. Dak.: f. **fasciculàtum** Munz.

Var. **laevicále** (Rydb.) Munz. [*E. l.* Rydb. *E. p.* f. *l.* St. John.] Pedicels glabrous or nearly so; fl.-tube 4–6 mm. long, subglabrous; sepals ca. 3 mm. long; petals 5–8 mm. long, rose to pink; caps. subglabrous.—Occasional with the sp. More common is f. **altíssimum** (Suksd.) Munz. [*E. a.* Suksd.] with pedicels, fl.-tubes and caps. glandular-puberulent.

Var. **jucúndum** (Gray) Trel. [*E. j.* Gray. *E. Hammondi* Howell.] Pedicels glandular-pubescent; fl.-tube 8–15 mm. long, glandular-pubescent; sepals 3–6 mm. long; petals 7–12 mm. long, purplish; caps. gl.-pubescent.—Occasional, n. Calif. to Wash., Ida. Plants with glabrous caps. and pedicels have been named f. **tubulòsum** Hausskn.

7. **E. minútum** Lindl. ex Hook. [*Crossostigma Lindleyi* Spach.] Annual, 0.5–3 dm. tall, simple or diffusely branched, the branches ± reddish, suberect, puberulent; lvs. mostly opposite, oblong-lanceolate to lanceolate or oblanceolate, entire to remotely denticulate, ± fleshy, 1–2 cm. long, on much shorter petiole; fls. in axils of upper somewhat reduced lvs.; pedicels 3–10 mm. long; fl.-tube less than 1 mm. long; sepals ca. 1.5 mm. long; petals rose-lavender to white, emarginate, 2–4 mm. long; stamens and style ca. half as long as petals; caps. subclavate, arcuate, 1.5–2.5 cm. long, beaked; seeds broadly obovoid, smooth, scarcely 1 mm. long.—Dry open disturbed places mostly below 5000 ft.; Foothill Wd., Chaparral, Yellow Pine F., Mixed Evergreen F., etc.; Coast Ranges from Santa Barbara Co. to Del Norte and Siskiyou cos., less common in Sierran foothills from Madera Co. n.; to B.C., Nev. April–Aug. A minor variant is var. *foliòsum* T. & G. [*E. m.* var. *Biolettii* Greene.] Lvs. narrow with some tendency to fascicles in axils; fls. smaller, petals scarcely 2 mm. long.—With the sp.

8. **E. glandulòsum** Lehm. Perennial with running rootstocks and large loosely formed turions; stems 3–9 dm. tall, rather thick, light-colored, simple or with few suberect branches above, glabrous below, crisp-pubescent and glandular above; lvs. mostly exceeding internodes, rather crowded above and not much decreased in size in infl., lance-ovate to ovate, prominently serrulate, 5–12 cm. long, subsessile; fls. erect, the infl. rather compact; fl.-tube narrow, 2–3 mm. long; sepals suberect, 3–5 mm. long; petals purple, 5–10 mm. long, not conspicuously spreading; pedicels 1–2 cm. long in fr.; caps. 4–7 cm. long, pubescent; seeds ca. 1.7 mm. long, with dingy coma.—Occasional in moist places, 5000–9000 ft.; Red Fir. to Subalpine F.; Sierra Nevada from Fresno Co. n. to Mt. Shasta; to Alaska, e. Asia, Saskatchewan. July–Aug.

9. **E. exaltàtum** E. Drew. [*E. californicum* var. *e.* Jeps. *E. glandulosum* var. *e.* Munz. *E. brevistylum* var. *e.* Jeps.] Perennial with large turions; stems 3–9 dm. tall, slender, ± pubescent, often freely branched above with very slender branches; lvs. lance-ovate, serrulate, subsessile, 5–12 cm. long, the uppermost much reduced; fls. near ends of glandular-pubescent branches; pedicels 5–10 mm. long in fr.; fl.-tube 2–3 mm. long, almost as broad; sepals suberect, 3–4 mm. long; petals pink to rose-purple, 5–10 mm. long, spreading; caps. 3–5 cm. long; seeds beaked, rugose, 1 mm. long, with white coma; $n = 18$ (Lewis et al., 1958).—Wet places, 4000–10,000 ft.; Montane Coniferous F.; San Bernardino Mts., Sierra Nevada, n. to Modoc Co., Coast Ranges from Lake and Tehama cos. n.; to Ore., Mont., Nev. June–Aug.

10. **E. brevístylum** Barb. [*E. concinnum* Congd.] Perennial with well formed compact turions, the dried scales of which persist at the base of the stem of the following season; stems erect, simple or subsimple, slender, 2–6 dm. tall, glabrous below, crisp-pubescent or ± glandular in infl.; lvs. ovate to lanceolate, denticulate, with rounded sessile base, mostly opposite, 2–4 cm. long, mostly shorter than internodes; fls. several; fl.-tube to ca. 1 mm. long; sepals 2–3 mm. long; petals purplish or paler, emarginate, 3–5 mm. long; fruiting pedicels 5–15 mm. long; caps. 4–6 cm. long; seeds ca. 1.5 mm. long, papillate, broad, with whitish coma.—Wet places, below 11,000 ft.; mostly Montane

Coniferous F.; N. Coast Ranges from Lake Co. n., San Jacinto and San Bernardino mts. n. through Sierra Nevada; to Wash., Mont., Colo. June–Aug.

Var. ursìnum (Parish) Jeps. [*E. u.* Parish.] Both lvs. and lower stems pilose with spreading white hairs.—With the sp. to Wash. & Ida.

11. **E. Pringleànum** Hausskn. [*E. brevistylum* var. *P.* Jeps. *E. ursinum* var. *subfalcatum* Trel.] Turions small, compact; stems mostly simple, slender, 0.5–2(–3) dm. high, very slender, pilose; lvs. oblong-linear to lanceolate, mostly much shorter than internodes, entire or nearly so, obtuse, sessile, 3–6(–8) pairs, 1–1.5(–3) cm. long; fls. few; fl.-tube to ca. 1 mm. long; sepals 1–1.5 mm. long; petals largely white, 2–2.5 mm. long; caps. mostly 3–4 cm. long, slender, ± glandular-pubescent; seeds papillose, ca. 1 mm. long.— Wet places, 5000–10,000 ft.; Red Fir F. to Subalpine F.; Coast Ranges, Trinity to Humboldt and Siskiyou cos., Sierra Nevada; to Ore., Ida., Nev. June–Aug.

Var. ténue (Trel.) Munz. [*E. delicatum* var. *t.* Trel. *E. brevistylum* var. *t.* Jeps.] Stems glabrous.—With the sp.

12. **E. Halleànum** Hausskn. Perennial with small turions; stems slender, erect, mostly simple, 1–4 dm. high, with lines of crisped hair on the ridges from the decurrent bases of the lvs., glandular-puberulent in the upper parts; lvs. lance-linear, erect, 1.5–4 cm. long, some with clasping base, acute apex, entire to serrulate; fls. small; fl.-tube 1–1.5 mm. long; sepals 2–3 mm. long; petals 2–4 mm. long, white to purplish; fruiting pedicels 3–5 mm. long; caps. 2–5 cm. long; seeds 1–1.5 mm. long, beaked, papillose.—Occasional, wet places, 5000–9000 ft.; Montane Coniferous F.; from Trinity and Mono cos. n. to B.C., Rocky Mts. July–Aug.

13. **E. oregonénse** Hausskn. [*E. o.* var. *gracillimum* Trel.] Perennial with subfiliform stolons; stems simple, slender, erect, 1–3 dm. high, glabrous except in the sparsely glandular-pubescent infl., often purplish above; lvs. oblong-linear to -ovate, suberect, entire to ± denticulate, obtuse, sessile, somewhat crowded on lower stem, remote above, 1–2.5 cm. long; fls. 1–few; sepals often purplish, 1–2 mm. long; petals cream to purplish, 4–7 mm. long, deeply emarginate; pedicels 1–3.5 cm. long in fr.; caps. erect, slender, 2–5 cm. long, often purplish; seeds smooth, scarcely 1 mm. long, with white coma.— Boggy places, 5000–11,500 ft.; Red Fir F. to Alpine Fell-fields; San Jacinto and San Bernardino mts., n. through the Sierra Nevada; Coast Ranges from Tehama Co. n.; to B.C., Ida., Nev. July–Aug.

14. **E. glabérrimum** Barb. [*E. pruinosum* Hausskn.] Perennial, with several stems from branching scaly tough rootstocks; stems simple or nearly so, slender, erect from ± decumbent base, glabrous and glaucous, sometimes slightly glandular-puberulent above, often purplish, mostly 3–6 dm. tall; lvs. glabrous, glaucous, ascending, oblong-lanceolate, obtuse, entire or minutely denticulate, sessile, 2–5 cm. long, gradually reduced up the stem; fls. erect or ± drooping; fl.-tube 1–2 mm. long; sepals 1–2 mm. long; petals 4–7 mm. long, purplish to almost white; fruiting pedicels 1–2 cm. long; caps. slender, sub-erect, 4–7 cm. long; seeds papillate, ca. 1 mm. long, with whitish coma; $n = 18$ (Lewis et al., 1958).—Stream banks and wet places, 3000–11,500 ft.; Montane Coniferous F., Alpine Fell-fields; San Jacinto Mts. n. through Sierra Nevada, N. Coast Ranges from Lake Co. n.; to Wash., Nev. July–Aug.

Var. fastigiàtum (Nutt.) Trel. [*E. affine* var. *f.* Nutt. *E. g.* var. *latifolium* Barb. *E. platyphyllum* Rydb.] Mostly 1–3 dm. tall; lvs. ovate, 1.5–2.5 cm. long, more crowded; petals 2–5 mm. long; $n = 18$ (Lewis et al., 1958).—With the sp. and to B.C., Alta., Mont., Utah.

15. **E. anagállidifòlium** Lam. [*E. alpinum* L., in part.] Densely cespitose perennial, stoloniferous, the stems simple, erect or decumbent at base, many, slender, sigmoidally bent, 0.3–1(–1.5) dm. long, glabrous or with pubescent lines, often purplish above; lvs. rather uniformly distributed, divergent, oblong-elliptic to -lanceolate, obtuse, entire or nearly so, 0.8–2 cm. long, obscurely petioled; fls. few, nodding in bud; sepals 2 mm. long; petals lilac to purple, 4–5 mm. long; pedicels 5–15 mm. long in fr.; caps. slender, linear, ca. 1 mm. thick, purplish, 2–4 cm. long; seeds smooth, obovoid, 1–1.3 mm. long, with dingy coma; $2n = 36$ (Bøcher & Larsen, 1950).—Moist rock slides and stony places, 8000–11,500 ft.; Subalpine F., Alpine Fell-fields; Sierra Nevada from Tulare and Inyo cos. n., Mt. Shasta; to Alaska, Rocky Mts., Maine, Labrador; Eurasia. July–Sept.

16. **E. clavàtum** Trel. Habit much like that of *E. anagallidifolium;* stems more

numerous, heavier, subglabrous to glandular-pubescent; lvs. broadly ovate, divergent, 1–2 cm. long, ± denticulate, short-petioled; fls. few, erect in bud; sepals 3–4 mm. long; petals purplish to rose, 5–6 mm. long; caps. purplish, ± subclavate, 1–2 mm. thick, 2–2.5 cm. long; seeds fusiform, papillose, 1.5–2 mm. long.—Moist rocky places, 7500–9000 ft.; Subalpine F.; Marble Mts. and Mt. Shasta, Siskiyou Co.; to B.C., Mont., Colo. July–Aug.

17. **E. Hornemánnii** Rchb. Perennial with subterranean scaly branches; stems slender, erect except at very base, simple, 1–4 dm. tall, glabrous except for the crisp pubescence on the decurrent lines below the lvs.; lvs. ovate to elliptic-ovate, 1.5–4 cm. long, mostly obtuse, subentire to remotely serrulate, short-petioled; fls. few, erect; sepals 3–4 mm. long; petals purplish or violet, 5–8 mm. long; pedicels 1–2 cm. long in fr.; caps. erect, slender, linear, less than 1 mm. thick, 4–5 cm. long; seeds papillose, 1 mm. long, with dingy coma.—Moist rocky and mossy places, 6000–11,500 ft.; Red Fir to Alpine Fell-fields; Sierra Nevada, Coast Ranges from Tehama Co. n.; to Alaska, Greenland, New England; Eurasia. July–Aug.

18. **E. lactiflòrum** Hausskn. [*E. alpinum* L., in part.] Size and habit of *E. Hornemannii*, but more glabrous on decurrent lines as well as in infl., the basal stolons epigaeous; lvs. delicate, pale, subentire or obscurely denticulate, elliptic to oblong-ovate, obtuse, 2–5 cm. long; fls. few; petals white or pink-tipped, ca. 3–4 mm. long; caps. slender, erect, linear, less than 1 mm. thick, 4–5 cm. long; seeds smooth, ca. 1 mm. long, beaked, with dingy coma.—Moist slopes and banks, 6000–11,500 ft.; San Jacinto and San Bernardino mts., Sierra Nevada, Coast Ranges from Glenn Co. n.; to Alaska, New England, Quebec; Eurasia. July–Aug.

19. **E. adenocáulon** Hausskn. [*E. glandulosum* var. *a.* Fern. *E. concinnum* Congd.] Perennial, flowering the first year, innovations by fleshy rosettes; stem erect, 3–10 dm. tall, glabrous below with some hair on decurrent lines below nodes, glandular-pubescent in infl., simple or weakly branched below, freely branched above; lvs. glabrous or nearly so, ovate- to elliptic-lanceolate, 3–6 cm. long, acute to obtuse, serrulate, rounded into very short flat petioles, upper lvs. gradually reduced, ± pubescent; sepals 2 mm. long; petals white or pale to ± reddish, 3–4 mm. long; fruiting pedicels 3–8 mm. long; caps. slender, ± reddish, 4–6 cm. long, with gland-tipped hairs, glabrate in age; seeds obovoid, abruptly short-beaked, ca. 1 mm. long, longitudinally ridged and rounded-papillose.—Moist places below 11,000 ft.; many Plant Communities; most of cismontane and montane Calif.; to Alaska and Atlantic states. July–Sept.

KEY TO VARIETIES

Infl. glandular-pubescent.
    Petals white to pale or reddish, 3–4 mm. long .... *E. adenocaulon*
    Petals rose to purple, 5–6 mm. long .... var. *occidentale*
Hairs of infl. not gland-tipped.
    Herbage green, the stems and lvs. scarcely pubescent .... var. *Parishii*
    Herbage grayish, the stems and lvs. densely soft-pubescent .... var. *holosericeum*

Var. **occidentàle** Trel. [*E. o.* Rydb. *E. glandulosum* var. *o.* Fern. *E. californicum* var. *o.* Jeps.] Lvs. lanceolate to lance-ovate; fls. purple or rose, the petals 5–6 mm. long.—With the sp., Coast Ranges from Santa Clara Co. n., less frequent in Sierra Nevada; to B.C., Rocky Mts.

Var. **Párishii** (Trel.) Munz. [*E. P.* Trel. *E. californicum* var. *P.* Jeps. *E. californicum* Hausskn. *E. Palmeri* Lév.] Aspect and stature of *E. adenocaulon* but the infl. with a whitish, ± appressed pubescence; petals white or pink, 2–4 mm. long; $n = 18$ (Lewis et al., 1958).—Wet places, mostly below 6000 ft.; many Plant Communities; common in cismontane s. Calif., and Coast Ranges to Del Norte Co., less so in Sierra Nevada, San Joaquin V. and deserts; to B.C.

Var. **holosericeum** (Trel.) Munz. [*E. h.* Trel. *E. californicum* var. *h.* Jeps.] Habit of the sp., but canescent throughout with soft subappressed hairs; petals pink to purple, 4–5 mm. long.—Occasional, in moist places, below 2500 ft.; many Plant Communities; much of cismontane Calif.

20. **E. ciliàtum** Raf. [*E. americanum* Hausskn. *E. adenocaulon* var. *perplexans* Trel.] Like *E. adenocaulon*, but more slender, mostly 1–3(–4) dm. high, scarcely glandular, but with crisped hairs in the infl., mostly simple in stature; lvs. thin, tapering at base

into narrow petioles 2–10 mm. long, the blades elliptic to ovate, faintly veined, 1–5 cm. long; fls. whitish, the petals 3–4 mm. long.—Occasional in wet places, 4000–8000 ft.; Sagebrush Scrub, Pinyon-Juniper Wd., etc.; e. side of Sierra Nevada, etc. from Inyo Co. to Modoc Co.; to e. Wash., Rocky Mts., Atlantic Coast. July–Oct.

21. E. Watsònii Barb. Rosuliferous perennial with rather coarse reddish stems 3–10 dm. tall, the herbage canescent with a short soft pubescence; lvs. many, mostly opposite, elliptic-lanceolate to ovate-lanceolate, obtuse, serrate, 3–6 cm. long, mostly rounded at base into very short broad petioles; fls. at first crowded, scarcely exceeding the somewhat reduced upper lvs.; fruiting pedicels mostly 5–10 mm. long; sepals 4–5 mm. long, reddish; petals usually red-purple, 6–10 mm. long, suberect, deeply emarginate; caps. slender, 5–8 cm. long; seeds 1 mm. long, half as wide, longitudinally ridged and papillate, with whitish coma.—Wet places, marshes, etc.; Mixed Evergreen F., Freshwater Marsh, etc.; Sonoma Co. May–July. Material from San Luis Obispo and Monterey cos. is intermediate with:

Var. franciscànum (Barb.) Jeps. [*E. f.* Barb.] Herbage subglabrous or slightly pubescent; $n = 18$ (Lewis et al., 1958).—Marshes, etc. at low elevs.; Freshwater Marsh, N. Coastal Scrub, Redwood F., Mixed Evergreen F.; immediate coast from San Luis Obispo Co. to Del Norte Co.; to Ore. May–Aug.

5. Boisduvàlia Spach.

Caulescent mostly erect annuals, sometimes decumbent. Lvs. generally alternate, sometimes opposite, simple, sessile. Fls. small, in leafy spikes or solitary in lf.-axils. Fl.-tube produced above the ovary, short in ours, funnelform. Sepals 4, erect. Petals 4, sessile, obovate, 2-lobed, purple to white. Stamens 8, those opposite the petals shorter; anthers basifixed, all fertile; pollen in tetrads. Stigma in ours with 4 cuneate lobes. Caps. 4-loculed, 4-valved, sessile, terete or ± angled. Seeds smooth, without a coma, having a thin margin at each end, mostly in a single row in each locule. Ca. 10 spp., w. N. Am., s. S. Am., Tasmania. (Named for J. A. *Boisduval*, French naturalist and physician.)

(Munz, P. A. A revision of the genus Boisduvalia. Darwiniana 5: 124–153, 1941.)
Caps. septifragal, the septa wholly adherent to the placental axis, making the latter 4-winged; lvs. lanceolate, dentate, the upper broader .... 1. *B. densiflora*
Caps. loculicidal, subterete, the septa adherent to the valves in dehiscence, or the caps. 4-sided and not dehiscent.
  The caps. coriaceous, 4-sided, tardily if at all dehiscent; lvs. narrowly lanceolate; ovules rather numerous, 10–14 in each row .... 2. *B. cleistogama*
  The caps. membranous, terete, usually dehiscent; ovules fewer except sometimes in B. glabella.
    Fl.-tube 0.5–1 mm. long; petals 1.5–4 mm. long.
      Fl.-bracts ovate or oblong, broader than the foliage lvs.; petals 2–4 mm. long; caps. 6–8 mm. long. Quite straight .... 3. *B. glabella*
      Fl.-bracts linear; petals 1–2 mm. long; caps. 8–10 mm. long, usually curved .... 4. *B. stricta*
    Fl.-tube 2–3 mm. long; petals 5–10 mm. long.
      Lvs. serrulate, crowded; petals 7–10 mm. long; caps. straight .... 5. *B. macrantha*
      Lvs. quite entire, not crowded; petals 5–8 mm. long; caps. curved .... 6. *B. pallida*

1. B. densiflòra (Lindl.) Wats. [*Oenothera d.* Lindl. *B. Douglasii* Spach. *B. bipartita* Greene.] Simple or branched (especially above), erect or nearly so, mostly 3–10 dm. tall, commonly villous and with some shorter gland-tipped hairs particularly in infl., the plant green to canescent, leafy throughout; lower lvs. lance-linear to lanceolate, entire to denticulate, acute, 2–5 cm. long, the fl.-bracts ovate, not conspicuously overlapping, mostly entire, 0.5–2 cm. long; infl. dense, long-spicate in fr.; sepals 2–4 mm. long, lanceolate; petals rose-purple, occasionally whitish, bilobed, 6–12 mm. long; caps. stout, straight, 8–10 mm. long, septifragal; seeds 4–6 in each locule, ovoid, angled, brown, paler at ends, concave on inner side, ca. 1.5 mm. long; $n = 10$ (Lewis et al., 1958).—Moist places, below 8500 ft.; many Plant Communities; cismontane and montane Calif. from San Diego Co. n.; to B.C., Ida., Nev. May–Aug. Plants with the spikes so dense that the bracts are imbricate and conceal the caps., are forma *imbricàta* (Greene) Munz. [*B. d.* var. *i.* Greene. *B. i.* Heller. *B. d.* var. *montana* Jeps.] With the sp.

Var. palléscens Suksd. Pubescence spreading, usually with some hairs gland-tipped; fl.-bracts often remote, broadly ovate, gradually acuminate; fls. pale; seeds 3–4 in each locule, 2 mm. long.—At 3000–7000 ft.; Montane Coniferous F.; Sierra Nevada from Madera Co. n.; to Wash.

Var. **salicìna** (Nutt. ex T. & G.) Munz. [*Oenothera s.* Nutt. *B. s.* Rydb. *B. sparsiflora* Heller.] Plant strigose-canescent with short pubescence, the hairs not gland-tipped; petals pale, 2.5–5 mm. long; caps. and seeds as in the sp.—Occasional below 7000 ft.; many Plant Communities; Colusa and Nevada cos. n.; to Wash., Ida., Nev.

2. **B. cleistógama** Curran. [*Oenothera c.* Lév.] Erect and simple or more usually branched from the base, 1–2 dm. tall, ± villous and glandular throughout, densely leafy; lvs. linear to lanceolate, 2–4.5 cm. long, 1.5–5 mm. wide, acute, remotely denticulate, pale, not much reduced up the stem; fls. axillary along the branches, the earliest cleistogamous, the later rose; sepals 1–2 mm. long; petals bifid, to 3 mm. long; caps. hard, 4-sided, sharply angled and with 4 nerves, pointed, slightly curved, 1 cm. long, 1.5 mm. thick, tardily if at all dehiscent; seeds light brown, linear, angled, 1–1.5 mm. long. —Dried beds of vernal pools, below 1000 ft.; largely V. Grassland, lower Foothill Wd.; in and about Cent. V. April–June.

3. **B. glabélla** (Nutt. in T. & G.) Walp. [*Oenothera g.* Nutt.] Mostly freely and decumbently branched from base, the stems 1–3 dm. long, subglabrous or pubescent on lf.-veins or throughout, rather uniformly leafy; lvs. lance-ovate to -oblong, serrulate, bright green, 1–1.5 cm. long; fls. axillary, sometimes even in lowest axils; sepals 2 mm. long; petals purplish, 2–4 mm. long; caps. straight, 6–8 mm. long, pointed at tip; seeds many, grayish brown, narrowly subfusiform, angled, 1 mm. long.—Dry mud flats and vernal pools, below 5000 ft.; many Plant Communities; cismontane and montane Calif. from San Diego Co. to Humboldt, Siskiyou, and Modoc cos.; B.C., Sask., Nev., Argentina. May–Aug.

Var. **campéstris** (Jeps.) Jeps. [*B. c.* Jeps.] Lvs. of upper branches densely overlapping and concealing the caps.—With the sp. from Monterey and Merced cos. to Modoc, Butte, and Glenn cos.

4. **B. strícta** (Gray) Greene. [*Gayophytum s.* Gray. *B. Torreyi* Wats.] Stems 1–5 dm. tall, simple and erect or virgately branched from near base, pilose and ± canescent throughout; lvs. linear to lance-linear, 1–4 cm. long, 2–4 mm. wide, acute, entire to sharply denticulate, the upper narrower than the lower; fls. axillary, often from near the base of the plant; sepals 1 mm. long; petals rose-purple or violet, 1.5–2 mm. long; caps. 8–10 mm. long, membranous, slender, tardily loculicidal, usually curved outwards and attenuate; seeds oblong-ovoid, brown, ca. 1 mm. long, cellular-pitted, ca. 6–8 in each locule; $n = 9$ (Lewis et al., 1958).—Moist places, below 8500 ft., many Plant Communities; from Santa Clara and Tulare cos. n.; to e. Wash., Ida., Nev. May–July.

5. **B. macrántha** Heller. Stems mostly erect, 1–10 dm. tall, simple or few-branched, glabrous below, villous above; lvs. rather crowded, 2–4 cm. long, 0.5–0.9 cm. wide, lanceolate to oblanceolate, the upper almost ovate, remotely serrulate, acute to acuminate; fls. solitary in upper axils; sepals linear-lanceolate, 3–6 mm. long; petals rose-purple when dry, divided ca. half their length, the lobes asymmetrically rounded at the tips; caps. straight, lance-linear, 1–2 cm. long, 2 mm. thick near base, with slender apical beak 2–3 mm. long; seeds 5–6 in each row, brownish, somewhat shining, 2 mm. long, microscopically cellular-punctate.—Moist gravelly or sandy places at low elevs.; Foothill Wd.; Butte, Shasta, Modoc, and Tehama cos. May–July.

6. **B. pállida** Eastw. Stems 1–4 dm. tall, mostly branched from base, sometimes simple, tomentulose and pilose, glabrescent below in age; lvs. not crowded, somewhat reduced above, 1.5–5 cm. long, 0.3–0.6 cm. wide, lanceolate, acute to acuminate, subentire, strigose to subglabrous; fls. axillary, even in lowermost axils; sepals 3–4 mm. long, pubescent; petals reddish, 5–8 mm. long; caps. 1.5–3 cm. long, 1.5–2 mm. thick at base, tapering gradually into a slender outcurved beak 2–4 mm. long; seeds ca. 6 in each locule, brownish, 1.5–2 mm. long, cellular-pitted.—Damp places, ca. 4000–6000 ft.; Montane Coniferous F.; Fresno Co. to Tehama, Shasta, and Modoc cos.; sw. Ore. June–July.

6. **Clárkia** Pursh.

Annual, with slender to stoutish stems, simple or branched, usually with epidermis exfoliating below. Lvs. simple, pinnately veined, linear to ovate, entire to denticulate, sessile or short-petioled. Infl. a leafy spike or raceme, the axis straight or recurved in bud; buds erect, deflexed or pendulous; fls. mostly showy, the fl.-tube obconic, campanulate or funnelform to slender and elongate, mostly with a ring of hairs within. Sepals 4, often colored, reflexed individually, in pairs, or united and deflexed to one side at anthesis. Petals 4, cuneate to obovate or oblanceolate, sometimes lobed, sessile to clawed, lavender to pink or purple or white, often variously spotted or flecked or blotched. Stamens 8 and in 2 series, or 4 and in 1 series; anthers basifixed. Stigma 4-lobed, the lobes short or elongate. Caps. clavate to cylindrical, terete or quadrangular, sessile to pedicelled, beakless or beaked, often 4- or 8-grooved; seeds brown to gray, minutely tubercled or scaly, usually angular, ± oblique, cubical or elongate, sometimes spindle-shaped, mostly crested. Ca. 33 spp., of temp. w. N. Am. and Chile. (Capt. William *Clark*, of Lewis and Clark Expedition.)

(Lewis, H., and M. E. Lewis. The genus Clarkia. Univ. Calif. Pub. Bot. 20[4]: 241–392, 1955.)
A. Petals with a conspicuously lobed limb.
    B. Stamens 8; fl.-tube less than 1.5 cm. long.
        C. Petals with 2 lobes, lacking a prominent tooth in the sinus . . . . 16. *C. biloba*
        CC. Petals with 3 lobes, the middle one merely a prominent tooth in the sinus
                                                      26. *C. Xantiana*
    BB. Stamens 4; fl.-tube slender, 1.5–3 cm. long.
        C. Fils. flattened but not club-shaped; petals ca. twice as long as broad, the middle lobe
            ca. as wide as the lateral ones . . . . 30. *C. concinna*
        CC. Fils. club-shaped toward tips; petals ca. as long as broad, the middle lobe narrower
            than the outer . . . . 31. *C. Breweri*
AA. Petals with an entire to erose or emarginate limb, not lobed.
    B. The petals constricted into a definite claw which is expanded into a pair of lateral lobes
        or teeth near the base.
        C. Axis of the infl. becoming erect as the fls. open, or at most 2 or 3 nodes above the
            uppermost fl.; lvs. ovate to lanceolate or elliptical; anthers lavender to red; petals
            mostly purplish-red or pink.
            D. Petals 6–12 mm. long, 3–7 mm. broad; widely distributed . . . . 27. *C. rhomboidea*
            DD. Petals 14–21 mm. long, 10–14 mm. broad. Butte, Plumas, and Shasta cos.
                                              28. *C. Mildrediae*
        CC. Axis of the infl. becoming erect in advance of anthesis and 4–6 or more nodes above
            the uppermost fl.; lvs. lanceolate to lance-linear; anthers usually purple. Foothills of
            cent. Sierra Nevada . . . . 29. *C. virgata*
    BB. The petals not constricted at base into a definite claw, or if so, the claw entire.
        C. Petals white to cream, fading pink, not flecked with purple, mostly less than 10 mm.
            long . . . . 20. *C. epilobioides*
        CC. Petals colored, or if nearly white, flecked with purple, mostly more than 10 mm. long.
            D. Buds, at least the older ones, pendulous or deflexed.
            E. The claw of the petals slender, equal to or longer than the limb.
                F. Petals usually more than 10 mm. long, the blade deltoid to semicircular;
                    style surpassing stamens; ovary at anthesis shorter than combined
                    lengths of fl.-tube and sepals. Widely distributed . . 24. *C. unguiculata*
                FF. Petals usually less than 10 mm. long, the blade rhombic; style equaling
                    stamens; ovary at anthesis equal to combined lengths of fl.-tube and
                    sepals. Tulare and Kern cos. 25. *C. exilis*
            EE. The claw of the petals short or lacking.
                F. Ring of hairs within the fl.-tube at or near the summit or lacking; fl.-tube
                  1–3, rarely 4 mm. long.
                    G. Sepals 6–10 mm. long; petals usually not more than 1.2 cm. long.
                      H. Petals oblanceolate, 2–3 times as long as broad, bright pink.
                          Mariposa Co. . . . . 17. *C. lingulata*
                      HH. Petals spatulate to obovate, not much longer than broad.
                        I. The immature caps. 4-grooved or the grooves obscure.
                          From Santa Clara and Amador cos. n.   5. *C. gracilis*
                        II. The immature caps. 8-grooved, -ribbed, or -striate.
                          J. Rachis of the infl. reflexed at the tip, becoming straight
                            as the buds mature.
                              K. Petals pink, often with reddish-purple flecks near
                                  the base. From Tulare and Santa Barbara cos. n.
                                  18. *C. modesta*
                            KK. Petals almost white or pale pink, flecked with red-
                                purple in lower half. San Benito Co. to San Diego
                              Co. 21. *C. similis*
                        JJ. Rachis of infl. straight, the buds deflexed. San Diego
                            Co. . . . . 23. *C. delicata*
                GG. Sepals mostly 10–20 mm. long; petals mostly 1.5–3 cm. long.
                    H. Rachis of infl. straight, only the buds deflexed. Orange Co. to
                      Monterey Co. . . . . 22. *C. deflexa*
                    HH. Rachis of the infl. reflexed at tip, becoming straight as the
                      buds mature.
                      I. Immature caps. 4-grooved or the grooves obscure.
                        Monterey Co. . . . . 14. *C. Bottae*

II. Immature caps. 8-grooved, -ribbed, or -striate.
    J. Petals mostly streaked with white. Tuolumne Co. to
        Riverside Co. .... 19. *C. Dudleyana*
    JJ. Petals not streaked with white. Eldorado Co. to Butte
        Co. .... 16. *C. biloba Brandegeae*
FF. Ring of hairs within the fl.-tube obviously below the upper margin;
  fl.-tube usually 3–7 mm. long.
    G. Anthers similar in size and color; pollen of the 2 series uniform in
      color, whitish to yellow; petals not flecked with red.
      H. Immature caps. 8-grooved or -ribbed, cylindrical or largest at
        the middle.
        I. The immature caps. bright green, shining, with a beak as
          long as 7 mm.; pedicels at maturity 5–15 mm. long; petals
          10–30 mm. long. Butte Co. to Mariposa Co.
                                          3. *C. arcuata*
        II. The immature caps. dull green, tapering to the summit
          or with a beak to 3 mm. long, sessile or on pedicels to
          3 mm. long; petals 8–16 mm. long. Plumas Co. to Modoc
          and Siskiyou cos. .... 4. *C. lassenensis*
      HH. Immature caps. 4-grooved, mostly largest above the middle.
        Widely distributed in n. half of Calif. .. 5. *C. gracilis*
    GG. Anthers of outer and inner series differing in color or size or both;
      pollen of at least the outer bluish, of the inner lighter; petals often
      flecked with red or purple.
      H. Petals oblanceolate, uniformly bright pink, not flecked. Mari-
        posa Co. .... 17. *C. lingulata*
      HH. Petals obovate to cuneate, pinkish-lavender to purple in upper
        half, lighter in lower, the base usually dark purple. Los
        Angeles to Monterey and Madera cos. .... 15. *C. cylindrica*
DD. The buds erect.
  E. Mature stigma well elevated above the stamens; petals usually more than
    15 mm. long.
    F. Immature caps. 4-grooved, the grooves sometimes obscure.
      G. Petals lavender to somewhat pinkish, usually with a bright red base.
        S. Marin Co. to Monterey Co. .... 2. *C. rubicunda*
      GG. Petals lavender to pink or nearly white, usually with a cent. red
        blotch or penciling, the base sometimes darker but not red. Marin
        Co. to Del Norte Co. .... 1. *C. amoena*
    FF. Immature caps. conspicuously 8-grooved or -ribbed.
      G. Petals with a red or purple spot.
      H. The petals with a wedge- or shield-shaped spot above the
        middle which often becomes diffuse or obscure near upper
        margin.
        I. Lvs. less than 5 times as long as broad, overlapping;
          fl.-tube ca. 15 mm. long, 10 mm. broad, conspicuously
          veined. Sonoma Co. .... 6. *C. imbricata*
        II. Lvs. at least 5 times as long as broad, not conspicuously
          overlapping; fl.-tube 5–13 mm. long, less than 10 mm.
          broad, not conspicuously veined.
          J. Buds mucronate, the sepal-tips free. Foothills of cent.
            Sierra Nevada .... 7. *C. Williamsonii*
          JJ. Buds not mucronate, the sepal-tips usually fused.
            Widely distributed .... 12. *C. purpurea*
      HH. The petals with an ovate, semicircular or crescent-shaped
        spot in the center or toward the base.
        I. Caps. conspicuously enlarged at the middle. Coastal
          Mendocino and Humboldt cos. .. 1. *C. amoena Whitneyi*
        II. Caps. not conspicuously enlarged at the middle.
          J. Infl. congested; ovary longer than the internode above.
            Foothills of cent. Sierra Nevada .... 8. *C. nitens*
          JJ. Infl. not congested; ovary usually shorter than the
            internode above. S. Sierra Nevada to S. Coast Ranges
                                    9. *C. speciosa*
      GG. Petals not spotted, sometimes darker in center.
      H. Ovary strigulose-puberulent with upwardly curved hairs.
        I. Petals dark red-purple; sepals reflexed individually. Cent.
          Sierra Nevada .... 7. *C. Williamsonii incerta*
        II. Petals bright pink or lavender, sometimes cream in
          lower half; sepals usually reflexed in pairs or remaining
          united.
          J. Fls. crowded; ovary longer than the internode above.
            Foothills of cent. Sierra Nevada ...... 8. *C. nitens*
          JJ. Fls. not crowded; ovary usually shorter than the
            internode above. S. Coast Ranges ....9. *C. speciosa*
      HH. Ovary usually densely spreading-pubescent . 12. *C. purpurea*

EE. Mature stigma not elevated above the stamens; stamens often adhering to
 stigma; petals usually less than 15 mm. long.
 F. Plants usually prostrate or decumbent; lvs. oblanceolate to elliptic or
  obovate, obtuse at tip. Along seacoast.
  G. Petals 5–11 mm. long, not spotted. Santa Cruz Co. to Humboldt Co.
                    10. *C. Davyi*
  GG. Petals 10–15 mm. long, usually with a reddish spot. Santa Cruz Co.
   to Santa Barbara Co. .... 11. *C. prostrata*
 FF. Plants usually erect; lvs. linear to elliptic or ovate, acute at tip. Away from
  the immediate coast.
  G. Immature caps. stout, not more than 8 times as long as broad,
   8-ribbed, tapering to the summit or with a beak to 2 mm. long;
   sepals mostly deflexed individually or in pairs .. 12. *C. purpurea*
  GG. Immature caps. slender, usually at least 10 times as long as broad,
   8-grooved, with a beak 3–7 mm. long at maturity; sepals usually
   united and deflexed to one side .... 13. *C. affinis*

1. **C. amoèna** (Lehm.) Nels. & Macbr. [*Oenothera a.* Lehm. *Godetia a.* G. Don. *G. Lehmanniana* Spach. *Oe. roseo-alba* Bernh. *G. vinosa* Lindl. *G. Nivertiana* Goujon.] Farewell-to-Spring. Erect or sprawling, mostly 3–10 dm. tall, ± puberulent above with upwardly curled hairs; lvs. lanceolate, 1–6 cm. long, 0.4–1.5 cm. wide, on petioles to ca. 0.6 cm. long; infl.-rachis erect; the fls. ± congested; buds erect; fl.-tube 6–10 mm. long, the ring of hairs below the middle; sepals lanceolate, mostly 1.5–2.5 cm. long, united and deflexed to one side at anthesis; petals obovate to fan-shaped, 2–3.5 cm. long, pink or lavender to white above, often pink or lavender at base, mostly penciled or blotched in center with bright red; fils. white to lavender, the anthers acute, lavender to yellow or with some red; mature style exceeding stamens; young caps. 4-grooved, slender, not much enlarged in middle, becoming 2–5 cm. long and subterete; seeds brownish, scaly, ca. 1.5 mm. long, the crest scarcely 0.1 mm. long; n = 7 (Håkansson). —On slopes and bluffs near the sea; N. Coastal Scrub, Coastal Prairie, etc.; Humboldt, Mendocino, Sonoma, and Marin cos. June–Aug.

KEY TO SUBSPECIES

Immature caps. 4-grooved, slender, less than 5 mm. broad, ± cylindrical.
 Plants coarse; infl. ± congested, the internodes usually shorter than the subtending ovary. Near
 immediate coast .... *C. amoena*
 Plants slender; infl. lax, the internodes longer than the subtending ovary. N. Coast ranges away
 from immediate coast .... ssp. *Huntiana*
Immature caps. 8-grooved, conspicuously enlarged in middle and often 10 mm. broad. Local in
coastal Mendocino and Humboldt cos. .... ssp. *Whitneyi*

Ssp. **Huntiàna** (Jeps.) Lewis & Lewis. [*Godetia a.* f. *H.* Jeps. *G. a.* var. *H.* Jeps.] Slender, usually much-branched, to 1 m. tall; lvs. lanceolate to linear, 0.2–0.5 cm. wide; infl. lax; fl.-tube 4–10 mm. long; petals 1.5–3 cm. long, lavender, usually with cent. red blotch; n = 7 (Håkansson, 1942).—Openings in forest, etc.; Redwood F., Mixed Evergreen F., N. Oak Wd., etc.; away from the coast from Del Norte, and Trinity cos. to Marin and Napa cos.; sw. Ore. June–Aug.

Ssp. **Whitneyi** (Gray) Lewis & Lewis. [*Oenothera W.* Gray. *Godetia W.* T. Moore. *Clarkia W.* Nels. & Macbr.] Coarse, erect or sprawling, simple or branched, 2–5 dm. tall; lvs. lanceolate to almost ovate, 3–6 cm. long; infl. congested; fl.-tube 6–11 mm. long; sepals 1.5–3 cm. long; petals cuneate to obovate, 4–6 cm. long, lavender with central red splotch; caps. broadly fusiform, 1.5–2.5 cm. long, 5–8 mm. thick, prominently 8-ribbed; seeds 1.5 mm. long, brown, minutely cellular-pubescent, fairly prominently crested; n = 7 (Lewis & Lewis, 1955).—Coastal bluffs and exposed places; largely N. Coastal Scrub; in a few localities in Humboldt and Mendocino cos. June–Aug.

2. **C. rubicúnda** (Lindl.) Lewis & Lewis. [*Godetia r.* Lindl. *Oenothera r.* H. & A.] Much like *C. amoena*, with stature, pubescence, lvs., erect buds, etc. as in that plant; lvs. 2–10 mm. broad; sepals 1.2–2.8 cm. long, united and deflexed at anthesis; petals subentire to erose, 1–3 cm. long, rose-pink to lavender, usually with a showy red or red-purple base; fils. and anthers red; stems slender, the infl. lax, with internodes longer than the subtending ovary; caps. terete, 2–4 cm. long, 2.5–3 mm. broad, mostly blunt; seeds much like those of *C. amoena;* n = 7 (Lewis & Lewis, 1955).—Openings in woods, etc.; N. Coastal Scrub, Mixed Evergreen F., etc.; from s. Marin and Alameda cos. to Santa Clara and Santa Cruz cos. May–July.

Ssp. **Blasdàlei** (Jeps.) Lewis & Lewis. [*Godetia B.* Jeps.] Stems coarse; infl. more crowded, the internodes shorter than the subtending ovary; larger lvs. 10–18 mm. broad; *n* = 7 (Håkansson, 1941).—N. Coastal Scrub, Coastal Strand; immediate coast from s. Marin Co. to n. San Luis Obispo Co.

C. **franciscàna** Lewis & Raven. Near to *C. rubicunda,* but more slender; the fl.-tube 1–3 mm. long; petals cuneate, 5–13 mm. long, erose; *n* = 7 (Lewis & Raven, 1958).— On serpentine, the Presidio, San Francisco. May–July.

3. **C. arcuàta** (Kell.) Nels. & Macbr. [*Oenothera a.* Kell. *Godetia a.* Jeps. *Oe. hispidula* Wats. *G. h.* Wats. *G. Hanseni* Jeps.] Erect, simple or branched from base or above, 1–7 dm. tall, glabrous or puberulent with spreading hairs; lvs. linear to lanceolate or oblanceolate, 1.5–6 cm. long, 1–5 mm. broad, sessile; rachis of infl. reflexed, becoming erect as fls. open; buds pendulous, ± glandular-puberulent; fl.-tube slender, 3–7 mm. long, with ring of hairs near the middle; sepals green to rose, lanceolate, 1–2.5 cm. long; petals fan-shaped, 1–3 cm. long, pink-lavender, with or without reddish spot near base; anthers dark purple; fils. lavender to dark red-purple; caps. 1–3 cm. long, terete to ± 4-sided, 1.5–2.5 mm. broad, with a beak to 7 mm. long and pedicel 5–15 mm. long; seeds brown, ca. 2 mm. long, conspicuously crested and cellular-pubescent; *n* = 7 (Håkansson, 1941). —Mostly grassy places, below 5000 ft.; Chaparral, Foothill Wd., Yellow Pine F.; w. base of Sierra Nevada from Butte Co. to Mariposa Co. March–June.

4. **C. lassenénsis** (Eastw.) Lewis & Lewis. [*Godetia l.* Eastw.] Erect, 1–8 dm. tall, mostly simple, puberulent above with ascending or spreading hairs; lvs. linear to lance-linear, 2–5 cm. long, 2–5 mm. broad, puberulent, on petioles 5–10 mm. long; infl.-axis recurved in bud, erect at anthesis; buds pendulous; fl.-tube 3–5 mm. long, the hair-ring below the middle; sepals lanceolate, 7–14 mm. long, united and deflexed to one side at anthesis; petals 8–16 mm. long, pinkish-lavender with red-purple base; caps. 8-ribbed when immature, 1–2.5 cm. long, sessile or short-pedicelled, subterete when mature and somewhat longer, with a beak 2 mm. long; seeds brownish or spotted, minutely tuberculate, 1.5 mm. long, the crest ca. 0.2 mm. long; *n* = 7 (Lewis & Lewis, 1955).—Open places, 4000–6900 ft.; Chaparral, Yellow Pine F., Red Fir F., Sagebrush Scrub, etc.; Trinity and Siskiyou cos. to Modoc and Plumas cos.; w. Nev. May–June.

5. **C. grácilis** (Piper) Nels. & Macbr. [*Godetia g.* Piper. *G. amoena* var. *g.* C. L. Hitchc. *G. a.* var. *concolor* Jeps. *G. a.* f. *pygmaea* Jeps.] Simple or branched, 3–9 dm. tall, subglabrous to silvery-pubescent above with hairs ± appressed; lvs. linear to lance-linear, 2–6 cm. long, 2–6 mm. broad, on petioles to 1 cm. long; infl.-axis recurved in bud, erect at anthesis; buds pendulous; fl.-tube 1.5–3 mm. long, with or without an inner ring of hairs; sepals lanceolate, mostly 8–10 mm. long; petals obovate to cuneate, entire to erose or apiculate, 0.6–2.2 cm. long, pink to lavender, sometimes darker near base; anthers cream; fils. lavender; ovary 4-grooved, puberulent; mature caps. 3–5 cm. long, 2–3 mm. broad, sometimes enlarged upward, with slender beak to 1 cm. long; seeds brown, often mottled, 1.5–2 mm. long, the crest to 0.2 mm. long; *n* = 14 (Håkansson, 1942).—Openings in woods, below 5000 ft.; Mixed Evergreen F., Foothill Wd., Yellow Pine F., etc.; from Santa Clara Co. to Siskiyou Co. and from Modoc Co. to Amador Co.; to Wash. April–July.

Ssp. **albicáulis** (Jeps.) Lewis & Lewis. [*Godetia amoena* var. *a.* Jeps. *G. cylindrica* var. *Tracyi* Jeps.] Fl.-tube 4–10 mm. long; sepals 1.5–2.5 cm. long; petals 2–4 cm. long, pink-lavender to light-purple shading to white with scarlet spot at base; *n* = 14 (Lewis & Lewis, 1955).—Mostly in Foothill Wd.; about the Sacramento V. in Butte, Colusa, Lake, Shasta, and Tehama cos.

Ssp. **sonoménsis** (C. L. Hitchc.) Lewis & Lewis. [*Godetia amoena* var. *s.* C. L. Hitchc.] Like ssp. *albicaulis* in fl.-size, but the petals with a cent. spot.—Openings in woods; Mixed Evergreen F., N. Oak Wd., Foothill Wd.; from Marin and Napa cos. to Del Norte Co.; sw. Ore.

6. **C. imbricàta** Lewis & Lewis. Erect, simple or branched, sparsely puberulent above, densely leafy; lvs. lanceolate, 2–2.5 cm. long, 4–7 mm. wide, entire to denticulate, ascending, overlapping; sessile or nearly so; infl. congested, erect; fl.-tube conspicuous, 1–1.5 cm. long, with rings of hairs near base; sepals lanceolate, 1–1.5 cm. long; petals fan-shaped, 2–2.5 cm. long, lavender above, paler downward but with light purple base and V-shaped purple spot in upper half; stamens lavender; young caps. ribbed, 1–1.5 cm. long, 4–5 mm. broad, sessile, deeply 8-ribbed; seeds brown or gray, 2 mm.

long, the crest less than 0.2 mm. long; $n = 8$ (Lewis & Lewis, 1955).—Chaparral near Pitkin Marshes, Sonoma Co. June–July.

7. **C. Williamsònii** (Dur. & Hilg.) Lewis & Lewis. [*Godetia W.* Dur. & Hilg. *G. viminea* vars. *Congdonii* and *incerta* Jeps.] Erect, 3–10 dm. high, simple or branched, puberulent above with short upwardly curled hairs, sometimes also pilose; lvs. linear to linear-oblanceolate, subentire, 2–6 cm. long, 2–5 mm. wide, not crowded, puberulent, sessile to short-petioled; infl. not congested, erect; buds erect; often pilose; fl.-tube 7–13 mm. long, the hair-ring in lower third; sepals lanceolate, 8–18 mm. long, reflexed individually or in pairs; petals cuneate, entire to erose, 1–3 cm. long, lavender near both ends, white near middle with a purple spot in upper half; anthers white to lavender; stigma ± purplish; immature caps. 1–3 cm. long, 4–5 mm. broad, 8-ribbed, subsessile, beak less than 2 mm. long; seeds brown or grayish, 1–1.5 mm. long, the crest short; $n = 9$ (Håkansson, 1941).—Dry places, below 5000 feet.; Foothill Wd., Yellow Pine F.; w. base of Sierra Nevada from Nevada to Fresno cos. May–Aug. Plants from Yosemite with dark wine-red fls. may be called forma *incérta* (Jeps.) Lewis & Lewis.

8. **C. nìtens** Lewis & Lewis. Much like *C. Williamsonii;* to ca. 4 dm. tall; lvs. ± crowded; infl. congested; sepals 10–12 mm. long, usually reflexed individually; petals 1–2 cm. long, lavender near upper margin, cream to pale yellow near base or cream throughout, not spotted or with a bright purple-red spot in lower half; immature caps. 1.5–2 cm. long, 8-ribbed; seeds 0.7–1 mm. long; $n = 9$ (Lewis & Lewis).—Foothill Wd., from San Joaquin Co. to Tulare Co. May–June.

9. **C. speciòsa** Lewis & Lewis. [*Oenothera viminea* var. *parviflora* H. & A. *Godetia p.* Jeps., not *C. p.* Eastw. *G. v.* var. *margaritae* Jeps. *G. p.* var. *luteola* C. L. Hitchc.] Erect to decumbent, the stems mostly basally branched, 1–6 dm. long, puberulent above with short upwardly curled hairs; lvs. linear to linear-lanceolate, 1–6 cm. long, 2–6 mm. wide, sessile or short-petioled; infl.-axis erect; buds erect; fl.-tube funnelform, mostly 8–15 mm. long, the hair-ring in lower half; sepals lanceolate, mostly 8–15 mm. long, united and deflexed at anthesis or in pairs or individually; petals fan-shaped, purplish-red to lavender or cream, light near base, with bright red spot near or below the center, or spot ± obscured; caps. ± 4-sided, 1–2.5 cm. long, 1–2 mm. broad, sometimes short-beaked; seeds brown or grayish, 0.7–1 mm. long, short-crested; $n = 9$ (Lewis & Lewis, 1955).—Dry open places, below 3000 ft.; Foothill Wd.; Coast Ranges from Monterey and San Benito cos. to Santa Barbara Co. May–July.

Ssp. **immaculàta** Lewis & Lewis. Petals 1.5–2.5 cm. long, unspotted, upper part pinkish or red-lavender streaking into the white or cream lower third; $n = 9$ (Lewis & Lewis, 1955).—Old sand dunes at edge of Chaparral and Oak Wd.; between Pismo and Arroyo Grande, San Luis Obispo Co.

Ssp. **polyántha** Lewis & Lewis. Erect, with wandlike many-fld. branches; petals 1–2.7 cm. long, purple or lavender, or yellowish, usually lighter near base, conspicuously spotted purple-red near the center; $n = 9$ (Lewis & Lewis, 1955).—At 2000–5000 ft.; Foothill Wd.; Sierran Foothills of Fresno, Tulare, and Kern cos.

10. **C. Dàvyi** (Jeps.) Lewis & Lewis. [*Godetia quadrivulnera* var. *D.* Jeps.] Prostrate to decumbent, the stems 2–8 dm. long, simple or branched, sparsely puberulent above with upwardly curled hairs; lvs. rather crowded, oblanceolate to elliptic or obovate, obtuse, subentire, 1–2 cm. long, 2–7 mm. broad, subsessile; infl.-axis straight; buds erect; fl.-tube 2–5 mm. long; sepals lanceolate, 5–7 mm. long, reflexed in pairs or individually at anthesis; petals obovate to fan-shaped, 6–11 mm. long, lavender-pink, the basal part cream to white, unspotted; stamens yellowish; caps. 1.5–2.5 cm. long, 2–3 mm. broad, 8-ribbed; seeds ca. 1 mm. long, ± obscurely crested; $n = 17$ (Lewis & Lewis, 1955).—Mostly sandy places; N. Coastal Scrub, Coastal Strand, etc.; coast from Del Norte to San Mateo cos. June–July.

11. **C. prostràta** Lewis & Lewis. Prostrate or decumbent, the stems simple or branched, 2–5 dm. long, puberulent above with upwardly curled hairs; lvs. elliptic to oblanceolate, 1–2.5 cm. long, 4–8 mm. broad, mostly obtuse, subsessile; infl.-axis straight; buds erect; fl.-tube 4–7 mm. long, the ring of hairs in lower third; sepals lanceolate, 6–8 mm. long, 2–3 mm. wide, usually reflexed in 2's; petals truncate-obovate, or fan-shaped, 10–15 mm. long, lavender-pink, ± cream or yellowish toward base, usually with deltoid red-purple blotch near middle; fils. greenish, anthers cream; caps. quadrangular, 2–3 cm. long, 2.5–3 mm. broad; seeds brown or grayish, 1–1.5 mm. long, short-crested; $n = 26$

(Lewis & Lewis, 1955).—Coastal bluffs, etc.; Closed-cone Pine F., V. Grassland, etc.; Santa Cruz Co. to San Luis Obispo Co., Santa Rosa Id. April–July.

12. **C. purpùrea** (Curt.) Nels. & Macbr. [*Oenothera p.* Curt. *Godetia p.* G. Don. *G. Willdenowiana* Spach. *G. p.* var. *lacunora* Jeps. *G. p.* var. *Elmeri* Jeps.] Erect, 1–5 dm. high, mostly simple, glabrous and often white-shining below, puberulent above with upwardly curled hairs; lvs. ovate or elliptic to broadly lanceolate, 1.5–4.5 cm. long, 5–20 mm. broad, usually less than 5 times as long as broad; infl. congested, erect; fl.-tube 3–9 mm. long, with inner ring of hairs in lower third, pilose without; sepals lanceolate, 8–15 mm. long, puberulent and pilose, reflexed individually or in pairs; petals obovate, 1–2.5 cm. long, lavender to purple or purplish-red, often with darker cent. spot; stigma lobed to subglobose; caps. quadrangular, 1–2 cm. long, 3–4 mm. broad, conspicuously long-pubescent, 8-ribbed, rounded at base, sometimes short-beaked; seeds brown or grayish, 1–2 mm. long, well crested; *n* = 26 (Lewis & Lewis, 1955).—Open places, at low elevs.; V. Grassland, Coastal Strand, N. Coastal Scrub, etc.; near the coast from Marin Co. to Santa Barbara Co., inland from Solano to Placer and Stanislaus cos. April–June.

Ssp. **vimínea** (Dougl.) Lewis & Lewis. [*Oenothera v.* Dougl. *Godetia v.* Spach. *Clarkia v.* Nels. & Macbr. *G. lepida* Lindl. *Oe.* Arnottii T. & G. *G. Goddardii* Jeps. *G. quadrivulnera* var. *Setchelliana* Jeps. *Oe. v.* var. *intermedia* Kell.] Stems erect or decumbent; lvs. linear to lance-linear, 3–7 cm. long, less than ⅙ as broad; infl. lax, petals 1.5–2.5 cm. long, lavender to purple, usually with dark spot in upper part; *n* = 26 (Lewis & Lewis, 1955).—Dry open places, mostly below 5000 ft.; Chaparral, Foothill Wd., Yellow Pine F., Mixed Evergreen F., etc.; scattered in much of cismontane Calif.; w. Ore. May–July.

Ssp. **quadrivúlnera** (Dougl.) Lewis & Lewis. [*Oe. q.* Dougl. in Lindl. *G. q.* Spach. *Oe. q.* var. *hirsuta* Kell. *G. q.* f. *flagellata* Jeps.; var. *apiculata* Jeps.; var. *Hallii* Jeps.; var *rubrissima* Jeps.; var. *vacensis* Jeps. *Oe. decumbens* Dougl. *G. d.* Spach. *Oe. lepida* var. *parviflora* Wats. *G. purpurea* var. *parviflora* C. L. Hitchc. *G. micropetala* Greene. *G. Goddardii* f. *capitata* Jeps.; var. *miguelita* Jeps. *G. purpurea* var. *procera* Jeps. *G. sparsifolia* Jeps. *Oe. pulcherrima* var. *Brauntonii* Lév.] Stems erect; lvs. linear to lanceolate, 1.5–5 cm. long, less than ⅛ as wide; infl. lax or congested; petals 5–15 mm. long, lavender to purple, often deep red, the paler fls. often with a wedge-shaped or shield-shaped purple spot; stigma-lobes short; *n* = 26 (Håkansson, 1941).—The most common ssp., in open spots, below 6000 ft.; many Plant Communities; throughout cismontane Calif.; to L. Calif., Ariz., Wash. April–July.

13. **C. affìnis** Lewis & Lewis. Resembling *C. purpurea quadrivulnera* in stature, foliage, erect infl.; fl.-tube obconic, 1.5–4 mm. long; sepals 5–12 mm. long, usually united and deflexed to one side at anthesis; petals obovate, entire to erose, 5–15 mm. long, pale-to lavender-pink, sometimes darker, often flecked or penciled with purple; ovary slender, 1–2.5 cm. long, the caps. shallowly 8-grooved, subterete or 4-sided, 1.5–3 cm. long, 1.5–2 mm. broad; seeds 1–1.5 mm. long, short-crested; *n* = 26 (Lewis & Lewis, 1955). —Openings in Chaparral, Foothill Wd., etc.; Coast Ranges from Lake and Napa cos. to Ventura Co. April–May.

14. **C. Bóttae** (Spach) Lewis & Lewis. [*Godetia B.* Spach. *Oenothera B.* T. & G.] Erect, usually branched, 2–5 dm. tall, puberulent above with upwardly curled hairs; lvs. lanceolate to lance-linear, subentire to denticulate, 2–5 cm. long, 3–6 mm. broad, with petioles to 7 mm. long; infl.-axis recurved in bud, erect at anthesis; buds pendent; fl.-tube 1.5–3 mm. long, the ring of hairs near summit; sepals lanceolate, 15–20 mm. long, united and deflexed to one side at anthesis; petals fan-shaped, entire to erose, 1–3 cm. long, pinkish-lavender distally, white basally, sometimes with darker flecks; anthers lavender, the outer mostly with blue, the inner with cream pollen; caps. subterete to quadrangular, 4-grooved when immature, 2–4 cm. long, 2–2.5 mm. broad, quite straight; seeds brown, ca. 1 mm. long, scarcely if at all crested; *n* = 9 (Håkansson, 1941).—Coastal Sage Scrub, Closed-cone Pine F., Mixed Evergreen F.; mostly near the coast, Monterey Co. May–July.

15. **C. cylíndrica** (Jeps.) Lewis & Lewis. [*Godetia Bottae* var. *c.* Jeps. *G. c.* Hitchc.] Much like *C. Bottae;* fl.-tube 3–5 mm. long, the ring of hairs near the middle; sepals 1–2.5 cm. long; petals 1–3.5 cm. long, purple to pinkish-lavender, shading to white near middle, usually flecked with red-purple and with a bright purple-red base; outer anthers

with bluish, the inner with yellow or bluish-gray pollen; caps. 2–5 cm. long, 1–2 mm. broad, usually with a beak (to 1 cm. long), 4-grooved when immature, often enlarged upward; seeds brown, 1–1.5 mm. long, the crest to 0.1 mm. long; $n = 9$ (Håkansson, 1943).—Dry slopes and flats, below 4000 ft.; V. Grassland, Foothill Wd., Chaparral; n. Los Angeles Co. to Madera, Mariposa and Monterey cos. April to July.

16. **C. bilòba** (Durand) Nels. & Macbr. [*Oenothera b.* Durand. *Godetia b.* Wats.] Erect, 3–10 dm. tall, simple or branched above, puberulent above with upwardly curled hairs; lvs. lanceolate to linear, entire to denticulate, 2–6 cm. long, 5–8 mm. broad, with petioles 5–15 mm. long; infl.-axis recurved in bud, straightening at anthesis; buds pendulous; fl.-tube 1–4 mm. long, with hair-ring near summit; sepals 6–17 mm. long, united and deflexed to 1 side at anthesis; petals cuneate, 1–2.5 cm. long, 0.6–1.8 cm. broad, pale pink to purplish-pink, deeply bilobed, the lobes up to ½ their total length; outer stamens with blue, the inner with blue or white pollen; caps. straight, 4-sided, 1–2.5 cm. long, 1.5–2 mm. broad, 8-ribbed, sessile or short-pedicelled; seeds 1 mm. long, brown, scarcely crested; $n = 8$ (Håkansson, 1941).—Open places, below 4000 ft.; Foothill Wd., Yellow Pine F.; Contra Costa Co., Eldorado to Mariposa cos. May–July.

Ssp. **austràlis** Lewis & Lewis. Lvs. 2–6 mm. broad; sepals 7–11 mm. long; petals narrowly cuneate, 1–2 cm. long, 0.3–1.2 cm. broad, bright pink to magenta, shallowly lobed, the lobes ca. ¼ the length; $n = 8$ (Lewis & Lewis, 1955).—Dry slopes, 1000–2000 ft.; Chaparral, Foothill Wd.; Merced R. drainage, Mariposa Co.

Ssp. **Brandègeae** (Jeps.) Lewis & Lewis. [*Godetia Dudleyana* f. *B.* Jeps.] Lvs. 5–8 mm. broad, denticulate; sepals 9–15 mm. long; petals broadly cuneate, lavender to pale pink, the lobes less than ⅓ the petal length; $n = 8$ (Lewis & Lewis, 1955).—Below 2500 ft.; Foothill Wd., Chaparral; Butte to Eldorado cos.

17. **C. lingulàta** Lewis & Lewis. Like *C. biloba* ssp. *australis;* sepals 7–10 mm. long; petals oblanceolate, obtuse, entire or minutely notched, 1–2 cm. long, 5–8 mm. broad, bright pink, sometimes flecked with red; $n = 9$ (Lewis & Lewis, 1955).—Known from 2 places on Merced R., Mariposa Co.

18. **C. modésta** Jeps. [*Phaeostoma m.* Heller. *Godetia epilobioides* var. *m.* Jeps.] Erect, 2–7 dm. tall, simple or branched, puberulent above with upcurled hairs; lvs. linear to lance-linear or elliptic, entire to denticulate, 2–4 cm. long, 2–7 mm. broad, on petioles 5–15 mm. long; infl.-axis reflexed in bud, erect at anthesis, the buds pendulous; fl.-tube 1–3 mm. long, the hair-ring near upper end; sepals 8–11 mm. long, 1.5–2 mm. broad, united and deflexed to one side at anthesis; petals rhomboid to oblanceolate, 8–12 mm. long, 3–7 mm. broad, bright pink, usually with some darker flecks, usually acute, with basal claw ca. 1 mm. long; pollen of outer anthers blue, of inner cream; caps. 1.5–3 cm. long, 1–2 mm. broad, subterete, sessile or short-pedicelled, shallowly 8-ribbed when young; seeds brown, 0.8–1.0 mm. long, obscurely or short-crested; $n = 8$ (Lewis & Lewis, 1955).—Dry slopes, below 2500 ft.; Foothill Wd.; Coast Ranges from Tehama and Lake cos. to Santa Barbara Co., Sierran foothills from Stanislaus Co. to Tulare Co. April–May.

19. **C. Dudleyàna** (Abrams) Macbr. [*Godetia D.* Abrams. *G. Bottae* var. *usitata* Jeps.] Erect, 3–7 dm. tall, simple or branched, puberulent above with upcurled hairs; lvs. lanceolate, subentire to denticulate, 1.5–7 cm. long, 3–15 mm. broad, with petioles 3–10 mm. long; infl.-axis recurved in bud, erect in anthesis; buds pendulous; fl.-tube slender, 1–3 mm. long, the inner hair-ring near summit; sepals 10–22 mm. long, united and deflexed to one side at anthesis; petals fan-shaped, entire to emarginate, 10–28 mm. long, lavender-pink, often flecked with red, lighter toward base; outer stamens with bluish or whitish pollen, inner with cream pollen; caps. 4- or 8-sided, ribbed, 1–3 cm. long, 2–2.5 mm. broad, sessile or nearly so; seeds brown, ca. 1 mm. long, short-crested; $n = 9$ (Lewis & Lewis, 1955).—Plants with petals white-streaked, on dry slopes below 4500 ft., Foothill Wd., Yellow Pine F., Sierran foothills from Tuolumne to Kern cos.; those with brighter color and usually not so streaked, below 6000 ft., Coastal Sage Scrub, Chaparral, Yellow Pine F., Los Angeles to w. Riverside cos. May–July.

20. **C. epilobioìdes** (Nutt.) Nels. & Macbr. [*Oenothera e.* Nutt. in T. & G. *Godetia e.* Wats.] Erect, 2–7 dm. tall, simple or branched, sparsely puberulent above with upwardly curled hairs; lvs. linear to narrowly lanceolate or oblanceolate, denticulate to subentire, 15–25 mm. long, 2–4 mm. broad, on short petioles; infl.-axis reflexed in bud, erect in

anthesis, buds pendulous; fl.-tube 0.5–1.5 mm. long, the ring of hairs in upper part or lacking; sepals 6–10 mm. long, united or in pairs at anthesis; petals white, ± pinkish in age, obovate, 5–10 mm. long, scarcely clawed; anthers white or cream; caps. slender, 4-sided, 1–3 cm. long, 1–1.5 mm. wide, on a pedicel to 1 cm. long, very short-beaked; seeds brown, 0.5–1 mm. long, obscurely or short crested; $n = 9$ (Lewis & Lewis, 1955). —Shaded places, below 2500 ft.; Chaparral, Coastal Sage Scrub, S. Oak Wd., etc.; cismontane Calif. from San Francisco Co. to San Diego Co.; Santa Cruz, Santa Catalina ids.; Baja Calif., Ariz. March–May.

21. **C. símilis** Lewis & Ernst. Resembling *C. epilobioides*, the lvs. 2–4 cm. long, 3–8 mm. broad; fl.-tube 1.5–2 mm. long; petals 6–12 mm. long, oblanceolate to rhombic or obovate, whitish to light pink, flecked with purple in lower half, with a basal claw ca. 1 mm. long; caps. 1.5–3 cm. long, 1–1.5 mm. broad, sessile or nearly so; seeds brown, ca. 1 mm. long, obscurely or short-crested; $n = 17$ (Lewis & Lewis, 1955).—Shaded places, below 3500 ft.; Chaparral, Foothill Wd., S. Oak Wd.; San Benito Co., w. Fresno Co., Los Angeles Co. to Riverside and San Diego cos. April–June.

22. **C. defléxa** (Jeps.) Lewis & Lewis. [*Godetia d.* Jeps. *G. Bottae* var. *d.* C. L. Hitchc.] Erect, 3–9 dm. tall, stout, mostly glabrous; lvs. lanceolate, 3–8 cm. long, 5–18 mm. wide, sparsely puberulent, short-petioled; infl.-axis erect, buds deflexed; fl.-tube 2–3 mm. long, the ring of hairs near summit; sepals 10–20 mm. long, united and deflexed to one side at anthesis; petals fan-shaped, entire or erose, 1–3 cm. long, pale lavender to pinkish-lavender, mostly white toward base, red-flecked, scarcely clawed; outer anthers lavender, inner yellowish, ± flecked with red; caps. subterete to quadrangular, 3–4 cm. long, 2 mm. broad, scarcely beaked, from sessile to having a pedicel 3 cm. long, obscurely 4-grooved when young; seeds brown or gray, 1.2–1.8 mm. long, the crest ± developed; $n = 9$ (Håkansson, 1941).—Dry openings, below 3000 ft.; Chaparral, Coastal Sage Scrub, Foothill Wd., S. Oak Wd.; Coast Ranges from s. Monterey Co. to Orange and w. Riverside cos. April–June.

23. **C. delicàta** (Abrams) Nels. & Macbr. [*Godetia d.* Abrams.] Erect, 2–7 dm. tall, the stems slender, simple or branched, subglabrous and glaucous above; lvs. lanceolate to ovate, 1.5–4 cm. long, 4–15 mm. wide, serrate or denticulate or upper entire, subglabrous, with petioles 5–10 mm. long; infl.-axis erect; buds deflexed; fl.-tube 2 mm. long, the inner ring of hairs near summit; sepals 6–9 mm. long, united and deflexed to one side at anthesis; petals spatulate to obovate, 8–12 mm. long, 4–8 mm. broad, pale pink to bright rose-lavender, lighter in basal half, short-clawed; outer stamens ± orange-red, inner cream; caps. quadrangular, 1.5–3.5 cm. long, ca. 2 mm. broad, subsessile, 8-grooved when young; seeds brown, 1–1.5 mm. long, the crest obscure; $n = 18$ (Lewis & Lewis, 1955).—Dry slopes, below 4000 ft.; mostly Chaparral, S. Oak Wd.; San Diego Co.; L. Calif. May–June.

24. **C. unguículàta** Lindl. [*Phaeostoma u.* Lilja. *C. elegans* Dougl., not Poir. *P. e.* Lilja. *Oenothera e.* Lév. *P. Douglasii* Spach. *C. Eiseniana* Kell.] Erect, 3–10 dm. high, the stems simple or branched, glabrous and glaucous throughout; lvs. lanceolate to elliptic or ovate, 1–6 cm. long, 0.5–2 cm. wide, with petioles 3–10 mm. long; infl.-axis erect; buds deflexed; fl.-tube obconic to campanulate, 2–5 mm. long, with broad band of inner hairs in upper half and other hairs at margin; sepals oblanceolate to spatulate, 10–16 mm. long, united and deflexed to one side at anthesis, subglabrous to densely pilose; petals deltoid to rhombic, 1–2 cm. long, lavender-pink to salmon or purplish or dark red-purple, with a slender claw almost half the total length; anthers 8, the outer red-orange to dull red, the inner lighter to whitish; caps. straight or curved, subterete to subquadrangular, 1.5–3 cm. long, 2–3 mm. broad, scarcely beaked, sessile or very short-petioled, 8-ribbed; seeds brown, 1–1.5 mm. long, minutely crested; $n = 9$ (Lewis, 1951).—Common on dry often shaded slopes, below 5000 ft.; Chaparral, Foothill Wd., V. Grassland, Coastal Strand, etc.; Coast Ranges from Mendocino and Lake cos. to n. San Diego Co., Sierran foothills from Butte to Kern cos. May–June.

25. **C. éxilis** Lewis & Vasek. Near to *C. unguiculata;* fl.-tube 1–3 mm. long, the ring of hairs at or above the middle; sepals 5–13 mm. long, mostly subglabrous; petals 5–15 mm. long, the slender claw as long as limb, the latter rhombic, lavender-pink to white, often with a darker red-purple spot at base; anthers white or outer sometimes reddish; seeds ca. 1 mm. long; $n = 9$ (Lewis & Lewis, 1955).—Mostly in shaded places, below 2000 ft.; Foothill Wd.; Sierran foothills of Tulare and Kern cos. April–May.

26. **C. Xantiàna** Gray. [*Phaeostoma X.* A. Nels. *C. parviflora* Eastw. *P. p.* A. Nels.] Erect, 2–8 dm. tall, the stems simple or branched above, glabrous and glaucous; lvs. linear to lance-linear, subentire, 2–6 cm. long, 1.5–8 mm. wide, sessile or nearly so; infl.-axis erect; buds deflexed; fl.-tube 2–5 mm. long, obconic to campanulate, with broad band of white hairs within at upper margin; sepals oblanceolate, 6–20 mm. long, united and deflexed to one side at anthesis; petals 6–20 mm. long, lavender to red-purple, 2-lobed with a subulate tooth 1–3 mm. long in the sinus, conspicuously clawed, upper petals often with red-purple spot; anthers 8, lavender to purple; caps. 1.5–2.5 cm. long, 1.5–2.5 mm. broad, 4-sided, usually with short slender beak, nearly or quite sessile, conspicuously 8-ribbed when young; seeds brown, 1.3–1.5 mm. long, minutely crested; n = 9 (Lewis & Lewis, 1955).—Dry slopes, 800–5700 ft.; Foothill Wd., Chaparral, etc.; e. base of Sierra Nevada in Inyo Co. to n. slope of San Gabriel Mts. in Los Angeles Co., n. to Tulare Co. at w. base of Sierra Nevada. May–June.

27. **C. rhomboìdea** Dougl. [*Oenothera r.* Lév. *Phaeostoma r.* A. Nels. *C. gauroides* Dougl. *Godetia latifolia* Nels. & Kenn. *P. atropurpureum* Heller.] Erect, simple or few-branched, 2–11 dm. tall, puberulent above with upcurled hairs; lvs. few, subopposite, lance-ovate to ovate-oblong or elliptic, 2–7 cm. long, 0.4–2.3 cm. wide, acute, entire or nearly so, with petioles 1–2.5 cm. long; infl.-rachis recurved, erect at anthesis; buds nodding; fl.-tube 1–3 mm. long, with white hairs at summit; sepals green, usually distinct at anthesis, 6–8 mm. long; petals 6–12 mm. long, pinkish-lavender, with or without darker flecks, often red at base, the limb deltoid-rhombic, often obscurely 3-lobed, the claw broad, with a pair of projections near the base; stamens 8, subtended by ciliated scales to 1.5 mm. long; anthers lavender to purple; caps. quadrangular, straight to curved, 1–2.5 cm. long, 2–2.5 mm. broad, with a beak to 3 mm. long, sessile or short-pedicelled; seeds brown or grayish, sometimes mottled, 1–1.5 mm. long, short-crested; n = 12 (Lewis & Lewis, 1955).—Rather dry slopes, below 8000 ft.; Red Fir F., Yellow Pine F., Foothill Wd., N. Oak Wd., Chaparral, etc.; cismontane Calif. from San Diego Co. to Del Norte and Modoc cos.; to L. Calif., Ariz., Wash., Mont., Utah. May–July.

28. **C. Mildrèdiae** (Heller) Lewis & Lewis. [*Phaeostoma M.* Heller.] Resembling *C. rhomboidea*, sepals 13–19 mm. long, reflexed separately or united and deflexed to one side at anthesis; petals 14–21 mm. long, red-purple to lavender, with purple flecks near base of limb; fils. having basal scales to 3 mm. long; caps. 2–3 cm. long, often tipped or streaked red; seeds 1.5–1.8 mm. long; n = 7 (Lewis & Lewis, 1955).—Coarse granite sand, at ca. 2000 ft.; Yellow Pine F.; N. Fork of Feather R. and Shasta Lake region, Butte, Plumas, and Shasta cos. June–July.

29. **C. virgàta** Greene. Like *C. rhomboidea;* lvs. lanceolate, 3–12 mm. broad; fl.-tube 3–4 mm. long; sepals 10–15 mm. long; petals 7–14 mm. long, rather narrow, purple to dark lavender, mottled or spotted with red-purple scales at base of the 8 stamens 1–2 mm. long; anthers mostly red-purple; stem becoming erect before the fls. open; anthers maturing ahead of stigmas; n = 5 (Lewis & Lewis, 1955).—Foothill Wd. and lower Yellow Pine F.; foothills of the Sierra Nevada from Eldorado to Mariposa cos. May–July.

30. **C. concínna** (F. & M.) Greene. [*Eucharidium c.* F. & M. *E. grandiflorum* F. & M. *C. g.* Greene.] Erect, freely branched to simple, strigulose above; lvs. lance-ovate to broadly elliptic, subentire, 1.5–5 cm. long, 0.6–2 cm. wide, with petioles an additional 0.5–2 cm. long; infl.-axis erect or bent to one side, straight at anthesis; fls. often crowded; fl.-tube slender, red, 1.3–2.5 cm. long; sepals linear to narrow-oblanceolate, 1–2 cm. long, 1–2 mm. broad, united at tips at anthesis; petals deep bright pink, often with white or purple streaks, 1–2.5 cm. long, 0.4–1.2 cm. wide, 3-lobed, the middle lobe spatulate, equal to or slightly longer than the lateral, claw ± conspicuous; anthers 4, red to lavender, ciliate, curling after dehiscence; caps. straight or ± curved, quadrangular, 1.5–2 cm. long, 2 mm. broad, 8-grooved when immature; seeds red-brown, ca. 2 mm. long, flattened, prominently crested; n = 7 (Schwemmle, 1926).—Dry loose soil below 4000 ft.; Mixed Evergreen F., N. Oak Wd., Douglas-Fir F., etc.; Coast Ranges from Santa Clara Co. to Humboldt and Siskiyou cos., Sierran foothills in Butte and Yuba cos. May–July.

31. **C. Brèweri** (Gray) Greene. [*Eucharidium B.* Gray. *Oenothera Eucharidium* Lév. *C. Saxeana* Greene.] Much like *C. concinna*, to ca. 2 dm. tall; lvs. 1–7 mm. broad; floral tube 2–3.5 cm. long; sepals 1.5–2.5 cm. long, united and deflexed to one side

at anthesis; petals pink, 1.5–2.6 cm. long, scarcely clawed, 3-lobed, the outer 2 lobes broad, the inner much longer, linear to spatulate; stamens 4, with clavate fils., anthers striped red and yellow and aging dark purple; caps. 1.5–4 cm. long; seeds 2–3 mm. long; $n = 7$ (Lewis & Lewis, 1955).—Talus and dry slopes, below 4000 ft.; Foothill Wd., Chaparral; inner Coast Ranges from Alameda to Fresno cos. April–May.

7. Oenothèra L. Evening-Primrose

Annual to perennial, caulescent or acaulescent herbs, sometimes suffruticose; lvs. alternate or basal. Fls. white to yellow or rose, often aging reddish or purplish. Fl.-tube prolonged beyond the ovary, deciduous after anthesis. Sepals 4, reflexed in anthesis. Petals 4. Stamens 8, equal, or if unequal, those opposite the petals shorter; anthers mostly versatile. Stigma with 4 linear lobes, or discoid or capitate. Caps. membranous to woody, straight to curved or coiled, 4-loculed, mostly 4-valved, and dehiscent. Seeds many, naked. Ca. 200 spp., of the New World, mostly of temp. regions. (Greek, meaning wine-scenting, a name given to some now unknown plant once used for that purpose.)

(Munz, P. A., pp. 190–206 in Abrams: Ill. Fl. Pacific States, vol. 3, 1951.)
A.   Stigma with 4 linear lobes; fls. opening in late afternoon.
    B.   Caps. terete or round-angled (winged above in *Oe. speciosa*); stems well developed and bearing lvs. and fls.
        C.   Fls. yellow, the buds erect.
            D.   The caps. angled, tapering from base toward apex; seeds prismatic-angled. (Subgenus Oenothera)
                E.   Petals 1–2 cm. long, not usually turning reddish in age. Extreme n. Calif.
                                                                                    1. *Oe. strigosa*
                EE. Petals 2.5–5 cm. long, turning reddish in age.
                    F.   Fl.-tube 3–5 cm. long.
                        G.   Cauline lf.-blades at least ⅓ as wide as long and strongly crinkled; petals 3.5–5 cm. long. N. Coast Ranges .... 3. *Oe. erythrosepala*
                        GG. Cauline lf.-blades usually less than ¼ as wide as long, mostly plane; petals 2.5–4 cm. long. Widespread ... 2. *Oe. Hookeri*
                    FF. Fl.-tube 8–12 cm. long. E. Mojave Desert .... 4. *Oe. longissima*
            DD. The caps. terete, not narrowed in upper part; seeds not angled. (Subgenus Raimannia)
                E.   Plants erect, leafy, biennial or perennial, 2–6 dm. tall; lvs. denticulate; sepals 12–20 mm. long; petals 12–20 mm. long .... 5. *Oe. stricta*
                EE. Plants mostly annual, semiprostrate, 1–7 dm. tall; lvs. sinuate-pinnatifid to subentire; sepals 6–12 mm. long; petals 5–18 mm. long .. 6. *Oe. laciniata*
        CC. Fls. white to rose, the buds nodding.
            D.   Caps. sterile and slender in lower part, thicker, fertile and ± winged in upper; seeds in more than 2 rows in each locule; fl.-tube 1–2 cm. long. Garden escape. (Subgenus Hartmannia) .... 7. *Oe. speciosa*
            DD. Caps. cylindric, sessile, not sterile in lower part; seeds in 1 row in a locule; fl.-tube 2–4 cm. long. Native. (Subgenus Anogra)
                E.   Plants annual (spring or winter), from a deep taproot; basal lvs. rhombic, 2–8 cm. long; caps. woody, with exfoliating epidermis .. 8. *Oe. deltoides*
                EE. Plants perennial, mostly from running underground rootstocks.
                    F.   Fl.-buds conspicuously long-pointed; lvs. 3–12 cm. long, runcinate-pinnatifid. Sand dunes, Contra Costa Co.
                                                                  8. *Oe. deltoides* var. *Howellii*
                    FF. Fl.-buds rather blunt; lvs. mostly shorter, not so deeply divided.
                        G.   Lvs. crowded, deltoid-ovate, 1–3 cm. long; old fl.-shoots becoming buried and resuming growth at their tips. Dunes, Eureka V., e. Inyo Co.   8. *Oe. deltoides* ssp. *eurekensis*
                        GG. Lvs. of stems mostly not crowded, oblong to lanceolate, 1–6 cm. long; old fl.-stems not resuming growth after being buried. Mostly cismontane .... 9. *Oe. californica*
    BB. Caps. crested or winged; plants acaulescent or nearly so, cespitose.
        C.   Caps. tapering toward the apex, not enlarged in upper half; seeds with a deep furrow along the raphe. (Subgenus Pachylophis)
            D.   Fls. white.
                E.   Petals 2–4 cm. long; fl.-tube 2–8 cm. long; heavy perennial
                                                                                    10. *Oe. caespitosa*
                EE. Petals 0.8–1.2 cm. long; fl.-tube 2.5–4 cm. long; winter annual.
                                                                                    13. *Oe. cavernae*
            DD. Fls. yellow.
                Caps. 3.5–6 cm. long, winged on angles on lower half; seeds with broad flat open depression along the raphe. Yellow Pine Belt .... 11. *Oe. xylocarpa*
                EE. Caps. 1.8–3.5 cm. long, not winged; seeds with narrow raphal groove. Desert plants .... 12. *Oe. primiveris*

    CC. Caps. enlarged in upper half, woody, winged especially above, 1–2 cm. long; fls.
        yellow; seeds cuneate-obovoid. (Subgenus Lavauxia) .... 14. *Oe. flava*
AA. Stigma discoid or capitate, not deeply lobed; fls. opening in morning.
    B.   Fl.-tube ca. 1 mm. long, orange and pubescent within and with a lobed disk; plants erect,
        annual, 2–10 dm. tall, cruciferlike. (Subgenus Eulobus) .... 15. *Oe. leptocarpa*
    BB. Fl.-tube not lined with a lobed disk.
        C.   Plants mostly acaulescent and cespitose; ovary fertile in lower portion only, the upper
            portion sterile, tubular, elongate, subfiliform and simulating a fl.-tube; fls. yellow.
            (Subgenus Taraxia)
            D.   Caps. broadly and truncately 4-winged, not over 1 cm. long; seeds obovoid;
                annuals.
                E.   Fls. small, the petals 2.5–3 mm. long; sterile portion of the ovary 10–15 mm.
                    long; epidermis of stems exfoliating .... 16. *Oe. Palmeri*
                EE. Fls. larger, the petals 8–12 mm. long; sterile portion of ovary 12–35 mm.
                    long; epidermis not exfoliating readily .... 17. *Oe. graciliflora*
            DD. Caps. somewhat cylindrical, at most angled, not winged, attenuate gradually at
                tip into sterile portion mostly over 1 cm. long; seeds not pointed at one end;
                perennials.
                E.   Lvs. entire or with few teeth; caps. glabrous.
                    F.   Plants glabrous or nearly so; sterile portion of mature ovary 2–6 cm.
                      long; caps. oblong-ovoid, 5 mm. or more thick; seeds 3 mm. long,
                      minutely pitted. Montane .... 18. *Oe. heterantha*
                  FF. Plants minutely pubescent, especially on veins and lf.-margins; sterile
                    portion of ovary 5–12 cm. long; caps. linear, not over 3 mm. thick;
                    seeds 2 mm. long, scurfy. Coastal .... 19. *Oe. ovata*
              EE. Lvs. deeply pinnatifid; caps. densely pubescent.
                  F.   Sterile portion of ovary 25–80 mm. long; petals 10–15 mm. long; caps.
                    ovoid, straight .... 20. *Oe. tanacetifolia*
                  FF. Sterile portion of ovary 10–25 mm. long; petals 6–8 mm. long; caps.
                    fusiform, frequently curved .... 21. *Oe. breviflora*
    CC. Plants caulescent; ovary fertile to near summit, not prolonged into a long sterile portion,
        though sometimes beaked.
        D.   Caps. nearly or quite sessile. (Subgenus Sphaerostigma; see also *Oe. kernensis* and
            cardiophylla)
            E.   Fls. white (yellowish in *minor* and 1 var. of *decorticans*), often drying
                pinkish, borne in terminal spikes.
                F.   Caps. cylindrical, terete, linear, not thickened in lower portion, scarcely
                  if at all coiled, not noticeably attenuate at tip.
                    G.   Petals 5–7 mm. long, suborbicular; style exceeding corolla, 10–13
                      mm. long; fl.-tube 4–6 mm. long; caps. refracted or spreading, oc-
                      casionally coiled .... 22. *Oe. refracta*
                    GG. Petals 3 mm. long, spatulate; style shorter than corolla, 3–4 mm.
                    long; fl.-tube 2.5–3 mm. long; caps. divaricately spreading
                      23. *Oe. chamaenerioides*
                FF. Caps. not strictly cylindrical, but somewhat enlarged at base and
                  attenuate toward tip.
                  G. Mature caps. usually distinctly contorted and coiled, not merely
                    bent and curved, quite slender, not subfusiform in shape (see
                    also *decorticans desertorum*).
                    H.   Fls. minute; petals 2 mm. long; fl.-tube 2 mm. long; fils.
                      distinctly unequal. Modoc Co. .... 24. *Oe. minor*
                    HH. Fls. larger; petals 3.5–5 mm. long; fl.-tube 3–8 mm. long;
                    fils, subequal. Lassen to Inyo cos. ...... 25. *Oe. alyssoides*
                GG. Mature caps. merely curved or bent, not distinctly contorted or
                  coiled, subfusiform in shape.
                    H.   Lvs. largely near base of plant, subglabrous, lance-ovate to
                      oblanceolate; stems glabrous or nearly so, with promptly ex-
                      foliating epidermis; caps. 15–25 mm. long; seeds ashy, linear-
                      obovoid .... 26. *Oe. decorticans*
                    HH. Lvs. well distributed, glandular-pubescent to glandular-
                    villous, ovate to oblong-ovate; stems glandular; epidermis ex-
                    foliating tardily if at all; caps. 10–15 mm. long; seeds brown-
                    ish, rhomboid-prismatic .... 27. *Oe. Boothii*
        EE. Fls. yellow, often drying greenish, borne in axils of foliage-lvs.
            F.   Caps. terete, cylindrical or subfusiform, but not quadrangular; lvs.
                 narrow, 1–4 mm. wide, usually linear-oblong.
                G.   Plant with several naked fine often capillary stems, each bearing
                    a leafy infl. at the tip; caps, subfusiform, almost straight, 5–8 mm.
                    long. Ne. Calif. .... 28. *Oe. andina*
                GG. Plants with stems leafy from the base; caps. terete, straight or
                  coiled, 15–40 mm. long.
                  H.   Fls. small; petals 2.5–3.5 mm. long; sepals 1.5–3.5 mm. long
                      29. *Oe. contorta*
                  HH. Fls. larger; petals 5–15 mm. long; sepals 3–12 mm. long
                      30. *Oe. dentata*
        FF. Caps. quadrangular; lvs. 5–20 mm. wide, lanceolate to ovate.

G. Fls. small; petals 1.5–7 mm. long.
    H. Mature caps. oblong-pyramidal, 10–12 mm. long, nearly
        straight. San Clemente Id. . . . . 32. *Oe. guadalupensis*
    HH. Mature caps. curved or contorted, 15–40 mm. long. Wide-
        spread . . . . 33. *Oe. micrantha*
GG. Fls. larger; petals 8–22 mm. long.
    H. Plants of sea bluffs and inland, greenish except in a desert
        form; cauline lvs. lanceolate to lance-ovate, acute, ± crisped,
        thin . . . . 34. *Oe. bistorta*
    HH. Plants of sea beaches, grayish to silvery (except in var.
        nitida); cauline lvs. lance-oblong to orbicular-ovate, obtuse,
        not crisped, thick in texture . . . . 35. *Oe. cheiranthifolia*
DD. Caps. definitely and plainly pedicelled. (Subgenus *Chylismia*)
    E. Seeds oblong and with an incurved wing so as to appear boat-shaped, cellular-
        pubescent; small slender plants, villous below, finely glandular-pubescent
        above with pinkish-white axillary fls. 4–5 mm. across. Inyo Co.
                                                                36. *Oe. pterosperma*
    EE. Seeds obovoid, rounded or angled, not winged; fls. not axillary but in
        terminal racemes or panicles.
        F. Lvs. orbicular-cordate, well distributed, not at all pinnatifid
                                                                37. *Oe. cardiophylla*
        FF. Lvs. ovate, oblong or lanceolate, commonly pinnatifid and near the base
            of the plant (except in *Oe. kernensis*).
            G. Caps. linear, elongate, usually over 2 cm. long.
                H. Lvs. not in basal rosette, the cauline secund; plant 8–15 cm.
                    high. Kern Co. . . . . 31. *Oe. kernensis*
                HH. Lvs. in basal rosette; plants 10–60 cm. tall.
                    I. Stems coarse, commonly branched at base only if at all;
                        pedicels 3–15 mm. long; caps. linear, widely spreading,
                        commonly 5–9 cm. long; anthers hairy
                                                                38. *Oe. brevipes*
                    II. Stems slender, commonly branched above; pedicels capil-
                        lary, 10–25 mm. long; caps. linear, 15–35 mm. long;
                        anthers glabrous . . . . 41. *Oe. multijuga*
            GG. Caps. somewhat clavate, usually less than 2 cm. long.
                H. Branches in well developed plants capillary and arising
                    at base of plant only, not capillary; caps. 10–25 mm. long;
                    anthers linear, beset with scattered white hairs; style longer
                    than petals.
                    I. Stems slender; fls. few, not crowded; lvs. ovate, sub-
                        entire; petals usually less than 4 mm. long
                                                                40. *Oe. scapoidea*
                    II. Stems fairly coarse; fls. crowded in close terminal
                        clusters; lvs. frequently with supplementary pinnules on
                        petioles; petals 4–7 mm. long . . . . . 39. *Oe. clavaeformis*
                HH. Branches in well developed plants capillary and arising
                    freely throughout the plant; anthers oblong, glabrous; style
                    not longer than petals . . . . 42. *Oe. heterochroma*

1. **Oe. strigòsa** Mkze. & Bush. [*Onagra s.* Rydb. *Oe. Rydbergii* House.] Biennial,
grayish-strigose throughout, erect, largely simple, 3–10 dm. tall, ± hirsute, sometimes
tinged with red and with red pustules at base of longer hairs; rosette-lvs. spatulate,
obtuse, 3–10 cm. long, 1–2 cm. wide, with petioles an additional 1–3 cm. long; cauline
lvs. lanceolate, acute, repand-denticulate, shorter-petioled; infl. with leafy lanceolate
subsessile bracts 1–5 cm. long; fls. many, vespertine, in elongate interrupted spikes;
bracts exceeding mature caps.; fl.-tube 3–4 cm. long, pubescent within, often hirsute
without; sepals strigose and hirsute, 10–15 mm. long, with free tips ca. 2 mm. long;
petals yellow, broadly obcordate, 1.2–2 cm. long; stamens subequal, ca. as long as petals,
glabrous; style pubescent near base; stigma-lobes 5–7 mm. long; caps. 2.5–3.5 cm. long,
tapering slightly; seeds red-brown, obtusely angled, irregular, 1–1.5 mm. long.—Moist
places; Mixed Evergreen F.; along the Klamath R., Siskiyou Co.; to e. Wash., Minn.,
Kans. July–Aug.

2. **Oe. Hóokeri** T. & G. [*Onagra H.* Small. *Oe. Jepsonii* Greene. *Oe. franciscana*
Bartlett.] Biennial, ± branched, 4–12 dm. high, the stems stout, usually red, erect to
ascending, abundantly muricate-hirsute and finely strigose; rosette-lvs. with rather little
red, elliptic-oblanceolate, sinuate-dentate, strigose, often with pilose midribs, the blades
plane to ± wrinkled, 10–18 cm. long, on petioles 5–10 cm. long; cauline lvs. lanceolate,
± crowded, many, acute, 6–12 cm. long, subsessile or short-petioled; infl. dense, simple
or branched, 1–4 dm. long, strigose, hirsute and glandular-pubescent; lowest bracts
broadly lanceolate, exceeding young buds, the upper narrower, shorter, spreading, hairy;

fl.-tube 3.5–4 cm. long; sepals 2–4 cm. long, the tips 3–6 mm. long; petals pale yellow, aging orange-red, 2.5–4 cm. long, retuse; anthers 12–14 mm. long; stigma-lobes 5–6 mm. long; caps. mostly 2.5–4.4 cm. long, 5–5.5 mm. thick at base, soft-hairy, strigulose and red-papillose; seeds angled, 1–1.4 mm. long; $n = 7$ (Cleland, 1935).—Moist places; Coastal Strand, N. Coastal Scrub, Mixed Evergreen F., etc.; Coast Ranges from Lake and Trinity cos. to San Luis Obispo Co. June–Sept.

KEY TO SUBSPECIES

A. Pubescence mostly appressed, not spreading or gland-tipped; stems scarcely or not muricate, 1–2.5 m. tall. S. Calif. . . . . ssp. *grisea*
AA. Pubescence largely loose, with some spreading hairs; stems muricate, hirsute.
    B. Sepals under a hand lens and in terminal half obviously papillose at base of some of longer hairs which are usually numerous and tend to conceal the short gland-tipped ones.
        C. Fls. small; sepals 2–3 cm. long; petals 2–2.5 cm. long. Nw. Calif. . . . . ssp. *Wolfii*
        CC. Fls. normally larger except in old infls.
            D. Sepal-tips 3–6 mm. long, the buds attenuate; sepals 3–3.5 cm. long. Trinity and Lake cos. to San Luis Obispo Co. . . . . *Oe. Hookeri*
            DD. Sepal-tips 1–2.5 mm. long, the buds blunt; sepals mostly 2–2.5 cm. long. Coast from San Mateo Co. to San Luis Obispo Co. . . . . ssp. *montereyensis*
    BB. Sepals under a lens slightly or not papillose, with few longer hairs and shorter gland-tipped ones, sometimes almost glabrous.
        C. Fl.-tube and sepals green; plant mostly less than 1 m. tall, simple or branched only below; lvs. plane. Montane . . . . ssp. *angustifolia*
        CC. Fl.-tube and sepals green; plant usually branched throughout and up to 2.5 m. tall; stem-lvs. crinkled. Below Yellow Pine F. . . . . ssp. *venusta*

Ssp. **montereyénsis** Munz. Bushy, spreading to horizontal to suberect, 3–15 dm. tall, muricate-hirsute, strigose and in infl. also glandular-pubescent; fl.-tube 2–3.5 cm. long; sepals 2–3 cm. long, reddish, the tips 1–2.5 mm. long; petals 2.5–3.5 cm. long; $n = 7$ (Cleland in Munz, 1949).—Springy places, mostly N. Coastal Scrub; mostly sea cliffs from San Mateo Co. to San Luis Obispo Co.

Ssp. **Wólfii** Munz. Branching, 5–15 dm. tall, the coarse stems green to red, muricate-hirsute, strigulose and in infl. with some gland-tipped hairs; fl.-tube 3.5–4.5 cm. long; sepals 2–3 cm. long, the tips 2–3 mm. long; petals 2–2.5 cm. long; seeds 1.5–1.9 mm. long.—Coastal Strand, Redwood F., Mixed Evergreen F., etc.; Marin to Del Norte and Siskiyou cos.; Ore.

Ssp. **venústa** (Bartlett) Munz. [*Oe. v.* Bartlett.] Mostly 1.5–2.5 m. tall, freely branched, gray-pubescent with long hairs with red basal papillae, also finely pubescent or strigose, upper parts also glandular-pubescent; stems mostly reddish; fl.-tube green, 4.5 cm. long; sepals 3–4.5 cm. long, the slender tips 3–5 mm. long; petals 3–4.5 cm. long; $n = 7$ (Cleland in Munz, 1949).—Moist places, mostly below 5000 ft.; Chaparral, Coastal Sage Scrub, Foothill Wd., etc.; Sierra Nevada foothills from Tuolumne Co. s., cismontane s. Calif.; to L. Calif.

Ssp. **angustifòlia** (Gates) Munz. [*Oe. H.* var. *a.* Gates.] Mostly 3–10 dm. high, erect to ascending, simple-stemmed, muricate-hirsute, strigose and glandular-pubescent (in upper parts); fl.-tube mostly red, 2.5–4.5 cm. long; sepals mostly red, 2.5–3.5 cm. long, the tips 3–5 mm. long; petals 2–4 cm. long; seeds 1–1.5 mm. long; $n = 7$ (Cleland in Munz, 1949).—Moist places, 3000–9000 ft.; Yellow Pine F. to Lodgepole F., Sagebrush Scrub, etc., from San Jacinto Mts. n. through the Sierra Nevada to Modoc, Siskiyou, and Trinity cos.; to Ore., Utah, Colo.

Ssp. **grísea** (Bartlett) Munz. [*Oe. venusta* var. *g.* Bartlett.] Plants 1–2.5 m. tall, freely branched, appressed-pubescent with fine white and occasional longer hairs; stems usually reddish; fl.-tube 2.5–4.2 cm. long; sepals 2–4.5 cm. long, not papillose or glandular, greenish, the tips 3–6 mm. long; petals 2.5–4.5 cm. long; $n = 7$ (Cleland in Munz, 1949).—Moist places below 4000 ft.; Coastal Sage Scrub, Chaparral, S. Oak Wd.; Ventura Co. to San Bernardino and San Diego cos.; L. Calif.

3. **Oe.** × **erythrosèpala** Borb. [*Oe. Lamarckiana* de Vries, not Ser. *Oe. Vrieseana* Lév.] Biennial to short-lived perennial, erect, bushy, 8–12 dm. tall; stems coarse, simple or branched, ± reddish, crisp-puberulent and with many longer spreading hairs (many red-pustulate at base); rosette well formed, the lvs. strongly crinkled, with blades 8–20 cm. long, 3–5 cm. wide, and petioles 5–10 cm. long; cauline lvs. gradually reduced up the stems; spikes dense, simple, in an open branched infl.; bracts ovate to ± oblong; fl.-tube red to green, glandular-pubescent and pilose, 3.5–5 cm. long; sepals ± reddish,

3–4 cm. long, the tips 5–8 mm. long; petals golden-yellow, aging orange-red, broadly obovate, emarginate, 3.5–5 cm. long; anthers 10–12 mm. long; style somewhat shorter than petals; stigma-lobes yellow-green, 5–7 mm. long; caps. green with median red bands, ± muricate, 2–2.5(–3) cm. long, 5–6 mm. wide; seeds dark, angled, 1–1.7 mm. long; n = 7 (Darlington 1931, as *Lamarckiana*).—Garden escape, near the coast from Alameda and San Francisco cos. to Del Norte Co.; to B.C. Of horticultural origin. July–Sept.

4. **Oe. longíssima** Rydb. ssp. **Clùtei** (A. Nels.) Munz. [*Oe. Clutei* A. Nels.] Biennial to short-lived perennial, simple to branched, erect, 1–3 m. tall, ± hirsute especially above, somewhat muricate on stems, hair mostly appressed, upper parts also glandular-pubescent; lvs. of rosette oblanceolate, the blades 1–2 dm. long, 1.5–3 cm. wide, with winged petioles 0.5–1 dm. long; cauline lvs. linear-lanceolate, plane, stiffly spreading-ascending, gradually reduced upward to sessile lanceolate bracts soon exceeded by buds; fl.-tube 8–12 cm. long, ± reddish; sepals 3.5–4.5 cm. long, the tips 3–5 mm. long; petals obovate, ca. 4 cm. long; anthers 14–18 mm. long; caps. subquadrangular, 3.5–4.5 cm. long, 4.5–5.5 mm. thick; seeds 1–1.5 mm. long; n = 7 (Cleland in Munz, 1949).— Moist places, 3400–5500 ft.; Creosote Bush Scrub, Pinyon-Juniper Wd.; Providence and New York mts., e. San Bernardino Co.; to Utah, Ariz. July–Sept.

5. **Oe. strícta** Ledeb. ex Link. [*Oe. arguta* Greene.] Decumbent biennial or perennial, simple or few-branched, 2–6 dm. tall; stems reddish, finely pubescent below, often villous above; foliage green, the lower lvs. oblanceolate to narrower, 5–8 cm. long, ± denticulate, the cauline rather remote, lanceolate, plane, sessile, 2–7 cm. long; uppermost becoming lance-ovate bracts, each with an axillary fl.; fl.-tube 1.5–2.5 cm. long; sepals 12–20 mm. long; petals yellow, aging reddish, 12–25 mm. long; caps. short-villous, enlarged in upper half, 2–2.5 cm. long; seeds brown, narrow-obovoid, smooth, 1.5 mm. long.—For some years reported from moist places near Monterey; native of Chile. April–Sept.

6. **Oe. laciniàta** Hill. [*Raimannia l.* Rose. *Oe. sinuata* L.] Annual or perennial, simple and erect or more commonly branched, semiprostrate to ascending, 1–7 dm. high, strigose and villous or hirsute; lvs. sinuate-pinnatifid to subentire, lance-oblong to oblanceolate, 2–10 cm. long, the lower petioled, the upper sessile or nearly so; fls. solitary in upper axils; fl.-tube 1.5–3.5 cm. long; sepals 6–12 mm. long, with free tips 1–2 mm. long; petals yellowish, 5–18 mm. long; caps. cylindrical, ± arcuate, 1–3.5 cm. long; seeds brownish, evenly pitted, ca. 1 mm. long.—Occasionally natur. in waste places; native of se. U.S. May–July.

7. **Oe. speciòsa** Nutt. [*Hartmannia s.* Small.] Perennial from a running rootstock, the stems 1–5 dm. high, erect to almost prostrate, mostly strigose, ± villous; lvs. oblanceolate to obovate, the cauline 3–8 cm. long, oblong, the lower ± deeply pinnatifid, the upper sinuate-dentate to subentire; fls. in uppermost axils, vespertine; buds nodding; fl.-tube 1–2 cm. long; sepals lance-linear, acuminate, 1.5–3 cm. long; petals white to pinkish, 2.5–4 cm. long; caps. stout, 10–15 mm. long, the basal part cylindrical, 1.5–2 mm. thick, sterile, the upper part 2–5 mm. thick, fertile, ± winged; seeds brown, asymmetrically obovoid, ca. 1 mm. long.—Occasional escape from gardens; native of s. cent. states. May–June. The common form in cult. is:

Var. **Chìldsii** (Bailey) Munz. [*Oe. tetraptera* var. *C.* Bailey.] Stems slender; fls. rose, the petals 2–2.5 cm. long.—Cult. as "Mexican Primrose" and escaping; native of s. Tex. and adjacent Mex. along the Gulf.

8. **Oe. deltoìdes** Torr. & Frém. Coarse spring or winter annual, simple or more often with cent. erect stem 0.5–3 dm. high and few to several decumbent branches naked at the base and 1–10 dm. long; stems pale, with exfoliating epidermis, glabrous below, spreading-pubescent in upper parts; lower lvs. in loose rosette, the blades rhombic-obovate to -lanceolate or oblanceolate, subentire to denticulate, 2–8 cm. long, with petioles ca. as long; cauline lvs. gradually somewhat reduced, the upper sessile and dentate; fls. solitary in upper axils, vespertine, the buds nodding, obtuse; fl.-tube slender, 2–4 cm. long, it and sepals with straight spreading hairs 1–1.5 mm. long; sepals 1.5–3.5 cm. long; petals white, aging pink, 2–4 cm. long; stamens subequal; stigma-lobes 3–6 mm. long; caps. spreading to ± reflexed, woody, with exfoliating epidermis, prismatic-cylindric, 4–6 cm. long, 2–3 mm. thick at base; seeds narrowly obovoid, 1.5–2 mm. long, light brown, usually with purple spots and rows of minute cellular pitting; n = 7 (Lewis

et al., 1958).—Common in sandy places, below 3500 ft.; Creosote Bush Scrub, Joshua Tree Wd.; Mojave and Colo. Deserts from Adelanto region e. and s.; to Ariz., L. Calif. March–May. Plants from w. Mojave Desert (Antelope V.) intergrade with var. *cognata*.

KEY TO VARIETIES

A. Hairs of sepals and infl. spreading, not at all appressed, the longer exceeding 1 mm.
  B. Plants annual. Growing in open sandy deserts or interior valleys.
    C. Petals 2–4 cm. long; uppermost lvs. rarely pinnatifid; caps. 2.5–7 cm. long.
      D. Hairs on buds straight, 1–1.5 mm. long; caps. 2–3 mm. thick at base. Deserts
            ...... *Oe. deltoides*
      DD. Hairs on buds curly, 2 or more mm. long; caps. 3–5 mm. thick at base. Cent. V.
            var. *cognata*
    CC. Petals mostly less than 2 cm. long; uppermost lvs. pinnatifid to deeply sinuate-dentate;
      buds with long curly hairs 2–2.5 mm. long. Siskiyou and Modoc cos. to Inyo Co.
            var. *Piperi*
  BB. Plants perennial; lvs. pinnatifid; buds long-pointed. Contra Costa Co. sand dunes
            var. *Howellii*
AA. Hairs of sepals and infl. appressed for most part, mostly less than 1 mm. long.
  B. Plant annual; lvs. 3–12 cm. long. Mostly Colo. Desert ....var. *cineracea*
  BB. Plant perennial; lvs. mostly less than 3 cm. long. Eureka V., Inyo Co. ....ssp. *eurekensis*

Var. **cognàta** (Jeps.) Munz. [*Oe. trichocalyx* var. *c.* Jeps.] Commonly 2–4 dm. high, branched from base; basal lvs. rhombic, upper coarsely sinuate-dentate; infl. with long curly hairs; buds obtuse; petals 2.5–3.5 cm. long; $n = 7$ (Lewis et al., 1958).—Sandy plains; V. Grassland, San Joaquin V.; Joshua Tree Wd., Antelope V.

Var. **Pìperi** Munz. Mostly not more than 1 dm. high, often unbranched; upper parts with soft curly hairs ca. 2 mm. long; buds blunt; petals barely 2 cm. long.—Dry flats, at 4000–5000 ft.; Sagebrush Scrub, etc.; Siskiyou Co. to Modoc and Lassen cos., Inyo Co., rare in e. San Bernardino Co.

Var. **Howéllii** Munz. Lvs. runcinate-pinnatifid, lanceolate, 3–12 cm. long, cinereous; buds acute, with few to many wavy hairs 1–3 mm. long; petals 2–3 cm. long.—Sand dunes near Antioch, Contra Costa Co.

Var. **cineràcea** (Jeps.) Munz. [*Oe. trichocalyx* var. *c.* Jeps.] Habit, foliage, fls. and fr. as in the typical form of the sp., but pubescence less than 1 mm. long and closely appressed.—Creosote Bush Scrub; Colo. Desert to Needles; Ariz., Sonora.

Ssp. **eurekénsis** Munz & Roos. With deep-seated perennial fleshy rootstocks; bushy, 3–6 dm. tall, densely white-strigose and ± spreading-villous; lvs. crowded, deltoid-ovate, 1–3.5 cm. long; petals 1.5–2.5 cm. long; old fruiting shoots becoming buried and resuming growth at tips.—Sand dunes; Creosote Bush Scrub; Eureka V., e. of Inyo Mts., Inyo Co.

9. **Oe. califórnica** (Wats.) Wats. [*Oe. albicaulis* var. *c.* Wats. *Anogra c.* Small. *Oe. pallida* var. *c.* Jeps.] Perennial from underground rootstocks, rather coarse-stemmed; stems mostly branched, 1–5 dm. tall, frequently decumbent, ashy with short appressed hairs and some longer spreading ones in upper parts, epidermis exfoliating; lvs. variable, lower blades oblanceolate to spatulate in outline, cauline oblong to lanceolate, all subentire to deeply and regularly sinuate-dentate, 1–6 cm. long, sessile or short-petioled; fls. several, vespertine; buds nodding; fl.-tube slender, 2–4 cm. long, strigulose and villous; sepals 1.5–2 cm. long, free tips wanting or very short; petals orbicular-obovate, 2–3 cm. long, white, aging pink; stamens subequal; stigma-lobes 4–6 mm. long; caps. terete, mostly divaricate, somewhat curved upward, 2–5 cm. long, ca. 3 mm. thick at base; seeds plump, obovoid, 1.5 mm. long, brown with dark spots; $n = 14$ (Lewis et al., 1958).—Mostly sandy dry places, below 8000 ft.; Coastal Sage Scrub, Chaparral, S. Oak Wd., Joshua Tree Wd., etc.; cismontane Calif. from Ventura Co. s.; desert edge from Mono Co. s.; to L. Calif., Nev. April–June. A glabrous form from cismontane Riverside, San Bernardino and Los Angeles cos. is var. *glabràta* Munz.

10. **Oe. caespitòsa** Nutt. var. **marginàta** (Nutt.) Munz. [*Oe. m.* Nutt. ex H. & A. *Pachylophis m.* Rydb.] Cespitose perennial with thick caudex, acaulescent or with stem to 2 dm. long, villous-hirsute throughout; lvs. linear-lanceolate, sinuate-pinnatifid, the blades 3–10 cm. long, on winged petioles almost as long; fls. fragrant, vespertine, white, aging pink; fl.-tube 5–8 cm. long; sepals 2.5–3.5 cm. long; petals 2.5–4 cm. long; stamens subequal, glabrous; caps. 3–4 cm. long, pedicelled, linear-cylindric, scarcely ridged, with low tubercles; seeds dark brown, ca. 3 mm. long, obovoid, minutely cellular-roughened,

conspicuously furrowed along the raphe.—Occasional on dry ± stony slopes, 3000–10,000 ft.; mostly Pinyon-Juniper Wd., Yellow Pine F., Shadscale Scrub; desert slopes from Santa Rosa Mts., Riverside Co. to White Mts., Inyo Co., and e.; to Utah, e. Wash. April–Aug.

Var. **longiflòra** (Heller) Munz. [*Anogra l.* Heller. *Pachylophis l.* Heller.] Plant subglabrous except for a few possible hairs along lf.-margins, on ovaries and sepals, which latter may be finely glandular-puberulent; fl.-tube 7–10 cm. long.—At 5000–7000 ft.; Pinyon-Juniper Wd.; Inyo-White Mts.

Var. **crinìta** (Rydb.) Munz. [*Pachylophis c.* Rydb.] Densely hirsute or pilose, the lvs. crowded, lanceolate, sinuate, the blades less than 2 cm. long; fl.-tube 2–4 cm. long; caps. 10–14 mm. long.—Dry rock-crevices, 7500–10,000 ft.; Pinyon-Juniper Wd., Bristle-cone Pine F.; Panamint Mts., Clark Mt.; to Utah. June–Sept.

11. **Oe. xylocárpa** Cov. [*Anogra x.* Small.] Acaulescent perennial with thick vertical root and caudex with crown of lvs.; lf.-blades pinnately parted, often spotted red, 2–7 cm. long, broadly oblanceolate to obovate in outline, densely soft pubescent, the terminal lobe much the largest, petioles ca. as long as blades; fls. vespertine; fl.-tube slender, almost villous, 2.5–4.5 cm. long; sepals 2–3 cm. long; petals bright yellow, aging salmon-red, 2.5–3 cm. long, with broad sinus 4–5 mm. deep; stamens subequal; stigma-lobes 4–5 mm. long; caps. somewhat woody, 3.5–6(–8.5) cm. long, the body proper 7–8 mm. thick at base, winged, tapering gradually into a slender wingless upper part; seeds brownish, 2–2.5 mm. long, narrow-obovoid, angled, roughened, minutely tubercled, with broad flat raphe.—Dry benches, 7000–9800 ft.; Yellow Pine F. to Lodgepole F.; Sierra Nevada from Mono Co. to Tulare Co.; w. Nev. July–Aug.

12. **Oe. primivèris** Gray. [*Lavauxia p.* Small. *Oe. bufonis* Jones.] Winter annual with long taproot, cespitose, acaulescent or nearly so, villous or pilose-pubescent throughout; lf.-blades oblanceolate in outline, 1–12 cm. long, usually deeply and regularly pinnatifid into lanceolate or ovate lobes which are in turn lobed or toothed, petioles shorter than blades; fls. vespertine; fl.-tube 2–6 cm. long; sepals 1.5–2.8 cm. long, without free tips; petals yellow, aging orange-red, cuneate-obovate, usually 2–3 cm. long, with a terminal sinus 4–5 mm. deep; stigma-lobes 6–8 mm. long; caps. pilose, quadrangular, with heavy rib down middle of each face, reticulate; gradually attenuate, 2–3.5 cm. long, 6–8 mm. thick at base; seeds brown, somewhat roughened-tuberculate, 2.5–3 mm. long, with narrow raphal groove.—Dry plains and slopes below 5000 ft.; Creosote Bush Scrub, Joshua Tree Wd., Pinyon-Juniper Wd.; deserts from Inyo and Kern cos. to n. Colo. Desert; to Utah, Texas. March–May.

13. **Oe. cávernae** Munz. Cespitose acaulescent winter annual; lvs. lyrate-pinnatifid, 3–12 cm. long; fl.-tube slender, 2.5–4 cm. long; sepals 5–9 mm. long; petals white, 8–12 mm. long; caps. woody, lance-ovoid, 1.5–3 cm. long, tubercular-costate; seeds 2 mm. long.—Limestone at 1500–4000 ft.; Creosote Bush Scrub; near state line in Clark Co., Nev. to n. Ariz. To be expected in e. San Bernardino Co. April–May.

14. **Oe. flàva** (A. Nels.) Garrett. [*Lavauxia f.* A. Nels.] Perennial, acaulescent, cespitose, subglabrous to finely glandular-pubescent about fls.; lf.-blades oblong-linear to oblanceolate in outline, 3–20 cm. long, deeply runcinate-pinnatifid, petiole shorter than blade; fl.-tube slender, 3–12 cm. long; sepals green, often drying purplish, 1–1.8 cm. long, the free tips an additional 1–5 mm. long; petals pale yellow, 1–2 cm. long, orbicular-obovate; stigma-lobes 3–4 mm. long; caps. indurate, ovoid, 1–2 cm. long, 4-winged, each wing reticulate-veined, 2–5 mm. wide, especially distally; seeds 2 mm. long, minutely granular, cuneate-obovoid, slightly concave with carinate ridge on ventral side and winglike margin around the obtuse summit.—About desiccating depressions; 3000–5000 ft.; Sagebrush Scrub, N. Juniper Wd., etc.; Sierra to Shasta and Modoc cos.; to Wash., Sask., Colo., Ariz., Mex. May–July.

15. **Oe. leptocárpa** Greene. [*Eulobus californicus* Nutt.] Erect, fairly coarse-stemmed annual, simple or with few short branches; stems subglabrous; lvs. few, mostly in basal rosette, these lanceolate in outline, pinnatifid, 5–15 cm. long, dying early; cauline smaller, remote, the upper pendulous; fls. not crowded; fl.-tube obconic, 1 mm. long, orange and pubescent within; sepals 5–8 mm. long, glabrous to pubescent; petals yellow or orange, drying pink, often with reddish spots near base, 6–14 mm. long, rhombic-obovate; stamens of 2 lengths; stigma globose; caps. linear, quadrangular, straight, commonly strongly refracted, 3–10 cm. long, ca. 1 mm. thick; seeds obovoid, light brown

with purplish dots, minutely cellular-pitted, 1 mm. long; $n = 7$, 14 (Lewis et al., 1958).—
Dry and disturbed places such as burns, below 5000 ft.; Chaparral, Coastal Sage Scrub,
Foothill Wd., V. Grassland, etc.; common in cismontane s. Calif., less so on w. side of
San Joaquin V. to San Benito Co., head of San Joaquin V., and deserts; to Ariz., Son.,
L. Calif. April–May.

16. **Oe. Pálmeri** Wats. [*Taraxia P.* Small.] Dwarf cespitose annual, finely strigulose
throughout, forming small acaulescent tufts 2–6 cm. tall or with a few horizontal
branches 2–4 cm. long; stems pubescent, exfoliating, tough and almost woody in age;
lvs. linear-lanceolate to -oblanceolate, subentire to minutely denticulate, 2–6 cm. long;
sterile part of ovary filiform, 8–18 mm. long; fl.-tube proper obconic, 1–2 mm. long;
sepals 2–3 mm. long; petals yellow, round-obovate, 3–5 mm. long, the fls. diurnal;
stamens of 2 unequal sets; stigma globose; caps. crowded, ovoid, 5–7 mm. long, coriaceous, 4-angled below, each angle growing into a thick obliquely truncate wing along the
upper edge of which is a line of dehiscence; seeds smooth, few, brownish, narrow-
obovoid, 1.5 mm. long, minutely cellular-pitted; $n = 7$ (Lewis et al., 1958).—Dry open
places, 2000–4000 ft.; Creosote Bush Scrub, Joshua Tree Wd., Foothill Wd.; Mojave
Desert from Barstow region w. and n. to Temblor Range, w. Kern Co.; e. Ore., Nev.
April–May.

17. **Oe. graciliflòra** H. & A. [*Taraxia g.* Raim.] Cespitose annual forming a single
acaulescent tuft or with several horizontal branches 1–3 cm. long, finely pubescent to
hirsute throughout; lvs. linear to linear-oblanceolate, entire or nearly so, 2–10 cm. long;
upper sterile portion of ovary filiform, 1.5–4 cm. long; fl.-tube proper 2 mm. long; sepals
6–8 mm. long; petals yellow, aging red, 8–14 mm. long, shallowly notched with an apical
tooth; stamens unequal; stigma globose; caps. ovoid-oblong, 8–12 mm. long, coriaceous,
4-angled near base, each angle expanding upward into a broad wing; seeds straw color
with grayish blotches, obovoid, 1.5–2 mm. long, minutely cellular-pitted.—Grassy slopes
and plains, below 3500 ft.; V. Grassland, Foothill Wd., Joshua Tree Wd.; Cent. V. and
borders to Ojai V. (Ventura Co.) and Antelope V. (Los Angeles Co.); to s. Ore. March–
May.

18. **Oe. heterántha** Nutt. [*Taraxia h.* Small.] Sun Cup. Acaulescent perennial, with
simple or branched crown with many lvs. and fls., essentially glabrous, except sometimes finely pubescent especially on lf.-margins; lvs. lanceolate to lance-ovate, entire to
repand-denticulate or sinuate-pinnatifid, especially at base, 3–15 cm. long, with winged
petioles almost as long; narrow sterile portion of ovary very slender, 3–10 cm. long;
fl.-tube proper 1–2 mm. long; sepals 5–8 mm. long; petals yellow, round-obovate, 8–10
mm. long; stamens unequal; stigma discoid; caps. oblong-ovoid, smoothish, coriaceous,
persistent, somewhat 4-angled, pointed above, 12–15 mm. long; seeds oblong, straw-
colored, 3 mm. long, minutely cellular-pitted.—Occasional, moist grassy places, 6000–
9100 ft.; Yellow Pine F. to Lodgepole F.; mts. from Tulare and Mono cos. to Modoc Co.;
to e. Wash. and Rocky Mts. May–Aug.

19. **Oe. ovàta** Nutt. in T. & G. [*Taraxia o.* Small.] Habit of *Oe. heterantha*, mostly
minutely pubescent, especially on lf.-margins and veins underneath; lf.-blades lanceolate
to lance-ovate, usually ± crisped, entire to denticulate or sinuate, glabrous above, 3–10
cm. long, with petioles to as long; sterile portion of ovary slender, 5–10(–18) cm. long;
fl.-tube 3 mm. long; sepals subglabrous to pubescent, lanceolate to lance-ovate, 7–12 mm.
long; petals yellow, obovate to roundish, 8–20 mm. long; stamens subequal; caps. linear-
ovoid, sessile to pedicelled, usually below the surface of the ground, torulose, 1–2 cm.
long, tardily dehiscent; seeds few, broadly oblong-ovoid, brownish or yellowish, shaggily
cellular-pubescent.—Open grassy slopes and fields; N. Coastal Scrub, Coastal Prairie,
Mixed Evergreen F., etc.; near the coast from San Luis Obispo Co. to s. Ore. March–
June.

20. **Oe. tanacetifòlia** T. & G. [*Oe. Nuttallii* T. & G., not Sweet. *Taraxia t.* Piper.]
Thick-rooted, acaulescent, cespitose perennial, with simple or branched crown, subglabrous or finely pubescent; lvs. lanceolate in outline, deeply sinuate-pinnatifid, the
blades 3–10 cm. long, with numerous unequal, acute, entire or toothed segms., petioles
slender, ca. as long as blades; sterile portion of ovary pubescent, slender, 2–10 cm. long;
fl.-tube proper 3 mm. long; sepals lance-ovate, acuminate, pubescent, 7–9 mm. long;
petals yellow, aging red, narrow-obovate, 10–15 mm. long; stamens unequal; stigmas
globular; caps. rarely developed, pubescent, narrow-ovoid, quadrangular, torulose, 17–20

mm. long, 5–6 mm. thick, relatively straight; seeds many, brown, oblong, slightly curved, carunculate, ca. 2 mm. long, finely pitted in longitudinal rows; $n = 21$ (Lewis et al., 1958).—Moist open places, 4000–8500 ft.; Sagebrush Scrub to Lodgepole F.; Mono and Butte cos. to Modoc and Siskiyou cos.; to e. Wash., Ida., Nev. May–July.

21. **Oe. breviflòra** T. & G. [*Taraxia b.* Nutt. ex T. & G. pro syn.] With appearance of *Oe. tanacetifolia*, sterile portion of ovary 1–2.5 cm. long; fl.-tube 2 mm. long; sepals ca. 4 mm. long; petals 6–8 mm. long; caps. fusiform, 1–3 cm. long, torulose; seeds 1.5 mm. long.—Wet meadow, at 7200 ft.; Yellow Pine F.; Warner Mts., Modoc Co.; to Assiniboia, Mont., Wyo. July–Aug.

22. **Oe. refrácta** Wats. [*Sphaerostigma r.* Small. *Oe. deserti* Jones.] Annual, 0.5–4 dm. tall, erect, with open divaricate branching, usually glandular-puberulent and ± strigulose, the stems slender, mostly red, with exfoliating epidermis; lvs. well distributed, but lower largest, oblanceolate to oblong-linear, entire to denticulate, 2–5 cm. long, sessile to short-petioled; infl. racemose, sometimes paniculate, 5–15 cm. long; fl.-tube 5–6 mm. long; sepals lance-oblong, 5–6 mm. long; petals white, roundish, 4–7 mm. long; stamens somewhat unequal; style exceeding corolla; stigma globose; caps. linear, commonly spreading or refracted, straight or curved to coiled, 3–5 cm. long, mostly not beaked; seeds pale, linear, 1 mm. long; $n = 7$ (Lewis et al., 1958).—Frequent in open sandy places, below 5500 ft.; Creosote Bush Scrub, Joshua Tree Wd.; deserts from Inyo to Imperial cos.; to Utah, Ariz. March–May.

23. **Oe. chamaenerioìdes** Gray. [*Sphaerostigma c.* Small. *S. erythrum* A. Davids. *Oe. e.* Macbr.] Erect, usually branched annual, 1–5 dm. tall, the stems slender, often reddish, glandular-puberulent below, the same above and strigulose; lf.-blades subglabrous, lance-ovate to lanceolate, 4–8 cm. long, with petioles 1–3 cm. long; infl. a corymbose raceme with linear bracts and becoming 2 dm. long in fr.; fl.-tube 2.5–3 mm. long; sepals 2.5 mm. long; petals white, often reddish in age, ca. 3 mm. long; stamens subequal; caps. terete, linear, divaricately spreading, 2.5–5 cm. long, scarcely beaked; seeds pale, linear, ca. 1 mm. long; $n = 7$ (Lewis et al., 1958).—Open desert, below 7500 ft.; Creosote Bush Scrub to Pinyon-Juniper Wd.; deserts from White Mts. to Imperial Co.; to Utah, Tex. March–June.

24. **Oe. mìnor** (A. Nels.) Munz. [*Sphaerostigma m.* A. Nels. *Oe. alyssoides* var. *minutiflora* Wats.] Annual, ± canescent-strigulose; stems mostly several, ascending, 0.5–3 dm. long; basal lvs. spatulate to elliptic-ovate, 0.5–2.5 cm. long, on shorter petioles; cauline narrower and smaller; fls. in axils of almost all lvs., the upper in spikes; fl.-tube 2 mm. long; sepals 2 mm. long; petals ochroleucous, 2 mm. long; caps. 1.8–2.5 cm. long, ± contorted, often beaked; seeds narrow-obovoid, grayish, ± angled, 1 mm. long.—Sand dunes in valley below Cedarville, Modoc Co.; to Ida., Wyo., Colo. May–June.

25. **Oe. alyssoìdes** H. & A. var. **villòsa** Wats. [*Sphaerostigma a.* var. *macrophyllum* Small.] Annual, mostly branched from base, cent. stem erect, the others ascending and curved at tip, grayish throughout with ± appressed or spreading hairs, 0.5–3.5 dm. tall; lvs. oblanceolate to lance-ovate, 1.5–4 cm. long, entire or nearly so, the lower with petioles ca. as long, the upper gradually reduced, subsessile; infl. a ± peduncled, spicate raceme; fl.-tube mostly 4–8 mm. long; sepals 4–5 mm. long; petals white, often drying yellowish, 4–5 mm. long; caps. 1.5–2.5 cm. long, gradually attenuate toward the beaklike tip, ± contorted or merely curved; seeds pale, linear-obovoid, minutely cellular-pitted; $n = 7$ (Lewis et al., 1958).—Dry gravelly and rocky places, 4000–8000 ft.; Sagebrush Scrub, Pinyon-Juniper Wd., etc.; Kingston and Panamint mts. in Inyo Co. to Lassen Co.; to Utah, n. Ariz. May–Aug. In s. part of its range intergrading with *Oe. decorticans* var. *desertorum*.

26. **Oe. decórticans** (H. & A.) Greene. [*Gaura d.* H. & A. *Sphaerostigma d.* Small. *Oe. gauraeflora* T. & G.] Erect annual, simple or branched below, subglabrous below, finely pubescent and glandular above, the shining epidermis readily exfoliating; lvs. largely near the base, bright green or tinged red, glabrous or finely pubescent, subentire, 2–8 cm. long, the petioles almost as long, upper lvs. reduced; infl. a fairly compact spike, to 3 dm. long in fr.; fl.-tube 4–6 mm. long; sepals 4–5 mm. long; petals white, 5 mm. long, not so wide, reddish in age only; stamens unequal; caps. subfusiform, thickest in lower half, 2 mm. thick, 1.5–2.5 cm. long, attenuate into slender beak, simply curved so that the beak spreads away from the stem; seeds ash-color, linear-obovoid, ± angled,

minutely pitted, 1 mm. long; $n = 7$ (Lewis et al., 1958).—Loose slopes and disturbed places, mostly below 3000 ft.; Foothill Wd., V. Grassland, etc.; away from the immediate coast, Monterey and San Benito cos., to n. Los Angeles Co., Kern Co. (Greenhorn and Tehachapi mts.). March–June.

KEY TO VARIETIES

Petals over 4 mm. long, usually whitish, if reddish, in age only.
  Caps. not more than 2 mm. thick at base, not conspicuously quadrangular or thickened and indurate at angles, scarcely woody; petals mostly distinctly longer than wide; plants fairly slender, 2–5 dm. tall.
    Frs. with simple curve ca. ⅓ from base, so that tips spread away from stem; base of caps. ca. 2 mm. thick; exfoliating epidermis of stem straw- or flesh-colored. Coast Ranges to Bakersfield and n. Los Angeles Co. .... *Oe. decorticans*
    Frs. often more contorted so that tips point down; base of caps. 1–1.5 mm. thick; epidermis of stem white. Deserts .... var. *desertorum*
  Caps. 3 mm. thick at base. conspicuously quadrangular and much thickened and indurate at angles, quite woody; plants coarse, rarely over 1.5–2 dm. high; exfoliating epidermis white; petals suborbicular. Deserts from Barstow and Victorville e. and s. .... var. *condensata*
Petals 3.5–4 mm. long, red, obovate, longer than wide; caps. curved with spreading tip. Mts. about w. edge of Mojave Desert .... var. *rutila*

Var. **desertòrum** Munz. More slender than sp., with whiter epidermis; caps. more slender; $n = 7$ (Lewis et al., 1958).—Open slopes and plains, below 6000 ft.; Creosote Bush Scrub, Joshua Tree Wd.; deserts from Kern and Inyo cos. to n. Colo. Desert; to Nev.

Var. **condensàta** Munz. Stems coarse, the epidermis white; caps. with prominent ridge down middle of each face; $n = 7$ (Lewis et al., 1958).—Open plains, below 3000 ft.; Creosote Bush Scrub; Death V. region, Inyo Co. to Imperial Co.; to sw. Utah.

Var. **rùtila** (A. Davids.) Munz. [*Oe. r.* A. Davids.] Stems slender, diffused with red; fls. small, red; caps. as in the sp.—Loose slopes, 4000–9000 ft.; Joshua Tree Wd. to Pinyon-Juniper Wd. and above; n. slope of San Gabriel Mts. to Mt. Pinos, Argus, Panamint, and Inyo mts.

27. **Oe. Boóthii** Dougl. ex Hook. [*Sphaerostigma B.* Walp. *S. Lemmonii* A. Nels.] Glandular-pubescent to -villous annual, erect, 1–4 dm. tall, usually with cent. stem more prominent than the ± widely spreading branches; lvs. ovate to oblong-ovate, fairly well distributed, subentire, 2–5 cm. long, with petioles 1–3 cm. more; infl. racemose-spicate, often quite congested, elongating in fr.; fl.-tube 4–8 mm. long; sepals 3–7 mm. long; petals white, aging pinkish, obovate, clawed, 3.5–9 mm. long; stamens subequal; caps. 10–15 mm. long, usually ascending in lower half and with terminal part spreading but not contorted, 1.5–2 mm. wide near the base; seeds brown, rhomboid-prismatic, minutely cellular-pubescent, 1 mm. long, gray when immature; $n = 7$ (Lewis et al., 1958).—Occasional on dry plains and slopes, 3000–8000 ft.; Pinyon-Juniper Wd., Joshua Tree Wd.; w. Mojave Desert from Victorville to Inyo and Mono cos.; to e. Wash., Ida., Utah, n. Ariz. June–Aug.

28. **Oe. andìna** Nutt. [*Sphaerostigma a.* Walp.] Low erect slender-stemmed annuals with spreading branches from near base or above, finely canescent, 0.2–1.5 dm. tall, ca. as broad, mostly leafless near base; lvs. linear to narrow-oblanceolate, entire, 1–3 cm. long, very short-petioled; fls. axillary in a crowded corymb that becomes racemose in fr.; fl.-tube 1 mm. long; sepals 1.5 mm. long; petals yellow, 1.5 mm. long; stamens unequal; caps. 5–6 mm. long, fusiform, ± quadrangular; seeds fusiform, smooth, brown, 0.7 mm. long.—Dry flats and plains, mostly 4000–7000 ft.; Coastal Sage Scrub to Yellow Pine F.; Lassen, Plumas, and Modoc cos.; to e. Wash., Assiniboia, Utah. May–July.

29. **Oe. contórta** Dougl. ex Hook. [*Sphaerostigma c.* Walp.] Slender-stemmed annual, 0.5–2 dm. high, usually branched from near base, subglabrous to finely pubescent; lvs. well distributed, linear to nearly, 0.5–2.5 cm. long, 1–2 mm. wide, subsessile, the lower often with fascicles of smaller lvs. in the axils, the upper reduced to lfy. bracts; fl.-tube 1–2 mm. long, subglabrous to strigulose or ± glandular; sepals lance-ovate, 1.5–2.5 mm. long; petals yellow, aging red, obcordate to narrowly obovate, 2.5–3 mm. long; stamens unequal; caps. linear, cylindrical, often torulose, sessile, curved to straight, 2.5–3.5 cm. long, beaked; seeds brown, obovoid, minutely cellular-pitted, less than 1 mm. long.—Dry loose soil, 2800–5000 ft.; Yellow Pine F., Sagebrush Scrub; Siskiyou and Shasta to Lassen and Modoc cos.; to e. B.C., Ida., w. Nev. May–June.

KEY TO VARIETIES

Plants low, commonly less than 1.5 dm. tall, with very slender subglabrous or finely pubescent lvs. less than 2 mm. wide.
  Caps. sessile, 2.5–3.5 cm. long, ending in a well defined beak. Ne. Calif. ...... *Oe. contorta*
  Caps. distinctly pedicelled, not beaked, 1.7–2.5 cm. long. Mojave Desert to Lassen Co.
                                                                              var. *flexuosa*
Plants taller or heavier, glabrous or pubescent.
  Stems densely pubescent with short appressed or incurved hairs; caps. 1.5–2.5 cm. long, not beaked. Sonoma to Monterey cos. .... var. *strigulosa*
  Stems glabrous or with spreading pubescence; caps. usually over 2.5 cm. long
    Plants spreading, coarse, grayish, with abundant spreading pubescence; lvs. over 2 mm. wide; caps. sessile or nearly so, not beaked. E. cent. Calif. .... var. *pubens*
    Plants tall, erect, glabrous or with some spreading hairs; caps. commonly beaked; lvs. usually 1–2 mm. wide. Cismontane .... var. *epilobioides*

Var. **flexuòsa** (A. Nels.) Munz. [*Sphaerostigma c.* var. *f.* A. Nels.] Stems and lvs. as in the sp.; caps. more slender, pedicelled, not beaked, frequently curved into a half-circle; *n* = 7 (Lewis et al., 1958).—Dry slopes, 4000–9000 ft.; Sagebrush Scrub, Joshua Tree Wd. to Red Fir F.; San Bernardino Mts., Kingston Mts., e. base of Sierra Nevada to Modoc Co.; to Wash., Wyo., Utah.
  Var. **strigulòsa** (F. & M.) Munz. [*Sphaerostigma s.* F. & M.] Stems 1.5–3 dm. tall; caps. short, not beaked; *n* = 14 (Lewis et al., 1958).—Sandy places, open fields, etc.; below 2500 ft.; Chaparral, N. Coastal Scrub, Coastal Strand, Mixed Evergreen F., etc.; near the coast from Humboldt to San Luis Obispo cos., occasional farther s., as Santa Rosa Id. and in w. Riverside Co.
  Var. **pùbens** (Wats.) Cov. [*Oe. s.* var. *p.* Wats.] Coarse-stemmed with abundant spreading pubescence; lvs. rather broad; caps. 2.5–3.5 cm. long, 1 mm. or more in diam.; sessile or subsessile, not beaked.—Occasional, mostly at 7000–9000 ft.; Pinyon-Juniper Wd. to Red Fir F., Sagebrush Scrub; w. Mojave Desert from Mt. Spring Canyon, Kern Co. n., to Lassen Co.; w. Nev.
  Var. **epilobioìdes** (Greene) Munz. [*Oe. strigulosa* var. *e.* Greene.] Commonly 2.5–4 dm. tall, subglabrous or ± spreading-pubescent; infl. often glandular; caps. 2.5–4 cm. long, sessile, slender, commonly beaked; *n* = 14, 21 (Lewis et al., 1958).—Dry disturbed or gravelly places, mostly below 5000 ft.; many Plant Communities; cismontane Calif. away from the immediate coast; to s. Ore., L. Calif., Chile.
  30. **Oe. dentàta** Cav. var. **campéstris** (Greene) Jeps. [*Oe. c.* Greene. *Sphaerostigma c.* Small.] Annual, usually bushy, freely branched from base, the stems subdecumbent to ascending, occasionally subsimple and erect, slender, even capillary, 0.5–2 dm. tall, short-villous below with spreading hair, the light-colored epidermis tending to exfoliate; lvs. well distributed, lance-linear, subsessile, often fascicled, remotely denticulate, 0.5–2.5 cm. long, gradually reduced to leafy bracts in infl.; fls. few, not crowded; fl.-tube 2–4 mm. long; petals bright yellow, with or without red basal dots, roundish, 5–8 mm. long; stamens unequal; caps. linear, terete, ± torulose, 2–4 cm. long, 0.5 mm. thick at base, usually beaked; seeds brown, linear-obovoid, somewhat angled and flattened, 0.5 mm. long, minutely cellular-punctate; *n* = 7 (Lewis et al., 1958).—Dry sandy plains, etc., below 3000 ft.; V. Grassland, Foothill Wd., etc.; Cent. V. and in interior valleys of Coast Ranges from Contra Costa Co. to Santa Barbara Co. March–May.

KEY TO VARIETIES

Whole plant not viscid with spreading gland-tipped hairs.
  Petals 5–8 mm. long; caps. slender, ca. 0.5 mm. thick; seeds 0.5 mm. long.
    Stems short-villous at least in lower part; infl. glandular; caps. with well defined beak. From Santa Barbara and Kern cos. n. .... *Oe. dentata* var. *campestris*
    Stems subglabrous or with appressed hairs; infl. canescent-strigulose or glandular; caps. not distinctly beaked. Mostly Santa Barbara and Tehachapi to w. Riverside Co. and sw. Mojave Desert .... var. *Parishii*
  Petals 8–15 mm. long; caps. thicker, ca. 1 mm. in diam.; seeds ca. 1 mm. long .... var. *Johnstonii*
Whole plant viscid with short spreading gland-tipped hairs. S. Death V. .... var. *Gilmanii*

Var. **Párishii** (Abrams) Munz. [*Sphaerostigma campestre* var. *P.* Abrams.] Lvs. and fls. of var. *campestris*, but pubescence appressed; infl. mostly strigulose; caps. not beaked; *n* = 7 (Lewis et al., 1958).—Sandy places, below 5000 ft.; Foothill Wd., V. Grassland,

Joshua Tree Wd., etc.; w. half of Mojave Desert to inner S. Coast Ranges (San Luis Obispo Co.), Sierran foothills to Mariposa Co., and cismontane Riverside Co.

Var. **Johnstònii** Munz. Fls. large, the petals 10–15 mm. long; caps. heavier; $n = 7$ (Lewis et al., 1958). Dry sandy places, below 5000 ft.; Joshua Tree Wd., Pinyon-Juniper Wd., V. Grassland, etc.; w. half of Mojave Desert to borders of San Joaquin V. (Kern and Kings cos.), occasional to w. Nev.

Var. **Gilmánii** Munz. Stems covered with short gland-tipped hairs; petals 6–12 mm. long.—Bradbury Well region, at s. end of Death V., Inyo Co.

31. **Oe. kernénsis** Munz. Erect annual with few spreading branches from base, 8–18 cm. tall, somewhat canescent throughout, minutely glandular-pubescent in infl.; lvs. well distributed, ± secund, the lower oblanceolate, 10–15 mm. long, 3–5 mm. wide, subsessile, subentire to denticulate, the cauline lance-linear, somewhat reduced; fls. few, solitary in upper axils in a lax raceme; floral tube 2–3 mm. long; sepals 5–6 mm. long; petals bright yellow, obovate, 10–14 mm. long; stamens unequal; caps. ascending, somewhat curved, cylindric-clavate, 2–2.5 cm. long, 1–1.5 mm. thick, not beaked, pubescent, on a pedicel 4–7 mm. long; seeds brownish, angled, narrow, ca. 1 mm. long, cellular-punctate; $n = 7$ (Raven, 1958).—Dry slopes of coarse sand and disintegrated granite, 3000–5000 ft.; Joshua Tree Wd., Pinyon-Juniper Wd.; Walker Pass, Red Rock Canyon, Kern Co. May.

32. **Oe. guadalaupénsis** Wats. Erect annual, much like forms of *Oe. micrantha,* strigulose throughout; stems leafy; petals 2.5–3 mm. long; caps. oblong-pyramidal, strongly angled, nearly straight, ca. 10–12 mm. long.—Sand dunes, San Clemente Id.; Guadalupe Id. May–June.

33. **Oe. micrántha** Hornem. ex Spreng. [*Sphaerostigma m.* Walp. *Oe. hirta* Link, not L.] Annual, simple or several-stemmed, prostrate, hairy, 0.5–5 dm. long, leafy throughout, the stems with readily exfoliating epidermis; basal lvs. forming a loose rosette, linear-lanceolate to oblanceolate, subentire, 2–10 cm. long, with equal or longer petioles; cauline lvs. shorter, sessile, denticulate, crisped; fls. small; petals yellow, often drying green, 2–4 mm. long; stamens unequal; caps. curved or contorted, quadrangular, 1.2–2 cm. long, gradually attenuate to apex, usually beaked; seeds brown, obovoid, 1 mm. long, finely cellular-pitted; $n = 7, 14$ (Lewis et al., 1958).—Common in dry disturbed places, burns, etc., below 3000 ft.; Coastal Strand, Coastal Sage Scrub, Chaparral, etc.; mostly near the coast from Marin Co. to L. Calif., Channel Ids. March–May. Integrading freely with the vars.

KEY TO VARIETIES

Fls. small, petals 2–3(–4) mm. long; stems and lvs. hirsutulous or villous, but not pallid.
  Stems semi-prostrate; cauline lvs. lance-oblong, obtuse, sessile but not clasping. Largely near the
  coast .... *Oe. micrantha*
  Stems erect or ascending; cauline lvs. oblong-ovate to broadly ovate, acute, with subcordate clasping
  base. Largely of interior valleys .... var. *Jonesii*
Fls. larger; petals 3–7 mm. long; foliage either subglabrous or pallid with close whitish pubescence.
  Plants subglabrous; stems often simple, erect. Interior cismontane valleys .... var. *ignota*
  Plants pallid with whitish pubescence; stems spreading, mostly several from the base. Deserts
                                             var. *exfoliata*

Var. **Jònesii** (Lév.) Munz. [*Oe. hirta* var. *J.* Lév. *Oe. hirtella* Greene. *Sphaerostigma arenicola* A. Nels.] Erect or ascending, densely villous-pubescent; cauline lvs. oblong-to broad-ovate, acute, often sessile with clasping base; $n = 7, 14, 21$ (Lewis et al., 1958).—Dry slopes, burns, etc., below 5500 ft.; Chaparral, Coastal Sage Scrub, Yellow Pine F., etc.; Mendocino and Glenn cos. s. to L. Calif. mostly in the Coast Ranges away from the immediate coast, occasional near Sierra Nevada foothills and on Channel Ids. March–Aug.

Var. **ignòta** Jeps. [*Oe. hirta* var. *i.* Munz.] Stems simple, erect, subglabrous; petals 5–7 mm. long; $n = 7, 14$ (Lewis et al., 1958).—Dry places below 8500 ft.; mostly Chaparral, Coastal Sage Scrub, Yellow Pine F.; interior valleys and mts. of s. Calif., occasional along w. base of Sierra Nevada to Madera Co. and in Coast Ranges to Yolo Co.; n. L. Calif.

Var. **exfoliàta** (A. Nels.) Munz. [*Sphaerostigma m.* var. *e.* A. Nels. *Oe. Abramsii* Macbr. *Oe. m.* var. *A.* Jeps. *S. pallidum* Abrams.] Pallid with dense whitish pubescence, usually ± erect; petals 3–6 mm. long; n = 7 (Lewis et al., 1958).—Dry sandy places, below 6000 ft.; Creosote Bush Scrub, Joshua Tree Wd.; deserts from Inyo Co. to e. San Diego and Imperial cos.; w. Ariz.

34. **Oe. bistórta** Nutt. ex T. & G. [*Sphaerostigma b.* Walp. *Oe. spiralis* var. *linearis* Jeps.] Annual, or of longer duration, occasionally simple, usually with several prostrate to ascending stems, these often with reddish tinge, the older epidermis exfoliating, stems rather slender, 0.5–8 dm. long; lvs. green, pubescent to pilose, subentire to denticulate, the basal 3–7 cm. long lance-linear and with petioles 1–4 cm. long, the cauline often secund, shorter, wider, the uppermost subsessile to cordate-clasping; fls. in lf.-axils, only a few in bloom at once; fl.-tube 3–5 mm. long; sepals 7–10 mm. long; petals yellow, often drying greenish, with or without dark spot at base, 8–14 mm. long; stamens unequal; caps. curved or contorted, 2–2.5 mm. thick, 12–15 mm. long, sharply quadrangular, without beak or this not more than 4–5 mm. long; seeds brown, obovoid, 1 mm. long, finely cellular-pitted; n = 7 (Lewis et al., 1958).—Beaches and bluffs; Coastal Strand, Coastal Sage Scrub; immediate coast from Los Angeles Co. to n. L. Calif. March–June. Intergrading with the vars.

KEY TO VARIETIES

Foliage in general green. Cismontane.
  Caps. short, 12–15(–20) mm. long, sharply quadrangular, usually 2–2.5 mm. thick, the beak
    none to 4 or 5 mm. long. Coast, largely in San Diego area . . . . *Oe. bistorta*
  Caps. longer, 20–40 mm. long, more slender, 1.5–2 mm. thick, and with beak 3–10 or more mm.
    long. General in cismontane s. Calif. . . . . var. *Veitchiana*
Foliage pallid with short appressed hair. Deserts . . . . var. *Hallii*

Var. **Veitchiàna** Hook. [*Sphaerostigma V.* Small.] Caps more slender, longer, beaked; n = 7 (Lewis et al., 1958).—Dry slopes and valleys, especially in disturbed places, below 4000 ft.; Chaparral, Coastal Sage Scrub, S. Oak Wd., etc.; Ventura and Kern cos. to San Diego Co.; n. L. Calif.

Var. **Hállii** (A. Davids.) Jeps. [*Sphaerostigma H.* A. Davids.] Pallid with appressed hairs; petals 7–14 mm. long; n = 7 (Lewis et al., 1958).—Sandy places; Creosote Bush Scrub; w. edge of Colo. Desert.

35. **Oe. cheiránthifòlia** Hornem. ex Spreng. [*Oe. spiralis* Hook. *Sphaerostigma s.* F. & M.] Perennial, apparently flowering the first year, with several prostrate to decumbent wiry stems radiating from a cent. rosette, these 1–6 dm. long; plant grayish-pubescent throughout; lvs. thick, those of the rosette oblanceolate, 1–7 cm. long, narrowed into petioles 1–2 cm. long; lower cauline lvs. lance-oblong, subsessile to short-petioled, obtuse, subentire, 2–4 cm. long, the upper still shorter and broader, oblong-ovate to orbicular-ovate; fls. in axils, mostly above the base of the stems; fl.-tube 2.5–5 mm. long; sepals 4–10 mm. long; petals bright yellow, with or without reddish spots near the base, drying green or red, 5–9 mm. long; stamens unequal; caps. coiled, distinctly quadrangular, short-beaked or not beaked, pubescent, 12–22 mm. long; seeds dark brown, obovoid, 1 mm. long, minutely cellular-pitted.—Coastal Strand; Point Conception, Santa Barbara Co. to Coos Co., Ore., Santa Rosa and San Miguel ids. April–Aug.

Var. **nítida** (Greene) Munz. [*Oe. n.* Greene.] Plants like the sp., but green and glabrous throughout.—Occasional with the sp., Monterey Co. to San Miguel Id.

Var. **suffrùticosa** Wats. [*Oe. viridescens* Hook. *Sphaerostigma v.* Walp.] Foliage silvery; plant usually suffrutescent; petals 13–22 mm. long.—Coastal Strand; Point Conception to n. L. Calif.

36. **Oe. pterospérma** Wats. [*Chylismia p.* Small.] Low annuals, 5–12 cm. tall, simple or few-branched, pilose below, finely glandular above; lvs. oblong to lance-ovate, often with a shoulder on each side near the tip, entire, subsessile, 5–20 mm. long; fls. axillary, pinkish-white; pedicels 5–8 mm. long; fl.-tube 1.2 mm. long; sepals 1.5–2.5 mm. long; petals obcordate, ca. as long, white; caps. cylindric-clavate, slightly curved, 10–16 mm. long; seeds oblong, brownish, flattened, bordered with a revolute winglike tubercled margin.—Rare, dry places 4500–6500 ft.; Pinyon-Juniper Wd.; Panamint and Inyo mts.; to e. Ore., Utah. May–June.

37. **Oe. cardiophýlla** Torr. [*Chylismia c.* Small. *Oe. c.* var. *petiolaris* Jones.] Annual to suffrutescent perennial, 1–5 dm. tall, erect, usually branched, mostly soft-pubescent

throughout, stems fairly coarse; lvs. round-cordate to ovate, irregularly dentate or denticulate, obtuse, subglabrous to white-villous, well distributed, 1–6 cm. long, on petioles ca. as long; fls. in upper axils or in dense terminal racemes; fl.-tube 5–10 mm. long; sepals ovate, 3–7 mm. long; petals yellow, aging red, broader than long, 6–8 mm. long; stamens unequal; caps. rather coarse, cylindrical, ± curved, 2–6 cm. long, on pedicels 2–10 mm. long; seeds obovoid, brown, ± irregularly angled, 0.6 mm. long; $n = 7$ (Lewis et al., 1958).—Desert mesas and canyons, below 5000 ft.; Creosote Bush Scrub; Argus and Panamint mts., Inyo Co., Colo. Desert from Riverside Co. to Ariz., L. Calif. March–May.

Var. **spléndens** M. & J. [*Oe. c.* var. *longituba* Jeps.] Fl.-tube 20–25 mm. long; petals 13–25 mm. long.—With the sp., from Mecca to Needles; Ariz., L. Calif.

38. **Oe. brévipes** Gray. [*Chylismia b.* Small.] Annual, frequently rather coarse, 1–few-stemmed from base, occasionally branched above, spreading-villous especially below, 1–4 dm. tall, erect with nodding tips; lvs. largely in basal rosettes with few scattering small ones on lower stem, the upper bracteate, lower petioled, subglabrous to villous, usually bicolored, red-veined beneath, ovate to oblong-cordate, subentire to pinnate; infl. racemose; pedicels short, 3–15 mm. long; fl.-tube 3–7 mm. long; sepals 6–10 mm. long, pilose and glandular; petals bright yellow, obovate, 7–15 mm. long; stamens subequal, the anthers scattered-hairy; caps. linear, spreading-divaricate, 5–9 cm. long, 2–3 mm. in diam.; seeds straw color, obovoid, 1–1.5 mm. long, ± angled; $n = 7$ (Lewis et al., 1958).—Dry slopes and washes, below 5000 ft.; Creosote Bush Scrub, Joshua Tree Wd.; Mojave Desert from Inyo Co. and w. San Bernardino Co. e. and s. to e. Imperial Co.; Nev., Ariz. March–May. Hybridizing with *Oe. clavaeformis,* the intermediates having various stages of pubescence, caps. length, etc., but usually not the spreading long hairs on the stem or the long spreading caps. of *brevipes.*

39. **Oe. clavaefórmis** Torr. & Frém. [*Oe. scapoidea* var. *c.* Wats. *Chylismia c.* Heller.] Annual, simple or with few simple stems from the base, 1–4 dm. tall, subglabrous to finely pubescent below, subglabrous about the infl.; lvs. mostly in a basal rosette, simple and irregularly dentate with ovate blades commonly 2–5 cm. long and petioles ca. as long, rarely pinnatifid; cauline lvs. much reduced; infl. racemose, somewhat peduncled, the fls. quite crowded at anthesis, pedicels 8–25 mm. long; fl.-tube and sepals glabrous, each ca. 5 mm. long; petals white, often drying reddish and often red-brown at base, 4–6 mm. long; stamens subequal, the anthers with white spreading hairs; caps. clavate, commonly ca. 2 mm. thick, 12–20 mm. long, generally curved and strongly ascending; seeds light brown, obovoid, ± angled, 1.2 mm. long; $n = 7$ (Lewis et al., 1958).—Sandy plains and washes, below 4000 ft.; Creosote Bush Scrub, Joshua Tree Wd.; desert from Inyo and Kern cos. to s. San Bernardino Co.; w. Nev. March–May.

KEY TO VARIETIES

Fls. whitish, often brown at throat; petals drying purplish.
  Sepals, ovaries and upper stems glabrous; lvs. usually not much pinnatifid . . . . *Oe. clavaeformis*
  Sepals, ovaries and upper stems strigulose; lvs. usually much pinnatifid . . . . var. *aurantiaca*
Fls. yellow.
  Lvs. scarcely or not pinnatifid; stems subglabrous to canescent-puberulent. Inyo Co. to Ore.
                                                         var. *purpurascens*
  Lvs. much pinnatifid; stems spreading-villous. Imperial and e. San Diego cos. . . . . var. *Peirsonii*

Var. **aurantìaca** (Wats.) Munz. [*Oe. scapoidea* var. *a.* Wats.] Stems subglabrous to finely pubescent; sepals and infl. finely strigulose; $n = 7$ (Lewis et al., 1958).—Creosote Bush Scrub; Inyo, San Bernardino, Riverside, and Imperial cos.; Ariz., Nev.

Var. **purpuráscens** (Wats.) Munz. [*Oe. scapoidea* var. *p.* Wats. *Oe. cruciformis* Kell. *Chylismia s.* var. *c.* Small. *C. lancifolia* Heller.] Stems closely and finely canescent-puberulent; fls. yellow, sometimes with basal reddish spots; lvs. scarcely or not pinnatifid; $n = 7$ (Lewis et al., 1958).—Dry slopes and flats, 4000–7000 ft.; Sagebrush Scrub, Pinyon-Juniper Wd.; Inyo, Mono, and Lassen cos.; e. Ore., w. Nev.

Var. **Pèirsonii** Munz. Stems spreading-villous; lvs. mostly much divided; fls. yellow; $n = 7$ (Lewis et al., 1958).—Creosote Bush Scrub; Imperial and e. San Diego cos.; Ariz., L. Calif.

40. **Oe. scapoìdea** Nutt. var. **seòrsa** (A. Nels.) Munz. [*Chylismia s.* var. *s.* A. Nels.] Annual, simple or branched from base, erect or spreading, subglabrous below, glandular-

puberulent above, 1–4 dm. tall; lvs. mostly below, prevailingly simple, ovate to ± oblong, 1–4 cm. long, on somewhat longer petioles; infl. mostly racemose; pedicels slender, 5–15 mm. long; fl.-tube 1.5–3 mm. long; sepals 2 mm. long; petals yellow, 2–3 mm. long; caps. quite erect, clavate, 1–2.5 cm. long, 2–2.5 mm. thick; seeds brownish, obovoid, 1.5–2 mm. long.—Dry slopes, at 2500 ft.; Creosote Bush Scrub; Funeral Mts., Inyo Co.; e. Ore., Ida., to Wyo., Colo. May–June.

41. **Oe. multijùga** Wats. var. **parviflòra** (Wats.) Munz. [*Oe. brevipes* var. *p.* Wats. *Chylismia p.* Rydb. *Oe. scapoidea* var. *tortilis* Jeps.] Annual, or of longer life, subglabrous to pubescent or villous, slender-stemmed, erect, simple or branched, 2–8 dm. tall; lvs. in basal rosette, pinnate with 5–8 pairs of major lateral pinnae and a larger terminal one, usually quite villous, red-veined beneath; upper parts of plant quite leafless; infl. of naked racemes, often in a panicle; pedicels capillary, 1–2 cm. long; fl.-tube 1–2.5 mm. long; sepals 3–4 mm. long; petals yellow, 3–5 mm. long; caps. linear, 1–1.5 mm. in diam., 1.5–2.5 cm. long; seeds many, light brown, obovoid, 1 mm. long.—Dry washes and loose slopes, below 6000 ft.; Creosote Bush Scrub, Joshua Tree Wd.; from Kelso, San Bernardino Co. to White Mts., Inyo Co. March–May.

42. **Oe. heterochròma** Wats. [*Chylismia h.* Small.] Annual, simple or branched at base and above, glandular-pubescent throughout, 2–5 dm. tall; lvs. in lower portion only, irregularly serrate, ovate, conspicuously veined beneath, villous, 2–6 cm. long, on petioles ca. as long; upper lvs. reduced; pedicels capillary, 2–5 mm. long; fl.-tube 2.5 mm. long; sepals 2.5 mm. long; petals 3–4 mm. long, purplish; caps. clavate, 8–13 mm. long, 2 mm. thick; seeds brown, obovoid, 1 mm. long; $n = 7$ (Lewis et al., 1958).—Dry loose soil, at 3400 ft., Grapevine Canyon above Scotty's, Death V.; w. Nev. May–June.

Var. **monoénsis** Munz. Stems subglabrous, ± glaucous; petals 3–5 mm. long.—Dry places, 5000–8000 ft.; Sagebrush Scrub, Pinyon-Juniper Wd.; Inyo and Mono cos.

8. Gayophỳtum Juss.

Slender-stemmed caulescent annuals. Lvs. alternate, entire, linear or lanceolate and subsessile, or the lower may be opposite, linear-oblanceolate and short-petioled. Fls. in upper axils. Fl.-tube not prolonged beyond the ovary. Sepals 4, usually reflexed in anthesis. Petals 4, small, rhomboid-spatulate to -obovate, white, frequently drying pink or red. Stamens 8, the alternate set much reduced and usually sterile. Stigma capitate. Caps. 2-loculed, 4-valved, linear or clavate. Seeds many, in 1 row in each locule, narrow-obovoid, not comose. Ca. 6 spp., of temp. w. N. Am. and Chile and Argentine. (C. *Gay*, author of Flora of Chile, and Greek word for plant.)

(Munz, P. A., *in* Am. Jour. Bot. 19: 768–778, 1932.)

Caps. torulose, pedicelled; plants freely openly branched above the base, repeatedly dichotomous, the upper lvs. bracteate. Dry slopes and banks.
  The petals 0.5 mm. long; caps. 2–5 mm. long, shorter than the deflexed pedicel; plants quite glabrous. Ne. Calif. . . . . 1. *G. ramosissimum*
  The petals 1–5 mm. long; caps. 5–12 mm. long, exceeding the erect or deflexed pedicel. Widespread
                                                                                    2. *G. Nuttallii*
Caps. not plainly torulose, subsessile; plants branched mostly at the base, not so much above; upper lvs. quite well developed. Plants of desiccating mud flats or other places where water has been.
  Seeds vertically placed in a very narrow caps., 5–10 in each locule . . . . 3. *G. racemosum*
  Seeds obliquely placed in a slightly broader caps., ca. 15–18 in each locule . . . . . . 4. *G. humile*

1. **G. ramosíssimum** T. & G. [*G. r.* var. *deflexum* Hook. *G. r.* var. *obtusum* Jeps.] Diffusely branched mostly above the base, 1–2(–3) dm. high, the ultimate branchlets filiform, quite glabrous, sometimes slightly strigulose among the fls.; lvs. lance-linear, mostly 1–2 cm. long, short-petioled, gradually reduced up the stems; pedicels capillary, 3–5 mm. long, mostly spreading-deflexed; fls. minute; sepals erect, 0.5 mm. long; petals 0.5 mm. long; stigma globose; caps. plump, 2–5 mm. long; seeds mostly 2–4 in a locule, glabrous, 0.6 mm. long; $n = 7$ (Lewis et al., 1958).—Dry slopes and ridges, 4500–8000 ft.; Pinyon-Juniper Wd., N. Juniper Wd.; Yellow Pine F.; Mono, Plumas, Lassen and Modoc cos.; to Wash., Wyo., Ariz. June–Aug.

2. **G. Nuttállii** T. & G. [*G. ramosissimum* var. *strictipes* Hook. *G. r.* var. *pygmaeum* Jeps.] Like *G. ramosissimum*, but usually more open and slightly larger plants, more evidently strigulose in upper parts; lvs. 1–3 cm. long; pedicels erect, 1–3(–5) mm.

long; sepals 1–1.5 mm. long; reflexed in anthesis; petals 1–1.5 mm. long; caps. 5–12 mm. long, erect, torulose, usually exceeding the pedicels; seeds glabrous, mostly 4–8 in a locule, 1–1.5 mm. long; *n* = 7, 14 (Lewis et al., 1958).—Common on dry slopes and ridges, mostly 3000–11,000 ft.; Montane Coniferous F.; mts. from San Diego and Orange cos. to Mt. Pinos, Sierra Nevada to Modoc, Humboldt, and Tehama cos.; Wash. to Dak., New Mex., Chile, Argentine. June–Aug. Variable; the following key distinguishes forms that have been named:

A. Petals 0.5–1.5 mm. long.
  B. Seeds glabrous.
    C. Pubescence of upper parts of plant appressed.
      D. Caps. erect .... *G. Nuttallii*
      DD. Caps. deflexed .... (1) *G. intermedium*
    CC. Pubescence spreading; caps. erect .... (2) *G.N.* var. *Abramsi*
  BB. Seeds appressed-pubescent.
    C. Pubescence of upper parts of plant appressed .... (3) *G. lasiospermum*
    CC. Pubescence of short spreading hairs .... *G. l.* var. *Hoffmannii*
AA. Petals 2–5 mm. long.
  B. Seeds glabrous .... (4) *G. diffusum*
  BB. Seeds pubescent .... (5) *G. eriospermum*

(1) *G. intermèdium* Rydb. Pedicels and caps. spreading or deflexed; pubescence appressed.—Occasional with *G. Nuttallii* in Calif.; to e. Wash., Wyo., Colo. (2) *G. N.* var. *Abramsi* Munz with short spreading puberulence.—Occasional with the sp. (3) *G. lasiospermum* Greene like *G. N.*, but with strigose-canescent seeds; not usually common; an occasional form with spreading hairs on upper parts of plant is var. *Hoffmannii* Munz. The next two have larger fls. (4) *G. diffusum* T. & G. is appressed-puberulent in upper parts; pedicels 2–8 mm. long, erect or divaricate; sepals 2–3 mm. long, reflexed in anthesis; petals 2–5 mm. long; caps. 5–12 mm. long; seeds glabrous.—Occasional with *G. N.*; plants with short spreading hairs are *G. d.* var. *villosum* Munz. (5) *G. eriospermum* Cov. [*G. lasiospermum* var. *e.* Jeps.] like *G. diffusum* but with canescent-strigose seeds; uncommon. The exact status of these various forms awaits further study.

3. **G. racemòsum** T. & G. Plants low, 1–2 dm. high, subsimple to repeatedly branched from the base, the ultimate branches leafy and relatively simple, subglabrous to strigulose; lvs. linear to linear-oblanceolate, 1–3 cm. long; pedicels from almost lacking to 2 mm. long, erect; sepals 0.5 mm. long; petals white, turning red, scarcely 1 mm. long; caps. subterete, narrowly linear, erect, not markedly torulose, 6–14 mm. long; seeds erect, ca. 5–10 in each locule, glabrous, 1 mm. long; *n* = 14 (Lewis et al., 1958). —Drying slopes and flats that have been moist, 5000–11,000 ft.; Montane Coniferous F.; San Bernardino Mts., Sierra Nevada to Modoc Co.; Wash. and Rocky Mts. July–Aug. Occasionally plants are found with spreading pubescence and are var. *caèsium* (T. & G.) Munz. [*G. caesium* T. & G.] Plants with short spreading hairs in the infl. and strigose seeds, are *G. Helleri* Rydb. [*G. H.* var. *erosulatum* Jeps.] Some plants are quite glabrous or ± strigulose and with strigose seeds; they have been called *G. H.* var. *glabrum* Munz.

4. **G. hùmile** Juss. [*G. pumilum* Wats.] Low, 0.5–1.5 dm. high, glabrous, branched from base and sometimes also above; lvs. linear to lance-linear, 1–3 cm. long, short-petioled, the upper ± reduced, but still leaflike; pedicels scarcely evident; sepals 1 mm. long; petals 1 mm. long; caps. flattened, erect, 10–15 mm. long; seeds 15–18 in each locule, ± overlapping so as to be obliquely placed, 0.6 mm. long; *n* = 7 (Lewis et al., 1958).—Drying flats, mostly 3000–9000 ft.; Montane Coniferous F.; Coast Ranges from Lake Co. to Siskiyou Co., Modoc Co. through Sierra Nevada to Santa Barbara Co.; to Wash., Ida., Chile. July–Aug.

9. Gáura L.

Annual to perennial caulescent herbs with alternate lvs. Fls. white or pink, in terminal spikes or racemes. Fl.-tube narrow, short; sepals 4, deciduous. Petals 4, clawed. Stamens 8, usually with scalelike appendage at base of each fil. Ovary 4-loculed, usually with 1 ovule in each locule. Stigma 4-lobed, with cuplike border at base. Caps. nutlike, obovoid, nearly or quite indehiscent, 1–4-seeded. Ca. 18 spp., temp. N. Am., Argentine. (Greek, *gauros*, proud, some spp. being showy.)

(Munz, P. A. Bull. Torrey Bot. Club 65: 105–122, 211–228, 1938.)
Caps. sessile or nearly so, not narrowed into a pedicellike base; petals 1.5–2 mm. long
                                                                    1. *G. parviflora*
Caps. narrowed into a distinct thick or slender stipelike base; petals 5–10 mm. long.
  Stipelike base of caps. very slender, almost filiform, 3–6 mm. long; the caps.-body being abruptly
    narrowed at base, plant soft-villous with long hairs .... 2. *G. villosa*
  Stipelike base of caps. stout, mostly shorter, the caps.-body more gradually narrowed at base.
    Fl.-tube 2.5–4 mm. long; fils. 8–10 mm. long; fr. 5–9 mm. long, 1–1.5 mm. wide . . 3. *G. sinuata*
    Fl.-tube 5–12 mm. long; fils. 3–6 mm. long; fr. 5–11 mm. long, 2–3 mm. wide.
      Sepals 10–13 mm. long; petals 7–8 mm. long; fl.-bracts lance-ovate, caducous; main cauline
        lvs. 5–25 mm. wide .... 4. *G. odorata*
      Sepals 5–9 mm. long; petals 3–6 mm. long; fl.-bracts lanceolate to linear, mostly persistent;
        main cauline lvs. mostly narrower .... 5. *G. coccinea*

1. **G. parviflòra** Dougl. ex Hook. var. **lachnocárpa** Weath. Biennial or winter annual, mostly 5–20 dm. tall, erect, rather simple below, freely and widely branched above, soft-villous throughout with spreading hairs and also glandular-pubescent; lvs. simple, those of basal rosette broadly oblanceolate, 5–15 cm. long, the cauline reduced upward to sessile lanceolate lvs. 3–10 cm. long; bracts lance-linear; spikes slender, nodding at tips, becoming 1–3 dm. long; fl.-tube 1.5–2.5 mm. long; sepals ca. as long; petals spatulate, 1.5–2 mm. long; caps. sessile, 6–10 mm. long, ca. 2 mm. thick, subfusiform, 4-nerved, puberulent; seeds 1–2, brown.—Occasional weed; native from Mo. and Utah to Mex.; Argentine. June–Aug.

2. **G. villòsa** Torr. Perennial, usually suffrutescent, the stems several from the base, ascending-erect, mostly 5–10 dm. tall, somewhat branched, ± villous; lvs. many, the lower spatulate to ovate or lanceolate, 3–7 cm. long, subsessile or short-petioled, the upper narrower, smaller; fl.-bracts ovate to lance-ovate, 1–3 mm. long, caducous; infl. 3–6 dm. long in fr., branched with very slender stems, subglabrous to strigulose; fl.-tube 2 mm. long; sepals cinereous-strigulose, 7–10 mm. long; petals ca. 8 mm. long; caps. cinereous-strigulose, the body oblong, wing-angled, 7–9 mm. long, somewhat narrowed toward summit.—Occasional weed; native from Kans. to New Mex. and Tex. July–Aug. More commonly introd. is var. **McKélveyae** Munz, with glabrous ovary.

3. **G. sinuàta** Nutt. ex Ser. in DC. Perennial, 3–8 dm. tall, simple or branched above base, subglabrous; basal lvs. oblanceolate to oblong-lanceolate, 3–8 cm. long, sinuate-dentate, short-petioled, cauline crowded, lanceolate to spatulate, 1–5 cm. long; fl.-bracts 1–3 mm. long, caducous; fl.-tube 2.5–3 mm. long, subglabrous to sparsely strigulose; sepals strigulose, 7–10 mm. long; petals 8–10 mm. long; caps. subglabrous, the body fusiform, obtusely 4-angled, 5–9 mm. long, 1–1.5 mm. wide, gradually tapered into the thick pedicellike base 2–5 mm. long; $n = 14$ (Lewis et al., 1958).—Occasional weed, in s. and cent. Calif.; native from Okla. to n. Mex. June–Sept.

4. **G. odoràta** Ses. ex Lag. Apparently biennial or winter annual, mostly branched at base, the stems ascending, 2–5 dm. high, slender, grayish-pubescent; lower lvs. oblanceolate, sinuate-dentate, 2–6 cm. long, upper narrower, shorter; bracts lance-ovate, 3–6 mm. long; fl.-tube ca. 3 mm. long; petals ca. 8 mm. long; caps. glabrous to strigulose, 8–11 mm. long, the lower third terete, gradually enlarged upward into an ovoid-pyramidal, 4-angled part, each face 2.5 mm. wide with median nerve; $n = 7$, 14 (Lewis et al., 1958).—Weed from San Diego to Ventura cos., Alameda Co.; native from Tex. to cent. Mex. May–Sept.

5. **G. coccínea** (Nutt.) Pursh. [*Malva c.* Nutt. ex Fraser.] Perennial, the stems several to many, branched so as to form a bushy plant 1–5 dm. tall, strigose-canescent; lvs. many, sessile, oblong-lanceolate to linear, entire to repand-dentate, sessile, acute to obtuse, 1–3 cm. long; fl.-bracts 3–6 mm. long; spikes short, 1–2 dm. long in fr.; fl.-tube 6–10 mm. long; sepals 6–9 mm. long; petals 5–8 mm. long; frs. canescent, short-obovoid, 4-angled in upper half, 5–7 mm. long, with stout terete base; $n = 7$ (Johansen, 1929).—Dry slopes, mostly near limestone, 3000–5000 ft.; Joshua Tree Wd., Pinyon-Juniper Wd.; mts. of e. Mojave Desert to S. Dak., Tex. April–June. Plants nearly or quite glabrous are the var. **glàbra** (Lehm.) T. & G. [*G. g.* Lehm.].—Natur. as a weed in Los Angeles and Orange cos.

10. Heterogáura Rothr.

Caulescent annuals with alternate lvs. Fls. pink, in terminal spicate racemes. Fl.-tube short, obconic. Sepals 4, deciduous. Petals 4, clawed. Stamens 8, erect, the 4 epipetalous sterile; fils. not appendaged. Stigma discoid, entire, without a basal cuplike border. Ovary 4-loculed, with 1 ovule in each locule. Fr. 2–4-celled, 1–2-seeded, indehiscent. One sp. (Greek, *heter*, different, and *Gaura*.)

1. **H. heterándra** (Torr.) Cov. [*Gaura h.* Torr. *H. californica* Rothr.] Erect, stem simple or paniculately few-branched, 1–4 dm. tall, minutely puberulent throughout; lvs. oblong-ovate to lanceolate, entire to remotely and shallowly denticulate, the blades 2–5 cm. long, ca. half as wide, on petioles 0.5–1 cm. long; pedicels 1–1.5 mm. long; fl.-tube 2–3 mm. long; sepals ca. the same; petals pink, aging lavender, spatulate, 3–5 mm. long; alternate stamens fertile, 2 mm. long, opposite sterile, 1 mm. long; caps. ridged, often triquetrous, 3 mm. long; seeds slender, 2 mm. long; *n* = 9 (Lewis et al., 1958).—Dry shaded places, 2000–5000 ft.; Foothill Wd., Yellow Pine F., Chaparral; w. base of Sierra Nevada from Placer to Kern cos., Siskiyou and Trinity cos., Los Angeles and San Bernardino cos. May–June.

11. Circaèa L. ENCHANTER'S NIGHTSHADE

Low slender perennial herbs with subterranean rootstocks. Lvs. opposite, thin, petioled. Fls. small, white to pinkish, paniculately arranged in racemes. Pedicels capillary. Fl.-tube short, deciduous and with a ringlike disk within. Sepals 2, reflexed. Petals 2, notched. Stamens 2, alternate with petals. Ovary 1–2-loculed, each locule 1-ovuled. Fr. nutlike, 1–2-seeded, obovoid, indehiscent, usually with hooked hairs. Ca. 6 spp., of N. Hemis. (*Circe*, the enchantress.)

1. **C. alpìna** L. var. **pacífica** (Asch. & Magnus) Jones. [*C. p.* Asch. & Magnus.] Rootstock tuberous-thickened; stem simple, 2–4 dm. tall, usually ± strigulose below the infl.; lf.-blades ovate to roundish, usually rounded at base, sometimes ± cordate, entire or minutely denticulate or obscurely repand-denticulate, 2–6 cm. long, acuminate; petioles 2–3 cm. long; racemes bractless; sepals and petals ca. 1 mm. long; caps. narrow-obovoid, 1-loculed, 1.5–2 mm. long, with hooked hairs; *n* = 11 (Lewis et al., 1958).—Deep woods, below 8000 ft.; Montane Coniferous F., Mixed Evergreen F., Redwood F.; San Bernardino Mts., Sierra Nevada, Coast Ranges from Marin Co. n.; to B.C., Rocky Mts. June–Aug.

88. Haloragàceae. WATER-MILFOIL FAMILY (Fig. 84)

Mostly aquatic perennial herbs with alternate or verticillate lvs., the blades of those submerged often pectinate-pinnatifid or pinnately divided into capillary divisions. Fls. perfect or imperfect, axillary, solitary or grouped or in interrupted spikes. Fl.-tube adherent to ovary, prolonged little or none beyond it. Sepals 2–4 or obsolete. Petals small, 2–4 or 0. Stamens 1–8. Ovary inferior, 1–4-loculed, styles 1–4. Fr. indehiscent, angular, ribbed or winged, with 2–4 1-seeded carpels. Endosperm fleshy; cotyledons minute. Ca. 100 spp., widely distributed.

Submerged lvs. pinnatifid, the emersed entire or toothed; stamens 4 or 8; ovary 2–4-loculed
1. *Myriophyllum*
Submerged lvs. simple, entire; stamen 1; ovary 1-celled .... 2. *Hippuris*

1. Myriophýllum L. WATER-MILFOIL

Perennial aquatics; lvs. often whorled, the submersed pinnately parted into capillary divisions. Fls. sessile, chiefly in the axils of the reduced upper lvs., usually above water in summer, the uppermost ♂. Calyx of the ♂ fls. of 4 sepals, of the ♀ of 4 teeth. Fr. nutlike, 4-loculed, deeply 4-lobed; stigmas 4, recurved; dehiscence into 4 indehiscent 1-seeded bony carpels. Ca. 20 spp.

Fls. in the axils of submersed lvs. .... 1. *M. brasiliense*
Fls. in spikes borne above the water.
 Bracts shorter than the fls. and frs., not lobed; rachis of lvs. threadlike and of almost same diam.
 throughout, the segms. not broadened toward their base .... 2. *M. spicatum*
 Bracts mostly longer than the fls., toothed or lobed; rachis of lvs. flat and broader toward the
 base, the segms. broadened at the base.
  Stamens 8; petals early fugacious; lvs. subtending fls. pinnate into linear segms. much like
  those of submerged lvs. .... 3. *M. verticillatum*
  Stamens 4; petals tardily deciduous; lvs. subtending fls. toothed but not pinnate into linear
  segms. .... 4. *M. hippurioides*

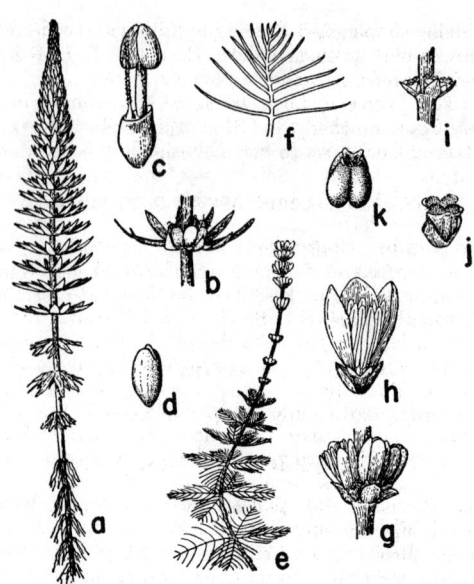

Fig. 84. HALORAGACEAE. *Hippuris vulgaris: a,* habit, × ½; *b,* whorl of lvs., × 2, with axillary fls.; *c,* fl., × 3, 1 stamen, stigmatic style; *d,* drupe, × 4. *Myriophyllum spicatum* var. *exalbescens: e,* habit, × ½, with dissected lvs. (see f, lf., × 1), terminal whorls of fls.; *g,* whorl of ♂ fls., × 2½; *h,* ♂ fl., × 5, 4 small sepals, 4 petals (1 removed), several stamens; *i,* whorl of ♀ fls., × 2½; *j,* ♀ fl., × 5, inferior ovary, 4-toothed calyx, 4 stigmas; *k,* fr., × 4, 4-lobed.

1. **M. brasiliénse** Camb. [*M. proserpinacoides* Gill.] PARROT'S-FEATHER. Stems weak, growing ca. 1–1.5 dm. out of water; lvs. all alike, feathery, in whorls of 4–6, 1.5–3 cm. long, with 10–25 capillary segms.; plants dioecious; fls. in axils of submerged lvs.— Occasional escape from cult., as in Kings R. near Hanford, Kings Co., and Mad River, Humboldt Co. Summer.

2. **M. spicàtum** L. ssp. **exalbéscens** (Fern.) Hult. [*M. e.* Fern. *M. s.* var. *e.* Jeps.] Stems simple or forked, to 1 m. long, purple, whitish on drying; lvs. in 3's or 4's, 1–3 cm. long, with 6–11 pairs of capillary segms.; fls. in emersed almost naked interrupted spikes, the lower ♀, the upper ♂; bracts rarely equaling the frs., spatulate-obovate to ± oblong, the lower serrate, the upper entire; petals fugacious, 2.5 mm. long; stamens 8; fr. subglobose, 2.5–3 mm. long, the carpels rounded on the back, smooth or rugulose; $2n = 14$ (Löve, 1954).—Our most common sp., in quiet water below 8000 ft.; many Plant Communities; scattered localities of most of cismontane Calif.; to Alaska, Atlantic Coast. June–Sept.

3. **M. verticillàtum** L. Much like the preceding, but lvs. in 4's and 5's, 1–4.5 cm. long, with 9–13 pairs of capillary flaccid divisions; spikes emersed, with pinnate bracts 2–8 mm. long; fr. 2–2.5 mm. long, deeply 4-furrowed, the carpels rounded on the back, smooth; $2n = 28$ (Scheerer, 1940).—Quiet water at low elevs.; Lake Co., Del Norte Co.; to B.C., Atlantic Coast; Eurasia. June–July.

4. **M. hippuroìdes** Nutt. ex T. & G. Lvs. in whorls of 4–6, the submerged 1.5–3 cm. long, with 7–10 pairs of divisions; the emersed linear to lanceolate, pectinate, 5–15 mm. long; petals greenish-white; frs. 2 mm. long, deeply sulcate, the carpels smooth, laterally compressed.—Scattered stations below 5000 ft., Fresno Co., San Joaquin Co., Plumas Co., Lake Co., Lassen Co., etc.; to Wash., N.Y. to Wis. May–July.

2. Hippùris L. MARE'S-TAIL

Stems simple, erect. Lvs. verticillate, simple, entire. Fls. small, axillary, mostly perfect. Sepals minute, entire. Petals 0. Stamens 1, inserted on the anterior rim of the fl.-tube. Style filiform, stigmatic its whole length along 1 side. Fr. 1-loculed, 1-seeded, becoming a nutlet. One sp. (Greek, *hippos,* horse, and *oura,* tail.)

1. **H. vulgàris** L. Stems simple, 2–5 dm. high, glabrous, usually ca. the upper half or fourth emersed, the lower part rooting at the nodes; lvs. ca. 7–10 in a whorl, lanceolate to linear, sessile, 0.5–2.5 cm. long, acute; anther ca. 1 mm. long; fr. ellipsoid-obovoid, ca. 2 mm. long; $2n = 32$ (Löve & Löve, 1948).—Uncommon, in quiet water, below 9000 ft.; many Plant Communities; San Bernardino Mts., Sierra Nevada and Coast Ranges; to Alaska, Atlantic Coast, Patagonia; Eurasia. July–Sept.

89. Myrtàceae. Myrtle Family

Aromatic trees and shrubs, mostly with opposite persistent thickish entire short-petioled exstipulate lvs., ± pellucid-punctate. Fls. bisexual, regular, solitary in axils or in racemes or corymbs, usually bracted. Fl.-tube ± adnate to the ovary, sometimes elongate. Sepals 4–5, usually persistent. Petals 4–5, imbricate, rarely lacking. Stamens usually numerous, often in fascicles opposite the petals, the fils. distinct or partly united. Ovary inferior, 1–many-loculed, with 1–many ovules in each locule, the placentation mostly axile; style simple. Fr. a berry, drupe, caps., or nut. Ca. 75 genera and 3000 spp., mostly trop. or subtrop., particularly in Australia and S. Am.

1. Eucalýptus L'Hér. Gum Tree

Lvs. mostly vertical, alternate, stiff, pinnate-veined, those of young shoots often opposite, sessile, horizontal. Fls. in umbels or heads. Fl.-tube turbinate or campanulate, adnate to ovary at base, the free part entire or 4-toothed. Sepals and petals united to form a lid or cap which is present on the bud and drops off at anthesis. Stamens many, in several series, white to highly colored. Ovary 3–6-loculed; ovules many. Fr. a caps., loculicidally dehiscent at top by 3–6 valves. Seeds minute. A large genus, largely Australasian, many spp. of importance for wood, oil, ornament. Many are grown in Calif. and some occasionally establish themselves outside of cult. In addition to those listed below, identification of others may be sought in the key by Eric Walther (Proc. Calif. Acad. Sci. ser. iv, 17: 67–87, 1928.)

Infl. a many-fld. panicle; lvs. round to ovate; fls. white, ca. 6 mm. long; lid conical, attenuate
   1. *E. polyanthemos*
Infl. umbellate or solitary; lvs. ± lanceolate.
   Fls. ca. 4 cm. across, borne singly on flattened stalks .... 2. *E. Globulus*
   Fls. smaller, in umbels, the stalks not flat .... 3. *E. tereticornis*

1. **E. polyánthemos** Schauer. Tree 10–40 m. high with brown or gray persistent bark; lvs. dull gray-green, rounded or obtuse; caps. goblet-shaped, ca. 6 mm. across.—Native of Australia. Jan.–April.

2. **E. Glóbulus** Labill. Blue Gum. Tree to 80 m. tall, the bark deciduous; lvs. of older branches 1.5–2 dm. long; fls. white, solitary in axils, ca. 4–5.5 cm. wide; lid of bud warty; fr. angular, 2–2.5 cm. across; $2n = 22$ (McAulay et al., 1936). Dec.–May.

3. **E. tereticórnis** Sm. Tree to 40 m., with smooth gray deciduous bark; lvs. 1–1.4 dm. long; fls. 4–8 in an umbel, 1.5 cm. across; lid of bud slender-conical, acuminate; caps. obovoid to subglobose, 6–8 mm. across. April–July.

90. Callítrichaceae. Water-Starwort Family (Fig. 85)

Mostly aquatic herbs, with very slender tufted stems. Lvs. opposite, entire, linear to obovate, estipulate. Fls. small, axillary, perfect or imperfect, with or without 2 saclike bracts. Perianth none. Stamen 1, the fil. elongate, anthers cordate, 4-loculed, dehiscent by longitudinal slits. Pistillate fl. a single 4-locular ovary of 2 carpels; styles 2, filiform, distinct. Fr. nutlike, compressed, 4-lobed, separating into 4 1-seeded mericarps; seeds pendulous; endosperm present. One genus with ca. 20 spp., widely distributed.

1. Callítriche L.

Characters of the family. (Greek, *kallos*, beautiful, and *trichos*, hair, because of the slender stems.)

(Fassett, N. C. Callitriche in the New World. Rhodora 53: 138 ff., 1951.)
A. Lf.-bases connected by a narrow wing; lvs. not uniform.

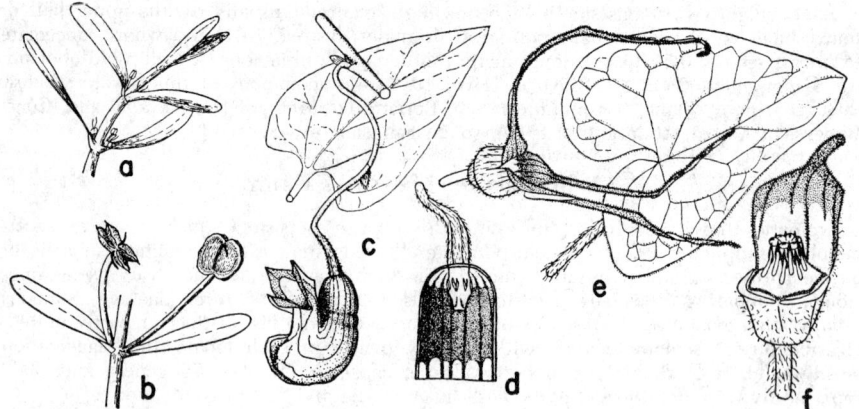

Fig. 85. CALLITRICHACEAE. *Callitriche: a,* habit, × 5, opposite lvs. with axillary ♂ and ♀ fls.;
b, compressed 4-lobed frs., × 5.
ARISTOLOCHIACEAE. *Aristolochia: c,* lf. and fl., × ½, inferior ovary, tubular curved pipelike
calyx; *d,* section of base of calyx, × 1, 6 united stamens adnate to style. *Asarum: e,* lf. and fl., × ½,
calyx 3-parted; *f,* calyx partly removed, × 1, 12 stamens with short fils.; style stout, stigmas 6.

   B.   Frs. pedicelled.
       C.   The fr. wider than high; lvs. spatulate, longer than wide, short-petioled
                                                                                   1. *C. marginata*
       CC. The fr. higher than wide; lf.-blade often wider than long, long-petioled
                                                                              5. *C. longipedunculata*
  BB. Frs. sessile or subsessile.
       C.   Carpels broadly winged all around the margin.
           D.   The fr. ca. as wide as high. N. Coast Ranges .... 4. *C. trochlearis*
           DD. The fr. higher than wide. Widely distributed .... 5. *C. longipedunculata*
       CC. Carpels wingless or winged at summit, the wing narrower or lacking on sides.
           D.   Fr. 0.2 mm. longer than wide; carpels winged at summit; reticulations on
               mericarps mostly in vertical rows .... 2. *C. verna*
           DD. Fr. scarcely if at all longer than wide; carpels wingless or obscurely falsely. so;
               reticulations on mericarps not in vertical rows .... 3. *C. heterophylla*
AA. Lf.-bases not connected by a wing; lvs. uniformly linear-lanceolate ...... 6. *C. hermaphroditica*

1. **C. marginàta** Torr. [*C. Nuttallii* Jeps., not Torr.] Stems slender, 0.5–1 dm. long;
lvs. nearly uniform, spatulate, the larger 3.5–8 mm. long, 0.8–2 mm. wide, 1–3-nerved;
fr. on ± inflated pedicels, 0.8–1.2 mm. wide, not quite so high, the wing 0.1–0.2 mm.
wide; stigmas to 1 mm. long, persistent or caducuous, sharply deflexed.—Drying mud
of vernal pools, below 2000 ft.; many Plant Communities; scattered stations from San
Diego Co. n. through cismontane Calif.; Ore., n. L. Calif. March–May.

    2. **C. vérna** L. [*C. palustris* many auth. *C. stenocarpa* Hegelm.] Slender perennial,
0.5–2.5 dm. long, the lvs. variable, the lower submersed, often linear, 1-nerved, to 1
mm. wide, the upper often dilated, the terminal in a floating rosette, obovate, ca.
4–6 mm. long, on petioles ca. as long; fr. 0.6–1.4 mm. wide, somewhat longer, sharply
reticulate on the face, the reticulations in ± well defined vertical rows, scarious-winged
at summit; $n = 10$ (Sokolovskaja, 1932).—Shallow water or on mud, to ca. 11,400 ft.;
mostly Montane Coniferous F.; mts. from San Diego Co. to Modoc Co.; N. Coast Ranges;
to Alaska, Atlantic Coast; Eurasia. May–Aug.

    3. **C. heterophýlla** Pursh. Habit of *C. verna;* fr. 0.6–0.8 mm. wide, ca. as high, the
carpels more broadly rounded at summit than at base, so that the outline of the fr. is ±
heart-shaped, the margins wingless or nearly so, the reticulations on the face not in
vertical rows.—Uncommon, quiet water, below 8000 ft.; many Plant Communities; San
Bernardino Mts., cent. Calif. n.; to Wash., Atlantic Coast, Central Am. April–Aug. The
more abundant form is:

    Var. **Bolánderi** (Hegelm.) Fassett. [*C. B.* Hegelm.] Fr. 0.9–1.2 mm. high and wide.
—From sea level to 8000 ft.; San Diego Co.; cent. Calif.; to B.C.

    4. **C. trochleàris** Fassett. Aquatic plants with linear 1-nerved lvs. below and broader
subpetiolate upper lvs.; stranded plants with spatulate lvs.; fr. nearly round, 1–1.2 mm.

in diam., distinctly winged, with fat or slightly plump faces and a shallow U-shaped commissural groove.—Ponds below 3000 ft.; Mixed Evergreen F. etc.; Sonoma to Humboldt cos. March–June.

5. **C. longipedúnculàta** Morong. [*C. marginata* var. *l.* Jeps.] Stems threadlike, forming mats; lvs. often crowded toward the ends of the branches, less than 1 cm. long, the blade often wider than long, abruptly narrowed to a long margined petiole; fr. oblong, with nearly parallel or slightly rounded sides, 0.8–1.2 mm. wide, 1–1.4 mm. long, olive-green when young, almost black when mature, distinctly pitted, the wing very narrow but equally wide all the way around, pale, slightly entering the lower sinus; commissural groove wide and flat-bottomed; pedicel 1–25 mm. long.—Water of vernal pools and later on mud, below 3000 ft.; various Plant Communities; San Diego Co., cent. Calif. March–April.

6. **C. hermaphrodítica** L. [*C. autumnalis* L. *C. a.* var. *bicarpellaris* Fenley ex Jeps. *C. h.* var. *b.* Mason. *C. bifida* Morong.] Submersed, the slender stems 1–2 dm. long; lvs. lance-linear, metallic green, narrowly white-margined, 5–15 mm. long, the bases not connected by a wing; frs. 1–2.5 mm. wide, ca. as high, obscurely and irregularly pitted, narrow-winged on margins and the outer part of the carpel itself strongly compressed and wing-like; $n = 3$ (Jorgensen, 1923).—Shallow water, below 5000 ft.; several Plant Communities; Lake, San Joaquin, Madera, Modoc cos.; to Alaska, Atlantic Coast; Eu. April–Aug.

91. Aristolochiàceae. Birthwort Family (Fig. 85)

Twining shrubs or low herbs. Lvs. alternate or basal, petioled, mostly cordate and entire. Fls. perfect, axillary or terminal, solitary or clustered, regular or irregular. Calyx-tube mostly adnate to the ovary, lurid to purple, valvate in bud, the lobes 3–6. Petals none. Stamens 5–12, ± united to the style; anthers 2-celled, adnate, extrorse. Ovary wholly or partly inferior, 6-locular, with parietal placentae, many-ovuled. Fr. a many-seeded caps. Seeds angled or compressed, smooth or wrinkled, mostly with a large raphe and minute embryo in fleshy endosperm. Ca. 200 spp., widely distributed.

Stemless herbs; calyx regular, 3-lobed; stamens 12 .... 1. *Asarum*
Twining vines; calyx irregular; stamens 6 .... 2. *Aristolochia*

1. Ásarum L. Wild-Ginger

Acaulescent perennial herbs, with creeping rootstocks. Lvs. basal, long-petioled, cordate or hastate, aromatic, mostly ovate or rounded. Fls. large, solitary, borne in lower axils. Calyx regular, 3-parted, campanulate. Stamens 12; fils. short, stout; anther-tips pointed. Caps. fleshy, globular, loculicidal or bursting irregularly. Seeds large, compressed. Ca. 20 spp., N. Temp. (Greek, *asaron*, of obscure derivation.)

Calyx-lobes long-attenuate, 25–85 mm. long.
   Rootstock rather closely scaly; styles nearly distinct; terminal appendages of anthers longer than
   pollen-sacs .... 1. *A. Hartwegii*
   Rootstock stolonlike, remotely scaly; styles united; terminal anther-appendages shorter than pollen-
   sacs .... 2. *A. caudatum*
Calyx-lobes acute or obtuse, not attenuate, 8–12 mm. long .... 3. *A. Lemmonii*

1. **A. Hartwègii** Wats. [*A. Hookeri* var. *majus* Ducharte in DC. *A. majus* Cov.] Rootstock rather closely scaly, stoutish; lvs. persistent, cordate, ovate, 4–10 cm. long, mostly acute, ± mottled above, ± pubescent beneath, glabrous above, sometimes pubescent on veins; petioles 5–15 cm. long; pedicels ca. 1–2.5 cm. long; calyx brownish-purple, hairy outside, pubescent within, the lobes 2.5–6.5 cm. long; styles shorter than stamens; seeds ca. 4 mm. long.—Shaded places, 2500–7000 ft.; Yellow Pine F., Red Fir F.; Sierra Nevada from Tulare Co. n., Siskiyou and Trinity cos.; s. Ore. May–June.

2. **A. caudàtum** Lindl. Rootstocks slender, elongate, remotely scaly; lvs. cordate-reniform, persistent, obtuse-rounded to subacute, 2–10 cm. long, pubescent below and above on veins; petioles glabrous to woolly, 3–15 cm. long; pedicels 1.5–3 cm. long, slender, pubescent; calyx-lobes 2.5–8.5 cm. long; styles united, as long as the stamens; seeds ca. 3 mm. long.—Deep shade, below 5000 ft.; Redwood F. to Yellow Pine F.; Coast Ranges from Santa Cruz Mts. n.; to B.C., Mont. May–July.

3. **A. Lemmònii** Wats. Rootstock stolonlike, remotely scaly; lvs. thin, cordate-reniform, 4–9 cm. long, rounded at apex, glabrous above, ± floccose beneath; petioles 5–12 cm. long, ± floccose; pedicel slender, 1–2 cm. long; calyx-lobes 8–12 mm. long, acute; appendage of anther shorter than anther-body; styles united.—Wet places, 3600–6000 ft.; Yellow Pine F., Red Fir F.; Sierra Nevada, from Tulare Co. to Plumas Co. May–June.

2. Aristolòchia L. PIPE VINE. DUTCHMAN'S PIPE

Perennial herbs or shrubs, twining or climbing. Lvs. cordate, entire or lobed. Fls. axillary, pendulous, irregular, greenish to lurid purple. Calyx tubular, strongly curved, pipelike, the tube usually inflated around the style and contracted at the throat, limb spreading or reflexed, entire to lobed or appendaged. Stamens mostly 6; anthers sessile, adnate to the short style or stigma. Stigma 3- to 6-lobed or -angled. Caps. septicidal, 6-valved. Seeds compressed. Ca. 180 spp., largely trop. (Greek, *aristos*, best, and *locheia*, parturition, from its supposed value for childbirth.)

1. **A. califórnica** Torr. Woody climber, 3–4 m. tall, ± pubescent, even silky; lvs. ovate-cordate, 3–15 cm. long, on petioles 2–5 cm. long; pedicels 1–2 cm. long, bracted near the middle; calyx greenish, with purple veining, 2–4 cm. long, strongly saccate, 1–1.5 cm. broad, the limb 2-lipped with 2 broad lobes in upper lip, the lower lip entire, the lobes lined with a disklike thickening forming a projection partly closing the tube; stigma-lobes 3, broad; caps. obovate, narrowed to a slender base, 3–4 cm. long, 6-winged; seeds cuneate-obovate, 6 mm. long, concave, with a spongy raphe in the concavity.—Stream banks, etc., growing over shrubs; mostly below 1500 ft.; Foothill Wd., Chaparral, Mixed Evergreen F.; Coast Ranges from Monterey Co. n., Sierra Nevada foothills from Sacramento and Eldorado cos. n. to head of Sacramento V. Jan.–April.

92. Rafflesiàceae. RAFFLESIA FAMILY

Dioecious or monoecious fleshy herbs, parasitic on roots or branches, the vegetative body reduced to myceliumlike tissues or thalloid; lvs. usually scalelike. Fls. minute to large, solitary, unisexual. Calyx of 4–10 ± distinct segms. Corolla 0. Stamens indefinite; anthers sessile, usually 2-celled. Pistillate fl. with single pistil, the ovary ± inferior, unilocular, of 4–6–8 carpels with parietal placentation and many ovules; style 1 or 0; stigma discoid or capitate or lobed. Fr. a berry; seed with endosperm. Seven genera and 27 spp., mostly of Old World.

1. Pilóstyles Guill.

Minute stem-parasite; lvs. reduced to fl.-bracts. Fls. brown, minute; bracts orbicular, imbricated; sepals 4–5, distinct, similar to the subtending bracts. Fls. with a thick cent. column expanded at apex into a fleshy disk. Anthers in ♂ fls. many, borne under margin of disk. Stigma in ♀ fls. ring-shaped. Ovary inferior; fr. a many-seeded caps. Several spp., mostly of trop. Am. (Latin, *pilus*, hair, and *stylus*, a pillar or stylus.)

1. **P. Thúrberi** Gray. Only the small brown fls. and their imbricated bracts visible on the outside of the host, the whole 1.5–2 mm. long.—On *Dalea*, especially *D. Emoryi*; Creosote Bush Scrub; Imperial Co.; to Ariz. Jan.

93. Celastràceae. STAFF-TREE FAMILY (Fig. 86)

Trees, shrubs or woody vines. Lvs. alternate or opposite, simple, deciduous or persistent. Stipules small and caducous or none. Fls. small, regular, mostly bisexual, commonly on jointed pedicels. Sepals 4–6, imbricate, persistent. Petals 4–6, spreading. Disk broad, flat or lobed, fleshy. Stamens usually as many as the petals, sometimes more, inserted under or on the disk. Ovary sessile, the base free from or adherent to the disk, 3–5-loculed. Style 1, short; stigma entire or 3–5-lobed. Ovules mostly 2 in locule. Fr. a caps., loculicidal to indehiscent. Seeds usually with an aril; embryo large. Ca. 450 spp., of temp. and trop. areas.

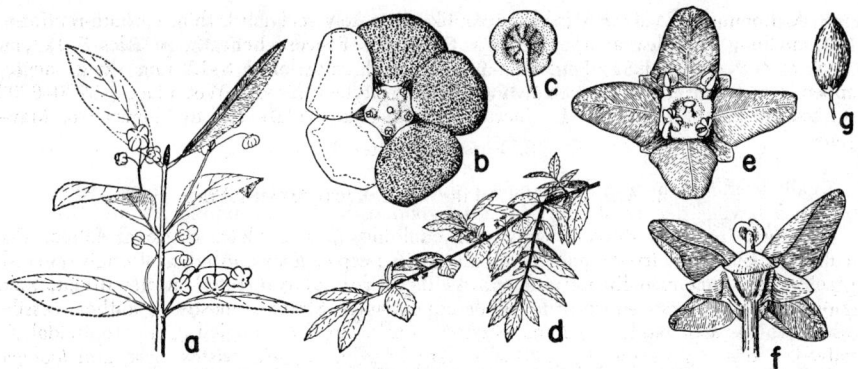

Fig. 86. CELASTRACEAE. *Euonymus occidentalis: a,* habit, × ¼, with opposite lvs. and small cymes of fls.; *b,* fl., × 3, 5 spreading sepals, cent. disk with inserted stamens and cent. style; *c,* disk from beneath. *Paxistima Myrsinites: d,* habit, × ½, opposite lvs. axillary fls.; *e* and *f,* fl. in 2 views, × 5, sepals 4, petals 4, stamens 4 and at edge of disk; style cent.; *g,* caps., × 2.

Lvs. opposite.
    Plants deciduous, 2–6 m. high; lvs. 3–9 cm. long .... 1. *Euonymus*
    Plants evergreen, to ca. 1 dm. high; lvs. 1–2.5 cm. long .... 2. *Paxistima*
Lvs. alternate.
    Twigs brownish or olive; lvs. spatulate to oblanceolate or elliptic; stamens often twice as many
        as the petals .... 3. *Forsellesia*
    Twigs yellowish, hispidulous; lvs. very thick, elliptic; stamens as many as the petals .. 4. *Mortonia*

1. Euónymus L. BURNING BUSH

Shrubs with opposite petioled lvs., evergreen or deciduous. Stipules minute or 0. Fls. greenish or purplish, small, in axillary few-fld. cymes, mostly 5- or 4-merous. Sepals spreading or recurved. Stamens inserted on the broad disk. Ovary 3–5-loculed, short; stigma 3–5-lobed. Caps. 3–5-loculed and 3–5-lobed or rounded. Seeds 1–2 in a locule, enveloped by a red aril. Ca. 175 spp., of N. Temp. regions, mostly Eurasian. (Greek, *eu,* good, and *onoma,* a name.)

1. **E. occidentàlis** Nutt. ex Torr. Deciduous shrub or arborescent, 2–6 m. high, with slender often straggling branches and smooth greenish angled branchlets; lvs. thin, glabrous, ovate, abruptly acuminate, serrulate, 3–9 cm. long; petioles 0.5–1.5 cm. long; peduncles slender, 1–5-fld., 2–6 cm. long; petals 5, rounded, brown-purple, finely dotted, scarious at edges, 3–4 mm. long; caps. depressed, smooth, deeply 3-lobed; seeds ca. 6 mm. long.—Damp wooded banks and canyons, below 5200 ft.; Redwood F., Mixed Evergreen F., Closed-cone Pine F., Yellow Pine F.; Monterey Co. to Del Norte Co., Siskiyou Co.; to Wash. Reported also from Plumas Co. April–June.

Var. **Párishii** (Trel.) Jeps. [*E. P.* Trel.] Branchlets whitish; cymes 3–5-fld.; lvs. usually obtusish.—Occasional in canyons, 4500–6500 ft.; Yellow Pine F.; San Jacinto Mts. to Cuyama and Palomar mts.

2. Paxístima Raf. MOUNTAIN LOVER. OREGON-BOXWOOD

Low evergreen glabrous shrubs with corky squarish branchlets. Lvs. opposite, coriaceous, serrulate, with minute caducous stipules. Fls. small, perfect, solitary or in few-fld. axillary cymes, 4-merous. Stamens inserted on the edge of the rounded disk. Ovary 2-loculed, 4-ovuled; style very short; stigma shallowly 2-lobed. Caps. oblong, compressed, 2-loculed, loculicidal. Seeds with a whitish lacerate aril at the base. A small N. Am. genus. (Greek, *pachus,* thick, and *stima,* stigma.)

1. **P. Myrsinìtes** (Pursh) Raf. [*Ilex M.* Pursh. *Myginda myrtifolia* Nutt.] Low densely branched very leafy shrub 3–10 dm. high, sometimes spreading and almost prostrate; lvs. oblong, ovate or oblanceolate, simple, rounded at ends, 1–2.5 cm. long, dark glossy green above, paler beneath, very short petioled; peduncles 2–3 mm. long; fls. 1–3; petals

reddish-brown, ovate, 1 mm. long; caps. 4–5 mm. long.—Abundant in ± shaded places, 2000–6000 ft.; Yellow Pine F., Red Fir F., Mixed Evergreen F.; Coast Ranges, Marin Co. to Del Norte and Siskiyou cos., Sierra Nevada from Mariposa Co. to Modoc Co.; to B.C., Rocky Mts. to New Mex. May–July.

3. Forsellèsia Greene

Small deciduous intricately branched shrubs with slender greenish angled spinescent branches having decurrent lines from the nodes; lvs. small, simple, entire, alternate, usually with minute stipules. Fls. usually solitary in axils, minute, mostly 5-merous. Petals white, narrow-oblanceolate, deciduous, inserted under the edge of the crenately lobed disk. Stamens equal or unequal, 4–10. Carpels 1–3, distinct, ovoid, attenuate to the stigma, sessile. Ovary superior, 1-loculed, 1–2-ovuled. Fr. a coriaceous follicle, striate, opening along the ventral suture.

(Ensign, M. A revision of the Celastraceous genus Forsellesia [Glossopetalon]. Am. Midl. Nat. 27: 501–511, 1942.)

Low matted shrubs, not spiny, 0.5–2 dm. high; lvs. elliptical, spine-tipped; fls. terminal, 5-merous
　　　　　　　　　　　　　　　　　　　　　　　　　　　　　　　　1. *F. pungens*
Intricately branched shrubs, ± spiny, 2–30 dm. high; lvs. oblong to oblanceolate, acute to acuminate; fls. axillary, 4–5-merous.
　　Young branches ca. 1 mm. in diam.; petals 4–7 mm. long. Death V. to Mojave Desert
　　　　　　　　　　　　　　　　　　　　　　　　　　　　　　2. *F. nevadensis*
　　Young branches ca. 0.5 mm. in diam.; petals 6–9 mm. long. White Mts., Inyo Co. and in
　　Trinity Co. .... 3. *F. stipulifera*

1. **F. púngens** (Bdg.) Heller var. **glàbra** Ensign. Stems and lvs. glabrous, the latter 6–10 mm. long, estipulate; pedicels 3–4 mm. long, with 3–4 scarious bracts at base; sepals 5, ovate, acuminate, 2- or 3-spinose-tipped, denticulate, hyaline-margined; petals 5, 6–8 mm. long; stamens 10, those opposite the sepals longer; carpels ovoid, less than 1 mm. long.—Limestone cliffs, ca. 5500–6500 ft.; Pinyon-Juniper Wd.; Clark Mts., e. San Bernardino Co. May–June.

2. **F. nevadénsis** (Gray) Greene. [*Glossopetalon n.* Gray. *G. spinescens* Calif. auth.?] Divaricately branched, the stems ribbed, pubescent; the young branches more than 1 mm. thick, yellowish in age; lvs. oblong to oblanceolate, 5–12 mm. long, pubescent; petioles 1 mm. long; stipules subulate, less than 1 mm. long, adnate to a persistent often thickened glandular base; fls. axillary, 4–5-merous; pedicels 3–5 mm. long, with several bracts; sepals entire, 1–3 mm. long; petals 4–7 mm. long; stamens 6–10, unequal; carpels 1–2, to 5 mm. long in fr.—At 3500–7000 ft., often on limestone; Joshua Tree Wd., Pinyon-Juniper Wd.; Mojave Desert, from n. base of San Bernardino Mts. to Inyo Mts., Death V. region; Nev., Ariz. April–May. Forma **glàbra** Ensign. Mostly glabrous; with the sp.

3. **F. stipulífera** (St. John) Ensign. [*Glossopetalon s.* St. John.] Freely branched, 1–3 m. tall, occasionally spinescent; young branches 0.5 mm. thick; lvs. glabrous, oblanceolate, 6–17 mm. long; petioles 1 mm. long; stipules mostly over 1 mm. long; pedicels 2–5 mm. long, axillary; sepals 2 mm. long; petals 6–9 mm. long; stamens 5–8, equal; carpels solitary, 3–5 mm. long.—At 5700 ft., White Mts., Inyo Co., also in Trinity R. Canyon, Trinity Co.; to Wash., Ida. May.

4. Mortònia Gray

Rather low evergreen much-branched shrubs. Lvs. alternate, crowded, subsessile, coriaceous, 1-nerved, revolute, with minute glandlike caducous stipules. Fls. small, white, in narrow terminal thyrsoid cymes. Calyx tubular at base, 10-ribbed, obconic, the lobes 5. Petals 5. Stamens 5, with short fils. Ovary 5-loculed; style columnar; stigmas 5; ovules 2 in a locule, erect, basal. Fr. dry, indehiscent, 1-seeded. Seed oblong, not arillate, with erect embryo. A small genus of sw. N. Am. (Dr. S. G. *Morton*, Am. naturalist of last century.)

1. **M. utahénsis** (Cov.) A. Nels. [*M. scabrella* var. *u.* Cov. ex Gray.] Low, 9–12 dm. high, yellow-green, hispidulous, densely leafy; lvs. broadly oval to roundish, 8–12 mm. long, scabrellous, with thickened margin; petioles almost lacking; infl. 3–6 cm. long; bracts lanceolate, 3–5 mm. long; calyx-lobes 2 mm. long, with scarious margins; petals

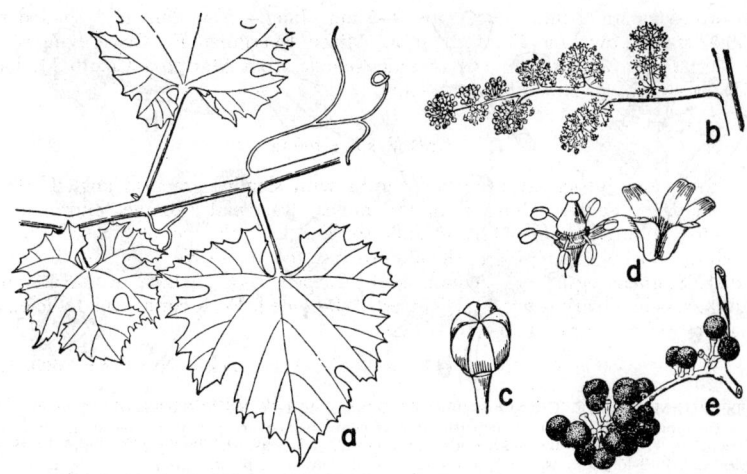

Fig. 87. VITACEAE. *Vitis: a,* habit, with lvs. and tendrils, × ¼; *b,* infl., × ½; *c,* bud, × 5, showing unexpanded petals; *d,* opened fl., × 5, the petals coherent and pushed off together; *e,* berries, × ½.

obovate, 3 mm. long; caps. oblong, 4 mm. long, glabrous.—On limestone, 3400–5500 ft.; mostly Joshua Tree Wd., Pinyon-Juniper Wd.; Clark, Kingston, and Panamint mts., e. Mojave Desert; to Utah, Ariz. March–May.

94. Vitàceae. Grape Family (Fig. 87)

Woody vines climbing by tendrils opposite the lvs. Lvs. alternate, petioled. Fls. small, paniculate-cymose, bisexual or unisexual. Calyx entire or 4–5-toothed. Petals 4–5, separate or coherent, valvate, deciduous without expanding. Stamens as many as the petals and opposite them. Ovary generally immersed in the disk, 2–6-loculed; ovules 1–2 in each locule. Fr. a berry, commonly 2-celled. Seeds with a bony testa and hard endosperm; embryo short. Ca. 500 spp., widely distributed in trop. and temp. regions.

1. Vìtis L. Grape

Lvs. palmately lobed or dentate, with small caducous stipules. Calyx minute, with entire limb. Petals hypogynous or perigynous, coherent in a cap. Ovary mostly 2-loculed; style short, conic; ovules 2 in each locule. Berry ovoid to round, pulpy. Ca. 50 spp., of N. Hemis. (Classical Latin name.)

(Bailey, L. H. The spp. of grapes peculiar to N. Am. Gentes Herb. 3: 150–244, 1934.)
Skin of fr. loose and easily separating from pulp; young growth ± tomentose. Native.
    Young growth arachnoid-pubescent or -tomentose, but green in color; berry purple, with a dense glaucous bloom. From San Luis Obispo and Kern cos. n. . . . . 1. *V. californica*
    Young growth densely white-tomentose, the lvs. permanently so beneath; berry black, scarcely or not glaucous. From Inyo and Santa Barbara cos. s. . . . . 2. *V. Girdiana*
Skin of fr. adhering and not easily separating from the pulp; young growth often glabrous. Escape
                                                        3. *V. vinifera*

1. V. califórnica Benth. Stems 2–15 m. long, with shreddy bark; lvs. roundish, pubescent or thinly arachnoid-tomentose, especially beneath, round-cordate to broadly ovate-cordate, the basal sinus usually deep, often narrow, blade 3-lobed or scarcely so, 7–14 cm. broad, usually with broad short-apiculate blunt teeth; petioles 3–12 cm. long; panicle ca. 5–15 cm. long; fls. greenish-yellow, fragrant, ca. 1.5 mm. long, 5-merous; pedicels of fr. rough or warty; berries globose, 6–10 mm. in diam., purplish, covered with whitish bloom; seeds pyriform, 4 mm. long; $2n = 38$ (Kobel, 1929).—Stream banks and canyons, below 4000 ft.; Mixed Evergreen F., N. Oak Wd., Foothill Wd., etc.; Coast Ranges from San Luis Obispo to Siskiyou cos., Cent. V. and Sierra Nevada foothills from Kern Co. n.; s. Ore. May–June.

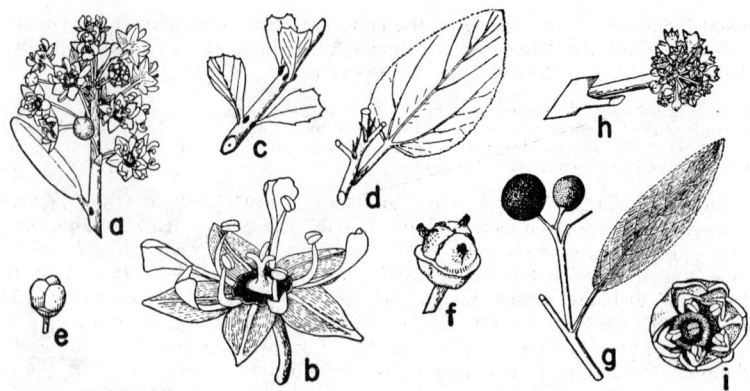

Fig. 88. RHAMNACEAE. *Ceanothus insularis: a,* infl., × 1; *b,* fl., × 5, sepals 5, petals 5, clawed, hooded, with opposite stamens from beneath edge of cent. disk, ovary partly immersed in disk. *C. gloriosus: c,* twig, × 1, opposite lvs. with corky stipules. *C. arboreus: d,* alternate lvs. and thin stipules. *C. integerrimus: e,* 3-lobed caps., × 1, without crests or horns. *C. prostratus: f,* caps., × 1, with horns. *Rhamnus: g,* lf. and drupes, × ½; *h,* axillary infl., × 1; *i,* fl., × 10, sepals 5, petals 5, stamens 5, ringlike disk with cent. pistil.

2. **V. Girdiàna** Munson. Stems 1–6 m. long; lf.-blades broadly cordate-ovate, 5–16 cm. wide, mostly with triangular apex and with rather deep and narrow sinus, obscurely or distinctly lobed as well as irregularly sharply dentate, ashy tomentose beneath; petioles mostly 3–8 cm. long; petals and stamens 6; pedicels of fr. smooth; berries globose, 3–6 mm. in diam., black, with little bloom.—Canyon bottoms along streams, etc., below 4000 ft.; S. Oak Wd., Coastal Sage Scrub, etc.; Santa Barbara to San Diego cos., occasional on desert edge from Inyo Co. s., Santa Catalina Id.; L. Calif. May–June.

3. **V. vinífera** L. The cultivated grape occasionally establishes itself as a wild plant; lvs. rather thin, 10–20 cm. wide; fr.-clusters large; berries more than 1 cm. long; $2n = 38$, 57, 76 (Olmo, 1937).—Native of Europe.

95. Rhamnàceae. BUCKTHORN FAMILY (Fig. 88)

Shrubs or small trees, sometimes climbers, often thorny. Lvs. simple, alternate, sometimes opposite, with small deciduous stipules or these sometimes thick, corky, persistent. Fls. small, regular, perfect or imperfect, usually in terminal or axillary cymes, corymbs or panicles made up of small umbels. Calyx ± tubular at base, 4–5-lobed, lined with a disk on edge of which are inserted the petals and stamens. Petals 4–5, usually clawed, hooded, sometimes lacking. Stamens 4–5, opposite the petals; anthers short. Ovary 2–3-loculed, partly immersed in the disk; ovules 1–2 in each locule. Styles and stigmas ± united. Fr. a caps. or drupe. Ca. 45 genera and 600 spp., of temp. and trop. regions.

Fruit fleshy, drupelike.
    Drupe with 1 nutlet; petals clawed or none; spinose desert shrubs . . . . 1. *Condalia*
    Drupe with 2–3 nutlets; petals clawless or none; plants not usually spinose . . . . . . 2. *Rhamnus*
Fruit dry, capsular.
    Calyx-tube joined to base of ovary, the sepals deciduous in fr.; style not jointed.
        Pedicels and calyx glabrous; calyx-lobes or sepals petaloid; petals showy, hooded and long-clawed, often spreading away from the stamens. Common, widespread . . . . . . 3. *Ceanothus*
        Pedicels and calyx tomentose; calyx-lobes not petaloid; petals small, sessile, surrounding the stamens. Rare, desert plant . . . . 4. *Colubrina*
    Calyx-tube free from lower part of ovary, the sepals persistent; style jointed. Sw. San Diego Co.
                                                              5. *Adolphia*

1. Condàlia Cav. CRUCILLO

Shrubs or small trees with divaricate spiny twigs. Lvs. alternate, entire, with minute stipules. Fls. in small axillary umbellike cymes. Calyx deeply 5-lobed. Petals hooded and clawed if present. Ovary free from the calyx and disk, incompletely 2-loculed; styles 2–3-notched or shallowly lobed. Fr. a drupe with a single seed. Ca. 10 spp., of warm parts of New World. (A. *Condal,* Spanish physician.)

Petals present; calyx-lobes deciduous.
   Plants canescent; infl. peduncled; drupe beakless, 6–10 mm. long .... 1. *C. lycioides*
   Plants glabrous; infl. sessile; drupe beaked, ca. 15 mm. long .... 2. *C. Parryi*
Petals none; calyx-lobes persistent .... 3. *C. globosa*

1. **C. lycioìdes** (Gray) Weberb. var. **canéscens** (Gray) Trel. in Gray. [*Zizyphus l.* var. *c.* Gray.] Rigid much-branched shrub 1–4 m. tall, with ± zigzag canescent spiny twigs; lvs. elliptical to oblong-ovate, 8–15 mm. long, ± canescent, short-petioled, soon deciduous; umbel short-peduncled, 2–6-fld.; fls. minute; drupe ellipsoid, 6–10 mm. long, deep blue to black, with ± bloom.—Uncommon in sandy places, below 1500 ft.; Creosote Bush Scrub; Colo. Desert; to Nev., Ariz., Son., L. Calif. April–June.

2. **C. Párryi** (Torr.) Weberb. [*Zizyphus P.* Torr.] Large rounded shrubs 1–4 m. high, glabrous, the ultimate branchlets spiny, flexuous; lvs. mostly fascicled, obovate to oblong-elliptical, 8–20 mm. long, coriaceous, bright green, on short slender petioles; fls. 1–2 mm. long; drupe ellipsoid-ovoid, 1–2 cm. long, usually distinctly beaked, with dry thin flesh, the pedicels 1–1.5 cm. long.—Local on dry slopes and in canyons, 1200–3500 ft.; Joshua Tree Wd., Pinyon-Juniper Wd., upper Creosote Bush Scrub; w. edge of Colo. Desert from Morongo Pass to L. Calif. Feb.–April.

3. **C. globòsa** Jtn. var. **pubéscens** Jtn. [*C. spathulata* of Calif. refs., not Gray.] Intricately branched spinose shrub with spreading twigs, puberulent, to 4 m. tall; lvs. narrow-spathulate to oblanceolate, 5–15 mm. long, puberulent, thick; umbels 1–2-fld.; pedicels 3–4 mm. long; sepals deciduous, 1 mm. long; drupe obliquely ovoid, 4–5 mm. long, black and juicy.—Uncommon, at low elevs.; Creosote Bush Scrub; e. Colo. Desert; to Ariz., n. L. Calif. March–April.

2. Rhámnus L. BUCKTHORN. CASCARA

Shrubs or small trees, evergreen or deciduous. Lvs. alternate, pinnately veined. Fls. small, greenish, bisexual or unisexual, in axillary clusters. Calyx 4–5-lobed or -toothed, the tube circumscissile after anthesis. Petals when present 4–5. Stamens 4–5, with short fils. Pistil 1, free from the disk; ovary 2–4-loculed. Fr. a berrylike drupe with 2–4 separate nutlets. Ca. 100 spp., of almost world-wide distribution, some of considerable medicinal value. (The ancient Greek name.)

(Wolf, C. B. The N. Am. spp. of Rhamnus. Rancho Santa Ana Bot. Gard. Mon. 1, 1938.)
Petals lacking; winter-buds with bud-scales.
   Lvs. deciduous, 4–10 cm. long; fr. black; fls. 5-merous .... 1. *R. alnifolia*
   Lvs. persistent, 1–4 cm. long; fr. red; fls. 4-merous .... 2. *R. crocea*
Petals present; winter-buds without bud-scales.
   Lvs. persistent, rather thick and coriaceous, mostly 3–8 cm. long; bark of branchlets usually grayish or brownish. Through most of Calif. .... 3. *R. californica*
   Lvs. deciduous, rather thin and not coriaceous. Cent. and n. Calif.
      Bark of branchlets cherry-red, rarely grayish; lvs. 3–7 cm. long; berry mostly 2-seeded
                                                     4. *R. rubra*
      Bark of branchlets grayish; lvs. 8–20 cm. long; berry mostly 3-seeded .... 5. *R. Purshiana*

1. **R. alnifòlia** L'Hér. [*Apetlorhamnus a.* Nieuwl.] Mostly 1–1.5 m. high, deciduous, with gray or reddish-brown branchlets; lvs. elliptical to ovate, 3–8 cm. long, acuminate, serrulate, glabrous or puberulent above and beneath; petioles 4–14 mm. long; umbels 1–3-fld., appearing with the lvs.; fls. unisexual, mostly 5-merous; pedicels 2–6 mm. long; calyx ca. 3 mm. long; petals none; berry black, 6–8 mm. in diam., spherical or obovoid, 3-seeded; seeds 5–6 mm. long, obovoid, with deeply notched apex.—Swampy or boggy places, 4500–7000 ft.; Red Fir F., Lodgepole F.; Sierra Nevada from Placer Co. to Plumas Co.; to Wash., Sask. and Atlantic Coast. May–June.

2. **R. cròcea** Nutt. in T. & G. BUCKTHORN. REDBERRY. Spreading gray-green much-branched shrub mostly 1–2 m. high, with rigid often spinescent branchlets, evergreen; lvs. often fascicled, glabrous or slightly puberulent, shining, rigidly coriaceous, elliptic to broadly ovate or obovate, 5–15 mm. long, usually glandular-serrulate; petioles 1–4 mm. long; fls. unisexual, 4(–5)-merous; pedicels 1–4 mm. long; petals 0; berry red, 5–6 mm. long, obovoid, 2-seeded; seeds 4 mm. long, finely reticulate, rounded at apex, the

outer side deeply grooved.—Dry washes and canyons, below 3000 ft.; Coastal Sage Scrub, Chaparral, Mixed Evergreen F., S. Oak Wd., etc.; Coast Ranges from Lake Co. to San Diego Co.; L. Calif. March–April.

KEY TO SUBSPECIES

Branchlets somewhat divaricate, ending in weak spines; lf.-blades 10–17 mm. long ...... *R. crocea*
Branchlets not divaricate, seldom ending in weak spines; lf.-blades 18–60 mm. long.
  Shrubs usually with many branches from the base; lvs. mostly serrate to dentate.
    Lvs. pilose on both surfaces, revolute. San Diego Co. .... ssp. *pilosa*
    Lvs. glabrous to pubescent, rarely slightly revolute. Widespread .... ssp. *ilicifolia*
  Small trees with distinct trunks; lvs. mostly crenate to entire, occasionally serrate. Insular
                                                           ssp. *pirifolia*

Ssp. **pilòsa** (Trel.) C. B. Wolf. [*R. crocea* var. *p.* Trel.] Upright sparsely branched, 1.5–2 m. high; young branchlets, lvs., floral parts densely pilose; lf.-blades 15–28 mm. long, orbicular to broadly ovate, mostly revolute.—Dry slopes, ca. 1500 ft.; Chaparral; San Diego Co.

Ssp. **ilicifòlia** (Kell.) C. B. Wolf. [*R. i.* Kell.] Branched from base or almost arborescent, 1.5–4 m. tall; petioles 2–8 mm. long; lf.-blades 18–40 mm. long, oval to roundish, glabrous, or slightly pubescent beneath, subentire to spinosely serrate; fr. to 8 mm. long; seeds to 6 mm. long.—Dry slopes mostly below 5000 ft.; Chaparral, N. Oak Wd., Foothill Wd., Yellow Pine F., etc.; Coast Ranges from Siskiyou Co. to San Diego Co.; foothills of Sierra Nevada; Providence Mts., San Bernardino Co.; Ariz., L. Calif. March–June.

Ssp. **pirifòlia** (Greene) C. B. Wolf. [*R. p.* Greene. *R. c.* var. *insularis* Sarg. *R. catalinae* A. Davids.] Treelike, to 10 m. high; petioles 5–10 mm. long; lf.-blades elliptical to almost round, to 4 cm. long.—Coastal Sage Scrub, Chaparral; Channel Ids.

3. **R. califórnica** Esch. [*Frangula c.* Gray. *R. laurifolia* Nutt. in T. & G.] COFFEEBERRY. Upright rounded shrub to low and spreading, 1–4 m. tall; bark of young twigs usually reddish; lvs. persistent, oblong to elliptic, plane or ± revolute, entire to serrate, acute to obtuse, 3–8 cm. long, usually shining and glabrous above, glabrous or with a few hairs beneath; fls. perfect, 5-, rarely 4-merous; umbels on peduncles 4–18 mm. long, 6–50-fld.; fls. 2–3 mm. long; berries green, black or red when ripe, subglobose or somewhat elongate, 10–12 mm. long; seeds mostly 2, green-brown, 7–9 mm. long, smooth; $2n = 24$ (Bowden, 1945).—Sandy and rocky places along the coast, hillsides and ravines farther back, below 3500 ft.; Coastal Strand, N. Coastal Scrub, Mixed Evergreen F., Redwood F., Chaparral, Coastal Sage Scrub, etc., outer Coast Ranges from w. Siskiyou Co. to Orange Co. May–July.

KEY TO SUBSPECIES

Lvs. glabrous or subglabrous beneath.
  The lvs. green beneath, entire or serrate; seeds mostly 2 .... *R. californica*
  The lvs. yellowish beneath, entire or crenate; seeds mostly 3 .... ssp. *occidentalis*
Lvs. pubescent or tomentose beneath.
  Lf.-margins entire or with blunt teeth; pubescence of short dense white hairs.
    Lf.-blades broadly elliptical. Inner N. Coast Ranges .... ssp. *crassifolia*
    Lf.-blades narrowly elliptical. Shasta and Trinity cos. to Mariposa and San Diego cos.
                                                 ssp. *tomentella*
  Lf.-margins with sharp teeth; pubescence largely of short white hairs mixed with long coarse ones.
    Margins of blades dentate. Madera and Mono cos. to Riverside and Orange cos. ... ssp. *cuspidata*
    Margins of blades serrate to almost entire. E. Mojave Desert .... ssp. *ursina*

Ssp. **occidentàlis** (Howell) C. B. Wolf. [*R. o.* Howell.] Rounded bushes to 2 m. high; young bark green; lvs. thick, 3–8 cm. long, oval or elliptical, glabrous or subglabrous, green above, yellow beneath.—Dry slopes, below 7500 ft.; Yellow Pine F., Red Fir F.; Del Norte to Trinity and Shasta cos.; sw. Ore. June–July.

Ssp. **crassifòlia** (Jeps.) C. B. Wolf. [*R. californica* var. *c.* Jeps.] Two to 3 m. high; young twigs gray-velvety; lvs. 3–10 cm. long, oval or broadly elliptical, gray-green above, whiter beneath, with dense short soft pubescence.—Dry slopes and ravines, below 2500 ft.; Chaparral; inner N. Coast Ranges from Trinity to Napa and Lake cos. May–July.

Ssp. **tomentélla** (Benth.) C. B. Wolf. [*R. t.* Benth.] Erect, to 5 m. high, densely pubescent on young growth; lvs. 3–7 cm. long, lanceolate to oblong, glabrous or pubescent above, tomentose beneath.—Dry slopes, below 3000 ft.; Chaparral, Foothill Wd.;

inner Coast Ranges from Trinity Co. s., Sierra Nevada foothills from Shasta Co. s. to Los Angeles Co.; L. Calif. May–June.

Ssp. **cuspidàta** (Greene) C. B. Wolf. [*R. c.* Greene. *R. californica* var. *viridula* (virida) Jeps. *Frangula v.* Grubov.] Upright, to 6 m. tall; young growth densely pubescent and with long hairs; blades 2–6 cm. long, oval to elliptical, bright or dull green above, white-tomentose beneath.—Dry slopes and canyons, 2000–7500 ft.; Creosote Bush Scrub to Pinyon-Juniper Wd., Foothill Wd. to Yellow Pine F.; e. and w. slopes of Sierra Nevada from Mono and Madera cos. s. to Orange and Riverside cos. April–July.

Ssp. **ursìna** (Greene) C. B. Wolf. [*R. u.* Greene.] Rounded shrubs to 5 m. high; lvs. 3–7 cm. long, elliptical or oval, dull or bright green and glabrous or pubescent above, white-tomentose beneath.—Canyons, 4000–7000 ft.; Joshua Tree Wd., Pinyon-Juniper Wd.; Providence, New York, and Clark mts., e. San Bernardino Co.; Ariz., New Mex. May–July.

√ 4. **R. rùbra** Greene. [*R. californica* var. *r.* Trel. *Frangula r.* Grubov.] Low shrubs 1–1.5 m. high, spreading or rounded, mostly with reddish subglabrous twigs, deciduous; lvs. rather thin, narrowly elliptic to oblong, 2–6 cm. long, mostly acute, green above and beneath, minutely and sharply serrulate; petioles 3–8 mm. long; fls. perfect, 5-merous in small peduncled umbels; pedicels 2–8 mm. long; calyx ca. 3 mm. long; petals 1 mm. long, notched at apex; berry obovoid to globose, black, mostly less than 1 cm. in diam., mostly 2-seeded; seeds 6–8 mm. long, broadest above the middle.—Dry slopes, 4000–7000 ft., as open thickets in Montane Coniferous F.; Siskiyou Co. to Calaveras Co. May–Aug.

KEY TO SUBSPECIES

Lf.-blades glabrous to merely puberulent along veins of lower surface, entire to serrulate.
 Blades acute at apex, bright green . . . . *R. rubra*
 Blades obtuse at apex.
  Lvs. scattered along the slender branchlets, not borne on stubby spurs; peduncles mostly over 2 mm. long. Eldorado Co. to Siskiyou and Modoc cos. . . . . ssp. *obtusissima*
  Lvs. ± fascicled at tips of short stubby spurs; peduncles mostly less than 2 mm. long. Modoc, Siskiyou, and Shasta cos. . . . . ssp. *modocensis*
Lf.-blades with both surfaces with fine soft hairs, sharply denticulate with minute teeth. Tuolumne, Mariposa, and Mono cos. . . . . ssp. *yosemitana*

Ssp. **obtusíssima** (Greene) C. B. Wolf. [*R. o.* Greene.] Lvs. oblong to obovate, rounded to obtuse at apex, 2.5–6 cm. long.—Dry open slopes, 2000–7000 ft.; Yellow Pine F.; Sierra Nevada from Tuolumne Co. to Shasta Co. May–Aug.

Ssp. **modocénsis** C. B. Wolf. Lvs. small, 1.5–4 cm. long, crowded at tips of short stubby spurs.—Dry places, 4000–5000 ft.; Sagebrush Scrub, N. Juniper Wd., etc.; Siskiyou, to Modoc and Lassen cos.

Ssp. **yosemitàna** C. B. Wolf. Lvs. soft-puberulent on both surfaces, 3–7 cm. long.—Dry open places, 4000–6500 ft.; Yellow Pine F., Sagebrush Scrub; Mono to Tuolumne and Mariposa cos. June–July.

5. **R. Purshiàna** DC. [*Frangula P.* Cooper. *R. anonaefolia* Greene. *F. a.* Grubov.] CASCARA SAGRADA. Large shrub to small tree 5–12 m. tall; bark gray; young growth sparingly pubescent; lvs. deciduous, tufted at ends of branches; petioles 6–23 mm. long; lf.-blades thin, 5–15 cm. long, broadly elliptical or obovate, obtuse to subtruncate at apex, green above, paler beneath, subglabrous to pubescent on both surfaces, mostly irregularly serrate or crenate; fls. perfect, 5-merous, in umbels of less than 25 fls.; peduncles to 25 mm. long; fls. 4–5 mm. long; petals notched at apex; berry black, spherical, ca. 10 mm. in diam., mostly 3-seeded; seeds 6 mm. long, narrowed slightly to basal notch.—Moist places, below 5000 ft.; Redwood F., Mixed Evergreen F., Yellow Pine F.; outer Coast Ranges from Lake and Mendocino cos. n., Sierra Nevada foothills from Placer Co. n.; to B.C., Mont. May–July.

3. Ceanòthus L. CALIFORNIA-LILAC

Shrubs or small trees, often with divaricate, sometimes spiny twigs. Lvs. alternate or opposite, deciduous or in most of ours persistent, frequently serrate, 3-nerved from the base or strictly pinnately veined, ± petioled. Stipules present. Fls. small but showy, white to blue or purplish, sometimes lavender or pinkish, borne in terminal or lateral panicles or umbellike cymes. Sepals 5, somewhat petallike, united at the base with the urn-shaped receptacle that is filled with a glandular disk in which the ovary is immersed. Petals 5, distinct, hooded and clawed. Stamens 5, opposite the petals, with elongate fils. Ovary 3-loculed, 3-lobed, with a short 3-cleft style. Fr. a 3-lobed caps., separating at maturity into 3 parts. Seeds smooth, convex on 1 side. Between 50 and 60 spp., of temp. N. Am., many of considerable horticultural value. Hybridizing freely. (Greek, *keanothus*, name used by Dioscorides for some spiny plant.)

McMinn, H. E. A systematic study of the genus Ceanothus. Part II of Ceanothus. Santa Barbara Bot. Gard., 1942.)

A. Stipules thin and early deciduous; lvs. alternate, with stomata not sunken; caps. without horns, but often with ridges or crests; fls. in simple or compound panicles. (Subgenus Ceanothus)
  B. Ultimate branchlets flexible, not rigidly divaricate and spinose.
    C. Lvs. with 3 distinct veins from the base, the lateral sometimes obscure.
      D. The lvs. entire, ± deciduous.
        E. Lvs. mostly more than 2 cm. long; peduncles ± leafy; fls. mostly white
                                                                  3. *C. integerrimus*
        EE. Lvs. mostly less than 2 cm. long; peduncles naked; fls. blue
                                                                  4. *C. parvifolius*
      DD. The lvs. serrate, serrulate or glandular-denticulate.
        E. Fls. white; lvs. 2.5–10 cm. long.
          F. Lvs. deciduous, thin, not varnished above, subglabrous. Trinity and Siskiyou cos. .... 2. *C. sanguineus*
          FF. Lvs. persistent, varnished above, often canescent beneath. From Tulare Co. n. .... 1. *C. velutinus*
        EE. Fls. blue or purple; lvs. persistent.
          F. Branchlets terete, not angled and striate.
            G. The lvs. tomentose beneath.
              H. Lf.-blades 3–8 cm. long. Insular .... 10. *C. arboreus*
              HH. Lf.-blades 1–2.5 cm. long. Mainland .... 13. *C. tomentosus*
            GG. The lvs. ± pubescent beneath, chiefly on the veins.
              H. Branchlets distinctly hispid or villous; lvs. usually pubescent above. San Luis Obispo Co. to San Diego Co.
                                          11. *C. oliganthus*
              HH. Branchlets subglabrous or slightly pubescent; lvs. usually glabrous above. Humboldt to Orange and Riverside cos.
                                        12. *C. sorediatus*
          FF. Branchlets angled or ridged and striate.
            G. The lvs. ± revolute.
              H. Lvs. usually grayish silky beneath, broadly ovate; panicles 2–5 cm. long. Santa Barbara to Mendocino cos.
                                        16. *C. griseus*
              HH. Lvs. ± cobwebby beneath, oblong to elliptic; panicles 5–15 cm. long. Napa to Humboldt cos. .... 17. *C. Parryi*
            GG. The lvs. not revolute.
              H. Veins on lower surface of lvs. very prominent, raised; branchlets lacking glandular tubercles. Santa Barbara Co. n.
                                        15. *C. thyrsiflorus*
              HH. Veins on lower surface of lvs. not prominent and raised; branchlets with small glandular tubercles. San Diego Co.
                                        14. *C. cyaneus*
    CC. Lvs. with 1 main vein from the base, even though the basal lateral pair may be longer than other pairs.
      D. The lvs. entire, mostly rather glabrous.
        E. Shrubs evergreen; lvs. rather firm and thick.
          F. Fls. white; caps. 5–7 mm. broad, conspicuously crested. Sierra Nevada foothills and in mts. of s. Calif. (where above 3000 ft.)
                                        5. *C. Palmeri*
          FF. Fls. blue to almost white; caps. 4–5 mm. broad, scarcely crested. San Luis Obispo Co. to n. L. Calif., from below 3000 ft. . . 6. *C. spinosus*
        EE. Shrubs deciduous; lvs. thinnish.
          F. The lvs. mostly more than 2 cm. long; peduncles ± leafy; fls. mostly white .... 3. *C. integerrimus*
          FF. The lvs. mostly less than 2 cm. long; peduncles naked; fls. blue
                                        4. *C. parvifolius*
      DD. The lvs. dentate to serrulate, usually pubescent at least beneath.
        E. Lf.-margins ± revolute.
          F. Infl. 7–20 cm. long; branchlets angled. Napa to Humboldt cos.
                                        17. *C. Parryi*
          FF. Infl. 1–5 cm. long; branchlets not angled.
            G. The lvs. distinctly glandular-papillate on upper surface, 1.5–5 cm. long. San Mateo to Orange cos. .... 18. *C. papillosus*
            GG. The lvs. not glandular-papillate on upper surface, mostly less than 1.5 cm. long.

H. Lvs. narrow-oblong to elliptical, not conspicuously furrowed above along the veins; caps. slightly crested laterally. San Luis Obispo Co. to Santa Cruz Co. .... 19. *C. dentatus*

HH. Lvs. suborbicular to elliptical, conspicuously furrowed on upper surface above the veins; caps. prominently crested laterally. Santa Barbara Co. .... 20. *C. impressus*

EE. Lf.-margins not revolute.

F. Low spreading shrubs to taller, but not strictly trailing or prostrate; lvs. rather deep green.

G. Lvs. glossy above, usually ± undulate, sometimes plane; caps. inconspicuously crested. Humboldt Co. to San Diego Co.
21. *C. foliosus*

GG. Lvs. mostly dull above, plane; caps. conspicuously crested. Sierra Nevada and inner Coast Ranges from Lake Co. n.
22. *C. Lemmonii*

FF. Prostrate or trailing shrubs forming extensive mats; lvs. pale bluish-green .... 23. *C. diversifolius*

BB. Ultimate branchlets rigidly divaricate and spinose; lvs. persistent, plane.

C. Lvs. with 1 main vein from base, mostly entire and glossy on both surfaces. San Luis Obispo Co. to n. L. Calif. .... 6. *C. spinosus*

CC. Lvs. with 3 main veins from the base, the 2 lateral sometimes obscure, gray-glaucous on both surfaces.

D. Lf.-blades 1–2.5 cm. long. Largely inner Coast Ranges and Sierra Nevada.

E. Shrubs 2–4 m. tall; fls. whitish to blue, the clusters 3–8 cm. long. Mostly below Yellow Pine F. .... 7. *C. leucodermis*

EE. Shrubs less than 2 m. high; fls. white, the clusters 1.5–3 cm. long. Mostly above Yellow Pine F. .... 9. *C. cordulatus*

DD. Lf.-blades 2–6 cm. long. Outer Coast Ranges from Santa Cruz Co. n.
8. *C. incanus*

AA. Stipules with thick corky persistent bases; lvs. alternate or opposite; the stomata in sunken pits; caps. usually with 3 horns; fls. in lateral umbels; lvs. firm-coriaceous, persistent. (Subgenus Cerastes)

B. Lvs. alternate at least in part.

C. The lvs. all alternate, broadly obovate to almost round; caps. 4–6 mm. in diam. W. San Diego Co. .... 24. *C. verrucosus*

CC. The lvs. both alternate and opposite, oblanceolate to broadly elliptic; caps. 8–12 mm. in diam.

D. Fr. scarcely or not at all horned; lvs. mostly opposite, 1.5–3.5 cm. long. Insular
25. *C. insularis*

DD. Fr. with conspicuous subapical horns; lvs. mostly alternate, 1–2 cm. long. Mostly mainland, n. San Diego Co. to Santa Barbara Co. .... 26. *C. megacarpus*

BB. Lvs. all opposite.

C. Fls. normally white.

D. Lvs. densely white-tomentose beneath, mostly revolute .... 27. *C. crassifolius*

DD. Lvs. not densely white-tomentose beneath, usually plane and minutely canescent.

E. The lvs. mostly concave on upper surface; horns of caps. lateral and spreading, usually small. Mojave Desert and inner cismontane s. Calif.
28. *C. Greggii*

EE. The lvs. not concave on upper surface; horns of caps. apical or subapical, conspicuous.

F. Lvs. entire, cuneate at base; caps. 5–6 mm. broad. Cismontane Calif.
29. *C. cuneatus*

FF. Lvs. mostly toothed, rounded at base.

G. Caps. 7–9 mm. broad; leaves short-toothed to subentire. Mt. Hamilton Range and Santa Cruz Mts. .... 30. *C. Ferrisae*

GG. Caps. ca. 6 mm. broad; lvs. hollylike, spine-toothed. Lake and Napa cos. .... 43. *C. Jepsonii* var. *albiflorus*

CC. Fls. blue or lavender, sometimes pale.

D. Plants prostrate, rooting along the branches.

E. Lvs. dull above, 0.6–1.2 cm. long, somewhat toothed at the ± rounded apex; horns of fr. subapical. Yellow Pine F., Fresno to Tuolumne cos.
31. *C. fresnensis*

EE. Lvs. mostly shining above.

F. The lf.-blades mostly not over 2 cm. long.

G. Lvs. almost round.

H. The lvs. entire or with few teeth near the apex; caps. 4–5 mm. broad, the horns near the top. Point Sal, Santa Barbara Co.
32. *C. ramulosus*

HH. The lvs. coarsely 6–9-toothed on each side; caps. ca. 7 mm. broad, the horns on the side. S. Sierra Nevada
42. *C. pinetorum*

GG. Lvs. definitely longer than wide.

H. The lvs. less than 5 mm. broad, narrowly oblanceolate or ± spatulate or elliptic, with 1–3 small teeth near the apex; horns on fr. short and stubby, not much wrinkled. N. Mendocino to Del Norte cos. .... 41. *C. pumilus*

> HH. The lvs. 6–12 mm. broad, oblanceolate to obovate; horns
> mostly wrinkled.
>> I. Lf.-blades white tomentose beneath, ± revolute, with 1–3
>> teeth on each side. Coastal San Luis Obispo Co.
>>> 38. *C. maritimus*
>> II. Lf.-blades at most canescent-tomentulose beneath.
>>> J. Erect stems not more than 1.5 dm. tall; caps. mostly
>>> 6–9 mm. in diam. Sierra Nevada to Trinity and
>>> Siskiyou cos. .... 40. *C. prostratus*
>>> JJ. Erect stems 2–6 dm. tall; caps. mostly ca. 5 mm. in
>>> diam. Sonoma Co. .... 39. *C. divergens* ssp. *confusus*
> FF. The lf.-blades mostly 2–5 cm. long, usually with numerous teeth.
> Coastal bluffs, Marin to Mendocino cos. .... 35. *C. gloriosus*
DD. Plants not prostrate or rooting along the branches.
> E. The lvs. with sharp spinulose teeth at least at the apex, often sinuate and
> hollylike.
>> F. Caps. ca. 6 mm. in diam., with large and conspicuous wrinkled horns
>> and prominent intermediate crests or ridges; lvs. 1–2 cm. long, infolded
>> lengthwise, with 4–5 coarse teeth on each side. Marin to Mendocino cos.
>>> 43. *C. Jepsonii*
>> FF. Caps. mostly 3–5 mm. in diam., with erect apical or subapical horns
>> or these almost obsolete.
>>> G. Lvs. mostly less than 12 mm. long, and 6 mm. broad, with 2–4
>>> spinulose teeth on each side; bark usually gray. Sonoma Co.
>>>> 37. *C. sonomensis*
>>> GG. Lvs. larger; bark brown.
>>>> H. The lvs. usually concave or troughlike above, undulate on
>>>> margins; stems stout and rigid. Napa Range
>>>>> 36. *C. purpureus*
>>>> HH. The lvs. usually plane, rarely undulate on margins; stems weak
>>>> and arching. Upper Napa V. near Calistoga
>>>>> 39. *C. divergens*
> EE. The lvs. usually dentate, but not hollylike, not with conspicuously spinose
> teeth.
>> F. Caps. 7–9 mm. broad, the horns prominent and wrinkled, with crests
>> or ridges between; teeth 6–8 on each side of lf.-blade. S. Sierra Nevada
>>> 42. *C. pinetorum*
>> FF. Caps. 4–5 mm. broad, the horns short or slender, the intervening sur-
>> face not crested.
>>> G. Lvs. entire or toothed at apex only; stipules less than 3 mm. long.
>>>> H. Branches usually stiff and straightish; lvs. crowded on short
>>>> lateral branchlets, mostly toothed, often retuse at tip; fls.
>>>> dark to light blue. Monterey Peninsula .... 33. *C. rigidus*
>>>> HH. Branches slender, elongate, arching; lvs. not crowded, entire
>>>> or with few teeth; truncate to obtuse; fls. lavender or pale
>>>> blue. Santa Barbara to Mendocino cos. .... 32. *C. ramulosus*
>>> GG. Lvs. toothed nearly all around; stipules ca. 4.5 mm. long.
>>>> H. Lvs. 0.6–1.8 cm. long; branches stout, erect, straightish.
>>>> Bolinas Ridge, Marin Co. .... 34. *C. Masonii*
>>>> HH. Lvs. 1.5–4 cm. long; branches lax, divergent, arching. Bolinas
>>>> Ridge, Marin Co. and Mendocino Co.
>>>>> 35. *C. gloriosus* var. *exaltatus*

1. **C. velutìnus** Dougl. ex Hook. Tobacco Brush. Mostly a spreading round-topped shrub 1–2 m. high, stout, much-branched, evergreen, puberulent on twigs and lower lf.-surfaces; lvs. broadly- to ovate-elliptical, rounded or subcordate at base, ± rounded to obtuse at apex, 2.5–8 cm. long, closely glandular-serrulate, dark green glabrous and ± varnished above, paler and finely canescent beneath; petioles 5–18 mm. long; fl.-clusters compound, in axils of last year's lvs., 5–10 cm. long, on ± angled peduncles 2–5 cm. long; fls. white; caps. subglobose to triangular, 3–4 mm. wide, 3-lobed at summit, mostly ± viscid, crestless or minutely crested; seeds brown, shining, obovoid, 2 mm. long; $n = 12$ (Nobs in McMinn, 1942).—Open wooded slopes, 3500–10,000 ft.; Montane Coniferous F., Chaparral; Sierra Nevada from Tulare Co. n., to Modoc and w. to Humboldt and Trinity cos.; to B.C., S. Dak., Colo. April–July. With strong balsamic odor.

Var. **laevigàtus** (Hook.) T. & G. [*C. l.* Dougl. ex Hook.] Larger, sometimes arborescent, to 6 m. tall, glabrous on twigs and lvs.—Woods below 3000 ft.; Mixed Evergreen F., Redwood F., etc.; outer Coast Ranges from Marin to Del Norte cos.; to B.C.

C. × **Lorenzènii** (Jeps.) McMinn. [*C. velutinus* var. *L.* Jeps.] Habit of *C. v.*, the lvs. 2–3 cm. long, entire to serrulate, not varnished, but dull above, ± silky beneath; fls. white.—Supposed to be a hybrid of *C. v.* and *C. cordulatus*, occurring in the Sierra Nevada from Kern Co. n., to Siskiyou Co.; w. Nev. June–July.

2. **C. sanguíneus** Pursh. [*C. oreganus* Nutt.] OREGON TEA-TREE. Erect shrub, 1.5–3 m. tall, with purple or reddish flexible glabrous twigs; lvs. deciduous, thin, broadly elliptical, to ovate or obovate, rounded or subcordate at base, acute to obtuse at apex, 3–10 cm. long, subglabrous, glandular-serrulate; petioles 1–2.5 cm. long; fls. white, the clusters compound, 5–10 cm. long, from last year's wood; peduncles almost as long; caps. obovoid, ca. 4 mm. long, slightly lobed above, inconspicuously crested.—Dry slopes, ca. 4000 ft.; Yellow Pine F., Douglas-Fir F.; Trinity and Siskiyou cos.; to B.C., Mont. May–July.

3. **C. integérrimus** H. & A. [*C. Andersonii* Parry.] DEER BRUSH. Loosely branched, ± deciduous, 1–4 m. tall, with green or yellowish branches, the twigs flexible, glabrous to strigose; lvs. broadly elliptical to ovate or suboblong, rounded at base, acute to obtuse at apex, 2.5–7 cm. long, light green and subglabrous to puberulent above, paler and mostly pubescent beneath, entire, sometimes denticulate near tips; petioles 6–12 mm. long; fl.-clusters mostly compound, 4–15 cm. long; peduncles ca. as long; fls. white to dark blue or pink; caps. globose to triangular, slightly depressed at summit, viscid, 4–5 mm. wide, often with small lateral crests; seeds dull brown, obovoid, ca. 2.5 mm. long.—Dry slopes and ridges, between 1000 and 5000 ft.; Yellow Pine F., Mixed Evergreen F.; mostly in mts. from Ventura Co. through inner Coast Ranges to Santa Cruz Co., occasional farther n. May–June. Hybrids have been observed with *C. cordulatus, C. tomentosus,* and *C. Lemmonii.* Variable, the following poorly defined vars. representing some of the forms.

KEY TO VARIETIES

Lvs. oblong to oblong-elliptical, mostly pinnate-veined. Ventura Co. to Santa Cruz Co.
                                               C. integerrimus
Lvs. oblong-ovate to lanceolate or elliptical, 3-veined from base.
    The lvs. usually glabrous above, glabrous or pubescent beneath. Sierra Nevada to N. Coast Ranges
                                         var. *californicus*
    The lvs. puberulent or pubescent on both surfaces.
        Fls. white. Chiefly s. Calif. . . . . var. *puberulus*
        Fls. mostly blue. Butte and Shasta cos. n. . . . . var. *macrothyrsus*

Var. **califórnicus** (Kell.) G. T. Benson. [*C. c.* Kell. *C. nevadensis* Kell. *C. myrianthus* Greene.] Lvs. ovate to broadly ovate, acutish at apex, roundish or subcordate at base, subglabrous at least above; $n = 12$ (Nobs, 1942).—At 2000–7000 ft.; Yellow Pine F., Mixed Evergreen F., N. Oak Wd., etc.; Sierra Nevada, N. Coast Ranges; to Wash.

Var. **pubérulus** (Greene) Abrams. [*C. p.* Greene.] Lvs. ovate-oval, obtuse, puberulent above, pubescent beneath; fls. white.—Yellow Pine F.; mts. from San Diego and Orange cos. to Kern Co.

Var. **macrothýrsus** (Torr.) G. T. Benson. [*C. thyrsiflorus* var. *m.* Torr. *C. m.* Greene.] Lvs. ovate to oval, ± rounded at apex, densely pubescent on both surfaces; fls. mostly blue.—Mixed Evergreen F., Yellow Pine F.; Butte and Shasta cos. to Del Norte and Mendocino cos.; sw. Ore.

4. **C. parvifólius** (Wats.) Trel. [*C. integerrimus* var. *p.* Wats.] A low spreading shrub 6–12 dm. high with slender flexible greenish or reddish twigs, glabrous throughout except sometimes on under lf.-surface, deciduous; lvs. oblong-elliptic to elliptic, acute to obtuse, 6–20 mm. long, 3-veined from base, green above, paler beneath, entire except sometimes near apex; petioles 2–5 mm. long; fl.-clusters simple, rarely compound, 3–7 cm. long, on peduncles ca. as long; fls. pale to deep blue; fr. globose, smooth, not or obscurely crested, 4–5 mm. long.—Wooded slopes, 4500–7000 ft.; Yellow Pine F., Red Fir F.; Sierra Nevada from Tulare to Plumas cos. May–July.

5. **C. Pálmeri** Trel. [*C. spinosus* var. *P.* K. Bdg.] Large spreading shrub 1–3.5 m. tall with gray-green bark and glabrous flexible twigs; lvs. oblong to oblong-ovate, mostly persistent, 1.5–3.5 cm. long, rounded to retuse at apex, rather firm and leathery, glabrous and light green above, pale and sometimes ± pubescent beneath on the midrib, 1-nerved, entire; petioles ca. 3–4 mm. long; fl.-clusters compound, 7–12 cm. long, on somewhat shorter peduncles; fls. white; caps. ± triangular, 5–7 mm. broad, 3-lobed, with glandular crests; seeds dark olive-brown, round-obovoid, ca. 3 mm. long.—Dry slopes, 3200–6000 ft.; Chaparral, Yellow Pine F., mts. of Riverside, Orange, and San Diego cos., n. L. Calif.; at 400–1500 ft., Chaparral, Foothill Wd., Sierra Nevada foothills of Amador and Eldorado cos. May–June.

6. **C. spinòsus** Nutt. in T. & G. RED-HEART. GREENBARK C. Large or arborescent, 2–6 m. tall, with smooth olive-green bark, the main branchlets flexible, ascending, glabrous, the ultimate short, stiff, divergent, spinescent; lvs. persistent, elliptic to oblong, obtuse to emarginate at apex, entire or sometimes toothed toward tips, 1.2–3 cm. long, glabrous or somewhat strigose on midrib of undersurface, bright green above, paler beneath; petioles 4–8 mm. long; fl.-clusters mostly large, compound, 4–15 cm. long, on somewhat leafy peduncles; fls. pale blue to almost white; caps. globose, 4–5 mm. broad, viscid, scarcely lobed or crested; seeds ca. 3 mm. long, dark olive-brown; n = 12 (Nobs, 1942).—Dry slopes, mostly below 3000 ft.; Chaparral, Coastal Sage Scrub; mts. mostly near the coast from San Luis Obispo Co. to n. L. Calif. Feb.–May. Hybrids with *C. thyrsiflorus* and *C. sorediatus* have been recorded.

7. **C. leucodérmis** Greene. [*C. divaricatus* auth., not Nutt. *C. d.* var. *eglandulosus* Torr., var. *laetiflorus* Jeps. and var. *grossesseratus* Torr.] Evergreen shrub, 2–4 m. high, with rigid divaricate branches with pale green smooth bark and short spreading subglabrous spinescent branchlets; lvs. elliptical-oblong to ovate, rounded or subcordate at base, obtuse to acute at apex, entire to serrulate, usually glabrous and glaucous on both surfaces, sometimes pubescent, 1–2.5 cm. long; petioles 2–3 mm. long; fl.-clusters mostly simple, 3–8 cm. long, the fls. white to blue; caps. globose, slightly depressed at top, 4.5–6 mm. wide, viscid, not crested; seeds shining, dark olive-brown, flattened-obovoid, 2.5–3 mm. long; n = 12 (Nobs, 1942).—Dry rocky slopes, below 6000 ft.; Chaparral, Foothill Wd.; mts. of cismontane s. Calif., n. along inner Coast Ranges to Alameda Co., and Sierra Nevada to Eldorado Co.; n. L. Calif. April–June.

8. **C. incànus** T. & G. Erect, evergreen, 2–4 m. high, with numerous glaucous branches and stout thornlike branchlets or some of the latter more slender and flexible; lvs. broadly ovate to elliptic, obtuse, 2–6 cm. long, 3-veined from base, puberulent or subglabrous on both surfaces, gray-green above, lighter beneath, entire or ± serrulate; petioles 3–12 mm. long; fl.-clusters compound, dense, 3–6 cm. long, short-peduncled; fls. white; caps. triangular, shallowly lobed at top, rugosely roughened, ca. 5 mm. broad; seeds shiny, dark, rounded-obovoid, ca. 3 mm. long; n = 12 (Nobs, 1942).—Open slopes and canyons, below 3000 ft.; Redwood F., Mixed Evergreen F.; outer Coast Ranges from Santa Cruz Co. to Humboldt and Siskiyou cos. April–May.

9. **C. cordulàtus** Kell. SNOW BUSH. Intricately branched, spreading, 1–2 m. high, with smooth whitish bark, the whole plant grayish-glaucous, the ultimate twigs stiff, divaricate, spinose; lvs. persistent, ovate to elliptic, mostly 1–2 cm. long, acute to obtuse, 3-veined, glabrous or ± puberulent, green and glaucous above, grayer beneath, mostly entire; petioles 3–6 mm. long; fls. white, the clusters dense, 1.5–3 cm. long; caps. triangular, mostly lobed, slightly crested, 4–5 mm. broad, ± viscid when young; seeds olive-brown, round-obovoid, almost 2 mm. long; n = 12 (Nobs, 1942).—Dry open flats and slopes, 3000–9500 ft.; Montane Coniferous F. mostly above Yellow Pine; San Jacinto Mts. n., Sierra Nevada, N. Coast Ranges from Lake Co. n.; to Ore., Nev., L. Calif. May–July. Hybridizing with several spp.

10. **C. arbòreus** Greene. [*C. velutinus* var. *a.* Sarg. *C. a.* var. *glabra* Jeps.] Tall shrub, ± arborescent, evergreen, 3–7 m. high, with smooth gray bark and soft-pubescent twigs; lvs. broadly ovate to elliptical, rounded at base, acute to obtuse at tip, 3–8 cm. long, 3-veined, dull green and puberulent above, paler and canescent-tomentose beneath, serrulate to serrate; petioles 8–25 mm. long; fl.-clusters compound, 5–12 cm. long, short-peduncled; fls. pale blue; caps. triangular, roughened all over, 6–8 mm. broad, ± crested on the backs of the lobes, almost black when mature; seeds round-obovoid, olive-brown, ca. 2 mm. long, shining; n = 12 (Nobs, 1942).—Brushy slopes; Chaparral; Santa Cruz, Santa Rosa, Santa Catalina ids. Feb.–May. Some subglabrous plants from the n. ids. are the var. *glàber*.

11. **C. oligánthus** Nutt. in T. & G. [*C. hirsutus* Nutt. *C. divaricatus* Nutt.] Shrub, often arborescent, 1–3 m. tall, younger branches ± reddish, hairy, not spinescent; lvs. persistent, ovate or oblong-ovate to elliptical, obtuse, rounded or acute, 1.5–4 cm. long, mostly 3-veined, dark green and sparingly pubescent above, paler and ± hirsute-pubescent beneath especially on veins, glandular-denticulate; petioles 3–8 mm. long; fl.-clusters mostly simple, rather loose, 1.5–5 cm. long; fls. mostly deep blue or purplish; frs. ± triangular, ca. 4 mm. broad, smooth or rough, crested, usually viscid; seeds shining, olive-brown, rounded, ca. 3 mm. long.—Dry slopes, below 4500 ft.; Chaparral, Foothill Wd.; San Luis Obispo Co. to Los Angeles and w. Riverside cos. Feb.–April.

Var. **Orcúttii** (Parry) Jeps. [*C. O.* Parry.] Fls. pale blue; caps. villous-pubescent, strongly wrinkled, more viscid.—Chaparral, Yellow Pine F.; Santa Ana Mts. (Orange Co.) to Cuyamaca Mts., San Diego Co.

12. **C. sorediàtus** H. & A. [*C. intricatus* Parry.] JIM BRUSH. Densely and rigidly branched, 1–3(–5) m. tall, with smooth gray-green bark and rigid ± villous stiff nearly spinescent branchlets, evergreen; lvs. elliptic to ovate, mostly acute, 1–4 cm. long, finely glandular-serrulate, 3-veined, dark glossy green and almost glabrous above, paler and strigose beneath especially along the veins; petioles 2–5 mm. long; fl.-clusters mostly simple, dense, 1–4 cm. long, on somewhat shorter peduncles; fls. light to deep blue; frs. globose, smooth, 4 mm. broad, scarcely lobed or crested, viscid; seeds dark, plump, shining, ca. 2 mm. long; $n = 12$ (Nobs, 1942).—Dry slopes, below 3500 ft.; Chaparral, Foothill Wd., Mixed Evergreen F.; Coast Ranges from Humboldt to Orange and w. Riverside cos. Feb.–May. It hybridizes with *C. oliganthus, C. papillosus, C. thyrsiflorus,* and *C. tomentosus.*

13. **C. tomentòsus** Parry. [*C. oliganthus* var. *t.* K. Bdg. *C. azureus* Kell., not Desf.] Evergreen shrub, 1–3 m. high, with grayish-brown or reddish bark, the branches long and slender, young growth rusty-tomentose; lvs. ovate to elliptic, glandular-serrulate, 1–2.5 cm. long, 3- or 1-veined, dark green and finely pubescent above, whitish- or brownish-tomentose beneath; petioles 1–5 mm. long; fl.-clusters compound, lateral or terminal, 2–5 cm. long, on shorter, 1 to 2-lvd. peduncles; fls. blue to almost white; frs. subglobose, slightly lobed at top, ca. 4 mm. broad, somewhat laterally crested, ± viscid when developing; $n = 12$ (Nobs, 1942).—Scattered on dry slopes, below 5000 ft.; Chaparral, Foothill Wd., Yellow Pine F.; foothills of Sierra Nevada from Placer to Mariposa cos. April–May.

Var. **olivàceus** Jeps. Lf.-blades glandular-denticulate, gray-green beneath; caps. more sticky and very dark in age.—Dry brush-covered slopes, below 3500 ft.; Chaparral; mts. from Redlands (San Bernardino Co.) and Santa Ana Mts. to San Diego Co.; L. Calif. March–May.

14. **C. cyàneus** Eastw. Arborescent shrub 1–5 m. high, with gray-green branches and glabrous to sparsely puberulent branchlets with scattered brownish glands; lvs. persistent, ovate-elliptical, rounded at base, acute or obtuse at apex, 2–4.5 cm. long, 3-veined, green and glabrous above, lighter and thinly puberulent beneath, glandular-serrulate to subentire; petioles 3–6 mm. long; fl.-clusters compound, 5–18 cm. long, terminal; fls. bright blue; frs. subglobose, shallowly 3-lobed, ca. 4 mm. broad, smooth, with small or no crests; $n = 12$ (Nobs, 1942).—Chaparral; San Diego Co. (Ramona, Lakeside, Alpine, El Capitan). April–June.

15. **C. thyrsiflòrus** Esch. [*C. bicolor* Raf. *C. elegans* Lem. *C. t.* var. *Chandleri* Jeps.] BLUE BRUSH. Large shrub, sometimes arborescent, 1–6 m. high, with green angled branchlets, evergreen; lvs. oblong-ovate to broadly elliptical, rounded to acute at both ends, 1–5 cm. long, prominently 3-veined, ± glandular-serrulate, dark green and glabrous above, paler beneath, sparsely hairy on the raised veins; petioles 3–12 mm. long; fl.-clusters compound, 3–8 cm. long, on peduncles ca. as long; fls. light to deep blue, rarely almost white; frs. subglobose, ca. 3 mm. broad, slightly lobed, glandular-viscid, dark in age; seeds rather dark, plump, ± dull, scarcely 2 mm. long; $n = 12$ (Nobs, 1942).—Wooded slopes and canyons below 2000 ft.; Chaparral, Redwood F., Mixed Evergreen F.; outer Coast Ranges from Santa Barbara Co. to s. Ore. March–June. Hybridizing with several other spp.

Var. **rèpens** McMinn. Plants prostrate.—N. Coastal Scrub; Marin to Monterey cos.

16. **C. gríseus** (Trel.) McMinn. [*C. thyrsiflorus* var. *g.* Trel.] Much like *C. thyrsiflorus;* lvs. broadly to roundish ovate, obtuse, dark green and glabrous above, gray-tomentulose beneath, the margins revolute and somewhat undulate between the teeth; fls. violet-blue in dense panicles 2–5 cm. long; fr. subglobose, glandular-viscid when young, black and shiny when mature, ca. 4 mm. broad.—Closed-cone Pine F., N. Coastal Scrub, etc.; outer Coast Ranges; Santa Barbara to Monterey cos., also Sonoma and Mendocino cos. March–May.

Var. **horizontàlis** McMinn. Low-spreading to prostrate.—N. Coastal Scrub; Yankee Point, Monterey Co.

C. × Veitchiànus Hook., possible hybrid between *C. griseus* and *C. rigidus,* collected long since near Monterey; with ovate to obovate lvs. and blue fls.

17. **C. Párryi** Trel. [*C. integerrimus* var. *P.* K. Bdg.] Evergreen, 1–5 m. tall, with grayish or reddish-brown bark, and tomentose angled young branchlets; lvs. oblong or ± elliptic, rounded at base, obtuse at tip, 1.5–4.5 cm. long, mostly 1-veined from base, subentire to glandular-serrulate, dark green above, lighter and cobwebby beneath, slightly revolute; petioles 3–10 mm. long; fl.-clusters mostly compound, 5–15 cm. long, on ± leafy peduncles; fls. deep blue; caps. subglobose, ca. 3–4 mm. broad, slightly lobed, smooth; seeds dark, plump, ca. 2 mm. long; *n* = 12 (Nobs, 1942).—Wooded canyons and slopes, below 2500 ft.; Redwood F., Mixed Evergreen F., Chaparral; outer and middle Coast Ranges from Humboldt to Napa and Sonoma cos. April–May.

18. **C. papillòsus** T. & G. [*C. dentatus* var. *p.* K. Bdg.] Evergreen, loosely branched, 1–5 m. tall, with densely tomentose-pubescent young branchlets; lvs. rather crowded, oblong-elliptical to elliptical, rounded to truncate at apex, 1.5–5 cm. long, 1–2 cm. broad, 1-veined from base, dark green, ± villous and glandular-papillose above, paler and hirsutulous to tomentose beneath, glandular-denticulate and somewhat revolute; petioles 4–6 mm. long; fl.-clusters mostly simple, narrow, dense, 2–5 cm. long, on naked peduncles ca. as long; fls. deep blue; fr. triangular, lobed, ca. 3 mm. wide, with low narrow subdorsal crests; seeds plump, dark, ca. 2 mm. long; *n* = 12 (Nobs, 1942).—Open or ± wooded slopes, below 3200 ft.; Redwood F., Mixed Evergreen F., Chaparral; outer Coast Ranges from San Mateo to San Luis Obispo cos. April–May.

Var. **Roweànus** McMinn. Lvs. linear to oblong, less than 1 cm. broad, ± retuse to truncate at apex.—Dry slopes, 2000–4000 ft.; largely Chaparral; San Benito and Monterey cos. to Santa Ana Mts., Orange Co. Feb.–June.

C. × **règius** (Jeps.) McMinn. [*C. papillosus* var. *r.* Jeps.] Lvs. like in *C. p.* but scarcely or not papillose and subglabrous above.—Known from King's Mt., San Mateo Co., near Ben Lomond, Santa Cruz Co.; Camp Lucia, Monterey Co. Probably a hybrid between *C. p.* and *C. thyrsiflorus.*

19. **C. dentàtus** T. & G. Densely branched, evergreen, 0.5–1.5 m. tall, with hairy slightly roughened branchlets; lvs. crowded, fascicled, elliptical to narrow-oblong or linear, rounded at base, usually truncate to retuse at apex, 0.5–1.2(–2) cm. long, 1-veined from base, dark green and hirsutulous above, paler and matted-hairy beneath, strongly revolute, the margins ± glandular-papillate; petioles 1–2 mm. long; fl.-clusters simple, subglobose, mostly less than 2 cm. long; peduncles small-lvd.; fls. deep blue; fr. globose, shallowly lobed, ca. 4 mm. in diam., slightly laterally crested.—Sandy and gravelly places, below 5000 ft.; Closed-cone Pine F., Chaparral, Mixed Evergreen F., Yellow Pine F.; near the coast from Santa Cruz Co. to n. San Luis Obispo Co. Feb.–June.

Var. **floribúndus** (Hook.) Trel. [*C. f.* Hook. *C. papillosus* var. *f.* Parry.] A poorly defined form with lvs. more than 6 mm. broad, less revolute, practically glandless on margins.—Occasional with the sp.

C. × **Lobbiànus** Hook. [*C. d.* var. *L.* Trel.] Lvs. 3-veined, dark above, paler and hairy beneath, glandular-dentate, revolute.—Probably hybrid of *C. dentatus* and *C. griseus;* occasional in Monterey region.

20. **C. impréssus** Trel. [*C. dentatus* var. *i.* Trel.] Low, evergreen, rather densely branched, the young growth densely hairy-pubescent; lvs. elliptical to suborbicular, 0.6–1.2 cm. long, 1-veined from base, loosely pubescent, upper surface deeply grooved over the veins, strongly revolute, sometimes slightly glandular, appearing crenate; petioles 1–3 mm. long; fl.-clusters simple, narrow, densely fld., 1–2.5 cm. long; fls. deep blue; caps. subglobose, ca. 4 mm. broad, with prominent lateral crests; seeds dark, shining, plump, ca. 2.5 mm. long.—Chaparral; Burton Mesa, near Lompoc, Santa Barbara Co. to San Luis Obispo Co. March–April.

Var. **nipoménsis** McMinn. To 3 m. tall; lvs. 1–2.5 cm. long, lighter green, not so deeply furrowed along the veins, scarcely revolute.—Chaparral; Nipomo Mesa and region, n. Santa Barbara Co. March–April.

21. **C. foliòsus** Parry. [*C. diversifolius* var. *f.* K. Bdg. *C. austromontanus* Abrams. *C. dentatus* var. *Dickeyi* Fosb.] Low shrub to ca. 1 m. high, evergreen, with flexuous pubescent glandular branchlets; lvs. oblong-elliptic to broadly elliptic, 0.5–1.5 cm. long, glandular-denticulate, 1-veined or faintly 3-veined from base, obtuse at both ends, dark green and somewhat strigose above, paler and with somewhat longer hairs beneath on the veins; petioles 1–3 mm. long; fl.-clusters simple, subglobose or somewhat elongate, 0.5–2.5 cm. long, sometimes compound and longer; peduncles as long or longer; fls.

blue; caps. subglobose, ca. 4 mm. broad, smooth, faintly lobed and crested; $n = 12$ (Nobs, 1942).—Dry slopes, below 5000 ft.; Chaparral, Yellow Pine F., Mixed Evergreen F., edge of Redwood F.; Coast Ranges from Humboldt to Santa Cruz cos., Cuyamaca Mts. of San Diego Co. March–May.

Var. vineàtus McMinn. Lvs. obovate to broadly elliptic, 1.3–1.9 cm. long, nearly plane on the margins.—Rolling hills, Sonoma and Mendocino cos.

Var. mèdius McMinn. More erect; lvs. more hairy between veins, 1–2 cm. long.— Chaparral; middle and inner Coast Ranges, Santa Clara, Monterey, and San Luis Obispo cos.

22. **C. Lemmònii** Parry. Low spreading shrubs ca. 3–8 dm. tall, with gray bark and gray or whitish pubescent glandular branchlets; lvs. persistent, elliptic to ± oblong, rounded to tapering at base, obtuse to acute at apex, 0.8–3 cm. long, 1-veined or indefinitely 3-veined from base, subglabrous to strigose above, paler and villous beneath, glandular-denticulate; petioles 3–8 mm. long; fl.-clusters simple, 1–3 cm. long, on peduncles almost as long; fls. pale blue; caps. somewhat triangular, 3–4 mm. broad, depressed and lobed at top, smooth, conspicuously crested; seeds brown, shining, rather flat, ca. 2.5 mm. long; $n = 12$ (Nobs, 1942).—Open slopes, 1200–3500 ft.; Yellow Pine F., Foothill Wd., etc.; w. base of Sierra Nevada from Tuolumne and Eldorado cos. n., inner Coast Ranges from Lake and Yuba cos. n., to Humboldt and Shasta cos. April–May.

23. **C. diversifòlius** Kell. [*C. decumbens* Wats.] PINE MAT. Low trailing shrub, 1–3 dm. tall, with long flexible villous branches forming extensive mats; lvs. persistent, ovate to orbicular or elliptical or ± oblong, obtusish, mostly 0.5–2 cm. long, 1-veined from base, pale bluish-green and villous above, paler and hairy-tomentose beneath, sparingly glandular-denticulate; petioles 3–12 mm. long; fl.-clusters simple, rather few-fld., ca. 1 cm. long, on longer peduncles; fls. blue to almost white; caps. subglobose, ca. 4 mm. broad, smooth, crested at apex.—Occasional, draws and flats in heavy forest, 3000–6000 ft.; Yellow Pine F.; from Greenhorn Mts., Kern Co. along w. side of Sierra Nevada to Shasta Co., Siskiyou Co. to n. Lake Co. May–June.

24. **C. verrucòsus** Nutt. in T. & G. Erect stiff-branched rounded evergreen shrub 1–3 m. tall with tomentulose young branchlets; lvs. alternate, crowded, round- to deltoid-obovate, tapering to rounded at base, obtuse to retuse at apex, plane, firm, entire or nearly so, 0.5–1.4 cm. long, dark green and glabrous above, minutely canescent beneath; petioles mostly 1–3 mm. long; fl.-clusters dense, axillary, few-fld., 1–2 cm. long; fls. white; caps. globose, 5 mm. broad, shallowly lobed, laterally short-horned; seeds black, shining, flattened, 2.5–3 mm. long; $n = 12$ (Nobs, 1942).—Dry hills and mesas; Chaparral; w. San Diego Co.; adjacent L. Calif. Jan.–April.

25. **C. insulàris** Eastw. [*C. megacarpus* var. *i*. Munz.] Erect rather stiffly branched shrub 1–3 m. tall, the ultimate branchlets tomentulose and rather gracefully arched; lvs. opposite or alternate, persistent, elliptical to cuneate-obovate, truncate to retuse at apex, entire, 1.2–3.5 cm. long, green and glabrous above, minutely canescent beneath; petioles 5–7 mm. long; fl.-clusters umbellate, short-peduncled, axillary; fls. white or with bluish centers; caps. globose, 8–10 mm. in diam., mostly without horns or crests; seeds shining, olive, ± flattened, ca. 4 mm. long.—Dry slopes; Chaparral; Santa Cruz, Santa Rosa, Santa Catalina ids. Jan.–March.

26. **C. megacàrpus** Nutt. [*C. macrocarpus* Nutt., not Cav. *C. cuneatus* var. *m*. K. Bdg.] Rather compact large shrub 1–4 m. tall, the branches reddish or grayish-brown, the young branchlets canescent-strigose; lvs. persistent, alternate, spatulate to obovate, cuneate at base, truncate or notched at apex, 1–2 cm. long, entire, firm, dull green and glabrous above, canescent beneath, ± revolute; petioles 2–3 mm. long; fl.-clusters on lateral branchlets, few-fld.; fls. white; caps. 8–12 mm. broad, scarcely lobed, laterally or dorsally horned; seeds olive, rather dull, biconvex, 4.5–5 mm. long.—Dry slopes, below 2000 ft.; Chaparral; near the coast from Santa Barbara Co. to n. San Diego Co. Jan.–April. A form with whiplike, ± pendulous branches has been called var. *péndulus* McMinn.

27. **C. crassifolius** Torr. [*C. verrucosus* var. *c*. K. Bdg.] Rather stiffly and openly branched, 2–3.5 m. high, with grayish, brownish, or whitish branches and ± stout tomentose twigs; lvs. opposite, persistent, broadly elliptic to ± elliptic-obovate, rounded to obtuse, 1.5–3 cm. long, revolute, thick and leathery, olive-green and glabrous above,

white-tomentose beneath, coarsely and pungently dentate, sometimes almost entire; petioles 4–7 mm. long; fl.-clusters umbellike, almost sessile, short; fls. white; caps. globose, 7–8 mm. broad, viscid, with short subdorsal horns, roughened at summit; seeds black, shining, ± flattened-oblong, 3.5–4 mm. long.—Common on dry slopes and fans, below 3500 ft.; Chaparral; Santa Barbara Co. to L. Calif. Jan.–April.

Var. **plànus** Abrams. Lvs. not or scarcely revolute, less tomentose beneath.—Santa Barbara and Ventura cos. and intergrading with the sp. farther s.

C. × **otàyénsis** McMinn. To ca. 1 m. high; lvs. broadly oval, tomentulose above, whiter beneath, very revolute, 6–12 mm. long.—Otay Mt., San Miguel Mt., San Diego Co. Probable hybrid between *C. crassifolius* and *C. Greggii* var. *perplexans.* Jan.–Feb.

28. **C. Gréggii** Gray var. **vestìtus** (Greene) McMinn. [*C. v.* Greene.] Erect rigidly branched, evergreen, 1–2 m. high with tomentulose branchlets; lvs. opposite, elliptic-ovate to oblanceolate, commonly concave above, entire to dentate, grayish-green above, gray beneath with a close tomentum (later ± glabrescent), rigid and firm, 0.7–1.5 cm. long; petioles mostly 1–2 mm. long; fls. cream-white, umbellate, on short axillary peduncles; caps. ca. 5 mm. broad, with spreading dorsal horns scarcely 1 mm. long.— Dry slopes, 3500–7500 ft.; Joshua Tree Wd., Pinyon-Juniper Wd., Sagebrush Scrub; desert slopes from Mono Co. to Kern, Los Angeles, and San Barnardino cos., w. into e. Santa Barbara and San Luis Obispo cos.; to Utah, Ariz. May–June.

Var. **perpléxans** (Trel.) Jeps. [*C. p.* Trel.] Lvs. roundish to broadly elliptical or broadly obovate, 1–2 cm. long, mostly conspicuously toothed, yellowish-green and glabrous above, finely white-tomentose beneath in the areolae; seeds shining, black, ± flattened, ca. 3 mm. long.—Dry slopes, below 7000 ft.; Chaparral, Pinyon-Juniper Wd., Yellow Pine F.; s. face of San Bernardino Mts. to n. L. Calif. March–April.

29. **C. cuneàtus** (Hook.) Nutt. [*Rhamnus c.* Hook.] Buck Brush. Rigid shrub much like the preceding, 1–3.5 m. tall; lvs. on spurlike branchlets, cuneate-obovate to spatulate, mostly obtuse, entire, finely tomentulose beneath, gray-green and glabrous above, firm, 0.5–1.5 cm. long, plane; fls. white, umbellate; caps. subglobose, 5–6 mm. broad with short erect horns near the top; seeds shining, black, round-oblong, ca. 4 mm. long; $n = 12$ (Nobs, 1942).—Common on dry slopes and fans, below 6000 ft.; Chaparral, Pinyon-Juniper Wd., Yellow Pine F.; cismontane Calif. to Ore., L. Calif. March–May. Variable; some of the variants with names are: var. *dùbius* J. T. Howell, with lvs. 2–4 cm. long, subsericeous beneath, and with much wrinkled base of fr.; Santa Cruz Mts.; var. *sub-montànus* (Rose) McMinn [*C. s.* Rose] with some lvs. concave above, and occurring where *C. cuneatus* and *C. Greggii* var. *vestitus* come together; and *C. oblanceolàtus* A. Davids., with narrowly oblanceolate lvs. 1.2–2 cm. long, plane, occurring in mts. bordering sw. corner of Mojave Desert. Covering so wide a range, *C. cuneatus* contacts many other spp.; some described hybrids are: *C.* × *fléxilis* McMinn, low, sometimes prostrate, with lvs. intermediate with *C. prostratus* and fls. white to bluish; Shasta, Lassen and Modoc cos. *C.* × *connìvens* Greene, low with weak flexible ± trailing branches; lvs. like in *C. cuneatus*, horns of caps. sometimes connivent; in Calaveras-Tuolumne area and possibly with genes of *C. fresnensis* and *C. prostratus.*

30. **C. Férrisae** McMinn. Erect shrub 1–2 m. high, with stiff divergent or arching branches and short lateral strigose branchlets; lvs. opposite, persistent, round to broadly elliptical, 1.5–2.5 cm. long, short-toothed to subentire, dark green and glabrous above, minutely canescent beneath; petioles 2–3 mm. long; fls. white, in short umbels; fr. globose, 7–9 mm. broad, with 3 dorsal or subdorsal horns, ± roughened.—Dry serpentine slopes, 400–1000 ft.; Foothill Wd., Chaparral; Mt. Hamilton Range and Santa Cruz Mts. Jan.–March.

31. **C. fresnénsis** Dudl. ex Abrams. [*C. rigidus* var. *f.* Jeps.] Forming mats to 6 m. broad, with some ± erect branchlets to 3 dm. tall; branchlets gray or brownish-purple, tomentulose; lvs. opposite, elliptical, oblanceolate to obovate, coriaceous, 0.6–1.2 cm. long, gray-green and minutely pubescent above, paler and ± tomentulose beneath, somewhat toothed at the ± rounded apex; fls. mostly blue, in few-fld., sessile or short-peduncled umbels; caps. globose, 5–6 mm. in diam., with slender subapical spreading to suberect horns.—Dry ridges, 3000–6500 ft.; Yellow Pine F.; w. slope of Sierra Nevada from Fresno to Tuolumne cos. May–June.

C. × **arcuàtus** McMinn. Rigidly branched, 3–6 dm. high with stiff arching branches and brownish tomentulose branchlets; lvs. oblanceolate to obovate or oval, 0.6–1.2 cm.

long, entire or few-toothed near apex, gray-green; fls. pale blue to whitish; caps. 4–6 mm. in diam., with slender horns.—At 2000–7500 ft.; w. slope of Sierra Nevada from Madera Co. to Plumas Co. Possibly hybrid between *C. fresnensis* and *C. cuneatus.*

32. **C. ramulòsus** (Greene) McMinn. [*C. cuneatus* var. *r.* Greene.] From 0.6–3 m. tall, with rather slender spreading or arching branches and slender ± strigose branchlets; lvs. opposite, roundish to elliptical or obovate, 0.6–2 cm. long, entire or with few teeth near the truncate or obtuse apex, glabrous above, canescent beneath; petioles 1–3 mm. long; fls. lavender, pale blue to whitish, in small peduncled umbels; caps. globose, ca. 4–5 mm. in diam., usually with 3 slender horns near top; seeds black, shining, somewhat flattened, ca. 3 mm. long; *n* = 12 (Nobs, 1942).—Dry rocky and sandy places, below 1500 ft.; Chaparral; outer Coast Ranges from n. Santa Barbara Co. to Mendocino Co. Jan.–April.

Var. **fascìculàris** McMinn. Lvs. narrowly oblanceolate, many apparently fascicled; horns of caps. obsolete or minute.—Chaparral; coastal mesas from Lompoc (Santa Barbara Co.) to San Luis Obispo Co. Feb.–April.

33. **C. rígidus** Nutt. [*C. verrucosus* var. *r.* K. Bdg. *C. r.* var. *pallens* Sprague.] Much-branched, 1–2 m. tall, the branchlets slender, divaricate, dark, pubescent-tomentose; lvs. opposite, cuneate- to rounded-obovate, 0.6–1.5 cm. long, toothed above the middle or near the rounded or notched apex, or entire, glabrous, dark green and shining above, coriaceous, paler beneath and minutely tomentose in the areoles; petioles 1–2 mm. long; fls. bright to pale blue, in few-fld. umbels; caps. globose, ca. 5 mm. broad, scarcely lobed, with short dorsal horns; *n* = 12 (Nobs, 1942).—Sandy hills and flats; Closed-cone Pine F., N. Coastal Scrub; Monterey Peninsula. Feb.–April.

34. **C. Masònii** McMinn. Erect or ± spreading, 0.6–2 m. tall, with stiff divaricate branches and dark tomentulose branchlets; lvs. opposite, broadly elliptical to roundish, 0.6–1.8 cm. long, obtuse to emarginate, dark green and glabrous above, grayish-white and minutely canescent beneath, short-toothed except near base; petioles 1–2 mm. long; fls. dark blue to purple, in many-fld. umbels; caps. globose, ca. 3.5 mm. broad, with short apical or subapical horns.—Dry rocky slopes; Chaparral; Bolinas Ridge, Marin Co. March–April.

35. **C. gloriòsus** J. T. Howell. [*C. rigidus* var. *grandifolius* Torr. *C. prostratus* var. *g.* Jeps.] Prostrate or decumbent, to 3 dm. tall and 2–3 m. across, the branches stout, ± strigose especially when young; lvs. opposite, broadly elliptical, broadly oblong, to roundish, rounded to retuse at apex, thick, leathery, dark green and glabrous above, paler and minutely canescent beneath, 2–5 cm. long, mostly conspicuously dentate; petioles 2–4 mm. long; fls. deep blue to ± violet, in many-fld. stout-peduncled umbels; caps. globose, viscid, 3.5–4 mm. in diam., with 3 small apical horns; seeds black, shining, 3–3.5 mm. long; *n* = 12 (Nobs, 1942).—Sandy places; Coastal Strand, Closed-cone Pine F., N. Coastal Scrub; Point Reyes Peninsula, Marin Co. to Mendocino Co. March–May.

Var. **exaltàtus** J. T. Howell. Stems erect, 1–2 m. tall; lvs. 1.5–4 cm. long.—Douglas-Fir F., Chaparral; Bolinas Ridge, Marin Co., Mendocino Co.

Var. **porréctus** J. T. Howell. Sprawling, 3–5 dm. tall; lvs. 1–2 cm. long.—Closed-cone Pine F.; Inverness Ridge and Ledum Swamp, Marin Co.

36. **C. purpùreus** Jeps. [*C. Jepsonii* var. *p.* Jeps.] Erect or spreading, 1–2 m. tall, with rigid divergent red-brown subglabrous branches; lvs. opposite, holly-like, broadly elliptical or roundish in outline, dark green and shining above, paler and minutely canescent beneath, 1.2–2.5 cm. long, with undulate and spine-toothed margins; petioles to ca. 2 mm.; fls. deep blue to purple, in several-fld. umbels; caps. globose, ca. 5 mm. in diam., with slender erect dorsal horns; seeds black, shining, ca. 3 mm. long.—Dry rocky hills, below 1800 ft.; Chaparral; Napa Range, Napa Co. Feb.–April.

37. **C. sonoménsis** J. T. Howell. Erect, ca. 1 m. high, with stiff gray or brown stems; branchlets spurlike, strigose; lvs. opposite, cuneate-obovate to roundish, 0.5–1.5 cm. long, 3-toothed at apex, often with 1–2 pairs of lateral teeth, coriaceous, glabrous above, gray-tomentulose beneath, subsessile; fls. blue to lavender, in short-peduncled to subsessile umbels; caps. globose, 5 mm. in diam., with subdorsal horns and low ridgelike intermediate crests.—Below 2200 ft.; Chaparral; Hood Mountain Range, Sonoma Co. Feb.–April.

38. **C. marítimus** Hoov. Stems rigid, prostrate, to 1 m. long; lvs. opposite, narrowly

to broadly obovate or oblong, mucronate, truncate to emarginate or obcordate, 0.8–2 cm. long, dark glossy green and glabrous above, white-tomentose beneath, revolute and with 1–3 teeth on each side; fls. light to deep blue; caps. with 3 short erect horns.— Coastal bluffs, San Luis Obispo Co. Jan.–March.

39. **C. divérgens** Parry. [*C. prostratus* var. *d.* K. Bdg.] Scrambling to erect, 6–15 dm. high, with rather weak arching to divergent subtomentose branches; lvs. oblong to obovate, 1.2–2.5 cm. long, bright green and glabrous above, grayish-tomentulose beneath, the ± revolute undulate margin with 5–8 coarse spinescent teeth, subsessile or short-petioled; fls. blue, racemosely arranged in small umbels; caps. subglobose, ca. 6 mm. in diam., with 3 prominent dorsal horns.—Chaparral and Foothill Wd.; upper Napa V., mostly near Calistoga. Feb.–March.

Ssp. **confùsus** (J. T. Howell) Abrams. [*C. c.* J. T. Howell.] Decumbent, the tips turned upward and 2–4 dm. high; lvs. 0.6–2 cm. long, denticulate to spinose with 3–11 teeth.—Chaparral; Rincon Ridge, ne. of Santa Rosa, Sonoma Co. to Bartlett Mt., Lake Co. and Mt. St. Helena, Napa Co. Feb.–April.

40. **C. prostràtus** Benth. [*C. p.* var. *laxus* Jeps.] SQUAW CARPET. MAHALA MATS. Prostrate, the branches rooting and forming mats 1–2.5 m. across; branchlets red-brown, ± strigose; lvs. cuneate-oblanceolate to -obovate, 0.8–2.5 cm. long, with 3 sharp teeth at apex or a few additional just below, coriaceous, glossy light green above, minutely canescent beneath; petioles 1–3 mm. long; fls. deep or light blue, in umbellate clusters on short peduncles; caps. subglobose, 5–9 mm. broad, with mostly erect horns and intermediate crests; seeds dark, shining, flattened, ca. 4 mm. long; $n = 12$ (Nobs, 1942).—Open flats in pine forests, 3000–6500 ft.; Yellow Pine F., Red Fir F.; Sierra Nevada from Calaveras and Alpine cos. n., to Modoc Co., and w. to Siskiyou and Trinity cos.; to Wash. April–June.

Var. **occidentàlis** McMinn. [*C. divergens* ssp. *o.* Abrams.] Lvs. cuneate-spatulate, often undulate and slightly troughed above; caps. with mostly spreading horns.—Chaparral, Yellow Pine F.; higher mts. of e. Sonoma and Mendocino and n. Napa and Lake cos. May.

C. × serrulàtus McMinn. A mat-former, with mostly alternate elliptical lvs. 1.2–2 cm. long, pale green above, minutely flocculent-canescent beneath; fls. white or pale blue.— Between Emerald Bay and Cascade Lake, Eldorado Co.; supposed hybrid between *C. prostratus* and *C. cordulatus.*

C. × rugòsus Greene. Prostrate plant; lvs. opposite, leathery, elliptic-obovate, 3-veined from base, tomentulose beneath, 1.5–4 cm. long; fls. pale, in short subsimple racemes.— Modoc to Nevada cos. June. Possible hybrid between *C. prostratus* and *C. velutinus.*

41. **C. pùmilus** Greene. [*C. prostratus* var. *profugus* Jeps.] Prostrate or decumbent, with stout rigid branches and tomentulose branchlets; lvs. opposite, cuneate-oblong, oblanceolate, or oblong-obovate, 0.4–1.5 cm. long, coriaceous, glabrous above, paler and white-tomentulose beneath, truncate and usually 3-toothed, slightly revolute, entire or few-toothed on sides; petioles 1–2 mm. long; fls. blue to white, in few-fld. umbels; caps. globose, 4–5 mm. in diam., with short lateral or subdorsal horns, usually without intermediate crests; seeds shining, flattened, ca. 3.5 mm. long.—Dry serpentine slopes and flats, 2000–5700 ft.; Chaparral, Yellow Pine F., Red Fir F.; Mendocino and Trinity cos. to Del Norte and Siskiyou cos.; sw. Ore. April–May.

42. **C. pinetòrum** Cov. [*C. prostratus* var. *p.* K. Bdg.] Erect or semiprostrate, 1.5–12 dm. tall, with divergent brownish or grayish branches that may root at nodes and become 4 m. long; lvs. opposite, often clustered at nodes, broadly obovate to roundish, 1.2–2.5 cm. long, glabrous and light green above, minutely canescent and paler beneath, slightly revolute, coarsely 6–9-toothed on each side; fls. whitish to blue, in few-fld. umbels; caps. globose, ca. 7 mm. broad, with wrinkled dorsal horns and intermediate crests; seeds dark, shining, 5 mm. long.—Dry slopes, 5400–9000 ft.; Yellow Pine to Lodgepole Pine F.; s. Sierra Nevada, Kern, Tulare, and Inyo cos. June–July.

43. **C. Jepsònii** Greene. Low, spreading, 3–6 dm. high, the branches stout, short, strigulose; lvs. opposite, rigidly coriaceous, elliptical, hollylike, rounded at ends, yellowish-green and glabrous above, 1–2 cm. long, white-canescent beneath, ± infolded lengthwise, with ± revolute undulate margins having 4–5 coarsely spinulose teeth on a side; petioles 1–3 mm. long; fls. blue to violet, few, on stout naked peduncles; caps. globose, ca. 6 mm. in diam., with large wrinkled dorsal horns and intermediate crests.—Dry slopes

of serpentine; Chaparral; Mt. Tamalpais, Marin Co. to Mendocino Co. March–April.
Var. **albiflòrus** J. T. Howell. Erect to ca. 1 m. tall; fls. white; caps. somewhat elongate.
—Chaparral, Lake and Napa cos.

4. Colubrìna Rich.

Evergreen or deciduous shrubs or trees with rigid or even spiny twigs. Lvs. alternate, entire or denticulate, pinnate or 3-nerved, usually with small stipules. Fls. inconspicuous, in sessile or pedunculate axillary umbels. Sepals tardily deciduous, united at the base with the urn-shaped receptacle which is lined with the disk and adherent to the base of the ovary. Petals minute, hooded and partly enclosing the anthers, sessile or clawed. Style short, 3-lobed nearly to the base. Caps. 3-loculed, ± 3-lobed, with 1 seed in each locule. Ca. 18 spp., of sw. N. Am. and of S. Am.; 1 in Old World. (Latin, *coluber*, a serpent, of uncertain application.)

1. **C. califórnica** Jtn. [*C. texensis* var. *c.* L. Benson.] Intricately branched shrub 1–2.5 m. tall, the branches ± spinescent, usually divaricate, finely grayish-tomentose; lvs. oblong-obovate, 0.8–2 cm. long, entire, rounded to obtuse, dull gray-green, ± tomentose; fls. in small axillary clusters; sepals deltoid, tomentose, ca. 1 mm. long; petals yellowish, ca. 1 mm. long; caps. globose, 6 mm. in diam.—Occasional, dry canyons below 3000 ft.; Creosote Bush Scrub; Joshua Tree National Monument, n. Riverside Co. to Eagle Mts., and Chuckawalla Mts.; to Ariz., L. Calif. April–May.

5. Adólphia Meissn.

Shrubs with opposite divaricate spinose twigs which are articulated with the stems. Lvs. opposite, small, mostly caducous, with stipules. Fls. inconspicuous, solitary or in few-fld. axillary clusters. Sepals 5. Petals 5, hooded. Caps. 3-loculed, 3-lobed, the lower third surrounded by but mostly free from the cuplike calyx-base. Style 3-cleft, articulate near the base. Seeds 1 in each locule, smooth, bony. Ca. 2 spp., of sw. N. Am. (*Adolphe* Brongniart, French botanist and student of Rhamnaceae.)

1. **A. califórnica** Wats. Intricately branched, ca. 1 m. high, the branches short-pubescent, green, stiff, divaricate, striate, the ultimate spinescent; lvs. oblong or obovate, entire or nearly so, 3–12 mm. long, short-petioled, entire, puberulent; fls. solitary or few, short-pedicelled; calyx pubescent, greenish-white; petals white, ca. 2 mm. long; caps. globose, 4–5 mm. in diam., 3-grooved; seeds dark, ca. 3.5 mm. long, rounded oblong, plump.—Dry canyons and washes; Chaparral; sw. San Diego Co.; n. L. Calif. Dec.–April.

96. Buxàceae. Box Family (Fig. 89)

Trees, shrubs, or perennial herbs, monoecious or dioecious. Lvs. mostly persistent, simple, alternate or opposite, exstipulate. Fls. regular, bracted, solitary or clustered. Calyx present or not. Petals none. Staminate fls. with 4–12 stamens. Pistillate fls. 2–4(mostly 3)-loculed, with 1–2 anatropous ovules in each locule. Styles as many as locules, simple. Fr. a caps. or drupe; endosperm fleshy or scanty. Seven genera and ca. 30 spp.; temp. and subtrop.

1. Simmóndsia Nutt. Goatnut. Jojoba

Dioecious shrubs, with opposite persistent leathery entire lvs. Fls. on short axillary peduncles, the ♂ clustered, the ♀ solitary. Sepals 5 (4–6). Stamens 10–12. Ovary mostly 3-loculed; styles 3. Fr. a caps. with a firm wall, partly enclosed by the enlarged sepals. Seed 1. One sp. (F. W. *Simmonds*, English naturalist.)

1. **S. chinénsis** (Link) C. K. Schneid. [*Buxus c.* Link. *S. californica* Nutt.] Shrub 1–2 m. tall, with stiff branches, pubescent on younger growth; lvs. oblong-ovate, subsessile, 2–4 cm. long, dull green, ± canescent-puberulent; peduncles 3–10 mm. long; sepals in ♂ fls. greenish, 3–4 mm. long, those of ♀ becoming 10–20 mm. long; caps. coriaceous, nutlike, ovoid, obtusely 3-angled, ca. 2 cm. long, filled by the large oily puberulent seed.—Locally common on dry barren slopes, below 5000 ft.; Creosote Bush Scrub,

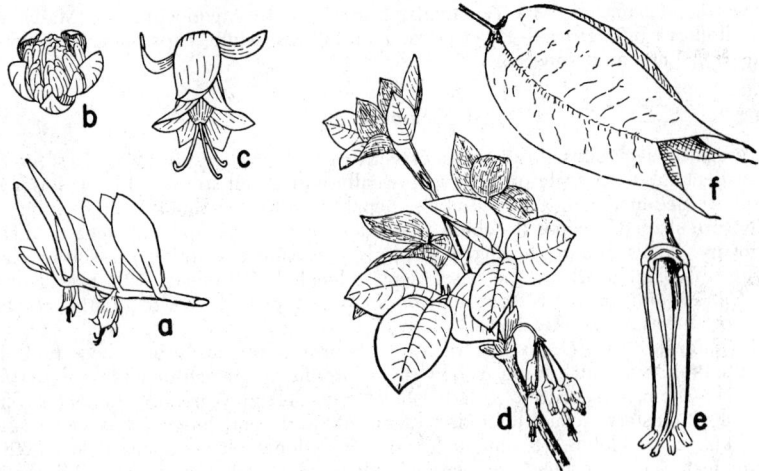

Fig. 89. BUXACEAE. *Simmondsia chinensis: a,* twig, × 1, with ♀ fls.; *b,* ♂ fl., × 2½, sepals 5, petals O, stamens many; *c,* ♀ fl., × 1½, sepals and pistil with 3 styles.
STAPHYLEACEAE. *Staphylea Bolanderi: d,* twig, × ½, lvs. 3-foliolate, fls. 5-merous; *e,* inner parts of fl., × 2½, pistil of 3 partly united carpels, stamens inserted outside the disk; *f,* 3-lobed bladdery caps., × ½.

Joshua Tree Wd.; Little San Bernardino Mts. and region of Twentynine Palms to Imperial Co.; also in Chaparral, interior cismontane s. Calif., w. Riverside and San Diego cos.; to Son., L. Calif. A form from Twentynine Palms to Beale's Well has quite glaucous foliage. The nuts have a useful oil and there is some interest in cultivating the sp. March–May.

97. Staphyleàceae. BLADDERNUT FAMILY (Fig. 89)

Trees or shrubs with opposite pinnately compound or 3-foliolate lvs. Stipules small, often caducous. Fls. regular, perfect, usually 5-merous, in axillary or terminal clusters. Stamens inserted outside a large disk. Pistil free from disk; ovary commonly 3-loculed. Fr. a 3-lobed bladdery caps., often dehiscent at apex. Five genera and ca. 20 spp.; N. Temp.

1. Staphylèa L. BLADDERNUT

Deciduous. Fls. white, on jointed pedicels in drooping axillary panicles. Calyx deeply 5-parted. Petals ca. as long as calyx-lobes. Pistil of 3 carpels united only at their axes; styles 3; ovules many. Caps. bladdery, deeply 3-lobed. Seeds globose. Ca. 6 spp. (Greek, *staphule,* a cluster.)

1. **S. Bolánderi** Gray. Erect, shrubby or arborescent, 2–6 m. high, glabrous; lvs. 3-foliolate, the lfts. broadly ovate to orbicular, 2.5–6 cm. long, serrulate, acutish; petiole slender, 2–6 cm. long; calyx white, 8–10 mm. long; petals white, 10–12 mm. long; caps. 2.5–5 cm. long, 3-horned, the carpels separating and dehiscing down the inner side of the free horn; seeds ± obovoid, light brown, smooth, 5–7 mm. long.—Uncommon, on canyon walls, 1000–4500 ft.; Chaparral, Foothill Wd., Yellow Pine F.; Sierra Nevada from Tulare Co. n., to Shasta and Siskiyou cos. April–May.

98. Thymelaeàceae. MEZEREUM FAMILY (Fig. 90)

Mostly woody plants with opposite or sometimes alternate simple exstipulate lvs. Fls. perfect or unisexual, in short racemes or spikes, rarely solitary. Calyx usually petaloid, 4–5-lobed. Petals none. Stamens as many or twice as many as calyx-lobes, inserted on

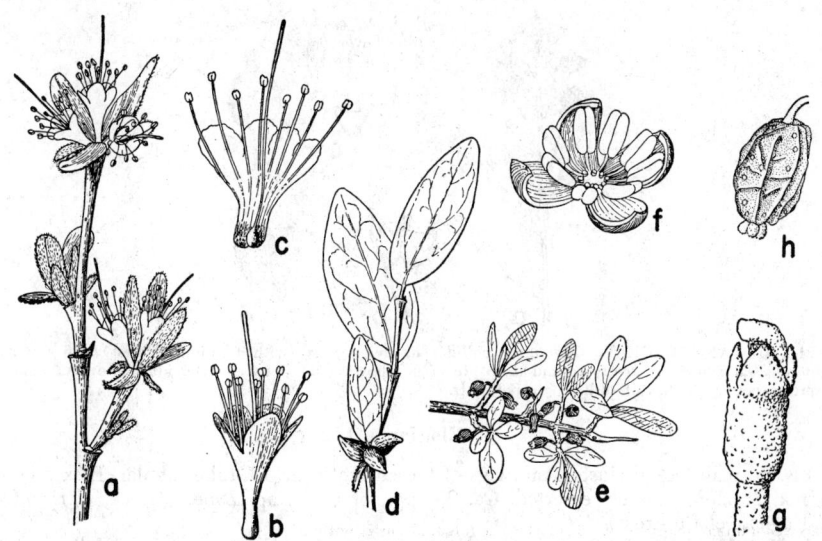

Fig. 90. THYMELAEACEAE. *Dirca occidentalis:* a, twig, × 1; b, fl. × 1½, disk wanting, calyx 5-lobed, stamens 10; c, section of fl., petals none; d, drupes, 3 in cluster.

ELAEAGNACEAE. *Shepherdia argentea:* e, twig, × ½; f, ♂ fl., × 5, sepals 4, petals O, stamens 8 and at edge of disk; g, ♀ fl., × 5, 4-lobed calyx and cent. style; h, wrinkled fr., × 3.

the tube. Disk hypogynous or lacking. Ovary superior, 1–2-loculed; style simple; stigma capitate or discoid. Ovules solitary in each locule, pendent. Fr. a berry or drupe. Ca. 40 genera and 400 spp., mostly trop.

1. Dírca L. LEATHERWOOD

Shrubs with smooth tough fibrous bark. Lvs. deciduous, alternate, entire, short-petioled. Fls. yellow, appearing before the lvs., 2–4 in fascicled clusters. Calyx 4-lobed. Stamens 8, exserted, the alternate longer. Disk wanting. Ovary subsessile, 1-loculed; style slender, elongate; stigma small, capitate. Fr. a drupe, oval-oblong. Two spp., 1 in e. N. Am. (Greek name of a fountain in Thebes, the plants from moist places).

1. **D. occidentàlis** Gray. Erect, 6–20 dm. high, with very pliable twigs and tough leathery bark; lvs. oval to obovate, rounded at ends, glabrous and light green above, paler and ± pubescent beneath, 3–6 cm. long, entire; petioles 3–6 mm. long; fls. lemon-yellow, regular, sessile, bent downward; calyx 8–10 mm. long; stamens mostly 8, exserted; drupe yellowish-green, ovoid, 8–10 mm. long.—Uncommon, wet slopes of rocky hills below 1500 ft.; Mixed Evergreen F., Chaparral; Sonoma, Marin, Contra Costa, Alameda, San Mateo cos. Jan.–March.

99. Elaeagnàceae. OLEASTER FAMILY (Fig. 90)

Shrubs or trees, with alternate or opposite, mostly silvery-scaly or stellate-pubescent simple entire lvs. Fls. perfect or imperfect, clustered or solitary, on nodes of previous year. Fl.-tube in ♂ fls. cup-shaped or saucer-shaped, in ♀ or perfect fls. tubular or urn-shaped and persistent. Sepals 4. Petals none. Stamens 4 or 8. Disk lobed or annular. Ovary superior, 1-loculed, 1-ovuled. Style slender. Fr. drupelike, the fleshy fl.-tube enclosing the ak. Ca. 3 genera and 20 spp., widely distributed.

Lvs. alternate; fls. perfect or the plants polygamous; stamens 4 .... 1. *Elaeagnus*
Lvs. opposite; fls. unisexual, the plants dioecious; stamens mostly 8 .... 2. *Shepherdia*

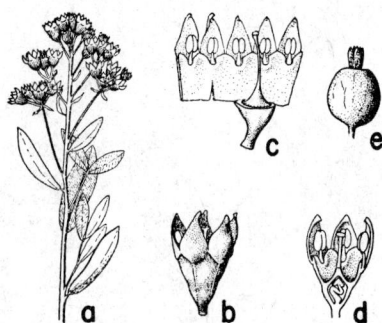

Fig. 91. SANTALACEAE. *Comandra pa'lida: a,* twig, × ½, with terminal cymes; *b,* fl., × 2½, 5-merous calyx, inferior ovary; *c* and *d,* inside of fl., showing floral tube, style, inferior ovary, stamen-insertion between lobes of disk; *e,* drupelike fr.

1. Elaeágnus L.

Fls. in small lateral clusters on twigs of the current year. Fl.-tube tubular. Disk present or not. Stamens scarcely exserted. Ca. 25 spp., of N. Temp. Zone. (Greek, *elais,* olive, and *agnos,* chaste-tree.)

1. **E. angustifòlia** L. RUSSIAN-OLIVE. OLEASTER. Shrub or tree, 3–7 m. tall, densely silvery; lvs. lanceolate, 3–10 cm. long; fls. 12–15 mm. long; fr. yellow with silvery scales.—Occasional escape in wet places, as at Victorville (San Bernardino Co.), Antioch (Contra Costa Co.), Lone Pine Creek (Inyo Co.); native of Eu. May–June.

2. Shephérdia Nutt.

Lvs. deciduous. Fls. small, clustered in axils or the ♀ solitary. Sepals 4, greenish-yellow within. Stamens 8, alternating with the lobes of the disk. Three spp., of N. Am. (John *Shepherd,* once curator of Liverpool Botanic Garden.)

1. **S. argéntea** Nutt. [*Lepargyraea a.* Greene. *Elaeagnus utilis* A. Nels.] BUFFALO-BERRY. Erect spiny shrub or small tree, 2–6 m. high, the branchlets brown or silvery-scurfy; lvs. opposite, oblong or lance-oblong, obtuse at apex, mostly cuneate at base, 2–6 cm. long, silvery-scurfy on both sides; petioles 4–12 mm. long; fls. subsessile, fascicled at the nodes, brown, 4–5 mm. broad; fr. broadly ellipsoid, 4–6 mm. long, red, sour, edible.—Along streams, 3500–6500 ft.; Sagebrush Scrub, Pinyon-Juniper Wd., N. Juniper Wd.; e. of Sierra Nevada from Inyo to Modoc cos., Cuyama V., Mt. Pinos region; to Canada and Rocky Mts. April–May.

100. Santalàceae. SANDALWOOD FAMILY (Fig. 91)

Herbs, shrubs or trees. Lvs. entire, estipulate. Fls. perfect or imperfect, axillary or terminal. Fl.-tube adnate to the base of the ovary. Calyx 4–5-merous, valvate in bud. Corolla none. Stamens as many as the sepals, opposite them, inserted on the fleshy disk. Style 1; ovary 1-loculed; stigma capitate. Ovules 2–4, suspended from the top of a free central placenta. Fr. a drupe or nut. Seed 1, round or ovoid, without testa; embryo small, axile at one end of the abundant endosperm. Ca. 26 genera and 250 spp.; mostly trop.

1. Comándra Nutt. BASTARD TOAD-FLAX

Smooth perennial herbs from rootstocks; stems striate. Lvs. alternate, entire, sub-sessile. Fls. greenish, in small terminal or axillary clusters, perfect. Fl.-tube campanulate or urn-shaped. Sepals 5. Anthers attached to sepals by a tuft of hairs. Fr. drupelike, crowned by the persistent calyx; seed globular. Four spp., 3 in Am., 1 in Eu. (Greek, *kome,* hair, and *ander,* man, referring to the hairy attachment of the stamens.)

1. **C. pállida** DC. [*C. umbellata* var. *p.* Jones. *C. californica* Eastw.] Stems many, 1–3 dm. high, leafy, glaucous, branched; lvs. lanceolate, 1–2.5 cm. long, acute to

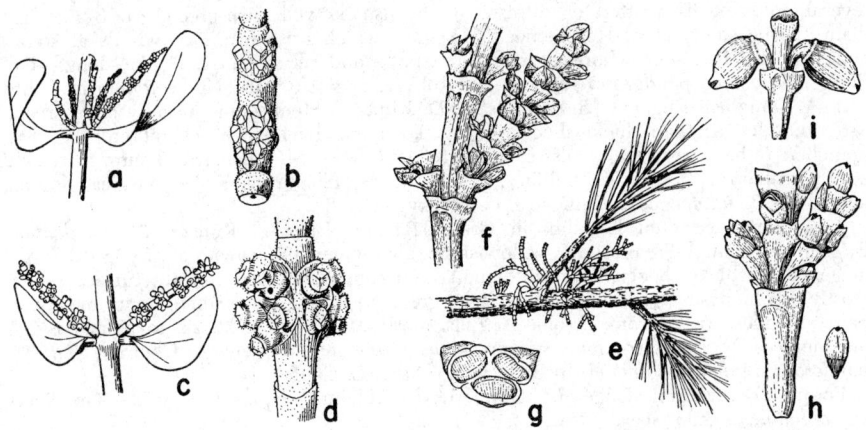

Fig. 92. LORANTHACEAE. *Phoradendron flavescens* var. *macrophyllum: a,* lvs. and ♂ spikes. × ½; *b,* ♂ spike, × 2½, fls. with 3 sepals; *c,* lvs. and ♀ spikes, × ½; *d,* ♀ fls. × 2½. *Arceuthobium campylopodum: e,* parasitic on pine, × ⅙; *f,* ♂ spike, × 2; *g,* ♂ fl., × 5, with 3 sepals and 3 anthers; *h,* ♀ fls. compressed, × 2, 2-toothed, shown in more mature stage in *i.*

cuspidate, cymes few- to several-fld., corymbosely clustered; pedicels 2–4 mm. long; fls. ca. 5 mm. long; sepals oblong, purplish-green; fl.-tube scarcely produced above the body of the fr., this 6–7 mm. long, oblong-ovoid.—Occasional, as root-parasite, on mostly dry slopes, 1000–9000 ft.; Mixed Evergreen F. and Sagebrush Scrub to Red Fir F.; Tehachapi Mts. and Death V. region, Sierra Nevada, Trinity, and Humboldt cos.; to B.C., Minn., New Mex. April–Aug.

101. Loranthàceae. Mistletoe Family (Fig. 92)

Evergreen shrubs or herbs, parasitic on woody plants. Branches dichotomous, swollen at the nodes. Lvs. opposite, exstipulate, simple, entire, sometimes reduced to scales. Ours dioecious, with greenish inconspicuous regular apetalous fls. Fl.-tube adnate to ovary. Calyx 2–5-merous. Petals 0. Stamens as many as the sepals and inserted on them; anthers 1- to 2-celled, sessile. Ovary inferior, 1-loculed. Fr. a berry with glutinous endocarp. Seed solitary, with copious endosperm. Ca. 20 genera and 500 spp., largely trop.

Berry on a recurved pedicel, compressed; ♀ sepals 2; anthers 1-celled . . . . 1. *Arceuthobium*
Berry sessile, globose; ♀ sepals 3; anthers 2-celled . . . . 2. *Phoradendron*

1. Arceuthòbium Bieb.

Yellowish or brownish plants with fragile jointed stems, the segms. glabrous, often ± 4-angled. Lvs. decussate, reduced to connate scales. Fls. solitary or several from the same axil, unisexual; the ♂ usually 3-merous, compressed, with anthers a single round locule opening by a circular slit. Pistillate fls. ovoid, compressed, 2-toothed. Berry fleshy, compressed, dehiscing on a short recurved pedicel. Parasitic on conifers. When mature the frs. discharge the viscid seeds with great force at slightest touch and seeds adhere to adjacent branches. Spp. several, of N. Hemis. (Greek, *arkeuthos,* juniper, and *bios,* life.)

(Gill, L. S. Arceuthobium in the U.S. Trans. Conn. Acad. Arts and Sci. 32: 111–245, 1935.)
Fls. blooming from April to June.
    Accessory branches in whorls, 2-many cm. long; ♂ fls. in panicles. On *Pinus* . . . . 1. *A. americanum*
    Accessory branches in fanlike arrangement, 1–2 cm. long; ♂ fls. in simple spikes. On *Pseudotsuga*
                                                                                            2. *A. Douglasii*
Fls. blooming in Aug. & Sept.; ♂ fls. in compound spikes. On *Pinus, Abies, Picea, Tsuga*
                                                                                            3. *A. campylopodum*

1. **A. americànum** Nutt. ex Engelm. in Gray. [*Razoumofskya a.* Kuntze.] Stem-segms. usually olive-green, 12–15 times as long as thick; shoots 3–6(–10) cm. long, with

several crops of fls., tufted or diffuse; ♂ fls. usually yellowish-green, ca. 3 mm. in diam., borne singly at end of pedicellike segms. which arise in 2's or whorls at stem-nodes; ♀ fls. in 2's or whorls at nodes or single and terminal; fr. 3 mm. long.—On *Pinus Murrayana, ponderosa, Jeffreyi,* etc.; to Wash., Rocky Mts. Not common in Calif.

2. A. Doúglasii Engelm. [*Razoumofskya D.* Kuntze.] Stem-segms. usually olive-green, 5–6 times longer than thick; shoots mostly 0.8–2 cm. long, with several crops of fls., branched rather sparingly; ♂ fls. usually green to yellowish-green, ca. 3 mm. in diam., mostly axillary in 2's; ♀ fls. axillary, mostly in 2's; fr. purplish, glaucous, ca. 4 mm. long.—On *Pseudotsuga;* n. Calif. to B.C., Rocky Mts.

3. A. campylópodum Engelm. in Gray. [*Razoumofskya c.* Kuntze. *A. occidentale* Engelm. in Wats.] Stem-segms. yellowish to olive-green or brown, mostly 5–10 times longer than thick; shoots 2–15 cm. long, often much-branched, mostly tufted; sepals mostly 4, oblong-ovate; ♂ fls. mostly light green to bright yellow or orange, mostly in pairs at nodes, ca. 2.5 mm. in diam.; ♀ fls. mostly axillary, the lateral in pairs; fr. 4–5 mm. long.—Our most common sp., on *Pinus attenuata, Murrayana, Coulteri, Jeffreyi, ponderosa, radiata, Sabiniana;* to B.C., Rocky Mts., L. Calif.

Forma divaricàtum (Engelm.) Gill. [*A. d.* Engelm.] Sepals mostly 3.—On *Pinus monophylla;* to Colo., Tex.

Forma abietìnum (Engelm.) Gill. [*A. a.* Engelm. *A. Douglasii* var. *a.* Engelm.] Sepals 7.—On *Abies;* to Wash., Mont., New Mex.

Forma tsugénsis (Rosend.) Gill. [*Razoumofskya t.* Rosend.] Sepals 4, orbicular.—On *Tsuga;* to Alaska, Ida.

Forma cyanocárpum (A. Nels.) Gill. [*Razoumofskya c.* A. Nels.] Sepals usually 3.—On *P. flexilis, albicaulis, aristata, Balfouriana.* Reported in Calif.; to Ore., Mont., New Mex.

Forma Blùmeri (A. Nels.) Gill. [*A. B.* A. Nels.] Shoots usually greenish.—On *Pinus Lambertiana* and *monticola;* to Canada, New Mex., Ariz.

Forma microcárpum (Engelm.) Gill. [*A. Douglasii* var. *m.* Engelm.] On *Picea;* Siskiyou Co.; to Mont., New Mex.

2. Phoradéndron Nutt. MISTLETOE

Parasitic woody plants, with much-branched rather brittle stems. Lvs. foliaceous, entire and faintly nerved, or reduced to connate scales. Fls. sunk in the jointed rachis, usually several in the axil of each bract. Staminate fls. mostly globose, the sepals 3 and with the sessile transversely 2-celled anthers at their base. Pistillate fls. with the fl.-tube adnate to the ovary. Berry sessile, ovoid to globose. A rather large genus of the Americas, largely trop. Largely spread from host to host by birds as they feed. (Greek, *phor,* a thief, and *dendron,* tree.)

(Trelease, William. The genus Phoradendron, 1–224; pls. 1–244, 1916.)
Pistillate spikes with 2 fls. at each joint.
  Lvs. reduced to connate scales.
    Plants canescent; ♀ spikes 3–4-jointed; berries usually red .... 1. *P. californicum*
    Plants glabrous; ♀ spikes 1-jointed; berries white .... 2. *P. juniperinum*
  Lvs. green, oblanceolate .... 3. *P. Bolleanum*
Pistillate spikes with 6 or more fls. at each joint; lvs. green .... 4. *P. flavescens*

1. P. califórnicum Nutt. MESQUITE M. Stems slender, terete, with ± pendulous clustered branches 1–5 dm. long and internodes 1–3(–4) cm. long, often reddish, puberulent when young; lvs. scalelike; spikes axillary, the ♂ of 2–5 joints, the ♀ of 3–4 joints; fr. globose, reddish, 3–4 mm. thick.—Mostly on *Prosopis,* occasional on *Acacia, Cercidium, Tamarix, Larrea,* etc.; deserts to Son., L. Calif. Plants with the internodes of the fruiting spikes unusually long are the var. *distans* Trel. and plants with white berries var. *leucocárpum* (Trel.) Jeps.

2. P. juníperinum Engelm. Stems glabrous, often bluntly squarish, rather stout; internodes rather short, to ca. 1 cm. long, microscopically granular; lvs. scalelike, spreading, 1–2 mm. long, not constricted at base; spikes solitary, glabrous, ca. 3 mm. long, with a single joint; ♂ fls. 6 or 8; ♀ 2; fr. straw- or wine-colored, ca. 3 mm. in diam.—On *Juniperus osteosperma,* e. Mojave Desert; to Colo., Tex., Chihuahua.

Var. ligàtum (Trel.) Fosb. [*P. l.* Trel.] Stem-scales mostly constricted-grooved at base, 1 mm. long.—On *Juniperus occidentalis,* etc.; to Ore., Mex.

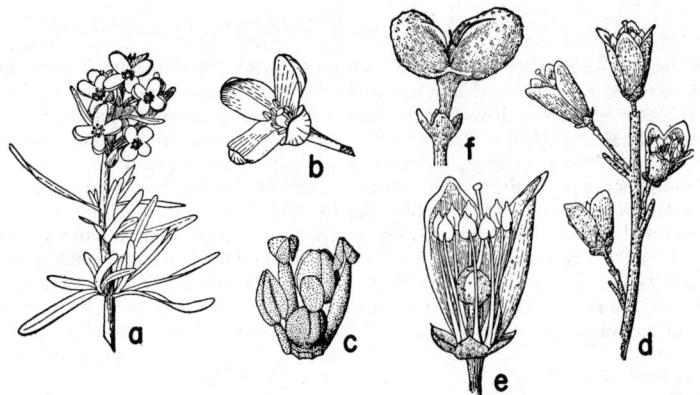

Fig. 93. RUTACEAE. *Cneoridium dumosum: a,* twig, × ¾, with 4-merous fils.; *b,* fl., × 1½, petals 4, stamens 4 long and 4 short, pistil cent.; *c,* fl. detail, × 5, 2 long and 2 short stamens shown, with ovary and lateral style. *Thamnosma montana: d,* part of infl., × 1; *e,* inside of fl., × 2½, with 4-lobed calyx, 4 petals, 8 stamens, stipitate 2-lobed ovary; *f,* fr., × 2.

Var. **Libocèdri** Engelm. in Wats. [*P. L.* Howell.] Scales obscurely constricted but not grooved at base.—On *Libocedrus;* to L. Calif. and Ore.

3. **P. Bolleànum** (Seem.) Eichler var. **dénsum** (Torr.) Fosb. [*P. d.* Torr. ex Trel. *P. d.* f. *Parishii* Trel.] Much-branched, dense, 1–2 dm. high, glabrous; lvs. oblanceolate, sessile, usually 1–1.5 cm. long, 3–4 mm. wide; spikes 3 mm. long, 1–2-jointed, the ♂ ca. 12-fld.; fr. 4 mm. thick, straw-colored, subglobose.—On *Juniperus californica,* etc. and *Cupressus;* Ore. to L. Calif., Ariz., Sonora.

Var. **pauciflòrum** (Torr.) Fosb. [*P. p.* Torr.] Plant more open, 2–3 dm. high; lvs. subpetiolate, 1.5–3 cm. long, 5–10 mm. wide.—On *Abies* and *Cupressus;* Sierra Nevada to L. Calif., Ariz., Coahuila.

4. **P. flavéscens** (Pursh) Nutt. var. **macrophýllum** Engelm. in Rothr. [*P. m.* Ckll. *P. coloradense* Trel. *P. longispicum* Trel. *P. l.* var. *cyclophyllum* Trel.] Plants stout, dioecious, 3–6 dm. high, the internodes 3–5 cm. long, minutely canescent to canescent-tomentose, later ± glabrate; lvs. thickish, elliptic-obovate to oblong-oblanceolate, obtuse, 1–5 cm. long, glabrous at least in age, cuneately petioled; spikes 1.5–5 cm. long, short-villous to glabrate, the ♂ 20–30-fld., the ♀ ca. 6- or 12-fld.; fr. white or tinged with pink, globose, 4–5 mm. in diam.—On *Platanus, Populus, Salix, Fraxinus, Juglans, Prosopis, Diospyros,* etc.; Sacramento V. to cismontane s. Calif., Colo. R. V.; to w. Tex.

Var. **villòsens** (Nutt.) Engelm. in Rothr. [*Viscum v.* Nutt. in T. & G. *P. v.* var. *rotundifolium* Trel.] Internodes and lvs. densely short-villous; lvs. narrow-obovate to roundish, 1.5–3(–4) cm. long.—Mostly on *Quercus,* sometimes on *Umbellularia, Arctostaphylos, Rhus, Populus,* etc.; from Ore. to L. Calif., Tex., Mex.

102. **Rutàceae.** RUE FAMILY (Fig. 93)

Aromatic herbaceous or mostly woody plants with punctate glands. Lvs. alternate or opposite, often compound, estipular. Fls. bisexual or unisexual, in axillary or terminal infl. Sepals and petals usually 3–5. Stamens as many or in ours twice as many, inserted on or at base of the hypogynous disk. Carpels 1–5, distinct or united; styles distinct or connate; stigma simple or lobed. Ovary superior; ovules usually 2 in each locule. Fr. a 2–5-celled caps. or drupelike. Seeds with or without endosperm. Ca. 110 genera and 900 spp., most abundant in trop. of S. Am. and Australia.

Lvs. simple.
   The lvs. alternate; ovary 2-loculed, forming a deeply 2-lobed caps. Deserts .... 1. *Thamnosma*
   The lvs. opposite; ovary 1-loculed, forming a round berry. Cismontane San Diego Co.
                                                                2. *Cneoridium*
Lvs. divided or compound.
   Perennial herb; fr. a 4–5-lobed caps. .... 3. *Ruta*
   Small tree; fr. a circularly winged samara .... 4. *Ptelea*

1. Thamnósma Torr. & Frém. in Frém. TURPENTINE-BROOM

Strong-scented undershrubs covered with glands. Lvs. narrow, alternate, early deciduous. Fls. in racemose cymes, bisexual. Calyx persistent, 4-lobed. Petals 4, erect. Stamens 8, inserted on the disk. Ovary usually 2-celled, 2-lobed, stipitate; style filiform; stigma capitate. Fr. a leathery 2-lobed caps. opening at the apex. Seeds 4–6, reniform. Two spp., of sw. N. Am. (Greek, *thamnos*, bush, and *osme*, odor.)

1. **T. montàna** Torr. & Frém. Stems branching, broomlike, yellowish-green, 3–6 dm. tall; lvs. oblance-linear, 0.5–1.5 cm. long; sepals ovate to round, 3–4 mm. long, greenish; petals ovate to oblong, 8–12 mm. long, purplish; stamens 4 long and 4 short with filiform fils.; style somewhat exceeding petals; caps. with 2 subglobose glandular lobes, each ca. 5 mm. thick; seeds 1–3, whitish, ca. 4 mm. long.—Frequent on dry slopes, below 5500 ft.; Creosote Bush Scrub to Pinyon-Juniper Wd.; deserts from Inyo Co. to Imperial and San Diego cos.; to Utah, New Mex., Son., L. Calif. March–May.

2. Cneorídium Hook. f. in Benth. & Hook.

Low evergreen shrub with grayish bark and slender twigs. Lvs. opposite, or fascicled, narrow, pellucid-punctate, rather heavily odorous. Fls. 1–3, usually on short, axillary peduncles. Calyx 4-lobed. Petals 4, white. Stamens 8, of 2 lengths, with dilated fils. Ovary 1-celled, sessile; ovules 2; style short, flattened, arising from near base of ovary on 1 side; stigma capitate. Fr. globose, fleshy, punctate. Seeds 1–2, subglobose, dark brown. One sp. (Resembling *Cneorum*, an Old World genus.)

1. **C. dumòsum** (Nutt.) Hook. f. [*Pitavia d.* Nutt. in T. & G.] BUSHRUE. Intricately branched, 1–2 m. tall; lvs. linear to oblong, glabrous, 1–2.5 cm. long, 1–3 mm. wide, gland-dotted, entire; calyx 1–1.5 mm. long; petals obovate, 5–6 mm. long; fr. 5–6 mm. thick, greenish to reddish, drying red-brown; seeds 5–6 mm. long.—Frequent on mesas and bluffs, below 2500 ft.; Chaparral, Coastal Sage Scrub; Orange and San Diego cos.; n. L. Calif. Nov.–March.

3. Rùta L. RUE

Perennial herbs and subshrubs, strong-scented. Lvs. alternate, pinnate, gland-dotted. Fls. small, in erect terminal clusters; sepals and petals 4–5. Calyx persistent. Petals toothed or fringed. Stamens 8 or 10. Ovary 4–5-loculed; style central; ovules several in each locule. Fr. a caps., 4–5-lobed, dehiscent or not. Seeds angled. Ca. 60 spp.; native Eurasia. (*Ruta*, the classical name.)

1. **R. chalepénsis** L. Glaucous perennial, the stems 4–8 dm. tall, puberulent; lvs. bipinnate or tripinnate, oblong in outline, to 1 or more dm. long, the segms. entire, narrow-elliptic, ca. 1–1.5 cm. long; infl. corymbose; sepals ovate, crenulate, 3–4 mm. long; petals yellow, 6–8 mm. long, involute, fringed on both sides; ovary and caps. with pointed lobes; $2n = 36$ (Negodi, 1939).—Occasional escape; native of Medit. region. Feb.–June.

R. gravèolens L. has also been reported from Calif.; its petals are undulate or denticulate on the margins, but not fringed; $2n = 72$ (Negodi, 1939), 81 (Revell, 1945).—Native of s. Eu.

4. Ptèlea L. HOP TREE

Shrubs or small trees, with a rather strong odor. Lvs. alternate, pinnately 3–5-foliolate; lfts. entire or serrulate, sessile. Fls. in corymbose or paniculate cymes, greenish-white, unisexual and bisexual. Sepals, petals and stamens 4–5; fils. hairy on the inner side, present in the ♀ fl., but sterile. Ovary 2-loculed; locules 2-ovuled, the lower abortive; style short; stigmas 2. Fr. a 2-celled, 2-seeded samara, winged all way around. A small genus of U.S. and Mex. (Greek name of the elm, which has a similar fr.)

1. **P. crenulàta** Greene. [*P. Baldwinii* var. *c.* Jeps. *P. brevistylis* Greene. *P. ovalifolia* Greene. *P. bullata* Greene. *P. cycloloma* Greene. *P. cinnamonea* Greene.] Shrub or small tree, 2–5 m. high, glabrous or slightly pubescent, the young twigs brown; lfts. 3, ovate

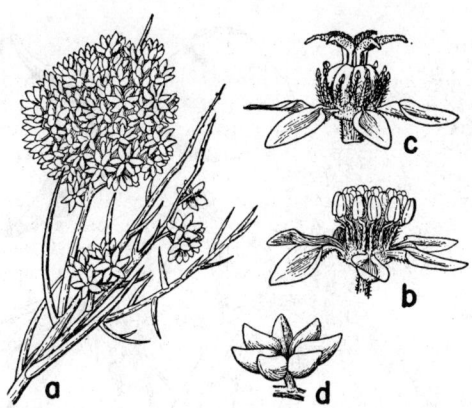

Fig. 94. SIMARUBACEAE. *Holacantha Emoryi: a,* thorny twigs, × ¼, with clusters of fr.; *b, ♂* fl., × 3, with 7–8-parted calyx, 7–8 spreading petals, many stamens with villous fils.; *c, ♀* fl., × 3, stamens abortive, the several styles divergent above; *d,* diverging drupes, × 1.

to lanceolate or obovate, acute to rounded at apex, cuneate to obliquely rounded at base, 2–7 cm. long, ± crenate, gland-dotted, glabrous above, ± pubescent beneath; petioles 2–5 cm. long; sepals minute; petals 4–5 mm. long; fils. hairy toward base; samara 1–2 cm. long, ca. as broad, the wing emarginate at both ends; style persistent.— Canyons and flats, below 2000 ft.; largely Foothill Wd., Yellow Pine F.; inner Coast Ranges from Shasta to Santa Clara Co. and foothills of Sierra Nevada s. to Tulare Co. April–May.

103. Simarubàceae. QUASSIA FAMILY (Fig. 94)

Shrubs or trees resembling *Rutaceae,* but the foliage without gland-dots; bark usually very bitter with contained oil-sacs. Lvs. alternate in ours. Fr. in ours a samara or drupe. Ca. 30 genera and 125 spp., of warmer regions.

Plant a tree, unarmed; lvs. large, odd-pinnate; fr. a samara .... 1. *Ailanthus*
Plant a shrub, thorny; lvs. scalelike; fr. drupaceous .... 2. *Holacantha*

1. Ailánthus Desf. TREE OF HEAVEN

Trees, polygamo-dioecious, with large odd-pinnate lvs. Fls. small, greenish, in large terminal panicles, the ♂ very strong-smelling. Calyx of 5 imbricate segms. Petals 5, spreading. Stamens 10, inserted at the base of the disk. Ovary in ♀ fls. 2–5-cleft, with flat 1-celled divisions. Ovule solitary in each cell. Fr. a linear or oblong samara. Three spp., of se. Asia. (*Ailanto,* a Moluccan name.)

1. **A. altíssima** (Mill.) Swingle. [*Toxicodendron a.* Mill. *A. glandulosa* Desf.] Tree 5–20 m. tall, spreading freely underground; lvs. 3–6 dm. long, 11–25-foliolate; lfts. lanceolate to oblong, acuminate, 7–18 cm. long; with 2–4 teeth near the base; samaras 3–5 cm. long, often ± reddish.—Rather widely natur. in cismontane Calif., especially about old dwellings; native of Asia. June. Introd. in early mining days by Chinese.

2. Holacántha Gray. CRUCIFIXION THORN

Rigid much-branched thorny shrubs. Lvs. alternate, deciduous, much-reduced. Fls. unisexual, solitary or glomerate on spinose branchlets. Calyx normally 7–8-parted. Petals 7–8. Stamens 12–16 in ♂ plants; fils. villous at base, 6–8 in ♀ fls. but sterile. Pistils 5–10, the bodies lightly cohering, tipped by divergent styles. Fr. of several dry diverging drupes. Seed ovoid. One sp. (Greek, *holos,* complete, and *akantha,* thorn.)

1. **H. Émoryi** Gray. Low and spreading and ca. 1 m. high; or taller, more erect and 2–2.5 m. high; appressed-canescent on younger branches; lvs. on mature plants scale-

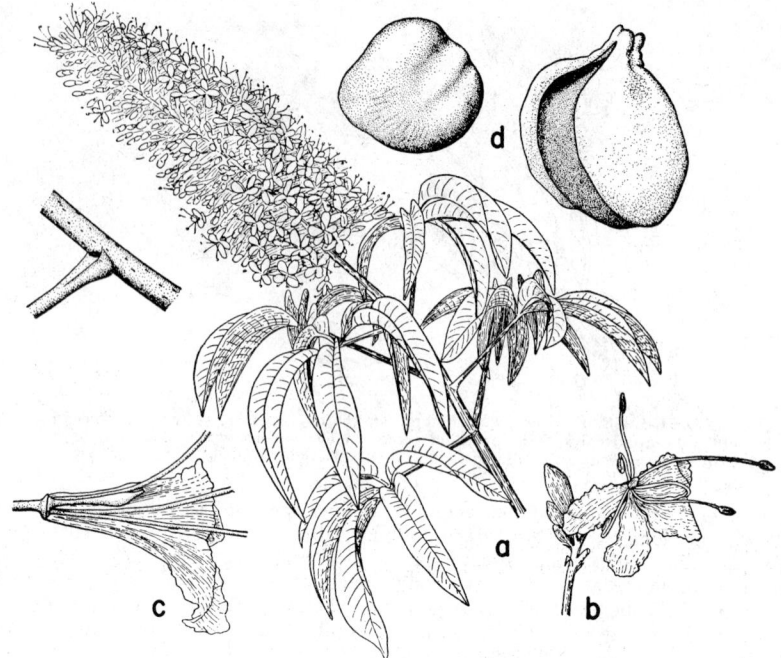

Fig. 95. HIPPOCASTANACEAE. *Aesculus californica: a,* twig, × ¼, with palmate lvs., terminal thyrse, petiole-base shown separately to left; *b,* fl., × 1, with 4 petals, 6 stamens, slender ovary (more detail shown in *c*); *d,* caps., × ⅓, splitting to reveal single large seed.

like, ovate to subulate, those of seedlings ca. 1 cm. long; fls. densely pubescent without, the ♂ ca. 3 mm. long, ♀ smaller; drupes 6–8 mm. long.—Occasional in gravelly places and on dry plains; Creosote Bush Scrub; s. Mojave Desert (Daggett, Ludlow, Amboy, Goffs, etc.) and Colo. Desert (Hayfields, Coyote Wells); to Ariz., Son. June–July.

104. Hippocastanàceae. BUCKEYE FAMILY (Fig. 95)

Trees or shrubs with opposite exstipulate palmate lvs. Fls. polygamous, irregular, showy, borne on jointed pedicels in a terminal panicle or thyrse. Calyx tubular or campanulate, 5-parted, the segms. unequal. Petals 4–5, clawed, unequal. Stamens 5–8; fils. long, slender. Ovary 3-loculed; ovules 2 in each cell; style slender. Fr. a leathery caps., globose or slightly 3-lobed, smooth or spiny, 3-celled or by abortion 1–2-celled and -seeded. Seeds large, shining; endosperm none; cotyledons large and thick. Three genera and ca. 18 spp., of N. Hemis.

1. Aésculus L. BUCKEYE. HORSE-CHESTNUT

Characters of family. Ca. 15 spp. (Latin name of some oak.)

1. **A. califórnica** (Spach) Nutt. [*Calothyrsus c.* Spach.] Large bush or tree to 7(–12) m. tall, with broad round top; lfts. 5–7, oblong-lanceolate, serrulate, acute to acuminate, 5–15 cm. long, subglabrous to finely pubescent; thyrse erect, 1–2 dm. long; fls. white or pale rose; calyx ca. 7–8 mm. long, 2-lobed, finely pubescent, the lobes shallowly toothed; petals ca. 13–15 mm. long; stamens 5–7, with orange anthers; fr. pear-shaped, smooth; seeds mostly 1, glossy brown, 2–3 cm. in diam., with large whitish hilum.— Common on dry slopes and in canyons, below 4000 ft.; largely Foothill Wd.; Coast Ranges and Sierra Nevada from Siskiyou and Shasta cos. s. to n. Los Angeles Co. and Kern Co. May–June. With various toxic principles; ground-up seeds at one time used by Indians for stupefying fish; fls. said to be poisonous to bees.

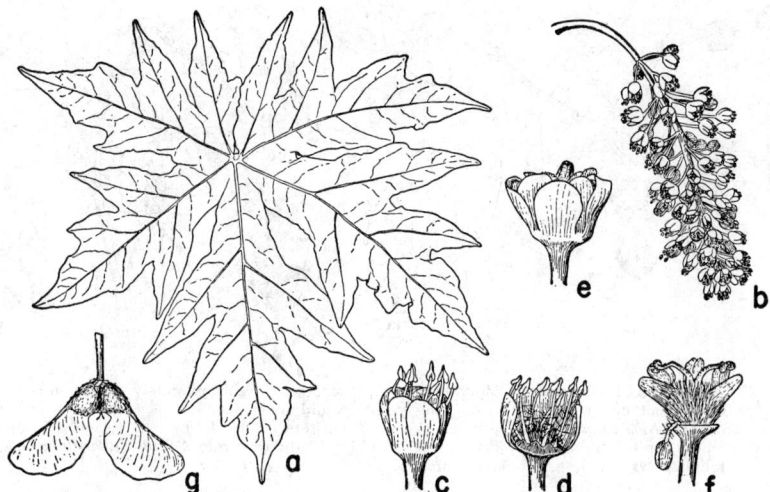

Fig. 96. ACERACEAE. *Acer macrophyllum: a,* lf., × ⅔; *b,* infl., × ½; *c,* ♂ fl., × 2, with erect sepals and petals; *d,* in section showing stamens from inside the disk; *e,* perfect fl., × 2, with sepals, petals, and stigmas; *f,* perfect fl. without perianth, with 2 stigmas, several stamens; *g,* 2 samaras, × ½.

105. Aceràceae. MAPLE FAMILY (Fig. 96)

Shrubs or trees, with opposite, simple or compound lvs. Plants polygamous to dioecious, the fls. small, borne in racemes, corymbs, or fascicles. Calyx generally 5-parted (sometimes 4–9). Petals as many as sepals or 0. Stamens 4–9, often 8, hypogynous or borne on the edge of the disk; fils. filiform. Pistil 1 with a 2-lobed 2-celled ovary and 2 styles. Fr. of 2 winged carpels, united below (samaras). Two genera and ca. 125 spp., of N. Hemis.

1. Acer L. MAPLE

Lvs. simple or pinnate. Fl.-clusters drooping. Ca. 120 spp., of N. Temp. regions. (Latin name of the maple.)

Lvs. simple, palmately lobed; petals present.
    Lvs. 2–4 cm. wide; fls. in corymbs; frs. glabrous . . . . 1. *A. glabrum*
    Lvs. usually much larger.
        The lvs. 5–12 cm. broad; samaras glabrous; fls. few, in corymbs . . . . 2. *A. circinatum*
        The lvs. 10–25 cm. broad; samaras hairy; fls. many, in long racemes . . . . . . 3. *A. macrophyllum*
Lvs. pinnate, with 3–5 lfts.; petals absent . . . . 4. *A. Negundo*

1. A. glàbrum Torr. var. **Tórreyi** (Greene) Smiley. [*A. T.* Greene.] MOUNTAIN MAPLE. Shrub or small tree, 2–6 m. high; twigs slender, usually reddish; lf.-blades broadly cordate to subreniform, usually 3-lobed, sometimes with an additional basal pair, rarely 3-parted, truncate to cordate at base, 2.5–5 cm. long, 3–5 cm. wide, the cent. lobe largest, cuneate, teeth few to many, usually obtuse; samara-pairs 3–6(–8), in corymbs; length of peduncle plus that of pedicel 1.7–4 cm.; samaras 2–3 cm. long, usually divergent at an angle of ca. 45°.—Moist to fairly dry slopes and canyons, mostly 5000–9000 ft.; Montane Coniferous F.; Sierra Nevada and N. Coast Ranges (Trinity to Del Norte and Siskiyou cos.); w. Nev. April–May.

Var. **diffùsum** (Greene) Smiley. [*A. d.* Greene. *A. bernardinum* Abrams.] Twigs whitish-gray; lf.-blades 1.5–2.5 cm. long, 1.2–2.8 cm. wide, with few blunt teeth on the lobes; peduncle plus pedicel 1–2 cm. long.—At 6500–9000 ft.; mostly Montane Coniferous F.; San Jacinto and San Bernardino mts., Panamint Mts., Inyo-White Range, e. slopes of Sierra Nevada, Clark Mt. (e. Mojave Desert); to Utah.

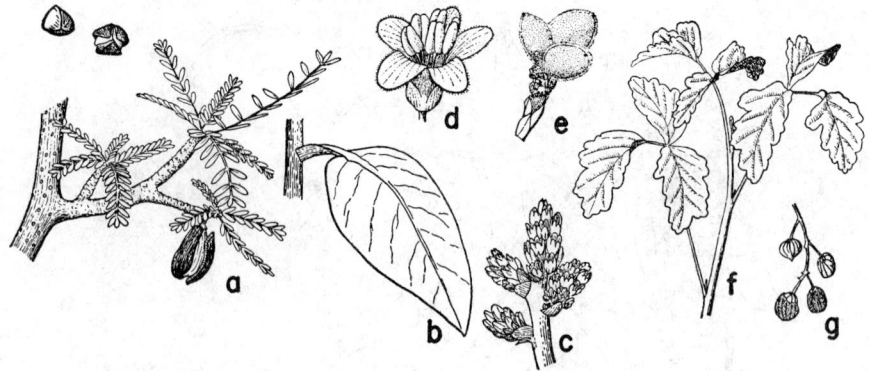

Fig. 97. BURSERACEAE. *Bursera microphylla:* a, twig, × 1, with pinnately compound lvs. and drupe (shown also at upper left, with dehiscing pericarp and inner seed).
ANACARDIACEAE. *Rhus ovata:* b, lf., × ½; c, fl.-clusters, × 1; d. fl., × 2½, with 5 sepals, 5 petals, 5 stamens (pistil hidden); e, drupes, × 1, with glandular surface. *Rhus diversiloba* (Poison-Oak): f, 3-foliolate lvs. × ¼; g, subglobose frs., × ½.

Var. **Greènei** Keller. Samaras not divergent, the wing of one overlapping the other.—Sierra Nevada of Tulare Co.

2. **A. circinàtum** Pursh. [*A. modocense* Greene.] VINE MAPLE. Shrub or small tree, 1–6 m. tall, or often vinelike and reclining; twigs slender, glabrous; lvs. round-cordate in outline, the blades thin, glabrous, palmately 5–11-lobed to near the middle, 5–12 cm. in diam. somewhat paler beneath than above; petioles 2–5 cm. long, grooved; fls. mostly 4–10 in a corymb, mostly ♂; calyx-lobes red-purple, villous, 4–6 mm. long; petals shorter, greenish-white; stamens 8, the fils. villous toward base; samaras glabrous, 2–3 cm. long, widely divergent, reddish when mature; $2n = 26$ (Foster, 1933).—Shaded stream banks, below 5000 ft.; Mixed Evergreen F., Redwood F., Yellow Pine F., etc.; from Mendocino, Yuba, and Butte cos. n.; to B.C. April–May.

3. **A. macrophýllum** Pursh. [*A. flabellatum, A. coptophyllum, A. platypterum, A. auritum, A. stellatum, A. hemionitis, A. dactylophyllum, A. leptodactylon* and *A. politum* Greene.] BIG-LEAF MAPLE. CANYON MAPLE. Round-topped tree 5–30 m. tall with coarse greenish to brownish glabrous twigs; lvs. roundish in outline, 1–2.5 dm. in diam., deeply 3–5-parted into coarsely irregularly few-toothed lobes, paler and more pubescent beneath; petioles commonly 5–12 cm. long; ♂ and perfect fls. in the same raceme; sepals and petals ca. 3 mm. long, greenish; stamens 7–9, villous near base; samaras variable, the body stiff-hairy, tawny, the wings 2–4 cm. long, diverging at an acute angle.—Common on stream banks and in canyons, below 5000 ft.; many Plant Communities; most of cismontane Calif. except valley plains; to Alaska. April–May.

4. **A. Negúndo** L. ssp. **califórnicum** (T. & G.) Wesmael. [*Negundo c.* T. & G. *A. N.* var. *c.* Sarg.] BOX ELDER. Round-headed tree 6–15(–20) m. tall; twigs rather slender, pubescent, greenish; lvs. pinnately 3-foliolate, the terminal lft. largest, 3–5-lobed, ovate, 5–12 cm. long, coarsely serrate, long-petiolulate, the lateral smaller, shorter-petiolulate, all densely pubescent especially beneath and when young; petioles mostly 2–8 cm. long; fls. unisexual, the ♂ fascicled, the ♀ drooping-racemose, appearing slightly before the lvs., greenish; pedicels filiform, stamens 4–5; samaras red when young, straw color when mature, finely pubescent, 2.5–3 cm. long, the pedicels 1.5–2 cm. long.—Along streams and bottom lands, below 6000 ft.; many Plant Communities; cismontane Calif. from stations in San Jacinto and San Bernardino mts., Santa Barbara and Kern cos. n.; to Shasta and Siskiyou cos. March–April.

106. **Burseràceae.** TORCHWOOD FAMILY (Fig. 97)

Deciduous aromatic shrubs or trees. Lvs. mostly alternate, usually pinnately compound or decompound, the rachis often winged, estipulate. Fls. bisexual or unisexual, solitary or in panicles. Sepals 3–5, ± connate basally. Petals 3–5, sometimes 0. Disc present, annular to cup-shaped. Stamens in 1–2 whorls; fils. naked. Pistil 1, the ovary superior, 2–5-loculed; ovules usually 2 in each locule. Fr. drupelike, containing 1–5 stones. Seeds without endosperm. Ca. 20 genera and 500 spp., largely trop.

1. Búrsera Jacq. ELEPHANT TREE

Petals inserted on the edge of the disk. Stamens 8–10. Ovary 3-loculed; stones covered by an aromatic pulp. Ca. 40 spp., of trop. Am. (J. *Burser*, 16th-century botanist.)

1. **B. microphýlla** Gray. [*Elaphrium m.* Rose.] Arborescent, 1–3 m. tall; older branches cherry-red; lvs. once pinnate, the pinnae 7–33, oblong-linear, 5–7 mm. long; fls. 5-merous; sepals ca. 1 mm. long; petals 4 mm.; stamens 10; drupes hanging, the exocarp splitting into 3 valves; stone yellow, 6 mm. long.—Local, rocky places; Creosote Bush Scrub; between Fish Creek and Carrizo Creek, w. edge of Colo. Desert; Ariz. to L. Calif. Early summer.

107. Anacardiàceae. SUMAC FAMILY (Fig. 97)

Trees or shrubs, with acrid, resinous or milky sap and alternate lvs. Polygamo-dioecious or with perfect fls., these small, regular, in axillary or terminal panicles. Calyx commonly 5-cleft, a glandular ring or cuplike disk lining its base. Petals commonly 5, the stamens as many or twice as many. Ovary 1-celled, 1-ovuled, free from calyx and disk; styles 3, terminal. Fr. a dry berrylike drupe; seed without endosperm. Ca. 70 genera and 600 spp., mostly from warm regions.

Stamens 10; introd. trees with pendent pinnate lvs. .... 1. *Schinus*
Stamens 5; native shrubs, mostly with entire or 3-foliolate lvs. .... 2. *Rhus*

1. Schìnus L. PEPPER TREE

Dioecious trees with alternate odd-pinnate resinous lvs. Fls. in bracteate panicles, small, whitish. Calyx short, 5-parted. Petals 5, imbricated. Ovary sessile forming a globose drupe. Ca. 15 spp., mostly from S. Am. (Greek name for Pistacia.)

1. **S. Mólle** L. Evergreen, 5–15 m. high, with slender pendulous twigs; lvs. 2–3 dm. long, pendulous, with numerous lance-linear lfts. 3–6 cm. long; fr. reddish, 6–8 mm. in diam.; $2n = 28$ (Schnack & Covas, 1947).—The Peruvian Pepper tree, commonly cult. in Calif., occasionally becomes natur.

2. Rhús L. SUMAC

Polygamous shrubs or small trees. Lvs. simple or compound. Fls. small, the sepals 4–6, mostly 5, persistent. Petals spreading in anthesis. Disk annular. Stamens 5. Drupe small, glabrous or pubescent, with persistent or deciduous exocarp. Seed 1, inverted. Ca. 120 spp., widely distributed. (Greek, *rhous*, ancient name for sumac.)

(Barkley, F. A. A monographic study of Rhus and its immediate allies in N. and Cent. Am. Ann. Mo. Bot. Gard. 24: 265–498, 1937.)
Lfts. 11–31. Rare .... 1. *R. glabra*
Lfts. 1–3. Common.
  Lvs. 3-foliolate, deciduous.
    Fls. whitish, in loose axillary panicles; fr. glabrous, whitish; lvs. subglabrous, shining; branches stiff, erect .... 2. *R. diversiloba*
    Fls. yellowish in sessile spikes; fr. hairy, red; lvs. usually pubescent; branches tending to be arched .... 3. *R. trilobata*
  Lvs. simple, persistent, coriaceous.
    Panicle with numerous slender branchlets; fr. glabrous, whitish; lvs. oblong-lanceolate, almost 3 times longer than wide .... 4. *R. laurina*
    Panicle with fewer stout branchlets; fr. pubescent, red.
      Lvs. oblong-elliptic, flat .... 5. *R. integrifolia*
      Lvs. ovate, somewhat folded along midrib .... 6. *R. ovata*

1. **R. glàbra** L. SMOOTH SUMAC. Few-branched straggling glabrous shrub, 1–3 m. tall; lvs. pinnate with 11–31 lfts., green above, white beneath, the lfts. lanceolate to ± oblong, acuminate, serrulate, 3–8 cm. long; fls. greenish, ca. 3 mm. long, in large terminal panicle; drupe compressed, ca. 4 mm. in diam., covered with short reddish

hairs.—Collected many years ago in Chino Canyon, San Jacinto Mts.; Ore. to B.C., Ariz. to Atlantic Coast.

2. **R. diversilòba** T. & G. [*Toxicodendron d.* Greene. *T. comarophyllum* Greene. *T. isophyllum* Greene. *T. oxycarpum* Greene. *T. dryophyllum* Greene. *T. vaccarum* Greene.] Erect shrub, bushy, 1–2(–3) m. tall, stiffly branched, pubescent or glabrous; lvs. pinnately 3-foliolate, the lfts. obtuse, usually crenulate or even lobed, roundish to ovate, 2–7 cm. long, bright green and shining above, paler beneath; panicles axillary, racemose; petals of ♂ fls. 3–4 mm. long, of ♀ 2–3 mm.; fr. whitish, glabrous, subglobose, 4–7 mm. thick; seeds flattened, 3–6 mm. long, irregularly roughened.—Common in low places and thickets and wooded slopes, below 5000 ft.; many Plant Communities; cismontane Calif.; to Wash., L. Calif., n. Mex. April–May. Causing a painful dermatitis for human beings. Plants climbing trees by means of aerial rootlets are forma *rádicans* McNair; with the sp. Occasional ones with 5 lfts. constitute the forma *quinquifòlia* McNair. The sp. is commonly called Poison-oak.

3. **R. trilobata** Nutt. ex T. & G. var. **malacophýlla** (Greene) Munz. [*Schmaltzia m.* Greene.] Squaw Bush. Diffusely branched shrub, 8–14 dm. tall, the branches spreading, often turned down at tips, strong-scented when crushed; ultimate branchlets rather coarse, ca. 2 mm. thick, heavily pubescent; lvs. trifoliate, pubescent, the lfts. rhombic-ovate to cuneate-obovate, obtuse, crenate, the terminal one 1–3 cm. long, ca. as wide, larger than the 2 lateral; fls. yellowish, in clustered spikes; petals ca. 2 mm. long; fr. viscid, pilose, 4–5 mm. in diam., reddish; seeds ± striate, 4–5 mm. long.—Common in canyons and washes especially of interior valleys, mostly below 3500 ft.; Coastal Sage Scrub, Chaparral, Foothill Wd., etc.; cismontane s. Calif., less common along base of Sierra Nevada to Butte Co. March–April.

Var. **quinàta** Jeps. [*Schmaltzia q.* Greene. *S. cruciata* Greene. *S. straminea* Greene.] Ultimate twigs slender, 1–1.5 mm. thick, very finely puberulent; lvs. sparsely and finely pubescent in maturity, the terminal lft. usually 3-lobed, so that the lf. appears ± 5-lobed or -foliolate.—Dry slopes and thickets; Foothill Wd., Chaparral, etc.; Coast Ranges from San Diego Co. to Siskiyou Co., occasional w. base of Sierra Nevada; L. Calif., Ore.

Var. **anisophýlla** (Greene) Jeps. [*Schmaltzia a.* Greene.] Twigs finely puberulent; lvs. green, subglabrous, the lfts. small, terminal 1–1.5 cm. long, not lobed, the lateral less than 1 cm. and often unequal; berries bright crimson.—Dry slopes, 3500–5500 ft.; Creosote Bush Scrub to Pinyon-Juniper Wd.; mts. of Mojave and n. Colo. deserts; to Utah, Ariz.

4. **R. laurìna** Nutt. in T. & G. [*Malosma l.* Nutt. ex Abrams.] Laurel Sumac. Large rounded evergreen shrub or small tree, 2–5 m. tall, glabrous, aromatic; twigs and veins reddish; lvs. lance-oblong, bicolored, 4–10 cm. long, 2–4 cm. wide, entire, mucronate, somewhat folded along the midrib; petioles 1–3 cm. long; panicle dense, intricately branched, 5–15 cm. long; fls. white, ca. 1 mm. long; drupe whitish, glabrous, 2–3 mm. in diam.; stone smooth, ca. 2 mm. long, flattish.—Dry slopes, below 3000 ft.; Chaparral, Coastal Sage Scrub; coastal slopes from n. L. Calif. to Santa Barbara Co., inland to w. Riverside Co. June–July.

5. **R. integrifòlia** (Nutt.) Benth. & Hook. [*Styphonia i.* Nutt. ex T. & G. *S. serrata* Nutt. *Neostyphonia i.* Shaf. *Schmaltzia i.* Barkl.] Lemonadeberry. Rounded shrub, aromatic, 1–3 m. tall, with finely pubescent, ± reddish, stoutish twigs; lvs. coriaceous, flat, subentire, or with shallow sharp teeth, oblong-ovate, rounded-obtuse at both ends, subglabrous, 2.5–5 cm. long, 2–3 cm. wide, sometimes ± lobed; petioles 3–6 mm. long; fls. in close panicles, white to rose, subtended by roundish hairy bracts; petals ciliolate, ca. 3 mm. long; drupe viscid, acid, pubescent, reddish, flattened, ca. 10 mm. in diam.; stone smooth, flat, ca. 7 mm. long.—Ocean bluffs and canyons, dry places below 2500 ft.; Coastal Sage Scrub, Chaparral; Santa Barbara Co. to n. L. Calif., inland to w. Riverside Co. Feb.–May.

6. **R. ovàta** Wats. [*Neostyphonia o.* Abrams. *Schmaltzia o.* Barkl. *R. o.* var. *Traskiae* Barkl.] Sugar Bush. Evergreen shrub, 1.5–3 m. high, with stout reddish glabrous twigs; lvs. coriaceous, ovate, acute, 4–8 cm. long, ± folded along midrib; infl. dense; bracts ovate, 1.5 mm. long; sepals and petals ciliate, the former 2.5, the latter 5 mm. long; drupes glandular, viscid, ± acid, reddish, 7–8 mm. in diam.; stone flat, smooth, ca. 4 mm. long.—Dry slopes, below 2500 ft.; mostly Chaparral, usually away from the

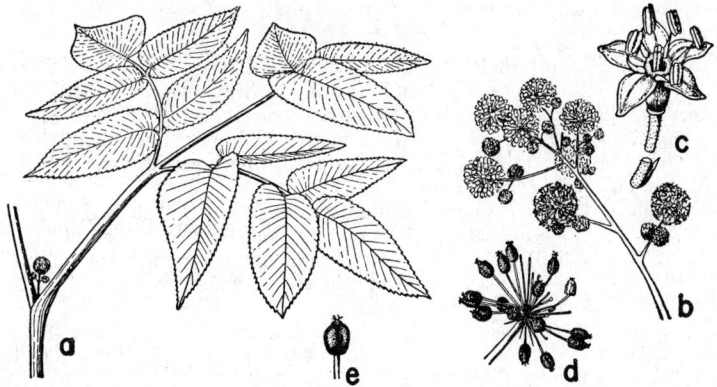

Fig. 98. ARALIACEAE. *Aralia californica: a*, lf., × ¹⁄₁₂; *b*, part of infl., × ¼, showing several umbels; *c*, fl., × 3, ovary inferior, sepals 5, hidden, petals 5, stamens 5, styles 5; *d*, umbel of frs., × ½; *e*, fr., × 1, with stigmas at top.

coast, Santa Barbara Co. to n. L. Calif., w. edge of Colo. Desert, Santa Cruz and Santa Catalina ids.; Ariz. March–May.

108. Meliàceae. MAHOGANY FAMILY

Trees or shrubs with hard often scented wood. Lvs. usually alternate, mostly pinnately compound or decompound, without pellucid dots, estipulate. Fls. bisexual, often in cymose panicles. Sepals 4–5, usually connate at base. Petals 4–5. Stamens 8–10, mostly monadelphous, hypogynous, outside the disk. Pistil 1, the ovary superior, mostly 2–5-loculed, usually with 2 ovules in a locule. Fr. a berry, caps., or drupe. Seeds often winged, with or without endosperm. Ca. 50 genera and 800 spp., largely trop.

1. Mèlia L.

Deciduous. Lvs. large, 1–3-pinnate. Fls. white or purple, in large axillary branched many-fld. panicles. Ca. 20 spp., from trop. Asia. (Ancient Greek name.)

1. **M. Azédarach** L. CHINABERRY TREE. Round-topped tree to 15 m. high, with furrowed bark and spreading branches; lvs. bipinnate, 3–8 dm. long, the many lfts. ovate to lanceolate, sharply serrate, 3–5 cm. long; fls. purplish, fragrant, in open panicles 1–2 dm. long; drupe round, 1–1.5 cm. in diam.; $2n = 28$ (Bowden, 1945).—Widely cult. in Calif. and sometimes becoming natur.; native of sw. Asia. More common in cult. is var. **umbraculifórmis** Berckmans. UMBRELLA TREE. Very compact and dense as to head with the many branches and drooping lvs. resembling a huge umbrella.

109. Araliàceae. GINSENG FAMILY (Fig. 98)

Herbs, shrubs or trees, sometimes climbing. Lvs. alternate to whorled, entire to compound or decompound. Fls. small, greenish or whitish, perfect or imperfect, mostly in umbels or umbellate heads. Fl.-tube adnate to the ovary. Calyx small, often absent. Petals usually 5, valvate or imbricate. Stamens as many as petals and alternate with them. Ovary 1, inferior, 1- to several-loculed, with 1 ovule in each locule; styles as many as the locules. Fr. a berry or drupelike. Seeds flattened or 3-angled; testa thin; embryo small in copious endosperm. Ca. 50 genera and 500 spp.; widely distributed in temp. and trop. regions.

Plant large perennial herb, with large compound lvs. .... 1. *Aralia*
Plant sprawling or climbing by aerial roots; lvs. simple, ± lobed .... 2. *Hedera*

1. Aràlia L. Spikenard

Lvs. large, pinnately or ternately compound or decompound; lfts. serrate, with sheathing petioles. Umbels 2 or more in an infl., radiating or laxly corymbose or in large compound racemes, with small bracts and bractlets. Pedicels articulate with calyx. Plants polygamo-monoecious, the fls. 5–6-merous, mostly glabrous. Fils. short. Disk small. Styles 4–6, free above or to the base. Fr. baccate, subglobose, sharply angled. Ca. 30 spp., of N. Am. and Asia. (Name unexplained.)

1. **A. califórnica** Wats. Roots large, with milky juice; stems 1–3 m. high; lvs. glabrous, ternate then pinnately 3–5-foliolate; petioles to 3 dm. long; lfts. ovate or ± oblong, serrate, subcordate at base, 0.5–2.5 dm. long, glabrous or sparsely puberulent on the nerves beneath; infl. ample, 3–4 dm. long, with numerous glandular-tomentulose many-fld. umbels; pedicels 1–2 cm. long; sepals minute; petals ca. 2 mm. long; fr. 3–5 mm. in diam., dark when mature; seeds almost 3 mm. long, light in color; $2n = 48$ (Bowden, 1945).—In ± moist and shaded spots, below 5000 ft.; many Plant Communities; cismontane Calif. from Orange Co. n.; to s. Ore. June–Aug.

2. Hédera L. Ivy

Plants with a sterile juvenile form which is climbing and with aerial roots and lobed lvs. and an adult form fertile, nonclimbing, and with nonlobed lvs. Lvs. alternate, persistent. Fls. bisexual, greenish, in umbels arranged in terminal racemes or panicles. Sepals 5; petals 5; styles united. Fr. a 3–5-seeded berry. Ca. 6 spp. of Eurasia, commonly cult. as English and Algerian Ivy. (The classical name.)

1. **H. Hèlix** L. English Ivy. Juvenile lvs. usually 3–5-lobed, triangular-ovate to subreniform, mostly 2–10 cm. wide, cordate to truncate at base; $2n = 48$ (Jacobsen, 1954).—Occasional escape and persisting for some time; native of Eurasia.

110. Umbellíferae. Carrot Family (Fig. 99)

Aromatic herbs, commonly with hollow stems and alternate compound or simple lvs., the petioles frequently dilated at the base. Fls. small, epigynous, perfect or sometimes imperfect, in compound or simple umbels or rarely a head, the umbels usually with an involucre of bracts and the umbellets with an involucel of bractlets. Sepals usually 5, sometimes obsolete. Petals 5, the tips often inflexed. Stamens 5, alternate with the petals; fils. filiform; anthers versatile. Ovary inferior, bicarpellate, 2-loculed; styles 2, usually borne on a stylopodium (swollen base). Ovule 1 in each locule, pendulous. Fr. dry, usually ribbed or winged, the 2 carpels separating at maturity along the plane of their contiguous faces (commissure), either flattened laterally (at right angles to the commissure) or dorsally (parallel with the commissure), sometimes terete. The 2 mericarps attached to a carpophore; pericarp usually containing oil-tubes between the ribs (in the intervals) and on the commissure. Seeds mostly adnate to the pericarp; endosperm cartilaginous; embryo small. Ca. 250 genera and 2000 spp., widely distributed and many of economic importance for food, flavoring, etc. Some are poisonous.

(Mathias, Mildred E., and L. Constance, *in* N. Am. Flora 28B: 43–295, 1944–1945.)
A. Infl. capitate, not umbellate.
    B. Fr. winged, not squamose .... 40. *Cymopterus*
    BB. Fr. not winged, ribless, variously scaly .... 41. *Eryngium*
AA. Infl. a distinct umbel, not capitate.
    B. Ovary and fr. bearing prickles, bristles, tubercles, or scales.
        C. Lvs. simple, entire, usually the basal with parallel veining and the upper broad and
            perfoliate .... 20. *Bupleurum*
        CC. Lvs. compound.
            D. Ovary and fr. armed with spines, hooked bristles, or tubercles.
                E. Plants perennial or biennial; fls. perfect and ♂ .... 4. *Sanicula*
                EE. Plants annual; fls. all perfect.
                    F. The plants glabrous; lf.-divisions ± elongate, filiform .... 5. *Apiastrum*
                  FF. The plants ± pubescent; lf.-divisions shorter.
                      G. Invol. of conspicuous foliaceous bracts; lvs. 3–4-pinnatisect; fr.
                        bristly on ribs only .... 6. *Caucalis*

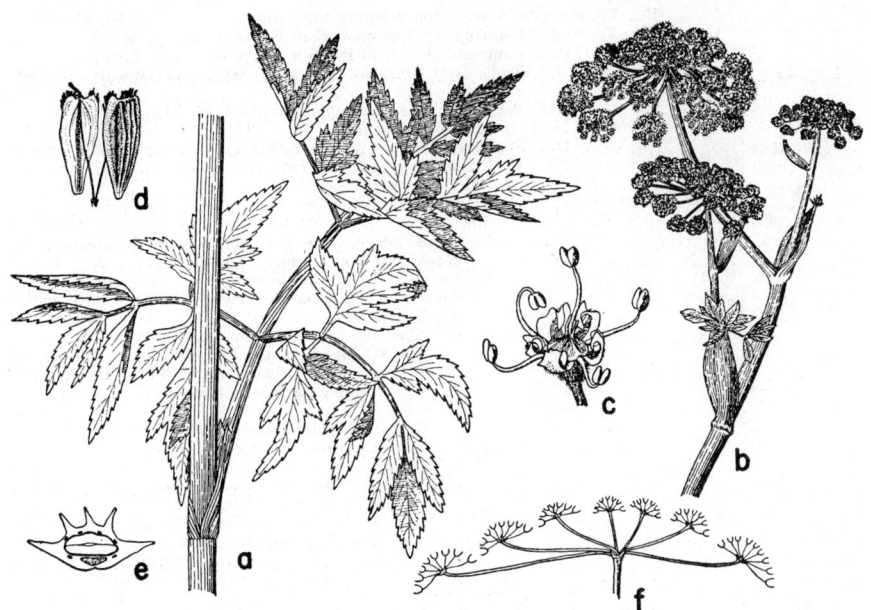

Fig. 99. UMBELLIFERAE. *Sphenosciadium capitellatum: a*, compound lf., × ¼, with inflated sheathing petiole; *b*, part of infl., × ¼, compound umbels; *c*, fl., × 3, with 5 incurved petals, 5 alternating stamens, 2 styles borne on a swollen base (stylopodium); *d*, mature fr., × 2½, the 2 carpels attached to a carpophore and pulling apart; *e*, diagram, × 5, of a cross section of 1 carpel with lateral wings, 3 winglike ribs and cent. seed with 2 cotyledons, oil-tubes solitary in intervals and 2 on commissural side. *Diagram of a compound umbel: f*, with invol. of bracts at base of main rays and smaller bractlets or involucel at base of each umbellet.

             GG. Invol. none or of linear bracts; lvs. pinnate to 3-pinnatisect; fr.
                bristly or tubercled throughout.
                H. Fr. not beaked; bractlets longer than the pedicels . . 7. *Torilis*
                HH. Fr. beaked; bractlets shorter than the pedicels . . 8. *Anthriscus*
      DD. Ovary and fr. armed with bristles which are never hooked.
         E. The fr. several times longer than broad; oil-tubes absent or obscure.
           F. Plants annual; the fr. with a beak longer than the body . . . . 9. *Scandix*
           FF. Plants perennial; fr. not beaked or if so, the beak shorter than the body
                                        10. *Osmorhiza*
        EE. The fr. not more than twice as long as broad; oil-tube evident.
         F. Foliage glabrous; fr. armed with unequal subulate bristles
                                     11. *Ammoselinum*
         FF. Foliage ± pubescent; fr. armed with barbed bristles. . . . . 12. *Daucus*
  BB. Ovary and fr. not bearing prickles, bristles, tubercles, or scales.
    C. Lvs. simple; umbels simple or proliferous.
      D. Ovary and fr. glabrous; lvs. glabrous.
        E. Lvs. with a definite ovate to roundish blade . . . . 1. *Hydrocotyle*
        EE. Lvs. reduced to hollow cylindrical jointed phyllodes . . . . . . 2. *Lilaeopsis*
      DD. Ovary and fr. covered with stellate hairs; lvs. ± stellate-pubescent . . 3. *Bowlesia*
  CC. Lvs. and umbels variously compound.
    D. Ribs of the fr. not prominently winged; fr. terete in cross section or flattened
      somewhat laterally.
      E. Fls. yellow.
        F. Involucel absent; lf.-divisions filiform; plants with anise odor
                                    28. *Foeniculum*
        FF. Involucel present; lf.-divisions linear to ovate; plants lacking anise odor.
          G. Stemmed biennial; stylopodium low, conical. Garden Parsley
                                     15. *Petroselinum*
         GG. Stemmed or stemless perennials; stylopodium none. Native
                                     29. *Tauschia*
    EE. Fls. not yellow.

F.   Fr. elongate, several times longer than wide .... 10. *Osmorhiza*
FF. Fr. roundish to oblong, not more than twice as long as broad.
    G.   Plants annual (if perennial, with celery odor and taste).
        H.   Petals conspicuously unequal; sepals prominent; fr. sub-
           globose .... 13. *Coriandrum*
        HH. Petals equal; sepals none; fr. ovoid to oblong .. 14. *Apium*
    GG. Plants perennial or biennial.
        H.   Plants caulescent, mostly rather tall; invol. usually present.
          I.   Bracts 3-parted to the middle into filiform divisions,
            closely reflexed .... 16. *Ammi*
          II. Bracts entire or toothed, spreading, rarely reflexed, some-
            times none.
            J.   Stems purple-dotted; oil-tubes none or obscure; lvs.
              decompound into small divisions ...... 17. *Conium*
            JJ. Stems not purple-dotted; oil-tubes present; lvs.
              pinnately or ternate-pinnately divided into mostly
              larger divisions.
              K.   The lvs. all once-pinnate.
                L.   Stylopodium conical; oil-tubes solitary in the
                  intervals; ribs of fr. filiform, the pericarp
                  forming a continuous corky covering.
                  18. *Berula*
               LL. Stylopodium depressed; oil-tubes 1–3 in the
                  intervals; ribs of fr. corky, equal
                    19. *Sium*
             KK. The lvs. pinnately or ternate-pinnately divided
               or the upper once-pinnate.
                L.   Stylopodium prominent; ribs not corky.
                  Meadows and dry ground.
                  M.   Lf.-divisions few, mostly entire; ribs
                    filiform; roots tuberous or fusiform
                      21. *Perideridia*
                  MM. Lf.-divisions many, incised or serrate;
                  ribs prominent to ± winged
                      22. *Ligusticum*
              LL. Stylopodium none or low; ribs corky. Marshes
               and stream banks.
                M.   Styles ⅕ to ⅓ as long as the fr.; fr.
                  broadly ovoid to roundish
                    23. *Cicuta*
                MM. Styles ca. ½ as long as fr.; fr. sub-
                  cylindric .... 24. *Oenanthe*
        HH. Plants acaulescent, low; invol. absent.
          I.   Plants pubescent; sepals evident.
             J.   Pedicels of the fls. subequal; sepals not rigid
               25. *Podistera*
             JJ. Pedicels of the sterile fls. longer than or equaling the
               fr.; sepals rigid .... 26. *Oreonana*
          II. Plants glabrous; sepals minute or evident.
             J.   Ribs unequal, the lateral conspicuously corky-
               thickened .... 27. *Orogenia*
             JJ. Ribs equal, filiform .... 29. *Tauschia*
DD. Ribs of the fr. (all or some of them) winged; fr. ± dorsally compressed.
    E.   Lateral ribs winged, the dorsal ribs filiform.
        F.   Corollas all alike; oil-tubes mostly as long as the fr.
          G.   Lvs. simply pinnate into ovate divisions.
             H.   Fls. white; dorsal and intermediate ribs apparently 5. Native
              aquatic plants .... 30. *Oxypolis*
            HH. Fls. yellow; dorsal and intermediate ribs 3. The garden
              Parsnip escaped .... 31. *Pastinaca*
         GG. Lvs. pinnately or ternate-pinnately divided into mostly linear or
          filiform divisions.
             H.   Plants annual, with leafy stems; lf.-divisions filiform; plants
              with anise odor .... 32. *Anethum*
            HH. Plants perennial, acaulescent or short-stemmed; lf.-divisions
             often wider; no anise odor .... 33. *Lomatium*
    FF. Corollas of marginal fls. of umbel larger than of cent.; oil-tubes reaching
      only halfway to base of fr.; tall ± woolly plants .... 34. *Heracleum*
EE. Lateral, dorsal and intermediate ribs winged or prominent.
    F.   Stems leafy, rather tall.
        G.   Umbellets not capitate.
          H.   Lf.-divisions large, mostly lanceolate to ovate, serrate to
             entire .... 35. *Angelica*
          HH. Lf.-divisions small, oblong, incised or deeply toothed
             36. *Conioselinum*
        GG. Umbellets capitate .... 37. *Sphenosciadium*
    FF. Stems scarcely or not developed above ground.

G. Lf.-divisions broad, 5–30 mm. wide. Maritime .... 38. *Glehnia*
GG. Lf.-divisions mostly less than 5 mm. wide. Interior.
  H. Sepals prominent; bractlets usually inconspicuous
    39. *Pteryxia*
  HH. Sepals not prominent; bractlets usually conspicuous
    40. *Cymopterus*

1. Hydrocótyle L. MARSH PENNYWORT

Low perennials with slender creeping stems or rootstocks, rooting at the nodes. Lvs. petioled, entire or parted to base. Petioles slender, not sheathing. Peduncles axillary, obsolete to longer than lvs. Invol. wanting or present. Fls. white, greenish to yellow; petals ovate, plane; calyx-teeth obsolete or minute. Fr. orbicular to ellipsoid, strongly flattened laterally. Carpel with 5 primary ribs. Ca. 75 spp., widely distributed. (Greek, *hudor*, water, and *cotule*, a low vessel or cup.)

Lvs. peltate; ribs of fr. thick and corky.
  Infl. a simple umbel .... 1. *H. umbellata*
  Infl. an interrupted spike .... 2. *H. verticillata*
Lvs. round-reniform, not peltate; ribs of fr. filiform .... 3. *H. ranunculoides*

1. **H. umbellàta** L. Branches of rootstock descending, with round tubers; stems creeping; petioles and peduncles 3–12 cm. long; lf.-blades round-peltate, crenate, mostly 0.5–5 cm. wide; umbels many-fld., simple, the pedicels 2–25 mm. long; fr. 1–2 mm. long, 2–3 mm. broad, strongly notched at base and apex; pericarp thin between corky thick ribs; $2n = 48$ (Wanscher, 1932).—Frequent in wet places and sluggish water, below 5000 ft.; many Plant Communities; cismontane Calif.; to Ore., Mex., Atlantic states, S. Am. March–July.

2. **H. verticillàta** Thunb. [*H. cuneata* Coult. & Rose. *H. v.* var. *c.* Jeps.] With habit of *H. umbellata*, but the infl. an interrupted spike, simple or bifurcate; fr. sessile or subsessile, shallowly notched at apex, narrowly rounded to abruptly cuneate at base, 1–3 mm. long.—Less frequent, cismontane cent. and s. Calif.; to Atlantic states, Mex., W.I. April–Sept.

Var. **triràdiata** (A. Rich.) Fern. [*H. polystachya* var. *t.* A. Rich. *H. prolifera* Kell.] Fr. with pedicels 1–10 mm. long, the spike mostly simple.—Sacramento and San Joaquin valleys to San Francisco Bay and s. Calif.; to Atlantic states, W.I., S. Am.

3. **H. ranunculoìdes** L. f. Floating or creeping on mud; lvs. 0.5–8 cm. broad, round-reniform with cordate base, 3–7-cleft with crenate lobes, on equal or longer petioles; peduncles shorter, recurved in fr.; umbels simple, capitate, 5–10-fld.; pedicels 1–3 mm. long; fr. suborbicular, 2–3 mm. broad, with thick pericarp and obscure filiform ribs.—Ponds and slow streams, below 5000 ft.; many Plant Communities; cismontane Calif.; to Wash., Penn., S. Am.; s. Eu. March–Aug.

H. sibthorpioìdes Lam. differs from no. 3 by having sessile, not pedicelled fr.—Reported from Westwood and San Mateo by Peter Raven; native of Eu.

2. Lilaeópsis Greene

Low tufted glabrous caulescent perennials from long creeping rhizomes. Lvs. reduced to fistulose septate phyllodes borne at the nodes. Infl. of axillary few-fld. umbels on slender peduncles. Invol. of few small bracts. Pedicels slender. Fls. white; calyx-teeth small; stylopodium depressed or obsolete. Fr. ovoid or round, the dorsal ribs filiform, the lateral thick and corky next to the commissure. Four to 5 spp., worldwide. (Resembling *Lilaea*.)

1. **L. occidentàlis** Coult. & Rose [*L. lineata* var. *o.* Jeps.] Lvs. linear, terete, 3–15 cm. long; peduncles 1–4 cm. long, weak; umbels 5–12-fld.; pedicels 2–8 mm. long, ascending to pendulous; fr. ovoid, 2 mm. long, the dorsal ribs obscure.—Marshes and ± brackish flats; Coastal Salt Marsh and environs; Solano and Marin cos. to B.C. June–Aug.

3. Bòwlesia R. & P.

Slender branching annuals with stellate pubescence. Lvs. opposite, simple, lobed, the stipules scarious, lacerate. Umbels on axillary pedicels, simple, few-fld. Fls. white, minute. Sepals rather prominent. Fr. broadly ovoid, stellate-pubescent, with narrow commissure and lacking ribs or oil-tubes, the dorsal part of each carpel turgid. Ca. 20 spp., mostly S. Am. (W. *Bowles,* Irish naturalist of the 18th century.)

1. **B. incàna** R. & P. [*B. lobata* auth., not R. & P. *B. septentrionalis* Coult. & Rose.] Delicate with weak trailing stems 1–5 dm. long, dichotomously branched; petioles slender, 2–8 cm. long; lf.-blades thin, reniform to cordate in outline, 0.5–3 cm. broad, 5–7-lobed, the lobes entire or toothed; umbels 1–6-fld., on short peduncles; fr. 1–1.5 mm. long, sessile or subsessile.—Shaded slopes, below 2500 ft.; S. Oak Wd., Coastal Sage Scrub, Foothill Wd., etc.; cismontane Calif. in Coast Ranges from Sonoma Co. s., Sierra Nevada from Amador Co. s. to San Diego Co., and desert edge, Channel Ids.; to La., L. Calif., also in S. Am. March–April.

4. Sanícula L. SANICLE. SNAKEROOT

Biennial or perennial herbs with nearly naked or few-lvd. stems. Lvs. palmately or pinnately divided, rarely entire. Umbels irregularly compound, few-rayed, bearing invols. and involucels. Sepals evident, persistent. Corolla greenish-yellow or purple. Fr. sub-globose, densely covered with tubercles or hooked bristles. Carpels not ribbed; oil-tubes usually several to many. Ca. 40 spp., widely distributed in n. temp. regions, S. Afr. and S. Am. (Diminutive, derived from Latin, *sanare,* to heal.)

(Shan and Constance. The genus S. in the Old World and the New. Univ. Calif. Pub. Bot. 25: 1–78, 1851. Bell, C. R. The S. crassicaulis complex. Univ. Calif. Pub. Bot. 27: 133–230, 1954.)
A. Basal lvs. ternately or palmately divided, rarely entire.
   B. Involucel-bractlets much longer than the umbellets, conspicuous; plants prostrate, with foliage greenish-yellow at anthesis .... 1. *S. arctopoides*
   BB. Involucel-bractlets shorter than the umbellets, inconspicuous.
      C. Frs. prickly to the base.
         D. Plants usually branched well above the base, the infl. leafy; lf.-margins mostly serrate or lobed.
            E. Involucel-bractlets 1–2 mm. long; calyx-lobes of ♂ fls. 0.5–0.7 mm. long; styles twice as long as calyx; frs. distinctly pedicellate; seed-face of mericarp deeply sulcate. Widely distributed .... 2. *S. crassicaulis*
            EE. Involucel-bractlets 3–5 mm. long; calyx-lobes of ♂ fls. 1–2.3 mm. long; styles ca. as long as calyx; frs. subsessile; seed-face of mericarp plane. Coastal San Luis Obispo and Santa Barbara cos., adjacent ids. .... 3. *S. Hoffmannii*
         DD. Plants usually branched from near base, the infl. nearly naked; lf.-margins laciniate .... 4. *S. laciniata*
      CC. Frs. prickly on upper part only.
         D. Frs. pedicellate or stipitate; basal lvs. ternately divided .... 5. *S. arguta*
         DD. Frs. sessile; basal lvs. entire or shallowly lobed .... 6. *S. maritima*
AA. Basal lvs. pinnatifid or pinnately or ternate-pinnately dissected or compound.
   B. Lvs. once or twice pinnatifid with a toothed rachis.
      C. Frs. usually 1–2 in each umbellet, shorter than the conspicuous ♂ fls.; frs. weakly prickly on upper part only. Nw. Calif. .... 7. *S. Peckiana*
      CC. Frs. several in each umbellet, longer than the inconspicuous ♂ fls.; frs. uncinate-prickly throughout. Widely distributed .... 8. *S. bipinnatifida*
   BB. Lvs. 1- to 3-pinnate or pinnately or ternate-pinnately decompound, without a margined or toothed rachis.
      C. Stems from fusiform taproots.
         D. Staminate fls. shorter than the frs.; frs. prickly on at least the upper half.
            E. Lvs. 1- to 3-pinnate; ♂ fls. few and inconspicuous .... 9. *S. bipinnata*
            EE. Lvs. ternate-pinnately decompound; ♂ fls. several, conspicuous
                                                          10. *S. graveolens*
        DD. Staminate fls. exceeding the frs.; frs. prickly at summit only, or not at all
                                                       11. *S. Tracyi*
      CC. Stems from globose or irregular tuberous roots.
         D. Fls. straw- or salmon-colored; frs. short-prickly on upper part ..12. *S. saxatilis*
         DD. Fls. yellow; frs. lacking prickles, but with inflated unarmed tubercles
                                                          13. *S. tuberosa*

1. **S. arctopoìdes** H. & A. YELLOW MATS. Perennial, with prostrate or decumbent stems 1–2 dm. long, glabrous or ± puberulent especially at nodes; lvs. deltoid in outline, deeply palmately 3-parted, 2–6 cm. broad, the divisions 1–2-laciniate-dentate, the whole margin mostly dissected into lanceolate acute segms.; lvs. of basal rosette with long flat petioles, upper cauline reduced, sessile; umbels terminal, 1–4-rayed; bracts 1–2, lfy.; bractlets mostly 8–12, 5–18 mm. long, exceeding heads; fr. short-pedicellate, 2–5 mm. long, bristly above, naked below; commissural face concave; $n = 8$ (Bell, 1954).—

Sandy flats and open hillsides, below 1000 ft.; N. Coastal Scrub, Mixed Evergreen F., etc.; mostly near the coast from Monterey Co. n.; to n. Ore. March–May.

2. **S. crassicáulis** Poepp. ex DC. [*S. Menziesii* H. & A. *S. nudicaulis* H. & A. *S. M.* vars. *foliacea* and *pedata* Jeps.] Perennial; stems solitary, erect, 3–8(–12) dm. high, branched above, from a stout taproot; lvs. near base, round-cordate in outline, 3–12 cm. broad, palmately and deeply 3–5-lobed into broad obovate sharply lobed or incised segms.; upper lvs. with narrower lobes; umbels with 3–4 slender rays and in an open panicle; invol. of 2–3 small foliaceous bracts; bractlets 6–8, small, entire; fls. yellow; the sterile short-pedicellate; fr. subglobose, 2–4 mm. long, distinctly stipitate, but not pedicellate; seed face deeply sulcate; n = 8, 12 or 16 (Bell, 1954).—Frequent on shaded and wooded slopes, below 4500 ft.; many Plant Communities; cismontane Calif. to B.C., also in S. Am. March–May.

3. **S. Hoffmánnii** (Munz) Shan & Const. (pro hybr.) [*S. bipinnatifida* var. *H.* Munz.] Stout, erect, 3–9 dm. high from a taproot; stem 1, mostly branched above; basal lvs. many, glaucous, rounded-deltoid in outline, 5–13 cm. broad, deeply 3–5-lobed, the segms. obovate, ± lobed, serrate; upper lvs. reduced; umbels in loose panicle, 3–4-rayed; involucel-bractlets 5–7, lanceolate, 3–5 mm. long; fls. greenish-yellow; frs. 4–10, ovoid, 3–5 mm. long, subsessile, covered with stout uncinate prickles, these reduced near base; seed-face plane; n = 8 (Bell, 1954).—Coastal Sage Scrub, etc.; near coast of San Luis Obispo and Santa Barbara cos., Santa Rosa and Santa Cruz ids. March–May.

4. **S. lacinìata** H. & A. [*S. serpentina* Elmer. *S. l.* var. *s.* Jeps.] Perennial, with a taproot, glabrous, the stems few, slender, branched from near base, 1–5 dm. high; lvs. ovate in outline, 1.5–4.5 cm. broad, 3-lobed or -parted, the divisions toothed to pinnately parted, with bristle-tipped teeth; cauline lvs. rather remote, reduced upward; umbels 3–6-rayed; bracts leafy, bractlets small; fls. yellow; fr. roundish, 2 mm. long, sessile, the lower bristles obsolete; seed-face sulcate; n = 8 (Bell, 1954).—Wooded slopes, below 1000 ft.; Closed-cone Pine F., Mixed Evergreen F., etc.; near coast from San Luis Obispo Co. to sw. Ore. March–May.

5. **S. argùta** Greene, ex Coult. & Rose. Stems erect from a taproot, glabrous, 1.5–5 dm. tall, ± branched; lvs. mainly basal, the blades deltoid in outline, 3–10 cm. long, palmately 3–5-parted, the middle divisions longer, all spinose-serrate to sublaciniate, decurrent with a broad toothed wing on the rachis, ± glandular-roughened above; bracts reduced, leaf-like; involucels with entire to 3-lobed bractlets; fertile rays 3–5, the umbellets globose; fls. yellow; fr. 4–6 mm. long, obovoid, stipitate, bristly above, almost naked basally; seed-face grooved; n = 8 (Bell, 1954).—Dry ± grassy slopes, below 2500 ft.; V. Grassland, Coastal Sage Scrub, etc.; cismontane s. Calif. from Monterey to San Diego cos., Channel Ids.; L. Calif. March–April.

6. **S. marítima** Kell. ex Wats. Taproot thickened; stems stout, 1.5–3 dm. high; basal lvs. many, long-petioled, the blades obovate to ovate-cordate, 3–8 cm. long, ± fleshy, entire to 3-parted, the median segm. large, the lateral smaller, entire to serrate; cauline lvs. reduced upward; bracts mostly 2; bractlets ca. 10, lanceolate, 1–4 mm. long; rays of umbel 1–4; umbellets 15–20-fld.; fls. yellow; frs. 3–8, obovoid, ca. 5 mm. long, subsessile, covered with stout inflated prickles; seed-face concave; n = 8 (Bell, 1954).—Heavy soil, dry or wet, San Francisco Bay region and San Luis Obispo Co. April–May.

7. **S. Peckiàna** Macbr. Taproot vertical; stems slender, erect, 1.5–4 dm. tall; basal lvs. many, forming a rosette, subcoriaceous, subglaucous beneath, oblong-ovate, the blades 5–10 cm. long, 1–2-pinnatisect with incised to pinnatifid ovate sessile primary divisions that are decurrent into a toothed winged rachis; bracts mostly 2; rays 3–4; involucel-bractlets ca. 6, acute, ca. 1 mm. long; fls. yellow; frs. 1–5, sessile, ovoid to globose, only the upper tubercles bearing prickles; seed-face sulcate; n = 8 (Bell, 1954).—Serpentine areas, below 2000 ft.; Yellow Pine F., Chaparral? Mixed Evergreen F.; Del Norte Co. to sw. Ore. May.

8. **S. bipinnatífida** Dougl. ex Hook. [*S. b.* var. *flava* Jeps.] PURPLE SANICLE. Stems rather stout, 1.5–7 dm. high, from a thickened often branched taproot-crown; basal lvs. several, polymorphic, the blades 3–7-parted, 5–12 cm. long, the divisions cleft to lobed, decurrent on the rachis forming a toothed wing; umbel 3–5-rayed, the bracts leaflike; involucel-bractlets 2.5 mm. long, lanceolate, 6–8; fls. purplish-red or yellow; frs. 5–10, sessile, ovoid to subglobose, 3–6 mm. long, covered with stout inflated prickles; seed-face broadly concave; n = 8 (Bell, 1954).—Frequent on open slopes,

below 3500 (6000) ft.; V. Grassland, Yellow Pine F., Chaparral, Foothill Wd., etc.; cismontane Calif. from L. Calif. to B.C. March–May. Plants from the Sierran foothills tend to have yellow fls. and are the var. *flàva* Jeps.

9. **S. bipinnàta** H. & A. [*S. pinnatifida* Torr.] POISON SANICLE. Taproot vertical, fusiform, slender; stems slender, erect, 1–6 dm. high, solitary, branched from near base; herbage quite aromatic, the lvs. largely near the base, the blades 4–10 cm. long, 2–3-pinnate, the divisions not decurrent, ovate to oblong, incisely toothed; cauline lvs. reduced, remote; umbels 2–5-rayed; bracts mostly two, 1–2 cm. long; involucel-bractlets 6–8, entire, 1–2 mm. long; fls. pale yellow; frs. 3–8, subglobose to subovoid, 2–3 mm. long, subsessile, cuneate, the bulbous tubercles mostly ending in hooked prickles; seedface sulcate; *n* = 8 (Bell, 1954).—Shaded or open slopes, below 5000 ft.; Foothill Wd., V. Grassland, etc.; interior cismontane Calif. from Butte and Mendocino cos. to Los Angeles Co. April–May.

10. **S. gravèolens** Poepp. ex DC. [*S. nevadensis* Wats. S. *n.* var. *glauca* Jeps. S. *nemoralis, septentrionalis, divaricata* Greene.] Root vertical, mostly fusiform; stems slender and erect or low and spreading, 0.5–4.5 dm. long, branched from near base or higher; lvs. ternate, ± ovate, 1.5–4 cm. long, the primary divisions oblong-ovate, petiolulate, 3–5-lobed or 1–2-pinnatisect, the segms. ± ovate, incised or lobed, with incised to serrate margins; cauline lvs. reduced; involucral bracts 2, pinnatifid, 5–10 mm. long; rays of umbel 3–5; bractlets 6–10, ca. 1 mm. long; fls. 10–15, yellow; frs. 2–5, ovoid-globose, 3–5 mm. long, subsessile, bristly throughout; seed-face concave; *n* = 8 (Bell, 1954.— Open forests, 4000–8000 ft.; Montane Coniferous F.; San Diego Co. to Siskiyou and Modoc cos.; to B.C., Mont., in Argentina. April–June.

11. **S. Tràcyi** Shan & Const. Taproot slender, vertical; stem solitary, purplish below, 3–6 dm. high, sparingly branched; basal lvs. few, the blades oblong- to deltoidovate, 2.5–3.5 cm. long, ternate, then pinnatisect, the first divisions ovate, petiolulate, 3–5-lobed or pinnatifid into ovate-cuneate or obovate incised or lobed segms.; cauline lvs. reduced upward; bracts 2, divided; rays of umbel 3–4; involucel-bractlets 5–8; fls. 6–10, yellow; frs. 1–3, subglobose to obovoid, 2–3 mm. long, sessile, the inflated tubercles mostly unarmed; seed face slightly concave; *n* = 8 (Bell, 1954).—In woods; Mixed Evergreen F.; Trinity and Humboldt cos.; sw. Ore.

12. **S. saxátilis** Greene. Stems solitary, low, stout, 1–2.5 dm. high, from a subglobose tuber; basal lvs. 3–5, the blades deltoid, 3–9 cm. long, ternate, then 1- to 2-pinnatisect, the primary and secondary divisions petiolulate, remote, the ultimate coarsely dissected with finely serrulate margins; bracts 2, foliaceous, ternate, 1–2 cm. long; rays of umbel 3; bractlets of involucels 3–5, ± ovate, 2–4 mm. long; fls. pale salmon to straw color; frs. 3–5, ovoid to subglobose, sessile, 2.5–3 mm. long, covered with inflated tubercles of which the upper bear short subulate prickles; seed-face almost plane; *n* = 8 (Bell, 1954).—At ca. 3000–3800 ft.; Chaparral, Foothill Wd.; Mt. Diablo and Mt. Hamilton. May–June.

13. **S. tuberòsa** Torr. Stems simple or divided near base, 1–8 dm. high, from a small globose tuber; lvs. mostly in lower part, the blades 3–12 cm. long, ternate then pinnatifid, or ternately bipinnatisect or tripinnatisect, the primary divisions petiolulate, remote, the segms. narrow, entire to dissected into fine divisions; cauline lvs. reduced upward; bracts 2, mostly pinnatifid; umbel-rays mostly 3; bractlets 6–10, ca. 1–3 mm. long; fls. yellow; frs. 3–5, ovoid to subglobose, 1.5–2 mm. long, mostly subsessile; seed-face nearly plane; *n* = 8 (Bell, 1954).—Open to wooded slopes, mostly below 8000 ft.; Montane Coniferous F., Foothill Wd., Chaparral, etc.; cismontane Calif. to L. Calif. and sw. Ore. March–July.

5. Apiástrum Nutt. in T. & G.

Smooth branching annuals with slender stems and finely dissected lvs. Umbels naked, unequally few-rayed. Sepals obsolete. Corolla white. Fr. ellipsoid-cordate, with obscure or obsolete ribs, ± papillose-roughened. Stylopodium depressed; styles short. Oil-tubes solitary in the intervals and beneath the ribs, 2 on the ± concave commissural side. One sp. (Latin, *apium*, celery, and *aster*, wild.)

1. **A. angustifòlium** Nutt. in T. & G. [*Helosciadium leptophyllum* var. *latifolium* H. & A.] WILD-CELERY. Stems 0.5–5 dm. long, erect; lvs. 2–5 cm. long, ternately finely

dissected; umbels sessile; rays unequal, 1–5 cm. long; pedicels 0–1.5 cm. long; fr. 1–1.5 mm. long, the ribs inconspicuous.—Common in dry sandy valleys and on slopes, below 3000 ft.; many Plant Communities; cismontane Calif. from San Diego Co. to Lake and Sacramento cos.; L. Calif. March–April.

6. Caúcalis L.

Hispid annuals with pinnately dissected decompound lvs. Umbels irregularly compound. Fls. white. Sepals evident. Fr. flattened laterally, oblong to ovoid. Carpel with 5 filiform bristly ribs and 4 prominently winged secondary ones with barbed or hooked bristles. Oil-tubes solitary in the intervals (under the secondary ribs), 2 on the commissural face. Stylopodium conical. Seed-face deeply grooved. Five spp., N. Temp. Zone. (Ancient classical name.)

1. **C. microcárpa** H. & A. [*Daucus brachiatus* Torr.] Erect, slender-stemmed, 1–3.5 dm. high; lvs. 2–3 times ternate, much dissected; umbels unequally 1–9-rayed; bracts foliaceous, pinnate; pedicels unequal; fr. oblong, 3–7 mm. long, armed with rows of hooked bristles.—Occasional, open and shaded slopes, below 5000 ft.; many Plant Communities; cismontane and sometimes deserts; to B.C., Ida., Utah, Ariz., Mex. April–June.

7. Tòrilis Adams. HEDGE-PARSLEY

Hispid or pubescent annuals with pinnately compound lvs. Fls. white, in compound umbels. Invol. of few small bracts or wanting. Involucels of several linear bractlets. Sepals triangular, acute, sometimes obsolete. Stylopodium thick, conic. Fr. ovoid or oblong, tuberculate or prickly; the primary ribs filiform, setulose; the lateral displaced on to the commissural surface, the secondary hidden by the many barbed or hooked bristles or tubercles; oil-tubes solitary under the secondary ribs, 2 on the commissural side. Ca. 20 spp., N. Hemis. (Derivation not known.)

Umbels sessile or short-pedunculate, capitate, opposite the lvs. .... 1. *T. nodosa*
Umbels usually long-pedunculate, spreading, terminal and lateral.
  Invol. of several bracts, 1 to each ray; bristles incurved-ascending, shorter than the width of the fr.
    2. *T. japonica*
  Invol. wanting or of 1 bract; bristles spreading almost at right angles, scarcely incurved, ca. as long as the width of the fr. .... 3. *T. arvensis*

1. **T. nodòsa** (L.) Gaertn. [*Tordylium n.* L.] Erect, few-branched, 1–7 dm. tall, the stems retrorsely scabrous; lvs. pinnately decompound, 0.5–2 dm. long, the lfts. bipinnately dissected; umbels short-peduncled, solitary, opposite the lvs.; invol. none or of 1 bract; involucel-bractlets longer than the pedicels; rays few; fr. ovoid, 3–5 mm. long, the outer carpel hooked-bristly, the inner smooth or with tubercles; $2n = 22$ (Gardé & Gardé, 1949).—Occasional on open shaded hills, below 2000 ft.; natur. from Eu. April–June.

2. **T. japónica** (Houtt.) DC. [*Caucalis j.* Houtt.] Plants rather stout, hispid throughout, 4–7 dm. high; lvs. 1–2-pinnate, the lfts. dentate to divided, 0.5–6 cm. long; peduncles 4–16 cm. long; rays 5–10, subequal; pedicels 1–3 mm. long; fr. ovoid, 1.5–4 mm. long, covered with uncinate bristles; $2n = 16$ (Melderis, 1930), $n = 6$ (Bell & Constance, 1957).—Occasionally natur. from Eurasia. April–July.

3. **T. arvénsis** (Huds.) Link. [*Caucalis a.* Huds.] Appressed-hispid throughout, 3–10 dm. high; fr. 3–5 mm. long, with long hooked bristles.—Natur. in nw. Calif.; from Eu.

8. Anthríscus Hoffm.

Annual or biennial, with ternately or pinnately compound lvs. Petioles sheathing. Infl. of loose compound umbels; peduncles terminal and lateral. Invol. mostly wanting. Involucel of several narrow usually reflexed bractlets. Rays few, spreading. Sepals none or minute. Fls. white. Stylopodium conic; styles short. Fr. ovoid to linear, beaked, laterally compressed; ribs and oil-tubes obsolete. Seed-face sulcate. Ca. 10 spp., of Old World. (Ancient Greek name.)

1. **A. scandicìna** (Weber) Mansf. [*Caucalis s.* Weber ex Wiggers. *Scandix A.* L.] BUR-CHERVIL. Annual, ± hispid throughout, stems rather slender, branched, 4–9 dm.

tall; lvs. pinnately decompound, with ciliate stipules; umbel-rays mostly 3–6; invol. none or of 1 bract; bractlets lanceolate, 2–5 mm. long; pedicels 2–10 mm. long; fr. ovoid, 4 mm. long, muricate with short hooked bristles.—Waste places, cent. Calif. to Ore.; natur. from Eu. April–June.

9. Scándix L.

Annual, with pinnately decompound lvs. with sheathing petioles. Infl. of loose compound or simple umbels. Involucral bracts usually none. Involucel-bractlets several, usually dissected. Sepals minute or obsolete. Petals white, mostly unequal. Stylopodium depressed; styles short. Fr. linear or linear-oblong, laterally flattened, prolonged into an elongate beak, prominently ribbed; oil-tubes solitary in the intervals or obsolete. Ca. 10 spp., of Old World. (Ancient Greek name.)

1. S. Pécten-Véneris L. SHEPHERD'S NEEDLE. Plants 1.5–3.5 dm. high, ± hispid, mostly branched from base; lvs. oblong in outline, the blades 3–12 cm. long, finely dissected; bractlets lanceolate, ciliate; pedicels 8 mm. or less long; fr. 6–15 mm. long with an additional 20–70 mm. in the beak; $2n = 16$ (Wanscher, 1931).—Occasional in waste places, etc. from s. Calif. to B.C.; natur. from Eurasia. April–June.

10. Osmorhìza Raf. SWEET-CICELY

Slender to stoutish caulescent perennials from thick fascicled roots. Lvs. petiolate, ternate or ternate-quinate, the lfts. lanceolate to roundish, serrate to pinnatifid, with mucronate teeth or lobes. Petioles sheathing. Infl. of loose compound umbels, on terminal or lateral peduncles. Invol. wanting or present. Involucels of several narrow reflexed bractlets or wanting. Rays few, slender, unequal. Fls. white, purple or greenish-yellow. Sepals obsolete. Stylopodium conic; styles slender to obsolete. Fr. linear to oblong, cylindric to clavate, obtuse, tapering, beaked or constricted at apex, rounded or tapering at base; ribs filiform, acute, often bristly. Oil-tubes obscure or wanting. Seed subterete in cross section, the face concave or grooved. Ca. 11 spp. of N. Am., w. S. Am., e. Asia. (Greek, *osme*, odor, and *rhiza*, root.)

(Constance, L., and R. H. Shan. The genus O. Univ. Calif. Pub. Bot. 23: 111–156, 1948.)
Fr. not attenuate at base, glabrous or rarely sparsely bristly toward base ...... 1. *O. occidentalis*
Fr. attenuate at base, conspicuously bristly at least at base.
  Involucels conspicuous, of several bractlets .... 2. *O. brachypoda*
  Involucels lacking, or rudimentary.
    Rays and pedicels divaricate; fr. clavate .... 3. *O. depauperata*
    Rays and pedicels spreading-ascending; fr. cylindrical, linear-oblong.
      Fr. 10–13 mm. long, constricted at apex; stylopodium depressed, the disk conspicuous; fls. purplish or greenish; styles 0.5–1 mm. long. Nw. Calif.   4. *O. purpurea*
      Fr. 12–27 mm. long, tapering at apex; stylopodium conic, the disk inconspicuous; fls. greenish-white or white; styles 0.2–0.5 mm. long. Widely distributed .... 5. *O. chilensis*

1. O. occidentàlis (Nutt.) Torr. [*Glycosma o.* Nutt. ex T. & G. *Washingtonia o.* C. & R. *Myrrhis Bolanderi* Gray. *O. B.* Jeps.] Rather stout, 3–12 dm. high, villous at nodes, subglabrous to hirtellous throughout; lf.-blades ovate to oblong, 1–2 dm. long, 1–3-ternate or ternate-pinnate, the lfts. lance-oblong to ovate, 2–10 cm. long, serrate and usually incised or lobed; petioles 0.5–3 dm. long; rays 5–12; involucels usually wanting; fls. yellow; fr. linear-fusiform, 12–20 mm. long, constricted below apex, obtuse at base, ± glabrous; $n = 11$ (Bell & Constance, 1957).—Wooded slopes, 2500–8500 ft.; Montane Coniferous F.; Mono and Madera cos. to Modoc and Siskiyou cos., then s. in Coast Ranges to Mendocino Co.; to B.C., Alta., Colo. May–July.

2. O. brachýpoda Torr. [*Washingtonia b.* Heller. *O. b.* var. *fraterna* Jeps.] Plants stoutish, 3–8 dm. high, hirsutulous to villous; lf.-blades deltoid or ovate, 0.8–2.5 dm. long, ternate-pinnate, the lfts. ovate, 2–6 cm. long, coarsely serrate to lobed; petioles 0.5–2 dm. long; involucels 2–10 mm. long, of several bractlets; rays 2–5; pedicels ascending, 1–3 mm. long; fls. greenish-yellow; fr. oblong-fusiform, 12–20 mm. long, with narrow beak and narrowed base; ribs bristly, prominent.—Shaded woods, below 8500 ft.; S. Oak Wd., Foothill Wd., Yellow Pine F., Red Fir F., etc.; Coast Ranges from Contra Costa Co. s., Sierra Nevada from Placer Co. s. to San Diego Co.; Ariz. March–May.

3. O. depauperàta Phil. [*O. obtusa* (Coult. & Rose) Fern. *Washingtonia o.* Coult. &

Rose.] Plants slender, 1.5–6 dm. high, hispid to subglabrous; lvs. roundish, the blades 4–11 cm. long, lfts. ± ovate, 1.5–5 cm. long; involucels wanting; rays 2–5; pedicels widely divergent, 1–3 cm. long; fls. greenish-white; fr. clavate, 8–15 mm. long, ± obtuse, densely hispid at the narrowed base.—At 6000 ft.; Yellow Pine F.; Deep Creek, Warner Mts., Modoc Co.; to Alaska, Atlantic Coast, New Mex., also in s. S. Am. May–July.

4. **O. purpùrea** (Coult. & Rose) Suksd. [*Washingtonia p.* Coult. & Rose. *W. Leibergii* Coult. & Rose.] Plants slender, 2–6 dm. high, subglabrous to ± hirsutulous; lf.-blades deltoid to roundish, 3–10 cm. long, 1–3-ternate, the lfts. lanceolate to ovate, 1.5–7 cm. long, serrate to lobed; petioles 5–12 cm. long; invol. none; involucels none; rays 2–6; fls. purplish to greenish-white; fr. linear-fusiform, constricted below the short-beaked apex, narrowed and bristly at base.—Woods; Redwood F.; Del Norte Co.; to Alaska, Mont. June–July.

5. **O. chilénsis** H. & A. [*O. nuda* Torr. *Washingtonia brevipes* Coult. & Rose.] Plants slender, 3–10 dm. high, hirtellous to villous; lf.-blades roundish, 5–15 cm. long, biternate, the lfts. lance-ovate to roundish, 2–6 cm. long, coarsely serrate to lobed; petioles 5–15 cm. long; peduncles 5–25 cm. long; invol. and involucel usually wanting; pedicels spreading-ascending, 5–30 mm. long; fr. linear-oblong, 10–20 mm. long, tapering toward the slender beak, narrowed and densely hispid at base; $n = 11$ (Bell & Constance, 1957).—Woods, mostly below 8000 (10,000) ft.; Yellow Pine F., Red Fir F., Lodgepole F., Mixed Evergreen F., etc.; San Diego Co. n., Sierra Nevada and Coast Ranges; to Alaska, Atlantic Coast, Ariz.; in s. S. Am. April–July.

11. Ammoselìnum T. & G.

Low mostly branched annuals, ± roughened. Lvs. ternately dissected into linear or spatulate ultimate divisions. Infl. of loose compound umbels. Invol. usually wanting. Involucel of several narrow entire or toothed bractlets. Rays few. Fls. white. Sepals obsolete. Stylopodium low-conic; styles short. Fr. oblong-ovoid to ovoid, laterally flattened; ribs prominent, glabrous to coarsely scabrous; oil-tubes 1–3 in the intervals, 2–4 on the commissure. Three spp., of sw. N. Am. (Greek, *ammos,* sand, and *Selinum,* an Old World genus of Umbelliferae.)

1. **A. gigantèum** Coult. & Rose. [*A. occidentale* M. & J.] Stems 1–several, 1–2 dm. high; lvs. 1.5–2.5 cm. long, glabrous, ternate-pinnately dissected into linear segms. 4–12 mm. long; umbels axillary and terminal; bractlets few, linear-lanceolate, 1–3 mm. long; pedicels 1–8 mm. long, unequal; fr. 3–5 mm. long, scabrous; oil-tubes 3 on the intervals, 2 on the commissure.—Occasional in basins; Creosote Bush Scrub; near Borrego and Hayfields, Colo. Desert; to Ariz., Coahuila. March–April.

12. Daúcus L.

Pubescent or bristly caulescent annuals or biennials. Lvs. pinnately decompound into small narrow ultimate divisions. Invol. of leafy pinnately divided bracts. Involucels of many entire or divided bractlets. Infl. compact in fr. Fls. usually white. Sepals obsolete or present. Fr. oblong to ovoid, flattened dorsally; primary ribs slender, bristly; secondary ribs winged and with a row of barbed bristles. Oil-tubes solitary in intervals, 2 on the commissure. Ca. 25 spp., widely distributed. (The ancient Greek name.)

Bracts pinnately divided into short linear or lanceolate divisions; rays usually 0.4–4 cm. long; carpel usually broadest below the middle; cent. fl. of the umbellet white . . . . 1. *D. pusillus*
Bracts pinnately divided into elongate filiform divisions; rays 3–7 cm. long; carpel broadest at middle; cent. fl. of the umbellet usually rose or purple . . . . 2. *D. Carota*

1. **D. pusíllus** Michx. RATTLESNAKE WEED. Annual, 3–8 dm. high, simple or few-branched, retrorsely papillate-hispid; lf.-blades 3–10 cm. long; peduncles 1–4 dm. long; rays few to many, unequal; pedicels unequal, 2–9 mm. long; fr. 3–5 mm. long, the commissure with 2 rows of stiff bristles.—Common on dry slopes, especially after fire and disturbance, below 5000 ft.; many Plant Communities; cismontane Calif. and occasional on desert; to B.C., S. Car., Fla., n. Mex. April–June.

2. **D. Caròta** L. WILD CARROT. QUEEN ANNE'S LACE. Biennial, with erect bristly stems 2–10 dm. high, from a fusiform root; lvs. finely dissected, large; rays many, unequal, the infl. compact, concave in fr.; pedicels unequal, 3–10 mm. long; fr. ovoid, 3–4 mm. long;

$2n = 18$ (Heiser & Whitaker, 1948).—Occasionally natur., especially in s. Calif.; native of Eu. May–Sept.

13. Coriándrum L. CORIANDER

Caulescent annual, with pinnately decompound lvs. Infl. of loose compound umbels. Invol. usually none. Involucel of few small narrow bractlets. Rays few, spreading-ascending. Pedicels spreading. Fls. white to rose; sepals prominent. Stylopodium conic; styles slender, spreading. Fr. subglobose, glabrous, hard, with slender ribs. Oil-tubes obscure. Two spp., of Old World. (Ancient Latin name.)

1. **C. sativum** L. Glabrous, 2–7 dm. high; lower lvs. ternately or pinnately divided into ovate or obovate segms., these 1–2 cm. long, variously toothed or incised; cauline lvs. decompound into linear segms.; rays 2–8, 1–2.5 cm. long; pedicels 2–5 mm. long; sepals lance-ovate, to 1 mm. long; fr. 2–5 mm. in diam.; $n = 11$ (Håkansson, 1953).—Occasional escape from gardens; native of s. Eu. May–July.

14. Apium L.

Glabrous annual to perennial caulescent herbs with pinnate to ternate-pinnately decompound lvs. Fls. white or greenish-yellow, in compound umbels. Invol. and involucel wanting to conspicuous. Sepals obsolete. Stylopodium short-conic to depressed; styles short. Fr. oblong-oval to subglobose or ellipsoid, mostly glabrous; ribs prominent, filiform; oil-tubes 1 in the intervals, 2 on the commissure. Ca. 30 spp., of Eurasia and S. Hemis. (The ancient Latin name.)

Plants perennial; lf.-divisions ovate to suborbicular or cuneate .... 1. *A. graveolens*
Plants annual; lf. divisions linear to filiform .... 2. *A. leptophyllum*

1. **A. graveolens** L. CELERY. Erect to ascending, 5–12 dm. tall; basal lvs. pinnate, 1–6 dm. long, petioled, the upper much reduced, subsessile; lfts. 5–9, 2.5–7 cm. long; petioles 3–25 cm. long; umbels sessile or short-peduncled; involucels and invol. none; rays 7–16; pedicels 1–6 mm. long; fr. subglobose to ellipsoid, ca. 1.5 mm. long; $2n = 22$ (Shah, 1953).—Common in wet places, at low elevs.; natur. from Eu. May–July.

2. **A. leptophýllum** (Pers.) F. Muell. ex Benth. & Muell. [*Pimpinella l.* Pers.] MARSH-PARSLEY. Prostrate to suberect, 1–6 dm. high; basal lvs. 3–4-pinnately decompound, 3–10 cm. long, long-petioled, the upper smaller, short-petioled; ultimate divisions 4–30 mm. long, linear to subfiliform; rays 3–5; bracts and bractlets none; fr. ovoid, 1.3–3 mm. long; $n = 7$ (Bell & Constance, 1957).—Occasional as weed in gardens, lawns, etc.; Riverside, Los Angeles, Ventura, Lompoc, Salinas, Berkeley, Humboldt Co., etc.; se. U.S., W.I., S. Am. April–Aug.

15. Petroselìnum Hoffm.

Slender erect glabrous biennials. Lvs. ternate-pinnately or pinnately decompound, the ultimate divisions linear to ovate, toothed to lobed. Invol. none or inconspicuous. Rays few to many. Bractlets several, linear. Fls. yellow or ± greenish. Sepals obsolete. Stylopodium low-conic; styles short. Fr. ovoid to oblong, glabrous, laterally flattened, with prominent filiform ribs. Three spp., of Eurasia. (Greek, *petra*, rock, and *selinon*, parsley.)

1. **P. críspum** (Mill.) Mansf. [*Apium crispum* Mill. *A. Petroselinum* L.] PARSLEY. Plants 3–10 dm. high; lvs. deltoid, divided into linear to lance-ovate segms. 2–5 cm. long; invol. of few inconspicuous bracts or none; involucel of 5–6 linear bractlets; rays 10–20; pedicels 2–5 mm. long; fr. ovoid-oblong, 2–4 mm. long; $n = 11$ (Håkansson, 1953).—Occasional escape into moist waste places; native of Eu.

16. Ammi L. BISHOP'S WEED

Slender erect caulescent branching essentially glabrous herbs. Lvs. petioled, ternate-pinnately or pinnately dissected into filiform or lanceolate ultimate segms. Infl. of loose compound umbels. Invol. of many entire to divided bracts. Rays many, spreading-ascending. Involucel of many entire bractlets. Fls. white. Sepals minute. Stylopodium depressed-conic; styles slender, elongate. Fr. oblong to ovoid, laterally flattened, glabrous; ribs acute; oil-tubes solitary in the intervals, 2 on the commissure. Six spp., of Medit. region. (Ancient Latin name.)

Umbels compact in fr.; infl. borne on a discoid receptacle .... 1. *A. Visnaga*
Umbels spreading in fr.; infl. not borne on a discoid receptacle .... 2. *A. majus*

1. **A. Visnága** (L.) Lam. [*Daucus V. L.*] Erect, glabrous, 2–8 dm. high; lvs. deltoid in outline, the blades 5–20 cm. long, with filiform or linear ultimate divisions 0.5–3 cm. long; petioles ca. 10 cm. long; peduncles 8–12 cm. long; bracts exceeding the rays; rays 60–100, 2–5 cm. long; pedicels 3–12 mm. long; fr. oblong-ovoid to ovoid, 2–2.5 mm. long; $2n = 20$ (Gardé & Gardé, 1951).—Occasional weed in waste places; native of Eurasia. June–July.

2. **A. màjus** L. Erect, 2–8 dm. high, scabrous in the infl.; lvs. oblong in outline, the blades 5–20 cm. long, ternate or pinnate, the lfts. lanceolate, 1–1.5 cm. long, setulose-serrate; petioles 3–12 cm. long; bracts exceeding rays; bractlets of involucel linear; rays 50–60, 2–7 cm. long; pedicels 1–10 mm. long; fr. oblong, 1.5–2 mm. long; $n = 11$ (Håkansson, 1953).—Occasional weed in fields and waste places; introd. from Eurasia. May–July.

17. Conìum L. Poison-Hemlock

Tall biennial glabrous herbs with spotted stems and pinnately decompound lvs. Infl. of compound, many-rayed umbels. Bracts of invol. many, lanceolate, inconspicuous. Bractlets of involucel many, shorter than the pedicels. Rays many, spreading-ascending. Pedicels spreading. Sepals obsolete. Fls. white. Stylopodium depressed-conic; styles short, reflexed. Fr. broadly ovoid, flattened laterally, glabrous; ribs prominent, obtuse, undulate; oil-tubes obscure, irregular. Two spp., 1 Eurasian, 1 S. African. (Greek, *coneion*, the ancient name.)

1. **C. maculàtum** L. Stems 5–30 dm. high; lower lvs. petioled, upper sessile, all finely dissected into segms. ovate in outline, dentate or incised; lower lf.-blades 15–30 cm. long; rays 1.5–4.5 cm. long; fr. 2–2.5 mm. long, the ribs prominent when dry; $2n = 22$ (Gardé & Gardé, 1949).—Common in low waste places, below 5000 ft., especially in cismontane Calif.; natur. from Eu. April–July.

18. Bérula Hoffm. in Bess. Water-Parsnip

Erect caulescent branched glabrous perennial with petioled pinnate variously cut lvs. and white small fls. Infl. of loose compound umbels. Invol. of conspicuous narrow entire or toothed bracts. Involucel of conspicuous narrow bractlets. Rays rather few. Sepals minute, subulate. Stylopodium conic; styles short. Fr. oval to orbicular, glabrous, laterally flattened; ribs filiform, obscure in the thick corky pericarp; oil-tubes many. One sp. of N. Temp. (Latin name of some aquatic plant.)

1. **B. erécta** (Huds.) Cov. [*Sium e.* Huds.] Stems 2–8 dm. high, stoutish; lfts. in 5–9 pairs, oblong, subentire to serrate or laciniate, 1.5–4 cm. long; rays 6–15; pedicels 2–5 mm. long; fr. scarcely 2 mm. long, with inconspicuous ribs; $2n = 18$ (Scheerer, 1940); $n = 6$ (Bell & Constance, 1957).—Marshes and sluggish water, below 8000 ft.; many Plant Communities; cismontane and rarely desert Calif.; to B.C., Ontario, New Mex., L. Calif.; Eurasia. July–Oct.

19. Sìum L. Water-Parsnip

Glabrous erect caulescent branching perennials with petioled pinnate to pinnately decompound lvs. Infl. of loose compound umbels. Invol. of subfoliaceous, entire or incised bracts. Involucel of conspicuous narrow bractlets. Rays rather few. Fls. white. Sepals minute to obsolete. Stylopodium mostly depressed; styles short, reflexed. Fr. oval to orbicular, glabrous, flattened laterally; ribs prominent, subequal, corky; oil-tubes 1–3 in the intervals, 2–6 on the commissure. Ca. 8 spp. of N. Temp. (Greek, *sion*, name of some marsh plant.)

1. **S. suàve** Walt. [*S. cicutaefolium* Schrank. *S. heterophyllum* Greene.] Erect, stout, 6–12 dm. tall; lower lvs. long-petioled, the upper subsessile; submerged lvs. with

pectinately dissected segms.; aerial lvs. with linear to lanceolate, sharply serrate divisions 1–9 cm. long; rays 10–20, angled, 1.5–3 cm. long; pedicels 3–6 mm. long; fr. 2–3 mm. long.—Wet places, below 6500 ft.; Freshwater Marsh, Sagebrush Scrub, Montane Coniferous F.; Tulare Co., San Joaquin Co., Butte to Modoc and Siskiyou cos.; to B.C., Atlantic Coast. July–Aug.

20. Bupleùrum L.

Annual to perennial herbs, rather low, erect or spreading, glabrous. Basal lvs. petioled, entire; cauline usually sessile and clasping, auriculate or perfoliate. Infl. of loose compound umbels. Invol. of conspicuous foliaceous bracts or wanting. Involucel of broad foliaceous often connate bractlets sometimes longer than the fls. and frs. Rays few. Fls. yellow. Sepals obsolete. Stylopodium depressed-conic; styles short. Fr. oblong to subglobose, slightly flattened laterally and constricted at the commissure; ribs filiform; oil-tubes many to wanting or obscure. Rather a large genus, chiefly of Old World. (Greek, *bous*, ox, and *pleuron*, a rib.)

1. **B. subovàtum** Link. Annual, 2–4 dm. high, glaucous and glabrous, divaricately branched; cauline lvs. broadly ovate, perfoliate, 2–10 cm. long; peduncles 1–8 cm. long; invol. none; involucel of 5–6 suborbicular ± united bractlets 1–1.5 cm. long; pedicels 10–20, half as long as fr.; fr. ovoid-globose, 3–5 mm. long, dark, tuberculate or rugose, the ribs prominent.—Occasional weed in gardens; introd. from Medit. region. May–June.

21. Periderídia Rchb.

Erect caulescent branching herbs from tuberous or fusiform fascicled roots. Lvs. petioled, ternately, pinnately or ternate-pinnately compound, the ultimate divisions linear to ovate. Infl. of loose compound umbels. Invol. of few to many entire narrow ± scarious bracts. Involucel of usually scarious or colored bractlets. Sepals evident. Fls. white to pinkish. Fr. flattened laterally; carpels with filiform ribs; stylopodium ± conical; oil-tubes 1–5 in intervals, 2–8 on commissure. Ca. 9 spp., of N. Am. The tuberous roots were important as food for the Indians. (Greek, *peri*, around, and *derris*, a leather coat.)

Styles erect or divaricate, usually less than 0.6 mm. long; plant coarse, from fascicles of numerous fibrous or slightly thickened roots.
    Lfts. ovate, 5–25 mm. broad; rays usually 20–25; lower petioles 6–25 cm. long .. 1. *P. Howellii*
    Lfts. linear, 1–6 mm. broad; rays 10–20; petioles 4–7 cm. long .... 2. *P. Kelloggii*
Styles filiform, reflexed, mostly 1–2 mm. long; plants usually slender, from solitary tubers of fascicles of few tuberous roots.
    Basal lvs. 1–2-pinnate or -ternate, the petioles and rachis not dilated, the ultimate divisions not dimorphic.
        Bractlets usually setaceous; fr. ca. as broad as long .... 3. *P. Gairdneri*
        Bractlets scarious or with scarious margins; fr. longer than broad.
            Fr. usually rounded at base and apex; oil-tubes solitary in the intervals. N. Calif.
                        4. *P. oregana*
            Fr. usually narrowed at base and apex; oil-tubes 2–4 in the intervals. Sierra Nevada to s. Calif.
                        5. *P. Parishii*
    Basal lvs. ternate-pinnately or pinnately decompound, the petioles and rachis dilated, the ultimate divisions usually dimorphic.
        Rays 10–20, short (1–2.5 cm: long), forming small compact umbels .... 6. *P. Bolanderi*
        Rays mostly 5–10, longer (3–8 cm.), forming larger looser umbels.
            Fr. 4–6 mm. long; oil-tubes several in the intervals .... 7. *P. Pringlei*
            Fr. 6–8 mm. long; oil-tubes solitary in the intervals .... 8. *P. californica*

1. **P. Howéllii** (Coult. & Rose.) Math. [*Carum H.* Coult. & Rose. *Taeniopleurum H.* Coult. & Rose] Stout, 6–12 dm. tall; lf.-blades oblong to oval in outline, 1–3 dm. long, 1–2-pinnate, the lfts. ± ovate, 2–4 cm. long, entire to serrate-lobed; peduncles 5–15 cm. long; bracts reflexed, narrow, 1–2 cm. long; bractlets 3–6 mm. long; rays 3–6 cm. long; pedicels 4–8 mm. long; fr. oblong, 3–6 mm. long; oil-tubes solitary in the intervals, 2 on the commissure.—Meadows, 2000–5000 ft.; Yellow Pine F., Mixed Evergreen F., etc.; Mariposa Co. to Nevada Co., Mendocino and Siskiyou cos.; to s. Ore. July–Aug.

2. **Kellóggii** (Gray) Math. [*Carum K.* Gray.] Stout, 7–15 dm. high; lf.-blades deltoid in outline, 15–30 cm. long, 1–2-ternate-pinnate, the ultimate divisions 3–10 cm. long; peduncles 5–18 cm. long; bracts narrow, reflexed, ± scarious, 5–15 mm. long; bractlets 2–6 mm. long; rays 1.5–6 cm. long; pedicels 2–6 mm. long; fr. oblong, 4–5 mm. long;

oil-tubes solitary in intervals, 2 on the commissure; $n = 20$ (Bell & Constance, 1957).—Moist places, open or wooded, below 4000 ft.; N. Coastal Prairie, Mixed Evergreen F., Chaparral, Foothill Wd., etc.; Coast Ranges from Santa Clara to Del Norte cos., Sierra Nevada foothills from Mariposa to Sierra cos. July–Aug.

3. **P. Gáirdneri** (H. & A.) Math. [*Ataenia G.* H. & A.] Squaw Root. Plants slender, 3–12 dm. tall, from a solitary fusiform tuber or a small cluster; lvs. mostly pinnate into linear or lanceolate, mostly entire divisions 2–12 cm. long; bracts mostly obsolete; bractlets several, linear, green or scarious, 1–4 mm. long; rays usually 8–20, 1.5–6 cm. long; pedicels 3–7 mm. long; fr. ± subglobose, 2–3 mm. long; oil-tubes solitary in the intervals, 2 on the commissure.—Wet places, below 11,000 ft.; many Plant Communities; cismontane and montane Calif. from San Diego Co. n.; to B.C., Alta., New Mex. June–July.

4. **P. oregàna** (Wats.) Math. [*Carum o.* Wats.] Slender, 3–6 dm. high, from a fascicle of tubers; lf.-blades deltoid to lanceolate in outline, 5–15 cm. long, 1–2-ternate or ternate-pinnate, the ultimate divisions linear to lance-linear, 1.5–6 cm. long; bracts of invol. lanceolate to linear, 2–8 mm. long; bractlets almost as long; rays 6–20, 1–3 cm. long; pedicels 2–5 mm. long; fr. oblong-ovoid, 2.5–3.5 mm. long; oil-tubes solitary in intervals, 2 on commissure.—Meadows and rocky slopes, 3000–7000 ft.; Yellow Pine F., Red Fir F., Lodgepole F.; Coast Ranges, from Glenn and Trinity cos. n.; to Wash. July–Aug.

5. **P. Paríshii** (Coult. & Rose) Nels. & Macbr. [*Pimpinella P.* Coult. & Rose. *Eulophus P.* Coult. & Rose. *E. simplex* Coult. & Rose. *Carum Lemmonii* Coult. & Rose.] Slender, from 1 or more tubers, 2–8 dm. tall; lvs. lanceolate to ovate in outline, the blades 5–15 cm. long, ternate or simple or biternate, the lfts. linear to lanceolate, 2–10 cm. long; petioles 3–7 cm. long; peduncle 6–15 cm. long; bracts mostly obsolete; involucel of several linear to obovate scarious to colored bractlets 2–4 mm. long; rays 8–15, ± unequal, 1–4 cm. long; pedicels 3–8 mm. long; fr. oblong to ovoid, 2.5–3.5 mm. long; oil-tubes 2–4 in the intervals, 6 on the commissure.—Damp meadows, etc., 3500–11,200 ft.; Montane Coniferous F.; mts. from San Diego Co. to Sierra Nevada. July–Sept.

6. **P. Bolánderi** (Gray) Nels. & Macbr. [*Podosciadium B.* Gray. *Eulophus B.* Coult. & Rose. *E. B.* var. *benignus* Jeps. *E. cuspidatus* Jeps.] Plants slender, 2–8 dm. tall, from fascicled tubers; lvs. deltoid in outline, 5–15 cm. long, ternate-pinnately dissected into ultimate filiform to oblong divisions 0.5–3 cm. long, or the terminal to 8 cm.; petioles 2–8 cm. long; peduncles 6–14 cm. long; bracts 1–several, scarious, 5–12 mm. long; bractlets 3–6 mm. long; rays 10–20, ca. 1–2.5 cm. long; pedicels 3–6 mm. long; fr. oblong, 3–5 mm. long; oil-tubes 2–5 in intervals, 6 on the commissure.—Meadows and surrounding slopes, 3000–10,500 ft.; Sagebrush Scrub to Subalpine F.; Sierra Nevada to Modoc Co.; e. Ore., Wyo., Utah. June–Aug.

7. **P. Prínglei** (Coult. & Rose) Nels. & Macbr. [*Eulophus P.* Coult. & Rose] Slender, 3–6 dm. high, from fascicled tubers; lf.-blades ovate to deltoid in outline, 5–10 cm. long, pinnately dissected into linear divisions 0.2–8 cm. long; petioles 4–9 cm. long; peduncles 3–10 cm. long; bracts 1–several, lanceolate or wanting; bractlets subulate, scarious, 2–4 mm. long; rays 5–8, ca. 3–8 cm. long; pedicels 5–10 mm. long; fr. oblong, 4–6 mm. long; oil-tubes 3–5 in intervals, 8 on the commissure.—Open slopes and canyons, 1000–3500 ft.; Coastal Sage Scrub, Chaparral, etc.; n. Los Angeles Co. to San Luis Obispo and Kern cos. April–June.

8. **P. califórnica** (Torr.) Nels. & Macbr. [*Chaerophyllum?* c. Torr. *Eulophus c.* Coult. & Rose. *E. c.* var. *sanctorum* Jeps.] Slender, 5–10 dm. high, from fascicled tubers; lf.-blades deltoid in outline, 15–25 cm. long, ternate-pinnately dissected, the ultimate divisions linear to ovate, the terminal being larger, to 8 cm. long; petioles 2–5 cm. long; peduncles 4–20 cm. long; bracts lanceolate, 3–12 mm. long; bractlets 1–4 mm. long; rays 5–10; pedicels 5–10 mm. long; fr. oblong, 6–8 mm. long; oil-tubes solitary in intervals, 4 on commissure; $n = 22$ (Bell & Constance, 1957).—Meadows and along streams, below 3000 ft.; Foothill Wd., etc.; Sierra Nevada foothills from Stanislaus to Mariposa cos., inner Coast Ranges from Contra Costa to San Luis Obispo cos. April–May.

22. Ligústicum L. Lovage

Erect scapose to caulescent perennials from fibrous root-crown. Lvs. petioled, ternate or ternate-pinnately decompound. Petioles sheathing. Infl. of loose compound umbels. Invol. wanting or inconspicuous. Involucel wanting or of several short narrow bractlets.

Rays few to many. Pedicels slender. Fls. white to pinkish. Sepals minute to evident. Stylopodium low-conic; styles short, spreading. Fr. ovoid to oblong, slightly flattened laterally to subterete; ribs prominent to narrowly thin-winged; oil-tubes small, 1–6 in intervals, 2–10 on commissure. Widely distributed genus with ca. 9 spp. in N. Am. (*Liguria*, Italian province, where lovage is endemic.)

Stems ± leafy. Coast Ranges.
  Fr. ribbed, not winged; infl. puberulent .... 1. *L. apiifolium*
  Fr. narrowly winged; infl. glabrous .... 2. *L. californicum*
Stems naked or with one much reduced cauline lf. Sierra Nevada .... 3. *L. Grayi*

1. **L. apiifòlium** (Nutt.) Gray. [*Cynapium a.* Nutt. ex T. & G. *L. apiodorum* Coult. & Rose.] Stout, branched, 3–15 dm. high, glabrous to pubescent; lvs. ovate to roundish in outline, the blades 1–3 dm. long, ternate-pinnate, lfts. ovate to oblong, 1–5 cm. long, coarsely toothed to deeply pinnatifid; petioles 1–3 dm. long; peduncles 1–3 dm. long; rays 12–20, slender, 2–5 cm. long; bractlets few, 3–7 mm. long; pedicels 5–10 mm. long; fr. oval to subglobose, 3.5–5 mm. long; ribs slender, not winged; oil-tube 3–6 in intervals, 6–8 on commissure.—Meadows and shaded banks, below 2500 ft.; Coastal Prairie, Redwood F., Mixed Evergreen F., N. Oak Wd., etc.; near coast from San Mateo Co. n.; to w. Wash. June–July.

2. **L. califórnicum** Coult. & Rose. Stout, glabrous throughout, 5–9 dm. tall; lf.-blades ovate-deltoid in outline, 1.5–3 dm. long, bipinnate to ternate-pinnate, the lfts. ovate, 2–4 cm. long, toothed or cleft into linear divisions; petioles 1–3 dm. long; peduncles 1–3 dm. long; rays 9–20; spreading-ascending, 3–7 cm. long; involucel of linear bractlets 3–5 mm. long; pedicels 5–11 mm. long; frs. oval, 4–6 mm. long; ribs narrowly winged; oil-tubes several in intervals and on commissure.—Open places, 2500–7200 ft.; Montane Coniferous F.; Glenn and Trinity to Del Norte and Siskiyou cos. June–Aug.

3. **L. Gràyi** Coult. & Rose. [*L. Pringlei* Coult. & Rose. *L. apiifolium* var. *minor* Gray.] Stems naked or nearly so, slender, 2–6 dm. tall, glabrous; lf.-blades ternate-pinnate, ovate to oblong in outline, 0.6–2 dm. long, the lfts. ovate to oblong, 1–2 cm. long, pinnatifid; petioles to 1 dm. long; peduncles 0.7–2 dm. long; invol. wanting or nearly so; involucel of 4–8 linear bractlets 2–5 mm. long; rays 5–14; pedicels 3–8 mm. long; frs. oval-oblong, 4–6 mm. long, ribs narrowly winged; oil-tubes 3–5 in intervals, 8 on commissure; $n = 22$ (Bell & Constance, 1957).—Meadows and slopes, 4000–10,500 ft.; Montane Coniferous F.; Sierra Nevada to Wash., Mont. June–Sept.

23. Cicùta L. WATER-HEMLOCK

Glabrous branching caulescent perennials from a tuberous base with fibrous to tuberous roots. Lvs. petiolate, broad, 1–3-pinnate or ternate-pinnate, the divisions serrate to incised. Petioles sheathing. Infl. of loose compound umbels. Invol. wanting or inconspicuous. Involucel of several narrow bractlets. Rays many, slender, spreading-ascending. Pedicels slender, spreading. Fls. white or greenish. Sepals evident. Stylopodium depressed or low-conic; styles spreading. Fr. oval to ovoid or subglobose, flattened laterally, often constricted at the commissure; ribs usually prominent, obtuse and corky; oil-tubes solitary in the intervals, 2 on the commissure. Ca. 8 spp.; circumboreal genus. (Ancient Latin name of poisonous hemlock.)

Oil-tubes large; seed evidently channeled under the oil-tubes. Plants of salt marshes
                                                                    1. *C. Bolanderi*
Oil-tubes small; seed terete or slightly sulcate under the oil-tubes. Mostly of fresh water
                                                                    2. *C. Douglasii*

1. **C. Bolánderi** Wats. Stout, 1–3 m. high; lf.-blades oblong to ovate in outline, 1.5–3.5 dm. long, 1–2-pinnate, the lfts. linear- to oblong-lanceolate, 5–9 cm. long, finely to coarsely serrate; petioles 1.5–4.5 dm. long; peduncles 4–13 cm. long; invol. present or absent; involucel-bractlets 2–5 mm. long; rays 2–5 cm. long; pedicels 3–4 mm. long; fr. oval, 3–4 mm. long, constricted at the commissure, the ribs low and corky, narrower than the broad darker intervals; seed very oily; $n = 22$ (Bell & Constance, 1957).—Salt marshes, Marin to Solano and Contra Costa cos. July–Sept.

2. **C. Doúglasii** (DC.) Coult. & Rose [*Sium? D.* DC. *C. californica* Gray. *C. occidentalis* Greene. *C. frondosa* Greene. *C. valida* Greene. *C. Sonnei* Greene.] Stout, 0.5–2

m. high; lf.-blades oblong to ovate in outline, 1.2–3.5 dm. long, 1–3-pinnate, the lfts. linear-lanceolate to broader, 3–10 cm. long, serrate to incised; petioles 1–8 cm. long; peduncles 5–15 cm. long; bracts of invol. 1–several or none; bractlets several, 2–15 mm. long; rays 2–6 cm. long; pedicels 3–8 mm. long; fr. ovoid to subglobose, 2–4 mm. long; ribs low, corky, broader than the red-brown or like-colored intervals; seed not very oily; $n = 22$ (Bell & Constance, 1957).—Mostly freshwater wet places, below 8000 ft.; many Plant Communities; most of Calif.; to Alaska, Alta., Ariz., n. Mex. June–Sept.

24. Oenánthe L.

Decumbent to erect, branching caulescent mostly glabrous perennials from fascicled or tuberous roots. Lvs. petioled, pinnate to pinnately decompound. Petioles sheathing. Infl. of loose compound umbels. Invol. wanting or reduced. Rays many, spreading, subequal. Involucels of many narrow bractlets. Pedicels spreading. Fls. white. Sepals lanceolate. Stylopodium conic; styles erect, elongate. Fr. in ours oblong, subterete, glabrous; ribs low, obtuse, subequal, corky, often ± confluent; oil-tubes usually solitary in intervals, 2 on commissure. Ca. 30 spp., mostly of Old World. (Ancient Greek name of some thorny plant.)

1. **O. sarmentòsa** Presl. [*Helosciadium californicum* H. & A. *O. c.* Wats. *O. s.* var. *c.* Coult. & Rose.] Stems succulent, 5–15 dm. long, decumbent and ascending; lf.-blades oblong to ovate in outline, 1–3 dm. long, bipinnate, the lfts. ovate, 1–5 cm. long, coarsely dentate to incised; petioles 1–3 dm. long; peduncles 5–13 cm. long; bracts few, linear, 5–15 mm. long; bracteoles 4–5 mm. long; rays 10–20, 1.5–3 cm. long; styles 2–3 mm. long; fr. oblong, 2.4–3.5 mm. long, often purplish, the prominent ribs broader than the intervals; $n = 22$ (Bell & Constance, 1957).—Marshes and sluggish water, below 3500 ft.; many Plant Communities; cismontane Calif.; to B.C. and Ida. June–Oct.

25. Podístera Wats.

Acaulescent cespitose perennials from a long thick root and crown of fibrous sheaths. Lvs. 1–2-pinnate, petioled. Infl. of compact compound umbels. Invol. of many narrow bracts or wanting. Involucel of evident, toothed or entire bractlets. Rays few, obsolete or short. Pedicels short or obsolete. Fls. yellow to purplish. Sepals ovate. Stylopodium conic; styles flattened. Fr. ovoid to ± oblong, slightly flattened laterally, glabrous; ribs filiform; oil-tubes 2–several in the intervals and on the commissure. Three spp., w. N. Am. (Greek, *podos*, foot, and *stereos*, solid, because of the compactness.)

1. **P. nevadénsis** (Gray) Wats. [*Cymopterus n.* Gray. *P. albensis* Jeps.] Caudex woody, with many short mostly subterranean branches; lvs. tufted, 3–10 mm. long, with 3–7 lanceolate divisions, sometimes bipinnate; peduncles 0.5–3 cm. long; bractlets 3–5-cleft; umbels congested, of 3–5 subsessile umbellets; fr. 1–2 mm. long.—Largely Alpine Fell-fields; mostly 10,000–13,000 ft.; White Mts., Sierra Nevada from Placer to Tuolumne and Mono cos., San Bernardino Mts. July–Sept.

26. Oreonàna Jeps.

Low tufted acaulescent plants from thickened taproot. Lvs. petioled, pinnately or ternately decompound, the segms. small, oblong, crowded. Infl. of subcapitate compound umbels. Invol. none. Involucel one-sided. Rays few to many, short, stout. Pedicels of sterile fls. longer than fr. and rays, those of fertile fls. obsolete. Fls. white or purplish. Sepals evident. Fr. ovoid, somewhat laterally compressed; ribs filiform; oil-tubes several in the intervals and on commissure. Two spp. (Greek, *oreos*, mountain and *nannos*, dwarf.)

Rays membranously winged; sterile pedicels equaling or little exceeding the fr. S. Sierra Nevada
                                                        1. *O. Clementis*
Rays not winged; sterile pedicels greatly exceeding the fr. Mts. of s. Calif. . . . . . . 2. *O. vestita*

1. **O. Cleméntis** (Jones) Jeps. [*Drudeophytum C.* Jones. *O. californica* Jeps.] Plants 3–8 cm. high, mostly gray-pubescent except on petioles and peduncles; umbels globose, slightly exserted from tuft of lvs.; rays 2–8 mm. long; fr. 3–4 mm. long, sessile; calyx of sterile fls. conspicuous, starlike.—Dry granitic gravel, mostly 8000–13,000 ft.; Sub-

alpine F., Alpine Fell-fields; s. Sierra Nevada of Tulare, Fresno, and Inyo cos. July–Aug.

2. **O. vestìta** (Wats.) Jeps. [*Deweya v.* Wats.] Plants 4–15 cm. high, densely white-woolly throughout; umbels dense, mostly above the lvs.; bractlets many; sterile pedicels 10–25 mm. long; fr. subsessile, 5–6 mm. long, pubescent; oil-tubes 3–4 in the intervals, 3 on the commissure.—Dry gravel or talus, 6500–11,000 ft.; upper Montane Coniferous F.; San Gabriel and San Bernardino mts. June–July.

27. Orogènia Wats.

Dwarf glabrous perennials with fleshy roots; stem largely underground and sheathed by large scarious bracts. Lvs. 1–3-ternate, with narrow divisions. Invol. none. Involucels none or of few linear bractlets. Fls. white, in compound umbels. Sepals minute. Fr. oblong to oval, slightly flattened laterally, with filiform dorsal ribs, the lateral corky, thickened. Oil-tubes minute, several in the intervals and on the commissure. Two spp., w. U.S. (Greek, *oros*, mountain, and *genos*, race, because of the habitat.)

1. **O. fusifórmis** Wats. Plants to ca. 1 dm. high, from a long fusiform root; lvs. deltoid-ovate in outline, ternate to triternate, the segms. to 6 cm. long, linear; peduncles 5–10 cm. high, slender; rays 2–10; fr. 3–4 mm. long.—Occasional, in wet sandy places, 4000–6000 ft.; Montane Coniferous F.; Sierra Nevada from Nevada to Plumas cos., Coast Ranges of Trinity and Humboldt cos.; to Ore., Mont. May–July.

28. Foenículum Adans.

Erect caulescent biennial or perennial herbs, with anise odor. Lvs. decompound into linear or capillary divisions. Fls. yellow, in large compound umbels. Invol. and involucels none. Sepals obsolete. Stylopodium conical. Fr. oblong, slightly flattened laterally, glabrous; ribs prominent; oil-tubes solitary in the intervals. Four spp., Old World. (Diminutive of Latin, *foenum*, hay, because of odor.)

1. **F. vulgàre** Mill. [*Anethum Foeniculum* L.] Sweet Fennel. Perennial, 1–2 m. high, with striate branching stems, glaucous; lf.-blades ovate to deltoid in outline, to ca. 3 dm. long, pinnately decompound into filiform divisions 4–40 mm. long; petioles broadly sheathing; rays 15–40, ± unequal, 1–6 cm. long; frs. oblong, 3.5–4 mm. long, the ribs acute; $2n = 22$ (Gardé & Gardé, 1949).—Common in waste places especially in s. and cent. Calif.; natur. from Eu. May–Sept.

29. Taúschia Schlecht.

Acaulescent or short-caulescent perennials from taproots or tubers. Lvs. petioled, entire, pinnate to ternately decompound. Infl. of loose compound umbels. Invol. usually wanting. Involucel dimidiate, usually of prominent bractlets. Fertile rays few to many; fertile pedicels short, spreading. Fls. yellow, white, or purplish. Sepals prominent to obsolete. Stylopodium obsolete; styles spreading. Fr. oblong to round or ellipsoid, slightly flattened laterally, glabrous; ribs prominent to filiform, not winged; oil-tubes large to small, solitary to several in intervals, 2–several on commissure. Ca. 20 spp., w. N. and Cent. Am. (I. F. *Tausch*, 19th-century botanist.)

Lvs. simply pinnate; fr. with conspicuous ribs. S. Calif. .... 1. *T. arguta*
Lvs. ternate-pinnate or ternately or pinnately decompound.
  Ultimate lf.-divisions 3–15 mm. long; plants glabrous and glaucous throughout.
    Sepals obsolete; rays 5–12; fr. 2–3 mm. long. N. Calif. .... 2. *T. glauca*
    Sepals evident.
      Rays 12–18; fr. 5–8 mm. long. S. Calif. .... 3. *T. Parishii*
      Rays 3–5; fr. 2–4 mm. long. N. Calif. .... 4. *T. Howellii*
  Ultimate lf.-divisions mostly 20–60 mm. long; plants ± scabrous, at least in infl.
    Involucels conspicuous, often exceeding the umbellet; fr. 4–7 mm. long ...... 5. *T. Hartwegii*
    Involucels inconspicuous; fr. 3–5 mm. long .... 6. *T. Kelloggii*

1. **T. argùta** (T. & G.) Macbr. [*Deweya a.* T. & G. *Velaea a.* Coult. & Rose] Short-caulescent, glabrous, 3–7 dm. high, from a crown from long taproot; lf.-blades oblong to ovate in outline, 8–16 cm. long, the lfts. oblong to oval, distinct, 3–8 cm. long, sharply serrate; petioles 6–20 cm. long; peduncles 1.5–4 dm. long; involucel of several linear

to lanceolate entire to lobed bractlets 2–10 mm. long; rays 15–25, 2–12 cm. long; pedicels 3–9 mm. long; sepals evident; fls. yellow; fr. oblong, 6–9 mm. long, 3–5 mm. broad, ribs prominent; oil-tubes 3–5 in intervals, 4–6 on commissure; $n = 11$ (Bell & Constance, 1957).—Locally common on dry fans and slopes, below 7000 ft.; Coastal Sage Scrub, Chaparral, etc.; cismontane s. Calif. from Santa Barbara Co. s.; to L. Calif. April–June. To desert edge.

2. **T. glaúca** (Coult. & Rose) Math. and Const. [*Velaea g.* Coult. & Rose. *V. g.* var. *purpurascens* J. T. Howell.] Short-caulescent, 2–4 dm. high, glabrous and glaucous; basal lv.-blades ternate-pinnate or biternate, 6–13 cm. long, the lfts. ovate to orbicular, 10–17 mm. long, coarsely serrate, or lobed toward the base; petioles 2–11 cm. long; peduncles 2–4 dm. long; involucel-bractlets lanceolate, 1–7 mm. long; rays 5–12, 1–6 cm. long; pedicels 1–3 mm. long; sepals minute; fls. yellow or purplish; fr. suborbicular, 2–3 mm. long, the ribs filiform; oil-tubes 2–3 in intervals, 4–5 on commissure.—Wooded slopes, 1000–3500 ft.; Mixed Evergreen F., Yellow Pine F.; Trinity and more n. cos.; sw. Ore. April–June.

3. **T. Párishii** (Coult. & Rose) Macbr. [*Velaea P.* Coult. & Rose. *Cymopterus owenensis* Jones.] Acaulescent, 1–4 dm. high, glabrous and glaucous throughout; lf.-blades oblong to ovate in outline, ternate-pinnate or bipinnate, 8–15 cm. long, the ultimate segms. ovate, cuspidate-toothed, 6–12 mm. long; petioles 5–15 cm. long; peduncles 1–3 dm. long; involucel-bractlets few, linear, entire, 5–12 mm. long; rays subequal, 3–6 cm. long; pedicels 2–7 mm. long; sepals evident; fls. yellow; fr. oblong to oval, 5–8 mm. long, the ribs filiform; oil-tubes 4–5 in intervals, 8–10 on commissure; $n = 11$ (Bell & Constance, 1957).—Frequent on dry slopes, 4000–9000 ft.; Chaparral, Yellow Pine F., Red Fir F., Pinyon-Juniper Wd.; mts. from San Diego Co. to Mt. Pinos and s. Sierra Nevada. May–July.

4. **T. Howéllii** (Coult. & Rose) Macbr. [*Velaea H.* Coult. & Rose.] Short-stemmed, 5–8 cm. high, glabrous; lf.-blades ovate in outline, 1.5–3 cm. long, pinnate or ternate, the oblong to ovate ± confluent lfts. 5–15 mm. long, dentate and lobed; petioles 2–3 cm. long; peduncles 1–2 cm. long; bractlets 1–2 cm. long, leafy; sepals prominent; frs. oblong, glabrous, 2–4 mm. long; oil-tubes several in intervals and on commissure.—Dry slopes, 2000–3000 ft.; upper Mixed Evergreen F., Yellow Pine F., etc.; Siskiyou Mts., Del Norte and Siskiyou cos.; sw. Ore. June–July.

5. **T. Hartwégii** (Gray) Macbr. [*Deweya H.* Gray.] Acaulescent, 3–10 dm. high, minutely scabrous; lf.-blades 1–2-ternate-pinnate, 1–2 dm. long, the lfts. oblong to ovate, confluent, 2–6 cm. long, mucronulate-serrate, rounded-obtuse at apex; petioles 5–25 cm. long; peduncles 2–8 dm. long; bractlets reflexed, 5–12 mm. long; sepals minute; rays 2–13 cm. long; pedicels 2–7 mm. long; fls. yellow; fr. suborbicular, 4–7 mm. long; oil-tubes 3–5 in intervals, 6–8 on commissure; $n = 11$ (Bell & Constance, 1957).—Occasional on wooded or brushy slopes, below 5000 ft.; Chaparral, Foothill Wd., Yellow Pine F.; from Santa Monica Mts. n. through Coast Ranges to Contra Costa Co. and Sierra Nevada foothills to Butte Co. March–May.

6. **T. Kellóggii** (Gray) Macbr. [*Deweya K.* Gray. *Velaea K.* Coult. & Rose.] Mostly acaulescent, 2–7 dm. high, minutely scabrous; lf.-blades ovate to roundish in outline, 8–20 cm. long, the lfts. oblong to ovate, confluent, 15–30 mm. long; petioles 5–15 cm. long; peduncles 2–5 dm. long; rays 2–12 cm. long; bractlets linear, few, 3–8 mm. long; pedicels 3–15 mm. long; fls. yellow; fr. suborbicular, 3–5 mm. long; oil-tubes 2–3 in intervals; ca. 6 on commissure.—Scattered stations mostly on shaded slopes, below 3500 ft., often on serpentine; Mixed Evergreen F., Yellow Pine F., Chaparral; Coast Ranges from Santa Cruz Co. n., Sierra Nevada from Tulare to Butte cos.; Ore. April–June.

30. Oxýpolis Raf.

Slender erect caulescent glabrous perennials from fascicled tubers. Lvs. petioled, simply pinnate or ternate. Petioles sheathing. Infl. of loose compound umbels. Invol. of few slender bracts or none. Involucel of lfts. like the bracts or none. Rays few to many. Pedicels slender, ascending to spreading. Sepals prominent or minute. Stylopodium conic; styles slender, spreading. Fr. oblong to obovoid, glabrous, strongly flattened laterally; ribs filiform, the lateral broadly thin-winged; the nerves resembling ribs and making the carpel appear 5-ribbed; oil-tubes large, solitary in the intervals, 2–6 on the commissure. N. Am. genus with ca. 7 spp. (Greek, *oxys*, sharp, and *polis*, city, of uncertain application.)

1. **O. occidentàlis** Coult. & Rose. Plants 6–12 dm. high, the stems simple or few branched; lf.-blades oblong in outline, 1–3 dm. long, pinnate, the lfts. 5–13, lance-linear to oblong-ovate, crenate to serrate or incised, 3.5–6.5 cm. long; petioles 1–3 dm. long; stem-lvs. few, ± reduced; peduncles 0.5–3 dm. long; rays 12–24, ± unequal, 2–8 cm. long; bracts mostly 1–2, 5–25 mm. long; bractlets 5–15 mm. long; fls. white; sepals evident; fr. oval to oblong, 5–6 mm. long; $n = 18$ (Bell & Constance, 1957).— Shallow water and wet places, 4000–8500 ft.; Montane Coniferous F.; San Bernardino Mts., Sierra Nevada to Siskiyou Co.; Ore. July–Aug.

31. Pastinàca L. Parsnip

Tall branching biennial to perennials. Lvs. pinnately compound. Invols. and involucels small or none. Fls. yellow or red. Sepals obsolete. Stylopodium depressed-conical. Fr. strongly flattened dorsally, oval to obovate. Carpels with winged lateral and filiform dorsal ribs; oil-tubes solitary in intervals, 2–4 on the commissure. Ca. 14 spp., Eurasian. (Latin name from *pastino*, to prepare the ground for planting of the vine.)

1. **P. satìva** L. Stout glabrous biennials, 3–10 dm. high; lvs. oblong to ovate, pinnate, the blades 1.5–2.5 dm. long, the lfts. 5–10 cm. long; petioles 1–1.5 cm. long; peduncles stout, 7–15 cm. long; rays 15–25, unequal, 2–10 cm. long; fls. yellow; fr. 5–6 mm. long; $2n = 22$ (Ogawa, 1929).—Locally natur. in damp places, escape from gardens; native of Eu.

32. Anèthum L. Dill

Tall branching annuals. Lvs. pinnately dissected into filiform divisions. Fls. yellow, in compound umbels. Invol. and involucels wanting. Rays many. Pedicels slender, spreading. Sepals obsolete. Stylopodium conic; styles short, reflexed. Fr. ovate, strongly flattened dorsally, glabrous; ribs narrowly winged, the lateral broader than the dorsal; oil-tubes solitary in intervals. Two spp., Old World. (The ancient Latin name.)

1. **A. gravèolens** L. Plants 4–15 dm. tall; lvs. oblong to obovate in outline, the blades 1–3 dm. long; petioles 5–6 cm. long; peduncles 7–15 cm. long; rays 10–45, spreading, 3–10 cm. long; pedicels 20–50, 6–10 mm. long; fr. ovoid, ca. 4 mm. long; $2n = 22$ (Tamamschjan, 1933).—Waste places, as a garden escape. June–Aug. Used in pickling.

33. Lomàtium Raf.

Acaulescent or short-caulescent perennials with slender or thick subfusiform to tuberous roots. Lvs. ternate, pinnate or decompound. Fls. in compound umbels, yellow to white or purple. Invol. mostly lacking; involucels present, sometimes none. Sepals small. Fr. strongly flattened dorsally; carpels with filiform dorsal ribs, the lateral winged, thin to corky; stylopodium lacking; oil-tubes 1–several in the intervals, sometimes obsolete, 2–10 on the commissure. Seed flattened dorsally, the face plane or ± concave. Ca. 80 spp., w. N. Am. (Greek, *loma*, a border, because of the wings on the fr.)

(Mathias, M. E. A revision of the genus Lomatium. Ann. Mo. Bot. Gard. 25: 225–297, 1938.)
A. Peduncles conspicuously inflated and swollen at the apex .... 35. *L. nudicaule*
AA. Peduncles not conspicuously swollen at the apex.
    B. Fr. ± deeply emarginate at each end, the wings distinct on each side of the body; lfts. mostly broad in outline.
        C. Lf.-divisions not pinnatifid, merely toothed or sometimes 3-lobed.
            D. Lvs. 1–2-ternate; wings thickened, much broader than the body; oil-tubes solitary in the intervals. Santa Barbara to San Diego cos. .... 1. *L. lucidum*
            DD. Lvs. ternate-pinnate; wings thin, ca. as broad as or broader than the body; oil-tubes 1–3 in intervals.
                E. Fr. broadly oval; plants mostly 1.5–2.5 dm. tall. Napa and Lake cos.
                                              2. *L. repostum*
                EE. Fr. suborbicular; plants mostly 2.5–4 dm. tall. Del Norte Co.
                                              3. *L. Howellii*
        CC. Lf.-divisions pinnatifid, usually incised.

    D. Lf.-blades longer than the petioles; fr. 12–15 mm. long. San Nicolas Id.

                                           5. *L. insulare*

   DD. Lf.-blades mostly shorter than or equal to petioles; fr. 7–10 mm. long.

       E. Ultimate lf.-divisions acerose-tipped; wings much narrower than body. Inyo Co.

                                        6. *L. rigidum*

       EE. Ultimate lf.-divisions not so tipped; wings broader than body. Monterey and
San Luis Obispo cos. .... 4. *L. parvifolium*

BB. Fr. not emarginate or scarcely so, the wings ± joined above and below the body; lfts. mostly narrow.

   C. Plants mostly 1–2 dm. high (more in *leptocarpum*), from round to elongate or ± irregular tubers. N. and ne. Calif.

     D. Fls. white; fr. 5–10 mm. long, ovate to oblong.

       E. Involucels none or inconspicuous; pedicels 0–2 mm. long; ultimate segms. of lvs. 5–30 mm. long .... 7. *L. Piperi*

      EE. Involucels evident; pedicels 8–12 mm. long; ultimate segms. of lvs. 4–5 mm. long .... 8. *L. Canbyi*

   DD. Fls. yellow; fr. 10–15 mm. long, narrow-oblong .... 9. *L. leptocarpum*

CC. Plants mostly stouter, often taller, from ± thickened elongate taproots.

   D. Lvs. decompound, dissected into many small divisions.

     E. Ovaries and young fr. variously pubescent or roughened.

       F. Bractlets of involucel oblanceolate to obovate.

         G. Plants scabrous or roughened; umbels 10–25-rayed; wings ca. as broad as body. Plumas Co. n.............. 10. *L. vaginatum*

        GG. Plants glabrous to pubescent, not scabrous or roughened; umbels 5–13-rayed; wings mostly broader than the body. Most of cismontane Calif. .... 11. *L. utriculatum*

      FF. Bractlets of involucels mostly linear, never obovate, sometimes reduced.

         G. Young fr. granulate-roughened, not pubescent, the wings less than half as wide as body. Siskiyou Co. ...... 18. *L. Peckianum*

        GG. Young fr. variously pubescent, not granulate-roughened.

          H. Bractlets of involucel with conspicuous scarious margin, not tomentose or villous. Alpine to Modoc and Siskiyou cos.

                                     19. *L. nevadense*

         HH. Bractlets not conspicuously scarious-margined, ± villous or tomentose.

           I. Plants acaulescent, 1–3 dm. tall.

             J. Fls. yellow or purple; lf.-blades oblong-ovate in outline, mostly 5–15 cm. long, the ultimate segms. not especially crowded.

               K. Plants ± villous throughout; petioles shorter than the lf.-blades. Inyo-White Mts.

                              21. *L. MacDougalii*

              KK. Plants hoary-pubescent, never villous; petioles as long as the lf.-blades. General Mojave Desert

                            24. *L. mohavense*

            JJ. Fls. white; lf.-blades narrow-ovate in outline, 1.5–5.5 cm. long, the ultimate segms. crowded. High Inyo Mts.

                            22. *L. inyoense*

         II. Plants mostly caulescent, 3–5 dm. tall.

           J. Petals glabrous; fr. narrow-oblong, sparingly pubescent with long hairs. Kern and Monterey cos. n.

                            25. *L. macrocarpum*

          JJ. Petals tomentose; fr. ovate-oblong to orbicular, densely pubescent.

             K. Pedicels mostly longer than the mature fr.; wings broader than the body, membranaceous, thinly pubescent to glabrate. San Diego to Humboldt cos.

                            23. *L. dasycarpum*

             KK. Pedicels mostly shorter than the mature fr.; wings narrower than to as broad as body, somewhat thickened, tomentose. Kern to Butte cos.

                            26. *L. tomentosum*

     EE. Ovaries and fr. glabrous.

       F. Bractlets of involucel absent.

         G. Foliage and peduncles pubescent (if glabrate, the fls. yellow); umbels 2–7-rayed. Siskiyou and Trinity cos. .. 29. *L. Engelmannii*

        GG. Foliage and peduncles glabrous (if scaberulous, not of n. Calif.); umbels 5–16-rayed.

          H. Fls. white; pedicels 6–15 mm. long; fr. 7–10 mm. long. Mariposa Co. foothills .... 27. *L. Congdonii*

         HH. Fls. yellow; pedicels 1–4 mm. long; fr. 10–16 mm. long. Pine belt, Tulare to Mariposa cos. .... 28. *L. Torreyi*

      FF. Bractlets of involucel present.

         G. Bractlets obovate, sometimes connate.

          H. Plants usually with several cauline lvs.; wings of fr. broader than body, the dorsal ribs obsolete.

    I. Fr. ovate to obovate, 9–15 mm. long; sepals prominent in young frs. From San Luis Obispo and Inyo cos. s. .... 12. *L. Vaseyi*

    II. Fr. oblong to ovate, 5–11 mm. long; sepals obsolete. General cismontane Calif. .... 11. *L. utriculatum*

HH. Plants with not more than 1 stem lf.; wings of fr. narrower than body, or if broader, the dorsal ribs evident.

    I. Plants glabrous or slightly pubescent; fls. yellow (or if purple, from w. Sierra Nevada).

        J. Lf.-blades broadly ovate to obovate; fertile rays 6–15. San Luis Obispo Co. to Mendocino Co. .... 16. *L. caruifolium*

        JJ. Lf.-blades oblong to ovate; fertile rays 2–5. Tehama Co. to Tulare Co. .... 17. *L. humile*

    II. Plants scaberulous to densely pubescent; fls. white or purple (if yellow, from N. Coast Ranges).

        J. Fls. purple or yellow; lvs. ternate, then 1–2-pinnate. N. Coast Ranges .... 13. *L. ciliolatum*

        JJ. Fls. white; lvs. 3-pinnate. Alpine to Modoc and Siskiyou cos. .... 19. *L. nevadense*

GG. Bractlets filiform to lance-linear, rarely oblance-acuminate.

    H. Bractlets ± tomentose or villous. Kern and Monterey cos. n. .... 25. *L. macrocarpum*

    HH. Bractlets glabrous or ± roughened.

        I. Fr. 12–16 mm. long, 6–10 mm. broad, with narrow corky-thickened wings. Mendocino Co. to Siskiyou and Modoc cos. and Sierra Nevada to San Bernardino Mts. .... 30. *L. dissectum*

        II. Fr. 5–13 mm. long, 3–7 mm. broad, with thin membranous wings.

            J. Plants ± pubescent. Sierra Nevada.

                K. Fls. yellow; plants less than 3 dm. tall .... 20. *L. Plummerae*

                KK. Fls. white; plants mostly 3–4.5 dm. tall .... 19. *L. nevadense*

            JJ. Plants glabrous or slightly scaberulous.

                K. Plants acaulescent.

                    L. Lf.-blades ovate to oblong in outline, 4–10 cm. long; fr. 3–5 mm. broad. Trinity and Humboldt cos. n. .... 14. *L. Tracyi*

                    LL. Lf.-blades obovate, 6–20 cm. long; fr. 5–7 mm. broad. From Tulare and Napa cos. to Shasta Co. .... 15. *L. marginatum*

                KK. Plants short-caulescent.

                    L. Lf.-divisions remote, mostly elongate, 5–80 mm. long; fr. apically obtuse .... 15. *L. marginatum*

                    LL. Lf.-divisions crowded, shorter, 3–7 mm. long; fr. apically acute. Sierra Nevada .... 20. *L. Plummerae*

DD. Lvs. with mostly few or large divisions, ternately or pinnately divided, the divisions mostly remote.

    E. Fls. whitish; plants 1–2 dm. tall; plants subglabrous. Del Norte Co. .... 32. *L. Martindalei*

    EE. Fls. yellow; plants mostly taller.

        F. Lf.-blades narrow-oblong in outline, 2–3-pinnate, the ultimate segms. linear, 2–9 mm. long. Mts. of Mojave Desert .... 31. *L. Parryi*

        FF. Lf.-blades ovate to obovate in outline, ternate-pinnately or quinate-pinnately divided.

            G. Plants puberulent to subglabrous, the lf.-blades 0.7–1.5 dm. long; oil-tubes 1 in intervals, 2 on commissure. Extreme n. Calif. .... 33. *L. triternatum*

            GG. Plants glabrous, glaucous; lf.-blades 1–3 dm. long; oil-tubes 3–4 in intervals, 6–10 on commissure, sometimes obscure. From Ventura and Kern cos. n. .... 34. *L. californicum*

1. **L. lùcidum** (Nutt.) Jeps. [*Euryptera l.* Nutt. ex T. & G. *Peucedanum E.* Gray. *P. Hassei* Coult. & Rose. *Cogswellia l.* and *H.* Jones.] Short-caulescent, 2–6 dm. tall, glabrous, from long slender taproot; lf.-blades ± ovate in outline, 5–9 cm. long, 1–2-ternate, the ultimate divisions deltoid to cuneate, 1.5–7 cm. long, dentate to lobed; petioles 3–13 cm. long; peduncles exceeding lvs.; rays 10–20, spreading, 2–8 cm. long; bractlets of involucel lance-linear, ca. as long as fls.; pedicels 7–17 mm. long; fls. yellow; fr. suborbicular to broadly elliptic, 6–15 mm. long, emarginate especially at base, the wings thick, broader than body; oil-tubes solitary in intervals, 2–4 on commis-

sure.—Occasional on brushy slopes, below 5000 ft.; mostly Chaparral; cismontane s. Calif., especially away from the coast, Santa Barbara Co. to San Diego Co. Jan.–April.

2. **L. repóstum** (Jeps.) Math. [*L. lucidum* var. *r.* Jeps.] Much like *L. lucidum*, lf.-blades 1–2-ternate or ternate-pinnate; bractlets lanceolate; fls. greenish-yellow; fr. broadly oval, emarginate at each end, 10–15 mm. long, the wings thin, equal to or broader than body; oil-tubes 1–3 in intervals, 4–6 on commissure; $n = 22$ (Bell & Constance, 1957).—Chaparral, Closed-cone Pine F., Foothill Wd.; inner Coast Ranges of n. Napa and s. Lake cos. April–May.

3. **L. Howéllii** (Wats.) Jeps. [*Peucedanum H.* Wats. *Cogswellia H.* Jones.] Sub-acaulescent, 2.5–4 dm. high, from slender taproot, glabrous; lvs. ternate, then 1–2-pinnate, the divisions deltoid, sharply toothed, sometimes lobed, 10–25 mm. long; rays 10–15, spreading, 2.5–5 cm. long; pedicels 8–12 mm. long; fls. yellow; fr. suborbicular, 7–11 mm. long, deeply emarginate at both ends, the wings ca. as broad as body; oil-tubes 2–3 in intervals, 9 on commissure.—Rocky serpentine slopes, to 3500 ft.; Chaparral, Mixed Evergreen F., Yellow Pine F.; Del Norte Co.; to sw. Ore. May–June.

4. **L. parvifòlium** (H. & A.) Jeps. [*Ferula p.* H. & A. *Euryptera p.* Coult. & Rose.] Short-caulescent, 1–4 dm. high, glabrous; lf.-blades oblong in outline, ternate then 1–2-pinnate, the ultimate segms. lanceolate to cuneate, 8–20 mm. long, sharply incised; petioles 3–15 cm. long, purplish; bractlets ca. as long as fls., narrow; rays 8–14, spreading, 1–2.5 cm. long; pedicels 3–6 mm. long; fls. yellow; fr. orbicular to oblong, 7–10 mm. long, emarginate at both ends, wings broader than body; oil-tubes 1–2 in dorsal intervals, 2–3 in lateral, 4–6 on commissure.—Largely Closed-cone Pine F.; Monterey Co. to San Luis Obispo Co. Feb.–May.

Var. **pállidum** (Coult. & Rose) Jeps. [*Euryptera p.* Coult. & Rose.] Foliage pale; rays 3–6 cm. long; pedicels 7–17 mm. long.—Santa Lucia Mts., Monterey and San Luis Obispo cos.

5. **L. insulàre** (Eastw.) Munz. [*Peucedanum i.* Eastw.] Acaulescent, 1–5 dm. tall; lf.-blades biternate to bipinnate, the divisions ± oblong, cuneate, irregularly pinnatifid, 4–14 mm. long; rays 15–20, subequal, 3–8 cm. long; bractlets filiform; pedicels 6–12 mm. long; fls. yellow; fr. oblong-ovate, 12–15 mm. long, emarginate at both ends; wings thick, ca. as broad as body; oil-tubes 2 in intervals, 4 on commissure.—Sea bluffs; San Nicolas Id.; Guadalupe Id. Feb.–April.

6. **L. rígidum** (Jones) Jeps. [*Cogswellia r.* Jones.] Acaulescent or nearly so, 1–4 dm. high, glabrous; lf.-blades oblong in outline, 8–12 cm. long, bipinnate into ovate to cuneate segms. 1–2 cm. long, sharply pinnatifid; petioles 5–10 cm. long; bractlets conspicuous, lanceolate, ca. 5–10 mm. long; rays 10–20, spreading, 2.5–5 cm. long; fls. yellow; sepals conspicuous; fr. ovate to oblong, 7–9 mm. long, emarginate at base, rounded at apex, the wings less than half as wide as body; oil-tubes 3 in intervals, 6 on commissure.—Rocky places, 4000–9000 ft.; Sagebrush Scrub, Pinyon-Juniper Wd.; e. slope of Sierra Nevada. April–May.

7. **L. Pìperi** Coult. & Rose. [*Cogswellia P.* Jones.] Acaulescent or nearly so, 1–2 dm. high, from a small rounded tuber, glabrous or ± puberulent; lf.-blades oblong-ovate in outline, 3–7 cm. long, ternate, then tripinnate into linear segms. 0.5–3 cm. long; petioles 3–10 cm. long; bractlets 0 or linear, few and inconspicuous; rays 3–20, spreading, 1–6 cm. long; pedicels 0–2 mm. long; fls. white, with purple anthers; fr. ovate to oblong, 5–9 mm. long, the wings half as wide as body; oil-tubes 1–8 in intervals, 2–4 on commissure.—Dry stony places, 2500–5000 ft.; Yellow Pine F., Red Fir F., etc.; Sierra to Siskiyou cos.; to Wash., Ida. March–May.

8. **L. Cánbyi** (Coult. & Rose) Coult. & Rose. [*Peucedanum C.* Coult. & Rose. *Cogswellia C.* Jones.] Acaulescent, glabrous, 1.5–2 dm. high, from a thick rootstock ending in a large tuber; lf.-blades oblong-ovate in outline, 7–9 cm. long, ternate then bipinnate into linear segms. 4–5 mm. long; petioles 4–6 cm. long, forming at base a purple-veined scarious sheath; bractlets linear; rays 12–17, spreading, 2.5–5.5 cm. long; pedicels 8–12 mm. long; fls. white; fr. oval-oblong, 7–10 mm. long, narrow-winged; oil-tubes 1–2 in intervals, 2–4 on commissure.—Sagebrush Scrub; Lassen Co. to e. Wash., w. Ida. April–May.

9. **L. leptocárpum** (T. & G.) Coult. & Rose. [*Peucedanum triternatum* var. *l.* T. & G. *L. ambiguum* var. *l.* Jeps.] Short-caulescent, 1.5–5 dm. high, glabrous to scaberulous, from elongate tuberous roots; lf.-blades broadly obovate in outline, 9–14 cm. long,

1–2-ternate, then 2–4-pinnate, the ultimate segms. linear to filiform, 7–40 mm. long; petioles 2–7 cm. long; bractlets several, linear; rays 4–15, suberect, 2–10 cm. long; pedicels 2–7 mm. long; fls. yellow; fr. narrow-oblong, 10–15 mm. long, the wings less than half as wide as body; oil-tubes 1 in intervals, 2–4 on commissure.—Dry rocky places, 3500–7000 ft.; Sagebrush Scrub to Red Fir F.; Lassen, Shasta, and Modoc cos.; to e. Ore., Colo. Ariz. April–June.

10. **L. vaginàtum** Coult. & Rose. [*L. Plummerae* var. *Helleri* Math.] Caulescent, 2–4.5 dm. tall, from an elongate ± thickened root, ± scabrous; lf.-blades ovate in outline, 5–12 cm. long, ternate, then 1–2-pinnate into crowded oblong segms. 1–5 mm. long; petioles 2–10 cm. long, with purplish sheaths completely surrounding the stem; bractlets conspicuous, oblanceolate to obovate; rays 6–15, ascending, 1–8 cm. long; pedicels 3–12 mm. long; fls. yellow; fr. broadly oval to obovate, 8–12 mm. long, granulate-roughened, the wings almost as broad as body; oil-tubes 1–4 in intervals, 4–5 on commissure.— Rocky slopes and ridges, ca. 3000–5000 ft.; Yellow Pine F.; Plumas Co. to Siskiyou and Modoc cos.; to cent. Ore. April–May.

11. **L. utriculàtum** (Nutt.) Coult. & Rose. [*Peucedanum u.* Nutt. ex T. & G. *Cogswellia u.* Jones. *L. u.* vars. *glabrum* and *anthemifolium* Jeps. *C. Chandleri* Jones.] Caulescent, 1–5 dm. tall, purplish below, glabrous to pubescent, from a long slender taproot; lf.-blades oblong in outline, 3–16 cm. long, tripinnate into linear segms. 4–25 mm. long; petioles 1–10 cm. long; bractlets obovate, entire to cleft, green to purplish, ± scarious on margins; rays 5–13, spreading to ascending, 2–12 cm. long; pedicels 2–9 mm. long; umbellets ca. 20-fld.; fls. yellow; fr. ovate to oblong, 5–11 long, puberulent when young, the wings thin, usually broader than body; oil-tubes 1–3 in dorsal intervals, 1–4 in lateral, 2–6 on commissure; $n = 11$ (Bell & Constance, 1957).—Grassy slopes and flats, below 5000 ft.; V. Grassland, Foothill Wd., Yellow Pine F., Sagebrush Scrub, etc.; cismontane Calif. to edge of desert (Argus Mts., etc.); to B.C. Feb.–May.

12. **L. Vàseyi** (Coult. & Rose) Coult. & Rose. [*Peucedanum V.* Coult. & Rose. *Cogswellia V.* Coult. & Rose.] Like the preceding vegetatively, the ultimate lf.-segms. oblong, 3–17 mm. long; rays 10–20, ascending, 2–7 cm. long; umbellets ca. 30-fld.; fls. yellow; fr. 9–15 mm. long, the thin wings usually broader than the body; oil-tubes 1 in intervals, 4 on commissure.—Dry slopes and mesas, below 5500 ft.; Chaparral, Coastal Sage Scrub, Foothill Wd., etc.; interior cismontane s. Calif. to San Luis Obispo and Inyo cos. March–April.

13. **L. ciliolàtum** Jeps. Acaulescent, 1–1.5 dm. high, hoary-pubescent, from a long slender taproot; lf.-blades ovate-oblong to -deltoid in outline, 3–7 cm. long, ternate, then 1–2-pinnate into oblong or ovate segms. 1–5 mm. long; petioles 2–3 cm. long; scarious-margined, wholly sheathing; bractlets purplish, entire, lanceolate or broader; rays 7–12, spreading-ascending, 1–4 cm. long, unequal; pedicels 2–4 mm. long; fls. purple or yellow; fr. oval, 7–9 mm. long, glabrous, the wings thickish, 1 mm. broad; oil-tubes obscure, 4–5 in intervals, 2 on commissure.—Rocky slopes, 6000–7000 ft.; Red Fir F.; inner Coast Ranges, Glenn and Lake to Trinity and Mendocino cos. June–July.

Var. **Hoòveri** Math. & Const. Taller, 1.5–3 dm. high, scaberulous; ultimate lf.-segms. linear, 1–10 mm. long; fr.-wings thin; $n = 11$ (Bell & Constance, 1957).—Serpentine slopes, ca. 2000–3000 ft.; Chaparral, etc.; inner Coast Ranges, Napa and Colusa and Lake cos. May.

14. **L. Tràcyi** Math. & Const. Acaulescent, 1–3 dm. tall, glabrous to sparsely scaberulous-puberulent; lf.-blades ternate, then 1–2-pinnate, 4–10 cm. long, the ultimate divisions linear to oblong, 1–7 mm. long; petioles 2–5 cm. long, scarious; rays 4–9, ascending, 1–8 cm. long, very unequal; bractlets linear to oblanceolate, scarious-margined; pedicels 1–5 mm. long; fls. yellow; fr. oblong-ovate to oval, 6–10 mm. long, acute at ends, glabrous, the thin wings much narrower than body; oil-tubes obscure.— Serpentine, 1500–6500 ft.; Yellow Pine F., Red Fir F.; Trinity and Humboldt cos.; sw. Ore. May–June.

15. **L. marginàtum** (Benth.) Coult. & Rose. [*Peucedanum m.* Benth. *Cogswellia m.* Jones. *L. alatum* var. *purpureum* Jeps. *L. caruifolium* var. *p.* Math.] Acaulescent or short-stemmed, 1.5–6 dm. high, scaberulous to glabrate; lf.-blades obovate in outline, 6–20 cm. long, 1–3-ternate or ternate then pinnate into segms. linear to filiform, 5–80 mm. long; petioles 3–12 cm. long, usually purplish, sheathing; bractlets several, narrow, scarious-margined; rays 3–15, 1.5–12 cm. long, unequal; pedicels 3–12 mm. long; fls.

yellow or purple; fr. oval to ± obovate, 9–12 mm. long, the wings thin, narrower to as wide as body; oil-tubes obscure; $n = 11$, 22 (Bell & Constance, 1957).—Rocky flats, 500–2250 ft.; Foothill Wd., Yellow Pine F., V. Grassland; Sierra Nevada foothills from Tulare Co. n., Coast Ranges from Napa Co. to Shasta Co. March–May.

16. **L. caruifòlium** (H. & A.) Coult. & Rose. [*Ferula c.* H. & A. *L. c.* vars. *solanense* and *erythropodum* Jeps.] Acaulescent or short-caulescent, 1.5–4.5 dm. high, glabrous to pubescent, from long slender taproot; lf.-blades ovate to obovate in outline, 5–30 cm. long, 1–3-ternate or 1-ternate then bipinnate, the ultimate segms. linear, 3–15 mm. long; petioles 4–7 cm. long; bractlets entire or toothed, green to purplish, scarious-margined; rays 6–15, unequal, 2–12 cm. long; pedicels mostly 2–8 mm. long; fls. yellow; fr. narrowly ovate to obovate, 8–12 mm. long, usually obtuse at both ends, glabrous, the wings thickish, narrower than body; oil-tubes obscure.—Moist grassy hills and meadows, below 4000 ft.; Foothill Wd., Mixed Evergreen F., etc.; Coast Ranges from San Luis Obispo Co. to Mendocino Co., Santa Cruz Id. March–May.

17. **L. hùmile** (Coult. & Rose) Hoov. ex Math. & Const. [*Leptotaenia h.* Coult. & Rose. *L. caruifolium* var. *denticulatum* Jeps. *Peucedanum erosum* Jeps.] Much like the preceding sp., lf.-divisions linear to filiform, 5–60 mm. long; bractlets round to lanceolate, conspicuously veined; pedicels 1–4 mm. long; fr. oval to round, 6–12 mm. long, the wings thick and corky, narrower than the body; $n = 11$ (Bell & Constance, 1957).—Grassy slopes, 1000–2000 ft.; V. Grassland, Foothill Wd.; foothills of Sierra Nevada from Tehama and Butte cos. to Tulare Co. April–May.

18. **L. Peckiànum** Math. & Const. Acaulescent, 1–3 dm. high, scaberulous to glabrous; lf.-blades ovate-oblong in outline, 3–10 cm. long, ternate, then 1–2-ternate, the segms. oblong to linear, 3–18 mm. long; petioles 2–4 cm. long; bractlets 0 or inconspicuous, linear; rays 1–5, ascending, obsolete to 5 cm. long, unequal; pedicels 2–7 mm. long; fr. oblong-oval, 2–15 mm. long, granulate-roughened to glabrate, the wings less than half as wide as body; oil-tubes several in intervals, 6–8 on commissure.—Dry hillsides; at ca. 2500 ft.; Yellow Pine F.; Siskiyou Co.; adjacent Ore. May.

19. **L. nevadénse** (Wats.) Coult. & Rose. [*Peucedanum n.* Wats. *Cogswellia n.* Jones.] Acaulescent or short-caulescent, 1–4 dm. tall, pubescent; lf.-blades oblong to obovate in outline, 5–6 cm. long, tripinnate, the segms. crowded, oblong, 2–3 mm. long; petioles 4–6 cm. long, sheathing to above the middle, purplish; bractlets conspicuous, linear and distinct or obovate and connate; rays 8–22, spreading, 1–2.5 cm. long; pedicels 3–10 mm. long; fls. white; fr. ovate to oblong-ovate, 6–8 mm. long, puberulent, the wings narrower than the body; oil-tubes 2–9 in intervals, 4–12 on commissure.—Dry hills, 4000–9500 ft.; mostly Montane Coniferous F., Sagebrush Scrub; Sweetwater Mts., Sierra Nevada, mostly on e. side from Mono Co. n., to Modoc and Siskiyou cos.; Ore., Nev., Utah, Ariz. April–July.

Var. **Párishii** (Coult. & Rose) Jeps. [*Peucedanum P.* Coult. & Rose.] Lf.-segms. sometimes longer, up to 3 cm. long; bractlets sometimes reduced to form a sheath; rays mostly 1.5–5.5 cm. long; pedicels 3–12 mm. long; ovaries glabrous; fr. 7–10 mm. long; oil-tubes 1–4 in intervals, 4–7 on commissure; $n = 11$ (Bell & Constance, 1957).—Dry slopes, 5000–9000 ft.; Sagebrush Scrub to Montane Coniferous F.; desert slopes of San Bernardino and San Gabriel mts., Sierra Nevada, to e. Ore., New Mex.

Var. **pseudorientàle** (Jones) Munz. [*Cogswellia n.* var. *p.* Jones. *L. n.* var. *holopterum* Jeps.] Near var. *Parishii*, but with more prominent scarious margins on petioles; wings ca. as broad as body, the dorsal ribs evident.—Dry slopes, 3000–6000 ft.; Creosote Bush Scrub to Pinyon-Juniper Wd.; mts. of e. Mojave Desert; to Nev., Ariz.

20. **L. Plúmmerae** (Coult. & Rose) Coult. & Rose. [*Peucedanum P.* Coult. & Rose. *Cogswellia P.* Jones.] Short-caulescent, 2–3.5 dm. high, glabrous, from long slender taproot; lf.-blades oblong in outline, 5–10 cm. long, ternate then bipinnate, the segms. linear to oblong, 3–7 mm. long; petioles 3–6 cm. long, wholly sheathing; bractlets dimidiate, linear-lanceolate, acute, ± connate; rays 10–25, unequal, 1–7 cm. long; pedicels 3–8 mm. long; fls. yellow or purplish; fr. oblong to oblong-ovate, mostly apically acute, 9–13 mm. long, glabrous, the wings narrower than the body; oil-tubes mostly 1 in intervals and 4–8 on commissure.—Sandy slopes and flats, 3000–5000 ft.; Sagebrush Scrub, N. Juniper Wd., Montane Coniferous F.; Sierra Nevada from Sierra to Shasta cos.; w. Nev. May–June.

Var. **Sónnei** (Coult. & Rose) Jeps. [*L. S.* Coult. & Rose.] Pubescent, the pedicels to

10 mm. long; fr. glabrous.—Sagebrush Scrub, Yellow Pine F.; e. Sierra Co.; w. Nev.
Var. **Austìniae** (Coult. & Rose) Math. [*Peucedanum A.* Coult. & Rose.] Plant pubescent; fr. ovate, not pointed apically, ca. 8 mm. long.—At 5000–7000 ft.; Yellow Pine F.; Plumas Co.

21. **L. MacDoùgallii** Coult. & Rose. [*Cogswellia M.* Jones. *L. Jonesii* Coult. & Rose.] Acaulescent, 1–3 dm. tall, short- and stiff-villous, from a long slender taproot; lf.-blades ternate, then tripinnate, 3–13 cm. long, the segms. crowded, linear to ovate, 1–5 mm. long; petioles 1–2.5 cm. long, shorter than blades, usually purplish, wholly sheathing; bractlets linear, usually distinct; rays 2–14, spreading, 1–6 cm. long; pedicels 3–10 mm. long; fls. yellow with some purplish tinge; fr. ovate to roundish, 6–11 mm. long, pubescent, the wings narrower than body; oil-tubes 1–4 in intervals, 4–6 on commissure.—At 6000–9000 ft.; Sagebrush Scrub, Pinyon-Juniper Wd.; White-Inyo Mts., Inyo Co. to Mono Co.; to Wyo., Ariz. April–June.

22. **L. inyoénse** Math. & Const. Acaulescent, cespitose, 2–10 cm. tall, hirtellous-puberulent, from a long slender taproot; lf.-blades narrow-ovate to oblong, 1.5–5.5 cm. long, bipinnate into crowded segms. 1–2 mm. long; petioles 1–2.5 cm. long, purplish, sheathing below; bractlets purplish, linear, acute; umbels appearing simple, usually only 1 ray developed, this up to 1.5 cm. long; fls. many, whitish; pedicels 1–5 mm. long; sepals evident in young frs.; fr. densely puberulent, ovate to ovate-oblong, 4–8 mm. long, the wings ca. ⅓ as wide as body; oil-tubes 3–7 in intervals, 7–8 on commissure.—Dry slopes and flats, 10,000–10,500 ft.; Bristle-cone Pine F.; Inyo Mts. June–July.

23. **L. dasycárpum** (T. & G.) Coult. & Rose. [*Peucedanum d.* T. & G. *Cogswellia d.* Jones. *P. Jaredii* Eastw. *L. d.* vars. *decorum* & *medium* Jeps.] Acaulescent or short-caulescent, 1–4 dm. tall, villous-tomentose to subglabrous, purplish, especially below, from a long taproot; lvs. quadripinnate, occasionally ternate, oblong to ovate in outline, the blades 3–12 cm. long, segms. crowded, linear, 1–3 mm. long; petioles 3–10 cm. long; bractlets linear-lanceolate, sometimes connate; rays 10–20, spreading, 1–8 cm. long; fls. greenish, appearing white because of the pubescence on petals, sometimes purplish; fr. orbicular to ovate-oblong, 8–15 mm. long, tomentulose to subglabrous, wings broader than body; oil-tubes 1–4 in intervals, 2–4 on commissure; n = 11 (Bell & Constance, 1957).—Common on dry ridges, below 5000 ft.; many Plant Communities; cismontane Calif. from San Diego to Humboldt cos.; n. L. Calif. March–June.

24. **L. mohavénse** (Coult. & Rose) Coult. & Rose. [*Peucedanum m.* Coult. & Rose *P. argense* Jones.] Acaulescent, 1–3 dm. tall, short hoary-pubescent; lf.-blades oblong in outline, 2–9 cm. long, 3–4-pinnate, the divisions crowded, linear 2–5 mm. long; petioles 3–12 cm. long, short-sheathing below; bracts linear; rays 10–16, subequal, 1–4 cm. long; pedicels 1–10 mm. long; fls. purple, rarely yellow; fr. ovate to roundish, 5–9 mm. long, pubescent, the wings narrower than or equal to the body; oil-tubes 1–4 in intervals, 4–6 on commissure.—Fairly abundant on dry plains, 2000–6000 ft.; Creosote Bush Scrub to Pinyon-Juniper Wd.; w. edge of Colo. Desert, Mojave Desert to Mt. Pinos, Independence and Death V.; w. Nev. April–May.

25. **L. macrocárpum** (H. & A.) Coult. & Rose. [*Ferula m.* H. & A. *L. m.* var. *Douglasii* Jeps.] Short-caulescent, 1–5 dm. tall, densely tomentose to villous or glabrate, purplish especially below; lf.-blades oblong to obovate in outline, 2.5–12 cm. long, ternate, then 2–3-pinnate, the segms. linear to oblong, 1–7 mm. long; petioles 2–7 cm. long, sheathing ca. to middle, subscarious; bractlets linear-lanceolate, equal to or exceeding fls.; rays 5–25, spreading, 1–8 cm. long; pedicels 1–14 mm. long; fls. many, white, yellow or purplish; fr. narrow-oblong, 1–2 cm. long, glabrous or glabrate in maturity, the wings narrower than the body; oil-tubes 1 (2–3) in intervals, 2–6 on commissure; n = 11 (Bell & Constance, 1957).—Dry stony places, below 8000 ft.; Red Fir F., Yellow Pine F., Sagebrush Scrub, N. Oak Wd.; from Kern and San Luis Obispo cos. n.; to B.C. April–June.

Var. **ellípticum** (T. & G.) Jeps. [*Peucedanum nudicaule* var. *e.* T. & G.] Pedicels often longer, up to 16 mm. long; fr. oblong-oval, 16–18 mm. long, glabrous, the wings twice as broad as body.—Foothills of n. Sierra Nevada.

26. **L. tomentòsum** (Benth.) Coult. & Rose. [*Peucedanum t.* Benth. *Cogswellia t.* Jones.] Short-caulescent, villous-tomentose, 3–5 dm. tall; lf.-blades pinnately decompound or ternate, then quadripinnate into crowded filiform segms. 2–6 mm. long; rays

12–20, subequal, 3–8 cm. long; bractlets lanceolate or broader, entire or cleft above; pedicels 5–20 mm. long; fls. greenish-white or purplish; fr. ovate-oblong, 16–22 mm. long, tomentulose; the wings ca. as broad as body; oil-tubes 1–3 in intervals, 3 on commissure.—Mostly rocky places, below 4000 ft.; V. Grassland, Foothill Wd.; from Tehachapi Mts. along base of Sierra Nevada to Butte Co. March–May.

27. **L. Cóngdoni** Coult. & Rose. [*Cogswellia C.* Jones.] Acaulescent or short-caulescent, 2–3.5 dm. high; lf.-blades broadly oblong in outline, 6.5–15 cm. long, ternate to quinate, then 2–3-pinnate, with scaberulous rachises and linear segms. 3–10 mm. long; petioles 2–6 cm. long, white-scarious, wholly sheathing; bractlets 0; rays 6–16, ascending, 3–13 cm. long; pedicels 6–15 mm. long; fls. white; fr. oblong to subovate, 7–10 mm. long, the wings half as wide as body; oil-tubes obscure, usually 1 in interval, 2–4 on commissure; $n = 11$ (Bell & Constance, 1957).—Rocky places, 1500–2500 ft.; Foothill Wd.; foothills of Sierra Nevada in Mariposa and Tuolumne cos. April–May.

28. **L. Tórreyi** (Coult. & Rose) Coult. & Rose. [*Peucedanum T.* Coult. & Rose.] Acaulescent or caulescent, 1–2.5 dm. high, glabrous to ± scaberulous; lf.-blades oblong in outline, 2–15 cm. long, ternate, then tripinnate, the segms. filiform, 3–8 mm. long; petioles 2–5 cm. long, sheathing throughout, the sheath with white scarious margin; involucel wanting; rays 5–9, erect, 1–4 cm. long, unequal; pedicels 1–4 mm. long; fls. yellow; fr. narrow-oblong, narrowed toward base, 10–16 mm. long, the wings less than half as broad as the body; oil-tubes solitary in intervals; $n = 11$ (Bell & Constance, 1957).—Granite rocks and slopes, 6000–11,000 ft.; upper Montane Coniferous F.; Sierra Nevada from Tulare to Tuolumne and Mariposa cos. May–Aug.

29. **L. Engelmánnii** Math. Acaulescent, 1–3 dm. high, pubescent to glabrate; lf.-blades oblong to ovate in outline, 3–15 cm. long, ternate or quinate, then 1–2-pinnate, the segms. linear-lanceolate to oblong-ovate, 1–15 mm. long; petioles purplish, wholly sheathing, 2–10 cm. long; rays 2–12, unequal, 1–10 cm. long; pedicels 2–12 mm. long; fls. yellow, few; fr. ovate-oblong, 9–14 mm. long, glabrous, the wings half as wide as body, the dorsal ribs filiform; oil-tubes 1–2 in intervals, 2–6 on commissure.—Gravelly slopes, 3000–6000 ft.; Yellow Pine F., Red Fir F.; n. Trinity Co., Siskiyou Co.; to sw. Ore. June–Aug.

30. **L. disséctum** (Nutt.) Math. & Const. [*Leptotaenia d.* Nutt. ex T. & G. *Ferula dissoluta* Wats.] Mostly caulescent, 8–14 dm. tall, puberulent to almost glabrous, from a stout caudex and thickened root; lf.-blades deltoid-roundish in outline, 1.5–3.5 dm. long, ternate, then 2–4-pinnate, the segms. linear-oblong, 2–8 mm. long, puberulent beneath; petioles 5–25 cm. long, broadly sheathing at base; cauline lvs. few, smaller; rays many, 3–13 cm. long, subequal; bractlets few, entire, linear; pedicels 1–3 mm. long; fls. purple or yellow; fr. oblong-oval, 12–16 mm. long, glabrous, the wings much narrower than the body, thick and corky; oil-tubes obscure.—Rocky places, below 3500 ft.; N. Oak Wd., Mixed Evergreen F.; Coast Ranges, Mendocino to Siskiyou cos.; to Wash., Ida. May–July.

Var. **multífidum** (Nutt.) Math. & Const. [*Leptotaenia multifida* Nutt. ex T. & G.] Lf.-segms. 2–22 mm. long; pedicels 4–20 mm. long.—Occasional on rocky slopes, mostly 2000–9500 ft.; Pinyon-Juniper Wd., Yellow Pine F., Red Fir F., Lodgepole F.; n. face of San Bernardino and San Gabriel mts., Tehachapi Mts., San Rafael Mts., Sierra Nevada to Siskiyou and Modoc cos.; to B.C., Alta., Colo., Ariz.

31. **L. Párryi** (Wats.) Macbr. [*Peucedanum P.* Wats.] Acaulescent, glabrous, 2–4 dm. tall, from a long stout taproot; lf.-blades narrow-oblong in outline, 1–2 dm. long, 2–3-pinnate, the segms. linear, 2–9 mm. long; petioles 6–10 cm. long, sheathing below; bractlets several, linear, acute, subscarious; rays ca. 15, suberect, 2–4 cm. long; pedicels 10–17 mm. long; fls. yellow, ca. 10 in no.; fr. oblong, 9–12 mm. long, the wings ca. as broad as body; oil-tubes 2–3 in intervals, 4 on commissure.—Occasional on rocky slopes, 4000–8500 ft.; mostly Pinyon-Juniper Wd.; Providence, New York, Kingston, Clark, Panamint mts., etc., of Mojave Desert; to Utah. May–June.

32. **L. Martindàlei** Coult. & Rose. Short-caulescent, 1–2 dm. high, purplish; lf.-blades ± oblong in outline, 2.5–5 cm. long, 1–2-pinnate, minutely roughened to glabrous on margins and veins beneath, the segms. oblong to cuneate, 5–10 mm. long, lobed to serrate; petioles 2.5–5 cm. long, sheathing below; bractlets few, filiform; rays 4–7, ascending, 1.5–5 cm. long; pedicels 7–10 mm. long; fls. whitish; fr. oblong, 12–16 mm. long, the wings ca. as wide as body; oil-tubes solitary in intervals, 2 on commissure.—

Infrequent on dry flats, 3500–4500 ft.; Yellow Pine F.; Siskiyou Mts., Del Norte Co.; sw. Ore. May–June.

33. **L. triternàtum** (Pursh) Coult. & Rose. [*Seseli t.* Pursh. *Cogswellia t.* Jones.] Caulescent or acaulescent, 1.7–8 dm. high, puberulent to subglabrous; lf.-blades broadly obovate in outline, 7–15 cm. long, ternate or quinate, then 1–2-pinnate into few narrow mostly linear segms. 1.5–12 cm. long; petioles 7–20 cm. long, purplish below, sheathing to ca. the middle; bractlets filiform, several; rays 10–20, unequal, 1.2–5.5 cm. long; pedicels 3–5 mm. long; fls. yellow, many; fr. oblong, 9–13 mm. long, glabrous, the wings narrower than body; oil-tubes 1 in intervals, 2 on commissure; $n = 11$ (Bell & Constance, 1957).—Stony ground, below 5000 ft.; Sagebrush Scrub, N. Juniper Wd., Yellow Pine F.; Trinity to Siskiyou, Modoc, and Placer cos.; to w. Wash., Alta., Wyo. April–July.

Var. **anómalum** (Jones) Math. [*L. a.* Jones ex Coult. & Rose. *L. giganteum* Coult. & Rose? *L. nudicaule* var. *puberulum* Jeps.?] Caulescent; lf.-segms. lance-ovate; pedicels 2–8 mm. long; ovaries glabrous; fr. 13–22 mm. long.—Nw. Calif.; to Ore., Ida.

Var. **macrocárpum** (Coult. & Rose) Math. [*Peucedanum t.* var. *m.* Coult. & Rose. *P. t.* var. *alatum* Coult. & Rose.] Lf.-blades 12–22 cm. long, ternate or quinate, then biternate to biquinate into linear to lance-ovate segms. 1.5–14 cm. long; rays 5–18, unequal, 1–10 cm. long; ovaries puberulent; fr. 8–20 mm. long.—N. Calif.; to B.C., Alta., Nev.

34. **L. califórnicum** (Nutt.) Math. & Const. [*Leptotaenia ? c.* Nutt. ex T. & G. *L. c.* var. *platycarpa* Jeps. and var. *dilatata* Jeps.] Caulescent, 3–12 dm. high, glabrous, glaucous; lf.-blades ovate to deltoid in outline, 1–3 dm. long, ternate or biternate, the segms. cuneate to obovate, usually 3-cleft and coarsely toothed, 2–5 cm. long; petioles 0.5–2.5 dm. long, sheathing at base; bractlets linear, scarious; rays many, ± subequal, 3–8 cm. long; pedicels 4–12 mm. long; fls. many, yellow; fr. oblong-oval, 10–15 mm. long, glabrous, the wings ± corky, narrower than body; oil-tubes 3–4 in intervals, 6–10 on commissure, or obscure; $n = 11$ (Bell & Constance, 1957).—Wooded or brushy slopes, below 5500 ft.; Foothill Wd., Yellow Pine F., Chaparral, etc.; Ventura and Kern cos. n. through Coast Ranges to Ore. April–June.

35. **L. nudicaúle** (Pursh) Coult. & Rose. [*Smyrnium n.* Pursh. *Peucedanum n.* Nutt. *P. robustum* Jeps. *P. leiocarpum* Jeps.] Acaulescent, or with 1 stem lf., 2.5–7 dm. high, glabrous; lf.-blades broadly ovate in outline, 5–18 cm. long, 1–2-ternate, then pinnate into distinct, lanceolate to broadly ovate, entire to toothed or lobed divisions 1.5–9 cm. long; petioles 4–25 cm. long, sheathing to above the middle; bractlets wanting; rays 10–20, ± swollen at apex, 1–20 cm. long; pedicels 3–15 mm. long; fls. yellow; fr. oblong, 10–14 mm. long, the wings narrower than body; oil-tubes 1–several in intervals, 4–7 on commissure.—Open slopes and plains, below 7000 ft.; Sagebrush Scrub, Yellow Pine F., Chaparral; from Ventura Co. n. through Coast Ranges, n. Sierra Nevada to B.C., Alta., Utah. April–June.

34. **Herácleum** L. Cow-Parsnip

Erect, tall, stout, biennial or perennial, from taproots or fascicled fibrous roots. Lvs. petioled, ternately or pinnately compound, the lfts. broad, serrate to lobed. Petioles sheathing, mostly dilated. Infl. of loose compound umbels on terminal and axillary umbels. Invol. usually none. Involucel of many narrow entire bractlets. Rays many. Fls. usually white; sepals obsolete; petals oval to obcordate, the outer of the marginal fls. radiant and often 2-cleft. Stylopodium conical. Fr. orbicular to obovate or elliptic, strongly flattened dorsally; dorsal ribs filiform, the lateral broadly thin-winged. Oil-tubes large, solitary in intervals, 2–4 on commissure, extending only part way to the base of the fr. Ca. 60 spp., circumboreal. (Named for *Hercules*, who was supposed to have first used it for medicine.)

1. **H. lanàtum** Michx. [*H. Douglasii* DC. *Spondylium l.* Greene.] Perennial, 1–3 m. high, tomentose; lf.-blades round to reniform in outline, 2–5 dm. long and broad, ternately compound, the lfts. ovate to roundish, 1.5–4 dm. long, cordate, coarsely serrate and ± lobed; petioles 1–4 dm. long; peduncles 0.5–2 dm. long; involucral bracts 5–10, deciduous, lanceolate-acuminate, 0.5–2 cm. long; bractlets similar; rays 15–30; pedicels 8–20 mm. long; petals white; fr. obovate to obcordate, 8–12 mm. long, ± pubescent; $n = 11$ (Bell & Constance, 1957).—Moist, ± shaded places, below 9000 ft.; many Plant Com-

munities; San Jacinto and San Bernardino mts., Sierra Nevada, Coast Ranges from Monterey Co. n.; to Alaska, Atlantic Coast. April–July.

35. Angélica L.

Stout or slender mostly erect perennials, caulescent, from stout taproots. Lvs. ternate-pinnately or pinnately compound, the lfts. broad, distinct, serrate, toothed or lobed. Petioles sheathing, the cauline sheaths often inflated and bladeless. Infl. of loose compound umbels. Invol. usually none; involucel of many entire bractlets or none. Sepals minute or obsolete. Fls. white, pink or purplish. Fr. strongly flattened dorsally, the carpels with filiform to narrowly or corky-winged dorsal ribs, the lateral broadly thin- or corky-winged; stylopodium low-conical; oil-tubes few to many. Ca. 50 spp., circumboreal. (Latin, *angelica*, angelic, because of medicinal properties.)

Ribs of fr. thick and corky, broader than the intervals; oil-tubes many, adhering to the seed which is free in the pericarp when mature. N. Calif. . . . . 1. *A. lucida*
Ribs of fr. thin; oil-tubes few, the seed adhering to the pericarp.
  Lf.-divisions linear to linear-oblong, less than 1 cm. broad. Sierra Nevada, Panamint Mts., Sweet-
  water Mts. . . . . 2. *A. linearíloba*
  Lf.-divisions lanceolate to oval, broader.
    Ovaries glabrous; pedicels conspicuously webbed at base. N. Calif. . . . . 8. *A. arguta*
    Ovaries pubescent or roughened.
      Petals glabrous; fr. roundish, 3–4 mm. long . . . . 7. *A. genuflexa*
      Petals pubescent or dorsally scabrous; fr. mostly narrower in proportion and larger.
        Rays 7–14; fr. 4–5 mm. long, 2–3 mm. broad; lvs. oblong. E. Calif. . . . . 3. *A. Kingii*
        Rays 25–45; fr. 7–14 mm. long, 4–9 mm. broad; lvs. ovate to deltoid.
          Lvs. white-tomentose beneath. Coastal bluffs . . . . 4. *A. Hendersonii*
          Lvs. scaberulous to villous, but not tomentose.
            The lvs. glaucous, villous with some hairs forked. S. Calif. and Coast Ranges
                                                    5. *A. tomentosa*
            The lvs. green, subglabrous to somewhat villous. Sierra Nevada . . . . 6. *A. Breweri*

1. **A. lùcida** L. [*Archangelica Gmelinii* DC.] Stout, 6–12 dm. tall, the foliage quite glabrous, infl. villous; lf.-blades ovate to deltoid in outline, 1–3 dm. long, 1–3-ternate, the lfts. ovate to ovate-lanceolate, 3–15 cm. long, 2–10 cm. broad; petioles 1–6 dm. long; bractlets many, narrow, villous, 5–15 mm. long; rays 20–45, subequal, 3–10 cm. long; petals white, oval, glabrous; ovaries glabrous; fr. oblong-oval, 4–9 mm. long, 2–5 mm. broad, the ribs ca. equally narrowly winged, corky-thickened; oil-tubes small, continuous about the seed.—High banks and bluffs, marshes along the coast; N. Coastal Scrub, Coastal Strand, etc.; Humboldt Co. to Alaska, Atlantic Coast. May–Aug.

2. **A. linearilòba** Gray. [*A. l.* var. *Culbertsonii* Jeps.] Stout, subglabrous to scabrous, 5–15 dm. tall; lf.-blades ternate-pinnately decompound, 1–3.5 dm. long, the segms. linear to linear-oblong, entire, 0.2–0.8 cm. broad; petioles 0.5–2.5 dm. long; involucel wanting; rays 20–40, subequal, 3–7 cm. long; pedicels 3–10 mm. long; fls. white or pinkish; ovaries glabrous to scabrous; fr. oblong to cuneate, 10–13 mm. long, 5–7 mm. broad, the dorsal ribs narrowly winged, the lateral broader, ca. equal to the body; oil-tubes 1–2 in intervals, 4 on commissure.—Moist places and gravelly or talus slopes, 6000–10,600 ft.; Red Fir F. to Subalpine F., Sagebrush Scrub; Sierra Nevada from Mono and Mariposa to Tulare and Inyo cos., Nelson, White, Sweetwater, and Panamint mts. June–Aug.

3. **A. Kíngii** (Wats.) Coult. & Rose. [*Selinum K.* Wats.] Stout, 3–9 dm. tall, glabrous to scaberulous; lf.-blades oblong in outline, 1.5–4 dm. long, ternate-pinnate, the lfts. lanceolate to lance-ovate, entire to ± serrate, 4–15 cm. long, mostly 1–3.5 cm. broad; petioles 1.5–3 dm. long; involucel wanting; rays 7–14, unequal, 1–10 cm. long; pedicels 1–6 mm. long, webbed; fls. white, the petals pubescent on back; ovaries hispid; fr. oblong, 4–5 mm. long, 2–3 mm. broad, the dorsal ribs narrowly winged, the lateral much narrower than body; oil-tubes 1–2 in intervals, 2 on commissure.—Damp banks, 7000–9500 ft.; Pinyon-Juniper Wd.; White Mts., e. Calif. to Nev. and Ida. June–Aug.

4. **A. Hendersònii** Coult. & Rose. Stout, 3–8 dm. tall, the lvs. green above, white-tomentose beneath, infl. tomentose; lf.-blades ovate to deltoid in outline, 1–2.5 dm. long, ternate-pinnately divided, the lfts. oval to lance-ovate, 4–8(–15) cm. long, 2.5–6 cm. broad, serrate; petioles 1–2 dm. long; involucel of several linear tomentose bractlets 5–7 mm. long; rays 30–45, subequal, 2–6 cm. long; pedicels 1–8 mm. long; petals

white, tomentose on back; ovaries tomentose; fr. oval, 7–10 mm. long, 6–9 mm. broad, the dorsal ribs scarcely winged, the lateral broader and ca. equal in width to the body; oil-tubes 1 in intervals, 4 on commissure.—Coastal bluffs and flats; largely N. Coastal Scrub, Coastal Strand; Monterey Co. to Del Norte Co.; to s. Wash. June–July.

5. **A. tomentòsa** Wats. [*A. californica* Jeps. *A. t.* var. *c.* Jeps. *A. t.* var. *elata* Jeps.] Stout, 6–18 dm. tall, the lvs. glaucous beneath and stiff-villous with occasional forked hairs; infl. villous; lf.-blades deltoid in outline, 1.5–4.5 dm. long, ternate-pinnately divided, the lfts. oval or oblong to lanceolate, 3–15 cm. long, 1–8 cm. broad; petioles 2–3 dm. long; involucel of several narrow villous bractlets 2–5 mm. long; rays 25–40, unequal, 3–12 cm. long; pedicels 2–12 mm. long; petals white, villous on back; ovaries densely villous; fr. oblong-oval, 8–10 mm. long, 6–7 mm. broad, the dorsal ribs narrowly winged, the lateral ca. as wide as the body; oil-tubes 1 in intervals, 4 on commissure.— Rather moist and ± shaded places, below 7000 ft.; Chaparral, Yellow Pine F., in mts. of s. Calif., and Coast Ranges from Santa Cruz Co. to Humboldt and Siskiyou cos., where it may be also in Mixed Evergreen F., Redwood F., etc. June–Aug.

6. **A. Brèweri** Gray. Stout, 9–12 dm. tall, subglabrous to ± villous; lf.-blades deltoid in outline, 1.5–3.5 dm. long, ternately to ternate-pinnately divided, the lfts. lanceolate, 4–12 cm. long, 0.8–3 cm. wide; petioles 2–3 dm. long; involucel-bractlets several, linear, villous, 4–6 mm. long; rays 25–40, unequal, 3–8 cm. long, webbed; pedicels 8–12 mm. long; petals white, villous on back; ovaries densely villous; fr. oblong to oval, 8–12 mm. long, 5–7 mm. broad, the dorsal ribs narrowly winged, the lateral ca. as broad as body; oil-tubes 1 in intervals, 2 on commissure; $n = 33$ (Bell & Constance, 1957).— Open wooded slopes, 3000–8600 ft.; Yellow Pine F. to Lodgepole F.; Sierra Nevada from Inyo to Shasta cos.; w. Nev. June–Aug.

7. **A. genufléxa** Nutt. ex T. & G. Stout, 4–18 dm. tall, glabrous to scaberulous, the infl. pilose to hispidulous; lf.-blades ovate to deltoid in outline, 1.5–3 dm. long, ternate-pinnate to bipinnate, the main divisions often reflexed, the lfts. broadly to narrowly ovate, 4–10 cm. long, 2–6 cm. broad; petioles 1–6 dm. long; involucel-bractlets many, narrow, 5–10 mm. long, hispidulous; rays 22–45, unequal, 2–7 cm. long; pedicels 5–15 mm. long; petals white to pinkish, glabrous; ovaries hispidulous; fr. roundish, 3–4 mm. long, the dorsal ribs filiform to narrowly winged, the lateral ca. as wide as body; oil-tubes 1 in intervals, 2 on commissure.—Stream banks, swampy places, etc., below 3100 ft.; Freshwater Marsh, Mixed Evergreen F., Yellow Pine F., etc., Humboldt and Del Norte cos.; to Alaska; Siberia. July–Aug.

8. **A. argùta** Nutt. ex T. & G. [*A. Lyallii* Wats.] Stout, 5–20 dm. tall, glabrous to scaberulous; lf.-blades ternate-pinnate or bipinnate, ovate to roundish in outline, 1.5–3 dm. long, the lfts. ovate to lanceolate, 5–15 cm. long, 2–5 cm. broad; petioles 1–3 dm. long; involucel wanting or of few filiform bractlets; rays 18–45, subequal, 1–8 cm. long, webbed; pedicels 2–10 mm. long, conspicuously webbed at base; petals white or pinkish, glabrous; ovaries glabrous; fr. oval to roundish or obovate, 4–7 mm. long, 4–5 mm. broad, the dorsal ribs narrowly winged, the lateral ca. as wide as body; oil-tubes 1 in intervals, several on commissure.—Moist or boggy places, below 7000 ft.; Mixed Evergreen F., Yellow Pine F., Red Fir F.; Del Norte, Humboldt, Shasta, and Siskiyou cos.; to B.C., Alta., Wyo. July–Aug.

36. Conioselìnum Hoffm.

Leafy caulescent perennials, mostly branched, glabrous except for the puberulent infl., from a taproot or cluster of fleshy roots. Lvs. ternate-pinnately decompound, the lfts. lobed or dissected. Petioles sheathing. Infl. of loose compound umbels. Invol. present or wanting. Involucel of narrow, scarious or scarious-margined, entire bractlets. Rays rather many, spreading-ascending. Pedicels slender. Fls. white; sepals obsolete; stylopodium conic, styles short. Fr. dorsally flattened, oblong-oval to oval, with prominent dorsal ribs sometimes winged, and lateral ribs broadly winged; oil-tubes 1–2 in intervals, 2–4 on commissure. Several poorly differentiated spp. of boreal and n. temp. regions. (Compounded from *Conium* and *Selinum*, umbelliferous genera which it resembles.)

1. **C. chinénse** (L.) BSP. [*Athamanta c.* L. *Selinum pacificum* Wats. *C. Gmelenii* Coult. & Rose.] Stout, 3–15 dm. high; lf.-blades ovate to deltoid in outline, 0.6–2 dm. long, the lfts. lanceolate to ovate, 1–4 cm. long; petioles 5–15 cm. long; invol. of

several narrow, entire or toothed, leafy bracts 0.5–2 cm. long; involucel-bractlets many, narrow, 5–15 mm. long, scarious-margined; rays 13–30, subequal, 1.5–4.5 cm. long; pedicels 5–8 mm. long; fr. oblong-oval to oval, 4–6 mm. long, 2–3.5 mm. broad, the dorsal ribs acute, the lateral broadly winged.—Ocean bluffs, cold marshes, etc., below 1500 ft.; N. Coastal Scrub, etc.; Mendocino Co. n.; to Alaska, Siberia, also on Atlantic Coast. June–Aug.

37. Sphenosciàdium Gray

Tall subsimple stout thick-rooted perennials, glabrous below the tomentose infl. Lvs. 1–2-pinnate or ternate-pinnate. Petioles inflated, sheathing. Infl. of loose compound umbels. Invol. wanting. Involucel of many linear-setaceous tomentose bractlets. Rays many, rather long. Fls. white or purplish, sessile in capitate umbellets. Stylopodium small, conical. Fr. strongly flattened dorsally, cuneate-obovate, ribbed at base, winged above. Oil-tubes 1 in intervals, 2 on commissure. One sp. (Greek, *sphenos,* a wedge, and *sciadios,* an umbrella, referring to the umbel.)

1. **S. capitellàtum** Gray. [*Selinum eryngifolium* Greene. *S. validum* Congd. *Sphenosciadium c.* vars. *scabrum, e.* and *v.* Jeps.] Ranger's Button. White Heads. Scabrous, 5–16 dm. high; lf.-blades oblong to ovate in outline, 1–4 dm. long, the lfts. linear-oblong to lance-ovate, 1–12 cm. long, 0.5–5 cm. broad, paler and scabrous beneath; petioles 1–4 dm. long; rays 4–18, subequal, 1.5–9 cm. long, tomentose; fr. 5–8 mm. long, 3–5 mm. broad, tomentose, the dorsal ribs prominent or narrowly winged, the lateral more broadly winged.—Swampy places, 3000–10,400 ft.; Yellow Pine F. to Subalpine F., San Jacinto and San Bernardino mts., Mt. Pinos, Tehachapi Mts., Sierra Nevada, White and Sweetwater mts., Glenn to Mendocino, Siskiyou, and Modoc cos.; to e. Ore., Ida., w. Nev., n. L. Calif. July–Aug.

38. Glèhnia F. Schmidt ex Miq.

Low, ± fleshy maritime herbs from a stout taproot. Lvs. 1–2-ternate or ternate-pinnate, the lfts. oblong-obovate or cuneate. Petioles sheathing. Infl. of subcompact compound umbels. Invol. none or of few bracts. Involucel-bractlets several, conspicuous, lanceolate. Rays few to many, spreading. Umbels capitate. Fls. white; sepals inconspicuous; stylopodium lacking; styles short. Fr. ovoid-oblong to roundish, flattened dorsally; lateral and dorsal wings conspicuous, thickened at base; oil-tubes many, large. Two spp., w. N. Am. and e. Asia. (Possibly in honor of P. von *Glehn,* botanist of last century.)

1. **G. leiocárpa** Math. [*Cymopterus? littoralis* Gray, not *G. l.* F. Schmidt.] Subacaulescent, prostrate or spreading, the sheathing petioles ± buried in the sand; lf.-blades broadly ovate in outline, 3–15 cm. long, the lfts. 1–5 cm. long, 0.5–3 cm. broad, hirtellous on veins above, mostly tomentose beneath; petioles 3–12 cm. long, hirtellous; infl. villous; umbel globose to spreading, the rays 5–13, and ca. 1–4 cm. long; fr. 4–12 mm. long, subglabrous, the lateral wings sometimes broader than the dorsal.—Coastal Strand; Mendocino Co. to Alaska. May–June.

39. Pterýxia Nutt. ex Coult. & Rose

Low cespitose perennials from a deep-seated root, caulescent or acaulescent. Lvs. bipinnate to pinnately or ternate-pinnately decompound, with ultimate divisions linear, oblong or subcuneate. Petioles sheathing. Infl. of loose to compact compound umbels. Invol. usually wanting. Involucel of narrow herbaceous bractlets. Fls. yellow to white or purple. Sepals evident. Fr. narrowly oblong to ovoid, flattened dorsally; lateral ribs winged, thin, some or all of the dorsal winged. Stylopodium none. Oil-tubes 1 to several in the intervals, several on the commissure. Five spp., w. N. Am. (Greek, *pteris,* fern, and *ixia,* the chameleon plant.)

Rays 7–24; fr. 5–10 mm. long; lvs. ovate to broadly oblong .... 1. *P. terebinthina*
Rays 3–7; fr. 4.5–7 mm. long; lvs. narrow-oblong in outline .... 2. *P. petraea*

1. **P. terebinthìna** (Hook.) Coult. & Rose var. **califórnica** (Coult. & Rose) Math. [*P. c.* Coult. & Rose.] Plants short-stemmed, 1–5 dm. high; lf.-blades gray-green, 3–15

cm. long, pinnately or ternate-pinnately decompound, the ultimate divisions not rigid, linear to subcuneate, 1–6 mm. long, ca. 1 mm. broad, ± confluent; petioles 2–15 cm. long; involucel dimidiate, the bractlets linear or broader, 2–6 mm. long; rays unequal, 1–8 cm. long; fls. yellow, on pedicels 1–8 mm. long; fr. 5–10 mm. long, ca. as broad, the wings plane, thickish, the dorsal usually equaling the lateral; oil-tubes 3–12 in intervals, 6–20 on commissure.—Dry brushy or rocky slopes, below 10,600 ft.; Mixed Evergreen F., Yellow Pine F. to Subalpine F.; Sweetwater Mts., Sierra Nevada from Tulare Co. n., N. Coast Ranges from Sonoma Co. to Siskiyou Co.; w. Nev. May–June.

2. **P. petraèa** (Jones) Coult. & Rose. [*Cymopterus p.* Jones.] Stems slender, 1.5–4 dm. tall; lf.-blades pale green, narrow-oblong in outline, 3–15 cm. long, ternate-bipinnate to tripinnate, the ultimate divisions linear, 1–8 mm. long, distinct; petioles 5–12 cm. long; involucel dimidiate, the bractlets linear, 1–3 mm. long; rays 3–7, unequal, to 5 cm. long; pedicels 1–6 mm. long; fls. yellow; fr. ovoid to ± oblong, 4.5–7 mm. long, 2–4 mm. broad, the wings plane, narrower than to as broad as the body, 1–3 of the dorsal ribs winged; oil-tubes 1–3 in intervals, 5–15 on commissure.—Rocky places, 5000–7000 ft., Pinyon-Juniper Wd., Inyo-White Mts.; and 8000–11,250 ft., Subalpine F. to Alpine Fell-fields, Sierra Nevada; to e. Ore., Ida., Nev. May–June.

40. Cymópterus Raf.

Perennial, from long slender, thickened or fusiform taproots, the stems mostly subterranean (pseudoscapes) bearing the tuft of lvs. and peduncles at the surface of the ground. Lvs. variously lobed, divided or compound, petioled, thin to subcoriaceous. Umbels congested and globose, or spreading; invol. present or absent; involucels usually present, with conspicuous, herbaceous to ± hyaline bractlets. Fls. yellow, purple or white; sepals small or obsolete; stylopodium wanting. Fr. ovoid to oblong, somewhat flattened dorsally, the ribs conspicuously winged or the dorsal sometimes wingless, the wings thin or thick and corky toward the outer edge; oil-tubes small, 1–many in intervals, 2–many on commissure. Ca. 30 spp. of w. N. Am. (Greek, *kuma*, wave, and *pteron*, wing, some spp. having undulate wings.)

(Mathias, M. Studies in Umbelliferae III. A monograph of Cymopterus and a critical study of related genera. Ann. Mo. Bot. Gard. 17: 213–476, 1930.)

Lvs. hirtellous to scaberulous.
  Umbels congested and globose; invol. conspicuous .... 1. *C. cinerarius*
  Umbels not globose, the rays evident; invol. wanting or of a few linear bracts .. 6. *C. aboriginum*
Lvs. glabrous.
  Umbels congested and globose, the rays obsolete.
    Fr. pubescent; fls. purple. San Bernardino Co. .... 2. *C. deserticola*
    Fr. glabrous; fls. white or purple. Mono Co. .... 3. *C. globosus*
  Umbels not globose, the rays evident.
    Involucral bracts mostly wanting, not scarious; bractlets of involucel inconspicuous or if conspicuous not hyaline.
      Lvs. simply ternate or pinnate, the divisions deltoid, confluent; fls. purplish .. 4. *C. Gilmanii*
      Lvs. ternate, then 2–3-pinnate, the divisions linear, distinct; fls. greenish-yellow
                          5. *C. panamintensis*
    Involucral bracts evident to conspicuous, scarious; bractlets conspicuous, hyaline.
      Bractlets white or whitish, few-nerved; pedicels 3–12 mm. long ...... 7. *C. purpurascens*
      Bractlets purple or greenish-white, many-nerved; pedicels less than 1 mm. long
                        8. *C. multinervatus*

1. **C. cineràrius** Gray. Acaulescent, the lvs. arising directly from the root-crown, to 7 or 8 cm. high; lf.-blades oblong-ovate in outline, 1.5–2.5 cm. long, cinereous-hirtellous, 2-pinnate; petioles 3–5 cm. long; umbel small, discoid; bracts united below the middle, triangular-lanceolate, scarious-margined; fls. white; fr. narrowly cuneate, 6 mm. long, glabrous, the wings a little constricted at the base, subacute at apex, the dorsal and lateral similar.—Dry open places, 9000–11,500 ft.; Bristle-cone Pine F., Lodgepole F., Subalpine F.; Sonora Pass region, Sierra Nevada, Sweetwater Mts., Mono Craters, Inyo-White Mts., w. Nev. June–July.

2. **C. deserticola** Bdg. Acaulescent, the lvs. arising from the root-crown, 1–1.5 dm. high, glabrous; lf.-blades broadly oblong-ovate, 2–6 cm. long, glaucous, glabrous, ternate-bipinnate, the ultimate divisions 1–4 mm. long; petioles ca. as long as blade; umbel globose, compact; invol. none; bractlets paleaceous, mostly aborted; fls. purple; fr. pubescent, 5–7 mm. long, oblong-ovoid to cuneate, the lateral wings narrower than the body,

the dorsal absent or reduced.—Rare, sandy or gravelly plains, 2500–3100 ft.; Creosote Bush Scrub, Joshua Tree Wd.; Mojave Desert from e. of Victorville to Kramer and Muroc. April.

3. **C. globòsus** (Wats.) Wats. [*C. montanus* var. *g.* Wats.] Lvs. and peduncles on long slender pseudoscape from the deep-seated root, glabrous and glaucous, the lf.-blades ternate-bipinnate or bipinnate, broadly ovate in outline, 2–7 cm. long, the lfts. incised or lobed, ultimate divisions 0.5–5 mm. long, apiculate; umbels globose, 2–3 cm. in diam.; invol. none; bractlets scarious, linear; sepals purplish; petals whitish; fr. glabrous, narrowly cuneate, 6–11 mm. long, the lateral wings broadest at summit, dorsal wings mostly 3, like the lateral.—Dry open flats, 4000–7000 ft.; Sagebrush Scrub; Benton, Mono Co.; to w. Utah. March–May.

4. **C. Gílmani** Mort. Subacaulescent, glabrous, 1–2 dm. high; lf.-blades round-reniform in outline, 2.5–4.5 cm. long, ternate, with deltoid confluent laciniately lobed lfts.; rays ca. 8, 1–2 cm. long; invol. none; bractlets several, distinct, linear-subulate, exceeding the purplish fls.; fr. broadly oval, 7–8 mm. long, the wings broader than the body, narrowed or broadened at the base.—Dry rocky slopes, 3500–6500 ft.; Creosote Bush Scrub; Inyo Co. from Last Chance Mts. to Death V. region; w. Nev. April–May.

5. **C. panaminténsis** Coult. & Rose. [*Aulospermum p.* Coult. & Rose.] Acaulescent, glabrous, 0.5–4 dm. high; lf.-blades broadly ovate-oblong, 3–12 cm. long, ternate, then 2–3-pinnate, the divisions linear, spinulose, distinct, 1–5 mm. long; rays 5–15, 1–6 cm. long; invol. wanting; bractlets several, ± united, linear-attenuate; fls. greenish-yellow; fr. oblong-ovoid, 6–10 mm. long, the wings equaling or exceeding the body, thin, enlarged at base.—Dry rocky places, 3000–8300 ft.; Creosote Bush Scrub to Pinyon-Juniper Wd.; Sweetwater Mts. to Funeral, Panamint, and Argus mts. March–May.

Var. **acutifòlius** (Coult. & Rose) Munz. [*Aulospermum p.* var. *a.* Coult. & Rose.] Lf.-divisions more remote, 3–20 mm. long, not spinulose.—At 2000–4700 ft.; Creosote Bush Scrub; Mojave Desert from Sheephole Mts. and Kelso, to Barstow, Newberry Mts., Eagle Mts. (Colo. Desert). March–April.

6. **C. aboríginum** Jones. [*Aulospermum a.* Math.] Acaulescent, the lvs. and peduncles from a short rootstock covered with persistent lf.-bases, 1–3.5 dm. high; lf.-blades oblong in outline, 3–10 cm. long, ternate-bipinnate or tripinnate, gray-green, hirtellous, ultimate divisions crowded, linear, acute, 2–8 mm. long; petioles 3–12 cm. long; invol. wanting or of few linear bracts; involucel-bractlets several, linear, slightly scarious; rays 3–10, spreading, 4–35 mm. long; fls. white; fr. ovoid to oblong, 6–11 mm. long, 5–8 mm. broad, the wings linear, ca. twice as wide as the body.—Dry rocky places, 4000–9300 ft.; largely Pinyon-Juniper Wd., Joshua Tree Wd.; Inyo-White Range, to Panamint Mts., Grapevine Mts.; Charleston Mts., Nev. April–June.

7. **C. purpuráscens** (Gray) Jones. [*C. montanus* var. *p.* Gray. *Phellopterus p.* Coult. & Rose.] Acaulescent or subacaulescent, 3–15 cm. high, from a slender taproot crowned with persistent lf.-bases; lf.-blades glabrous, ovate-oblong in outline, 1.5–5 cm. long, bipinnate or pinnate, pale, the ultimate lobes rounded to acute, 1–8 mm. long, confluent; petioles 1–4 cm. long; bracts of invol. white, mostly connate below the middle; bractlets white with 1–5 green or white nerves; umbels globose, compact; fertile rays 3–5, only 4–10 mm. long; fls. purplish; fr. usually broadly ovoid, 8–18 mm. long, 8–16 mm. broad, the wings thin, 2–3 times as broad as body.—Rocky limestone hills, 4500–7000 ft.; largely Pinyon-Juniper Wd.; Cottonwood Mts. (Death V. region), New York Mts., Clark Mt., e. Mojave Desert; to Ida., Utah, Ariz. March–May.

8. **C. multinervàtus** (Coult. & Rose) Tides. [*Phellopterus m.* Coult. & Rose.] Acaulescent or with a pseudoscape, glabrous, 0.5–2 dm. high; lf.-blades ovate-oblong in outline, 1–9 cm. long, pinnate, bipinnate, or ternate-pinnate, pallid, the lfts. entire or lobed, the lobes acute to obtusish, 1–6 mm. long; petioles 2–7 cm. long; invol. a low scarious sheath or of 1–2 conspicuous nerved bracts or a purplish connate lobed cup; bractlets conspicuous, ± ovate, many-nerved, subconnate; umbels compact, the fertile rays mostly 1–5, glabrous, 5–25 mm. long; fls. purplish; fr. ovoid to ovoid-oblong, 8–17 mm. long and broad, often purplish, the wings long, slender, 2–3 times as broad as body.—Dry slopes and plains, 3500–6000 ft.; Joshua Tree Wd., Pinyon-Juniper Wd.; Cushenbury Springs, n. base of San Bernardino Mts., New York Mts., e. Mojave Desert; to Utah, w. Tex., Son. March–April.

41. Erýngium L.

Biennials or perennials, usually glabrous, from taproots or fibrous clustered roots. Lvs. often prickly, entire to lobed or divided, the blades somewhat obsolete; petioles sheathing, sometimes septate. Infl. capitate, the heads solitary, racemose or cymose; invol. of entire or lobed bracts subtending the head. Bractlets of involucel entire or lobed. Fls. white, blue, or purple. Sepals evident, entire to spinescent. Rays and pedicels lacking. Stylopodium lacking. Fr. globose to ovoid, slightly flattened laterally if at all, variously covered with thin scales or with tubercles, the ribs obsolete. Oil-tubes none or obscure. Ca. 200 spp., of temp. and trop. regions. (Greek name used by Dioscorides.)

(Mathias, M. E., and L. Constance. A synopsis of the N. Am. spp. of Eryngium. Am. Midl. Nat. 25: 361–387, 1941.)

Heads blue or bluish.
  Petioles of basal lvs. septate, much lcnger than the blades; bracts and bractlets not callous-margined.
    Lower San Joaquin V. to Siskiyou and Modoc cos. .... 1. *E. articulatum*
  Petioles of basal lvs. not septate, not longer than the blades; bracts and bractlets callous-margined.
    Coast from San Luis Obispo Co. n. .... 2. *E. armatum*
Heads greenish.
  Bracts and bractlets ± callous-margined.
    Plants diffusely branched from base; lvs. serrate or incised. Coastal .... 2. *E. armatum*
    Plants branched above; lvs. twice pinnately parted. Sierra Nevada ...... 3. *E. pinnatisectum*
  Bracts and bractlets not callous-margined.
    Bractlets mostly scarious-margined at the base; fr. with subequal scales.
      Lvs. nearly entire to serrate, incised or pinnatifid, the teeth or lobes entire to spinose.
        Lf.-blades 3–15 cm. long, 1–3 cm. broad; infl. cymose. Ne. Calif. .... 4. *E. alismaefolium*
        Lf.-blades 2–3 cm. long, 0.4–0.6 cm. broad; infl. falsely racemose. San Joaquin Delta
                                    5. *E. racemosum*
      Lvs. deeply pinnatifid, the lobes mostly remote, spinulose-lobed or pinnatifid. Salinas and Central valleys .... 7. *E. Vaseyi*
    Bractlets scarious-lobed at base; fr. with unequal scales. Coast Ranges, Sierra Nevada foothills, and San Diego Co. .... 6. *E. aristulatum*

1. **E. articulàtum** Hook. [*E. a.* var. *Bakeri* Jeps. *E. Harknessii* Curran.] Erect, stout, 3–10 dm. tall; basal lvs. lanceolate to ovate, the blades 4–9 cm. long, entire or spinose with petioles 1–3(–6) dm. long; cauline lvs. similar, sessile, often laciniate at base; infl. cymose, the heads many, pedunculate, bright blue, ovoid, 1–2 cm. long; bracts 10–15, rigid, reflexed, linear-lanceolate, 1–2 cm. long, spinose-ciliate and scarious-dilated at base; bractlets tricuspidate at tip; sepals usually entire, linear-lanceolate, 3–5 mm. long; petals 1.5–2 mm. long; fr. ovoid, 2–3 mm. long, covered with white appressed lanceolate scales to ca. 1 mm. long; $n = 16$ (Bell & Constance, 1957).—Wet places below 400 ft.; Coastal Salt Marsh, Freshwater Marsh; lower San Joaquin V. to Trinity, Siskiyou, and Modoc cos.; to Ida. May–Sept.

2. **E. armàtum** (Wats.) Coult. & Rose. [*E. petiolatum* var. *a.* Wats. *E. longistylum* Coult. & Rose. *E. Harmsianum* H. Wolff.] Stems prostrate or ascending, 0.5–4 dm. long, diffuse; basal lf.-blades oblanceolate, 0.5–3 dm. long, 0.5–3 cm. broad, remotely serrate to coarsely spinose-incised; petioles short, broad; cauline lvs. reduced; heads many, in a cyme, short-pedunculate, yellowish, many-fld.; bracts 8–10, rigid, lanceolate, 10–20 mm. long, mostly entire; bractlets 5–10 mm. long, scarious-winged at base, mostly entire; sepals ovate-lanceolate, 1–2 mm. long, acuminate, entire; petals 0.5–1 mm. long, oblong to oblanceolate; fr. ovoid, 1.5–3 mm. long, covered with flat white or brown scales, the upper larger; $n = 16$ (Bell & Constance, 1957).—Vernal pools, etc., open fields and flats, below 500 ft.; N. Coastal Scrub, Coastal Strand, Coastal Prairie, Mixed Evergreen F., etc.; near coast from Santa Barbara Co. to Humboldt Co. May–Aug.

3. **E. pinnatiséctum** Jeps. Stout, erect, branched above, 1–4 dm. high; basal lvs. lanceolate, 1–3 dm. long, pinnatifid to the midrib, the lobes oblong to lance-linear, callous-margined, entire or with few spinose teeth; petioles short and broad; cauline lvs. similar, but nearly sessile; infl. cymose, the heads globose, pedunculate, 8–15 mm. long; bracts 8–12, rigid, lance-linear, 1–3 cm. long, entire or with few spines near base; bractlets similar, 5–8 mm. long; sepals lanceolate, 3–4 mm. long, entire; petals oblanceolate, 0.8–1 mm. long; fr. ovoid, 3 mm. long, covered with appressed white lanceolate scales 1–1.5 mm. long.—Summer beds of winter pools, 1000–3000 ft.; Foothill Wd.,

Yellow Pine F.; Sierran foothills from Amador to Sacramento and Tuolumne cos. June–Aug.

4. **E. alismaefòlium** Greene. [*E. petiolatum* var. *minimum* Coult. & Rose. *E. m.* Coult. & Rose. *E. articulatum* var. *microcephalum* Coult. & Rose.] Spreading, caulescent, 0.5–3 dm. high, the stems many; basal lvs. lanceolate to ovate, 3–15 cm. long, 1–3 cm. broad, coarsely spinose-serrate, incised or pinnatifid, with short broad petioles; cauline lvs. reduced; infl. cymose, heads many, short-pedunculate, 5–10 mm. long; bracts few, rigid, 6–15 mm. long, pungent, often spinose-ciliate; bractlets 5–8 mm. long, densely spinose to entire, broadly scarious-winged at base; sepals ovate-lanceolate, 1–3 mm. long, pungent; petals oblong, 1.5 mm. long; fr. ovoid, ca. 2 mm. long, covered with flat white subequal scales 0.5–1 mm. long.—Seasonally moist places, 4000–6100 ft.; N. Juniper Wd., Yellow Pine F., Red Fir F.; Nevada to Modoc cos. then to e. Ore. July–Aug.

5. **E. racemòsum** Jeps. Stems slender, decumbent or prostrate, 1–3 dm. long; basal lvs. lanceolate, 2–5 cm. long, 4–8 mm. broad, subentire to spinulose-serrate; petioles septate, slender, 1–4 dm. long; cauline lvs. similar but sessile and reduced above; infl. falsely racemose, the heads many, pedunculate, rather few-fld., 4–8 mm. in diam.; bracts ca. 8, linear, 8–10 mm. long, spinose-serrate near base; bractlets similar, 5–10 mm. long; sepals ovate, 1–1.5 mm. long, scarious-margined; petals oblanceolate, 1 mm. long; fr. ovoid, 1.5 mm. long, densely covered with short white or tawny lanceolate subequal scales 0.5 mm. long.—Low wet places below 100 ft.; Freshwater Marsh, etc.; San Joaquin Delta. Aug.–Oct.

6. **E. aristulàtum** Jeps. [*E. Jepsoni* Coult. & Rose. *E. oblanceolatum* Coult. & Rose. *E. californicum* Jeps. *E. laxibracteum* Math.] Stems stout or slender, erect to prostrate, 1–8 dm. long; basal lvs. oblanceolate to obovate, 3.5–25 cm. long, 0.5–2.5 cm. broad, spinulose-serrate to incised or lobed, or blades obsolete; petioles slender, 5–25 cm. long; cauline lvs. reduced upward; infl. cymose, the heads few to many, 5–12 mm. in diam.; bracts rigid, spreading, linear to ± lanceolate, 5–25 mm. long, spinose on margins, ± scarious-winged at base; bractlets 5–15 mm. long, with 1–3 pairs of lateral spines and sometimes dorsal ones, or spineless, the scarious-winged margin at the base enfolding the fr.; sepals lanceolate, 1.5–3.5 mm. long; petals oblanceolate, 1.5 mm. long; fr. ovoid, 1.5–2.5 mm. long, covered with lanceolate scales, the upper to 1 mm. long, the lower much shorter.—Salt marshes and summer flats of vernal pools, below 3000 ft.; Coastal Salt Marsh to Mixed Evergreen F., Foothill Wd., Yellow Pine F., etc.; Coast Ranges, Humboldt to Ventura and San Benito cos., Sierran foothills of Plumas Co. May–Aug.

Var. **Párishii** (Coult. & Rose) Math. & Const. [*E. P.* Coult. & Rose. *E. Jepsonii* var. *P.* Jeps.] Sepals ovate, usually puberulent; fr. with subequal scales.—Vernal pools, Chaparral; w. San Diego Co.; n. L. Calif.

7. **E. Vàseyi** Coult. & Rose. Rather stout, 1.5–4 dm. high, erect or ascending, branching above; basal lvs. oblong-lanceolate to ovate, 1–2.5 dm. long, deeply pinnatifid, the unequal segms. spinulose-lobed or again pinnatifid; petioles dilated, 1–4 cm. long; cauline lvs. similar, reduced upward; infl. corymbose, the heads many, 5–10 mm. in diam.; bracts ca. 8, rigid, linear-subulate, 5–15(–25) mm. long, densely spinose with 1–5 pairs of lateral spines, but no dorsal ones; bractlets similar, 5–15 mm. long; sepals lanceolate to ovate, 1–3 mm. long, scarious-margined; petals oblong, 1–1.5 mm. long; fr. ovoid, 2–3 mm. long, covered with subequal white lanceolate scales 0.5–1 mm. long.—Summer beds of vernal pools, below 800 ft.; V. Grassland; Salinas and San Joaquin valleys. June–Aug.

Var. **vallícola** (Jeps.) Munz. [*E. castrense* var. *vallicolum* Jeps. *E. castrense* Jeps. *E. V.* var. *c.* Hoov. *E. globosum* var. *medium* Jeps.] Bracts and bractlets usually with dorsal spines; sepals spinose, usually not pinnatifid; $n = 16$ (Bell & Constance, 1957).—Vernal pools, below 1500 ft.; V. Grassland, Foothill Wd.; plains and Sierran foothills from San Joaquin and Mariposa cos. to Butte and Shasta cos.

Var. **globòsum** (Jeps.) Hoov. ex Math. & Const. [*E. g.* Jeps. *E. spinosepalum* Math.] Bracts 1–3 cm. long, with few dorsal spines; sepals pinnatifid with 3–8 spiny teeth.—Vernal pools, below 1200 ft.; V. Grassland; base of Sierra Nevada from Tulare to San Joaquin cos.

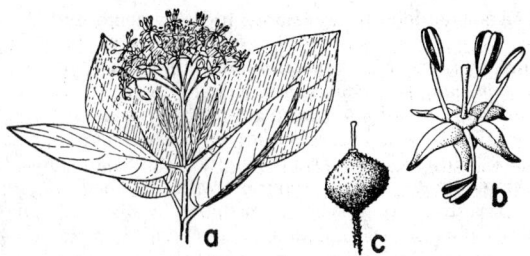

Fig. 100. CORNACEAE. *Cornus occidentalis:* *a,* twig, × ½, lvs. opposite, infl. cymose; *b,* fl., × 2½, petals 4, stamens 4, style from inferior ovary; *c,* drupe, × 2.

111. Cornàceae. Dogwood Family (Fig. 100)

Trees, shrubs, or herbs, usually with opposite, whorled or alternate entire lvs. and perfect or imperfect fls. in capitate or cymose infl. Sepals 4–5, minute. Petals 4–5, rarely more, inserted at base of epigynous disc. Stamens as many as petals or more. Ovary inferior, 1–2-loculed; style 1; stigma terminal. Ovule solitary in each locule, pendulous. Fr. a drupe; stone 1–2-celled; endosperm present. With 16 genera and 80 spp., mostly of N. Hemis.

1. Córnus L. Dogwood

Fls. small, 4-merous. Ca. 25 spp., of n. temp. regions and s. to Peru. (Latin, *cornu,* horn, because of the hard wood.)

(Rickett, H. W. Cornaceae. N. Am. Fl. 28B: 299–316, 1945.)
A. Fls. in a bractless cyme, appearing with or after the lvs.
    B. Veins 3–4 on each side of midrib; lf.-blades 3–5 cm. long . . . . 1. *C. glabrata*
    BB. Veins 4–7 on each side of midrib; lf.-blades mostly 5–10 cm. long.
        C. Style less than 2.5 mm. long; petals 2–3 mm. long; cyme and lower surfaces of lvs.
            variously pubescent but not hirsute; endocarp smooth, usually at least as long as broad
                                                                          2. *C. stolonifera*
        CC. Style 2.5 mm. long or more; petals 3–4 mm. long; cyme and lower surface of lvs.
            densely hirsute; endocarp ridged, mostly broader than long . . . . . . 3. *C. occidentalis*
AA. Fls. in a head or umbel subtended by persistent bracts, usually appearing before the lvs.
    B. Stems woody; drupes ellipsoid.
        C. Bracts of invol. 1–1.5 cm. long, scarcely petaloid; infl. umbelliform; drupes on slender
            hairy pedicels . . . . 4. *C. sessilis*
        CC. Bracts of invol. 5 cm. or longer, petaloid; infl. a head; drupes sessile . . 5. *C. Nuttallii*
    BB. Stems herbaceous from woody rhizomes; drupes globose . . . . 6. *C. canadensis*

1. **C. glabràta** Benth. [*Svida g.* Heller. *S. catalinensis* Millsp.] Brown Dogwood. Shrubs to small trees, 1.5–6 m. tall, often forming thickets by means of underground shoots, the twigs nearly or quite glabrous, brownish to reddish-purple, slender; lvs. lanceolate to elliptic, acute at tip, cuneate at base, almost glabrous, gray-green, slightly paler beneath, the blades mostly 2–5 cm. long, 1.5–2.5 cm. wide, on petioles 3–8 mm. long; infl. 2.5–4.5 cm. across, ± strigulose; pedicels 2–3 mm. long; sepals 0.8 mm. long; petals 4.5–5 mm. long; style 3.5 mm. long; drupes white to bluish, subglobose, 9 mm. in diam., the stone 5–6 mm. broad, almost smooth.—Moist places, below 5000 ft.; many Plant Communities; cismontane Calif. from San Diego Co. n. to Ore., uncommon in s. May–June.

2. **C. stolonífera** Michx. [*Svida s.* Rydb. *C. alba* var. *s.* Wanger. *C. sericea* ssp. *s.* Fosb.] American Dogwood. Spreading shrubs 2–5 m. tall, often rooting at tips of branches; twigs strigulose, bright red-purple; lf.-blades commonly 5–9 cm. long, 1.5–5 cm. broad, lanceolate to elliptic or ovate, acute to acuminate, subglabrous above, strigulose beneath, villous-tufted in axils of veins, minutely papillose; petioles 5–8(–12) mm. long; infl. 3–6 cm. across, rather flat, strigulose to hirtellose; pedicels mostly 1–5 mm. long; sepals 0.5 mm. long; petals 2–3 mm. long; styles ca. 2 mm. long; drupes white, 7–9 mm. in diam., subglobose, the stone smooth on the faces, furrowed laterally, 4–5 mm.

broad.—Moist places below 9000 ft., mostly in Montane Coniferous F.; Sierra Nevada to Siskiyou, Del Norte, and Trinity cos.; to Alaska, Nfld., N.Y., Mex. May–July.

3. **C. occidentàlis** (T. & G.) Cov. [*C. sericea* var.? *o.* T. & G. *C. pubescens* Nutt., not Willd.] Much like *C. stolonifera*, but the twigs ± hirsute; lf.-blades mostly 6–10 cm. long, 3–5 cm. broad, sparsely strigulose above, paler and densely hirsute beneath; infl. ± hirsute; sepals 0.7 mm. long; petals 3–4.5 mm. long; style 2.5–3 mm. long; drupes white, ca. 8 mm. in diam., the stone with 3 often broad and low ridges on each face and furrowed laterally.—Moist places, below 8000 ft.; many Plant Communities; San Diego Co. through cismontane Calif., n. to B.C. May–July.

C. × califórnica C. A. Mey. [*C. pubescens* var. *c.* Coult. & Evans. *C. Torreyi* Wats. *C. c.* var. *nevadensis* Jeps.] A series of segregates from hybrids of *C. stolonifera* and *C. occidentalis* and showing most intermediate stages between the two.

4. **C. séssilis** Torr. ex Durand. Shrub or small tree, 1.5–4 m. tall, with twigs glabrous, pale; lf.-blades commonly 4.5–9 cm. long, 2–3.5 cm. broad, elliptic, acute or short-acuminate, cuneate at base, glabrous above or nearly so, strigulose beneath and ± tomentose at vein axils; petioles 5–10 mm. long; infl. subtended by 2 pairs of deciduous bracts 1 cm. long, these brown, often with yellow edges; pedicels ca. 1 cm. long, white-villous; sepals 0.5 mm. long; petals 3 mm. long; style 1 mm. long; drupes at first whitish, turning yellow and red to very dark, 1–1.5 cm. long, ellipsoid, acute at ends.—Stream banks, 500–5000 ft.; Redwood F. to Yellow Pine F.; Sierra Nevada from Calaveras Co. n., Coast Ranges from Tehama, Trinity, and Humboldt cos. to Siskiyou Co. March–April.

5. **C. Nuttállii** Aud. [*Cynoxylon N.* Shafer. *Benthamidia N.* Mold.] Mountain Dogwood. Arborescent bush or tree, 4–25 m. tall; twigs at first green, later dark red to almost black, strigulose; lf.-blades commonly 6–12 cm. long, 3–7 cm. broad, elliptic to obovate, cuneate at base, short-acuminate, minutely strigulose above, paler beneath and pubescent with appressed and looser hairs; petioles 5–10 mm. long; infl. appearing in autumn, subtended by 2 lvs. and 2 bracts that persist until spring; at anthesis the head of fls. subtended by 4–7 white or ± pinkish petaloid bracts 4–6 cm. long, 3–6 cm. broad, also by smaller bracts; calyx 2.5 mm. high; petals 4 mm. long; style 2 mm. long; drupes red, 1–1.5 cm. long, ± ellipsoid (often angled by mutual pressure), the stone smooth.— Mountain woods, below 6000 ft.; Mixed Evergreen F., Yellow Pine F., etc.; occasional in mts. from San Diego Co. to Los Angeles Co., more common in Coast Ranges from Santa Lucia Mts. n., Sierra Nevada from Tulare Co. n.; to B.C. and Ida. April–July.

6. **C. canadénsis** L. [*Cornella c.* Rydb.] Bunchberry. Rootstocks ± horizontal; flowering stems mostly 0.7–2 dm. high, each usually with 4–6 apparently whorled lvs. near the summit, these 2.5–7 cm. long, oval to obovate, glabrous or strigose; peduncle 1, slender, 2–4 cm. long; invol. of 4 ovate bracts 0.8–1.6 cm. long, white; infl. a head; sepals 0.4 mm. long; petals 1.5 mm. long, yellowish or ± purplish; style 1.5–2 mm. long; drupes globose, red, ca. 8 mm. in diam., the stone smooth.—Moist places, below 3500 ft.; Redwood F. to Yellow Pine F.; Mendocino Co. n.; to Alaska, Atlantic Coast, ne. Asia. May–July.

112. Garryàceae. Silk-Tassel Family (Fig. 101)

Evergreen dioecious shrubs or small trees with opposite simple rather leathery short-petioled lvs. Fls. small, apetalous, imperfect, borne in pendulous catkinlike clusters. Staminate fls. pedicelled, in 3's in axil of each of the decussately connate bracts; calyx 4-parted, the tips often connate; stamens 4; fils. distinct. Pistillate fls. subsessile, borne 1 in each bract-axil; calyx with 2 lobes or obsolete; ovary inferior, 1-celled; styles 2, persistent, stigmatic on inner side. Fr. a berry, the bitter pulp surrounding the 1–2 seeds, dark purple to black, enclosed in and soon free from the dry brittle epicarp. Seeds with thin testa and horny endosperm; embryo minute.

1. Gárrya Dougl. Silk-Tassel Bush

One genus, with the characters of the family. Ca. 15 spp. of w. N. Am., many of which intergrade freely where they come in contact with each other. (Named for N. *Garry* of the Hudson Bay Company, friend of David Douglas.)

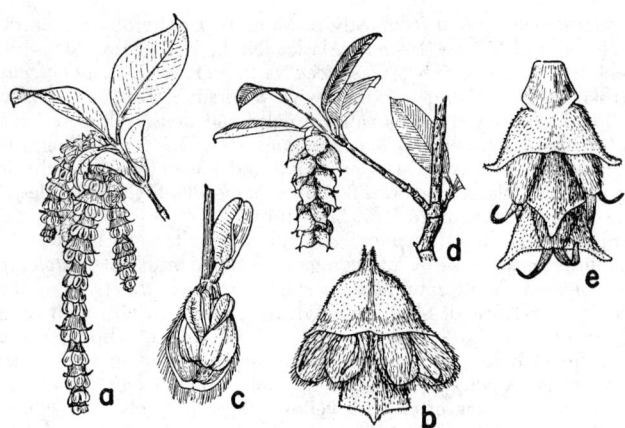

Fig. 101. GARRYACEAE. *Garrya Congdonii:* *a*, ♂ infls., × ½, the fls. in 3's in axil of each of connate bracts; *b*, ♂ fl., × 2; *c*, ♂ fl., × 4, calyx 4-parted, stamens 4; *d*, ♀ infl., × ½; *e*, ♀ fls., × 2, each single in bract-axil, with 2 styles.

Lower surface of lvs. glabrous or nearly so; lvs. plane; fr. glabrous or pubescent when young
　　　　　　　　　　　　　　　　　　　　　　　　　　　　　　　　1. *G. Fremontii*
Lower surface of lvs. usually covered with hairs (glabrate in age in no. 3.)
　　Hairs of lower lf.-surface straight, upwardly appressed.
　　　　Lvs. glossy and bright green above or olive-green; fr. glabrate. N. Coast Ranges
　　　　　　　　　　　　　　　　　　　　　　　　　　　　　　　2. *G. buxifolia*
　　　　Lvs. gray-green or yellowish-gray above; fr. densely strigose. From Fresno and Alameda cos. s.
　　　　　　　　　　　　　　　　　　　　　　　　　　　　　　　3. *G. flavescens*
　　Hairs of lower lf.-surface curly or wavy.
　　　　Tomentum of under surface of lvs. of very short curled hairs and forming a dense felt.
　　　　　　Lvs. oval or broadly elliptical, rounded or obtuse at apex, strongly undulate on margins.
　　　　　　Coast Ranges from Ventura Co. n. .... 4. *G. elliptica*
　　　　　　Lvs. lanceolate to ovate, ± acuminate, plane or nearly so. From San Luis Obispo Co. s.
　　　　　　　　　　　　　　　　　　　　　　　　　　　　　　　5. *G. Veatchii*
　　　　Tomentum of under surface of lvs. of long wavy hairs which are mostly ascending. Sierran foot-
　　　　hills and inner Coast Ranges .... 6. *G. Congdoni*

1. **G. Fremóntii** Torr. [*G. F.* var. *laxa* Eastw. *G. rigida* Eastw.] Erect shrub 1.5–3 m. tall, the young twigs strigose, soon glabrate and red-brown; lvs. oblong-elliptical to -ovate, 2–5 cm. long, coriaceous, usually tapered at ends, glabrous and shining above and yellow-green in age, paler and glabrous to sparingly pilose beneath, plane on margins; petioles to ca. 1 cm.; ♂ catkins solitary or clustered, simple, lax, 7–20 cm. long yellowish; ♀ 4–5 cm. long or more in fr.; fr. ca. 6 mm. in diam., globose, sub-glabrous, almost black to purplish.—Dry brushy slopes, mostly below 7500 ft.; Chaparral, Mixed Evergreen F., Yellow Pine F., Red Fir F.; Sierra Nevada n. to Modoc Co., Coast Ranges s. to Monterey Co., mts. of Orange and San Diego cos.; to Wash. Jan.–April.

2. **G. buxifòlia** Gray. [*G. flavescens* var. *b.* Jeps.] Low, 0.5–2(–3) m. tall, ± strigose on young twigs; lf.-blades oblong-elliptic to roundish, 1–4 cm. long, glossy and ± glabrous above, silvery-gray beneath with dense appressed pubescence; petioles 3–6 mm. long; ♂ catkins 5–7 cm. long, in clusters of 2–4; fruiting 3–9 cm. long; fr. sub-glabrous, bluish-black, 4–6 mm. in diam.—Rocky slopes at 1500–4600 ft.; Chaparral, Mixed Evergreen F., Yellow Pine F.; Mendocino to Del Norte and Siskiyou cos.; sw. Ore. Feb.–April.

3. **G. flavéscens** Wats. var. **pállida** (Eastw.) Bacig. ex Ewan. [*G. p.* Eastw.] Erect, 1.5–3.5 m. tall, grayish; young twigs cinereous-strigose above, densely so beneath; petioles 3–10 mm. long; ♂ catkins 3–4 cm. long; fruiting catkins compact, densely silky, 3–5 cm. long; fr. broadly ovoid, 6–8 mm. broad, densely silky.—Dry slopes, 3000–8000 ft.; Chaparral, Yellow Pine F., Pinyon-Juniper Wd.; Coast Ranges from San Diego Co. to Alameda Co., Sierra Nevada of Tulare, Fresno and Inyo cos., mts. of e. Mojave Desert where it approaches *G. flavescens* of the area to the east in its more yellowish less glaucous appearance. Feb.–April.

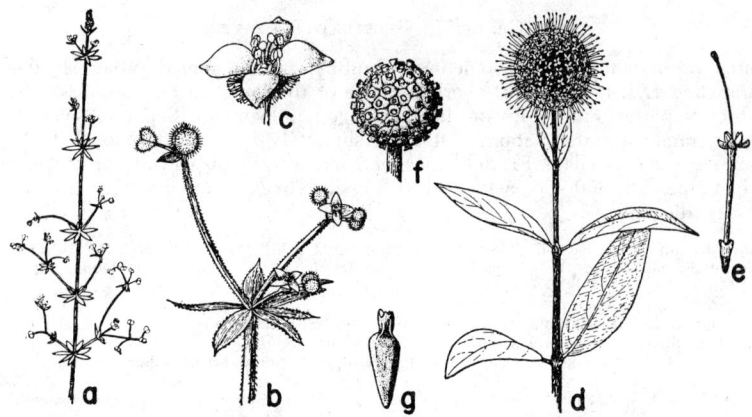

Fig. 102. RUBIACEAE. *Galium Aparine: a,* habit, × ¼; *b,* whorl of lvs., × 2½, retrorse hairs on stem-angles; *c,* fl., × 14, sepals obsolete, corolla 4-parted, stamens 4, syles 2, ovary inferior. *Cephalanthus occidentalis* var. *californicus: d,* twig, × ½, with head of fls.; *e,* fl., × 2 to show inferior ovary, 4 sepals, tubular corolla, exserted style; *f,* head in fr., × 1; *g,* single akenelike fr., × 2½.

4. **G. elliptica** Dougl. Shrub or small tree to 8 m. high, the young twigs densely short-villous; lf.-blades elliptic to oval, 6–8(–10) cm. long, green and subglabrous above, densely felty-woolly beneath with short ± curly hairs, mostly strongly crisped-undulate on margins; petioles 6–12 mm. long; ♂ catkins largely 8–15 cm. long; fruiting catkins 8–15 cm. long; fr. globose, 7–11 mm. in diam., white-tomentose, but ± glabrate in age.—Dry slopes and ridges, below 2000 ft.; N. Coastal Scrub, Chaparral, Mixed Evergreen F., etc.; outer Coast Ranges from Ventura Co. to Ore., Santa Cruz Id. Jan.–March.

5. **G. Veàtchii** Kell. [*G. V.* vars. *Palmeri* and *undulata* Eastw.] Shrub 1–2 m. tall, the young twigs densely white-tomentose; lf.-blades lanceolate to ovate or ovate-elliptic, ± acuminate, 2.5–6 cm. long, green and glabrous above, felty-tomentose beneath, plane or slightly undulate on margins; petioles 6–14 mm. long; ♂ catkins 5–10 cm. long; fruiting 2.5–6 cm. long; fr. ovoid to rounded, 7–8 mm. in diam., buff to purple-brown, ± glabrescent in age.—Dry slopes, below 7000 ft.; Chaparral, Foothill Wd.; San Luis Obispo Co. to Baja Calif. Feb.–April.

6. **G. Cóngdoni** Eastw. [*G. flavescens* var. *venosa* Jeps.] Shrub, 1–2 m. tall of yellowish-green appearance; young twigs silky-pubescent; lf.-blades narrowly oval to lance-ovate or elliptic, 2.5–6 cm. long, thinly puberulent to subglabrous above, densely hairy beneath with rather long wavy hairs which ± intertwine; petioles 4–8 mm. long; ♂ catkins 3–8 cm. long; fruiting 2–5 cm. long; fr. roundish to broadly ovoid, densely pubescent or ± glabrate at base, 6–8 mm. in diam.—Dry canyons and ridges, below 2750 ft.; probably largely on serpentine; Chaparral, Foothill Wd.; inner Coast Ranges from Tehama to San Benito cos., foothills of Sierra Nevada from Mariposa Co. n. Feb.–April.

113. **Rubiàceae.** MADDER FAMILY (Fig. 102)

Herbaceous or woody plants with opposite entire lvs. connected by interposed stipules or in whorls without apparent stipules. Fls. perfect or imperfect, mostly 4-, sometimes 3- or 5-merous. Calyx sometimes obsolete. Corolla regular, the stamens as many as the corolla-lobes and inserted on the tube. Ovary in ours 2-loculed, splitting when dry into 2 or 4 indehiscent 1-seeded carpels; styles 1 or 2. Embryo in fleshy or horny endosperm. A large family, largely trop., many of great economic importance (*Coffea, Cinchona*), others ornamental (*Gardenia, Bouvardia*).

Lvs. in whorls mostly of 4 or more; plants herbs or low shrubs.
  Fls. solitary or in cymes, pedicelled; corolla rotate .... 1. *Galium*
  Fls. in involucrate heads; corolla funnelform .... 2. *Sherardia*
Lvs. opposite, or if whorled, on large shrubs.
  Fls. in cymes; plants low perennial herbs of montane pine belt .... 3. *Kelloggia*
  Fls. in heads; plants large shrubs or small trees, below the montane pine belt .... 4. *Cephalanthus*

1. Gàlium L. Bedstraw. Cleavers

Annual or perennial herbs, sometimes shrubby, with 4-angled rather slender stems and branches. Lvs. in apparent whorls because of the large, leaflike stipules. Fls. small, perfect or imperfect, in cymes or these arranged in panicles. Sepals obsolete. Corolla rotate, 4-, rarely 3-parted. Stamens 4 or 3, short. Ovary 2-lobed, 2-loculed, 2-ovuled; styles 2; stigmas capitellate. Fr. didymous, of 2 indehiscent carpels, dry or fleshy, separating when ripe. Ca. 300 spp., widely distributed. (Greek, *gala*, milk, certain spp. being used to curdle milk.)

(Hilend, M. and J. T. Howell. The genus Galium in s. Calif. Leafl. W. Bot. 1: 145–168, 1935. Ehrendorfer F. Survey of the G. multiflorum complex in w. N. Am. Contr. Dudley Herb. 5[1]: 1–36, 1956.)

A. Lvs. 5–8 in a whorl.
  B. Ovary and fr. glabrous or at most granulate, not bristly-hairy.
    C. Stems ± retrorsely scabrous, at least when young.
      D. Lvs. blunt or rounded at tip, without terminal bristle; plants perennial
                                                                                9. *G. trifidum*
      DD. Lvs. sharply acute, ending in a bristle or mucronate at tip; plants annual.
        E. Frs. not more than 1 mm. in diam.; lf.-margins upwardly spinose-edged
                                                                                2. *G. divaricatum*
          EE. Frs. 3–4 mm. in diam.; lf.-margin retrorsely prickly-edged .. 3. *G. tricorne*
      CC. Stems and lvs. not retrorsely scabrous; fls. in loose ample almost leafless panicles;
        plants perennial .... 6. *G. Mollugo*
  BB. Ovary and fr. bristly-hairy.
    C. Fls. several on each side-branch, the whole upper part of the plant forming an open panicle; fr. ca. 1 mm. in diam.
      D. Stems slender, with filiform diffuse branches; lvs. 5–12 mm. long; plant annual
                                                                                1. *G. parisiense*
      DD. Stems coarser, mostly simple; lvs. 12–25 mm. long; plant perennial
                                                                                8. *G. asperrimum*
    CC. Fls. mostly 2–3 in axillary cymules, rarely 6; fr. larger.
      D. Plants perennial; stems not readily clinging to other vegetation; lvs. ovate-oblong to broadly obovate .... 7. *G. triflorum*
      DD. Plants annual; stems easily clinging to other vegetation; lvs. linear to linear-oblong or -oblanceolate.
        E. Fr. 4–5 mm. long; corolla 2 mm. broad, white. Common .... 4. *G. Aparine*
          EE. Fr. 1.5–3 mm. long; corolla 1 mm. broad, green. Rare .... 5. *G. spurium*
AA. Lvs. mostly 4 in a whorl, occasionally 2 or 5.
  B. Plants herbaceous, not at all woody at the base.
    C. Annuals; frs. with hooked hairs.
      D. The plants to ca. 5 cm. high, filiform-stemmed; lvs. 2–3 mm. long
                                                                                13. *G. murale*
      DD. The plants 5–20 cm. high, coarser; lvs. 5–20 mm. long.
        E. Lvs. approximately equal in the whorl; fls. almost sessile between a pair of bracted lvs.; herbage usually hispidulous .... 12. *G. proliferum*
          EE. Lvs. very unequal in the whorl; fls. on naked peduncles recurved in fr.; herbage glabrous .... 11. *G. bifolium*
    CC. Perennials; fr. glabrous.
      D. Fls. 1.5–2 mm. broad, 1–3 in upper axils or on branchlets; pedicels strongly arcuate in age .... 9. *G. trifidum*
      DD. Fls. 2.5–3.5 mm. broad, in small cymes; peduncles often curved, but pedicels straight .... 10. *G. cymosum*
  BB. Plants woody at the base, at least below the current year's growth.
    C. Mature fr. glabrous or very slightly pubescent, the hairs if present much shorter than the body of the fr. .
      D. Lvs. linear; plants forming low dense mats, the stems densely leafy.
        E. The lvs. linear-subulate, glabrous, prickly-tipped. Lake Co. to San Diego Co.
                                                                                14. *G. Andrewsii*
          EE. The lvs. narrowly linear, acute but not prickly, often pubescent.
            F. Lvs. 6–12 mm. long; fls. greenish-white. From Tehama and Lake Cos. n.
                                                                                15. *G. ambiguum*
              FF. Lvs. 3–6 mm. long; fls. yellowish. Monterey Co. .... 16. *G. Clementis*
      DD. Lvs. lance-linear or broader; plants taller, more open.
        E. Stems mostly retrorse-scabrous on the angles, woody; lvs. 2–6(–10) mm. long, oval to linear-oblong .... 22. *G. Nuttallii*
          EE. Stems not retrorse-scabrous, scarcely woody; lvs. mostly longer.
            F. Mature fr. dry; main cauline lvs mostly 15–30 mm. long.
              G. Fls. solitary or in few-fld. cymes, terminal or in uppermost axils of very leafy compacted branchlets; lvs. 1-nerved. Insular.
                H. Main lvs. ⅓–¼ as wide as long, with inconspicuous lateral veins; hairs of fr. slender, spreading, to ca. 1 mm. long. Catalina and San Clemente ids. .... 23. *G. catalinense*
                  HH. Main lvs. ½–⅓ as wide as long, with conspicuous lateral

veins; hairs of fr. coarse, appressed, less than 0.5 mm. long.
Santa Cruz and San Miguel ids. ...... 24. *G. buxifolium*
GG. Fls. many in a compact terminal panicle; lvs. 3-nerved. Extreme
n. Calif. .... 25. *G. boreale*
FF. Mature fr. fleshy; main cauline lvs. mostly 6–15 mm. long.
G. Lvs. glabrous on surfaces, ciliate on margins, oblong to lance-
oblong, 5–10 mm. long. Sierra Nevada from Mariposa Co. to
Siskiyou, Humboldt, and Solano cos. .... 19. *G. Bolanderi*
GG. Lvs. ± stiff-pubescent on surfaces.
H. Plants tufted, erect, 6–18 cm. tall; fr. quite glabrous. Widely
distributed.
I. Lvs. dull, 5–15 mm. long. Los Angeles to Humbolt cos.
17. *G. californicum*
II. Lvs. shining, 5–9 mm. long. Mendocino Co.
18. *G. muricatum*
HH. Plants diffuse, 30–60 cm. tall.
I. Lvs. 8–16 mm. long, 6–10 mm. broad, cuspidate-
pointed; fr. glabrous. From above 6000 ft., Sierra Nevada,
Tulare to Plumas cos. .... 20. *G. sparsiflorum*
II. Lvs. 6–12 mm. long, 2–6 mm. broad, mostly not cuspidate-
pointed; fr. pubescent at least when young. Mostly below
5000 ft., Sierra Nevada, San Gabriel, and San Bernardino
mts. .... 21. *G. pubens*
CC. Mature frs. densely hispid with hairs usually almost as long as the body of the fr.
D. Corolla-lobes mostly glabrous on the back; lvs. oblong-linear to subovate.
E. Hairs on fr. ascending and subappressed. At 7000 ft. or above, San Gabriel
and San Bernardino mts. .... 29. *G. Jepsonii*
EE. Hairs on fr. spreading.
F. Corolla reddish, the lobes lance-acuminate; style slender. E. Mojave
Desert .... 26. *G. Wrightii*
FF. Corolla greenish-white, the lobes mostly acute; style stout.
G. Lvs. lance-linear to linear-oblong, 5–30 mm. long; corolla 2 mm.
in diam. Largely cismontane and montane .. 27. *G. angustifolium*
GG. Lvs. broadly ovate, 2–8 mm. long.
H. Corolla 4–5 mm. in diam.; plant glabrous, 1–3 dm. tall.
Mono Co. .... 33. *G. multiflorum*
HH. Corolla 2.5–3 mm. in diam.; plant minutely cinereous-sca-
brous, 0.3–1.5 dm. tall. Inyo to Plumas cos.
37. *G. hypotrichium*
DD. Corolla-lobes mostly hairy externally; lvs. lanceolate to oblong or round-ovate.
E. Lvs. acerose-acute or -acuminate, the midveins prominent, lateral veins mostly
lacking, margin often strongly revolute. Deserts, mostly below 4500 ft.
30. *G. stellatum*
EE. Lvs. acute but not acerose-acuminate, the lateral veins usually present, sometimes as prominent as the midvein.
F. Fl.-clusters drooping; mature frs., including hairs, 4–5 mm. in diam.;
corolla-lobes silky-hairy without. Mts. about w. end of Mojave Desert
31. *G. Hallii*
FF. Fl.-clusters not drooping; frs. mostly smaller; corolla-lobes mostly
bristly-hairy.
G. Lvs. linear-oblong to oblong, 1–2 mm. wide; plants tufted, 0.5–2
dm. tall, cinerous-pubescent. Coniferous F. from the San Gabriel
to San Jacinto mts. .... 28. *G. gabrielense*
GG. Lvs. lance-ovate to roundish, mostly wider.
H. Vegetative parts of plant glabrous; lvs. remote, 6–8 mm. long,
the upper cuspidate-acute. Mts. of Inyo Co.
34. *G. Matthewsii*
HH. Vegetative parts of plant hispid-pubescent or -puberulent.
I. Lvs. of a given whorl unequal, 3–7 mm. long.
J. Stems 1–3 dm. long; fls. mostly ca. 1.5 mm. in diam.;
fr. including hairs, ca. 3 mm. in diam. San Gabriel
Mts. to Santa Rosa and Kingston mts.
35. *G. Parishii*
JJ. Stems 0.3–1(–1.5) dm. long; fls. 2–3 mm. in diam.;
fr. including hairs, 5–6 mm. in diam. Panamint Mts.
to Sierra Nevada .... 37. *G. hypotrichium*
II. Lvs. of a given whorl subequal, 5–12 mm. long.
J. Lvs. not more than 2½ times as long as broad.
K. Stems 1–3.5 dm. tall; corolla 1.5–2 mm. broad.
Mts. of Mojave Desert and e. slope of Sierra
Nevada .... 32. *G. Munzii*
KK. Stems mostly less than 1 dm. tall; corolla ca.
3.5 mm. in diam. Placer to Siskiyou Cos.
36. *G. Grayanum*
JJ. Lvs. 3–3½ times as long as broad. Modoc and Siskiyou
cos. .... 38. *G. Watsonii*

1. **G. parisiénse** L. Slender-stemmed diffusely branched annual, mostly 1.5–4 dm. high, minutely scabrous; lvs. chiefly 5–7 at a node, linear to oblanceolate, 5–12 mm. long, spinulose-ciliate, bristle-tipped; infl. a lax panicle of small several-fld. cymules, the upper bracts reduced or wanting; fls. ca. 1 mm. broad, greenish-white, glabrous; fr. 0.7–1 mm. long, hispid-pubescent; $2n = 22$, 44, 66 (Fagerlind, 1937).—Becoming rather widely distributed in waste and open places, as from Fresno, Butte, and Lake to Humboldt cos.; native of Eu. April–Aug.

2. **G. divaricàtum** Lam. [*G. parisiense* var. *d.* (Lam.) Koch. *G. anglicum* Huds.] Much like *G. parisiense*, but more tufted, with finer more capillary branches, less scabrid stems; frs. granulate, not hispid.—Natur. in grassy places, etc., in Mendocino, Lake, Sonoma, Yuba, Marin, Napa, and Eldorado cos.; native of Eu. May–July.

3. **G. tricórne** Stokes. Rather coarse annuals, 3–6 dm. high, with subsimple stems bearing stoutish recurved prickles on the angles; lvs. 6–8 at a node, 1.5–2.5 cm. long, narrowly oblanceolate, strongly recurved-prickly on margins; fls. white, mostly in 3's, glabrous; fr. tuberculate-granulate, 3–4 mm. in diam., on short recurved pedicels; $2n = 44$ (Fagerlind, 1937).—Weed in grainfields, etc., San Luis Obispo to Sonoma and Tuolumne cos.; native of Eu. April–May.

4. **G. Aparìne** L. Annual, the stems weak, reclining or scrambling over other plants, retrorsely hispid at the angles, hairy at the nodes, 1–10 dm. long, rather slender; lvs. 6–8 in a whorl, mostly linear-oblanceolate, mostly 1.5–7 cm. long, bristle-tipped, with coarse divergent or reflexed marginal setae and ± hispidulous on surfaces; fls. 2 mm. in diam., whitish, 2–5 in cymes in upper axils, the peduncles with a whorl of leaflike bracts; frs. bristly, 3–5 mm. in diam., the bristles with tuberculate bases.—Common on shaded banks to ca. 7500 ft.; many Plant Communities; cismontane Calif., Channel Ids., occasional on deserts; to Alaska, e. Coast; said to be introd. from Eu. March–July.

5. **G. spùrium** L. var. **echinospérmum** (Wallr.) Hay. [*G. Vaillantii* DC.] Like *G. Aparine* but with narrower lanceolate long-mucronate lvs. 0.5–2 cm. long; fls. greenish, 1 mm. across, in axillary cymes of 3–9 fls. with only 2 leaflike bracts; frs. 1.5–3 mm. in diam., the hairs nontuberculate at base.—Occasional in grassy and wooded places, where introd.; native of Eu. March–June. Commonly mistaken for *G. Aparine*.

6. **G. Mollùgo** L. Perennial, decumbent or ascending, with diffusely branched glabrous or pubescent weak stems 2–12 dm. long; lvs. 6–8 at a node, mostly oblanceolate, to obovate or almost linear, 1–2.5 cm. long, rough on margins with forwardly directed prickles; infl. a terminal panicle with spreading branches, almost leafless; fls. white, ca. 3 mm. in diam., the corolla-lobes cuspidate; fr. glabrous, rugulose, 1 mm. long.—Occasionally natur. as in Humboldt and San Mateo cos.; native of Eu. June–July.

7. **G. triflòrum** Michx. Sweet-Scented Bedstraw. Perennial, with weak simple or remotely forking smooth or ± scabrous herbaceous stems 2–8 dm. long; lvs. mostly in 6's, oblong-ovate to ± obovate, thin, cuspidate, 1.5–6 cm. long, the surfaces mostly glabrous, margins and midribs beneath minutely scabrous; axillary peduncles 3-fld. or 3-forked, the fls. all pedicelled; corolla greenish-white; fr. densely bristly, scarcely 2 mm. broad.—Moist shaded canyons and wooded places, below 8000 ft.; many Plant Communities; cismontane and montane Calif., uncommon in s. more frequent n.; to Alaska, Atlantic Coast; Eurasia. May–July.

8. **G. aspérrimum** Gray. [*G. a.* var. *asperulum* Gray.] Perennial from slender rootstocks; stems weak, 3–8 dm. long, freely branched above, retrorse-hispid on angles; lvs. mostly in whorls of 6 or 8, oblanceolate to broadly linear, acute to obtuse, abruptly mucronulate, 1.5–4 cm. long, hispid on margins and midrib; fls. in a diffuse terminal ± leafy infl., on capillary pedicels; corolla white, 2.5–3.5 mm. across; fr. small with very short hooked bristles, the pedicels ± clavate at summit.—Shaded places, 1500–7300 ft.; Foothill Wd., Yellow Pine F., Red Fir F.; Coast Ranges from Santa Clara Co. n., Sierra Nevada from Mariposa Co. n., to Siskiyou and Modoc cos.; to Wash., Rocky Mts. June–Aug.

9. **G. trífidum** L. var. **subbiflòrum** Wieg. [*G. t.* var. *pacificum* Wieg. *G. s.* Rydb. *G. tinctorium* var. *s.* Fern. and var. *submontanum* Wight.] Slender-stemmed perennial from slender rootstocks, ascending, branched and intertangled, 1.5–4 dm. long, the angles ± retrorse-scabrous; lvs. elliptic-oblong to linear-spatulate, obtuse, not bristle-pointed, in 4's, 5's, or 6's, thin, obscurely scabrous on margins and midribs, 0.5–2 cm. long; fls. mostly 1.5–2 mm. broad, whitish, 3–4-merous, 1, 2, or 3 on axillary peduncles, the

pedicels strongly arcuate in age, ± indistinctly scabrous, slender, mostly 5–18 mm. long; fr. glabrous, dry, each carpel 1–1.5 mm. in diam.—Wet places, below 8000 ft.; many Plant Communities; most of cismontane and montane Calif.; to Alaska, Atlantic Coast. June–Aug.

Var. **púsillum** Gray. [*G. Brandegeei* Gray.] More glabrous throughout, ± matted, the weak stems mostly 0.5–1.8 dm. long, simple or forked; lvs. in 4's, oblanceolate, 4–8 mm. long, fleshy; fls. 1–2, on short stocky glabrous pedicels that thicken in fr.; fr. 1–2 mm. in diam.—Wet places, 5000–10,500 ft.; Yellow Pine F. to Subalpine F.; San Jacinto Mts., San Bernardino Mts., Sierra Nevada from Fresno Co. n., to Modoc Co.; to Wash., Nfld., Rocky Mts., New Mex. June–Sept.

10. **G. cymòsum** Wieg. Perennial, with slender weak erect or ascending stems 3–8 dm. high, 4-angled, minutely roughened on angles; lvs. in 4's to 6's, unequal, oblanceolate, 1–2 cm. long, minutely scabrous on margins and midrib; fls. abundant, in small cymes at ends of upper branches; bracts foliaceous, small; pedicels short, straight; corolla white, 2.5–3.5 mm. in diam., 3–4-lobed; fr. glabrous.—Moist places, below 5000 ft.; Yellow Pine F., Red Fir F.; Humboldt to Siskiyou cos.; to B.C. July–Aug.

11. **G. bifòlium** Wats. Slender erect glabrous annuals, 0.5–1.5 dm. high, simple or slightly branched; lvs. 2–4, often quite unequal, 1–2.5 cm. long, lanceolate to sublinear, mostly acute; pedicels axillary and terminal, 1-fld., recurved and equaling lvs. in fr.; fls. whitish, ca. 1 mm. wide; fr. uncinate-hispid, ca. 2 mm. in diam.—Moist partly shaded places, 5000–10,500 ft.; Montane Coniferous F.; San Bernardino Mts. to Sierra Nevada, then to Trinity, Humboldt, and Siskiyou cos.; to Wash., Rocky Mts. June–Sept.

12. **G. prolíferum** Gray. Erect branching glabrous to hispidulous annual, 1–2 dm. tall; cotyledons conspicuous, persistent, oblong, paired; stems many from base, slender; lvs. 4 at a node, unequal, lanceolate to ovate, acute, 5–8 mm. long; fls. whitish, solitary, subsessile in axil of 2 leafy bracts on axillary peduncles that may have 1–2 prolifications; fr. 1.5–3 mm. in diam., uncinate-hairy.—Dry limestone slopes, 3500–4000 ft.; Creosote Bush Scrub, Joshua Tree Wd.; Kingston and Providence mts., e. Mojave Desert; to Nev., Utah, Tex. April–May.

13. **G. muràle** (L) All. [*Sherardia m.* L.] Delicate slender-stemmed, few-branched glabrous annual, 0.3–0.5 dm. high; lvs. 4 or 6, lance-ovate, 2–3 mm. long, bristly-ciliate; fls. 2 at a node, short-pedicelled, the pedicels recurved in fr.; frs. elongate, bristly-hairy at the apex, ca. 1 mm. long.—Established in grassy places in hills about San Francisco Bay; native of Eu. April–May.

14. **G. Andrèwsii** Gray. Depressed-cespitose or matted prickly plants, with branching underground creeping stems from which arise tufted leafy slender erect flowering stems 3–8 cm. tall, glabrous or slightly scabrous; lvs. in 4's, crowded, subulate, rigid, cuspidate, 4–10 mm. long; ♀ fls. solitary, the ♂ in few-fld. terminal cymes; corolla greenish-white, 1.5 mm. wide, glabrous, the lobes acute; fr. glabrous, 2–3 mm. wide, dark, baccate.— Dry benches and ridges, 1000–6500 ft.; Chaparral, Foothill Wd., Yellow Pine F., Pinyon-Juniper Wd., etc.; Coast Ranges from Lake and Tehama cos., occasional along base of Sierra Nevada, Mariposa to Kern cos. to L. Calif. April–June.

15. **G. ambíguum** Wight. Habit of *G. Andrewsii*, but soft to the touch and not at all prickly, low, tufted, from a branched creeping rootstock; stems leafy, lax, 4–10 cm. long, they and the lvs. grayish with short spreading hairs; lvs. 6–12 mm. long, linear, 1.5–2 mm. wide, hairy on both surfaces, acute, ± revolute, with prominent midrib beneath; fls. greenish-white, ca. 3 mm. across, solitary in the upper lf.-axils, or in short-peduncled cymes; fr. fleshy, glabrous to hairy, dark when dry.—Dry open wooded rocky places, 4000–7000 ft.; Yellow Pine F., Red Fir F.; Log Spring Ridge, Tehama Co., Bartlett Mt., Lake Co. to sw. Ore. June–July.

Var. **siskiyouénse** Ferris. More diffuse, the stems lax, to 16 cm. long; lvs. dark green, shining, glabrous or nearly so, except often ciliate on margins.—At 1400–5500 ft.; Yellow Pine F., Red Fir F.; Trinity and Humboldt cos. to Siskiyou Mts.

16. **G. Cleméntis** Eastw. Densely matted soft perennial with branched stems, spreading-pubescent throughout; lvs. in 4's, 3–6 mm. long, 1–1.5 mm. wide, elliptic to oblong, but seemingly linear by inrolling of margins, gray-green; fls. yellowish, in upper lf.-axils, solitary or in short-peduncled cymes; fr. fleshy, 1–2 mm. in diam., pubescent.— Rocky dry places, 3200–5800 ft.; Yellow Pine F., Red Fir F.; Santa Lucia Mts., Monterey Co. June–July.

17. **G. califórnicum** H. & A. [*G. flaccidum* Greene. *G. occidentale* McClat. *G. chartaceum* Wight.] Herbaceous from slender creeping rootstocks, tufted, the stems slender, rather stiff-pubescent, 0.5–2(–3) dm. high; lvs. in 4's, elliptic to ovate, 5–15 mm. long, subacuminate, apiculate, dull, short-hairy to hispidulose, mostly not longer than the internodes; ♂ fls. largely 2–3 in cymes, the ♀ solitary; corolla yellowish, ca. 2 mm. across, glabrous or pu escent externally; fr. fleshy, glabrous or hairy, 3–4 mm. in diam., whitish when mature and dark when dried.—Open hills and woods, below 3500 ft.; Redwood F., Closed-cone Pine F., Mixed Evergreen F., etc.; Coast Ranges from Humboldt to Los Angeles cos., Santa Cruz Id. March–July. Along the coast sometimes approaching:

Var. **miguelénse** (Greene) Jeps. [*G. m.* Greene.] Hairs of stem short and curved or almost none; lvs. ovate to roundish, crowded at ends of branchlets, glabrous and shining above.—Santa Rosa and San Miguel ids.

18. **G. muricàtum** Wight. Near to *G. californicum,* but lvs. shining, 5–9 mm. long, broadly to narrowly elliptic, sparsely short-stiff-hairy on upper surface, ciliate on margins with ascending curved hairs.—Uncommon, Mendocino Co.; w. Ore.

19. **G. Bolánderi** Gray. [*G. arcuatum* Wieg. *G. margaricoccum* Gray.] Stems erect or diffusely spreading, tufted, from slender woody underground branches from a cent. root, 1–4 dm. high, slender, angled, glabrous except sometimes on the angles; lvs. in 4's, oblong to lance-oblong, acute, glabrous or hispid-ciliate, 0.6–1.5 cm. long; ♂ fls. in small terminal cymes, ♀ solitary in upper axils; corolla purplish-red, 4-merous, ca. 2 mm. in diam.; fr. fleshy, glabrous, dark purplish, 3–3.5 mm. in diam.—Dry rocky places, 600–6000 ft.; Mixed Evergreen F., but mostly in Yellow Pine F., Red Fir F.; Coast Ranges from Solano Co. n., Sierra Nevada from Fresno Co. n.; to s. Ore. May–Aug.

20. **G. sparsiflòrum** Wight. [*G. pubens* var. *scabridum* Gray. *G. subscabridum* Wight.] Erect or spreading perennial from a woody root with subterranean rootstocklike branches, ± scabrous, 3–5 dm. long; lvs. in 4's, glabrous to hispidulous, oval or ovate, 8–15 mm. long; ♂ fls. rather few, in leafy cymules, ♀ solitary; corolla greenish, ca. 2 mm. across; fr. fleshy, glabrous, ca. 4 mm. in diam., on recurved pedicels.—Dryish places, largely 5500–8000 ft.; Montane Coniferous F.; Sierra Nevada from Plumas to Kern cos., but mostly in Fresno and Tulare cos. June–Aug.

21. **G. pùbens** Gray. [*G. Culbertsonii* Greene.] With much the same habit as *G. sparsiflorum,* cinereous with soft and stiff pubescence; lvs. mostly gray-green, linear-oblong to ovate-oblong, 0.5–1.5 cm. long, short-hispidulous on surfaces and edges, mostly not cuspidate-pointed; ♂ fls. in small cymules in paniculate arrangement, ♀ solitary in axils; corolla purplish or greenish; fr. fleshy, grayish to dark, pubescent, later sometimes glabrate.—At 2500–5000 ft.; Chaparral, Foothill Wd., Yellow Pine F.; Sierra Nevada from Plumas Co. s., Coast Ranges of Napa and Lake cos. April–June.

Var. **gránde** (McClat.) Gray. [*G. g.* McClat.] Finely and densely cinereous-pubescent. —Rare, 2000–4000 ft.; Chaparral; s. face of San Gabriel Mts.

22. **G. Nuttállii** Gray. [*G. suffruticosum* T. & G., not H. & A.] Woody at the base, with slender branches 2–20 dm. long, varying from low and tufted to elongate and clambering over other plants, usually retrorse-scabrous on the stem-angles and lf.-margins; lvs. in 4's, generally dark green, oval to linear-oblong, 2–6(–10) mm. long, with harsh pubescence, variable but mostly with hairs on both surfaces; plants dioecious, the ♂ fls. in small cymules, the ♀ solitary; corolla yellowish-green, ca. 1 mm. wide; fr. glabrous, baccate, 2–3 mm. in diam.—Common on dry slopes, mostly below 3000 ft.; Chaparral, Coastal Sage Scrub, Foothill Wd., etc.; cismontane Calif. to s. Ore., n. L. Calif. March–June.

Ssp. **insulàre** Ferris. Stems rather stout; secondary branches almost glabrous; lvs. ± linear, revolute on margins.—Santa Catalina, Santa Rosa, and Santa Cruz ids.

23. **G. catalinénse** Gray. Stems woody, stout, erect, 5–12 dm. high, quadrangular, mostly ± scabrous, with a swollen ring at the nodes and bearing above numerous leafy branches with short internodes and thus ± tufted; main lvs. mostly in 4's, narrow-elliptical, 1.5–2.5 cm. long, ca. ⅕–⅙ as wide, ± retrorse-scabrous on under surface, the lateral veins inconspicuous; infl. tufted, leafy, cymosely branched; pedicels short, glabrous or pubescent; corolla whitish, 4-lobed, 2–3 mm. across; fr. dry, ca. 2 mm. in diam., mostly with slender spreading hairs 0.5–1 mm. long, sometimes glabrate.—Dry rocky canyons; Coastal Sage Scrub, Chaparral; Santa Catalina and San Clemente ids. April–July.

24. **G. buxifòlium** Greene. Like *G. catalinense*, but nodes less conspicuously swollen; main cauline lvs. ca. ⅓–½ as wide as long, the lateral veins conspicuous and on under side retrorse-scabrous, the intervening surface ± glabrous; fr. with short coarse appressed hairs less than 0.5 mm. long; $2n = 22$, 44 (Fagerlind, 1937).—Dry rocky places; Chaparral, Closed-cone Pine F.; Santa Cruz and San Miguel ids. March–July.

25. **G. boreàle** L. Perennial with slender woody rootstocks and erect stems 3–6 dm. high, branching, mostly glabrous except puberulent at nodes; lvs. in 4's, narrowly to broadly lanceolate, 3-nerved, mostly obtuse, 1.5–4 cm. long, glabrous or ± scabrous, with ciliate margins; infl. a large rather dense showy panicle of white fls.; corolla 3–4 mm. wide; fr. glabrous or ± hairy, scarcely 2 mm. in diam.; $n = 53$ (Löve & Löve, 1954).—Dry slopes, below 7000 ft.; largely Yellow Pine F., Red Fir F.; Humboldt and Shasta cos. to Alaska, Atlantic Coast; Eurasia. June–Aug.

26. **G. Wrìghtii** Gray ssp. **Rothróckii** (Gray) Ehrend. [*G. R.* Gray.] Suffrutescent, many-stemmed, glabrous, not at all scabrous, the stems slender, 1.5–6 dm. tall; lvs. in 4's, narrowly linear, rigid, 0.8–1 cm. long, uppermost bracteate; plant polygamo-monoecious, the fls. in open paniculate infls. of small cymes; corolla usually red, 1–2 mm. wide, the lobes long-acuminate; fr. sparsely hairy, including the bristles ca. 2 mm. broad, the bristles straight, scarcely as long as the body of the fr.—Dry, rocky places, 5000–6000 ft.; Pinyon-Juniper Wd.; mts. of e. Mojave Desert; to Ariz., New Mex. May–June, Aug.–Sept.

27. **G. angustifòlium** Nutt. in T. & G. Erect, shrubby, 3–12 dm. tall, the branches stiff, subglabrous not scabrous on angles, often ± stiff-pubescent at nodes; lvs. linear to lance-linear, 0.5–2 cm. long, glabrous or pubescent on surface and edges, acute; plants polygamo-dioecious, the fls. greenish-white, 1.5–2.5 mm. wide, numerous in elongate panicles of cymes; body of fr. 1–2 mm. in diam., the hairs 1–2 mm. long, spreading.— Dry slopes and bushy places, below 6000 ft.; Chaparral, Coastal Sage Scrub, Foothill Wd., etc.; cismontane Calif. from Monterey and Kern cos. to L. Calif. April–June. Intergrading freely with the vars.

KEY TO VARIETIES

Pistillate fls. many, generally congested at ends of the branches of the infl.
  Body of fr. 1.5–2 mm. in diam., the hairs 1–2 mm. long; ♂ infl. broad; plant 3–12 dm. tall.
  Cismontane, mostly below 4000 ft.
    Internodes glabrous . . . . *G. angustifolium*
    Internodes cinereous-pubescent . . . . var. *siccatum*
  Body of fr. 1.5 mm. in diam., the hairs 1 mm. long; ♂ infl. generally narrow, plant 1.5–3 dm.
  tall. Montane, mostly above 5000 ft. . . . . var. *bernardinum*
Pistillate fls. few, or if many, ± scattered in an open cymose panicle.
  Staminate fls. in broad diffuse panicles; the whole plant laxly and diffusely branched. Desert
                                                             var. *diffusum*
  Staminate fls. in a narrow infl.
    Pistillate infl. conspicuously leafy-bracted, the fls. short-pedicelled or subsessile; branches nearly
    equally leafy throughout. Insular . . . . var. *foliosum*
    Pistillate infl. not conspicuously leafy, the fls. frequently on long pedicels; lvs. congested at base
    of plant, the upper branches sparsely leafy. Pine belt . . . . var. *pinetorum*

Var. **siccàtum** (Wight) Hilend & Howell. [*G. s.* Wight.] Like the sp., but cinereouspuberulent.—Occasional with the sp., usually near the coast; San Diego Co. to San Benito Co.

Var. **bernardìnum** Hilend & Howell. Plants low, compact, 1.5–3 dm. high; lvs. 0.5–1 cm. long; ♂ infl. long and narrow; ♀ infl. pyramidal with divaricate branches and numerous congested fls.; fr. 1.5 mm. in diam.—Dry places, 4500–8000 ft.; Pinyon-Juniper Wd., Yellow Pine F.; San Bernardino Mts. to Laguna Mts. (San Diego Co.). May–July.

Var. **diffùsum** Hilend & Howell. Lax and diffusely branched, mostly 3–6 dm. tall; lvs. mostly 0.5–1 cm. long; fls. on very slender pedicels.—Dry places, below 5500 ft.; Creosote Bush Scrub, Joshua Tree Wd., etc.; desert slopes along w. edge of Colo. Desert and San Bernardino Mts. to Providence Mts. April–June.

Var. **foliòsum** Hilend & Howell. Dark green, compact, mostly 1–3 dm. tall; lvs. mostly crowded, horizontal, mucronulate, 0.5–0.8 cm. long.—Exposed rocky places; Coastal Sage Scrub, Chaparral; Santa Cruz, Santa Rosa, Anacapa, Santa Barbara ids. March–July.

Var. **pinetòrum** M. & J. Herbaceous, 1–2 dm. tall; upper internodes much exceeding lvs.;

lvs. 1–3 cm. long; panicle lax, few-fld., 3–8 cm. long.—Occasional on dry slopes, 5000–7500 ft.; Yellow Pine F.; San Gabriel, San Bernardino, and San Jacinto mts. June–July.

28. **G. gabrielénse** M. & J. [*G. siccatum* var. *anotinum* Jeps.] Dioecious perennial 5–15(–30) cm. tall, erect, tufted, from woody base of branched caudex, cinereous-pubescent with short spreading hairs; lvs. in 4's, linear to narrow-elliptic, mostly 5–10 mm. long, 1–2 mm. wide, acute, cinereous-pubescent; infl. cylindrical, 3–8 cm. long; corolla yellowish, 2.5–3 mm. in diam., bristly-hairy without; fr. including hairs ca. 3 mm. in diam., the body 1.5–2 mm. in diam.—Dry ridges and slopes, 4000–8500 ft.; Chaparral to Red Fir F.; s. face of e. half of San Gabriel Mts. June–Aug.

29. **G. Jepsònii** Hilend & Howell. [*G. angustifolium* var. *subglabrum* Jeps.] Perennial, the stems several and tufted from the ends of underground stems, erect, 1–2 dm. tall, glabrous, smooth; lvs. linear-oblong, or the lower broader, 0.5–1.5 cm. long, bristly-ciliate but glabrous on surfaces; plants dioecious; the fls. 1–3 in a cyme, the cymes few in a narrow infl.; corolla glabrous, ca. 2 mm. in diam.; body of fr. 2.5 mm. in diam., subpilose with appressed hairs 1–1.5 mm. long.—Rare, in dry rocky and gravelly places, 7000–8000 ft.; Red Fir F.; San Gabriel and San Bernardino mts. July–Aug.

30. **G. stellàtum** Kell. ssp. **erèmicum** (Hilend & Howell.) Ehrend. [*G. s.* var. *e.* Hilend & Howell.] Bushy, much-branched above the woody base, 2–7 dm. high, stems and lvs. ± scabrous-pubescent, the younger branches with shining white exfoliating epidermis; old lvs. whitish, persistent; lvs. of the current year lanceolate to almost linear, ± revolute, 0.4–1 cm. long, acuminate-cuspidate, pale green; corolla greenish-yellow, 2–2.5 mm. in diam., bristly-hairy outside; ♂ fls. in crowded panicles, ♀ on short pedicels and solitary at ends of leafy-bracted branchlets; fr. including the soft hairs ca. 4 mm. in diam., the body ca. 2 mm.—Common on dry rocky slopes, below 5000 ft.; Creosote Bush Scrub, Joshua Tree Wd.; deserts from Inyo to Imperial cos.; to Nev., Ariz. March–April.

31. **G. Hàllii** M. & J. Dioecious; stems many from a branched root-crown, woody below with white exfoliating bark, erect and 2–6 dm. tall or ± spreading, cinereous-pubescent throughout with spreading hairs; lvs. oblong-elliptic to broadly ovate, 0.5–1.5 cm. long, 0.3–1 cm. broad, ± revolute, with strong midvein and weakish lateral veins; panicles loose, leafy, the branches recurved at tips; corolla yellowish, 2–2.5 mm. in diam., pilose without; body of fr. ca. 3 mm. in diam., the dense white hairs 2 mm. long.—Dry rocky slopes and canyons, 4000–7000 ft.; Pinyon-Juniper Wd., Yellow Pine F.; desert slopes from San Bernardino Mts. to Mt. Pinos and Santa Ynez Mts. May–Aug.

32. **G. Múnzii** Hilend & Howell. [*G. M.* var. *carneum* Hilend & Howell. *G. Matthewsii* var. *scabridum* Jeps.] Erect or decumbent, many-stemmed, from a ± cespitose woody base, 1.5–3.5 dm. tall, short-stiff-pubescent throughout, the stems slender, exfoliating in age; lvs. in 4's, subequal, broadly to narrowly ovate, 5–12 mm. long, mucronulate, mostly 1-nerved; panicles 5–10 cm. long, erect, not densely fld.; plants dioecious; corolla greenish to brownish, hispid without, 1.5–2 mm. wide; fr. including hairs, 2.5–3 mm. wide, the whitish hairs 1–1.5 mm. long.—Occasional in dry rocky places, 3200–10,000 ft.; Pinyon-Juniper Wd., Sagebrush Scrub, etc.; desert slopes of Sierra Nevada and Inyo-White Mts., from s. Plumas Co. to n. slope San Bernardino Mts., Providence Mts., Kingston Mts., etc.; w. Nev. May–July. Plants with glabrous lvs. are f. **glàbrum** Ehrend.

33. **G. multiflòrum** Kell. [*G. Bloomeri* Gray.] Many-stemmed glabrous plant from woody caudex, the stems slender, 1–3 dm. high, scarcely or not scabrous; lvs. in 4's, broadly ovate, mostly 0.4–0.8′ cm. long, coriaceous, cuspidate; plants dioecious, the fls. in narrow panicles; corolla greenish, glabrous externally, ca. 4 mm. broad; fr. including hairs ca. 4.5–5.5 mm. in diam., the body ca. 1.5 mm. in diam.—Dry rocky hills, 4700–8000 ft.; Pinyon-Juniper Wd., Yellow Pine F., Sagebrush Scrub; e. slope of Sierra Nevada, Sweetwater Mts., Last Chance Mts., etc., Mono to Inyo cos.; adjacent Nev., Utah. May–Aug.

34. **G. Matthèwsii** Gray. Stems erect, branched, slender, somewhat tufted from a woody caudex, glabrous, 1.5–3 dm. high; lvs. in 4's, remote, lance-ovate, 1-nerved, cuspidate-acute, 3–8 mm. long, the upper much reduced; panicle lax, 5–18 cm. long; fls. greenish-white, 1.5 mm. broad, stiff-pubescent externally; fr. including hairs, 3 mm. wide, the hairs ca. 1 mm. long.—Dry rocky places, 4000–8500 ft.; Sagebrush Scrub, Pinyon-Juniper Wd., Shadscale Scrub; Argus Mts., Kingston Mts., Panamint Mts., Inyo Mts., e. slope of Sierra Nevada in Inyo Co. May–Aug.

35. **G. Párishii** Hilend & Howell. [*G. multiflorum* var. *parvifolium* Parish. *G. p.* Jeps., not Gaud.] Stems many, slender, scabrous-puberulent, 1–3 dm. long, from a branched

woody root-crown, erect or spreading, tufted; lvs. in 4's, the lowermost scalelike, the main ones ovate, acute, abruptly mucronate, 3–7(–10) mm. long, scabrous-puberulent; infl. spicate-paniculate, slender, with congested cymules; plants dioecious; fls. mostly reddish, ca. 1.5 mm. in diam., short-bristly externally; fr., including hairs, ca. 3 mm. broad, the body 1.5–2 mm. across.—Common in dry rocky places, 6000–10,000 ft.; Pinyon-Juniper Wd. to Lodgepole F.; San Gabriel Mts. to Santa Rosa Mts., New York Mts., Kingston Mts., etc.; s. Nev. June–Aug.

36. **G. Grayànum** Ehrend. Stems tufted, slender, herbaceous, 0.5–1.5 dm. tall, from woody underground rootstocklike branches from a cent. caudex; plant cinereous with short stiff puberulence throughout; lvs. in 4's, subequal, broadly ovate, 0.5–1 cm. long, abruptly short-cuspidate; plants dioecious; infl. narrow, leafy, 3–10 cm. long; corolla greenish-white, ca. 3.5 mm. in diam., mostly hispid-puberulent externally; fr., including hairs, ca. 6–7 mm. in diam., the hairs soft, tawny, ca. 3 mm. long.—Dry rocky slopes, 6000–10,000 ft.; Montane Coniferous F.; Sierra Nevada from Placer Co. n., Coast Ranges from Snow Mt. and Yolla Bolly Mts. to Siskiyou Mts. July–Aug. Ssp. **glabréscens** Ehrend. Herbage subglabrous.—Trinity and Siskiyou cos. to Lake Tahoe; sw. Ore.

37. **G. hypotríchium** Gray. Tufted perennial with underground rootstocks from a woody caudex; stems herbaceous, 0.3–1.5 dm. high, slender, mostly simple, leafy; herbage minutely cinereous-scabrous; lvs. in 4's, round-ovate, apiculate, mostly 3-nerved, 2–7 mm. long, almost as broad; plants largely polygamo-dioecious; infl. a narrow leafy panicle, 2–5 cm. long; corolla ± yellowish to pinkish, ca. 2.5–3 mm. across, mostly glabrous externally; fr. including hairs, ca. 6 mm. across, the hairs soft, ca. 2–2.5 mm. long.—Dry places about rocks, etc., 7300–11,200 ft.; upper Montane Coniferous F.; e. slope of Sierra Nevada from Inyo to Plumas cos., Sweetwater Mts.; to Utah. May–Aug.

Ssp. **subalpìnum** (Hilend & Howell) Ehrend. [*G. Munzii* var. *s.* Hilend & Howell.] Lvs. ovate to lance-ovate, subacuminate; corolla mostly ± pubescent externally.—About rocks, 8000–12,000 ft.; Lodgepole F. to Alpine Fell-fields; Sierra Nevada of Inyo, Tulare, and Fresno cos.; s. Nev. June–Aug.

Ssp. **tomentéllum** Ehrend. Herbage very closely cinereous-pubescent; lvs. broadly ovate, 3–5 mm. long; corolla reddish, ca. 2 mm. in diam., ± pubescent externally.—About rocks, 10,500–11,000 ft.; Bristle-cone Pine F.; Telescope Peak, Panamint Mts., Inyo Co. June–Aug.

38. **G. Watsònii** (Gray) Heller f. **scàbridum** Ehrend. Herbaceous, ± cespitose, 1.5–3 dm. tall, minutely scabrous; lvs. lanceolate, 1–2 cm. long, 0.3–0.5 cm. wide; infl. narrow, elongate; fls. 3–4 mm. across; frs. 6–8 mm. in diam., densely beset with whitish hairs.—Sagebrush Scrub to Yellow Pine F.; Modoc and Siskiyou cos.; to Wash., Ida. June–July.

2. Sherárdia L. FIELD MADDER

Slender many-stemmed annual, procumbent, the stems square, subglabrous to hispidulous-roughened, 1–2 dm. long. Lvs. in whorls of 4–6, lance-ovate, pungent, mostly retrorse-hispid on margins and midrib beneath, 5–10 mm. long. Fls. surrounded by a gamophyllous invol., small, blue or pinkish, 2–3 in a head. Sepals persistent. Corolla funnelform, 4–5-lobed. Stamens 4–5. Style filiform, slightly 2-cleft. Fr. dry, didymous, forming 2 indehiscent 1-seeded carpels. One sp. (Named for Dr. W. *Sherard*, a patron of Dillenius.)

1. **S. arvénsis** L. Stems 1–2 dm. long, tufted; invol. deeply 6–8-lobed.—Natur. widely but usually not abundantly, in fields, orchards, lawns, meadows, etc.; cismontane Calif.; native of Eu. Jan.–July.

3. Kellóggia Torr.

Slender perennial herbs with few stems from slender underground parts. Stems slender, simple ,or branched, subglabrous. Lvs. opposite, lanceolate, with interposed stipules. Fls. in loose forking terminal cymes, pinkish-white, small. Corolla funnelform, with exserted stamens. Ovary inferior, 2-celled; style filiform; stigmas 2, linear-clavate. Fr. dry, uncinate-hispid, separating into 2 closed carpels. One sp. (Named for Dr. A. *Kellogg*, pioneer Calif. botanist.)

1. **K. galioìdes** Torr. Stems 1–2.5 dm. tall; lvs. 1–2.5 cm. long, glabrous; pedicels

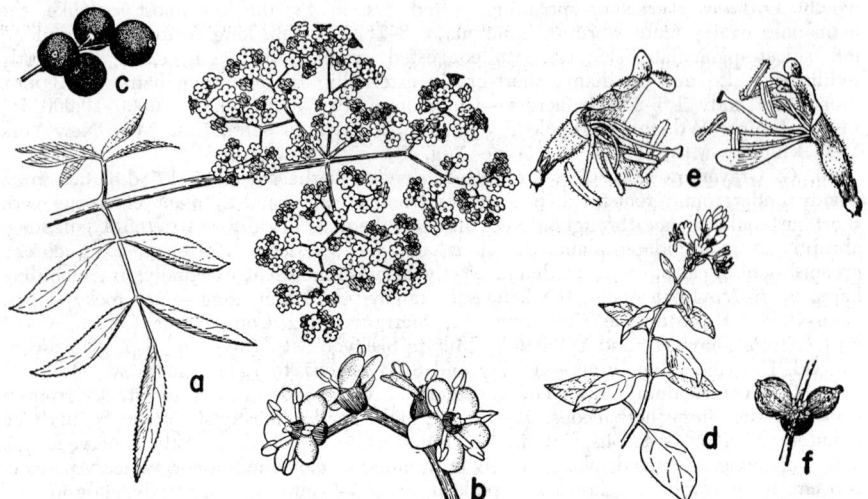

Fig. 103. CAPRIFOLIACEAE. *Sambucus mexicana: a,* infl. and compound lf., × ½; *b,* fls., × 2½, ovary inferior, sepals 5, corolla rotate, stamens 5, stigmas 3; *c,* frs., × 1. *Lonicera subspicata* var. *denudata: d,* twig, × ¼; *e,* fls., × 2; ovary inferior, sepals 5, corolla bilabiate, stamens 5; *f,* fr., × 1.

filiform, 1–12 cm. long; corolla 3–4 mm. long, mostly 4-lobed; fr. broadly clavate, 3–4 mm. long.—Dry benches and slopes, 3000–9600 ft.; Montane Coniferous F.; San Jacinto and San Bernardino mts.; Sierra Nevada to Siskiyou and Modoc cos., Humboldt to Lake and Trinity cos.; to Wash., Ida., Utah, Ariz. May–Aug.

4. Cephalánthus L. Buttonbush

Shrubs or small trees. Lvs. opposite or whorled, with short intervening stipules. Fls. densely aggregated in round peduncled heads. Sepals 4, short. Corolla narrow-funnelform, slender, the limb small, 4-cleft. Style filiform, exserted; stigma capitate. Fr. small, dry, at length splitting from the base upward into 2–4 closed 1-seeded ak.-like portions. Ca. 6 spp., of Am., Asia, and Afr. (Greek, *cephale,* a head, and *anthos,* fl.)

1. **C. occidentàlis** L. var. **califórnicus** Benth. Two to 9 m. high, with gray bark and reddish younger twigs; lvs. essentially glabrous, oblong- to lance-ovate, abruptly short-acuminate, 5–15 cm. long, entire, on petioles ca. 1 cm. long; peduncles in upper axils and terminal, 2–8 cm. long; heads 1–2.5 cm. in diam.; calyx greenish; corolla white, 7–8 mm. long, with obtuse dark-tipped lobes; fr. 3–4 mm. long; seed ca. 2 mm. long, flat, acute.—Along lake shores and streams, below 3000 ft.; Foothill Wd., V. Grassland, Mixed Evergreen F., etc.; Cent. V. and bordering ranges from Kern Co. n., inner Coast Ranges from Napa Co. to Siskiyou Co.; w. U.S. June–Sept.

114. Caprifoliàceae. Honeysuckle Family (Fig. 103)

Shrubs or trees or vines. Lvs. opposite, mostly without stipules. Fls. bisexual, regular or irregular, 4–5-merous, with inferior ovaries. Corolla gamopetalous, rotate to tubular; stamens alternate with corolla-lobes, inserted on the tube. Ovary 1–5-loculed, each locule 1–many-ovuled; placentation axile; style 1 or obsolete; stigmas 1–5. Fr. an ak., berry, caps., or drupe. Seed-coat adherent to the fleshy endosperm; embryo small. Ca. 14 genera and 400 spp. mostly of N. Temp. Zone; many of ornamental use.

A. Corolla rotate or nearly so, regular; style short, 3–5-lobed.
    B. Lvs. pinnately compound .... 1. *Sambucus*
    BB. Lvs. simple .... 2. *Viburnum*
AA. Corolla tubular to funnelform, usually irregular; style elongate, mostly with capitate stigma.
    B. Fr. a 1-seeded ak.; plants low trailing herbs .... 3. *Linnaea*

BB. Fr. fleshy; plants more woody.
    C. Berry white, 1–2-seeded; corolla subregular, open-campanulate or tubular-funnelform
                                                               4. *Symphoricarpos*
   CC. Berry red or black, few-seeded; corolla mostly irregular, tubular, often gibbous at base
                                                                       5. *Lonicera*

1. Sambùcus L. ELDERBERRY

Shrubs or small trees with opposite odd-pinnate lvs. and serrate lfts. Twigs with large pith. Fls. small, mostly whitish, in compound cymes, jointed with their pedicels. Sepals 5. Corolla regular, rotate, 5-lobed. Stamens 5, inserted at the base of the corolla. Ovary 3–5-loculed; style short; stigmas 3–5. Fr. a berrylike drupe with 3–5 cartilaginous nutlets. Ca. 20 spp., of temp. and subtrop. regions. (Greek, *sambuke,* a musical instrument made of Elder wood.)

Berry blue or white; cymes flat-topped, the axis not or seldom extended beyond the lowest branches.
  Lfts. mostly 5–9, oblong-lanceolate to lanceolate, 6–15 cm. long, gradually short-acuminate, sharply and often rather deeply serrate; infl. 0.5–2 dm. across. Mostly Montane Coniferous F.
                                                                  1. *S. caerulea*
  Lfts. mostly 3–5, roundish to ovate or oblong-lanceolate, rather abruptly acuminate, mostly 1.5–6 cm. long, finely serrate; infl. mostly 0.3–1 dm. across. Generally below Montane Coniferous F.
                                                                  2. *S. mexicana*
Berry bright red or black; cymes dome-shaped or thyrsoid, the axis extended beyond the lowest branches.
  The berries red at maturity; branchlets and often lvs. glabrous or glabrate; lfts. mostly 7; corolla yellowish.
    Plants 2–6 m. high; lvs. stiff-pubescent along veins beneath; fruiting infl. 6–10 cm. across. Coastal .... 3. *S. callicarpa*
    Plants 0.5–2 m. high; lvs. glabrous or nearly so; fruiting infl. 3–6 cm. across. Montane
                                                                    4. *S. microbotrys*
  The berries black at maturity; branchlets and lower lf.-surfaces usually scurfy-puberulent or sparsely villous; lfts. commonly 5; corolla whitish .... 5. *S. melanocarpa*

1. **S. caerùlea** Raf. [*S. glauca* Nutt. *S. fimbriata* Greene.] Large shrub or small tree 2–8 m. tall; lvs. glabrous to pubescent or hispidulose beneath, often ± persistent, the lfts. mostly quite asymmetrical at the base; fls. white, 5–6 mm. wide; berries nearly black but densely glaucous, thus appearing bluish, 5–6 mm. in diam.—Open places, up to 10,000 ft.; largely Montane Coniferous F.; mts. from San Diego Co. n. through Sierra Nevada, N. Coast Ranges; to B.C., Alta., Ida. June–Sept.

2. **S. mexicàna** Presl. [*S. caerulea* var. *m.* L. Benson. *S. coriacea* Greene. *S. orbiculata* Greene. *S. velutina* Dur. & Hilg.] Much like the preceding, the lfts. mostly fewer, smaller, often quite deciduous in the dry season; infl. smaller; berries either blue or white under the white bloom, often dryer at maturity.—Open flats and cismontane valleys and canyons, below 4500 ft.; many Plant Communities, like Foothill Wd., S. Oak Wd., etc.; cismontane Calif. from Lake and Glenn cos. s. to L. Calif., Ariz., etc. March– Sept. Occasional in Pinyon-Juniper Wd., mts. of Mojave Desert. Exceedingly variable and in need of study; plants from Monterey to Santa Barbara cos., with roundish lfts. have been called *S. orbiculata;* and from the head of the San Joaquin V., with rather large quite pubescent lfts., *S. velutina.* (This is near *S. caerulea.*)

3. **S. callicárpa** Greene. [*S. racemosa* var. *c.* Jeps. *S. maritima* Greene.] Woody, 2–6 m. high; lvs. 5–7-foliolate, stiff-pubescent especially along the veins beneath, the lfts. lanceolate to oblong-ovate, 5–16 cm. long, sharply and finely serrate; infl. mostly 6–10 cm. across in fr., rounded or pyramidal; fls. whitish; fr. bright scarlet, 4–5 mm. thick, without bloom.—Damp woods and flats, at low elevs.; N. Coastal Scrub, Redwood F., Douglas-Fir F., etc.; along the coast from San Mateo Co. to B.C. March–July.

4. **S. microbòtrys** Rydb. Low shrub, 0.5–1(–2) m. high, with rank odor; lvs. thin, glabrous or nearly so; lfts. 5–7, oval to elliptic, acute to short-acuminate, 3–8 cm. long, coarsely serrate; fls. cream-color; fruiting infl. mostly 3–6 cm. across, almost as long as broad; frs. red, 4–5 mm. in diam., without bloom.—Common in moist places, 6000– 11,000 ft.; Red Fir F. to Subalpine F.; San Bernardino Mts., Sierra Nevada, N. Coast Ranges from Yolla Bolly Mts. n.; to Rocky Mts. June–Aug.

5. **S. melánocarpa** Gray. [*S. racemosa* var. *m.* McMinn.] Shrub 1–2 m. tall, ± pubescent, especially on young lvs. which generally remain scurfy beneath when older; lfts. 5–7, oval to lance-ovate, coarsely serrate, abruptly acuminate, 4–12 cm. long; infl. 4–7

cm. in diam.; corolla white, turning brownish on drying; fr. purplish-black, 4–5 mm. in diam., without a bloom.—Occasional in moist places, 6000–12,000 ft.; Lodgepole F. to Alpine Fell-fields; e. Sierra Nevada and Siskiyou Co.; to Canada, Utah, New Mex. July–Aug.

2. Vibúrnum L.

Shrubs mostly with opposite simple lvs. Fls. in terminal compound cymes, white. Sepals 5. Corolla open-campanulate, 5-lobed. Stamens 5. Ovary inferior, 1-loculed, with 1 ovule. Fr. a drupe with a flattened stone. Ca. 100 spp., widely distributed in temp. and subtrop. regions of N. Hemis. (Classical Latin name of uncertain meaning.)

1. **V. ellípticum** Hook. Slender-stemmed shrub 1–4 m. tall; young growth pubescent and ± glandular; lvs. elliptical to roundish, coarsely dentate except at base, 2–6 cm. long, 3–5-veined from base, glabrous or somewhat pubescent above, paler and more pubescent beneath; petioles an additional 6–12 mm. long; fls. 6–8 mm. broad, in peduncled cymes; fr. ellipsoidal, ca. 10–12 mm. long, the stone 5-grooved.—Occasional at 1000–4500 ft.; Yellow Pine F., Chaparral, etc.; N. Coast Ranges (Contra Costa, Sonoma, Mendocino, Humboldt, and Glenn cos.) and in Fresno and Eldorado cos. of Sierra Nevada; to Wash. May–June.

3. Linnaèa L. Twin Flower

Slender-stemmed creeping herbaceous or slightly woody perennial with opposite lvs. Fls. in pairs, on long erect slender terminal peduncles. Sepals 5, subulate. Corolla pinkish, funnelform to slender-campanulate, regular, 5-lobed. Stamens 4, of 2 lengths. Ovary 3-loculed, with 2 abortive and 1 fertile ovule. Fr. a dry 1-seeded pod. One circumpolar sp. (Named for *Linnaeus* by Gronovius.)

1. **L. boreàlis** L. ssp. **longiflòra** (Torr.) Hult. [*L. b.* var. *l.* Torr. *L. l.* Rydb.] Trailing stems to 1 m. long, finely pubescent; lvs. elliptical, acutish at apex, with scattered longish hairs on margins and sometimes on surface, somewhat toothed in upper half, ± cuneate at base, mostly 1–2 cm. long, the petioles 2–4 mm. long; peduncles commonly 4–8 cm. long; pedicels ca. 1 cm.; calyx-lobes 3–4 mm. long, glandular-puberulent; corolla 12–15 mm. long, funnelform, the tube exceeding calyx; fr. ovoid, glandular, ca. 3 mm. long.—Dense woods, 400–8000 ft.; Mixed Evergreen F. to Subalpine F.; Humboldt and Trinity cos. to Del Norte, Siskiyou, Plumas, and Modoc cos.; to Ida., Alaska. June–Aug.

4. Symphoricárpos Duhamel. Snowberry

Shrubs with simple short-petioled exstipulate, entire or ± toothed or lobed, opposite deciduous lvs. Fls. white or pink, 2-bracteolate, in small terminal or axillary clusters, sometimes solitary, 4–5-merous. Corolla campanulate or salverform or long-funnelform, regular or nearly so. Stamens inserted on the corolla. Ovary inferior, 4-loculed, 2 of the locules with several abortive ovules, the other 2 each with 1 pendulous ovule. Fr. a berrylike drupe, in ours white, with 2 nutlets. Ca. 17 spp. of N. Am. and 1 in China. Some in cult. (Greek, *sumphoreo*, to bear together, and *karpos*, fr., because of the clustered frs.)

(Jones, G. N. A monograph of the genus Symphoricarpos. Jour. Arnold Arb. 21: 201–252, 1940.)
A.   Corolla short-campanulate, the lobes at least as long as the tube.
    B.   Plants erect; corolla 5–6 mm. long; nutlets 4–6 mm. long; lvs. mostly glabrous or ± pilose
        beneath. Coast Ranges s. to Monterey Co., n. Sierra Nevada . . . . 1. *S. rivularis*
    BB.   Plants trailing or low and spreading; corolla 3–5 mm. long; nutlets 2.5–3 mm. long; lvs.
        pubescent, at least beneath.
        C.   Young twigs closely puberulent with short curved hairs.
            D.   Lvs. firm, round-oval, rounded at each end, copiously short-pilose beneath, some-
                what so above, young twigs closely tomentulose-puberulent. Widely distributed
                                                    2. *S. mollis*
           DD.   Lvs. thin, oval, mostly acutish at ends, glabrous or nearly so above, sparsely
               short-pilose beneath, young twigs subglabrous to sparingly short-pilose. From
               Trinity Co. . . . . 3. *S. hesperius*
        CC.   Young twigs with short spreading hairs; lvs. oval, often sinuate, prominently veined,
        copiously pubescent, definitely bicolored. Sierra Nevada and N. Coast Ranges
                                                    4. *S. acutus*

AA. Corolla elongate-campanulate, to salverform, the lobes shorter than the tube.
B. The corolla elongate-campanulate, 6–9 mm. long.
 C. Plant erect; corolla 7–9 mm. long; lvs. scarcely glaucous; young twigs tomentulose-puberulent. Sierra Nevada from Fresno and Mono cos. n. to Modoc and Siskiyou cos.
                           5. S. *vaccinoides*
 CC. Plant low, spreading, often rooting at tips of branches; corolla 6–7 mm. long; lvs. glaucous; young twigs with scattered short spreading hairs or glabrous .. 6. S. *Parishii*
BB. The corolla salverform, 11–13 mm. long. Desert slopes .... 7. S. *longiflorus*

1. **S. rivulàris** Suksd. [S. *albus* auth., not (L.) Blake.] Erect, branched, 3–20(–30) dm. tall, with slender branches, the young twigs glabrous, the older with gray bark; lvs. oval, mostly 2–3 cm. long, 0.7–1.5 cm. wide, acute to obtusish, dark green and glabrous above, slightly paler and glabrous or with scattered hairs beneath, entire, or sinuate or lobed and larger on young shoots; fls. often many, in short-peduncled racemes 1–2.5 cm. long; bracts and bractlets glabrous; sepals triangular, 0.5 mm. long; corolla rose-pink to white, villous within, 5–7 mm. long, the lobes as long as the tube, obtuse; style glabrous, 2 mm. long; fr. subglobose to ellipsoid, 8–12 mm. in diam., white; nutlets plano-convex, 4–6 mm. long, 3–3.5 mm. wide.—Banks and flats in canyons and near streams, below 4000 ft.; Mixed Evergreen F., Foothill Wd., Yellow Pine F., etc.; Coast Ranges from Monterey Co. n., n. Sierra Nevada; to Alaska, Mont. May–July.

2. **S. móllis** Nutt. in T. & G. [S. *ciliatus* Nutt. S. *nanus* Greene.] Low, trailing, diffusely branched, the stems 3–9 dm. long; twigs closely tomentulose-puberulent with short curved hairs; lvs. thin, mostly entire, round-oval, 1–4 cm. long, 0.7–3 cm. wide, short-pubescent above, paler and more densely pubescent beneath, obtuse; fls. in pairs or small clusters; sepals 0.5–0.8 mm. long, deltoid, ciliate; corolla pink, 3–5 mm. long, the lobes 2–3 mm. long, sparsely pilose within near base; style glabrous, 2 mm. long; frs. white, globose, 4–6 mm. in diam.; nutlets 2.5–3 mm. long.—Common on shaded slopes, below 3000 (5000) ft.; Chaparral, S. Oak Wd., Mixed Evergreen F., etc.; Coast Ranges from Mendocino Co. to n. L. Calif. April–June.

3. **S. hespérius** G. N. Jones. Prostrate, the branches 1–3 m. long; young twigs sparsely short-pubescent to glabrous; lvs. oval, 1–3 cm. long, 0.5–2 cm. wide, tapering toward the ends, dark green and subglabrous above, pale, prominently reticulate beneath, with short-pilose veins; sepals ovate, ciliolate, 1 mm. long; corolla pink, 3–5 mm. long; fr. subglobose, 5–6 mm. in diam.; nutlets 2.5–3 mm. long.—Open slopes and ridges, 3200–5500 ft.; Red Fir F., Yellow Pine F.; Trinity and Humboldt to Siskiyou cos.; to B.C., Ida. June–Aug.

4. **S. acùtus** (Gray) Dieck. [S. *mollis* var. a. Gray. S. *racemosus* var. *trilobus* Durand.] Procumbent or trailing, diffuse, the branches 4–8 dm. long; young twigs soft-pubescent with spreading hairs; lvs. oval to ovate, 1–3 cm. long, 0.6–1.8 cm. wide, obtuse to acutish, dark green and ± pubescent above, somewhat paler and densely pubescent beneath with prominent veins, often sinuate on margins; fls. 1 or 2 in upper axils; sepals 4–5, ciliate, pubescent on back, less than 1 mm. long; corolla bright pink, campanulate, 4–5 mm. long, the obtuse lobes villous within; style glabrous, 2–2.5 mm. long; fr. subglobose, 4–5 mm. in diam.; nutlets 4 mm. long, 2–2.5 mm. wide.—Rather damp places, 3500–8000 ft.; Yellow Pine F. to Lodgepole F.; Sierra Nevada from Tulare Co. n., Coast Ranges from Lake and Glenn cos. n.; s. Ore., w. Nev.; apparently also in Cuyamaca Mts. June–Aug.

5. **S. vaccinoides** Rydb. [S. *rotundifolius* auth., not Gray. S. *Austinae* Eastw.] With underground runners; stems branched, to ca. 1.5 m. tall; the twigs puberulent or pubescent with short curved hairs; lvs. dark green above, ± pale beneath, puberulent, oval, acutish at ends, 1–2 cm. long, entire, ± revolute; fls. 1–2 in uppermost axils; sepals triangular, ca. 1 mm. long; corolla pink, 7–9 mm. long, elongate-campanulate, the lobes rounded, ca. ⅓ as long as tube; style glabrous, 4 mm. long; fr. ellipsoid, ca. 1 cm. long, 6–7 mm. thick; nutlets 5–6 mm. long.—Dry stony slopes, 5000–10,500 ft.; mostly Lodgepole F. to Subalpine F.; Sierra Nevada from Fresno Co. n., to Modoc Co.; to B.C., Mont., Colo. June–Aug.

6. **S. Párishii** Rydb. [S. *parvifolius* Eastw.] Low, spreading, the declined branches 5–10 dm. long, often rooting at tips; young twigs short-pubescent, or internodes sometimes glabrous, glaucous; lvs. glaucous, oval to narrow-elliptic, usually acutish, 1–2 cm. long, 0.5–1.3 cm. wide, grayish-green, thickish, short-pubescent above and beneath, sometimes almost glabrous, frequently lobed on young shoots; fls. 2 to several; sepals

ca. 1 mm. long; corolla pink, elongate-campanulate, 6–7 mm. long, the lobes ca. ½ as long as tube; fr. rounded to somewhat ellipsoid, 6–8 mm. in diam.; nutlets 3.5–4.5 mm. long.—Dry rocky slopes and ridges, 4000–11,000 ft.; Pinyon-Juniper Wd. to Subalpine F.; mts. from San Diego Co. to San Rafael Mts. (Santa Barbara Co.) and Sierra Nevada of Mono and Tulare cos.; to Nev., Ariz. June–Aug.

7. **S. longiflòrus** Gray. Low, spreading, the branches somewhat declined, 5–10 dm. long; young twigs glaucous, glabrous or sparsely pubescent; lvs. lanceolate to elliptical, acute to obtuse, entire, 0.6–1.5 cm. long, 0.3–0.5 cm. wide, pale, glabrous to sparsely pubescent; fls. 1–2 in upper axils or in small racemes; sepals deltoid, ca. 1 mm. long; corolla salverform, pink, 11–13 mm. long, the lobes ⅓–⅕ as long as the tube; style 5–7 mm. long, usually pilose; frs. ellipsoid, 8–10 mm. long; nutlets 4.5–5 mm. long.—Dry slopes, often on limestone, 4500–7000 ft.; mostly Pinyon-Juniper Wd.; desert mts. from Modoc and Mono cos. to Argus Mts. and e. San Bernardino Co.; to se. Ore., Colo., Tex. May–June.

5. Lonícera L. Honeysuckle

Erect or twining shrubs, with simple entire lvs., the uppermost pair often connate-perfoliate. Fls. in terminal spikes or heads or in small axillary clusters. Sepals 5 or obsolete. Corolla campanulate to tubular, ± gibbous at base, the limb somewhat irregular to strongly bilabiate with 4 lobes in upper lip and 1 in lower. Stamens 5, adnate to corolla-tube. Ovary 2–3-loculed. Fr. a fleshy, few-seeded berry. Ca. 100 spp., mostly N. Temp. (Named for A. *Lonitzer*, latinized Lonicerus, a German herbalist.)

A. Fls. in peduncled axillary pairs.
  B. Stems twining; fls. 3–4 cm. long. Introd. plant . . . . 9. *L. japonica*
  BB. Stems erect, fls. 0.6–2 or 2.5 cm. long. Native plants.
    C. Peduncles 2–5 mm. long; the ovaries enclosed in saclike juicy bractlets, the whole appearing as one fleshy fr. Sierra Nevada . . . . 1. *L. cauriana* Fern.
    CC. Peduncles 12–25 mm. long; berries not enclosed in the juicy saclike bractlets.
      D. Fls. subtended by large broad bracts that usually are red in age; corolla yellow or reddish, subregular . . . . 2. *L. involucrata*
      DD. Fls. subtended by minute bractlets or the bractlets obsolete; corolla bilabiate.
        E. Fls. purplish-black; ovaries of a pair connate . . . . 3. *L. conjugialis*
        EE. Fls. white to salmon; ovaries of a pair distinct . . . . 4. *L. utahensis*
AA. Fls. sessile, in whorls; plants ± twining or trailing.
  B. Fls. mostly in 1 whorl, sometimes 2–3; lvs. glabrous except for the ciliate margins. N. Calif. . . . . 5. *L. ciliosa*
  BB. Fls. in several whorls in spikes or panicles; lvs. not ciliate.
    C. Lvs. all distinct, the uppermost pair rarely slightly connate; infl. glandular-pubescent. S. Calif. . . . . 6. *L. subspicata*
    CC. Lvs. mostly with the uppermost connate. Widely distributed.
      D. Lvs. without stipules; corolla yellow, glabrous; infl. glabrous . . . . 7. *L. interrupta*
      DD. Lvs. mostly with stipules; corolla pinkish to purplish; infl. glandular-pubescent 8. *L. hispidula*

1. **L. cauriàna** Fern. [*L. caerulea* auth., not L.] Undershrub, 2–9 dm. high with shreddy light brown bark and ± pubescent young twigs; lvs. membranous, oval or elliptic to oblong-obovate, apically rounded, 1.5–3.5 cm. long, glabrous or pubescent above, paler and pilose beneath, entire and ciliate; petioles 2–5 mm. long; fls. in pairs on peduncles 2–5 mm. long; bracts at base of ovaries green, oblong-linear, 6–7 mm. long; bractlets surrounding the ovaries of the 2 fls. forming a saclike cup; corolla yellow, funnelform, 9–13 mm. long, pilose, slightly 2-lipped, the lobes almost as long as the tube; stamens slightly exserted; style filiform, ca. 13–15 mm. long; fr. blue-black, oval, ca. 6 mm. in diam., fleshy, consisting of the fused bractlets and containing the 2 ovaries. —Moist banks, 5000–10,500 ft.; Red Fir F. to Subalpine F.; Sierra Nevada from Tulare to Nevada cos.; to Wash., Wyo. May–July.

2. **L. involucràta** (Richards.) Banks. [*Xylosteon i.* Richards.] Twinberry. Upright shrub, 6–30 dm. high, deciduous, glabrous to ± pilose and glandular-pubescent; lvs. obovate to ovate or oval, acutish, darker and more glabrous above, paler and more pubescent beneath, often ciliate, 3–12 cm. long, on petioles 5–12 mm. long; fls. in axillary pairs, the peduncle 1.5–2.5 cm. long, rather coarse, with 2 bracts at its summit, ovate to oblong, 1–1.3 cm. long, often turning reddish or purplish; bractlets united, broad, resembling the bracts; corolla yellow or yellowish, subcylindric, ca. 12–16 mm.

long, viscid-pubescent, the lobes subequal, ca. ⅓ as long as the tube; ovaries not united; fr. black, ovoid to globose, ca. 8 mm. in diam., almost enclosed in the bractlets; $2n = 18$ (Jan Ammal & Saunders, 1952).—Moist places, 6000–10,000 ft.; Red Fir to Subalpine F.; Sierra Nevada from Tulare Co. n., to Modoc Co. and below 1000 ft.; Coastal Strand, Closed-cone Pine F., etc., Santa Barbara to Del Norte cos.; to Alaska, Quebec, Mex. The montane form usually less than 1 m. high, with corolla rarely tinged red and anthers ± exserted, is sometimes called *L. involucràta* and sometimes var. *flavéscens* (Dippel) Rehd.; June–Aug. The coastal plant 1–3 or more m. high, with corolla often tinged red and stamens included, is sometimes called var. *Ledeboùrii* (Esch.) Zabel. March–April.

3. **L. conjugiàlis** Kell. Slender-stemmed straggling deciduous shrub 6–15 dm. tall, with ± strigulose branchlets; lvs. oblong-ovate to -obovate, 2–6 cm. long, thin, light green, subglabrous above, lighter and more pubescent beneath, on petioles 2–3 mm. long; fls. in pairs on slender peduncles 2–2.5 cm. long; bracts minute or none; ovaries of the 2 fls. well united; corolla dull dark purple, 6–8 mm. long, bilabiate; fils. and style and corolla-throat white hairy; fr. bright red, ca. 5–6 mm. long.—Wooded slopes, 4000–10,200 ft.; Red Fir F. to Subalpine F.; Sierra Nevada from Tulare Co. n., to Modoc, Humboldt, and Trinity cos.; Wash., w. Nev. June–July.

4. **L. utahénsis** Wats. Deciduous, 6–15 dm. high, with slender glabrous branches; lvs. oblong-ovate to slightly -obovate, rounded at tip, 2–5 cm. long, thin, glabrous, sometimes scattered-pilose beneath, on petioles 2–5 mm. long; fls. in pairs on a slender peduncle 10–20 mm. long; the bracts narrow, ca. 1.5 mm. long; corolla white to light yellow, ca. 18–20 mm. long, somewhat saccate on one side of base of tube, the subequal lobes almost half as long as tube; fr. red, globular, ca. 6 mm. in diam.—Wooded slopes, 2700 ft.; Yellow Pine F.; near Yreka, Siskiyou Co.; to B.C., Rocky Mts. May–June.

5. **L. ciliòsa** (Pursh) Poir. [*Caprifolium c.* Pursh.] Trailing or climbing deciduous shrub, stems 1–5 m. long; lvs. broadly elliptic to oval, mostly obtusish, 3–8 cm. long, glabrous and green above, glaucous and subglabrous or ± pilose beneath, ciliate, on petioles 3–5 mm. long; fls. mostly in a single terminal whorl, which is sessile or short-peduncled; corolla yellow to red-orange, narrow-funnelform, 2–3 cm. long, swollen on one side at base, slightly 2-lipped, the lobes ca. half as long as tube; fr. red, ca. 5–6 mm. in diam.—Dry slopes, 2000–5000 ft.; mostly Yellow Pine F.; mts. of Humboldt, Trinity, and Siskiyou cos. to Shasta and Butte cos.; to B.C., Mont., Ariz. May–June.

6. **L. subspicàta** H. & A. Clambering shrub, evergreen, 1–2.5 m. high, with puberulent twigs; lvs. linear-oblong to oblong, 1–3 cm. long, 0.6–1 cm. wide, rounded at ends, ± revolute, entire, coriaceous, bicolored, pubescent especially beneath; petioles 1–5 mm. long; fl.-whorls several, compact, in short leafy spikes 2–12 cm. long; sepals broadly lanceolate, 1 mm. long; corolla yellowish or cream-color, 8–10 mm. long, glandular-pubescent, with a 2-lipped often recurved limb; fils. pubescent at base; berry 5–7 mm. long, yellowish or red, ellipsoid.—Dry slopes, below 3000 ft.; Chaparral; near Santa Barbara. June–July.

Var. **Johnstònii** Keck. [*L. J.* McMinn.] Lvs. whitish beneath, oblong-ovate to suborbicular, less than twice as long as wide; corolla 10–14 mm. long.—Common on dry slopes, below 5000 ft.; Chaparral; mts. from San Diego Co. to Santa Monica Mts. and to Kern Co., occasional, n. to Mt. Diablo, and Mt. Hamilton Range. April–June.

Var. **denudàta** Rehd. [*L. d.* Davids. & Mox.] Lvs. of width of var. *Johnstonii*, but yellowish and subglabrous beneath.—Chaparral; near San Diego to Temecula, Riverside Co. April–June.

7. **L. interrúpta** Benth. Evergreen bushy intricate shrub with the branches twining or leaning on other vegetation; twigs glabrous, glaucous; lvs. orbicular to elliptical, entire, 1.5–3.5 cm. long, glabrous or puberulent, glaucous beneath, green above, the uppermost pair usually connate; spikes interrupted, 3–16 cm. long, in an open panicle, subglabrous; corolla yellowish, funnelform, 10–14 mm. long, glabrous without; fils. pubescent; fr. red, subglobose, ca. 5 mm. in diam.—Dry slopes, 1000–6000 ft.; Chaparral, Foothill Wd., Yellow Pine F., etc.; middle and inner N. Coast Ranges, Sierra Nevada to San Bernardino Mts. May–July.

8. **L. hispídula** Dougl. var. **vácillans** Gray. [*L. h.* var. *californica* (T. & G.) Rehd. *L. catalinensis* Millsp.] Climbing shrubs 2–6 m. high, with long glabrous twigs; lvs.

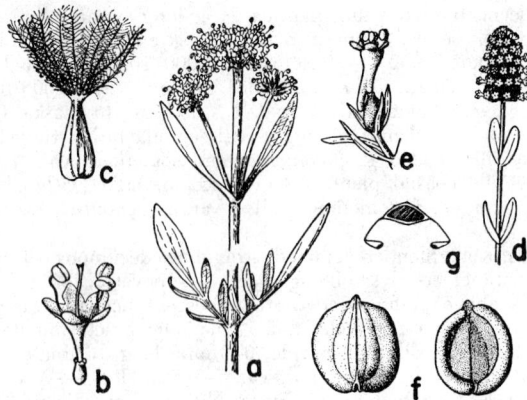

Fig. 104. VALERIANACEAE. *Valeriana capitata* ssp. *californica: a,* lvs. and terminal infl., × ½; *b,* fl., × 2, ovary inferior, calyx small, corolla 5-lobed, stamens 3, stigma 3-lobed; *c,* fr., × 2, ovary surmounted by plumose setae developed from calyx. *Plectritis macrocera: d,* habit, × ½; *e,* fl., × 3, ovary inferior, calyx obsolete, corolla ± tubular, 5-lobed and with basal spur; *f,* fr. × 5, in dorsal view (left) and ventral view (right) with inrolled lateral wings shown diagrammatically in g.

elliptic to oblong-ovate, obtusish, 3.5–8 cm. long, puberulent and glaucous beneath, green and mostly glabrous above, the several upper pairs usually connate-perfoliate, lower with petioles 2–3 mm. long; fls. in many whorls, forming spikes or loose panicles, the infl. glandular-pubescent; corolla purplish or pink, funnelform, 12–18 mm. long, glandular-pubescent, bilabiate; fr. red, subglobose, 5–6 mm. in diam.—Along streams and on wooded slopes, below 2500 ft.; Mixed Evergreen F., Redwood F., Douglas-Fir F., Foothill Wd., etc.; Sierra Nevada, Coast Ranges s. to w. Riverside Co. (rare in s. Calif.); Santa Cruz and Santa Catalina ids.; to s. Ore. April–July.

9. **L. japónica** Thunb. Half evergreen climber to 6 or 8 m., the twigs mostly hairy when young; lvs. ovate to oblong, 3–8 cm. long, pubescent especially beneath; fls. white, changing to yellow, sometimes purplish on outside, borne in axillary pairs on short peduncles with 2 ovate bracts; bractlets half as long as ovary; corolla 3–4 cm. long, pubescent; fr. black; $2n = 18$ (Jan Ammal & Saunders, 1952).—Occasional escape from cult. in Calif.; commonly so in se. states; native of Asia. Spring and summer.

115. Valeriànceae. Valerian Family (Fig. 104)

Herbs, sometimes shrubs, with opposite estipulate lvs. Fls. small, bisexual or unisexual, in cymose or capitate infl. Calyx annular or variously toothed, often inrolled in fl. and forming a feathery pappus in fr. Corolla funnelform to almost salverform, the base often saccate or spurred on one side. Stamens 1–4, inserted near the base of the corolla-tube. Ovary inferior, mostly 3-loculed, one locule fertile, the other 2 sterile or almost none. Ovule 1, pendulous. Fr. dry, indehiscent. Ca. 9 genera and 300 spp., mostly of N. Hemis.

Plants perennial.
   Corolla saccate at base; stamens 3; lvs. mostly ± pinnatifid or pinnate. Natives ...... 1. *Valeriana*
   Corolla with a tubular spur; stamen 1; lvs. entire. Escape from cult .... 4. *Centranthus*
Plants annual.
   Stem forked at summit; fls. in a flat-topped infl. .... 2. *Valerianella*
   Stem not forked at summit; fls. in a head or interrupted spike .... 3. *Plectritis*

1. Valeriàna L. Valerian

Perennial herbs from thickened strong-scented roots or rhizomes and leafy or sub-scapose stems. Lvs. opposite, estipulate, entire to pinnate. Infl. determinate, the cymes clustered or paniculate. Calyx initially involute, later spreading, the limb sessile, hyaline, membranaceous, becoming setose distally, the setae plumose or the limb short-cupuliform, ± toothed or lobed. Corolla rotate to funnelform, the tube straight or gibbous, throat ± hairy, the 5 lobes subequal. Stamens 3 (4), anthers sessile or on fils., 2- or 4-lobed. Ovary inferior, basically 3-loculed, maturing 1 fertile adaxial carpel with 1 pendulous ovule, the other 2 abaxial carpels vestigial. Style 1; stigma 3-lobed. Fr. an ak. Ca. 175 spp., largely of temp. and colder parts of N. Hemis. (Mediaeval Latin name.)

(Meyer, F. G. Valeriana in N. Am. and the W. I. Ann. Mo. Bot. Gard. 38: 377–503, 1951.)
Corolla 3–9 mm. long, funnelform to salverform, the tube distinctly gibbous; lvs. mostly ovate to rounded or ± spatulate.
    Lower cauline lvs. definitely petioled, essentially glabrous. Mendocino to Humboldt and Siskiyou cos.
                                                 1. *V. sitchensis*
    Lower cauline lvs. practically sessile, mostly puberulent. Sierra Nevada to Humboldt and Siskiyou cos. . . . . 2. *V. capitata*
Corolla ca. 3 mm. long, rotate to subrotate, the tube not gibbous; lvs. largely ± oblong. Modoc Co.
                                              3. *V. occidentalis*

1. **V. sitchénsis** Bong. [*V. Hookeri* Shuttl. *V. anomala* Eastw.] Plants 3.5–12 dm. tall, leafy, from stout rhizomes; lvs. mostly cauline, 1–2 dm. long, pinnate to pinnatifid, the lobes crenate to repand-dentate or entire, membranaceous, glabrous to hirtellous, the terminal lobe obovate to suborbicular, 2.5–4.5 cm. broad, the lateral lobes 1–4 pairs; basal lvs. obovate to ovate-elliptic, 1–4 dm. long, glabrous, the terminal lobe 5–10 cm. long; infl. in anthesis 2–8 cm. wide, later diffuse, the nodes pilosulous; bracts ca. 5 mm. long, glabrous or ciliate; calyx-limb 11–20-fid; fls. mostly perfect; corolla 4.5–7 mm. long, white to pinkish, sometimes pilosulous toward base, the limb scarcely ½ as long as the tube; anthers 3–6 mm. long, glabrous; aks. ovate to oblong-ovate, 3–6 mm. long, tawny or purplish, often spotted.—Woods and meadows, 4000–6000 ft.; largely Red Fir F. and above; mts. of Siskiyou, Humboldt cos.; to Alaska, Mont. July–Aug.

Ssp. **Scoùleri** (Rydb.) F. G. Mey. [*V. S.* Rydb. *V. Adamsiana, Follettiana,* and *humboldtiana* Eastw.] More slender, 3–7 dm. high; lvs. predominantly basal, 3–10 cm. long, entire to pinnate, the lobes mostly entire, glabrous, the terminal lobe or the undivided blade obovate, ovate-rhombic to roundish, 1–3 cm. wide, the lateral lobes 2–5 pairs; calyx 12–18-fid; corolla 5–9 mm. long, glabrous without; aks. 5–6 mm. long, tawny, rarely purplish.—Moist places, below 4000 ft.; Mixed Evergreen F., Douglas-Fir F., etc.; Mendocino to Del Norte cos.; to B.C. April–June.

2. **V. capitàta** Pall. ex Link ssp. **califórnica** (Heller) F. G. Mey. [*V. c.* Heller. *V. Whiltonae* Eastw. *V. sylvatica* var. *glabra* Jeps.] Plants 2–6 dm. tall from a rather stout rhizome, the stem puberulent to pilosulous, glabrescent above; lvs. mostly basal, usually forming a loosely tufted rosette, mostly undivided, sometimes pinnate or pinnatifid, 5–15 cm. long (including petioles), ± spatulate, entire to 3–7-toothed, puberulent or ciliate or glabrous; cauline lvs. mostly pinnate or pinnatifid, 3–8 cm. long; infl. 1.5–2.5 cm. wide at anthesis; calyx-limb 12–17-fid; corolla ca. 3 mm. long, glabrous or pilosulous without toward base; aks. elliptic to ovate-oblong, 4–15 mm. long, glabrous or pilosulous. —Moist or dryer places, 5000–10,500 ft.; Red Fir F. to Subalpine F.; Sierra Nevada from Tulare Co. n., to Trinity, Humboldt, Siskiyou, and Modoc cos.; Ore., w. Nev. July–Sept.

3. **V. occidentàlis** Heller. Plants 4–9 dm. tall, rather stout, glabrous or glabrescent, the nodes with tufts of white hairs; basal lvs. in a loose rosette, simple to pinnate, oblong to narrowly ovate, rarely roundish, 1–3 dm. long (including petiole), glabrous, the undivided blade and terminal lobe 2–10 cm. long, 1.3–6 cm. wide, the lateral pairs 1–2 pairs; cauline lvs. 2–4 pairs, shorter; infl. 3.5–5 cm. wide at anthesis, with bracts 5–6 mm. long; calyx-limb 11- to 16-fid; corolla 3–3.5 mm. long, rotate, white, glabrous without, scattered-pilosulous in the throat; aks. linear- to ovate-oblong, 3–5 mm. long, tawny.—Moist to damp places; Yellow Pine F.; Willow Creek, Modoc Co.; to Ore., Mont., Colo., S. Dak., Ariz. June–Aug.

2. **Valerianélla** Mill. CORN SALAD

Annual or biennial herbs, erect, with stems forking above. Lvs. ± succulent, opposite, mostly entire. Fls. small, white or pale blue, cymose-clustered, bracted. Calyx-limb obsolete or toothed. Corolla funnelform or salverform, equally or unequally 5-lobed. Stamens 3, rarely 2. Fr. 3-locular, with 2 of the locules empty and sometimes confluent, the 3d 1-seeded. Ca. 50 spp., of the N. Hemis. (Latin diminutive of *Valeriana*.)

(Dyal, Sarah C. Valerianella in N. Am. Rhodora 40: 185–212, 1938.)
Frs. almost round in lateral view, with a well developed corky mass on the back of the fertile locule
                                                                    1. V. olitoria
Frs. ± oblong in lateral view, much longer than thick, lacking a dorsal corky mass
                                                                    2. V. carinata

1. **V. olitòria** (L.) Poll. [*Valeriana Locusta* var. *o*. L. *Valerianella L.* Am. auth.] Stem simple below, 1–2.5 dm. high, pubescent; lower lvs. petioled, spatulate, connate, entire, the upper oblong-ovate, sessile, with few teeth near base, all ciliate, 1–5 cm. long; bracts ciliate, foliose, 3–6 mm. long; infl. corymbosely cymose; corolla white, often with bluish lobes, funnelform, 1.5 mm. long; fr. yellowish, mostly ca. 2, sometimes to 4 mm. long, laterally compressed, glabrous or finely pubescent, the cent. locule with a thick dorsal corky mass, the lateral locules sterile, narrow, with a shallow groove between them.—Shaded and often damp places, below 4000 ft.; Amador, Calaveras, Trinity, and Siskiyou cos. as natur. plant; native of Eu. April–May.

2. **V. carinàta** Lois. Vegetatively much like the preceding; corolla 1.5–2 mm. long, white with purplish-blue limb; frs. oblong, 1.5–2 mm. long, finely pubescent, the fertile locule smaller than the combined width of the sterile lateral cells which are ventrally incurved forming a deep cavity with prominent nerve down the middle; $2n = 18$ (Elvers, 1932).—Collected in Foothill Wd., on banks of Jackson Creek, 3 miles ne. of Jackson, Amador Co.; native of Eu. April–May.

3. **Plectrìtis** DC.

Subglabrous annuals, with erect to sprawling simple or few-branched stems, having tufts of hairs at nodes. Lvs. entire or with a few teeth, obovate-oblong, the lower short-petioled, the upper sessile and oblong to linear. Infl. capitate or interruptedly spicate; bracts linear-subulate, the base of several being fused to give a palmate appearance. Calyx obsolete. Corolla small, 5-lobed, sometimes strongly bilabiate, funnelform, usually spurred near the base. Stamens 3. Stigma-lobes mostly 2, flat-reniform. Ovary inferior, 1-loculed, the fr. a wingless or winged ak., the wings being an outgrowth from the dorsal angles of the fertile locule. Ca. 11 spp., w. N. Am. and Chile; much in need of study. (Latin, *plecto*, to plait, because of the complex infl.)

(Nielsen, Sarah Dyal. Systematic studies in the Valerianaceae. Am. Midl. Nat. 42: 480–501, 1949.)
A. Corolla rose to bright pink, 5–8 mm. long (exclusive of spur).
    B. Wings of fr. definitely incurved.
        C. Spur short and thick, much shorter than the tube and throat; keel of the convex side
            of the fr. acute, not grooved. N. Calif. . . . . 1. *P. congesta*
        CC. Spur long and slender, equal to the tube and throat; keel on the convex side of the
            fr. broad with a groove down the center. Cent. Calif.
            D. Wings of fr. short, thick, forming a narrow elliptic opening . . . . . . 9. *P. ciliosa*
            DD. Wings of fr. broad and thin, forming a round or broadly elliptic opening
                                                                    10. *P. californica*
    BB. Wings of fr. scarcely incurved, but flat . . . . 11. *P. macroptera*
AA. Corolla pale pinkish to white, 2–4 mm. long.
    B. Fr. wingless or nearly so.
        C. Corolla not spurred, but with a saclike swelling at its base .. 4. *P. anomala* var. *gibbosa*
        CC. Corolla with a short definite spur . . . . 5. *P. samolifolia*
    BB. Fr. with a definite, often inrolled wing on each side.
        C. Corolla spurless, merely with a saclike gibbosity at its base . . . . 4. *P. anomala*
        CC. Corolla with a definite spur at the base.
            D. Fr. keeled on the back, not grooved, the infolded wings of the concave side
                usually connivent at the base and diverging at the apex.
                E. Fr. very shining, obtusely keeled on the back . . . . 2. *P. magna*
                EE. Fr. not shining, acutely keeled on the back.
                    F. Lvs. not reduced in size toward the infl.; corolla-lobes more than half
                        as long as tubular part . . . . 3. *P. aphanoptera*
                    FF. Lvs. reduced in size toward the infl.; corolla-lobes less than half as long
                        as tubular part . . . . 5. *P. samolifolia* var. *involuta*
            DD. Fr. obtusely angled or rounded on the back with a narrow groove down the
                center, the margins of the wings ca. equally connivent at base and apex.
                E. Corolla white or pale, weakly bilabiate; spur thick.
                    F. Fr. glabrous or with stiff clavate hairs, if glabrous with wide thin wings
                                                                    6. *P. macrocera*
                    FF. Fr. woolly or with long cylindrical hairs, if glabrous with narrow thick
                        wings.

G.   Wings thick and narrow; fr. woolly or glabrous . . 7. *P. Jepsonii*
GG. Wings thin and wide, pubescent on the convex side or glabrous
                                                     8. *P. Eichleriana*
EE. Corolla pink, strongly bilabiate; spur slender.
   F.   Wings of fr. thick and short, forming a narrow elliptic opening on the
      concave side of the fr. . . . . 9. *P. ciliosa*
   FF. Wings of fr. broad and thin, forming a round or broad elliptic opening
      on the concave side of the fr.
      G.   Wings of fr. strongly incurved, forming a round deep cup-shaped
         opening on the concave side of the fr. . . . . 10. *P. californica*
      GG. Wings of fr. spreading, only slightly incurved, forming an elliptic
         or saucer-shaped opening on the concave side . . 11. *P. macroptera*

1. **P. congésta** (Lindl.) DC. [*Valerianella c.* Lindl.] Stems stoutish, 2–4.5 dm. tall; lvs. mostly entire, glabrous, obtuse, the lower obovate, the upper obovate to oblong-ovate, 1–5 cm. long; infl. mostly capitate, the bracts linear-subulate, slightly glandular-ciliate; corolla pink to rose, 5–8 mm. long, strongly bilabiate, the limb ca. half as long as tube; spur very short; fr. 2–4 mm. long, glabrous or pubescent, with wide wings that spread at the apex, keel acute.—Open slopes and grassy places; largely Yellow Pine F.; Humboldt and Trinity cos.; to B.C. April–June.

Var. **màjor** (F. & M.) Dyal. [*Betckea m.* F. & M.] Frs. wingless.—With the sp.

2. **P. mágna** (Greene) Suksd. [*Valerianella m.* Greene.] Plants 1–4 dm. high, erect or sprawling, simple or rather freely branched; lvs. largely oblong-ovate, 1–5 cm. long, the lower obovate-spatulate, entire, obtuse; infl. capitate to spicate; corolla white to pale pink, 3–4 mm. long, strongly bilabiate, with a short spur, the limb ca. ⅓ the length of the tube; fr. shining, 2–2.5 mm. long, glabrous or pubescent, yellow-brown, rounded on the convex side, wings wide, diverging at apex.—Grassy or brushy places; Mixed Evergreen F., N. Coastal Prairie, etc.; Coast Ranges from Santa Clara Co. to Humboldt and Siskiyou cos. April–May.

Var. **nítida** (Heller) Dyal. [*P. n.* Heller.] Frs. wingless.—Grassy fields, San Mateo Co.

3. **P. aphanóptera** (Gray) Suksd. [*Valerianella a.* Gray.] Stems 1–6 dm. tall; lvs. glabrous, oblong-ovate or the lower spatulate-obovate, obtuse; infl. capitate; corolla pale pink, 2.5–3.5 mm. long, strongly bilabiate, the limb ca. ⅓ as long as tube; spur very short; fr. 2–3 mm. long, glabrous or pubescent, narrow-winged, the keel on the convex side acute.—Grassy slopes and about brush; Mixed Evergreen F., N. Coastal Scrub, Coastal Prairie, etc.; Coast Ranges from Santa Barbara Co. n.; to Wash. March–May.

4. **P. anómala** (Gray) Suksd. [*Valerianella a.* Gray.] Freely branched, 1–5 dm. tall; lvs. oblong to oblong-ovate, obtuse to acutish; infl. capitate to interrupted-spicate; corolla whitish or light pink, 2–3 mm. long, obscurely bilabiate, the limb ⅔ as long as the gamopetalous part; spur reduced to a small sac; fr. 3–4 mm. long, glabrous or pubescent, the keel acute, the wings broad, conspicuous, spreading toward apex.—Reported from region of San Francisco Bay; Wash., B.C.

Var. **gibbòsa** (Suksd.) Dyal. [*P. g.* Suksd.] Frs. wingless.—San Mateo Co.; Ore. to B.C.

5. **P. samolifòlia** (DC.) Hoeck. [*Betckea s.* DC.] Stems 1–6 dm. high; upper lvs. acute; corolla 2–4 mm. long, white to pink, strongly bilabiate, the limb scarcely half as long as the tube; spur short; fr. yellow-brown to brown, glabrous or pubescent, 3-cornered, the wings almost abortive.—Dampish places, below 4500 ft.; many Plant Communities; Santa Clara Co. to Humboldt, Shasta, and Siskiyou cos.; B.C. March–June.

Var. **involùta** (Suksd.) Dyal. [*P. i.* Suksd.] Fr. winged with a concave side.—With the sp. and reported also from San Diego Co.

6. **P. macrócera** T. & G. [*Aligera m.* Suksd.] Slender or stoutish, 1–6 dm. tall, sometimes glandular-puberulent in infl.; upper lvs. remotely serrate toward base; infl. often with 2–3 whorls; corolla white or pale pink, 2–3.5 mm. long, bilabiate or almost regular, the limb ca. ⅓ the length of the tube; spur well developed, thick; fr. yellowish, 2–3 mm. long, pubescent or glabrous, ca. as long as wide, the wings broad, ca. equally connivent at both ends, with a narrow groove at the outer margin and with the opening on the concave side wider than long.—Moist grassy and wooded places, below 4000 ft.; Foothill Wd., V. Grassland, Oak Wd., Yellow Pine F., etc. cismontane Calif. from Ventura Co. n. to Wash. April–May.

Var. **collìna** (Heller) Dyal. [*P. c.* Heller.] Fr. smaller, more than twice as long as wide, almost wingless; upper lvs. conspicuously serrate. Grassy places, ca. 4300 ft., Mt. Hamilton Range.

Var. **mamillàta** (Suksd.) Dyal. [*Aligera m.* Suksd.] Fr. intermediate between *macrocera* and var. *collina*, hardly twice as long as wide, the wings covering ca. ⅔ of the concave side of the fr.—With the typical form.

Var. **Gràyii** (Suksd.) Dyal. [*Aligera G.* Suksd.] Fr. ca. as long as wide, the wings without a marginal groove and the opening on the concave side at least as long as wide.—At 4500–6500 ft.; Sagebrush Scrub, N. Juniper Wd., etc.; Plumas Co. to Modoc Co.; e. Wash., Mont., Utah. May–July.

7. **P. Jepsònii** (Suksd.) Davy. [*Aligera J.* Suksd. *P. glabra* Jeps.] Stout, 1.5–5 dm. tall; upper lvs. remotely toothed at base; corolla white, 4–5 mm. long, bilabiate; spur short, thick; fr. 2–2.3 mm. long, glabrous or woolly with long cylindrical hairs, the wings narrow and thick with a groove along the outer margin, strongly incurved to form a narrow elliptic opening.—Moist places in Vaca Mts., Napa and Colusa cos. April–May.

8. **P. Eichleriàna** (Suksd.) Heller. [*Aligera E.* Suksd.] Like the preceding, but fr. 2–4 mm. long, ca. as wide, with broad thin wings, the opening on the concave side wider than long.—Near San Francisco Bay region, as at Antioch, n. to Colusa Co. March–May.

9. **P. ciliòsa** (Greene) Jeps. [*Valerianella c.* Greene. *Aligera c.* Suksd.] Slender, 1–4 dm. tall; lvs. obovate to narrow-oblong and 3–4 times as long as wide; corolla 5–7 mm. long, deep pink, bilabiate with a deep red dot at the base on either side of middle lobe; spur slender, at least as long as tube; fr. 2–3 mm. long, pubescent, the wings sharply inrolled, thick, narrow, forming a narrow-elliptic opening on the concave side.—Grassy places, mostly below 4000 ft.; Foothill Wd., V. Grassland, etc., w. base of Sierra Nevada, Coast Ranges from Mendocino Co. s., to Los Angeles Co. March–May.

Var. **Davyàna** (Jeps.) Dyal. [*P. D.* Jeps.] Fls. smaller, 2–3 mm. long, short-spurred. —With the sp. especially in s. part of its range.

10. **P. califórnica** (Suksd.) Dyal. [*Aligera c.* Suksd.] Much like the preceding, the corolla 5–7 mm. long; fr. 3–4 mm. long, pubescent, the keel broad with a groove down the center and wings thin and incurved to form an oval opening on the concave side. —Damp or shaded places, below 6000 ft.; Yellow Pine F., Chaparral, Foothill Wd., etc.; Napa and Calaveras cos. to Ore. April–June.

Var. **rùbens** (Suksd.) Dyal. [*Aligera r.* Suksd.] Corolla 2–3 mm. long; fr. 2–2.5 mm. long.—San Diego Co. through cismontane Calif. to s. Wash. The common form in the s. part of the state.

11. **P. macróptera** (Suksd.) Rydb. [*Aligera m.* Suksd.] Corolla 5–8 mm. long, dark pink, bilabiate, with long slender spur; fr. pubescent, 3–4 mm. long and wide, the wings thin, wide, extending beyond the apex of the fertile locule and weakly inrolled, the keel broad with cent. groove.—Shaded or damp places, mostly below 3000 ft.; many Plant Communities; cent. Calif. to Wash. March–May.

Var. **patellifórmis** (Suksd.) Dyal. [*Aligera p.* Suksd.] Corolla 2–3 mm. long, spur short; frs. 3–3.5 mm. in diam.—At 5000 ft.; Yellow Pine F.; Cuyamaca Mts., Laguna Mts. (San Diego Co.), Kern and Monterey cos. to Wash. April–June.

4. Centránthus DC.

Annual or perennial herbs, sometimes suffruticose. Lvs. entire, dentate, or pinnatisect. Fls. small, red to white, in dense terminal clusters. Calyx cut into 5–15 narrow divisions infolded at anthesis, enlarging and spreading in fr. Corolla with a slender tube, the limb 5-parted; spur basal. Stamen 1. Fr. 1-loculed, narrow, crowned with a pappuslike calyx, 1-seeded. Ca. 12 spp., Old World. (Greek, *kentron*, a prickle or spur, and *anthos*, fl.)

1. **C. rùber** (L.) DC. [*Valeriana r.* L.] RED VALERIAN. Smooth, ± glaucous, 3–8 dm. high; lvs. lance-ovate to ovate, sessile and broad at base or short-petioled, mostly entire, 3–10 cm. long; fls. many, fragrant, red to pink or occasionally white, rather crowded; corolla-tube ca. 10 mm. long, the limb spreading, spur slender; stamen exserted; fr. an elongate narrow nut, 3–4 mm. long; $2n = 14$ (Poucques, 1949).—Occasionally established as wild plant; escape from cult. April–Aug.

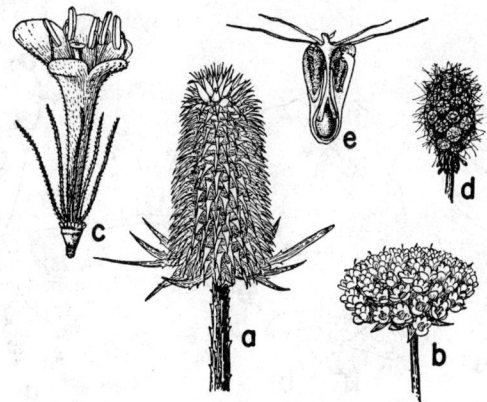

Fig. 105. DIPSACACEAE. *Dipsacus fullonum:* a, oblong head, × ½, with basal invol. of bracts and floriferous bracts above. *Scabiosa atropurpurea:* b, head of fls., × ½; c, fl., × 2, calyx of 5 setae, ovary inferior, corolla oblique, 5-lobed, stamens 5; d, fruiting head, × ½; e, section of fr., × 2, with seed below and calyx-setae above.

116. Dipsacàceae. Teasel Family (Fig. 105)

Annual to perennial herbs with lvs. opposite or whorled, entire, toothed or pinnatifid, estipulate. Fls. small, in dense bracted and involucrate heads or interrupted spikes. Calyx cup-shaped, discoid or divided into spreading bristles. Corolla limb 4-5-lobed. Stamens 2-4, inserted on throat of corolla. Ovary inferior, 1-loculed; style filiform; stigma simple. Ovule 1. Fr. an ak. Ca. 11 genera and 160 spp., of Old World.

Scales of the elongate receptacle prickly pointed; plant ± thistlelike .... 1. *Dipsacus*
Scales of the receptacle not prickly, but herbaceous, inconspicuous and concealed among the fls.; plant not thistlelike .... 2. *Scabiosa*

1. Dípsacus L. Teasel

Tall stout herbs with opposite lvs., the cauline united at the base. Fls. pinkish, in dense terminal peduncled oblong heads surrounded by invol. of elongated bracts. Bracts subtending the fls. (i.e. the receptacular) shorter, rigid and spinelike in fr. Calyx and corolla 4-lobed. Stamens 4. Ca. 15 spp., of Eurasia and N. Afr. (Greek, *dipsa*, thirst, since the connate lf.-bases in some spp. hold water.)

Bracts of invol. curving upwards, the longest equaling or exceeding the head; bracts of receptacle exceeding the fls., ending in long straight slender spine .... 1. *D. sylvestris*
Bracts of invol. spreading ± horizontally, often not overtopping the head; those of receptacle almost equal to fls., ending in a stiff recurved spine .... 2. *D. fullonum*

1. **D. sylvéstris** Huds. Biennial, prickly, 0.5-2 m. tall; lvs. lance-oblong, toothed and often prickly on the margin, to 2.5 dm. long; involucral bracts slender, unequal; heads 3-8 cm. long, conical, blunt, erect; receptable-bracts exceeding the fls., ciliate, ending in a long straight spine; corolla mostly rose-purple, 7-8 mm. long; fr. 5 mm. long; $2n = 16$ (Poucques, 1949).—Occasionally natur., especially in the Bay Region and in Humboldt Co.; native of Eurasia. April-Aug.

2. **D. fullònum** L. Fuller's Teasel. Heads 5-8 cm. long; involucral bracts spreading ± horizontally; receptacle-bracts almost equaling the fls., spinose-ciliate, ending in a stiff recurved spine; corolla-tube 10-13 mm. long; $2n = 18$ (Kachidze, 1929).—Locally natur. ± throughout cismontane Calif.; grown in Eu. and used for fulling in textile mills. May-July.

2. Scabiòsa L. Scabiosa. Mourning-Bride

Fairly large herbs with opposite lvs., no prickles, and blue or white or purplish fls. in peduncled involucrate heads. Bracts of invol. herbaceous, separate. Scales on receptacle

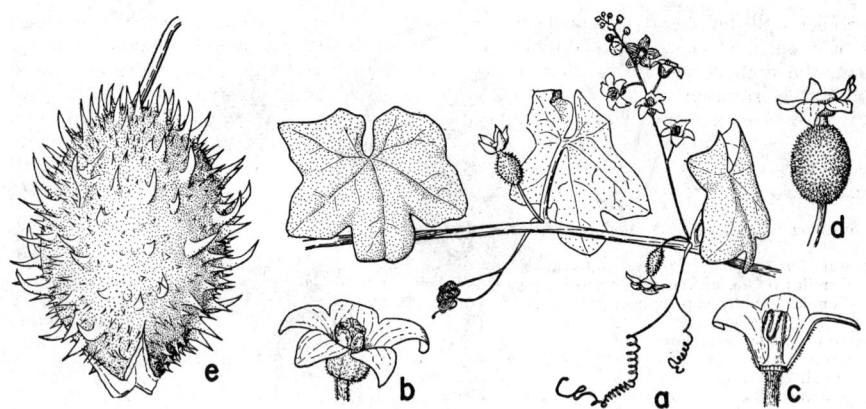

Fig. 106. CUCURBITACEAE. *Marah macrocarpus: a,* branch, × ½, with lvs., raceme of ♂ fls. and individual ♀ fls.; *b,* ♂ fl., × 2½, corolla 5-merous, stamens united; *c,* same in section; *d,* ♀ fl., × 1, ovary inferior, corolla 5-lobed, stigma round; *e,* fr., × ½.

small or lacking, not sharp-pointed. Calyx of 5 setae united at base. Corolla oblique, 5-lobed, the marginal fls. larger with the upper lobes smaller than the lower. Stamens 4, rarely 2. Ca. 80 spp., mostly temp., Eurasia and Afr. (Latin, name meaning scurfy, the plants having been used for skin diseases.)

Fls. blue; plants 1.5–4 dm. tall .... 1. *S. stellata*
Fls. purple or white; plants 4–6 dm. tall .... 2. *S. atropurpurea*

1. **S. stellàta** L. Hairy annual; lvs. cut or lyrate, with large terminal obovate lobe; fls. blue.—Rare escape; native of Eu.

2. **S. atropurpùrea** L. Pincushions. Branching annual ca. 5–6 dm. tall, the stems subglabrous; basal lvs. oblong-spatulate, simple or lyrate, coarsely dentate; cauline lvs. pinnately parted, the lobes oblong, dentate or cut; fls. dark purple, rose or white, in heads to 5 cm. across; inner corollas ca. 1 cm. long; calyx-awns 5–8 mm. long; $2n = 16$ (Braun, 1937).—Natur. as escape from gardens; native of Eu. Much of year.

117. Cucurbitàceae. Gourd Family (Fig. 106)

Annual or perennial herbs, mostly with soft stems. Lvs. alternate, broad, palmately veined or lobed, glabrous to scabrous or hairy, sometimes compound. Tendrils lateral, simple or branched, opposite the lvs. Fls. regular, unisexual, the plants mostly monoecious. Calyx 5-lobed. Corolla 5-lobed or almost polypetalous; stamens seemingly 3, but really 5 with 2 pairs united, mostly monadelphous by contorted anthers; fils. free or connate. Ovary inferior, with parietal placentation and many ovules; carpels commonly 3. Fr. mostly an indehiscent pepo (fleshy berrylike structure with rind and spongy interior), sometimes a papery bladdery ± spiny pod. Ca. 90 genera and 700 spp., largely trop.

A. Fr. fleshy, with hard or firm rind, indehiscent, often dry when mature; fls. yellow.
    B. Corolla campanulate, distinctly gamopetalous, 5-lobed, 5–12 cm. long; plants perennial from large fleshy root .... 3. *Cucurbita*
    BB. Corolla almost rotate, almost polypetalous, less than 2 cm. long; plants annual
                                                       4. *Citrullus*
AA. Fr. not fleshy, ± spinose, dry; fls. white or cream.
    B. Fr. more than 2 cm. long, dehiscent, 2–many-seeded; corolla 5–30 mm. in diam. Cismontane
                                                   1. *Marah*
    BB. Fr. less than 1 cm. long, indehiscent, 1-seeded; corolla 1.5 mm. in diam. Deserts
                                                   2. *Brandegea*

1. Màrah Kell. Wild Cucumber. Big-Root

Climbing or trailing herbs with annual stems arising from large fusiform to subglobose perennial tubers. Lvs. suborbicular, ± cordate and 5–7-lobed or -cleft. Tendrils 1–3-fid;

petioles well developed. Staminate fls. in axillary racemes or panicles, tardily deciduous; sepals small, obsolete; corolla rotate to campanulate, mostly 5(4–8)-merous; stamens 3–4, the anthers forming a globose head. Pistillate fl. solitary, from same axil as ♂ infl.; style short or obsolete; stigma discoid or subglobose. Ovary 4(2 or 8)-loculed; ovules 1–4 in a locule. Fr. a turgid caps., almost globose to ± ellipsoid or fusiform, pendent, with many large spines or smaller, sometimes almost lacking, irregularly dehiscent near apex. Seeds large, brownish-gray to olive or tan; cotyledons large; seed-coat thick-walled. Seven spp., of Pacific Coast. (An aboriginal name.)

(Stocking, K. M. Some taxonomic and ecological considerations of the genus Marah. Madroño 13: 113–137, 1955.)
Mature corolla rotate or only slightly cup-shaped.
  Corollas rotate, cream; fr. globose; ovules 4 or fewer. Ventura and Kern cos. n. .... 1. *M. fabaceus*
  Corollas slightly cup-shaped, white; fr. oblong-cylindrical; ovules more than 4. Santa Barbara Co. s.
                                                              2. *M. macrocarpus*
Mature corollas campanulate.
  Staminate fls. less than 8 mm. in diam.; ovary and fr. globose; lvs. glaucous beneath, 3–8 cm. broad, deeply dissected. Sonoma and Calaveras cos. to Shasta Co. .... 3. *M. Watsonii*
  Staminate fls. more than 8 mm. in diam.; ovary and fr. elongate; lvs. not glaucous beneath, 8–30 cm. broad.
    Ovaries and fr. ovoid, the spines inconspicuous; seeds discoid. Coast Ranges from Santa Clara Co. n. .... 4. *M. oreganus*
    Ovaries and fr. cylindrical, with conspicuous spines; seeds cylindrical. N. Los Angeles to Tuolumne cos., along Sierran foothills .... 5. *M. horridus*

1. **M. fabàceus** (Naud.) Greene. [*Echinocystis f.* Naud. *Megarrhiza californica* Torr. in Wats. *Micrampelis f.* Greene.] Stems 3–7 m. long, subglabrous to ± pubescent; lvs. ,suborbicular, 5–10 cm. broad, ± deeply 5–7-lobed, the lobes less than half the lf.-length, acute to obtuse at apices, the basal sulcus 1–3 cm. deep; petioles 3–6 cm. long; ♂ fls. 8–18(–25) in a raceme; pedicels 3–6 mm. long; ♂ corolla rotate, 7–10(–12) mm. in diam., mostly cream; ♀ 10–15 mm. in diam., the lobes unequal; fr. globose below, tapering to a tip, 4–5 cm. in diam., with rigid spines 5–25 mm. long; mature seeds usually 4, asymmetrical, laterally flattened, 18–24 mm. long, 12–15 mm. thick, brownish tan; n = 16 (McKay, 1931).—Banks and slopes, below 2500 ft.; Coastal Strand, Mixed Evergreen F., etc.; near the coast, Marin to Monterey cos. Feb.–April.

Var. **agréstis** (Greene) Stocking. [*Micrampelis f.* var. *a.* Greene. *E. inermis* Congd. *E. scabrida* Eastw.] Spines of fr. soft, less than 5 mm. long; seeds mostly 1–3, seldom flattened laterally.—Below 5000 ft.; Mixed Evergreen F., Foothill Wd., Chaparral, etc.; inner Coast Ranges and Sierran foothills; Mendocino and Butte cos. s. to Kern Co., to the coast from San Luis Obispo to Ventura cos.

2. **M. macrocárpus** (Greene) Greene. [*Echinocystis m.* Greene. *Micrampelis leptocarpa* Greene.] Vegetatively much like the preceding sp.; lf.-blades 5–10 cm. broad; ♂ corollas 8–13 mm. in diam., white, the lobes ovate, 3–12 mm. long; ♀ 15–20 mm. in diam.; fr. cylindrical, mostly 8–12 cm. long, 6–9 cm. in diam., beaked, densely spiny, the spines flattened, 5–30 mm. long, 1–3 mm. wide at base; seeds oblong, somewhat flattened, 15–20 mm. long, 12–18 mm. wide, 11–14 mm. thick, brown to tan; n = 16 (McKay, 1931), 32 (Whitaker, 1950).—Dry places, mostly below 3000 ft.; Coastal Sage Scrub, Chaparral, S. Oak Wd.; cismontane s. Calif., Santa Barbara to San Diego cos.; n. L. Calif. Jan.–April.

Var. **màjor** (S. T. Dunn) Stocking. [*M. m.* S. T. Dunn.] Lf.-blades mostly 15–25 cm. broad; ♂ fls. 15–30 mm. broad; seeds 28–33 mm. long, 21–25 mm. broad, 12–14 mm. thick.—Chaparral etc.; Channel Ids.

3. **M. Watsònii** (Cogn.) Greene. [*Echinocystis W.* Cogn. *Micrampelis W.* Greene. *E. muricata* Kell., not *Marah m.* Kell.] Stems 1–3 m. tall, subglabrous with a few scattered hairs; lvs. 3–8 cm. broad, orbicular, 5-cleft, the lobes mostly obtuse; petioles 2–6 cm. long; ♂ fls. 3–12 in a raceme; pedicels 10–15 mm. long; corolla greenish, campanulate, the ♂ 5–7 mm., ♀ 8–12 mm. in diam.; fr. globose, 2–3 cm. in diam., with weak spines 1–2 mm. long; seeds 11–14 mm. in diam., globose, 2, grayish-brown with some mottling in a reticulum of dark lines.—Below 2500 ft.; Chaparral, Foothill Wd.; about the Sacramento V. from Shasta to Sonoma, Solano, and Calaveras cos. March–April.

4. **M. oregànus** (T. & G.) Howell. [*Sicyos o.* T. & G. *Echinocystis o.* Cogn. *Micrampelis o.* Greene. *Marah muricatus* Kell. *E. Marah* Cogn.] Stems 1–7 m. long, sparsely

pubescent to subglabrous; lvs. suborbicular, 8–30 cm. broad, with broad rounded sinuses between the 5–7 ± shallow lobes; petioles 4–10 cm. long; ♂ fls. 5–20 in a raceme; corolla campanulate, 12–15 mm. in diam., white; ♀ 15–17 mm. in diam.; fr. ovoid, attenuate at both ends, 4.5–7 cm. long, 3–4 cm. thick, with alternating longitudinal dark and light bands, the spines few to almost none, weak, to 6 mm. long; seeds discoid, 16–20 mm. in diam., 8–12 mm. thick, dark reddish-brown; n = 16 (McKay, 1931).— Hilly places, below 6000 ft.; Redwood F., Mixed Evergreen F. to Yellow Pine F.; Coast Ranges from Santa Clara to Siskiyou and Del Norte cos.; B.C. March–May.

5. **M. hórridus** (Congd.) S. T. Dunn. [*Echinocystis h.* Congd.] Stems 1–4 m. long, sparsely puberulent; lvs. roundish, 10–15 cm. in diam., usually rather deeply 5–7-lobed; petioles 3–8 cm. long; ♂ fls. 5–12 in a raceme; pedicels 5–15 mm. long; ♂ corolla campanulate, 10–12 mm. in diam., white; ♀ 13–17 mm. in diam.; fr. oblong-ellipsoidal, 9–15 cm. long, 6–9 cm. in diam., very spinose, the spines 5–35 mm. long, 3–7 mm. wide at base; seeds lenticular, oblong-obovoid, 26–32 mm. long, 15–18 mm. wide, 13–15 mm. thick, light olive.—Dry slopes below 2500 ft.; Foothill Wd., etc.; n. Los Angeles Co. and Tehachapi Mts. along Sierran foothills to Tuolumne Co. March–April.

2. Brandègea Cogn.

Perennial herbs with large thick roots. Lvs. 3–5-parted. Fls. small; corolla rotate, 5-parted. Ovary 1-loculed, oblique, 1-ovuled. Fr. narrowly obovoid, asymmetrical, indehiscent, sparsely echinate, less than 1 cm. long, thin-walled. Four or 5 spp. of sw. N. Am. (Named for T. S. *Brandegee*, pioneer California botanist.)

1. **B. Bigelòvii** (Wats.) Cogn. [*Elaterium B.* Wats. *Echinocystis parviflora* Wats.] Stems slender, trailing or climbing over shrubs, 1–2 m. long, glabrous; lf.-blades ± round in outline, 1–5 cm. across, shallowly or deeply 3–5-lobed, the upper lobes lance-oblong to triangular, the upper surface closely covered with disklike pustules; petioles shorter than blades; ♂ fls. few, in small axillary clusters; corolla 1.5 mm. broad, ± cup-shaped; body of fr. 5–6 mm. long; style 5 mm. long.—Locally common in washes and canyons, below 2500 ft.; Creosote Bush Scrub; Colo. Desert to Sheep Hole Mts., s. Mojave Desert; Ariz., L. Calif. March–April.

3. Cucúrbita L. GOURD. MELON. PUMPKIN

Annual and perennial scandent or trailing herbs, with fibrous to tuberous roots; usually monoecious. Lvs. entire to lobed; tendrils opposite lvs., well developed. Fls. large, yellow to yellowish, campanulate, solitary in axils, the ♂ long-, the ♀ short-peduncled. Fils. distinct; anthers linear, contorted. Fr. a pepo, morphologically 3-loculed with 3–5 placentae often in 1 cavity; stigmas 3–5, each 2-lobed. Seeds many, ovate or oblong-ovate, white to tawny or black, flat. Spp. ca. 25, of the warmer parts of Am.; many grown for food, ornaments, etc. (Latin name of the gourd.)

(Bailey, L. H. Species of Cucurbita. Gentes Herbarum 6: 267–322, 1943.)
Lf.-blades triangular-ovate, 1–2.5 dm. long; fr. 3-celled, on a slender pedicel without thickened ridges . . . . 1. *C. foetidissima*
Lf.-blades palmately 5-lobed or -cleft, less than 1 dm. long; fr. seemingly 5-celled, with thickened ridges near summit of pedicel.
    Lvs. lobed to near the base, the lobes very narrow . . . . 2. *C. digitata*
    Lvs. lobed ca. half way, the lobes ± deltoid . . . . 3. *C. palmata*

1. **C. foetidíssima** HBK. [*Pepo f.* Britton.] CALABAZILLA. Coarse rough strong-smelling perennial with an immense fusiform root; stems mostly trailing, 2–4 m. long; lvs. erect, triangular-ovate, somewhat cordate at base, 1–2.5 dm. long, on somewhat shorter petioles; ♂ fls. 10–12 cm. long, rough-pubescent, the corolla ribbed, veiny, with broad lobes; ♀ fls. 9–10 cm. long, pubescent; calyx-lobes narrow, 8–10 mm. long; fr. slightly oblong-globose, 6–9 cm. high, dull light green with 5–6 main cream-white stripes and a few intermediate ones as well as narrow mottling; peduncles angled; seeds oblong-ovate, ca. 12 mm. long, obtusely edged; 2n = 40 (McKay, 1931).—Common in sandy and gravelly places, below 2000 ft., Coastal Sage Scrub, Coastal Strand, V. Grassland, etc., cismontane Calif. from San Joaquin V. to San Diego Co.; to ca. 4000 ft. in Shadscale Scrub, etc., Mojave Desert; to Nebr., Tex. June–Aug.

2. **C. digitàta** Gray. Perennial with a deep fleshy fusiform root; stems trailing, slender, sparsely pilose, to ca. 1 m. long; lvs. 5-parted to the base, gray-green, with usually lighter midribs, the lobes linear-lanceolate, 5–10 cm. long, asperulate with whitish points and some short stiff setae; calyx-lobes lance-linear, ca. 1 cm. long; corollas 4–5 cm. long; fr. subglobose, ca. 8 cm. in diam., vivid green with narrow whitish-green stripes; peduncle shallowly ridged; seeds white, ovate, 10–11 mm. long; $2n = 40$ (McKay, 1931).—Occasional in sandy places; Creosote Bush Scrub at Whitewater, Colo. Desert; Chaparral, Buckman Springs, e. San Diego Co.; Coastal Sage Scrub, Santa Ana River Canyon, Orange Co.; possibly introd. from Ariz., ranging to New Mex., Son. Aug.–Oct.

3. **C. palmàta** Wats. [*C. californica* Torr. ex Wats.] Stems 3–12 dm. long; lvs. grayish, palmately 5-lobed, ca. 3–9 cm. long and broad, with low bulbate prickles and some longer stiff short setae, the lobes triangular to deltoid-lanceolate; corolla 4–6 cm. long; fr. globose, 8–9 cm. long, dull light green with broad, ill-defined bands and splashes of greenish-white; seeds white, narrow-ovate, 10–14 mm. long; $2n = 40$ (McKay, 1931).—Occasional in dry sandy flats, below 4000 ft.; Creosote Bush Scrub, Colo. and Mojave deserts; Coastal Sage Scrub, V. Grassland, interior cismontane s. Calif. and San Joaquin V.; Ariz., L. Calif. April–Sept. Plants with triangular lf.-lobes rather than lanceolate, have been called *C. californica*.

4. Citrúllus Neck.

Monoecious annuals or perennials, ± hairy, with long trailing stems bearing branched tendrils. Lvs. deeply pinnatifid, the lobes again lobed. Fls. of medium size, light yellow, solitary in axils; corolla 5-parted nearly to base. Ovary with 3 placentae and many ovules; stigmas 3. Ca. 4 spp., of trop. Old World. (Latin diminutive of *citrus*, some frs. having similar odor and flavor.)

1. **C. vulgàris** Schrad. var. **citroìdes** Bailey. CITRON. MELON. Hairy annual; lf.-blades ovate in outline, 3–8 cm. long, pinnately divided into 3–4 pairs of lobes, these again lobed and toothed, the segms. broad at apex; fr. hemispherical, hard, smooth, green with white spots, small, white-fleshed; seeds greenish or tan.—What seems to be this has been found a number of times as a garden escape; native of Afr. *C. vulgaris* itself is the Watermelon.

118. Campanulàceae. BELLFLOWER FAMILY (Fig. 107)

Plants herbaceous or rarely suffrutescent, usually with milky juice. Lvs. simple, alternate, estipulate. Fls. mostly perfect, usually 5-merous except as to carpels. Ovary inferior or partly so, a fl.-tube mostly not developed to any extent above it. Sepals or calyx-lobes mostly 5, persistent. Corolla sympetalous, regular or irregular. Stamens distinct or united. Style 1; ovary 2–5-loculed, sometimes without internal septum. Fr. a caps. with many minute seeds. Ca. 60 genera and 1600 spp., temp. and trop. regions; many of ornamental use.

A. Corolla regular; anthers and fils. distinct. (Campanuloideae)
  B. Caps. dehiscent on the side; lvs. broad to roundish (except the cauline lvs. of *Campanula rotundifolia*).
    C. The caps. opening by lateral openings formed by uplifting of small lids.
      D. At least the terminal fls. pedicelled; all fls. normally opening and expanding; ovary short and broad; annual or perennial . . . . 1. *Campanula*
      DD. At least the terminal and upper fls. sessile; lower fls. mostly cleistogamous with vestigial corollas, upper expanded; ovary somewhat elongate; annuals
                                                        2. *Triodanis*
    CC. The caps. opening by irregular fissures; ovary short and broad; slender annual with roundish sessile lvs. and cleistogamous as well as normal fls. . . . . 3. *Heterocodon*
  BB. Caps. dehiscent at the apex; lvs. narrow, oblong . . . . 4. *Githopsis*
AA. Corolla irregular. (Lobelioideae)
  B. Anthers all alike, distinct; fls. minute, inconspicuous.
    C. Caps. dehiscent by apical valves or irregularly; lvs. mostly in a basal rosette; fls. racemose . . . . 5. *Nemacladus*
    CC. Caps. circumscissile; lvs. basal and cauline; fls. capitate . . . . 6. *Parishella*
  BB. Anthers united into a tube, 2 shorter than the others; fls. mostly showier.
    C. Fls. sessile in the axils of foliaceous bracts, the ovary linear and simulating a pedicel; caps. elastically dehiscent by long slits on the sides . . . . 7. *Downingia*

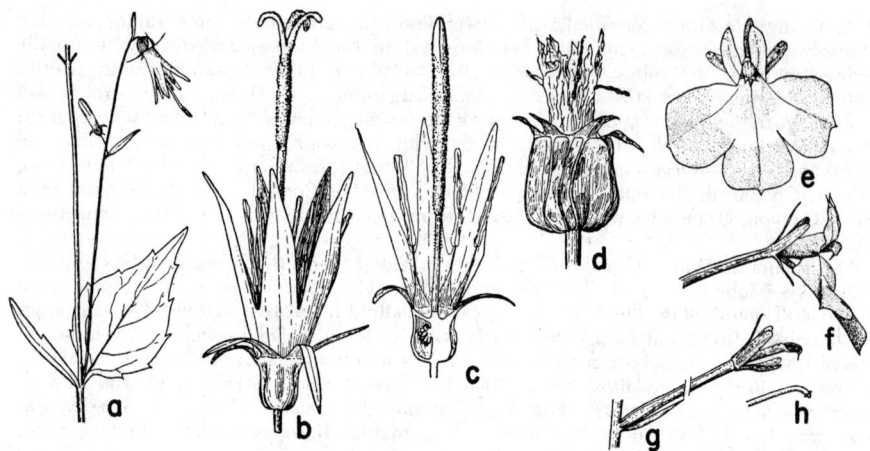

Fig. 107. CAMPANULACEAE. *Campanula prenanthoides: a,* upper lf. and fl., × 1; *b,* fl., × 2, ovary inferior, calyx spreading, corolla erect, stamens 5 (dilated at base as seen in *c*), style long-exserted, papillate; *d,* fr., × 2. *Downingia: e,* fl., × 2, as seen from front, *f,* from side, with inferior ovary, 5 narrow sepals, bilabiate corolla, the upper lip of 2 short lobes, the lower of 3 broad ones; *g,* corolla removed to show united stamens; *h,* style and stigma.

> CC. Fls. pedicelled; caps. round to fusiform, indehiscent or opening by valves.
>   D. Corolla wanting or not more than 4 mm. long, cleft nearly to the base on the
>   upper side; fr. with 1 cavity .... 8. *Legenere*
>   DD. Corolla 6–40 mm. long; caps. bilocular.
>   E. Plants annual; corolla 6–9 mm. long .... 9. *Porterella*
>   EE. Plants perennial; corolla 20–40 mm. long .... 10. *Lobelia*

1. Campánula L. BELLFLOWER

Annual or perennial herbs with lower lvs. oblanceolate, spatulate or ovate, petioled, the upper narrower. Fls. solitary to paniculate, blue or violet, sometimes white, showy; fl.-tube turbinate-obconic. Calyx mostly 5-merous. Corolla campanulate or nearly so. Fils. dilated at the base. Stigmas 3–5. Ovary ± inferior, 3-loculed; fr. a caps., opening near the base or the apex; seeds many, minute. Ca. 250 spp., largely of N. Hemis. and in Old World. Many used for ornamentals. (Diminutive of Latin, *campana,* a bell.)

A. Plants annual.
  B. Corolla showy, ca. twice as long as calyx; stamens 4–6 mm. long .... 1. *C. exigua*
  BB. Corolla inconspicuous, ca. as long as calyx; stamens ca. 2 mm. long .... 2. *C. angustiflora*
AA. Plants perennial.
  B. Style longer than and well exserted from corolla; cauline lvs. serrate.
    C. Lvs. sessile or nearly so; corolla bright blue, the lobes narrowly lanceolate. Monterey
      and Tulare cos. .... 3. *C. prenanthoides*
    CC. Lvs. on margined petioles at least half as long as the blade; corolla pale blue to white,
      the lobes ovate-oblong. Humboldt and Sierra cos. n. .... 4. *C. Scouleri*
  BB. Style not longer than corolla.
    C. Plant finely scabrous-puberulent throughout, 0.3–1.2 dm. high. Siskiyou Co.
                                   5. *C. scabrella*
    CC. Plant essentially glabrous, the stems often 1.5–4 dm. long.
      D. Upper stem-lvs. entire to crenate; stems mostly more than 1.5 dm. high.
        E. Upper cauline lvs. linear or lance-linear, entire; corolla bright blue, 1.5–2 cm.
          long. From above 4000 ft., Trinity and Siskiyou cos. .... 6. *C. rotundifolia*
        EE. Upper cauline lvs. oval or elliptic-ovate, crenate; corolla pale blue, 1.2–
          1.4 cm. long. Coastal swamps .... 7. *C. californica*
      DD. Upper stem-lvs. serrate; stems less than 1.5 dm. high ...... 8. *C. Wilkinsiana*

1. **C. exígua** Rattan. Slender-stemmed annual, 0.5–1.6 dm. high, mostly branched, stiff-pubescent especially below; lvs. obovate to linear, passing upward into subulate bracts; fls. scattered in upper parts, erect; calyx-lobes subulate, 4–6 mm. long; corolla light blue, 8–10 mm. long; stamens 4–6 mm. long, anthers linear, longer than fils.; fils.

with a broad bearded base; style linear-clavate, papillate above the middle, with 3 short terminal branches; caps. ± urn-shaped, with 3 valvelike openings near the middle. —Open places like talus, 2000–4000 ft.; Chaparral; Mt. Diablo, Mt. Hamilton Range, San Carlos Range of inner S. Coast Ranges. May–June.

2. **C. angustiflòra** Eastw. Resembling the preceding, but stouter, at least when young; lvs. oblong-ovate, sessile, serrate; the upper bracteate; sepals 3–4 mm. long; corolla ca. as long; stamens ca. 2 mm. long; anthers lance-linear, ca. as long as the sparsely ciliate fils.; style tapering from the base, not papillate on sides, branched to ca. ¼ its length. —Uncommon, burns and disturbed places; Chaparral; mts. from Sonoma and Lake cos. to Marin Co. and Boulder Creek, Santa Cruz Co. May–June. A more slender form with narrower subentire lvs. is var. *éxilis* J. T. Howell; Mt. Tamalpais and the Pinnacles (San Benito Co.).

3. **C. prenanthoìdes** Durand. CALIFORNIA HAREBELL. Perennial with slender rootstocks, stems slender, erect, simple to branched, 2–8 dm. tall, angled, sparsely stiff-pubescent; lvs. oblong-ovate to lanceolate, sessile or short-petioled, 1.5–4 cm. long, serrate, ± stiff-puberulent; infl. long, narrow, the fls. mostly in scattered clusters of 2–5; pedicels 2–6 mm. long; sepals subulate, 2.5–5 mm. long; corolla bright blue, 8–12 mm. long, the linear-lanceolate lobes 2–3 times as long as the tube; style long-exserted, curved, papillate and slightly enlarged upward, with 3 short lobes; caps. turbinate or short-oblong.—Dryish wooded places, below 6000 ft.; Redwood F., Mixed Evergreen F. to Red Fir F.; Coast Ranges from Monterey Co. n., Sierra Nevada from Tulare Co. n.; s. Ore. June–Sept.

4. **C. Scoùleri** Hook. Perennial with slender rootstocks, the stems slender, mostly simple, 1–3 dm. high, glabrous or ± puberulent; lvs. ovate to lanceolate, serrate, the blades 1–4 cm. long, narrowed into a margined petiole at least half as long; infl. mostly racemose, the few fls. solitary in upper lf.- or bract-axils; pedicels slender, 0.5–2 cm. long; sepals subulate, 5–8 mm. long; corolla ca. 8–10 mm. long; the lobes almost as long as the tube; style well exserted, straight, swollen and papillose upward.—Wooded places, 1500–5000 ft.; Mixed Evergreen F. to Yellow Pine F.; Humboldt, Trinity, and Sierra cos. n.; to Alaska. June–Aug.

5. **C. scabrélla** Engelm. From elongate branching rootstocks, the stems clustered, 0.3–1.2 dm. tall, scabrous-puberulent; lvs. lanceolate to oblanceolate, 2–5 cm. long, puberulent; fls. solitary and terminal or 2–4; sepals subulate, 3–8 mm. long; corolla light blue, 7–10 mm. long; caps. opening near summit.—Rocky places, 9000 ft.; Alpine Fell-fields; Mt. Eddy, Siskiyou Co.; to Wash., Ida. Aug.

6. **C. rotundifòlia** L. HAREBELL. BLUEBELL. From slender rootstocks, the stems diffuse or erect, 1–6 dm. high, slender; some of basal lvs. ovate to roundish, few-toothed, slender-petioled; most lvs. linear to lance-linear, sessile, entire, 3–6 cm. long; infl. loosely paniculate, the fls. 1–several, nodding; pedicels slender, mostly 1–2 cm. long; sepals subulate, 4–6 mm. long; corolla bright purplish-blue, campanulate, 12–20 mm. long, the lobes shorter than the tube; caps. short-ovoid to sub-cylindric, opening by basal pores; $2n = 34$, 55–56, 68 (Guinochet, 1942).—Moist slopes, 4500–8000 ft.; Yellow Pine F. to Lodgepole F.; Trinity and Siskiyou cos.; to Alaska, Atlantic Coast; Eurasia. July–Sept. This sp. has been called *C. petiolata* A. DC. by some auth.

7. **C. califórnica** (Kell.) Heller. [*Wahlenbergia c.* Kell. *C. linnaeifolia* Gray.] Slender-stemmed perennial, the stems simple or few-branched, 1–3(–4) dm. long, leafy, ± retrorsely scabrous; lvs. ovate-oblong to ± elliptic, crenulate, sessile or nearly so, 1–2.5 cm. long; fls. few, solitary; sepals lanceolate, 3–4 mm. long; corolla pale blue, campanulate, 10–14 mm. long, the lobes ca. as long as the tube; caps. globular.—Freshwater Swamp; near the Coast, Marin to Mendocino cos. June–Sept.

8. **C. Wilkinsiàna** Greene. Glabrous perennial, with erect leafy stems 7–15 cm. high, from slender rootstocks; lvs. from obovate-cuneiform and toothed at apex only, to lance-oblong and serrate on margins; fls. 1–3, on slender erect peduncles; calyx obpyramidal, the lobes lanceolate, entire, erect, longer than the tubular part; corolla deep blue-purple, funnelform, erect, cleft almost to middle, the lobes ± spreading; style ca. as long as corolla.—Moist places, 6000–8500 ft.; Red Fir F. to Subalpine F.; Mt. Shasta (Siskiyou Co.) and Trinity Mts. July–Sept.

2. Triodànis Raf. Venus Looking-Glass

Low annuals, simple or with ascending branches; stems leafy. Lvs. well distributed, gradually passing upward into leafy bracts. Fls. solitary, axillary, sessile, those from lower nodes normally cleistogamous, at least some of the upper open with well developed corollas. Fils. abruptly dilated and ciliate at base. Caps. ovoid to clavate or linear, opening at apex or near middle, 3-loculed, dehiscing by small valvelike openings in ours. Eight spp., 1 Medit., the others Am. (Greek, *treis*, three, and *odons*, tooth.)

(McVaugh, R. The genus Triodanis Rafinesque, and its relationships to Specularia and Campanula. Wrightia 1: 13–52, 1945.)

Pores at or very near the apex of the caps.; bracts usually longer than broad, not prominently veined beneath; open corollas terminal or scattered ....1. *T. biflora*
Pores well below the caps.-apex; bracts as broad or broader than long, often with 1–2 pairs of veins prominent beneath; several upper nodes usually with open corollas ....2. *T. perfoliata*

1. **T. biflòra** (R. & P.) Greene. [*Campanula b.* R. & P. *Specularia b.* F. & M.] Stems slender, simple or branched, 0.5–4 dm. long, ± sparsely hirsute below, scabrous to retrorsely hispid above; lvs. glabrous to hispid beneath along the veins, ovate to elliptic, all but the lowermost sessile, 1–3 cm. long, crenate; axillary fls. mostly 1–4 at a node, the cleistogamous scarcely 0.5 mm. wide, the open blue to violet or lilac, with corollas 5–10 mm. long; caps. glabrous or ± hispid, ellipsoid to ovoid, 4.5–7 mm. long, with oval or roundish pores; style 4–6 mm. long; sepals deltoid to deltoid-ovate; seeds brown, smooth, shining.—Open disturbed and burned places, below 6500 ft.; many Plant Communities; cismontane Calif. to Ore., Atlantic Coast, S. Am. April–June.

2. **T. perfoliàta** (L.) Nieuwl. [*Campanula p.* L. *Specularia p.* A. DC.] Stems usually somewhat stouter; lvs. round-cordate, clasping, crenate, 1–3 cm. long; bracts to ca. 3 cm. wide, scarcely as long, crenate; corollas of cleistogamous fls. less than 0.5 mm. long, of upper fls. bluish-lavender, 8–12 mm. long; style 5–7 mm. long; sepals 5–8 mm. long; caps. obovoid to oblong, 5–10 mm. long, the pores round to broadly elliptical, 1–2 mm. below the apex of the caps.; seeds ellipsoid, brown, often ± muriculate.—Open disturbed places, below 3000 ft.; Mixed Evergreen F., etc.; Humboldt Co. to B.C., Atlantic Coast, Mex., n. S. Am. June–July.

3. Heterocòdon Nutt.

Delicate annual, simple or few-branched. Lvs. roundish, dentate, sessile. Fls. solitary, axillary, of 2 kinds: the lower cleistogamous with inconspicuous corollas and the uppermost with showy corollas; sepals round-ovate, few-toothed; corolla open-campanulate. Caps. short and broad, mostly 3-loculed, bursting by irregular fissures. One sp. (Greek, *heteros*, different, and *kodon*, bell, the campanulate fls. of 2 kinds.)

1. **H. rariflòrum** Nutt. [*Specularia r.* McVaugh.] Delicate, the stems 5–30 cm. long, scattered bristly-pubescent; lvs. 5–8 mm. long; sepals of lower fls. ca. 1.5–2 mm. long, of upper ca. 3 mm.; corolla of upper fls. blue, 3–4 mm. long; caps. ca. 2.5 mm. in diam.—Damp grassy or shaded places, below 7500 ft.; many Plant Communities; cismontane Calif. to B.C., Ida., Mont., Nev. April–July.

4. Githópsis Nutt.

Small annuals, simple or diffusely branched, the stems commonly angled. Lvs. mostly narrow, inconspicuous, scattered, sessile or nearly so. Fls. in upper axils or terminal or both. Sepals usually conspicuous, lanceolate to acicular, firm. Corolla minute to exserted and salverform or campanulate. Ovary usually enlarged upward, elongate, 3-loculed, lineate or ribbed. Caps. clavate, opening by a terminal perforation. Seeds minute, smooth, shining. Ca. 6 spp., of w. N. Am. (Resembling *Githago*.)

(Ewan, J. A review of the genus Githopsis. Rhodora 41: 302–313, 1939.)
A. Stems elongate-filiform, the branches few, widely spreading, almost twining; fls. minute, 3–6 mm. long; ovary 3–5 mm. long. San Diego .... 6. *G. filicaulis*
AA. Stems stout to very slender, but not filiform, few- to many-branched; fls. over 4 mm. long; ovary 4–12 mm. long.

B. Middle cauline lvs. oblong-ovate, to ca. 1.5 cm. wide; corolla with broad obtuse lobes. Plumas Co. .... 1. *G. latifolia*
BB. Middle cauline lvs. linear to oblong, narrower, mostly inconspicuous; corolla with rounded to acute lobes.
    C. Corolla 1½ times as long as calyx; plants subglabrous above. W. base of cent. Sierra Nevada .... 2. *G. pulchella*
    CC. Corolla inconspicuous or if showy less than 1½ times as long as sepals (if longer, then the plant pubescent above); plants hairy to subglabrous.
        D. The corolla 5–15 mm. long; caps. distended in middle, ± strongly ribbed.
            E. Sepals erect or turbinate-spreading in fr., linear-subulate, not outwardly sickle-curved; corolla ½ to 1 times as long as calyx. Widely distributed
                        3. *G. specularioides*
            EE. Sepals lance-subulate, outwardly sickle-curved in fr.; corolla 1–1½ times as long as sepals. Cent. and n. Calif. .... 4. *G. calycina*
        DD. The corolla 3–5 mm. long; caps. linear, slightly enlarged upward, not ribbed
                        5. *G. diffusa*

1. **G. latifòlia** Eastw. Branching with weak slender stems, with a few hairs on the angles; cauline lvs. oblong-ovate, sessile, entire, 1–2 cm. long; fls. terminal; sepals linear-lanceolate, thin, widely spreading, 1 cm. wide; corolla open-campanulate, light purple, 1 cm. across; young caps. lightly ribbed.—Known from a collection at Big Meadows, Plumas Co. Aug.–Sept.

2. **G. pulchélla** Vatke. [*G. specularioides* var. *glabra* Jeps.] Stems slender, simple or divergently branched, 1–2 dm. high, glabrous or setulose along angles; upper lvs. oblong-linear, serrulate, ca. 6–12 mm. long, the lower withered by anthesis; sepals linear or subulate, 8–15 mm. long; corolla campanulate, 15–20 mm. long, the lobes deep blue, shortly acute; caps. broadest above the middle, ribbed, glabrous or retrorse-pubescent; seeds brown, shining, ca. 1 mm. long.—Dry rocky places, burns, etc., 500–3500 ft.; Foothill Wd., Yellow Pine F.; Sierran foothills from Amador to Mariposa cos. May–June.

3. **G. specularioìdes** Nutt. Stems stiff, usually branched slightly above base, angled below, 0.5–1.5 dm. tall, gray-hispidulose; lowermost lvs. ovate to broader, small, withering early; middle cauline lvs. cuneate-obovate, few-toothed, ca. 3–6 mm. long; fls. in upper axils and terminal; sepals linear-subulate, 6–10 mm. long at anthesis, later 8–14 mm.; corolla ± salverform, 4–7 mm. long, bright blue, the lobes acute; caps. ribbed, mostly 5–9 mm. long, inflated above the middle; seed 0.5–0.8 mm. long, ± angled, smooth, shining, light brown.—Open places, burns, etc., below 4000 ft.; many Plant Communities; Mariposa and Santa Barbara cos. n. to Wash.; Santa Cruz Id. May–June.

Ssp. cándida Ewan. Plants 15–20 cm. tall; corolla white, 4–5 mm. long.—Burns and disturbed places; Chaparral; San Diego and w. Riverside cos.

4. **G. calycìna** Benth. Somewhat bushy spreading plants 1–2.5 dm. tall, the stems angled; herbage at first downy-pubescent, later more scabrous-hispidulose; middle and upper lvs. narrow-oblong to ± ovate, 8–12 mm. long, serrulate; fls. both axillary and terminal; sepals becoming 12–20 mm. long in fr.; corolla cylindro-campanulate, 6–10 mm. long, bright dark blue; caps. slender-conic, ribbed, 6–12 mm. long; seeds 0.8 mm. long, smooth, chestnut-brown.—Grassy slopes, etc., below 2000 ft.; Chaparral, Foothill Wd., Yellow Pine F., etc.; Contra Costa and Tuolumne cos. to sw. Ore. April–May. Plants from Butte Co. have especially large fls.

5. **G. diffùsa** Gray. [*G. gilioides* Ewan.] Slender strict plants, simple or few-branched, glabrous except for slightly ciliate bases of sepals or bushier and setulose-pubescent throughout; cauline lvs. obovate, ± serrulate, 4–7 mm. long; fls. few, scattered, in upper axils; sepals linear, erect, 4–6 mm. long in fr.; corolla subcylindrical, light purple, 3–5 mm. long, the lobes acute; caps. 5-nerved, slightly enlarged upward, ca. 6–10 mm. long; seeds 0.6–0.8 mm. long, subterete, smooth, shining, 0.7 mm. long.—Banks and disturbed places, 1500–4700 ft.; Chaparral; hills, Santa Barbara to w. Riverside cos. April–June. Some plants are more tufted and pubescent and constitute *G. gilioides* Ewan.

6. **G. filicáulis** Ewan. Stems filiform, 1–2.5 dm. long, hirsutulous on the angles; lvs. inconspicuous, ovate-oblong, serrulate, subglabrous, 1–3 mm. long; sepals minute, long, linear; corolla shorter than sepals; caps. 3–5 mm. long; seeds 0.6–0.9 mm. long, shining, red-brown.—Mission Canyon, San Diego. May.

5. Nemácladus Nutt.

Small annuals with capillary diffusely branched stems. Basal lvs. in compact rosette, cauline largely reduced to subulate bracts. Fls. minute, loosely racemose, borne on capillary pedicels from most axils. Fl-tube ± evident above the ovary. Sepals triangular to ovate, entire. Corolla irregular, ± bilabiate, the lower lip 2-, the upper 3-lobed. Stamens monadelphous, the fils. separate below and sometimes above, the staminal tube curved near the tip and surmounted by the stellately spreading distinct anthers. Ovary with 3 flattened rounded glands near the base of the free part, these opposite the 3 lobes of the upper corolla-lip. Fils. between these glands with small stipelike appendages, each of which bears 1 or more terminal transparent rodlike cells. Caps. fundamentally bilocular, dehiscent by 2 valves and these splitting, so that caps. appears 4-valved. Seeds many, cylindrical, longitudinally ridged and with transverse lines forming pits. Ca. 10 spp., of sw. U.S. (Greek, *nemos*, thread, and *clados*, branch, because of the capillary stems.)

(McVaugh, R. Some realignments in the genus Nemacladus. Amer. Midl. Natur. 22: 521–550, 1939.)
A. Corolla-tube mostly 2–5 mm. long, much exceeding the calyx; caps. superior, free from the calyx its entire length, and exceeding it . . . . 1. *N. longiflorus*
AA. Corolla-tube 1.5 mm. long or less, scarcely or not exceeding the calyx; caps. ca. half inferior, the free part not or scarcely longer than the calyx.
    B. Corolla-lobes united at base so that the tube is almost or ca. as long as the free lobes.
        C. Stems straight, not at all zigzag; seeds subglobose, with ca. 6 pits in each row. Monterey Co. to L. Calif. . . . . 2. *N. ramosissimus*
        CC. Stems ± zigzag; seeds longer than thick, with 8–12 pits in a longitudinal row.
            D. Basal lvs. usually pinnatifid; bract at base of pedicel sublinear, not or scarcely enfolding the base of the pedicel; mature caps. 3–4 mm. long. Los Angeles Co. to L. Calif. . . . . 3. *N. pinnatifidus*
            DD. Basal lvs. entire or toothed; bract at base of pedicel ovate to lanceolate, enfolding and concealing the base of the pedicel; caps. 1.5–2.5 mm. long.
                E. Basal lvs. rhombic-ovate to elliptic, entire or crenate-dentate; flowering branches of large plants often repeatedly and intricately branched; pedicels mostly finely capillary and forming a double curve. Deserts
                                        5. *N. sigmoideus*
                EE. Basal lvs. narrower (oblong-oblanceolate or ± spatulate), crenate-dentate to subpinnatifid; flowering branches simple or sparingly branched. Inner S. Coast Ranges.
                    F. Racemes distichous; corolla-tube scarcely longer than calyx; pedicels finely capillary, commonly forming a double curve . . 4. *N. gracilis*
                    FF. Racemes often strongly secund; corolla-tube equal to or twice as long as calyx; pedicels slightly stoutish, straight to somewhat sigmoid (in age)
                                      6. *N. secundiflorus*
    BB. Corolla-lobes scarcely united at base, the tube very short.
        C. Calyx and ovary much enlarged in fr., ca. 2.5 times as long as at anthesis; plant compact, very robust. Near the state line from Inyo Mts. to Modoc Co.
                                        7. *N. rigidus*
        CC. Calyx and ovary not notably enlarged in fr.; plant diffuse, with very slender stems.
            D. Lobes of upper lip of corolla ciliate on margin; stamens appearing well exserted. Deserts.
                E. Anthers 0.4–0.8 mm. long; fils. 2–3 mm. long . . . . 8. *N. rubescens*
                EE. Anthers 0.1–0.3 mm. long; fils. 1.3–2 mm. long . . . . . . 10. *N. glanduliferus*
            DD. Lobes of upper lip of corolla not ciliate; stamens mostly not much exserted.
                E. Base of fr. long-turbinate; seeds few, 5–12; caps. swollen above, exceeding calyx; fls. ca. 2 mm. long. Mostly from Merced and Santa Clara cos.
                                      11. *N. capillaris*
                EE. Base of fr. rounded; seeds more numerous; fls. ca. 3 mm. long.
                    F. Seeds ca. 0.9 mm. long, each broad ridge with ca. 15–30 transverse lines. Lake and Napa cos. . . . . 12. *N. montanus*
                    FF. Seeds ca. 0.5–0.6 mm. long, the narrow ridges separated by rows of 10–12 pits. Along the base of the Sierra Nevada from Kern Co. n.
                                        9. *N. interior*

1. **N. longiflòrus** Gray. Simple to diffusely branched from the base, 0.3–2 dm. high, usually ± pubescent at least below; rosette-lvs. few to many, oblanceolate to obovate or spatulate, the blades 0.5–1 cm. long, mostly entire; pedicels spreading to ascending, 1–2.5 cm. long; fls. 4–7 mm. long; fl.-tube almost none; sepals 1–3 mm. long; corolla white to pale blue, 5–8 mm. long, tubular for more than half its length, the upper lip bearded at base; fils. 3.5–7.5 mm. long; caps. fusiform, 3–4.5 mm. long; seeds broadly ellipsoid, 0.4–0.5 mm. long, with 9–10 weak ridges and ca. 10–12 cells in each row.—Dry

sandy and disturbed places like burns, below 6000 ft.; Chaparral, Coastal Sage Scrub; Santa Lucia Mts. and Los Angeles Co. to L. Calif. April–June.

Var. **breviflòrus** McVaugh. Corolla 3–3.5 mm. long; fils. 2–2.8 mm. long.—Creosote Bush Scrub, Joshua Tree Wd., etc.; desert edge from Little San Bernardino Mts. to e. San Diego Co.; Ariz.

2. **N. ramosíssimus** Nutt. [*N. tenuissimus* Greene.] Stems not at all zigzag, glabrous or puberulent, 0.5–2.5 dm. long; rosette-lvs. oblanceolate, dentate, the blades 0.5–2 cm. long; pedicels spreading, 8–20 mm. long, glabrous; bracts of infl. linear, 3–6 mm. long; ovary half inferior, the sepals lanceolate, 0.6–1 mm. long; corolla white, glabrous, campanulate, 1.5–2.5 mm. long; stamens pubescent near the anthers; caps. exceeding calyx; seeds ca. 0.5 mm. in diam., with ca. 10 rows, each with ca. 6 pits.—Sandy places, burns, etc., below 5000 ft.; Chaparral, Coastal Sage Scrub; Monterey Co., Los Angeles Co. to L. Calif. April–May.

3. **N. pinnatífidus** Greene. [*N. ramosissimus* var. *p.* Gray.] Habit of *N. ramosissimus* but the stems slightly zigzag; basal lvs. mostly deeply pinnatifid with toothed lobes; pedicels 7–12 mm. long; sepals 1.2–2 mm. long, linear to deltoid; corolla 1.6–2 mm. long, campanulate; caps. 3–4 mm. long, ca. half inferior; seeds ellipsoid, ca. 0.6 mm. long, with 8–10 rows each of 8–10 broad pits.—Burns, disturbed places, etc., 1000–4000 ft.; Chaparral, Coastal Sage Scrub; s. base of San Gabriel and San Bernardino mts. to e. San Diego Co.; L. Calif. May–June.

4. **N. grácilis** Eastw. [*N. ramosissimus* var. *g.* Munz.] Stems ± zigzag, diffuse, pubescent, 0.3–1 dm. tall; rosette-lvs. spatulate, coarsely dentate, 5–8 mm. long including petioles; bracts of infl. involute, falcately recurved, fleshy, entire, 2–4 mm. long, scarcely clasping the pedicels; pedicels capillary, sigmoidally recurved-spreading, 8–12 mm. long; sepals ca. 1 mm. long, lanceolate; corolla campanulate, 1.5 mm. long, the ovate-oblong obtuse divisions twice as long as the tube; anthers 0.3–0.4 mm. long; caps. not exceeding calyx, acute, half inferior; seeds many, red-brown, oblong, ca. 0.4 mm. long, with ca. 10 rows of ca. 13 pits.—V. Grassland, Foothill Wd.; w. Merced to w. Kern cos. March–April.

5. **N. sigmoìdeus** Robbins. Like *N. gracilis*, but coarser; pedicels largely 12–14 mm. long, the subtending bracts more clasping; sepals lance-deltoid; corolla ca. 2–2.5 mm. long, the 3 lower lobes white with pinkish or yellowish tips and bases; anthers ca. 0.3 mm. long; seeds broadly ellipsoid.—Common in sandy and gravelly places, 1500–7500 ft.; mostly Joshua Tree Wd., Pinyon-Juniper Wd., desert borders and deserts from n. Inyo Co. to Mt. Pinos (Ventura Co.) and San Bernardino Co. and e. San Diego Co.; L. Calif., Nev., Ariz. April–June.

6. **N. secundiflòrus** Robbins. Plants 2.5–12.5 cm. high; stems solitary, simple or dichotomously branched just above the base, dull reddish-brown below, glabrous or ± puberulent; basal lvs. 3–6 mm. long, narrowed to a broad petiolar base; pedicels capillary, spreading at right angles to the branches; floral bracts conduplicate, 2–3 mm. long; calyx-tube campanulate, the lobes acutish; corolla 3.5–5 mm. long, often 1½ to 2 times as long as calyx, the 2 upper lobes concolorous, widely spreading, the 3 lower each with a yellow blotch at base, all lobes hispidulous on their inner surfaces, outer surfaces lavender or pinkish, sparingly hispidulous; staminal appendages 2, each with ca. 12–15 radially spreading rods; seeds subglobose or broadly ellipsoid.—Dry slopes and flats, 1000–5000 ft.; Foothill Wd., etc.; Coast Ranges from Monterey, San Benito and San Luis Obispo cos., Greenhorn Range in Kern Co. April–May.

7. **N. rígidus** Curran. Compact, coarse-stemmed, glabrous or finely pubescent, the stems strongly zigzag, 0.5–1.5 dm. long, spreading-decumbent, somewhat shining, purple; rosette-lvs. elliptic, 5–10 mm. long; pedicels spreading, stiff, straight, 10–12 mm. long, subtended by bracts 2–3 mm. long, broad elliptic; caps. ca. half inferior, 3–4 mm. long; the sepals lanceolate to ovate, unequal, 1–2.5 mm. long; corolla deeply divided, ± purplish, 1–1.5 mm. long; seeds ellipsoid, ca. 0.7 mm. long, with 8–10 longitudinal flattened ridges and rows of ca. 15 narrow pits.—Dry caked adobe, etc., 4000–7500 ft.; Sagebrush Scrub, Pinyon-Juniper Wd., N. Juniper Wd.; e. Mono and Inyo to e. Modoc cos.; adjacent Ore., Nev. May–June.

8. **N. rubéscens** Greene. [*N. rigidus* var. *r.* Munz. *N. adenophorus* Parish.] Mostly diffusely branched, 0.5–2 dm. tall, the stems slightly pubescent to glabrous, the lower parts silvery-gray; rosette lvs. elliptic to oblanceolate, with winged petiole, entire or

± toothed, 1–3 cm. long, ½–⅓ as wide; pedicels ± ascending, slender, 8–15 mm. long, glabrous, subtended by bracts lanceolate to ovate, 1–2.5 mm. long, ± conduplicate; caps. half-inferior, ca. 1.5–2 mm. high; sepals elliptic to deltoid, 0.7–1.3 mm. long; corolla yellow with purple-brown markings, at least 2–3 mm. long, with long spreading lobes, the upper 3 ciliate; fils. 2–3 mm. long; anthers 0.6 mm. long; terminal cells of staminal appendage longer than the stipe; seeds 0.5–0.7 mm. long, broadly ellipsoid, each with 8–10 ridges and obscure pits between.—Dry sandy to rocky places, below 4000 ft.; Creosote Bush Scrub, Joshua Tree Wd.; deserts from Kern Co. to L. Calif., Nev., Ariz. April–May.

Var. ténuis McVaugh. Lvs. oblanceolate, pinnatifid-toothed, ⅕–⅛ as wide as long; pedicels more finely capillary; terminal cells on staminal appendage shorter than the stipe.—Dry sandy places, mostly below 1000 ft.; Creosote Bush Scrub; Colo. Desert to Inyo Co.

9. **N. intèrior** (Munz) Robbins. [*N. rigidus* var. *i.* Munz. *N. rubescens* var. *i.* McVaugh.] Plants rather strict, 1.5–3 dm. tall, the stems brown or purplish; rosette-lvs. ca. half as wide as long; fl.-bracts scarcely enfolded, 1–4 mm. long; caps. 2–3.5 mm. long; corolla pale lilac with red-violet blotches or whitish with pink tinge, the lobes not ciliate.—Dryish slopes and disturbed places, 1500–5600 ft.; largely Yellow Pine F.; w. base of Sierra Nevada, Kern to Butte cos. May–July.

10. **N. glandulíferus** Jeps. Plants erect, pubescent below to glabrous, the stems brownish or purplish below, dull, 1–2.5 dm. tall, with ascending somewhat flexuous branches; basal lvs. mostly oblanceolate, toothed or pinnatifid, 1–2.5 cm. long; pedicels spreading, 7–13 mm. long, glabrous, slender, usually curved near the tip, subtended by linear to lanceolate bracts 2–5 mm. long, flat or slightly folded; caps. ca. half-inferior, 2–3 mm. long; sepals linear to narrowly triangular, 1.5–3 mm. long; corolla 1.5–2.5 mm. long, white, with the lobes ± purplish at tip, at least the 3 upper ciliate; fils. 1–2 mm. long, bearing appendages; anthers 0.2–0.35 mm. long, seeds cylindrical to ± ellipsoid, 0.5–0.6 mm. long, with 6–8 longitudinal ridges divided by 15–20 fine transverse lines.—Open sandy places, below 2000 ft.; Creosote Bush Scrub; w. Colo. Desert to Sheephole Mts., Mojave Desert; L. Calif. March–May.

Var. orientàlis McVaugh. Pedicels stiffly spreading-ascending, not or scarcely curved near the tip; sepals 0.8–1.8 mm. long.—More common than the sp., sandy washes, etc., below 5000 ft.; Creosote Bush Scrub, Joshua Tree Wd., etc.; deserts from Inyo Co. to L. Calif., Son., Utah. March–May.

11. **N. capillàris** Greene. [*N. rigidus* var. *c.* Munz.] Stems glabrous or minutely pubescent, ± lustrous, brownish or purplish, forked, 0.5–1.6 dm. tall; rosette-lvs. few, ovate, entire or nearly so, 0.3–1.5 cm. long; pedicels spreading to ascending, mostly straight, 8–12 mm. long; fl.-bracts elliptic to lance-ovate, glabrous, 1–3 mm. long, flat or slightly folded; caps. half-inferior, 1.5–2.5 mm. long; sepals elliptic to ± triangular, 0.6–1.3 mm. long; corolla white, 0.7–1.3 mm. long, almost tubeless; fils. 0.8–1.2 mm. long; anthers scarcely 0.2 mm. long; seeds broadly ellipsoid, 0.5–0.7 mm. long, with 8–10 narrow ridges and rows of 9–12 pits between.—Dry slopes and burns, 2000–4500 ft.; Chaparral, Yellow Pine F.; Coast Ranges, s. Ore. to Santa Clara Co., w. base of Sierra Nevada s. to Kern Co. May–July.

12. **N. montànus** Greene. [*N. rigidus* var. *m.* Munz.] Diffusely stiff-branched, 1–2 dm. tall, strongly zigzag, glabrous or minutely puberulent; rosette-lvs. oblanceolate, dentate, glabrous; pedicels strongly ascending, 12–15 mm. long; fl.-bracts linear to lanceolate, 1–3 mm. long, scarcely or not enfolded; caps. ca. half inferior, 2.5–3 mm. long; sepals blunt, 1.2–1.6 mm. long; corolla white or purplish, 1.5–2.5 mm. long; the lobes almost separate; fils. 2–2.5 mm. long, glabrous; seeds ellipsoid, 0.9 mm. long, with 10–12 longitudinal ridges divided by ca. 30 transverse lines.—Serpentine; Chaparral, Foothill Wd.; Napa and Lake cos. May–July.

6. Parishélla Gray

Low spreading annual with lvs. and fls. in subcapitate tufts, the first at the base of the plant, the others at the ends of short naked proliferous branches. Ovary half inferior. Sepals spatulate. Corolla rotate, almost equally 5-parted. Stamens with appendages and ovary with glands as in *Nemacladus*. Caps. turbinate, circumscissile just above the sepals. One sp. (Named for S. B. and W. F. *Parish*, early s. Calif. botanical collectors.)

1. **P. califórnica** Gray. Stems 2–5 cm. long; lvs. oblanceolate to spatulate, 2–5 mm. long, the basal ones petioled; corolla white, 3–4 mm. long; seeds many, brown, oblong-cylindrical, ca. 0.4 mm. long, with ca. 10 rows of ca. 10 pits.—Local, gravelly places, 2500–5000 ft.; largely Joshua Tree Wd.; Victorville and Barstow regions, sw. Mojave Desert; and Caliente Mt., s. San Luis Obispo Co. April–May.

7. Downíngia Torr.

Low glabrous soft-stemmed annuals, rather succulent and tender. Lvs. sessile, lanceolate to subulate, or the uppermost broader, mostly entire, the upper passing into leafy bracts. Fls. perfect, 5-merous, sessile in the axils of bracts, but appearing pedicelled because of the long inferior twisted ovaries which "invert" the fls. Sepals rather narrow, ± unequal. Corolla mostly blue, blotched on lower lip, bilabiate, the 2 upper lobes usually smaller than the 3 fused lower. The 2 smaller anthers each with a terminal tuft of bristles and often with a terminal hornlike process also. Caps. linear to fusiform, many times longer than thick, with or without a cent. septum, opening elastically by slits. Seeds many, fusiform, light brown, small, mostly smooth. Ca. 12 spp., w. N. Am. and s. S. Am. (Named for A. J. *Downing*, Am. horticulturist.)

(McVaugh, R. A monograph of the genus Downingia. Memoirs Torr. Bot. Club. 19(4): 1–57, 1941.)
A. Anther-tube strongly incurved so as to be almost vertical to the fils.; lower corolla-lip not forming a sharp angle with the corolla-tube.
  B. Ovary and caps. with 2 internal cavities separated by a longitudinal septum which bears the ovules; anthers almost uniformly dirty white, granular-roughened; fls. from March to June
    1. *D. insignis*
  BB. Ovary and caps. without an internal septum, the ovules on the ovary-wall; anthers ± striped longitudinally with blue, not granular-roughened; fls. from June–Sept. . . 2. *D. elegans*
AA. Anther-tube not strongly incurved, but almost in line with the fils.
  B. Corolla short, scarcely or not exceeding the sepals; lower lip of corolla ascending.
    C. The corolla 2–4 mm. long; fils. 1–1.8 mm. long; anthers 0.6–1 mm. long; seeds appear as if twisted, the cellular lines ± oblique. Cent. V. . . . . 3. *D. pusilla*
    CC. The corolla 4–7 mm. long; fils. 1.8–2.8 mm. long; anthers 1.3–2 mm. long; seeds not twisted. E. of Sierra Nevada . . . . 4. *D. laeta*
  BB. Corolla definitely exceeding sepals, its lower lip forming a sharp angle with the corolla-tube.
    C. Pair of bristles at apex of anther-tube usually tightly twisted together; base of lower lip of corolla with dark purple nipplelike projections, the tube bearded within on lower side
      5. *D. bicornuta*
    CC. Pair of bristles, if present on apex of anther-tube, not twisted together; base of the lower lip of corolla without purple nipplelike projections, the tube usually not bearded within.
      D. Anther-tube 2.5–3.5 mm. long, the fil.-tube long and prominently exposing the anthers; lower corolla-lip with 3 basal purple spots . . . . 6. *D. pulchella*
      DD. Anther-tube 1.5–2.5 mm. long, usually not prominently exserted, the fil.-tube shorter; lower corolla-lip often not with 3 basal purple spots.
        E. Dorsal sinus of corolla (between the lobes of upper lip) usually produced backward and outward into a short hornlike process; anthers well exserted; corolla-tube ± pilose within . . . . 7. *D. ornatissima*
        EE. Dorsal sinus of corolla without projecting horn or fold; anthers not much exserted; corolla-tube glabrous within.
          F. Ovary and caps. with a longitudinal septum to which the ovules are attached (this can be seen by splitting ovary lengthwise with a needle).
            G. Lower lip of corolla without any purple areas; seeds appearing twisted, the cellular lines oblique . . . . 8. *D. cuspidata*
            GG. Lower lip of corolla with 1 or more basal purple spots; seeds not appearing twisted.
              H. Two upper corolla-lobes smooth on margins, the lower lip with 3 or 2 basal purple spots. Vernal pools, Cent. V.
                9. *D. bella*
              HH. Two upper corolla-lobes minutely ciliate-scabrous on margins toward tip, the lower lip with 1 large basal purple spot. Coast Ranges . . . . 10. *D. concolor*
        FF. Ovary and caps. lacking a cent. septum, the ovules borne in 2 lines on the ovary wall.
          G. Three larger anthers mostly abundantly pilose near apex; seeds not shining. Mts. from Shasta Co. to Tuolumne Co.
            11. *D. montana*
          GG. Three larger anthers glabrous or sparsely pilose near tip; seeds shining. From Humboldt and Siskiyou cos. n. . . . . 12. *D. Yina*

1. **D. insígnis** Greene. Stems slender, 1–2.5 dm. long; lvs. 5–15 mm. long, 1–2 mm. wide; infl. 4–18 cm. long, with bracts 6–20 mm. long, 1–5 mm. wide; sepals elliptic, 3–10 mm. long, 1–2 mm. wide; corolla 9–15 mm. long, sky-blue with darker veins, the lower lip with cent. light area and 2 yellow folds in a field of dark purple or with this divided into 3 purple spots, the tube ° 3.5–5 mm. long, the 2 upper lobes 6–10 mm. long, the 3 lobes of the lower lip 3–7 mm. long; fil.-tube 9–10 mm. long; anther-tube 2.5–3 mm. long, 1–1.3 mm. in diam., the 2 shorter anthers white-tufted at tip, each also with a short hornlike process; caps. 4–8 cm. long, 1–1.5 mm. in diam.—Clay mud of vernal pools; largely V. Grassland, Great V. and adjacent foothills, Stanislaus Co. n.; Sagebrush Scrub, Lassen, and Modoc cos.; w. Nev. March–May.

2. **D. elegáns** (Dougl.) Torr. [*Clintonia e.* Dougl. *Bolelia e.* Greene.] Mostly 1–4 dm. high, stout-stemmed; lvs. 5–25 mm. long, 0.5–4 mm. wide; infl. 5–25 cm. long, with bracts 8–25 mm. long, 2–9 mm. wide; sepals 4–10 mm. long, 1–2.5 mm. wide; corolla mostly blue, the lower lip with cent. white spot and 2 low yellow ridges, the tube with purple veins and often with 3 oblong purple blotches, 2 upper lobes 4–12 mm. long, 1–2.5 mm. wide, acute, the 3 lower 2–4 mm. long; the tube 2–3 mm. long; fil.-tube glabrous, 6–8 mm. long; anther-tube 0.6–1 mm. in diam., 2.5–3.5 mm. long; caps. 2.5–4.5 cm. long, with papery easily ruptured walls when dry; seeds shining.—Mud flats, vernal pools, etc., below 5000 ft.; Mixed Evergreen F., Sagebrush Scrub, Yellow Pine F., etc.; Humboldt and Sierra cos. n.; to Wash., Ida., Nev. June–Sept.

Var. **brachypétala** (Gand.) McVaugh. [*D. b.* Gand.] Corolla 5–9 mm. long, the tube 1.5–2.5 mm. long; fils. 3–4 mm. long; anther-tube 2–2.5 mm. long.—Mixed Evergreen F., etc.; Mendocino and Lake cos., Lassen Co. n. to Wash., Ida.

3. **D. pusílla** (G. Don) Torr. [*Clintonia p.* G. Don. *Bolelia humilis* Greene. *D. h.* Greene.] Stems rather stout, 2–12 cm. high; lvs. 4–7 mm. long, 0.5–1 mm. wide; infl. 2–5 cm. long; bracts 2–8 mm. long, 1–1.5 mm. wide; sepals 3–8 mm. long, blunt; corolla 2.5–4 mm. long, white or the lower lip blue-tipped, with white center and yellowish patch near base, the tube 1–2 mm. long; fil.-tube 1–1.8 mm. long; anthers white-apiculate, the 2 shorter with a few bristles and each with a blunt hornlike process; caps. 2–2.5 cm. long, 1 mm. in diam., the walls firm; seeds twisted.—Vernal pools; V. Grassland, Foothill Wd.; Cent. V. and adjacent hills to Sonoma V.; Chile. March–May.

4. **D. laèta** (Greene) Greene. [*Bolelia l.* Greene.] Plants 5–25 cm. tall; lvs. 5–20 mm. long, 0.5–2 mm. wide; infl. 2–10 cm. long, with bracts 7–20 mm. long, 1–4 mm. wide; sepals elliptic, 3–7 mm. long, 1–2 mm. wide; corolla 4–7 mm. long, light blue or purplish, the lower lip with cent. white or yellow area and basal transverse band of purple, the yellowish tube 1.3–1.6 mm. long; fil.-tube 1.8–2.5 mm. long; anthers ± ciliate on backs or glabrous, the 2 shorter white-tufted at apex and each with a hornlike process; caps. 2–4 cm. long, 1–2 mm. in diam.; seeds not or scarcely twisted.—Muddy places, 4000–5000 ft.; Sagebrush Scrub, etc.; Lassen and Modoc cos.; to Sask., Wyo., Nev., Utah. May–July.

5. **D. bicornùta** Gray. [*Bolelia b.* Greene. *D. sikota* Appleg.] Stems 5–25 cm. long; lvs. 6–16 mm. long, 0.5–1 mm. wide; infl. 4–15 cm. long, with bracts 5–20 mm. long, 1–4 mm. wide; sepals 3–8 mm. long, usually plainly unequal, often rotate-spreading; corolla 9–19 mm. long, bearded within on lower side, the lower lip deep purplish-blue with a cent. whitish or yellowish area with 2 yellow or green spots, the upper lip purplish-blue, corolla-tube 3–4 mm. long, cut very deeply, upper lobes 4–8 mm. long; fil.-tube 3–4 mm. long; anther-tube 2–2.5 mm. long, anthers minutely tufted at tip, each of the 2 shorter with a recurved elongate process from the tip, these 0.6–1.5 mm. long, twisted together; caps. 4–6.5 cm. long, with tough walls; seeds not twisted.—Moist places and drying mud, below 6000 ft.; V. Grassland, Foothill Wd., Yellow Pine F.; Cent. V. (mostly n. of the San Joaquin) and surrounding hills from Nevada and Stanislaus cos. n.; to se. Ore., sw. Ida., w. Nev. April–July.

Var. **pícta** Hoov. Corolla 7–10 mm. long, the tube 2–2.5 mm. long; caps. 3.5–5.5 cm. long; twisted anther-bristles 1.6–2.7 mm. long.—V. Grassland, etc., of Cent. V. from Shasta Co. s., but especially in San Joaquin V.

6. **D. pulchélla** (Lindl.) Torr. [*Clintonia p.* Lindl. *Bolelia p.* Greene.] Stems 5–25 cm. long; lvs. 4–12 mm. long, 1–2 mm. wide; infl. 5–18 cm. long, with bracts 8–20 mm. long, 2–7 mm. wide; sepals 3–8 mm. long, 0.5–2 mm. wide, subequal; corolla 8–13 mm.

° Tube-length measurements are from dorsal sinus to base.

long, deep blue, the lower lip with a cent. white area with 2 yellow spots passing into narrow yellow folds at the base, which alternate with 3 dark purple spots, the tube purple, 2–3 mm. long, the 2 upper lobes divergent, 6–8 mm. long, lower lip reflexed, the lobes 4.5–6 mm. long; fil.-tube 3–4.5 mm. long; anther-tube prominently exserted, 2.5–3.5 mm. long, pointed at tip, 2 shorter anthers minutely white-tufted at tip and each with a slender hornlike process; caps. 3–7 cm. long, the walls tough; seeds shining, not twisted.—Muddy places, vernal pools, etc.; Coastal Salt Marsh, Mixed Evergreen F., V. Grassland, Foothill Wd., etc.; Monterey and Merced cos. to Colusa Co. April–June.

7. **D. ornatíssima** Greene. [*Bolelia o.* Greene.] Plants 5–20 cm. high, mostly 1–few-stemmed; lvs. 5–12 mm. long, 0.5–2 mm. wide; infl. 4–12 cm. long, with bracts 6–12 mm. long, 1–4 mm. wide; sepals 2–7 mm. long, subequal or the 2 lower shorter; corolla 8–13 mm. long, the lower lip blue to lilac or almost white, with squarish white center having 2 yellowish spots passing into folds at angle of throat, tube whitish, sparsely pilose within on lower side, 2.5–4 mm. long; the 2 upper lobes 2.5–6 mm. long, divergent with the tip of each curled outward and backward into a ring or at least strongly recurved, glabrous within; lower lobes oblong; fil.-tube 3–4.5 mm. long; anther-tube 1.7–2.5 mm. long, wholly exserted, anthers white-tufted at tip and each also with a hornlike process; caps. 2.5–6.5 cm. long, tough; seeds not twisted.—Heavy clay mud; V. Grassland, Foothill Wd.; Sacramento V. s. to Merced Co. April–May.

Var. **exímia** (Hoov.) McVaugh. [*D. mirabilis* var. *e.* Hoov. *D. m.* J. T. Howell.] Two upper corolla-lobes minutely pubescent within near tips, divergent but not curled into rings.—V. Grassland, Foothill Wd.; San Joaquin V. to Butte Co. April–May.

8. **D. cuspidàta** (Greene) Greene. [*Bolelia c.* Greene. *D. pulchella* var. *arcana* Jeps. *D. immaculata* M. & J. *D. pallida* Hoov.] Plants 5–25 cm. high; lvs. 3–13 mm. long, 0.2–2 mm. wide; infl. 3–12 cm. long, with bracts 4–10 mm. long, 1–3 mm. wide; sepals 3–8 mm. long, ca. 1 mm. wide, subequal; corolla 7–15 mm. long, from bright or pale blue to lavender or almost white, lower lip with a cent. white area with a yellow spot or 2 ± confluent ones; the 2 upper lobes 3–6 mm. long, often darker than lower and with purplish veins; tube 3–4 mm. long, lower lobes oblong to broadly ovate; fil.-tube 2.5–4 mm. long; anther-tube 1.4–2.5 mm. long, all minutely white-tufted at tip, the 3 shorter anthers also with a short hornlike process; caps. 2–4 cm. long, the valves easily separated; seeds shining, twisted.—Wet and drying clay soil, below 1600 ft.; Foothill Wd., Mixed Evergreen F., etc.; Coast Ranges from Humboldt Co. and Shasta Co. to San Luis Obispo Co., Sierran foothills from Calaveras to Madera cos., Egg Lake, Modoc Co. and w. Riverside and San Diego cos. March–June.

9. **D. bélla** Hoov. Stems spreading, 5–20 cm. long, ± fistulous; lvs. 5–12 mm. long, 1–1.5 mm. wide; infl. 4–8 cm. long, the bracts 7–18 mm. long, 1–2.5 mm. wide; sepals rotately spreading to ascending, 3–6 mm. long, blunt; corolla 10–12 mm. long, deep bright blue, the lower lip with a cent. white area with yellow center and 2 basal yellow ridges that alternate with 3 small purple spots; tube 3–4 mm. long, bluish; 2 upper lobes 3–4.5 mm. long, ± parallel; lower lobes divergent; fil.-tube 2.7–3.5 mm. long; anther-tube 1.6–2.4 mm. long; caps. 3.5–5 cm. long, tough-walled; seeds not twisted.—Vernal pools of alkaline plains; V. Grassland; Cent. V. from Colusa to Tulare cos. March–May.

10. **D. concólor** Greene. [*Bolelia c.* Greene. *B. tricolor* Greene. *B. c.* var. *t.* Jeps.] Stems 5–20 cm. long; lvs. 5–18 mm. long, 0.4–2 mm. wide; infl. 3–15 cm. long, with bracts 5–16 mm. long, 1–3 mm. wide; sepals ascending or rotately spreading, 3–8 mm. long, subequal; corolla 7–13 mm. long, blue, the lower lip with a quadrate or 2-lobed purple spot at base of cent. sometimes obsolete white area and with 2 low ridges or nipples at the base; tube 3–4 mm. long; 2 upper lobes 3.5–5 mm. long, mostly recurved, minutely ciliate-scabrous on edges; fil.-tube 2.5–4 mm. long; anther-tube 1.8–2.3 mm. long, glabrous to pubescent on back, minutely tufted at tips, the 2 shorter anthers also each with a hornlike process; caps. 3–5 cm. long, tardily dehiscent; seeds not twisted.—Vernal pools, muddy banks, etc., below 3000 ft.; Mixed Evergreen F., Foothill Wd., Yellow Pine F., etc.; Coast Ranges, Lake and Trinity cos. to Monterey Co. April–July.

Var. **brévior** McVaugh. Caps. 1.2–2.5 cm. long, the valves soon opening.—Mud, below 5000 ft.; Yellow Pine F., Chaparral; Cuyamaca Mts. and coastal San Diego Co. May–July.

11. **D. montàna** Greene. [*Bolelia m.* Greene. *D. bicornuta* var. *m.* Jeps.] Stems 3–15 cm. long; lvs. 5–13 mm. long, 0.3–1 mm. wide; infl. 3–12 cm. long, the bracts 8–16 mm.

long, 1–1.5 mm. wide, often toothed; sepals 4–8 mm. long, the 2 lower usually shorter than others; corolla 9–12 mm. long, the lobes of the lower lip light blue or violet, the cent. part of lip white, the base dark blue-purple, with 2 prominent purple folds at the angle; the tube 3–5 mm. long, blue to violet, with 2 greenish-yellow ridges; 2 upper lobes erect, 4–5.5 mm. long; fil.-tube 3–4 mm. long; anther-tube 1.7–2.2 mm. long, anthers prominently bearded at tip, the 2 shorter each with a short hornlike process also; caps. 1.5–3.5 cm. long, with 3 evident hyaline valves apparent as longitudinal impressed lines; seeds not twisted.—Open grassy meadows, mostly 3000–5500 ft.; Yellow Pine F.; Sierra Nevada from Tuolumne to Shasta cos. May–Aug.

12. **D. Yìna** Appleg. Stems 3–10 cm. long; lvs. 6–10 mm. long; infl. 2–5 cm. long, with bracts 8–13 mm. long; sepals 5–6 mm. long, 0.5–1 mm. wide; corolla 8–10 mm. long, dark blue or purplish; the lower lip with cent. yellow area and 2 low yellow ridges at base; the tube 3.5–4.5 mm. long; upper lobes erect; fil.-tube 2.5–3.5 mm. long; anther-tube 1.6–2 mm. long, the shorter anthers white, tufted at tip and each also with a short hornlike process; caps. 2–2.5 cm. long, with thin walls and impressed lines of dehiscence; seeds not twisted.—Boggy places and vernal pools, below 5500 ft.; Yellow Pine F.; Siskiyou Co. to s. Ore. July–Aug.

Var. **màjor** McVaugh. [*D. willamettensis* Peck. *D. pulcherrima* Peck.] Corolla usually 7–12 mm. long, the tube 3.5–5.5 mm. long; fil.-tube 2–4 mm. long; anther-tube 2–2.5 mm. long, the 2 short anthers white-tufted at tips and each with a short hornlike process; caps. 2–4 cm. long, with no impressed lines; seeds more shining.—Drying pools, etc., below 3500 ft.; Yellow Pine F. to Redwood F.; Siskiyou and Humboldt cos. to w. Wash. June–July.

8. Legénere McVaugh.

Small annuals. Fl.-bracts elliptic. Fls. inverted, both petaliferous and apetalous; corollas in the former cleft dorsally. Fils. connate in a tube. Anthers connate, 2 shorter than the others. Ovary 1-locular. Fr. capsular, dehiscent apically, with parietal placentae. Seeds many, small, smooth. One sp. (Anagram of *E. L. Greene.*)

1. **L. limòsa** (Greene) McVaugh. [*Howellia l.* Greene.] Stems weakly erect, 1–3 dm. long, simple or branched, sometimes forming mats, smooth, glabrous; lvs. few, entire, sessile, lanceolate to ca. 2.5 cm. long, early deciduous; infl. 8–14 cm. long, loosely 6–12-fld.; pedicels filiform, 1–3 cm. long; sepals subulate to deltoid, 1–1.5 mm. long; corolla yellowish, 3.5–4 mm. long in petaliferous fls.; fil.-tube ca. 1 mm. long; caps inferior; seeds ca. 0.9 mm. long.—Dried beds of vernal pools; V. Grassland; Sacramento V. May–June.

9. Porterélla Torr.

Erect annual, simple or few-branched to diffuse, glabrous. Cauline lvs. few to many, soft, lax, early deciduous, sessile, linear-subulate, entire to sinuate. Infl. loosely 1–25-fld., with spreading-ascending slender pedicels, straight or arcuate. Fl.-bracts linear to ovate. Ovary essentially inferior. Sepals linear to narrowly triangular or elliptic, entire. Corolla blue with yellow or whitish eye, strongly bilabiate, with narrow tube ca. as long as the limb. Stamens free from corolla, coherent the whole length. Seeds fusiform, light brown, smooth, slightly lustrous. One sp. (T. C. *Porter*, 19th-century Am. botanist.)

1. **P. carnòsula** (H. & A.) Torr. [*Lobelia c.* H. & A. *Laurentia c.* Benth. & Hook.] Plants 0.5–2(–3) dm. high; cauline lf.-blades 1–2.5 cm. long, acute to acuminate; infl. 6–10(–20) cm. long; pedicels 5–25 mm. long; bracts 4–18 mm. long; caps. 7–10 mm. long; sepals 3–9 mm. long; corolla 6–9 mm. long, the 2 upper lobes erect, elliptic, 3.5–5.5 mm. long, the lower elliptic to obovate, 5–9 mm. long; fils. 3–6 mm. long; seeds ca. 1 mm. long; $n = 11$ (Carlquist in 1956).—Wet places, 5000–10,000 ft.; Yellow Pine F. to Subalpine F.; Sierra Nevada from Tulare to Lassen cos.; to se. Ore., nw. Wyo., Utah, Ariz. June–Aug.

10. Lobèlia L.

Annual or perennial herbs, or sometimes somewhat woody. Lvs. alternate, the cauline sometimes reduced to bracts. Fls. blue, red, yellowish, or white, in terminal racemes, spikes or panicles, even solitary. Fl.-tube short or globular. Fls. inverted in anthesis, the pedicel twisted. Sepals 5, short. Corolla-tube often split to the base on one side, the limb 5-lobed, the upper 3 lobes forming the "upper" lip, the other 2 (1 on each side of the cleft) erect or recurved. Anthers united into a tube or ring around the style, 2 or all hairy at tips. Ovary largely inferior, 2-loculed, many-ovuled. Fr. a 2-valved caps. Ca. 250 spp., a widely distributed genus. (Named for de *Lobel* or *L'Obel*, a 16th-century Flemish botanist.)

(McVaugh, R. A revision of "Laurentia" and allied genera in N. Am. Bull. Torr. Club 67: 778–798, 1940.)

Corolla bright red, 2.5–4 cm. long, the tube split to the base on the upper side ...... 1. *L. Cardinalis*
Corolla blue, less than 2 cm. long, the tube not cleft, at least at the upper end .... 2. *L. Dunnii*

1. **L. Cardinális** L. ssp. **gramínea** (Lam.) McVaugh. [*L. g.* Lam. *L. splendens* Willd.] SCARLET LOBELIA. Perennial by means of slender offsets; stems simple, erect, 3–10 dm. high, glabrous to pubescent; lvs. lanceolate to linear-lanceolate (or the lower oblanceolate), 5–12 cm. long, irregularly serrulate, the lowermost petioled; raceme 1–3 dm. long in fr.; pedicels 5–15 mm. long; sepals subulate, ca. 8–10 mm. long; anther-tube bluish-gray, 3–4 mm. long; caps. papery, ca. 4 mm. long, 5–6 mm. thick, ca. 10-ribbed; seeds many, ± obpyramidal, golden-brown, deeply and regularly pitted.—Occasional in boggy places, below 6000 ft.; Chaparral, Coastal Sage Scrub, etc.; Los Angeles Co. to San Bernardino and San Diego cos.; to Tex., Mex., Panama. Aug.–Oct. Our plants have been referred to var. *pseudospléndens* McVaugh when with lvs. 8–14 times as long as wide and the infl. short, or to var. *multiflòra* (Paxt.) McVaugh when with lvs. 5–8 times as long as wide and the infl. ample, often leafy.

2. **L. Dúnnii** Greene var. **serràta** (Gray) McVaugh. [*Palmerella debilis* var. *s.* Gray. *Laurentia d.* var. *s.* McVaugh.] Perennial, decumbent or erect, 3–5 dm. tall; lvs. linear-lanceolate to oblanceolate, glabrous, 3–7 cm. long, serrate, sessile, the lower petioled; infl. sometimes puberulent, densely few- to several-fld.; pedicels 4–10 mm. long; bracts linear, 1–2 cm. long, or the lower longer; sepals linear-subulate, 8–14 mm. long; corolla pubescent, the tube split incompletely on the dorsal side from near the base; anther-tube bluish-gray, 2–3 mm. long, pilose; caps. 6–10 mm. long, campanulate to ellipsoid; seeds ellipsoid-lenticular, smooth, shining, light brown, ca. 0.5 mm. long.—Occasional in moist canyons, below 3000 ft.; Coastal Sage Scrub, Chaparral, etc.; cismontane Calif. from Monterey Co. to n. L. Calif. July–Sept.

Compósitae. SUNFLOWER FAMILY (Fig. 108)

Genera 1 through 95 were prepared by David D. Keck.

Herbs or shrubs; in the tropics, sometimes trees. Lvs. opposite or alternate, rarely whorled, entire to dissected, estipulate. Fls. borne in a head, on a common receptacle, surrounded by an invol. of phyllaries (bracts), perfect, polygamous, monoecious or dioecious. Heads usually many-fld., sometimes few-fld., or even 1-fld. Receptacle usually flattened, sometimes conical or even columnar, with bracts (*chaff*), scales or bristles subtending the fls., or without (*naked*). Corolla gamopetalous, either regular, tubular, with a usually 5-toothed limb, valvate in bud, the veins bordering the margins of the lobes, or bilabiate, or ligulate, with the limb (*ray*) strap-shaped and toothed at apex, the corolla rarely wanting in the ♀ fls. Stamens 5 (rarely 4 or 3), usually united by the elongate anthers into a tube (*syngenesious*), inserted on the corolla. Style normally 2-branched, the branches stigmatiferous inside, in hermaphrodite (functionally ♂) fls. usually entire; ovary inferior, 1-celled, maturing into an ak., with a single seed without endosperm, usually bearing a persistent pappus (representing the calyx-limb) of bristles, scales, or paleae, or even a crown or ring. The tubular ♂ or hermaphrodite fls. are called *disk-fls.*, and the head is *discoid* when composed only of these. The ligulate (ray) fls. are commonly ♀, sometimes perfect or neutral, and the head is called *radiate* when composed of cent. tubular fls. and marginal ray-fls.; if all are strap-shaped, it is called *ligulate*. The largest family of vascular plants, with possibly 950 genera and 20,000 spp., chiefly herbaceous and world-wide in distribution.

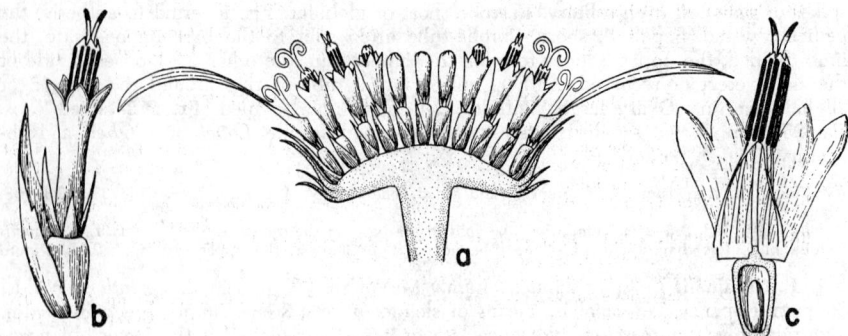

Fig. 108. COMPOSITAE. *Generalized floral stucture: a,* section through head showing enlarged receptacle with invol. of overlapping bracts (phyllaries), outer ring of ray-fls. or rays (corolla tubular only at base, with exserted pistil, no stamens); disk-fls. cent., tubular, regular, with 5-lobed corolla, exserted stamens and pistil; *b,* disk-fl. enlarged to show subtending receptacular bract (chaff), inferior ovary, pappus (in this case of 4 paleae or scales and representing modified calyx); *c,* disk-fl. with opened corolla to reveal 5 stamens united by the anthers (syngenesious) and 2-lobed style, basal ak. in section and with 1 ovule.

KEY TO THE TRIBES

A. Fls., or some of them, tubular or elongate; juice watery, very rarely milky. Subfamily
I. TUBULIFLORAE.
  B. Corollas all regular (the heads then discoid) or with the marginal ones ligulate (the heads then radiate).
    C. Style without any ring of hairs or distinct thickened ring below the branches; anthers (except in Inuleae) not tailed; plants seldom prickly; receptacle chaffy or naked, rarely somewhat bristly.
      D. Style-branches ± flattened, commonly stigmatic to middle or beyond, the stigmatic portion often conspicuously defined; heads discoid or generally radiate, at least some of the fls. often yellow.
        E. Anthers truncate to sagittate at base but scarcely tailed.
          F. Receptacle with chaffy bracts .... 1. *Heliantheae*   p. 1082
          FF. Receptacle naked (setaceous in Gaillardia).
            G. Style-branches appendaged at tip, not truncate.
              H. Phyllaries in few series, little imbricated; disk-fls. commonly yellow; rays, when present, usually yellow; style-branches various, the appendage, when present, hairy on both faces .... 2. *Helenieae* p. 1129
              HH. Phyllaries usually imbricated; disk-fls. commonly yellow; rays, when present, of the same or different color; style-branches with the appendages glabrous on inner face .... 3. *Astereae* p. 1162
           GG. Style-branches mostly truncate, rarely if at all appendaged but usually penicillate.
              H. Phyllaries mostly several seriate, usually well imbricated, often dry with broad hyaline margins; pappus usually none, sometimes paleaceous, never of capillary bristles .... 4. *Anthemideae* p. 1227
              HH. Phyllaries mostly equal and uniseriate, chiefly herbaceous, often with much reduced secondary bractlets.
                I. Pappus of capillary bristles; aks. homomorphic 5. *Senecioneae* p. 1238
                II. Pappus none; aks. heteromorphic 6. *Calenduleae* p. 1256
        EE. Anthers tailed at base; style-branches rounded or truncate, unappendaged at tip; corollas all tubular (or in Inula some of them with yellow rays); herbage ± white-woolly (except in Pluchea) 7. *Inuleae* p. 1257
      DD. Style-branches terete or clavate, strongly flattened, papillate, not hairy, stigmatic only near base; fls. all tubular and perfect, never yellow; receptacle naked .... 8. *Eupatorieae* p. 1266
    CC. Style with a ring of hairs, or sometimes merely with a thickened ring, below the papillate branches, the branches generally connate to near tip; plants often prickly.
      D. Anthers tailed at base; rays none; receptacle densely bristly or sometimes naked .... 9. *Cynareae* p. 1271
      DD. Anthers not tailed at base; rays present in ours; receptacle glabrous or alveolate in ours, never bristly .... 10. *Arctotideae* p. 1284

BB. Corollas all bilabiate, the fls. perfect .... 11. *Mutisieae* p. 1284
AA. Fls. all ligulate and perfect; juice milky or colored. Subfamily II. LIGULIFLORAE
12. *Cichorieae* p. 1286

ARTIFICIAL KEY TO THE GENERA OF COMPOSITAE

A. Anthers tailed or sagittate at their base.
  B. Some or all of the corollas appearing definitely bilabiate ...... Tribe *Mutiseae* p. 1284
  BB. Corollas not at all or obscurely bilabiate; plants thistlelike ...... Tribe *Cynareae* p. 1271
AA. Anthers not tailed or sagittate at their base.
  B. Heads with fls. all perfect, the corollas strap-shaped, 5-toothed at apex (true ligules) .... Tribe *Cichorieae* p. 1286
  BB. Heads with fls. tubular when perfect, and ± regular; marginal fls. often ♀ or neutral and often strap-shaped, 2–3-toothed (ray-fls. or rays).
    C. Pappus none or vestigial, sometimes a mere crown.
      D. Rays none or vestigial .... **Group A** p. 1075
      DD. Rays evident .... **Group B** p. 1076
    CC. Pappus evident on some or all of the aks.
      D. Rays none or vestigial.
        E. Aks. with a pappus of scales or awns or both, these sometimes united into a low crown .... **Group C** p. 1077
        EE. Aks. with a pappus of capillary bristles, rarely with additional outer scales .... **Group D** p. 1078
      DD. Rays present, evident.
        E. Aks. with a pappus of firm awns or chaffy scales.
          F. Rays yellow to orange or brown .... **Group E** p. 1079
          FF. Rays white to pink, red or purplish .... **Group F** p. 1080
        EE. Aks. with a pappus of capillary bristles, rarely with additional outer scales .... **Group G** p. 1081

Group A. (Rays none or vestigial; pappus none or vestigial)

A. Heads unisexual, both ♂ and ♀ on the same plant; invol. of ♀ heads closed and ± burlike; ♂ heads in a raceme or spike, with open invols.
  B. Phyllaries of ♂ heads united.
    C. Fruiting invol. winged with broad scarious scales; lvs. or their divisions filiform
          24. *Hymenoclea*
    CC. Fruiting invol. not so winged; lvs. toothed or pinnatifid.
      D. The fruiting invol. with a ring of tubercles or short spines below the single beak
          25. *Ambrosia*
      DD. The fruiting invol. with several rows of spines below the 1–4 beaks
          26. *Franseria*
  BB. Phyllaries of ♂ heads distinct; invol. of ♀ heads becoming a stout spiny bur
          27. *Xanthium*
AA. Heads not unisexual; ♀ invol. not becoming a bur.
  B. Phyllaries in a single series or lacking.
    C. Lvs. opposite; low annual herbs.
      D. Proper invol. present; plant green .... 51. *Baeria*
      DD. Proper invol. none; plant white-woolly .... 129. *Psilocarphus*
    CC. Lvs. mostly alternate.
      D. Fls. all perfect; low small-lvd. perennial. White Mts. .... 45. *Laphamia*
      DD. Fls. imperfect, ♂ and ♀ in same head.
        E. Plant leafy only at base, the large lvs. deltoid-ovate, soon glabrate above, white-woolly beneath; phyllaries reflexed .... 107. *Adenocaulon*
        EE. Plant not leafy only at base; lvs. not as above; phyllaries not reflexed.
          F. Lvs. or their lobes linear-filiform; aks. long-villous; plant woody
            21. *Oxytenia*
          FF. Lvs. or their lobes not linear-filiform; aks. not villous; plant not woody.
            G. Plants not white-woolly, rather coarse.
              H. Aks. obovoid or pyriform, not winged; perennial herbs; phyllaries ca. 5 .... 22. *Iva*
              HH. Aks. flattened, pectinate-winged; annual herbs; phyllaries 6–7, the 1 or 2 inner enlarged .... 23. *Dicoria*
           GG. Plants low white-woolly annuals.
               H. Fr.-bearing bracts conduplicate or sacklike, completely enclosing the ♀ fl. and falling away with the ak.
                I. Receptacle chaffy throughout; proper invol. below the sacklike bracts none .... 127. *Stylocline*
                II. Receptacle naked in the center, the innermost fls. bractless; invol. below the sacklike bracts of ca. 5 phyllaries
                  130. *Micropus*
              HH. Fr.-bearing bracts open, merely subtending the ♀ fls., not falling away with the aks. .... 131. *Evax*
  BB. Phyllaries in 2 or more usually imbricated series.
    C. Heads in spikes, racemes or panicles .... 106. *Artemisia*
    CC. Heads solitary or in capitate clusters or in corymbs.
      D. Aks., especially the marginal, stipitate .... 105. *Cotula*

DD. Aks. sessile.
  E.   Receptacle convex or hemispheric, not chaffy .... 103. *Matricaria*
  EE. Receptacle flat.
    F.   Receptacle chaffy; plant woody, 4–6 dm. high, with tomentose pinnately
      divided lvs. .... 98. *Santolina*
    FF. Receptacle not chaffy.
      G.   Perennials with capitate to corymbiform infl. .. 102. *Tanacetum*
      GG. Annuals with burlike heads sessile in forks of stems
                                          104. *Soliva*

Group B. (Rays evident; pappus none or vestigial)

A.   Receptacle chaffy with bracts at base of at least some of the fls., or with stiff bristles.
  B.   Phyllaries usually imbricated, or at least in more than 1 series.
    C.   Lvs. opposite. Introd. weeds.
      D.   Ray-ak. completely enclosed by the subtending inner phyllary .. 19. *Melampodium*
      DD. Ray-ak. not so enclosed.
        E.   Heads (including rays) barely 1 cm. in diam.; rays 1–5, white
                                        11. *Eclipta*
        EE. Heads (including rays) ca. 3 cm. in diam.; rays yellow .... 15. *Guizotia*
    CC. Lvs. alternate (at least in upper ⅔ of the plant) or largely basal.
      D.   Rays white.
        E.   Heads relatively large, solitary or few .... 96. *Anthemis*
        EE. Heads very small, many, in dense flattish or rounded panicles
                                      97. *Achillea*
      DD. Rays yellow.
        E.   Invol. of 2 very unlike series of phyllaries; herbs or ± woody and with
          fleshy stems .... 14. *Coreopsis*
        EE. Invol. of 2 or more series of similar phyllaries; often ± suffrutescent.
          F.   Disk-aks. thickened, angled, not or scarcely compressed.
            G.   Receptacle ± columnar; rays sterile .... 5. *Rudbeckia*
            GG. Receptacle flat or merely convex; rays fertile .... 2. *Balsamorhiza*
          FF. Disk-aks. strongly flattened.
            G.   Plants scapose, herbaceous, with very large solitary heads; aks.
              with a cartilaginous border .... 7. *Enceliopsis*
            GG. Plants leafy-stemmed, ± shrubby, the heads medium-sized, several;
              aks. ciliate .... 8. *Encelia*
  BB. Phyllaries in a single series.
    C.   Receptacle bristly, honeycombed; phyllaries not enclosing the ray-aks.; plant ± tomen-
      tose; annuals.
      D.   Rays orange. Garden escape .... 148. *Venidium*
      DD. Rays yellow. Desert annual .... 56. *Eriophyllum ambiguum*
    CC. Receptacle not bristly.
      D.   Ray-aks. enclosed in the obcompressed phyllaries, of which the infolded lateral
        margins meet.
        E.   Disk-fls. 6; ray-aks. obcompressed .... 29. *Lagophylla*
        EE. Disk-fl. 1 and the ray-ak. obcompressed, or more than 1 and the ray-aks.
          not obcompressed .... 32. *Madia*
      DD. Ray-aks. not completely enclosed, the phyllaries not obcompressed.
        E.   Rays 3-toothed or 3-lobed, the lobes subparallel; herbage without tack-
          shaped glands.
          F.   Upper lvs. and phyllaries not terminated by open pit glands
                                 33. *Hemizonia*
          FF. Upper lvs. and phyllaries terminated by open pit glands
                                34. *Holocarpha*
        EE. Rays 3-cleft or -parted into palmately spreading lobes; herbage mostly with
          tack-shaped glands .... 35. *Calycadenia*
AA. Receptacle naked, without bristles or chaffy bracts.
  B.   Phyllaries graduated in , several unequal series, closely imbricate, at least the inner ±
    scarious.
    C.   Rays well developed, the heads solitary to corymbose .... 101. *Chrysanthemum*
    CC. Rays short, inconspicuous, the heads in capitate to crowded-corymbose infls.
                                        102. *Tanacetum*
  BB. Phyllaries in 1–2 series, or at least not imbricate.
    C.   Rays white or pink. Low introd. daisy .... 83. *Bellis*
    CC. Rays yellow or orange.
      D.   Phyllaries scarious-margined, incurved; ray-aks. strongly incurved; lvs. entire.
        Garden escape .... 121. *Calendula*
      DD. Phyllaries and ray-aks. not strongly incurved.
        E.   Lvs. opposite.
          F.   Rays persistent, becoming papery; plant perennial, ± white-woolly. Pine
            belt .... 40. *Whitneya*
          FF. Rays not persistent or becoming papery.
            G.   Phyllaries in 1 series; plants mostly annual.
              H.   Phyllaries distinct .... 51. *Baeria*
              HH. Phyllaries united into a toothed cup.
                I.   Aks. 2–4-angled, usually minutely papillate under mag-
                  nification. Widely distributed .... 52. *Lasthenia*

          II. Aks. flat, hispidulous on the callous margin. Cent. V.
                         53. *Crockeria*
        GG. Phyllaries in 2 or more series; plant perennial, succulent
                         42. *Jaumea*
    EE. Lvs. alternate.
      F. Phyllaries broad, in 2 distinct series (the outer ± reflexed, the inner
          erect); sunflowerlike perennial .... 41. *Venegasia*
      FF. Phyllaries not as above.
          G. Rays persistent and becoming papery; phyllaries many; woolly per-
              ennial of the deserts .... 39. *Baileya*
          GG. Rays not as above.
             H. Rays with a tooth or lobe on the inner side of the throat
                opposite the ligule .... 55. *Monolopia*
             HH. Rays without a lobe on the inside.
                I. Herbage puberulent; annual herbs 3–6 dm. high, the lvs.
                  1–3 times ternately divided. San Bernardino and Santa
                  Rosa mts. .... 54. *Bahia*
               II. Herbage white-woolly, at least when young.
                  J. Aks. ± flattened; corolla-tube glabrous with a ring of
                    hairs at junction with throat. Cent. V.
                         57. *Pseudobahia*
                JJ. Aks. angled; corolla-tube glandular or pubescent.
                  K. Rays yellow or white. Widely distributed
                         56. *Eriophyllum*
                  KK. Rays ± purplish. Desert slopes, San Gabriel and
                    San Bernardino mts. ...... 58. *Syntrichopappus*

**Group C. (Rays none or vestigial; aks. with a pappus of paleae or flattened scales or stiff bristles
or awns or both)**

A. Receptacle chaffy.
    B. Pappus of plumose awns; plant a low almost leafless shrub .... 17. *Bebbia*
    BB. Pappus not of plumose awns.
      C. Aks. with pectinately toothed wings. Annuals .... 23. *Dicoria*
      CC. Aks. not with pectinately toothed wings.
        D. Awns or teeth of pappus not retrorsely barbed. Shrubby.
          E. Lvs. ovate to broadly lanceolate, often 3-nerved from base. Deserts
                         8. *Encelia*
             EE. Lvs. linear, 1-nerved. About the Cent. V. .... 18. *Eastwoodia*
        DD. Awns or teeth retrosely barbed. Herbaceous .... 12. *Bidens*
AA. Receptacle not chaffy.
    B. Lvs., at least all the lower, opposite.
      C. Plant woody; lvs. deltoid, small, on long petioles. Deserts .... 134. *Hofmeisteria*
      CC. Plants herbaceous.
        D. Perennial; lvs. hastate-triangular, 3–7 cm. long; plants 1–1.5 m. tall
                         46. *Pericome*
        DD. Annuals; lvs. mostly narrower, or if broad, shorter; plants smaller.
          E. Lvs. not auricled at base; phyllaries ca. 5–15 .... 51. *Baeria*
          EE. Lvs. auricled at base, ± oblong; phyllaries ca. 15–30 .. 132. *Trichocoronis*
    BB. Lvs. mostly alternate.
      C. Pappus a mere crown; phyllaries imbricate, ± scarious or scarious-margined.
        D. Perennials with heads in capitate or corymbiform infl. .... 102. *Tanacetum*
        DD. Annuals with heads solitary or somewhat corymbed .... 103. *Matricaria*
      CC. Pappus of well developed paleae or awns or both.
        D. Lvs. and invol. punctate with translucent oil-glands.
          E. Pappus of 2–6 entire paleae with or without awns; phyllaries united into
              a cup or tube .... 68. *Tagetes*
          EE. Pappus of 10–15 paleae or these dissected into bristles; phyllaries ± 2-seriate
                         69. *Dyssodia*
        DD. Lvs. and invol. not punctate with translucent oil-glands.
          E. Pappus of bristles, or these alternating with minute paleae or awns.
             F. The pappus of 2–8 deciduous awns; plants perennial, usually strongly
                viscid. Mostly cismontane .... 71. *Grindelia*
             FF. The pappus of ca. 4 bristles alternating with ca. 4 minute paleae;
                annual, not glutinous. Colo. Desert .... 133. *Malperia*
          EE. Pappus of flat or awnlike paleae.
             F. Pappus of 20–30 silvery awns; plant woody. Deserts
                         74. *Acamptopappus*
             FF. Pappus of mostly fewer parts.
               G. Phyllaries in a single series.
                H. Pappus-paleae erose or fimbriate.
                  I. Pappus-paleae 11–17, deciduous in a ring. Subalpine
                    annual   .... 62. *Orochaenactis*
                  II. Pappus-paleae fewer, deciduous separately. Woolly annuals.
                    J. Lvs. 3-lobed at apex; plants 3–8 cm. high
                         56. *Eriophyllum Pringlei*
                  JJ. Lvs. linear, entire to sinuate-dentate; plants 8–30 cm.
                  high   .... 43. *Eatonella Congdonii*

      HH. Pappus-paleae nearly or quite entire.
        I. Pappus not of subulate awns.
          J. Fls. whitish or purplish.
            K. Pappus-paleae linear, with a strong midrib; lvs.
               entire .... 63. *Palafoxia*
           KK. Pappus-paleae nerveless; lvs. toothed to pinnatifid
               61. *Chaenactis*
        JJ. Fls. yellow.
          K. Pappus-paleae 4; lvs. dissected
              61. *Chaenactis glabriuscula*
          KK. Pappus-paleae more numerous.
            L. Plants annual, viscid; lvs. mostly linear, en-
               tire. Coastal .... 64. *Amblyopappus*
            LL. Plants perennial, not viscid; lvs. mostly dis-
               sected. Interior .... 56. *Eriophyllum*
        II. Pappus of 3–5 subulate awns; slender annuals with entire
          linear lvs. .... 59. *Rigiopappus*
    GG. Phyllaries in 2 or more series.
      H. Pappus-paleae 5, dissected into bristles; white-woolly annual.
        Colo. Desert .... 60. *Trichoptilium*
      HH. Pappus-paleae 10–22, not so dissected.
        I. The pappus-paleae obtuse; lvs. in a basal rosette
          99. *Hymenopappus*
        II. The pappus-paleae lanceolate, each ending in an awn;
          lvs. both basal and cauline .... 100. *Hymenothrix*

Group D. (Rays none or vestigial; pappus of capillary bristles)

A. Phyllaries completely scarious or hyaline; herbage ± white-woolly.
  B. Plants taprooted, annual or perennial; heads all with outer ♀ and cent. perfect or func-
    tionally ♂ fls.
    C. Receptacle chaffy except in the center; plants small, to ca. 3 dm. high
        128. *Filago*
    CC. Receptacle naked; plants mostly larger .... 124. *Gnaphalium*
  BB. Plants fibrous-rooted, perennial, often with rhizomes or stolons, dioecious or nearly so.
    C. Plants mostly less than 3 dm. high, with a basal rosette of lvs., strictly dioecious
        125. *Antennaria*
    CC. Plants mostly 3–9 dm. high, without a basal rosette; ♀ plants commonly with a few
      ♂ fls. in center of each head .... 126. *Anaphalis*
AA. Phyllaries herbaceous or only partly scarious or hyaline and then the herbage usually not white-
  woolly.
  B. Heads unisexual; plants dioecious, ± woody at base .... 95. *Baccharis*
  BB. Heads not strictly unisexual.
    C. Plants well developed shrubs.
      D. Phyllaries proper 4–7, in a single series, of equal length; lvs. narrow, entire
        119. *Tetradymia*
      DD. Phyllaries more numerous.
        E. Lvs. scalelike, at least on flowering branches.
          F. Aks. sericeous; heads solitary at ends of small branchlets
            88. *Aster intricatus*
          FF. Aks. glabrous; heads closely placed in large clusters
            120. *Lepidospartum*
        EE. Lvs. not scalelike.
          F. Phyllaries arranged in ± distinct vertical ranks .... 82. *Chrysothamnus*
          FF. Phyllaries not so arranged.
            G. Fls. not yellow.
              H. Fls. purplish; plant a shrub 1–4 m. high; phyllaries not striate
                123. *Pluchea*
             HH. Fls. whitish; plants mostly not so tall; phyllaries ± striate.
               I. Pappus double, the inner of bristles, the outer of short
                  paleae; petioles longer than the small lf.-blades
                    134. *Hofmeisteria*
               II. Pappus simple, of capillary bristles only; petioles mostly
                 shorter than lf.-blades .... 136. *Brickellia*
           GG. Fls. yellow.
             H. Phyllaries merely of 1 series, subequal, subulate; desert shrub
               with terete resinous-punctate lvs. ...... 118. *Peucephyllum*
             HH. Phyllaries of more than 1 series, unequal; widely distributed
               78. *Haplopappus*
    CC. Plants herbaceous.
      D. Plants annual.
        E. Herbage white-woolly, aromatic; lvs. broad, rounded; heads solitary in forks
        of branches. Deserts .... 114. *Psathyrotes*
        EE. Herbage and lvs. not as above.

   F. Phyllaries and fls. only 2–3 in each head; small annual with broad lvs.
                  108. *Dimeresia*
  FF. Phyllaries and fls. more numerous.
    G. Phyllaries in a single series, equal or nearly so.
      H. Plants 1–3 dm. high, with filiform branches; lvs. linear, entire
                  86. *Tracyina*
      HH. Plants much taller, coarser; lvs. broader, dentate to lobed
                 117. *Erechtites*
    GG. Phyllaries in more than 1 series, unequal.
      H. Outer corollas enlarged, more deeply cleft on inner side;
        herbage often strong-smelling .... 93. *Lessingia*
      HH. Outer corollas not enlarged, very slender; herbage not usually
        strong-smelling .... 94. *Conyza*
  DD. Plants biennial to perennial.
    E. Principal phyllaries essentially equal and in 1 series, although some basal
      outer reduced bracts may be present.
      F. Lvs. opposite .... 109. *Arnica*
      FF. Lvs. alternate or basal.
        G. Pappus-bristles plumose; plants ± scapose ...... 110. *Raillardella*
        GG. Pappus-bristles barbellate, not truly plumose.
          H. Lvs. large, palmately cleft or parted, mostly basal; heads
            corymbose .... 115. *Cacaliopsis*
          HH. Lvs. entire to toothed or pinnate or pinnatifid.
            I. Fls. white or pinkish .... 116. *Luina*
            II. Fls. yellow to orange .... 111. *Senecio*
  EE. Principal phyllaries ± imbricate in 2–several series.
    F. Phyllaries and lvs. with conspicuous translucent oil-glands
                 70. *Porophyllum*
    FF. Phyllaries and lvs. not having translucent oil-glands.
      G. Fls. yellow.
        H. Style-appendages very short, 0.5 mm. long or less; phyllaries
          scarcely imbricate .... 92. *Erigeron*
        HH. Style-appendages mostly 0.7 or more mm. long; phyllaries
          plainly imbricate.
          I. Plant a biennial or tap-rooted perennial with squarrose
            phyllaries .... 90. *Machaeranthera*
          II. Plant not as above.
            J. Aks. ± flattened; invol. 9–12 mm. high; plants
              herbaceous .... 75. *Chrysopsis*
            JJ. Aks. terete or angled; invol. mostly 12 mm. or higher;
              plants mostly definitely woody at base
                78. *Haplopappus*
    GG. Fls. not yellow.
      H. Aks. 5-angled or -ribbed; lvs. often opposite
                135. *Eupatorium*
      HH. Aks. 10-ribbed; lvs. alternate .... 136. *Brickellia*

Group E. **(Rays well-developed, yellow to orange or brown, at least when fresh;**
    pappus of well-developed paleae or scales or of stiff awns)

A. Receptacle chaffy or bristly throughout or with a circle of chaffy bracts surrounding the disk-fls.
  B. Invol. of 1–several series of phyllaries, none enfolding the ray-aks.
    C. Receptacle bristly .... 48. *Gaillardia*
    CC. Receptacle chaffy.
      D. Phyllaries in 2 distinct dissimilar series; aks. flattened at right angles to the radius
        of the head.
        E. Pappus of 2–6 firm mostly retrorsely barbed awns; lvs. entire to pinnately
          dissected.
          F. Aks. strongly flattened, not beaked .... 12. *Bidens*
          FF. Aks. not much flattened, strongly beaked .... 13. *Cosmos*
        EE. Pappus of 2 minute teeth; lvs. pinnately dissected .... 14. *Coreopsis*
      DD. Phyllaries in 2 or more similar series.
        E. Rays sessile and persistent on their aks., becoming papery; disk-aks. 4-angled.
          E. Mojave Desert .... 20. *Sanvitalia*
        EE. Rays not persistent or papery.
          F. Plants scapose or nearly so, with broad entire ± silvery-pubescent lvs. and
            large long-peduncled heads .... 7. *Enceliopsis*
          FF. Plants not scapose.
            G. Rays ♀.
              H. Disk-aks. strongly compressed, 2-winged; lvs. mostly opposite
                10. *Verbesina*
              HH. Disk-aks. compressed-quadrangular; lvs. alternate
                1. *Wyethia*
            GG. Rays neutral, the aks. not developing.
              H. Aks. ± thickened.
                I. Pappus of 2 awns and several short squamellae
                  3. *Viguiera*

II. Pappus of 2, rarely many, deciduous paleaceous awns
                                                             4. *Helianthus*
HH. Aks. very flat.
    I. Aks. with 2 white wings; lvs. opposite, at least below
                                                             10. *Verbesina*
    II. Aks. not winged; lvs. usually alternate.
      J. Plants annual; awns 2, strong; ak.-body conspicuously
          white-margined .... 9. *Geraea*
      JJ. Plants perennial.
        K. Low rounded shrubs; awns 1–2, weak
                                                             8. *Encelia*
        KK. Herbs; awns 2, short, with intermediate scales
                                                             6. *Helianthella*
BB. Invol. with 1 series of phyllaries, each partly or completely enclosing an ak.
  C. Ray-aks. completely enfolded by the phyllaries.
    D. Fls. red-brown; pappus silvery-scarious, the longer paleae longer than the aks.;
        aks. 10-costate and tuberculate-scabrous .... 28. *Achyrachaena*
    DD. Fls. yellow; pappus and aks. not as above .... 31. *Layia*
  CC. Ray-aks. half enclosed by the phyllaries .... 33. *Hemizonia*
AA. Receptacle not chaffy or bristly.
  B. Invol. and lvs. with translucent oil-glands.
    C. Phyllaries uniseriate, united almost to apex; pappus of 5–6 unequal paleae . 68. *Tagetes*
    CC. Phyllaries ± 2-seriate, if 1-seriate, then quite free; pappus not as above.
      D. Aks. pubescent; lvs. lacking stiff spreading basal bristles; phyllaries ± 2-seriate
                                                             69. *Dyssodia*
      DD. Aks. glabrous; lvs. with a few stiff spreading bristles at base; phyllaries 1-seriate
                                                             66. *Pectis*
  BB. Invol. and lvs. lacking translucent oil-glands.
    C. Invol. of several series of graduated phyllaries.
      D. Pappus of 2–8 slender deciduous awns; heads rather large, gummy; coarse herbs
                                                             71. *Grindelia*
      DD. Pappus not as above; heads small.
        E. Pappus of disk-aks. of 1 series of straight paleae; lvs. mostly linear
                                                             72. *Gutierrezia*
        EE. Pappus of disk-aks. of many ± twisted narrow paleae or flat bristles; lvs.
          elliptic to obovate .... 73. *Amphipappus*
    CC. Invol. of ca. 1–2 equal or subequal but not graduated series of phyllaries.
      D. Lvs. opposite; mostly low annuals.
        E. Phyllaries distinct .... 51. *Baeria*
        EE. Phyllaries united into a toothed cup .... 52. *Lasthenia*
      DD. Lvs. mostly alternate.
        E. Rays persistent and becoming papery, few, broad, conspicuous; herbage
          white-woolly. Deserts ... . 38. *Psilostrophe*
        EE. Rays not persistent or becoming papery.
          F. Phyllaries in 1 series .... 56. *Eriophyllum*
          FF. Phyllaries in more than 1 series.
            G. Lvs. entire to pinnately cut, green above, white-woolly beneath;
              heads 3–8 cm. across; subacaulescent escape from gardens
                                                             149. *Gazania*
            GG. Lvs. various; heads mostly smaller. Natives.
              H. Phyllaries reflexed; rays short, broad ...... 50. *Helenium*
              HH. Phyllaries not reflexed.
                I. Receptacle ± rounded; aks. turbinate, mostly 5-angled;
                  pappus-paleae mostly 5 .... 49. *Hymenoxys*
                II. Receptacle flat; aks. linear, flattened or somewhat 3-
                  angled; pappus-paleae mostly 4 .... 47. *Hulsea*

Group F. (Rays well-developed, white to pink, red, or purplish; pappus of paleae or scales or of stiff awns)

A. Heads 5–7 cm. across; peduncles 1.5–3 dm. long. Garden escapes.
  B. White-woolly perennial; lvs. toothed .... 147. *Arctotis*
  BB. Green annual; lvs. dissected into linear segms. .... 13. *Cosmos*
AA. Heads smaller.
  B. Outermost aks. ± enfolded by their subtending phyllary.
    C. Ray-aks. half enclosed by the phyllaries.
      D. Disk-aks. sterile, undeveloped, glabrous, epappose .... 33. *Hemizonia*
      DD. Disk-aks. fertile, 10-ribbed, hairy, pappose with plumose paleae
                                                             36. *Blepharizonia*
    CC. Ray-aks. completely enclosed by the phyllaries .... 31. *Layia*
  BB. Outermost aks. not so enfolded.
    C. Lvs. opposite; rays 4–5, short; aks. black. Introd. weed .... 16. *Galinsoga*
    CC. Lvs. alternate or basal.
      D. Phyllaries dotted or striped with oil-glands; rays purplish; perennial
                                                             67. *Nicolletia*
      DD. Phyllaries not bearing conspicuous oil-glands; mostly annuals.
        E. Receptacle naked.

> F. Aks. compressed, 2-edged or 2-nerved.
> > G. Pappus of several to many paleae or flattened bristles; phyllaries scarious-margined .... 85. *Townsendia*
> > GG. Pappus of 1–2 bristlelike awns or a crown of scales or both; phyllaries not scarious-margined .... 44. *Perityle*
> FF. Aks. ± thickened, 4–5-angled.
> > G. Herbage hispid-pilose; upper lvs. subtending the heads
> > > 84. *Monoptilon*
> > GG. Herbage woolly; upper lvs. not subtending the heads
> > > 56. *Eriophyllum*
> EE. Receptacle chaffy throughout; slender annual 1–3 dm. tall. N. Calif.
> > 37. *Blepharipappus*

Group G. (Rays well-developed; pappus of soft capillary bristles, mostly without any paleae)

A. Rays yellow.
> B. Phyllaries ca. 5, partly enclosing the ray-aks.; pappus of 35–40 white bristles. Mojave Desert
> > 58. *Syntrichopappus*
> BB. Phyllaries not at all enclosing the ray-aks.
> > C. Phyllaries many, usually ± imbricated; style-tips not truncate.
> > > D. Plants annual.
> > > > E. Lvs. entire, narrow, not bristle-tipped .... 77. *Chaetopappa*
> > > > EE. Lvs. pinnatifid, the lobes bristle-tipped .... 78. *Haplopappus gracilis*
> > > DD. Plants perennial.
> > > > E. Ray-aks. without pappus; tall leafy herbs with broad lvs. . 76. *Heterotheca*
> > > > EE. Ray-aks. with pappus.
> > > > > F. Pappus in 1 series.
> > > > > > G. Pappus-bristles unequal; invol. mostly 6–18 mm. high, if shorter, mostly on shrubby plants.
> > > > > > > H. Disk-fls. fertile, their aks. pubescent; pappus-bristles 12–40 or more, at least as long as invol.; outer phyllaries lacking a prominent capitate gland .... 78. *Haplopappus*
> > > > > > > HH. Disk-fls. sterile, their undeveloped aks. glabrous; pappus-bristles 1–8, shorter than invol.; outer phyllaries tipped by a prominent capitate gland .... 79. *Benitoa*
> > > > > > GG. Pappus-bristles equal; invol. mostly 4–5 mm. high, or if larger, on herbaceous plants.
> > > > > > > H. Heads subcylindric in dense flattish fastigiate cymes; lvs. coriaceous; stems low, cespitose from a branched caudex
> > > > > > > > 81. *Petradoria*
> > > > > > > HH. Heads neither cylindric nor in dense fastigiate cymes; lvs. not definitely coriaceous; stems not cespitose from a branched caudex .... 80. *Solidago*
> > > > > FF. Pappus double, in 2 series, the outer much shorter than the inner and of short linear scales or bristles.
> > > > > > G. Lvs. oblong-spatulate to oblanceolate or elliptic, well distributed
> > > > > > > 75. *Chrysopsis*
> > > > > > GG. Lvs. linear, mostly basal .... 92. *Erigeron*
> > CC. Phyllaries fewer, or if more, at least not strongly imbricated, mostly subequal and in 1 series.
> > > D. Lvs. mostly opposite .... 109. *Arnica*
> > > DD. Lvs. alternate or basal.
> > > > E. Pappus plumose .... 110. *Raillardella*
> > > > EE. Pappus not plumose.
> > > > > F. Receptacle flat or nearly so; plants mostly perennial .. 111. *Senecio*
> > > > > FF. Receptacle strongly conic; delicate spring annual .... 112. *Crocidium*
AA. Rays white to pink, blue, or purple.
> B. Ray-aks. completely enfolded by their phyllaries; plant perennial; heads solitary
> > 30. *Holozonia*
> BB. Ray-aks. not so enclosed.
> > C. Lvs. basal, large, palmately lobed, appearing after the scapelike flowering stems with bractlike lvs.; heads in a dense corymb .... 113. *Petasites*
> > CC. Lvs. not as above and not appearing after the flowering stems.
> > > D. Pappus of ray-fls. much reduced.
> > > > E. Style-appendages of disk-fls. ± comose; perennial. Montane and cismontane
> > > > > 91. *Corethrogyne*
> > > > EE. Style-appendages of disk-fls. not comose; annual. Deserts .... 87. *Psilactis*
> > > DD. Pappus well developed in both rays and disk-fls.
> > > > E. Heads nodding in bud; lvs. narrow; annuals with slender stems
> > > > > 77. *Chaetopappa*
> > > > EE. Heads erect in bud.
> > > > > F. Rays very numerous, filiform, scarcely exceeding the disk-fls.; annual weeds with numerous sublinear lvs. .... 94. *Conyza*
> > > > > FF. Rays much surpassing the disk.
> > > > > > G. Phyllaries subequal or ± imbricate, neither leafy nor with chartaceous base and herbaceous green tip; style-appendages 0.5 mm. long or less, lanceolate or broader .... 92. *Erigeron*

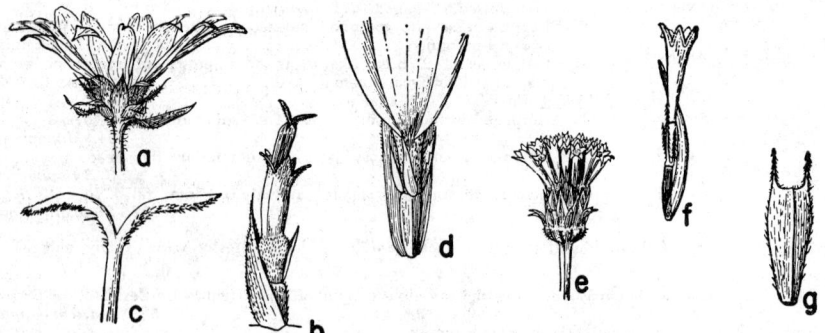

Fig. 109. HELIANTHEAE. *Helianthus annuus: a,* head, × ½, with invol. of foliaceous phyllaries, and ray-fls.; *b,* disk-fl., × 2½, with subtending receptacular bract, inferior ovary, pappus (1 palea shown), swollen base of corolla-tube, 5 corolla-lobes, exserted stamens and stigmatic branches (latter enlarged, × 7, in *c*); *d,* base of ray-fl., × 2½, in axil of phyllary and with corolla having tubular and expanded portions. *Bidens pilosa: e,* discoid head, × 1, with basal phyllaries different from upper; *f,* discoid fl., × 2½, with subtending bract, pappus of ca. 4 retrorsely barbed awns. *B. frondosa: g,* ak., × 2½, with 2 awns.

> GG. Phyllaries either subequal and the outer leafy, or usually imbricate, with chartaceous base and green tip; style-appendages lanceolate or narrower, usually more than 0.5 mm. long.
>> H. Plants annual or biennial or perennial from a distinct taproot; lvs. mostly toothed; phyllaries chartaceous or coriaceous toward base, often spreading or recurved at the herbaceous tips .... 90. *Machaeranthera*
>> HH. Plants mostly perennial and if so, rhizomatous or fibrousrooted; phyllaries mostly herbaceous.
>>> I. Style-appendages lanceolate to subulate, acute; lvs. mostly 3–20 cm. long; plants often tallish. Mostly montane and cismontane .... 88. *Aster*
>>> II. Style-appendages ovate or oblong, obtuse; lvs. to ca. 1 cm. long; plants low, tufted. Of desert mts.
>>>> 89. *Leucelene*

Subfamily I. Tubuliflòrae

Tribe 1. Heliántheae. SUNFLOWER TRIBE (Fig. 109)

Herbs or shrubs with mostly yellow fls. and at least the lower lvs. frequently opposite, the herbage often glandular. Heads heterogamous, with ♀ fls. radiate (rarely with the ligule wanting, in which case some are monoecious), or sometimes homogamous and discoid. Receptacle paleaceous, with chaffy bracts subtending all the disk-fls. or the marginal ones only. Pappus various or none but never of truly capillary bristles.—An almost wholly Am. tribe of over 150 genera.

A. Phyllaries in 1–several series, none enfolding ray-aks.; receptacle very chaffy.
> B. Anthers united; heads not unisexual, the corollas not much reduced.
>> C. Ligule of ray-fls. not becoming papery and persistent on the fr.
>>> D. Disk-fls. fertile.
>>>> E. Pappus cupulate or coroniform, or of teeth or awns from the 2–4 principal angles, or of some scales, or of a few stout bristles, or none.
>>>>> F. Disk-aks. various, but never obcompressed. (Subtribe Verbesininae)
>>>>>> G. Rays yellow or none.
>>>>>>> H. Disk-aks. thickened, angled, only slightly if at all compressed.
>>>>>>>> I. Receptacle flat or convex.
>>>>>>>>> J. Ray-fls. fertile; herbs with lvs. usually basal.
>>>>>>>>>> K. Lvs. mostly entire, the cauline well developed; pappus usually present .... 1. *Wyethia*
>>>>>>>>>> KK. Lvs. pinnatifid to entire, the cauline strongly reduced or none; pappus none . 2. *Balsamorhiza*
>>>>>>>>> JJ. Ray-fls. neutral.
>>>>>>>>>> K. Pappus persistent, of 2 awns and several short scales, or none .... 3. *Viguiera*
>>>>>>>>>> KK. Pappus caducous, of 2 paleaceous awns and

rarely a few shorter scales . . . . . . **4.** *Helianthus*
 II. Receptacle conic or columnar . . . . **5.** *Rudbeckia*
HH. Disk-aks. strongly compressed or flattened.
 I. Aks. not winged; ray-fls. neutral or sterile.
  J. Aks. not ciliate; pappus squamellae present.
   K. Herbage green; stems leafy. Ours cismontane
    6. *Helianthella*
   KK. Herbage silvery-velutinous; stems scapose. Mo-
    jave Desert . . . . **7.** *Enceliopsis*
  JJ. Aks. ciliate; pappus squamellae absent.
   K. Shrubs; aks. without crown, usually epappose
    8. *Encelia*
   KK. Herbs; aks. with strong white margin, and
    thickened crown produced into 2 awns
    9. *Geraea*
 II. Aks. winged; ray-fls. usually fertile . . . . . . **10.** *Verbesina*
 GG. Rays white, very short . . . . **11.** *Eclipta*
FF. Disk-aks. ± obcompressed, the subtending chaffy bracts flat or scarcely
 concave. (Subtribe Coreopsidinae)
 G. Receptacle flat or nearly so; pappus present.
  H. Pappus of 2–6 barbed awns.
   I. Aks. not beaked, strongly flattened; rays yellow
    12. *Bidens*
   II. Aks. beaked, not much flattened; rays rose or lilac in ours
    13. *Cosmos*
  HH. Pappus of scales or teeth, never barbed . . **14.** *Coreopsis*
 GG. Receptacle conic or convex; pappus none . . . . **15.** *Guizotia*
EE. Pappus of many paleae or bristles. (Subtribe Galinsoginae)
 F. Annual herb; ray-fls. present, very short . . . . **16.** *Galinsoga*
 FF. Xerophytic shrubs; heads discoid.
  G. Pappus of 15–20 plumose bristles . . . . **17.** *Bebbia*
  GG. Pappus of 5–6 narrow paleae . . . . **18.** *Eastwoodia*
DD. Disk-fls. sterile; ray-aks. completely enclosed in their phyllaries. (Subtribe
 Melampodiinae) . . . . **19.** *Melampodium*
CC. Ligule of ray-fls. becoming papery and persistent on the fr. (Subtribe Zinniinae)
 20. *Sanvitalia*
BB. Anthers free or nearly so; heads mostly unisexual, the corollas reduced; plants mostly wind-
 pollinated. (Subtribe Ambrosiinae, p. 1100)
AA. Phyllaries mostly uniseriate, each partly or completely enclosing its ray-ak., except in *Layia*
 discoidea. (*Melampodium* and *Rigiopappus* might be sought here) (Subtribe Madiinae, p. 1106)

1. Wyèthia Nutt.

Coarse perennial herbs with erect or ascending usually unbranched stems from a branching crown surmounting a thick fusiform taproot. Lvs. cauline and often in basal rosettes, alternate, linear, lanceolate, or deltoid-ovate, entire or toothed. Heads usually large and radiate, many-fld., yellow, solitary or several, terminal and axillary. Invol. 2–4-seriate, the herbaceous or coriaceous phyllaries linear-lanceolate to ovate, subequal, or the outer series enlarged and exceeding the disk. Paleae of the convex receptacle firm, conduplicate, persistent. Ray-fls. ♀ and fertile, yellow in ours. Disk-fls. numerous, perfect, fertile. Aks. glabrous or pubescent, those of the ray dorsally compressed and trigonous, those of the disk quadrangulate or rhomboidal in x-section. Pappus a crown of unequal laciniate persistent scales, united at base and often awn-tipped, or lacking. Fourteen spp. of w. U.S. (For Capt. Nathaniel J. *Wyeth*, who discovered it in 1833.)

(Weber, Wm. A. A taxonomic and cytological study of the genus Wyethia. Amer. Midl. Nat. 35: 400–452, 1946.)
Basal lvs. absent, reduced, or similar in size to the cauline. (Section Agnorhiza)
 Invol. campanulate; phyllaries few, foliaceous; heads small; plants 3 dm. or less high.
  Heads clustered in upper lf.-axils; lvs. densely tomentose at first; pappose . . . . . . **1.** *W. ovata*
  Heads solitary, terminal; lvs. glabrous; epappose . . . . **2.** *W. Bolanderi*
 Invol. hemispheric; phyllaries numerous, linear, graduate; heads large; plants 5 dm. or more high.
  Plants scabrous and sparsely hispid; aks. 6 mm. long; pappus minute . . . . . . **3.** *W. reticulata*
  Plants uniformly pubescent.
   Heads radiate; aks. 8–12 mm. long; pappus conspicuous; plants densely short-pubescent
    4. *W. elata*
   Heads discoid; aks. 7–8 mm. long, epappose; plants pilose-hispid . . . . **5.** *W. invenusta*
Basal lvs. present, conspicuous.
 Outer phyllaries foliaceous, greatly exceeding the disk; heads very large, solitary, terminal.
 (Section Álarconia)
  Plants shining, sparsely short-pilose to glabrous; aks. 10–12 mm. long . . . . **6.** *W. glabra*
  Plants densely white-tomentose, becoming glabrate in age; aks. 12–15 mm. long
   7. *W. helenioides*

Outer phyllaries narrow, bractlike, or, if foliaceous, scarcely exceeding the disk. (Section Wyethia)
    Plant densely white-tomentose, becoming glabrate in age; phyllaries and rays few
                                                8. W. mollis
    Plant not tomentose; phyllaries and rays numerous.
      Phyllaries ciliate; heads usually solitary .... 9. W. angustifolia
      Phyllaries puberulent, not ciliate; heads 1–4 .... 10. W. longicaulis

1. **W. ovàta** T. & G. [*W. coriacea* Gray.] Stems few, stout, 1–3 dm. high, much exceeded by the lvs.; herbage silky-villous when young; lvs. broadly ovate to suborbicular, 7–20 cm. long, coarsely reticulate, entire, petiolate; heads somewhat hidden by the lvs., 1.5–2 cm. wide; invol. biseriate, the outer phyllaries 4–6, erect, foliaceous, with dilated tips usually exceeding the 5–8 short rays, the inner linear-oblong, equaling the disk; paleae narrow, silky-tipped; disk-fls. 12–20, glabrous; pappus coroniform, projected into short awns on the angles.—Grassy, openly wooded hillsides, 1200–6000 ft.; Chaparral, Yellow Pine F.; s. Sierra Nevada (Tulare Co.); mts. of coastal s. Calif. (Los Angeles Co. to San Diego Co.); adjacent L. Calif. May–Aug.

2. **W. Bolánderi** (Gray) W. Weber. [*Balsamorhiza B.* Gray.] Stems few, slender, 1.5–3 dm. high, not exceeded by the lvs., glandular above; basal lvs. bractlike; cauline lvs. ovate, 4–12 cm. long, obliquely truncate to cordate at base, green, glutinous when young, coarsely reticulate with age, entire, petiolate; heads solitary, 2–2.5 cm. wide; invol. biseriate, the outer phyllaries 4–6, erect, foliaceous, ovate to ovate-lanceolate, exceeded by the large rays, the inner shorter and narrower, all tomentose-ciliate when young; paleae linear, densely ciliate; disk-fls. usually more than 20, glabrous; aks. ca. 7 mm. long, epappose.—Sierran foothills, in grassy places, 1000–3000 ft.; V. Grassland, Chaparral; Butte Co. to Mariposa Co. March–May.

3. **W. reticulàta** Greene. Stem erect, 4–7 dm. high, leafy throughout, sparsely hirsute and densely glandular-puberulent; lvs. ovate-lanceolate to deltoid, the largest 12–15 cm. long, 5–8 cm. wide, acute at apex, truncate or cordate at base, pustulate-scabrous on upper surface, viscous beneath, entire to serrulate, petiolate; heads 1–4, 1–2.5 cm. wide; outer phyllaries unequal, ca. equaling the disk, their dilated tips squarrose, ciliate, viscid; rays showy; aks. ca. 6 mm. long; pappus coroniform, minute.—Stony clay openings, 1200–1500 ft.; Foothill Wd., Chaparral; Eldorado Co. May–July.

4. **W. elàta** Hall. [*W. ovata* Gray, not T. & G.] Stems 5–10 dm. high, leafy throughout, densely hirsute and glandular-puberulent; lvs. ovate-lanceolate to deltoid, the largest 15–20 cm. long, 8–12 cm. wide, acute at apex, truncate or subcordate at base, moderately hispidulous above, densely canescent and glandular beneath, entire or shallowly toothed, short-petiolate; heads 1–4, 2.5–4 cm. wide; outer phyllaries unequal, equaling or exceeding the disk, their tips sometimes dilated and squarrose; rays showy; aks. 8–12 mm. long; pappus coroniform, projected into short stout awns on the angles.—Dry open slopes, 3000–4600 ft.; Foothill Wd., Yellow Pine F.; Sierran foothills from Mariposa Co. to Tulare Co. June–Aug.

5. **W. invenùsta** (Greene) W. Weber. [*Helianthus (?) i.* Greene. *Balsamorhiza i.* Cov.] Stems 3–8 dm. high, leafy throughout, hirsute and becoming densely glandular-puberulent above; lvs. as in *W. elata*, entire or crenate; heads usually solitary, 2–3 cm. wide, discoid or with 1–3 short rays; outer phyllaries unequal, lanceolate, foliaceous, ca. equaling the disk, their dilated tips erect or spreading; aks. 7–8 mm. long, epappose.—Open wds., 3800–6000 ft.; Yellow Pine F.; Sierra Nevada, from Fresno Co. to Kern Co., Greenhorn Mts. July–Aug.

6. **W. glàbra** Gray. MULE-EARS. Stems 2–6 dm. high; herbage green, beset with minute ferruginous glands, glabrous or very sparingly hispidulous, at length ± scabrid; basal lvs. oblong-lanceolate to elliptic-ovate, 30–50 cm. long, 10–12 cm. wide, abruptly short-petiolate, entire or sometimes irregularly serrate-dentate; cauline lvs. similar but smaller; heads usually solitary, 3.5–6 cm. wide; outer phyllaries ovate, up to 8 cm. long and 3 cm. wide, usually exceeding the ray-fls., the inner ones reduced; paleae glabrous; pappus a crown of short broad awns.—Frequent on shady, brushy slopes, up to 2700 ft.; Mixed Evergreen F., Chaparral; Coast Ranges, Mendocino and Lake cos. to San Luis Obispo Co. March–May.

7. **W. helenioìdes** (DC.) Nutt. [*Alarconia h.* DC. *Melarhiza inuloides* Kell.] Similar to *W. glabra* except herbage tomentose, becoming glabrate in age, resin-dotted; lobes of disk-fls., paleae and pappus pubescent; aks. 12–15 mm. long, pubescent; $n = 19$

(Weber, 1946).—Open fields and sunny woodland borders, up to 6000 ft.; V. Grassland, Foothill Wd., Yellow Pine F.; Sierran foothills from Eldorado Co. to Mariposa Co., Sutter Buttes, Coast ranges from Tehama and Mendocino cos. to San Luis Obispo Co. March–July.

8. **W. móllis** Gray. Stems 4–10 dm. high; herbage densely tomentose when young, becoming glabrate, resin-dotted; basal lvs. lanceolate to oblong-ovate, 2–4 dm. long, 6–17 cm. wide, narrowed to an ample petiole, entire; cauline lvs. similar but reduced; heads 1–4, 2–3 cm. wide; outer phyllaries 4–6, lance-ovate, erect, slightly exceeding the disk; paleae pubescent; aks. 8–11 mm. long, pubescent above; pappus a crown of long rigid awns; $2n = 38$ (Weber, 1946).—Dry wooded slopes and rocky openings, 5000–10,600 ft.; Montane Coniferous F.; both slopes of Sierra Nevada from Fresno Co. n., to Siskiyou and Modoc cos.; adjacent Ore. and Nev. May–Aug.

9. **W. angustifòlia** (DC.) Nutt. [*Alarconia* (?) *a.* DC. *Helianthus longifolius* Hook. *H. Hookerianus* DC. *W. robusta* Nutt. *W. foliosa* Congd. *W. a.* var. *f.* Hall. *W. a.* var. *solanensis* Jeps.] Stems 2–6 dm. high, subscapose or leafy, appressed-pubescent to hirsute and scabrous; basal lvs. ca. equaling the stems, linear-lanceolate, tapering to base and apex, long-petiolate, the cauline reduced, often becoming ovate-lanceolate and sessile, entire or sometimes serrate or undulate; heads 1(–3), 3–4 cm. wide, on long slender peduncles; phyllaries numerous, erect, lanceolate, prominently coarse-ciliate, often anthocyanous at base; paleae pubescent; aks. 7–9 mm. long, puberulent at summit; pappus of ± united coroniform scales, usually prolonged into 1–4 stout scabrous awns at the angles; $2n = 38$ (Weber, 1946).—Open grassy slopes, sea level to 5500 ft.; many Plant Communities up to Yellow Pine F.; cismontane Calif. from Tulare and San Luis Obispo cos. n.; to Wash. April–July. Variable in pappus and lf.-size and the most common sp. The phyllaries become particularly foliose in the San Francisco Bay region.

10. **W. longicaúlis** Gray. Stems rather slender, erect or ascending, 2–5 dm. high; herbage glandular-punctate, smooth except for the scabrous lf.- and bract-margins; basal lvs. oblong-lanceolate, ca. equaling the stems, entire or serrate, tapering to a sometimes-winged petiole, the cauline sessile, much smaller; heads 1–4, 1–3 cm. wide, on slender peduncles; outer phyllaries 4–10, foliaceous, broadly lanceolate, erect, slightly exceeding the disk; paleae sparsely scabrid; aks. 7–8 mm. long; pappus inconspicuous, the lacerate scales ca. 1 mm. long.—Open woods and exposed ridges, 2500–5000 ft.; Yellow Pine F., Douglas-Fir F., Mixed Evergreen F., Coastal Prairie; N. Coast Ranges of Humboldt, Trinity and n. Mendocino cos. May–July.

2. Balsamorhìza Nutt. BALSAM ROOT

Perennial herbs with erect or ascending scapose stems from the usually multicipital caudex surmounting a thick fusiform rough-barked taproot. Lvs. almost all in basal rosettes, opposite, lanceolate to deltoid or sagittate, entire, coarsely toothed, pinnately divided to bipinnatifid. Heads radiate, many-fld., yellow, usually solitary. Invol. 2–4-seriate, the herbaceous or coriaceous phyllaries linear-lanceolate to oblong, the outer series sometimes enlarged and foliaceous. Paleae of the convex receptacle firm, conduplicate, persistent. Ray-fls. ♀, fertile. Disk-fls. perfect, fertile. Aks. trigonous (ray) or quadrangular (disk), usually glabrous and without pappus. Ca. 12 spp. of w. U.S. (Greek, *balsamos,* balsam, and *rhiza,* root.)

(Sharp, Ward M. A critical study of certain epappose genera of the Helantheae-Verbesininae of the natural family Compositae. Ann. Mo. Bot. Gard. 22: 51–153, 1935.)

Lvs. deltoid or sagittate, entire or merely crenate-toothed; root woody, deep-seated, surmounted by a multicipital caudex. (Section Artorhiza)

    Herbage rather densely hairy, the lvs. tomentose beneath at least when young; invol. persistently woolly tomentose . . . . 1. *B. sagittata*

    Herbage sparsely hairy, the lvs. green on both surfaces, their hairs fewer and coarser; invol. not truly woolly . . . . 2. *B. deltoidea*

Lvs. lanceolate to oblong-lanceolate, sharply toothed to incised or dissected; root smaller, usually unbranched, or sometimes with a few-branched crown. (Section Eubalsamorhiza)

    Lf.-blades sharply serrate-dentate and sometimes incised, the teeth prickle-tipped; phyllaries ± caudate-tipped . . . . 5. *B. serrata*

    Lf.-blades incised to pinnatifid or bipinnatifid, the teeth or lobes not prickle-tipped; phyllaries with shorter blunter tips.

        Phyllaries not exceeding the disk, entire . . . . 3. *B. Hookeri*

        Phyllaries exceeding the disk, often dentate or incised toward apex . . . . 4. *B. macrolepis*

1. **B. sagittàta** (Pursh) Nutt. [*Buphthalmum* s. Pursh. *Espeletia* s. Nutt. *E. helianthoides* Nutt. *Balsamorhiza* h. Nutt.] Stems scapose or bracteate, usually several, 2–6 dm. high; lvs. long-petiolate, the blade triangular-hastate to cordate-ovate, 15–30 cm. long, 5–15 cm. wide, entire, silvery velutinous especially on under surface, greener above and becoming less hairy in age; heads solitary, 6–10 cm. across the rays, like the top of the peduncle usually lanate; phyllaries broadly lanceolate, acuminate, the outer series longest; paleae with prominent strongly hirsute glandular subulate tips extending above the disk; ligules up to 4 cm. long; aks. glabrous; $n = 19$ (Weber, 1946). —Deep sandy soils, plains and forest openings, 4300–8300 ft.; Sagebrush Scrub, N. Juniper Wd., Yellow Pine F.; both slopes of the Sierra Nevada from Kern and Inyo cos. n., to Tehama and Modoc cos.; e. of the Cascades to Canada and the Rocky Mts. May–July.

2. **B. deltoìdea** Nutt. [*B. glabrescens* Benth.] Habit of *B. sagittata;* stems 2–8 dm. high, scapiform, but usually with several greatly reduced lvs.; blades of basal lvs. triangular-hastate to cordate-ovate, 10–30 cm. long, 5–20 cm. wide, green, moderately hispidulous and glandular, becoming sparingly scabrous and reticulate-coriaceous in age, usually entire or crenate, sometimes irregularly dentate; heads solitary or few, the summit of peduncle and base of invol. often densely hirsute and glandular, but hardly woolly; aks. glabrous.—Deep sandy soil, mostly in the mts., 600–7000 ft.; many Plant Communities; San Rafael Mts. e. to n. Los Angeles Co., n. on w. slope of Sierra Nevada to Siskiyou Co.; inner N. Coast Ranges from Lake Co. to Del Norte Co.; n. mostly w. of the Cascades to Vancouver Id. April–June.

3. **B. Hoòkeri** Nutt. [*Heliopsis Balsamorhiza* Hook. *B. b.* Heller. *B. platylepis* Sharp. *B. hirsuta* var. *neglecta* Sharp. *B. Hookeri* var. *lanata* Sharp. *B. H.* var. *n.* Cronq.] Woody taproot covered with black rugose bark, its upper portion often densely clothed in old lf.-bases; stems 1–3 dm. tall, pilose and glandular; lvs. 1–3 dm. long, lanceolate, pinnatifid, the divisions often cleft into spreading lobes, subsericeous to velutinous or lanate; heads solitary, 4–7 cm. across the rays; phyllaries linear to lance-ovate, lanate to sericeous, long-ciliate; ray-fls. pubescent and glandular dorsally; aks. glabrous.— Grassy, often rocky and subalkaline flats and scablands, 2500–5000 ft.; Sagebrush Scrub, N. Juniper Wd., Yellow Pine F.; occasional in Calif., n. Mono Co. to Modoc Co., w. through the Scott and Siskiyou mts.; to Wash., Utah. April–June. Highly variable, possibly with well-marked sspp. in the Great Basin, but some Calif. material seems indistinguishable from the typical form from the Columbia R.

4. **B. macrolèpis** Sharp. Habit of *B. Hookeri* but larger, the usually scapose stems 2–6 dm. tall, pilose and glandular-puberulent; lvs. 2–5 dm. long, pinnatifid to bipinnatifid, the lanceolate divisions 3–8 cm. long, 5–20 mm. wide, entire to pinnatifid with acute or rounded lobes, sericeous to tomentulose and glandular; heads solitary, 4–8 cm. across the rays; phyllaries of outer series foliaceous, oblong, twice as long as the disk, 2–3.5 cm. long, subvelutinous; aks. glabrous.—Fields and rocky hillsides, up to 2000 ft.; V. Grassland, Foothill Wd.; interior slopes near San Francisco Bay; Sierran foothills from Mariposa Co. to Butte Co. March–June.

5. **B. serràta** Nels. & Macbr. Habit of *B. Hookeri;* lvs. green, scabrous and strongly reticulate-veiny, lanceolate to oblong-ovate, sharply serrate but also sometimes pinnatifid, 4–10 cm. long, 2–8 cm. wide.—Locally common in open scrub country, 4800 ft.; Sagebrush Scrub, n. Modoc Co.; adjacent Nev., to Wash. May–June.

The spp. of *Balsamorhiza* hybridize rather freely where their ranges overlap. One that has been collected frequently, because of its intermediate appearance between the entire deltoid-lvd. parent, *B. deltoidea,* and the pinnatifid-lvd. parent, *B. Hookeri,* has received the name *B. terebinthàcea* (Hook.) Nutt. [*Heliopsis* (?) *terebinthacea* Hook.] Plants of the parentage *B. sagittata* × *B. Hookeri* also bear this name, which should be written *B.* × *terebinthacea.*

3. Viguiéra HBK.

Herbs or shrubs with leafy usually branching stems. Lvs. opposite (at least below), linear to ovate. Heads medium-sized, on slender peduncles, solitary or cymose, yellow. Invol. 2–3-seriate, the phyllaries linear-lanceolate. Paleae of the convex receptacle clasping the aks., persistent. Ray-fls. neutral, the ligules dorsally pubescent. Disk-fls.

numerous, fertile. Aks. laterally compressed, 4-angled. Pappus of 2 awns and several intervening scales, or none. Ca. 150 spp., distributed from s. U.S. to temp. S. Am. (Honoring L. G. A. *Viguier*, physician and botanist of Montpellier.)

(Blake, S. F. A revision of the genus Viguiera. Contr. Gray Hb. n.s. 54: 1–205, 1918.)
Shrubs or subshrubs; phyllaries indurated at base; pappose; aks. pubescent.
    Lvs. ovate; invol. canescent.
        Lvs. tuberculate-hispidulous on both surfaces, the blade 1–2.5 cm. long ...... 1. *V. deltoidea*
        Lvs. silvery appressed-canescent on upper surface, crisped-pubescent and prominently reticulate
            on lower surface, the blade 3–8 cm. long .... 2. *V. reticulata*
    Lvs. lanceolate; invol. green .... 3. *V. laciniata*
    Herbs; phyllaries herbaceous; epappose; aks. glabrous .... 4. *V. multiflora*

1. **V. deltoìdea** Gray var. **Paríshii** (Greene) Vasey & Rose. [*V. P.* Greene.] Rounded subshrub 3–10 dm. high, multibranched, the stems densely and harshly tuberculate-strigillose; lvs. deltoid-ovate, strongly toothed and reticulate, the fine spreading hairs from prominent tuberculate bases; heads on elongated peduncles, solitary at branch tips or in small cymes; invol. 5–9 mm. high, densely canescent, the phyllaries broadly lanceolate, narrowed to an acuminate tip; paleae much thinner, medianly hairy, hyaline margined, several-nerved, abruptly acute; ray-fls. ca. 8, the narrow ligules 12–15 mm. long; aks. appressed-pilose; pappus awns 2–3 mm. long, the fimbriate squamellae less than 1 mm. long.—Sandy desert canyons and mesas, sea level to 4800 ft.; Creosote Bush Scrub; Colo. Desert and e. Mojave Desert (Providence Mts., etc.), w. to coastal San Diego Co.; s. Nev., Ariz., L. Calif. (where the genuine var. occurs), Son. Feb.–June; Sept.–Oct.

2. **V. reticulàta** S. Wats. Suffrutescent, the several pallid pilose stems 5–15 dm. high; lvs. broadly ovate, often subcordate, entire or undulate, coriaceous, strongly reticulate dorsally, short-petiolate, densely white-pilose with appressed hairs ventrally and spreading hairs dorsally; heads well overtopping the lvs. in a nearly naked cymose panicle; invol. 4–6 mm. high, densely canescent, the phyllaries lance-oblong, thickened; ray-fls. 10–15, the ligules 8–12 mm. long; pappus awns 1 or 2, 1.5–2.7 mm. long, stoutish, the squamellae coherent at base, less than 1 mm. long.—Gravelly washes and rocky canyons, 10–5000 ft.; Creosote Bush Scrub; desert ranges of Inyo Co. Feb.–June.

3. **V. laciniàta** Gray. Multibranched round-topped shrub 6–12 dm. high, the herbage ± resinous and harshly hispidulous; lvs. numerous, lanceolate from a broad often subhastate base, 2–5 cm. long, pinnatifid to subentire, acute, coriaceous, strongly veiny beneath, short-petiolate, with axillary fascicles; heads in rather compact corymbs; invol. 5–8 mm. high, the outer phyllaries ovate with abruptly acuminate tips; ray-fls. 8–13, the ligules 10–15 mm. long; pappus paleae lanceolate, awn-tipped, deciduous, 2.5 mm. long, the squamellae coherent at base, 0.5 mm. long.—Dry slopes, up to 2500 ft.; Coastal Sage Scrub, Chaparral; sw. San Diego Co.; L. Calif. Feb.–June.

4. **V. multiflòra** (Nutt.) Blake var. **nevadénsis** (A. Nels.) Blake. [*Gymnolomia n.* A. Nels.] Perennial herb, sometimes suffrutescent at crown, the several slender erect stems 3–9 dm. high; herbage strigose to scabrous-puberulent and gland-dotted; lvs. linear to linear-lanceolate, entire, revolute, 2–5 cm. long, 2–5 mm. wide; heads few, loosely paniculate; invol. 6–7 mm. high, the narrowly lanceolate phyllaries exceeding the paleae; ray-fls. 8–13, the ligules 10–15 mm. long; aks. glabrous; pappus none.—Canyons, 4000–7500 ft.; Sagebrush Scrub, Pinyon-Juniper Wd.; desert ranges of Inyo and ne. San Bernardino cos.; to Utah, Ariz. May–Sept. Genuine *V. multiflora* occurs n. and e. through the Great Basin to the Rocky Mts.

V. CILIATA (Rob. & Greenm.) Blake, a slender annual sp. from the Arizona region, has been reported in the manuals as an introd. at Santa Monica and Los Angeles in 1890. Apparently it is no longer a member of the Calif. flora.

4. Heliánthus L. SUNFLOWER

Coarse annual or perennial herbs with simple lvs., at least the lowermost ones opposite. Heads radiate, large, solitary at branch-tips or in corymbs. Phyllaries imbricate or subequal, ± herbaceous. Receptacle plano-convex, the persistent paleae clasping the aks. Ray-fls. conspicuous (in ours), yellow, neutral. Disk-fls. yellow or reddish, fertile. Aks. narrowly obovate, quadrangular but obcompressed, generally glabrous. Pappus of 2 principal awns paleaceous at base, rarely with additional short squamellae, all readily caducous. Some 60 spp. of temp. N. and S. Am. (Greek, *helios*, sun, and *anthos*, fl.)

Plants annual; disk-corollas reddish.
    Outer phyllaries ovate, abruptly caudate, strongly ciliate .... 1. *H. annuus*
    Outer phyllaries lanceolate, gradually attenuate, not obviously ciliate.
        Cent. paleae of the disk densely white-hirsute at tip .... 2. *H. petiolaris*
        Cent. paleae of the disk not hirsute-tipped .... 3. *H. Bolanderi*
Plants perennial; disk-corollas yellow or brownish.
    Plants strongly glaucous and rhizomatous .... 4. *H. ciliaris*
    Plants not strongly glaucous and rhizomatous.
        Foliage white with strigose pubescence; subshrub .... 5. *H. tephrodes*
        Foliage greenish; perennial herbs.
            Plants with thickened taproot. Ne. Calif. .... 6. *H. Cusickii*
            Plants with thickened, often fleshy roots or short rhizomes, not taprooted.
                Outer phyllaries longer than the disk. Wet places.
                    Phyllaries 2–3 mm. wide at base, slightly longer than the disk, erect ... 7. *H. Nuttallii*
                    Phyllaries 3–4 mm. wide at base, considerably longer than the disk, reflexed at tip
                                          8. *H. californicus*
                Outer phyllaries shorter than the disk. Dry ground .... 9. *H. gracilentus*

1. **H. ánnuus** L. ssp. **lenticuláris** (Dougl.) Ckll. [*H. l.* Dougl. ex Lindl.] COMMON SUNFLOWER. Stem usually stout, 3–20 or more dm. high, often openly branching, very hispid; herbage rough-hairy; lvs. petiolate, the blade 6–15 cm. long, narrowly to broadly ovate, truncate to cordate at base, mostly serrate, the uppermost often entire; heads large, the low-convex usually reddish disk 2–3.5 cm. across; phyllaries narrowly to broadly ovate, abruptly and slenderly acuminate, densely scabrous and usually at least medianly hispid-hirsute dorsally, strongly hispid-ciliate; paleae often 3-toothed, not conspicuously hairy; $2n = 34$ (Tahara, 1915).—Roadsides and waste places, frequent in lowlands, occasional up to 5000 ft.; a ruderal growing almost throughout the state, excepting the N. Coast ranges and the Mojave Desert; s. Canada to n. Mex. Feb.–Oct. The typical form of the sp., which tends to be larger in all its parts, occurs in cent. U.S. The common cult. sunflower, with disk 6–25 cm. across, and widely grown for its oily seeds, is *H. a.* var. **macrocárpus** (DC.) Ckll. [*H. m.* DC.]

Ssp. **Jaègeri** (Heiser) Heiser. [*H. J.* Heiser.] Lvs. lance-ovate, cuneate to truncate at base, serrulate; heads with reddish disk 1.5–2.5 cm. across; phyllaries lance-ovate; $n = 17$ (Heiser, 1948).—Wet alkaline areas; Alkali Sink; Owens V. (Rock Creek, Lone Pine), e. Mojave Desert (Tecopa Hot Springs, Soda Dry Lake, etc.); w. Nev. July–Oct.

H. Maximiliánii Schrad., a scabrous-hispidulous perennial from a short rhizome, with stems to 2.5 m. high, lanceolate often infolded lvs. to 30 cm. long, linear-lanceolate canescent phyllaries exceeding the yellow disk, which is 1.5–2.5 cm. across, is reported as a garden escape in Fresno Co. It is native in the Great Plains; $n = 17$ (Geisler, 1931).

2. **H. petioláris** Nutt. Annual; stems 3–20 dm. high, glabrate or sparingly hispid; lf.-blades 3–8 cm. long, lanceolate, entire or nearly so, the strigulose hairs from pustulate bases not harsh as in *H. annuus;* heads with reddish disk ca. 2 cm. across; phyllaries lanceolate, gradually acuminate, hispidulous, not prominently hispid-ciliate; cent. paleae white-bearded at tip; $n = 17$ (Heiser, 1947).—Ruderal of waste places, rare in s. Calif. (San Bernardino V. and Mts., Hemet V.); common to Tex., Mo., Sask. May–Sept.

Var. **canéscens** Gray. [*Gymnolomia encelioides* Gray. *H. petiolaris* var. *canus* Britton. *H. canus* Woot. & Standl.] Densely canescent-strigose throughout; $n = 17$ (Heiser, 1948).—Open desert; Creosote Bush Scrub; Colo. Desert, Borrego V.; to Tex., Mex. March–June.

3. **H. Bolánderi** Gray [*H. scaberrimus* Benth., not Ell. *H. exilis* Gray.] Annual; stems usually slender, 3–12(–25) dm. high, often freely branching, herbage ± glandular, strigulose and hirsute; lvs. linear- to ovate-lanceolate, cuneate (rarely truncate) at base, entire to obscurely serrate; heads with red-purple or yellowish convex disk 1.5–2.5 cm. across; phyllaries lanceolate or oblong, gradually attenuate, scabrous and hirsute, but not obviously ciliate; paleae (scabrous above) tipped with a glabrous acerose awn; $n = 17$ (Heiser, 1948).—On serpentine outcrops in the foothills up to 5000 ft., and a ruderal on v. floors; V. Grassland up to Yellow Pine F.; sw. Ore. (Rogue R. V.), through cismontane Calif. to Tulare and Santa Barbara cos., also Plumas and Sierra cos. (Sagebrush Scrub). July–Nov. The serpentine form received the name *H. éxilis.* It is not as

large as the otherwise indistinguishable v. ruderal (typical *Bolanderi*), which Heiser believes may have arisen through the introgression of genes of *H. annuus* into the *exilis* race.

4. **H. ciliàris** DC. BLUEWEED. Perennial herb spreading by strong rhizomes; stems 2–5 dm. high, longitudinally striate; herbage blue-glaucous, glabrate, or especially the lf.-margins sparingly strigose; lvs. linear-lanceolate to broadly lanceolate, 3–8 cm. long, 5–15 mm. wide, entire to crisped-undulate; heads on often long naked peduncles, the disk 12–16 mm. across; phyllaries ovate to oblong, obtusish, white-ciliolate, otherwise glabrous, shorter than the disk; paleae canescent at tip; *n* = 51 (Heiser, 1955).—A pernicious weed introd. within the last two decades into the s. half of the state, from San Luis Obispo Co. to San Diego Co., and in the Cent. V. Its complete eradication from our flora is sought. Native from Ariz. to Tex. June–Nov.

5. **H. tephròdes** Gray. [*Viguiera nivea* Gray. *V. t.* Gray.] Woody and decumbent at base, 3–20 dm. high, silvery white with dense strigillose pubescence; lvs. deltoid-ovate, acute, entire to crenulate, the blade 2–7 cm. long; heads several, with reddish disk 1.5–2.5 cm. across; phyllaries lanceolate, not exceeding the disk; disk-corollas very slender, puberulent at bulbous base of tube and on lobes; pappus of 2 long awns and several squamellae.—Sandy desert, rare; Creosote Bush Scrub; Yuma sand-hills, Imperial Co.; s. to Son. March–May, Oct.–Jan.

6. **H. Cusíckii** Gray. Herbaceous perennial; stems several, usually erect, from the summit of a stout fleshy taproot, 2–10 dm. high, often branched; lvs. narrowly lanceolate, tapering to base and apex, subsessile, entire, 6–15 cm. long, 1–2 cm. wide, hirsutulous and usually gland-dotted; heads usually solitary, the yellow disk 12–25 mm. across; phyllaries lance-linear, rather loose, often caudate-tipped, hirsute (especially marginally) and hispidulous; rays showy, 2–3 cm. long; *n* = 17 (Heiser, 1955).—Dry rocky or sandy slopes, 4000–5500 ft.; Sagebrush Scrub, N. Juniper Wd.; Lassen Co. to Modoc Co.; to Wash., Nev. May–Aug.

7. **H. Nuttállii** T. & G. Herbaceous perennial with short rhizomes and tuberous roots; stems to 4 m. high, glabrous, glaucescent; lvs. scabrous and glandular-dotted, linear-lanceolate, tapering to base and apex, up to 15 cm. long and 30 mm. wide, entire to serrate; heads loosely corymbose, the yellow disk 15–23 mm. across; phyllaries linear-lanceolate, 2–3 mm. wide at base, scabro-puberulent, the narrow tip erect, slightly exceeding the disk; rays 1.5–3 cm. long; n = 17 (Heiser, 1955).—Moist meadows, 4000–8000 ft.; Pinyon-Juniper Wd.; Modoc Co., e. Inyo and ne. San Bernardino cos.; to N. Mex., Black Hills, Canada. July–Sept.

Ssp. **Paríshii** (Gray) Heiser. [*H. P.* Gray. *H. Oliveri* Gray. *H. P.* f. *O.* Ckll. *H. californicus* var. *P.* Jeps. *H. c.* var. *O.* Blake.] Lvs. canescent, the soft ± erect hairs scarcely at all pustulate; summit of peduncles and invols. densely white-villous; *n* = 17? (Heiser).—Wet ground; 1000–1500 ft.; Los Angeles, San Bernardino and Orange cos. Aug.–Oct.

8. **H. califórnicus** DC. [*H. c.* var. *mariposianus* Gray.] Herbaceous perennial with stout smooth erect stems up to 5 m. high from somewhat tuberous woody roots, branching above; lvs. lanceolate, tapering to apex and usually to base, short-petiolate, usually entire, up to 20 cm. long and 50 mm. wide, hispidulous and gland-dotted; heads in loose corymbs, the yellow disk 18–25 mm. across; phyllaries lanceolate, the white-ciliolate basal portion narrowed to a densely hispidulous spreading attenuate tip exceeding the disk; rays 2–3 cm. long; *n* = 51 (Heiser, 1955).—Boggy meadows, stream banks and moist ground, 10–1500 ft.; V. Grassland, Foothill Wd., Freshwater Marsh, etc.; Sierran foothills, Calaveras Co. to Mariposa Co., Coast Ranges, Napa Co. to Santa Clara Co., s. Calif. (up to 6000 ft.), Los Angeles Co. to San Diego Co.; L. Calif. June–Oct.

9. **H. graciléntus** Gray. Herbaceous perennial with ascending or erect tufted stems from a woody caudex, up to 15 dm. high; lvs. lanceolate to lance-ovate, mostly acuminate and entire, sometimes serrate, 3–12 cm. long, 1–3 cm. wide, short-petiolate, or the upper subsessile, sometimes with fascicles in the axils, coarsely hispid on both faces; heads solitary or corymbose on long nearly naked peduncles, the yellow disk 1–2.5 cm. across; phyllaries lance-oblong to ovate, acuminate, apiculate, densely hispidulous-puberulent, not exceeding the disk; rays 1.5–2 cm. long; *n* = 17 (Heiser, 1955).—Dry hillsides, 200–6000 ft.; Chaparral, Yellow Pine F., etc.; mostly outer S. Coast Ranges, from Contra Costa Co. to San Diego Co.; n. L. Calif. May–Oct.

5. Rudbéckia L.

Tall erect annual or usually perennial herbs. Lvs. alternate, simple or compound, entire to pinnatifid. Heads large, radiate or discoid, on terminal peduncles. Invol. 2–3-seriate, herbaceous or subfoliaceous; phyllaries equal or irregular, mostly spreading or reflexed. Receptacle conic or columnar, its bracts clasping the aks. Ray-fls. neutral, yellow, orange, or with red-purple base. Disk-fls. fertile. Aks. quadrangular, or if laterally compressed, with a salient angle on each face. Pappus a short irregular crown, or none. Ca. 25 spp. of temp. N. Am. (For the Olaf *Rudbeck*'s, father and son, professors of botany and predecessors of Linnaeus at Uppsala.)

Heads radiate.
    Pappus none; disk globose or ovoid .... 1. *R. hirta*
    Pappus present; disk ± columnar .... 2. *R. californica*
Heads discoid .... 3. *R. occidentalis*

1. **R. hírta** L. var. **pulchérrima** Farw. [*R. serotina* Nutt.] BLACK-EYED SUSAN. Biennial or short-lived perennial, rarely flowering the first year; stems simple or sparingly branched, 3–8 dm. high; herbage rough-hirsute throughout; lvs. oblanceolate to oblong or linear-lanceolate, the lower petiolate, entire; heads long-pedunculate, the hemispheric or ovoid disk 1–2 cm. wide, generally black-purple or brown; rays 8–21, golden, 1.5–3 cm. long; paleae of the receptacle linear, acutish, hispid and ciliate at tip; pappus none; $n = 19$ (Battaglia, 1947).—Introd. in mid-altitude meadows of the Sierra Nevada from Amador Co. to Mariposa Co. and occasional in moist land in the Cent. V.; native in the Miss. V. June–Aug. Typical *R. hirta* is an Appalachian plant.

2. **R. califórnica** Gray. CALIFORNICA CONE-FLOWER. Leafy and unbranched perennial 6–18 dm. high, with a single showy head on a long smooth peduncle, lvs. broadly lanceolate to elliptic, firm, glabrous on the upper, pubescent on the lower surface, the lower with blades 10–25 cm. long, on long slender petioles, entire or irregularly dentate or incised, the upper becoming sessile and entire; rays 8–21, ligulate, yellow, 2.5–6 cm. long; disk ovoid-oblong to columnar, 3–5 cm. high; paleae of the receptacle with densely pubescent acute tips; aks. flattish; pappus an irregularly 4-toothed cup.— Occasional in moist meadow slopes, 5500–7800 ft.; Red Fir F.; Sierra Nevada from Kern Co. to Eldorado Co. July–Aug.

Var. **glaúca** Blake. [*R. glaucescens* Eastw.] Lvs. glaucous, ample, mostly lanceolate and tapering to base and apex, entire or subdentate, glabrous on both surfaces, the margin hispidulous.—Stream banks and springy places, often on serpentine outcrops or in *Darlingtonia* swamps, 400–4000 ft.; N. Coastal Coniferous F., Douglas-Fir F.; Del Norte Co.; to Douglas Co., Ore. July–Sept.

Var. **intermèdia** Perdue. Similar to var. *glauca*, but lvs. somewhat thinner and more coarsely and crenately dentate; heads somewhat smaller, the disk only 1.5–3 cm. high.— At 4000–5000 ft.; Yellow Pine F.; Mt. Eddy and other mts. of Siskiyou Co. Aug.–Sept.

3. **R. occidentàlis** Nutt. Simple-stemmed perennial 6–15 dm. high, with the columnar head surmounting a stout striate peduncle; lvs. broadly ovate or elliptic, sometimes acuminate, mostly rounded or subcordate at base, subglabrous to scabrous-margined, or particularly the lower surface hirsutulous, entire or irregularly toothed; heads discoid, the columnar black disk 2–6 cm. long; pappus a short crown of teeth or quadrate scales.—Wet ground in woods, 4000–6000 ft.; Yellow Pine F., Red Fir F.; Sierra Nevada from Placer Co. to Plumas and Butte cos.; to Wash., Mont., Utah. June–Aug.

6. Helianthélla T. & G.

Perennial herbs with several leafy stems from a woody caudex surmounting a tap-root, the lvs. simple, entire, broadly linear to ovate-lanceolate, the lower entire. Heads radiate, large, usually solitary. Phyllaries linear-lanceolate and subequal (in ours), ± herbaceous. Receptacle plano-convex, the persistent paleae clasping the aks. Ray-fls. conspicuous, yellow, neutral. Disk-fls. yellow or purple. Aks. laterally compressed, usually thin-edged, cuneate-obovate. Pappus of 2 slender persistent awns plus short lacerate squamellae, or none. 8 spp. of w. N. Am. (Diminutive of *Helianthus*.)

(Weber, Wm. A. The genus Helianthella (Compositae). Amer. Midl. Nat. 48: 1–35, 1952.)
Invol. 1.5–2(–2.5) cm. wide; outer phyllaries rarely enlarged, not incurved; aks. strongly compressed
.... 1. *H. californica*
Invol. 2.5–4 cm. wide; outer phyllaries usually enlarged, leaflike, curved over the disk at maturity;
aks. plump .... 2. *H. castanea*

1. **H. califórnica** Gray. Caudex becoming 2.5 cm. thick, the stems slender, 2–6 dm. high; herbage moderately hirsutulous when young, becoming glabrate; lvs. linear-lanceolate, up to 25 cm. long and 25 mm. wide, tapering to a slender petiole, often mostly basal; heads solitary on long naked peduncles, the yellow disk 1.5–2.2 cm. across; phyllaries narrowly lanceolate, pubescent to ciliate or glabrate, occasionally with a few tips ± foliaceous; aks. cuneate-obovate, narrowly margined, 6–10 mm. long, deeply emarginate apically; pappus none; $n = 15$ (Weber, 1952).—Grassy slopes, 200–3000 ft.; V. Grassland, Foothill Wd.; Coast Ranges from Mendocino and Glenn cos. to Santa Clara Co. April–June.

Var. **nevadénsis** (Greene) Jeps. [*H. n.* Greene.] Similar except having a pappus of 2 short marginal awns 1–2 mm. long and usually with several low squamellae between.— Dry openings on wooded slopes, 800–7000 ft.; Foothill Wd., Chaparral, Yellow Pine F., Red Fir F.; Cascade-Sierra Nevada axis, Kern Co. to Shasta and Modoc cos., inner N. Coast Range, Lake Co. to Trinity Co.; s. Ore., w. Nev. May–Sept.

Var. **shasténsis** W. Weber. Stems 1–3 dm. high; lvs. principally basal and only 5–8 mm. wide; heads solitary, the disk mostly 1–1.5 cm. across; pappus of 2 awns 1 mm. long.—At 3000–5500 ft.; Yellow Pine F., Red Fir F.; Siskiyou, Shasta, and Trinity cos. May–July.

2. **H. castánea** Greene. [*H. Cannonae* Eastw. *H. c.* var. *C.* Jeps.] Caudex slender to stout, the several stems 2–4.5 dm. high, almost equalled by the basal lvs.; cauline lvs. oblong to oblanceolate, up to 15 cm. long and 50 mm. wide, petiolate; heads large, solitary, the yellow disk 2.5–3.5 cm. across; phyllaries linear-acuminate, conspicuously hispid-ciliate, a few of the outermost becoming enlarged and leaflike; aks. broadly obovate, not margined, the outer ones plump but thin-edged, ca. 8 mm. long; pappus of 2 short awns only, or none; $n = 15$ (Weber, 1952).—Grassy hillsides, 500–4000 ft.; V. Grassland, Foothill Wd.; San Francisco Bay region. April–May.

7. Enceliópsis (Gray) A. Nels.

Scapose xerophytic perennial herbs, with stout woody taproot and often much branched caudex, the herbage silvery-velutinous or canescent. Lvs. in basal tufts, thick, oval or rhombic, entire. Heads large, many-fld., solitary on scapiform peduncles, radiate in ours, yellow. Invol. 2–3-seriate. Bracts of the ± convex receptacle soft and scarious, embracing the aks. with which they fall. Ray-fls. pistillate, sterile. Disk-aks. strongly compressed, blackish, with a narrow white cartilaginous border and crown. Pappus of 2 short subulate awns, with or without minute ± confluent squamellae between, or none. 3 spp. of arid regions of sw. U.S. (*Encelia*, and Greek, *opsis*, likeness, habitally similar to *Encelia*.)

(Blake, S. F. A revision of Encelia and some related genera. Proc. Am. Acad. 49: 351–355, 1913.)
Lvs. obtuse, obovate or suborbicular, the blade usually shorter than the slender petiole; pubescence
dull .... 1. *E. nudicaulis*
Lvs. acute, rhombic-ovate, the blade usually longer than the winged petiole; pubescence silvery
.... 2. *E. argophylla*

1. **E. nudicaúlis** (Gray) A. Nels. [*Encelia n.* Gray. *Helianthella n.* Gray.] Cespitose, 1–4 dm. high, the herbage densely tomentose-canescent, often with flocs of wool in the lf.-axils; lvs. ovate to orbicular, the blade 2–6 cm. long, abruptly narrowed to a petiole 1–3 times its length; scapes ascending-erect; disks 2–3.5 cm. across; invol. broadly hemispheric, the phyllaries lance-subulate from a broad base, 1–2 cm. long; rays ca. 21, 2–2.5 cm. long; aks. cuneate, silky-villous; pappus awns usually short, connected by a fimbriate crown of very short fused squamellae, or none.—In sand, rocky clays or compacted arid soils, up to 6000 ft.; Sagebrush Scrub, Shadscale Scrub, Creosote Bush Scrub; Death V. region, Inyo Co.; to Ariz., Utah, Ida. May.

2. **E. argophýlla** (D. C. Eat.) A. Nels. var. **grandiflòra** (Jones) Jeps. [*Encelia g.* Jones, not Hemsl. *Enceliopsis g.* A. Nels. *Helianthella Covillei* A. Nels. *Enceliopsis C.* Blake.]

Habit of the preceding but tending to be stouter throughout; lvs. broadly rhombic-oval or orbicular, the blade 4–10 cm. long, gradually tapering to a winged petiole not as long, silvery-velutinous; scapes 3–5 dm. high, stout; disks 3.5–5 cm. across; rays 20–34, 3.5–4 cm. long; aks. broadly obovate, sparingly puberulent or glabrate; pappus as in the last.—Rocky or clayey, subalkaline canyon sides and sandy washes, 1200–4000 ft.; Creosote Bush Scrub; w. side of Panamint Mts., Inyo Co. April–June. The typical var., which occurs in s. Nev., Utah, and Ariz., differs only in having rays ca. 2 cm. long and silky-villous aks. It may be found in Calif.

8. Encèlia Adans.

Low branching shrubs. Lvs. alternate, entire or remotely toothed. Heads medium-sized, solitary or panicled, peduncled, radiate or discoid, yellow, or the disk purple. Invol. 2–3-seriate. Bracts of the convex receptacle soft and scarious, embracing the aks. and falling with them. Ray-fls. neutral. Disk-aks. flat, obovate, villous-ciliate, ± pubescent on the faces; pappus 0, or with 2 slender awns. Ca. 14 spp. from sw. U.S. to Mex., Peru, Chile, and Galapagos Ids. (Christopher *Encel*, published on oak-galls in 1577.)

(Blake, S. F. A revision of Encelia and some related genera. Proc. Am. Acad. 49: 358–376, 1913.)
Heads in cymose panicles; peduncles glabrous; lvs. white-tomentulose . . . . 1. *E. farinosa*
Heads solitary at the tips of pubescent peduncles; lvs. not tomentulose.
    Disks yellow. Interior and deserts.
        Rays none; lvs. scabrous with scattered pustulate-based hairs. E. deserts . . . . 2. *E. frutescens*
        Rays present, 5–15 mm. long; lvs. finely appressed-pubescent, sometimes with slender scabrous
        hairs between. Desert borders . . . . 3. *E. virginensis*
    Disks purple; rays 18–30 mm. long. Coastal . . . . 4. *E. californica*

1. **E. farinòsa** Gray. BRITTLE-BUSH. INCIENSO. Roundish bush 3–8(–16) dm. high, fragrant, the brittle stems arising from a woody trunk and bearing dense clusters of lvs. of the season apically; lvs. narrowly to broadly ovate, obtuse to acute, entire or ± repand-toothed, silvery-tomentose, 3–8 cm. long, on shorter petioles; peduncles much exceeding the lvs., cymosely branched with several heads or simple and 1-headed; heads often nodding in fr.; disk 1–1.5 cm. across; invol. 4–7 mm. high, ± villous and gland-dotted; rays showy, 8–12 mm. long, orange-yellow; disks yellow; aks. narrowly obovate, emarginate, epappose.—Dry stony slopes, up to 3000 ft.; Creosote Bush Scrub, Coastal Sage Scrub; Death V. region, through e. Mojave Desert to Colo. Desert (where abundant), Kern R. Canyon, San Gabriel Mts. and San Bernardino V. to San Diego Co.; to sw. Utah, Ariz., L. Calif., Sinaloa. March–May.

Var. **phenicodónta** (Blake) Jtn. [*E. f. f. p.* Blake.] Disk-corollas purple.—E. and s. Colo. Desert; adjacent Ariz., L. Calif.

2. **E. frutéscens** (Gray) Gray. [*Simsia f.* Gray.] Rounded much-branched shrub 8–15 dm. high; stems whitish; herbage very scabrous with pustulate-based hairs; lvs. green, scattered, lance-oblong to ovate, obtuse or acute, mostly truncate at base, entire or with an obscure pair of teeth toward base, mostly 1–2 cm. long and less than 1 cm. wide; heads on long naked peduncles; disk 1–2.3 cm. across; invol. 6–10 mm. high, whitish hispid-scabrous and sometimes glandular; rays 0; disk-fls. yellow.—Dry slopes and mesas; Creosote Bush Scrub; e. and s. Mojave Desert, Colo. Desert to e. San Diego Co.; Ariz. Feb.–May.

3. **E. virginénsis** A. Nels. [*E. frutescens* f. *v.* Hall. *E. f.* var. *v.* Blake.] Similar to *E. frutescens;* stems scabrous; lvs. green, finely canescent and with sessile or stalked glands, with stouter tuberculate-based hairs interspersed, usually 1–2.5 cm. long, 6–16 mm. wide; rays 5–10 mm. long.—Washes, gravelly mesas and canyonsides, 1000–5000 ft.; Creosote Bush Scrub; e. Mojave Desert (Providence Mts., Needles, etc.); to s. Nev., sw. Utah, nw. Ariz. Dec.; April–May.

Ssp. **áctoni** (Elmer) Keck. [*E. a.* Elmer. *E. frutescens* f. *a.* Hall. *E. f.* var. *a.* Blake.] Lvs. often larger, commonly broadly ovate, up to 4 cm. long and 3 cm. wide, ± densely canescent or velutinous, often yellowish-green; rays 10–16 mm. long.—Desert slopes, up to 5000 ft.; w. Colo. and Mojave deserts, from San Diego, Riverside (also cismontane) and Los Angeles cos. to Kern R. Canyon, Owens V., White Mts., Death V. region and adjacent Nev. Feb.–July.

4. **E. califòrnica** Nutt. Broad rounded clumps 6–15 dm. high (on San Clemente Id.

to 30 dm.), shrubby at base; stems and peduncles canescent; lvs. green, lanceolate to ovate, acute, appressed-pubescent, 3–6 cm. long, 1–3 cm. wide; heads showy; disk 1.5–2.5 cm. across; invol. 10–12 mm. high, densely villous; rays oblong, 15–30 mm. long; disks purple.—Coastal bluffs and open or brushy slopes, up to 1,600 ft.; Coastal Sage Scrub, Chaparral; Casmalia, Santa Barbara Co. to San Diego, inland to San Bernardino V. and Riverside; Channel Ids.; L. Calif. Feb.–June.

9. Geraèa T. & G.

Herbs with alternate dentate lvs. Heads rather few, paniculate, showy, yellow-fld. Invol. hemispheric, 2–3-seriate. Bracts of the low-convex receptacle softly scarious, clasping the aks. and falling with them. Ray-fls. neutral. Disk-aks. flat, narrowly cuneate, villous-ciliate, the sides villous medially, black with whitish narrow margin and thickened crown produced into 2 slender persistent awns. 2 spp. (Greek, *geraios,* old, the aks. white-villous.)

(Blake, S. F. A revision of Encelia and some related genera. Proc. Am. Acad. 49: 355–357, 1913.)
Heads radiate; phyllaries densely ciliate .... 1. *G. canescens*
Heads discoid; phyllaries densely glandular .... 2. *G. viscida*

1. **G. canéscens** T. & G. [*Simsia c.* Gray. *Encelia eriocephala* Gray.] Desert-Sunflower. Annual, often with several stems from the base, 2–6 dm. high, white-hirsute and glandular, asperous; lvs. lanceolate or oblanceolate to broadly ovate, acute, entire or few-toothed, narrowed to a margined petiole, 1–7 cm. long, the upper ones reduced to bracts; heads solitary or paniculate, peduncled; invol. 7–12 mm. high, the green lance-acuminate phyllaries very prominently white-villous-ciliate; rays 10–21, golden, oblong, up to 20 mm. long; aks. 6–7 mm. long; pappus-awns 3 mm. long.—Sandy desert floors, up to 3000 ft.; Creosote Bush Scrub; mostly e. Mojave and Colo. deserts, Inyo Co. to L. Calif.; to Utah, Ariz., Son. Feb.–May, Oct.–Nov.

2. **G. víscida** (Gray) Blake. [*Encelia v.* Gray.] Short-lived perennial with few coarse leafy stems, 5–8 dm. high, hirsute and densely glandular-puberulent throughout; lvs. thin, ovate-oblong, acute to obtuse, clasping at base, irregularly dentate, 3–9 cm. long; heads corymbose-paniculate, discoid, 1.5–4 cm. across; phyllaries lance-oblong; aks. 7–10 mm. long; pappus-awns 3–4 mm. long, villous.—Dry hillsides, 2000–4000 ft.; Chaparral; se. San Diego Co.; n. L. Calif. May–June.

10. Verbesìna L. Crown-Beard

Ours annual or perennial branching herbs. Lvs. opposite or alternate, usually toothed. Heads corymbose on long peduncles, medium-sized and radiate in ours, yellow. Invol. hemispheric to campanulate, its linear to ovate phyllaries 2–4-seriate. Receptacle conical, its bracts thin, clasping the aks. Disk-aks. flat, 2-winged, each wing running up into an awn in ours. Over 100 spp. of warm temp. to trop. N. and S. Am. (From *Verbena,* because of a foliage resemblance.)

Lvs. sessile, green on both faces; phyllaries broadly oblong .... 1. *V. dissita*
Lvs. petioled, grayish at least beneath; phyllaries lance-linear, densely canescent .. 2. *V. encelioides*

1. **V. díssita** Gray. Suffrutescent, 5–10 dm. high, glabrate, scabrid; lvs. mostly opposite, rather remote, lance-ovate, acute, remotely serrate to entire, 4–8 cm. long, 1.5–4 cm. wide; heads several in a terminal corymb; invol. 10–12 mm. high, 15–20 mm. wide, graduate, the phyllaries spatulate and rounded (outer) to oblong and acute (inner), sparsely strigillose; rays 13–18 mm. long, neutral; aks. glabrate, the thin wings broadening above.—Once adv. in Mill Creek, San Bernardino Mts. and Arch Beach, Orange Co.; n. L. Calif. May.

2. **V. encelioìdes** (Cav.) Benth. & Hook. var. **exauriculàta** Rob. & Greenm. [*Ximenesia exauriculata* Rydb. *V. australis* of Calif. auth.] Much-branched erect annual 3–12 dm. high, with a taproot; stems canescent; lvs. narrowly lanceolate to deltoid-ovate, acute or acuminate, the blade 4–10 cm. long, saliently dentate, white-strigose, greener on upper surface, on slender petioles; invol. 7–12 mm. high, 10–15 mm. wide, scarcely graduate, densely strigose, the phyllaries slender; rays 10–15 mm. long, fertile; disk-aks. when fully mature turning from black to whitish-olive and the canescent corky wing

of uniform width becoming more prominent; $n = 17$ (Carlquist, 1954).—A weed of field borders, etc.; V. Grassland; Salinas V., upper San Joaquin V., Ventura Co. to Riverside Co.; Ariz. to Kans., Mex. May–Dec. The typical var., with lvs. auriculate at base, ranges from the Mississippi V. to Fla. and Mex.

11. Eclípta L.

Low annual herbs with leafy procumbent or ascending stems. Lvs. opposite, narrow, entire or toothed. Heads radiate but inconspicuous, white-fld., peduncled in the upper axils. Invol. hemispheric, 2-seriate, the outer phyllaries somewhat longer and broader than the inner. Receptacle flat, its bracts bristlelike. Rays short. Disk-fls. perfect, fertile. Aks. short, thick, 3- or 4-angled, becoming corky-rugose on the angles, hairy at summit. Pappus an obscure fimbriate crown or none. Small genus of riparian herbs, chiefly trop. (Greek, *ekleipo*, to be deficient, because lacking in pappus.)

1. **E. álba** (L.) Hassk. [*Verbesina a.* L.] Stem 1–10 dm. long; lvs. narrowly to broadly lanceolate, widest near the middle, sparingly serrate or entire, 2–10 cm. long, ± short-strigose, sessile; rays equaling disk.—Shores and wet banks in lowlands; Freshwater Marsh; San Francisco Bay region, Los Angeles Co. to L. Calif.; adv. from e. U.S. All months.

12. Bìdens L. Bur-Marigold

Annual or perennial herbs with opposite simple to ternately or pinnately compound lvs. Heads radiate or discoid, many-fld., usually yellow, solitary or paniculate. Invol. mostly 2-seriate, the inner phyllaries membranous, often striate, broader than the herbaceous outer ones. Paleae narrow, subplane. Ray-fls. neutral (or ♀). Disk-fls. perfect. Aks. obcompressed or 3–4-angled, usually bearing a pappus of 2–4 rigid retrorsely barbed persistent awns. Ca. 200 spp. of all warm regions, chiefly Am. (Latin, *bidens*, 2-toothed, in reference to the persistent pappus.)

(Sherff, Earl E. The genus Bidens. Field Mus. Nat. Hist., Bot. Ser. 16: 1–709, 1937.)

Aks. narrow, linear-tetragonal, often attenuate above; rays inconspicuous or lacking; lvs. compound
                                                                              1. *B. pilosa*
Aks. flat, cuneate to obovate.
  Lvs. simple.
    Rays 1.5–3 cm. long, golden-yellow .... 2. *B. laevis*
    Rays mostly less than 1.5 cm. long or lacking, pale yellow .... 3. *B. cernua*
  Lvs. compound; rays inconspicuous.
    Outer phyllaries 5–8, the heads ca. 1 cm. broad .... 4. *B. frondosa*
    Outer phyllaries 10–16, the heads 1.5–2.5 cm. broad .... 5. *B. vulgata*

1. **B. pilòsa** L. [*B. californica* DC. *Kerneria p.* Lowe. *B. p.* var. *minor* Sherff.] Beggar-ticks. Rather weak-stemmed usually branched annual 3–15 dm. high, the tetragonous stem glabrate to sparingly pilose; lvs. pinnate with 3–5 lfts., these ovate, serrate, 1–3 cm. long, the terminal one petiolulate, the lateral ones sessile or decurrent; heads discoid or inconspicuously radiate, the rays 2–3 mm. long, yellowish-white; invol. 5–7 mm. high, the inner phyllaries lance-ovate, hyaline-margined; aks. accrescent, linear, tetragonal, tuberculate-strigose, with 2–4 awns.—Natur. weed in lowlands of s. Calif.; native in Am. tropics (Mex. to Chile) and found in most warm countries. May–Nov.

2. **B. laèvis** (L.) BSP. [*Helianthus l.* L. *B. expansa* Greene. *B. speciosa* Parish, not Gardn.] Bur-Marigold. Erect or decumbent glabrous perennial 5–20 dm. high; stem stout, smooth; lvs. narrowly to broadly lanceolate, tapering to the sessile ± connate base, 7–16 cm. long, serrate; heads rather few, large, radiate; invol. with herbaceous spreading oblong ciliolate outer phyllaries somewhat exceeding the broadly ovate thin brownish yellow-hyaline-margined smooth inner ones; paleae reddish at tip; rays deep yellow, 1.5–3 cm. long; aks. narrowly cuneate, retrorsely hispidulous, 2–4-awned.—Wet lowlands, river-bottoms and in sluggish streams, below 1500 ft.; Freshwater Marsh; Cent. V. from lower Sacramento V. to upper San Joaquin; S. Coast Ranges; cismontane s. Calif.; e. to Fla., whence n. to New England. Aug.–Nov.

3. **B. cérnua** L. [*B. Kelloggii* Greene.] Nodding Bur-Marigold. Closely resembling *B. laevis*, 2–10 dm. high; lvs. as in *B. laevis*, to 20 cm. long and 4 cm. wide; heads commonly nodding in age, radiate or discoid; invol. with rather foliaceous spreading outer

phyllaries plainly exceeding the broad membranous inner ones; paleae yellowish at tip; rays light yellow, up to 1.5 cm. long, or wanting; aks. narrowly cuneate, compressed-quadrangular, the angles becoming cartilaginous-thickened and pale at maturity; $n = 12$ (Lewitzky, 1937).—Infrequent, wet lowlands and sloughs; up to 5000 ft.; Freshwater Marsh; San Francisco; Plumas Co. to Modoc and Siskiyou cos.; to B.C., New Mex., N. Carolina, Nova Scotia; Old World. June–Oct.

4. **B. frondòsa** L. STICK-TIGHT. Annual; stem erect, branching, 2–12 dm. high, nearly glabrous; lvs. slenderly petiolate, sparsely hispidulous, with 3–5 lanceolate acuminate serrate lfts. up to 8 cm. long and 25 mm. wide, the terminal one slenderly petiolulate; heads discoid or nearly so, *ca.* 1 cm. in diameter at anthesis; outer phyllaries commonly 8, ± foliaceous, prominently ciliate toward the narrowed base; disk orange; aks. cuneate, sparsely hispidulous, papillate-rugose in age, 2-awned.—Infrequent, damp ground, in waste places, up to 5200 ft.; natur. in Calif.; Riverside, San Bernardino Mts., Cent. V., Sonoma Co., Owens V.; to Wash., Nfld., La.; introd. in Eu. Aug.–Oct.

5. **B. vulgàta** Greene. Very similar to *B. frondosa*, but somewhat more robust; heads larger, 12–15 mm. in diameter at anthesis; outer phyllaries commonly 13, otherwise similar; disk yellow.—Rare, adv.; Sacramento V.; to Wash., Atlantic Coast; introd. in Eu. Aug.–Nov.

13. Cósmos Cav.

Annual or perennial herbs. Lvs. opposite, entire or lobed or mostly pinnately cut. Heads long-peduncled, solitary or in open corymbose panicles. Phyllaries in 2 series, connate at base. Receptacle flat, chaffy. Rays showy, neutral. Disk-fls. perfect, fertile. Aks. linear, mostly beaked, the beak ending in 2–4 retrorsely barbed awns. Ca. 25 spp. of warmer parts of Am., several spp. cult. for their showy fls. (Greek, *kosmos*, an ornament.)

Lf.-divisions linear to filiform; fls. white to crimson .... 1. *C. bipinnatus*
Lf.-divisions lanceolate to mucronate; fls. yellow .... 2. *C. sulphureus*

1. **C. bipinnàtus** Cav. Annual 1–3 m. high, openly branched, glabrous or ± pubescent; lvs. 4–10 cm. long, bipinnately cut into remote linear or filiform entire lobes; heads mostly long-peduncled, 4–7 cm. across; rays white, pink, or crimson, truncate or ± toothed; disk yellow; aks. glabrous, 8–11 mm. long, with an abrupt beak much shorter than the body; $2n = 24$ (Sugiura, 1936).—Occasional escape from gardens; native from Mex. to Brazil. Aug.–Nov.

2. **C. sulphùreus** Cav. Annual 1–2 m. tall, branched, pubescent; lvs. 5–30 cm. long, 2–3 times pinnately cut into lanceolate or elliptic mucronate lobes; heads 4–7 cm. across; rays strongly 3-toothed, yellow; disk yellow; anthers black with orange tips; aks. hispid, 2–2.5 mm. long including the slender beak; $2n = 24$ (Sugiura, 1936).—Occasional escape from gardens; native of Mex. Autumn.

14. Coreópsis L. COREOPSIS. TICKSEED

Annual or perennial herbs, or sometimes shrubby. Lvs. usually opposite, entire, dentate, tripartite or 1–4 times pinnately dissected. Heads showy, long-peduncled, radiate. Invol. 2-seriate in ours, the outer phyllaries 5–8, spreading, herbaceous, the inner ones 8–12, erect, membranous. Paleae of the flat receptacle thin, scarious. Aks. obcompressed, oblong to orbicular, the margin smooth or ciliate, thin-winged or thickened, with pappus a small cup, or of 2 teeth or narrow paleae over the angles, or obsolete. Ca. 100 spp. of warm and temp. regions, chiefly Am. (Greek, *koris*, a tick, and *opsis*, resemblance, from the form of the ak.)

(Sherff, Earl E. Revision of the genus Coreopsis. Field Mus. Nat. Hist., Bot. Ser. 11: 277–475, 1936. Sharsmith, Helen K. The native Calif. spp. of the genus Coreopsis. Madroño 4: 209–231, 1938.)
Ray-fls. sterile. Garden escapes.
  Lvs. entire. (Section Coreopsis) .... 1. *C. lanceolata*
  Lvs. 1–2-pinnate. (Section Calliopsis)
    Outer phyllaries much shorter than the inner; aks. narrowly oblong, thin, subplane
                                                            2. *C. tinctoria*
    Outer phyllaries not much shorter than the inner; aks. obovate, thick, convex .... 3. *C. basalis*

Ray-fls. fertile. Native spp.
  Perennials; stems stout; heads 5–8 cm. across; coastal and insular. (Section Tuckermannia)
    Heads few, on naked peduncles 15–50 cm. long .... 4. *C. maritima*
    Heads numerous, cymosely clustered on shorter leafy peduncles .... 5. *C. gigantea*
  Annuals; stems slender; heads 2–5 cm. across; not maritime.
    Aks. monomorphic, nonciliate; pappus-paleae none. (Section Leptosyne)
      Lvs. 1–2 pinnate into spatulate lobes 1–3 mm. wide, the terminal one usually broadest; annulus
        of disk-corolla mostly glabrous; outer phyllaries linear-spatulate, not gibbous at base
                                                                                6. *C Stillmanii*
      Lvs. entire or with 1–2 filiform pinnae, the terminal lobe not broader; annulus bearded;
        outer phyllaries linear-lanceolate, gibbous at base.
          Aks. shining, smooth or nearly so, the corky wing thin .... 7. *C. Douglasii*
          Aks. dull, beset with clavellate hairs, the corky wing thick and rugose .... 8. *C. californica*
    Aks. dimorphic, disk-aks. ciliate, with 2 pappus paleae, ray-aks. glabrous, epappose. (Section
      Pugiopappus)
        Outer phyllaries linear, obtuse; stems essentially scapose.
          Paleae adherent to and falling with disk-aks.; pappus 2 mm. long; rays horizontal in
            anthesis .... 9. *C. Bigelovii*
          Paleae not adherent to or falling with disk-aks.; pappus 1 mm. long; rays strongly reflexed
            in anthesis .... 10. *C. hamiltonii*
        Outer phyllaries broadly ovate, acute; stems leafy below; pappus 4 mm. long
                                                                                11. *C. calliopsidea*

1. **C. lanceolàta** L. [*Coreopsoides l.* Moench. *Leachia l.* Cass. *Chrysomelea l.* Tausch.]
Perennial herb with 1–several erect or ascending few-branched stems 2–6 dm. high
from the small caudex; lvs. several, in the lower half of the stem, linear-lanceolate to
oblanceolate, the lowermost slender-petioled, all entire; heads 3–6 cm. across, solitary
on long slender peduncles, yellow; phyllaries all broad; tips of style-branches caudate;
aks. strongly concavo-convex, orbicular, 2.5–3 mm. long, the body black, papillate, with
broad thin wing; pappus of 2 small scales.—Occasional garden escape; native in e. U.S.
May–July.
2. **C. tinctòria** Nutt. [*Diplosastera t.* Tausch. *Calliopsis t.* DC.] CALLIOPSIS. Slender
erect glabrous annual 4–10 dm. high, dichotomously or trichotomously branching above;
lower lvs. bipinnately divided into linear divisions, 5–10 cm. long, the uppermost pinnately few-lobed or entire; heads subcorymbose, 2–3 cm. across; outer phyllaries inconspicuous, narrow, ca. 2 mm. long, the inner ones ovate, 5–6 mm. high; rays yellow, or
red-brown at base or throughout; disks dark reddish; style-branches obtuse; aks. linear-
oblong, black, wingless, awnless.—Occasional garden escape; native from Minn. and
B.C. to La. and New Mex. June–Aug.
  C. **Atkinsoniàna** Dougl., resembling *C. tinctoria*, but with narrowly winged aks., native
in the NW., has been reported as introd. in Stanislaus Co.
3. **C. basális** (Otto & Dietr.) Blake var. **Wrìghtii** (Gray) Blake. [*C. Drummondii*
var. *W.* Gray.] Erect branching annual 2–5 dm. high, glabrous or nearly so, the petioles
sometimes ciliate with multicellular hairs; lvs. 2–3-pinnatifid, the lower with linear segments, the upper with filiform; heads numerous, subcorymbose, 1–3 cm. across; outer
phyllaries linear, spreading, 5–8 mm. long, scarcely shorter than the ovate scarious-
margined inner ones; rays yellow, orange-red at base; disks dark reddish; aks. obovate,
thickish.—Introd. below 1000 ft., V. Grassland, in San Diego Co. (Pechstein Reservoir);
native to Tex.; May.
4. **C. marìtima** (Nutt.) Hook. f. [*Tuckermannia m.* Nutt. *Leptosyne m.* Gray.] Robust
erect glabrous perennial 3–8 dm. high, from a tuberous taproot; stems fistulous; lvs.
alternate, fleshy, slenderly petiolate, 5–25 cm. long, 2–3-pinnate into widely divaricate
linear segments mostly 2–3 mm. wide; heads few, on stout naked peduncles 15–50 cm.
long at branch-tips, 6.5–8 cm. or more across, yellow; phyllaries all broad, the inner
scarcely exceeding the outer; rays 15–20; style-branches triangular-acute; aks. oblong
to obovate, plane, the thin wing up to 1 mm. wide, rarely with 1–2 awns.—Coastal
bluffs and dunes; below 200 ft.; Coastal Strand, Coastal Sage Scrub; s. San Diego Co.;
n. L. Calif. and adjacent ids. March–May.
5. **C. gigantèa** (Kell.) Hall. [*Leptosyne g.* Kell. *Tuckermannia g.* Jones.] Stout erect
fleshy few-branched arborescent shrub 4–15(–30) dm. high, glabrous, the main trunk
up to 12 cm. thick; lvs. in dense tufts at the ends of the primary branches, alternate,
5–25 cm. long, 3–4-pinnate into nearly filiform segments; heads in cymose clusters on
long scapiform peduncles, 4–8 cm. across, yellow; rays 10–16; aks. oblong to obovate,
plane, narrowly winged; pappus none.—Rocky sea cliffs and exposed dunes, up to 150 ft.;

Coastal Strand, Coastal Sage Scrub; San Luis Obispo Co. to Los Angeles Co. and most of the Channel Ids.; Guadalupe Id., L. Calif. March–May.

6. **C. Stillmánii** (Gray) Blake. [*Leptosyne S.* Gray.] Delicate erect glabrate annual 10–15(–30) cm. high; stems several from the base; lvs. chiefly basal, opposite or alternate, 2–10 cm. long, pinnately divided into few remote linear-spatulate lobes; heads solitary on long peduncles, 1.5–3.5 cm. across, orange-yellow; outer phyllaries oblong, nearly as long as the ovate inner ones, bearded at base with brownish glandular hairs; rays 5–8; aks. obovate, ± plane, the dorsal face smooth, the ventral face sometimes papillate-setose, the wing (and midrib) whitish, corky-thickened, rugose; pappus reduced to a cupule or rarely 1–2 short awns.—Grassy slopes, sometimes serpentine, 100–3000 ft.; V. Grassland, Foothill Wd., Chaparral; foothills surrounding Cent. V. from Butte Co. to Tulare Co. and from Contra Costa Co. to Stanislaus Co. March–May.

7. **C. Douglásii** (DC.) Hall. [*Leptosyne D.* DC. *C. Stillmanii* var. *Jonesii* Sherff.] Delicate erect glaucescent annual, with few scapose monocephalous stems 5–25 cm. high; lvs. basal or nearly so, nearly filiform, grooved above, rounded beneath, remotely and sparingly glandular-ciliolate, otherwise glabrous, 2–8 cm. long, entire or with 1–2 pinnae; heads 1–2.5 cm. across, golden; outer phyllaries narrowly lanceolate, shorter than the lance-ovate apically tufted inner ones, glandular-hairy at base; aks. obovate, concavo-convex, essentially glabrous, smooth, brown, the wing yellowish, not rugose and scarcely thickened; pappus reduced to a cupule.—Gravelly or rocky open slopes, 500–2000 ft.; Foothill Wd., Chaparral; inner S. Coast ranges, from Santa Clara Co. to Santa Barbara Co. March–April.

8. **C. califòrnica** (Nutt.) H. K. Sharsm. [*Leptosyne c.* Nutt. *L. Newberryi* Gray. *L. Douglasii* of many auth., not DC.] Similar in habit, herbage and heads to *C. Douglasii*, but often larger; scapes 5–45 cm. high; lvs. terete, 2–10(–15) cm. long; heads 1–3.5 cm. across, yellow; aks. obovate, dull and minutely papillate, adorned with stubby hyaline clavellate or capitate hairs on wing and body, brownish, the lighter-colored wing conspicuously spongy-thickened, ± ciliolate; pappus reduced to a cupule.—Sandy or gravelly soils of valley floors and canyon sides, 100–3000 ft.; S. Oak Wd., V. Grassland, Joshua Tree Wd., Creosote Bush Scrub; cismontane s. Calif., from Los Angeles Co. to San Diego Co., both deserts from s. Inyo Co. to Imperial Co.; Ariz., L. Calif. March–May.

9. **C. Bigelòvii** (Gray) Hall. [*Pugiopappus B.* Gray. *P. Breweri* Gray. *Leptosyne Bigelovii* Gray.] Erect glabrous annual, with few monocephalous scapes 1–6 dm. high, stouter than in the 2 preceding spp.; lvs. basal or nearly so, 4–12 cm. long, ovate in outline, 1–2-pinnate into linear lobes as wide as the 1–2 mm. wide rachis; heads 2–4.5 cm. across, golden; outer phyllaries linear, shorter or longer than the dark oblong-ovate inner ones, glabrous; palea adherent to the disk-ak. and falling with it; ray-aks. oblong-obovate, glabrous, narrowly winged, epappose; disk-aks. oblanceolate, dorsally glabrous, ventrally pilose along midrib, the thin margin densely ciliate with silky ascending hairs, their pappus of two bright white linear-lanceolate fimbriate paleae ca. 2.5 mm. long.— Dry gravelly hillsides, 1000–5000 ft.; Creosote Bush Scrub, Joshua Tree Wd., Pinyon-Juniper Wd., Foothill Wd.; slopes around both sides of the upper San Joaquin V., from s. Monterey and Tulare cos. s. to the San Rafael and Santa Monica mts., w. part of the Mojave Desert from Death V. region to San Gorgonio Pass. March–May.

10. **C. hamiltònii** (Elmer) H. K. Sharsm. [*Leptosyne h.* Elmer.] Closely related to *C. Bigelovii* and much confused with that sp.; scapes somewhat leafy below, 8–25 cm. high; lvs. 2–5 cm. long, bipinnate into lobes 1 mm. wide; heads 1–2 cm. across; outer phyllaries linear, always shorter than the broad inner ones, glabrous; paleae not adherent to the disk-aks.; ray-aks. oblong-oblanceolate, concavo-convex, blackish, narrowly winged, epappose; disk-aks. linear-oblanceolate, convex and moderately silky sericeous about the midnerve on both faces, the thin margin densely ciliate with similar hairs, the white pappus paleae ca. 1 mm. long.—Exposed, dry rocky slopes, 1800–4250 ft.; Foothill Wd.; Mt. Diablo and the Mt. Hamilton Range, Contra Costa, Stanislaus, and Santa Clara cos. March–May.

11. **C. calliopsidèa** (DC.) Gray. [*Agarista c.* DC. *Pugiopappus c.* Gray. *Leptosyne c.* Gray. *L. c.* var. *nana* Gray.] Stout erect glabrous annual, the monocephalous stems 1–5 dm. high; lvs. on lower half of stem, the lower ones 4–8 cm. long, ovate in outline, 1–2-pinnate into linear lobes 0.5–2 mm. wide; heads 2–9 cm. across, golden; outer phyllaries deltoid-ovate, shorter than the ovate inner ones; palea adherent to the disk-ak. and fall-

ing with it; ray-aks. oval, glabrous, broad-winged, epappose; disk-aks. lance-oblong, dorsally glabrous and shining, ventrally rather copiously silky-sericeous, the margin densely ciliate with similar but longer hairs, their pappus of two yellowish channeled linear-lanceolate fimbriate awns ca. 4 mm. long.—Dry, open gravelly ground, 500–3200 ft.; V. Grassland, Foothill Wd., Joshua Tree Wd., Creosote Bush Scrub; inner Coast Ranges facing the San Joaquin V., from Corral Hollow, Alameda Co., to the Cuyama R., Santa Barbara Co., Calabasas, Los Angeles Co., e. to w. Mojave Desert in Kern, Los Angeles, and San Bernardino cos. March–May.

15. Guizòtia Cass.

Annual herbs with opposite lvs. or the upper alternate. Heads peduncled, axillary and terminal; ray-fls. ♀, 1-seriate, fertile, yellow; disk-fls. perfect, fertile. Invol. campanulate; phyllaries 2-seriate, the outer subfoliaceous, the inner like the paleae. Receptacle convex or conic. Paleae flat, scarious. Disk-corollas pubescent without at base. Aks. glabrous, dorsally compressed, rounded at tip; pappus none. A few spp., of trop. Afr.

1. **G. abyssínica** (L. f.) Cass. [*Polymnia a.* L. f. *G. oleifera* DC.] RAMTILLA. Stout, erect, leafy, smooth or scabrid, 3–9 dm. high; lvs. sessile, 7–16 cm. long, linear or lance-oblong, partly amplexicaul, obtuse, serrate; heads 1–2 cm. in diam.; peduncles naked, 3–5 cm. long; outer phyllaries broadly elliptic or ovate, green; ligules few, broad; $2n = 30$ (Richharia & Kalamar, 1938).—Occasional weed as at Santa Barbara, Loma Linda, etc.; native of Afr. Aug.–Sept. Cult. in India, etc., for the oil.

16. Galinsòga R. & P.

Annual herbs with opposite lvs. and small cymose heads. Phyllaries few, relatively broad, rather membranous, with several green nerves, each subtending a ray. Receptacle conic, the paleae membranous, flat and narrow. Ray-fls. 4 or 5, only slightly surpassing the disk, fertile, white in ours. Disk-fls. perfect, yellow. Aks. 4-angled, the outer somewhat obcompressed. Pappus paleae fimbriate, sometimes reduced or none in the rays. Ca. 6 habitally similar spp. from U.S. to Argentina, some widely distributed as weeds. (For M. M. *Galinsoga*, Spanish physician and botanist.)

1. **G. parviflòra** Cav. Slender annual 2–7 dm. high, leafy and often freely branching throughout, sparsely hairy; lvs. ovate, acute, short-petiolate, 2–5 cm. long, 1–3 cm. wide, crenulate or entire, thin; heads broadly campanulate, 3–4 mm. high; pappus of the disk-fls. of 8–14 conspicuously fimbriate blunt paleae, ca. as long as the corolla, that of the ray-fls. obsolete.—Locally common weed in irrigated orchards, etc., cismontane Los Angeles, Orange, and San Bernardino cos., Inyo Co.; native from Mex. to S. Am., but now a cosmopolitan weed. May–Nov.

17. Bébbia Greene

Rounded odorous xerophytic shrub with green nearly leafless multibranched stems. Heads solitary, terminating the twigs, or loosely cymose, discoid, yellow. Invol. hemispheric, ca. 3-seriate, the phyllaries imbricate, striate, shorter than the disk. Receptacle umbonate, the paleae scarious, lanceolate, partially enfolding the aks. Corolla linear, the tube glandular, the limb hairy. Aks. slender, pubescent with appressed hairs. Pappus of 15–20 plumose bristles much longer than the ak. 2 spp. of the N. Am. desert region. (For Michael S. *Bebb*, Illinois student of w. willows.)

1. **B. júncea** (Benth.) Greene. [*Carphephorus j.* Benth. *B. j* var. *aspera* Greene. *B. a.* A. Nels.] SWEET BUSH. Diffuse shrub 5–10 dm. high, glabrate or minutely hispid, or toward the heads somewhat canescent, the often leafless slender branches junciform, glaucescent; lvs. remote, early deciduous, linear (to oblanceolate), 1–5 cm. long, often bearing medianly 1–2 salient teeth; invol. 4–8 mm. high; phyllaries lance-ovate to lanceolate, the outer usually canescent, the inner often becoming anthocyanous.—Gravelly fans, rocky washes and canyon-sides, 100–3900 ft.; Creosote Bush Scrub; both deserts, but only in the e. Mojave (rare), from White Mts. s. to L. Calif., cismontane s. Calif. from San Bernardino s.; Nev. to New Mex., Son. April–July. Typical *juncea*, from L.

Calif., has broader, blunter phyllaries than almost all U.S. material, which would accordingly be referable to var. *aspera,* but the distinction is difficult to maintain.

18. Eastwoódia Bdg.

Xerophytic shrub, somewhat glutinous. Stem white-barked, striate, glabrous, brittle. Lvs. alternate, essentially linear, entire. Heads discoid, many-fld., yellow, solitary at ends of the often cymosely arranged branches. Invol. hemispheric, ca. 4-seriate, graduate, the phyllaries appressed, narrow, firm, 1-nerved, smooth, narrowly scarious-margined. Receptacle broad, paleaceous throughout, its bracts firm, boat-shaped, caducous. Corollas funnelform, the 5 lobes long. Style-branches linear, the acuminate appendages equaling the stigmatic portion. Aks. ± quadrangular, narrowly obpyramidal, silky-pilose especially on the ribs. Pappus of 5–6 free persistent linear acute or acuminate firm paleae ca. ⅔ as long as the corolla. Monotypic. (In honor of Alice *Eastwood,* California botanist.)

1. **E. élegans** Bdg. Rounded desert shrub 3–10 dm. high, erect-branched; herbage minutely and sparingly hispidulous, rather glutinous, pallid; lvs. linear to linear-oblanceolate or the upper subulate, up to 2.5 cm. long, to 3 mm. wide; heads depressed-subglobose, 1–1.5 cm. thick; invol. impressed at base, 5–7 mm. high; corollas ca. 5.5 mm. long; aks. ca. 2 mm. long; pappus ca. 4 mm. long.—Hot dry hillsides, up to 2500 ft.; V. Grassland; foothills on w. and s. side of San Joaquin V. from Alameda Co. to Santa Barbara Co., e. to Tehachapi Mts., Kern Co. April–July.

19. Melampòdium L.

Herbs with opposite ample mostly sessile lvs. Heads radiate, small or medium-sized, pedunculate, in leafy cymes. Invol. 2-seriate, dimorphous, the outer phyllaries few, loose, often foliaceous, the inner ones smaller, completely enfolding the marginal aks. and deciduous with them as a thickened covering. Rays white or yellow, ♀, fertile. Disks yellow, perfect, sterile. Pappus none. Less than a dozen spp. of chiefly subtrop. Am., especially Mex. (Greek, *melas,* black, and *pous,* foot, meaninglessly applied to these plants.)

1. **M. perfoliàtum** HBK. [*Alcina p.* Cav. *Wedelia p.* Willd.] Coarse widely branched annual 2–12 dm. high; lvs. deltoid-ovate or hastate to oval-oblong, membranaceous, 5–15 cm. long, scabrid, dentate, narrowed to a winged petiole, these connate in pairs; outer phyllaries 4 or 5, ovate, herbaceous, scabro-ciliate, 9–13 mm. long, united at base; rays ca. 10, yellow, shorter than the invol. and scarcely longer than the disk.—Rare weed of waste lands in the Los Angeles area (Compton, Whittier, etc.); adv. from Mex. May–Nov.

20. Sanvitália Lam.

Low mostly branching annual herbs. Lvs. opposite, ± petioled, entire. Heads small, terminating the branches or subsessile in the upper leaf-axils, radiate. Invol. 1–3-seriate, the phyllaries rather firm and dry. Receptacle strongly conical in fr. in ours, its chaffy bracts cuspidate-tipped. Ray-fls. ♀, fertile, the ligule sessile and persistent on the ak. Disk-fls. perfect, fertile. Ray-aks. apically bearing 3–5 divaricate horny cusps or awns. Disk-aks. 4-angled, tuberculate, epappose or nearly so. Ca. 4 spp., of sw. U.S. and n. Mex. (Name honoring *Sanvitali,* a noble Italian family.)

1. **S. Àberti** Gray. Erect or rather diffuse annual 1–3 dm. high, the opposite ascending or divaricate branches slender, with rather long internodes; herbage scabrid-hispidulous with white curly appressed hairs to glabrate; lvs. linear-lanceolate, up to 7 cm. long and 10 mm. wide, short-petiolate; heads conical when ripe; ray-fls. 6–11, the ligule yellow, drying stramineous, firm, bidentate, green-veined dorsally; disk-fls. greenish-yellow; ray-aks. 3–4 mm. long, thick, narrowly 4-sulcate, smooth to ± papillate, white, 3-cuspidate-awned; disk-aks. mostly 4-angled or -ribbed, conspicuously and irregularly warty, dark-colored.—Rocky slopes and dry washes, 5400 ft.; Pinyon-Juniper Wd.; Clark Mt., e. San Bernardino Co.; to Tex., n. Mex. Aug.–Sept.

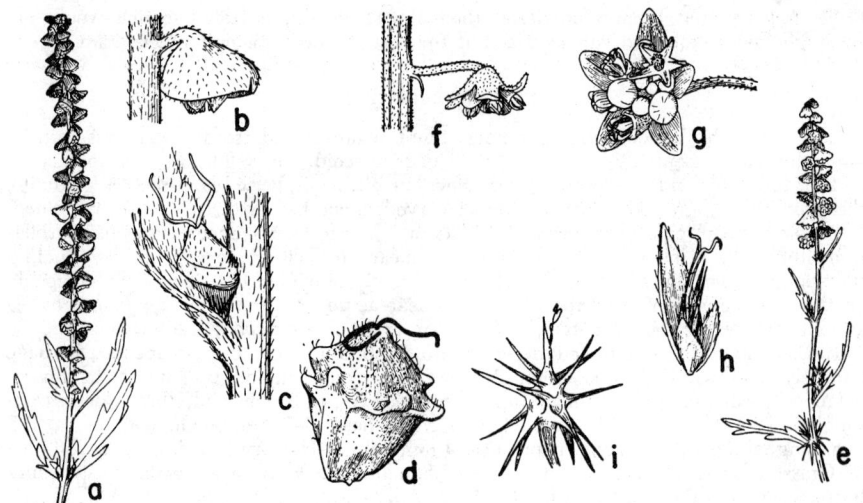

Fig. 110. AMBROSIINAE. *Ambrosia psilostachya: a,* infl., × 1, ♂ heads in raceme, ♀ in upper lf.-axil; *b,* ♂ head, × 7, cuplike invol.; *c,* ♀ head, × 7, in lf.-axil; *d,* fr., × 7, tubercled invol. adnate to and inclosing the ak. *Franseria acanthicarpa: e,* infl., × 1; *f,* ♂ head, × 2½, with lobed invol.; *g,* same, × 5, with ♂ fls.; *h,* ♀ head, with 1 fl.; *i,* ♀ head in fr., × 2½, the invol. spiny.

Tribe **Heliántheae**. Subtribe **Ambrosìinae** (Fig. 110)

Coarse homely plants, often weedy. Heads small, greenish or whitish. Lvs. alternate or the lowest opposite. Fls. unisexual, the ♂ and the ♀ in separate heads, the former in a raceme or spike above the latter, which are few and axillary; or sometimes ♂ and ♀ fls. in the same axillary head. Receptacle of the ♂ or perfect heads with chafflike bracts. Rays none. Corolla of ♀ fls. none or rudimentary. Anthers distinct or nearly so, not caudate. Pappus none. Fr. commonly a bur.

A.   Heads containing both ♂ and ♀ fls., the latter at the margin.
   B.   Aks. densely long-villous; lvs. or their lobes linear-filiform .... 21. *Oxytenia*
   BB. Aks. not long-villous; lvs. or their lobes not linear-filiform.
      C.   Aks. obovoid or pyriform, not winged .... 22. *Iva*
      CC. Aks. flattened, with pectinate or toothed wings .... 23. *Dicoria*
AA. Heads unisexual, monoecious; ♀ heads with 1–4 fls. enclosed in a nutlike or burlike invol., only the style-tips exserted.
   B.   Phyllaries of the ♂ heads united.
      C.   Pistillate invol. with several transverse scarious wings; lvs. or their lobes linear-filiform
                          24. *Hymenoclea*
      CC. Pistillate invol. without transverse wings; lvs. and their lobes not linear-filiform.
         D.   Fruiting invol. unarmed or with a single row of short prickles below the single beak .... 25. *Ambrosia*
         DD. Fruiting invol. armed with several rows of spines below the 1–4 beaks
                          26. *Franseria*
   BB. Phyllaries of the ♂ heads free; fruiting invol. burlike, covered with barbed prickles
                         27. *Xanthium*

21. Oxytènia Nutt.

Virgate perennial herb with erect branches, *Artemisia*-like in habit, the stems almost woody at base. Lvs. alternate, 3–5-parted into filiform divisions, or the upper ones sparse and entire. Heads numerous, in dense panicles. Phyllaries 5, orbicular, mucronate-tipped. Paleae of receptacle slender, thin, with dilated villous tips. Fls. unisexual, the outer ca. 5, ♀, lacking corolla, the inner numerous, ♂, with funnelform corolla. Aks. obovoid, with wide-beaked areola, densely villous, epappose. Monotypic. (Greek, *oxytenes*, pointed, referring to the slender acerose lvs.)

1. **O. aceròsa** Nutt. Suffrutescent strong-scented perennial 0.5–2 m. high; lvs. canescent, 5–10 cm. long, pinnatifid with linear-filiform subterete divisions 0.5 mm. wide,

or the upper entire; heads hemispheric, 5 mm. wide, the fls. whitish, fragrant; aks. 2 mm. long.—Saline stream beds and plains, 500–2000 ft.; Alkali Sink; Amargosa Desert, Funeral Mts., Death V. region; e. to sw. Colo. and New Mex. Aug.–Dec. Often nearly leafless and rushlike; said to cause human dermatitis.

22. Iva L.

Herbs or low shrubs. Lvs. alternate (or opposite), entire in ours, with small nodding heads mostly racemosely disposed in the axils. Phyllaries few, roundish, green, sometimes with a shorter inner scarious series. Paleae of the receptacle narrowly linear, membranaceous. Fls. greenish-white, inconspicuous, the marginal ones ♀, fertile, 1–5, their corollas tubular or obsolete, the inner ones (disk) perfect, sterile, their corollas funnelform, their anthers almost distinct. Aks. obovoid in ours, glabrous, epappose. Ca. 15 spp., all N. Am. (After the mint, *Ajuga iva,* because of the similar odor.)

Plants perennial.
  Phyllaries distinct .... 1. *I. Hayesiana*
  Phyllaries united into a cup .... 2. *I. axillaris*
Plants annual .... 3. *I. nevadensis*

1. **I. Hayesiàna** Gray. Frutescent perennial with several virgate stems from base, 5–10 dm. high, sometimes racemosely branching; lvs. oblong-oblanceolate, obtuse, entire, subsessile, 3–6 cm. long, 4–12 mm. wide, thickish, strigose-puberulent and often glandular-atomiferous, reduced gradually upwards, the uppermost scarcely exceeding the heads; invol. hemispheric, 5–6 mm. wide, the phyllaries ovate to obovate, distinct.—Alkaline places, below 1000 ft.; Alkali Sink; s. San Diego Co.; L. Calif. April–Sept.

2. **I. axilláris** Pursh. [*I. a.* var. *pubescens* Gray.] POVERTY WEED. Less woody, or almost wholly herbaceous, spreading from slender rhizomes, the several stems erect from an often decumbent or prostrate base, mostly 2–6 dm. high; herbage strigose or hirsutulous, red-glandular-punctate; lvs. oblanceolate, obtuse, entire, subsessile, 1–4 cm. long, 2–12 mm. wide, thickish; heads solitary in the axils, 5–6 mm. wide, the phyllaries united into a deeply lobed to subentire cup; $2n = 32$–34 (Heiser & Whitaker, 1948).—Alkaline plains or saline marsh borders, sea level to 6700 ft.; Coastal Salt Marsh, Alkali Sink, and elsewhere; scattered throughout Calif., from Shasta V., through the Sierra Nevada to w. Mojave Desert and San Bernardino Mts., and through the Coast Range valleys to Ventura and Orange cos.; to Nebr., Canada. May–Sept. Weedy, but more troublesome in other states.

3. **I. nevadénsis** Jones. [*Chorisiva n.* Rydb.] Diffusely branched annual 0.5–3 dm. high; stems yellow; herbage strigosely pubescent with curved multicellular blunt white hairs and somewhat glandular; lvs. pinnately parted, the ultimate segms. thickish, obtuse; aks. obcompressed, with concave inner face, 2 mm. long, at length muricate.—Sandy alkaline plains, 4700–6500 ft.; Sagebrush Scrub, Pinyon-Juniper Wd.; w. foothills of White Mts., Mono and Inyo cos.; w. Nev. June–Oct.

23. Dicòria T. & G.

Diffusely branched annual herbs. Upper lvs. alternate, petioled, entire or toothed. Heads small, numerous, nodding, spicately arranged in loose panicles, heterogamous and also some ♂. Phyllaries strongly dimorphic, the outer ones ca. 5, small, herbaceous, the inner ones subtending the 1–2 ♀ fls. subscarious, accrescent, much surpassing the outer at maturity. Paleae narrow, tardily deciduous. Pistillate fls. without corolla. Staminate fls. with funnelform corolla, the stamens with coherent filaments and free anthers. Aks. plano-convex, keeled on each face, pectinately wing-margined, epappose. Ca. 4 spp. of arid sw. U.S. (Greek, *dis,* twice, and *koris,* a bug, from the aspect of the 2 aks. of the original sp.)

1. **D. canéscens** T. & G. [*D. oblongifolia* Rydb.] Widely branching plants 3–9 dm. high, white-strigose-canescent or -hirsute and often also somewhat hispid and glandular; juvenile lvs. deltoid-lanceolate, dentate, 3–5 cm. long, floral lvs. broadly ovate to subrotund, denticulate to subentire, ca. 1 cm. long; outer phyllaries elliptic, 2–3 mm. long, reflexed in age, the inner ones glandular-puberulent, orbicular, becoming deeply concave and accrescent in fr., to 6–8 mm. long; ak. 4.5–5.5 mm. long, the wing shallowly

toothed to deeply pectinate-lobed.—Open sandy deserts, below sea level to 2000 ft.; Creosote Bush Scrub; Colo. Desert (common), Mojave Desert (rare); sw. Ariz., nw. Son. Sept.–Jan. Variable in the development of the fruiting inner phyllaries and the wing of the ak. The following sspp. appear to have regional validity.

Ssp. hispídula (Rydb.) Keck. [*D. hispidula* Rydb.] Stems white-strigose-canescent, also with scattered spreading pustulate-based hispid hairs; inner phyllaries scarcely accrescent, becoming 4–5 mm. long; ak. 3.5–4.5(–5) mm. long, the margin relatively narrow with ± remote teeth.—Sandy places; Creosote Bush Scrub, Joshua Tree Wd.; w. Mojave Desert (frequent), Colo. Desert (uncommon); w. Ariz. Sept.–Dec.

Ssp. Clárkae (Kenn.) Keck. [*D. C.* Kenn.] Very similar to the typical ssp. except the inner phyllaries strongly accrescent, becoming 10–13 mm. long in fr.; ak. 5.5–6 mm. long, its wing a somewhat interrupted row of shorter but firmer ± corneous teeth.— Sand dunes, 3000–4000 ft.; Alkali Sink, Creosote Bush Scrub; Inyo Co.; Nev., sw. Utah. Sept.–Dec.

24. Hymenoclèa T. & G.

Low diffusely much-branched monoecious xerophytic shrubs. Lvs. alternate, filiform, entire or the lower pinnately parted into few lobes. Heads small, very numerous, scattered or glomerate-paniculate, both sexes usually found in the same lf.-axils, the ♂ above the ♀. Staminate heads several-fld.; invol. shallow, 4–6-lobed. Pistillate heads 1-fld.; invol. ovoid or fusiform, becoming indurate, pericarplike and beaked in fr., the phyllaries persistent as scarious horizontal wings near the middle. Two spp., sw. U.S. and Mex. (Greek, *humen,* membrane, and *kleio,* to enclose.) The pollen of these plants is said to be an important cause of hay fever.

Wings of the fruiting invol. reniform-orbicular, spirally arranged, 3.5–8 mm. long .... 1. *H. Salsola*
Wings of the fruiting invol. cuneate or obovate, in a single whorl of 7–12, 1–2 mm. long
2. *H. monogyra*

1. **H. Sálsola** T. & G. [*H. fasciculata* A. Nels. & var. *patula* A. Nels.] Erect or spreading twiggy bush 6–10(–15) dm. high, the herbage yellowish-green, resinous, minutely canescent or glabrous; lvs. sparse, 2–5 cm. long; scales of the ♀ invol. spirally arranged from the base to near the middle, reniform, parallel-veined, mucronate-tipped, narrowed to a clawlike base, often with a deep pit in the axils, sometimes imbricated like a cone, sometimes all spreading.—Common in sandy washes and rocky uplands in the desert area, often in alkaline soils, below sea level to 5000 ft.; Creosote Bush Scrub, Shadscale Scrub, Pinyon-Juniper Wd.; n. Inyo Co. to Colo. Desert, w. to Cuyama R. and head of San Joaquin V.; Nev., s. Utah, Ariz., L. Calif. March–June.

Var. **pentalèpis** (Rydb.) L. Benson. [*H. p.* Rydb. *H. hemidioica* A. Nels.] Very similar, but the scales of the ♀ invol. in a single radiating whorl of usually 5 or 6 members.—E. Colo. Desert; s. Ariz., Son., L. Calif.

2. **H. monogỳra** T. & G. More strictly erect and more leafy than *H. Salsola;* shrub 1–3 m. high; scales of the ♀ invol. 7–12, spreading in a single whorl around the middle of the considerably less indurate invol., obovate to suborbicular, neither prominently veined, mucronate-tipped, nor with pits in their axils.—Sandy washes, up to 1500 ft.; Chaparral; Rialto, around San Diego; more common e. to Tex. and Coahuila, s. to L. Calif. Aug.–Nov.

25. Ambròsia L. Ragweed

Annual or perennial coarse monoecious weedy herbs. Lvs. opposite or alternate, mostly lobed or dissected. Heads small, numerous, greenish, inconspicuous. Staminate heads nodding in bractless terminal racemes or spikes; phyllaries united into a 5–12-lobed broadly turbinate cup; paleae of the receptacle subtending at least the outer fls. Pistillate (fertile) heads borne in lf.-axils below the ♂ racemes, 1-fld., their invols. ± turbinate, short-beaked, enclosing the ak., at length armed with a single row of prickles above the middle; their fls. without corolla or pappus. Ca. 15 spp., mostly Am. The windborne pollen is a common cause of hay fever. (Ancient Greek, and also Latin, name of several plants.)

Principal lvs. once or twice pinnatifid.
Annuals .... 1. *A. artemisiifolia*
Perennials from creeping rootstocks.
  Lvs. once-pinnatifid into coarse lobes, greenish, rough-pubescent .... 2. *A. psilostachya*
  Lvs. twice-pinnatifid into minute obtuse lobes 1 mm. wide, very canescent with soft hairs
                                                                     3. *A. pumila*
Principal lvs. 3–5-lobed or undivided; annual .... 4. *A. trifida*

1. **A. artemisiifòlia** L. [*A. elatior* L. *A. e.* var. *a.* Farw.] LOW RAGWEED. Erect annual, branching at least above, the stem glabrate to strigose and often hirsute, 2–10 dm. high; lvs. alternate above, usually bipinnatifid, strigillose, ovate or elliptic in outline, 3–8 cm. long; ♂ heads in tight terminal racemes; ♀ heads in sessile clusters, their invols. 3–5 mm. long, short-beaked, with 5–7 rather small spines.—Uncommon introd. weed with us; Gasquet, Del Norte Co., Penryn, Placer Co.; native in e. U.S. Aug.–Oct.

2. **A. psilostáchya** DC. [*A. coronopifolia* T. & G. *A. californica* Rydb. *A. p. californica* Blake.] WESTERN RAGWEED. Perennial with running rootstocks; herbage rather harshly pubescent with short stiff appressed or spreading hairs, glandular, aromatic when rubbed; stem simple, 5–12 dm. high; lvs. mostly only once pinnatifid, 4–12 cm. long, thickish, often subsessile; ♂ invols. scabrous-hirtellous; ♀ invols. merely tuberculate above or quite unarmed; $n = 50$–52 (Heiser & Whitaker, 1948).—One of the very common weeds of roadsides and uncultivated land, sea level to 2500 ft.; throughout cismontane Calif. but especially in coast valleys, Colo. Desert; to Wash., Sask., Ill., n. Mex. July–Nov.

3. **A. pùmila** (Nutt.) Gray. [*Franseria p.* Nutt. *Hemiambrosia heterocephala* Delpino.] Herbaceous stems erect from a small branching rhizomelike caudex, 0.5–3(–5) dm. high; herbage grayish silky-canescent; lvs. crowded, ± oblanceolate in outline, 2–3-pinnatifid into narrowly oblong-oblanceolate obtuse crowded lobes, with long slender petioles; ♀ invols. in fr. unarmed, pubescent, obovoid, 2 mm. long.—Dry, sunny places, along roadsides, etc., 100–600 ft.; V. Grassland; sw. San Diego Co., Lake Hodges to National City; n. L. Calif. June–Sept.

4. **A. trífida** L. GIANT RAGWEED. Stout annual up to 2 m. high, ± hispid-hirsute; lvs. ample, elliptic to broadly ovate, palmately 3–5-cleft into broadly lanceolate serrate lobes (or lobeless, but this form not yet detected in Calif.); ♀ invol. 5–7-ribbed, each rib in fr. terminated by an acute tubercle; fr. 6–10 mm. long.—Introd. weed of moist ground, rare in Calif. (Orange Co.); to B.C., Atlantic Coast; adv. in Eu. June–Sept. Pollen of this sp. is perhaps the most serious cause of hay fever in e. U.S.

26. Fransèria Cav.

Annual or perennial herbs or subshrubs, very similar to *Ambrosia*. Lvs. opposite or alternate, toothed or dissected. Staminate heads nodding in terminal catkinlike narrow naked spiciform racemes; invols. bowl-shaped or turbinate; corolla funnelform, 5-toothed. Pistillate (fertile) heads borne in lf.-axils below the ♂ racemes or rarely mixed, 1(–4 or more)-fld., their invols. armed with spines or prickles in several series and in fr. becoming a bur, ending in a 2-toothed beak for each floret. Ca. 20–25 spp. of warm temp. N. and S. Am. (For Antonio *Franseri*, Spanish physician and botanist.)

Plants shrubby, or at least woody at base.
  Lvs. petioled, the blades not spinose.
    Bur with straight spines.
      Lvs. deeply toothed to pinnatifid; fr. fusiform, 8 mm. long, 1-beaked, the spines densely long-villous .... 1. *F. eriocentra*
      Lvs. once to thrice pinnately divided into very small obtuse lobes; fr. globular, 4–6 mm. long, 2-beaked, the spines not villous .... 2. *F. dumosa*
    Bur with hooked spines; lvs. not pinnatifid or divided.
      Body of the bur lanate like the base of the spines .... 3. *F. chenopodiifolia*
      Body of the bur essentially glabrous, the spines prominently glandular-puberulent
                                             4. *F. ambrosioides*
  Lvs. sessile and cordate-clasping, coarsely spinose-toothed; fr. ovoid, 10–20 mm. long
                                             5. *F. ilicifolia*

Plants herbaceous.
  Bur 2–4 mm. long, its spines mostly hooked .... 6. *F. confertiflora*
  Bur 7–10 mm. long, its spines not hooked but flattened and channeled.
    Plants decumbent littoral perennial herbs .... 7. *F. Chamissonis*
    Plants erect weedy annuals .... 8. *F. acanthicarpa*

1. **F. eriocéntra** Gray. [*Gaertneria e.* Kuntze.] Pallid aromatic multibranched shrub or subshrub 3–12 dm. high; branchlets white-barked, slender, spinescent; herbage finely gray-tomentulose, the lvs. becoming glabrate on upper surface; lvs. lance-oblong, acute, subsessile, often fascicled, sparingly incised-dentate to sinuately lobed or pinnatifid, 2–4 cm. long; ♂ heads ± glomerate, the invol. shallow, glandular, tomentulose, with 5–7 deltoid teeth; ♀ heads 1-fld., sessile; fr. fusiform, glandular and densely villous, the hairs tufted toward the ends of the 12–20 flattened-subulate straight spines.—Dry sandy or gravelly washes, sometimes on rocky slopes, 2500–5000 ft.; Creosote Bush Scrub, Joshua Tree Wd., Pinyon-Juniper Wd.; mts. of e. Mojave Desert, San Bernardino Co.; to sw. Utah, Ariz. March–May.

2. **F. dumòsa** Gray. [*F. albicaulis* Torr. *F. d.* var. *a.* Gray. *Gaertneria d.* Kuntze.] BURRO-WEED. BUR-SAGE. Low intricately branched rounded shrub 2–6 dm. high; stems white, becoming spinescent; herbage densely cinereous-strigose; lvs. mostly fascicled, 5–20 mm. long, ovate in outline, 1–3-pinnatifid with very short and obtuse lobes; ♂ heads spicate-racemose, rather few, not glomerate, with ♀ heads often scattered between, ♂ invol. very shallow, often becoming explanate, moderately cinereous and glandular; fr. obovoid, 2-beaked, glandular-puberulent or glabrate, the 20–35 lance-subulate flattened channeled scattered spines 2–2.5 mm. long, not hooked.—Very abundant on well drained soils throughout, mostly up to 3500 ft.; Creosote Bush Scrub, Joshua Tree Wd.; Mojave and Colo. deserts, Inyo Co. s.; to Utah, Ariz., Son., L. Calif. Feb.–June; Sept.–Nov.

3. **F. chenopòdiifòlia** Benth. [*Gaertneria c.* Abrams.] Rounded shrub 3–10 dm. high, the many stems white-tomentose when young, the herbage gray-yellow-green; lvs. greenish on upper surface, tomentose beneath, deltoid-ovate, petiolate, the blade 1–6 cm. long, dentate and often obscurely 3–5-lobed; ♂ heads spicate, rather few, the invol. rotate, shallowly 7–10-lobed, tomentulose; fr. globose, 2–3-beaked, lanate between the 20–30 glandular spines, the spines subulate, flattened and sulcate below, hook-tipped, 2–3 mm. long.—Dry sunny hillsides, up to 500 ft.; Coastal Sage Scrub; sw. San Diego Co.; L. Calif. April–June.

4. **F. ambrosioìdes** Cav. [*Xanthidium a.* Delpino. *Gaertneria a.* Kuntze.] Spreading shrubby perennial 6–15 dm. high; stems hispid-hirsute and glandular; lvs. gray-green, hirsutulous and scabrous, simple, lanceolate, acuminate, ± truncate at base, 4–12 cm. long, 1.5–3 cm. wide, ascending, on petioles 1–2 cm. long, irregularly dentate; ♂ heads few, 6–8 mm. wide; fr. 2–4-fld., 10–15 mm. long, the 70–90 slender subulate hook-tipped spines glandular-puberulent, 2–4 mm. long.—Hillsides and waste places, below 1000 ft.; Coastal Sage Scrub; near San Diego; s. Ariz., n. Mex. March–June.

5. **F. ilicifòlia** Gray. [*Gaertneria i.* Kuntze.] Hemispherical much-branched shrub 5–10 dm. high; stems very leafy, white-barked, hispid-hirsute and glandular; lvs. elliptic to oval, clasping at base, 4–8 cm. long, 2–6 cm. wide, rigidly coriaceous, shiny, scabrous, coarsely and unevenly spinose-dentate, strongly reticulate-veiny on under surface; ♂ heads peduncled, large, the shallow glandular-pubescent invol. with 10–12 lanceolate spine-tipped lobes; fr. appearing globose or ovoid, 2-beaked, very glandular-pubescent, the spines 40–70, shallowly grooved but subterete, prominently hooked.—Rare, forming mats in sandy washes and rocky canyon sides, up to 1000 ft.; Creosote Bush Scrub; Colo. Desert e. of Salton Sea, Riverside and Imperial cos.; w. Ariz., n. L. Calif. Feb.–April.

6. **F. confertiflòra** (DC.) Rydb. [*Ambrosia c.* DC. *F. tenuifolia* Gray. *F. strigulosa* Rydb. *Gaertneria t.* Kuntze.] Herbaceous perennial, with 1–several erect stems 3–12 dm. high from slender rootstocks; herbage gray-green, strigose-hispidulous; lvs. bipinnately parted into oblong-linear acute segments, the terminal one often elongate, short-petioled, 4–10 cm. long, lance-oblong to obovate in outline; ♂ heads narrowly racemose, campanulate, 2–2.5 mm. wide; fr. obovoid, glomerate, 1–2-beaked, glandular-puberulent, 2.5–4 mm. long, reticulate-ridged between the 10–20 broad-based hook-tipped spines ca. 0.5 mm. long.—Dry plains and waste land, up to 1000 ft.; Coastal Sage Scrub, S. Oak Wd.; cismontane s. Calif.; San Francisco (introd.); to Kans., Tex., Mex. May–Nov.

7. **F. Chamissònis** Less. [*Gaertneria C.* Kuntze. *Ambrosia C.* Greene. *F. cuneifolia* Nutt.] Perennial herb with radiating procumbent branching stems, forming loose mats 1–3 m. across and 1.5–3 dm. high; herbage silvery-canescent with silky hairs, the stems more hirsute; lvs. simple, the blade ovate, rhombic, or oval-oblanceolate, usually obtuse,

crenate-serrate to bluntly toothed or lobed or even incised, 2–5 cm. long, tapering to a petiole nearly as long; ♂ heads in congested terminal spikes, 7–8 mm. wide; fr. 1-fld., sparsely hirsute and glandular-atomiferous, 8–10 mm. long, the spines 15–25, dilated at base, boat-shaped; $n = 16$ (Wiggins & Stockwell, 1937).—Coastal beaches and dunes, sea level to 50 ft.; Coastal Strand; San Clemente, Santa Cruz and San Miguel ids., Monterey Co. to Del Norte Co.; to B.C. July–Nov.

Ssp. **bipinnatisécta** (Less.) Wiggins & Stockw. [*F. C.* var. *b.* Less. *F. bipinnatifida* Nutt. *F. Lessingii* Meyen & Walp. *Gaertneria bipinnatifida* Kuntze. *Ambrosia bipinnatifida* Greene. *F. bipinnatifida* var. *dubia* Eastw. *F. villosa* Rydb.] Lvs. once to thrice pinnatifid into oblong or obovate segments, less silvery; fr. glandular but not hairy, the spines often more slender and less sulcate; $n = 16$ (Wiggins & Stockwell, 1937).—Continuous along the coast from B.C. to L. Calif., Channel Ids.; Chile. March–Sept. The 2 sspp. hybridize and recombine their characters freely, with no loss of fertility, yet the 2 main types persist side by side.

8. **F. acanthicárpa** (Hook.) Cov. [*Ambrosia a.* Hook. *F. Hookeriana* Nutt. *F. montana* Nutt. *Gaertneria a.* Britt. *G. H.* Kuntze. *F. californica* Gand. *F. Palmeri* Rydb.] SAND-BUR. Rather strict or corymbosely branching annual 1–7 dm. high, rather densely appressed-canescent and sparingly hispid; lvs. bipinnatifid, the blade ovate in outline, with oblong obtusish divisions, 2–6 cm. long, petiolate; ♂ heads very numerous, 2–4 mm. wide; fr. 1-fld., glabrous or sparsely villous, 5–10 mm. long, the spines 9–18, much flattened, shallowly sulcate.—Common weed of sandy plains, river bottoms, etc.; below 2500 (6500) ft.; V. Grassland, Foothill Wd., Coastal Sage Scrub, etc.; San Benito and Fresno cos. to San Diego Co., w. Mojave Desert n. to Mono Co.; to Wash., Sask., Tex.; adv. eastward. Aug.–Nov.

27. Xánthium L. COCKLEBUR

Coarse weedy annuals with stout branching stems. Lvs. alternate, usually lobed or toothed, petioled. Staminate heads borne above the ♀ in terminal and axillary clusters, many-fld.; invols. subglobose, the phyllaries free, in 1–3 series; filaments monadelphous; anthers free, very slender. Invol. of ♀ heads enclosing the 2 florets and becoming an indurated bur, 2-beaked, covered with rigid hooked prickles; corolla and pappus none; style-branches exserted through the beaks. Possibly only 3 spp., of which the 2 here treated are cosmopolitan weeds of doubtful origin. (Greek, *xanthos*, yellow, from the ancient name of some plant, the fr. of which was used to dye the hair that color.)

(Millspaugh & Sherff. Revision of the N. Am. Spp. of Xanthium. Field Mus. Pub., Bot. Ser. 4: 9–49, 1919. Widder, F. J. Die Arten der Gattung Xanthium, *in* Fedde, Repert. Beih. 20: 1–221, 1923.)

Lvs. lanceolate, tapering at both ends, short-petiolate, with conspicuous 3-forked spines in their axils
  1. *X. spinosum*
Lvs. broadly ovate or cordate, long-petiolate, without spines . . . . 2. *X. strumarium*

1. **X. spinòsum** L. SPINY CLOTBUR. SPANISH-THISTLE. Much-branched annual 3–10 dm. high, to 15 or more dm. wide; lvs. lanceolate, entire or few-toothed or -lobed, 3–8 cm. long, 6–26 mm. wide, glabrate or strigillose above, silvery-tomentulose beneath, armed with basally 3-forked stout yellow spines 1–2.5 cm. long in their axils; ♂ fls. rusty-pubescent; burs numerous but not very crowded, cylindric, ca. 1 cm. long, puberulent, the 2 beaks inconspicuous and unarmed, the body puberulent and beset with numerous uncrowded slender sharply hooked prickles; $n = 18$ (Heiser & Whitaker, 1948).—Common weed of old pastures and waste places at low elevs., cismontane Calif.; e. to the Atlantic and now cosmopolitan in warm and temp. regions. July–Oct.

2. **X. strumàrium** L. var. **canadénse** (Mill.) T. & G. [*X. c.* Mill. *X. italicum* Moretti. *X. campestre, californicum, acutum* and *palustre* Greene. *X. canadense* var. *p.* Jeps.] Coarse branching annual 5–15 dm. high; lf.-blades 3–12 cm. long and equally wide on slender petioles as long, deltoid-ovate, ± cordate at base, dentate or serrate and somewhat lobed, thickish, scabridulous; burs narrowly to broadly ovoid, greenish- or yellowish-brown, mostly 2–3.5 cm. long, the ± crowded often very stout prickles hispid with stout gland-tipped hairs in lower half, the beaks short and very stout, with short incurved tips.—The common form of this cosmopolitan weed in Calif., often abundant in moist v. floors throughout. July–Oct.

Var. **glabràtum** (DC.) Cronq. [*X. macrocarpum* var. *g.* DC. *X. chinense* Mill. *X.*

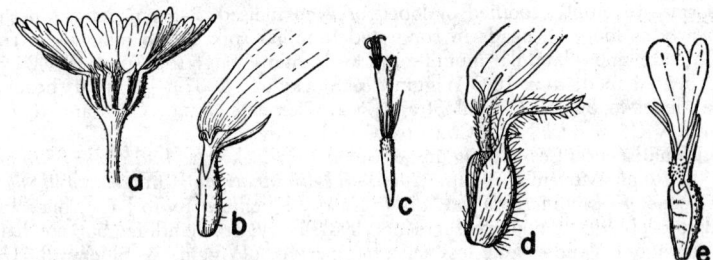

Fig. 111. MADIINAE. *Layia platyglossa: a,* head, × 1, the phyllaries wholly enveloping the ray-aks.; *b,* ray-fl., × 2½, the ak. obcompressed; *c,* disk-fl., × 2½, pappus of bristles. *Madia elegans: d,* ray-fl., × 3, the ak. laterally compressed, quite enfolded by the phyllary. *Hemizonia fasciculata: e,* ray-fl., × 5, ak. half enclosed by the phyllary.

pennsylvanicum Wallr. X. *calvum* Millsp. & Sherff.] Burs mostly less than 2 cm. long, the prickles merely glandular-puberulent to glabrate, not hispid; $n = 18$ (Heiser & Whitaker, 1948).—Rare; low fields, San Francisco Bay region and occasionally else-where; more abundant in e. U.S.

This pantropic weed is an anthropophyte, first described from Eu., but probably of Am. origin. As an apparent facultative apomict it has developed innumerable local races based largely on variations in the bur. Innumerable intergrades prevent satisfactory taxonomic separation of these, although various authors have tried it by proposing dozens of inconsequential "species." Typical *strumarium,* with a straight-beaked, greenish, puberulent, small bur about 1.5 cm. long, is a rare waif in Calif. (Cameron L., Colo. Desert) but widespread in the Am. tropics and in Eu.

Tribe **Heliantheae.** Subtribe **Madiinae** (Fig. 111)

Mostly annuals with glandular, viscid or heavy-scented herbage. Lvs. alternate or opposite. Phyllaries mostly in a single series, each partly or completely enclosing its ray-ak. (except in *Layia discoidea*). Bracts of the receptacle commonly in a single series between ray- and disk-fls. and often united into a cup, or sometimes scattered among the disk-fls. Rays usually present, showy or not. Ray-aks. fertile, mostly without pappus; disk-aks. fertile or sterile, their pappus paleaceous, awnlike or none.

A. Ray-aks. obcompressed, each hidden within the obcompressed phyllary, the abruptly infolded
    lateral margins of which ± overlap.
    B. Disk-fls. 6 or more.
        C. Aks. 10-costate and tuberculate-scabrous; pappus oblong, obtuse . . 28. *Achyrachaena*
        CC. Aks. not costate or scabrous; pappus, when present, at least attenuate to apex.
            D. Disk-aks. sterile, undeveloped; heads closing during midday.
                E. Pappus none; annuals . . . . 29. *Lagophylla*
                EE. Pappus present; perennial . . . . 30. *Holozonia*
            DD. Disk-aks. fertile; heads open continuously . . . . 31. *Layia*
    BB. Disk-fl. 1 . . . . 32. *Madia*
AA. Ray-aks. not obcompressed or each hidden by the enveloping lateral margins of the subtending
    phyllary.
    B. Style of disk-fls. glabrous below the subulate branches.
        C. Disk-aks. not prominently ribbed; ray-aks. usually glabrous and epappose.
            D. Ray-ligules 3-toothed or 3-lobed, the lobes subparallel; herbage without tack-
                shaped glands; lvs. not narrowly linear or grasslike.
                E. Ray-aks. usually laterally compressed and finely longitudinally striated, each
                    completely enfolded by the deeply sulcate subtending phyllary; basal lvs.
                    subentire . . . . 32. *Madia*
                EE. Ray-aks. broader, not laterally compressed or striated, each only partially
                    enfolded by the subtending phyllary; basal lvs. pinnatifid or toothed, rarely
                    subentire.
                    F. Upper lvs. and phyllaries not terminated by open pit glands
                            33. *Hemizonia*
                    FF. Upper lvs. and phyllaries terminated by open pit glands; ligules yellow;
                      each disk-fl. subtended by a bract; pappus none . . . . 34. *Holocarpha*
        DD. Ray-ligules 3-cleft or parted into palmately spreading lobes; tack-shaped glands
        present (excepting *C. tenella*); lvs. narrowly linear and grasslike
                             35. *Calycadenia*

CC. Disk-aks. prominently ribbed; ray-aks. pubescent and pappose; pappus plumose
   36. *Blepharizonia*
BB. Style of disk-fls. hairy below the 2-cleft tip; pappus plumose .... 37. *Blepharipappus*

28. Achyrachaèna Schauer

Vernal mesophytic annuals. Lvs. opposite and clasping below, alternate above, linear. Heads solitary, long-peduncled, terminating the stem and few ascending branches, heterogamous, radiate, all fls. fertile. Invol. oblong-campanulate, nearly equaling the inconspicuous fls. Ray-fls. 1-seriate; ligules yellow, turning crimson-red. Receptacular bracts in a single row between ray and disk, free. Aks. 10-costate, tuberculate-scabrous; pappus of silvery scales, spreading on ripe aks. to form a globose head. One sp. (Greek, *achuron*, chaff, and Latin, *achaenium*, an ak., in ref. to the very chaffy pappus.)

1. **A. móllis** Schauer. [*Lepidostephanus madioides* Bartl.] BLOW-WIVES. Stem simple or few-branched, 1–4 dm. high, the herbage villous, becoming moderately glandular above; lvs. entire or remotely serrulate, up to 13 cm. long and 7 mm. wide, mostly much less; heads in fl. 1.5–2 cm. high, in fr. forming a globose cluster 3 cm. across; phyllaries completely investing the ray-aks.; ray-fls. ca. 8, the ligules inconspicuous; disk-fls. ca. 15–35; aks. clavate, black, ca. 5 mm. long, those of the ray epappose and smooth-ribbed, those of the disk pappose and scabrous with brown teeth; pappus of 10 oblong blunt scales, the 5 outer half as long as and alternate with the inner, the inner 7–9 mm. long; self-fertile; $n = 8$ (Johansen, 1933).—Frequent in moist grassy fields with heavy soils, below 1000 ft.; V. Grassland, Foothill Wd., S. Oak Wd.; cismontane valleys from Douglas Co., Ore. to n. L. Calif. April–May.

29. Lagophýlla Nutt.

Mesophytic or xerophytic annuals. Basal lvs. serrate-dentate to subentire; cauline lvs. entire, readily caducous. Invol. turbinate to hemispheric, the phyllaries completely enfolding the obcompressed ray-aks. and caducous with them. Fls. yellow, the heads opening toward evening and closing in the morning. Receptacle penicillate-pubescent centrally, its bracts in a single row between ray and disk, slightly united. Ray-fls. 5, fertile. Disk-fls. 6, sterile. Pappus none. Five spp. of w. U.S. (Greek, *lagos*, a hare, *phyllon*, lf., in allusion to the copious sericeous pubescence of the upper lvs. of the original sp.)

A. Ligules 8–13 mm. long, conspicuous, bright yellow.
   B. Stem dichotomously branched; invol. hemispheric. Chiefly Coast Range.
      C. Ray-ak. broadly oblanceolate, glossy, smooth, midnerve evident; phyllaries pilose-ciliate; low slender flexuous habit; vernal .... 1. *L. minor*
      CC. Ray-ak. obovate, dull, striate, midnerve obscure; phyllaries barbellate-ciliate; taller and stricter habit; late vernal and early aestival .... 2. *L. dichotoma*
   BB. Stem simple or paniculately branched above; invol. turbinate; late vernal to autumnal. Sierra Nevada .... 3. *L. glandulosa*
AA. Ligules 3–5.5 mm. long, inconspicuous, pale lemon-yellow; vernal to autumnal.
   B. Heads scattered, or at least not densely glomerate .... 4. *L. ramosissima*
   BB. Heads densely congested in compact glomerules .... 5. *L. congesta*

1. **L. mìnor** (Keck) Keck. [*L. dichotoma* ssp. *m.* Keck.] Slender herb 1–3 dm. high, the stem repeatedly forking, at least in upper half, to form an open corymbose crown, the ultimate twigs very dark, strigose; lvs. linear, some tapering to base, up to 5.5 cm. long and 3 mm. wide, entire except for the basal, narrowly involute, soon deciduous, strigulose-puberulent, becoming hirsute-ciliate toward the heads; infl. bearing few or no stalked glands, if present, not extending down the peduncles; invol. 4–5 mm. high, pilose, the phyllaries long-ciliate on the angles with hairs almost as long as the bract is wide, acuminate; ray-aks. smooth and glossy black, 0.8–1.3 mm. wide; self-sterile; $n = 7$ (Keck).—Occasional (or locally abundant) on gravelly clay or serpentine foothill slopes; Foothill Wd.; inner N. Coast Range from Glenn Co. to Napa Co.; Eldorado Co. April–May.

2. **L. dichótoma** Benth. Stouter, 1.5–6 dm. high; infl. bearing few to many evident stalked glands, these extending down the peduncles; invol. 4.5–5.8 mm. high, hispid-hirsute, the hairs much shorter than the width of the acute bracts; ray-aks. blackish or lightly mottled, often with cellular surface, with 20–30 striae on each face, 1.4–1.9 mm.

wide; self-sterile; $n = 7$ (Johansen, 1933).—Colonies rare, up to 2000 ft.; V. Grassland, Foothill Wd.; e. side of Cent. V. from Butte Co. to Tulare Co.; inner S. Coast Range from San Benito Co. to Monterey Co. April–May.

3. **L. glandulòsa** Gray. Stem 1–10 or 15 dm. high, simple or paniculately branched, the plant slender or bushy with uneven crown; herbage densely appressed hirtellous and distinctly glandular; lower lvs. linear-oblanceolate to spatulate, lost before flowering, 3–12 cm. long, 5–12 mm. wide, the cauline bearing fascicles or short sterile shoots in their axils, readily caducous, reduced to bracts upwards; heads on short peduncles; invol. turbinate, 5–7 mm. high; ligules 8–13 mm. long, 6–9 mm. wide, obcordate, the lateral lobes rounded, much wider than the cent. one; ray-aks. clavate-obovate, smooth, 2.7–3.9 mm. long; $n = 7$ (Johansen, 1933).—Common in the Sierran foothills, up to 2000 ft.; V. Grassland, Foothill Wd., Chaparral; Shasta Co. to Kern Co., occasional in the upper Sacramento V.; e. Mendocino Co. July–Oct.

Ssp. **serràta** (Greene) Keck. [*L. s.* Greene.] Inconspicuous herb 1–6 dm. high, simple or with few ascending branches, almost glandless below the crown; lvs. without fascicles or sterile shoots in their axils, the lower often persistent after time of flowering; heads few, on short or elongated very slender peduncles; invol. 4.75–6 mm. high; ray-aks. 2.5–3.2 mm. long; $n = 7$ (Keck).—A seasonal ecotype growing with the sp. May–July.

4. **L. ramosíssima** Nutt. [*L. minima* Kell.] Stiffly erect, 2–10(–15) dm. high, racemosely or paniculately branched above, sometimes from the base, rarely simple; herbage grayish or dull green, densely white-hirtellous or white-sericeous, the prominent yellow stipitate glands confined to upper lvs. and heads; lvs. as in *L. glandulosa;* heads short-peduncled or subsessile, racemosely disposed along the branchlets and in small clusters at their ends; invol. 4.4–6.7 mm. high, the lanceolate phyllaries densely villous-ciliate on the angles; ray-aks. clavate and slightly arcuate, 2.5–4 mm. long; self-fertile; $n = 7$ (Johansen, 1933).—Abundant in open places, often with hard dry soils, from sea level up to 5200 ft.; many Plant Communities; throughout cismontane Calif., mostly away from the immediate coast, San Diego Co. to Del Norte and Modoc cos.; to Ore., e. Wash., Ida., Nev. May–Oct.

5. **L. congésta** Greene. [*L. ramosissima* var. *c.* Jeps.] Similar to *L. ramosissima* except more robust, the heads densely congested in compact glomerules 1.5–6 cm. thick, the glomerules simply terminal or in interrupted spikes or, more rarely, on short lateral branches racemosely arranged, the stems usually simple; invol. 5–7.5 mm. high; self-fertile; $n = 7$ (Keck).—With *L. ramosissima,* from Humboldt Co. to Santa Cruz Co. and from Eldorado Co. to Stanislaus Co. May–Sept. Maintained as a sp. because it is cohabitant without intergradation with, and occurs almost to the exclusion of, *L. ramosissima* over wide areas, as in Lake Co.

30. Holozònia Greene

Perennial herb, hemicryptophyte, with elongated fleshy rootstocks. Stems erect, usually simple and densely leafy below, diffusely branched at the nearly leafless apex, frequently with short leafy spurs or peduncles in the lf.-axils. Lvs. opposite and connate below, alternate above, ligulate, reduced to small ovate bracts on peduncles. Heads scattered, solitary on long slender peduncles. Invol. broadly turbinate. Ray-fls. 1-seriate, fertile; ligules white, purplish dorsally. Receptacular bracts united into a cup surrounding the sterile disk-fls. Ray-aks. clavate, finely striate, the cupulate areola entire- or apiculate-margined. Disk-aks. with or without a fragile pappus of 1–5 caducous palaceous bristles nearly as long as the corolla. One sp. (Greek, *holos,* whole, and *zone,* girdle, from the wholly enclosed ray-aks.)

1. **H. fílipes** (H. & A.) Greene. [*Hemizonia f.* H. & A. *Lagophylla f.* Gray.] Stems 3–10 dm. high, stramineous, the mature herbage cinereous and villous, beset with stalked glands towards the infl.; lower lvs. remotely serrate to subentire, 3–10 cm. long, 4–8 mm. wide, progressively smaller upward; invol. 3.5–5 mm. high, as broad, the linear-lanceolate (as folded) phyllaries prominently hirsute; ray-fls. 4–8, the ligule 3.5–4.5 mm. long, deeply 3-cleft into linear lobes; disk-fls. 10–20(–28); self-sterile; $n = 14$ (Johansen, 1933).—Occasional, in dry alkaline clays of gulches, or among rocks in beds of intermittent streams, mostly below 2000 ft.; V. Grassland, Foothill Wd.; Shasta Co. to Mariposa Co.; Napa Co. to Marin and Santa Clara cos.; sw. Monterey Co. June–Oct.

31. Làyia Hook. & Arn.

Vernal annuals with chiefly alternate subentire or toothed to pinnatifid narrow lvs. Heads many-fld., on usually naked terminal peduncles, heterogamous and radiate (except in *L. discoidea*), both ray- and disk-fls. fertile. Invol. campanulate to broadly hemispheric, the thin dilated lower margins of the phyllaries abruptly infolded and enclosing the ray-aks., the tip flat (except in *L. discoidea*). Receptacle broad, chaffy marginally or throughout. Ray-fls. 8–24; ligules 3-lobed or -toothed, white, yellow, or yellow with white tip. Disk-fls. numerous, yellow; anthers black or yellow. Ray-aks. obcompressed, commonly smooth and glabrous, with prominent terminal areola, epappose. Disk-aks. pubescent and pappose, rarely glabrous and epappose; pappus of numerous bristles, awns or paleae, the bristles often plumose below. All spp. occur in Calif., only two extend beyond the state; in massed stands, among our showiest wild-flowers. (G. Tradescant *Lay*, botanist to Capt. Beechey's voyage in the "Blossom," which visited Calif. in 1827.)

A. Bracts of receptacle subtending each disk-fl.; invol. (and herbage) not glandular, the phyllaries with prominent pustulate processes at base of hairs.
    B. Pappus of rigid subulate awns; disk-aks. obviously obcompressed, cuneate; phyllaries hispid-ciliate or pectinate along the folded edge, otherwise nearly smooth. Coast Ranges
                                    1. *L. chrysanthemoides*
    BB. Pappus of ovate-lanceolate thin paleae; disk-aks. plump, clavate; phyllaries uniformly papillate-hispid dorsally. Cent. V. . . . . 2. *L. Fremontii*
AA. Bracts of receptacle limited to a ring between ray- and disk-fls.; invol. (and herbage) glandular, the phyllaries pubescent but not as above.
    B. Pappus paleaceous, glabrous except for capillary hairs radiating from very base.
        C. Ray-fls. in 1 series, their ligules obovate to flabelliform, their aks. dull, glabrous or pubescent, flattened at least ventrally; invol. hemispheric, green, the phyllaries not inflated, with prominent tips; pappus 2–3.5 mm. long.
            D. Ligules white; anthers yellow; ray-aks. sericeous; lvs. mostly entire
                                      3. *L. leucopappa*
            DD. Ligules yellow with long white tips; anthers black; ray-aks. glabrous, or lightly pubescent toward apex; lvs. mostly pinnately lobed . . . . 4. *L. Munzii*
        CC. Ray-fls. usually in 2 series, their ligules oblong-spatulate, their aks. polished, glabrous, plump; invol. broadly urceolate, dark-dotted, the phyllaries inflated, with inconspicuous tips; pappus 1–2 mm. long . . . . 5. *L. Jonesii*
    BB. Pappus setaceous, or if paleaceous, plumose.
        C. Ligule yellow with white tip (rarely yellow throughout); pappus commonly merely scabrid, if plumose, also with inner wool, of 18–30 bristles; anthers black; tips of phyllaries longer than basal portion . . . . 6. *L. platyglossa*
        CC. Ligule yellow or white (none in 8) but not as above (except in 11); pappus plumose, with inner woolly hairs only in 7 and 9; tips of phyllaries shorter than basal portion (except in 15).
            D. Pappus persistent; stem not prominently fistulous, pubescent below (glabrate in 14); disk-aks. strigose.
                E. Ligules more than 5 mm. long, conspicuous (none in 8); self-sterile.
                    F. Anthers yellow; invol. not urceolate; stems not dark-dotted; pappus white.
                      G. Invol. ovoid, 7–12 mm. high; ligules golden yellow; disk-corollas 5.5–8 mm. long; ray-aks. 3.8–5.2 mm. long; pappus 3.5–6.2 mm. long, the slender bristles 16–21, densely plumose, with inner wool. Lake and Colusa cos. . . . . 7. *L. septentrionalis*
                      GG. Invol. turbinate or campanulate to hemispheric, 5–9.5 mm. high; disk-corollas 3–6 mm. long.
                          H. Heads discoid; pappus paleaceous, 1–1.5 mm. long, fulvous. San Benito Co. . . . . 8. *L. discoidea*
                          HH. Heads radiate; pappus setaceous, 2–5 mm. long, white.
                            I. Stems hispid; basal lvs. dentate or lobed, the cauline entire; ray-aks. 3–4 mm. long; pappus 3–5 mm. long, the bristles 10(–12), flattened, linear-attenuate, densely plumose, usually with inner wool . . . . . 9. *L. glandulosa*
                            II. Stems pilose or hirsute; basal lvs. pinnatifid to bipinnatifid; ray-aks. 2.5–4 mm. long; pappus 2–4 mm. long, the bristles 10–18 (rarely less), filiform, moderately plumose, without wool . . . . 10. *L. pentachaeta*
                  FF. Anthers black; invol. urceolate; stems dark-dotted; pappus rufous
                                    11. *L. gaillardioides*
            EE. Ligules 2–4 mm. long, inconspicuous; self-fertile; pappus rufous to whitish; anthers black.
                F. Ligules yellow; pappus bristles 11–15; stem rigidly erect, 20–100 cm. tall, dark-dotted; pungently odorous; ray-fls. 8–16.
                  G. Lower lvs. laciniate-dentate, broadly oblanceolate or oblong; stems mostly stout. San Francisco Bay to Santa Cruz
                                    12. *L. hieracioides*

GG. Lower lvs. pinnatifid, if merely dentate narrowly linear, narrower;
    stems mostly slender. Mt. Diablo to Santa Barbara
                                                            13. *L. paniculata*
  FF. Ligules white; pappus bristles 25–32; stem freely branched, to 15 cm.
      tall, without dots; not odorous; ray-fls. 5–7(–9). Coastal dunes
                                                            14. *L. carnosa*
DD. Pappus deciduous; stem prominently fistulous, glabrate below; disk-aks. sericeous;
    ligules creamy white to pale yellow .... 15. *L. heterotricha*

1. **L. chrysanthemoìdes** (DC.) Gray. [*Oxyura c.* DC. *Blepharipappus c.* Greene.
Calliglossa Douglasii H. & A. *Callichroa D. T.* & G. *Layia Calliglossa* Gray. *L. C.* var.
oligochaeta Gray. *B. D.* Greene and var. *o.* Greene.] Stem erect, corymbosely branched,
1–4 dm. high; lvs. scabro-ciliate, otherwise smooth, the basal pinnately parted into
linear or oblong blunt lobes; invol. 6–11 mm. high; ligules yellow with white lobes,
7–15 mm. long, 5–9 mm. wide; disk-fls. 30–100, the corolla 3–5 mm. long; ray-aks.
2.7–3.9 mm. long, strongly compressed, glabrous; disk-aks. 2.5–4 mm. long, densely
strigose, sometimes lacking pubescence and pappus; pappus of 9–17 tawny rigid subulate
scabrous awns, the lateral ones 1.5–3.5 mm. long, the intervening ones usually shorter
or even reduced to rudiments; self-sterile; *n* = 7 (Johansen, 1933).—Frequent in
vernally moist ground, mostly below 1000 ft., Foothill Wd., V. Grassland; Coast Range
valleys, Mendocino and Glenn cos. to Monterey Co. March–May.

Ssp. **marítima** Keck. Stem prostrate, even the cent. peduncle horizontal; lvs. succulent,
the lower with broad rounded lobes; tips of phyllaries dilated, rounded; late-flowering;
n = 7 (Clausen et al., 1936).—Rare, on the immediate coast; N. Coastal Scrub;
Mendocino and Sonoma cos. May.

2. **L. Fremóntii** (T. & G.) Gray. [*Calliachyris F.* T. & G. *Blepharipappus F.* Greene.]
Stem 1–4 dm. high; lvs. scabrous or short-hispid, not at all viscid, the basal pinnately
parted; invol. 6–11 mm. high; ligules yellow, the outer half white, 9–18 mm. long,
7–12 mm. wide; disk-fls. 40–100, the corolla 3.6–5.3 mm. long; ray-aks. 2.5–3.7 mm.
long, strongly compressed, glabrous; disk-aks. 2–3.5 mm. long, densely strigose and
with a row of long capillary hairs on the areola; pappus of 10 tawny white ovate-
lanceolate paleae with attenuate tips, 2–4.5 mm. long; self-sterile; *n* = 7 (Clausen et al.,
1934).—Frequent in open fields; V. Grassland, Foothill Wd.; Cent. V. and Sierran
foothills, Tehama Co. to Tulare Co. March–May.

3. **L. leucopáppa** Keck. Stem branched from base upward, to 5 dm. high; herbage
± glaucescent; lvs. hispidulous-ciliate, villous on upper surface, viscidulous and with
scattered black glands, the basal pinnately toothed or lobed, the cauline mostly entire;
invol. 6–9 mm. high, the phyllaries papillate, short-hirsute and black-glandular on the
back, the infolded margin lanuginous-ciliate; ligules white, 8–12 mm. long, 8–14 mm.
wide; disk-fls. 50–100, the corolla 3.7–5.2 mm. long; ray-aks. 2.5–2.9 mm. long; disk-aks.
2.5–4 mm. long, densely white-sericeous and with a row of long capillary hairs on the
areola; pappus of 10(–13) bright white lanceolate acuminate paleae 2–3.5 mm. long;
self-sterile; *n* = 7 (Clausen et al., 1941).—Colonial on moist benches; V. Grassland;
limited area along Tejon Hills, head of San Joaquin V., s. of Bakersfield. March–April.

4. **L. Múnzii** Keck. Stem branched from base upward, 2–3.5 dm. high; lvs. mostly
glabrous except for scabrous or ciliate margin, the upper sparsely villous and viscid, the
basal and lower cauline pinnately lobed to parted; invol. 7–8.5 mm. high, the bracts
glabrate to sparsely hispid; ligules yellow tipped with white for ⅓–⅔ their length,
9–12 mm. long, 6–9 mm. wide; disk-fls. 50–100, the corolla 3.6–5 mm. long; ray-aks.
2.8–3.5 mm. long; disk-aks. 2.5–4 mm. long, densely strigose and with a row of long
capillary hairs on the areola; pappus of 9–12 dirty white linear-lanceolate attenuate
paleae, usually with a spot of anthocyanin at base, 2.3–3.4 mm. long; self-sterile; *n* = 7
(Clausen et al., 1941).—Widely scattered colonies on alkaline flats of heavy clay,
500–2000 ft.; V. Grassland, Alkali Sink; west side of San Joaquin V. and adjacent inner
Coast Range; Merced Co. to Kern and San Luis Obispo cos. March–April.

5. **L. Jònesii** Gray. [*Blepharipappus J.* Greene.] Stem usually corymbosely branched,
up to 3.5 dm. high; lvs. glandular-puberulent and moderately short-hispid, ± pustulate,
the basal and lower cauline pinnately parted; invol. 6–8 mm. high; ligules yellow, with
short white tips, 6–9 mm. long, 4–5.5 mm. wide; disk-fls. 35–75(–100), the corolla
3.1–5.2 mm. long; ray-aks. 2.3–3.3 mm. long; disk-aks. 2.4–4 mm. long, moderately
strigose and with a row of capillary hairs on the areola; pappus of 10–14 dirty-white

ovate to oblong-ovate imbricate acuminate paleae 1–2 mm. long; self-sterile; $n = 7$ (Johansen, 1933).—Pastures and grassy slopes, sea level to 500 ft.; V. Grassland; locally frequent along coast near San Luis Obispo. March–May.

6. **L. platyglóssa** (F. & M.) Gray. [*Callichroa p.* F. & M. *Blepharipappus p.* Greene.] TIDY TIPS. Stem prostrate or decumbent, stout, succulent, corymbosely branched, 1–3 dm. long; lvs. short-hirsute or pilose, the basal and lower cauline dentate to pinnatifid with salient rotund short lobes; invol. 6–12 mm. high, the phyllary-tips rounded; ligules 6–15 mm. long, 5–10 mm. wide; disk-fls. 30–100, the corolla 4–6 mm. long; ray-aks. 3–3.8 mm. long; disk-aks. 2.8–5 mm. long, strigose; pappus of 18–32 white or tawny scabrid bristles; self-sterile; $n = 7$ (Johansen, 1933).—Grassy flats on the immediate coast, below 300 ft.; N. Coastal Scrub, V. Grassland; Mendocino Co. to Santa Cruz Id. May–June.

Ssp. **campéstris** Keck. [*Madaroglossa elegans* Nutt. *L. e.* T. & G. *Blepharipappus e.* Greene. *L. p.* var. *breviseta* Gray.] Stem erect, the side branches ascending to erect, slender, less succulent, to 5 dm. high, the herbage a grayer green; lf.-lobes narrower and longer; phyllary-tips not dilated; pappus as in the sp. or the bristles densely plumose and interlaced with woolly hairs within; $n = 7$ (Johansen, 1933).—Grassy slopes and openings, sea level to 3000 (4000) ft.; several Communities; mostly from Bay Region s. to cent. L. Calif., occasional n. to Mendocino and Butte cos.; abundant on the coastal plains, infrequent in the Cent. V. March–May.

7. **L. septentrionális** Keck. Stem often corymbosely branched, 1.5–4 dm. high; lvs. hirsute, glandular, the lower pinnately toothed to parted; invol. 7–12 mm. high, rather densely pilose and glandular; ligules golden-yellow, 8–16 mm. long, 7–9 mm. wide; disk-fls. 35–60, the corolla 5.5–8 mm. long; ray-aks. 3.8–5.2 mm. long; disk-aks. 4.5–7.5 mm. long; pappus of 16–21 glistening white bristles, not appreciably flattened, plumose to above middle with capillary hairs outside and tangled woolly hairs inside; $n = 8$ (Clausen et al., 1941).—Scattered colonies in fields and grassy slopes in light or serpentine soils; Foothill Wd.; inner N. Coast Ranges from Colusa Co. to Sonoma Co., Sutter Buttes. April–May.

8. **L. discoìdea** Keck. Stem simple or corymbosely few-branched, 5–12(–15) cm. high; lvs. white-hispid, all but those of the rosette moderately black-glandular, the basal pinnately short-lobed, the cauline few-toothed or mostly entire; invol. turbinate-campanulate, anthocyanous, 5–6 mm. high, moderately villous and rather densely beset with stalked black glands; fls. 10–25, the corolla 3.5–4 mm. long; aks. 3.5–4.5 mm. long; pappus of (8–)11–15 lanceolate acuminate entire or deeply lacerate (and then often truncate) fulvous paleae 1(–1.5) mm. long, plumose with short capillary hairs; $n = 8$ (Clausen et al., 1941).—Very rare, serpentine slopes, 3000–3500 ft.; Foothill Wd.; vicinity of New Idria, San Benito Co. May.

9. **L. glandulòsa** (Hook.) H. & A. [*Blepharipappus g.* Hook. *Layia Douglasii* H. & A. *Madaroglossa angustifolia* DC. *L. g.* var. *rosea* Gray. *L. hispida* Greene. *L. g.* var. *h.* Jeps. *B. h.* Greene.] Stem commonly corymbosely branched, 1–4(–6) dm. high, often anthocyanous; lvs. hispid, often densely strigose on upper surface, the basal dentate or lobed, the cauline mostly entire; invol. 6.5–9.5 mm. high, short-hispid and usually stipitate-glandular; ligules white, often fading rose-purple, 6–15 mm. long, 5–9(–12) mm. wide; disk-fls. 25–50(–100), the corolla 4–6 mm. long; ray-aks. 3–4 mm. long; disk-aks. 3.5–6 mm. long; pappus of 10(–12) glistening white flattened linear-attenuate paleae, plumose to above middle with capillary hairs outside and tangled woolly hairs inside; $n = 8$ (Clausen et al., 1934).—Common in sandy soil, sea level to 7800 ft.; several Communities; Great Basin deserts n. to Wash. and Ida., e. to Utah and New Mex., s. through Mojave Desert to L. Calif., in cismontane Calif. in Coast Ranges from Contra Costa Co. to San Diego. March–June.

Ssp. **lùtea** Keck. Ligules golden yellow; $n = 8$.—Sandy soil; V. Grassland, Chaparral, Foothill Wd.; inner S. Coast Range, San Benito Co. to Monterey Co., local in foothills of Santa Lucia Mts. and mts. of Ventura Co. April–May.

10. **L. pentachaèta** Gray. [*Blepharipappus p.* Greene. *L. p.* var. *Hanseni* Jeps.] Stem branching above, 2–8 dm. high; herbage with acrid odor; lvs. hirsute and finely glandular, pinnatifid to bipinnatifid; invol. broad, 6–9 mm. high, pustulate-hirsute and moderately glandular; ligules 5–18 mm. long, 4–9 mm. wide; disk-fls. 20–125, the corolla 3–5.5 mm. long; ray-aks. 2.5–4 mm. long; disk-aks. 2.5–4(–4.7) mm. long;

pappus of 10–18 (rarely 5 or less) filiform bristles, moderately plumose at base with rather short straight capillary hairs; $n = 8$ (Clausen et al., 1934).—Frequent on grassy slopes, up to 3200 ft.; Foothill Wd., Chaparral, V. Grassland; Sierran foothills from Placer Co. to Kern Co., plains of Tulare Co. in a robust form. March–May.

Ssp. **álbida** Keck. Ligules white; $n = 8$.—Frequent on plains and hills below 3000 ft.; Foothill Wd., V. Grassland; about the head of the San Joaquin V., from w. Fresno Co. to the Tehachapi foothills, Kern Co. Readily separable from *L. glandulosa*, with which it is often confused, by the pappus, odor, lf.-cut, etc. March–May.

11. **L. gaillardioìdes** (H. & A.) DC. [*Tridax?* g. H. & A. *Blepharipappus* g. Greene. *B. nemorosus* Greene. *L. n.* Jeps. *L. g.* var. *n.* Jeps.] Stem simple or freely branched, 2–6 dm. high; herbage pungently odorous; lvs. hirsute to hispid, pustulate, somewhat viscid or stipitate-glandular or both, remotely serrate or dentate to pinnatifid (or bipinnatifid); invol. 5–8.5 mm. high, the phyllaries ventricose, villous and stipitate-glandular; ligules golden-yellow throughout to pale yellow with whitish tip, 7–12 mm. long, 4–7 mm. wide; disk-fls. (20–)40–100, the corolla 3.5–5.2 mm. long; ray-aks. 3–4.4 mm. long; disk-aks. 3.2–5 mm. long; pappus of 17–21 slender reddish-brown to white bristles 1–4 mm. long, plumose to above middle with straight capillary hairs; $n = 8$ (Johansen, 1933).—Occasional, grassy places, below 3800 ft.; several Communities; both outer and inner Coast Ranges from Humboldt and Colusa cos. to n. Monterey Co. April–June. Especially variable in fl.-color.

12. **L. hieracioìdes** (DC.) H. & A. [*Madaroglossa h.* DC. *Blepharipappus h.* Greene.] Stem stout, rigidly erect, corymbosely branched toward apex, 2–8(–10) dm. high; herbage pungently odorous; lvs. hispid, usually pustulate, viscid, laciniate-dentate (or pinnatifid); heads terminating short leafy branches; invol. 5–8 mm. high, villous and stipitate-glandular; ligules 2–4 mm. long, 1.5–2.5 mm. wide; disk-fls. 15–75, the corolla 2.3–3.3 mm. long; ray-aks. 3–3.3 mm. long; disk-aks. 3–4 mm. long; pappus of 11–15 slender purplish to white bristles 2–4 mm. long, plumose to middle with straight capillary hairs; $n = 8$ (Johansen, 1933).—Occasional in grassy openings, up to 2800 ft.; Mixed Evergreen F., Foothill Wd.; Mt. Diablo, Mt. Hamilton, Berkeley Hills, Santa Cruz Mts. April–June.

13. **L. paniculàta** Keck. Similar to the preceding except stems more slender; lvs. narrower and more cut; disk-corolla 3–4.3 mm. long; ray-aks. 3.5–4 mm. long; disk-aks. 3.8–5 mm. long; $n = 16$ (Clausen et al., 1941).—Occasional among scattered shrubs in sandy or light loamy soils; Coastal Sage Scrub, Chaparral, Mixed Evergreen F.; both S. Coast Ranges from San Benito Co. and from Monterey Bay to Santa Barbara. April–May.

14. **L. carnòsa** (Nutt.) T. & G. [*Madaroglossa c.* Nutt. *Blepharipappus c.* Greene.] Stem freely branched, erect, the branches divergent to decumbent, up to 15 cm. high and to 4 dm. across, usually much less; lvs. scabro-ciliate, glabrous beneath, villous above, the uppermost sparingly black-glandular, the basal sinuate-pinnatifid, the cauline mostly entire; invol. 5–7.5 mm. high; ligules white (or fading pink), 2–2.4 mm. long, 1.5 mm. wide; disk-fls. 12–26, the corolla 2.5–3.5 mm. long; ray-aks. 3.7–4.5 mm. long; disk-aks. 3.8–5.5 mm. long; pappus of 25–32 slender bristles 2.5–3.7 mm. long, plumose for ⅔ their length with rather short straight capillary hairs; $n = 8$ (Johansen, 1933).—Widely scattered stations on coastal sand dunes; Coastal Strand; Humboldt Co. to near Point Conception; San Diego(?). April–June.

15. **L. heterotrícha** (DC.) H. & A. [*Madaroglossa h.* DC. *Layia h.* var. *major* Gray. *L. graveolens* Greene. *Blepharipappus h.* Greene. *B. g.* Greene. *B. glandulosus* var *h.* Jeps. *L. glandulosas* var. *h.* Hall, as to synonymy.] Stem stout, simple or with few sharply ascending branches, 2–9 dm. high; herbage pallid, ± succulent, heavily odorous; lvs. short-hispid, strigulose and glandular, crenate or short-dentate, the upper entire; invol. 8–12 mm. high, densely hispid and glandular; ligules creamy white to pale yellow, not tipped with white, often fading rose-purple, 6–15(–22) mm. long, 4.5–12(–15) mm. wide; disk-fls. 35–85, the corolla 4.5–7 mm. long; ray-aks. 4.2–5 mm. long; disk-aks. 4.3–6.3 mm. long; pappus of 14–19 very slender glistening white deciduous bristles 3.5–6 mm. long, plumose to above middle with straight capillary hairs nearly equaling the bristles; $n = 8$ (Clausen et al., 1941).—Scattered, from sea level to 5000 ft.; V. Grassland, Foothill Wd.; w. side of San Joaquin V. and adjacent inner Coast Range (also Santa Lucia Mts.) to Mt. Pinos and Tehachapi Pass. March–June.

32. Màdia Mol. TARWEED

Mesophytic or xerophytic herbs, usually very glandular and heavy-scented. Lvs. linear or oblong, the basal entire to remotely denticulate. Invol. ± angled by its deeply sulcate phyllaries which completely enclose the ray-aks. Receptacle bearing between ray and disk a single row of chaffy bracts, these usually ± united into an often persistent prismatic cup. Ray-fls. few to many, fertile. Disk-fls. fertile or sterile. Ray-aks. usually laterally compressed, with flat sides, narrow back, and sharp ventral angle (see also *M. minima*), longitudinally striate. Pappus usually none in ray-aks., sometimes present in disk-aks. 18 spp. of Pacific N. and S. Am. (*Madi*, the Chilean name of the original sp.)

A. Pappus present; heads not closing. (Section Anisocarpus)
  B. Tall perennials of dense woods; lvs. opposite well up stem, those of the rosette wider; pappus of soft lanceolate paleae.
    C. Disk-aks. fertile; ray-aks. without beak; invol. 10–12 mm. high; anthers black
                                                1. *M. Bolanderi*
    CC. Disk-aks. sterile; ray-aks. beaked; invol. 5–6 mm. high; anthers yellow
                                             2. *M. madioides*
  BB. Low vernal annuals; lvs. alternate above first basal pairs, those of the rosette narrow.
    C. Disk-aks. fertile; pappus of wider paleae; anthers yellow; ligules flabelliform; plants 5–25 cm. tall.
      D. Branches flexuous; buds and fruiting heads nodding; pappus only on disk-aks., lance-attenuate, fimbriate, 2.5–3.7 mm. long .... 3. *M. nutans*
      DD. Branches strict, subumbellate; heads strictly erect; pappus of ray and disk similar, oblong or quadrate, erose-fimbriate, 0.2–0.3 mm. long .... 4. *M. Hallii*
    CC. Disk-aks. sterile; pappus of long flexuous awns; anthers black; ligules obovate.
      D. Heads medium; ray-fls. 7–10, ligules 5–9 mm. long; disk-fls. 20–65; plants 20–60 cm. high .... 5. *M. Rammii*
      DD. Heads small; ray-fls. 4–7, ligules 2.5–3 mm. long; disk-fls. 2–7; slender plants 7–20 cm. high .... 6. *M. yosemitana*
AA. Pappus none; annuals. (Section Madia)
  B. Disk-fls. more than 1 (except rarely in *M. glomerata*), pubescent; phyllary with empty flat tip.
    C. Ray-aks. prominently beaked; receptacular bracts deciduous; heads not closing, the ligules golden-yellow; disk-aks. fertile .... 7. *M. radiata*
    CC. Ray-aks. not prominently beaked; receptacular bracts ± persistent; heads mostly closing at midday.
      D. Disk-aks. sterile; receptacle pubescent; ray-fls. conspicuous (except *M. citriodora*).
        E. Ray-aks. compressed, with narrow backs (except in ssp. *Wheeleri*); ligules bright yellow, conspicuous; herbage not lemon-scented ..... 8. *M. elegans*
        EE. Ray-aks. scarcely compressed, with broad backs; ligules greenish-yellow, inconspicuous; herbage lemon-scented .... 9. *M. citriodora*
      DD. Disk-aks. fertile; receptacle glabrous; ray-fls. inconspicuous.
        E. Ray-fls. 6–15; ray-aks. broadest toward top with small areola.
          F. Aks. obovoid, black and shining .... 10. *M. anomala*
          FF. Aks. strongly laterally compressed, dull.
            G. Glandular above the middle, slender and flexuous; heads small (to 9 mm. high); early. Mostly interior.
              H. Heads in a spikelike raceme; herbage yellow-green. Foothills, Butte Co. to Mariposa Co. .... 11. *M. subspicata*
              HH. Heads paniculate (or openly racemose); herbage grayer or deeper green.
                I. Side branches not surpassing main axis; cauline lvs. narrow at base; herbage coarsely pilose with spreading hairs
                                12. *M. gracilis*
                II. Side branches surpassing main axis; cauline lvs. broad at base; herbage densely villous with rather appressed hairs
                                13. *M. citrigracilis*
           GG. Glandular well down toward base; stem stout; heads usually large; late. Mostly near the coast; often a ruderal.
              H. Heads short-peduncled, scattered or subglomerate, the subtending bracts narrowly lanceolate, short; invol. 7–12 mm. high; stems rather slender but tall, the branches, if any, usually ascending .... 14. *M. sativa*
              HH. Heads congested in terminal glomerules, very large, the subtending (mature) bracts deltoid-lanceolate and usually exceeding them; invol. 8–15 mm. high; stems stout, not tall, the branches, if any, divaricate .... 15. *M. capitata*
        EE. Ray-fls. 0–3; ray-aks. broadest at middle, truncate at both ends with broad areola .... 16. *M. glomerata*
  BB. Disk-fls. 1 (or 2), glabrous, phyllary embracing the ak. and sulcate to tip; tiny herbs.
    C. Ray-aks. laterally compressed, with sharp ventral angle, glabrous; invol. strongly stipitate-glandular, the glands yellow .... 17. *M. exigua*

CC. Ray-aks. obcompressed, not angled, pubescent; invol. minutely stipitate-glandular, the
glands black .... 18. *M. minima*

1. **M. Bolánderi** (Gray) Gray. [*Anisocarpus B.* Gray.] Perennial herb with woody
rootstocks; stem simple, 5–12 dm. high, densely hirsute to glabrate below, strongly
stipitate-glandular above; lvs. entire, hirsute, linear, attenuate, the basal crowded
into an erect tuft, the lower pairs connate, 10–30 cm. long, 4–12 mm. wide, the cauline
often abruptly reduced in size and frequency; heads few, openly corymbose, terminating
peduncles up to 25 cm. long; invol. campanulate to hemispheric; ray-fls. 8–12; disk-fls.
30–65; ray-aks. 5.7–6.5 mm. long, broadly linear, obscurely 5-nerved, glabrous (when
pappose somewhat hispid); disk-aks. 6–8 mm. long, 5-angled, tawny hispid; pappus of
ray none or rudimentary, of disk of 5–10 lance-attenuate to lance-ovate ciliate stramineous
paleae 1.8–5 mm. long; self-sterile; $n = 6$ (Clausen et al., 1934).—Damp mountain
meadows or along streamways, 3500–6700 ft.; Yellow Pine F., Red Fir F.; Sierra Nevada
and Cascades from Tulare Co. to Lane Co., Ore., Scott and Marble mts., Trinity Summit.
July–Sept.

2. **M. madioìdes** (Nutt.) Greene. [*Anisocarpus m.* Nutt. *M. Nuttallii* Gray.] Perennial
herb with a short usually simple rootstock; stem simple 3–6.5 dm. high, hirsute below,
glandular-pubescent above; lvs. coarsely strigose, linear to linear-oblong, the basal horizontal in the rosette, then erect, 6–12 cm. long, 5–15 mm. wide; heads few, racemose
or cymose; invol. globose-urceolate; ray-fls. 8–15; disk-fls. 10–30; ray-aks. 3.5–5 mm.
long, semilunar; disk-aks. 3–4 mm. long, undeveloped, pubescent; ray-pappus none or
vestigial, the disk-pappus of 5–8 lanceolate to quadrate discrete or somewhat united
unequal fimbriate paleae 1 mm. or less long; self-fertile; $n = 7$ (Johansen, 1933).—
Moist coniferous woods, below 4000 ft.; Redwood F., N. Coastal Coniferous F., Mixed
Evergreen F.; outer (rarely inner) Coast Ranges, Monterey Co. to Del Norte Co.; to
Vancouver Id. May–Sept.

3. **M. nùtans** (Greene) Keck. [*Callichroa n.* Greene. *Blepharipappus n.* Greene.
Layia n. Jeps.] Stem erect, commonly with few ascending or decumbent branches from
base, 1–2.5 dm. high, hirsute below, puberulent and beset with minute stipitate glands
above; lvs. mostly entire, linear, the lower 2–3 cm. long, up to 2.5 mm. wide; heads
solitary or loosely paniculate; invol. 4–7 mm. high; ray-fls. 5–8; disk-fls. 7–30; ray-aks.
2.7–4.1 mm. long, narrowly clavate, slightly arcuate, without prominent nerve; disk-
aks. 2.3–4.5 mm. long, tapering to a slender base, the stipe dilated; self-sterile; $n = 9$
(Clausen et al. 1936).—Uncommon, in volcanic ash, 600–3200 ft.; Mixed Evergreen F.,
Foothill Wd.; Napa and Sonoma cos. from Mt. St. Helena to Napa Range, Hoods Peak,
etc. April–May.

4. **M. Hállii** Keck. Principal stem short, very leafy, hirsute, topped by a dense rosette
of lvs., then 2–5-ramified into assurgent leafless glandular-pubescent branches terminating
in solitary heads or in leafy rosettes from which similar peduncles arise, 5–18 cm. high;
lvs. hirsute, linear, the margins thickened, 0.5–3 cm. long, 0.7–1.7 mm. wide; heads
solitary on black-glandular pedicels 1–5 cm. long, or ± congested in corymbs terminating
peduncles 5–9 cm. long; invol. obovoid, 4.5–5.2 mm. high; ray-fls. 3–6; disk-fls. 8–20;
ray-aks. 2.7–3.1 mm. long, narrowly clavate, arcuate, slightly laterally compressed; disk-
aks. 2.8–3.2 mm. long, prismatic, the stipe scarcely dilated; self-sterile; $n = 9$
(Johansen, 1933).—Rare, on stony serpentine ridges, 1500–3000 ft.; Foothill Wd.,
Chaparral; s. Lake and n. Napa cos. May.

5. **M. Rámmii** Greene. [*Anisocarpus R.* Greene.] Stem usually simple below, openly
branching above, slender, pilose below, glandular above, 2–6 dm. high; lvs. scattered,
essentially entire, linear, strigose to hirsute, 1.5–6(–10) cm. long, 1–4 mm. wide;
heads cymose, terminating long naked peduncles; invol. broadly urceolate, 4.8–5 mm.
high; ray-aks. 2.4–3 mm. long, laterally flattened, ventral angle ± sharp, slightly concave, the back rounded, with ca. 12 longitudinal striae per face, the stipe none, the
areola on a short stout eccentric beak; ray-pappus a microscopic ring of fimbriate
rudimentary paleae less than 0.3 mm. long, the disk-pappus equaling or exceeding the
disk-fls., the exposed portions anthocyanous, of 5–7 ciliate flexuous awns; self-sterile;
$n = 8$ (Clausen et al., 1934).—Frequent in sunny grassy places, 1400–5000 ft.; Foothill Wd., Yellow Pine F.; Sierran foothills from Butte Co. to Calaveras Co. May–July.

6. **M. yosemitàna** Parry ex Gray. [*Anisocarpus y.* Greene.] Stem usually simple below,
dichotomously or trichotomously branching above, very slender, 6–15(–24) cm. high;

habit as in *M. Rammii;* invol. broadly turbinate, 3.2–4.2 mm. high; ray-aks. 2.7–3.8 mm. long, the beak slender, otherwise these and the pappus similar to *M. Rammii;* self-fertile; *n* = 8 (Johansen, 1933).—Uncommon, grassy seeps or swales, 4000–7500 ft.; Yellow Pine F.; Eldorado Co. (near Fallen Leaf L.), Tuolumne Co. to Tulare Co. May–July.

7. **M. radiàta** Kell. [*Anisocarpus r.* Greene.] Stem commonly branched above or throughout, 1.5–9 dm. high, fistulous, yellowish, glandular-pubescent; lvs. linear-lanceolate, acuminate, glandular and often hirsute, the lower 4–10 cm. long, 4–15 mm. wide, the upper gradually reduced in size and broadest at base; heads many, in an irregular compound cyme; invol. depressed hemispheric, 4.5–6.5 mm. high, bearing glands on pustulate stipes and additionally crisped-pubescent or pilose; ray-fls. 8–16, the obovate ligules 6–16 mm. long; disk-fls. 20–65, the anthers yellow; ray-aks. 3.3–4.2 mm. long, flattened, semilunar; disk-aks. 3.2–4.4 mm. long; self-sterile; *n* = 8 (Clausen et al., 1934).—Frequent on grassy slopes, up to 2500 ft.; V. Grassland, Foothill Wd.; inner S. Coast Range from e. Contra Costa Co. to w. Kern Co. March–May.

8. **M. élegans** D. Don. [*Madaria e.* DC. *Madaria e.* var. *depauperata* Gray. *Madia hispida* and *polycarpa* Greene. *M. villosa* Eastw. *M. e.* var. *h.* Hall.] COMMON MADIA. Stem commonly corymbosely branching above, 2–8 dm. high, villous below, inconspicuously to densely glandular above; lvs. linear to broadly lanceolate, the basal few or in a small rosette, the lower cauline often crowded, the upper well spaced; invol. campanulate to hemispheric, to 10 mm. high, hirsute and stipitate-glandular (sometimes hispid-hirsute and eglandular), the attenuate tips of the phyllaries equaling the basal portion; ray-fls. 8–16, the ligules 6–15 mm. long, yellow or with maroon blotch at base; disk-fls. 25 or more, yellow or maroon; anthers purple-black; self-sterile; *n* = 8 (Johansen, 1933).—Common on rather dry slopes, meadow edges, etc., 3000–8000 ft.; many Plant Communities; throughout cismontane Calif. except in the Cent. V. and rare in the Coast Ranges below 4000 ft., rare e. of Sierra Nevada; Ore. June–August. This summer-flowering montane ecotype is replaced at low elevs. by the following two ecotypes:

Ssp. **vernális** Keck. Stem simple to openly branching throughout, 3–8 dm. high, the herbage sparingly pubescent to densely hispid or pilose, sparingly glandular below the infl.; basal rosette scarcely developed, the cauline lvs. scattered; heads large, remaining open longer through the day than the other sspp.; spring-flowering; fls. all yellow, rarely with maroon blotch at base of ligules; anthers black; self-sterile; *n* = 8.—Common in valleys and foothills, usually below 3000 ft.; many Plant Communities; cis-montane Calif. from Kern and San Luis Obispo cos. n. to Ore. March–June.

Ssp. **densifòlia** (Greene) Keck. [*Madaria corymbosa* DC. *Madia c.* Greene. *M. d.* Greene. *M. e.* var. *d.* Jeps.] Stem stout, usually branching well above the middle to form a corymbose panicle, to 25 dm. high, usually strongly glandular-pubescent above; basal lvs. forming a large rosette, the lower cauline closely imbricated, densely villous or hirsute, the upper scattered, strongly glandular and pubescent; invol. broad, to 12 mm. high, the tips of the phyllaries often exceeding the basal portion; ray-fls. 12–20, the ligules 10–20 mm. long, all yellow or usually with maroon blotch at base; disk-fls. always yellow; anthers purple-black; self-sterile; *n* = 8.—Dry soils of valley floors and foothills, up to 3000 ft., throughout cismontane Calif., n. to Ore. Aug.–Nov.

Ssp. **Wheeleri** (Gray) Keck. [*Hemizonia W.* Gray. *Madia tenella* Greene. *M. W.* Keck.] Stem slender, simple or commonly divaricately branching from the middle, 1–4.5 dm. high, moderately villous, sparingly glandular; basal rosette small or none, the cauline lvs. scattered; heads small, scattered and solitary on leafy peduncles or in leafy cymules, not closing; invol. 4.5–5.5 mm. high, the phyllaries ¾ enclosing ray-aks., villous, inconspicuously yellow-glandular, the tip shorter than the basal portion; ligules 4–5 mm. long, flabelliform, yellow; disk-fls. yellow; anthers yellow, with rounded ovate appendages; ray-aks. much less laterally compressed, triquetrous; self-sterile; *n* = 8.—Conifer belt, 5000–10,000 ft.; Yellow Pine F., Red Fir F.; s. Sierra Nevada, Mt. Pinos se. to San Jacinto Mts.; Sierra Juarez, L. Calif. June–Aug. Looking very distinct from *elegans* in the Sierra Nevada, but intergrading strongly in s. Calif.

9. **M. citriodòra** Greene. [*Hemizonia c.* Gray.] Stem usually simple below, corymbosely or subumbellately branching above, rather slender, 2–7 dm. high, villous-hirsute, especially below, with black stipitate glands above; lvs. not crowded, linear, the

lower 4–9 cm. long, 3–7 mm. wide; heads in a corymbose panicle; invol. broadly turbinate to hemispheric-urceolate, 6–8 mm. high, villous-hirsute, scarcely glandular;
ray-fls. 5–12, disk-fls. 15–50; anthers black; ray-aks. 3.3–4.4 mm. long, the lateral angles
rounded; self-fertile; n = 8 (Clausen et al., 1937).—Not common, dry open slopes,
1400–5200 ft.; Yellow Pine F., N. Juniper Wd.; Modoc and Siskiyou cos. to Amador Co.,
inner N. Coast Range from Lake Co. to Napa Co.; to Wash. May–July.

　　10. **M. anómala** Greene. [*Hemizonia a.* Gray. *M. dissitiflora* var. *a.* Jeps.] Stem slender,
usually flexuously branching from the middle, 2–5 dm. high, glandular only in upper
half; herbage rather anthocyanous; lvs. not very crowded, villous, the upper ones somewhat glandular, linear, up to 7 cm. long and 7 mm. wide; heads paniculate to subracemose, not congested; invol. globose, ca. 6–7 mm. high, the phyllaries dilated but
usually entirely enclosing the ak., thickly beset with stout gland-tipped hairs and viscid-
puberulent; bracts of the receptacle united only toward base, ciliate; ray-fls. 3–7; disk-fls.
3–6; ray- and disk-aks. similar; self-fertile; n = 16 (Johansen, 1933).—Not common,
grassy slopes, from sea level to 500 ft.; Mixed Evergreen F., Foothill Wd.; N. Coast
Ranges from Humboldt Co. to Mt. Diablo, Sacramento Co. May–June.

　　11. **M. subspicàta** Keck. Stem strict, slender, usually simple, or sparingly short-
branching above, 1–1.5 dm. high, the herbage yellow-green, pilose and viscid-puberulent
with prominent stipitate glands; lvs. linear, 2–7 cm. long, to 3 mm. wide; heads
subspicate on very short peduncles, overtopped by the subtending lvs.; invol. ovate,
6–7 mm. high; disk-fls. 5–15, the anthers black; ray-aks. ca. 3 mm. long, minutely
striate, purple-spotted, the disk-aks. similar; self-fertile; n = 8 (Clausen et al., 1945).—
Rare, in grassy places, 150–1300 ft.; V. Grassland, Foothill Wd.; Sierran foothills from
Butte Co. to Mariposa Co. May–June.

　　12. **M. grácilis** (Sm.) Keck. [*Sclerocarpus g.* Sm. *Madorella racemosa* and *dissitiflora*
Nutt. *Madia r.* and *d.* T. & G. *M. sativa* var. *r.* and var. *d.* Gray. *Lagophylla Hillmani*
A. Nels. *M. s.* ssp. *d.* Keck.] GUMWEED. Stem usually slender, simple or flexuously
branching from the middle, the branches not overtopping the main stem, 1–10 dm. high;
herbage not as viscid as *M. sativa*, resinously fragrant; lvs. not very crowded even at
base, mostly linear, sessile by a narrow base, to 10 cm. long and 5 mm. wide; heads
paniculate to racemose, not congested, the leafy bracts rarely prominent; invol. ovoid to
depressed-globose, 6–9 mm. high; phyllaries thickly beset with stout gland-tipped hairs,
the acuminate tips short; ray-fls. 8–12; ligules 3–8 mm. long; disk-fls. 15–35; anthers
included, black; ray-aks. 2.8–5 mm. long, gibbously obovate, often mottled; disk-aks.
similar but straighter; self-fertile; n = 16, 24 (Clausen et al., 1945).—Abundant on
wooded hillsides and in open places, sea level to 7800 ft.; many Plant Communities;
throughout cismontane Calif.; to B.C., Mont., Utah, L. Calif.; Chile, Argentina. April–
Aug.

　　Ssp. **collìna** Keck. More robust, usually branching near the top, the heads spicate or
glomerulate, the infl. very viscid, the stem not glandular below; heads large, the
globose-urceolate invols. 8–10 mm. high, the phyllaries with elongated tips.—Below
3000 ft.; Foothill Wd.; Sierra Nevada of Amador and Calaveras cos. May–June.

　　Ssp. **pilòsa** Keck. Like the sp., but moderately pilose throughout, especially on
peduncles and invols., rather sparsely glandular, but usually so well down toward base;
lvs. few, large, the upper ones rather broad; heads solitary or racemose on elongated
peduncles, large, the invols. depressed-globose, contracted above the aks., the phyllaries
somewhat dilated, lightly holding the aks., the erect tips elongated.—Mixed Evergreen
F., N. Oak Wd.; local in Humboldt Co.

　　13. **M. citrigrácilis** Keck. Stem sparsely branched throughout, the branches strict,
slender, often exceeding the main stem, 2.5–5 dm. high, hispid-hirsute below, villous
and viscid-puberulent above with prominent stipitate glands; lvs. linear-oblong, often
subamplexicaul, the lower 4–8 cm. long, to 6 mm. wide; heads racemose or terminating
leafy peduncles; invol. obovate, 6–8 mm. high; phyllaries somewhat hirsute and densely
stipitate-glandular, villous-ciliate over the face of the ak.; ray-fls. 5(–14); disk-fls.
3–10(–30); ray-aks. ca. 4 mm. long, broadly lanceolate in cross section; self-fertile;
n = 24 (Clausen et al., 1945).—Uncommon, 4000–6000 ft.; Yellow Pine F., N. Juniper
Wd.; Modoc, Shasta and Lassen cos. July–August.

　　14. **M. satíva** Mol. CHILE TARWEED. COAST TARWEED. Stem usually stout, often

rigidly branched above, glandular well down toward base, 5–10(–20) dm. high; herbage strongly odorous; lvs. rather crowded, sessile by a broad base, up to 18 cm. long and 12 mm. wide; heads paniculate, racemose or subspicate, often approximate along the branchlets, not foliose-bracted; ray-aks. falcate-oblanceolate; disk-aks. oblanceolate, the sides sometimes 1-nerved; self-fertile; $n = 16$ (Johansen, 1933).—Close to the coast, up to 1000 ft.; many Plant Communities; behaving as a ruderal, Los Angeles Co. n.; to Alaska; native of Chile, Argentina. May–Oct.

15. **M. capitàta** Nutt. [*M. sativa* ssp. *c.* Piper. *M. s.* var. *congesta* T. & G.] Stem stout, simple or rigidly corymbosely branched above, glandular to base, more viscid than *M. sativa*, 3–6 dm. high; strongly odorous; lvs. sessile by a broad base; heads in congested spikes or in terminal or lateral glomerules, often with foliose bracts exceeding the heads; ray-aks. narrower and longer than in *M. sativa*, the sides usually 1-nerved; self-fertile; $n = 16$ (Johansen, 1933).—Low fields, near the coast, sometimes weedy, Santa Barbara Co. n.; to B.C. May–Oct., but usually earlier than *M. sativa*.

16. **M. glomeràta** Hook. [*M. ramosa* Piper. *M. g.* var. *r.* Jeps.] Stem rigid, very leafy, simple or with assurgent or fastigiate branches, 1.5–8 dm. high, villous to hispid, with yellow stipitate glands above; strongly and unpleasantly odorous; lvs. narrowly linear, often with fascicles in their axils, the lower 3–9 cm. long, 2–7 mm. wide; heads in dense terminal glomerules of 5–30, or in more open cymes and panicles; invol. narrowly ovoid, 5.5–9 mm. high; ligules inconspicuous, greenish-yellow to purplish, 1.5–2.5 mm. long; disk-fls. 1–10; aks. 4–6 mm. long, 5-nerved; self-fertile; $n = 14$ (Clausen et al., 1934).— Grassy places in coniferous woods, 3500—8800 ft.; Yellow Pine F., Red Fir F.; Modoc and Siskiyou cos. to Tulare Co., San Bernardino Mts.; to Alaska, S. Dak., New Mex. July–Sept. Sometimes adv., as at Eureka and Santa Cruz Mts.

17. **M. exígua** (Sm.) Gray. [*Sclerocarpus e.* Sm. *Harpaecarpus madarioides* Nutt. *H. e.* Gray. *Madia filipes* Gray. *H. californicus* Gand.] Stem simple, corymbosely branched above, or paniculately branched throughout, very slender, 0.5–3(–4) dm. high, hirsute, prominently stipitate-glandular above; aromatic; lvs. strigose, linear, 1–4 cm. long, 2 mm. or less wide; heads on filiform divaricate often elongated peduncles in corymbose panicles; invol. depressed-globose, 2.5–4.8 mm. high; phyllaries early deciduous with the mature fr., linear dorsally, lunate in outline laterally, covered with prominent glands on thick often pustulate stalks; ray-fls. 5–8, the ligule 1 mm. long; ray-aks. 1.8–2.8 mm. long, crescentic, stoutly beaked; disk-ak. fertile; self-fertile; $n = 16$ (Johansen, 1933).— Abundant at low to moderate elevs.; many Plant Communities; throughout cismontane Calif.; to adjacent L. Calif., B.C., Mont. May–July.

18. **M. mínima** (Gray) Keck. [*Hemizonia m., parvula* and *Durandi* Gray. *Hemizonella m., p.* and *D.* Gray. *Harpaecarpus m.* and *p.* Greene. *Melampodium m.* and *D.* Jones. *Hemizonella m.* var. *p.* Hall.] Stems 1 to several from near the base, divaricately branched to form often a hemispheric plant, villous below, glandular-puberulent above, 2–15 cm. high; lvs. often in little clusters at the nodes, otherwise usually scattered, linear-oblong, 1–2 cm. long; heads solitary or in small terminal glomerules, napiform, 2–3 mm. high; phyllaries loosely appressed, becoming arcuate with the ripening aks., rounded on the back, their glands tiny, on prominent pustulate processes; ray-aks. incurved, ± beaked; self-fertile; $2n = 32$ (Keck, 1949).—Gravelly slopes, 3500–8600 ft.; mostly Yellow Pine F., Red Fir F.; common in the Sierra Nevada, Modoc Co. to Kern Co., inner N. Coast Ranges s. to Lake Co., Mt. Hamilton, Mt. Pinos to Laguna Mts.; to B.C. May–July.

33. Hemizònia DC. TARWEED

Annual or perennial herbs or shrubs, usually very glandular and aromatic, mostly fall-flowering xerophytes, very few spring-flowering. Basal lvs. variously lobed, rarely subentire; upper lvs. and bracts not terminated by open pit glands. Phyllaries half enclosing the ray-aks. Receptacular bracts in a single row (and often ± united) outside the disk-fls. or scattered. Fls. yellow or white. Ray-aks. beaked (except in *Euhemizonia*), triquetrous, the odd angle posterior, epappose, fertile. Disk-aks. usually bearing a paleaceous pappus. Self-sterile. 31 spp., all in Calif. and L. Calif. (Greek, *hemi*, half, and *zone*, girdle, the phyllaries but half-enclosing the ray-aks.)

A. Shrubs; lvs. small, crowded. (Section 1. Zonamra)
    B. Flocs present in axils of older lvs.; phyllaries scarcely keeled; anthers black; insular
                                                                   1. *H. clementina*
    BB. Flocs absent; phyllaries strongly keeled; anthers yellow. Santa Susana Mts.
                                                                   2. *H. Minthornii*
AA. Herbs; lvs. larger.
    B. Ray-aks. obviously beaked; ligules yellow; inner receptacular bracts, when present, not adnate or deliquescent.
        C. Lvs. not spine-tipped; receptacular bracts confined to a row surrounding the outer disk-fls. and united into a cup; pappus of quadrate or oblong paleae. (Section 2. Deinandra)
            D. Disk-aks. sterile.
                E. Phyllaries not keeled; anthers yellow (except in *corymbosa*).
                    F. Heads not glomerate.
                      G. Ray-fls. 5; disk-fls. 6. Inner Coast Ranges to s. Calif.
                                              3. *H. Kelloggii*
                    GG. Ray-fls. 8 or more; disk-fls. 10 or more.
                      H. Lvs. pubescent throughout; stems not strongly fistulous.
                          I. Ray-fls. 8–12; disk-fls. 10–20; herbage pallid. Inland.
                            J. Pappus obvious; radical lvs. lobed; herbage villous below; ligules pale yellow. Cismontane .. 4. *H. pallida*
                          JJ. Pappus none or vestigial; radical lvs. entire or obscurely toothed; herbage hispid-hirsute below; ligules deep yellow. Transmontane .... 5. *H. arida*
                        II. Ray-fls. 18–32; disk-fls. 28–48; herbage bright green; radical lvs. pinnatifid or bipinnatifid. Coastal
                                              6. *H. corymbosa*
                    HH. Lvs. merely ciliolate, mostly entire; stems strongly fistulous; pappus usually none. Inner Coast Range .... 7. *H. Halliana*
                FF. Heads glomerate, small; ray-fls. 5; disk-fls. 6. Mojave Desert
                                            8. *H. mohavensis*
            EE. Phyllaries keeled; anthers black. Coast Ranges.
                F. Ray-fls. 3; disk-fls. 3; herbage hispid and gray-green; plants low and divaricate. Bay region to s. Monterey Co. .... 9. *H. Lobbii*
                FF. Ray- and disk-fls. more than 3.
                    G. Ray-fls. 5; disk-fls. 6.
                      H. Herbage bearing pustulate hairs, gray-green; plants low, intricately branched. S. Monterey Co. to San Luis Obispo Co.
                                            10. *H. pentactis*
                    HH. Herbage bearing simple filamentous hairs, yellow-green; plants taller, less intricately branched. S. Calif.
                        I. Heads paniculate; stems tall, scarcely hispid, flexuous and much-branched. San Luis Obispo Co. to San Diego Co.
                                          11. *H. ramosissima*
                      II. Heads in dense glomerules; stems lower, rather hispid above, with strict divaricate branches. Coastal plain to L. Calif. .... 12. *H. fasciculata*
                  GG. Ray-fls. 8; disk-fls. 13–21. Sw. San Diego Co. .. 13. *H. conjugens*
        DD. Disk-aks. mostly fertile.
            E. Radical lvs. pinnatifid or bipinnatifid; herbage grayish-hirsute; ray-fls. 8–13, the ligules as broad as long. Common, cent. Calif. to L. Calif.
                                              14. *H. paniculata*
            EE. Radical lvs. merely lobed; herbage bright green, soft-pubescent; ray-fls. 13–20, the ligules half as broad as long. Rare; s. San Diego Co.
                                            15. *H. floribunda*
    CC. Lvs. tipped with a rigid spine or apiculation; receptacular bract subtending each disk-fl., free, persistent; pappus none, or of very narrow paleae. (Section 3. Centromadia)
        D. Pappus none; anthers yellow.
            E. Receptacular bracts pungent. N. and cent. Calif. .... 16. *H. pungens*
            EE. Receptacular bracts obtuse or ± acute, not cuspidate. S. Calif.
                                              17. *H. laevis*
        DD. Pappus present.
            E. Receptacular bracts fleshy at tip, not long-villous; pappus-paleae 3(–5), sparsely ciliolate; herbage bearing small yellowish glands, if any, of mild odor.
                F. Anthers yellow. Cent. Calif. .... 18. *H. Parryi*
                FF. Anthers black; infl. very glandular. S. Calif. .... 19. *H. australis*
            EE. Receptacular bracts long-villous at tip; pappus-paleae 8–12, densely fimbriate at tip; herbage bearing large stalked black glands, of rank odor; anthers black
                                              20. *H. Fitchii*
BB. Ray-aks. not obviously beaked; ligules white or yellow; inner receptacular bracts adnate, forming a cell about each disk-fl., deliquescent. (Section 4. Hemizonia)
    C. Ligules yellow, dorsally veined with purple.
        D. Spring-flowering (May–June); lvs. prominent at time of flowering, elongated, not crowded into a basal rosette; herbage glandular only above .. 21. *H. multicaulis*
        DD. Fall flowering (Aug.–Oct.); taller; lvs. inconspicuous at time of flowering, very short, crowded into a basal rosette; densely glandular throughout
                                            22. *H. lutescens*

CC. Ligules white, dorsally veined with purple.
    D. Heads paniculate or terminal; herbage puberulent to sericeous or villous with soft hairs; lower lvs. broad.
       E. Peduncular bracts short, not overtopping the heads.
          F. Herbage green, merely puberulent; heads scarcely glandular; spring-flowering .... 23. *H. Tracyi*
          FF. Herbage gray or silvery, pubescent; heads obviously glandular; spring-to fall-flowering.
              G. Infl. widely paniculate, the heads scattered; herbage villous and copiously dark glandular, or sericeous; phyllaries 3.5–6 mm. long. Mostly inland, n. and cent. Calif. .... 24. *H. luzulaefolia*
              GG. Infl. corymbosely paniculate, the heads usually glomerate; herbage shaggy and with few pale glands; phyllaries 6–9 mm. long. Coastal valleys, n. of San Francisco Bay .... 25. *H. congesta*
       EE. Peduncular bracts long, overtopping the heads, calyculate; herbage prominently glandular. Mendocino and Lake cos. .... 26. *H. calyculata*
    DD. Heads spicate, at least on the fully developed side branches, obviously glandular; herbage pilose; lower lvs. narrow; basal rosette not obvious. N. Calif. to Ore. 27. *H. Clevelandii*

1. **H. clementína** Bdg. [*Zonanthemis c.* Davids. & Mox.] Shrub 3–8 dm. high, with many erect, ascending or decumbent branches from base, densely leafy above, the older lvs. deciduous; herbage sparsely hirsute, viscid; lower lvs. opposite, narrowly linear, 3–8 cm. long, 1.5–6 mm. wide, remotely dentate; upper lvs. alternate, entire, with fascicles or leafy shoots in their axils; heads in compound cymes; invol. broadly campanulate to hemispheric, 5–8 mm. high; ray-fls. 13–14, the ligules yellow, 4.5–6.5 mm. long; receptacular bracts in 2 series; disk-fls. 18–30, the cent. ones without bracts; ray-aks. 2–3 mm. long, transversely rugose, the prominent recurved beak 1.5–3 times as long as thick; disk-aks. sterile; pappus of 7–10(–15) linear-lanceolate attenuate fimbriate unequal paleae 1–3 mm. long; n = 12 (Johansen, 1933).—Heavy soils of Anacapa, S. Barbara, S. Nicholas, S. Catalina and S. Clemente ids. May–Aug.

2. **H. Minthórnii** Jeps. Shrub 6–10 dm. high, 10–30 dm. wide, with up to 500 stiff woody ascending stems from base; herbage short-hirsute, viscid, fragrantly resinous; lvs. alternate, somewhat thickened, like those of *clementina;* heads mostly solitary on long peduncles, racemosely or corymbosely paniculate; invol. 5.5–6 mm. high, 4–6 mm. broad; ray-fls. 8, the ligules yellow, 5.5–6.5 mm. long; receptacular bracts in ca. 3 series; disk-fls. 18–23, each subtended by a bract; ray-aks. 2.5–3 mm. long, smooth, the beak scarcely longer than thick; disk-aks. developed but sterile; pappus of 8–12 linear ± fimbriate subequal paleae 1.2–2.6 mm. long; n = 12 (Clausen et al., 1934).—Rare; Chaparral; Santa Susana Mts. July–Oct.

3. **H. Kellóggii** Greene. [*H. Wrightii* Gray. *Deinandra K.* and *W.* Greene. *H. W.* var. *K.* Jeps.] Stem 2–10 dm. high, corymbosely branching above or intricately branching throughout, soft-villous at base, hispid-hirsute and densely glandular above; lower lvs. narrowly oblong, remotely sharp-toothed or pinnatifid, 3–8 cm. long, 3–10 mm. wide; upper lvs. entire; heads pedunculate, in an open panicle or thyrse; invol. 4–5.5 mm. long, 3.5–5 mm. wide, densely stipitate-glandular, usually hirsute; ligules 4–7 mm. long, 3.5–5.5 mm. wide; pappus of 6–12 white linear to oblong paleae; n = 9 (Johansen, 1933).—Abundant (rare above 2000 ft.); V. Grassland, S. Oak Wd.; San Francisco Bay region s. through the Cent. V. to San Diego Co.; adv. in Mendocino, Colusa and Imperial cos. April–July.

4. **H. pállida** Keck. Stem 2–8 dm. high, branching above or throughout with ascending or divergent branches, whitish or reddish; herbage villous-hirsute and hispidulous, lightly glandular and mildly odorous; lower lvs. linear to oblanceolate, remotely sharp-toothed or cleft, 5–10 cm. long, 3–6(–10) mm. wide, the upper ones becoming entire; heads numerous, cymose; invol. 4.5–6.5 mm. high, 5–8 mm. wide, hispid-hirsute and minutely stipitate-glandular; ligules 6–10 mm. long, 3–5 mm. wide; pappus of 4–8 narrow distinct paleae 0.8 mm. or less long; n = 9 (Johansen, 1933).—Common on plains and hills up to 2200 ft.; V. Grassland; head of San Joaquin V. April–May.

5. **H. árida** Keck. Stem 2–4 dm. high, intricately corymbosely branching throughout; herbage hispid-hirsute and hirsutulous, lightly glandular and mildly odorous; lvs. ca. as in *pallida;* heads numerous, cymose-paniculate; invol. 4 mm. high, 5 mm. wide, hispid-hirsute and stipitate-glandular; ligules 5–6 mm. long, 3–4 mm. wide; pappus vestigial or none; n = 12 (Clausen et al., 1935).—Very rare; Creosote Bush Scrub; Red Rock Canyon, Kern Co., at 2500 ft. May–Nov.

6. **H. corymbòsa** (DC.) T. & G. [*Hartmannia. c.* DC. *Hemizonia angustifolia* DC. *H. decumbens* Nutt. *H. balsamifera* Kell.] Stem erect or decumbent, 2–10 dm. long, branching from base upward or from above the middle, the branches ascending or spreading; herbage densely villous and glandular, strongly and pleasantly odorous; lower lvs. linear to oblanceolate, deeply pinnatifid, 3–10 cm. long, 3–20 mm. wide; upper lvs. becoming entire in older plants; heads cymose or paniculate; invol. 5–9 mm. high, 6–12 mm. wide; ligules 4–7 mm. long, 2.5–4 mm. wide; pappus of narrow distinct unequal entire or laciniate paleae mostly 1 mm. or less long, or wanting; $n = 10$ (Johansen, 1933).—Frequent along the coast, up to 500 ft.; N. Coastal Scrub; Mendocino Co. to Monterey Co. May–Oct.

Ssp. **macrocéphala** (Nutt.) Keck. [*H. m.* Nutt. *H. angustifolia* ssp. *m.* Keck.] Stouter denser; heads densely glomerate at the apices of the branches, cymose, larger; ray-fls. 20–35; disk-fls. 40–70; $n = 10$.—Coastal s. Monterey and n. San Luis Obispo cos.

7. **H. Halliàna** Keck. Stem 2–12 dm. high, up to 16 mm. thick, simple and glabrous below, corymbosely branching (only near summit or to below the middle) and becoming pilose and viscid-puberulent above; pungently odorous; lvs. linear-lanceolate, sessile, glabrous except for the scabrous margin, the lower remotely short-dentate, 5–8 cm. long, 3–9 mm. wide; invol. 5–7 mm. high, 8–10 mm. wide; ray-fls. 10–14, the ligule 5.5 mm. long; disk-fls. 30–60; ray-aks. 3.5–4 mm. long; disk-aks, rarely bearing a rudimentary paleaceous pappus; $n = 10$ (Clausen et al., 1934).—Rare, adobe flats and serpentine, 1200–2900 ft.; Foothill Wd.; inner S. Coast Range, San Benito Co. to San Luis Obispo Co. April–May.

8. **H. mohavénsis** Keck. Stem 1.5–3 dm. high, subsimple or divaricately branched above; herbage very soft-pubescent and viscid, pleasantly odorous; lower lvs. oblanceolate, subentire, the upper oblong-lanceolate, entire, amplexicaul, much reduced toward the expanded-corymbose infl.; heads sessile in glomerules at ends of the branches; invol. 5–6 mm. high, 5 mm. wide; ligules 5–6 mm. long; pappus of 6–8 quadrate ± connate paleae 0.5 mm. long; $n = 11$ (Clausen et al., 1934).—Very rare, 3000–4000 ft.; Joshua Tree Wd., Chaparral; Mojave R. at Deep Creek; Mt. San Jacinto. July–Sept.

9. **H. Lóbbii** Greene. [*H. fasciculata* var. *L.* Gray. *Deinandra L.* Greene.] Stem erect, slender, 2–5 dm. high, rigidly branched above or throughout, the branches divaricate and again very freely branching toward tips; herbage glabrate to sparingly hispid-hirsute, often beset with sessile glands above, mildly odorous; lvs. small, linear and entire at anthesis, those of peduncles often crowded, the deciduous lower lvs. laciniately sharp-toothed or pinnatisect; heads very numerous, subsessile to short-peduncled; invol. 4–6 mm. long, 2.5–4 mm. wide, beset with sessile glands and sometimes ± hispid with pustulate hairs; ray- and disk-fls. 3 each; pappus of 6–8 lanceolate or oblong entire or toothed paleae equaling the corolla-tube; $n = 11$ (Johansen, 1933).—Dry interior hills; Foothill Wd., V. Grassland; Solano Co. to s. Monterey Co. Introd. in Lassen and Shasta cos. May–Nov.

10. **H. pentáctis** (Keck) Keck. [*H. Lobbii* subsp. *p.* Keck.] Having the low intricately branched habit, grayish herbage and pinnatisect basal lvs. of *Lobbii*, the stems and heads more definitely pustulate-hirsute and hispid, the lvs. whiter pilose; ray-fls. 5; disk-fls. 6; otherwise similar to *Lobbii*; $n = 11$ (Johansen, 1933).—Dry hills; Foothill Wd., V. Grassland; s. Monterey and San Luis Obispo cos.; introd. about Stanford University. May–Oct.

11. **H. ramosíssima** Benth. [*H. fasciculata* var. *r.* Gray. *H. f.* ssp. *r.* Keck. *Deinandra f.* var. *r.* Davids. & Mox.] Stem erect, 2–10 dm. high, simple below and corymbosely branching above or with ascending branches from the base upward; herbage moderately hirsute to glabrate, the glands (most frequent on invols.) sessile, yellow; lower lvs. (mostly missing at anthesis) linear-oblanceolate, remotely dentate, 3–15 cm. long, 3–30 mm. wide; upper lvs. becoming entire, small and bractlike; heads pedunculate, numerous, ± remote and solitary, sometimes in 2's at ends of branches; ray-fls. 5; disk-fls. 6; pappus as in *Lobbii*; $n = 12$ (Johansen, 1933).—Dry fields and hillsides, up to 1500 ft.; Coastal Sage Scrub, S. Oak Wd., V. Grassland; Santa Barbara Co. to Orange Co., largely replaced by the next in San Diego Co., Channel Ids., L. Calif. May–Aug.

12. **H. fasciculàta** (DC.) T. & G. [*Hartmannia f.* DC. *H. glomerata* Nutt. *Deinandra f.* Greene.] Stem 1–10 dm. high, mostly branching from above the middle, the branches rigid, sharply ascending and with comparatively few twigs; heads subsessile in glomer-

ules of 3 to many, the glomerules terminating short leafy branches and sometimes a few solitary heads below; otherwise similar to *ramosissima;* n = 12 (Johansen, 1933).—Dry coastal plains, up to 1000 ft.; Coastal Sage Scrub, S. Oak Wd., V. Grassland; frequent from San Luis Obispo Co. to Riverside Co., abundant thence to L. Calif. May–Sept.

13. **H. conjùgens** Keck. Stem 2–5 dm. high, branching as in *fasciculata;* foliage and invols. soft-hirsute and sometimes hispidulose, especially the invol. bearing large flat and small capitate sessile or subsessile glands; heads solitary on short peduncles or subsessile in few-headed glomerules; ray-fls. 8–10; disk-fls. 13–21.—Mesas; Coastal Sage Scrub; sw. San Diego Co. May–June.

14. **H. paniculàta** Gray. [*Deinandra p.* Davids. & Mox.] Stem 2–8 dm. high, the central shaft replaced ca. midway by numerous ascending branches which ramify to form a twiggy infl.; herbage moderately hispid-hirsute (especially below) and stipitate-glandular (especially above), fragrant; lower lvs. persistent until anthesis, linear to oblanceolate, sharp-toothed or pinnatifid with entire or dentate lobes, 1.5–8 cm. long, up to 25 mm. wide; heads obviously peduncled; invol. 5–7 mm. high, 5–7 mm. wide, the phyllaries densely glandular-pubescent; ray-fls. 8; disk-fls. 8–13; disk-aks. sparsely pubescent, some fertile and dark-colored, the majority sterile and pale; pappus of 6–12 white oblong fimbriate paleae; n = 12 (Johansen, 1933).—Dry hills and mesas, low elevs.; V. Grassland; w. Riverside Co. to San Diego Co.; n. L. Calif. May–Nov.

Ssp. **incrèscens** Hall ex Keck. Plants erect or somewhat spreading, deep green, even ultimate twigs rather rigid (never capillary); ray-fls. 8–13; disk-fls. 14–30; n = 12 (Johansen, 1933).—Abundant in fields near coast; V. Grassland; Monterey Co. to Santa Barbara Co. and Santa Rosa Id. May–Nov.

15. **H. floribúnda** Gray. [*Deinandra f.* Davids. & Mox.] Stem 3–10 dm. high, corymbosely branched above or throughout with strict ascending branches; herbage moderately pilose and densely stipitate-glandular throughout, mildly odorous; lvs. mostly entire, the lower oblanceolate, the middle cauline linear-lanceolate, 1.5–3 cm. long, 2–3.5 mm. wide; heads short-peduncled, in racemosely compound cymes or panicles, not glomerate; invol. 5–6 mm. high, 6–7.5 mm. wide, soft-pubescent and very glandular; ray-fls. 13–20; disk-fls. ca. 28, their aks. all or mostly fertile; pappus of 6–9 oblong or elliptic closely fimbriate rufous and flecked paleae; n = 12 (Johansen, 1933).—Hillsides and dry valleys; Coastal Sage Scrub; s. San Diego Co.; n. L. Calif. Aug.–Oct.

16. **H. púngens** (H. & A.) T. & G. [*Hartmannia ? p.* H. & A. *Centromadia p.* Greene. *Hemizonia p.* ssp. *interior* Keck.] COMMON SPIKEWEED. Stem 1–12 dm. high, divergently and rigidly branched above or throughout, leafy, sparsely to copiously hirsute, not glandular; herbage yellowish-green, inodorous; basal lvs. linear-lanceolate in outline, 5–15 cm. long, 10–40 mm. wide, bipinnately parted into segments; upper cauline lvs. with fascicles in their axils, mostly entire, spine-tipped, stiffly ciliate and scabrous-puberulent; heads subsessile in upper axils and terminal; invol. hemispheric, 3–6 mm. high, usually overtopped by bracts; phyllaries keeled, pungent, scabrous, persistent; ray- and disk-fls. numerous, the ligules only 2-toothed; ray-aks. angular, often rugose; disk-aks. in part sterile, the fertile ones ± fusiform; n = 9 (Johansen, 1933).—Interior dry valleys and foothills, below 1000 ft.; V. Grassland; San Joaquin V. from San Joaquin Co. to Kern Co.; Salinas V.; introd. Los Angeles, San Bernardino, and San Diego cos. and n. Ore. April–Oct.

Ssp. **marítima** (Greene) Keck. [*Centromadia m.* Greene.] Plants fairly low, the branches usually divaricate and strict; lvs. and bracts not scabrous; heads in glomerules of 5–40 members, large; n = 9.—Salt marshes and valley land around San Francisco Bay, s. to n. Monterey Co., ne. to San Joaquin and Yolo cos.

Ssp. **septentrionàlis** Keck. Plants large, the branches elongated and often lax; lvs. and bracts not scabrous; heads scattered, small; n = 9.—Sacramento V. and Shasta V.; introd. n. Ore., Wash.

17. **H. laèvis** (Keck) Keck. [*H. pungens* ssp. *l.* Keck.] Similar to *pungens* but of somewhat lower habit; upper lvs. and bracts sparsely setose-ciliate, otherwise very glabrous; heads small, scattered, or approximate in loose glomerules; receptacular bracts obtuse or slightly acute, sometimes minutely but weakly mucronate, not cuspidate; n = 9 (Keck).—Grassy fields, low elevations; V. Grassland; San Diego Co. to San Bernardino V. and Los Angeles region, away from immediate coast; sw. Kern Co. April–Sept.

18. **H. Párryi** Greene. [*Centromadia P.* Greene. *C. pungens* var. *P.* Jeps. *Hemizonia p.*

var. *P.* Hall.] Stem 1–7 dm. high, erect, corymbosely branching above, the branches rigid or lax, or rather prostrate, ± leafy, sparsely to copiously hirsute; basal lvs. linear-to lance-oblong, 5–20 cm. long, 10–30 mm. wide, pinnatifid to bi-pinnatisect, the lower cauline ± hirsute, their lobes ± spine-tipped, the upper mostly entire, linear-subulate, spine-tipped, with fascicles (incipient shoots or heads) in their axils; heads sessile or short-peduncled along the branches and terminal, or in small clusters; peduncular bracts often exceeding the invol.; invol. hemispheric, 5–10 mm. high, the phyllaries glandular-pubescent, copiously ciliate, the glands subsessile; ray-fls. 9–20, the ligules 3–6 mm. long, yellow, not fading saffron, 2-lobed; receptacular bracts with resinous-thickened soft obtuse to acute (not cuspidate) tip; disk-fls. 40–60 or more; disk-aks. in part sterile; pappus of 3–5 linear-subulate usually fimbriate paleae ca. equaling the corolla, often united at base and contorted; $n = 12$ (Johansen, 1933).—Grassy often alkaline fields, low elevs.; V. Grassland, Coastal Salt Marsh; sw. Colusa Co. to n. San Mateo Co.; n. Monterey Co. June–Oct.

KEY TO SUBSPECIES

Infl. glandular; invol. 5–10 mm. high; herbage not scabrid. Valleys mostly n. of San Francisco Bay
                                                                                            H. Parryi
Infl. not glandular.
    Lvs. and bracts very scabrid, herbage grayish-pilose; invol. 3.5–4 or 5 mm. high. Sacramento V.
                                                                                            ssp. *rudis*
    Lvs. and bracts ± pilose but not scabrid or glandular, herbage yellow-green; invol. 4.5–8 mm. high.
    San Francisco Bay to San Luis Obispo Co. . . . . ssp. *Congdonii*

Ssp. **rùdis** (Greene) Keck. [*Centromadia r.* Greene.] Stem erect, corymbosely or diffusely branched, not prostrate; herbage pale green; lvs. and bracts scabrid-puberulent; infl. obscurely if at all glandular; peduncular bracts not exceeding the invol.; heads small; ray-fls. often fading saffron; $n = 11$ (Clausen et al., 1934).—Common. low fields; V. Grassland; Merced Co. n. through the w. side of the Sacramento V. to Glenn Co.

Ssp. **Congdónii** (Rob. & Greenm.) Keck. [*H. C.* Rob. & Greenm. *Centromadia C. C. P.* Sm. *C. pungens* var. *C.* Jeps.] Stem erect with ascending or horizontal branches or all branches closely prostrate; herbage yellow-green, sparsely to copiously hirsute but smooth between the long hairs; peduncular bracts exceeding the invol.; ray-fls. not fading saffron; $n = 12$ (Johansen, 1933).—Locally common; V. Grassland; Salinas V., Monterey Co., s. end of San Francisco Bay, mostly Alameda Co., and San Ramon V., Contra Costa Co. June–Oct.

19. **H. austràlis** (Keck) Keck. [*H. Parryi* ssp. *a.* Keck.] Similar to *Parryi;* stem erect, the central leader of medium length and with long and lax divaricate branches, not prostrate, the twiggery above dense; herbage dark green; lvs. and bracts densely glandular-puberulent and villous; peduncular bracts not exceeding the invol.; heads small; phyllaries 4–5.5 mm. long; ray-fls. often fading saffron; anthers black; $n = 11$ (Venkatesh, 1956).—Lowlands near the coast; V. Grassland; Santa Barbara Co. to San Diego Co.; adjacent L. Calif. June–Sept.

20. **H. Fítchii** Gray. [*Centromadia F.* Greene.] FITCH'S SPIKEWEED. Stem 2–8 dm. high, rigid, erect, above diffusely branched; herbage dark, densely villous and beset with prominent stalked glands, unpleasantly heavy-scented; lvs. crowded, at least in the rosette, very slender, the lower pinnatisect, up to 15 cm. long, the upper entire, rigid, linear-subulate, acerose-tipped, often with fascicles in their axils, the uppermost in-volucroid, spreading, radiating around the head which they overtop; invol. hemi-spheric, densely villous, with few prominent glands, the phyllaries rigidly 1-nerved, with subulate tip slightly surpassing the rays; receptacular bracts soft, pointless, hairy, enfolding the disk-fls.; ray-fls. 10–20, the short light yellow ligules bifid; pappus of 8–12 glistening white soft oblong paleae united at base and long-fimbriate above, equaling the disk-corolla; $n = 13$ (Clausen et al., 1934).—Abundant but not gregarious, dry hills and plains, up to 3000 ft.; V. Grassland, Foothill Wd.; Sierran foothills from Siskiyou Co. to Fresno Co., Cent. V., Coast Range valleys from Mendocino Co. to San Luis Obispo Co., n. base San Bernardino Mts.; s. Ore. May–Nov.

21. **H. multicaúlis** H. & A. [*H. citrina* Greene.] Stem erect, ascending or decumbent, 1–3(–5) dm. high, subsimple or openly branching above or throughout, pilose and glandular-pubescent; lvs. not crowded into a basal rosette, the lower remotely serrulate,

linear-lanceolate, 6–15 cm. long, 6–12 mm. wide, deep green; heads corymbose; invol. hemispheric, 5–6.5 mm. high, the tips of the phyllaries ca. equaling the body; ray-fls. 8–13, showy, clear yellow (sometimes drying greenish); *n* = 14 (Johansen, 1933).— Seaward bluffs along the immediate coast; N. Coastal Scrub; Sonoma and Marin cos. May–July.

Ssp. **vernàlis** Keck. Stem erect, corymbosely branching above, slender; lvs. grasslike, narrower, usually silvery sericeous; *n* = 14.—Mostly away from the immediate coast, Sonoma and Marin cos. April–June.

22. **H. lutéscens** (Greene) Keck. [*H. luzulaefolia* var. *l.* Greene. *H. congesta* ssp. *l.* Babc. & Hall. *H. c.* var. *l.* Jeps.] Stem erect, robust, 2–7 dm. high, usually much-branched from the base upward; herbage dark-glandular within infl. or throughout, heavily scented; basal rosette at first prominent; lvs. short and narrow (up to 12 cm. long and 5 mm. wide), often silvery-villous with long cobwebby hairs; heads small; invol. 4.5–5 mm. high, the tips of the phyllaries shorter than the body; ray-fls. deep yellow; *n* = 14 (Clausen et al., 1937).—Abundant in grassy valleys and hillsides away from the immediate coast; V. Grassland; s. Mendocino Co. to Marin and Alameda cos. July–Nov.

23. **H. Tràcyi** (Babc. & Hall) Keck. [*H. congesta* ssp. *T.* Babc. & Hall. *H. c.* var. *T.* Jeps.] Stem slender, 1.5–6 dm. high, subsimple or openly few-branched above, some-what villous or merely densely puberulent; herbage soft-pubescent, glandular only around the heads; lvs. not crowded into a basal rosette, grasslike, very narrow, the lower up to 15 cm. long, and 5 mm. wide, bright green; heads corymbose, relatively few; invol. 5–7 mm. high, the tips of the phyllaries shorter than the body; ray-aks. ½–⅗ as broad as long; *n* = 14 (Clausen et al., 1937).—Bald hills and forest openings, up to 3000 ft.; Coastal Prairie, Redwood F.; s. Humboldt and n. Mendocino cos. May–July.

24. **H. luzulaefòlia** DC. [*H. sericea* H. & A. *H. l.* var. *fragarioides* Kell. *H. grandiflora* Abrams. *H. congesta* ssp. *l.* Babc. & Hall. *H. c.* var. *l.* Jeps.] Stem slender, 2–5 dm. high, openly branching above or throughout, mostly viscid-pubescent well down toward base, often also ± villous; basal rosette none or obscure, lower lvs. prominent at anthesis, subentire, up to 15 or more cm. long and to 13 mm. wide, green and chiefly puberulent, or silky or silvery with cobweb-tomentum, not glandular; heads showy; invol. 5–6 mm. high, the tips of the phyllaries shorter than the body; ray-aks. ca. ½ as broad as long; *n* = 14 (Johansen, 1933).—Coastal valleys and hills, up to 1000 ft.; V. Grassland; Lake Co. to San Luis Obispo Co. April–June.

Ssp. **rùdis** (Benth.) Keck. [*H. r.* Benth.] HAYFIELD TARWEED. More robust, 2–8 dm. high, branching and densely glandular-viscid throughout; basal rosette well developed, the shorter and narrower lvs. often silvery villous; heads smaller but very numerous; invol. 3.5–5 mm. high; *n* = 14.—The most widespread and abundant member of the section, in stubblefields, pastures, and other open ground, up to 3000 ft.; V. Grassland, Foothill Wd.; Sacramento V. from Tehama Co. to San Joaquin Co., Coast Range valleys from Lake Co. to Santa Barbara Co., very rare n. of Sonoma Co. and in Monterey Co. July–Nov.

25. **H. congésta** DC. Stem virgate, 1–5 dm. high, with rather few strict or lax branches only above or from the base upward; herbage ± densely villous throughout, the hairs sometimes appressed but never silky, somewhat viscid toward heads but the glands usually obscure; basal rosette none or obscure; lower lvs. usually prominent at anthesis, remotely denticulate; heads in small terminal clusters or solitary at the ends of the short branches and often a few scattered lateral heads; invol. 6–9 mm. high, the tips of the phyllaries equaling to much exceeding the body; ray-aks. mostly ⅗ as broad as long; *n* = 14 (Johansen, 1933).—Fields near the coast; N. Coastal Scrub; Del Norte Co. to San Mateo Co. but frequent only in Sonoma and Marin cos. May–Oct.

26. **H. calyculàta** (Babc. & Hall) Keck. [*H. congesta* ssp. *c.* Babc. & Hall. *H. c.* var. *c.* Jeps.] Stem 2–8 dm. high, divaricately branching above; herbage moderately villous and densely glandular-pubescent, the dark stalked glands abundant in the infl.; lower lvs. not prominent at anthesis, linear-lanceolate, 4–10 cm. long, 2–5 mm. wide; heads terminating short stiff peduncles, the fl.-bracts crowded, tending to surpass the heads; invol. 6–12 mm. high, the tips of the phyllaries exceeding the body; ray-aks. ½–⅗ as broad as long; *n* = 14 (Johansen, 1933).—Fields and open hillsides; V. Grassland, Foothill Wd.; Lake and s. Mendocino cos. July–Oct.

27. **H. Clevelándii** Greene. [*H. congesta* ssp. *C.* Babc. & Hall. *H. c.* var. *C.* Jeps.]

Stem 2–6 dm. high, simple to divaricately branched above or throughout, or with several ascending branches from base; herbage of early flowering forms usually densely villous especially toward base of stem and rosette and moderately glandular above, of late flowering forms moderately villous or merely tomentose but strongly glandular-pubescent especially toward the infl.; basal rosette small or none, basal lvs. narrow, to 15 cm. long, to 7 mm. wide; heads terminal on the branches in early forms, also spicate-paniculate in late forms; invol. 4–7 mm. high, the tips of the phyllaries usually shorter than the body; ray-aks. ca. ⅔ as broad as long; $n = 14$ (Johansen, 1933).—Dry interior grassy hills, 100–4500 ft.; N. Oak Wd., Foothill Wd., V. Grassland; Sonoma and Napa cos. n. through both Coast Ranges to s. Ore. June–Oct.

34. Holocárpha (DC.) Greene. TARWEED

Annual herbs, mostly very glandular and aromatic. Lvs. linear, the basal remotely serrulate to dentate with slender teeth, the upper reduced to entire bracts usually bearing fascicles or peduncles in their axils, truncated at apex by a crateriform gland. Phyllaries half-enclosing the ray-aks., bearing on back and at apex stoutly stalked pit glands. Receptacular bracts subtending each disk-fl., enclosing the ak. and corolla-tube, free, persistent. Fls. yellow, the rays 3-lobed. Ray-aks. laterally beaked, obcompressed, triquetrous, the odd angle anterior. Disk-aks. sterile or outermost fertile. Pappus none. Self-sterile. Four spp., all in Calif. (Greek, *holos*, whole, and *karphos*, chaff, the entire receptacle chaffy.)

A. Anthers black; herbage not puberulent above.
   B. Disk-fls. 40–90; ray-fls. 8–16; heads very large, densely glomerate; low, with robust divaricate branches. Rare, near the coast .... 1. *H. macradenia*
   BB. Disk-fls. 10–25; ray-fls. 3–7; heads small, fusiform to subglobose, racemose; tall with virgate ascending branches. Interior .... 2. *H. virgata*
AA. Anthers yellow; medium tall. Interior.
   B. Herbage not densely puberulent above; infl. corymbose-paniculate; invol. obconical, the phyllaries bearing 5–15(–20) stout gland-tipped processes; receptacle broad . 3. *H. obconica*
   BB. Herbage densely puberulent above; infl. racemose; invol. subglobose, the phyllaries bearing 25–50 very slender gland-tipped processes; receptacle narrow .... 4. *H. Heermannii*

1. **H. macradènia** (DC.) Greene. [*Hemizonia m.* DC.] Stem 1–5 dm. high, rigid, divaricately branched above the base, the few branches arising near together and usually again similarly branched, rather densely leafy; basal lvs. usually lost before anthesis, broadly linear, up to 12 cm. long and 8 mm. wide including the slender remote teeth; upper lvs. becoming bractlike with revolute margin, conspicuously to obscurely glandular-pubescent and often hirsute, the apical gland sometimes occurring only on the uppermost; heads mostly in terminal glomerules; invol. subglobose, 5.5–8 mm. high, the phyllaries bearing ca. 25 stout terete gland-tipped processes; ray-aks. 2.5–3 mm. long; $n = 4$ (Clausen et al., 1934).—Colonial in heavy soils on grassy flats along the coast; N. Coastal Scrub; Marin, Alameda, and Santa Cruz cos., now possibly extinct. June–Oct.

2. **H. virgàta** (Gray) Keck. [*Hemizonia v.* Gray.] Stem 2–12 dm. high, rigid, usually simple below, with few to several strict sharply ascending branches above; herbage sparingly pilose below and glandular-puberulent, canescent or hispidulous above and strongly resinous and odorous; basal lvs. 6–15 cm. long, 3–10 mm. wide, mostly lost before anthesis, the upper reduced to bracts, often fasciculate, those of the peduncles crowded, heathlike, spreading or recurved, the apex terete, prominently gland-tipped; heads short-peduncled, spicately or racemosely disposed along the virgate branches; invol. 5–6 mm. high, the phyllaries bearing 5–20 stout terete ascending gland-tipped processes, otherwise glabrous; ray-aks. 2.4–3.5 mm. long; $n = 4$ Clausen et al., 1934).— Very abundant in hard-baked soils, up to 2500 ft.; V. Grassland, Foothill Wd.; Cent. V. from Glenn Co. to Fresno Co., Sierran foothills, inner Coast Ranges (Lake, Napa, Santa Clara, and San Benito cos.). June–Nov.

Ssp. **elongàta** Keck. Stem slender, 5–12 dm. high, multibranched, the ultimate branchlets gracefully curving, elongated, the heads more scattered on less leafy peduncles up to 15 cm. long.—Sw. San Diego Co.

3. **H. obcónica** (Clausen & Keck) Keck. [*Hemizonia o.* Clausen & Keck. *H. vernalis* Keck.] Stem 1.5–8(–12) dm. high, the central shaft replaced by several divaricate

branches below the middle, the branches paniculately compound to form a dense twig-gery; herbage strongly resinous and odorous; lvs. like *virgata* except the basal more hispid and the peduncular often well spaced; heads paniculate on long spreading peduncles; invol. 4–5 mm. high, the phyllaries bearing relatively few gland-tipped processes, the terminal one less prominent than in *virgata,* the surface otherwise glabrous or minutely glandular, rarely ± hispid but never puberulent, often failing to cover com-pletely the ripe aks.; $n = 6$ (Clausen et al., 1934).—Scattered colonies in open fields and hillsides; V. Grassland, Foothill Wd.; inner S. Coast Range from Alameda Co. to w. Fresno Co., Sierran foothills of Fresno and Tulare cos. April–Oct.

Ssp. **autumnàlis** Keck. Cent. shaft quite concealed by short sterile shoots densely clothed with bractlike lvs. and progressively shorter from the base upward, so that before anthesis the plants form tapering green columns; principal lvs. exceptionally nar-row; flowering branches relatively short; $n = 6$.—Plains n. and e. of Mt. Diablo. Sept.–Nov.

4. **H. Heermánnii** (Greene) Keck. [*Hemizonia H.* Greene. *H. parvifolia* Greene. *Deinandra H.* Greene. *Hemizonia virgata* var. *H.* Jeps.] Stem 2–12 dm. high, simple below and virgately few-branched above to diffusely branched throughout, pilose or hispid below, cinereous above, with interspersed glandular-pubescence throughout; strongly resinous and odorous; lvs. canescent, the upper also densely stipitate-glandular, otherwise like *virgata,* the peduncular bracts usually overlapping each other but not the base of the invol., appressed to moderately recurved, flattened to apex, 2–5 mm. long; heads paniculate or openly racemose; invol. 5–6 mm. high, puberulent and viscid in addition to bearing many gland-tipped processes; $n = 6$ (Clausen et al., 1934).—Common in hard-baked soils, up to 4000 ft.; V. Grassland, Foothill Wd.; foothills sur-rounding the San Joaquin V. from Contra Costa Co. to the Mt. Hamilton Range and from Calaveras Co. to Tehachapi Pass and Ventura Co.; local in Monterey and San Luis Obispo cos. May–Oct.

35. Calycadènia DC. Rosin Weed

Xerophytic annuals with linear to filiform entire revolute grasslike lvs., those at the base of the rigid stem crowded into an erect rosette, more scattered above and often fasciculate, those of the fascicles and the uppermost usually bearing a prominent tack- or saucer-shaped gland terminally and often similar glands along their margins, these bractlike lvs. frequently bearing also pectinate cilia. Ray-fls. few, 1–5(–8), with very broad palmately 3-lobed or -parted ligules. Receptacular bracts united into a cup surrounding the few disk-fls. Ray-aks. with nearly cent. terminal areola. Disk-aks. like-wise fertile, angular, usually bearing a paleaceous pappus. Self-sterile. Eleven spp., nearly confined to Calif. (Greek, *kalux,* cup, and *adenos,* gland, in reference to the glands of the infl.)

A. Tack-shaped or saucer-shaped glands none; lobes of ray-fls. linear; ray-aks. distinctly beaked. S. Calif. (Section 1. Osmadenia) One sp.; pappus paleae 10, alternating long and short
1. *C. tenella*
AA. Tack-shaped or saucer-shaped glands present on bracts; lobes of ray-fls. oblanceolate or broader; ray-aks. not distinctly beaked but areola sometimes elevated. S. Ore. to cent. Calif. (Section 2. Calycadenia)
  B. Ray-aks. rugose.
    C. Stems glabrous throughout or scabrid toward summit; pappus of ca. 10–12 broad blunt fimbriate paleae 0.5–1.8 mm. long or none   2. *C. truncata*
    CC. Stems softly pubescent, never scabrid; pappus of 5 long-attenuate paleae 4–5 mm. long, sometimes alternating with 1–3 short blunt ones .... 3. *C. molis*
  BB. Ray-aks. smooth; stems pubescent, at least above; herbage scabrous.
    C. Bracts subtending the heads nerved quite to apex, i.e., ± flattened throughout, the rounded or truncate apex not impressed, irregularly shorter or longer than the invol.
      D. Tack-shaped glands confined to the one terminating most of the peduncular bracts; ray-aks. villous .... 4. *C. villosa*
      DD. Tack-shaped glands more numerous and on the heads (cf. *Fremontii*); ray-aks. glabrous to moderately hairy.
        E. Stem simple or openly branched; heads medium-large, crowded and often glomerate; ray-fls. 2–8.
          F. Ligules with very broad lateral lobes; late flowering.
            G. Lobes of ligule equal; bracts much shorter than the invol., densely and finely hispidulous; pappus much shorter than the disk-ak.
5. *C. ciliosa*

GG. Lobes of ligule unequal, the central one only half as wide as the laterals; bracts equaling or exceeding invol., coarsely hispid; pappus equaling or exceeding disk-ak.
H. Pappus paleae 2.7–3.5 mm. long, alternating long and short; peduncular bracts nearly straight, rigidly ascending or erect, rarely longer than invol.; infl. ± anthocyanous
6. C. multiglandulosa
HH. Pappus paleae 4–6.5 mm. long, all long; peduncular bracts recurved from heads, usually much longer than invol. (except in ssp. reducta); herbage not anthocyanous, more finely and densely pubescent .... 7. C. hispida
FF. Ligules equally 3-parted to base with linear-oblanceolate divaricate lobes; early flowering.
G. Lvs. all opposite, the uppermost thrice as long as the invols.; heads in verticillate glomerules .... 8. C. oppositifolia
GG. Lvs. alternate above, the uppermost but slightly exceeding the invols.; heads racemose .... 9. C. Fremontii
EE. Stem densely and often dichotomously branched above; heads small, never glomerate; ray-fls. usually 1, white .... 10. C. pauciflora
CC. Bracts subtending the heads not nerved to apex, the outer third terete, the apex truncate and impressed, strongly imbricated, regularly equaling the invol.
11. C. spicata

1. **C. tenélla** (Nutt.) T. & G. [*Osmadenia t.* Nutt. *Hemizonia t.* Gray.] Stem 1.5–5 dm. high, divaricately branched, often intricately twiggy, the ultimate branchlets subcapillary, leafy and sparsely villous, becoming somewhat viscid-puberulent within the infl.; lvs. 1.5–5 cm. long, scabrous and somewhat hirsute, the floral bracts whiteciliate; heads scattered; invol. 3.5–4.8 mm. high, ovoid; ray-fls. 3–5, the ligules white, occasionally with crimson blotch at center of each narrow lobe, the lateral lobes opposite; disk-fls. 3–10, white, like the rays fading roseate, the lobes medially streaked with crimson; ray-aks. 1.8–2.7 mm. long, the cent. areola strongly beaked; pappus of 4 or 5 rufous-flecked aristiform-tipped paleae 3.5 mm. long, alternating with acute paleae 1 mm. long; n = 9 (Clausen et al., 1934).—Abundant, in light soils, below 2000 ft.; Coastal Sage Scrub, S. Oak Wd.; Los Angeles Co. to L. Calif. May–July.

2. **C. truncàta** DC. [*Hemizonia t.* Gray.] Stem 3–11 dm. high, arcuately few-branched above, reddish, smooth and glabrous, or minutely scabrous toward summit; herbage glaucescent, odorous; lvs. sessile by a widened base, firm, 2–9 cm. long, 1–4.5 mm. wide, minutely scabrous, the lower sparsely ciliate, the upper tipped with a broad saucer-shaped gland, the floral bracts scabrous and pectinate-ciliate; heads spicately scattered and terminal; invol. 4–7 mm. high, campanulate; ray-fls. 3–8, the ligules light to deep yellow, sometimes with small red "eye" at base or suffused with red dorsally, flabelliform-orbicular, 7–12 mm. long; disk-fls. 10–25, bearing short quadrate brownish pappus; n = 7 (Johansen, 1933).—Dry soils in sunny places, up to 3800 ft.; Foothill Wd., Chaparral; Sierran foothills from Plumas Co. to Mariposa Co., inner Coast Ranges from Mendocino Co. to Santa Clara Co., Santa Lucia Mts. June–Oct.

Ssp. **scabrélla** (E. Drew) Keck. [*Hemizonia s.* E. Drew. *C. s.* Greene. *C. truncata* var. *s.* Jeps. *Lagophylla s.* Jones. *H. speciosa* Congd.] Stem more slender, usually less branched or even simple, nearly always glabrous throughout; fl.-bracts smooth, neither glandular nor ciliate; heads smaller, fewer-fld., terminal, neither scabrous nor glandular; epappose; n = 7.—Dry ground, up to 4400 ft.; Chaparral, Yellow Pine F.; interior foothills from s. Ore. to Placer and Lake cos. June–Oct.

Ssp. **microcéphala** Hall ex Keck. Slender habit of ssp. *scabrella*, the branching stem glabrous; bracts smooth or sparsely scabrous, some with a few cilia; heads 5 mm. high, spicate or nearly all terminal, the phyllaries smooth or with a few stiff bristles dorsally; receptacular bracts 3(–6); ray-fls. 3, the ligules 2–3 mm. long; disk-fls. 3–4, epappose.— Inner N. Coast Range, from Trinity Co. to Lake Co., local in the Santa Lucia Mts., s. Monterey Co. May–Aug.

3. **C. móllis** Gray. [*Hemizonia m.* Gray.] Stem 2.5–7 dm. high, rigidly erect, virgate, usually simple, or with 2–6 main stems from base and few short divergent branches, arachnoid-villous at base, villous and viscid above; herbage gray-green, the odor acidulous; lvs. 2–7 cm. long, not rigid, appressed-pilose and beset with granular glands, the fl.-bracts also white-ciliate and bearing 1–10 saucer-shaped glands; heads congested in axillary and terminal glomerules of 5–50; invol. 4–5.5 mm. high, the phyllaries often reduced to 1; ray-fls. 1–3, the ligules deep yellow, white, or deep rose, with or without

crimson "eye," 5–6 mm. long, 8–10 mm. wide; disk-fls. 5–7(–10); $n = 7$ (Johansen, 1933).—Frequent in meadow borders, 1500–4500 ft.; Yellow Pine F.; Sierra Nevada from Tuolumne Co. to Tulare Co. June–Sept. The type was white-fld. Colonies of it extend over the range of the sp. Colonies with golden yellow fls. (f. *aurea* Keck) occur in pure stands from Mariposa Co. s., most abundantly in Madera Co., and colonies with pure rose to deep claret fls. (f. *rosea* Keck) are best developed in Tuolumne Co.

4. **C. villòsa** DC. [*Hemizonia Douglasii* Gray.] Stem 1–3 dm. high, commonly with several ascending branches, often somewhat zigzag above, densely leafy at base and moderately so to summit, scabrid-puberulent and ± hirsute, sometimes viscidulous; lvs. rigid, 2–5 cm. long, 1 mm. wide, scabrous, long-ciliate with stiff bristles; peduncular bracts ca. as long as, and closely investing, the invol., terete toward apex, truncate, tipped by a large solitary T-gland, scabrous, densely ciliate and glutinous from pellucid glands; heads 1(–3) at a node along the interrupted spike; invol. 5–6 mm. high; ray-fls. 1–4, the ligules white (or pinkish), 6 mm. long, 8–9 mm. wide; disk-fls. 6–14; pappus of 10–11 linear-lanceolate awn-pointed paleae 4–6 mm. long; $n = 7$ (Johansen, 1933).— Uncommon, dry slopes; Foothill Wd.; inner slopes of the outer Coast Range, from Monterey Co. to Santa Barbara Co. June–Oct.

5. **C. ciliòsa** Greene. Stem 2–8 dm. high, usually corymbosely few-branched above, densely leafy toward base, hirsute and viscidulous-puberulent; lvs. rigid, 2–10 cm. long, up to 2 mm. wide, scabrous, sparsely bristly ciliate; peduncular bracts not crowded, tipped with a T-gland and often bearing a subapical pair or so, scabrous, copiously white-ciliate and viscid with pellucid glands; heads spicate; invol. 5–6 mm. high; ray-fls. (1–)3–6, the ligules yellow to white, with or without red "eye" at base, 6–7 mm. long; disk-fls. 7–18; pappus of 5 awn-tipped paleae 2.2–3.2 mm. long alternating with much shorter blunt erose paleae, variable; $n = 6$ (Johansen, 1933).—Stony barren soils, up to 4000 ft.; Foothill Wd., N. Oak Wd., Yellow Pine F.; s. Ore. to Butte and Lake cos. June–Sept. Most frequently found with yellow fls. as originally described, but white-fld. colonies (f. *alba* Keck) occur throughout the range of the sp.

6. **C. multiglandulòsa** DC. [*Hemizonia m.* Gray.] Stem 2–7 dm. high, strict, usually simple, leafy throughout, glabrous or puberulent below, puberulent to villous and often ± crisp-pubescent above, scarcely viscid; herbage yellowish-green to anthocyanous, pungently odorous; lvs. rigid, 3–9 cm. long, to 2 mm. wide, scabrous, hispid-ciliate toward base; heads spicate or terminally glomerate; invol. 5–6.5 mm. high, prominently glandular; ray-fls. 2–5, the ligules white, often tinged with rose, with or without a red eye-spot at base, 5–7 mm. long, 9–12 mm. wide; disk-fls. 4–12. A variable sp. separable into 3 geographic sspp., but the type cannot with surety be included in any one of them, so nomenclaturally it stands alone.

KEY TO SUBSPECIES

Heads simply spicate-racemose; upper lvs. usually short; stems slender. Sierra Nevada foothills
    ssp. *bicolor*
Heads glomerate at least apically on the stem; upper lvs. usually long. Coast Ranges.
    Heads principally in a terminal cluster, prominently anthocyanous; upper lvs. ascending; stems short and slender. Humid forest borders n. of San Francisco Bay .... ssp. *cephalotes*
    Heads spicate-thyrsoid, glomerate apically and at each node well down the stem, obscurely anthocyanous; upper lvs. arcuate; stems stout. Arid hillsides s. of San Francisco Bay . ssp. *robusta*

Ssp. **bícolor** (Greene) Keck. [*C. b.* Greene.] Stems slender; heads dark-anthocyanous and black-glandular, usually in spicate racemes well down the stem, sometimes glomerate, especially in Eldorado Co., but in this case the glomerules rarely limited to the apex of the stem; $n = 6$ (Johansen, 1933).—Abundant on dry stony soils, 500–3000 ft.; V. Grassland, Foothill Wd., Chaparral; Sierran foothills from s. Butte Co. to Tulare Co. May–Aug.

Ssp. **cephalòtes** (DC.) Keck. [*C. c.* DC. *C. m.* var. *c.* Jeps. *Hemizonia c.* Greene. *H. m.* var. *c.* Gray.] Stems rather slender, mostly 1–3 dm. high; lvs. long, the uppermost usually well exceeding the heads, bright green; heads congested at apex of stem in a terminal glomerule, their bracts anthocyanous apically and black-glandular; $n = 6$ (Clausen et al., 1934).—Grassy valleys and ridges, below 1000 ft.; Foothill Wd., Mixed Evergreen F.; Coast Ranges from s. Mendocino Co. to Napa and San Mateo cos. May–Oct.

Ssp. **robústa** Keck. Stems stout, often tall; lvs. long, the floral ones tending to exceed the heads and recurve, yellowish-green; heads aggregated into apical and axillary or subaxillary glomerules, their bracts yellow-green and bearing honey-colored glands; $n = 6$.—Dry chaparral openings and open savannah, 1000–3000 ft.; Foothill Wd., Chaparral, Mixed Evergreen F.; outer and inner Coast Ranges, Santa Clara Co. June–Sept.

7. **C. híspida** (Greene) Greene. [*Hemizonia h*. Greene. *C. campestris* Greene.] Stem 2–7 dm. high, usually robust, simple or with several short branches above, leafy throughout and with conspicuous basal rosette, puberulent and long-villous or hispid, usually visciculous; herbage yellowish-green, with fragrant acetic odor; lvs. 2–7 cm. long, mostly 1 mm. wide, considerably reduced in length upward, almost all tipped with a small T-gland, the upper ones also margined with several, viscid with pellucid glands, scabrous, long-ciliate; heads spicate, racemose or paniculate, not glomerate; invol. 6–7 mm. high; ray-fls. 3–8, the ligules deep yellow, pale yellow, or white, without red "eye," 7–8 mm. long, 9–10 mm. wide; disk-fls. 10–20, their corollas 7–9 mm. long; $n = 6$ (Clausen et al., 1934).—Open ground, low elevs.; V. Grassland, Foothill Wd.; Sierran foothills and adjacent plains from Placer Co. to Kern Co., Diablo Range, s. Monterey Co. July–Sept. The type form was yellow; the form with white rays that turn pink in age (f. *albiflora* Keck) occurs on Parkfield Grade.

Ssp. **redúcta** Keck. More slender, without conspicuous rosette; heads exceeding their subtending bracts; ligules short, 1–4, white; disk-fls. 4–8, their corollas 6–6.8 mm. long; pappus 4–4.8 mm. long; $n = 6$.—Grassy slopes, below 3000 ft.; Foothill Wd.; Contra Costa Co. to Santa Clara Co. July–Sept.

8. **C. oppositifòlia** (Greene) Greene. [*Hemizonia o*. Greene.] Stem 1–3 dm. high, simple or corymbosely branched, cinereous-strigose, equally leafy throughout; lvs. opposite, 1–5 cm. long, up to 2 mm. wide, increasing in length up the stem, scabrous, covered with pellucid punctate glands, rigidly ciliate near base; peduncular bracts short, crowded, with terminal and often a few lateral T-glands; heads in verticillastrate glomerules from ca. midway up stem to apex; invol. 5–6 mm. high; ray-fls. 2–4, the ligules white, fading deep rose, 6–9 mm. long; disk-fls. 6–20; pappus paleae brownish, alternating awn-pointed to 3 mm. long and shorter blunt ones; $n = 7$ (Clausen et al., 1934).—Dry plains and hillsides, up to 2000 ft.; V. Grassland, Foothill Wd.; Butte Co. April–July.

9. **C. Fremóntii** Gray. [*Hemizonia F*. Gray.] Stem 1.5–2.5 dm. high, usually with several ascending branches above, strigose, not viscid; lvs. alternate above, 2–4 cm. long, up to 2 mm. wide, scabrous, sparsely bristly ciliate near base; peduncular bracts with or without a T-gland, not otherwise glandular; heads paniculate, few; invol. 5–6 mm. high; ray-fls. 3–6, the ligules white tinged with rose, 8 mm. long; disk-fls. 10–12; pappus paleae subequal or alternating short and medium long.—Plains; V. Grassland; near Chico, Butte Co. May.

10. **C. pauciflòra** Gray. [*Hemizonia p*. Gray. *C. elegans* and *ramulosa* Greene. *C. p.* var. *e.* Jeps.] Stem 1.5–5 dm. high, corymbosely branched above or divaricately branched throughout with dichotomously forking twigs ascending in a ± zigzag manner, villous at base, often glabrate below but always strigose-puberulent above, scarcely at all viscid; lvs. 1–5 cm. long, 1–2 mm. wide, scabrous, sparsely ciliate, the upper often gland-tipped; peduncular bracts more ciliate, with terminal and often several lateral T-glands; heads spicately scattered, never clustered, very small; ray-fls. 1 (or 2), the ligules white, fading pink, 4.5–6.5 mm. long, 10–13 mm. wide; disk-fls. 3–6; pappus paleae unequal, up to 2.3 mm. long; $n = 5$ (Johansen, 1933).—Exposed rocky soils, at low elevs.; Foothill Wd.; inner N. Coast Range from Glenn Co. to Sonoma Co., Sierran foothills in Calaveras and Stanislaus cos. May–Sept.

11. **C. spicàta** (Greene) Greene. [*Hemizonia s*. Greene.] Stem 1–6 dm. high, often unbranched, or with several rigidly ascending branches, leafless at base at anthesis, densely leafy bracteate above, densely puberulent and ± strigose; herbage gray-green, with strong but pleasant odor; lvs. 2–6 cm. long, 1.5 mm. or less wide; peduncular bracts closely investing and concealing invol., glutinous from pellucid glands and tipped with a T-gland; ray-fls. 1–5, the ligules white, fading old rose, 6–9 mm. long, 7–9 mm. wide; disk-fls. 5–9(–12); pappus paleae rather unequal, all awn-tipped, the longer 4–5.5 mm. long; $n = 4$ (Clausen et al., 1934).—Rolling hills, below 2000 ft.; V. Grassland, Foothill Wd.; e. border of Cent. V. from Butte Co. to Tulare Co. June–Sept.

36. Blepharizònia Greene

Stout fall-flowering xerophytic odorous annuals with white densely hirsute stems and pallid herbage. Uppermost lvs. copiously covered with prominent yellow tack-shaped glands. Heads large, scattered. Phyllaries concave, not surrounding the subtended ray-aks. Ray-fls. 5–11, white, dorsally veined with rose. Ray- and disk-aks. similar, 10-ribbed, densely rufous-hirtellous, all fertile. Pappus of ray very minute, of disk of 15–20 plumose narrow paleae united at base. Monotypic. (Greek, *blepharis*, eyelash, and *zone*, girdle, from the short plumose pappus-paleae of the typical form.)

1. **B. plumòsa** (Kell.) Greene. [*Calycadenia p.* Kell. *Hemizonia p.* Gray. *H. p.* var. *subplumosa* Gray. *B. laxa* Greene. *B. p.* var. *s.* Jeps.] Stem 3–18 dm. high, paniculately branching above, not rigid; herbage gray-green, densely pilose, sparingly glandular below the heads; lvs. crowded at base, less congested above, the basal linear, 4–15 cm. long, 5–8 mm. wide, sharply incised-serrate to subentire, the cauline entire, broadest at base, mostly with fascicles in their axils, the fascicular lfts. tipped with several prominent T-glands; heads subracemose to loosely paniculate; invol. 5.5–7.5 mm. high, subglobose, canescent, with large T-glands and sometimes also small granular glands; disk-fls. 10–35; pappus half as long as the disk-aks. or longer; n = 14 (Johansen, 1933).—Occasional in dry ground, low elevs.; V. Grassland; Solano and San Joaquin cos. to Alameda and Stanislaus cos. July–Oct.

Ssp. víscida Keck. Herbage yellowish-green, moderately to densely hispid and very conspicuously glandular; invol. hispid; pappus reduced to less than 1 mm. in length and frequently obsolete; n = 14.—Dry openings, below 3000 ft., Foothill Wd., Chaparral; frequent through the inner Coast Range from San Benito Co. to Kern Co.; local in the Santa Lucia Mts., Monterey Co. July–Oct.

37. Blepharipáppus Hook.

Late vernal xerophytic annuals, corymbosely branched with alternate narrowly linear entire lvs. Heads many-fld., terminating the branchlets, heterogamous, radiate, all fls. fertile. Invol. 1–2-seriate, the outer phyllaries concave, lightly holding the ray-aks. Ray-fls. 1-seriate; ligules broad, 3-lobed. Disk-fls. each subtended by a distinct ± scarious bract, the style linear, hairy, bearing glabrous extremely short truncate branches; anthers black. Aks. all alike and fertile, turbinate, silky villous; pappus present or absent. One sp. (Greek, *blepharis*, eyelash, and *pappos*, pappus.)

1. **B. scàber** Hook. [*Ptilonella s.* Nutt. *B. s.* var. *subcalvus* Gray.] Stem erect, corymbosely branched from ca. the middle or throughout, sometimes proliferous, 1–3 dm. high, puberulent, becoming glandular upward; lvs. spreading or ascending, up to 3 cm. long, rarely more than 1 mm. wide, rapidly reduced to small bracts on the peduncles, scabrous, margins becoming revolute; invol. turbinate to hemispheric, 4–6 mm. high, glandular-puberulent and somewhat hispid; ray-fls. 3–6, the ligules white with 3 purple dorsal veins; disk-fls. 8–20; pappus of 12–18 linear hyaline densely fimbriate scales up to 2 mm. long, or frequently very short, rarely lacking; self-sterile; n = 8 (Clausen et al., 1941).—Rather common on arid plains and slopes, 3000–6000 ft.; N. Juniper Wd.; Sierra Nevada from Sierra Co. n. to Siskiyou Mts.; to Wash., Ida, Nev. May–Aug.

Ssp. laèvis (Gray) Keck. [*B. s.* var. *l.* Gray. *B. l.* Gray. *Ptilonella l.* Greene.] Branching more subumbellate; lvs. all small, becoming very small closely appressed bracts on the peduncles; peduncles sharply ascending, wiry, nearly filiform, surmounted by unusually small heads. The usual absence of pappus in Ore. material does not hold for Calif. material. Dry ground, often with yellow pine; Yellow Pine F.; Scott and Siskiyou mts., Siskiyou Co., n. on w. side of Cascades to Lane Co., Ore. June–Aug.

Tribe 2. Helenìeae. SNEEZEWEED TRIBE (Fig. 112)

Herbs or occasionally subshrubs. Lvs. alternate or opposite, in one subtribe punctate with oil glands. Heads radiate or discoid. Phyllaries mostly herbaceous and green, in only .1–3 series. Receptacle not paleaceous (sometimes a few paleae near the margin in *Helenium* and *Rigiopappus*), but sometimes bristly or hairy. Ray-fls. usually fertile and ligulate. Pappus of paleae, awns, or bristles, or often wanting.—An almost ex-

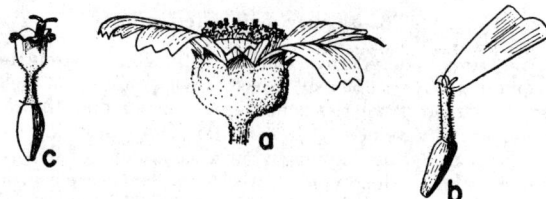

Fig. 112. HELENIEAE. *Monolopia major: a,* head, $\times$ 1, phyllaries almost united; *b,* ray-fl., $\times$ 2½, inferior ovary, no pappus, corolla-tube with 1 or 2 small terminal appendages; *c,* disk-fl., $\times$ 2½.

clusively Am. tribe, centering in the sw. U.S. and Mex. Perhaps an artificial assemblage of polyphyletic origin within the preceding tribe, from which it differs principally in the absence of chaff on the receptacle, but retained in the customary sense here for convenience.

A. Invol. not dotted or striped with oil glands.
   B. Ray-corollas marcescent and becoming papery, falling off with the aks.; herbage ± woolly. (Subtribe Riddelliinae)
      C. Pappus paleaceous; ligules broad and mostly 3–5 .... 38. *Psilostrophe*
      CC. Pappus wanting; ligules usually numerous.
         D. Receptacle naked; lvs. alternate; invol. hemispheric, 2–3-seriate. Deserts
                                                          39. *Baileya*
         DD. Receptacle villous; lvs. opposite; invol. campanulate, uniseriate. Sierra Nevada
                                                             40. *Whitneya*
   BB. Ray-corollas deciduous from the aks. or none.
      C. Invol. of broad phyllaries in 3–4 series; pappus none in ours. (Subtribe Jaumeinae)
         D. Lvs. alternate; tall perennial herb .... 41. *Venegasia*
         DD. Lvs. opposite; low succulent perennial .... 42. *Jaumea*
      CC. Invol. of narrower subequal phyllaries in only 1–2 series (2–3-seriate in *Hulsea* and *Rigiopappus*).
         D. Aks. flat, with only marginal ribs or veins. (Subtribe Peritylinae)
            E. Phyllaries mostly in 2 equal series, distinct.
               F. Aks. ciliate on the narrow callous margin; at least some of the pappus squamellate or paleaceous
                  G. Lvs. alternate; pappus of few persistent paleae, sometimes awn-tipped .... 43. *Eatonella*
                  GG. Lvs. opposite; pappus a crown of squamellae plus 1–2 bristles
                                              44. *Perityle*
               FF. Aks. hairy but not ciliate; pappus of 1–2 bristles or wanting
                                             45. *Laphamia*
            EE. Phyllaries uniseriate, loosely united into a cup .... 46. *Pericome*
         DD. Aks. terete or 3–5-angled, not strongly flattened or prominently 2-nerved. (Subtribe Heleniinae)
            E. Phyllaries mostly 20 or more, 2–3-seriate though seldom much imbricate, usually not individually subtending the rays; heads mostly radiate; lvs alternate or all basal.
               F. Aks. linear, at least 4 times as long as wide; receptacle flat or nearly so.
                  G. Plants aromatic biennial or perennial herbs; heads relatively large, the invol. 10 mm. or more high .... 47. *Hulsea*
                  GG. Plants seldom annual; heads small, the invol. 8 mm. high or less, the inconspicuous ray-fls. scarcely surpassing the disk
                                          59. *Rigiopappus*
               FF. Aks. turbinate or obpyramidal, only 2–3 times as long as wide; receptacle convex to conic or subglobose.
                  G. Receptacle bristly; style-branches with a subulate appendage
                                         48. *Gaillardia*
                  GG. Receptacle naked; style-branches truncate or nearly so.
                    H. Phyllaries appressed, permanently erect; lvs. entire (in which case all basal) or in ours mostly pinnatifid, not decurrent
                                          49. *Hymenoxys*
                  HH. Phyllaries loose, at length deflexed; lvs. entire or merely toothed, usually decurrent; stems not scapose . 50. *Helenium*
            EE. Phyllaries mostly 5–13, sometimes 15 (or more in *Chaenactis*), mostly subequal and subuniseriate, tending to subtend the rays individually when the heads are radiate.
               F. Lvs. all opposite; rays present, yellow; receptacle conic.
                  G. Phyllaries distinct or only slightly united at base .. 51. *Baeria*
                  GG. Phyllaries united at least halfway into a cup.
                      H. Aks. not prominently fringed on the margin .. 52. *Lasthenia*

HH. Aks. prominently fringed with clavate glandular hairs on the margin .... 53. *Crockeria*
FF. Lvs. alternate, at least above.
  G. Phyllaries wholly herbaceous or ± chartaceous at base, without prominent or colored scarious margins.
    H. Heads radiate.
      I. Rays obvious or showy.
        J. Plants not white-woolly .... 54. *Bahia*
        JJ. Plants white-woolly, at least when young.
          K. Ray-corolla with a lobe at summit of the tube opposite the ligule; pappus none . 55. *Monolopia*
          KK. Ray-corolla without a lobe opposite the ligule.
            L. Pappus paleaceous, or if none the rays yellow.
              M. Aks. 4–5-angled, not flattened; pappus usually present .... 56. *Eriophyllum*
              MM. Aks. ± flattened; pappus usually wanting .... 57. *Pseudobahia*
            LL. Pappus of barbellate bristles, or if none the rays white .... 58. *Syntrichopappus*
      II. Rays filiform, very inconspicuous; annual with very slender wiry stems .... 59. *Rigiopappus*
    HH. Heads discoid.
      I. Pappus-paleae deeply fimbriate into numerous bristles; aks. obpyramidal .... 60. *Trichoptilium*
      II. Pappus-paleae not fimbriate into bristles.
        J. Aks. linear-clavate, terete.
          K. Phyllaries many; pappus-scales distinct or rarely wanting; lvs. usually compoundly dissected 61. *Chaenactis*
          KK. Phyllaries 4; pappus-scales united at base, deciduous as a ring; lvs. entire 62. *Orochaenactis*
        JJ. Aks. 3–4-angled.
          K. Heads relatively large, the invol. 13–20 mm. high. Deserts .... 63. *Palafoxia*
          KK. Heads small, the invol. 3–5 mm. high. Coastal dunes and marshes .... 64. *Amblyopappus*
  GG. Phyllaries with thin ± scarious and colored margins; lvs. parted or dissected into very narrow lobes.
    H. Aks. narrowly obovoid, papillate ...... 65. *Blennosperma*
    HH. Aks. obpyramidal, 4–5-angled, glabrous to pubescent.
      I. Pappus-paleae not awn-tipped; fls. yellow 99. *Hymenopappus*
      II. Pappus-paleae awn-tipped; fls. white or purplish 100. *Hymenothrix*
AA. Invol. dotted or striped with prominent oil glands as is the foliage; herbage strong-scented. (Subtribe Tagetininae)
  B. Pappus at least partly squamellate.
    C. Phyllaries uniseriate, sometimes with accessory bractlets.
      D. Lvs. all opposite, ciliate with few stiff setae toward base, entire. Ours annual 66. *Pectis*
      DD. Lvs. usually alternate at least above, glabrous, toothed or lobed.
        E. Pappus of 5 awn-tipped paleae surrounded by many capillary bristles 67. *Nicolletia*
        EE. Pappus of few unequal paleae and 1 or 2 subulate bristles .. 68. *Tagetes*
    CC. Phyllaries biseriate, usually with accessory bractlets .... 69. *Dyssodia*
  BB. Pappus of numerous scabrous bristles .... 70. *Porophyllum*

38. Psilostròphe DC. PAPERFLOWER

Perennial herbs or low shrubs from a ligneous taproot, ± woolly. Lvs. alternate, linear or spatulate, entire, or the lower ones pinnatifid. Heads yellow, radiate, corymbosely arranged. Invol. cylindric to campanulate; phyllaries subequal, distinct, obscured by the thick tomentum. Receptacle small, naked. Ray-fls. ♀, the prominent persistent ligules becoming papery and paler in age. Disk-fls. perfect, somewhat hispidulous and glandular, the throat narrowly cylindric. Style-branches truncate-capitellate. Aks. slender, terete, obscurely angled. Pappus of 4–6 hyaline entire or laciniate paleae. Six spp. of sw. U.S. and n. Mex. (Greek, *psilos*, naked, and *strophe*, to turn.)

(Heiser, Charles B., Jr. Monograph of Psilostrophe. Ann. Mo. Bot. Gard. 31: 279–300, 1944.)

1. **P. Coóperi** (Gray) Greene. [*Riddellia C.* Gray.] Shrubby, with many ascending stems from the woody caudex, 2–5 dm. high, white-lanose particularly on younger parts:

lvs. entire, linear, 3–7 cm. long; heads terminating the upper branches, pedunculate; invol. subcylindric, 5–8 mm. high; phyllaries 10–20, the outer much firmer than the inner; rays 4–8, broadly oval, 8–16 mm. long, shallowly 3-lobed at the broad apex; disk-fls. 5–20, often much exserted; aks. glabrous; pappus paleae various, lance-oblong to ovate, fulvous, shorter than the aks.—Rocky desert mesas and sandy fans, 2000–5000 ft.; Creosote Bush Scrub, Joshua Tree Wd.; e. Mojave Desert, San Bernardino Co., from Kingston and Clark mts. to Little San Bernardino Mts. and n. edge of Colo. Desert; to sw. Utah, w. New Mex., n. Mex. April–June; Oct.–Dec.

39. Bàileya Harv. & Gray. Desert-Marigold

Annual or perennial densely floccose-woolly herbs, branching from or above the base. Lvs. alternate, the lower once or twice pinnatifid, the upper entire or nearly so. Heads solitary or cymose, long-peduncled, yellow, radiate. Invol. hemispheric; phyllaries numerous, distinct, densely floccose. Receptacle naked. Ray-fls. ♀, numerous; ligules large, persistent, accrescent, becoming deflexed and papery in age. Disk-fls. many, fertile. Style-branches short, truncate. Aks. oblong or clavate, striate, epappose. Ca. 3 or 4 spp. of sw. U.S. and n. Mex. (Honoring Jacob Whitman *Bailey*, early American microscopist.)

Ray-fls. 5–7; heads relatively small, loosely cymose; invol. ca. 6 mm. broad . . . . 1. *B. pauciradiata*
Ray-fls. 20–50; heads larger, mostly solitary on elongated peduncles; invol. 10 mm. or more broad.
    Stem leafy to above the middle or nearly to apex, the medium-sized heads on peduncles 10 cm. or
    less long . . . . 2. *B. pleniradiata*
    Stem leafy only below the middle, the large heads on peduncles 10–20 cm. long
                                                             3. *B. multiradiata*

1. **B. pauciradiàta** Harv. & Gray. Densely floccose-lanate much branched annual from a slender or woody taproot, sometimes persisting, 2–6 dm. high, leafy throughout; lvs. linear or linear-lanceolate and nearly all entire, or the lower irregularly pinnatifid or bipinnatifid with 2–5 pairs of short linear lobes, 3–10 cm. long; peduncles 2–5 cm. long; invol. 5–6 mm. high; ligules 5–7, lemon yellow, oval, 5–8 mm. long; aks. pale, muriculate.—In sand, −100–2200 ft.; Creosote Bush Scrub; rather common in Colo. Desert, occasional in e. Mojave Desert; sw. Ariz., adjacent Mex. Feb.–June; Oct.

2. **B. pleniradiàta** Harv. & Gray. [*B. multiradiata* (var.) *p.* Cov. *B. nervosa* Jones.] Annual or sometimes perennial, white-floccose with ± appressed wool, with many branching stems, 2–5 dm. high, leafy to above the middle; lvs. spatulate to linear-oblong, the lower ones (soon withering) 4–7 cm. long, pinnately few-lobed, tapering to a petiole, the upper smaller, mainly entire and sessile; peduncles mostly less than 10 cm. long and leafy below, slender; invol. 6–8 mm. high, 8–14 mm. wide; ligules 20–40, golden to pale yellow, oblong, 8–10 mm. long; aks. pale, sparsely glandular-atomiferous.—Sandy soils, up to 5000 ft.; Creosote Bush Scrub, Joshua Tree Wd.; common on Mojave and ne. Colo. deserts, Inyo Co. to Riverside Co.; to sw. Utah, w. Tex. and n. Mex. March–June; Oct.–Nov. Not always easy to separate from no. 3.

3. **B. multiradiàta** Harv. & Gray. [*B. m.* var. *nudicaulis* Gray.] Biennial or perennial, much like no. 2, but the basal lvs. persisting usually as a rosette and in the lower half of the stem; peduncles bractless, 1–3 dm. long, often stout; invol. 7–8 mm. high, 11–16 mm. wide; ligules 25–50, 11–15 mm. long; $n = 16$ (Carlquist, 1956).—Sandy plains and rocky slopes, 2000–5200 ft.; Creosote Bush Scrub, Joshua Tree Wd.; infrequent in e. Mojave Desert, San Bernardino Co.; to Utah, Tex., n. Mex. April–July; Oct.

40. Whítneya Gray

Perennial herb, with one or few erect stems from a black slender usually branched woody rootstock. Lvs. whitish from a close dense tomentum, opposite, oblanceolate to obovate, entire or nearly so, apiculate, 3-nerved, the basal rather numerous, tapering into the petiole, the cauline relatively few, soon reduced, sessile. Heads radiate, solitary or few in a cyme, long-peduncled, yellow-fld. Invol. broadly campanulate, uniseriate. Ray-fls. ♀, fertile; ligules showy, becoming papery. Disk-fls. perfect, sterile; corolla rather slender, hairy and glandular, persistent. Fertile (ray-) aks. narrowly obovoid, ±

compressed, hirsutulous, epappose. Monotypic. (For Josiah D. *Whitney*, Director of Calif. Geol. Survey, 1860–1870.)

1. **W. dealbàta** Gray. *Arnica*-like plant with somewhat striate stems 15–35 cm. high; basal lvs. 5–10 cm. long including petiole; peduncles 5–15 cm. long, somewhat glandular toward summit; invol. 8–11 mm. high; phyllaries erect, narrowly ovate to broadly lanceolate, acute, thin, softly pubescent and finely glandular; ligules ca. 2 cm. long, spreading. —In light, moist soil of open hillsides and forest openings, uncommon, 4000–7000 ft.; Yellow Pine F., Red Fir F.; Sierra Nevada, from Shasta Co. to Fresno Co. June–Aug.

41. Venegàsia DC.

Tall leafy branching perennial herbs. Lvs. alternate, deltoid-ovate, cordate or truncate at base, acuminate, crenate-dentate to subentire. Heads large, radiate, yellow-fld., in the upper lf.-axils and terminal. Invol. hemispheric, the broadly oval membranaceous phyllaries 3-seriate, the intermediate ones widest. Receptacle nearly flat, hairy. Ray-fls. ♀, numerous, showy, fertile. Corolla-tube of ray and disk glandular-hairy. Aks. clavate, muricate, ca. 10-striate, epappose. Monotypic. (For Michael *Venegas*, early missionary writer upon Calif.)

1. **V. carpesioìdes** DC. [*Parthenopsis maritimus* Kell. *V. deltoidea* Rydb.] Coarse widely branched herbs from a somewhat woody base, 1–2.5 m. high, glabrous or above becoming sparsely canescent; lvs. thin, 5–15 cm. long, gland-dotted beneath; heads usually solitary, on slender peduncles 2–6 cm. long; outer phyllaries foliaceous, loose, the erect inner ones nearly circular, thin, puberulent toward base, the innermost narrower; rays 13–21, 15–20 mm. long, obscurely toothed at the blunt tip; aks. ca. 6 mm. long.—Rocky canyon walls, shaded slopes, and moist stream banks, not far from the coast, 15–2700 ft.; Coastal Sage Scrub, Chaparral, S. Oak Wd.; s. Monterey Co. to San Diego Co.; Channel Ids.; L. Calif. Feb.–Sept.

42. Jaúmea Pers.

Perennial herbs or subshrubs. Lvs. opposite, narrow, entire. Heads radiate or rarely discoid, terminal or axillary, pedunculate. Invol. of rather few broad herbaceous graduated phyllaries in 3–4 series. Receptacle flat or conic, naked. Ray-fls. (when present) ♀, fertile, yellow, deciduous. Disk-fls. fertile, yellow, glabrous. Style-branches with very short blunt minutely papillate appendages. Aks. oblong, 10-nerved. Pappus of scales or bristles, or none. Half a dozen or so spp., one in the Pacific States, the rest scattered from Mex. to S. Am. and s. Afr. (Honoring the French botanist, I. H. *Jaume* St. Hilaire.)

1. **J. carnòsa** (Less.) Gray. [*Coinogyne c.* Less. *Kleinia c.* Kuntze.] Glabrous perennial herb with creeping branched rhizomes, and numerous mostly simple decumbent or ascending stems, 1–3 dm. high; lvs. linear-oblanceolate, fleshy, slightly connate at base, 2–4 cm. long, 2–5 mm. wide; heads mostly solitary; invol. 8–12 mm. high, the phyllaries rounded and often pinkish above; rays 6–10, narrow, inconspicuous, sharply 3-toothed, 3–5 mm. long; receptacle conic; aks. glabrous, ca. 3 mm. long, epappose.— Coastal salt marshes and tidal flats, near sea level; Coastal Strand, Coastal Salt Marsh; at scattered stations along the coast from n. L. Calif. to the Puget Sound region and Vancouver Id.; Channel Ids. May–Oct.

43. Eatonélla Gray

Rather floccose spring annuals. Lvs. alternate. Heads small, in the upper axils or at the ends of the branches, sessile or nearly so, radiate or discoid. Invol. uniseriate; phyllaries free to base, at length reflexed. Receptacle flat, naked. Marginal fls. ♀, usually with inconspicuous ligules, sometimes with flaring-campanulate throat. Disk-fls. yellow, fertile, with glandular tube, 4–5-toothed limb. Style-branches short, truncate. Aks. shiny black, flattened, densely silky-hirsute on the narrow callous margin. Pappuspaleae few, erose, sometimes awn-tipped. Two spp., both in Calif. (For Daniel Cady *Eaton*, botanist at Yale.)

Plants 2–5 cm. high, subacaulescent. Great Basin .... 1. *E. nivea*
Plants 10–25 cm. high, caulescent. San Joaquin V. .... 2. *E. Congdonii*

1. **E. nívea** (D. C. Eat.) Gray. [*Burrielia n.* D. C. Eat. *Actinolepis n.* Gray.] Dwarf cespitose annual, very leafy and much branched, 2–5 cm. high, densely and persistently woolly; lvs. linear-oblanceolate, rounded, entire, 8–18 mm. long; heads usually sessile, sometimes on short filiform peduncles; invol. campanulate, 5–6 mm. high; phyllaries 8–12, oblong-lanceolate; ray-fls. purplish or yellow, narrow, scarcely exceeding the disks; aks. all alike, linear-oblanceolate, compressed but the faces convex, the narrow white callous margin silky-ciliate, ca. 3 mm. long; pappus-paleae 2, quadratish, folded on the angle of the ak., erose, awn-tipped.—Sandy soil among shrubs, 4500–9400 ft.; Sagebrush Scrub, N. Juniper Wd.; Inyo and Mono cos.; to se. Ore., sw. Ida., cent. Nev. May–June.

2. **E. Congdònii** Gray. [*Lembertia C.* Greene.] Annual 5–30 cm. high, with several decumbent sometimes succulent branches from the base, loosely floccose-woolly; lvs. narrowly oblong, repand or sinuate-dentate, or sometimes entire, 1–3.5 cm. long; heads clustered towards the ends of the branches, sessile or short-pedunculate; invol. hemispheric, ca. 4.5 mm. high; phyllaries 4–7, broadly oval or obovate, with deciduous tomentum, the tips often black-tomentose; marginal fls. not ligulate but modified-tubular, not unlike the yellow disk-fls.; aks. flattened, those of the marginal fls. 3-angled, densely pubescent on the faces, those of the disk 4-angled from a low ridge on each face, sparsely pubescent or glabrate on the faces, all ca. 3 mm. long, with a densely hirsute-ciliate callous margin; pappus of ca. 5–7 broad thin hyaline erose scales up to 1 mm. long, muticous; *n* = 10 (Carlquist, 1954).—Sandy or clayey often alkaline plains, below 1200 ft.; Alkali Sink, V. Grassland; San Joaquin V., from Fresno Co. to Kern Co., Carrizo Plain, San Luis Obispo Co. March–April.

44. Perítyle Benth.

Annual herbs or perennial undershrubs, almost glabrous to tomentose, often glandular and aromatic. Lvs. opposite at least below, petioled, toothed, or dissected. Heads small, corymbosely arranged, radiate or discoid. Invol. hemispheric to turbinate, 1–2-seriate; phyllaries ± boat-shaped. Receptacle flat, naked, alveolate. Ray-fls. ♀, fertile, white or yellow. Disk-fls. fertile, numerous, yellow, the tube glandular-puberulent, the limb 4-toothed. Aks. strongly compressed, ciliate on the narrow callous margin. Pappus of a crown of squamellae and usually 1–2 barbellate bristles. Genus of 25 spp., almost wholly confined to sw. U.S. and Mex. (Greek, *peri*, around, and *tyle*, a callus; the aks. callous-margined.)

(Everly, M. L. A taxonomic study of the genus *Perityle* and related genera. Contr. Dudl. Herb. 3: 375–396, 1947.)

1. **P. Emòryi** Torr. in Emory. [*P. nuda* Torr. *P. E.* var. *n.* Gray. *P. californica* Gray, not Benth. *P. E.* var. *Orcuttii* Rose. *P. Greenei* Rose.] Winter annual 1–4 dm. high with ascending brittle branches, glandular-puberulent and sparsely hirsute; lvs. mostly alternate, broadly cordate to ovate, deeply toothed to palmately lobed, the lobes laciniate-toothed, 1–4 cm. long with petioles as long; invol. 5–6 mm. high, the oblanceolate carinate thin-margined glandular-puberulent phyllaries ciliate at tip; rays 10–13, white, rather inconspicuous; aks. arcuate-oblanceolate, flat, black, those of the ray puberulent on the faces, the scarcely thickened margin densely short-ciliate with stiff tawny hairs, 2.5–3 mm. long; pappus a crown of very small lanceolate squamellae and a slender awn ca. as long as the disk-corolla, or the awn wanting.—Crevices of cliffs and among boulders on dry desert slopes, sea level (Death V.) to 3000 ft.; Creosote Bush Scrub; e. Mojave Desert and throughout Colo. Desert, from s. Inyo Co. to Imperial and San Diego cos.; Coastal Sage Scrub; Santa Monica Mts. near the coast, Ventura Co. and Los Angeles Co.; Channel Ids.; to s. Nev., sw. Ariz., Sinaloa, L. Calif. Feb.–June.

45. Laphàmia Gray

Low corymbosely branched subshrubs. Lvs. small, toothed, lobed, or parted, sometimes entire, petioled, the lower opposite, the upper usually alternate, usually scabrous-puberulent. Heads radiate or discoid, small or medium-sized, yellow or the rays white. Invol. hemispheric, 2-seriate. Receptacle low-convex, alveolate, naked. Disk-fls. with slender glandular tube, glabrous rather slender throat, 4-toothed. Aks. flattish, sometimes 4-angled, not ciliate. Pappus of 1 or 2 barbellate awns, or wanting. Nearly 20 spp., of sw. U.S. and n. Mex. (For Dr. Increase A. *Lapham,* Wisconsin amateur botanist.)

Stems loosely villous .... 1. *L. villosa*
Stems merely scabrous-puberulent .... 2. *L. megacephala*

1. **L. villòsa** Blake. Plant suffruticulose with a thick woody caudex, 1–2.5 dm. high; herbage densely and laxly villous and ± glandular, odoriferous; lvs. relatively thin and obscurely nerved, broadly ovate, obtuse at apex, rounded, truncate or cordate at base, coarsely dentate to hastately 3-lobed or subentire, 5–15 mm. long, nearly as wide, short-petioled or the upper sessile; invol. 3.5–7 mm. high; rays none; disk-fls. numerous, yellow or reddish; aks. 3 mm. long, hispidulous, epappose.—Cliff crevices and rocky places, 5800–8500 ft.; Pinyon-Juniper Wd.; Inyo and Panamint ranges, Inyo Co. July–Sept.

2. **L. megacéphala** Wats. [*Monothrix m.* Rydb. *Perityle m.* Macbr.] Rounded aromatic shrub from a thick woody taproot, 1.5–3.5 dm. high, intricately or diffusely branched; herbage scabrid-puberulent to densely hirtellous and beset with small sessile or stalked glands; lvs. oval to rotund-ovate, subentire, 4–6 mm. long, firm, short-petiolate; invol. 4–7 mm. high; rays none; disk-fls. many, yellow; aks. 3–3.5 mm. long, hispidulous, epappose.—Dry cliff crevices in granite or limestone, 5000–9000 ft.; Pinyon-Juniper Wd.; White, Inyo, Panamint, and Cottonwood mts., Inyo Co.; adjacent Nev. n. to the Wassuk Range. June–Nov.

Ssp. **intricàta** (Bdg.) Keck. [*L. i.* Bdg. *Monothrix i.* Rydb.] Lvs. narrower, linear-spatulate, 5–8 mm. long, dilated above to 1.5 mm. wide; invol. 3.5–6.5 mm. high; otherwise similar.—At 6000–8500 ft.; Creosote Bush Scrub, Pinyon-Juniper Wd.; Mazourka Canyon, Inyo Mts., Tin Mt., Cottonwood and Grapevine mts., Inyo Co.; s. Nev. (Pahrump V., Sheep Mt.). July–Aug.

46. Perícome Gray

Tall branching perennial herbs with minutely puberulent and glandular-punctate herbage. Lvs. opposite, petiolate, hastate or deltoid-lanceolate, often caudate-acuminate. Heads in terminal corymbiform cymes, discoid, yellow, the fls. much exserted. Phyllaries thin, uniseriate, loosely united into a cup. Receptacle small, low-conic. Fls. perfect, the slender cylindric corolla ± glandular throughout, 4-toothed. Anthers much exserted. Style-branches slender, elongated. Aks. black, flattened, hirsute on the callous margin. Pappus a crown of lacerate-ciliate squamellae, sometimes with a pair of short marginal awns. Two spp., native of sw. U.S. and Mex. (Greek, *peri,* around, and *come,* a tuft of hairs, referring to the ak.-margin.)

1. **P. caudàta** Gray. Widely spreading aromatic herb up to 2 m. high, suffrutescent at base; lvs. deltoid-lanceolate or subhastate, usually long-caudate, 3–5 cm. long, on petioles 1–5 cm. long; invol. broadly campanulate, 5–6 mm. high; phyllaries 20–25, linear, their tips tomentulose within; corollas 4–4.5 mm. long; aks. linear-oblanceolate, glabrous except for the hirsute-ciliate whitish callous margin, 4–5 mm. long; pappus-paleae ca. 1 mm. long.—Dry stony canyons, 4000–7500 ft.; Pinyon-Juniper Wd.; White and Inyo mts., Sierra Nevada (mostly e. face), from s. Mono Co. to Tulare Co.; to Okla., Tex., Chihuahua. July–Oct.

47. Húlsea T. & G.

Annual, biennial or perennial rather fleshy viscid-pubescent aromatic herbs from a taproot. Lvs. alternate, the usually very numerous basal ones broadly petiolate, the cauline sessile, entire to pinnatifid. Heads many-fld., radiate, yellow or purple. Invol. hemispheric, 2–3-seriate, the herbaceous subequal phyllaries persistent and at length reflexed. Receptacle convex, naked. Ray-fls. numerous, ♀, fertile. Disk-fls. perfect, fertile. Aks. linear, 2–3-angled, compressed, softly villous especially on the ribs. Pappus of 4 hyaline nerveless paleae united at base. Ca. 8 spp. of w. U.S. (G. W. *Hulse,* U.S. Army surgeon and collector.)

Stems ± leafy.
  Plants viscid-pubescent, not at all woolly.
    Heads solitary, 4–6.5 cm. across .... 7. *H. algida*
    Heads corymbose, 2–3 cm. across.
      Stem-lvs. oval to lance-ovate, mostly 15–35 mm. wide, strongly dentate; rays 25–60, purplish
                                                                         1. *H. heterochroma*
      Stem-lvs. narrowly oblong, mostly 5–10 mm. wide, remotely and ± obscurely denticulate; rays
      10–21, yellow .... 2. *H. brevifolia*
  Plants ± woolly at least below.
    All upper cauline lvs. green, not woolly even when young .... 3. *H. californica*
    All lvs. woolly when young .... 4. *H. callicarpha*
Stems scapose.
  Lvs. broadly spatulate, entire or dentate, pannosely tomentose .... 5. *H. vestita*
  Lvs. linear-oblanceolate, pinnatifid, glabrate or tomentulose .... 6. *H. nana*

1. **H. heterochròma** Gray. Robust perennial, with several erect leafy stems from the crown, viscid-villous and heavy-scented, 4–12 dm. high; lvs. oblong, saliently dentate, up to 10 cm. long and 35 mm. wide, sessile or clasping; heads racemose or corymbose; invol. 12–15 mm. high; phyllaries long-acuminate; rays narrowly linear, reddish-purple to yellow, hirsute and glandular; aks. 6–7 mm. long; pappus-paleae strongly unequal, lacerate.—Infrequent in forest openings or chaparral, 3000–8000 ft.; Chaparral, Yellow-Pine F.; Sierra Nevada from Alpine Co. to Tulare and Inyo cos., S. Coast Ranges from Santa Clara and San Benito cos. to San Jacinto Mts.; Pinyon-Juniper Wd.; Panamint Mts., Inyo Co.; Nev. June–Aug.

2. **H. brevifòlia** Gray. Slender perennial, with several usually erectly branched stems from the base, moderately villous and glandular, 3–5 dm. high; lvs. relatively small, ± obscurely repand-dentate; heads terminating leafy stems; invol. 11–12 mm. high; phyllaries acuminate; rays linear, yellow, glabrate or glandular-puberulent; aks. and pappus about as in *H. heterochroma*.—Occasional in forest openings, 6000–8000 ft.; Red-Fir F.; w. slope of Sierra Nevada, Tuolumne Co. to Fresno Co. June–Aug.

3. **H. califórnica** T. & G. Robust perennial, with few erect branching stems from the base, 5–10 dm. high; herbage loosely woolly when young, becoming moderately hirsute and glandular; lower lvs. ample, 6–10 cm. long, 2–5 cm. wide, sinuate-dentate, the upper ones much reduced; heads several, racemose or corymbose; invol. 15 mm. high; phyllaries short- or long-acuminate; rays 20–35, broadly linear, 10 mm. long, yellow, glabrate; aks. 5 mm. long; pappus-paleae unequal, those over the angles 2–3 mm. long. —Open places, up to 6000 ft.; V. Grassland, Yellow Pine F.; San Diego Co.; L. Calif. (*H. mexicana* Rydb.) April–June.

Ssp. **inyoénsis** Keck. Phyllaries broadly lanceolate, acuminate; aks. 7.5 mm. long; pappus-paleae subequal, quadrate, 1 mm. long.—On rocky slopes or talus, 6500 ft.; Pinyon-Juniper Wd.; Inyo and Panamint mts., Inyo Co. June.

4. **H. callicárpha** (Hall) S. Wats. ex Rydb. [*H. vestita* var. *c.* Hall.] Biennial or short-lived perennial; stems branching above or ± throughout, 2–10 dm. high, leafy principally at base or in lower half; lvs. white-tomentose, or the upper ones greenish, the basal crowded, spatulate, winged-petiolate, 3–10 cm. long, to 35 mm. wide, entire or dentate, the upper sessile, becoming bractlike; heads on long scapiform peduncles; invol. 10–13 mm. high; phyllaries acute or acuminate; rays yellow or purplish; aks. 5–7 mm. long; pappus-paleae broadly oblong.—Infrequent on open dry stony slopes or meadow borders, 4000–10,000 ft.; Chaparral, Yellow Pine F.; San Gabriel Mts., Little San Bernardino Mts. to Cuyamaca Mts. May–Aug. Not sharply set off from *H. vestita*, but typically a much larger, leafier plant.

5. **H. vestita** Gray. [*H. Parryi* Gray.] Subscapose perennial 8–30 cm. high, with dense rosettes of white-tomentose spatulate entire or dentate lvs. 2–5 cm. long surmounting the branches of the subterranean caudex; stems several, bearing a few linear bracts, monocephalous, like the invols. viscid-pubescent; invol. 10–13 mm. high; phyllaries acuminate; rays linear, yellow tinged with reddish, or purple; aks. 5 mm. long; pappus-paleae subequal, broadly oblong, lacerate.—Sand flats or gravelly soils, 6000–11,000 ft.; Sagebrush Scrub, Yellow Pine F., Red-Fir F.; Mono Lake and high s. Sierra Nevada, Mono Co. to Kern Co., Frazier Mt. to San Jacinto Mts.; w. Nev. June–Aug.

Var. **pygmaèa** Gray. Depressed compact alpine dwarf mostly less than 6 cm. high, the few heads rising but slightly above the dense basal tufts of lvs. only 1.5–2.5 cm.

long, the lvs. sometimes green on upper surface.—High peaks, 8000–11,400 ft.; Subalpine F., Alpine Fell-fields; s. Sierra Nevada in Tulare Co., San Bernardino Mts. June–Aug.

6. **H. nàna** Gray. [*H. n.* var. *Larseni* Gray. *H. L.* Rydb.] Cespitose scapose perennial, less than 15 cm. high; lvs. in dense rosettes on the branching caudex, oblanceolate, pinnately lobed, 2–7 cm. long, glandular-hirsute and green, or ± white-villous and tomentose; invol. 11–15 mm. high; phyllaries acuminate; rays yellow, 15–25; pappus-paleae often dissected into several linear segms.—In volcanic ash, gravels, or talus, 8000–10,500 ft.; Lodgepole F., Alpine Fell-fields; peaks of the Cascades, from Mt. Lassen, Mt. Eddy, Mt. Shasta and Goosenest; n. to Mt. Rainier; Wallowa Mts. July–Aug.

7. **H. álgida** Gray. Similar to *H. nana*, but more robust throughout; rather succulent strongly glandular-pubescent disagreeably odorous perennial with few to many erect unbranched leafy monocephalous stems 1–4 dm. high from the branching subterranean caudex; lvs. to 15 cm. long including the margined petiole, not woolly; invol. 13–20 mm. high, usually lanate; phyllaries narrower, attenuate; rays 25–50.—Rocky peaks above timber line, 10,000–14,000 ft.; Alpine Fell-fields; Sierra Nevada from Mt. Whitney to Mt. Rose, White Mts.; to Nev., Mont., e. Ore., Ida. July–Aug.

48. **Gaillárdia** Foug. BLANKET FLOWER

Annual or perennial herbs. Lvs. alternate (or all basal), entire to pinnatifid, sometimes resinous-punctate. Heads solitary, radiate, rather large, mostly long-pedunculate. Invol. 2–3-seriate, herbaceous, reflexed in fr. Receptacle convex to subglobose, with soft or firm setae that do not individually subtend the disk-fls. Ray-fls. neutral, or sometimes ♀ and fertile; ligules yellow or partly or wholly reddish-purple. Disk-fls. perfect, fertile. Aks. broadly obpyramidal, villous at least at base. Pappus of 5–10 scarious often awned paleae. Ca. a dozen spp., chiefly e. N. Am., one S. Am. (For *Gaillard* de Merentonneau, French botanist.)

1. **G. pulchélla** Foug. Leafy-stemmed annual 2–4 dm. high, harsh-puberulent; lower lvs. oblanceolate, short-petioled, 4–8 cm. long, entire, dentate or sinuately pinnatifid, the upper lvs. sessile, lanceolate, acute or acuminate, entire; invol. ca. 10 mm. high, the phyllaries lanceolate, acuminate, hirsute, herbaceous, chartaceous at base, soon spreading; ligules 12–20 mm. long, yellow with purple base or wholly purple; *n* = 18 (Morinaga et al., 1929).—Locally natur. as an escape from gardens; native from se. Ariz. to Colo., Nebr., La. June–Aug.

49. **Hymenóxys** Cass.

Annual or perennial herbs with basal or alternate entire to pinnatifid or ternate lvs. Heads mostly radiate; rays ♀ and fertile, broad, yellow, mostly 5–35. Phyllaries partly or wholly herbaceous, in 2–3 similar or sharply differentiated series. Receptacle naked, ± rounded. Disk-fls. ± numerous, yellow, perfect, fertile. Aks. turbinate, hairy, mostly 5-angled. Pappus of a few, mostly 5, hyaline often aristate or awned scales. Ca. 20 spp. of w. N. Am. and S. Am. (Greek, *humen*, membrane, and *oxus*, sharp, because of the pointed pappus scales.)

Lvs. entire, mostly basal .... 1. *H. acaulis*
Lvs. ± dissected, mostly or partly cauline.
Plants perennial; ultimate lobes of lvs. 2–5 mm. wide. N. Calif. .... 2. *H. Lemmonii*
Plants annual to perennial; ultimate lobes of lvs. mostly ca. 1 mm. wide. S. Calif.
Invol. ca. 12 mm. broad, hemispheric; outer phyllaries 12–14; rays 12–14, ca. 12–25 mm. long
3. *H. Cooperi*
Invol. ca. 6–9 mm. broad; outer phyllaries ca. 8; rays 8–10, ca. 7–10 mm. long . . 4. *H. odorata*

1. **H. acáulis** (Pursh) Parker var. **arizónica** (Greene) Parker. [*Tetraneuris a.* Greene. *Actinea a.* var. *a.* Blake.] Tufted perennial from a branched caudex; stems scapose, 1–3 dm. high, subglabrous; lvs. gray-silky, basal, linear-oblanceolate, 2–5 cm. long; heads solitary; invol. hemispheric, 6–8 mm. high, white-villous; rays bright yellow when young, ca. 1 cm. long; aks. silky-villous; pappus of 5–7 hyaline paleae ca. 2 mm. long, abruptly contracted into a very short awn; *n* = 30 (Parker, 1946).—Rocky largely limestone ridges, 4000–8000 ft.; Pinyon-Juniper Wd.; Chuckawalla, Providence, New York, Clark

mts.; to Colo., Ariz. April–June. Widespread variable sp. that has been known under 8 generic names.

2. **H. Lemmònii** (Greene) Ckll. [*Picradenia L.* Greene. *Actinea L.* Blake.] Perennial with a branched root-crown; stems leafy, 3–6 dm. high, simple below, branched above, glabrous; lvs. well distributed, 6–16 cm. long, pinnately divided into linear 3–5 entire or divided segms.; petioles ca. as long as blades; heads in ± corymbose infl.; invol. hemispheric, 7–8 mm. high; outer phyllaries lanceolate, punctate; inner ovate, somewhat erose; rays ca. 12, 8–12 mm. long; disk-fls. densely glandular; aks. silky; pappus-paleae lanceolate, gradually acuminate.—Open places, 2000–5000 ft.; Sagebrush Scrub, Yellow Pine F.; e. Siskiyou Co. to Plumas Co.; Nev., Ariz. June–Aug.

3. **H. Coòperi** (Gray) Ckll. [*Actinella C.* Gray. *A. biennis* Gray. *H. b.* Hall.] Biennial or short-lived perennial, with 1–few stems 3–7 dm. high, puberulent; lvs. 3–10 cm. long, puberulent and glandular-punctate, divided into 3–5 linear divisions, gradually reduced up the stem; heads ± open-corymbose, few to many; disk to ca. 15 mm. wide; invol. turbinate, 5–7 mm. high; outer phyllaries ca. 10–14, keeled, pointed, ± woolly, the inner more rounded, subglabrous; ligules yellow, 8–14 mm. long; pappus-paleae 5, oblong or ovate, obtuse, erose to ± acuminate; *n* = 15 (Parker, 1947).—Dry slopes, 4000–7500 ft.; largely Pinyon-Juniper Wd.; Little San Bernardino, Providence, Clark, and New York mts.; to s. Utah, n. Ariz. May–June, Sept.–Oct.

Var. **canéscens** (D. C. Eat.) Parker. [*Actinella Richardsonii* var. *c.* D. C. Eat. *Actinea c.* Blake.] Plants more canescent; stems 1.5–2.5 dm. high; heads larger, the disk 15–25 mm. wide.—Rocky places, 9000–10,100 ft.; Bristle-cone Pine F.; White Mts.; to Ida., nw. Ariz. July–Aug.

4. **H. odoràta** DC. [*H. chrysanthemoides* var. *excurrens* Ckll.] Annual; stem slender, 3–6 dm. high, branched mostly above or at base, ± pubescent; lvs. 3–6 cm. long, pinnately dissected into linear divisions; heads many, corymbose; invol. ca. 6 mm. high; outer phyllaries ca. 8, lanceolate, united at the thickened base; pappus-paleae ovate, acuminate, awn-pointed; *n* = 11 (Parker, 1947).—Lowlands along Colo. R., Parker Dam to Yuma; to Kans., Mex. Feb.–May.

50. Helènium L. SNEEZEWEED

Annual or perennial herbs, with simple or branched stems. Lvs. alternate, glandular-punctate, usually decurrent. Heads solitary or corymbose, pedunculate, usually radiate, golden-yellow, or the disk-fls. purple-brown. Invol. 1–3-seriate, the phyllaries subequal and usually soon deflexed. Receptacle naked, convex to subglobose. Ray-fls. ♀ or neutral. Disk-fls. very many, perfect. Aks. turbinate or obpyramidal, 4–5-angled, with as many intermediate ribs. Pappus of 5–10 scarious often awn-tipped scales. Ca. 40 spp., restricted to the New World. (A Greek name for some plant, said by Linnaeus to be named for Helen of Troy.)

Lvs. essentially filiform, mostly in fascicles and densely crowded, not decurrent; annual
    1. *H. amarum*
Lvs. wider, not fascicled or densely crowded.
  The lvs. not decurrent on the stems; mature invols. ascending or spreading but not obviously deflexed; rays narrow, showy .... 2. *H. Hoopesii*
  The lvs. decurrent on the stems; mature invols. obviously deflexed; rays reflexed, with flabellate or cuneate tips.
    Lvs. ovate-lanceolate, mostly less than 8 cm. long, crowded, scarcely reduced in size upwards, denticulate or subentire; heads several, mostly short-peduncled .... 3. *H. autumnale*
    Lvs. linear to lanceolate, up to 15 cm. long, fewer, reduced in size upwards, always entire; heads solitary or few, long-peduncled.
      Rays inconspicuous, much shorter than the disk; stems openly branched above
        4. *H. puberulum*
      Rays conspicuous, equaling to much exceeding the disk; stems simple or few-branched above.
        Pappus ca. half as long as the 3.5–4 mm. long disk-corolla; disk 1.5–2 cm. wide, subglobose; peduncles slender .... 5. *H. Bigelovii*
        Pappus nearly as long as the 4.5–5 mm. long disk-corolla; disk 2–3 cm. wide, hemispheric; peduncles stout and enlarged at summit .... 6. *H. Bolanderi*

1. **H. amárum** (Raf.) Rock. [*Gaillardia a.* Raf. *H. tenuifolium* Nutt.] Taprooted glabrous slender annual, 2–7 dm. high; fastigiately branched, densely covered with linear-filiform sessile usually fascicled lvs. 2–4 cm. long, these not decurrent; heads many, golden-yellow, on very slender peduncles, the disk depressed-globose, ca. 1 cm.

wide; rays commonly 8, ca. 1 cm. long, spreading to reflexed; aks. tawny-hirsute; pappus-paleae ovate or orbicular, the slender bristle-tip as long as the body.—A weed of open places near Hamilton Field, Marin Co., and near Modesto, Stanislaus Co.; native in se. U.S., whence it is spreading rather widely as a ruderal.

2. **H. Hoópesii** Gray. [*Heleniastrum H.* Kuntze. *Dugaldia H.* Rydb.] Rather stout perennial herb 4–10 dm. high, the few to several striate but wingless stems arising from a thick aromatic rhizome or caudex; herbage yellow-green, ± villous-tomentose but soon glabrate; basal lvs. narrowly to broadly oblanceolate, up to 3 dm. long, tapering to the broadly clasping base, the cauline gradually reduced in size upwards, lanceolate, all entire; heads orange or yellow, in rather loose terminal corymbs of 3–8, the disk hemispheric, 2–2.5 cm. wide; rays 13–21, 15–25 mm. long, narrowly ligulate, scarcely reflexed; aks. ca. 3 mm. long, densely tawny-hirsute; pappus-paleae hyaline, broadly lanceolate, the thicker midrib excurrent into an awn-tip, nearly as long as the corolla.— Meadows, stream sides and wet slopes, 7500–10,800 ft.; Lodgepole F., Subalpine F.; Sierra Nevada from Tulare and Inyo cos. to Tuolumne and Mono cos., Warner Mts.; to New Mex., Wyo., s. Ore. July–Sept. The cause of much sheep poisoning in the Rocky Mts.

3. **H. autumnàle** var. **montànum** (Nutt.) Fern. [*H. m.* Nutt.] Perennial, with the usually single narrowly angled stem 2–7 dm. high, corymbosely branched at summit, rather densely puberulent and glandular, very leafy; lvs. elliptic to ovate-lanceolate, acuminate, 3–8 cm. long, entire or obscurely serrulate, the lowermost usually withering before anthesis, the uppermost not much reduced; heads yellow, 3-many in a terminal corymb, the disk globose, 1–2 cm. wide; rays 10–20, ca. 1 cm. long, soon reflexed; aks. 1.5 mm. long, tawny-hirsute on the angles and glandular; pappus-paleae lance-ovate, with slender awn-tip as long as the body, much shorter than the corolla.—Damp meadows and bottom lands, rather low elevs.; Sagebrush Scrub, Alkali Sink; Modoc and Siskiyou cos.; e. of the Cascades to B.C., e. to Rocky Mts. July–Aug.

Var. **grandiflòrum** (Nutt.) T. & G. [*H. g.* Nutt.] Stems 4–12 dm. high; herbage brighter green, sparingly puberulent or glabrate; lvs. thinner, more obviously denticulate but verging to subentire; heads larger, the rays 1.5–2.5 cm. long.—Damp places; N. Coastal Coniferous F.; Del Norte Co.; w. of the Cascades to Vancouver Id. Aug. Not a sharply defined var.

4. **H. pubérulum** DC. [*Cephalophora decurrens* Less. *H. d.* Vatke, not *Helenia d.* Moench. *H. pubescens* H. & A., not Ait. *H. californicum* Link. *H. Rosilla* Turcz.] Biennial or short-lived perennial blooming the first year, 3–15 dm. high, divaricately branching above or throughout; herbage puberulent; basal rosette and lower cauline lvs. petiolate, gone before flowering, the cauline lvs. lance-oblong to linear, 3–15 cm. long, 5–40 mm. wide, sessile, prominently decurrent; heads yellow or reddish-brown, terminating the branches on long slender peduncles, the disk globose, 1–1.5 cm. wide; rays 5–10, 3–8 mm. long, reflexed and often nearly concealed; aks. 1.4–2 mm. long, sparsely hairy; pappus-paleae 5, broadly or narrowly ovate, awn-tipped, less than half as long as the disk-corolla.—Moist meadows, marshes, slough-banks, etc., mostly below 2000 ft. (up to 4200 ft. in s. Calif.); many Plant Communities; occasional, Sierra Nevada foothills, Butte Co. to Tuolumne Co., frequent, N. Coast Ranges, Humboldt Co. s., infrequent, outer S. Coast Range to cismontane s. Calif.; L. Calif., Ore. June–Sept.

5. **H. Bigelòvii** Gray. [*Heleniastrum B.* Kuntze. *H. occidentale* Greene. *H. rivulare* Greene. *Helenium r.* Rydb.] Habit of *H. puberulum*, single stems or clumps 4–8 dm. high, simple or few-branched above; lower lvs. often persisting until anthesis, petiolate, the cauline becoming sessile, decurrent, oblanceolate to linear-lanceolate, up to 22 cm. long and 4 cm. wide, the upper ones rapidly reduced; heads mostly solitary on very long naked peduncles, the depressed-globose disk usually deep yellow like the rays, sometimes red-purple, 1.5–2 cm. wide; rays 13–30, showy, 8–22 mm. long, reflexed; aks. 2 mm. long, sparsely hairy; pappus-paleae 6–8, lanceolate, attenuate into the awn-tip, ca. half the length of the disk-corolla.—Common in moist meadowy places; many (especially f. and wd.) Plant Communities; both slopes of Sierra Nevada (mostly 3000–10,000 ft.), Tulare Co. n., mts. of s. Calif. (5000–8500 ft.), mostly N. Coast Ranges, 10–5000 ft.; s. Ore. June–Aug. Intergrading to some extent with *H. puberulum* where their ranges overlap.

6. **H. Bolánderi** Gray. [*Heleniastrum B.* Kuntze. *Dugaldia grandiflora* Rydb. *Helenium*

Bigelovii var. *festivum* Jeps.] Clumps 3–7 dm. high, with several stout stems from a thick rhizome or caudex; herbage villous-tomentose, more densely so about the heads, becoming glabrescent in age; basal lvs. oblanceolate to obovate, narrowed to a petiole, broadly clasping, 8–16 cm. long, to 4 cm. wide, the cauline lanceolate, oblong or ovate, sessile with a clasping base, obscurely decurrent; heads terminating leafless fistulous often ± inflated peduncles, usually solitary, the depressed-globose or hemispheric disk purplish or yellow, 2–3 cm. wide; rays yellow, 15–30, showy, 15–25 mm. long; aks. ca. 2 mm. long, tawny-hirsute on the 9–10 nerves; pappus-paleae 6–8, lance-subulate, gradually tapering to the awn tip, ± laciniate, nearly as long as the corolla-throat.— Frequent in moist meadows and coastal swamps and bluffs, sea level to 500 ft.; N. Coastal Scrub; Mendocino Co. n. to Coos Co., Ore. June–Aug.

51. Baèria F. & M.

Slender vernal annuals or short-lived perennials, glabrous or pubescent; lvs. opposite, sessile, mostly linear, entire or pinnatifid. Heads small, terminal, yellow, radiate, mostly many-fld., pedunculate. Invol. cylindric to hemispheric; phyllaries 1–2-seriate, distinct or slightly united, herbaceous, plane or ± carinate. Receptacle subulate, conic or hemispheric, muricate or scrobiculate. Disk-fls. with slender glandular tube, broad throat, and 5-lobed limb. Style-branches truncate-capitate to ovate, sometimes apiculate. Aks. ± 4-angled or somewhat flattened, slender, all fertile. Pappus of awns or scales or both, sometimes wanting. Genus of 9 spp., largely confined to Calif. (Honoring Karl Ernst von *Baer*, Russian zoologist.)

(Hall, H. M. *Baeria*, in N. Am. Fl. 34 [1]: 76–80, 1914.)
Receptacle subglobose to conic; phyllaries 8–18 (or fewer in #4).
  Herbage ± glandular or viscid. (Section Ptilomeris) .... 1. *B. californica*
  Herbage neither glandular nor viscid.
    Lvs. entire; pappus, when present, of awns or squamellae. (Section Baeria)
      Biennials or short-lived perennials; maritime plants .... 2. *B. macrantha*
      Annuals; not maritime plants .... 3. *B. chrysostoma*
    Lvs., at least the lower ones, pinnately lobed to divided, the lobes sometimes minute.
      Pappus, when present, all aristate; invol. turbinate, the phyllaries ± carinate. (Section Platycarpha) .... 4. *B. platycarpha*
      Pappus, when present, of awns and squamellae intermixed; invol. hemispheric, the phyllaries not keeled. (Section Dichaeta)
        Aks. 2–2.5 mm. long; pubescence of woolly hairs; receptacle conic, beset with terete processes, glabrous .... 5. *B. minor*
        Aks. 1–1.5 mm. long; pubescence of short straight hairs; receptacle globose-conic, beset with rounded papillae, usually pubescent .... 6. *B. Fremontii*
Receptacle subulate; phyllaries 3–6. (Section Burrielia)
  Rays 2–5 mm. long, invol. turbinate or campanulate.
    Stems wiry, simple or few-branched; lvs. nearly filiform; anther-tips subulate .. 7. *B. leptalea*
    Stems weak, freely branching; lvs. linear or lanceolate; anther-tips acute ...... 8. *B. debilis*
  Rays vestigial or wanting; invol. cylindric .... 9. *B. microglossa*

1. **B. califórnica** (Hook.) Chamb. [*Hymenoxys c.* Hook. *Ptilomeris aristata, coronaria, mutica, anthemoides, tenella*, and *affinis* Nutt. *H. coronaria, mutica*, and *calva* T. & G. *Actinolepis coronaria, anthemoides, mutica*, and *tenella* Gray. *B. affinis, tenella, coronaria, mutica*, and *anthemoides* Gray. *B. Parishii* Wats. *B. aristata*, vars. *affinis* and *Parishii* Hall.] Annual 1–3 dm. high, the stems slender, simple to diffusely branched, the herbage densely beset with short- but thick-stalked tiny glands and often also ± villous especially toward the heads; lvs. mostly once or twice pinnately parted into filiform lobes, 1–4 cm. long; peduncles slender; invol. hemispheric, the phyllaries 3–5 mm. long, ovate, acute, thin, with prominent midrib; rays 8–15, 5–10 mm. long; receptacle conic, usually scrobiculate and pilosulous; aks. 2–2.5 mm. long, linear, obscurely nerved, densely scabridulous; pappus-paleae 8–12 or none, quadrate and awnless to lanceolate and awn-tipped, fimbriate, extremely variable, and many colonies exhibiting several combinations, sometimes variable in the same head, as between ray- and disk-fls.; $n = 5$ (Chambers, 1957).—Sunny, sandy spots, often abundant, 100–3000 ft.; Coastal Sage Scrub, V. Grassland, Chaparral, Creosote Bush Scrub; cismontane s. Calif., Los Angeles Co. to San Diego Co. w. edge of Colo. Desert; L. Calif. March–May.

2. **B. macrántha** (Gray) Gray. [*Burrielia chrysostoma* var. *m.* Gray. *Baeria c.* var. *m.* Gray. *Lasthenia m.* Greene. *Baeria m.* var. *littoralis* Jeps.] Short-lived perennial, appar-

ently often blooming the first year, the stems erect or ascending, 1–4 dm. high, simple or few-branched, leafy at least below, from a cluster of thickened roots or short stolons, ± villous with ascending hairs; lvs. almost always entire, narrowly linear to linear-oblong, acute or obtuse, 3–10 cm. long, ± short-hirsute, hispid-ciliate below; invol. broadly hemispheric, 7–12 mm. high, strigose, the phyllaries 2-seriate, ovate, flat; rays 10–20, 8–18 mm. long; receptacle conic, muricate, glabrous; aks. 2.5–3 mm. long, clavate, gray-green, sparingly papillate-scabrid with reddish hairs or glabrate; pappus of 1–4 flattened bristles, or often absent.—Grassy slopes on hills near the sea, below 500 ft.; N. Coastal Scrub; scattered stations, Marin Co. n., to Curry Co., Ore. March–Aug.

Var. **pauciaristàta** Gray. [*B. m.* var. *thalassophila* J. T. Howell.] Low subcespitose plant, spreading by short rhizomes, the stems often subscapose, ascending, 5–20 cm. high; lvs. shorter, broader, thicker, often more crowded; heads equally large, the rays usually 7–12 mm. long; $n = 24$ (Chambers, 1956).—Grassy sea bluffs and sandy beaches; Coastal Strand, Coastal Sage Scrub, N. Coastal Scrub; San Miguel and Santa Rosa ids., San Luis Obispo Co. n., to Curry Co., Ore. Feb.–Aug.

Var. **Bàkeri** (J. T. Howell) Keck. [*B. B.* J. T. Howell.] Stems more slender and rather strictly erect, sometimes few-branched, 1.5–4 dm. high, from a cluster of thickened roots; lvs. of basal rosette linear-filiform, mostly gone before flowering time, those of the stem linear, to 2 mm. wide; heads smaller, the invols. 5–7 mm. high, the rays 7–9 mm. long.—Grassy forest-openings, below 200 ft.; Closed-cone Pine F.; Mendocino Co. coast. May–June.

3. **B. chrysóstoma** F. & M. [*Burrielia tenerrima* DC. *Baeria gracilis* var. *aristosa* Gray. *B. g.* var. *t.* Gray. *B. a.* Howell.] GOLDFIELDS. Slender annual, unbranched, 5–15 cm. high in poor or dry soils, stouter, corymbosely branched, to 25 cm. high in good soils, strigulose or sparsely villous, more densely hairy on the peduncles; lvs. nearly filiform, entire, hirsute-ciliate toward the clasping base; peduncles very slender; invol. broadly hemispheric, the phyllaries 3–6 mm. long, glabrate to moderately strigose, ovate, rather thin; rays 5–10 mm. long, as many as the phyllaries; receptacle narrowly conic, muricate; aks. 2–3 mm. long, linear-clavate, glabrous to rather densely scabrid with vesicular hairs; pappus none, or of 2–7 brownish paleae attenuate from a narrow base, often nearly as long as the corollas; $n = 8$, 16 (Chambers, 1956).—Abundant on open grassland in clay or sand, up to 3000 ft.; N. Coastal Scrub, Coastal Sage Scrub, Coastal Prairie, N. Oak Wd., V. Grassland, Foothill Wd.; near the coast and in the outer Coast Ranges, in cent. Calif. also in the inner Coast Ranges, San Diego Co. to Del Norte Co., Channel Ids., Curry Co., Ore. March–May.

Ssp. **hirsùtula** (Greene) Ferris. [*Lasthenia h.* Greene. *B. h.* Greene.] Rather stout, often branching, erect or depressed, typically short-hirsute and ± strigose; lvs. linear-oblong to oblong-spatulate, obtuse, entire or essentially so; phyllaries 5–7 mm. long, broadly ovate; otherwise similar; $n = 8$ (Chambers, 1956).—The maritime ecotype, on coastal headlands and dunes, below 500 ft.; N. Coastal Scrub, Coastal Sage Scrub; Mendocino Co. to San Luis Obispo Co., Channel Ids., Santa Barbara Co. March–June.

Ssp. **grácilis** (DC.) Ferris. [*Burrielia g.* DC. *Burrielia longifolia, hirsuta* and *parviflora* Nutt. *Baeria g.* and var. *paleacea, curta, Clevelandii,* and *Palmeri* var. *clementina* Gray. *B. c.* var. *gracilis* Hall.] Erect, 1–2 dm. high, strigose throughout and somewhat villous, the lvs. obscurely if at all hirsute-ciliate toward base; phyllaries lance-ovate, obviously hairy; pappus usually present, of 2–5 whitish ovate-lanceolate paleae attenuate into a long slender awn; $n = 8$ (Chambers, 1956).—The most abundant composite in the state, coloring sunny plains and slopes, up to 4000 ft.; V. Grassland, Foothill Wd., Chaparral, etc.; Cent. V. and bordering foothills, inner S. Coast Range, cismontane s. Calif., w. edge of deserts; sw. Ore., Ariz., Son., L. Calif. March–May.

4. **B. platycárpha** (Gray) Gray. [*Burrielia p.* Gray. *Baeria carnosa* Greene. *Lasthenia p.* and *c.* Greene.] Erect annual, usually with several reddish stems from the base, 1–3 dm. high, the herbage slightly succulent, sparingly woolly, most densely so on the peduncles, becoming glabrate; lvs. linear or filiform, with ca. 2 short filiform lobes, or entire, 2–6 cm. long; invol. broadly turbinate, the 5–9 phyllaries 6–9 mm. high, elliptic to ovate, rather firm, 3-nerved, usually ± keeled; rays elliptic, 6–8 mm. long; receptacle narrowly conic, strongly muricate, glabrous; aks. ca. 3 mm. long, narrowly clavate, puberulent; pappus of 4–7 stramineous narrowly ovate scales abruptly tapering to a

narrow awn, or none.—Alkaline grassy places, below 500 ft.; V. Grassland; Cent. V. from Tehama Co. to Merced and Fresno cos. March–May.

5. **B. minor** (DC.) Ferris. [*Monolopia m.* DC. *Dichaeta tenella* and *uliginosa* Nutt. *B. u.* and vars. *tenella* and *tenera* Gray. *B. tenella* Greene. *Lasthenia u.* and *tenella* Greene. *Eriophyllum minus* Rydb.] Annual 1–3.5 dm. high, the stems erect or ascending, often diffusely branched, rather lax, sometimes fistulous, the herbage ± woolly-villous above when young, becoming glabrate; lvs. ligulate, thin, 2–15 cm. long, with rachis 2–10 mm. wide, and few or several salient linear lobes 2–15 mm. long, or linear and quite entire in slender plants; heads small, on short peduncles, 10–13 mm. across; invol. depressed-hemispheric, the phyllaries 8–13, ovate, thin, hairy on the margin, midrib and within the tip; rays usually 8–11, oblong, 4–8 mm. long, pale yellow; receptacle broadly conic, densely beset with terete pedicellike processes that bear the florets, glabrous; aks. 2–2.5 mm. long, narrow, angled, ± strigose; pappus, when present, of 4–6 truncate fimbriate paleae, with 2–4 awns nearly as long as the corolla interspersed; $n = 4$ (Chambers, 1956).—Forming colonies in damp soil of bottom lands, stream banks, seepages and slopes, 100–2000(–3300) ft.; V. Grassland, Foothill Wd., Coastal Sage Scrub; San Joaquin V. from Amador and San Joaquin cos. to Kern Co., Coast Range valleys from Sonoma Co. to Santa Barbara; n. Los Angeles Co. March–April.

Ssp. **marítima** (Gray) Ferris. [*Burrielia m.* Gray. *Baeria m.* Gray.] Decumbent or prostrate succulent plant with relatively stubbier, short-lobed lvs. and larger heads, these 15–20 mm. across, commonly with 13 rays only 2–4 mm. long; $n = 4$ (Chambers, 1956).—The maritime ecotype, occasional along the immediate coast, on grassy headlands, etc., 10–500 ft.; N. Coastal Scrub, Coastal Sage Scrub; Del Norte Co. to Monterey Co.; to B.C. March–May.

6. **B. Fremóntii** (Torr.) Gray. [*Dichaeta F.* Torr. *Burrielia F.* Benth. *Baeria Burkei* Greene. *B. F.* var. *heterochaeta* Hoov. *Lasthenia conjugens, F.* and *B.* Greene.] Slender erect annual with 1 to several stems from the base, sometimes freely branching, 1–3 dm. high, glabrous below, softly and finely appressed-pubescent above, particularly on the peduncles; lvs. 2–6 cm. long, filiform or narrowly linear, at least the lower pinnately parted with 2–4 divergent lobes, rarely all entire; heads on elongated peduncles, obscurely glandular; invol. broadly hemispheric, the phyllaries ovate, ± strigose, canescent on inner face, 4–5 mm. high; rays 8–13, 3–6 mm. long, elliptic; receptacle conic-globose, papillate, pilosulous; aks. 1–1.4 mm. long, clavate, scabrous; pappus of (1–)4 slender awns gradually widening to the base, interspersed between 4–8 minute subquadrate lacerate scales scarcely wider, rarely none; $n = 6$ (Chambers, 1956).—Low sunny flats, drying borders of vernal pools, etc., up to 700 ft.; V. Grassland; Cent. V. from Butte and Tehama cos. to Kern Co., Sierran foothills, drier inner Coast Range valleys in Mendocino, Solano, Contra Costa, and Santa Barbara cos. April–May. Pappus variable, occasionally almost all members being minute squamellae (*B. Burkei*), sometimes, flattened awns (*B. F.* var. *heterochaeta*).

7. **B. leptàlea** (Gray) Gray. [*Burrielia l.* Gray.] Very slender erect annual 7–15 cm. high, usually unbranched, glabrous or somewhat pubescent toward heads; lvs. filiform, entire; invol. campanulate to turbinate, the 4–6 phyllaries broadly oblong, 3–5 mm. high, ciliate; rays few, 3–4 mm. long, oblong; receptacle subulate; anther-tips filiform; aks. 2 mm. long, sparsely puberulent apically; pappus of 1–4 flattened awns tapering from a somewhat broadened base.—Dry mt. slopes, 700–2000 ft.; Foothill Wd., V. Grassland, etc.; e. Monterey and San Luis Obispo cos. April.

8. **B. débilis** Greene. Weak-stemmed annual 1–3 dm. high, openly branching, sparingly downy or villous with multicellular hairs; lvs. thin, linear or linear-oblong, 1–6 cm. long, 1–5 mm. wide, entire or sometimes obscurely few-toothed or -lobed; invol. campanulate to broadly turbinate, the 5–6 phyllaries narrowly ovate to broadly obovate, acute or obtuse, thin, 4–6 mm. high; rays 5–6, 3–5 mm. long, elliptic; receptacle subulate, 0.5 mm. wide at base, foveolate; anther-tips ovate-lanceolate; aks. 2.5–4 mm. long, nearly linear, somewhat flattened, scabrid with vesicular hairs; pappus of 2–4 ovate-lanceolate paleae tapering to a slender awn; $n = 4$ (Chambers, 1956).—Grassy places, often under the protection of bushes or boulders, 900–3300 ft.; V. Grassland, Foothill Wd., Chaparral; e. side of San Joaquin V. and adjacent Sierran foothills, Mariposa Co. to Kern Co. March–May.

9. **B. microglóssa** (DC.) Greene. [*Burrielia m.* DC. *Lasthenia m.* Greene. *Pentachaeta laxa* Elmer.] Ephemeral usually much-branched annual 5–20 cm. high, sparsely pilose; lvs. thin, linear, entire, 2–4 cm. long, 1–2.5 mm. wide; heads usually numerous on short peduncles; invol. cylindric, the 3–5 phyllaries oblong, acute or obtuse, more densely hairy on the inner face, 4–8 mm. high; rays 1–4, shorter than the disks, very inconspicuous; receptacle short-subulate, ca. 1 mm. long, papillate, glabrous; aks. 3.5–5 mm. long, twice as long as the very slender yellow florets, linear, angular, scabrous; pappus of 2–4 rigid flattened paleae, tapering from the narrow base to a slender awn, equaling the fls.; $n = 12$ (Chambers, 1956).—Frequent on drying sunny or shaded banks, 200–3200 ft.; V. Grassland, Foothill Wd., Chaparral, Creosote Bush Scrub; Tehama Co., inner S. Coast Ranges from Contra Costa Co. to Kern Co., Greenhorn Range, w. borders of Mojave and Colo. deserts. March–May.

52. Lasthènia Cass.

Slender rather succulent winter annuals, glabrous to moderately puberulent especially below the heads. Lvs. opposite, narrow, entire or sparsely denticulate, sessile and somewhat sheathing at base. Heads yellow-fld., pedunculate, terminating the stems. Invol. hemispheric, externally glabrous, the phyllaries 1-seriate, united to form a 5–15-toothed or -lobed herbaceous cup. Receptacle high-conic, muricate. Ray-fls. ♀, fertile. Disk-fls. hermaphrodite, fertile, with slender glandular tube, widely campanulate throat, and 4–5-lobed spreading limb. Aks. oblanceolate in outline, somewhat compressed and obtusely 2–4-angled. Pappus paleaceous or none. Genus of 3 spp., 2 in Calif. and Ore., 1 in Chile. (Named for a Greek girl who attended the lectures of Plato in the garb of a man.)

Ligules shorter than the invol., the head appearing discoid; pappus conspicuous .... 1. *L. glaberrima*
Ligules conspicuous; pappus none .... 2. *L. glabrata*

1. **L. glabérrima** DC. Stems slender, lax, frequently simple or few-branched, rooting from the lower nodes, 1–3.5 dm. high, essentially glabrous throughout; lvs. narrowly to broadly linear, entire, obtuse, 3–7 cm. long; peduncles up to 15 cm. long, enlarged below the heads; invol. 5–7 mm. high, the usually deltoid teeth pubescent within; rays shorter than the invol.; disk-corollas 1.5–2 mm. long, the limb essentially glabrous; aks. sublinear, compressed, 4 mm. long, gray, with a sparse ferrugineous spiculate appressed pubescence; pappus of 8–10 brown firm laciniate paleae, some awn-tipped, to 2 mm. long.—Wet meadowy places, mucky clay soils, 10–1000 ft.; N. Coastal Scrub, N. Oak Wd., V. Grassland; Sacramento V., Coast Range valleys from Alameda Co. n.; to the Columbia R. April–June.

2. **L. glabràta** Lindl. [*L. californica* DC. ex Lindl. *Hologymne g.* and *c.* Bartl. *Monolopia g.* and *c.* F. & M. *Xantho g.* Remy. *L. g.* var. *Coulteri* Gray. *L. C.* Greene. *L. g.* var. *californica* Jeps.] Plant 2–6 dm. high, the stem simple or corymbosely branched, often fistulous, glabrous or becoming canescent toward the heads; lvs. linear or the upper ones lanceolate, entire or obscurely toothed, 3–10 cm. long; peduncles up to 12 cm. long; invol. depressed-hemispheric, 6–8 mm. high, the usually deltoid teeth pubescent within; rays 5–10 mm. long, orange-yellow; aks. narrowly oblong-oblanceolate, slightly compressed, obscurely angled, 1.8–2.3 mm. long, gray, smooth or ± rusty-glandular-muriculate.—Colonial in heavy soils, vernal pools, low alkaline fields, hillsides, etc., 10–4500 ft.; V. Grassland, Alkali Sink, N. Oak Wd., Coastal Salt Marsh, etc.; Cent. V., Butte Co. to Kern Co., Coast Range valleys, Humboldt Co. to San Diego; Santa Rosa Id. March–May.

53. Crockéria Greene ex Gray

Slender low vernal annuals. Lvs. opposite, sessile, linear, entire. Heads solitary or loosely cymose, peduncled, radiate, many-fld., golden yellow. Invol. hemispheric; phyllaries thin, united more than half their length into a cup, the tips deltoid and pilose within. Receptacle conic. Ray-fls. ♀, fertile. Disk-fls. perfect, fertile, the very slender glandular-atomiferous tube equaling the campanulate glabrous throat and 5-toothed limb. Anther-appendages obcordate. Style-branches very short. Aks. flat, striate, obovate, glabrous to ± glandular-scabrous on the faces, strongly crested with a fringe of clavate glandular hairs on the margin. Pappus none. Monotypic. (Honoring Charles *Crocker*, San Francisco patron of Calif. botany.)

1. **C. chrysántha** Greene. Stem erect, 10–25 cm. high, simple or with sharply ascending branches, the herbage glabrous except for puberulent peduncles; lvs. 1–5 cm. long, 1–2 mm. wide, ± connate at base; peduncles 2–7 cm. long; invol. ca. 5 mm. high, the phyllaries 8–12; rays 6–10, narrowly oblong, 5–7 mm. long, entire or irregularly 2–3-toothed; aks. 2 mm. long, dull black, microscopically muriculate.—Low alkaline fields and grassy plains, below 500 ft.; Alkali Sink, V. Grassland; San Joaquin V. from Stanislaus Co. to King and Tulare cos. March–April.

54. Bahía Lag.

Annuals or perennials, rarely suffrutescent, pubescent but not woolly. Lvs. alternate or opposite, entire to dissected. Heads radiate, or rarely discoid, yellow, terminating the branches. Invol. obconic to hemispheric; phyllaries not punctate, 1–2-seriate, subequal, broader above the middle. Receptacle flat. Ray-fls. when present ♀, fertile. Disk-fls. perfect, fertile. Aks. narrow, tapering to base, 4-angled. Pappus paleaceous, with callous-thickened base or midrib, or rarely none. Ca. 15 spp., of sw. U.S., Mex., and w. S. Am. (Honoring Juan Francisco *Bahi*, Barcelona botany professor.)

Biennial; ray-fls. present; pappus wanting .... 1. *B. dissecta*
Annual; ray-fls. wanting; pappus present .... 2. *B. neomexicana*

1. **B. dissécta** (Gray) Britton. [*Amauria d.* Gray. *Villanova chrysanthemoides* Gray. *B. c.* Gray. *V. d.* Rydb. *Amauriopsis d.* Rydb.] Biennial or short-lived perennial 3–8 dm. high; stems minutely hirtellous and glandular, often anthocyanous, openly branching above; lvs. once to thrice ternately divided into linear or oblong rounded lobes, 2–7 cm. long, strigillose; peduncles 1–6 cm. long, densely glandular-pubescent; invol. hemispheric, 5–6 mm. high, the phyllaries ± glandular-hairy, narrowly obovate, abruptly narrowed to a short caudate tip; rays mostly 10–13, ca. 6 mm. long; aks. black, longitudinally striate, minutely hirtellous, epappose.—Gravelly open-timbered slopes and dry rocky ridges, 6000–8600 ft.; Yellow Pine F.; San Bernardino Mts., chiefly Santa Ana R. system, and Santa Rosa Mts.; to s. Nev., Wyo., w. Tex., n. Mex. Aug.–Sept.

2. **B. neomexicàna** (Gray) Gray. [*Schkuhria n.* Gray. *S. multiflora* H. & A., not *B. m.* Nutt. *Amblyopappus n.* Gray. *Achyropappus n.* Gray ex Rydb. *Cephalobembix n.* Rydb.] Slender often multi-branched annual 1–2 dm. high, the herbage hirtellous; lvs. opposite below, alternate within the corymbose infl., linear-filiform, usually medianly 3-divided, impressed-punctate, 2–3 cm. long; heads broadly obconic, discoid; invol. short-pilose and glandular-puberulent, ca. 6 mm. high, the phyllaries narrowly to broadly oblanceolate, obtuse, the thin outer margin purplish; fls. pale yellow or whitish, very small; aks. quadrangular, slender, tapering to the sericeous base, strigillose above, 2–5 mm. long; pappus of 8 obovate scarious paleae 1–1.5 mm. long.—Sandy slopes and washes, 5000–5500 ft.; Pinyon-Juniper Wd.; Clark Mt., e. Mojave Desert; e. to Colo., Tex., n. Mex.; w. S. Am. Sept.

55. Monolòpia DC.

Erect floccose-woolly annual herbs. Lvs. opposite below, the upper alternate, sessile. Heads rather large, pedunculate, terminating the branches. Invol. hemispheric or campanulate; phyllaries distinct to base or connate into a cup, the tips usually bearing some black hairs. Receptacle high-conic. Ray-fls. ♀, fertile, the corolla with a minute posterior lobe. Disk-fls. hermaphrodite, fertile, the tube glandular-hairy, the throat dilated. Aks. oblanceolate, sharply 3–4-angled, somewhat compressed, or the outer obcompressed, epappose. Four spp., all in Calif. (Greek, *mono*, single, and *lopos*, covering, in reference to the uniseriate invol.)

(Crum, Ethel. A revision of the genus Monolopia. Madroño 5: 250–270, 1940.)
Phyllaries distinct to base.
  Disk-aks. ca. as broad as thick; rays mostly 4–9 mm. long, rounded apically, subentire.
    Branches widely divergent; aks. essentially glabrous, 2 mm. long. Coast Ranges, Contra Costa and

San Mateo cos. to San Luis Obispo Co. . . . . 1. *M. gracilens*
Branches strict; aks. densely strigose, 2.5 mm. long. W. side of San Joaquin V. and adjacent
foothills . . . . 2. *M. stricta*
Disk-aks. obcompressed; rays mostly 10–20 mm. long, truncate, dentate. San Joaquin V., inner
S. Coast Ranges, s. Calif. . . . . 3. *M. lanceolata*
Phyllaries united half their length into a cup. Cent. V. and Coast Ranges, Tehama Co. to Monterey co.
4. *M. major*

1. **M. grácilens** Gray. [*M. major* var. *g.* Macbr.] Stems 1.5–8 dm. high, usually simple below, divergently several-branched in upper half, lanate or arachnoid-tomentose, this partially deciduous in age; lvs. crowded below, scattered above, oblanceolate to linear-oblong, dentate to entire, 2–10 cm. long; heads terminating slender divergent peduncles up to 10 cm. long; invol. hemispheric, 5–7 mm. high; phyllaries 7–11, free, thin, ovate or rhombic, acute, somewhat black-pilose and glandular; rays golden-yellow, 5–10 mm. long; aks. sparsely strigillose to glabrous, scarcely at all compressed, 1.7–2 mm. long; *n* = 12 (Carlquist, 1956).—Well-drained openings in woods or brush, 500–3800 ft.; Mixed Evergreen F., Redwood F., Chaparral; mostly outer S. Coast Range, Mt. Diablo and Santa Cruz Mts. to Santa Lucia Mts., San Luis Obispo Co. April–June.

2. **M. strícta** Crum. Stems 1–6 dm. high, simple to branched ± throughout, the branches strict, the herbage sparsely floccose to glabrate; lvs. evenly distributed, oblanceolate or narrowly oblong, entire or sometimes undulate or remotely dentate, 3–8 cm. long; heads scattered, on peduncles 3–5 cm. long; invol. hemispheric, 5–7 mm. high; phyllaries usually 8, free, ovate-lanceolate to rhombic, the short-acuminate tips black-lanate; rays 8, bright yellow, 4–7 mm. long; aks. uniformly gray-strigose, scarcely at all compressed, 2.5–2.8 mm. long; *n* = 10 (Carlquist, 1956).—Abundant on barren hills and valley floors, 150–2000 ft.; V. Grassland; inner S. Coast Ranges and adjacent San Joaquin V., from San Benito Co. to w. Kern Co. March–April. On the e. side of the San Joaquin V. from Tulare Co. to the Tehachapi foothills occur occasional recombinations between this sp. and *M. lanceolata*.

3. **M. lanceolàta** Nutt. [*M. major* var. *l.* Gray.] Stems 1–6 dm. high, slender to stout and fistulous, simple to diffusely much branched, densely lanate but becoming glabrate on lower parts; lower lvs. broadly lanceolate, 3–10 cm. long, the upper ones lanceolate to linear, entire or undulate-dentate; heads mostly terminal, on divergent peduncles 1–12 cm. long; invol. broadly campanulate to depressed-hemispheric, 6–10 mm. high, the 8 or so free lanceolate to ovate-lanceolate phyllaries densely white-lanate below and black-lanate on the acute or acuminate tips; rays mostly 8, bright yellow, 10–20 mm. long; aks. mostly uniformly gray-strigose, obcompressed, 2.2–3.8 mm. long, those of the ray tending to be triquetrous, those of the disk quadrangular; *n* = 10 (Carlquist, 1956).—Locally abundant on grassy slopes and valley floors, 500–4000 ft.; V. Grassland, Chaparral, Foothill Wd., S. Oak Wd.; S. Coast Range valleys and San Joaquin V. from Salinas V. and San Joaquin Co. to Riverside Co.; on the e. side of San Joaquin V. from Fresno Co. to Tehachapi; rare e. to Mojave, Kern Co. March–May.

4. **M. màjor** DC. Very similar in aspect to *M. lanceolata*, but tending to be larger in all its parts; stems 1–6 dm. high, relatively few-branched; invol. 8–13 mm. high, the usually 8 phyllaries united for half their length into a cup; aks. glabrate or ± strigillose toward apex, more strongly obcompressed, 2.5–4 mm. long; *n* = 12 (Carlquist, 1956).—Locally abundant on rolling hills and plains, 100–1500 ft.; V. Grassland; inner Coast Ranges and adjacent Cent. V., from Tehama Co. to San Benito Co., San Francisco Bay region to n. Monterey Co. March–May.

56. Eriophýllum Lag.

Annual to perennial ± permanently woolly herbs or subshrubs. Lvs. mostly alternate, usually toothed or divided, sometimes entire. Heads solitary or cymosely or corymbosely clustered at the ends of the branches. Invol. narrowly campanulate to hemispheric; phyllaries 1-seriate, firm and permanently erect, distinct or somewhat united at base, the base sometimes cartilaginous-thickened. Receptacle strongly convex to nearly flat, usually naked. Ray-fls. (0–)4–15, fertile, yellow (white in 2 spp.). Disk-fls. fertile, almost always with glandular or hairy tube, terete to broadly campanulate throat, 5-lobed limb. Anther-appendages triangular-ovate to linear-subulate. Style-branches truncate to obtuse or conical. Aks. linear, 4(5)-angled. Pappus of firm nerveless often fimbriate hyaline or opaque paleae, rarely obsolete. (Greek, *erion,* wool, and *phyllon,* lf., the herbage woolly.)

(Constance, Lincoln. A systematic study of the genus Eriophyllum. Univ. Calif. Publ. Bot. **18:** 69–136, 1937.)
A. Perennials (or biennials), somewhat woody at base.
  B. Heads 1.5–3 cm. broad, long-peduncled (3 cm. or more).
    C. Plants woody only at base; heads solitary or very loosely corymbose; invol. 5–12 mm. high. Widespread .... 1. *E. lanatum*
    CC. Plants woody above the base; heads in loose corymbiform clusters. Restricted endemics.
      D. Invol. 5–7 mm. high; aks. hispidulous. .... 2. *E. Jepsonii*
      DD. Invol. 4–5 mm. high; aks. glabrous .... 3. *E. latilobum*
  BB. Heads 1.5 cm. or less broad, sessile or short-peduncled.
    C. Phyllaries 5–6; peduncles supporting infls. slender .... 4. *E. confertiflorum*
    CC. Phyllaries 8–12; peduncles supporting infls. stout.
      D. Lvs. glabrate on upper surface, 3–7 cm. long, entire to bipinnatifid; rays 3–5 mm. long, 2–3 mm. wide; pappus to 1 mm. long .... 5. *E. staechadifolium*
      DD. Lvs. white-tomentose on both surfaces, 5–25 cm. long, 2–3-pinnatifid; rays 2 mm. long, 1 mm. wide; pappus to 2.2 mm. long .... 6. *E. Nevinii*
AA. Small annuals, 2–20 cm. high.
  B. Heads pedunculate; pappus-paleae, if present, essentially entire.
    C. Rays present; plants 3–20 cm. high; receptacle conical or convex.
      D. Anther-appendages ovate or lanceolate.
        E. Pappus 0.5–2 mm. long. Cent. Sierra Nevada .... 7. *E. nubigenum*
        EE. Pappus less than 0.4 mm. long or none. S. Sierra Nevada and deserts
                                                                                 8. *E. ambiguum*
      DD. Anther-appendages linear-subulate.
        E. Rays white; pappus of unequal paleae .... 9. *E. lanosum*
        EE. Rays yellow; pappus of equal paleae, or reduced to a short crown
                                                                                 10. *E. Wallacei*
    CC. Rays absent; plants 1–2 cm. high .... 11. *E. mohavensis*
  BB. Heads sessile in leafy-bracteate terminal clusters; receptacle nearly flat.
    C. Heads radiate .... 12. *E. multicaule*
    CC. Heads discoid .... 13. *E. Pringlei*

1. **E. lanàtum** (Pursh) Forbes. [*Actinella l.* Pursh. *Trichophyllum l.* Nutt. *Actinea l.* Steud. *Helenium l.* Spreng. *Bahia l.* DC.] Herbaceous perennial, erect or decumbent from a woody base, 1–8 dm. high, with a persistent or partially deciduous tomentum throughout; stems few to numerous, simple or few-branched; lvs. variable, usually toothed or divided, 1–8 cm. long; heads solitary or loosely corymbose, on long naked peduncles, showy; invol. campanulate to hemispheric, 5–12 mm. high, 6–15 mm. broad; ray-fls. 8–13, yellow; aks. 2.5–4 mm. long, 4-angled, tapering to base, glabrous, hairy, or glandular; pappus of 4–12 scales of diverse shape and length, or reduced to a toothed crown, or none.—Rather common in brushy places, 10–10,000 ft.; many Plant Communities; s. Calif. n.; to B.C., Great Basin, Rocky Mts. The typical var., marked by robust plants with large showy heads solitary at the ends of stems and usually pinnatifid lvs., ranges from nw. Mont. to cent. Wash. and ne. Ore. Constance referred here smaller-headed material from w. of the Cascades which we refer to var. *leucophyllum.*

KEY TO VARIETIES

A. Tube of disk-corolla conspicuously glandular or glandular-hairy.
  B. Lvs. more densely tomentose on the under surface, the upper surface usually glabrate and green.
    C. Upper lf.-surface darkening on drying; pappus squamellate, but much reduced.
      D. Lf.-blades narrowly dissected, the lobes ± narrow and acute. Del Norte Co. n.
                                                                              var. *leucophyllum*
      DD. Lf.-blades relatively prominent, the lobes mostly shallow, broad, ± blunt. Redwood Belt .... var. *arachnoideum*
    CC. Upper lf.-surface not turning dark.
      D. Lvs. oblanceolate to obovate, cuneate and entire in lower half, coarsely serrate or lobed in upper; pappus reduced to a ring without squamellae, or nearly so. Sierran .... var. *croceum*
      DD. Lvs. narrowly dissected; herbage whiter.
        E. Heads numerous, corymbose, 7–10 mm. across the disk; lvs. mostly 1–2-pinnately or -ternately divided. Mostly Coast Ranges.
          F. Heads radiate .... var. *achillaeoides*
          FF. Heads discoid .... var. *aphanactis*
        EE. Heads fewer, longer- and stouter-peduncled, 10–15 mm. across the disk; lvs. less divided. Mostly Sierran foothills .... var. *grandiflorum*
  BB. Lvs. ca. equally tomentose on both surfaces, not pinnatifid.

C. Plants 1–2 dm. high, cespitose; rays mostly 6–8; invol. campanulate. High Sierran
var. *monoense*
CC. Plants 1.5–5 dm. high, less compact.
    D. Invol. loosely tomentose, broadly campanulate, 6–8 mm. high; rays mostly 8;
    plants mostly less than 2 dm. high. Sierran . . . . var. *integrifolium*
    DD. Invol. densely and permanently woolly, hemispheric (or nearly so), 7–12 mm.
    high; rays 10–15; plants mostly more than 2 dm. high.
        E. Aks. hairy; herbage ± pannose; stems relatively stout. Klamath region
var. *lanceolatum*
        EE. Aks. glabrous; herbage moderately tomentose; stems relatively slender.
        Mts. of s. Calif. . . . . var. *obovatum*
AA. Tube of disk-corolla glabrous. Tehachapi Mts.   var. *Hallii*

Var. leucophýllum (DC.) W. R. Carter. [*Bahia l.* DC. *E. caespitosum* Dougl. *E. c.* var. *l.* Gray. *E. superbum* Rydb.] Stems 2–6 dm. high, erect or decumbent at short woody base; herbage loosely tomentose, the upper surface of lvs. becoming glabrous and darkening on drying; lvs. usually pinnatifid or pinnatisect, the linear lobes with involute margin, or the lower lvs. sometimes entire and the others with shallower and broader lobes; heads loosely corymbose or solitary on the stems; invol. 5–8 mm. high; rays 8–12 mm. long; aks. 2.5–3.5 mm. long, glabrous; pappus variable, usually of 8–10 lanceolate or quadrate toothed paleae ca. 0.5 mm. long.—Moist uplands or dry canyonsides, below 1000 ft.; Douglas-Fir F.; Del Norte Co.; w. of the Cascades to B.C. June–July. Most of the vars. intergrade freely.

Var. arachnoidèum (Fisch. & Avé-Lall.) Jeps. [*Bahia a.* Fisch. & Avé-Lall. *B. latifolia* Benth. *B. lanata* var. *brachypoda* Gray. *E. a.* Greene.] Stems 3–6 dm. long, erect, decumbent, or prostrate, densely leafy to the relatively short peduncles; herbage loosely floccose, the lvs. green and glabrate on upper surface; lvs. relatively broad and thin, few-lobed or pinnatifid, the lobes entire or sinuately dentate, the margin usually revolute; heads few, on terminal or axillary peduncles 3–10 cm. long; invol. 8–10 mm. high, loosely to densely floccose; rays 8–15, 8–10 mm. long; aks. 3–4 mm. long, glabrous; pappus a mere crown of obtuse erose teeth, typically shorter than areole, sometimes none; $2n = 16$ (Carlquist, 1956).—Coastal bluffs or shady banks near the coast, often in dense stands of perennials, mostly below 1000 ft.; N. Coastal Scrub, Redwood F.; Del Norte Co. to Monterey Co. April–Aug.

Var. cròceum (Greene) Jeps. [*E. c.* Greene. *E. chrysanthum* Rydb.] Stems mostly simple, 1–6 dm. high, from a decumbent ± cespitose base or from slender stolons; herbage arachnoid-woolly, the lvs. becoming glabrate on upper surface; lvs. narrowly oblanceolate to obovate in outline, coarsely serrate or lobed in outer half, the short broad usually obtuse lobes mostly entire; heads terminating peduncles 3–8 cm. long; invol. 5–8 mm. high; rays usually 13, 8–10 mm. long; aks. and pappus as in var. *arachnoideum.*—Infrequent, on partially shaded slopes, 2500–5500 ft.; Yellow Pine F.; w. slope of Sierra Nevada, from Butte Co. to Tulare Co. May–July.

Var. achillaeòides (DC.) Jeps. [*Bahia a.* DC. *E. a.* Greene. *E. ternatum* Greene. *E. idoneum* Jeps. *E. Greenei* Elmer. *E. Cineraria* Rydb.] Biennial or perennial, often forming shrubby clumps 1.5–7 dm. high; stems corymbosely branching above; herbage loosely tomentose; lvs. ± ternate-pinnatifid, the segments rather narrow; heads usually numerous, relatively small, on slender peduncles; invol. 5–8 mm. high; rays 8–13, 6–9 mm. long; aks. 2.5–3 mm. long, hispidulous or glabrous; pappus usually of 6–10 oblong white paleae ca. 0.5 mm. long; $n = 8$, 16 (Carlquist, 1956).—Frequent in light or rocky soils on dry slopes with scattered trees or brush, 100–5400 ft.; Mixed Evergreen F., Douglas-Fir F., Yellow Pine F., N. Juniper Wd., N. Oak Wd., Chaparral; common in the Coast Ranges from Klamath Mts. to Santa Clara and Santa Cruz cos.; rarer in the Cascade-Sierran axis s. to Mariposa Co.; s. Ore. May–July.

Var. aphanáctis J. T. Howell. Differing from var. *achillaeoides* only in the absence of ray-fls.—Common subshrub in Chaparral; local in the inner N. Coast Range of Glenn, Colusa, and adjacent Lake cos. May–July.

Var. grandiflòrum (Gray) Jeps. [*Bahia l.* var. *g.* Gray. *E. caespitosum* var. *g.* Gray. *E. speciosum* Greene. *E. g.* Greene. *E. Bolanderi* Rydb.] Short-lived perennial from a taproot; stems erect almost from base, 3–10 dm. high, relatively stout, sparsely leafy above; herbage rather densely tomentose, at length ± deciduous from upper surface of lvs.; lvs. linear and entire or lanceolate and laciniately toothed, occasionally pinnatifid; heads large, terminating stout long naked peduncles; invol. persistently tomentose,

8–11 mm. high; rays 10–15, 9–20 mm. long; aks. 3.5–4 mm. long, appressed-hispidulous to glabrate; pappus of 8–12 oblong or lanceolate paleae 0.5–1.5 mm. long; $n = 8$, 16 (Carlquist, 1956).—Dry sunny hillsides, 300–5000(–9000) ft.; mostly V. Grassland, Foothill Wd., Chaparral, Yellow Pine F.; common in Sierra Nevada foothills from Mariposa Co. to Shasta Co., Sacramento V., occasional from Mendocino Co. to Siskiyou and Del Norte cos.; sw. Ore. May–July.

Var. **monoénse** (Rydb.) Jeps. [*E. m.* Rydb. *E. lutescens* Rydb.] Densely cespitose, the numerous stems 8–15 cm. high from a spreading woody caudex, densely leafy, persistently tomentose; lvs. spatulate, entire or apically 3-toothed or -lobed, 5–20 mm. long, 3–5 mm. wide; heads usually solitary; invol. 6–8 mm. high, often somewhat reddish; rays 6–8, 6–8 mm. long; aks. ca. 4 mm. long, glabrous or sparingly hairy or glandular; pappus variable, the white paleae 0.5–1.2 mm. long.—Infrequent, rocky slopes or meadow borders, 6500–12,000 ft.; Yellow Pine F. up to Alpine Fell-fields; Sierra Nevada, from Nevada Co. to Tulare Co.; Sweetwater and White mts.; w. Nev., s. Ore. (Crater L. to Siskiyou Mts.). July–Aug.

Var. **integrifòlium** (Hook.) Smiley. [*Trichophyllum i.* Hook. *Bahia cuneata* Kell. *E. i.* Greene. *E. c.* Rydb. *E. nevadense* Gand. *E. l.* var. *c.* Jeps.] Stems numerous from a short cespitose caudex, 1.5–2 dm. high; herbage, lvs. and heads similar to var. *monoense* but the parts larger and less compact; rays commonly 8, 7–10(–16) mm. long; aks. and pappus as in var. *monoense.*—Forest openings, rocky plateaus, sagebrush slopes, 4000–7500 ft.; Sagebrush Scrub, Yellow Pine F., Red Fir F.; Sierra Nevada, Mariposa Co. n., to Modoc Co.; e. of the Cascades to Canadian boundary, sw. Mont., Nev. July–Aug.

Var. **lanceolàtum** (Howell) Jeps. [*E. l.* Howell. *E. Rixfordii* Eastw.] Stems few to several, 2–5.5 dm. high, usually relatively stout, leafy up to the solitary heads, or the heads on long peduncles; herbage persistently white- or yellowish-tomentose; lvs. thick, oblanceolate, entire to coarsely serrate, 2–4 cm. long; invol. densely woolly, 8–12 mm. high; rays 10–15, 7–13 mm. long; aks. 2.6–3.8 mm. long, sparsely hispidulous; pappus of 8–12 lance-oblong white erose paleae ca. 1 mm. long.—Forest openings, 400–7600 ft.; Mixed Evergreen F., Douglas-Fir F., Yellow Pine F., Red Fir F.; Klamath Mts. region of Humboldt, Trinity, Siskiyou, and Del Norte cos.; sw. Ore. May–Aug.

Var. **obovàtum** (Greene) Hall. [*E. o.* Greene. *E. brachylepis* Rydb.] Stems several, slender, leafy, 2–4 dm. high; herbage moderately white-tomentose; lvs. oblanceolate, entire or shallowly toothed, 1.5–4 cm. long; invol. 7–10 mm. high; rays 10–13, 8–10 mm. long; aks. 3–4 mm. long, glabrous; pappus of 8–10 lance-oblong erose paleae up to 1 mm. long.—Openings under pines, 4000–7600 ft.; Yellow Pine F., Red Fir F.; uncommon, s. Sierra Nevada (Tulare Co.), Greenhorn Mts., Kern Co., frequent, San Bernardino Mts. June–July.

Var. **Hállii** Const. Stems numerous, 3–4 dm. high; herbage thinly floccose; lvs. broad and thin, 2.5–5 cm. long, ovate in outline, pinnately incised or pinnatifid, the oblong divisions entire or toothed; heads solitary or few, relatively small; invol. campanulate, 8–12 mm. high; rays ca. 8, 10–13 mm. long; tube of disk-corolla glabrous (thus differing from all other vars.); aks. 4–5 mm. long, sparsely hispidulous; pappus of 8–12 unequal or subequal oblong and/or lanceolate erose paleae up to 2 mm. long.—Known only from the type region, 3500 ft.; Foothill Wd.; near Ft. Tejon, Kern Co. June–July.

2. **E. Jepsònii** Greene. Bushy perennial 3–8 dm. high, to 1 m. in diameter, the stems woody below, persistently white-pannose; lvs. ovate in outline, 3–6 cm. long, pinnatifid into 5–7 linear lobes, the tomentum soon deciduous from upper surface; heads few, loosely corymbose at ends of erect branches, the peduncles 5–10 cm. long; invol. broadly campanulate, 5–7 mm. high, with deciduous tomentum; rays 5–9, golden, 6–10 mm. long; aks. 4-angled, 3 mm. long, hispidulous; pappus paleae unequal, lanceolate to ovate-oblong, up to 1 mm. long; $n = 16$ (Carlquist, 1956).—Dry rocky often serpentine slopes, 1000–3500 ft.; Foothill Wd.; inner S. Coast Ranges, Contra Costa Co. to San Benito Co. April–June.

3. **E. latilòbum** Rydb. Bushy perennial 3–6 dm. high, loosely tomentose, the wool at length ± deciduous from stems and upper surface of lvs.; lvs. as in *Jepsonii,* but the divisions broader; heads several to many in loose corymbiform clusters on peduncles 2–6 cm. long; invol. broadly campanulate, 4–5 mm. high; rays 6–8, golden, 6–9 mm. long; aks. 2.5–3 mm. long, glabrous or nearly so; pappus paleae unequal, linear-oblong to lanceolate, up to 1 mm. long; $n = 16$ (Carlquist, 1956).—Grassy or rocky sparsely

wooded slopes, below 500 ft.; Foothill Wd.; restricted endemic of the Crystal Springs Lake region, San Mateo Co. May–June. This sp. and the closely related *Jepsonii* possibly arose independently from combined *lanatum* and *confertiflorum* parentage.

4. **E. confertiflòrum** (DC.) Gray. [*Bahia c.* DC. *B. tenuifolia* DC. *B. trifida* Nutt. *E. c.* vars. *trifidum* and *laxiflorum* Gray. *E. c.* var. *discoideum* Greene. *E. c.* var. *latum* Hall. *E. trifidum, cheiranthoides, tenuifolium, biternatum* and *tridactylum* Rydb. *E. confertiflorum* var. *tridactylum* Munz.] Shrub 2–6 dm. high, with many erect herbaceous stems from the woody base; lvs. cuneate to obovate in outline, 1–4 cm. long, deeply 3–5-lobed with oblong subentire divisions to bipinnatifid with narrowly linear divisions, persistently tomentose beneath, usually green and glabrate above, the margin revolute; heads numerous, subsessile in small terminal clusters, or distinctly peduncled in rather tight corymbs; invol. campanulate, tomentose, 3–6 mm. high; phyllaries 5–6, oval, obtuse; rays (0–)4–6, deep yellow, 2–5 mm. long; aks. 2–3 mm. long, linear-clavate, 4-angled, hispidulous; pappus of 5–12 oblong unequal paleae up to 1 mm. long; $n = 8$, 16 (Carlquist, 1956).—Common on brushy slopes, sea level to 8000 ft. or more; many Plant Communities; Coast Ranges (particularly in the outer) from Mendocino and Tehama cos. to San Diego Co., Sierra Nevada foothills from Calaveras Co. s., Inyo Co. (rare), mts. of s. Calif., Channel Ids.; L. Calif. April–Aug. Highly variable in compactness, cut and size of foliage, proximity of heads, etc., but unsatisfactorily divisible into regional taxa except for the following.

Var. **tanacetiflòrum** (Greene) Jeps. [*E. t.* Greene.] Plant 4–6 dm. high, the stems pannose and persistently woolly; lvs. thin, spatulate to obovate in outline, 2–5 cm. long, pinnatifid with few oblong divisions, becoming glabrate above; heads somewhat larger, several in corymbiform clusters; invol. 5–7 mm. high, loosely tomentose; rays (0–)4–6, 4–5 mm. long; aks. 3.5 mm. long, hispidulous to glabrate; $n = 32$ (Carlquist, 1956).—Hillsides, 1000–3000 ft.; Foothill Wd.; Calaveras and Mariposa cos. May–July.

5. **E. staechadifòlium** Lag. [*Bahia s.* DC. *B. s.* var. *californica* DC.] Shrubby much-branched perennial 3–15 dm. high, the stems decumbent to erect, tomentose but often glabrate in age; lvs. linear or linear-oblanceolate, entire or with few linear lobes, subcoriaceous, 3–7 cm. long, persistently tomentose beneath, glabrous above, the margin revolute; heads numerous, short-peduncled, in rather dense corymbs; invol. campanulate, 5–7 mm. high; phyllaries 8–11, oblong, obtuse, carinate-thickened at base; rays (0–)6–9, yellow, 3–5 mm. long; aks. 3–4 mm. long, linear-oblong, 4-angled, ± hispidulous; pappus of 8–12 oblong or oblanceolate unequal paleae up to 1 mm. long; $n = 16$ (Carlquist, 1956).—In seaside vegetation, on bluffs or beaches, up to 200 ft.; Coastal Strand, Coastal Sage Scrub; Santa Cruz Co. to Monterey Co. April–Sept.

Var. **artemisiaefòlium** (Less.) Macbr. [*Bahia a.* Less. *B. a.* var. *Douglasii* DC. *E. a.* Kuntze.] Lvs. always 1–2-pinnatifid, with linear to oblong lobes, the rachis commonly broader than in the preceding; $n = 16$ (Carlquist, 1956).—The commoner and more widely distributed form, shores and fields near the coast; Coastal Strand, N. Coastal Scrub, Coastal Sage Scrub; Del Norte Co. to Point Conception, Santa Barbara Co., Santa Barbara Ids.; to Coos Co., Ore. March–Aug.

Var. **depréssum** Greene. Stems stout, 1–2 dm. high, depressed; lvs. pinnately lobed, with broad rachis up to 10 mm. wide, densely clothing the stem and sometimes overtopping the heads.—Santa Barbara Co. coast near Surf; Santa Rosa, Santa Cruz and Anacapa ids. March–July.

6. **E. Nevínii** Gray. Shrubby branching perennial 5–20 dm. high; stems stout, with large pith, tomentose, the old lvs. deciduous, the shoots of the season densely leafy, terminated by a glabrate nearly leafless portion bearing the corymbiform much-branched many-headed compact infl.; lvs. 5–25 cm. long, broadly ovate in outline, 2–3-pinnatifid into numerous linear-oblong segments, tomentose on both faces; invol. narrowly campanulate, 6–7 mm. high; phyllaries 8–12, oblong, obtuse; rays 4–8, inconspicuous, 2 mm. long; aks. 3–4 mm. long, hispidulous on the angles; pappus-paleae 4–6, often ± fused, erose, very unequal, the longest to 2.2 mm.; $n = 16$ (Carlquist, 1956).—Rocky coastal bluffs, 5–100 ft.; Coastal Sage Scrub; Santa Catalina and San Clemente ids. April–Aug.

7. **E. nubígenum** Greene. [*Actinolepis n.* Greene.] Low loosely tomentose annual 5–15 cm. high; stem simple or branched, leafy; lvs. spatulate or oblanceolate, mostly

entire, 1–2 cm. long; heads solitary at the ends of the branches; invol. narrowly campanulate, 5–6 mm. high, the 4–6 oblong phyllaries distinct to base; rays 4–6, yellow, scarcely exceeding disk, ca. 1 mm. long; aks. 2.5–3 mm. long, appressed-hirsute; pappus of 8–10 oblong subequal paleae 0.5–1.5 mm. long.—Rare endemic of forest openings, 5000–9000 ft.; Red Fir F. to Subalpine F.; Yosemite National Park. June–July.

Var. **Congdònii** (Bdg.) Const. [*E. C.* Bdg.] Larger in most of its parts, somewhat less woolly; stems 1–3 dm. high, freely branched; lvs. 1–4 cm. long, the larger often few-lobed or -toothed; invol. campanulate, 5–9 mm. high; phyllaries 8–10, oblong, free to base; rays 6–10, yellow, sometimes tinged with purple, 3–7 mm. long; pappus paleae unequal, alternately lanceolate-acute and shorter oblong-obtuse, 1–2 mm. long; $n = 7$ (Carlquist, 1956).—Gravelly slopes, 1500–3000 ft.; Foothill Wd.; Sierran foothills of Mariposa Co. April–June.

8. **E. ambíguum** (Gray) Gray. [*Lasthenia a.* Gray. *Bahia a.* Gray. *B. parviflora* Gray. *E. paleaceum* Bdg. *E. parviflorum* Rydb.] Loosely tomentose annual 5–30 cm. high, openly branching; lvs. linear-oblong to spatulate, entire or shallowly few-toothed or -lobed toward apex, 1–4 cm. long; heads solitary on peduncles 1–5 cm. long; invol. campanulate, 4–7 mm. high, the 6–9 oblong-lanceolate rigid phyllaries free or partially united, with indurated base; rays 5–10, yellow, 2–9 mm. long; receptacle conical, naked, or with few linear scales toward apex; aks. 2.2–3 mm. long, linear-clavate, strigillose; pappus of 6–10 truncate erose paleae 0.2–0.3 mm. long, or reduced to a fused ring, or obsolete; $n = 7$ (Carlquist, 1956).—Gravelly slopes or washes, 200–7500 ft.; Foothill Wd., Pinyon-Juniper Wd., Joshua Tree Wd., Creosote Bush Scrub; s. Sierra Nevada and Greenhorn Mts., Kern Co. to Ft. Tejon, scattered across the deserts from Palm Springs to cent. Inyo Co.; Nye Co., Nev. April–June. The most variable annual sp.

9. **E. lanòsum** (Gray) Gray. [*Burrielia l.* Gray. *Actinolepis l.* Gray. *Antheropeas l.* Rydb. *A. tenuifolium* Rydb.] Loosely floccose annual 3–15 cm. high, openly branching, the reddish stems sometimes decumbent; lvs. linear to linear-oblanceolate, mostly entire, 1–2 cm. long; heads solitary on slender peduncles 1–5 cm. long; invol. turbinate, 5–7 mm. high; phyllaries 8–10, narrowly oblong, acute, carinate, distinct or nearly so; rays 5–10, white, usually veined with red, 3–5 mm. long; aks. 2.5–4.5 mm. long, narrowly linear, sparsely strigillose; pappus of ca. 5 hyaline slender awn-tipped paleae up to 2.5 mm. long, alternating with oblong obtuse ones of half the length; $n = 4$ (Carlquist, 1956).—Sandy deserts, 500–3500 ft.; Creosote Bush Scrub; rare in Calif., e. Mojave and Colo. deserts, San Bernardino to Imperial cos.; to Nev., sw. Utah, Ariz., L. Calif. Feb.–May.

10. **E. Wàllacei** (Gray) Gray. [*Bahia W.* Gray. *Actinolepis W.* Gray. *E. aureum* Bdg. *Antheropeas W.* Rydb. *A. australe* Rydb. *E. W.* var. *calvescens* Blake. *E. W.* ssp. *australe* Wiggins.] Persistently tomentose annual 1–8(–15) cm. high, well branched, often appearing tufted; lvs. spatulate to obovate, entire or rarely 3-lobed at apex, 7–15 mm. long; heads scattered, on short peduncles; invol. obconic, 5–7 mm. high; phyllaries 6–10, ovate, carinate, distinct; rays 5–10, golden or yellow, 3–4 mm. long; aks. ca. 2 mm. long, narrowly linear, sparsely strigillose or glabrous; pappus of 6–10 opaque oblong obtuse or truncate regular paleae mostly ca. 0.5 mm. long, or none; $n = 5$ (Carlquist, 1956).—Sandy washes and fans, 200–5000(–6500) ft.; Creosote Bush Scrub, Joshua Tree Wd., Chaparral; common on both deserts, from Mono Co. to San Diego Co., adjoining ranges to the w.; to sw. Utah, Ariz., L. Calif. March–May. The pappus variations are too randomized for formal recognition.

Var. **rubéllum** (Gray) Gray. [*Bahia r.* Gray.] Rays roseate or purplish, or rarely whitish.—W. end of the Colo. Desert in the San Felipe region, Riverside and San Diego cos.

11. **E. mohavénse** (Jtn.) Jeps. [*Eremonanus m.* Jtn.] Woolly-villous cespitose dwarf annuals 1–2.5 cm. high, 2–3 cm. wide; lvs. spatulate, entire and mucronate, or sharply 3-toothed at apex, attenuate to base, 5–10 mm. long; heads solitary at ends of the branches, discoid, 3–4-fld.; invol. cylindric, 3–4 mm. high; phyllaries 3–4, linear-oblong, obtuse, concave, at length embracing the ak.; disk-fls. ca. 2 mm. long; aks. terete, 5-nerved, 2–2.5 mm. long, strigose; pappus of 12–14 oblong to obovate subequal paleae ca. 1.5 mm. long.—Sandy or rocky places, 2000–3000 ft.; Creosote Bush Scrub; known only from within 30 mi. of Barstow, Mojave Desert. April–May.

12. **E. multicaùle** (DC.) Gray. [*Actinolepis m.* DC. *A. m.* var. *papposa* Gray.]

Loosely woolly annual 2–15 cm. high, becoming glabrate below, diffusely branched from the base upward; lvs. mostly cuneate, bluntly toothed or lobed at apex, ca. 1 cm. long; heads subsessile in leafy-bracteate terminal clusters; invol. campanulate, 3–4 mm. high; phyllaries 5–7, ovate-oblong, distinct, boat-shaped, loosely enclosing the ray-aks.; rays 3–7, obovate, yellow, 2 mm. long; aks. 1.6–2 mm. long, sparsely strigillose to glabrous; pappus of ca. 12 brownish attenuate fimbriate unequal paleae 1 mm. or less long; n = 7 (Carlquist, 1956).—Sandy fields, old dunes, and chaparral openings, 20–2000 ft.; Coastal Sage Scrub, Foothill Wd., Chaparral; Monterey and San Benito cos. to San Diego Co. March–May.

13. **E. Prínglei** Gray. [*Actinolepis P.* Greene.] Densely and persistently woolly annual 1–5 cm. high, branched from the base and often densely tufted; lvs. oblong to broadly spatulate, thickened, entire or usually 3-lobed at apex, 5–10 mm. long; heads in terminal leafy clusters, very woolly, discoid, 10–25-fld.; invol. campanulate, 3–4 mm. high; phyllaries 6–8, oblanceolate, distinct, boat-shaped; corolla 2 mm. long, the slender glandular tube equaling the widely campanulate throat and limb; aks. 1.5–2 mm. long, silvery-sericeous; pappus of 8–12 tawny hyaline oblong-lanceolate fimbriate paleae ca. 1 mm. long; n = 7 (Carlquist, 1956).—Sandy desert, 1200–6600 ft.; Creosote Bush Scrub, Sagebrush Scrub; w. borders of both deserts, Mono Co. to San Diego Co.; w. and s. Ariz. April–June.

57. Pseudobahía (Gray) Rydb.

Erect floccose-woolly spring annuals. Lvs. alternate, entire, 3-lobed or 1–2-pinnatifid. Heads radiate, solitary at ends of the branches. Invol. broadly campanulate; phyllaries 1-seriate, herbaceous, usually united at base. Receptacle conic, naked. Ray-fls. ♀, fertile, yellow, not at all bilabiate. Disk-fls. many, fertile, with prominently hairy slender tube, dilated smooth throat and limb. Anther-appendages ovate. Style-branches truncate. Aks. oblanceolate in outline, ± obcompressed, epappose or sometimes with vestigial pappus-paleae. Three spp., all in Calif. (Greek, *pseudo,* false, and *Bahia,* another helenioid genus.)

Lvs. entire or 3-lobed. Very rare .... 1. *P. bahiaefolia*
Lvs. 1–2-pinnatifid.
    Invol. 4–5 mm. high; phyllaries united half their length, often with callous processes at their
    sinuses; lvs. 1–2 cm. long, mostly pinnatifid .... 2. *P. Heermannii*
    Invol. 6–9 mm. high; phyllaries free to base, without callous processes at their sinuses; lvs. 2–6 cm.
    long, mostly bipinnatifid .... 3. *P. Peirsonii*

1. **P. bahiaefòlia** (Benth.) Rydb. [*Monolopia b.* Benth. *Eriophyllum b.* Greene.] Slender tomentose annuals 5–20 cm. high, the stem simple or few-branched from below; lvs. 1–2 cm. long, linear-spatulate to linear, all entire or some with 3 blunt teeth at apex; heads on slender peduncles 2–5 cm. long; invol. hemispheric, becoming urceolate when ripe, 5–6 mm. high; phyllaries 6–8, narrowly ovate, united below the middle, the basal portion becoming indurated; rays 6–8, oblong-elliptic, bright yellow, 5–10 mm. long; aks. 2–2.5 mm. long, brown-strigose; pappus none or vestigial-paleaceous; n = 4 (Carlquist, 1956).—Widely scattered on dry gravelly soil or grassy slopes, 100–1000 ft.; V. Grassland, Foothill Wd.; e. side of Cent. V. from Placer Co. to Madera Co. March–May.

2. **P. Heermánnii** (Durand) Rydb. [*Monolopia H.* Durand. *M. bahiaefolia* var. *pinnatifida* Gray. *Eriophyllum H.* Greene.] Loosely floccose annuals 1–3 dm. high, usually openly branching, the rather straggly slender reddish stems at length glabrate; lvs. 1–3 cm. long, pinnately lobed with linear divisions 0.5–1.5 mm. wide, or the upper sometimes entire; peduncles 2–5 cm. long, often curved at maturity; invol. 5–6 mm. high, persistently tomentose; phyllaries ca. 8, ovate, united ca. half their length, the basal portion becoming indurated and often developing callous processes between the lobes; rays 6–10 mm. long; aks. 2–2.5 mm. long, brown-strigillose; pappus none or vestigial-paleaceous; n = 3 (Carlquist, 1956).—Infrequent, in colonies in sun or half-shaded grassy slopes, 800–3000(–5000) ft.; Foothill Wd., Yellow Pine F.; Sierran foothills, Butte Co. to Kern Co.; Santa Lucia Mts., Monterey Co. (rare). March–May. Possibly insufficiently distinct from the preceding.

3. **P. Peirsònii** Munz. Stouter floccose annual 2–7 dm. high, usually rather strictly

branching above the base; lvs. 2–6 cm. long, triangular-ovate in outline, bipinnatifid, the segments 1–5 mm. wide, the flattened petiolar portion usually shorter than the divided portion; invol. 6–9 mm. high, persistently tomentose; phyllaries ca. 8, lance-ovate, free essentially to base; rays broadly oblong, 5–10 mm. long; aks. ca. 3 mm. long, glabrous or with few appressed hairs; pappus none or vestigial; $n = 8$ (Carlquist, 1956). —Grassy floors and rolling hills, in adobe, up to 1000 ft.; V. Grassland; e. side of San Joaquin V., Tulare and n. Kern cos. March–April.

58. Syntrichopáppus Gray

Low branched floccose-woolly annuals, with mostly alternate spatulate to linear lvs., often 3-lobed at apex. Heads many, small, radiate, solitary. Invol. narrowly campanulate, 1-seriate; phyllaries 5–8, oblong, concave and partly enfolding the ray-aks. Receptacle flat. Ray-fls. ♀, fertile, as many as the phyllaries. Disk-fls. fertile, yellow. Anther-appendages very slender, acuminate. Aks. clavate, 5-angled. Pappus of numerous barbellate bristles united at base into a ring, falling off together, or none. Genus of 2 spp., natives of the Mojave Desert and adjacent borders. (Greek, *syn*, together, *thrix*, hair, and *pappos*, pappus, from the united pappus bristles.)

Rays yellow; pappus of numerous bristles .... 1. *S. Fremontii*
Rays white flushed with rose, red-veined; pappus none.................... 2. *S. Lemmonii*

1. **S. Fremóntii** Gray. Low floccose annual 2–10 cm. high, diffusely much-branched throughout; lvs. 5–20 mm. long, linear-spatulate to cuneate, 3-toothed at apex or entire; heads on short terminal or axillary peduncles. Invol. narrowly campanulate to sub-cylindric, 5–7 mm. high, the phyllaries becoming boat-shaped, firm, scarious-margined; rays mostly 5, golden-yellow, 3–5 mm. long; aks. 2.2–3.2 mm. long, hoary; pappus bright white, of 30–40 barbellate bristles united at base, deciduous; $n = 6$ (Carlquist, 1956).—Frequent, sandy or gravelly washes or hillsides, 2000–7500 ft.; Creosote Bush Scrub, Joshua Tree Wd.; Mojave Desert, Death V. region, Inyo Co. to Kern, Los Angeles, and San Bernardino cos.; to sw. Utah, w. Ariz. April–June.

2. **S. Lemmònii** (Gray) Gray. [*Actinolepis L.* Gray. *Microbahia L.* Ckll.] Thinly floccose to glabrate annual 2–8 cm. high, the thin reddish stem corymbosely branching above the middle or throughout; lvs. linear, rounded at apex, entire, 3–8 mm. long; invol. campanulate to narrowly turbinate, 4–5 mm. high, the 6–8 phyllaries scarious-margined; rays 6–8, white with cream-yellow base ventrally, rose with bright red veins dorsally, 2–3 mm. long; aks. 2.5 mm. long, sparsely strigillose, epappose; $n = 7$ (Carlquist, 1956).—Sandy slopes, 3000–4500 ft.; Joshua Tree Wd., Chaparral; sw. border of Mojave Desert and adjoining slopes of the San Gabriel and San Bernardino mts., Los Angeles and San Bernardino cos. April–May.

59. Rigiopáppus Gray

Slender wiry annuals with linear alternate entire lvs. Heads solitary at the ends of the simple stems or the filiform corymbosely proliferous branches, radiate, many-fld., all fls. fertile. Invol. irregularly 2–3-seriate, the outer phyllaries concave and half-embracing the ray-aks., the inner narrower and shorter and sometimes encroaching upon the receptacle. Ray-fls. 1-seriate, very narrow, scarcely exceeding the disk; disk-corollas linear, 3–5-toothed. Aks. linear, transversely wrinkled, pubescent. Pappus of 3–5 rigid tapering awnlike scales, rarely wanting. One sp. (Greek, *rigios*, stiffened, and *pappos*, pappus.)

1. **R. leptócladus** Gray. [*R. l.* var. *longiaristatus* Gray. *R. longiaristatus* Rydb.] Stems erect, 1–3 dm. high, sparsely pubescent, leafy, the branches naked below and over-topping the main stem; lvs. 1–3 cm. long, mostly less than 1 mm. wide; invol. 5–8 mm. high, turbinate; fls. yellow, usually suffused with reddish, the rays ca. 5–15, the disk ca. 5–35; aks. all alike, beset with short appressed hairs with swollen tips; pappus scales ochroleucous, often with a spot of purple at the ± united bases; $n = 9$ (Raven).—Dry, open, hard or rocky ground and grassy slopes up to 5000 ft.; several Plant Communities: San Gabriel Mts., Los Angeles Co., through the Coast Ranges to sw. Ore., Sierra Nevada foothills from Kern Co. to Shasta Co.; to Wash., Ida., Utah. April–June.

60. Trichoptílium Gray

Low diffusely branched floccose-woolly annual. Lvs. crowded, mostly alternate, oblanceolate, incised-dentate with spinescent teeth. Heads solitary on long slender peduncles in the upper axils, yellow, discoid, the outer florets slightly enlarged. Invol. hemispheric, 2-seriate. Receptacle hemispheric, naked. Aks. turbinate, hairy. Pappus of 5 thin paleae almost as long as the corolla, palmately deeply fimbriate into slender bristles. Monotypic. (Greek, *trichos*, hair, and *ptilon*, feather, referring to the dissected pappus-paleae.)

1. **T. incìsum** (Gray) Gray. [*Psathyrotes i.* Gray.] Stems several to many from the taproot, dichotomously branching, forming plants 5–20 cm. high and as broad or much broader, white-woolly; lvs. mostly 1–3 cm. long, 3–9 mm. wide, petiolate, at last glabrate; peduncles not woolly but glandular-puberulent, much overtopping the lvs.; heads depressed-globose, 7–11 mm. high; phyllaries lanceolate, tomentose and glandular; florets 35–80; pappus bright white to fulvous.—Common on desert pavements or sand, up to 2200 ft.; Creosote Bush Scrub; Colo. Desert and s. Mojave Desert; s. Nev., w. Ariz., L. Calif. Feb.–May, Oct.–Nov.

61. Chaenáctis DC.

Annual, biennial or low perennial herbs. usually ± floccose, at least when young. Lvs. alternate, entire to compoundly dissected. Heads discoid, the fls. all perfect, but the marginal ones often enlarged and subligulate, pedunculate. Invol. campanulate, turbinate or hemispheric; phyllaries herbaceous, in 1–3 series, free to base. Receptacle flat, alveolate, naked, or in one sp. with bristles subtending some fls. Fls. yellow, white, or pink, perfect; limb 5-lobed. Aks. linear-clavate, terete or compressed. Pappus of 4–20 hyaline linear-lanceolate to obovate equal or unequal erose-denticulate scales, or obsolete. Ca. 25 spp., of w. U.S. and adjacent borders. (Greek, *chaino*, to gape, and *aktis*, ray, the marginal fls. in many spp. flaring and raylike.)

(Stockwell, Palmer. A revision of the genus Chaenactis. Contr. Dudl. Herb. 3: 89–168, 1940.)
Pappus-paleae 10 or more in 2 series; perennials or biennials.
  Lvs. linear-oblong, with 10–15 pairs of small crispate pinnae. Mts. of s. Calif. .. 1. *C. santolinoides*
  Lvs. broader, with fewer segments.
    Segments of lvs. linear, mostly entire; suffruticose perennials.
      Pappus of 14–18 unequal scales. San Jacinto and Santa Rosa mts. .... 2. *C. Parishii*
      Pappus of 10 equal scales. N. Calif. .... 3. *C. suffrutescens*
    Segments of lvs. shorter and broader, often again toothed or lobed.
      Plants alpine or subalpine dwarfs, up to 10 cm. high.
        Lvs. flabelliform to ovate, few-lobed, the deep-set taproot slender-stoloniferous.
          Invol. glandular; lvs. bipinnate, gray-tomentose, 10–25 mm. wide. N. Sierra Nevada
                                                  4. *C. nevadensis*
          Invol. eglandular, tomentose; lvs. palmately to pinnately lobed, yellow-tomentose, 3–10
          mm. wide. S. Sierra Nevada .... 5. *C. alpigena*
        Lvs. oblong, with 4–12 pairs of segments; caudex more compact .... 6. *C. alpina*
      Plants of lower elevs., mostly over 20 cm. high .... 7. *C. Douglasii*
Pappus-paleae 4–5 in one series or with outer reduced paleae, or obsolete; annuals.
  Fls. yellow .... 8. *C. glabriuscula*
  Fls. white or purplish tinged.
    Pappus of persistent entire or erose paleae; aks. subterete.
      Stamens included; corollas pink, hoary, much exceeding the invol. .... 9. *C. macrantha*
      Stamens exserted.
        Phyllaries attenuate into a bristle-tip .... 10. *C. carphoclinia*
        Phyllaries obtuse or acute.
          Lvs. entire or once-parted into few slender lobes, green; plants soon glabrate.
            Phyllaries 8–10 mm. long, essentially glabrous including the short straight tip; marginal corollas much enlarged; pappus uniseriate .... 11. *C. Fremontii*
           Phyllaries 12–16 mm. long, at least the curving tip densely puberulent; marginal corollas scarcely enlarged; pappus biseriate .... 12. *C. Xantiana*
          Lvs. bipinnatifid with short thick segments, grayish; plants ± persistently tomentulose;
          phyllaries 6–9 mm. long, glandular-puberulent .... 13. *C. stevioides*
    Pappus none or rudimentary; aks. flattened .... 14. *C. artemisiaefolia*

1. **C. santolinoìdes** Greene. [*C. s.* var. *indurata* Stockw.] Perennial with a cespitose often branching crown surmounting a woody taproot; stems subscapose, sometimes branched, 1–3.5 dm. high, ± glandular; lvs. principally in dense basal rosettes, permanently white-tomentose and punctate, 3–10 cm. long, 4–7 mm. wide, linear-oblong, pinnatisect with many short oblong or rounded crispate segments; invol. broadly

turbinate, 8–13 mm. high, densely glandular; phyllaries oblong, obtuse; corollas cream-white or pink, 6–7 mm. long, ± pilose; aks. equaling corollas, densely hirsute; pappus of 12–16 very unequal oblanceolate scales shorter than the corolla.—Open pine woods and dry ridges, 4500–8000 ft.; Yellow Pine F., Red Fir F.; Greenhorn Mts. and Pah Ute Peak, Kern Co., Mt. Pinos, Ventura Co. to San Bernardino Mts. June–July.

2. **C. Paríshii** Gray. Suffruticose perennial 2–4 dm. high; stems several, erect, sub-scapose, from a branched decumbent woody base, simple or erectly branched above to bear 2–4 long-peduncled heads, the sterile shoots short, pannose; lvs. lanulose and punctate, 2–5 cm. long, oblong in outline, pinnate with relatively few linear obtuse mostly entire lobes, petiolate; invol. turbinate, 11–13 mm. high, minutely glandular; phyllaries linear, loose, unequal; corolla cream-white or pinkish, 7–9 mm. long, at least the lobes hispidulous; aks. densely hirsute; pappus of 14–18 unequal linear-lanceolate scales shorter than the corolla.—Infrequent, dry rocky slopes, 5000–7000 ft.; Chaparral; San Jacinto and Santa Rosa mts., Riverside Co.; n. L. Calif. June–July.

3. **C. suffrutéscens** Gray. [*C. s.* var. *incana* Stockw.] Suffruticose perennial 2–3.5 dm. high; stems few to many, erect from a branched woody crown or terminating slender woody stolons, densely leafy and pannose in lower half, terminating in a stout densely glandular peduncle 10–20 cm. long; lvs. ± white-pannose and punctate, 5–10 cm. long, ovate in outline, pinnate or partially bipinnate with linear obtuse lobes up to 2 cm. long, petiolate; invol. broadly turbinate, 12–17 mm. high; phyllaries unequal, densely glandular-puberulent; corolla cream-white, glandular-hairy; aks. densely tawny-hirsute; pappus of 10 equal oblong-oblanceolate scales nearly as long as the corolla.—Clayey meadows or gravelly talus or washes, 2800–6500 ft.; V. Grassland, Yellow Pine F.; valleys around Mt. Shasta, Siskiyou Co., w. to Trinity Co. May–June.

4. **C. nevadénsis** (Kell.) Gray. [*Hymenopappus n.* Kell. *C. pumila* Greene.] Cespitose or mat-forming perennial 5–10 cm. high, the deep-set woody taproot giving off several slender woody rootstocks terminating in dense rosettes of lvs.; lvs. densely to loosely gray-tomentose, puncticulate, 2.5–4.5 cm. long, 10–25 mm. wide, ovate or roundish in outline, bipinnatifid, the lobes spatulate, irregular, 6–12 mm. long, the petiole exceeding the blade; heads solitary on glandular-puberulent peduncles; invol. broadly turbinate, 10–12 mm. high, often purplish; phyllaries broadly linear, obtuse, densely glandular-puberulent; corolla whitish tinged with pink apically; pappus of 10–16 subequal linear-oblanceolate paleae nearly as long as the corolla; $n = 6$ (Gillett, 1954).—In gravelly soils or talus, 8300–10,200 ft.; Red Fir F. to Subalpine F.; Cascade-Sierran axis, from Mt. Lassen region to Placer and Tuolumne cos. (Sonora Pass); Mt. Rose, Nev. July–Aug.

5. **C. alpigèna** C. W. Sharsm. Densely cespitose or mat-forming perennial 2–7 cm. high, the rootstocks from the small woody taproot very slender; lvs. densely yellow-tomentose, 1–3 cm. long, 3–10 mm. wide, flabelliform to ovate in outline, palmately to pinnately few-lobed, the lobes 1.5–3 mm. long, the petiole ca. equaling the blade; peduncles short, tomentose; invol. turbinate, 8–14 mm. high; phyllaries relatively few (8–11), rather unequal, densely to sparsely tomentose, eglandular or nearly so; corolla cream-white; pappus of 8–20 unequal to subequal linear-oblanceolate paleae nearly as long as the corolla.—Occasional in sandy or gravelly flats or slopes, 8600–12,500 ft.; Lodgepole F. to Alpine Fell-fields; Sierra Nevada from Eldorado Co. to Tulare Co.; Mt. Rose, Nev. July–Aug.

6. **C. alpina** (Gray) Jones. [*C. Douglasii* var. *a.* Gray.] Perennial 5–10 cm. high, with usually condensed caudex of several short branches each surmounted by a dense rosette of lvs.; lvs. thinly gray-tomentose or glabrate, 2–6 cm. long, oblong to elliptic, with 4–12 pairs of rounded crenate often again lobed pinnae; heads solitary on glandular peduncles 1–5 cm. long; invol. broadly campanulate, 10–14 mm. high, densely glandular-puberulent; corolla whitish; pappus of 10 subequal linear-oblong rounded paleae ⅔ as long as the corolla.—Stony slope, 9500 ft.; Lodgepole Pine F.; Mud Creek Canyon, Mt. Shasta, Siskiyou Co., apparently this; ne. Ore. to sw. Mont. and nw. Colo. Aug.

7. **C. Douglásii** (Hook.) H. & A. [*Hymenopappus D.* Hook. *C. achilleaefolia* H. & A. *Macrocarphus D.* and *a.* Nutt. *C. D.* var. *a.* A. Nels.] Biennial or short-lived perennial with basal rosette and erect stem 2–4 dm. high, floccose when young, usually simple below, corymbosely branching above with sharply ascending branches; lvs. glandular-punctate, loosely tomentose, at length glabrate, 3–10 cm. long, 1–4 cm. wide, 2–3-pinnatifid with 4–8 pairs of pinnae; invol. broadly turbinate, 10–14 mm. high, densely glandular-puberulent; corolla whitish or pinkish; pappus of usually 10 linear-oblong

paleae, alternating long and somewhat shorter, ½–¾ as long as the corolla.—Sandy plains, or open gravelly or serpentine slopes, 4000–7000 ft.; N. Juniper Wd., Chaparral, Yellow Pine F., Red Fir F.; mostly w. side of Sierra Nevada from Fresno Co. n.; both sides of Cascades n. to Modoc and Siskiyou cos.; inner N. Coast Ranges from Lake Co. to Trinity Co.; to B.C., Mont., Colo., n. Ariz. June–July. Variable in size, habit, and duration, but separable with some difficulty into definable segregates.

Var. **rubricaúlis** (Rydb.) Ferris. [*C. r.* Rydb. *C. panamintensis* Stockw.?] Plants 1–3 dm. high, soon glabrate, the stems often several from the base, often reddish, relatively slender, corymbosely branching from below the middle, the branches divaricate.—Dry woods, open gravelly slopes, etc., 3700–10,600 ft.; Sagebrush Scrub, Pinyon-Juniper Wd., Coniferous F.; mostly e. side of Sierra Nevada and adjacent ranges from Tulare and Inyo cos. n., to Siskiyou and Trinity cos.; s. Ore., w. Nev. June–Aug.

8. **C. glabriúscula** DC. Winter annual 1–4 dm. high, openly branching above, thinly floccose, soon glabrate; lvs. scattered throughout, 3–8 cm. long, once or twice pinnatifid with few to several pairs of irregular narrowly linear lobes usually 2–8 mm. long, the uppermost lvs. entire; peduncles 3–10 cm. long; invol. broadly turbinate to hemispheric, 7–10 mm. high, the phyllaries linear-oblong, plane, 1-nerved; marginal corollas with flaring palmate limb surpassing the disk; pappus-paleae 4, those of marginal fls. quadrate or the inner scale longer, those of cent. fls. equal, ¾ as long as corolla; $n = 6$ (Stockwell, 1940).—Sandy valleys and foothills, 300–4000 ft.; Foothill Wd., Chaparral, V. Grassland; inner S. Coast Ranges, Monterey Co. to Santa Barbara and Los Angeles cos., coming to the coast in the s. March–May.

KEY TO VARIETIES

Peduncles very long, subscapose and slender or from short leafy stem and stout.
  Heads on long delicate scapose peduncles; plants branching only at base, woolly; lvs. in a basal
    rosette .... var. *lanosa*
  Heads on stout often fistulous peduncles; plant branching in lower half, glabrate; leafy below
                                                      var. *denudata*
Peduncles relatively short, usually much less than half the length of the stem.
  Phyllaries oblong, plane, with thick obtuse tips; marginal corollas mostly with palmate limb.
    Pappus-paleae 4, in one series; lvs. pinnatifid to subentire.
      Phyllaries 8–10 mm. high, the heads medium large. Lowland .... *C. glabriuscula*
      Phyllaries 5–7 mm. high, the heads small. Montane .... var. *curta*
    Pappus-paleae 4–5 in one series, with an incomplete outer whorl of much shorter paleae; some
    lvs. bipinnatifid.
      Invol. hemispheric, ca. 10 mm. high. Bordering Cent. V. .... var. *megacephala*
      Invol. campanulate, 6–7 mm. high. N. Coast Ranges .... var. *gracilenta*
  Phyllaries boat-shaped, narrowly linear; marginal corollas only slightly enlarged, seldom with
  palmate limb. S. Calif.
    Lf.-segs. very narrow, rather short and recurved .... var. *tenuifolia*
    Lf.-segs. succulent, short, obtuse .... var. *Orcuttiana*

Var. **lanòsa** (DC.) Hall. [*C. l.* DC.] Leafy only at the branching base, 2–3 dm. high, with many scapose peduncles; herbage floccose with tardily deciduous wool; lvs. mostly pinnatifid with few narrowly linear lobes or the upper entire; invol. ca. 8 mm. high; marginal corollas not very ampliate; pappus-paleae of inner fls. 4 (or 5), subequal, acutish, nearly as long as the corolla; $n = 6$ (Stockwell, 1940).—Common in dry sandy places, below 2000 ft.; V. Grassland, Chaparral, Foothill Wd.; inner S. Coast Ranges, San Francisco Bay to Los Angeles Co. March–June.

Var. **denudàta** (Nutt.) Munz. [*C. d.* Nutt.] Larger and coarser than the preceding, 3–4 dm. high, the numerous peduncles becoming fistulous; heads larger, the invol. 9–10 mm. high; marginal corollas prominently enlarged; pappus-paleae of inner fls. 4–5, subequal, nearly as long as the corolla; $n = 6$ (Stockwell, 1940).—Old dunes and sandy places usually near the coast, below 1000 ft.; Coastal Sage Scrub, Foothill Wd., S. Oak Wd.; San Luis Obispo Co. to Orange Co.; San Gabriel and San Bernardino valleys. April–July.

Var. **cúrta** (Gray) Jeps. [*C. heterocarpha* var. *c.* Gray. *C. g.* var. *h.* f. *c.* Hall. *C. aurea* Greene. *C. g.* var. *a.* Stockw.] Heads small, the invols. 5–7 mm. high; phyllaries narrow; pappus-paleae of inner fls. 4–5, usually half as long as the corolla, sometimes longer or reduced to a mere crown.—Sandy or gravelly soils, up to 6500 ft.; V. Grassland, Chaparral, Yellow Pine F.; head of San Joaquin V., s. mostly through the mts. from Santa Barbara Co. to Tehachapi region to n. San Diego Co. April–June.

Var. **megacéphala** Gray. [*C. heterocarpha* Gray. *C. g.* var. *h.* Hall. *C. h.* var. *m.*

Stockw.] Often strict, rather robust, branching above, 1.5–4 dm. high; lvs. scattered, pinnatifid or bipinnatifid, with narrowly linear lobes, lanate to glabrate; heads large, campanulate to hemispheric; invol. 7–11(–15) mm. high; phyllaries lance-oblong, acute; pappus tawny, of inner fls. of 4 prominent obtuse scales nearly as long as the corolla and 1–4 very short roundish outer ones or these obsolete.—Sandy places, 100–4000 ft.; V. Grassland, Foothill Wd., Chaparral, Yellow Pine F.; Cent. V. and surrounding slopes, from Shasta and Trinity cos. to Tehachapi Mts., Kern Co. March–June.

Var. **gracilénta** (Greene) Keck. [*C. g.* Greene. *C. tanacetifolia* Gray. *C. t.* var. *g.* Stockw. *C. glabriuscula* var. *filifolia* sensu Jeps., not *C. filifolia* Gray.] Low openly branching throughout, 1–2.5 dm. high, more delicate than var. *megacephala;* lvs. pinnatifid or bipinnatifid, with very narrow thickish segms.; heads relatively small, campanulate; invol. 7–8 mm. high; pappus whitish, of inner fls. of 4–5 narrow scales ¾ as long as the corolla and 2–6 minute quadratish outer ones; $n = 6$ (Stockwell, 1940).—Gravelly places or serpentine soil, up to 4000 ft.; Foothill Wd., Chaparral; inner Coast Ranges, but w. of var. *megacephala,* Colusa and Mendocino cos. to San Benito Co. May–June.

Var. **tenuifòlia** (Nutt.) Hall. [*C. t.* Nutt. *C. filifolia* Gray. *C. g.* var. *t. f. f.* Hall.] Tall and slender, 2–6 dm. high, simple or much branched; lvs. pinnatifid, their divisions filiform, short or elongated, ± succulent and viscid, not very lanose; heads small; invol. 5–7 mm. high; phyllaries ± keeled, narrow; pappus-paleae of inner fls. uniseriate, somewhat shorter than the corolla; $n = 6$ (Stockwell, 1940).—Open, sandy ground, up to 5000 ft.; Coastal Sage Scrub, Chaparral, S. Oak Wd.; cismontane s. Calif., Los Angeles Co. to San Diego; L. Calif. April–June.

Var. **Orcuttiána** (Greene) Hall. [*C. tenuifolia* var. *O.* Greene. *C. O.* Parish.] Similar to var. *tenuifolia* but stouter and more succulent; lvs. sometimes crowded toward the base, 2–3-pinnatifid, fleshy, the ultimate lobes very short and blunt; infl. viscid; invol. 6–7 mm. high; pappus-paleae of inner fls. 4, ¾ as long as the corolla; $n = 6$ (Stockwell, 1940).—Dunes near the coast, below 500 ft.; Coastal Strand; San Diego Co.; n. L. Calif. April–July.

9. **C. macrántha** D. C. Eat. Winter annual 5–20 cm. high, commonly widely branched from base, floccose-tomentose when young; lvs. once or twice pinnatifid with linear or oblong lobes; invol. campanulate, 12–15 mm. high, ± tomentose; corollas vespertine, white to flesh-color, hoary, much exceeding the invol., the marginal ones not much enlarged; stamens included; pappus of 4 linear-oblong paleae half as long as the corolla and 2–4 very short outer ones, or these missing.—Gravelly desert plains or hills, 2000–5000 ft.; Creosote Bush Scrub, Shadscale Scrub, Sagebrush Scrub, Pinyon-Juniper Wd.; s. Mono Co. s. throughout the Mojave Desert but uncommon; to Ariz., Utah, sw. Ida. May.

10. **C. carphoclínia** Gray. Diffusely branched winter annual, the branching tending to be zigzag, the stems very slender, 1–4 dm. high, farinose; lvs. scattered, 1–4 cm. long, 1–2-pinnatifid, with filiform rather remote segments; heads numerous, on short slender peduncles; invol. broadly campanulate, 6–8 mm. high; phyllaries 15–30, linear-lanceolate, tapering into pungent awnlike reddish tips; bracts of the receptacle subtending all but the innermost fls., filiform, bristle-tipped, exceeding the unopened fls.; fls. whitish, the marginal not much enlarged; pappus variable, paleae of inner fls. usually 4, lance-acuminate, almost as long as the corollas, those of outer fls. sometimes short, broad and obtuse.—Common on hot desert pavements, −100–2500 ft.; Creosote Bush Scrub, Alkali Sink; s. Inyo Co. across both deserts to n. L. Calif.; e. to Utah, Ariz. March–May.

Var. **attenuàta** (Gray) Jones. [*C. a.* Gray.] Pappus-paleae of even the cent. fls. very short and obtuse, or some of them acute but not more than ⅓ the length of the corolla. —Deserts, up to 4800 ft.; n. Inyo Co. to Imperial and San Diego cos.; Nev., w. Ariz., n. L. Calif. Largely replacing the typical var. in s. Mojave and Colo. deserts. March–June.

Var. **Peirsònii** (Jeps.) Munz. [*C. P.* Jeps.] Lvs. almost all gathered in a dense basal rosette, even tripinnatifid, with short congested segments; pappus-paleae all very short. —Local, e. end of Santa Rosa Mts., Colo. Desert, Riverside and San Diego cos. March–April.

11. **C. Fremóntii** Gray. Winter annual with few ascending branches, 1–4 dm. high,

glabrate, or slightly woolly when young; lvs. scattered, 2–5 cm. long, pinnatifid with few linear-filiform lobes or commonly entire; heads several, on long glandular-puberulent peduncles; invol. 8–10 mm. high; phyllaries linear-oblong, acute, essentially glabrous, thickish; corollas white or flesh-color, 5–6 mm. long, the marginal ones usually much enlarged, often with palmatifid limb; aks. hirsute; pappus-paleae of inner fls. 4, equal, lance-acuminate, almost as long as the corollas, those of outer fls. in part shorter and obtuse; $2n = 10$ (Stockwell, 1940).—Common on sandy mesas and open desert slopes, 100–3500 ft.; Creosote Bush Scrub, Joshua Tree Wd., V. Grassland, Chaparral; Mojave and Colo. deserts, Inyo Co. to San Diego Co., w. to Kettleman Hills and head of San Joaquin V.; to s. Nev., sw. Utah, w. Ariz. March–May.

12. **C. Xantiána** Gray. [*C. X.* var. *integrifolia* Gray.] Rather stout annual with few ascending branches, 1–4 dm. high, sometimes anthocyanous toward base, early glabrate except around the heads; lvs. somewhat fleshy, narrowly linear, entire or pinnatifid into 3–7 linear lobes; peduncles often fistulous; invol. 12–16 mm. high; phyllaries variable, linear-oblong, the strongly 1-nerved body glabrous, the acute tips shorter than the disk, or somewhat loosely spreading and exceeding the disk, but always densely puberulent; corollas white or flesh-color, the marginal ones scarcely enlarged; pappus double, the 4 long paleae surrounded by 4 very short outer squamellae; $n = 7$ (Stockwell, 1940). —Slopes of desert ranges, 1400–7000 ft.; Chaparral, Pinyon-Juniper Wd., Sagebrush Scrub; n. base of San Gabriel Mts. and Mojave R. to Mt. Pinos, Ventura Co., n. in inner S. Coast Range to w. Merced Co., and along e. face of Sierra Nevada to Lassen Co.; to se. Ore., w. Nev. Possibly only adv. in nw. Ariz. April–June.

13. **C. stevioìdes** H. & A. [*C. latifolia* Stockw.] Annual 5–25 cm. high, often freely and openly branching, sparingly tomentose; lvs. scattered, grayish, once or twice pinnatifid into numerous short linear lobes, or some of the upper entire; heads on short slender peduncles; invol. 6–9 mm. high; phyllaries linear, obtusish, glandular-puberulent and sometimes sparingly arachnoid-tomentose; corollas white, 5 mm. long, the marginal ones moderately enlarged; pappus-paleae 4, oblong-lanceolate, in the typical form ca. ⅔ the length of the corolla and acute.—Sandy desert floors and ranges, 500–5000 ft.; mostly Creosote Bush Scrub; Mojave and especially Colo. deserts, from White Mts., Inyo Co. to San Diego Co.; adjacent Mex., e. to New Mex., n. to Wyo. and Ida. March–May.

Var. **brachypáppa** (Gray) Hall. [*C. b.* Gray.] Pappus reduced to ⅓ or less the length of the corolla, obtuse.—Desert slopes, 2000–7000 ft.; Creosote Bush Scrub, Alkali Sink, Joshua Tree Wd., Pinyon-Juniper Wd.; mts. around head of San Joaquin V., to w. and n. Mojave Desert, n. Colo. Desert; s. Nev. April–June.

14. **C. artemisiaefòlia** (Harv. & Gray) Gray. [*Acarphaea a.* Harv. & Gray. *Acicarphaea a.* Walp.] Relatively stout erect annual 3–15 dm. high, openly branching above to form a leafless often paniculate cyme; herbage farinose below, glandular-hirsute above; lvs. 5–20 cm. long, broadly lanceolate to deltoid-ovate in outline, 2–3-pinnatifid into numerous irregularly linear or oblong divisions; heads large, hemispheric; invol. 9–12 mm. high, densely glandular-pubescent; phyllaries linear-lanceolate, acute; corollas nearly uniform, white or pinkish, the tube densely glandular-puberulent; aks. compressed, glabrous or nearly so; pappus none or rudimentary; $n = 8$ (Stockwell, 1940).—Dry canyonsides and sandy fields, often in disturbed places, 10–5600 ft.; Coastal Sage Scrub, Chaparral; near the coast, Ventura Co. to Santa Monica and San Gabriel mts., along the foothills to San Diego Co.; L. Calif. April–July.

62. Orochaenáctis Cov.

Low slender annuals, branched from the base. Lvs. linear, entire, alternate or the lowest opposite. Heads discoid, sessile singly or in glomerules at the ends of the branches or in the upper axils. Invol. narrowly turbinate, of ca. 4 equal herbaceous distinct phyllaries. Florets 4–9, greenish-yellow, hermaphrodite; corolla 5-toothed. Anthers sagittate. Style-branches linear, without appendages. Aks. terete, linear-clavate, closely striate. Pappus of several spatulate hyaline erose-fimbriate paleae narrowed toward the united base, deciduous as a ring. Monotypic restricted endemic. (Greek, *horos*, mountain, and *Chaenactis*, the related genus.)

1. **O. thysanocárpha** (Gray) Cov. [*Chaenactis t.* Gray. *Bahia Palmeri* Wats.] Plant

1–25 cm. high, openly and slenderly branching, ± arachnoid-tomentose and glandular-atomiferous, at length glabrate; lvs. narrowly linear, obtuse, tapering to base, well spaced, 5–25 mm. long; heads 1–5 in a glomerule, 6–8 mm. high; invol. 4–5 mm. high; phyllaries narrowly oblanceolate, obtuse, glandular-puberulent; corolla glandular-granuliferous, 3–4 mm. long; aks. minutely glandular; pappus a crown of 10–20 spatulate hyaline irregularly laciniate paleae deciduous in a ring.—In well-drained granitic soils, 7000–11,500 ft.; Red Fir F., Subalpine F.; s. Sierra Nevada, Inyo, Tulare and Kern cos. June–Aug.

63. Palafóxia Lag.

Annual or perennial herbs. Lvs. mostly alternate, entire. Heads corymbose, narrow, discoid. Phyllaries narrow, subequal, mainly uniseriate, herbaceous. Receptacle small, flat. Disk-fls. all alike or the outer with very unequal lobes; corolla with short tube, long throat, narrow-lobed limb. Aks. linear-tetragonal, nearly as long as the invol., pubescent. Pappus-paleae 4–8, slender, unequal, with an excurrent nerve, nearly as long as the ak. and hence well exserted. Less than a dozen spp., natives of s. U.S. and adjacent Mex. (Honoring José *Palafox*, noted Spanish general, defender against the armies of Napoleon.)

(Ammerman, E. A monographic study of the genus *Palafoxia* and its immediate allies. Ann. Mo. Bot. Gard. 31: 249–278, 1944.)

1. P. lineàris (Cav.) Lag. [*Ageratum l.* Cav. *Stevia l.* Cav. *Paleolaria carnea* Cass.] SPANISH NEEDLES. Erect branching annual, 2–7 dm. high, hispid or scabrous, glandular above; lvs. ± canescent, linear or linear-lanceolate, tapering to base and apex, strongly 1-nerved, 2–6 cm. long; invol. narrowly turbinate, 13–18 mm. high; phyllaries linear, acuminate; fls. 10–20, the corolla white, the exserted styles pink; aks. strigose, the cent. ones longer, equaling the invol.; pappus-paleae usually 4, composed of a thickened midrib bordered by a narrow fragile hyaline wing, ca. ⅔ as long as the ak., or the paleae of outer fls. sometimes reduced.—Sand flats and washes, −100–2500 ft.; Creosote Bush Scrub, Alkali Sink; Mojave and Colo. deserts, but more abundant on the latter, s. Inyo Co. to Imperial Co.; to sw. Utah, w. Ariz., n. Mex. Jan.–Sept.

Var. gigantèa Jones. [*P. l.* var. *arenicola* A. Nels.] Annual, or becoming perennial and woody at base, 1–2 m. high, the herbage green and scabrous; lvs. lanceolate, 3-nerved; invol. 15–20 mm. high, eglandular; fls. 25 or more; pappus-paleae 8, 4 long and 4 short.—Common in sand hills w. of Yuma, Imperial Co. March–May.

64. Amblyopáppus H. & A.

Small annual herbs. Lvs. narrowly linear, all but the lowermost alternate. Heads small, apparently discoid, cymose-paniculate. Invol. broadly obconic, 1–2-seriate, the phyllaries obovate to obovate-oblong, herbaceous, rather thin, concave. Receptacle short-conic. Marginal fls. ♀; corolla tubular, 2–3-toothed. Disk-fls. perfect, fertile; corolla 5-lobed. Aks. short-clavate, 3–4-angled. Pappus of 7–12 irregular oblong obtuse paleae. Monotypic. (Greek, *amblus*, blunt, and *pappos*, pappus.)

1. A. pusíllus H. & A. [*Aromia tenuifolia* Nutt. *Infantea chilensis* Remy.] Plant yellow-green, balsamic glandular-granuliferous; stem 1–4 dm. high, leafy, erect, corymbosely but strictly branching above; lvs. entire or the lower pinnately divided into 3–5 lobes; heads many, on short bracteate peduncles; invol. 3–5 mm. high, nearly equaling the 10–30 inconspicuous yellowish fls.; aks. 1.5–2 mm. long, black, sparsely glandular and hispidulous; pappus paleae white or reddish, 0.5 mm. long.—Beaches, old dunes, coastal bluffs and salt-marsh borders, below 500 ft.; Coastal Strand, Coastal Salt Marsh, Coastal Sage Scrub; coast, San Luis Obispo Co. to San Diego, rare inland (Glendora), Channel Ids.; L. Calif.; Peru, Chile. March–June.

65. Blennospérma Less.

Low annual herbs with alternate usually pinnately parted lvs. Heads radiate, many-fld., terminating the branches. Invol. hemispheric or depressed, the phyllaries uniseriate, thin, with membranous margin, united at base. Receptacle flat, naked. Ray-fls. fertile. Disk-fls. perfect but functionally ♂; tube slender; throat broadly campanulate. Anthertips broadly obcordate. Style of ray-fls. with flat linear-oblong branches; that of disk-fls. undivided. Aks. narrowly obovoid, densely covered with minute papillae which become mucilaginous when wet; epappose. Genus of 3 spp., of Calif. and Chile. (Greek, *blenna*, mucus, and *sperma*, seed.)

Lower lvs. 7–15-lobed, the lobes mostly 5 mm. or less long .... 1. *B. nanum*
Lower lvs. entire or 3-lobed, the lobes 10–15 mm. long .... 2. *B. Bakeri*

1. **B. nànum** (Hook.) Blake. [*Chrysanthemum ? n.* Hook. *Coniothele californica* DC. *B. c.* T. & G.] Slender-stemmed rather succulent herb 7–20 cm. high, often branching from the base and upward, glabrous except for a scanty microscopic glandularpuberulence and very sparse flocs of wool in the lf.-axils and beneath the heads; lower lvs. 2–5 mm. long, pinnately parted into 7–15 subfiliform lobes; invol. depressedhemispheric, 5–6 mm. high, the 5–12 phyllaries elliptic, purple-tipped and with an apical tuft of hairs; rays as many as the phyllaries, oblong, entire, yellow with purplishbrown backs; aks. 2.5–3 mm. long; *n* = 7 (Heiser, 1947).—Locally common at scattered stations in low, wet, sunny, clay flats or slopes, 25–1500 ft.; V. Grassland, Foothill Wd., etc.; Cent. V. from Butte and Glenn cos. to Tulare Co., Coast Ranges from Mendocino and Lake cos. to San Luis Obispo Co., Los Angeles and San Diego (4600 ft.) cos. Feb.–April. One of the first fls. of spring.
Var. **robústum** J. T. Howell. More robust, the stouter fistulous stems 15–30 cm. long; invol. 7 mm. high; aks. 3–4.5 mm. long, sometimes without papillae and glabrous. —Apparently a maritime ecotype of sandy soil, known only from Pt. Reyes Peninsula, Marin Co. May.

2. **B. Bàkeri** Heiser. Somewhat succulent herb up to 30 cm. high; stems fistulous, branching from the base and above; pubescence as in no. 1; lower lvs. 8–15 mm. long, linear, entire or 3-lobed, the upper lvs. shorter, 3–5-lobed; invol. 6–8 mm. high, the 6–8 phyllaries curved over the aks. at maturity; stigmas of ray-fls. red; aks. 3–4 mm. long; *n* = 9 (Heiser, 1947).—In vernal pools, 100 ft.; V. Grassland; Sonoma, Sonoma Co. March–April.

66. Péctis L.

Low slender-stemmed aromatic annual or perennial herbs. Lvs. opposite, dotted with oil-glands, entire, usually ciliate with stiff setae toward the base. Heads small, radiate, yellow, solitary or cymose. Invol. 1-seriate. Ray-fls. ♀, fertile, the rays often tinged purplish dorsally. Style-branches short, obtuse, hispidulous. Aks. slender, terete or ± angled. Pappus various, of scales, awns, or bristles, or reduced to a mere crown. Ca. 70 spp., from sw. U.S. and Fla., through Mex., to S. Am. (Greek, *pecteo*, to comb, the lvs. of most spp. pectinately ciliate.)

1. **P. pappòsa** Harv. & Gray ex Gray. CHINCH WEED. Slender dichotomously branched yellowish-green annual 1–2.5 dm. high, heavy scented; lvs. narrowly linear, 1–6 cm. long, 1–2 mm. wide, with 2–5 pairs of marginal bristles near base, the marginal oil-glands prominent; heads ± clustered in leafy cymes, yellow-fld.; invol. 5 mm. high; phyllaries 7–10, linear, strongly round-keeled and clasping the ray-aks., gibbous at base, prominently dotted with oil-glands; aks. narrowly linear, sparsely hispidulous; pappus of disk-fls. of 15–20 short weak tawny plumose bristles, of ray-fls. coroniform from very much reduced and ± united bristles, or obsolete.—Sandy and clayey flats, sea level to 5000 ft.; Creosote Bush Scrub, Joshua Tree Wd.; infrequent, Colo. Desert from San Jacinto Mts. and San Diego Co. e., e. Mojave Desert, Death V. region to Needles; to sw. Utah, New Mex., n. Mex. Usually appearing after summer rains, June, Sept.–Nov.

67. Nicollètia Gray

Perennial herbs, with slender woody rootstocks atop a deep-seated taproot; herbage very glabrous, glaucous, somewhat succulent. Lvs. alternate, pinnately parted into linear gland- and bristle-tipped lobes. Heads large, terminating the branches. Invol. turbinate; phyllaries oblong, abruptly acute, with a prominent gland near tip, all equal and uniseriate, or often with a smaller and shorter calyculate series of subtending bracts. Receptacle convex. Ray-fls. few, ♀, fertile, purplish or pinkish. Disk-fls. many, fertile, yellow, aging pink. Aks. linear, rusty-pubescent. Pappus double, the outer of many capillary bristles, the inner of 5 attenuate hyaline-margined awn-tipped paleae. Five spp., of sw. U.S. and Mex. (For J. N. *Nicollet*, early Am. astronomer and explorer.)

1. N. occidentàlis Gray. Stout ill-scented perennial with several erect leafy stems 1–4 dm. high arising from the deep-seated root-crown; lvs. 3–5 cm. long; heads nearly sessile among the upper lvs.; phyllaries 8–12, 11–15 mm. long, hyaline-margined; rays normally 8, 5–7 mm. long, narrow, lurid purple, striped dorsally with pink.—Sandy washes, 2500–4500 ft.; Creosote Bush Scrub, Joshua Tree Wd.; w. borders of the Mojave and Colo. deserts, from ne. Kern Co. to Little San Bernardino Mts., occasional to San Diego Co. April–June.

68. Tagètes L. MARIGOLD

Annual or sometimes perennial strong-scented herbs. Lvs. opposite, or the upper sometimes alternate, pinnate or pinnatifid, or sometimes simply serrate, conspicuously gland-dotted. Heads of various sizes, solitary at the ends of branches or in leafy cymes. Invol. of united phyllaries forming a cylindric or campanulate tube, naked at base. Ray-fls. usually present, ♀, fertile. Disk-fls. fertile, orange or yellow like the rays. Aks. angled or flattened. Pappus of few entire mostly unequal often ± united paleae, with 1 or 2 subulate bristles longer than the rest. Ca. 20 spp., from New Mex. and Ariz. to Argentina. (Perhaps from *Tages*, an Etruscan God.)

Fls. obscure, extending only 1–2 mm. out of the invol. . . . . 1. *T. minuta*
Fls. showy, extending several mm. above the invol. . . . . 2. *T. patula*

1. **T. minùta** L. Leafy branched glabrous annual 2–8 dm. high; lvs. pinnate, 5–15 cm. long, the lfts. 11–17, linear-lanceolate, sharply serrate, 2–4 cm. long; heads numerous in congested cymes; invol. ca. 10 mm. long, 2.5–3 mm. broad; phyllaries 5, with rounded free lobes; rays usually 3, yellow, 1 mm. long.—Natur. at Riverside in 1921; native of S. Am.

2. **T. pátula** L. FRENCH MARIGOLD. Glabrous annual 2–5 dm. high, with branching often purplish stem; lvs. pinnate, 5–10 cm. long, the lfts. linear-lanceolate, sharply toothed, 1–3 cm. long; heads cymosely arranged on long peduncles; invol. campanulate, ca. 12 mm. long, 6–7 mm. broad; phyllaries 5–7, with deltoid-acuminate or acute free lobes; rays ca. 5, orange-yellow, 12–15 mm. long; $2n = 48$ (Eyster, 1939).—Natur. in vacant lots and dumps, as at Santa Barbara; native of Mex.

69. Dyssòdia Cav.

Strong-scented annual or perennial herbs or subshrubs, the herbage marked with conspicuous translucent oil-glands, the stems striate. Lvs. opposite or alternate, entire to pinnatisect. Heads small to rather large, terminal, radiate, yellow or orange. Invol. of 2 equal series of quite distinct to ± united phyllaries, these usually subtended by an outer series of much shorter calyculate bracts. Receptacle convex, puberulent. Ray-fls. ♀, fertile. Disk-fls. fertile. Style-branches with a short conic appendage. Aks. terete or angled, striate. Pappus of 10–15 bristle-tipped paleae, or these usually dissected into numerous bristles. Possibly 40 spp., natives of sw. U.S. and Mex. (Greek, *dysodia*, a disagreeable odor.)

Invol. 12–16 mm. high; lvs. alternate, relatively inconspicuous, mostly shorter than the internodes.
   Stems glabrous; lvs. 3–5-parted into narrow lobes . . . . 1. *D. porophylloides*
   Stems scabridulous; lvs. merely spinulose-dentate . . . . 2. *D. Cooperi*
Invol. 5–7 mm. high; lvs. opposite, conspicuous, crowded, mostly longer than the internodes.
   Pappus-paleae each dissected into 5–12 capillary bristles; heads sessile or subsessile
                                                            3. *D. papposa*
   Pappus-paleae tipped with only 1–3 bristles; heads on long slender peduncles . . 4. *D. Thurberi*

1. **D. porophylloìdes** Gray. [*Lebetina p.* A. Nels. *Clomenocoma p.* Rydb.] Shrubby very glabrous perennial from a woody base, 3–6 dm. high, with numerous branching virgate glaucescent stems; lvs. narrow, thick, 1–2 cm. long, the lower petiolate, parted

into lance-linear acerose-tipped divisions, the upper often entire or merely incised; invol. turbinate, 12–15 mm. high, the principal phyllaries 14–20, indurate, linear-oblong, abruptly acute, free or nearly so, with oval terminal gland and several linear marginal ones, subtended by attenuate calyculate bractlets; rays orange-yellow, sometimes turning purplish, few and inconspicuous or none.—Sandy washes, mesas and rocky slopes, 600–3500 ft.; Creosote Bush Scrub; s. Mojave and Colo. deserts, from the w. borders of the latter in Riverside and San Diego cos. e.; to s. Ariz., Son., L. Calif. March–June.

2. **D. Coóperi** Gray. [*Lebetina C. A. Nels. Clomenocoma C.* Rydb.] Habit of the preceding, 3–5 dm. high, the stems rather stout from branching woody rootstocks, puberulous or hispidulous; lvs. lanceolate to ovate, 1–2 cm. long, spinulose-dentate, rarely with small lobes; invol. turbinate, 14–18 mm. high, the principal phyllaries 20–30, strongly nerved, linear-lanceolate, with subulate-acuminate tip, free or nearly so, the glands, if present, terminal only; rays 8–12, bright orange or yellow, often turning saffron, 1 cm. long.—Gravelly washes, open mesas and slopes, 2100–5100 ft.; Creosote Bush Scrub, Joshua Tree Wd.; e. and s. Mojave Desert, from Death V. region to Victorville and n. Colo. Desert; s. Nev., nw. Ariz. May–June; Sept.–Nov.

3. **D. pappòsa** (Vent.) Hitchc. [*Tagetes p.* Vent. *D. glandulosa* Cav. *Boebera chrysanthemoides* Willd. *B. g.* Pers. *B. p.* Rydb. *D. c.* Lag.] Much-branched annual 1–4 dm. high, moderately canescent or glabrate; lvs. 2–5 cm. long, pinnatifid or bipinnatifid into linear divisions; heads sessile or nearly so; invol. campanulate, 6–7 mm. high, the principal phyllaries 8–10, lance-elliptic, obtuse, hyaline-margined, with several conspicuous oblong brownish glands, the subtending calyculate bracts herbaceous, ⅔ as long; rays yellow, scarcely exceeding the disk; pappus-paleae dissected into many bristles.—Waste places and cult. ground; introd. at Loma Linda, San Bernardino Co.; widespread as a weed from Ariz. to La., and from Mont. to Ohio. May–June.

4. **D. Thúrberi** (Gray) Rob. [*Hymenatherum T.* Gray. *D. cupulata* A. Nels. *Thymophylla T.* Woot. & Standl.] Low pubescent perennial 1–2 dm. high, the slender densely leafy striate branching stems ascending directly from the wiry taproot or from a suffrutescent caudex; lvs. sessile, pinnately parted into 3–7 rigid filiform acerose spreading lobes 1–2 cm. long; peduncles filiform, glabrous, 2–6 cm. long, terminating the branchlets; invol. ca. 5 cm. high, the principal phyllaries ± united, their free margins ciliate and beset with small oval glands; rays orange, 3 mm. long; pappus of 10 linear-lanceolate ± laciniate barbellate-awn-tipped unequal paleae.—Washes or rocky ground, often on limestone, 3000–5600 ft.; Creosote Bush Scrub, Joshua Tree Wd.; Clark, Kingston, Ivanpah, and Providence mts. and Mid Hills of e. Mojave Desert, San Bernardino Co.; to Tex., Mex. April–June.

70. Porophýllum [Vaill.] Adans.

Herbs or subshrubs with glabrous herbage. Lvs. alternate or opposite, simple, with translucent oil-glands near the margin. Heads discoid. Invol. cylindric or campanulate, uniseriate, the phyllaries 5–9, equal, free or united only at base, usually with 2 rows of oil-glands, without accessory bractlets. Fls. perfect, fertile, purplish or yellow. Style-branches elongated, subulate, hirtellous. Aks. slender, striate, often tapering to apex. Pappus of numerous scabrous bristles. Ca. 30 spp., native from sw. U.S. and Mex. to Panama and S. Am. (Greek, *poros*, a passage or pore, and *phyllon*, lf., the translucent glands making the lvs. appear punctate.)

1. **P. grácile** Benth. [*P. junciforme, Vaseyi* and *caesium* Greene.] Multibranched perennial 3–7 dm. high, from a lignescent base; herbage dark purplish-green, glaucous, with rank odor; stems slender, erect, striate, sparingly leafy, rushlike; lvs. filiform to narrowly linear, entire, 1–5 cm. long; heads solitary on the branches; invol. cylindraceous, 10–16 mm. high; phyllaries 5, oblong, obtuse, purplish, hyaline-margined, the oil-glands linear, few; corolla 7–9 mm. long, purplish or whitish with purple veins; aks. linear, attenuate to a beaklike apex, hispidulous; pappus of very numerous fuscous capillary bristles.—Dry rocky slopes, washes, mesas, sometimes alkaline, 200–4800 ft.; Creosote Bush Scrub, Coastal Sage Scrub; Clark Mt., San Bernardino Co., s. through e. Mojave Desert to L. Calif., w. through Colo. Desert to Little San Bernardino Mts., San Jacinto Mts., San Diego Co., near coast in Orange Co.; to s. Nev., Ariz., Son. Oct.–June.

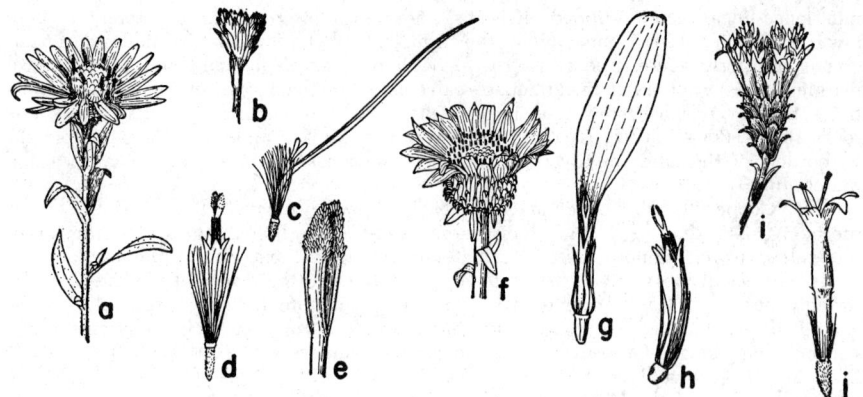

Fig. 113. ASTEREAE. *Aster: a*, head and upper lvs., × 1; *b.* invol., × 1, of overlapping phyllaries; *c*, ray-fl., × 2½; *d*, disk-fl., pappus of many capillary bristles; *e*, stigmatic branches, × 10, with "collecting hairs" grouped at tips. *Grindelia arenicola: f*, head, × ½, with squarrose invol.; *g*, ray-fl.; *h*, disk-fl., × 2½, pappus of flattened paleae. *Lessingia germanorum: i*, discoid head, × 2½, with gland-margined phyllaries; *j*, disk-fl., × 5.

Tribe 3. **Astèreae.** ASTER TRIBE (Fig. 113)

Herbs or shrubs with alternate lvs. Heads commonly radiate and insect-pollinated. Phyllaries usually well imbricated (nearly equal in *Erigeron* and *Conyza*). Receptacle naked in ours. Disk-fls. mostly yellow, perfect in all ours except *Baccharis*. Style-branches of perfect fls. flattened, conspicuously margined by the stigmatic lines, usually prominently tipped with a hairy appendage. Pappus mostly of awns or bristles, paleaceous in Calif. only in *Gutierrezia* and *Amphiachyris*.—Over 100 genera, of world-wide distribution.

A. Heads all with perfect disk-fls., commonly radiate; not dioecious.
 B. Ray-fls. yellow, only exceptionally wanting; disk-fls. yellow. (Homochromeae; Subtribe Solidagininae)
  C. Pappus of paleae, scales, or flattened awns.
   D. Lvs. usually serrate and more than 5 mm. wide; aks. glabrous; pappus of 2–8 firm deciduous awns .... 71. *Grindelia*
   DD. Lvs. entire, mostly much less than 5 mm. wide; aks. hairy; pappus of 10 or more persistent members. Low shrubs of desert areas.
    E. The lvs. filiform to narrowly oblanceolate; pappus of 10–12 sordid oblong free scales; phyllaries not prominently scarious-margined .. 72. *Gutierrezia*
    EE. The lvs. broader in outline; pappus of 15–40 bright white narrow paleae; phyllaries broad, rounded, with prominent scarious erose margin.
     F. Heads glomerate; disk-fls. 4–7; plant spinescent .. 73. *Amphipappus*
     FF. Heads solitary but sometimes approximate; disk-fls. mostly 20 or more; plant not spinescent .... 74. *Acamptopappus*
  CC. Pappus of capillary bristles, sometimes with some short outer scales as well.
   D. Ray-fls. present.
    E. Pappus of disk-fls. double, the inner series of capillary bristles, the outer much shorter.
     F. Ray-fls. with copious pappus and pubescent aks. ...... 75. *Chrysopsis*
     FF. Ray-fls. epappose or the pappus reduced to a deciduous crown and with essentially glabrous aks. .... 76. *Heterotheca*
    EE. Pappus of all fls. simple, generally unequal but not divided into 2 lengths.
     F. Disk-fls. fertile.
      G. Plants taprooted, never rhizomatous.
       H. Heads nodding in bud; very slender usually diffuse annuals with slender or filiform stems; phyllaries thin with prominent scarious margin .... 77. *Chaetopappa*
       HH. Heads erect or at least not nodding in bud; coarser mostly perennial herbs or shrubs; heads usually few and relatively large, if small and panicled, then the plants shrubby
                                                                                78. *Haplopappus*
      GG. Plants with numerous fibrous roots arising from a rhizome or caudex, without a taproot; heads usually small and very numerous; always herbaceous .... 80. *Solidago*

FF. Disk-aks. sterile.
  G. Plant annual; phyllaries not in vertical ranks, with recurved tip bearing a prominent gland .... 79. *Benitoa*
  GG. Plant a tufted perennial herb; phyllaries in vertical ranks, the erect tip not bearing a prominent gland .... 81. *Petradoria*
DD. Ray-fls. none.
  E. Plants shrubby.
    F. Phyllaries in ± distinct vertical ranks .... 82. *Chrysothamnus*
    FF. Phyllaries not in vertical ranks .... 78. *Haplopappus*
  EE. Plant a diffusely branched annual herb .... 93. *Lessingia germanorum*
BB. Ray-fls. white, pink, purple, or blue, sometimes very inconspicuous, occasionally lacking; disk-fls. yellow, white, pink, or reddish-purple. (Heterochromeae)
  C. Pappus none; receptacle conic; plants scapose. (Subtribe Bellidinae) .... 83. *Bellis*
  CC. Pappus present; receptacle not conic; habit various (Subtribe Asterinae)
    D. Pistillate fls. few to numerous, in one or more series, usually with conspicuous rays; disk-fls. few to generally numerous.
      E. Pappus of paleae or scales, sometimes with additional bristles.
        F. The pappus of unequal bristles alternating with short paleae or of a scarious cup and 1 bristle; diminutive desert annuals .. 84. *Monoptilon*
        FF. The pappus of numerous rigid narrow awns or awns and scales or scales only; ours mostly montane cushion plants .... 85. *Townsendia*
      EE. Pappus of mostly capillary bristles.
        F. Ray-fls. present, sometimes very inconspicuous.
          G. Pappus of 5 firm bristles; phyllaries with broad scarious margins; small annual with white or purplish rays
                                 77. *Chaetopappa bellidiflora*
          GG. Pappus usually of numerous bristles, in *Erigeron* very rarely as few as 5 and then plants perennial.
            H. Aks. beaked; phyllaries deciduous; erect annual with filiform branches .... 86. *Tracyina*
            HH. Aks. not beaked; phyllaries persistent.
              I. Pappus in ray-fls. obsolete or none; lvs. mostly laciniate-pinnatifid; slender desert annual .... 87. *Psilactis*
              II. Pappus present in both ray- and disk-fls.
                J. Phyllaries usually obviously graduated and imbricated in 3 or more series.
                  K. Style-branches of disk-fls. not tipped with a prominent tuft of yellow hairs; ray-fls. fertile; pappus not reddish.
                    L. Perennials, usually rhizomatous or with fibrous roots, if annuals then lvs. essentially entire.
                      M. Appendages of style-branches lanceolate to subulate, acute; phyllaries lacking an even scarious margin .. 88. *Aster*
                      MM. Appendages of style-branches ovate or oblong, obtuse; low tufted desert perennial with bristle-tipped subulate lvs.
                                     89. *Leucelene*
                    LL. Annuals, biennials, or short-lived perennials with a taproot; lvs. spinulose-tipped, entire or usually spinulose-dentate to pinnatifid or dissected .... 90. *Machaeranthera*
                  KK. Style-branches of disk-fls. densely tufted with rigid yellow hairs; ray-fls. neutral; pappus reddish .... 91. *Corethrogyne*
                JJ. Phyllaries only slightly or not at all graduated, in 1 or 2 series; style-branches of disk-fls. with lanceolate and acute or usually with triangular and obtuse appendages .... 92. *Erigeron*
        FF. Ray-fls. absent, but the marginal disk-corollas sometimes enlarged to resemble rays.
          G. Marginal corollas enlarged, becoming palmately 5-lobed; pappus usually brownish to reddish .... 93. *Lessingia*
          GG. Marginal corollas not enlarged, regular; pappus not brownish or reddish. (Rayless spp. of) .... 88. *Aster*
    DD. Pistillate fls. very numerous, several-seriate, tubular-filiform, eligulate, or the inconspicuous ligule barely equaling the pappus .... 94. *Conyza*
AA. Heads unisexual, discoid; dioecious; pappus of ♂ fls. of clavellate bristles. (Subtribe Baccharidinae) .... 95. *Baccharis*

71. Grindélia Willd. GUM-PLANT

Annual, biennial, or usually perennial herbs, mostly with a taproot, rarely suffrutescent at base, ± resinous particularly on the invol. Lvs. alternate, punctate, usually serrate and sessile, often clasping. Heads medium to large, yellow, usually radiate, solitary at branch-tips. Invol. multiseriate, imbricate, the phyllaries thickish, with pale appressed base and narrow often squarrose or revolute herbaceous tip. Receptacle flattish, foveolate. Ray-fls. 10–45, uniseriate, fertile. Disk-fls. usually fertile. Style-branches with slender hispidulous appendages. Aks. compressed to subquadrangular, few-angled, glabrous; pappus of 2–8 stiff often curved deciduous corneous or paleaceous awns. Ca. 50 spp. of w. N. and S. Am. (Named for Professor David Hieronymous *Grindel*, 1776–1836, botanist of Dorpat and Riga.)

(Steyermark, J. A. A monograph of the N. Am. spp. of the genus Grindelia. Ann. Mo. Bot. Gard. 21: 433–608, 1934.)
A. Tips of phyllaries erect or spreading, some gradually curved but not sharply reflexed.
  B. Fruticose. Salt marshes of San Francisco Bay .... 1. *G. humilis*
  BB. Herbaceous, ± woody at caudex.
    C. Invol. pubescent. Coast Ranges .... 2. *G. hirsutula*
    CC. Invol. glabrous.
      D. Stems 8–18 dm. high, erect, strictly branching above. Interior valleys
          3. *G. procera*
      DD. Stems 3–8 dm. high, ascending, openly branched. San Francisco .. 4. *G. maritima*
AA. Tips of phyllaries (at least of some middle and outer ones) sharply reflexed or looped.
  B. Coastal succulent plants.
    C. Woody below; nearly erect .... 5. *G. stricta*
    CC. Herbaceous to caudex; ± decumbent or prostrate.
      D. Lvs. obovate-cuneate or rounded. Mostly n. of San Francisco Bay
          5. *G. stricta*
      DD. Lvs. acutely or obtusely pointed at tip. Mostly s. of San Francisco Bay
          6. *G. latifolia*
  BB. Interior non-succulent plants; stems ± erect.
    C. Lvs. callous-serrulate. Rare introd. .... 7. *G. squarrosa*
    CC. Lvs. sharply toothed or entire, not callous-serrulate.
      D. Invols. mostly 1 cm. in diameter; ligules 5–8 mm. long. San Diego Co.
          8. *G. Hallii*
      DD. Invols. 12–25 mm. in diameter; ligules (when present) mostly 8–15 mm. long.
        E. Tips of phyllaries spreading or reflexed (often some sharply reflexed) but
            rarely looped. Great V. .... 9. *G. camporum*
        EE. Tips of phyllaries looped back to form a tight ring.
          F. Heads relatively small, the invol. 7–10 mm. high, 9–15 mm. thick.
              N. Calif. .... 10. *G. nana*
          FF. Heads larger, the invol. 8–15 mm. high, 12–25 mm. thick. S. Calif.
              11. *G. robusta*

1. **G. hùmilis** H. & A. [*G. robusta* var. *angustifolia* Gray. *G. cuneifolia* of most auth., not Nutt.] Frutescent, to 1.5 m. high, stout, the woody stems giving rise to herbaceous shoots of the season, the branches subcorymbose, glabrous to villous, few-headed; lvs. coriaceous, scarcely resinous, remotely serrulate, cuneate-oblanceolate to lance-oblong, 2–8 cm. long, 5–15 mm. wide, narrowed to a sessile base or amplexicaul; heads 3–5 cm. across; phyllaries lanceolate, erect, with short flat appressed or slightly reflexed tip, or tip more subulate and strongly reflexed; rays 16–34, 12–18 mm. long.—Coastal Salt Marsh; San Francisco, San Pablo, and Suisun bays. Aug.–Oct.

  G. paludòsa Greene. [*G. cuneifolia* var. *p.* Jeps.] Stems herbaceous to the crown, stout, freely branching, glabrous, 5–15 dm. high; lvs. subcoriaceous, not resinous, remotely serrate, serrulate, or entire, lance-ovate, oblong, or oblanceolate, strongly clasping, the lower mostly up to 8 cm. long and 1.5 cm. wide, those near the heads much reduced; heads 3–4 cm. across, very resinous, the prominent green tips of the phyllaries erect or spreading, the outer ones strongly recurved; rays 20–35, 10–12 mm. long.—Salt marshes, Suisun Bay, Solano Co. In all probability this is a stabilized hybrid population of the parentage *G. humilis* and *G. camporum*. July–Sept.

2. **G. hirsùtula** H. & A. [*G. pacifica* Jones. *G. h.* vars. *brevisquama, calva* and *subintegra* Steyerm.] Herbaceous perennial 3–8 dm. high; stems erect, slender, simple or commonly corymbosely branching above, ± villous with crisped hairs, especially on peduncles; lvs. chartaceous, gray-green, the basal oblanceolate or spatulate, obtuse, remotely serrate or lobed, sometimes merely shallowly crenate, tapering to a narrowly margined petiole, 10–22 cm. long, 1–2.5 cm. wide, the cauline much reduced in size upwards, from oblanceolate and sessile below to narrowly oblong, entire, bract-like and clasping above; heads 2.5–4 or 5.5 cm. across; invol. ± pubescent, the tips of the phyllaries erect or nearly so, not caudate; rays 20–35, 1–2 cm. long.—Arid slopes;

Coastal Sage Scrub, Foothill Wd.; Coast Ranges from Napa Co. to Monterey Co., also Ventura Co. April–July.

Ssp. **rubricaùlis** (DC.) Keck. [*G. r.* DC. *G. patens* Greene. *G. robusta* var. *p.* Jeps.] Stems usually densely villous above; peduncles prominently bracteate, the uppermost linear bracts crowded, retrorse, usually lanate, equaling or surpassing the disk; outer phyllaries often spreading with recurved tips.—Morphologically well marked but sympatric with the typical ssp.; also in the Sierran foothills of Fresno Co.

3. **G. prócera** Greene. [*G. camporum* var. *parviflora* Steyerm.] Herbaceous perennial 8–18 dm. high; stems erect, pallid, unbranched below, rather strictly much-branched above, bearing numerous heads; lvs. chartaceous, dark green, principally cauline, oblong-obovate to lance-oblong, obtuse, clasping, shallowly toothed, 3–7 cm. long, up to 2 cm. wide; heads 2–3.5 cm. across; invol. slightly resinous, mostly under 8 mm. high, the short green phyllaries erect or slightly squarrose at tip; rays 21–45, 8–10 mm. long.— Deep soils; V. Grassland; bottom lands of the San Joaquin R., Sacramento Co. to Kern Co., also Marin Co. July–Oct.

4. **G. marítima** (Greene) Steyerm. [*G. rubricaulis* var. *m.* Greene.] Stems erect or ascending from a ligneous caudex, 3–8 dm. high, slender, glabrous or rarely villous, loosely branching; lvs. thickish, mostly serrulate, the basal tufted, narrowly oblanceolate, up to 18 cm. long, tapering to slender petioles, the cauline oblong to lanceolate, mostly 3–7 cm. long, 8–15 mm. wide, clasping; heads 2.5–4 cm. across; invol. ± glandular, the phyllaries with mostly erect green relatively short tips; rays 30–40, 10–13 mm. long.— Ocean bluffs and open hillsides; N. Coastal Scrub; San Francisco. Aug.–Sept.

5. **G. stricta** DC. ssp. **venulòsa** (Jeps.) Keck. [*G. v.* Jeps. *G. arenicola* Steyerm. *G. a.* var. *pachyphylla* Steyerm. *G. s.* var. *procumbens* Steyerm.] Procumbent to decumbent herbaceous perennial up to 3 or 4 dm. high and 1 m. across, often subligneous at base, usually glabrous, rarely villous, the stems usually whitish or yellowish; lvs. thickish, resinous-punctate, serrulate, mostly spatulate and rounded at apex, the basal narrowed to a margined petiole, the upper cauline amplexicaul; heads 3–4.5 cm. across, often leafy-bracted, very resinous, the stout or slender herbaceous tips of the phyllaries squarrose.—Coastal salt marshes and seaside bluffs; Coastal Salt Marsh, Coastal Strand, N. Coastal Scrub; Marin Co. n. to Coos Co., Ore.; Monterey Co. (rare). June–Sept.

Ssp. **Blàkei** (Steyerm.) Keck. [*G. B.* Steyerm.] Suffrutescent, the stout red-brown stems of the season ascending from a woody base, up to 1 m. high; lvs. subcoriaceous, usually entire; heads to 5.5 cm. across, the tapering herbaceous outer phyllaries with prominently reflexed tips.—Restricted to the salt marshes bordering Humboldt Bay. July–Sept. Despite its woody axis and essentially erect branches, this plant is doubtless more closely related to forms of *G. stricta* from the Puget Sound area than to *G. humilis* of the San Francisco Bay area, which it also mimics.

6. **G. latifòlia** Kell. [*G. robusta* var. *l.* Jeps. *G. rubricaulis* var. *l.* Steyerm.] Very succulent and leafy herbaceous perennial 4–6 dm. high, the stout decumbent or ascending branches monocephalous or topped by a close cluster of 2 or 3 large heads, the heads often partially enveloped by subtending lvs.; lvs. principally cauline, thick, irregularly serrate to regularly and sharply dentate, scabro-ciliate, lance-ovate to broadly oblong, amplexicaul to subcordate, rounded at apex, 3–8 cm. long, 1.5–4 cm. wide; heads 3–5 cm. across, milky-resinous, the outermost phyllaries foliaceous, their green tips and usually those of many inner ones squarrose; rays 30–45, 10–15 mm. long.— Salt marshes and dunes; Coastal Salt Marsh, Coastal Strand, Coastal Sage Scrub; Watsonville coast, Surf to Pt. Conception and Channel Ids. May–Sept.

Ssp. **platyphýlla** (Greene) Keck. [*G. robusta* var. *p.* Greene. *G. rubricaulis* vars. *p.* and *permixta* Steyerm.] Neither so succulent nor so leafy, the many simple or branched decumbent or ascending usually glabrous stems radiating from the crown, forming plants up to 6 dm. high and 1 m. in diam., the milky-resinous heads sometimes subtended by several small leafy bracts but never enveloped by these.—Beaches and low ground near the coast; Coastal Strand, Coastal Sage Scrub; the common form of the species from Marin Co. to Santa Barbara Co. June–Sept.

7. **G. squarròsa** (Pursh) Dunal. [*Donia. s.* Pursh.] Erect biennial or short-lived perennial 2–10 dm. high, openly branched above and bearing many heads; lvs. regularly callous-serrulate, sometimes sharply toothed or even entire, mostly oblong, 2–5 cm. long, the upper clasping; heads 2–3 cm. across, strongly resinous, the green tips of the

phyllaries strongly rolled back; rays 25–40, 7–15 mm. long.—Dry fields and waste places; Joshua Tree Wd.; sparingly introd. in Antelope V. Native of the Great Plains. July–Sept.

An adventive discoid form has been collected in Clark Mts., e. Mojave Desert. This keys to *G. s.* var. *nuda* f. *angustior* Steyerm., but it would seem more logically to be a discoid form of *G. s.* var. *serrulata*.

8. **G. Hállii** Steyerm. Several glabrous herbaceous stems from the woody crown, 3–6 dm. high, corymbosely branching above; lvs. mostly sharply and regularly serrate, subcoriaceous, prominently resinous-punctate, the basal in often persistent rosettes, oblanceolate, 5–7 cm. long, the cauline oblong, much smaller; heads numerous, terminating the branches, 2–3 cm. across, strongly resinous, the green tips of the outer phyllaries strongly recurved or hooked; rays 13–21, 5–8 mm. long.—Dry flats and grassy mesas; Pinyon-Juniper Wd., Yellow Pine F.; Cuyamaca Mts., San Diego Co. July–Oct.

9. **G. campòrum** Greene. [*G. robusta* var. *rigida* Gray. *G. robusta* var. *Davyi* Jeps. *G. rubricaulis* var. *interioris* Jeps. *G. c.* vars. *D.* and *i.* Steyerm.] Several erect herbaceous stems from the subligneous caudex, 5–12 dm. high, simple or openly branched, glabrous or nearly so; lvs. glabrous or scabrous, rarely more hairy, subcoriaceous, very resinous, saliently dentate, narrowly oblong to broadly oblanceolate, the cauline 2–8 cm. long, 7–15 mm. wide; heads terminating the branches, 2.5–4 cm. across, strongly and translucently resinous, the green tips of the multiseriate elongated phyllaries strongly recurved or hooked; rays 18–35, 8–15 mm. long; *n* = 12 (Heiser & Whitaker, 1948).—Dry banks, rocky fields and plains, low alkaline ground; V. Grassland, Alkali Sink; the common Grindelia of the Great V. and adjacent foothills, s. to n. Los Angeles Co., w. to San Francisco Bay. May–Oct.

10. **G. nàna** Nutt. [*G. n.* var. *integrifolia* Nutt. *G. n.* vars. *altissima* and *turbinella* Steyerm.] Glabrous perennial with erect usually branching stems, 2–8 dm. high; lvs. spinulose-toothed to entire, the lower oblanceolate, up to 15 cm. long and 3 cm. wide, tapering to slender petioles, the upper much smaller and clasping; heads 2–3 cm. across; invol. rarely very resinous, the green tips of the phyllaries strongly rolled back; rays mostly 12–25, 5–15 mm. long.—Open ground, dry hillsides, roadsides and subalkaline flats; Yellow Pine F., N. Oak Wd.; inner N. Coast Range from Napa Co. to Humboldt Co., e. to Modoc Co.; n. to Ida. June–Oct.

11. **G. robústa** Nutt. [*G. cuneifolia* Nutt. *G. camporum* var. *australis* Steyerm. *G. rubricaulis* vars. *elata* and *robusta* Steyerm.] Stems few, erect from a subligneous crown, stout, usually corymbosely branching above, glabrous, 5–12 dm. high; lvs. sharply toothed to finely and remotely serrate or often entire, the basal oblanceolate, up to 18 cm. long (including the margined petiole) and 3 cm. wide, the cauline much reduced, ovate-lanceolate to linear-oblong, broadly clasping; heads 3–5 cm. across, often strongly and translucently resinous, the long green tips of the phyllaries rolled back in a loop; rays 25–45, 8–15 mm. long.—Dry fields and banks, also borders of salt marshes and shores; Coastal Sage Scrub, Chaparral; not far inland, Santa Barbara Co. to n. L. Calif. March–Sept.

Var. **bracteòsa** (J. T. Howell) Keck. [*G. b.* J. T. Howell. *G. rubricaulis* var. *b.* Steyerm.] Heads discoid.—Dry hills and canyons; Coastal Sage Scrub; away from the immediate coast, intermittent, Puente Hills, Los Angeles Co. to Pine Hills, San Diego Co. May–July.

72. Gutierrèzia Lag. MATCHWEED. SNAKEWEED

Perennial herb or subshrub, glutinous, glabrous to hirtellous. Lvs. alternate, entire, filiform to narrowly oblanceolate, usually punctate-glandular. Heads very small, radiate, yellow, numerous, scattered or crowded in cymes or panicles. Invol. cylindric to turbinate-subglobose, the imbricated phyllaries coriaceous, appressed, whitish. Receptacle foveate, sometimes hairy. Ray-fls. ♀, fertile. Disk-fls. perfect, sometimes sterile. Aks. obovoid or oblong, pubescent; pappus of 10–12 oblong unequal free scales, shorter on the ray-aks. Ca. 25 spp., native to w. N. and S. Am. (Named for the Spanish nobleman, Pedro Gutierrez.)

Heads flaring, more than 4-fld.
  Invol. campanulate, 3–5 mm. thick; ray-fls. 5–10; disk-fls. 6–16; cismontane . . . . 1. *G. californica*

Invol. narrowly turbinate, 2–3 mm. thick; ray- and disk-fls. each 3–6; transmontane
 .... 2. *G. Sarothrae*
Heads cylindric, ca. 1 mm. thick, 2–4-fld. .... 3. *G. microcephala*

1. **G. califòrnica** (DC.) T. & G. [*Brachyris c.* DC. *G. bracteata* Abrams. *G. c.* var. *b.* Hall. *Xanthocephalum c.* Greene.] Subshrub 3–6 dm. high, nearly glabrous to densely hirtellous, the branches stiff and erect or sometimes divergent or spreading, very twiggy above; lvs. spreading to deflexed, up to 5 cm. long, mostly 1 mm. wide; heads usually solitary at tips of branchlets, rarely glomerulate; invol. broadly turbinate-obovoid or campanulate, 5–7 mm. high, 3–5 mm. thick, the usually broad blunt phyllaries with definite green tips; ray-fls. 5–10; disk-fls. 6–16.—Dry hills and plains, mostly below 1000 ft.; V. Grassland, Foothill Wd., Chaparral; inner Coast Range valleys from Yolo Co. to San Francisco Bay and upper San Joaquin V., cismontane s. Calif. to San Diego Co.; to s. Ariz. and Chihuahua. May–Oct.

2. **G. Saròthrae** (Pursh) Britt. & Rusby. [*Solidago. S.* Pursh. *Brachyris Euthamiae* Nutt. *G. E.* T. & G. *G. divergens* Greene, in large part. *Xanthocephalum S.* Shinners.] Subshrub 1.5–6 dm. high, mostly hirtellous, the numerous slender stems cymosely paniculate above; lvs. punctate, 2–5 cm. long, 1–2 mm. wide; infl. flat-topped, the numerous resinous heads scattered or usually in small glomerules; invol. narrowly turbinate, 4–5 mm. high, 2–3 mm. thick, the phyllaries with often obscurely thickened herbaceous tips; ray- and disk-fls. 3–8 each, the ligule ca. 3 mm. long; aks. subsericeous-pilose; pappus of ray ca. 0.7 mm. long, of disk ca. 1.5 mm. long.—Dry plains, ascending the desert ranges to 8000 ft.; Shadscale Scrub, Sagebrush Scrub; sw. Mojave Desert, Kern Co., to Los Angeles and San Diego cos.; n. to se. Wash., e. to Great Plains, Sask., n. Mex. May–Oct.

3. **G. microcéphala** (DC.) Gray. [*Brachyris m.* DC. *G. Euthamiae* var. *m.* Gray. *Xanthocephalum lucidum* Greene. *G. l.* Greene. *G. Sarothrae* var. *m.* L. Benson. *X. m.* Shinners.] Many-stemmed, 3–6 dm. high, strongly resinous, essentially glabrous, the slender stems striate-angled, much branched above; lvs. 2–5 cm. long, 0.5–2 mm. wide, spreading-deflexed; infl. cymose-paniculate, the tiny heads ca. 3 mm. high, 1 mm. wide, in small terminal glomerules of 2–5, or a few solitary; phyllaries yellowish-white without green tips, the hyaline margin prominent, the inner row of only 2 members, as long as the disk-fls.; ray- and disk-fls. 1–2 each, the ligule up to 2.5 mm. long; ray-aks. fertile, appressed-pilose, ca. 2 mm. long; ray-pappus of ca. 8 oblong paleae ca. 0.8 mm. long; disk-pappus of ca. 12 paleae 1 mm. long.—Open desert, up to 7000 ft.; Shadscale Scrub, Creosote Bush Scrub, Joshua Tree Wd.; Mojave Desert from White Mts., Inyo Co. to Palmdale and Little San Bernardino Mts.; e. to Colo., Tex., and n. Mex. July–Oct.

73. Amphipáppus T. & G. CHAFF-BUSH

Low shrub with divaricate spinescent branchlets. Lvs. alternate, small, entire, oval to elliptic or obovate, short-petioled. Heads radiate, few-fld., small, glomerate at tips of branchlets. Invol. obovoid, ca. 3-seriate, strongly graduate, pale, the 7–12 broad rounded phyllaries thin, with scarious erose margin. Receptacle fimbrillate. Ray-fls. 1–2, pale yellow, small, ♀, fertile. Disk-fls. 4–7, perfect, sterile. Anthers narrowly lance-tipped. Style branches thick, short-subulate. Ray-aks. broadly oblanceolate, compressed, pilose, their pappus of ca. 15–20 short basally united unequal white paleae. Disk-aks. undeveloped, glabrous or sparingly pilose, their pappus of ± 25 tortuous hispidulous irregularly basally united white paleae of unequal width, equaling the corolla. Monotypic. (Greek, *amphi*, both [kinds of], and *pappos*, pappus.)

(Porter, C. L. The genus, Amphipappus Torr. & Gray. Am. Jour. Bot. 30: 481–483, 1943.)

1. **A. Fremóntii** T. & G. [*Amphiachyris F.* Gray.] Much-branched white-barked shrub 3–6 dm. high, with yellow-green cast, glabrous throughout, slightly glutinous around heads; lvs. 5–12 mm. long, 2–4 mm. wide, obtuse or acute, 1-nerved; heads 4–5 mm. high, the phyllaries closely appressed.—Open desert and alkaline flats, 500–5200 ft.; Creosote Bush Scrub; Death V. region from Inyo and Argus ranges to Funeral Mts. and Pahrump V., Inyo and adjacent San Bernardino cos.; Nev. April–May.

Ssp. **spinòsus** (A. Nels.) Keck. [*Amphiachyris F.* var. *s.* A. Nels. *Amphipappus s.* A. Nels. *A. F.* var. *s.* C. L. Porter.] Herbage densely scabro-hispidulous.—Similar sites, e. Mojave Desert farther s., Kelso, Goffs., etc., San Bernardino Co.; to Utah, Ariz.

74. Acamptopáppus Gray. GOLDENHEAD

Low much-branched desert shrubs with white bark. Lvs. alternate, small, usually spatulate or oblanceolate, entire, 1-nerved. Heads yellow, radiate or discoid, subglobose, solitary or cymosely arranged at tips of branches, the fls. all fertile. Invol. broad, ca. 4-seriate, strongly graduate, the phyllaries broad, rounded, whitish with greenish tip, firm with prominent thin scarious erose margin. Receptacle convex, alveolate-fimbrillate. Style-branches linear, the narrowly lanceolate appendages equaling the stigmatic portion. Aks. subturbinate, short, densely villous. Pappus persistent, of ca. 30–40 silvery flattened paleae and bristles of different widths, the broader ones usually somewhat dilated apically. Two known spp. (Greek, *akamptos*, stiff, and *pappos*, pappus.)

Heads discoid, small, gathered in small cymes .... 1. *A. sphaerocephalus*
Heads radiate, large, solitary at tips of branches .... 2. *A. Shockleyi*

1. **A. sphaerocéphalus** (Harv. & Gray) Gray. [*Aplopappus s.* Harv. & Gray.] Round-topped shrub 2–9 dm. high, corymbosely branched, often densely twiggy, glabrous throughout or sparsely hispidulous on lf.-margins; lvs. linear to spatulate, 5–20 mm. long, 1.5–5 mm. wide, obtuse to acute, mucronulate, firm, sessile, gray-green; heads subglobose, 7–10 mm. high, mostly solitary but approximate at tips of cymosely arranged branchlets; invol. 5–6 mm. high.—Open desert, 200–4000 ft.; Creosote Bush Scrub, Joshua Tree Wd.; e. Mojave Desert and w. border of Colo. Desert, Inyo Co. to San Diego Co.; to Nev., Utah, Ariz. April–June.

Var. **hirtéllus** Blake. Stems and lvs. densely hirtellous.—W. borders of Mojave Desert to Lone Pine, Inyo Co.: w. Nev. May–June.

2. **A. Shóckleyi** Gray. Rounded shrub 1.5–5 dm. high, spinescent-branched, the herbage finely hirtellous or hispidulous; lvs. spatulate, oblanceolate, or elliptic, 5–15 mm. long, usually mucronulate; heads globose, radiate, 1.5–3 cm. wide, solitary at ends of nearly naked peduncles; invol. 8–11 mm. high; rays 8–13, the oblong ligule ca. 1 cm. long.—Desert plains and washes, 3400–6200 ft.; Creosote Bush Scrub, Joshua Tree Wd.; White Mts., Inyo Co. to Clark Mt., San Bernardino Co.; s. Nev. April–June.

75. Chrysópsis (Nutt.) Ell. GOLDEN-ASTER

Low perennial (rarely annual) pubescent herbs, sometimes suffrutescent. Lvs. alternate, usually entire. Heads radiate or discoid, yellow, medium-sized, terminating the stems and branches. Invol. campanulate to hemispheric, the phyllaries numerous, narrow and imbricated. Receptacle low-convex, foveolate. Ray-fls. ♀, fertile; ligules narrow. Disk-fls. perfect, fertile. Style-branches flattened, with elongated hairy appendages much longer than the stigmatic portion. Aks. ± flattened; pappus usually double, brownish, the inner of numerous capillary bristles, the outer (when present) of short linear scales (or bristles). Ca. 20 spp., native in temp. N. Am. (Greek, *chrysos*, golden, and *opsis*, appearance, from the color of the heads.)

Heads radiate; outer pappus linear-squamellate .... 1. *C. villosa*
Heads discoid; outer pappus none or indistinct.
   Corollas filiform; pappus exceeding the invols. by most of its length. Mostly below 3000 ft.
                                                                                2. *C. oregona*
   Corollas funnelform; pappus mostly included in the invols. Above 4500 ft. .... 3. *C. Breweri*

1. **C. villòsa** (Pursh) Nutt. [*Amellus v.* Pursh. *Diplopappus v.* Hook. *Diplogon v.* Kuntze.] Gray-green perennial herb, with several erect stems from an often ± woody base surmounting a taproot, 1–5 dm. high; herbage canescently strigose and often somewhat glandular; lvs. oblong-spatulate, entire, 1–5 cm. long, the aestival lvs. smaller, firmer and more sessile than the early-deciduous vernal ones; heads paniculate or cymose in a short infl.; invol. 7–10 mm. high; phyllaries linear, acuminate, hirsutulous; rays 10–16, becoming revolute from the tip; aks. oblong-ovate, villous; outer pappus evident.—Open, well-drained slopes, below 6000 ft.; Yellow Pine F., Sagebrush Scrub; rare in Calif., reported in the Sierra Nevada from Placer Co. n.; to B.C., e. to the Great Plains. July–Aug.

KEY TO VARIETIES

Heads large, solitary or few in a cluster, closely subtended by leafy bracts; herbage obviously glandular. Mostly coastal.
    Upper lvs. mostly 1 cm. long, broadly oblong, not tapering to base, densely villous
                                                     var. *sessiliflora*
    Upper lvs. mostly larger, more ample, less crowded, 2–4 cm. long, usually petiolate, hirsute
                                                     var. *Bolanderi*
Heads more numerous, cymose or paniculate, not subtended by leafy bracts. More inland.
    Lvs. pubescent with silky appressed hairs.
        Outer pappus usually lacking. S. Calif. .... var. *fastigiata*
        Outer pappus conspicuous. Sierra Nevada .... *C. villosa*
    Lvs. pubescent with spreading hairs.
        Herbage densely canescent and hispid-hirsute .... var. *echioides*
        Herbage moderately hispid, rarely canescent.
            Plant moderately, if at all, glandular. Cismontane .... var. *hispida*
            Plant prominently glandular. Coast Ranges .... var. *camphorata*

Var. **sessiflòra** (Nutt.) Gray. [*C. s.* Nutt. *C. californica* Elmer. *Heterotheca s.* Shinners.] Herbage mostly grayish, villous-canescent and glandular; lvs. oblong or spatulate, crowded, 1–2 cm. long; heads mostly large and solitary, foliose-bracteate; outer pappus evident, squamellate.—Brushy sand flats, washes, etc., sea level to 5300 ft. (Palomar Mt.); Coastal Sage Scrub, Chaparral; rare, usually near the coast, Santa Barbara to San Diego; reported n. to Mendocino Co.; L. Calif. July–Sept.

Var. **Bolánderi** (Gray) Gray ex Jeps. [*C. B.* Gray. *C. sessiliflora* var. *B.* Gray. *C. arenaria* Elmer.] Stems decumbent or erect, 1–3 dm. high, several from the branching woody root-crown, the lvs. of lower half often soon shaded out by other vegetation; herbage densely hirsute and ± densely beset with sessile glands; lvs. oblanceolate or spatulate, rounded, mucronate, the lower petiolate, 2–4 cm. long, 5–10 mm. wide; heads large, mostly solitary, foliose-bracteate; invol. 10–15 mm. high; phyllaries linear-lanceolate, acuminate, villous and glandular; outer pappus of very slender scales.—Grassy slopes, mostly overlooking the coast, below 1500 ft.; N. Coastal Scrub; Mendocino Co. to San Francisco Bay. June–Nov.

Var. **fastigiàta** (Greene) Hall. [*C. f.* Greene.] Stems 3–10 dm. high, subsimple to multibranched, usually closely beset with small ascending linear-oblong to elliptic obtuse or acute mucronate lvs., these silky-tomentose, especially dorsally, 1–2 or 3 cm. long; outer pappus absent, or present as short bristles, not as squamellae.—Dry washes, rocky canyons, and plains, 250–6500 ft.; Coastal Sage Scrub, Chaparral, Foothill Wd.; w. foothills of the Sierra Nevada, Tulare and Kern cos., cismontane s. Calif. from Ventura Co. to Orange Co. July–Nov.

Var. **echioìdes** (Benth.) Gray. [*C. e.* Benth. *C. sessiliflora* var. *e.* Gray. *Heterotheca e.* Shinners.] Stems erect, 3–8 dm. high, virgate, simple or branching throughout; lvs. crowded, spreading, densely hirsute-canescent, firm, the upper 1–2 cm. long, sessile; heads numerous, paniculate to cymose; phyllaries numerous, very slender, hispidhirsute; outer pappus of short narrow squamellae.—Dry sandy soils, open fields, sea level to 2000 ft. (to 5300 ft. in San Jacinto Mts.); Coastal Sage Scrub, V. Grassland, Foothill Wd., Yellow Pine F.; Coast Ranges, somewhat away from the coast, from Sonoma and Solano cos. to n. San Diego Co.; s. edge of Mojave Desert. July–Nov.

Var. **híspida** (Hook.) Gray ex D. C. Eat. [*Diplopappus h.* Hook. *C. h.* DC.] Stems slender, virgate, 2–4 dm. high; herbage moderately hispid or villous, somewhat glandular; lvs. linear-oblanceolate, 8–20 mm. long, 2–5 mm. wide; heads numerous, rather small; phyllaries minutely glandular and sparsely hispid; outer pappus of short narrow squamellae.—Rocky ground, 3000–6400 ft.; Red Fir F., Joshua Tree Wd.; Lake Tahoe region (July), and Little San Bernardino Mts., San Bernardino and Riverside cos. (April–May; Oct.–Nov.); n. to B.C., e. to Rocky Mts.

Var. **camphoràta** (Eastw.) Jeps. [*C. c.* Eastw.] Stems slender, from a slender branching rootstock, 3–8 dm. high; herbage moderately to rather densely hirsute and conspicuously glandular-pubescent; lvs. lance-oblong, not very crowded; heads usually many, cymose-paniculate, rather small, usually subtended by small foliaceous bracts; phyllaries sparsely hispid and ciliate or merely ciliolate from the fine dissection of the narrow hyaline margin, prominently glandular.—Open sandy slopes and wooded banks, up to

2000 ft.; mostly Mixed Evergreen F.; Santa Cruz Mts., Mt. Hamilton Range, n. Santa Lucia Mts., Santa Clara Co. to Monterey Co. July–Sept.

2. **C. oregòna** (Nutt.) Gray. [*Ammodia o.* Nutt. *Diplogon o.* Kuntze. *Heterotheca o.* Shinners.] Stems clustered from a woody base, much branched, erect, 3–6 dm. high; herbage glabrate or sparsely hirsute, rather viscid, green; lvs. lanceolate to elliptic-ovate, sessile, ascending, 2–4 cm. long, 4–10(–15) mm. wide; heads corymbosely paniculate, discoid; invol. 9–12 mm. high; phyllaries 4-seriate, regularly imbricate, glandular-atomiferous but scarcely hairy, with white hyaline margin, acuminate; corollas nearly filiform; outer pappus setulose and obscure.—Gravel bars of dry stream beds, 200–2000 ft.; Redwood F., Mixed Evergreen F.; outer Coast Range valleys, Marin Co. to Del Norte Co.; n. to Wash. June–Sept. The following vars. intergrade.

KEY TO VARIETIES

Lvs. glabrate or sparsely hirsute, without harsh hairs. Near the coast . . . . *C. oregona*
Lvs. more densely and harshly pubescent.
    Upper lvs. crowded and conspicuous.
        Herbage hispid-hirsute but greenish. Widespread inland . . . . var. *rudis*
        Herbage densely hirsute-canescent, grayish or whitish. Bordering the Cent. V. . . . . var. *compacta*
    Upper lvs. much reduced, scalelike. Inner S. Coast Range and Sierra Nevada foothills
                                                            var. *scaberrima*

Var. **rùdis** (Greene) Jeps. [*C. r.* Greene.] Stems 3–8 dm. high, often virgate, brittle; herbage hispid-hirsute and ± viscid; lvs. linear-lanceolate to broadly oblong or oblanceolate, mostly sessile, erect or spreading, 2–5 cm. long; invol. 8–11 mm. high, glandular and somewhat hispid.—Dry gravelly places, 500–3500 ft.; Redwood F., N. Oak Wd., Foothill Wd., Yellow Pine F.; N. Coast Ranges from Siskiyou and Humboldt cos. to San Francisco Bay, n. Sierra Nevada, Tehama Co. July–Oct.

Var. **compácta** Keck. Low rounded bushes 1.5–4 dm. high, with many stems from the woody base; herbage gray-green, densely hispid- or hirsute-canescent and glandular-atomiferous, the hairs from prominent pustulate bases; lvs. typically small (1–2 cm.), densely crowded, spreading, lance-oblong; heads crowded toward the ends of the branches; invol. 8–10 mm. high, glandular but usually sparingly hairy.—Dry sandy places, below 1000 ft.; Foothill Wd., Chaparral; Cent. V. from Tehama Co. to Fresno Co. June–Oct.

Var. **scabérrima** Gray. Tall, rather openly branching, the heads commonly corymbosely paniculate on slender peduncles; lvs. in upper part of plant becoming bractlike; herbage viscidulous and scabrous from the pustulate bases of the fractured hairs.—Sandy roadsides and dry stream-ways, below 2000 ft.; Coast Range valleys and foothills, Alameda and Santa Clara cos. to Monterey Co., Sierran foothills, Tulare Co. June–Oct.

3. **C. Bréweri** Gray. [*C. Wrightii* Gray. *Heterotheca B.* Shinners.] Few to 20 or so erect stems from a woody caudex, 2–8 dm. high, simple or more commonly racemosely branching throughout, the slender leafy branches terminated by a solitary head; herbage moderately hirtellous and glandular-puberulent; lvs. lanceolate or narrowly oblong to lance-ovate, mucronate, sessile, thin, 1–3 cm. long; heads broadly campanulate, discoid; invol. 7–11 mm. high, 2–3-seriate; phyllaries lance-attenuate; corollas slender-funnel-form; outer pappus of numerous short fine bristles.—Open rocky slopes and coniferous forests, 4500–10,900 ft.; Montane Coniferous F.; Sierra Nevada, both slopes, from Shasta Co. to Tulare Co., San Gorgonio Peak, San Bernardino Mts. July–Sept.

Var. **multibracteàta** Jeps. [*C. gracilis* Eastw.] Herbage sparingly arachnoid-villous; phyllaries somewhat thicker, narrowly ovate to lanceolate, acute or acuminate, sometimes purple-margined, 4–5-seriate.—Dry rocky slopes, 6000–7500 ft.; Red Fir F.; Siskiyou, Trinity, and Tehama cos. July–Aug.

76. Heterothèca Cass. TELEGRAPH WEED

Coarse erect herbs, with yellow-fld. heads disposed in terminal corymbose panicles. Lvs. alternate. Invol. hemispheric; phyllaries narrow, closely imbricate in several series, appressed. Ray-fls. 1-seriate, fertile, the ak. triangular-compressed, its pappus none or caducous; disk-fls. many, fertile, the ak. cuneiform, its pappus double, the copious inner bristles capillary, long, the outer setose, short. Three or more spp. in s. U.S. and Mex.

(Greek, *heteros,* different, and *theke,* case or ovary, from the unlike aks. of ray and disk.)

Upper lvs. narrowed to a sessile base; heads relatively large; invol. 7–10 mm. high, glandular-pubescent but not also canescent . . . . 1. *H. grandiflora*
Upper lvs. subcordate-clasping at base; heads smaller; invol. 6–8 mm. high, glandular-puberulent and also canescent . . . . 2. *H. subaxillaris*

1. **H. grandiflòra** Nutt. [*H. floribunda* Benth.] Annual or biennial, the stout stem simple below, 5–20 dm. high, hirsute, the ample infl. glandular-pubescent and heavy-scented; lvs. thickish, densely villous, ovate to oblong or oblanceolate, 2–7 cm. long, obtuse, serrate, the lower petiolate, commonly with a pair of stipulelike lobes at base; heads medium large; invol. 7–9 mm. high; ray-fls. 25–35, the corolla 6–8 mm. long, 1 mm. wide, revolute from the tip, with hairy tube; disk-fls. 50–65, very slender, the stubby style-branches scarcely exserted; pappus brick-red, the outer series inconspicuous; $n = 9$ (Heiser & Whitaker, 1948).—Sandy, open, coastal valleys, below 2000 ft., behaving as a weed; Coastal Sage Scrub, S. Oak Wd., Foothill Wd.; apparently native throughout cismontane s. Calif. from Santa Barbara to San Diego, spreading n. through the Salinas, Santa Clara, and Cent. valleys to Butte Co. Jan.–Dec.

2. **H. subaxillàris** (Lam.) Britt. & Rusby. [*Inula s.* Lam. *I. scabra* Pursh. *Chrysopsis scabra* Nutt. *H. Lamarckii* Cass. *H. scabra* DC.] Annual or biennial, the moderately stout stem simple below or openly branching, 5–20 dm. high, hispid-hirsute, glandular above; lvs. rather coarsely hirsute, glandular, ovate to lance-oblong, serrate-dentate or subentire, the lower petiolate, the upper subcordate-clasping; heads relatively small, numerous; invol. 6–8 mm. high, glandular and somewhat canescent; ray-fls. ca. 20–28; disk-fls. 40–60; pappus rufous, the outer series usually conspicuous.—Sandy roadsides and ditches; Creosote Bush Scrub; easternmost Mojave and Colo. deserts, rare; introd. at San Gabriel; common e. to Fla. and Del. Aug.–Oct.

77. Chaetopáppa DC.

Ours low very slender annuals with simple or diffusely branched stems and alternate entire chiefly linear lvs. Heads small, few- to many-fld., terminating very slender peduncles, radiate, disciform, or discoid, all florets potentially fertile, yellow, white, or reddish. Invol. turbinate to hemispheric, the phyllaries 2–5-seriate, graduate or equal, thin, green-centered, prominently scarious-margined, usually setulose-tipped, persistent. Receptacle convex, naked. Pistillate florets 1–3-seriate, ligulate or tubular; hermaphrodite florets very slender, 3–5-toothed. Aks. linear-fusiform, often compressed, pubescent. Pappus of 3-many fragile slender bristles, sometimes dilated and ± joined at very base, or wanting. Ca. 15 spp. of sw. U.S. and Mex. (Greek, *chaite,* bristle, and *pappos,* pappus.)

Ray-fls. present, conspicuous; invol. broadly hemispheric.
  Ligules golden-yellow. S. Calif.
    Invol. pubescent . . . . 1. *C. Lyonii*
    Invol. glabrous.
      Phyllaries in strongly graduated series, tapering to the short-caudate tip . . . . 2. *C. aurea*
      Phyllaries in subequal series, obtuse, mucronulate . . . . 3. *C. fragilis*
  Ligules white or purplish. San Francisco Bay region . . . . 4. *C. bellidiflora*
Ray-fls. reduced to a filiform tube, or absent; invol. turbinate.
  Stem simple or with few erect branches; disk-corollas dilated at throat, contracted at orifice
     5. *C. exilis*
  Stem diffusely branched; disk-corollas narrowly linear . . . . 6. *C. alsinoides*

1. **C. Lyònii** (Gray) Keck. [*Pentachaeta L.* Gray.] Stem 1–5 dm. high, simple or branched, hirsute chiefly on the lf.-margins; lvs. narrowly linear or spatulate-linear, 2–5 cm. long; invol. ca. 5 mm. high, the phyllaries hirsute, subequal, lance-linear, acuminate, narrowly scarious-margined; ray-fls. ca. 30–50; disk-fls. ca. 80–100; aks. of ray and disk similar, dark brown, moderately strigose with short hairs; pappus of 10–12 very fragile filiform bristles, flared at very base and forming a rudimentary corona.—In clayey soil; V. Grassland; coastal part of Los Angeles Co., and Santa Catalina Id. March–April.

2. **C. aúrea** (Nutt.) Keck. [*Pentachaeta a.* Nutt.] Usually diffusely branched, 8–30 cm. high, entirely glabrous, or the lf.-margins usually ciliate; lvs. mostly narrowly linear,

the lower 1–3.5 cm. long, up to 2 mm. wide, the upper much shorter and up to 1 mm. wide; heads solitary at tips of branches, 1–2.5 cm. wide; invol. broad, 4–7 mm. high, the phyllaries ca. 4–5-ranked, lance-ovate (outer) to oblong, from setaceous-acuminate to obtuse and apiculate, the greenish cent. portion scarcely wider than the shining scarious margin; ray-fls. 0–70, usually 10–40, the ligule 5–12 mm. long; disk-fls. usually numerous; aks. mahogany brown, ca. 1 mm. long, sparsely and finely short-strigose; pappus of 5(–7) scabrous filiform bristles slightly thickened toward apex, flared at very base and often united, ca. equaling the corolla.—Dry open ground and grassy slopes, up to 6000 ft.; V. Grassland, S. Oak Wd.; Los Angeles and San Bernardino cos. to San Diego; n. L. Calif. April–July.

3. **C. frágilis** (Bdg.) Keck. [*Pentachaeta f.* Bdg.] Very slender, wiry, diffusely branched, glabrous except for the hirsutulous lf.-margins, the branches 4–10 cm. long; lvs. spatulate to linear-oblanceolate, the basal 8–15 mm. long, 2–3 mm. wide, obtuse, the cauline 2–6 mm. long, 1–1.8 mm. wide; invol. ca. 4 mm. high, the phyllaries subequal, oblong, subtruncate, rather broadly scarious-margined, glabrous or sparsely pilose, lacerate-ciliate at apex; ray-fls. 10–25, the ligule 5 mm. long; aks. sparsely pilose; pappus of ca. 20 very fragile filiform bristles scarcely enlarged toward apex, not at all dilated at base, slightly shorter than the disk-corolla.—Dry grassy slopes, below 4000 ft.; Foothill Wd.; s. Sierra Nevada and Greenhorn Mts., Kern Co., San Joaquin plains and inner S. Coast Range, Merced and San Luis Obispo cos. March–June.

4. **C. bellidiflòra** (Greene) Keck. [*Pentachaeta b.* Greene.] Stems 6–20 cm. high, simple or with few erect branches, glabrous or nearly so; lvs. narrowly linear, 8–35 mm. long, ca. 1 mm. wide; heads 1–1.7 cm. wide; invol. 3–4 mm. high, the phyllaries subequal, often purplish, oblong, short-acuminate or apiculate from the truncate lacerate-ciliate apex, broadly scarious-margined, glabrous dorsally; ray-fls. 5–16, white or purplish-tinged, the ligule 5 mm. long; disk-fls. numerous, yellow, the ample throat not contracted at orifice; aks. densely tawny hirsute, rarely glabrous; pappus of 5 relatively firm scabrous bristles, not dilated at base, shorter than the disk-corolla.—Open dry rocky slopes; N. Coastal Scrub, Coastal Prairie; Marin, San Mateo, and Santa Cruz cos. March–May.

5. **C. éxilis** (Gray) Keck. [*Aphantochaeta e.* Gray. *Pentachaeta e.* Gray. *P. e.* var. *Aphantochaeta* Gray. *P. A.* Greene.] Stem 3–18 cm. high, simple or with few erect branches, often purplish, sparsely pubescent; lvs. filiform or nearly so, up to 25 mm. long, ciliate-pubescent toward base; peduncle white-villous beneath the head; invol. 3–5 mm. high, glabrous, often purplish, the phyllaries subequal, few, broad, oblong, weakly short-bristle-tipped at the obtuse or truncate ± laciniate apex, moderately scarious-margined; ♀ fls. 0–5, reduced to a short filiform tube; disk-fls. 4–10(–50), purplish, distinctly widened at throat, contracted at orifice; aks. brown, moderately villous or rarely glabrous; pappus of 3–5 slender scabrid bristles, not dilated at base, or these reduced to triangular scales, or often entirely wanting.—Grassy slopes, 100–1500 ft.; V. Grassland, Foothill Wd.; from Placer Co. to Mariposa Co., and from Mendocino and Colusa cos. to Monterey Co. April–May.

6. **C. alsinoìdes** (Greene) Keck. [*Pentachaeta exilis* var. *discoidea* Gray, in part. *P. a.* Greene.] Diffusely branched, 3–12 cm. high and wide, somewhat villous; lvs. narrowly linear or filiform, 1 mm. wide or less; heads tiny, not strictly solitary; invol. 2.6–3 mm. high, glabrous or somewhat hirsute, the phyllaries subequal, few (6–7), oblong or oval-oblong, obtuse or shortly setaceous-acuminate, green, narrowly scarious-margined, lacerate toward apex; ♀ fls. ca. 4–6, capillary, tubular or with minute involute erect ligule, not exceeding the 3–5 very similar slightly thicker reddish-tinged disk-fls., these commonly with imperfect anthers; aks. brownish, lightly to moderately villous; pappus usually of 3 capillary bristles somewhat exceeding the florets, scarcely at all dilated at base.—Grassy slopes of foothills; N. Coastal Scrub, Coastal Sage Scrub, V. Grassland, Coastal Prairie; mostly near the coast, Humboldt Co. to Santa Barbara Co., less common, Eldorado Co. to Tulare Co. Inconspicuous and probably overlooked. April–May.

78. Haplopáppus Cass.

Herbs or shrubs, very varied in habit, often glandular. Lvs. alternate, entire to bipinnatifid, often thickish, sometimes glandular-punctate. Heads radiate or discoid, large or small, solitary to numerous and cymose or paniculate, yellow, rarely creamy white. Invol. cylindric or turbinate to hemispheric, the phyllaries numerous, subequal to strongly graduate, usually narrow and indurate or chartaceous, at least below. Receptacle ± alveolate. Ray-fls. ♀, rarely sterile; disk-fls. fertile, their style-branches ovate to subulate. Aks. terete or angled, linear-fusiform to turbinate, glabrous to silky pilose. Pappus of numerous capillary subequal or graduate bristles, usually persistent. Some 150 spp., all Am., chiefly w. U.S., Mex., and Chile. (Greek, *haploos*, simple, and *pappos*, pappus.)

(Hall, H. M. The genus Haplopappus. Carnegie Inst. Wash. Publ. 389: 1–391, 1928.)
A. Aks. turbinate, 2–3 mm. long; lvs. dentate to bipinnatifid, the teeth spinulose- or bristle-tipped. (Section Blepharodon)
    B. Annual; herbage, including invol., hirsute with strigose hairs, the phyllaries also minutely glandular-puberulent . . . . 1. *H. gracilis*
    BB. Perennials; invol. not strigose.
        C. Tufted with several slender erect stems from a suffrutescent base; lvs. very narrow; heads radiate.
            D. Phyllaries prominently glandular-puberulent and scabrous; lvs. 2–5 cm. long
                                                    2. *H. Gooddingii*
            DD. Phyllaries beset with granular glands; lvs. 1–2 cm. long . . . . 3. *H. junceus*
        CC. Rigidly branched shrub; lvs. mostly oval, merely dentate; heads discoid
                                                  4. *H. brickellioides*
AA. Aks. nearly prismatic, subcylindric, or fusiform, 3 mm. or more long; lvs. various, but if toothed, the teeth not as above.
    B. Perennial herbs with shoots of the season arising from prominent leafy rosettes surmounting a deep fusiform taproot. (Section Pyrrocoma)
        C. Heads large, the invol. 12–18 mm. high; ray-fls. inconspicuous or lacking
                                                  5. *H. carthamoides*
        CC. Heads smaller, the invol. up to 12 mm. high (higher in *H. lucidus*); ray-fls. showy.
            D. Heads solitary or cymose, or if racemose not numerous in a long narrow infl.
                E. Stipitate-glandular as well as hirsute or villous . . . . 6. *H. hirtus*
                EE. Not stipitate-glandular.
                    F. Heads usually solitary, terminating long peduncles.
                        G. Phyllaries typically herbaceous, thin, obscurely graduate; aks. pubescent . . . . 7. *H. uniflorus*
                        GG. Phyllaries green only toward tip, firm, evidently graduate; aks. glabrous . . . . 8. *H. apargioides*
                  FF. Heads corymbose or cymose-paniculate, rarely solitary; phyllaries firm toward base, green-tipped; aks. sericeous . . . . 9. *H. lanceolatus*
             DD. Heads usually numerous, spicate or racemose.
                E. Plants not strongly if at all glandular; phyllaries coriaceous and distinctly graduate, obtuse to acute . . . . 10. *H. racemosus*
                EE. Plants vernicose from sessile glands; phyllaries coriaceous-herbaceous, obscurely graduate, acuminate to attenuate . . . . 11. *H. lucidus*
    BB. Shrubs or subshrubs.
        C. Plants cespitose or tufted, usually mat-forming, with much-branched caudex; heads mostly solitary at ends of branches (usually several in *H. Whitneyi*).
            D. Stems relatively leafy; herbage glandular-puberulent; lvs. toothed in ours. Plants of high mts. (Section Tonestus)
                E. Aks. glabrous; lvs. sharply serrate; invol. campanulate-oblong; phyllaries linear-lanceolate; stems to 50 cm. high . . . . 12. *H. Whitneyi*
                EE. Aks. densely pubescent; lvs. saliently dentate in outer half; invol. hemispheric; stems to 15 cm. high.
                    F. Invol. 7.5–10 mm. high, 12–15 mm. wide; disk-corolla 6–7 mm. long
                                                  13. *H. eximius*
                  FF. Invol. 14–18 mm. high, 20–30 mm. wide; disk-corolla 9–10 mm. long
                                              14. *H. Peirsonii*
            DD. Stems sparsely leafy or subscapose; herbage scarcely if at all glandular; lvs. entire. Mid-altitude plants. (Section Stenotus)
                E. Lvs. linear, 7–18 mm. long, up to 1 mm. wide . . . . . 15. *H. stenophyllus*
                EE. Lvs. linear-oblanceolate to spatulate, 1–6 cm. long, 1.5–7 mm. wide
                                            16. *H. acaulis*
      CC. Plants not at all cespitose or mat-forming, but sometimes forming low rounded bushes.
        D. Appendages of style-branches at least twice as long as the stigmatic portion; low intricately branched shrubs mostly under 3 dm. high. (Section Macronema)
            E. Twigs closely white-tomentose; heads discoid; lvs. densely glandular
                                             17. *H. Macronema*
             EE. Twigs glabrous or glandular, if rarely loosely tomentose, then heads radiate.
                F. Heads relatively large, 20–45-fld.; invol. broadly campanulate, ca. 2-seriate, the phyllaries not obviously ciliate or scarious-margined; lvs. undulate-margined, stipitate-glandular . . . . 18. *H. suffruticosus*
                FF. Heads smaller and narrower; invol. more than 2-seriate; lvs. not crisped, very rarely stipitate-glandular.

G. Phyllaries villous-ciliate; lvs. without axillary fascicles.
  H. Invol. narrowly campanulate, 3–6-seriate, the outer phyllaries not squarrose-tipped; heads radiate, 6–25-fld.
    I. Lvs. oblanceolate, 1.5–3.5 cm. long, 3–7 mm. wide; heads solitary or few, cymose .... 19. *H. Greenei*
    II. Lvs. mostly linear, 2–6 cm. long, 0.5–3 mm. wide; heads several or many in a raceme or thyrsoid panicle
                                                                20. *H. Bloomeri*
  HH. Invol. cylindric, 5–8-seriate, the outer phyllaries squarrose-tipped; heads discoid, mostly 5-fld., solitary or few, openly cymose .... 21. *H. ophitidis*
GG. Phyllaries glabrous (obscurely ciliate in *H. Gilmanii*); lvs. often with axillary fascicles.
  H. Lvs. mostly spatulate, spaced; outer phyllaries squarrose; disk-fls. 15–18, white .... 22. *H. Gilmanii*
  HH. Lvs. mostly linear, crowded; outer phyllaries erect; disk-fls. 5–10, yellow .... 23. *H. nanus*
DD. Appendages of style-branches only equaling or shorter than the stigmatic portion.
  E. Invol. hemispheric, 10–18 mm. wide; herbage glandular-punctate; lvs. entire. (Section Stenotopsis) .... 24. *H. linearifolius*
  EE. Invol. turbinate or subcylindric.
    F. Disk-corolla abruptly dilated from narrow tube to much wider throat; heads discoid. (Section Isocoma)
      G. Phyllaries with green but thin tips; stems brownish. Coastal
                                                                25. *H. venetus*
      GG. Phyllaries with thickened subepidermal resin-pocket near tip, not green; stems whitish. Deserts .... 26. *H. acradenius*
    FF. Disk-corolla only slightly ampliate upward.
      G. Heads large, the 8–15 mm. high invols. tightly imbricate, mostly 6–8-seriate; herbage without distinct resin pits. (Section Hazardia)
        H. Herbage tomentose. Insular .... 27. *H. canus*
        HH. Herbage not tomentose. Mainland ...... 28. *H. squarrosus*
      GG. Heads small, the 3–8 mm. high invols. loosely imbricate, 2–6-seriate; herbage with distinct resin pits. (Section Ericameria)
        H. Ray-fls. present; lvs. filiform; heads ± paniculate.
          I. Outer phyllaries ± caudate-tipped.
            J. Lvs. 10–35 mm. long, with shorter fascicles in the axils; aks. pilose .... 29. *H. pinifolius*
            JJ. Lvs. 4–12 mm. long, scarcely exceeding the axillary fascicles, ericoid; aks. glabrous (except in ssp.)
                                                                30. *H. ericoides*
          II. Outer phyllaries acute to obtuse.
            J. Heads solitary or few; invol. campanulate, 7–8 mm. high .... 31. *H. Eastwoodae*
            JJ. Heads many; invol. turbinate, 5–6.5 mm. high
                                                                32. *H. Palmeri*
        HH. Ray–fls. reduced or wanting (see also *H. laricifolius*).
          I. Lvs. filiform to linear, less than 3 mm. wide.
            J. Heads solitary or racemose-paniculate, discoid. San Diego Co. s. .... 33. *H. propinquus*
            JJ. Heads cymose.
              K. Lvs. 0.5–2 cm. long; broad rounded shrubs seldom more than 1 m. high. Deserts.
                L. Herbage glabrous; lvs. subterete, without persistent fascicles in the old axils; ray-fls. 3–11 .... 34. *H. laricifolius*
                LL. Herbage hairy; lvs. flat, with persistent fascicles in the old axils; ray-fls. 0–2
                                                                35. *H. Cooperi*
              KK. Lvs. 3–6 cm. long; arborescent shrubs 1–3 m. high. Cismontane .... 36. *H. arborescens*
          II. Lvs. oblanceolate to obovate, 3–10 mm. wide.
            J. Heads 9–12-fld., discoid; erect shrub 2–5 m. high; lvs. 2–6 cm. long, mostly tapering to base and apex
                                                                37. *H. Parishii*
            JJ. Heads 16–30-fld., sometimes radiate; spreading subshrub mostly less than 1 m. high; lvs. cuneate
                                                                38. *H. cuneatus*

1. H. grácilis (Nutt.) Gray. [*Dieteria g.* Nutt. *Aster D.* Kuntze. *Eriocarpum g.* Greene. *Sideranthus g.* A. Nels.] Annual herb mostly 6–25 cm. high, usually divaricately branching, hirsute throughout with strigose hairs; lvs. numerous, the lower oblanceolate, pinnatifid or bipinnatifid, 1.5–3 cm. long, 3–7 mm. wide, the upper linear, appressed, much reduced, serrate-dentate to subentire, teeth, lobes and apex tipped with a prominent white bristle; heads cymose or solitary; invol. hemispheric, 6–7 mm. high, 8–12 mm. wide; phyllaries linear-lanceolate, well imbricated, green with hyaline margin, cinereous or strigose and minutely glandular-puberulent, with appressed bristle-tip; ray-fls. 16–28, the ligules 7–12 mm. long; pappus of numerous tawny unequal bristles slightly dilated below; *n* = 2 (Jackson, 1957).—Sandy or rocky flats and hillsides, up to 5,000 ft.; Creosote Bush Scrub, Joshua Tree Wd.; e. Mojave Desert from Clark Mt. to Providence Mts., San Bernardino Co.; to Colo., Tex., Mex. April–June.

2. **H. Gooddíngii** (A. Nels.) M. & J. [*Sideranthus G.* A. Nels. *H. spinulosus G.* Blake. *H. s.* ssp. *G.* Hall.] Taprooted perennial with several stiffly erect or ascending slender stems 2–6 dm. high, glabrate to harshly glandular-puberulent, sometimes also canescent; lvs. scattered, lanate ventrally, scabrid dorsally, pinnatifid, the axis and remote lobes linear and bristle-tipped, 2–5 cm. long, the upper becoming entire and much reduced; heads terminating long branches; invol. depressed-hemispheric, 6–9 mm. high, 10–18 mm. wide; phyllaries linear-lanceolate, well imbricated, greenish, prominently glandular-puberulent and scabrous, with short apical bristle; ray-fls. 30–45, the ligules 6–16 mm. long; pappus of numerous tawny unequal bristles.—Rocky mesas and canyon-sides, 450–1900 ft.; Creosote Bush Scrub; e. Mojave Desert from Needles to Vidal, San Bernardino Co.; s. Nev. to cent. Ariz. Feb.–May.

3. **H. júnceus** Greene. [*Eriocarpum j.* Greene. *Sideranthus j.* Davids. & Mox.] Stems tufted, 4–10 dm. high, suffrutescent at base, slender, branching, sparingly strigose, slightly glandular near heads; lvs. chiefly linear, pinnatifid or serrate with bristle-tipped teeth, 1–2 cm. long, the upper reduced, entire, bristle-tipped; heads solitary on long slender scaly-bracted branches or in open cymes; invol. hemispheric, 5–8 mm. high, the closely imbricated phyllaries linear, covered with granular glands, bristle-tipped; ray-fls. 15–25, the ligules 5–6 mm. long; pappus of numerous tawny unequal bristles.— Dry brushy hillsides, 500–3000 ft.; Chaparral, Coastal Sage Scrub; s. San Diego Co.; n. L. Calif., Son. June–Oct.

4. **H. brickellioìdes** Blake. Rigidly branched rounded shrub 2–8 dm. high, the older stems white-barked and ± pilose, the branches yellowish, densely pilosulous, some hairs thickened and tipped with yellow glands; lvs. oval, elliptic or obovate-cuneate, 1–3.5 cm. long, 5–25 mm. wide, acute, spinescent-tipped, dentate with 1–4 pairs of spinescent teeth, firm, triplinerved, the midnerve prominent, pilose and yellow-glandular; heads discoid, ca. 12-fld., rather small, sessile or subsessile in 1's–3's toward tips of leafy branchlets; invol. 6–7 mm. high, the phyllaries 4–5-seriate, lanceolate, 1-ribbed, hispidulous and glandular, the tip greenish; aks. oblong, hispidulous; pappus sparse.—Rare in rocky canyons, 3000–6500 ft., Creosote Bush Scrub; Inyo Co., Last Chance Mts. to Death V. vicinity and adjacent Nev. April–Sept.

5. **H. carthamoìdes** (Hook.) Gray ssp. **Cusíckii** (Gray) Hall. [*H. c.* var. *C.* Gray. *Pyrrocoma C.* Greene.] Stems few, decumbent or sometimes erect, from a stout taproot, 1–2(–4) dm. high; lvs. tufted, mostly spatulate or oblanceolate, the basal with blades 5–12 cm. long, 5–15 mm. wide, with slender petiole nearly as long, the cauline mostly sessile or clasping, entire to sparingly spinulose-serrate, glabrous or hirtellous (especially beneath); heads solitary or racemose or subspicate, turbinate-campanulate; invol. 12–18 mm. high, often subtended by a few leafy bracts, the phyllaries graduate or subequal, mostly oblong-lanceolate with indurate base and elongated herbaceous tip; ray-fls. usually lacking, sometimes few, inconspicuous, seldom exceeding the pappus; aks. glabrous; pappus stiff, sordid.—Open, barren, often rocky desert slopes, 4000–6000 ft.; Sagebrush Scrub, N. Juniper Wd.; Shasta and Lassen cos. to Siskiyou and Modoc cos.; n. to Blue Mts., Wash., e. to n. Nev. and cent. Ida. June–Aug. Typical *H. carthamoides* grows farther n., from Columbia R. gorge to w. Mont.

6. **H. hírtus** Gray. [*Aster Grayanus* Kuntze. *Pyrrocoma h.* Greene. *Hoorebekia h.* Piper.] Stems several, erect or ascending, often decumbent at base, 1.5–3 dm. high, from a woody crown topping a taproot, ± equably leafy throughout; herbage sparingly to rather densely villous with jointed hairs, stipitate-glandular at least above; basal lvs. tufted,⋅ elliptic-lanceolate, the blade 3–8 cm. long, 8–25 mm. wide, short-petiolate, sharply pectinate-serrate or doubly serrate, rarely subentire, the cauline smaller, sessile; heads few or several, loosely racemose or subcymose; invol. broadly campanulate, 9–12 mm. high; phyllaries ca. 3-seriate, subequal, linear-lanceolate, acuminate, herbaceous with loose tip, stipitate-glandular and villous; ray-fls. 13–25, the ligules 7–10 mm. long;

aks. silky; pappus sordid.—Dry rocky meadows or open woods; N. Juniper Wd.; Lassen Co.; to Wallowa Co., Ore., n. Nev. July–Aug.

Ssp. **lanulòsus** (Greene) Hall. [*Pyrrocoma l.* Greene. *P. turbinella* Greene. *H. hirtus* var. *l.* Peck.] Cauline lvs. more obviously reduced in size, the herbage tending to be more obviously woolly and less glandular; invol. 8–11 mm. high, the phyllaries imbricate, merely green-tipped, the tip not elongated or spreading.—Moist or dry soil of sagebrush flats or coniferous woods; Sagebrush Scrub, N. Juniper Wd., Yellow Pine F.; Lassen Co. to Modoc Co.; n. to Grant Co., Ore. June–Aug.

7. **H. uniflòrus** (Hook.) T. & G. [*Donia u.* Hook. *Homopappus u.* Nutt. *Pyrrocoma u.* Greene. *Hoorebekia u.* Jones.] Stems 1–3 dm. high, from a fibrous-coated fusiform crown, decumbent to ascending-erect, usually anthocyanous, glabrate to tomentulose; lvs. mostly basal, narrowly to broadly lanceolate, tapering to base and apex, the blade 5–12 cm. long, 6–15 mm. wide, entire to laciniate-dentate, the cauline much reduced and sessile; heads solitary, terminating rather long peduncles; invol. hemispheric, 8–10 mm. high, the phyllaries 2–3-seriate, obscurely graduate, typically herbaceous from apex to base at least medianly, not thickened, the margin scarious, glabrous to tomentose; ray-fls. 18–32, the ligules 6–9 mm. long; aks. sericeous.—Moist alkaline meadows, often in aspen belt, 6000–7000 ft.; Sagebrush Scrub, N. Juniper Wd.; Mono Co., Modoc Co.; to se. Ore., n. Colo., Sask. June–Aug.

Ssp. **gossypìnus** (Greene) Hall. [*Pyrrocoma g.* Greene. *Haplopappus g.* Hall.] Stems 1–3 dm. high, decumbent, floccose-tomentose like the lvs.; basal lvs. lanceolate, the blade 2–8 cm. long, 7–20 mm. wide; heads racemose or usually solitary; invol. 10–13 mm. high, the thin linear-oblong acuminate phyllaries rather loose, of unequal lengths but scarcely imbricate, somewhat arachnoid-pilose.—Dry meadows; Yellow Pine F.; Bear V., San Bernardino Mts. July–Sept.

8. **H. apargioìdes.** Gray. [*Aster a.* Kuntze. *Pyrrocoma a.* Greene. *P. demissa* Greene.] Stems several, decumbent to ascending-erect, 0.5–1.5(–2.5) dm. high, from a thick taproot that is sometimes branched at the crown, glabrate or somewhat villous, few-lvd. or scapiform; basal lvs. tufted, linear-lanceolate to oblanceolate, acuminate, petiolate, 3–10 cm. long, 3–10 mm. wide, laciniate with spinescent teeth to entire, typically ciliate toward base with scabrous or rather coarse hairs, otherwise glabrous, coriaceous; heads usually solitary, long-peduncled; invol. subhemispheric, 8–12 mm. high, the phyllaries loosely imbricated in few ranks, usually narrowly oblong and pungently acute, sometimes obtuse, firm, green toward tip with pale margin becoming hyaline below, glabrous; ray-fls. 13–34, the ligules 6–9 mm. long; aks. flattened, glabrous, striate, 3–7 mm. long; pappus sordid.—Open, rocky slopes and meadows, 7500–12,000 ft.; Lodgepole F., Subalpine F., Alpine Fell-fields; Sierra Nevada from Plumas Co. to Tulare Co., Sweetwater and White mts.; adjacent Nev. July–Sept.

9. **H. lanceolàtus** (Hook.) T. & G. [*Donia l.* Hook. *Aster l.* Kuntze. *Pyrrocoma l.* Greene. *Hoorebekia l.* Jones.] Stems few to several from a taproot and simple or forked fibrous-coated crown, decumbent to ascending-erect, 1–5 dm. high, glabrous or slightly villous-tomentulose; lvs. spiny-toothed or entire, rather thin, glabrous to lanulose, the basal tufted, oblanceolate, acuminate at each end, the blade 5–15 cm. long, 5–15 mm. wide, much exceeding the slender petiole, the cauline much reduced; heads several to many in an open corymb or racemose panicle, rarely solitary; invol. subhemispheric, 7–10 mm. high, the phyllaries 3–4-seriate, distinctly to obscurely graduate, typically herbaceous toward tip, firm toward base, linear-lanceolate to linear-oblong, glabrate to loosely villous; ray-fls. 13–34, the ligules 5–10 mm. long; aks. sericeous; pappus sordid.—Infrequent, meadows and alkaline flats, 4000–6000 ft.; Sagebrush Scrub, N. Juniper Wd.; Mono Co., Plumas Co. to Siskiyou Co.; to e. Ore., Sask., Nebr., Colo. June–Aug.

Ssp. **tenuicaùlis** (D. C. Eat.) Hall. [*H. t.* D. C. Eat. *H. l.* var. *t.* Gray. *Pyrrocoma t.* Greene. *P. solidagineus* Greene. *H. l.* ssp. *s.* Hall.] Heads smaller, the invol. 5–7 mm. high; stems typically slender, flexuous, decumbent at base.—Intergrading with the sp., Sierra Co. n. to e. Ore., where the common form of the sp., e. to n. Nev.

10. **H. racemòsus** (Nutt.) Torr. [*Homopappus r.* Nutt. *Pyrrocoma r.* T. & G. *P. elata, longifolia* and *balsamitae* Greene. *Aster P.* Kuntze. *Hoorebekia r.* Piper. *Haplopappus l.* Jeps. *H. r.* ssp. *l.* Hall.] Stems few to several, erect from a stout taproot and short sometimes branched caudex, 3–10 dm. high; herbage essentially glabrous; basal lvs. tufted,

oblanceolate to elliptic, 10–30 cm. long (including slender petiole), 10–25 mm. wide, entire to shallowly serrate, rarely spinulose-toothed, the cauline reduced, sessile or clasping; heads several to many, racemose or spicate (rarely solitary), not glomerate; invol. ± hemispheric, 8–12 mm. high; phyllaries 4–5-seriate, firm, herbaceous throughout or green-tipped, the narrow margin whitish or hyaline, sometimes ciliate, otherwise glabrous, sharply acute to obtusish; ray-fls. 13–30, the ligules 5–12 mm. long; aks. densely villous; pappus sordid, ca. 7 mm. long.—Coastal valleys, in neutral or saline soils, mostly below 1000 ft.; Coastal Salt Marsh, Coastal Prairie, V. Grassland; San Benito Co. to Napa Co., n. and e. to Shasta V. and Willamette V., Ore. June–Oct.

Ssp. **congéstus** (Greene) Hall. [*Pyrrocoma c.* Greene. *Haplopappus r.* var. *c.* Peck.] Stems slender, curving upward from base, 2–7.5 dm. high, glabrous or sparingly tomentose; lvs. green, relatively thin, the lateral veins visible, entire or sparsely denticulate; heads sessile or rarely short-peduncled in a typically glomerate-spicate infl.; invol. 6–7.5 mm. high; phyllaries firm but not thick, oblong, abruptly narrowed to a short acute tip, bright yellow-green, glandular-ciliolate at apex, sparsely tomentulose throughout or at least on margin below.—Dry or moist serpentine slopes, 1000–2500 ft.; Chaparral, Yellow Pine F., Douglas-Fir F.; Del Norte Co. to Josephine Co., Ore. Aug.–Sept.

Ssp. **glomerátus** (Nutt.) Hall. [*Homopappus g., argutus* and *paniculatus* Nutt. *Pyrrocoma g.* and *p.* T. & G. *Haplopappus g. virgatus* and *stenocephalus* Gray. *H. r.* vars. *glomerellus, s.* and *v.* Gray. *Pyrrocoma ciliolata*, and *microdonta* Greene. *P. p. v.* Davids. & Mox. *H. r.* var. *glomeratus* Peck.] Habitally similar to ssp. *congestus*, with stems 2–7 dm. high, but herbage gray-green or glaucous, the foliage stiff and thickish, the lateral veins obscure, entire to sharply denticulate-serrate, even the basal lvs. usually less than 15 mm. wide; heads sessile or short-pedunculate, sometimes ± glomerate, racemose or spicate; invol. 6–8 mm. high; phyllaries firm, thickish, lance-oblong to oblong, acute to obtuse, pale below the green tip, glabrous or ciliate, rarely more hairy.—Alkaline plains and meadows, 400–6700 ft.; N. Juniper Wd., Sagebrush Scrub, Alkali Sink; upper San Joaquin V.; thence transmontane from n. Inyo Co. to Modoc Co.; to n. Ore., Utah, s. Ida. July–Oct.

Ssp. **sessiflórus** (Greene) Hall. [*Pyrrocoma s.* Greene.] Stems slender, nearly prostrate to ascending-erect or erect, 2–5 dm. long; herbage blue-glaucous; lvs. thick, mostly entire, ciliolate, the basal tufts often grasslike, 5–10 cm. long, or oblanceolate and even up to 20 cm. by 28 mm., the small amplexicaul ± recurving cauline lvs. rather crowded; infl. spicate, the heads often in clusters of 2's or 3's; invol. 5–7 mm. high; phyllaries thick, broad, pale, the prominent green tip often squarrose, ciliate.—Moist saline flats and meadows, 1000–7000 ft.; Sagebrush Scrub, Alkali Sink; s. Mono Co. to s. Inyo Co., e. to cent. Nev. July–Oct.

Ssp. **pinetòrum** Keck. [*H. r.* var. *p.* J. T. Howell. *H. r.* var. *praticola* J. T. Howell.] Stems erect or decumbent at base, 2–6 dm. high, red-brown, sparingly arachnoid-villous; lvs. entire or spinulose-serrate, pilosulous to ± densely arachnoid-villous, the basal narrowly lanceolate, up to 20 cm. long and 16 mm. wide; infl. spicate-racemose or sometimes paniculate; invol. campanulate to hemispheric, 10–15 mm. high, multiseriate, rather densely villous but not glandular.—Rocky forested slopes and meadows, 2800–5400 ft.; Yellow Pine F., Red Fir F.; cent. Siskiyou Co. to Scott Mts., Trinity Co. July–Aug.

11. **H. lùcidus** (Keck) Keck. [*H. racemosus* ssp. *l.* Keck.] Stems few, slender or stout, erect from a taproot, 2–10 dm. high, the herbage very sticky from numerous punctate glands, not pubescent; lvs. deep green, entire (usually) or serrate-dentate, often scabrid-ciliolate, the basal (long- or short-petiolate) with narrowly to broadly lanceolate blades 6–18 cm. long, 5–30 mm. wide; infl. spiciform-paniculate, the heads solitary or clustered; invol. campanulate, 10–15 mm. high, the phyllaries 2–3-seriate, obscurely imbricate, coriaceous-herbaceous, vernicose, linear-lanceolate, acuminate to attenuate, the tip at length spreading; aks. sericeous.—Alkaline flats and f. openings, 2500–5000 ft.; Yellow Pine F.; Plumas, Sierra, and Yuba cos. July–Sept.

12. **H. Whítneyi** Gray. [*Aster W.* Kuntze. *Hazardia W.* Greene.] Perennial herb with several ascending simple stems from a woody root-crown, 2–5 dm. high, moderately pilose with septate hairs and stipitate-glandular; lvs. broadly oblong or slightly spatulate, the lower narrowed at base, the upper subamplexicaul, 2.5–5 cm. long, 7–16 mm. wide,

mucronate-serrate, firm, glandular-scabrid, sometimes ± pilose; heads solitary, spicate, racemose, or cymose-clustered, leafy-bracteate; invol. campanulate, 11–13 mm. high; phyllaries 4–6-seriate, loosely graduate, linear-lanceolate, acuminate, chartaceous, granular-glanduliferous, the herbaceous portion decreasing and the hyaline often roseate margin increasing on the inner ones; ray-fls. 5–18; disk-fls. 15–30; aks. glabrous, 8–14-ribbed; pappus copious, brownish.—Rocky, open forested slopes, 4000–10,000 ft.; Red Fir F., Lodgepole F.; occasional throughout the Sierra Nevada from Plumas Co. to Tulare Co. July–Sept.

Ssp. discoidèus (J. T. Howell) Keck. [*H. W.* var. *d.* J. T. Howell.] Heads discoid; otherwise the same.—Similar habitats, 3000–7000 ft.; Red Fir F.; inner N. Coast Ranges from Snow Mt., Lake Co., to Mt. Eddy and Siskiyou Mts.; adjacent Ore.

13. **H. exímius** Hall. [*Tonestus e.* Nels. & Macbr.] Perennial herb with few to several erect leafy stems 3–15 cm. high from a subterranean branched slender caudex or deep rhizomes, forming a loose mat up to 8 dm. across, the herbage glandular-puberulent; lvs. cuneate or spatulate, saliently dentate above middle, obtuse, mucronate, firm, 2–5 cm. long, 7–15 mm. wide; heads solitary; invol. hemispheric, 7.5–10 mm. high, 12–15 mm. wide (pressed); phyllaries shorter than the disk, scarcely graduate, unlike, the outer ca. 2-seriate, obovate, oblong or oblanceolate, obtuse, herbaceous and glandular, the inner 2-seriate, subequal, lanceolate, attenuate, ciliolate, scarious, reddish above; ray-fls. 15–20, the ligules 8 mm. long; aks. densely pubescent; pappus sordid; $n = 9$ (Stebbins, 1950).—Granitic soils near treeline, 8600–9600 ft.; Subalpine F.; e. flank of Sierra Nevada from s. Washoe Co., Nev., to Eldorado Co. July–Aug.

14. **H. Peirsònii** (Keck) J. T. Howell. [*H. eximius* ssp. *P.* Keck.] Habit of *H. eximius* but more robust, the caudex branching but not producing elongate rhizomes, the stems to 20 cm. high; lvs. 3–8 cm. long, 10–25 mm. wide, less deeply toothed; heads larger, the invol. 14–18 mm. high, 20–30 mm. wide (pressed); phyllaries equaling the disk, the outer lanceolate or lance-oblong, mostly acute; $n = $ ca. 45 (Stebbins, 1950).—Rocky summits near treeline, 9600–12,000 ft.; Subalpine F., Alpine Fell-fields; Sierra Nevada of Mono and Inyo cos., from Mammoth Lakes Basin to Sawmill Pass, and just entering Fresno Co. July–Aug.

15. **H. stenophýllus** Gray. [*Aster s.* Kuntze. *Stenotus s.* Greene. *Hoorebekia s.* Piper.] Densely cespitose from a much dissected rather slender matlike woody caudex surmounting a taproot, the mats up to 45 cm. across, the numerous stems densely leafy in lower half, scapiform above and monocephalous, 3–8 cm. high, like the lvs. densely hispidulous or hirtellous-scabrous and sometimes glandular; lvs. crowded, linear-spatulate to nearly filiform, 7–18 mm. long; invol. hemispheric, 5–9 mm. high, the linear to oblanceolate acute to acuminate phyllaries 2-seriate, subequal, herbaceous, densely glandular-scaberulous; ray-fls. 8–12, the ligules 7–11 mm. long; aks. appressed-villous; pappus whitish, copious.—Open flats or slopes, often with sagebrush, in volcanic ash, adobe, or rocky soils, 3000–8500 ft.; Sagebrush Scrub; Lassen and Modoc cos.; to cent. Wash., occasional e. to cent. Ida. and n. and w. Nev. May–July.

16. **H. acaúlis** (Nutt.) Gray. [*Chrysopsis a.* Nutt. *C. caespitosa* Nutt. *Stenotus a.* Nutt. *S. a.* var. *Kennedyi* Jeps. *Haplopappus nevadensis* Kell. *H. a.* var. *glabratus* D. C. Eat. *H. a.* ssp. *g.* Hall.] Stems scapiform, numerous, monocephalous, cespitose from a much-branched woody caudex, 5–10 cm. high, densely clothed at base with the marcescent lvs., the whole mat sometimes up to 5 dm. across; lvs. linear-oblanceolate to spatulate, mostly erect, entire, obtuse to usually acuminate, cuspidate-tipped, veiny, pale green, densely hispidulous to glabrous except for the scabrid margin, 1–6 cm. long, 2–7 mm. wide; invol. hemispheric, 7–10 mm. high; phyllaries 2–3-seriate, broad, acute or acuminate, pallid; ray-fls. 6–10, the ligules 6–10 mm. long; aks. densely sericeous to glabrous.—Dry, often rocky ridges and slopes; 5300–10,500 ft.; Sagebrush Scrub, Yellow Pine F., Red Fir F., Lodgepole F.; Inyo and White mts., e. side of the Sierra Nevada, n. Inyo Co. to Modoc Co.; to n. Ore., Utah, n. Colo., Sask. May–Aug.

17. **H. Macronèma** Gray. [*Macronema discoideum* Nutt. *Aster M.* Kuntze. *Bigelovia M.* Jones. *Haplopappus d.* Hall & Hall, not DC.] Undershrub 1–4 dm. high, with numerous erect pannose-tomentose twigs of the season from a multibranched woody base; lvs. numerous, oblong to oblanceolate, sessile, entire or the margin crisped, 1–3 cm. long, 3–6 mm. wide, green, densely stipitate-glandular; heads discoid, yellow, solitary and terminal or few subracemose, 10–26-fld.; invol. turbinate-campanulate, 11–

15 mm. high, glandular-puberulent; phyllaries subequal, 2–3-seriate, the outer broader, more herbaceous, the inner acuminate to attenuate, thin, dry; aks. appressed-pilose; pappus brownish.—Rocky, mostly open slopes, often on talus above timber line, 9000–12,000 ft.; Subalpine F., Alpine Fell-fields; Sierra Nevada from Tulare Co. to Nevada Co., Sweetwater Mts., Warner Mts.; to se. Ore., Utah, Colo., w. Wyo. July–Sept.

18. **H. suffruticòsus** (Nutt.) Gray. [*Macronema s.* Nutt. *M. imbricatum* Nels. & Macbr. *Aster s.* Kuntze. *H. s.* ssp. *tenuis* Hall.] Low compact subshrub up to 2 or even 4 dm. high, with densely stipitate-glandular fragrant herbage; lvs. very numerous on the brittle twigs, linear-oblanceolate to spatulate-oblong, 1–3 cm. long, 1.5–5 mm. wide, entire, usually crisped and with axillary fascicles; heads 1–4 at branch tips, mostly solitary, broadly campanulate; invol. with several outer foliaceous oblong bracts often longer than the 2-seriate chartaceous lanceolate acuminate obscurely ciliolate inner phyllaries, 10–14 mm. high, stipitate-glandular; ray-fls. 3–6, showy; disk-fls. 18–40; aks. villous, somewhat flattened; pappus stramineous.—Open rocky slopes and ridges, 7800–12,000 ft.; Red Fir F. to Alpine Fell-fields; high Sierra Nevada from Tulare Co. to Nevada Co., White and Sweetwater mts.; Wallowa Mts., Ore.; e. across n. Nev. and cent. Ida. to w. Wyo. and sw. Mont. July–Sept.

19. **H. Greènei** Gray. [*H. mollis* Gray. *H. G.* var. *m.* Gray. *H. G.* ssp. *m.* Hall. *Aster G.* Kuntze. *Macronema G.* Greene. *M. m.* Greene. *M. G.* var. *m.* Jeps. *Hoorebekia G.* ssp. *m.* Piper.] Undershrub 1–3 dm. high, glabrate to ± densely tomentose, eglandular to ± resinous from punctate or sessile (very rarely stalked) glands; lvs. very numerous, oblanceolate, 1.5–3.5 cm. long, 3–7 mm. wide, plane or rarely slightly crisped; heads clustered at ends of twigs; invol. 8–12 mm. high, subtended by a few leafy bracts; phyllaries 3–4-seriate, subequal to somewhat imbricate, the outer with herbaceous ligulate tips 3–4 times as long as the body, the inner with proportionately shorter caudate tips or acuminate or even acute, the body prominently scarious-margined, villous-ciliate; ray-fls. 1–7, showy; disk-fls. 6–20.—Rocky flats and sparsely wooded slopes, 5500–6500 ft.; Red Fir F.; inner N. Coast Range, from Snow Mt., Lake Co., to Trinity Summit and Siskiyou Mts., Warner Mts., rare in n. Sierra Nevada; n. through the Cascades to cent. Wash.; mts. of e. Ore. and cent. Ida. July–Sept.

Closely related to and intermediate between *H. suffruticosus* and *H. Bloomeri*, behaving as a subalpine ecotype of the latter. Fairly well marked morphologically and, considering the considerable series of closely related taxa here, best treated as a sp. The random occurrence of tomentose forms, not ecologically separable, in *H. Greenei* and *H. Bloomeri* precludes their recognition as natural units.

20. **H. Bloómeri** Gray. [*H. B.* var. *angustatus* Gray. *H. B.* var. *Sonnei* Greene. *H. B.* sspp. *a.* and *S.* Hall. *Ericameria erecta* Klatt. *Aster B.* Kuntze. *Chrysothamnus B.* and var. *a.* Greene. *Ericameria B.* Macbr.] Low compact shrub broader than tall, the woody trunk up to 3 cm. thick, 1.5–4(–8) dm. high, glabrate to sometimes ± tomentose, eglandular to pruinose-glandular or occasionally glutinous from sessile glands; lvs. numerous, nearly filiform to narrowly oblanceolate, 2–6 cm. long, 0.5–3 mm. wide, plane (rarely twisted); heads in small terminal clusters or commonly more numerous in subracemose spikes or panicles, narrowly campanulate; invol. 7–12 mm. high; phyllaries 3–6-seriate, clearly imbricate, linear-lanceolate to oblong, stramineous, the outer with caudate herbaceous tips, at least the inner prominently scarious-margined, villous-ciliate; ray-fls. 1–5, wanting in some heads, not very showy; disk-fls. 4–13.—Sandy or rocky soils, open flats or slopes near coniferous woods, 3500–9500 ft.; Yellow Pine F. to Lodgepole F., N. Juniper Wd.; Sierra Nevada from Tulare Co. to Siskiyou and Modoc cos.; adjacent Nev., n. through cent. Ore. to cent. Wash. July–Oct.

21. **H. ophitìdis** (J. T. Howell) Keck. [*H. Bloomeri* var. *o.* J. T. Howell.] Low matlike undershrub up to 20 cm. high, the stout woody trunk (up to 12 mm. thick) soon intricately and compactly branched, black-barked, the herbage resinous from sessile yellow glands; lvs. numerous on the very slender stems, narrowly linear, apiculate, falcate-recurved, sulcate, very sparsely arachnoid-ciliate, 5–15 mm. long, 0.5–1 mm. wide; heads discoid, terminating leafy twigs, solitary or in small cymes, cylindric at anthesis; invol. 12–14 mm. high; phyllaries 5–8-seriate, strongly imbricate, the outermost linear-lanceolate, herbaceous, the inner broadly oblong with wide hyaline margin, truncate to base of the lanceolate to deltoid herbaceous thickened often squarrose tip, glutinous, somewhat ciliate; fls. mostly 5 (4–6), light yellow.—Serpentine soil, 5000

ft.; Yellow Pine F.; known only from the summit of Mt. Tedoc, nw. Tehama Co. July–Aug.

22. **H. Gilmánii** Blake. Low rounded intricately branched aromatic shrub 2–3.5 dm. high, the old stems pallid like the new, the very slender twigs rather rigid, sulcate, sticky; lvs. numerous, often with axillary fascicles, vernicose (often in droplets), spatulate, 6–12 mm. long, 2–3 mm. wide, often conduplicate and recurved at tip; heads solitary or cymose, narrowly campanulate; invol. 7–9 mm. high, resinous; phyllaries 4–6-seriate, imbricate, the outer ovate-lanceolate, often with thickened appendagelike green squarrose or reflexed tip, the inner oblong, chartaceous, with prominent hyaline ciliolate margin, appressed; ray-fls. 4–6, white or pale yellow (as also the disk); disk-fls. 15–18; aks. silky-sericeous; pappus ochroleucous.—Ridges and walls of peaks, often in limestone, 7100–11,000 ft.; Pinyon-Juniper Wd., Bristle-cone Pine F.; Inyo and Panamint ranges, Inyo Co. Aug.–Sept.

23. **H. nànus** (Nutt.) D. C. Eat. [*Ericameria n.* Nutt. *Chrysoma n.* Greene. *Chrysothamnus n.* Howell.] Intricately branched rounded compact aromatic shrub 1–3 dm. high, the old gnarled wood blackish; lvs. very numerous, narrowly linear-spatulate to linear, or involute and linear-filiform, to 20 mm. long, subulate-mucronate, ± resinous but not punctate, sometimes minutely scaberulous, often with budlike axillary fascicles; heads in small leafy cymes, yellow; invol. turbinate, 6–8.5 mm. high, glabrous; phyllaries 4–6-seriate, chartaceous and indurate at base, only the outermost somewhat herbaceous, the tips obtuse to sharply acute or even aristate-caudate, not squarrose; ray-fls. (0–)2–8; disk-fls. 5–10; pappus stramineous, fragile, scanty.—Dry rocky plains, cliffs and crevices, 6000–7000 ft. in Calif.; Pinyon-Juniper Wd.; se. Mono Co. and n. Inyo Co.; e. and n. across the Great Basin to Utah and sw. Mont. July–Nov. Hall and Goodspeed reported that this species contains 5–10% rubber, a higher percentage than that known at the time (1919) for any other N. Am. shrub except guayule (*Parthenium argentatum*).

24. **H. linearifòlius** DC. [*Stenotus l.* T. & G. *Aster l.* Kuntze. *H. interior* Cov. *H. l.* var. *i.* Jones. *H. l.* ssp. *i.* Hall. *Stenotus l.* T. & G. S. *i.* Greene. *S. l.* var. *i.* Hall. *Stenotopsis l.* Rydb. *S. i.* Rydb. *S. l.* var. *i.* Macbr.] Much-branched shrub 4–15 dm. high, essentially glabrous but usually puberulent below heads, the twigs fastigiate, resinous; lvs. crowded, sometimes fasciculate, prominently glandular-punctate, nearly linear, narrowed toward base, entire, 1–4 cm. long, 1–2.5 mm. wide, flat to subterete; heads numerous, terminal on nearly leafless peduncles; invol. hemispheric, 8–14 mm. high, the phyllaries 2–3-seriate, scarcely graduate, lance-oblong to linear, acuminate, beset with granular glands, with greenish center and lacerate-ciliate scarious margin; ray-fls. 13–18, the ligules 8–15 mm. long; aks. silky-pilose; pappus white, deciduous.—Arid slopes and banks, rocky or sandy soils, mostly below 6000 ft.; Chaparral, Creosote Bush Scrub, Pinyon-Juniper Wd., Joshua Tree Wd.; Marysville Buttes and Lake Co. s. through the inner S. Coast Range to interior cismontane s. Calif. along desert borders, e. across Mojave D. to Inyo Co.; s. Utah, w. Ariz., L. Calif. March–May. Continued recognition of a desert ssp. *interior,* based on the smaller size of heads and lvs., seems scarcely warranted, as this moderate reduction seems to be due to environmental modification rather than to genetically governed variation.

25. **H. venètus** (HBK.) Blake ssp. **vernonioìdes** (Nutt.) Hall. [*Pyrrocoma Menziesii* H. & A. *Isocoma vernonioides* Nutt. *I. veneta* var. *v.* Jeps. *I. leucanthemifolia, microdonta* and *villosa* Greene. *Haplopappus M.* T. & G.] Shrub 4–12 dm. high, with erect ascending or decumbent stems from a branched suffrutescent base, usually simple below the infl.; somewhat resinous, from nearly glabrous to pilose or tomentose, very leafy; lvs. linear to spatulate-oblong, 1–4 cm. long, 2–8 mm. wide, spinulose-dentate throughout or only near apex, sometimes even lobed or essentially entire, usually with axillary fascicles; heads discoid, turbinate, 15–30-fld., in round terminal usually compact cymes; invol. 5–7 mm. high, strongly graduate, the phyllaries oblong, obtuse or acute, firm, pale, with short usually granulose greenish appressed tips; aks. silky; pappus brownish.—Coastal valleys and headlands, often in sandy soils or subsaline places, mostly below 1000 ft.; Coastal Strand, Coastal Salt Marsh, Coastal Sage Scrub; infrequent from San Francisco to Santa Barbara (inland as far as San Benito Co. and upper Salinas V.), abundant s. to San Diego and the ids. April–Dec. Typical *H. venetus* occurs in cent. Mex.; its subdivisions are found in Calif. and L. Calif. The more noteworthy of these, all of which intergrade with the widespread ssp. *vernonioides,* are as follows:

Ssp. **oxyphýllus** (Greene) Hall. [*Isocoma o.* Greene. *H. v.* var. *o.* Munz.] Robust shrub 1–2 m. high, loosely villous to glabrate; lvs. oblanceolate to narrowly spatulate, acute or acuminate, entire, 3–5 cm. long; cymes openly paniculate.—Largely replacing the preceding in San Diego Co. and L. Calif. June–Nov.

Ssp. **furfuràceus** (Greene) Hall. [*Bigelovia f.* Greene. *Isocoma decumbens* Greene. *I. v.* var. *d.* Jeps. *H. venetus* vars. *d.* and *f.* Munz.] Stems slender, decumbent or curved, 3–5 dm. long; lvs. crowded, narrow, small, few-toothed or entire, with prominent fascicles in the axils, mostly glabrous, sometimes woolly; heads loosely cymose, not in large glomerules.—Dry, sandy mesas, s. San Diego Co., L. Calif. and the s. ids. April–Nov.

Var. **sedoìdes** (Greene) Munz. [*Bigelovia v.* var. *s.* Greene. *Isocoma s.* and *latifolia* Greene. *I. v.* var. *s.* Jeps.] Prostrate or decumbent, almost glabrous, stout; lvs. succulent, obovate, obtuse, toothed; heads large, in a capitate cluster. Santa Rosa, Santa Cruz, San Miguel and Anacapa ids. and adjacent coast from Mono to Pt. Mugu.

Var. **argùtus** (Greene) Keck. [*Isocoma a.* Greene. *I. v.* var. *a.* Jeps.] Low bush with erect stems 1.5–4 dm. high; lvs. pinnately cleft into acute lobes or only saliently but pungently dentate.—Both sides of Carquinez Straits, Solano and Contra Costa cos.

26. **H. acradènius** (Greene) Blake. [*Bigelovia a.* Greene. *Aster a.* Kuntze. *Isocoma a.* Greene. *I. veneta* var. *a.* Hall.] Similar to *H. venetus;* shrub 3–10 dm. high, the stems erect, more woody, brittle, white-barked, striate, glabrous; lvs. linear-spatulate to oblong, 1–4 cm. long, 1–5 mm. wide, thick, entire, mostly mucronate and glabrous, minutely impressed-punctate, fewer in axillary fascicles; heads in smaller cymes, nearly sessile, 6–13-fld.; invol. 5–6.5 mm. high; phyllaries with conspicuous thick subepidermal resin-pocket near the rounded or blunt tip, the very narrow hyaline margin fimbrillate. —Subalkaline or sandy flats, 1500–3000 ft.; Shadscale Scrub, Alkali Sink, Joshua Tree Wd.; sw. Mojave Desert (common in Antelope V.), rare in the e. Mojave s. of Death V.; s. Nev., w. Ariz. Aug.–Nov.

Ssp. **eremóphilus** (Greene) Hall. [*Isocoma e.* Greene. *H. a.* var. *e.* Munz.] Lvs. denticulate to dentate or even saliently lobed, with some entire, 2–5 cm. long, to 7 mm. wide, often scabrid-hirtellous; heads 15–25-fld.; invol. 6–8 mm. high.—Open desert, —200–3000 ft.; Alkali Sink, Creosote Bush Scrub; s. border of the Mojave Desert, Colo. Desert, extending slightly into Ariz. and L. Calif.

Ssp. **bracteòsus** (Greene) Hall. [*Isocoma b.* Greene. *H. a.* var. *b.* McMinn.] Lvs. mostly denticulate, the lowermost 2–3 cm. long, the remainder mostly 6–10(–15) mm. long, horizontal or reflexed; heads 15–22-fld.; invol. 6–9 mm. high.—Both sides of San Joaquin V. from Panoche and Merced s., Cuyama V.

27. **H. cànus** (Gray) Blake. [*Diplostephium c.* Gray. *Corethrogyne detonsa* Greene. *C. c.* Greene. *Hazardia c.* Greene. *H. d.* Greene. *H. serrata* Greene. *Haplopappus Traskae* Eastw.] Openly branched shrub 6–25 dm. high, moderately to densely lanate-tomentose throughout, the lvs. often glabrate on upper surface; lvs. obovate to oblanceo-late, obtuse, entire to sharply serrate, petiolate or sessile, 4–12 cm. long, 1–5 cm. wide, thick; heads numerous, in large panicles or cymes, pedunculate or sessile; invol. broadly turbinate, 10–13 mm. high; phyllaries 6–8-seriate, appressed, oblong, acute, tomentose, or the inner becoming glabrous; ray-fls. 6–14, inconspicuous, not exceeding disk, like the 20–54 disk-fls. turning from yellow to purplish; aks. nerved, pilose; pappus tawny or brown.—Dry bluffs above the ocean; Coastal Sage Scrub; Santa Rosa, Santa Cruz, and San Clemente ids.; Guadalupe Id., L. Calif. June–Dec.

28. **H. squarròsus** H. & A. [*Hazardia s.* Greene.] Multi-stemmed shrub, woody at base, 3–10 dm. high, glabrate to hirtellous, or, especially above, pilose; lvs. many, oblong to cuneate-obovate, obtuse, sharply serrate throughout with mucronulate teeth, clasping at base, firm, resinous from punctate glands, essentially glabrous, 1.5–4 cm. long, 1–2 cm. wide; heads discoid, 15–30-fld., spicate or racemosely paniculate; invol. turbinate, 11–15 mm. high; phyllaries 8–10-seriate, strongly graduate, prominently glandular-scurfy at the green obtuse to acute usually squarrose tips; aks. glabrous or sparsely pilose; pappus yellow-tawny.—Coastal bluffs and montane canyons and ridges, up to 2200 ft.; Coastal Sage Scrub, Chaparral; Monterey s. through the Santa Lucia Mts. to Santa Ynez Mts., Santa Barbara, also e. side of Salinas V. Aug.–Oct.

Ssp. **grindelioìdes** (DC.) Keck. [*Pyrrocoma g.* DC. *Aster g.* Kuntze.] More strongly hairy, the stems often tomentulose near the heads, even the upper surface of lvs. often

± hairy; heads smaller; invol. mostly 8–11 mm. high; phyllaries prominently cinereous on both faces of the green usually squarrose tip, but glandular only marginally if at all; pappus red-brown.—Cismontane s. Calif. from Santa Barbara to San Diego and the ids.; n. L. Calif. July–Oct.

Ssp. **obtùsus** (Greene) Hall. [*Hazardia o.* Greene. *H. s.* var. *o.* Jeps. *Haplopappus s.* var. *o.* McMinn.] Heads relatively large, 18–25-fld.; invol. broadly turbinate, 13–15 mm. high; phyllaries relatively broad, very blunt, mucronate, the resinous-granular pallid glabrous tips appressed.—Canyons up to 4000 ft., mts. of Kern Co. w. of Tejon Pass, and Ventura Co. s. to Ojai V. Sept.–Nov.

Ssp. **stenolèpis** Hall. [*H. s.* var. *stenolepis* McMinn.] Usually woodier at base, very branched, forming dense shrubs to 1 m. high and 2 m. across, often much smaller, the twigs scabrid or hirsutulous; lvs. relatively small; heads 4–8-fld., densely spicate; invol. very narrowly turbinate, 13–17 mm. high; phyllaries 6–7-seriate, regularly but loosely imbricate, linear-acuminate, at length spreading but not squarrose, sparingly dotted with granular glands on the small herbaceous portion near tip, somewhat viscid but glabrous; pappus red-brown.—Serpentine and shaly soils, 500–3500 ft.; Foothill Wd., Chaparral; inner S. Coast Range from Parkfield Grade, Fresno and Monterey cos. to Cuyama R., Santa Barbara Co. Sept.–Nov.

29. **H. pinifòlius** Gray. [*Aster pityphyllus* Kuntze. *Chrysoma pinifolia* Greene. *Ericameria pinifolia* Hall. *H. illinitus* Eastw.] Stout shrub 6–25 dm. high, the main stems trunklike, fastigiately branched, resinous, glandular-punctate, sometimes slightly pilose; lvs. linear-filiform, 1–3.5 cm. long, mucronate, subterete, with shorter lvs. fascicled in the axils; heads solitary and few in spring flowering, large, terminating leafy twigs, often surpassed by subtending lvs., in short racemes or apical cymose clusters and numerous in autumn flowering, smaller; invol. of latter form turbinate, 6–8 mm. high, overlapped by leafy bracts; phyllaries loosely 3–5-seriate, lance-acuminate to oblong, the outer often ± caudate-tipped, the inner short-tipped or merely acute, the tips green, otherwise pale with scarious ciliate margin, the costa sometimes glandular-thickened above; ray-fls. 5–10 (15–30 in vernal heads); disk-fls. 12–18; aks. sparsely pilose; pappus buff or reddish.—Cismontane dry slopes and washes away from the coast, 500–5400 ft.; S. Oak Wd., Chaparral; n. Los Angeles Co. to s. San Diego Co. April–July, Sept.–Jan.

30. **H. ericoìdes** (Less.) H. & A. [*Diplopappus e.* Less. *Ericameria microphylla* Nutt. *E. e.* Jeps. *Chrysoma e.* Greene. *Aster ericina* Kuntze.] Broad compact fastigiately branched shrub 3–8(–15) dm. high; herbage sparsely pilosulous, somewhat resinous; lvs. very numerous, nearly filiform, divaricate, 4–12 mm. long, subterete, with dense fascicles of scarcely shorter lvs. in the axils; heads cymose-paniculate, terminating leafy branches; invol. turbinate, 5–6 mm. high; phyllaries loosely 3–5-seriate, villous-ciliate, the outer ovate-lanceolate with short-caudate greenish tip, the inner broadly oblong, acute, the costa thickened above into a filiform gland; ray-fls. 2–6; disk-fls. 8–14; aks. glabrous.—Sand dunes on and near the coast; Coastal Strand; Point Reyes, Marin Co. to El Segundo coast, Los Angeles Co.; reported from San Miguel Id. Aug.–Nov.

Ssp. **Blàkei** C. B. Wolf. Aks. moderately sericeous.—Sand hills away from the immediate coast, up to 1500 ft.; Coastal Sage Scrub, Mixed Evergreen F.; Santa Cruz Mts., Santa Cruz and Monterey cos., Santa Maria R. V., San Luis Obispo Co. scattered s. to Ventura Co. Sept.–Nov. The lvs. are usually, but not always, longer and the appearance less ericoid in the s. material.

31. **H. Eastwoòdae** Hall. [*Chrysoma fasciculata* Eastw. *Ericameria f.* Macbr.] Stout dense fastigiately branched shrub 5–10 dm. high, glabrous, very resinous, glandular-punctate, densely leafy; lvs. linear, 8–20 mm. long, terete or ± flattened, mucronate, nearly all with axillary fascicles of smaller lvs.; heads solitary or usually several in terminal cymes or close racemes; invol. campanulate, 7–8 mm. high; phyllaries ca. 5-seriate, strongly graduate, scarious, pale yellow, the outer lance-ovate, sharply acute, the inner oblong, obtusish, villous-ciliate on the narrow hyaline margin, the costa ± glandular-thickened above; ray-fls. 1–6; disk-fls. 18–22; aks. densely silky pilose.—Coastal sand dunes; Coastal Strand; Monterey and Carmel bays. July–Oct.

32. **H. Pálmeri** Gray. [*Aster Nevinii* Kuntze. *Chrysoma P.* Greene. *Ericameria P.* Hall.] Stout shrub 1–4 m. high, with numerous ascending stems, puberulous, resinous, glandular-punctate, very leafy; lvs. filiform, 1.5–4 cm. long, subterete, fasciculate; heads

many, in thyrsoid panicles; invol. turbinate, 5–6.5 mm. high; phyllaries 30–40, loosely 4–5-seriate, linear or nearly so, blunt, glabrous, or the outer glandular-atomiferous, ciliolate at tip, the hyaline margin narrow, the costa thickened for most of its length into a linear-oblong gland; ray-fls. 4–10; disk-fls. 8–20; aks. moderately sericeous.— Dry plains and foothills, 300–2000 ft.; Coastal Sage Scrub; s. San Diego Co.; L. Calif. Sept.–Nov.

Ssp. **pachylèpis** Hall. [Var. *p.* Munz.] Smaller, 5–15 dm. high; lvs. mostly 8–12 mm. long; invol. cylindro-turbinate, 6–7 mm. high; phyllaries 16–25, somewhat thicker, 4-seriate, broadly lanceolate to oblong, the bullate costal gland in apical half only; ray-fls. 1–4; disk-fls. 5–10; aks. densely sericeous.—Rather common on sandy or clayey plains, up to 2300 ft.; Coastal Sage Scrub; s. Ventura Co. across cismontane s. Calif. to desert borders, Riverside Co.; Santa Catalina Id. Aug.–Dec.

33. **H. propínquus** Blake. [*Bigelovia brachylepis* Gray. *Aster b.* Kuntze. *Chrysoma b.* Greene. *Ericameria b.* Hall. *H. b.* Hall, not Phil.] Shrub 1–2 m. high, rigidly branched, glabrous, ± resinous, glandular-punctate; lvs. crowded, linear-filiform, 1–2 cm. long, 0.5–1 mm. wide, flattish or subterete, often mucronate, sharply ascending, with axillary fascicles; heads discoid, 9–14-fld., yellow, terminal or racemose-paniculate; invol. 4.5–5.5 mm. high; phyllaries 3–4-seriate, strongly graduate, ovate to linear-oblong, the outer grading into the scaly bracts of the peduncles, the costa prominently glandular-thickened throughout its length; aks. densely villous.—Arid brushy slopes, up to 4500 ft.; Chaparral; s. San Diego Co.; n. L. Calif. Sept.–Dec.

34. **H. laricifòlius** Gray. [*Aster l.* Kuntze. *Chrysoma l.* Greene. *Ericameria l.* Shinners.] Turpentine-Brush. Fastigiately branched broadly rounded shrub 3–10 dm. high, the herbage resinous, prominently impressed-punctate, glabrous; lvs. linear, 1–2 cm. long, 1–2 mm. wide, usually subterete, mucronate, sometimes with much smaller ones in axillary fascicles; heads in small leafy cymes; invol. broadly turbinate, 3–5 mm. high; phyllaries loosely 3–4-seriate, lance-acuminate, rather firm, the tip soft and ciliolate, the costa thickened for most of its length into an olive-brown gland; ray-fls. 3–11; disk-fls. 10–16, much exceeding the invol.; aks. densely pilose.—Rocky slopes and mesas, 3500–6500 ft.; Creosote Bush Scrub, Pinyon-Juniper Wd.; e. Mojave Desert (Clark Mt., New York Mts., Hackberry Mt.); to w. Tex. and adjacent Mex. Sept.–Oct. Contains rubber, especially in the roots.

35. **H. Coóperi** (Gray) Hall. [*Bigelovia C.* Gray. *Haplopappus monactis* Gray. *Aster C.* and *m.* Kuntze. *Ericameria m.* McClat. *E. C.* Hall. *Chrysoma C.* Greene. *Acamptopappus microcephalus* Jones. *Chrysothamnus corymbosus* Elmer. *Tumionella monactis* Greene.] Low flat-topped fastigiately branched woody shrub 2.5–6(–15) dm. high, glutinous, puberulent; lvs. linear-spatulate, 6–15 mm. long, up to 1.5 mm. wide, with much smaller glandular-punctate ones fasciculate in the lower axils and persistent after the fall of the primaries; heads prominently pedicelled in small cymes; invol. 4–5 mm. high; phyllaries 2–3-seriate, 9–15, the outer ovate, acute, the inner broadly oblong, obtuse, ± puberulent, the glandular thickening of the costa obscure; ray-fls. 0–2; disk-fls. 4–7(–11), much exceeding the invol.; aks. silky-pilose.—Common in rocky basins and mesas, 2600–5700 ft.; Creosote Bush Scrub, Joshua Tree Wd.; Mojave Desert from s. Mono Co. and adjacent Nev. s. to Lancaster V. and Little San Bernardino Mts.; rare in interior cismontane s. Calif. March–June.

36. **H. arboréscens** (Gray) Hall. [*Linosyris a.* Gray. *Bigelovia a.* Gray. *Aster Chrysothamnus* Kuntze. *Ericameria a.* Greene. *Chrysoma a.* Greene.] Golden Fleece. Stout erect shrub 0.6–3 m. high, fastigiately branched, glabrous, resinous, prominently glandular-punctate; lvs. narrowly linear to filiform, 3–6 cm. long, up to 2 mm. wide, thick, crowded, rarely with axillary fascicles; heads discoid, 18–23-fld., many, in rounded terminal cymes or cymose panicles; invol. turbinate, 4–5 mm. high; phyllaries 4-seriate, graduate, lanceolate to linear, acuminate to acute, thin and chaffy except for the glandular-thickened costa; aks. turgid, obscurely 5-angled, less than 2 mm. long, finely sericeous; pappus very fragile.—Dry foothills, 300–4000 ft. (up to 9350 ft. in s. Sierra Nevada); Chaparral, Foothill Wd.; cismontane Sierra Nevada from Nevada Co. to Tulare Co., outer Coast Ranges from Oregon line to Ventura Co. Aug.–Nov.

37. **H. Paríshii** (Greene) Blake. [*Bigelovia P.* Greene. *Aster P.* Kuntze. *Chrysoma P.* Greene. *Ericameria P.* Hall.] Erect shrub 2–5 m. high, arborescent, glabrous, resinous, prominently glandular-punctate, the shoots densely leafy; lvs. linear-oblanceolate to

lance-elliptic, 2–6 cm. long, 3–10 mm. wide, flat, coriaceous; heads discoid, 9–12-fld., in compact rounded cymes, the short pedicels with scalelike bracts; invol. ca. 5 mm. high; phyllaries 4-seriate, lanceolate to lance-oblong, acutish to acuminate, whitish, firm, carinate by the glandular-thickened costa; aks. appressed-pilosulous; pappus copious, fragile.—Locally frequent on outwash fans and exposed hillsides, 1500–7000 ft.; Chaparral; s. slopes of San Gabriel and San Bernardino mts., San Jacinto and Santa Ana mts., through San Diego Co. to L. Calif. July–Oct.

38. **H. cuneàtus** Gray. [*Bigelovia spathulata* Gray. *Aster c.* Kuntze. *Ericameria c.* McClat. *E. c.* var. *s.* Hall. *Chrysoma c.* and var. *s.* Greene. *C. Merriami* Eastw. *H. c. s.* Blake.] Deep green spreading much-branched shrub 1–5(–12) dm. high, glabrous, balsamic-resinous, glandular-punctate; lvs. crowded, cuneate to suborbicular-obovate, entire, often undulate, apiculate at the obtuse or broadly rounded or retuse apex, 5–20 mm. long, 3–10 mm. wide, thick; heads compactly cymose; invol. turbinate, 5–7 mm. high; phyllaries 4–6-seriate, regularly imbricate, ± glandular-thickened along the costa, linear-oblong to lance-ovate, the outer passing into minute ovate thick peduncular bracts; ray-fls. 1–5 (mostly in n. Sierra Nevada) or usually wanting; disk-fls. 16–28; aks. densely appressed-pilose.—Cliffs and slopes, frequently in crevices of granitic rocks, 2500–9000 ft.; Yellow Pine F., Pinyon-Juniper Wd.; Sierra Nevada from Plumas Co. to Tulare Co., ranges surrounding the Mojave Desert from s. Mono Co. to New York, Tehachapi and Little San Bernardino mts., s. Calif. mts. (mostly desert slopes) from e. San Luis Obispo Co. to San Diego Co.; L. Calif., Nev., Ariz. Sept.–Nov.

79. **Benitòa** Keck

Xerophytic annual with erect stem, cymose-paniculate above, the heads terminating the branchlets, heavy scented from a harsh glandular pubescence. Lvs. entire, reticulate-veiny. Fls. yellow, tinged with red. Invol. cylindro-turbinate, 5–6-seriate, the 30–50 phyllaries corneous, linear-attenuate, glandular-atomiferous, the outer ones with recurved tips bearing a prominent gland. Ray-fls. fertile, 5–14, the ak. 3-angled; disk-fls. sterile, 9–25, the corolla constricted between tube and throat, its style-branches included, appressed. Pappus of ray and disk of 2–8 brownish very fragile caducous bristles as long as the aks. One sp. (Named for San Benito Co.)

1. **B. occidentàlis** (Hall) Keck. [*Haplopappus o.* Hall.] Rather rigid, 3–15 dm. high, viscid throughout with rough septate gland-tipped hairs interspersed between a denser glandular puberulence, the lvs. often additionally somewhat hispid-hirsute, often ± anthocyanous; lvs. linear-lanceolate, obtuse or acute, narrowed to a clasping base, 5–9 cm. long, 6–15 mm. wide, becoming bractlike and apiculate in upper half of plant; invol. 8–10 mm. high, 3–5 mm. wide; ray-aks. olive, brown-maculate, finely sericeous, 3.5 mm. long; $n = 5$ (Raven, 1958).—Hot dry exposed serpentine or clay hillsides, 500–2500 ft.; Foothill Wd.; Diablo Range, in San Benito, Monterey, and Fresno cos. June–Nov.

80. **Solidàgo** L. GOLDENROD

Perennial herbs with leafy usually simple stems arising from rhizomes or a caudex. Lvs. alternate, entire or toothed. Heads numerous, radiate, yellow (in ours), small, campanulate to subcylindric, panicled, racemose, or cymose. Invol. few-seriate, graduate or subequal, the phyllaries usually with obscurely herbaceous tips. Receptacle usually alveolate. Ray-fls. small, fertile; disk-fls. perfect, fertile, their anthers subentire at base, their style-branches with mostly lanceolate appendages. Aks. short, pubescent (in ours), usually few-nerved. Pappus copious, setose, whitish. (Latin, *solidus*, whole, and th. suffix *-ago*, from its reputed medicinal value.)

Plants with well developed creeping herbaceous rhizomes; stems rather equably leafy, the lowest lvs. not prominently different from the upper and at length deciduous (except in S. *missouriensis*).
    Lvs. glandular-punctate, lance-linear; infl. ample, copiously bracteate, the heads in terminal cymose clusters; ray-fls. 15–25 . . . . 1. S. *occidentalis*
    Lvs. not punctate, broader; infl. more compact, not interrupted, the heads not glomerate in cymes; ray-fls. 8–13.
        Stems puberulent, at least above the middle.
            Lvs. densely puberulent on both faces, the middle and upper usually elliptic and entire
                                                       2. S. *californica*

Lvs. puberulous mostly on nerves beneath or usually glabrous, the middle and upper lanceolate, chiefly sharply toothed .... 3. *S. canadensis*
Stems glabrous below the infl. .... 4. *S. missouriensis*
Plants with rather short woody rhizomes or a branched caudex; stems with lower and basal petiolate lvs. much larger than the upper reduced sessile ones; infl. without recurving branches.
Plants of low or mid-elevations, usually more than 4 dm. high.
Infl. not glutinous; invol. 3.5–5 mm. high; lvs. entire (rarely remotely serrulate in *S. spectabilis*).
Panicle subracemose, few-headed; stems slender. Cent. Calif. .... 5. *S. Guiradonis*
Panicle usually oblong and very dense; stems rather stout.
Phyllaries acute or acuminate; rays mostly 8. S. Calif. .... 6. *S. confinis*
Phyllaries obtusish; rays mostly 13. Great Basin .... 7. *S. spectabilis*
Infl. glutinous; invol. 5–6 mm. high; lvs. crenate-serrate. Coastal .... 8. *S. spathulata*
Plants of high altitudes, mostly less than 3 dm. high .... 9. *S. multiradiata*

1. **S. occidentàlis** (Nutt.) T. & G. [*Euthamia o.* Nutt. *E. californica* Gand.] WESTERN GOLDENROD. Stem stout, from creeping rhizome, much-branched, 6–20 dm. high, glabrous, often glutinous above; lvs. lance-linear, acuminate, sessile, entire, 3–5-nerved, 4–10 cm. long, 3–9 mm. wide, glandular-punctate, the margin often scabrid; infl. ample, leafy-bracteate, interrupted-elongate or rounded, the heads in small cymose clusters; invol. 4 mm. high; phyllaries firm, lance-oblong to lance-linear, acute; ray-fls. 15–25, 1.5–2.5 mm. long; disk-fls. 7–14; aks. pilose.—Wet meadows, banks of rivers, lakes, and sloughs, usually below 2000 ft.; Coastal Salt Marsh, Freshwater Marsh, V. Grassland, Coastal Prairie, Sagebrush Scrub; frequent throughout cismontane Calif., rare e. of the mts., Inyo and Modoc cos.; to B.C., Alta., Nebr., Tex., L. Calif. July–Nov.

2. **S. califórnica** Nutt. [*Aster c.* Kuntze. *S. c.* var. *nevadensis* Gray.] CALIFORNIA GOLDENROD. Stems from creeping rhizomes, 2–12 dm. high, like the lvs. densely cinereous-puberulent; basal and lower cauline lvs. spatulate to obovate or oval, obtuse to acute, attenuate to base, firm, crenate or serrate, 5–12 cm. long, 1–3.5 cm. wide, the upper cauline usually much reduced, elliptic, entire, sessile; infl. a narrow dense thyrse, or sometimes broadened into a pyramidal panicle with spreading branches; invol. 3–4.5 mm. high; phyllaries lance-linear to narrowly oblong, sharply acute to obtuse, puberulent or glabrous; ray-fls. 8–13; disk-fls. 4–12; aks. hispidulous.—Common in dry or moist fields, clearings and forest openings, 200–7500 ft.; several Plant Communities from the Coastal Scrubs up to Yellow Pine F.; from sw. Ore. almost throughout cismontane Calif. to L. Calif.; rare e. of the Sierra Nevada (Inyo Co.). July–Oct.

3. **S. canadénsis** L. ssp. **elongàta** (Nutt.) Keck. [*S. e.* Nutt. *Aster e.* Kuntze. *S. lepida* var. *e.* Fern.] Stems from creeping rhizomes, 3–10(–15) dm. high, puberulent or pilosulous up towards (and always including) the infl. or throughout, densely leafy; lvs. nearly uniform, lanceolate or oblong-lanceolate, 5–12 cm. long, 1–2 cm. wide, tapering to base and apex, triplinerved, sharply serrate to entire, scabrid-margined, from essentially glabrous to scabrid-puberulent on both faces; panicle 5–20 cm. long, dense, usually rhombic and broad or oblong, the branches mostly erect; invol. 3.5–5 mm. high; phyllaries thin, linear-lanceolate; ray-fls. ca. 13, little exceeding the disk; aks. hispidulous.—Meadows and moist openings in woods, from sea level to 8500 ft.; Coastal Prairie and V. Grassland to Yellow Pine F.; Monterey Co. to Del Norte Co., N. Coast Range valleys, Sierra Nevada from Tulare Co. to Modoc Co.; n. on both sides of the mts. to s. B.C., e. to Rocky Mts., L. Calif. May–Sept.

S. altíssima L., 1–2 m. high, scabrous-pubescent throughout, with triple-nerved lvs. and dense pyramidal panicle, is occasionally established; native of e. U.S. Oct.–Nov.

4. **S. missouriénsis** Nutt. [*Aster m.* Kuntze.] Stems from creeping rhizomes, or clustered from a simple or branched caudex, 2–5(–9) dm. high, essentially glabrous throughout, or sparsely puberulent and viscidulous within the infl.; basal lvs. oblanceolate, 5–20 cm. long, 5–20 mm. wide, entire or toothed above, tapering to the petiole, scabrid-ciliolate, sometimes lost early, the cauline becoming strongly reduced above, the upper mostly linear or subulate, entire; panicle usually oblong or rhombic, with erect branches; invol. 3–5 mm. high; phyllaries linear-oblong to broadly lanceolate, obtuse to acutish, rather firm; ray-fls. 8(–13), usually distinctly exceeding the 15–28 disk-fls.; aks. hispidulous.—Meadowlands and upland valleys, 4000–5000 ft.; Sagebrush Scrub; rare in Calif. (Sierra V.); frequent n. to B.C., and e. to the Great Plains and Ariz. Aug.–Oct.

5. **S. Guiradònis** Gray. Stems slender, erect, from a woody rhizome, 8–10 dm. high, glabrous throughout; lvs. entire, the basal lanceolate, tapering to the petiole, 12–18 cm. long, 5–10 mm. wide, the cauline elongate but reduced, linear, the uppermost becoming

linear-subulate bracts; panicle erect, slender, subracemose, few-headed, not secund, 10–20 cm. long; invol. ca. 4–5 mm. high, the phyllaries lance-linear, the midvein broad; ray-fls. 8–10, little exceeding the 10–12 disk-fls.; aks. puberulent.—Moist slopes, rare, 100–4500 ft.; V. Grassland, Foothill Wd.; inner S. Coast Range, from San Francisco Bay s., and southernmost Sierra Nevada. Sept.–Oct.

6. **S. confinis** Gray. [*S. c. f. luxurians* Hall. *S. c.* var. *l.* Jeps.] Stems usually stout, terminating rather short woody rhizomes or sometimes clustered on a short caudex, 3–14 dm. high, glabrous throughout; lvs. thick, pale green, entire, glabrous or only the margin scabrid, the basal spatulate or oblanceolate to obovate, tapering to the petiole, 15 cm. or less long, the cauline gradually reduced, lance-elliptic to linear, sessile; panicle usually oblong and very dense, up to 25 cm. long, the branches erect, rarely spreading; invol. 3.5–4.5 mm. high, the phyllaries only slightly imbricate, linear-lanceolate, acute or acuminate, rather firm; ray-fls. 6–10, scarcely surpassing the 11–21 disk-fls.; aks. sparsely hispidulous to canescent.—Dry or moist banks, 500–8200 ft.; Coastal Sage Scrub, Chaparral, Yellow Pine F.; w. San Luis Obispo Co. to Ventura Co., San Gabriel and San Bernardino mts. and desert borders, to San Diego Co. and L. Calif. July–Oct.

7. **S. spectábilis** (D. C. Eat.) Gray. [*S. Guiradonis* var. *s.* D. C. Eat.] Stem rather stout, terminating a rather short woody rhizome or caudex, 4–13 dm. high, glabrous throughout, or becoming ± hispidulous within the infl.; lvs. usually entire, the basal oblanceolate, acute or obtuse, tapering to a long winged clasping petiole, 9–28 cm. long, 1.3–4 cm. wide, the upper cauline linear-lanceolate and often much reduced, scabrid-ciliolate, otherwise glabrous; panicle usually oblong and very dense, mostly less than 10 cm. long; invol. ca. 4 mm. high; phyllaries linear-oblong, obtusish; ray-fls. 11–15, short; disk-fls. 15–22; aks. puberulent.—Alkaline meadows or bogs, sea level to 7500 ft.; Alkali Sink, Shadscale Scrub, Sagebrush Scrub; Mono and Inyo cos.; to se. Ore., Utah. July–Sept. A Great Basin sp.

8. **S. spathulàta** DC. [*Homopappus s.* Nutt. *S. spiciformis* T. & G. *Aster Candollei* Kuntze.] Stems stout, from a caudex or woody rhizome, 2–6 dm. high, usually glabrous throughout, somewhat glutinous above; basal lvs. broadly obovate to spatulate-oblanceolate, mostly blunt or rounded, crenate-serrate, tapering to the petiole, the cauline similar but reduced, the uppermost acute and subsessile; heads in a simple or compound sometimes racemiform thyrse 6–25 cm. long; invol. 5–6 mm. high, the phyllaries firm, oblong, very blunt; ray-fls. 7–9, scarcely exceeding the 10–16 disk-fls.; aks. densely pubescent.—Sandy coastal hills and dunes, up to 600 ft.; Coastal Strand, N. Coastal Scrub; Monterey coast, n. to Coos Co., Ore. May–Nov.

9. **S. multiradiàta** Ait. [*S. corymbosa* Nutt., not Poir. or Ell. *S. m.* var. *scopulorum* Gray. *Aster m.* Kuntze. *S. ciliosa* Greene.] Stems erect from a woody rhizome or branching caudex, 0.5–4 dm. high, hairy at least above; basal lvs. oblanceolate to elliptic, obtuse or acute, entire to crenate-serrulate, mostly 2–10 cm. long, 5–18 mm. wide, scabrid-margined, tapering to a ciliate petiole, the cauline spatulate to lanceolate, usually acute, sessile, mostly entire, ciliate at least toward base; heads few to rather numerous, in a loose or usually dense terminal corymb, or with 1 or 2 loose axillary clusters; invol. 4–6.5 mm. high, the phyllaries linear to lance-linear, acute to acuminate, not much imbricate, thin, ± ciliolate; ray-fls. commonly 13, distinctly exceeding the 13–34 disk-fls.; aks. hispidulous.—Sunny, rocky or grassy places, (5200–)8000–12,500 ft.; Lodgepole F., Subalpine F., Alpine Fell-fields; White and Sweetwater mts., Sierra Nevada from Tulare Co. n. through the Cascades to Alaska and adjacent Siberia; e. to Labrador, s. through the Rocky Mts. to New Mex. and Ariz. June–Sept.

81. Petradòria Greene. ROCK GOLDENROD

Low tufted perennial herb, with a short branched caudex surmounting a stout black scaly taproot. Lvs. narrow, rigid, veiny, sharp-pointed, entire. Heads yellow, small, numerous, in terminal corymbs. Invol. cylindric, 4-seriate, strongly graduate, the phyllaries corneous, stramineous, in vertical ranks. Ray-fls. ♀, fertile, 1–3, the ak. somewhat flattened, 5–10-striate, glabrous; disk-fls. sterile, 3–5. Pappus of numerous very slender stramineous bristles. One sp. Seeming to combine the characters of *Chrysothamnus* and *Haplopappus*. (Greek, *petra*, a rock, and *Doria*, an early name for the Goldenrod.)

1. **P. pùmila** (Nutt.) Greene. [*Chrysoma p.* Nutt. *Solidago p.* T. & G., not Crantz. *Aster p.* Kuntze. *Solidago Petradoria* Blake.] Rather rigid flat-topped plant with numerous simple erect leafy stems from the caudex, 1–2.5 dm. high, glabrous, resinous, light green; lvs. crowded in basal rosettes, fewer above, obscurely punctate, 3–5-nerved, the basal nearly linear to oblanceolate, slenderly petiolate, 5–8 cm. long, 3–7 mm. wide, the cauline sessile, becoming bractlike within the infl.; invols. 5.5–8 mm. high, ca. 2.5 mm. wide; phyllaries lance-oblong to lanceolate, obtusish to acuminate, apiculate, glabrous, ± carinate, sometimes with small green tip; ligules 2 mm. long.—Dry, stony, often limestone hillsides, 3500–7000 ft.; Pinyon-Juniper Wd.; Clark, New York and Providence mts., e. Mojave Desert, San Bernardino Co.; to Wyo., Tex. July–Oct.

82. Chrysothámnus Nutt. RABBIT-BRUSH

Shrubs or subshrubs, usually much-branched with erect stems. Lvs. alternate, entire, narrow, not fasciculate, sometimes glandular-punctate. Heads numerous, discoid, narrow, mostly 5-fld., in cymes, racemes or panicles. Invol. cylindraceous; phyllaries strongly imbricate, usually in 5 vertical ranks, chartaceous or coriaceous, sometimes herbaceous-tipped, mostly carinate and often with a costal gland. Appendages of style-branches usually longer than the stigmatic portion. Aks. slender, terete or angled or flattened, densely sericeous to glabrous. Pappus copious, of soft capillary bristles. Thirteen spp. in w. N. Am. (Greek, *chrysos,* gold, and *thamnos,* shrub, the infls. very showy in most.)

(Hall, H. M., and F. E. Clements. Carnegie Inst. Wash. Pub. 326: 157–234, 1923.)
A. Lvs. resinous-punctate, terete.
    B. Fls. yellow.
        C. Phyllaries thin, not glandular-thickened apically .... 1. *C. paniculatus*
        CC. Phyllaries tipped with a conspicuous thick green glandular spot .... 2. *C. teretifolius*
    BB. Fls. white .... 3. *C. albidus*
AA. Lvs. not resinous-punctate.
    B. Twigs not tomentose.
        C. Plants definitely shrubby; lvs. not prominently ribbed; phyllaries not cuspidate at a
            truncate or retuse apex.
            D. Phyllaries in obscure vertical ranks; aks. hairy.
                E. Fls. white; phyllaries with slender tapering tip .... 3. *C. albidus*
                EE. Fls. yellow; phyllaries obtuse to acute .... 4. *C. viscidiflorus*
            DD. Phyllaries in 5 sharply defined vertical ranks.
                E. Stems glabrous; lvs. terete, 0.5 mm. wide; invol. 5–6 mm. high; aks. sericeous
                        5. *C. axillaris*
                EE. Stems scabrid-puberulent; lvs. oblanceolate, 1.5–4 mm. wide; invol. 9–12
                    mm. high; aks. glabrate .... 6. *C. depressus*
        CC. Plants largely herbaceous above a branched woody caudex; lvs. longitudinally
            ribbed; phyllaries broad, cuspidate at the truncate or retuse apex .... 7. *C. gramineus*
    BB. Twigs densely pannose-tomentose.
        C. Infl. mostly racemose or spicate; phyllaries very attenuate .... 8. *C. Parryi*
        CC. Infl. mostly cymose; phyllaries obtuse to moderately attenuate .... 9. *C. nauseosus*

1. **C. paniculàtus** (Gray) Hall. [*Bigelovia p.* Gray. *Aster Asae* Kuntze. *Ericameria p.* Rydb.] Loosely branched broadly rounded shrub 6–20 dm. high, the herbage glabrous, resinous, strongly glandular-punctate, ± glaucescent; lvs. terete, mucronate-tipped, 1–3 cm. long, 0.5 mm. wide; heads in profuse panicles, 5–8-fld.; invol. subcylindric, 5.5–6.5 mm. high; phyllaries 4-seriate, graduate, the vertical ranks not sharply defined, oblong, obtuse, whitish, indurate, the costa narrow, obscure, scarcely if at all glandular-thickened above; aks. appressed-villous or sericeous; pappus brownish-white.—Stony open deserts and washes, mostly below 4000 ft. (to 8000 ft. in Inyo Co.); Creosote Bush Scrub; w. and n. borders of Colo. Desert, local across Mojave Desert from Tehachapi to e. Inyo and San Bernardino cos.; to sw. Utah, Ariz. June–Dec.

2. **C. teretifòlius** (Dur. & Hilg.) Hall. [*Linosyris t.* Dur. & Hilg. *Bigelovia t.* Gray. *Aster Durandii* Kuntze. *Chrysoma t.* Greene. *Ericameria t.* Jeps.] Fastigiately branched globose or spreading shrub 2–15 dm. high, the herbage glabrous, densely glandular-punctate, balsamic-resinous, dark green; lvs. terete, obtuse, not mucronate, 1–2.5 cm. long, 0.5–1 mm. wide; heads in short terminal spikes, 4–6-fld.; invol. subcylindric, 6–8 mm. high; phyllaries 4–5-seriate, strongly graduate in rather definite vertical ranks, oblong, obtuse, indurate, stramineous, obscurely carinate, tipped with a conspicuous green glandular spot; aks. appressed-villous; pappus stramineous.—Canyon walls and rocky slopes, 3200–8000 ft.; Sagebrush Scrub, Creosote Bush Scrub, Joshua Tree Wd.,

Pinyon-Juniper Wd.; deserts from Santa Rosa Mts. n. to se. Mono Co., w. to Tehachapi and Liebre Mts.; to sw. Utah, nw. Ariz. Sept.–Nov.

3. **C. álbidus** (Jones) Greene. [*Bigelovia a.* Jones ex Gray.] Shrubs 3–15 dm. high, fastigiately branched, white-barked, glabrous, resinous-viscid, aromatic; lvs. filiform, 1.5–3 cm. long, 0.5–1 mm. wide, terete, mucronate, impressed-punctate, crowded, with axillary fascicles; heads in small compact terminal cymes, 4–6-fld.; invol. 7–9 mm. high; phyllaries ca. 4-seriate, graduate, in obscure vertical ranks, the outer lanceolate to ovate, herbaceous-thickened in outer half, with subulate-attenuate curved tip, the inner oblong, acuminate-tipped, the narrow hyaline margin ± erose; corollas white, 7–8 mm. long, the lobes ca. 2 mm. long; aks. pilose, glandular above; pappus copious.—Dry, alkaline, sandy or silty plains, 1000–6400 ft.; Shadscale Scrub, Alkali Sink; very rare in Calif., Owens V.(?), Saline V., Amargosa Desert se. of Death V.; across Nev. to Salt Lake Desert, Utah. Aug.–Nov.

4. **C. viscidiflòrus** (Hook.) Nutt. [*Crinitaria v.* Hook. *Bigelovia v.* DC. *Linosyris v.* T. & G. *B. Douglasii* Gray. *B. D.* var. *tortifolia* Gray. *Aster v.* Kuntze. *Chrysothamnus v.* var. *t.* Greene. *C. t.* Greene. *C. D.* Clem. & Clem.] Rounded fastigiately branched white-barked shrub usually less 1 m. high, the twigs brittle; lvs. linear or linear-lanceolate, flat or twisted, erect, spreading, or reflexed, 2–5 cm. long, 2–5 mm. wide, 1–3-nerved, glabrous, sometimes scabrous-ciliolate, viscid, sometimes with punctate glands ventrally; heads ca. 5-fld., in terminal broad cymes; invol. 5–7 mm. high; phyllaries linear-oblong to lanceolate, obtuse (but commonly mucronate) to acute, not keeled, strongly graduate but in obscure vertical ranks, chartaceous; aks. densely villous; pappus brownish-white.—Dry open places, 4000–7500 ft.; Yellow Pine F.; San Jacinto and San Bernardino mts., e. of the Sierran crest from Mono Co. to Modoc Co.; n. to B.C., e. to New Mex. and Mont. July–Sept. The typical ssp. of this highly polymorphic sp. is most frequent in the s. Rocky Mts.

KEY TO SUBSPECIES

A. Lvs. glabrous (sometimes viscid or with punctate glands ventrally) at least on the faces, the
    margin sometimes scabro-ciliate.
    B. Shrubs mostly 4–12 dm. high (shorter in some wide-lvd. plants); lvs. mostly 2 mm. or more
        wide.
        C. Lvs. linear to linear-lanceolate, tapering to base and apex, 2–5 mm. wide, 1–3-nerved,
            acute or acuminate; widespread .... *C. viscidiflorus*
        CC. Lvs. broadly elliptic to elliptic-oblong, 6–12 mm. wide, 3–5-nerved, obtuse but
            mucronate; shrubs often only 2–4 dm. high; nearly limited to ne. Nev.
                                                                            (a) ssp. *latifolius*
    BB. Shrubs mostly 1–3.5 dm. high; lvs. linear or linear-filiform, 0.5–2 mm. wide
                                                                            (b) ssp. *pumilus*
AA. Lvs. ± densely puberulent, 1–2 mm. wide, 1-nerved .... (c) ssp. *puberulus*

(a) Ssp. **latifòlius** (D. C. Eat.) Hall & Clem. [*Linosyris v.* var. *l.* D. C. Eat. *Chrysothamnus v.* var. *l.* Greene. *C. l.* Rydb. *Bigelovia Douglasii* var. *l.* Gray.] Mts. of ne. Nev. and adjacent Ida. Reported by Hall and Clements from Modoc Co., but doubtfully a member of our flora.

(b) Ssp. **pùmilus** (Nutt.) Hall & Clem. [*C. p.* Nutt. *Linosyris p.* Gray. *Bigelovia Douglasii* var. *stenophylla* Gray. *B. D.* var. *p.* Gray. *C. s.* Greene. *C. v.* var. *s.* Hall. *C. v.* ssp. *s.* Hall & Clem. *C. v.* var. *p.* Jeps.] Dry sagebrush-covered plains, often in alkaline soils, 5000–11,000 ft.; Sagebrush Scrub, N. Juniper Wd.; San Bernardino Mts., Inyo Co. to Modoc Co.; to s. Wash., Mont., Utah.

(c) Ssp. **pubérulus** (D. C. Eat.) Hall & Clem. [*C. pumilus* (var.) *euthamioides* Nutt., apparently this. *Linosyris v.* var. *puberula* D. C. Eat. *Bigelovia Douglasii* var. *p.* Gray. *C. p.* Greene. *C. humilis* Greene. *C. v.* ssp. *h.* Hall & Clem. *C. v.* var. *h.* Jeps. *C. v.* var. *p.* Jeps.] Sagebrush-covered plains and slopes, up to 10,500 ft.; Sagebrush Scrub, N. Juniper Wd., Pinyon-Juniper Wd.; San Bernardino, Inyo, and White mts., e. flank of Sierra Nevada; n. to s. Wash., e. to Yellowstone Park; abundant in Nev. and s. Ida.

5. **C. axilláris** Keck. Low rounded shrub 3–6 dm. high, intricately branched, the branches slender, white-barked, glabrous, with small axillary buds in the axils of the oldest lvs.; lvs. tightly involute, terete, green, spreading, filiform, 0.5 mm. wide, acicular-tipped; heads in rounded cymose panicles, 3–5-fld.; invol. turbinate, 5–6 mm.

high; phyllaries 4-seriate, strongly graduate in 4 or 5 sharply defined vertical ranks, spreading at maturity, broadly linear, sharply acute or apiculate, the costa glandular-thickened above, the small herbaceous tip ciliolate; corolla ca. 5 mm. long, the lobes 1–1.2 mm. long; aks. sericeous; pappus tawny, not very long or copious.—Desert slopes in granitic sand and limestone canyons, 5000–6000 ft.; Shadscale Scrub, Sagebrush Scrub; head of Deep Springs V. and n. Inyo Mts., Inyo Co.; also Esmeralda Co., Nev. July–Oct.

6. **C. depréssus** Nutt. [*Linosyris d.* Torr. *Bigelovia d.* Gray.] Depressed subshrub with many erect herbaceous stems from a much branched spreading woody crown, 1–3 dm. high, cinereous with a dense scabrid puberulence; lvs. oblanceolate or spatulate, the lowermost rounded or obtuse, the upper becoming sharply apiculate, 7–20 mm. long, 1.5–4 mm. wide, firm; heads in compact terminal cymes, 5-fld.; invol. 9–12 mm. high; phyllaries 5-seriate, in 5 sharply defined vertical ranks, lance-acuminate, drawn to a soft mucro, strongly keeled, the outer herbaceous and minutely puberulent, the inner broader, scarious, with hyaline margin; aks. glabrate; pappus brownish-white.—Dry canyons, 3500–6000 ft.; Joshua Tree Wd., Pinyon-Juniper Wd.; Clark, New York and Providence mts., e. Mojave Desert; to s. Colo. and n. New Mex. Aug.–Oct.

7. **C. gramíneus** Hall. Many-stemmed from a branched woody caudex, 2.5–6 dm. high, light green, essentially glabrous, the striate-angled stems simple or erect-branched above; lvs. narrowly linear-lanceolate, 3–7 cm. long, 3–8 mm. wide, acuminate, 3–5-ribbed, coriaceous; heads mostly subsessile in small terminal clusters, pale yellow; invol. 10–13 mm. high; phyllaries 4–6-seriate, stramineous, oblong, abruptly cuspidate at the truncate or retuse ciliolate apex; corolla ca. 10 mm. long; aks. glabrous, ca. 6 mm. long.—Rocky, wooded slopes, 7500–9500 ft.; Pinyon-Juniper Wd., Bristle-cone Pine F.; Inyo and Panamint mts., Inyo Co.; Charleston Mts., Nev. July–Aug.

8. **C. Párryi** (Gray) Greene. [*Linosyris P.* Gray. *Bigelovia P.* Gray.] Shrub up to 5 dm. high, the numerous pliable branches erect or spreading, densely white-, gray-, or greenish-yellow-pannose, very leafy; lvs. narrowly linear to elliptic, 1–8 cm. long, 0.5–8 mm. wide, 1–3-nerved; heads in short leafy racemes or racemiform panicles, yellow; invol. 9–14 mm. high; phyllaries 4–6-seriate, in ± obscure vertical ranks, acuminate or attenuate, the outer often with herbaceous tip, ciliate and often ± tomentose; corolla 8–11 mm. long; aks. densely appressed-pilose; pappus brownish-white.—Mountainsides and flats, Calif. to Nebr. and New Mex. July–Sept. Variable in habit, pubescence, foliage, infl., and characters of head. The typical ssp., having 10–20-fld. heads., ranges from e. Nev. to Colo. and Wyo.

KEY TO SUBSPECIES

A. Lvs. 4 mm. or more wide.
   B. Invol. 9–10 mm. high; heads 7–11-fld., in short racemes .... ssp. *Bolanderi*
   BB. Invol. 12–15 mm. high; heads 5–7-fld., in racemiform panicles .... ssp. *latior*
AA. Lvs. 3 mm. or less wide.
   B. Stems spreading at base, ca. 1 dm. high; heads 11–15-fld. .... ssp. *imulus*
   BB. Stems usually taller; heads 4–8(–10)-fld.
      C. Herbage finely stipitate-glandular .... ssp. *asper*
      CC. Herbage sometimes viscid, but not obviously stipitate-glandular.
         D. Plants low, rigidly much-branched; heads very few .... ssp. *monocephalus*
         DD. Plants taller, the stems mostly erect; heads several.
            E. Tips of the phyllaries erect. S. Sierra Nevada .... ssp. *vulcanicus*
            EE. Tips of phyllaries tending to recurve. N. Calif. .... ssp. *nevadensis*

Ssp. **Bolánderi** (Gray) Hall & Clem. [*Linosyris B.* Gray. *Bigelovia B.* Gray. *Aster B.* Kuntze. *Chrysothamnus B.* Greene. *Macronema B.* Greene. *C. Parryi* var. *B.* Jeps.] Lvs. 3–4 cm. long, 4–5 mm. wide, green, viscid-glandular; heads 7–11-fld., in short sometimes branched racemes; invol. 9–10 mm. high; phyllaries ca. 11–15, in obscure ranks.—Known only from type loc., Mono Pass, Sierra Nevada, 9000–10,000 ft.

Ssp. **látior** Hall & Clem. [*C. Parryi* var. *l.* Jeps.] Lvs. elliptic or oblanceolate, 2–4 cm. long, 4–8 mm. wide, 3-nerved, green, scabro-glandular; heads 5–7-fld., in racemiform panicles; invol. 12–15 mm. high; phyllaries 12–15, in rather distinct vertical ranks, nearly smooth, with attenuate rather pungent tips; corolla 11–12 mm. long.—Slopes, 3000–4500 ft.; Yellow Pine F.; mts. of Siskiyou and Modoc cos., Lake and Plumas cos.

Ssp. **imùlus** Hall & Clem. [*C. Parryi* var. *i.* Jeps.] Spreading at base, the shoots ca. 1

dm. high; lvs. narrowly spatulate, 1–1.5 cm. long, 2–3 mm. wide, gray-tomentose; heads few, 11–15-fld.; invol. 11–12 mm. high; phyllaries ca. 16, obscurely ranked, the outer tomentose.—San Bernardino Mts.

Ssp. **ásper** (Greene) Hall & Clem. [*C. a.* Greene. *C. Parryi* var. *a.* Munz.] Stems 1.5–4 dm. high, white-tomentose; lvs. linear, 2–5 cm. long, 1–3 mm. wide, green, finely stipitate-glandular and roughish; heads several, 5–10-fld.; invol. 11–15 mm. high; phyllaries 9–13, usually glandular-puberulent as well as arachnoid-ciliate below.—Mountainsides rimming the desert, 7000–8500 ft.; Pinyon-Juniper Wd., Yellow Pine F.; Mono Co. to Ventura Co. and San Bernardino Mts.; e. to Nev.

Ssp. **monocéphalus** (Nels. & Kenn.) Hall & Clem. [*C. m.* Nels. & Kenn. *C. nevadensis* var. *m.* Smiley. *C. Parryi* var. *m.* Jeps.] Usually low, rigidly much-branched; lvs. 1–3 cm. long, 0.8–1.5 mm. wide, tomentulose and viscidulous; heads 1–4, crowded, 5–6-fld.; invol. 9–11 mm. high; phyllaries 8–12, thinly tomentose and arachnoid-ciliate, sometimes purplish.—Dry rocky slopes, 10,000–11,300 ft.; Alpine Fell-fields; crest of the Sierra Nevada, Mono, Tuolumne, and Madera cos.; w. Nev.

Ssp. **vulcánicus** (Greene) Hall & Clem. [*C. v.* Greene. *C. Parryi* var. *v.* Jeps.] Very similar to ssp. *asper;* lvs. viscid but not obviously stipitate-glandular.—Mountain flats, 4600–10,600 ft.; Montane Coniferous F.; s. Sierra Nevada, Mono and Mariposa cos. to Tulare Co.

Ssp. **nevadénsis** (Gray) Hall & Clem. [*Linosyris Howardii* var. *n.* Gray. *Bigelovia H.* var. *n.* Gray. *B. n.* Gray. *C. n.* Greene. *C. Parryi* var. *n.* Jeps.] Up to 6 dm. high; branches white-, gray-, or greenish-tomentose; lvs. linear to linear-oblanceolate, 1.5–4 cm. long, 0.5–3 mm. wide, gray-tomentulose to green and glandular; heads few to numerous, 4–6-fld.; invol. 12–15 mm. high; phyllaries 13–18, thinly tomentose and arachnoid-ciliate, the attenuate herbaceous tips tending to recurve.—Dry mountainsides, 4600–9000 ft.; Yellow Pine F., Red Fir F.; e. Calif. from Modoc Co. to Alpine Co.; to Nev. and n. Ariz.

9. **C. nauseòsus** (Pall.) Britton. [*Chrysocoma n.* Pall. *Chondrophora n.* Britton.] Shrub 3–20 dm. high, usually with several fibrous-barked main stems from the base, these much branched, the often ill-smelling erect usually densely leafy twigs clothed with a persistent feltlike gray, white, or greenish tomentum; lvs. variable, linear-filiform to narrowly linear-oblanceolate, 2–7 cm. long, 0.5–4 mm. wide, tomentose to subglabrous; heads in terminal rounded cymose clusters; invol. 6–13 mm. high; phyllaries usually 3–4-seriate, strongly graduate, in rather definite ranks, mostly lanceolate or linear-lanceolate, not green-tipped, usually with resinous-thickened costa; fls. usually 5, yellow, the corolla 7–12 mm. long; pappus copious, dull white.—Widespread in open, dry places on plains and mountainsides, often in alkaline soils; mostly e. Calif., n. to B.C., e. to Sask., Tex. and n. Mex. The typical form occurs mostly e. of the Continental Divide and on the Great Plains.

KEY TO SUBSPECIES

A.   Aks. densely pubescent.
  B.   Phyllaries, at least the outer, ± pubescent or tomentose (sometimes only ciliate).
    C.   Corolla-lobes lanceolate, 1.3–2.5 mm. long; style-appendage longer than the stigmatic portion.
      D.   Invol. 7–10(–13) mm. high; phyllaries rarely glandular, the margin not prominently hyaline or fimbriate (but often ± ciliate); corolla 8–11 mm. long
                                                ssp. *albicaulis*
      DD. Invol. 10–13 mm. high; phyllaries ± glandular-atomiferous, the margin prominently hyaline and slashed-fimbriate (or merely ciliate); corolla 9–12 mm. long
                                                  ssp. *bernardinus*
    CC. Corolla-lobes ovate, 0.5–1 mm. long; style-appendage shorter than the stigmatic portion; invol. 6–7 mm. high.
      D.   Herbage mostly grayish or whitish, not obviously glandular . . . . ssp. *hololeucus*
      DD. Herbage mostly yellow-green, the infl., including the invols., glandular-pubescent
                                                 ssp. *viscosus*
  BB. Phyllaries not hairy, but sometimes glandular or viscid; lvs. not white-tomentose; corolla-tube glabrous.
    C.   Phyllaries without a recurved mucronate tip; corolla 7 mm. or more long.
      D.   Invol. 6.5–9 mm. high, not sharply angled, the phyllaries slightly if at all keeled
                                                  ssp. *consimilis*
      DD. Invol. 9–10 mm. high, sharply angled, the strongly keeled phyllaries in very distinct vertical rows . . . . ssp. *mohavensis*

CC. Phyllaries with a very slender recurved mucronate tip; corolla less than 7 mm. long
ssp. *ceruminosus*
AA. Aks. glabrous or with merely a few hairs on the prominent ribs . . . . ssp. *leiospermus*

Ssp. **albicáulis** (Nutt.) Hall & Clem. [*C. speciosus* and var. *a.* Nutt. *Linosyris a.* T. & G. *Bigelovia graveolens* var. *a.* Gray. *C. californicus* and var. *occidentalis* Greene. *C. o.* Greene. *C. nauseosus* var. *a.* Rydb. *C. orthophyllus* Greene. *C. n.* vars. *occidentalis, s.* and *c.* Hall. *C. n.* sspp. *s.* and *o.* Hall & Clem. *C. n.* var. *macrophyllus* J. T. Howell.] Shrub 5–20 dm. high; foliage gray or white with a rather copious tomentum; lvs. 1(–4) mm. wide; invol. 7–13 mm. high; phyllaries mostly acute or acuminate, at least the outer ± tomentose, but even these sometimes merely obscurely ciliate; corolla 8–11 mm. long, the tube loosely arachnoid-villous to puberulent, the lobes mostly 1.3–2 mm. long.—Open slopes and flats, 3000–8000 ft.; Chaparral, Sagebrush Scrub, Yellow Pine F., N. Juniper Wd.; inner N. Coast Range (Lake Co. to sw. Ore.), both slopes of Sierra Nevada, Fresno and Inyo cos. n. to Siskiyou and Modoc cos.; to B.C., w. Mont. and nw. Colo. The most variable ssp., but apparently not naturally divisible.

Ssp. **bernardínus** (Hall) Hall & Clem. [*C. n.* var. *b.* Hall.] Branches gray- or white-pannose; lvs. usually green, 1–2 mm. wide; invol. 10–13 mm. high; phyllaries sharply acuminate, with margin broadly hyaline and cut-fimbriate or merely ± ciliate, the outer often puberulent or rarely tomentose, together with the peduncles ± glandular-atomiferous; corolla 9–12 mm. long, the lobes 1.5–2.5 mm. long, the tube puberulent.—Dry benches, 6000–9500 ft.; Yellow Pine F., Red Fir F.; San Gabriel, San Bernardino and San Jacinto mts.

Ssp. **hololeùcus** (Gray) Hall & Clem. [*Bigelovia graveolens* var. *h.* Gray. *C. speciosus* var. *gnaphalodes* Greene. *C. gnaphalodes* Greene. *C. n.* vars. *h.* and *g.* Hall. *C. n.* ssp. *g.* Hall & Clem.] Shrub 5–20 dm. high, the twigs white, gray or yellowish-green; herbage fragrant; lvs. 0.5–1.5 mm. wide, gray- or white-tomentose; invol. 6–7 mm. high; phyllaries rather obtuse, keeled; corolla 6.5–8 mm. long, the lobes 0.5–1 mm. long.—Sandy, neutral soils of high deserts, 3000–9000 ft.; Alkali Sink, Sagebrush Scrub, Pinyon-Juniper Wd.; w. ends of Colo. and Mojave deserts, w. to Cuyama V., n. to Mono Co.; throughout Nev.

Ssp. **viscòsus** Keck. Dense round shrub 5–20 dm. high, yellowish- or grayish-green; lvs. spreading or recurved, thickish, 1–2 mm. wide, permanently pannose, together with the mostly divaricate flowering branchlets, peduncles and invols. yellow-glandular; invol. 5–8 mm. high, the phyllaries narrow but obtuse, ± glandular-tomentose, the costal gland prominent; corolla 7–8 mm. long, the lobes 0.4–0.6 mm. long.—Sandy washes and flats, 4000–8000 ft.; Pinyon-Juniper Wd.; s. White Mts., Inyo Co., and adjacent Nev., s. along Sierra Nevada to Walker Pass.

Ssp. **consímilis** (Greene) Hall & Clem. [*C. c. tortuosus* and *angustus* Greene. *C. oreophilus* and var. *artus* A. Nels. *C. n.* vars. *o., c.* and *viridulus* Hall. *C. n.* ssp. *v.* Hall & Clem. *C. n.* var. *artus* Cronq.] Shrub 5–30 dm. high; tomentum gray, greenish-yellow, or whitish; lvs. mostly linear-filiform, less than 1 mm. wide, green or gray-tomentulose; infl. tending to be narrow and elongate rather than flat-topped; invol. 6.5–8.5 mm. high, glabrous; phyllaries acute or obtuse, not sharply keeled; corolla 7–9.5 mm. long, the lobes 1–2.5 mm. long.—Alkaline open valleys; Sagebrush Scrub; a Great Basin plant, not abundant in Calif., skirting the deserts from San Diego Co. n. to Shasta V. and e. Ore., e. to Ida. and Utah.

Ssp. **mohavénsis** (Greene) Hall & Clem. [*Bigelovia m.* Greene. *C. m.* Greene. *C. n.* var. *m.* Hall.] Shrub 6–20 dm. high, often fastigiately branched, the branches often leafless and rushlike, closely gray- or greenish-yellow-tomentose; lvs. narrowly linear, 1 mm. or less wide, tomentulose or nearly glabrous; infl. a rounded or somewhat elongate thyrse; invol. narrow, 8–10.5 mm. long, glabrous, sharply 5-angled; phyllaries keeled, mostly obtuse, in very distinct vertical ranks, the costa usually dilated at apex into an oblong gland; corolla 8–10 mm. long, the lobes 1.5–2.5 mm. long.—Well-drained, scarcely alkaline soils, 2500–6000 ft.; Foothill Wd., Joshua Tree Wd., Creosote Bush Scrub; S. Coast Ranges (Mt. Hamilton, Tassajara region, etc. to Ventura Co.), southernmost Sierra Nevada (Tulare and Kern cos.), head of San Joaquin V. and w. Mojave Desert; e. to Nev.

Ssp. **ceruminòsus** (Dur. & Hilg.) Hall & Clem. [*Linosyris c.* Dur. & Hilg. *Bigelovia c.* Gray. *C. c.* Greene. *C. n.* var. *c.* Hall.] Shrub 5–12 dm. high; tomentum yellowish-green;

lvs. linear-filiform, less than 1 mm. wide, early deciduous; invol. 7–9 mm. high, glabrous; phyllaries with abrupt filiform recurved tip ca. 1 mm. long; corolla 6.5 mm. long, the lobes 1.5–2 mm. long.—Creosote Bush Scrub, Chaparral; rare, s. Mojave Desert, Tejon Pass to Little San Bernardino Mts.

Ssp. **leiospérmus** (Gray) Hall & Clem. [*Bigelovia l.* Gray. *C. l.* Greene. *C. n.* var. *l.* Hall.] Shrub 3–12 dm. high, with many short leafy branches, sometimes nearly leafless, white- (usually) or yellowish-green-tomentose; lvs. short and very narrow, essentially glabrous; heads in small terminal cymes; invol. 6–9 mm. high, glabrous; phyllaries obtuse, not obviously keeled; corolla 5–8 mm. long, the lobes 0.5 mm. long or less; ak. glabrous or essentially so.—Very arid slopes; Joshua Tree Wd., Pinyon-Juniper Wd.; mts. of e. Mojave Desert; to Nev., s. Utah, n. Ariz.

83. Béllis L. DAISY

Low annual or perennial herbs, with rosulate lvs. and medium-sized solitary heads on scapelike peduncles. Invol. hemispheric; phyllaries herbaceous, blunt, biseriate, equal. Receptacle conic, naked. Ray-fls. ♀, white to pink or purple. Disk-fls. perfect, greenish-yellow; style-branches flattened, with short ovate puberulent appendage scarcely longer than broad. Aks. compressed, mostly 2-nerved; pappus none. Ca. half a dozen spp., native to Eu. (Latin, *bellus*, pretty.)

1. **B. perénnis** L. ENGLISH DAISY. Fibrous-rooted perennial, moderately hirsute throughout except the upper part of the scapes becoming densely silky-strigose; lvs. broadly oblanceolate to obovate or orbicular, the dentate or denticulate thin blade tapering into a slender or margined petiole as long, together 2–7 cm. long, 7–20 mm. wide; scape 5–20 cm. high; invol. 4–7 mm. high; rays 30–80, 1 cm. or less long; $2n = 18$ (Negodi, 1935).—Natur. in lawns and in grassy places near the coast, particularly in the white-fld. form, from Santa Barbara Co. n. to Vancouver Id.; introd from Eu. April–Sept.

84. Monóptilon T. & G. DESERT STAR

Diminutive depressed desert annuals, branched from base, with hispid herbage. Lvs. alternate, linear to spatulate, entire. Heads radiate, daisylike, terminal on the branchlets, subtended by the upper lvs. Invol. hemispheric to campanulate, nearly uniseriate; phyllaries equal, linear, firm, ± keeled below, herbaceous. Receptacle flat, naked. Ray-fls. ♀, showy; lamina elliptic, white to pinkish. Disk-fls. fertile, the corolla yellow or sometimes purplish. Aks. oblong-obovate, compressed, marginally nerved, pubescent. Pappus alike in ray and disk, of numerous unequal bristles alternating with short paleae, or of a short scarious cup and one apically plumose bristle. Two spp., of deserts of the sw. U.S. and n. Mex. (Greek, *monos*, single, and *ptilon*, feather, referring to the pappus of the original sp.)

Pappus a cup of minute scales and a single apically plumose bristle; disk-corollas densely pilose below
.... 1. *M. bellidiforme*
Pappus of 1–12 nonplumose bristles, alternating with shorter lacerate paleae; disk-corollas glabrate to sparsely pilose below .... 2. *M. bellioides*

1. **M. bellidifórme** T. & G. ex Gray. Several decumbent or ascending slender branching stems from the base, white-hirsute throughout, the plant 2–5 cm. high, 3–12 cm. across; lvs. 4–10 mm. long; invol. 4–5 mm. high; phyllaries linear, acuminate or attenuate, hirsute and microscopically glandular, herbaceous, 1-ribbed toward base; rays 12–20, ca. 5 mm. long, with pilose tube; ak. 2 mm. long; pappus of one long apically plumose bristle and several very short ± united scales, or the scales lacking, or very rarely the bristle lacking.—Rare, sandy desert, 2000–4000 ft.; Creosote Bush Scrub; n. border of Colo. Desert, Riverside Co., across Mojave Desert to Inyo Co.; Nev. (to Reno), sw. Utah, nw. Ariz. March–June.

2. **M. bellioides** (Gray) Hall. [*Eremiastrum b.* Gray. *E. Orcuttii* Wats. *E. b.* var. *O.* Cov.] Very similar to *M. bellidiforme* but often larger, the plant up to 25 cm. across, the lvs. up to 2 cm. long; invol. 4–6 mm. high; rays up to 7 mm. long; pappus of few to many unequal tawny merely scabrid bristles and as many deeply and finely lacerate white squamellae half as long.—Abundant on sandy or stony desert plains, —100–

3000 ft.; Creosote Bush Scrub; both deserts from Death V. region, Inyo Co. to San Diego and Imperial cos.; Ariz., nw. Mex. Feb.–May; Sept.

85. Townséndia Hook.

Low taprooted many-stemmed perennial or rarely annual or biennial herbs. Lvs. alternate, linear to spatulate, entire. Heads medium to large, Asterlike, with white to rosy or violet rays and a yellow disk. Invol. broad, few- to several-seriate, ± graduate; phyllaries appressed, mostly lanceolate, with green center and a narrow colored scarious usually ciliate margin. Aks. obovate or oblong, flattened, usually thick-margined, pubescent with 2-forked or glochidiate hairs. Pappus of rather numerous rigid narrow barbellate awns, often united at base, or of few awns and several squamellae, or of squamellae only, often reduced on ray-aks. Ca. 20 spp., centering in the Rocky Mt. region; 1 in Mex. (For David *Townsend*, amateur botanist of West Chester, Pa.)

(Larsen, E. L. A revision of the genus Townsendia. Ann. Mo. Bot. Gard. 14: 1–46, 1927. Beaman, J. H. Chromosome numbers, apomixis, and interspecific hybridization in the genus Townsendia. Madroño 12: 169–180, 1954.)

Aks. at or near maturity with readily deciduous pappus; plants villous-woolly with long hairs
             1. *T. condensata*

Aks. with persistent pappus; plants not villous-woolly.
  The aks. glabrous .... 2. *T. leptotes*
  The aks. pubescent with duplex bifurcate hairs .... 3. *T. scapigera*

1. **T. condensàta** D. C. Eat. in Parry. Depressed stemless perennial (or sometimes biennial) cushion-plant 2–3 cm. high, with a taproot and branched caudex, densely woolly-villous throughout with long loose hairs; lvs. narrowly to broadly spatulate, 1–1.5 cm. long, 2–4 mm. wide, with the narrow petiole longer than the blade, crowded around the large sessile or subsessile many-fld. heads; invol. 9–12 mm. high, the numerous slender subequal thin scarious-margined phyllaries ± attenuate; rays many, lavender to bluish-violet, 1–2 cm. long; pappus of ray and disk similar; $2n = 18$ (for *T. anomala* Heiser, a synonym) (Beaman, 1954).—Gravelly ridges, 10,500–11,500 ft.; Subalpine F., Alpine Fell-fields; Sweetwater and White mts., Mono Co.; to cent. Ida., Wyo., Mont., Alta. July–Aug.

2. **T. leptòtes** (Gray) Osterh. [*T. sericea* var. *l.* Gray.] Depressed stemless perennial 1.5–3 cm. high, with a taproot and often branched caudex, the herbage strigillose when young, often glabrate in age; lvs. linear-oblanceolate to spatulate-obovate, 1–2 cm. long, 1.5–3 mm. wide (larger in Rocky Mtn. material); heads solitary, sessile or nearly so; invol. 7–9 mm. high in ours, the phyllaries relatively broad, acutish or obtuse; rays blue or violet, up to 1 cm. long; aks. glabrous; ray-pappus usually much shorter than that of the disk; $2n = 18$ (Beaman, 1954).—Alpine summits, 11,500–12,000 ft.; Alpine Fell-fields; White Mts., Mono Co.; to cent. Ida., w. Mont., New Mex. July–Aug.

3. **T. scapìgera** D. C. Eat. [*T. s.* var. *caulescens* D. C. Eat.] GROUND-DAISY. Subscapose perennial from a taproot, densely cespitose, with numerous simple stems from the caudex, 2–8 cm. high, strigose-hirsute throughout; basal lvs. spatulate, 1–4 cm. long including the long slender petiole, 2–5 mm. wide; heads on scapiform 1–2-lvd. stems; invol. 8–13 mm. high, broadly campanulate to hemispheric; phyllaries 4–5-seriate, graduate, lance-oblong, acuminate; rays 8–35, pink or lavender to almost white, 6–12 mm. long, densely stipitate-glandular dorsally; aks. and pappus (of awns) of ray and disk similar.—Rare, rocky ridges and flats, 4600–11,200 ft.; Sagebrush Scrub, Pinyon-Juniper Wd., Lodgepole F., Subalpine F., Bristle-cone Pine F.; Warner Mts., Modoc Co., Sweetwater, White, and Inyo mts., Mono and Inyo cos.; Nev. May–Aug.

86. Tracyína Blake

Slender annuals with narrow alternate lvs. Heads many-fld., terminating slender ascending or erect peduncles, heterogamous, the tiny rays erect. Invol. turbinate, the phyllaries 3–4-seriate, plane, linear, acuminate or attenuate, appressed, narrowly scarious-margined. Receptacle small, naked. Ray-fls. 1-seriate, the corollas filiform, slightly reddish-tinged; disk-fls. even more inconspicuous. Aks. linear-fusiform, subterete, 5-nerved, tapering above to a short sterile beak, dilated at very tip. Pappus of persistent but very fragile capillary bristles. One sp. (Joseph P. *Tracy*, California botanist.)

1. **T. rostràta** Blake. Habit of *Rigiopappus leptocladus,* the plant 15–30 cm. high; stem solitary, simple, or occasionally with few erect branches from base, usually with 2–4 filiform ascending branches subterminally, glabrous; lvs. appressed, narrowly lance-linear, up to 2 cm. long, ciliate-margined, callous-apiculate; peduncle loosely pilose beneath the head; invol. 6–7 mm. high, the phyllaries glabrous, deciduous; ray-fls. ca. 12–20, the corollas 4–5 mm. long; disk-fls. ca. 15–25, the corollas very slender, shorter than the ray; aks. brown, 5–5.5 mm. long, hirsutulous; pappus of ca. 36–38 graduated bristles, the majority longer than and nearly concealing the disk-corollas, not quite as long as the rays.—Dry grassy slopes, 400–1000 ft.; Coastal Prairie; locally frequent in s. Humboldt Co., w. Lake Co. May–June.

87. Psiláctis Gray

Slender somewhat glandular desert annuals. Lvs. alternate, entire to pinnatifid. Heads small, in lax panicles, radiate. Invol. hemispheric, ca. 3-seriate; phyllaries graduate, with pale dry base and herbaceous tip. Ray-fls. ♀, numerous, uniseriate; ligule white to purplish. Disk-fls. fertile; corollas 5-toothed, yellow; style-branches short, with triangular appendages. Aks. nearly linear, somewhat compressed. Pappus in ray-aks. obsolete or none, in disk-aks. of fine capillary bristles shorter than the fls. Five spp. of sw. U.S. and n. Mex. (Greek, *psilos,* naked, and *aktis,* ray, the ray-ak. lacking pappus.)

1. **P. Còulteri** Gray. Divaricately branching throughout, 5–30 cm. high, stipitate-glandular throughout and usually hispidulous or pilose; lvs. oblanceolate or oblong, mostly laciniate-pinnatifid, the teeth spinescent-tipped, the lower petiolate, 2–3 cm. long, the upper closely sessile and appressed, becoming much reduced, sometimes entire; invol. ca. 4 mm. high; phyllaries lance-oblong, acute, granular-glanduliferous except on the whitish chartaceous basal portion; rays 20–35, white to lavender, 5 mm. long; aks. pubescent.—Rare; dry open sand flats, playa margins, etc., 300–3000 ft.; Creosote Bush Scrub, Alkali Sink; cent. and e. Mojave Desert from Inyo Co. s., Whipple Mts., Colo. Desert; s. Nev., s. Ariz., Son. March–June.

88. Aster L. ASTER

Summer- or fall-flowering herbs, usually rhizomatous or fibrous-rooted perennials, rarely shrubs. Lvs. alternate, entire or toothed. Heads usually numerous and radiate, in panicles, corymbs or racemes, rarely solitary. Invol. turbinate, campanulate or hemispheric; phyllaries usually imbricated and herbaceous throughout, sometimes coriaceous with green tips. Ray-fls. usually present, occasionally without ligule, ♀, shades of purple or blue, more rarely white. Disk-fls. yellow (sometimes pale) or reddish-purple. Style-branches flattened, with lanceolate to subulate appendages. Aks. ± compressed, hairy or glabrous; pappus of subequal ± numerous capillary bristles, occasionally with a few short outer bristles or scales. Some 250 or more species, centered in N. Am. but widely distributed in temp. regions. (Greek, *aster,* a star, from the radiate heads of fls.)

(Cronquist, A. W. N. Am. Spp. of Aster centering about Aster foliaceus Lindl. Am. Midl. Nat. 29; 429–468, 1943; Revision of the Oreastrum group of Aster. Leafl. West. Bot. 5: 73–82, 1948.)
A. Plants perennial.
  B. Plants herbaceous; stems leafy or subscapose, the lvs. obvious.
    C. Pappus distinctly double, the outer series of bristles very short; heads solitary; lvs. numerous, narrow, uniform. (Section Ianthe) .... 1. *A. scopulorum*
    CC. Pappus simple or occasionally ± double; habit various but not as in Ianthe.
      D. Plants with creeping rhizomes or if with woody caudex, this ± horizontal and with numerous fibrous roots, without a taproot or taprootlike caudex; style-appendages in most spp. shorter than the stigmatic portion.
        E. Rays usually 5 (4–7), white. (Section Sericocarpus) .... 2. *A. oregonensis*
        EE. Rays more numerous, or colored, or both (sometimes lacking).
          F. Creeping rhizomes lacking or poorly developed; phyllaries coriaceous toward base, tending to be keeled; lower cauline lvs. reduced, scalelike. (Section Eucephalus)
            G. Heads radiate; rays 6–15 .... 3. *A. ledophyllus*
            GG. Heads discoid; rays 0 (1–5) .... 4. *A. brickellioides*
          FF. Creeping rhizomes present.
            G. Plants with lvs. various, often toothed or petioled and if at all glandular not sessile and grasslike. (Section Aster)
              H. Invol. and peduncles glandular.

      I. Lvs., at least the lower, obovate, with clasping base; heads few, 2–4.5 cm. wide .... 5. *A. integrifolius*

      II. Lvs. linear to linear-spatulate; heads usually many, 1.5–2 cm. wide .... 6. *A. campestris*

    HH. Invol. and peduncles not glandular.

      I. Phyllaries often with purple tip or margin; disk-corollas with tube equaling or surpassing the limb

                                           7. *A. radulinus*

      II. Phyllaries sometimes purple-tipped, but not purple-margined (except in some forms of *A. foliaceus*); disk-corollas with tube distinctly shorter than the limb.

        J. Pubescence of stem in lines below the lf.-bases, commonly neither uniform under the heads nor confined to the infl. .... 8. *A. hesperius*

        JJ. Pubescence of stem uniform, or if in lines, uniform under the heads or scanty and confined to the infl.

                                           (go to K.)

K. Invol. not strongly graduated, or, if so, the phyllaries markedly acute; phyllaries acute, or, if obtuse, enlarged and foliaceous.

    L. Outer phyllaries with evident scarious margin toward base; lvs. usually toothed. Coast Ranges .... 9. *A. subspicatus*

    LL. Outer phyllaries essentially without scarious margin; lvs. usually entire.

      M. Infl. a narrow leafy panicle with usually numerous heads; rays usually pink or white. Sierra Nevada .... 10. *A. Eatonii*

      MM. Infl. few-headed, or if many-headed, shorter, more open, cymose-paniculate, often with much-reduced lvs.

        N. Middle stem lvs. mostly less than 1 cm. wide and more than 7 times as long as broad.

          O. Plant cespitose, decumbent, less than 2 dm. high; phyllaries purple-tipped and -margined .... 11. *A. foliaceus* var. *apricus*

          OO. Plant rhizomatous, mostly over 2 dm. high; phyllaries often purple-tipped but rarely purple-margined .... 13. *A. occidentalis*

        NN. Middle stem lvs. 1 cm. wide or more, mostly less than 7 times as long as broad.

          O. Lvs. entire, relatively narrow .... 11. *A. foliaceus*

          OO. Lvs. frequently toothed, the middle ones very large, thin, mostly over 2.5 cm. wide .... 12. *A. Greatai*

KK. Invol. strongly graduated, at least the outer phyllaries obtuse, markedly shorter than the inner, not foliaceous.

    L. Infl. conspicuously bracteate with small lvs.; innermost phyllaries usually obtusish; mostly cismontane.

      M. Herbage, including phyllaries, glabrate, or if pubescent the hairs mostly short or appressed; infl. an open often divergent panicle. Santa Barbara Co. n.

                                        14. *A. chilensis*

      MM. Herbage, including phyllaries, obviously cinereous with soft spreading hairs; infl. a close panicle or raceme. Los Angeles Co. s. .... 15. *A. bernardinus*

    LL. Infl. not closely beset with small uniform bracts; innermost phyllaries usually acuminate; mostly transmontane .... 16. *A. adscendens*

      GG. Plants with sessile grasslike lvs., glabrous except for the glandular infl. (Section Orthomeris) .... 17. *A. pauciflorus*

DD. Plants with taproot or erect taprootlike caudex, without rhizomes or numerous fibrous roots; style-appendages equaling or longer than the stigmatic portion; alpine or subalpine long-lived subscapose perennials. (Section Oreastrum)

    E. Stems and invols. ± woolly to glabrous, not glandular.

      F. Stems and invols. ± woolly-pubescent; plants mostly 2–30 cm. high

                                        18. *A. alpigenus*

      FF. Stems and invols. glabrous; plants 30–70 cm. high .... 19. *A. elatus*

    EE. Stems and invols. ± glandular-scaberulous .... 20. *A. Peirsonii*

BB. Plants woody toward base; stems rushlike, almost leafless, the lvs. inconspicuous at anthesis. (Section Leucosyris)

    C. Plant essentially herbaceous, not glaucous, often spinose in the lf.-axils; heads radiate, white; aks. glabrous .... 21. *A. spinosus*

    CC. Plant shrubby, glaucous, not spiny; head discoid, yellowish; aks. pubescent

                                        22. *A. intricatus*

AA. Plants annual.

  B. Rays evident, surpassing the pappus; phyllaries distinctly graduate, acuminate with a prominent scarious margin. (Section Oxytripolium) .... 23. *A. exilis*

  BB. Rays present but very inconspicuous, equaling or shorter than the mature pappus, several-seriate or more numerous than the disk-fls. (Section Conyzopsis) ...... 24. *A. frondosus*

1. A. scopulòrum Gray. [*Chrysopsis alpina* Nutt. *Diplopappus a.* Nutt. *Ionactis a.* Greene.] Stems numerous, tufted, from a branching woody caudex, erect or ascending, simple, monocephalous, 5–12 cm. high, sparsely pilose to subtomentose; lvs. crowded, sessile, entire, firm, spatulate to elliptic or lance-linear, spinulose-tipped, gray-green, scabrous, with narrow white margin, 5–15 mm. long, 1–3 mm. wide; peduncles 1–4 cm. long; invol. 7–11 mm. high, broadly turbinate; phyllaries 3–4-seriate, linear-lanceolate, glandular-puberulent, at least the inner with narrow scarious margin; rays 8–15, blue or deep violet, 7–12 mm. long; aks. silky-pilose; outer pappus as much as 1 mm. long.— Dry, rocky hillsides, 5000–10,000 ft.; Sagebrush Scrub, Pinyon-Juniper Wd.; widely scattered in Calif., transmontane, White Mts., Inyo Co., to Warner Mts., Modoc Co.; to e. Nev., e. Ore., Wyo., Mont. May–July.

2. **A. oregonénsis** (Nutt.) Cronq. [*Sericocarpus o.* Nutt. *S. rigidus* var. *o.* Ferris.] Stems from a horizontal thick caudex or an elongated woody rhizome, erect, 4–12 dm. high, leafy throughout, branching above; herbage scabrous-puberulent; lowermost lvs. somewhat reduced and soon lost, the largest ones just above, oblanceolate, obtuse or acute, subpetiolate, 4–8 cm. long, 7–10 mm. wide, the middle and upper cauline lance-elliptic or oblong, sessile; heads numerous, in close terminal clusters in a leafy cymose panicle; invol. 7–8 mm. high, turbinate-campanulate; phyllaries 3–4-seriate, graduate, slightly keeled, with white chartaceous base and short broad green often loose or spreading tip; rays usually 5 (4–7), white, 4–7 mm. long; disks ca. 13–21, ochroleucous, with purple anthers; aks. hairy; pappus unequal, with some broader flattened bristles.— Scattered in open woods, below 5000 ft.; mostly Yellow Pine F.; mts. of Humboldt, Del Norte, Shasta, and Siskiyou cos.; to Wash. July–Sept.

Ssp. **califórnicus** (Durand) Keck. [*Sericocarpus c.* Durand. *S. rigidus* var. *c.* Blake.] Stem ± densely hirsute or pilose; lvs. ± canescent, larger, to 9 cm. long and 17 mm. wide, sometimes ± undulate-margined.—In dry woods, 3500–6500 ft.; Yellow Pine F., Red Fir F.; Sierra Nevada from Plumas and Butte cos. to Eldorado Co., infrequent s. to Tulare Co. July–Sept.

3. **A. ledophýllus** (Gray) Gray. [*A. Engelmannii* var. *l.* Gray. *Eucephalus l.* Greene. *A. E.* of Calif. auth., not Gray.] Stems several, erect from a stout often horizontal woody caudex, simple below the corymbose infl., 3–6(–10) dm. high; lowermost lvs. reduced to scales, the others uniform, numerous, narrowly lanceolate to elliptic-oblong, sessile, entire or nearly so, 3–6 cm. long, 5–20 mm. wide, callous-apiculate, green and glabrate above, pilosulose on veins to ± generally cinereous-tomentose beneath; heads several on divergent bracteate often glandular and tomentulose peduncles to occasionally solitary; invol. 7–10 mm. high, broadly campanulate; phyllaries 4–5-seriate, narrow, with indurate base and elongated sharp ± herbaceous tip, ± finely glandular and ciliate, sometimes anthocyanous-margined; rays 6–15, lavender-purple, 10–15 mm. long; aks. moderately hairy.—Open woods and grassy places, 5000–7000 ft.; Montane Coniferous F.; rare in mts. of Siskiyou, Humboldt, and Trinity cos.; Cascade Mts. to n. Wash. July–Sept.

4. **A. brickellioìdes** Greene. [*Sericocarpus tomentellus* Greene, not *A. t.* Hook. *Eucephalus t.* Greene. *E. bicolor* Eastw.] Stems erect from a stout scaly branching caudex, 3–8 dm. high, ± tomentulose, leafy, often branched well down from apex; lvs. mostly spreading, linear-oblong to oval, sessile, entire, 3–6 cm. long, 5–20 mm. wide, acute or usually obtuse, apiculate, subcoriaceous, glabrous or nearly so above, tomentulose beneath, reticulate-veined; heads cymose-clustered at ends of branches, variable in size; invol. 7–10 mm. high, turbinate to campanulate, ± gray-tomentose; phyllaries 5–6-seriate, linear-oblong to lance-ovate, acute or obtusish, with indurate base, narrow fimbro-ciliate sometimes purplish margin, and ± herbaceous tip; rays none (or 1–5); aks. sparsely villous.—Dry rocky ridges and wooded slopes, 2000–4000 ft.; Douglas-Fir F., Yellow Pine F.; Siskiyou Mts., Del Norte Co.; adjacent Ore. July–Oct.

Var. **glabràtus** Greene. [*A. g.* Blake ex Peck, not Kuntze. *A. siskiyouensis* Nels. & Macbr. *Eucephalus g.* Greene. *E. glandulosus* Eastw.] Lvs. glabrous on both surfaces or sparingly pilosulose, not tomentulose.—Dry forested slopes, 5000–7000 ft.; Yellow Pine F., Red Fir F.; inner N. Coast Range, from Glenn Co. to Siskiyou Co.; s. Ore. July–Oct. Intergrading freely with the sp.

5. **A. integrifòlius** Nutt. Stems several, rather stout, ascending from a short woody densely fibrous-rooted rhizome, 2–7 dm. high, glabrate to densely villous, also glandular-pubescent above; lvs. thickish, entire, glabrate to white-pilose, the basal broadly oblanceolate or elliptic, 4–12 cm. long with slender petiole as long, the lower cauline oblanceolate or obovate, narrowed to a clasping base, the upper cauline reduced, elliptic to lance-linear; infl. bracteate, cymose or often elongate, the heads short-pedunculate; invol. 8–14 mm. high, broadly campanulate, not much imbricate; phyllaries narrowly oblong, the outer sometimes leafy, the inner much narrower and deeply anthocyanous,

densely stipitate-glandular; rays 10–21, purple or violet, 10–15 mm. long; aks. short-pilose.—Dry meadow borders and open woods, 5500–10,500 ft.; Red Fir F. to Sub-alpine F.; Sierra Nevada, from Tulare Co. to Plumas Co.; n. to Wash., e. to Rocky Mts. July–Sept.

6. **A. campéstris** Nutt. var. **Bloómeri** (Gray) Gray. [*A. B.* Gray. *A. argillicola* Peck.] Stems slender, erect from slender rhizomes, 1–3 dm. high, simple or much branched, the herbage ± densely hispidulous with short spreading hairs from pustulate bases, also ± stipitate-glandular toward heads; lvs. linear or linear-spatulate, obtuse to acuminate, entire, firm, 1–5 cm. long, 2–7 mm. wide, rather crowded; heads usually numerous, sometimes solitary, small; invol. 5–6 mm. high, glandular; phyllaries scarcely to evidently imbricate, broadly linear, acuminate, with green midnerve and herbaceous tip; rays 13–25, violet or purplish, 5–10 mm. long.—Dry meadowy slopes and open timber, 6000–8000 ft.; N. Juniper Wd., Yellow Pine F.; e. side of the Sierra Nevada, Lake Tahoe region n.; to s. Wash., Nev. July–Sept.

7. **A. radulìnus** Gray. [*A. Torreyi* Porter.] BROAD-LEAF ASTER. Stems 1 or more from a rather stout woody rhizome, 2–6 dm. high; herbage sparsely to densely canescent, hirtellous, or hirsute, or sometimes glabrate below, not glandular; lvs. firm, mostly sharply dentate or serrate, the lowest broadly oblanceolate to obovate, narrowed to a margined ciliate petiole, usually reduced, the lower cauline the largest, obovate to oval or oblanceolate, 4–10 cm. long, 2–5 cm. wide, sessile, the upper cauline gradually reduced and relatively wide, becoming entire clasping bracts; heads usually numerous in a short flat-topped infl., the pedicels densely hirsutulous or crisped-pilose; invol. 6–9 mm. high; phyllaries 5–7-seriate, strongly graduate, linear-oblong, acute, ciliate and hirsutulous, with pale indurate base, prominent midrib and short herbaceous tip, often the inner purple-tipped or -margined; rays 10–15, white to pale violet, 6–12 mm. long.—Dry forest floors and wooded slopes, 300–5000 ft.; Foothill Wd., Mixed Evergreen F., Douglas-Fir F., Yellow Pine F.; Sierra Nevada from Mariposa Co. n., Coast Ranges from San Luis Obispo Co. n., Santa Cruz Id.; w. of the Cascades of Vancouver Id. July–Oct.

8. **A. hespérius** Gray. [*A. coerulescens* of auths., not DC. *A. ensatus* Greene. *A. foliaceus* var. *h.* Jeps.] Stem from a creeping rhizome, 1–2 m. high, with numerous ascending branches, pubescent strictly in lines; lvs. numerous, lance-linear, entire or the lower serrate, thickish, scabro-ciliolate, 5–15 cm. long, 3–15 mm. wide; infl. ample and leafy, with many heads in narrow or spreading panicles; invol. 5–8 mm. high, broadly turbinate; phyllaries 4–5-seriate, irregularly imbricate, linear, acuminate, ciliolate, erect or loose, the green tip extending as a midnerve well into the indurate lower portion; rays 21–35, white or pink to bluish-purple, 6–12 mm. long.—Stream banks and marshy meadows, up to 5500 ft.; S. Oak Wd., Chaparral, Sagebrush Scrub; cismontane s. Calif., n. through Owens V. to Mono Co.; to Alta., Wis., Mo., Tex. Aug.–Oct.

9. **A. subspicàtus** Nees. [*A. Douglasii* Lindl. in Hook.] Stems from a fibrous-rooted caudex, 3–10 dm. high, with erect or ascending branches, pubescent in lines; lvs. lanceolate to oblanceolate, acute or acuminate, scabrid-ciliolate, often ± serrate, the lower narrowed to a winged clasping petiole, 5–15 cm. long, 5–25 mm. wide; heads few to many, in a leafy cymose panicle; invol. 6–7 mm. high; phyllaries subequal to ± graduate, rather loose, linear, acute or the outer spatulate, with obvious green tip and usually conspicuous yellowish thickened basal portion with evidently scarious and fimbriolate margin; rays 20–30, violet; pappus often pinkish; aks. sparsely hairy.—Moist places, often in wds., near the coast, up to 5000 ft.; several Plant Communities; Coast Ranges, Monterey Co. to Del Norte and Siskiyou cos.; n. to coastal Alaska, less common e. to Mont. July–Sept. One of the most variable spp.

10. **A. Eatònii** (Gray) Howell. [*A. foliaceus* var. *E.* Gray. *A. oreganus* of auths., not Nutt.] Stems from a stout creeping rhizome, 4–10 dm. high, with numerous erect or ascending branches, minutely and ± uniformly pubescent below the infl., if in lines, usually with scattered hairs between; lvs. lance-linear, numerous, usually entire, scabrous-margined, the lower narrowed to a petiole, early deciduous, the middle and upper cauline sessile with a broad base, 5–15 cm. long, 4–18 mm. wide; heads numerous, in a narrow leafy panicle, crowded toward tips of the branches; invol. 5–7.5 mm. high; phyllaries subequal or loosely graduate, sometimes subtended by foliose bracts, thin, oblanceolate, acute, ciliolate but otherwise glabrous, the herbaceous tip exceeding the pale base; rays

20–35, lavender or violet, 5–10 mm. long.—Wet often shady places along streams, etc., 2000–6000 ft.; Foothill Wd., Sagebrush Scrub, Yellow Pine F.; Cascade-Sierran axis, chiefly e. side, from Siskiyou Co. to Mono Co., White Mts., Inyo Co. (8000 ft.); to B.C., Alta., New Mex. July–Sept.

11. **A. foliàceus** Lindl. Stems erect from a creeping rhizome or caudex, 2–9 dm. high, glabrate to pubescent especially above; lvs. glabrous to soft-pubescent, mostly entire, the lower oblanceolate to obovate, usually petiolate, often early deciduous, the middle ones sessile or clasping, 5–12 cm. long, 1–4 cm. wide; heads usually several, medium-sized to large, in a cymose panicle; phyllaries subequal, imbricate, white-margined at base, ciliolate, the outer sometimes foliaceous and conspicuous, oblong, obtuse or broadly acute; rays 15–50, rose-purple to blue or violet, 1–2 cm. long.— The typical var., marked by solitary heads and very leafy invol., occurs from Alaska s. to cent. Wash. Several other vars. can be distinguished in this variable sp., which also hybridizes with its relatives. The vars. that occur in Calif. may be keyed as follows:

Middle and upper stem lvs. strongly auriculate-clasping; outer foliaceous phyllaries conspicuous; upper part of plant tending to be soft hairy. Siskiyou Co. . . . . var. *Lyallii*
Middle and upper lvs. not strongly auriculate-clasping; outer foliaceous phyllaries not conspicuous; upper part of plant inconspicuously pubescent.
  Plants 1.5–2.5 dm. high, often monocephalus (or 2–6 heads); subalpine or alpine. N. Calif.
                                                                 var. *apricus*
  Plants usually over 4 dm. high; heads few to several; not subalpine or alpine. Sierra Nevada
                                                                 var. *Parryi*

Var. **Lyállii** (Gray) Cronq. [*A. Cusickii* var. *L.* Gray.] Stems 3–6 dm. high, frequently pubescent with spreading hairs; lower lvs. sometimes deciduous before flowering, the middle and upper cauline ones strongly auriculate-clasping; phyllaries often puberulent on the back, the outer foliaceous, when present, linear-lanceolate, acuminate.—Damp meadows, 5000–6000 ft.; Montane Coniferous F.; Marble and Siskiyou mts., Siskiyou Co.; to s. B.C., Ida., Mont. July–Aug.

Var. **aprìcus** Gray. [*A. a.* Rydb.] Stems 2.5 dm. or less high, cespitose, decumbent, pubescent above; lvs. persistent to base of stem, small, the middle cauline 3–8 cm. long, 5–12 mm. wide, scabrous-ciliate, often slightly pubescent on upper surface; heads usually solitary, sometimes few and subcymose; phyllaries glabrous or nearly so on backs, obviously ciliate, puberulent within.—Subalpine or alpine ridges and meadows, Warner Mts., Modoc Co.; to s. B.C., Mont., Colo. Aug.

Var. **Párryi** (D. C. Eat.) Gray. [*A. adscendens* var. *P.* D. C. Eat. *A. f.* var. *frondeus* Gray.] Stems mostly 3–10 dm. high, with rather large thin lvs., the basal ones usually persistent; heads several or many, rarely 1; phyllaries as in *apricus*.—Moist wd. or meadows, 5000–8000 ft.; Montane Coniferous F.; Sierra Nevada, from Tuolumne Co. n.; to Wash., Rocky Mts., Alta. to New Mex. July–Aug.

12. **A. Greàtai** Parish. Stems from a creeping rhizome, 5–12 dm. high, leafy, sparsely but rather uniformly coarse-pubescent above, often glabrous below; lvs. ample, thin, entire or sometimes serrulate above, ± pilosulous, especially on under surface; basal lvs. oblanceolate, petioled, lost before flowering, the lower cauline lance-elliptic, clasping, acuminate, 6–15 cm. long, 2–4 cm. wide, those of the infl. much reduced; heads medium-small, few to many in a cymose panicle; invol. 5.5–6.5 mm. high; phyllaries subequal to graduate, linear-lanceolate, ciliolate, with slightly enlarged acuminate apiculate green often loose tip, sometimes hispidulous on back; rays 25–50, light purple, 8–12 mm. long.—Moist or dry sites in canyons, 2000–4000 ft.; Chaparral, S. Oak Wd.; s. face of San Gabriel Mts., Los Angeles and San Bernardino cos. Aug.–Oct.

13. **A. occidentális** (Nutt.) T. & G. [*Tripolium o.* Nutt. *A. Fremontii* var. *Parishii* Gray. *A. delectabilis* Hall. *A. adscendens* var. *d.* Jeps.] Stems from a slender branching caudex, 2–4 or 5 dm. high, often reddish, glabrous below, scantily pubescent above, often in lines, but uniformly toward summit of peduncles, occasionally pubescent throughout; lvs. entire, persistent, linear-oblanceolate, usually less than 10 cm. long, the basal with narrowly winged ciliate petiole, the middle cauline 3–10 mm. wide, sessile; heads 1–several, in a nearly naked cyme or cymose panicle; invol. 5–7 mm. high, slightly if at all graduate; phyllaries linear, acute or obtuse, often purple-tipped but seldom purple-margined, glabrous on the backs, ciliolate; rays 25–35, lavender or violet, 7–10 mm. long; $n = 16$ (Clausen et al. 1940).—Moist meadows, lake margins, etc., 4000–10,500

ft.; Montane Coniferous F.; throughout the Sierra Nevada, Siskiyou Co. to Tulare Co., White, Inyo, San Bernardino, and San Jacinto mts.; L. Calif., n. to Alaska, e. to Rocky Mts. July–Sept.

Var. **intermèdius** Gray. Mostly 4–8 dm. high; basal lvs. sometimes enlarged, entire or serrulate, often withering early, upper lvs. much reduced, rather numerous and bractlike; heads comparatively numerous and small in an open cymose panicle; invol. ca. 5 mm. high.—Occasional, mid-altitudes; Montane Coniferous F.; Sierra Nevada, Tulare Co. to Siskiyou Co., Del Norte Co.; n. to s. B.C. and n. Ida. July–Sept.

Var. **yosemitànus** (Gray) Cronq. [*A. adscendens* var. *y.* Gray. *A. y.* Greene. *A. Copelandii* Greene. *A. paludicola* Piper.] Stems slender, leafy throughout; lvs. mostly linear, usually less than 5 mm. wide, the lower not appreciably wider than the upper, gradually reduced upwards; heads solitary or several, on erect peduncles.—Not very common; meadows, 4000–7000 ft.; Yellow Pine F., Red Fir F.; Sierra Nevada, Tulare Co. to Siskiyou Co.; s. Ore., Nev. July–Sept.

14. **A. chilénsis** Nees. [*A. Menziesii* Lindl. in Hook. *Diplopappus incanus* Lindl. *A. Chamissonis* Gray ex Torr. *A. militaris* Greene.] Stems erect or ascending, 4–10 dm. high, paniculately branching above, uniformly pubescent at summit of peduncles, often ± in lines below; lvs. usually entire, sometimes serrulate, with scabrous margin and sometimes rough above, glabrous beneath, the lower oblanceolate or broader, narrowed to a winged petiole, 8–12 cm. long, 1–2.5 cm. wide, the mid-cauline linear-lanceolate, sessile, the upper crowded, becoming bractlike; infl. mostly paniculate; heads small to medium-sized, usually numerous; invol. 5–7 mm. high, 4–5-seriate, strongly graduate; phyllaries ciliolate, glabrous on back, green-tipped, narrowly oblong, often slightly broadened above, the outer mostly obtuse or rounded, the innermost obtuse or abruptly acute; rays 20–35, violet or purplish to whitish, 6–12 mm. long.—Dry banks, grassy fields, etc., sea level to 4500 ft.; many Plant Communities; mostly outer Coast Ranges, s. to Santa Barbara Co. and Santa Rosa Id.; sw. Ore. June–Oct. Nees erroneously thought the plant came from Chile, hence the name. Similarly glabrate plants with more numerous smaller lvs. and more divaricately branched infl., occurring in the Sierran foothills from Eldorado Co. to Tuolumne Co. and in the Cent. V. to the w., are referable to the weakly defined var. *medius* Jeps.

Var. **invenùstus** (Greene) Jeps. [*A. i.* Greene.] Similar to var. *chilensis* but the herbage obviously canescent to short-hirsute.—In moist ground, up to 4500 ft.; Mixed Evergreen F., Foothill Wd.; Siskiyou Co. s. through the inner Coast Ranges to Santa Clara Co., through the Sierran foothills to Tulare Co. July–Nov.

Var. **léntus** (Greene) Jeps. [*A. l.* Greene.] Glabrous or nearly so, slightly succulent, up to 2 m. high; lvs. linear-lanceolate; infl. ample, widely branching, the branches conspicuously bracteate; heads large.—Coastal Salt Marsh; marshes around Suisun Bay, Solano Co.

Var. **sononénsis** (Greene) Jeps. [*A. s.* Greene.] Only 3 or 4 dm. high, glabrous, glaucescent; basal lvs. rather few, the cauline reduced, narrowly lanceolate; infl. open-cymose, the heads few and solitary at the tips of the bracteate peduncles.—Coastal Salt Marsh; saline ground around San Francisco Bay, Sonoma, Napa, and Santa Clara cos.

15. **A. bernardínus** Hall. [*A. deserticola* Macbr. *A. chilensis* ssp. *adscendens* var. *b.* Cronq. Possibly *A. defoliatus* Parish.] Stems erect from a woody rhizome, 4–12 dm. high, very leafy with fascicles developing in the axils; herbage densely pubescent with soft straight hairs; primary lvs. linear or narrowly lanceolate, the lower ones promptly deciduous, 3–7 cm. long, 3–6 mm. wide; heads numerous, medium-sized, terminating densely bracteate twigs in racemose panicles; invol. 5.5–6.5 mm. high; phyllaries pubescent on back and ciliate, the outer usually very obtuse or rounded.—Damp meadows, 100–3500 ft.; Freshwater Marsh, Coastal Sage Scrub, S. Oak Wd.; Los Angeles Co. to San Bernardino V. and Orange Co., w. border of Mojave Desert, mts. of San Diego Co. July–Nov.

16. **A. adscéndens** Lindl. in Hook. [*A. chilensis* ssp. *a.* Cronq.] Stems slender, 2–7(–10) dm. high, with erect or ascending branches above, uniformly pubescent at summit of peduncles, often in lines and becoming glabrous below; lvs. usually entire, with scabrous margin, otherwise glabrous to ± pubescent, the lower narrowly oblanceolate, the middle cauline lanceolate to linear, amplexicaul, mostly less than 1 cm. wide; heads few to many, in a nearly naked cyme or cymose panicle; invol. 4–7 mm. high,

strongly graduate; phyllaries ciliolate, usually glabrous on back, green-tipped, linear or linear-oblong, the innermost acuminate or sharply acute; rays 20–35, violet or purplish, 6–10 mm. long; $n = 8$ (Clausen et al. 1940).—Moist or dry soil of meadows, etc., 200–7500 ft.; N. Juniper Wd., Foothill Wd. up to Red Fir F.; Modoc and Siskiyou cos. s. to Tulare and Inyo cos., mostly e. of the crest, San Gabriel, San Bernardino, and San Jacinto mts.; to New Mex., Sask., Alta., se. Wash. July–Oct.

17. **A. pauciflòrus** Nutt. Stems erect from creeping rootstocks, wiry, simple or freely branched, 3–12 dm. high; herbage glaucescent, glabrous below the glandular-atomiferous infl., ± fleshy; lvs. linear or nearly so, acuminate, sessile, entire, 5–10 cm. long, 3–6 mm. wide, reduced above and bractlike within the open corymbiform infl.; heads small, solitary at the ends of the branchlets; invol. 5–7 mm. high, very viscid; phyllaries linear-lanceolate, thick, mostly herbaceous, rather loose, 2–3-seriate, obscurely graduate; rays 5–8 mm. long, lavender to whitish.—Moist saline soils, 800–2300 ft.; Alkali Sink; Amargosa Desert region, se. Inyo Co.; to Ariz., Tex., Sask. June–Oct.

18. **A. alpígenus** (T. & G.) Gray ssp. **Andersònii** (Gray) Onno. [*Erigeron A.* Gray. *A. A.* Gray. *Oreastrum A.* Greene. *Oreostemma A.* Greene.] Subscapose herb from a short stout simple or somewhat branched caudex surmounting a fleshy taproot, 5–40 cm. high, very thinly tomentulose below, more densely and permanently so toward heads; basal lvs. tufted, grasslike, persistent, linear or oblance-linear, tapering to base and apex, entire, 4–15(–25) cm. long, 2–10 mm. wide; cauline lvs. few, much reduced, sessile by a broad base; heads solitary, showy; invol. broadly hemispheric, 2–3-seriate, 6–10 mm. high; phyllaries subequal or ± graduate, the inner with subscarious often purplish margin; rays 7–15 mm. long, purple, pilosulous dorsally; aks. pilose to base.— Moist or boggy meadows, 4000–11,500 ft.; Montane Coniferous F., Alpine Fell-fields; Siskiyou Co. to Humboldt Co., s. through the Sierra Nevada to Tulare Co., Sweetwater, White, and San Jacinto mts.; adjacent Nev., sw. Ore. June–Sept. Typical *alpigenus* occurs from cent. Ore. to Wash.

19. **A. elàtus** (Greene) Cronq. [*Oreastrum e.* Greene. *Oreostemma e.* Greene.] Very similar to *A. alpigenus* ssp. *Andersonii*, but in many respects larger and coarser; stem 3–7 dm. high; herbage glabrous throughout or ± tomentulose; basal lvs. 15–35 cm. long, 5–10 mm. wide; invol. 9–14 mm. high, the phyllaries very firm, the yellow base coriaceous; aks. glabrous to hirsute throughout.—Meadows, 5000–7000 ft.; Red Fir F., Lodgepole Pine F.; Sierra Nevada, Lassen Co. to Sierra Co.; Josephine Co., Ore. June– Aug. Further study is needed to determine whether this can stand as a sp.

20. **A. Peirsònii** C. W. Sharsm. Dwarf subscapose herb with an erect black branching caudex on a taproot; stems 1 to a rosette, 2–7 cm. high, monocephalous, like the invol. and often also the lvs. ± densely glandular-puberulent, not otherwise pubescent; basal lvs. tufted, persistent, linear, ± involute, sharply acute, 2–5 cm. long, 1–3 mm. wide, 3-nerved; cauline lvs. few and reduced; invol. purplish, turbinate to hemispheric, 7–11 mm. high; phyllaries imbricate, linear-lanceolate, herbaceous, often with mucronate tip; rays 8–20, purple or violet, 12–16 mm. long; aks. glabrous or scantily pubescent. —Subalpine and alpine meadows and moist gravelly slopes, 11,000–12,250 ft.; Sub-alpine F., Alpine Fell-fields; Sierra Nevada of Tulare, Inyo and s. Fresno cos. July– Sept.

21. **A. spinòsus** Benth. [*Leucosyris s.* Greene.] MEXICAN DEVIL-WEED. Broomlike clumps of much-branched almost leafless erect stems from a deep-seated woody rhizome, 6–28 dm. high, glabrous, with axillary or supra-axillary subulate thorns up to 1.5 cm. long on lower parts; lvs. thickish, entire or remotely and obscurely toothed, the basal linear to linear-spatulate, 2–4 cm. long, 2–5 mm. wide, the cauline rapidly reduced to minute remote scales; heads small, solitary at ends of the branches, a few sometimes axillary; invol. 4–6 mm. high, hemispheric, essentially glabrous; phyllaries 3–5-seriate, imbricate, lanceolate, mostly acuminate with merely acute tip, herbaceous with prominent scarious margin; rays 15–30, white, scarcely exceeding the disk; aks. glabrous.—Troublesome weed in low subalkaline places, below 500 ft.; Coastal Sage Scrub, Alkali Sink, Creosote Bush Scrub; s. San Diego Co., Imperial and Palo Verde valleys, Needles; adjacent Nev. and L. Calif. to Tex., La., Mex., Cent. Am. June–Dec.

22. **A. intricàtus** (Gray) Blake. [*Bigelovia i.* Gray. *Linosyris ? carnosa* Gray. *A. c.* Gray ex Hemsl., not Gilib. *Leucosyris c.* Greene.] Rounded bush 5–9 dm. high, the slender glaucescent almost leafless stems rigidly and divaricately much-branched, woody

below, essentially glabrous; lower lvs. linear, fleshy, 1–2 cm. long, 1–2 mm. wide, those of the branches reduced to small subulate appressed scales; heads numerous, small, few-fld., discoid, yellowish, solitary at tips of the branches; invol. 5.5–6.5 mm. high, turbinate, glabrous; phyllaries 4–5-seriate, imbricate, lanceolate (the inner becoming linear), acute, chartaceous with greenish midline and narrow scarious margin; aks. slender, striate, ± sericeous; pappus copious, reddish.—Occasional, silty saline soils of meadow borders and seeps, below 3500 ft.; Alkali Sink, Coastal Sage Scrub, Creosote Bush Scrub; Mono Co., s. through the Mojave Desert to Los Angeles and San Bernardino cos., San Joaquin V. from Madera Co. to Kern Co., Colton; s. Nev., w. and s. Ariz. June–Oct.

23. **A. éxilis** Ell. Slender erect glabrous annual 3–15 dm. high, openly much branched above; lvs. narrowly linear to oblanceolate, 5–12 cm. long, 2–10 mm. wide, entire or nearly so, fleshy, 1-nerved, the uppermost more numerous and reduced to subulate bracts; panicle diffuse, rather narrow; heads numerous, small; invol. 4–6 mm. high, turbinate, 3–4-seriate; phyllaries lance-linear, acuminate or attenuate, herbaceous, with translucent midnerve and hyaline margin; rays light pink to purple, numerous, scarcely exceeding the rather scanty pappus.—Common in wet, often alkaline places, mostly below 500 ft.; Coastal Salt Marsh, V. Grassland; Cent. V. and San Francisco Bay region to coastal s. Calif.; e. to Fla. and S. Car., s. to Mex., S. Am. July–Oct.

24. **A. frondòsus** (Nutt.) T. & G. [*Tripolium f.* Nutt. *Brachyactis f.* Gray.] Annual, erect or decumbent, 1–5(–10) dm. high, nearly glabrous; lvs. oblanceolate with margined petiole to linear and sessile, 2–6(–10) cm. long, 2–10 mm. wide, ciliolate in lower half; heads usually very numerous in spicately paniculate infls.; invol. 5–8 mm. high; outer phyllaries linear-oblong, obtuse, ciliolate, the inner with subscarious margin below; rays very numerous, in several ranks, pinkish, filiform, ca. 2 mm. long; pappus copious, surpassing the disk-corollas.—Not common in Calif., moist alkaline flats, marshes and lake borders, 2600–7000 ft.; Alkali Sink, Sagebrush Scrub, etc.; Shasta V., Siskiyou Co. and e. side of Sierra Nevada s. to Inyo Co., Death V., s. Calif. (Baldwin L., Elsinore, Arrastre Ck.); to e. Wash., Wyo., New Mex. May–Oct.

89. Leucelène Greene

Low perennial herb, with numerous tufted stems from a buried cordlike rootstock. Lvs. alternate, linear or subulate, or the lower narrowly spatulate. Heads solitary at ends of the branches, radiate. Invol. turbinate; phyllaries ca. 3-seriate, imbricate, herbaceous with narrow scarious even margin. Ray-fls. ♀, fertile; ligules white, often turning reddish. Disk-fls. perfect, fertile, yellowish. Style-branches with obtuse appendages shorter than the stigmatic portion. Aks. linear. Pappus of long subequal bristles. Monotypic. (Etymology unknown.)

(Shinners, L. H. Revision of the genus Leucelene Greene. Wrightia 1: 82–89, 1946.)

1. **L. ericoìdes** (Torr.) Greene. [*Inula ? e.* Torr. *Diplopappus e.* var. *hirtellus* Gray. *Aster ericaefolius* Rothr., not *A. ericoides* L. *A. bellus, Leucelene* and *hirtifolius* Blake.] Tufted, heathlike, the numerous slender very leafy crowded stems from the cespitose caudex 5–15 cm. high; herbage hoary-strigose, ± glandular; lvs. hispid-ciliate, bristle-tipped, 4–12 mm. long; invol. 5–6 mm. high; phyllaries strigulose, slightly glandular-puberulent, lance-oblong, acuminate, green, thickened, with white fimbrillate scarious margin; rays 12–21, 5 mm. long.—Dry rocky slopes, 4400–9400 ft.; Joshua Tree Wd., Pinyon-Juniper Wd., Bristle-cone Pine F.; e. face of Sierra Nevada, Mono Co., Inyo Mts., Inyo Co., Clark, New York, and Providence mts., San Bernardino Co.; e. to Nebr., Tex., Mex. April–Aug. Variable in pubescence, some plants being more densely strigose and nonglandular (*Aster bellus*); others from the same locality less hairy but glandular (*A. hirtifolius*). This genus lies between *Aster* and *Erigeron*.

90. Machaeránthera Nees

Annual, biennial or perennial herbs or shrubs from a distinct taproot that is often surmounted by a branching caudex. Lvs. alternate, spinulose-tipped, entire or more often spinulose-dentate to pinnatifid or pinnately dissected. Heads solitary to numerous, small to large. Invol. several-seriate; phyllaries narrow or broad, chartaceous or coriaceous toward base, herbaceous (or at least greenish) toward tip, appressed to loose or squarrose. Ray-fls. usually present, ♀, fertile, shades of blue or purple to white. Style-branches flattened, with narrowly acute to obtuse hairy appendage longer or shorter than the stigmatic portion. Aks. ± compressed, mostly several-nerved, hairy or glabrous; pappus of ± numerous markedly unequal barbellate often brownish bristles. Ca. 25–30 spp. of temp. w. N. Am. (Greek, *machaira*, sword, and *anthera*, anther, from the lanciform anther-tips, which, however, do not distinguish this genus from its close relatives.)

(Cronquist, A., and D. D. Keck. A reconstitution of the genus Machaeranthera. Brittonia 9: 231–239, 1957.)
A. Plants herbaceous; heads usually several or numerous, small to medium-sized; biennials or short-lived perennials. (Section Machaeranthera)
  B. Phyllaries narrowly linear, not glandular, the long narrow loose herbaceous tips longer than the indurated base .... 1. *M. tephrodes*
  BB. Phyllaries not narrowly linear, the erect or squarrose tips shorter than the indurated base.
    C. Stems viscid-glandular to base; cauline lvs. ovate, mostly reduced to small bracts
                                                                         2. *M. leucanthemifolia*
    CC. Stems often glandular above but not to base; cauline lvs. mostly better developed and more elongate.
      D. Invol. 12–15 mm. high; phyllaries 6–10-seriate, their short broad canescent tips not squarrose; herbage and invol. not noticeably glandular .... 3. *M. lagunensis*
      DD. Invol. 6–10 mm. high; phyllaries often squarrose; herbage very often stipitate-glandular above.
        E. Phyllaries mostly 1–1.6 mm. wide, 3–4(–5)-seriate, somewhat or not at all squarrose; usually low perennials .... 4. *M. shastensis*
        EE. Phyllaries up to ca. 1 mm. wide, 4–8 seriate, strongly squarrose; usually coarse biennials .... 5. *M. canescens*
AA. Plants shrubby or woody at base; heads usually solitary at branch-tips, large, the disk mostly 2–4 cm. wide, the invol. 13–19 mm. high. (Section Xylorhiza)
  B. Stem glandular and hispid, often also tomentose; lvs. chiefly lanceolate or lance-linear, pubescent or rarely merely ciliate; phyllaries glandular-hirtellous or pilose; plants scarcely shrubby .... 6. *M. tortifolia*
  BB. Stems glabrous or slightly glandular toward apex; lvs. chiefly oblong, glabrous or sparsely villous-ciliate; phyllaries essentially glabrous dorsally, usually stipitate-glandular on margin.
    C. Outer phyllaries linear-attenuate, equaling or exceeding the inner; pedicels glandular above .... 7. *M. cognata*
    CC. Outer phyllaries lanceolate, acuminate, much shorter than the inner; pedicels glabrous
                                                                                  8. *M. Orcuttii*

1. **M. tephròdes** (Gray) Greene. [*Aster canescens* var. *t.* Gray. *A. t.* Blake.] Erect biennial, often paniculately branched, 2–7 dm. high; herbage cinereous-puberulent or canescent, with occasional gland-tipped hairs; lvs. entire to moderately spinose-dentate, the lower oblanceolate, winged-petioled, 4–10 cm. long, 4–12 mm. wide, the middle and upper nearly linear, reduced, sessile; heads solitary at ends of cymose-paniculate branches; invol. 8–10 mm. high, hemispheric; phyllaries 5–7-seriate, graduate, linear, with whitish chartaceous base and a somewhat longer herbaceous attenuate ± mucronate soft and spreading or squarrose tip, ciliate, finely canescent; rays 25–40, lavender to violet, 10–15 mm. long; aks. linear-fusiform, ± flattened, striate, velutinous.—Low ground, river bottoms, etc., 100–2000 ft.; Coastal Sage Scrub, Creosote Bush Scrub; Mojave and Colo deserts and borders, e. San Bernardino Co. to San Diego and Imperial cos.; to s. Utah, Tex., Son., L. Calif. April–May; Sept.–Dec.

2. **M. leucanthemifòlia** (Gřeene) Greene. [*Aster l.* Greene.] Short-lived pallid perennial from a deep woody taproot; stems few, erect, rather intricately and rigidly divaricately branching above, 2–4 dm. high, densely cinereous-puberulent and beset throughout with stout- but short-stalked yellow glands; lvs. thick, grayish with a dense almost feltlike fine canescence, ± glandular, prominently spinulose-serrate-dentate, the basal rosulate, spatulate or elliptic, 8–12 mm. wide, 2–3 cm. long including the broadly emarginate petiolar portion, often deciduous before anthesis, the cauline amplexicaul, rapidly reduced, the lower oblong, the upper ovate bracts only 3–4 mm. long but remotely spinulose-dentate and -tipped; heads terminating the branchlets, inconspicuous; invol. 6–7 mm. high, broadly campanulate, canescent and glandular-viscid; phyllaries ca. 5-seriate, imbricate, broadly linear, acute to acuminate, mucronate-tipped, the white base longer than the erect or spreading rarely squarrose herbaceous tip; rays 8–15, pale purple, 6–8 mm. long, fertile; aks. striate, pubescent, ca. 3.2 mm. long; pappus-bristles very slender, soft, brownish-white.—In sand dunes, 2200 ft.; Creosote Bush Scrub; Kelso, Mojave Desert, San Bernardino Co.; sw. Nev. May–June.

3. **M. lagunénsis** Keck. Rather stout-stemmed perennial from a woody root, 4–7 dm. high, corymbosely branching above; herbage ± finely cinereous-puberulent throughout, not glandular; lvs. linear-oblong to linear, firm, remotely spinose-dentate or entire, the lower up to 8 cm. long and 8 mm. wide including the emarginate petiole, the upper narrow, reduced, sessile by a broad base; heads large, several to numerous, solitary at tips of the branches; invol. 12–15 mm. high, broadly campanulate to hemispheric; phyllaries 6–10-seriate, conspicuously imbricate, lance-oblong, the larger up to 2 mm. wide, with chartaceous base and shorter green triangular or short-acuminate appressed or somewhat spreading tip, cinereous-puberulent but scarcely at all glandular; rays 13–30, deep lavender, 10–17 mm. long, ca. 2 mm. wide; aks. pubescent; pappus copious, reddish.—Dry slopes under pines, 5000–7000 ft.; Yellow Pine F.; Laguna Mts., San Diego Co., and in somewhat smaller form, Big Bear Lake, San Bernardino Mts. July–Aug.

4. **M. shasténsis** Gray. [*Aster s.* Gray.] Biennial or short-lived perennial from a woody taproot topped by a slender branched caudex; stems cespitose, ascending or decumbent, usually branching above, 1–4 dm. high; herbage finely and often densely canescent-puberulent with crisped hairs, glandular above; lvs. entire or few-toothed, the basal oblanceolate to spatulate, 3–6 cm. long including petiole, 3–8 mm. wide, the cauline gradually reduced; heads solitary to numerous in panicles or cymes; invol. 6.5–10 mm. high, campanulate to hemispheric, canescent but rarely glandular; phyllaries 4–5-seriate, conspicuously imbricate, broadly linear to oblong, acuminate or acute, often purplish, the indurate base longer than the erect or spreading rarely regularly squarrose herbaceous tip; rays 8–15, violet, 6–8 mm. long, sterile, their pappus greatly reduced or obsolete.—Dry open gravelly places, 3500–6200 ft.; Chaparral, Yellow Pine F., Red Fir F.; Mt. Shasta n. in the Cascades to cent. Ore. July–Sept.

KEY TO VARIETIES

Rays present.
  Phyllaries 4–5-seriate, canescent but sparingly glandular.
    Rays fertile, with well developed pappus .... var. *glossophylla*
    Rays sterile, their pappus reduced or obsolete .... *M. shastensis*
  Phyllaries ca. 3-seriate, ± canescent but abundantly glandular .... var. *montana*
Rays wanting .... var. *eradiata*

Var. **glossophýlla** (Piper) Cronq. & Keck. [*Aster g.* Piper. *A. s.* var. *g.* Cronq.] Stems erect or stiffly ascending, scarcely cespitose, 1–5 dm. high; herbage canescent-puberulent, sparingly glandular only above; lvs. narrow, linear-oblanceolate, sometimes toothed, the basal usually gone before anthesis; heads radiate, the rays often few but mostly fertile; phyllaries 4–5-seriate, puberulent but sparingly glandular, the inner erect, purplish, the outer green-tipped, scarcely squarrose.—Sandy plains and montane slopes, 4000–7000 ft.; Sagebrush Scrub, N. Juniper Wd., Yellow Pine F., Red Fir F.; Cascade-Sierran axis and e., Eldorado Co. to Siskiyou and Modoc cos.; cent. and e. Ore., Nev. July–Oct. Variable, tending to intergrade with *A. canescens*.

Var. **montàna** (Greene) Cronq. & Keck. [*M. m.* Greene.] Cespitose low perennial from a woody taproot beneath a stout branched caudex, the many decumbent to ascending stems mostly 5–20 cm. high; herbage densely to moderately canescent-puberulent (or glabrate), conspicuously glandular toward the heads; lvs. spatulate, often deeply incised-dentate, 2–4 cm. long, 3–8 mm. wide, crowded below; heads radiate; phyllaries usually 3-seriate, firm, glandular, often squarrose; rays radiate.—Dry gravelly timbered or open slopes, 9500–11,500 ft.; Subalpine F., Bristle-cone Pine F.; Sierra Nevada, Sweetwater, and White mts., Tuolumne, Mono, and Inyo cos.; high mts. e. across Nev. July–Aug.

Var. **eradiàta** (Gray) Cronq. & Keck. [*Aster shastensis* var. *e.* Gray. *M. e.* Howell.] Similar in habit to var. *glossophylla;* stems few, 1.5–3 dm. high, ascending, from a woody root or small caudex; lvs. oblong-spatulate, occasionally toothed; phyllaries variable in vesture, usually prominently glandular, the crisped puberulence moderate to lacking, the tips erect or squarrose; rays none.—Light dry soils of timbered slopes and ridges, up to 7000 ft.; Sagebrush Scrub, Yellow Pine F.; peaks of inner N. Coast Ranges from Lake and Glenn cos. to Siskiyou Co.; in the Cascade region to cent. Ore. June–Aug.

5. **M. canéscens** (Pursh) Gray. [*Aster c.* Pursh. *Dieteria c.* Nutt. *A. inornatus* Greene, a discoid form. *M. pinosa* Elmer.] Taprooted biennial or short-lived perennial; stems several, paniculately or racemosely branched, 2–7 dm. high; herbage cinereous-

puberulent, often also stipitate-glandular above; lvs. spinulose-toothed, the basal linear-oblanceolate, 2–8 cm. long, 4–12 mm. wide, persistent or usually lost early, the cauline smaller, often linear, the upper often entire and becoming bractlike; heads numerous; invol. 6–10 mm. high, turbinate or campanulate, canescent or glandular or both; phyllaries 4–8-seriate, conspicuously imbricate, relatively narrow, with short green ± squarrose tips, rarely anthocyanous; rays 8–25 (rarely wanting), pale bluish-purple, 5–10 mm. long.—Dry gravelly slopes, meadow borders, etc., 4500–9000 ft.; Sagebrush Scrub, N. Juniper Wd., Pinyon-Juniper Wd., Yellow Pine F.; infrequent in Calif., San Jacinto, San Bernardino, and San Gabriel mts., Mt. Pinos, Providence, Clark, and Grapevine mts., n. along e. side of Sierra Nevada to Modoc and Siskiyou cos.; common n. to B.C., e. to Sask., Colo., Ariz. July–Oct.

6. **M. tortifòlia** (Gray) Cronq. & Keck. [*Aplopappus t.* T. & G. *Aster t.* Gray, not Michx. *A. mohavensis* Cov., not Kuntze. *A. t.* var. *funereus* Jones. *A. abatus* Blake. *Xylorhiza t.* Greene.] MOJAVE ASTER. Suffruticose, the leafy sparingly branched stems rather numerous, erect from a thick woody crown, 3–7 dm. high; herbage tomentose at least when young, sometimes merely stipitate-glandular and very sparingly hairy; lvs. linear to lanceolate or oblong, pungently acute to attenuate, sessile, spinose-toothed or rarely subentire, 3–6 cm. long, 3–15 mm. wide, coriaceous; heads 3.5–6.5 cm. in diameter, solitary at ends of long nearly naked glandular peduncles; invol. 12–16 mm. high, hemispheric, usually canescent or tomentose and ± glandular-puberulent; phyllaries 4–5-seriate, imbricate, linear-attenuate, with yellowish indurate base and long loose herbaceous tip; rays ca. 40–60, blue-violet or lavender to almost white, 12–25 mm. long; aks. silky villous; pappus of ± unequal flattened tawny rigid scabrid bristles.—Dry rocky slopes and washes, 2000–5500 ft.; Creosote Bush Scrub; Mojave and n. Colo. deserts, from White Mts., Inyo Co. to Riverside Co.; to sw. Utah, n. and w. Ariz. March–May, Oct.

7. **M. cognàta** (Hall) Cronq. & Keck. [*Aster c.* Hall. *A. Standleyi* A. Davids. *Xylorhiza S. A.* Davids.] Rounded openly branched shrub 5–15 dm. high, the stems white-barked; herbage with scattered mostly minute stipitate glands; lvs. lance-ovate or elliptic, sessile, ± coarsely spinose-dentate, 2–4 cm. long, 8–17 mm. wide, coriaceous, veiny, extending nearly to the large solitary terminal heads; invol. 15–25 mm. high, hemispheric, glandular-atomiferous; phyllaries ca. 5-seriate, not regularly graduate, the green outer ones often elongated and exceeding the inner, narrowly lance-linear, attenuate, the inner ones broader, indurate at base, scarious-margined and fimbriate; rays 21–35, pink-lavender or violet, 15–20 mm. long; aks. densely long-villous; pappus unequal in length and width, ± flattened, reddish.—Local, in sandstone or clay crevices, below 500 ft.; Creosote Bush Scrub; Painted, Hidden Springs, and Box canyons, Mecca Hills, Colo. Desert; Cucopa Mts., n. L. Calif. Jan.–June.

8. **M. Orcúttii** (Vasey & Rose) Cronq. & Keck. [*Aster O.* Vasey & Rose. *Xylorhiza O.* Greene.] Stems many, erect, from a woody base, forming a white-barked glabrous bush 4–9 dm. high, leafy up to the heads; lvs. closely spinulose-toothed, the lower oblanceolate, cuneate at base, the middle and upper lance-elliptic to oblong, obtuse to sharply acute, sessile by a broad base, 2–5 cm. long, 8–17 mm. wide, coriaceous, veiny; heads terminal, solitary, very large, 5.5–7 cm. in diameter; invol. 15–20 mm. high, hemispheric; phyllaries ca. 6-seriate, strongly graduate, the outer narrowly, the inner broadly lanceolate, indurate, pallid with greenish tip and mid-line, sparingly glandular-ciliolate or fimbriolate, otherwise glabrous; rays 30–45, lavender or purple, ca. 2 cm. long; aks. obscured by a dense silky tomentum; pappus of very unequal (in length and width) flattened tawny rigid scabrid bristles.—Occasional in gypsum soils of canyons, below 1000 ft.; Creosote Bush Scrub; sw. side of Colo. Desert, San Diego and Imperial cos.; n. L. Calif. March–April.

91. Corethrógyne DC.

Perennial herbs, sometimes suffrutescent at base, leafy, clothed with a soft white at length ± deciduous wool, resembling *Aster.* Heads many-fld., solitary, or cymose, or paniculate. Invol. turbinate to hemispheric; phyllaries imbricate in several series, narrow, with erect to squarrose green tips. Ray-fls. neutral, numerous, 1-seriate; ligule linear, violet. Disk-fls. yellow. Style-branches linear, the short blunt appendage with a dense tuft of rigid yellow hairs. Aks. of ray none or rudimentary, those of the disk cuneiform or linear-turbinate, pubescent. Pappus of numerous slender fulvous bristles. Three spp. (Greek, *korethron*, a brush of twigs for sweeping, and *gune*, style, from the brushlike style appendages.)

(Canby, Margaret L. The genus Corethrogyne in southern California. Bull. So. Calif. Acad. 26: 8–16, 1927.)
Heads usually solitary at ends of decumbent stems, with hemispheric invol.
  The heads 20–35 mm. across; stems relatively stout; style-branches exserted, prominently tufted. Mostly n. coast .... 1. *C. californica*
  The heads 12–18 mm. across; stems relatively slender and straggly; style-branches mostly included, less prominently tufted. Monterey coast .... 2. *C. leucophylla*
Heads more numerous on virgate erect or ascending stems, with narrower invol. Mostly s. half of Calif. .... 3. *C. filaginifolia*

1. **C. califórnica** DC. [*C. caespitosa* Greene.] Stems decumbent from spreading rootstocks, somewhat woody at base, densely leafy, simple or ± branched apically, 1.5–4 dm. long; herbage below the heads ± permanently white-tomentose; lvs. mostly linear-oblanceolate, 2–4 cm. long, 2–6 mm. wide, entire or sparingly toothed toward apex, the lower distinctly petiolate, the uppermost often bractlike; invol. 8.5–11 mm. high, ca. 4–5 seriate, densely glandular-puberulent and sometimes somewhat tomentose; phyllaries narrowly oblong, acute or apiculate, ± loose; rays 30–40, glabrous or sparsely pubescent dorsally with short multicellular hairs, violet-purple to lilac-pink, 10–13 mm. long.— Scattered colonies, grassy slopes, often in serpentine, below 1000 ft.; Mixed Evergreen F.; San Francisco Peninsula to Monterey Peninsula. April–July.

KEY TO VARIETIES

Invol. 8–12 mm. high, the linear-oblong phyllaries erect. On or near the coast.
  Principal lvs. linear-oblanceolate. S. of the Golden Gate .... *C. californica*
  Principal lvs. spatulate or obovate. N. of the Golden Gate .... var. *obovata*
Invol. mostly 12–18 mm. high, the lance-oblong broader phyllaries more lax or squarrose. Inner S. Coast Ranges .... var. *Lyonii*

Var. **obováta** (Benth.) Kuntze. [*C. o.* Benth. *C. spathulata* Gray. *C. c.* var. *s.* Kuntze.] Similar to the typical var. but the stems decumbent or trailing, up to 6 dm. long, the peduncles often more glandular and less tomentose; lvs. with spatulate-oblanceolate to obovate blade.—Grassy slopes or dunes near the coast, below 500 ft.; N. Coastal Scrub; Marin Co. to Coos Co., Ore. June–Aug.

Var. **Lyònii** Blake. Stems decumbent or ascending-erect, 1.5–2.5 dm. long; lvs. obovate, to 4.5 cm. long and to 15 mm. wide; invol. 12–18 mm. high, the phyllaries subequal or slightly graduated, 2–2.5 mm. wide, rather lax or ± squarrose.—Sunny slopes, 2000–3000 ft.; probably Chaparral; inner S. Coast Ranges from San Benito Co. to Merced Co. May–June.

2. **C. leucophýlla** Jeps. [*Diplopappus l.* Lindl. *C. californica* var. *l.* Kuntze.] Stems ascending from a decumbent base, often very slender, simple or sparingly branched, 2–5 dm. high, monocephalous; herbage ± persistently tomentose; lower lvs. ovate-spatulate to narrowly oblanceolate, obscurely to prominently serrate, 2–6 cm. long including the narrow petiole, the upper becoming very narrow and bractlike; invol. broadly obconic, 8–9 mm. high, with some deciduous tomentum and obscurely viscid.—Local on sand dunes and in pine woods by the sea, below 200 ft.; Coastal Strand, Closed-cone Pine F.; Año Nuevo Pt., San Mateo Co., to Monterey Peninsula. July–Oct. This imperfectly known sp. recombines the characters of the other two and needs further study.

3. **C. filaginifòlia** (H. & A.) Nutt. [*Aster ? f.* H. & A. *C. tomentella* (H. & A.) T. & G.? *C. californica* var. *f.* Kuntze.] Suffrutescent, slender, erect or ascending, paniculately few-branched above, 4–8 dm. high; herbage with tardily deciduous tomentum, scarcely at all glandular above; lvs. lanceolate to oblanceolate, acute to obtuse, often sharply serrate toward apex, 2–6 cm. long including petiole, the upper reduced and sessile; heads not very numerous; invol. broadly campanulate or turbinate, 7–9 mm. high, glabrate; rays violet or purple.—Grassy or brushy slopes, below 1000 ft.; Coastal Strand, Coastal Sage Scrub; along the coast, Monterey to Santa Barbara. June–Dec.

KEY TO VARIETIES

A. Invol. smooth or glandular but not tomentose at anthesis.
  B. Infl. scarcely if at all glandular. Coastal .... *C. filaginifolia*

BB. Infl. glandular.
  C. Invol. under 9 mm. high, the glands sessile or short-stipitate.
    D. Plants ± tomentose, at least below.
      E. Stems usually under 4 dm. high. Montane or maritime.
        F. Tomentum only about base, not more than halfway up the stem, the upper glandular portion bright green. San Gabriel Mts. . . var. *pinetorum*
        FF. Tomentum at least halfway up the stem.
          G. Invol. 7–9 mm. high; stems very stout, ± depressed. Morro to Santa Barbara Ids. . . . . var. *robusta*
          GG. Invol. 6–7 mm. high; stems fairly slender, quite erect. Montane var. *brevicula*
      EE. Stems usually over 4 dm. high. Low elevations.
        F. Tomentum extending up to invol. Inland valleys . . . . . . var. *bernardina*
        FF. Tomentum not extending up to invol. but upper parts glandular.
          G. Invol. turbinate to broadly campanulate. Coastal . . . . var. *virgata*
          GG. Invol. cylindric, 6–8-seriate; phyllaries squarrose, very glandular. San Fernando to base of Mt. Pinos . . . . var. *Peirsonii*
    DD. Plants ± green and glandular from base to apex, very obscurely if at all tomentose. Monterey region . . . . var. *viscidula*
  CC. Invol. over 9 mm. high, ± hemispheric, with long-stalked glands. Coastal San Diego Co. var. *incana*
AA. Invol. tomentose at anthesis.
  B. Plants of cent. Calif. . . . . var. *hamiltonensis*
  BB. Plants of s. Calif.
    C. Lvs. linear. San Diego region . . . . var. *linifolia*
    CC. Lvs. not linear.
      D. Invol. 9–14 mm. high, campanulate; lvs. ovate to oblong to spatulate. San Bernardino Mts. . . . . var. *sessilis*
      DD. Invol. 7–8 mm. high, turbinate; lvs. mostly broadly oblong. Coastal Ventura Co. var. *latifolia*

Var. **pinetòrum** Jtn. Herbaceous, 1–5 dm. high; lower lvs. and lower part of stem permanently tomentose, the upper half of plant oily-green, with dense stipitate glands; heads few; invol. turbinate, 5–7 mm. high; phyllaries squarrose.—Frequent in dry rocky ground, 1400–5500 ft.; San Gabriel Mts. July–Oct.

Var. **robústa** Greene. Suffrutescent; stems stout, somewhat depressed or ascending, ca. 3 dm. long; herbage tomentose up to the rather dense corymbose infl., the peduncles then glandular; lvs. numerous and conspicuous up to infl., broadly obovate to spatulate, 2–4 cm. long; invol. hemispheric, 7–9 mm. high, scarcely glandular.—Coastal Sage Scrub; Morro, San Luis Obispo Co., Surf and Lompoc, and San Miguel and Santa Rosa ids., Santa Barbara Co.

Var. **brevícula** (Greene) Canby. [*C. b.* Greene. *C. racemosa* Greene.] Stems stiff, erect, 2–4(–8) dm. high; herbage tomentose up to the infl. at anthesis, the infl. then glandular; lvs. spatulate to obovate, yellow-green; heads corymbosely arranged, relatively few; invol. turbinate, 6–7 mm. high, with generally recurved phyllaries.—Scattered in openings on wooded slopes, (2000–)3500–8200 ft.; mostly Yellow Pine F.; Sierra Nevada from Mariposa Co. to Kern Co., Tehachapi Mts., San Bernardino Mts. to Laguna Mts. July–Oct.

Var. **bernardìna** (Abrams) Hall. [*C. virgata* var. *b.* Abrams. *C. f.* var. *glomerata* Hall.] Stems slender, mostly 4–9 dm. high, often paniculately branching above, usually persistently white-tomentose except for the upper parts of peduncles and the invols. which are glandular; lvs. variable, 1–7 cm. long, commonly narrow, sometimes obovate and up to 3 cm. wide; heads terminating the rather long divaricate branches, sometimes spicate and ± glomerulate (the var. *glomerata*); invol. turbinate, 5–7 mm. high, the phyllaries often squarrose.—Sandy plains and rocky canyons, 1000–3500 ft.; cismontane s. Calif. toward the base of the mts., Ventura Co. to Riverside Co. July–Oct.

Var. **virgàta** (Benth.) Gray. [*C. v.* Benth. *C. f.* var. *rigida* Gray. *C. californica* vars. *v.* and *r.* Kuntze. *C. r.* Heller. *C. floccosa* and *lavandulacea* Greene.] Slender, erect, 4–12 dm. high, ± gray- or white-tomentose below, becoming green and short-stipitate-glandular above; lvs. oblanceolate to linear-lanceolate, entire or with serrate tips; infl. a diffuse many-headed panicle; invol. 5–8 mm. high, the phyllaries sometimes squarrose. —The commonest var.; open wasteland or brushy slopes, often in sand, below 2000 ft.; Coastal Strand, Coastal Sage Scrub, Foothill Wd., S. Oak Wd.; along the coast and in outer Coast Ranges, Alameda and Santa Cruz cos. to n. L. Calif.; Santa Catalina Id. July–Oct. (May).

Var. **Peirsònii** Canby. Stout, 3–9 dm. high, tomentose except in the paniculate to virgate dark green densely glandular infl.; lvs. oblanceolate to obovate, 1–5 cm. long, 8–20 mm. wide; invol. cylindric, dark-glandular, 7–8 mm. high, the numerous phyllaries with prominently squarrose tips.—Sandy flats and canyons, below 5000 ft.; Coastal Sage Scrub, Chaparral, S. Oak Wd.; mts. of n. Ventura Co. to Tehachapi Mts., w. end of San Gabriel Mts., and San Fernando V. Aug.–Nov.

Var. **viscídula** (Greene) Keck. [*C. v.* Greene. *C. scabra* Greene?] Erect, 4–7 dm. high; herbage tomentose when young, the tomentum soon lost, the plant yellowish-green and very viscid essentially throughout at anthesis, often much-branched.—Sandy hills, near sea level; Coastal Sage Scrub, Chaparral; Monterey region, probably Los Angeles. June–Dec.

Var. **incàna** (Nutt.) Canby. [*C. i.* Nutt. *C. californica* var. *i.* Kuntze. *C. f.* var. *pacifica* Hall.] Stout, erect, 5–8 dm. high, tomentose below; lvs. linear to narrowly lanceolate or oblanceolate, mostly entire, up to 7 or 8 cm. long; infl. an open panicle with many large heads, or simply cymose with few heads; invol. hemispheric, 10–12 mm. high, together with the peduncles densely beset with stout-stalked prominent glands.—Common on sandy slopes facing the sea; Coastal Sage Scrub; sw. San Diego Co. June–Aug.

Var. **hamiltonénsis** Keck. Slender, erect, 4.5–6.5 dm. high, the tomentum (like the lower lvs.) early deciduous, but persisting on the invols. at anthesis; lower lvs. narrowly oblanceolate, with serrate tips, the upper lance-linear, much reduced; infl. a narrow or open panicle, few-headed; invol. obconic, 7.5–11 mm. high, 6–7-seriate, scarcely at all glandular.—Dry slopes, 900–3000 ft.; Foothill Wd.; Mt. Hamilton Range, principally on w. side, Santa Clara Co., to San Benito Co. Aug.

Var. **linifòlia** Hall. Erect, 2–8 dm. high, permanently but loosely tomentose throughout; lvs. all narrowly linear, sometimes revolute, 1–2(–5) mm. wide, crowded toward base, scattered above; heads few, terminating slender divaricate branches, many-fld.; invol. broadly campanulate, 7.5–11 mm. high, 6–7-seriate, prominently cobwebby-tomentose and microscopically viscid; rays lavender, 6–8 mm. long.—Common on bluffs or brushy slopes near the sea; Coastal Sage Scrub, Chaparral; sw. San Diego Co. July–Sept.

Var. **séssilis** (Greene) Canby. [*C. s.* Greene.] Stout, erect, 2–7.5 dm. high, densely and rather permanently tomentose including the invols.; lvs. spatulate to ovate, 5–20 mm. wide, rather equally distributed; heads cymose-paniculate to subspicate, many-fld.; invol. campanulate, 9–14 mm. high, 7–9-seriate; phyllaries broad, the tips occasionally squarrose; rays violet, 7–11 mm. long.—Common in dry forest openings, 4000–7500 ft.; Yellow Pine F.; San Bernardino Mts. July–Oct.

Var. **latifòlia** Hall. [*C. flagellaris* Greene.] Suffrutescent, stout, 3–6 dm. high, tomentose throughout including the invols.; principal lvs. broadly oblong, 1–4 cm. long, 5–12 mm. wide; heads few, in prominently bracteate panicles; invol. turbinate, 7–8 mm. high, ca. 5-seriate; phyllaries with slightly spreading tips; rays showy, 10 mm. long.—Slopes overlooking the coast; Coastal Sage Scrub; Pt. Sal, Santa Barbara Co. to Redondo, Los Angeles Co., Anacapa Id. Aug.–Dec.

92. Erígeron L. Wild Daisy. Fleabane

Annual, biennial or perennial herbs. Lvs. alternate (sometimes all basal), usually narrow, entire to toothed or pinnatifid, often sessile. Heads solitary to numerous and corymbose or paniculate, many-fld. Invol. campanulate to hemispheric; phyllaries only slightly or not at all graduated, narrow, not herbaceous-tipped, but herbaceous throughout to scarcely herbaceous throughout. Receptacle flat, naked. Ray-fls. ♀, fertile, numerous, bearing evident often narrow white, pink, purple, or bluish ligules, or the ligules or even the ray-fls. occasionally lacking. Disk-fls. ± numerous, yellow, rarely reddish; style-branches with lanceolate and acute or usually triangular and obtuse appendages. Aks. flattened, usually pubescent and 2-nerved, sometimes 4–10-nerved; pappus of capillary often fragile bristles, sometimes with an outer series of short minute bristles or scales. Largely Am. genus of nearly 200 spp., some in the Old World, nearly all of temp. or boreal regions. (Greek, *eri*, early, and *geron*, old man, the ancient name of an early-fl. plant with hoary pubescence.)

(Cronquist, Arthur. Revision of the N. Am. Spp. of Erigeron, N. of Mexico. Brittonia 6: 121–302, 1947.)

A. Pistillate corollas few to numerous (rarely absent), the tube generally cylindrical; rays well developed and spreading, or sometimes reduced or absent, but not short, narrow, and erect.

   B. Pappus of ray- and disk-fls. alike, of bristles, sometimes also with outer setae or scales; plants mostly perennial, a few spp. biennial or casually annual, seldom weedy except for *E. philadelphicus* and *E. divergens*.

   C. Internodes not excessively numerous or usually very short; lvs. variously shaped, sometimes linear, but then the basal obviously larger than the cauline; phyllaries equal or imbricate.

   D. Aks. in most spp. 2-nerved, if more-nerved, in not at all silvery-strigose plants with essentially equal phyllaries of montane, woodland, or maritime habitats. (Section Erigeron)

   E. Plants maritime and submaritime; heads large, hemispheric, the disk 14–35 mm. wide; disk-corollas 4.5–7 mm. long.

   F. Heads radiate; cauline lvs. ample .... 5. *E. glaucus*
   FF. Heads discoid; cauline lvs. narrow .... 6. *E. supplex*

   EE. Plants not maritime or submaritime; heads mostly smaller except in some tall spp.

   F. Cauline lvs. ample, usually lanceolate or broader, never trilobed; aks. 2–7-nerved.

   G. Plants normally tall and erect, ± Asterlike; aks. 2–7-nerved.

   H. Rays 2–4 mm. wide (1–2 mm. in *E. Coulteri*, which has hairs on invol. with black cross-walls near base).

   I. Hairs of invol. without black cross-walls; rays mostly 2–4 mm. wide, white or colored; lvs. hairy or glabrous.

   J. Aks. 4–7-nerved; lvs. usually glabrous; phyllaries mostly merely glandular, sometimes hirsute or ± ciliate, but not hirsute below and glandular above
   1. *E. peregrinus*

   JJ. Aks. 2–4-nerved; lvs. hirsute; phyllaries hirsute below, glandular above .... 2. *E. Aliceae*

   II. Hairs of invol. with black cross-walls near base; rays 1–2 mm. wide, white; lvs. hairy .... 3. *E. Coulteri*

   HH. Rays less than 1 mm. wide; hairs of invol. without black cross-walls .... 4. *E. philadelphicus*

   GG. Plants low and often spreading or ascending, 0.5–3 dm. high, scarcely Asterlike; aks. 2-nerved.

   H. Lvs. rather equably distributed, the middle cauline about as large as the lower, the upper gradually reduced
   8. *E. delicatus*

   HH. Lvs. inequably distributed, the basal obviously larger than the few progressively reduced cauline ...... 9. *E. cervinus*

   FF. Cauline lvs. mostly not very well developed, commonly linear or oblanceolate, sometimes linear- or lance-oblong; aks. 2-nerved.

   G. Lvs., or many of them, trilobed or 2–4 times ternate.

   H. Caudex divided into several or many long slender rhizomelike branches; lvs. mostly merely trilobed, the lobes short, broad, rounded .... 10. *E. vagus*

   HH. Caudex stout, the branches, if present, stout and short; lvs. trifid or more often 2–4 times ternate, mostly with relatively slender lobes .... 11. *E. compositus*

   GG. Lvs. all entire or nearly so.

   H. Pubescence of stem widely spreading, sometimes scanty (appressed except under the heads in *E. multiceps*).

   I. Plants alpine or subalpine with solitary or rarely 2 heads, radiate.

   J. Stems ± foliose.

   K. Stem viscid or glandular as well as hirsute; phyllaries subequal; taproot poorly or scarcely developed .... 12. *E. petiolaris*

   KK. Stem not viscid or glandular; phyllaries ± graduate; taproot well developed .... 13. *E. Clokeyi*

   JJ. Stems scapose or very nearly so.

   K. Basal lvs. oblanceolate, tapering to the petiole, up to 2.5 mm. wide; rays blue or purple. Sierra Nevada .... 14. *E. pygmaeus*

   KK. Basal lvs. subrotund to broadly oblanceolate, rather abruptly contracted to and much shorter than the petiole, up to 8 mm. wide; rays pink or white. Desert ranges of Death V. region
   15. *E. uncialis*

   II. Plants of low elevs., or if higher in mts., discoid; heads 1-many.

   J. Basal lvs. not triplinerved.

   K. Rays, if present, not yellow.

   L. Heads evidently radiate.

   M. Pappus simple or nearly so; disk-corollas

5–6 mm. long. San Luis Obispo and Santa Barbara cos.
                                    7. *E. sanctarum*
MM. Pappus distinctly double.
  N. Perennials; disk-corollas (except in *E. multiceps*) 3–5
       mm. long.
    O. Pubescence of stem all spreading; disk-corollas
         3.5–5 mm. long. N. and e. Calif.
                                    17. *E. pumilus*
    OO. Pubescence of stem appressed except spreading un-
         der the heads; disk-corollas 2.5 mm. long. S. Sierra
         Nevada .... 18. *E. multiceps*
  NN. Biennial or nearly so; disk-corollas 2–3 mm. long. In-
       terior .... 19. *E. divergens*
LL. Heads disciform or discoid.
  M. The heads disciform, the ♀ fls. present but essentially rayless.
    N. Biennials.
      O. Inner pappus of 5–12 fragile bristles; disk-corollas
           2–3 mm. long .... 19. *E. divergens*
      OO. Inner pappus of ca. 15–20 firm bristles; disk-
           corollas 4–5 mm. long. Inyo Mts. .. 20. *E. calvus*
    NN. Perennials.
      O. Outer pappus of evident scales; stems leafy (except
           in var. from San Bernardino Mts.)
                                    21. *E. aphanactis*
      OO. Outer pappus setose and obscure, or wanting; stems
           subscapose. Ne. Calif. ...... 22. *E. chrysopsidis*
  MM. The heads discoid, the ♀ fls. absent. Var. of
                                    23. *E. Bloomeri*
  KK. Rays yellow ....*..................... 22. *E. chrysopsidis*
  JJ. Basal lvs. ± strongly triplinerved .... 28. *E. lassenianus*
HH. Pubescence of stem and lvs. ± closely appressed, or wanting, never definitely spread-
     ing.
  I. Heads discoid; fls. all tubular and perfect. Nevada Co. n. .... 23. *E. Bloomeri*
  II. Heads evidently radiate.
    J. Base of stem conspicuously enlarged, shining and ± indurated, stramineous
         or purplish.
      K. Rays yellow; pappus-bristles 10–20. N. Sierra Nevada .. 24. *E. linearis*
      KK. Rays blue, purple, or pink; pappus-bristles mostly 20–50.
        L. Lvs. linear, 1 mm. or less wide; invol. 3.5–5 mm. high; disk 6–11
             mm. wide. Lassen Co. n. .... 25. *E. elegantulus*
        LL. Lvs. oblanceolate, the basal 1.5–5 mm. wide; invol. 5.5–9 mm.
             high; disk 13–18 mm. wide. Lassen Co. s. ... 26. *E. barbellulatus*
    JJ. Base of stem not conspicuously enlarged, shining, and indurated (alpine speci-
         mens of *E. peregrinus* would be sought here, except for the 4–7-nerved aks.).
      K. Basal lvs. narrow, linear or rather narrowly oblanceolate, the blade, if
           distinguishable, tapering very gradually to the petiole.
        L. Bases of basal lvs. neither enlarged nor of different texture from the
             blades; lvs. linear or linear-filiform; stem more densely hairy toward
             base. Sierra Co. n. .... 27. *E. filifolius*
        LL. Bases of basal lvs. ± enlarged, membranous or indurated; lvs. ob-
             lanceolate or sometimes linear; stems not more densely hairy to-
             ward base.
          M. Plants neither pulvinate-cespitose nor scapose; basal lvs.
               mostly triplinerved (except in *E. flexuosus*).
            N. Disk-corollas 5–7 mm. long; style-appendages 0.3–0.4
                 mm. long; pappus coarse and copious, the inner of 25–
                 40 bristles; invol. 7.5–10 mm. high. E. side, Sierra Ne-
                 vada .... 29. *E. nevadincola*
            NN. Disk-corollas 2.5–5 mm. long; style-appendages 0.1–
                 0.25 mm. long; pappus ± scanty and fragile, the inner of
                 10–25 bristles; invol. 3.5–8 mm. high.
              O. Cauline lvs. abruptly reduced; rays white. E. side,
                   Sierra Nevada .... 30. *E. Eatonii*
              OO. Cauline lvs. gradually reduced; rays blue, lilac, or
                   pink. Cismontane.
                P. Basal lvs. ± evidently triplinerved; heads few
                     or solitary on ± naked peduncles; rays 20–50;
                     invol. hairy but scarcely glandular, 6–8 mm.
                     high. N. Calif. .... 31. *E. decumbens*
                PP. Basal lvs. obscurely if at all triplinerved; heads
                     rather numerous on leafy branches; rays 12–
                     20; invol. glandular and hairy, 4–5 mm. high.
                     Trinity Alps. .... 32. *E. flexuosus*
          MM. Plants pulvinate-cespitose, scapose; lvs. not triplinerved. Inyo
               Mts. e. .... 33. *E. compactus*
      KK. Basal lvs. broadly oblanceolate or usually broader, the well defined blade
           ± abruptly contracted to the petiole.
        L. Lvs. glabrous; phyllaries subequal, not hairy but finely glandular.

Siskiyou Mts. .... 9. *E. cervinus*
    LL. Lvs. strigose; phyllaries graduate, hairy and glandular. Sierra Ne-
        vada n. .... 16. *E. tener*
  DD. Aks. 4–8(–10)-nerved; phyllaries evidently graduate; lvs. silvery-strigose. Desert
    ranges. (Section Wyomingia)
    E. Aks. mostly 4-nerved, rarely 6-nerved; basal lvs. often withered by fl.-time, not
        forming a persistent tuft.
        F. Outer pappus of evident narrow scales; stems silvery pubescent. N. base,
            San Bernardino Mts. .... 34. *E. Parishii*
        FF. Outer pappus of inconspicuous setae; stem merely gray-green, except near
            base. Providence Mts. e. .... 35. *E. utahensis*
    EE. Aks. mostly 6–8(–10)-nerved; basal lvs. tufted and persistent, the cauline
        reduced. Inyo Co. e. .... 36. *E. argentatus*
  CC. Internodes very numerous and short; lvs. linear or narrowly oblong, uniform from base
    to near top of plant; phyllaries markedly graduate. (Section Pycnophyllum)
    D. Heads evidently radiate.
      E. Root-crown subterranean, giving rise to slender rhizomatous stems rarely
        more than 1.5 mm. thick, which become aerial stems; heads solitary or few
        on slender stems; pubescence of stem spreading or retrorse.
        F. Stem finely glandular and spreading-hirsute. Tulare Co.
            37. *E. aequifolius*
        FF. Stem not glandular, though the hairs may be a little viscid; pubescence
          of stem short or wanting. Shasta Co. to Riverside Co. e.
            38. *E. Breweri*
      EE. Root-crown superficial; base of stem, if rhizomatous, very short or stout or
        both; heads often numerous; pubescence of stem, when present, strigose to
        hirtellous or hirsute. Coast Ranges and Sierra Nevada ...... 39. *E. foliosus*
    DD. Heads discoid.
      E. Stem and lvs. spreading-villous or -villosulous.
        F. Larger lvs. 2–4 cm. long; outer pappus obscure or wanting. Coast
          Ranges n. .... 40. *E. petrophilus*
        FF. Larger lvs. 1–2 cm. long; outer pappus evident. Donner Pass region
            41. *E. miser*
      EE. Middle and upper part of stem not spreading-villous, glabrous to glandular
        and sometimes ascending-hirtellous or appressed-hairy; lvs. not villous but
        sometimes short-hairy. Coast Ranges and n. Sierra Nevada .. 42. *E. inornatus*
  BB. Pappus of ray and disk unlike, the disk-pappus composed of bristles and short outer setae,
    the ray-pappus lacking the bristles; weedy mostly annual plants. (Section Phalacroloma)
    C. Foliage ample; plants mostly 6–15 dm. high; pubescence of stem long and spreading
            43. *E. annuus*
    CC. Foliage sparse; plants mostly 3–7 dm. high; pubescence various .... 44. *E. strigosus*
AA. Pistillate corollas very numerous, filiform, with very narrow short erect rays sometimes not ex-
  ceeding the disk (the inner ones sometimes rayless). (Section Trimorphaea)
  B. Rayless ♀ fls. wanting; infl. racemiform, the peduncles erect or nearly so, or the head
    solitary .... 45. *E. lonchophyllus*
  BB. Rayless ♀ fls. present between the ray- and disk-fls.; infl. corymbiform, the peduncles
    arcuate or obliquely ascending, or the head solitary .... 46. *E. acris*

1. **E. peregrìnus** (Pursh) Greene ssp. **calliánthemus** (Greene) Cronq. [*E. c.* Greene. *E. salsuginosus* of many auths., not *Aster s.* Rich.] Fibrous-rooted perennial from a short rhizome or short stout caudex, up to 7 dm. high, amply leafy, essentially glabrous except for the closely villous peduncles; lower lvs. well developed, entire, oblanceolate or broader, tapering to the petiole, 5–10 cm. long, 8–20 mm. wide, the cauline lance-ovate or ovate and not greatly reduced, sessile and often slightly clasping; heads solitary or few, the disk 10–25 mm. wide; invol. 7–11 mm. high, the phyllaries linear, attenuate, loose, densely glandular, subequal; rays 30–80, 8–25 mm. long, 2–4 mm. wide, mostly rose-purple; disk-corollas mostly 4–6 mm. long; aks. usually 5(4–7)-nerved; pappus of 20–30 bristles.—Montane meadows, 5500–10,500 ft.; Red Fir F., to Subalpine F.; Sierra Nevada, from Tulare Co. n., Trinity Co. to Modoc Co.; to B.C., Alta., Colo., New Mex. July–Sept. This glabrate broad-lvd. ssp. is relatively uncommon in Calif. but the abundant form to the n. and e. In Calif. it is largely replaced by the following vars., with which it intergrades. Typical *E. peregrinus,* with villous instead of glandular phyllaries, extends from the n. Cascades to Alaska.

Var. **hirsùtus** Cronq. Lvs. ± hirsute on the surfaces; peduncle spreading-hirsute; otherwise resembling var. *angustifolius,* though sometimes with broader lvs.—Sierra Nevada, from Mono and Tuolumne cos. to Tulare Co.; w. Nev. June–Aug.

Var. **angustifòlius** (Gray) Cronq. [*Aster salsuginosus* var. *a.* Gray. *E. s.* var. *a.* Gray. *E. a.* Rydb.] Stems slender or stout, often reddish at base, 2–5 dm. high; lvs. all narrow, the basal oblanceolate or narrower, the cauline reduced, linear (in small forms) or lanceolate (in large forms), acute.—The common form in the Sierra Nevada, from Tulare Co. n., Trinity Co. to Modoc Co., Sweetwater Mts.; adjacent Nev., n., less commonly, to s. B.C. June–Aug.

2. **E. Alíceae** Howell. Fibrous-rooted perennial from a rather short rhizome or woody caudex, 3–8 dm. high, amply leafy, ± hirsute throughout; lvs. entire or coarsely toothed, thin, the lower broadly oblanceolate, the blade (to 2 or 3 cm. wide) tapering to a slender petiole as long, the whole up to 20 cm. long, the middle cauline sessile, ovate to oblong or lanceolate; heads solitary or few, the disk 12–20 mm. wide; invol. 7–10 mm. high, the phyllaries linear, attenuate, loose, conspicuously white-hirsute at and near base, finely glandular-puberulent but sparsely or not at all hirsute toward the tip; rays 45–80, 8–14 mm. long, 1.5–2.5 mm. wide, whitish to rose-purple; disk-corollas 3–4 mm. long; aks. 2(or 4)-nerved; pappus of 20–30 bristles.—Meadows or moist forest openings, 4500–5500 ft.; Yellow Pine F., Red Fir F.; N. Coast Ranges, Humboldt, Trinity, and Siskiyou cos.; n. in the Cascades to Mt. Hood and the Olympics. June–Aug.

3. **E. Coùlteri** Porter. [*E. frondeus* Greene.] Perennial from a slender rhizome, or this shortened into a branching caudex; stem 1–6 dm. high, amply leafy, spreading-hirsute at least above; lvs. ± hirsute, entire, or some of the lower dentate, lance-oblong, the basal petiolate, the cauline sessile, ± clasping at base, up to 8 cm. long and to 2.5 cm. wide; heads solitary to several, the disk 10–15 mm. wide; invol. 7–11 mm. high, the phyllaries thin, green, attenuate, villous with spreading black-crosswalled hairs, obscurely viscidulous; rays 50–100, 10–20 mm. long, ca. 1.5 mm. wide, white; disk-corollas 3–4.4 mm. long; aks. 2-nerved; pappus of 20–25 bristles.—Stream banks and wet meadows, 6200–11,000 ft.; Red Fir F. to Subalpine F.; Sierra Nevada, from Tulare and Inyo cos. to Nevada Co., Warner Mts.; e. to the Rocky Mts. July–Aug.

4. **E. philadélphicus** L. Biennial or short-lived perennial, 4–10 dm. high; stem slender or rather stout, simple below the middle; herbage hispid-hirsute; basal lvs. spatulate to oblong-obovate, crenate-serrate to sharply dentate or sometimes entire, 8–15 cm. long, 1–3 cm. wide, the cauline becoming sessile and ± clasping, mostly ample; heads few to many, corymbose, the disk 6–15 mm. wide; invol. 4–6 mm. high, the phyllaries subequal, the brownish midvein hirsute or subglabrate, the broad hyaline margin occasionally purplish; rays 150–400, 5–10 mm. long, 0.2–0.6 mm. wide, deep pink to white; aks. 2-nerved.—Moist, open or protected grassy places, 10–4000; many Plant Communities, mostly below Yellow Pine F.; widely but often locally distributed, cismontane s. Calif. through the outer Coast Ranges to Del Norte Co. and through the Sierran foothills to Siskiyou Co.; to Canada and e. U.S. April–July.

5. **E. glaùcus** Ker. [*Aster californicus* Less. *Stenactis glauca* Nees. *Woodvillea calendulacea* DC. *E. hispidus* and *maritimus* Nutt.] SEASIDE DAISY. Stems erect from a decumbent base, 1–4 dm. high, leafy, hirsute and ± viscid, arising from a basal rosette of lvs. on the fleshy caudex and often also from offsets terminating stout prostrate branches; basal lvs. broadly spatulate to obovate, entire or sometimes toothed above the middle, up to 12 cm. long and 3 cm. wide, tapering to a winged petiole, pilosulous-ciliate; upper lvs. not abruptly reduced; heads terminating long branches, the disk 1.5–3.5 cm. wide; invol. 8–12 mm. high, the phyllaries equal, acuminate or attenuate, sparsely to densely shaggy or long-villous and ± viscid; rays ca. 100, 9–15 mm. long, pale violet to lavender; aks. 2–6-nerved.—Common on coastal bluffs, sandhills and beaches, mostly below 500 ft.; Coastal Strand, N. Coastal Scrub, Coastal Sage Scrub; Santa Barbara Ids. and San Luis Obispo coast n.; to Clatsop Co., Ore. April–Aug.

6. **E. súpplex** Gray. Perennial with a taproot and short branched caudex; stems decumbent, 1.5–4 dm. high, moderately hirsute; lvs. moderately villous or mostly ciliate, entire, the lower oblanceolate, petiolate, up to 8 cm. long and 7 mm. wide, the upper lance-linear, sessile, more crowded; heads solitary on subnaked peduncles 2–10 cm. long, the disk 14–20 mm. wide; invol. 7–11 mm. high, the slender attenuate phyllaries villous-hirsute and ± glandular; rays wanting; disks orange-yellow; aks. 2-nerved.—Occasional on the seacoast of Humboldt and Mendocino cos. May–June.

7. **E. sanctárum** Wats. Stems erect or ± decumbent from a very slender branching rootstock, 5–30 cm. high, sparsely pubescent with spreading or retrorse hairs; lvs. entire, .minutely rough-hairy at least on upper surface and ± ciliate, oblanceolate, 2–5 cm. long, 3–10 mm. wide, evidently reduced and becoming linear and sessile up the stem; heads usually solitary on a subnaked peduncle 2–10 cm. long, the disk 12–17 mm. wide; invol. 6–9 mm. high, the narrow attenuate phyllaries densely hirsute; rays 45–90, 8–12 mm. long, 1.3–1.9 mm. wide, rose-purple.—Hills near the coast, up to 1000 ft.; Coastal Sage Scrub, Chaparral, Foothill Wd.; Cambria coast, San Luis Obispo Co. to Santa Ynez Mts., Santa Barbara Co.; Santa Rosa Id. March–June.

8. **E. delicàtus** Cronq. Fibrous-rooted perennial from a slender caudex, glabrous below, finely stipitate-glandular above; stems slender, leafy, 2–3 dm. high; lvs. entire, the lowermost oblanceolate, obtuse or rounded, petiolate, 2–7 cm. long, 4–8 mm. wide, the cauline gradually reduced and becoming sessile; heads 1 or 2, the disk 9–13 mm. wide; invol. 5–6 mm. high, the glandular or glabrous phyllaries greenish-stramineous, with acute or acuminate sometimes purplish tips; rays ca. 40, 7–10 mm. long, 1.5–2 mm. wide, blue.—At lower elevs.; probably Douglas-Fir F.; Smith R., at mouth of Mill Creek, Del Norte Co. (type locality); adjacent Ore. June–July.

9. **E. cervìnus** Greene. Fibrous-rooted perennial from a branching caudex, glabrous only up to near the summit of the peduncles, where usually finely glandular, 1–3 dm. high; lvs. entire, largely basal, broadly oblanceolate to elliptic or obovate, petiolate, up to 10 cm. long and 12 mm. wide, the cauline scattered and reduced, sessile; heads solitary (or 2–4), the disk 10–15 mm. wide; invol. 5–7 mm. high, the finely glandular acute or acuminate phyllaries herbaceous; rays 20–50, 7–10 mm. long, 1.2–1.7 mm. wide, pale blue or lavender to whitish.—Open rocky slopes of meadows, 3000–6000 ft.; Douglas-Fir F.; Salmon-Trinity Alps, Trinity Co. to Siskiyou Mts., Siskiyou Co.; s. Ore. June–Aug.

10. **E. vàgus** Pays. Perennial from a diffuse slenderly branching caudex eventually connecting with a taproot, ± coarsely hirsute with white multicellular hairs and usually also sparsely glandular, or more densely glandular and sparsely hirsute; lvs. crowded in basal rosettes, mostly digitately 3-lobed, 8–20 mm. long (including the margined petiole 2–3-times as long as the blade), 2–5 mm. wide; heads solitary on short scapiform peduncles up to 5 cm. long, the disk 8–16 mm. wide; invol. 5–7 mm. high, the glandular and villous phyllaries purple-tinged; rays 25–35, 4–7 mm. long, rose-color or whitish.— Screes and rock-crevices, 11,000–14,100 ft.; Alpine Fell-fields; Sierra Nevada from Mono and Tuolumne cos. to Tulare Co., White Mts.; to Nev., Utah, sw. Colo., Wallowa Mts., Ore. July–Aug.

11. **E. compósitus** Pursh var. **glabràtus** Macoun. [*E. multifidus* Rydb. *E. c.* var. *m.* Macbr. & Pays.] Dwarf cushion-plant from a stout compactly branched caudex atop a taproot; herbage ± hispid-hirsute and glandular; lvs. crowded at base, 2–6 cm. long, long-petiolate, the fanlike blade 2–3-times ternate, the lobes commonly crowded and linear-oblong; stems monocephalous, subscapose, with few reduced often entire lvs., 5–15 cm. high; invol. 5–8 mm. high, the thin slender phyllaries purple-tinged; ♀ fls. 20–50, the ligules white, pink, or bluish, up to 10 mm. long, usually shorter or reduced and inconspicuous or essentially lacking.—Rocky slopes and ridges, 8000–13,000 ft.; Lodgepole F. to Alpine Fell-fields; Modoc and Siskiyou cos., s. through the Sierra Nevada to Inyo and Tulare cos., Sweetwater and White mts., San Gorgonio Peak; to Alaska, Greenland, s. in the Rocky Mts. to n. Ariz. July–Aug. The typical variety, a somewhat larger plant with 3–4-times ternate lvs. and these with long linear divisions, occurs in w. Ida. and e. Wash. and Ore.

Var. **discoìdeus** Gray. [*E. trifidus* Hook. *E. c.* var. *t.* Gray.] Plant commonly very compact and small; lvs. mostly only once ternate; otherwise similar to var. *glabratus.*— Range and environments of var. *glabratus* but much less common in Calif. and w. of the Rocky Mts.

12. **E. petiolàris** Greene. [*E. algidus* Jeps.] Fibrous-rooted to weakly taprooted tufted perennial with a branching caudex, 5–25 cm. high, ± hirsute and glandular, or occasionally glabrate; lvs. crowded at base, narrowly oblanceolate to obovate, entire, rather abruptly contracted to the petiole, up to 7 cm. long and 12 mm. wide, those of the scapiform stems few, reduced, linear; heads solitary, the disk 5–18 mm. wide; invol. 5–8 mm. high, glandular and villous-hirsute, somewhat purplish; rays 25–120, 5–13 mm. long, lavender-purple to pink or bluish.—Stony slopes and dry meadows, 9500–12,000 ft.; Subalpine F., Alpine Fell-fields; Sierra Nevada from Mt. Whitney to Tahoe region, Sweetwater Mts.; Mt. Rose, Nev. July–Aug.

13. **E. Clòkeyi** Cronq. Cespitose perennial with woody taproot and usually branched caudex; stems slender, decumbent or suberect, 3–20 cm. long; herbage moderately or densely hirsute with curved hairs, not glandular below the summits of the peduncles; lvs. crowded at base (the old bases persistent on the caudex), narrowly oblanceolate, entire, slenderly petiolate, up to 8 cm. long and 6 mm. wide, the cauline rather numerous, mostly linear or lance-linear; heads solitary (sometimes 2), the disk 8–11 mm. wide; invol. 4–7 mm. high, glandular and ± hirsute; rays 20–50, 6–10 mm. long, lavender.— Sandy or rocky slopes, 8000–10,500 ft.; Sagebrush Scrub, Bristle-cone Pine F., Subalpine

F.; s. Sierra Nevada, Tulare Co., Sweetwater and White mts., Mono Co. to Inyo Mts., Inyo Co.; Wassuk Range to Charleston Mts., Nev., e. to sw. Utah. June–Aug.

14. **E. pygmàeus** (Gray) Greene. [*E. nevadensis* var. ? *p.* Gray.] Dwarf compact cushion-plant from a woody taproot and a stout branched caudex; herbage rather densely canescent or hispidulous and finely stipitate-glandular; lvs. almost all basal, in dense rosettes, linear-oblanceolate, 1–2.5 cm. long, 1–2.5 mm. wide; heads solitary on erect subscapose peduncles 2–6 cm. high, the disk 6–13 mm. wide; invol. 4–7 mm. high, the phyllaries glandular and usually only sparsely hirsute, blackish-purple or green with purple tips; rays 15–35, 4–7 mm. long, purple or lavender.—Rocky slopes, 10,000–12,000 ft.; Subalpine F., Alpine Fell-fields; Sweetwater Mts., Mono Co., Sierra Nevada from Mt. Whitney region n. to Mt. Rose, Nev. July–Aug.

15. **E. unciàlis** Blake. Habit of *E. pygmaeus;* lvs. all basal, hirsute or hirsute-strigose, sometimes glabrate on one or both faces, the broadly oblanceolate to subrotund blade abruptly contracted to a longer petiole, 1–4 cm. long, 1.5–8 mm. wide; scapes 1–5 cm. high, hirsute with spreading hairs, densely so just below the solitary head; disk 6–11 mm. wide; invol. 4.5–5 mm. long, villous-hirsute and obscurely glandular, ± purplish; rays 15–40, 4–6 mm. long, pale lavender or rose to white.—Crevices of limestone cliffs, 8000–9500 ft.; Sagebrush Scrub, Bristle-cone Pine F.; Inyo Mts., Tin Mt., Death V. region, Inyo Co., Clark Mt., San Bernardino Co. June–July. The var. *conjugans* Blake, with petioles and lower part of stem pubescent with appressed or ascending hairs, of high mts. in Nye and Clark cos., Nev., may occur in Calif.

16. **E. téner** (Gray) Gray. [*E. caespitosus* var. *t.* Gray. *E. Copelandii* Eastw.] Cespitose from a taproot and branched caudex covered with persistent old lf.-bases; stems slender, ascending or erect, 3–15 cm. high; herbage white-strigose, more strongly so on younger parts; basal lvs. spatulate or lanceolate, acute or nearly so, long-petiolate, up to 7 cm. long and 7 mm. wide, the cauline few and linear; heads 1–3, the disk 6–12 mm. wide; invol. 3.5–5 mm. high, the phyllaries finely glandular and sparsely hirsute, the outer somewhat thickened, mostly greenish brown with darker midrib; rays 15–40, 4–8 mm. long, purple to lavender-pink.—Cliff-faces and open stony places, 8000–11,000 ft.; Sagebrush Scrub and Yellow Pine F. to Alpine Fell-fields; rare in Calif., Mt. Eddy, Siskiyou Co., Thompson Peak, Lassen Co., Ebbetts Pass, Alpine Co., Rock Creek Lake Basin, Inyo Co.; to Nev. (frequent), Utah, Wyo., Ida., e. Ore. July–Aug.

17. **E. pùmilus** Nutt. ssp. **intermèdius** Cronq. [*E. p.* ssp. *i.* var. *gracilior* Cronq.] Stems erect from a multicipital caudex surmounting a taproot, 1–5 dm. high, leafy throughout, from rather stout and polycephalous to slender and monocephalous; herbage densely hirsute and sometimes a little glandular; lvs. oblanceolate, up to 10 cm. long and 8 mm. wide; heads hemispheric, the disk 7–15 mm. wide; invol. 4–7 mm. high; phyllaries finely glandular and villous, with thickened brown midrib; rays 50–100, 6–15 mm. long, blue, pink, or white; disk-corollas glabrous or nearly so; pappus all bristly.—Valley floors and open slopes, often among sagebrush, 4000–7000 ft.; Sagebrush Scrub, N. Juniper Wd., Yellow Pine F.; mostly transmontane, from Placer Co. to Siskiyou and Modoc cos.; n. to B.C., e. to Mont., Utah. June–Aug.

Ssp. **concinnoìdes** Cronq. [*Distasis concinna* H. & A. *E. concinnus* T. & G. *E. concinnus* var. *eremicus* Jeps.] Usually more compact and sometimes a little more robust, more hispid-hirsute; throat of disk-corollas obviously scabrous-puberulent; inner pappus of bristles, the outer of relatively broad (even subquadrate) scales.—Dry, gravelly slopes, 4000–6000 ft.; Joshua Tree Wd., Pinyon-Juniper Wd.; e. Mojave Desert, s. Inyo Co., e. San Bernardino Co. May–June.

18. **E. múlticeps** Greene. Stems ascending from a stout multicipital caudex surmounting a woody taproot, 1.5–2 dm. long, slender, branched; herbage cinereous-strigose; lvs. principally basal, oblanceolate to narrowly obovate, up to 4 cm. long, the blade shorter than the slender petiole, the cauline nearly linear; heads several, on bracteate peduncles, the disk 8–10 mm. wide; invol. 4–5 mm. high, like the summit of the peduncle finely glandular and villous with spreading hairs; rays 75–125, 5–7 mm. long, pale purple; outer pappus bristles ± squamately united.—Gravelly spots near river banks, N. Fork of Kern R., near Kernville. June.

19. **E. divérgens** T. & G. [*E. divaricatus* Nutt., not Michx. *E. californicus* Jeps. *E. tephrodes* Greene.] Biennial or short-lived perennial from a slender taproot, 1–5 dm. high, the slender stems decumbent to erect, branched; herbage rather densely pubescent with short spreading hairs; basal lvs. oblanceolate or spatulate, petiolate, up to 6 cm. long, rosulate but soon deciduous, the cauline numerous, smaller, linear-oblong

or linear; heads numerous, the disk 7–11 mm. wide; invol. 4–5 mm. high, finely glandular and hairy; rays 75–150, 5–10 mm. long (rarely much reduced), light blue to lavender-pink or whitish; inner pappus of 5–12 very fragile bristles, the outer of very short narrow scales.—Moist or dry sandy soils of valleys, ravines, ridges and meadow borders, 200–5000(–7700) ft.; mostly Foothill Wd. up to Lodgepole F.; mts. of n. Calif. (Trinity and Siskiyou cos.) through the Sierra Nevada to Kern Co., San Gabriel Mts. to San Bernardino and Laguna mts., rare on both deserts; to L. Calif., Tex., Okla., Mont., B.C. March–Nov.

20. **E. cálvus** Cov. Taprooted hirsute biennial ca. 1 dm. high, widely branching from the base, the stems finely glandular toward the heads; basal lvs. numerous, with oblong to obovate blade 1–1.5 cm. long tapering to a petiole twice as long, the cauline reduced, spatulate; heads terminating the branches, the disk ca. 13 mm. wide; invol. 5 mm. high, finely glandular and coarsely hirsute; ♀ fls. present, but the rays wanting or inconspicuous and shorter than the style; inner pappus of 15–20 firm sordid bristles, the outer of several relatively long firm setae.—Foot of Inyo Mts., ca. 4 miles n. of Keeler, Inyo Co. May.

21. **E. aphanáctis** (Gray) Greene. [*E. concinnus* var. *a.* Gray.] Cespitose perennial with a taproot and branching caudex, the stems mostly erect, 1–3 dm. high, ± branched, the herbage densely short-hirsute; basal lvs. crowded, linear-oblanceolate to spatulate, long-petiolate, up to 8 cm. long and 6 mm. wide, often deciduous, the cauline numerous but reduced; heads several or solitary, yellow, often brownish in age; invol. 4–6 mm. high, obscurely glandular and short-hirsute; ♀ fls. eradiate, or sometimes with rays shorter than the disk; inner pappus of 7–20 bristles, the outer of evident sometimes narrow scales.—Dry stony slopes and mesas, 5000–8000 ft.; Sagebrush Scrub, Yellow Pine F., Joshua Tree Wd., N. Juniper Wd.; largely transmontane, San Bernardino Mts. (rare), Providence, Grapevine, Argus, and Inyo mts., s. Inyo Co. to Sierra and Lassen cos. May–Sept.

Var. **congéstus** (Greene) Cronq. [*E. c.* Greene.] Plant essentially scapose; corolla-lobes usually turning reddish or purplish, at least in age.—Dry places, 6000–8500 ft.; Yellow Pine F.; the common form in the San Bernardino Mts. June–Aug.

22. **E. chrysópsidis** Gray ssp. **Aústinae** (Greene) Cronq. [*E. A.* Greene.] Cespitose perennial with a taproot and compact branching caudex, the many scapiform peduncles erect, 5–15 cm. high; herbage rather densely and coarsely short-hirsute; lvs. densely tufted, nearly all basal, linear-oblanceolate to linear, 2–8 cm. long, 1–3 mm. wide; heads solitary, the disk 10–17 mm. wide; invol. 5–7 mm. high, hispid-hirsute and obscurely glandular; ♀ fls. 20–50, eradiate or usually with yellowish rays present but no longer than the orange-yellow disk; pappus of 15–25 slender bristles, usually with a few short outer setae.—Dry openings among brush, 4000–5500 ft.; Sagebrush Scrub, N. Juniper Wd.; Lassen Co. to Modoc Co.; se. Ore., sw. Ida. May–July. The sp. in typical form has prominent yellow rays and occurs in n. Ore.

23. **E. Bloòmeri** Gray. [*E. filifolius* var. *B.* A. Nels.] Cespitose perennial with a taproot and much-branched caudex, 5–15 cm. high; herbage finely white-strigose; lvs. densely tufted at or very near the base, linear, 2–7 cm. long, 0.8–2 mm. wide; heads solitary, the disk 7–20 mm. wide; invol. 5–10 mm. high, strigose to villous, especially toward base, obscurely glandular, the phyllaries broad; ♀ fls. wanting; disk-corollas yellow, 4.5–7 mm. long; style-appendages mostly lanceolate, 0.3–0.5 mm. long; pappus essentially uniseriate, of 25–40 bristles, sometimes with a few small outer setae.—Dry often rocky ground, 4000–6500 ft.; Sagebrush Scrub, N. Juniper Wd., Yellow Pine F.; Nevada Co. to Trinity, Siskiyou and Modoc cos.; to Wash., Ida., Nev. June–Aug.

Var. **nudàtus** (Gray) Cronq. [*E. n.* Gray.] Stem and lvs. glabrous or finely and sparsely strigose; invol. glabrous or very nearly so; style-appendages mostly deltoid, 0.2–0.3 mm. long.—Serpentine slopes and rocky ridges, 2000–7500 ft.; Coniferous F.; Siskiyou and Klamath ranges, Del Norte Co.; sw. Ore. June–July.

Var. **pùbens** Keck. Herbage soft-pilose with spreading hairs; otherwise like the typical var.—At 6000–6200 ft.; Red Fir F.(?); Marble Mts., Siskiyou Co. July–Aug.

24. **E. lineàris** (Hook.) Piper. [*Diplopappus l.* Hook.] Cespitose perennial with a taproot and much-branched caudex, forming clumps 0.5–2 dm. in diameter and 1–3 dm. high; herbage finely white-strigose; stem- and basal lf.-bases conspicuously indurated and somewhat enlarged, stramineous (or purplish); lvs. linear or nearly so,

1.5–9 cm. long, 0.5–3 mm. wide, principally basal or sometimes also cauline; heads solitary or few on erect stems, the disk 8–13 mm. wide; invol. 4–7 mm. high, strigose-villous and sometimes finely glandular, the phyllaries subequal or imbricate, relatively broad, with oil-filled midrib; rays 15–45, 4–11 mm. long, yellow; pappus of 10–20 fragile inner bristles and ca. as many short usually narrow outer scales.—Rocky slopes, 4500–10,000 ft.; N. Juniper Wd., Sagebrush Scrub, up to Subalpine F.; largely trans-montane, Modoc and Siskiyou cos. through the Sierra Nevada to Tuolumne Co.; to Nev., Mont., B.C. May–Aug.

25. **E. elegántulus** Greene. [*E. linearis* var. *e.* J. T. Howell.] Habitally very similar to *E. linearis*, 3–15 cm. high; herbage rather sparsely gray-strigose; lvs. narrowly linear, not over 1 mm. wide, almost all in basal tufts; heads solitary, small, the disk 7–11 mm. wide; invol. 3.5–5 mm. high, strigose and viscidulous, as in *E. linearis;* rays 15–25, 6–9 mm. long, blue or pink to almost white; pappus of 20–30 fragile bristles, occasionally with a few slender outer short setae.—Open places in sagebrush plains, often among basaltic rocks, 3500–7000 ft.; Sagebrush Scrub, N. Juniper Wd.; Lassen, Modoc, and Siskiyou cos.; s. and e. Ore. June–July.

26. **E. barbellulàtus** Greene. Loosely cespitose perennial 5–15 cm. high, from a much-branched caudex, the taproot often poorly developed; herbage finely strigose; stem- and basal lf.-bases conspicuously smooth, shining, ± enlarged and indurated, usually purplish; lvs. oblanceolate, mostly 1.5–4 cm. long and 1.5–5 mm. wide, principally basal; heads solitary, the disk 13–18 mm. wide; invol. 5.5–9 mm. high, loosely hirsute and ± finely strigose, somewhat glutinous; phyllaries subequal or slightly imbricate, long-acuminate, the back ± thickened; rays 15–35, 7–12 mm. long, purplish; pappus of 25–40 firm bristles, with very few short outer setae.—Gravelly or rocky slopes, 7000–9000 ft.; Red Fir F. to Subalpine F.; scattered from Lassen Co. through the Sierra Nevada to Fresno Co. June–Aug.

27. **E. filifòlius** Nutt. [*Diplopappus f.* Hook. *Chrysopsis canescens* DC.] Perennial from a taproot and branching woody caudex, 1–5 dm. high; herbage white-strigose, the stems densely so toward base; lvs. linear-oblanceolate to filiform, 1–8 cm. long, up to 2 mm. wide, well distributed but the basal usually more crowded and longer; heads 1–several per stem, the disk 5–15 mm. wide; invol. 4–6 mm. high, villous-strigose or finely glandular-puberulent or commonly both; phyllaries subequal or imbricate, firm, ± thickened on the back, acuminate; rays 15–50, 4–10 mm. long, lavender or bluish; pappus of 20–30 slender bristles, sometimes with a few short slender outer setae.—Dry sandy soils, often among sagebrush, 4000–6500 ft.; Sagebrush Scrub, N. Juniper Wd., Yellow Pine F.; e. side of the Sierra Nevada, from Sierra Co. to Modoc and Siskiyou cos.; to w. Nev., n. Utah, Mont., B.C. June–July.

28. **E. lasseniànus** Greene. [*E. l.* var. *deficiens* Cronq.] Perennial from an evident taproot, the weak spreading slender decumbent stems 6–20 cm. long, forming a mat; herbage finely hirsute with spreading hairs; basal lvs. linear-oblanceolate, ± triplinerved, acute or obtusish, tapering to the slender petiole, up to 10 cm. long and 7 mm. wide, the cauline rather numerous, reduced, linear; heads usually several (up to 20), the disk 6–10 mm. wide; invol. 5–6 mm. high, white-hirsute and glandular; phyllaries ± evidently imbricate, acuminate; rays 10–25, 5–6 mm. long, blue or pink to sometimes white, rarely wanting; pappus of 15–25 very slender bristles with several very short and slender outer setae.—Dry sandy soils, as meadow borders, 4000–6500 ft.; Yellow Pine F., Red Fir F.; Sierra Nevada from Eldorado Co. to Plumas and Butte cos. June–Aug.

29. **E. nevadíncola** Blake. [*E. nevadensis* Gray, not Wedd.] Coarse perennial with a stout taproot and a thickened simple or forked crown; stems usually decumbent at base, 1–3 dm. high; herbage strigose-canescent or appressed-hirsute; basal lvs. linear-oblanceolate, triplinerved, acute or obtuse, tapering to the petiole, up to 20 cm. long and 10 mm. wide, the cauline few and abruptly reduced; heads usually solitary, on rather naked peduncles, the disk 10–20 mm. wide; invol. 7.5–10 mm. high, villous-hirsute, scarcely or not at all glandular; phyllaries subequal, indurate toward base, the thin tips scarious-margined, acute or acuminate; rays 20–40, 5–11 mm. long, white to blue-lavender; disk-corollas mostly 5–7 mm. long; style-appendages acute, 0.3–0.4 mm. long; pappus of 25–40 rather coarse bristles and a few slender inconspicuous short outer setae. —Rocky slopes, often with sagebrush, 6000–7500 ft.; Sagebrush Scrub, N. Juniper Wd.;

mostly transmontane, Plumas Co. to Mono Co.; to Elko and Nye cos., Nev. June–July.

30. **E. Eatònii** Gray ssp. **plantaginèus** (Greene) Cronq. [*E. p.* Greene. *E. Sonnei* Greene. *E. nevadensis* var. *S.* Smiley. *E. E.* var. *p.* Cronq.] Perennial from a taproot and a woody compact or ± branched caudex, the decumbent stems 5–25 cm. long; herbage finely strigose or rather loosely strigose-canescent; lvs. crowded at base, narrowly linear-oblanceolate, tapering to apex and the petiolate base, 3–12 cm. long, 1–5 mm. wide, triplinerved, the cauline several, ± strongly reduced; heads solitary or few, the disk 9–13 mm. wide; invol. 5–7 mm. high, sparsely villous-hirsute, sparsely if at all glandular; phyllaries subequal or slightly imbricate, with dark midvein, often purple-tipped; rays 15–35, 5–8 mm. long, usually white, sometimes lavender; disk-corollas 3.5–5 mm. long; style-appendages obtusish, 0.1–0.2 mm. long; pappus of 15–25 fragile bristles and a few very fine short outer setae.—Sand or clay flats or rocky places, 4000–9000 ft.; Sagebrush Scrub up to Lodgepole F.; mostly e. of the Sierran crest, Mono Co. n., Salmon Mts., Trinity Co., Warner Mts., Modoc Co.; s. Ore., w. Nev. May–July. Typical *E. Eatonii*, with more distinctly imbricate and glandular invol. and more numerous florets, occurs in the Rocky Mts.

31. **E. decúmbens** Nutt. ssp. **robústior** Cronq. Taprooted perennial with stems decumbent and ± anthocyanous at base, simple or branching above, 1.5–4 dm. high; herbage rather sparsely strigose; lvs. well distributed, the basal linear-lanceolate, tapering to the apex and the slender petiole, to 15 cm. long and 8 mm. wide, triplinerved, the cauline gradually reduced up to base of the naked peduncles; heads solitary to several, the disk 10–15 mm. wide; invol. 6–8 mm. high, moderately hirsute and obscurely viscid; phyllaries subequal, rather thin and broad, acuminate; rays ca. 20–40, up to 20 mm. long and 3 mm. wide, blue-lavender; pappus of 12–16 slender fragile bristles, sometimes with a few very fine and short outer setae.—Openings in woods, 3000–5000 ft.; Yellow Pine F., Douglas-Fir F.; Humboldt, Trinity, and Plumas cos.; s. Ore. June–July. Typical *E. decumbens*, with invol. 3.5–5 mm. high, occurs in w. Ore.

32. **E. flexuòsus** Cronq. Taprooted perennial with a short slenderly branched caudex, the several lax slender stems 1–3 dm. high, branching above; herbage strigose, the hairs under the heads ± spreading; lvs. well distributed, linear, 1-nerved, the basal not conspicuously tufted, long-petiolate, up to 10 cm. long and 3 mm. wide, the cauline extending up to the several or rather numerous heads; disk 5–10 mm. wide; invol. 4–5 mm. high, moderately villous and obviously glandular-viscid, the phyllaries subequal, short-acuminate, with dark thickened midvein; rays 12–20, 5–7 mm. long, pink; pappus of 15–25 very fragile bristles and few fine short outer setae.—Forest openings, 2000–5000 ft.; Yellow Pine F.; Trinity Alps Resort, Minersville, and Weaverville, Trinity Co. June–July.

33. **E. compáctus** Blake. [*E. pulvinatus* Rydb., not Wedd.] Pulvinate-cespitose perennial with a short multicipital caudex surmounting a shreddy-barked stout taproot; lvs. basal, densely clustered, white-strigose, linear, 4–20 mm. long, up to 1 mm. wide; scapes 2–6 cm. high; heads solitary, the disk 7–14 mm. wide; invol. 5–6 mm. high, finely strigose and glutinous; phyllaries subequal or ± imbricate, pallid with dark midrib, firm, acute; rays 15–30, 6–9 mm. long, white or pink; pappus of 30–40 slender firm sordid bristles and a few slender outer setae.—Dry gravelly, often calcareous soils, 6000–7500 ft.; Pinyon-Juniper Wd.; Westgaard Pass, Inyo Mts.; to ne. Nev., nw. Utah. June.

34. **E. Paríshii** Gray. Stems erect, somewhat tufted on the stout woody short-branched caudex above the taproot, 1–3.5 dm. high, mostly unbranched; herbage silvery with a dense villous-strigose pubescence; lvs. well distributed, linear or linear-oblanceolate, the lower ones up to 6 cm. long and 4 mm. wide, sometimes withered before anthesis, the cauline gradually reduced upward; heads solitary or few, the disk 10–15 mm. wide; invol. 5–7 mm. high, silvery strigose near base and densely glandular-puberulent; phyllaries imbricate, acuminate, olivaceous; rays 30–50, 6–13 mm. long, rose to lavender; pappus of 18–26 firm sordid bristles and several conspicuous whitish outer setae.—Dry slopes, 3500–5000 ft.; Joshua Tree Wd.; Cushenbury Canyon, s. Mojave Desert, n. base San Bernardino Mts. May–June.

35. **E. utahénsis** Gray. Habit of *E. Parishii*, from a heavy root, the erect stems 1–5 dm. high, sometimes branching above, silvery strigose, densely so at base, ± viscidulous; lvs. well distributed, strigose and gray-green, the lower linear-oblanceolate, up to 10 cm. long and 6 mm. wide, commonly much narrower, the upper gradually reduced; heads

1–10, the disk 8–15 mm. wide; invol. 4–6 mm. high, ± silvery strigose and glandular-puberulent; phyllaries strongly imbricate; rays 10–40, 9–18 mm. long, deep to pale lavender or whitish; pappus of 35–45 firm bristles, the outer smaller setae often obscure; aks. 4–6-nerved, densely hairy.—Dry limestone slopes, 5000 ft.; Joshua Tree Wd.(?); rare, Providence Mts., Mojave Desert; to Nev., Utah, sw. Colo., n. Ariz. May–June.

36. **E. argentàtus** Gray. [*Wyomingia a.* A. Nels.] Densely cespitose perennial with a taproot and compact branching caudex; stems rather numerous, erect, sparsely leafy, simple or branching above, 1–4 dm. high; herbage silvery strigose; basal lvs. densely tufted, linear-oblanceolate, up to 6 cm. long and 6 mm. wide, mostly smaller; heads usually solitary, the disk 10–18 mm. wide; invol. 6–9 mm. high; phyllaries strongly imbricate, the outer silvery strigose, all obscurely glandular, acute or acuminate; rays 20–50, 9–15 mm. long, lavender-purple or paler; pappus of 25–40 slender white bristles and as many fine short outer setae; aks. 6–8-nerved, rather densely coarse-hairy.—Rocky slopes, 6000–8500 ft.; Pinyon-Juniper Wd.; White and Inyo mts., Inyo Co.; Nev. and Utah. June–July.

37. **E. aequifòlius** Hall. Root-crown deep-seated, giving rise to very slender branching rhizomes that terminate in erect stems 1–2 dm. high, these sometimes branching at base; herbage rather rough-hirsute and densely glandular-puberulent; lvs. evenly distributed, much longer than the internodes, oblanceolate, rounded at apex, tapering to a narrow base, 5–20 mm. long, 1.5–3 mm. wide, only gradually reduced upwards; heads 1(–3), the disk 8–13 mm. wide; invol. 4–5 mm. high, rather densely glandular-puberulent and sometimes also sparsely hirsute; phyllaries imbricate, acuminate, with brown midvein; rays 20–40, 5–6 mm. long, lavender or light blue; pappus of 20–35 slender white bristles and a few well developed outer setae; aks. 2-nerved, sparsely hairy.—Dry ridges, 6000–7000 ft.; Yellow Pine F., Red Fir F., upper Kern R., Tulare Co. July.

38. **E. Bréweri** Gray. Root-crown usually deep-seated, giving rise to slender rhizomes that terminate in trailing to erect stems 1–3 dm. or more high, these often branched at base; herbage rather densely retrorse-hirtellous or partly spreading-hirsute to glabrate; lvs. numerous, rather uniform, longer than the internodes, linear to oblong or oblong-oblanceolate, acute to rounded, 0.5–4 cm. long, 1–6 mm. wide; heads several or occasionally solitary, the disk 7–15 mm. wide; invol. 4–6 mm. high, rather densely glandular-atomiferous; phyllaries relatively broad, abruptly acute to acuminate, usually with rather obviously greenish tips; rays 10–45, 4–10 mm. long, generally blue; pappus of 20–50 slender bristles and a few inconspicuous outer setae.—Dry rocky places, 5000–10,500 ft.; Yellow Pine F. to Lodgepole F.; Sierra Nevada from Shasta and Lassen cos. to Kern Co., San Gabriel and San Bernardino mts.; w. Nev. July–Aug. A glabrate form of the cent. Sierra Nevada has been given the name var. *Élmeri* (Greene) Jeps. [*Aster E.* and *E. E.* Greene.]

Var. **porphyrèticus** (Jones) Cronq. [*E. p.* Jones. *E. foliosus* var. *p.* Compton.] Invol. generally hispidulous as well as glandular, 5–8 mm. high, the phyllaries relatively narrow and long-acuminate, without evident green tips; lvs. rarely more than 2 mm. wide.— Largely transmontane, 5000–9500 ft.; Pinyon-Juniper Wd., Yellow Pine F.; e. flank of the Sierra Nevada (uncommon), Sweetwater, White, Inyo, and San Bernardino ranges; to cent. and n. Nev. May–Aug.

Var. **jacínteus** (Hall) Cronq. [*E. j.* Hall.] Stems 1 dm. or less long; pubescence longer and finer; lvs. not over 12 mm. long, crowded.—Rocky ridges and talus, 6000–11,480 ft.; Yellow Pine F. to Subalpine F.; San Gabriel, San Antonio, San Bernardino, and San Jacinto mts. July–Aug.

39. **E. foliòsus** Nutt. [*E. Douglasii* T. & G. *E. mariposanus* Congd. *E. striatus* Greene.] Perennial with a stout taproot and nearly superficial root-crown, not clearly rhizomatous; stems erect, sometimes suffrutescent at base, 2–10 dm. high, glabrous to ± strigose; lvs. numerous, ± crowded, rather uniform, linear to narrowly oblong, 2–6 cm. long, the larger ones mostly over 1.5 mm. wide, hairy, often with pustulate hairs; heads arranged in a usually naked corymbiform infl., the disk 10–18 mm. wide; invol. 4–7 mm. high, glabrate to strigose or spreading-hairy or glandular or both; phyllaries imbricate, acuminate, the inner stramineous, with brown midrib; rays 20–60, 5–12 mm. long, blue or purple; pappus of 20–30 slender tawny bristles and a few inconspicuous outer setae; aks. 2-nerved, sparsely hairy.—Grassy or brushy slopes, up to 3500 (or even 5500) ft.;

Foothill Wd., Chaparral, Yellow Pine F., Oak Wd.; widespread in cismontane Calif., occasional in the Sierran foothills and N. Coast Ranges, common from San Francisco Bay s., Channel Ids.; n. L. Calif. May–July.

KEY TO VARIETIES

A. Plant 25 cm. high or usually less; lvs. very crowded, the central internodes usually less than 3 mm. long. Klamath region .... var. *confinis*
AA. Plants taller; lvs. less crowded, the central internodes usually more than 3 mm. long.
    B. Stems glabrous to strigose, or the hairs near base of stem spreading; lvs. glabrous to variously hairy.
        C. Lvs. linear to narrowly oblong, the larger ones mostly over 1.5 mm. wide, if less, then flat and without pustulate hairs.
            D. Herbage ± hairy, often with pustulate hairs. Chiefly from Eldorado Co. and San Francisco s. .... var. *foliosus*
           DD. Herbage glabrate or sparsely strigose, without pustulate hairs. Chiefly n. Sierra Nevada and N. Coast Ranges .... var. *Hartwegii*
        CC. Lvs. linear-filiform or narrowly linear, 1.5 mm. wide or less, either folded (or involute) or with some of the marginal hairs pustulate. Fresno and Monterey cos. s.
                                      var. *stenophyllus*
    BB. Stems and lvs. densely puberulent or hirtellous with spreading hairs.
        C. Aks. glabrous; stem canescent. Subcoastal dunes, San Luis Obispo and Santa Barbara cos. var. *Blochmanae*
        CC. Aks. ± hairy; stem hirtellous. Mojave Desert, extending into the Sierra Nevada var. *Covillei*

Var. **confinis** (Howell) Jeps. [*E. c.* Howell.] Low plants with several stems, these often monocephalous or with few heads; herbage usually strigose-hirsute, sometimes spreading-hirsute or glabrate; lvs. very crowded, linear-filiform to linear-oblanceolate, 0.5–2.5 mm. wide and less than 4 cm. long; rays comparatively showy; aks. sparsely hairy.—On brushy slopes and rocky ridges, 3000–6000 ft.; Yellow Pine F., Douglas-Fir F.; Trinity, Humboldt, w. Siskiyou, and Del Norte cos.; to Linn Co., Ore. June–Aug.

Var. **Hartwégii** (Greene) Jeps. [*E. H.* Greene. *E. Blasdalei* and *mendocinus* Greene.] Stems 2.5–5(–8) dm. high; herbage subglabrous or sparsely strigose; lvs. 1–4 mm. wide, flat, without pustulate hairs; rays comparatively showy; aks. hairy.—Open or brushy, usually rocky slopes, mostly below 3000 ft.; Foothill Wd., Coniferous F., etc.; Sierran foothills from Merced Co. n., Coast Ranges from Monterey Co. n.; sw. Ore. May–Aug.

Var. **stenophýllus** (Nutt.) Gray. [*E. s.* Nutt. *E. f.* var. *tenuissimus* Gray. *E. t.* Greene. *E. Nuttallii* Heller. *E. Setchellii* Jeps. *E. fragilis* Greene.] Generally tall and stout, rarely less than 4 dm. high; lvs. approaching filiform, not over 1.5 mm. wide, either folded or bearing some marginal pustulate hairs or commonly both, the surfaces commonly glabrate, sometimes short-hairy.—Dry places, up to 6200 ft.; Coastal Sage Scrub, S. Oak Wd., Chaparral, Yellow Pine F.; from the coast to the borders of the desert, San Luis Obispo Co. s., Channel Ids.; L. Calif. May–Aug.

Var. **Blóchmanae** (Greene) Hall. [*E. B.* Greene.] Tall and stout; herbage finely but rather densely canescent with crinkled hairs; infl. densely corymbiform; aks. glabrous or nearly so.—Limited to sand dunes near sea level; Coastal Strand, Coastal Sage Scrub; coast of San Luis Obispo and n. Santa Barbara cos. July–Aug.

Var. **Covíllei** (Greene) Compton. [*E. C.* Greene.] Habit of var. *foliosus* but herbage finely and usually densely hirtellous, the hairs of the lvs. loose or spreading, those of the stem spreading or retrorse.—Dry places, up to 6000 ft.; Joshua Tree Wd., Pinyon-Juniper Wd., Yellow Pine F., etc.; scattered across the Mojave Desert from n. base of San Bernardino Mts. to Inyo Co. and e. flank of Sierra Nevada, also occasional on w. flank of Sierras from Mariposa Co. to Tulare Co. May–Aug.

40. **E. petróphilus** Greene. Stems clustered from a branching suffrutescent root-crown, 1–3 dm. high, rigid, very leafy up to the single or few loosely cymose heads; herbage rather densely hirsute-pubescent, often also ± glandular; lvs. much longer than the internodes, similar, broadly linear to narrowly oblanceolate, the larger 2–4 cm. long and 2–5 mm. wide; disk 9–15 mm. wide; invol. 5–9 mm. high, densely glandular but not hirsute; phyllaries strongly imbricate, firm, acuminate, greenish with brown midrib; ♀ fls. wanting; disk corollas 4.5–6.5 mm. long, yellow to reddish; pappus of 20–40 firm sordid bristles and few inconspicuous outer setae.—Rock outcrops, 1500–7500 ft.; Mixed Evergreen F., Foothill Wd., Chaparral, Yellow Pine F.; scattered through the

Coast Ranges, from Santa Lucia Mts., Monterey Co. to Siskiyou Mts., Siskiyou Co.; sw. Ore. July–Sept.

41. **E. miser** Gray. Stems numerous, rather tufted on the short-branched woody caudex or more slender root-crown above the taproot, ascending or erect, densely leafy, 5–25 cm. high; herbage ± densely villous-canescent and obscurely glandular; lvs. linear-oblong or oblanceolate, similar, crowded, 1–2 cm. long, 2–3 mm. wide, the lower ones subpetiolate; heads solitary or few in rather close cymes, the disk 7–14 mm. wide; invol. 4–6 mm. high, densely glandular-atomiferous like the summit of the peduncle, sometimes with a few stiff hairs; phyllaries evidently imbricate, firm, acuminate; ♀ fls. wanting; disk-corollas 3.7–5 mm. long; pappus double, the inner of 15–30 slender bristles, the outer of evident short setae.—Clefts in the granite, 6500–7500 ft.; Red Fir F.; Donner Pass vicinity, Nevada and Placer cos. July–Aug.

42. **E. inornàtus** (Gray) Gray. [*E. foliosus* var. *i.* Gray.] Stems several, from a woody root-crown above a taproot, erect, leafy throughout, 2–8 dm. high; herbage usually glabrous above and ± hispid-hirsute toward base, sometimes sparsely short-strigose almost throughout; lvs. numerous, linear to narrowly oblong, up to 5 cm. long and 6 mm. wide; heads few to many in a broad corymb, the disk 9–15 mm. wide; invol. 3–6(–7) mm. high, glabrous (or very inconspicuously glandular), occasionally with a few short stiff hairs; phyllaries strongly imbricate, linear-lanceolate, stramineous with brown midrib; ♀ fls. wanting; disk-corollas 4.2–7 mm. long; pappus of 30–50 firm sordid bristles and ± evident short outer setae.—Forest openings and dry rocky places, 1100–7350 ft.; Mixed Evergreen F., Foothill Wd., Yellow Pine F., Red Fir F.; Sierra Nevada from Tulare Co. n. (frequent only from Amador Co. n.), inner N. Coast Ranges from Lake Co. n.; to se. Wash., w. Nev. June–Sept.

KEY TO VARIETIES

A. Invol. not glandular; herbage usually glabrous in upper part of plant .... var. *inornatus*
AA. Invol. glandular-atomiferous.
  B. Principal lvs. mostly over 2 mm. wide; stem commonly either hirsute at base or glandular or both .... var. *viscidulus*
  BB. Principal lvs. 1–1.5 mm. wide; stem subglabrous, neither glandular nor spreading-hairy at base.
    C. Plants 2–10 dm. high; longer lvs. mostly over 2.5 cm. long. Coast Ranges
      var. *angustatus*
    CC. Plants 5–15(–20) cm. high; longer lvs. less than 2.5 cm. long. N. Sierra Nevada
      var. *reductus*

Var. **viscídulus** Gray. [*E. v.* Greene. *E. Biolettii* Greene. *E. i.* var. *B.* Jeps. *E. decumbens* Eastw., not Nutt.] Stems 1–9 dm. high; herbage rough-hairy or glandular-atomiferous or both; lvs. usually linear-oblong or linear-oblanceolate (narrower in a rare form in the s. Sierras); invol. 5–8 mm. high.—At 50–8000 ft.; many Communities from N. Coastal Scrub to Red Fir F.; Mt. Shasta region to Humboldt Co., thence s. to Pt. Reyes, Marin Co.; rare in the Sierra Nevada, Tuolumne Co. to Fresno Co. July–Aug.

Var. **angustàtus** Gray. [*E. a.* Greene.] Stems relatively tall, slender or stout; lvs. generally glabrous, needlelike; characters as in the key.—Low elevs.; Coast Ranges from Trinity Co. to San Mateo Co. July–Aug.

Var. **redúctus** Cronq. Stems short and slender; lvs. short and narrow; heads small, the invol. 4–7 mm. high.—At 5000–7000 ft.; Yellow Pine F., Red Fir F.; Sierra Nevada from Plumas Co. to Placer Co. July–Sept.

43. **E. ánnuus** (L.) Pers. [*Aster a.* L.] Stout leafy annual 6–12 dm. high; stem moderately hirsute, becoming strigose toward the corymbosely disposed heads; basal lvs. elliptic to broadly ovate, coarsely serrate, often abruptly contracted to the long petiole, the blade up to 10 cm. long and 5 cm. wide; cauline lvs. numerous, ample, lanceolate or wider, acute, becoming entire, hirsute or the upper merely ciliate; heads numerous, the disk 6–10 mm. wide; invol. 3–5 mm. high, the phyllaries subequal, sparsely hirsute; rays 80–120, 5–8 mm. long, white or bluish; pappus of 10–15 very fragile bristles (wanting in the ray-fls.) and outer slender squamellae; $2n = 27$ (Okabe, 1934).— Rare, Humboldt and Plumas cos.; an apomict introd. from n. and e. U.S. where widespread. June–July.

44. **E. strigòsus** Muhl. ex Willd. [*Doronicum ramosum* Walt. *E. r.* B.S.P., not Raf.

Diplemium s. Raf. *Stenactis* s. DC. *E. s.* var. *gracilis* Nutt. *Tessenia* r. Lunell. S. r.
Domin.] Annual (or biennial), slender, erect, simple below, often much-branched above,
3–8 dm. high, sparingly leafy, hirsute below, moderately strigose above, sometimes
glabrate; basal lvs. oblanceolate or elliptic, entire or toothed, tapering to a long slender
petiole, the blade 3–6 cm. long, mostly less than 15 mm. wide, the cauline becoming
reduced and entire upwards; heads small, usually numerous, the disk 5–10 mm. wide;
invol. 2–5 mm. high; phyllaries subequal, yellowish with narrow brown midrib, ±
hairy and obscurely glandular; rays 50–100, 5–6 mm. long, white or sometimes lilac;
pappus of 10–15 very fragile bristles (wanting in the rays fls.) alternating with prominent
but very short setose scales.—Moist places, below 5000 ft.; Humboldt Co. to Siskiyou
Co. and s. in the mts. to Mariposa Co. Adv. from n. and e. U.S. where a widespread
apomictic weed. June–July.

45. **E. lonchophýllus** Hook. [*E. racemosus* Nutt.] Biennial or perennial with solitary
or tufted erect slender stems 1–6 dm. high, hirsute, or the lvs. glabrate; basal lvs.
narrowly oblanceolate, entire, tapering to long petioles, the blade 3–6 cm. long, the
cauline mostly linear; heads usually several, borne on erect peduncles, the disk 8–15
mm. wide; invol. 4–9 mm. high, ± hirsute, not glandular; phyllaries usually imbricate,
thin, green, acuminate; rays 25–50, 2–3 mm. long, inconspicuous, white or pale lavender;
pappus of 20–30 long slender bristles, sometimes with few inconspicuous outer setae.
—Moist meadows, often alkaline, 6000–11,700 ft.; Sagebrush Scrub and Yellow Pine F.
up to Subalpine F. and Bristle-cone Pine F.; both slopes of Sierra Nevada from
Tuolumne and Mono cos. s., White, Inyo, and San Bernardino mts.; e. to Rocky Mts., n.
to Alaska, Sask., Que. July–Aug.

46. **E. àcris** L. var. **débilis** Gray. [*E. d.* Rydb.] Biennial or perennial, the stems
ascending-erect from a short caudex, 5–25 cm. high, spreading-hirsute or subglabrous,
becoming also glandular-puberulent in the infl.; basal lvs. spatulate or oblanceolate,
entire, up to 8 cm. long and 8 mm. wide, ± tufted, the cauline much reduced; heads
few in rather open corymbs to solitary, the disk 9–17 mm. wide; invol. 5–8 mm. high,
glandular-puberulent and hirsute, purplish; ♀ fls. several-seriate, of 2 kinds, the outer
with generally pinkish rays 3–4.5 mm. long, a few of the inner becoming rayless; pappus
of 25–35 slender bristles, sometimes with few short outer setae.—Lava outcrop, at
9500 ft.; Subalpine F.; Mt. Shasta; e. to Rocky Mts., n. to Alaska. July–Aug. Typical
E. acris is circumpolar in distribution.

93. **Lessíngia** Cham.

Erect or depressed branching summer annuals with flocculent-woolly to glabrate
generally glandular herbage. Lvs. alternate, sessile, entire or serrate or deeply lobed,
the basal rosulate and petioled, the upper sessile and often reduced to scalelike bracts.
Infl. paniculate, glomerate, or spicate, the heads small, turbinate to campanulate, 3–38-
fld. Phyllaries imbricated in several series, the green tips appressed or recurved. Fls. all
discoid, perfect, yellow, lavender, rose, or white; outer corollas enlarged, more deeply
cleft on the inner side, the ligulelike limb palmately 5-lobed, ± reflexed. Style-branches
flattened, the appendage very short, obtuse, penicillate-tufted. Aks. turbinate, silky-
villous. Pappus of numerous unequal usually distinct scabrous bristles, usually brownish
to red. Seven spp., almost restricted to Calif. (Honoring the *Lessing* family of Germany,
G. E., eminent author, Karl, the artist, and Christian F., author of a work on the
Compositae.)

(Howell, J. T. A systematic study of the genus Lessingia Cham. Univ. Calif. Publ. Bot. 16: 1–44,
1929.)

A. Glands spherical, short-stipitate; corollas yellow, sometimes with purplish band in throat
   (sometimes white or pinkish in var. *tenuis*) .... 1. *L. germanorum*
AA. Glands tack-shaped, or sessile, or none; corollas white to lavender or rose-color, never yellow.
    B. Plants of depressed dwarf habit, mostly less than 8 cm. high; tips of inner phyllaries white,
       crustaceous .... 2. *L. nana*
    BB. Plants or erect habit, mostly more than 15 cm. high; tips of inner phyllaries herbaceous or
       scarious, not crustaceous.
       C. Cauline lvs. with punctate glands.
          D. Phyllaries ± glandular but not hairy; heads 3–8-fld.......... 3. *L. nemaclada*
          DD. Phyllaries arachnoid-tomentose.
             E. Infl. cymose-paniculate; heads not borne sessile in a lf.-axil . 4. *L. leptoclada*

EE. Infl. spicate, virgate (glomerate in var.); heads borne sessile in lf.-axils
                         5. *L. virgata*
CC. Cauline lvs. without punctate glands.
   D. Lvs. with tack-shaped glands; outer corollas only slightly enlarged, not palmately
       spreading . . . . 6. *L. ramulosa*
   DD. Lvs. not glandular; outer corollas conspicuously enlarged, palmately lobed and
       spreading . . . . 7. *L. hololeuca*

1. **L. germanòrum** Cham. Divaricately branching throughout, 1.5–3 dm. high, gray-green with matted tomentum when young, obscurely glandular, the adult stems becoming very dark; basal lvs. narrowly oblanceolate, up to 3 cm. long, pinnately cleft into 4–5 pairs of oblong acute lobes, the cauline reduced, broadly sessile, pinnatifid, only the uppermost subentire; heads 25–38-fld.; invol. narrowly campanulate, 5–7 mm. high; corolla deep lemon-yellow, often with brown band in throat, the outer palmate; aks. long turbinate; pappus bristles uniseriate, 21–31.—Coastal dunes, 100–300 ft.; Coastal Strand, Coastal Sage Scrub; San Francisco and adjacent San Mateo cos. Aug.–Nov.

KEY TO VARIETIES

A. Glands very few, confined to the phyllaries, inodorous. San Francisco peninsula . *L. germanorum*
AA. Glands usually numerous in the infl., odorous.
  B. Outer phyllaries ± tomentose.
    C. Plants low, openly branched, spring-flowering. Cismontane.
      D. Basal lvs. toothed to pinnately divided; corolla yellow.
        E. Phyllaries broadly oblong, purple-tipped. Inner S. Coast Ranges
                                      var. *parvula*
        EE. Phyllaries broadly linear, green-tipped. Tehachapi region . . var. *Peirsonii*
      DD. Basal lvs. entire or rarely few-toothed; corollas white, yellow, or rose-color.
        Mt. Pinos region . . . . var. *tenuis*
    CC. Plants low, compact, fall-flowering. W. Colo. Desert, San Diego Co. . . var. *tomentosa*
  BB. Outer phyllaries glandular and often puberulent but not tomentose.
    C. Phyllaries purple-tipped, lacking large glands; plants spring-flowering. See vars. *parvula*
      and *tenuis* above.
    CC. Phyllaries green-tipped, often with large marginal glands; plants summer- or fall-flowering.
      D. Style-branch appendages truncate or nearly so, rarely ± subulate, mostly less than
        0.5 mm. long. Cismontane.
        E. Lvs. pectinate-pinnatifid; herbage ± glabrate, the stems often dark, rather
          spreading. Coastal sand hills, Monterey and San Luis Obispo cos.
                                       var. *pectinata*
        EE. Lvs. entire or pinnately lobed, or the lower divided; herbage prominently
          glandular or tomentose, the stems pallid, ± erect. Widespread in interior valleys
                                     var. *glandulifera*
      DD. Style-branch appendages subulate, 0.7–1 mm. long. Mojave Desert and Cuyama V.
                                        var. *Lemmonii*

Var. **párvula** (Greene) J. T. Howell. [*L. p.* Greene. *L. tenuis* var. *Jaredii* Jeps.] Stems slender, diffusely branching from the base or above or rarely simple, the ultimate branch-lets filiform, 5–15 cm. high; lvs. tomentose, the basal deeply pinnately divided, 2–3 cm. long, the cauline with salient acute lobes or entire; invol. glandular-granuliferous, larger marginal glands rare or none; corolla yellow with purplish ring in throat; pappus bristles free to base or several united at base, rarely reduced to few paleaceous awns. —Hillside openings, 1000–3000 ft.; Foothill Wd., Chaparral; inner Coast Ranges from Santa Clara and Stanislaus cos. to San Luis Obispo Co. Mostly May–June.

Var. **Peirsònii** J. T. Howell. Divaricately branching above or from the base, 3–15 cm. high, matted-woolly throughout or the tomentum somewhat deciduous; glands on margins of upper lvs. and phyllaries; lower lvs. entire, or the larger pinnately divided, 1–4.5 cm. long, the linear to oblanceolate entire to pinnately toothed cauline ones passing into the bractlike upper ones; phyllaries permanently tomentose and with only few large marginal glands.—Open slopes, 3000–4000 ft.; Foothill Wd., Chaparral; w. end of Mojave Desert and surrounding hills, n. Los Angeles Co., s. and e. Kern Co. July.

Var. **ténuis** (Gray) J. T. Howell. [*L. ramulosa* var. *t.* Gray. *L. t.* Cov. *L. heterochroma* Hall.] Habit and tomentum of var. *parvula*, laxly branched from the base, the ultimate peduncles very slender; basal lvs. spatulate, mostly entire, 1–3.5 cm. long; invol. glandular-granuliferous, larger marginal glands essentially none; outer corollas pinkish or whitish, the inner ones yellow; pappus bristles free or ± united.—Dry open slopes, 4000–7000 ft.; Yellow Pine F.; Mt. Pinos region, Ventura Co. May–July.

Var. **tomentòsa** (Greene) J. T. Howell. [*L. t.* Greene.] Low and compact, densely leafy, 10–15 cm. high; lvs. persistently white-tomentose, glandular only along the margin, thick, oblong, entire, apiculate, less than 1 cm. long; phyllaries with tack-shaped marginal glands apically, erect or subsquarrose; fls. yellow.—W. edge of Colo. Desert near Warners Ranch, San Diego Co. Oct.

Var. **pectinàta** (Greene) J. T. Howell. [*L. p.* Greene. *L. glandulifera* var. *p.* Jeps.] Plants erect or usually spreading, diffusely branching from near base, 1–3 or more dm. high, tomentose only at base, sparingly glandular and glabrate above; lower lvs. pectinate-pinnatifid, the segments cuspidate; phyllaries usually beset with large glands, the tips ± squarrose; fls. yellow.—Sand dunes near the coast; Coastal Strand, Coastal Sage Scrub; Monterey and San Luis Obispo cos. June–Sept.

Var. **glandulifera** (Gray) J. T. Howell. [*L. g.* Gray. *L. germanorum* vars. *vallicola* and *tenuipes* J. T. Howell.] Plants erect, 1–7 dm. high, cymose-paniculately much-branched above, tomentose at least at base, usually glabrate but with prominent yellow marginal glands throughout above, sometimes tomentose well up toward peduncles; phyllaries with marginal glands; corolla yellow, often with a brown band in throat.—Frequent in open ground, 50–5500 ft.; V. Grassland, Foothill Wd., S. Oak Wd., Coastal Sage Scrub, Yellow Pine F.; away from the immediate coast, from San Francisco Bay s. through the Coast Range valleys and the Cent. V., and from Placer Co. through the Sierran foothills to L. Calif. May–Nov.

Var. **Lemmònii** (Gray) J. T. Howell. [*L. L.* Gray. *L. ramulosìssima* A. Nels. *L. g.* var. *r.* J. T. Howell.] Low and spreading, branching from the base or below the middle, often very intricately so, 0.6–3 dm. high; spring-flowering plants tomentose almost throughout, by summer the tomentum restricted to base and plant finely hirtellous above, usually with glands restricted to bract- and phyllary-margins; corollas yellow without purplish band in throat; style-appendages slender-subulate.—In scattered stands, 1000–6000 ft.; Creosote Bush Scrub to Pinyon-Juniper Wd., Yellow Pine F.; Mojave Desert n. to Owens V., w. to Cuyama V., Santa Barbara Co., and Mt. Pinos, Ventura Co., s. to San Gabriel Mts., e. to nw. Ariz. May–Oct.

2. **L. nàna** Gray in Benth. [*L. n.* var. *caulescens* Gray. *L. Parryi* Greene.] Acaulescent, 1–2 cm. high, or caulescent with stout decumbent to erect stems 2–10 cm. long, gray-green with long thick densely matted wool, at length deciduous, the upper lvs. punctate-glandular beneath; basal lvs. linear-oblanceolate, petiolate, subentire to coarsely serrate, 1–5 cm. long; heads 12–18-fld.; invol. 8–10 mm. high; outer phyllaries herbaceous, acute, woolly, mucronate, the inner with broad scarious margin below and shiny white crustaceous attenuate tips exceeding the rose-color fls.; style-branch appendage short, blunt; pappus-bristles deep red, biseriate, 50–70.—Open clay hills, below 3000 ft.; V. Grassland, Foothill Wd.; Sierra foothills and adjacent plains, from Tehama Co. to Kern Co. June–Sept.

3. **L. nemáclada** Greene. [*L. leptoclada* var. *microcephala* Gray. *L. ramulosa* var. *m.* Jeps. *L. mendocina, cymulosa, fastigiata* and *paleacea* Greene. *L. n.* var. *mendocina* J. T. Howell.] Plants 1–8 dm. high, divaricately branching above or throughout, the ultimate branchlets filamentous; basal lvs. oblanceolate, sharply toothed, 3–4 cm. long, deciduous before anthesis, the cauline entire sessile, with tomentum sometimes persistent on the upper surface, the upper scalelike, punctate- and stipitate-glandular; heads 3–8-fld., solitary at ends of the branchlets, rarely 2–4-clustered; invol. slenderly turbinate or fusiform, 5–6 mm. high; phyllaries erect, ca. 4-seriate, linear-oblong, acuminate, stipitate-glandular; corollas all tubular or 1 or 2 palmate, purplish; style-branch appendage subulate; pappus of ca. 25 partly united bristles or of 5 bristly paleaceous awns.—Common on dry gravelly slopes, 800–3500 ft.; Foothill Wd., Yellow Pine F.; Sierran foothills, Fresno Co. to Siskiyou Co. (to Klamath R.), w. to Humboldt Co., s. to Mendocino and Lake cos. July–Oct.

Var. **albiflòra** (Eastw.) J. T. Howell. [*L. a.* Eastw. *L. glandulifera* var. *a.* Jeps.] Invol. sometimes more glandular; corollas white, often with purple throat, or the limb suffused with lavender.—Barren hills, in clay or serpentine, below 2500 ft.; Foothill Wd.; plains and hills w. and s. of the San Joaquin V., from Stanislaus Co. to Kern Co. May–Sept.

4. **L. leptóclada** Gray. [*L. l.* var. *tenuis* Gray.] Plants 3–9 dm. high, ± fastigiately

branched above the base and in upper half, the divergent branches slender-virgate; herbage tomentose or becoming glabrate, the upper lvs. glandular-punctate; basal lvs. spatulate, with few salient teeth, 2–5 cm. long, deciduous before anthesis, the upper broadly sessile, much reduced; heads 12–22-fld., solitary or 2–5-glomerate; invol. 6–10 mm. high, silvery woolly, less glandular than the lvs.; corollas lavender to blue-purple, the outer palmate; style-branch appendage very short, tufted; pappus 1–2-seriate, of 18–40 brownish distinct bristles or rarely united at base into sets.—Abundant on open dry ground, 1000–6200 ft.; Foothill Wd. to Yellow Pine F.; Sierra Nevada from Eldorado Co. to Kern Co. July–Oct.

5. **L. virgàta** Gray. [*L. subspicata* Greene.] Plants 1.5–6 dm. high, fastigiately few-branched above the base or in upper part, the divergent or divaricate branches rigidly virgate, rather closely and uniformly clothed with broadly ovate to oblong appressed lvs., these 1–2.5 cm. long, acute, usually entire, ± concave, densely but loosely woolly, or the lower side becoming glabrate, disclosing punctate glands; basal lvs. linear-oblanceolate, petiolate, to 3 cm. or so long, deciduous before anthesis; heads 3–6-fld., spicately sessile in the axil of a lf. of nearly the same length; invol. 5–7 mm. high, woolly; corollas lavender.—Open plains and bordering foothills, up to 1000 ft.; V. Grassland, Foothill Wd.; e. side of the Sacramento V., from Tehama Co. to San Joaquin and Stanislaus cos. June–Oct. Altitudinally below *L. nemaclada* and *L. leptoclada* with both of which it may hybridize. Var. *glomerata* (Greene) J. T. Howell (*L. g.* Greene), with a tendency to glomerate heads and having the habit of *L. nemaclada,* may be hybrid derivatives with that sp.

6. **L. ramulòsa** Gray in Benth. [*L. bicolor* and *adenophora* Greene. *L. r.* var. *a.* Gray.] Plants 1.5–5 dm. high, with slender ascending branches usually from the base upwards, tomentose to glabrate, rather densely glandular-pubescent above; basal lvs. oblanceolate, serrate, 3–5 cm. long, the cauline round-ovate to oblong, carinate, acute, all with copious persistent tomentum on upper side; heads 6–15-fld., solitary at tips of filamentous branchlets; invol. turbinate, 5–7 mm. high, glandular-puberulent; corollas pink or rose-purple, the outer enlarged but not palmate-reflexed; style-branch appendage subtruncate; pappus of 25–35 mostly free bristles or these ± united into several bristle-tipped or simple and paleaceous awns.—Clay soils, open ground, below 3000 ft.; Mixed Evergreen F., N. Oak Wd., Foothill Wd.; Mendocino, Lake, and Colusa cos. to Sonoma and Solano cos. July–Oct.

Var. **micradènia** (Greene) J. T. Howell. [*L. m.* Greene.] Simulating *L. nemaclada* in its slender habit and narrow few-fld. heads; divaricately and paniculately much branched from near the middle; lvs. with persistent copious tomentum on upper side; heads narrowly cylindro-turbinate, glandular-puberulent, 5–10-fld.; outer corollas not palmate; pappus often reduced in length and number of bristles or paleae.—Dry gravelly or serpentine slopes, 200–1500 ft.; Chaparral, Mixed Evergreen F.; n. of Mt. Tamalpais, Marin Co. Aug.–Oct.

Var. **glabràta** Keck. Resembling var. *micradenia,* 4–7 dm. high, glabrate and scarcely glandular at anthesis, paniculately much-branched, the heads terminating capillary branchlets; lower lvs. deciduous before anthesis, the upper rapidly reduced upwards to tiny appressed scalelike bracts, permanently tomentose on the upper face, soon glabrous on the lower face; heads narrowly cylindro-turbinate, 3–5-fld.; invol. 6–8-seriate, strongly imbricate, the phyllaries glabrous or the outer tipped with a sessile gland and often with a few small sessile glands on and near the margin, otherwise glabrous; marginal corollas not palmate.—Dry open gravelly slopes, in serpentine or clay, below 1000 ft.; Mixed Evergreen F., Foothill Wd.; e. side of Santa Cruz Mts., Santa Clara Co. Aug.–Sept.

7. **L. hololeùca** Greene. [*L. Bakeri* and *imbricata* Greene. *L. leptoclada* var. *h.* Jeps.] Stems slender to rather rigid, solitary or several from the base, with few to numerous rigid virgate branches above, 2–4 dm. high; herbage persistently white-tomentose or becoming glabrate, with glands essentially none or limited to minute granules on the phyllaries; basal lvs. ± persistent, oblanceolate, mostly entire, 2–5 cm. long, even the uppermost ± tomentose; heads 13–18-fld., cymose-paniculate; invol. turbinate, 9–12 mm. high, arachnoid-tomentose, the phyllaries with dark acuminate apiculate tips; marginal corollas palmately liguloid, pinkish to lavender; pappus biseriate, of 35–55 unequal free bristles.—Open grassland, below 1000 ft.; Coastal Sage Scrub, Mixed Evergreen

F.; mostly outer Coast Ranges, Napa and Marin cos. to Santa Clara Co., e. to Yolo Co. June–Oct.

Var. **arachnoìdea** (Greene) J. T. Howell. [*L. a.* Greene. *L. leptoclada* var. *a.* Blake.] Stems 3–8 dm. high, widely paniculately branching above; herbage glabrate, the lower lvs. deciduous before anthesis; heads 8–18-fld.; invol. 6–9 mm. high, arachnoid-tomentose; pappus of 14–27 bristles, shorter than in the sp. and sometimes ± united; n = 5 (Raven, 1958).—Grassy slopes, 300–2000 ft.; Coastal Sage Scrub; outer Coast Ranges, San Mateo Co. to Santa Cruz Co. July–Sept.

94. Conỳza Less.

Annual or perennial herbs. Lvs. alternate, pinnatifid or bipinnatifid to subentire. Heads small, several to numerous, disciform or minutely radiate. Invol. few-seriate, the subequal phyllaries narrow, subherbaceous, strongly reflexed in age. Pistillate fls. very numerous, several-seriate, their corollas tubular-filiform, in ours usually with an in-conspicuous white or purplish ligule barely equaling the style and pappus. Disk-fls. perfect, few, their narrowly tubular corollas (in ours) enlarged above. Style-branches with short ovate hispidulous appendages. Aks. small, oblong, compressed; pappus of rather few and fragile capillary bristles. More than 50 spp., chiefly trop. and subtrop., in both hemis. (Name used by Dioscorides and Pliny for some kind of Fleabane, presumably from the Greek, *konops*, a flea.)

(Cronquist, A. The separation of Erigeron from Conyza. Bull. Torr. Club 70: 629–632, 1943.)
A.   Upper lvs., at least, entire; rays-fls. ligulate.
    B.   Invol. glabrous or nearly so; phyllaries with translucent oil-filled midvein; aks. less than
        1 mm. long . . . . 1. *C. canadensis*
    BB. Invol. copiously hairy; aks. more than 1 mm. long.
        C.   Lateral branches often overtopping the main axis; phyllaries often whitish on inner
            face when reflexed; heads 1 cm. across when pressed . . . . 2. *C. bonariensis*
        CC. Lateral branches shorter than the main axis, the infl. pyramidal-compound; phyllaries
            usually red-brown on inner face when reflexed; heads less than 1 cm. across
                                                           3. *C. floribunda*
AA. Upper and lower lvs. coarsely dentate; rays-fls. eligulate; herbage glandular-viscid
                                                                      4. *C. Coulteri*

1. **C. canadénsis** (L.) Cronq. [*Erigeron c.* L.] Horseweed. Strict leafy annual up to 2 m. high, sometimes with erect branches in upper half, green, rather densely hirsute to nearly glabrous; lvs. many, the lower oblanceolate, petiolate, up to 10 cm. long and 10 mm. wide, entire or serrate, often deciduous before anthesis, the upper narrower, sessile, entire; heads small, numerous, paniculate; invol. 3–4 mm. high, the phyllaries linear and glabrous or nearly so, with conspicuous oil-filled central area and narrow subscarious margin; rays 25–40, white, very inconspicuous, scarcely exceeding the pappus; disks 7–12; $2n = 18$ (Okabe, 1934).—Common weed of waste ground, at low elevs.; throughout the U.S. and s. Canada, trop. and S. Am.; natur. in the Old World. June–Sept. Much of the Calif. material, with essentially glabrous stems, is referable to the var. *glabrata* (Gray) Cronq. [*Erigeron c.* var. *g.* Gray.]

2. **C. bonariénsis** (L.) Cronq. [*Erigeron b.* L. *E. linifolius* Willd. *C. ambigua* DC. *Conyzella l.* Greene.] Annual up to 1 m. high, subsimple or with erect leafy branches often overtopping the main axis; herbage grayish-green, ± densely strigose and hirsute; lvs. as in *C. canadensis;* heads rather numerous; invol. 4–5 mm. high, the phyllaries densely hirsutulous, whitish on the inner face when reflexed in fr.; rays 125–180, as in *C. canadensis;* disks 10–20; pappus whitish or stramineous; $2n = 54$ (Holmgren, 1919). —Frequent weed in waste ground, from the Cent. V. and San Francisco Bay to s. Calif.; to the Atlantic Coast; introd. from S. Am. June–Aug.

3. **C. floribúnda** HBK. [*Erigeron f.* Sch. Bip. *C. albida* Willd. ex Spreng.] Closely related to *E. bonariensis,* the lateral branches not surpassing the main axis, the whole forming a pyramidal compound infl.; stems to 1 m. high; heads slightly smaller, the phyllaries chestnut-brown on the inner face when reflexed in fr.; pappus stramineous or rufous.—Adv. at San Diego; widely established elsewhere, as in se. U.S. and in the Old World; native of trop. Am. June.

4. **C. Còulteri** Gray. [*Erigeron discoidea* Kell. *Conyzella C.* Greene. *Eschenbachia C.* Rydb.] Erect annual herb 2–10 dm. high, glandular-pubescent and villous or hirsute

throughout, the rigid very leafy stem usually simple below the much-branched paniculate infl.; lvs. narrowly oblong, coarsely toothed, the lower tapering to a margined petiole, the cauline sessile and 2–6 cm. long; heads small, very numerous; invol. 2–3 mm. high, the linear-attenuate phyllaries hirsute and glandular; ray-fls. 125–250, eligulate; disk-fls. ca. 5–15; pappus soft, whitish, well exceeding the invol.—Occasional in moist often alkaline flats, sometimes weedy, below 1000 ft.; V. Grassland, Foothill Wd., Coastal Sage Scrub, Creosote Bush Scrub; Santa Clara and Amador cos. s. to San Diego and Imperial cos.; Channel Ids.; L. Calif. e. to Colo., Tex., Mex. May–Oct.

95. Báccharis L.

Dioecious shrubs or undershrubs, or some perennial herbs, often resinous. Lvs. alternate, entire to toothed. Heads discoid, relatively small, many-fld., whitish or yellowish, corymbose or paniculate. Invol. imbricated, of chartaceous whitish phyllaries. Pistillate heads composed of tubular-filiform truncate or obscurely toothed ♀ fls. bearing copious pappus of capillary bristles. Staminate heads composed of tubular slightly dilated 5-toothed ♂ fls. with abortive ovary and scabrous often scanty and tortuous pappus of clavellate bristles. Aks. small, somewhat compressed, 5–10-ribbed. Complex and diverse Am. genus of some 300 spp., best developed in e. S. Am. (After the god *Bacchus*.)

A. Herbage puberulent or pubescent; aks. puberulent.
  B. Lvs. entire, 0.5–1.2 cm. long, mostly lost before anthesis . . . . 1. *B. brachyphylla*
  BB. Lvs. acutely serrate, 2–5 cm. long, persistent . . . . 2. *B. Plummerae*
AA. Herbage not hairy.
  B. Stems herbaceous to base; lvs. ovate-lanceolate; phyllaries thin, viscid-ciliate; aks. puberulent
                                                  3. *B. Douglasii*
  BB. Stems woody below; plants shrubby; phyllaries not viscid-ciliate; aks. glabrous.
    C. Plants persistently leafy, not broomlike (or sometimes ± so in *B. Emoryi*).
      D. Lvs. linear or linear-lanceolate, entire or evenly toothed, 5–15 cm. long; aks. 5-nerved.
        E. Heads in a cymose panicle at apex of stem; lvs. usually toothed
                                        4. *B. glutinosa*
        EE. Heads in cymose panicles at tips of numerous short lateral branches; lvs. mostly entire . . . . 5. *B. viminea*
      DD. Lvs. cuneate to oblong-lanceolate, few-toothed above the middle or entire; aks. 10-nerved.
        E. Lvs., at least the upper, narrowly oblong; infl. sparsely leafy . . 6. *B. Emoryi*
        EE. Lvs. all ovate or broadly cuneate; infl. densely leafy . . . . . 7. *B. pilularis*
    CC. Plants with lvs. mostly deciduous before anthesis, broomlike with numerous erect strongly sulcate fastigiate branches; heads scattered; aks. 10-nerved.
      D. Pistillate pappus very short, ca. 3 mm. long in fr.; larger lvs. obovate
                                         8. *B. sergiloides*
      DD. Pistillate pappus elongating in fr. (to 10 mm. or more); larger lvs. linear
                                         9. *B. sarothroides*

1. **B. brachyphýlla** Gray. Intricately branched shrub 5–10 dm. high, woody at base, the numerous slender stems sharply ascending, sulcate, densely hirtellous toward the heads; lvs. linear, acute, entire, sessile, scabrous, mostly scalelike on the branchlets; heads loosely paniculate; invol. of ♀ heads 5–6 mm. high, 3–4-seriate, the phyllaries lance-oblong, with green hispidulous midrib and whitish scarious margin; ripe aks. ca. 2.5 mm. long, puberulent, their brownish pappus 7 mm. long.—Local, in dry rocky washes, below 2500 ft.; Creosote Bush Scrub; Morongo Wash, Little San Bernardino Mts., to San Diego Co.; to Ariz., Son. Aug.–Nov.

2. **B. Plúmmerae** Gray. Rounded bush 6–10 dm. high; stems herbaceous above the woody base, rather stout, openly branching in the infl.; herbage crisped-pilose and a little viscid; lvs. elliptic-oblong to linear, acute to obtuse, sessile, acutely serrate, 1.5–5 cm. long, 3–13 mm. wide, becoming glabrate on upper surface; heads openly or closely and corymbosely paniculate; invol. of ♀ heads 6–8 mm. high, ca. 5-seriate, the linear-lanceolate phyllaries green except for the ciliate scarious margin; ♂ heads somewhat smaller, the pappus notably shorter, stouter, and more hairy than in the ♀ fls.; ripe aks. 2.7–3.2 mm. long, viscid-puberulent.—Brushy canyons and mountainsides near the sea, below 1000 ft.; Coastal Sage Scrub; coast of Santa Barbara and Los Angeles cos., Santa Cruz Id. Aug.–Oct.

3. **B. Douglásii** DC. [*B. Haenkei* DC.] Stems several from a suffrutescent base, 1–2 m. high, simple or with ascending branches above, green, glutinous; lvs. narrowly to

broadly lanceolate, short-petiolate, entire or serrulate, obviously glandular-punctate, 3–10 cm. long, 6–25 mm. wide, 3-nerved; heads numerous in terminal often compound corymbose clusters; invol. of ♀ and ♂ heads ca. 5 mm. high, 4–5-seriate, the phyllaries narrowly lanceolate, acuminate, thin, with narrow greenish center and whitish scarious margin becoming viscid-ciliate toward apex; ripe aks. 0.8 mm. long, minutely viscid-puberulent, 4–5-nerved.—Moist ground, near streams, below 1500 ft.; Coastal Salt Marsh, Coastal Scrub, Redwood F., Foothill Wd., Yellow Pine F.; Sierran foothills, Amador Co. to Tulare Co., Cent. V., more common along the coast, Humboldt Co. to San Bernardino and San Diego cos., Santa Rosa, Santa Cruz, and Santa Catalina ids.; Curry Co., Ore. July–Oct.

4. **B. glutinòsa** Pers. [*B. coerulescens* DC.] SEEP-WILLOW, WATER-WALLY. Willowlike shrub 1–3 m. high, the virgate stems simple up to the infl. or branched; herbage green, glutinous, the granular scurfy glands not strongly impressed-punctate; lvs. lance-linear, 5–15 cm. long, 7–18 mm. wide, acuminate, tapering into the short petiole, remotely serrate to entire, the midnerve much stronger than the 2 lateral nerves; heads numerous, in terminal compound corymbs; invol. ca. 4 mm. high, 3–4-seriate, the phyllaries ovate to lance-oblong, acutish, chartaceous, with obscure green midnerve above and narrow scarious margin ± fimbrillate toward tip, not at all viscid; ripe aks. less than 1 mm. long, glabrous, olivaceous, 5-nerved.—Common along watercourses and often forming thickets, up to 2500 ft.; several Communities; Owens V. s. through the Mojave Desert to the Colo. Desert, Riverside, San Diego, Imperial V.; to Colo., Tex., Mex.; S. Am. April–Oct.

5. **B. viminèa** DC. MULE FAT. More woody than *B. glutinosa*, which it simulates, 2–4 m. high, the willowlike stems with numerous short lateral fl.-branches; lvs. usually entire, sometimes denticulate, obscurely glutinous, somewhat smaller; heads in close cymose clusters at the ends of the lateral branches; invol. ca. 5 mm. high, 4–5-seriate, the phyllaries broad, more evidently ciliolate, not viscid; otherwise similar to *B. glutinosa*.—Dry stream beds, ditch-banks, etc., mostly below 1500 ft.; V. Grassland, Foothill Wd., Coastal Sage Scrub, etc.; Sacramento V. and bordering foothills, from Tehama and Butte cos. s. through the Coast Ranges to San Diego Co. and the Channel Ids.; n. L. Calif.; less common e. to sw. Utah, Ariz. Mostly March–July, but some fl. all year.

6. **B. Emòryi** Gray. Erect loosely branched shrub 1–3(–4) m. high, with tough stems and ascending sometimes subfastigiate branchlets, ± glutinous; lvs. cuneate or oblong oblanceolate, obtuse or acute, tapering to base, commonly few-toothed in outer half, 2–5 cm. long, 5–20 mm. wide, 3-nerved, the upper linear, 1-nerved, entire; heads numerous, glomerate in ample pyramidal panicles; invol. of ♀ heads 7–9 mm. high, ca. 6-seriate, the outer phyllaries ovate, obtuse, thick, whitish, with narrow scarious subciliate margin, the inner lance-linear, elongate, acuminate; invol. of ♂ heads broader, ca. 6 mm. high, 5-seriate, the innermost narrower phyllaries not prominent; ripe aks. 1.5–2 mm. long, glabrous, 10-nerved, their pappus 1 cm. long.—Mostly along streams, low elevs.; Coastal Sage Scrub, Creosote Bush Scrub, etc.; Bakersfield and Los Angeles to San Diego, Colo. Desert, Amargosa Desert, Inyo Co., Santa Catalina Id.; to n. L. Calif., s. Utah, w. Tex. Aug.–Dec.

7. **B. pilulàris** DC. Prostrate shrubs, forming dense mats 10–15 cm. high and 1–4 m. across; older stems very woody; lvs. very numerous, oval or obovate, very obtuse, cuneate at base, with 4–6 coarse teeth usually in outer half, occasionally entire, mostly 5–15 mm. long, thick, resinous, 1-nerved; heads numerous, in small axillary and terminal glomerules on the leafy branchlets; invol. 3–5 mm. high, ca. 5-seriate; phyllaries ovate (outer) to lance-oblong (inner), obtuse, stramineous, scurfy-glandular, indurate, with narrow scarious fimbrillate margin; ripe aks. 1.3–1.5 mm. long, glabrous, 10-nerved, their pappus 6–10 mm. long.—Windswept dunes and headlands along the coast, low elevs.; Coastal Strand, N. Coastal Scrub, Coastal Sage Scrub; Russian R., Sonoma Co. to Point Sur, Monterey Co. Aug.–Dec.

Ssp. **consanguínea** (DC.) C. B. Wolf. [*B. c.* DC. *B. p.* var. *c.* Kuntze. *B. congesta* DC.] CHAPARRAL BROOM. COYOTE BRUSH. Much-branched erect or rounded shrub 1–4 m. high; lvs. usually with 5–9 teeth, mostly 1.5–4 cm. long and 5–15 mm. wide; otherwise like the typical ssp., with which it intergrades, and much more widespread.—Common on hillsides and in canyons, below 2000 ft.; Coastal Scrub, Chaparral, Foothill Wd., Closed-cone Pine F., Mixed Evergreen F.; outer Coast Ranges throughout to San Diego Co., Channel Ids., local in cent. Sierran foothills from Butte Co. to Tuolumne Co.; to Tillamook Co., Ore. Aug.–Dec.

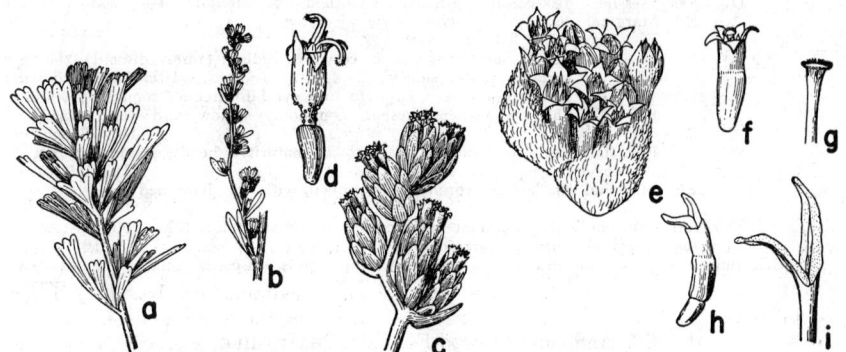

Fig. 114. ANTHEMIDEAE. *Artemisia tridentata: a,* vegetative shoot, × 1, with tridentate lvs.; *b,* flowering shoot, × ½, with entire lvs.; *c,* discoid heads, × 3, with imbricate phyllaries; *d,* fl., × 5, no pappus. A. *pycnocephala: e,* head, × 5, phyllaries woolly; *f,* ♂ fl., × 5, with reduced stigmatic area (see *g*); *h,* fertile fl., × 5, with normal stigmatic branches (shown also in *i*).

8. **B. sergilòides** Gray. Squaw Waterweed. Green glabrous rounded often nearly leafless shrub 7–20 dm. high, the broomlike branches strongly striate-angled; larger lvs. spatulate or obovate, entire or rarely few-toothed, 1–2.5 cm. long, 3–10 mm. wide, obtuse, apiculate, thick, very obscurely punctate; heads numerous, densely panicled, not glomerate; invol. 3 mm. high, 4–5-seriate, stramineous; phyllaries oval to lance-oblong, obtuse, firm, scurfy-glandular, with narrow whitish margin; ripe aks. 1.5 mm. long, glabrous, 10-nerved, their pappus not exceeding the styles, 2.5–3 mm. long.— Washes and canyon-bottoms, up to 4500 ft.; Creosote Bush Scrub to Pinyon-Juniper Wd.; both deserts from the Death V. region, Inyo Co. to San Diego Co.; to sw. Utah, Ariz., Son. April–Sept.

9. **B. sarothròides** Gray. Broom Baccharis. Erect glabrous glutinous green shrub 2–4 m. high, nearly leafless, with broomlike sharply angular-sulcate branches; lvs. all nearly linear, entire, rigid, up to 2 cm. long; heads mostly solitary at the tips of the numerous branchlets; invol. of ♀ heads 6–8 mm. high, ca. 6-seriate, cream-color, the outer phyllaries broadly ovate, the inner linear-oblong, obtuse, indurate; invol. of ♂ heads 3–4 mm. high, the broad blunt phyllaries with small green apical spot; ripe aks. 1.7–2.2 mm. long, glabrous, 10-nerved, their pappus much exceeding the styles, 6–11 mm. long.—Sandy washes, below 1000 ft.; Coastal Sage Scrub, Creosote Bush Scrub; San Diego region, e. to Colo. R.; to L. Calif., Ariz., sw. New Mex., Sinaloa. June–Oct.

Tribe 4. **Anthemídeae.** Mayweed Tribe (Fig. 114)

Strongly scented annual or perennial herbs or shrubs. Lvs. alternate, mostly ± dissected, if entire, then mostly small. Phyllaries subequal or usually graduate, imbricate, commonly dry and scarious or with scarious margins. Receptacle chaffy or naked. Fls. white, yellow, or greenish, either all perfect or the outer ones ♀ or neutral; rays present or none. Style-branches of perfect fls. obtuse or truncate and penicillate. Pappus usually none, sometimes a small scarious or chaffy crown or even paleaceous.—Cosmopolitan tribe, most abundant in the Medit. area, of about 50 genera.

A. Receptacle chaffy; pappus none or a short crown. (Subtribe Anthemidinae)
  B. Heads radiate.
    C. The heads solitary at branch-tips; receptacle chaffy at least centrally; aks. terete or ribbed, scarcely ever flattened .... 96. *Anthemis*
    CC. The heads many in close terminal corymbs; receptacle chaffy throughout; aks. obcompressed, callous-margined .... 97. *Achillea*
  BB. Heads discoid; ours a tomentose shrub .... 98. *Santolina*
AA. Receptacle naked. (Subtribe Chrysantheminae)
  B. Pappus prominently paleaceous (in ours).
    C. Pappus-paleae not awn-tipped; fls. yellow .... 99. *Hymenopappus*
    CC. Pappus-paleae awn-tipped; fls. white or purplish .... 100. *Hymenothrix*
  BB. Pappus none or a small crown.
    C. Infl. corymbiform to capitate or the heads solitary; annual or perennial herbs.

D. Aks. sessile; disk-corollas usually 5-toothed, occasionally 4-toothed.
  E. Marginal fls. with corolla; style deciduous.
    F. Receptacle flat or low-convex.
      G. Heads borne singly or in corymbs, radiate (rarely discoid), the rays
          usually white, showy .... 101. *Chrysanthemum*
      GG. Heads borne in a capitate to corymbiform infl., sometimes solitary,
          the rays, when present, commonly yellow and very short.
                                                        102. *Tanacetum*
    FF. Receptacle high-conic, at least at maturity; heads peduncled, not in
        corymbs .... 103. *Matricaria*
  EE. Marginal fls. lacking corolla; aks. tipped with the hardened persistent style
                                                        104. *Soliva*
DD. Aks., especially the marginal ones, conspicuously stipitate; disk-corollas 4-toothed,
    the marginal ones lacking corolla .... 105. *Cotula*
CC. Infl. spiciform, racemiform or paniculiform; heads discoid, small; pappus none
                                                        106. *Artemisia*

96. Ánthemis L. Dog-Fennel. Chamomile

Annual or perennial usually aromatic herbs with alternate incised-dentate to pinnately
dissected lvs. Heads medium-sized, radiate, rarely discoid. Rays elongate, white or
yellow, ♀ or neutral. Phyllaries subequal or imbricate, dry, ± hyaline on margins. Disk-
fls. many, perfect. Receptacle convex to conic, ± chaffy, at least near the summit. Aks.
terete or 4–5-angled, sometimes ± compressed. Pappus a short crown or none. Ca. 60
spp., of Old World. (Ancient Greek name of the Chamomile.)

Phyllaries quite brown, the summit largely membranous, dark; stems very slender, ± reddish.
Sonoma Co. .... 1. *A. fuscata*
Phyllaries greenish, the summit not broadly membranous; stems rather coarse, green. Common weed
                                                        2. *A. Cotula*

1. **A. fuscàta** Brot. Annual, mostly 1–2 dm. high, commonly several-stemmed from
base; lvs. pinnately dissected into minute segms., ca. 1–4 cm. long, scattered; heads
peduncled, ca. 2.5–3.5 cm. across, the phyllaries rounded-obtuse, dark-membranous,
imbricate; rays white, ca. 1.5 cm. long; aks. finely striate.—Locally commonly natur.
in vineyards about Asti, Sonoma Co.; native Medit. region. March–April.

2. **A. Cótula** L. Mayweed. Ill-smelling annual, 1–5 dm. high, ± branched, subglabrous;
lvs. mostly 2–6 cm. long, 2–3-pinnatifid into very narrow segms.; heads rather many,
short-peduncled, 1.5–2 cm. across; phyllaries ± pointed; rays white, 5–11 mm. long,
sterile; aks. subterete, ca. 10-ribbed; pappus none; $2n = 18$ (Harling, 1950).—Common
weed in waste places, fields, etc.; much of Calif.; native of Eu. April–Aug.

97. Achillèa L. Yarrow

Perennial aromatic herbs with alternate subentire to pinnately dissected lvs. Heads
several to many, rather small, in terminal corymbs, radiate (rarely discoid). Rays mostly
3–12, ♀ and fertile, white, sometimes pink or yellow, short, broad. Phyllaries imbricate
in several series, dry, with scarious or hyaline margins and often greenish midrib.
Receptacle conic or convex, chaffy throughout. Disk-fls. ca. 10–75, perfect and fertile.
Aks. compressed parallel to the phyllaries, callous-margined, glabrous. Pappus none.
Ca. 75 spp., N. Hemis. The spp. variable; chromosome counts often needed for exact
determination. (In honor of *Achilles*.)

(Clausen, J., D. D. Keck, and W. M. Hiesey. Experimental studies on the nature of spp. Carnegie
Inst. Wash. Pub. 520: 296–324, 1940.)
Invol. largely 5–6 mm. high; larger lvs. mostly 1.5–3 cm. broad; ligules 3–4 mm. long. Mostly below
2500 ft. elevs. .... 2. *A. borealis*
Invol. mostly 4–4.5 mm. high; ligules 2.5–3 mm. long.
    Lvs. 1–2.5 cm. wide; rachis winged, 1–2 mm. broad. Natur. weed in lawns, etc. .. 1. *A. Millefolium*
    Lvs. 1–1.5 cm. wide; rachis scarcely margined. Native, mostly above 3000 ft. .... 3. *A. lanulosa*

1. **A. Millefòlium** L. Rootstock creeping; stem simple, 3–10 dm. high, arachnoid to
glabrescent; cauline lvs. many, smooth to loosely pubescent, bipinnately parted and
dissected into fine segms.; lower lvs. 1–2 dm. long, petioled, the upper sessile, linear,
oblong; primary divisions ovate in outline, divaricate; rachis winged, 1–2 mm. broad;
ultimate divisions linear-lanceolate, spinulose-tipped; heads many, in corymbiform

panicles; invol. 4–5 mm. high, villous; phyllaries ca. 20, in 4 series, the outer ovate, obtuse, ca. half as long as the innermost, margins light brown; rays mostly 5, white to rose, round, 2.5–3 mm. long; aks. 2 mm. long, with thick wing-margins; $2n = 54$ (Turesson, 1938).—To be expected in lawns, etc.; natur. in e. U.S.; native of Eurasia. Summer.

2. **A. borèàlis** Bong. ssp. **califórnica** (Pollard) Keck. [*A. c.* Pollard, *A. puberula* Rydb., *A. pacifica* Rydb.] Much like the preceding, 5–10 dm. high, usually branched above, villous especially above; lvs. many, bipinnatifid or tripinnatifid, the lower petioled, 10–15 cm. long, ca. 1.5–3 cm. wide, the upper sessile and clasping; primary divisions ovate in outline, the ultimate linear or lance-linear, spinulose-tipped; heads many; invol. 5–6 mm. high, rather densely villous; phyllaries ca. 20, in 4 series, all obtuse or rounded at tip; margins brown; rays 5–6, white, the ligules 3–4 mm. long; disk-fls. 25–30, corollas 3 mm. long; aks. 2 mm. long, thick-margined; $2n = 54$ (Clausen et al., 1950).—Open and grassy places, below ca. 2500 ft.; many Plant Communities; Coast Ranges, w. foothills of Sierra Nevada, cismontane s. Calif.; to L. Calif., n. to Wash. March–June.

Ssp. **arenícola** (Heller) Keck. [*A. a.* Heller.] Stem 3–5 dm. high; lf.-segms. more numerous, thicker; $2n = 54$ (Clausen et al., 1940).—Coastal Strand; Santa Barbara Co. to Del Norte Co. June–July.

3. **A. lanulòsa** Nutt. Much like the preceding, to ca. 1 m. high, ± densely villous; lvs. 5–10 cm. long, rarely more than 1.5 cm. wide, the ultimate divisions linear; rachis scarcely margined; invol. 4–4.5 mm. high; phyllaries ca. 20, the outer lance-ovate, obtusish, the inner elliptic or oblong, obtuse; margins light brown; rays with ligules 2.5–3.5 mm. long, roundish; disk-fls. ca. 20, corollas 2.5 mm. long; aks. 2 mm. long, with thick margins; $2n = 36$ (Clausen et al., 1940).—Meadows and dampish places, largely 2500–8000 ft.; Yellow Pine F. to Lodgepole F.; scattered in mts. of s. Calif., more frequent in Sierra Nevada, to Yolla Bolly and Siskiyou mts., etc.; to Cascade Range and Rocky Mts. June–Aug.

Ssp. **alpícola** (Rydb.) Keck. [*A. l. a.* Rydb. *A. subalpina* Greene.] Plants mostly 1–2 dm. high, simple; lvs. to ca. 1 cm. wide; phyllaries often with darker almost black margins; $2n = 36$ (Clausen et al., 1940).—Borders of meadows, etc., largely 9000–11,300 ft. (lower northward); Subalpine F., Alpine Fell-fields; Sierra Nevada; Cascade Range, Rocky Mts. July–Aug.

98. Santolìna L.

Shrubs, sometimes herbs with aromatic herbage. Lvs. alternate, pinnately toothed or lobed or finely divided. Heads many-fld., discoid, yellow or rarely white, solitary on long peduncles. Invol. mostly campanulate; phyllaries appressed, imbricated. Receptacle with chaffy bracts. Aks. angled, without pappus. Ca. 8 spp., mostly of Medit. region. (Derivation of name uncertain.)

1. **S. Chamaecyparíssus** L. LAVENDER-COTTON. Much-branched, evergreen, woody, 4–6 dm. high; herbage silvery-gray and tomentose; lvs. 1.5–3.5 cm. long, pinnately divided into small ovate-oblong segms.; heads yellow, subglobose, 12–15 mm. in diam.; phyllaries ± carinate, subtomentose.—Occasionally established as in Sonoma, Monterey, and Los Angeles cos.; native of Medit. region. May–June.

99. Hymenopáppus L'Hér.

Biennial and perennial, subscapose to leafy-stemmed herbs from a single taproot or woody branched crown on a taproot. Stems slender to stout, erect, angled and sulcate. Lvs. in a basal rosette, mostly bipinnately dissected, to entire, reduced up the stem, minutely impressed-punctate. Infl. a cymose panicle. Heads discoid or radiate, subturbinate to broadly campanulate, on slender peduncles. Invol. of 6–14 subequal phyllaries in 2–3 series, membranous at apex or rarely throughout. Receptacle dome-shaped to flattish. Rays when present ♀ and fertile with white conspicuous ligules. Disk-fls. regular, yellow to white or red-purple, the tube mostly glandular, lobes reflexed after anthesis. Aks. 4-sided, obpyramidal to incurved, glabrous to pubescent. Pappus 0 or of 12–22 hyaline scales, the medial nerve completely included. Ca. 10 spp. of w. N. Am. (Greek, *humen*, membrane, and *pappos*, pappus, because of the hyaline paleae.)

(Turner, Billie L. A cytotaxonomic study of the genus H. Rhodora 58: 163–186, 208–242, 250–269, 295–308, 1956.)

1. **H. filifòlius** Hook. var. **eriópodus** (A. Nels.) Turner. [*H. e.* A. Nels.] Perennial, 4–8 dm. high, ± tomentose below, glabrate above; rosette-lvs. 1–2 dm. long, 3–7 cm. wide, bipinnately dissected into filiform divisions 1–2 cm. long; stem-lvs. 3–7, green, glabrate; heads discoid, 3–8 on a stem, 30–60-fld., on peduncles 9–16 cm. long; phyllaries 7–10 mm. long, sparsely tomentose to glabrate, yellow-membranous at tip; corollas ochroleucous, 4–5 mm. long, the tube glandular, ca. 2 mm. long, throat campanulate; aks. ca. 6 mm. long, pubescent; pappus of 12–16 linear-oblong scales 1.5–2 mm. long.—Dry limestone slopes, ca. 4900–7000 ft.; Pinyon-Juniper Wd.; New York Mts., Clark Mt., e. San Bernardino Co.; to sw. Utah. May–June, Oct.

KEY TO VARIETIES

Corolla 4–7 mm. long; anthers 3–4 mm. long; aks. 5–7 mm. long.
  Fls. whitish; peduncles 8–16 cm. long; stem-lvs. 2–7; aks. with hairs 0.5–1 mm. long
                                                  var. *eriopodus*
  Fls. yellow; peduncles 2–12 cm. long; stem-lvs. 0.6; ak.-hairs 1–2 mm. long.
    Stem-lvs. 2–7; tips of basal lvs. 5–30 mm. long; phyllaries 8–12 mm. long. Providence Mts.
                                              var. *megacephalus*
    Stem-lvs. 0–3; tips of basal lvs. 3–15 mm. long; phyllaries 6–10 mm. long. San Bernardino
    Mts. to Cuyamaca Mts. ....var. *lugens*
Corolla 3–4 mm. long; anthers 2–3 mm. long; aks. 4.5–5.5 mm. long ....var. *nanus*

Var. **megacéphalus** Turner. Ultimate lf.-divisions 1–3 cm. long; heads 3–14 on a stem; peduncles 2–10 cm. long; phyllaries 8–12; corollas yellowish, 4–6 mm. long; aks. 5–7 mm. long; pappus of 14–18 scales 1–3 mm. long.—Reported from Providence Mts.; to Ariz., Colo.

Var. **lùgens** (Greene) Jeps. [*H. l.* Greene.] Rosette-lvs. 0.5–1.4 dm. long, the ultimate divisions 3–10 mm. long; stem-lvs. mostly 0–2; heads 3–8 per stem; peduncles 2–12 cm. long; corolla mostly yellow, 4–6 mm. long; aks. 5–6 mm. long; pappus-scales 16–18, 1.2–2.5 mm. long; *n* = 17, 34 (Turner, 1956).—Dry slopes, 4000–7500 ft.; largely Yellow Pine F.; mts. from San Diego Co. to San Bernardino Mts.; to Utah, New Mex., L. Calif. June–Aug.

Var. **nànus** (Rydb.) Turner. [*H. n.* Rydb.] Mostly evenly sparsely tomentose throughout; basal lvs. 2–12 cm. long, the ultimate divisions 5–15 mm. long; stem-lvs. 0–2; heads 1–5 on a stem; peduncles 3–15 cm. long; corollas light yellow, 3–4 mm. long; aks. 4.5–5.5 mm. long; pappus of 14–18 scales, 1.5–3 mm. long; *n* = 17 (Turner, 1956).—Dry rocky places, 5500–10,000 ft.; Pinyon-Juniper Wd., Bristle-cone Pine F.; Inyo-White Mts.; to Utah, Ariz. July–Aug.

100. Hymenòthrix Gray

Slender-stemmed annual or perennial herbs. Lvs. alternate, biternately or triternately dissected into linear divisions. Heads few to many, cymose-panicled, radiate or discoid, yellow, white, or purple. Disk-corollas zygomorphic, the lobes unequal, erect or merely spreading. Aks. narrowly obpyramidal, 4–5-angled; pappus-scales with medial nerve extending to the apex or excurrent into a distinct awn. A genus of several spp., of sw. N. Am. (Greek, *humen*, membrane, and *thrix*, bristle.)

Invol. ca. 10 mm. high; pappus ca. 4 mm. long ....1. *H. Wrightii*
Invol. 4–6 mm. high; pappus ca. 2 mm. long .... 2. *H. Loomisii*

1. **H. Wrìghtii** Gray. [*Hymenopappus* W. Hall. *H. W.* var. *viscidulus* Jeps.] Perennial with a branched root-crown; stems leafy, few to several, 3–8 dm. high, glandular and hirsute; lf.-blades ovate to roundish in outline, 1.5–2.5 cm. long, the lower on equally long petioles, gradually somewhat reduced up the stems; heads few, in open infl.; invol. ca. 1 cm. high, subglabrous, the phyllaries membranous on edges; fls. white or purplish, ca. 7 mm. long, the narrow tube glandular; aks. villous, ca. 4 mm. long, the pappus ca. 4 mm. long.—Dry slopes, at ca. 5000 ft.; lower edge of Yellow Pine F.; Laguna and Cuyamaca mts.; to New Mex., Son., L. Calif. June–Sept.

2. **H. Loomísii** Blake. Annual or biennial 3–6 dm. high, finely incurved-puberulous throughout; lvs. alternate, the deltoid blades 1–8 cm. long; heads several to many; invol. 4–6 mm. high; corollas whitish, 5–6 mm. long; aks. 3–5 mm. long; pappus ca. 2 mm.

long.—Reported as natur. along Santa Fe Ry. tracks (Riverside and e. Los Angeles Co.);
cent. Ariz. Nov.–Dec.

101. Chrysánthemum L.

Annual or perennial herbs with alternate entire or toothed to pinnatifid lvs. Heads 1
to many, smallish to quite large, radiate or rarely discoid. Rays, when present, ♀ and
fertile, white to yellow or pink. Phyllaries ± imbricate in 2–4 series, dry, becoming
scarious or hyaline at least on margin and tip, sometimes with greenish midrib. Receptacle
flat or convex, naked. Disk-fls. tubular and perfect, the corolla 5(4)-lobed. Aks. subterete or angular, 5–10-ribbed, or those of rays with 2–3 wing-angles. Pappus a short
crown or none. Ca. 100 spp., chiefly of N. Hemis. and Old World. (Greek, *chrusos*,
gold, and *anthemon*, fl.)

Plants annual.
    Heads with a tricolored effect, the disk purple, the rays white or red or purple with a yellow or
    purple ring; phyllaries broad, papery-margined .... 1. *C. carinatum*
    Heads not tricolored.
        Rays light yellow; lvs. linear-lobed and incised, the main lower part narrowed almost to a
        petiolelike base; ak. strongly angled, but not with intermediate ribs .... 2. *C. coronarium*
        Rays golden yellow; lvs. not reduced to linear lobes, the main part usually ca. 6 mm. broad; ak.
        with many deep ribs .... 3. *C. segetum*
Plants perennial; rays white.
    Heads borne singly.
        Cauline lvs. dentate or simply lobed; plant not woody at base .... 4. *C. Leucanthemum*
        Cauline lvs. 2-pinnatifid into linear segms.; plant woody below .... 5. *C. anethifolium*
    Heads borne in dense flat-topped clusters .... 6. *C. Parthenium*

1. **C. carinàtum** L. Tricolor Chrysanthemum. Glabrous simple or forked annual
5–9 dm. high; lvs. ± succulent, remotely twice pinnatifid into linear lobes; heads solitary,
long-peduncled, 4–6 cm. across; outer phyllaries keeled; rays differently colored at base
so as to make a ring in the fl.-head; disk purple; aks. flat, winged.—Escape in San Diego
region, El Segundo, etc.; native of Morocco.

2. **C. coronàrium** L. Garland C. Ca. 2–10 dm. tall, annual, stiff; lvs. not succulent,
twice pinnatifid into oblong to subovate divisions; heads ca. 2–4 cm. across, yellow or
yellowish-white; phyllaries not keeled; aks. ± prismatic, strongly angled but not winged,
with minor ribs between; $2n = 18, 36$ (Shimotomai & Huziwara, 1935).—Occasionally
natur.; native of Medit. region. April–Aug.

3. **C. segètum** L. Corn C. Erect annual 3–6 dm. high, much branched, glabrous; lvs.
incised or pinnatifid, the lobes broad-linear, ± obtuse; base usually clasping; phyllaries
broad, obtuse, hyaline-margined; heads 2.5–5 cm. across, mostly golden-yellow; ray-aks.
broad, conspicuously notched at apex; disk-aks. prismatic, with prominent ribs; $2n = 18,
36$ (Dowrich, 1952).—Natur. in fields and waste places, especially in coastal cent. and
n. Calif.; native of Medit. region. April–Aug.

4. **C. Leucánthemum** L. [*L. vulgare* Lam.] Ox-eye Daisy. Perennial with ± of a
rhizome; stems mostly 2–8 dm. high, ± simple, glabrous or sparsely hairy; lower lvs.
spatulate to oblanceolate, petioled, 4–15 cm. long, ± deeply crenate, the cauline reduced, subsessile, mostly blunt-toothed to subentire; heads solitary, naked-pedunculate,
2.5–5 cm. across; disk 1–2 cm. in diam.; phyllaries narrow, with a dark brown submarginal area; rays mostly 15–30, 1–2 cm. long; aks. terete, ca. 10-ribbed; pappus none;
$2n = 36, 54$ (Dowrich, 1952).—Natur. in waste places, fields, etc. especially in n. and
cent. Calif., less commonly in the s.; native of Old World. June–Aug. A form with lvs. ±
lobed or cleft is occasionally found and is the var. *pinnatifidum* Lec. & Lam.

5. **C. anethifòlium** Brouss. Much-branched bush, woody at base, 5–9 dm. high,
glaucous, glabrous; lvs. thickish, oblong to ovate in outline, glaucous, finely cut into
linear segms.; heads solitary, 3–5 cm. in diam.; rays white.—Reported from near San
Ysidro, San Diego Co.; native of Canary Ids. July.

6. **C. Parthènium** Pers. (L.) Bernh. [*Matricaria P.* L.] Perennial with a caudex; stems
leafy, nearly glabrous, 3–8 dm. high; lvs. finely puberulent at least beneath, pinnatifid,
with rounded to pinnate segms., blade to ca. 7 cm. long; heads several to many,
corymbose; disk 5–9 mm. across; phyllaries narrow, the inner hyaline at tip; rays ca.
10–20, 4–8 mm. long, white; aks. subterete, ca. 8–10-nerved; pappus a minute crown
or obsolete; $2n = 18$ (Harling, 1951).—Occasional in waste places; native of Eu.
June–Sept.

102. Tanacètum L. Tansy

Aromatic annual or perennial herbs, sometimes suffrutescent. Lvs. alternate, entire to more commonly pinnately dissected or ternate. Heads small or medium-sized, globular, discoid or shortly radiate, in a capitate to corymbiform infl. Outer fls. ♀, with short tubular corolla which may have a short ray, or ♀ fls. sometimes wanting. Invol. of many imbricate phyllaries, the margins and tips of at least the inner usually ± scarious. Receptacle flat or low-conic, naked. Disk-fls. perfect, the tubular yellow corolla 5-toothed. Aks. mostly 5-ribbed or -angled, commonly glandular; pappus a short crown or none. Ca. 50 spp., mostly of N. Hemis. of Old World. Near to *Chrysanthemum*. (Name obscure.)

Large robust leafy-stemmed plants, mostly 4–20 dm. tall; lvs. pinnately dissected, 5–20 cm. long.
  Lvs. glabrous or nearly so, punctate; pinnae with broad winged rachis. Introd. plants
                                                                 1. *T. vulgare*
  Lvs. ± pubescent, scarcely or not punctate; pinnae with rachis scarcely winged. Coastal natives.
    Rays short, but ± evident; plant ± villous but not whitish. From Humboldt Co. n.
                                                                 2. *T. Douglasii*
    Rays not at all evident; plant villous-tomentose, quite whitish on younger parts. San Francisco Bay
    area .... 3. *T. camphoratum*
Smallish slender-stemmed plants mostly 0.5–3 dm. high. High montane.
  Lvs. dissected; heads 6–8 mm. in diam. E. Sierra Nevada .... 4. *T. potentilloides*
  Lvs. entire to 3–4 toothed; heads 3–6 mm. wide. E. Sierra Nevada to Panamint Mts.
                                                                 5. *T. canum*

1. **T. vulgàre** L. Common Tansy. Coarse aromatic perennial, the stems erect, 0.5–1 m. high, subglabrous; lfts. and wings of petiole cut-toothed or deeply cut, the lvs. many, 1–2 dm. long, sessile to short-petioled; heads in a flat corymb, disciform, many (20–200); disk ca. 5–10 mm. wide; ♀ fls. terete, with oblique 3-toothed limb; pappus a minute 5-lobed crown; $2n = 18$ (Shimotomai, 1937).—Occasionally natur. from old gardens, especially in cent. and n. Calif.; native of Old World. Aug.–Oct.

2. **T. Douglàsii** DC. [*Chrysanthemum D.* Hult.] Stout rhizomatous perennial 2–6 dm. tall, ± villous; basal lvs. often well developed, larger than the cauline; cauline 0.5–2 dm. long, bipinnatifid or tripinnatifid, the pinnae without much winging of rachis, ultimate segms. mostly blunt or rounded; heads several, ca. 5–20, corymbose; disk ca. 8–15 mm. in diam.; phyllaries many, the outer acute, the inner oblong-rounded; rays short but ± evident; pappus a minute crown; aks. ± clavate, glandular, ca. 3–4 mm. long.— Coastal Strand; Humboldt and Del Norte cos.; to B.C. May–Aug.

3. **T. camphoràtum** Less. Like no. 2, but larger, 3–8 dm. high, more tomentose especially on young parts; lvs. mostly bipinnatifid, the ultimate segms. much crowded, oblong to oval, ± revolute; disk mostly 12–17 mm. in diam.; rays not evident.—Coastal Strand; San Francisco Bay region. June–Sept.

4. **T. potentilloìdes** (Gray) Gray. [*Artemisia p.* Gray. *Sphaeromeria p.* Heller.] Perennial from a short caudex, silky-tomentose; stems slender, 1–2.5 dm. long, moderately leafy; basal lvs. tufted, ovate in outline, 2–3-times pinnately dissected, the blade 2–3 cm. long, the petiole ca. as long; cauline lvs. gradually reduced upward; heads 1–few, pedunculate, disciform, the disk ca. 5–10 mm. in diam.; receptacle convex, white-hairy; phyllaries short- to round-ovate, villous; rays obsolete; disk yellow; pappus obsolete.— Semialkaline meadows, etc., 5000–7500 ft.; Sagebrush Scrub; e. Sierra Nevada, Mono Co. to Lassen Co.; Ore., Ida., Nev. May–July.

5. **T. cànum** D. C. Eat. Stems several from a woody root-crown, mostly 1–2 dm. long, slender; herbage closely tomentose-canescent; lvs. mostly 1–2 cm. long, well distributed, subsessile, linear-oblong and entire to narrowly obovate and with 3 linear-oblong segms.; corymbs compact, 2–several-headed; heads 3–6 mm. in diam.; phyllaries ovate, canescent, round-oblong; disk-fls. lemon-yellow.—Dry rocky places, 9000–12,000 ft.; Subalpine F., Bristle-cone Pine F., Alpine Fell-fields; Panamint Mts., Sierra Nevada n. to Madera and Mono cos.; Nev., se. Ore. July–Aug.

103. Matricària L.

Annual or perennial herbs with alternate pinnatifid or finely pinnately dissected lvs. Heads small, solitary or somewhat corymbed, radiate or discoid; rays white, ♀, usually fertile, or wanting. Phyllaries dry, of ca. 2–3 series, ± imbricated, with scarious margins. Receptacle naked, ± conic to hemispheric. Disk-corollas yellow, 4–6-toothed. Aks. 3–5-ribbed, wingless, generally nerved on the margin and ventrally, nerveless dorsally, glabrous or roughened. Pappus a short crown or none. Ca. 35 spp., of N. Hemis. and S. Afr. (Latin *matrix,* because used medicinally.)

1. **M. matricarioìdes** (Less.) Porter. [*Artemisia m.* Less. *M. suaveolens* (Pursh) Buch., not L. *M. discoidea* DC. *M. occidentalis* Greene.] PINEAPPLE WEED. Branched erect aromatic annual 1–3 dm. high; lvs. 1–5 cm. long, 1–3-pinnatifid into linear or filiform segms.; heads several to many, rayless, the disk mostly 4–10 mm. wide; phyllaries with broad hyaline margin, the invol. 5–7 mm. high, much shorter than the conical disk; aks. with 2 marginal and 1 rather weak ventral nerves; pappus a short crown, sometimes ± unequally lobed at summit; $2n = 18$ (Rutland, 1941).—Exceedingly common weed in waste places through most of Calif.; to Alaska, Rocky Mts., L. Calif.; apparently introd. in e. states. May–Aug.

104. Solìva R. & P.

Rather small depressed annuals with short stiff leafy branches. Lvs. petioled, pinnately dissected. Fls. greenish, in small discoid burlike heads sessile in the forks of the stems. Invol. of 5–12 subequal phyllaries in not more than 2 series. Receptacle flat. Outer series of florets ♀, apetalous; innermost fls. perfect but sterile, the corolla 4-toothed. Aks. obcompressed, membranous, callous-margined or -winged, pointed with the hardened persistent style. Pappus none. Several spp. of temp. N. and S. Am. (Dr. Salvador *Soliva,* of Spain.)

(Crampton, B. Observations on the genus Soliva in California. Leafl. W. Bot. 7: 196–8, 1954.)
Aks. with broad membranous wings, each wing projecting above the body of the ak. as a tooth 1–1.5 mm. long.
    Wings entire . . . . 1. S. sessilis
    Wings conspicuously notched toward the base . . . . 2. S. pterosperma
Aks. with wings reduced to a hardened marginal callus, or if toothed, the teeth minute and little or not projecting above the ak.-body . . . . 3. S. daucifolia

1. **S. séssilis** R. & P. Plants 5–25 cm. across, pubescent to villous; lvs. ca. 1–3 cm. long, with 3–5 primary divisions, these parted into 3–5 narrowly lanceolate lobes; first heads sessile at very base of plant, the stems radiating out from beneath them; invol. ca. 2.5–3 mm. high; ♀ fls. 9–12; each wing of the ak. terminating above in an incurved tooth; ak.-body 3–3.5 mm. long; disk-fls. ca. 7–9, their styles stout, conspicuously exserted from the subglobose head.—Moist open places at low elevs.; various Plant Communities; Coast Ranges from Del Norte Co. to Santa Barbara Co., San Diego; Chile. April–July.
2. **S. pterospérma** (Juss.) Less. [*Gymnostyles p.* Juss.] Much like *S. sessilis* in habit; aks. with broad membranous wings (ca. 1 mm. wide) having a conspicuous notch indented 0.5–0.7 mm. in the lowest third above the base and with an apical tooth projecting above the ak.-body and usually curving outward.—Sierran foothills from Eldorado to Tuolumne cos.; s. Ore.; Argentina. Spring.
3. **S. daucifòlia** Nutt. Of same general habit and stature; aks. wingless but sometimes with a minute apical tooth on each margin, little or not projecting above the 2–2.3 mm. long body of the ak.—Near the immediate coast, as on bluffs above the beaches, etc.; Humboldt Co. to Santa Barbara Co.; also a lawn weed in Ventura, Riverside, and Escondido. April–July.

105. Cótula L.

Low mostly diffuse rather strong-scented herbs, with alternate toothed, lobed or dissected lvs. Heads pedunculate, disciform. Fls. yellow. Receptacle flat or nearly so. Phyllaries greenish, subequal, in 1 or 2 ranks. Outer series of fls. ♀ only, fertile. Disk-fls. tubular, 4-toothed, fertile or not. Mature aks. ± pedicelled, compressed in ours and spongy-margined or narrow-winged. Pappus none. Ca. 50 spp., of wide distribution in S. Hemis. (Greek, *kotule,* a small cup, referring to a hollow at the base of the amplexicaul lvs.)

Annual; lvs. finely pinnate; heads 2–5 mm. broad .... 1. *C. australis*
Perennial; lvs. entire or pinnatifid; heads 8–10 mm. broad .... 2. *C. coronopifolia*

1. **C. austràlis** (Sieber) Hook. f. [*Anacyclus a.* Sieber.] Slender-stemmed, branched from base, spreading, 0.3–2 dm. high, sparsely pubescent with spreading hairs; lvs. 1–3 cm. long, 1–2-pinnate into linear lobes; phyllaries brownish at apex, round-oblong, with scarious margins; marginal aks. minutely glandular on both faces, but glabrous on margins; slightly over 1 mm. long.—Very common and troublesome weed about gardens, city lots, etc.; from Australia. Jan.–May.

2. **C. coronopifòlia** L. BRASS-BUTTONS. Perennial, decumbent or repent, fleshy, quite glabrous, branched; stems 2–3 dm. long; lvs. 2–7 cm. long, linear-oblong to -oblanceolate, entire to coarsely and deeply few-toothed, with broad rachis; heads depressed, bright yellow; phyllaries oblong, 3–5-veined, ± anthocyanous; ♀ fls. in 1 row, on pedicels as long as invol., without corolla; disk-fls. on much shorter pedicels, numerous; aks. winged, almost 2 mm. long; $2n = 20$ (Castro & Fontes, 1946).—Common on mud and moist banks, about salt marshes, etc.; many Plant Communities; cismontane Calif.; natur. from S. Afr. March–Dec.

106. Artemísia L. SAGEBRUSH. WORMWOOD

Herbs or shrubs, usually aromatic. Lvs. alternate, entire to toothed or dissected. Infl. spiciform, racemiform, or paniculate. Heads small, discoid or disciform, sometimes only with perfect fls., sometimes the outer ♀ and the cent. then sometimes sterile. Phyllaries dry, imbricate, at least the inner scarious or scarious-margined. Receptacle flat or hemispheric, naked or with long hairs. Style-branches flattened, truncate, penicillate. Aks. obovoid or oblong, usually glabrous; pappus a very short crown or more often none. Over 100 spp., of N. Hemis. and S. Am. (*Artemisia*, wife of Mausolus, king of Caria.)

(Keck, D. D. A revision of the A. vulgaris complex in N. Am. Proc. Calif. Acad. Sci. IV. 25: 421–468, 1946. Ward, G. H. Artemisia, section Seriphidium, in N. Am. Contr. Dudley Herb. 4: 155–205, 1953.)

A. Fls. all perfect; shrubs (except No. 6).
    B. Ray-fls. present in at least some of the heads. Mojave Desert .... 1. *A. Bigelovii*
    BB. Ray-fls. absent, all fls. perfect and fertile.
        C. Receptacle naked; plants mostly ± canescent.
            D. Lvs. mostly cuneate or flabelliform with 3–7 teeth at apex, or sometimes entire and truncate or rounded.
                E. Plants not root-sprouting, seldom layering; outer phyllaries not acute or acuminate; inner phyllaries usually canescent; fls. mostly 3–11 in a head.
                    F. Lvs. 3 or more times as long as wide, 0.5–5 cm. long; plants mostly more than 4 dm. high .... 2. *A. tridentata*
                    FF. Lvs. 1–3 times as long as wide, 0.5–1.5 cm. long; plants 1–4 dm. high
                                            3. *A. arbuscula*
                EE. Plants often root-sprouting or layering; outer phyllaries acute or acuminate; inner phyllaries usually glabrous; fls. 8–20 .... 5. *A. Rothrockii*
            DD. Lvs. mostly linear or oblanceolate, entire, acute or acuminate, or lvs. deeply divided into 3 or more narrow lobes.
                E. Lvs. mostly entire; shrub 1–4 dm. high .... 4. *A. cana*
                EE. Lvs. mostly deeply divided into 3 or more linear lobes; shrub 1–4 dm. high
                                            3. *A. arbuscula*
        CC. Receptacle with chaff, most fls. subtended by bracts; lvs. green above. San Diego region
                                              6. *A. Palmeri*
AA. Fls. not all perfect, the marginal ♀, the cent. ones fertile or sterile.
    B. Plants definitely woody at base or above, not rhizomatous.
        C. Plant with long naked spines; stems 1–3 dm. long; lvs. with spatulate lobes. Deserts
                                            8. *A. spinescens*
        CC. Plant not spiny; stems taller; lvs. with linear lobes.
             D. Shrub 6–15 dm. high; herbage strigulose; invol. 2–3 mm. high. Mostly back of the coastal beaches .... 7. *A. californica*
                DD. Suffrutescent, 4–7 dm. high; herbage silky-tomentose; invol. 3.5–4.5 mm. high. Sandy coastal beaches and dunes .... 9. *A. pycnocephala*
    BB. Plants not woody at base.
        C. Herbage glabrous, rarely pubescent, never tomentose.
            D. Annual or biennial; lvs. (at least the lower) 2–3-pinnatifid. Weed in waste places
                                            14. *A. biennis*
            DD. Perennial; lvs. linear, mostly entire. Native .... 13. *A. Dracunculus*
        CC. Herbage ± pubescent, mostly white-tomentose beneath.

D.   Fls. all fertile; style 2-cleft; plants with branched caudex; stems mostly loosely
     villous; invol. 4–7 mm. high. High montane .... 15. *A. norvegica*
DD. Fls. of center of disk sterile; plants from horizontal rhizomes.
     E.   Principal lvs. narrow, 1 cm. or less wide exclusive of lobes when present,
          tomentose on both sides or green above; stems rarely more than 1 m. tall
          10. *A. ludoviciana*
     EE. Principal lvs. 1–5 cm. wide exclusive of lobes when present, frequently
          entire, lobes if present few, entire, discolored; stems more frequently more
          than 1 m. tall.
          F.   Invol. campanulate, 2–3 mm. wide, ± tomentose; ray-fls. 6–10; disk-fls.
               10–25. Widely distributed .... 11. *A. Douglasiana*
          FF. Invol. terete or narrow-ovoid, less than 2 mm. wide, yellow-green and
               shining; ray-fls. 3–7; disk-fls. 2–8. Mendocino Co. to Del Norte Co.
               12. *A. Suksdorfii*

1. **A. Bigelòvii** Gray in Torr. Low evergreen shrub 2–4 dm. high, many-stemmed, with grayish-brown bark; twigs and lvs. canescent; lvs. of vegetative shoots sessile or short-petioled, narrowly cuneate, 1–2 cm. long, 2–5 mm. wide, with a truncate sharply 3-toothed apex, sometimes entire; lvs. of infl. mostly entire; infl. narrowly paniculate, dense, with several heads on each short recurved panicle-branch; invol. turbinate, 2–4 mm. long; phyllaries densely tomentose, 8–12, the outer ovate, the inner oblong; rays 0–2, ♀, the tubular corolla 2-toothed, 1–2 mm. long; disk-fls. 1–3, perfect, fertile, funnelform, 5-toothed, 1.5–3 mm. long; aks. ellipsoid, 4–5-ribbed, glabrous; $2n = 18$ (Ward, 1953).—Dry limestone slopes, 5000–6000 ft.; mostly Pinyon-Juniper Wd.; Inyo, Clark, and New York mts., e. Mojave Desert; to Colo., w. Tex. Aug.–Sept.

2. **A. tridentàta** Nutt. [*A. t.* var. *angustifolia* Gray.] BASIN SAGEBRUSH. Rounded evergreen shrub mostly 0.5–3 m. high, with a short thick trunk or few branches from base, ± silvery canescent throughout; lvs. of vegetative shoots sessile or short-petioled, cuneate with 3 blunt teeth or sometimes 4–9-toothed or entire, 1–4 cm. long, 0.2–1.3 cm. wide, those of fl.-shoots mostly entire and linear to oblanceolate; infl. mostly with erect branches 1.5–4 dm. long; invol. 3–3.8 mm. high; phyllaries 8–15, the outer short, round-ovate, the inner elliptic; fls. 3–6(–12) in a head, perfect; corolla narrow-funnelform, 5-toothed, 2–3 mm. long; aks. resinous-granuliferous, rarely sparsely short-villous; $2n = 18, 36$ (Ward, 1953).—Dry slopes and plains, 1500–10,600 ft.; Sagebrush Scrub, Yellow Pine F., N. Juniper Wd., Pinyon-Juniper Wd., etc.; Laguna Mts., San Diego Co. along w. edge of deserts to Sierra Nevada, then to Modoc and Siskiyou cos.; to B.C., Rocky Mts., L. Calif. Aug.–Oct. Plants from s. Calif. tend to have linear to linear-cuneate lvs. and have been referred to var. *angustifòlia*, probably incorrectly.

Ssp. **Paríshii** (Gray) Hall & Clem. [*A. P.* Gray. *A. t.* var. *P.* Jeps.] Plants erect; lvs. mostly linear, entire or shallowly notched; infl. often with drooping branches; aks. glandular and sparingly short-villous to arachnoid-hairy; $2n = 36$ (Ward, 1953).—Dry valleys, 1000–2500 ft.; Coastal Sage Scrub, Joshua Tree Wd., etc.; Santa Clara R. V., Ventura Co., n. Los Angeles Co. to w. Mojave Desert (Antelope V.). Oct.–Nov.

3. **A. arbúscula** Nutt. [*A. tridentata* var. *a.* McMinn.] Low spreading evergreen shrub 1–4 dm. high with light to dark brown bark; twigs densely canescent or later glabrate; vegetative lvs. canescent, broadly cuneate to flabelliform, 0.5–1.5 cm. long, 0.3–1 cm. wide, usually 3–5-toothed, sometimes lobed; lvs. of infl. linear-oblanceolate and entire to cuneate and 3-toothed at apex; infl. narrow, with few mostly erect branches; invol. campanulate, 3–4 mm. long; phyllaries 10–15, mostly canescent; fls. 6–11, perfect; corolla funnelform, 5-toothed, 2–3 mm. long; aks. resinous-granuliferous; $2n = 18, 36$ (Ward, 1953).—Dry slopes and ridges, 4000–9500 ft.; Sagebrush Scrub, N. Juniper Wd., etc.; Sierra Nevada from Alpine Co. n., to Modoc and Siskiyou cos.; to Wash., Rocky Mts. July–Aug.

Ssp. **nòva** (A. Nels.) Ward. [*A. n.* A. Nels. *A. tridentata* ssp. *n.* Hall & Clem.] Phyllaries 8–12, the inner mostly glabrous; fls. mostly 3–5, perfect; aks. glabrous; $2n = 18, 36$ (Ward, 1953).—Dry rocky places, 5000–11,000 ft.; Sagebrush Scrub, Pinyon-Juniper Wd., etc.; desert slopes of San Bernardino Mts., New York, Clark, Panamint, and Inyo-White mts., Benton Range; to Ida., Wyo., Ariz. Sept.–Nov.

4. **A. càna** Pursh. Low rounded evergreen shrub 4–9(–12) dm. high, often root-sprouting or layering, freely branched, old branches with brown bark; twigs with dense tomentum; lvs. mostly 2–6 cm. long, linear or ± oblanceolate, entire, acute or acuminate, rarely with 1–2 teeth, silvery-canescent; infl. mostly loosely paniculate to subspicate,

1–4 dm. long; invol. campanulate, 3.5–5 mm. long; phyllaries 8–14, the outer broadly ovate, the inner elliptic to linear-obovate, canescent to subglabrous; fls. 8–20, perfect; corolla funnelform, 5-toothed, 2–3.5 mm. long; aks. granuliferous; $2n = 18$, 36 (Ward, 1953).—Dry gravelly to rocky places, 6000–10,600 ft.; Montane Coniferous F., Sagebrush Scrub; Mono Co. along e. side of Sierra Nevada to Modoc Co.; to Ore., Alta., N. Dak., Nebr., Colo. July–Sept.

Ssp. **Bolánderi** (Gray) Ward. [*A. B. Gray. A. tridentata* ssp. *B.* Hall & Clem.] Shrub 2–5 dm. high; lvs. 1–3 cm. long, often lobed; corolla 1.8–2.5 mm. long; $2n = 18$ (Ward, 1953).—Grassy flats and hollows among sandy or rocky hills, 6500–9000 ft.; Sagebrush Scrub; Mono Co. Sept.–Oct.

5. **A. Rothróckii** Gray. [*A. tridentata* ssp. *R.* Hall & Clem.] Low spreading evergreen shrub 2–6 dm. high, often root-sprouting, with light gray bark; twigs densely tomentose at least when young; lvs. of vegetative shoots 0.5–5 cm. long, mostly broadly cuneate or flabelliform, 3-toothed or -lobed, sometimes lanceolate, entire, densely canescent, sometimes glabrate in age; infl.-lvs. similar or lanceolate, entire; infl. spicate or narrowly paniculate, 0.5–4 dm. long; invol. campanulate, 4–5.5 mm. long; phyllaries 10–14, the outer broadly ovate, acute or acuminate, ± tomentose, the inner elliptic, less hairy; fls. mostly 10–16, perfect; corolla 5-toothed, 2.3–3.5 mm. long; aks. granuliferous; $2n = 36$, 54, 72 (Ward, 1953).—Dry to wet rocky slopes, mostly 6500–11,500 ft.; Red Fir F. to Alpine Fell-fields; Sierra Nevada from Eldorado Co. s., White, Panamint, and San Bernardino mts. Aug.–Sept.

6. **A. Pálmeri** Gray. [*Artemisiastrum P.* Rydb.] A ± suffrutescent perennial with long wandlike stems 5–8 dm. high, grayish-puberulent above over a reddish color; lvs. 5–12 cm. long, pinnately parted into 3–9 linear or lance-linear lobes, or linear and entire, subglabrous above, finely tomentose beneath; infl. terminal, broadly paniculate, 1.5–5 dm. long; heads numerous; invol. hemispheric, 2.5–4 mm. long; phyllaries 8–12, ovate-acutish, sparingly pubescent or glabrous, the inner little longer than the outer, mostly subglabrous; fls. 12–30, perfect, usually each subtended by a bract; corolla 1.5–2.5 mm. long; aks. 4-angled, granuliferous; $2n = 18$ (Ward, 1953).—Lower slopes; Coastal Sage Scrub; sw. San Diego Co.; n. L. Calif. July–Sept.

7. **A. califórnica** Less. CALIFORNIA SAGEBRUSH. Grayish shrub mostly 6–15 dm. high, usually broader; lvs. numerous, strigulose, the lower 1–5 cm. long, palmately once or twice parted into linear-filiform segms. less than 1 mm. wide, the upper sometimes entire and fascicled; heads many, in long racemose panicles; invol. 2–3 mm. high; fls. rather numerous; aks. with a minute squamellate crown.—Dry slopes and fans, below 2500 ft.; Coastal Sage Scrub, Coastal Strand, etc.; Marin and Napa cos. to cismontane s. Calif.; L. Calif. Aug.–Dec. A form on San Clemente and San Nicolas ids. has lf.-segms. 1–3 mm. wide and has been named var. *insuláris* (Rydb.) Munz. [*Crossostephium i.* Rydb.]

8. **A. spinéscens** D. C. Eat. Woody, intricately branched, 1–3 dm. high, spiny, white-tomentose on young branches and ± so or stiff-pubescent on the grayish foliage; lvs. crowded, 5–8 mm. long, pedately 5–7-parted and the divisions 3-lobed; heads few, in short lateral spikes which become the spines the next year; fls. 5–12; invol. globose, 3 mm. long; phyllaries 5–8, obovate, tomentose; corolla and ak. villous-cobwebby.—Semi-alkaline flats, 2000–6000 ft.; Shadscale Scrub, Creosote Bush Scrub, etc.; w. Mojave Desert to Owens V. and Lassen Co.; to Ore., Mont., Colo., New Mex. April–May.

9. **A. pycnocéphala** DC. [*A. campestris* ssp. *p.* Hall & Clem.] Suffrutescent perennial, with stout subsimple stems 4–7 dm. high, densely leafy to the infl.; herbage densely whitish or grayish silky-tomentose; lvs. with blades 1–2.5 cm. long, 1–3-pinnate into linear lobes, the lower petioled, the upper subsessile and subsimple; heads many, erect, in spicate dense thyrsuslike panicles; invol. densely villous, subglobose, ca. 3.5–4.5 mm. in diam.; corolla slender, 2–2.5 mm. long; aks. brown, ca. 1.5 mm. long.—Coastal Strand; Monterey Co. to Humboldt Co.; Ore. June–Aug.

10. **A. ludoviciàna** Nutt. [*A. gnaphalodes* Nutt. *A. l.* var. *g.* T. & G. *A. vulgaris* sspp. *l.* and *g.* Hall & Clem. *A. Purshiana* Bess.] Perennial rhizomatous herb; stems mostly 3–10 dm. tall, slender to moderately stout, simple up to the infl., ± white-tomentose at least above; lvs. many, linear to lanceolate, oblanceolate or elliptic, sometimes cuneate, mostly 3–11 cm. long, entire or few-toothed or -lobed especially apically, the lobes mostly entire, permanently and densely white-tomentose on both sides or loosely floccose

to green and glabrate above, plane; infl. ample or narrow, an elongate usually compact panicle 1.5–5 cm. broad; heads erect or nodding, often in glomerules; invol. ovoid to campanulate, 3–4 mm. high, usually densely tomentose, sometimes glabrate; phyllaries 7–13; ray-fls. 5–12; disk-fls. 6–20; $n = 18$ (Clausen et al., 1940).—Dry open places, below 8500 ft.; Sagebrush Scrub, Montane Coniferous F.; San Jacinto and San Bernardino mts., e. slope of Sierra Nevada, n. to Modoc Co.; to Wash., Alta., Ontario, Ark., New Mex. July–Sept.

Ssp. **álbula** (Woot.) Keck. [*A. a.* Woot. *A. microcephala* Woot., not Hillebr.] Lvs. mostly 1–2 cm. long, obovate to elliptic and with forward-projecting teeth or lobes, or lance-linear and entire, often ± revolute; invol. ca. 3 mm. high; phyllaries 11–16; ray-fls. 8–11; disk-fls. 8–13.—Mostly dry slopes, below 7000 ft.; Creosote Bush Scrub to Pinyon-Juniper Wd., Yellow Pine F.; w. edge Colo. Desert, Mojave Desert, s. Sierra Nevada; to Colo., w. Tex., Son., L. Calif. May–Oct.

Ssp. **incómpta** (Nutt.) Keck. [*A. i.* Nutt. *A. vulgaris* var. *discolor* of Jeps., not *A. d.* Dougl.] Stems usually entirely herbaceous; herbage commonly ± green, the lvs. glabrate above, white-tomentose beneath; lower lvs. 2–8 cm. long, parted into linear or lanceolate lobes, some of these again toothed or lobed; upper lvs. less cut to entire; invol. 3–3.5 mm. high, silky-tomentose to shining-glabrate; phyllaries 9–14; ray-fls. 6–10; disk-fls. 15–30; $n = 18$ (Clausen et al., 1940).—Dry places, 2600–11,600 ft.; Joshua Tree Wd., Pinyon-Juniper Wd., Montane Coniferous F., etc.; Santa Rosa Mts. to San Gabriel Mts., Kingston Mts., Sierra Nevada, n. to Modoc Co.; to Rocky Mts. July–Sept.

11. **A. Douglasiàna** Bess. in Hook. [*A. vulgaris* var. *californica* Bess. *A. heterophylla* auth., not Bess. *A. v.* var. *h.* Jeps. *A. Kennedyi* A. Nels. *A. v.* var. *Lindleyana* of Jeps., not *A. L.* Bess.] Stout rhizomatous perennial, 5–15(–30) dm. tall; stems simple or sometimes branched above; lvs. commonly 7–15 cm. long, lanceolate to elliptic and entire, or oblanceolate to obovate in outline and coarsely few-toothed or -lobed toward apex, the lobes entire, mostly lanceolate, strongly discolored, glabrous and green to sparsely tomentulose above, densely gray-tomentose beneath, mostly plane; infl. leafy, open or dense, elongate, paniculate; heads erect or nodding; invol. campanulate, 3–4 mm. high, ± tomentose; phyllaries 8–14; ray-fls. 6–10; disk-fls. 10–25; $n = 27$ (Clausen et al., 1940).—Mostly low waste places, up to 6000(9500) ft.; many Plant Communities; cismontane Calif., Sierra Nevada e. to w. Nev.; to Wash., Ida., L. Calif. June–Oct.

12. **A. Suksdórfii** Piper. [*A. vulgaris* var. *litoralis* Suksd.] Much like *A. Douglasiana* in habit; lvs. broadly lanceolate to elliptic, mostly entire, 8–15 cm. long, bright deep green above, silvery-tomentose beneath; infl. dense; heads small, commonly erect; invol. terete or narrow-ovoid, yellow-green, shining, 3–4 mm. high; phyllaries 6–9; ray-fls. 3–7; disk-fls. 2–8; $n = 9$ (Clausen et al., 1940).—Along the immediate coast, below 500 ft.; N. Coastal Scrub; Sonoma Co. n.; to Vancouver Id. May–Aug.

13. **A. Dracúnculus** L. [*A. glauca* Pall. *A. dracunculoides* Pursh.] Almost odorless perennial with hard horizontal rhizome; stem erect, 5–15 dm. tall, glabrous to villous-puberulent; lvs. linear or nearly so, 3–8 cm. long, entire or sometimes cleft, the lower mostly deciduous; panicle with elongate leafy ascending branches; heads many, soon spreading or nodding; invol. subglobose, 2–3 mm. broad; phyllaries 7–12; scarious-margined; fls. ca. 20–30, the outer ♀ and fertile, the inner sterile; aks. ellipsoid, glabrous, not ribbed; $2n = 18$ (Weinedel, 1928).—Dry ± disturbed or dry places, below 9000 ft.; Coastal Sage Scrub to Montane Coniferous F., cismontane s. Calif. to cent. Calif.; Joshua Tree Wd. and above, Mojave Desert n.; to B.C., Wis., Tex., etc. Aug.–Oct.

14. **A. biénnis** Willd. Erect glabrous annual or biennial, 3–8 dm. high, leafy; lvs. 2-pinnately parted or the upper pinnatifid, the lobes linear, acute, mostly cut-toothed; heads subglobose, crowded, erect, subsessile; invol. 2–3 mm. high; larger glabrous phyllaries roundish, broadly scarious on margin, with green midrib; disk-fls. 15–40, fertile; corolla campanulate, ca. 1 mm. long; aks. glabrous, 4–5-nerved.—Weed in moist waste places, below 7000 ft.; several Plant Communities; cismontane Calif.; native in nw. U.S. Aug.–Oct.

15. **A. norvègica** Fries var. **saxátilis** (Bess. in Hook.) Jeps. [*A. Chamissoniana* var. *s.* Bess. *A. arctica* auth.] Perennial from a branched caudex; stems 2–6 dm. tall, mostly loosely villous; lvs. ovate in outline, the basal with a dissected blade 2–10 cm. long and ultimate acute narrow segms.; cauline lvs. progressively reduced, becoming sessile; infl. loosely racemose or racemose-paniculate; heads many-fld.; invol. 4–7 mm. high;

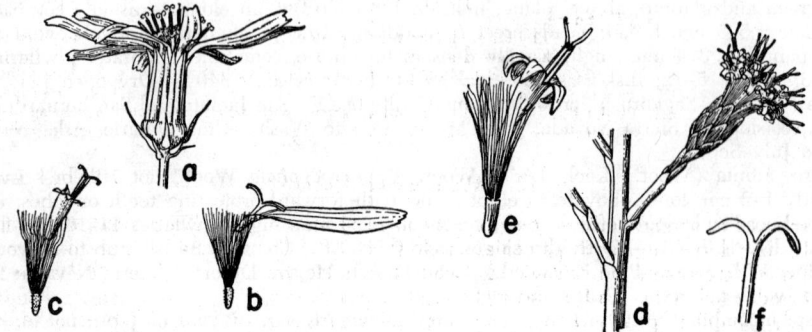

Fig. 115. SENECIONEAE. *Senecio: a,* rayed head, × 1, cylindrical invol. with 1 series of phyllaries and smaller basal bractlets; *b,* ray-fl., × 1½, pappus capillary; *c,* disk-fl., × 1½, the stigmatic-branches truncate. *Lepidospartum: d,* discoid head, × 1, invol. of imbricate phyllaries; *e,* fl., × 2½; *f,* stigmatic branches, × 5.

phyllaries ± woolly villous to glabrous, dark-margined; disk-corollas long-hairy near the junction with the mostly glabrous ak.—Talus slopes, rocky places, 5000–13,600 ft.; Red Fir F. to Subalpine Fell-fields; Sierra Nevada, White Mts.; to the Arctic, Rocky Mts. July–Sept.

Tribe 5. **Senecióneae.** GROUNDSEL TRIBE (Fig. 115)

Herbs or shrubs, or in the tropics, sometimes trees. Lvs. mostly alternate or basal, rarely opposite (as in *Arnica*). Heads radiate or rarely discoid. Phyllaries herbaceous, mostly subequal and 1-seriate, sometimes with an outer calyculate series, rarely imbricated in several series (*Lepidospartum*). Receptacle almost always naked. Style-branches of perfect fls. usually flat and truncate, often penicillate, lacking a well-defined sterile appendage. Pappus of numerous capillary bristles, very rarely lacking (*Adenocaulon*).—A cosmopolitan tribe with over 50 genera; *Senecio* itself is one of the largest known world-wide genera, with over 1000 spp.

A. Pappus none; disk–fls. sterile; fls. whitish .... 107. *Adenocaulon*
AA. Pappus of many bristles.
   B. Lvs. opposite or the uppermost sometimes alternate.
      C. Fls. 2–3, each embraced by a phyllary; pappus deciduous in a ring; diminutive annual
                                                        108. *Dimeresia*
      CC. Fls. more numerous, not individually embraced by a phyllary; pappus not deciduous
         in a ring; perennial herbs .... 109. *Arnica*
   BB. Lvs. alternate or basal.
      C. Pappus plumose, the bristles ± flattened toward base .... 110. *Raillardella*
      CC. Pappus not plumose, mostly of numerous soft capillary bristles.
         D. Heads radiate.
            E. Disk-fls. fertile.
               F. Style-branches truncate or nearly so, penicillate; receptacle flat or
                   merely convex .... 111. *Senecio*
               FF. Style-branches with deltoid appendages; receptacle conic; diminutive
                   annual .... 112. *Crocidium*
            EE. Disk-fls. sterile; style undivided or nearly so; fl.-stems preceding the true lvs.
                                                       113. *Petasites*
        DD. Heads discoid.
            E. Phyllaries 1- or at most 2-seriate.
               F. Herbs, sometimes with a woody caudex (*Luina*).
                   G. Corolla-throat elongate, much exceeding the proper tube; low
                       compact desert annuals .... 114. *Psathyrotes*
                   GG. Corolla-throat shorter than the tube, slender. Coast Ranges.
                       H. Plants tall annuals; lvs. pinnately lobed or pinnatifid;
                         herbage glabrate to deciduously tomentulose
                                                117. *Erechtites*
                  HH. Plants biennial to perennial.
                     I. Lvs. palmately cleft or parted, mostly basal; heads
                       corymbose .... 115. *Cacaliopsis*
                     II. Lvs. entire to toothed or pinnate or pinnatifid.
                       J. Fls. yellow to orange .... 111. *Senecio*

JJ. Fls. white to pinkish .... 116. *Luina*
FF. Shrubs.
　G. Phyllaries linear-subulate, 2-seriate; lvs. terete, resinous-punctate; herbage glabrous .... 118. *Peucephyllum*
　GG. Phyllaries oblong, 4–6, uniseriate; lvs. not as above; herbage white-tomentose .... 119. *Tetradymia*
EE. Phyllaries 3–4-seriate; broomlike shrub with small numerous heads in large panicles .... 120. *Lepidospartum*

107. Adenocáulon Hook.

Perennial or annual herbs with slender stems leafy at base. Lvs. large, alternate. Heads small, in a large subnaked paniculate infl. Invol. of a single row of subequal green phyllaries. Receptacle flat, naked. Fls. whitish, tubular, the outer 3–7, ♀, the inner ♂ with undivided style. Anthers strongly sagittate. Aks. elongate, clavate, large, slightly nerved, glandular; pappus none. Ca. 4 spp., of w. N. Am., e. Asia, Guatemala, Chile. (Greek, *aden*, gland, and *kaulos*, stem.)

1. **A. bìcolor** Hook. Perennial from slender rootstocks, 5–9 dm. tall, floccose-woolly below, glandular above; lvs. deltoid-ovate, thin, ± cordate, the blades 5–12 cm. long, ca. as wide, subglabrous above, closely white-woolly beneath, mostly ± coarsely sinuate-dentate; petioles equal to or longer than blades; infl. forming upper ⅔ or more of plant, the branches very slender, glandular; phyllaries ca. 1.5–2 mm. long, reflexed in fr. and then deciduous; aks. 5–8 mm. long, coarsely glandular above.—Moist shaded woods, below 6000 ft.; Red Fir F., Yellow Pine F., Mixed Evergreen F., Douglas-Fir F., Redwood F.; Coast Ranges from Santa Cruz Co. n., Sierra Nevada from Tulare Co. n.; to B.C., Mont., Mich., Minn. June–Aug.

108. Dimerèsia Gray

Small tufted annual with opposite entire lvs. Heads in close terminal clusters. Invol. of 2–3 oblong phyllaries ± united at base, broadly rounded on the back and each embracing a fl. Fls. 2–3, tubular, perfect. Anthers sagittate, but scarcely caudate. Style-branches flattened, papillate-puberulent. Aks. clavate, striate; pappus of ca. 20 plumose bristles united at base and deciduous together. One sp. (Greek, *di*, two, and *meres*, number, from the 2-flowered heads.)

1. **D. Howéllii** Gray. [*Ereminula H.* Greene.] Stems stout, fleshy; plants less than 1 dm. across, ± arachnoid below, more glandular upwards; lvs. oblanceolate to spatulate or ovate, 1–3 cm. long including the petiole, crowded about the compact clusters of heads; invol. 4–6 mm. high; corolla-tube purple, lobes white to pinkish or purplish; aks. 3 mm. long.—Dry gravelly places, at ca. 5000 ft.; e. slope of Warner Mts., Modoc Co., Siskiyou Co.; n. Nev., se. Ore. June–July.

109. Árnica L.

Perennial herbs, ± glandular or aromatic, from a rhizome or caudex and with fibrous roots. Lvs. simple, opposite or the uppermost sometimes alternate. Heads rather large, turbinate to hemispheric, solitary to rather many, radiate or discoid. Phyllaries green, ± biseriate, subequal and connivent. Receptacle flat or convex, naked. Rays, when present, ♀ and fertile, yellow or orange, rather few and broad. Disk-fls. perfect and fertile, yellow or orange. Anthers entire to minutely sagittate. Style-branches ± flattened, truncate, with an inner longitudinal groove, the outer surface papillose. Aks. subterete, 5–10-nerved; pappus white to tawny, the bristles many, capillary, barbellate to subplumose. Ca. 30 spp., circumboreal, but running s. in the mts. Some spp. apparently of hybrid origin and confused; some with apomixis. (Origin of name obscure.)

(Maguire, B. A monograph of the genus Arnica. Brittonia 4: 386–510, 1943.)
A. Heads characteristically radiate (rayless plants sometimes occurring with normal rayed ones).
　B · Cauline lvs. mostly 5–12 pairs.
　　C. Phyllaries obtusish, with a tuft of long hairs just within the tip; rhizomes conspicuously elongate, almost naked; stems solitary .... 1. *A. Chamissonis*
　　CC. Phyllaries ± sharply acute, the tip not markedly more hairy than the rest.
　　　D. Cauline lvs. entire, sessile; plants densely tufted; rhizomes mostly shortened into a branching caudex .... 2. *A. longifolia*

        DD. Cauline lvs. ± toothed; plants mostly not much tufted; rhizomes mostly elongate
                                                    3. *A. amplexicaulis*
  BB. Cauline lvs. mostly 2–4 pairs.
    C. Rays mostly 7–15 mm. long; young heads nodding; lower petioles and lower stem
       densely and conspicuously long-pubescent .... 6. *A. Parryi* var. *Sonnei*
    CC. Rays mostly 15–30 mm. long, or if shorter, the plants not having long conspicuous hairs;
       heads erect.
      D. Pappus ± tawny, plainly subplumose; rhizomes rooting freely but often short.
        E. Heads broad, ± hemispheric; lower lvs. generally largest .. 4. *A. mollis*
        EE. Heads narrower, ± turbinate; middle lvs. commonly largest
                                                 5. *A. diversifolia*
     DD. Pappus whitish, mostly merely barbellate; rhizomes nearly naked or rooting freely.
        E. Lf.-blades mostly 3–10 times as long as wide; rhizomes mostly short and
          rooting freely.
          F. Lower cauline lvs. sessile or subsessile; ligules denticulate, the teeth
             less than 0.5 mm. long or subentire; rays ca. 8 .... 9. *A. Rydbergii*
          FF. Lower cauline lvs. petioled; ligules obviously toothed, the teeth 0.5–1
             mm. long; rays ca. 13 or ca. 21.
             G. Phyllaries broadly lanceolate to elliptic-oblong, obtusish, ciliolate,
                the tips pilose within; old lf.-bases with dense tufts of long tawny
                hair in axils; disk-corollas pilose .... 7. *A. fulgens*
             GG. Phyllaries narrowly lanceolate, acute, scarcely ciliolate, the tips
                sparingly or not pilose within; old lf.-bases without axillary tuft of
                hair, or with a scant one; disk-corollas glandular, not pilose
                                        8. *A. sororia*
      EE. Lf.-blades mostly 1–2.5 times as long as broad; rhizomes mostly elongate and
       subnaked.
        F Aks. mostly glabrate below the middle; basal lvs. (those on separate short
          shoots) seldom cordate when present, the cauline ones rarely cordate;
          lvs. generally ± toothed; phyllaries with few or no long hairs.
          G. Pappus merely barbellate; stems mostly 2–6 dm. tall; heads
             mostly 1–3 .... 10. *A. latifolia*
          GG. Pappus subplumose; stems 1–3 dm. tall; heads mostly 1
                                  11. *A. cernua*
      FF. Aks. mostly short-hairy or glandular nearly or quite to the base; basal lvs.
       often cordate.
        G Phyllaries ± obtusish, with a subapical tuft of hairs within; heads
          mostly 5–7; lvs. not cordate. Rare. .... 12. *A. tomentella*
        GG. Phyllaries ± sharply acute, mostly lacking a subapical tuft of hairs;
          heads mostly 1–3, sometimes more in plants with cordate lvs.
          H. Phyllaries with few or no long hairs; lvs. entire or subentire,
             the lower broadly rounded to subcordate at base
                              13. *A. nevadensis*
          HH. Phyllaries with some to many long white hairs especially
             below; lvs. mostly toothed, the basal and usually the lower
             cauline ± cordate .... 14. *A. cordifolia*
AA. Heads characteristically discoid, the marginal corollas sometimes enlarged, rarely short-radiate.
  B. Lowermost lvs. ± petioled.
    C. Young heads nodding; phyllaries gradually and slenderly acute .... 6. *A. Parryi*
    CC. Young heads erect; phyllaries acutish.
      D. Lowermost lvs. with blades mostly ovate to cordate, ± abruptly contracted to the
        narrowly or scarcely winged petiole, rarely spatulate and wing-petiolate; aks.
        usually hairy as well as stipitate-glandular. Widespread ...... 15. *A. discoidea*
      DD. Lowermost lvs. ± spatulate, usually with broadly winged often poorly marked
        petiole; aks. stipitate-glandular. Siskiyou and Del Norte cos. .. 16. *A. spathulata*
  BB. Lvs. all broadly sessile.
    C. Lvs. dentate, strongly 3–5-nerved; disk-corollas copiously soft pilose; aks. canescently
       villous. Shasta Co. .... 17. *A. venosa*
    CC. Lvs. entire, not strongly 3–5-nerved; disk-corollas sparingly long-stipitate-glandular; aks.
       moderately long-stipitate-glandular. Siskiyou Co. .... 18. *A. viscosa*

1. A. Chamissònis Less. ssp. **foliòsa** (Nutt.) Maguire. [*A. f.* Nutt. *A. Bruceae* Rydb.] Perennial from long nearly naked rhizomes; stem solitary, 3–8 dm. tall; herbage ± villous-puberulent to -hirsute, ± glandular or viscid above; cauline lvs. mostly 5–10 pairs, not much reduced upwards, lanceolate to oblanceolate, 5–20 cm. long (including petioles which in lower lvs. may equal the blades), uppermost sessile, blades mostly entire or merely denticulate; heads 5–15, hemispheric-campanulate, 15–18 mm. high; invol. 8–12 mm. high; phyllaries obtuse to acutish, with hairs at base ± septate; ligules pale yellow, 12–18 mm. long; disk-corollas 7.5–9 mm. long; aks. 4–5 mm. long; pappus stramineous, barbellate, the bristles ca. 0.2 mm. long.—Meadows and moist places, 5000–11,000 ft.; Montane Coniferous F.; San Bernardino Mts., Sierra Nevada to Alaska, Alta., Mont. July–Aug. Exceeding variable, the following forms recognizable in Calif.

A. Phyllaries moderately to broadly obtuse; lvs. not thin, mostly entire to remotely and minutely denticulate.
B. Phyllaries very blunt and rounded at tips; herbage often tomentose. San Bernardino Mts., White Mts. .... **var. *bernardina***
BB. Phyllaries narrower, more pointed.
    C. Herbage hairy, but not silvery. Mostly from dryish habitats. San Bernardino Mts., Sierra Nevada n. .... ssp. *foliosa*
    CC. Herbage conspicuously silvery-tomentose. Mostly in very wet places. Sierra Nevada n. var. *incana*
AA. Phyllaries acutish; herbage minutely puberulent; lf.-blades thin, remotely but sharply denticulate. White Mts., Truckee, etc. .... var. *Jepsoniana*

Var. **bernardina** (Greene) Maguire. [*A. b.* Greene. *A. foliosa* var. *b.* Jeps.] Wet places, 7000–9000 ft.; Red Fir F., San Bernardino Mts.; Bristle-cone Pine F., White Mts. July–Aug.

Var. **incana** (Gray) Hult. [*A. foliosa* var. *i.* Gray. *A. i.* Greene.] Range of spp. *foliosa*, but from wetter places.

Var. **Jepsoniana** Maguire.—E. slope of Sierra Nevada and White Mts.

2. **A. longifòlia** D. C. Eat. ssp. **myriadènia** (Piper) Maguire. [*A. m.* Piper.] Tufted perennial from a branching caudex or with short rhizomes; floriferous stems mostly 3–6 dm. high, the sterile lower, leafy; herbage scantily puberulent; lvs. mostly 5–7 pairs, sessile or shortly connate-perfoliate, only gradually reduced upwards, lanceolate to lance-elliptic, mostly acuminate, entire to slightly denticulate, mostly 5–12 cm. long; heads several to many, campanulate; invol. 7–10 mm. high; phyllaries acuminate to sharply acute, glandular-puberulent and often with some longer septate hairs; rays ca. 8–13, ca. 10–20 mm. long; aks. subglabrous or glandular and hairy; pappus ± tawny, somewhat subplumose.—Wet places, 5000–11,000 ft.; Montane Coniferous F.; Sierra Nevada to Modoc Co.; to e. Ore., Mont., Alta. July–Aug.

3. **A. amplexicáulis** Nutt. Rhizomes rather coarse, rooting freely; stems 3–7 dm. high, ± glandular and hairy especially upwards, leafy; lvs. all cauline, mostly 5–12 pairs and mostly sessile, lanceolate or elliptic-lanceolate, mostly irregularly serrate-dentate, 4–12 cm. long; heads generally several, 12–16 mm. high; invol. campanulate; phyllaries lanceolate, sharply acute or acuminate, stipitate-glandular at base; rays 8–14, pale yellow, 10–20 mm. long; aks. sparsely hirsute, sometimes glandular; pappus tawny, subplumose.—Moist places, 7000–10,000 ft.; Lodgepole F., Subalpine F.; Sierra Nevada; to Alaska, w. Mont. July–Aug.

4. **A. móllis** Hook. [*A. Merriami* and *scaberrima* Greene. *A. m.* var. *s.* Smiley.] Rhizomes dark brown, 2.5–4 mm. thick, short, rooting freely; stems 2–6 dm. high, simple, scantily scabrid-puberulent to conspicuously pilose and glandular; cauline lvs. mostly 3–4 pairs, the lower commonly the largest, all often sessile, lanceolate to ovate or obovate, entire to irregularly denticulate; heads 1–few, hemispheric-campanulate, the disk to ca. 3 cm. wide; invol. 10–16 mm. high; phyllaries ± acuminate, long-hairy at base, glandular above; rays mostly 12–18, yellow, 15–25 mm. long; pappus tawny, sub-plumose; aks. hirsute.—Moist places, ca. 7500–11,500 ft.; Red Fir F. to Subalpine F.; Sierra Nevada n.; to B.C., Alta., Rocky Mts. July–Sept.

5. **A. diversifòlia** Greene. [*A. latifolia* var. *viscidula* Gray.] Rhizomes 2–3 mm. in diam., rooting freely; stems 1 or in loose tufts, 1.5–4 dm. high, simple, subglabrous to scabrid-granular, more glandular-puberulent upward; cauline lvs. 3 pairs, ovate to elliptic, obtuse to subcordate at base, serrate-dentate, scabrid-ciliate, the blades 4–8 cm. long, almost sessile to petioled; heads mostly 3–5, campanulate-turbinate, 12–16 mm. high, stipitate-glandular and pilose at base; phyllaries obscurely 2-ranked, 10–14 mm. long, acute to acuminate, stipitate-glandular; rays 12–15, pale yellow, 18–20 mm. long; aks. strongly angled, short-hispidulous; pappus ± tawny, subplumose.—Wet places, ca. 7000–11,000 ft.; Lodgepole F., Subalpine F.; Sierra Nevada from n. Inyo Co. n.; to Alaska, Rocky Mts. July–Aug.

6. **A. Párryi** Gray. Rhizomes horizontal, 1.5–2.5 mm. in diam., dark brown, with slender roots; stem single, 2–6 dm. tall, moderately or scantily lanate-villous at base, glandular at least above; cauline lvs. mostly 2–4 pairs, strongly reduced upwards, the lowermost with blades 5–20 cm. long, lanceolate, short-petioled, denticulate to entire, sparingly villous; infl. cymose, the heads generally several, nodding in bud, campanulate, discoid; invol. mostly 10–14 mm. long; phyllaries sharply acute or acuminate; aks. glabrous to hairy; pappus tawny, strongly barbellate to weakly subplumose.—Occasional

in open damp or dryish places; Montane Coniferous F.; Placer Co.; to Wash., July–Aug. More common in Calif. is the

Ssp. **Sónnei** (Greene) Maguire. [*A. S.* Greene. *A. foliosa* var. *S.* Jeps.] Stem-base densely lanate-pilose; heads normally radiate; rays mostly 7–15 mm. long.—Moist places, 7000–11,000 ft.; upper Montane Coniferous F.; Sierra Nevada from n. Inyo Co. to Eldorado and Nevada cos.; w. Nev. July–Aug.

7. **A. fúlgens** Pursh. [*A. alpina* in part for Calif. refs.] Rhizomes short, 2–7 mm. in diam., rooting profusely, thickly clothed with bases of previous lvs. in axils of which are dense tufts of long tawny hair; stems 2–6 dm. high, stipitate-glandular and often also hairy, more densely so upwards; lvs. 3–5-nerved, mostly ca. 5 pairs, ± hairy and glandular, the lower petioled, oblanceolate to elliptic, mostly 3–12 cm. long; heads 1(–3), broadly hemispheric; invol. 10–15 mm. high, glandular and hairy; phyllaries elliptic, ± ciliate upwards; rays ca. 10–23, 10–15 mm. long; aks. densely hairy; pappus whitish or straw color, barbellate.—Rare, ± moist open places, ca. 6000–9500 ft.; Montane Coniferous F.; Alpine Co. to Modoc Co.; to B.C., Rocky Mts. May–July.

8. **A. sorória** Greene. Near to *A. fulgens;* rhizomes more slender; stems slender; lvs. narrower, the basal without or with white axillary wool; phyllaries narrowly lanceolate, more acute at tip; pappus whitish.—Dryish places; Sagebrush Scrub, Montane Coniferous F.; Sierra Nevada to B.C., Rocky Mts. May–Aug.

9. **A. Rydbérgii** Greene. [*A. sulcata* Rydb.] Rhizomes horizontal, brown, moderately long; stems ± clustered, 1–3 dm. tall; herbage ± glandular and short-hairy to subglabrous; basal lvs. petioled, oblanceolate to spatulate, blades 3–5-nerved, 4–7 cm. long; cauline lvs. 3–4 pairs, sessile or nearly so, oblanceolate to spatulate, subentire, 3–5-nerved, ± bunched on lower stem; heads 1–few, turbinate-campanulate; invol. 9–13 mm. high; phyllaries glandular, ± hairy, ± ciliate, acute; rays ca. 8, 10–20 mm. long, entire to minutely toothed at tip; aks. short-villous, the upper hairs longest; pappus white, barbellate.—Dry meadows and open slopes; Subalpine F.; Scott Mts., Siskiyou Co.; to B.C., Alta., Colo. July–Aug.

10. **A. latifòlia** Bong. Rhizomes slender, horizontal, light brown, moderately rooting; stems 1–few, 1–6 dm. tall; basal lvs. long-petioled, blades 8–25 cm. long, ovate to lanceolate; cauline lvs. 2–4 pairs, 2–12 cm. long, mostly sessile, toothed, obtusish, subglabrous to thinly pilose; heads mostly 1–3, turbinate to subhemispheric; invol. mostly 10–18 mm. high, ± glandular, sometimes pilose; rays 8–12, yellow, 12–25 mm. long; aks. glabrous to pubescent especially upwards; pappus white, barbellate.—Moist places, 5500–7000 ft.; Subalpine F.; Eldorado and Humboldt cos. n.; to Alaska, Rocky Mts. July–Aug.

11. **A. cérnua** Howell. [*A. Chandleri* Rydb.] Rhizomes slender, subnaked; stems 1–few, 1–3 dm. tall; basal lvs. ± ovate-cordate, long-petioled, the blades 2–4 cm. long; cauline lvs. mostly 3–4 pairs, glabrous or scabrous-puberulent, obovate to lance-ovate or the lower even subcordate, up to ca. 5 cm. long; heads mostly 1, turbinate-campanulate; invol. 10–18 mm. high, puberulent, often glandular; phyllaries ciliolate; rays 7–10, 15–25 mm. long, strigose; aks. hirsute above; pappus white, subplumose.—Dryish serpentine slopes, 4500–5500 ft.; Yellow Pine F.; Humboldt, Del Norte, and Siskiyou cos.; sw. Ore. April–July.

12. **A. tomentélla** Greene. Rhizomes slender, long, subnaked; stems solitary, 2–5 dm. high, glandular especially above, hairy; lvs. cauline, mostly 3–4 pairs, grayish tomentulose-puberulent, the lower petioled, ovate to ovate-elliptic, the blade 3–7 cm. long, mostly toothed; heads mostly 5–7, campanulate-hemispheric; invol. 10–13 mm. high; phyllaries acutish to obtuse 9–12, biseriate, 3–5 mm. wide, tomentulose; rays 12–20 mm. long; aks. short-hairy, sparsely glandular; pappus white to straw color, mostly subplumose.—Open slopes and woods, ca. 5000–7000 ft.; mostly Red Fir F. to Subalpine F.; Sierra Nevada in Tulare Co., Placer Co. n. to Siskiyou Co.; s. Ore. July–Aug.

13. **A. nevadénsis** Gray. Rhizomes slender, almost naked; stems solitary or few, 1–2.5 dm. high; herbage ± glandular and puberulent; cauline lvs. mostly 2–3 pairs, the lower larger, petioled, broadly ovate, the blade 3–8 cm. long, mostly entire; heads 1–3, broadly turbinate to campanulate; invol. 10–18 mm. high; phyllaries ± sharply acute, stipitate-glandular; rays ca. 12–20 mm. long; aks. glandular and/or short-hairy; pappus white or straw color, ± subplumose.—Open rocky banks near streams, etc., 6600–11,900 ft.; Red Fir F. to Alpine Fell-fields; Sierra Nevada to Wash., Nev. July–Aug.

14. **A. cordifòlia** Hook. [*A. Austinae* Rydb.] Rhizomes extensive, nearly naked; stems mostly solitary, 2–6 dm. high, loosely white-hairy to glandular-puberulent; basal lvs. cordate, long-petioled, mostly on separate short shoots; cauline lvs. 2–3 pairs, the lower petioled, the blades 2–9 cm. long, lanceolate to round-ovate, cordate to truncate at base, ± puberulent or with some longer hairs, the upper blades sessile, narrower; heads mostly 1–3, broadly turbinate, 1.8–2.5 cm. high; phyllaries 10–15, biseriate, 14–18 mm. high, pilose at base, puberulent above; rays 9–14, yellow, elliptic-oblong, 15–30 mm. long; aks. mostly short-hairy or glandular or both; pappus whitish, barbellate.—Dry or moist open or wooded places, 3500–10,000 ft.; Montane Coniferous F.; Cuyamaca and Santa Ana mts., Sierra Nevada to Modoc Co., Coast Ranges from Monterey Co. n.; to Alaska, Rocky Mts. to New Mex. May–Aug. Dwarfed plants 1–2 dm. high with lvs. mostly 2–5 cm. long and less hairy invol., distributed with the sp., have been called var. *pùmila* (Rydb.) Rydb. [*A. p.* Rydb.]

15. **A. discoìdea** Benth. Rhizomes long and naked or forming an approximate crown; stems mostly solitary, 3–6 dm. tall, glandular-puberulent and often also ± long-hairy, commonly branched above or even lower; cauline lvs. 3–several pairs, mostly crowded toward base, there opposite, and long-petioled; lvs. ± long-hairy, the blade ovate to deltoid or subcordate, 3–8 cm. long, sharply or undulately dentate; upper lvs. sessile, reduced; heads several to many, turbinate-campanulate to subhemispheric, discoid; invol. 9–13 mm. high, glandular and spreading-villous; phyllaries mostly obtusish, 9–11 mm. long; aks. glandular and hairy throughout or glabrate above; pappus white or straw color, strongly barbellate.—Open woods, 1500–6000 ft.; Mixed Evergreen F. to Yellow Pine F.; outer Coast Ranges from n. Ventura Co. to Mendocino Co. May–July.

Var. **eradiàta** (Gray) Cronq. [*A. cordifolia* var. *e.* Gray. *A. parviflora* Gray.] Petioles not or scarcely winged; lvs. less apt to be cordate; cauline lvs. 2–3 pairs; heads smaller, the invol. 10–12 mm. high; pappus barbellate.—Mostly open woods; Mixed Evergreen F. to Yellow Pine F.; inner Coast Ranges from Alameda Co. n.; to Wash. May–Aug.

Var. **alàta** (Rydb.) Cronq. [*A. a.* Rydb. *A. sanhedrensis* Rydb.] Basal lvs. mostly cordate; cauline lvs. 2–4 pairs, the middle and upper with broadly winged petioles; invol. 12–15 mm. high; pappus subplumose.—Mixed Evergreen F. to Red Fir F.; Coast Ranges from Monterey Co. to Del Norte and Siskiyou cos., Sierra Nevada from Mariposa Co. to Shasta Co.; Ore. June–Aug.

16. **A. spathulàta** Greene. Much like *A. discoidea;* plants pilose and conspicuously long- and short-stipitate-glandular; lower cauline lvs. spatulate to oblanceolate, the petioles mostly broadly wing-margined; cauline lvs. 3–5 pairs; invol. 12–17 mm. high, more coarsely and conspicuously glandular and spreading-hairy; phyllaries acute; aks. stipitate-glandular throughout; pappus barbellate.—Dry open woods; Siskiyou Co.; s. Ore. April–July.

Ssp. **Eastwoòdiae** (Rydb.) Maguire. [*A. E.* Rydb.] Plants villous only, not conspicuously glandular; stems more apt to be branched.—At ca. 2000–3000 ft.; Mixed Evergreen F., Yellow Pine F.; Del Norte and Siskiyou cos.; sw. Ore. June–July.

17. **A. venòsa** Hall. Rhizomes 10–14 mm. thick; stems angled and ribbed, glabrous at base, viscid and with weak septate hairs above; cauline lvs. 6–8 pairs, 4–6 cm. long, broadly sessile, lanceolate to ovate-elliptical, coarsely cuspidate-dentate, 3–5-ribbed from base, stipitate-glandular and pubescent; heads solitary on each branch, 18–22 mm. high; invol. turbinate-campanulate; aks. canescent and hispid; pappus white, barbellate to subplumose.—Rare, on hot dry slopes, ca. 1300–2500 ft.; Foothill Wd.; Sierran foothills, Shasta Co. May–June.

18. **A. viscòsa** Gray. [*Chrysopsis shastensis* Jeps.] Much like the preceding, stems 2–5 dm. high, coarse, angled; cauline lvs. 8–12 pairs, 2–5 cm. long, broadly sessile, entire, 3–5-nerved; heads 10–20; phyllaries 10–15, lanceolate, 8–11 mm. long, long-stipitate-glandular and pilose; aks. long-stipitate-glandular; pappus white or creamy, strongly barbellate.—Rocky places, ca. 8000 ft.; Subalpine F.; Mt. Shasta, Siskiyou Co.; Mt. Mazama, Ore. Aug.

110. Raillardélla Gray in Benth. & Hook.

Perennial herbs with simple entire or subentire alternate or basal lvs. and solitary or few heads. Heads discoid or rayed. Phyllaries subequal, subherbaceous, uniseriate or nearly so. Receptacle naked, flat. Anthers subtruncate at base. Style-branches flattened, with introrsely marginal stigmatic lines and a slender hispidulous appendage. Aks. linear, subcompressed, several-nerved. Pappus of ± numerous plumose bristles somewhat flattened toward base. Ca. 5 spp. (Diminutive of *Raillardia,* a shrubby genus of Hawaii.)

Stems leafy, usually branched; lvs. linear; herbage hirsute-pubescent.
  Lvs. acute, apparently all alternate, none sheathing; heads discoid. S. Sierra Nevada
                                                                 1. *R. Muirii*
  Lvs. ± obtuse, the lower opposite, somewhat scalelike, ± connate-sheathing at base; heads ordinarily with some rays. Inner N. Coast Ranges . . . . 2. *R. scabrida*
Stems simple and monocephalous, ± scapose; lvs. mostly basal.
  Lvs. green, not tomentose.
    Heads with at least 20 phyllaries and 6–10 rays; lvs. glabrous. Siskiyou Co. . . . . 3. *R. Pringlei*
    Heads with ca. 8–13 phyllaries, discoid or with mostly 1–3 rays; lvs. glandular and sometimes inconspicuously hairy. Sierra Nevada to Ore. . . . . 4. *R. scaposa*
  Lvs. gray, silky-tomentose. San Bernardino Mts., Sierra Nevada n. . . . . 5. *R. argentea*

1. **R. Muirii** Gray. [*Raillardiopsis M.* Rydb.] Stems several from the branched caudex, 2–3 dm. high; herbage roughish-hirsute, the stems glandular-villous; lvs. linear or lance-linear, acute, sessile, none sheathing; heads terminal and with 1–2 lateral, discoid; invol. with ca. 13 phyllaries, campanulate, hirsute and glandular-villous; pappus-bristles ca. 10–12, ca. 8 mm. long.—Rare, open slopes, 4000–7000 ft.; Yellow Pine F.; Fresno Co., Calif. July.

2. **R. scabrida** Eastw. [*Raillardopsis s.* Rydb.] Stems several from the branched root-crown, 1–4 dm. long, often branched above and with several heads; lvs. well distributed, many, mostly alternate, linear, sessile, obtuse, ca. 1–3 cm. long, hirsute and stipitate-glandular, the lowest ± scalelike, opposite, sheathing; heads pedunculate, discoid or with 2–3 broad rays 3–5 mm. long; invol. 9–12 mm. high; phyllaries ca. 8–10, stipitate-glandular; rays without pappus, the aks. glabrous or ± hairy; pappus of disk-fls. of ca. 15 or more plumose parts.—Open stony places, 6500–7500 ft.; Red Fir F.; n. Lake Co. to Tehama and Mendocino cos. July–Aug.

3. **R. Pringlei** Greene. [*R. scaposa* var. *P.* Jeps.] Scapose from a branched caudex; lvs. glabrous, linear-oblanceolate, 3–10 cm. long, firm, entire or shallowly toothed, basal or on the lower stems; stems scapose, 2–4 dm. high, stipitate-glandular; head solitary, ca. 2 cm. across; invol. ca. 1 cm. high, shorter than the disk; phyllaries ca. 20–30, stipitate-glandular, ± ciliate; rays ca. 6–10, orange to salmon, ca. 1–2 cm. long, deeply trifid; aks. sparsely hairy above; pappus of ca. 15 stout flattened bristles.—Boggy and wet places, at ca. 5500 ft.; Red Fir F.; near Mt. Eddy, Siskiyou Co. July–Aug.

4. **R. scaposa** (Gray.) Gray. [*Raillardia s.* Gray. *Raillardella s.* var. *Eiseni* Gray.] Perennial, scapose or nearly so, from branched caudex; lvs. linear-oblanceolate, mostly 3–15 cm. long, entire, green, ± stipitate-glandular; scape 0.5–4 dm. high, stipitate-glandular; head discoid or with 1–3(–5) short rays; invol. 12–16 mm. high; phyllaries mostly ca. 8 or ca. 13, stipitate-glandular and sometimes with eglandular hairs; pappus of ca. 15 flattened plumose bristles; aks. pubescent.—Dry stony places and edge of meadows, etc., 6500–11,100 ft.; mostly Lodgepole F., Subalpine F.; Sierra Nevada to w. Nev., Ore. July–Aug.

5. **R. argentea** (Gray) Gray. [*R. minima* Rydb. *Raillardia a.* Gray.] Scapose perennial 1–10 cm. high, conspicuously silky-tomentose; lvs. mostly 1–6 cm. long; heads strictly discoid; invol. ca. 9–14 mm. high; phyllaries linear, acuminate; fls. yellow; pappus-bristles flattened.—Dry open rocky places, mostly 9000–12,000 ft.; Subalpine F., Alpine Fell-fields; San Gorgonio Mt., Sierra Nevada, n. to Ore. July–Aug. Hybrids have been found between this sp. and *R. scaposa* in Yosemite National Park.

111. Senècio L. GROUNDSEL. RAGWORT

Herbs, shrubs or vines, sometimes arborescent. Lvs. alternate or basal, entire to toothed or divided. Heads solitary to many, cylindric to campanulate or hemispheric, mostly many-fld., usually radiate; ray ♀ and fertile, or none. Phyllaries essentially equal, mostly in 1 series, often with smaller bracteoles at the base. Receptacle flat or convex. Disk-fls. perfect, fertile. Anthers entire to minutely sagittate. Style-branches ± flattened, truncate. Aks. subterete, 5–10-nerved; pappus of many usually white soft bristles. Ca. 1000 spp., of very wide distrib. The N. Am. spp. very much need study. (Latin, *senex,* old man, because of the white pappus.)

(Greenman, J. M. Monograph of the N. and Cent. Am. spp. of the genus Senecio. Ann. Mo. Bot. Gard. 2: 573–626, 1915; 3: 85–194, 1916; 4: 15–36, 1917; 5: 37–108, 1918 [not completed].)
A. Plants perennial.
    B. Stems climbing; lvs. ivylike .... 38. *S. mikanioides*
    BB. Stems not climbing; lvs. not ivylike.
        C. Lvs. pinnatifid into narrowly linear segms. or sometimes entire and linear; plants ±
            woody at base. Mostly below Yellow Pine belt.
            D. Lf.-lobes obtuse, numerous; rays 6–8 mm. long; invol. 6–7 mm. high. Insular
                                                                                1. *S. Lyonii*
            DD. Lf.-lobes acute, mostly few; rays 8–15 mm. long; invol. mostly 7–11 mm. high.
                Mostly mainland.
                E. Phyllaries mostly ca. 21; bracteoles at base of invol. rather long and con-
                    spicuous .... 2. *S. Douglasii*
                EE. Phyllaries mostly ca. 8 or 13; bracteoles short, inconspicuous.
                    F. Lvs. or segms. linear-filiform, mostly ca. 1 mm. wide; aks. 4 mm.
                        long. Coastal Strand .... 3. *S. Blochmanae*
                    FF. Lvs. or segms. wider, ca. 2 mm. wide; aks. 2.5–3 mm. long. Slopes of
                        high mts. bordering Mojave Desert .... 4. *S. spartioides*
        CC. Lvs. not narrowly linear or divided into linear segms.; plants largely herbaceous.
            Mainly from Yellow Pine belt or above.
            D. Lvs. well distributed along the stems, only slightly or gradually reduced up the
                stem.
                E. Stems 0.5–3 dm. long; lvs. 1–3.5 cm. long.
                    F. Lvs. broadly oblanceolate to obovate, mostly 1–2 cm. wide; plants from
                        a caudex with taproot, the branches sometimes ± rhizomelike. Widely
                        distributed in high mts. .... 5. *S. Fremontii*
                    FF. Lvs. narrower, 0.3–0.5 cm. wide; plants from slender rhizomes, without
                        a taproot. Mts. of Mono Co. .... 6. *S. pattersonianus*
                EE. Stems 5–15 dm. long; lvs. 4–20 cm. long.
                    F. Lvs. 2–3 times pinnatifid. Introd. weed in Mendocino Co.
                                                                        31. *S. Jacobaea*
                    FF. Lvs. once pinnatifid to toothed. Native plants.
                        G. Some lvs. pinnatifid to pinnatilobate; phyllaries not black-tipped.
                            Sierra Nevada .... 7. *S. Clarkianus*
                        GG. Lvs. mostly toothed, not deeply cut; phyllaries often with black tips.
                            H. Lower lvs. with deltoid to subcordate base. Widely distributed
                                in higher mts. .... 8. *S. triangularis*
                            HH. Lower lvs. tapering at base. Sierra Nevada to Modoc Co.
                                                                        9. *S. serra*
            DD. Lvs. largely basal, the cauline rapidly reduced up the stem.
                E. Stems mostly solitary, simple, 3–20 dm. tall, arising from a short erect short-
                    lived crown and with numerous fibrous roots.
                    F. Herbage ± villous with multicellular hairs, later glabrate.
                        G. Rays 5–12; invol. mostly 8–12 mm. high. Yellow Pine belt and
                            above .... 10. *S. integerrimus*
                        GG. Rays 0–2; invol. 5–8 mm. high. Often below Yellow Pine belt
                                                                        11. *S. aronicoides*
                    FF. Herbage subglabrous from the beginning.
                        G. Plants not glaucous, 3–10 dm. tall; lvs. toothed. Ne. Calif.
                                                                        12. *S. foetidus*
                        GG. Plants glaucous, 6–20 dm. tall; lvs. mostly entire. Cent. and n.
                            Calif. .... 13. *S. hydrophilus*
                EE. Stems few to several, often shorter, arising from a ± horizontal caudex or
                    creeping rhizome.
                    F. Lvs. toothed, not lobed, crenate or pinnatifid.
                        G. Heads discoid; invol. 9–10 mm. high. San Bernardino Mts. to
                            San Luis Obispo Co. .... 14. *S. astephanus*
                        GG. Heads mostly rayed; invol. 4–6 mm. high. Sierra Nevada to Mt.
                            Lassen .... 15. *S. Scorzonella*
                    FF. Lvs. lobed, crenate or pinnatifid, sometimes entire.
                        G. Basal lvs. entire or toothed or few-lobed, not pinnatifid or lyrate.
                            H. Herbage ± white-tomentose at anthesis. Plants of dry places.
                                I. Heads several to many.
                                    J. Lvs. nearly all in a basal tuft. San Bernardino Mts.
                                                                        18. *S. bernardinus*
                                    JJ. Lvs. not all in a basal tuft.
                                        K. Stems 1–3 dm. high; lvs. round to elliptic, 1–3
                                            cm. long; rays 6–12 mm. long. Sierra Nevada
                                            and White Mts. to Siskiyou and Modoc cos.
                                                                        16. *S. canus*
                                        KK. Stems 3–4.5 dm. high; lvs. linear-oblong, 3–7 cm.
                                            long; rays ca. 15 mm. long. Siskiyou Mts.
                                                                        17. *S. ligulifolius*
                                II. Heads ca. 1–6.

J. Invol. 8–14 mm. high; disk ca. 15–20 mm. wide.
    K. Rays yellow. San Bernardino and San Gabriel mts.
                                  19. *S. ionophyllus*
    KK. Rays orange or flame. N. Coast Ranges
                                 20. S. *Greenei*
JJ. Invol. 6–9 mm. high; disk ca. 7–15 mm. wide. Sierra
    Nevada .... 21. *S. werneriaefolius*
HH. Herbage subglabrous from the first. Plants of ± moist places.
  I. Invol. 8–10 mm. high; heads 1-few. S. Calif.
                              19. *S. ionophyllus*
  II. Invol. often shorter. Sierra Nevada and N. Coast Ranges.
    J. Heads mostly solitary and radiate; rootstock very
       slender .... 22. *S. subnudus*
    JJ. Heads mostly several to many, or if few, they are
       discoid; rootstock rather coarse.
       K. Heads discoid, usually orange or reddish
                          23. *S. pauciflorus*
       KK. Heads normally radiate.
         L. Some of basal lvs. ± cordate
                23. *S. pauciflorus* var. *jucundulus*
         LL. Basal lvs. not ± cordate
           M. Herbage not glaucous; rays yellow.
              Sierra Nevada to Siskiyou Co.
                   24. *S. cymbalarioides*
           MM. Herbage strongly glaucous; rays orange.
              Napa and Lake cos.
                 25. *S. Clevelandii*
GG. Basal lvs. dissected or pinnatifid to lyrate.
  H. Stems mostly 3–8 dm. high.
    I. Invol. 5–8 mm. high, mostly rather hairy. Coastal Strand,
      Mendocino Co. n. .... 26. *S. Bolanderi*
    II. Invol. 8–12 mm. high, glabrous.
      J. Herbage glabrous except for axillary tufts of wool; lvs.
       pinnatifid, the segms. globulate-dentate. S. Coast
       Ranges to Greenhorn Mts. .... 27. *S. Breweri*
      JJ. Herbage tomentulose at least while young; lvs. more
       deeply dissected. Sonoma and Colusa cos. to Siskiyou
       and Modoc cos. .... 28. *S. eurycephalus*
  HH. Stems mostly 1–3 dm. high.
    I. Herbage thinly tomentulose, the lvs. not usually with a
      larger terminal segm. Death V. region to Mono Co.
                    29. *S. multilobatus*
    II. Herbage quite glabrous except for axillary tufts of wool;
      lvs. usually with terminal lobe large and rounded. E.
      Mojave Desert .... 30. *S. stygius*
AA. Plants annual.
  B. Rays purplish-red, rarely white; plant ± viscid-pubescent. Escape from cult. along the
    coast .... 32. *S. elegans*
  BB. Rays yellow or none; plant not viscid-pubescent.
    C. The rays well developed, conspicuous. Native, from Monterey and Tulare cos. s.
                         33. *S. californicus*
    CC. The rays none or very inconspicuous.
      D. Bracteoles at base of invol. well developed, black-tipped. Common introd. weed
                       36. *S. vulgaris*
      DD. Bracteoles at base of invol. none or poorly developed, not black-tipped.
        E. Phyllaries ca. 8; slender-stemmed plant. From near the coast, Solano Co.
         to San Diego Co. .... 34. *S. aphanactis*
        EE. Phyllaries ca. 13.
         F. Lvs. ± pinnatifid and irregularly toothed, not much broadened at their
          base. Introd. weed, nw. coast .... 35. *S. sylvaticus*
        FF. Lvs. coarsely dentate, cordate-clasping at the broad base. Desert native
                      37. *S. mohavensis*

1. **S. Lyònii** Gray. Rather shrubby, 6–12 dm. high, branched, tomentose on younger parts, soon glabrate but with persistent tufts of wool in the axils; lvs. well spaced along the stems, 2–12 cm. long, 1–2-pinnate into linear or broadly linear segms. and lobes, green above, ± white-tomentose beneath, sessile and auriculate or petioled; infl. corymbiform on a long bracteate peduncelike stem; invol. 6–7 mm. high, broadly turbinate; phyllaries linear; rays yellow, 6–8 mm. long; aks. canescent, linear, almost 3 mm. long.—Coastal bluffs; Coastal Sage Scrub; Santa Catalina and San Clemente ids. March–June.

2. **S. Douglásii** DC. Shrubby, branched, bushy, mostly ca. 1–1.6 m. high, leafy up to the infl.; stems striate; herbage white-tomentose when young, the lvs. glabrate

and gray-green above, whitish tomentose beneath; most lvs. 3–10 cm. long, pinnate into mostly 5–9 linear revolute lobes generally 2–6 cm. long, 1–2.5 mm. wide, or the upper lvs. 3-lobed or entire; principal lvs. often with axillary fascicles of smaller ones; infl. corymbiform, of several to many heads; invol. broadly turbinate, 8–10 mm. high, the longer phyllaries mostly 16–21, with quite well developed shorter bracteoles at their base; rays 10–13, yellow, showy, ca. 10–15 mm. long; aks. canescent, ca. 4 mm. long; $n = 20$ (Vasek).—Common in washes and dry stream beds, below 5000 ft.; Foothill Wd., Coastal Sage Scrub, Chaparral, V. Grassland, etc.; Coast Ranges from Mendocino Co. to L. Calif., less abundant in Sierran foothills and w. parts of desert; Santa Cruz and Santa Catalina ids. June–Oct.

Var. tularénsis Munz. Lvs. mostly 2–4 cm. long, rather evenly pinnate into segms. 0.5–1.5 cm. long.—Sandy and rocky places below 2500 ft.; Foothill Wd.; Sierran foothills, Tuolumne Co. to Tulare Co. May–Oct.

Var. monoénsis (Greene) Jeps. [*S. m.* Greene.] Less shrubby, often scarcely woody even at base, 3–10 dm. high, more yellowish-green, glabrous or rarely slightly tomentulose; lvs. from scarcely dissected with long linear segms. to more so and with many shorter segms.; invol. mostly ca. 7–8 mm. long, the outermost bracteoles quite conspicuous; aks. ca. 5 mm. long.—Dry slopes and washes, 2000–6500 ft.; Creosote Bush Scrub to Pinyon-Juniper Wd.; n. Colo. Desert, Mojave Desert n. to Mono Co.; to Utah, Ariz. March–May, sometimes also in fall.

3. **S. Blóchmanae** Greene. Undershrub 5–12 dm. high, glabrous or with some tomentum, leafy to the summit of the branches; lvs. linear-filiform, entire, mostly 3–7 cm. long, persistent; heads several to many in a corymbose arrangement; invol. 7–10 mm. high, the principal phyllaries ca. 13, the outer bracteoles few and short; rays yellow, ca. 1 cm. long; aks. scabrous-canescent, ca. 4 mm. long, linear.—Coastal Strand; San Luis Obispo and n. Santa Barbara cos. May–Oct.

4. **S. spartioìdes** T. & G. [*S. serra* var. *sanctus* Hall. *S. multicapitatus* of Jeps., not Greenm.] Perennial with a heavy woody branched caudex, 2–6 dm. high, severalstemmed, leafy to the infl.; herbage glabrous or occasionally ± pubescent; lvs. many, mostly linear and entire, the lower small, the others 3–10 cm. long, 1.5–5 mm. wide, often spreading or ± recurved; infl. paniculate-corymbiform, leafy-bracteate; invol. 6–11 mm. high, subcylindric, the principal phyllaries 8–13, the outer bracteoles inconspicuous; rays 8–15 mm. long, yellow; aks. strigulose to subglabrous, ca. 2.5–3 mm. long.—Dry stony slopes and edge of meadows, 7800–10,400 ft.; Red Fir F., Bristle-cone Pine F., etc.; San Bernardino Mts., e. slope of Sierra Nevada, Panamint Mts., Inyo-White Mts., n. to Mono Co.; to Colo., New Mex., L. Calif. July–Aug. The pubescent plants occur widely scattered over the range of the sp.

5. **S. Fremóntii** T. & G. Perennial from a branching caudex, the stems decumbent at base, glabrous, mostly ca. 1–1.5 dm. long, leafy, lvs. ± succulent, glabrous, obovate to spatulate or almost ovate, rather deeply and fairly sharply toothed, petiolate at base, mostly 1.5–2.5(–4) cm. long; heads solitary at ends of branches, with short naked peduncles; invol. ca. 8–12 mm. high, principal phyllaries ca. 13; shorter bracteoles few; rays yellow, 6–10 mm. long; aks. mostly glabrous.—Occasional, about rocks, 9000–12,400 ft.; Subalpine F., Alpine Fell-fields; Sierra Nevada in n. Inyo Co., ne. Tulare Co., Alpine Co., Warner Mts.; to B.C., n. Rocky Mts. July–Aug. Intergrading with

Var. occidentàlis Gray. [*S. o.* Greene.] Stems more slender and flexuous, longer, the basal connections more rootstocklike; lvs. round-ovate to suborbicular, with blunter shallower teeth.—About rocks, 8500–12,000 ft.; Subalpine F., Alpine Fell-fields; San Bernardino Mts., Sierra Nevada n. to Alpine Co.; w. Nev. July–Aug.

6. **S. pattersoniànus** Hoov. [*S. revolutus* Hoov., not Kirk.] Near to *S. Fremontii*, but with slender creeping rhizomes; stems 7–10 cm. tall, purplish near base; lvs. almost linear, 1–3.5 cm. long, sinuate, dentate, or subentire, strongly revolute; heads solitary; phyllaries 12–18, scarious-margined, 7–8 mm. long; rays yellow, 5–6 mm. long; aks. light brown, 3.5 mm. long.—Open rocky places, 9500–10,600 ft.; Alpine Fell-fields; Sweetwater Mts. and e. slope of Sierra Nevada, Mono Co. July–Aug.

7. **S. Clarkiànus** Gray. Perennial with fibrous roots; stems largely single, 6–12 dm. high, subglabrous to ± scaberulous-strigulose; lvs. well distributed, lanceolate to oblong in outline, 5–20 cm. long, gradually reduced upward, the lower petioled, others sessile, laciniate-dentate to pinnatifid; infl. corymbiform, of several heads; invol. 7–11 mm.

high; phyllaries ca. 15–21, not black-tipped; rays ca. 8–12 yellow, 8–11 mm. long; aks. glabrous.—Moist places, mostly 7000–9000 ft., sometimes much lower; Red Fir F., Lodgepole F.; Sierra Nevada from Mariposa Co. to Tulare Co. Mostly July–Aug.

8. **S. triangulàris** Hook. [*S. subvestitus* Howell. *S. t.* var. *Hanseni* Greene. *S. trigonophyllus* Greene.] Perennial herb from short stout rootstocks and fibrous roots; stems several, simple, often coarse, mostly 5–15 dm. tall, glabrous, rarely villous-puberulent; lvs. many, rather well distributed and only gradually reduced upward, the lower triangular-ovate to -lanceolate, truncate or cordate at base, long-petioled, the upper shorter petioled or sessile, often narrower; lf.-blades mostly ca. 4–20 cm. long, generally sharply or sinuately toothed; heads several to many in a flat-topped infl.; invol. 7–10 mm. high; phyllaries ca. 9–13, often black-tipped, outer bracteoles few, narrow, ± elongate; rays 6–12, ca. 7–13 mm. long; aks. glabrous, ca. 4 mm. long; $n = 20$ (Afzelius, 1949).—Common in wet meadows and on stream banks, 4000–11,150 ft.; Montane Coniferous F.; San Jacinto, San Bernardino, and San Gabriel mts., Sierra Nevada, Coast Ranges n. of Trinity Co.; to Alaska, Rocky Mts. July–Sept.

9. **S. sérra** Hook. [*S. s.* var. *integriusculus* Gray and var. *altior* Jeps. *S. Millikeni* Eastw.] Much like *S. triangularis* in stature, but the lvs. tapering or contracted toward base, sharply toothed to subentire; heads often more numerous and in a larger more branched infl.; invol. 6–8 mm. high; phyllaries ca. 8–13, often black-tipped; bracteoles rather elongate; rays ca. 5–8 in no., 5–8 mm. long; aks. glabrous.—Meadows and moist banks, 4700–10,600 ft.; Montane Coniferous F.; Sierra Nevada, to Modoc Co.; to Wash., Wyo., Utah. July–Aug. Plants from Calif. tend to have minutely denticulate or subentire ± lanceolate lvs. and are the basis of var. *áltior*.

10. **S. integérrimus** Nutt. var. **màjor** (Gray) Cron. [*S. eurycephalus* var. *m.* Gray. *S. m.* Heller. *S. mendocinensis* and *Whippleanus* Gray. *S. caulanthifolius* Davy. *S. Sonnei* and *fodinarum* Greene. *S. lugens* var. *megacephalus* Jeps.] Stout perennial from a short erect crown with fibrous roots; stems solitary, leafy to or below the middle, 3–9 dm. tall, thinly arachnoid or ± floccose, later ± glabrate; lower lvs. entire to denticulate or dentate, oval or ovate to lance-oblong, the blades ca. 3–13 cm. long, on petioles from much shorter to ca. as long; cauline lvs. gradually reduced upward; infl. of rather few heads in a flat-topped open cluster; invol. 8–12 mm. high; phyllaries ca. 13–21, pale or sometimes ± purplish at tip; rays 5–12, yellow, mostly 1–2 cm. long; aks. glabrous or hispidulous.—Woods and slopes, mostly 2500–10,600 ft.; Montane Coniferous F.; N. Coast Ranges from Trinity and Mendocino cos. n., Sierra Nevada especially on the w. side, to Siskiyou and Modoc cos., White Mts.; May–Aug.

Var. **exaltàtus** (Nutt.) Cronq. [*S. e.* Nutt. *S. lugens* var. *e.* D. C. Eat.] Heads more numerous, shorter, the phyllaries ca. 5–10 mm. long, black-tipped; rays yellow, 6–15 mm. long.—Montane Coniferous F.; Sierra Nevada, largely on e. slope; to B.C., Rocky Mts. May–July.

Var. **ochroleùcus** (Gray) Cronq. [*S. lugens* var. *o.* Gray.] Basal lvs. ± deltoid to subcordate; rays white or creamy.—Rather moist places, ca. 4000 ft.; Yellow Pine F.; Siskiyou Co.; to Wash., Mont. May–July.

11. **S. aronicoìdes** DC. [*S. exaltatus* var. *uniflosculus* Gray. *S. Rawsonianus* and *leptolepis* Greene.] Habit much like that of the preceding sp., 3–10 dm. high, leafy chiefly below the middle, loosely woolly when young, later glabrate; lowest lvs. often deltoid or subcordate, mostly ovate to oblong, irregularly and coarsely dentate to subentire, the blades 7–25 cm. long, on somewhat shorter petioles; cauline lvs. often narrower, reduced; infl. of many heads in flat-topped cluster; invol. 5–8 mm. high; phyllaries ca. 8–13, with pale, purplish or blackish tips; rays none to 1 or 2; aks. ca. 3 mm. long, glabrous.—Scattered over burns, in brushy and wooded places, etc., below 8000 ft.; Chaparral, Montane Coniferous F., Sagebrush Scrub, etc.; Coast Ranges from San Mateo Co. n., Sierra Nevada, to Modoc Co.; s. Ore. April–July.

12. **S. foètidus** Howell. Perennial with fibrous roots, glabrous, 3–10 dm. high; stems 1–several, often clustered; lvs. ± fleshy, the blades elliptic to broadly oblanceolate, dentate, 6–20 cm. long, the lower long-petioled; middle and upper lvs. few, much reduced; heads several to mostly many, in a congested infl.; invol. 6–9 mm. high; phyllaries ca. 9–13, minutely black-tipped; outer bracteoles few, narrow; rays to ca. 8 mm. long, yellow, ca. 5–7, or none; aks. glabrous.—Wet meadows, ca. 4500–5500 ft.; Yellow Pine F.; ne. Calif.; to B.C., Mont. May–July.

13. **S. hydróphilus** Nutt. [*S. h.* var. *pacificus* Greene.] Fibrous-rooted perennial from a short erect crown; stem simple, erect, ± glaucous, often purplish, hollow, 6–20 dm. tall; lvs. thick, firm, the lower oblong-oblanceolate to elliptic, entire to callous-toothed, the blades ca. 10–20 cm. long, with long winged petiole; middle and upper lvs. few, progressively reduced, sessile; heads rather many, crowded; invol. 5–8 mm. high; phyllaries ca. 8–13, often black-tipped, with few short narrow basal bracteoles; rays few, 4–8 mm. long, or 0; aks. glabrous.—Marshes and swamps or stream banks, below 7200 ft.; Freshwater Swamp to Red Fir F.; region of San Francisco Bay to Siskiyou Co., Mono Co. to Modoc Co.; to B.C., Rocky Mts. May–Aug.

14. **S. astéphanus** Greene. [*S. ilicetorum* A. Davids.] Erect coarse perennial herb, 4–10 dm. high, floccose-woolly; lvs. well distributed, sharply dentate, elliptic to broadly oblanceolate, thin, 1–3 dm. long, the lower petioled, the uppermost sessile and passing into leafy bracts; heads several to rather many, in compact or more often open infl., discoid, yellow; invol. ca. 9–10 mm. high; phyllaries ca. 19–23, green-tipped; outer bracteoles few, short, narrow; aks. glabrous.—Steep rocky slopes, 2500–5000 ft.; Chaparral; s. face of San Bernardino and San Gabriel mts. to San Luis Obispo Co. May–July.

15. **S. Scorzonélla** Greene. [*S. Covillei* Greene. *S. C.* var. *S.* Jeps.] Stems 1–3 from a short horizontal or ascending rootstock, erect, simple, 2–5 dm. high; herbage thinly tomentose or ± glabrate later; lower lvs. tufted, oblong-oblanceolate to oblong, subentire or more usually saliently toothed, broadly short-petioled; cauline lvs. few, poorly developed; heads mostly ca. 10–30 in a compact cymelike infl.; invol. 4–6.5 mm. high; phyllaries mostly 9–15, minutely black-tipped; outer bracteoles few; rays mostly 5–7, ca. 7–9 mm. long, yellow, sometimes wanting; aks. glabrous, ca. 3 mm. long.—Meadows and moist places, 6000–11,500 ft.; Montane Coniferous F.; Sierra Nevada, n. to Mt. Lassen. July–Aug.

16. **S. cànus** Hook. [*S. Howellii* Greene. *S. kernensis* and *oreopolus* Greenm.] Perennial from a ± branching caudex; stems several, slender, 1–3 dm. high; herbage white-tomentose, sometimes ± glabrate in age especially on upper surface of lvs.; lower lvs. ± tufted, round to elliptic, entire or rarely toothed, the blades mostly 1–4 cm. long, on petioles shorter to as long; middle and upper lvs. few, reduced upward; heads several in flat terminal clusters; invol. 4–8 mm. high; phyllaries ca. 13(–17), green-tipped; bracteoles none or inconspicuous; rays mostly 5–8, yellow, 6–12 mm. long, rarely none; aks. glabrous, ca. 3 mm. long.—Dry ± rocky places, 4200–11,750 ft.; Montane Coniferous F., N. Juniper Wd., etc.; Sierra Nevada, White Mts., to Modoc and Siskiyou cos.; B.C., Sask., Rocky Mts. May–Aug. Occasional discoid plants are the f. *aphanáctis* Greenm.

17. **S. ligulifòlius** Greene. [*S. Howellii* of Jeps., not Greene.] Perennial from a caudex with long rootstocklike branches; stems several, erect, 3–4.5 dm. high, leafy mainly below; herbage canescently tomentulose, later ± glabrate on upper lf.-surfaces; lower lvs. tufted, linear-oblong, obtuse, entire, ± revolute, the blades 3–7 cm. long, on mostly somewhat shorter petioles; cauline lvs. few, rapidly reduced upward; heads few to rather many, corymbosely arranged; invol. narrow-subcampanulate, 7–9 mm. high; phyllaries ca. 13–15, green-tipped; bracteoles ± tomentose; rays ca. 7–10, to ca. 15 mm. long, yellow; aks. glabrous.—Dry often serpentine slopes, 3500–5000 ft.; Yellow Pine F., etc.; Siskiyou Mts. June–July.

18. **S. bernardìnus** Greene. [*S. ionophyllus* var. *b.* Hall. *S. neomexicanus*, Calif. refs.] Perennial from a caudex, the stems 1–several, 1–3 dm. high; herbage mostly permanently white-tomentose; lvs. in a basal tuft, numerous, the blades round-ovate to spatulate-cuneate, 1–2 cm. long, subentire to ± toothed, on longer slender petioles; cauline lvs. few, much reduced; heads ca. 3–20, in flat-topped rather open clusters; invol. campanulate, tomentose, 6–8 mm. long; rays yellow, 8–10, ca. 7–10 mm. long; aks. glabrous or hispidulous.—Locally abundant on dry rocky slopes, 6400–7500 ft.; Yellow Pine F.; San Bernardino Mts., chiefly in Bear V. May–July.

19. **S. ionophýllus** Greene. [*S. sparsilobatus* Parish. *S. i.* var. *s.* Hall. *S. i.* var. *intrepidus* Greenm.] Perennial from a root-crown, the stems several, 1.5–3.5 dm. tall, subglabrous to thinly tomentulose, leafy at base, sparingly so above; lower lvs. 2–5 cm. long, obovate to round or flabellate, few-toothed to lyrately pinnatifid; heads 1–few; invol. 8–10 mm. high; phyllaries ca. 15–21; bracteoles 0 to few; rays yellow, 6–9 mm. long; aks. glabrous, 5–6 mm. long.—Dry slopes, 5000–9000 ft.; Montane Coniferous F.; San Bernardino and

San Gabriel mts. June–July. Plants with pinnatifid lvs. have been called var. *sparsilobàtus* and occur with the others.

20. **S. Greènei** Gray. Perennial with slender running rootstocks; stems 1–2(–3) dm. tall; herbage thinly floccose-tomentose, ± glabrate; basal lvs. ± tufted, roundish, 2–5 cm. long, ± dentate, often purplish especially beneath, on rather long slender petioles; cauline lvs. few, reduced; heads 1–3(–few), on short naked peduncles, erect; invol. 10–14 mm. high; phyllaries ca. 13(–21); bracteoles 0 or few; rays deep orange, 1–2.5 cm. long; aks. glabrous.—Local on brushy or wooded slopes, often on serpentine, below 5000 ft.; Yellow Pine F., Mixed Evergreen F., etc.; N. Coast Ranges n. to Trinity Co. May–June.

S. Làyneae Greene. Described as 2–5 dm. high, subglabrous, with oblanceolate mostly basal sharp-serrate lvs. and heads 10–14 mm. high, may be near *S. integerrimus.* Known from a single collection from foothills in Eldorado Co. at 800 ft. May.

21. **S. werneriaefòlius** Gray. [*S. Muìrii* Greenm., not Bolus. *S. speculicola* J. T. Howell. *S. petrocallis* Calif. refs.] Perennial with a loosely branched root-crown; stems several, 0.5–1.5 dm. tall, almost naked, thinly tomentulose when young; lvs. basal, tufted, oblanceolate to lance-elliptic, white-tomentose to glabrate and greenish, entire or nearly so, acute to obtuse, gradually tapering into a petiole, the blade and petiole ca. 1.5–7 cm. long; heads 1–6; invol. 7–9 mm. high; phyllaries ca. 21 or ca. 13; bracteoles 0 or poorly developed; rays 9–13, yellow, 5–8 mm. long; aks. glabrous.—Dry rocky places, 10,400–13,000 ft.; mostly Alpine Fell-fields; Sierra Nevada from Tulare and Inyo cos. to Fresno and Mono cos.; Rocky Mts. July–Aug.

22. **S. subnùdus** DC. [*S. pauciflorus* var. *s.* Jeps.] Perennial from a short slender rhizome, fibrous-rooted, glabrous; stems solitary, 0.5–3 dm. high, slender; basal lvs. roundish, crenate, the blade 1.5–2.5 cm. long, on longer petiole; cauline lvs. few, reduced, sometimes ± pinnatifid; heads 1–2; invol. 5–8 mm. high; phyllaries ca. 21, sometimes 13, sometimes purplish toward ends; rays yellow, ca. 7–12 mm. long; aks. glabrous.—Wet meadows, 6500–10,500 ft.; Lodgepole F., Subalpine F.; Sierra Nevada; to Wash., Mont. July–Aug.

23. **S. pauciflòrus** Pursh. Perennial from a subsimple caudex with fibrous roots; stem 1.5–3.5 dm. high, glabrous or slightly floccose when young; lvs. chiefly basal, rather thin, roundish to elliptic-ovate, abruptly contracted to a ± truncate base, subentire to crenulate, petioled, 1.5–4 cm. long including petiole; stem-lvs. reduced, toothed to pinnatifid, subsessile; heads mostly several, orange or reddish, mostly discoid; invol. 6–8 mm. high; phyllaries often ± red-purple; bracteoles 0 or small; aks. glabrous.—Meadows and moist places, 8250–11,150 ft.; largely Subalpine F.; Sierra Nevada to Alaska, Labrador. July–Aug.

Var. **fállax** Greenm. ex Jeps. [*S. p.* ssp. *f.* Greenm. *S. indecorus* Greene.] Like the sp., but with thinner lvs.; heads more numerous (6–40), yellow, mostly discoid; invol. 7–10 mm. high.—Moist places; Pine Creek, Lassen Co.; to Alaska, Wyo. July–Aug.

Var. **jucúndulus** Jeps. [*S. pseudaureus* Rydb.] Lower lf.-blades ± subcordate; heads several to many; invol. 5–8 mm. high; phyllaries mostly greenish; rays yellow, ca. 6–10 mm. long.—Moist places, 8000–11,000 ft.; Subalpine F.; Sierra Nevada; to B.C., Rocky Mts. July–Aug.

24. **S. cymbalarioìdes** Nutt. [*S. laetiflorus* Greene.] Perennial from a caudex or short rhizome; stems few, 1–4 dm. high; herbage glabrous or ± floccose when young; lvs. thickish, ± succulent, the lower roundish to broadly elliptic, entire to coarsely crenulate, the blades mostly 2–4 cm. long, on somewhat longer petioles; cauline lvs. few, reduced, becoming sessile and mostly pinnatilobed; heads several or more; invol. 7–9 mm. high, mostly greenish; rays 6–12 mm. long; aks. glabrous.—Wet places, mostly 6000–10,000 ft.; Lodgepole F., Subalpine F.; Sierra Nevada, Siskiyou Co.; to w. Canada, Rocky Mts. July–Aug.

25. **S. Clevelándii** Greene. Perennial with a short caudex and several stems 3–7 dm. high; herbage glabrous, glaucous; lvs. in a basal tuft and on lower stems, thick and firm, entire, oblanceolate to oblong, the blades 2–6 cm. long, on petioles as long or longer, the upper gradually reduced to bracts; heads many in a compound corymbiform infl.; invol. 6–7 mm. high; phyllaries ca. 13 or 21; bracteoles small; rays 5–6, deep orange, 5–7 mm. long; aks. glabrous.—Moist places mostly on serpentine; Chaparral, Foothill Wd.; Napa and Lake cos. June–July.

Var. heterophýllus Hoov. Some of upper cauline lvs. ± pinnatifid.—Moist places in serpentine areas; Sierran foothills, Tuolumne Co. June–July.

26. S. Bolánderi Gray. Perennial, from rootstocks; stems slender, 2–6 dm. tall; herbage glabrous or nearly so; basal lvs. round-cordate or subcordate, ca. 1–3 cm. broad, rather firm, mostly palmately lobulate with the lobes in turn toothed, sometimes with a pair of small lobes below the main blade; petioles long, slender; cauline lvs. several, much reduced upwards; heads several, in a ± cymose infl.; invol. 5–8 mm. high; phyllaries ca. 13–21, mostly rather hairy; rays 5–10, yellow, 6–12 mm. long; aks. glabrous.—Coastal Strand, N. Coastal Scrub; Mendocino Co. to sw. Ore. June–July.

27. S. Brèweri Davy. [*S. B.* var. *contractus* Greenm.] Perennial from a short caudex; stems largely solitary, 3–8 dm. high; herbage mostly glabrous or with tufts of tomentum in axils; lvs. thin, pinnatifid, mostly with 1–9 pairs of lateral segms., the terminal segm. enlarged, lobulate-dentate, 4–8 cm. long and almost as broad; petioles to as long as the blades; cauline lvs. ± reduced up the stem; heads several to many, in a flat-topped infl.; invol. ca. 8–11 mm. high; phyllaries ca. 13 or ca. 21, rather wide; bracteoles small or 0; rays yellow, 8–15 mm. long; aks. glabrous.—Open or ± wooded or brushy slopes, below 5500 ft.; Foothill Wd., Yellow Pine F., Chaparral, etc.; Coast Ranges from Contra Costa Co. to n. Los Angeles Co., Tehachapi and Greenhorn mts. n. to Tulare Co. April–June.

28. S. eurycéphalus T. & G. [*S. Austinae* Greene. *S. e.* var. *A.* Jeps. *S. Lewisrosei* J. T. Howell.] With much the same habit as *S. Breweri*, but stems more numerous and caudex more woody; herbage mostly tomentulose at least when young, later ± glabrate; lvs. mostly more deeply dissected, lyrate to sharply toothed; rays 10–15 mm. long.— Dry open places, 1500–5000 ft.; Foothill Wd., N. Oak Wd., etc.; inner Coast Ranges from Sonoma and Colusa cos. to Siskiyou, Shasta, Butte, Lassen, and Modoc cos.; Ore. April–June. Plants with unusually long branches to the caudex, finely dissected lvs. and from Feather R. region, are *S. Lewisròsei*.

29. S. multilobàtus T. & G. [*S. Nelsonii* var. *uintahensis* Nels. *S. u.* Greenm.] Perennial from a short taproot; stems mostly several, suberect, 1.5–3.5 dm. high; herbage thinly tomentulose, ± glabrous in age; lower lvs. obovate to oblong-oblanceolate in outline, mostly lyrately pinnatifid into many segms., the terminal segm. small, all rather thickish in texture, mostly 3–10 cm. long including the petiole; upper lvs. sessile, reduced; heads rather many, in a corymbose cyme; invol. campanulate, mostly 6–8 mm. high; bracteoles poorly developed; rays ca. 8, yellow, 6–10 mm. long; aks. glabrous to hirtellous.—Dry stony slopes, 6000–10,500 ft.; Pinyon-Juniper Wd., Bristle-cone Pine F.; mts. from Death V. region n. to Mono Co.; to Utah, Colo., Ariz. May–July.

30. S. stýgius Greene. [*S. prolixus* Greenm.] Much like the preceding, but quite glabrous except for small tufts of wool in lf.-axils; lower lvs. lyrate-pinnatifid, 3–8 cm. long, the terminal segm. much larger than others, ± rounded, dentate; infl. loose; invol. 5–6 mm. high; rays ca. 13, yellow, 8–11 mm. long; aks. glabrous.—Frequent on dry slopes, 4000–6500 ft.; Joshua Tree Wd., Pinyon-Juniper Wd.; mts. of e. Mojave Desert; sw. Nev., nw. Ariz. April–May.

31. S. Jacobaèa L. Biennial or short-lived perennial; stems 1–few, coarse, erect, simple except above, 3–10 dm. tall, ± purplish-red; herbage thinly floccose-tomentose, later ± glabrate; lvs. well distributed, mostly 2–3-pinnatifid, ca. 5–20 cm. long, only the lower petioled; infl. short, broad, of several to many heads; invol. ca. 4 mm. high; phyllaries ca. 13, mostly with dark tips; bracteoles narrow, sometimes quite evident; rays mostly ca. 13, yellow, 5–10 mm. long; aks. of disk-fls. pubescent, of rays glabrous; $n = 20$ (Afzelius, 1924).—Creek bottom lands, pastures and waste places, Mendocino Co.; introd. from Eu. July–Sept. Poisonous to livestock.

32. S. élegans L. PURPLE RAGWORT. Annual, viscid-pubescent, ± diffuse, 3–6 dm. high; lvs. mostly oblong in outline, 3–8 cm. long, pinnatifid into toothed rather broad lobes with rounded sinuses; heads in open infl.; invol. ca. 1 cm. high; rays purple or reddish, 10–15 mm. long; aks. ribbed; $n = 10$ (Afzelius, 1924).—Occasional escape from cult., mostly near the coast; introd. from Afr. May–July.

33. S. califórnicus DC. [*S. Coronopus* Nutt. *S. ammophilus* Greene.] Annual; stems simple or branched from base, glabrous or ± arachnoid-villous, 1–5 dm. high; lvs. narrowly oblong-lanceolate to -linear in outline, acute or obtuse, remotely toothed to

pinnatifid, all but the lower with auriculate base, 1–6 cm. long; heads several in an open infl., rarely solitary; invol. 5–7 mm. high, quite naked at base; phyllaries usually black-tipped; rays ca. 1 cm. long, usually several; aks. canescent.—Mostly dry open places, below 3500 ft.; Coastal Strand, Coastal Sage Scrub, Chaparral, etc.; cismontane s. Calif. n. to Monterey, Kern, and Tulare cos., w. edge Colo. Desert; L. Calif. March–May.

34. **S. aphanáctis** Greene. Slender annual, simple or branched, 1–2.5 dm. high; glabrous except for some woolly pubescence in infl.; lvs. small, sessile, 1–4 cm. long, coarsely toothed to pinnately lobed; heads several, narrow, especially near top; invol. ca. 5 mm. high, with ca. eight 2–4-nerved phyllaries not black-tipped; rays inconspicuous; aks. densely canescent.—Dry open places; Chaparral, Coastal Sage Scrub, etc.; near the coast from Solano Co. to San Diego Co.; L. Calif. Feb.–March.

35. **S. sylváticus** L. Erect annual 2–8 dm. tall, leafy throughout, ± pubescent; lvs. ± pinnatifid and irregularly toothed, mostly 3–12 cm. long; heads several to many; invol. cylindric, 5–7 mm. high; phyllaries 1–2-nerved, not black-tipped; rays much reduced, inconspicuous; aks. canescent.—Weed in disturbed places, Marin Co. to Humboldt Co.; introd. from Eu. June–Sept.

36. **S. vulgáris** L. Common Groundsel. Annual 1–5 dm. high, leafy throughout, simple or branched from base, sparsely strigulose to glabrous; lvs. rather coarsely pinnatifid and then toothed or simply coarsely toothed, ca. 3–10 cm. long, the lower petioled, the upper clasping; heads discoid, the disk 5–10 mm. wide; invol. ca. 5–8 mm. high; phyllaries mostly ca. 21 and black-tipped; bracteoles well developed; aks. strigulose-hirtellous; $n = 20$ (Afzelius, 1924).—Common weed in gardens and waste places; introd. from Eu. Fls. most months.

37. **S. mohavénsis** Gray. Slender erect annual, mostly freely branched, 1.5–4 dm. tall, entirely glabrous; lvs. well distributed, broadly oblong to oblong-ovate, obtuse, 2–6 cm. long, coarsely dentate, mostly with broad clasping base; heads in loose corymbs; invol. ca. 5–7 mm. high; phyllaries mostly ca. 13, linear; rays inconspicuous or wanting; aks. canescent.—Occasional in washes and canyons, below 2500 ft.; Creosote Bush Scrub; Colo. Desert, Death V. region; to Ariz., Son. March–May.

38. **S. mikanioídes** Otto. German-Ivy. Glabrous perennial with slender twining stems to 6 m. long; lvs. roundish-cordate, sharply 5–7-angled, 2–8 cm. long, ca. as wide; petioles as long or longer; infl. condensed, pedunculate, axillary toward the summit of stems; invol. 3–4 mm. high; principal phyllaries ca. 8; rays 0; aks. glabrous.—Natur. in canyons and gullies along coast of s. and cent. Calif.; native of S. Afr. Dec.–March.

112. Crocídium Hook.

Small annual; lvs. alternate and basal, small, entire or few-toothed. Heads solitary at ends of pedunclelike branches. Herbage flocculent or glabrate above. Heads radiate; rays ♀, often fertile, yellow. Invol. of 1 series of broad herbaceous equal phyllaries. Receptacle conic, naked. Disk-fls. perfect, fertile, yellow. Anthers with deltoid-ovate acute tips. Style-branches short, broad, with deltoid externally papillate-hairy appendage. Aks. covered with thick papillalike hairs, becoming mucilaginous when wet. Pappus of rather many fragile deciduous white bristles. One sp. (Diminutive of Greek, *croce*, loose tufts, because of persistent wool in lf.-axils.)

1. **C. multicáule** Hook. Stems slender, several, 1–2 dm. high, glabrous or glabrate except for the loose tufts of axillary wool; lvs. slightly fleshy, those of the basal tuft oblanceolate, 1–2.5 cm. long, to 1 cm. wide; cauline lvs. rather few, reduced, on lower half of stems; rays 8–13, fertile or sterile, 4–10 mm. long, individually subtended by thin membranous phyllaries 3–7 mm. long.—Open dry places, below 5000 ft.; Sagebrush Scrub, Foothill Wd., N. Oak Wd., etc.; Mariposa Co., Mt. Hamilton, Mendocino Co. to Siskiyou Co., Modoc Co.; to B.C. March–May.

113. Petasìtes Mill.

Perennial herbs with creeping rootstocks. Flowering stems in early spring, with large leaflike bracts and a terminal dense corymb of heads. Lvs. later, large, basal, long-petioled, deltoid to reniform, coarsely toothed to deeply lobed. Heads discoid or radiate, subdioecious; fls. in ♀ heads mostly fertile, tubular or ligulate, those in ♂ heads mostly bisexual but sterile. Invol. a single series of ± equal phyllaries. Receptacle flat, naked. Corolla of ♀ fl. filiform, of others tubular. Aks. linear, 5–10-ribbed; pappus of many capillary bristles. Ca. 12 spp., of cooler parts of N. Hemis. (Greek, *petasos*, broad-brimmed hat, because of the large lvs.)

1. **P. palmàtus** (Ait.) Gray. [*Tussilago p.* Ait. *P. frigidus* var. *p.* Cronq.] WESTERN COLTSFOOT. Stem 1–5 dm. tall, sparingly arachnoid, glandular, ± inflated; cauline bracts lance-elliptic to lance-linear, parallel-veined, mostly 3–6 cm. long; basal lvs. ± roundish or broader in outline, 1–4 dm. wide, palmately lobed and veined, the lobes extending at least halfway to the base and coarsely few-lobed or toothed, green and subglabrous above, densely white-tomentose beneath; petioles 1–4 dm. long; infl. rather crowded; invol. campanulate, 5–9 mm. high, glabrous or slightly strigose; fls. whitish, the ♀ with short rays; pappus copious, white.—Deep shade in wooded places, below 1000 ft.; Mixed Evergreen F., Redwood F., etc.; Coast Ranges from Santa Lucia Mts. n. to Siskiyou and Del Norte cos.; to Alaska, Mass. March–April.

114. Psathyròtes Gray

Low herbs, aromatic, ours scurfy or tomentose, with numerous spreading-decumbent branches. Lvs. alternate, petioled, with broad rounded rather small blades. Heads solitary in the forks of the branches, peduncled, discoid; fls. all tubular, perfect, yellow, often turning purple in age. Phyllaries biseriate, the outer more herbaceous and often a little shorter than the inner. Receptacle flat, naked. Corolla-tube short; throat elongate, cylindric; lobes short, woolly or glandular. Aks. oblong-obconic, obscurely 10-striate, appressed-hirsute. Pappus of many scabrous capillary bristles, white to reddish. Four spp., of sw. N. Am. (Greek, *psathurotes*, brittleness, referring to the stems.)

Plant lanate-tomentose as well as scurfy; outer phyllaries much broader than the inner, the obtuse tips spreading or recurved .... 1. *P. ramosissima*
Plant scurfy-tomentose; outer phyllaries not very different from the inner, all with erect tips
2. *P. annua*

1. **P. ramosíssima** (Torr.) Gray. [*Tetradymia r.* Torr.] Compact round and rather flat plant 5–12 cm. high, 5–30 cm. broad, white-tomentose, largely annual, with strong turpentinelike odor; lvs. numerous, thick, coarsely and irregularly toothed, 6–20 mm. long, suborbicular to reniform; invol. 6–8 mm. high, the outer phyllaries 5 in number, the inner more membranous; corollas yellow to purplish, long-hairy above; aks. ca. 2 mm. long; pappus tawny, ca. 3 mm. long.—Common on hard dry soil of flats and ledges, largely below 3000 ft.; Creosote Bush Scrub; both deserts, especially in e. part; to Utah, Ariz., nw. Mex. March–June, sometimes in winter.

2. **P. ánnua** (Nutt.) Gray. [*Bulbostylis a.* Nutt.] Much like *P. ramosissima*, less lanate, more greenish; lvs. thinner; invol. 5–6 mm. high.—Dry sandy often ± alkaline places, below 6500 ft.; Shadscale Scrub, Alkali Sink, etc.; s. Mojave Desert to Mono Co.; to Utah, Ariz. June–Oct.

115. Cacaliópsis Gray

Stout floccose-woolly perennial herbs from stout rootstocks. Lvs. largely basal, round-cordate, deeply palmately cleft or parted. Heads few, corymbosely placed at summit of almost leafless stem. Fls. yellow, many; all tubular. Invol. broadly campanulate; phyllaries many, lance-linear, acuminate, stiff, subequal. Anthers entire at base. Style puberulent below the slightly flattened branches. Receptacle flat. Aks. 10-nerved; pappus of many soft white bristles. One sp. (Greek, *kakalia*, ancient name of some plant, and *opsis*, resemblance.)

1. **C. Nardósmia** Gray. SILVER-CROWN. Stem erect, 3–8 dm. high; loosely floccose; basal lvs. 5–20 cm. broad, glabrous above, tomentose beneath, deeply palmately cleft into 7–9 lobes, these irregularly cleft and toothed, the apices acute; petioles 1–2 dm. long; cauline lvs. few, small; heads mostly 5–10, ± racemosely arranged; invol. 10–13 mm. high, mostly ± white-woolly; phyllaries carinate; corollas yellowish, with conspicuous

lanceolate lobes; aks. glabrous, brown, slender, ca. 5 mm. long; pappus copious, white. —Wooded or bushy slopes, below 6000 ft.; Yellow Pine F., Mixed Evergreen F., etc.; Coast Ranges from Lake Co. n.; w. Ore. April–May.

116. Luìna Benth.

Low perennials from a woody caudex, with many erect simple stems. Lvs. alternate, entire to deeply cleft. Heads discoid, 10–19-fld., in terminal corymbs. Fls. perfect, fertile, mostly yellow. Invol. oblong-campanulate; phyllaries 8–12, in 1 series, equal, linear, rigid, 1-nerved. Receptacle naked. Corolla with a narrow tube and somewhat enlarged throat. Anthers sagittate at base. Style-branches flattened, papillate externally, with a short blunt papillate appendage. Pappus soft, white. Four spp. of Pacific Nw.

1. **L. hypoleùca** Benth. [*L. h.* var. *californica* Gray.] Stems 1.5–4 dm. high, white-tomentose, leafy throughout; lvs. elliptic to oblong-ovate, sessile, entire or subentire ca. 2–6 cm. long, greenish and thinly tomentose to glabrate above, white-tomentose beneath; heads dull yellow, 10–17-fld.; invol. thinly tomentose, 5–8 mm. high; aks. glabrous, brown, ca. 2.5 mm. long.—Uncommon, dry rocky places, below 5000 ft.; Mixed Evergreen F., Redwood F., etc.; Santa Cruz Mts., Mendocino Co. to Del Norte and Siskiyou cos.; to B.C. June–Sept.

117. Erechtìtes Raf. Fireweed

Erect annual or perennial herbs with rank odor. Lvs. alternate, entire to pinnately dissected. Heads cylindric to ovoid, discoid, dull yellow or whitish. Invol. a single series of narrow equal ± herbaceous phyllaries, sometimes with a few minute bracteoles at base. Receptacle flat, naked. Outer fls. ♀, filiform-tubular, in 2–several series; cent. fls. perfect but sometimes sterile, the corolla narrowly tubular. Anthers entire or slightly sagittate at base. Style-branches flat, minutely penicillate near the tip. Aks. 5-angled or 10–20-nerved; pappus-bristles elongate, many, soft, fine. Ca. 12 spp., of Am. and Australasia. (Greek name of a perhaps related plant.)

Lvs. ± lobed or pinnatifid; heads scarcely calyculate at base .... 1. *E. arguta*
Lvs. denticulate, not lobed or pinnatifid; heads markedly calyculate .... 2. *E. prenanthoides*

1. **E. argùta** (A. Rich.) DC. [*Senecio a.* A. Rich.] Annual or short-lived perennial, 6–20 dm. high, white-tomentulose, early ± glabrate, the stems stout, erect, branched above; lvs. many, well distributed, oblong-ovate to lanceolate in outline, 5–15 cm. long, the lower ± petioled, the upper sessile, reduced; heads in loose or crowded cymes in a subcorymbose infl.; invol. 5–7 mm. high, subcylindric, ± woolly; fls. dull yellow; aks. strigose, dark, scarcely 2 mm. long; pappus ca. 5 mm. long, white.—A weed in woods, at low elevs.; Redwood F., etc.; from San Mateo Co. near the coast to Ore.; native of Australasia. June–Aug.

2. **E. prenanthoìdes** (A. Rich.) DC. [*Senecio p.* A. Rich.] Habit and size of the preceding sp., but subglabrous to obscurely puberulent; lvs. arachnoid beneath, linear-lanceolate, evenly and finely dentate; infl. large, to 3 dm. broad; pappus ca. 6–7 mm. long.—Waste places as a weed; Mixed Evergreen F., Redwood F.; Marin Co. to Humboldt Co.; Ore.; native of Australasia. July–Sept.

118. Peucephýllum Gray. Pigmy-Cedar

Aromatic shrub, rather openly branched and with crowded alternate terete resinous-punctate lvs. near the ends of the branches. Heads solitary, discoid; invol. broadly campanulate; phyllaries in 2 series, linear-subulate, the outer subterete, the inner flattened, slightly scabrous-margined. Receptacle flat, naked. Fls. perfect, fertile. Corolla-tube short; throat cylindric; lobes short. Aks. oblong-obconic, obscurely 10-striate, appressed-hirsute. Pappus of many scabrous capillary bristles or some of them flattened and with hyaline margins. One sp. (Greek, *peuke*, the fir, and *phyllon*, lf.)

1. **P. Schóttii** (Gray) Gray. [*Psathyrotes S.* Gray. *Inyonia dysodioides* Jones.] Plant 0.5–2.5 m. high, with trunklike stem; lvs. bright green, 0.5–2.5 cm. long; heads scattered, subsessile; invol. ca. 8 mm. high, bright green; fls. yellow; aks. slender, ca. 4 mm.

long, dark brown; pappus 3–4 mm. long, tawny.—Rocky places, canyons, etc., below 3000 ft.; Creosote Bush Scrub; Death V. region through e. Mojave Desert to Colo. Desert; Nev., Ariz., L. Calif. Dec.–May.

119. Tetradỳmia DC.

Low rigid shrubs with dense matted or floccose wool which may be deciduous. Lvs. alternate, entire, solitary or fascicled, the primary ones often modified into spines. Heads discoid, 4–9-fld., usually in small corymbs or racemes. Invol. cylindric to oblong; phyllaries 4–6, firm, concave, overlapping, often enlarged and thickened at base. Receptacle flat, small. Fls. yellow, the lobes longer than the throat. Anthers strongly sagittate, almost caudate. Aks. terete, obscurely 5-nerved, glabrous to densely long-hairy; pappus of numerous white or whitish, soft capillary bristles. Several spp., w. N. Am. (Greek, *tetra*, four, and *dymos*, together, the first known sp. with 4 fls.)

Aks. with numerous long ascending hairs which almost equal and conceal the pappus.
   Heads peduncled, scattered or loosely racemose; primary lvs. transformed into stiff slender spines
   1. *T. axillaris*
   Heads nearly sessile in a close terminal corymb; primary lvs. foliose, often spine-tipped
   2. *T. comosa*
Aks. glabrous or with much shorter hairs than the pappus which is evident and conspicuous.
   Heads 5-fld., terminal; primary lvs. modified into spines.
      Aks. canescent; spines 2–3 cm. long, perpendicular to the twigs; tomentum on young twigs complete, not separated into ridges . . . . 3. *T. stenolepis*
      Aks. glabrous; spines 0.8–1.5 cm. long, ascending; tomentum on young twigs separated into ridges . . . . 4. *T. argyraea*
   Heads 4-fld.; primary lvs. not changed into spines.
      Primary lvs. erect, slender-subulate, softly spinescent, not over 1 mm. wide, 5–10(–20) mm. long
      5. *T. glabrata*
      Primary lvs. linear to oblanceolate, 1–4 mm. wide, 20–30 mm. long . . . . 6. *T. canescens*

1. **T. axillàris** A. Nels. [*T. spinosa* var. *longispina* Jones.] COTTON-THORN. Rigidly branched, densely white-tomentose, 6–12 dm. high, the foliage deciduous in the dry season; primary lvs. modified into slender rigid spines mostly 2–3 cm. long; secondary fascicled lvs. ca. 1 cm. long, linear, green; heads on short stout axillary peduncles; invol. short-woolly, sometimes green, 8–10 mm. high, usually 6–7-fld.; phyllaries 5–6, tomentose, the outer oblong, the inner broader, all obtuse; aks. ca. 5 mm. long, with long soft white hairs nearly equaling the pappus.—Common on dry slopes and flats, 2000–6400 ft.; mostly Joshua Tree Wd., Pinyon-Juniper Wd.; Mojave Desert n. to Mono Co.; to Utah, Ariz. April–May. The closely related *T. spinòsa* H. & A., with shorter ± recurved spines, may occur in ne. Calif.

2. **T. comòsa** Gray. Erect bush with many stems and virgate branches, 5–12 dm. high, permanently and densely white-tomentose; primary lvs. linear, 2.5–5 cm. long, white-tomentose, the early ones soft, the later rigid and spine-tipped; secondary fascicled lvs. when present short and narrow; heads in close terminal corymbs; invol. 8–10 mm. high, 6–9-fld.; phyllaries oblong, woolly, obtuse, 5–6 in number; aks. long- and soft-woolly.—Dry places, below 5000 ft.; Coastal Sage Scrub, Chaparral; interior cismontane s. Calif. n. to Newhall (Los Angeles Co.), occasional on Mojave Desert; w. Nev. June–Sept.

3. **T. stenólepis** Greene. Four–8 dm. high, much-branched, permanently white-tomentose; lower primary lvs. oblanceolate, ca. 2 cm. long, mucronate, most other primary lvs. modified into stout spreading spines 2–3 cm. long, with oblanceolate secondary fascicled lvs. ca. 1 cm. long; heads in close terminal corymbs; invol. 10–12 mm. high, 5-fld.; phyllaries 5, oblong, thick, obtuse; aks. canescent, ca. 9 mm. long, glabrate; pappus rather stiff, whitish, ca. 1 cm. long.—Occasional in dry places, 2000–5000 ft.; Creosote Bush Scrub, Joshua Tree Wd.; Mojave Desert. May–Aug.

4. **T. argyraèa** Munz & Roos. Six–15 dm. tall, the younger branches with a densely matted silvery-white wool forming firm ridges on the internodes; primary lvs. linear, becoming acicular spines 8–15 mm. long, at first woolly, later glabrate and yellowish; fascicles of secondary lvs. 5–12 mm. long, sometimes present in the axils; heads 1–few, in close clusters at ends of branchlets; phyllaries 5, equal, 6–8 mm. long, rigid, thick, obtuse, silvery-white with dense matted wool; fls. 5, ca. 8 mm. long; aks. greenish, glabrous, ca. 4.5 mm. long; pappus copious, yellowish-white, 7–8 mm. long.—Dry rocky

slopes, 5000–7000 ft.; Pinyon-Juniper Wd.; Kingston and Clark mts., e. Mojave Desert. Aug.–Sept.

5. **T. glabràta** Gray. Branched, the stems ascending or arched, 3–9 dm. high, ± white-tomentose; lvs. early glabrate and greenish, the primary rigid, subulate, cuspidate, 5–10 mm. long; secondary lvs. fascicled, soft, persistent; heads in small terminal clusters; invol. 7–8 mm. high, tomentose to green, 4-fld.; phyllaries 4–5, oblong, keeled; aks. villous, ca. 3 mm. long.—Dry open places, 2300–6800 ft.; Joshua Tree Wd., Sagebrush Scrub, etc.; w. Mojave Desert to Lassen Co. etc.; to Ore., Utah. May–July.

6. **T. canéscens** DC. Stems 1–3 (or more) dm. high, freely branched, sometimes curved downward toward the tips and with numerous short erect flowering branches; lvs. and heads white-tomentose; lvs. linear or lance-linear, numerous, rarely fascicled, 1–2 cm. long; heads in short-peduncled close corymbs; invol. 8 mm. high, 4-fld.; phyllaries 4–5, carinate, oblong; aks. subglabrous to villous, 7–8 mm. long; pappus yellowish or sordid, ca. 1 cm. long.—Dry slopes, 4000–8000(–10,000) ft.; Sagebrush Scrub, Pinyon-Juniper Wd., Yellow Pine F., etc.; San Bernardino Mts. n. along e. slope of Sierra Nevada and through higher desert mts. to Modoc and Siskiyou cos.; to B.C., Mont., Utah. July–Aug. Passing by imperceptible degrees especially in Mono and Inyo cos. to var. *inérmis* Gray, with primary lvs. oblanceolate, less than 2 cm. long and usually with fascicles of oblanceolate to obovate lvs. in the axils; this var. extends eastward.

120. Lepidospártum Gray

Rigid broomlike shrubs with alternate entire lvs. Heads numerous, in small close terminal racemes arranged in large panicles. Phyllaries imbricated in 3–4 series, chartaceous, oblong, obtuse. Receptacle naked. Heads discoid. Fls. yellow, with long tube, short campanulate throat and rather long lobes. Anthers exserted. Aks. terete; pappus of copious minutely scabrous capillary bristles in several series. Two spp. (Greek, *lepis*, scale, and *sparton*, the broom-shrub.)

Branches glabrate in maturity, not much striate; heads 10–15-fld.; invol. 4–6 mm. high, subglabrous
                                                                    1. *L. squamatum*
Branches persistently tomentose, plainly striate; heads 4–5-fld.; invol. 9–10 mm. high, usually tomentose .... 2. *L. latisquamum*

1. **L. squamàtum** (Gray) Gray. [*Linosyris s.* Gray. *L. s.* var. *Breweri* Gray. *Tetradymia s.* Gray. *Baccharis sarathroides* var. *pluricephala* Jeps.] A ± round-topped shrub 1–2 m. tall, with many ascending virgate branches; young growth white-tomentose, densely leafy with entire obovate lvs. ca. 1 cm. long; older shoots glabrate, green, with discrete mere lf.-scales; heads many; invol. 4–6 mm. high, campanulate; phyllaries not extending down on to peduncles or grading into scaly lvs., 1–1.5 mm. wide; aks. narrow-clavate, ca. 3.5 mm. long, glabrous; pappus whitish, ca. 5 mm. long.—Common in washes and gravelly places, below 5000 ft.; Coastal Sage Scrub, Chaparral, Joshua Tree Wd., etc.; drier cismontane Calif. from Santa Clara and Tulare cos. s., Mojave and Colo. deserts; L. Calif. Aug.–Oct.

Var. **Pálmeri** (Gray) Wheeler. [*Linosyris s.* var. *P.* Gray. *Lepidospartum s.* var. *obtectum* Jeps.] Phyllaries extending 5–8 mm. onto peduncles and grading into scaly lvs.; scaly lvs. crowded on fertile stems.—Washes, 500–2000 ft.; Creosote Bush Scrub; head of Coachella V. (Palm Canyon to Whitewater and Morongo washes), Riverside Co. and near Twentynine Palms.

2. **L. latisquàmum** Wats. [*L. striatum* Cov.] Taller more narrow shrub 1–2.5 m. high; stems strongly striate; lvs. numerous, needlelike, 2–3 cm. long; invol. 8–10 mm. high, tomentose, subcylindric; phyllaries 2–2.5 mm. wide; aks. white-villous, ca. 5 mm. long; pappus whitish, 8–9 mm. long.—Dry slopes and benches, 5000–8000 ft., mostly on limestone; Pinyon-Juniper Wd., Shadscale Scrub, Sagebrush Scrub; Inyo-White Mts., n. slope of San Gabriel Mts., Clark Mts. (e. Mojave Desert); Nev. June–Oct.

Tribe 6. Calendùleae. CALENDULA TRIBE

Herbs or shrubs. Lvs. commonly alternate. Heads radiate. Phyllaries often with dry scarious margins, subequal, 1–2-seriate. Receptacle naked, or very rarely with a few

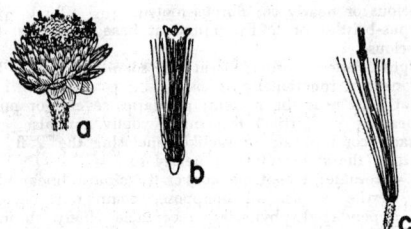

Fig. 116. INULEAE. *Anaphalis margaritacea: a,* head, × 2, discoid, with several series of phyllaries; *b,* ♂ fl., × 5, ovary abortive; *c,* ♀ fl., × 5, pappus capillary.

bristles. Ray-fls. ♀ and fertile, commonly ligulate. Disk-fls. mostly sterile. Anthers mostly sagittate at base, or sometimes short-caudate. Style branches bifid or entire. Aks. usually large, mostly heteromorphic. Pappus none, sometimes of a few deciduous hairs. —An Old World, principally African, tribe of 8 genera.

One genus .... 121. *Calendula*

121. Caléndula L.

Annual or perennial herbs with simple alternate lvs. Heads mostly large. Invol. broad; phyllaries usually scarious-margined, incurved, in 1–2 rows. Receptacle naked, plane. Ray-fls. yellow or orange; ray-aks. glabrous, incurved. Disk-fls. infertile. Pappus none. Ca. 15 spp., Medit. region to Iran, Canary Ids. (Latin, *calendae,* throughout the months.)

Lvs. linear-lanceolate; heads nodding in fr., 1–2 cm. broad in fl. .... 1. *C. arvensis*
Lower lvs. spatulate, the upper lanceolate to linear; fruiting heads erect; flowering heads 2–5 cm. broad .... 2. *C. officinalis*

1. **C. arvénsis** L. Annual; stems slender, 1–3 dm. high, leafy throughout, finely glandular-pubescent; lower lvs. short-petioled, the middle and upper sessile, lanceolate to lance-oblong, 2–7 cm. long, entire or remotely denticulate; heads peduncled, solitary at ends of branches; invol. broadly campanulate, ca. 8–10 mm. high, the phyllaries uni-seriate, lanceolate to linear; ligules mostly yellow, 7–12 mm. long, 3-toothed; aks. strongly incurved, strongly muricate on back; $2n = 36$ (Negodi, 1936).—Occasionally established in fields and waste places; native of Eurasia. March–April.

2. **C. officinàlis** L. Pot-Marigold. Annual 3–6 dm. high, much-branched; lvs. oblong to oblong-obovate, 5–15 cm. long, entire or remotely denticulate, ± clasping; heads solitary on stout peduncles; phyllaries lanceolate to ovate, 12–20 mm. high, ciliate; rays pale yellow to deep orange, 2–3.5 cm. long; aks. strongly incurved, thorny-muricate on back; $2n = 28$ (Negodi, 1936), 32 (Weddle, 1941).—Occasional waif, escaping from gardens; introd. from Eu. March–May.

Tribe 7. Inùleae. Everlasting Tribe (Fig. 116)

Herbs, shrubs, or rarely trees, with usually woolly or glandular herbage. Lvs. commonly alternate or basal (opposite in *Psilocarphus*), entire or nearly so. Heads radiate or discoid (in all ours), homogamous or heterogamous, sometimes dioecious. Phyllaries commonly dry and scarious and several-seriate, sometimes foliaceous or petaloid. Receptacle naked or scaly. Anthers tailed at base. Style-branches various, mostly obtuse or truncate, not appendaged. Pappus (in all ours) capillary or none.—A cosmopolitan tribe of about 150 genera.

A. Heads with conspicuous yellow rays .... 122. *Inula*
AA. Heads rayless.
  B. Receptacle naked; phyllaries many, not saclike; often perennials.
    C. Phyllaries dry but not scarious; plants not tomentose; large herbs or shrubs
      123. *Pluchea*
    CC. Phyllaries scarious; herbage woolly.
      D. Fls. all fertile, perfect and ♀ in the same head; taprooted .... 124. *Gnaphalium*

DD. Fls. dioecious or nearly so; fibrous-rooted.
   E. Pappus-bristles of ♀ fls. united at base; basal lvs. tufted, persistent; strictly
     dioecious .... 125. *Antennaria*
   EE. Pappus-bristles distinct; basal lvs. soon deciduous; ♀ plants commonly with
     few central functionally ♂ fls. .... 126. *Anaphalis*
BB. Receptacle chaffy, at least near the margin; phyllaries several or none, when present saclike
   (open in *Evax*), bearing ♀ (fertile) fls.; small woolly annuals.
   C. Fr.-bearing bracts conduplicate or saclike, enclosing the ♀ fl., falling with the ak.
     D. Aks. straight, the style apical or nearly so.
       E. Lvs. alternate; receptacle convex to almost linear.
         F. Fertile ♀ fls. all epappose, completely enclosed in woolly hyaline-
           appendaged phyllaries; receptacle chaffy throughout; true invol.
                      127. *Stylocline*
         FF. Fertile ♀ fls. of 2 sorts, the outer epappose and bract-enclosed, the inner
           pappose but not bract-subtended; true invol. scanty .... 128. *Filago*
       EE. Lvs. opposite; receptacle subglobose to obpyriform; true invol. and pappus
        none; inner fls. bractless .... 129. *Psilocarphus*
     DD. Aks. gibbous, the style lateral; true invol. below the fertile bracts of ca. 5 scarious
      scales .... 130. *Micropus*
   CC. Fr.-bearing bracts open, merely subtending the ♀ fls., persistent .... 131. *Evax*

122. Ínula L. ELECAMPANE

Coarse ± glandular or hairy herbs or subshrubs, with alternate simple lvs. and large yellow heads. These many-fld., radiate or rarely disciform. Rays ♀ and fertile. Invol. imbricated, hemispherical or campanulate, the phyllaries narrow, scarious or the outer-most broader and often herbaceous. Receptacle flat or convex, naked. Disk-fls. tubular, perfect, yellow; anthers caudate; stigmatic-branches flattened, slightly broader upward. Pappus of few to many capillary bristles; aks. prominently to obscurely 4–5-ribbed or -angled. Ca. 100 spp., of Old World. (The ancient Latin name.)

1. **I. Helènium** L. Stout perennial 1–1.5 m. high, the stem spreading-hairy; lvs. large, irregularly dentate, densely velvety beneath, the lower long-petioled and elliptic, the upper ovate, sessile, cordate-clasping; heads few, peduncled; disk 3–5 cm. in diam.; rays many, narrow, yellow, over 1 cm. long; aks. slender, glabrous, ca. 4-angled; $n = 10$ (Rutland, 1941).—Natur. in nw. and e. U.S., from Eu.; to be expected in nw. Calif. June–Sept.

123. Plùchea Cass.

Tall leafy shrubs or herbs with alternate lvs. Heads many, in terminal corymbose infl., discoid, with numerous purplish fls. Marginal fls. ♀ and perfect, the corolla narrowly tubular, truncate, entire or 2–3-toothed; style 2-cleft. Cent. fls. few, perfect, but some-times sterile, with 5-cleft corolla. Invol. imbricate. Receptacle flat, naked. Aks. grooved; pappus of 1 series of capillary bristles. A genus of some size, from warmer parts of World. (N. A. *Pluche*, Parisian 18th-century naturalist.)

Plant a glandular-puberulent herb; lvs. narrowly ovate .... 1. *P. purpurascens*
Plant a sericeous shrub; lvs. lance-linear to elliptic .... 2. *P. sericea*

1. **P. purpuráscens** (Sw.) DC. [*Conyza p.* Sw. *P. camphorata*, Calif. refs.] MARSH-FLEABANE. Annual to perennial, erect, branched above, 3–12 dm. high, green; lvs. ovate to lanceolate, glandular-dentate, 5–10 cm. long, 2–3 cm. wide, the lower petioled; corymbs large; invol. chartaceous, campanulate, ca. 5 mm. high; phyllaries lance-ovate; pappus of all the fls. similar, setaceous, without dilated tips.—Occasional in ± alkaline wet places, at low elevs.; Coastal Salt Marsh, Freshwater Marsh, etc.; cis-montane Calif. n. to San Francisco Bay and Cent. V.; deserts from Inyo Co. s.; to Atlantic Coast. July–Nov.

2. **P. serícea** (Nutt.) Cov. [*P. borealis* Gray. *Polypappus s.* Nutt.] ARROWWEED. Slender-stemmed willowlike shrub 1–4 m. high; herbage silvery-silky; lvs. entire, lanceo-late, 1–4 cm. long, 0.3–0.6 cm. wide, acute, sessile; invols. 5–6 mm. high; outer phyllaries ovate, coriaceous, the inner narrower, less firm; pappus of perfect fls. with clavellate tips.—Frequent in wet places like river bottoms; Coastal Sage Scrub, Creosote Bush Scrub; cismontane s. Calif. from n. Santa Barbara Co. s., deserts; to L. Calif., Tex. March–July.

124. Gnaphàlium L. Cudweed. Everlasting

Annual to perennial woolly herbs, with alternate entire lvs. Heads disciform, white, yellowish or tinted rose, arranged in panicles, corymbs or spikes. Invol. ovoid or campanulate; phyllaries slightly to evidently imbricated, scarious at tip or almost throughout. Receptacle naked. Numerous outer fls. slender, ♀, the few inner ones coarser and perfect. Style-branches flattened, truncate, without sharply differentiated stigmatic part. Pappus of capillary bristles, sometimes ± thickened at summit, sometimes united at base. Aks. small, nerveless. Over 100 spp., widely distributed. (Greek, *gnaphalon*, a lock of wool, these plants floccose-woolly.)

A. Pappus-bristles united at base, falling away in a ring.
  B. Lower lf.-surface closely white-pannose, the subappressed hairs tightly enmeshed; invol. densely woolly at base only, 4–6 mm. high .... 1. *G. purpureum*
  BB. Lower lf.-surface loosely villose-lanate; invol. nearly buried in wool, 3–4 mm. high
                                                             2. *G. peregrinum*
AA. Pappus-bristles not or only partly united at base, deciduous separately or in small groups.
  B. Heads small, the invol. mostly 2–4 mm.; glomerules of heads leafy-bracted; plants rarely as much as 3 dm. high, usually much branched.
    C. Plants annual.
      D. Height 3–5 dm.; heads in dense globose clusters with almost linear foliar subtending bracts. Nw. Calif. .... 3. *G. japonicum*
      DD. Height 0.5–2 dm.; heads in small rather loose clusters with ± spatulate bracts. Widespread .... 5. *G. palustre*
    CC. Plants perennial with a tufted or creeping rhizome; heads subtended by narrow lvs.
                                                       4. *G. collinum*
  BB. Heads larger, mostly 4–7 mm. high; glomerules of heads not leafy-bracted; plants mostly 2–10 dm. high, the stems often simple below the infl.
    C. Lvs. green in age, at least on the upper surface.
      D. Mature lvs. usually green on both surfaces; plants with rather strong balsamlike odor; annual to biennial.
        E. Infl. corymbose; invol. rounded, white; lvs. lanceolate to oblong. Common, widespread .... 6. *G. californicum*
        EE. Infl. paniculate; invol. narrower, often pinkish; lvs. lance-linear. Occasional, Orange Co. to cent. Calif. .... 7. *G. ramosissimum*
      DD. Mature lvs. green above, white beneath; plants not strongly scented, perennial.
        E. Lvs. oblong to broadly linear, broadly auriculate at base; stems tending to branch, 6–9 dm. high .... 8. *G. bicolor*
        EE. Lvs. narrowly linear, with short-decurrent base; stems mostly simple, 3–6 dm. high .... 9. *G. leucocephalum*
    CC. Lvs. permanently tomentose.
      D. The lvs. not or scarcely decurrent.
        E. Phyllary-tips pearly white or ± straw color. Native plants.
          F. Middle phyllaries ca. twice as long as broad, rounded-obtuse at apex. Cismontane and montane .... 10. *G. microcephalum*
          FF. Middle phyllaries ca. three times as long as broad, acutish. Desert mts.
                                       (11.) *G. Wrightii*
        EE. Phyllary-tips greenish or brownish. Introd. weed ...... 13. *G. luteo-album*
      DD. The lvs. strongly decurrent.
        E. Stems usually branched above; heads in rather open corymbose panicles, at least the inner phyllaries abruptly pointed .... 11. *G. beneolens*
        EE. Stems usually simple above; heads in dense terminal clusters; phyllaries all decidedly obtuse .... 12. *G. chilense*

1. **G. purpùreum** L. [*G. ustulatum* Nutt.] Annual or biennial, simple or branched, 1–5 dm. high, closely woolly-canescent; lower lvs. spatulate to oblanceolate, petioled, 2–5(–9) cm. long, ± bicolored, closely white-pannose beneath; cauline lvs. gradually reduced up the stem; spike terminal, dense or ± interrupted, leafy-bracted in lower part; invols. crowded, lanate at base only, 4–6 mm. long, brown to chestnut or purple; pappus-bristles united at base.—Dry, often disturbed or waste places, Santa Catalina, Santa Rosa, and Santa Barbara ids., near the coast from Santa Barbara Co. to Wash., in e. and cent. U.S., S. Am. April–July.

2. **G. peregrìnum** Fern. [*G. spathulatum* auth.] Like no. 1, simple or loosely branched; lvs. spatulate, 2–4 cm. long, greenish, loosely villous-lanate; heads nearly buried in dense gray wool; invol. 3–4 mm. high.—Occasional weed; to Atlantic Coast. March–May.

3. **G. japónicum** Thunb. Erect annual mostly 2–4 dm. high, ± cottony-white, slender; lvs. from oblong-spatulate and narrowed into a petiole to linear and sessile, glabrate above, cottony-white beneath, to ca. 7 cm. long; heads small, in dense globose clusters, surrounded by a few floral lvs.; invol. oblong, white-woolly at base, the phyllaries brown

or straw color, all but the inner obtuse; pappus-bristles scarcely cohering at base.—Locally abundant weed, Humboldt and Trinity cos.; native of Australasia. Aug.–Oct.

4. **G. collìnum** Labill. Near to *G. japonicum,* but perennial, smaller, with closer indumentum; lvs. more acute, the lower persistent, glabrous above; heads not in so compact a cluster; invol. broader, brown.—Weed in Humboldt and Del Norte cos.; native of Australasia. March–Aug.

5. **G. palústre** Nutt. Annual, commonly branched at base, 5–20(–30) cm. high, floccose-tomentose, especially upwards and on the stems; lvs. spatulate, 1–3 cm. long, the uppermost oblong or lanceolate and subtending the glomerules of heads; invol. 3–3.5 mm. high, densely woolly below, the bracts not much imbricate, brown, usually with whitish tips; pappus-bristles distinct, falling separately.—Frequent on damp banks, in stream beds, etc., below 9500 ft.; many Plant Communities; cismontane and montane Calif.; to B.C., Alta., New Mex. May–Oct.

6. **G. califórnicum** DC. [*G. decurrens* var. *c.* Gray.] Biennial, stoutish, 4–8 dm. high, corymbosely branched at summit, leafy, green, glandular, strongly scented; lower lvs. oblong-lanceolate, 4–10 cm. long, cauline gradually reduced and more lanceolate; infl. a large terminal paniculate corymb; invol. 6–7 mm. high, rounded, phyllaries mostly white, blunt to broadly rounded.—Dry wooded hills, disturbed places, etc., below 4000 ft.; Chaparral, Foothill Wd., Mixed Evergreen F., etc.; cismontane s. Calif. from San Diego Co. n.; to Ore. Jan.–July.

7. **G. ramosíssimum** Nutt. Biennial, with 1–several slender erect stems 5–12 dm. high, glandular, sweet-scented, early greenish and glabrate; lvs. lance-linear, mostly 2–6 cm. long; infl. a large terminal ± oblong panicle; invol. 4–5 mm. high, turbinate; phyllaries often pinkish, ± rounded to pointed.—Occasional, dry slopes, below 1500 ft.; Mixed Evergreen F., Coastal Strand, Chaparral, etc.; near the coast from San Diego Co. to Marin Co. March–Sept.

8. **G. leucocéphalum** Gray. Perennial, with few erect leafy permanently tomentose stems mostly 3–6 dm. high; lvs. narrowly linear (or lowermost oblance-linear), 2–8 cm. long, green above, white-tomentose beneath, attenuate; infl. a corymbose panicle; invol. pearly white, 6–7 mm. high; phyllaries papery, obtuse, ovate.—Occasional, dry disturbed places and hillsides, at low elevs.; Coastal Sage Scrub, Chaparral; cismontane Los Angeles Co. s.; to Tex. Aug.–Sept.

9. **G. bícolor** Bioletti. Biennial or perennial, with several stout branched stems 4–9 dm. high, very leafy, white-tomentose; lvs. glabrate and becoming green above, white-tomentose beneath, lance-oblong, 2–7 cm. long, closely sessile by a broad auriculate base, ± crisped along margin; heads in rather loose corymbs; invol. campanulate, ca. 6 mm. high; phyllaries whitish, the outer ovate, the inner narrowly oblong.—Common in dry open places, mostly below 2000 ft.; Chaparral, Coastal Sage Scrub, Mixed Evergreen F., etc.; coastal Calif. from Monterey Co. to L. Calif. Jan.–May.

10. **G. microcéphalum** Nutt. [*G. albidum* Jtn.] Biennial or short-lived perennial 5–10(–15) dm. high, permanently densely white-tomentose throughout, loosely branched; lvs. oblanceolate to spatulate, 2–5 cm. long, 4–10 mm. wide, subdecurrent; panicle corymbose, the heads in small clusters at the ends of the branches; invol. 5–6 mm. high; phyllaries whitish, the outer ovate, the inner oblong.—Dry slopes and open places, below ca. 4000 ft.; several Plant Communities; n. L. Calif. to Marin Co. July–Oct. Plants from Montane Coniferous F., from San Bernardino Mts. to n. Calif. and B.C. tend to have the tomentum somewhat more compact, invols. slightly shorter and infl. narrower; they have been described as var. *thermále* (E. Nels.) Cronq. [*G. t.* E. Nels.]

11. **G. benèolens** A. Davids. Perennial, persistently woolly, the stems rather simple, erect, 5–8 dm. high, often with a slightly greenish-yellow cast; lvs. lance-linear, 2–7 cm. long, the lowest spatulate-oblong; panicle lax, with small clusters of heads at ends of branches; invol. usually slightly yellowish, 5–6 mm. long, the inner phyllaries pointed.—Dry places, below 5000 ft.; Chaparral, Coastal Sage Scrub; cismontane s. Calif., Santa Cruz Id. July–Nov. A closely related plant with base of invol. more woolly occurs at ca. 4000 ft. in the Little San Bernardino Mts. and may be *G. Wrìghtii* Gray, which ranges from Ariz. to Tex. and Mex.

12. **G. chilénse** Spreng. Annual or biennial, several-stemmed, 2–6 dm. high, usually quite leafy and with greenish-yellow tomentum; lvs. lanceolate to narrowly spatulate, 2–5 cm. long, gradually reduced upward; heads usually in a single rather close glomerule

at end of each stem, sometimes paniculate; invol. 5–6 mm. high, greenish-yellow; phyllaries all obtuse.—Rather moist often waste places, below 6000 ft.; many Plant Communities; cismontane Calif., occasional on desert; to B.C., Mont., Tex. June–Oct. An occasional plant lower, densely leafy to the compact cluster of heads, is var. *confertifòlium* Greene.

13. **G. lùteo-álbum** L. Permanently tomentose annual weed 2–6 dm. high, the stems erect from a decumbent base, leafy, loosely branched; lvs. linear-oblanceolate, 2–4 cm. long, 2–4 mm. wide, the upper cauline becoming linear-lanceolate and acute or acuminate; heads in usually several cymosely-arranged terminal clusters; invol. 4–5 mm. high, the hyaline-tipped phyllaries greenish or brownish-stramineous, the inner ones lance-oblong; $2n = 14$ (Wulff, 1937).—Frequent along roadsides, etc., in coastal hills and valleys; Sonoma Co. to Monterey Co.; introd. from Old World. Flowering all year.

125. **Antennària** Gaertn. Pussytoes

Low woolly perennials with simple entire alternate and basal lvs., the latter mostly tufted and conspicuous, the cauline mostly ± reduced. Plants dioecious; heads 1 to many in a usually congested infl., disciform or discoid. Fls. many. Phyllaries imbricate in several series, scarious at least in upper part, often whitish or colored. Receptacle convex or flattish, naked. The ♂ fls. with filiform corolla, entire or merely notched style, and a scanty pappus of usually clavate bristles. The ♀ fls. with tubular 5-toothed corolla, 2-cleft style and a copious pappus of capillary naked sometimes barbellate bristles. Aks. terete or slightly compressed. Taxonomy confused; some spp. having both bisexual and parthogenetic races. Perhaps 25–30 spp., largely of N. Am., some circumpolar, some in S. Am. (Latin, *antenna*, because of the resemblance of pappus in ♂ fls. to insect antennae.)

Heads solitary.
  Stems ca. 5–12 cm. high; lvs. green above .... 1. *A. suffrutescens*
  Stems 1–4 cm. high; lvs. silky-tomentose .... 2. *A. dimorpha*
Heads several to many.
  Upper surface of basal lvs. soon green, the basal lvs. in tufts at ends of stolons .. 9. *A. neglecta*
  Upper surface of basal lvs. mostly remaining tomentose, or lvs. not in tufts at ends of stolons.
    Plants forming mats because of the presence of stolons.
      Terminal part of outer and middle phyllaries brownish to dirty green.
        Phyllaries with their terminal portion a dirty blackish-green throughout, usually sharp-
          pointed .... 3. *A. alpina*
        Phyllaries with their terminal part white or pale brown .... 10. *A. umbrinella*
      Terminal part of outer and middle phyllaries pink or white, sometimes with a basal dark spot.
        Phyllaries lacking a basal dark spot .... 6. *A. rosea*
        Phyllaries with a basal dark spot .... 7. *A. corymbosa*
    Plants lacking stolons and not forming mats.
      Invol. scarious and glabrous nearly to the base, pale.
        Pubescence of stems closely silky-lanate; plants with a branched caudex or short stout
          rhizomes; middle cauline lvs. mostly linear .... 4. *A. luzuloides*
        Pubescence of stems rather loosely woolly; plants with vigorous rhizomes; middle cauline
          lvs. linear-oblanceolate .... 5. *A. argentea*
      Invol. densely pubescent in lower part, mostly reddish .... 8. *A. Geyeri*

1. **A. suffrutéscens** Greene. Low shrubby evergreen; lvs. spatulate, 5–12 mm. long, green above, white-tomentose beneath; peduncles erect, slender, glandular, 7–12 cm. high, with scattered reduced lvs.; heads solitary; invol. 7–10 mm. high, hemispheric, woolly below; phyllaries marked with brown and reddish.—At 3500–4000 ft.; Yellow Pine F., Red Fir F.; Siskiyou Mts., Del Norte Co.; sw. Ore. June.

2. **A. dimórpha** (Nutt.) T. & G. [*Gnaphalium d.* Nutt.] Dwarf matlike perennial from a compact much-branched caudex, the stems freely branched but not stoloniferous; herbage silky-tomentose; lvs. many, linear to oblanceolate, to ca. 3 cm. long and 5 mm. wide; heads solitary, terminal on leafy stems 1–4 cm. high; ♂ invol. ca. 5–7 mm. high, the phyllaries marginally and apically colorless, subhyaline, otherwise dingy brown or bluish; ♂ pappus-bristles barbellate upwards but scarcely clavate; ♀ invol. mostly 10–15 mm. high, the phyllaries narrow, pointed, ± tinged with brown or reddish-brown; pappus-bristles ca. twice as long as invol.—Dry slopes and ridges, 4700–9700 ft.; Sagebrush Scrub to Lodgepole F., Pinyon-Juniper Wd.; San Bernardino and San Gabriel

mts., Mt. Pinos, Panamint Mts., Sierra Nevada, n. to Modoc and Siskiyou cos.; to B.C., Mont., Nebr., Colo. May–July.

3. **A. alpìna** (L.) Gaertn. var. **mèdia** (Greene) Jeps. [*A. m.* Greene. *A. densa* Greene.] Mat-forming stoloniferous perennial with a branching root-crown and erect or ascending stems to 1 (1.5) dm. high and short leafy-tufted basal shoots; basal lvs. oblanceolate to ± spatulate, ± white-tomentose, sometimes glabrate; heads mostly 3–6 in a subcapitate group; ♀ invol. 4–7 mm. high, woolly below, the scarious tips of the phyllaries mostly blackish green, the inner slender and pointed; ♂ invol. often with striate white tips to the scarious portion of the phyllaries; aks. glandular; pappus-bristles dilated, often dentate above.—Rocky and gravelly often ± moist places, 7600–12,500 ft.; Subalpine F., Alpine Fell-fields; San Bernardino Mts., White Mts., Sierra Nevada, Yolla Bolly Mts.; n. to B.C., Rocky Mts. July–Aug. Plants from the White Mts. with the tomentum promptly deciduous and exposing a dense glandular-scabrous green surface are the var. *scàbra* (Greene) Jeps. [*A. s.* Greene.]

4. **A. luzuloìdes** T. & G. [*A. microcephala* Gray.] Stems clustered, slender, from a branched rather woody caudex, 1–2(–2.5) dm. high, rather leafy; herbage rather permanently gray-tomentulose; basal lvs. erect, linear-oblanceolate, with a petiolar base, ca. 3–8 cm. long, 2–8 mm. wide; cauline lvs. more narrow, linear, gradually reduced up the stem; heads several to many in a subcapitate to subcorymbiform group; ♂ and ♀ invols. alike, 4–5 mm. high, glabrous; lower part of phyllaries pale greenish-brown, the upper part whiter; ♂ pappus-bristles with dilated tips, ± serrulate.—Open slopes and valleys, ca. 5000–6000 ft.; Yellow Pine F., Sagebrush Scrub; n. Sierra Nevada to Lassen and Modoc cos.; to B.C., Mont., Colo. May–July.

5. **A. argéntea** Benth. [*A. luzuloides* var. *a.* Gray.] Rhizomatous perennial, ± loosely tomentose throughout, the stems from a branched root-crown, scantily leafy above the middle, 2–4 dm. high; lvs. largely basal, oblanceolate or broader, 2–5 cm. long including the petiole, up to 1.5 cm. wide; lower cauline gradually reduced; heads as in *A. luzuloides,* but a little larger and whiter or even pinkish on the invol.—Dryish open slopes, 2500–6700 ft.; Yellow Pine F., Red Fir F.; Sierra Nevada from Fresno Co. n., Coast Ranges from Lake Co. n.; Ore. June–July.

6. **A. ròsea** Greene. [*A. dioica* var. *r.* D. C. Eat. ex Ckll. *A. angustifolia* Rydb. *A. speciosa* E. Nels.] Cespitose, forming leafy mats, closely and persistently gray-tomentose, or some of lvs. finally glabrate and greenish above; stems mostly 0.5–2.5 dm. high, scatteringly leafy; basal lvs. oblanceolate or spatulate, mostly ca. 1–3 cm. long, 2–7 mm. wide; stem-lvs. narrower, gradually reduced upward; heads several, in a dense or ± loose cyme; ♀ invol. ca. 4–7 mm. high, woolly below, the scarious part of the phyllaries ± oblong, deep pink to bright or dull white, ± striate under magnification; inner phyllaries more acutish; pappus-bristles linear, ± barbellate.—Dryish to moist ± wooded places, 4500–12,000 ft.; upper Montane Coniferous F. to Alpine Fell-fields; San Bernardino Mts., Sierra Nevada, White Mts., N. Coast Ranges from Yolla Bolly Mts. n.; to Alaska, Ontario, Rocky Mts. June–Aug. Variable; ♂ plants apparently infrequent, and with clavate pappus.

7. **A. corymbòsa** E. Nels. [*A. dioica* var. *c.* Jeps.] Near to *A. rosea* and differing chiefly in having a dark spot at the base of each whitish phyllary.—Meadows and moist places, 7200–10,600 ft.; upper Montane Coniferous F.; Sierra Nevada from Fresno Co. n.; to Ore., Mont., Colo. July–Aug.

8. **A. Géyeri** Gray. Plant with a branched ± woody base, densely and persistently tomentose, the several stems 5–15 cm. high; lower stem-lvs. oblanceolate, 1–2 cm. long, somewhat reduced upward but stems quite leafy; infl. rather dense; ± paniculate; invol. woolly to above the middle, the ♀ narrow, mostly cylindric-turbinate, 7–9 mm. high, with short mostly pinkish scarious tips; ♂ invol. slightly shorter and broader, often white-tipped; ♂ pappus barbellate, scarcely clavate.—Dry places, ca. 5000–7600 ft.; Sagebrush Scrub to Red Fir F.; Eldorado and Tehama cos. to Modoc and Siskiyou cos.; to e. Wash., Nev. June–Aug.

9. **A. neglécta** Greene var. **Howéllii** (Greene) Cronq. [*A. H.* Greene.] Stems scape-like, 1.5–4 dm. high, arising from stolons terminating slender rootstocks, bearing few lvs.; lvs. mostly basal, obovate to spatulate, glabrate and green above, ± 3-nerved, white-woolly beneath, 2.5–4 cm. long, petioled; cauline lvs. sessile, narrow; heads several, in a usually dense cyme; ♀ invol. mostly 6–8 mm. long, the phyllaries narrow, pointed, the

scarious part dingy white; pappus surpassing the stigmas; ♂ plants rare.—Woods, at ca. 4000 ft.; Yellow Pine F.; Siskiyou Co.; to Wash. May–July.

10. **A. umbrinélla** Rydb. [*A. dioica* var. *kernensis* Jeps.] Densely cespitose, forming leafy mats; stems 0.5–2 dm. high; herbage tomentose; basal lvs. spatulate to oblanceolate, 5–20 mm. long, the cauline gradually reduced; heads few, in capitate clusters; invol. ca. 4–5 mm. high, woolly at base, the scarious part of at least the outer phyllaries brownish or dirty green, the inner whitish; ♂ plants uncommon.—Borders of meadows, etc., 9000–12,000 ft.; Subalpine F. to Alpine Fell-fields; s. Sierra Nevada; to most of Canada, Colo., Ariz.

126. Anáphalis DC. Pearly Everlasting

White-woolly perennials with simple erect equably leafy stems. Lvs. entire, alternate. Heads mostly numerous, in terminal compound corymbs. Plants dioecious or polygamo-dioecious, the ♀ heads sometimes with a few cent. ♂ fls. Receptacle naked, flat or convex. Heads discoid or disciform, many-fld. Phyllaries imbricate in several series. Staminate fls. tubular, generally with undivided style; ♀ fls. tubular-filiform, with bifid style. Pappus of distinct capillary bristles, not conspicuously barbellate. Ca. 25 spp., of N. Temp. (Ancient Greek name of some Everlasting.)

1. **A. margaritàcea** (L.) Benth. ex Clarke [*Gnaphalium m. L. A. m.* var. *occidentalis* Greene. *A. o.* Heller. *A. sierrae* Heller. *A. m.* var. *subalpina* Gray.] Perennial with slender running rootstocks; stems commonly 2–9 dm. high, loosely white-woolly or somewhat rusty in age; lvs. lanceolate to linear or linear-oblong, 2–8(–12) cm. long, sessile, obtuse to acuminate, often greener above than beneath, sometimes ± revolute; heads to 1 cm. wide; invol. ca. 5–7 mm. high; phyllaries pearly white, ovate, sometimes with a basal dark spot; aks. papillate; $2n = 28$ (Maude, 1939).—Openings in woods, talus, etc., below 8500 ft.; many Plant Communities; Coast Ranges from Monterey Co. n., San Bernardino Mts., Sierra Nevada n.; to Alaska, Atlantic Coast, Eurasia. June–Aug.

127. Stylocline Nutt.

Low woolly annuals resembling *Micropus*. Receptacle longer than broad, oblong to nearly linear, the membranous bracts attached spirally and the winged margins spirally imbricate in a subglobose slightly conic head. True invol. absent. Perfect cent. fls. usually subtended by plane bracts; sterile cent. aks. generally with a few pappus-bristles. The ♀ marginal fls. completely enclosed in woolly nonindurate phyllaries. A small genus of Eurasia and N. Am. (Greek, *stulos*, column, and *kline*, bed.)

Bracts of the ♂ (sterile) fls. thick, with conspicuous terminal hooklike cusps; fertile fls. 5–9
1. *S. filaginea*
Bracts of the ♂ (sterile) fls. thin, not ending in hooks; fertile fls. more.
The ♀ fls. enclosed in the cent. portion of wholly hyaline bracts which have winglike margins. Cismontane.
Bracts surrounding the ♀ fls. obviously imbricate and not much enveloped in wool. Monterey and San Joaquin V. to w. edge Colo. Desert and San Diego Co. . . . . 2. *S. gnaphalioides*
Bracts surrounding the ♀ fls. not plainly much imbricate and quite enveloped in loose wool. Alameda Co. to Sonoma and Lake cos. . . . . 3. *S. amphibola*
The ♀ fls. enfolded by the basal part of the bracts which do not have winglike margins. Deserts
4. *S. micropoides*

1. **S. filagínea** (Gray) Gray. [*Ancistrocarphus f.* Gray.] Plant branched at or above the base, mostly 3–10 cm. high, appressed woolly-canescent; lvs. linear to oblance-linear, 5–12 mm. long, the uppermost broader; fertile fls. 5–9, their boat-shaped bracts ca. 3 mm. long, firm except at the hyaline tip; sterile fls. surrounded by 5 empty bracts, these 4–5 mm. long and each ending in a rigid incurved hook; sterile fls. without pappus. —Dry often brushy places, below 5500 ft.; Coastal Sage Scrub, Chaparral, Foothill Wd., etc.; w. Riverside Co. to Kern Co., Coast Ranges to Mendocino Co., Mariposa Co. April–May.

2. **S. gnaphalioìdes** Nutt. More or less branched from base, 5–18 cm. high; lvs. oblong- to spatulate-linear, 5–12 mm. long; fr.-bearing bracts many, barely 3 mm. long, ovate, nearly plane on outer surface, the basal part firm, forming a sac enclosing the ak., the rest hyaline; ak. without pappus; sterile fls. little shorter than their bracts and

with scanty pappus.—Dry slopes, burns, etc., below 5500 ft.; Chaparral, Coastal Sage Scrub, Yellow Pine F., etc.; cismontane Calif. from Lake and Marin cos. and Monterey Co. and San Joaquin V. to San Diego Co., w. edge Colo. Desert. March–May.

3. **S. amphíbola** (Gray) J. T. Howell. [*Micropus a.* Gray.] Slender, erect, floccose-woolly; lvs. oblanceolate, ± appressed; fruiting bracts 9–10, rather thin and soft, somewhat imbricate on the elevated receptacle; ♂ fls. subtended by linear thin chafflike bracts and with a pappus of a few soft bristles.—Not common, shallow soil in rocky places; Mixed Evergreen F?; Sonoma, Lake, Marin, Contra Costa, and Alameda cos. April–May.

4. **S. micropoìdes** Gray. Branched from base, 2–8 cm. high; lvs. linear, 5–8 mm. long, those below the heads lance-linear; ♀ fls. many, with oblong woolly bracts lacking hyaline margins and ending in a small hyaline tip, the whole body enveloping the ak.; sterile fls. naked or slightly subtended by glabrous oblong paleae and bearing a pappus of 3–4 slender deciduous bristles.—Occasional in sandy washes and on dry slopes, below 4000 ft.; Creosote Bush Scrub, Joshua Tree Wd.; n. Colo. Desert, Mojave Desert; to sw. Utah, New Mex. March–May.

128. Filàgo L.

White woolly annuals with entire alternate lvs. Heads small, discoid, in small capitate clusters. Invol. scanty, the phyllaries resembling the bracts of the cylindric to obconic or convex receptacle. Outer fls. ♀, fertile, with tubular-filiform corolla, placed in several series, the outer epappose and subtended by concave partly enclosing bracts, the inner bractless and with pappus of capillary bristles; cent. fls. 2–5, often sterile, bractless, with capillary pappus. Aks. subterete. Ca. 12 spp., of temp. and warm-temp. Eurasia, Afr., Am. (Latin, *filum*, thread, referring to the hairs.)

```
Uppermost lvs. much longer than the heads.
  Lvs. subulate with broadish base; receptacle low, nearly flat ................. 1. F. gallica
  Lvs. linear; receptacle ± obconic or convex .............................. 2. F. arizonica
Uppermost lvs. scarcely or not longer than the heads.
  Plants erect, 6–25 cm. high ............................................. 3. F. californica
  Plants depressed, spreading, the stems 3–10 cm. long ..................... 4. F. depressa
```

1. **F. gállica** L. Erect, simple or branched, 2–3 dm. high; lvs. appressed, subulate, 1–2.5 cm. long, those below the heads involucrate, divaricate; receptacle almost flat; heads 2–5, obconic, ca. 4 mm. high; marginal aks. completely enclosed in the subtending triangular pointed woolly bract; aks. 0.5 mm. long.—Burns and waste places, San Diego Co. to Humboldt Co.; natur. from Eu. April–June.

2. **F. arizónica** Gray. Stem branched at or above the base, erect or diffuse, 4–15 cm. long, with slender naked internodes and few lvs. except at upper part; lvs. linear to ± oblong, mostly ca. 0.5–1.5 cm. long, the uppermost involucrate around and exceeding the heads; marginal ♀ fls. 10–15, their bracts firm, ovate, those within the inner circle of paleae ca. 4–5, all perfect; aks. slightly curved, smooth.—Occasional in dry and disturbed places; Chaparral, Coastal Sage Scrub, Creosote Bush Scrub; Channel Ids., cismontane and desert s. Calif.; s. Ariz., L. Calif. March–May.

3. **F. califórnica** Nutt. Erect, simple or branched, 0.5–3.5 dm. high, leafy; lvs. 0.8–2 cm. long, oblong-linear to subspatulate, sessile; heads ovoid, 3–4 mm. high, scarcely exceeded by the involucrate lvs.; bracts of outer ♀ fls. 8–10, woolly, boat-shaped, with hyaline tip, the inner ones thinner and less woolly, the inner florets ca. 12–20, only ca. 2–4 often perfect; inner aks. papillose.—Common in dry open places, on burns, etc., below 3500 ft.; Foothill Wd., Chaparral, Coastal Sage Scrub, V. Grassland, etc.; cismontane Calif. from Mendocino and Butte cos. s., desert edge; to Utah, s. Ariz., L. Calif. March–June.

4. **F. depréssa** Gray. Branched from base, depressed or spreading, the stems 4–12 cm. long; lvs. oblong to narrow-obovate, 3–9 mm. long; heads 2–3 mm. high, the involucrate lvs. short; outer bracts with a hyaline appendage ca. as long as the body, the marginal ♀ fls. 5–6; aks. all smooth.—Open places, below 2500 ft.; Creosote Bush Scrub, Joshua Tree Wd.; both deserts to s. Ariz. March–May.

129. Psilocárphus Nutt.

Small woolly annuals with opposite entire lvs. Heads disciform, small, round, terminal, commonly subtended by a pair of branches. True invol. lacking, although the heads are commonly subtended by foliage lvs. Receptacle subglobose to obpyriform. Pistillate fls. in several series, with short filiform-tubular corolla, each loosely enclosed in a saccate woolly bract which bears a hyaline appendage below the summit. Cent. fls. few, bractless, functionally ♂, sterile; pappus none. Aks. small, smooth, nerveless, turgid or compressed. Five spp. (1 in w. S. Am.). (Greek, *psilos*, bare, and *karphos*, chaff.)

(Cronquist, A. A review of the genus P. Res. Studies State Coll., Wash. 18: 71–89, 1950.)
Well developed receptacular bracts ca. 3 mm. long at maturity . . . . 1. *P. brevissimus*
Well developed receptacular bracts ca. 2 mm. long.
    Lvs. linear or linear-oblanceolate, mostly 6–12 times as long as wide; aks. narrowly oblong or elliptic-oblong . . . . 2. *P. oregonus*
    Lvs. spatulate, oblong or oblanceolate, mostly 1.5–6 times as long as wide; aks. broadly oblanceolate to narrowly obovate . . . . 3. *P. tenellus*

1. **P. brevíssimus** Nutt. [*P. globiferus* Nutt. *P. oregonus* var. *b.* Jeps.] Plants usually prostrate and branching, the stems seldom up to 1 dm. long; herbage loosely and copiously woolly; lvs. lance-oblong to ± linear, 5–25 mm. long, generally broadest near their base and well surpassing heads; ♀ fls. 8–80; receptacular bracts mostly 2.5–4 mm. long; aks. obliquely oblanceolate, flattened, 1–2 mm. long.—Dried beds of vernal pools, below 3000 ft.; various Plant Communities, especially V. Grassland; cismontane Calif., Modoc Co. to near Lake Tahoe; to Wash., Mont. April–June.
Var. **multiflòrus** Cronq. Plants ± erect; lvs. mostly linear-oblong; wool rather thin; heads with more than 100 ♀ fls.—Sites of vernal pools about San Francisco Bay. May–June.

2. **P. oregònus** Nutt. Slender, at first erect, becoming much-branched and prostrate or decumbent later; stems to ca. 15 cm. long; tomentum fine, short and close, silvery, generally persistent; lvs. linear to linear-oblanceolate, 6–20 mm. long, usually conspicuously surpassing the heads; ♀ fls. ca. 20–80 in a head; aks. 0.6–1.2 mm. long, turgid, ± oblong.—Dried vernal pools, mostly from Santa Barbara Co. n. through Coast Ranges and Cent. V. to Wash., Ida. May–June.

3. **P. tenéllus** Nutt. Slender annual, at first erect, later branched and prostrate; tomentum generally thin and rather loose; lvs. spatulate to oblong, 2–6 times as long as wide, 4–15 mm. long; ♀ fls. 25–46 in a head; receptacular bracts ca. 2 mm. long; aks. 0.6–1.2 mm. long, moderately compressed, turgid, ± obovate.—Dried vernal cismontane and montane pools, below 6000 ft.; n. L. Calif. through cismontane and montane Calif. to B.C. April–June.
Var. **ténuis** (Eastw.) Cronq. [*P. t.* Eastw.] Lvs. subtending the heads elliptic-ovate to ovate, 1.5–3 times as long as wide; plants less branched and more erect.—Dried vernal pools, largely of Salinas and San Joaquin valleys.

130. Micròpus L.

Rather small floccose-woolly annuals. Lvs. entire, alternate. Heads terminal, usually clustered, discoid, several-fld. Receptacle usually broader than long, rather flat. Phyllaries open, scarious, surrounding the fl.-bearing bracts of the receptacle; these woolly, conduplicate, each enclosing a fertile ♀ fl. of which only the corolla-tube and style are exserted through a slit. Cent. perfect fls. sterile and mostly naked. Aks. with 1 lateral corolla and style. Pappus wanting in sterile cent. aks.; ak. obovate, laterally compressed. Several spp. of Eu., N. Am. (Greek, *micros*, small, and *pous*, foot.)

1. **M. califórnicus** F. & M. [*M. angustifolius* Nutt.] Stem slender, erect, usually simple, 0.5–3.5 dm. high, leafy; lvs. linear-oblong, short-acuminate, 5–15 mm. long; phyllaries commonly ca. 5, rounded to ovate, scarious, with a cent. green spot; receptacular bracts loose-woolly, 4–6, semiobovate, indurate in age, ca. 3.5 mm. long, with a scarious lateral beaklike appendage; sterile fls. ca. 3, the corolla filiform but expanded toward the throat.—Common in low dry open places, below 5000 ft.; many Plant Communities; cismontane Calif. to Ore. April–June. Plants with wool of the bracts short and wholly appressed, from the general range, have been called var. *subvestìtus* Gray.

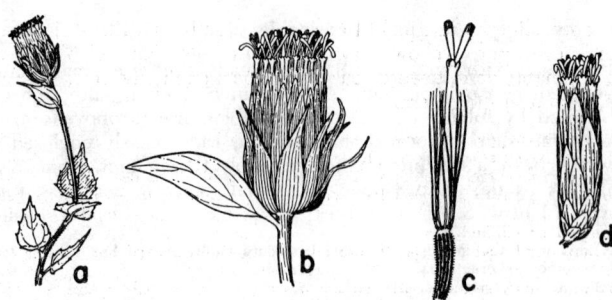

Fig. 117. EUPATORIEAE. *Brickellia arguta: a,* head and upper lvs., × ½; *b,* discoid head, × 1½, phyllaries in several series, striate; *c,* fl., × 2½, ovary nerved, pappus setose, stigmatic-branches with papillate appendages. *B. californica: d,* head, × 2, with phyllaries more erect.

131. Eváx Gaertn.

Ours low woolly annuals with entire alternate lvs. and small discoid heads with circles of bractlike lvs. beneath. Phyllaries very closely imbricate, covering but not enclosing the ♀ fls. and forming a short-cylindric head. Receptacle columnar, villous, tipped with a cuplike whorl of mostly 5 bracts at summit, the cup bearing mostly 2–4 ♂ fls. Phyllaries and bracts of ♀ fls. indurate, persistent, the upper or sterile bracts deciduous. The ♀ fls. with filiform corollas. Pappus none. A genus of some size, sometimes perennial in the Old World. (Supposedly the name of an Arabian chief.)

Lf.-blades ca. 7–15 mm. wide; heads in terminal clusters if stems developed. Mostly dry beds of vernal pools, Butte Co. to Merced and Contra Costa cos. . . . . 1. *E. caulescens*
Lf.-blades ca. 3–5 mm. wide; heads solitary in axils of cauline lvs. if stems developed. Dry grassy slopes and ridges.
    Plants mostly caulescent; lf.-blades ca. ½ as wide as long. Outer Coast Ranges . . 2. *E. sparsiflora*
    Plants mostly acaulescent; lf.-blades ca. ⅓ as wide as long. Foothills of Sierra Nevada
                                                                      3. *E. acaulis*

1. **E. cauléscens** (Benth.) Gray. [*Psilocarphus c.* Benth. *E. involucrata* Greene.] Stems 1–several, simple, mostly 1–20 cm. high; heads crowded into a dense terminal hemispherical cluster 1–2 cm. across and surrounded by a whorl of spatulate-obovate lvs. 1–3 cm. long on somewhat longer petioles; cauline lvs. scattered, petiolate; receptacle slender-columnar, the ♀ fls. and bracts crowded at base, the summit with mostly 3–7 broad bracts subtending a few ♂ fls.; aks. pyriform, obovate, ± compressed, smooth.—Dry beds of vernal pools; V. Grassland; Sacramento V. s. to Merced Co. April–June. Plants with axis shortened or not developed and lvs. broad have been called var. *hùmilis* (Greene) Jeps. [*Hesperevax h.* Greene.]

2. **E. sparsiflòra** (Gray) Jeps. [*E. caulescens* var. *s.* Gray.] Erect, simple or branched, mostly 3–10 cm. high, grayish; lf.-blades spatulate, 1–1.5 cm. long, on slender petioles ca. as long; heads axillary, often ± crowded toward top; receptacle-bracts woolly on back, long-hirsute at base; cent. ♂ fls. ca. 4.—Dry open places, below 2500 ft.; many Plant Communities; Coast Ranges from w. Ore. and Humboldt Co. to San Luis Obispo Co., w. San Diego Co., Santa Rosa Id. March–May. Plants with stems 0.5–ca. 3 or 4 cm. high occur at scattered stations and have been called var. *brevifòlia* (Gray) Jeps. [*E. caulescens* var. *b.* Gray.]

3. **E. acaúlis** (Kell.) Greene. [*Stylocline a.* Kell. *Hesperevax a.* Greene. *E. caulescens* var. *minima* Gray.] Much like no. 2, but mostly quite acaulescent, sometimes with ± prostrate stems 1–3 cm. long; lvs. spatulate, the blades mostly 2–6 mm. long and 1–3 mm. wide.—Open places; V. Grassland, Foothill Wd.; w. base of Sierra Nevada from Calaveras Co. to Kern Co. March–May.

Tribe 8. Eupatorìeae. EUPATORY TRIBE (Fig. 117)

Herbs or shrubs with mostly opposite but sometimes alternate or whorled usually undivided lvs. Heads always discoid. Phyllaries usually imbricate in several series or in

1–2 series and only slightly unequal. Receptacle usually flattish and naked. Corollas tubular, regular, never pure yellow. Style-branches commonly elongate, ± thickened above, obtuse, stigmatic only toward base. Pappus commonly setose.—Principally of the Am. tropics, with over 40 genera.

A. Annuals; lvs. sessile; pappus of awns alternating with paleae.
    B. Lvs. opposite; receptacle convex; phyllaries equal, nerveless . . . . 132. *Trichocoronis*
    BB. Lvs. alternate; receptacle flat; phyllaries very unequal, several-nerved . . . . 133. *Malperia*
AA. Perennial herbs or shrubs; lvs. petioled; pappus of bristles (in *Hofmeisteria* some of these paleaceous at base).
    B. Aks. 5-angled or -ribbed.
        C. Phyllaries imbricated, striately nerved; intricately branched shrub with scattered heads
                                                134. *Hofmeisteria*
        CC. Phyllaries in 2 series, nearly equal, obscurely nerved; stems simple or few-branched, the heads in compact cymes . . . . 135. *Eupatorium*
    BB. Aks. 10-nerved; phyllaries well imbricated, striate . . . . 136. *Brickellia*

132. Trichocorònis Gray

Small annuals, the several stems creeping at base or spreading, branching, leafy. Lvs. opposite, sessile. Heads peduncled, terminal. Receptacle convex, naked. Phyllaries 12–18, greenish, subequal, nerveless. Corolla abruptly much dilated above the narrow tube, flesh-color to rose-purple. Aks. grayish-black, ± 5-angled, short-hispidulous toward the summit; pappus forming a sort of crown of unequal paleae and awns. Perhaps 3 spp., of sw. N. Am. (Greek, *trichos*, hair, and *koronis*, crown.)

1. **T. Wrìghtii** (T. & G.) Gray. [*Ageratum?* W. T. & G. *Biolettia ripària* Greene. *T. r.* Greene.] Stems 1–2 dm. long, ascending, rooting at lower nodes; lvs. oblong-ovate to narrowly lanceolate, entire or serrate, auricled at base, 0.8–2 cm. long; heads 4–5 mm. broad; aks. ca. 1 mm. long, with barbellate bristles alternating with small fimbriate paleae.—Mud flats and shores; Mystic Lake near Moreno, Riverside Co., occasional in Cent. V.; s. Tex., n. Mex. May–Sept.

133. Malpèria Wats.

Erect slender annual with narrow alternate mostly sessile lvs. Heads loosely cymose; invol. turbinate; phyllaries narrow, thin, unequal, several-nerved, scarious-margined. Receptacle flat, naked. Style-branches thickened, exserted only in age. Aks. slender, 5-angled. Pappus of 3 hispidulous setae as long as the corolla and alternating with minute truncate erose paleae. (Etymology obscure.)

1. **M. ténuis** Wats. Stems branching above, 1–3 dm. high, minutely scabrous; lvs. linear, acuminate, 1.5–5 cm. long, mostly entire; invol. 8–10 mm. high; phyllaries linear, strigulose; fls. brownish; aks. somewhat hispid on the angles.—Sandy places, Creosote Bush Scrub; Colo. Desert (Signal Mt., Split Mt., Fish Mt.); n. L. Calif. March–April.

134. Hofmeistèria Walp.

Bushy suffrutescent plants, the lower lvs. opposite, the upper alternate. Heads medium-sized, long-peduncled to almost sessile, many-fld. Invol. campanulate; phyllaries narrow, striate, imbricated. Receptacle naked. Fls. tubular. Aks. 5-angled, callous-thickened at base. Pappus in ours of 10–12 scabrous bristles with additional short thin paleae. Two or 3 spp., of sw. N. Am. (W. *Hofmeister*, German botanist.)

1. **H. plurisèta** Gray. ARROW-LEAF. Low rounded shrubs mostly 3–8 dm. high, somewhat wider, intricately much branched; older stems with white shreddy bark; young growth green, glandular-puberulent; lf.-blades 4–10 mm. long, deltoid-lanceolate, entire or few-toothed; petioles 2–4 cm. long; heads few, in small terminal corymbs; invol. 7–9 mm. high; phyllaries 3-striate, acuminate, often with recurved tips; corollas whitish; stigmas purplish at tips; aks. ca. 2–3 mm. long.—Common in rocky places, below 4000 ft.; mostly Creosote Bush Scrub; both deserts; to Utah, Ariz. March–May.

135. Eupatòrium L.

Mostly perennial herbs or shrubs. Lvs. mostly opposite, sometimes alternate or whorled, entire to toothed or even dissected, often glandular-punctate. Heads large to small, mostly in a corymbiform infl. rarely paniculate or solitary. Fls. all tubular, perfect, few–many in a head. Phyllaries ± imbricate to subequal. Receptacle flat to conic, naked. Corollas pink or purple to blue or white. Anthers obtuse, entire at base or minutely sagittate. Style-branches with short stigmatic lines and an elongate papillate obtuse often clavate appendage. Aks. mostly 5-angled; pappus of ± numerous capillary bristles. Perhaps 500 spp., almost cosmopolitan, most numerous in Am. tropics. (*Eupator* Mithridates, king of Pontus, 132–63 B.C.)

Lvs. opposite; heads 5–8 mm. high.
  Lf.-blades mostly 4–10 cm. long; petioles mostly 2–4 cm. long. Escape from cult
    1. *E. adenophorum*
  Lf.-blades mostly 1.5–3.5 cm. long; petioles mostly 0.5–1 cm. long. Native in e. Mojave Desert
    2. *E. herbaceum*
Lvs. mostly alternate; heads 8–10 mm. high. Sierra Nevada and N. Coast Ranges .. 3. *E. occidentale*

1. **E. adenóphorum** Spreng. [*E. pasadense* McClat.] Stems purplish, erect, simple or few-branched, mostly 5–15 dm. high, glandular-puberulent; lvs. deltoid-ovate, serrate, ± cuneate at base, ± acuminate at apex, strongly bicolored; cymes small, compact, in open panicles with many heads; heads ca. 5 mm. high; invol. glandular-puberulent; fls. white, sometimes tinged pink.—Occasional escape from gardens, San Diego Co. to San Francisco Bay region; native of Mex. Most months.

2. **E. herbàceum** (Gray) Greene. [*E. ageratifolium* var.? *h.* Gray. *E. arizonicum* (Gray) Greene.] Perennial from a woody caudex; stems several, 5–7 dm. high, leafy, branching above; herbage light green, cinereous-puberulent; lvs. mostly opposite, ovate, commonly cordate, crenate-serrate, acute; heads 6–8 mm. high, numerous, in dense ± corymbose cluster of cymes; invol. puberulent; corollas white; aks. black, ca. 2 mm. long.—Among rocks, 5000–6400 ft.; Pinyon-Juniper Wd.; New York Mts., Clark Mt., e. Mojave Desert; to Utah, New Mex. Sept.–Oct.

3. **E. occidentàle** Hook. Perennial from a woody ± rhizomatous base; stems tufted, 1.5–7 dm. tall, greenish or purplish, crisped-puberulent; lvs. mostly alternate, deltoid to deltoid-ovate, ± toothed, mostly atomiferous-glandular, acutish at tip, mostly ovate or truncate at base; heads rather compactly clustered at tips of stems and branches, 8–10 mm. high; phyllaries puberulent; fls. pink, red-purple to whitish; aks. black, 3–3.5 mm. long; $2n = 34$ (Grant, 1953).—About rocks, 6500–11,100 ft.; Montane Coniferous F.; Sierra Nevada; below 7500 ft., Redwood F. to Red Fir F.; Coast Ranges from Tehama Co. n.; to Wash., Ida., Utah. July–Sept. Plants from Sierra Nevada tend to be lower, with smaller lvs. than in the n. form.

136. Brickéllia Ell.

Perennial to ± woody or annual; lvs. alternate or opposite, simple, veiny, mostly resinous-dotted. Fls. white to creamy or pink-purple, all perfect and tubular, 3–many in discoid mostly narrow heads. Invol. cylindric to campanulate, the phyllaries striate, chartaceous to membranous, imbricated. Receptacle naked, mostly flat. Anthers minutely rounded-sagittate at base; style-branches with short stigmatic lines and an elongate papillate appendage. Aks. 10-nerved; pappus of 10–80 barbellate or nearly smooth to rarely subplumose bristles. Almost 100 spp., of N. and S. Am., chiefly in warm regions. (Dr. J. *Brickell*, early physician and botanist of Georgia.)

(Robinson, B. L. A monograph of the genus Brickellia. Mem. Gray Herb. 1: 1–151, 1917.)
A. The heads clustered, usually racemose-paniculate, not in corymbs.
    B. Heads 3–7-fld.
        C. Phyllaries 10–12; lvs. linear .... 1. *B. longifolia*
        CC. Phyllaries ca. 20; lvs. lanceolate to ovate.
            D. Lvs. entire; phyllaries subglabrous .... 2. *B. multiflora*
            DD. Lvs. dentate-serrate; phyllaries puberulent .... 3. *B. Knappiana*
    BB. Heads 8–26-fld.
        C. Phyllaries with recurved or spreading tips; lvs. sessile or nearly so.
            D. Intermediate phyllaries entire; lvs. white-lanate .... 4. *B. Nevinii*
            DD. Intermediate phyllaries mostly 3-toothed; lvs. greenish.

E. Stems glandular-villous; aks. 4–4.5 mm. long .... 5. *B. microphylla*
EE. Stems finely woolly; aks. 3.5 mm. long .... 6. *B. Watsonii*
CC. Phyllaries erect; lvs. petioled, the petioles at least ⅕ the length of blades.
D. Heads small, 9–10 mm. high; invol. puberulous.
E. Lf.-blades mostly 4–8 mm. long; heads 8–12-fld. S. Calif.
7. *B. desertorum*
EE. Lf.-blades 25–60 mm. long; heads 20–38-fld. Sierra Nevada and n.
9. *B. grandiflora*
DD. Heads larger, ca. 14 mm. high; invol. essentially glabrous .... 8. *B. californica*
AA. The heads solitary or at the ends of corymbosely arranged branchlets.
B. Stems leafy up to the heads. N. Calif. .... 10. *B. Greenei*
BB. Stems not leafy up to the heads, or if so, plants not from n. Calif.
C. Plants green; heads not over 17 mm. high.
D. Outer phyllaries broadly ovate, leaflike; lvs. ovate, usually sharply toothed
11. *B. arguta*
DD. Outer phyllaries linear to narrowly oblong, not leaflike.
E. Lvs. spatulate, those of the divaricate branchlets not over 8 mm. long
12. *B. frutescens*
EE. Lvs. linear-oblong, those of the ascending branchlets 12–20 mm. long
13. *B. oblongifolia*
CC. Plants white-tomentose; heads ca. 22 mm. high .... 14. *B. incana*

1. **B. longifòlia** Wats. [*Coleosanthus l.* Kuntze.] Glabrous much-branched shrub 1–1.5 m. high; lvs. alternate, linear, entire, attenuate, 3–10 cm. long, 0.2–1 cm. wide, almost sessile; heads 3–5-fld. in small cymes which are racemosely arranged; outer phyllaries ovate, obtuse, the inner lance-oblong, 3-nerved; aks. 1.8 mm. long.—Rare, on dry stream terraces, etc., 3000–5000 ft.; Creosote Bush Scrub, Joshua Tree Wd.; Inyo Co. (Owens Lake, Argus Mts., Panamint Mts.); to sw. Utah, Ariz.

2. **B. multiflòra** Kell. [*Coleosanthus m.* Kuntze.] Erect branched shrub 1–2 m. tall; lvs. glabrous, alternate, lanceolate, acute, subentire, 3-nerved, gummy, 3–8 cm. long, 1–2.5 cm. wide, on very short petioles; heads 3–5-fld., ca. 7 mm. high; phyllaries 3-nerved; aks. 1.7 m. long.—Common, dry washes and stony places, 1800–7000 ft.; Creosote Bush Scrub to Pinyon-Juniper Wd.; Clark Mt. (e. San Bernardino Co.), mts. of Inyo Co.; w. Nev. Sept.–Nov.

3. **B. Knappiàna** E. Drew. [*Coleosanthus K.* Greene.] Slender willowlike shrub 1–2 m. tall, somewhat viscid and hispidulous; lvs. alternate, lanceolate or narrow-ovate, 2.5–3.5 cm. long, 1–1.5 cm. wide, usually serrate; petioles 4–5 mm. long; panicle leafy with a few heads at tip of each branch; heads ca. 7 mm. high, 5–7-fld., the phyllaries obtuse, 4-nerved; aks. 2.5 mm. long, minutely hispidulous.—Rare, at ca. 2500–3500 ft.; Joshua Tree Wd.; Mojave R., and Panamint Mts.

4. **B. Nevínii** Gray. [*Coleosanthus N.* Heller.] Dense white-tomentose shrub, the stems several, erect, 3–5 dm. tall, loosely branched; lvs. alternate, ovate, sometimes subcordate, sessile, 0.6–1.5 cm. long; panicle open, with 1–few heads at ends of short branchlets; heads ca. 23-fld., 15 mm. tall; phyllaries woolly, 3–4-nerved, the outer with spreading or recurved tips; aks. ca. 4 mm. long, hispidulous; $n = 9$ (Gaiser, 1953).—Dry slopes and washes, 1000–5500 ft.; Chaparral, etc.; s. face of San Gabriel Mts., Santa Monica Mts., Newhall to Santa Ynez Mts. and San Emigdio Canyon, sw. Kern Co. Sept.–Nov.

5. **B. microphýlla** (Nutt.) Gray. [*Bulbostylis m.* Nutt. *Coleosanthus m.* Kuntze.] Shrubby, branched at base, glandular-villous, 3–6 dm. high, the stems erect, paniculately much-branched above, the short branchlets with 1–3 heads; lvs. green, round-ovate, 0.7–2 cm. long, subentire to denticulate, short-petioled; heads ca. 22-fld., 10–11 mm. high; phyllaries green-tipped; aks. 4–4.5 mm. long, hispidulous; $n = 9$ (Gaiser, 1953).— Occasional, dry rocky places, 3000–8000 ft.; Joshua Tree Wd., Pinyon-Juniper Wd., Yellow Pine F., etc.; San Gabriel Mts., w. Mojave Desert, Inyo Mts., e. slope of Sierra Nevada n. to Mono Co., Lake Tahoe, etc.; to Ore., Nev. Aug.–Nov.

6. **B. Watsònii** Rob. Shrubby at base, intricately branched, 2–3 dm. high, aromatic, tomentulose-puberulent; lvs. ovate, 3–10 mm. long, sparingly dentate, light green; heads 1–3 on ends of branchlets of rather a small panicle, 15–18-fld., 9–11 mm. high; phyllaries green-tipped; aks. 3.5 mm. long.—Dry rocky ledges and walls, 4500–7000 ft.; largely Pinyon-Juniper Wd.; Clark, Funeral, Panamint and Inyo mts.; to Utah. Sept.–Oct.

7. **B. desertòrum** Cov. [*B. californica* var. *d.* Parish. *Coleosanthus d.* Cov.] Shrubby, intricately branched, 1–1.5 m. tall, puberulent; lvs. opposite or alternate, ovate, crenate-serrate, obtuse, 3–12 mm. long, short-petioled, tomentulose; heads in glomerules at ends of short lateral branchlets, 8–9 mm. high, 8–12 fld.; phyllaries 3–4-striate, greenish or

yellowish, tomentulose, acutish; aks. 2.3 mm. long, pubescent; $n = 9$ (Gaiser, 1953).—Occasional in dry rocky places, 800–4000 ft.; Creosote Bush Scrub, Coastal Sage Scrub, Joshua Tree Wd.; w. edge of Colo. Desert, cismontane Riverside and San Bernardino cos., to Eagle and Panamint mts.; Nev., Ariz. Aug.–Nov.

8. **B. califórnica** (T. & G.) Gray. [*Bulbostylis c.* T. & G. *Coleosanthus c.* Kuntze.] Woody at base, 5–10 dm. high, with many stems, ± branched, puberulent to thinly tomentose; lvs. alternate, deltoid-ovate, crenate-serrate, 1–4 cm. long, subtruncate or subcordate at base, short-petioled, the uppermost reduced; panicle leafy, with heads in small clusters at ends of lateral branchlets which may be quite short; heads 12–14 mm. high, 8–18-fld., cylindrical; phyllaries 3–5-striate, green or purplish, obtuse, subglabrous; aks. 3 mm. long; $n = 9$ (Gaiser, 1953).—Common in washes and on dry slopes, below 8000 ft.; Chaparral, Coastal Sage Scrub; cismontane s. Calif.; many Plant Communities, Coast Ranges n. to Humboldt and Siskiyou cos., foothills of Sierra Nevada, occasional on Mojave Desert; to Colo., Tex., n. Mex. Aug.–Oct. A form from Solano Co. to Lake Co., with heads on peduncles 0.8–5 cm. long and more acutish phyllaries, is var. *Jepsònii* Rob.

9. **B. grandiflòra** (Hook.) Nutt. [*Eupatorium ? g.* Hook. *Coleosanthus g.* Kuntze. *C. gracilipes* Greene.] Perennial herb from a long fusiform root, the stems 3–7 dm. high, usually simple up to the infl.; herbage finely puberulent to ± cinereous; lvs. usually opposite, deltoid-ovate to lanceolate, with truncate or cordate base, 3–11 cm. long, dentate, acuminate, the petiole 0.5–3 cm. long; heads ± nodding, usually sub-umbellately corymbose at ends of paniculately arranged branchlets; fls. 20–38, greenish or yellowish-white; invol. 12–14 mm. high, green, ± pubescent; phyllaries 30–37, 5–7-striate, loosely imbricated; aks. 4 mm. long, brown, ± hirtellous; $n = 9$ (Gaiser, 1953). —Dry rocky slopes, mostly 4500–8000 ft.; Yellow Pine F., Red Fir F.; Sierra Nevada from Mariposa Co. n., Coast Ranges from Tehama Co. n.; to Wash., Colo., Ariz. July–Sept. A form from Sonora Pass, Mono Co. to Colo., New Mex., with axils often having leafy shoots and petioles ca. as long as blade, is var. *petiolàris* Gray.

10. **B. Greènei** Gray. [*Coleosanthus G.* Kuntze.] Perennial herb with many stems from a woody base, simple or somewhat branched above, 2–4.5 dm. high, very leafy; herbage viscid and glandular-pilose up to the heads; lvs. alternate, ovate, mostly serrate, 1.5–3 cm. long, 1–2 cm. wide, very short-petioled; heads terminal and solitary or corymbosely arranged, subtended by leafy bracts; invol. 12–14 mm. high; phyllaries linear, acuminate, 2–4-nerved; aks. hispidulous, ca. 7 mm. long; $n = 9$ (Gaiser, 1953).—Mostly dry open rocky places, 2700–8000 ft.; Montane Coniferous F.; from Tehama and Placer cos. n.; to Ore. July–Sept.

11. **B. argùta** Rob. [*B. atractyloides* var. *a.* Jeps.] Much-branched shrub 2–4 dm. high, glandular-pilose; stems zigzag; lvs. bright green, rigid-coriaceous, alternate, ovate, saliently toothed to entire, acute or acuminate, granular-scabrous, 1–2 cm. long, sub-sessile; heads solitary, slender-pedunculate, 13–15 mm. high, the invol. campanulate; outer phyllaries lance-ovate, entire or nearly so, the inner linear-attenuate; aks. 4 mm. long.—Frequent in rocky places, below 4500 ft.; Creosote Bush Scrub, Joshua Tree Wd.; from Darwin and Last Chance Range (Inyo Co.) s. to n. L. Calif. April–May.

Var. **odontolèpis** Rob. Outer phyllaries conspicuously dentate.—W. Colo. Desert.

12. **B. frutéscens** Gray. [*Coleosanthus f.* Kuntze.] Intricately branched aromatic rigid shrub 3–6 dm. high, with divaricate ± spiny cinereous-pubescent slender branches; lvs. alternate, spatulate-oblong, obtuse, entire, pale green, 3–12 mm. long, 1–3 mm. wide, short-petioled; heads solitary, terminal, 13–14 mm. high; phyllaries oblong-linear, 4-striate, obtusish; aks. hispidulous-scabrous, ca. 3.3 mm. long.—Occasional, dry rocky slopes, 2000–3500 ft.; Creosote Bush Scrub; w. edge Colo. Desert; L. Calif. April–June.

13. **B. oblongifòlia** Nutt. var. **linifòlia** (D. C. Eat.) Rob. [*B. l.* D. C. Eat. *B. mohavensis* Gray. *Coleosanthus l.* Kuntze.] Stems many from a branching woody base and forming rounded clumps 2–4 dm. high; plant cinereous-pubescent and somewhat glandular; lvs. ovate-oblong to linear, acute, entire or 1–2-toothed, 10–22 mm. long, sessile or nearly so; heads solitary, 12–16 mm. high, cylindrical, pedunculate; phyllaries linear, 2–4-nerved, imbricate, acute, the outer short and lance-oblong; aks. pilose, eglandular.—Occasional in dry stony places, below 8500 ft.; Joshua Tree Wd., Pinyon-Juniper Wd.; w. Colo. Desert, Mojave Desert, n. to Mono Co.; to Colo., New Mex. May–July.

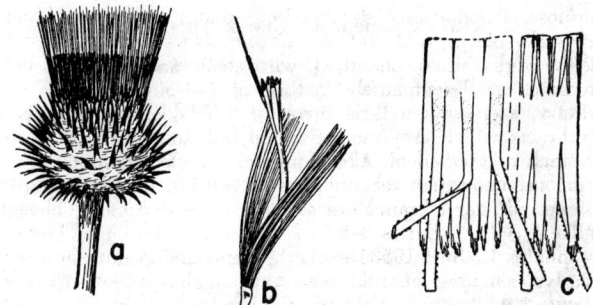

Fig. 118. CYNAREAE. *Cirsium occidentale: a,* discoid head with several series of spine-tipped phyllaries imbedded in loose wool; *b,* fl., × 1, ovary short, pappus capillary, tubular corolla narrow, stamens and style exserted; *c,* stamens, × 10, from outside (left) and inside (right), with papillose fils., the connate anthers caudate at the base.

14. B. incàna Gray. [*Coleosanthus i.* Kuntze.] Globose white-tomentose bush 4–10 dm. high, woody at base; lvs. alternate, sessile, ovate, serrulate to entire, obtuse to acute, 1–3 cm. long; heads solitary, pedunculate, ca. 2.2 cm. high; invol. campanulate; phyllaries imbricated, ovate-oblong, tomentose, the inner narrower; aks. 1 cm. long, cinereous; $n = 9$ (Gaiser, 1953).—Sandy washes and flats, below 5100 ft.; Creosote Bush Scrub, Joshua Tree Wd., Shadscale Scrub; n. Colo. Desert, Mojave Desert n. to Tecopa; w. Nev., Ariz. April–Oct.

Tribe 9. Cynàreae. Thistle Tribe (Fig. 118)

Thistles or thistlelike herbs with alternate often spinose-toothed or -lobed lvs. Heads large, homogamous, the fls. all perfect and regular, or the marginal fls. ♀ or neutral and raylike in a few genera. Phyllaries imbricated in many series, often spinose or scarious or foliaceous at tip. Receptacle mostly bristly, sometimes chaffy, rarely naked. Corollas mostly tubular with 5 long narrow lobes. Anthers with elongated appendage at tip, caudate at base. Style with a thickened often hairy ring below the branches, thence papillate to the obtuse tips, the branches commonly connate below. Pappus bristly or plumose, rarely paleaceous or wanting.—Ca. 35 genera of all continents except S. Am., but concentrated in the Medit. region.

137. Echinóps L. GLOBE THISTLE

Coarse thistlelike herbs, many perennial, with stems and lower lf.-surfaces ± white-woolly; lvs. alternate, usually pinnately toothed or 2–3-pinnatifid, the lobes and teeth prickly. Fls. solitary, each with a little invol. of bristlelike outer bracts and linear to lanceolate inner bracts which are free or united, all united into a dense round head subtended by a small reflexed invol. Aks. elongate, 4-angled to subterete, mostly villous, crowned by an inconspicuous pappus of many short scales. Ca. 100 spp. of Medit. region to cent. Asia; some cult. for their foliage and fls. (Greek, *echinos,* hedgehog, and *ops,* eyed, hedgehoglike.)

1. **E. spherocéphalus** L. Bushy, 1–2 m. high; stems and peduncles gray-woolly and ± viscid-glandular; lvs. sinuate-pinnatifid into ± triangular spinose-dentate lobes, green and roughish above, ± white-arachnoid beneath; heads ca. 5 cm. across, blue or whitish; aks. gray-silky; pappus of ciliate scales united to middle; $2n = 32$ (Poddubnaja, 1927).—Natur. in 1952 at Tulelake, n. Modoc Co.; Eurasian. Summer.

138. Árctium L. BURDOCK

Coarse biennials, branched, with large unarmed rounded or ovate mostly cordate petioled lvs. Heads several to many, many-fld. Fls. all tubular, perfect, similar, pink or purplish. Invol. subglobose. Phyllaries imbricate, coriaceous, appressed at base, attenuate to long stiff points with hooked tips. Receptacle bristly. Aks. oblong, flattened, transversely wrinkled. Pappus short, of many rough bristles, separate and deciduous. Four spp., Eurasian. (Greek, *arction,* a plant name from *arctos,* a bear, because of the rough invol.)

Heads corymbose; larger lf.-blades mostly rounded apically; petioles angled, solid .... 1. *A. Lappa*
Heads subracemose; larger lf.-blades tapering apically; petioles scarcely angled, hollow
2. *A. minus*

1. **A. Láppa** L. Plant 1–2.5 m. tall; petioles to 4 or 5 dm. long, woolly; lf.-blades 1–5 dm. long, 0.5–3 dm. wide, subglabrous above, ± white-tomentose beneath; infl. corymbiform, with leafy bracts; heads 3–4.5 cm. in diam.; invol. glabrous, green, the outer phyllaries ca. 1 cm. long; aks. 6–7 mm. long, fawn-color, often mottled, slightly rugose below the summit; $2n = 32$ (Sugiura, 1936), 36 (Nakajima, 1936).—Occasional weed in widely scattered cismontane Calif.; natur. from Eu. June–Aug.

2. **A. mìnus** (Hill) Bernh. [*Lappa m.* Hill.] Mostly less than 1.5 m. tall; lf.-blades to ca. 3 dm. long, 2.5 dm. wide, often eventually glabrate beneath; branches of infl. ± racemiform; heads 1.5–2.5 cm. in diam., glabrous to glandular or ± arachnoid; aks. 5–6 mm. long, gray or ashy, usually mottled; $2n = 32$ (Wulff, 1937).—Introd. in scattered locations, cent. and n. Calif.; from Eu. June–Oct.

139. Saussurèa DC.

Perennial herbs with alternate entire to toothed or pinnatifid lvs., not spiny. Heads discoid, rather small, the fls. blue or purple, all tubular, perfect. Phyllaries imbricate in several series. Receptacle flat or convex, densely paleaceous-bristly or rarely naked. Aks. glabrous, variously nerved, flattened or thick. Pappus-bristles plumose, united at base and falling together, usually with a short outer nonplumose series. Over 50 spp., largely Eurasian. (Named for Theodore and Horace *Saussure,* Swiss naturalists.)

1. **S. americàna** D. C. Eat. Coarse, rather stout, the stems simple, 5–15 dm. tall, loosely tomentose when young; lvs. thin, 6–15 cm. long, lance-ovate, the lower petioled and ± cordate, the upper narrower and sessile, all dentate or denticulate; infl. corymbiform, rather dense; invol. 10–14 mm. high, narrow; phyllaries firm, pale, with dark edges and tips; fls. ca. 13, violet-purple or rarely white; outer pappus shedding separately, brownish.—Mostly moist places, ca. 5700 ft.; upper Montane Coniferous F.; Siskiyou Mts., Siskiyou Co.; to Alaska. July–Aug.

140. Sílybum Adans. MILK THISTLE

Stout annual or biennial herbs with large prickly sinuate-lobed or pinnatifid mottled lvs. Heads large, solitary, terminal, nodding. Fls. purple. Invol. broadly subglobose, the phyllaries in many rows, stiff, spiny-margined and -tipped. Receptacle fleshy, bristly. Aks. glabrous, shining, flattened; pappus of several series of minutely barbed bristles and falling away from the ak. together because of the united base. Two spp., of Old World. (Old Greek name applied to thistlelike plants.)

1. **S. Mariànum** (L.) Gaertn. [*Carduus M.* L.] Erect, branched, 1–2 m. tall; lvs. 3–7 dm. wide, with clasping bases and wavy or lobed margins bearing many yellow prickles; upper lf.-surface ± mottled with white blotches; heads 2.5–5 cm. broad; phyllaries leathery, the spine 1–2 cm. long, or outer phyllaries mucronate; aks. glabrous, ca. 6 mm. long, spotted brown; pappus shining white; $2n = 34$ (Heiser & Whitaker, 1948). —Common weed in cismontane pastures and waste places; natur. from Medit. region. May–July.

141. Carthàmus L.

Annuals with alternate spinose lvs. Heads terminal, solitary or corymbose. Invol. with spreading leafy outer phyllaries and ± spiny inner ones. Receptacle chaffy. Corolla expanded above the tube. Aks. glabrous, mostly 4-ribbed; pappus lacking or scalelike. Ca. 20 spp., Eurasia. One cult. for the dye from the fls. (Arabic name, alluding to fl. color.)

Fls. yellow; aks. brownish.
  Heads broadly ovoid; phyllaries shorter or little longer than fls. . . . . 1. *C. lanatus*
  Heads narrowly ovoid or oblong, the outer phyllaries attenuate and much longer than fls.
                                          2. *C. baeticus*
Fls. red-orange; aks. white . . . . 3. *C. tinctorius*

1. **C. lanàtus** L. DISTAFF THISTLE. Leafy spiny plant 4–10 dm. high; lvs. rigid, pinnatifid, clasping at base, to ca. 1 dm. long, coarsely sinuate-toothed or pinnatifid with long stout marginal spines; outer phyllaries ending in leaflike appendage; inner rigid, appressed, spine-tipped; fls. yellow with red veins; aks. obpyramidal, 5–6 mm. long, straw color with some deeper brown; pappus-scales of 2 sorts, some of outer short, the inner to ca. 1 cm. long, rigid; $2n = 64$ (Poddubnaja, 1931).—Occasionally adventive, San Francisco, Marin, San Mateo, and Santa Clara cos.; native of Medit. region. July–Aug.

2. **C. baèticus** (Boiss. & Reut.) Nym. [*Kentrophyllum b.* Boiss & Reut. *C. nitidus,* Calif. refs., not Boiss.] Much like *C. lanatus,* but with narrower heads, longer less arachnoid phyllaries.—Occasional weed, from San Diego to San Joaquin and Tuolumne cos.; native of w. Medit. July–Aug.

3. **C. tinctòrius** L. SAFFLOWER. Lvs. oblong to lance-ovate, the upper clasping, minutely and softly spinose-toothed; fls. red-orange; aks. white, shining, ca. 6 mm. long, with a slightly notched scar near base; pappus of numerous narrow scales; $2n = 24$ (Poddubnaja, 1931).—Reported in Antelope V. and Kings, Tulare, San Benito and Sacramento cos.; native of Eurasia. Sometimes cult. as oil seed crop.

142. Cynàra L.

Stout thistlelike perennial herbs, mostly coarse and prickly, with large pinnatifid or bipinnatifid lvs. and very large solitary globose heads. Invol. broad or nearly globose; phyllaries in many series and ± enlarged at base. Receptacle fleshy, plane, bristly. Corolla slender-tubed, violet, blue, or white. Aks. thick, glabrous, compressed or 4-angled, with truncate apex. Pappus of many series of plumose bristles. Ca. 12 spp., of Medit. region and Canary Ids.; 2 grown as vegetables. (Greek, *kuon,* dog, the invol.-spines likened to dog's teeth.)

Lvs. prominently spiny; phyllaries spinose-tipped . . . . 1. *C. Cardunculus*
Lvs. scarcely spiny; phyllaries unarmed . . . . 2. *C. Scolymus*

1. **C. Cardúnculus** L. CARDOON. Robust, freely branched, 5–10(–20) dm. high; stems and under side of lvs. white-tomentulose, the upper side of lvs. grayish-green, ± arachnoid; lvs., 3–6 dm. long, deeply pinnatifid, prominently spiny; invol. 3–4 cm. high; phyllaries spine-tipped; heads purple-fld.; $2n = 34$ (Covas & Schnack, 1947).—Natur. about Benicia and Cordelia, occasional elsewhere; native of Medit. region. Grown for food. May–July.

2. **C. Scòlymus** L. ARTICHOKE. Stem fleshy, branched above, to ca. 1 m. high; lvs. grayish, less pinnatifid, hardly at all spiny; invols. ca. 5 cm. high, the phyllaries unarmed; fls. bluish; $2n = 34$ (Janaki-Ammal.).—Occasional escape from artichoke fields near the coast; native of Medit. region. May–July.

143. Círsium Mill. THISTLE

Annual, biennial or perennial herbs, spiny, with alternate toothed or more usually pinnatifid lvs. and solitary to clustered discoid heads. Receptacle flat to subconic, densely bristly. Invol. with several series of phyllaries, these subequal or usually ± imbricate, generally some or most of them spine-tipped, in some spp. with a thickened glutinous dorsal ridge. Corolla tubular, its segms. linear, white or yellowish to red or purple. Fils. usually papillose-hairy; anthers caudate at base. Style with a thickened minute hairy ring, the branches nearly connate; stigmatic lines marginal, extending nearly to the free tips. Aks. glabrous, flattened or quadrangular, 4–many-nerved; pappus of bristles mostly plumose, sometimes barbellate, deciduous in a ring. Ca. 200 spp., of N. Hemis. Our spp. much needing study; as here presented they seem to be real units but apparently hybridize rather freely where 2 spp. come together, and much introgression occurs. (Greek, *kirsion*, a kind of thistle.)

A. Heads usually unisexual, the plants partly dioecious; perennial from creeping rootstocks; invol.
    1–2.5 cm. high. Introd. weed .... 30. *C. arvense*
AA. Heads with bisexual fls.
    B. Stem spinose-winged by decurrent lf.-bases almost as long as internodes; lvs. scabrous-hispid
        above. Introd. weed .... 1. *C. vulgare*
    BB. Stems not spinose-winged by decurrent lf.-bases or only shortly so; lvs. not scabrous-hispid
        on upper surface.
        C. Middle or outer phyllaries with a conspicuous glutinous dorsal ridge.
            D. Heads many, rather small, the invol. mostly ca. 2–2.5 cm. high.
                E. Spines on the phyllaries mostly not much over 5 mm. long; lvs. not or
                    scarcely decurrent. Mostly cismontane.
                    F. Upper lvs. and bracts fimbriate-spinulose along the margins. Suisun
                        Marshes, Solano Co. .... 16. *C. hydrophilum*
                    FF. Upper lvs. and bracts not closely fimbriate-spinulose.
                        G. Stems and lvs. densely white-tomentose; fls. dark red-purple. N.
                            Coast Ranges to n. Sierra Nevada .... 17. *C. Breweri*
                        GG. Stems and lvs. (at least upper surface of upper lvs.) glabrous; fls.
                            light purplish-rose. Marin Co. .... 18. *C. Vaseyi*
                EE. Spines on the phyllaries mostly ca. 1 cm. long; upper lvs. quite strongly
                    decurrent. Mojave Desert .... 19. *C. mohavense*
            DD. Heads mostly solitary, larger, the invol. ca. 2.5–4 cm. high.
                E. Stems 1–4 dm. high; spines of the phyllaries not more than 5–6 mm. long.
                    Occasional introd. in cismontane s. Calif. .... 15. *C. undulatum*
                EE. Stems 5–15 dm. high; spines of the phyllaries ca. 1 cm. long. At 5000–
                    10,000 ft., n. Inyo Co. .... 20. *C. ochrocentrum*
        CC. Middle or outer phyllaries not bearing a conspicuous glutinous dorsal ridge (except
            possibly in *C. neomexicanum*).
            D. Heads leafy-bracteate (the spiny foliaceous subtending bracts usually more than
                3), usually ± clustered.
                E. Middle phyllaries conspicuously recurved, broadened above the middle; corolla
                    unequally lobed.
                    F. Stems mostly single from the caudex, green; heads permanently nodding;
                        outer phyllaries green, 2–3 cm. long, the spines 3–5 mm. long. Mt.
                        Hamilton Range .... 3. *C. campylon*
                    FF. Stems mostly 3–4 from a caudex, reddish; heads usually erect in ma-
                        turity; outer phyllaries reddish, 1.5–2 cm. long, tipped with a spine
                        1–2 mm. long. San Mateo and San Luis Obispo cos. ... 2. *C. fontinale*
                EE. Middle phyllaries not as above; corolla equally lobed.
                    F. Outer and sometimes middle phyllaries pinnately spinose.
                        G. Phyllaries not dilated above.
                            H. Plant covered with a close feltlike white tomentum; lf.-lobes
                                broad, obtuse, few-spined. Coast at Surf
                                                                        4. *C. rhothophilum*
                            HH. Plant not covered with a feltlike tomentum; lf.-lobes sharper,
                                more spiny.
                                I. Wool on phyllaries, if present, not confined to margins.
                                    Plants of low elevs.
                                    J. Heads mostly arachnoid-woolly. Coast Ranges.
                                                                        11. *C. Andrewsii*
                                    JJ. Heads subglabrous. Lower San Joaquin V.
                                                                        10. *C. crassicaule*

II. Wool on phyllaries confined to margins. Subalpine
23. *C. Eatonii*
GG. At least some of the phyllaries dilated above.
H. Invol. broadly turbinate; phyllaries mostly lanceolate, thin and loose .... 5. *C. remotifolium*
HH. Invol. campanulate; phyllaries mostly dilated above the middle, the outer oblong, imbricated ...... 6. *C. callilepis*
FF. Phyllaries not pinnately spinose.
G. Plants acaulescent, to ca. 3 dm. high.
H. Outer phyllaries brownish-red with yellow margin and base, coriaceous, cuspidate or not. Below 2000 ft., Coast Ranges
7. *C. quercetorum*
HH. Outer phyllaries without a conspicuous yellow margin, chartaceous, prickle-tipped. Mostly above 4000 ft.
I. Stem well developed, with a terminal cluster of few sessile heads exceeded by upper sublinear lvs. Plumas and Siskiyou cos. .... 12. *C. foliosum*
II. Stems low or scarcely developed, with a tuft of rather loosely arranged heads; lvs. mostly wider.
J. Phyllaries mostly striate or longitudinally veined, ending in rather weak spines mostly less than 5 mm. long. Mostly below 8000 ft. Siskiyou and Modoc cos. through Sierra Nevada, s. to San Diego Co.
13. *C. Drummondii*
JJ. Phyllaries not striate or veined, ending in well developed spines ca. 5 mm. long. At 9500–11,000 ft., Sierra Nevada .... 14. *C. tioganum*
GG. Plants mostly 5–20 dm. high.
H. Heads solitary at ends of rather long peduncles, glabrous; plant from a horizontal rootstock ...... 8. *C. Walkerianum*
HH. Heads clustered, ± arachnoid; plants from a taproot
9. *C. brevistylum*
DD. Heads not subtended by leafy bracts, usually not much clustered.
E. Phyllaries essentially glabrous or arachnoid along margins only, not squarrose or reflexed.
F. Outer phyllaries spinose-fimbriate.
G. Invol. broadly turbinate; phyllaries mostly lanceolate, thin and loose
5. *C. remotifolium*
GG. Invol. campanulate; phyllaries mostly dilated above the middle, the outer oblong, imbricated .... 6. *C. callilepis*
FF. Outer phyllaries not spinulose-fimbriate.
G. Stems 3–9 dm. high.
H. Spines of phyllaries well developed, 1.5–2 cm. long. E. Mojave Desert .... 22. *C. nidulum*
HH. Spines smaller.
I. Invol. 2.5–3 cm. high; plants from a horizontal rootstock. Alameda and Sonoma cos. .... 8. *C. Walkerianum*
II. Invol. 3–4.5 cm. high; plants from a taproot. Siskiyou and Trinity cos. to Sierra Nevada .... 21. *C. Andersonii*
GG. Stems 0.3 dm. high, from taproots.
H. Outer phyllaries brownish-red with yellow margin and base, coriaceous, cuspidate or not. From below 2000 ft., Coast Ranges .... 7. *C. quercetorum*
HH. Outer phyllaries without a conspicuous yellow margin, chartaceous, prickle-tipped. From mostly above 4000 ft.
I. Phyllaries mostly striate or longitudinally veined, the weak spines mostly less than 5 mm. long. From below 8000 ft.
13. *C. Drummondii*
II. Phyllaries not longitudinally striate with veins, the spines well developed, ca. 5 mm. long. Subalpine
14. *C. tioganum*
EE. Phyllaries densely arachnoid-tomentose on back, sometimes ± glabrate in age, the middle ones ± squarrose, the outer reflexed.
F. Plants with horizontal underground rootstocks; phyllaries slender, rather loosely imbricated; fls. sordid white to buff. Coast Ranges from San Benito and Monterey cos. n., Sierra Nevada from Plumas Co. n.
24. *C. cymosum*
FF. Plants from a taproot; phyllaries mostly wider.
G. Stems coarse, the peduncles mostly 3–5 mm. in diam.
H. Outer phyllaries much shorter than the inner, their base appressed, the tips spreading or reflexed, ± glabrate; fls. conspicuously exceeding the invol.; herbage ± glabrate in age. Widespread .... 25. *C. Coulteri*
HH. Outer phyllaries not much shorter than the inner, spreading to erect, permanently densely white-tomentose with a loose wool; herbage ± permanently hoary-tomentose.

1. **C. vulgàre** (Savi) Ten. [*Carduus v.* Savi. *Cirsium lanceolatum* auth., not Hill.]
BULL THISTLE. Coarse biennial, spreading, 6–12 dm. tall; rosette-lvs. oblanceolate to
elliptic, coarsely toothed; cauline lvs. lanceolate, to ca. 3 dm. long, deeply pinnatifid
into lanceolate lobes, green and hirtellous above, armed with long fierce prickles, tomentose beneath; lf.-base decurrent on stem as long interrupted prickly wings; heads 1–
few; invol. ovoid to subglobose, 3–4 cm. high, the phyllaries mostly lanceolate to
linear, attenuate to subulate-acerose, spreading; fls. purple, well exserted; $2n = 68$
(Poddubnaja, 1931).—An aggressive weed becoming common in waste places, below
5000 ft., especially in n. and cent. Calif.; from Eu. June–Sept.

2. **C. fontinàle** (Greene) Jeps. [*Cnicus f.* Greene.] Perennial 3–6 dm. high, with
widely spreading branches; stems and upper lf.-surfaces ± glandular-pubescent, lower
surface ± tomentose; basal lvs. 1–2 dm. long, deeply pinnatifid, petioled, the lobes broadly
deltoid, spine-tipped; cauline somewhat reduced, the upper sessile; heads clustered, ±
paniculate, ca. 2.5–3 cm. high, roundish, a little nodding; phyllaries glandular-puberulent,
very broad below, ciliate-fimbriate, spreading or recurved from near the middle and
then drawn out into slender lanceolate tips; fls. dull white, well exserted; aks. smooth,
shining, brown, ± oblong, ca. 5 mm. long.—In wet spots in clay overlying serpentine,
near Crystal Springs Lake, San Mateo Co. June–Oct.
Var. **obispoénse** J. T. Howell. More obviously tomentose especially on under side of
lvs.; phyllaries less pubescent and the cent. ones rarely drawn out into lanceolate tips;
fls. ± pinkish; aks. more turgid, sparsely roughened near apex.—Boggy places near
serpentine, Chorro Creek, San Luis Obispo Co.

3. **C. campÿlon** H. K. Sharsm. Coarse pale green erect perennial 6–20 dm. tall, the
stems arachnoid in age, leafy throughout; lvs. lanceolate, lanate beneath, more glandular-
papillate and less arachnoid above to subglabrate in age; basal lvs. 6–7 dm. long,
somewhat reduced up the stems, shallowly to deeply pinnatifid into irregularly 3–4-
lobed segms., with rounded sinuses and with terminal spines 5–15 mm. long; lvs. sessile
and decurrent with prickly auriculate bases; infl. paniculate, many-headed; heads
nodding, 2.5–3 cm. long; outer phyllaries strongly recurved, narrow-ovate, broad at
base, ciliate or erose near middle; fls. white, ± exserted; aks. 4 mm. long, dark brown,
shining, truncate at apex.—Moist sandy places along streams in serpentine areas, 1000–
2500 ft.; Foothill Wd., Chaparral, etc.; Mt. Hamilton Range, Stanislaus and Santa Clara
cos. May–July.

4. **C. rhothophìlum** Blake. [*Carduus maritimus* Elmer. *Cirsium m.* Petr., not Makino.]
Deep-rooted, apparently not of long duration, with a densely matted white wool and
forming bushy plants 5–8 dm. high; lvs. ± fleshy, 1–3 dm. long, deeply pinnatifid to
spinosely broadly lobed, ± elliptical in outline, only gradually reduced upward, the
uppermost auriculate-clasping; heads 1–few at ends of branches, subtended by leafy
bracts; invol. 3–4 cm. high, densely white-woolly, the phyllaries lanceolate, straight,
long-spined, spinose-ciliate above the middle; fls. white, ± exserted; aks. oblong, brown,
ca. 4 mm. long.—Dunes; Coastal Strand; Surf, Santa Barbara Co. April–June.

5. **C. remotifòlium** (Hook.) DC. [*Carduus r.* Hook. *Cirsium r.* ssp. *pseudocarlinoides*
and var. *odontolepis* Petr. *C. mendocinum* Petr.] Short-lived perennial 5–15 dm. high,
openly but sparsely cymose-branched; stem arachnoid-villous, the branches slender; lvs.
glabrate above to thinly arachnoid-villous, ± tomentose beneath, weakly spiny, lanceolate
to oblong, 1–4 dm. long, pinnately lobed or divided, the lower with divisions divergently 2–3-lobed; heads 1–few at ends of subnaked branches; invol. 2–3 cm. high, the
phyllaries ± arachnoid-villous at margins, loose or partly spreading, narrow, tapering
gradually, the middle and outer spine-tipped, the inner not, but often dilated and
fringed-scarious at the apex; fls. cream to pale brownish; aks. brownish, 5–6 mm. long,

shining.—Dry brushy slopes and open woods, below 5000 ft.; Mixed Evergreen F., Yellow Pine F., etc.; Marin Co. n.; to Wash. June–Sept.

6. **C. callilèpis** (Greene) Jeps. [*Carduus c.* Greene. *Cirsium americanum* and *C. Centaureae* Calif. refs. *C. remotifolium* ssp. *pseudocarlinoides* Petr.] Short-lived perennial 3–12 dm. tall, the stem ± arachnoid; lvs. green, glabrate above, thinly floccose beneath, 1–3 dm. long, well distributed, ± spinose, sinuately toothed to deeply pinnatifid, the upper auriculate-clasping; heads ± rounded, solitary or few at ends of branches; invol. mostly 1.5–2.5 cm. high; phyllaries subglabrous, the outer and middle oblong, ± dilated, scarious and laciniate-fringed toward tip, the outer also spiny at tip; fls. mostly cream to pink; aks. oblong, brownish, ca. 4 mm. long, shining.—Dry ± open slopes, below 4000 ft.; Mixed Evergreen F., Yellow Pine F., etc.; Bay Region to Ore. May–Sept. Apparently hybrids of this sp. with *C. quercetorum* have been described as *C. amblylèpis* Petr.

7. **C. quercetòrum** (Gray) Jeps. [*Cnicus q.* Gray. *Cirsium q.* var. *xerolepis* Petr. *C. q.* var. *mendocinum* Petr.] Low and almost stemless to ca. 3 dm. high, perennial; herbage ± arachnoid when young, usually glabrate in age; lvs. oblong in outline, 1–2.5 dm. long, pinnately parted into lanceolate or oblong often cleft divisions; heads solitary at ends of short branches; invol. campanulate, 2.5–4 cm. high, the phyllaries glabrous, many-ranked, the outer ovate, the inner lanceolate, coriaceous but scarcely lacerate on the edges, all mostly with a short spiny tip; fls. purplish or whitish; aks. pale brown, 3–4 mm. long, shining.—Open places, below 2000 ft.; Mixed Evergreen F., Coastal Prairie, etc.; outer Coast Ranges from Bay Region n. to Mendocino Co. March–July.

8. **C. Walkeriànum** Petr. [*C. quercetorum* var. W. Jeps.] From a horizontal rootstock; stems 3–9 dm. high, ± purplish, subglabrous; lvs. rather permanently whitish tomentose beneath, ± glabrate above, deeply cut into lance-linear ± revolute spine-tipped lobes; heads solitary at ends of rather long peduncles, ovoid or round-ovoid, the invol. 2.5–3 cm. long; phyllaries many ranked, glabrous, scarcely spine-tipped, the inner thin and ± twisted at apex; fls. purplish or rose; aks. brown, ca. 5 mm. long, shining.—Dry slopes, largely Mixed Evergreen F.; Berkeley Hills, Alameda Co., and in a modified form in Sonoma Co. June–July.

9. **C. brevístylum** Cronq. [*C. edule* Calif. refs., not Nutt.] Plant robust short-lived perennial with taproot; stem mostly 1–2 m. high, ± crisp-arachnoid, leafy to the top; lvs. green above, ± arachnoid beneath, oblong to oblanceolate, mostly 5–15 cm. long, shallowly sinuate-pinnatifid, the upper auriculate-clasping, all ± weakly spinulose-ciliate; heads clustered at summit of branches, often exceeded by subtending lvs.; invol. 2–4 cm. high, ± arachnoid, not much imbricate; phyllaries slender, tapering gradually, all but the innermost with short erect spines; fls. dull purple-red or sometimes whitish, not much exserted; aks. 3.5–4 mm. long, finely undulate-marked, ± shiny, yellowish at tip. —Brushy and wooded slopes; Mixed Evergreen F., N. Coastal Scrub, Coastal Sage Scrub, etc.; occasional, coastal s. Calif., more common along the Coast Ranges to B.C., Mont., Ida. April–Aug.

10. **C. crassicáule** (Greene) Jeps. [*Carduus c.* Greene.] Possibly annual, 1–2 m. tall, stout, strongly striate, branched above, with several paniculately disposed heads; herbage permanently hoary-tomentose or glabrescent above; lvs. lanceolate, sinuate-pinnatifid with ovate lobes, spinulose-ciliate; invol. glabrate in age, 1.5–2 cm. high; phyllaries lanceolate, often pinnately spinose above the middle and ending in a long prickle; fls. whitish or pinkish, well exserted; aks. dark, 3.5–4 mm. long.—Shallow water and wet places in fields near San Joaquin R., San Joaquin Co. June–Aug.

11. **C. Andrèwsii** (Gray) Jeps. [*Cnicus A.* Gray. *C. amplifolius* Greene.] Perennial to ca. 1 m. tall, the stems striate, branched above; tomentum mostly deciduous except from invols. and lower surface of lvs.; lower lvs. oblong, 1–4 dm. high, deeply sinuate-pinnatifid into 3-cleft lobes, spinulose on margins; cauline lvs. laciniate-pinnatifid, with lanceolate lobes ending in long spines; heads hemispheric, solitary or ± clustered, sessile to peduncled, subtended by ± reduced lvs.; invol. arachnoid-woolly, 1–2.5 cm. high; phyllaries sometimes pinnately spinose, abruptly contracted upward into an awnlike spine; corollas bright red, sometimes whitish, exceeding the invol.; aks. 4.5–5 mm. long. —Uncommon, moist places; N. Coastal Scrub, Mixed Evergreen F., etc.; bluffs and canyons near the coast from San Mateo Co. to Sonoma Co. June–July.

12. **C. foliòsum** (Hook.) DC. [*Carduus f.* Hook.] Perennial from a taproot, ±

arachnoid-villous; stem ± fleshy, simple, 2–6 dm. high, with a terminal cluster of a few sessile heads; lvs. mostly light green, ± arachnoid above, tomentose beneath, subentire to ± pinnatifid, mostly lance-linear, 0.5–2 dm. long, with rather weak yellow spines, the uppermost sublinear lvs. far surpassing the heads as rather simple erect leafy bracts; heads 3–5 cm. high, the invol. glabrous; phyllaries smoothish, puberulent-ciliate, the outer ovate and with short weak spine, the innermost much elongate, spineless and ± twisted and fimbriate-expanded terminally; corollas pale or white, not exceeding the invol.—Occasional in meadows and similar places; Montane Coniferous F.; Plumas and Siskiyou cos. n.; to B.C., Sask., Rocky Mts. July–Aug.

13. **C. Drummóndii** T. & G. [*Cnicus D.* Gray. *C. D.* var. *acaulescens* Gray. *Cirsium a.* Jeps. *C. D.* ssp. *latisquamum* Petr.] Much like *C. foliosum* and doubtfully specifically distinct; from acaulescent with a tuft of heads centered in a lf.-rosette to ca. 3 dm. high with loosely arranged heads; herbage ± arachnoid-tomentose especially beneath, oblong to oblanceolate, mostly 1–3 dm. long, subentire to deeply sinuate-pinnatifid, spinulose; heads 2.5–4 cm. high; phyllaries thin, mostly plainly longitudinally veined, the outer ovate, tipped with bristlelike spines, the inner ± twisted and innocuous; fls. white to lavender, well exserted.—Common in meadowy and damp places, mostly below 8000 ft.; Montane Coniferous F., Sagebrush Scrub, S. Oak Wd.; cismontane s. Calif., Sierra Nevada to Modoc and Siskiyou cos.; to Canada, Rocky Mts. June–Aug. Variable. Descending below the pine belt mostly in s. Calif.

14. **C. tiogànum** Congd. [*C. Drummondii* ssp. *lanatum* Petr.] Resembling acaulescent forms of *C. Drummondii*, but seemingly distinct in the more deeply pinnatifid and more heavily spinose lvs., often more persistently tomentose; phyllaries more chartaceous and not longitudinally veined when dry, ending in spines 4–5 mm. long.—Edges of meadows, etc., 9500–11,000 ft.; Subalpine F.; Sierra Nevada from Inyo and Tulare cos. to Mono and Eldorado cos. July–Aug.

15. **C. undulàtum** (Nutt.) Spreng. [*Carduus u.* Nutt.] Our plants from a well developed underground rootstock with tuberous thickenings; stems 1–4 dm. high, mostly simple, persistently white-tomentose, as are lower surfaces of lvs.; lvs. coarsely toothed to pinnatifid with ovoid or deltoid lobes; heads mostly several, at branch-ends; invol. 2.5–4 cm. high, the phyllaries ± tomentose on margins, with a dorsal glutinous ridge, well imbricate, the inner attenuate and often crisped at tip; fls. mostly pink-purple, well exserted.—Apparently occasional introd. in fields as at Beaumont, Riverside Co., W. Covina and Puente, Los Angeles Co., and at Otay Mesa, San Diego Co.; native Ariz. and New Mex. to B.C. and Minn. July–Sept.

16. **C. hydróphilum** (Greene) Jeps. [*Carduus h.* Greene. *Cirsium Vaseyi* var. *h.* Jeps.] Perennial, rather slender, freely branched above, 1–1.5 m. high, thinly arachnoid when young, later green and glabrate; lower lvs. to ca. 6 dm. long, deeply pinnatifid, the broad segms. with several spine-bearing lobes; upper cauline lvs. much reduced, deeply laciniate; heads mostly solitary at ends of slender peduncles; invol. ca. 2–3 cm. high, round-ovoid, the phyllaries appressed-imbricate, with a green glutinous ridge on the back and ending in a slender ± spreading spine; fls. pale, well exserted.— Brackish marshes about Suisun Bay, Solano Co. July–Sept.

17. **C. Brèweri** (Gray) Jeps. [*Cnicus B.* Gray. *Cirsium B.* vars. *Wrangelii, canescens,* and *lanosissimum* Petr. *C. triacanthum* Petr?] Apparently biennial, slender, simple below, branched above, 1–2 m. high, mostly ± white-tomentose throughout; lower lvs. oblong, 2–5 dm. long, spinose-dentate to sinuate-pinnatifid, the broad lateral segms. and long terminal segm. tipped with strong spines; upper lvs. mostly elongate-lanceolate, shorter, subentire to deeply pinnatifid; heads solitary to ± crowded; invol. subglobose, 1–2 cm. high, with appressed ± woolly coriaceous phyllaries, each with a glandular spot or ridge and terminal weak spine; fls. dark red-purple to almost white.—Wet places often in serpentine areas, below 5000 ft.; N. Coastal Scrub to Yellow Pine F., etc.; Coast Ranges from Monterey Co. to Del Norte Co., thence to Siskiyou, Modoc, and Inyo cos.; w. Nev., s. Ore. June–Aug.

18. **C. Vàseyi** (Gray) Jeps. [*Cnicus Breweri* var. *V.* Gray.] Like *C. Breweri* but stems and upper surface of lvs. glabrous and light green; corollas light purplish-rose.— Moist places in serpentine areas; Mixed Evergreen F., etc.; Mt. Tamalpais, Marin Co. May–July.

19. **C. mohavénse** (Greene) Petr. [*Carduus m.* Greene.] Short-lived perennial or

biennial, the stem rather stout, 6–15 dm. high, simple or paniculately branched above; herbage white-tomentose or the wool deciduous on upper surface of lvs.; lvs. rather narrow, 1–2.5 dm. long, evenly sinuate-pinnatifid, the lobes tipped with numerous long yellow spines; upper lvs. rather strongly decurrent; invol. oblong, becoming hemispheric, 1.5–3 cm. high, glabrate; phyllaries with dorsal glutinous ridge, straw color and ending in stout spines; fls. pinkish or light red.—Moist alkaline places, 1400–5000 ft.; Creosote Bush Scrub, Alkali Sink, etc.; Mojave Desert. July–Oct.

20. **C. ochrocéntrum** Gray. Stem stout, 5–15 dm. high, from a perennial base, tomentose throughout; lvs. pinnatifid with rather crowded segms. and armed with long yellowish spines; upper lvs. quite decurrent; heads mostly solitary at ends of branches, 4–6 cm. high; invol. of many imbricate lanceolate phyllaries with conspicuous glutinous ridges and yellow spines 1–2 cm. long; fls. red-purple, well exserted.—Dry slopes, 5000–10,000 ft.; Sagebrush Scrub, Pinyon-Juniper Wd., Yellow Pine F.; Inyo Co.; to Nebr., Tex. April–July.

21. **C. Andersònii** (Gray) Petr. [*Cnicus A.* Gray.] Biennial or short-lived perennial, the stem slender, purplish-red, simple or sparsely branched, 4–9 dm. high, nearly naked above; herbage loosely woolly to glabrate; lvs. lanceolate to oblong, mostly 1–2 dm. long, spinose-toothed to deeply pinnatifid with toothed or cleft spiny lobes; heads broadly turbinate, rather few, mostly well peduncled; invol. 3–4.5 cm. high, the phyllaries glabrous save for the ciliate sometimes arachnoid margins, the outer pinnately spinose, the innermost mostly purplish-red, acuminate or tipped with a short spine; corollas rose-purple, exserted.—Dry slopes, 4000–10,500 ft.; Montane Coniferous F.; Sierra Nevada from Tulare Co. n. to Siskiyou and Trinity cos.; to Nev., Ida. June–Oct.

22. **C. nìdulum** (Jones) Petr. [*Cnicus n.* Jones.] Root perennial; stems erect, branched above, ca. 5–8 dm. high, arachnoid-tomentose; basal lvs. 2–3.5 dm. long, deeply and regularly sinuate-pinnatifid, the short lanceolate lobes with very long yellow spines; upper lvs. reduced, sessile, very spiny; heads solitary, ovoid; invol. ca. 3.5 cm. high, the phyllaries glabrous except for the arachnoid margins, with a rather faint glutinous ridge, the lower phyllaries spreading-reflexed, they and the middle ending in yellow spines to 2.5 cm. long, the innermost acuminate, not spine-tipped; fls. light red-purple, well exserted.—Rocky places, 5000–6000 ft.; Pinyon-Juniper Wd.; New York Mts., e. Mojave Desert; n. Ariz. July–Oct.

23. **C. Eatònii** (Gray) Rob. [*Cnicus E.* Gray. *Cirsium Rothrockii*, Calif. refs.] Apparently short-lived perennial, the stem 2–5 dm. high, ± arachnoid, angled; lvs. linear in outline, pinnatifid, with short ovate lobes ending in yellow spines, ± arachnoid or almost glabrate in age especially on upper surface, 5–15 cm. long; heads 1–few, rather crowded, leafy-bracted; invol. 3–3.5 cm. high, ± arachnoid at edges of phyllaries; outer phyllaries spinulose-ciliate, ending in stout yellow spines ca. 5–10 mm. long, the inner lance-subulate; fls. rose-purple, well exserted.—Dry stony slopes, 10,000–11,500 ft.; Subalpine F.; Sierra Nevada of Tulare, Alpine, and Inyo cos.; to Utah, Colo. July–Aug.

24. **C. cymòsum** (Greene) J. T. Howell. [*C. Vaseyi* Jeps., in part *Carduus c.* Greene.] Perennial from underground rootstocks; stem stout, simple or branched above, leafy, 3–12 dm. high; herbage thinly tomentulose to glabrate; lvs. lanceolate to broadly oblanceolate, 0.5–3.5 dm. long, the lower petioled, the upper reduced and sessile to decurrent, most ± pinnately lobed into broad segms. which are ± toothed and end in rather stout spines; heads at ends of rather short branches, sometimes ± paniculately clustered; invol. greenish, arachnoid-tomentose, 3–3.5 cm. high, the subtending bracts spinulose-ciliate, they and outer phyllaries spine-tipped, the inner lance-acuminate with ± spreading scarcely spinose tips; fls. buff to sordid white, well exserted.—Dry slopes, below 5000 ft.; Mixed Evergreen F., Chaparral, Foothill Wd., Yellow Pine F., etc.; Coast Ranges from Monterey and San Benito cos. n. to Siskiyou Co., and from Butte and Plumas cos. to Modoc Co.; Santa Catalina Id.; Ore. May–Aug.

25. **C. Còulteri** Harv. & Gray. [*C. occidentale* var. *C.* Jeps. *C. C.* var. *venustum* (Greene) Jeps. *Carduus v.* Greene.] Stout, from a taproot, mostly 6–12 dm. high; herbage permanently arachnoid-tomentose; stems simple below, ± cymosely few-branched above; lvs. mostly lanceolate, rather shallowly pinnatifid, prickly, the basal 1–3 dm. long, petioled, the cauline gradually reduced up the stems, sessile but scarcely decurrent; heads solitary at ends of long pedunculate branches, large, subglobose, not subtended by lvs.; invol. arachnoid-woolly to glabrate, 4–5 cm. high, the outer phyllaries

much shorter than the inner, the base appressed and the tips spreading or reflexed, outer and middle weakly spine-tipped, innermost scarcely so; fls. conspicuously exceeding the invol., mostly dark crimson, occasionally pale, especially in s. Calif.—Common on dry slopes, below 8500 (10,000) ft.; many Plant Communities; cismontane Calif. n. to Mendocino and Butte cos. and e. slope of Sierra Nevada to Inyo Co.; May–July.

26. **C. occidentàle** (Nutt.) Jeps. [*Carduus o.* Nutt.] Much like *C. Coulteri*, but more heavily coated with a cottony wool; invol. more white-woolly, 4–6 cm. high, the outer phyllaries not so much shorter than the inner and less appressed at base, then spreading; fls. scarcely exceeding invol., red-purple for most part, to ± lavender. —Chiefly sandy places; Coastal Strand; but running into the adjacent hills and apparently hybridizing with *C. Coulteri;* Mendocino Co. to Monterey Co.; San Miguel, Santa Cruz and Santa Rosa ids. April–July.

27. **C. pastòris** J. T. Howell. [*Carduus candidissimus* Greene. *C. occidentalis* var. *c.* Hall. *Cirsium c.* Davids. & Mox., not Damner.] Much like *C. occidentale,* but herbage more snowy white with a feltlike persistent tomentum; heads mostly longer than thick; outer and middle phyllaries rigid, long; fls. long-exserted.—Dry slopes, below 5600 ft.; many Plant Communities; Humboldt and Del Norte cos. to Modoc, Plumas and Nevada cos.; s. Ore., w. Nev. June–Sept.

28. **C. califórnicum** Gray. [*Carduus c.* Greene. *C. lilacinus* and *neglectus* Greene.] Mostly biennial, slender, 5–18 dm. high, leafy near the base, cymosely branched above, white-woolly, ± glabrate in age; lvs. narrow, 1–3.5 dm. long, sinuately to deeply pinnatifid, moderately spinose; heads mostly solitary on the long slender peduncles; invol. hemispherical, ± arachnoid-woolly, 2–3 cm. high; outer phyllaries with spreading terminal spines, the inner erect and more herbaceous; fls. lavender to whitish, well exceeding invol.—Common on dry slopes, in washes, etc., below 7000 ft.; Coastal Sage Scrub, S. Oak Wd., Foothill Wd., Chaparral, Yellow Pine F., etc.; cismontane and montane s. Calif. to Placer and Contra Costa cos. April–July. Rather low less spinose plants from s. Calif. have been called var. *bernardinum* (Greene) Petr. [*Carduus b.* Greene. *Cirsium c.* ssp. *pseudoreglense* Petr.]

29. **C. neomexicànum** Gray. Biennial, stout, 1.5–2 m. tall, white-woolly; basal lvs. narrow to ± elliptic, deeply and regularly sinuate-pinnatifid, strongly spiny, to ca. 4 dm. long; cauline lvs. gradually reduced up the stem, quite strongly decurrent; heads 4–5 cm. broad; invol. 2.5–4 cm. high, subglobose, sparsely arachnoid, glabrate; outer phyllaries reflexed, the middle squarrose, strongly spinose, innermost attenuate, scarcely spine-tipped; fls. whitish, well exserted.—Dry rocky places, 3500–6000 ft.; Pinyon-Juniper Wd., Joshua Tree Wd.; mts. of e. Mojave Desert; to Colo., New Mex. April–May.

30. **C. arvénse** (L.) Scop. [*Serratula a.* L.] CANADA THISTLE. Perennial with creeping rootstocks; stems 3–9 dm. high, slender; lvs. oblong or lanceolate, subrigid, smooth or slightly woolly beneath, at length green on both sides, strongly sinuate-pinnatifid, prickly margined, 5–12 cm. long; heads imperfectly dioecious, rather numerous, the invol. 1.5–2 cm. high; outer phyllaries appressed, ovate, subulate-tipped, the inner with thin attenuate tips; fls. mostly pinkish-purple, not much exserted; $2n = 34$ (Ehrenberg, 1945).—Natur. in low places; native of Eu. June–Sept.

144. Cárduus L. PLUMELESS THISTLE

Much like *Cirsium,* but pappus-bristles merely barbellate or smoothish. Plants biennial. Lvs. conspicuously decurrent, spiny. Large genus of Eurasia and n. Afr. (Ancient Latin name.)

Heads 1–2 cm. broad; invol. 1.5–2 cm. high.
 The heads usually 1–5 at ends of branches; phyllaries with small rough hairs on margin and back
                            1. *C. pycnocephalus*
 The heads usually 5–20 at the ends of the branches; phyllaries glabrous or subciliate
                            2. *C. tenuiflorus*
Heads 4–6 cm. broad; invol. 2.5–3 cm. high .... 3. *C. nutans*

1. **C. pycnocéphalus** L. ITALIAN THISTLE. Annual 3–18 dm. high, the stems slender, narrowly spiny-winged especially below; lvs. pinnatifid, to ca. 12 cm. long, the lobes and teeth spine-tipped, white-tomentose beneath, greenish but ± arachnoid above; heads

subcylindric; phyllaries not membranous-margined, ± persistently floccose-tomentose; corolla-lobes ca. 3 times as long as the throat, rose-purple; aks. light tan or buff, ca. 20-nerved, 5–6 mm. long; pappus 15–20 mm. long, sordid.—Weed along roadsides, waste places, etc.; Sonoma Co. to Santa Barbara Co., San Diego Co.; natur. from s. Eu. May–July.

2. **C. tenuiflòrus** Curt. Like the preceding, but stems more definitely spiny-winged in the upper parts; phyllaries ± membranous-margined, very scantily tomentose; corolla-lobes 1.5–2.5 times as long as the throat; aks. gray-brown, usually 10–13-nerved; pappus 1–1.5 cm. long.—Natur. from Humboldt Co. to San Benito Co.; from s. Eu. May–July.

3. **C. nùtans** L. var. **leiophýllus** (Petrovic) Arènes. [*C. l.* Petrovic.] MUSK THISTLE. Biennial 4–10 dm. high; lvs. bipinnately lobed, to 2 dm. long; heads solitary, hemispherical, nodding, on long naked peduncles; outer phyllaries reflexed; fls. purplish; aks. shining, ca. 5 mm. long; pappus ca. 2 cm. long; $2n = 16$ (Poddubnaja, 1931).—Walnut, e. Los Angeles Co. as a weed; native of Eu. June–July.

145. Centaurèa L. STAR THISTLE

Annual to perennial herbs with alternate entire dentate incised or pinnatifid lvs. and large or middle-sized heads of tubular purple, violet, pinkish, white, or yellow fls., the outer sometimes enlarged. Invol. ovoid, globose to subcylindric, the phyllaries imbricated, appressed, fimbriate or dentate to entire, sometimes spine-tipped. Receptacle flat, bristly. Marginal fls. usually sterile; cent. perfect and fertile. Aks. ± compressed, usually smooth and shining, attached obliquely or laterally at or near the base. Pappus setose or partly chaffy or none. Ca. 500 spp., mostly of Old World, many cult. for the fls. (*Centaurie*, ancient Greek name, without clear application.)

A. Phyllaries fringed but not definitely spiny at tip.
  B. Plant annual with flocculent pubescence and simple linear or lanceolate lvs. .. 1. *C. Cyanus*
  BB. Plants perennial or biennial.
    C. Lvs. pinnatifid into narrow segms.
      D. Plant hoary with close white wool; middle phyllaries round-ovate, almost as wide
        as long .... 2. *C. Cineraria*
      DD. Plant greenish, ± flocculent; middle phyllaries oblong, much longer than broad
                                      3. *C. maculosa*
    CC. Lvs. entire.
      D. Invol. whitish green, ca. 1 cm. high, the outer phyllaries entire, the inner long-
        pubescent; pappus 5–7 mm. long .... 4. *C. repens*
      DD. Invol. brownish to darker at least near the tips, mostly 1.5–2 mm. high;
        phyllaries with lacerate margins near their apex; pappus to 1 mm. long.
        E. Innermost phyllaries irregularly toothed or lacerate; pappus lacking
                                      5. *C. Jacea*
        EE. Innermost phyllaries regularly pectinate or toothed; pappus present, to ca.
           1 mm. long .... 6. *C. nigra*
AA. Phyllaries ending in definite terminal, but sometimes short, spines.
  B. Terminal spine of phyllary less than 5 mm. long; fls. purplish to white.
    C. Phyllaries entire, not fringed.
      D. Invol. glabrous; pappus-parts in 2 series .... 7. *C. salmantica*
      DD. Invol. villous; pappus-parts in several series, the outer paleaceous .. 8. *C. muricata*
    CC. Phyllaries deeply fringed toward apex.
      D. Invol. cylindric, less than 1 cm. long .... 9. *C. virgata*
      DD. Invol. round-ovoid, 1.5–2 cm. long .... 10. *C. diluta*
  BB. Terminal spine of phyllary more than 5 mm. long.
    C. Fls. purplish or pinkish; stems wingless.
      D. Aks. with definite pappus .... 11. *C. iberica*
      DD. Aks. without pappus or with merely a vestige .... 12. *C. Calcitrapa*
    CC. Fls. yellow; stems winged with decurrent lf.-bases.
      D. Pappus black, copious; spines of phyllaries blackish at base.
        E. Invol. nearly glabrous; cent. spine of middle phyllaries 2–2.5 cm. long
                                      13. *C. sulphurea*
        EE. Invol. arachnoid-tomentose; cent. spine of middle phyllaries 1–2 cm. long
                                      14. *C. eriophora*
      DD. Pappus whitish.
        E. Spines slender, purplish, 1 cm. long or less; plants branched mostly above
          the base; corolla glandular .... 15. *C. melitensis*
        EE. Spines stout, yellow, 1–2 cm. long; plants branched from the base;
          corolla without glands .... 16. *C. solstitialis*

1. **C. Cyànus** L. BACHELOR'S BUTTON. CORNFLOWER. Slender-stemmed annual 3–6 dm. high, with long ascending branches ending in solitary heads; lightly flocculent-

tomentose when young; invol. ovoid, ca. 1.5 cm. long, of ca. 4 unequal series of pale scarious-fimbriate phyllaries; fls. deep purplish-blue to pink or white, the marginal enlarged and raylike; aks. metallic pale blue, ca. 4–5 mm. long, notched on 1 side at base, the pappus-bristles unequal, shorter than the ak.; $2n = 24$ (Fritsch, 1935).—Escaped from gardens and natur. at widely scattered stations, cismontane Calif.; native of Eu. May–Aug.

2. **C. Cinerària** L. DUSTY MILLER. Erect branching perennial 3–10 dm. high, closely white-tomentose throughout; lvs. pinnately parted into narrow obtuse lobes; heads rather large, the invol. round-ovoid, ca. 1.5–2 cm. long; phyllaries with a membranous black margin, long-ciliate, the apical bristle thicker than the others; fls. purple, the marginal slightly enlarged; pappus copious, white, ca. 7–8 mm. long.—Occasional escape from gardens; from Eu. June–Sept.

3. **C. maculòsa** Lam. Biennial with slender stems 3–9 dm. high; lvs. canescent, the lower pinnately divided into linear or lanceolate segms., the upper simpler; heads 1.5–2.5 cm. broad, terminal; invol. pale, 1–1.4 cm. high, the phyllaries with 5–7 pairs of cilia and a dark tip; fls. whitish to pink or purplish, the marginal somewhat enlarged; pappus whitish, copious.—Occasional escape; native of Eu. Summer.

4. **C. rèpens** L. [*C. Picris* Pall.] RUSSIAN KNAPWEED. Perennial with creeping rootstocks, branched, 3–10 dm. tall, arachnoid-tomentose; lvs. firm, the basal oblong, sinuate-pinnatifid, 4–10 cm. long, the cauline numerous, linear-lanceolate to -oblong, entire, 2–3 cm. long; heads peduncled, oblong-cylindric; invol. 10–14 mm. high, the phyllaries roundish, entire, scarious-margined, the inner hairy at tip; fls. blue to pink or white, the marginal enlarged; pappus-bristles in several ranks, deciduous, barbellulate; $2n = 26$ (Heiser & Whitaker, 1948).—Becoming increasingly widely natur.; from the region of the Caucasus. May–Sept.

5. **C. Jàcea** L. BROWN KNAPWEED. Perennial 3–10 dm. high, rather simple; lower lvs. lanceolate, entire to pinnatifid, petioled, to ca. 1 dm. long; cauline scattered, lanceolate, entire to coarsely few-toothed; heads showy, the subglobose invol. 1.5–2.5 cm. high; phyllaries with brown scarious concave entire to lacerate margins; fls. rose-purple, the outer enlarged; $2n = 44$ (Roy, 1937).—Occasionally natur.; introd. from Eu. Summer.

6. **C. nìgra** L. BLACK KNAPWEED. Coarse perennial; stems 5–8 dm. high, branching; rosette-lvs. lance-elliptic to oval, ± sinuate, to ca. 1.5 dm. long; cauline gradually reduced up the stem; invol. subglobose, 1.5–2 cm. high, the outer and middle phyllaries deep brown to black and fringed on upper margins, the innermost paler, lacerate; corollas rose-purple; pappus blackish, scarcely 1 mm. long.—Occasionally natur.; introd. from Eu. July–Aug.

7. **C. salmántica** L. [*Microlonchus s.* DC.] Perennial or annual 2–8 dm. high, ± villous below, subglabrous above; basal lvs. lance-ovate, lyrate-pinnatifid into dentate lobes, the terminal lobe much the largest; upper lvs. entire, reduced; stem-branches long, slender, loosely paniculate, ending in single heads; invol. ovoid, ca. 1.5 cm. high, glabrous, the phyllaries imbricate in many ranks, ovate, entire, each with a spine to ca. 1 mm. long; fls. violet or white; aks. fuscous, shining, oblong, transversely rugose, the pappus forming a short crown.—Long ago collected in Sonoma Co.; from Medit. region. Aug.–Sept.

8. **C. muricàta** L. Much like *C. salmantica* in habit, but the invol. villous-pubescent; terminal spine on phyllary ca. 3–4 mm. long; aks. dark, ca. 3 mm. long, striate and with transverse pits between, the pappus of outer imbricate scales and inner white bristles.—Occasional escape at Santa Barbara; native of Spain. May–June.

9. **C. virgàta** Lam. var. **squarròsa** (Willd.) Boiss. [*C. s.* Willd.] Perennial, 5–9 dm. high, virgately branched above; basal lvs. petioled, parted into linear lobes; cauline lvs. sessile, pinnately lobed; uppermost small, entire; heads on short peduncles, few-fld.; invol. slender urn-shaped, 8–10 mm. long; phyllaries with a terminal spine ca. 1.5–2 mm. long and 5–6 shorter lateral ones; fls. pink, with slender lobes; aks. ca. 3 mm. long, olive-drab to dull brown, longitudinally striate with straw-colored lines; pappus persistent, of bristles to ca. ½ the ak.-length.—Fields and waste places, Big V., Lassen Co.; native of Eurasia. June–Aug.

10. **C. dilùta** Ait. Lvs. with scabrous margins, the lower oblong, dentate, the upper lance-oblong, entire, decurrent; heads on well formed peduncles; invol. conic-ovoid, 1.5–2 cm. long, subglabrous; phyllaries lacerate on margins, the outer and middle with

a stout terminal spine 1–2 mm. long; fls. pinkish or pale violet, ca. 2 cm. long, or the outer much longer; pappus white, capillary.—Occasionally natur. as at East Whittier and Watts in Los Angeles Co. and on dunes, San Francisco; from Medit. region. May–July.

11. **C. ibèrica** Trev. Perennial 5–10 dm. high, ± cobwebby-pubescent; basal lvs. deeply sinuate-lobed, 1.5–2.5 dm. long, the cauline remote, reduced, few- and narrow-lobed; heads in open corymbose panicle; invol. ovoid, ca. 1.5 cm. high, the phyllaries green-chartaceous, ovate at base, abruptly narrowed into widespread stout spines 1–2 cm. long with traces of short spines near their base; fls. purplish-pink toward the tips, occasionally whitish; aks. 3–4 mm. long, straw color to grayish or mottled with dark brown, with ca. 3 rows of flattened barbellulate bristles; $2n = 16$ (Poddubnaja, 1931).—Occasionally natur. as near Santa Rosa, Solvang, San Diego; native of Asia Minor. Aug.–Nov.

12. **C. Calcítrapa** L. Annual or biennial, divaricately branched, 3–6 dm. high; lvs. pubescent, the basal rosulate, pinnately divided into linear-lanceolate toothed or incised segms.; cauline reduced, simpler; heads sessile or nearly so, on leafy branches; invol. ovoid, smooth, ca. 1.5 cm. high without the spines; several of the subcoriaceous phyllaries tipped with divergent to reflexed stiff spines 1–2.5 cm. long; fls. purplish; aks. ca. 3 mm. long, straw color, mottled with dark brown, without pappus; $2n = 20$ (Vignoli, 1945).—In waste places and uncult. lands, especially in Solano Co., but at scattered stations from San Diego Co. to Humboldt Co.; native of Eu. July–Oct.

13. **C. sulphùrea** Willd. [*C. sicula* Calif. refs.] Erect annual 3–8 dm. high, leafy below; lvs. 1–2 dm. long, on lower parts of plant, irregularly shallowly toothed, ± pubescent, decurrent; uppermost much reduced; peduncles long; invol. 1.5–2.5 cm. long, conic-ovoid, straw color except for the blackish-purple base of the spines, these palmately arranged at the apex of the phyllary, the cent. one to almost 2 cm. long, spreading to reflexed, the lateral ones ca. 3 on a side, much shorter; fls. yellow; ak. ca. 10 mm. long, shiny, dark brown, with a bristly black pappus.—Known for the past 30 years from near Folsom, Sacramento Co.; from Medit. region. June–July.

14. **C. erióphora** L. Plant ± arachnoid; appendages on middle phyllaries pinnately spinose.—Collected in 1909 in Highland Park, Los Angeles; native of Medit. region. May.

15. **C. meliténsis** L. Tocalote. Annual 3–8 dm. high, erect, commonly much-branched, grayish-pubescent, the stems winged by the decurrent lvs.; basal lvs. lyrate, 5–12 cm. long, with obtuse lobes; upper lvs. narrow, entire; heads solitary or 2 or 3 together; invol. ovoid, ca. 1 cm. high, arachnoid, the phyllaries rigid, the outer with palmatifid spining, the inner and middle with a rigid spine 4–8 mm. long; fls. yellow; aks. ca. 2.5 mm. long, grayish, with pappus-bristles in ca. 3 rows; $2n = 22$ (Covas & Schnack, 1947).—Common weed in grain fields, pastures, along roadsides, etc.; much of Calif.; introd. from Eu. May–June.

16. **C. solstitiàlis** L. Barnaby's Thistle. Cottony-pubescent annual, branched from the base, 3–7 dm. high; basal lvs. 5–8 cm. long, deeply lobed; the upper to ca. 3 cm. long, entire, narrow, decurrent on the thus winged stems; invol. globose-ovoid, 14–18 mm. high; lower phyllaries with 3-pronged spines, middle with simple stout spines 6–20 mm. long, uppermost spineless; fls. bright yellow; aks. light-colored and with white pappus-bristles or dark and without pappus; $2n = 16$ (Heiser & Whitaker, 1948).—Widely distrib. weed in Calif.; native of Eu. May–Oct.

146. Cnìcus L. Blessed Thistle

Annual loosely pubescent herb with sinuate-dentate or pinnatifid lvs. having spiny teeth or lobes. Heads large, sessile, of yellow tubular fls., solitary, terminal, subtended by the upper lvs. Phyllaries imbricated in several series, the outer ovate, the inner lanceolate, tipped by long pinnately branched spines. Receptacle bristly. Aks. terete, striate, laterally attached, 10-toothed at apex. Pappus of 2 series of awns, the outer long, naked, yellow, the inner white, hispidulous. One sp. (Latin name of the Safflower, from the Greek *cnecos*.)

1. **C. benedíctus** L. Branched, 2–6 dm. high; lvs. oblong-lanceolate in outline, conspicuously veiny, 6–15 cm. long; invol. 3–4 cm. high; $2n = 22$ (Vaarama, 1947).—

Occasional weed in waste places, fields, etc.; cismontane Calif.; natur. from Eu. April–July.

Tribe 10. Arctotídeae. ARCTOTIS TRIBE

Herbs or shrubs with lvs. mostly alternate or basal and often ± spiny. Heads usually radiate. Phyllaries imbricated in several series, separate or united at base, broadly scarious to very acute or spinescent at apex. Receptacle naked or sometimes chaffy. Disk-corollas tubular, regular. Anthers not tailed at base. Style-branches generally connate to near tip, papillate, not appendaged, the style thickened just below. Aks. often thick, without pappus, or the pappus paleaceous or coroniform.—Ca. 11 genera, centering in S. Afr.

A. Phyllaries free; plants without latex.
   B. Aks. usually villous; pappus of hyaline often convolute scales .... 147. *Arctotis*
   BB. Aks. glabrous; pappus none or a crown of minute scales .... 148. *Venidium*
AA. Phyllaries grown together at base; plants with latex .... 149. *Gazania*

147. Arctòtis L.

Perennial herbs with ± white-woolly herbage. Lvs. alternate. Heads of large or medium size, solitary, long-peduncled, with rays. Phyllaries imbricate in several series, the inner scarious. Receptacle honeycombed, mostly bristly. Aks. grooved, usually villous, ours with a tuft of silky hairs at the base; pappus crownlike or composed of scales. Ca. 30 spp., mostly of S. Afr.; a few grown as ornamentals. (Greek, *arktos,* bear, and *otis,* ear, referring to the pappus-scales.)

1. **A. stoechadifòlia** Berg. var. **grándis** (Thunb.) Less. [*A. g.* Thunb.] AFRICAN DAISY. Perennial but often grown as an annual, bushy, 4–7 dm. high; stems leafy; lvs. 8–15 cm. long, obovate-oblong, toothed, concolorous; peduncles 1.5–3 dm. long; heads 5–7 cm. across; outer phyllaries tomentose; rays white, but violet on outside; $2n = 18$ (Bilquez, 1951).—Occasional escape from cult.; native of S. Afr. June–Aug.

148. Venídium Less.

Annual or perennial herbs, ± tomentose, much like *Arctotis,* but pappus lacking or of 4 minute scales. Ca. 20 spp., of S. Afr. (Name of obscure application.)

1. **V. fastuòsum** (Jacq.) Stapf. [*Arctotis f.* Jacq.] MONARCH-OF-THE-VELD. Annual 3–8 dm. high, branched from base, cobwebby when young; lvs. oblanceolate or obovate, 5–8 cm. long, irregularly lobed to sublyrate, the lower on petioles to 5 cm. long; heads 7–10 cm. across; rays bright orange with dark purplish-brown zone at base; disk yellowish-brown; $2n = 18$ (Bilquez, 1951).—Occasional escape from cult.; native S. Afr. March–May.

149. Gazània Gaertn.

Mostly perennial herbs with short leafy stems or lvs. crowded in a basal tuft, entire or pinnatifid. Heads radiate, fairly large, solitary on long peduncles. Phyllaries in 2–several rows, united and cuplike at base. Rays white, yellow, orange, or scarlet, closing at night. Receptacle pitted. Aks. villous, the pappus of 2 series of delicate scarious toothed scales, often hidden in the ak.-wool. Ca. 25 spp., of S. Afr. (Named for T. *Gaza,* 15th-century Italian scholar.)

1. **G. longiscàpa** DC. Subacaulescent; lvs. variable, lanceolate, entire to pinnately cut, subglabrous and green above, white-tomentose beneath, ± revolute; heads 3–8 cm. across, on glabrous pedicels exceeding lvs.; rays yellow, often with black spot at base; invol. glabrous, the phyllaries finely pointed, equal to or longer than the tube, the outer ones ciliate.—Occasional escape from cult.; native of S. Afr. Many months.

Tribe 11. Mutisìeae. MUTISIA TRIBE (Fig. 119)

Herbs, shrubs, lianas or small trees with alternate lvs. Heads homogamous, the corollas all regular and deeply lobed, or more commonly some or all bilabiate, or the

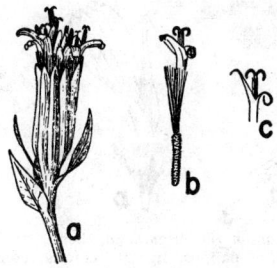

Fig. 119. MUTISEAE. *Trixis californica: a,* head, × 1, discoid; *b,* fl., × 1, with 2-lipped corolla shown also in *c.*

heads heterogamous with ligulate ray-fls. Phyllaries usually many-seriate, imbricate. Receptacle usually naked. Anthers long-tailed at base, their tips also elongated. Style-branches not appendaged, usually short and blunt, without node below. Pappus of simple or plumose bristles or narrow paleae.—Ca. 60 genera, chiefly of Mex. and S. Am.

A. Heads many-fld., not capitate-clustered.
  B. Plants shrubby; corollas yellow; invol. in only 2 distinct rows . . . . 150. *Trixis*
  BB. Plants herbaceous; corollas pink or whitish; invol. well imbricated . . . . . . 151. *Perezia*
AA. Heads 1-fld., capitate-clustered and surrounded by spiny-toothed semitranslucent bracts
        152. *Hecastocleis*

150. Tríxis P. Br.

Shrubs or herbs with alternate lvs. that are lanceolate, entire to denticulate, densely sessile-glandular beneath. Heads 9–12-fld., yellow, solitary or not, in ours in corymbosely branched leafy clusters. Invol. double, the outer series of phyllaries several, linear to elliptic, herbaceous; inner series ca. 8, linear, acuminate, corky-thickened at base. Corollas all 2-lipped, the outer lip 3-toothed, the inner 2-cleft. Aks. densely hispidulous, tapering or ± beaked, 5-ribbed. Pappus of many straw-colored soft bristles. Ca. 40 spp., ranging from sw. U.S. to Chile. (Greek, *trixos,* 3-fold, because of the 3-cleft corolla-lip.)

1. **T. califórnica** Kell. Erect, bushy, to ca. 1 m. high, leafy to the heads, green, minutely puberulent; lvs. lanceolate, narrow at base, acute, 2–5 cm. long, remotely shallowly denticulate, subsessile; invol. ca. 15 mm. high, subtended by leafy bracts; corolla ca. 13–17 mm. long; aks. brown, glandular, ca. 11–12 mm. long, narrowed and beaklike at summit; pappus sordid, ca. 1 cm. long.—Frequent in canyons, washes, etc., below 3000 ft.; Creosote Bush Scrub; Colo. Desert, n. to Sheephole Mts., e. San Bernardino Co.; to w. Tex., n. Mex. Feb.–April.

151. Perèzia Lag.

Perennial herbs, ours branched and with leafy stems. Lvs. alternate, sessile and often clasping, usually spinulose-toothed. Heads many, cymose-paniculate in ours. Receptacle flat, usually naked. Invol. imbricated. Fls. all perfect, bilabiate, the upper lip deeply 2-lobed, the lower broad, 3-toothed at apex. Aks. subcylindric or fusiform, densely glandular or glandular-hispidulous. Pappus of many scabrous bristles. Perhaps 70 spp. of warmer parts of Am. (L. Perez, 16th-century medical botanist of Toledo.)

1. **P. microcéphala** (DC.) Gray. [*Acourtia m.* DC.] Rather stout, ca. 1 m. high; lvs. scabrous-puberulent and minutely glandular, 1–2 dm. long, 3–8 cm. broad, sessile by broad clasping base, denticulate; panicle 3–5 dm. long, 1–3 dm. broad; invol. 7–9 mm. high, the phyllaries oblong, abruptly acuminate or mucronate; corollas rose to white, ca. 1 cm. long; pappus white, soft, ca. 8 mm. long.—Common on dry slopes, below 4000 ft.; Chaparral; San Luis Obispo Co. to San Diego Co.; Channel Ids. June–Aug.

Fig. 120. CICHORIEAE *Stephanomeria virgata: a,* head, $\times$ 1, phyllaries few, fls. all ligulate; *b,* fl., $\times$ 2½, inferior ovary, plumose pappus, ligulate corolla, connate (syngenesious) anthers, style; *c,* ak., $\times$ 2½.

152. Hecastocleis Gray

Low rounded subshrub, glabrous, with rigid branches and rigid lvs., the cauline broadly linear, cuspidate and with spinose teeth especially in the primary ones, less spinose in the axillary fascicled lvs.; uppermost lvs. oval or ovate and forming a loose envelope around the heads, reticulate, semitranslucent, sparsely beset with slender marginal prickles. Heads 1-fld., in a fascicle. Invol. cylindraceous, of several linear-lanceolate rather rigid cuspidate-acuminate phyllaries. Fl. perfect; corolla with linear spreading lobes. Aks. glabrous. Pappus coroniform, laciniate-dentate, corneous. One sp. (Greek, *ekastos,* each, and *kleio,* to shut up, each fl. in its own invol.)

1. **H. Shóckleyi** Gray. Plants 4–6 dm. high and somewhat broader; cauline lvs. 1.5–3.5 cm. long, the primary longer and spinier than the secondary fascicled ones; floral envelope 1.5–2.5 cm. long, chartaceous; invol. ± woolly; fl. reddish-purple within before opening, greenish-white at anthesis.—Dry slopes, washes, etc., 4000–7000 ft.; Creosote Bush Scrub, Shadscale Scrub, etc.; Death V. region and e. of Inyo Mts.; w. Nev. May–June.

Subfamily II. Liguliflòrae

Tribe 12. Cichorìeae. Chicory Tribe (Fig. 120)

Herbs (seldom shrubs or trees) with milky juice and alternate or basal lvs. Heads ligulate, the fls. all perfect and with a 5-toothed commonly yellow ligule. Phyllaries subequal, 1–2-seriate, often calyculate at base, less often imbricate in several series. Anthers with thin short appendage at apex, sagittate at base, not tailed. Style-branches semicylindric, very narrow, essentially smooth, not appendaged, stigmatic throughout their length on the inner face.—Widely distributed, especially in n. temp. regions, with ca. 65 genera.

A. Plants thistlelike; oil-ducts present; receptacle with broad chaffy paleae enclosing the aks.; pappus of few coarse bristles. (Subtribe Scolyminae) .... 153. *Scolymus*
AA. Plants not thistlelike; oil-ducts absent.
    B. Herbage usually coarsely hairy; fls. blue (in ours) or yellow; aks. short, turbinate or columnar, truncate at apex; pappus paleaceous or none; pollen pale yellow; style-branches mostly elongate. (Subtribe Cichoriinae) .... 154. *Cichorium*
    BB. Not in all respects as above.
        C. Mostly low acaulescent glabrous plants, rarely glandular, never hirsute; fls. yellow; pappus of paleae, awns, setae, or absent; pollen bright orange; style-branches short and blunt, rather broad. (Subtribe Microseridinae)
            D. Pappus present.
                E. Aks. not beaked; pappus (at least in part) broad at base, tapering into bristlelike awns (bristlelike and brown in *Microseris borealis*).
                    F. Scapose perennials; heads erect; invol. white-villous (if hairy); phyllaries subequal .... 155. *Nothocalais*
                    FF. Annuals, or if perennial, then scapose or branched; heads nodding in bud; invol. appressed-black-tomentose; outermost phyllaries much shorter than the inner .... 156. *Microseris*
                EE. Aks. beaked; pappus of white capillary bristles, never plumose
                                        157. *Agoseris*
            DD. Pappus none; aks. not beaked .... 158. *Phalacroseris*
        CC. Not in all respects as above.
            D. Annuals to shrubs, mostly glabrous, some appressed-tomentose, none hirsute with

spreading hairs; fls. pink, white, or sometimes yellow; aks. cylindrical, fusiform, or beaked, never flattened; pappus setose (none in *Atrichoseris*); style-branches mostly short. (Subtribe Stephanomeriinae)

E. Aks. pappose.
    F. Plants nonscapose, usually much branched; lvs. ± reduced in size; heads narrow, few-fld. (except in *Rafinesquia*).
        G. Pappus not plumose.
            H. The pappus of 5 rigid tapering awns; perennial
                159. *Chaetadelpha*
            HH. The pappus of many capillary bristles; annual or perennial
                160. *Lygodesmia*
        GG. Pappus plumose at least above.
            H. Aks. beakless; fls. rose or flesh-colored; annual or usually perennial .... 161. *Stephanomeria*
            HH. Aks. beaked; fls. white or nearly so; fistulous-stemmed annual
                162. *Rafinesquia*
    FF. Plants scapose or subscapose; heads many-fld.; annuals (except a few spp. of *Malacothrix*).
        G. Pappus of capillary bristles, never plumose, quickly deciduous.
            H. Aks. beakless, truncate .... 163. *Malacothrix*
            HH. Aks. beaked.
                I. Infl. not glandular; aks. abruptly short-beaked, truncate
                  164. *Glyptopleura*
                II. Infl. with tack-shaped glands; aks. tapering into the shallowly cup-tipped beak .... 165. *Calycoseris*
        GG. Pappus of plumose bristles .... 166. *Anisocoma*
EE. Aks. epappose; glaucous scapose annual with broad spinulose-denticulate lvs. .... 167. *Atrichoseris*
DD. Not in all respects as above.
    E. Annual, biennial, or perennial herbs, glabrous or somewhat tomentose, never hirsute; lvs. mostly linear and entire; aks. cylindric, fusiform, or beaked, often more than 2 cm. long, pale; pappus setae coarse, strongly plumose; style-branches elongate. (Subtribe Scorzonerinae)
        F. Phyllaries uniseriate .... 168. *Tragopogon*
        FF. Phyllaries multiseriate .... 169. *Scorzonera*
    EE. Not in all respects as above.
        F. Annual or perennial herbs with coarse spreading hirsute pubescence, the hairs often forked; fls. almost always yellow; aks. mostly fusiform or beaked; pappus usually of coarse plumose setae; style-branches elongate. (Subtribe Leontodontinae)
            G. Receptacle chaffy-bracted; at least the inner aks. beaked; lvs. mostly basal .... 170. *Hypochoeris*
            GG. Receptacle naked.
                H. Aks. terete, truncate, the outer ones enveloped by the hardening phyllaries; not scapose ...... 171. *Hedypnois*
                HH. Aks. beaked.
                  I. Stems leafy; outer phyllaries broader than the inner (in ours) .... 172. *Picris*
                II. Stems scapose; outer phyllaries very small
                173. *Leontodon*
        FF. Habit, invols. and aks. various; pappus of numerous coarse or fine nonplumous setae (none in *Lapsana*); style-branches elongate. (Subtribe Crepidinae)
            G. Aks. flattened (obscurely so in *Sonchus arvensis*); leafy-stemmed herbs; pappus of soft and copious bristles.
            H. Invol. campanulate; aks. not beaked ...... 174. *Sonchus*
            HH. Invol. cylindrical; aks. beaked .... 175. *Lactuca*
            GG. Aks. not obviously flattened; lvs. mostly basal or subbasal; pappus various or none.
              H. Pappus present; phyllaries usually 2-seriate.
                I. Aks. truncate; pappus fragile, usually sordid or fuscous; caulescent .... 176. *Hieracium*
                II. Aks. beaked; pappus mostly bright white.
                  J. Heads in panicles or cymes; pappus early deciduous
                  177. *Crepis*
                JJ. Heads solitary on fistulous naked scapes; pappus not early deciduous .... 178. *Taraxacum*
            HH. Pappus none; stems branched, ± leafy; phyllaries uniseriate
                179. *Lapsana*

153. Scólymus L.

Erect thistlelike herbs with alternate coriaceous stiff sinuate-dentate to pinnatifid decurrent lvs. having spinescent lobes. Heads rather large, sessile, terminal and lateral. Fls. ligulate, yellow. Phyllaries in few rows, scarious-margined, spinescent-tipped, passing below into leafy bracts. Receptacle chaffy, the chaff ± embracing the beakless aks. Pappus a crown of unequal scarious paleae. (Old Greek name.)

1. **S. hispánicus** L. GOLDEN-THISTLE. Biennial or perennial 2–8 dm. tall, usually pubescent; stems interruptedly spinose-winged; lvs. spiny, white-veined, sinuate-pinnatifid, 5–8 cm. long; heads ca. 3 cm. high, axillary, few-fld.; phyllaries lance-linear, acuminate; aks. with 2 or 4 deciduous unequal paleae; $2n = 20$ (Stebbins et al., 1953). —Natur. at Los Gatos in 1894 and 1896; native of Medit. region. July. Cult. for its salsifylike taproot.

154. Cichòrium L. CHICORY

Annual or perennial herbs, the lvs. mostly basal. Stems branching, with reduced lvs. Fls. mostly blue, in sessile heads, all ligulate and perfect. Invol. with 2 series of phyllaries, the outer shorter. Receptacle naked. Aks. glabrous, striate-nerved, 5-angled, truncate, beakless. Pappus in 2–3 series of short blunt paleae. Ca. 9 spp., largely of Medit. region. (Name altered from the Arabic.)

1. **C. íntybus** L. Stems erect from a deep perennial taproot, glabrous or hairy, 3–10 dm. high; lower lvs. oblanceolate, toothed or pinnatifid, 1–2 dm. long, reduced and sessile upward; heads 1–3 in axils of much-reduced upper lvs., the branches nearly naked, racemiform; heads ca. 4 cm. broad in anthesis, blue, occasionally white; invol. 10–15 mm. high, the outer phyllaries loose, fewer and ca. half as long as the inner, callous-thickened at base; aks. 2–3 mm. long; pappus-scales minute, narrow; $2n = 18$ (Stebbins et al., 1953).—Natur. in waste places in much of cismontane Calif.; from Eu. June–Oct. The root used in many parts of the world as an adulterant or substitute for coffee.

155. Nothocalàis Greene

Near to *Agoseris*, perennial, with the root and caudex as in perennial spp. of that genus; lvs., scapes and phyllaries glabrous or villous with white-opaque hairs (not translucent or glandular); hairs on lvs. restricted to midrib and/or margins, the lvs. often crisped on edges. Infl. strictly scapose from a basal rosette or the scapes sometimes bearing a single lf. Heads erect; phyllaries subequal. Fls. conspicuously longer than invol. at anthesis. Aks. truncate-columnar or slender-fusiform, not narrowed into a beak, although the upper portion may be empty. Pappus of 15–90 silvery nonplumose awns, some or all of which are compressed and paleaceous toward the base, or if capillary throughout they number less than 40. A small genus of w. N. Am. (Greek, *nothos*, spurious, and *Calais*, of Greek mythology, who had scales on his back).

Aks. 5–8 mm. long; pappus-bristles unequal, 30–50, essentially capillary . . . . 1. *N. alpestris*
Aks. 8–10 mm. long; pappus-bristles subequal, 10–30, distinctly flattened . . . . . . 2. *N. troximoides*

1. **N. alpéstris** (Gray) Chamb. [*Troximon a.* Gray. *Agoseris a.* Greene. *Microseris a.* Q. Jones. *A. barbellulata* Greene.] Scapose perennial from a stout root, 5–20 cm. high, essentially glabrous; lvs. sometimes ± ciliate, 3–12 cm. long, remotely toothed or pinnatifid, sometimes subentire; heads solitary; invol. 10–20 mm. high, the phyllaries subequal, acuminate, often with minute dark dots, the outer phyllaries broader; ligules well exserted; aks. linear-fusiform, 5–8 mm. long, not beaked; pappus-bristles unequal, 30–50, barbellate, capillary, ± flattened toward the base.—Meadows and open slopes, 7000–11,500 ft.; Lodgepole F. to Subalpine F., Alpine Fell-fields; White Mts., Sierra Nevada, to Modoc and Siskiyou cos.; to Wash. July–Aug.

2. **N. troximoìdes** (Gray) Greene. [*Microseris t.* Gray. *Scorzonella t.* Jeps.] Stems few from a scaly root-crown, 0.5–3 dm. high, glabrous or puberulent; lvs. glabrous, basal, linear to lanceolate, 5–15 cm. long, with entire ± crisped margins; heads solitary; invol. 15–25 mm. high, the phyllaries subequal, lanceolate, acuminate, usually with dark midrib; aks. ca. 8–10 mm. long, puberulent upward; pappus of 10–30 gradually attenuate white narrowly linear paleae tapering into short awns; $2n = 18$ (Stebbins et al., 1953).—Dry places, 4000–6400 ft.; Sagebrush Scrub, N. Juniper Wd., N. Oak Wd.; Lassen and Siskiyou cos.; to B.C., Rocky Mts. May–June.

156. Micróseris D. Don

Annual to perennial herbs from a taproot or rhizomatous or vertical caudex. Lvs. principally basal and rosulate, sometimes cauline, entire to pinnatifid. Infl. of 1 to many 1-headed naked or bracteate scapes. Heads 5–300-fld., usually nodding in bud. Invol. of inner subequal lanceolate phyllaries and sometimes outer imbricate shorter ones. Receptacle alveolate. Corollas ligulate, yellow to white, often striped red. Aks. truncate-columnar to slender-fusiform, not beaked, ca. 10-ribbed. Pappus of 2–60 silvery to sordid mostly persistent paleae tapering into bristlelike awns. Ca. 15–20 spp., of w. N. Am., Chile, Australasia. (Greek, *micros*, small, and *seris*, a lettucelike plant.)

(Chambers, K. L. A biosystematic study of the annual spp. of Microseris. Contr. Dudley Herb. 7[4]: 207–312, 1955.)
A. Plants perennial with pale taproots; pappus-parts 6–60, the awns spiculate to subplumose; fls. conspicuous.
  B. Plants usually with 1–2 taproots; caudex short, vertical, unbranched; infl. bracteate below; pappus-awns paleaceous at base.
    C. Pappus-bristles ± plumose.
      D. Pappus-bristles white, 15–20 .... 4. *M. nutans*
      DD. Pappus-bristles brownish or yellowish, 5–10.
        E. Outer phyllaries lanceolate, glabrous .... 3. *M. paludosa*
        EE. Outer phyllaries ovate, scurfy .... 2. *M. sylvatica*
    CC. Pappus-bristles not plumose, white, ca. 10 .... 1. *M. laciniata*
  BB. Plants with several taproots; caudex rhizomatous; infl. not bracteate below; pappus-awns capillary .... 5. *M. borealis*
AA. Plants annual; pappus-parts 2–5.
  B. Heads always erect; pappus-paleae deciduous; awns filiform, white ...... 6. *M. Lindleyi*
  BB. Heads nodding or semierect, at least in the bud; pappus-paleae persistent; awns stouter, brown or yellow.
    C. Paleae narrowly lanceolate, ± bifid at apex; one or more of outer phyllaries lanceolate after anthesis.
      D. Aks. blue- or red-violet to brown, tawny or gray, never white-pruinose between the ribs, the tips of the ribs thickened and flared outward. Alameda and Mariposa cos. to L. Calif. ...... 7. *M. heterocarpa*
      DD. Aks. brown to black, often minutely white-pruinose between the ribs, the tips of the ribs not thickened or flared outward. Marin, Santa Cruz, and Monterey cos. 8. *M. decipiens*
    CC. Paleae various, but if narrowly lanceolate then acute at apex; outer phyllaries ovate or deltoid after anthesis.
      D. Pappus-parts 5, the paleae smooth or minutely scabrous, straight and flat at maturity at least in the upper half; awns hairlike.
        E. Aks. 1.5–3.5 mm. long, not dark-spotted, the flared apex as broad as or broader than the body; paleae 0.2–2 mm. long .... 12. *M. elegans*
        EE. Aks. 3–5.2 mm. long, often dark-spotted, the apex scarcely as broad as the body; paleae 1–4 mm. long .... 11. *M. Bigelovii*
      DD. Pappus-parts 2–5, the paleae often scabrous to villous, curved, the margins usually incurved or convolute at maturity; awns stouter.
        E. Paleae linear-lanceolate, flat at maturity, the broad stout midrib forming ⅓–⅕ of the maximum palea-width. From Alameda and Mariposa cos. n. 9. *M. acuminata*
        EE. Paleae round to lanceolate or almost obsolete, ± incurved or convolute at maturity, the stout or slender midrib forming less than ⅕ of the maximum palea-width in those paleae which approach the ak. in length.
          F. Paleae smooth or only minutely scabrous; aks. gray or pale brown. Alameda and Tuolumne cos. to San Luis Obispo and Kern cos. 10. *M. campestris*
          FF. Paleae villous or scabrous; aks. black, brown or tawny. Widely distributed .... 13. *M. Douglasii*

1. **M. lacinìata** (Hook.) Sch.-Bip. [*Hymenonema l.* Hook. *Scorzonella l.* Nutt. *Calais l.* Gray. *S. B.* Greene. *M. B.* Gray. *M. procera* Gray. *S. p.* Greene. *S. maxima* Bioletti. *S. pratensis* Greene. *S. arguta* E. Drew.] Perennial, from a thickened root, the stems glabrous or nearly so, several, 2–8 dm. tall, leafy in lower part, simple or branched above; lvs. entire to laciniate-pinnatifid, 5–25 cm. long; heads solitary at ends of branches; invol. 12–25 mm. high, variable, the inner phyllaries mostly lance-linear, attenuate, the outer lanceolate to round-ovate; ligules ± exserted; aks. 4–6 mm. long, narrowed upward; pappus white, 7–8 mm. long, mostly of ca. 10 barbellate awns flattened at base; $2n = 18$ (Stebbins et al., 1953).—Mostly in meadowy or woody places, below 4000 ft.; Mixed Evergreen F., N. Oak Wd., Yellow Pine F., etc.; Coast

Ranges from Napa and Sonoma cos. to Siskiyou Co., e. to Lassen and Modoc cos.;
to Wash. May–July. Rather distinct is a slender form, to ca. 3 dm. high, with ± linear
lvs., invol. ca. 12–13 mm. high and with the phyllaries narrow; from near the coast
of Mendocino and Humboldt cos. to Wash.; it has been called *M. leptosèpala* (Nutt.)
Gray and *M. laciniàta* ssp. *leptosèpala* Chamb. [*Calais Bolanderi* Gray.]

2. **M. sylvática** (Benth.) Gray. [*Scorzonella s.* Benth.] Perennial with 1–few stems
3–6 dm. high arising from a tuft of lvs. above the root-crown; herbage glabrous to
densely mealy-puberulent; lvs. erect, linear to lanceolate, mostly laciniate-pinnatifid,
5–22 cm. long; heads solitary at ends of branches; invol. 16–20 mm. long, the phyllaries
in 2–3 series, the inner acuminate, the outer lanceolate to ovate and shorter; ligules
light yellow, with a median pinkish line; aks. linear, ca. 10-ribbed, 6–9 mm. long; pappus
brownish or sordid, 12–14 mm. long, of 5–10 linear-attenuate parts paleaceous below
and with a plumose awn above; $2n = 18$ (Stebbins et al., 1953).—Open grassy or
wooded places, below 4500 ft.; Foothill Wd., V. Grassland, etc.; Tehachapi Mts. and
hills bordering San Joaquin and Sacramento valleys. March–May.

3. **M. paludòsa** (Greene) J. T. Howell. [*Scorzonella p.* Greene. *M. sylvatica* var.
Stillmanii Gray. *S. p.* var. *integrifolia* Jeps.] Resembling *M. sylvatica;* invol. ca. 15
mm. high; outer phyllaries lanceolate, acuminate; aks. 4–6 mm. long, the lanceolate
part of the palea ca. 2 mm., the barbellate (not plumose) awn ca. 8–10 mm. long.—
Grassy and wooded places; N. Coastal Scrub, Closed-cone Pine F., etc.; near the coast,
Monterey Co. to Sonoma Co. May–June.

4. **M. nùtans** (Hook.) Sch.-Bip. [*Scorzonella n.* Hook. *Calais n.* Gray. *M. major*
(Gray) Sch.-Bip. *M. m.* var. *laciniata* Gray. *C. gracililoba* Kell.] Stems 1–several from
the thickened perennial roots, slender, erect, 1–4.5 dm. high; herbage glabrous or
slightly scurfy; lvs. on lower part of stem, lance-linear to oblanceolate or wider, 1–2.5
dm. long, entire or toothed or saliently pinnatifid; heads solitary on ends of branches;
invol. 12–14(-20) mm. high, outer phyllaries much shorter than the inner; aks. fusiform,
5–7 mm. long, glabrous or puberulent; pappus of 15–20 narrow scales, each bearing
a long white terminal plumose bristle; $2n = 18$ (Stebbins et al., 1953).—Mostly in
woods or moist places, 4000–9500 ft.; Montane Coniferous F.; Sierra Nevada to Modoc,
Siskiyou, Humboldt, and Trinity cos.; to B.C., Rocky Mts. June–Aug.

5. **M. boreàlis** (Bong.) Sch.-Bip. [*Apargia b.* Bong. *Apargidium b.* T. & G.] Essentially
glabrous; lvs. entire or remotely denticulate, 7–25 cm. long; scapes 1–2, mostly 1–4
dm. high; heads ± campanulate, nodding in bud, 10–12 mm. broad; phyllaries lanceolate, acuminate, in more than 1 series; receptacle naked; aks. linear, truncate, 10–12-
nerved; pappus of ca. 30–40 persistent brownish capillary barbellate bristles 5–6 mm.
long; $2n = 18$ (Stebbins et al., 1953).—Boggy places, 3000–4000 ft.; Yellow Pine F.;
Bald Mt., Humboldt Co.; to Alaska. June–Sept.

6. **M. Líndleyi** (DC.) Gray. [*Calais L.* DC. *C. linearifolia* DC.] Subacaulescent
glabrous or villous annual 1–6 dm. tall; lvs. basal and lower-cauline, 3–25 cm. long,
linear to narrow-elliptic, acuminate, entire or with a few retrorse teeth or pinnatifid;
heads terminal on scapose peduncles; invol. 15–30 mm. high, the inner phyllaries 3–18,
somewhat unequal, lanceolate to lance-ovate, green or striped red, glabrous or with a
few black hairs; outer phyllaries 2–8, imbricate; aks. 8–15 mm. long, black or dark
brown, fusiform, narrowed upward; pappus of 5 narrow scales 6–20 mm. long, each
2-cleft at apex and slender-awned from between the lobes; $2n = 18$ (Stebbins et al.,
1953).—Grassy places, open woods, etc., below 6000 ft.; common in many Plant Communities through most of cismontane Calif., occasional on Mojave Desert; to se. Wash.,
Ida., Utah, New Mex., n. L. Calif. April–June.

7. **M. heterocárpa** (Nutt.) Chamb. [*Uropappus h.* Nutt. *Calais Parryi* Gray. *C. Kelloggii, Clevelandii,* and *pluriseta* Greene. *M. Lindleyi* auth. *U. leucocarpus* Greene.]
Caulescent or acaulescent annual, glabrous or furfuraceous; lvs. 5–30 cm. long, linear
to narrow-elliptic, entire to pinnatifid; scapes 1–6 dm. high; heads 10–125-fld.; invol.
1–3 cm. high, the inner phyllaries subequal, lanceolate or lance-ovate, often with
thickened midrib, green or striped with red; outer phyllaries imbricate or very short;
aks. 5–10 mm. long, brown to straw color, violet or gray, plain or dotted, the ribs
lightly scabrous; pappus-parts 5, persistent, the broad paleaceous part 5–10 mm. long,
apically toothed, ending in an awn 4–8 mm. long; $n = 18$ (Chambers, 1955).—Grassy
places, below 2000 ft.; V. Grassland, etc.; cismontane s. Calif. n. to Napa and Butte
cos.; Channel Ids.; L. Calif. March–May.

8. **M. decípiens** Chamb. Lvs. cauline and/or basal, 5–15 cm. long; heads 10–35-fld.; invol. 8–18 mm. long, the inner phyllaries 6–12, subequal; outer imbricate or mostly basal; aks. 6–8 mm. long, brown or purplish-brown with darker flecks; pappus-parts 5, persistent, 7–10 mm. long, the basal paleae 2.5–5 mm. long, lacerate or cleft at apex, the awn 4–5 mm. long, minutely spiculate, straw color; $2n = 36$ (Chambers, 1955).— Grassy places; Coastal Prairie, V. Grassland, etc.; near the coast, Marin, Santa Cruz, and Monterey cos. April–May.

9. **M. acumináta** Greene. [*M. cognata* Greene.] Acaulescent annual, the lvs. 3–20 cm. long, mostly pinnatifid; scapes 1–3 dm. high, furfuraceous; heads 5–50-fld., the inner phyllaries 4–12, subequal, lanceolate, thickened along the midrib, 8–18 mm. long; outer phyllaries 6–12, calyculate; aks. 5–7 mm. long, uniformly brown; pappus-parts 5, persistent, 9–18 mm. long, the paleaceous base 4–10 mm. long, arcuate, awn 4–7 mm. long, barbellulate, white or brown; $2n = 36$ (Chambers, 1955).—Grassy flats and depressions; V. Grassland, etc.; Cent. V. from Mariposa and Alameda cos. n.; s. Ore. March–May.

10. **M. campéstris** Greene. [*M. aphantocarpha* var. *palealis* Gray.] Acaulescent annual, the lvs. all basal, 5–20 cm. long, usually toothed or pinnatifid; scapes 0.5–5 dm. high, mostly slender, furfuraceous; heads 5–120-fld., the inner phyllaries 5–20, subequal, lanceolate to lance-ovate, acute or acuminate, carinate along the midrib, 6–12 mm. long, the outer phyllaries 5–12, calyculate; aks. 3–5 mm. long, gray or pale brown, often spotted; pappus-parts 5, persistent, 5–8 mm. long, the basal paleae 1–4 mm. long, deltoid to lanceolate, usually not apically lobed, the awn 3–5 mm. long, spiculate or barbellulate; $2n = 36$ (Chambers, 1955).—Grassy places; V. Grassland; San Joaquin V. from Kern Co. to Alameda and Tuolumne cos., Livermore V. March–April.

11. **M. Bigelòvii** (Gray) Sch.-Bip. [*Calais B.* Gray, *M. intermedia, castanea, obtusata,* and *maritima* Greene.] Acaulescent annual, the lvs. 5–25 cm. long, glabrous or furfuraceous, entire to pinnatifid; scapes 1–6 dm. high, mostly slender, furfuraceous; heads 10–110-fld., the inner phyllaries 5–18, subequal, lance-ovate, acute or acuminate, carinate along the midrib, 5–14 mm. long; outer phyllaries 4–14, calyculate; aks. 2.5–5 mm. long, brownish, sometimes spotted with black; pappus-parts 5, persistent, 5–10 mm. long, the paleaceous base 1–4 mm. long, narrow-lanceolate, tapering to an awn 3–8 mm. long; $2n = 18$ (Chambers, 1955).—Coastal bluffs and flats; Coastal Strand, Closed-cone Pine F., etc.; Santa Barbara Co. to Ore., B.C. April–June.

12. **M. élegans** Greene ex Gray. [*M. stenocarpha* Greene. *M. aphantocarpha* vars. *mariposana* and *elegans* Jeps.] Acaulescent annual; lvs. 5–20 cm. long, glabrous or furfuraceous, mostly toothed or pinnatifid; scapes 1–3.5 dm. high, mostly slender; heads 10–100-fld., inner phyllaries 5–15, lance-ovate, acute to acuminate, keeled on midrib, 4–8 mm. long, the outer 6–12, calyculate; aks. 1.5–3.5 mm. long, gray-brown to brown or almost black, not spotted; pappus-parts 5, persistent, 3.5–7 mm. long, the paleaceous base to 2 mm. long, tapering into a hairlike awn 3–5 mm. long; $2n = 18$ (Chambers, 1955).—Grassy flats and hills; largely V. Grassland; much of cismontane Calif. especially in valleys away from the coast except in s. Calif. where near the coast; n. L. Calif. April–May.

13. **M. Douglásii** (DC.) Sch.-Bip. [*Calais D.* DC. *C. cyclocarpha* Gray. *C. eriocarpha* Gray. *M. attenuata, atrata, Aliciae, callicarpa, conjugens, insignis, leucocarpha, leiosperma, melanocarpa, parvula, proxima, picta, Parishii* and *tenuisecta* Greene.] Acaulescent annual; lvs. 5–25 cm. long, glabrous or furfuraceous, entire to pinnatifid; scapes 1–6 dm. high; heads 5–100-fld., the inner phyllaries 5–18, subequal, lanceolate to lance-ovate, acute or acuminate, carinate, 5–15 mm. long; outer phyllaries 4–12, calyculate; aks. 4–10 mm. long, gray to straw color, brown or blackish, often dark-spotted; pappus-parts 5–2, rarely none, persistent, 5–13 mm. long, the basal paleaceous part 1–6 mm. long, round to ovate or lanceolate, tapering into a stout or slender barbellulate awn 3–8 mm. long; $2n = 18$ (Chambers, 1955).—Grassy places; V. Grassland, Coastal Prairie, etc.; much of cismontane Calif.; s. Ore. March–April.

KEY TO SUBSPECIES

Palea mostly more than 1 mm. long, conspicuous; ak.-ribs usually broadened and flared outward at the tip.

    The paleae scabrous to villous, 1–6 mm. shorter than the ak., or if not, then the ak. more than 4.5 mm. long . . . . *M. Douglasii*

The paleae scabrous, 0.5–2 mm. shorter than the ak.; ak. 4.5 mm. long or less .... ssp. *platycarpha*
Palea mostly 1 mm. long or less, sometimes almost obsolete; ak.-ribs usually not broadened and flared
outward at the tip, but linear .... ssp. *tenella*

Ssp. **platycárpha** (Gray) Chamb. [*Calais p.* Gray. *M. p.* Sch.-Bip. *M. breviseta* Greene.] Heads 5–50-fld.; inner phyllaries 5–15, 5–12 mm. long, the outer 4–10, calyculate; aks. 3–4.5 mm. long, straw color to dark brown, commonly spotted black; pappusparts 5 to 1, rarely 0, 4–10 mm. long, the paleae 2.5–6 mm. long, round to lanceolate tapered into an awn 1–4 mm. long; $2n = 18$ (Chambers, 1955).—Grassy places; V. Grassland, Coastal Sage Scrub; Los Angeles Co. to L. Calif.; Santa Catalina and San Clemente ids. March–May.

Ssp. **tenélla** (Gray) Chamb. [*Calais t.* Gray. *C. aphantocarpha* Gray. *M. furfuracea, indivisa, pulchella* Greene.] Heads 10–300-fld., the inner phyllaries 5–30, 5–14 mm. long, outer 4–13, calyculate; aks. 3–6.5 mm. long, light to dark brown, sometimes spotted with black; pappus-parts 5–0, fragile, 3.5–9 mm. long, the paleae to 1 mm. long, deltoid, the awns 3–8.5 mm. long, slender; $2n = 18$ (Chambers, 1955).—Grassy places; V. Grassland, Coastal Prairie, etc.; Santa Barbara Co. to Bay region, occasional in Cent. V. April–May.

157. Agóseris Raf. Mountain Dandelion

Annual or perennial herbs with strong taproots with blackish periderm; vertical caudex short or long, single or multicipital. Lvs. mostly in radical tufts or a few scattered on lower stem, glabrous to villous with the hairs not restricted to midrib or margins. Infl. scapose, the head erect; invol. campanulate to subcylindric, the phyllaries subequal or imbricate. Fls. yellow to orange, often drying darker, usually longer than phyllaries at anthesis. Aks. narrowed above into a stout or slender beak ⅙ to almost 4 times as long as the body; pappus of 50 or more white capillary bristles, never plumose. Ca. 8–9 spp., of w. N. Am. and 1 in S. Am. (Greek, *aix*, goat, and *seris*, chicory.)

A. Plants annual; ak.-beak mostly 2–3 times as long as ak.-body; invol. 12–18 mm. high.
                                               1. *A. heterophylla*
AA. Plants perennial; ak.-beak variable; invol. mostly higher.
    B. Ak.-beak mostly not more than half as long as ak.-body, stout, ± striate. Sierra Nevada and ranges to the e. .... 2. *A. glauca*
    BB. Ak.-beak more than half as long as ak.-body, slender, not plainly striate.
        C. Beak less than twice as long as body of ak.
            D. Fls. yellow, sometimes drying pinkish.
                E. Ak.-body mostly 8–10 mm. long; invol. in fr. 20–30 mm. high. Sierra Nevada, above 5000 ft. .... 3. *A. elata*
                EE. Ak.-body mostly 4–6 mm. long; invol. in fr. 10–18 mm. long. Near the coast, below 2500 ft. .... 4. *A. apargioides*
            DD. Fls. burnt-orange, often drying purplish. High montane .... 5. *A. aurantiaca*
        CC. Beak 2–4 times as long as ak.-body; fls. yellow, sometimes pinkish on drying.
            D. Ak.-body truncate at apex, abruptly beaked; lf.-segms. retrorse; phyllaries in 2 quite unlike sets .... 6. *A. retrorsa*
            DD. Ak.-body tapering at apex, gradually beaked; lf.-segms. not retrorse; phyllaries mostly in 3 sets .... 7. *A. grandiflora*

1. **A. heteróphylla** (Nutt.) Greene. [*Macrorhyncus h.* Nutt. *Troximon h.* Greene. *A. h.* var. *kymapleura* Greene. *A. h.* var. *cryptopleura* Greene. *T. h.* var. *californicum* Hall and f. *idiale,* f. *crenulatum* and f. *turgidum* Hall. *A. h.* var. *turgida* Jeps.] Subglabrate to pubescent annual 0.3–4 dm. tall, often with several scapes from the base; lvs. oblong to spatulate or linear, entire to denticulate or sinuate, 5–15 cm. long, in a basal cluster, but not all strictly basal; invol. 5–20 mm. long, sparsely villous, the phyllaries lance-acuminate; ligules scarcely to conspicuously exserted, yellow; ak.-body 3–5 mm. long, several-ribbed, the beak ca. twice as long, slender; $2n = 18$, 36 (Stebbins et al., 1953). —Open grassy places and flats, below 7500 ft.; many Plant Communities; cismontane Calif. to B.C., Ida., Utah, Ariz. April–July. Variable; some of the forms that have been recognized for Calif. are: var. *kymapleùra* with ribs of outer aks. corky-thickened and undulate, var. *crenulàta* with similar ribs and longer ligules, var. *túrgida* with outer aks. much inflated.

2. **A. gláuca** (Pursh) Greene var. **montícola** (Greene) Q. Jones. [*A. m.* Greene.] Perennial with 1 to several stems from the short branches of the root-crown, the scapes stout, mostly 1–3 dm. high, pubescent just below the heads and toward the base; lvs.

largely basal, linear to oblanceolate, 5–20 cm. long, acute or acuminate, glabrous or short-hairy, not arachnoid; invol. 15–25 mm. high, subglabrous, the phyllaries in ca. 3 lengths, the outer conspicuously shorter, acute or acutish; fls. yellow; aks. 5–10 mm. long, with a short stout striate beak; pappus 10–12 mm. long; $2n = 18$, 36 (Stebbins et al., 1953).—Edge of meadows to dry rocky slopes, 5000–10,500 ft.; Sagebrush Scrub, Montane Coniferous F.; Inyo-White Range, Sierra Nevada, to Modoc and Siskiyou cos.; to Wash., Nev. July–Aug.

Var. **laciniàta** (D. C. Eat.) Smiley. [*Macrorhyncus g.* var. *l.* D. C. Eat.] Petioles and lower part of lvs. ± arachnoid; phyllaries mostly in 2 ranks; pappus-bristles 14–18 mm. long.—At 4000–8500 ft.; Sagebrush Scrub, N. Juniper Wd.; Mono, Nevada, Lassen, and Modoc cos.; to Rocky Mts. May–July.

3. **A. elàta** (Nutt.) Greene. [*Stylopappus e.* Nutt. *Troximon e.* Nels.] Glabrous or short-villous perennial, the scapes rather stout, 3–6 dm. tall; lvs. oblanceolate, entire or remotely denticulate to pinnatifid, 1–3 dm. long; fls. yellow, or drying pinkish; invol. 2–2.5 cm. high, the phyllaries well imbricate, the outer broader and blunter; ak.-body 3–10 mm. long, with a slender scarcely striate beak ca. as long; pappus ca. 8–10 mm. long.—Uncommon, near meadowy places, 5200–10,500 ft.; Yellow Pine F. to Subalpine F.; Sierra Nevada from Tulare and Inyo cos. n.; to Wash. July–Sept.

4. **A. apargioìdes** (Less.) Greene. [*Troximon a.* Less. *Leontodon hirsutum* Hook. *A. h.* Greene.] Perennial, the herbage thinly hirsute to hirsutulose; lvs. oblong-lanceolate in outline, entire to pinnatifid, 7–25 cm. long, the segms. from lance-oblong to linear; scapes several to many, 1–4 dm. high; heads 14–20 mm. high, the phyllaries oblong to linear, in ca. 2 series, but not very unequal, long-pubescent, often a little glandular; ligules exserted, yellow with a little brown on the back; aks. strongly ribbed, ca. 3 mm. long, the beak slightly longer; pappus 4–5 mm. long; $2n = 18$ (Stebbins et al., 1953).—Grassy or ± wooded slopes, often in sand or dried mud; Mixed Evergreen F., Coastal Scrub, etc.; coastal slopes from n. Santa Barbara Co. to Humboldt Co.; Santa Rosa Id? Mostly March–June.

Ssp. **marítima** (Sheld.) Q. Jones. [*A. m.* Sheld.] Plant ± arachnoid at least when young; phyllaries well imbricate; ak.-body mostly 4–6 mm. long, the slender beak mostly ca. as long or longer.—Sandy places, to ca. 1000 ft.; Coastal Strand; Mixed Evergreen F., etc.; Humboldt Co. to sw. Wash. June–Aug.

Var. **Eastwoòdae** (Fedde) Munz [*A. E.* Fedde. *A. maritima* Eastw.] Herbage with some ± crisped hairs, glabrate; phyllaries well imbricate; beak ca. half as long as ak.-body; $2n = 36$ (Stebbins et al., 1953).—Coastal Strand, Closed-cone Pine F., etc.; Monterey Co. to Mendocino Co. April–Aug.

5. **A. aurantìaca** (Hook.) Greene. [*Troximon a.* Hook. *A. gracilens* (Gray) Kuntze. *A. g.* var. *Greenei* Jeps.] Perennial, with few to several scapes from the root-crown; herbage glabrous or ± villous, 1–5 dm. high; lvs. oblong-lanceolate, subentire to remotely divergently pinnatifid, 5–25 cm. long; fls. burnt-orange, mostly turning purple to deep pink in age; invol. 18–24 mm. high; inner phyllaries lanceolate, the outer oblong, ca. half as long as inner; aks. linear-fusiform, ribbed, the body 4–8 mm. long, rather abruptly tapered into the slender beak which is half as long to longer than body.—Moist mostly grassy places, 6000–11,500 ft.; Montane Coniferous F.; Sierra Nevada to Modoc Co., Coast Ranges from Yolla Bolly Mts. n.; to B.C., Rocky Mts. July–Aug.

6. **A. retròrsa** (Benth.) Greene. [*Macrorhynchus r.* Benth. *Troximon r.* Gray.] Perennial, the stout scapes 1.5–5 dm. tall; herbage woolly-pubescent when young, ± glabrate in age; lvs. lanceolate to oblong-lanceolate, 8–30 cm. long, pinnately parted into linear or lanceolate retrorse and rather scattered segms., the terminal one very long; invol. 2.5–4 cm. high, the phyllaries in 2 sets; ligules somewhat exceeding invol., yellow, often pinkish in drying; ak.-body 5–7 mm. long, ± truncate at apex and narrowed suddenly into the slender nonstriate beak 2–4 times as long; pappus-bristles 10–16 mm. long.—Common on dry ridges and slopes, mostly 2500–8000 ft.; Montane Coniferous F.; mts. of s. Calif., Sierra Nevada, Coast Ranges; to Wash., Nev. May–Aug.

7. **A. grandiflòra** (Nutt.) Greene. [*Stylopappus g.* Nutt. *Troximon g.* Gray. *T. plebeium* Greene. *A. p.* Greene. *A. laciniata* (Nutt.) Greene. *A. intermedia* Greene.] Scapes robust, few to several from the branched root-crown, stout, 1.5–6 dm. high; herbage hirsute-pubescent to glabrate, or the invol. and lower parts ± tomentose; lvs. oblanceolate to lance-elliptic, 10–25 cm. long, entire to laciniate or with spreading lobes; heads 2.5–4

cm. high, the invol. with broad short outer phyllaries and narrow much more elongate inner ones; ligules exserted, yellow; ak.-body ca. 4–7 mm. long, tapering to a beak 2–4 times as long; $2n = 18$ (Stebbins et al., 1953).—Dry or ± moist places, below 6200 ft.; many Plant Communities; cismontane Calif.; to B.C., Mont., Utah. May–July.

158. Phalacróseris Gray

Glabrous perennial with naked 1-headed scapes from a tuft of narrow lvs. coming from the root-crown. Invol. campanulate, of 12–16 subequal lanceolate phyllaries and naked or unibracteate at the base. Ligules yellow. Receptacle naked. Aks. short-oblong, slightly incurved, obscurely quadrangular. Pappus none. One sp. (Greek, *phalakros*, bald-headed, and *seris*, some lettucelike plant.)

1. **P. Bolánderi** Gray. Lvs. narrow- to oblong-lanceolate, entire, slightly fleshy, 6–30 cm. long; heads 2–2.5 cm. broad at anthesis; invol. 8–10 mm. high; ligules conspicuous; aks. truncate at both ends, 3–4 mm. long; $2n = 18$ (Stebbins et al., 1953).—Wet meadows, 7000–9300 ft.; Red Fir F., Lodgepole F.; Sierra Nevada from Tuolumne Co. to Tulare Co. June–Aug. Plants having aks. with a short crown have been named var. *coronàta* Hall.

159. Chaetadélpha Gray

Diffuse much-branched broomlike perennial herb from a heavy underground branching rootstock. Lvs. alternate, entire, linear to lance-linear, thickish, acuminate, scarious-margined. Heads solitary at ends of branches; invol. with 5 principal phyllaries and a few short outer ones. Fls. 5 to a head, pale lavender, almost whitish. Aks. 5-angled, narrow; pappus of 5 rigid upwardly tapering awns bearing on each side toward the base 3–5 shorter slender rigid bristles. One sp. (Greek, *chaete*, bristle, and *adelphe*, sister, referring to the united bristles in the pappus.)

1. **C. Wheèleri** Gray. Glabrous, mostly 1–3 dm. high, the branches ridged; lvs. 2–5 cm. long, deciduous; invol. 12–15 mm. high; aks. 8–10 mm. long; pappus light brown, almost as long; $2n = 18$ (Stebbins et al., 1953).—Sand dunes, etc., 3000–5000 ft.; Creosote Bush Scrub, Sagebrush Scrub; base of Inyo-White Range; s. Ore., w. Nev. May–Sept.

160. Lygodésmia D. Don

Herbs or ± woody, with lvs. largely near the base. Heads terminal on the branchlets, erect, 3–12-fld. Receptacle naked. Invol. slender, of a few equal phyllaries and outer minute ones. Fls. pink or rose. Aks. slender, few-ribbed, truncate at apex; pappus-bristles many, capillary, stiffish or soft, unequal, not plumose, deciduous separately. A small N. Am. genus. (Greek, *lugos*, a pliant twig, and *desme*, bundle, the type sp. with fascicled stems.)

Perennial, the branches rigid, spine-tipped; plant with mats of wool at base . . . . 1. *L. spinosa*
Annual, not spiny, not woolly at base . . . . 2. *L. exigua*

1. **L. spinòsa** Nutt. [*Pleiacànthus s.* Rydb.] Perennial from woody branched root-crown, with numerous slender rigid zigzag glabrous to puberulent branches 3–4 dm. high, woolly-matted at base, almost leafless and with spiny branchlets; lower lvs. linear, entire, 0.5–4 cm. long, the upper reduced to scales; heads 3–5-fld., subsessile or on very short branchlets; invol. 8–9 mm. high, the inner phyllaries lanceolate, the outer broad, calyculate; ligules well exserted; aks. tapering slightly, 4 mm. long, ± 5-angled; pappus 4 mm. long.—Dry sandy or rocky places, 5000–9400 ft.; Sagebrush Scrub, Pinyon-Juniper Wd., Yellow Pine F.; desert slopes of San Bernardino and San Gabriel mts., Panamint Mts., Inyo-White Range, e. slope of Sierra Nevada, Inyo Co. to Modoc Co.; to B.C., Mont., n. Ariz. July–Sept.

2. **L. exigua** Gray. Diffusely branched glabrous annual 1–4 dm. high; lvs. oblanceolate, mostly basal, entire or runcinate-dentate or -pinnatifid, 2–4 cm. long, the upper few and reduced; heads ca. 3–4-fld., terminal on very slender branchlets; invol. 4–5 mm. long, the inner phyllaries lanceolate, the outer few, minute; ligules inconspicuous;

aks. 3–4 mm. long, the pappus bright white, conspicuous, 3–4 mm. long.—Well distributed but not common, open places, rocky hills, etc., below 5000 ft.; Creosote Bush Scrub, Joshua Tree Wd., Pinyon-Juniper Wd.; both deserts; to Colo., Tex. March–May.

161. Stephanomèria Nutt.

Annual or perennial herbs with slender strict or paniculately branched stems. Lvs. alternate, linear to oblong, entire to pinnatifid, the uppermost usually much reduced. Heads usually paniculate, small or middle-sized, rose or flesh-colored. Invol. of several equal phyllaries, with some small outer bracts, sometimes more regularly graduated. Aks. columnar, 5-angled; pappus-bristles 1-seriate, plumose at least above, often paleaceous toward the base, sometimes connate into groups. A dozen or more spp., of w. N. Am. (Greek, *stephane*, wreath, and *meros*, division; of uncertain application.)

A. Plants annual or at most biennial.
  B. Pappus-bristles clear white. Widely distributed in Calif.
    C. Pappus-bristles plumose almost throughout, the base scarcely thickened; plants mostly
      5–30 dm. high . . . . 1. *S. virgata*
    CC. Pappus plumose only above, naked or merely hirsutulous below the middle, the base
      thickened or paleaceous; plants mostly below 5 dm. high . . . . 3. *S. exigua*
  BB. Pappus-bristles tawny or brownish, plumose nearly or quite to the base. Extreme n. Calif.
      2. *S. paniculata*
AA. Plants perennial.
  B. Invol. not imbricated, but calyculate at base with minute bractlets; receptacle naked; herbage
    mostly glabrous.
    C. Stems from slender creeping rootstocks; lvs. mostly linear, thin, entire or with a few
      salient teeth, few if any of the lvs. bractlike. Pine belt of cent. and n. Calif.
      4. *S. lactucina*
    CC. Stems from the crown of a taproot; lvs. of branchlets commonly small and minute or
      bractlike.
      D. Invol. 10–14-fld., 12–15 mm. high; lvs. callous-margined. Deserts . . 5. *S. Parryi*
      DD. Invol. 3–9-fld., less than 12 mm. high; lvs. not callous-margined.
        E. Stems herbaceous, erect, very slender; lvs. mostly filiform and entire; pappus-
          bristles bright white, plumose to the base. Sierra Nevada . . 6. *S. tenuifolia*
        EE. Stems ± woody below, coarser, with ± spreading branches; lvs. linear-subulate,
          mostly runcinate; pappus brownish.
          F. Pappus-bristles plumose essentially to the base; invol. 5–8 mm. high.
            Occasional . . . . 7. *S. myrioclada*
          FF. Pappus-bristles naked and merely scabrous toward the base, plumose
            above; invol. 8–10 mm. high. Common . . . . 8. *S. pauciflora*
  BB. Invol. strongly imbricated; receptacle pitted; young herbage woolly . . . . 9. *S. cichoriacea*

1. **S. virgàta** Benth. [*Ptiloria v.* Greene. *P. pleurocarpa* Greene.] Erect stiff annual 5–20(–30) dm. tall, glabrous or puberulent, virgate or usually with virgate branches from ca. the middle; lower lvs. oblong or spatulate, 1–2 dm. long, often sinuate, dying before anthesis; upper lvs. small, linear, quite entire; heads subsessile along the elongate naked branches, 4–15-fld.; invol. 6–8 mm. high; phyllaries obtusely narrowed at tip; ligule purplish on back, pinkish to white above, ca. 7–9 mm. long; aks. ribbed, linear-clavate to oblong, ± rugose, ca. 3–4 mm. long; pappus-bristles clear white, 4–5 mm. long, plumose almost throughout; $2n = 16$ (Stebbins et al., 1953).—Common late summer annual in deserted fields and disturbed places, below 6000 ft.; many Plant Communities from Coastal Strand and Coastal Sage Scrub to Yellow Pine F.; widely distrib. in cismontane Calif.; Ore., Nev., L. Calif. July–Oct.

Var. **tomentòsa** (Greene) Munz. [*S. t.* Greene. *Ptiloria t.* Greene. *P. canescens* Greene.] Near to the preceding and doubtfully distinct; herbage white-tomentose at least when young; heads on short bracted peduncles; phyllaries subtruncate at apex.—Santa Cruz Id. and Morro Bay. July–Sept.

2. **S. paniculàta** Nutt. [*Ptiloria p.* Greene.] Mostly annual, usually 1-stemmed from base, 2–9 dm. high, ± branched above with short branches; basal lvs. toothed, oblanceolate, ± auriculate, upper lvs. narrower, reduced; infl. ± paniculiform, the heads short-pedunculate; invol. 6–9 mm. high; phyllaries and fls. ca. 5–8; pappus-bristles 15–20, ± brownish, plumose almost throughout; aks. 4–5 mm. long, pitted and rugose-tuberculate. —Dry open places, 3000–4500 ft.; Sagebrush Scrub, etc.; Siskiyou Co.; to Wash. July–Sept.

3. **S. exígua** Nutt. [*Ptiloria e.* Greene.] Annual 2–6 dm. tall, diffusely branched, glabrous below, minutely glandular-pubescent above; lower lvs. pinnatifid with linear

or lanceolate divisions, oblong in outline, auriculate-clasping, 2–5 cm. long; upper lvs. small, bractlike; heads many, scattered or paniculate; invol. 6–7 mm. high; fls. mostly 5; aks. 5-angled, with a double row of tubercles between the angles; pappus-bristles 10–18, naked on lower third, commonly united into 4–5 groups by the thickened bases. —Low dry places, below 8500 ft.; many Plant Communities; cismontane and desert Calif. n. to Monterey and Mono cos.; to Wyo., New Mex. May–Sept. Variable; a no. of forms have been recognized: (1) var. *Dèanei* Macbr., an intricately branched conspicuously glandular form with slender twigs and a pappus in which the bristles are deciduous above the paleaceous base so as to leave a crown of setose scales; (2) var. *coronària* (Greene) Jeps. [*S. c.* Greene.] Plant glabrous, not glandular; pappus-bristles 10 or more, tending to be deciduous above the base.—With the sp.; and; (3) var. *pentachaèta* (D. C. Eat.) Hall. [*Ptiloria p.* D. C. Eat.] Herbage not glandular; pappus-bristles ca. 5, all distinct and dilated at base.—Occasional with the sp.

4. **S. lactucìna** Gray. [*Ptiloria l.* Greene.] Stems single, 1–3 dm. high, simple or branched at or near the base, glabrous or puberulent below; lvs. mostly linear, 3–10 cm. long, the lower with a few divergent teeth, upper lvs. not greatly reduced; heads few, terminal on the ± corymbosely arranged peduncles; invol. 12–15 mm. high; ligules bright pink, 8–11 mm. long; aks. ± angled, not rugose, 5–6 mm. long; pappus-bristles ca. 20, white, completely plumose, thickened and ± united at base.—Dry flats and ridges, 4000–7600 ft.; Yellow Pine F.; Red Fir F.; Sierra Nevada from Mariposa Co. n., to Modoc and Siskiyou cos., thence s. in Coast Ranges to Lake Co.; Ore. July–Aug.

5. **S. Párryi** Gray. Perennial, with 1–few weak branching herbaceous stems 2–4(–6) dm. high, glabrous; lvs. runcinate-pinnatifid, thickish, 2–8 cm. long, the lobes turned downward, somewhat spinulose-tipped; heads short-peduncled, 10–14-fld.; invol. ca. 15 mm. high; ligules whitish, 1.5–2 cm. long; aks. 3–4 mm. long, ribbed, not rugose; pappus-bristles sordid, naked at base, often united basally into 2's and 3's.—Occasional in dry open places, 2000–7000 ft.; Creosote Bush Scrub, Joshua Tree Wd., etc.; w. Mojave Desert to Mono Co.; to Utah, Ariz. May–June.

6. **S. tenuifòlia** (Torr.) Hall. [*Prenanthes? t.* Torr. *Ptiloria t.* Raf. *Lygodesmia t.* Shinners. *S. minor* (Hook.) Nutt.] Stems few to several from a woody root-crown, erect, with many slender ascending branches, 1–5 dm. tall; herbage pale, glabrous; lvs. mostly erect, filiform, 1.5–5 cm. long, the lowermost ± runcinate; heads terminal, ca. 5-fld.; invol. 8–10 mm. high, mostly with 5 principal phyllaries; aks. striate, ca. 2 mm. long; pappus-bristles 15–25, bright white, plumose throughout; $2n = 16$ (Stebbins et al., 1953).—Dry slopes, 4000–11,000 ft.; Sagebrush Scrub to Subalpine F.; Sierra Nevada especially on the e. side, from Tulare and Inyo cos. n.; to Wash., Mont., Colo., Ariz. July–Aug.

7. **S. myrioclàda** D. C. Eat. in King. [*Ptiloria tenuifolia m.* Blake. *S. runcinata* var. *m.* Jeps. *S. pauciflora* var. *m.* Munz. *S. t.* var. *m.* Cronq.] From a woody root-crown, the branches very many, slender, often flexuous, 3–6 dm. high; lvs. linear, inconspicuous, occasionally 1–2-toothed; heads mostly at ends of branches, often only 3–4-fld.; invol. 5–8 mm. high, the phyllaries 3–4–5; ligules pink; aks. striate, not at all tuberculate, ca. 4–5 mm. long; pappus-bristles plumose essentially to base and only slightly thickened there.—Occasional in dry places, 3000–6500 ft.; Creosote Bush Scrub to Pinyon-Juniper Wd.; w. Mojave Desert to Owens V.; Nev. May–Aug.

8. **S. pauciflòra** (Torr.) Nutt. [*Prenanthes? p.* Torr. *Ptiloria p.* Raf. *S. runcinata* Calif. refs. *Lygodesmia p.* Shinners.] Stems woody at base, rigid, intricately and divaricately branched, forming rounded bushes 3–6 dm. high, pale, glabrous; basal lvs. 3–7 cm. long, runcinate-pinnatifid with narrow divisions, the upper lvs. entire, often reduced to scales; heads solitary, short-peduncled, 3–5-fld.; invol. 8–10 mm. high, principal phyllaries 5; aks. 5-angled, striate, often ± rugulose; pappus-bristles tinged brownish, not plumose to very base; $2n = 16$ (Stebbins et al., 1953).—Common in washes and open places, below 6000 ft.; Creosote Bush Scrub to Pinyon-Juniper Wd.; both deserts and adjacent mts.; occasional in innermost cismontane valleys in s. Calif. (Mint C., Claremont, Lytle Creek, etc.); to Kans., Tex., L. Calif. May–Sept.

Var. **Paríshii** (Jeps.) Munz. [*S. runcinata* var. *P.* Jeps. *S. cinerea* (Blake) Blake.] Plant densely white-tomentulose.—Occasional on the deserts; to Nev.

9. **S. cichoriàcea** Gray. [*Ptiloria c.* Greene.] Perennial, woody at base, with erect simple or virgately branched stout stems 4–12 dm. high; herbage woolly when young,

later ± glabrate; lvs. oblong to oblanceolate, 5–18 cm. long, sessile, acute, subentire to remotely and saliently toothed; heads on short bracteate peduncles along the branches, ca. 12-fld.; invol. 12–15 mm. high, the phyllaries imbricate; receptacle with hirsute pits; ligules pink, 15–20 mm. long; aks. faintly 5-angled, smooth, ca. 3 mm. long; pappus-bristles 18–22, plumose throughout, brownish; $2n = 16$ (Stebbins et al., 1953).— Rocky slopes and canyons, below 6000 ft.; Chaparral, Mixed Evergreen F., etc.; from San Bernardino and Santa Ana mts. n. near the coast to Monterey Co., inland to Tejon Pass; Santa Cruz Id. Aug.–Oct.

162. Rafinésquia Nutt.

Glabrous or slightly puberulent ± fistulous or stoutish branching annuals with toothed or pinnatifid lvs. and ± cymosely branched panicles. Heads rather large, solitary at ends of branchlets, 15–30-fld.; invol. cylindro-conical, of ca. 7–15 equal lanceolate acuminate scarious-margined phyllaries and of some shorter unequal outer ones. Receptacle flat, naked. Ligules white or slightly rose in tinge, unequal; fls. fragrant. Aks. subfusiform, tapering into a beak; pappus of 10–15 capillary bristles, long-plumose from base to tip. Two spp. (C. S. *Rafinesque*, early Am. naturalist.)

Rays white, exceeding the invol. by ca. 5 mm.; ak.-beak slender, ca. as long as the body; pappus dull or brownish, the bristles plumose to the tip with straight hairs . . . . 1. *R. califórnica*

Rays white but veined outside with rose-purple, exceeding the invol. by 10–15 mm.; beak of ak. stout, shorter than the body; pappus bright white, the bristles not plumose near the tips, with soft, sub-arachnoid hairs . . . . 2. *R. neomexicana*

1. **R. califórnica** Nutt. [*Nemoseris c.* Greene.] Plants 2–15 dm. high, branched especially above; lvs. oblong in outline, 5–20 cm. long, subentire to pinnatifid, the lower petioled, others auriculate-clasping, the uppermost much reduced; heads scattered, the invol. 15–18 mm. high; ligules 5–8 mm. long, white; $2n = 16$ (Stebbins et al., 1953).— Most frequent in burns and disturbed places, at low elevs.; Coastal Sage Scrub, Chaparral, etc.; cismontane Calif. n. to Humboldt and Kern cos.; Channel Ids.; occasional in Creosote Bush Scrub, Joshua Tree Wd., etc., deserts; to Utah, Ariz., L. Calif. April–July.

2. **R. neomexicàna** Gray. Stems rather weak; invol. 18–22 mm. high; ligules 15–18 mm. long; $2n = 16$ (Stebbins et al., 1953).—Common in shade of shrubs, in canyons, etc.; Creosote Bush Scrub, Joshua Tree Wd.; deserts; to Utah, Tex. Feb.–May.

163. Malacòthrix DC.

Annual to perennial herbs, sometimes suffruticose. Lvs. usually basal and stems scapose, sometimes leafy. Lvs. toothed to pinnatisect. Heads small to medium in size, yellow or white, solitary or panicled, never sessile, commonly nodding in bud. Receptacle bristly or naked. Invol. of subequal inner phyllaries and much shorter outer ones, or strongly graduated, the phyllaries scarious-margined. Aks. columnar, truncate, ribbed. Pappus-bristles soft, scabrous, deciduous ± in a ring, 1–8 of them (rarely 0) stiffer and persistent, the ak. also often crowned with a ring of minute teeth. Ca. 15 spp., of w. N. Am. (Greek, *malakos*, soft, and *thrix*, hair.)

A. Outer phyllaries mostly round to ovate, with broad scarious margins.
    B. Outer phyllaries orbicular, 4–5 mm. wide; persistent pappus-bristles 1–4 .. 1. *M. Coulteri*
    BB. Outer phyllaries ovate, acute, ca. 2 mm. wide; persistent pappus none . . . . 9. *M. squalida*
AA. Phyllaries lanceolate to linear, acute, with narrow pale margins.
    B. Annuals.
        C. Lf.-segms. linear-filiform, elongate; persistent pappus-bristles 2 or more.
            D. Stems simple, ending in solitary heads and without lvs.; invol. 12–15 mm. high
                                        2. *M. californica*
            DD. Stems few-branched, each branch ending in a head; stems with a few lvs.; invol. 5–12 mm. high . . . . 3. *M. glabrata*
        CC. Lf.-segms. lanceolate to oblong, short, usually toothed; persistent pappus-bristles 1 or 0.
            D. Ligules ca. 1 cm. long; aks. 15-ribbed . . . . 4. *M. sonchoides*
            DD. Ligules shorter or, if as long, the lvs. with tufts of white wool on the margins; aks. often only 5-ribbed.
                E. Stems leafy at base only.
                    F. Fls. white to pink; pappus all deciduous; aks. 5-ribbed . 5. *M. floccifera*

        FF. Fls. yellow; 1 pappus-bristle and a crown of setulose teeth persistent;
            aks. minutely 15–striate .... 6. *M. Clevelandii*
    EE. Stems leafy throughout. Insular only.
        F. Invol. 10–12 mm. high; lf.-lobes acute; plants erect .... 7. *M. foliosa*
        FF. Invol. 6–7 mm. high; lower lvs. with obtuse lobes; plants often depressed
                                                                    8. *M. indecora*
  BB. Perennials.
    C. Suffrutescent herbs; lvs. linear to lanceolate or oblanceolate.
        D. Herbage densely white-tomentose when young; inner phyllaries obtuse
                                                                    10. *M. incana*
        DD. Herbage essentially glabrous; inner phyllares attenuate .... 11. *M. saxatilis*
    CC. Shrub; lvs. obovate .... 12. *M. Blairii*

1. **M. Còulteri** Gray. [*M. C.* var. *cognata* Jeps.] SNAKE's-HEAD. Erect annual, simple
or branched above, glabrous and rather pale, leafy, 1–5 dm. high; lvs. oblong, the lower
sinuate-pinnatifid to dentate, 3–6(–10) cm. long, the upper shorter, usually less deeply
toothed; heads short-peduncled, subglobose; invol. 12–16 mm. long, the phyllaries much
imbricated, the outer orbicular; ligules light yellow, 5–8 mm. long; aks. 15-ribbed,
4–5-angled; 1–4 pappus-bristles persistent; $2n = 14$ (Stebbins et al., 1953).—Occasional
on dry slopes, about bushes, in washes, etc., below 3500 ft.; Creosote Bush Scrub, V.
Grassland, Coastal Sage Scrub, etc.; cismontane and desert s. Calif., upper San Joaquin
V. and bordering hills; Santa Cruz and Santa Rosa ids.; to Utah, Ariz.; Argentina.
March–May.
2. **M. califórnica** DC. Annual with dense rosette of radical lvs. which are conspicuously woolly when young, 6–12 cm. long, pinnatifid into linear-filiform lobes; scapes
several, simple, not leafy, 1-headed, 1.5–3.5 dm. high; invol. 12–15 mm. high, the wool
often persistent, the phyllaries in ca. 3 ranks; ligules pale yellow, 12–16 mm. long; fls.
opening only in sunlight, fragrant; aks. narrow, lightly striate; pappus white, of 2
persistent bristles and some minute intervening teeth; $2n = 14$ (Stebbins et al., 1953).—
Dry sandy places, below 6000 ft.; V. Grassland, Coastal Sage Scrub, Joshua Tree Wd.,
etc.; cismontane s. Calif. to Sacramento V., out to coast in Monterey and San Luis
Obispo cos., extreme w. edge of desert; Santa Cruz and Santa Catalina ids. March–May.
3. **M. glabràta** Gray. [*M. californica* var. *g.* Gray ex D. C. Eat.] DESERT-DANDELION.
General habit of *M. californica,* but the stems few-branched, bearing 2 or more heads
each and few lvs.; plants quite glabrous, 1–4 dm. high; lvs. pinnatifid into linear lobes;
invol. 5–12 mm. high, mostly glabrous; $2n = 14$ (Stebbins et al., 1953).—Dry sandy
plains and washes, below 6000 ft.; Creosote Bush Scrub, Joshua Tree Wd., Shadscale
Scrub, etc.; both deserts, inner cismontane valleys from San Diego Co. to Santa Barbara
Co.; to Ida., Ariz. March–June.
4. **M. sonchoìdes** (Nutt.) T. & G. [*Leptoseris s.* Nutt.] Freely branched annual,
glabrous or early glabrate, the stems several from the base, 3–5 dm. high, freely branching; lvs. mostly near the base, 3–10 cm. long, oblong in outline, pinnatifid into short
callus-toothed lobes; uppermost lvs. much reduced; heads short-peduncled; invol. 7–8
mm. high, the phyllaries linear-acuminate; ligules bright yellow, ca. 10–12 mm. long;
aks. 15-striate, 5 of the ribs best developed; persistent bristles none, but the ak. with
a finely denticulate whitish crown; $2n = 18$ (Stebbins et al., 1953).—Occasional in
open ± sandy places, 2000–5000 ft.; Creosote Bush Scrub, Joshua Tree Wd.; w. and n.
Mojave Desert to e. Modoc Co.; to Ida., Nebr., Ariz. April–June.
5. **M. floccífera** (DC.) Blake. [*Senecio f.* DC. *M. obtusa* Benth. *M. parviflora* Benth.]
Annual, the stems 1–several from the base, 1–4 dm. tall, almost leafless; lvs. basal, oblong
in outline, usually with tufts of wool near the margin and in the axils, 2–8 cm. long,
dentate-pinnatifid with short teeth on the lobes; heads many; invol. 4–6 mm. long, the
main phyllaries linear, subequal, acute, often purplish at tip; ligules 3–6(–10) mm.
long, white or pale-yellow, often tinged pinkish; aks. obovate-oblong, ca. 2 mm. long,
entire at top, rather conspicuously 5-ribbed; persistent pappus-bristles none; $2n = 14$
(Stebbins et al., 1953).—Damp to mostly dry sandy and rocky places, below 5000 ft.;
Foothill Wd., Coastal Sage Scrub, N. Oak Wd., Mixed Evergreen F., etc.; Coast Ranges
from Ventura Co. to Siskiyou Co., then s. in Sierra Nevada to Mariposa Co. April–June.
6. **M. Clevelándii** Gray. Annual, usually branched from the base, 1–5 dm. high,
glabrous, slender; lvs. mostly basal, many, 3–10 cm. long, pinnatifid into lobes 1–3 mm.
wide; stems paniculately branched above, with many heads; invol. 6–7 mm. high, the

principal phyllaries in 1 series; corollas yellow, ca. 2–3 mm. long; aks. oblong-linear, finely 15-ribbed, with 1 (2) persistent pappus-bristles and a crown of minute white setulose teeth; $2n = 14$, 28 (Stebbins et al., 1953).—Burns and other disturbed places, below 5000 ft.; Chaparral, Coastal Sage Scrub, Foothill Wd., Coastal Strand, etc.; cismontane Calif. n. to Tehama and Tuolumne cos., occasional in Creosote Bush Scrub on the deserts; Ariz. April–June.

7. **M. foliòsa** Gray. Annual 2–5 dm. high, glabrous, branched above, leafy nearly to the heads; lvs. lanceolate, mostly laciniate-pinnatifid, 3–10 cm. long, the uppermost linear-attenuate; heads many, short- and slender-peduncled; invol. ca. 8–10 mm. high, the outer phyllaries in 1–2 series, lance-oblong, much shorter than the inner ones, all with narrow scarious margin and ± obtuse; fls. yellow, 4–6 mm. long; aks. subequally 15-ribbed; pappus wholly deciduous.—Occasional in dryish places; Coastal Sage Scrub; Santa Barbara, San Clemente, and Coronado ids. March–July.

8. **M. indécora** Greene. Near to *M. foliosa* and possibly intergrading with it, diffuse, depressed or erect, the lvs. more succulent and more obtusely lobed; invol. largely 6–7 mm. high.—Santa Cruz, San Miguel, and San Nicolas ids.

9. **M. squàlida** Greene. Much like the 2 preceding, but with the phyllaries much more imbricated, the outer in 3–4 series, round to ovate and with conspicuous scarious margins.—Santa Cruz, Santa Rosa, Santa Barbara, and Anacapa ids. Near to *M. insulàris* Greene of Coronado Ids.

10. **M. incàna** (Nutt.) T. & G. [*M. succulenta* Elmer. *Malacomeris i.* Nutt.] Perennial with stout root and woody crown; herbage white-woolly, ± glabrate in age; stems several, 1.5–2.5 dm. high; lvs. 5–10 cm. long, pinnately lobed to subentire, the lobes short, broad, blunt; peduncles slightly exceeding the lvs., mostly 1–2-headed; invol. ca. 2 cm. long, much imbricated; ligules lemon-yellow, ca. 10–12 mm. long; aks. 15-striate, ca. 1.5 mm. long; pappus-bristles all deciduous; $2n = 14$ (Stebbins et al., 1953). —Coastal Strand; Ventura Co. to San Luis Obispo Co.; Santa Rosa and San Miguel ids. Most months. Plants from about Surf, Santa Barbara Co., tend to be more glabrous and constitute *M. succulénta* Elmer.

11. **M. saxátilis** (Nutt.) T. & G. [*Leucoseris s.* Nutt.] Perennial, suffrutescent, with several stems 3–9 dm. long, diffuse or decumbent, branched, minutely tomentulose when young; lvs. many, well distributed, succulent, 3–10 cm. long, lanceolate to spatulate, obtuse, entire to dentate; infl. rather few-headed; invol. 10–16 mm. high, the phyllaries in 3–4 series; ligules ca. 15–18 mm. long, white with some rose; aks. 10–15-costate, ca. 5 of the striae riblike, the ak. crowned with a minute denticulate white border; none of pappus persistent; $2n = 18$ (Stebbins et al., 1953).—Sea bluffs, road-cuts, etc.; Coastal Sage Scrub; nw. of Santa Barbara to Gaviota. March–Sept.

KEY TO VARIETIES

A. Lvs. obtuse, fleshy, entire to ± toothed; heads few, the infl. condensed. Coast of Santa Barbara Co.
   M. saxatilis
AA. Lvs. acute, the lower often pinnatifid; heads many, the infl. open, diffuse.
   B. Herbage quite glabrous. Plants of s. Calif.
      C. Lower lvs. toothed or once-pinnatifid. Santa Barbara Co. to Orange and w. Riverside cos. . . . . Var. *tenuifolia*
      CC. Lower lvs. twice pinnatifid into narrowly linear segms. N. Channel Ids. . . Var. *implicata*
   BB. Herbage not glabrous.
      C. Plant ± canescent, Monterey and San Luis Obispo cos. . . . . Var. *commutata*
      CC. Plant hoary-tomentose. Carmel V. . . . . Var. *arachnoidea*

Var. **tenuifòlia** (Nutt.) Gray. [*Leucosyris t.* Nutt. *M. s.* var. *tenuissima* Munz. *M. altissima* Greene.] Lower lvs. acute, not succulent, often toothed or pinnately lobed with broad rachis, the upper linear-filiform, entire; infl. open, diffuse, slender-stemmed, many-headed.—Dry slopes and ridges, below 2000 ft.; Coastal Strand, Coastal Sage Scrub, etc.; San Luis Obispo and Kern cos. to Laguna Beach, Santa Ana Mts., Redlands; Santa Catalina Id. Most months.

Var. **implicàta** (Eastw.) Hall. [*M. i.* Eastw.] Lvs. twice pinnatifid into linear or filiform segms.; fls. purplish-tinged.—Sandy and rocky places; San Nicolas, San Miguel, Santa Rosa, and Santa Cruz ids.; Gaviota Canyon, Santa Barbara Co. Most months.

Var. **commutàta** (T. & G.) Ferris. [*M. c.* T. & G. *Hieracium ? californicum* DC.]

Plants with habit and lvs. of var. *tenuifolia* but more canescent with a light tomentum and lvs. more definitely lanceolate, less deeply toothed.—Chaparral, Foothill Wd.; Santa Lucia Mts. March–Sept.

Var. **arachnoìdea** (McGreg.) E. Williams. [*M. a.* McGreg.] Plants with habit of var. *commutata* but completely densely white-woolly except on the ± glabrous invols. and upper parts of peduncles.—Rocky open banks; Mixed Evergreen F.; Carmel V., Monterey Co. June–Dec.

12. **M. Blaìrii** (M. & J.) M. & J. [*Stephanomeria B*. M. & J.] Coarse straggly shrub 1–1.5 m. tall; flowering branches woody, 8–10 mm. thick; lvs. crowded, obovate to oblong-obovate, 5–13 cm. long, glabrate, coarsely sinuate or lobulate; infl. paniculate, leafless, 1–2 dm. long; heads many, 9–12-fld.; invol. ca. 6 mm. high; ligules rose, 1 cm. long; aks. 3–3.5 mm. long; pappus-bristles many, ca. 4 mm. long, plumose.—Rocky canyon-walls; Coastal Sage Scrub; San Clemente Id. July–Sept.

164. Glyptopleùra D. C. Eat.

Depressed small winter annuals with compact rosette of pinnatifid lvs. having a toothed white crustaceous edge. Heads many, short-peduncled, white or pale yellow, often pinkish in drying; invol. of ca. 7–12 equal lanceolate scarious-margined phyllaries and a basal group of ± spatulate bractlets which are pinnatifid to toothed on their crustaceous margins. Aks. oblong or columnar, obtusely 5-angled, each face with 2 rows of tubercles, the ak.-apex abruptly beaked; pappus-bristles white, in several series, the outer falling separately. Two spp. (Greek, *glyptos*, carved, and *pleura*, side, referring to the sculptured aks.)

Ligules conspicuous, well exserted; outer short bractlets lacerate-fringed at the conspicuously dilated apex and naked or sparsely setose below it; lvs. with narrow white margin, much narrower than the acuminate teeth . . . . 1. *G. setulosa*
Ligules inconspicuous, scarcely exserted; outer short bractlets pinnatifid above, lacerate-fringed most of way to base; lvs. with broad crustaceous margin and short teeth . . . . 2. *G. marginata*

1. **G. setulòsa** Gray. [*G. marginata* var. *s.* Jeps.] Stems to ca. 5 cm. long; lvs. oblong, sinuately pinnatifid into rounded or oblong lobes with the scarious margin acicular-toothed, 2–5 cm. long; peduncles very short; invol. ca. 10–12 mm. high; ligules broad, creamy to yellowish, to over 2 cm. long; aks. 4 mm. long; pappus ca. 8 mm. long.— Locally common, sandy flats, 2000–3500 ft.; Creosote Bush Scrub; w. Mojave Desert from Inyo Co. to s. Utah, nw. Ariz. April–May.

2. **G. marginàta** D. C. Eat. Much like no. 1; lvs. with a more crustaceous margin; outer short bractlets of invol. different; ligules short, little exserted; $2n = 18$ (Stebbins et al., 1953).—Creosote Bush Scrub; Mojave Desert (Nelson Range, Keeler, Box "S" Springs, Barstow, Kelso, Fenner, Goffs); to Ore., Nev., Utah. April–June.

165. Calycóseris Gray

Annual, branched from the base, glabrous below, sprinkled with tack-shaped glands above. Lvs. mostly basal, pinnately parted into narrow divisions. Heads rather showy, peduncled. Invol. many-fld., of many narrow scarious-margined equal phyllaries and of an outer series of very short loose ones. Receptacle with capillary bristles. Aks. fusiform, 5–6-ribbed, tapering into a short beak, this expanded apically into a shallow denticulate cup. Pappus abundant, white, of hispidulous white bristles falling away together. Two spp. (Greek, *kalux*, cup, and *seris*, a cichoriaceous genus.)

Fls. yellow; aks. smooth, not rugulose, ca. 8 mm. long, including the beak . . . . 1. *C. Parryi*
Fls. white to rose; aks. rugulose, ca. 6 mm. long including the beak . . . . 2. *C. Wrightii*

1. **C. Párryi** Gray. Plant mostly 1–3 dm. high, subsimple to divaricately branched; lvs. 3–12 cm. long, pinnately parted into short linear lobes or the upper subentire; stipitate glands dark; invol. 10–15 mm. high; phyllaries linear, attenuate at tip; ligules 15–25 mm. long; pappus longer than ak.—Common on open desert flats and on slopes, up to 6000 ft.; Creosote Bush Scrub, Joshua Tree Wd.; Calif. deserts; to Utah, Ariz. April–May.

2. **C. Wrìghtii** Gray. [*C. W.* var. *californica* Bdg.] Habit of no. 1; glands pale; ligules

white with rose or purplish dots or streaks on back; aks. rugulose on sides; $2n = 14$ (Stebbins et al., 1953).—Less frequent; Creosote Bush Scrub; Eureka V. and Death V. region, e. Mojave Desert, w. edge Colo. Desert; to Utah, w. Tex. March–May.

166. Anisócoma T. & G.

Low scapose annual with basal rosette of pinnately parted or toothed lvs. and several ascending 1-headed scapes. Invol. cylindric, strongly graduated, the inner phyllaries linear and acute, the outer successively shorter and very obtuse, appressed, all with brownish-green to purplish midrib and broad scarious margins. Receptacle flat with scarious linear bracts. Ligules pale yellow. Aks. oblong, truncate, 10–15-nerved, pubescent. Pappus bright white, of 10–12 plumose bristles in 2 unequal series. One sp. (Greek, *anisos*, unequal, and *kome*, tuft of hair, because of the 2 unlike sets of pappus-bristles.)

1. **A. acáulis** T. & G. Lvs. many, ± tomentulose, 3–5 cm. long, oblong in outline, with toothed segms.; scapes glabrous, 5–20 cm. high; invol. 2–3 cm. long, the phyllaries often edged with red toward the tips and with reddish dots; aks. 4 mm. long; longer pappus-bristles ca. 2–3 cm. long, the others much shorter; $2n = 14$ (Stebbins et al., 1953).—Common in washes and sandy places, above 2000 ft.; Creosote Bush Scrub, Joshua Tree Wd., etc.; Mojave and Colo. deserts, reaching 7500 ft. in adjacent mts.; Nev., nw. Ariz. April–June.

167. Atrichóseris Gray. Tobacco-Weed

Glabrous annual with basal rosette of broad thick lvs. and 1–few scapose stems corymbosely branched above. Heads on slender peduncles; invol. of ca. 15 equal linear acute phyllaries and several small outer ones. Receptacle scrobiculate. Ligules white. Aks. oblong with corky-thickened ribs. Pappus none. One sp. (Greek, *athrix*, without hair, and *seris* a cichoriaceous genus.)

1. **A. platyphýlla** Gray. Lvs. flat on the ground, often spotted, ± glaucous, round-oblong, 3–10 cm. long, short-petioled, spinulose-denticulate; scapes 3–18 dm. high, minutely bracteate; invol. 6 mm. high, the phyllaries with scarious margins; ligules oblong, 1–2 cm. long; fls. fragrant; aks. white, 4 mm. long; $2n = 18$ (Stebbins et al., 1953).—Common in sandy washes; Creosote Bush Scrub; Colo. Desert e. of Mecca and Imperial V., Mojave Desert e. of Ord Mts. and in Death V. region; to Utah, Ariz. Mostly March–May.

168. Tragopògon L. Goat's Beard

Stout nearly glabrous biennial or perennial herbs with entire elongate grasslike clasping lvs. and large solitary heads of many yellowish or purple fls. Invol. of 8–13(–17) lanceolate equal acuminate phyllaries. Aks. long-beaked, muricate; pappus of ± flattened plumose bristles. Ca. 45 spp., of Eu. and w. Asia. (Greek, *tragos*, goat, and *pogon*, beard.)

Fls. purple; phyllaries longer than the corollas .... 1. *T. porrifolius*
Fls. yellowish or greenish-yellow.
  Peduncles strongly inflated toward the apex; phyllaries usually 10–13, much longer than the lemon-yellow corollas .... 2. *T. dubius*
  Peduncles not or scarcely inflated toward apex; phyllaries usually 8–9, not longer than the chrome-yellow corollas .... 3. *T. pratensis*

1. **T. porrifòlius** L. Salsify. Oyster Plant. Stems from a stout taproot, leafy below, 6–12 dm. high; lvs. linear-lanceolate, long-acuminate, 1–3 dm. long, tapering uniformly from base to apex, not crisped on margins or curled backward at tip; peduncles strongly inflated toward apex; fls. averaging ca. 90 in a head; phyllaries mostly 8–9, not margined with purple; aks. 3–3.5 cm. long (including beak), tapering abruptly to the slender beak which exceeds the body; pappus brownish; $2n = 12$ (Poddubnaja et al., 1935).—Occasional weed in orchards, fields and waste places; natur. from Eu. April–June.

2. **T. dùbius** Scop. Much like no. 1; heads averaging over 100 fls.; phyllaries usually 13, sometimes up to 17, or as few as 8, long and narrow, not margined with purple;

ligules pale lemon-yellow, shorter than phyllaries; aks. slender, 2.5–3.5 cm. long, outer pale brown, inner straw-colored, the beak not strongly differentiated; pappus whitish; $2n = 12$ (P. et al., 1935).—Occasionally reported; native of Eu. May–July.

3. **T. praténsis** L. Lvs. narrowed more abruptly below, the margins concave and crisped, the tips recurved, not glaucous; heads averaging ca. 75 fls.; phyllaries mostly 8–9, margined with purple; aks. thickish, 2–2.5 cm. long, tapering abruptly into a slender beak; $2n = 12$ (Winge, 1926).—Rare in Calif.; from Eu. June–July.

169. Scorzonèra L.

Herbs closely resembling *Tragopogon*, with copious latex. Lvs. simple, entire, linear to lance-ovate. Heads yellow or purple, the invol. with many rows of imbricate phyllaries. Pappus of several rows of hairs, the outermost simple or all plumose. Ca. 90 spp., of Eurasia, especially in Medit. region. (Old French, *scorzon*, serpent; used against snake bite.)

1. **S. hispánica** L. VIPER'S GRASS. Perennial 4–12 dm. high, from a thick taproot; stems erect, branched above, leafy; lvs. elliptic-ovate to linear, entire, acuminate, the lower petioled, the upper reduced and sessile; heads solitary, long-peduncled; invol. cylindric-ovoid, 2.5–4 cm. long, the phyllaries ± arachnoid at base; fls. lemon-yellow, twice as long as invol.; aks. nearly white, ca. 15 mm. long, beaked, the inner smooth, the outer 5-angled, rough; pappus dirty white, ca. as long as ak., plumose; $2n = 14$ (Poddubnaja et al., 1935).—Reported as occasional escape in Mendocino and Napa cos.; native of Eu.

170. Hypochoèris L. CAT'S EAR

Annual or perennial herbs with lvs. in radical rosette or cluster and naked stems bearing a solitary head or a somewhat corymbose cluster of long-peduncled heads. Fls. yellow. Invol. cylindric or campanulate, with rather few lanceolate erect imbricated phyllaries. Receptacle flat, the bracts scarious, chaffy, thin. Aks. glabrous, upwardly scabrous, the body 10-ribbed, narrow-oblong or fusiform, truncate or beaked. Pappus-bristles plumose or some of outer shorter and simple. Ca. 70 spp., 12 in Eu., the others S. Am. (Greek name used by Theophrastus for this or some other genus.)

Annual; glabrous; outer aks. truncate, inner beaked .... 1. *H. glabra*
Perennial; pubescent; aks. all beaked.. 2. *H. radicata*

1. **H. glàbra** L. Lvs. usually not deeply lobed, spatulate-oblong, 2–10 cm. long; stems 1–4 dm. high; heads campanulate; invol. 12–16 mm. high; ligules scarcely exceeding invol.; aks. dark brown, the outer almost 4 mm. long; pappus ca. 9 mm. long, tinged yellowish or brownish; $2n = 10$ (Stebbins et al., 1953).—Weed in fields and waste places; natur. from Eu. March–June.

2. **H. radicàta** L. Perennial; lvs. hispid, pinnatifid, 6–14 cm. long; stems 4–15 dm. high; ligules longer than the invol.; aks. brown, the body ca. 3.5 mm. long; $2n = 8$ (Stebbins et al., 1953).—Weed about lawns, waste places, etc., scattered stations in most of cismontane Calif., more abundant northward; from Eu. May–Nov.

171. Hedýpnois Schreb.

Annual, low, simple or branched from base. Heads rather small, mostly peduncled, yellow-fld.; invol. with short outer phyllaries, the principal ones in one row, hardened in fr., enveloping the outer aks. Aks. cylindrical, ribbed, not beaked; pappus of outer aks. crownlike, of the inner usually in 2 series and consisting of free or basally connate scales ending in bristles and of short outer scales. A small genus of Medit. region. (Name given by Pliny to a kind of wild endive.)

1. **H. crètica** (L.) Willd. [*Rhagadiolus* H., Calif. refs.] Branches diffuse or spreading, 1–3 dm. long; basal lvs. petioled, often lobed, the cauline oblanceolate, 3–10 cm. long, serrate; peduncles rather long, mostly naked; invol. 8–10 mm. high; inner phyllaries strigose with short stiff bristles; ligules yellowish with purple tip; aks. cylindric, 6–7 mm. long; pappus-paleae 2 mm. long, the bristles 5 mm. long; $2n = 8$ (Stebbins et al., 1953).—Local but rather widely natur. weed; from e. Medit. April–May.

172. Pícris L. Ox Tongue

Coarse rough-bristly annuals or biennials with numerous yellow fls. in the terminal heads. Stems leafy. Outer phyllaries loose or spreading. Aks. with 5–10 rugose ribs; pappus of 1–2 rows of plumose bristles. Perhaps 50 spp., of Old World. (Greek, *picros,* bitter, the name of some allied herb.)

1. **P. echioìdes** L. Three–8 dm. high; lvs. oblong to oblanceolate, 5–20 cm. long, sessile, coarsely toothed; heads short-peduncled, or along the branches; outer phyllaries ovate, subcordate, the inner long-acuminate, ± spinose-pinnatifid toward tip, becoming thickened below, 15–20 mm. long; aks. beaked, the body oblong, brownish, ca. 3 mm. long, plainly rugose; pappus densely plumose; $2n = 10$ (Schnack & Covas, 1947).— Well established weed in waste places, fields, etc., especially in heavy soil; from Medit. region. June–Dec.

173. Leóntodon L. Hawkbit

Low acaulescent perennials with toothed or pinnatifid basal lvs. and simple or forked scapes bearing 1 or more yellow many-fld. rather large heads. Invol. oblong, scarcely imbricated, but with several bractlets at the base, the principal phyllaries subequal. Aks. finely striate, narrowed at summit or beaked; pappus persistent, of plumose bristles enlarged and flattened at base or the outer sometimes paleaceous. Ca. 45 spp., of Old World. (Greek, *leon,* lion, and *odous,* tooth, because of the toothed lvs.)

1. **L. Leýsseri** (Wallr.) G. Beck. [*Thrincia* L. Wallr. *L. nudicaulis* (L.) Banks.] Lvs. and scapes densely clustered; the former narrowly oblanceolate, toothed to pinnatifid, hispid, 5–15 cm. long; scapes slender, 1–3 dm. high, simple; heads solitary, the invol. 7–9 mm. long; phyllaries 6–12, narrow-lanceolate, the outer very small; outer aks. not roughened and with a pappus of short scales, the inner aks. roughened, with a plumose pappus with an outer ring of fine bristles; $2n = 8$ (Stebbins & al., 1953).— Open fields and hills, as a weed; Santa Clara Co. to Humboldt Co.; from Eu. May–Aug. A variable sp.

174. Sónchus L. Sow-Thistle

Leafy-stemmed mostly smooth and glaucous rather coarse herbs, with corymbed or umbellate heads. Invol. campanulate, the phyllaries few, thin, with many shorter ones at the base; these becoming callus-thickened. Fls. many, yellow. Aks. obcompressed, ribbed or striate, not beaked; pappus copious, of very white exceedingly soft and fine bristles mainly falling together. Ca. 45 spp., of Old World. (The ancient Greek name.)

Plant perennial, with running rootstocks; heads large, 15–25 mm. high; aks. oblong, but little compressed .... 1. *S. arvensis*
Plant annual; heads smaller, not more than 15 mm. high.
  Lvs. pinnately parted into narrow lobes; plant slender; aks. narrow, thickish .... 2. *S. tenerrimus*
  Lvs. serrate or with broad lobes; plants stout; aks. broad, flat, thin-edged.
    Basal auricles of the lvs. acute; aks. striate, transversely wrinkled .... 3. *S. oleraceus*
    Basal auricles of the lvs. rounded; aks. 3-nerved on each side, otherwise smooth .. 4. *S. asper*

1. **S. arvénsis** L. Extensively creeping by underground rootstocks; flowering stems 4–12 dm. high, leafy below; lvs. many, runcinate-pinnatifid to spinulose-dentate, short-petioled, 5–15 cm. long, the upper sessile, clasping, often not divided; heads few to many, on glandular-hispid peduncles; invol. 1.5–2.5 cm. long, the phyllaries linear-lanceolate, in ca. 3 series; aks. ca. 3 mm. long, transversely wrinkled on the ribs; $2n = 64$ (Wulff, 1937).—At one time natur. in peat lands near Wintersburg, Orange Co., Sacramento Co.; from Eu.

2. **S. tenérrimus** L. Much-branched annual 3–10 dm. tall, leafy to the infl.; lvs. oblong, 1–2 dm. long, with linear to narrowly lanceolate lobes; aks. narrow, longitudinally striate and transversely rugose, 2 mm. long; $2n = 14$ (Stebbins et al., 1953).—Weed in Tulare Co., w. San Diego Co.; Channel Ids.; L. Calif.; from Eu. April–June.

3. **S. oleràceus** L. Rather simple, 5–25 dm. high; lower lvs. petioled, 1–2 dm. long, lyrate-pinnatifid, the lobes lanceolate, spinulose-dentate, the upper lobe large and deltoid; upper lvs. reduced; auricles acute; invol. 9–15 mm. high; aks. longitudinally striate on

each face, transversely rugose, 2.5 mm. long; $2n = 32$ (Ishikawa, 1916).—Common weed in waste places, gardens, etc.; from Eu. Most months.

4. **S. ásper** L. Like *S. oleraceus* but the auricles of the clasping base rounded; aks. more plainly 3-nerved on each side, otherwise smooth; $2n = 18$ (Barber, 1941).— Common weed; natur. from Eu. Most months.

175. Lactùca L. LETTUCE

Leafy-stemmed herbs with mostly panicled heads. Heads several–many-fld. Invol. cylindrical, or in fr. conical; the phyllaries imbricated in 2 or more sets of unequal length. Rays 5-toothed at summit. Aks. contracted into a beak which is dilated at apex and bears copious early deciduous pappus of soft capillary bristles falling away separately. Perhaps 90 spp.; widely distributed. *L. sativa* L. is Garden Lettuce. (Ancient name from Latin, *lac*, milk, because of the milky sap.)

(Stebbins, G. L., Jr. Notes on Lactuca in w. N. Am. Madroño 5: 123–126, 1939.)

A. Aks. narrowed above but tapering to a thick summit or stout or firm beak, 3-ribbed on each face; fls. blue or blue-purple to creamy; pappus tawny to brownish.
  B. Plant perennial with long underground rhizome; heads 2–3 cm. broad at anthesis. Sierra Nevada n. to Modoc Co. . . . . 1. *L. tatarica*
  BB. Plant biennial; heads smaller. Marin Co. to Humboldt Co. . . . . 2. *L. biennis*
AA. Aks. narrowed into a slender beak ca. as long as the body; fls. yellow, sometimes drying bluish.
  B. Plant mostly biennial; aks. black, rugose.
    C. Main cauline lvs. not lobed, cordate-amplexicaul, 2–4 cm. broad. Bay Region
                                         3. *L. virosa*
    CC. Main cauline lvs. with linear-falcate lobes, sagittate-clasping, 4–7 cm. broad. N. Calif.
                                         6. *L. canadensis*
  BB. Plant annual; aks. gray or brownish, minutely rugulose on the ribs below, spiculate or setose toward the apex, not wing-margined.
    C. Lvs. oblong or elliptic in outline, the margins conspicuously spinulose-denticulate; panicle open, with widely spreading branches; fls. 14–20   4. *L. Serriola*
    CC. Lvs. linear-lanceolate or pinnatifid with linear-lanceolate lobes, the margins entire or remotely denticulate; panicle spiciform, with short ascending branches; fls. ca. 8–10
                                         5. *L. saligna*

1. **L. tatárica** (L.) C. A. Mey. ssp. **pulchélla** (Pursh) Steb. [*Sonchus p.* Pursh. *L. p.* DC.] Stems stoutish, simple, erect, very leafy, 3–10 dm. high, glabrous, ± glaucous; lvs. 5–15 cm. long, linear to lance-oblong, entire to pinnatifid, sessile or with a winged petiole; invol. 1.4–1.8 cm. long, the phyllaries 3–4-ranged; aks. oblong with a very short beak, ca. 3 mm. long; pappus 8 mm. long; $2n = 18$ (Stebbins et al., 1953).— Valleys, 3500–6500 ft.; Sagebrush Scrub, N. Juniper Wd., etc.; Sierra Nevada especially on e. side from Mono and Tuolumne cos. to Modoc Co.; to Alaska, e. Can., Okla., New Mex., Ariz. June–Sept.

2. **L. biénnis** (Moench) Fern. [*Sonchus b.* Moench. *L. spicata* Hitchc.] Subglabrous biennial, coarse, 5–40 dm. high, leafy; lvs. irregularly pinnatifid, coarsely toothed, the upper cauline sessile and auriculate; fls. pale blue to cream; aks. ± oblong, thickish, mottled, practically beakless; pappus brownish.—Rare or local; Redwood F.; Marin Co. to Humboldt Co.; to B.C., Atlantic Coast. June–Aug.

3. **L. viròsa** L. Usually biennial, the stem leafy, 6–20 dm. high, glabrous or prickly below; lvs. ovate-oblong in outline, undivided to ± deeply pinnatifid, the lower petioled, the upper sessile and cordate-amplexicaul with appressed auricles; panicle pyramidal; invol. 8–12 mm. long, glabrous, the phyllaries numerous, glaucous, with white margin and reddish tip; fls. pale greenish-yellow; ak.-body ca. 4 mm. long, ± rugulose, blackish, the beak white and ca. as long; $2n = 18$ (Thompson et al., 1941).—Brushy and wooded slopes; Mixed Evergreen F., Coastal Prairie, etc.; San Francisco Bay region; from Eu. Aug.–Oct.

4. **L. Serriòla** L. [*L. Scariola* L.] PRICKLY LETTUCE. Annual, ± prickly-bristly especially below, 6–15 dm. tall; lvs. spinose-denticulate, pinnatifid, sessile or sagittate-clasping, 4–16 cm. long, soft-prickly beneath along midrib; heads many, 9–14-fld.; invol. 10–12 mm. high, with phyllaries in ca. 4 lengths; fls. yellow, often drying purplish; aks. oblanceolate, with muriculate apex, 5–7-ribbed on each face, the beak filiform; pappus white; $2n = 18$ (Thompson et al., 1941).—Common and widespread weed; from Eu. May–Sept.

Var. **integràta** Gren. & Godr. Lvs. oblong, entire; plant usually smaller than in the typical form.—Occasional weed.

5. **L. salígna** L. Plants 5–18 dm. high, more slender than *L. Serriola*; lvs. linear to

lanceolate, mostly entire, 5–20 cm. long, not prickly on under side; infl. virgate, narrow; $2n = 18$ (Thompson et al., 1941).—Waste places, fields, etc.; San Francisco Bay Region, Lake Co.; from Eu. Aug.–Nov.

6. **L. canadénsis** L. Weedy, mostly biennial, 0.5–2.5 m. tall, mostly glabrous; principal lvs. sagittate-clasping, mostly 1–3 dm. long, their linear-falcate lobes mostly entire; heads many; invol. 10–15 mm. long in fr.; fls. mostly 13–22, yellow; aks. ± elliptic, blackish, cross-rugulose, 5–6 mm. long including the pale beak; pappus white, 5–7 mm. long; $2n = 34$ (Whitaker & Jagger, 1939).—Waste places and wds.; Siskiyou, Shasta, Plumas, and Amador cos.; to B.C. and Atlantic Coast. July–Sept. Our plants are largely the var. *longifòlia* (Michx.) Farw.

176. Hieràcium L. HAWKWEED

Perennial herbs, often hairy, sometimes glandular, mostly caulescent. Lvs. entire or dentate, never deeply lobed. Heads variously panicled, rarely solitary. Receptacle flat, usually naked. Invol. cylindric or campanulate, the phyllaries in 1–3 series, subequal or ± imbricate, not thickened on the back, with a few small calyculate ones at the base. Ligules white or yellow in ours, truncate and 5-toothed at apex. Aks. oblong or columnar, truncate, 10–15-ribbed. Pappus of 1–2 series of brownish or sordid fragile capillary bristles. More than 700 spp., mostly of Eu. and S. Am.; variable. (Greek, *hierax*, a hawk; the ancients believed that hawks used the sap to sharpen their eyesight.)

Ligules white; or ochroleucous; stem (except at base) and invol. glabrous, or invol. with a few
long hairs .... 1. *H. albiflorum*
Ligules yellow; stem and invol. pubescent or hirsute.
  Stems densely leafy at least below.
    Herbage densely stellate-pubescent, plant also villous-hirsute below .... 2. *H. Greenei*
    Herbage not or sparsely stellate-pubescent, but covered with long mostly brown shaggy hairs.
      Lvs. entire; pappus brownish .... 3. *H. horridum*
      Lvs. toothed; pappus almost white .... 4. *H. argutum*
  Stems sparsely or not at all leafy.
    Plants mostly over 3 dm. high; invol. copiously glandular-viscid, not hirsute
                                             5. *H. cynoglossoides*
    Plants mostly less than 3 dm. high; invol. either ± hirsute or not very glandular-viscid.
      Invol. black-hirsute and glandular; basal lvs. glabrous. Subalpine .... 6. *H. gracile*
      Invol. subglabrous; basal lvs. ± hirsute. N. Coast Ranges .... 7. *H. Bolanderi*

1. **H. albiflòrum** Hook. [*H. occidentale* Eastw.] Stems 1–several, slender, erect, 4–8 dm. high, leafy below, quite naked above, the lower parts mostly densely hirsute with whitish or tawny hairs; lvs. mostly basal, oblong to oblanceolate, 8–15 cm. long, 2–4 cm. wide, the lower with winged petiole, the upper sessile, reduced; heads in a loose panicle, white-fld.; invol. 9–10 mm. long, glabrous or often glandular and with black hairs; phyllaries linear-subulate, the outer very short; ligules 3–4 mm. long; aks. red-brown, 2–3 mm. long; pappus dull white, or tawny; $2n = 18$ (Stebbins et al., 1953).— Dry open wooded slopes, below 9700 ft.; Redwood F. to Lodgepole F.; montane and cismontane Calif. from San Diego Co. n.; to Alaska, Colo. June–Aug.

2. **H. Greènei** Gray. Stem mostly 2–4 dm. high, from a tuft of subbasal lvs., densely stellate-tomentose and below villous-hirsute, simple or paniculately branched; lvs. oblong-oblanceolate, ± petioled, irregularly low-dentate to subentire, 5–10 cm. long, the upper much reduced, densely gray- or rusty-stellate and ± hirsute; heads narrow, few-fld.; invol. 8–10 mm. long, stellate but not glandular; inner phyllaries 5–7, narrow, 7–9 mm. long, the outer shorter; ligules yellow, ca. 1 cm. long; aks. dark brown, 10-ribbed, 4–7 mm. long; pappus deep tawny.—Dry slopes in woods, 4000–9200 ft.; Montane Coniferous F.; n. Lake Co. to Humboldt Co. and Mt. Shasta; s. Ore. July–Aug.

3. **H. hórridum** Fries. [*H. Breweri* Gray.] Stems few to several, paniculately branched above, 1–3.5 dm. high; herbage densely shaggy with whitish or brownish long soft hairs; lvs. spatulate to oblong, entire, 3–10 cm. long, not densely crowded; infl. open, with many small heads; invol. narrow, 6–9 mm. long, the phyllaries lanceolate with medial part ± glandular and long-hairy; ligules ca. 15, bright yellow, ca. 1 cm. long; aks. 10-ribbed, red-brown, ca. 3 mm. long; pappus brownish; $2n = 18$ (Stebbins et al., 1953).—Common in dry ± rocky places, 5000–11,000 ft.; Montane Coniferous F.; Santa Rosa and San Jacinto mts. n. through Sierra Nevada, Mt. Shasta to Modoc Co.; Ore. July–Aug.

4. **H. argùtum** Nutt. [*H. Brandegei* Greene.] Stem simple, leafy below, slender, 3–10

dm. high, shaggy below with brown hairs, subglabrous above, terminating in a long racemose panicle; lvs. oblong to oblanceolate, remotely dentate, the lower 8–16 cm. long, 1–4 cm. wide, the upper reduced; invol. 7–10 mm. long, with dark stipitate glands, the inner phyllaries linear-attenuate, the outer shorter; ligules yellowish, ca. 8 mm. long; aks. dark brown, ca. 2.5 mm. long; pappus sordid.—Dry slopes; S. Oak Wd., Closed-cone Pine F.; coastal Santa Barbara and San Luis Obispo cos.; Santa Cruz and Santa Rosa ids. June–Aug.

Var. **Paríshii** (Gray) Jeps. [*H. P. Gray. H. Grinnellii* Eastw.] Invol. with light-colored stipitate glands.—Below 4000 ft.; s. face of San Bernardino and San Gabriel mts. to Ventura Co., also in Santa Lucia Mts. June–Oct.

5. **H. cynoglossoìdes** Arv.-Touv. ex Gray. Stems 1 to few, simple, rather slender, 3–7 dm. high, subglabrous to yellowish-hirsute; lvs. mostly near the base, lanceolate to oblanceolate, entire, 1–2 dm. long, ± copiously villous-hirsute; infl. a broad cymose panicle, glabrous to somewhat glandular or short-bristly; invol. 8–10 mm. high, glandular-pubescent, the inner phyllaries linear-attenuate, the outer much shorter; ligules bright yellow, ca. 8–10 mm. long; aks. dark brown, ca. 3 mm. long, striate; pappus sordid.— Open woods and rocky places, 1500–6500 ft.; largely Chaparral, Yellow Pine F.; Yolla Bolly Mts. and Sierra Nevada n.; to B.C., Rocky Mts. May–July.

Var. **nudicáule** Gray. Stems quite naked.—Trinity Co. to Del Norte and Shasta cos.

6. **H. grácile** Hook. Stems 1–several from a nearly basal cluster of lvs., simple or ± branched, slender, puberulent, mostly 1–3 dm. high; lvs. spatulate to oblong-oblanceolate, subentire, mostly glabrous, 2–8 cm. long; heads few, in an open infl.; invol. campanulate, 7–8 mm. high, black-hirsute and glandular; aks. dark or red-brown, 2 mm. long, strongly ribbed; pappus light tawny; $2n = 18$ (Stebbins et al., 1953).—Woods and rocky places, 8000–11,000 ft.; Lodgepole F., Subalpine F.; Sierra Nevada from Inyo and Tulare cos. to Tuolumne Co., lower in Siskiyou and Trinity cos.; to B.C., Rocky Mts. July–Aug. Some plants with the involucral hairs light-colored are the var. *detónsum* Gray.

7. **H. Bolánderi** Gray. Stems 1 or few, simple or loosely branched, glabrous or puberulent, 1–3 dm. high; lvs. basal, oblanceolate, to narrowly obovate, mostly subentire, sparingly hirsute on upper surface, sometimes on lower, 2–7 cm. long; infl. narrow to open, with rather few heads; invol. narrow, 8–9 mm. long, glabrous or nearly so; ligules 4–5 mm. long, yellow; aks. linear, dark brown, striate, ca. 4 mm. long; pappus fuscous.—Dry wooded places, 1000–9000 ft.; Mixed Evergreen F., Yellow Pine F., Red Fir F.; Mendocino and Trinity cos. to Del Norte, Siskiyou, and Shasta cos.; sw. Ore. June–Aug.

177. Crèpis L. HAWKSBEARD

Annual to perennial herbs, usually ± caulescent. Lvs. largely ± basal, entire to toothed or deeply lobed. Heads in panicles or cymose clusters. Invol. cylindric or campanulate, the inner phyllaries equal, in a single series, with ± thickened midribs; outer calyculate phyllaries very small or wanting. Receptacle naked. Fls. yellow. Aks. columnar or fusiform, narrowed toward the summit, 10–35-ribbed. Pappus of many soft mostly white bristles. Many polyploid apomicts occur that are ± intermediate between the diploid spp. and cannot be accounted for in a brief treatment. Ca. 200 spp.; widely distrib. in N. Hemis. (Greek, *krepis*, a sandal, an ancient plant name.)

(Babcock, E. B. The genus Crepis. Univ. Calif. Publ. Bot. 21–22, pp. 1–1030, 1947.)
A. Native perennials.
  B. Plants glabrous, densely tufted, 2–7 cm. high; lvs. entire .... 1. *C. nana*
  BB. Plants ± hairy or tomentose, taller; lvs. toothed or pinnatifid.
    C. Stems and lvs. glabrous or ± hispidulous; invols. glandular. Sierra Co. to Inyo Co.
                                                                          9. *C. runcinata*
    CC. Stems and lvs. at least slightly tomentose, often also otherwise hairy.
      D. Herbage and invols. slightly tomentose and shaggy-hirsute with long (1–3 mm.) glandular hairs; inner phyllaries long attenuate, the tips not folded over the florets in the buds; outer phyllaries lance-linear. N. Calif. ...... 2. *C. monticola*
      DD. Herbage and invols. sometimes setose or glandular-pubescent, but hairs short; inner phyllaries somewhat attenuate toward tip, folded over the florets in the buds; outer phyllaries mostly lanceolate or ovate-lanceolate.
        E. Invols. densely beset with curved or crisped setae, or if not, the basal part of the stems and petioles conspicuously setose; aks. weakly ribbed or merely striate .... 5. *C. modocensis*

EE. Invols. glabrous, tomentose, glandular-pubescent, or, if with a few setae, these straight and the stems and petioles not setose; aks. strongly ribbed.
  F. Largest heads of the infl. with 5–7 inner phyllaries; heads 5–10-fld.
    G. Phyllaries with a glabrate median part and white-tomentose near the scarious margins; aks. reddish or brownish, shorter than the pappus, strongly ribbed; cauline lvs. much reduced or none
          6. *C. pleurocarpa*
    GG. Phyllaries glabrous or sparingly and evenly tomentulose, the inner yellow-green, shading indistinctly into the scarious margins; aks. yellow, buff or tawny, equaling or longer than pappus, finely ribbed; cauline lvs. at least 1–3, well developed . . 7. *C. acuminata*
  FF. Largest heads of the infl. with 8–13 inner phyllaries; heads 9–40-fld.
    G. Plants mostly 1–3 dm. high, with a cymose infl. of 2–25 heads; invols. broadly cylindric, 5–9 mm. wide at anthesis, mostly 12–25-fld.; longest outer phyllaries mostly ⅓–⅔ as long as inner.
      H. Lvs. grayish-tomentose, not glandular-pubescent; peduncles not expanded toward the apex . . . . 3. *C. occidentalis*
      HH. Lvs. green, glandular-pubescent, in fresh specimens with a reddish midrib and petiole; peduncles expanded toward apex
          4. *C. Bakeri*
    GG. Plants mostly 2.5–6 dm. high, with a narrowly panicle of 20–60 heads; invols. narrowly cylindric, 3–5.5 mm. wide at anthesis, mostly 8–10-fld.; longest outer phyllaries mostly ⅕–⅓ as long as inner bracts.
      H. Basal lvs. glabrate; phyllaries glabrate on median part, strongly tomentose near the scarious margin
          6. *C. pleurocarpa*
      HH. Basal lvs. grayish tomentose; phyllaries evenly tomentose or tomentulose . . . . 8. *C. intermedia*
AA. Introduced annuals or biennials.
  B. Invols. 5–8 mm. long; aks. 2–3.5 mm. long . . . . 10. *C. capillaris*
  BB. Invols. 8–12 mm. long; aks. 3.2–5 mm. long.
    C. Aks. 6–8 mm. long; invols. not strongly setose . . . . 11. *C. vesicaria*
    CC. Aks. 3–5 mm. long; invols. strongly setose . . . . 12. *C. setosa*

1. **C. nàna** Richards. Plants tufted, 0.2–0.7 dm. high, the stems many from short branches from the root-crown, slender; lvs. mostly basal, glabrous, obovate-spatulate to ± elliptical, the blades 1–5 cm. long on petioles ca. as long, ± purplish, entire or few-lobed; branches congested, 2–4-headed, the heads borne among the lvs.; invol. 10–13 mm. long, cylindrical; outer phyllaries 5–8, unequal, short; the inner 10, equal, oblong, narrowed near the obtuse purplish ciliate apex, scarious-margined; corolla 7–8 mm. long, yellow with purple tinge on outer face; aks. 4–6(–8) mm. long, golden-brown, subterete, columnar, 10–13-ribbed, the ribs rounded, smooth or rugulose; pappus white, 4–6 mm. long; $2n = 14$ (Babcock, 1947).—Stony or gravelly scree and talus, 8000–13,000 ft.; Alpine Fell-fields, Subalpine F., Lodgepole F., Bristle-cone F.; San Gabriel, Panamint, and Sweetwater mts., Sierra Nevada; to B.C., Alaska, Labrador; Asia. July–Aug.

Ssp. **ramòsa** Babc. Taller, with more elongate branches (to 1.8 dm. high); stems leafy; heads borne well above the basal lvs.; aks. with broad nerves.—Sierra Nevada to B.C.

2. **C. montícola** Cov. Sparsely tomentulose and densely hirsute with long glandular hairs, from a vertical woody elongate taproot, the stems 1–3 dm. high, leafy; lvs. 1–2.5 dm. long, 2–4 cm. wide, pinnately parted into oblong lobes, sometimes merely dentate; stems several-branched, the branches 1–6-headed; fls. 16–20; invol. campanulate, 16–24 mm. long, 5–10 mm. wide at middle; outer phyllaries 3–10, acuminate, ½–¾ as long as inner; inner 7–12, long-acuminate, with yellowish, spongy-thickened keel; receptacle glabrous; corolla 16–21 mm. long; aks. reddish-brown, 6–9 mm. long, fusiform, ca. 13-ribbed; pappus white or pale cream, 9–13 mm. long; $2n = 22$, 33? 44, 55? 77? or 88? (Babcock, 1947).—Dry stony places, 3000–7100 ft.; largely Yellow Pine F., Red Fir F.; Sierra and Santa Clara cos. to s. Ore. June–July.

3. **C. occidentàlis** Nutt. Mostly 1.5–2.5(–4) dm. high, with a close gray tomentum and infl. also ± glandular, arising from a deep thick taproot; lvs. mostly 1–3 dm. long, 2–6 cm. wide, sinuately dentate or runcinately or deeply pinnatifid with linear or lanceolate toothed lobes; lvs. gradually reduced up the stems; stems 1–3, erect, stout, several-branched, each main branch with ca. 10–30 heads in a corymbiform infl.; larger heads 18–30-fld., the invols. broadly cylindric or cyathiform; outer phyllaries 6–8, to ca. ½ as long as inner; inner 8–13, ± carinate toward outer base; corolla ca. 22 mm. long; aks. mostly medium brown, 6–10 mm. long, moderately strongly 10–18-ribbed;

pappus whitish, 10–12 mm. long; $2n = 22$, 33, 44, 55? 66? 77, 88? (Babcock, 1947).—
Dry ± rocky or stony places, 4000–9000 ft.; Sagebrush Scrub, Yellow Pine F. to Lodgepole F.; cent. Sierra Nevada and Argus Mts. to se. Wash., Wyo., New Mex. June–Aug.

KEY TO SUBSPECIES

Invols. with at least some glandular pubescence.
    The invols., peduncles and upper lvs. ± glandular but not setose; largest heads 18–30-fld., and
      with 10–13 inner phyllaries . . . . *C. occidentalis*
    The invols., peduncles and usually upper lvs. with conspicuous gland-tipped setae; larger heads
      12–14-fld., with 8 inner phyllaries . . . . ssp. *costata*
Invols. without glandular pubescence, or if with a few glands, the invols. with 8 inner phyllaries
and less than 15 florets.
    Stems well developed, 1–4 dm. high, with a definite primary axis; invols. mostly with 8 inner
      phyllaries and with outer ca. ⅓ as long as inner; lvs. if pinnatifid, with closely spaced strongly
      toothed or pinnatifid lobes . . . . ssp. *pumila*
    Stems low, 0.5–2 dm. high; infl. branching from near base of stem, bearing heads on long divergent
      peduncles; invols. with 8–12 inner phyllaries and with outer ⅓–½ as long as inner; lvs. deeply
      pinnatifid, with remotely spaced lanceolate entire or coarsely few-toothed lobes . . . . ssp. *conjuncta*

Ssp. **costàta** (Gray) Babc. & Steb. [*C. o.* var. *c.* Gray. *C. grandifolia* Greene.]—Lassen
and Siskiyou cos. to Alta., S. Dak.

Ssp. **pùmila** (Rydb.) Babc. & Steb. [*C. p.* Rydb.] With the sp. from Kern and
Ventura cos. n.; to s. Ore., Wash., Mont., Nev.

Ssp. **conjúncta** (Jeps.) Babc. & Steb. [*C. o.* var. *c.* Jeps.] Sierra Nevada to n. Calif.;
se. Wash., nw. Wyo.

4. **C. Bàkeri** Greene. One–3 dm. high, with dark green glandular-pubescent lvs.,
stems, and invols., often with some red-purple on midveins, etc.; lvs. 1–2 dm. long,
2–5 cm. wide, elliptic, mostly deeply pinnatifid with lanceolate or narrowly elliptic
dentate segms.; stems 1–3, erect, remotely several-branched, each branch with 1–4 heads,
these 11–40-fld.; invol. 14–21 mm. long; outer phyllaries lanceolate, 8–10, the longest
ca. ½ as long as inner; inner 10–14, lanceolate, narrowly carinate dorsally; corolla ca.
20 mm. long; aks. 8–10 mm. long, mostly only slightly attenuate at apex, ca. 13-ribbed,
brown to yellowish; pappus 9–12 mm. long; $2n = 22$, 33? 44, 55? (Babcock, 1947).—
Dry places, 4000–8000 ft.; Sagebrush Scrub, N. Juniper Wd. to Lodgepole F.; Alpine
and Trinity cos. to Wash. June–Aug.

Ssp. **Cusíckii** (Eastw.) Babc. & Steb. [*C. C.* Eastw.] Invols. 10–17 mm. long; pappus
6–9 mm. long; mature aks. somewhat contracted at apex.—Lake Co. to Lassen and
Siskiyou cos.; s. Ore.

5. **C. modocénsis** Greene. Stems 1–4.5 dm. high, always with a well developed primary
axis, at least the lower part with spreading glandless setae or hairs; basal lvs. 0.7–2.5
dm. long, glabrate or tomentose and with some setae, deeply pinnatifid with lanceolate
toothed or pinnatifid segms.; infl. with 1–10 heads; heads 10–60-fld., 11–16 mm. long,
the phyllaries setose throughout, the inner 8–18, the outer ca. half as long as inner;
corolla 13–22 mm. long; aks. 7–12 mm. long, weakly if at all striate, greenish black to
deep red-brown; pappus 5–10 mm. long; $2n = 22$, 33? 44, 55? 66? 88? (Babcock, 1947).
—Dry places, 4500–7000 ft.; Sagebrush Scrub, N. Juniper Wd., Yellow Pine F.; Lassen
and Modoc cos.; to Ore., Mont., Colo. June–Aug.

Ssp. **subacáulis** (Kell.) Babc. & Steb. [*C. o.* var. *s.* Kell. *C. s.* Cov.] Mostly 0.6–2
dm. high and branched from near the base, with 1–5 heads; invols. 13–21 mm. high,
less setose; aks. blackish to brownish or reddish, more strongly ribbed, the pappus as
long as or longer than ak.—At 5000–10,000 ft.; San Bernardino Mts., n. Sierra Nevada,
Warner Mts.; s. Ore.

6. **C. pleurocárpa** Gray. [*C. acuminata* var. *p.* Jeps.] Root slender; stem 1.5–4(–6)
dm. high, mostly greenish-tomentulose; lower lvs. 0.7–2.8 dm. long, 1–7 cm. wide,
elliptic or oblanceolate, denticulate to pinnately divided with deltoid or lanceolate lobes,
the terminal segm. acuminate, 1–7 cm. long; stems branched below or above middle,
each branch ± cymose toward apex; heads 7–40, 4–12-fld.; invol. cylindric-campanulate,
8–16 mm. long; outer phyllaries 5–6 with 1–3 subtending ones, the longest 1.5–4 mm.
long; inner mostly 5, lanceolate, deep green or blackish when dry, glabrate in middle,
densely floccose toward the scarious margins; corollas 15–20 mm. long; aks. deep
chestnut-brown, 5–8 mm. long, 10-ribbed; pappus 6–12 mm. long; $2n = 22$, 33? 44?

55? 77? 88? (Babcock, 1947).—Largely 3000–7000 ft.; Yellow Pine F. to Lodgepole F.; Lake and Eldorado cos. to Wash. June–Aug.

7. **C. acumìnàta** Nutt. Stems 1–2, 2–6 dm. high, with gray-green foliage; basal lvs. 1–4 dm. long, 1–10 cm. wide, pinnately lobed with 5–10 pairs of lateral entire or dentate segms., the apical part 3–8 cm. long, acuminate; cauline lvs. few, remote, reduced upward; infl. corymbiform, usually ca. 30–100-headed; heads 5–10-fld.; invol. cylindric-campanulate, 9–15 mm. long; outer phyllaries 5–7, ciliate; inner 5–8, lanceolate, ciliate at apex; corollas 10–18 mm. long; aks. pale yellow or brownish, 6–9 mm. long, ca. 12-ribbed; pappus 6–9 mm. long; $2n = 22, 33, 44, 55?$ 88? (Babcock, 1947). —Dry places, 2600–10,700 ft.; Sagebrush Scrub, Montane Coniferous F.; San Bernardino Mts. n. through Sierra Nevada to Modoc and Siskiyou cos., then s. to Trinity Co.; to Wash., Mont., Colo. June–Aug.

8. **C. intermèdia** Gray. [*C. acuminata* var. *i.* Jeps.] Mostly 3–7 dm. high, canescent-tomentose; lower lvs. 1.5–4 dm. long, 2–9 cm. wide, elliptic-lanceolate, pinnatifid, the apical part long-attenuate, the lateral lobes lanceolate, entire or dentate; infl. corymbiform, with 10–60 heads, these 7–12(–16)-fld.; invol. 10–16 mm. long, tomentose to glabrous; outer phyllaries 6–8, lance-deltoid, the inner mostly 7–8, lanceolate, obtuse, becoming carinate; corollas 14–30 mm. long; aks. yellow to brown, 6–9 mm. long, 10–12-ribbed; pappus 7–10 mm. long; $2n = 33?$ 44? 55, 88? (Babcock, 1947).—Dry slopes, 1500–10,500 ft.; Sagebrush Scrub, N. Oak Wd., Montane Coniferous F., etc.; cent. Sierra Nevada and Inyo-White Mts. to Warner Mts. and N. Coast Ranges; to Wash., Alta., Colo., New Mex. June–Aug.

9. **C. runcinàta** T. & G. ssp. **Andersònii** (Gray) Babc. & Steb. [*C. A.* Gray.] Stems 1–several, 2.5–5 dm. high, nearly leafless, glabrous to hispidulous; basal lvs. oblong-obovate, 0.5–2.5 dm. long, 1–8 cm. wide, laciniately dentate to pinnatifid, narrowed to a winged base; heads mostly 6–20; invols. 19–21 mm. long, glandular-pubescent; outer phyllaries 5–12, 2–3 mm. wide, unequal; inner 10–16, lanceolate, attenuate; corollas 10–18 mm. long; aks. pale yellow to red-brown, 6–8 mm. long, short-beaked, 10–13-ribbed; pappus 6–9 mm. long; $2n = 22$ (Babcock, 1947).—Alkaline meadows, 4500–5500 ft.; Sagebrush Scrub, etc.; Sierra Co.; w. Nev. June–July.

Ssp. **Hállii** Babc. & Steb. Heads 5–14; invol. 9–13 mm. high, glandular-puberulent; aks. chestnut-brown, 4.5–6.5 mm. long; pappus 6–7 mm. long.—Alkaline places, 5000–7000 ft.; Sagebrush Scrub, Pinyon-Juniper Wd.; Inyo and Mono cos.

10. **C. capillàris** (L.) Wallr. [*Lapsana c.* L.] Annual or biennial, 1–several-stemmed, 1–9 dm. high; stems simple below, slender, paniculate above; herbage glabrous to glandular-pubescent; lvs. mostly basal, thinnish, oblong or spatulate in outline, dentate to pinnately parted, 4–10 cm. long; upper smaller, amplexicaul; infl. ± corymbiform, 20–60-fld.; invol. cylindric in anthesis, 5–8 mm. long; outer phyllaries 8, linear, glabrous to tomentose or sparsely gland-hairy; inner 8–16, lanceolate, membranous-margined; corolla 8–10 mm. long; aks. brownish, 1.5–2.5 mm. long, 10-ribbed; pappus 3–4 mm. long; $2n = 6$ (Babcock, 1947).—Weed at widely scattered stations; introd. from Eu. June–Aug.

11. **C. vesicària** L. ssp. **taraxacifòlia** (Thuill.) Thell. [*C. t.* Thuill.] Annual or biennial 1–8 dm. high; stems 1–few, paniculately branched above; herbage subglabrous, but with short scattered bristles; lower lvs. pinnately parted to dentate, 4–10 cm. long; heads many-fld.; invol. 8–12 mm. long; outer phyllaries 6–12, ± lanceolate; inner 9–13, often shortly gland-pubescent, spongy-thickened in fr.; corolla 10–12 mm. long; aks. pale brown, 6–8 mm. long, 10-ribbed; pappus 4–6 mm. long; $2n = 8, 16$ (Babcock, 1947).—Introd. weed, Mendocino and Los Angeles cos.; native of Eu. June–Aug.

12. **C. setòsa** Haller f. Annual 1–8 dm. high; lower lvs. 0.5–3 dm. long, dentate to pinnately parted, finely hispid, with a winged petiole; stems ± hispid; heads many-fld.; invol. cylindric-campanulate, the outer phyllaries 10–14, linear, carinate; inner 12–16, lanceolate, acuminate; aks. tawny, 3–5 mm. long; pappus white, 1-seriate; $2n = 8$ (Babcock, 1947).—Natur. weed in Humboldt Co.; native of Medit. region. July–Aug.

178. Taráxacum Zinn. DANDELION

Perennials or biennials, with lvs. in a basal tuft and with heads solitary and terminal on naked hollow slender scapes. Lvs. pinnatifid to toothed. Fls. many, yellow; invol.

double, the outer phyllaries short, in several series, often reflexed or spreading, the inner in a single row, erect, linear. Receptacle naked. Aks. fusiform to oblong-ovoid, 4–5-ribbed, the ribs roughened; apex prolonged into a very slender beak and bearing copious soft pale capillary pappus. Inner invol. closing after anthesis and the slender beak elongating and exposing the pappus, the whole invol. finally reflexing and exposing the globular head of frs. to the wind. A rather large genus according to many, but many of the taxa representing apomictic races. (Name possibly of Arabian origin.)

Outer phyllaries erect, not reflexed, lance-ovate; lvs. toothed, usually not lobed to near the midrib. Native, San Bernardino Mts. . . . . 1. *T. californicum*
Outer phyllaries reflexed or spreading, ± linear; lvs. mostly lobed to near the midrib. Introd. weeds.
   All or some of the phyllaries corniculate-appendaged; aks. ± reddish at maturity; lvs. tending to be deeply cut their whole length . . . . 2. *T. laevigatum*
   All or nearly all of the phyllaries unappendaged; aks. pale gray-brown to olive-brown when mature; lvs. tending to have a petiolar base . . . . 3. *T. officinale*

1. **T. califórnicum** M. & J. [*T. ceratophorum* var. *bernardinum* Jeps.] Usually glabrous, 0.5–2 dm. high; root rather thick; lvs. ascending to spreading, light green, oblanceolate, 0.5–1.2 dm. long, 1–2(–3) cm. wide, subentire or sinuate-dentate, rarely incised; heads stout, broadly cylindrical, 10–15 mm. long, the outer phyllaries many, erect, ovate-lanceolate, 5–7 mm. long, the inner lance-linear, 12–15 mm. long; fls. many; aks. pale brown, the body 3 mm. long, the beak 7–9 mm. long; pappus 5 mm. long.—Moist meadows, 6500–8000 ft.; Yellow Pine F.; San Bernardino Mts. May–July.

2. **T. laevigàtum** (Willd.) DC. [*Leontodon l.* Willd. *T. erythrospermum,* auth.] Red-seeded Dandelion. Slender, the scapes 0.2–2.5 dm. high; lvs. narrowly oblanceolate, slender-petioled, 0.5–2.5 dm. long, 1–5 cm. broad, deeply cleft into remote divergent to reflexed narrow lobes with many smaller intermediate shorter slender lobes; invol. 1.3–2.5 cm. long, the inner phyllaries mostly appendaged near the tip; ligules sulphur-yellow above; aks. narrow-oblanceolate, red or red-purple, smooth below, sharply muricate upward, 2.8–3.5 mm. long; beak 6–8 mm. long; pappus sordid; $2n = 24$ (Poddubnaja & Dianowa, 1934).—Occasional as weed at widely scattered stations; native of Old World. May–July.

3. **T. officinàle** Wiggers. [*T. vulgare* (Lam.) Schrank.] Common Dandelion. Lvs. scarcely petioled, 5–30 cm. long, oblong or spatulate, sinuate-pinnatifid to subentire, the longer marginal lobes toothed and with intermediate small teeth; heads 2–5 cm. broad, orange-yellow; phyllaries green to fuscous, mostly not appendaged; aks. 2–4 mm. long, drab or olivaceous, tubercled at summit; pappus whitish; $2n = 16, 24, 48$ (Woess, 1949).—Damp often low places in scattered localities in much of cismontane and montane Calif.; natur. from Eu. Most months.

179. Lápsana L.

Slender branching annuals with alternate lvs. Fls. yellow, the heads with slender peduncles and paniculately arranged. Invol. subcylindric, the 8 phyllaries erect, in 1 series, subequal, with a few minute outer ones at base. Receptacle naked. Aks. oblong, ± flat, narrowed below, rounded at tips, 20–30-nerved; pappus none. (Greek name used by Dioscorides for some plant.)

1. **L. commùnis** L. Nipplewort. Erect, nearly smooth, 3–8 dm. tall, paniculately branched above; lower lvs. ovate, dentate, the blades 2–5 cm. long, commonly with 1–2 pairs of small supplementary lfts. on the petiole which is 3–8 cm. long; invol. pale green, glabrous, glaucous, 4–5 mm. long; ligules ca. 6–7 mm. long; aks. 2.5 mm. long; $2n = 12$ (Marchal, 1920), 14 (Stebbins et al., 1953).—Natur. weed in coastal cent. and n. Calif.; to Wash., e. U.S.; native of Eu. June–Aug.

Class II. Monocotylèdoneae

Stem mostly with vascular bundles closed and scattered irregularly throughout the pith, sometimes the whole stem woody, but not exhibiting xylem, then cambium, then phloem and cortex (bark). Lvs. usually parallel-veined, mostly alternate and entire, commonly sheathing the stem at the base and often with little or no distinction into blade and petiole. Fls. mostly 3-merous or 6-merous. Embryo with 1 cotyledon; the first lvs. of the germinating plantlet alternate. Perhaps ¼ of the angiosperms fall here, many in the tropics.

KEY TO FAMILIES

A. Plants small, floating on water or stranded on mud, not differentiated into stem and lf.,
 sometimes with 1 to few simple roots .... *Lemnaceae* p. 1358
AA. Plants with stem and lvs., the latter sometimes scalelike.
  B. Perianth wanting or reduced, its parts often bristles or mere scales, not petallike in color and texture.
    C. Fls. in the axils of chaffy or husklike scales and ± concealed by them, or stamens and styles protruding; fls. in spikes, spikelets or heads.
      D. Plants either with cauline lvs. or with sheathing scales only or with several to many spikes, spikelets or heads of fls. on each stem.
        E. Lf.-sheaths split lengthwise on the side opposite the blade; lvs. usually 2-ranked; stems mostly hollow and terete; anthers versatile

                                                                  Gramineae p. 1462
        EE. Lf.-sheaths continuous around the stem or ruptured in age; lvs. mostly 3-ranked; stems often triangular in cross section, usually with a pith; anthers basifixed .... *Cyperaceae* p. 1414
      DD. Plants with linear basal lvs. only and a terminal compact head at tip of each scape .... *Eriocaulaceae* p. 1325
    CC. Fls. not in the axils of chaffy bracts or, if subtended by bracts, exceeding or equaling them and not concealed.
      D. Plants floating or submersed; fls. floating or submersed, or barely raised above the surface of the water.
        E. Marine plants; fls. on 1 side of a flattened axis ...... *Zosteraceae* p. 1322
        EE. Freshwater plants or of brackish water.
          F. Fls. in spikes or heads.
            G. Lvs. in dense rosettes; fls. on a spadix adnate to and shorter than the inclosing spathe .... *Araceae* (*Pistia*) p. 1357
            GG. Lvs. not in rosettes; fls. not on a spadix.
              H. Fls. perfect, in peduncled axillary spikes; sepals 4
                                     Potamogetonaceae p. 1315
              HH. Fls. unisexual, in globose heads, the lower ♀, the upper ♂ .... *Sparganiaceae* p. 1364
          FF. Fls. axillary and solitary or very few in a group.
            G. Lvs. alternate; frs. in umbelliform clusters

                                               Ruppiaceae p. 1319
            GG. Lvs. opposite.
              H. Carpels 2 or more; lvs. mostly 3–10 cm. long.
                                   Zannichelliaceae p. 1323
              HH. Carpel 1; lvs. mostly 1–3 cm. long .... *Naiadaceae* p. 1324
      DD. Plants of land or shallow water; usually lvs. and fls. well emersed.
        E. Fls. on a spadix (fleshy, spikelike) surrounded by a yellow or white spathe .... *Araceae* p. 1357
        EE. Fls. not on a spadix.
          F. Infl. a dense elongate spike.
            G. Plants 1–4 dm. high; ♂ and ♀ fls. intermingled, or fls. perfect.
              H. Perianth present; stamens 3 or 6 ... *Juncaginaceae* p. 1320
              HH. Perianth none; stamen 1 .... *Lilaeaceae* p. 1321
            GG. Plants 10–20 dm. high; stamens usually 3, the ♂ fls. in a separate upper portion of the spike .... *Typhaceae* p. 1366
          FF. Infl. of subglobose heads, racemes or open clusters.
            G. Fls. unisexual, the lower heads ♀ ...... *Sparganiaceae* p. 1364
            GG. Fls. bisexual; perianth-segms. distinguishable.
              H. Ovaries 3 or 6, separating at maturity.
                I. Raceme slender, bractless; seed 1 in each carpel
                                     Juncaginaceae p. 1320
                II. Raceme bracted; seeds 2 .... *Scheuchzeriaceae* p. 1314
              HH. Ovary 1, forming a caps. .... *Juncaceae* p. 1400
  BB. Perianth well developed, at least the inner segms. petaloid in color and texture.
    C. Carpels ± free, 1-loculed, mostly 1-ovuled, maturing into a bunch or whorl of aks. .... *Alismataceae* p. 1312
    CC. Carpels united into a mostly 3–12-loculed ovary maturing into a caps. or berry.
      D. Lvs. large, plicately folded in bud; plant arborescent, unbranched; fls. in a large panicle with a spathaceous bract .... *Palmae* p. 1364
      DD. Lvs. not as above and plant branched if arborescent.
        E. Plants quite woody with long stiff ± swordlike lvs. .... *Agavaceae* p. 1360
        EE. Plants herbaceous or, if woody, climbing vines.
          F. Ovary superior.
            G. Infl. with a spathe; submersed or floating aquatic plants with perianth-segms. partly connate .... *Pontederiaceae* p. 1356

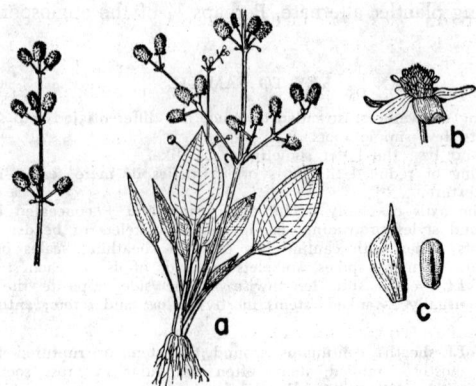

Fig. 121. ALISMATACEAE. *Echinodorus cordifolius: a,* habit, × ¼, with basal lvs. and terminal infl., top of which is shown detached at the left; *b,* fl., × 1½, 3 sepals, 3 petals, several stamens, many cent. pistils; *c,* beaked ak. to left, × 6, seed to right.

GG. Infl. without a spathe; plants not floating or submersed.
    H.   Sepals green; petals colored; fls. in umbels; prostrate
         introd. plant .... *Commelinaceae*   p. 1324
    HH. Sepals and petals concolored, or if unlike, the fls.
         not in umbels.
         I.   Fls. in a scapose umbel subtended by ± mem-
           branous spathelike bracts .... *Amaryllidaceae*   p. 1367
         II. Fls. not in a scapose umbel with spathelike
           bracts .... *Liliaceae*   p. 1326
  FF. Ovary inferior.
    G.   Aquatic plants with mostly submersed lvs.
                                       Hydrocharitaceae   p. 1325
    GG. Land plants, the lvs. not submersed.
        H.   Fls. regular; ovary 3-loculed; stamens 3; lvs.
          equitant .... *Iridaceae*   p. 1388
        HH. Fls. irregular; ovary 1-loculed; stamens 1–2; lvs. not
          equitant .... *Orchidaceae*   p. 1394

1. Alismatàceae. WATER-PLANTAIN FAMILY (Fig. 121)

Annual or perennial aquatic or marsh plants with scapose stems. Lvs. basal, long-petioled, sheathing; the blades linear to ovate-rounded, longitudinally nerved. Fls. whorled, subtended by a whorl of bracts, perfect or unisexual, in racemes or panicles. Receptacle flat to globose. Sepals 3, green, imbricate, persistent. Petals 3, imbricate, deciduous, white or pink. Stamens hypogynous, 6 or more, free; anthers 2-celled, extrorse. Carpels ± free, 1-celled, mostly 1-ovuled; style persistent. Fr. a whorl or bunch of aks.; seeds curved, with horseshoe-shaped embryo and no endosperm. A family of 10 genera, mostly of N. Hemis.

Carpels in fr. spreading stellately, long-beaked; petals toothed .... 1. *Machaerocarpus*
Carpels not spreading stellately or long-beaked in fr.; petals entire.
  Lowest fls. with carpels only; staminate fls. above; stamens many .... 2. *Sagittaria*
  Lowest fls. perfect.
    Upper fls. with stamens only; frs. winged; stamens 9–15 .... 2. *Sagittaria*
    Upper fls. bisexual; frs. not winged.
      Frs. in a ring; stamens 6 .... 3. *Alisma*
      Frs. in a dense head; stamens 6–30 .... 4. *Echinodorus*

1. Machaerocárpus Small

Perennial scapose herbs from a short erect cormlike rootstock. Lvs. erect, ascending or floating; blades ovate to linear-oblong, 3–5-veined, long-petioled. Fls. perfect, in a simple panicle of several whorls. Sepals 3, persistent, broad, ribbed. Petals 3, white or pink, sharply and unevenly toothed, spreading, deciduous. Stamens 6, 2 opposite each sepal; fils. flattened; anthers elongate. Carpels few, in 1 whorl, free from each other,

1-ovuled. Aks. ribbed on back, the faces depressed; beak erect, at least as long as ak.-body. One sp. of w. U.S. (Greek, *machaira*, dagger, and *karpos*, fr.)

1. **M. califórnicus** (Torr.) Small. [*Damasonium c.* Torr.] Lf.-blades 2.5–8 cm. long, long-petioled; scapes usually 2 or more, 2–4 dm. high; pedicels spreading to recurved in fr., 2–5 cm. long; sepals 4–5 mm. long; petals suborbicular, 8–10 mm. long; aks. 7–14, strongly divergent, 7–12 mm. long.—Shallow water or mud, below 5000 ft., V. Grassland, Foothill Wd., Sagebrush Scrub; Petaluma (Sonoma Co.), Sacramento V., adjacent Sierra Nevada foothills, Modoc Co.; se. Ore., w. Nev.? April–Aug.

2. Sagittària L. ARROWHEAD

Mostly perennial aquatic or marsh herbs, stoloniferous or tuber-bearing, with milky juice. Lvs. with sheathing bases, at least the outermost lvs. without distinct blades, the lf.-blades arrow-shaped or lanceolate. Scapes erect or lax. Fls. conspicuous, pedicelled, in whorls of 3, the plants usually monoecious; the ♀ fls. below, ♂ fls. above. Sepals loosely spreading or reflexed in fr.; petals white, exceeding sepals. Stamens many. Pistils many, crowded on a large globose receptacle, forming flat aks. in fr. Ca. 20 spp., of temp. and trop. regions, but largely Am. (Latin, *sagitta*, arrow, because of the lf. shape.)

(Bogin, Clifford. Revision of the genus Sagittaria. Mem. N.Y. Bot. Gard. 9: 179–233, 1955.)
Sepals of the ♀ fls. erect and accrescent .... 1. *S. montevidensis*
Sepals of ♀ fls. reflexed or spreading.
  Pedicels of ♀ fls. thick, recurved in fr.; lvs. entire .... 2. *S. Sanfordii*
  Pedicels of ♀ fls. slender, ascending; lvs. with basal lobes.
    Beak of ak. laterally inserted, horizontal, well developed .... 3. *S. latifolia*
    Beak of ak. apically inserted, erect, minute.
      Basal lobes of lvs. not longer than terminal lobe; ak. not winged .... 4. *S. cuneata*
      Basal lobes of lvs. longer than terminal lobe; ak. wing-margined .... 5. *S. longiloba*

1. **S. montevidénsis** Cham. & Schlecht. ssp. **calycìna** (Engelm.) Bogin. [*S. c.* Engelm. *Lophotocarpus c.* J. G. Sm. *L. californicus* J. G. Sm.] Plants to 2 m. tall, glabrous; lf.-blades hastate or sagittate, 0.5–2 dm. long, with basal lobes long, acute, sharply divergent; scapes stout; bracts connate, membranous; sepals erect, obtuse, becoming 1–1.5 cm. long in fr.; petals white, ca. 1.5 cm. long; fils. glandular-pubescent; fruiting heads 1.5–3 cm. in diam.; aks. 2–3 mm. long, winged, the beak short, slender, oblique-horizontal; $2n = 20$ (Taylor, 1925).—Sloughs and slowly moving water; Freshwater Marsh; Cent. V.; Miss. V. drainage. July–Aug. Typical *montevidensis* is native in S. Am. and has a purple spot at the base of each petal.

2. **S. Sanfórdii** Greene. Plants 9–15 dm. high, glabrous; lvs. entire, 5–9 dm. long, the blades 5–15 cm. long, lanceolate or wider; scapes shorter than the lvs., simple; bracts connate, membranous; fertile pedicels 15–20 mm. long, recurved in fr.; sepals spreading, ovate, 4–6 mm. long; petals somewhat longer; fils. glabrous, dilated; fruiting head 12–14 mm. in diam.; aks. 2 mm. long, winged on both margins, beak short, oblique, triangular. —Sloughs and sluggish streams; Freshwater Marsh; Cent. V., Ventura Co. (Mirror Lake). May–June.

3. **S. latifòlia** Willd. [*S. variabilis* Engelm.] WAPPATO. TULE-POTATO. Plants 2–12 dm. tall, glabrous; lvs. variable, 1.5–6 dm. long, the blades 1–4 dm. long, the basal lobes acuminate, divaricate, the terminal lanceolate to broadly ovate, obtuse to acute; scape simple or branched; bracts boat-shaped, nearly free, rather firm; fertile pedicels spreading-ascending; sepals ovate, 5–11 mm. long; petals ca. 10–20 mm. long; fils. glabrous, not dilated; fruiting heads 1.5–3 cm. in diam.; aks. obovate, ca. 3 mm. long, broadly winged, the beak subhorizontal, slender, well developed; $n = 11$ (Oleson, 1941). —Edge of ponds or slow streams, meadows, below 7000 ft.; largely Freshwater Marsh; most of Calif.; to B.C., Atlantic states, Mex. July–Aug. Exceedingly variable.

4. **S. cuneàta** Sheld. [*S. arifolia* Nutt.] Mostly emersed and 2–4 dm. high, or submerged and longer; lvs. hastate-sagittate, the blades ovate, acute, 3–17 cm. long; scapes simple or branched; bracts linear to lance-attenuate, ± membranous; fertile pedicels ascending; sepals ovate, 4–8 mm. long; petals 6–10 mm. long; fils. glabrous, not dilated; fruiting heads 1–1.5 cm. in diam.; aks. obovate, 2–2.5 mm. long, with thickened margins, the beak minute, erect; $n = 11$ (Brown, 1946).—Shallow ponds and swampy places, below 7500 ft.; Montane Coniferous F.; San Bernardino Mts., Sierra Nevada and N. Coast Ranges; to B.C., Nova Scotia, New Mex. June–Aug.

5. **S. longíloba** Engelm. [*S. Greggii* J. G. Sm.] Plants 5–14 dm. high, stout; lf.-blades 1–3 dm. long, the terminal lobe ovate to lanceolate, acute to subacuminate, the basal lobes equaling it or longer, lanceolate, widely divergent; scape paniculately branched; bracts linear to lance-attenuate, ± membranous; fertile pedicels ascending; sepals ovate, 4–7 mm. long; petals 8–12 mm. long; fils. glabrous, not dilated; fruiting heads 8–15 mm. in diam.; aks. 1.2–2.3 mm. long, broadly obovate, thin-margined, the beak minute, subapical.—Sloughs and sluggish water; Freshwater Marsh; Cent. V.; Ariz. to Nebr., Tex., Mex. May–June.

3. Alísma L. WATER-PLANTAIN

Ours perennial herbs. Lvs. basal, erect or floating, several-ribbed and with transverse veinlets. Fls. many, small, perfect, in pyramidal panicles of whorled branches, each of which has a simple or compound umbel. Sepals 3, broad, usually ribbed, persistent. Petals 3, white or pinkish, entire, deciduous, spreading. Stamens 6, two opposite each petal; fils. slender, short; anthers short. Carpels few–many, in 1 whorl, distinct; aks. 2–3-ribbed and curved on the back. Ca. 8 spp., widely distributed in temp. and trop. regions. (Alisma, the Greek name.)

(Fernald, M. L. N. Am. Representatives of A. Plantago-Aquatica. Rhodora 48: 86–88, 1946.)

Aks. longer than wide, 1-grooved on back; petals usually white .... 1. *A. triviale*
Aks. not longer than wide, 2-grooved on back; petals pink .... 2. *A. Geyeri*

1. **A. triviàle** Pursh. [*A. Plantago-aquatica* auth., not L. *A. P-a.* var. *t.* Farw.] Scapes 6–12 dm. high, exceeding lvs.; lf.-blades oblong to ovate, cuneate to rounded at base, 3–15 cm. long, long-petioled; pedicels slender, erect-spreading, 2–4 cm. long; sepals scarious-margined, 2.5–4 mm. long in anthesis; petals 3–6 mm. long; fruiting heads 4–7 mm. in diam., the aks. 2–3 mm. long; $2n = 14$ (Brown, 1946).—Margins of ponds, in swales and marshy shores, below 5000 ft.; Mixed Evergreen F., V. Grassland, etc.; Ventura Co. (Mirror Lake), n. in Coast Ranges, Cent. V., Sierra Nevada; to B.C., Nova Scotia, N.Y. June–July.

2. **A. Gèyeri** Torr. [*A. gramineum* var. *G.* Samuels.] Scapes 1–5 dm. high, not exceeding lvs.; lf.-blades oblong to almost linear, 5–9 cm. long; pedicels recurved; sepals ca. 2.5 mm. long; petals 2–4 mm. long; fruiting heads 4.5–5.5 mm. broad, the aks. suborbicular, 2 mm. wide.—Wet places; Sagebrush Scrub; Modoc Co. (Likely, Alturas); to Alta., Quebec, N.Y. June.

4. Echinódorus Rich. BUR HEAD

Annual or perennial herbs, often with runners. Lvs. long-petioled, elliptic-ovate to lanceolate, often cordate, 3–9-ribbed and punctate with dots or lines. Scapes often exceeding lvs., bearing a paniculate or racemose infl. Fls. perfect, in umbellike whorls. Petals white. Stamens 6–30. Pistils many; style obliquely apical, persistent; aks. turgid, ribbed, beaked, forming spinose heads. Ca. 14 spp., widely distributed in Am., Eu., Afr. (Greek, *echinos*, hedgehog, and *doros*, utricle, leather bottle, referring to the fr.)

1. **E. Berteròi** (Spreng.) Fassett. [*Alisma B.* Spreng. *E. cordifolius* of Calif. references. *E. rostratus* Engelm.] Lvs. basal, broadly ovate to lanceolate, with cordate or truncate base, obtuse, the blade 2–15 cm. long; scape 1–5 dm. high; umbels remote, 3–12-fld., proliferous; sepals ovate, 4–5 mm. long; petals 6–10 mm. long; fruiting heads sub-globose, 4–10 mm. in diam.; aks. 2.5–3 mm. long, the beak 1.2–2 mm. long, ½–⅔ as long as the cuneate, ribbed body; $n = 11$ (Heiser & Whitaker, 1948).—Occasional in wet places, at low elevs.; Freshwater Marsh; scattered stations, Calif.; to Kans., Tex., Mex.

Var. **lanceolàtus** (Engelm.) Fassett. [*E. rostratus* var. *l.* Engelm.] Beak of ak. 0.5–0.8 mm. long, ⅓–½ as long as the body.—With the sp. in Calif.; Kans. to Ohio, Tex.

2. Scheuchzeriàceae. ARROW-GRASS FAMILY

Perennial marsh herbs with creeping rootstock and erect leafy stems. Lvs. linear, sheathing at base, ligulate at junction of sheath with blade. Fls. perfect, in terminal few-fld. bracteate racemes. Perianth-segms. 6, persistent, alike, free. Stamens 6, free;

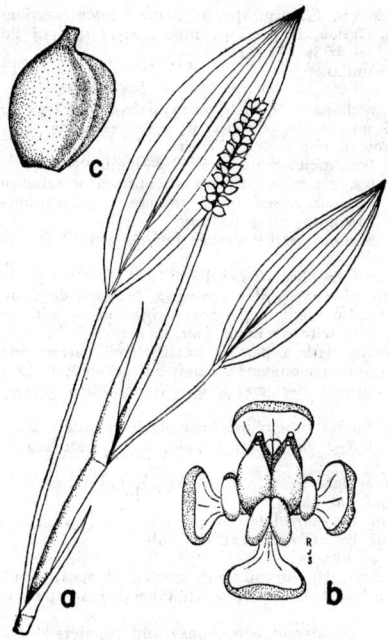

Fig. 122. POTAMOGETONACEAE. *Potamogeton: a,* flowering branch, × 1, with floating lvs., stipules, spike of frs.; *b,* fl., perianth with 4 segms., stamens 4, carpels 4, sessile; *c,* fr.

anthers linear, basifixed, extrorse. Carpels 3–6, ± fused at base; stigmas sessile; ovules 2 to few. Fruiting carpels divaricate, 1–2-seeded, inflated, folliclelike. Seeds ellipsoid, smooth, without endosperm. One genus and 1 sp. of circumpolar distribution.

1. Scheuchzèria L.

Characters of family. (J. J. *Scheuchzer,* 1672–1733, Swiss botanist.)

1. **S. palústris** L. var. **americàna** Fern. Stems 1 to several, 1–3 dm. long; lvs. 1–4 dm. long, striated, the upper reduced to bracts; pedicels spreading in fr., 6–20 mm. long; perianth-segms. 3 mm. long, 1-nerved; follicles 4–8 mm. long.—Bogs, in cold places; found long ago in Sierra and Plumas cos.; to Alaska, Pa., Labrador; Eurasia. July.

3. Potamogetonàceae. Pondweed Family (Fig. 122)

Perennial herbs of fresh waters, growing from rhizomes. Stems jointed, leafy, fibrous-rooted at lower nodes. Lvs. 2-ranked, alternate or opposite, the submersed thin, the floating thicker, all sheathing at base, the sheath free or partially adnate to the petiole. Fls. perfect, small, on pedunculate axillary spikes; the peduncle surrounded by a basal sheath formed by stipules. Bracts lacking. Perianth of 4 free, rounded, short-clawed, valvate segms. Stamens 4, inserted on the claws; anthers 2-celled, sessile, extrorse. Carpels 4, free, sessile, 1-celled and 1-ovuled; stigmas on short styles or sessile. Fr. drupelike when fresh, ± compressed, with bony endocarp; seed 1, without endosperm. One widely distributed genus. Many of the spp. of great importance as food for wild fowl and various mammals; in some cases the whole plant, in others the starchy rhizomes or nutlets, are eaten.

1. Potamogèton L. Pondweed

Characters of the family. A genus of perhaps 70 spp. of temp. regions. (Greek, *potamos,* a river, and *geiton,* a neighbor, because of the habitat.)

(Fernald, M. L. The linear-leaved N. Am. spp. of Potamogeton, section Axillares. Mem. Am. Acad.
Arts & Sci. 17: pt. 1, 1932. Ogden, E. C. The broad-leaved spp. of Potamogeton of N. Am. n. of
Mex. Rhodora 45: 57–104, 119–163, 171–213, 1943.)

A.   Lvs. all submerged and similar.
  B.   The lvs. linear.
    C.   Stipules fused with base of lf. to form a sheath at least 1 cm. long.
      D.   Lvs. 4–8 mm. wide, auricled at base, serrulate .... 4. *P. Robbinsii*
      DD. Lvs. narrower, not auricled, entire.
        E.   The lvs. fascicled terminally, 2–4 mm. wide .... 2. *P. latifolius*
        EE. The lvs. generally distributed, filiform or setaceous.
          F.   Stigmas raised on a minute style, capitate; lvs. gradually acuminate;
            rhizomes tuber-bearing .... 3. *P. pectinatus*
          FF. Stigmas sessile, broad; lvs. not acuminate; rhizomes not tuber-bearing
            1. *P. filiformis*
    CC. Stipules free from the lvs. or nearly so.
      D.   Lvs. 9- to 35-nerved; stems winged, ca. as wide as lvs. ...... 6. *P. zosteriformis*
      DD. Lvs. 1- to 7-nerved; stems not winged or as wide as lvs.
        E.   The lvs. without basal glands; nerves 3–5 .... 7. *P. foliosus*
        EE. The lvs. with a pair of basal glands; nerves mostly 3.
          F.   Spikes elongate, strongly interrupted, 6–12 mm. long; peduncles filiform,
            usually 3–8 cm. long; stipules when young with edges partly connate
            8. *P. pusillus*
          FF. Spikes subglobose, scarcely interrupted, 2–8 mm. long; peduncles stouter,
            0.5–3 cm. long; stipules flat or convolute, not connate
            9. *P. Berchtoldii*
  BB. The lvs. broad (lanceolate to elliptic or ovate), never linear.
    C.   Lvs. not clasping at base.
      D.   Margin of lvs. serrulate throughout .... 5. *P. crispus*
      DD. Margin of lvs. serrulate at tip only .... 17. *P. illinoensis*
    CC. Lvs. clasping at base.
      D.   Blades of lvs. 10–30 cm. long, hooded at apex; stem whitish .. 18. *P. praelongus*
      DD. Blades of lvs. 2–12 cm. long, not hooded at apex; stem green
        19. *P. Richardsonii*
AA. Lvs. of 2 kinds, floating (broad and coriaceous) and submerged (broad or narrow, but thin).
  B.   Submerged lvs. linear or threadlike, not differentiated into petiole and blade.
    C.   Floating lvs. oblong, gradually narrowed at base; submerged lvs. with median fine
      reticulate veining .... 11. *P. epihydrus*
    CC. Floating lvs. elliptical, subcordate at base; submerged not so veined .... 15. *P. natans*
  BB. Submerged lvs. linear to lanceolate or ovate, with true blades.
    C.   The submerged lvs. ovate to lanceolate, more than 4 mm. wide.
      D.   Floating lvs. with 30–55 nerves, the submerged with 30–40 nerves
        13. *P. amplifolius*
      DD. Floating and submerged lvs. with fewer nerves than above.
        E.   Blades of submerged lvs. petioled and not sharp-pointed at apex
          14. *P. nodosus*
        EE. Blades of submerged lvs. sessile (except sometimes the lowermost), sharp-
        pointed.
          F.   Submerged lvs. sessile; floating lvs. usually lacking, if present delicate,
            translucent, tapering gradually into petiole .... 12. *P. alpinus*
          FF. Submerged lvs. sessile or petioled; floating lvs. coriaceous and opaque,
            distinct from petiole.
            G.   Stem 1.5–5 mm. thick; submerged lvs. 1.5–4 cm. wide, the nerves
              mostly 9–17 .... 17. *P. illinoensis*
            GG. Stem 0.5–1 mm. thick; submerged lvs. 0.2–1.2 cm. wide, the
              nerves mostly 3–9 .... 16. *P. gramineus*
    CC. The submerged lvs. filiform to linear, not over 4 mm. wide.
      D.   Spikes all alike, cylindric; stipules all free from the lf.-bases .. 16. *P. gramineus*
      DD. Spikes of 2 kinds, the emersed cylindric with many fls., the submerged capitate,
        with few fls.; stipules fused with the lf.-base .... 10. *P. diversifolius*

1. **P. filifórmis** Pers. var. **Macoùnii** Morong. Rootstock creeping; stem to 3 dm. long,
usually short and much-branched; lvs. all submerged; blades narrow, rather blunt, 5–12
cm. long, 0.8–2.0 mm. wide; stipules fused with lf., forming a short clasping sheath;
peduncle ca. as long as lvs.; spike short, 1–2.5 cm. long, the whorls mostly approximate;
fr. obovoid, 2 mm. long, 1.5 mm. wide, beakless, the stigma broad and sessile.—Shallow
pools, chiefly in calcareous areas; cent. and n. Calif.; to Canada, Atlantic Coast. May–
Aug.

2. **P. latifòlius** (J. W. Robbins) Morong. [*P. pectinatus* var. *l.* J. W. Robbins.] Stem
1–2 mm. thick, whitish, freely branched above, 2–6 dm. long; lvs. all submerged,
fascicled terminally, linear, obtuse or short-apiculate, 2–8 cm. long, 2–4 mm. wide, 3-
to 5-nerved, with many cross-veins; sheaths loose, swollen, 1–2.5 cm. long, much thicker
than stem; peduncles 2–10 cm. long; spikes 1–3 cm. long, interrupted; fr. obovoid, 3
mm. long, 2 mm. wide, the style slender, incurved, forming a beak.—In slightly alkaline

ponds; Merced Co.; Sagebrush Scrub, Mono Co. to Modoc Co.; Ore., Nev. July–Aug.

3. **P. pectinàtus** L. Rootstock creeping, branched, slender, with terminal thickened tubers; stems filiform, ca. 1 mm. thick, much-branched 5–20 dm. long; lvs. submerged, setaceous, mostly 1-nerved, attenuate at apex, 3–15 cm. long, scarcely 1 mm. wide; sheaths 2–5 cm. long, only slightly thicker than the stem, often bleached; peduncles filiform, 5–25 mm. long; whorls 2–6, widely spaced; fr. obliquely obovoid, 2.5–4 mm. long, 2–3 mm. wide, the style slender, incurved, forming a beak; $2n = 78$ (Harada, 1942).—Brackish or fresh water, widely distributed in Calif., below 7000 ft.; to B.C., Nfld.; temp. and trop. regions of world. The small tubers of great importance to wild ducks. May–July.

4. **P. Robbínsii** Oakes. Stems stout, branched, to 5 dm. long; lvs. submerged, linear, auricled at base, borne in 2 ranks and stiffly divergent, 4–10 cm. long, 4–8 mm. wide, the margins thickened, finely serrate; sheaths 1–1.5 cm. long, white, free portion of the stipules longer; peduncles reddish, 3–10 cm. long; spikes 1–2 cm. long, loosely fld.; fr. obliquely obovoid, 4 mm. long, 3 mm. wide, 3-keeled, the beak subapical, 1 mm. long.—Quiet water, Subalpine F., Big Pine Lakes, Inyo Co.; Sagebrush Scrub, Honey Lake, Lassen Co.; to B.C., Atlantic Coast.

5. **P. críspus** L. Stems compressed, branching; lvs. all submerged, sessile, half-clasping, oblong, 2–6 cm. long, 5–12 mm. wide, obtuse, crisped and serrulate, with prominent midrib and at least 2 fine, lateral veins; stipules scarious, deciduous, splitting; propagating buds prominent in axils of decayed lvs.; spikes cylindric, 1–1.8 cm. long; fr. ovoid, 3-keeled, almost 3 mm. long, with beak ca. 2 mm. long; $n = 26$ (Scheerer, 1939).—Natur. in Santa Ana R. system, Cent. V., etc.; to Atlantic Coast; native of Eu. July–Sept.

6. **P. zosterifórmis** Fern. [*P. compressus* auth., not L. *P. zosterifolius* auth., not Schumacher.] EELGRASS PONDWEED. Stems flat or winged, 1–3 mm. wide, branched, 3–10 dm. long, often with propagating buds; lvs. submerged, linear, without basal glands, 5–20 cm. long, 2–5 mm. wide, 9- to 35-nerved, abruptly pointed; stipules free, firm, 1.5–3.5 cm. long, soon deciduous; peduncles 2–6 cm. long; spikes 1.5–3 cm. long, rather densely fld.; fr. oblong or suborbicular, 4–5 mm. long, 3-keeled, the middle keel prominent, usually toothed.—Still or slow water, below 5000 ft.; Lake Co., San Joaquin R. delta., Lassen and Modoc cos.; to B.C., Atlantic Coast. June.

7. **P. foliòsus** Raf. [*P. f.* var. *californicus* Morong. *P. c.* Piper. *P. f.* var. *niagarensis* Gray.] LEAFY PONDWEED. Stems flattened, leafy, slender, freely branched, 2–10 dm. long; winter-buds on very short branches; lvs. submerged, numerous, linear, 3–10 cm. long, 1.5–2.5 mm. wide, 3–5-nerved, without basal glands, the midrib with 1–3 rows of loosely cellular tissue on each side near the base; stipules at first connate, later free, 15–25 mm. long, early deciduous; peduncles 5–20 mm. long; spikes many, few-fld. and short; frs. pitted, broadly ovoid, 2–2.5 mm. long, strongly 3-keeled, the middle keel winged and denticulate.—Hard or brackish water, below 5000 ft.; well distributed in Calif.; to B.C., Atlantic Coast, Cent. Am. July–Oct. Variable as to lf.-width, a form with lvs. 0.3–1.5 mm. wide, 1–3-nerved, the midrib with little or no loosely cellular tissue, winter-buds on long branches, has been called var. *macéllus* Fern.—With the sp.

8. **P. pusíllus** L. [*P. panormitanus* Biv.] Stems filiform, branching, 3–10 dm. long; lvs. submerged, linear, pointed or rounded at tip, 2–6 cm. long, 0.5–3 mm. wide, with a translucent gland on each side of the base; stipules hyaline, 6–15 mm. long, free from the lvs. except at very base, edges connate so as to form a tube; peduncles filiform, 5–8 cm. long; spikes 6–12 mm. long, few-whorled, interrupted; frs. obliquely obovoid, 2–3 mm. long, rounded on back, obscurely keeled.—Ponds and slow streams below 8000 ft., only occasional in Calif., but widely distributed; to Alaska, Atlantic Coast; Eurasia. May–June. Very variable as to lf.-width.

9. **P. Berchtòldii** Fieb. [*P. pusillus* auth., not L. *P. p.* var. *tenuissimus* Mert. & Koch. *P. B.* var. *t.* Fern.] Much like *P. pusillus*, but the stipules flat, not connate; the peduncles stouter, 0.5–3 cm. long; spikes denser, subcapitate, 2–8 mm. long; frs. rugulose, 2–2.5 mm. long, rounded on back, with obscure and very low keel.—Cold waters, Del Norte Co., and in a narrow-lvd. form (var. *tenuissimus*) at 5000–9000 ft. in the Sierra Nevada, Tulare to Butte cos.; to Atlantic Coast, circumpolar. July–Aug. The collection from Del Norte Co. is said to have fibrously disintegrating stipules and may represent *P. fibrillòsus* Fern.

10. **P. diversifòlius** Raf. [*P. dimorphus* of Calif. refs.] Stems freely branched, flattened

to subterete, 2–10 dm. long; submerged lvs. narrowly linear, acute, 0.5–1.5 mm. wide, 2.5–7 cm. long, faintly 3-nerved; their stipules 6–10 mm. long, fused ca. half their length; floating lvs. elliptic to narrowly obovate, the blades 1–4 cm. long, 3–20 mm. wide, 7–15-nerved, on usually shorter petioles; their stipules usually becoming fibrous; submerged spikes subglobose, on peduncles to 5 mm. long, emersed spikes cylindric, 5–20 mm. long, on peduncles 5–15 mm. long; frs. suborbicular, compressed, pitted, 3-keeled, 1–1.5 mm. long, the beak obtuse.—Shallow ponds and slow streams, Lake Surprise, San Jacinto Mts. at 9000 ft., Cent. V. to Ore., Atlantic Coast, Mex.

11. **P. epihỳdrus** Raf. [*P. Claytonii* auth., not Tuckerm.] Stems compressed, sometimes forking above, 2–20 dm. long; submerged lvs. ribbonlike, 5–20 cm. long, 5–10 mm. broad, 7–13-nerved, loosely cellular-reticulate in lower half and near midvein; floating lvs. coriaceous, oblance-oblong to elliptic, tapering into flattened petioles, the blades 3–8 cm. long, 1.5–3.5 cm. broad, 19–41-nerved; stipules free, 2–4.5 cm. long; peduncles 2–16 cm. long, thickish; spikes cylindric, 1.8–4 cm. long in fr.; frs. flattened, pitted on sides, round-obovoid, 3–4.5 mm. long, 3–3.6 mm. broad, the middle keel prominent, 0.6–1.2 mm. broad.—Cool waters, Covelo, Mendocino Co.; to Wash., Atlantic Coast. July–Aug.

Var. **Nuttállii** (Cham. & Schlecht.) Fern. [*P. N.* Cham. & Schlecht.] Submersed lvs. 2–8 mm. wide, 5–7-nerved; fruiting spikes 0.8–3 cm. long; frs. 2.5–3.5 mm. long.— Yosemite V., Sierra V.; Little Hot Springs V. (Modoc Co.); to Alaska, Atlantic Coast. July–Aug.

12. **P. alpìnus** Balbis var. **tenuifòlius** (Raf.) Ogden. [*P. t.* Raf. *P. rufescens* Schrad.] Plant reddish; stems 1–2 mm. thick, terete, but often flat when pressed, 3–10 dm. long; submerged lvs. oblong-linear to lance-linear, 7–25 cm. long, 8–18 mm. wide, sessile, obtusish; stipules 1.5–3 cm. long, 2–8 mm. broad, 7-nerved, oblong, obtuse; floating lvs. thin, translucent, often absent, the blades elliptic to obovate or linear-oblong, 4–6.5 cm. long, 1–2 cm. wide, tapering gradually into the petiole, 9–13-nerved; stipules 15-nerved, 1.5–3 cm. long, 3–10 mm. broad; peduncles 3–15 cm. long, as thick as the stems; spikes 5–9-whorled, cylindric and crowded in fr., 1.5–3.5 cm. long; frs. obovoid, hard, smooth, tawny-olive, 3–3.5 mm. long, with added 1–1.3 mm. in the linear curved beak.—At 3000–8500 ft., San Gabriel Mts. (Los Angeles Co.), Sierra Nevada from Tulare Co. n.; to Alaska, Atlantic Coast. July–Sept.

13. **P. amplifòlius** Tuckerm. Stems terete, 1–3 mm. thick, simple or short-branched, 4–12 dm. long; lower submerged lvs. lanceolate, dark green, 19–25-nerved, upper broadly lanceolate to ovate, 23–27-nerved, arcuate, petioled, the blades 8–20 cm. long, 2.5–7.5 cm. wide; stipules somewhat persistent, fibrous, 3.5–11 cm. long, triangular; floating lvs. coriaceous, ovate to elliptic, cuneate to rounded at base, the blades 5–10 cm. long, 2.5–5 cm. wide, 29–41(–51)-nerved; stipules 5–12 cm. long; peduncles 4–30 cm. long, somewhat thickened upward; spikes 9–16-whorled, 4–8 cm. long in fr.; frs. obovoid, cuneate at base, 3.5–4.5 mm. long, reddish or orange-brown when mature, the beak often prominent and to 1 mm. long.—Rather deep water, mostly 6000–9000 ft., Sierra Nevada, N. Coast Ranges; to Wash., Atlantic Coast. July–Aug.

14. **P. nodòsus** Poir. [*P. americanus* Cham. & Schlecht. *P. lonchites* auth. *P. fluitans* auth.] Stems terete (often flat when pressed), branched, 1–1.5 mm. thick, 1–2 m. long; submerged lvs. thin, 7–15-nerved, lance-linear to lance-elliptic, the blades 9–20 cm. long, 1–3.5 cm. wide, gradually tapering into long petioles; stipules brownish, often decaying early, linear, 3–6(–9) cm. long; floating lvs. coriaceous, elliptical, cuneate or rounded at base, 5–9 cm. long, 2–4 cm. wide, 13–21-nerved; stipules 3–6 cm. long, linear, somewhat 2-keeled; peduncles thicker than stem, 3–15 cm. long; spikes 10–15-whorled, 3–6 cm. long in fr.; frs. obovoid, 3.5–4 mm. long, prominently keeled, brown or reddish when mature, with erect linear beak to 1 mm. long.—Ponds and streams, from sea-level to 6600 ft.; from Orange Co. n. mostly w. of the Sierra Nevada, occasional on deserts (Victorville) and Bridgeport, etc.; to B.C., Atlantic Coast, S. Am. May–Aug.

15. **P. nàtans** L. Stem simple or nearly so, terete, 1–2 mm. thick; submerged lvs. coriaceous, linear, 10–20 cm. long, 1–2 mm. wide, obscurely nerved; stipules clasping the stem, whitish, fibrous, persistent, linear to lanceolate, 4–9 cm. long, 2-keeled; floating lvs. coriaceous, ovate to oblong-ovate, usually subcordate at base, the blades 4–9 cm. long, 2.5–6 cm. wide, 23–37-nerved; stipules broader than those of submerged lvs.; peduncles scarcely thickened, 3–8 cm. long; spikes 8–14-whorled, 3–5 cm. long

in fr.; frs. obovoid, 3.5–5 mm. long, rounded on back, beak short and broad.—At 4000–7500 ft., San Bernardino Mts. and Sierra Nevada; lower in Coast Ranges, San Mateo Co. n.; to Alaska, Atlantic Coast; Eurasia. June–Aug.

16. **P. gramíneus** L. [*P. heterophyllus* Schreb.] Stems much-branched, terete to 1 mm. thick; submerged lvs. linear to lance-elliptical, 1.5–4.5 cm. long, 2–10 mm. wide, tapering gradually to a sessile base, 5–7-nerved; stipules persistent, 5–30 mm. long, 1–2 mm. wide, 10–30-nerved; floating lvs. coriaceous, ovate to elliptical, the blades 2–6 cm. long, 1–2 cm. wide, 13–17-nerved; stipules broader than in submerged lvs.; peduncles thickened upward, 2–10 cm. long; spikes 5–10-whorled, dense, 1–2.5 cm. long in fr.; frs. obovoid, 2–2.5 mm. long, usually keeled, greenish, with short beak; $2n = 52$ (Palmgren, 1939).—At 6000–9000 ft., Sierra Nevada from Tulare Co. n., Modoc Co.; to Alaska, Atlantic Coast, Mex.; Eurasia. July–Aug.

Var. **máximus** Morong. Principal submerged lvs. 6–9 cm. long, 7–9-nerved.—Sierra Nevada (Eldorado Co., Butte Co.); to Alaska, Labrador. July–Aug.

17. **P. illinoénsis** Morong. [*P. lucens* auth., not L.] Stem simple or branched, terete when fresh, 1.5–5 mm. thick, 3–6 dm. long; submerged lvs. thin, elliptic to lanceolate, often somewhat arcuate, the blades 5–20 cm. long, 1.5–4 cm. wide, sessile or with petioles to 4 cm. long; stipules persistent, obtuse, 2.5–8 cm. long, 0.5–1.2 cm. wide, 15–35-nerved; floating lvs. absent, or if present, subcoriaceous, elliptic to oblong, the blades 4–13 cm. long, 2–6.5 cm. wide, cuneate or rounded at base, 13–29-nerved; stipules broader than for submerged lvs.; peduncles ± thickened, 4–15 cm. long; spikes cylindric, 8–15-whorled, dense and 3–6 cm. long in fr.; frs. obovoid to orbicular, 2.5–3.5 mm. long, prominently and acutely keeled, the beak deltoid, 0.5 mm. long.—Below 5000 ft., San Diego Co. to B.C., Quebec, Fla., L. Calif. June–Aug.

18. **P. praelóngus** Wulf. Stem simple or branched, 1–2 m. long, whitish, 1.5–4 mm. thick; lvs. all submersed, ovate-oblong, 10–30 cm. long, 1–3 cm. wide, 13–25-nerved, cordate or rounded at base and clasping, hooded at apex; stipules white, oblong-linear to ovate-lanceolate, 5–10 cm. long, usually persistent; peduncles clavate, 15–60 cm. long; spikes 6–12-whorled, 3–5 cm. long in fr.; frs. obovoid, rounded on back, cuneate at base, 4–5 mm. long, with prominent short thick beak, the dorsal keel acute and well developed.—Deep cold water, Sierra, Plumas, and Lassen cos.; to Alaska, Atlantic Coast; Eurasia. July–Aug.

19. **P. Richardsònii** (Benn.) Rydb. [*P. perfoliatus* var. *R.* Benn.] Stem often branched, 1–2.5 mm. thick, 1–3 m. long; lvs. all submerged, mostly lance-ovate, 3–10 cm. long, 1–2 cm. wide, 17–29-nerved, cordate and clasping at base; stipules whitish, lanceolate to ovate, 1–2 cm. long, disintegrating into whitish fibers; peduncles enlarged upward, 1.5–20 cm. long; spikes 6–12-whorled, 1.5–4 cm. long in fr.; fr. obovoid, rounded on back, 2.5–3 mm. long, the beak up to an additional mm.; $n = 13$ (Löve, 1954).—At 4000–7000 ft., Sierra Nevada (Plumas Co., Sierra Co.), Modoc Co.; at lower elevs., Humboldt, and Siskiyou cos.; to Alaska, Atlantic Coast. July–Aug.

4. Ruppiàceae. DITCH-GRASS FAMILY

Aquatic herbs. Stems capillary, widely branched. Lvs. submerged, slender, attenuate, opposite or alternate, membranous-sheathing at base. Fls. perfect, few, small, arranged in terminal spikes at first enclosed by the sheathing lf.-bases, later elongated; bracts absent. Perianth absent. Stamens 2, with short broad fils. and extrorse anthers. Carpels 4 or more, stigmas peltate; ovule 1. Fruiting carpels long-stipitate, indehiscent. Seeds without endosperm. One genus of 2 spp. in saline and alkaline marshes of temp. and subtrop. regions.

1. Rúppia L. DITCH-GRASS

Characters of the family. (H. B. *Rupp*, an 18th-century German botanist.)

(Setchell, W. A. The genus Ruppia. Proc. Calif. Acad. Sci., IV, 25: 469–478, 1946.)

Peduncles 0.2–2.5 cm. long, not spiraling; lvs. obtusish .... 1. *R. maritima*
Peduncles 3–30 cm. long, spiraling or flexuous; lvs. acute .... 2. *R. spiralis*

1. **R. marítima** L. [*R. m.* vars. *intermedia* (Thed.) Asch. & Graebn. and *rostrata* Ag. *R. rostellata* Koch.] Stems branching, 6–10 dm. long, with internodes 2–8 cm. long;

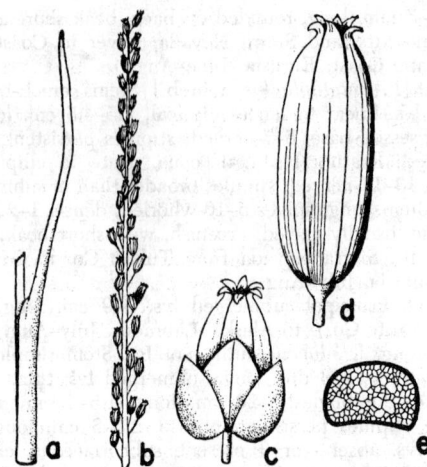

Fig. 123. JUNCAGINACEAE. *Triglochin: a,* sheathing lf. with ligule at base of blade; *b,* spike, × ½; *c,* young fr. with perianth at base (perianth-segms. in 2 series of 3 each); *d,* older fr.; *e,* cross section of subterete lf.

lvs. threadlike, 2–10 cm. long, 0.3 mm. wide, blunt at apex, with basal stipular sheath 6–10 mm. long; peduncles 2–25 mm. long; stipes of carpels 1–2.5 cm. long in fr.; fr. ovoid, ca. 2 mm. long; *n* = 8 (Wulff, 1937).—At low elevs., salt marshes along the coast and inland, as at Panamint V., Mojave Desert; widely distributed over the world. April–July. The var. *intermèdia* has ovoid, slightly oblique carpels, not conspicuously beaked, while var. *rostràta* has them strongly excentric, curved, and slenderly beaked.

2. **R. spiràlis** L. ex Dumort. [*R. marítima* auth., not L. *R. m.* var. *longipes* Hagström.] Much like no. 1, but the peduncles becoming 5–30 cm. long in fr. and spiraling or flexuous.—More common; brackish water at low elevs. throughout Calif.; cosmopolitan. March–Aug. Uncertainly distinct from no. 1.

5. Juncáginaceae. ARROW-WEED FAMILY (Fig. 123)

Annual or perennial marsh herbs, with rhizomes and scapes. Lvs. mostly basal, linear, sheathing at base. Fls. small, in terminal bractless spikes or racemes, perfect or imperfect, anemophilous. Perianth-segms. 3 or 6, in 1 or 2 series, greenish or reddish. Stamens 6, 4, or 3; anthers subsessile, 2-celled, extrorse. Carpels 6, 4 or 3, superior, free or connate; style short to absent; ovule 1, basal. Fr. cylindric to obovoid; seeds basal, erect, without endosperm. Four genera, mostly in cooler parts of both hemis.

1. Triglóchin L. ARROW-GRASS

Lvs. fleshy, ligulate at junction of sheath and the semiterete or compressed blade. Scapes long, naked, smooth. Perianth-segms. concave. Carpels united, 1-celled; style short; stigmas as many as carpels, plumose. Fr. a cluster of 3 to 6 1-seeded carpels, separating when mature from base upward and dehiscing. Seeds compressed or angular. Spp. ca. 12; temp. and cooler parts of both hemis. (Greek, *tri,* three, and *glochis,* a point, referring to fr. of some spp.)

Carpels and stigmas 3.
   Fr. nearly globose; perianth-segms. 3 .... 1. *T. striata*
   Fr. elongate; perianth-segms. 6 .... 2. *T. palustris*
Carpels and stigmas 6; fr. oblong or ovoid.
   Ligules simple, entire or emarginate; blade obcompressed .... 3. *T. maritima*
   Ligules 2-parted to base; blade nearly terete .... 4. *T. concinna*

1. **T. striàta** R. & P. [*T. triandra* Michx.] Scapes 1 or 2, ± angular, 1.5–2 dm. high; lvs. ca. as long as scapes, 1.5–2 mm. wide, ligules not divided; racemes 3–12 cm. long; pedicels 1–2 mm. long, not elongating in fr.; perianth-segms. 3; carpels and stigmas

3; fr.-subglobose, 2 mm. thick; carpels 3-ribbed on back.—Coastal Salt Marsh; Mendocino to Ventura cos.; Md. to La., Mex., S. Am. May–Sept.

2. **T. palústris** L. Lvs. linear, 5–30 cm. long, shorter than scapes; ligule very short, 2-parted; scapes slender, 1–6 dm. high; racemes 5–30 cm. long; pedicels appressed, becoming 4–6 mm. long in fr.; perianth-segms. 6; carpels and stigmas 3; fr. linear to clavate, 5–7 mm. long; carpels 2-ribbed on back; $2n = 24$ (Löve & Löve, 1944).— Mud flats and springy places, 7500–11,500 ft.; Subalpine. F. and above; Sierra Nevada (Center Basin, Whitney Meadows, Rattlesnake Creek, Mineral King, in Tulare Co. and Rock Creek Lake Basin in Inyo Co.); to Wash., Atlantic Coast; Eurasia; S. Hemis. July–Aug.

3. **T. marítima** L. Densely tufted, from short thick rootstocks; lvs. strongly obcompressed, ca. 2 mm. wide, with indurate-corky persistent bases; ligules entire or emarginate, but not 2-parted; scapes 3–7 dm. high; racemes 1–4 dm. long; pedicels ascending, 2–3 mm. long, 4–5 mm. in fr.; perianth-segms. 6; carpels and stigmas 6; fr. oblong-ovoid, 3–4 mm. long, each carpel with 2 dorsal winged angles and groove between; $2n = 12$ (Löve & Löve, 1944), 48 (LaCour, 1952).—Coastal Salt Marsh, San Francisco Bay n.; alkaline flats and boggy places, below 7500 ft., Montane Coniferous F., San Bernardino Mts., Sierra Nevada; to Alaska, New Mex., N.J.; Eurasia. April–Aug.

4. **T. concínna** Davy. From slender elongate rootstocks; lvs. subterete, 1–2 mm. wide, with bases of old lvs. evanescent; ligules 2-parted; scapes slender, 1–4 dm. high; racemes 5–15 cm. long; pedicels 2–3 mm. long; perianth-segms. 6; carpels and stigmas 6; fr. oblong-obovoid, 3–4 mm. long, shallowly grooved.—Coastal Salt Marsh; L. Calif. to B.C. March–Aug.

Var. **débilis** (Jones) J. T. Howell. [*T. maritima* var. *d.* Jones.] Rootstock fibrous-coated; scapes 1.5–6 dm. high.—Wet saline places especially near hot springs, below 6500 ft.; Creosote Bush Scrub, Sagebrush Scrub, Pinyon-Juniper Wd.; Mojave Desert (Rosamond, Tecopa), e. of Sierra Nevada to Modoc Co.; to Ore., N. Dak., Colo. May–Oct.

6. Lilaeàceae. Flowering Quillwort Family (Fig. 124)

Aquatic acaulescent annual herbs, with fibrous roots and basal spongy narrowly linear lvs. with sheathing base. Infls. of two types, some of the fls. basal and some in spikes on slender scapes. Basal fls. ♀, borne singly in the axils of the lvs., consisting of a single sessile erect carpel with a long filiform style and hemispheric oblique penicillate floating stigma; fr. indehiscent, divergent, longitudinally 25–30-ribbed, obliquely beaked by the persistent sharply recurved base of the style. Fls. of the scape usually perfect, tiny, irregularly imbedded in the fleshy axis of the spike, consisting of a single caducous bracteate perianth-segm., a single subsessile 2-celled extrorse anther opening by lateral slits, and a sessile closely appressed pistil; some spikes wholly ♀ and ebracteate, the styles of the lower florets up to 1 mm. long; fr. similarly ribbed, but with 2–4 ribs drawn into narrow undulate wings, the short beak erect. Ovule solitary, basal, anatropus, the straight embryo without endosperm. One genus.

1. Lilaèa Humb. & Bonpl.

A monotypic genus with the characters of the family, occurring in Pacific N. and S. Am. (Named for the French botanist, Alire Raffeneau-*Delile*, 1778–1850, who wrote on the plants of Egypt.)

1. **L. scilloìdes** (Poir.) Haum. [*Phalangium s.* Poir. *Lilaea subulata* Humb. & Bonpl. *Anthericum s.* Schult. f. *Heterostylus gramineus* Hook.] Flowering Quillwort. Tufted; lvs. 5–20(–35) cm. long, 1–5 mm. wide, subulate-tipped, the hyaline sheaths 3–10 cm. long; basal fls. with styles 6–20 cm. long; scapes 6–20 cm. long, the short dense erect spike borne at water level; fl.-spike 4–7 mm. long, in fr. up to 35 mm. long; perianth-segm. oblong, recurved, purplish; anther purplish; fr. in the lf.-axils 5–7 mm. long, 2.5 mm. wide, lance-oblong, compressed, light brown; fr. in the spike broadly lanceolate, more tapering, green.—In muddy and marshy places (sometimes brackish), below 5000 ft.; many Plant Communities; scattered through most of cismontane Calif.; to B.C., Alta., Ida., Mex., S. Am. March–Oct.

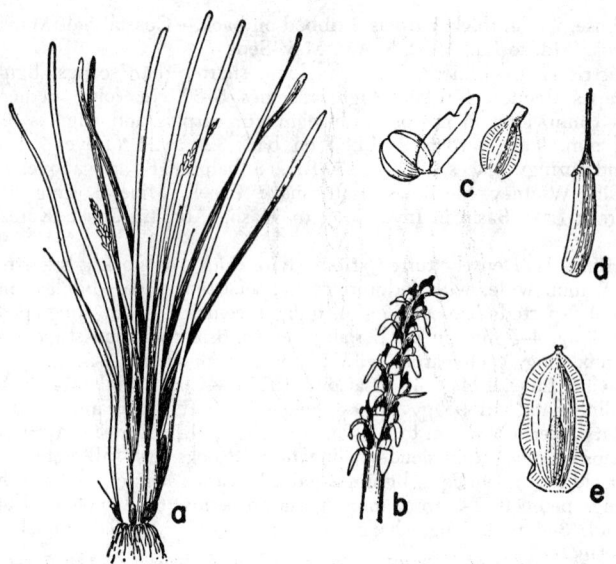

Fig. 124. LILAEACEAE. *Lilaea scilloides:* *a*, habit, × ½, acaulescent, lvs. terete with adnate stipules, each axil with a spike of fls. and with basal ♀ fls. with long filiform styles; *b*, spike, × 2½, with ♂ and ♀ fls., shown above in *c* (to the left a ♂ fl. with 1 anther and a perianthlike appendage arising from its connective, to the right a ♀ fl. with short style); *d*, fr. of a basal ♀ fl. with remnant of long style (see also bottom of *a*); *e*, fr. of ♀ fl. from a spike, × 3, angled, short-styled.

7. Zosteràceae. EEL-GRASS FAMILY (Fig. 125)

Submerged marine perennials, with creeping rootstocks. Stems flattened, slender, simple or branched. Lvs. in 2 rows, linear, sheathing at base. Plants monoecious or dioecious, the fls. arranged on one side of a flattened axis or spadix, at first enclosed in the lf.-sheath; bracts lacking. Perianth absent or represented by a row of bractlike lobes on each side of the axis. Staminate fls. reduced to a single anther arranged alternately in 2 rows on the spadix. Pistillate fl. an ovary with 2 stigmas and 1 ovule. Fr. indehiscent or bursting irregularly. Seed without endosperm. Family of 2 widely distributed genera.

Plants monoecious; fr. ovoid .... 1. *Zostera*
Plants dioecious; fr. cordate at base .... 2. *Phyllospadix*

1. Zostèra L. EEL-GRASS

Rootstocks slender. Monoecious; lvs. to 2 m. long. Fls. imperfect, the ♂ and ♀ alternating, in 2 rows in a 1-sided spike. Anther 1-celled; pollen threadlike. Style elongate; stigmas capillary. Mature carpels flasklike, beaked, forming a utricle. Seed ribbed. A genus of ca. 5 spp. from seas of both hemis. Of great importance as food for waterfowl. (Greek, *zoster*, a girdle or band, because of the ribbonlike lvs.)

1. **Z. marìna** L. [*Z. pacifica* Wats. *Z. oregona* Wats.] Stems branched, 1–3 m. long; lvs. 3–15 dm. long, 2–8 mm. wide, with 3–7 main nerves; spadix 2.5–6 cm. long, bearing ca. 10–20 of each kind of fls.; anthers escaping from spathe and discharging pollen into water; stigmas protruding through the spathe and dropping off before anthers of same spadix open; seed 20-ribbed, ca. 3 mm. long; $n = 6$ (Wulff, 1937).—Shallow water of bays, San Diego to Alaska; Eurasia. June–Sept.

Var. **latifòlia** Morong. [*Z. l.* Morong.] Lvs. 6–12 mm. wide; seed not ribbed.—Santa Barbara to Puget Sound.

2. Phyllospàdix Hook. SURF-GRASS

Rootstocks thickened. Plants dioecious, with the infl. at or near the summit of the slender stems, the spikes (spadices) solitary or 2 or 3 within a spathe. Staminate fls. in

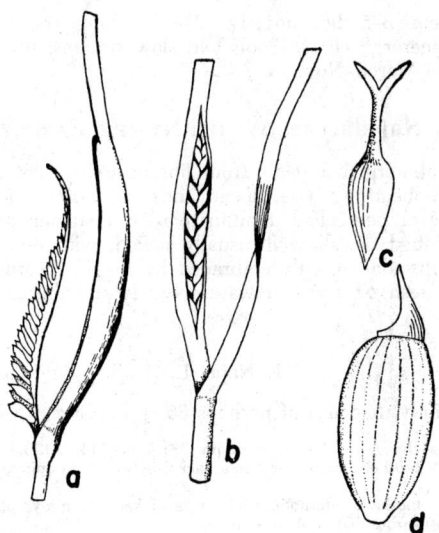

Fig. 125. ZOSTERACEAE. *Phyllospadix: a,* spathe and spadix, ca. ½. *Zostera: b,* spathe and spadix with lf. below; *c,* pistil; *d,* mature beaked fr.

2 rows, many, each of 1 sessile stamen. Pistillate of simple sessile ovaries; style short, with 2 filiform stigmas. Fr. beaked, cordate-sagittate. Seed globose, not ridged. Few spp., of the n. Pacific. (Greek, *phullon,* lf., and *spadix,* spicate infl.)

Pistillate spadices 2–6, borne on flowering stems 3–10 dm. long .... 1. *P. Torreyi*
Pistillate spadix usually 1, on flowering stems 1–2 dm. long .... 2. *P. Scouleri*

1. **P. Tórreyi** Wats. Stems flat, 3–10 dm. long; lvs. linear, flat to almost terete, ca. 1.5 mm. wide, 1–2 m. long; spadices cauline, in 2–3 pairs, ♀ usually ca. 5 cm. long, ♂ shorter; fr. flask-shaped, 2–3 mm. long, beaked by the persistent style, cordate-sagittate at base.—Usually below low-tide level in quiet waters, Ensenada (L. Calif.) to Ore. May–Nov.

2. **P. Scoùleri** Hook. Stems 1–2(–4) dm. long; lvs. flat, mostly 2–4 mm. wide; ♂ spadix ca. 6 cm. long; ♀ 2–5 cm. long; fr. ca. 3.5 mm. long.—On surf-beaten rocky shores, Santa Monica to B.C.; Japan. May–July.

8. Zannichelliàceae. HORNED PONDWEED FAMILY

Submerged aquatic monoecious or dioecious herbs with slender creeping rhizome. Stems capillary, branched. Lvs. alternate or opposite or crowded at nodes, linear, sheathing at base, the sheaths usually ligulate. Fls. minute, axillary, solitary or cymose. Perianth of 3 small free scales or lacking. Stamens 3, 2 or 1; anthers 2–1-celled, opening lengthwise. Carpels free, 1–9; style short or long, simple or 2–4-lobed; stigma capitate, peltate or spathulate. Fruiting carpels indehiscent, 1-seeded, sessile or stipitate. Seed lacking endosperm. Ca. 6 genera, widely distributed, in fresh, brackish or salt water.

1. Zannichéllia L. HORNED PONDWEED

Lvs. opposite, filiform but flat, 1-nerved. Staminate and ♀ fls. in the same axil, enclosed in the bud by a hyaline spathelike envelope; ♂ solitary, without perianth. stamen 1. Pistillate fls. 2–5 (usually 4) almost sessile carpels in a hyaline invol., each with short style and peltate stigma. Frs. flattish, usually toothed down one side. One or 2 spp. of fresh or brackish waters; cosmopolitan. (G. G. *Zanichelli,* 1662–1729, Italian botanist.)

1. **Z. palústris** L. Stems 3–5 dm. long; lvs. 2–8 cm. long; frs. 2–2.5 mm. long, short-stipitate; $2n = 28$ (Scheerer, 1940).—Pools and slow streams, below 7000 ft., throughout Calif.; cosmopolitan. March–Nov.

9. Najadàceae. WATER-NYMPH FAMILY

Slender branching submerged annuals from fibrous roots. Lvs. small, subopposite or verticillate, sessile with sheathing base, linear, entire or toothed. Fls. unisexual, minute, axillary, solitary, sessile or pedicelled. Staminate with 1 stamen and 2-lipped perianth, the latter entire or 4-lobed. Pistillate fl. usually naked, with one, 1-ovuled carpel. Fr. sessile, indehiscent, ellipsoidal, usually embraced by the lf.-sheath. Seed without endosperm. One genus of fresh or brackish water, widely distributed in temp. and warm parts of the world.

1. Nàjas L.

Characters of the family. A genus of perhaps 35 spp. (Greek, *Naias*, a water-nymph.)

(Clausen, R. T. Studies in the genus Najas. Rhodora 38: 333–344, 1936.)
Lvs. coarsely and definitely toothed; internodes and backs of lvs. often spiny; plants dioecious
                                                                          1. *N. marina*
Lvs. almost entire or finely toothed; internodes and backs of lvs. not spiny; plants monoecious.
  The lf.-sheaths truncate or rounded, not auriculate.
    Seed smooth and shining, with 30–40 rows of areolae; styles 1 mm. or more long; lvs. finely and
    closely spined .... 2. *N. flexilis*
    Seed dull, with 10–20 rows of areolae; styles 0.5 mm. long; lvs. finely but remotely spined
                                                                          3. *N. guadalupensis*
  The lf.-sheaths auriculate .... 4. *N. graminea*

1. **N. marìna** L. [*N. m.* var. *californica* Rendle.] Stems rather stout, often prickly, 1–4 dm. long; lvs. stiff, narrowly linear, 1–4.5 cm. long, 0.5–3 mm. wide with coarse teeth on margins to 1 mm. long and often spiny on the back; sheath almost entire; ♂ fls. 3–4 mm. long; anther 4-celled; ♀ fl. 3–4 mm. long; stigmas 3; frs. 4–5 mm. long, finely reticulate; $n = 6$ (Lewitzky, 1931).—Occasional in quiet fresh water, Lake Co. to L. Calif.; to Ariz., Utah, Wis., N.Y., Fla.; Eurasia, Afr. July–Aug.

2. **N. fléxilis** (Willd.) Rostk. & Schmidt. [*Caulinia f.* Willd.] Stems 1–2 m. long, unarmed; lvs. 1.5–3.5 cm. long, 0.5–1 mm. wide, with 20–30 minute teeth on each side; sheath serrulate; ♂ fls. 2.5–3 mm. long; anther 1-celled; ♀ fls. 1–2.5 mm. long; stigmas 2–4, subulate; fr. reticulate under a lens, 2–3 mm. long; $2n = 12, 24$ (Chase, 1947).—Occasional in ponds at low elevs.; Orange Co. to San Francisco; to B.C., Atlantic Coast; Eu. July.

3. **N. guadalupénsis** (Spreng.) Morong. [*Caulinia g.* Spreng.] Stems 3–6 dm. long, unarmed; lvs. 1–2.5 cm. long, 0.5 mm. wide, unarmed or remotely finely spined; ♂ fls. 2–3 mm. long; anthers 4-celled; ♀ fls. 2–3 mm. long; stigmas 2–3; fr. dull, finely reticulate under a lens, 2 mm. long; $2n = 12, 36, 42, 48, 54, 60$ (Chase, 1947).—Wet places in much of cismontane Calif.; to Ore., Atlantic Coast, S. Am. July.

4. **N. gramínea** Raffeneau-Delile. Lateral branches often very short, with tufted lvs. 1–2.5 cm. long, thin and translucent and tipped with 1 or more short spines; lf.-sheaths usually with 2 prominent auricles that are minutely spine-tipped.—Rice fields, Colusa Co. to Butte Co.; native of Old World.

10. Commelinàceae. SPIDERWORT FAMILY

Annual or perennial herbs, with knotty and leafy stems. Lvs. alternate, entire, lance-ovate to linear, amplexicaul or vaginate. Fls. bisexual, usually almost regular. Sepals 3, sometimes slightly connate, green. Petals 3, free or united into a tube, alternate with sepals, colored. Stamens usually 6, in 2 series; fils. often hairy. Style 1; stigma simple or obscurely lobed; ovary 2–3-celled, few–many-ovuled. Fr. a caps. A family of 34 genera and ca. 500 spp., of wide distribution in warmer regions.

1. Tradescántia L. SPIDERWORT. WANDERING JEW

Plants prostrate or erect. Lvs. variable. Peduncles solitary, fascicled or paniculate. Fls. mostly umbellate. More than 30 spp., temp. to trop. regions. (J. *Tradescant*, English gardener, who died about 1638.)

1. **T. fluminénsis** Vell. WANDERING JEW. Prostrate, rooting at the nodes; lvs. oblong to oblong-ovate, 3–7 cm. long, acute, glabrous; sheath hairy at summit; umbels many-fld., subtended by 2 lance-ovate bracts; petals white, 6 mm. long; fils. hairy; $2n = 60$ (Anderson & Sax, 1935).—Common in cult. and becoming natur. in damp places.; S. Am.

11. Eriocauláceae. PIPEWORT FAMILY

Annual or perennial, usually marsh or aquatic herbs, with mostly linear lvs. in radical tufts or rarely crowded on an elongated stem. Scapes usually exceeding lvs., simple, usually enclosed at the base in a sheathing scale and bearing a single terminal head. Fls. very small, usually many, bracteate, the outer mostly ♀, the inner chiefly ♂. Staminate fls. with calyx of 2 or 3 keeled sepals and tubular corolla of 2 or 3 lobes; stamens mostly 4 or 6. Pistillate fls. with calyx of 2 or 3 sepals often attached considerably below the corolla; corolla of 2 or 3 narrow petals, rarely wanting; stamens 0; pistil with stalked or sessile ovary and 2 or 3 slender stigmas. Ovary superior, 2- or 3-celled and -lobed, with 1 ovule in each cell. Fr. a caps., each lobe 2-valved. Seeds globular or ovoid, with mealy endosperm. A family of several genera of temp. and warmer regions.

1. Eriocaùlon L. PIPEWORT. BUTTON RODS

Characters as above. A genus of perhaps 200 spp., of Old and New Worlds. (Greek, *erion*, wool, and *kaulos*, stem, many spp. being woolly.)

1. **E. cinèreum** R. Br. Lvs. linear-filiform, 1–5 cm. long; scapes filiform, 3–12 cm. high; heads hemispherical, 2–3 mm. in diam.; bracts scarious, glabrous; petals of ♂ fls. lanceolate, fringed, with small glands; stamens usually 3; ♀ fls. stipitate, with 3 filiform sepals and almost obsolete petals; $2n = 32$ (Erlandsson, 1942).—Introd. in rice fields near Modesto, Stanislaus Co., *Markos* in 1947; s. China to Australia. Aug.–Sept.

12. Hydrocharitáceae. FROGBIT FAMILY

Dioecious, or rarely monoecious or polygamous, perennial, submerged or floating, aquatic herbs. Lvs. opposite or clustered. Fls. regular, sessile or on scapelike peduncles in a spathe. Calyx usually of 3 sepals or lobes; petals 3 or 0. Staminate fls. with 3 to 21 distinct or fused stamens. Pistillate fls. with 3–12 stigmas and compound pistil fused with the tubular perianth; ovary inferior, maturing into an indehiscent submerged fr. A family of ca. 14 genera and 40 spp., widely distributed in fresh and salt waters.

1. Elodèa Michx. WATERWEED. ELODEA

Plants submerged, forming large masses of branching elongate stems with occasional slender roots. Lvs. whorled, sometimes opposite, 1-nerved, thin, elongate, entire or minutely serrulate. Plants dioecious or polygamous; fls. 1–3 in axillary tubular spathes. Staminate fls. with 3 almost separate sepals, 3 petals, 3–9 stamens, reaching the surface or breaking off and floating. Pistillate fls. solitary, sessile, with 6-parted limb at the end of an elongate tube and with capillary style, the 3 stigmas thus on the surface. Fr. oblong or fusiform, with 1–5 seeds. A genus of ca. 12 spp. of temp. regions of New World. (Greek, *elodes*, marshy.)

Fls. usually 3 in a spathe; lvs. 2–3 cm. long .... 1. *E. densa*
Fls. solitary in the spathe; lvs. 0.5–1.5 cm. long.
    Lvs. averaging 2 mm. wide (1.2–4 mm.), obtuse; spathe of ♂ fl. 10–12 mm. long
                                                  2. *E. canadensis*
    Lvs. averaging 1.3 mm. wide (0.7–1.8 mm.), acute; spathe of ♂ fl. 2–3 mm. long
                                                  3. *E. Nuttallii*

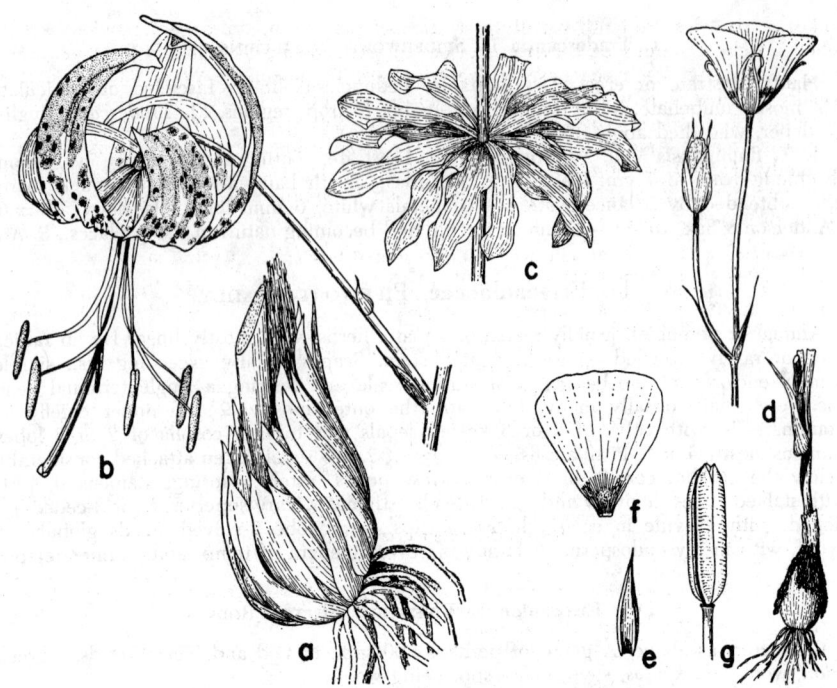

Fig. 126. LILIACEAE. *Lilium Humboldtii: a*, bulb, × ⅓, with unjointed fleshy scales; *b*, fl., × ½, 6 similar perianth-segms, 6 versatile anthers, long cent. style; *c*, whorl of lvs., × ¼. *Calochortus catalinae: d*, habit, × ⅓, to show base and apex of plant; *e*, green sepal, × ½; *f*, colored petal, × ½, with basal gland and hairs above it; *g*, caps. from superior ovary, × ½, 3-loculed.

1. **E. dénsa** Planch. [*Anacharis d.* Victorin.] BRAZILIAN WATERWEED. Stems several dm. long, densely leafy; lvs. lance-linear 2.5–3 cm. long, 3–5 mm. wide; petals 7–8 mm. long; $n = 24$ (Matsuura & Suto, 1935).—Used in aquaria and sparingly natur. in sloughs and ponds at low elevs., as at Robert's Id., San Joaquin Co.; S. Am. July–Aug.

2. **E. canadénsis** Michx. [*Anacharis c.* Planch. *Philotria c.* Britton. *E. Planchonii* Casp. *A. P.* Peck.] Stems 2–5 dm. long; lvs. narrowly oblong-ovate, 6–15 mm. long, 1.2–4 mm. wide, averaging 2 mm., obtuse; sepals of ♂ fl. 4–5 mm. long, this remaining attached on its long pedicel; sepals of ♀ fl. ca. 2.5 mm. long; petals 2.5–3 mm. long; fr. 10–15 mm. long; $n = 12$ (Heppell, 1945).—Occasional in ponds and sluggish waters, probably as an escape, as near San Dimas, Stockton, Mill V.; Rocky Mts. to Quebec, Ky., Va. July–Aug.

3. **E. Nuttállii** (Planch.) St. John. [*Anacharis N.* Planch. *Philotria N.* Rydb. *E. occidentalis* St. John.] Stems 3–10 dm. long; lvs. linear-oblong, 7–15 mm. long, 0.7–1.8 mm. wide, averaging 1.3 mm., acute; ♂ fl. sessile, breaking free and floating at anthesis, with sepals 2–2.5 mm. long; ♀ fl. with sepals 1–1.8 mm. long; petals 1.5–2 mm. long. —Rare, in slow streams and ponds, below 9200 ft.; Montane Coniferous F. and below; San Bernardino Mts., Sierra Nevada, N. Coast Ranges, Modoc Co.; to Wash., Atlantic Coast. July–Aug.

13. **Liliàceae.** LILY FAMILY (Fig. 126)

Leafy-stemmed or scapose perennial herbs, sometimes somewhat woody or climbing; from bulbs, corms, or rootstocks which may be somewhat tuberous. Lvs. mostly alternate, sometimes opposite or whorled, sometimes all basal, broad or grasslike, parallel-veined, the veins often connected by cross veinlets. Fls. perfect or imperfect, mostly showy and with colored parts, sometimes small and greenish, but then usually many in racemes, spikes or panicles. Perianth usually of 6 distinct parts, sometimes gamophyllous and 6-lobed, rarely 4-merous or 3-merous, often all parts petaloid. Stamens 6, sometimes 4 or 3, inserted at or on the base of the segms. or borne on the perianth-tube; anthers usually 2-celled, rarely 1-celled. Ovary mostly superior or nearly so, usually 3-celled, sometimes 1–2-celled; ovules few to many in each locule with axile placentation; styles 1 or 3, stigma capitate to lobed. Fr. a caps. or a berry, mostly several- to many-seeded. A family of 2000 or more spp., widely distributed, most abundant in temp. and subtrop. regions. Many of great importance as food or ornamental plants, as the source of fibers, drugs, etc.

A. Plant not bulbous, producing a rootstock or thickened tuberous ± branched underground parts.
  B. Stem short, subterranean.
    C. The lvs. linear; fls. white .... 6. *Leucocrinum*
    CC. The lvs. broad; fls. greenish with red veins .... 23. *Scoliopus*
  BB. Stem evident above ground.
    C. Lvs. all basal, or the basal much larger than the reduced cauline lvs.
      D. The lvs. linear and grasslike.
        E. Lvs. 5–10 dm. long, not equitant, dry; pedicels 2–5 cm. long.
                                                  1. *Xerophyllum*
        EE. Lvs. 0.5–3 dm. long, equitant, not dry; pedicels less than 2 cm. long.
          F. Perianth 6–10 mm. long; fls. racemose .... 2. *Narthecium*
          FF. Perianth 3.5–5.5 mm. long; fls. in a subcapitate panicle .. 3. *Tofieldia*
      DD. The lvs. broader, not grasslike.
        E. Fls. nodding; fr. a caps. .... 17. *Erythronium*
        EE. Fls. not nodding; fr. a berry .... 12. *Clintonia*
    CC. Lvs. not all basal, but well distributed on the stem.
      D. Green "foliage" represented by needlelike branchlets borne in the axils of scalelike lvs. .... 16. *Asparagus*
      DD. Green foliage of true lvs. not borne in the axils of scales.
        E. Cauline lvs. in a whorl of 3 .... 24. *Trillium*
        EE. Cauline lvs. alternate.
          F. Stems climbing or straggling, vinelike, bearing tendrils; plant dioecious
                                              25. *Smilax*
          FF. Stems not vinelike, or tendril-bearing.
            G. Fls. in terminal panicles or racemes.
              H. Panicles mostly 2–6 dm. long; lvs. 1.5–5 dm. long; fr. a caps.
                                        15. *Veratrum*
              HH. Panicles or racemes less than 2 dm. long; lvs. less than 2 dm. long; fr. a berry.
                I. Perianth-segms. 6; lvs. sessile or nearly so .. 8. *Smilacina*
                II. Perianth-segms. 4; lvs. well petioled .. 9. *Maianthemum*
            GG. Fls. solitary or umbellate.
              H. The fls. terminal, solitary or in a small umbel
                                        10. *Disporum*
              HH. The fls. axillary, 1 or 2 together .... 11. *Streptopus*
AA. Plant with a tunicated or scaly bulb, or a corm.
  B. The stems leafy, sometimes branched above.
    C. Styles 3, distinct to the base.
      D. Fls. nodding; perianth-segms. glandless .... 13. *Stenanthium*
      DD. Fls. erect; perianth-segms. with prominent gland .... 14. *Zigadenus*
    CC. Style 1, ± lobed at summit.
      D. Perianth-segms. unlike, the outer often sepallike, the inner petaloid
                                        20. *Calochortus*
      DD. Perianth-segms. alike, petaloid.
        E. Anthers versatile, attached to fils. at middle of the back ...... 19. *Lilium*
        EE. Anthers often seemingly basifixed.
          F. Lvs. several to many; plants from scaly bulbs .... 18. *Fritillaria*
          FF. Lvs. 2; plants from funicated corms .... 17. *Erythronium*
  BB. The stems scapelike, with lvs. mostly at the base.
    C. Perianth-segms. united at base into a tube.
      D. The perianth 4–6 cm. long, funnelform .... 7. *Hesperocallis*
      DD. The perianth ca. 1 cm. long, salverform .... 22. *Odontostomum*
    CC. Perianth-segms. not united at base into a tube.
      D. Fls. bractless, nodding; lvs. 2 .... 17. *Erythronium*
      DD. Fls. with scarious bracts; lvs. more numerous.
        E. Style short, perianth becoming scarious .... 4. *Schoenolirion*
        EE. Style filiform, long; perianth not scarious.
          F. Infl. a simple raceme .... 21. *Camassia*
          FF. Infl. a widely branching panicle .... 5. *Chlorogalum*

1. Xerophýllum Michx. Bear-Grass. Indian Basket-Grass

Tall perennial herbs from short thick woody rootstocks. Stem stout, simple, leafy. Lvs. many, rigid, grasslike, linear, dry, rough-margined. Fls. many, whitish, in a dense terminal raceme; perianth-segms. distinct, persistent, without glands, oblong to ovate, 5–7-nerved, spreading. Stamens 6; fils. subulate; anthers oblong. Ovary 3-lobed; styles 3, distinct. Caps. ovoid, 3-grooved; seeds 2–4 per cell, oblong, scarcely if at all appendaged. Two or 3 spp., N. Am. (Greek, *xeros,* dry, and *phullon,* lf.)

1. **X. ténax** (Pursh) Nutt. [*Helonias t.* Pursh.] Stem 3–15(–18) dm. high; basal lvs. densely clustered, 5–10 dm. long, 2–4 mm. wide above the base; cauline lvs. passing into long linear bracts; raceme 1–6 dm. long, dense; pedicels slender, 2–5 cm. long, erect in fr.; perianth-segms. linear-oblong, 6–10 mm. long; stamens somewhat longer; caps. acute, 5–7 mm. long; seeds linear-oblong, 4 mm. long.—Open dry slopes and ridges, below 6000 ft.; Mixed Evergreen F., Montane Coniferous F.; Coast Ranges s. to Monterey Co.; w. slope of Sierra Nevada s. to Placer Co.; to B.C., Rocky Mts. May–Aug.

2. Narthècium Huds. Bog-Asphodel

Perennial herbs with creeping rootstocks. Stems slender, erect, simple. Basal lvs. linear, grasslike, equitant; cauline few, reduced. Fls. racemose, small, greenish-yellow. Pedicels bracteate at base and near middle. Perianth-segms. distinct, glandless. Stamens 6; fils. subulate, woolly. Ovary attenuate upward to the scarcely lobed stigma. Caps. oblong, loculicidal, many-seeded. Seeds linear, with bristlelike tail at each end. A N. Temp. genus of 4 spp. (*Narthex,* Greek name.)

1. **N. califórnicum** Baker [*Abama c.* Heller. *A. occidentalis* Heller. *N. ossifragum* var. *o.* Gray.] Stem solitary, 4–5 dm. high; basal lvs. 1–3 dm. long, 3–6 mm. wide; raceme 8–15 cm. long; pedicels 6–15 mm. long; perianth-segms. lance-linear, 6–10 mm. long, spreading; stamens 3–4 mm. long; caps. ovoid-lanceolate, 8–12 mm. long; seeds, including tails, 6–8 mm. long.—Wet meadows and banks, 2500–5500 ft., Montane Coniferous F.; N. Coast Ranges s. to Mendocino Co.; up to 8500 ft., Sierra Nevada s. to Fresno Co. July–Aug.

3. Tofièldia Huds.

Perennial herbs from slender rootstocks. Stem solitary, slender, erect, scapose. Lvs. mostly basal, linear, equitant. Fls. small, green to white, in ours borne in 3's in a subcapitate panicle. Pedicels bracted at base and with 3 membranous, somewhat united bractlets below each fl. Perianth-segms. oblong-obovate, alike, persistent, glandless, distinct. Stamens 6; fils. filiform; anthers ovate, marginally dehiscent. Ovary 3-lobed at summit; styles 3, short, recurved. Caps. 3-lobed, 3-beaked, septicidal, many-seeded. Seeds with a membranous testa, in ours tailed at free end. A genus of 12 or more spp., of n. temp. regions and the Andes. (*Tofield,* an English botanist and correspondent of Hudson.)

(Hitchcock, C. L. The Tofieldia glutinosa complex. Am. Midl. Nat. 31: 487–498, 1944.)

1. **T. glutinòsa** (Michx.) Pers. ssp. **occidentàlis** (Wats.) C. L. Hitchc. [*T. o.* Wats.] Stems 2–7 dm. tall, glandular; lvs. 5–20 cm. long, 3–6 mm. wide; pedicels 2–10 mm. long; perianth 3.5–5.5 mm. long, light yellow, the outer segms. broader than inner; styles 1–2.5 mm. long; caps. 5–9 mm. long; seeds with spongy free testa and ca. 3 mm. long including the appendage.—Boggy places and meadows; below 5500 ft., Montane Coniferous F., N. Coast Ranges s. to Sonoma Co., and to 10,200 ft. in Sierra Nevada s. to Tulare Co.; s. Ore. July–Aug.

4. Schoenolírion Durand

Erect herb with a coated bulb and simple stem. Lvs. linear, flat. Fls. many in a dense terminal spicate raceme, this sometimes with 1 or 2 short racemose branches at base. Bracts small, scarious. Pedicels short, jointed. Perianth white or greenish, persistent and becoming scarious, divided to base, the segms. 6, oblong, 3-nerved. Stamens 6, adnate to base of segms.; fils. subulate; anthers versatile. Ovary ovoid, short-stipitate, with 2 ovules in a locule; style short, persistent. Caps. ovoid, loculicidal. Seeds black, oblong. Spp. 2. (Greek, *schoinos,* a rush, and *lirion,* lily.)

Perianth-segms. 3–6 mm. long, linear-oblong .... 1. *S. album*
Perianth-segms. 10–12 mm. long, lanceolate .... 2. *S. bracteosum*

1. **S. álbum** Durand. [*Hastingsia a.* Wats.] Bulb 2 or more cm. long; scape 2.5–15 dm. high; lvs. 15–60 cm. long, 4–12 mm. wide; bracts lance-acuminate, 3–8 mm. long; pedicels mostly 1–4 mm. long; perianth white, tinged with green or pink, 3–6 mm. long; stamens 3–5 mm. long; caps. 6–7 mm. long; seeds 4 mm. long.—Meadows and swampy places, 1500–8000 ft.; Redwood F., Mixed Evergreen F., Yellow Pine F.; Sierra Nevada, Nevada Co. to s. Ore., Coast Ranges s. to Mendocino Co. June–July.

2. **S. bracteòsum** (Wats.) Jeps. [*Hastingsia b.* Wats.] Much like the preceding, lvs. 3–7 mm. wide; perianth-segms. 10–12 mm. long, lanceolate, acuminate; stamens ½–⅔ as long; caps. 8–10 mm. long; seeds ca. 3 mm. long.—Moist places, Mixed Evergreen F., Yellow Pine F.?; s. Ore. to Siskiyou and Del Norte cos. June.

5. Chlorógalum Kunth. Soap Plant. Amole

Herbs with coated bulbs and tall almost leafless paniculate infl. Basal lvs. several, tufted, linear; cauline reduced, passing upward into scarious bracts. Pedicels scattered or crowded, jointed at apex. Perianth white to pink or blue, the segms. 6, distinct, linear to oblong, persistent, twisted together above the caps., spreading, each with 3 approximate nerves. Stamens 6, inserted on bases of segms.; fils. filiform; anthers versatile. Style long-filiform, slightly 3-cleft at apex. Caps. subglobose, loculicidal, 3-valved. Seeds black, rounded, 1 or 2 in each locule. Spp. 5, mainly in Calif. (Greek, *chloros*, green, and *gala*, milk or juice.)

(Hoover, R. A. Monograph of the genus Chlorogalum. Madroño 5: 137–146, 1940.)
Fls. vespertine; style not exceeding perianth; perianth-segms. 8–30 mm. long.
    Lvs. strongly undulate, 4–25 mm. wide; perianth-segms. 15–30 mm. long.
      Pedicels 5–35 mm. long; style 10–15 mm. long; bulb-coats usually with coarse fibers
                                       1. *C. pomeridianum*
      Pedicels 2–5 mm. long; style 18–28 mm. long; bulb-coats membranous .. 2. *C. grandiflorum*
    Lvs. plane or nearly so, 2–5 mm. broad; perianth-segms. 8–12 mm. long .... 3. *C. angustifolium*
Fls. diurnal; style exceeding perianth; perianth-segms. 5–8 mm. long.
    Perianth white or pink; pedicels mostly shorter than fls. .... 4. *C. parviflorum*
    Perianth blue; pedicels mostly longer than fls. .... 5. *C. purpureum*

1. **C. pomerídianum** (DC.) Kunth. [*Scilla p.* DC. *Antheridium p.* Ker. *Phalangium p.* Sweet. *Laothoe p.* Raf.] Bulb 7–15 cm. long, heavily coated with persistent dark brown fibers of old coats; lvs. 2–7 dm. long, 6–25 mm. wide, very wavy; stem glaucous, stout, 6–25 dm. tall, freely branched above; pedicels slender, 5–25(–35) mm. long; perianth-segms. linear, white with green or purple midvein, 15–23 mm. long, spreading and recurved at anthesis; stamens ca. ⅔ as long; anthers 2 mm. long; style 10–15 mm. long; caps. short-stipitate, 5–7 mm. long; $n = 18$ (Cave, 1949).—Dry open hills and plains, sometimes in woods, below 5000 ft., chiefly V. Grassland, Coastal Scrub, Foothill Wd.; s. Ore. to San Diego Co. The bulbs were roasted and eaten by the Indians; uncooked they have lather-producing properties in water. Crushed material of the plant was used in streams to stupefy fish. May–Aug.

Var. **divaricàtum** (Lindl.) Hoov. [*Ornithogalum d.* Lindl. *C. d.* Kunth. *Laothoe d.* Greene.] Stem with widely divaricate branches from base and not over 3 dm. high.— Along immediate coast; Closed-cone Pine F., Mixed Evergreen F.; Sonoma Co. to Monterey Co. May–July.

Var. **mìnus** Hoov. Bulb-coats membranous, with few fibers; stems 3–4 dm. tall.— Serpentine rocks; Chaparral; Tehama Co.

2. **C. grandiflòrum** Hoov. Bulb 5–7 cm. long, with membranous coats and few delicate fibers; lvs. 1–3 dm. long, 4–12 mm. wide; stem 3–6 dm. tall, slender; pedicels 2–5 mm. long; perianth 2–3 cm. long, the segms. white, with purple midvein; stamens slightly shorter than perianth; anthers 3 mm. long; caps. stipitate, 5–8 mm. long.— Among serpentine rocks, on open brushy hills; Foothill Wd.; Tuolumne Co. May–June.

3. **C. angustifòlium** Kell. [*Laothoe a.* Greene.] Bulb 3–5 cm. long, with membranous coats and very delicate fibers; lvs. 1–3 dm. long, 2–5 mm. wide, not strongly undulate; stem rather slender, 3–6 dm. tall; pedicels slender, 2–3 mm. long; perianth-segms. white with greenish-yellow midvein, divaricate, oblong, 8–12 mm. long; stamens ca. as long; style 3–4 mm. long; caps. 4–5 mm. long; seeds 2 mm. long.—Heavy soil, valleys and hills, below 1500 ft.; V. Grassland, Foothill Wd.; interior Calif., Shasta Co. to Lake Co. in inner Coast Ranges, and to Fresno Co. in Sierran foothills. April–July.

4. **C. parviflòrum** Wats. [*Laothoe p.* Greene.] Bulb 4–7 cm. long, with brown membranous coats; lvs. 1–2 dm. long, 3–9 mm. wide, undulate; stem 3–8 dm. tall; pedicels 2–8 mm. long; fls. white to pink, with dark midvein, 7–8 mm. long, sub-rotate; stamens 3–4 mm. long; style 7–9 mm. long; caps. ca. 4 mm. long.—Dry open places, below 2000 ft.; V. Grassland, Coastal Sage Scrub; e. Los Angeles Co. (San Dimas) to Riverside and San Diego cos.; L. Calif. May–June.

5. **C. purpùreum** Bdg. [*Laothoe p.* Greene.] Bulb 2.5–3 cm. long, outer coat brown, inner white; lvs. undulate, ca. 10 cm. long, 2–5 mm. wide; stem 2.5–4 dm. tall; pedicels 4–10 mm. long; perianth deep blue, 5–7 mm. long, recurved; stamens almost as long; style 5–6 mm. long; caps. 3 mm. long.—Plains at ca. 1000 ft.; Foothill Wd.; Jolon, Monterey Co. June.

6. Leucocrìnum Nutt. SAND-LILY. STAR-LILY

Acaulescent herb with short rootstock, fleshy roots and tufted narrowly linear lvs. surrounded at base by membranous bracts. Fls. in a cent. sessile cluster on the ground and arising from the rootstock, white, fragrant, showy. Perianth persistent, salverform, with elongate slender tube and lance-oblong segms. Stamens 6, inserted near summit of tube; fils. filiform; anthers linear, attached near base. Ovary below ground; style filiform, elongate, persistent, tubular, expanded and slightly 3-lobed at orifice; ovules several in each locule. Caps. triangular-obovate, loculicidal. Seeds black, angled, obovate. A single sp. of w. U.S. (Greek, *leuco*, white, and *krinon*, lily.)

1. **L. montànum** Nutt. Lvs. several, 1–2 dm. long, 2–6 mm. wide, flat, several-nerved, the underground part 3–7 cm. long; fls. several; pedicels 5–30 mm. long; perianth-tube 5–10 cm. long, segms. 14–20 mm. long; fils. 7–10 mm. long, anthers yellow, 4–5 mm. long; caps. 6–8 mm. long, somewhat wrinkled; $2n = 28$ (Cave, 1948).—Sandy loam about meadows and flats, 4000–5000 ft.; Sagebrush Scrub, N. Juniper Wd.; Siskiyou and Modoc cos. to Sierra Co.; Ore. to Nebr. April–May.

7. Hesperocállis Gray. DESERT-LILY

Bulb tunicated, deep-set; stems stout, straight, simple, leafy at base, with elongated terminal raceme. Lvs. linear, mostly strongly crisped on the white margins. Bracts subtending the fragrant fls., conspicuous, scarious. Pedicels jointed at apex. Perianth united below middle into a tube, the segms. 6, spatulate, withering-persistent, 5–7-nerved. Stamens 6, inserted on throat; fils. filiform; anthers linear, versatile. Ovary oblong, sessile, 3-celled; style white, filiform, persistent, equaling segms.; stigma discoid. Caps. subglobose, deeply 3-lobed, loculicidal. Seeds many, black, horizontal, flattened. A single sp. of the deserts of Calif. and Ariz. Desert Indians used the bulbs for food. (Greek, *hesperos*, western, and *kallos*, beauty.)

1. **H. undulàta** Gray. Bulb ovoid, 4–6 cm. long; stem 3–18 dm. high; basal lvs. 2–5 dm. long, 8–15 mm. wide, blue-green; cauline reduced, few; raceme 1–3 dm. long; bracts broadly ovate, 1–1.5 cm. long, scarious; pedicels ca. 1 cm. long; perianth-tube 1.5–2 cm. long, the segms. 3–4 cm. long, 6–10 mm. wide, divergent, white within, with a silvery-greenish lineate band without; fils. 2–2.5 cm. long; anthers golden, 7 mm. long; caps. 12–16 mm. long; seeds ca. 5 mm. long and wide; $2n = 48$ (Cave, 1948).— Locally common on dry sandy flats and gentle slopes, below 2500 ft.; Creosote Bush Scrub; Mojave Desert e. of Yermo, Colo. Desert; w. Ariz. March–May.

8. Smilacìna Desf. FALSE SOLOMON'S-SEAL

Perennial herbs with creeping rootstocks. Stem simple, scaly below, leafy above. Lvs. alternate, short-petioled or sessile, lanceolate to ovate. Fls. small, usually white, spreading, in a terminal panicle or raceme; perianth of 6 equal segms. Stamens 6, inserted at base of segms.; fils. slender; anthers small, ovate. Ovary subglobose, superior, 3-celled, with 2 ovules in each cell; style columnar; stigma trifid to 3-grooved. Fr. a berry; seeds subglobose. Ca. 25 spp., N. Am. and Asia. (Diminutive of *Smilax*.)

(Galway, D. H. The N. Am. spp. of Smilacina. Am. Midl. Nat. 33: 644–666, 1945.)
Fls. panicled, numerous .... 1. *S. racemosa*
Fls. racemose, few to several .... 2. *S. stellata*

1. **S. racemòsa** (L.) Desf. [*Convallaria r.* L. *Vagnera r.* Morong.] Rootstock stout; stem erect, 3–9 dm. high, ± pubescent above; lvs. several, ovate to lance-oblong, broadest near middle, 7–20 cm. long, 3–10 cm. wide, long-acuminate, pubescent beneath, distinctly petioled; peduncle naked or with 1–2 bracts; infl. 3–18 cm. long, many-fld., pubescent; perianth-segms. 1–2 mm. long; stamens 1.5–3 mm. long; berry ca. 5 mm. long, mostly red or with small purple spots; seeds ca. 4 mm. long.—Reported from Marin Co. and Amador Co.; common from B.C. to Atlantic Coast. The usual form in Calif. is:

Var. **amplexicaùlis** (Nutt.) Wats. [*S. a.* Nutt. ex Baker. *Vagnera a.* Greene. *V. pallescens* Greene. *Unifolium a.* Greene.] Lvs. acute or short-acuminate, clasping and sessile, or with short dilated clasping petiole, broadest near base; n = 18 (Rattenbury, 1948).—Shaded woods, below 6000 ft.; mostly Mixed Evergreen F., Montane Coniferous F.; through the Coast Ranges and Sierra Nevada; to B.C., Rocky Mts. March–May.

Var. **glàbra** (Macbr.) St. John. [*S. amplexicaulis* var. *g.* Macbr. *Vagnera a.* ssp. *g.* Abrams.] Plant glabrous and slightly glaucous.—At 5000–8000 ft.; Montane Coniferous F.; San Jacinto and San Bernardino mts., Sierra Nevada, Humboldt Co. June–July.

2. **S. stellàta** (L.) Desf. [*Convallaria s.* L. *Unifolium s.* Greene. *Vagnera s.* Morong. *Unifolium liliaceum* Greene. *Vagnera l.* Rydb.] Rootstock 2–4 mm. thick; stem 3–6 dm. high, erect, rather strict, glabrous to puberulent; lvs. oblong-lanceolate to lanceolate, 5–6 times as long as wide, acuminate, 5–15 cm. long, 15–30 mm. wide, puberulent beneath, sessile or somewhat clasping, ascending, rather evenly many-veined, often somewhat folded; raceme sessile or short-peduncled, 3–15-fld., puberulent; pedicels ascending, 5–15 mm. long; perianth-segms. oblong, obtuse, 5–7 mm. long; stamens somewhat shorter; berry red-purple, becoming black, 7–10 mm. long; seeds 3–4 mm. long; n = 18 (Stenar, 1935).—Wet places, often in brush, mostly 4000–8000 ft.; Montane Coniferous F. and below; Sierra Nevada and e., s. to San Jacinto Mts.; n. to B.C., Atlantic Coast. April–June. Variable.

Var. **sessilifòlia** (Baker) Henders. [*Tovaria s.* Baker. *S. s.* Nutt. *Vagnera s.* Greene.] Stems usually flexuous above; lvs. lance-ovate, flat and spreading, ca. 2–4 times as long as wide with 3 veins more prominent than others; n = 18 (Rattenbury, 1948).—Moist slopes and streambanks; Chaparral, Redwood F., Mixed Evergreen F.; Coast Ranges, Santa Barbara Co. n.; to B.C., Mont. March–May.

9. Maiánthemum Weber. FALSE LILY-OF-THE-VALLEY

Low perennial herbs with creeping rootstock. Stem erect, simple, few-lvd. Lvs. broad, alternate, cordate-ovate. Fls. small, white, in terminal raceme. Perianth of 4 distinct spreading segms. Stamens 4, inserted at base of segms.; fils. filiform; anthers short. Ovary globose, 2-celled, with 2 ovules in each cell; style equal to ovary; stigma 2-lobed. Fr. a 1–2-seeded berry; seeds subglobose. Spp. 3 or 4; N. Temp. Zone. (Greek, *maios*, May, and *anthemon*, fl., referring to time of flowering.)

1. **M. dilatàtum** (Wood) Nels. & Macbr. [*M. bifolium* var. *d.* Wood. *Unifolium d.* Howell. *M. b.* var. *camtschaticum* (Gmel.) Jeps.] Stem 1.5–3.5 dm. high, glabrous; lvs. 2–3, the blades 5–20 cm. long, 5–10 cm. wide, on petioles 5–15 cm. long; peduncle 2–5 cm. long; raceme ca. the same; perianth white, 2–3 mm. long; stamens shorter; berry red, globose, 6 mm. long; seeds brown, 3 mm. long; n = 16 (Matsuura & Sato, 1935).—Moist shaded banks near the coast; Redwood F., Mixed Evergreen F., Closed-cone Pine F.; Coast Ranges, Marin Co. n.; to Alaska, Ida. May–June.

10. Dísporum Salisb. FAIRY BELLS

Perennial, ± pubescent herbs with slender rootstocks. Stems branched, scaly below, leafy above. Lvs. alternate, sessile or clasping, somewhat inequilateral, with 3–5 main longitudinal veins. Fls. greenish or white, terminal, drooping, solitary or a few in an umbel. Perianth-segms. 6, narrow, equal, distinct, deciduous. Stamens 6; fils. filiform or flat; anthers oblong or linear. Ovary 3-celled, ovules mostly 2 in each cell; style slender; stigma 3-cleft or entire. Fr. a berry, ovoid, obtuse; seeds subovoid. About 15 spp., of N. Am. and Asia. (Greek, *dis*, double, and *spora*, seed, because of 2 ovules per locule in some spp.)

(Jones, Q. A cytotaxonomic study of the genus Disporum in N. Am. Contr. Gray Herb. 173: 1–39, 1951.)
Fls. cylindrical, truncate or slightly swollen below, or small with subsessile anthers and an abortive ovary; fr., if formed, usually more than 6-seeded; stigma usually 3-cleft.
   The fls. large, mostly 15–25 mm. long; ovary normal; stems usually more than 4 dm. tall; lvs. usually more than 5 cm. long. Santa Cruz Co. n. .... 1. *D. Smithii*
   The fls. smaller, 10 mm. or less long; anthers subsessile; ovary abortive; stem usually less than 3 dm. tall; lvs. usually less than 4 cm. long. Del Norte Co. .... 2. *D. parvifolium*
Fls. turbinate, narrowing at base; fr. never more than 6-seeded; stigma entire .... 3. *D. Hookeri*

   1. **D. Smíthii** (Hook.) Piper. [*Uvularia S.* Hook. *Prosartes Menziesii* D. Don.] Stem pubescent to subglabrous, 3–9 dm. high; lvs. ovate to lance-ovate, rounded or subcordate at base, acute to acuminate at apex, 5–12 cm. long; fls. 1–5 in a cluster, mostly whitish, subtruncate at base, the segms. acute, suberect, 1.5–2.5 cm. long; stamens ⅔ as long; style pubescent throughout, 3-lobed at apex; berry obovoid, light orange to red, 12–15 mm. long; seeds ca. 4 mm. long; $2n = 16$ (Q. Jones, 1951).—Moist shaded woods near coast; Redwood F., Mixed Evergreen F.; Santa Cruz Co. to B.C. March–May.
   2. **D. × parvifòlium** (Wats.) Britton. [*Prosartes p.* Wats.] Stoutish, branched, 1.5–4 dm. tall, woolly-pubescent; lvs. ovate to ovate-lanceolate, 2–3 cm. long, acuminate to acute, the lower with cordate clasping base; fls. in 2's or 3's, creamy-white, 5–10 mm. long, narrow-campanulate, the perianth-segms. oblong-lanceolate, acute; anthers subsessile, lanceolate, acute, ca. 4 mm. long; ovary abortive; style glabrous; stigma entire; plants completely sterile.—Siskiyou Mts., n. Calif.; s. Ore. May–June. A probable hybrid between *D. Smithii* and *D. Hookeri*.
   3. **D. Hoòkeri** (Torr.) Nichols. [*Prosartes H.* Torr. *P. H.* var. *oblongifolia* Wats.] Stems mostly 3–8 dm. tall, usually with spreading hairs; lvs. ovate to oblong-ovate, usually acute, scabridulous beneath, 3–10 cm. long, at least the lower cordate-clasping; fls. 1–3, creamy-white to greenish-white, 10–15 mm. long, turbinate, the perianth-segms. oblanceolate; stamens included or slightly exserted; anthers glabrous; style glabrous or pubescent; stigma entire; ovary glabrous or sometimes lanulose, ellipsoid; berry scarlet, ca. 8 mm. long; seeds commonly 4–6, 6 mm. long; $n = 9$ (Rattenbury, 1948).—Shaded woods away from immediate coast; Mixed Evergreen F., Redwood F.; Coast Ranges from Monterey Co. to s. Ore. March–May.
   Var. **trachyándrum** (Torr.) Q. Jones. [*Prosartes t.* Torr. *P. lanuginosa* var. *t.* Baker. *D. t.* Britton.] Fls. mostly 9–12 mm. long; anthers hispidulous; $2n = 18$ (Q. Jones, 1951). —Dry shaded slopes and benches below 5000 ft.; Yellow Pine F., Red Fir F.; Sierra Nevada from Tulare Co. n., N. Coast Ranges from Lake Co. n.; to s. Ore. April–June.

11. Stréptopus Michx. Twisted Stalk

   Perennial herbs, with horizontal rootstocks. Stem leafy, simple to several-times forking. Lvs. alternate, elliptic to ovate, thin, many-veined, ± tapering at apex, sessile to clasping at base. Fls. 1 or 2 together, on slender, supra-axillary, bent or twisted peduncles; perianth campanulate or rotate, of 6 distinct deciduous segms., the outer flat, the inner keeled. Stamens 6; fils. short, flat; anthers sagittate, apiculate to aristate at apex. Ovary 3-celled, with many ovules; style slender, entire to 3-cleft. Fr. a berry; seeds elongate. Seven spp., N. Hemis. (Greek, *streptos*, twisted, and *pous*, foot or stalk, referring to peduncles.)
   1. **S. amplexifòlius** (L.) DC. var. **denticulàtus** Fassett. Rootstock stout, with thick fibrous roots; stem 3–8 dm. high, glabrous; lvs. ovate to ovate-lanceolate, acuminate at apex, cordate-clasping, glaucous beneath, 5–15 cm. long, 2.5–5 cm. wide, with minute marginal teeth; peduncles 1.5–3 cm. long, sharply bent and twisted at the joint with the 1 or 2 pedicels; fls. greenish or yellowish, narrow-campanulate, 1–1.5 cm. long, the segms. lanceolate, spreading or recurved above; anthers subulate-pointed; stigma entire; berry oval to elliptic, 1–1.6 cm. long; seeds longitudinally grooved, 3–3.5 mm. long; $n = 16$ (Matsuura & Sato, 1935).—Moist woods, 1000–5500 ft.; Redwood F., Montane Coniferous F.; Mendocino and Plumas cos. n. to Alaska; Great Lakes. May–June.

12. Clintònia Raf.

   Scapose perennial herbs with slender rootstocks. Stems erect, simple. Lvs. few, largely basal, broad, petioled, sheathing. Fls. bractless, terminal, 1 to many, if the latter, in one or more umbellate clusters. Perianth-segms. 6, distinct, subequal, ascending, deciduous. Stamens 6, inserted at bases of segms.; fils. filiform; anthers oblong, basifixed. Ovary 2–3-celled, with 2–several ovules in each cell; style stout or slender; stigma obscurely 2–3-lobed. Fr. a berry, in ours blue. Seeds few, rounded on back, angled at edges, with 2 or 3 inner faces. Ca. 6 spp., N. Am. and E. Asia. (Named for De Witt *Clinton*, 1769–1828, naturalist and governor of N.Y.)

Fls. 1 or 2, white .... 1. *C. uniflora*
Fls. several to many, rose-purple .... 2. *C. Andrewsiana*

1. **C. uniflòra** (Schult.) Kunth. [*Smilacina borealis* var. *u.* Schult. S. *u.* Menz.] Brides's Bonnet. Queen Cup. Lvs. usually 2–3, obovate to oblanceolate, 7–15 cm. long, 2.5–6 cm. wide, ± pubescent; peduncle slender, pubescent, naked or with 1–2 bracts; fls. erect, white, the segms. pubescent, oblanceolate, spreading, 18–22 mm. long, 5–7 mm. wide; fils. pubescent, 10–12 mm. long; anthers ca. 4 mm. long; berry 8–12 mm. long; seeds 3–4 mm. long; $2n = 28$ (Walker, 1944).—Shaded woods, 3500–6000 ft.; Montane Coniferous F.; Tulare Co. to Lassen, Siskiyou, and Humboldt cos.; to B.C., Mont. May–July.

2. **C. Andrewsiàna** Torr. Lvs. usually 5 or 6, oblanceolate to broadly elliptic, 15–25 cm. long, 5–12 cm. wide, sparsely pubescent on margins; peduncle stout, 2.5–5 dm. high, with 1–3 bracts; infl. pubescent, with a terminal umbel and often 1 or more lateral ones; pedicels unequal, 1–3 cm. long; fls. deep rose-purple, the segms. gibbous at base, 10–15 mm. long; fils. pubescent, 8–10 mm. long; anthers ca. 2 mm. long; berry ovoid; 8–12 mm. long; seeds 3–4 mm. long; $2n = 28$ (Walker, 1944).—Shaded damp woods; Redwood F.; Del Norte Co. to Monterey Co.; sw. Ore. May–July.

13. Stenánthium Gray

Perennial herb from a tunicated bulb. Stem slender, solitary. Lvs. mostly basal, narrow, grasslike. Fls. racemose or paniculate, nodding; perianth-segms. purplish-green, distinct, linear-lanceolate, withering-persistent with reflexed tips. Stamens 6, inserted on base of perianth; fils. dilated toward base; anthers reniform, 1-celled, explanate. Ovary ovoid; styles 3. Caps. septicidal, oblong-ovoid, 3-beaked. Seeds oblong, winged. A N. Am. and Asiatic genus of ca. 5 spp. (Greek, *steno*, narrow, and *anthos*, fl.)

1. **S. occidentàle** Gray. [*Stenanthella o.* Rydb.] Bulb ovoid, 1–2 cm. long; stem 3–6 dm. high; lvs. 1.5–3 dm. long, 6–20 mm. wide, with sheathing base; infl. 1–2 dm. long; bracts lance-linear; pedicels 5–30 mm. long; perianth narrow-campanulate, 1.5–2 cm. long; stamens 7–8 mm. long; caps. 12–16 mm. long; seeds 3–4 mm. long.—Moist banks and seeps, 5000–6000 ft.; Montane Coniferous F.; Trinity and Siskiyou cos.; to B.C., Alta. July–Aug.

14. Zigadènus Michx. Zygadene

Perennial herbs with tunicated bulbs in our spp. Stem simple, leafy below. Lvs. linear, glabrous. Fls. greenish or yellowish-white, in terminal racemes or panicles, perfect or unisexual, subtended by long-acuminate bracts. Perianth withering-persistent, free or adnate to lower part of ovary; segms. ovate to lanceolate, with 2 or (in our spp.) 1 gland just above the base. Stamens 6, free from the segms.; anthers reniform, explanate after dehiscence. Ovary 3-celled, styles 3, persistent; caps. 3-lobed, loculicidal to the base. Seeds many, subrhomboid, irregularly angled. Ca. 18 spp., N. Am. and N. Asia. At least some of the spp. are poisonous to cattle and sheep, but said to be less so to hogs. (Greek, *zugon*, yoke, and *aden*, gland, referring to the paired glands of the first known sp. *Zygadenus*, a later spelling.)

Stamens definitely longer than perianth in mature fls.
  Fls. normally racemose; perianth-segms. obtuse .... 1. *Z. venenosus*
  Fls. paniculate; perianth-segms. acute .... 4. *Z. paniculatus*
Stamens not thus exceeding perianth.
  Perianth 8–14 mm. long; stamens ca. half as long .... 5. *Z. Fremontii*
  Perianth smaller, or if 8 mm. long, the stamens more than half as long.
    Fls. normally racemose, 4–5 mm. long. Contra Costa Co. n. .... 2. *Z. micranthus*
    Fls. normally paniculate, ca. 6–8 mm. long.
      Sepals cordate at base, Mendocino Co. to Marin Co. .... 3. *Z. fontanus*

Sepals not cordate.
  Lvs. 8–30 mm. wide, 5–7 dm. long. Sierra Nevada .... 6. *Z. exaltatus*
  Lvs. 6–8 mm. wide, 1.5–2 dm. long. S. Calif. deserts and San Luis Obispo Co.
                                                                    7. *Z. brevibracteatus*

1. **Z. venenòsus** Wats. [*Toxiscordion v.* Rydb. *T. arenicola* Heller. *Z. v.* var. *ambiguus* Jones. *Z. diegensis* A. Davids.] DEATH-CAMAS. Bulbs oblong-ovoid, 1.5–2.5 cm. long, with dark outer coats; stems 2.5–6 dm. high, glabrous, slender; basal lvs. 1.5–3 dm. long, 4–10 mm. wide, usually folded, slightly scabrous on margins; lowest cauline lvs. with scarious sheaths; raceme usually simple, 5–20 cm. long; bracts membranous, lanceolate to setaceously acuminate; pedicels ascending in fr., 1–1.5(–2) cm. long; perianth-segms. whitish, ovate, obtuse at apex, subcordate at base, 3–4 mm. long, distinctly clawed, the gland with a thick well-defined toothed upper margin; stamens ca. 5 mm. long; caps. cylindric, 1–1.5 cm. long; seeds brown, ca. 3 mm. long.—Moist grassy places, below 8200 ft.; N. Coastal Scrub, Coastal Sage Scrub, etc. to Montane Coniferous F.; San Diego Co., Kern and San Luis Obispo cos. n. through Sierra Nevada and Coast Ranges; to B.C., Utah, L. Calif. Mostly May–July.

2. **Z. micránthus** Eastw. [*Z. venenosus* var. *m.* Jeps. *Toxicoscordion m.* Heller.] Bulbs 2–2.5 cm. long; stems 2–5 dm. tall, scabrous; basal lvs. 1.5–4 dm. long, 4–10 mm. wide, folded, scabrous; raceme usually simple, 5–25 cm. long; bracts long-attenuate; pedicels wide-spreading in fr., becoming 3–6 cm. long; perianth-segms. 4–5 mm. long, obtuse, the petals definitely clawed; gland toothed on upper margin; stamens 4–5 mm. long; caps. 8–12 mm. long, subcylindric; seeds 4.5 mm. long.—Dry slopes and flats, below 3000 ft.; Mixed Evergreen F., Chaparral; Coast Ranges, Lake Co. to Ore. May–June.

3. **Z. fontànus** Eastw. Bulb 3–4 cm. long; stems 4–10 dm. high, smooth; basal lvs. often longer than stem, 8–20 mm. wide; panicles large, with horizontal-divaricate branches; bracts linear-attenuate; pedicels horizontal-divaricate, 2.5–4 cm. long; perianth-segms. white, 5–7 mm. long, cordate at base, obtuse at apex, clawed; gland yellow, truncate and dentate at apex; stamens 5–6 mm. long; caps. 1.5–2 cm. long; seeds 4–5 mm. long.—Wet places, often on serpentine; Chaparral, Mixed Evergreen F.; Mendocino Co. to Marin Co., San Benito Co. April–June.

4. **Z. paniculàtus** (Nutt.) Wats. [*Helonias p.* Nutt. *Toxicoscordion p.* Rydb.] SAND-CORN. Bulb ovoid, 3–4 cm. long; stems 2–6 dm. high, smooth; basal lvs. 3–5 dm. long, 5–20 mm. wide, ± scabrous; infl. usually paniculate, 5–25 cm. long, the lower branches elongate, ascending; bracts long-acuminate; pedicels slender, ascending, 1–2 cm. long; perianth-segms. yellowish-white, broadly ovate, acute to acuminate, 4 mm. long, the sepals sessile, the petals subcordate at base and clawed; gland indefinitely margined, green; stamens 4–5 mm. long; caps. 10–12 mm. long; seeds 4–5 mm. long.—Dry places, 4000–7000 ft.; largely Sagebrush Scrub; Nevada Co. to Siskiyou Co.; to Wash., Mont. May–June.

5. **Z. Fremóntii** Torr. [*Anticlea F.* Torr. *Toxicoscordion F.* Rydb. *Z. speciosus* of Greene.] STAR-LILY. Bulb 3–6 cm. long; stems 3–10 dm. high, smooth; basal lvs. 2–6 dm. long, 8–25 mm. wide, folded and arched, with scabrous margins; infl. racemose or paniculate, 1–4 dm. long; bracts long-acuminate; pedicels spreading and to 4 cm. long in fr.; perianth-segms. yellowish-white, 10–12 mm. long, lance-ovate, obtusish, the sepals subsessile, the petals clawed; gland greenish-yellow, toothed on upper margin; stamens 5–7 mm. long; caps. 1.5–3 cm. long, oblong; seeds ca. 5 mm. long; *n* = 11 (Miller, 1931).—Dry grassy or bushy slopes, below 3500 ft.; V. Grassland, Chaparral, Mixed Evergreen F.; Coast Ranges from San Diego Co. n., Butte Co.; s. Ore., n. L. Calif. March–May.

Var. **mìnor** (H. & A.) Jeps. [*Z. chloranthus* var. *m.* H. & A.] Plants 1–2 dm. high; racemes corymbose, 3- to few-fld.; fls. white with greenish glands.—Open moist fields near the coast; Mendocino to San Luis Obispo cos., San Diego. Feb.–March.

Var. **sálsus** Jeps. Stout, 4–6 dm. high; basal lvs. in conspicuous sheathing tuft; gland pale.—Alkaline flats; V. Grassland; Solano Co.

Var. **ineziànus** Jeps. Perianth-segms. more distinctly rotate; gland green.—Santa Ynez Mts.

6. **Z. exaltàtus** Eastw. [*Toxicoscordion e.* Heller.] Bulb 5–8 cm. long; stem stout, 6–10 dm. high, smooth; basal lvs. 4–7 dm. long, 1.5–3 cm. wide, with scabrous margins;

infl. paniculate, 2–4 dm. long, the lower branches with ♂ fls. only; pedicels ascending; perianth-segms. oblong-elliptic, 7–8 mm. long, obtuse, the sepals subsessile, petals distinctly clawed; stamens 5–6 mm. long; caps. 2–2.8 cm. long; seeds 6–7 mm. long.— Wooded slopes, 2000–4000 ft.; Foothill Wd., Yellow Pine F.; w. slope Sierra Nevada, Butte Co. to Tulare Co. May–July.

7. **Z. brevibracteàtus** (Jones) Hall [*Z. Fremontii* var. *b.* Jones.] Bulbs 2–4.5 cm. long; stems slender, 3–5 dm. tall, glabrous; basal lvs. 1.5–3 dm. long, 5–10 mm. wide, slightly scabrous on margins, folded; infl. usually paniculate, 1–2.5 dm. long; pedicels spreading, 1–3 cm. long; perianth-segms. 5–7 mm. long, yellowish, elliptic, the outer broader, the inner more distinctly clawed; stamens 4–6 mm. long; caps. oblong, 1.5 cm. long; seeds 3–4 mm. long.—Sandy flats and mesas, 2000–5000 ft.; Creosote Bush Scrub, Joshua Tree Wd., w. and s. Mojave Desert and Whitewater Creek; Foothill Wd., inner Coast Ranges, San Luis Obispo Co. April–May.

15. Veràtrum L. False-Hellebore

Tall stout leafy perennial herbs from short thick rootstocks. Lvs. broad, clasping, strongly veined and plaited. Fls. greenish-white to purple, rather large, bisexual, unisexual, or together on same plant, on short pedicels, in large terminal panicles. Perianth-segms. 6, oblong-ovate, obtuse, glandless or nearly so, sometimes adnate to the base of the ovary. Stamens 6, opposite the segms. and free from them, almost as long; fils. filiform; anthers cordate, their sacs confluent, ovoid. Ovary 3-celled; styles 3, persistent. Caps. septicidal, 3-lobed, each cell several-seeded. Seeds compressed, the body narrow, surrounded by a broad margin or wing. Ca. 12–14 spp. of N. Temp. Zone. Plants often reported as poisonous to stock. (Ancient name of hellebore.)

Ovary densely woolly; perianth erose .... 1. *V. insolitum*
Ovary glabrous.
  Perianth-segms. deeply fringed .... 2. *V. fimbriatum*
  Perianth-segms. entire or serrulate.
    Fls. whitish; branches of panicle erect .... 3. *V. californicum*
    Fls. green; branches of panicle drooping .... 4. *V. viride*

1. **V. insolìtum** Jeps. Stems 1–1.5 m. tall, pubescent; lf.-blades elliptic, acute, 15–25 cm. long, pubescent beneath and on sheathing bases, the upper lanceolate and somewhat smaller; panicle 2.5–6 dm. long, lanate-tomentose; bracts linear-lanceolate; pedicels 6–15 mm. long, perianth-segms. whitish, obovate, 6–8 mm. long, irregularly and shallowly fringed, with 2 dark glandular spots near base; stamens 3–4 mm. long; ovary densely woolly; caps. ovoid-lanceolate, 2–2.5 cm. long, somewhat glabrate; seeds straw-colored, ca. 1 cm. long including the broad wings.—Openings in thickets in red clay; Mixed Evergreen Forest; Del Norte Co.; also at ca. 3500 ft., Castle Crag, Siskiyou Co.; sw. Ore. June–July.

2. **V. fimbriatum** Gray. Stems 1–2 m. tall; lvs. lanceolate, 1.5–5 dm. long, 2.5–5(–10) cm. wide, acute to acuminate, pubescent; panicle 1.5–5 dm. long, tomentose; pedicels 6–12 mm. long; perianth-segms. rhombic-ovate, 6–10 mm. long, irregularly fimbriate, with 2 oblong subglandular spots below the middle; stamens 4 mm. long; caps. obovoid, depressed at apex, 8 mm. long; seeds green, 6 mm. long, scarcely margined.—Wet meadows; N. Coastal Scrub; Mendocino and Sonoma cos. July–Sept.

3. **V. califórnicum** Durand. Corn-Lily. Skunk-Cabbage. Plants 1–2 m. tall, tomentose above; lvs. ovate to broadly elliptic, 2.5–4 dm. long, 1–2 dm. wide, loose-pubescent beneath, the uppermost lanceolate; panicles 2–5 dm. long, short-tomentose; pedicels 2–6 mm. long; perianth-segms. dull white, oblong-ovate, 9–14 mm. long, with a Y-shaped green gland near the base; stamens 6–8 mm. long; caps. 2–3.5 cm. long; seeds 10–12 mm. long, strongly winged.—Wet meadows and banks, below 11,000 ft.; Montane Coniferous F.; Sierra Nevada to Palomar Mts. (San Diego Co.), N. Coast Ranges; to Wash., Rocky Mts., L. Calif. July–Aug.

4. **V. víride** Áit. Similar to *V. californicum;* the lower branches of the panicle drooping; bracts mostly leaflike; perianth-segms. greenish, oblong or oblanceolate, 8–10 mm. long, woolly-pubescent, the margin ± erose.—Subalpine F. meadows; Salmon Mts., Siskiyou Co.; to Alaska, Atlantic Coast. July–Aug.

16. Aspáragus L. Asparagus

Perennial herbs, with thick matted rootstocks or tuberous roots. Stems much-branched, ending in filiform or flattened green branchlets borne in the axils of scalelike lvs. Fls. small, greenish yellow, usually in racemes or umbels. Perianth-segms. alike. Stamens 6; anthers versatile. Ovary 3-celled, 2- or few-ovuled; stigmas 3. Fr. a globose berry. Ca. 150 spp.; Old World. (The ancient Greek name.)

1. A. officinàlis L. Garden Asparagus. Young stems stout, edible, later branching and becoming 1–2 m. high; perianth bell-shaped, ca. 6 mm. long; berry red; $n = 10$, (Nagao, 1938).—Escape from gardens in low subsaline places; native of Eu. May–June.

17. Erythrònium L. Adder's-Tongue. Fawn-Lily

Perennial herbs from deep-seated elongate corms with membranous coats and arising in succession on a short rhizome. Lvs. 2, basal or nearly so. Stem low, simple, producing a solitary fl. or 2 to several in a simple raceme. Fls. nodding, showy, bractless; perianth-segms. separate, lanceolate to oblanceolate, deciduous, with nectar-bearing groove, and sometimes (particularly the inner segms.) with 2 or 4 saclike appendages near base, the lateral pair often adjacent to a pair of auricles. Stamens 6, hypogynous, shorter than the segms. and often in 2 sets, the alternate shorter; anthers * oblong-linear, not versatile. Ovary sessile; style 3-cleft into stigmatic lobes or subentire. Caps. oblong to obovoid, somewhat 3-angled, loculicidal; seeds compressed to swollen, in 2 rows in each locule. A genus of ca. 25 spp., mostly of temp. N. Am., many of great beauty and excellent for gardens. (Greek, *eruthros,* red, the color of the fls. of some spp.)

(Applegate, E. I. The genus Erythronium: A taxonomic and distributional study of the w. N. Am. spp. Madroño 3: 58–113, 1935.)

Lvs. not mottled. Mostly above the yellow pine belt.
  Stigma entire or nearly so.
    Style 4–5 mm. long, clavate; fls. white to cream.
      Perianth-segms. without basal appendages, 2–4 mm. wide . . . . 1. *E. purpurascens*
      Perianth-segms. with appendages, and 5 mm. wide . . . . 2. *E. klamathense*
    Style ca. 10 mm. long, slender-clavate; fls. golden-yellow . . . . 3. *E. tuolumnense*
  Stigma definitely lobed; style usually not strongly clavate; fls. golden-yellow . . . . 4. *E. grandiflorum*
Lvs. mottled. Mostly below or in yellow pine belt.
  Stigma entire or nearly so; style 6–8 mm. long, strongly clavate.
    Fls. white or cream.
      Perianth-segms. without appendages . . . . 5. *E. Howellii*
      Perianth-segms. with basal appendages . . . . 6. *E. citrinum*
    Fls. lavender, the segms. purple near base . . . . 7. *E. Hendersonii*
  Stigma plainly lobed; style 8–15 mm. long, not or moderately clavate.
    Fils. strongly dilated toward base.
      The fls. rose-pink . . . . 8. *E. revolutum*
      The fls. white to cream . . . . 9. *E. oregonum*
    Fils. more filiform.
      Anthers golden-yellow; style 12–15 mm. long. Mt. St. Helena . . . . 10. *E. helenae*
      Anthers white; style 8–10 mm. long.
        Fils. 8–10 mm. long; lvs. 2.5–5 cm. wide; stigma-lobes stubby . . . . . . 11. *E. californicum*
        Fils. 5–7 mm. long; lvs. 1–2.5 cm. wide; stigma-lobes long and filiform
                                           12. *E. multiscapoideum*

1. **E. purpuráscens** Wats. [*E. p.* var. *uniflorum* Wats. *E. grandiflorum* var. *multiflorum* Torr.] Corm 3–4 cm. long, 5–6 mm. thick; lvs. narrow- to oblong-lanceolate, 10–15 cm. long, 1–2.5 cm. wide, crisped along margin, not mottled, yellow-green; scape 8–20(–35) cm. high, 1–8-fld.; pedicels unequal, 5–40 mm. long; perianth-segms. white with yellow base, tinged purple with age, lance-linear, 10–15 mm. long, 2–4 mm. wide, spreading, slightly recurved, the inner not appendaged; fils. filiform, 4–5 mm. long; anthers white, 3–6 mm. long; styles clavate, shorter than stamens; caps. oblong-obovoid, 2–3 cm. long, 7–8 mm. thick.—Meadows and along streams, brushy or forested slopes, 4000–8000 ft.; Montane Coniferous F.; Sierra Nevada, Shasta Co. to Tulare Co. May–Aug.

2. **E. klamathénse** Appleg. Corm oblong, 3–5 cm. long, 7–10 mm. thick; lvs. lanceolate, acute, yellow-green, 7–15 cm. long, 1.5–2.5 cm. wide; scape 7–20 cm. high, 1–3-several-fld.; perianth-segms. lanceolate-acuminate to acute, 2–3 cm. long, 5 mm. wide, white in

* Measurements for anthers are from dried material, hence shorter than in fresh fls.

upper part, yellow in lower, with well-developed globular basal processes and auricles; fils. slender, 5–8 mm. long; anthers yellow, 2–3 mm. long; style clavate, 5 mm. long; stigma subentire; caps. obovoid, 3 cm. long, 2 cm. thick.—Glades and openings in woods, 5000–6000 ft.; Red Fir F., Lodgepole F.; Siskiyou Co.; s. Ore. June–July.

3. **E. tuolumnénse** Appleg. Corm oblong, 6 cm. long, 2 cm. thick; lvs. not mottled, lanceolate to broadly oblanceolate, 2–3 dm. long, 4–8 cm. wide, yellow-green; scape 1–several-fld., ca. as long as lvs.; perianth-segms. broadly lanceolate, deep pure yellow with pale greenish-yellow base, 3–3.5 cm. long, 8–10 mm. wide, median basal processes well developed, lateral scarcely so; stamens ca. half the length of the segms., in 2 distinct sets; fils. slender; anthers yellow; style scarcely clavate, ca. 10 mm. long; caps. obovoid, retuse.—Open woods, 1000–2000 ft., Foothill Wd.; base of Sierra Nevada in Tuolumne and Stanislaus cos. March–May. Reproduction mostly vegetative.

4. **E. grandiflòrum** Pursh var. **pállidum** St. John. [*E. g.* var. *parviflorum* Wats., in part. *E. parviflorum* Goodd., in part.] Corm slender, 4–5 cm. long, 5–8 mm. thick; lvs. not mottled, 10–20 cm. long, 1–4.5 cm. wide, oblong-elliptic, acutish; scape 1.5–3 dm. high, 1–several-fld.; perianth-segms. lanceolate, acuminate, 2–3.5 cm. long, 4–7 mm. wide, golden-yellow, lighter within at base, streaked green without, basal appendages globular, auricled; fils. slender, unequal, ca. 8 and 12 mm. long; anthers white, 3–4 mm. long; style slender, 9–13 mm. long; stigma plainly 3-lobed; caps. obovoid, 3 cm. long, 1 cm. thick; $n = 12$ (F. H. Smith, 1955).—Open woods, to 6000 ft. elev.; Montane Coniferous F.; Humboldt, Trinity, and Siskiyou cos.; to Vancouver Id., Alta., Mont. June–July.

5. **E. Howéllii** Wats. Corm narrow-oblong, 3–4 cm. long; lvs. mottled, lance-oblong to lanceolate, 8–15 cm. long, 15–40 mm. wide; scape 1–2 dm. high, 1–4-fld.; perianth-segms. white, proximally yellow-barred, or with orange base, tinged pink in age, lanceolate, 20–22 mm. long, lacking basal appendages; fils. filiform, 8–10 mm. long; anthers white, 4–6 mm. long; style clavate, 6–8 mm. long; caps. narrow-obovoid.—Woods, 3500 ft.; Yellow Pine F.; Del Norte Co.; s. Ore. April–May.

6. **E. citrìnum** Wats. Corm slender, 4–7 cm. long, 1–1.5 cm. thick; lvs. mottled, lanceolate to oblong, obtuse, 6–15 cm. long, 1.5–5 cm. wide, crisped on margins; scape 1–3(–7)-fld., 12–30 cm. high; perianth-segms. creamy-white, greenish-yellow at base, sometimes pinkish without, lance-oblong, acuminate, 2.5–4.0 cm. long, 5–9 mm. wide, the inner with median basal sacs and smaller lateral auricles; fils. in 2 unequal sets, slender, cream-color, 10–15 and 8–12 mm. long; anthers white, 4–5 mm. long; style clavate; caps. narrow-obovoid, 2.5–3 cm. long.—Wooded and brushy slopes, 2000–3500 ft.; Yellow Pine F., Chaparral; Siskiyou and Del Norte cos.; Josephine Co., Ore. March–April.

7. **E. Hendersònii** Wats. Corm somewhat curved and flattened, 3–6 cm. long, 1–1.5 cm. thick; lvs. mottled, lance-oblong, 10–20 cm. long, 2–4.5 cm. wide, crisped on margins; scape 1–3 dm. high, purplish, 1–4-fld.; perianth-segms. lanceolate, lavender with dark purple base surrounded by yellowish or whitish zone, strongly recurved, 2.5–4.0 cm. long, 6–8 mm. wide, the inner segms. with median pair of inflated sacs and lateral pair covering the auricles; fils. purple, slender, 6–12 mm. long; anthers brownish, 3–5 mm. long; style clavate; caps. narrowly obovoid, 2.5–4 cm. long, 2 cm. thick; $n = 12$ (La Cour in Darlington, 1945).—Wooded slopes, 1400–5000 ft.; Yellow Pine F.; Siskiyou Co.; sw. Ore. April–July.

8. **E. revolùtum** Sm. [*E. grandiflorum* var. *Smithii* Hook. *E. S.* Orcutt. *E. r.* var. *Bolanderi* Wats. *E. r.* var. *Johnsonii* Purdy.] Corm 3–5 cm. long, 5 mm. thick; lvs. mottled, lance-ovate, mostly acute, 15–20 cm. long, 3–6 cm. wide, crisped on margins; scape 15–40 cm. high, 1–3-several-fld.; perianth-segms. lance-linear, acuminate to acute, often with involute margins, 3.5–4.5 cm. long, 7–10 mm. wide, rose-pink without, lighter within and with transverse yellow bands near base, with conspicuous elongate median appendages and reduced or no lateral ones; fils. 10–15 mm. long, 3–4 mm. broad near base; anthers yellow, 4–6 mm. long; style filiform, 8–10 mm. long; stigma deeply lobed and lobes recurved; caps. oblong, 3–4 cm. long, 5–8 mm. thick; $n = 12$ (LaCour, in Darlington, 1945).—Margins of swamps and bogs and along wooded streams near the coast, to 3500 ft.; Redwood F., Mixed Evergreen F.; Mendocino Co. to Vancouver Id. March–June.

9. **E. oregònum** Appleg. ssp. **leucándrum** Appleg. Corm ca. 5 cm. long, 1–1.5 cm.

thick; lvs. mottled, lanceolate to lance-oblong, acute, 12–15 cm. long, 3–6 cm. wide; scape brownish, 1.5–3 dm. high, 1–few-fld.; perianth-segms. broadly lanceolate, 3.5–5 cm. long, 8–12 mm. wide, with twisted tips, white or pink, basally reddish or brown without, yellow within; the appendages 4-saccate; fils. 8–10 mm. long, 1.5–2.5 mm. wide at base; anthers white, 5–8 mm. long; style slender, 10–12 mm. long, the stigma-lobes recurved; caps. narrowly obovoid, 3.5 cm. long, 8 mm. thick; *n* = 12 (F. H. Smith, 1955).—Openings in woods, at 700 ft.; Mixed Evergreen F.; Mendocino Co.; sw. Ore. April–May.

10. **E. helènae** Appleg. Corms 6–8 cm. long, 10–15 mm. thick; lvs. strongly mottled, ovate to lanceolate, obtuse to acute, 5–15 cm. long, 7–22 mm. wide; scape 2–3 dm. high, 1–several-fld.; perianth-segms. broadly lanceolate, 3.5–4 cm. long, 0.5–1.5 cm. wide, white with golden-yellow base, without spots or bands, with median appendages slightly elongate, inflated and auricles transversely folded; fils. slender, 7–10 mm. long; anthers golden, 3–4 mm. long; style scarcely clavate, declined, 12–15 mm. long; stigma-lobes short, stout; caps. 15–20 mm. long, 8–10 mm. thick.—Well watered volcanic soil and leaf mold, in brush and woods, 1500–2200 ft.; Chaparral, Foothill Wd.; about Mt. St. Helena, Napa, Lake and Sonoma cos. March–May. With much vegetative reproduction.

11. **E. califórnicum** Purdy. Corm rather large, 3.5–5 cm. long; lvs. strongly mottled, 10–15 cm. long, 2.5–5 cm. wide, from lance- or ovate-oblong to oblong; scape 1–2.5(–3.5) dm. high, often reddish, 1–3–several-fld.; perianth-segms. broadly lanceolate, 2.5–3.5 cm. long, 7–10 mm. wide, white to cream, with pale greenish-yellow base, and transverse band of yellow, orange or brown; fils. slender, 8–10 mm. long; anthers white, 5–7 mm. long; style moderately clavate, 8–10 mm. long; stigma with erect or spreading lobes; caps. narrow-obovoid; *n* = 12 (La Cour, in Darlington, 1945).—Openings on brushy slopes and in woods, up to 3200 ft.; Chaparral, Foothill Wd., Yellow Pine F.; Coast Ranges, Humboldt and Shasta cos. to Sonoma and Colusa cos. March–April.

12. **E. multiscapoìdeum** (Kell.) Nels. & Kenn. [*Fritillaria m.* Kell. *E. grandiflorum* var. *m.* Wood. *E. Hartwegii* Wats.] Corms 1–2 cm. long, 5 mm. thick, oblong-ovoid; lvs. mottled, oblanceolate, 4–10 cm. long, 1–2.5 cm. wide; scape 1–2 dm. high, usually 1-fld.; perianth-segms. lanceolate to lance-oblong, 2.5–4 cm. long, 7–12 mm. wide, white to whitish, with pale greenish-yellow base; appendages lacking or with 4 small sacs; fils. 5–7 mm. long, slender; anthers white, 3–5 mm. long; style slender, 8–10 mm. long; stigma deeply cleft, the lobes often recurved; caps. oblong-ovoid, 1 cm. long, 5 mm. thick.—Rich loam, brushy and wooded hillsides, ca. 1800 ft.; Yellow Pine F., Foothill Wd.; foothills of Sierra Nevada, Tehama and Butte cos. to Mariposa Co. March–May.

18. Fritillària L. FRITILLARY

Perennial; bulb of one or more fleshy scales, with or without rice-grain bulblets. Stem erect, simple. Lvs. alternate or whorled, sessile, linear to almost ovate. Fls. 1 to several in a terminal raceme; perianth campanulate to funnelform, deciduous, of 6 distinct segms. in 2 series; segms. with ± evident gland or nectary above the base. Stamens 6, included, inserted on base of segms.; fils. slender; anthers extrorse, ± versatile. Ovary sessile or subsessile; style 1, entire or trifid; caps. membranaceous, 6-angled or -winged, 3-valved, loculicidal; seeds many, flat, in 2 rows in each locule, brownish. Spp. ca. 100, of N. Temp. Zone; a few of horticultural value, the bulbs of others eaten by the Indians. (Latin, *fritillus*, a dicebox, because of the shape of the caps.)

(Beetle, D. E. Monograph of N. Am. spp. of Fritillaria. Madroño 7: 133–158, 1944.)

A. Style not or barely cleft.
  B. Perianth-segms. yellow to orange, 15–19 mm. long .... 1. *F. pudica*
  BB. Perianth-segms. white to pink or purple.
    C. Fls. 25–35 mm. long.
      D. The fls. uniformly pinkish-purple; style exceeding stamens ...... 2. *F. pluriflora*
      DD. The fls. white to pink, often with red stripes; style not exceeding stamens
                                                        3. *F. striata*
    CC. Fls. 12–17 mm. long, pink to purplish; style exceeding stamens .... 4. *F. Brandegei*
AA. Style obviously 3-cleft.
  B. Lvs. on lower part of stem just above ground.
    C. Fls. not mottled or checkered.

D. The fls. white with green striations and 10–16 mm. long ...... **5.** *F. liliacea*
DD. The fls. not white, 20–35 mm. long.
    E. Color of fls. greenish-yellow outside.
        F. Stems 3–6 dm. high; lvs. 5–12; fls. 25–30 mm. long .... **6.** *F. agrestis*
        FF. Stems 1–2 dm. high; lvs. 2–4; fls. 15–25 mm. long ...... **7.** *F. glauca*
    EE. Color of fls. dark brown to greenish-purple; fls. 20–35 mm. long
                                                  8. *F. biflora*
CC. Fls. mottled or checkered.
    D. Lvs. ovate; fls. white with purple spots .... **9.** *F. Purdyi*
    DD. Lvs. broadly linear; fls. rust-brown with yellow .... **10.** *F. falcata*
BB. Lvs. on upper stem, the lower part above ground naked.
C. Fls. faintly or not mottled.
    D. Color of fls. purplish or greenish-white; gland on perianth-segm. lance-oblong
                                                11. *F. micrantha*
    DD. Color of fls. greenish to reddish-yellow; gland on perianth-segm. indistinct
                                               12. *F. phaeanthera*
CC. Fls. plainly mottled.
    D. Color of fls. scarlet, checkered yellow; perianth campanulate-funnelform
                                               13. *F. recurva*
    DD. Color of fls. not red; perianth campanulate to bowl-shaped.
        E. Perianth-segms. mostly 2–4 cm. long; lvs. 5–30 mm. wide
                                            14. *F. lanceolata*
        EE. Perianth-segms. mostly 1–2 cm. long; lvs. 2–6 mm. wide.
            F. Stem slender; bulb without "rice-grain" bulblets; caps. angled
                                       15. *F. atropurpurea*
            FF. Stem fistulose in region of lvs.; bulb with rice-grain bulblets; caps. with
                curved hornlike process at base and top of each wing
                                       16. *F. pinetorum*

1. **F. pùdica** (Pursh) Spreng. [*Lilium ? p.* Pursh. *Amblirion p.* Raf. *Theresia p.* Klatt. *Ochrocodon p.* Rydb.] Bulb cream, 1–3 cm. long, of a few fleshy scales and many "rice-grain" bulblets; stem 7–30 cm. high; lvs. alternate, 3–8, linear to lanceolate, 6.2–20 cm. long, 2–11 mm. wide; fls. 1–3, nodding, bell-shaped, yellow to orange, sometimes with brown veins outside, aging brick-red; perianth-segms. 15–19 mm. long, 4–7 mm. wide, the basal gland small, green; stamens half the length of segms.; anthers 3–5 mm. long; style equaling segms.; stigma knobbed; caps. obovoid-oblong, 17 mm. long; seeds light brown; $n = 12, 13, 39/2$ (Tischler, 1938).—Grassy and brushy or wooded slopes, below 5000 ft.; Sagebrush Scrub, Yellow Pine F.; Siskiyou and Modoc cos. to Sierra Co.; to B.C., Mont., Wyo. April–June. A variable sp.

2. **F. pluriflòra** Torr. in Benth. ADOBE-LILY. Bulb yellowish, 1.3–2.5 cm. long, of several large scales; stem 2–4.5 dm. long; lvs. clustered near base, alternate, elliptic to obovate-oblong, 6–12 cm. long, 7–15 mm. wide; fls. 1–3(–7), nodding, bell-shaped, uniformly pinkish-purple; perianth-segms. obovate, 25–35 mm. long, 7–15 mm. wide, the gland continuing as a green vein; stamens half as long as segms.; style exceeding stamens; stigmas 3-lobed; caps. truncate, obtusely angled, equally long and broad; $n = 12$ (Beetle, 1944).—Adobe soil of interior foothills below 1500 ft.; Foothill Wd.; Mendocino, Glenn, Solano, and Butte cos.; to s. Ore. Feb.–April.

3. **F. striàta** Eastw. Bulb with thick scales; stem 2.5–3.8 dm. tall; lvs. on lower half of stem, oblong-ovate, somewhat glaucous, alternate, 6–7 cm. long, 10–15 mm. wide; fls. 2–3(–7), nodding, white to pink, often with red stripes, fragrant; perianth-segms. 2.5–3.5 cm. long, 7–10 mm. wide, usually recurved; gland obscure; style slightly 3-parted at tip, equal to shorter than stamens; caps. quadrate, 2 cm. long, not winged.—Adobe soil, up to 3000 ft.; V. Grassland, Foothill Wd.; Sierra Nevada foothills, Greenhorn Mts., Tulare and Kern cos. March–April.

4. **F. Brandègei** Eastw. [*F. Hutchinsonii* A. Davids.] Bulbs with "rice-grain" bulblets; stem 4–10 dm. long; lvs. on upper stem, in whorls of ca. 5, lanceolate, 5–10 cm. long, 4–20 mm. wide; fls. 4–12, nodding, pink to purplish; perianth-segms. oblong-lanceolate, 12–17 mm. long, 2–3 mm. wide, spreading, involute in age; gland small, subdeltoid; fils. 2 mm. wide, 6 mm. long; pistil ca. as long as perianth-segms., scarcely lobed; caps. winged, truncate.—Granitic soils, open forests, 5000–7000 ft.; Yellow Pine F.; Tulare and Kern cos. April–June.

5. **F. liliàcea** Lindl. [*Liliorhiza lanceolata* Kell.] Bulbs 15–20 mm. thick, of few round scales; stem 1–3.5 dm. high; lvs. just above ground level, alternate, ovate to linear, 5–10 cm. long, 4–15 mm. wide; fls. 1–5, campanulate, white with green striations; perianth-segms. 10–16 mm. long, 5–6 mm. wide, with apical tuft of short white hairs;

gland green, with purple dots; stamens ca. half as long as perianth-segms.; pistil ca. the same, style cleft halfway; caps. 12–15 mm. long, stipitate, obtusely angled; $n = 12$ (Beetle, 1944).—Heavy soil, open hills and fields near the coast; N. Coastal Scrub, Redwood F.; Sonoma Co. to Monterey Co. Feb.–April.

6. **F. agréstis** Greene. STINK BELLS. Bulb deep-seated, with several large fleshy scales; stem 3–6 dm. high; lvs. 5–12, crowded near lower center of part above ground, lance-oblong to -linear, 5–11 cm. long, 8–20 mm. wide, glaucous; fls. 1–5(–8), nodding, campanulate, greenish-white outside, purplish-brown inside, with obnoxious odor; perianth-segms. 2.5–3 cm. long, 7–10 mm. wide; gland green, continuing as cent. stripe; fils. ca. 10 mm. long; anthers 5 mm.; style cleft halfway; caps. 17–20 mm. long, quadrate.— Low heavy soil; V. Grassland, Foothill Wd.; interior valleys, Mendocino Co. to San Luis Obispo Co., Cent. V. March–April.

7. **F. glaùca** Greene. Bulb small, with few fleshy scales; stem 8–18 cm. long; lvs. 2–4, glaucous, alternate, oblong-lanceolate, 3.5–9 cm. long, 5–14 mm. wide; fls. 1–3, nodding, purplish or greenish marked with yellow; perianth-segms. lance-oblong, 15–25 mm. long, 6–9 mm. wide; gland not obvious, appearing as a small vein; stamens 12 mm. long; style cleft to middle; caps. 2 cm. long, broadly winged; seeds 5–6 mm. long, tan-colored.—Barren serpentine slopes, 2000–7000 ft.; Yellow Pine F., Red Fir F.; s. Ore. to Humboldt, Lake, and Tehama cos. April–July.

8. **F. biflòra** Lindl. [*F. Grayana* Rchb. f. and Baker. *F. b.* var. *agrestis* Greene. *F. b.* vars. *Ineziana* and *inflexa* Jeps. *F. succulenta* Elmer.] CHOCOLATE-LILY. MISSION BELLS. Bulb 15–20 mm. thick, of a few fleshy scales; stem 1.5–4 dm. high; lvs. 3–7, alternate, often somewhat crowded just above ground level, oblong to ovate-lanceolate, 5–12 cm. long, 1–2.5(–4) cm. wide; fls. 1–7, nodding, dark brown or greenish-purple, campanulate; perianth-segms. lance-oblong, 2–3.5 cm. long, 5–12 mm. wide; gland appearing as a green band; stamens 8–10 mm. long; style cleft ½–⅔ its length; caps. 15–25 mm. long, not winged.—Heavy soil on grassy slopes and mesas, below 3000 ft.; V. Grassland, Foothill Wd.; Coast Ranges, Mendocino and Napa cos., San Mateo to Riverside and San Diego cos. Feb.–June.

9. **F. Púrdyi** Eastw. Bulb 1–1.3 cm. long, with few large fleshy scales; stem 1–4 dm. long; lvs. few, pale green, ovate, alternate, but crowded just above ground level, 2.5–6(–9) cm. long, 1–3 cm. wide; fls. 1–2(–7), campanulate, horizontal or nodding, white with purple spots and lines and pink shadings; perianth-segms. 2–2.5 cm. long, 7–9 mm. wide, recurved at apex and with tuft of short whitish hairs; gland obscure; stamens nearly equaling segms.; style cleft halfway; caps. 1.5 cm. long, not winged; $n = 12$ (Beetle, 1944).—Serpentine ridges, 1400–6900 ft.; Chaparral, Foothill Wd., Yellow Pine F.; inner Coast Ranges, Humboldt and Trinity cos. to Napa Co. March–June.

10. **F. falcàta** (Jeps.) D. E. Beetle. [*F. atropurpurea* var. *falcata* Jeps.] Bulb of several fleshy scales; stem 7–20 cm. long; lvs. 2–6, alternate, fleshy, near base of stem, folded, broadly linear, 3.5–8.5 cm. long, 7–15 mm. wide; fls. 1–4, erect, campanulate, greenish without, mottled rusty-brown and yellow within; perianth-segms. obovate, 15–22 mm. long, 5–7 mm. wide, with pinkish terminal hairs; gland yellow, with brown spots; style 3-parted; caps. 2 cm. long, acutely angled with hornlike processes at top and base of angle; $n = 12$ (Beetle, 1944).—Serpentine talus, 1000–3000 ft., Chaparral, Foothill Wd.; inner S. Coast Ranges (Stanislaus, Santa Clara, and San Benito cos.). March–May.

11. **F. micrántha** Heller. [*F. parviflòra* Torr., not Mart. *F. multiflòra* Kell., nom. prov.] BROWN BELLS. Bulb with few scales and numerous "rice-grain" bulblets; stem 4–9 dm. long, light green; lvs. on upper part of stem, in whorls of 4–6, linear to lance-linear, 5–15 cm. long, 3–10 mm. wide; fls. nodding, broadly campanulate, 4–10, purplish or greenish-white, occasionally faintly mottled; perianth-segms. 12–20 mm. long, 4–5 mm. wide, apically white-tufted; gland lance-oblong, on lower third of segm.; stamens slightly more than half as long as segm.; style cleft ⅓–⅔ its length; caps. broadly winged, slightly wider than long.—Dry benches and slopes, 1000–6000 ft.; Montane Coniferous F.; Sierra Nevada, Plumas to Tulare cos. April–June.

12. **F. phaeánthera** Eastw. Bulb of thick scales and many "rice-grain" bulblets; stem 3–4.5 dm. long; lvs. linear to linear-lanceolate, 4–6 cm. long, 5–9 mm. wide, in whorls of 3–5; fls. 3–7, nodding, open-campanulate, pale greenish-yellow to speckled red-

purple; perianth-segms. narrow-ovate, sometimes partially recurved, 1–1.5 cm. long, 3 mm. wide, white-tufted at apex; gland indistinct; stamens 7–8 mm. long; style cleft halfway; caps. truncate, winged; $n = 12$, 18 (Beetle, 1944).—Dry slopes; Chaparral, Foothill Wd.; Butte and Napa cos. March–June.

13. **F. recúrva** Benth. SCARLET FRITILLARY. Bulb 2–2.5 cm. long, of thick scales and "rice-grain" bulblets; stem 3–9 dm. long; lvs. linear to lance-linear, usually 8–10, in 2–3 whorls of 2–5, near the middle of stem, 3–10 cm. long, 3–14 mm. wide; fls. nodding, campanulate-funnelform, 1–4(–9), scarlet, checkered with yellow within, and tinged purple without; segms. recurved at tips, oblanceolate, 2–3.5 cm. long, ·5–7 mm. wide; gland prominent, oval, yellow with red spots; stamens almost as long as segms.; style cleft ¼–⅕ its length; caps. winged, 9–11 mm. long; $n = 12$, 18 (Beetle, 1944).—Dry hillsides in brush or woods, 2000–6000 ft.; Chaparral, Foothill Wd., Yellow Pine F.; s. Ore. to inner Coast Ranges of Lake Co. and Sierra Nevada of Nevada Co.; w. Nev. March–July.

Var. **coccínea** Greene [*F. c.* Greene.] Perianth-segms. not recurved at apex, more brilliantly scarlet, mottled yellow.—Inner N. Coast Ranges, Mendocino Co. to Napa Co. March–June.

14. **F. lanceolàta** Pursh. [*Amblirion l.* Sweet. *F. mutica* Lindl. *F. l.* var. *floribunda* Benth. *F. l.* var. *gracilis* Wats. *F. l.* var. *tristulis* Grant in Jeps. *F. viridea* Kell. *Liliorhiza v.* Kell. *F. esculenta* Nutt. ex Baker. *F. m.* var. *gracilis* Jeps. *F. ojaiensis* A. Davids. *F. eximia* Eastw.] CHECKER-LILY. Bulbs of few scales and many "rice-grain" bulblets; stem 3–8(–12) dm. long; lvs. in several whorls of 3–5 on upper stem, ovate- to linear-lanceolate, 4–16 cm. long, 5–30 mm. wide; fls. 1–4(–12), nodding, deeply bowl-shaped, brown-purple mottled with yellow, to pale greenish-yellow and faintly mottled with purple; perianth-segms. 1.5–4 cm. long, 4–11 mm. wide, ovate to oblong; gland yellow-green with purple dots; stamens half the length of the segms.; style cleft to middle; caps. 1.5–2.5 cm. long, broadly winged; $n = 12$, 18, 24 (Beetle, 1944).—Mostly in brush and among oaks and pines, below 2500 ft.; N. Coastal Scrub, Mixed Evergreen F., Chaparral, Foothill Wd., Yellow Pine F.; Coast Ranges from Ventura to Del Norte and Siskiyou cos.; to B.C., Ida. Feb.–May. Exceedingly variable in coloration and size of fls.

15. **F. atropurpùrea** Nutt. [*F. gracillima* Smiley.] Bulb of few fleshy scales; stem slender, 1.5–6 dm. long; lvs. 7–14, linear to lanceolate, alternate or ± whorled, on upper half of stem, 5–9 cm. long, 2–6 mm. wide; fls. 1–4(–12), open-campanulate, nodding, purplish-brown spotted with yellow and white; perianth-segms. oblong to rhombic, 1–2 cm. long, 4–8 mm. wide, with yellowish apical tuft; gland indistinct, brownish-yellow; stamens ca. ⅔ the length of segms.; style cleft more than half its length; caps. broadly obovoid, 10–17 mm. long, acutely angled.—Leaf mold under trees, 6000–10,500 ft., Montane Coniferous F.; Sierra Nevada from Tulare Co. n., to Trinity, Siskiyou, and Modoc cos.; to Ore., N. Dak., New Mex. April–July.

16. **F. pinetòrum** A. Davids. [*F. atropurpurea* var. *p.* Jtn.] Bulb 1–2 cm. thick, with many "rice-grain" bulblets; stem glaucous, 1–3(–5) dm. high, somewhat fistulous; lvs. glaucous, 12–20, somewhat whorled, linear, 5–15 cm. long, 2–7 mm. wide; fls. 3–9, erect or nearly so, purplish, mottled with greenish-yellow; segms. 14–19 mm. long, 2–6 mm. wide; gland indefinite; stamens ca. ⅔ the length of the segms.; style cleft to near base; caps. 12–15 mm. long, angled, with short hornlike process at base and summit of each wing; seeds 3–4 mm. long.—Somewhat shaded granitic slopes, 6000–10,500 ft.; Montane Coniferous F.; Alpine Co. to San Bernardino Co.; Nev. May–July.

19. Lílium L. LILY

Perennial herbs with scaly bulbs or scaly rootstocks. Stems simple, tall and leafy. Lvs. narrow, sessile, alternate to whorled. Fls. showy, large, solitary to many in a terminal raceme. Perianth funnelform or campanulate, deciduous, the segms. 6, spreading or recurved, each with a nectar-bearing gland near base. Stamens 6; fils. slender, ca. ⅔ as long as perianth-segms.; anthers ° linear, versatile. Ovary 3-loculed; style long, deciduous, with 3-lobed stigma. Caps. loculicidal; seeds many, flat, horizontal, in 2 rows in each cell. A genus of almost 100 spp. of the temp. N. Hemis.; of great horticultural interest. (Greek, *lirion*, the classical name.)

° Anther measurements as given in this treatment are for before the pollen is shed.

Plants of rather dry places; bulbs ovoid, the scales mostly over 25 mm. long, not jointed.
    Fls. white to pink or purplish or red.
        Perianth-segms. 8–10 cm. long .... 1. *L. Washingtonianum*
        Perianth-segms. 3–5(–6.5) cm. long.
            The perianth-segms. revolute to the base, pink .... 3. *L. Kelloggii*
            The perianth-segms. revolute only in upper part.
                Fls. erect or ascending, at first white with purple spots .... 2. *L. rubescens*
                Fls. horizontal or nodding, deep crimson with purple spots .... 4. *L. Bolanderi*
    Fls. yellow to orange.
        Perianth-segms. 7–9 cm. long; anthers 12–15 mm. long .... 5. *L. Humboldtii*
        Perianth-segms. 3.5–6 cm. long; anthers 4–6 mm. long .... 6. *L. columbianum*
Plants of wet places; bulbs rhizomatous, the scales mostly less than 2.5 cm. long and jointed.
    Fls. horizontal to ascending.
        Perianth 3–4 cm. long, bell-shaped; anthers 3–4 mm. long.
            The fls. dark red; caps. 2–4 cm. long. Coastal .... 7. *L. maritimum*
            The fls. orange-red to -yellow; caps. 1.2–1.6 cm. long. Sierra Nevada ...... 9. *L. parvum*
        Perianth 6–10 cm. long, trumpet-shaped, yellow; anthers 6–8 mm. long. S. Calif. .. 13. *L. Parryi*
    Fls. nodding.
        Anthers 4–6 mm. long; perianth 3.5–5 cm. long.
            Perianth-segms. recurved to near base; caps. 2.5 cm. long ....10. *L. Kelleyanum*
            Perianth-segms. recurved on outer half, caps. 1.5 cm. long .... 8. *L. occidentale*
        Anthers 7–15 mm. long; perianth 5–8 cm. long.
            Rhizomes unbranched, scales 3–4-jointed; anthers 7–10 mm. long .... 11. *L. Vollmeri*
            Rhizomes branched, scales usually 1-jointed; anthers 10–15 mm. long .... 12. *L. pardalinum*

1. **L. Washingtònianum** Kell. WASHINGTON LILY. Bulb oblique, semirhizomatous, 10–20 cm. long, with lanceolate imbricated nonjointed scales 25–50 mm. long; stem 6–18(–24) dm. tall; lvs. light green, mostly horizontal, in several whorls of 6–12, some solitary, oblanceolate, 3.5–14 cm. long, undulate or plane; fls. up to 20 or more, trumpet-shaped, fragrant, 8–10 cm. long, segms. 5–10 mm. wide and with a space between them, white, with a few reddish dots; stamens a little shorter than segms.; anthers 9–12 mm. long; style 4.5–5.5 cm. long; caps. oblong or obovoid, 2.5–3 cm. long; $n = 12$ (Stewart, unpub.).—Among bushes, dry granitic and loamy slopes and flats, 4000–7000 ft.; Chaparral, Montane Coniferous F.; Sierra Nevada, Fresno Co. n., to Siskiyou Co. July–Aug.

Var. **mìnus** Purdy. SHASTA LILY. Bulb ovoid, 2–5 cm. long; fls. fewer, ca. 8 cm. long, segms. overlapping.—Volcanic soil, 4000–5000 ft.; Chaparral; base of Mt. Shasta. July.

Var. **purpuráscens** Stearn. [*L. W.* var. *purpureum* auth., not Baker.] CASCADE LILY. Bulb subrhizomatous; fls. 10–15; perianth-segms. 8–10 cm. long, the inner 12–18 mm. wide, segms. overlapping, white to pink or lavender at first and finely dotted, later deep wine-purple.—Dry open, granitic and loamy slopes, 2500–6000 ft.; Chaparral, Montane Coniferous F.; Humboldt and Siskiyou cos.; to Mt. Hood, Ore. July.

2. **L. rubéscens** Wats. [*L. Washingtonianum* var. *purpureum* Baker.] CHAPARRAL LILY. REDWOOD LILY. CHAMISE LILY. Bulbs ovoid, somewhat asymmetrical, the broad "crown" 1–2.5 cm. in diam., 3.5–5 cm. long, the scales not jointed, 2.5–3.5 cm. long; stem 5–15(–30) dm. tall; lvs. glaucous beneath, oblanceolate, 4–8 cm. long, 12–25 mm. wide, in whorls of 5–10, the lower scattered; fls. 3–8, or many, erect or ascending, fragrant, trumpet-shaped, 3.5–5(–6.5) cm. long, at first white with purple spots, later wine color, segms. papillose at apex, ca. 1 cm. wide; anthers 5–6 mm. long; caps. 2.5–3.5 cm. long; seeds broadly obovate, 5–7 mm. long, 4–5.5 mm. wide, often straight along one side; $n = 12$ (Stewart, unpub.).—Brushy and wooded slopes and ridges, below 5000 ft.; Chaparral, N. Coastal Coniferous F., Mixed Evergreen F.; Coast Ranges, s. to Santa Cruz Co. June–July.

3. **L. Kellóggii** Purdy. Bulbs 4–6 cm. long, broadly ovoid-conical the scales to 5 cm. long, lance-ovate, not jointed; stem 6–9(–30) dm. high; lvs. dark green above, paler beneath, whorled, 6–10 cm. long, 1–2 cm. wide; fls. 1–15 (or many), fragrant, the perianth-segms. revolute to the base, 3–5 cm. long, pink to paler, with cent. yellow band, sometimes dotted purple, aging rose-purple; anthers 8–9 mm. long; caps. cylindrical, 4–5 cm. long; seeds obovate, 5–6 mm. long; $n = 12$ (Stewart, unpub.).—Dry rocky places or shaded deeper soil, below 3500 ft.; Redwood F., Mixed Evergreen F.; Del Norte and Humboldt cos. June–July.

4. **L. Bolánderi** Wats. [*L. Howellii* Jtn.] Bulb ovoid, 3–5 cm. long, the lanceolate scales not jointed, 2.5–5 cm. long; stem 3–10(–12) dm. high; lvs. oblanceolate to obovate, 3.5–4.5(–6) cm. long, 10–18 mm. wide, glaucous on both surfaces, but more so beneath, in 3–4 whorls, with 1–3 solitary lvs. below each whorl; fls. 2–3–7, horizontal

or somewhat nodding, funnelform, the upper third somewhat spreading, deep crimson with purple spots, the segms. 3–4 cm. long, 8–9 mm. wide; anthers 5–6 mm. long; caps. 3–3.5 cm. long; seed obovate, 5–6 mm. long; $n = 12$ (Sansome & La Cour, 1934).—Dry reddish clay soil, 500–2500 ft.; openings in Chaparral, Mixed Evergreen F.; s. Ore. to Del Norte Co. June–July.

5. **L. Humbòldtii** Roezl & Leichtl. [*L. canadense* var. *puberulum* Torr.] HUMBOLDT LILY. (Incorrectly called Tiger Lily). Bulbs ovoid, oblique, 5–15 cm. long, the scales fleshy, lance-ovate, 4–7 cm. long, not jointed, white; stems 9–20(–30) dm. high, stout, ± puberulent; lvs. in 4–8 whorls of 10–20 each, 9–12 cm. long, 15–30 mm. wide, oblanceolate, bright green to purplish, puberulent on margins and on larger veins beneath; fls. nodding, few to many, orange-yellow, spotted maroon or purple, the segms. 7–9 cm. long, 12–25 mm. wide, revolute to near the base; stamens and style ca. 5 cm. long; anthers 12–15 mm. long; caps. obovoid, acutely angled, 2.5–5 cm. long; $n = 12$ (Sansome & La Cour, 1934).—Dry rather heavy soil, open places and forests, below 4500 ft.; Chaparral, Yellow Pine F.; Sierra Nevada, Butte to Fresno cos. June–July. Bulb-scales very bitter.

Var. **ocellàtum** (Kell.) Elwes. [*L. Bloomerianum* var. *o.* Kell. *L. o.* Beane. *L. H.* var. *magnificum* Purdy.] Scales purplish; stems to 36 dm. high; fls. orange-red, with the maroon spots or blotches margined with red.—Gravelly soil, gulleys and canyons, below 3000 ft.; Chaparral; Santa Barbara Co. (including ids.) to San Jacinto Mts. and Santa Ana Mts. June–July.

Var. **Bloomeriànum** (Kell.) Jeps. [*L. B.* Kell. *L. ocellatum* ssp. *B.* Beane. *L. Fairchildii* Jones.] Stems 9–15 dm. high; lvs. 5–15 mm. wide; perianth-segms. with dark spots usually with red on margins.—Open flats, 4000–5500 ft., mts. of e. San Diego Co. June–July.

6. **L. columbiànum** Hanson ex Baker. [*L. canadense* var. *parviflorum* Hook. *L. canadense* vars. *minus* and *Walkeri* Wood. *L. lucidum* Kell. *L. parviflorum* Holz. *L. Bakeri* Purdy. *L. Purdyi* Waugh.] COLUMBIA LILY. OREGON LILY. Bulb ovoid, 3–5 cm. long, with unjointed lanceolate whitish scales ca. 3 cm. long; stems slender, 6–12(–20) dm. high; lvs. light green, lanceolate to oblanceolate, 5–10 cm. long, 12–30 mm. wide, the lower usually in whorls of 5–9, the upper scattered; fls. few to many, recurved, lemon-yellow to golden to deep red with yellow center, usually spotted maroon; perianth-segms. 3.5–6 cm. long, 8–12 mm. wide, recurved ca. half their length; anthers 4–6 mm. long; caps. 3–6 cm. long, oblong, acutely angled; $n = 12$ (Sansome & La Cour, 1934).—Among ferns and brush, in cut-over or virgin forests, at low elevs.; N. Coastal Scrub, Redwood F.; Humboldt Co. n. to B.C., Ida. June–July.

7. **L. marítimum** Kell. COAST LILY. Bulb rhizomatous, 2.5–3.5 cm. thick, the scales 15–25 mm. long, often nearly as wide, nonjointed or 1-jointed; stems 3–12(–20) dm. tall; lvs. usually scattered, dark green, oblanceolate to linear, 3–12 cm. long, 5–15 mm. wide; fls. 1–6(–12), horizontal, bell-shaped, dark red, spotted maroon; perianth-segms. 3–4 cm. long, 6–8 mm. wide, the upper third recurved; anthers 3–4 mm. long; caps. oblong to subglobose, 2–4 cm. long, wing-angled.—Sometimes in sandy soil, usually on raised hummocks in bogs, also in brush and woods, at low elevs.; N. Coastal Scrub, N. Coastal Coniferous F.; Marin to Mendocino cos. June–July.

8. **L. occidentàle** Purdy. WESTERN LILY. Bulb rhizomatous, 4–5 cm. long, the scales simple or jointed, 12–18 mm. long; stems slender, 6–18(–25) dm. tall; lvs. dark green, narrowly oblanceolate, 6–15 cm. long, 5–25 mm. wide, usually only the cent. whorled; fls. 1–10(–25), nodding, green at center, then dark orange and usually dotted maroon, crimson at outside, the segms. recurved on outer half, 3.5–5.5 cm. long; anthers 5–6 mm. long, standing close to pistil; caps. broadly ellipsoid, 1.5–2.5 cm. long; seeds rounded, 6–7 mm. wide; $n = 12$ (Stewart, 1947).—Sandy loam or peat, thickets and among ferns; N. Coastal Scrub; Humboldt Co. to s. Ore. June–July.

L. pitkinénse Beane & Vollmer. A plant from the Pitkin Marsh, Sonoma Co., closely related to *L. occidentale*, but occasionally with short stolons between the rhizome sections.

9. **L. párvum** Kell. [*L. canadense* var. *p.* Baker.] ALPINE LILY. Bulb rhizomatous, 2.5–3.5 cm. long, the scales 1–2 cm. long and 3-jointed; stems 4–15(–20) dm. tall, slender; lvs. light green, broadly lanceolate to linear, 5–12 cm. long, 5–30 mm. wide, mostly scattered but usually a few in whorls; fls. erect or ascending, few to many, orange

to dark red, spotted maroon, bell-shaped, the segms. 3.5–4 cm. long, recurved in upper third; anthers 3 mm. long; caps. ovoid, 12–16 mm. long; $n = 12$ (Stewart, unpub.).— Boggy places at edge of swamps or streams, often among alders and willows, 6500–9000 ft.; Montane Coniferous F.; throughout Sierra Nevada. July–Sept. Forma *crocàtum* Stearn. [*L. p.* var. *luteum* auth., not Purdy.] Fls. orange-yellow, spotted maroon.—At 4000–6500 ft., Sierra Nevada.

10. **L. Kelleyànum** Lemmon. [*L. nevadense* Eastw. *L. pardalinum* var. *n.* Stoker. *L. parviflorum* of many references. *L. n.* vars. *fresnense, monense,* and *shastense* Eastw. *L. s.* Beane. *L. f.* Eastw. *L. inyoense* Eastw. *L. parvum* var. *luteum* Purdy.] Bulb rhizomatous, unbranched, 5–6 cm. long, with numerous jointed scales ca. 1 cm. long; stems 6–18 dm. high; lvs. oblong-lanceolate, 5–15 cm. long, 1–4 cm. wide, the lower alternate, middle and upper in whorls of 3–8; fls. fragrant, few to 25, nodding, orange toward tips or yellow throughout, with minute maroon dots, segms. 3–5 cm. long, lanceolate, recurved to near base; anthers 4–6 mm. long; caps. oblong, 2.5 cm. long.— Wet banks and boggy places, 4000–10,500 ft.; Montane Coniferous F.; Sierra Nevada, Tulare Co. n., to Siskiyou and Trinity cos. July–Aug.

11. **L. Vóllmeri** Eastw. [*L. Roezlii* auth., not Regel.] Rhizomes unbranched, the scales 3–4-jointed, thick, ovoid, acute or obtuse; stems one, rarely 2, ca. 1 m. tall; lvs. glabrous, linear-lanceolate, light green on both surfaces, 10–15 cm. long, 5–10 mm. wide, the middle often whorled; fls. few to many, nodding, on long erect peduncles; perianth-segms. yellow to reddish-orange, with reddish tinge along the margins near the apex and with dark spots on lower half of inner surface, 6–8 cm. long, 5–10 mm. wide, recurved from near base; fils. divaricate; anthers 7–10 mm. long; capsules ellipsoid, obtuse at both ends; $n = 12$ (Stewart, 1947).—Hillside bogs; Redwood F., N. Coastal Coniferous F.; Humboldt and Del Norte cos.; Josephine Co., Ore.

L. Wígginsii Beane & Vollmer. Differing from *L. Vollmeri* in the shorter and broader perianth-segms., clear yellow spotted with purple, and anthers 11 mm. long.—From Siskiyou Mts. Apparently representing a recombination-type between *L. Vollmeri* and *L. pardalinum.*

12. **L. pardalìnum** Kell. [*L. p.* var. *angustifolium* Kell. *L. p.* var. *pallidifolium* Baker. *L. Roezlii* Regel. *L. californicum* Lindl. *L. Harrisianum* Beane & Vollmer.] LEOPARD LILY. PANTHER LILY. Bulbs branching-rhizomatous, 6–10 cm. long, the scales usually 1-jointed, 1.5–3 cm. long; stems stout, 10–25 dm. high; lvs. linear to lanceolate, 1–2 dm. long, 6–25 mm. wide, in 3–4 whorls of 9–15 and some scattered, pale to deep green; fls. nodding, often not fragrant, 1–several, yellow with some red to dark red, with maroon spots, the segms. recurved to the middle or below, 5–8 cm. long, 12–18 mm. wide; anthers 10–15 mm. long; caps. oblong, 3 cm. long; $n = 12$ (Sansome & La Cour, 1934).—Forming large colonies, stream banks and springy places, up to ca. 6000 ft.; N. Coastal Coniferous F., Mixed Evergreen F., Coast Ranges, Santa Barbara to Humboldt cos.; Yellow Pine F., Sierra Nevada s. to Kern Co.; mts. of e. San Diego Co. May–July. Exceedingly variable.

13. **L. Párryi** Wats. LEMON LILY. Bulb rhizomatous, 2.5–3 cm. long, the scales jointed, 12–20 mm. long; stems slender, 6–15 dm. high; lvs. scattered, sometimes the lower whorled, linear-oblanceolate or lanceolate, 8–15 cm. long, 6–15 mm. wide; fls. 1– few(–25), horizontal, fragrant, clear lemon-yellow, sometimes with maroon spots, trumpet-shaped, the segms. recurved or spreading in upper third, 6–10 cm. long, 8–12 mm. wide; anthers brown, 6–8 mm. long; caps. narrow-oblong, 3.5–5 cm. long; seeds obovate, 3–4 mm. long; $n = 12$ (Stewart, 1947).—Springy places and wet banks, 4000– 9000 ft.; Montane Coniferous F.; San Gabriel Mts. to San Diego Co.; Ariz. Plants from Little Rock Creek, San Gabriel Mts., with lvs. 4 cm. wide, are the var. *Késsleri* A. Davids. July–Aug.

20. Calochórtus Pursh. MARIPOSA-LILY. STAR-TULIP. BUTTERFLY-TULIP

Glabrous perennial herbs from tunicated bulbs, with membranous or fibrous-reticulate coats. Stems scapiform or leafy, simple or branched, frequently bulbiferous in axils of lowest lf. or lvs., sometimes of upper lvs. Lvs. usually linear, the basal solitary, often very large, the cauline reduced up the stem. Infl. with cent. axis or subumbellate by its suppression, the bracts usually as long as pedicels and opposite them. Fls. conspicuous, globose to open-campanulate, erect or nodding, white, yellow, red, lavender, purple, bluish or brownish, often variously tinged. Perianth-segms. well differentiated, the sepals ovate to lanceolate, ± colored, usually naked; the petals usually larger and broader, cuneate or clawed, usually bearded on inner face, variously spotted and patterned and bearded, with a gland near base. Stamens 6, inserted on the base of the segms.; fils. basally dilated; anthers basifixed, linear to oblong. Ovary 3-celled, with many ovules in 2 rows in each locule; stigmas 3, sessile, persistent. Caps. linear to orbicular, 3-angled to -winged, septicidal, erect or nodding. Seeds irregular or flattened, usually with hexagonally reticulate coats. Ca. 60 spp. of temp. w. N. Am., from B.C. to the Dakotas and Guatemala; many of great beauty and horticultural interest. The bulbs of many were eaten by the Indians. (Greek, *kalos,* beautiful, and *chortos,* grass, referring to the fls. and lvs.)

(Ownbey, M. A. A monograph of the genus Calochortus. Ann. Mo. Bot. Gard. 27: 371–560, 1940.)
A. Fls. closed-campanulate to subglobose, mostly nodding.
  B. The fls. white to purple; petals not conspicuously fringed.
    C. Lower membranes of glands extending entirely across the base of the petals; petals deep rose, drying purple . . . . 1. *C. amoenus*
    CC. Lower membranes extending only ⅓–⅔ the breadth of the petals; petals white to rose
                      2. *C. albus*
  BB. The fls. yellow; petals conspicuously fringed.
    C. The petals sparsely hairy within to near the apex, overlapping by 7–9 mm.
                      3. *C. pulchellus*
    CC. The petals nearly naked or with few hairs near the gland, and overlapping by 3–5 mm.
                      4. *C. amabilis*
AA. Fls. broadly campanulate to bowl-shaped, erect or spreading.
  B. Surface of gland naked or nearly so, although the bordering membranes may be fringed and extend over the gland (see also nos. 19 and 20).
    C. Petals glabrous or nearly so, not ciliate or bearded.
      D. Stems leafy, the lower internodes sometimes very short.
        E. Internodes elongate; stems branched, usually not bulbiferous
                      9. *C. umbellatus*
        EE. Internodes very short; stems usually unbranched, bulbiferous near base
                      10. *C. uniflorus*
      DD. Stems scapiform.
        E. Petals acute; frs. nodding . . . . 11. *C. minimus*
        EE. Petals rounded; frs. erect . . . . 12. *C. nudus*
    CC. Petals ± densely bearded, often ciliate or fimbriate.
      D. Infl. subumbellate; fr. elliptic to orbicular in outline.
        E. Petals ciliate and mostly densely bearded.
          F. The petals yellow . . . . 5. *C. monophyllus*
          FF. The petals white, cream, or with some purple or rose.
            G. Stems usually branched; petals inconspicuously fringed, bearded to tip . . . . 6. *C. Tolmiei*
            GG. Stems rarely branched; petals rather conspicuously fringed, bearded just above the gland.
              H. Anthers acute; petals not papillose on inner face
                      7. *C. coeruleus*
              HH. Anthers long-apiculate; petals papillose on inner face
                      8. *C. elegans*
        EE. Petals not ciliate, rather sparsely bearded.
          F. Anthers ca. half as long as fils.; lower internodes very short; stems forming bulblets near the base . . . . 13. *C. longebarbatus*
          FF. Anthers nearly as long as fils.; stems not forming basal bulblets.
            G. Stems 1–3 dm. tall; anthers obtuse or acute; caps. erect
                      14. *C. Greenei*
            GG. Stems ca. 1 dm. tall; anthers apiculate; caps. nodding
                      15. *C. persistens*
      DD. Infl. with an axis; fr. linear.
        E. Petals obovate, scarcely if at all shorter than the sepals.
          F. The petals rarely fimbriate, glabrous at apex, pinkish or purplish
                      35. *C. Plummerae*
          FF. The petals fimbriate, bearded nearly or quite to the apex, orange to brown or purplish . . . . 36. *C. Weedii*
        EE. Petals oblong, half as long as sepals . . . . 37. *C. obispoensis*
  BB. Surface of gland densely hairy, often with peculiar thickened hairs.
    C. Glands not depressed or but slightly so and not surrounded by a membrane.
      D. Infl. with a distinct axis, even though its internodes be short. S. Calif.
        E. Glands covered with linear hairs or processes.
          F. Petals with dark blotch near gland; stems erect.
            G. Processes on gland dark; petal-blotch purplish; caps. oblong, obtuse. Coastal drainage, Orange Co. to Santa Barbara Co.
                      16. *C. catalinae*

GG. Processes on gland yellow; petal-blotch red-brown; caps. linear,
         acute. Mts. of San Diego Co. .... 18. *C. Dunnii*
   FF. Petals without dark blotch; stem sinuous. Deserts ...... 17. *C. flexuosus*
EE. Glands with clavate or fungoid processes.
   F.   Gland-hairs or processes clavate; anthers white ...... 19. *C. Palmeri*
   FF. Gland-hairs or processes fungoid with stellate tips, or lacking; anthers
         purplish or blue .... 20. *C. splendens*
DD. Infl. subumbellate. Mostly cent. and n. Calif.
   E.   Gland oblong, with long slender gland-processes.
      F.   Petals conspicuously striate, not spotted. Mojave Desert .. 21. *C. striatus*
      FF. Petals not striate, with red spot above the gland. N. Calif.
                                                       22. *C. monanthus*
   EE. Glands not oblong, with short thick processes.
      F.   Gland quadrate, each petal with dark red blotch .... 23. *C. venustus*
      FF. Gland not quadrate, with or without cent. blotch.
         G.   The glands shaped like an inverted V; petal-blotch surrounded by
               bright yellow .... 24. *C. superbus*
         GG. The glands not as above.
            H.   Shape of gland doubly lunate; petal-blotch very large and
                  surrounded by pale yellow .... 25. *C. Vestae*
            HH. Shape of gland not doubly lunate.
               I.   Fls. yellow; glands simply lunate; anthers not sagittate
                                                       26. *C. luteus*
               II. Fls. white or smoke-colored; glands triangular-ovate;
                     anthers sagittate .... 27. *C. Leichtlinii*
CC. Glands ± depressed, surrounded by a membrane.
   D.   Sepals usually longer than petals; anthers linear .... 28. *C. macrocarpus*
   DD. Sepals usually not longer than petals; anthers lanceolate to oblong.
      E.   Fls. white to purplish.
         F.   Petals with a dark spot above the gland .... 29. *C. Nuttallii*
         FF. Petals without such dark spot.
            G.   Basal lvs. withering before anthesis. Plants of dry places.
               H.   Petals broadly obovate, 20–32 mm. long. Panamint Mts.
                                                       29. *C. Nuttallii*
               HH. Petals cuneate-obovate, 15–35 mm. long. Sierra Nevada to San
                     Jacinto Mts. .... 30. *C. invenustus*
            GG. Basal lvs. persistent at anthesis. Moist meadows .. 31. *C. excavatus*
      EE. Fls. red to yellow.
         F.   Hairs on face of petal not enlarged distally. San Bernardino Co. to
               L. Calif. .... 32. *C. concolor*
         FF. Hairs on face of petals distally enlarged.
            G.   Glands with simple or somewhat branched processes; petal-hairs
                  few, somewhat thickened distally. Deserts ...... 33. *C. Kennedyi*
            GG. Glands with much-branched fungoid processes; petal-hairs clavate.
               W. base of the Sierra Nevada and in S. Coast Ranges
                                                       34. *C. clavatus*

1. **C. amoènus** Greene. [*C. albus* var. *a.* Purdy.] Stem slender, erect, somewhat
flexuous, 2–5 dm. tall, branched; basal lf. 2–5 dm. long, 5–25 mm. wide; cauline lvs.
2–5, lanceolate, 5–15 cm. long; bracts paired, lanceolate, acuminate, 2.5–4 cm. long;
fls. deep rose, drying purple, narrow-campanulate, nodding to suberect; sepals lance-
ovate, acute, glabrous, 1–1.5 cm. long; petals elliptic-obovate, obtuse to acute, somewhat ciliate, with slender hairs above gland on inner face, 1.6–2.5 cm. long; gland
broad, maroon, slightly depressed, with 4–5 wide transverse membranes fringed with
papillae, the lowest membrane extending across the petal-base; fils. dilated at base, 3–4
mm. long; anthers oblong, obtuse, 3–4 mm. long, whitish; caps. broadly elliptic, narrowly
3-winged, nodding, 2–3 cm. long; seeds irregular, dark brown, reticulate; $n = 10$ (Beal
& Ownbey, 1943).—Leafy loam of grassy slopes, in partial shade, 1800–4500 ft.; Foothill Wd., Yellow Pine F.; w. foothills of Sierra Nevada (Madera Co.) to Greenhorn
Mts. (Kern Co.) April–June.

2. **C. álbus** Dougl. ex Benth. [*Cyclobothra a.* Benth. *Cyclobothra paniculata* Lindl.
Calochortus a. var. *paniculatus* Baker. *C. lanternus* A. Davids.] Fairy Lantern. Globe-
Lily. Stem rather slender, erect, 2–8 dm. tall, branched; basal lf. 3–7 dm. long, 1–5
cm. wide; cauline lvs. 2–6, lanceolate to linear, 0.5–2.5 dm. long; bracts often paired,
lanceolate, 1.5–5 cm. long; fls. white, globose to globose-campanulate, nodding; sepals
ovate to lanceolate, glabrous, 1–1.5 cm. long; petals elliptic or wider, ciliate, and with
slender hairs above the gland, 2–2.5 cm. long; gland ⅓–⅔ as wide as petal, depressed,
with several transverse fringed membranes; fils. dilated at base, 4–5 mm. long; anthers
oblong, 4 mm. long; caps. elliptic-oblong, 3-winged, nodding, 2.5–4.0 cm. long; seeds

irregular, dark brown, reticulate; $n = 10$ (Beal & Ownbey, 1943).—Shaded often rocky places in open woods or brush, below 5000 ft.; Footland Wd., Chaparral, Yellow Pine F.; Sierran foothills, Butte to Madera cos., Coast Ranges, San Francisco to Cuyamaca Mts., Santa Barbara Ids. April–June. Material from the Santa Cruz Mts. to the Santa Lucia Mts. tends to rose-tinged fls. and is var. **rubéllus** Greene.

3. **C. pulchéllus** Dougl. ex Benth. [*Cyclobothra p.* Benth.] Stem stoutish, erect, 1–3 dm. tall, often branched; basal lf. 1–4 dm. long, 1–3 cm. wide; cauline lvs. 2–3, linear to lanceolate, 0.5–2.5 dm. long; bracts opposite, lanceolate, 4–8 cm. long; fls. greenish-yellow, globose, nodding; sepals ovate to lanceolate, 2–3 cm. long; petals lance-ovate, obtuse, clawed, fringed, with short thick hairs on inner face, 2.5–3.3 cm. long, overlapping by 7–9 mm.; gland deeply depressed with transverse band above of long slender processes; fils. dilated at base, 6–8 mm. long; anthers oblong, 3–5 mm. long; caps. elliptic-oblong, broadly 3-winged, nodding, 2–3 cm. long; seeds irregular, dark brown, reticulate; $n = 10$ (Beal & Ownbey, 1943).—Frequent on wooded and brushy slopes, above 700 ft.; Foothill Wd., Chaparral; Mt. Diablo, Contra Costa Co. April–June.

4. **C. amábilis** Purdy. [*C. pulchellus* var. *a.* Jeps. *C. p.* var. *maculosus* Wats.] Stem stoutish, erect, 2–5 dm. tall, usually branched; basal lf. 2–5 dm. long, 5–40 mm. wide; cauline lvs. 2–4, lanceolate to linear, 2–20 cm. long; bracts opposite, lanceolate, 2–10 cm. long; fls. deep clear yellow, triangular in outline, nodding; sepals ovate to lanceolate, glabrous, 15–20 mm. long; petals lance-ovate, obtuse, fringed, inner face naked or nearly so, clawed, 16–20 mm. long, overlapping by 3–5 mm.; gland brown on back, deeply depressed, with transverse band above of slender processes; fils. dilated basally, ca. 5 mm. long; anthers 3–4 mm. long; caps. oblong, 3-winged, nodding, 2–3 cm. long; seeds irregular, dark brown, reticulate; $n = 10$ (Beal & Ownbey, 1943).—Loamy to rocky soil, dry slopes in brush or woods, up to 3000 ft.; Chaparral, Foothill Wd., Mixed Evergreen F.; N. Coast Ranges, Humboldt Co. to Solano and Marin cos. April–June.

5. **C. monophýllus** (Lindl.) Lem. [*Cyclobothra m.* Lindl. *Cyclobothra elegans* var. *lutea* Benth. *Calochortus Benthamii* Baker. *C. B.* var. *Wallacei* Purdy. *C. nitidus* var. *cornutus* Wood. *C. pulchellus* var. *parviflorus* Regel. *C. maculatus* Eastw.] YELLOW STAR-TULIP. Stem flexuous, simple or branched, 8–20 cm. high; basal lf. 1–3 dm. long, 3–15 mm. wide; cauline lvs. 0–3, lanceolate to linear, much-reduced; bracts paired, lanceolate to linear, 2–6 cm. long; fls. deep yellow, often with red-brown spot on claw of each petal, open-campanulate, weakly erect; sepals lance-oblong, 16–20 mm. long, glabrous; petals narrow-obovate, acute to obtuse, clawed, fringed, densely bearded above gland with clavate hairs; gland transverse, arched upward, naked, bordered below by a fringed membrane, above by short processes; fils. dilated below, 4–5 mm. long; anthers apiculate, 3–4 mm. long, yellowish; caps. broadly elliptic, 3-winged, nodding, 12–20 mm. long; seeds irregular, dark brown, reticulate; $n = 10$ (Beal & Ownbey, 1943).—Clay loam of wooded slopes, 1200–3600 ft.; upper Foothill Wd., lower Yellow Pine F.; Sierran foothills, Shasta Co. to Tuolumne Co. April–May.

6. **C. Tólmiei** H. & A. [*C. Maweanus* Leichtl. *C. coeruleus* var. *M.* Jeps. *C. elegans* var. *Lobbii* Baker. *C. glaucus* Regel. *C. Purdyi* Eastw.] PUSSY EARS. Stem simple or branched, 1–4 dm. high, somewhat flexuous; basal lf. 1–4 dm. long, 2–30 mm. wide; cauline lf. usually present; bracts 2 or more, lanceolate to linear, 1–7 cm. long; fls. white or cream, or tinged with rose or purple, erect or spreading, open-campanulate; sepals lance-oblong, acute or acuminate, glabrous, 10–15 mm. long; petals obovate, cuneate, acute or obtuse, fringed laterally, bearded on most of inner surface, 12–25 mm. long; gland transverse, arched upward, naked, depressed, bordered below with a fringed membrane and above with short processes; fils. dilated at base, 6–8 mm. long; anthers lanceolate, acute to apiculate, white to purple, 3–5 mm. long; caps. elliptic-oblong, 3-winged, nodding, 2–3 cm. long; seeds irregular, dark brown, reticulate; $n = 10$ (Beal & Ownbey, 1943).—Dry, often rocky soil, on open grassy slopes or in woods, below 6000 ft., Redwood F., Chaparral, Mixed Evergreen F., Yellow Pine F.; Wash. and Ore. through Coast Ranges to Santa Cruz Co. and to upper Sacramento V. April–July.

7. **C. coerùleus** (Kell.) Wats. [*Cyclobothra c.* Kell.] BEAVERTAIL-GRASS. Stem usually simple, slender, erect or flexuous, 3–15 cm. high; basal lf. 1–2 dm. long, 2–10 mm. wide; bracts 2–several, lanceolate to linear, acuminate, 5–10 mm. long; fls. 1–8(–10), subumbellate, bluish, erect or ascending, open-campanulate; sepals lance-oblong, ca. 10 mm. long; petals obovate, clawed, with smooth inner face, but laciniate-fringed, bearded

above gland, 8–12 mm. long; gland transverse, arched upward, naked on surface, slightly depressed, bordered below with fringed membrane and above with short processes, these and membrane papillose; fils. basally dilated, ca. 3.5 mm. long; anthers oblong, acute to short-apiculate, ca. 4 mm. long; caps. broadly elliptic, nodding, 1–1.5 cm. long; seeds reticulate.—Open gravelly places in woods, 3500–7500 ft.; Montane Coniferous F.; w. Sierra Nevada, Lassen and Tehama cos. to Amador Co. May–July.

Var. **fimbriàtus** Ownbey. [*C. elegans* var. *nanus* Ownbey, not Wood.] Anthers lanceolate, short-apiculate.—Dry woods at 2000–6000 ft.; Montane Coniferous F.; inner N. Coast Ranges, Siskiyou Co. to Lake Co. May–July.

Var. **Wéstoni** (Eastw.) Ownbey. [*C. W.* Eastw.] Petals lanceolate, laciniate-fringed only on sides; anthers apiculate.—Greenhorn Mts., Kern Co. May.

8. **C. élegans** Pursh. var. **nànus** Wood. [*C. n.* Piper. *C. e.* var. *oreophilus* Ownbey.] Stem slender, flexuous, usually simple, 5–15 cm. tall; basal lf. 1–2 dm. long, 2–10 mm. wide; bracts 2–several, lanceolate to linear, 15–30 mm. long; fls. 1–3(–7), open-campanulate, erect or spreading, greenish-white, often with purplish crescent above gland and on each sepal; sepals lance-ovate, acuminate, 12–16 mm. long; petals oblanceolate or wider, clawed, papillose on inner face, fringed laterally, bearded above gland; gland transverse, arched upward, with naked surface, bordered below with erose or shallowly fringed membrane, and above with short processes, both these and membrane papillose; fils. dilated basally, 3–4 mm. long; anthers long-apiculate, 3–4 mm. long; caps. nodding, broadly elliptic, winged, 1.5–2 cm. long; seeds light brown, reticulate; $n = 10$ (Beal & Ownbey, 1943).—Open woods, 5000–7500 ft.; Montane Coniferous F.; Siskiyou Co.; Ore. May–July.

9. **C. umbellàtus** Wood. [*C. collinus* Lemmon.] Stem 8–25 cm. high, usually 2-branched, each branch subumbellately 2–6-fld.; basal lf. 2–4 dm. long, 5–15 mm. wide; cauline lf. 1, linear; bracts 2–several, linear, 1–6 cm. long; fls. white or pale lilac, often with purple spot on each petal near gland and on each sepal, suberect, open-campanulate; sepals lance-elliptic, 10–14 mm. long; petals broadly obovate, cuneate, rounded-erose above, naked except for a few hairs near the gland, 12–18 mm. long; gland convex basally, truncate above, naked on surface, covered with fringed membrane and bordered above by row of short processes; fils. dilated basally, 5 mm. long; anthers oblong, 2–3 mm. long; caps. oblong to orbicular, nodding, winged, 10–14 mm. long; seeds irregular, dark brown, reticulate; $n = 10$ (Beal & Ownbey, 1943).—Openings on dry wooded or barren hills, often on serpentine, 1000–2200 ft.; Chaparral, Mixed Evergreen F.; Coast Ranges, Lake Co. to Santa Clara Co. March–May.

10. **C. uniflòrus** H. & A. [*Cyclobothra u.* Kunth. *Calochortus lilacinus* Kell.] Stem usually simple and barely reaching above the surface of the ground; basal lf. 1–4 dm. long, 5–20 mm. wide; cauline lvs. 1–3, linear, bulbiferous in axils; bracts 2–several, linear, 1–4 cm. long; fls. subumbellate, 1–5, lilac, often with purple spot above the gland; sepals lance-elliptic, acuminate, 12–16 mm. long; petals broadly obovate, cuneate, rounded and erose-denticulate above, 15–28 mm. long, with few hairs near gland; gland convex basally, truncate above, covered with broad fringed membrane and bordered above by slender processes, but its own surface naked; fils. dilated basally, 7–8 mm. long; anthers oblong, 3–4 mm. long; caps. oblong-elliptic, narrowly winged, 15–25 mm. long; seeds irregular, dark brown, reticulate; $n = 20$ (Beal & Ownbey, 1943).—Wet, low, often alkaline places, below 500 ft.; N. Coastal Scrub, Coastal Coniferous F., Mixed Evergreen F.; Coast Ranges, Monterey to Trinity and Mendocino cos.; sw. Ore. April–June.

11. **C. mínimus** Ownbey. [*C. elegans* var. *subclavatus* Baker.] Stem scapiform, barely rising above the surface of the ground; basal lf. 1–2 dm. long, 3–10 mm. wide; bracts usually 2, opposite, lanceolate, 1–2(–6) cm. long; fls. 1–3(–10), subumbellate, white, suberect, open-campanulate; sepals lanceolate, 8–10 mm. long; petals obovate, cuneate, erose-denticulate above, not ciliate, naked or few-haired near gland, 10–14 mm. long; gland transverse, straight, naked but widely bordered by a laciniate membrane; fils. dilated basally, 4–5 mm. long; anthers bluish, linear-oblong, acute, 3–4 mm. long; caps. elliptic, winged, nodding, 15–20 mm. long; seeds reticulate.—Moist grassy places, in open woods, 4000–9500 ft.; Montane Coniferous F.; Sierra Nevada, e. Eldorado Co. to Tulare Co. May–Aug.

12. **C. nùdus** Wats. [*C. shastensis* Purdy. *C. n.* var. *s.* Jeps.] Stem erect, 10–25 cm. high; basal lf. shorter, 5–15 mm. wide; bracts usually 2, 1–2(–5) cm. long; fls. erect, white to pale lavender, open-campanulate; sepals lance-ovate, 10–12 mm. long; petals broadly obovate, rounded, erose-denticulate, not ciliate, naked or nearly so, 14–16 mm. long; gland transverse, straight or arched upward, naked, bordered below with broad fringed membrane; fils. dilated below, 4–5 mm. long; anthers linear-oblong, obtuse or acute, 5–6 mm. long; caps. elliptic, winged, erect, 15–20 mm. long; seeds irregular, light brown, reticulate; $n = 10$ (Beal & Ownbey, 1943).—Low meadows and damp places, 4000–7500 ft.; Montane Coniferous F.; Siskiyou Co. to w. Eldorado Co. May–July.

13. **C. longebarbàtus** Wats. Stem erect, 1–3 dm. tall, bulbiferous in axil of nearly basal lf., this 2–3 dm. long, 5–10 mm. wide; bracts opposite, 2–6 cm. long; fls. 1–4, subumbellate, campanulate, lavender-pink, with purplish-red spot above each gland, drying purple, erect; sepals lance-ovate, 15–20 mm. long; petals broadly obovate, rounded to acute, with few hairs within above the gland, 2–3 cm. long; gland transversely oblong, naked but bordered below by broad deeply fringed membrane and above by short processes, these and fringe papillose; fils. dilated at base, 7–10 mm. long; anthers obtuse to short-apiculate, 3–4 mm. long; caps. broadly elliptic, winged, erect, 2–2.5 cm. long; seeds irregular, light brown, reticulate; $n = 10$ (Beal & Ownbey, 1943).—Infrequent in moist meadows, 4000–6000 ft.; Sagebrush Scrub, N. Juniper Wd.; Shasta and Modoc cos.; to se. Wash. June–Aug.

14. **C. Greènei** Wats. Stem erect, 1–3 dm. tall; basal lf. ca. 2 dm. long, 1–2 cm. wide; cauline lvs. 1–2, ca. 2 dm. long; bracts 2–4, linear-lanceolate, 2–4 cm. long; fls. 1–5, subumbellate, erect, open-campanulate, purplish with darker purple crescent on each petal above gland; sepals ovate, acuminate, 2.5–3 cm. long; petals broadly obovate, rounded, loosely bearded above gland, 3–4 cm. long; gland lunate, deeply impressed, naked on surface, bordered below with broad fringed membrane, above with short processes, these and membrane papillose; fils. 10–14 mm. long; anthers lance-oblong, obtuse or acute, 8–12 mm. long; caps. elliptic, winged, erect, 2–2.5 cm. long.—Rare, brushy hillsides, etc.; Siskiyou Co.; Jackson Co., Ore. June–July.

15. **C. persístens** Ownbey. Stem erect, ca. 1 dm. tall; basal lf. ca. 2 dm. long, 15–20 mm. wide; cauline lf. 1, bractlike; bracts 2, lanceolate, 2–3 cm. long; fls. 2, purplish, erect, with persistent perianth; sepals lance-elliptic, acuminate; petals obovate, obtuse, laterally fringed, yellow-haired above gland, 3.5–4 cm. long; gland transverse, sublanate, depressed, naked, bordered below with broad fringed membrane, above with short processes, these and fringe papillose; fils. ca. 10 mm. long; anthers lanceolate, apiculate, ca. 10 mm. long; caps. elliptic, winged, nodding; $n = 10$ (Beal & Ownbey, 1943).—Known only from type locality, mts. near Yreka, Siskiyou Co. June–July.

16. **C. catalìnae** Wats. [*Mariposa c.* Hoov. *C. Lyoni* Wats.] Stem erect, somewhat zig-zag, usually branched above, 2–6 dm. high, bulbiferous near base; lvs. linear, 1–2.5 dm. long, 3–6 mm. wide, the basal usually withered at anthesis, the cauline reduced upward; bracts opposite the pedicels, 2–10 cm. long; fls. 1–several, bowl-shaped, white tinged with lilac or light purple, with purple spot near base of each sepal and petal; sepals lanceolate, acuminate, 2–3 cm. long; petals obovate, cuneate, rounded and obtuse, naked except for few hairs near base, 2–5 cm. long; gland not depressed, oblong, densely covered with slender processes; fils. 8–10 mm. long; anthers lilac, oblong, 4–5 mm. long; caps. narrow-oblong, obtuse, erect, 2–5 cm. long, 8–10 mm. thick; seeds elliptic, flattened, light-colored, reticulate; $n = 7$ (Beal & Ownbey, 1943).—Heavy soil on open grassy slopes and openings in brush, below 2000 ft.; V. Grassland, Chaparral; s. San Luis Obispo Co. to San Diego Co.; Santa Barbara Ids. March–May.

17. **C. flexuòsus** Wats. Stem branched, usually sinuous and intertwined with other plants, or straggling over the ground, 2–4 dm. long; lvs. linear, attenuate, reduced upward, 1–2 dm. long; bracts 1–3 cm. long; fls. 1–4, campanulate, erect, white with lilac tinge and having purple spot with yellow band on each segm.; sepals lanceolate to lance-ovate, 2–2.5 cm. long; petals obovate, cuneate, rounded above, with few short hairs near gland, 3–4 cm. long; gland not depressed, transverse-lunate or wider, with dense short processes; fils. 6–10 mm. long; anthers oblong, 5–7 mm. long; caps. lanceolate, erect, angled, 2.5–3.5 cm. long, 8–10 mm. thick; seeds flat, light-colored, reticulate;

$n = 7$ (Beal & Ownbey, 1943).—Dry stony slopes, desert hills and mesas, 2000–5000 ft.; Creosote Bush Scrub, Sagebrush Scrub; e. Mojave Desert to Chuckawalla Mts., Riverside Co.; to sw. Colo. April–May.

18. **C. Dúnnii** Purdy. [*C. Palmeri* var. *D.* Jeps. & Ames.] Stem erect, usually branched, slender, 2–6 dm. high; basal lvs. 1–2 dm. long, 6–10 mm. wide, channelled; cauline reduced; bracts 1–2 cm. long; fls. open-campanulate, erect, white or flushed with pink, with red-brown spot above the gland; sepals lance-ovate, 1.5–2 cm. long; petals obovate, cuneate, usually rounded above, 2–3 cm. long, with yellowish hairs near the gland; gland rounded, not depressed, covered with linear yellow processes; fils. 5–6 mm. long; anthers oblong, white, acute, 4–5 mm. long; caps. linear, angled, erect, 2.5–3 cm. long. —Dry stony ridges, 4500–5000 ft.; Chaparral, Yellow Pine F.; Julian and Otay Mt., San Diego Co. to Guadalupe Mts., L. Calif. June.

19. **C. Pálmeri** Wats. [*C. splendens* var. *montanus* Purdy. *C. invenustus* var. *m.* Parish. *C. m.* A. Davids. *C. paludicola* A. Davids. *C. Palmeri* var. *p.* Jeps. & Ames.] Stem erect, bulbiferous, often branched, 3–6 dm. high; basal lvs. 15–20 cm. long, 4–5 mm. wide; cauline reduced; bracts 1–2 cm. long; fls. broadly campanulate, erect, white to lavender, sometimes with a brownish spot above each gland and near base of each sepal; sepals lanceolate, acuminate, 2.5 cm. long; petals obovate, cuneate, rounded above, 2–2.5 cm. long, with some yellow hairs near gland; gland not depressed, rounded, with many, short, thick, distally knobbed, yellowish or purplish processes or sometimes quite naked; fils. 7–8 mm. long; anthers oblong, white, acutish, 5–7 mm. long; caps. erect, linear, angled, 4.5–5 cm. long; $n = 7$ (Beal & Ownbey, 1943).—Meadows and places moist in early spring, 3500–6500 ft.; Yellow Pine F., Chaparral; San Bernardino Mts. to Tehachapi Mts. and e. San Luis Obispo Co. May–July.

Var. **Múnzii** Ownbey. Stem not bulbiferous at base; pedicels paired; bracts opposite; glands purple or none.—Yellow Pine F.; San Jacinto Mts. June.

20. **C. spléndens** Dougl. ex Benth. [*C. s.* vars. *atroviolaceus, major* and *ruber* Purdy. *C. Davidsonianus* Abrams. *Mariposa s.* Hoov.] Stem erect, branched, rarely bulbiferous, 2–6 dm. high; basal lvs. 1–1.5 dm. long, 5–6 mm. wide; cauline reduced; bracts 2–5 cm. long; fls. erect, campanulate, narrow at base, deep lilac, often with purple spot on each sepal and sometimes each petal; sepals lance-ovate, acuminate, 2–2.5 cm. long; petals obovate, cuneate, rounded and irregular above, 2.5–5 cm. long, with scattered hairs below middle; gland not depressed, usually with many, branched, fungoid processes; fils. 7–8 mm. long; anthers dark, obtuse to short-apiculate, 5–7 mm. long; caps. linear, erect, angled, 5–7 cm. long, 5–8 mm. thick; seeds flat, light-colored, cellular; $n = 7$ (Beal & Ownbey, 1943); $n = 14$ for *Davidsonianus* (Beal & Ownbey, 1943).—Dry slopes in heavy or granitic soil, up to 6000(–8500) ft.; Chaparral, V. Grassland, Yellow Pine F.; Colusa Co. through the Coast Ranges to n. L. Calif.; Santa Catalina Id. May–June.

21. **C. striàtus** Parish. Stem erect, not bulbiferous, 1–4.5 dm. high; basal lvs. 1–2 dm. long, 6–8 mm. wide; bracts linear, 1–2.5 cm. long; fls. subumbellate, campanulate, with narrow base, erect, lavender with purple veins; sepals lanceolate, 1.5–2 cm. long, acuminate; petals obovate, cuneate, rounded and erose above, 2–2.5 cm. long, sparsely hairy about the gland; gland not depressed, oblong, covered with linear processes; fils. 5–7 mm. long; anthers oblong, 4–6 mm. long; caps. linear, angled, erect, 4.5–5 cm. long; seeds flat, light-colored, reticulate.—Alkaline meadows and springy places, 2500–4300 ft.; Creosote Bush Scrub; Mojave Desert at n. base of San Gabriel and San Bernardino mts. to Las Vegas, Nev. April–June.

22. **C. monánthus** Ownbey. Stem erect, simple, bulbiferous; lvs. 7–10 mm. wide; bracts opposite, linear-attenuate; fl. 1, turbinate-campanulate, pinkish, with a ∧-shaped, dark red spot above each gland; sepals lanceolate, attenuate, ca. 4 cm. long; petals obovate, cuneate, rounded and erose above, ca. 4.5 cm. long, with few flexuous hairs near the gland; gland not depressed, oblong, densely covered with slender processes; fils. shorter than the lance-linear, short-apiculate anthers; ovary linear; fr. and seeds not known.—Known from a single collection from meadow along Shasta R., near Yreka, Siskiyou Co., *Greene* in June, 1876.

23. **C. venústus** Dougl. ex Benth. [*C. v.* var. *purpureus* Baker. *C. v.* var. *purpurascens* Wats. *C. purpurascens* Purdy. *C. v.* var. *Caroli* Ckll. *C. v.* var. *roseus* Reuthe. *C. v.* vars. *eldorado* and *sulphureus* Purdy. *Mariposa v.* Hoov. *M. simulans* Hoov.] Stem erect,

usually branched, bulbiferous; basal lvs. linear, 1–2 dm. long, 2–5 mm. wide; cauline somewhat reduced; bracts 2–8 cm. long; fls. 1–3, erect, campanulate, white to yellow, purple or dark red, each petal with dark red median blotch and often with a second paler blotch above the first; sepals lanceolate, curled back at tip, 2.5–3 cm. long; petals obovate, cuneate, 3–4.5 cm. long, scattered-hairy on lower part; gland not depressed, subquadrate, covered with short processes; fils. 7–10 mm. long; anthers linear-oblong, 7–10 mm. long; caps. linear, angled, erect, 5–6 cm. long, 5 mm. thick; $n = 7$ (Beal & Ownbey, 1943).—Light sandy soil, often decomposed granite, 1000–8000 ft.; V. Grassland, Foothill Wd., Yellow Pine F.; Sierra Nevada from Eldorado to Kern cos., Coast Ranges from San Francisco Bay to n. Los Angeles Co. May–July.

24. **C. supérbus** Purdy ex J. T. Howell. [*C. venustus* var. *citrinus* Baker, as to type. *C. luteus* var. *c.* Wats. *C. l.* var. *robustus* Purdy. *C. v.* var. *lilacinus* Baker. *C. s.* var. *pratensis* Purdy ex J. T. Howell. *C. p.* Hoov. *Mariposa s.* Hoov. *M. argillosa* Hoov.] Stem erect, bulbiferous, 4–6 dm. high; lvs. linear, 1.5–2.5 dm. long, 4–6 mm. wide, reduced upward; bracts 2–8 cm. long; fls. 1–3, subumbellate, erect, campanulate, white to yellowish or lavender, usually penciled with purple toward base and each segm. with median brown or purple blotch surrounded by zone of bright yellow; sepals lanceolate, attenuate, 2–3.5 cm. long; petals obovate, cuneate, rounded, obtuse, with few short hairs near gland, 2.5–4 cm. long; gland not depressed, linear, ± Λ -shaped, densely covered with short hairlike processes; fils. 7–9 mm. long; anthers lance-linear to -oblong, acute to obtuse, 8–10 mm. long; caps. linear, acute, angled, erect, 5–6 cm. long; $n = 6$, 7 (Beal & Ownbey, 1943).—Open slopes, dry meadows or wooded places up to 5000 ft.; V. Grassland, Foothill Wd., Yellow Pine F.; Sierra Nevada from Shasta to Kern cos. Coast Ranges at scattered stations, Shasta to San Diego cos. May–July.

25. **C. Véstae** Purdy (spelled "*Vesta*".) [*C. luteus* var. *oculatus* Wats.] Stem erect, bulbiferous, 3–5 dm. high; lvs. linear, 12–20 cm. long; bracts 2–5 cm. long; fls. 1–3, subumbellate, erect, campanulate, white to purplish, the petals penciled with red to purple below, and with median red-brown blotch, surrounded by pale yellow zone; sepals linear-lanceolate, attenuate, 2.5–3 cm. long; petals cuneate-obovate, rounded above, with few short hairs near gland, 3–4 cm. long; gland not depressed, transverse, doubly lunate, with dense short processes; fils. 7–10 mm. long; anthers oblong-linear, obtuse or acute, 7–10 mm. long; caps. linear, angled, erect; $n = 14$ (Beal & Ownbey, 1943).—Heavy clay soils, 1500–2500 ft.; Mixed Evergreen F., Yellow Pine F.; N. Coast Ranges, Humboldt to Napa and Sonoma cos. May–July.

26. **C. lúteus** Dougl. ex Lindl. [*C. venustus* var. *citrinus* auth., not Baker. *C. l.* var. *c.* of Calif. references. *Mariposa l.* Hoov.] Stem erect, slender, bulbiferous, 2–5 dm. high; lower lvs. linear, 1–2 dm. long, 2–5 mm. wide, reduced upward; bracts 1.5–8 cm. long; fls. 1–4, subumbellate, erect, campanulate, deep yellow, the petals usually penciled below, with red-brown lines, often with median red-brown blotch; sepals lance-oblong, attenuate, 2–3 cm. long; petals cuneate-obovate, with few slender hairs near the gland, 2.5–3.5 cm. long; gland not depressed, transverse, sublunate, with matted short hairlike processes; fils. 7–9 mm. long; anthers linear-oblong, obtuse or acute, 4–6 mm. long; caps. lance-linear, angled, erect, 3.5–6 cm. long; seeds flat, reticulate; $n = 7$, 10 (Beal & Ownbey, 1943).—Usually in heavy soils in open places, below 2000 ft.; V. Grassland, Foothill Wd., Mixed Evergreen F.; Coast Ranges, Mendocino to Santa Barbara cos., Santa Cruz Id., Sierra Nevada from Tehama Co. to Kern Co. April–June. Coastal material largely triploid, interior largely diploid.

27. **C. Leichtlínii** Hook. f. [*Mariposa L.* Hoov. *C. Nuttallii* var. *L.* Smiley. *C. N.* var. *subalpinus* Jones.] Stem erect, simple, bulbiferous, 2–4(–6) dm. high; lvs. linear, the lower 1–1.5 dm. long, 2–3 mm. wide, the cauline shorter; bracts paired, 1–5 cm. long; fls. 1–5, subumbellate, erect, campanulate, white to smoky-blue, often tinged pink, each petal with red to dark spot above the gland; sepals lance-ovate, 15–20 mm. long; petals cuneate-obovate, rounded above, 15–40 mm. long, with few short hairs near gland; gland slightly depressed, subovate, densely covered with short hairlike processes; fils. ca. 5 mm. long; anthers linear-oblong, sagittate at base, 5–7 mm. long; caps. lance-linear, erect, 3–6 cm. long; seeds flat, inflated, reticulate; $n = 7$ (Beal & Ownbey, 1943).—Open gravelly places, 4000–11,000 ft.; Montane Coniferous F.; mts. from Modoc Co. to Tulare Co.; w. Nev. June–Aug.

28. **C. macrocárpus** Dougl. [*Mariposa m.* Hoov.] Stem stout, usually simple, often

bulbiferous, 2–5 dm. high; lvs. several, linear, involute, curled at tip, 5–10 cm. long, reduced up the stem; bracts 3–5 cm. long; fls. 1–3, campanulate, subumbellate, erect, purple, each petal with median green stripe and sometimes a purple band above the gland; sepals lanceolate, acuminate, 4–5 cm. long; petals narrow-obovate, acuminate, bearded just above gland, 4–6 cm. long; gland slightly depressed, triangular-oblong, with sagittate base, densely covered with slender processes; fils. 8–9 mm. long; anthers lance-linear, obtuse, ca. 1 cm. long; caps. lance-linear, acuminate, angled, erect, 4–5 cm. long; seeds flat, with inflated reticulate coats; $n = 7$ (Beal & Ownbey, 1943).—Dry plains and slopes, usually in volcanic soil, 4000–6000 ft.; Sagebrush Scrub to Yellow Pine F.; Shasta, Siskiyou, Modoc, and Lassen cos.; to B.C., Mont. July–Aug.

29. **C. Nuttállii** Torr. var. **bruneaúnis** (Nels. & Macbr.) Ownbey. [*C. b.* Nels. & Macbr. *C. discolor* A. Davids.] Stem erect, usually simple, often bulbiferous, 2–4 dm. high; lvs. linear, reduced upward, becoming involute, the lower 10–15 cm. long, cauline reduced; bracts 2–4 cm. long; fls. 1–4, subumbellate, erect, campanulate, white, tinged lilac, the petals with median green stripe and dark red or purple spot above gland; sepals lanceolate, 1.5–2.5 cm. long; petals narrowly obovate, glabrous or nearly so, 2.5–3.5 cm. long; gland circular, depressed, surrounded by a conspicuous fringed membrane and densely covered with short simple or distally branched processes; fils. 5–6 mm. long; anthers yellow to maroon, 5–7 mm. long; caps. linear-lanceolate, acuminate, angled, erect, 3–5.5 cm. long; seeds flat, with loose reticulate coats.—Dry brushy and grassy slopes and flats, 5000–9000 ft.; Sagebrush Scrub, Pinyon-Juniper Wd.; e. of the Sierra Nevada, Lassen Co. to Inyo Co.; to Ore., Mont., Nev. May–Aug.

Var. **panaminténsis** Ownbey. Stem 4–6 dm. high; petals not spotted; anthers bluish or reddish.—Dry slopes, 7500–9500 ft.; Pinyon-Juniper Wd.; Panamint Mts., Inyo Co. June–July.

30. **C. invenústus** Greene [*C. Nuttallii* var. *australis* Munz.] Stem slender, erect, mostly simple, bulbiferous, 1.5–5 dm. tall; lvs. linear, 1–2 dm. long, 2–4 mm. wide, becoming involute, reduced upward; bracts 2–5 cm. long; fls. 1–5, subumbellate, erect, campanulate, white or dull lavender to purplish, sometimes with purplish spot below the gland; sepals lance-ovate, acuminate, 1.5–2.5 cm. long; petals cuneate-obovate, obtuse to apiculate, with few short hairs near gland, 2–3.5 cm. long; gland small, circular, slightly depressed, surrounded by a fringed membrane, densely covered with short distally branched processes; fils. 6–7 mm. long; anthers oblong, obtuse, purplish or yellowish, 7–8 mm. long; caps. lance-linear, acute, angled, erect, 5–7 cm. long; seeds flat, with loose reticulate coats.—Dry soil, mostly granitic, usually in pine woods, 4500–9000 ft.; Montane Coniferous F.; Sierra Nevada, Tuolumne Co. to Tulare Co., S. Coast Ranges, Santa Clara Co. to Monterey Co., Tehachapi Mts. to Laguna Mts. May–Aug.

31. **C. excavàtus** Greene. [*C. campestris* A. Davids.] Stem slender, erect, simple, often bulbiferous, 1–3 dm. high; lvs. linear, 1–2 dm. long, reduced upward, the basal green at anthesis; bracts 3–8 cm. long, paired; fls. 1–4, subumbellate, erect, open-campanulate, lavender, with or without spots; sepals lanceolate, 2–2.5 cm. long, acuminate; petals broadly cuneate-obovate, 2.5–3.5 cm. long, with few short hairs near gland; gland circular, depressed, surrounded by fringed membrane, densely covered with short distally branched processes; fils. 6–8 mm. long; anthers oblong, obtuse, red-brown, 7–10 mm. long; caps. lance-linear, angled, erect.—Grassy meadows, 4000–6000 ft.; Shadscale Scrub; e. base of Sierra Nevada, Inyo Co. April–May.

32. **C. cóncolor** (Baker) Purdy. [*C. luteus* var. *c.* Baker.] GOLDENBOWL MARIPOSA. Stem rather stout, erect, sparingly branched, rarely bulbiferous, 3–6 dm. high; lvs. glaucous, linear, becoming involute, 1–1.5 dm. long, reduced up the stem; bracts opposite, 4–8 cm. long; fls. 1–4, subumbellate, erect, campanulate, yellow within, often tinged purple in drying, usually with dark red blotch near base of each sepal and petal; sepals lance-ovate, 2.5–3 cm. long; petals cuneate-obovate, with few long yellow hairs near the gland, 3–5 cm. long; gland usually small, roundish, depressed, surrounded by fringed membrane, densely covered with slender unbranched processes; fils. 9–10 mm. long; anthers oblong, yellowish, obtuse, 8–10 mm. long; caps. lance-linear, acuminate, angled, erect, 5–8 cm. long.—Dry slopes, frequently in decomposed granite, 2000–7500 ft.; Chaparral, Yellow Pine F.; s. face of San Bernardino Mts. to n. L. Calif. May–July.

33. **C. Kénnedyi** Porter. Stem erect, rather simple, 1–2(–5) dm. high, sometimes

twisted, rarely bulbiferous; lower lvs. glaucous, linear, channeled, 1–2 dm. high, cauline reduced; bracts 2–4 cm. long, with dilated base; fls. 1–6, subumbellate, vermilion to orange, campanulate, often with brown-purple spots near base of each segm.; sepals ovate, acute, 1.5–2.5 cm. long; petals cuneate-obovate, with few slightly enlarged hairs near the gland, 2.5–5 cm. long; gland round, depressed, surrounded by fringed membrane, densely covered with simple or cleft processes; fls. 4–5 mm. long; anthers lanceolate, purplish, 5–8 mm. long; caps. lance-linear, acuminate, angled, longitudinally striped, 4–6 cm. long; seeds flat, with loose reticulate coats; $n = 8$ (Beal & Ownbey, 1943).—Heavy soil of open or brushy flats, or dry rocky slopes, 2000–6500 ft.; Creosote Bush Scrub, Pinyon-Juniper Wd., Joshua Tree Wd.; deserts of Inyo, Ventura, Kern, and San Bernardino cos.; Nev., Ariz. April–June. Plants from below 6000 ft. and from the Panamint Mts. to Cajon Pass have vermilion fls., while those in the e. Mojave have largely orange, but those from higher elevs. in the Panamint, Clark, and Providence mts. have yellow and constitute the var. *Múnzii* Jeps.

34. **C. clavàtus** Wats. [*C. c.* var. *avius* Jeps. *Mariposa c.* Hoov.] Stem coarse, zigzag, simple or branched, rarely bulbiferous, 5–10 dm. high; lower lvs. 1–2 dm. long, linear, upper reduced; bracts 4–8 cm. long, dilated at base; fls. 1–6, subumbellate, erect, cup-shaped, lemon-yellow, sometimes with red-brown markings on sepals and transverse line on petals above the glands; sepals lance-ovate, acute, 2.5–3.5 cm. long; petals broadly cuneate-obovate, with clavate hairs about the gland, 3.5–5 cm. long; gland circular, deeply impressed, surrounded by fringed membrane, densely covered with short processes with branched fungoid tips; fls. ca. 1 cm. long; anthers brownish, oblong, 8–10 mm. long; caps. lance-linear, acuminate, angled, 6–9 cm. long; seeds flattened; $n = 8$ (Beal & Ownbey, 1943).—Dry often rocky slopes, below 4000 ft.; Chaparral, Foothill Wd.; Sierra Nevada, Eldorado Co. to Mariposa Co., Coast Ranges, Stanislaus Co. to Los Angeles Co. April–June.

Var. **grácilis** Ownbey. Stem slender, 2–3 dm. high; petals sparsely bearded, 3–4 cm. long; anthers 4–7 mm. long.—Lower canyons, San Gabriel Mts., Los Angeles Co. May–June.

35. **C. Plúmmerae** Greene. [*C. Weedii* var. *purpurascens* Wats.] Stem rather slender, usually branched, not bulbiferous, 3–6 dm. high; basal lf. 2–4 dm. long, 10–15 mm. wide, usually withered at anthesis; cauline lvs. reduced upward; bracts like upper lvs.; fls. 2, sometimes 4, erect, broadly campanulate, pink to rose, drying purplish; sepals lanceolate, long-acuminate, 3.5–4 cm. long; petals broadly cuneate-obovate, erose-dentate, rarely fimbriate, conspicuously bearded with long yellow hairs in transverse band; gland circular, slightly depressed, nearly naked, bordered with ring of orange hairlike processes; fls. 9–11 mm. long; anthers lance-linear, acute to subapiculate, 10–14 mm. long; caps. linear, acute, angled, erect, 4–8 cm. long; seeds flat, finely reticulate; $n = 9$ (Beal & Ownbey, 1943).—Dry rocky slopes, often in brush, below 5000 ft.; Chaparral, Yellow Pine F.; Santa Monica Mts. to San Jacinto Mts. May–July.

36. **C. Weèdii** Wood. [*C. luteus* var. W. Baker. *C. citrinus* Baker.] Stem slender, usually branched, not bulbiferous, 3–6(–9) dm. high; basal lf. 2–4 dm. long, 10–15 mm. wide, withering before anthesis; cauline lvs. reduced upward, passing into bracts; fls. 2, sometimes 3–4, erect, open-campanulate, orange-yellow, flecked and often margined with red-brown; sepals ovate-lanceolate, attenuate, 2.5–3 cm. long; petals broadly cuneate-obovate, dentate to fimbriate, bearded on most of inner face with long yellow hairs; gland round, slightly depressed, almost naked, bordered with ring of long yellow hairs; fls. 8–12 mm. long; anthers lance-oblong, acutish, 10–14 mm. long; caps. linear, acute, angled, erect, 4–5 cm. long; seeds flat, finely reticulate; $n = 9$ (Beal & Ownbey, 1943).—Dry often heavy or rocky soil, below 5000 ft.; Chaparral; hills from Santa Ana Mts. to San Diego Co. May–July.

Var. **véstus** Purdy [*C. Weedii* var. *purpurascens* auth.] Petals subtruncate, red-brown to purplish, fringed with brown hairs; anthers apiculate; $n = 9$ (Beal & Ownbey, 1943). —Dry slopes; Chaparral; Monterey Co. to Ventura Co. July–Aug.

Var. **intermèdius** Ownbey. Petals purplish, rounded, fringed with dark or yellow hairs.—Hills, Orange Co. June–July.

37. **C. obispoénsis** Lemmon. [*C. Weedii* var. *o.* Purdy.] Stem slender, erect, branched, 3–6 dm. high; basal lf. 2–3 dm. long, 2–3 mm. wide; cauline linear, reduced upward; fls. 2–several, erect, opening flat; sepals reflexed, lanceolate, 1.5–2 cm. long; petals

oblong-ovate, deep orange to yellow, tipped purple-brown, 1–1.5 cm. long, fimbriate, with hairy tuft at apex; gland round, naked, slightly depressed, surrounded with ring of slender, hairlike processes; fils. 7–8 mm. long; anthers oblong, acutish, 3–5 mm. long; caps. linear, acute, angled, erect, 2.5–3.5 cm. long; seeds flat, finely reticulate.— Dry stony hills and canyons; Chaparral; San Luis Obispo Co. May–June.

21. Camássia Lindl. CAMAS

Perennial herbs with tunicated bulbs, a basal whorl of linear keeled lvs., and slender scapose stem. Infl. racemose, dense or lax. Pedicels jointed at apex and in axils of lance-acuminate, 5- to 10-veined, scarious bracts. Fls. rather large, blue, blue-violet, or white; perianth-segms. 6, somewhat spreading, 3- to 9-veined, persistent. Stamens 6, on the base of the segms.; fils. filiform, ca. ¾ the length of the perianth-segms.; anthers versatile, oblong-linear. Ovary 3-celled; style filiform, slightly surpassing stamens; stigma 3-lobed. Caps. subglobose to oblong, loculicidal. Seeds several in each cell, obpyriform, blackish, smooth or wrinkled. A N. Am. genus of 5 spp. The bulbs were much eaten by the Indians. Some spp. have horticultural merit. (*Camas*, or *quamash*, the name used by the nw. Indians.)

(Gould, F. A systematic treatment of the genus Camassia Lindl. Am. Midl. Nat. 28: 712–742, 1942.)
Fls. regular, usually 1–3 in bloom at one time; caps. not appressed to axis of raceme
                                                                                    1. *C. Leichtlinii*
Fls. somewhat irregular, with 5 perianth-segms. curving upward and the 6th downward, many fls. blooming at one time; caps. closely appressed to raceme-axis .... 2. *C. Quamash*

1. **C. Leichtlìnii** (Baker) Wats. ssp. **Suksdórfii** (Greenm.) Gould. [*C. S.* Greenm. *Quamasia S.* Piper.] Bulb 1.5–3.5 cm. thick; lvs. 2–6 dm. long, 5–25 mm. wide; scape 2–13 dm. tall; fls. 4–many; pedicels 1.5–5 cm. long, usually spreading-erect; fls. deep blue-violet to bright blue, rarely albino, the segms. 2–4 cm. long, twisting together over caps. after anthesis, but soon deciduous; anthers 4–7 mm. long; caps. oblong or ovoid, 1–3 cm. long; seeds 6–12 per locule, 2–4 mm. long; *n* = 15 (Gould, 1942).—Mt. meadows, usually 2000–8000 ft.; Montane Coniferous F.; Coast Ranges, Napa Co. n., Sierra Nevada, Mono and Tulare cos. n.; to B.C. May–Aug.

2. **C. Quamash** (Pursh) Greene. ssp. **lineáris** Gould. [*C. Q.* var. *l.* J. T. Howell.] Bulbs 1–3 cm. thick; lvs. 2–5 dm. long, 6–15 mm. wide, green on both surfaces; scape 2–8 dm. tall; pedicels 1–2.5 cm. long, incurving-erect in fr.; perianth-segms. 2–3.5 cm. long, deep blue-violet, connivent after anthesis and quite persistent; anthers 4–7 mm. long; caps. ovoid, 8–25 mm. long; seeds 5–10 per locule, 2–4 mm. long.—Wet meadows, below 2500 ft.; Coastal Coniferous F., Mixed Evergreen F.; N. Coast Ranges, Del Norte Co. to Marin Co. May–June. Coastal plants have pedicels to 2.5 cm. long and perianth-segms. quite thick when dry, while those from inland have pedicels to 1.5 cm. long and the dry segms. thinner.

Ssp. **breviflòra** Gould. Lvs. 1.5–3 dm. long, 10–17 mm. wide, glaucous above; scape 2–5 dm. tall; pedicels 5–15 mm. long; perianth 15–20 mm. long; anthers 4–5 mm. long. —Wet meadows, at 2500–7000 ft.; Montane Coniferous F.; Sierra Nevada, El Dorado Co. n., to Siskiyou and Modoc cos.; to s. Wash. May–July.

22. Odontóstomum Torr.

Erect herb with subglobose, deep-seated corm and flexuous stem. Lvs. mostly basal, linear, sheathing the stem. Fls. in bracteate racemes. Perianth tubular, with 6 spreading or reflexed oblong segms., white or yellowish. Stamens 6, inserted on the throat and alternating with as many short staminodia; fils. short, subulate; anthers basifixed, dehiscing apically. Ovary globose, sessile, with 2 ovules in each locule. Style filiform, deciduous; stigmas 3. Caps. obovoid, 3-lobed, loculicidal. Seed 1 in each cavity, obovoid, dark brown, subrugose. One sp. (Greek, *odous*, tooth, and *stoma*, mouth, referring to the staminodia.)

1. **O. Hartwégii** Torr. Corm ca. 2.5 cm. thick; stem 12–40 cm. high, simple or branched; lvs. 8–22 cm. long, 5–10 mm. wide, the upper reduced; bracts scarious, subulate, 3–10 mm. long; pedicels 3–5 mm. long, with a bractlet near middle; fls. numerous, tube nerved, 4–6 mm. long, segms. ca. as long, 5–6-nerved; stamens and

style ca. 1 mm. long, the stamen opposite the lower outer perianth-segm. longer and facing the others; caps. ca. 4 mm. long; $n = 10$ (Cave, 1949).—Locally frequent, in clay soil, sometimes rocky and later drying, up to 2000 ft.; Foothill Wd.; Sierran foothills, Butte Co. to Mariposa Co., Coast Ranges, Tehama and Napa cos. April–May.

23. Scoliopus Torr.

Low glabrous perennial herbs from a short slender rootstock. Stem short, underground. Lvs. mostly 2, broad, many-nerved with transverse veinlets and with purplish dots. Fls. ill-scented, in a sessile umbel, on long sharply angled pedicels; perianth-segms. 6, distinct, deciduous, the sepals broad, spreading, the petals narrow, erect. Stamens 3, attached to base of sepals; fils. filiform-subulate; anthers oblong, 2-celled, extrorse. Ovary narrow, 3-angled, 1-celled; style short; stigmas 3, linear, recurved, persistent. Caps. oblong-lanceolate, thin-walled, irregularly dehiscent. Seeds oblong, slightly curved, longitudinally sulcate-striate. Two spp. of the Pacific states. (Greek, *skolios*, crooked, and *pous*, foot, because of the tortuous pedicels.)

1. **S. Bigelòvii** Torr. Slink Pod. Brownies. Lvs. elliptic to oblong, 1–2 dm. long, 0.5–1 dm. wide, obtuse to acute, sheathing at base; pedicels 3–12, 1–2 dm. long, 3-angled, erect at anthesis, tortuous-recurved in fr.; sepals lance-ovate, 14–17 mm. long, greenish, with reddish veins; petals ca. as long; stamens 5–6 mm. long; stigmas 5–6 mm. long, spreading-recurved; caps. 15–18 mm. long; seeds 3 mm. long; $n = 7$ (Johansen, 1932).—Moist shaded slopes, below 1500 ft.; Redwood F.; Coast Ranges, Humboldt Co. to Santa Cruz Co. Feb.–March.

24. Tríllium L. Wake Robin. Trillium

Erect glabrous unbranched perennial herbs with short rootstocks scarred by the abscission of the old sheaths. Stem with scarious basal sheaths. Lvs. in a whorl of 3, subtending the sessile or peduncled, solitary, bractless, perfect fl. Perianth-segms. distinct; the sepals usually green, persistent; the petals white, pink, purplish, or yellow, deciduous or withering. Stamens 6; fils. short; anthers linear. Ovary 3-celled, with several to many ovules in each cell; styles 3, stigmatic along inner surface. Fr. a globose or ovoid berry. Seeds ovoid. A genus of ca. 30 spp., from Asia and N. Am. (Latin name referring to the 3 lvs. and 3-parted fls.)

Fls. sessile .... 1. *T. chloropetalum*
Fls. pedicelled.
  Lvs. sessile or subsessile; ovary winged .... 2. *T. ovatum*
  Lvs. petioled; ovary scarcely angled .... 3. *T. rivale*

1. **T. chloropétalum** (Torr.) Howell. [*T. sessile* var. *c.* Torr. *T. giganteum* var. *c.* Gates. *T. s.* var. *californicum* Wats.] Stems stout, 1 or more from same root, 3–5 dm. high; lvs. rhombic-ovate, 7–15 cm. long, nearly or quite as wide, often mottled with dark blotches, obtuse to sharply acute at apex, rounded at sessile base; fls. sessile; sepals oblong-lanceolate, 3–5.5 cm. long; petals mostly green or greenish-yellow, sometimes whitish or with some purple, especially in age, mostly oblanceolate, 4–9 cm. long, 1–3 cm. wide; anthers 15–25 mm. long; berry winged, becoming reddish, 2.5–3 cm. long; seeds ca. 3 mm. long; $n = 5$ (Warmke, 1937).—Brushy or wooded slopes, below 3500 ft.?; N. Coastal Scrub, Redwood F., Mixed Evergreen F.; outer Coast Ranges mostly from San Mateo and Napa cos. n., Sierra Nevada n. of Placer Co.; to Wash. April–May. Exceedingly variable and with very poorly marked:
Var. **gigantèum** (H. & A.) Munz. [*T. sessile* var. *g.* H. & A. *T. g.* Heller.] Petals obovate to oblanceolate, mostly deep red or lilac.—Chaparral; hills e. and sw. of San Francisco Bay to Monterey Co. Feb.–July.
Var. **angustipétalum** (Torr.) Munz. [*T. sessile* var. *a.* Torr. *T. giganteum* var. *a.* Gates.] Lvs. narrowed to petiolelike base; petals linear, mostly 2–10 mm. wide.—Below 4500 ft.; Montane Coniferous F.; Sierra Nevada, Fresno Co. to Siskiyou Co. (May–July); Foothill Wd., Lake, San Luis Obispo and Santa Barbara cos. (March–April).
2. **T. ovàtum** Pursh. [*T. californicum* Kell. *T. crassifolium* Piper. *T. Scouleri* Rydb. *T. venosum* Gates. *T. o.* var. *stenosepalum* Gates.] Stem 2–5 dm. high; lvs. rhombic-ovate, abruptly acuminate, subsessile, 5–15 cm. long, 3–14 cm. wide; pedicel erect, 2–7

cm. long; sepals green, lanceolate-oblong, 2.5–5 cm. long; petals white, turning rose, ascending, ovate to oblong, 2.5–5.5 cm. long; anthers 8–15 mm. long; berry somewhat winged, 1–2 cm. long; seeds 3–4 mm. long; *n* = 5 (Warmke, 1937).—Moist wooded slopes; Redwood F., Mixed Evergreen F.; Coast Ranges, Monterey Co. to Siskiyou and Shasta cos.; to B.C., Mont. Feb.–April. Variable.

3. **T. rivàle** Wats. Stems 1–2.5 dm. high; lvs. lance-ovate, acute to acuminate, the blades 2.5–5 cm. long, rounded or subcordate at base, on petioles 5–25 mm. long; pedicel erect or later nodding, 2.5–8 cm. long; sepals green, elliptic-lanceolate, 12–15 mm. long; petals white, or marked with rose-carmine, ascending, ovate-cordate, acute. 15–25 mm. long; anthers 4–5 mm. long; berry globose, scarcely angled; *n* = 5 (Warmke, 1937).—Rocky stream banks; Yellow Pine F.; Del Norte and Siskiyou cos.; sw. Ore. March–April.

25. Smìlax L. Greenbrier

Shrubby or herbaceous, perennial, climbing or straggling dioecious plants, with a usually tuberous rootstock. Stems often prickly. Lower lvs. reduced to scale, the upper entire or lobed, broad, 3–7-nerved, deciduous to persistent, commonly with paired tendrils near base. Fls. greenish or yellowish, regular, small, in peduncled axillary umbels; the segms. distinct, deciduous. Stamens 6, inserted on segm.-bases; anthers 1-celled, basifixed. Ovary 3-celled, with 1–2 ovules in each cell; style short or none; stigmas 1–3. Fr. a berry with 1–6 seeds. Ca. 200 spp., generally distributed. (Ancient Greek name of uncertain application.)

1. **S. califórnica** (A. DC.) Gray. [*S. rotundifolia* var. *c.* A. DC.] Stems woody, ± prickly, 1–3 m. long; lvs. ovate, subcordate, 5–10 cm. long, on petioles 10–15 mm. long; peduncles 2–5 cm. long; fls. 8–20, pedicels 6–10 mm. long; perianth ca. 5 mm. long; berries black, 6 mm. long; seeds 3–4 mm. long.—Thickets and stream banks, below 5000 ft.; Yellow Pine F., Mixed Evergreen F.; Napa Co., Trinity Co. to Siskiyou and Butte cos.; s. Ore. May–June.

14. Pontedèriàceae. Pickerel-Weed Family

Perennial aquatic or bog plants. Lvs. petioled, with rounded to elongate blades. Fls. perfect, 1 to many and spicate, regular to irregular, without bracts, subtended by leaflike spathes. Perianth with a tube, 6-lobed or -parted, petaloid, free from the ovary. Stamens 3 or 6, inserted on the perianth; anthers 2-celled, introrse. Ovary superior, 1- or 3-celled; style 1; stigma 3-lobed or 6-toothed. Fr. a caps. or ak.; endosperm copious. Six genera and ca. 20 spp., mostly in warmer parts of world.

Perianth ± tubular; stamens essentially alike.
    Lvs. grasslike; submerged aquatics .... 1. *Heteranthera*
    Lvs. rounded; floating plants .... 2. *Eichhornia*
Perianth almost completely parted; 1 stamen enlarged .... 3. *Monochoria*

1. Heteranthèra R. & P. Mud-Plantain

Ours a submerged plant with slender forked stems and linear sessile lvs. having thin sheaths tipped with small stipulelike appendages. Spathe 1-fld. Fl. small petaloid; perianth-lobes linear, subequal. Stamens 3, inserted on the throat. Ovary fusiform, ± 3-celled; ovules many; stigma 3-lobed. Caps. ovoid, enclosed in persistent withered perianth-tube. Seeds ovoid, ribbed. Ca. 10 spp., mostly trop. (Greek, *heteros*, different, and *anthera*, anther, since some spp. have unequal stamens.)

1. **H. dùbia** (Jacq.) MacM. [*Commelina d.* Jacq.] Stems flexuous, 2–10 dm. long; lvs. 5–15 cm. long, 1–5 mm. wide; fls. terminal on the branches; perianth light yellow, the tube 2–3 cm. long, segms. 5–6 mm. long; 2*n* = 30 (Bowden, 1945).—Reported from still water in Mendocino Co. and along the Pitt R. in Modoc, Lassen, and Shasta cos.; to Wash., Atlantic Coast.

2. Eichhórnia Kunth. Water-Hyacinth

Floating or rooting at nodes on mud. Lvs. floating or emersed, obovate, cordate to lanceolate. Fls. spicate, rarely paniculate; perianth funnelform with long or short tube.

Stamens 6, unequally inserted, some of them exserted. Ovary sessile, 3-celled, many-ovuled; style filiform. Six spp., in tropics. (Named for J. A. F. *Eichhorn,* 1779–1856, German statesman.)

1. **E. crássipes** (Mart.) Solms. [*Pontederia c.* Mart. *Piaropus c.* Britton.] Lvs. 1–12 cm. wide, ovate to rounded, slightly scabrous above; petioles inflated at base; scape 1–4 dm. high, sheathed near middle; fls. many, showy; perianth ca. 5 cm. long, 6-lobed, violet, the upper lobe enlarged and with patch of blue having yellow center; *n* = 16 (Taylor, 1925).—Occasionally natur. in sloughs and sluggish water, Sacramento and San Joaquin valleys, Santa Ana R. system; native of trop. Am., where it may be a serious pest interfering with navigation. June–Oct.

3. Monochòria Presl

Perennial herbs with short or creeping rootstock. Lvs. radical and solitary at the top of the emerged stem or branches. Perianth campanulate, 6-parted. Stamens 6, 1 usually largest with the fil. toothed on 1 side; anthers basifixed, slit terminal at last elongating. Ovary 3-loculed, many-ovuled. Ca. 6 spp., Old World trop. (Greek, *mono,* one, and *choris,* apart, referring to the enlarged stamen.)

1. **M. vaginàlis** (Burm.f.) Presl. [*Pontederia v.* Burm.f.] Rootstock short, spongy; lvs. variable, long-petioled, the blades linear to ovate, 5–10 cm. long, sometimes cordate at base, acuminate, 7–9-veined; peduncle emerging from the sheath of uppermost lf., bearing few to many fls. in a spiciform, at first short, raceme; terminal fl. opening first; pedicels 3–6 mm. long; perianth ca. 12 mm. across, blue, sprinkled with red, the 3 larger segms. obovate, the 3 smaller oblong; fil. of long stamen spurred; large anther dark blue, the others yellow; 2*n* = 52 (Morinaga & Fukushima, 31).—Weed in rice-fields, Butte Co.; introd. from se. Asia. Sept.

15. Aràceae. ARUM FAMILY

Herbs, but stems sometimes hard and woody, erect, prostrate or climbing. Lvs. large, mostly basal, simple or compound. Fls. reduced, often without perianth, crowded on a ± fleshy spadix which is usually surrounded by a large, often colored bract or spathe. Fls. usually imperfect, the ♂ above, the ♀ below, or perfect, or plants dioecious. Perianth of 4–6 scalelike segms. or lacking. Stamens 4–10; fils. short; anthers 2-celled. Ovary 1–several-celled; ovules 1–several per cell; style short or none; stigma terminal, usually minute and sessile. Fr. a berry or utricle. Seeds with or without endosperm. Ca. 100 genera and 1500 spp., mostly trop.

Plant floating; lvs. in dense rosettes .... 1. *Pistia*
Plant not floating; lvs. not in rosettes.
    Spathe white; lvs. cordate-ovate to sagittate-ovate .... 2. *Zantedeschia*
    Spathe yellowish; lvs. lance-oblong .... 3. *Lysichiton*

1. Pístia L. WATER-LETTUCE

Free-floating or sometimes rooting in mud. Lvs. simple, sessile, entire, oblong to cuneate with rounded ends and several ribs, green, velvety; spathes small, sessile or subsessile, axillary, white; spadix adnate to and shorter than spathe; plant monoecious; fls. unisexual, without perianth; stamens 2; ovary 1-celled, several-ovuled; fr. a berry. One sp. of trop. Am. (Greek, *pistos,* liquid, in reference to aquatic habitat.)

1. **P. Stratiòtes** L. Stoloniferous; lvs. 5–15 cm. long, plaited; spathe ca. 1 cm. long, constricted near middle; 2*n* = 28 (Blackburn, 1933).—Drainage canal on the Calif. side of the Colo. R., near Yuma, collected by *Peebles & Noble;* Tex. to Fla.

2. Zantedéschia Spreng. CALLA-LILY

Perennial herbs from thick rhizomes. Lvs. basal, long-petioled, the blade hastate to cordate-ovate in ours. Spathe showy, on peduncles equaling or exceeding lvs., appearing with the lvs., not deciduous. Spadix ♀ below, ♂ above. Perianth lacking. Stamens 2–3, free. Pistil 1- to more-celled, with ca. 4 ovules in each cell. Fr. berrylike. Eight spp., of S. Afr. (Named for F. *Zantedeschi,* writer on Italian botany, 1825.)

1. **Z. aethiòpica** (L.) Spreng. [*Calla a.* L.] Robust, to 12 dm. high, with shining dark green lvs.; fls. fragrant; spathe creamy or white, 1–2.5 dm. long; spadix yellow; $2n = 32$ (Ito, 1942).—Much cult. in Calif. and occasionally natur. March–May.

3. Lysichìton Schott.

Acaulescent herb, from a thick horizontal rootstock, the sap ill-scented. Lvs. large, oblong to elliptic, basal. Peduncle stout, shorter than lvs.; spathe sheathing at base, expanded and yellowish or whitish above. Spadix greenish; fls. perfect, crowded; perianth 4-lobed. Stamens 4. Carpels 2, united; ovary conical; ovules 2. Fr. berrylike, green. Two spp., 1 of w. N. Am. and 1 of e. Siberia. (Greek, *lusis,* loose, and *chiton,* a covering, alluding to the spathe.)

1. **L. americànum** Hult. & St. John. [*L. camtschatcense* Am. auth., not Schott.] YELLOW SKUNK-CABBAGE. Lvs. 3–15 dm. long, narrowed at base into a short petiole; fls. malodorous, appearing before the lvs.; spathe yellow to greenish-yellow, the expanded part 1–2 dm. long; peduncle to 3 or 5 dm. long; spadix cylindric, 5–14 cm. long in fr.—Open swamps and wet woods, near the coast, N. Coastal Scrub, Redwood F.; Santa Cruz Mts. to Del Norte Co.; to Alaska, Mont. April–June.

Àcorus cálamus L. SWEET-FLAG. Aromatic, with creeping rootstock and 2-ranked cauline lvs.; spathe a leaflike continuation of the scape.—Reported from Blue Lake, Humboldt Co.; transcontinental.

16. Lemnàceae. DUCKWEED FAMILY

Minute floating perennial aquatics, consisting of a fleshy or membranaceous, loosely cellular, thalluslike stem or frond, rootless or with 1 or more simple rootlets. Lvs. lacking. New stems produced from 2 lateral vegetative pouches or 1 terminal, the new frond being attached to the old by a short slender stalk and usually soon separating. Fls. very rare, the infl. consisting of 1 ♀ and 1 or 2 ♂ fls., borne in a saclike spathe in a flowering pouch on the edge of the upper surface of the frond. Perianth absent. Staminate fl. consists of 1 stamen, with 2–4 pollen sacs. Pistillate fl. of 1 flasklike pistil, with 1–several ovules. Fr. a 1–6-seeded utricle. Ca. 4 genera and 30 spp.

Rootlets present.
Frond 7–15-nerved, with several rootlets .... 1. *Spirodela*
Frond 1–5-nerved, with 1 rootlet .... 2. *Lemna*
Rootlets absent.
Frond thin, sickle-shaped to elongate .... 3. *Wolffiella*
Frond thick, globose or ellipsoid .... 4. *Wolffia*

1. Spirodèla Schleid. GREATER DUCKWEED

Fronds flattened, discoid, 5–15-nerved, 2–several-rooted. Vegetative pouches 2, triangular, opening as clefts in either margin of the basal portion of the frond. Two ♂ fls. and 1 ♀ borne in a flowering pouch and subtended by a saclike spathe. Fils. curving upward from the margin of the frond; anthers 2-celled. Pistil 1, naked, 2-ovuled. Fr. round-lenticular, with wing-margin. Spp. few, from all continents. (Greek *speira,* a cord, and *delos,* evident, referring to the roots.)

1. **S. polyrhìza** (L.) Schleid. [*Lemna p.* L.] Fronds solitary or in groups of 2–5, 3–8 mm. long, 3–4.5 mm. wide, flat and green above, subconvex and purplish beneath, usually sterile, with 4–several roots; $2n = 40$ (Blackburn, 1933).—Occasional in quiet water, below 5000(?) ft.; throughout Calif.; to B.C., Atlantic Coast; cosmopolitan.

S. oligorrhìza (Kurz) Hegelm. [*Lemna o.* Kurz.] Fronds with 2–3 roots.—Reported from near Berkeley; S. Hemis.

2. Lémna L. DUCKWEED

Fronds flattened, 1–5-nerved, each with 1 root and 2 vegetative pouches on the margin near the base. Fls. 3, borne in the flowering pouch, 2 ♂, each of 1 stamen, and 1 ♀, consisting of a single naked pistil. Ovary 1-celled, 1–several-ovuled. Fr. ribbed. Spp. ca. 8, all continents. (Possibly Greek *limnos,* lake, referring to living in swamps.)

Fronds oblong, long-stalked, 6–12 mm. long, mostly submerged ....1. *L. trisulca*
Fronds oblong-ovate to elliptical, not stalked, 2–5 mm. long, floating.

Outline of frond symmetrical or nearly so; frond usually with row of papillae along the midnerve.
The frond oblong-elliptical, 1-nerved or nerveless .... 2. *L. minima*
The frond oblong-obovate, ± faintly 3–5-nerved.
  Upper surface uniformly green.
    Air-chambers within the frond in 2 layers .... 3. *L. minor*
    Air-chambers within the frond in 1 layer .... 4. *L. perpusilla*
  Upper surface mottled or streaked .... 5. *L. gibba*
Outline of front oblique.
  The frond obliquely obovate, 3–5-nerved, with papillae .... 5. *L. gibba*
  The frond narrow, obscurely 1-nerved, without papillae .... 6. *L. valdiviana*

1. **L. trisúlca** L. Fronds oblong to almost lanceolate, 6–10 mm. long, 2–3 mm. wide, remaining connected and forming chainlike colonies, dentate toward tips, obscurely 3-nerved, usually submerged, with or without rootlets; $2n = 44$ (Blackburn, 1933).—Occasional in cold springs, streams and ponds, below 7500 ft., Santa Ana R. system and Bluff Lake (San Bernardino Mts.), Victorville, e. of Sierra Nevada to Modoc Co.; to B.C., Atlantic Coast; all continents.

2. **L. mínima** Phil. Fronds solitary, or in 2's or 4's, oblong-elliptical, 1.5–4 mm. long, 1–1.5 mm. wide, 1-nerved or nerveless, with row of papillae along nerve on upper surface, flat or slightly convex beneath.—Occasional but well distributed, pools below 5000 ft., San Diego Co. n. to Trinity Co.; to Ore., Wyo., Ohio, Fla., S. Am.

3. **L. mìnor** L. Fronds 1 to few together, round to oblong-obovate, thickish, with convex, green or purplish surfaces, 2–4 mm. long, ± faintly 3-nerved and sometimes with upper row of papillae, the apical one prominent; $2n = 40$ (Blackburn, 1933).—Stagnant pools and quiet water below 3500 ft., San Diego Co. n. to Wash.; all continents except S. Am.

4. **L. perpusílla** Torr. Fronds more plainly 3-nerved and with 1 layer of air spaces within.—Widely distributed in Calif.; to Atlantic Coast.

5. **L. gíbba** L. Fronds rounded to obovate, 1–4 in a group, 2–5 mm. long, 2–4 mm. wide, usually 3–5 nerved, convex above, pale beneath and ± gibbous; $2n = 64$ (Blackburn, 1933).—Fairly common in pools below 7000 ft., throughout Calif.; to Nebr., Tex., Mex.; Eurasia.

6. **L. valdivIàna** Phil. [*L. minor* var. *cyclostasa* Ell. *L. c.* auth.] Fronds thin, 1–2–8 in a group, oblong, subfalcate, asymmetrical, 2.5–4 mm. long, 1–1.5 mm. wide, without papillae, obscurely 1-veined.—Quiet water, below 7000 ft., San Diego Co. to Lake Co.; to Nebr., Atlantic Coast, S. Am.

Var. **abbreviàta** Hegelm. Fronds broadly elliptic or oval, 1.5–2.3 mm. wide, veinless. —S. Calif. to Va., S. Am.

3. Wolffiélla Hegelm.

Fronds minute, rootless, sickle-shaped or strap-shaped, thin, asymmetrical, curved, brown-punctate on both surfaces. Vegetative pouch 1, triangular, opening as a cleft in the basal margin of the frond. Fls. 2 in a pouch, the ♂ a single stamen, the ♀ a single pistil. Ca. 7 spp., subtrop. and trop.

Fronds sickle-shaped, 3–6 times as long as wide .... 1. *W. oblonga*
Fronds straight, 1–4 times as long as wide .... 2. *W. lingulata*

1. **W. oblónga** (Phil.) Hegelm. [*Lemna o.* Phil.] Fronds 1–2, rarely 3, tapering from the obliquely rounded base to the slightly narrower bluntly rounded apex, 2–4 mm. long, 0.5–1 mm. wide.—San Bernardino, *Parish* in 1881; Mex. to S. Am.

2. **W. lingulàta** (Hegelm.) Hegelm. [*Wolffia l.* Hegelm.] Fronds 1, rarely 2, oblong, tongue-shaped, 2.5–6 mm. long, 1.5–3 mm. wide.—Ponds and sloughs, below 2000 ft., San Mateo and San Joaquin cos. s.; to Mex. June.

4. Wólffia Horkel

Fronds floating, globular, ellipsoid, rootless, proliferating from a funnel-shaped pouch, but soon pulling apart, mostly 1–2 mm. long. Fls. 2 in a pouch from the upper surface, ♂ consisting of 1 stamen, ♀ of 1 pistil. Several spp., mostly of warmer waters. (J. F. *Wolff*, 1778–1806, German physician and botanist.)

1. **W. columbiàna** Karst. [*Bruneria c.* Nieuwl.] Frond globose, 0.5–1 mm. long, only

slightly emersed and this portion with 1–several stomates.—Dune Lakes, San Luis Obispo Co.; e. and cent. U.S. The smallest known flowering plants. Growing mixed with *Lemna*. Material from Oso Flaco L., San Luis Obispo Co., may be **W. arrhìza** (L.) Wimm.; it has a diam. of 0.7–1.5 mm. and a flat top to the fronds. In the Cent. V. occur plants with narrow fronds, a much flattened top and air-spaces bounded by cells with thicker walls; they may be **W. cylindràcea** Hegelm.

17. Agavàceae. AGAVE FAMILY

Mostly xerophytic plants, with a rhizome. Stem short to arborescent. Lvs. fibrous, narrow, often thick or fleshy, usually crowded on or near the base of the stem, entire or prickly on the margin. Fls. bisexual or unisexual, regular or somewhat irregular, racemose or paniculate, they and the branches subtended by bracts. Perianth-segms. usually partially united into a short or long tube, segms. subequal to unequal, petaloid, often fleshy. Stamens 6, on the tube or segm.-bases; fils. filiform to thickened, free; anthers linear, usually dorsifixed. Ovary superior or inferior, 3-celled; style slender, often very short; stigma more or less evidently 3-lobed. Fr. a berry or loculicidal caps. A family of ca. 19 genera and 500 spp., formerly partly placed in *Liliaceae* and partly in *Amaryllidaceae*.

Ovary superior; fls. white to purplish.
  Perianth 3–13 cm. long; fls. all perfect .... 1. *Yucca*
  Perianth less than 1 cm. long; fls. both perfect and imperfect .... 2. *Nolina*
Ovary inferior; fls. yellow to greenish .... 3. *Agave*

1. Yúcca L. SPANISH BAYONET. YUCCA

Trees or shrubs, simple or branched, caulescent to subacaulescent. Lvs. narrow, usually stiff, linear, with expanded base, and terminal spinose tip, ± persistent. Fls. large, pendent, numerous, in dense, elongate, terminal panicles, the perianth-segms. subequal, fleshy, withering-persistent, distinct, or somewhat united at base, lanceolate to ovate, the tips tending to curve inward. Stamens 6, much shorter than perianth; fils. fleshy. Ovary superior, 3-loculed. Style short with 3 stigmatic lobes or subcapitate stigma. Fr. a loculicidal or septicidal caps. or fleshy. Seeds obovoid or compressed, black, in 2 rows in each locule. Pollination by small Pronuba moths which work at night and deposit eggs in the ovary of a given flower after pollinating it. Ca. 40 spp., largely from the arid parts of N. Am. Many have yielded economic products such as fibers, material for splints, and protectors about base of young fr. trees. The fleshy frs. and the seeds are eaten by cattle and the frs. by man. The young fl.-stalks are sometimes roasted for food. (Haitian name for manihot.)

(McKelvey, Susan D. Yuccas of the southwestern U.S. 1: 1–150, 1938; 2: 1–192, 1947. Webber, J. M. Yuccas of the southwest. U.S.D.A., Agriculture Mon. 17: 1–97, 1953.)
Lvs. with free marginal fibers; fr. fleshy, eventually pendent. (Section Sarcocarpa)
  Pistil 6–8 cm. long; perianth 5–13 cm. long; plants acaulescent or nearly so ...... 1. *Y. baccata*
  Pistil 2–3 cm. long; perianth 3–5 cm. long; plants with definite trunk .... 2. *Y. schidigera*
Lvs. lacking free marginal fibers; fr. spongy or dry, spreading to erect.
  Plant with definite trunk, arborescent; style stout with 3 stigmas. (Section Clistocarpa)
                                          3. *Y. brevifolia*
  Plant practically stemless; style slender with capitate stigma. (Section Hesperoyucca)
                                          4. *Y. Whipplei*

1. Y. baccàta Torr. Mostly simple, stemless, sometimes in clumps with 2–6 short procumbent stems; fl.-stem 3–10 dm. long; lf.-cluster rather open, 6–7.5 dm. high, about twice as wide; lf.-base ca. 1 dm. long, ca. half as wide at middle, reddish, the blade 5–7.5 dm. long, 5–7 cm. wide if flattened, 2.5–4 cm. wide across concavity, often glaucous, with broad coarse recurved marginal fibers and thickened acute to acuminate apex ending in stout spine 1.5–7 mm. long; infl. ca. 6–7.5 dm. high, the scape 1–1.5 dm. long, 2.5–4 cm. thick, fl.-cluster proper fleshy, heavy, subglabrous, with red-purple tinge and with ca. 15 branches; bracts subtending pedicels 2.5–5 cm. long; pedicels 7–40 mm. long; perianth campanulate, red-brown outside, cream-white inside, 6–13 cm. long, the segms. united at base for 7–12 mm., lanceolate, thickened at center, the inner wider than the outer; fils. 3–4.5 cm. long, united at base, fleshy-pubescent; anthers

5–7 mm. long; pistil 6–8 cm. long; the style 5–7 mm. long; fr. ellipsoid, fleshy, 15–17 cm. long; seeds obovoid to compressed, thick, somewhat ridged, 8–12 mm. long.—Uncommon on dry slopes, 3000–4000 ft.; Joshua Tree Wd.; Clark, New York, and Providence mts., e. San Bernardino Co.; to Utah, Tex. May–June.

Var. **vespertìna** McKelvey. Dense, cespitose, in larger colonies, acaulescent, blue-green; infl. short, 3.5–4 dm. high, with 10–12 branches, red-purple.—With the sp., but more common at 4000–6000 ft., mts. of e. San Bernardino Co.; to Utah and Ariz. April–May.

2. **Y. schidígera** Roezl ex Ortgies. [*Y. mohavensis* Sarg.] MOHAVE YUCCA. Trunk 1–4.5 m. high, simple or branched; head of lvs. 6–13 dm. long, 1–2 m. wide; lf.-base light, 2.5–7.5 cm. long, 4–11 cm. wide at insertion; lvs. 3–15 dm. long, 2.5–4 cm. wide, yellow-green, with few coarse marginal fibers, the spine 7–12 mm. long; infl. 6–12 dm. high, the scape 1.5–5 dm. long, fl.-cluster with 15–25 branches; bracts subtending pedicels 5 cm. or less long; pedicels 1–3 cm. long; perianth globose, 3–4.5(–6) cm. long, cream or with purplish tinge; fils. 2–2.5 cm. long; anthers 3 mm. long; pistil 2–3 cm. long; style 1.5–3 mm. long; fr. 5–7(–10) cm. long, capsular; seeds thick, obovoid to compressed, somewhat ridged, 5–8 mm. long.—Dry rocky slopes and mesas below 5000 (7800) ft.; Chaparral, Creosote Bush Scrub; and mesas, coastal and desert areas, San Diego Co., Riverside and San Bernardino cos.; Nev., Ariz., L. Calif. April–May. Morongo V. region, San Bernardino Co. has plants with blue-green lvs.; the character is heritable.

3. **Y. brevifòlia** Engelm. in Wats. [*Y. Draconis* var. *arborescens* Torr. *Y. a.* Trel. *Clistoyucca a.* Trel. *C. b.* Rydb.] JOSHUA TREE. Tree 5–12(–15) m. high, branched mostly at 1–3 m. above ground; stem stout, red-brown or gray, the bark checked into small squarish plates; lf.-clusters near ends of branches, 0.3–1.5 m. long, lf.-bases whitish, 4–5 cm. wide at insertion, 2.5–4 cm. long, the blade 2–3.5 dm. long, rigid, denticulate, not fibrous on margin, with apical spine 7–12 mm. long; infl. 3–5 dm. long, the scape 1–1.5 dm. long; fl.-cluster 3–3.8 dm. thick, dense, heavy, with many branches; bracts subtending pedicels ca. 2.5 cm. long; pedicels 7–10(–20) mm. long; perianth 4–7 cm. long, fleshy, waxen, cream to greenish-white, the segms. united half its length; fils. ca. 1.2 cm. long; anthers 3.2 mm. long; pistil 2.5–3 cm. long; style 1.5 mm. long; fr. 6–10 cm. long, plump, but drying in age; seeds thin, compressed, smooth, 7–10 mm. long.—Dry mesas and slopes, 2000–6000 ft.; Joshua Tree Wd.; Mojave Desert to Owens V.; sw. Utah, w. Ariz. April–May.

Var. **Jaegeriàna** McKelvey. [*Y. b.* var. *Wolfei* Jones.] Plant 3–3.5 m. tall, branching at 7–10 dm. from the ground; lvs. mostly 1 dm. (up to 2 dm.) long; infl. ca. 3 dm. long.—Clark Mt. to New York Mts., e. Mojave Desert; s. Nev., sw. Utah. March–May.

Var. **Herbértii** (J. M. Webber) Munz. [*Y. b.* f. *H.* J. M. Webber.] Plants with many stems, 1–5 m. tall, arising from scaly underground rootstocks, forming clumps up to 10 m. in diam.—W. end of Antelope V., Los Angeles Co. to Monolith, and Walker Pass, Kern Co. April–May.

4. **Y. Whípplei** Torr. [?*Y. californica* Groenl. *Y. Ortgesiana* Roezl. *Hesperoyucca W.* Trel.] OUR LORD'S CANDLE. QUIXOTE PLANT. Plant quite acaulescent, simple, bulbous, with single fl.-stalk, the whole plant dying after fruiting; lf.-bases subrectangular, whitish to greenish, 4–7 cm. wide; lvs. in dense basal rosette, gray-green, 3–8(–10) dm. long, mostly 8–10 mm. wide, ± serrulate, flattened, rigid, the slender terminal spine 1–2 cm. long; fl.-stalk 1.5–2.5(–3.5) m. tall, the panicle compact, 5–12(–18) dm. long, averaging 3 dm. thick; branches numerous; bracts subtending pedicels 2–4 cm. long; pedicels 1–3 cm. long; fls. pendent, 2.5–3.5 cm. long, often tinged purple; fils. ca. 1.3 cm. long; anthers 3 mm. long; pistil 1–1.3 cm. long, style ca. 2 mm. long; caps. oblong, 3–4 cm. long; seeds compressed, very thin, smooth, 6–8 mm. long; $n = 30$ (5 long, 25 short) (McKelvey & Sax, 1933).—Dry often stony slopes, 1000–4000 ft.; mostly Chaparral, sometimes Coastal Sage Scrub, Creosote Bush Scrub; San Jacinto and Santa Ana mts. to n. L. Calif. April–May.

KEY TO SUBSPECIES

A. Crown branched, only the flowering branch dying after fruiting; both dead and living rosettes usually present.
  B. Stem branching by means of rhizomes 6–18 dm. long, forming colonies. Monterey and Santa Barbara cos. . . . . ssp. *percursa*

BB. Stem not forming rhizomes.
    C. The stem branching to form a clump of rosettes prior to flowering, with several infls. possible in one season. Desert edge, w. base of s. Sierra . . . . ssp. *caespitosa*
    CC. The stem branching to form new rosettes only after first fl.-stalk formed, with one infl. in a season. Near coast, Ventura and Los Angeles cos. . . . . ssp. *intermedia*
AA. Crown simple, the whole plant dying after fruiting.
    B. Fl.-stalk averaging 2–2.5 m. tall; lvs. 8–10 mm. wide. Riverside and Orange cos. s.
                                         Y. Whipplei
    BB. Fl.-stalk averaging 4 m. tall; lvs. 12–20 mm. wide. Coastal slopes of San Gabriel and San Bernardino mts. . . . . ssp. *Parishii*

Ssp. **percúrsa** Haines. Stem 3–6 dm. long, with rhizomes 6–18 dm. long; fl.-stalk averaging 3 m.; lvs. averaging 4.8 dm. long; fls. creamy-white.—Below 2000 ft. elev.; Chaparral, Coastal Sage Scrub; Monterey Co. to Santa Barbara Co. May–June.

Ssp. **caespitòsa** (Jones) Haines. [*Y. W.* var. *c.* Jones.] Stem branching above ground to form crowded cespitose clump; fl.-stalk averaging 2.5 m.; lvs. triquetrous, averaging 8 dm. long; fls. creamy-white.—Dry slopes, mostly 2000–4000 ft.; Joshua Tree Wd., Pinyon-Juniper Wd., Chaparral; edge of Mojave Desert from San Bernardino Mts. to Walker Pass, and in a slender form in the region of Kern and King rivers. May–June.

Ssp. **intermèdia** Haines. Stem branching by axillary buds only after it is well matured; fl.-stalk averaging 3 m.; lvs. flat, averaging 9 dm. long.—Below 2000 ft.; Coastal Sage Scrub, Chaparral; Santa Monica and Santa Susana mts. April–June.

Ssp. **Paríshii** (Jones) Haines. [*Y. Whipplei* var. *P.* Jones. *Y. graminifolia* Wood. *Y. W.* var. *g.* Trel.] Stem solitary; fl.-stalk averaging 4 m.; lvs. flat, averaging 1 m. long; fls. creamy-white.—On slopes and fans, 1000–8000 ft.; Chaparral, Coastal Sage Scrub, Yellow Pine F.; s. front San Gabriel and San Bernardino mts. May–July.

2. Nolìna Michx. NOLINA

Perennial with a yuccalike aspect, the stem forming a thick woody trunk or underground. Lvs. many, narrowly linear with greatly distended rigid bases. Fls. numerous, imperfect, borne on a nearly naked stem in an elongate congested, compound panicle, the main branches subtended by long white bracts. Pedicels jointed, subtended by minute scarious bracts. Perianth small, usually whitish, persistent, the 6 segms. distinct, 1-nerved. Stamens 6, usually reduced in the ♀ fls.; fils. short and slender. Ovary sessile, 3-lobed, with 2 ovules per locule; style short; stigmas 3. Caps. membranous, broadly 3-winged, bursting irregularly; seeds subglobose to ovoid, light-colored, often solitary. Ca. 25 spp. of sw. N. Am. (C. P. *Nolin*, joint author of an agricultural essay, 1755.)

Lvs. mostly 8–15 mm. wide above the expanded base.
  Stems forming underground platforms bearing many rosettes; lvs. flat . . . . 1. *N. interrata*
  Stems erect or ascending, each ending in a rosette; lvs. concave above . . . . 2. *N. Parryi*
Lvs. mostly 15–30 mm. wide above the expanded base.
  The lvs. hispid-serrulate on the margins, not fibrous; perianth 5 mm. long . . 2. *N. Parryi* ssp. *Wolfii*
  The lvs. smooth on margins and with shredding fibers; perianth 2–2.5 mm. long . . 3. *N. Bigelovii*

1. **N. interràta** Gentry. Plant with subterranean caudex, 2–3 m. long, and forming a platform bearing many aerial rosettes of 10–20 or more scabrous glaucous lvs. 7–9 dm. long, 8–15 mm. wide, the margins with minute teeth of 2 sizes; fl.-stalk 1.5–2 m. tall; scape-internodes 5–12 cm. long with bracts 2–4 dm. long; pedicels jointed above middle; perianth-segms. 3 mm. long, puberulent at the swollen tips; caps. 12–15 mm. wide; seeds yellowish, wrinkled, 5 mm. long, 4 mm. thick.—Dry slope; Chaparral; w. of Dehesa School, 8 miles e. of El Cajon, San Diego Co. June–July.

2. **N. Párryi** Wats. [*N. Bigelovii* var. *P.* L. Benson.] Caudex several-branched, erect, 0.3–1 m. tall, ca. 3–4 dm. thick; lvs. in dense crown, thickish, ± gray-green, concave, scarcely rigid, 6–10 dm. long, 8–16(–20) mm. wide, with serrulate margins, not becoming fibrous; total flowering stalk 1–1.5 m. long, the bracts 1–2 dm. long; infl. very dense, 5–6 dm. long; branches to 3 dm. long; fls. 5 mm. long; caps. 12–15 mm. wide, scarcely as long; seeds light brown, wrinkled, ca. 4 mm. long, 3 mm. thick; $n = 19$ (Lenz, 1950).—Dry slopes, below 3000 ft.; Chaparral, Coastal Sage Scrub, San Diego, Orange, Riverside, and Ventura cos.; and 3500–5500 ft., Pinyon-Juniper Wd., Chaparral, desert slopes of Santa Rosa and San Jacinto mts. April–June.

Ssp. **Wólfii** Munz. Flowering plant 4–5 m. tall, the caudex simple or few-branched,

8–20 dm. high, 6–9 dm. in diam.; lvs. green, flat, stiff, conspicuously hispid-serrulate on margins, 12–15 dm. long, 2.5–3.5 cm. wide; fl.-stalk 3–4 m. high, the base bare for 1–2 m. and 12–15 cm. thick, infl. proper 2–2.5 m. long, 0.8–1 m. thick; lower bracts 3–3.5 dm. long; perianth-segms. 5 mm. long, oblong; caps. 12–14 mm. wide, 10–11 mm. long; seeds light brown, oblong-spherical, 3–3.5 mm. long, slightly wrinkled.—Dry slopes, 3500–6000 ft.; Pinyon-Juniper Wd., Joshua Tree Wd.; Kingston, Eagle, and Little San Bernardino mts. (Mojave Desert), se. San Jacinto Mts. May–June.

3. **N. Bigelòvii** (Torr.) Wats. [*Dasylirion B.* Torr.] Caudex many-branched, 6–10 dm. high; lvs. flat, 8–12 dm. long, 1.5–3 cm. wide, glaucous, not or scarcely serrulate on margins, but with shredding brown fibers; fl.-stalks 1–3 m. high, the panicle dense, 6–15 dm. long; perianth 2–2.5 mm. long; caps. 8–10 mm. wide, somewhat shorter; seeds grayish-white, oblong-ovoid, 3 mm. long.—Dry slopes, below 3000 ft.; Creosote Bush Scrub; w. and n. edges of Colo. Desert, Old Woman Mts., Sheephole Mts., Eagle Mts. (Mojave Desert); to Ariz., L. Calif., Son. May–June.

3. Agáve L. Century Plant. Maguey

Acaulescent or sometimes short-stemmed perennials, from a thick fibrous-rooted crown. Lvs. fleshy, persistent, mostly in basal rosettes and spine-edged and -tipped. Fl.-stems tall, arising from the center of the rosette and forming panicles or spikes. Fls. numerous, relatively small, fleshy; perianth ± funnelform, with short tube, the 6 segms. narrow, subequal. Stamens inserted at throat or in tube; fils. filiform, exserted; anthers versatile. Ovary inferior, 3-loculed; style 1, awl-shaped; stigma capitate but 3-lobed. Caps. oblong, loculicidal; seeds many, superposed in 2 rows, obovate, black, thin, flat. A genus of perhaps 300 spp., from warmer dry parts of W. Hemis.; many of horticultural interest. Various fibers such as sisal and henequen are from Agave. The young fl.-stalks yield a liquid which is fermented for drinks like pulque and, if distilled, tequila. (Greek, *agave*, noble or admirable.)

(Berger, A., *in* Die Agaven, 1–288, 1915.)
Fls. ca. 4 in a cluster, the panicle subracemose and narrow .... 1. *A. utahensis*
Fls. many, in terminal clusters on the branches of an open panicle.
  Trunk subterranean; lvs. glaucous; prickles pale. Deserts .... 2. *A. deserti*
  Trunk rising above ground; lvs. deep green; prickles red. Coastal .... 3. *A. Shawii*

1. **A. utahénsis** Engelm. var. **nevadénsis** Engelm. ex Greenm. & Roush. [*A. n.* Hester.] Acaulescent, with subterranean trunk, usually cespitose; lvs. glaucous, curved inward at tips, 10–25 cm. long, 2–3 cm. wide, margined by ca. 8–10 lateral white teeth on each side, and with a slender terminal spine 3–8.5 cm. long; fl.-stalk 1.5–2.5 m. high, with narrow subracemose panicle; fls. yellow, 2.5–3.5 cm. long; perianth abruptly expanded above the tube, thus 1–2 mm. long, the segms. 12–15 mm. long; fils. 2–3 cm. long; anthers ca. 8 mm. long; caps. fusiform, 1.8–2 cm. long, 9–12 mm. thick; seeds 3.5–4 mm. long; $n = 30$ (Lenz, 1951).—Dry stony limestone slopes, 3000–5000 ft.; Shadscale Scrub, Joshua Tree Wd.; Ivanpah, Potosi, Clark, and Kingston mts., e. Mojave Desert. May–July.

2. **A. desérti** Engelm. [*A. consociata* Trel.] Densely cespitose, acaulescent, forming large colonies; lvs. gray-green, glaucous, triangular-lanceolate, 1.5–4 dm. long, edged with straight or curved pale prickles 3–7 mm. long, with dark terminal spine 1–3 cm. long; scape 2–5 m. high, the infl. slender; fls. yellow, 3.5–5 cm. long, tube 6–8 mm. long, segms. 12–18 mm. long; fils. 2–3 cm. long; anthers 10–15 mm. long; caps. 3–4.5 cm. long, short-pointed at apex; seeds ca. 5 mm. long; $n = 30$ (5 long, 25 short) (McKelvey & Sax, 1933).—Washes and dry rocky slopes, below 5000 ft.; Creosote Bush Scrub, Shadscale Scrub; w. edge Colo. Desert to Providence, Old Dad, Granite, and Whipple mts. of s. Mojave Desert; n. L. Calif., Ariz. May–July.

3. **A. Shàwii** Engelm. Cespitose, the trunks 1–3 dm. long, leafy; lvs. ovate to lance-ovate, green, glossy, openly concave, 2–5 dm. long, with red hooked prickles 3–10 mm. long and terminal spine 2–2.5 cm. long; fl.-stems 1–3.5 m. long, the panicle congested; bracts and buds purplish-brown; fls. greenish-yellow, 7–10 cm. long; perianth-tube 14–18 mm. long, the segms. 1.5–2 cm. long; fils. dilated in middle and toward base, 6–7 cm. long; anthers 2–3 cm. long; caps. 4–6 cm. long, oblong; seeds 6–8 mm. long;

$n = 30$ (5 long, 25 short) (Lenz, 1950).—Dry coastal bluffs and slopes; Coastal Sage Scrub; near Mexican boundary, where now quite extinct; n. L. Calif. Sept.–May.

18. Pálmae. PALM FAMILY

Ours trees with fibrous roots and cylindrical unbranched trunks. Lvs. in ours large, fan-shaped, long-petioled, plaited in early stages, borne in a large apical tuft. Plants commonly monoecious, the fls. small, on large compound axillary spadices, surrounded by large spathes. Perianth of 6 segms. in 2 series, usually firm in texture. Stamens 6; fils. dilated, ± united at base. Carpels 3, ± united, each 1-ovuled. Fr. a drupe or berry. Ca. 140 genera and some thousands of spp., of the warmer parts of the world, many of great economic and horticultural value.

1. Washingtònia Wendl. FAN PALM

Tall, columnar, the trunks commonly clothed with a dense shag or thatch of dead drooping lvs. which may be burned away. Lvs. large and heavy, variously filiferous, the petioles stout, flattish, ± spiniferous on the margins and projected into the blade with an "arrowhead" point or hastula. Spadix long, projecting from a flattened spathe 1 m. or so long. Fls. whitish, ca. 8 mm. long including the exserted stamens; calyx persistent, tubular; corolla-lobes reflexed, later deciduous. Style straight, erect, elongate, apically 3-toothed. Fr. hard, short-oblong, 8–10 mm. long, blackish; seed brown. A genus of sw. N. Am.; 2 spp. (Named in honor of George *Washington*.)

(Bailey, L. H. Washingtonia. Gentes Herb. 4: 52–82, 1936.)
    1. **W. filífera** (Lindl.) Wendl. [*Pritchardia f.* Lindl. ex André. *P. filamentosa* Fenzi. *Neowashingtonia filamentosa* Sudw. *W. filifera* var. *robusta* Parish, not *W. r.* Wendl.] Tree 10–15(–25) m. high, the trunk 6–10 dm. thick; lvs. gray-green; the petioles 1–2 m. long, ca. 5 cm. wide, armed along lower half or farther; blades 1–2 m. long, with 40–60 folds, segmented ca. to the middle, copiously fibrous; spadices to 3 or 4 m. at maturity; $2n = 36$ (Sato, 1952).—In groves in moist alkaline spots about seeps, springs, and streams, below 3500 ft.; Creosote Bush Scrub; w. and n. edge of Colo. Desert, Turtle Mts., region of Twentynine Palms on Mojave Desert; w. Ariz., n. L. Calif. Much planted in subtrop. regions. June.

19. Sparganiàceae. BUR-REED FAMILY (Fig. 127)

Perennial herbs from a creeping rhizome. Stems simple or branched, leafy. Lvs. elongate, alternate, sheathing at base, floating and flaccid or erect and stiffer. Fls. unisexual, densely crowded in globose heads on the upper stem and branches, the ♂ heads above the ♀. Perianth of a few chaffy elongate scales. Staminate fls. of 3 or more stamens; fils. free or partly united; anthers oblong, basifixed. Pistillate fls. with a sessile mostly 1-celled ovary narrowed at base; style simple or forked; ovule 1. Frs. indehiscent, crowded, narrowed at base, nutlike. Seed with mealy endosperm. One genus.

1. Spargànium L. BUR-REED

Characters of family. Ca. 15 spp., of aquatic habitats, temp. and cold regions of N. Hemis. and Australasia. (Greek, *sparganion*, a swaddling band, referring to the long narrow lvs.)

Stigmas 2; mature aks. rather truncate at apex .... 1. *S. eurycarpum*
Stigma 1; mature aks. tapered at apex.
    Stipes and beaks less than 1 mm. long; ♂ head 1 .... 2. *S. minimum*
    Stipes and beaks 2 mm. or more long; ♂ heads 2–several.
        Lvs. rounded on back, 1.5–4 mm. wide, floating, the middle and upper enlarged and sheathing at the base .... 3. *S. angustifolium*
        Lvs. mostly flat, 5–12 mm. wide, usually erect, scarcely inflated at base . 4. *S. multipedunculatum*

    1. **S. eurycárpum** Engelm. [*S. Greenei* Morong. *S. e.* var. *G.* Graebn. *S. californicum* Greene.] Stems 5–25 dm. high, erect; lvs. equaling or shorter than the stem, flat, some-

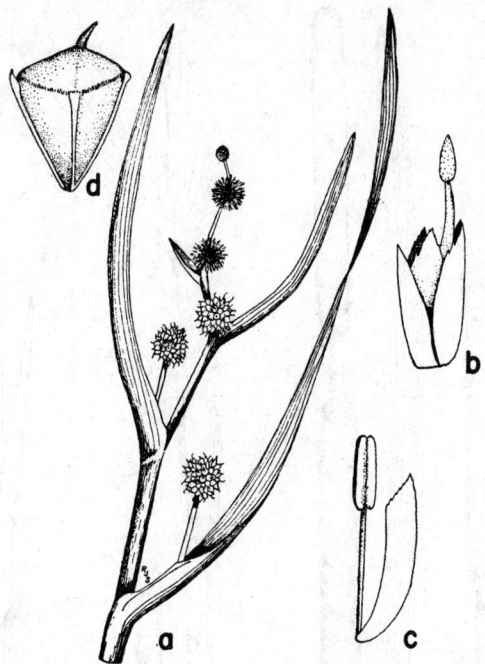

Fig. 127. SPARGANIACEAE. *Sparganium: a,* upper part of stem, $\times$ ca. ½, with heads of ♂ fls. above and of ♀ fls. below; *b,* ♂ fl. with perianth of subtending bracts; *c,* single stamen; *d,* nutlike fr.

what keeled beneath, 7–12 mm. wide; infl. branched; ♂ heads 5–many; ♀ heads 1–4 on a branch, sessile or peduncled, 2–3 cm. in diam. when mature; perianth-scales almost as long as aks.; aks. sessile, obpyramidal, subtruncate at summit, 6–9 mm. long, the beak ending in 2 stigmas.—Swamps and along streams, at low elevs., w. of the desert regions; L. Calif. to B.C., Atlantic Coast. April–Aug. The material near the coast seems often to have the aks. more rounded at the summit and is the basis of the var. *Greènei.*

2. **S. mínimum** Fries. Stems floating and elongate to 8 dm., or emergent and shorter; lvs. dark green, flat, mostly floating, ca. as long as stem, 3–7 mm. wide; infl. simple; ♂ head solitary; ♀ heads 1–3 in upper axils, ca. 1 cm. in diam. when mature; aks. short-stipitate, ca. 6 mm. long, the stipe and beak each ca. 1 mm. long, with 1 stigma; $2n = 30$ (Wulff, 1938).—Doubtful for Calif.; Ore to Alaska, Atlantic Coast; Eurasia.

3. **S. angustifòlium** Michx. [*S. simplex* var. *a.* Torr.] Stems floating, slender, 2–10 dm. long; lvs. flaccid, largely floating, usually longer than stem, 2–4(–5) mm. wide, with the nerves 0.2–0.8 mm. apart, rounded on back, the upper ones with enlarged sheathing bases; infl. simple; ♂ heads 2–6; ♀ 2–4, the lower on supra-axillary peduncles, 1.2–2 cm. thick when mature; aks. fusiform, 5–6 mm. long, stipitate, the beak ca. 1 mm. long; $2n = 30$ (Löve & Löve, 1942).—Ponds and slowly moving water, 4000–11,200 ft.; Montane Coniferous F.; San Bernardino Mts., Sierra Nevada, Salmon Mts.; to Alaska, Atlantic Coast. July–Aug.

4. **S. multipedunculàtum** (Morong) Rydb. [*S. simplex* var. *m.* Morong. *S. s.* auth., not Huds.] Stems 4–10 dm. high, erect; lvs. flat, commonly exceeding stem, 5–12 mm. wide, with nerves 0.8–2 mm. apart, lvs. scarcely inflated at base; infl. simple; ♂ heads 4–8; ♀ 2–5, the lower usually supra-axillary and peduncled, 18–25 mm. in diam. when mature; aks. fusiform, stipitate, 4–5 mm. long, the beak ca. 1.5 mm. long, with 1 stigma.—Swamps and ponds, below 8000 ft., Sierra Nevada (Eldorado Co.) and n., Mendocino Co. to Del Norte Co.; to Alaska, Atlantic Coast. June–Aug.

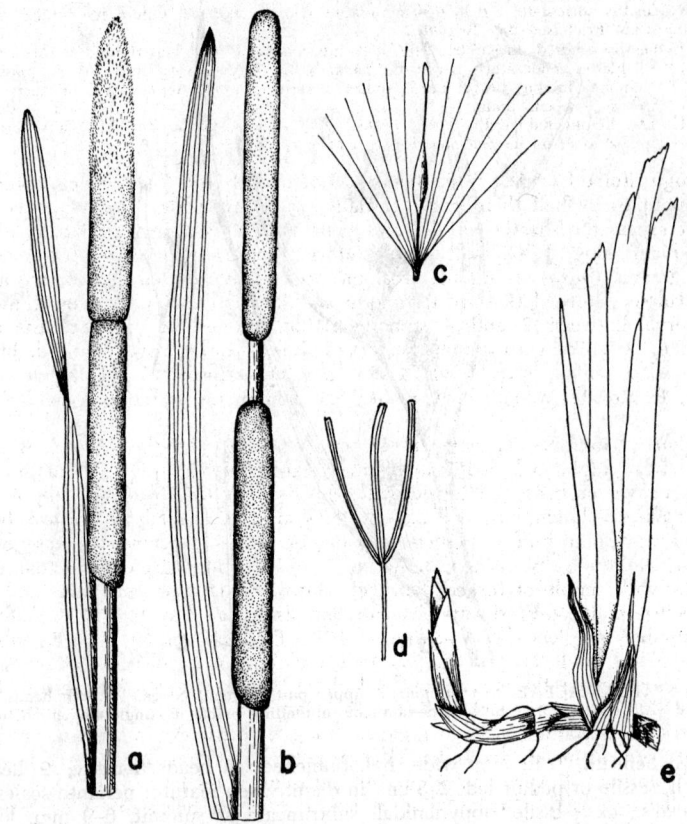

Fig. 128. TYPHACEAE. *Typha: a,* sp. with upper ♂ spike contiguous with lower ♀ spike; *b,* sp. with interval between ♂ and ♀ spikes; *c,* ♀ fl. with stipitate ovary and basal hairs; *d,* ♂ fl. with 3 stamens; *e,* part of creeping rhizome and bases of lvs.

20. Typhàceae. Cat-Tail Family (Fig. 128)

Tall perennial herbs from creeping rhizomes. Stems simple, submerged at base, cylindrical, jointless. Lvs. mostly radical, elongate-linear, alternate, rather thick and spongy. Fls. unisexual, anemophilous, very numerous, crowded in a terminal elongate spike, the ♂ above, ♀ below. Perianth of many slender jointed threads. Staminate fls. with 2–5 stamens; fils. free or connate; anthers linear, basifixed. Fertile ♀ fls. with 1-celled stipitate ovary, narrowed into a style with narrow or ligulate stigma. Many ♀ fls. sterile and with abortive terminal ovaries and perianth hairs beneath. Fr. dry, tardily dehiscent. Seed striate, with mealy endosperm. One genus.

1. Tỳpha L. Cat-Tail

Characters of family. Ca. 10 spp. of fresh water and marshy places, temp. and trop. regions. (The ancient Greek name.)

(Hotchkiss, N., & H. L. Dozier. Taxonomy and distribution of N. Am. cat-tails. Am. Midl. Nat. 41: 237–253, 1949.)

A. Lvs. mostly 5 mm. wide, strongly convex on back, dark green; ♀ spikes much overtopped by lvs., and with bracts very dark brown, opaque and firm; lf.-sheaths auriculate. Cent. Calif.
                                                                                        1. *T. angustifolia*
AA. Lvs. 6–15 mm. wide; ♀ spikes not or moderately overtopped by lvs. and with bract light brown and translucent, or lacking.

B. Lv.-sheaths auriculate, the blades moderately convex on back, blue-green; bracts of ♀ spikes none. San Francisco Bay Region .... 2. *T. glauca*
BB. Lf.-sheaths tapered to subtruncate, not auriculate. Widely distributed.
   C. Lf.-blades moderately convex on back, light yellow-green; bracts of ♂ spikes cuneate, laciniate, brown; bracts of ♀ spikes ovate and apiculate; interval between ♂ and ♀ spikes 1–4 cm. long .... 3. *T. domingensis*
   CC. Lf.- blades flat, light green; bracts of ♂ spikes simple, hairlike, white; bracts of ♀ spikes none; no interval between ♂ and ♀ spikes .... 4. *T. latifolia*

1. **T. angustifòlia** L. NAIL ROD. Stems 1–1.5 m. tall; lvs. 7–13, much exceeding ♀ spikes, sometimes by half their length, usually ca. 5 mm. wide, strongly convex on the back, dark green, the sheaths usually auriculate at their summit; ♀ portion of mature spike 8–20 cm. long, 1.3–2 cm. thick, uniform, dark brown or red-brown in color, becoming greenish-brown as stigmas wear off, and finally mottled dark brown and buff; bracts spatulate, blunt, dark brown, opaque and firm; stigmas dark brown, linear, not fleshy; interval between ♀ and ♂ spikes variable, but usually ca. twice the diam. of the ♀ spike; ♂ spike with simple or forked, linear, brown bracts and light lemon-yellow, 1-celled pollen; *n* = 15 (S. Galen Smith).—Mostly in subalkaline water at low elevs.; Freshwater Marsh; cent. Calif.; common in ne. states, e. Canada; Eurasia. June.

2. **T. glaùca** Godron. [*T. angustifolia* var. *elongata* (Dudl.) Wieg.] BLUE-FLAG. Plants 2–3.5 m. high; lvs. 8–12, moderately exceeding ♀ spikes, 6–12 mm. wide, moderately convex on back, medium to dark bluish-green, the sheaths usually auriculate; mature ♀ spike 10–25 cm. long, 1.8–2.5 cm. thick, uniform, usually red-brown, becoming mottled red-brown and buff as stigmas wear off; bracts usually none; stigmas red-brown, lance-linear, slightly fleshy; interval between ♀ and ♂ spikes usually less than diam. of ♀; ♂ spike with simple or forked, hairlike, whitish to light brown bracts and golden-yellow 1-celled pollen.—Freshwater Marsh; San Francisco Bay region; e. states. June.

3. **T. domingénsis** Pers. [*T. truxillensis* HBK. *T. bracteata* Greene. *T. angustifolia* many references.] Plant 2–3 m. tall; lvs. 6–9, equaling or slightly exceeding ♀ spikes, 6–12 mm. wide, moderately convex on back, light yellowish-green, the sheaths tapering into the blade; mature ♀ spike 15–25 cm. long, 1.5–2.2 cm. thick, uniform, light cinnamon-brown, becoming buffy or grayish; bracts light brown, translucent and spongy; stigmas light brown, linear, not fleshy; interval between ♀ and ♂ spikes usually ca. the diam. of the former; ♂ spikes with cuneate, laciniate, brown bracts and golden-yellow 1-celled pollen.—Widely distributed in subsaline habitats below 5000 ft.; Freshwater Marsh; to L. Calif., Ore., Utah, Kans., s. Atlantic Coast, S. Am.; Eu. June–July.

4. **T. latifòlia** L. SOFT-FLAG. Plant 1–2.5 m. high; lvs. 12–16, moderately exceeding ♀ spikes, 8–15 mm. wide, nearly flat, light green; the sheaths tapering into the blade or truncate; mature ♀ spike 10–18 cm. long, 1.8–3 cm. thick, often thickened upward, dark green-brown to red-brown, becoming whitish as stigmas wear off; bracts none; stigmas medium to dark brown, lance-ovate, fleshy; no interval between ♀ and ♂ spikes; ♂ spikes with simple, hairlike, white bracts, and deep orange-yellow, 4-celled pollen; 2*n* = 30 (Harada, 1947).—Freshwater Marsh; throughout Calif. below 5000 ft.; to Alaska, Atlantic Coast; Eu. June–July.

21. Amaryllidàceae. AMARYLLIS FAMILY (Fig. 129)

Perennial herbs, with bulbs, corms, or rhizomes. Stems well developed or not. Lvs. radical or cauline, alternate, mostly linear, entire. Fls. in scapose umbels, subtended by ± spathaceous bracts. Fls. usually perfect, the perianth of 6 segms. or lobes, usually petaloid. Stamens usually 6 (sometimes 3 sterile or lacking) inserted in the throat or on the base of the segms.; fils. filiform to thickened; anthers introrse. Ovary superior or inferior, 3-celled, ovules usually many, on axillary placentation; style 1; stigma usually 3-lobed. Fr. a caps. or berry, usually loculicidal. Ca. 90 genera and 1200 spp., of temp. and warm regions. Many are valuable horticulturally and economically.

Perianth-segms. distinct or nearly so; anthers versatile.
  Fils. not appendaged; pedicels not jointed.
    Pedicels not subtended by bractlets within the main spathelike bracts; plants with onionlike odor and taste .... 1. *Allium*
    Pedicels subtended by bractlets above the main spathelike bracts; plants lacking onionlike odor and taste .... 2. *Muilla*

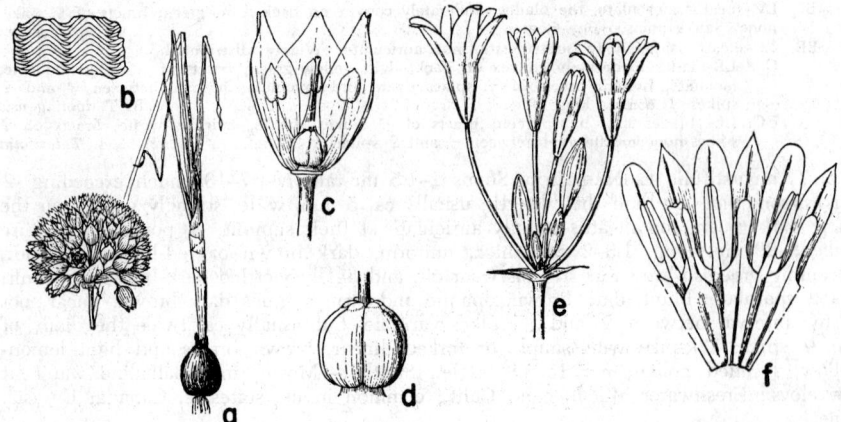

Fig. 129. AMARYLLIDACEAE. *Allium serratum: a,* habit, × ½, for bulb, stem, lvs., umbellate infl.; *b,* bulb-coat, × 15, with transverse-serrate reticulation; *c,* fl., × 3, with 3 perianth-segms. removed, showing superior ovary, fils. dilated at base; *d,* ovary, × 8, with small cent. crests. *Brodiaea coronaria: e,* umbellate infl. with basal spathelike bracts, × ½; *f,* detail of inside of fl., × 1, with 3 outer and 3 inner perianth-segms. different, fertile stamens 3, staminodia or sterile stamens 3, pistil with superior ovary and 3-lobed stigma.

Fils. surrounded at base by cup-shaped winged appendages; pedicels jointed ...... **3. Bloomeria**
Perianth-segms. united into definite basal tube; anthers often basifixed.
    Fils. separate, not united to form a corona. Common and widely distributed .... **4. Brodiaea**
    Fils. united to form a tubular corona with erect bifid segms. between the anthers. Rare, e. desert
    region .... **5. Androstephium**

1. Állium L. WILD ONION

Prepared with the collaboration of Marion Ownbey.

Scapose plants from tunicated bulbs. Lvs. mostly basal, linear, flat or terete, sometimes hollow, sometimes convolute-filiform. Herbage with taste and smell of onions. Fls. variously colored, small, few to many in a terminal umbel, which is subtended by a scarious, sometimes colored, 1-, 2-, 3- or more-lvd. sheath of ± connate bracts. Pedicels rather slender, not jointed. Perianth persistent, its segms. nearly or quite distinct, 1-nerved, erect to spreading. Stamens 6, usually attached to base of perianth; fils. often dilated at base. Pistil with 3-celled superior ovary, slender style, and entire to 3-lobed stigma. Fr. a loculicidal caps., tending to be obovoid-globose, obtusely 3-lobed, often with terminal crests. Seeds obovoid, wrinkled, black, finely cellular-punctate under a lens. A large genus of perhaps 500 spp., widely spread in the N. Hemis. Many from the Old World grown in the garden for food, as onions, garlic, chives, shallots, leeks, etc.; others as ornamentals. Some of our Am. spp. are well-flavored. (Latin for garlic.) The so-called reticulations on the bulb-coats characteristic of some spp. are the cells of the persistent inner epidermis and become evident only if the outer layers of the bulb-coat have been removed by decay or otherwise. The chromosome nos. indicated are from the unpublished work of Dr. Hannah C. Aase.

A. Lvs. mostly 2–several, if solitary, strongly flattened, or the ovary not prominently crested; crests
    various.
    B. Lvs. concave-convex to subterete, never falcate, if flattened, more than 2 in no.; scape
      persistent at maturity.
      C. Bulbs oblong, long-necked, clustered on a ± well developed rhizome.
        D. Scapes 1–4 dm. high; stamens included. Low elevs. .... 1. *A. haematochiton*
        DD. Scapes 5–10 dm. high; stamens exserted. Montane .... 2. *A. validum*
      CC. Bulbs round or ovoid, mostly separate or enclosed only by bulb-coats in common.
        D. Ovary prominently 6-crested; reticulations on bulb-coats with thin minutely sinuous
        walls.
          E. Lvs. 3–10 mm. broad, usually green at anthesis; perianth-segms. papery in fr.,
          tips with neither strongly involute margins nor a pronounced keel.

  F. Perianth-segms. mostly 9–10 mm. long, elliptic-ovate, acute, thin; pedicels once to twice the length of the fls. ...... 3. *A. membranaceum*

  FF. Perianth-segms. mostly 7–8 mm. long, lance-ovate, acuminate, thicker; pedicels twice to thrice the length of the fls. .... 4. *A. bisceptrum*

 EE. Lvs. 1–5 mm. wide, usually withering at anthesis; perianth-segms. becoming rigid, the tips strongly involute and with a pronounced keel.

                5. *A. campanulatum*

DD. Ovary crestless or less prominently 3- or 6-crested; reticulations on bulb-coats thick-walled, prominent under a lens.

 E. Reticulations quadrate-polygonal, ± isodiametric.

  F. Perianth-segms. 8–12 mm. long, the inner serrulate at tip, the outer recurved; bracts 1.5–2 cm. long. N. Calif. .... 36. *A. acuminatum*

  FF. Perianth-segms. 5–8 mm. long, not serrulate; bracts 8–12 mm. long. S. Calif.

   G. Pedicels 5–10 mm. long; scape 10–15(–20) cm. tall. Marin Co. to Santa Barbara Co. .... 37. *A. lacunosum*

   GG. Pedicels 10–20 mm. long; scape 15–30 cm. tall. Greenhorn Mts., Kern Co. to Mojave R., San Bernardino Co. .. 38. *A. Davisiae*

 EE. Reticulations transversely elongate, commonly in regularly serrate rows.

  F. Ovary without crests or very obscurely crested; reticulations of bulb-coats rather irregularly arranged.

   G. Perianth 5 mm. long; stamens almost as long. Monterey region.

                15. *A. Hickmannii*

   GG. Perianth 5–14 mm. long; stamens definitely shorter.

    H. Bulbs arising from slender lateral rootstocks; perianth-segms. oblong-ovate, 10–15 mm. long. Coast Ranges n. of Monterey.

                16. *A. unifolium*

    HH. Bulbs not arising as above; perianth-segms. lance-ovate.

     I. Perianth 5–9 mm. long, pale, becoming thin and membranous. Sierra Nevada to Tehachapi Mts.

              17. *A. hyalinum*

     II. Perianth 9–12 mm. long, rose to rose-purple, less papery. S. Calif. to L. Calif. .... 18. *A. praecox*

  FF. Ovary with ± evident mostly cent. crests; reticulations of bulb-coats usually very regularly serrate and in vertical rows.

   G. Perianth 6–7 mm. long, becoming papery; fls. many in subglobose umbel; ovary with 6 lateral crests .... 19. *A. amplectens*

   GG. Perianth usually 8–14 mm. long, not papery in age (except in *A. serratum*); fls. in open umbels; ovary with 3 cent. crests.

    H. Inner perianth-segms. undulate-crisped or serrulate.

     I. Outer and inner perianth-segms. similar in width, linear-lanceolate, ± attenuate, the inner commonly serrulate; scape sometimes laterally attached to bulb. Mostly Lake Co. n. .... 20. *A. Bolanderi*

     II. Outer perianth-segms. much wider than inner, oblong-ovate, the inner usually undulate-crisped; scape terminal on bulb. Mostly from Santa Clara Co. to Kern Co.

              21. *A. crispum*

    HH. Inner perianth-segms. plane and entire.

     I. The perianth-segms. lance-ovate, acuminate. Spp. of the interior.

      J. Pedicels 15–30 mm. long; umbel open; perianth 10–13 mm. long, not papery ..... 22. *A. peninsulare*

      JJ. Pedicels 8–14 mm. long; umbel congested; perianth 7–9 mm. long, papery .... 23. *A. serratum*

     II. The perianth-segms. ovate-oblong, blunt; pedicels mostly less than 10 mm. long and umbel congested. Outer Coast Ranges .... 24. *A. dichlamydeum*

BB. Lvs. flat, ± falcate, usually much longer than the short scape, 1–2 in no., breaking off with the scape at the surface of the ground at maturity; ovary crestless or rather obscurely 3- or 6-crested.

 C. Lvs. 2.

  D. Perianth-segms. linear, linear-lanceolate or linear-oblong, more than 4 times as long as broad; stamens ⅘ as long as the perianth or longer.

   E. Segms. connivent above the ovary, narrowly lanceolate, the long-attenuate tips infolded and widely spreading; lvs. 8–15 mm. wide .. 6. *A. platycaule*

   EE. Segms. not connivent above the ovary, tips not strongly infolded; lvs. mostly less than 5 mm. wide.

    F. Perianth-segms. narrowly lanceolate, long-attenuate; bracts usually 2. E. of Sierra Nevada .... 7. *A. anceps*

    FF. Perianth-segms. narrowly oblong, not attenuate; bracts 3. W. of Sierra Nevada .... 8. *A. yosemitense*

  DD. Perianth-segms. lanceolate, ovate or oblong, less than 4 times as long as broad, if narrower and 4 or more times as long as broad, the stamens much shorter.

   E. Segms. oblong or oblong-lanceolate, rarely elliptic-ovate, mostly obtuse, tips usually not involute.

    F.   Perianth-segms. narrowly oblong to lanceolate, thin, plane, midnerve not thickened; bracts often 3. Mt. Shasta, Sierra Nevada and s.
                                                                9. *A. tribracteatum*
  FF. Perianth-segms. elliptic-oblong to oblong-lanceolate, sometimes broader, thicker, keeled, the midnerve thickened; bracts 2. Mostly e. of Sierra Nevada or n. of Mt. Shasta . . . . 10. *A. parvum*
  EE. Segms. lanceolate to lance-ovate, tips usually involute or at least infolded, thus seemingly acute, acuminate or attenuate.
    F.   Fls. 6–10 mm. long, white to deep rose; segms. acute or acuminate; lvs. mostly less than 5 mm. wide.
      G. Stamens ca. the length of the very short (6–8 mm.) perianth; scape tall, little shorter than the lvs.; bracts usually 3.
                                                          11. *A. Lemmoni*
      GG. Stamens much shorter than the perianth; scape low, much shorter than the lvs.
        H. Perianth-segms. lanceolate with strongly seemingly acuminate tips, entire, mostly white or whitish, rarely rose-tinged. Lassen Co. n. . . . . 12. *A. Tolmiei*
        HH. Perianth-segms. ovate or ovate-lanceolate, not strongly involute, often denticulate, mostly pale to deep rose, rarely white. Siskiyou Mts. . . . . 13. *A. siskiyouense*
    FF. Fls. 9–15 mm. long, mostly deep rose to purple, rarely white; segms. often denticulate, commonly attenuate; lvs. mostly 5–9 mm. wide. Coast Ranges . . . . 14. *A. falcifolium*
CC. Lf. 1.
    D.   Perianth-segms. linear or lance-linear; stamens ca. equal to perianth or exserted.
      E. Fils. minutely papillose. N. Coast Ranges . . . . 25. *A. Hoffmanii*
      EE. Fils. smooth. S. Calif. . . . . 26. *A. Burlewii*
    DD. Perianth-segms. elliptic or rarely lanceolate; stamens included.
      E. Perianth 5–8(–11) mm. long, mostly whitish; crests obscure. Sierra Nevada
                                                         27. *A. obtusum*
      EE. Perianth 8–14 (usually 10–12) mm. long, pinkish or purplish; crests usually prominent. Sierra Nevada foothills and N. Coast Ranges . . 28. *A. cratericola*
AA. Lvs. solitary and terete; ovary prominently 6-crested.
    B.   Scape 2–4 dm. tall. Sierran foothills, Mariposa to Shasta cos. . . . . 29. *A. Sanbornii*
    BB. Scape usually much shorter. Not Sierran, n. of Madera Co.
      C. Stigma entire, capitate, at most 3-lobed, not distinctly trifid.
        D.   Perianth-segms. 8–12 mm. long; pedicels slender, longer than the fls.
          E. Outer bulb-coats usually with ± distinct contorted reticulations; stamens usually less than ⅔ the perianth-length. E. San Bernardino Co.
                                                        32. *A. nevadense*
          EE. Outer bulb-coats without reticulations; stamens relatively longer. N. San Bernardino to Mono cos. . . . . 33. *A. atrorubens*
        DD. Perianth-segms. 12–20 mm. long; pedicels stout, mostly shorter than the fls. S. Calif.
         E. Perianth-segms. elliptic-lanceolate, acute, becoming papery and abruptly spreading at tips in fr. . . . . 34. *A. Parishii*
         EE. Perianth-segms. linear-lanceolate, attenuate, becoming rigid and widely divergent apically in fr. . . . . 35. *A. monticola*
      CC. Stigma distinctly trifid, the divisions often slender and recurved. See also *A. Parishii*.
        D.   Stamens well included . . . . 31. *A. fimbriatum*
        DD. Stamens ca. equal to perianth-length or exserted . . . . 30. *A. Howellii*

   1. **A. haematochìton** Wats. [*A. Marvinii* A. Davids.] Bulbs oblong, 2–3 cm. long, usually clustered, from a short rootstock, the coats membranous, deep red to white, with fine vertical striations; scape 1–4 dm. high, slightly compressed; lvs. several, flat, 1–2 dm. long, 1–4 mm. wide, with much wider sheath; bracts 2–4, connate, obtuse; fls. 10–30; pedicels 1–2 cm. long; perianth-segms. white to rose, with darker midvein, broadly ovate to lance-ovate, acute, 6–8 mm. long; stamens and style included; fils. dilated at base; anthers yellow; caps. 4 mm. long, obcordate, with 6 short rounded crests; seeds broadly obovoid, ca. 2 mm. long; $n = 7$.—Dry slopes and ridges of clay or stony soil, below 2500 ft.; Chaparral, V. Grassland, Coastal Sage Scrub; San Luis Obispo Co. to L. Calif. March–May.

   2. **A. válidum** Wats. Bulbs oblong-ovoid, 3–5 cm. long, commonly clustered, from a short rootstock and with membranous white to reddish ribbed coats, with fine vertical striations and persistent fibers; scape stout, angled, somewhat compressed, 5–10 dm. high; lvs. 3–6, flat or slightly keeled, almost as long as scape, 5–12 mm. wide; bracts 2–4, united at base, pointed; fls. many; pedicels 10–18 mm. long; perianth-segms. rose to almost white, lanceolate, acuminate, 6–10 mm. long, saccate at base; stamens and style usually exserted; fils. subulate; anthers dark; caps. subglobose, 5–7 mm. long, not crested; seeds narrow, 4–5 mm. long; $n = 14, 28$.—In wet meadows, 4000–11,000 ft.;

Montane Coniferous F.; Sierra Nevada, Coast Ranges s. to Lake Co.; to B.C. and Ida. An onion of good flavor. July–Sept.

3. **A. membranàceum** Ownbey. Bulb ovoid, 1–1.6 cm. long, outer coats dark, inner whitish or reddish, the quadrate reticulations with strongly sinuous vertical lines; scapes 1–2, 1.5–3.5 dm. high; lvs. flat, commonly 2, equaling scape, 2–9 mm. wide; bracts 2, broadly lanceolate, 1–2 cm. long, setose-acuminate; fls. mostly 15–35, in globose umbels; pedicels slender, 5–15 mm. long; perianth-segms. whitish to pink, elliptic-ovate, acute, thin, 7–12 mm. long, slightly saccate at base; stamens half as long; fils. slightly dilated at base; anthers yellow; stigma capitate; caps. ca. 2.5 mm. high, with 6 short triangular crests; seeds narrow-obovoid, almost 2 mm. long; *n* = 7.—Wooded slopes, 500–4500 ft.; mostly Yellow Pine F.; w. slope of Sierra Nevada, Plumas, and Butte to Mariposa cos. and in Humboldt and Trinity cos. May–July.

4. **A. biscéptrum** Wats. Bulb round-ovoid, 1–1.5 cm. long, with supplementary bulblets in a tight basal cluster, outer coats grayish, inner white, with very irregular reticulations, the vertical walls very sinuous; scapes 1–3, terete, 1–3 dm. high; lvs. 2–3, flat, 3–10 mm. wide, ca. as long as scape; bracts 2, lance-ovate, acuminate, 5–15 mm. long; fls. 15–40, in open globose umbel; pedicels stoutish, 1–2 cm. long; perianth-segms. rose-purple, lance-ovate, acuminate, 6–10 mm. long, divaricate, greenish at base; stamens ¾ as long as perianth; fils. dilated at base; anthers dark; stigma subcapitate; caps. 3–4 mm. long, the 6 crests triangular, away from the style; seeds broadly obovoid, ca. 1.5 mm. long; *n* = 7.—Meadows and aspen groves or moist banks, 6500–9500 ft.; Montane Coniferous F.; s. Ore. along e. slope of Sierra Nevada to Inyo Co., White Mts.; Nev., Utah, Ida. May–July.

5. **A. campanulàtum** Wats. [*A. Bidwelliae* Wats. *A. acuminatum* var. *B.* Jeps. *A. Austinae* Jones. *A. tenellum* A. Davids. in part. *A. Bullardii* A. Davids.] Bulb ovoid, 1–2 cm. long, with supplementary bulblets either in a tight basal cluster or on filiform rhizomes up to 1 dm. long, outer bulb-coats brown, inner reddish to white, with minute quadrate reticulation, or the horizontal walls faint and vertical ones prominent and sinuous; scapes 1–2, 1–3 dm. high; lvs. 2–3, flat, 1–5 mm. wide, shorter than to longer than scape; bracts 2, ovate, abruptly acuminate, 7–15 mm. long; fls. 15–40, loosely arranged; pedicels slender, 1–2 cm. long; perianth-segms. pale rose, spreading, lance-ovate, acute, 5–8 mm. long; stamens ca. ¾ as long; fils. slender, basally dilated; anthers reddish; stigma subcapitate; caps. ca. 3 mm. high, with 6 low cent. triangular crests; seeds obovoid, ca. 2 mm. long; *n* = 7, 14.—Dry slopes in woods, 2000–8500 ft.; Pinyon-Juniper Wd., Montane Coniferous F.; s. Ore., along Sierra Nevada to mts. of San Diego Co., N. Coast Ranges s. to Monterey Co.; w. Nev. May–July.

6. **A. platycaùle** Wats. Bulb obliquely ovoid, 2–3.5 cm. long, outer coats gray, inner white without definite reticulations; scape 4–12 cm. long, flattened, 3–5 mm. wide; lvs. usually 2, falcate, thick, 8–15 mm. wide, much longer than scape; bracts 3–5, lance-ovate, acuminate; fls. 30–many; pedicels 12–20 mm. long; perianth-segms. deep rose, with pale tips, lance-linear, long-attenuate, infolded, connivent above the ovary, then widely spreading, 10–12 mm. long; stamens ca. as long; fils. widened at base; anthers dark; caps. 3–4 mm. high, not crested; seeds 2–3 mm. long, obovoid; *n* = 7.—Gravelly slopes and knolls, 4000–9000 ft.; Sagebrush Scrub, Montane Coniferous F.; s. Ore. through Modoc Co. and the Sierra Nevada to Placer Co.; w. Nev. May–Aug.

7. **A. ánceps** Kell. Bulb ovoid, 15–18 mm. long, the outer coats brownish-gray, the inner paler, faintly and transversely rectangular-reticulate; scape 6–10 cm. high, flattened; lvs. 2, somewhat falcate, exceeding scapes, 2–5 mm. wide; bracts 2, broadly ovate, obtuse and mucronate, rose-purple; fls. 15–25; pedicels stout, 8–15 mm. long; perianth-segms. pale rose with purplish midveins, lance-linear, attenuate, 8–10 mm. long; stamens as long or longer; fils. flat, with dilated base; anthers whitish; style 6 mm. long; caps. subglobose, 4 mm. long, usually with 6 low broad crests; seeds angled-obovoid, 2.5–3 mm. long; *n* = 7.—Dry flats and slopes, 4000–5000 ft.; Sagebrush Scrub; e. base of Sierra Nevada; Carson City, Nev. to s. Ore. April–June.

8. **A, yosemiténse** Eastw. Bulb ovoid, 2–3 cm. long, the outer coats dark, without definite reticulations; scape 5–10 cm. long, reddish, somewhat flat, 1–3 mm. wide; lvs. 2, flat, 2–3 dm. long, 1–3 mm. wide; bracts 3, ovate, purplish, 1 cm. wide, abruptly apiculate, the narrow tip 5 mm. long; fls. many; pedicels purplish, 1–2 cm. long; perianth-segms. pale rose, linear-oblong, acute, ca. 1 cm. long; stamens ca. as long, dilated at base;

caps. 5 mm. long, with 3 or 6 low obtuse crests; seeds obovoid, 3 mm. long; $n = 7$.—Montane Coniferous F.; Sierra Nevada. June–July.

9. **A. tribracteàtum** Torr. [*A. parvum* var. *jacintense* Munz.] Bulbs ovoid, 1–2 cm. long, with thin white coats with faint mostly oblong reticulations; scape 3–12 cm. long; lvs. 2, longer than scape, 2–6 mm. wide; bracts 2–3, abruptly acuminate; fls. 10–20; pedicels 4–16 mm. long; perianth-segms. pale rose with dark purplish midveins, narrowly oblong to lanceolate, 7–11 mm. long, obtuse or acute; stamens and styles less than ⅓ perianth-length; fils. dilated at base; anthers yellow or purple; caps. 4 mm. long, 3-lobed, crestless or obscurely 3-crested; seeds ca. 2 mm. long; $n = 7$.—Dry rocky ridges and slopes, 4000–8000 ft.; Montane Coniferous F.; San Jacinto Mts. (Riverside Co.) to Sierra Nevada and Mt. Shasta. April–July.

10. **A. párvum** Kell. [*A. tribracteatum* var. *p.* Jeps. *A. t.* var. *Andersoni* Wats. *A. modocense* Jeps.?] Bulb ovoid, 10–15 mm. long, the coats grayish to brownish, without definite reticulations; scape 3–5 cm. long, mostly underground; lvs. 2, ca. twice as long as scape, somewhat curved, 2–4 mm. wide; bracts 2, round-ovate, obtusish to sub-acuminate; fls. 8–120; pedicels 5–10 mm. long; perianth-segms. rose-purple or pink with purplish midrib, elliptic- to lance-oblong, obtuse, 7–10 mm. long; stamens and style included; fils. not much dilated at base; anthers yellow or purple; caps. rounded, 3–4 mm. long, obscurely 3-crested; seeds obovoid, 2 mm. long; $n = 7$.—Gravelly and stony soil, 4500–8000 ft.; Sagebrush Scrub, Montane Coniferous F.; Siskiyou and Modoc cos. to Mono Co.; e. Ida., Utah, Nev. April–July.

11. **A. Lemmònii** Wats. [*A. anceps.* var. *L.* Jeps.] Bulb ovoid, 15–22 mm. long, the outer coats gray, inner tawny, with fine transversely oblong reticulations in vertical rows; scape 1–2 dm. high, narrowly 2-edged; lvs. 2, flat, 2–5 mm. wide, ca. equaling scape; bracts 2–4, ovate, abruptly acuminate, 10–17 mm. long; fls. numerous; pedicels slender, 10–16 mm. long; perianth-segms. whitish to pale rose, ovate-lanceolate, acuminate, erect, 6–8 mm. long; stamens ca. as long; fils. dilated at base; anthers yellow; stigma capitate; caps. 3–4 mm. high, obscurely crested; seeds obovoid, ca. 2 mm. long; $n = 7$.—Heavy soil, at about 5000 ft.; Yellow Pine F.; Modoc Co. to Sierra Co.; w. Nev. to se. Ore., sw. Ida. May–June.

12. **A. Tólmiei** Baker. Bulb ovoid, 1–2 cm. long, the outer coats grayish or brownish, mostly without definite reticulations; scape flattened, 5–12 cm. long, 3–5 mm. wide; lvs. 2, much longer than the scape, falcate, 2–8 mm. wide; bracts 2, lance-ovate, acuminate; fls. 15–30; pedicels 10–20 mm. long; perianth-segms. white with broad pink midveins, lanceolate, 7–10 mm. long, tips strongly involute, appearing acuminate; stamens ca. ⅔ as long, united at the dilated bases into a scalloped cup; caps. 3–4 mm. long, crestless to obscurely 3- or 6-crested; seeds obovoid, 2 mm. long; $n = 7$.—Stony clay flats, 5000–7000 ft.; Sagebrush Scrub; Lassen Co. to Wash., Ida. April–June.

13. **A. siskiyouénse** Ownbey. [*A. Watsoni* auth., not Howell.] Bulb ovoid, ca. 15 mm. long, the outer coats grayish-brown, the inner pink or white, without definite reticulations; scape 3–7 cm. high, slightly compressed, 1–2 mm. wide; lvs. 2, somewhat falcate, much longer than scape, 2–3 mm. wide; bracts 2, round-ovate, connate at base, abruptly pointed; fls. 10–20; pedicels 6–12 mm. long; perianth-segms. rose with darker midveins, lance-ovate, 7–10 mm. long, often denticulate, acutish, erect; stamens ca. ⅔ as long; fils. dilated below; caps. 3–4 mm. long, with 3 cent. entire crests; seeds obovoid, 2 mm. long; $n = 7$.—Rocky slopes; Montane Coniferous F.; Siskiyou Mts. to s. Ore. June–July.

14. **A. falcifòlium** H. & A. [*A. f.* var. *demissum* Jeps. and var. *Breweri* Jones. *A. B.* Wats.] Bulb ovoid, 1.5–2.5 cm. long, with brownish outer coats lacking definite reticulations; scapes 3–12 cm. long, ca. half underground, flattened, 2.5–6 mm. wide, 2-winged above; lvs. 2, flat, thick, falcate, exceeding scape, 4–9 mm. wide; bracts 2, lance-ovate to ovate, acuminate or abruptly acute, 1.5–2 cm. long; fls. 10–30; pedicels 6–16 mm. long; perianth-segms. deep rose to purple, or greenish-white tinged with rose, lanceolate, acute to attenuate, 9–15 mm. long, often minutely glandular-serrate and spreading above; stamens ca. half the length of the segms.; fils. dilated at base; anthers yellow; caps. 4–5 mm. long, rounded, with 3 cent. narrow crests; seeds obovoid, 2–3 mm. long; $n = 7$.—Heavy or rocky soil, often on serpentine outcrops, 500–7000 ft.; openings in Chaparral, Foothill Wd., Mixed Evergreen F., Yellow Pine F.; Coast Ranges, s. Ore. to Santa Cruz Co. March–July.

15. **A. Hickmánii** Eastw. [*A. hyalinum* var. *H.* Jeps.] Bulb round-ovoid, 8–10 mm.

long, outer coats gray, inner white, the reticulations transversely and rather irregularly undulate-serrate; scapes 1–2, slender, 1–1.5 dm. tall; lvs. 2–3, filiform, exceeding scapes; bracts round-ovate, obtuse, mucronate, scarcely 1 cm. long; fls. 8–22; pedicels filiform, 5–10 mm. long; perianth-segms. white to pink, ovate, acuminate, 5 mm. long; stamens almost as long; fils. dilated at base; anthers white; caps. without crests; *n* = 7.—Grassy places; Closed-cone Pine F.; Monterey Peninsula and near Jolon, Monterey Co. April.

16. **A. unifòlium** Kell. [*A. u.* var. *lacteum* Greene. *A. grandisceptrum* A. Davids.] Bulb ovoid, 10–15 mm. long, arising terminally on a stout lateral rootstock, the old bulb not persisting; outer bulb-coats pale, with obscure narrow horizontal undulating reticulations; scape stout, 2–6 dm. high; lvs. 2–4, flattish, 2–8 mm. wide, shorter than scape; bracts 2, lance-ovate, acuminate, 2–3 cm. long; fls. many, in loose umbel; pedicels stoutish, 2–3.5 cm. long; perianth-segms. rose-pink to lilac, sometimes white, spreading, oblong-ovate, 10–15 mm. long, subacuminate; stamens scarcely ⅔ as long, wide at base; anthers yellow or purplish; stigma subentire; caps. 4–5 mm. high, crestless; seeds 3 mm. long, 1.5 mm. thick; *n* = 7.—Occasional in moist often heavy soil below 3500 ft.; Closed-cone Pine F., Mixed Evergreen F., Chaparral; Coast Ranges, Del Norte to Monterey cos. April–June.

17. **A. hyalìnum** Curran. Bulbs ovoid, 6–10 mm. long, the coats gray, thin, with horizontal undulate reticulation in vertical rows; scape slender, 1.5–3 dm. high; lvs. 2, sometimes 1 or 3, 1–5 mm. wide, ca. as long as scape, often convolute; bracts 2, lance-ovate, 1–2 cm. long, acuminate; fls. 6–15, in open umbel; pedicels slender, spreading, 2–2.5 cm. long; perianth-segms. white or pinkish, lance-ovate, acute, 5–9 mm. long, thin and membranous in age; stamens ⅔ as long, dilated at base; anthers pale; stigma subentire; caps. ca. 3 mm. high, crestless; seeds broadly obovoid, 1.5–2 mm. long; *n* = 7.—Rather moist places, grassy and rocky slopes, 500–5000 ft.; V. Grassland, Foothill Wd.; w. base of Sierra Nevada, Eldorado Co. to Tehachapi Mts., Kern Co. March–June.

18. **A. praècox** Bdg. [*A. hyalinum* var. *p.* Jeps.] Bulbs round-ovoid, 7–15 mm. long, outer coats gray-brown, inner paler, all with horizontal somewhat irregular and somewhat undulating reticulation; scape stout, 2–5 dm. high; lvs. 2–4, almost or quite as long as scape, flat, 1–5 mm. wide; bracts 2, lance-ovate, acuminate, 1–2.5 cm. long; fls. 6–30, in open umbel; pedicels stoutish, spreading, 1.5–3 cm. long; perianth-segms. rose-purple or lighter, with dark midveins, lance-ovate, 9–12 mm. long, acuminate, the inner somewhat narrower; stamens ⅔ as long as perianth, ± dilated at base; anthers yellowish or reddish; stigma slightly lobed; caps. ca. 4–5 mm. high, minutely crested; seeds broadly obovoid, 3–4 mm. long, almost as thick.—Shaded slopes and canyons, below 2500 ft.; Chaparral, S. Oak Wd.; Ventura, e. Los Angeles and San Bernardino cos. to L. Calif.; Santa Cruz, Santa Rosa, Santa Catalina and San Clemente ids. March–April.

19. **A. ampléctens** Torr. [*A. attenuifolium* Kell. *A. occidentale* Gray. *A. monospermum* Jeps. *A. attenuifolium* var. *m.* Jeps.] Bulb ovoid, 1–1.5 cm. long, the outer coats brownish, with transverse broadly V-shaped regular reticulations in vertical rows, the inner coats reddish or whitish; scape 2–5 dm. high; lvs. 2–4, shorter than scape, narrow, flattened, but convolute-filiform in age; bracts 2–3, broadly ovate, abruptly acuminate, 7–10(–19) mm. long; fls. many, in a subglobose umbel; pedicels slender, 6–15 mm. long; perianth-segms. white to pinkish, lance-ovate, acute, 6–7 mm. long, papery after anthesis; stamens equal to or shorter than perianth; fils. slightly dilated at base; anthers yellow or purplish; stigma capitate; caps. ca. 3 mm. high, with 6 very low crests; seeds round-obovoid, ca. 2 mm. long; 2*n* = 14?, 21, 28 (Levan, 1940).—Dry slopes and fields, mostly below 6000 ft.; Yellow Pine F., Foothill Wd.; B.C. s. through Coast Ranges to Alameda and Santa Clara cos., w. base of Sierra Nevada, Cuyamaca Mts. March–June.

20. **A. Bolánderi** Wats. [*A. stenanthum* Drew. *A. B.* var. *s.* Jeps.] Bulb round-ovoid, 10–12 mm. long, outer coats gray or brown, inner white, with close transverse-serrate reticulation; sometimes reproducing by more elongate short-stalked lateral offsets, the old bulb disappearing in the process; scape slender, 1–2.5 dm. high; lvs. mostly 2, ca. as long as scape, 1–2.5 mm. wide; bracts 2, lance-ovate, acuminate, 10–15 mm. long; fls. 6–25, in open umbel; pedicels slender, 1–2 cm. long; perianth-segms. rose-purple to white, 8–15 mm. long, narrowly lanceolate, acuminate-attenuate, mostly involute, saccate at base, inner segms. generally serrulate; stamens ca. half as long; fils. dilated at base; anthers yellow; stigma slightly lobed; caps. ca. 3 mm. high, obscurely crested;

seeds ca. 2 mm. long; $n = 7$.—Heavy soil, openings in brush and woods, below 3000 ft.; Chaparral, Foothill Wd., Yellow Pine F.; inner Coast Ranges, s. Ore. to Lake Co., Mt. Hamilton, e. to Siskiyou, Shasta and Modoc cos. May–July.

21. **A. críspum** Greene. [*A. peninsulare* var. *c.* Jeps.] Bulb round-ovoid, 8–12 mm. long, outer coats gray-brown, inner whitish, all closely and regularly horizontal-serrate; scape 1–3 dm. high; lvs. 2, ca. 1.5 mm. wide, mostly shorter than scape; bracts 2, ovate, 1–2 cm. long, abruptly short-acuminate; fls. 10–40, in open umbels; pedicels stout, spreading, 1–3 cm. long; perianth-segms. bright red-purple to orchid, 9–12 mm. long, the outer oblong-ovate, acute, spreading at apex, inner narrower, undulate-crisped; stamens ½–⅔ as long, broadly dilated at base; anthers yellowish; stigma slightly lobed to distinctly trifid; caps. ca. 4 mm. high, the crests cent. usually evident; seeds 3 mm. long; $n = 7$.—Heavy soil on rolling hills, below 2500 ft.; V. Grassland, Foothill Wd.; inner Coast Ranges, San Joaquin and Contra Costa cos. to Figueroa Mt. (Santa Barbara Co.), and s. Sierran foothills in Kern Co. March–May.

22. **A. peninsulàre** Lemmon. [*A. montigenum* A. Davids.] Bulb round-ovoid, 1–1.5 cm. long, outer coats gray-brown, inner whitish, all with close regular horizontal-serrate reticulation; scape 2–4 dm. high; lvs. 2–4, almost as long as scape, 1–6 mm. wide; bracts mostly 2, lance-ovate, acuminate, 1–2 cm. long; fls. 6–25, in open umbel; pedicels slender, 1.5–3 cm. long, spreading; perianth-segms. deep rose-purple, 10–13 mm. long, the outer ovate-lanceolate, acuminate and with spreading apex, the inner narrower, erect or spreading, plane; stamens scarcely ⅔ as long; fils. broad at base; anthers yellowish; stigma ± distinctly 3-lobed; caps. ca. 4 mm. high, the crests very minute; seeds broad-obovoid, ca. 2 mm. long; $n = 7$.—Dry open or wooded slopes, below 3000 ft.; V. Grassland, Foothill Wd.; Butte Co. s. along the base of the Sierra Nevada, Coast Ranges around s. end of San Francisco Bay to n. L. Calif.; Santa Catalina and San Clemente ids. March–June.

23. **A. serràtum** Wats. Bulb round-ovoid, 8–12 mm. long, the outer coats gray-brown, inner lighter, all horizontally and regularly serrate-reticulate; scape 2–3.5 dm. high, slender; lvs. 2–4, shorter than scape, 1.5–3 mm. wide; bracts mostly 2, ovate-lanceolate, acuminate, 1–1.5 cm. long; fls. many, in crowded umbels; pedicels slender, 8–14 mm. long; perianth-segms. rose-pink to purplish, papery, ovate-lanceolate, acuminate, ± spreading at apex, 7–9 mm. long; stamens ca. ⅔ as long, with dilated bases; anthers yellowish to reddish; stigma slightly 3-lobed; caps. 2.5–3 mm. high, the crests minute, cent., narrow, 2-lobed; seeds thick, 1.5–2 mm. long; $n = 7$.—Grassy slopes and meadows, usually on serpentine, below 3000 ft.; Foothill Wd., V. Grassland; inner Coast Ranges, Lake and Colusa cos. to Merced Co. March–May.

24. **A. dichlamýdeum** Greene. [*A. serratum* var. *d.* Jones.] Bulb round-ovoid, 1–1.5 cm. long, outer coats gray-brown, inner whiter, all with regular close transverse-serrate reticulations in vertical rows; scape rather stout, 1–3 dm. high; lvs. 1–3, flat, 1–2 mm. wide, equaling or exceeding the scape; bracts 2, ovate, subacuminate, 10–20 mm. long; fls. few–many, in congested umbels; pedicels stout, 1–1.5 cm. long; perianth-segms. deep rose-purple, ovate-oblong, acutish, 9–11 mm. long; stamens ⅔ as long; fils. dilated at base; anthers yellow; stigma capitate, entire or slightly 3-lobed; caps. 3–4 mm. high, minutely crested; seeds obovoid, 1.5–2 mm. long; $n = 7$.—Dry heavy soil, or rocky places; N. Coastal Scrub, Mixed Evergreen F.; outer Coast Ranges, Mendocino to Monterey cos. May–July.

25. **A. Hoffmánii** Ownbey. Bulb ovoid, 1.5–2.5 cm. long, the outer coats grayish or brownish, without definite reticulations, the inner whitish or pinkish; scape 5–10 cm. long, not strongly flattened; lf. 1, flat, 4–8 mm. wide, exceeding scape; bracts 2–5 (mostly 3–4), ovate, acuminate; fls. 10–40; pedicels slender, 8–15 mm. long; perianth-segms. purplish-pink with green midveins, linear-lanceolate, attenuate, 8–10 mm. long; stamens exserted; fils. little dilated and adnate to perianth-segms. 1–2 mm., strongly papillose-glandular at 20X; anthers purple; caps. ca. 4 mm. long, with 6 low crests; seeds obovoid, ca. 2.5 mm. long; $n = 7$.—Serpentine outcrops at ca. 5000 ft.; probably Yellow Pine F.; extreme sw. Shasta and adjacent Tehama and Trinity cos. (head of Beegum Creek) and near Mt. Lassic on Trinity-Humboldt co. line. June–July.

26. **A. Burlèwii** A. Davids. [*A. Johnstonii* Jones in Jeps. in synon.] Bulb ovoid, 1.5–2 cm. long, the outer coats grayish or brownish, usually without reticulations, the inner white, obscurely cellular; scape 2–8 cm. long, somewhat flattened; lf. 1, flat, 1–10 mm.

wide, exceeding scape; bracts 2–3, ovate, obtuse or abruptly pointed; fls. 8–18; pedicels stout, 1–2 cm. long; perianth-segms. pinkish-purple with dark midveins, 7–10 mm. long; stamens ca. equal to perianth or exserted; fils. dilated basally, smooth; anthers purple or yellowish; caps. ca. 4 mm. long, with 6 low crests or crestless; seeds obovoid, 2.5–3 mm. long; $n = 7$.—Dry granitic slopes and ridges, 6000–9000 ft.; Montane Coniferous F.; s. Sierra Nevada, Kern Co. to San Jacinto Mts. and at lower elev. in San Benito Co. May–July.

27. **A. obtùsum** Lemmon. [*A. ambiguum* Jones, not Sibth. & Sm. *A. concinnum* K. Bdg. ex Jones, pro synon.] Bulb ovoid, 12–20 mm. long, the outer coats grayish, with or without fine rectangular reticulations, the inner whitish, not reticulate; scape 4–8 cm. high, largely underground, little flattened; lf. 1, flat, much exceeding scape, 1–3(–5) mm. wide; bracts 2–3(–4), ovate, acuminate, 5–8 mm. long; fls. several to many; pedicels 3–5 mm. long; perianth-segms. greenish-white with purplish midribs and base, oblong-ovate, obtuse, rarely lanceolate and acute, 5–8(–11) mm. long; stamens and style included; anthers yellowish or purplish; stigma subentire; caps. 3 mm. long, with 6 low crests, or crestless; seeds 1.5 mm. long; $n = 7$.—Sandy or gravelly slopes and benches, 7000–12,000 ft.; Lodgepole F. to Subalpine Fell-fields; Sierra Nevada, Plumas Co. to Kern Co. May–July.

28. **A. cratericola** Eastw. [*A. parvum* var. *Brucae* Jones.] Bulb ovoid, ca. 15 mm. long, outer coats dark, inner whitish, scarcely or faintly oblong-reticulate; scape 2–7 cm. long, somewhat flattened, reddish, partly underground; lf. 1, flat, falcate, 1–2 dm. long, 3–5(–12) mm. wide; bracts (2–)3–4, dark, ovate, setose-acuminate; fls. many; pedicels 5–10 mm. long; perianth-segms. pale to dark purple, linear-oblong, obtuse, 8–14 mm. long; fils. half to ¾ as long, broad at base; anthers purplish or yellowish; stigma capitate; caps. 4–5 mm. long, with 3 cent. broad low crests; seeds ca. 2.5 mm. long; $n = 14$.—Volcanic and serpentine soils mostly below 2500 ft.; Chaparral, Foothill Wd., etc.; N. Coast Ranges from Trinity to Napa cos., and lower foothills of Sierra Nevada from Butte to Tuolumne cos. March–May.

29. **A. Sanbórnii** Wood. Bulb ovoid, 15–25 mm. long, the outer coats brownish, the inner pinkish, obscurely cellular; scape 2–5 dm. high; lf. 1, terete above the tubular sheath, 2–3 mm. thick; bracts usually 4, lanceolate, attenuate, ca. 15 mm. long; fls. many, rather crowded; pedicels slender, 6–15 mm. long; perianth-segms. deep pink, 5–7 mm. long, ovate, acute or cuspidate, erect, inner series ¼ longer than outer; stamens slightly exserted; fils. filiform, slightly dilated at base; anthers yellow; stigma capitate or irregularly branched, not definitely trifid; caps. 2.5 mm. long, the crests 6, entire, conspicuous; seeds 2 mm. long.—Serpentine, 1200–4000 ft.; mostly Yellow Pine F.; w. slope of Sierra Nevada, Shasta Co. to Calaveras Co. June–Sept.

Var. **Congdònii** Jeps. [*A. intactum* Jeps.] Lf. ca. as long as the scape; bracts usually 4; perianth-segms. 5–7 mm. long, white or faintly pinkish, erect, ovate, acuminate, inner series ¼ longer than outer, both erose-dentate; stamens exserted; anthers mostly purple; stigma distinctly trifid; $n = 7$.—Placer to Mariposa cos. June–Oct.

Var. **Jepsònii** Ownbey & Aase. Lf. ca. as long as scape; bracts 3–4; perianth-segms. 7–8 mm. long, pale with deep pink midribs, spreading, broadly ovate, acute, inner series only slightly longer than outer, both obscurely erose; stamens well included; stigma distinctly trifid; $n = 7$.—Butte and Tuolumne cos. May–June.

Var. **tuolumnénse** Ownbey & Aase. Bulb round-ovoid, with reddish inner coats; lf. shorter than the scape; bracts 3; perianth-segms. 6–8 mm. long, elliptic-lanceolate, obtusish, entire, inner and outer series ca. equally long, papery in fr.; stamens ¾ as long as perianth; stigma trifid; crests laciniate.—Rawhide Hill, Tuolumne Co. May.

30. **A. Howéllii** Eastw. Bulb round-ovoid, 7–12 mm. long, outer coats red-brown, inner paler, obscurely cellular; scape 1–3 dm. high, slender, terete; lf. 1, terete above the tubular sheath, ca. equaling scape; bracts usually 3, reddish, broad-ovate, abruptly pointed, 4–15 mm. long; fls. 10–30; pedicels 5–15 mm. long; perianth-segms. pale lilac-violet to deep rose, with darker midveins, oblong or ovate, obtuse, 5–8 mm. long; stamens and style usually exserted; fils. dilated at base; anthers yellow; stigma trifid; caps. ca. 3 mm. long, with 6 red acuminate dentate-laciniate to entire crests; seeds 2 mm. long; $n = 7$.—Heavy or granitic soil, 700–2000 ft.; Foothill Wd., V. Grassland; about the head of San Joaquin V. from Kern Co. to inner S. Coast Ranges, Santa Clara Co. April–May.

Var. **Clòkeyi** Ownbey & Aase. Scape 1.5–3 dm. high, stout; fls. many, white, sometimes with reddish midveins; stamens ca. equal to perianth; $n = 7$.—Heavy soil, 4500–6000 ft.; Sagebrush Scrub, etc.; region of Mt. Pinos, Ventura Co. April–June.

Var. **sanbeniténse** (Traub) Ownbey & Aase. [*A. s.* Traub. *A. robustum* Eastw., not Kar. & Kir.] Scape 2.5–4 dm. high, robust; lf. shorter than scape; bracts 2–4, ca. 15 mm. long; fls. many; perianth-segms. pale rose with red midveins; stamens exserted 2–4 mm.; $n = 7$.—S. Coast Ranges, San Benito Co. April–May.

31. **A. fimbriàtum** Wats. [*A. f.* var. *aboriginum* Jeps.] Bulb globose-ovoid, 12–16 mm. long, outer coats dark, inner brownish, obscurely cellular; scape 3–10(–16) cm. high, slender; lf. 1, ca. twice as long as the scape, terete above the tubular sheath; bracts 2–3, setaceous at apex, 6–12 mm. long; fls. 8–40; pedicels 4–15 mm. long, stoutish; perianth-segms. purple to rose, with darker midveins, lanceolate, entire, obtuse to acuminate, 7–15 mm. long, with erect or spreading tips; stamens ½–¾ as long, broad at base; anthers yellow; stigma with 3 linear lobes; caps. 3–4 mm. high, 3-lobed, the 6 erect crests 1 or more mm. high, usually fimbriate; seeds obovoid, ca. 2 mm. long; $n = 7$.— Dry slopes and flats, often in heavy soil, 2000–8000 ft.; Creosote Bush Scrub, Joshua Tree Wd., Pinyon-Juniper Wd.; w. edge of Colo. Desert, w. Mojave Desert n. to Olancha and to Mt. Pinos; serpentine talus, inner Coast Ranges n. to Napa and Lake cos.; L. Calif. March–July.

KEY TO VARIETIES

A. Perianth purple to rose.
  B. Perianth-segms. not conspicuously recurved-spreading at tips; ovary-crests usually fimbriate or toothed.
    C. Scape 0.3–1(–1.5) dm. long, slender; pedicels 4–15 mm. long. Napa and Lake cos. to Mojave and Colo. deserts .... *A. fimbriatum*
    CC. Scape 1–3 dm. long, stout; pedicels 15–20 mm. long. Lake-Colusa co. line
        var. *Purdyi*
  BB. Perianth-segms. conspicuously recurved-spreading at tips; ovary-crests entire or nearly so.
    C. Plant not glaucous; inner perianth-segms. erose-dentate or crisped.
      D. Scape mostly less than 1 dm. long, very slender. From above 4500 ft., Sierra Nevada from Tulare Co. n. .... var. *Abramsii*
      DD. Scape 1–2 dm. long, stout. At 3000–5000 ft., Kern Co. to Little San Bernardino Mts. .... var. *denticulatum*
    CC. Plants glaucous; inner perianth-segms. not erose-dentate. From below 1700 ft., w. Stanislaus Co. .... var. *Sharsmithae*
AA. Perianth mostly pale pink to white.
  B. Ovary-crests coarsely dentate-laciniate; scape 0.7–2.5 dm. long. Serpentine, w. Stanislaus to n. Santa Barbara cos. .... var. *diabolense*
  BB. Ovary-crests entire to ± toothed.
    C. Scape 1–2.5 dm. long. Below 4000 ft.
      D. Perianth-segms. oblong-lanceolate, acuminate. W. Mojave Desert .. var. *mohavense*
      DD. Perianth-segms. oblong-ovate, ± rounded at apex. W. Riverside Co.
        var. *Munzii*
    CC. Scape mostly 0.6–1.5 dm. long. From 4000–7000 ft., San Bernardino Mts. to n. L. Calif.
      var. *Parryi*

Var. **mohavénse** Jeps. [*A. m.* Tides.] Bracts often 3; perianth-segms. pale pink to white with pink midveins, lance-oblong, 7–10 mm. long, not recurved-spreading at tip; ovary-crests toothed to subentire.—At 2500–4000 ft.; Creosote Bush Scrub; w. Mojave Desert. April–May.

Var. **Abrámsii** Ownbey & Aase. Inner bulb-coats whitish or pinkish, outer grayish, sometimes with distinct cellular markings; perianth-segms. narrow, conspicuously recurved-spreading, deep rose-purple, inner series erose-dentate or crisped; crests entire or nearly so; $n = 7$.—Granitic gravel, 4500–10,000 ft.; Montane Coniferous F.; Sierra Nevada, Madera to Tulare cos. May–July.

Var. **denticulàtum** Ownbey & Aase. Inner perianth-segms. minutely denticulate; crests entire or toothed; $n = 7$.—At 3000–5000 ft.; largely Joshua Tree Wd.; Piute Mts., Walker Pass, Kern Co. to Little San Bernardino Mts. April–May.

Var. **Púrdyi** (Eastw.) Ownbey & Aase. [*A. P.* Eastw.] Scape 1–3 dm. high, rather stout; lf. little longer; bracts 3, 1–2 cm. long; pedicels 15–20 mm. long; perianth-segms. rose-purple, ca. 10 mm. long, scarcely recurved at tip; crests 1.5 mm. long, subacuminate, toothed; $n = 7$.—Serpentine, below 1000 ft.; Foothill Wd.; Lake-Colusa co. line. May.

Var. **diabolénse** Ownbey & Aase. Scape 7–25 cm. high, rather slender; lf. 1–2 times as long; perianth-segms. 7–9 mm. long, pale with deep pink midribs, elliptic-lanceolate, obtuse or acute, erect, not recurved at tip; crests coarsely dentate-laciniate; $n = 7$.—Serpentine, 1700–4000 ft.; Chaparral, etc.; inner S. Coast Ranges from Arroyo del Puerto, Stanislaus Co. to Figueroa Mts., Santa Barbara Co. April–June.

Var. **Sharsmíthae** Ownbey & Aase. Glaucous; scape 5–10 cm. high, stout, half as long as the lf.; pedicels stout, shorter than fls.; perianth-segms. 10–15 mm. long, rose-purple, lanceolate-attenuate, recurved at tip; crests entire, or nearly so; $n = 7$.—Serpentine gravel slides, 1300–1600 ft.; Red Mts., Stanislaus Co. March–May.

Var. **Múnzii** Ownbey & Aase. Scape slender, 1–2.5 dm. high; bracts 7–10 mm. long; pedicels 6–12 mm. long; perianth-segms. pink, 7–9 mm. long, oblong-ovate, usually rounded at apex, erect, tip not recurved; crests ca. 1 mm. high, triangular-lanceolate, entire or coarsely toothed.—At 1000–2000 ft., w. end of Riverside Co. April–May.

Var. **Párryi** (Wats.) Ownbey & Aase. [*A. P.* Wats. *A. tenellum* A. Davids. in part. *A. Kessleri* A. Davids.] Scape slender, 6–16 cm. high; bracts 8–10 mm. long; pedicels 10–15 mm. long; perianth-segms. 6–8 mm. long, pink or with pink midveins, lanceolate, acuminate; crests erect, 1–1.5 mm. high, entire or erose; $n = 7$.—Dry slopes, 4000–7000 ft.; Yellow Pine F.; San Bernardino Mts. to n. L. Calif. June–July.

32. **A. nevadénse** Wats. Bulb ovoid, 12–15 mm. long, with basal bulblets, outer coats grayish or brownish, the inner pinkish, with narrow intricately contorted reticulations; scape 8–14 cm. long, mostly underground; lf. 1, terete above the tubular sheath, exceeding scape; bracts 2–3, lance-ovate, acuminate, 8–14 mm. long; perianth-segms. pink with red midveins, ovate-lanceolate, acute to attenuate, 8–12 mm. long; stamens included; fils. dilated at base; anthers yellow or purple; stigma subentire; caps. ca. 3 mm. long, with 6 erect triangular entire or toothed crests ca. 1 mm. high; seeds ca. 2 mm. long; $n = 7, 14$.—Stony soils, 2500–7000 ft.; Sagebrush Scrub, Pinyon-Juniper Wd.; s. Nev., possibly adjacent Calif. to nw. Ariz., se. Ore., Ida. and w. Colo. April–June.

Var. **cristàtum** (Wats.) Ownbey. [*A. c.* Wats.] Differing in its obscure or imperceptible markings on the bulb-coats.—Dry stony slopes at ca. 5000 ft.; Pinyon-Juniper Wd.; Providence, New York, Ivanpah, and Clark mts.; to Utah, Ariz. April–May.

33. **A. atrorùbens** Wats. Bulb ovoid, ca. 15 mm. long, with sessile bulblets, or these on slender rhizomes, the coats red-brown with indistinct quadrate reticulations; scape 6–15 cm. high, terete; lf. 1, exceeding scape, terete above the tubular sheath, coiled above; bracts usually ovate, 1.2–1.8 cm. long, long-pointed; fls. 20–25; pedicels 10–15 mm. long; perianth-segms. red-purple, rarely white, stiff, spreading, lanceolate, attenuate, 9–12 mm. long; stamens and style included; fils. slender, dilated at base; anthers yellow or purple; stigma entire; caps. 3–4 mm. high, the 6 crests 1 mm. high, entire or toothed; seeds ca. 2 mm. long; $n = 7$.—Dry hillsides, 5000–7000 ft.; Sagebrush Scrub; near Bridgeport, Mono Co., Avawatz Mts., Death V.; to Nev. May–June.

Var. **inyònis** (Jones) Ownbey & Aase. [*A. i.* Jones. *A. decipiens* Jones, not Fisch.] Perianth pale with dark midveins, broader, not attenuate, 8–12 mm. long.—At 4000–7000 ft.; Sagebrush Scrub, Pinyon-Juniper Wd.; e. of Sierra Nevada, Mono and Inyo cos.; adjacent Nev. May–June.

34. **A. Paríshii** Wats. Bulb round-ovoid, 10–15 mm. long, with pinkish nonreticulate coats, sometimes with basal bulblets; scape stout, 1–2 dm. long, largely underground; lf. 1, terete above the tubular sheath, much longer than scape; bracts 2–3, ovate, 1–1.5 cm. long, acuminate; fls. 10–many; pedicels 6–12 mm. long; perianth-segms. pale pink, 12–15 mm. long, elliptic-lanceolate, acute, with spreading tips; fils. half as long, lanceolate; anthers yellow; stigma entire or slightly lobed, rarely trifid; caps. 2–3 mm. long, with 6 triangular entire or toothed crests ca. 1 mm. long; $n = 7$.—Open rocky slopes mostly at 3000–4000 ft.; Joshua Tree Wd.; n. base of San Bernardino Mts., Little San Bernardino Mts.; w. Ariz. April–May. *A. anserìnum* Jeps. was based on a specimen typical of *A. Parishii* except for its trifid stigma and deeply fimbriate crests. No similar specimens having since been collected in n. Calif., it may be suspected that the type locality (Goose Lake V., Modoc Co.) is in error and that the type actually came from s. Calif. Until specimens exactly matching the type are again found, however, the identity and distribution of *A. anserinum* must remain in doubt.

35. **A. montícola** A. Davids. [*A. Peirsonii* Jeps.] Bulb ovoid, 1–2 cm. long, some-

times with basal bulblets, the outer coats dark gray-brown, the inner whitish or pinkish, obscurely quadrate-reticulate; scape 5–15 cm. long, largely underground, stout; lf. 1, terete above the tubular sheath, exceeding scape, glaucous; bracts 3 or 2, broadly ovate, 1–1.8 cm. long, subacuminate; fls. 6–many; pedicels 5–12 mm. long, stoutish; perianth-segms. pale to deep pink, 12–20 mm. long, linear-lanceolate, attenuate, with spreading tips; fils. ½–⅔ as long as segms., scarcely dilated at base; anthers purple; stigma entire; caps. ca. 5 mm. high, the 6 crests deltoid-lanceolate, 1–2 mm. high, entire or irregularly toothed; seeds obovoid, 3 mm. long; $n = 7$.—Loose rock or talus, mostly 5000–10,000 ft.; Montane Coniferous F.; San Gabriel and San Bernardino mts. May–July.

Var. **Kéckii** (Munz) Ownbey & Aase. [*A. Parishii* var. *K.* Munz.] Crests linear, 2–3 mm. high, papery; fls. deep purple.—Summits, 4000–5500 ft.; Chaparral, Yellow Pine F.; Santa Ynez Mts., Topatopa Mts., Santa Ana Mts. June.

36. **A. acuminàtum** Hook. Bulb broad-ovoid, 1–1.5 cm. long, the outermost coats brownish, strongly reticulate with irregular heavy-walled polygonal meshes, the inner thin, white, obscurely cellular; scape 1–3 dm. high, slender; lvs. 2–3, shorter than scape, 2–3 mm. wide; bracts 2 or 1, ovate, acuminate, 1.5–2 cm. long; pedicels slender, 1–2 cm. long; perianth-segms. bright rose, saccate at base, 8–12 mm. long, lanceolate, acuminate, the outer with recurved tips, the inner serrulate; stamens and style somewhat shorter than perianth; fils. dilated at base; anthers yellow; stigma subcapitate; caps. 4–5 mm. high, with 3 inconspicuous deltoid cent. crests; seeds 2.5 mm. long; $n = 7$.—Rather loose, often stony, sometimes alluvial soil, about shrubs, below 6000 ft.; mostly Yellow Pine F.; interior Humboldt to Modoc and Lassen cos.; Mt. Diablo, Contra Costa Co.; to B.C., Ida., Colo., Ariz. May–June.

37. **A. lacunòsum** Wats. Bulb ovoid, often with remnants of older bulbs in a row above, outermost coats gray-brown, with quadrate heavy-walled reticulations, inner coats thin, pale, the reticulations with thinner walls; scape 1–2 dm. high; lvs. 1–2, ca. as long as scape, terete, slender; bracts 2, ovate, short-acuminate, 8–12 mm. long; fls. 8–25; pedicels 5–10 mm. long; perianth-segms. white to pinkish, with green or red midveins, lance-oblong, acute or subacuminate, 5–7 mm. long; stamens almost as long; fils. dilated at base; anthers yellow; stigma capitate; caps. 3–4 mm. long, with an obtuse thickened ridge at summit of each side; seeds 2 mm. long; $n = 7$.—Open slopes and flats, often on serpentine hills, below 3000 ft.; Foothill Wd., Mixed Evergreen F.; Coast Ranges from Marin Co. s. to Santa Barbara Co.; Santa Rosa Id. April–May. Plants with 3 bracts and perianth 4 mm. long, from San Benito and San Luis Obispo cos., are var. *micránthum* Eastw.

38. **A. Davísiae** Jones. [*A. pseudobulbiferum* A. Davids.] Like *A. lacunosum,* but tending to be taller, mostly 2–4 dm. high; lvs. 2–3, ca. as long as scape, coarser than in the preceding; bracts 2, broadly ovate, 8–12 mm. long, abruptly acuminate; pedicels 1–2(–2.5) cm. long, stoutish; perianth-segms. pale with red midveins, lance-ovate, acuminate, 6–8 mm. long; stamens ¾ as long; fils. dilated at base; caps. ca. 4 mm. high, 3-lobed, scarcely crested; $n = 7$ (Greenhorn Mts.).—Dry flats, 2000–3500 ft.; Creosote Bush Scrub, Joshua Tree Wd.; Mojave R. basin w. to Palmdale, Mojave, Pilot Knob, etc. and in San Gabriel Mts. to 5600 ft. April–May. Specimens from the Greenhorn Mts., Kern Co., having smaller fls. with greenish midveins, more slender pedicels, often 3-bracted spathe and thinner bulb-coats are provisionally placed here.

2. Muílla S. Wats. MUILLA

Scape from a corm with fibro-membranous coats. Lvs. few, narrow, subterete. Umbel subtended by scarious acuminate bracts; pedicels slender, not jointed. Perianth subrotate, persistent, of 6 subequal slightly united lance-oblong segms., whitish or greenish, with dark 2-nerved midribs. Stamens 6, inserted near the base; fils. filiform to dilated; anthers versatile. Ovules 8–10 per locule; style short, clavate, persistent and finally splitting. Caps. globose, 3-angled, loculicidal. Seeds compressed, irregularly angled, black. An alliumlike genus of 5 spp., from the sw. U.S. and Mex., but lacking the odor and taste of onions. (Anagram of *Allium.*)

(Ingram, J. A monograph of the genera Bloomeria and Muilla. Madroño 12: 19–27, 1953.)

Fils. filiform or subulate. Cismontane . . . . 1. *M. maritima*

Fils. greatly dilated. Transmontane.
Scapes 3–15 cm. high; perianth 3–4 mm. long. W. Mojave Desert .... 2. *M. coronata*
Scapes 15–50 cm. high; perianth 6–8 mm. long. Lassen to Mono cos. .... 3. *M. transmontana*

1. **M. marítima** (Torr.) Wats. [*Hesperocordium* (?) *m.* Torr. *Allium m.* Benth. *Nothoscordum m.* Hook. f. *Bloomeria m.* Macbr. *M. tenuis* Congd. *M. serotina* Greene. *B. m.* var. *s.* Macbr.] Corm 1.2–2.0 cm. thick; scapes 1–5 dm. high; lvs. equaling or exceeding scape, ± scabrous; bracts 3–6, lanceolate, mostly 3-nerved; fls. 4–70; pedicels 1–5 cm. long; perianth-segms. greenish-white with brownish midnerve, 3–6 mm. long, the inner generally wider; fils. subulate, dilated toward base; anthers usually purplish; caps. 5–8 mm. high; seeds 2–3 mm. long.—On subalkaline flats and granitic or serpentine slopes, below 7500 ft.; Coastal Sage Scrub, Chaparral, V. Grassland, Foothill Wd., Yellow Pine F.; principally Coast Ranges, Glenn to San Diego cos., rarer in Cent. V. and cent. Sierran foothills; L. Calif. March–June.

2. **M. coronàta** Greene. Scapes 3.5–15 cm. high; lvs. longer, scabrous; fls. 3–10; perianth-segms. 3–4 mm. long, bluish or whitish within, green without; fils. petaloid, hyaline, conspicuously dilated, retuse at summit, their broad margins overlapping to form a cylindrical crown; anthers yellow.—Infrequent, in heavy soil, 3000–5000 ft.; Joshua Tree Wd., Shadscale Scrub; Lone Pine to Antelope V. March–April.

3. **M. transmontàna** Greene. [*Bloomeria t.* Macbr.] Corm 2.5–3 cm. thick; scape 1.5–5 dm. high; fls. 12–30; pedicels 2–3 cm. long; perianth white, with lilac tinge in age, 6–8 mm. long; fils. dilated at base to form a cuplike corona; anthers yellow, 1 mm. long; caps. 8–10 mm. long; seeds 4–5 mm. long.—Sagebrush Scrub, Yellow Pine F.; Lassen to Mono cos.; w. Nev. June.

3. Bloomèria Kell. GOLDEN STARS

Corm fibrous-coated; stem scapose. Lvs. basal, few, linear, carinate. Fls. yellow, many, in a loose umbel subtended by membranous bracts; pedicels jointed at summit. Perianth persistent, subrotate, of 6 subequal oblong-linear segms. Stamens 6; fils. filiform, each with a cup-shaped winged basal appendage; anthers versatile, but attached near the base. Ovules several in each locule; style filiform-clavate, persistent and splitting with the caps. Caps. subglobose, angular, loculicidal. Seeds black, subovoid, angular and wrinkled. Spp. 2, Calif. (H. G. *Bloomer,* early San Francisco botanist.)

(Ingram, J. A monograph of the genera Bloomeria and Muilla. Madroño 12: 19–27, 1953.)
Lf. one, 4–14 mm. wide; stamen-appendages papillose, bicuspidate at apex .... 1. *B. crocea*
Lvs. several, 1–2 mm. wide; stamen-appendages smooth, obtuse .... 2. *B. Clevelandii*

1. **B. cròcea** (Torr.) Cov. [*Allium c.* Torr. ?*B. gracilis* Borzi.] Corms ca. 15 mm. thick; scape 1.5–6 dm. high; lf. ca. half as long; bracts lanceolate; pedicels many, 2–6 cm. long; perianth-segms. orange-yellow, with median dark lines, 8–12 mm. long; fils. 6 mm. long, the appendages shallowly bicuspidate; style 5 mm. long; caps. 5–6 mm. long; seeds 2 mm. long; $n = 9$ (Burbanck, 1944).—Common, dry flats and hillsides, often in heavy soil, up to 5000 ft.; Chaparral, Coastal Sage Scrub, V. Grassland, S. Oak Wd.; S. Coast Ranges, Santa Barbara and w. Kern cos. to L. Calif.; Channel Ids. April–June.

Var. **aúrea** (Kell.) Ingram. [*B. a.* Kell. *Nothoscordum a.* Hook. f.] Perianth-segms. yellow, 11–12 mm. long; fil.-appendages with acute cusps.—Largely Foothill Wd.; Coast Ranges, San Benito and Monterey cos. to n. Santa Barbara Co.

Var. **montàna** (Greene) Ingram. [*B. m.* Greene.] Perianth-segms. yellow, 11–13 mm. long; fil.-appendages with linear-attenuate cusps ca. as long as the body.—Largely Chaparral, Yellow Pine F.; mts., w. Santa Barbara Co. to Tehachapi.

2. **B. Clevelándii** S. Wats. Scape 8–30 cm. tall; lvs. 6–15 cm. long; pedicels 2–3.5 cm. long; perianth-segms. 6–10 cm. long, yellow with greenish stripe; fils. 3–5 mm. long, appendages 3 mm. long, oblong, entire; style 1.5 mm. long; caps. 4–5 mm. long.—Dry mesas and hillsides; Chaparral, Coastal Sage Scrub; sw. San Diego Co. May.

4. Brodiaèa Sm. BRODIAEA

Perennial herbs with tunicated underground corms bearing few grasslike lvs. and scapose erect or occasionally twining stems. Fls. in loose or subcapitate umbels subtended by usually scarious bracts. Pedicels jointed beneath the perianth. Perianth-tube from short and saccate to elongate and funnelform; segms. similar or the outer narrow. Stamens 6, or the alternate transformed into dilated sterile fils. or staminodia. Fils. slender to broad; anthers versatile to basifixed. Ovary sessile or stipitate; style developed; stigma subentire to 3-lobed. Fr. a caps.; seeds black, rounded to elongate, variously punctate to striate, and longitudinally ridged. As here recognized, a genus of over 40 spp., of temp. w. N. Am. and w. S. Am. (especially Chile). Especially well developed in Calif., containing spp. of horticultural merit and possibility. The corms of some were eaten by the Indians. Many spp. are exceedingly variable and polyploidy is apparently common. (J. J. *Brodie,* Scotch cryptogamic botanist.)

(Hoover, R. F. A definition of the genus Brodiaea. Bull. Torr. Bot. Club 66: 161–166, 1939; A revision of the genus B. Am. Midl. Nat. 22: 551–574, 1939; The genus Dichelostemma. *Ibid.* 24: 463–476, 1940; A systematic study of Triteleia. *Ibid.* 25: 73–100, 1941.)

A. Anthers versatile, 6; corm-coats straw color; stigma not evidently lobed. (Subgenus Triteleia).
   B. Stamens alternately attached at 2 different levels.
      C. Perianth-tube obtuse and rounded at the base .... 1. *B. Douglasii*
      CC. Perianth-tube acute or narrow at the base.
         D. Anthers 1.5 mm. long; fils. dilated toward base. Ids. ...... 3. *B. clementina*
         DD. Anthers 2–5 mm. long; fils. not or scarcely dilated. Mainland.
            E. Pedicels 2–5 times the length of the perianth; perianth-tube acute at base, not attenuate .... 2. *B. peduncularis*
            EE. Pedicels not more than twice the length of the perianth; perianth-tube attenuate at base.
               F. Fils. 3–6 mm. long; stipe 2–3 times the ovary at anthesis .. 4. *B. laxa*
               FF. Fils. alternately 3 mm. and 1.5 mm. long; stipe ca. as long as ovary at anthesis .... 5. *B. crocea*
   BB. Stamens all attached at same level.
      C. Fils. forked at apex.
         D. Perianth-segms. abruptly spreading; stamens not more than half as long as segms.; anthers 1–2 mm. long .... 8. *B. lutea*
         DD. Perianth-segms. spreading, but at an angle to the tube; lower stamens ⅔ the segms.; anthers 2.5–3.5 mm. long .... 9. *B. versicolor*
      CC. Fils. not forked at apex.
         D. Perianth-tube more than ⅔ the length of the segms.
            E. The perianth-tube 8–12 mm. long; fils. of 2 lengths ...... 6. *B. Dudleyi*
            EE. The perianth-tube 17–25 mm. long; fils. subequal ...... 13. *B. Bridgesii*
         DD. Perianth-tube ⅓ to ½ the length of the segms.
            E. Longer fils. almost twice the length of the shorter .... 7. *B. lugens*
            EE. Longer fils. scarcely if at all longer than the others.
               F. Perianth-tube open-campanulate, 2–4 mm. long .... 10. *B. hyacinthina*
               FF. Perianth-tube attenuate at base, 4–7 mm. long.
                  G. Fls. 12–15 mm. long; ovary-stipe equaling the body at anthesis 11. *B. gracilis*
                  GG. Fls. 19–23 mm. long; ovary-stipe almost twice as long as body at anthesis .... 12. *B. Hendersonii*
AA. Anthers erect and appressed to style, attached near their base, 3 (6 in pulchella); inner corm-coats brown; stigma ± 3-lobed.
   B. Lvs. narrow, rounded, not keeled, on lower side; stigma 3-parted, with elongate recurved-spreading lobes; seeds obtusely angled. (Subgenus Brodiaea)
      C. Fils. with dorsal appendages extending back of the anthers.
         D. Staminodia 8–12 mm. long, linear; perianth-tube thin-membranous, splitting in fr. 21. *B. appendiculata*
         DD. Staminodia 4–6 mm. long, almost as wide as perianth-segms.; perianth-tube firm, not splitting in fr. .... 19. *B. stellaris*
      CC. Fils. not so appendaged.
         D. Perianth-segms. 3 times as long as tube, the latter splitting in fr. 20. *B. californica*
         DD. Perianth-segms. 1–2 times the length of the tube.
            E. Perianth-tube firm, not splitting in fr.; staminodia obtuse or acute.
               F. The perianth-tube strongly constricted above ovary, half as long as segms. .... 17. *B. minor*
               FF. The perianth-tube not or scarcely constricted above the ovary and more than half as long as the segms.
                  G. Earliest fls. of umbel with shortest pedicels; staminodia as wide as perianth-segms. .... 18. *B. pallida*
                  GG. Earliest fls. of umbel with longest pedicels; staminodia narrower than perianth-segms.
                     H. Fils. 2–6 mm. long, dilated at base only if at all; staminodia not incurved at apex.
                       I. Staminodia plane, acute, usually shorter than stamens; perianth-tube funnelform .... 14. *B. elegans*

II. Staminodia usually involute and obtuse, longer than stamens; perianth-tube subcampanulate to ovoid
15. *B. coronaria*
HH. Fils. 1 mm. long, broadly triangular; staminodia with incurved apex .... 16. *B. jolonensis*
EE. Perianth-tube thin-membranous and splitting in fr.; staminodia acuminate or lacking.
F. The fils. 1 mm. long; staminodia present .... 22. *B. filifolia*
FF. The fils. 4–6 mm. long; staminodia lacking .... 23. *B. Orcuttii*
BB. Lvs. flat, keeled on lower surface; stigma with 3 small lobes which run into wings on upper style; seeds sharply angled. (Subgenus Dichelostemma)
C. Anthers 6 .... 24. *B. pulchella*
CC. Anthers 3.
D. Fls. blue or violet, sometimes white; pedicels ascending in fr.
E. Perianth strongly constricted at throat; pedicels free from each other
25. *B. multiflora*
EE. Perianth scarcely constricted at throat; pedicels joined near base
26. *B. congesta*
DD. Fls. red or pink; pedicels spreading or recurved in fr.
E. Scape flexuous, twining; perianth-tube not much longer than segms.
27. *B. volubilis*
EE. Scape stiff, erect; perianth-tube 2–3 times as long as segms.
F. Perianth rose-red throughout; tube 14–17 mm. long .... 28. *B. venusta*
FF. Perianth-tube red, segms. green; tube 20–25 mm. long
29. *B. Ida-Maia*

1. **B. Douglásii** Wats. var. **Howéllii** (Wats.) Peck. [*B. H.* Wats. *Triteleia H.* Greene. *Hookera H.* Piper. *T. grandiflora* var. *H.* Hoov. *B. bicolor* Suksd. *B. g.* var. *H.* Peck.] Scape 2–7 dm. tall, glabrous; lvs. 1–3 dm. long, 4–10 mm. wide; pedicels 1–4 cm. long, ascending; perianth bluish-purple to almost white, 17–30 mm. long, the tube rounded at base, 8–18 mm. long; segms. divergent, 9–13 mm. long; fils. broadly deltoid, the upper row 3–4 mm. long, lower scarcely 2 mm.; anthers 3–4 mm. long; ovary-stipe half as long as body; seeds ca. 2 mm. long; $n = 16$ (Burbanck, 1941).—Open, moist to dry places, Hornbrook, Siskiyou Co.; to B.C., mostly w. of Cascade Range. April–May.

2. **B. pedunculàris** (Lindl.) Wats. [*Triteleia p.* Lindl. *Milla p.* Baker. *Hookera p.* Kuntze.] Scape 1–8 dm. tall, smooth; lvs. 2–4 dm. long, 5–15 mm. wide; pedicels 2–16 cm. long; perianth white or with some lilac, 15–26 mm. long, the tube acute at base, 7–10 mm. long; segms. divergent, 8–16 mm. long; fils. narrow-subulate, the longer 2–3 mm. long, the shorter 1–1.5 mm.; anthers white, 2–4 mm. long; ovary-stipe about as long as the yellow body at anthesis; $n = 7$, 14 (Burbanck, 1941).—Not common, low fields and in places wet during the growing season; Closed-cone Pine F., Mixed Evergreen F., Foothill Wd.; Coast Ranges, Humboldt and Tehama cos. to Monterey Co. May–July.

3. **B. clementìna** (Hoov.) Munz. [*Triteleia c.* Hoov.] Scape 3–4 dm. high, smooth; lvs. 3–5 dm. long, 4–10 mm. wide; pedicels 3–8 cm. long; perianth light blue, 17–25 mm. long, the tube acute at base, 7–10 mm. long; segms. erect, 10–15 mm. long; fils. 2 mm. long, triangular; anthers 1.5 mm. long; stipe of ovary ca. as long as body at anthesis; seeds 3–4 mm. long.—Damp clefts on rocky walls; Coastal Sage Scrub; San Clemente Id. March–April.

4. **B. láxa** (Benth.) Wats. [*Triteleia l.* Benth. *Seubertia l.* Kunth. *Milla l.* Baker. *Hookera l.* Kuntze. *T. candida* Greene. *B. c.* Baker. *B. laxa* var. *c.* Jeps. *B. l.* var. *nimia* Jeps. *B. l.* var. *Tracyi* Jeps. *T. angustiflora* Heller.] Grass Nut. Ithuriel's Spear. Scapes 1–7 dm. high, smooth to scabrous or retrorse-pubescent below; lvs. 2–4 dm. long, 4–25 mm. wide; pedicels 2–9 cm. long, usually slightly bent at apex; perianth blue to white, 20–45 mm. long, subhorizontal, with pistil on lower side and fils. curved upward; tube attenuate at base, 12–25 mm. long, with narrow membranous appendages within from the adnate fil.-base; segms. divergent, 8–20 mm. long; fils. of both rows 3–6 mm. long; anthers 2–5 mm. long; ovary-stipe 2–3 times the length of the body at anthesis; seeds 1.5–2 mm. long; $n = 14$, 21, 24 (Burbanck, 1941).—Common in heavy soils, below 4600 ft.; mostly Mixed Evergreen F., Foothill Wd., Chaparral; Coast Ranges, Curry Co., Ore. to San Bernardino Co., and Sierran foothills, Tehama Co. to Kern Co. April–June. Variable and with the following poorly defined tendencies: (1) plants from s. Sierra Nevada foothills and adjacent San Joaquin V. with large fls. (over 4 cm.), large lvs., fls. often white, are the var. *cándida* (Greene) Jeps.; (2) some of

those from around the Sacramento V. are often retrorsely hairy on lower scapes and with sky-blue fls.; (3) those from Marin Co. coast, with deep blue fls. ca. 2 cm. long and very long stipes, are var. *nimia* Jeps.; and (4) the n. coast form flowering in July and Aug. is var. *Tràcyi* Jeps.

5. **B. cròcea** (Wood) Wats. [*Seubertia c.* Wood. *Milla c.* Baker. *Triteleia c.* Greene. *Hookera c.* Kuntze.] Scape 1–3 dm. high, smooth to slightly scabrous below; lvs. 1.5–2.5 dm. long, 4–10 mm. wide; pedicels 7–20 mm. long; perianth bright yellow, 14–19 mm. long, the tube attenuate at base, 5–8 mm. long; segms. divergent, 9–11 mm. long; fils. slender, slightly dilated below, the upper 3 mm. long, the lower 1.5 mm. long; anthers 1.5–2 mm. long; ovary-stipe ca. as long as body; $n = 8$ (Burbanck, 1941).—Dry slopes, 4000–7000 ft.; Yellow Pine F.; Josephine and Jackson cos., Ore. to Trinity Co., Calif. May–June.

Var. **modésta** (Hall) Munz. [*B. m.* Hall. *Triteleia m.* Abrams. *T. c.* var. *m.* Hoov.] Plant smaller; fls. blue.—Similar places in n. Trinity Mts., Calif.

6. **B. Dúdleyi** (Hoov.) Munz. [*Triteleia D.* Hoov.] Scape 1–3 dm. tall, smooth; lvs. ca. as long, 4–8 mm. wide; pedicels slender, 15–35 mm. long; perianth pale yellow, drying purplish, 18–24 mm. long, the tube slender-funnelform, 8–12 mm. long; segms. spreading, 10–12 mm. long; fils. narrow-triangular, inserted at one level, but alternately long and short, ca. 3.5 and 2 mm. long; anthers 1 mm. long, lavender; ovary-stipe ca. as long as body.—Rare, black soil at 9500–11,500 ft.; Subalpine F.; Upper Tule R. region, Tulare Co. July.

7. **B. lùgens** (Greene) Baker. [*Triteleia l.* Greene. *Calliprora l.* Greene. *Hookera ixioides* var. *l.* Jeps. *B. i.* var. *l.* Jeps. *Calliprora i. l.* Abrams. *B. lutea* var. *l.* Mort.] Scape 1–4 dm. tall, smooth or somewhat scabrous; lvs. ca. as long, 3–10 mm. wide; pedicels 1–2.5 cm. long; perianth dull yellow, 12–14 mm. long, the tube funnelform, 4–5 mm. long; segms. spreading, with dark midvein and 8–9 mm. long; fils. alternately long and short, very broad, 3–4 and 1–2 mm. long, apically rounded; anthers 1.5–2 mm. long; stipe scarcely as long as body of ovary at anthesis.—Rare, apparently along streams; N. Coast Ranges, Sonoma Co. to Solano Co., n. of Mt. Waterman, Los Angeles Co.; Guadalupe Id.

8. **B. lùtea** (Lindl.) Mort. [*Calliprora l.* Lindl. *Ornithogalum ixioides* Ait. f. *Themis i.* Salisb. *Milla i.* Baker. *Triteleia i.* Greene. *B. i.* Wats., not Hook. *Hookera i.* Kuntze.] GOLDEN BRODIAEA. PRETTY FACE. Scape 2–8 dm. tall, smooth or slightly scabrous below; lvs. 2–4 dm. long, 3–14 mm. wide; pedicels 1.5–7 cm. long, curved upward; perianth golden-yellow with dark midveins, or purplish when dry, 15–24 mm. long, the tube acute at base, 6–8 mm. long; segms. divergent, 10–16 mm. long; fils. broad, flat, apically forked, 4–5 and 2.5–3 mm. long; anthers 1.5–2 mm. long, yellow to blue, borne between forks of fils.; ovary-stipe shorter than body.—Sandy soil; Closed-cone Pine F., Foothill Wd.; near the coast from San Mateo Co. to San Luis Obispo Co. May–Aug.

Var. **scàbra** (Greene) Munz. [*Calliprora s.* Greene. *B. s.* Baker. *B. ixioides* var. *s.* Smiley. *Triteleia i.* var. *s.* Hoov. *Calliprora aurantea* Kell. *B. a.* Mort.] Scape retrorsely pubescent or scabrous near base; perianth cream or straw color to deep yellow, the tube 4–6 mm. long; segms. spreading-deflexed, 10–17 mm. long; fils. 5–7 and 4–5 mm. long; anthers 1–2 mm. long; seeds 1.5 mm. long; $n = 5$ (Burbanck, 1941).—Frequent on dry grassy slopes or in open woods, heavy to granitic soils, 500–7000 ft.; V. Grassland, Foothill Wd., Yellow Pine F.; w. side of Sacramento V. in Tehama Co., and along Sierra Nevada foothills, Butte Co. to Kern Co. March–May.

Var. **analìna** (Greene) Munz. [*Calliprora scabra* var. *a.* Greene. *B. s.* var. *a.* Baker. *Calliprora a.* Heller. *Triteleia ixioides* var. *a.* Hoov.] Scape smooth or scabrous; perianth dull yellow, or bluish when dry, the tube 4–7 mm. long; segms. divergent-spreading, 7–12 mm. long; fils. rather narrow, 3.5–5 and 2.5–3.5 mm. long; anthers ca. 1 mm. long; $n = 25$ (Burbanck, 1941).—Frequent in open or shady places, sandy gravelly or loamy soil, 2000–10,000 ft.; Chaparral, Yellow Pine F., Lodgepole F.; Jackson Co., Ore. and Siskiyou Co., Calif., e. and w. slopes of Sierra Nevada s. to Tulare Co., Calif. May–July.

9. **B. versicólor** (Hoov.) Munz. [*Triteleia v.* Hoov.] Scape 3–6 dm. tall, scabrous at base; lvs. 7–11 mm. wide; pedicels spreading, 2–4 cm. long; perianth at first pale yellow, later white or even purple, the tube open-campanulate, 6–7 mm. long; segms. divaricate, 8–10 mm. long; fils. flat, 5–6 and 3.5–4 mm. long, the terminal forks sub-

parallel; anthers 2.5–3.5 mm. long; stipe half as long as ovary.—Rare, Closed-cone Pine F.; Point Lobos State Park, Monterey Co.

10. **B. hyacinthìna** (Lindl.) Baker. [*Hesperoscordum h.* Lindl. *Milla h.* Baker. *Triteleia h.* Greene. *Hookera h.* Kuntze. *Hesperoscordum lacteum* Lindl. *Allium l.* Benth. *Milla h.* var. *l.* Baker. *B. l.* Wats. *B. h.* var. *l.* Baker. *Hookera h.* var. *l.* Jeps. *Triteleia l.* Davids. & Mox. *Hesperoscordum Lewisii* Hook. *Veatchia crystallina* Kell. *Allium Tilingi* Regel. *B. lactea* var. *lilacina* Wats. *B. hyacinthina* var. *l.* Jeps. *B. dissimulata* Peck.] WHITE BRODIAEA. Scape 3–6 dm. tall, smooth or scabrous; lvs. 1–4 dm. long, 4–22 mm. wide; pedicels 5–50 mm. long; fls. usually white, or bluish with green midveins, the tube bowl-shaped, 2–4 mm. long; segms. spreading, 7–12 mm. long; fils. dilated toward base, 2–4 mm. long; anthers 1–2 mm. long, white to blue; stipe ca. half as long as ovary at anthesis; seeds ca. 1.5 mm. long; $n = 14$ (Burbanck, 1941).—Common in low moist places as meadows, vernal pools, and along streams, occasional on dryer slopes; ascending to 5500 ft.; Closed-cone Pine F., V. Grassland, Foothill Wd.; Coast Ranges, interior valleys and w. slope of the Sierra Nevada, from Monterey and Tulare cos. to Del Norte and Modoc cos.; to Vancouver Id., Idaho. An old collection reported from Camp Baldy, w. San Bernardino Co. A variable sp. with: (1) a large, white-fld. form from wet places and with large corms forming offsets; (2) a smaller rigid scabrous form from dry places in the Great V. and without offsets; and (3) a form with rather long pedicels (var. *láctea*). The above seem too indefinite to merit names.

Var. **Greènei** (Hoov.) Munz. [*Triteleia lilacina* Greene. *B. l.* Baker. *Hesperoscordum l.* Heller ex Abrams. *T. hyacinthina* var. *Greenei* Hoov. Not *B. lactea* var. *l.* Wats. or *B. h.* var. *l.* Jeps.] Perianth-tube with internal hyaline vesicles evident only in fresh material; fils. filiform; anthers blue.—Volcanic soil, plains and foothills; V. Grassland, Foothill Wd.; from Tehama Co. to Tuolumne Co.

11. **B. grácilis** Wats. [*Triteleia g.* Greene, not Phil. *Hookera g.* Kuntze. *T. montana* Hoov.] Scape 5–25 cm. tall, ± scabrous; lvs. 1–3 dm. long, 2–5 mm. wide; pedicels 5–30 mm. long; perianth yellow, or purplish in age, the tube slender, 4–5 mm. long; segms. somewhat spreading, 8–10 mm. long, the midveins brown outside; fils. filiform, 5–6 mm. long; anthers cream to bluish, 1–1.5 mm. long; stipe ca. as long as ovary.— Locally rather plentiful on gravelly plains and granitic ridges, 4000–9800 ft.; Yellow Pine F., Red Fir F., Lodgepole F.; Sierra Nevada from Plumas to Mono and Mariposa cos. June–July.

12. **B. Héndersonii** Wats. [*Triteleia H.* Greene.] Scapes 1–3 dm. high, smooth or somewhat scabrous; lvs. 2–3.5 dm. long, 3–12 mm. wide; pedicels 1.5–4 cm. long; perianth yellow, often tinged blue when dry, the tube narrow-funnelform, 6–7 mm. long; segms. wide-spreading, 13–16 mm. long, with dark midvein; fils. slender, 3–4 mm. long; stipe nearly twice the ovary at anthesis; $n = 16$ (Burbanck, 1941).—Dry slopes in s. Ore.; reported from "Northwest California." May–July.

13. **B. Brídgesii** Wats. [*Triteleia B.* Greene. *Hookera B.* Kuntze.] Scapes 1–5 dm. high, smooth or somewhat scabrous or retrorse-pubescent below; lvs. 2–5.5 dm. long, 3–10 mm. wide; pedicels 2–9 cm. long; perianth lilac or blue, sometimes pale or pinkish, the tube 17–25 mm. long, with very slender base; segms. 10–20 mm. long, spreading; fils. triangular, 3–4 mm. long; anthers blue, 3.5–4.5 mm. long; stipe 3–4 times the ovary at anthesis; $n = 8$ (Burbanck, 1941).—Dry bluffs and hillsides, often at edge of woods; Mixed Evergreen F.; Humboldt Co. to s. Ore.; and below 3000 ft., in Foothill Wd., Yellow Pine F., Shasta Co. to Mariposa Co. April–June.

14. **B. élegans** Hoover. [*Hookera coronaria* auth., not Salisb. *B. c.* auth.] HARVEST BRODIAEA. Scape mostly 1–4 dm. tall; lvs. ca. same length, 2 mm. wide, apparently largely withered at anthesis; pedicels 1–8 cm. long; perianth violet to deep blue-purple, rarely pink, 25–42 mm. long, the tube funnelform, 9–17 mm. long; segms. 17–26 mm. long, ascending-recurved; staminodia erect, 9–11 mm. long, distant from the stamens; fils. angular, 4–6 mm. long; anthers 7–10 mm. long; stipe shorter than the obovoid ovary; seeds 2–3 mm. long; $n = 16$ (Burbanck, 1941).—Common, usually in heavy soils of open or wooded plains and foothills, or even in wet meadows, up to 7000 ft.; V. Grassland, Foothill Wd., and Yellow Pine F.; Monterey and Tulare cos. to Ore.; occasional in San Gabriel and San Bernardino mts. at ca. 5000 ft.; rare near the coast. April–July, or even later near the coast.

Var. **múndula** (Jeps.) Hoov. [*B. coronaria* var. *m.* Jeps.] Scapes 4–6 dm. high; pedicels

6–15 mm. long; staminodia 7–8 mm. long; fils. 3 mm. long.—Tuolumne Co. at ca. 2500 ft.

15. **B. coronària** (Salisb.) Engler. [*Hookera c.* Salisb. *B. grandiflora* Sm. *B. synandra* Jeps., not *H. s.* Heller. *B. s.* var. *insignis* Jeps. *B. Howellii* Eastw., not Wats.] Scape 5–25 cm. tall; lvs. as long or longer; pedicels mostly few, 2–10 cm. long; perianth lilac to violet, 20–23 mm. long, the tube rounded at base, 7–11 mm. long; segms. spreading, 13–22 mm. long; staminodia usually leaning toward and sometimes covering stamens, involute, 6–11 mm. long; fils. 2–4 mm. long, flat, dilated at base; anthers 4–7 mm. long; caps. ovoid; seeds 1.5–2 mm. long; $n = 21$ (Johansen, 1932).—Dry adobe or clay soil or gravelly places, on grassy hillsides, alkaline plains and wooded slopes, below 5000 ft.; V. Grassland, Foothill Wd., Yellow Pine F.; most common about the Sacramento V., n. to Shasta and Modoc cos. and B.C.; less frequent e. of San Joaquin V. and in Coast Ranges s. to Cuyamaca and Laguna mts. May–July. In addition to the vars. named below, variants occur such as lavender-fld. forms in dry hard-packed soil and deep violet forms in more moist places.

Var. **ròsea** (Greene) Hoov. [*Hookera r.* Greene. *B. r.* Baker.] Fls. pale lavender, drying pink; staminodia folded around the stamens, abruptly constricted at base; fils. triangular.—Indian V., Lake Co.

Var. **kérnensis** Hoov. Scape 3–7 cm. tall, about as long as pedicels; staminodia erect, not bent toward stamens.—V. Grassland, Foothill Wd.; Kern Co. April–May.

Var. **macropòda** (Torr.) Hoov. [*B. grandiflora* var. *m.* Torr. *Hookera m.* Kuntze. *B. terrestris* Kell., originally spelled *terrestria. Hookera t.* Britten. *B. Torreyi* Wood.] Scape to 5 cm. high, the portion above ground usually shorter than pedicels, these 3–20 cm. long; staminodia not closely folded about anthers.—Heavy or lighter soils; Closed-cone Pine F., Mixed Evergreen F., Foothill Wd.; San Luis Obispo Co. to Coos Co., Ore. April–July.

16. **B. jolonénsis** Eastw. Scape 3–20 cm. high; lvs. usually longer; pedicels 2–12 cm. long; perianth violet, 18–27 mm. long, the tube rounded at base, 8–13 mm. long; segms. spreading, 10–14 mm. long; staminodia violet, involute, 4–6 mm. long, incurved at apex; fils. 1 mm. long, broader than anthers; these 3–5 mm. long, emarginate at apex; caps. ovoid.—Clay depressions on mesas and gentle slopes, below 1500 ft., V. Grassland, Coastal Sage Scrub, Chaparral; Monterey Co. to San Diego Co.; Santa Cruz Id. April–June.

17. **B. mìnor** (Benth.) Wats. [*B. grandiflora* var. *m.* Benth. *Hookera m.* Britten. *B. Purdyi* Eastw. *H. P.* Heller.] Scape 1–3 dm. tall; pedicels 2–6 cm. long; perianth violet, 15–25 mm. long, tube rounded at base, much constricted at throat; segms. oblong, divaricate-recurved, 10–18 mm. long; staminodia 7–10 mm. long, 2–2.5 mm. wide, involute, curved outward at tips; fils. 1–2 mm. long; anthers 4–6 mm. long, emarginate; caps. short-ovoid; seeds 1.5–2 mm. long; $n = 16$ (Burbanck, 1941).—Heavy to gravelly soil of dry open flats, below 3500 ft.; V. Grassland, Foothill Wd.; Sacramento V. and adjacent foothills, Shasta Co. to Amador Co. May–June.

Var. **nàna** (Hoov.) Hoov. [*B. n.* Hoov.] Scape 1–5 cm. above ground; perianth-segms. bluish-lilac, 8–15 mm. long, the inner almost obovate; staminodia 5–7 mm. long; anthers 3–4 mm. long; $n = 6$ (Burbanck, 1941).—With the sp., Tehama Co. to Sacramento Co., Butte Co. to Merced Co.

18. **B. pállida** Hoov. Like *B. minor,* but the first fls. with pedicels scarcely 1 cm. long, the later ones 2–6 cm. long; perianth-tube cylindric, 7–10 mm. long; segms. 10–15 mm. long; staminodia 7–10 mm. long, 3.5–4 mm. wide, erect; fils. with strongly reflexed margins; anthers covered with fleshy hairlike processes.—Known only from Chinese Camp, Tuolumne Co.

19. **B. stellàris** Wats. [*Hookera s.* Greene.] Scape 1–3 dm. tall; pedicels 3–15 cm. long; perianth violet-purple, 17–23 mm. long, the tube greenish, campanulate, 7–10 mm. long, firm; segms. spreading, 10–13 mm. long; staminodia nearly as wide as perianth-segms., 4–6 mm. long, white, involute, retuse, loosely folded around the stamens; fils. 2 mm. long, broad, each with 2 appendages extending along the back of the anther; anther 4–5 mm. long, deeply notched; caps. fusiform; $n = 6$ (Johansen, 1932).—Clay soil or rocky slopes, open places, below 3000 ft.; Redwood F., Mixed Evergreen F.; inner portion of outer Coast Ranges, Humboldt Co. to Sonoma Co. May–July.

20. **B. califórnica** Lindl. [*Hookera c.* Greene. *B. grandiflora* vars. *elatior* and *major*

Benth.] Corm without offsets; scape 2–7 dm. tall, often scabrous; pedicels 2–10 cm. long; perianth violet to lilac-purple, rarely pink, 28–45 mm. long, the tube rounded at base, 8–12 mm. long, firm and brittle in fr.; segms. 20–23 mm. long, ascending-recurved; staminodia 15–28 mm. long, folded around stamens; fils. angular, 6–12 mm. long; anthers 9–13 mm. long, notched; caps. ovoid, ca. 1 cm. long; seeds ca. 3.5 mm. long; $n = 5$ (Johansen, 1932), 6 (Burbanck, 1941).—Wooded hills and open plains, below 2500 ft.; V. Grassland, Foothill Wd.; Shasta Co. to Butte and Nevada cos. May–July.

Var. **leptándra** (Greene) Hoov. [*Hookera l.* Greene. *B. l.* Baker. *H. synandra* Heller. *B. s.* Jeps., as to name only.] Corm with offsets; caps. smaller.—Similar situations; Sonoma and Napa cos.

21. **B. appendículata** Hoov. Scape 1–4.5 dm. tall; perianth deep violet-purple, the tube 8–10 mm. long, brittle in fr.; segms. not recurved, 15–20 mm. long; staminodia linear, 8–12 mm. long, approximate around the anthers; fils. 4–5 mm. long, flat, with 2 filiform appendages from apex, 3–5 mm. long and extending back of the anthers, these 7–8 mm. long, notched; caps. subglobose; seeds ca. 2 mm. long.—Heavy soil of low open hills and plains, below 1800 ft.; V. Grassland, Foothill Wd.; Sutter Co. to Madera Co., occasional in Coast Ranges, Napa Co. to Santa Clara Co. April–May.

22. **B. filifòlia** Wats. [*Hookera f.* Greene.] Scape 2–4 dm. high; lvs. several, shorter than or nearly as long as scapes, 1–2 mm. wide; pedicels 2–5 cm. long; perianth violet, the tube greenish, narrow-campanulate, 6–7 mm. long, membranous and splitting in fr.; segms. spreading, 9–12 mm. long; staminodia plane, linear, 6–7 mm. long when fresh, 2–3 mm. in dry fls., curved outward above; fils. 1 mm. long, triangular; anthers 4 mm. long, broad, notched; caps. short-ovoid; seeds 2–2.5 mm. long.—Local, in heavy clay soil, below 2000 ft.; Coastal Sage Scrub, Chaparral; Glendora, San Bernardino V., Perris, Vista, San Clemente Id. May–June.

23. **B. Orcúttii** (Greene) Baker. [*Hookera O.* Greene. *B. filifolia* var. *O.* Jeps. *H. multipedunculata* Abrams.] Scape 1–4 dm. high; pedicels 2–8 cm. long; perianth violet, the tube 5–8 mm. long, narrow-campanulate, thin-membranous and brittle in fr.; segms. 10–18 mm. long, ascending; staminodia lacking; fils. slender, 4–6 mm. long; anthers 5–6 mm. long, subentire at apex.—Near streams, about vernal pools and seeps, up to 5500 ft.; Chaparral, Yellow Pine F.; San Diego Co. April–July.

24. **B. pulchélla** (Salisb.) Greene. [*Hookera p.* Salisb. *Dichelostemma p.* Heller. *Dipterostemon p.* Rydb. *B. parviflora* Torr. *Hookera p.* Torr. *B. capitata* Benth. *Dichelostemma c.* Wood. *Milla c.* Baker. *Hookera c.* Kuntze. *Dipterostemon c.* Rydb. *B. insularis* Greene. *Dichelostemma i.* Burnham. *Dipterostemon i.* Greene. *B. c.* var. *i.* Macbr.] BLUE DICKS. WILD-HYACINTH. Scapes 3–6(–9) dm. high, smooth; lvs. 1.5–4 dm. long, 5–12 mm. wide; bracts purple, usually ovate; pedicels 2–15 mm. long; perianth-tube pale, 4–8 mm. long, cylindro-campanulate; the segms. violet, rarely white, ascending, 7–11 mm. long; fils. opposite the outer segms. dilated, 2 mm. long, bearing anthers 2–3 mm. long, those opposite the inner segms. adnate but extending beyond the anthers (which are 3.5–4.5 mm. long) as 2 lanceolate appendages; style 4–6 mm. long; caps. ovoid, 4–6 mm. long; sessile; seeds 2.5–4 mm. long; $n = 36$ (Johansen, 1932), 18 (Lenz, unpub.), 9 Burbanck, 1941).—Common on plains and hillsides below Yellow Pine F., in most portions of Calif. w. of Sierra Nevada; to Douglas Co., Ore., L. Calif.; more uncommon e. of the Sierra, in Pinyon-Juniper Wd. and Yellow Pine F., to s. Utah and n. Ariz. Mostly March–May.

Var. **pauciflòra** (Torr.) Mort. [*B. capitata* var. *p.* Torr. *Milla c.* var. *p.* Baker. *Dichelostemma p.* Standl. *Dipterostemon p.* Rydb. *Hookera p.* Tides.] Bracts lanceolate, not or scarcely colored; pedicels 6–35 mm. long; perianth segms. pale blue, spreading.—Dry open places; Creosote Bush Scrub; deserts of s. Calif.; New Mex., Son. Intergrading freely with the sp.

25. **B. multiflòra** Benth. [*Hookera m.* Britten. *Dichelostemma m.* Heller. *B. grandiflora* var. *brachypoda* Torr.] Scape 3–8 dm. high, usually scabrous; lvs. 3 or more, equaling or exceeding scape, 3–8 mm. wide; bracts ovate, purplish; pedicels stiff, 3–15 mm. long; umbels subglobose; perianth-tube pale, 8–10 mm. long, inflated, cartilaginous in fr., constricted at throat; segms. violet to lilac, 8–10 mm. long, spreading at anthesis; staminodia white to violet, 5–6 mm. long, entire; fils. adnate, not appendaged; anthers 4–5 mm. long; caps. sessile, ovoid, 8 mm. long; seeds 2.5–3.5 mm. long; $n = 15$ (Johansen, 1932).—Clay soil, open and wooded slopes, below 5000 ft.; Foothill Wd., Yellow

Pine F.; s. Ore to Trinity and Sacramento cos., along the Sierra Nevada to Mariposa Co., also in San Mateo and Santa Clara cos. May–June.

26. **B. congésta** Sm. [*Dichelostemma c.* Kunth. *Hookera c.* Jeps. *B. pulchella* auth., not Greene.] Ookow. Scapes 4–10 dm. high; lvs. 2 or more, 6–12 mm. wide, almost as long as scape; pedicels joined near base, the free part 1–6 mm. long; perianth-tube slightly constricted at throat, 6-angled, 8–10 mm. long; segms. ca. as long, blue-violet, ascending; staminodia 5–6 mm. long, bifid; fils. adnate; anthers 4–5 mm. long; caps. subsessile, ovoid; *n* = 18 (Burbanck, 1941).—Open hills; Foothill Wd., Mixed Evergreen F.; Santa Clara Co. and Sacramento Co. to Island Co., Wash. April–June.

27. **B. volùbilis** (Morière) Baker. [*Rupalleya v.* Morière. *Macroscapa v.* Kell. *Dichelostemma v.* Heller. *Hookera v.* Jeps. *Stropholirion californicum* Torr. *Dichelostemma c.* Wood. *B. c.* Jeps., not Lindl.] Snake-Lily. Twining Brodiaea. Scape 5–15 dm. long, contorted and twining over bushes; lvs. 3–7 dm. long, 8–14 mm. wide; pedicels 1–4 cm. long, flexuous and spreading or drooping at anthesis, later curved upward; perianth-tube inflated, 6-angled, 5–7 mm. long, equally wide; segms. rose to pink, 5–7 mm. long, spreading at anthesis, later erect; staminodia white, narrow-oblong, 2.5–3 mm. long, bifid; fils. adnate, apically bearing 2 oblong appendages ca. as long as anthers; anthers 3–4 mm. long; caps. subsessile, ovoid; seeds ca. 3 mm. long; *n* = 18 (Johansen, 1932), 9 (Burbanck, 1941).—On clay or granite, bushy or open slopes, below 2500 ft., Butte Co. to Kern Co., inner Coast Ranges, Tehama Co. to Solano Co.

28. **B. venústa** (Greene) Greene. [*Brevoortia v.* Greene. *Dichelostemma v.* Hoov.] especially on recent burns; Chaparral, Foothill Wd.; foothills of Sierra Nevada from Scape 4–9 dm. tall; pedicels 1–2 cm. long; fls. rose, the perianth-tube distended-cylindric, 14–17 mm. long; segms. spreading, 7–8 mm. long; staminodia 3–3.5 mm. long, ca. as wide, rounded at tip; fils. obsolete; anthers joined to staminodia, 5–6 mm. long; caps. short-stipitate.—Rare, Del Norte Co. to Trinity Co.

29. **B. Ìda-Màia** (Wood) Greene. [*Brevoortia I.-M.* Wood. *Dichelostemma I.-M.* Greene. *B. coccinea* Gray. *Brevoortia c.* Wats.] Fire-Cracker Fl. Scape 3–9 dm. tall; lvs. mostly 3, 3–5 dm. long, 4–8 mm. wide; pedicels 1–5 cm. long, somewhat pendulous at anthesis, curved upward in fr.; perianth-tube bright red, 20–25 mm. long, broadly tubular; segms. greenish, lance-ovate, 7–9 mm. long, recurved at anthesis; staminodia white, involute, apically rounded, 3–4 mm. long, somewhat wider; fils. very short; anthers 6–8 mm. long; caps. stipitate, triangular-ovoid; seeds ca. 4 mm. long; *n* = ca. 20 (Johansen, 1832), 25 (Burbanck, 1941).—Grassy slopes in open places, 1000–4000 ft.; Redwood F., Mixed Evergreen F.; Curry Co., Ore. to Mendocino, Lake and Shasta cos., Calif. May–July.

5. Androstèphium Torr.

Flowering stem scapose, from fibrous-coated corm. Lvs. linear, channeled. Bracts several, scarious. Umbel few-fld., pedicels not jointed. Perianth funnelform, segms. 6, narrow-oblong, united below. Stamens 6, in one row in the throat; fils. partly united into a tube with erect bifid lobes between the versatile anthers. Ovary sessile, with persistent style. Caps. subglobose, obtusely 3-angled. Seeds several in each locule, black. Three spp. of sw. U.S. (Greek, *andros*, stamen, and *stephanos*, crown, referring to united fils.)

1. **A. breviflòrum** Wats. Scape 1–3 dm. high, scabrous below; lvs. several, ca. as long as scape, 2 mm. wide, scabrous; fls. 3–12; pedicels 15–30 mm. long; perianth whitish to light violet-purple, drying brownish-yellow, 15–20 mm. long; segms. longer than tube; anthers 3 mm. long; caps. 10–15 mm. long, deeply 3-lobed; seeds 8–9 mm. long, flat.— Dry slopes and plains below 5000 ft.; Creosote Bush Scrub; supposedly e. Calif.; to w. Colo.

22. Iridàceae. Iris Family (Fig. 130)

Perennial herbs, mostly low with simple or branching stems. Lvs. mostly basal and equitant, parallel-veined, linear or sword-shaped. Fls. terminal, showy, bisexual, issuing from a spathe of herbaceous or scarious bracts. Perianth of 6 parts in 2 series, the outer sometimes calyxlike, all united into a tube that is adnate to the ovary. Stamens 3;

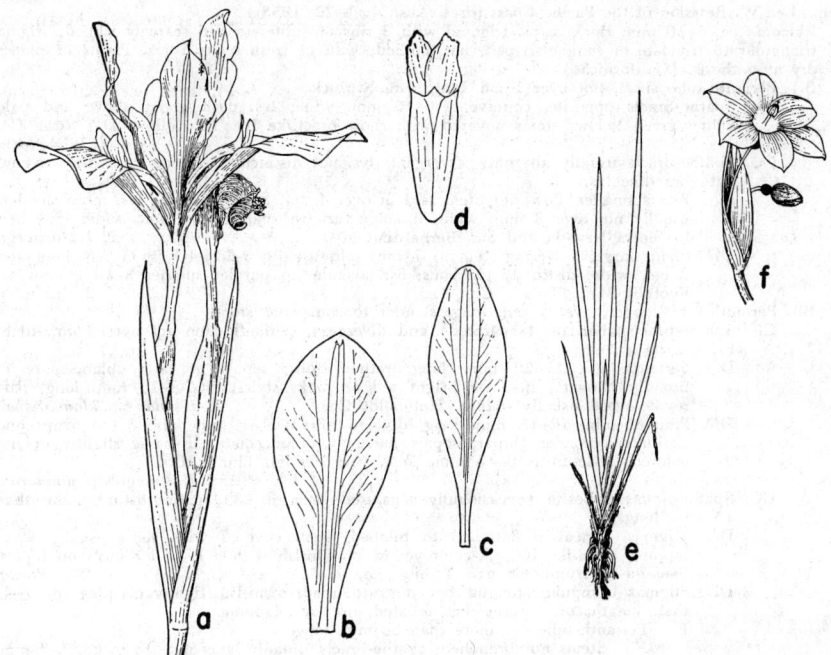

Fig. 130. IRIDACEAE. *Iris Douglasiana: a*, fl., × ½, in axil of 2 spathe-valves, and with inferior ovary, 3 spreading perianth-segms. (sepals or falls) and 3 erect ones (petals or standards); *b*, fall, × ⅔; *c*, standard, × ⅔; *d*, petallike branch of style, seen from beneath, with a 2-lobed crest and a flat median scale with the stigmatic surface. *Sisyrinchium bellum: e*, habit, × ¼, to show fan of equitant flat lvs.; *f*, umbellate infl. subtended by a spathe of 2 bracts, the fl. with a regular perianth.

anthers extrorse, opposite the outer perianth-segms. Pistil 1; ovary inferior, mostly 3-loculed and with axile placentation; style single; stigmas 3, sometimes expanded and petallike or divided. Fr. a few- to many-seeded loculicidal caps. Ca. 1200 spp., a cosmopolitan family with many plants used for ornamental purposes (gladioli, irises, freesias, crocuses, ixias, etc.).

Infl. umbellate; perianth-segms. all alike, the perianth not more than 2 cm. long .... 2. *Sisyrinchium*
Infl. not umbellate; perianth-segms. not all alike, or if so, more than 2 cm. long.
  Perianth with 3 erect and 3 spreading or drooping perianth-segms.; fls not in elongate spikelike infl.
                                                  1. *Iris*
  Perianth not as above; infl. spikelike.
    The perianth irregular, the 3 upper segms. larger than the 3 lower .... 3. *Gladiolus*
    The perianth regular or essentially so .... 4. *Watsonia*

1. Íris L. Iris. Fleur-de-lis

Perennial herbs with creeping ± tuberous rhizomes or bulblike base. Stems erect, simple or branched, with 1–several fls. at the top and equitant mostly radical and basal lvs. (these sometimes flexuously grasslike). Fls. in axils of bracts (or spathe) which may be opposite or alternate. Perianth of 6 clawed segms. united below into a tube, the 3 outer broad, spreading or reflexed (the sepals or falls), the 3 inner usually narrower and erect (the petals or standards). Stamens inserted at the base of the falls; anthers linear or oblong. Ovary 3-loculed; style with 3 petallike branches that arch over the stamens and bear a stigmatic surface beneath on a flat scale just below the usually 2-lobed tip (crest); base of style adnate to perianth-tube. Caps. oval or oblong, 3–6-angled or -lobed, many-seeded. Seeds vertically compressed, in 1 or 2 rows in each locule. Perhaps 150 spp., mostly in N. Temp. Zone. (Greek, the rainbow, because of variegated color in fls.)

(Lenz, Lee W. Revision of the Pacific Coast irises. Aliso 4: 1–72, 1958.)

A.   Rhizome ca. 5–10 mm. thick; caps. trigonal with 3 ribs or subterete and scarcely ribbed; stigma triangular to truncate or tongue-shaped, not toothed; wall of fresh ovary thick. Plants of places dry at anthesis. (Californicae)

    B.   Perianth-tube short, not over 1 cm. long, usually thick.

        C.   Spathe-bracts opposite, connivent, 6–10 mm. wide; lvs. dark green above and pale yellow-green below; stems covered with short bractlike lvs.; fls. yellow. Del Norte Co.
                                                                     8. *I. bracteata*

        CC.   Spathe-bracts usually alternate, divergent; lvs. not distinctly 2-sided; stems not covered with bractlike lvs.

            D.   Plants smaller, flowering stem seldom over 4 dm. high, 1–3-fld.; lvs. more slender, usually not over 8 mm. wide; fl.-color various; spathes 3–9 mm. wide. Pine belt in Sierra Nevada and San Bernardino Mts. . . . . 2. *I. Hartwegii*

            DD.   Plants large, flowering stem to 7 dm. tall, usually 3-fld.; lvs. to 5 dm. long and 2 cm. wide, distinctly glaucous; fls. lavender to purple; spathes 8–14 mm. wide. Foothill Wd., Tulare Co. . . . . 3. *I. Munzii*

    BB.   Perianth-tube longer, over 1 cm. long, slender to somewhat stout.

        C.   Spathe-bracts alternate (separated) and divergent; perianth-tube not over 2 cm. long; fls. 1–2 on a stem.

            D.   Perianth-tube 11–20 mm. long, rather stout; sepals or falls oblanceolate to broadly obovate; lf.-bases bright red or pink; style-crests 8–16 mm. long, narrowly ovate, usually obtuse. Humboldt Co. . . . . 1. *I. tenax* ssp. *klamathensis*

            DD.   Perianth-tube 12–15 mm. long, divided into a short tube above the ovary and a dilated broader throat; sepals narrowly oblanceolate; lf.-bases slightly or not colored; style-crests 9–13 mm. long, very slender. Plumas Co.
                                                  2. *I. Hartwegii* ssp. *pinetorum*

        CC.   Spathe-bracts opposite (occasionally separated, then fls. 3); perianth-tube more than 15 mm. long.

            D.   Stigmas truncately flattened to bilobed; stem covered with short usually overlapping bractlike lvs.; fl.-color yellow or whitish with a lavender flush on sepals. Sonoma to Humboldt and Trinity cos. . . . . 9. *I. Purdyi*

            DD.   Stigmas triangular, tongue-shaped or rounded; stem-lvs. tightly clasping for ½–⅓ their length, or if somewhat inflated, not over-lapping.

                E.   Perianth-tube not more than 30 mm. long.

                    F.   Stems not branched; spathe-bracts broadly lanceolate to ovate, 5–7 mm. wide, 33–60 mm. long; perianth-tube 15–30 mm. long; fl.-stem 1–2-fld.; lvs. narrow, grasslike. Del Norte Co. . . . . 10. *I. innominata*

                    FF.   Stems usually branched, each branch with 2–3 fls.; spathe-bracts lanceolate, acuminate, 7–12 mm. wide, 60–120 mm. long; perianth-tube 15–28 mm. long; ovary triangular in cross-sect. with nipplelike projection at apex; lvs. to 2 cm. wide, deep green. Santa Barbara Co. to Ore.
                                                         11. *I. Douglasiana*

              EE.   Perianth-tube over 30 mm. long, usually slender.

                  F.   Style-crests long and very slender, occasionally as long as the style-branches; perianth-parts usually very narrow, wide-spreading, fragile; fl.-color whitish to cream, with dark veining.

                    G.   Perianth-tube lacking distinct throat, upper part often with a short bowllike enlargement. Del Norte Co. . . 12. *I. chrysophylla*

                    GG.   Upper part of perianth-tube dilated to form a conspicuous throat. Plumas and Sierra cos., Glenn and Trinity cos. to Butte Co.
                                                       5. *I. tenuissima*

                  FF.   Style-crests short, rounded, much shorter than the style-branches.

                    G.   Spathe-bracts 4–9 mm. wide, linear-lanceolate; perianth-tube never with a distinct throat; fl.-stem to 2.5 dm. tall; lvs. to 5 mm. wide; lf.-bases usually colorless; fl.-color deep purple to golden-yellow. Both sides of Cent. V. . . . . 6. *I. macrosiphon*

                    GG.   Spathe-bracts 6–11 mm. wide, broadly lanceolate; perianth-tube sometimes showing a distinct throat; fl.-stems 2–4 dm. tall; lvs. to 7 or 8 mm. wide, drying a peculiar gray-green color, entire plant often intensely colored with red; fls. creamy-yellow, sometimes veined darker. Lake and Napa cos. to Santa Cruz Co.
                                                         7. *I. Fernaldii*

AA.   Rhizome ca. 20–30 mm. thick; caps. hexagonal or each of the 3 angles 2-ribbed; stigma 2-lobed; wall of fresh ovary paper-thin. Plants of meadows or places wet at anthesis. (Longipetalae)

    B.   Lvs. light green, mostly 3–6 mm. wide; spathes largely scarious; fls. usually 2–3 on a stem. Plants mostly of Montane Coniferous F. . . . . 12. *I. missouriensis*

    BB.   Lvs. dark green, ca. 6–10 mm. wide; spathes largely herbaceous; fls. up to 8 on a stem. Plants from near coast . . . . 13. *I. longipetala*

1. **I. tenáx** Dougl. ssp. **klamathénsis** Lenz. Rhizome slender, 3–5 mm. in diam.; lvs. 3–5 mm. wide, to 4 dm. long, usually exceeding the fl.-stem; lf.-bases brilliantly colored; fl.-stem simple, 1(2)-fld., to 4 dm. long, with 1–several clasping lvs. free ca. ⅓ their length; spathe-bracts linear to narrowly lanceolate, divergent, widely separated on stem, the outer 3–6 mm. wide, 4.7–6.7 cm. long; pedicel 4–15 mm. long at anthesis;

ovary 1.3–3.0 cm. long, stoutish, ± funnelform; sepals oblanceolate to broadly obovate, 5.3–7.3 cm. long, 1.5–2.4 cm. wide; petals narrowly lanceloate, 3.7–6.2 cm. long, 0.7–1.1 cm. wide; fl.-color pale buff-yellow, sepals usually distinctly marked with deep maroon or brown veins; style-branches 2–3 cm. long; style-crests narrowly ovate, usually obtuse, 8–16 mm. long; stigmas triangular; fils. ca. 7 mm. long; anthers 11–14 mm. long; caps. oblong, 3–4 cm. long.—Wooded hillsides; Mixed Evergreen F.; near Orleans, Humboldt Co. May. Natural hybrids with *I. chrysophylla* occur.

2. **I. Hartwègii** Baker. Rhizome 5–8 mm. in diam., usually covered with remains of old lvs.; lvs. few, 2–6 mm. wide, up to 4.5 dm. long, usually lacking anthocyanin pigments at base; fl.-stem slender, simple, 0.5–3 dm. tall, with 1–several sheathing lvs. free ca. ½ their length; fls. 1–2; spathe-bracts linear to lance-linear, the outer 4–7 mm. wide, 5–11.5 cm. long, usually divergent and as much as 4 cm. apart on the stem, herbaceous; pedicels 1.7–8.5 cm. long; ovary subcylindrical, 10–20 mm. long; perianth-tube 5–10 mm. long; sepals 4–7 cm. long, 1.4–2 cm. wide; petals narrowly oblanceolate, 3.5–6 cm. long, 5–11 mm. wide; fl.-color variable, usually pale yellow or cream, sometimes lavender or deep yellow; style-branches 1.6–3 cm. long; style-crests obtusely rounded, 0.5–1.1 cm. long; anthers ca. 1 cm. long; caps. oblong-oval, 2–3 cm. long, rather abruptly tapered at both ends.—Wooded slopes, 2000–6000 ft.; Yellow Pine F.; w. base of Sierra Nevada, Butte to Kern cos. May–June. Natural hybrids with *I. macrosiphon* are found.

KEY TO SUBSPECIES

Perianth-tube 5–10 mm. long.
  Fls. 1–2 on a stem; anthers 10–15 mm. long; lvs. 3–8 mm. wide.
    Spathe-bracts 4–7 mm. wide; fl.-stem to 3 dm. tall; fl.-color pale yellow to golden-yellow and lavender. Sierra Nevada .... *I. Hartwegii*
    Spathe-bracts 6–9 mm. wide; fl.-stem to 4 dm. tall; fl.-color lavender to purple. San Bernardino Mts. and adjacent ranges .... ssp. *australis*
  Fls. 3 on a stem; anthers ca. 18 mm. long; lvs. to 10 mm. wide; fls. pale yellow with golden-yellow veining. Tuolumne Co. .... ssp. *columbiana*
Perianth-tube 12–15 mm. long; lvs. pale green, ca. 5 mm. wide; fls. 2 on a stem; fls. pale yellow with deep gold veining. Plumas Co. .... ssp. *pinetorum*

Ssp. **austràlis** (Parish) Lenz. [*I. H.* var. *a.* Parish. *I. tenax* var. *a.* Foster.] Rhizome 6–9 mm. thick; lvs. 3–8 mm. wide, to 4 or 5 dm. long, usually pinkish at base; fl.-stem simple, 1–4 dm. tall, with 1–several sheathing lvs. free ca. ½ their length, 2-fld.; outer spathe-bract 6–9 mm. wide, 7–12.5 cm. long; perianth-tube 7–13 mm. long; sepals broadly oblanceolate, 5–6.5 cm. long, 1.5–2.7 cm. wide; petals 5–6 cm. long, 0.8–1.4 cm. wide; fls. purple to bluish-violet; style-branches 2–3.2 cm. long; style-crests 11–15 mm. long; $2n = 40$ (Lenz, 1950).—Dry woods, 5000–7500 ft.; Yellow Pine F.; e. San Gabriel, San Bernardino, San Jacinto mts. May–June.

Ssp. **columbiàna** Lenz. Rhizome 7–9 mm. thick; lvs. to 9 dm. long and 1 cm. wide, bright green, the bases slightly red or colorless; fl.-stem to 3.5 dm. tall, usually 3-fld., with ca. 3 sheathing lvs. free ca. ½ their length; outer spathe-bract 5–8 mm. wide; perianth-tube ca. 8 mm. long; sepals 5–6.5 cm. long, 1.7–2.4 cm. wide; petals 4.5–5.9 cm. long, to 1.4 cm. wide; fl.-color pale clear creamy-yellow with golden veining; caps. ca. 4 cm. long.—Dry slopes on border line between Foothill Wd. and Yellow Pine F.; Italian Bar and near Columbia, Tuolumne Co. May–June.

Ssp. **pinetòrum** (Eastw.) Lenz. [*I. p.* Eastw.] Rhizomes 5–6 mm. thick; lvs. pale green, to 4 dm. long and 5 mm. wide, usually faint pink at bases; fl.-stem to 3 dm. tall, with 2–4 closely clasping cauline lvs. free ¼–⅓ their length; spathe-bracts separated, the outer 3–6 mm. wide, 5–7 cm. long; ovary 12–15 mm. long; perianth-tube 12–15 mm. long; sepals narrowly oblanceolate, 4.8–6.3 cm. long, 1.2–2 cm. wide; petals 4–6 cm. long, 0.6–1.2 cm. wide; color pale creamy-yellow with sepals conspicuously and deeply gold-veined; style-branches ca. 2.7 cm. long; style-crests 0.9–1.3 cm. long.— Dry shaded slopes, 3000–4000 ft.; Yellow Pine F.; Plumas Co. May–June.

3. **I. Múnzii** Foster. Rhizomes ca. 1 cm. in diam., covered with bases of old lvs.; lvs. to 5 dm. long and 2 cm. wide, gray-green and glaucous, without anthocyanin pigments at base; fl.-stalk to ca. 7 dm. high, simple, 3(2–4)-fld.; lower spathe-bract 8–14 mm. wide, 6.5–11 cm. long; pedicels 1–5 or more cm. long; ovary roundish in cross section, 1.4–3 cm. long, gradually tapered into pedicel, abruptly into perianth-tube; this stout, 7–10 mm. long, funnelform; sepals oblong-ovate to broadly oblanceolate, 6–9

cm. long, 1.8–3.7 cm. wide; petals oblanceolate to spatulate, 5–9.5 cm. long, 1.2–2.1 cm. wide; fl.-color pale lavender to blue-violet; style-branches ca. 3 cm. long; style-crests 1.1–2 cm. long, subquadrate, obscurely lobed to entire; anther ca. 2.5 cm. long; caps. to ca. 5 cm. long; $2n = 40$ (Lenz, 1950).—Dry partly wooded slopes, 1800–2600 ft.; Foothill Wd.; Forks of Tule R., Tulare Co. March–April.

4. **I. chrysophýlla** Howell. Rhizome slender; lvs. exceeding stems, 3–5 mm. wide, light green, the bases pink to red; fl.-stem simple, to 2 dm. tall, usually 2-fld., usually with 1–3 cauline lvs.; spathe-bracts opposite, lanceolate to narrower, the outer 6–10 mm. wide, 5–8.6 cm. long, often reddish or purplish; pedicels 5–15 mm. long; ovary 11–18 mm. long, tapering gradually toward apex and abruptly at base; perianth-tube slender, 4.3–9 cm. long; sepals oblanceolate, 4.6–6.7 cm. long, 1–2 cm. wide; petals lanceolate, 3–5.5 cm. long, 0.6–1.2 cm. wide; fl.-color pale creamy-yellow to almost white, sometimes with faint bluish tinge, usually veined darker; style-branches 1.7–2.5 cm. long, the crests 1.5–2.2 cm. long, usually subentire; stigmas triangular; caps. oblong, 2–3 cm. long; $2n = 40$ (Clarkson, 1955).—Dry slopes; Yellow Pine F.; Sanger Peak Road, Del Norte Co.; to w. Ore. May–June. Natural hybrids with *I. bracteata*.

5. **I. tenuíssima** Dykes. [*I. citrina* Eastw. *I. humboldtiana* Eastw.] Rhizome slender; lvs. to 4 dm. long and 6 mm. wide, gray-green, sometimes slightly glaucous, the bases often pinkish or red; fl.-stem simple, with 2–3 cauline lvs. free most of their length, usually 2-fld.; spathe-bracts opposite, lanceolate, 5–10 mm. wide, 4–8 cm. long, often pinkish or reddish; pedicels 8–18 mm. long; ovary 1–2 cm. long, equally tapered at both ends; perianth-tube 3–5.8 cm. long, slender below, abruptly dilated into a broader throat ¼–⅓ the length of the tube; sepals lanceolate or narrower, 4.7–7.5 cm. long, 1.1–1.8 cm. wide; petals 4.4–6.4 cm. long, 0.6–1.4 cm. wide; color usually pale cream with distinct lavender or brownish or red-brown veining; style-branches 2–3 cm. long; style-crests 1.1–2.3 cm. long; stigmas triangular to tongue-shaped; caps. oblong, 3–4 cm. long; $2n = 40$ (Clarkson, 1955).—Dry sunny woods; Mixed Evergreen F., N. Oak Wd., Foothill Wd., Yellow Pine F.; Butte to Siskiyou and Humboldt cos. April–June. Natural hybrids with *I. Purdyi* and *I. tenax* ssp. *klamathensis*.

Ssp. **purdyifórmis** (Foster) Lenz. [*I. p.* Foster.] Stem-lvs. 3–4, closely clasping, free only at their tips; spathe-bracts broadly lanceolate, somewhat inflated; stigmas broadly triangular to rounded.—Shaded places; Yellow Pine F.; Plumas and Sierra cos. April–June.

6. **I. macrosìphon** Torr. [*I. amabilis* Eastw. *I. californica* Leichtl. *I. m.* var. *elata* Eastw.] Rhizome to 8 mm. in diam.; lvs. linear, exceeding stems, to 5 mm. wide, the bases colorless; fl.-stem simple, to ca. 2.5 dm. long, mostly 2-fld., with 2–4 cauline lvs. free ca. half their length; spathe-bracts opposite, the outer 4–9 mm. wide, 3.9–4.5 cm. long; pedicel 3–20 mm. long; ovary ovoid, 1.2–2.4 cm. long; perianth-tube slender, 3.6–8.6 cm. long, ± bowl-shaped above; sepals narrowly oblanceolate to broadly obovate, 4–6.8 cm. long; petals slightly shorter; fl.-color deep golden-yellow, cream, pale lavender, or deep blue-purple, usually distinctly veined; style-branches 2–3.3 cm. long; style-crests 8–18 mm. long, subquadrate to semi-ovate, erose on margins; stigma triangular; caps. oblong to ovoid, 2.5–3 cm. long; $2n = 40$ (Clarkson, 1955).—Sunny grassy to woody slopes, below 3000 ft.; Foothill Wd., N. Oak Wd., Mixed Evergreen F., Yellow Pine F.; Sierran foothills, Tuolumne to Butte cos. and inner Coast Ranges from Santa Clara to Tehama and Glenn cos. April–May. Natural hybrids are with: *Hartwegii*, *Douglasiana*, *Purdyi* and *Fernaldii*, and one involves both *Douglasiana* and *Purdyi*.

7. **I. Fernáldii** Foster. Rhizome ca. 6 mm. in diam.; lvs. to 7 or 8 mm. wide and 4 dm. long, usually brightly colored at base, otherwise gray-green and glaucous; fl.-stem 2–4 dm. tall, 2-fld., with 2–several cauline lvs. free ca. half their length; spathe-bracts opposite, rather broadly lanceolate, often flushed with anthocyanin pigment, 6–11 mm. wide, 5–9 cm. long; pedicels 1–2.2 cm. long; ovary ellipsoid, 1.5–2.3 cm. long; perianth-tube 3–6 cm. long, usually rather abruptly dilated above into a conspicuous throat; sepals oblanceolate to spatulate, 4.7–6.8 cm. long, 1.2–2.1 cm. wide; petals narrowly oblanceolate, 4.3–6 cm. long, 0.6–1.4 cm. wide; fl.-color soft creamy-yellow, often variously veined; style-branches 2.2–3 cm. long; style-crests linear to narrowly lunate, 1–1.7 cm. long; stigma triangular; caps. oblong, beaked, 2.5–3.5 cm. long.—Rather shaded slopes; Mixed Evergreen F.; Santa Cruz Co. to Sonoma, Lake, and Solano cos. April–May. Natural hybrids with: *Douglasiana, macrosiphon*.

8. **I. bracteàta** Wats. Rhizome 6–9 mm. in diam.; lvs. rather few, deep glossy green on upper surface, yellow-green on lower, to 10 mm. wide, 6 dm. long, usually pink or

reddish at base; fl.-stalk usually shorter than lvs., simple, with 3–6 cauline lvs., the upper ones free ca. ⅓ their length, all ± pinkish or reddish; fls. 2; spathe-bracts opposite, subequal, 6–10 mm. wide, 5–9 cm. long, lanceolate; pedicels 3–6 cm. long; ovary tapered abruptly above, gradually below; perianth-tube funnelform, 6–10 mm. long; sepals oblanceolate to obovate, 4.3–7.8 cm. long, 1.7–3 cm. wide; petals oblanceolate, 4.8–7.4 cm. long, 0.8–2 cm. wide; fl.-color yellow, veined with deep maroon or brown; style-branches 2.2–3 cm. long; style-crest 0.9–1.7 cm. long; stigmas triangular; caps. 2–3 cm. long; $2n = 40$ (Foster, 1937).—Rather shady places; Yellow Pine F.; Del Norte Co. to sw. Ore. May–June. Natural hybrids with: *chrysophylla* and × *Thompsonii* Foster (pro sp.).

9. **I. Púrdyi** Eastw. [*I. macrosiphon* var. *P.* Jeps. *I. Lansdaleana* Eastw.] Rhizome 4–6 mm. in diam.; lvs. few, to 4.5 dm. long and 8 mm. wide, somewhat 2-sided, dark green above, gray-green beneath, pinkish or reddish at base; fl.-stems 1.5–3.5 dm. long, covered with many overlapping bracts that are often flushed with anthocyanin and free only at tips; spathe-bracts opposite, ± inflated, the outer broadly lance-ovate, 8–13 mm. wide, 5–7.7 cm. long; fls. usually 2; pedicel 1.1–2.2 cm. long; ovary narrow, 1.3–2 cm. long; perianth-tube 2.8–4.8 cm. long; sepals oblanceolate, 5.5–8.4 cm. long; petals lanceolate, 5–7 cm. long, 0.9–2 cm. wide; fl.-color pale cream-yellow, conspicuously veined with brown-purple, or often pale creamy-yellow or whitish with light lavender tinge to sepals; style-branches 2–3.2 cm. long; style-crests 0.9–2.1 cm. long, narrowly ovate, laciniate on margins; stigmas truncate, broadly rounded or bilobed; caps. oblong-ovoid, 2–3 cm. long; $2n = 40$ (Foster, 1937).—Wooded places; Redwood F., N. Coastal Coniferous F., Mixed Evergreen F.; Sonoma to Humboldt and Trinity cos. May–June. Natural hybrids with: *tenuissima, Douglasiana, macrosiphon,* and 1 involving *Douglasiana* and *macrosiphon.*

10. **I. innomináta** Henders. Rhizome slender, 3–4 mm. in diam.; lvs. abundant, 2–4 mm. wide, to 3.5 dm. long, dark shining green above, lighter beneath, the bases pinkish to deep purplish-red; fl.-stem 1–2-fld., to 2 dm. tall, with 2–4 cauline lvs. free ca. ⅓ their length; spathe-bracts opposite, subequal, broadly lanceolate to ovate, with scarious margins, 3.3–6 dm. long, 5–7 mm. wide; pedicels 4–13 mm. long; perianth-tube 1.5–3 cm. long; sepals broadly lanceolate to oblanceolate, 4.4–6.3 cm. long, 1.7–3 cm. broad; petals slightly shorter and 0.9–1.6 cm. wide; fl.-color variable, deep golden-yellow, ± veined with darker colors, or occasionally clear yellow with no veining, to lavender or deep purple; style-branches 1.9–2.6 cm. long; style-crests 0.9–1.4 cm. long, subquadrate to semiovate, irregularly toothed; stigma triangular; caps. oblong to oval, 2–3 cm. long; $2n = 40$ (Lenz, 1958).—Sunny or partly shaded slopes; Mixed Evergreen F., etc.; n. Del Norte Co.; to sw. Ore. May–June. Natural hybrids with *Douglasiana* are *I.* × *Thompsonii* Foster (pro sp.).

11. **I. Douglasiàna** Herb. [*I. Beecheyana* Herb. *I. D.* var. *bracteata* and var. *nuda* Herb. *I. D.* var. *major* Torr. *I. D.* var. *alpha* Dykes, var. *altissima* Purdy and var. *mendocinensis* Eastw. *I. Watsoniana* Purdy.] Rhizome to 9 mm. in diam.; lvs. to 2 cm. wide and 10 dm. long, prominently ribbed, yellow-green to very deep green, with pinkish or reddish bases; fl.-stalk 1.5–8 dm. tall, usually with 1–4 side branches, each usually 3-fld., cauline lvs. 1–3; spathe-bracts usually opposite, sometimes separated, lanceolate-acuminate, 7–12 mm. wide, 6–12 cm. long; pedicels 2–5.3 cm. long; ovary elliptic-ovoid, strongly triangular in cross sect., tapering to either end, 2.4–4.8 cm. long; perianth-tube 1.5–2.8 cm. long; sepals oblanceolate to obovate, 5–8.7 cm. long, 1.4–3 cm. wide; petals oblanceolate, 4.5–7 cm. long, 0.9–1.8 cm. wide; fl.-color varying from pale cream through light and dark lavender to deep red-purple; style-branches 1.7–3.5 cm. long; style-crests 1–2 cm. long, subquadrate, coarsely toothed; stigma triangular; anthers 1–1.5 cm. long; caps. 2.5–5 cm. long, sharply triangular in cross section; $2n = 40$ (Simonet, 1934).—Abundant on grassy slopes and open places; Coastal Prairie, Mixed Evergreen F., etc.; Santa Barbara Co. to Ore. March–May. Natural hybrids with: *tenax, innominata* [*I.* × *Thompsonii* Foster (pro sp.)], *Purdyi, macrosiphon, Fernaldii,* and one involving both *Douglasiana* and *macrosiphon.*

12. **I. missouriénsis** Nutt. Rhizome 2–3 cm. in diam., clothed with dark remnants of old lvs.; lvs. rather light green, glaucous, to ca. 4.5 dm. long and 3–6(–9) mm. wide, sometimes purplish at base; fl.-stem rather slender, 2–5 dm. high, branched, sometimes simple, with 1–2 fls. in a cluster; spathe-bracts opposite, scarious with herbaceous parts at base and along the keel, lanceolate or ovate, 4–7 cm. long; pedicel to

20 cm. long; ovary narrow, trigonal, 1.5–3 cm. long; perianth-tube funnel form, to ca. 1 cm. long; sepals ca. 6 cm. long and 2 cm. wide, obovate, largely pale lilac to whitish with lilac-purple veins; petals shorter, 1 cm. wide, oblanceolate to spatulate; style-branches to 2.5 cm. long; style-crests ca. 8 mm. long, subquadrate, incised; stigma ± bilobed; anthers ca. 15 mm. long; caps. 3–5 cm. long, oblong, trigonal; $2n = 38$ (Foster, 1937).—Flats and meadows that are moist until flowering time, mostly 3000–11,000 ft.; Montane Coniferous F.; mts. of San Diego Co., San Bernardino Mts., Mt. Pinos, Sierra Nevada (largely on e. slope), to Modoc Co. and inner Coast Ranges of Mendocino and Solano cos. where occasional at lower levels; to B.C., S. Dak., Coahuila. May–June. Plants from n. Calif. with ± narrow spathes have been called forma *angustispàtha* Foster.

13. **I. longipétala** Herb. in H. & A. Rootstock 2–2.5 cm. in diam., covered with unsplit bases of old lvs.; lvs. to ca. 7 dm. long, ca. 6–9 mm. wide, dark green, ± glaucous; fl.-stem mostly simple, stout, 3–6 dm. high, 3–6-fld., sometimes with 1–2 reduced cauline lvs.; spathe-bracts herbaceous, sometimes scarious in upper part, linear-lanceolate, 7–12(–15) cm. long, the outer distant from the inner by 1–10 cm.; pedicels 3–9 cm. long; ovary trigonal, to 2.5 cm. long, ridged in middle of each side; perianth-tube short, funnelform, 0.5–1.3 cm. long; sepals 6–8(–10) cm. long, 5 cm. wide, the blade obovate, with lilac-purple veining on a lighter ground; petals 5–9 cm. long, 1.5–2 cm. wide, oblong, emarginate, style-branches cuneate, to 4 cm. long; style-crests subquadrate, irregularly incised, to 1.5 cm. long; stigma obscurely to prominently bilobed, with crenate edges; anthers to 2 cm. long; caps. oblong-ovoid, 6-ribbed, 5–9 cm. long; $2n = 86–88$ (Simonet, 1932).—Moist open places; Coastal Prairie, Mixed Evergreen F., etc.; near coast from Mendocino Co. to Monterey Co. April–May.

2. Sisyrínchium L.

Tufted perennials from short rootstocks. Stems slender, compressed, ± winged. Lvs. narrow, grasslike. Fls. ephemeral, opening in sun, umbellate, subtended by a spathe of 2 bracts. Perianth blue, violet, purplish, or yellow, sometimes white, the segms. all alike. Fils. ± united. Caps. globose; seeds subglobose or ovoid, smooth or pitted. A fairly large genus, no. of spp. uncertain; N. Am. and W. I. (Name used by Theophrastus for a plant related to *Iris*.)

Fls. mostly blue to purplish-red.
  Perianth 1.5–2 cm. long, reddish-purple; fils. united at base only . . . . 1. *S. Douglasii*
  Perianth less than 1.5 cm. long, bluer; fils. united almost to top.
    Stem bearing 1 or more lf.-bearing nodes, each node with 2 or more peduncles . . . . 2. *S. bellum*
    Stem leafless, simple, with a single terminal spathe.
      Pedicels glabrous; perianth 10–15 mm. long; caps. brownish. High mt. meadows
                                                                                   3. *S. idahoense*
      Pedicels pubescent; perianth 9–10 mm. long; caps. not brownish. In ± alkaline places e. of
      Sierra Nevada . . . . 4. *S. halophilum*
Fls. yellow.
  Lvs. 2–6 mm. wide, drying almost black; perianth 12–18 mm. long. Coastal . . 5. *S. californicum*
  Lvs. 1–3 mm. wide, not drying dark; perianth 8–12 mm. long. Montane . . . . 6. *S. Elmeri*

1. **S. Douglásii** A. Dietr. [*S. grandiflorum* Dougl., not Cav. *Olsynium* g. Raf. *O. D.* Bickn.] PURPLE-EYED-GRASS. Stems 1.5–3 dm. high, not winged, ± flattened, with 2–3 lvs., basal lvs. bractlike, with much reduced blades; cauline with blades to 1 dm. or more long; spathe solitary, terminal; outer bract 5–12 cm. long, inner much shorter; pedicels 2–5 cm. long; perianth-segms. oblong-obovate, 5-nerved, obtuse or apiculate; anthers orange, 3–4 mm. long; caps. 6–8 mm. long, often with anthocyanin; seeds dark, finely pitted, ca. 2 mm. long.—Moist hollows and open slopes, 3000–6000 ft.; Sage-brush Scrub, N. Juniper Wd., N. Oak Wd., Foothill Wd., Yellow Pine F.; Humboldt to Modoc and Lassen cos.; to B.C., Nev. March–May.

2. **S. béllum** Wats. [*Bermudiana b.* Greene. *S. maritimum* Heller.] BLUE-EYED-GRASS. Tufted, from 1–4(–6) dm. high, green to glaucescent; stems 1–4 mm. wide, stout or slender, narrowly firm-margined to winged, smooth or denticulate on edges; lvs. mostly basal, soft to firm, shorter than to almost as tall as stems, mostly 2–4(–6) mm. wide; cauline lvs. 1–3, each commonly bearing in its axil 2–4 peduncles 2–15 cm. long; spathe 2–6 mm. wide when pressed, often ± purplish, the bracts acute to obtuse, subequal or very unequal, the outer mostly ca. 2–4.5 cm., the inner 1.5–2.5 cm. long; pedicels ±

exserted, slender, glabrous; perianth-segms. blue, violet, lilac, rarely white, emarginate and often aristulate, 12–15 or more mm. long; caps. dark or pale brown, 2–7 mm. long; seeds 1–few in a locule, dark, pitted, ca. 1.5 mm. in diam.; $2n = 32$ (Bowden, 1945).—Widely distributed, the more typical form largely in open grassy places, below 3000 ft.; many Plant Communities; near the coast principally from Ventura Co. or somewhat farther s. to Humboldt Co. March–May.

Exceedingly variable and in a complex much in need of study. Some of the variants which represent tendencies of uncertain taxonomic worth are: (1) *S. Greènei* Bickn., a smaller more delicate plant, paler green in color, drying paler, outer bract of the spathe scarcely 2 cm. long; perianth-segms. scarcely 1 cm. long; Foothill Wd. and Montane Coniferous F. of lower w. slopes of Sierra Nevada and Mt. Shasta to Yolla Bolly Mts.; April–July. (2) *S. Eastwoòdiae* Bickn. Similarly slender, with narrow lvs. and short spathes, but tending to be taller and more grasslike; wet meadows, largely of Montane Coniferous F.; mts. mostly of s. Calif. May–July. (3) *S. hespérium* Bickn. Slender, near to typical *bellum*, but with shorter spathe and smaller fls.; grassy slopes and flats often away from the coast; Coastal Sage Scrub, etc.; San Luis Obispo Co. to L. Calif. March–April. (4) *S. funèreum* Bickn. Stems rather tall; pale with very little anthocyanin; fls. 12–14 mm. long; ± alkaline places; Death V. region and Mojave Desert. Feb.–April.

3. **S. idahoénse** Bickn. [*S. oreophilum* Bickn. *S. leptocaulon* Bickn?] Plants pale green, glaucous, 1–4.5 dm. high; stems mostly leafless, slender, 1–4 dm. high, 1–2 mm. wide, winged; lvs. basal, ca. half as long as stems, 1–3 mm. wide, smooth on edges; spathe 1, narrow, green or faintly purplish; outer bract 2–6 cm. long, the inner somewhat shorter; pedicels glabrous, 1.5–3 cm. long; perianth-segms. 10–15 mm. long, scarcely emarginate, mucronulate-aristulate; caps. 3–6 mm. in diam., brown, sparsely puberulent; seeds black, ca. 1 mm. long, rugulose.—Wet meadows 4500–10,800 ft.; Montane Coniferous F.; San Jacinto Mts., San Bernardino Mts., Sierra Nevada to Siskiyou Co.; to Wash., Ida. July–Aug.

4. **S. halóphilum** Greene. Habit of no. 3; outer bract of spathe 1.5–2 cm. long, broadly scarious-margined; inner somewhat shorter; pedicels ± glandular-pubescent; perianth violet-purple, 9–10 mm. long, prominently aristulate at apex; caps. 3–4 mm. high, pubescent, pale grayish-green; seeds ca. 1.5 mm. in diam., rugulose-pitted.—In ± alkaline meadows and wet places, ca. 5000–7000 ft.; Sagebrush Scrub, etc.; e. base of Sierra Nevada in Inyo and Mono cos.; to Nev. May–June.

5. **S. califórnicum** (Ker) Dryand. [*Marica c.* Ker. *Hydastylus c.* Salisb. *S. flavidum* Kell. *Bermudiana c.* Greene.] GOLDEN-EYED-GRASS. Stems 1.5–4(–6) dm. high, dull green, glaucescent, turning dark on drying and staining paper a deep purple; lvs. all basal, thin, ca. half as high as stem, sometimes almost as long, ca. 2–5 mm. wide; stems 2–5 mm. wide, mostly broadly winged; outer bract of spathe 2–5 cm. long, mostly surpassing the inner; fls. bright yellow, each segm. with 5–7 dark nerves, oblong, obtuse or acutish, 12–18 mm. long; caps. 7–12 mm. long; seeds black, pitted, 1.3–1.5 mm. in diam.; $2n = 36$ (Bowden, 1945).—Moist places at low elevs.; Freshwater Marsh, etc.; near the coast from Monterey Co. to Ore. May–June.

6. **S. Élmeri** Greene. [*Hydastylus E.* Bickn. *H. rivularis* Bickn.] Slender, mostly less than 2 dm. high, the stems 1–2 mm. wide; lvs. ca. 1–3 mm. wide, half as long as stems; outer bract of spathe 1.2–3.5 cm. long, scarcely longer than inner; perianth-segms. 8–12 mm. long, orange-yellow, with 5 dark veins, obtusish; caps. ca. 6–7 mm. long.—Boggy and wet places, mostly 4000–8500 ft.; Montane Coniferous F.; San Bernardino Mts., Sierra Nevada n. to Plumas Co., Trinity Co. July–Aug.

3. Gladìolus L.

Plants with tunicated corms and simple ± leafy stems with large usually herbaceous spathes (bracts) each yielding 1 sessile fl. Lvs. mostly sword-shaped, broad, many-nerved, sometimes linear or terete. Fls. mostly several cm. across, commonly showy, with dilated funnel-shaped tube, curved upward in most spp. Perianth-segms. usually oblong, obtuse or acute, the 3 upper larger than the 3 lower. Stamens inserted below the throat. Style slender. Caps. oblong to obovoid, loculicidal, shorter than the lanceolate outer spathe-valve, but sometimes ca. equal to the inner one. Seeds mostly flattened or winged. Ca. 250 spp. of Medit. region to S. Afr. Many in cult. (Latin *gladiolus*, small sword.)

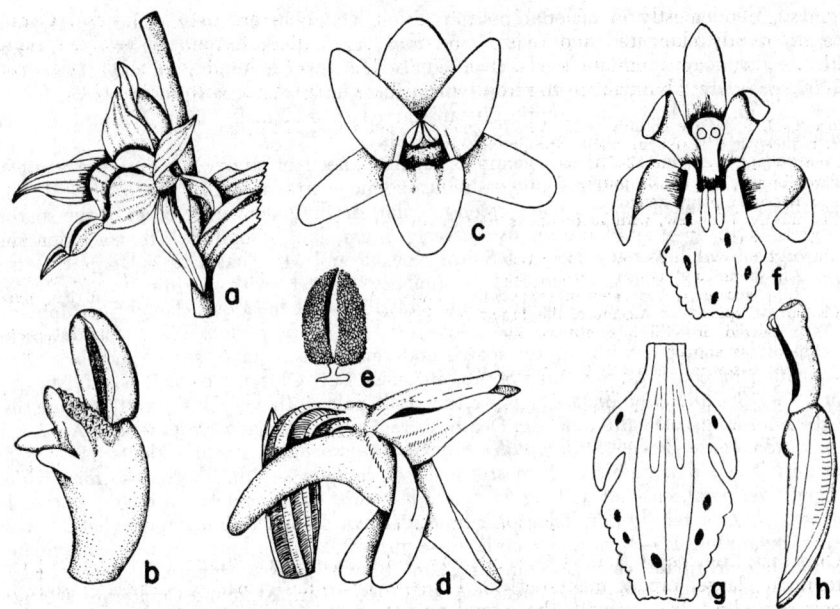

Fig. 131. ORCHIDACEAE. *Epipactis gigantea: a,* fl., × 1, with 3 sepals, 2 similar petals and 3d forming a lip, column-tip evident, ovary inferior; *b,* column in side view, anther above, stigma broad. *Habenaria unalaschensis: c,* front view of fl., × 5, with 3 sepals, 3 petals, 2 anther-cells; *d,* fl., from side, showing inferior ovary and large spur from lip; *e,* pollinium. *Corallorhiza maculata: f,* fl. in front view, × ca. 4, lower petal much modified (lip), column prominent; *g,* lip 3-lobed; *h,* column and inferior ovary from side.

1. **G. segètum** Ker. Plants ca. 1 m. high, the stems slender, several-lvd.; lvs. 1–2 dm. long, ca. 1 cm. wide; fls. several, the infl. elongating and becoming ca. 3 dm. long in fr.; outer bract of spathe 3–3.5 cm., inner 1.5–2.5 cm. long; fls. purplish-pink, ca. 5 cm. long; $2n = 120$ (Bamford, 1935).—Occasionally natur. in walnut groves (Orange Co.) and in fields (Sonoma Co.); native of Medit. region. April–May.

4. Watsònia Mill.

Gladioluslike, but with regular perianth having a more evident and curved tube; style-branches divided. Ca. 20–30 spp. of S. Afr. (Named for Sir William *Watson,* 1715–1787, English botanist.)

1. **W. angústa** Ker. Corm globose, ca. 3–5 cm. in diam.; basal lvs. 4–6, ensiform, rigid, 3–6 dm. long, 2.5–6 cm. wide; stems usually branched, ca. 1 m. high, with a few reduced lvs.; spikes lax, becoming ca. ·4 dm. in fr., the axis ± purplish; outer spathe-bract lance-oblong, ca. 2–2.5 cm. long, brown; perianth bright scarlet, 5–6 cm. long, the tube strongly curved, upper half narrow funnelform; perianth-segms. oblance-oblong, cuspidate, ca. 2 cm. long.—Natur. in large areas in fields in Mendocino Co.; native of S. Afr. May–July.

23. Orchidàceae. ORCHID FAMILY (Fig. 131)

Ours perennial herbs, terrestrial, from a short or elongate, rarely coralloid rhizome, with fibrous to fleshy roots. Lvs. sheathing, often reduced to scales. Fls. in ours perfect, irregular, bracted, solitary or in spikes or racemes. Sepals 3, similar. Petals 3; 2 alike, usually resembling the sepals, the third forming the lip which is sometimes saccate and often spurred. Fils. 1 or 2, united with the style to form the column; the perfect anther 1 (2 in *Cypripedium*), on the apex of the column and above or behind the sticky stigmas. Pollen mostly in masses (pollinia), which in ours are usually 2 or 4. Ovary inferior, mostly elongated and twisted, 3-celled to 1-celled, forming a 3-valved caps. with very numerous minute seeds. The family has several hundred genera and over 15,000 spp.; most abundant in the trop., where largely epiphytic with aerial roots.

(Correll, D. S. Native orchids of N. Am. [Chronica Botanica Co., 1950], pp. 1–399.)

Lvs. reduced and scalelike; plant saprophytic, not green.
  Plant white; perianth 12–15 mm. long .... 5. *Eburophyton*
  Plant brown, yellow, or purple; perianth mostly shorter .... 9. *Corallorhiza*
Lvs. foliaceous; plant green.
  Fls. mostly 1 to few, with leafy bracts if more than 1.
    Lf. 1, basal; fl. 1 .... 8. *Calypso*
    Lvs. 2 to several; fls. often more than 1.
      Lip an inflated pouch; fertile anthers 2 .... 1. *Cypripedium*
      Lip saccate at base, but not pouchlike; anther 1 .... 6. *Epipactis*
  Fls. many, spicate or racemose, the bracts not foliaceous.
    Lip spurred or with a scrotiform sac .... 2. *Habenaria*
    Lip not as above.
      Lvs. 3-several; perianth 6–8 mm. long.
        Spike twisted, glabrous .... 3. *Spiranthes*
        Spike not twisted, glandular-hairy .... 7. *Goodyera*
      Lvs. 1–2, perianth 2–5 mm. long.
        The lvs. 2, subopposite .... 4. *Listera*
        The lf. 1 .... 10. *Malaxis*

1. Cypripèdium L. LADY-SLIPPER

Ours terrestrial herbs from short rootstocks and fibrous roots. Stems in ours leafy. Lvs. 2 to many, large, broad, many-nerved. Fls. few or solitary, usually large and showy. Sepals spreading, free or with the lateral pair ± united; petals spreading, free, usually smaller and narrower than the sepals; lip inflated, large and saccate or pouch-shaped. Column declined, with 2 laterally placed fertile stamens, each bearing a 2-celled anther, and a dorsally placed sterile petallike staminode; pollen granular. Stigma terminal and somewhat 3-lobed; ovary 1-celled; caps. obovoid to ellipsoid. A genus of 100 or more spp. of boreal, temp. and trop. Eurasia, Oceania and Am. Many of horticultural interest. (Greek, *Cypris*, Venus, and *pedilon*, shoe; the lip moccasinlike.)

Lvs. 2, opposite; lip greenish-yellow .... 1. *C. fasciculatum*
Lvs. several, alternate; lip white or pinkish.
  Sepals 4–6 cm. long, dark brown .... 2. *C. montanum*
  Sepals ca. 1.5 cm. long, greenish-yellow .... 3. *C. californicum*

1. **C. fasciculàtum** Kell. [*C. pusillum* Rolfe. *C. f.* var. *p.* Hook. f.] Stems woolly-pilose, 0.5–3.5 dm. high, with 1 or 2 basal scarious bracts; lvs. sessile, elliptic to elliptic-orbicular, 7–10 cm. long, 4–7 cm. wide, obtusish, 3-ribbed; fls. 1–4, clustered, subtended by leafy bracts, the stem above the lvs. peduncielike and with 1 or 2 smaller additional lanceolate bracts near its middle; sepals lanceolate, acuminate, 1.5–2.5 cm. long, greenish with brown-purple veins, the lateral sepals largely united; petals similar to sepals; lip ovoid, 8–12 mm. long, greenish-yellow with purple-brown veins; sterile stamen 3 mm. long; caps. oblong, 2 cm. long.—Open rocky woods; Redwood F. to Yellow Pine F.; Santa Cruz Co., Plumas Co. to Del Norte Co.; to Wash. April–May.

2. **C. montànum** Dougl. ex Lindl. [*C. occidentale* Wats.] Stems stout, 3–6 dm. high, glandular-puberulent; lvs. 4–7, elliptic-ovate or narrower, 8–15 cm. long, 4–7 cm. wide, clasping; fls. 1–3, in axils of leafy bracts; sepals brown-purple, lanceolate, acuminate, 4–6 cm. long, the lateral quite united; petals similar to sepals, but crisped and narrower; lip obovoid, white with purple veins, 2–3 cm. long; sterile stamen 8 mm. long; caps. ascending, oblong, 2 cm. long.—Moist woods, below 5000 ft.; Mixed Evergreen F., Yellow Pine F.; Santa Cruz Mts. to Siskiyou Co., thence s. in Sierra Nevada to Mariposa Co.; to B.C., Wyo. May–Aug.

3. **C. califórnicum** Gray. Plant puberulent throughout, 3–6 dm. high; lvs. several, lance-oblong to round-ovate, acute to acuminate, 7–12 cm. long, clasping, reduced upward to leafy bracts; fls. 3–10, solitary in axils of upper bracts; sepals broadly ovate, greenish-yellow to brown, 13–16 mm. long, the lateral united; petals oblong-linear, obtusish, equaling the sepals; lip obovoid, white or pinkish, spotted brown, 1.6–2 cm. long; sterile stamen round or arched, 4 mm. long; caps. reflexed, oblong, 1.5–2.5 cm.

long.—Wet rocky ledges and hillsides, below 5000 ft.; Mixed Evergreen F.; Marin Co., Mendocino Co. to Lassen Co.; s. Ore. May–June.

2. Habenària Willd. REIN ORCHID

Perennial terrestrial herbs with fleshy tuberlike roots. Stems erect, simple, ± leafy at least at the base. Lvs. basal or cauline, essentially sessile, with the basal part sheathing. Fls. in ours small, in terminal spike or raceme. Sepals free, similar or dissimilar. Petals slightly smaller, free, erect; lip entire in ours, with a basal scrotiform sac or a spur. Column short; stigmas with or without papillose processes; anther-cells 2, separate, divergent; pollen granular, attached to exposed glands. Caps. narrow-cylindric to ellipsoid. Ca. 500 spp., mostly of warmer parts of world. (Latin, habena, rein of a horse, because of the shape of the spur in some spp.)

(Correll, D. S. Habenaria in w. N. Am. Leafl. W. Bot. 3: 233–247, 1943.)
Sepals 1-nerved; lvs. usually near base of stem and withering before anthesis.
 Spur not much longer than the lip; raceme loosely fld., usually not more than 1 cm. in diam.
   1. H. unalascensis
 Spur at least twice as long as lip; raceme densely fld., 1.5–2.5 cm. in diam. ...... 2. H. elegans
Sepals 3-nerved; lvs. usually scattered and green at anthesis.
 Lip abruptly dilated at base; fls. mostly white .... 3. H. dilatata
 Lip not prominently dilated at base; fls. mostly greenish.
   Spur usually less than ⅔ as long as lip.
     The spur saccate; raceme loosely fld. .... 4. H. saccata
     The spur cylindrical to clavate; raceme densely fld. .... 5. H. hyperborea
   Spur as long as or longer than lip, filiform .... 6. H. sparsiflora

1. **H. unalascénsis** (Spreng.) Wats. [*Spiranthes u.* Spreng. *Piperia u.* Rydb. *H. Cooperi* Wats. *P. C.* Rydb.] Stem stiff, 3–10 dm. high; lvs. 2–4, basal, lanceolate to broadly oblanceolate, obtuse to acute, 7–20 cm. long, 1–5 cm. wide; spike slender, mostly loosely fld., 1–3 dm. long; bracts shorter than the greenish fls.; sepals oblong, 3–4 mm. long, almost alike, obtuse; lateral petals as long as sepals, lanceolate; lip oblong-ovate, slightly longer; spur equal to or slightly longer than lip, but shorter than ovary; caps. 12–15 mm. long.—Dry to moist soil, flats and slopes, below 8000 ft., Chaparral, V. Grassland, Oak Wds., Mixed Evergreen F.; Sierra Nevada to Montane Coniferous F.; Sierra Nevada to the coast; Alaska to Alta., Utah, L. Calif. April–Aug.

2. **H. élegans** (Lindl.) Boland. [*Platanthera e.* Lindl. *Piperia e.* Rydb. *H. Michaelii* Greene. *P. M.* Rydb. *P. elongata, leptopetala, multiflora,* and *longispica* Rydb. *Gymnadenia l.* Durand. *H. longispicata* Parish. *H. elegans* var. *elata* Jeps. *H. unalascensis* var. *elata* Correll.] Much like *H. unalascensis,* but the cylindrical spike up to 6 dm. long, rather densely fld., 1.5–2 cm. in diam.; fls. greenish-white; sepals 4–5 mm. long; lip 5 mm. long; spur 10–18 mm. long, filiform, not shorter than ovary.—Mostly in dry woods, below 8000 ft., Chaparral, Foothill Wd., Montane Coniferous F.; Sierra Nevada to the coast, San Diego Co. to B.C., Ida. May–Sept.

Var. **marítima** (Greene) Ames. [*H. m.* Greene, not Raf. *P. m.* Rydb. *H. unalascensis* var. *m.* Correll. *H. Greenei* Jeps.] Spike very dense, conic-cylindric, 4–10 cm. long, commonly 2–2.5 cm. thick; fls. white.—Dry hills and bluffs along the immediate coast; N. Coastal Scrub, Closed-cone Pine F.; Monterey Co. to Del Norte Co.; Santa Rosa Id.; Ore. July–Sept.

3. **H. dilatàta** (Pursh) Hook. [*Orchis d.* Pursh. *Limnorchis d.* Rydb.] Stems 4–8 dm. high, leafy; lvs. lanceolate, narrowly acute, 8–20 cm. long, gradually reduced upward; spike usually dense, 1–2 dm. long, the bracts usually incurved against the rachis, the lower ones equaling the fls.; fls. white, 10–14 mm. long; upper sepal obtuse, ovate, 4–6 mm. long, the lateral slightly longer and narrower; lateral petals lanceolate, 5–6 mm. long; lip rhombic-lanceolate, prominently dilated at base, 5–6 mm. long; spur ca. as long as lip, subclavate; $n = 21$ (Humphrey, 1934).—Moist places such as bogs and meadows; Montane Coniferous F.; Humboldt to Siskiyou and Placer cos.; to Alaska, Rocky Mts. June–Sept.

Var. **leucóstachys** (Lindl.) Ames. [*Platanthera l.* Lindl. *H. l.* Wats. *Limnorchis l.* Rydb. *H. pedicellata* Wats. *H. flagellans* Wats.] Lip 7–8 mm. long; spur 10–20 mm. long, more filiform.—Wet and springy places, below 11,000 ft., Montane Coniferous

F.; mts. from San Diego Co. to Sierra Nevada, White Mts., Panamint Mts.; N. Coastal Scrub to Mixed Evergreen F. and Montane Coniferous F.; Coast Ranges and adjacent coast, San Luis Obispo Co. to B.C. and Mont. May–Aug.

4. **H. saccàta** Greene [*Limnorchis stricta* Rydb., not *Platanthera s.* Lindl. *H. gracilis* Wats., not Colebrook or Lindl.] Stem 2–6 dm. high; lower lvs. oblanceolate, obtuse, 5–10 cm. long, upper lanceolate, acute, moderately reduced; spike usually slender, laxly fld., 1–2.5 dm. long, bracts erect; fls. greenish, 12–14 mm. long; upper sepal broadly ovate, 4–5 mm. long, the lateral lance-oblong, 5–6 mm. long; lateral petals 4–5 mm. long; lip pendent, oblong-linear, 5–7 mm. long, sometimes purple; spur saccate, shorter than lip, purplish; caps. ca. 10 mm. long.—Moist places, bogs, meadows, etc., below 8000 ft.; Montane Coniferous F.; Del Norte and Humboldt cos. to Modoc Co.; to B.C., Alaska, Mont., New Mex. June–Aug.

5. **H. hyperbòrea** (L.) R. Br. [*Orchis h.* L. *Limnorchis h.* Rydb. *Habenaria borealis* var. *viridiflora* Cham. *L. v.* Rydb.] Stem 2–5 dm. high; lvs. lanceolate to oblanceolate, 6–15 cm. long, the lower obtuse, the upper acute; spike dense, 8–20 cm. long, with divergent bracts; fls. light green, 10–12 mm. long; upper sepal broadly ovate, erect, 3–4 mm. long; the lateral lanceolate, slightly shorter; petals lanceolate, equaling lateral sepals; lip lanceolate, obtuse, 4–5 mm. long; spur clavate to cylindric, curved, usually definitely shorter than lip; caps. ca. 1 cm. long.—Boggy and moist places, up to 9000 ft.; Montane Coniferous F.; Inyo and Mono cos.; to Rocky Mts., Alaska, Atlantic Coast; Iceland. June–Aug.

6. **H. sparsiflòra** Wats. [*L. s.* Rydb. *L. laxiflora* Rydb. *H. l.* Parish. *H. s.* var. *l.* Correll. *H. leucostachys* var. *virida* Jeps.] Stem 3–6 dm. high, leafy; lower lvs. oblanceolate, obtuse, 1–2 dm. long, upper reduced, acuminate; spike slender, sparsely fld., 1–3 dm. long; fls. greenish, 10–12 mm. long; upper sepal ovate, 4–5 mm. long; the lateral lance-oblong, 5–6 mm. long; petals deltoid-lanceolate, ca. 4 mm. long; lip linear to lance-linear, 6–8 mm. long; spur 6–8 mm. long; caps. ca. 12–14 mm. long.—Mostly along streams or boggy places, 4000–11,000 ft.; Montane Coniferous F.; mts., San Diego Co. to Sierra Nevada, N Coast Ranges; to Wash., Ariz., Rocky Mts. June–Aug.

3. **Spiránthes** Rich. LADIES' TRESSES

Roots clustered, tuberous in ours. Stem erect, leafy only below, bracted above. Lvs. linear, lanceolate, or even ovate. Fls. in a terminal twisted spike, small, whitish in ours, spurless, fragrant. Sepals narrow, erect, free, the dorsal sepal and the petals coherent. Lip sessile or short-clawed, the base concave and embracing the column, each side with a small callosity. Column short; anther 1, 2-celled, with 2 powdery pollinia. Caps. erect, ellipsoid to ovoid. A genus of ca. 200 spp., of very wide distribution. (Greek, *speira*, spiral, and *anthos*, fl., alluding to the spiral character of the twisted infl.)

Lip definitely narrowed below the tip, its basal callosities obsolete . . . . 1. *S. Romanzoffiana*
Lip not narrowed below the tip, its basal callosities developed . . . . 2. *S. porrifolia*

1. **S. Romanzoffiàna** C. & S. [*Gyrostachys R.* MacM. *Ibidium R.* House *Orchiastrum R.* Greene.] Stem stout, glabrous, 1–5 dm. high; lower lvs. 3–5, linear to lance-linear, 0.5–3 dm. long, upper reduced to sheathing lance-linear bracts; spike 5–12 cm. long; bracts lanceolate, scarious, appressed, exceeding the fls.; perianth-segms. greenish-white, 6–8 mm. long, oblong; lip oblong, constricted below the dilated crisped apex.—Wet banks and meadows, below 10,000 ft., Montane Coniferous F., Mixed Evergreen F., Redwood F.; Sierra Nevada, Coast Ranges, occasional in mts. of s. Calif.; to Alaska, Atlantic Coast. June–Aug.

2. **S. porrifòlia** Lindl. [*Gyrostachys p.* Kuntze. *Orchiastrum p.* Greene. *Ibidium p.* Rydb. *S. Romanzoffiana* var. *p.* Ames & Correll.] Like *S. Romanzoffiana*, but the perianth-segms. narrower, creamy-yellowish; lip lance-oblong, scarcely or not constricted below the apex and with 2 nipplelike callosities at the base.—Wet springy places below 8000 ft.; Redwood F., Mixed Evergreen F., Montane Coniferous F.; San Bernardino Mts., Coast Ranges from Santa Cruz Mts. n., Sierra Nevada from Tulare Co. n.; to Wash., Rocky Mts. July–Aug.

4. Lístera R. Br. Twayblade

Roots fibrous. Stems low, slender, usually more or less glandular-pubescent above the two opposite lvs. Lvs. sessile, broad, at middle of stem. Fls. small, in a terminal raceme, greenish or purplish, spurless. Sepals and petals free, similar, subequal, spreading or reflexed. Lip longer than sepals, bilobed or notched, flat, dilated at base. Column slender, with 1 posterior anther; pollinia 2, powdery. Caps. small, pedicelled. Genus of ca. 20 spp., in temp. and colder parts of N. Hemis. (Named for M. *Lister,* 17th-century English naturalist.)

Lvs. somewhat cordate; lip oblong, 2-cleft to near middle .... 1. *L. cordata*
Lvs. not cordate; lip wedge-shaped.
    Lip 2-lobed at tip, 7–10 mm. long .... 2. *L. convallarioides*
    Lip rounded or retuse at tip, 5–6 mm. long .... 3. *L. caurina*

1. **L. cordàta** (L.) R. Br. [*Ophrys c.* L.] Stem 1–2 dm. high; lvs. cordate, ovate, 1.5–3 cm. long, mucronate; raceme 4–20-fld., puberulent to subglabrous; bracts scarcely 1 mm. long; pedicels 2–3 mm. long; sepals and petals greenish, 2–3 mm. long, oblong-linear; lip 4–5 mm. long, deeply cleft into 2 narrow lobes; caps. glabrous, ovoid, ca. 4 mm. long; $n = 21$ (Blackburn, 1936).—Floor of rich woods; Mixed Evergreen F.; Humboldt Co.; to Alaska, Atlantic Coast; Eurasia. June.

2. **L. convallarioìdes** (Sw.) Torr. [*Epipactis c.* Sw. *Ophrys c.* W. Wight.] Stem 1–2 dm. high; lvs. broadly ovate, 3–5 cm. long, rounded at base; raceme glandular-pubescent, 3–15-fld.; bracts 2–4 mm. long; pedicels 5–8 mm. long; sepals and petals yellow-green, reflexed in flowering, 4–5 mm. long; lip cuneate, 7–10 mm. long, shallowly 2-lobed at tip; caps. glandular-puberulent, ca. 6 mm. long.—Springy places in shade, below 8000 ft.; Montane Coniferous F.; San Jacinto and San Bernardino mts., Sierra Nevada from Fresno Co. n., N. Coast Ranges from Lake Co. n.; to Alaska, Atlantic Coast. June–Aug.

3. **L. caurìna** Piper. [*Ophrys c.* Rydb. *L. retusa* Suksd.] Stem 1–3 dm. high; lvs. broadly ovate, 3–6 cm. long, rounded at base; racemes 5–40-fld., glandular-pubescent; bracts 6–8 mm. long; pedicels slightly longer; sepals and petals greenish, 4–5 mm. long, erect; lip 5–6 mm. long, spatulate, entire to notched; caps. glabrous, ovoid, 5–6 mm. long.—Damp woods, 4000–5000 ft.; Montane Coniferous F.; Humboldt Co.; to B.C., Mont. June–July.

5. Eburophỳton Heller. Phantom Orchid

Saprophytic perennial herb, with erect stem from a branched creeping rootstock with fleshy roots. Lvs. reduced to long sheathing bracts. Fls. spicate. Lateral sepals keeled, spreading, subconcave; upper sepal and petals erect and somewhat connivent. Lip free, concave, the base saccate, the middle articulate. Column slender, with 1 stipitate anther; pollinia 2, granulose. Monotypic genus of w. U.S. (Latin, *ebur,* ivory, and *phyton,* plant, because not green).

1. **E. Aústinae** (Gray) Heller. [*Chloroea* A. Gray. *Cephalanthera* A. Heller. *Serapias* A. A. A. Eat. *C. oregana* Rchb. f.] Plant white, stout, 2–5 dm. high; lvs. 2–4 cm. long; raceme 5–15 cm. long; perianth white, with yellowish tinge, 12–15 mm. long; lip shorter, 3–5-nerved; column 4 mm. long.—Dry woods, below 6000 ft.; Montane Coniferous F.; San Bernardino Mts., Sierra Nevada from Fresno Co. n., Coast Ranges from Monterey Co. n.; to Wash., Ida. May.

6. Epipáctis Sw. Stream Orchis. Helleborine

Stem simple, leafy, from a short creeping rootstock; roots fibrous. Lvs. lance-linear to suborbicular, plicate-venose. Fls. racemose; fl.-bracts foliaceous. Sepals and petals almost equal, spreading. Lip saccate at base, flattened above, constricted at middle. Column short, with 1 sessile, 2-celled anther and broad truncate stigma. Pollinia 4, mealy-granulose, becoming attached above to the gland capping the rounded beak of the stigma. Caps. ellipsoid, pendent. Ca. 20 spp., of temp. Eurasia and N. Am. (Greek, *epipegnuo,* the name for hellebore and referring to a milk-curdling property ascribed to some spp.)

1. **E. gigantèa** Dougl. ex Hook. [*Peramium* g. Coult. *Serapias* g. A. A. Eat. *Helleborine* g. Druce. *Amesia* g. Nels. & Macbr.] Stout, 3–9 dm. tall, somewhat pubescent; lower lvs. ovate, 5–15 cm. long, upper lanceolate and gradually reduced; fls. 3–15, on pedicels 4–8 mm. long; sepals 12–18 mm. long, greenish, deeply concave, the dorsal erect, elliptic-lanceolate, the lateral ovate to lance-ovate, spreading, petals shorter, purplish or reddish; lip ovate-lanceolate, strongly veined and marked with purple or red, unequally 3-lobed; caps. 2–2.5 cm. long.—Moist stream banks, below 7500 ft.; almost all Plant Communities; throughout Calif.; to L. Calif., B.C., S. Dak., Tex. May–Aug.

7. Goodyèra R. Br. RATTLESNAKE-PLANTAIN

Low plants from somewhat fleshy rootstock with thickened fibrous roots. Lvs. all basal, usually white-reticulate in ours. Stem few-bracted and with fls. in terminal spike-like raceme. Fls. whitish, glandular-hairy; lateral sepals free, the dorsal adnate to the petals forming a galea. Lip saccate, sessile, entire, spurless. Column short, the anther on the back; pollinia 2, granulose, attached to small glandular disk. Caps. ellipsoid, ascending. Ca. 25 spp., of N. Temp. Zone, trop. Asia and to New Caledonia. (Named for John *Goodyear*, 17th-century English botanist.)

1. **G. oblongifòlia** Raf. [*Spiranthes decipiens* Hook. *Peramium d.* Piper. *G. d.* St. John & Const. *G. Menziesii* Lindl.] Scape stout, 2–4 dm. high; lvs. oblong-ovate, 3–6 cm. long, the winged petioles ca. half as long; perianth ca. 8 mm. long; column with awl-shaped beak above stigma; caps. ca. 1 cm. long.—Dry forest floor, below 5500 ft.; Yellow Pine F. and Red Fir F.; Coast Ranges from Marin Co. n. and Sierra Nevada from Mariposa Co. n.; to B.C., Quebec. July–Aug.

8. Calýpso Salisb. CALYPSO

Scapose, from a cormlike rootstock with coralloid roots. Lf. 1, basal, petioled, ovate. Bracts sheathing. Fl. 1, terminal, drooping, showy. Sepals and petals alike, lance-linear. Lip saccate, inflated, large, 2-parted below, with 3 rows of yellowish, translucent hairs in front and a translucent apronlike appendage, as well as 2 horns below the expanded apex. Column dilated or winged, petaloid, with 1 anther just below apex; pollinia 2, sessile on thickened square gland. Sp. 1, temp. Old and New Worlds. (*Calypso*, a Greek goddess.)

1. **C. bulbòsa** (L.) Oakes. [*Cypripedium b.* L. *Cytherea b.* House. *Calypso occidentalis* (Holz.) Heller.] Stem 8–25 cm. high, the sheathing scale 2–5 cm. long; lf. 3–6 cm. long, on a petiole 1–3 cm.; sepals and petals magenta-crimson, rarely white, ca. 11–14 mm. long, purple-striped; lip spotted madder-purple with sac whitish and marked with purple, the hairs yellowish; caps. many-nerved, ca. 1 cm. long; $2n = 28$ (Hagerup, 1944).—Bogs or leaf mold in rich woods; Redwood F., Mixed Evergreen F., Yellow Pine F.; Marin Co. to Del Norte and Siskiyou cos.; to Alaska, Atlantic Coast; Eurasia. March–July.

9. Corallorhìza Chat. CORAL ROOT

Saprophytic scapose herbs with rather short rhizomes which are much branched and coralloid. Stem brown or yellow to purple. Lvs. reduced to several sheathing scales. Infl. a loose terminal raceme. Fls. brown, yellowish or purplish, sometimes with some white or green. Sepals subequal, ascending spreading or connivent, the lateral united at base to form a short spurlike structure more or less adnate to ovary. Petals ca. equal to sepals, 1–3-nerved. Lip 1–3-ridged, simple to 3-lobed. Column compressed; anther terminal; pollinia 4, waxy, free. Caps. pendent, ovoid to ellipsoid. Genus of ca. 12 spp., N. and Cent. Am.; 1 in Eurasia. (Greek, *korallion*, coral, and *rhiza*, root.)

Perianth-parts longitudinally striped, 8–15 mm. long .... 1. *C. striata*
Perianth-parts not striped.
  Lip 3-lobed, spotted; spur wholly attached to ovary .... 2. *C. maculata*
  Lip entire, not spotted; spur free at tip .... 3. *C. Mertensiana*

1. **C. striàta** Lindl. [*C. Bigelovii* Wats.] Stems glabrous, erect, 1.5–5 dm. tall; lvs. tubular sheaths, whitish to purplish; fls. few to many, laxly arranged, the bracts ca. 2

mm. long; fls. pinkish-yellow or whitish, tinged and striped with purple, on pedicels ca. 2 mm. long; sepals oblong-elliptic to linear-lanceolate, 3- to 5-nerved, 6.5–16 mm. long; petals linear-oblong to obovate-elliptic, 3- to 5-nerved, 6–15 mm. long; lip white with purple veins, broadly elliptic, 6–12 mm. long; column slender, curved, 3–5 mm. long; caps. 1.2–2 cm. long.—Rich woods, below 7500 ft., Montane Coniferous F., Mixed Evergreen F., Redwood F.; Coast Ranges from Santa Cruz Mts. n., Sierra Nevada from Sierra Co. n.; to Wash., Quebec, Tex., Nuevo Leon. May–July.

2. **C. maculàta** Raf. [*C. multiflora* Nutt. *C. m.* var. *occidentalis* Lindl.] Stems 2–7 dm. tall, glabrous, the sheaths whitish; fls. laxly arranged, few to many; bracts 1.5–3 mm. long; pedicels 2–3 mm. long; sepals and petals crimson-purple to greenish, 3-nerved; dorsal sepal linear, 7–9 mm. long, the lateral lance-oblong to linear, 7–9 mm. long; petals lance-oblong to oblong-elliptic, 6–7.5 mm. long; lip white, spotted and veined with crimson, unequally 3-lobed, 6–8 mm. long; column yellow, curved, compressed, 4–5 mm. long; caps. nodding, 1.5–2.5 cm. long.—Montane woods, below 9000 ft.; Montane Coniferous F.; mts. from San Diego Co. n. through the Sierra Nevada and Coast Ranges; to B.C., Nfld., N. Car., Guatemala. June–Aug.

3. **C. Mertensiàna** Bong. Stems 2–5 dm. high, the sheaths scarious; raceme lax, 10- to 40-fld.; pedicels 2–3 mm. long; sepals and petals broadly linear, pinkish with purple tinge or veins, 3-nerved, 7–9 mm. long; sepals reflexed; lip broadly oblong, concave, red-purple, not spotted, clawed at base; column slender, almost equal to petals; free part of spur 1–5 mm. long; caps. 12–16 mm. long.—Rich woods, 4000–5000 ft.; Montane Coniferous F.; Humboldt Co. to Alaska, Mont. June–July.

10. Maláxis Soland. ex Sw. ADDER'S MOUTH

Small plants with simple stems from cormlike base. Lf. 1 in ours and ovate. Fls. minute, in terminal raceme. Sepals spreading, free or the lateral connate, oblong in ours. Petals narrower, spreading. Lip often auriculate, entire or lobed, concave, spurless. Column short, terete, 2-toothed at apex in ours; anther terminal; pollinia 4, waxy. Caps. small, ovoid to subglobose. A genus of ca. 150 spp., of Eurasia, Oceania, N. Am. (Greek, *malakos*, a softening, perhaps in reference to tender nature of the plant.)

1. **M. brachýpoda** (Gray) Fern. [*Microstylis b.* Gray. *Microstylis monophyllos* auth., not Lindl. *Malaxis m.* var. *b.* Morris & Eames.] Scape slender, 5–12 cm. high; lf. 3–5 cm. long, with sheathing petiolelike base; raceme spicate; bracts minute; pedicels 2–3 mm. long; sepals and petals greenish, 2–3 mm. long; lip rounded at base, with long slender tip; caps. 3–4 mm. long.—Rare, silty humps in wet meadows, 7300–8700 ft., Montane Coniferous F.; S. Fork of Santa Ana R. (San Bernardino Mts.), Tahquitz V. (San Jacinto Mts.); Rocky Mts. to Atlantic Coast. July–Aug.

24. Juncàceae. RUSH FAMILY (Fig. 132)

Perennial, sometimes annual, herbs, usually of moist places. Stems frequently from creeping sometimes erect rootstocks, often tufted, mostly simple, terete or compressed. Lvs. alternate, sheathing, grasslike, flat to terete. Infl. usually compound or decompound, paniculate, corymbose or umbelloid, rarely reduced to 1 fl. Fls. borne singly and each subtended by a bractlet, or in headlike or spikelike clusters and not individually subtended; infl. usually subtended by 1 or more bracts. Perianth small, regular, usually 6-parted into 3 outer and 3 inner glumaceous segms. Stamens usually 3 or 6; anthers adnate, introrse, 2-celled. Ovary superior, 3- or 1-celled, with 3–many ovules, 1 style, 3 filiform stigmas. Fr. a loculicidal caps., 3–many-seeded; seeds small, round to elongate, with fleshy endosperm. Eight genera, perhaps 300 spp., widely distributed.

Lf.-sheaths open; lvs. stiff (terete to flat); stems usually with spongy pith . . . . 1. *Juncus*
Lf.-sheaths closed; lvs. soft (flat); stems hollow . . . . 2. *Luzula*

1. Júncus L. RUSH. WIRE-GRASS

Perennial or annual, glabrous. Stems leafy or scapose. Lf.-sheaths with free margins; blades stiff, terete, flat, channeled, or compressed and gladiate. Infl. paniculate, corymbose, capitate or 1-fld. Fls. greenish to brownish or purplish. Stamens 3 or 6. Ovary

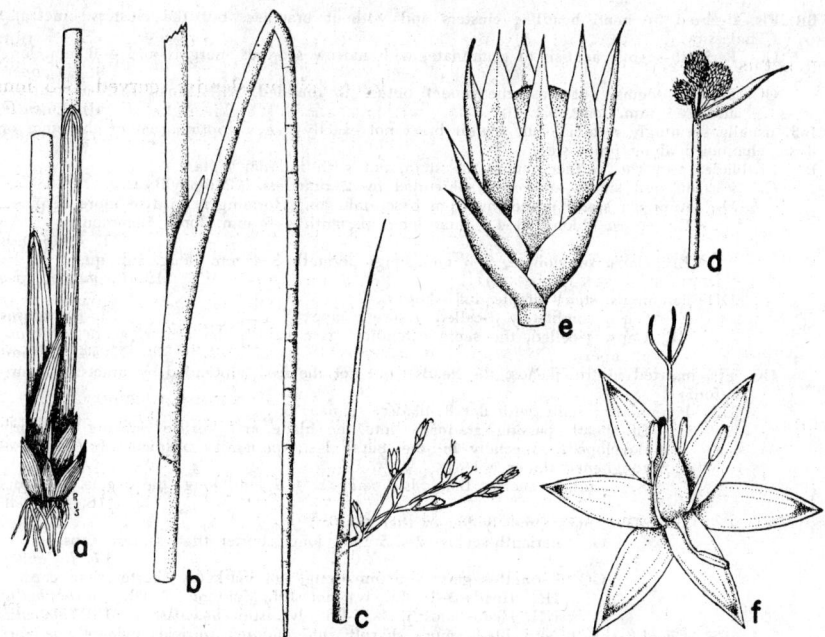

Fig. 132. JUNCACEAE. *Juncus: a*, base of plant of a sp. with basal lf.-sheaths bladeless; *b*, a sp. with cauline lf. showing sheath, auricle at its summit, and septate blade; *c*, infl. of seemingly lateral type with lowest bract terete and like a continuation of the stem; *d*, infl. seemingly terminal and fls. in heads; *e*, fl. with 6-parted perianth and subtending bracts; *f*, detail of opened fl., the perianth in an outer and an inner series, stamens 6, ovary superior, stigmas 3.

1-celled or 3-celled by intrusion of the parietal placentae. Seeds several to many, usually reticulate or ribbed, sometimes tailed. A genus of over 200 spp., most numerous in temp. zones. Many growing in meadows and swales mixed with grasses and sedges and of some economic importance for hay and pasture. (Latin name for rush, perhaps from *jungere*, to bind, the stems used for binding.)

A. Infl. seemingly lateral, the lowest bract terete and exactly like a continuation of the stem.
   B. Fls. inserted singly on the branches, each subtended by a pair of bractlets. (Genuini)
      C. Fls. 1–3, rarely 4–5; tufted subalpine plants; seeds conspicuously tailed.
         D. Uppermost lf.-sheath bearing filamentous rudiment of a blade; caps. retuse
                                                                          1. *J. Drummondii*
         DD. Uppermost lf.-sheath with well developed blade; caps. acute ...... 2. *J. Parryi*
      CC. Fls. many; plants usually from creeping rootstocks; seeds not tailed.
         D. Stamens 3, opposite the outer perianth-segms.; anthers shorter than fls.
                                                                          3. *J. effusus*
         DD. Stamens 6.
            E. Anthers ca. as long as fils.; stems in dense tufts; caps. subglobose
                                                                          4. *J. patens*
            EE. Anthers much longer than fils.; stems not conspicuously tufted; caps. longer
                than thick.
               F. Upper lf.-sheaths usually with well developed blades; stems compressed,
                   frequently twisted .... 5. *J. mexicanus*
               FF. Upper lf.-sheaths without blades; stems usually terete.
                  G. Perianth 5–6 mm. long. Coastal .... 7. *J. Lesueurii*
                  GG. Perianth mostly 3.5–4.5 mm. long. More generally distributed.
                     H. Stems smooth, or irregularly ridged when dry (diagonal sec-
                         tion of stem showing 1–2 rows of vascular bundles and no
                         fascicles of subepidermal strengthening cells); caps. acute
                                                                          6. *J. balticus*
                     HH. Stems finely and evenly ridged (with 3–4 rows of vascular
                         bundles and evident fascicles of subepidermal strengthening
                         cells); caps. obtuse .... 8. *J. textilis*

BB. Fls. inserted in small headlike clusters and without bractlets, but the clusters bracteate.
  (Thalassici)
  C. Perianth-segms. acutish to acuminate, with narrow scarious margins and 4–6 mm. long
                                                                                                  9. *J. Cooperi*
  CC. Perianth-segms. (at least the inner) obtuse to rounded, with broad scarious margins
        and 2–4 mm. long .... 10. *J. acutus*
AA. Infl. usually seemingly terminal, the lowest bract not exactly like a continuation of the stem or
      if so, channeled along inner side.
  B. Lf.-blades flat, the surface facing the stem, not with internal septa.
    C. Fls. inserted singly and each subtended by 2 bractlets. (Poiophylli)
      D. Annuals; stems branching from base; infl. long, forming a third or more of plant.
        E. Caps. oblong, 3–4.5 mm. long; perianth 4–6 mm. long. Common
                                                                                              11. *J. bufonius*
        EE. Caps. subglobose, 2–3 mm. long; perianth 3–4 mm. long. Infrequent
                                                                                        12. *J. sphaerocarpus*
      DD. Perennials; stems simple; infl. shorter.
        E. Caps. completely 3-celled, retuse at apex .... 13. *J. confusus*
        EE. Caps. 1-celled, the septa extending part way to the center, acute to obtuse
              at apex .... 14. *J. tenuis*
    CC. Fls. inserted in true heads, the heads (but not the fls.) subtended by bracts. (Grami-
        nifolii)
      D. Perennials usually with flat lf.-blades; stamens 6.
        E. Lf.-sheath passing gradually into the blade and auricles lacking or poorly
              developed, especially in the basal lvs.; perianth conspicuously papillose-
              roughened under a lens.
          F. Anthers shorter than fils.; perianth 4–5 mm. long, the segms. subequal
                                                                                                16. *J. Regelii*
          FF. Anthers mostly longer than fils.
            G. Perianth-segms. 2–3.5 mm. long, shorter than mature caps.
                                                                                              17. *J. Covillei*
            GG. Perianth-segms. 5–6 mm. long, not markedly shorter than caps.
              H. Heads 5–10-fld.; lvs. not stiff. Montane   18. *J. orthophyllus*
              HH. Heads mostly 8–25-fld.; lvs. stiff. Seacoast .. 19. *J. falcatus*
        EE. Lf.-sheath and blade more sharply differentiated, auricles evident; perianth
              smoother.
          F. Outer perianth-segms. distinctly shorter than the inner; lf.-blades chan-
                neled .... 20. *J. macrophyllus*
          FF. Outer perianth-segms. longer than or as long as inner; lf.-blades flat.
            G. Perianth-segms. brown, the inner longer than the outer; caps.
                  slightly shorter than perianth; seeds not tailed .. 21. *J. longistylis*
            GG. Perianth-segms. brown with green center, the inner shorter than
                  the outer; caps. much shorter than perianth; seeds tailed
                                                                                              15. *J. Howellii*
      DD. Annuals with ± setaceous lf.-blades; stamens 3.
        E. Lowest bract of infl. leaflike and extending beyond the head; caps. less than
              half as long as perianth .... 22. *J. capitatus*
        EE. Lowest bract not as above; caps. longer.
          F. Anthers longer than fils.; plants 3–12 cm. high; style 1.5–3 mm. long.
            G. Caps. equaling or exceeding perianth; perianth-segms. subequal
                                                                                            23. *J. leiospermus*
            GG. Caps. shorter than perianth; inner perianth-segms. longer than the
                  outer.
              H. Heads 5–7-fld.; bracts wide-spreading, blunt
                                                                                              24. *J. triformis*
              HH. Heads 2–4-fld.; bracts appressed or erect, pointed
                                                                                            25. *J. megaspermus*
          FF. Anthers shorter than fils.; plants 0.5–4 cm. high; style 0–1 mm. long.
            G. Heads often 2-several-fld.; bracts 2, subequal, 1–1.5 mm. long.
              H. Caps. 2 mm. long; seeds with prominent longitudinal ribs;
                    perianth-segms. not squarrose .... 26. *J. Kelloggii*
              HH. Caps. less than 2 mm. long; seeds with longitudinal ridges no
                    more prominent than lateral; perianth-segms. squarrose
                                                                                              27. *J. capillaris*
            GG. Heads 1-fld.; bract 1, 0, or if 2, these unequal and shorter.
              H. Bracts 1–2; lvs. much shorter than peduncles.
                I. Perianth 1.5–2 mm. long; caps. shorter than perianth
                                                                                              28. *J. bryoides*
                II. Perianth 2–3.5 mm. long; caps. equaling or exceeding
                    perianth.
                J. The perianth 2–3. mm. long; caps. much longer
                                                                                            29. *J. hemiendytus*
                JJ. The perianth 3–3.5 mm. long; caps. equaling perianth
                                                                                              30. *J. uncialis*
              HH. Bracts none; lvs. ca. as long as peduncles .. 31. *J. abjectus*
  BB. Lf.-blades terete, or if flat, with the edge toward the stem (gladiate), with ± evident
      internal septa.

C. The lf.-blades terete or somewhat compressed but not gladiate, the septa usually complete. (Nodosi)
  D. Plants growing in ponds, the early lvs. capillary, submerged, the later lvs. emerged. N. Coast .... 32. *J. supiniformis*
  DD. Plants not with 2 kinds of lvs. (i.e., submerged and erect).
    E. Stamens 3.
      F. Heads 1–few, usually congested, many-fld. Coast Ranges
                                                          33. *J. Bolanderi*
      FF. Heads 6–50, in open infl., few-fld. Cent. V. and n. .. 34. *J. acuminatus*
    EE. Stamens 6.
      F. Epidermis of stems and lvs. conspicuously transversely rugose under a lens .... 35. *J. rugulosus*
      FF. Epidermis smooth.
        G. Anthers definitely shorter than fils.; auricles 1–2 mm. high.
          H. Perianth 2–3 mm. long; heads few-fld. .... 37. *J. articulatus*
          HH. Perianth 3.5–5 mm. long; heads many-fld.
            I. Stems 4–10 dm. high; perianth 4–5 mm. long, brownish to greenish; caps. subulate .... 39. *J. Torreyi*
            II. Stems mostly 1–3 dm. high; perianth 3–4 mm. long, dark brown to blackish; caps. oblong-obovoid
                                                          41. *J. Mertensianus*
        GG. Anthers equaling or exceeding length of fils.; heads and auricles various.
          H. Caps. narrowed into long beak and well exserted when mature.
            I. Auricles less than 1 mm. high; perianth 3–4 mm. long. Rare in Calif. .... 38. *J. nodosus*
            II. Auricles mostly 4–6 mm. high; perianth 2.5–3 mm. long. Well distributed .... 36. *J. dubius*
          HH. Caps. abruptly contracted and not well exserted.
            I. Perianth-segms. light green, 4 mm. long, obtusish; auricles 4–6 mm. high .... 40. *J. chlorocephalus*
            II. Perianth-segms. brownish to blackish, 3–3.5 mm. long, acuminate; auricles 1–3 mm. high.
              J. Head single, purplish-black, with spathaceous bract; lvs. and perianth-segms. flaccid .. 41. *J. Mertensianus*
              JJ. Heads if single, brown, with narrow bract; lvs. and perianth-segms. stiffish.
                K. Rootstocks slender and creeping; stems not tufted; auricles mostly pointed, 2–3 mm. high; heads 1–many. General montane
                                                          42. *J. nevadensis*
                KK. Rootstocks short, subvertical; stems tufted; auricles mostly rounded, ca. 1 mm. high; heads usually solitary. S. Calif. ...... 43. *J. Duranii*
CC. The lf.-blades gladiate and the septa usually incomplete. (Ensifolii)
  D. Anthers mostly longer than fils.; style at least as long as ovary at anthesis.
    E. Perianth 4–5 mm. long; caps. acute below the beak and almost as long as perianth .... 44. *J. phaeocephalus*
    EE. Perianth 3–4 mm. long.
      F. Caps. shorter than perianth, obtuse below beak; perianth purple-brown
                                                          45. *J. macrandrus*
      FF. Caps. exceeding perianth, tapering into beak; perianth greenish or with brownish tinge .... 46. *J. oxymeris*
  DD. Anthers much shorter than fils.; style short.
    E. Auricles of lvs. small but evident; perianth-segms. somewhat unequal, exceeding caps. .... 47. *J. saximontanus*
    EE. Auricles lacking; perianth-segms. equal, shorter than caps.
      F. Stamens usually 6; lvs. 3–12 mm. wide; caps. tapering gradually into elongate beak .... 48. *J. xiphioides*
      FF. Stamens usually 3; lvs. 2–5 mm. wide; caps. tapering abruptly into very short mucro .... 49. *J. ensifolius*

1. J. Drummóndii E. Mey. [*J. D.* var. *humilis* Engelm. *J. compressus* var. *subtriflorus* E. Mey. *J. s.* Cov.] Rootstocks matted; stems in dense tufts, 5–30(–40) cm. high, slender, terete; lf.-sheaths basal, straw-colored, 2–7 cm. long, the inner often tipped with a bristlelike rudimentary blade; lowest bract of infl. mostly 1.5–3 cm. long; fls. 1–3(–5), inserted singly, each subtended by a pair of membranous ovate bractlets; perianth-segms. lanceolate, acuminate, 6–7 mm. long, green in center, with broad brownish membranous margins, mostly subequal, inner sometimes slightly shorter; anthers 6, much longer than fils.; caps. oblong, retuse at top, ca. as long as perianth; seeds narrow-obovoid, finely striate, long-tailed at each end.—Moist, steep, rocky or gravelly slopes, 6000–11,600 ft.; Montane Coniferous F.; Sierra Nevada from Tulare Co. n.; Siskiyou Co.; to Alaska, Alta., New Mex. July–Aug.

2. **J. Párryi** Engelm. Cespitose, mostly 1–3 dm. high; stems very slender; sheaths slightly brownish, mostly 1–3 cm. long, the uppermost bearing a blade; lf.-blade grooved at base, terete above, 3–6(–8) cm. long, very slender; auricles low, rounded, membranous; lowest bract of infl. 1.5–5 cm. long; fls. 1–3, inserted singly and subtended by 2 membranous ovate bractlets; perianth-segms. lanceolate, acuminate, 5–7 mm. long, sometimes with green in the center, mostly tinged brown, with broad scarious margins, the inner segms. somewhat shorter; anthers 6, much longer than fils.; caps. narrow-oblong, acute, usually exceeding perianth; seeds narrow-ovoid, finely striate, long-tailed at both ends.—Rather dry rocky places, above 9000 ft., Montane Coniferous F., San Bernardino Mts.; 6000–12,500 ft., Subalpine F., Alpine Fell-fields, Sierra Nevada, to Siskiyou and Trinity cos.; to B.C., Rocky Mts. July–Aug.

3. **J. effúsus** L. var. **pacíficus** Fern. & Wieg. Rootstocks stout, profusely branched; stems in dense tufts, 6–13 dm. tall, 2–3.5 mm. in diam. at base, terete, stiff, faintly many-striate; lf.-sheaths basal, chocolate-brown, dull, 5–15 cm. long, the uppermost mostly 6–15 cm. long, subtruncate or emarginate at the top, with strongly converging veins, and edges overlapping nearly or quite to the summit, often with terminal filamentous blade-rudiment; bract of infl. 6–20 cm. long; infl. lax, fastigiate, 2.5–15 cm. long; perianth-segms. 2.5–3.5 mm. long, subequal, rather soft, pale greenish-brown, with pale brownish membranous margins; anthers 3, shorter than fils.; caps. obovoid, ca. as long as perianth, brownish, obtuse to somewhat retuse, slightly apiculate; seeds ellipsoid, reticulate in ca. 16 longitudinal rows.—Rather common in moist places, below 8000 ft. in s. Calif., at lower elevs. farther n.; many Plant Communities; well distributed through the state except on the deserts; L. Calif. to B.C. June–Aug.

Var. **brúnneus** Engelm. [*J. e.* var. *hesperius* Piper.] Stems slender, 1–2 mm. in diam.; sheaths loose, membranous, dull, red-brown at base, the inner pale and somewhat greenish toward the rounded summit, mostly 10–15 cm. long; infl. 1–3 cm. in diam.; perianth-segms. 2.5–3 mm. long, green at center and bordered by 2 dark brown bands extending almost to margin.—Damp places near the coast; Coastal Salt Marsh, Coastal Strand; Santa Barbara Co. to B.C. June–July.

Var. **grácilis** Hook. Much like var. *brunneus*, but the sheaths usually rich red-brown to the summit and shining, coriaceous and rather tight.—Occasional, in wet places, Mt. Sanhedrin, Lake Co.; to B.C. July.

Var. **exíguus** Fern. & Wieg. Culms 1–1.5 mm. thick at base; sheaths red-brown at base, the uppermost greenish above and 6–11 cm. long; perianth 2 mm. long, pale brown with greenish center and broad scarious margins; caps. definitely shorter than perianth.—Occasional in wet places, mostly 3000–7000 ft., Montane Coniferous F.; Mt. Shasta, Sierra Nevada, Yolla Bolly Mts., and Plaskett Meadows (Glenn Co.); Ariz. July–Aug.

4. **J. pàtens** E. Mey. Perennial, tufted, but from short stout creeping rootstocks; stems blue-green, slender, 4–8 dm. high, terete; basal sheaths with or without awnlike blade-rudiments, brown; cauline lvs. none; infl. usually many-fld., open-paniculate to dense, 2.5–7 cm. long, the basal bract 5–20 cm. long; perianth segms. lanceolate, subequal, acuminate, light brown with green midrib, 2.5–3 mm. long; anthers 6, ca. as long as fils.; caps. subglobose, slightly angled, almost as long as perianth, obtuse and apiculate; seeds oblique-oblong, obscurely and irregularly reticulate.—Moist places such as meadows and stream banks, below 5000 ft.; most Communities from the shore to Montane Coniferous F.; Coast Ranges from Santa Rosa Id. and Santa Barbara Co. n.; to sw. Ore.; Ensenada, L. Calif. June–July.

5. **J. mexicànus** Willd. [*J. compressus* HBK. not Jacq. *J. balticus* var. *m.* Kuntze.] Rootstocks creeping, stout; stems not or scarcely tufted, slender, smooth, usually compressed and twisted, grasslike, mostly 2–6 dm. high; lf.-sheaths basal, brown to straw-colored, at least some usually with stemlike blades 5–20 cm. long; auricles short, rounded, somewhat cartilaginous; infl. mostly loosely 5–many-fld., 2–8 cm. high, the lowest bract 3–15 cm. long; perianth-segms. greenish or straw-colored, 4–5 mm. long, lanceolate, acuminate, subequal, with broad scarious margins; anthers 6, much longer than fils.; caps. ovoid, brown, mucronate, ca. as long as perianth; seeds oblong-obovoid, irregularly reticulate.—Moist, usually alkaline spots, or at higher elevs. in rather dry gravelly flats, sea level to 11,000 ft.; most Plant Communities; s. from Lake, Tuolumne, and Inyo cos.; to Tex., Mex. May–Aug. Possibly better treated as a var. of the next.

6. **J. bálticus** Willd. [*J. b.* var. *eremicus* Jeps. *J. Breweri* Engelm.] Rootstock creeping, stout; stems terete, wiry, grasslike, 3–8 dm. high, rather slender; sheaths basal, rather loose, straw-colored to brown, without blades except sometimes with a filament-like rudiment; panicle dense to diffuse, many-fld., 2–7 cm. high, the bract 3–20 cm. high; perianth-segms. purplish-brown, often with greenish center, 3.5–5 mm. long, lanceolate, acuminate, subequal, scarious-margined; anthers 6, much longer than fils.; caps. ovoid, brownish, acute, mucronulate, from nearly as long as to slightly longer than perianth; seeds oblong-obovoid, oblique, many-striate.—Moist places, mostly below 5000 (9000) ft.; many Plant Communities and well distributed in Calif.; variable and widespread in N. Am. and Old World. May–Aug.

Var. **montànus** Engelm. [*J. ater* Rydb.] Stems 1–2 dm. high, slender, wiry; infl. compact.—Up to 10,500 ft., Montane Coniferous F.; San Bernardino Mts., Sierra Nevada; to Alaska, Rocky Mts. July–Aug.

7. **J. Lesueúrii** Bol. [*J. L.* var. *Tracyi* Jeps. *J. balticus* ssp. *pacificus* Engelm.] Rootstocks creeping, long, stout; stems rather stout, smooth, terete, 3–9 dm. high, arising singly; lf.-sheaths basal, bladeless, straw-colored to brown; lowest lf. of infl. sometimes almost as long as stem; infl. many-fld., dense and subglobose to lax and 3–4-rayed; perianth-segms. lanceolate, acuminate, 5–6 mm. long, green in center, purplish-brown in the membranous margins, outer somewhat longer than inner; anthers 6, much longer than fils.; caps. oblong-ovoid, acute, brown, somewhat shorter than perianth; seeds ovoid, finely reticulate, the meshes broader than long.—Borders of marshes and near dunes, along the coast; Coastal Strand, Coastal Salt Marsh; San Luis Obispo Co. n.; to B.C., S. Am. May–Aug.

8. **J. téxtilis** Buch. [*J. Lesueurii* var. *elatus* Wats.] Rootstock stout, creeping; stems stout, terete, stiff, 1–2 m. high, 3–5 mm. thick, pale and distinctly striate and with low flat ridges (under a lens); lf.-sheaths basal, brown, bladeless, the upper sometimes with a filamentous rudiment; lowest bract of infl. 5–15 cm. long, spinescent; infl. lax, many-fld., 5–15 cm. high; perianth-segms. ca. 4 mm. long, lanceolate, acuminate, subequal or the inner slightly shorter, greenish or pale brownish, scarious-margined; anthers 6, much longer than fils.; caps. oblong-ovoid, obtuse, brownish, beaked, equal to or shorter than perianth; seeds oblong, finely reticulate.—Wet places, below ca. 6000 ft.; various Plant Communities; Ventura Co. to Orange and San Bernardino cos. May–June.

9. **J. Coòperi** Engelm. Perennials in large tufts, 4–8 dm. high; stems rigid; lvs. basal, terete, stiff, spinescent, much shorter than stems; auricles not developed; lowest bract of infl. 6–10 cm. long, spinescent; infl. paniculate, with uneven branches, 2–10 cm. long; fls. 2–several in a cluster; perianth-segms. lanceolate, greenish or straw-colored, 4–6 mm. long, acutish to acuminate, with narrow scarious margins; anthers 6, longer than fils.; caps. narrowly ovoid, acute to acuminate, ca. as long as perianth; seeds narrow-obovoid, short-appendaged, reticulate.—Alkaline flats, below 2000 ft.; Alkali Sink; Mojave and Colo. deserts; Nev. May.

10. **J. acùtus** var. **sphaerocárpus** Engelm. Perennial, in large tufts, 6–12 dm. high, with stout, rigid, pungent stems; lvs. all basal, terete, nearly as long as stems, the sheaths inflated, brownish; auricles from scarcely developed to several mm. high, cartilaginous; lowest bract of infl. foliose, 5–15 cm. long, stout, spinescent; infl. paniculate, with unequal branches, 5–20 cm. long; fls. 2–4 in small clusters; perianth-segms. lanceolate, pale brown, 2–4 mm. long, broadly scarious-margined, brown, shining, indurate, the outer acutish, the inner obtuse to rounded; anthers 6, longer than fils.; caps. subglobose, obtuse, mucronate, ca. 5 mm. long; seeds obliquely obovoid, finely reticulate, acute or slightly tailed at each end.—Moist saline places; Coastal Salt Marsh, San Luis Obispo Co. to L. Calif.; alkaline seeps, Alkali Sink, Colo. Desert; Ariz. May–June.

11. **J. bufònius** L. Toad Rush. Tufted annual, branching from base; stems slender, 3–20(–30) cm. high; lvs. 1–3 per stem, short, flat, to 1 mm. wide, often involute; auricles low, rounded, hyaline; infl. cymose, forming the upper third to more than half of the plant, the lower nodes with lvs. having short blades and conspicuous sheaths; fls. borne singly on the branches; perianth-segms. light green, 4–6 mm. long, lanceolate, acuminate, scarious-margined, the outer somewhat longer than inner; anthers 6 (sometimes 3), ca. as long as fils.; caps. oblong, shorter than perianth, obtuse, mucronate; seeds oblong-obovoid, minutely reticulate with 30–40 rows; $2n =$ ca. 60, 120 (Wulff, 1937).—Common in moist especially open places; dried pools, etc., below 8000 ft., in s. Calif.

and 5000 in the Sierra Nevada; all Plant Communities, even occasional on the desert;
San Diego Co. n.; cosmopolitan except polar regions and tropics. April–Sept. Variable.

Var. **congéstus** Wahl. [*J. bufonius* var. *fasciculiflorus* Boiss. *J. Congdoni* Wats.] Fls.
congested in small headlike clusters; inner perianth-segms. long-pointed.—Largely
halophytic, near the coast; occasional in the interior. April–Aug.

Var. **halóphilus** Buch. & Fern. Fls. ± congested; inner perianth-segms. obtuse.—Rare
in Calif., brackish spots mostly near the coast; Atlantic Coast. May–Aug.

12. **J. sphaerocárpus** Nees. Much like *J. bufonius,* tending to be smaller; perianth
3–4 mm. long; anthers shorter than fils.; caps. subglobose to broadly ovoid.—Occasional
on mud flats and in damp places, below 7500 ft., various Plant Communities throughout Calif.; to Ore., Ida., Ariz.; Old World. April–Aug.

13. **J. confùsus** Cov. Perennial, in small tufts; stems very slender, simple, 3–5 dm.
high; lvs. basal, ½ to ⅔ as long as stems; the blades almost filiform, channeled; auricles
white, rounded, scarcely 1 mm. long; infl. small, compact, mostly 1–1.5 cm. across, the
lowest bract filiform, 2–8 cm. long; perianth-segms. with greenish center, lateral brownish stripe and broad scarious margins, ovate-lanceolate, acute, 3.5–4 mm. long, subequal, acuminate; anthers 6, shorter than fils.; caps. broadly oblong, slightly shorter than
perianth, triangular and retuse at summit, completely 3-celled; seeds oblique-obovoid,
apiculate at both ends, coarsely and shallowly areolate.—Occasional, wet grassy places,
ca. 4000–6000 ft., Montane Coniferous F.; Trinity and Tuolumne cos. and n.; to Wash.,
Rocky Mts. July–Aug.

14. **J. ténuis** Willd. [*J. macer* S. F. Gray.] Tufted perennial, the stems slender,
2–6 dm. high, bright green; lvs. basal, with rather close sheaths, blades flat to involute,
ca. 1 mm. wide, commonly from half as long as stem to much longer; auricles white,
scarious, 1–3.5 (or more) mm. long; panicles 1–7 cm. long, usually rather many-fld., the
bracts 2–3, leaflike and much exceeding infl.; bracteoles blunt; perianth-segms. green
with white scarious margins, lanceolate, acuminate; anthers 6, much shorter than fils.;
caps. oblong-ovoid, shorter than perianth, rounded at summit, 1-celled, the placentae
extending only halfway to axis; seeds oblique-obovoid, short-apiculate, obscurely oblong-
reticulate; $2n = 30$ (Löve & Löve, 1948).—Occasional in damp grassy places, mostly
below 5000 ft.; largely Chaparral, Foothill Wd., Yellow Pine F.; San Bernardino Co. n.;
to Wash., Nfld., Fla.; S. Am. May–Aug. Intergrading with:

Var. **congéstus** Engelm. [*J. t.* var. *occidentalis* Cov. *J. o.* Wieg.] Stems and lvs. stiffer;
lvs. mostly less than half as long as stems; sheaths looser, auricles mostly less than 1 mm.
long; infl. subcapitate to open; perianth-segms. 4–5 mm. long, with brown on each side
of cent. green and onto the hyaline part; caps. oblong-ovoid, tending to be ca. ¾ the
length of the perianth.—Mostly in the Coast Ranges from Santa Barbara Co. to Ore.;
also in Sierra Nevada below 7500 ft. and intergrading with the sp. March–July.

Var. **Dúdleyi** (Wieg.) F. J. Herm. [*J. D.* Wieg.] Auricles cartilaginous, yellow, stiff,
glossy, to ca. 1 mm. long.—Wet places; Montane Coniferous F.; Trinity and Siskiyou
cos.; to Wash., Maine, Ariz. July–Aug.

15. **J. Howéllii** F. J. Herm. Rootstocks stout, stoloniferous; stems erect, subterete,
1.5–6 dm. high; lvs. grasslike, linear-subulate, 2–4 mm. wide, 2–20 cm. long; stem-lvs.
mostly 2; auricles acute, 1–3 mm. long; infl. of 2–10 heads, these 7–15 mm. across,
3–8-fld., sessile to well-peduncled; peduncles quite smooth; perianth-segms. lanceolate,
short-acuminate, dark brown with broad green center, smoothish, outer 5–6 mm. long,
with narrow hyaline margins, inner 4–5 mm. long, wider-margined; anthers 6, much
longer than fils.; caps. obovoid, shorter than perianth, obtuse, mucronate, pale; seeds
subellipsoid, caudate, conspicuously reticulate.—Meadows, 3000–7500 ft.; Montane
Coniferous F.; Butte and Plumas cos. to Del Norte and Siskiyou cos.; s. Ore. July–Aug.

16. **J. Regèlii** Buch. Rootstocks stout, stoloniferous; stems single or tufted, slender,
3–6 dm. high; lvs. flat and grasslike, 1–3 mm. wide, the blades 10–15 cm. long;
stem-lvs. 1–3; auricles not developed or poorly so; heads 1–5, globose to hemispherical,
10–30-fld., 8–15 mm. across; perianth-segms. 4–5 mm. long, minutely roughened, dark
brown with lighter middle, subequal, broadly lanceolate to subovate, the inner broader
and more blunt; anthers 6, slightly shorter than fils.; caps. oblong-ovoid, truncate to
retuse, mucronate, almost as long as perianth; seeds linear, long-tailed at both ends.—
Wet places, 6000 ft.; Montane Coniferous F.; Siskiyou Co.; to Wash. and Rocky Mts.
Aug.

17. **J. Covíllei** Piper. [*J. falcatus* var. *paniculatus* Engelm. *J. latifolius* var. *p.* Buch. *J. obtusatus* Engelm., not Kit.] Rootstocks creeping; stems in large tufts, 1–2.5 dm. high, slender; lvs. grasslike, mostly basal, flat, 2–3 mm. wide, ca. as long as stems; cauline lvs. 1–2; auricles none; heads few, in small panicles, commonly 3–5-fld.; bracts and peduncles roughened; perianth-segms. 2–3.5 mm. long, ovate, obtuse, subequal, the outer pointed, the inner rounded at tip, brown, hyaline-margined, the cent. roughened portion brown or green, narrow; anthers 6, slightly longer than fils.; caps. oblong-ovoid, brown, obtuse but depressed at apex, mucronate, exceeding perianth; seeds minutely reticulate, ribbed, somewhat truncate.—Not common, moist sandy banks, below 8000 ft.; mostly Redwood F. and Montane Coniferous F.; Sierra Nevada, San Gabriel, and San Bernardino mts., N. Coast Ranges s. to Mendocino and Lake cos.; to Wash. June–July.

18. **J. orthophýllus** Cov. [*J. longistylis* var. *latifolius* Engelm. *J. latifolius* Buch., not Wulf.] Rootstocks creeping, stout; stems compressed, 2–4 dm. high, leafless or with 1–3 lvs.; lvs. mostly basal, the blades grasslike, short to almost as long as stems, 2–6 mm. wide; auricles none or poorly developed; heads 1–several, mostly 6–10-fld., hemispheric, 8–15 mm. across; perianth-segms. minutely roughened, brown with green midrib, with scarious brown margins, lance-ovate, 5–6 mm. long, subulate-tipped, the outer shorter than inner; anthers 6, longer than fils.; caps. oblong-ovoid, obtuse, mucronate, slightly shorter than perianth; seeds oblique-obovoid, short-apiculate.—Wet places, such as meadows and stream banks, 4000–11,000 ft.; Montane Coniferous F.; San Bernardino Mts., Sierra Nevada, N. Coast Ranges from Yolla Bolly Mts. n.; to Wash., Ida. July–Aug.

19. **J. falcàtus** E. Mey. Rootstocks scaly, stoloniferous; stems 1–2(–3) dm. high, slender, leafless or with 1–3 lvs.; lf.-blades thick, stiff, flat, nearly or quite equaling the stems, 1.5–3 mm. wide; auricles usually lacking; heads 1, sometimes 2–3, 5–25-fld., hemispheric; perianth-segms. 5–6 mm. long, subequal, dark brown or with some green near center, papillose-roughened, lance-ovate, the outer subacuminate, the inner obtuse; anthers 6, longer than fils.; caps. brown, short-oblong to obovoid, ca. as long as perianth, retuse; seeds oblong-ovoid, obtuse, reticulate with elongate meshes.—Moist sandy places near the shore; Coastal Strand, N. Coastal Scrub; Santa Barbara Co. n.; to B.C.; Japan, Australia. May–Aug.

20. **J. macrophýllus** Cov. [*J. canaliculatus* Engelm., not Liebm.] Tufted perennials; stems 2–8(–10) dm. high, rather stiff, subterete; lvs. pale green, somewhat channeled; the basal striate, from ½ to ca. as long as stems, 1.5–3 mm. wide; cauline lvs. 1–3, the blades mostly 8–15 cm. long; auricles 1.5–3 mm. long; infl. loosely paniculate, with 8–25 heads; lowest sheath foliaceous, shorter than infl.; heads 3–5-fld.; perianth-segms. green, with reddish or brownish tinge, 5–6 mm. long, ovate, hyaline-margined, acute to obtuse, the inner distinctly longer than the outer; anthers 6, red-brown, much longer than fils.; caps. short-obovoid, short-beaked, not nearly as long as perianth; seeds obliquely obovoid, ca. 20-ribbed.—Wet banks and damp slopes of the interior, below 7500 ft.; mostly Chaparral, Montane Coniferous F.; Ventura Co. to n. L. Calif. and e. across the desert to Ariz. May–Aug.

21. **J. longistýlis** Torr. Loosely tufted perennial; stems slender, 2–5 dm. high; lvs. flat, grasslike, 1–3 mm. wide, the basal ⅓–½ as long as stem; cauline 1–3; sheaths with distinct, obtuse to truncate auricles 0.5–1.5 mm. long; heads 1–8, mostly approximate, 3–12-fld.; perianth-segms. 5–6 mm. long, broadly lanceolate, brown with scarious, usually whitish margins, acute to acuminate, the inner segms. longer than the outer; anthers 6, cream to pale yellow, longer than fils.; caps. oblong, truncate, a little shorter than perianth, slender-beaked; seeds oblong, 14–20-ribbed.—Occasional, meadows and moist banks, 5000–9500 ft.; Montane Coniferous F.; Sierra Nevada from Inyo and Mariposa cos. n.; to B.C., Nebr., New Mex. July–Aug.

22. **J. capitàtus** Weigel. Annual, 4–8 cm. high; stems very slender; lvs. ⅓–½ as long as stems, linear, flat or channeled; heads 1–2(–3), 4–8-fld., the lowermost bract leafy, exceeding the head; perianth-segms. ca. 3.5 mm. long, unequal, the outer longer and acuminate-aristate, hyaline-margined, the inner ovate, almost entirely membranous; anthers 3, shorter than fils.; caps. ovoid, obtuse, much shorter than perianth; seeds obovoid, transversely reticulate.—Vernal pool, 2.5 mi. e. of Fair Oaks, Sacramento Co.; Graham Hill between Fulton and Santa Cruz, Santa Cruz Co.; introd. from Old World. April.

23. **J. leiospérmus** F. J. Herm. Annual, 3.5–6 cm. high; lvs. ca. ⅓ as long as peduncles, the blades ca. 0.5 cm. long, ± canaliculate; peduncles 3–7, rather stout; heads mostly 2–4-fld., 6–10 mm. in diam.; bracts 2, broadly ovate, rounded at apex, 1–1.5 mm. long, hyaline, spreading; pedicels 1 mm. long; perianth-segms. appressed-ascending, subequal, 3–4 mm. long, lance-oblong, translucent with broad opaque midribs; anthers three, 3–4 times as long as fils.; caps. ca. length of perianth, broadly oblong, obtuse to truncate at apex, red to almost black; style 2–2.5 mm. long; seeds ovoid, short-apiculate, scarcely if at all reticulate.—Low place in grain field; V. Grassland; near Red Bluff, Tehama Co. April.

24. **J. trifórmis** Engelm. [*J. t.* var. *stylosus* Engelm.] Annual, cespitose, 5–12 cm. high; lf.-blades 1–2 cm. long, setaceous; peduncles 3–12, slender; heads 5–9 mm. in diam., 4–7-fld., dark; bracts ca. 1.2 mm. long, ovate, hyaline, blunt, spreading; pedicels ca. 1 mm. long; perianth-segms. appressed-ascending, 2.5–3.5 mm. long, slightly unequal, lance-linear, membranaceous, brownish-red with hyaline margins and green midribs, acute to acuminate; anthers 3, twice as long as fils.; caps. shorter than perianth, oblong, obtuse or truncate, somewhat apiculate; style 2–2.5 mm. long; seed subglobose to broadly obovoid, apiculate, longitudinally ribbed.—Moist sandy places, below 5000 ft., Montane Coniferous F., Shasta Co. to Madera Co., and clay depressions, Chaparral, San Diego Co. May–July.

25. **J. megaspérmus** F. J. Herm. Annual, 5–9 cm. high; lf.-blades 2–15 mm. long, channeled; peduncles 1–4, capillary; heads 5–7 mm. in diam., 2–4-fld.; bracts ca. 1.5 mm. long, ovate, acutish to acuminate, brown-tinged, erect to appressed; pedicels 0.5–1.5 mm. long; perianth-segms. appressed-ascending, 2–3.5 mm. long, unequal (inner longer), linear to lance-elliptic, acuminate, reddish at apex, membranous with opaque midribs; anthers 3, ca. twice the length of the fils.; caps. much shorter than perianth, broadly oblong to ovoid, truncate to obtuse at apex, with mucro; style 1.5–2 mm. long; seed elliptic-obovoid, longitudinally ribbed.—Moist sandy places, 3400–8100 ft.; Montane Coniferous F.; Placer Co. to Tulare Co. June–July.

26. **J. Kellóggii** Engelm. [*J. triformis* var. *brachystylus* Engelm. *J. b.* Piper.] Annual, 1.5–4 cm. high; lf.-blades 5–20 mm. long, linear-setaceous; peduncles 5–many; heads 2–5 mm. in diam., 1–3-fld.; bracts 2, ca. 1.5 mm. long, ovate, usually acutish to acuminate, hyaline, often tinged red, usually ascending to appressed; fls. nearly or quite sessile; perianth-segms. 4 or 6, erect to appressed, 2.5–3.5 mm. long, subequal, lanceolate, ± acute and acicular, hyaline-margined, reddish with green opaque midribs; anthers 3, ca. ⅓ as long as fils.; caps. shorter than to slightly longer than perianth, oblong to elliptic-ovoid, blunt to emarginate, brown to reddish, style 0.25 mm. long, becoming a beak on the caps.; seeds ± obovoid, apiculate, ribbed lengthwise and lineolate crosswise.—Moist banks and clay depressions, below 9000 ft.; most Plant Communities; San Diego Co. to Modoc and Del Norte cos.; L. Calif., Ore. to B.C. May–July.

27. **J. capilláris** F. J. Herm. Annual, 1–3 cm. high; lf.-blades linear to filiform, 3–6 mm. long; peduncles 1–15, setaceous; heads 1–4 mm. in diam., 1–2-fld.; bracts 2, ca. 1 mm. long, ovate, generally acute to acuminate, appressed or ascending; fls. subsessile; perianth-segms. erect to divergent, 2–2.5 mm. long, subequal, lanceolate, acicular, broadly hyaline-margined with some red, the midribs green; anthers 3, ca. ⅓ as long as fils.; caps. oblong-obovoid, shorter than perianth, obtuse to truncate; style ca. 0.1 mm. long, persistent; seed elliptic-obovoid, apiculate, longitudinally and transversely marked. —Moist places, below 11,000 ft.; many Plant Communities; Monterey and Tulare cos. to s. Ore. May–Aug.

28. **J. bryoides** F. J. Herm. Annual, 5–15 mm. high; lf.-blades 1–3 mm. long, triquetrous or setaceous-canaliculate; peduncles 1–25, filiform; head 0.8–1.2 mm. across, 1-fld.; bracts 2, 0.5–0.9 mm. long, ovate to lanceolate, blunt to pointed, hyaline, usually appressed; perianth-segms. appressed, somewhat unequal, 1.5–2 mm. long, ellipticoblong, red at center, incurved at tips; anthers 3, ca. half as long as fils.; caps. ellipticoblong to subspherical, shorter than perianth, obtuse; style none; seed turbinate, smooth, minutely apiculate.—Damp flats and sandy places, 4200–11,000 ft.; mostly Montane Coniferous F.; mts. from San Diego Co. to Alpine Co.; Utah. May–Aug.

29. **J. hemiendytus** F. J. Herm. Annual, 1–2.3 cm. high; lf.-blades 3–9 mm. long, linear-setaceous; peduncles 3–40, fairly stout; heads 1.7–2.5 mm. across, 1-fld.; bracts 1–2, 0.5–1.25 mm. long or obsolete, elliptic, usually blunt, appressed; perianth-segms.

erect or ascending, 2–3 mm. long, subequal, linear-lanceolate, acute, with broad hyaline margin and reddish midrib; anthers 3, ca. half as long as fils.; caps. longer than perianth, narrowly oblong, truncate to somewhat retuse; style very short, indurate; seed broadly ellipsoid, smooth, conspicuously apiculate.—Moist mud flats, mostly 2000–10,600 ft.; Montane Coniferous F.; San Bernardino Mts., Sierra Nevada to Modoc Co., N. Coast Ranges from Napa Co. n.; to Wash., Nev. June–July.

30. **J. unciàlis** Greene. Annual, 1.5–3 cm. high; lf.-blades 0.8–1.2 cm. long, linear, canaliculate; peduncles 4–35; heads 2–3 mm. wide, 1-fld.; bract 1, spathiform, 0.5 mm. long, shorter than wide; perianth-segms. divergent, 3–3.5 mm. long, subequal, lanceolate, abruptly acute to obtuse, with hyaline margins and broad green midribs; anthers 3, ca. ⅓ as long as fils.; caps. oblong-ovoid, blunt to subretuse; style very short, somewhat persistent; seeds obovoid, gibbous, conspicuously apiculate, faintly reticulate.—About rainpools, below 4500 ft.; V. Grassland, Foothill Wd.; e. San Luis Obispo Co. to Shasta Co.; Ore. April–May.

31. **J. abjéctus** F. J. Herm. Annual, 1–2 cm. high; lf.-blades 5–8 mm. long, linear-setaceous; peduncles 6–20, stoutish; heads 2–3 mm. wide, 1-fld.; bracts none; perianth-segms. usually 4, erect-ascending, 2.5–3 mm. long, subequal, oblong-lanceolate, obtusish, with broad hyaline margins and opaque green midribs; anthers 3, ca. half as long as fils.; caps. oblong-obovoid, equaling or slightly surpassing perianth, truncate to retuse; style 0.2 mm. long, somewhat indurate; seeds elliptic-obovoid, apiculate, faintly reticulate.—Mud flats, at 11,200 ft.; upper edge Subalpine F.; Center Basin, Tulare Co.; Harney Co., Ore. July.

32. **J. supinifórmis** Engelm. Rootstalks much-branched and matted; stems closely tufted, 8–40 cm. high, wiry; early lvs. mostly submerged, capillary, flaccid, up to 3 dm. long; stem-lvs. terete, much longer than stems, ca. 1 mm. thick, not conspicuously septate; auricles 1–1.7 mm. high, blunt to acutish, membranaceous; infl. loose, of few heads, these 3–9-fld.; perianth-segms. light brown, 4 mm. long, linear-lanceolate, acute, 3–4-nerved; anthers 3, shorter than fils.; caps. oblong, short-beaked, longer than perianth; seeds obovoid, pointed at each end, reticulate in ca. 16 linear rows.—Ponds near the coast; Closed-cone Pine Forest?; Mendocino Co. to Coos Co., Ore. July–Aug.? Fl.-stems appear only as the season advances and the ponds become dry.

33. **J. Bolánderi** Engelm. Rootstocks stout, creeping, with rather close, slender solitary stems 3–8 dm. high; basal sheaths bladeless; stem-lvs. 3–4, strongly septate, subcompressed, 1–2 dm. long, 1–1.5 mm. wide; auricles 3–5 mm. high; infl. of 1–few heads, usually surpassed by lowest bract; heads subglobose, mostly 10–12 mm. in diam.; perianth-segms. brown, 3–3.5 mm. long, subequal, lance-linear, setaceously acuminate, narrow-margined; anthers 3, shorter than fils.; caps. brown, clavate-oblong, beaked, slightly shorter than perianth; seeds obovoid, apiculate at both ends, reticulate.—Swampy ground, low elevs.; various Plant Communities; from the coast to inner Coast Ranges, Santa Clara Co. n. to B.C. June–Aug.

34. **J. acuminàtus** Michx. [*J. Bolanderi* var. *riparius* Jeps.] Rootstocks short, inconspicuous; stems tufted, 2–8 dm. high; lvs. 1–3 on a stem, terete, septate, 1–2 mm. thick, the lower 1–3 dm. long, upper reduced; auricles 1.5–5 mm. long; lowest bract shorter than the loose infl.; heads 5–20-fld., turbinate to subglobose, 6–50; perianth-segms. light brown to greenish, 3–3.5 mm. long, subequal, narrowly acuminate, narrow-margined; anthers 3, shorter than fils.; caps. narrow-ovoid, mucronate, ca. as long as perianth; seeds oblique-obovoid, reticulate with ca. 20 longitudinal rows.—Apparently in irrigated places, lower Sacramento and San Joaquin valleys; reported also from Trinity and Mendocino cos.; Ore. to B.C., Ariz., Atlantic Coast. May–June?

35. **J. rugulòsus** Engelm. Rootstocks stout, creeping, with rather close, single, erect, fairly stout stems 4–10 dm. high; stems, lvs. and sheaths minutely transverse-rugose; basal sheaths bladeless; cauline lvs. 2–4, terete, 1–4 dm. long, septate; auricles commonly 4–6 mm. high; panicle diffuse, 1–2.5 dm. long, many-headed; heads 4–8-fld.; perianth-segms. light greenish-brown, 2.5–3 mm. long, lance-acuminate, subequal, hyaline-margined to near apex; anthers 6, equaling fils.; caps. narrow, beaked, exceeding the perianth; seeds reticulate, brown, fusiform, apiculate at each end.—Frequent in wet places; all Plant Communities below 6500 ft. and to the desert edge; San Diego Co. to San Luis Obispo Co.

36. **J. dùbius** Engelm. Much like *J. rugulosus* vegetatively but lacking the transverse

ridges on the epidermis; perianth brownish, 2.5–3 mm. long; anthers longer than fils.—
Wet places, below 10,000 ft., Montane Coniferous F., cent. Sierra Nevada; at low elevs.
chiefly in Chaparral, San Diego Co. n. along Coast Ranges; to Ore., L. Calif. April–Aug.

37. **J. articulàtus** L. Rootstalks branching; stems tufted, erect or ascending, 2–6 dm.
high; basal lvs. 1–2, bearing blades, withering early; stem-lvs. 5–10 cm. long, with
loose sheaths and conspicuously septate terete blades; auricles 1–1.5 mm. high, cartilaginous; infl. 5–10 cm. high, with divaricate branches; heads rather numerous, turbinate
to hemispherical, 3–12-fld.; perianth 2–3 mm. long, the segms. subequal, lanceolate,
acuminate, red-brown with green midrib or all green; anthers 6, shorter than fils.;
caps. dark brown, 3-angled, sharply acute, exceeding perianth, 1-celled; seeds oblong-
obovoid, reticulate in 16–20 rows; 2*n* = ca. 60 (Wulff, 1937), 80 (Timm & Clapham,
1940).—Moist places; Montane Coniferous F.; China Sam Gulch, Tuolumne Co., Trinity
Co.; to B.C., Atlantic Coast; Eurasia. July.

38. **J. nodòsus** L. Rootstocks slender, creeping, with tuberlike thickenings from which
arise the slender stems; these solitary, 1–4 dm. high; lvs. erect, terete, septate, ca. 1 mm.
thick, the cauline 2–4 in number, the uppermost equaling or exceeding infl.; auricles less
than 1 mm. long, tawny; infl. open, of 1–15 heads, sometimes surpassed by the lowest
bract; heads 6–30-fld.; perianth-segms. 3–4 mm. long, light brown in greenish, subequal,
lance-acuminate, plainly striate, the membranous margin running toward the tips;
anthers 6, equaling fils.; caps. lance-subulate, 3-sided, 1-celled, longer than perianth;
seeds oblong, narrowed at base, reticulate in 20–30 longitudinal rows.—Rare in Calif.;
wet places, at 1800 ft., Middle Fork of Tule R. and 6500 ft., Wild Rose Spring, Panamint Mts.; Nev. to B.C., Nova Scotia, Va. July–Sept.

39. **J. Tórreyi** Cov. [*J. nodosus* var. *megacephalus* Torr. *J. m.* Wood, not Curtis.]
Rootstock slender, creeping, with tuberlike thickenings giving rise to stout stems, these
4–10 dm. high; lvs. septate, terete, divaricate, 2–5 mm. in diam., the upper equaling
or exceeding infl.; auricles 1–2 mm. high; infl. congested, 1–20-headed, surpassed by
lowest bract; heads 10–15 mm. in diam., round, many-fld.; perianth-segms. 4–5 mm.
long, brownish to greenish, long-subulate, the inner shorter than the outer; anthers 6,
shorter than fils.; caps. subulate, slightly exceeding perianth; seeds acute at both ends,
reticulate in ca. 20 longitudinal rows.—Wet places, sea level to ca. 6000 ft.; all Plant
Communities; s. Calif. from the coast across the desert, e. of the Sierra Nevada to
Wash., Atlantic Coast. July–Sept.

40. **J. chlorocéphalus** Engelm. Rootstocks slender, forming mats; stems slender, 2–5
dm. high; basal sheaths bladeless; stem lvs. 3–4, inconspicuously septate, somewhat
compressed, 4–10 cm. long, to 1 mm. wide; auricles ca. 4–6 mm. high; heads 1 to few,
8–15 mm. in diam., depressed-globose, few–many-fld.; perianth- segms. 4 mm. long,
light green, with broad scarious margins, subequal, oblong-lanceolate, obtusish; anthers
6, longer than fils.; caps. oblong-obovoid, obtuse, much shorter than perianth.—Meadows
and moist woods, 4000–10,000 ft.; Montane Coniferous F.; Sierra Nevada, Nevada Co.
to Tuolumne and Mono cos.; w. Nev. June–Aug.

41. **J. Mertensiànus** Bong. Rootstocks short; stems tufted, very slender, 1–3(–4) dm.
high; lvs. 1–3 per stem, obscurely septate, laterally compressed, 0.5–1.5 mm. wide, 5–12
cm. long; auricles round, opaque, 1–2 mm. long; heads usually 1, sometimes 2–3, subglobose or somewhat hemispherical, usually many-fld., subtended by bract broad and
spathelike at its base; perianth-segms. dark brown to almost black, 3–4 mm. long,
lanceolate, subequal, narrow-margined, subulate at tip; anthers 6, mostly much shorter
than fils.; caps. oblong-obovoid, obtuse or retuse at apex, almost equal to perianth; style
ca. as long as ovary at anthesis; seeds lance-ovoid, reticulate.—Wet banks and meadows,
4000–11,500 ft.; Montane Coniferous F.; Sierra Nevada from Tulare and Inyo cos. n.,
to Modoc and w. to Trinity and Del Norte cos.; to Alaska, Rocky Mts. July–Aug.

42. **J. nevadénsis** Wats. [*J. phaeocephalus* var. *gracilis* Engelm.] Rootstocks slender
to stoutish, creeping; the stems arising singly, slender, 1.5–3(–5) dm. high, somewhat
compressed; lvs. subterete, septate, subfiliform to (rarely) 2 mm. thick, blades 5–15(–20)
mm. long; ligules translucent, pointed to rounded, mostly 2–3 mm. high; infl. varying
from 1–2-headed and crowded to several heads in a loose narrow panicle, the lowest
bract short, narrow; heads few-fld., flattened-hemispherical to turbinate, 4–10 mm.
wide when pressed; perianth-segms. brown, lanceolate, acuminate, 3–3.5 mm. long,
hyaline-margined, subequal; anthers 6, linear, longer than fils.; caps. oblong, dark

brown, ca. as long as perianth, abruptly contracted to a short beak; seeds obliquely obovoid, apiculate at each end, reticulate, the meshes in ca. 15 rows.—Wet banks and meadows, 4000–10,700 ft.; Montane Coniferous F.; San Jacinto and San Bernardino mts. through Sierra Nevada, to B.C., Mont., Wyo. July–Aug.

43. **J. Duránii** Ewan. [*J. Mertensianus* of s. Calif. refs.] Rootstock short, subvertical; stems tufted, capillary, 1–2 dm. high; lvs. compressed, subfiliform, 7–15 cm. long; auricles round to pointed translucent, ca. 1 mm. long; bract of infl. subulate; heads usually solitary, 7–9 mm. across, flattened-hemispherical, 5–15-fld.; perianth-segms. pale brown, lance-linear, hyaline-margined, 3 mm. long; anthers 6, equal to or slightly longer than fils.; caps. obtuse-obovoid, strongly beaked, ca. as long as perianth; seeds brown, narrow-lanceolate, obscurely longitudinally lineate.—Occasional in wet places, 6000–9000 ft.; Montane Coniferous F.; San Gabriel, San Bernardino, and San Jacinto mts. July–Aug.

44. **J. phaeocéphalus** Engelm. [*J. p.* var. *glomeratus* Engelm.] Rootstocks stout, creeping; stems flat, 1–5 dm. high; lvs. compressed, equitant, without auricles, the blades 6–20 cm. long, 1.5–4 mm. wide, often with complete septa; infl. of 1 to several heads, the lowest exceeded by the bracts; heads spherical, many-fld., mostly 1–1.5 cm. in diam.; perianth-segms. dark-brown, 4.5–5 mm. long, subequal, broadly lanceolate; anthers 6, longer than fils.; caps. oblong-ovoid, shorter than to ca. as long as perianth, abruptly acute below the prominent beak; seeds ovoid, reticulate, subtruncate at apex.—Beaches, dunes and meadows near the coast; mostly Coastal Strand, N. Coastal Scrub, Coastal Sage Scrub; Los Angeles Co. and Santa Rosa Id. to Ore. May–July. Intergrading freely with:

Var. **paniculàtus** Engelm. Stems 3–9 dm. high; heads many in a loose panicle, few-fld., mostly 5–10 mm. in diam.; perianth 4–5 mm. long.—Swampy places, stream banks and meadows, at low elevs., but to 4600 ft. at Cuyamaca Lake, most inland Plant Communities w. of the Cent. V., San Diego Co. to Napa Co. May–Aug.

45. **J. macrándrus** Cov. Much like *J. phaeocephalus* var. *paniculatus,* the lvs. 2–3 mm. wide; panicle diffuse; heads many, few-fld.; perianth 3–3.5 mm. long, brownish; caps. obtuse below the beak, definitely shorter than perianth.—Moist places, 4000–9500 ft.; Montane Coniferous F.; San Jacinto Mts. to n. Sierra Nevada. July–Aug.

46. **J. oxýmeris** Engelm. Rootstocks creeping, stout; stems 3–6 dm. high, slender, compressed; lvs. equitant, flat, with obscure auricles, blades 5–20 cm. long, 3–6 mm. wide; panicles well developed, compound, the sheaths short; heads many, few-fld.; perianth-segms. subequal, green or with brownish tinge, 3–4 mm. long, acuminate and with broad scarious margins; anthers 6, as long as to much longer than fils.; caps. narrow-ovoid, longer than perianth, tapering into slender beak; seeds narrow-obovoid, reticulate. —Occasional, wet places, below 7000 ft.; most Plant Communities; coastal to montane, San Diego Co. n., mostly in the mts.; to Wash. June–Aug.

47. **J. saximontànus** A. Nels. [*J. xiphioides* var. *montanus* Engelm.] Rootstocks rather stout, creeping; stems flat, 4–6 dm. high, lf.-blades equitant, flat, 1–2.5 dm. long, 2–5 mm. wide; with small but definite auricles; panicle open, the lowest bract rather short; heads few, 7–10 mm. thick, 15–25-fld.; perianth-segms. light brown, ca. 3 mm. long, lanceolate, the outer slightly longer than the inner; anthers 6, shorter than fils.; caps. brown, oblong, slightly shorter than perianth, obtuse below the beak; seeds reticulate, subfusiform.—Occasional on wet banks, 6000–7500 ft.; Montane Coniferous F.; San Bernardino Mts., Sierra Nevada; Ore. to B.C., Alta., New Mex., Mex. July–Aug.

Forma **brunnéscens** (Rydb.) F. J. Herm. [*J. b.* Rydb.] Heads many, 5 mm. in diam., 5–12-fld.—Wet places, 6900–7500 ft.; Big Pines, San Gabriel Mts., Camp Tulakes, San Bernardino Mts., Santa Rosa Mts.; Ariz. to Colo., Mex. July–Aug.

48. **J. xiphioìdes** E. Mey. [*J. x.* vars. *auratus* and *littoralis* Engelm.] Rootstock creeping, rather stout; stems flat, 5–9 dm. high; lvs. equitant, the blades flat, 1–4 dm. long, 3–12 mm. wide, without auricles; bracts less than half as long as infl.; heads many, in a compound panicle, 3–20-fld.; perianth-segms. equal, lance-acuminate, light brown, 3–3.5 mm. long, narrow and revealing the caps.; anthers 6, a little shorter than fils.; caps. slightly exceeding perianth, oblong, acute, gradually contracted below the beak; seeds lance-ovoid, reticulate.—Well distributed in wet places, sea level to ca. 7000 ft.; all Plant Communities from the coast to the desert; s. Calif. to s. Ore., Ariz., L. Calif. May–Oct.

49. J. ensifòlius Wikstr. [*J. xiphioides* var. *triandrus* Engelm.] Rootstocks slender, creeping; stems compressed, 2-edged, 2–5 dm. tall; lf.-blades distinctly equitant, flat laterally, 7–15 cm. long, 2–5 mm. wide, without auricles; lowest bract of infl. ensiform, sometimes more than half as long as infl.; heads few and rather large or many and smaller, dark to pale brown or greenish; perianth ca. 3 mm. long, with subequal lance-acuminate segms.; stamens 3, anthers shorter than fils.; caps. oblong, contracted abruptly to the beak and slightly exceeding the perianth; seeds broadly fusiform, reticulate.—Wet meadows and marshy places, from sea level to ca. 9000 ft.; largely Redwood, Mixed Evergreen and Montane Coniferous forests; Coast Ranges, Napa and Mendocino cos. n., Sierra Nevada n. of Fresno Co.; to Alaska, Alta., Utah, Ariz. July–Aug.

2. Lúzula DC. Wood Rush

Perennial with slender hollow simple stems. Lvs. flat, grasslike, soft, the sheaths with united margins. Infl. umbellate, paniculate or congested, bracteolate. Perianth-segms. 6; stamens 6 in ours. Ovary 1-celled, the 3 ovules inserted basally. Seeds 1–3, indistinctly reticulate, sometimes carunculate at base or apex. Spp. ca. 80, of temp. and colder regions mostly in N. Hemis., less so in S. (Latin, *lucus*, a wood or thicket, in allusion to the habitat.)

Fls. borne singly at ends of branches of infl., or at most in 2's or 3's.
    Panicle-branches drooping; lvs. somewhat hairy at base.
        Lvs. thin, shining; fls. pale green; seeds brown .... 1. *L. parviflora*
        Lvs. thickish, dull; fls. dark brown; seeds yellow .... 2. *L. glabrata*
    Panicle-branches divaricate; lvs. not hairy at base .... 4. *L. divaricata*
Fls. borne in clusters.
    Lvs. channeled; infl. dense, nodding, spicate .... 5. *L. spicata*
    Lvs. flat.
        Infl. not of distinctly capitate clusters; perianth 1.5 mm. long .... 3. *L. subcongesta*
        Infl. of 1 to many distinctly capitate clusters; perianth 2.5–5 mm. long.
            The infl. distinctly rayed (except in a few coastal populations); culms and infls. predominantly green or pallid.
                Perianth 2.5–3.5 mm. long, ca. as long as caps. Montane .... 6. *L. comosa*
                Perianth mostly 3.5–6 mm. long, much exceeding caps. Low elevs. .... 7. *L. subsessilis*
            The infl. very compact and subcapitate; culms and infls. predominantly blackish. Alpine
                                            8. *L. orestera*

1. L. parviflòra (Ehrh.) Desv. [*Juncus p.* Ehrh. *Juncoides p.* Cov.] Plants tufted, with short rootstocks, the stems 2–7 dm. high, slender; lvs. thin, mostly 2–5 per stem, glabrous except for the pilose summit of the sheath, blades 4–15(–20) cm. long, 4–12 mm. wide; panicles decompound, open, 5–10 cm. long, with long capillary drooping branches; bracts mostly fimbriate, but those immediately subtending the fls. not or scarcely ciliate; perianth 1.5–2 mm. long, greenish to pale brown, slightly shorter than or equal to the ovoid caps.; anthers longer than fils.; seeds dark brown, ellipsoid, scarcely caruncled, ca. 1 mm. long; $2n = 24$ (Nordenskiöld, 1949).—Moist places in woods and meadows, on stream banks, 3500–11,000 ft.; Montane Coniferous F.; Sierra Nevada from Kern Co. n., N. Coast Ranges from Mendocino Co. n.; to Alaska, Labrador; Eurasia. June–Aug. Variable; plants with caps. dark brown to almost black have been called var. *melanocárpa* (Michx.) Buch. [*Juncus m.* Michx.]—With the sp.

2. L. glabràta (Hoppe) Desv. [*Juncus g.* Hoppe. *Juncoides Piperi* Cov. *L. P.* Jones.] Densely tufted, with short rootstocks, 1–3.5 dm. high; lvs. thickish, dull, often pilose on margins, 4–12 cm. long, 3–10 mm. wide; panicles 3.5–8 cm. long, nodding; bracts brown, lacerate; perianth 1.5–3 mm. long, dark brown; anthers shorter than to almost as long as fils.; caps. ovoid, longer than perianth, dark; seeds amber to buff, ca. 1.2 mm. long, narrowed and carunculate at each end.—Moist slopes and walls, at 7000 ft.; Subalpine F.; Upper Canyon Creek, above Dedrick, Trinity Co. to B.C., Alta., Rocky Mts. July–Aug.

3. L. subcongésta (Wats.) Buch. [*L. spadicea* var. *s.* Wats. *Juncoides s.* Cov. *L. parviflora* var. *s.* Buch.] Plants tufted, 2–5 dm. tall; stems 3–4-lvd.; basal lvs. flat, 0.8–2 dm. long, 4–8 mm. wide; infl. ± congested; fls. in subcapitate clusters at ends of drooping branches of infl.; bracts immediately subtending the fls. markedly ciliate; perianth dark brown, ca. 1.5 mm. long; caps. brown to blackish, almost as long as perianth, apiculate; seeds brown, darker at apex, oblong-ovoid; $2n = 24$ (Nordenskiöld, 1951).—At 7000–11,200 ft.; Lodgepole F. to Alpine Fell-fields; Sierra Nevada from Tulare Co. n.; to s. Ore. July–Aug.

4. **L. divaricàta** Wats. [*Juncoides d.* Cov. *L. parviflora* ssp. *d.* Hult.] Tufted, 2–5 dm. high; lvs. rather soft, 5–25 cm. long, 3–10 mm. wide, shining, sometimes with few long hairs near base; infl. decompound, diffuse, 8–15 cm. long, the branches and branchlets divaricate; bracts entire to somewhat lacerate; perianth 2–2.5 mm. long, light brown; anthers ca. as long as fils.; caps. scarcely as long as perianth, greenish; seeds light brown, ca. 1 mm. long, carunculate at base.—Rather dry ridges and slopes, 7000–11,000 ft.; Lodgepole F., Subalpine F.; Sierra Nevada from Tulare Co. n.; to Alaska. July–Aug.

5. **L. spicàta** (L.) DC. [*Juncus s.* L. *Juncoides s.* Kuntze. *L. s.* var. *nova* Smiley.] Densely tufted, 1–3.5 dm. high; lvs. mostly basal, thickish, 3–15 cm. long, 1–4 mm. wide, stiffly erect, pilose-ciliate toward base; infl. nodding, spikelike, often interrupted, 1–2.5 cm. long; bracts pilose-fringed; perianth 2–2.5 mm. long, brown with hyaline margins, the segms. aristate-acuminate; anthers ca. as long as fils.; caps. ca. ⅔ the perianth, brown; seeds brown, oblique-ellipsoid, ca. 1 mm. long, short-carunculate; $2n = 12, 14, 24$ (Nordenskiöld, 1951).—Moist gravelly lake banks and open places, 9000–12,500 ft.; Subalpine F., Alpine Fell-fields; Sierra Nevada from Tulare Co. n.; to Alaska, Atlantic Coast; Eurasia. July–Aug.

6. **L. comòsa** E. Mey. [*Juncoides c.* Sheld. *L. multiflora* ssp. *c.* Hult. and *L. campestris* and *L. multiflora,* Calif. refs.] Tufted, 1–6 dm. tall; lvs. thin, light green, 5–15 cm. long, 3–6 mm. wide, sparsely long-pilose especially at junction of sheath and blade; infl. umbellate, the rays 0.5–5 cm. long; spikes subglobose to short-cylindric, ca. 5–10(–15) mm. long, 5–7 mm. thick; lowest bract foliose, 2–5 cm. long; bractlets hyaline, subentire to lacerate, ± ciliate especially in terminal half; perianth 2.5–3.5 mm. long, light brown with broad hyaline margins; anthers much longer than fils.; caps. light brown, nearly or quite as long as perianth; seeds red-brown, 1.5 mm. long including the paler and large caruncle; $2n = 24$ (Nordenskiöld, 1951).—Meadows and open woods, 3000–10,500 ft.; Montane Coniferous F.; mts. of s. Calif., through Sierra Nevada, N. Coast Ranges, Glenn to Humboldt and Siskiyou cos.; to Alaska, Rocky Mts. May–Aug. Variable.

Confused with *L. campestris* (L.) DC., a 12-chromosome sp., and *L. multiflora* (Retz.) Lejeune, predominantly a 36-chromosome sp., both native to the Old World.

7. **L. subséssilis** (Wats.) Buch. [*L. comosa* var. *s.* Wats. *L. c.* var. *macrantha* Wats. *Juncoides c.* var. *m.* Howell. *L. campestris* var. *m.* Fern. & Wieg.] Tufted, 1–3 dm. high; lvs. mostly basal, dull green, loosely long-ciliate, 5–15 cm. long, 3–7 mm. wide; infl. umbellate, sometimes subcapitate; heads usually separate, rarely clustered, globose to subcylindric; lowest bracts foliaceous, from shorter than to exceeding infl.; bractlets ± lacerate-ciliate, brownish (rarely pale) with hyaline margin, 3–6 mm. long; anthers ca. as long as fils.; caps. greenish, much shorter than perianth; seeds brown, ca. 1.25 mm. long, the light caruncle well developed; $2n = 12$ (Nordenskiöld, 1951).—Grassy and wooded slopes, up to 4000 ft.; Closed-cone Pine F., Foothill Wd., Mixed Evergreen F., etc.; Coast Ranges, Santa Barbara Ids. to Humboldt and Trinity cos., w. base of Sierra Nevada from Madera Co. to Plumas Co.; probably to B.C. March–May.

8. **L. oréstera** C. W. Sharsm. [*L. congesta, L. campestris* var. *c.* and var. *sudetica,* and *L. multiflora* var. *c.* of Calif. refs.] Densely tufted, 5–18(–25) cm. high, deep green with purplish culms and infls.; lvs. mostly basal, thickish, 3–8 cm. long, 2–3(–5) mm. wide, often rather long-ciliate; infl. subcapitate, the 2–4 heads scarcely rayed and forming a pyramidal cluster 5–12 mm. high and subtended by 1–2 longer leafy bracts; bractlets subentire to somewhat lacerate, naked or somewhat ciliate; perianth-segms. purplish-brown, with hyaline margin, 2–3 mm. long; anthers somewhat shorter than fils.; caps. purplish-brown, shorter than perianth; seeds dark brown, 1–1.5 mm. long, the caruncle prominent; $2n = 20, 22$ (Nordenskiöld, 1951).—Moist meadows and near lakes, 9000–11,500 ft.; Subalpine F., Alpine Fell-fields; Sierra Nevada, Tulare and Inyo cos. to Tuolumne and Mono cos. July–Aug. A pallid, leafier, more robust form occurs sparingly on the e. flank of the Sierra Nevada.

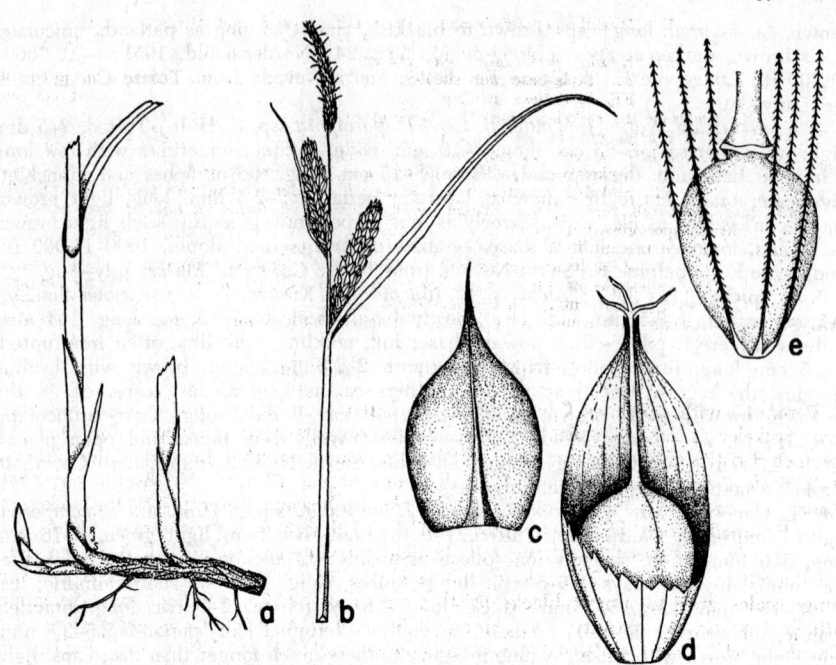

Fig. 133. CYPERACEAE. *Carex: a,* base of plant with creeping rhizome, basal lf.-sheaths, upper lf.-sheath of closed type and with blade; *b,* infl. with culm triangular in cross section, 3 spikes (upper one ♂, 2 lower ♀); *c,* bract from spike; *d,* perigynium of ♀ fl. partly cut away revealing pistil. *Heleocharis: e,* fl. from which stamens have fallen, perianth-bristles 6, downward-barbed, style-base forming a tubercle on summit of ak.

25. Cyperaceae. Sedge Family (Fig. 133)

Text except *Carex* contributed by Verne E. Grant

Grasslike or rushlike herbs, perennials with rhizomes or annuals with fibrous roots. Culms (stems) solid or rarely hollow, terete to variously angled. Lvs. mainly basal, alternate, commonly 3-ranked, the blades narrow, and the sheaths closed. Infl. commonly subtended by 1 to several involucral lvs. Infl. a simple or compound umbel or raceme or capitate cluster of spikelets, or a terminal spikelet. Spikelets composed of a series of scales (bracts) subtending individual fls. Scales spirally imbricated or 2-ranked, persistent or deciduous. Fls. perfect or unisexual. Perianth represented by several bristles or by an inner hyaline member, or absent. Stamens 1–3. Pistil 1, the ovary superior, 1-celled, with 1 ovule and a single 2- or 3-fid style. Fr. a triangular or lenticular ak. Embryo minute. Endosperm mealy. A large cosmopolitan family of ca. 60 genera and 2600 spp.

A. Fls. perfect, or perfect and ♂; ak. naked.
    B. All the scales of the spikelet bearing aks., or empty basal scales not more than 2; fls. all perfect. (Scirpoideae)
        C. Scales of the spikelet spirally imbricated. (Scirpineae)
            D. Bristles present.
                E. Bristles much exserted beyond the scales.
                    F. Bristles smooth and white .... 1. *Eriophorum*
                    FF. Bristles barbed .... 2. *Scirpus criniger*
                EE. Bristles included within the scales.
                    F. Style-base deciduous from the summit of the ak., tubercle none; 1 or more involucral lvs. present (except 1 sp.) .... 2. *Scirpus*
                    FF. Style-base persistent as a tubercle on the summit of the ak.; involucral lvs. none .... 3. *Heleocharis*
        DD. Bristles absent.

E. Hyaline perianth-member present between fl. and rachis of spikelet
                      5. *Hemicarpha*
  EE. Hyaline perianth-member absent.
    F. Style-base not swollen .... 2. *Scirpus setaceus* complex
    FF. Style-base swollen .... 4. *Fimbristylis*
CC. Scales of the spikelet 2-ranked. (Cyperineae)
  D. Bristles present; ak. beaked .... 6. *Dulichium*
  DD. Bristles absent; ak. not beaked.
    E. Spikelets elongated, many-fld., in ± loose umbels or heads ...... 7. *Cyperus*
    EE. Spikelets minute, with 1 perfect fl., in dense glomerate heads .. 8. *Kyllinga*
BB. Basal 3-several scales of the spikelet empty; spikelet consisting of both ♂ and perfect fls.
  (Rhynchosporoideae)
  C. Bristles present; infl. a capitate cluster or close cyme.
    D. Style-base deciduous from the summit of the ak.; ak. triangular .... 9. *Schoenus*
    DD. Style-base persistent as a tubercle on the ak.; ak. lenticular .... 10. *Rhynchospora*
  CC. Bristles absent; infl. a loose compound umbel. .... 11. *Cladium*
AA. Fls. all unisexual; ak. surrounded by a saclike bractlet (*perigynium*). (Caricoideae).. 12. *Carex*

1. Erióphorum L. COTTON-GRASS

Perennial with rhizomes. Culms erect, triangular to subterete, leafy. Invol. 1–several-lvd. Spikelets several, sessile, terminal or subterminal. Scales spirally imbricated. Fls. perfect. Bristles numerous, smooth, soft, white, much exserted beyond scales. Stamens 1–3. Style 3-fid. Ak. triangular. A small genus of ca. 13 spp., N. Temp. and Arctic Zones. (Greek, *erion*, cotton or wool, and *phoros*, bearing, referring to the spikelets.)

1. **E. grácile** Koch. [*E. g.* var. *caurianum* Fern.] SLENDER COTTON-GRASS. Culms slender, smooth, subterete, 30–60 cm. tall; lf.-blades triangular, channeled, 2–30 cm. long; involucral lf. solitary, 1–2 cm. long; infl. a cluster of 2–5 spikelets; spikelets 6–8 mm. long; scales gray to almost black; bristles numerous, white, 1–2 cm. long, remaining attached to ak. at maturity; ak. oblong, 2–3 mm. long; *n* = 38, 30 (Hagerup, 1944; Tanaka, 1942, 1948).—Mt. meadows and bogs, up to 7000 ft.; Yellow Pine F., Red Fir F., Mixed Evergreen F.; Sierra Nevada from Calaveras Co. n., N. Coast Ranges, from San Francisco and Sonoma cos. n.; to Alaska, New England; Eurasia. May–July.

2. Scírpus L. BULRUSH. TULE

Perennial or rarely annual herbs. Culms erect, triangular to terete, leafy or lvs. reduced to basal sheaths. Invol. of several blade-bearing lvs., or reduced to a solitary lf., or rarely absent. Infl. an open umbel or close cluster of numerous spikelets, or a solitary terminal spikelet. Scales spirally imbricated. Fls. perfect. Bristles (0–)1–6, barbed or ciliate, usually included within the scales, in one sp. exserted. Stamens 2–3. Style 2- or 3-fid. Ak. lenticular or triangular. A large cosmopolitan genus of ca. 200 spp. (Latin, *scirpus*, the classical name.)

(Beetle, A. A., Scirpus, in N. Am. Fl. 18[8]: 481–504, 1947.)
A. Bristles much exserted, upwardly barbed .... 1. *S. criniger*
AA. Bristles, where present, included within scales and usually downwardly barbed.
  B. Infl. subtended by involucral lvs.
    C. Involucral lvs. 2–5, usually exceeding the infl.; culms leafy.
      D. Spikelets small, 0.3–0.6 cm. long.
        E. Ak. lenticular; style 2-fid; stamens 2 .... 2. *S. microcarpus*
        EE. Ak. triangular; style 3-fid; stamens 3 .... 3. *S. Congdoni*
      DD. Spikelets larger, 1–4 cm. long.
        E. Lf. commonly 8–10 mm. broad; secondary rootlets conspicuous.
          F. Infl. loose-umbellate; ak. triangular; style 3-fid; bristles nearly equaling
            ak. .... 4. *S. fluviatilis*
          FF. Infl. capitate with 1–several elongated rays; ak. lenticular; style usually
            2-fid; bristles half as long as ak. .... 5. *S. robustus*
        EE. Lf. normally not over 6 mm. broad; secondary rootlets not conspicuous
                    6. *S. tuberosus*
  CC. Involucral lf. solitary, often appearing as a continuation of the culm.
    D. Culm leafy, triangular or subterete; spikelets few, 1–12.
      E. Perennials with rhizomes; culms not filiform; bristles present.
        F. Culm subterete; scales not awned .... 8. *S. nevadensis*
        FF. Culm sharply triangular; scales short-awned.
          G. Lf. blades convolute; involucral lf. 3–10 cm. long; plant small, to
            1.1 m. tall .... 7. *S. americanus*
          GG. Lf. blades flat; involucral lf. 1–3 cm. long; plant larger, to 2.2 m.
            tall.

H. Aks. light brown or gray, the surface minutely pitted; common, native .... 9. *S. Olneyi*
HH. Aks. dark brown, the surface horizontally rugose; rare, rice-field weed .... 10. *S. mucronatus*
EE. Annuals with fibrous roots, or rarely perennials; culms filiform; bristles absent.
　F. Annuals; aks. punctate.
　　G. Scales sharply keeled, acute or acuminate; involucral lf. to 2.5 cm. long .... 13. *S. koilolepis*
　　GG. Scales only slightly keeled at tip, obtuse; involucral lf. less than 0.5 cm. long .... 12. *S. cernuus*
　FF. Perennials; aks. longitudinally ribbed .... 11. *S. setaceus*
DD. Lvs. of culm reduced to basal sheaths with short blades, to 8 cm. long; culm stout and terete; spikelets numerous in umbels.
　E. Bristles filiform, barbed.
　　F. Scales well exceeding the ak.; common .... 14. *S. acutus*
　　FF. Scales equaling or only very slightly exceeding the ak.; rare
　　　　　　　　　　　　　　　　　　　　　　　　　　　15. *S. validus*
　EE. Bristles broad, often dark red, ciliate or plumose but never barbed
　　　　　　　　　　　　　　　　　　　　　　　16. *S. californicus*
BB. Infl. subtended by the long-awned lowermost scale of the spikelet, true involucral lvs. absent
　　　　　　　　　　　　　　　　　　　　　　　17. *S. Clementis*

1. **S. crìniger** Gray. [*Eriophorum c.* Beetle.] Perennial with rhizomes; culms slender, lightly scabrous above, erect, triangular, striate, (1–)2–10 dm. tall; lvs. flat, linear-lanceolate, scabrous, 4–10 cm. long; involucral lvs. 2–5, short, 4–8 mm. long; infl. a capitate head composed of 5–10 spikelets; spikelets 3–10 mm. long; scales dark to light brown, lanceolate, fimbriate; bristles ca. 6, upwardly barbed, much exserted, 3–7 mm. long, caducous; fils. elongated, 6–7 mm. long; style 3-cleft; ak. triangular, oblong-obovate, mucronate, 2–3 mm. long, dark brown.—Meadows, 7000–11,500 ft.; Red Fir F., Lodgepole F., Subalpine F.; Sierra Nevada from Tulare Co. n., N. Coast Ranges from Mendocino Co. n., to s. Ore. June–Aug.

2. **S. microcárpus** Presl. Perennial with stout rhizomes; culms stout, erect, leafy, subterete, 7–17 dm. tall; lvs. flat, broad, 1–2 cm. wide, margins scabrous, blades acuminate, often overtopping the stem; involucral lvs. 2–5, exceeding the panicle; infl. a loose spreading compound umbel, the primary rays to 10 cm. long; scales green to brown, acute, ovate, with a prominent midrib, not awned; bristles 4, downwardly barbed, somewhat surpassing the ak.; stamens 2; style 2-cleft; ak. whitish, ovate, lenticular with an obscure dorsal crest, mucronate, small, 1 mm. long.—Along streams and about springs in mts., up to 9000 ft.; many Plant Communities; throughout cismontane Calif. from San Diego Co. n.; to Alaska, Rocky Mts., New Mex. May–Aug.

3. **S. Cóngdoni** Britton. Similar to *S. microcarpus* but culms shorter, to 5 dm. tall; culm and involucral lvs. shorter, the latter not exceeding infl.; spikelets somewhat larger, 5–6 mm. long, acuminate; scales acuminate; bristles slender and curled at ends; stamens 3; style 3-fid; ak. triangular, mucronate.—Mt. meadows, 4000–9000 ft.; Yellow Pine F., Red Fir F., Lodgepole F.; Sierra Nevada from Fresno Co. to Plumas Co., n. and w. to Siskiyou and Humboldt Cos.; to Ore. June–Aug.

4. **S. fluviátilis** (Torr.) Gray. [*S. maritimus* var. ? *f.* Torr.] Rɪᴠᴇʀ Bᴜʟʀᴜsʜ. Perennial with horizontal tuber-forming rhizomes; culms stout, sharply triangular, erect, 10–15 dm. tall; lvs. 8–16 mm. wide; involucral lvs. 3–5, unequal, to 20 cm. long; infl. umbellate; spikelets acute, 1.6–2.5 cm. long; bristles 6, unequally long, nearly equaling ak.; style 3-fid; ak. sharply triangular, angled on back, 4 mm. long; $n = $ ca. 55 (Hicks, 1928).— In tule lands, below 4000 ft.; Freshwater Marsh; widely scattered in interior, Sacramento V., ne. Calif., Lassen Co. to Modoc Co., rare in Coast Ranges, San Mateo and Lake Cos.; throughout temp. N. Am.; Asia, New Zealand. June–July.

5. **S. robústus** Pursh. [*S. campestris* Britton, not Roth. *S. paludosus* A. Nels., *S. pacificus* Britton ex Parish.] Perennial with horizontal tuber-forming rhizomes; culms erect, sharply triangular, 5–15 dm. tall; lvs. typically 4–6 mm. wide, sometimes to 15 mm. wide; involucral lvs. 2–5, unequal, to 30 cm. long; infl. capitate, or with one to several elongated rays; spikelets ovate, 1.0–2.5 cm. long, sometimes cylindric, to 4 cm. long; scales reddish-brown to pale straw-brown; bristles 1–6, half as long as ak.; style usually 2-fid, sometimes 3-fid; ak. lenticular, 3–4 mm. long; $n = 53, 55, 57$ (Hicks, 1928).— Common, Freshwater Marsh, Coastal Salt Marsh, Alkali Sink; mostly below 1000 ft., ascending to 2000 ft. in s. Calif. and to 5000 ft. in Great Basin; throughout Calif. and N. Am., W. I., Argentina. April–Aug. Exceedingly variable, the spikelets reddish-brown

near the coast, becoming pale to straw-colored toward the interior. Apparently hybridizes with S. *fluviatilis*.

6. **S. tuberòsus** Desf. Culms slender, 2–5 dm. tall, from swollen nodes on black rhizomes; culms sheathed to at least ½ their length, trigonous, very leafy, blades mostly less than 5 mm. wide; bracts of infl. unequal, the outer to 1 dm. long; infl. of many spikelets, 1–2 cm. long; scales persistent, 7 mm. long, brownish; bristles 6, unequal; stamens 3; ak. 2.5 mm. long.—Introd. in Glenn, Colusa and Butte Cos.; native of Old World.

7. **S. americànus** Pers. [*S. a.* var. *polyphyllus* (Boeck.) Beetle.] THREE-SQUARE. Perennial with horizontal rhizomes; culms erect or arched, sharply triangular, stiff and slender, 0.3–1.1 m. tall; lf.-blades to 18 cm. long, keeled, convolute, narrow, 2–3 mm. wide; involucral lf. solitary, 3–10 cm. long; infl. a capitate cluster of 1–7 spikelets; spikelets oblong, acuminate, 8–12 mm. long; scales pale to chocolate brown, cleft at apex, short-awned; bristles 2–6, unequal, from slightly surpassing to half the length of ak.; style 2–3-cleft; ak. lenticular or obtusely triangular, mucronate, 3 mm. long; $n = 38$, ca. 40 (Hicks, 1928; Wulff, 1937).—Widespread up to 7000 ft.; many Plant Communities; in wet ground throughout Calif.; e. to Atlantic coast, S. Am.; Eu., New Zealand. May–Aug.

8. **S. nevadénsis** Wats. in King. Perennial with slender rhizomes; culms slender, erect or sometimes arched, subterete, 6–45 cm. tall; lf. blades narrow, ca. 1 mm. wide, convolute, to 15 cm. long; involucral lf. solitary, 1–10 cm. long; infl. a capitate cluster of 1–8 spikelets; spikelets narrow, acuminate, 0.5–2.0 cm. long; scales shining chestnut-brown, obtuse, not awned; bristles unequal, from very minute to nearly half the length of ak.; style 2-cleft; ak. convex on one surface, cuneate, shining, olive-green and punctate, 2 mm. long.—Wet alkaline ground, 4000–8000 ft.; N. Juniper Wd., Sagebrush Scrub; Mono Co. to Modoc Co.; Great Basin to B.C., Sask., Wyo. June–Aug.

9. **S. Ólneyi** Gray. [*S. chilensis*, Calif. refs.] OLNEY BULRUSH. Perennial with long rhizomes; culms stout, sharply triangular, the sides concave, 5–22 dm. tall; lf.-blades short, 2–13 cm. long; involucral lf. short, erect, 1–3 cm. long; infl. a capitate cluster of 5–12 spikelets; spikelets ovoid, 5–8 mm. long; scales flecked with brown, short-awned; bristles 4–6, unequal, from shorter than to equaling ak.; style 3-cleft; ak. lenticular, mucronate, light brown or gray, minutely pitted, 2.5 mm. long; $n = 39$ (Hicks, 1928).—Widespread in marshes and about springs up to 6700 ft.; Coastal Salt Marsh, Freshwater Marsh; throughout Calif.; throughout temp. N. Am., W. I., Cent. Am., S. Am. June–Aug.

10. **S. mucronàtus** L. Similar to S. *Olneyi*, but with longer spikelets, 8–10 mm. long; scales longitudinally striate, buff-colored, ovate, obtuse, mucronate; ak. dark brown, mucronulate, horizontally rugose; $n = 21$ (Morinaga & Fukushima, 1931).—Introd. as a ricefield weed in Sacramento V.; native in Eurasia. July–Aug.

11. **S. setàceus** L. Perennial with very slender horizontal rootstocks; culms filiform, erect, 6–12 cm. tall; lf. blades filiform, ca. 0.5 mm. wide, to 8 cm. long; involucral lf. solitary, 4–10 mm. long; spikelets solitary or 2, if 2, the lower one often larger; spikelets narrow-ovate, 2–4 mm. long; scales dark brown with a broad green midvein, obtuse, strongly keeled near apex, not awned; aks. somewhat angled on one side, prominently longitudinally ribbed, with fine horizontal striations between the ribs, brown, 1 mm. long; $n = 14$ (Scheerer, 1940).—Introd. in moist places at low or mid-elevs. in Humboldt, Del Norte, Nevada, and Placer Cos.; native in Eu. July–Aug.

12. **S. cérnuus** Vahl. var. **califórnicus** (Torr.) Beetle. [*Isolepis pygmaea* var. *c.* Torr.] Annual with fibrous roots; culms cespitose, nearly filiform, 4–20 cm. high; one basal lf.-sheath bearing a convolute blade 2–5 cm. long, the others without blades or with short blades to 3 mm. long; involucral lf. solitary, appearing as a continuation of the culm, 2–5 mm. long; spikelet solitary, ovoid, 2–5 mm. long; scales broad, obtuse, keeled at tip, pale to deep brown with green midrib; bristles none; stamens 2–3; style 3-fid; ak. triangular, punctate, white turning brownish at maturity, 1 mm. long.—Tidal flats and marshes; Coastal Salt Marsh, Freshwater Marsh, N. Coastal Scrub; L. Calif. to Del Norte County; n. to Wash. The sp. is represented by other vars. in w. S. Am. and Eu. May–Aug.

13. **S. koilolèpis** (Steud.) Gleason. [*Isolepis k.* Steud. *I. carinatus* H. & A. S. *carinatus* Gray, not Sm.] Annual with fibrous roots; culms cespitose, 4–20 cm. tall,

filiform; lf.-blades to 4 cm. long, obtuse; involucral lf. solitary, appearing as a continuation of the culm, to 2.5 cm. long, obtuse; spikelets 1–3 on a culm, ovate, 2–5 mm. long; scales strongly carinate, greenish-brown, acute to acuminate, with broad hyaline margins and a broad green midrib produced into a short mucro; bristles none; style 3-fid; ak. sharply triangular, punctate, light to dark brown, 1.5 mm. long.—Coastal marshes; Freshwater Marsh; Mendocino Co. to L. Calif.; through s. U.S. to Georgia. April–June.

14. **S. acùtus** Muhl. [*S. occidentalis* (Wats.) Chase, *S. rubiginosus* Beetle. *S. a.* var. *o.* Beetle.] COMMON TULE. Perennial with stout rootstocks and thick brown rhizomes; culms stout, erect, terete throughout to 5 m. tall and 2 cm. thick; lvs. reduced to basal sheaths with blades to 8 cm. long; involucral lf. solitary, terete, shorter than infl., appearing as a continuation of culm; infl. densely capitate to lax and umbellate; spikelets ovate or cylindric, 8–18 mm. long; scales ovate, carinate, short-awned, flecked with red so that the spikelet varies from pale brown to reddish-brown; bristles slender, usually 6, varying in length from ⅓ to equal the length of ak.; style 2-fid or sometimes 3-fid; aks. 2–3 mm. long, well enclosed by scales, lenticular or sometimes triangular; $n = 20$ (Hicks, 1928).—Abundant below 5000 ft. (to 8500 ft. in Mariposa Co.); Freshwater Marsh; throughout Calif.; n. to B.C., e. to Atlantic Coast. May–Aug.

15. **S. válidus** Vahl. TULE. GREAT BULRUSH. Similar to *S. acutus,* but with slender scaly rhizomes; spikelets ovate, 5–10 mm. long; scales equaling or only very slightly exceeding the mature aks.; $n = 21$ (Hicks, 1928).—Apparently rare; Freshwater Marsh; Humboldt, Trinity, and Siskiyou cos.; throughout N. Am., and nearly around the Pacific. Aug.–Sept.

16. **S. califórnicus** (C. A. Mey.) Steud. [*Elytrospermum c.* C. A. Mey.] CALIFORNIA BULRUSH. Stout perennial; culms subterete to triangular, to 4 m. tall; lvs. reduced to basal sheaths; involucral lf. solitary, erect, shorter than infl.; infl. loosely umbellate; spikelets narrow, acute, 5–10 mm. long; scales ovate, reddish-brown; bristles 2–4, dark red or sometimes pale red, broad and ciliate or plumose, not barbed; style 2-fid; ak. lenticular, 2 mm. long.—Abundant; Freshwater Marsh; along coast from Marin Co. to L. Calif., Cent. V.; to S. Carolina, Fla.; temp. S. Am., Hawaiian Ids. June–Sept.

17. **S. Cleméntis** Jones. [*S. yosemitanus* Smiley.] Dwarf perennial with densely cespitose slender grooved culms 3–10 cm. tall; basal lf.-sheath bearing a linear blade 0.5–3 cm. long; true involucral lf. wanting, but lowermost scale of spikelet simulating an invol. by the extension of its green midrib into an awn 2–4 cm. long; spikelet ovate, 2–4-fld.; scales ovate, obtuse, brown; bristles 6, shorter than ak.; stamens 3; style 3-fid; ak. trigonous, the surface minutely apiculate, ca. 1–5 mm. long.—High mt. meadows, 8000–12,000 ft.; Lodgepole F., Subalpine F., Alpine Fell-fields; Sierra Nevada from Tulare and Inyo cos. to Tuolumne and Mono cos. July–Aug.

3. Heleócharis R. Br. SPIKE-RUSH

Annual or perennial herbs with rhizomes, stolons, or fibrous roots. Culms simple, terete or subterete, usually striate. Lvs. reduced to basal lf.-sheaths, sometimes with an apiculate tip. Spikelets solitary, terminal, erect, several- to many-fld., not subtended by an involucral lf. Scales ovate to lanceolate, spirally imbricated. Bristles 1–8 or wanting, downwardly barbed. Fls. perfect. Stamens 2–3. Style 2- or 3-fid. Aks. lenticular or triangular. Style-base persistent, forming a tubercle on the apex of the achene. A large cosmopolitan genus of about 140 spp., inhabiting bogs, shallow ponds and salt marshes, mostly in trop. and warm-temp. regions. (Greek, *helos,* marsh, and *charis,* grace, many spp. found in marshes.)

(Svenson, H. K. Monographic studies in Rhodora, vols. 31–41, 1929–1939.)
A. Aks. plump or triangular; style 3-fid.
    B. Aks. with several longitudinal ridges and many fine horizontal lines ...... 1. *H. acicularis*
    BB. Aks. not longitudinally ribbed.
        C. Spikelets 2–7-fld.
            D. Culms 7–14 cm. tall .... 2. *H. pauciflora*
            DD. Culms 2–7 cm. tall .... 3. *H. parvula*
        CC. Spikelets 10–many fld.
            D. Scales spirally arranged; tubercle not 3-lobed.
                E. Tubercle long-subulate, continuous with apex of ak. .... 4. *H. rostellata*
                EE. Tubercle short-conic or pyramidal or apiculate, not continuous with apex
                    of ak. .... 5. *H. montevidensis*

DD. Scales arranged in 2 ranks; tubercle with 3 lobes decurrent on base of ak.
6. *H. pachycarpa*
AA. Aks. lenticular; style 2-fid.
B. Culms terete or subterete but not 4-angled.
C. Aks. shining black.
D. Apex of lf.-sheath conspicuously white-membranous .... 9. *H. flavescens*
DD. Apex of lf.-sheath firm, not membranous.
E. Aks. 0.5 mm. long .... 7. *H. atropurpurea*
EE. Aks. 1 mm. long .... 8. *H. geniculata*
CC. Aks. brown to yellow.
D. Annuals with fibrous roots; tubercle lamelliform .... 10. *H. obtusa*
DD. Perennials with rhizomes; tubercle pyramidal, constricted at base
11. *H. palustris*
BB. Culms 4-angled .... 12. *H. quadrangulata*

1. **H. aciculàris** (L.) R. & S. [*Scirpus a.* L. *H. a.* var. *occidentalis* Svens.] Perennial with filiform stolons; culms matted, capillary, furrowed, 2–20 cm. tall; basal lf.-sheaths truncate, inconspicuous; spikelets ovate to linear, acute, 2–7 mm. long, 5–10-fld.; scales ovate-lanceolate with brown sides, green midrib and hyaline margins; bristles 3–4 or wanting, equaling ak.; stamens 3; style 3-fid; ak. obovoid-oblong with ca. 40 close transverse lines; tubercle compressed-conical; *n* = 10, 15–19, 25–29 (Tanaka, 1937; Hicks, 1929).—Muddy riverbanks, meadows, vernal pools and marshes, sea level to 8000 ft. throughout the state except in the desert regions; throughout N. Am. and Eurasia. May–Aug.

KEY TO VARIETIES

Perennial with stolons or rhizomes; plants 2–20 cm. tall; bristles 3–4 or sometimes wanting.
Culms capillary; stamens 3; tubercle compressed-conical .... *H. acicularis*
Culms spongy; stamens 2; tubercle narrowly conical, not compressed .... var. *radicans*
Annual with fibrous roots; plants 2–6 cm. tall; bristles wanting .... var. *bella*

Var. **rádicans** (Poir.) Britton. [*Scirpus r.* Poir. *H. r.* Kunth. *H. Lindheimeri* Svens.] Perennial with long-creeping rhizomes; culms soft and spongy, striate, 3–8 cm. tall; basal lf.-sheaths closely investing the culm; spikelets 2–4 mm. long; scales with scarious straw-colored sides; bristles 4, exceeding ak.; stamens 2; tubercle narrowly conical, not compressed.—Moist places, below 4500 ft.; V. Grassland, Coastal Sage Scrub, etc.; cismontane Los Angeles and San Bernardino cos. to Fresno Co.; through s. U.S., n. Mex., W. I., temp. S. Am.

Var. **bélla** Piper [*H. b.* Svens.] Dwarf annual with fibrous roots and cespitose culms, often forming dense round tufts 5–10 cm. in diam.; culms capillary, furrowed, 2–6 cm. tall, light green; basal lf.-sheaths loose, obliquely truncated; spikelets 1–3 mm. long, (2–3–)8–10-fld.; scales with purplish-brown sides; bristles none; stamens 2; ak. with ca. 30 fine transverse lines; tubercle compressed-conical.—Mt. marshes, 3000–8000 ft.; Sagebrush Scrub to Lodgepole F.; widely distributed in montane Calif. except desert ranges; to Wash., Ida., New Mex.

2. **H. pauciflóra** (Lightf.) Link. [*Scirpus p.* Lightf.] Perennial with filiform rhizomes bearing small leafy tubers; culms capillary, grooved, erect, 7–14(–40) cm. tall; basal lf.-sheaths 2–3 cm. long, truncate; spikelet 4–7 mm. long, ovate, 2–7-fld.; scales lanceolate, acuminate, purplish-brown; bristles 2–6, shorter than to equaling or exceeding ak.; style 3-fid; ak. triangular, the surface finely reticulate, yellowish-brown, 2 mm. long; tubercle a subulate beak merging into the dark base of the style.—High mt. meadows, 5000–12,000 ft.; Montane Coniferous F.; n. to Modoc Co., 7000–12,000 ft.; San Bernardino and San Jacinto mts., through Sierra Nevada and N. Coast Ranges to B.C., Atlantic Coast; Eurasia, Chile. May–Aug.

KEY TO VARIETIES

Spikelets 2–7-fld.
Culms erect, not densely cespitose .... *H. pauciflora*
Culms arching, densely cespitose .... var. *bernardina*
Spikelets 9–12-fld. .... var. *Suksdorfiana*

Var. **bernardìna** (M. & J.) Svens. [*Scirpus b.* M. & J. *H. b.* M. & J.] Culms arching and densely cespitose, forming large, dense, grayish-green turfs.—At 7500–9250 ft.; Red Fir F., Lodgepole F.; San Bernardino Mts., Mt. Pinos.

Var. **Suksdorfiàna** (Beauverd) Svens. [*H. S.* Beauverd.] Culms compressed and conspicuously grooved, 1 mm. wide; spikelets 9–12-fld.—Meadowy places below 8000 ft.; many Plant Communities; Marin and Lake cos. n., Lassen and Modoc cos.; to Wash.

3. **H. párvula** (R. & S.) Link. [*Scirpus p.* R. & S.; *S. nanus* Spreng., not Poir.] Cespitose with fibrous roots, often with minute tuberous stolons; culms capillary, 2–7 cm. tall, greenish or straw-colored; spikelets 2.0–3.5 mm. long, broadly ovate, 2–9-fld.; scales ovate, obtuse or acute, green to yellowish, often reddish-brown on the sides; bristles 6, equaling or exceeding the ak.; stamens 3; style 3-fid; aks. obovate, 1.0–1.4 mm. long, straw-colored, smooth and shining, sometimes lightly striate under high magnification; tubercle very small, triangular, confluent with apex of ak., greenish; $n = 5$ (Wulff, 1937).—Coastal Salt Marsh; local, San Luis Obispo and Humboldt cos.; to B.C., Atlantic Coast; Eu., N. Afr. July–Aug.

Var. **coloradoénsis** (Britton) Beetle. [*Scirpus c.* Britton. *H. c.* Gilly. *H. leptos* Svens., not Clarke; *H. l.* var. *Johnstonii* Svens. *H. p.* var. *anachaeta* Svens. in part.] Dwarf sedge with somewhat arched culms 2–3 cm. tall; spikelets 4–6-fld.; scales with brown or purple sides, green midrib and hyaline margins; bristles 3 or none, short; aks. subtriangular, the convex surface bearing an obscure keel, the surface papillose; tubercle pyramidal.—Muddy places, up to 7000 ft.; several Plant Communities; widely scattered inland locs., Imperial and Los Angeles cos. to Modoc Co., to e. Ore., S. Dak., New Mex.

4. **H. rostellàta** (Torr.) Torr. [*Scirpus r.* Torr. *H. r.* var. *occidentalis* Wats. *H. r.* var. *Congdonii* Jeps.] Perennial from a short caudex; culms wiry, coarse, grooved, erect or somewhat arched, 0.25–1.5 m. tall, certain culms often procumbent and rooting at tips; basal lf.-sheaths 2–7 cm. long, obliquely truncate; spikelets oblong, 6–12 mm. long, 10–20-fld.; scales ovate-lanceolate; bristles 4–8, downwardly barbed, about equaling ak.; style 3-fid; ak. obtusely triangular, plump, 1.2–2.0 mm. long, the surface finely reticulate; tubercle a subulate beak continuous with the apex of the ak.—Saline marshes; mostly Alkali Sink; cismontane San Bernardino Co. to Death V., Owens V., Mono Co., Sacramento V., Marin and Sonoma cos.; throughout N. Am., W. Indies, Andes Mts. May–Aug.

5. **H. montevidénsis** Kunth. [*H. montana* auth., not R. & S. *H. arenicola* Torr.] Perennial with slender creeping reddish rhizomes; culms slender, striate, 1–4 dm. tall, tufted; sheaths reddish-brown at base, usually becoming straw-colored toward apex, the apex truncate, often with a minute tooth; spikelets narrowly ovoid to oblong, obtuse, 4–15 mm. long, many-fld.; scales ovate, obtuse, brownish or yellowish with a hyaline margin; bristles 4–6, pale brown or whitish, equaling or shorter than the ak.; style 3-fid; ak. triangular, obovoid, yellowish-brown to brown, minutely pitted or reticulate; the tubercle conic, short.—Moist ground, up to 6500 ft.; many Plant Communities; cismontane s. Calif. to Lake Co., w. edge of deserts; to Fla., Mex., temp. S. Am. May–Sept.

KEY TO VARIETIES

Tubercle conic to subulate; bristles long; spikelets linear-lanceolate to oblong.
  Spikelets ovoid to oblong, obtuse; bristles 2–6.
    Culms from slender rhizomes; mostly s. Calif. .... *H. montevidensis*
    Culms from stout rootstocks; n. Calif., rare .... var. *decumbens*
  Spikelets linear-lanceolate, acute; bristles 6–7.
    Perennial; not rare .... var. *Parishii*
    Annual; rare .... var. *disciformis*
Tubercle short-apiculate; bristles short; spikelets ovate .... var. *Bolanderi*

Var. **Paríshii** (Britton) V. Grant. [*H. P.* Britton.] Perennial; spikelets linear-lanceolate, acute, 10–15 mm. long; scales chestnut or dark brown with a short hyaline tip; bristles 6–7, sometimes exceeding the ak.; tubercle short-subulate to conic.—Moist ground, below 7000 ft.; many Plant Communities; Colo. and Mojave deserts n. to Mono Co., Coast Ranges, Ventura to Trinity cos.; Ariz., New Mex.

Var. **disciformis** (Parish) V. Grant. [*H. d.* Parish.] Similar to var. *Parishii* but annual.—E. base of San Jacinto Mts.

Var. **decúmbens** (Clarke) V. Grant. [*H. d.* Clarke.] Perennial with a stout rootstock; culms 1–6 dm. tall; spikelets oblong, 5–7 mm. long; scales obtuse; bristles 2–6, subequaling ak.; ak. and tubercle similar to var. *Parishii*.—Rare, Yellow Pine F., Red Fir F.;

Sierra Nevada, Mariposa to El Dorado cos., and probably in the N. Coast Ranges in Trinity and Siskiyou cos.

Var. **Bolánderi** (Gray) V. Grant. [*H. B.* Gray.] Perennial with short woody rhizomes; culms cespitose, glaucous green; basal lf.-sheaths slightly inflated at the indurated purplish summit; spikelets ovate, 3–8 mm. long; scales acute, dark brown; bristles 3–4, ½–¾ the length of ak.; tubercle flat, shortly apiculate.—Meadows, 4000–6000 ft.; Yellow Pine F.; Sierra Nevada from Madera to Plumas cos. to s. Ore. May–Aug.

6. **H. pachycárpa** Desv. Perennial with thick rootstocks; culms cespitose, slender to nearly filiform, quadrangular, grooved, 10–40 cm. tall, frequently proliferous; basal lf.-sheaths straw-colored, brownish or purplish-tinged; spikelets ovate, 5–10 mm. long; the scales arranged loosely in 2 ranks; scales ovate-lanceolate, purplish-brown; bristles equaling ak. or frequently absent; style 3-fid; ak. obtusely triangular, ovoid to obovoid, truncate, yellowish-white, the surface smooth or very finely reticulate; tubercle pyramidal, acute to acuminate, the base 3-lobed and decurrent on the angles of the ak.—Introd. in Coastal Salt Marsh in Humboldt Co. and in vernal pools of Foothill Wd. in El Dorado and Amador cos.; native in Chile.

7. **H. atropurpùrea** (Retz.) Kunth. [*Scirpus a.* Retz.] Annual with fibrous roots and cespitose culms; culms filiform, striate, 3–12 cm. tall; basal lf.-sheaths loose, obliquely truncate, with an attenuated tooth; spikelets ovoid, many-fld.; scales ovate, obtuse, with purple-brown sides, a green midrib and a very narrow scarious margin; bristles 2–4 or none, slender, shorter than ak.; stamens 2–3; style 2-fid; ak. lenticular, obovoid, 0.5 mm. long, shining black, the surface smooth; tubercle minute, depressed, constricted at base, conic, white.—Weed in low wet places, cismontane San Bernardino Co., Fresno Co., ricefields of Sacramento V.; Wash., cent. and s. U.S., trop. S. Am.; Eu., India, Afr.

8. **H. geniculàta** (L.) R. & S. [*Scirpus g.* L. *H. capitata* R. Br.; *H. caribaea* (Rottb.) Blake.] Similar to *H. atropurpurea;* culms sub-filiform, 5–25(–40) cm. tall; scales pale brown with a scarious margin; bristles 6–8, downwardly barbed, equaling ak.; ak. 1 mm. long; tubercle spongy.—Freshwater Marsh; San Bernardino, Riverside, Orange, and Imperial cos.; to Ind., Fla., trop. Am.; Old World. March–Dec.

9. **H. flavéscens** (Poir.) Urban. [*Scirpus f.* Poir. *S. flaccidus* Rchb. *H. f.* Urb.] Similar to *H. atropurpurea;* culms light green, 5–40 cm. tall; apex of lf.-sheath conspicuously membranous, usually consisting of 2 white inflated and somewhat emarginate lobes; spikelets 2–6 mm. long, ovate; scales elliptic to oblong-lanceolate, purplish-brown with paler midvein; bristles 6–7, white, equaling or slightly longer than the ak.; stamens 3; ak. shining, purplish-brown at maturity, the surface minutely punctate, obovate, ca. 1 mm. long; tubercle green, conic, acute.—Introd. in San Joaquin V. (Merced Co.); native to Atlantic Coastal Plain of N. Am., W. I., S. Am.

10. **H. obtùsa** (Willd.) Schult. [*Scirpus o.* Willd. *H. ovata* auth., not R. & S.] Annual with fibrous roots; culms slender, erect, 10–50 cm. tall; basal lf.-sheaths obliquely truncate with a minute tooth; spikelets ovoid to ovoid-oblong, obtuse, many-fld., 4–12 mm. long; scales obovate to nearly orbicular, obtuse, brown, scarious-margined; bristles 6–8, usually well exceeding the ak.; style 2-fid; ak. lenticular, obovate, ca. 1 mm. long, shining brown, the surface smooth; tubercle deltoid, lamelliform, broad, covering the top of the ak. and ⅓–½ its length; n = 5 (Hicks, 1929).—Marshes and ponds; Mixed Evergreen F., Yellow Pine F.; N. Coast Ranges (from Marin to Humboldt Cos.; to B.C., e. N. Am., Hawaii. June–Sept.

Var. **Engelmánni** (Steud.) Gilly. [*H. E.* Steud. *H. monticola* Fern.] Spikelets brownish, oblong-cylindric, 5–16 mm. long, obtuse to acute; scales ovate, obtuse to acute; bristles ca. 6 to rudimentary or wanting, when developed about equaling the ak.; ak. shining brown or greenish; tubercle less than ¼ the length of the achene.—Meadows and ponds; largely Yellow Pine F.; Sierra Nevada, Mariposa to Plumas cos., Modoc Co., Cent. V., Lake Co.; throughout temp. N. Am. Aug.–Oct.

11. **H. palústris** (L.) R. & S. [*Scirpus p.* L. *H. perlonga* Fern & Brack. *H. mamillata* Lindb. f.; *H. macrostachya* Britton.] Perennial with long, creeping rhizomes; culms loosely or densely cespitose, terete or oval in cross section, stout or slender, striated, erect, 0.5–1.0 m. tall, pale green to dark green; lvs. reduced to basal sheaths, obliquely truncated or sometimes mucronate; terminal spike 0.5–2.5 cm. long, subtended by 2–3 empty scales or 1 sterile clasping scale; fertile scales lanceolate, brown to purple or green, with a green midrib and scarious margin; bristles 4 or sometimes wanting,

downwardly barbed, ca. equaling the ak.; stamens 2–3; style mostly 2-fid; ak. lenticular, yellowish-brown, 1.0–1.5 mm. long; tubercle broadly to narrowly pyramidal or cone-shaped, constricted at base; $n = 5, 8, 16, 18, 19, 21, 23$ (Piech, 1924, 1928; Håkansson, 1928; Hicks, 1929; Kostrionkoff, 1930; Lewitzky, 1940; Pfeiffer, 1942; Walters, 1949). —Common and widespread in marshes, ponds and ditches up to 8000 ft.; many Plant Communities; throughout the state; throughout temp. N. Am.; Eurasia. A polyploid complex of great variability which has not as yet been satisfactorily analyzed into its elementary spp.

12. **H. quadrangulàta** (Michx.) R. & S. [*Scirpus q.* Michx.] Perennial, often bearing tubers; culms sharply 4-angled, coarse, 3–4 mm. wide on a side, 0.5–1.0 m. tall; basal lf.-sheaths membranous, reddish-brown; spikelets cylindric, 3–4 mm. in diam., 2–5 cm. long; the scales in 4 ranks; scales elliptic, rounded or acutish, not keeled, 5 mm. long, straw-colored; bristles downwardly barbed, equaling the ak.; stamens 3; ak. lenticular, obovate, yellowish-brown, shining, the surface finely marked with numerous rows of transverse linear cells; tubercle elongated, triangular, constricted at base.—Introd. in Merced Co.; native to Atlantic Coastal Plain of N. Am.

4. Fimbrístylis Vahl

Perennial or annual sedges. Culms erect, leafy at base. Involucral lvs. 1 to several. Infl. a loose umbel or capitate cluster of spikelets. Scales spirally imbricated. Fls. all perfect. Bristles none. Stamens 1–3. Style 2–3-fid, pubescent or glabrous, the style-base swollen and in one sp. persistent as a tubercle. Ak. triangular or lenticular.—A large genus of ca. 200 spp., trop. and temp. regions. (Greek, *fimbria*, fringe, and *stylus*, style, referring to the fringed style of some spp.)

Ak. triangular, bearing a tubercle .... 1. *F. capillaris*
Ak. lenticular, mucronate but not bearing a tubercle.
    Dwarf annuals; infl. a simple capitate cluster of 3–8 spikelets .... 2. *F. Vahlii*
    Taller perennials; infl. a loose umbel of several to many spikelets .... 3. *F. thermalis*

1. **F. capillàris** (L.) Gray. [*Scirpus c.* L. *Stenophyllus c.* Britton.] Annual with fibrous roots and cespitose culms; culms filiform, grooved, 5–25 cm. tall; lvs. basal, the blades filiform, shorter than the culm, the sheaths pubescent with long hairs; involucral lvs. 1–3, short and bristly; infl. a terminal umbel of (1–)2–several spikelets; spikelets oblong, 5–8 mm. long; scales oblong, obtuse, keeled, puberulent, with deep brown sides and green midrib; stamens 2; style 3-fid, glabrous; ak. triangular, truncate, ca. 1 mm. long, white to light brown, the surface transversely wrinkled; tubercle conic, constricted at base.—Meadows, 4000–6000 ft.; Yellow Pine F.; Sierra Nevada in Mariposa and Tuolumne cos.; n. to Ore., Atlantic Coast, W. Indies, Cent. Am., S. Am. June–Aug.

2. **F. Váhlii** (Lam.) Link. [*Scirpus V.* Lam.] Annual with fibrous roots and cespitose culms; culms filiform, striate, 2–10 cm. tall; lvs. basal, the blades nearly filiform, as long as or exceeding the culms; involucral lvs. 3–5, much exceeding the infl.; infl. a simple capitate cluster of 3–8 spikelets; spikelets many-fld., 4–8 mm. long; scales lanceolate, acuminate, pale greenish-brown; stamen 1; style 2-fid, papillose above, glabrous below; ak. lenticular, mucronate, minute, 0.5 mm. long, white, cancellate with ca. 14 vertical rows each with 15–20 rectangular pits.—Occasional wet banks, low elevs.; San Joaquin V., Colo. R.; Tex. to N. Car. and Fla., Cent. Am., S. Am. July–Nov.

3. **F. thermàlis** Wats. Perennial with rhizomes; culms slender, grooved, erect, 20–70 cm. tall; lvs. basal, the blades flat, 1.0–2.5 mm. wide, shorter than the culm, somewhat pubescent near the sheath; involucral lvs. few, short, narrow; infl. a loose simple or compound umbel, the rays to 6 cm. long; spikelets oblong, 8–15 mm. long, many-fld.; scales oblong, mucronate, light brown; style 2-fid, pubescent along its length, more heavily so above; ak. lenticular, obovate, mucronate, ca. 1.5 mm. long, slate-colored, the surface with numerous rows of very fine pits.—Marshes and hot springs; Freshwater Marsh; San Bernardino, Kern, and Inyo cos.; Nev. Aug.–Sept.

5. Hemicárpha Nees & Arn.

Dwarf annuals with fibrous roots and cespitose culms. Culms erect, filiform, grooved. Lvs. basal, the blades convolute, shorter than the culms. Involucral lvs. 1–3, one much exceeding the infl. Infl. a solitary spikelet or capitate cluster of 2–3 spikelets. Scales spirally imbricated. Fls. all perfect. Bristles absent. A single hyaline perianth-member developed between the fl. and the rachis of the spikelet, or sometimes wanting. Stamens 1–3. Style 2-fid, short, not swollen at base. Ak. oblong, finely pitted. A small genus of about 5 spp., trop. and temp. regions. (Greek, *hemi,* half, and *carphos,* chaff, referring to the perianth member.)

(Friedland, S. The Am. spp. of Hemicarpha. Am. Jour. Bot. 28: 855–861, 1941.)
Scales short-awned, awns 0.2–0.5 mm. long .... 1. *H. micrantha*
Scales long-awned, awns ca. 1 mm. long .... 2. *H. occidentalis*

1. **H. micrántha** (Vahl) Pax. [*Scirpus m.* Vahl.] Culms erect or somewhat recurved, 1–3 cm. tall or sometimes 10 cm. tall; lf.-sheaths weakly united below, open above, with loose hyaline margins, the blade filiform and grooved; involucral lvs. 2–3, the blades convolute, 0.5–2 cm. long; spikelets ovoid, obtuse, minute, ca. 2 mm. long, many-fld.; scales obovate, acuminate, with brown sides and a green midrib produced into a short recurved awn 0.2–0.5 mm. long, except the lowermost scale which possesses an awn 1–4 mm. long and simulates an involucral lf.; perianth-member hyaline, broad and fimbriate-tipped, or capillary, 0.5 mm. long, persistent on the rachis; stamen 1; style short; ak. obtuse, mucronulate, 0.5–0.8 mm. long, white to brown or black.—Calif. plants fall into two vars.

Var. **aristulàta** Cov. [*H. a.* Smyth.] Perianth-scale equaling or exceeding ak., margin entire.—Moist places; Yellow Pine F.; Coast Ranges from Tehama Co. n. to Wash. and e. to Miss. V. Aug.–Sept.

Var. **mìnor** (Schrad.) Friedl. [*Isolepis subsquarrosa* var. *m.* Schrad.] Perianth-scale shorter than ak., often vestigial or absent, margin bifid.—Moist places; V. Grassland, Foothill Wd., Yellow Pine F.; Cent. V. and Sierran foothills, from Fresno to Nevada cos.; to Wash., Atlantic Coast, Mex., Brazil. Aug.–Oct.

2. **H. occidentàlis** Gray. Similar to *H. micrantha;* culms 0.3–4.0 cm. tall; scales ovate-lanceolate, with brown sides and a green midrib produced into a long recurved awn 1.0–1.5 mm. long, subequaling the body of the scale.—Moist places, 4000–6000 ft.; Yellow Pine F.; San Bernardino and San Jacinto mts., cent. Sierra Nevada, N. Coast Ranges; n. to Wash. June–Aug.

6. Dulíchium Pers.

Perennial with rhizomes. Culms tall, leafy, terete, hollow, jointed. Culm-lvs. bearing racemes of ca. 6 spikelets in their axils. Spikelets flat, 2-ranked, jointed, the lowermost scale of each spikelet sterile. Scales lanceolate, decurrent on the proximal joint. Fls. perfect. Bristles 6–9, rigid, downwardly barbed. Stamens 3. Style 2-fid, the style-base persisting on the apex of the ak. as a beak. Ak. linear-oblong, long-beaked. Monotypic. (Name said to be from Latin, *dulcichimum,* sedge.)

1. **D. arundinàceum** (L.) Britton. [*Cyperus a.* L. *D. spathaceum* (L.) Pers.] Culms stout, erect, 3–10 dm. tall; lvs. numerous, the basal lvs. reduced to brown sheaths, the culm-lvs. green, flat, linear, 2–8 cm. long, 4–8 mm. wide, spreading or ascending; peduncles of the axillary racemes 0.4–2.5 cm. long; spikelets linear, spreading, 1.0–2.5 cm. long, 6–12-fld.; scales carinate, strongly nerved, greenish-brown; bristles ca. twice as long as the body of the ak.; ak. lenticular, smooth, ca. 3 mm. long, the beak long, subulate, ca. 3 mm. long; *n* = 16 (Hicks, 1929).—Occasional in swamps, 5000–7500 ft.; Yellow Pine F., Red Fir F.; Sierra Nevada from Fresno to Plumas cos.; N. Coast Ranges in Humboldt Co.; to B.C., Atlantic Coast. July–Oct.

7. Cypèrus L. Umbrella-Sedge

Perennial or annual herbs. Culms in ours simple, triangular, leafy at base, striate. Involucral lvs. 1 to several, much exceeding the infl. Infl. umbellate or capitate, the rays commonly bearing divaricate clusters of spikelets. Spikelets flat or subterete, many-fld. Spikelets falling away from the head, or persistent and then the scales deciduous. Rachis straight, offset or zigzag, unwinged or frequently bearing a pair of wings at each node, these being the decurrent bases of the next distal scale. Scales 2-ranked, keeled, all fertile or the lower ones empty. Fls. perfect. Bristles none. Stamens 1–3. Style 2–3-fid; the style-base deciduous from the summit of the ak. Ak. triangular or lenticular, naked or clasped by the rachis-wings. A large genus of about 600 spp., trop. and temp. regions. (Greek, *cypeiros,* the classical name.)

(McGivney, Sister. Subgenus Eucyperus. Cath. Univ. Am. Biol. Ser. No. 26, 1938. Corcoran, Sister. Subgenus Pycreus, Cath. Univ. Am. Biol. Ser. No. 37, 1941.)
A.   Style 2-fid; ak. lenticular.
  B.   Ak. laterally flattened.
    C.   Spikelets 2.0–2.5 mm. wide .... 1. *C. niger*
    CC.   Spikelets ca. 5 mm. wide .... 2. *C. unioloides*
  BB.   Ak. dorsally flattened .... 3. *C. laevigatus*
AA.   Style 3-fid; ak. triangular.
  B.   Spikelets persistent on the spike; scales deciduous.
    C.   Rachis not winged.
      D.   Perennial with short rhizomes.
        E.   Spikelets 10–20 mm. long.
          F.   Culms bluntly triangular, smooth; common .... 4. *C. Eragrostis*
          FF.   Culms sharply triangular, scaberulous; rare .... 5. *C. virens*
        EE.   Spikelets 4–7 mm. long .... 15. *C. alternifolius*
      DD.   Annual with fibrous roots.
        E.   Scales acuminate or awned, the tip recurved.
          F.   Low plants with celery-scented herbage; scales awned, several-nerved
                                                              6. *C. aristatus*
          FF.   Taller plants without a marked odor; scales sharply acute, 3-nerved
                                                              7. *C. acuminatus*
        EE.   Scales obtuse .... 8. *C. difformis*
    CC.   Rachis winged with a pair of inner hyaline appendages at each node.
      D.   Perennials; aks. obtuse or mucronulate.
        E.   Stoloniferous with tubers; lvs. about equaling culms.
          F.   Scales dull brown or yellow-brown .... 9. *C. esculentus*
          FF.   Scales shining reddish-brown .... 10. *C. rotundus*
        EE.   Short-rhizomatous; lvs. much shorter than culm. .... 11. *C. Parishii*
      DD.   Robust annual; aks. distinctly mucronate .... 12. *C. erythrorhizos*
  BB.   Spikelets disarticulating above the basal pair of scales, or breaking up into 1-fruited joints; scales persistent.
    C.   Spikelets disarticulating above the sterile basal pair of scales; rachis-wings thin and hyaline.
      D.   Culm swollen at base into a corm; rhizomes absent .... 13. *C. strigosus*
      DD.   Culm not swollen at base; short rhizomes present .... 11. *C. Parishii*
    CC.   Spikelets breaking up into 1-fruited joints; rachis-wings firm and brown .. 14. *C. ferax*

1. **C. niger** R. & P. var. **capitatus** (Britton) O'Neill. [*C. diandrus* var. *c.* Britton. *C. d.* var. *castaneus* of Wats., not Torr. *C. melanostachyus* HBK. *C. n.* var. *castaneus* Kükenth.] Perennial with short rhizomes, fibrous roots, and cespitose culms; culms slender, smooth, erect, 15–50 cm. tall; lvs. 2 on a culm, shorter than the culm, narrow, smooth, the sheaths reddish-brown; involucral lvs. 3, to 15 cm. long; infl. a capitate cluster of spikelets; spikelets lanceolate, acutish, 5–12 mm. long, 2.0–2.5 mm. wide, the rachis zigzag; scales ovate, acutish to obtuse, keeled, 3–5-nerved, ochre-brown, deciduous; stamens 2; style 2-fid; ak. lenticular, oblong, short-stipitate, short-mucronate, 1.0–1.3 mm. long, brown or gray, the surface puncticulate.—Marshes, ditches, and wet places, up to 3000 ft.; Foothill Wd., S. Oak Wd., Coastal Scrub, etc.; Coast Ranges from San Diego to Sonoma cos., Cent. V., Inyo Co.; to Tex., Mex. July–Nov. The typical var. of the sp. occurs in trop. Cent. and S. Am.

Var. **rivularis** (Kunth.) V. Grant. [*C. r.* Kunth.] Annual; culms 4–40 cm. tall; infl. umbellate; scales deep reddish-brown.—Marshy meadows and wet ground, up to 5000 ft.; largely Mixed Evergreen F., Foothill Wd., Yellow Pine F.; mostly N. Coast Ranges; to Atlantic Coast. Aug.–Sept.

2. **C. unioloides** R. Br. [*C. bromoides* auth., not Link.] Perennial with slender rootstocks; culms slender, 50–80 cm. tall; lvs. mostly shorter than the culms, 2–4 mm. wide; involucral lvs. 3–5, one of them much exceeding the infl.; infl. a capitate or loose cluster of spikelets; spikelets ovate-lanceolate, flat, acute, 12–18 mm. long, many-fld.; scales yellow, acute; ak. obovate, dark.—Reported long ago from a cienega near Los Angeles; Mex., W. I., Cent. Am., S. Am.; S. Afr., Asia, Australia.

3. **C. laevigatus** L. Perennial with horizontal rhizomes; culms smooth, 1.5–5 dm. tall; lvs. 2–5 on a culm, shorter than the culm, sometimes reduced to basal sheaths;

involucral lvs. 1–2, very unequal; infl. a capitate cluster of 1 to several spikelets, appearing lateral; spikelets linear-oblong, compressed, 6–12 mm. long, 2–3 mm. wide; scales ovate, obtuse, ca. 1.5 mm. long and 1 mm. wide, whitish on base and keel and chocolate-brown on sides; stamens 2; style 2-fid; ak. lenticular, compressed parallel to rachis, elliptic, obtuse, stipitate, the surface minutely cellular.—Wet places, up to 2500 ft.; Alkali Sink, Creosote Bush Scrub, Coastal Sage Scrub, etc.; throughout desert and coastal s. Calif.; to Mex.; trop. S. Am.; Old World trop. July–Dec.

4. **C. Eragróstis** Lam. [*C. vegetus* Willd. *C. serrulatus* Wats.] Perennial with short thick rhizomes and coarse fibrous roots; culms triangular, smooth, erect, 1–9 dm. tall, slightly swollen at base; lvs. basal, 6–10, about equaling culm, scaberulous on the margin and sometimes on midrib; involucral lvs. 5–8, unequal, to 50 cm. long, scaberulous on margins and midrib; infl. a compact globose head, rays 0–12 cm. long; spikelets flat, 10–20 mm. long, 3.0–3.5 mm. wide, rachis straight; scales ovate, acute, keeled, 3-nerved, straw-colored, falling off with the ak. enclosed; stamen 1; style 3-fid; aks. sharply triangular, obovoid, stipitate, the stipe broadened at base, mucronate, brown, puncticulate.—Common in shallow water and on moist ground, at low elevs.; many Plant Communities; most of cismontane Calif., Modoc Co.; to Ore., Mex., temp. S. Am. May–Nov.

5. **C. vìrens** Michx. Similar to *C. Eragrostis,* but with culms sharply triangular and upwardly scaberulous; aks. oblong-elliptic, short-stipitate, the stipe not broadened at base.—Rare, at low elevs., Fresno County; s. U.S., Mex., W. I., Cent. Am., S. Am. July–Oct.

6. **C. aristàtus** Rottb. [*C. inflexus* Muhl. *C. a.* var. *i.* Kükenth.] Small annual with fibrous roots and celery-scented herbage; culms cespitose, slender, smooth, 1–20 cm. tall; lvs. 2–3 on a culm, longer or shorter than culm, flat, 0.5–3.0 mm. wide, involucral lvs. much exceeding infl.; infl. umbellate, the rays lacking or to 2 cm. or more long, bearing capitate clusters of spikelets; spikelets linear-oblong, 4–10 mm. long, compressed; rachis straight, deciduous after the scales; scales lanceolate, awned, the awn recurved, keeled, strongly several-nerved, green to light brown, deciduous; stamen 1; style 3-fid; ak. triangular, obovoid or oblong, obtuse, mucronulate, 0.7–1.0 mm. long, brown, the surface puncticulate.—Wet ground, up to 8500 ft.; many Plant Communities; almost throughout the state; to B.C., Atlantic Coast, Mex., Cent. Am., temp. S. Am.; Asia, Afr. June–Nov.

7. **C. acuminàtus** Torr. & Hook. Annual with fibrous roots and slender cespitose culms 7–40 cm. tall; lvs. 2–4 on a culm, about equaling culms, flat, 0.5–2.5 mm. wide, scaberulous on margins; involucral lvs. 3–4, unequal, to 18 cm. long; infl. umbellate with 2–5 unequal rays, bearing globose heads of spikelets; spikelets ovate-oblong, obtuse, compressed, 4–10 mm. long; rachis straight, unwinged; scales ovate, acuminate, with a recurved tip, 3-nerved, pale green to light brown, the surface reticulate; stamen 1; ak. triangular, oblong, 0.6–0.9 mm. long, short-stipitate, short-mucronate.—Uncommon, wet ground, at low elevs., several Plant Communities; at scattered stations, Ventura to Siskiyou cos.; to Wash., Ill., Ga. June–Oct.

8. **C. diffórmis** L. [*C. lateriflorus* Torr.] Annual with fibrous roots and smooth cespitose culms 15–50 cm. tall; lvs. 2–4 on a culm, ca. equaling culms, 1–4 mm. wide, minutely scaberulous on margins near apex; involucral lvs. 2–3, unequal; infl. umbellate, the rays varying from 0.7 cm. long, bearing globose heads of linear obtuse subcompressed spikelets 4–8 mm. long; rachis straight, unwinged; scales roundish, obtuse, 0.6–0.8 mm. long, membranous, green with brown sides, readily deciduous; stamen 1; ak. triangular, obovate, minutely mucronate, 0.5 mm. long, pale greenish-brown, the surface minutely cellular.—Common weed in wet spots, low elevs.; Cent. V. (serious pest in ricefields), Sierran foothills, San Diego; Mex., Va.; native in Asia. July–Nov.

9. **C. esculéntus** L. [*C. e.* var. *Hermanni* (Buckl.) Britton.] CHUFA, YELLOW NUT-GRASS. Perennial with scaly stolons terminating in edible tubers; culms stout, smooth, 15–50 cm. tall; lvs. numerous, ca. equaling the culms, flat, 3–10 mm. wide, smooth; involucral lvs. 2–6; infl. umbellate, with 5–10 rays, 0–12 cm. long, bearing numerous remotely divaricate spikelets; spikelets linear, 6–30 mm. long, 2–3 mm. wide, flat; rachis winged with narrow, hyaline, persistent members; scales ovate, obscurely mucronulate, 7–9-nerved, light brown; stamens 3; style 3-fid; ak. triangular, oblong, obtuse, 1.3–2.0

mm. long, the surface puncticulate; $n =$ ca. 54 (Hicks, 1929).—Noxious weed of cult. fields and wet places, at low elevs.; many Plant Communities; cismontane Calif.; to Alaska, Atlantic Coast, trop. S. Am.; Old World. June–Oct.

10. **C. rotúndus** L. Purple Nut-Grass. Similar to *C. esculentus,* but with scales shining reddish-brown; $n = 54$ (Tanaka, 1937).—Noxious garden weed in coastal s. Calif. and San Joaquin V.; se. N. Am., W. I., S. Am.; natur. from Eurasia. July–Nov.

11. **C. Paríshii** Britton. [*C. sphacelatus* auth., not Rottb.] Perennial with short rhizomes and fibrous roots; culms subtriangular, smooth, 10–25 cm. tall; lvs. several, much shorter than the culm, 3–5 mm. wide, minutely scaberulous on the margins and midrib; involucral lvs. 3–4, scaberulous; infl. umbellate, the rays 0.5–5.0 cm. long; spikelets linear, acute, 12–20 mm. long, ca. 2 mm. wide; rachis winged with a pair of hyaline members at each node; scales ovate, acute, 2–3 mm. long, strongly several-nerved, keel green and sides reddish-brown; stamens 3; style 3-fid; ak. triangular, obovoid-ellipsoid, 1–1.2 mm. long, mucronulate, nearly black.—Coastal Sage Scrub; San Bernardino and Riverside cos.; to New Mex.

12. **C. erythrorhízos** Muhl. [*C. occidentalis* Torr.] Annual with fibrous roots and cespitose culms; culms bluntly triangular, smooth, 1–10 dm. tall; lvs. several, ca. equaling culm, flat, 2–10 mm. wide, margins and midrib scaberulous, sheaths loosely enveloping culm; involucral lvs. 4–10, unequal, margins scaberulous; infl. a compound umbel or sometimes simple, rays 1–30 cm. long, bearing numerous divaricate linear spikelets 3–10 mm. long, 1.0–1.5 mm. wide; rachis winged with a pair of inner hyaline members at each node; scales oblong-obovate, mucronulate, keeled, with green midrib and satiny golden-brown sides; stamens 3; style 3-fid; ak. sharply triangular, oblong, not stipitate, distinctly mucronate, grayish-white, the surface smooth or very finely cellular under magnification.—Marshy ground and river beds, up to 5300 ft.; cismontane Calif., Imperial V.; to Wash., Atlantic Coast. July–Oct.

13. **C. strigòsus** L. [*C. Hanseni* Britton. *C. s.* var. *H.* Kükenth.] Perennial with swollen corms at the base of the culms; culms 1–9 dm. tall, smooth; lvs. 1–several, longer or shorter than culm, flat, scaberulous on margins and midrib; involucral lvs. 3–several, scaberulous; infl. umbellate, the rays unequal, to 20 cm. long, terminating in loose divaricate clusters of spikelets; spikelets linear, flat, 5–25 mm. long, 1–2 mm. wide, disarticulating above the sterile basal pair of scales when ripe; rachis somewhat zigzag, winged at each node with a pair of wide hyaline members which clasp the ak.; scales oblong-lanceolate, subacute, strongly several-nerved, straw-colored to golden-brown; stamens 3; style 3-fid; ak. triangular, linear-oblong, 1.5–2.0 mm. long, 0.5 mm. wide, mucronate, purplish-brown, the surface puncticulate.—Wet ground; V. Grassland, Foothill Wd.; Cent. V., Sierra Nevada foothills from Mariposa to Butte cos., n. to Wash., Atlantic Coast. July–Oct.

14. **C. fèrax** Rich. [*C. californicus* Wats.; *C. speciosus* auth., not Vahl.] Annual; culms 1–several, 2–5 dm. tall, stout, smooth, bluntly triangular; lvs. several to many, equaling or exceeding culms, blades folded, channeled above, 3–8 mm. wide, margin beset with minute stout prickles; involucral lvs. several, 5–30 cm. long, very unequal, folded; infl. simulating a compound spike, rays arising from the top of the culm and umbellately spreading, branching into secondary rays each in the axil of an involucel-lf., rays becoming enlarged in age by a gibbous succulent tissue filling the base of the involucral and involucel-lvs.; spikelets linear, subterete, 10–25 mm. long, ca. 2 mm. wide, disarticulating above the sterile basal pair of scales when ripe, the rachis succulent, becoming corky on ripening and breaking up into 1-seeded joints enclosing the ak.; fl. deeply recessed in the joint of the rachis and almost enclosed by its winglike margins; scales ovate, obtuse, sheathing the rachis, strongly 7–9-nerved, yellowish-brown; ak. triangular, obovoid, obtuse, 1.0–1.2 mm. long, brown, the surface puncticulate.—Wet ground, often in grain fields, low elevs.; many Plant Communities; Imperial V., w. Mojave Desert, cismontane s. Calif. from Riverside to Los Angeles cos., Cent. V.; to se. U.S., trop. Am.; Asia. July–Oct.

15. **C. alternifòlius** L. Umbrella-Plant. Cespitose perennial from tough cordlike roots; rhizome short, thick; culms 3–15 dm. tall, stout, angled, striate, clothed below with leafless brown sheaths; umbel terminal in an invol. of many long firm spreading leaflike bracts 1–3 dm. long; rays as many as the bracts, 2–10 cm. long; spikelets oblong, 5–10 mm. long, pale; 10–30-fld.; glumes mostly deciduous, 1.6–2 mm. long, subacute;

stamens 3; $2n = 32$ (Tanaka, 1937).—Common in cult. and occasionally natur.; native of Afr. Most of year.

8. Kyllínga Rottb.

Perennial or annual herbs. Culms slender, triangular, leafy. Involucral lvs. 2–several, much exceeding infl. Infl. a dense head of numerous small sessile spikelets. Spikelets flat, falling away from head at maturity, consisting of 3–4 scales in 2 ranks, only the middle one subtending a perfect fl. Bristles none. Stamens 1–3. Style 2-fid; the style-base deciduous from the summit of the lenticular ak. A small genus of ca. 40 spp., trop. and warm temp. regions of both hemis. (Named for Peter *Kylling*, a Danish botanist of the 17th century.)

1. **K. brevifòlia** Rottb. Perennial with rhizomes; culms 25–75 cm. tall, leafy; involucral lvs. 2–4, 1–4 cm. long; head nearly globose, 5–7 mm. long; spikelets 3 mm. long; empty outer 2 scales ovate, acuminate, strongly several-nerved, enclosing the fertile scale and ak. at maturity; ak. lenticular, much flattened, obtuse; $2n = 120$ (Tanaka, 1941).— Introd. weed (San Diego, Pasadena, San Francisco); se. N. Am.; native Am. trop.; Old World. July–Sept.

9. Schoènus L.

Perennial with stiff quill-like culms and lvs. Lvs. basal, the blades convolute. Involucral lvs. 2, stiff, sharp-pointed, one of them elongated to 8 cm. Infl. a capitate cluster of spikelets. Spikelets flattened, the scales in 2 series, the 3–5 basal scales empty. Fls. 5–8, perfect, ♀, and ♂. Bristles 6. Stamens 3. Style 3-fid. Ak. triangular, without a tubercle. A genus of ca. 60 spp., trop. and warm temp. regions, largely Australia, New Zealand. (Greek, *schoinos,* a rush.)

1. **S. nìgricans** L. BLACK SEDGE. Culms 20–70 cm. tall; some of the lvs. nearly as long as the culms; basal lf.-sheaths shining chestnut-brown; scales of the spikelet lanceolate, carinate, acuminate, chestnut-brown; bristles 6, plumose, exceeding the ak.; ak. triangular, white, ca. 2 mm. long; $2n = 54$ (Rodrigues, 1953).—Marshes and hot springs, to 5000 ft.; San Bernardino and San Gabriel mts., Death V.; Tex., Fla., W. I., Eu. Aug.–Sept.

10. Rhynchóspora Vahl. BEAKED-RUSH

Ours perennial with rhizomes and cespitose culms. Culms erect, triangular to subterete, leafy. Lvs. flat or involute. Infl. cymose or capitate, axillary, containing 6–20 spikelets. Scales spirally imbricated, thin, with 1 midvein usually produced as a mucro, the basal 2–several scales empty, the middle 1–5 scales perfect, the apical scales subtending ♂ florets. Bristles 5–12 in ours, downwardly or upwardly barbed. Stamens 3. Style 2-fid; the style-base persistent on the ak. as a tubercle. Ak. lenticular, the surface wrinkled or smooth, detaching from the rachis of the spikelet with a persistent gynophore to which the bristles remain attached. Tubercle attenuate and subulate to compressed and deltoid. A large genus of ca. 200 spp., trop. and warm temp. regions. (Greek, *rhynchos,* snout, and *spora,* seed, from the beaked aks.)

(Gale, Shirley, Section Eurhynchospora. Rhodora 46: 89–134, 159–197, 207–249, 255–278, 1944. Kükenthal, G. Vorarbeiten zu einer Monographie der Rhynchosporoideae. Bot. Jahrb. 74: 375–509, 1949; 75: 90–195, 273–314, 1950–1951.)

Ak. obovate with a subulate tubercle; bristles downwardly barbed except in *R. californica.*
Scales whitish; bristles 10–12 .... 1. *R. alba*
Scales brown; bristles 5–7
    Aks. smooth .... 2. *R. glomerata*
    Aks. rugulose .... 4. *R. californica*
Aks. subglobose with a compressed tubercle; bristles upwardly barbed .... 3. *R. globularis*

1. **R. álba** (L.) Vahl. [*Schoenus a.* L.] Culms triangular, slender, erect, 1–7 dm. tall, leafy; lvs. linear, shorter than the culm; infl. consisting of 1–4 axillary capitate clusters of 6–20 spikelets; spikelets oblong, acute at both ends, 4–6 mm. long, bearing 1–2 aks.; scales ovate to ovate-lanceolate, mucronate, whitish; bristles 10–12, stiff, downwardly barbed, about equaling the ak.; ak. lenticular with an obscure margin,

obovate, faintly rugulose, ca. 1.7 mm. long, light brown; tubercle subulate, 0.6–1.2 mm. long; *n* = 13, 21 (Scheerer, 1940; Löve & Löve, 1942).—Bogs; Freshwater Marsh, Yellow Pine F., Mixed Evergreen F., N. Coastal Scrub; Yosemite V., Pitkin Marsh in Sonoma Co., Mendocino Co., Del Norte Co.; to Alaska, Atlantic Coast; Eurasia. July–Aug.

2. **R. glomeràta** (L.) Vahl var. **capitellàta** (Michx.) Kükenth. [*Schoenus c.* Michx. *R. c.* Vahl.] Culms subtriangular, slender, erect, 2–9 dm. tall; lvs. short, flat, 1.5–3.5 mm. wide; infl. consisting of 1–5 axillary capitate clusters of from 6–20 spikelets; spikelets oblong, acute at both ends, 3.5–5.0 mm. long, bearing 2–5 aks.; scales ovate-lanceolate, mucronulate, brown, caducous; bristles 6, downwardly barbed, ca. equaling the ak.; ak. lenticular with a narrow margin, the surface smooth, brown; tubercle subulate, 1–1.6 mm. long.—Rare, boggy places, below 3300 ft.; Freshwater Marsh, Yellow Pine F.; Sonoma, Nevada, and Trinity cos.; sw. Ore.; Miss. V. to Atlantic Coast. July–Aug.

3. **R. globulàris** (Chapm.) Small. [*R. cymosa* var. *g.* Chapm. *R. g.* var. *recognita* Gale.] Robust with triangular to subtriangular culms 1.5–10 dm. tall; lvs. basal, the blades flat, shorter than the culms; infl. consisting of 1–4 axillary somewhat open cymes of from 6–20 beadlike spikelets; spikelets subrotund, plump, 2.5–4.0 mm. long, bearing 1–3 aks.; scales suborbicular, obtuse, mucronate, splitting at the apex as the aks. ripen; bristles 5–6, ca. half the length of the ak.; upwardly barbed; aks. subglobose, ca. 1.5 mm. long, nearly as broad as long, transversely rugose, reticulate, brown, the persistent gynophore shorter than in *R. alba* and *R. glomerata;* tubercle compressed, conical, 0.5 mm. long.—Bogs, Pitkin Marsh, Sonoma Co.; lower Miss. V. to Atlantic Coast. July–Aug.

4. **R. califórnica** Gale. In habit similar to *R. glomerata;* spikelets ovate; scales ovate, mucronate, caducous; bristles 6–7, upwardly barbed, exceeding the tubercle; ak. lenticular with an obscure margin, obovate, light brown, lightly rugulose, 2.0 mm. long; tubercle deltoid, 1 mm. long.—Bogs, Ledum Swamp, Point Reyes, and Pitkin Marsh, Sonoma Co. July.

11. Clàdium R. Br. Saw-Grass

Perennial with hollow leafy culms. Infl. a compound umbel of numerous spikelets. Spikelets small, few-fld., the lower scales empty or imperfect. Scales spirally arranged. Perfect fl. one. Bristles none. Stamens 2. Style 2–3-fid. A genus of ca. 40 spp., trop. and temp. regions, mostly in Australia and New Zealand. (Greek, *klados,* branch, referring to the branched infl.)

1. **C. Maríscus** R. Br. var. **califórnicum.** Wats. [*Mariscus c.* Fern.; *C. c.* O'Neill.] Culms stout, 1–2 m. tall, subtriangular to subterete; lvs. 1–2 m. long, flat, 7–10 mm. wide, the margins serrate; infl. an axillary compound umbel with the spikelets borne in glomerules of 3–6; spikelets oblong, acute, 3 mm. long, reddish-brown, bearing one terminal perfect fl. and several ♂ fls., the lowermost scales empty; ak. ovoid, smooth, ca. 2 mm. long, without a tubercle; *n* = 18 (Pfeiffer, 1942).—Uncommon; Freshwater Marsh, Alkali Sink; deserts from Riverside to Inyo cos., cismontane San Bernardino Co. to San Luis Obispo Co.; Nev., Ariz., Mex. June–Sept. The typical form of the sp. occurs in e. N. Am., Cent. Am.; Old World.

12. Cárex L.

Treatment by John Thomas Howell

Perennial herbs, cespitose or developing shorter or longer rootstocks. Culms mostly solid, sharply triangular to nearly terete, leafy at the base (phyllopodic) or the lowest lvs. scalelike (aphyllopodic). Lvs. 3-ranked, sheath closed, a small inconspicuous ligule generally present on the ventral side of the blade at the junction of the sheath, the uppermost lvs. (bracts) subtending the spikelets either scalelike or foliaceous. Fls. unisexual, the plants monoecious or sometimes dioecious, the fls. borne in the axils of generally scalelike bracts (scales) and ± congested into short or elongate, few- to many-flowered spikelets (spikes of Mackenzie), spikelets solitary or more often several to many and disposed in spikes, heads, racemes, or panicles, the spikelets unisexual and either ♂ or ♀, or bisexual and either androgynous (with ♂ fls. above and ♀ fls. below) or gynaecandrous (with ♀ fls. above and ♂ fls. below). Fls. without perianth, each ♂ fl. consisting of 3 (or 2) stamens, each ♀ fl. consisting of a 2- or 3-(or rarely 4-) stigmatic pistil enclosed in an urn- or flask-shaped bractlet (perigynium) through the small orifice of which the style protrudes, the ♀ fl. rarely accompanied within the perigynium by a prolongation of the floral axis (rachilla). Ak. lenticular or triangular, contained by and falling with the perigynium at maturity.

Carex, one of the largest genera of plants in the world with more than 1000 spp., is the largest genus of flowering plants in Calif. The following treatment is based largely on the monograph of the N. Am. spp. by Kenneth Kent Mackenzie in the N. Am. Flora (vol. 18, pp. 9–478, 1931–1935) and upon studies of the Calif. spp. by John William Stacey (1934–1939). Hybridization has been recognized as rather common in the genus in many parts of its range, but here in Calif. instances of hybridization have been rarely observed or reported. Of the 144 spp. treated here, 3 have been introd. into Calif., 15 extend beyond N. Am., and 21 are endemic to Calif.

KEY TO THE SECTIONS

A. Spikelet 1 on each culm, androgynous or unisexual.
  B. Styles 2; aks. lenticular; lvs. filiform; perigynia glabrous .... 1. *Capitatae*
  BB. Styles 3 (or rarely 4); aks. triangular.
    C. Perigynia lightly to densely pubescent.
      D. Spikelets androgynous; lvs. filiform, less than 1 mm. wide .... 15. *Filifoliae*
      DD. Spikelets mostly unisexual, the plants dioecious; lvs. flat or canaliculate, 2–3 mm. wide .... 17. *Scirpinae*
    CC. Perigynia glabrous.
      D. Perigynia distended and completely filled by the aks. .... 20. *Firmiculmes*
      DD. Perigynia scarcely distended, loosely enclosing the aks.
        E. Perigynia beakless, finely many-striate, appressed-ascending, the spikelet only 2–3 mm. wide .... 14. *Polytrichoideae*
        EE. Perigynia with short beak, smooth, ascending to deflexed, the spikelet usually more than 3 mm. wide.
          F. Scales persistent; perigynia ascending or spreading; lvs. involute-filiform, 1 mm. wide or less .... 2. *Inflatae*
          FF. Scales deciduous; perigynia at length deflexed; lvs. flat, 1.5–2 mm. wide
                                         3. *Callistachys*
AA. Spikelets several to many, rarely only 2, on each culm.
  B. Styles 2; aks. lenticular.
    C. Spikelets sessile, short, with relatively few perigynia.
      D. Spikelets androgynous or sometimes the spikelets unisexual and the plants ± dioecious.
        E. Infl. simple, the spikelets in spikes or heads.
          F. Plants open or loosely cespitose, with long-creeping rootstocks or stolons.
            G. Spikelets in dense globose or subglobose heads; margin of perigynium and beak smooth .... 4. *Foetidae*
            GG. Spikelets in elongate heads or spikes.
              H. Spikelets generally unisexual, the plants tending to be dioecious; perigynia usually with a prominent serrulate beak 5. *Divisae*
              HH. Spikelets androgynous.
                I. Perigynia with very short smooth beak: *C. disperma* in 10. *Heleonastes*
                II. Perigynia with conspicuous serrulate beak: *C. tumulicola* in .... 6. *Bracteosae*
          FF. Plants cespitose, rootstocks short or short-prolonged, not long-creeping.
            G. Culms firm; perigynium-body ± contracted into the beak 6. *Bracteosae*
            GG. Culms with a spongy texture; perigynium-body tapering into the beak .... 9. *Vulpinae*
        EE. Infl. ± compound, some of the spikelets on short-condensed or elongate branches.
          F. Perigynia plano-convex.
            G. Culms firm; infl. forming compact subglobose to oblongish heads, or if the infl. is looser (in *C. alma*) then the ♀ scales broadly white-hyaline-margined .... 7. *Multiflorae*
            GG. Culms with a spongy texture; infl. somewhat loose or openly capitate, the scales not conspicuously white-hyaline: *C. stipata* in 9. *Vulpinae*
          FF. Perigynia thick-plano-convex or unequally biconvex; infl. ± paniculate; ♀ scales brownish .... 8. *Paniculatae*
    DD. Spikelets gynaecandrous.

E.   Margin of perigynia winged, the wing narrow or broad; perigynia not spongy-
thickened at base .... 13. *Ovales*
EE. Margin of perigynia rounded or sharp-edged but not winged; perigynia ±
spongy-thickened at base.
   F.   Perigynia puncticulate (cellular-dotted), nearly or quite filled by the aks.
10. *Heleonastes*
   FF. Perigynia not puncticulate, not filled by the aks.
      G.   Perigynia spreading or ascending in the spikelets . 11. *Stellulatae*
      GG. Perigynia appressed .... 12. *Deweyanae*
CC. Spikelets stalked, the stalks usually elongate and conspicuous, or if the lateral spikelets
are sessile, then the spikelets elongate, linear, or oblong with many perigynia.
   D.   Lowest bract long-sheathing, the blade leaflike; perigynia beakless
21. *Bicolores*
   DD. Lowest bract sheathless or nearly so, the blade bractlike or leaflike; perigynia
beaked but the beak usually very short.
      E.   Aks. not constricted .... 32. *Acutae*
      EE. Aks. constricted in the middle .... 33. *Cryptocarpae*
BB. Styles 3 (or 4); aks. triangular (or quadrangular).
   C.   Perigynia pubescent.
      D.   Plants dioecious; spikelets usually 1, rarely 2 or 3 .... 17. *Scirpinae*
      DD. Plants monoecious; spikelets generally unisexual and more than 2.
         E.   Aks. with sides convex above.
            F.   Perigynium-beak shallowly to deeply bidentate; bract of lowest nonbasal
spikelet sheathless or nearly so .... 16. *Montanae*
            FF. Perigynium-beak entire or nearly so; bract of lowest nonbasal
spikelet short-sheathing, its blade rudimentary or lacking
18. *Digitatae*
         EE. Aks. with sides flat or concave.
            F.   Perigynium-beak ca. 0.3 mm. long, bidentulate; bracts foliaceous, long-
sheathing .... 19. *Triquetrae*
            FF. Perigynium-beak 1 mm. long or more, bidentulate or bidentate.
               G.   Ak. continuous with the style, the two with ca. the same texture
36. *Paludosae*
               GG. Ak. and style jointed, not continuous.
                  H.   Leaves glabrous.
                     I.   Perigynia 6–8 mm. long: *C. obispoensis* in
25. *Sylvaticae*
                     II. Perigynia 2.5–5 mm. long .... 28. *Hirtae*
                  HH. Leaves pubescent.
                     I.   Lowest bract long-sheathing .... 25. *Sylvaticae*
                     II. Lowest bract short-sheathing .... 28. *Hirtae*
   CC. Perigynia glabrous.
      D.   Style and ak. not continuous, the style withering and deciduous.
         E.   Lowest bract of infl. long-sheathing.
            F.   Plants with conspicuous long-creeping stolons or rootstocks; perigynia
beakless or with subentire beak .... 22. *Paniceae*
            FF. Plants densely cespitose, the rootstocks very short or short-prolonged.
               G.   Perigynia bidentate, spreading or somewhat deflexed . . 26. *Extensae*
               GG. Perigynia bidentate or bidentulate, ascending or appressed.
                  H.   Rootstocks and base of plants conspicuously fibrous-coated
27. *Ferrugineae*
                  HH. Rootstocks and base of plants ± scaly, not fibrous-coated.
                     I.   Base of plant brownish, not reddish; lvs. and bracts not
rough- or slender-tipped .... 23. *Laxiflorae*
                     II. Base of plant strongly reddish-tinged; lvs. and bracts
drawn out to a rough slender tip: *C. mendocinensis*
in .... 25. *Sylvaticae*
         EE. Lowest bract of infl. short-sheathing or sheathless.
            F.   Lf.-blades ± pubescent; plants cespitose without prolonged rootstocks or
stolons .... 24. *Longicaules*
            FF. Lf.-blades glabrous, smooth or roughened.
               G.   Roots covered with a yellow felt; spikelets containing rather few
perigynia, drooping on long slender stalks ...... 30. *Limosae*
               GG. Roots not covered with a yellow felt; spikelets containing few to
very many perigynia, the stalks (sometimes very short) usually
erect or ascending, rarely elongate and drooping.
                  H.   Plants stoloniferous, rank, with blades mostly 1–2 cm. broad;
♀ spikelets elongate, linear, with very many spreading
perigynia .... 29. *Anomalae*
                  HH. Plants loosely to densely cespitose, strongly stoloniferous only
in *C. Buxbaumii*; lf.-blades less than 1 cm. broad; ♀ spikelets
linear to oblong, with few to many perigynia . . 31. *Atratae*
      DD. Style and ak. continuous, the style indurate and persistent.
         E.   Perigynia 3–4.5 mm. long with a short merely emarginate beak
34. *Hispidae*

EE. Perigynia 3.5–10(–12) mm. long with an elongate strongly bidentate beak, the teeth 0.5–3 mm. long.
    F. Lf.-sheaths and lower side of lf.-blades ± hairy .... 36. *Paludosae*
    FF. Lf.-sheaths and lf.-blades glabrous.
        G. Pistillate scales with awn much longer than the broadened base; perigynia closely and finely many-ribbed .. 35. *Pseudo-Cypereae*
        GG. Pistillate scales acute or awned, the scale-body broader and longer; perigynia coarsely fewer-ribbed .... 37. *Vesicariae*

KEYS TO THE SPECIES

1. Capitatae

Represented in Calif. by .... 1. *C. capitata*

2. Inflatae

Pistillate scales 1-nerved, generally covering the mature perigynia; perigynia 3.5–4 mm. long
    2. *C. subnigricans*
Pistillate scales 3-nerved, much smaller than the mature perigynia; perigynia strongly inflated, 5–6.5 mm. long .... 3. *C. Breweri*

3. Callistachys

Represented in Calif. by .... 4. *C. nigricans*

4. Foetidae

Culms mostly 0.5–3 dm. tall; lf.-blades 2–4 mm. wide, flattish; perigynia 3.5–4.5 mm. long
    5. *C. vernacula*
Culms 2–4 cm. long; lf.-blades folded-involute above the base; perigynia 3.25 mm. long
    6. *C. incurviformis* var. *danaensis*

5. Divisae

Beak of perigynium from half as long to nearly as long as the body.
    Culms generally smooth and less than 2 dm. long; lvs. folded-involute near the apex; perigynia finely nerved ventrally .... 7. *C. Douglasii*
    Culms generally roughened above and more than 2 dm. long; lvs. flat or canaliculate; perigynia nerveless ventrally .... 8. *C. praegracilis*
Beak of perigynium a third as long as the body or less.
    Perigynia 3.5–4.5 mm. long .... 9. *C. pansa*
    Perigynia 1.75–3 mm. long.
        Lvs. folded-involute near the apex, 1–1.5 mm. wide; perigynia 2.5–3 mm. long
        10. *C. Eleocharis*
        Lvs. flattish, 2–4 mm. wide; perigynia 1.75–2.25 mm. long .... 11. *C. simulata*

6. Bracteosae

Perigynia not conspicuously corky at base, the beak serrulate.
    Pistillate scales much smaller than the perigynia; perigynia 2–3.5 mm. long.
        Beak of perigynium bidentulate, the teeth very short .... 12. *C. vallicola*
        Beak of perigynium bidentate, the teeth conspicuous.
            Margin of perigynium raised ventrally .... 14. *C. cephalophora*
            Margin of perigynium not raised, flat ventrally .... 15. *C. Leavenworthii*
    Pistillate scales ca. as large as the perigynia, ± concealing them; perigynia 2.5–5 mm. long.
        Spikelets in a compact oblong to roundish-ovate capitate infl. .... 13. *C. Hoodii*
        Spikelets in an elongate, narrowly oblong or linear infl.
            Infl. not conspicuously bracteate; perigynia loose and ± spreading at maturity
            16. *C. occidentalis*
            Infl. conspicuously bracteate; perigynia appressed .... 17. *C. tumulicola*
Perigynia corky ⅓–½ length of body, the margin and beak smooth .... 18. *C. texensis*

7. Multiflorae

Infl. to 20 cm. long, from closely to loosely paniculate, the branches ± apparent; ♀ scales strongly hyaline, covering the perigynia .... 19. *C. alma*
Infl. to 5 cm. long, capitate-paniculate, the branches closely aggregated and usually not very apparent; ♀ scales scarcely hyaline, generally shorter than the perigynia (or the hyaline-margined scales with the awn longer than the perigynia in *C. Dudleyi*).
    Pistillate scales acute or cuspidate, not prominently awned, shorter than the perigynia.
        Ventral side of perigynia smooth and nerveless .... 21. *C. vicaria*
        Ventral side of perigynia nerved.
            Ligule conspicuous, as long as wide; perigynia 3.5–4.5 mm. long .... 20. *C. densa*
            Ligule inconspicuous, much shorter than wide; perigynia 3.25–3.75 mm. long
            22. *C. breviligulata*
    Pistillate scales conspicuously awned, the awn generally exceeding the perigynium
    23. *C. Dudleyi*

8. Paniculatae

Lf.-blades 1–2.5 mm. wide; perigynia 2–2.75 mm. long .... 24. *C. diandra*
Lf.-blades 2.5–6 mm. wide; perigynia 3–4 mm. long .... 25. *C. Cusickii*

9. Vulpinae

Spikelets in an unbranched simple head; perigynia 3–4 mm. long, the beak ⅓ to ½ the length of
the perigynium-body.
 Lvs. mostly near the base of the culm, the culms inconspicuously aphyllopodic, the lf.-sheaths not
 green-and-white-mottled dorsally .... 26. *C. Jonesii*
 Lvs. not all near the base, the culms conspicuously aphyllopodic, the lf.-sheaths green-and-white-
 mottled dorsally.
  Lf.-sheaths cross-rugulose ventrally; beak of perigynium bidentulate .... 27. *C. neurophora*
  Lf.-sheaths smooth ventrally; beak of perigynium conspicuously bidentate ...... 28. *C. nervina*
Spikelets in a branched infl.; perigynia 4–5 mm. long, the beak ca. as long as the perigynium-body
                         29. *C. stipata*

10. Heleonastes

Perigynium widest near the middle, the beak very short, entire or emarginate, smooth or slightly
serrulate.
 Spikelets androgynous .... 30. *C. disperma*
 Spikelets gynaecandrous.
  Spikelets densely aggregated, the infl. capitate; scales brownish, hyaline only on margin;
  perigynia 1.5–1.75 mm. long .... 31. *C. praeceptorum*
  Spikelets mostly discrete, the infl. spicate; scales pale, whitish or brownish, hyaline; perigynia
  1.75–3 mm. long .... 32. *C. canescens*
Perigynium widest at the base, the beak conspicuous, ⅓–½ the length of perigynium-body, serrulate,
bidentate .... 33. *C. arcta*

11. Stellulatae

Beak ¼–⅓ the length of perigynium-body, entire or bidentulate.
 Perigynium narrowly obovate, widest at or above the middle, 2.5–4 mm. long, the beak scarcely
 serrulate .... 34. *C. laeviculmis*
 Perigynium ovate, widest near the base, 2.25–3.25 mm. long, the beak serrulate .. 35. *C. interior*
Beak ½ the length of perigynium-body or more, bidentate.
 Pistillate scales obtuse; perigynia many-nerved dorsally and ventrally, 3.5–4.5 mm. long.
  Spikelets separate; scales ca. as long as the perigynia-bodies .... 36. *C. ormantha*
  Spikelets approximate, ± capitate-congested; scales ca. ⅔ as long as the perigynia-bodies
                       37. *C. phyllomanica*
 Pistillate scales short-cuspidate; perigynia obscurely nerved or nerveless, 2.5–3.5 mm. long
                        38. *C. angustior*

12. Deweyanae

Perigynia 3.5–4 mm. long, scarcely nerved, the beak bidentulate .... 39. *C. leptopoda*
Perigynia 4–4.5 mm. long, strongly nerved dorsally, the beak deeply bidentate .... 40. *C. Bolanderi*

13. Ovales

A. Perigynia visible in the spikelets, the subtending scales shorter and usually narrower.
 B. Infl. not conspicuously leafy-bracteate (except in *C. amplectens*); perigynia plano-convex
  or flat (except where distended by the aks.).
  C. Lf.-sheaths white-hyaline ventrally.
   D. Lf.-sheaths not conspicuously prolonged ventrally.
    E. Perigynia strongly flattened, thin except where distended by the aks.
     F. Perigynium-margin undulate, beak flattened and serrulate-margined to
      the tip .... 69. *C. straminiformis*
     FF. Perigynium-margin not undulate, beak terete and smooth at the tip or
      rarely somewhat flattened and margined nearly or quite to the tip.
      G. Perigynia lanceolate to lanceolate-ovate, the margin very narrow
                   42. *C. microptera*
      GG. Perigynia usually ovate, rarely lanceolate, the margin broad to the
       base.
       H. Lf.-blades flat, 1.5–6 mm. wide; infl. densely capitate.
        I. Culms erect, 3–10 dm. tall; scales dark brown
                  41. *C. festivella*
        II. Culms decumbent or spreading, rarely erect, 1–3 dm. tall;
         scales blackish .... 43. *C. Haydeniana*
       HH. Lf.-blades folded-canaliculate, 0.5–1.5 mm. wide; infl. openly
        spicate to capitate .... 44. *C. proposita*
   EE. Perigynia plano-convex, thickish.
    F. Perigynia small, 2.25–3.5 mm. long.
     G. Perigynia strongly nerved on the ventral side
                 48. *C. montereyensis*
     GG. Perigynia smooth and nearly or quite nerveless on the ventral side.

H. Perigynium-margin very narrow, smooth or nearly so.
    I. Infl. roundish, blackish; perigynia 3 mm. long
          49. *C. illota*
    II. Infl. oblongish, light brownish; perigynia less than 3 mm.
        long .... 50. *C. integra*
HH. Perigynium-margin wide and conspicuous, serrulate above the middle.
    I. Culms slender; lf.-blades 1–2.5 mm. wide; perigynia ±
        spreading-ascending .... 51. *C. teneraeformis*
    II. Culms stiffish; lf.-blades 2–3.5 mm. wide; perigynia
        appressed .... 52. *C. subfusca*
FF. Perigynia larger, 3.5–8 mm. long.
  G. Perigynia narrowly to broadly ovate, 3.5–5.5 mm. long.
    H. Perigynia strongly nerved on the ventral side.
      I. Infl. generally densely capitate; perigynia 3.75–4.25 mm.
        long, the beak hyaline-tipped.
        J. Lf.-blades 1.5–2.5 mm. wide; perigynia membrana-
          ceous, 3.75–4 mm. long .... 45. *C. abrupta*
        JJ. Lf.-blades 2.5–4.5 mm. wide; perigynia coriaceous,
          4.25 mm. long .... 47. *C. Harfordii*
      II. Infl. generally openly spicate below, subcapitate above;
        perigynia 3.5–5 mm. long, the beak scarcely hyaline
        46. *C. mariposana*
    HH. Perigynia with ventral side smooth or very finely nerved or
      nerved only near the base.
      I. Infl. generally densely capitate, oblong to roundish.
        J. Perigynium-beak not hyaline at tip or only slightly so.
          K. Scales with a conspicuous white-hyaline margin;
            perigynia many-nerved dorsally, the beak flat-
            tened, serrulate-margined to the tip or nearly
            68. *C. multicostata*
          KK. Scales with a narrow hyaline margin; perigynia
            ± few-nerved dorsally, the beak terete or some-
            what flattened, margined nearly to the tip or the
            tip slender and smooth.
            L. Perigynia coriaceous, green at first, becoming
              yellowish-brown, strongly wing-margined,
              beak bidentate .... 53. *C. Preslii*
            LL. Perigynie membranaceous, copper-color at
              maturity, narrowly wing-margined, beak
              bidentulate .... 57. *C. pachystachya*
        JJ. Perigynium-beak conspicuously hyaline at the tip.
          K. Culms 1–2.5 dm. long; lf.-blades 3–7 cm. long;
            perigynia many-nerved dorsally
            54. *C. paucifructa*
          KK. Culms 3–12 dm. long; lf.-blades 10–30 cm. long;
            perigynia several-nerved dorsally
            56. *C. subbracteata*
      II. Infl. generally elongate, capitate above, spicate-interrupted
        below.
        J. Lf.-blades 1–2 mm. wide; lowest bracts not dilated
          at the base and enclosing the subtended spikelets; tip
          of perigynium-beak hyaline .... 55. *C. gracilior*
        JJ. Lf.-blades 2.5–4 mm. wide; lowest bracts dilated at
          base and enclosing the subtended spikelets; tip of
          perigynium-beak not hyaline .... 58. *C. amplectens*
  GG. Perigynia lanceolate, 5–8 mm. long.
    H. Culms 3–8 dm. tall; perigynia 5–6 mm. long, tip of beak
      reddish-tinged, not hyaline .... 67. *C. specifica*
    HH. Culms 2.5–3.5 dm. tall; perigynia 6–8 mm. long, tip of beak
      hyaline .... 66. *C. Davyi*
DD. Lf.-sheaths conspicuously prolonged ventrally into a fragile oblongish hyaline
  appendage .... 59. *C. fracta*
CC. Lf.-sheaths green-striate to the top ventrally .... 70. *C. feta*
BB. Infl. usually densely capitate, conspicuously leafy-bracteate, the bracts frequently exceed-
  ing the head; perigynia flattened.
  C. Lowest bract not erect; perigynium-beak hyaline at the tip ...... 71. *C. athrostachya*
  CC. Lowest bract erect, simulating a continuation of the culm; perigynium-beak not hyaline
    72. *C. unilateralis*
AA. Perigynia covered and nearly or quite concealed by the scales.
  B. Perigynium-beak reddish-brown at the tip, not hyaline .... 60. *C. Tracyi*
  BB. Perigynium-beak hyaline at the tip.
    C. Perigynia 1 mm. wide or less, very narrowly margined .... 63. *C. leporinella*
    CC. Perigynia 1.5–2.25 mm. wide, usually conspicuously margined.
      D. Perigynia many-nerved on both sides.
        E. Perigynia 6–8 mm. long .... 65. *C. petasata*

     EE. Perigynia 5–6 mm. long .... 62. *C. tahoensis*
  DD. Perigynia scarcely nerved or nerveless ventrally.
    E.   Perigynium-beak very short, ca. ¼ the length of the perigynium-body
                                                       61. *C. phaeocephala*
    EE. Perigynium-beak longer, ⅓–½ the length of the perigynium-body
                                                    64. *C. praticola*

14. Polytrichoideae

The only species .... 73. *C. leptalea*

15. Filifoliae

Pistillate scales with very broad white-hyaline margins, nearly or quite concealing the perigynia; perigynia 3–3.5 mm. long, with a short beak .... 74. *C. filifolia*
Pistillate scales with narrower sordid margins, much shorter than the perigynia; perigynia about 2.5 mm. long, almost beakless .... 75. *C. exserta*

16. Montanae

Pistillate spikelets approximate or a little discrete, close to the terminal ♂ spikelet .... 76. *C. inops*
Pistillate spikelets widely separate, the uppermost near the terminal ♂ spikelet, the lowest nearly basal.
  Perigynia many-nerved as well as 2-keeled; ♀ scales usually as long as the perigynia or longer.
    Perigynium-body globose or nearly so; the basal spikelets on elongate lax capillary stalks
                                                   77. *C. globosa*
    Perigynium-body oval; the basal spikelets on short stout stalks .... 78. *C. Brainerdii*
  Perigynia 2-keeled, otherwise nerveless or nearly so; ♀ scales mostly shorter than the perigynia.
    Perigynia 2.5–3 mm. long, the beak 0.25–0.75 mm. long .... 79. *C. brevipes*
    Perigynia 3–4.5 mm. long, the beak 0.75–1.5 mm. long.
      Beak strongly bidentate; bract of lowest nonbasal spikelet scarcely colored at the base
                                                   80. *C. Rossii*
      Beak bidentulate; bract of lowest nonbasal spikelet reddish-brown at the auriculate base
                                                 81. *C. brevicaulis*

17. Scirpinae

Rhizomes extended, the plants loosely cespitose; ♀ scales obtuse .... 82. *C. pseudoscirpoidea*
Rhizomes very short, the plants densely cespitose; ♀ scales acute .... 83. *C. gigas*

18. Digitatae

Represented in Calif. by .... 84. *C. concinnoides*

19. Triquetrae

Represented in Calif. by .... 85. *C. triquetra*

20. Firmiculmes

Culms terete or obtusely triangular; lf.-blades 1–1.5 mm. wide, channeled or involute; lowest ♀ scales bractlike or leaflike, prolonged 1–4 cm. .... 86. *C. multicaulis*
Culms sharply triangular; lf.-blades 2–3.5 mm. wide, flat or channeled; lowest ♀ scales short-awned
                                                      87. *C. Geyeri*

21. Bicolores

Perigynia short-tapering and a little constricted at the apex, granular .... 88. *C. salinaeformis*
Perigynia obtuse at apex, granular or smooth.
  Perigynia granular, whitish-pulverulent at maturity, not fleshy .... 89. *C. Hassei*
  Perigynia not granular or pulverulent, fleshy and golden-yellow or brownish at maturity
                                                    90. *C. aurea*

22. Paniceae

Lf.-blades glaucous; perigynia glaucous, beakless .... 91. *C. livida*
Lf.-blades green; perigynia greenish, with a beak 0.75 mm. long .... 92. *C. californica*

23. Laxiflorae

Represented in Calif. by .... 93. *C. Hendersonii*

24. Longicaules

Lf.-blades flat, 2–6 mm. wide; ♀ scales spreading; perigynia ovate, the beak 0.5–1 mm. long
                                                     94. *C. Whitneyi*
Lf.-blades flat or revolute, 2–3 mm. wide; ♀ scales appressed-ascending; perigynia elliptic-ovate to elliptic-obovate, the beak 0.25–0.5 mm. long .... 95. *C. Jepsonii*

25. Sylvaticae

Pistillate spikelets oblong-cylindric, 5–9 mm. wide, with closely packed perigynia, erect, not flexuous or nodding; lf.-blades pubescent.
  Staminate spikelet sessile or short-stalked; perigynia 4.5–5.5 mm. long ...... 96. *C. gynodynama*

Staminate spikelet long-stalked; perigynia 3.5–4 mm. long .... 97. *C. hirtissima*
Pistillate spikelets linear, 3.5–6 mm. wide, with loosely placed but approximate perigynia, flexuous-erect or nodding; lf.-blades mostly glabrous or sparsely pubescent.
  Perigynia glabrous, 3.5 cm. long, the beak ca. ⅙ the length of perigynium-body
                                                               98. *C. mendocinensis*
  Perigynia appressed-pubescent, 6–8 mm. long, the beak ⅓–½ as long as the perigynium-body
                                                               99. *C. obispoensis*

26. Extensae

Represented in Calif. by .... 100. *C. viridula*

27. Ferrugineae

Perigynium-beak hyaline at apex.
  Terminal spikelet ♂, much exceeding the uppermost lateral ♀ spikelet; ♀ scales reddish-brown;
    perigynia 3–4 mm. long .... 101. *C. Lemmonii*
  Terminal spikelet ♂ or frequently ± ♀, little exceeding the uppermost lateral ♀ spikelet; ♀ scales
    pale, broadly hyaline; perigynia 3–4.5 mm. long .... 102. *C. albida*
Perigynium-beak purplish or blackish at apex, not hyaline.
  Perigynia triangular but somewhat compressed; ♀ scales finely ciliate, obtuse to acutish.
    Perigynium shortly stipitate, the beak ¼–⅓ the length of the perigynium-body; ♀ scales dark
      purplish, the midrib not extending to the apex .... 103. *C. ablata*
    Perigynium sessile, the beak very short, ⅐–⅕ the length of the perigynium-body; ♀ scales red-
      dish-brown, the midrib extending to the apex or nearly so ...... 104. *C. luzulina*
  Perigynia strongly flattened; ♀ scales acute to cuspidate.
    Pistillate scales glabrous and smooth; perigynia glabrous .... 105. *C. luzulaefolia*
    Pistillate scales ciliate and hispidulous; perigynia sparsely hispid-ciliate above
                                                               106. *C. fissuricola*

28. Hirtae

Perigynium-beak strongly bidentate; ♀ scales glabrous or ciliate.
  Lowest bract long-sheathing; perigynia 4–5 mm. long .... 107. *C. Halliana*
  Lowest bract scarcely sheathing; perigynia 2.5–3.5 mm. long .... 108. *C. lanuginosa*
Perigynium-beak bidentulate, hyaline-tipped; ♀ scales pubescent.
  Perigynia 2.5–3.5 mm. long, oblong-obovate to obovate, abruptly contracted to the beak, at ma-
    turity greenish or pale brown and completely filled by the ak. .... 109. *C. Sartwelliana*
  Perigynia 3.5–4 mm. long, ovate-lanceolate to oblanceolate, tapering to the beak, at maturity dark
    or purplish-brown and loosely enveloping the ak. .... 110. *C. Congdonii*

29. Anomalae

Represented in Calif. by .... 111. *C. amplifolia*

30. Limosae

Represented in Calif. by .... 112. *C. limosa*

31. Atratae

Perigynium or perigynium-beak granular-roughened, or the beak bidentate.
  Perigynia puncticulate, the beak roughened, bidentate .... 120. *C. serratodens*
  Perigynia granular-roughened or papillose, the beak bidentulate.
    Pistillate scales broadly ovate, with midvein mostly not evident .... 118. *C. albo-nigra*
    Pistillate scales lanceolate, with light midvein evident to the apex .... 121. *C. Buxbaumii*
Perigynium mostly smooth or puncticulate, not papillate, the beak mostly emarginate or bidentulate.
  Terminal spikelet staminate.
    Perigynia strongly flattened, nearly nerveless; scales tipped by excurrent midrib
                                                               113. *C. spectabilis*
    Perigynia turgid, nearly round in cross section, strongly several-nerved; midrib of scales not
      excurrent .... 114. *C. Raynoldsii*
  Terminal spikelet gynaecandrous.
    Spikelets closely approximate, the infl. subcapitate; ♀ scales longer than the perigynia
                                                               115. *C. Helleri*
    Spikelets discrete or separate, or at least the lowest a little separate; ♀ scales equaling the
      perigynia or shorter.
      Spikelets 5–10, ± drooping on slender stalks; perigynia 4.5–5 mm. long .. 119. *C. Mertensii*
      Spikelets 3–6, erect or ascending; perigynia 2.5–4.5 mm. long.
        Pistillate scales with midvein almost obsolete; width of the ak. ca. ⅓ the width of the
          perigynium; perigynia 3–4.5 mm. long .... 116. *C. epapillosa*
        Pistillate scales with midvein usually conspicuous to the tip; width of the ak. usually more
          than ⅓ the width of the perigynium; perigynia 2.5–3.5 mm. long .... 117. *C. heteroneura*

32. Acutae

Lf.-sheaths not becoming filamentose when breaking.
  Lowest bract of infl. shorter than the infl., not conspicuously foliaceous.
    Dried lvs. of previous season conspicuous at base of culm, all the lvs. of the current season
      at base of culm bearing well-developed blades; perigynia papillose ...... 122. *C. scopulorum*
    Dried lvs. of previous season not conspicuous at base of culm, the lowest lvs. of the current

season at base of culm reduced to bladeless or short-bladed sheaths; perigynia granular
                                                                    123. *C. gymnoclada*
Lowest bract of infl. equaling or exceeding the infl., foliaceous.
Beak of perigynium 0.5–1 mm. long, bidentate; perigynia coriaceous, many-ribbed
                                                                    127. *C. nebrascensis*
Beak of perigynium 0.1–0.5 mm. long, entire or nearly so; perigynia membranaceous, nerveless, nerved, or ribbed.
Perigynia obviously nerved or ribbed.
The perigynia roundish or nearly so, short-stipitate .... 124. *C. paucicostata*
The perigynia ovate, long-stipitate.
Perigynia glaucous-green, finely nerved; ♀ scales persistent; plants of the mts.
                                                                    125. *C. Kelloggii*
Perigynia yellowish-green, coarsely ribbed; ♀ scales deciduous; plants of coastal meadows
                                                                    126. *C. Hindsii*
Perigynia nerveless or nearly so.
Spikelets, at least the lowest, drooping; plants of lowland hills and valleys, generally near the coast .... 128. *C. sitchensis*
Spikelets erect; plants of mountain meadows .... 129. *C. aquatilis*
Lf.-sheaths breaking and becoming filamentose.
Lowest bract of infl. not leaflike, usually small and much shorter than the infl.; plants of rocky stream beds .... 133. *C. nudata*
Lowest bract of infl. leaflike, shorter to longer than the infl.; plants of moist slopes, meadows, and stream beds.
Beak of perigynium 0.5 mm. long, bidentate, hispidulous between the teeth
                                                                    130. *C. barbarae*
Beak of perigynium ca. 0.25 mm. long (or 0.2–0.5 mm. in *C. eurycarpa*), emarginate or entire.
Lf.-sheaths strongly carinate dorsally; blades 6–12 mm. broad; culms 10–15 dm. tall; perigynia slightly granular .... 131. *C. Schottii*
Lf.-sheaths slightly carinate or rounded dorsally; blades 2–5 mm. broad; culms 3–10 dm. tall; perigynia granular-roughened.
Culms roughened on angles throughout, arising from dried lvs. of preceding season (phyllopodic); blades 3–5 mm. wide .... 132. *C. senta*
Culms slightly roughened above, aphyllopodic; blades 2–3 mm. wide .... 134. *C. eurycarpa*

33. Cryptocarpae

Perigynia slightly to strongly nerved, minutely granular and dull .... 135. *C. Lyngbyei*
Perigynia nerveless, smooth and shining .... 136. *C. obnupta*

34. Hispidae

Represented in Calif. by .... 137. *C. spissa*

35. Pseudo-Cypereae

Perigynia inflated and suborbicular; teeth of perigynium-beak erect, 0.5 mm. long
                                                                    138. *C. hystricina*
Perigynia scarcely inflated, flattened-triangular; teeth of perigynium-beak divergent, awnlike, 1–2 mm. long .... 139. *C. comosa*

36. Paludosae

Perigynia glabrous, 7–10 mm. long .... 140. *C. atherodes*
Perigynia pubescent, 5–6 mm. long .... 141. *C. Sheldonii*

37. Vesicariae

Culms arising oblique to the ground; perigynia appressed-ascending even in fr., rather loosely arranged in the spikelets in a few rows.
Perigynia 3.5–8 mm. long, ± contracted into the beak, the beak 1–2 mm. long, the teeth 0.5–1 mm. long .... 142. *C. vesicaria*
Perigynia 7–10 mm. long, tapering into the beak, the beak 2–3 mm. long, the teeth 0.75–1.5 mm. long .... 143. *C. exsiccata*
Culms strictly erect; perigynia appressed-ascending when young, squarrose-spreading in fr., more closely arranged in the spikelets in more numerous rows .... 144. *C. rostrata*

1. C. capitàta L. Loosely to densely cespitose, the rootstocks short or a little prolonged; culms 2–40 cm. long, erect or somewhat spreading, shorter to much longer than the lvs.; blades filiform, about 0.5 mm. wide, folded along the middle, pale green, stiffish; spikelet solitary, androgynous, ovoid to nearly orbicular, 4–10 mm. long; ♀ scales smaller than the perigynia or frequently nearly covering them; perigynia spreading or ascending, ovoid, 2–3 mm. long, pale greenish or tan, thin, sharp-edged, nerveless or nearly so, abruptly contracted into a short beak, the beak brownish, smooth, entire or bidentulate.—Marshy meadows and alpine slopes, 6300–12,900 ft.; Red Fir F. to Alpine Fell-fields; Sierra Nevada from Tulare Co. to Nevada Co., Tehama and Shasta cos.; n. to Alaska, e. to Greenland; Eu.; Asia; s. S. Am.

2. **C. subnìgricans** Stacey. Rootstock well developed but usually only short-creeping, the culms and lvs. closely to loosely cespitose; culms 0.5–2 dm. tall, stiffly erect, usually much longer than the lvs.; blades nearly filiform, folded along the middle, erect; spikelet solitary, androgynous, lanceolate to oblong-lanceolate, 8–12 mm. long; ♀ scales 1-nerved, mostly covering the perigynia even at maturity; perigynia ascending, subappressed, little inflated but frequently distended by the ak., lanceolate to ovate-lanceolate, 3.5–4 mm. long, nerveless, brownish, stipitate, tapering into a short beak, the beak hyaline and obliquely cut.—Meadows and moist rocky slopes, 8500–12,500 ft.; Red Fir F. to Alpine Fell-fields; Sierra Nevada from Tulare Co. to Eldorado Co., White and Sweetwater mts.; Nev. n. to Ore., e. to Utah and Ida.

3. **C. Brèweri** Boott. Rootstocks long-creeping, culms and lvs. loosely cespitose in scattered tufts; culms 1–3 dm. tall, erect, usually longer than the lvs.; blades very narrow, folded along the middle, stiffly erect; spikelet solitary, androgynous, widely or narrowly ovoid, 1–2 cm. long; ♀ scales 3-nerved, much smaller than the mature perigynia; perigynia ascending or in age spreading, strongly inflated, broadly ovate, 5–6.5 mm. long, generally nerveless, pale brown, subsessile, abruptly short-beaked, the beak smooth, white-hyaline, becoming bidentulate.—Open gravelly slopes of high mts., 7700–12,300 ft.; Subalpine F. and Alpine Fell-fields; Sierra Nevada from Tulare Co. to Tuolumne Co., Mt. Lassen, Mt. Shasta, Mt. Eddy; n. to Wash.

4. **C. nìgricans** C. A. Mey. Rootstocks well developed but usually not long-creeping, the plants loosely cespitose; culms 0.5–3 dm. high, erect or spreading, usually longer than the lvs.; blades generally flat, 1.5–2 mm. wide; spikelet solitary, androgynous (or rarely entirely ♀), lanceolate to ovate, 8–15 mm. long; ♀ scales shorter than the perigynia, deciduous; perigynia appressed in anthesis, spreading-reflexed in fr., early deciduous, little-inflated, lanceolate to narrowly ovate, 3.5–4 mm. long, nerveless, stipitate, tapering into a short beak, the beak smooth, hyaline, obliquely cut.—Moist rocky slopes and meadows, 7500–12,000 ft.; Subalpine F. and Alpine Fell-fields; Sierra Nevada from Tulare Co., to Nevada Co., Mt. Lassen, Mt. Shasta, Coast Ranges in Siskiyou Co.; n. to Alaska, e. to Colo. and Alta.

5. **C. vernácula** Bailey. Culms and lvs. loosely cespitose from long-creeping rootstocks; culms 0.5–3 dm. tall, erect, slender, usually longer than the lvs.; blades flat or nearly so, 2–4 mm. wide; spikelets numerous, androgynous, closely compacted into an ovoid or roundish head about 1 cm. in diam.; ♀ scales ca. the size of the perigynia and nearly concealing them; perigynia flattened, ovate-lanceolate, 3.5–4.5 mm. long, membranaceous, finely or obscurely nerved, stipitate, contracted into a conspicuous smooth beak, the beak obliquely cut, becoming bidentulate.—Moist meadowy slopes, 6000–12,300 ft.; Red Fir F. to Alpine Fell-fields; Sierra Nevada from Tulare Co. to Plumas Co.; Mt. Lassen, Modoc Co., Sweetwater and White mts.; Nev. to Wash., e. to the Rocky Mts.

6. **C. incurvifórmis** Mkze. var. **danaénsis** (Stacey) F. J. Herm. [*C. d.* Stacey.] Loosely to compactly cespitose from elongate slender rootstocks; culms 2–4 cm. long, erect or recurved, mostly shorter than the lvs.; blades flattened at base, folded-involute above, 1.5 mm. wide; spikelets few, androgynous, densely aggregated into a globose-ovoid head 5–6 mm. in diam.; ♀ scales shorter than the perigynia; perigynia ovate, 3.25 mm. long, with several distinct ribs on each face, not stipitate, tapering above into a smooth dark brown beak ⅛ the length of the body.—Open gravelly or rocky slopes, rare; Alpine Fell-fields; Sierra Nevada in Tulare, Inyo, and Tuolumne cos., 12,000–12,880 ft.; Colo.

7. **C. Douglásii** Boott. Plants open, the lvs. and culms arising in small scattered tufts from long-creeping rootstocks; culms 0.5–3 dm. tall, erect, mostly longer than the lvs.; blades flattened below, folded along the middle above, erect or spreading, 1–2.5 mm. wide; spikelets several to numerous in a rather compact head, the heads 1.5–5 cm. long, usually unisexual and the plants dioecious, the ♂ heads narrowly oblong, the ♀ heads oblong to suborbicular; ♀ scales larger than the perigynia; perigynia ovate-lanceolate, 3.5–4 mm. long, buff, ± nerved on each face, stipitate, tapering into a beak nearly as long as the body, the beak serrulate, becoming bidentulate, hyaline at the apex.—Alkaline plains, sagebrush flats, and gravelly mt. slopes, widespread and often common, 900–12,500 ft.; Sagebrush Scrub to Alpine Fell-fields; mts. of s. Calif., Sierra Nevada n. to Modoc Co., rare in the Coast Ranges, White Mts.; n. to B.C., e. to Manitoba, Neb., and New Mex.

8. **C. praegrácilis** W. Boott. [*C. Douglasii* var. *brunnea* Olney. *C. usta* Bailey.] Culms and lvs. loosely cespitose or in scattered tufts, arising from blackish long-creeping rootstocks; culms 2–7.5 dm. high, stiffly erect or laxly spreading, usually longer than the lvs.; blades 1.5–3 mm. wide, flat or canaliculate; spikelets few to numerous, androgynous or ± unisexual, closely aggregated to form an elongate oblong or ovate head 1–5 cm. long, or the lower spikelets ± discrete; ♀ scales mostly covering the perigynia; perigynia ovate-lanceolate to ovate, 3–4 mm. long, dull and dark brown at maturity, nerveless ventrally, finely nerved dorsally, serrulate on the margin above the middle, substipitate, tapering into a serrulate beak ca. half the length of the body, the beak becoming bidentulate, hyaline at the apex.—Moist places in valleys, hills, and mts., widespread, sea level to 9000 ft.; Coastal Strand, V. Grassland, Coastal Sage Scrub and Creosote Bush Scrub to Red Fir F.; in Calif. from coast to transmontane desert oases and volcanic plateaus; n. to Yukon, e. to Manitoba and Mich., s. to L. Calif. and cent. Mex.; S. Am.

9. **C. pánsa** Bailey. Culms and lvs. arising in scattered tufts from long-creeping rootstocks; culms 1.5–3 dm. tall, stiff, suberect or curving, longer than the lvs.; blades 1–3 mm. wide, flattish or canaliculate; spikelets few to several, androgynous, clustered to form a rather dense ovate head 1.5–2.5 cm. long; ♀ scales larger than the perigynia and completely covering them; perigynia ovate-lanceolate or elliptic, 3.5–4.5 mm. long, shining and dark brown at maturity, scarcely nerved ventrally, many-nerved dorsally, margin serrulate above the middle, stipitate, tapering into a serrulate beak ca. ⅓ the length of the body, the beak becoming bidentulate, scarcely hyaline at the apex.— Maritime dunes; Coastal Strand; Santa Rosa and San Miguel ids., San Luis Obispo Co., Monterey Peninsula, Humboldt and Del Norte cos.; n. to Wash. A doubtful collection is labeled as from San Francisco.

10. **C. Eleócharis** Bailey. Lvs. and culms in small scattered tufts that arise from slender long-creeping rootstocks; culms 2.5–20 cm. tall, erect, shorter or longer than the lvs.; blades narrow, 1–1.5 mm. broad, flattened below, folded-involute above; spikelets few or several, rather closely capitate in narrowly oblong to ovate heads 0.5–2 cm. long, the lower spikelets generally androgynous, the upper spikelets tending to be ♂; the ♀ scales ca. covering the perigynia; perigynia broadly ovate, 2.5–3 mm. long, becoming blackish at maturity, nerved dorsally, scarcely nerved ventrally, substipitate, contracted above into a beak ¼–⅓ the length of the body, the beak serrulate, becoming bidentulate, hyaline at apex.—Elevated sagebrush flats and dry coniferous slopes; in Calif. known only from the White Mts. at 11,500 ft.; Bristle-cone Pine F.; n. to Yukon, e. to Manitoba, Ia., and New Mex.

11. **C. simulàta** Mkze. Culms and lvs. loosely cespitose or in scattered tufts, arising from long-creeping rootstocks; culms 3–5 dm. tall, erect or spreading, usually longer than the lvs.; blades 2–4 mm. wide, flat or canaliculate; spikelets few to numerous, androgynous or ± unisexual, closely aggregated into a narrowly oblong to broadly ovate head 1–2.5 cm. long, the heads tending to be unisexual, the ♂ more slender, the ♀ broader; ♀ scales covering the perigynia; perigynia broadly ovate, 1.75–2.25 mm. long, brown and shining at maturity, few-nerved dorsally, almost nerveless ventrally, short-stipitate and frequently subtruncate at base, contracted above into a very short beak less than ⅓ the length of the body, the upper part of the body and the beak serrulate, the beak slightly hyaline at the apex.—Moist soil of meadows, bogs, and stream banks, sea level to 10,800 ft.; Coastal Prairie and N. Coastal Scrub to Subalpine F.; Sierra Nevada from Tulare Co. to Plumas Co., n. to Modoc Co., local in the Coast Ranges; n. to Wash., e. to Mont. and New Mex.

12. **C. vallícola** Dewey. Plants cespitose, the rootstocks short; culms erect, 2.5–6 dm. tall, longer than the lvs.; blades narrow, ca. 1 mm. wide below the long-attenuate apex; spikelets few to numerous, androgynous, closely aggregated to form a linear or oblong head 1.5–3 cm. long; ♀ scales much shorter than the perigynia; perigynia oblong-elliptic, 3.5 mm. long, dark green, nerveless ventrally, somewhat nerved dorsally, margin serrulate above the middle, abruptly contracted into a serrulate beak, the beak bidentulate, less than half the length of the body.—In Calif. known only from Upper Deep Creek in the Sweetwater Mts., Mono Co.; n. to Ore., e. to Mont. and S. Dak.

13. **C. Hoòdii** Boott. [*C. H.* var. *nervosa* Bailey.] Plants densely cespitose, the rootstocks very short; culms slender, erect, 2.5–8 dm. tall, longer than the lvs.; blades

1.5–3.5 mm. wide, flat; spikelets androgynous, few to several, forming a compact oblong or roundish-ovate head 1–2 cm. long, rarely the lower spikelets tending to be discrete; ♀ scales ca. the size of the perigynia and nearly covering them; perigynia somewhat spreading, ovate-elliptic, 3.5–5 mm. long, greenish or brownish, nerveless or nearly so, or nerved dorsally, margin serrulate above the middle, tapering or contracted into a serrulate beak, the beak bidentate, ca. ⅓ the length of the body.—Rocky gravelly slopes and meadow borders in the mts., 4000–11,000 ft.; Yellow Pine F. to Subalpine F.; San Jacinto and San Bernardino mts., Sierra Nevada from Tulare Co. to Plumas Co., n. to Mt. Lassen and Modoc Co., Coast Ranges s. to Glenn Co.; n. to B.C., e. to Alta., S. Dak., and Colo.

14. **C. cephalóphora** Muhl. Plants densely cespitose, the rootstocks very short; culms erect, 2–5 dm. tall, ca. as long as the lvs.; blades flat, 2–4.5 mm. wide; spikelets androgynous, few to several, aggregated to form a compact ovate or roundish head 1–2 cm. long; ♀ scales narrower and much shorter than the perigynia; perigynia narrowly to broadly ovate, 2.5 mm. long, nerveless or finely few-nerved dorsally, margins raised ventrally, contracted to a serrulate beak, the beak bidentate, ¼–½ the length of the body.—A lawn weed in Pasadena, Los Angeles Co.; widely distributed in woods and dry meadows in e. N. Am.

15. **C. Leavenwórthii** Dewey. Plants cespitose with short-prolonged rootstocks, forming a loose turf; culms erect or decumbent, slender, 1–6.5 dm. long, mostly exceeding the lvs.; blades lax, flat, 1–3 mm. wide; spikelets androgynous, few to several in a dense oblong to ovate head 1–2 cm. long; ♀ scales shorter and narrower than the perigynia; perigynia broadly ovate, 2–3.5 mm. long, nerveless ventrally, obscurely nerved dorsally, margins flat, serrulate only at base of beak, the beak bidentate, serrulate, ca. ¼ the length of the body.—A lawn weed in Arcadia, Santa Barbara, and Berkeley; widespread in woods and meadows in e. N. Am.

16. **C. occidentàlis** Bailey. Plants loosely cespitose, the lvs. and culms from short-creeping rootstocks; culms slender, erect or spreading, 2.5–7 dm. high, exceeding the lvs.; blades flat, 1.5–2.5 mm. wide; spikelets androgynous, several to numerous, aggregated into an elongate-oblong head 1.5–3 cm. long, the lower spikelets ± separate; ♀ scales ca. the size of the perigynia and nearly concealing them; perigynia elliptic-oblong, 2.5–3.5 mm. long, slightly nerved on both faces, margin serrulate only at base of beak, the beak serrulate, bidentate, ca. ⅓ the length of the body.—Dry woods and meadows, San Bernardino Mts., 6300 ft.; Yellow Pine F.; e. to Ariz. and New Mex., n. to Wyo.

17. **C. tumulícola** Mkze. Plants not cespitose or loosely so, the culms and lvs. from somewhat prolonged rootstocks; culms erect or spreading, 2–8 dm. long, generally longer than the lvs.; blades 1.5–2.5 mm. wide, flat, or sometimes folded along the middle; spikelets several to numerous, the lower ± discrete, the upper approximate, the spike slender, 2–5 cm. long; ♀ scales ca. the size of the perigynia and usually concealing them; perigynia elliptic to ovate, 3.5–5 mm. long, nerved dorsally and ventrally or nerveless ventrally, serrulate on the margin above the middle, stipitate, contracted into a serrulate beak, the beak bidentate, ⅓–½ the length of the body.—Meadows and grassy slopes of open wd., 100–4000 ft.; Coastal Prairie and Mixed Evergreen F. to Yellow Pine F.; Santa Cruz Id., Coast Ranges from Monterey Co. to Del Norte Co., Sierra Nevada from Madera Co. to Tuolumne Co.; n. through w. Ore. to nw. Wash.

18. **C. texénsis** (Torr.) Bailey. [*C. rosea* var. *t.* Torr.] Plants cespitose with short-prolonged rootstocks, forming a loose turf; culms erect, slender, 1.5–3 dm. tall, longer than the lvs.; blades flat, 0.75–1.5 mm. wide; spikelets androgynous, few to several in a narrow head or interrupted spike 1–3 cm. long; ♀ scales ca. equaling and ± concealing the perigynium-body, early deciduous; perigynia ascending at first, spreading or reflexed in age, lanceolate or narrowly ovate, 3 mm. long, nerveless, margins a little raised, smooth, beak smooth, bidentate, ca. ⅓ the length of the body.—Adventive in gardens and waste ground in Arcadia, Hollywood, and Santa Barbara; widely distributed in woods in cent. and se. U.S.

19. **C. álma** Bailey. [*C. vitrea* Holm.] Lvs. and culms cespitose, arising from short rootstocks; culms 3–12 dm. tall, erect, generally exceeding the lvs.; blades flat or ± folded along the middle, 3–6 mm. wide, the margins strongly scabrous; infl. paniculately branching, 2.5–20 cm. long, the spikelets androgynous, congested in clusters along the branches; ♀ scales conspicuously white-hyaline, mostly covering the perigynia;

perigynia ovate or oblong-ovate, 3.5–4 mm. long, slightly nerved on both sides, margin serrulate above the middle, narrowed to a serrulate beak, the beak bidentate, ca. ⅓ the length of the body.—About springs and along stream courses in the hills and mts., 400–8000 ft.; Coastal Sage Scrub and Creosote Bush Scrub to Yellow Pine F. and Pinyon-Juniper Wd.; San Diego Co. n. in the Coast Ranges to Monterey Co. and in the Sierra Nevada to Fresno Co., e. to desert oases and mts.; s. to L. Calif., e. to Nev. and Ariz.

20. **C. dénsa** (Bailey) Bailey. [*C. Brongniartii* var. *d.* Bailey. *C. chrysoleuca* Holm.] Plants densely cespitose, the lvs. and culms from short rootstocks; culms 3–7 dm. tall, erect, usually exceeding the lvs.; blades flat, 3–6 mm. wide; ligule conspicuous, as long as wide; infl. paniculate, the branches closely aggregated to form an oblong to ovate head 2–5 cm. long, the spikelets androgynous, closely congested; ♀ scales acute to cuspidate, ca. as wide as the perigynia but shorter; perigynia ovate, 3.5–4.5 mm. long, ± nerved dorsally and ventrally, margin serrulate above the middle, narrowed to a serrulate beak, the beak half as long as the body or longer, bidentate.—About springs, along streams, and in low fields, sometimes abundantly gregarious, from sea level to 4600 ft.; V. Grassland and N. Coastal Scrub to Yellow Pine F.; Coast Ranges, Great V., and Sierra Nevada s. to Ventura and Tulare cos.; San Diego Co.; n. to s. Ore.

21. **C. vicària** Bailey. Plants densely cespitose with very short rootstocks; culms 3–6 dm. tall, erect, longer than the lvs.; blades flat, 3–4.5 mm. wide; ligule very short; infl. paniculate-capitate, oblong, 1.5–3.5 cm. long, the branches congested, spikelets androgynous; ♀ scales ovate, acute or cuspidate, ca. as wide as the perigynia but somewhat shorter; perigynia ovate, 3–3.5 mm. long, ventrally flat and nerveless, dorsally convex and few-nerved, serrulate on upper part of margin, narrowed into a serrulate beak, the beak ca. half as long as the body, bidentate.—Moist or wet places in the hills and valleys, 100–2000 ft.; Mixed Evergreen F., Coastal Prairie and Yellow Pine F.; Santa Cruz Co. to Humboldt Co., local in Eldorado Co.; n. to Ore. and Wash.

22. **C. breviligulàta** Mkze. Densely cespitose, the culms and lvs. closely clustered on very short rootstocks; culms 3–6 dm. tall, erect, generally much longer than the lvs.; blades flattish, 3–4.5 mm. wide; ligule very short; infl. paniculate, condensed into an oblong head 1.5–3.5 cm. long, the androgynous spikelets closely congested; ♀ scales ovate, acute to cuspidate, as wide as the perigynia but a little shorter; perigynia ovate, 3.25–3.75 mm. long, few- to several-nerved dorsally and ventrally, serrulate on the margin above the middle, abruptly narrowed into a serrulate beak, the beak bidentulate, half the length of the body or longer.—Wet hillsides and marshy flats, 50–4800 ft.; Coastal Prairie and Mixed Evergreen F. to Yellow Pine F. and N. Juniper Wd.; Coast Ranges from Marin Co. to Del Norte Co., Fresno Co., Lassen Co.; n. to Ore. This sp. particularly in the s. part of its range, is not readily separable from *C. densa.*

23. **C. Dúdleyi** Mkze. Cespitose, the culms and lvs. ± clustered on short rootstocks; culms 3–7 dm. long, stiffish, erect or somewhat spreading, longer than the lvs.; blades flat or a little channeled, 4–7 mm. wide; ligule short; infl. ovate or oblong, 2–3.5 cm. long, paniculate, the branches capitate-congested, the spikelets androgynous; ♀ scales ovate or ovate-lanceolate, ca. as wide as the perigynia, conspicuously awned, the awns usually longer than the perigynia; perigynia broadly ovate-lanceolate, 2.5–3 mm. long, several-nerved dorsally, scarcely nerved ventrally, serrulate on the margin above the middle, contracted into a serrulate beak ca. as long as the body, the beak bidentulate.— Moist slopes and flats in the Coast Ranges, Great V., and Sierra Nevada, 20–2500 ft.; Mixed Evergreen F., Coastal Prairie, and V. Grassland to Yellow Pine F.; Monterey and Fresno cos. n. to Lake and Humboldt cos.

24. **C. diándra** Schrank. [*C. bernardina* Parish.] Loosely cespitose in large clumps, the culms and lvs. from somewhat prolonged rootstocks; culms 3–7 dm. tall, erect, usually longer than the lvs.; blades flat or folded along the middle, 1–2.5 mm. wide; infl. 2.5–5 cm. long, slender, the numerous androgynous spikelets arranged in an elongate, ± interrupted panicle; ♀ scales scarcely concealing the perigynia, acute or cuspidate; perigynia ovate, 2–2.75 mm. long, somewhat biconvex, turgid, nerved dorsally, nerved only at base ventrally, serrulate on the margin above the middle, spongy at the rounded base, narrowed above into the serrulate beak, the beak nearly as long as the body, bidentulate.—Rather rare in swampy meadows, 500–7800 ft.; N. Coastal Coniferous F. and Coastal Sage Scrub to Red Fir F.; at scattered localities from

San Bernardino Co. to Butte and Humboldt cos.; n. to Yukon, e. to the Atlantic; Eu.; Asia.

25. **C. Cusíckii** Mkze. Plants loosely to densely cespitose, forming large clumps, the rootstocks short; culms 7–12 dm. tall, much longer than the lvs.; blades flat, revolute on the margins, 2.5–6 mm. wide; infl. 4–8 cm. long, paniculately branched, the branches separate or aggregated, the spikelets congested on the branches, androgynous; ♀ scales ovate, acute to short-acuminate, nearly concealing the perigynia; perigynia broadly ovate, 3–4 mm. long, turgid and strongly biconvex, nearly smooth ventrally, few-nerved dorsally, spongy at base, serrulate above at base of beak, the beak strongly serrulate, bidentulate, shorter than the perigynium body.—Occasional, marshes and wet meadows, 20–6500 ft.; Coastal Sage Scrub and Mixed Evergreen F. to Yellow Pine F.; Coast Ranges s. to San Luis Obispo Co. and Sierra Nevada s. to Tuolumne Co.; n. to B.C., e. to Sask. and Mont.

26. **C. Jònesii** Bailey. Culms and lvs. cespitose, arising from ± elongate rootstocks; culms erect or somewhat spreading, 2–6 dm. long, exceeding the lvs.; blades flat, 1.5–2.5 mm. wide, sheaths not green-and-white-mottled dorsally and not cross-rugulose ventrally; infl. densely capitate, the head oblongish to nearly round, 8–18 mm. long, spikelets androgynous; ♀ scales broadly ovatish, as wide as the perigynia but a little shorter; perigynia ovate-lanceolate, 3–4 mm. long, nerved dorsally and ventrally, margin entire, tapering into a smooth or slightly serrulate bidentate beak, the beak ca. ⅓ as long as the perigynium-body.—Common in moist soil in meadows and along streams and lakes, 3000–10,600 ft.; Yellow Pine F. to Subalpine F.; San Bernardino Mts., Sierra Nevada, n. to Mt. Lassen, Mt. Shasta, and Modoc Co., Coast Ranges s. to Glenn Co.; n. to Wash., e. to the Rocky Mts.

27. **C. neuróphora** Mkze. Plants cespitose, the rootstocks usually short; culms 3–7 dm. tall, slender, erect, longer than the lvs.; blades flat, to 3.5 mm. wide, sheaths green-and-white-mottled dorsally, cross-rugulose ventrally; infl. capitate, the androgynous spikelets densely congested into an oblong to ovate head 1.5–2.5 cm. long; ♀ scales ovate, much shorter than the perigynia but ca. as wide; perigynia lanceolate, 3.5–4 mm. long, finely nerved on both sides, margin smooth, attenuate into a slightly serrulate beak, the beak ca. half the length of the perigynium-body, bidentulate.— Rather rare, marshy meadows and stream banks, 5200–7000 ft.; Yellow Pine and Red Fir forests; N. Coast Ranges from Glenn Co. to Humboldt Co.; n. to Wash., e. to the Rocky Mts.

28. **C. nervìna** Bailey. Plants loosely cespitose with short rootstocks; culms 3–9 dm. tall, rather stout, erect, usually longer than the lvs.; blades flat, 3.5–5 mm. wide, sheaths smooth ventrally, green-and-white-mottled dorsally; infl. capitate, the androgynous spikelets densely aggregated into an oblong to ovate head 1.5–3 cm. long; ♀ scales ovate, shorter than the perigynia but ca. as wide; perigynia lanceolate to narrowly ovate, 3.5–4 mm. long, strongly nerved dorsally and ventrally, margin smooth, tapering into a smooth or slightly serrulate beak, the beak ca. half the length of the perigynium-body, bidentate.—Moist or wet soil along streams or in meadows, 4000–10,000 ft.; Yellow Pine F. to Subalpine F.; Sierra Nevada from Tulare Co. to Plumas Co., Mt. Lassen, Mt. Shasta, n. Coast Ranges from Glenn Co. to Siskiyou Co.; n. to s. Ore., e. to w. Nev.

29. **C. stipàta** Muhl. in Willd. Culms and lvs. densely cespitose, the rootstocks short; culms 3–12 dm. tall, thickish but spongy, erect, shorter or longer than the lvs.; blades flat, 4–8 mm. wide, sheaths green dorsally, cross-rugulose ventrally; infl. branching, elongate or broader, 3–10 cm. long, rather dense, the spikelets congested on the branches, androgynous; ♀ scales hyaline, broadly ovate, narrower and shorter than the perigynia; perigynia lanceolate, 4–5 mm. long, nerved both dorsally and ventrally, margin smooth, tapering into a serrulate beak, the beak bidentate, ca. as long as the perigynium-body.—Wet ground near streams and lakes and in marshy meadows, 100–5700 ft.; Mixed Evergreen F. and Foothill Wd. to Yellow Pine F.; s. in Coast Ranges to Sonoma Co. and in Sierra Nevada to Tuolumne Co.; n. to Alaska, e. to the Atlantic; Japan.

30. **C. dispérma** Dewey. Loosely cespitose, the culms and lvs. arising from elongate slender rootstocks; culms lax and spreading, 1.5–6 dm. long, slender, longer than the lvs.; blades 0.75–2 mm. wide, flat; infl. spicate-capitate, 1.5–2.5 cm. long, the spikelets

androgynous, the lower separate, the upper congested; ♀ scales broadly ovate, hyaline, narrower and shorter than the perigynia; perigynia thick and ± biconvex, puncticulate, narrowly ovate, 2.5–3 mm. long, nerved dorsally and ventrally, margin entire, abruptly contracted into a smooth beak, the beak entire, very short, ca. ⅒ the length of perigynium-body.—Rather rare or else overlooked; wet ground near streams and lakes, 3700–11,000 ft.; Yellow Pine F. to Subalpine F.; Sierra Nevada from Tulare Co. to Plumas Co.; White Mts., Nev.; n. to Yukon, e. to the Atlantic; Eu.; Asia.

31. **C. praeceptòrum** Mkze. Culms and lvs. cespitose, sometimes densely so, from short rootstocks; culms 1–1.5 dm. tall, erect or somewhat spreading, usually exceeding the lvs.; blades 1.25–2 mm. wide, channeled; infl. loosely capitate, ovate or oblongish, 1–1.5 cm. long, the spikelets gynaecandrous; ♀ scales ovate, as wide as the perigynia and nearly as long; perigynia plano-convex, thick, puncticulate, ovatish, 1.5–1.75 mm. long, margin entire, nerved on both sides, abruptly contracted into a short beak, the beak entire or nearly so, serrulate.—Occasional in boggy places and wet soil around ponds and lakes, 8000–11,500 ft.; Subalpine F.; Sierra Nevada from Tulare Co. to Tuolumne Co.; n. to Wash., e. to Ida.

32. **C. canéscens** L. Plants densely cespitose with very short rootstocks; culms 1–8 dm. tall, shorter than the lvs. or longer, erect; blades glaucous-green, 2–4 mm. wide, flat; infl. spicate, slender, 2–15 cm. long, spikelets discrete or approximate, gynaecandrous; ♀ scales broadly ovate, hyaline, nearly covering the perigynia; perigynia plano-convex, thick, puncticulate, ovatish, 1.8–3 mm. long, usually nerved on both faces but obscurely so, margin smooth or serrulate at base of beak, the beak very short, entire or slightly emarginate.—Rare in swamps or wet soil of meadows, 3500–10,500 ft.; Red Fir F. to Subalpine F.; Sierra Nevada from Inyo Co. to Sierra Co., Humboldt Co.; n. to Alaska, e. to the Atlantic; Eurasia; S. Am.; Australia.

33. **C. árcta** Boott. Culms and lvs. densely clustered on very short rootstocks; culms 1.5–8 dm. tall, erect, exceeded by the lvs. or the lvs. somewhat shorter; blades 2–4 mm. wide, flat; infl. capitate or nearly so, the head oblong to ovate, 1.5–3 cm. long; spikelets gynaecandrous, the lower sometimes a little separate; ♀ scales ovate, as wide as the perigynia but shorter; perigynia plano-convex, puncticulate, ovate, 2–3 mm. long, conspicuously nerved dorsally, inconspicuously ventrally, tapering into a conspicuous flattened beak ⅓–½ the length of the perigynium-body, the beak bidentate, serrulate.—Rare and local, wet soil of bogs and marshes, frequently in sphagnum bogs, 200–4600 ft.; N. Coastal Coniferous F., Douglas-Fir F.; Mendocino Co. to Del Norte Co.; n. to B.C., e. to the Atlantic.

34. **C. laevicúlmis** Meinsh. Culms and lvs. densely cespitose, arising from short rootstocks; culms 3–7 dm. long, very slender and lax, longer than the lvs.; blades lax, 1–2 mm. wide, flat; infl. 2–6 cm. long, spicate with the spikelets all separate, or the uppermost approximate and capitate, the lower spikelets ♀, the uppermost gynaecandrous with a slender ♂ base; ♀ scales ovate, scarcely covering the perigynia; perigynia narrowly obovate, widest above the middle, 2.5–4 mm. long, ± few-nerved on both sides, margin entire, tapering above into a scarcely serrulate beak, the beak ¼–⅓ the length of the body, entire or emarginate.—Rare, moist soil of woods, 2300–6000 ft.; Yellow Pine F., Red Fir F.; Sierra Nevada in Eldorado and Butte cos., Coast Ranges in Humboldt, Trinity, Del Norte, and Siskiyou cos.; n. to Alaska, e. to the Rocky Mts.

35. **C. intèrior** Bailey. Plants densely cespitose with very short rootstocks; culms 1.5–5 dm. tall, generally longer than the lvs., erect; blades 1–3 mm. wide, flat or somewhat canaliculate; infl. spicate-capitate, 1–2 cm. long, the spikelets generally approximate, the lower spikelets ♀ or gynaecandrous, the terminal gynaecandrous, usually with an elongate ♂ base; ♀ scales broadly ovate, obtuse, as wide as the perigynia but much shorter; perigynia narrowly ovate to ovate, 2.25–3.25 mm. long, few-nerved on the back, nerveless or nearly so ventrally, serrulate at base of beak, abruptly narrowed into the serrulate beak, the beak bidentulate, ¼–⅓ the length of the perigynium-body.— Rare, bogs and wet soil of meadows, 3500–5000 ft.; Yellow Pine F.; Plumas and Siskiyou cos.; n. to B.C., s. to Nev., Ariz., and n. Mex., e. to the Atlantic.

36. **C. ormántha** (Fern.) Mkze. [*C. echinata* var. *o.* Fern.] Culms and lvs. cespitose, from very short rootstocks; culms 1.5–4 dm. tall, slender and rather lax, longer than the lvs.; blades 1.5–2 mm. wide, flat or channeled; infl. spicate, 2–6 cm. long, the several spikelets separate, the terminal gynaecandrous with a conspicuous elongate ♂

base, the lower either ♀ or gynaecandrous; ♀ scales ovate, obtuse, ca. as long and wide as the bodies of the perigynia; perigynia lanceolate, 3.5–4 mm. long, many-nerved dorsally and ventrally, smooth on the margin except at the base of the beak, tapering into a bidentate beak, the beak serrulate, over half the length of the perigynium-body.— Occasional in marshy meadows, 300–10,500 ft.; N. Coastal Coniferous F. and Yellow Pine F. to Subalpine F.; San Bernardino Mts., Sierra Nevada from Tulare Co. to Butte Co., Coast Ranges in Humboldt, Trinity, Del Norte, and Siskiyou cos., Mt. Lassen, Mt. Shasta; n. to Wash.

37. **C. phyllománica** W. Boott. Plants densely cespitose, the culms and lvs. congested on short rootstocks; culms 2.5–6 dm. long, slender and becoming rather lax, shorter or longer than the lvs.; blades 1.75–2.75 mm. wide, flat or canaliculate; infl. capitate-spicate, 1.5–3.5 cm. long, the spikelets approximate or somewhat discrete, the terminal gynaecandrous, narrowed at the ♂ base, the lower spikelets ♀ or gynae-candrous; ♀ scales ovate, obtuse, as wide as the perigynia-bodies but only ca. ⅔ as long; perigynia ovate-lanceolate, 3.75–4.5 mm. long, striate on both faces, entire on the margin, tapering above into a serrulate beak, the beak bidentate, ca. half as long as the perigynium-body.—Occasional, cold bogs and marshes, sea level to 600 ft.; N. Coastal Scrub, Coastal Prairie, and N. Coastal Coniferous F.; Coast Ranges from Santa Cruz Co. to Del Norte Co.; n. to Alaska.

38. **C. angústior** Mkze. Plants closely cespitose, the rootstocks not at all elongate; culms 1–6 dm. tall, erect, ca. as tall as the lvs.; blades 0.75–2 mm. wide, flat or channeled; infl. spicate-capitate, 1–2 cm. long, the spikelets approximate or somewhat separate, the terminal gynaecandrous, the lower ♀; the ♀ scales ovate, short-cuspidate, narrower than the perigynia-bodies but nearly as long; perigynia lanceolate, 2.5–3.5 mm. long, nerveless or obscurely nerved, margins somewhat raised, smooth, tapering or contracted into a slightly serrulate beak, the beak bidentate, more than half to as long as the perigynium-body.—Rare, swampy mountain meadows, 3500–7000 ft.; Douglas-Fir F. to Red Fir F.; Humboldt Co., Sierra Nevada from Mariposa Co. to Plumas Co.; Washoe Co., Nev., n. to B.C., e. to the Atlantic.

39. **C. leptópoda** Mkze. Culms and lvs. not closely cespitose, arising from elongate slender rootstocks; culms slender, erect or lax, 2–8 dm. long, generally ca. as long as the lvs.; blades flat or margins revolute, 2.5–5 mm. wide; infl. spicate or spicate-capitate, 2–5 cm. long, the spikelets separate or the uppermost approximate, gynaecandrous or the lateral often ♀; the ♀ scales ovate or oblongish, hyaline, covering the perigynia-bodies; perigynia oblong-lanceolate, 3.5–4 mm. long, scarcely nerved, margin serrulate at base of beak, tapering into a serrulate beak, the beak somewhat shorter than the perigynium-body, bidentulate.—Moist soil of wooded slopes and flats, from near sea level to 8000 ft.; Coastal Sage Scrub and Coastal Prairie to Red Fir F.; Coast Ranges from Santa Cruz Co. to Del Norte and Siskiyou cos., Sierra Nevada from Tulare Co. to Butte Co.; n. to B.C., e. to Ariz. and Alta.

40. **C. Bolánderi** Olney. Loosely cespitose, the clustered culms and lvs. from short-prolonged rootstocks; culms 1.5–9 dm. long, slender, lax and frequently spreading, longer than the lvs.; blades 2–5 mm. wide, flat; infl. ± spicate-capitate, the lower spikelets sometimes discrete, the upper approximate and capitate-congested, spikelets gynaecandrous; ♀ scales ovate or somewhat narrower, covering the perigynia-bodies; perigynia lanceolate, 4–4.5 mm. long, strongly nerved dorsally, lightly nerved ventrally, serrulate on upper margin, tapering above into a serrulate beak, the beak more than half as long as the perigynium-body, strongly bidentate.—Common, wet soil in meadows, along streams, or about springs, 150–8100 ft.; Coastal Prairie and Mixed Evergreen F. to Red Fir F.; San Bernardino and San Gabriel mts., Sierra Nevada from Tulare Co. to Plumas Co., Coast Ranges from Monterey Co. to Del Norte Co., Mt. Shasta; n. to B.C., e. to New Mex. and Mont.

41. **C. festivélla** Mkze. Plants densely cespitose, the culms and lvs. in rather large clumps from very short rootstocks; culms 3–10 dm. tall, erect, longer than the lvs.; blades 2–6 mm. wide, flat; infl. densely capitate, the head ovoid or oblongish, 1–2.5 cm. long, spikelets gynaecandrous; ♀ scales ovate, dark brown, narrower and shorter than the perigynia; perigynia thin, distended by the ak., ovate, 3.75–5 mm. long, ± lightly nerved dorsally and ventrally, strongly margined to base, serrulate above the middle, tapering into a terete-tipped beak, the beak not serrulate above, ca. half the length

of the perigynium-body.—Open slopes and flats, occasionally in woods, 5000–11,500 ft.; Red Fir F., Subalpine F.; s. Calif. and desert mts., Sierra Nevada from Tulare Co. to Plumas Co., Warner Mts.; n. to B.C., e. to the Rocky Mts., s. to Chihuahua.

42. **C. micróptera** Mkze. Plants densely cespitose, the rootstocks very short; culms 3–10 dm. tall, erect, much longer than the lvs.; blades 2–4.5 mm. wide, flat; infl. capitate, the head roundish or ovate, 1–2 cm. long, the spikelets gynaecandrous; ♀ scales ovate-lanceolate, brownish, narrower and a little shorter than the perigynia; perigynia thin, distended by the ak., lanceolate to lanceolate-ovate, 3.5–4.5 mm. long, lightly nerved on both faces, very narrowly margined to the base, serrulate above the middle, tapering into a terete-tipped beak, the beak not serrulate above, ca. ⅓–½ the length of the perigynium-body.—Meadows and open slopes in the mts., 5000–11,000 ft.; Red Fir F., Subalpine F.; Sierra Nevada from Tulare Co. to Placer Co., high N. Coast Ranges, White, Sweetwater, and Warner mts., Mt. Shasta; n. to B.C., e. to the Rocky Mts.

43. **C. Haydeniàna** Olney. [*C. nubicola* Mkze.] Plants densely cespitose, the rootstocks short; culms 1–4 dm. long, equaling or exceeding the lvs., erect or frequently recurving and spreading obliquely; blades 1.5–4 mm. wide, flat; infl. capitate, ovate or roundish, 1–2 cm. long, the spikelets gynaecandrous; ♀ scales ovate, blackish, shorter and narrower than the perigynia; perigynia thin, distended by the ak., lanceolate to broadly ovate, 4.5–6 mm. long, obscurely nerved or nerveless, strongly wing-margined to the base, strongly serrulate upward from below the middle, contracted above into a terete-tipped beak, the beak ca. half the length of the body.—Rare, rocky slopes and flats kept moist by melting snow, 8000–13,600 ft., Subalpine F. and Alpine Fell-fields; Sierra Nevada from Tulare Co. to Alpine Co., Warner Mts.; n. to Ore., e. to the Rocky Mts.

44. **C. propòsita** Mkze. Cespitose, the culms and lvs. forming large dense clumps, the rootstocks very short; culms 1–3 dm. tall, erect, longer than the lvs.; blades 0.5–1.5 mm. wide, folded-canaliculate; infl. openly spicate to densely capitate, flexuous or stiffly compact, 1.5–2 cm. long, the spikelets gynaecandrous; ♀ scales lanceolate to ovate, light brown, shorter and narrower than the perigynia; perigynia thin, distended by the ak., 3.5–4.5 mm. long, broadly ovate, scarcely nerved, broadly wing-margined to the base, serrulate upward from below the middle, contracted into a terete-tipped beak, the beak less than half as long as the perigynium-body, smooth at the tip.—Among rocks and on open scree slopes, 10,000–13,400 ft.; Subalpine F. and Alpine Fell-fields; Sierra Nevada in Tulare, Inyo, Mono, and Fresno cos.; Ida.

45. **C. abrúpta** Mkze. Plants densely cespitose, the culms and lvs. from very short rootstocks; culms 4–6 dm. high, erect, much longer than the lvs.; blades 1.5–2.5 mm. wide, flat; infl. capitate, the head roundish, 9–17 mm. long, the spikelets gynaecandrous; ♀ scales ovate, narrower and shorter than the perigynia; perigynia broadly lanceolate to ovatish, 3.75–4 mm. long, plano-convex, membranaceous, conspicuously nerved dorsally and ventrally, wing-margined, serrulate on margins above, contracted into a slender beak ¼ to ⅓ the length of the body, the beak terete, smooth, and white-hyaline-tipped.—Common, mt. meadows and open slopes, usually in dry soil, from near sea level to 11,500 ft.; Coastal Prairie and Yellow Pine F. to Alpine Fell-fields; s. Calif. mts., Sierra Nevada from Tulare Co. to Butte Co., n. Coast Ranges s. to Glenn Co., Modoc Co., Mt. Lassen, Sweetwater Mts., Panamint Mts.; n. to Ore., e. to Nev.

46. **C. mariposàna** Bailey. Culms and lvs. densely cespitose, the rootstocks very short; culms 2.5–6 dm. tall, slender, erect but not strictly so, longer than the lvs.; blades 2–3 mm. wide, flat; infl. 2–3.5 cm. long, spicate to capitate, the spikelets discrete or the uppermost closely approximate, gynaecandrous; ♀ scales ovate, much shorter and narrower than the perigynia; perigynia ovate, 3.5–5 mm. long, plano-convex, membranaceous, conspicuously nerved on both faces, narrowly winged, serrulate above the middle, tapering into a terete-tipped smooth beak ca. ¼ the length of the body, the beak scarcely hyaline at the apex.—Drier parts of mt. slopes and meadows, 4000–11,400 ft.; Yellow Pine F. to Subalpine F.; s. Calif. mts., Sierra Nevada from Tulare Co. to Nevada Co., n. to Shasta Co., w. to Trinity Co.; Washoe Co., Nev.

47. **C. Harfórdii** Mkze. Rather loosely cespitose, the culms and lvs. from short-creeping rootstocks; culms 2.5–8 dm. long, erect or lax, frequently growing up through supporting shrubs, longer than the lvs.; blades 2.5–4.5 mm. wide, flat; infl. oblong to nearly round, 1.5–2.5 cm. long, the spikelets generally closely aggregated, gynaecandrous; ♀ scales

ovate, nearly as wide as the perigynia but shorter; perigynia ovate, 4.25 mm. long, plano-convex, coriaceous, with slender nerves dorsally and ventrally, narrowly margined, the margin serrulate above the middle, tapering into a slender terete beak ¼ the length of the body, the beak smooth, hyaline-tipped.—Coastal marshes and springy places in the hills, sea level to 1000 ft.; N. Coastal Scrub, Closed-cone Pine F., and N. Coastal Coniferous F.; Coast Ranges from San Luis Obispo Co. to Humboldt Co.

48. **C. montereyénsis** Mkze. Loosely to densely cespitose, the rootstocks short or a little prolonged; culms 8–10 dm. tall, erect or rather lax, longer than the lvs.; blades 2.5–3 mm. wide, flat; infl. capitate, 1.5–2.5 cm. long, ovate or oblongish, the spikelets closely aggregated, gynaecandrous; ♀ scales lanceolate-ovate, nearly as long as the perigynia-bodies but somewhat narrower, cuspidate or short-awned; perigynia ovate, 3.25 mm. long, plano-convex, membranaceous, nerved dorsally and ventrally, narrowly wing-margined, serrulate above the middle, contracted into a terete-tipped beak half the length of the body, the beak slightly hyaline-tipped.—Occasional, open or brushy slopes in coastal wds. up to 2000 ft.; Closed-cone Pine F., Redwood F., Mixed Evergreen F.; Santa Barbara Co. to Sonoma Co.

49. **C. illòta** Bailey. Culms and lvs. cespitose, sometimes in rather broad patches, from short-creeping matted rootstocks; culms erect to spreading, 1–3.5 dm. long, longer than the lvs.; blades 1.5–3 mm. wide, flat or channeled; infl. capitate, nearly round, 6–15 mm. long, the spikelets closely aggregated, gynaecandrous; ♀ scales broadly ovate, obtuse, blackish, wider than the perigynia and ca. as long as the body; perigynia lanceolate-ovate, 3 mm. long, plano-convex, not conspicuously nerved, very narrowly margined or only sharp-edged, smooth on the margins, tapering into a terete-tipped beak ca. ⅓ the length of the body, the beak smooth or minutely serrulate.—Wet meadows or moist soil along streams, 7000–11,000 ft.; Red Fir F. to Subalpine F.; San Bernardino Mts., Sierra Nevada from Tulare Co. to Nevada Co., Mt. Lassen, Modoc Co.; n. to B.C., e. to the Rocky Mts.

50. **C. intégra** Mkze. Loosely cespitose, the lvs. and culms on short rootstocks; culms 1.5–5 dm. long, erect or generally spreading or recurving, longer than the lvs.; blades 1–3 mm. wide, flat; infl. spicate-capitate, the spikelets approximate or aggregated into an oblong to ovate-oblong head 1–3 cm. long, spikelets gynaecandrous; ♀ scales ovate, nearly as wide as the perigynia but a little shorter; perigynia lanceolate, 2.25–2.75 mm. long, plano-convex, nerveless ventrally, lightly nerved dorsally, very narrowly wing-margined, margins smooth or nearly so, gradually or abruptly contracted into a slender beak ½ to ¾ the length of the perigynium-body, the beak smooth or nearly so.—Drier places in meadows and on open forest borders, 4000–11,000 ft.; Yellow Pine F. to Subalpine F.; San Bernardino Mts., Sierra Nevada from Tulare Co. to Plumas Co., Coast Ranges from Glenn Co. to Siskiyou Co., Mt. Lassen, Mt. Shasta, Modoc Co.; w. Nev., n. to Ore.

51. **C. teneraefórmis** Mkze. Plants cespitose, the rootstocks short; culms 3–4.5 dm. tall, erect, slender, longer than the lvs.; blades flat, 1–2.5 mm. wide; infl. 1.5–2.5 cm. long, slender, spicate with the gynaecandrous spikelets discrete or approximate; ♀ scales ovate, narrower and shorter than the perigynia; perigynia loosely appressed or somewhat spreading in the spikelets, 2.75–3.25 mm. long, plano-convex, nerveless ventrally, nerved dorsally, the margin winged, serrulate above the middle, tapering above into a slender beak ca. ½ the length of the perigynium-body, the tip of the beak terete and nearly smooth.—Common, meadows and open forests, 200–9500 ft.; Mixed Evergreen F. and Foothill Wd. to Red Fir F.; s. Calif. mts., Sierra Nevada from Kern Co. to Butte Co., Coast Ranges n. to Siskiyou Co., Mt. Lassen, Mt. Shasta, Modoc Co.; w. Nev., n. to Ore.

52. **Ç. subfúsca** W. Boott. [*C. stenoptera* Mkze.] Plants densely cespitose, the rootstocks very short; culms 2–6.5 dm. high, erect, stiffish, smooth; blades 2–3.5 mm. wide, flat; infl. elongate-capitate, the head oblong or ovate, 1–3.5 cm. long, the spikelets aggregated, gynaecandrous; ♀ scales ovate, narrower and shorter than the perigynia; perigynia appressed in the spikelets, narrowly to broadly ovate, 3–3.5 mm. long, plano-convex, nearly nerveless ventrally, lightly nerved dorsally, the margin winged, serrulate, gradually or abruptly narrowed into a serrulate beak half as long as the body, the beak somewhat flattened and serrulate at the tip.—Dry meadows and forest borders, 1000–11,500 ft.; S. Oak Wd. and Mixed Evergreen F. to Subalpine F.; San Diego Co., n.

through the Coast Ranges and Sierra Nevada, Panamint Mts., Mt. Lassen, Mt. Shasta; n. to B.C., s. to L. Calif., e. to Ariz. and Nev.

53. **C. Préslii** Steud. Culms and lvs. densely cespitose, the rootstock short and much thickened; culms 2.5–7.5 dm. high, erect, longer than the lvs., roughened above; blades flat, 1.5–4 mm. wide; infl. capitate or subcapitate, oblong to roundish, 1–2 cm. long, the spikelets gynaecandrous, usually aggregated or the lowest sometimes discrete; ♀ scales ovate, narrower and shorter than the perigynia; perigynia ovate, 3.5–4 mm. long, plano-convex, scarcely nerved, wing-margined, serrulate, contracted into a serrulate beak ⅓ to ½ as long as the body of the perigynium, the tip of the beak terete or a little flattened, not hyaline, bidentate.—Dryish, generally open slopes, 5500–10,800 ft.; Yellow Pine F. to Subalpine F.; Sierra Nevada from Madera Co. to Eldorado Co., n. Coast Ranges from Tehama Co. to Siskiyou Co., Mt. Lassen, Mt. Shasta, Warner Mts.; n. to B.C., e. to the Rocky Mts.

54. **C. paucifrúcta** Mkze. Plants densely cespitose, the rootstocks short; culms 1–2.5 dm. tall, smooth, longer than the lvs.; blades 1.5–3 mm. wide, flat or canaliculate; infl. capitate, ovate or oblong, 1–2 cm. long, the spikelets approximate, gynaecandrous; ♀ scales ovate, margins conspicuously hyaline, ca. as wide as the perigynia but shorter; perigynia ovate, 4 mm. long, plano-convex, coriaceous, nearly smooth ventrally, finely many-nerved dorsally, wing-margined, serrulate above, narrowed into a serrulate beak ⅓ the length of the perigynium-body, the beak terete and smooth at the tip, the apex hyaline.—Rare and little known; Red Fir and Subalpine forests, 6500–8300 ft.; Eldorado and Sierra cos.

55. **C. gracílior** Mkze. Loosely cespitose, the lvs. and culms from short rootstocks; culms 3–6 dm. high, erect, longer than the lvs.; blades 1–2 mm. wide, flat; infl. narrowly capitate-spicate, 1–2 cm. long, the gynaecandrous spikelets aggregated above but generally somewhat separate below; ♀ scales ovate, ca. as wide as the perigynia but shorter, margin white-hyaline; perigynia narrowly ovate, 3.5–4.5 mm. long, plano-convex, coriaceous, nerveless ventrally, few-nerved dorsally, narrowly wing-margined, serrulate above, contracted into a slender beak half the length of the perigynium-body, the beak smooth and hyaline at the tip.—Grassy and wooded slopes and flats that are wet in the spring, 100–2000 ft.; S. Oak Wd., Mixed Evergreen F., and Foothill Wd.; Coast Ranges from Santa Barbara Co. to Humboldt Co., local in n. Sierra Nevada foothills.

56. **C. subbracteàta** Mkze. Loosely cespitose, the culms and lvs. from short rootstocks; culms 3–12 dm. tall, taller than the lvs., frequently growing up through supporting shrubs; blades 2.5–4 mm. wide, flat; infl. capitate, ovate or roundish, 1.5–2.5 cm. long, the spikelets gynaecandrous; ♀ scales ovate, narrower and shorter than the perigynia, with white-hyaline margins; perigynia narrowly ovate, 3.5–4.5 mm. long, plano-convex, coriaceous, nerveless ventrally, finely nerved dorsally, narrowly margined, serrulate above, contracted into a slender serrulate beak ⅓ the length of the body, the tip hyaline at apex.—Moist soil of grassland, brush, or open woods, sea level to 5200 ft.; S. Oak Wd. to N. Coastal Coniferous F. and Coastal Prairie; Coast Ranges from Santa Barbara Co. and ids. n. to Humboldt Co.

57. **C. pachystáchya** Cham. Densely cespitose, the lvs. and culms arising from very short rootstocks; culms 3–10 dm. tall, erect, longer than the lvs.; blades 2–4 mm. wide, flat; infl. capitate, oblong or nearly orbicular, 1–2.5 cm. long, the 4–12 spikelets gynaecandrous; ♀ scales ovate, nearly as wide as the perigynia and only a little shorter; perigynia ovate, 3.5–5 mm. long, plano-convex, submembranaceous, smooth ventrally, nerved dorsally, narrowly margined, serrulate above, rather abruptly contracted into a slender serrulate beak less than half the length of the body, the tip slightly hyaline, bidentulate.—Moist meadows and open woods, 100–6800 ft.; N. Coastal Coniferous F. to Yellow Pine F.; Coast Ranges from Mendocino Co. to Del Norte and Siskiyou cos., Mt. Shasta; n. to Alaska, e. to the Rocky Mts. A slender form with narrower leaves and only 3–6 spikelets in the head is var. *grácilis* (Olney) Mkze. [*C. festiva* var. *g.* Olney.]

58. **C. ampléctens** Mkze. Plants cespitose with short rootstocks; culms 5–8 dm. tall, stiff, exceeding the lvs.; blades 2.5–4 mm. wide, flat; infl. capitate above, spicate-interrupted below, oblongish or ovatish, 2–5 cm. long, the spikelets gynaecandrous, closely subtended by the conspicuous dilated bracts; ♀ scales ovate, acute or shortly

cuspidate, a little shorter and narrower than the perigynia; perigynia ovate, 3.5–4 mm. long, plano-convex, several-nerved on both sides, wing-margined, serrulate from below the middle, contracted into a slender beak ⅓ to ½ the length of the body, the tip smooth, reddish-brown.—Rare and but little known; meadows, 4000–6500 ft.; Yellow Pine and Red Fir forests; Sierra Nevada from Tulare Co. to Eldorado Co., n. to Mt. Shasta.

59. **C. frácta** Mkze. Loosely cespitose, the culms and lvs. from short rootstocks; culms 5–12 dm. tall, erect, much longer than the lvs.; blades 3–6 mm. wide, flat or the margins a little revolute, sheaths hyaline ventrally and breaking easily, prolonged at the mouth into a fragile scarious and often evanescent appendage; infl. slender, oblong, spicate or the spikelets ± approximate, 2.5–7.5 cm. long; spikelets gynaecandrous; ♀ scales lanceolate, long-acuminate or aristate, a little shorter and narrower than the perigynia, hyaline; perigynia lanceolate-ovate, 3–4.5 mm. long, plano-convex but somewhat distended by the ak., nerved on each side, narrowly winged, serrulate above, contracted into a serrulate beak ca. ⅓ the length of the body, tip tawny, terete, smooth.— Common, meadows and moist slopes in open forests, 2600–10,800 ft.; Yellow Pine F. to Subalpine F.; s. Calif. mts., Sierra Nevada from Kern Co. to Eldorado Co., n. to Mt. Shasta, Coast Ranges from Lake Co. to Siskiyou Co.; n. to Wash.

60. **C. Tràcyi** Mkze. Plants densely cespitose, the rootstocks very short; culms 2–8 dm. high, erect or somewhat spreading, longer than the lvs.; blades 2–4 mm. wide, flat or ± channeled; infl. loosely capitate or somewhat spicate, oblongish or ovate, 1.5–4 cm. long, the spikelets gynaecandrous; ♀ scales ovate, ca. as long and as wide as the perigynia, margins white-hyaline; perigynia broadly ovate, 4–5 mm. long, flattened and distended by the ak., nerved on both sides, wing-margined, serrulate above the middle, contracted into a serrulate beak somewhat shorter than the body, the tip reddish-brown, smooth.—Boggy meadows or wet soil bordering marshes, from near sea level to 3500 ft.; Mixed Evergreen F., Coastal Prairie, and N. Oak Wd.; Coast Ranges from Marin Co. to Humboldt Co.; n. to B.C.

61. **C. phaeocéphala** Piper. Plants densely cespitose, frequently developing large clumps, the rootstocks short and matted; culms 1–3 dm. tall, erect or spreading, longer than the lvs.; blades 1.5–2 mm. wide, channeled or somewhat involute; infl. subcapitate, oblongish, 1–2.5 cm. long, the spikelets ± aggregated or the lower somewhat separate, gynaecandrous; ♀ scales ovate, the length and width of the perigynia, dark or reddish-brown with broad white-hyaline margins; perigynia oblong-ovate, 4–6 mm. long, thin, slightly nerved or nearly nerveless ventrally, nerved dorsally, conspicuously margined, serrulate above the base, contracted into a short beak ca. ¼ the length of the body, the tip white-hyaline.—Rocky ridges and slopes, 9000–12,800 ft.; Subalpine F., Bristle-cone F. and Alpine Fell-fields; Sierra Nevada from Tulare Co. to Eldorado Co., White and Sweetwater mts., Mt. Lassen, Mt. Shasta; n. to B.C., e. to the Rocky Mts.

62. **C. tahoénsis** Smiley. Plants densely cespitose, the culms and lvs. from short rootstocks; culms 1.5–3 dm. tall, firm and stout, obtusely triangular, longer than the lvs.; blades folded along the middle, 1 mm. wide; infl. 1.5–3.5 cm. long, subspicate to capitate, the lowest spikelets a little separate, closely approximate above, spikelets gynaecandrous; ♀ scales lanceolate-ovate, ca. as wide as the perigynia but somewhat shorter, margins broadly hyaline; perigynia oblong-lanceolate, 5–6 mm. long, many-nerved on both sides, narrowly wing-margined, serrulate above the middle, tapering into a serrulate beak ca. ⅓ to ½ the length of the body, the tip terete, smooth and hyaline above.—Open dry rocky slopes, 9500–12,000 ft.; Subalpine F. and Alpine Fell-fields; Sierra Nevada, chiefly along easterly slopes near the crest in Inyo, Mono, Fresno, and Eldorado cos.

63. **C. leporinélla** Mkze. Culms and lvs. densely cespitose, forming large clumps, the rootstocks very short; culms 1.5–3 dm. tall, stiffly erect, longer than the lvs.; blades 0.75–1.5 mm. wide, involute; infl. subcapitate or somewhat spicate, slender, 1.5–3 cm. long, the spikelets gynaecandrous; ♀ scales oblong-ovate, a little longer and wider than the perigynia, the margins white-hyaline; perigynia linear-oblanceolate, 3.5–4 mm. long, scarcely 1 mm. wide, ± nerved on both sides, very narrowly wing-margined, serrulate down to the middle, tapering into a short beak ca. ¼ the length of the body,

the tip white-hyaline.—Rocky slopes and flats, 7000–13,150 ft.; Red Fir F. to Alpine Fell-fields; Sierra Nevada from Tulare Co. to Placer Co., Sweetwater Mts., Mt. Lassen, n. Coast Ranges in Siskiyou Co.; n. to Wash., e. to Utah and Wyo.

64. **C. praticola** Rydb. Plants cespitose, the culms and lvs. in small clusters from short rootstocks; culms 2–7 dm. tall, ± flexuous, longer than the lvs.; blades 1–3.5 mm. wide, flat; infl. spicate, slender, erect or nodding, 1.5–5 cm. long, the spikelets approximate or frequently separate, gynaecandrous; ♀ scales ovate, as large as the perigynia, broadly white-margined; perigynia ovate-lanceolate, 4.5–6.5 mm. long, finely nerved dorsally, scarcely nerved ventrally, margined, serrulate above the base, narrowed to a serrulate beak ⅓ to ½ the length of the body, the tip white-hyaline.—Rare, meadows and open woods, near sea level to 2000 ft., Coastal Prairie and N. Coastal Coniferous F.; Humboldt Co.; n. to Alaska, e. to Quebec and Greenland.

65. **C. petasàta** Dewey. Plants cespitose, the culms and lvs. arising from short rootstocks; culms 3–8 dm. tall, erect, smooth or nearly so, longer than the lvs.; blades flat, 2–3 mm. wide; infl. capitate, oblongish or narrowly ovate, 2–4 cm. long, the spikelets gynaecandrous; ♀ scales ovate, ca. as wide and long as the perigynia, margins white-hyaline; perigynia oblong-lanceolate, 6–8 mm. long, many-nerved on both sides, narrowly margined, serrulate to below the middle, tapering into a serrulate beak half as long as the body or shorter, the tip terete, smooth, hyaline.—Dry meadows and open woods, in California known only from Patterson Flat, Lassen Co., 5700 ft.; n. to B.C., e. to the Rocky Mts.

66. **C. Dàvyi** Mkze. Plants densely cespitose, the rootstocks very short; culms 2.5–3.5 dm. tall, longer than the lvs., smooth; blades flat or channeled, 1.5–2.5 mm. wide; infl. narrowly to broadly oblong, subcapitate, 2.5 cm. long, the spikelets approximate, gynaecandrous; ♀ scales ovate, obtuse, narrower than the perigynia and only ca. half as long; perigynia oblong-lanceolate, 6–8 mm. long, nerved on both sides, narrowly margined, serrulate from below the middle, tapering into a serrulate beak ⅓–½ as long as the body, the tip terete, smooth, hyaline.—Rare, dry meadows and open woods, 4800–10,600 ft.; Red Fir and Subalpine forests; Sierra Nevada from Tuolumne Co. to Nevada Co.

67. **C. specífica** Bailey. [*C. scoparia* var. *fulva* W. Boott. *C. lancifructa* Mkze.] Plants cespitose, the culms and lvs. rather loose on short rootstocks; culms 3–8 dm. tall, much exceeding the lvs., smooth or nearly so; blades flat or channeled, 2–5 mm. wide; infl. capitate, oblong or roundish, 1.5–4 cm. long, the spikelets ± aggregated, gynaecandrous; ♀ scales lanceolate-ovate, acute, a little shorter than the perigynia but nearly as wide; perigynia lanceolate, 5–6 mm. long, nerved on both faces, narrowly margined, serrulate above, tapering into a serrulate beak ca. ⅓ the length of the body, the tip slender, smooth, reddish-tinged.—Common, dry soil of meadows and open forests, 4000–11,200 ft.; Yellow Pine F. to Subalpine F.; Sierra Nevada from Tulare Co. to Nevada Co.; w. Nev.

68. **C. multicostàta** Mkze. [*C. adusta* var. *congesta* W. Boott. *C. pachycarpa* Mkze.] Plants densely cespitose, the rootstocks short; culms 3–9 dm. tall, exceeding the lvs.; blades flat, 2.5–6 mm. wide; infl. usually compact, the spikelets aggregated into an oblong to roundish head, 1.5–4 cm. long, the spikelets gynaecandrous; ♀ scales ovate, margins hyaline, a little shorter and narrower than the perigynia; perigynia narrowly to broadly ovate, 3.5–5.5 mm. long, conspicuously nerved dorsally, finely nerved or smooth ventrally, wing-margined, serrulate from below the middle, abruptly contracted into a short beak ¼ to ⅓ the length of the body, the beak broad and flat, winged to the tip or nearly so.—Common, dry soil of meadows and open woods, 500–11,400 ft.; Yellow Pine F. to Subalpine F.; s. Calif. mts., Sierra Nevada from Tulare Co. to Plumas Co., Coast Ranges from Glenn Co. to Siskiyou Co., Mt. Lassen, Mt. Shasta; n. to Ore., e. to w. Nev. and Ida.

69. **C. straminifórmis** Bailey. [*C. straminea* var. *congesta* Boott.] Plants densely cespitose, the culms and lvs. from very short rootstocks and forming large clumps; culms 1–4 dm. tall, erect or curving; blades flat, 2–3.5 mm. wide; infl. capitate, the spikelets in an ovate to roundish head 1.5–2.5 cm. long, the spikelets gynaecandrous; ♀ scales ovate-lanceolate, shorter and much narrower than the perigynia; perigynia flattened, broadly ovate, 4.5–5 mm. long, nerved dorsally, smooth or slightly nerved ventrally, strongly winged to the base, the wing undulate, serrulate above the middle, abruptly

narrowed into a serrulate beak ca. ⅓ the length of the body, flat and margined at the reddish-brown tip.—Common, rocky and gravelly slopes, 6500–12,500 ft.; Red Fir F. to Alpine Fell-fields; Sierra Nevada from Tulare Co. to Plumas Co., n. Coast Ranges in Siskiyou Co., Sweetwater Mts., Mt. Lassen, Mt. Shasta, Modoc Co.; n. to Wash., e. to Utah and Mont.

70. **C. fèta** Bailey. [*C. straminea* var. *mixta* Bailey. *C. f.* var. *multa* Bailey.] Plants densely cespitose, the rootstocks short; culms 5–12 dm. long, smooth or nearly so, longer than the lvs.; blades flat, 2.5–4 mm. wide, sheath green-striate ventrally to the top; infl. rather openly spicate to subcapitate, oblongish to nearly round, 2–8 cm. long, spikelets gynaecandrous; ♀ scales ovate, narrower and shorter than the perigynia; perigynia ovate, 3–3.5 mm. long, ± nerved dorsally and ventrally, margined to the base, serrulate above the middle, narrowed above into a flat serrulate beak ½ the length of the body, the margined tip reddish.—Common in moist or wet soil of meadows and stream banks, 100–7800 ft.; Foothill Wd. and Coastal Prairie to Yellow Pine F.; San Bernardino Mts., Sierra Nevada from Kern Co. to Plumas Co., Coast Ranges from Santa Clara Co. to Siskiyou Co., Mt. Lassen; n. to B.C.

71. **C. athrostáchya** Olney. [*C. a.* var. *minor* Olney.] Plants cespitose in small or large clumps, the rootstocks very short; culms 0.5–6 dm. tall, longer or shorter than the lvs., smooth or a little roughened; blades flat, 1.5–4 mm. wide; infl. capitate or the lowest spikelets a little separate, the head mostly ovate or rounded, 1–2 cm. long, the spikelets gynaecandrous, the lowest subtended by foliaceous bracts that usually strongly exceed the infl.; ♀ scales oblong-ovate, a little narrower and shorter than the perigynia; perigynia flattened, lanceolate-ovate, 3–4.5 mm. long, nerved dorsally, slightly nerved or nerveless ventrally, margined, serrulate above, tapering into a subterete or little-flattened beak ca. ⅓ as long as the body.—Common in marshes, dry meadows, and open woods, 400–10,500 ft.; Coastal Prairie and N. Oak Wd. to Subalpine F.; San Bernardino Mts., Sierra Nevada from Kern Co. to Plumas Co., Mt. Lassen, Coast Ranges from Marin Co. to Del Norte Co., e. to Modoc Co.; n. to Alaska, e. to the Rocky Mts.

72. **C. unilateràlis** Mkze. Plants cespitose, the stems and lvs. clustered on short rootstocks; culms 2–9 dm. long, longer than the lvs., smooth or a little roughened above; blades flat, 2–4 mm. wide; infl. densely capitate, short-oblong to roundish, 1–3 cm. long, spikelets gynaecandrous, the lowest bract foliaceous, strictly erect and simulating a continuation of the culm, the head appearing as if laterally attached; ♀ scales ovate, narrower and a little shorter than the perigynia; perigynia flat, ovate, 4–5 mm. long, nerved on both sides, winged, serrulate above, narrowed into a serrulate beak from ⅛ to nearly as long as the body, the tip reddish-brown.—Wet soil of meadows and thickets, near sea level to 3100 ft.; N. Coastal Coniferous F. and Coastal Prairie; Humboldt and Trinity cos.; n. to B.C., e. to Ida.

73. **C. leptàlea** Wahl. Cespitose in small or large clumps with short rootstocks; culms very slender, 1.5–6 dm. high, erect but lax, usually much longer than the lvs.; blades 0.5–1.25 mm. wide, flat or folded; spikelet solitary, androgynous, slender, 5–15 mm. long, 2–3 mm. wide; ♀ scales yellowish-green, lanceolate, obovate, or ovate-orbicular, acute or obtuse, or the lowest sometimes attenuate-cuspidate, usually ca. half the length of the perigynia, or sometimes much longer; perigynia oval-elliptic, 2.5–5 mm. long, finely many nerved, the apex rounded, entire or emarginate, beakless.—Rare, wet ground of swamps and low meadows, 100–2300 ft.; Coastal Prairie to Yellow Pine F.; Coast Ranges in Marin, Humboldt, and Trinity cos.; n. to B.C., e. to the Atlantic.

74. **C. filifòlia** Nutt. Plants densely cespitose, the rootstocks very short; culms 8–30 cm. tall, erect, very slender, tough, ca. equaling the lvs. or a little longer; blades filiform, involute, ca. 0.25 mm. wide; spikelet solitary, androgynous, linear, 1–3 cm. long; ♀ scales broadly obovate or roundish, obtuse, with very broad white-hyaline margins, nearly as long and wide as the perigynia; perigynia obovate or roundish, 3–3.5 mm. long, faintly 2-ribbed, thinly hairy above, closely enclosing the ak. and rounded on the angles, rounded-truncate below the very short (0.2–0.4 mm.) beak.—Rare, dry rocky slopes and flats in the mts. at ca. 10,500 ft.; Subalpine F.; Sierra Nevada, Inyo Co.; n. to Yukon, e. to Tex. and Manitoba.

75. **C. exsérta** Mkze. [*C. filifolia* var. *erostrata* Kük.] Densely cespitose with very short rootstocks, the plants tufted or forming rings; culms 5–25 cm. long, erect, very slender, wiry; blades filiform, channeled, 0.25–0.5 mm. wide; spikelet solitary, androgy-

nous, narrowly oblong-lanceolate, 0.7–2 cm. long; ♀ scales ovate to obovate and orbicular, obtuse, with hyaline margins, shorter than the perigynia; perigynia obovate or roundish, 2.5 mm. long, obtuse, faintly 2-ribbed, short-puberulent, almost beakless. —Common, dry open slopes and flats, 5000–12,000 ft.; Yellow Pine F. to Alpine Fell-fields; San Bernardino Mts., Sierra Nevada from Tulare Co. to Lassen Co.; w. Nev., s. Ore.

76. **C. ínops** Bailey. [*C. verecunda* Holm.] Culms and leaves clustered in small or large clumps, arising from long stout rootstocks; culms 1.5–4 dm. tall, erect, much longer than the lvs.; blades flat or somewhat channeled, 1.5–2.5 mm. wide, pale green; ♂ spikelet terminal, solitary, linear, 0.7–2.5 cm. long, sessile or short-stalked; ♀ spikelets 1–3, approximate or discrete, oblong to roundish, 5–14 mm. long, the uppermost sessile or subsessile, the lower short-stalked; ♀ scales oblong-ovate, cuspidate or acuminate, larger or smaller than the perigynia; perigynia 3–4.5 mm. long, short-hispid, the body roundish or a little elongate, almost round in cross section, 2-keeled, abruptly stipitate and beaked, the beak ⅓–¾ the length of the perigynium-body, deeply bidentate, the teeth erect, hyaline.—Rare, dry rocky slopes and flats, 3200–6500 ft.; Yellow Pine F.; Mt. Shasta and vicinity, Siskiyou Co.; n. to B.C.

77. **C. globòsa** Boott. Culms and lvs. clustered, arising from somewhat extended rootstocks; culms 1.5–4 dm. long, slender, spreading, longer or shorter than the lvs.; blades flat, or channeled below, 1.5–2.5 mm. wide; ♂ spikelet terminal, solitary, short-stalked, 0.7–2 cm. long; ♀ spikelets 4–6, the uppermost 2 or 3 approximate, sessile or shortly stalked, the lowest nearly basal and widely separated, on filiform stalks, the spikelets broadly oblong to suborbicular, 5–10 mm. long; ♀ scales ovate, ca. the size of the perigynia; perigynia 4–5 mm. long, globose or nearly so with prominent stipe and beak, short-pubescent, 2-keeled and finely several-ribbed, the beak 0.75–1.25 mm. long, bidentate.—Wooded and brushy slopes and flats in well-drained or rocky soil, near sea level to 4200 ft.; common in the Redwood belt; S. Oak Wd. to N. Coastal Scrub and Mixed Evergreen F.; San Diego Co. n. to Humboldt Co.

78. **C. Braínerdii** Mkze. Culms and lvs. clustered, arising from prolonged rootstocks; culms to 3 dm. tall, shorter than the lvs.; blades 1.5–3 mm. wide, ± channeled throughout; ♂ spikelet terminal, sessile or short-stalked, 5–12 mm. long; ♀ spikelets 4–6, subglobose or broadly elliptic, the upper 2 or 3 approximate and close to the ♂, the others nearly basal and widely separated; ♀ scales ovate or ovate-lanceolate, cuspidate or long-awned, usually a little longer and narrower than the perigynia; perigynia 4.5 mm. long, pubescent, the body oval, strongly 2-keeled, the outer side many-ribbed, stipe and beak prominent, the beak nearly half as long as the body, bidentate.—Dry rocky slopes and open woods, 3000–7200 ft.; Yellow Pine and Red Fir forests; Sierra Nevada from Mariposa Co. to Plumas Co., Mt. Lassen, Mt. Shasta, n. Coast Ranges from Glenn Co. to Siskiyou Co.; s. Ore.

79. **C. brévipes** W. Boott. [*C. deflexa* var. *Boottii* Bailey.] Densely cespitose, the clustered lvs. and culms arising from short rootstocks; culms to 18 cm. long, longer or shorter than the lvs.; blades 1.5–2.5 mm. wide, flat, or channeled at the base; ♂ spikelet terminal, sessile or short-stalked, 4–12 mm. long; ♀ spikelets 3–5, the upper 1 or 2 approximate, sessile or stalked, the others widely separated, nearly basal; ♀ scales ovate, acute to cuspidate, shorter than the perigynia but ca. as wide; perigynia 2.5–3 mm. long, pubescent, the body roundish, 2-keeled, stipitate, contracted into a bidentate beak 0.25–0.75 mm. long.—Dry meadows and partially shaded forest slopes, 2400 to 12,300 ft.; Yellow Pine F. to Alpine Fell-fields; San Bernardino and San Gabriel mts., Sierra Nevada from Tulare Co. to Plumas Co. and n. to Tehama Co., N. Coast Ranges in Humboldt, Trinity, and Siskiyou cos.; n. to Wash., e. to Nev. and Ida.

80. **C. Róssii** Boott. Plants loosely to densely cespitose, the clustered lvs. and culms from short rootstocks; culms 5–30 cm. long, mostly equaling or longer than the lvs.; blades 1–2.5 mm. wide, flat or folded; ♂ spikelet terminal, sessile or short-stalked, 3–15 mm. long; ♀ spikelets 3–5, the uppermost 1 or 2 approximate or a little separate, sessile or short-stalked, the others widely separated, nearly basal, the spikelets 3–5 mm. long, short-oblong or roundish; ♀ scales ovate, acute to acuminate or awned, wider but shorter than the perigynia; perigynia 3–4.5 mm. long, pubescent, the body nearly round, 2-keeled, prominently stipitate and beaked, the bidentate beak 0.75–1.5 mm. long.—Dry meadows and forests, from near sea level to 12,400 ft.; Coastal Prairie and

Yellow Pine F. to Alpine Fell-fields; San Bernardino and San Gabriel mts., Sierra Nevada from Tulare Co. to Nevada Co., N. Coast Ranges from Glenn Co. to Del Norte Co., Mt. Lassen, Modoc Co.; n. to Yukon, e. to Mich. and Colo.

81. **C. brevicaùlis** Mkze. Densely cespitose in small to large clumps, with somewhat prolonged rootstocks; culms 3–15 cm. tall, usually shorter than the lvs.; blades flat or ± channeled, 1.5–3.5 mm. wide; ♂ spikelet terminal, 6–18 mm. long, short- or long-stalked; ♀ spikelets 2–4, the uppermost near the ♂ and sessile, the second, if present, a little separate, the others widely separate and nearly basal; ♀ scales ovate, acute to cuspidate, narrower and shorter than the perigynia; perigynia 4 mm. long, pubescent, the body roundish, 2-keeled, prominently stipitate and beaked, the bidentulate beak ca. half as long as the perigynium-body.—In shallow soil around rocks or on sandy or grassy flats and hills near the coast, near sea level to 2500 ft.; Coastal Sage Scrub and Coastal Prairie to Mixed Evergreen F.; Coast Ranges from San Luis Obispo Co. to Del Norte Co.; n. to B.C.

82. **C. pseudoscirpoìdea** Rydb. Plants loosely cespitose, the lvs. and culms arising from long-creeping rootstocks; culms 1.5–3.5 dm. tall, longer than the lvs., phyllopodic; blades flat, or channeled at the base, 2–3 mm. wide; spikelet solitary, unisexual, linear, 1–3.5 cm. long; ♀ scales broadly ovate, ± pubescent and ciliate, margin white-hyaline, broader and longer than the perigynia; perigynia obovate, (2.5–)3–4 mm. long, pubescent, 2-ribbed and faintly nerved, abruptly contracted above into a beak ca. ¼ the length of the body, the beak hyaline-tipped and bidentulate.—Rare, dry rocky slopes, 11,500–12,000 ft.; Subalpine F. and Alpine Fell-fields; Rock Creek Lake Basin in Inyo Co. and Dana Plateau in Mono Co.; n. to Wash., e. to Colo. and Mont.

83. **C. gìgas** (Holm) Mkze. [*C. scirpoidea* var. *g.* Holm.] Densely cespitose, the rootstocks short-creeping; culms 3–4.5 dm. tall, exceeding the lvs., phyllopodic; blades flat, or channeled near the base, 2–3 mm. wide; spikelet solitary (or rarely 2 or 3), unisexual, linear or linear-oblong, 1.5–2.5 cm. long; ♀ scales oblong-ovate, glabrous on back, margin finely ciliate and narrowly hyaline, narrower than the perigynia and ca. as long; perigynia obovate or oblong, 3–4 mm. long, hirsutulous-puberulent above, ± 2-ribbed and obscurely to strongly nerved, abruptly contracted into a short beak ca. ⅕ the length of the body, the beak hyaline and ± bidentulate at apex.—Rare, meadows and rocky slopes, 2800 to 6000 ft.; Yellow Pine and Red Fir forests; Plumas Co. in the Sierra Nevada and Trinity and Siskiyou cos. in the Klamath Mts.

84. **C. concinnoìdes** Mkze. Plants very loosely cespitose, the culms and lvs. from long-creeping rootstocks; culms spreading or erect, 1.5–3.5 dm. long, longer than the lvs.; blades flat or somewhat channeled, 2–4 mm. wide; spikelets 2–4, the terminal ♂, linear, 1–2 cm. long, the lateral spikelets ♀, approximate, sessile or short-stalked, linear or oblong, 5–10 mm. long; ♀ scales ovate-lanceolate or obovate, narrower and shorter than the perigynia, margins ciliate and wide-white-hyaline; perigynia oblong-ovate, 2.5–3 mm. long, pubescent, 2-keeled and several-nerved, abruptly short-beaked, the hyaline-tipped beak ⅕–¼ the length of the perigynium-body.—Clay soil of flats or rocky slopes, 200–3000 ft.; Coastal Prairie, N. Coastal Scrub, and Yellow Pine F.; Coast Ranges from Mendocino Co. to Del Norte Co.; n. to B.C., e. to Alta. and Mont.

85. **C. triquètra** Boott. [*C. monticola* Dewey.] Plants densely cespitose, the rootstocks very short and not at all prolonged; culms 3–6 dm. tall, longer than the lvs.; blades 2.5–6 mm. wide, flat but with margins revolute; spikelets usually 3 or 4, the terminal ♂, linear, 1–3 cm. long, the lateral spikelets ♀ or with a few terminal ♂ flowers, the uppermost approximate, the others ± distant from the ♂ spikelet, oblong or narrower, 1–4.5 cm. long, sessile or short-stalked above, long-stalked below; ♀ scales broadly ovate, ca. as wide as the perigynia but shorter; perigynia ovate or obovate, 4–4.5 mm. long, greenish, pubescent, obscurely nerved, abruptly very short-beaked, the beak bidentulate and less than ¹⁄₁₀ the length of the perigynium-body.—Clay soil of flats and rocky slopes, 600–3000 ft.; Coastal Sage Scrub and Chaparral; San Luis Obispo Co. s. to San Diego Co.; n. L. Calif.

86. **C. multicaùlis** Bailey. Loosely to densely cespitose, the rootstocks short or somewhat prolonged; culms terete or obtusely triangular, 2–6 dm. tall, green, much exceeding the short lvs.; lvs. clustered near the base, usually only 1 or 2 well developed, the blades 1–1.5 mm. wide, canaliculate or involute; spikelet solitary, androgynous, elongate, the ♂ part linear and 1–2.5 cm. long, the ♀ part consisting of 1–6 approxi-

mate or discrete perigynia; ♂ scales oblong-obovate, obtuse, with broad white-hyaline margin; ♀ scales bractlike or leaflike below, the lower prolonged into a blade 1–4 cm. long above a widened hyaline-margined base, the upper lanceolate, cuspidate or awned; perigynia 5–7 mm. long, oblong-ovate, triangular, glabrous, 2-ribbed and obscurely nerved, short-stipitate, minutely beaked.—Common, dry soil of dense or open forests, 500–7300 ft.; Yellow Pine and Red Fir forests; s. Calif. mts. n. through the Sierra Nevada and Coast Ranges to Siskiyou and Modoc cos.; n. to Ore., e. to Nev.

87. **C. Gèyeri** Boott. Loosely cespitose or the lvs. and culms more densely clustered, the rootstocks generally elongate; culms sharply triangular, 1–4 dm. tall, stiffly erect, longer or shorter than the lvs.; blades 2–3.5 mm. wide, flat or channeled; spikelet solitary, androgynous, elongate, the ♂ part linear and 0.5–2.5 cm. long, the ♀ part consisting of 1–3 perigynia, approximate or separate; ♂ scales oblong-obovate, obtusish, hyaline-margined; ♀ scales usually longer and wider than the perigynia, the lower short-awned, the upper acute or obtusish; perigynia 6 mm. long, oblong-obovate, triangular, glabrous, 2-ribbed, nerveless, short-stipitate, minutely beaked.—Rare, dry slopes and open woods, ca. 5000 ft.; Sagebrush Scrub and Yellow Pine F.; Siskiyou and Humboldt cos. and Sierra V.; n. to B.C., e. to Alta. and Colo., and Pa. (where probably introd.).

88. **C. salinaefórmis** Mkze. Loosely cespitose, the rootstocks long and slender; culms 5–15 cm. long, erect or somewhat spreading, shorter than the lvs.; blades flat, 2–5 mm. wide; spikelets 4 or 5, the terminal ♂, linear, 1–1.5 cm. long, stalked, the lateral spikelets ♀, oblong, 0.5–1.5 cm. long, short- or long-stalked, the stalks exserted from the sheath of the subtending bracts; ♀ scales ovate, appressed, ca. as wide as the perigynia but usually a little shorter; perigynia oblong-ovate, 2.5–3.75 mm. long, nerved, densely finely granular, beakless but short-tapering and a little constricted at the apex.—Uncommon, open coastal slopes and mesas in moist or wet soil, below 400 ft.; Coastal Prairie and N. Coastal Scrub; Santa Cruz Co. to Humboldt Co.

89. **C. Hàssei** Bailey. [*C. aurea* var. *celsa* Bailey.] Loosely cespitose, the rootstocks long and slender; culms 0.5–7 dm. long, erect or lax, usually exceeding the lvs.; blades flat (or channeled below), 2–4 mm. wide; spikelets 4–6, the terminal ♂ or gynaecandrous, 6–20 mm. long, short-stalked, the lateral spikelets ♀, linear-oblong, 0.7–2.5 cm. long, approximate or the lowest ± distant, the uppermost sessile or short-stalked, the lower long-stalked, the stalks exserted from the sheaths of the subtending bracts; ♀ scales ovate-suborbicular, obtuse to acuminate, appressed, usually a little shorter and narrower than the perigynia; perigynia elliptic-ovate, 2.5–3 mm. long, obscurely ribbed, membranous, not fleshy, densely finely granular, whitish-pulverulent, rounded and beakless at apex.—Coastal marshes to montane meadows and seepages, 20–9000 ft.; Coastal Prairie and Pinyon-Juniper Wd. to Red Fir F.; s. Calif. mts. n. through Coast Ranges and Sierra Nevada, Panamint Mts.; s. to L. Calif., n. to Yukon, e. to Utah and Ariz.

90. **C. aùrea** Nutt. Usually loosely cespitose, rarely more tufted and turf-forming, the rootstocks usually elongate; culms 0.5–5.5 dm. long, erect or spreading, much shorter to much longer than the lvs.; blades 2–4 mm. wide, flat, or channeled at the base; spikelets 4–6, the terminal ♂ or with a few ♀ fls., 3–10 mm. long, sessile or short-stalked, the lateral spikelets ♀, oblong or linear-oblong, 0.5–2 cm. long, the upper approximate, the lower ± widely separate, the lowest frequently basal, the stalks of the lowest exserted from the sheaths of leaflike bracts; ♀ scales ovate to roundish, obtuse to cuspidate, spreading at maturity; perigynia roundish-obovate, 2–3 mm. long, coarsely ribbed, golden-yellow or brownish and fleshy at maturity, rounded and beakless at apex.—Wet meadows and springy places in the mts., 3500–10,900 ft.; Yellow Pine F. to Subalpine F.; s. Calif. mts., Sierra Nevada from Tulare Co. to Butte Co., White Mts., Mt. Shasta, Modoc Co., Coast Ranges from Glenn Co. n.; n. to B.C., e. to the Atlantic.

91. **C. lívida** (Wahl.) Willd. [*C. limosa* var. *livida* Wahl.] Loosely cespitose, the small clusters of culms and lvs. arising from elongate rootstocks; culms 0.5–6 dm. tall, erect, longer or shorter than the lvs.; blades glaucous, 0.5–3.5 mm. wide, channeled; spikelets 2–4, the terminal ♂ or rarely gynaecandrous, linear, 1–3 cm. long, the lateral spikelets ♀, oblong, 1–2 cm. long, the uppermost approximate, the lowest sometimes distant and subbasal, long-stalked; ♀ scales ovate, ca. as wide as the perigynia but

shorter; perigynia oblong-ovate, 2–4.5 mm. long, glaucous-green, puncticulate, 2-keeled, finely many-nerved, tapering to the rounded or somewhat pointed beakless apex.— Very rare, coastal swamps near sea level; Coastal Prairie; Mendocino Co.; n. to Alaska, e. to the Atlantic; Eu.

92. **C. califórnica** Bailey. Scarcely cespitose, the culms and lvs. from long rootstocks; culms 2–7 dm. tall, erect, much longer than the lvs.; blades green, flat, 1.5–5 mm. wide; spikelets 2–6, the terminal ♂, linear, 1.5–3.5 cm. long, usually long-stalked, the lateral spikelets ♀, linear-oblong, 1–4 cm. long, discrete, the lowest often nearly basal, short- to long-stalked, the stalks exserted from the sheathing bases of leaflike bracts; ♀ scales ovate, narrower and shorter than the perigynia; perigynia ovate, 3.5–4 mm. long, 2-keeled and obscurely several-nerved, puncticulate and minutely granulose, abruptly beaked, the beak ca. ¼–⅓ the length of the perigynium-body, the tip white-hyaline, entire or nearly so.—Rare, coastal flats below 300 ft.; Coastal Prairie; Mendocino Co.; n. to Wash. and Ida.

93. **C. Hendersònii** Bailey. Loosely cespitose, the rootstocks short; culms 4–9 dm. tall, erect or spreading, longer than the lvs.; blades flat, 3–10 mm. wide; spikelets 3–5, the terminal ♂, linear, 1.5–3 cm. long, the lateral ♀, linear, 1–4 cm. long, the upper approximate and subsessile, the lower one or two separate and long-stalked, the stalks exserted from the sheath-based bracts; ♀ scales broadly obovate, as wide as the perigynia but shorter; perigynia narrowly obovate, triangular, 5–6 mm. long, 2-keeled and many-nerved, tapering above into a bidentulate beak, the beak ⅕–¼ the length of the perigynium-body, hyaline-tipped.—Coastal woods, mostly below 500 ft.; Coastal Prairie and N. Oak Wd.; Sonoma Co. to Humboldt Co.; n. to B.C., e. to Ida.

94. **C. Whítneyi** Olney. [*C. flaccifolia* Mkze.] Densely cespitose; culms 6–9 dm. tall, sparingly pubescent, sharply but not roughly triangular, longer than the lvs.; blades flat, 2–6 mm. wide, pubescent on both sides; spikelets 3 or 4, the terminal ♂, linear, 1–2.5 cm. long, the lateral spikelets ♀, oblong, 1–3 cm. long, 6–10 mm. wide, approximate, sessile or short-stalked, the lowest bract short-sheathing; ♀ scales ± spreading, 3–5-nerved, ovate, ca. as wide as the perigynia and either shorter or longer; perigynia ovate, 4–5 mm. long, 2–2.3 mm. wide, triangular, strongly several-nerved on each side, contracted above into a short beak, the beak 0.5–1 mm. long, hyaline and bidentulate or oblique at the tip.—Dry sandy flats on the edge of meadows or in open forests, 4000–6000 ft.; Yellow Pine F.; Sierra Nevada in Mariposa and Tuolumne cos. The type of *C. flaccifolia*, said to have come from "dry plains in southwestern California," was probably collected in Yosemite V., the type locality of *C. Whitneyi*.

95. **C. Jepsònii** J. T. Howell. [*C. Whitneyi* of Mackenzie in part.] Densely cespitose, the rootstocks short and stout; culms 2.5–6 dm. tall, longer than the lvs.; blades flat or ± revolute, 2–3 mm. wide, pubescent above and below; spikelets 3 or 4, the terminal ♂, linear, 0.5–2 cm. long, short-stalked, the lateral spikelets ♀, oblong or linear-oblong, 1–2.5 cm. long, 5–7 mm. wide, approximate or the lowest separate, sessile or short-stalked, the bracts with short sheaths; ♀ scales ovate, appressed-ascending, 3-nerved, about as wide as the perigynia and either longer or shorter; perigynia elliptic-ovate or -obovate, 3.75–4.5 mm. long, 1.5–2 mm. wide, triangular, glabrous, conspicuously nerved on each face, tapering above into a short bidentulate beak, the beak somewhat hyaline, about ⅛–⅙ the length of the perigynium-body.—Dry sandy or gravelly flats in meadows or open forests, 5000–11,000 ft.; Yellow Pine F. to Subalpine F.; Sierra Nevada from Tulare Co. to Plumas Co., Mt. Lassen, Mt. Shasta, Warner Mts., and possibly the Trinity Alps, Siskiyou Co.; s. Ore., w. Nev.

96. **C. gynodynàma** Olney. [*C. Blankinshipii* Fern.] Rather loosely cespitose, the rootstocks short and the tufts of lvs. and culms not large; culms 2–9 dm. long, erect or somewhat spreading, much longer than the lvs.; blades 3–9 mm. wide, flat, soft-hairy; spikelets 3–5, the terminal ♂ or with a few ♀ fls., oblongish, 1–2 cm. long, the lateral spikelets ♀, oblong-cylindric, 1–3.5 cm. long, the uppermost sessile or short-stalked and approximate, the lowest widely separate on stalks long-exserted from the long-sheathing bracts; ♀ scales broadly ovate or obovate, ± hairy, wider but shorter than the perigynia; perigynia oblong-ovate, 4.5–5.5 mm. long, short-hairy above, rounded and abruptly beaked at apex, the beak 0.5–1 mm. long, hyaline and bidentulate at the tip.—Moist slopes and meadows, near sea level to 2000 ft.; Coastal Prairie, N. Coastal

Scrub, and Mixed Evergreen F.; Coast Ranges from Monterey Co. to Del Norte Co.; s. Ore.

97. **C. hirtíssima** W. Boott. Rather loosely cespitose, the small clumps from short rootstocks; culms 3–6 dm. tall, erect, much longer than the lvs.; blades flat, 3–7 mm. wide, soft-hairy; spikelets 3 or 4, the terminal ♀ or gynaecandrous, linear-oblong, 1.5–2.5 cm. long, the lateral spikelets ♀, oblong-cylindric, 0.8–2.5 cm. long, all ± separate, the lowest distant on stalks long-exserted from conspicuously sheathing bracts; ♀ scales ovate or obovate, slightly pubescent, as wide as the perigynia but shorter; perigynia obovate, 3.5–4 mm. long, pubescent, abruptly beaked, the beak ⅓–½ the length of the perigynium-body, hyaline and bidentulate at the tip.—Rare, wet springy or boggy slopes and meadows, 200–4000 ft.; Mixed Evergreen F. and Foothill Wd. to Yellow Pine F.; Lake and Mendocino cos. in Coast Ranges, Tuolumne Co. to Butte Co. in the Sierra Nevada.

98. **C. mendocinénsis** Olney. [*C. cinnamomea* Olney. *C. debiliformis* Mkze.] Rather densely cespitose, the clumps small or large, arising from elongate rootstocks; culms 5–8 dm. high, erect or somewhat spreading, generally exceeding the lvs.; blades flat or channeled, 1.75–2.5 mm. wide, glabrous or sparsely hispidulous; spikelets 3 or 4, the terminal ♂ or with a few ♀ fls., linear, 1.5–3.5 cm. long, long- or short-stalked, the lateral spikelets ♀, linear, ± flexuous, 1.5–5 cm. long, the upper approximate or all separate, at least the lower on slender stalks long-exserted from the long-sheathed bracts; ♀ scales ovate, finely ciliate, shorter and a little narrower than the perigynia; perigynia oblong-oblanceolate, 3.5 cm. long, glabrous, abruptly contracted into a short bidentulate beak ca. ⅙ the length of the perigynium-body.—Moist meadows and springy slopes, frequently in serpentine areas, 500–5300 ft.; Coastal Prairie and Mixed Evergreen F.; Coast Ranges from Marin Co. to Del Norte Co.; s. Ore.

99. **C. obispoénsis** Stacey. Densely cespitose, the lvs. and culms in large clumps; culms 6–18 dm. long, erect or spreading, longer than the lvs.; blades 3–8 mm. wide, channeled; spikelets 4–9 or sometimes even more, the terminal and usually 1 or 2 subsidiary ones ♂, linear, 2–5 cm. long, long-stalked, the lateral spikelets ♀, linear, 2.5–8 cm. long, flexuous or nodding, the uppermost approximate, the lower distant and on slender stalks long-exserted from sheathing leaflike bracts; ♀ scales ovate, shorter than the perigynia; perigynia lanceolate-ovate, 6–8 mm. long, appressed-puberulent, tapering above into a stout bidentulate or bidentate beak ⅓–½ as long as the body.—Restricted to the vicinity of brooks and springs on serpentine below 2000 ft.; Chaparral; San Luis Obispo Co.

100. **C. virídula** Michx. Plants densely cespitose in small clumps, the rootstock very short; culms 0.6–3 dm. tall, erect, shorter or longer than the lvs.; blades channeled, 1–3 mm. wide; spikelets 3–7, the terminal usually ♂, linear, 3–15 mm. long, the lateral spikelets ♀, oblong to roundish, 5–10 mm. long, the upper approximate, the lower separate or approximate, the upper subsessile, the lower with short stalks; ♀ scales obovate, narrower and much shorter than the perigynia; perigynia obovate, 2–3 mm. long, yellowish-green, few- to many-ribbed, abruptly beaked, the beak bidentulate, ca. ⅛ the length of the perigynium-body.—Low wet ground near the coast, near sea level; N. Coastal Coniferous F.; Mendocino Co. to Del Norte Co.; n. to Alaska, e. to the Atlantic and Greenland; Japan.

101. **C. Lemmònii** W. Boott. [*C. Abramsii* Mkze.] Plants in loose or dense clumps with short rootstocks; culms 2–8 dm. tall, slender and somewhat spreading, longer than the lvs.; blades flat, 1.5–4 mm. wide; spikelets 3–5, the terminal ♂, linear, 0.5–2.5 cm. long, exceeding the uppermost ♀ spikelet, the lateral spikelets ♀, linear-oblong, 0.5–2 cm. long, the upper approximate, the lower widely separate and on stalks long-exserted from the bract-sheaths; ♀ scales broadly ovate, a little narrower and shorter than the perigynia; perigynia narrowly ovate, 3–4 mm. long, 2-ribbed and finely nerved, narrowed into a bidentulate hyaline-tipped beak ⅛ the length of the perigynium-body.—Moist meadows and springy slopes in the mts., 2300–10,000 ft.; Yellow Pine F. to Subalpine F.; San Jacinto and San Bernardino mts., Sierra Nevada from Tulare Co. to Plumas Co., Mt. Lassen, N. Coast Ranges from Lake Co. to Siskiyou Co.

102. **C. álbida** Bailey. [*C. sonomensis* Stacey.] Plants loosely cespitose with short rootstocks; culms 4–6 dm. tall, erect, longer than the lvs.; blades flat, 3–5 (or 6) mm. wide; spikelets 4–7, the terminal spikelet generally ♂, sometimes androgynous or with middle fls. ♀ and uppermost and lowest ♂, oblanceolate to ovate or obovate, 8–

10 mm. long, little exceeding the uppermost lateral ♀ or androgynous spikelet, the lateral (♀) spikelets oblong, 5–20 mm. long, the uppermost approximate, the lowest 1 or 2 distant and on stalks short- or long-exserted from leaflike long-sheathing bracts; ♀ scales ovate, ca. as wide as the perigynia but shorter; perigynia lanceolate, 3–4.5 mm. long, strong-nerved on both faces, tapering into a bidentulate, hyaline-tipped beak, the beak ¼–⅓ as long as the body of the perigynium.—Open marshy places, below 300 ft.; Mixed Evergreen F.; Sonoma Co.

103. **C. ablàta** Bailey. Densely cespitose, the rootstocks short; culms 2.5–6 dm. tall, erect, much longer than the lvs.; blades flat, 3–4.5 mm. wide; spikelets 4–7, the terminal ♂ or with a few perigynia, linear, 1–2 cm. long, the lateral spikelets ♀ or a few uppermost flowers ♂, narrowly oblong, 1–3 cm. long, the uppermost close to the terminal spikelet and not much shorter, the lower widely separated and on stalks long-exserted from long-sheathing bracts; ♀ scales ovate, somewhat ciliate, as wide as the perigynia but much shorter; perigynia lanceolate, 3.5–5 mm. long, 2-ribbed and finely nerved, somewhat hispid-ciliate, gradually tapering into a bidentulate beak ¼–⅓ the length of the perigynium-body, the beak dark purplish at the tip.—Rare, wet meadows and about springs, 3800–7200 ft.; Yellow Pine and Red Fir forests; N. Coast Ranges in Humboldt, Trinity, and Siskiyou cos., Mt. Shasta, possibly also in the Sierra Nevada in Nevada Co.; n. to B.C., e. to Mont. and Wyo.

104. **C. luzulìna** Olney. Plants densely cespitose, the rootstocks short; culms 1.5–9 dm. tall, erect, much longer than the lvs.; blades flat, 3–8 mm. wide; spikelets 3–5, the terminal ♂ or with a few perigynia, linear, 1–1.5 cm. long, the lateral spikelets ♀, oblong, 0.7–2 cm. long, the uppermost close to the terminal spikelet and scarcely any shorter, the lower widely separated, on stalks long-exserted from long-sheathing leaflike bracts; ♀ scales ovate, somewhat ciliate, ca. as wide as the perigynia but a little shorter; perigynia lanceolate, 4–5 mm. long, 2-ribbed and obscurely few-nerved, somewhat hispid-ciliate, tapering into a short bidentulate beak, the beak ⅐–⅕ as long as the perigynium-body, the tip dark purplish.—Coastal bogs and mt. meadows, 100–7900 ft.; Coastal Prairie to Red Fir F.; Coast Ranges from Marin Co. to Siskiyou Co., Mt. Shasta; n. to Wash. and Ida.

105. **C. luzulaefòlia** W. Boott. [*C. l.* var. *strobilantha* Holm. *C. pseudo-japonica* Clarke.] Plants densely cespitose with short rootstocks; culms 4–10 dm. tall, erect, much longer than the lvs.; blades flat, 5–15 mm. wide, strongly nerved, leathery; spikelets 4–7, the terminal ♂ or with a few perigynia, oblong, 1–2 cm. long, the lateral spikelets ♀, oblong, 1–2.5 cm. long, the uppermost approximate, the lower distantly separated, on stalks long-exserted from long-sheathing bracts; ♀ scales ovate, as wide as the perigynia but shorter; perigynia ovate, 4.5–5 mm. long, much flattened, obscurely nerved, glabrous, contracted into a bidentate beak nearly ½ the length of the perigynium-body, the tip of the beak purplish-black.—Mt. meadows and rocky slopes, 7000–9300 ft.; Red Fir F.; Sierra Nevada from Tulare Co. to Sierra Co., n. to Shasta Co.

106. **C. fissurícola** Mkze. [*C. ablata* var. *luzuliformis* Bailey.] Densely cespitose, with very short rootstocks; culms 5–8 dm. long, much longer than the lvs.; blades flat, 3–8 mm. wide; spikelets 5 or 6, the terminal ♂ or with a few perigynia, oblongish, 1–1.5 cm. long, the lateral spikelets ♀, oblong, 1–3 cm. long, the upper close to the terminal spikelet, the lower distant and separate, on stalks ± exserted from long-sheathing bracts; ♀ scales narrowly ovate, narrower and shorter than the perigynia; perigynia ovate, 4.5–5 mm. long, much flattened, obscurely nerved, sparsely hispid-ciliate above, contracted to a bidentate beak, the beak ¼–⅓ length of perigynium.—Common, meadowy slopes and flats among rocks, 5000–11,150 ft.; Red Fir and Subalpine forests; Sierra Nevada from Tulare Co. to Nevada Co., Mt. Shasta; w. Nev., e. to Utah.

107. **C. Halliàna** Bailey. Plants open, scarcely cespitose, the lvs. and culms in small clumps from elongate rootstocks; culms 1–5 dm. long, equaling the lvs. or longer; blades 3–5 mm. long, flat above, channeled near the base; spikelets 5–7, the upper 2 or 3 ♂, linear-clavate, 1–2.5 cm. long, approximate, ♀ spikelets separate, subsessile above, stalked below, bracts leaflike, the lowest bract long-sheathing; ♀ scales ovate, usually somewhat smaller than the perigynia; perigynia ovate or obovate, 4–5 mm. long, hispid, many-ribbed, contracted into a bidentate beak, ¼–⅓ the length of perigynium.—Loose soil of dry meadows, Medicine Lake, Siskiyou Co., at ca. 6000 ft., the only known station in California; N. Juniper Wd.; n. to Wash.

108. **C. lanuginòsa** Michx. Loosely cespitose, the culms and lvs. from long-creeping

rootstocks; culms 3–10 dm. tall, frequently exceeded by the lvs.; blades 1.5–5 mm. wide, flat; spikelets usually 4–6, the upper 2 usually ♂, linear or linear-oblong, 2–6 cm. long, the lower spikelets ♀, distant, sessile or short-stalked, oblong, 1.5–5 cm. long, the bracts sheathless or nearly so, the lowest leaflike and exceeding the culm; ♀ scales lanceolate, longer or shorter than the perigynia; perigynia roundish-ovate or obovate, 2.5–3.5 mm. long, hairy, ribbed, contracted to a bidentate beak ¼–⅓ the length of the perigynium-body.—Moist or dry slopes and meadows, 200–10,750 ft.; Coastal Prairie and Yellow Pine F. to Subalpine F.; s. Calif. mts., Sierra Nevada from Tulare Co. to Sierra Co., White and Sweetwater mts., n. to Siskiyou and Modoc cos., Coast Ranges from Ventura Co. to Del Norte Co., garden weed in Santa Barbara; n. to B.C., e. to the Atlantic.

109. **C. Sartwelliàna** Olney. [*C. yosemitana* Bailey.] Plants densely cespitose, frequently forming large clumps, the culms and lvs. from stout rootstocks; culms 3–9 dm. long, much longer than the lvs.; blades flat, 3–7 mm. wide, softly pubescent; spikelets 4 or 5, the terminal ♂ or with a few perigynia, linear, 1–3 cm. long, the ♀ spikelets oblong to linear, 1–4 cm. long, approximate or separate, the bracts sheathless or nearly so, the lowest leaflike and about equaling the infl.; ♀ scales ovate or ovate-lanceolate, hairy, longer or shorter than the perigynia; perigynia oblong-obovate to obovate, 2.5–3.5 mm. long, pilose, 2-ribbed and obscurely nerved, contracted into a bidentulate beak, the beak ⅓–½ the length of the perigynium-body, hyaline-tipped.—Meadows and open slopes in coniferous forests, 4000–8600 ft.; Yellow Pine and Red Fir forests; San Jacinto Mts., Sierra Nevada from Kern Co. to Tuolumne Co.

110. **C. Congdònii** Bailey. Densely cespitose, the culms and lvs. frequently in large clumps from stout rootstocks; culms 4–9 dm. long, usually longer than the lvs.; blades flat or channeled, 3–8 mm. wide, hairy or glabrate; spikelets 4–6, the terminal one or two ♂ or with a few perigynia, linear or wider, 1.5–3.5 cm. long, the ♀ spikelets linear-oblong or oblong, 1.5–5 cm. long, approximate or separate, the bracts sheathless or nearly so, the lowest leaflike, shorter or longer than the infl.; ♀ scales ovate, hairy, narrower than the perigynia; perigynia ovate-lanceolate to oblanceolate, 3.5–4 mm. long, pilose, gradually or rather abruptly tapering into the beak, the hyaline bidentulate beak only ⅕–⅓ the length of the perigynium-body.—Rocky slopes and taluses, 8500–12,800 ft.; Subalpine F. and Alpine Fell-fields; Sierra Nevada from Tulare Co. to Tuolumne Co.

111. **C. amplifòlia** Boott. Plants cespitose, the culms and lvs. in large clumps from stout long rootstocks; culms 5–10 dm. tall, frequently exceeded by the lvs. and leaflike bracts; blades flat, 8–18 mm. wide; spikelets 4–7, the terminal ♂, linear, 4–9 mm. long, ♀ spikelets linear to oblong, 3.5–14 cm. long, approximate or separate, the bracts scarcely sheathing; ♀ scales lanceolate and awned below to ovate and acute in the upper part of the spikelets, shorter or longer than the perigynia; perigynia ovate, 3 mm. long, 2-ribbed, nerveless, glabrous, contracted into a bidentulate beak, the beak ca. as long as the body, hyaline-tipped.—Wet soil along streams and in meadows, sea level to 8000 ft.; Coastal Prairie and Mixed Evergreen F. to Red Fir F.; Sierra Nevada from Tulare Co. to Plumas Co., Coast Ranges from San Mateo Co. to Del Norte Co., Modoc Co.; n. to B.C., e. to Ida.

112. **C. limòsa** L. Scarcely cespitose, the culms and lvs. in small clumps on elongate rootstocks; culms 2–6 dm. long, longer than the lvs.; blades channeled, 1–3 mm. wide, frequently glaucous; spikelets 2–4, the terminal ♂, linear, 1–3 cm. long, erect or drooping, the lateral spikelets ♀ (or with a few ♂ fls. at the top), oblong, 1–2.5 cm. long, drooping on slender stalks, the lowest bract leaflike, nearly or quite sheathless; ♀ scales ovate to nearly round, ca. the size of the perigynia or a little longer; perigynia broadly ovate, 2.5–4 mm. long, prominently nerved, glaucous-green, papillate, obtuse, minutely beaked, the orifice entire or emarginate.—Rare, in swamps, usually associated with *Sphagnum*, 4000–8700 ft.; Yellow Pine and Red Fir forests; Sierra Nevada from Fresno Co. to Plumas and Tehama cos.; n. to Alaska, e. to the Atlantic; Eu.; Asia.

113. **C. spectábilis** Dewey. [*C. invisa* Bailey.] Plants loosely cespitose, the culms and lvs. in small or large clumps from short rootstocks; culms 2.5–9 dm. tall, usually much longer than the lvs.; blades flat, 2–5 mm. wide; spikelets 3–7, the terminal ♂ (or rarely the upper two ♂), linear-oblong, 8–20 mm. long, the ♀ linear-oblong to oblong, 1–3 cm. long, mostly separate, subsessile or shortly peduncled, the lowest bract leaflike; ♀ scales oblong-ovate, the whitish midvein extending beyond the tip as a small point,

usually a little narrower and shorter than the perigynia; perigynia elliptic-oblong or ovate-oblong, 4–5 mm. long, 2-ribbed, nearly nerveless, contracted into an emarginate or bidentulate beak $\frac{1}{10}$–$\frac{1}{7}$ the length of the perigynium-body.—Wet soil of meadows or rocky slopes where snow banks linger, 5800–12,000 ft.; Red Fir F. to Alpine Fell-fields; Sierra Nevada from Tulare Co. to Plumas Co., Mt. Lassen, Mt. Shasta, N. Coast Ranges in Trinity and Siskiyou cos.; n. to Yukon, e. to Mont. and Alta.

114. **C. Raynóldsii** Dewey. Plants loosely cespitose, with short-creeping rootstocks; culms 2–7.5 dm. tall, mostly longer than the lvs.; blades flat, 3–8 mm. wide; spikelets 3–5, the terminal ♂, linear, 1–2 cm. long, the ♀ spikelets oblong, 1–2 cm. long, approximate or the lowest separate, sessile or stalked, the lowest bract leaflike; ♀ scales broadly ovate, shorter than the perigynia; perigynia oblong to oblong-obovate, 3.5–4.5 mm. long, ± inflated, 2-ribbed and several-nerved, contracted into an entire or emarginate beak, the beak $\frac{1}{8}$–$\frac{1}{6}$ the length of the perigynium-body.—Meadows and flats in rather dry soil, 6000–10,300 ft.; Red Fir and Subalpine forests; Sierra Nevada from Tulare Co. to Plumas Co., n. to Mt. Lassen and Siskiyou and Modoc cos.; n. to B.C., e. to the Rocky Mts.

115. **C. Hélleri** Mkze. Densely cespitose, the smaller or larger clumps of culms and lvs. from very short rootstocks; culms 0.5–3 dm. tall, longer than the lvs.; blades flat, 2–3.5 mm. wide; spikelets 3–5, closely aggregated, the terminal gynaecandrous, the lateral ♀, oblong, 1–2 cm. long, sessile or nearly so, the lowest bract leaflike; ♀ scales lanceolate-ovate or narrower, acuminate, longer and narrower than the perigynia; perigynia oval to obovate or roundish, 2.5–3.5 mm. long, 2-ribbed, nerveless, smooth, contracted into a very short bidentulate beak $\frac{1}{12}$–$\frac{1}{8}$ the length of the perigynium-body. —Rocky and gravelly slopes, 8500–13,600 ft.; Subalpine F. and Alpine Fell-fields; Sierra Nevada from Tulare Co. to Eldorado Co., Mt. Lassen, White and Sweetwater mts.; e. to Elko Co., Nev.

116. **C. epapillòsa** Mkze. Densely cespitose, the culms and lvs. usually forming large clumps, rootstocks short; culms 1.5–6 dm. tall, longer than the lvs.; blades flat, 3.5–7 mm. wide; spikelets 3–6, approximate, the terminal gynaecandrous, the lateral ♀, oblong, 1–2.5 cm. long, the lower ± strongly stalked, the lowest bract leaflike; ♀ scales lanceolate to ovate, ca. as long as the perigynia but narrower; perigynia oval to roundish, 3–4.5 mm. long, strongly flattened, nerveless or nearly so, smooth, contracted into a short bidentate beak $\frac{1}{7}$–$\frac{1}{5}$ the length of the perigynium-body.—Meadows and rocky slopes, 6000–12,500 ft.; Red Fir F. to Alpine Fell-fields; Sierra Nevada from Inyo Co. to Eldorado Co., Medicine Lake in Siskiyou Co., Sweetwater and Warner mts.; n. to Wash., e. to the Rocky Mts.

117. **C. heteroneùra** W. Boott. [*C. quadrifida* Bailey. *C. q.* var. *lenis* Bailey. *C. q.* var. *caeca* Bailey.] Densely or loosely cespitose, the rootstocks usually short; culms 2.5–10 dm. tall, longer than the lvs.; blades flat, 2–4 mm. wide; spikelets approximate, 3–6, the terminal gynaecandrous or rarely ♂, the lateral ♀, oblong or somewhat narrower, 0.7–2.5 cm. long, sessile or short-stalked, the lowest bract leaflike; ♀ scales lanceolate-ovate or wider, narrower and shorter than the perigynia or ca. as long as the perigynia; perigynia oval or obovate to nearly round, 2.5–3.5 mm. long, strongly flattened, nerveless or nearly so, smooth, contracted into a very short bidentulate beak ca. $\frac{1}{10}$ the length of the perigynium-body.—Meadows, forest opens, and rocky slopes, 6000–11,500 ft.; Yellow Pine F. to Alpine Fell-fields; San Jacinto and San Bernardino mts., Sierra Nevada from Tulare Co. to Plumas Co., Mt. Lassen, N. Coast Ranges in Glenn and Trinity cos., Medicine Lake in Siskiyou Co., White, Sweetwater, and Warner mts.; e. to Nev. and Wyo.

118. **C. albonìgra** Mkze. Loosely cespitose, the culms and lvs. usually arising in small clumps from short-creeping rootstocks; culms 1–3 dm. tall, longer than the lvs.; blades flat, 2.5–5 mm. wide; spikelets usually 3, approximate or the lowest a little separate, the terminal gynaecandrous, the lateral ♀, oblong, 8–10 mm. long, sessile or short-stalked, the lowest bract leaflike; ♀ scales broadly ovate, ca. as long and as wide as the perigynia; perigynia broadly ovate to obovate, 3–3.5 mm. long, flattened, nerveless, finely papillate-granular, contracted above into a bidentulate beak ca. $\frac{1}{6}$–$\frac{1}{5}$ the length of the perigynium-body.—Rocky meadows and slopes in rather dry soil, 10,000–12,880 ft.; Subalpine F. and Alpine Fell-fields; Sierra Nevada from Tulare Co. to Mt. Dana, Mono Co.; n. to Wash., e. to Alta. and Colo.

119. **C. Merténsii** Prescott. Plants densely cespitose with short rootstocks; culms 3–10

dm. tall, usually much longer than the lvs.; blades flaccid, flat, 4–7 mm. wide; spikelets 5–10, approximate, ± drooping on slender stalks, the terminal gynaecandrous or nearly entirely ♂, the lateral ♀ or with some basal ♂ fls., oblong or somewhat narrower, 1–4 cm. long, the lower bracts leaflike; ♀ scales ovate-lanceolate, narrower and shorter than the perigynia; perigynia ovate or obovate to roundish, 4.5–5 mm. long, flattened, nearly smooth, finely few-nerved, tapering above into a short entire or emarginate beak 0.25–0.5 mm. long.—Rare, open rocky slopes or forests, 5000 ft.; Red Fir F.; N. Coast Ranges in Trinity and Siskiyou cos.; n. to Yukon, e. to Mont. and Ida.

120. C. serrátodens W. Boott. [C. bifida Boott, not Roth. C. aequa Clarke.] Plants loosely cespitose with short rootstocks; culms 3–13 dm. tall, longer than the lvs.; blades flat, 1.5–4 mm. wide, glaucous-green; spikelets 3–6, the terminal ♂ or with some perigynia, linear-oblong or narrower, 1.5–3 cm. long, the ♀ spikelets oblong or broader, 6–18 mm. long, ± separate, the lowest rather distant, sessile or short-stalked, the lowest bract leaflike; ♀ scales ovate, a little shorter and narrower than the perigynia; perigynia oblong-ovate or ovate, 3–5 mm. long, ca. 10-nerved, puncticulate, somewhat contracted into a roughened bidentate beak 0.5–1 mm. long, the teeth hispidulous.— Moist meadows and rocky places near streams and seepages, frequently on serpentine, 100–6000 ft.; Coastal Prairie and Foothill Wd. to Mixed Evergreen F. and Yellow Pine F.; Tehachapi Mts. and Ventura Co., n. in Coast Ranges to Del Norte Co., rare in the Sierra Nevada from Tuolumne Co. to Butte Co.; s. Ore.

121. C. Buxbàumii Wahl. Plants loosely cespitose with long-creeping rootstocks; culms 2.5–10 dm. tall, much longer than the lvs.; blades flat or channeled at the base, 1.5–4 mm. wide, glaucous-green; spikelets 2–5, approximate or the lower separate, the terminal gynaecandrous, the lateral ♀, oblong-ovate to ovate, 0.5–2 cm. long, sessile or nearly so, lowest bract shorter than the infl. or equaling it; ♀ scales lanceolate, narrower than the perigynia but usually longer; perigynia elliptic or obovate, 2.5–4 mm. long, densely papillose, finely many-nerved, contracted into a very short bidentulate beak only 0.2 mm. long.—Bogs and wet meadows, 20–10,700 ft.; Coastal Prairie and Yellow Pine F. to Subalpine F.; Sierra Nevada from Inyo Co. to Tuolumne Co., Coast Ranges from Marin Co. to Humboldt Co.; n. to Alaska, e. to the Atlantic; Eu.; Asia.

122. C. scopulòrum Holm. Loosely cespitose, the culms and lvs. arising from elongate rootstocks; culms 1–4 dm. tall, longer than the lvs.; blades flat, 3–7 mm. wide; spikelets 3–7, the terminal ♂ or with a few ♀ fls., linear, 1–2.5 cm. long, the lateral spikelets ♀ or the uppermost fls. ♂, aggregated or the lowest somewhat separate, oblong, 1–2.5 cm. long, the lowest bract much shorter than the infl.; ♀ scales obovate, narrower and shorter than the perigynia, or nearly as long; perigynia broadly obovate or roundish, 2.5–3.5 mm. long, nerveless, papillose, the beak entire, 0.2–0.5 mm. long.—Meadows and marshy banks and strands, 6900–10,600 ft.; Red Fir and Subalpine forests; Sierra Nevada from Tulare Co. to Eldorado Co., Mt. Lassen, Sweetwater and Warner mts.; n. to Wash., e. to the Rocky Mts.

123. C. gymnoclàda Holm. [C. vulgaris var. bracteosa Bailey.] Loosely cespitose, the rootstocks elongate; culms 2–6 dm. tall, longer than the lvs.; blades flat, 2.5–5 mm. wide; spikelets 2–4, the terminal ♂, linear-clavate, 1–3 cm. long, the ♀ spikelets approximate or a little separate, often with ♂ fls. at the apex, oblong, 0.5–2.5 cm. long, the lowest bracts shorter than the infl.; ♀ scales ovate, longer or shorter than the perigynia; perigynia broadly obovate or nearly round, 2.25–3.5 mm. long, nerveless, granular, the beak entire, 0.1–0.25 mm. long.—Common, wet meadows and swampy places, 5200–11,200 ft.; Yellow Pine F. to Subalpine F.; Sierra Nevada from Tulare Co. to Plumas Co., Coast Ranges from Glenn Co. to Siskiyou Co., Sweetwater Mts., Mt. Lassen, Mt. Shasta; n. to B.C., e. to Colo.

124. C. paucicostàta Mkze. [C. interrupta var. impressa Bailey.] Cespitose, forming dense clumps, the culms and lvs. arising from very short rootstocks; culms 2.5–5 dm. tall, shorter or longer than the lvs.; blades flat or somewhat channeled toward the base, 2–4 mm. wide; spikelets 5–7, the terminal ♂ or with a few perigynia, linear, 2–3 cm. long, the lateral spikelets ♀, separate or the uppermost approximate, linear, 1–4 cm. long, the lowest bract leaflike, exceeding the infl.; ♀ scales oblong or oblong-lanceolate, a little shorter and usually much narrower than the perigynia; perigynia broadly ovate or obovate to roundish, 2 mm. long, granular and resinous, coarsely nerved dorsally and ventrally, short-stipitate at base, above abruptly contracted into a subentire beak

0.1–0.25 mm. long.—Wet meadows and shallows of lakes and streams, 4000–11,500 ft.; Yellow Pine F. to Subalpine F.; Sierra Nevada from Tulare Co. to Plumas Co., Coast Ranges from Tehama Co. to Siskiyou Co., Mt. Lassen; n. to Ore.

125. **C. Kellóggii** W. Boott. Plants cespitose in dense clumps, the rootstocks short or elongate; culms 1–6 dm. tall, shorter or rarely longer than the lvs.; blades flat above and channeled near the base, 1.5–2.5 mm. wide; spikelets 4–6, the terminal ♂ or with a few perigynia, linear, 1–4 cm. long, the lateral spikelets ♀, approximate or a little separate, linear, 1.5–3.5 cm. long, the lowest bract leaflike, longer than the infl.; ♀ scales oblong-ovate, narrower than the perigynia, shorter or ca. as long; perigynia ovate, 1.5–3 mm. long, granular, nerved dorsally and ventrally, slender-stipitate at base, abruptly beaked, the beak 0.1–0.25 mm. long, entire.—Common in shallows of streams and lakes and in wet meadows, 1000–11,200 ft.; Foothill Wd. to Subalpine F.; Sierra Nevada from Kern Co. to Plumas Co., Mt. Lassen, Mt. Shasta, N. Coast Ranges from Glenn Co. to Siskiyou Co.; n. to Alaska, e. to the Rocky Mts.

126. **C. Hìndsii** Clarke. Plants cespitose in medium-sized or large clumps with short to long rootstocks; culms 1–5 dm. tall, shorter or longer than the lvs.; blades flat near the tip, channeled near the base, 1.5–3 mm. wide; spikelets 4–7, the terminal ♂, linear, 1.5–3.5 cm. long, the lateral spikelets ♀, separate or approximate above, linear, 1–4.5 cm. long, the lowest bract leaflike, longer than the infl.; ♀ scales oblong-oblanceolate or oblong-lanceolate, much shorter and narrower than the perigynia; perigynia ovate, 2–3.5 mm. long, nerved dorsally and ventrally, papillate, slender-stipitate at base, tapering or contracted into a subentire beak 0.1–0.25 mm. long.—Coastal swamps and bogs near sea level; N. Coastal Coniferous F.; Mendocino Co. to Del Norte Co.; n. to Alaska.

127. **C. nebrascénsis** Dewey. [*C. jacintoensis* Parish.] Plants cespitose, the clustered lvs. and culms from elongate rootstocks; culms 2.5–12 dm. long, shorter or longer than the lvs.; blades flat or somewhat channeled below, 3–8 mm. wide, frequently ± glaucous; ♂ spikelets 1 or 2, broadly linear, 1.5–4 cm. long, the ♀ spikelets 2–5, approximate or separate, oblongish, 1.5–6 cm. long, the lowest bract leaflike, longer or shorter than the infl.; ♀ scales lanceolate, longer or shorter than the perigynia and narrower; perigynia oblong-obovate, 3–3.5 mm. long, granular and coriaceous, many-ribbed, sessile or subsessile at base, abruptly beaked, the beak 0.5–1 mm. long, bidentate, the teeth somewhat ciliate within.—Wet or dry meadows or swamps, near sea level to 10,500 ft.; Creosote Bush Scrub and Yellow Pine F. to Subalpine F.; mts., meadows, and oases from San Diego Co. to Modoc Co., local in the N. Coast Ranges, garden weed in Pasadena; n. to B.C., e. to Kans.

128. **C. sitchénsis** Prescott. Plants cespitose with short rootstocks; culms 2.5–12 dm. tall, equaling or exceeding the lvs.; blades flat or somewhat channeled near the base, 3–9 mm. wide, light- or glaucous-green; ♂ spikelets 1–4, linear, 2–8 cm. long, ♀ spikelets 3–5, the upper often androgynous, separate, drooping to erect on long stalks, linear, 2–9 cm. long, the lowest bract leaflike, usually longer than the infl.; ♀ scales lanceolate to ovate, narrower than the perigynia and either longer or shorter; perigynia narrowly to broadly ovate, 2.5–3.5 mm. long, nearly or quite nerveless, abruptly beaked, the beak 0.2–0.5 mm. long, entire.—Rare, swampy places, usually near the coast, sea level to 3500 ft.; Mixed Evergreen F. to Yellow Pine F.; Coast Ranges from Santa Cruz Co. to Del Norte Co., Butte and Siskiyou cos.; n. to Alaska, e. to Ida.

129. **C. aquátilis** Wahl. Plants cespitose, the lvs. and culms in small or large clumps from elongate rootstocks; culms 2–8 dm. tall, exceeding the lvs.; blades flat or channeled at the base, 2.5–5 mm. wide; ♂ spikelets 1 or 2, linear, 1–2.5 cm. long, ♀ spikelets 2–4, the upper sometimes androgynous, separate below, approximate above, linear or oblong, 1–4 cm. long, the lowest bract leaflike, usually as long as the infl. or longer; ♀ scales oblong-ovate to ovate, narrower than the perigynia and shorter or longer; perigynia oval to obovate, 2.5–3 mm. long, nerveless or nearly so, abruptly beaked, the beak 0.1–0.3 mm. long, entire.—Wet places and marshy meadows, 5000–10,500 ft.; Yellow Pine F. to Subalpine F.; Sierra Nevada from Fresno and Inyo cos. to Sierra Co., Coast Ranges in Glenn Co., Warner Mts.; n. to Alaska, e. to the Atlantic; Eu.; Asia.

130. **C. bárbarae** Dewey. [*C. Wilkesii* Olney. *C. lacunarum* Holm.] Plants cespitose, with long stout rhizomes; culms 3–10 dm. tall, usually longer than the lvs.; blades flat above, channeled near the base, 3.5–9 mm. wide, the lf.-sheaths breaking and becoming

filamentose; ♂ spikelets 1 or 2, linear, up to 6 cm. long, ♀ spikelets 2–5, the upper often androgynous, usually separate, linear to oblong, 2.5–8 cm. long, the lowest bract leaflike, shorter or longer than the infl.; ♀ scales ovate-lanceolate to ovate, narrower than the perigynia and from shorter to longer; perigynia oblong-obovate to roundish, 3–4.5 mm. long, weakly to strongly nerved both dorsally and ventrally, abruptly beaked, the beak 0.5 mm. long, bidentate, the teeth hispidulous.—Open or brushy slopes and valley flats that are wet in the spring, sea level to 3000 ft.; V. Grassland, Foothill Wd. and Coastal Prairie to Mixed Evergreen F. and Yellow Pine F.; coastal s. Calif. n. through the Coast Ranges and Great V., occasional in Sierra foothills; s. Ore.

131. **C. Schóttii** Dewey. Plants cespitose in large clumps, producing long yellowish-felted stolons; culms 1–1.5 m. tall, longer than the lvs.; blades flat, 6–12 mm. wide, the lf.-sheaths breaking and becoming filamentose; ♂ spikelets frequently 3, linear, 8–14 cm. long, ♀ spikelets 3, usually androgynous, separate, linear, 5–20 cm. long, the lowest bract leaflike, usually, but not always, longer than the infl.; ♀ scales linear-lanceolate to oblong, narrower than the perigynia but usually longer; perigynia oval or obovate, 3–3.5 mm. long, few-nerved both dorsally and ventrally, shortly beaked, the beak 0.25 mm. long, ± emarginate.—Stream banks in foothills and lower mts., near sea level to 2500 ft.; Coastal Sage Scrub and S. Oak Wd.; San Diego Co. n. to Santa Clara Co.

132. **C. sénta** Boott. [*C. auriculata* Bailey. *C. austromontana* Parish. *C. Bishallii* Clarke. *C. nudata* f. *sessiliflora* Kük. *C. Bolanderi* Gand., not Olney.] Plants cespitose in large clumps, producing long stout stolons; culms 3–10 dm. tall, longer than the lvs.; blades flat, 3–5 mm. wide, the lf.-sheaths breaking and becoming filamentose; ♂ spikelets 2 or 3, linear, 3–4.5 cm. long, often with a few perigynia near the base, the ♀ spikelets 1 or 2, rarely androgynous, linear to oblong, 2.5–5 cm. long, the lowest bract leaflike, shorter or longer than the infl.; ♀ scales oblong-lanceolate to oblong-ovate, shorter and narrower than the perigynia; perigynia broadly ovate or obovate, 3–3.5 mm. long, granular, few-nerved on both faces, abruptly beaked, the beak 0.25 mm. long, entire.—Along streams and in swampy places and meadows, 200–9500 ft.; Coastal Sage Scrub and Foothill Wd. to Yellow Pine F. and Red Fir F.; s. Calif. lowlands and mts. n. through the Coast Ranges to Solano Co. and the Sierra Nevada to Butte Co.; s. to L. Calif., e. to Ariz.

133. **C. nudàta** W. Boott. Plants densely cespitose with oblique rootstocks; culms 3–8 dm. long, slender, erect or curving outward, longer than the lvs.; blades 1.75–3.5 mm. wide, flat, channeled towards the base, the lf.-sheaths breaking and becoming filamentose; spikelets 3 or 4, the terminal ♂, linear, 1.5–3.5 cm. long, the ♀ spikelets approximate, linear, 1–4 cm. long, often androgynous, the lowest bract only occasionally leaflike, much shorter than the culms; ♀ scales oblong to oblong-ovate, narrower than the perigynia but ca. as long; perigynia oblong-obovate to obovate, 2.5–4 mm. long, smooth or slightly granular above, nerved dorsally and ventrally, abruptly beaked, the beak 0.2 mm. long, entire.—Common, rocky stream beds, 150–5000 ft.; Mixed Evergreen F. and Foothill Wd. to Yellow Pine F.; Coast Ranges from Santa Barbara Co. to Del Norte Co., Sierra Nevada from Mariposa Co. to Plumas Co.; n. to Ore.

134. **C. eurycárpa** Holm. Plants ± cespitose with short-creeping rootstocks and stolons; culms 4–9 dm. tall, longer than the lvs.; blades flat, 2–3 mm. wide, the lf.-sheaths breaking and becoming filamentose; ♂ spikelets 2 or 3, the terminal largest, linear, 3–4 cm. long, the ♀ spikelets often separate, usually androgynous, linear, 2.5–4.5 cm. long, lowest bract leaflike and longer than the infl.; ♀ scales lanceolate, narrower than the perigynia and shorter to longer; perigynia obovate to roundish, 3 mm. long, finely ribbed dorsally and ventrally, granular-roughened, abruptly beaked, the beak 0.2–0.5 mm. long, entire or emarginate.—Meadows and marshy places, 2000–7500 ft.; Foothill Wd. to Yellow Pine F.; Sierra Nevada from Tulare Co. to Plumas Co., Mt. Lassen, Mt. Shasta, N. Coast Ranges in Siskiyou Co.; n. to Wash. and Ida.

135. **C. Lýngbyei** Hornem. Plants strongly stoloniferous, forming large beds with the culms and lvs. rather openly spaced; culms 2–9 dm. tall, longer than the lvs.; blades flat, 2–12 mm. wide; ♂ spikelets 2 or 3, linear, 1.5–4 cm. long, ♀ spikelets 2–4, usually androgynous, somewhat separate, pendulous on slender stalks, mostly linear-oblong, 1–8 cm. long, the lowest bract leaflike and generally longer than the infl.; ♀ scales ovate-lanceolate to ovate, from narrower to wider and shorter to longer than the perigynia; perigynia oblong-ovate or oblong-obovate to roundish, 2.5–3.5 mm. long,

slightly to strongly nerved, somewhat granular, abruptly beaked, the beak 0.2 mm. long, entire; aks. constricted in the middle.—Rare, brackish soil and tidal flats at sea level; Coastal Salt Marsh; Marin Co. to Humboldt Co.; n. to Alaska, e. to the Atlantic, Greenland, Iceland; Eurasia.

136. **C. obnúpta** Bailey. Plants long-stoloniferous and forming beds, the culms and lvs. generally cespitose in large clumps; culms 3–15 dm. long, shorter or longer than the lvs.; blades channeled, 2–5 mm. wide; ♂ spikelets 1–3, linear, 1.5–6 cm. long, the ♀ spikelets 2–4, usually androgynous, erect, spreading, or cernuous, sessile or long-stalked, oblong to linear, 3–14 cm. long, the lowest bract leaflike, usually longer than the infl.; ♀ scales narrowly ovate, longer than the perigynia and either narrower or wider; perigynia elliptic-ovate, 2.5–3.5 mm. long, smooth and shining, abruptly beaked, the beak 0.3 mm. long, entire or subentire; aks. constricted in the middle.—Common; dunes, flats, marshes in fresh or saline soils, generally near the coast, sea level to 2000 ft.; Coastal Strand, Coastal Prairie, Closed-cone Pine F., and N. Coastal Coniferous F.; San Luis Obispo Co. to Del Norte Co.; n. to s. Alaska.

137. **C. spíssa** Bailey. Plants large, loosely cespitose, with stout rootstocks; culms 1–2 m. tall, longer than the lvs.; blades flat above, 7–14 mm. wide; ♂ spikelets 3 or 4, linear, 4–10 cm. long, the ♀ spikelets 3–7, androgynous, approximate or the lower somewhat separate, linear, 6–14 cm. long, the bracts leaflike, the lowest longer than the infl.; ♀ scales lanceolate to ovate-lanceolate, narrower than the perigynia but longer; perigynia broadly obovate, 3–4.5 mm. long, smooth and only obscurely nerved, abruptly beaked, the beak 0.5 mm. long, frequently bent, emarginate; style indurate, persistent, continuous with the ak.—Watercourses in hills and canyons, near sea level to 2000 ft.; Coastal Sage Scrub, Chaparral, and S. Oak Wd.; San Luis Obispo Co. s. to San Diego Co.; s. to L. Calif., possibly e. to Ariz.

138. **C. hystricìna** Muhl. Plants cespitose with short rootstocks and few elongate stolons; culms 1.5–10 dm. tall, shorter or longer than the lvs.; blades flat, 2–10 mm. wide; spikelets 2–5, the terminal ♂, linear, 1–5 cm. long, ♀ spikelets approximate or separate, the lower nodding on long stalks, oblong, 1–6 cm. long, the bracts leaflike; ♀ scales strongly awned from a small obovate or oblanceolate base, the base narrower and shorter than the perigynia; perigynia narrowly ovate, 5–7 mm. long, suborbicular and inflated, shining, many-ribbed, tapering into a slender beak about 2 mm. long, the beak bidentate with slender erect teeth 0.5 mm. long; ak. continuous with the indurate persistent style.—In Calif. known only from Rush Creek, Trinity Co.; n. to Wash. and Alta., e. to the Atlantic.

139. **C. comòsa** Boott. Plants densely cespitose, forming large clumps with short rootstocks; culms 5–15 dm. tall, usually shorter than the lvs.; blades flat, 6–16 mm. wide; spikelets 4–7, the terminal ♂, linear, 3–7 cm. long, the ♀ approximate or a little separate, the lower stalked and ± nodding, oblong, 1.5–7.5 cm. long, the lower bracts leaflike and longer than the infl.; ♀ scales long-awned from a small lanceolate to ovate base, the base much shorter and narrower than the perigynia; perigynia lanceolate, 5–7 mm. long, shining, strongly ribbed, tapering into a slender beak, 1.5–2 mm. long, the beak bidentate with spreading awnlike teeth 1–2 mm. long; ak. continuous with the indurate persistent style.—Swamps and marshes, sea level to 1400 ft.; V. Grassland, Coastal Prairie, Mixed Evergreen F.; San Bernardino V., cent. Coast Ranges from Santa Cruz Co. to Lake Co., San Joaquin and Shasta cos.; n. to Wash., e. to the Atlantic.

140. **C. atheròdes** Spreng. Plants with long slender stolons, loosely cespitose; culms 3–15 dm. tall, nearly as long as the lvs. or longer, smooth; blades flat, glabrous above, sometimes sparsely hairy below, 3–12 mm. wide; ♂ spikelets 2–6, rarely with a few perigynia, linear, 4–10 cm. long, ♀ spikelets 2–4, erect, separate, narrowly cylindric, 5–12 cm. long, the lowest bracts leaflike and longer than the infl.; ♀ scales ovate, abruptly narrowed to an awnlike tip, narrower than the perigynia; perigynia lanceolate or a little broader, 7–10 mm. long, subinflated, glabrous, strongly many-ribbed, tapering into a smooth beak with teeth 1.2–3 mm. long; ak. continuous in texture with the persistent style.—Low wet ground and ditches, 4300–4500 ft.; N. Juniper Wd.; in Calif. known only from the Pit River V., Modoc Co.; n. to Yukon, e. to N.Y.; Eu.; Asia.

141. **C. Sheldònii** Mkze. Plants loosely cespitose, with long stout stolons; culms 5–10 dm. tall, glabrous, smooth, longer than the lvs.; blades flat, ± pubescent, 3.5–6 mm. wide; ♂ spikelets 2 or 3, linear, 2–3.5 cm. long, ♀ spikelets 2 or 3, erect, separate,

oblong, 2–5 cm. long, lowest bract leaflike, longer than the infl.; ♀ scales lanceolate-ovate, narrower and shorter than the perigynia; perigynia lanceolate, 5–6 mm. long, short-pubescent, strongly ribbed, tapering into a bidentate beak 2 mm. long, the teeth nearly 1 mm. long, ± spreading; ak. continuous with the indurate persistent style.— Rare, wet meadows and marshes, 4000–5000 ft.; Yellow Pine F.; Placer and Modoc cos.; n. to Wash., e. to Utah.

142. **C. vesicària** L. [*C. v.* var. *obtusisquamis* Bailey. *C. monile* var. *pacifica* Bailey.] Plants cespitose with short-creeping rootstocks; culms 3–10 dm. long, arising oblique to the ground, usually longer than the lvs.; blades flat, 2–7 mm. wide; ♂ spikelets 2–4, linear, 2–4 cm. long, ♀ spikelets 1–3, generally separate, erect, oblongish, 2.5–7.5 cm. long, perigynia in rather few rows, ascending even in fr., lowest bract leaflike and exceeding the infl.; ♀ scales lanceolate to ovate, narrower and shorter than the perigynia or ca. as long; perigynia ovate, 4–8 mm. long, inflated, shining, strongly nerved, narrowed ± gradually into a slender beak 2 mm. long, the beak bidentate with slender erect teeth 0.5–1 mm. long; ak. continuous with the indurate persistent style.—Common; marshes, wet meadows, and strands, near sea level to 10,800 ft.; Mixed Evergreen F. and Yellow Pine F. to Subalpine F.; Sierra Nevada from Tulare Co. to Plumas Co., Mt. Lassen, Modoc Co., N. Coast Ranges; n. to B.C., e. to the Atlantic; Eurasia.

143. **C. exsiccàta** Bailey. Plants loosely cespitose, the rootstocks short-creeping; culms 3–10 dm. long, arising oblique to the ground, shorter than the upper lvs.; blades flat, 3–7 mm. wide; ♂ spikelets 2 or 3, linear, 2–4.5 cm. long, ♀ spikelets 2 or 3, usually separate, erect, oblong, 2–7.5 cm. long, perigynia in rather few rows, ascending even in fr., the bracts leaflike, exceeding the infl.; ♀ scales lanceolate, narrower and shorter than the perigynia; perigynia lanceolate, 7–10 mm. long, somewhat inflated, shining, strongly ribbed, tapering into a slender beak 2–3 mm. long, the beak bidentate with erect teeth 0.75–1.5 mm. long; ak. continuous with the indurate persistent style.— Marshes and strands, 100–6000 ft.; Coastal Prairie and Mixed Evergreen F. to Red Fir F.; Coast Ranges from Santa Clara Co. to Del Norte Co.; n. to Alaska, e. to Mont.

144. **C. rostràta** Stokes. [*C. utriculata* var. *globosa* Olney.] Plants loosely or scarcely cespitose, the lvs. and culms arising from short rootstocks and elongate stolons; culms strictly erect, 3–12 dm. tall, shorter than the uppermost lvs.; blades flat or ± channeled at base, 2–12 mm. wide; ♂ spikelets 2–4, linear, 1–6 cm. long, ♀ spikelets 2–5, separate, erect, oblong, 1–15 cm. long, perigynia in many rows, ascending at first, spreading-squarrose at maturity, the bracts leaflike, longer than the infl.; ♀ scales lanceolate to ovate, narrower than the perigynia and longer or shorter; perigynia elliptic-ovate to ovate, 3.5–8 mm. long, ± inflated, shining, strongly nerved, gradually or abruptly contracted into the bidentate beak 1–2 mm. long, the teeth erect, 0.5–0.75 mm. long; aks. continuous with the indurate persistent style.—Common; marshes, wet meadows, shallow ponds, and strands, sea level to 11,000 ft.; Coastal Prairie and Yellow Pine F. to Subalpine F.; San Bernardino Mts., Sierra Nevada from Tulare Co. to Plumas Co., local in Coast Ranges from Santa Cruz Co. to Siskiyou Co., Mt. Lassen; n. to Alaska, e. to the Atlantic, Greenland; Eurasia.

26. Gramíneae. GRASS FAMILY (Fig. 134)

Herbs, rarely shrubs or trees, usually with hollow stems (culms) solid at the nodes, and with 2-ranked, parallel-veined lvs. consisting of 2 parts; the *sheath* and the *blade*, the former enveloping the culm with the margins overlapping or rarely grown together; at the junction of the sheath and blade, on the inside, is a membranaceous hyaline or hairy appendage, the *ligule*. Fls. perfect, sometimes unisexual, minute, without a distinct perianth, arranged in spikelets consisting of a shortened axis (*rachilla*) and 2-many distichous bracts, the 2 lowest (*glumes*) being empty; rarely 1 or both of them obsolete; in the axil of each succeeding bract (*lemma*) is borne a single fl., subtended and usually enveloped by a 1–2-nerved bract (*palea*) with its back to the rachilla; at the base of the fl., between it and the lemma, are usually 2 very small hyaline scales (*lodicules*). Stamens usually 3, with delicate fils. and 2-celled versatile anthers. Pistil 1, with a 1-celled, 1-ovuled ovary, usually 2 styles and plumose stigmas. Fr. a caryopsis with starchy endosperm and usually enclosed at maturity in the lemma and palea, free or adnate to the latter. The lemma with its palea and fl. constitute a floret. Spikelets ar-

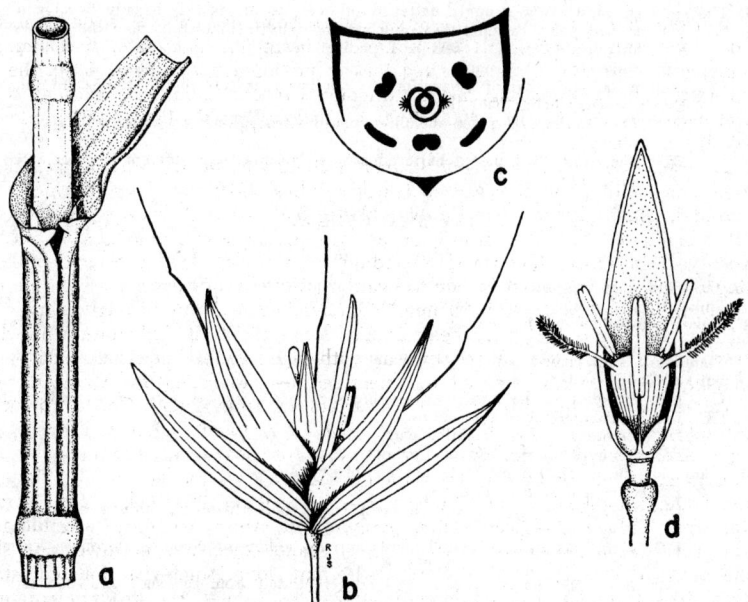

Fig. 134. GRAMINEAE. *Vegetative features: a,* diagram to show solid nodes below and above, with hollow internode between, largely enclosed by lf.-base (sheath of open type) passing into blade at upper right and with erect ligule at junction of sheath and blade. *Spikelet: b,* diagram, the 2 lower bracts are glumes subtending group of 3 florets, the lower right floret showing a lemma awned from its back, a palea above and stamen of floret emerging between; *c,* cross section of floret with lower lemma, upper palea, cent. pistil, 3 stamens, and 2 lodicules (1 on either side of lower anther). *Flowers: d,* diagram with palea in background, 3 stamens, ovary with 2 feathery stigmas, 2 basal lodicules.

ranged in spikes, racemes or panicles with bractless branches. Ca. 450 genera and 4500 spp.; from all parts of world.

(Hitchcock, A. S. Manual of the grasses of the U.S. ed. 2, revised by Agnes Chase. U.S. Dept. Agric., Misc. Pub. 200: 1–1051, 1951. [This volume has been the basis of the treatment of the family presented in this Calif. manual.])

KEY TO TRIBES

A. Spikelets with the glumes persistent, the rachilla articulated above them, 1–many-fld.; upper lemmas frequently empty; rachilla often prolonged beyond the upper lemma.
  B. The spikelets borne in an open or spikelike raceme or panicle, usually upon distinct pedicels.
    C. Spikelets 1-fld.
      D. The spikelets with 2 sterile or ♂ lemmas below the fertile lemma; palea 1-nerved
                                                                                    7. *Phalarideae*
      DD. The spikelets without sterile lemmas below the fertile lemma; palea 2-nerved
                                                                                    4. *Agrostideae*
    CC. Spikelets 2–many-fld.
      D. Lemma usually shorter than the empty glumes; the awn dorsal and usually bent
                                                                                    3. *Aveneae*
      DD. Lemma usually longer than the empty glumes; the awn terminal and straight or none (rarely dorsal in *Bromus*) .... 1. *Festuceae*
  BB. The spikelets borne in 2 rows, sessile or nearly so.
    C. Spikelets on one side of the continuous axis, forming 1-sided spikes; spikes usually more than 1 .... 6. *Chlorideae*
    CC. Spikelets alternately on opposite sides of the axis which is often articulated; spike terminal, single .... 2. *Hordeae*
AA. Spikelets falling from the pedicels entire, articulate below the glumes, naked or enclosed in bristles or burlike invols., 1-fld. or, if 2-fld., the lower fl. ♂; no lemmas empty; rachilla not extending beyond the upper lemma.
  B. The spikelets very flat, strongly laterally compressed, paniculate; glumes reduced or wanting; lemma and palea subequal, both keeled .... 8. *Oryzeae*

  BB. The spikelets mostly not strongly flattened; glumes, or at least 1, usually developed.
     C. Lemma and palea hyaline, thin, much more delicate in texture than the glumes.
        D. Spikelets in pairs, 1 sessile, 1 pedicellate . . . . 10. *Andropogoneae*
        DD. Spikelets not in pairs . . . . 4. *Agrostideae*
    CC. Lemma, at least that of the perfect fl., similar in texture to the glumes, or thicker and
       firmer, never hyaline and thin.
       D. The lemma and palea membranous; the first glume usually the larger
                                                     5. *Zoysieae*
       DD. The lemma and palea chartaceous to coriaceous, very different in color and appear-
         ance from the glumes . . . . 9. *Paniceae*

KEYS TO GENERA

Tribe 1. Festùceae

A. Lemmas divided at top into 5 or more awns or awnlike teeth or lobes.
  B. Spikelets several-fld.; awnlike lobes 5, or in 1 sp. of 7–11 short teeth . . . . . . 26. *Orcuttia*
  BB. Spikelets 3-fld.; awns 9, plumose . . . . 28. *Enneapogon*
AA. Lemmas awnless, or 1- or 3-awned.
  B. Plants tall stout reeds with large plumelike panicles; lemmas or rachilla with long silky
    hairs as long as the lemmas.
    C. Lvs. crowded at the base of the culms . . . . 20. *Cortaderia*
    CC. Lvs. distributed along the culms.
      D. Lemmas naked; rachilla hairy . . . . 21. *Phragmites*
      DD. Lemmas hairy; rachilla naked . . . . 19. *Arundo*
  BB. Plants low or to ca. 1.5 m. high, not reedlike.
    C. Plants dioecious, perennial.
      D. The plants densely tufted, erect from short rhizomes; lemmas scabrous. Of dry
        mountain slopes . . . . 8. *Hesperochloa*
      DD. Plants not densely tufted, spreading by stolons or creeping rhizomes; lemmas gla-
        brous. Of saline or alkaline places.
        E. Spikelets obscure, scarcely differentiated from the short crowded rigid lvs.
                                             14. *Monanthochloe*
        EE. Spikelets in narrow simple exserted panicles . . . . 15. *Distichlis*
    CC. Plants not dioecious, except in some spp. of *Poa* (with villous lemmas).
      D. Spikelets of 2 forms, sterile and fertile intermixed; panicle dense, ± secund.
        E. Fertile spikelets 2–3-fld.; sterile spikelets with numerous rigid awn-tipped lem-
         mas; panicle dense, spikelike . . . . 17. *Cynosurus*
        EE. Fertile spikelets with 1 perfect floret, long-awned; sterile spikelets with many
         obtuse sterile lemmas; panicle-branches short, nodding . . . . 18. *Lamarckia*
      DD. Spikelets all alike in the same infl.
        E. Lemmas 3-nerved, the nerves prominent.
          F. Infl. a few-fld. woolly capitate panicle overtopped by the lvs.; lemmas
          toothed or cleft. Low desert plants.
            G. Lemmas cleft either side of the midnerve to near the base, the
            lower 2 sterile, the 3d floret fertile, the 4th reduced to a 3-awned
            rudiment . . . . 27. *Blepharidachne*
           GG. Lemmas 2-lobed but not deeply cleft, all fertile but the uppermost
                                         24. *Tridens*
         FF. Infl. an exserted open or spikelike panicle.
          G. Lemmas pubescent on the nerves or callus, obtuse . . 24. *Tridens*
          GG. Lemmas not pubescent on the nerves or callus, but sometimes on the
           internerves, acute or acuminate.
            H. Glumes 6–7 mm. long, unequal; infl. a dense panicle, ±
            secund . . . . 12. *Cutandia*
            HH. Glumes 0.5–4 mm. long, subequal or unequal.
              I. Glumes longer than the lemmas; lateral nerves of lemma
               marginal, the internerves pubescent . . 13. *Dissanthelium*
              II. Glumes shorter than the lemmas; lateral nerves of the
               lemma not marginal, the internerves glabrous
                                         11. *Eragrostis*
        EE. Lemmas 5- to many-nerved, the nerves sometimes obscure.
          F. Lemmas flabellate; glumes wanting; infl. dense, cylindric; low annual
                                            25. *Neostapfia*
        FF. Lemmas not flabellate; glumes present; infl. not cylindric.
          G. Lemmas as broad as long, with outspread margins; florets closely
           imbricate, horizontally spreading . . . . 10. *Briza*
          GG. Lemmas longer than broad, the margins clasping the palea; florets
           not horizontally spreading.
            H. The lemmas keeled on the back (sometimes somewhat
            rounded in *Poa*).
             I. Spikelets strongly compressed, crowded in 1-sided clusters
              at the ends of the stiff naked panicle-branches
                                            16. *Dactylis*
            II. Spikelets not strongly compressed, not crowded in 1-sided
            clusters.

J. Lemmas awned from a minutely bifid apex, or nearly
    awnless; spikelets large .... 1. *Bromus*
JJ. Lemmas awnless; spikelets small .... 9. *Poa*
HH. The lemmas rounded on the back.
  I. Glumes papery; lemmas firm, strongly nerved, scarious-
    margined; upper florets sterile, often reduced to rudiments
    infolded by the broad upper lemmas; spikelets tawny or
    purplish ....,.......... .. 22. *Melica*
  II. Glumes not papery; upper florets like the others.
    J. Nerves of lemma parallel, not converging at the
      summit or but slightly so.
      K. Spikelets in long loose racemes .. 7. *Pleuropogon*
      KK. Spikelets in open or contracted panicles.
        L. Sheaths of lvs. usually connate; nerves of
          upper empty glumes single; styles present
          6. *Glyceria*
        LL. Sheaths of lvs. open; nerves of upper empty
          glumes 3 .... 5. *Puccinellia*
    JJ. Nerves of lemma converging toward the summit, the
      lemmas narrowed at apex.
      K. Lemmas awned or awn-tipped from a minutely
        bifid apex (rarely awnless); palea adhering to the
        caryopsis.
        L. Spikelets in open to contracted panicles;
          stigmas borne at the sides of the summit of
          ovary .... 1. *Bromus*
        LL. Spikelets nearly sessile in strict racemes;
          stigmas terminal on the ovary
          2. *Brachypodium*
      KK. Lemmas entire, pointed, awnless or awned from
        the tip.
        L. Spikelets awned except in a few perennial
          spp.; lemmas pointed .... 3. *Festuca*
        LL. Spikelets awnless.
          M. Second glume 5–11-nerved; spikelets
            ca. 1 cm. or more long; lemmas broad.
            Sand dunes e. of Inyo Mts.
            23. *Ectosperma*
          MM. Second glume 1- to 3-nerved; spikelets
            smaller.
            N. Spikelets on slender pedicels in
              compound panicles; perennials
              9. *Poa*
            NN. Spikelets on thick short pedicels
              in simple panicles; annual
              4. *Scleropoa*

Tribe 2. Hórdeae

A. Spikelets solitary at each node of the rachis (rarely 2 in ssp. of *Agropyron*, but never throughout).
  B. The spikelets 1-fld., sunken in hollows in the rachis; spikes slender, cylindric; plants low annuals.
    C. Lemmas with awns; florets lateral to the rachis. Mts. .... 40. *Scribneria*
    CC. Lemmas not awned; florets dorsiventral to the rachis. Coastal marshes.
      D. First glume wanting; spikes 1–2 dm. long .... 38. *Monerma*
      DD. First glume present; spikes 0.7–1 dm. long .... 39. *Parapholis*
  BB. Spikelets 2–several-fld., not sunken in the rachis.
    C. Spikelets placed edgewise to the rachis, the lateral ones with a single glume
      37. *Lolium*
    CC. Spikelets placed flatwise to the rachis.
      D. Plants perennial .... 29. *Agropyron*
      DD. Plants annual.
        E. Spikelets turgid or cylindric .... 34. *Aegilops*
        EE. Spikelets compressed.
          F. Glumes ovate, 3-nerved .... 33. *Triticum*
          FF. Glumes subulate, 1-nerved .... 35. *Secale*
AA. Spikelets normally more than 1 at each node of the rachis.
  B. Spikelets 3 at each rachis-node, 1-fld., the lateral pair pedicelled, usually reduced to awns
    36. *Hordeum*
  BB. Spikelets 2 or more (sometimes 1 in *Elymus*) at each node of rachis, alike, 2–6-fld.
    C. Glumes lacking or reduced to 2 short bristles; spikelets horizontally spreading or ascending at maturity; spikes very loose .... 32. *Hystrix*
    CC. Glumes usually as long as florets; spikelets appressed or ascending.
      D. Rachis continuous, disarticulating only tardily, and that rarely; glumes not greatly elongate .... 30. *Elymus*

DD. Rachis articulating into joints at maturity; glumes usually setaceous and greatly elongate .... 31. *Sitanion*

Tribe 3. Avèneae

A. Florets 2, one perfect, the other ♂.
   B. Lower floret ♂, the awn twisted, geniculate, exserted .... 48. *Arrhenatherum*
   BB. Lower floret perfect, awnless; upper floret awned .... 49. *Holcus*
AA. Florets 2 or more, all alike except the reduced upper ones.
   B. Articulation below the glumes, the spikelets falling entire.
      C. Lemmas, at least the upper, with a conspicuous bent awn; glumes nearly alike
                         44. *Trisetum*
      CC. Lemmas awnless; second glume much wider than the first .... 43. *Sphenopholis*
   BB. Articulation above the glumes, the glumes similar in shape.
      C. Lemmas bifid at apex, awned or mucronate between the lobes; spikelets several-fld.
         D. Awns conspicuous, flat, bent; spikelets 1 cm. or more long ...... 51. *Danthonia*
         DD. Awns minute to obsolete.
            E. Spikelets 8–12 mm. long .... 50. *Sieglingia*
            EE. Spikelets not more than 5 mm. long; awns when present, slender, rounded
                        41. *Schismus*
      CC. Lemmas toothed but not bifid and awned or mucronate between the lobes.
         D. Glumes 2–3.5 cm. long, 7–9-nerved; spikelets 2-fld., or with a rudimentary 3d one, pendulous; annual .... 47. *Avena*
         DD. Glumes not more than 1 cm. long, 1–5-nerved; spikelets not pendulous; mostly perennial.
            E. Lemmas keeled, the awn, when present, from above the middle.
               F. Rachilla-joints very short, glabrous or short-pubescent; lemmas awnless or with a straight awn from a toothed apex .... 42. *Koeleria*
               FF. Rachilla-joints slender, villous; lemmas with a dorsal geniculate awn in most spp. .... 44. *Trisetum*
            EE. Lemmas convex, awned from below the middle.
               F. Rachilla prolonged behind the upper floret; lemmas truncate and erose-dentate at summit .... 45. *Deschampsia*
               FF. Rachilla not prolonged; lemmas tapering into 2 slender teeth
                        46. *Aira*

Tribe 4. Agrostídeae

A. Rachilla articulate below the glumes, these falling with the spikelet.
   B. Glumes long-awned .... 57. *Polypogon*
   BB. Glumes awnless.
      C. Panicle dense; rachilla not prolonged behind the palea.
         D. Glumes united toward the base, ciliate on the keel; infl. not capitate and bracteate
                        56. *Alopecurus*
         DD. Glumes not united, glabrous; infl. capitate, in the axils of broad bracts
                        63. *Crypsis*
      CC. Panicle narrow or open, not dense; rachilla prolonged behind the palea .. 55. *Cinna*
AA. Rachilla articulate above the glumes.
   B. Lemma indurate when mature and very closely embracing the grain; callus usually well developed, bearded.
      C. Awn 3-branched, the lateral branches sometimes obsolete (then no line of demarcation between lemma and awn) .... 67. *Aristida*
      CC. Awn simple, a line of demarcation between the awn and lemma.
         D. The awn twisted and bent, persistent, several to many times longer than the fr.
                        66. *Stipa*
         DD. The awn not twisted, deciduous, rarely more than 3–4 times as long as the plump fr. .... 65. *Oryzopsis*
   BB. Lemma usually hyaline or membranaceous at maturity; callus not well developed.
      C. Glumes mostly longer than the lemma (sometimes not much longer in *Agrostis*).
         D. Panicle feathery, capitate, nearly as broad as long; spikelets woolly .. 60. *Lagurus*
         DD. Panicle not feathery; spikelets not woolly.
             E. Glumes compressed-carinate, stiff-ciliate on keel; panicle dense, cylindric or ellipsoid .... 58. *Phleum*
            EE. Glumes not compressed-carinate, not ciliate.
                F. Glumes saccate at base; lemma long-awned; panicle contracted, shining
                        59. *Gastridium*
                FF. Glumes not saccate at base; lemma awned or not; panicle open or contracted.
                   G. Floret bearing a tuft of hairs at the base from the short callus; palea well developed, the rachilla prolonged behind the palea as a hairy bristle .... 52. *Calamagrostis*
                   GG. Floret without hairs at base or with short hairs; palea usually small or obsolete .... 54. *Agrostis*
      CC. Glumes not longer than the lemma, usually shorter (except sometimes for the awn-tips).
         D. Lemma awned from the tip or mucronate, usually 3- to 5-nerved
                        61. *Muhlenbergia*
         DD. Lemma awnless or awned from the back.

    E. Floret with a tuft of hairs at the base from the short callus; lemma and palea chartaceous, awnless .... 53. *Ammophila*

  EE. Floret without hairs at base.

      F. Fr. at maturity falling from the lemma and palea; seed loose in the pericarp; lemma 1-nerved; panicle open or contracted . 62. *Sporobolus*

    FF. Fr. not falling from the lemma and palea, but permanently enclosed in them; seed adnate to the pericarp; panicle spikelike.

        G. Panicle short, partly enclosed in the sheath; annual

                  64. *Heleochloa*

      GG. Panicle elongate; perennial .... 61. *Muhlenbergia*

Tribe 5. Zoÿsieae

One genus in our flora .... 68. *Hilaria*

Tribe 6. Chlorídeae

A. Spikelets with more than 1 perfect floret.

  B. Spikes many, slender, racemose on an elongate axis .... 69. *Leptochloa*

  BB. Spikes few, digitate or nearly so.

    C. Rachis of spike not prolonged beyond the spikelets .... 70. *Eleusine*

    CC. Rachis of spike extending beyond the spikelets .... 71. *Dactyloctenium*

AA. Spikelets with only 1 perfect floret, though sometimes with additional imperfect florets above or below.

  B. The spikelets without additional modified florets, the rachilla sometimes prolonged.

    C. Rachilla articulate below the glumes, the spikelets falling entire; spikes not digitate, but arranged on an elongate axis.

      D. Glumes, narrow, unequal .... 74. *Spartina*

      DD. Glumes broad and boat-shaped, equal .... 73. *Beckmannia*

    CC. Rachilla jointed above the glumes; spikes digitate .... 72. *Cynodon*

  BB. The spikelets with 1 or more modified florets above the perfect one.

    C. Spikes digitate or nearly so .... 75. *Chloris*

    CC. Spikes racemose along a main axis .... 76. *Bouteloua*

Tribe 7. Phalarídeae

A. Lower floret ♂; spikelets brown, shining .... 77. *Hierochloe*

AA. Lower florets neuter; spikelets green or yellowish.

  B. Lower florets not reduced to small scalelike lemmas.

    C. Lower florets consisting of awned hairy sterile lemmas exceeding the fertile floret

                  78. *Anthoxanthum*

    CC. Lower florets consisting of awnless glabrous sterile lemmas enclosing the fertile floret

                  79. *Ehrharta*

  BB. Lower florets reduced to small awnless scalelike lemmas much smaller than the fertile florets

                  80. *Phalaris*

Tribe 8. Orÿzeae

One genus .... 81. *Leersia*

Tribe 9. Paníceae

A. Spikelets sunken in the cavities of the flattened corky rachis; coarse creeping grass escaping from cult .... 83. *Stenotaphrum*

AA. Spikelets not sunken in the rachis.

  B. The spikelets subtended or surrounded by 1 to many distinct or ± connate bristles forming an invol.

    C. Bristles persistent, the spikelets deciduous .... 88. *Setaria*

    CC. Bristles falling with the spikelets at maturity.

      D. The bristles not united at the base, slender, often plumose ...... 89. *Pennisetum*

      DD. The bristles united into a burlike invol., the bristles retrorsely barbed

                  90. *Cenchrus*

  BB. The spikelets not subtended by bristles.

    C. Glumes awned or mucronate; apex of palea not enclosed by the lemma

                  87. *Echinochloa*

    CC. Glumes awnless; apex of palea usually enclosed by the lemma.

      D. Spikelets in panicles .... 86. *Panicum*

      DD. Spikelets in 1-sided spikelike racemes.

        E. First glume and the rachilla-joint forming a swollen ringlike callus below the spikelet; racemes several along the main axis .... 84. *Eriochloa*

        EE. First glume present or lacking, not forming a ringlike callus; racemes aggregate at the summit of the culm.

          F. Racemes slender, 3–12 .... 82. *Digitaria*

          FF. Racemes stout, in pairs .... 85. *Paspalum*

Tribe 10. Andropogòneae

A. Spikelets all alike, fertile, surrounded by copious soft hairs; infl. a narrow panicle .. 91. *Imperata*

AA. Spikelets unlike, the sessile perfect, the pedicellate ♂ or neuter.

  B. Racemes of several joints, silky-villous .... 92. *Andropogon*

BB. Racemes reduced to 1 or few joints, these in a compound panicle, not silky villous
93. *Sorghum*

1. Bròmus L. Bromegrass

Low or rather tall annuals or perennials with closed sheaths, usually flat blades and open or contracted panicles of large spikelets. These several- to many-fld., the rachilla disarticulating above the glumes and between the florets; glumes unequal, acute, the 1st 1- to 3-nerved, the 2d usually 3- to 5-nerved; lemmas convex on the back or keeled, 5- to 9-nerved, 2-toothed, awned from between the teeth or awnless; palea usually shorter than the lemma, ciliate on the keels. Ca. 100 spp. of the temp. regions of the world. (Ancient Greek name for the oat.)

(Wagnon, H. Keith. A revision of the genus Bromus, section Bromopsis, of N. Am. Brittonia 7: 415–480, 1952.)
A. Plants annual, largely of introd. weedy type.
    B. Spikelets strongly flattened, the lemmas compressed-keeled and with terminal teeth not more than 0.5 mm. long.
        C. Lemmas awnless or the awn less than 3 mm. long.
            D. Lemmas usually with 13 veins .... 1. *B. catharticus*
            DD. Lemmas usually with 9 veins .... 2. *B. Haenkeanus*
        CC. Lemmas with awns 7–15 mm. long.
            D. Spikelets 6–10-fld.; 2d glume shorter than the lowest lemma.
                E. Panicle with spreading or drooping branches; lemma 1.6–2 mm. wide. Native
                                                                                        4. *B. carinatus*
                EE. Panicle with stiff ascending branches; lemma 2.5 mm. wide. Introd.
                                                                                        5. *B. stamineus*
            DD. Spikelets 5–7-fld.; 2d glume almost equal to the lowest lemma .. 6. *B. arizonicus*
    BB. Spikelets terete or somewhat flattened, the lemmas not compressed-keeled, their terminal teeth mostly 0.6–5 mm. long.
        C. Awn usually geniculate, twisted at base; lemma-teeth aristate .... 36. *B. Trinii*
        CC. Awn straight or spreading, sometimes minute, not twisted at base.
            D. Lemmas broad, rounded apically, not acuminate, the teeth mostly less than 1 mm. long; 1st glume 3–5-nerved.
                E. The lemmas awnless or the awn not more than 1 mm. long; spikelets ovate, 0.8–1.3 cm. broad .... 21. *B. brizaeformis*
                EE. The lemmas with awns 3–15 mm. long; spikelets narrower.
                    F. Panicles compact, the spikelets mostly longer than the pedicels; lemmas membranaceous or scarious, prominently nerved, unequal, the lower longer than the upper.
                        G. Lemmas glabrous.
                            H. Culms mostly 3–8 dm. high; lf.-blades pubescent; awns straight   . ... 26. *B. racemosus*
                            HH. Culms mostly 2–3 dm. high; lf.-blades glabrous to sparingly pilose; awns becoming divaricate .... 27. *B. scoparius*
                        GG. Lemmas pubescent.
                            H. Culms mostly 3–8 dm. tall; panicle 5–10 cm. long; spikelets turgid; lemmas ca. ¾ as wide as long, the awn straight
                                                                                        24. *B. mollis*
                            HH. Culms mostly 1–2 dm. tall; panicle 2–4 cm. long; spikelets compressed; lemmas ca. half as wide as long, the awn ± divaricate   . ... 25. *B. molliformis*
                    FF. Panicles more open, the spikelets mostly not longer than the pedicels; lemmas ± coriaceous, not prominently veined.
                        G. Foliage glabrous .... 22. *B. secalinus*
                        GG. Foliage pubescent.
                            H. Awns straight; lemmas not conspicuously inrolled at maturity and not exposing rachilla to view.
                                I. Lower lemmas 9–11 mm. long.
                                    J. The lemmas ca. equal; awn 10–16 mm. long; glumes pilose .... 30. *B. arenarius*
                                    JJ. The lemmas unequal, the lower much longer than the upper; awn 4–8 mm. long; glumes glabrous
                                                                                        23. *B. commutatus*
                                II. Lower lemmas 7–8 mm. long; awn 7–10 mm. long
                                                                                        29. *B. arvensis*
                            HH. Awns flexuous, usually ± divergent; lemmas inrolled at maturity and exposing to view the rachilla .... 28. *B. japonicus*
            DD. Lemmas narrow, elongate, tapering at the tip, the teeth 2–5 mm. long; 1st glume 1-nerved.
                E. Panicle erect, contracted; awn 1–2 cm. long.
                    F. Culms pubescent below the dense panicle; sheaths pubescent. Common
                                                                                        33. *B. rubens*

FF. Culms glabrous below the slightly open panicle; sheaths mostly smooth. occasional .... 34. *B. madritensis*
EE. Panicle open, with spreading or drooping branches.
   F. Awn 12–14 mm. long; lemmas ca. 1 cm. long; spikelets 1–2 cm. long
                35. *B. tectorum*
   FF. Awns 2–5 cm. long; lemmas 1.7–3 cm. long; spikelets 2.5–4 cm. long.
     G. Lemmas 25–30 mm. long; awns 3.5–5 cm. long; 1st glume ca. 16–20 mm. long .... 31. *B. rigidus*
     GG. Lemmas 17–20 mm. long; awns 2–3 cm. long; 1st glume 7–9 mm. long .... 32. *B. sterilis*
AA. Plants perennial, largely native members of our Plant Communities.
  B. First glume 3–5-nerved.
    C. Spikelets strongly flattened, the lemmas compressed-keeled.
      D. Blades canescent, densely short-pilose, 2–5 mm. wide, often involute; panicle narrow. Montane and n. .... 3. *B. breviaristatus*
      DD. Blades not canescent, glabrous to puberulent or sparsely pilose; blades mostly 4–12 mm. wide.
        E. Awns 2–3 mm. long; spikelets 3–4 cm. long; panicles strict, the short branches erect. Immediate coast .... 8. *B. maritimus*
        EE. Awns 4–15 mm. long.
          F. The awns 7–15 mm. long; spikelets 2–3 cm. long.
            G. Panicle with spreading or drooping branches; lemma 1.6–2 mm. wide. Native .... 4. *B. carinatus*
            GG. Panicle with stiff ascending branches; lemma 2.5 mm. wide, more conspicuously white-margined. Introd. .... 5. *B. stamineus*
          FF. The awns 4–7 mm. long; spikelets 2.5–3.5 cm. long; panicles rather narrow with ascending branches (except in forms of *marginatus*).
            G. Sheaths usually pilose; lemmas mostly pubescent; spikelets 7–8-fld. Widespread .... 7. *B. marginatus*
            GG. Sheaths glabrous; lemmas glabrous; spikelets 7–11-fld. Yosemite Nat. Park .... 9. *B. polyanthus*
    CC. Spikelets terete before anthesis or somewhat flattened, but the lemmas not compressed-keeled.
      D. Panicle narrow, 2 cm. or less broad at anthesis, erect; nodes 2–3, glabrous; sheaths glabrous and smooth. Montane .... 14. *B. Suksdorfii*
      DD. Panicle more than 2 cm. broad at anthesis, or if less, with pilose sheaths, erect or nodding.
        E. Second glume 5-nerved.
          F. Ligule of the culm-lvs. 2–4 mm. long; blades glabrous; glumes glabrous; lemma with hairs on the marginal parts only, or with scattered hairs over the back. Cismontane and montane .... 13. *B. laevipes*
          FF. Ligule of culm-lvs. 1 mm. long or less; blades mostly pilose; glumes usually with scattered hairs; lemma usually with hairs across the back. Coast Ranges .... 11. *B. pseudolaevipes*
        EE. Second glume 3-nerved.
          F. Spikelets 2–4 cm. long; sheaths retrorsely pubescent.
            G. Panicle 1–1.5 dm. long, erect or pyramidal; awn 5–8 mm. long. Montane Coniferous F. .... 15. *B. Orcuttianus*
            GG. Panicle 1.5–2 dm. long, with slender drooping branches; awn 3–6 mm. long. Below Yellow Pine F. .... 16. *B. grandis*
          FF. Spikelets 1.3–1.5 cm. long; sheaths mostly glabrous .... 19. *B. Porteri*
  BB. First glume 1-nerved.
    C. Creeping rhizomes present; lemma awnless or with an awn up to 3 mm. long; lvs. usually glabrous. Introd. .... 17. *B. inermis*
    CC. Creeping rhizomes wanting; lemma awned.
      D. Panicles not more than 2 cm. broad at anthesis; nodes 2–3; awns 5–7 mm. long. Occasional waif .... 18. *B. erectus*
      DD. Panicles broader; nodes mostly more.
        E. Awns 7–11 mm. long; ligule 3–5 mm. long. From Monterey and Butte cos. n. 12. *B. vulgaris*
        EE. Awns 3–5 mm. long; ligule mostly less than 2 mm. long.
          F. Anthers 1–1.3 mm. long; blades pubescent above. Moist places 10. *B. ciliatus*
          FF. Anthers 2–3.5 mm. long; blades glabrous above. Dry places 20. *B. Richardsoni*

1. **B. cathárticus** Vahl. [*Festuca unioloides* Willd. *B. u.* HBK.] Rescue Grass. Annual or biennial; culms erect or spreading, 6–10 dm. tall; sheaths pilose or glabrous; blades glabrous or sparsely pilose, ca. 3–6 mm. wide; panicle open (or narrow in depauperate plants), 1–2 dm. long, the branches naked at the base; spikelets 2–3 cm. long, 0.5–0.9 cm. broad, 6–12-fld.; glumes smooth, the 1st 5-nerved, 7–10 mm. long, the 2d 7-nerved, 10–12 mm. long; lemmas acute, subcoriaceous, glabrous or scabrous, 12–16 mm. long, ca. 13-nerved; awn 0–2.5 mm. long; palea ½–¾ as long as lemma; $2n = 42$

(Stebbins & Tobgy, 1944).—Occasional as introd. in waste places in cismontane Calif., especially in the cent. part; s. U.S. Apparently native of S. Am. April–Nov.

2. **B. Haenkeànus** (Presl.) Kunth. [*Ceratochloa H.* Presl.] Resembling *B. catharticus,* panicle 1–1.3 dm. long, the branches stiff, in 2's or 3's, bearing 1–3 sessile spikelets; spikelets much compressed, 2–3-fld., 1–1.3 cm. long; glumes 5–7-nerved; lemmas 10–12 mm. long, 9–11-nerved, finely scabrous-pubescent; awn very short or none.—Occasional as a weed in s. Calif.; from S. Am.

3. **B. breviaristàtus** Buckl. [*B. subvelutinus* Shear. *B. carinatus* var. *linearis* Shear.] Perennial, tufted, erect, 2.5–5 dm. tall; sheaths canescent to densely retrorse-pilose; blades narrow, becoming involute, mostly erect or ascending, canescent and pilose, mostly 1–3 mm. wide; panicle narrow, erect, 5–15 cm. long, with short appressed branches often bearing only 1 spikelet; spikelets 2–3 cm. long; glumes puberulent, the 1st 3–5-nerved, 8–10 mm. long, the 2d 7-nerved, 10–12 mm. long; lemmas appressed-puberulent, 12–14 mm. long; awn 3–10 mm. long; $n = 28$ (Staehlin, 1929).—Occasional in dry places, mostly 4000–8000 ft.; Montane Coniferous F.; mts. s. Calif. to B.C., Nev. May–July. Sometimes lower along n. coast and as a street weed in cent. Calif.

4. **B. carinàtus** H. & A. CALIFORNIA BROME. Erect annual or biennial, 5–10(–12) dm. tall; sheaths scabrous to sparsely pilose; blades flat, mostly 2–3 dm. long, 3–10 mm. broad, scabrous or sparsely pilose; panicle 1.5–3 dm. long, with spreading or drooping branches; spikelets 2–3 cm. long (without the awns), mostly 6–10-fld., the florets scarcely or not overlapping at anthesis; 1st glume 6–9 mm. long, 3-nerved, the 2d 5-nerved, 10–15 mm. long; awns 7–15 mm. long; palea acuminate, nearly equaling the lemma; $n = 28$ (Stebbins & Love, 1941).—Frequent in dry open places, below 10,500 ft.; many Plant Communities; cismontane Calif. to B.C., Ida., New Mex., L. Calif. April–Aug. Variable; plants with sheaths smooth have been called var. *califórnicus* Shear and those with spikelets 3–4 cm. long, var. *Hookeriànus* (Thurb.) Shear.

5. **B. stamíneus** Desv. in Gay. Resembling *B. carinatus,* lf.-blades ± pilose, 2–7 mm. wide, scabrous; panicle 1–2 dm. long, with stiff ascending branches; spikelets 1.8–2.4 cm. long, greenish, 4–6-fld.; glumes lance-ovate, acuminate, pubescent, the lower 7–8 mm. long (without awnlike tip), 5–7-nerved, the upper longer, 9-nerved; lemmas 10–12 mm. long, keeled, broadest above the middle, weakly 9-nerved, pubescent, with prominent whitish or slightly purplish hyaline margin; awn 8–10 mm. long.—Weed, cent. Calif.; introd. from S. Am. May–July.

6. **B. arizónicus** (Shear) Steb. [*B. carinatus* var. *a.* Shear.] Annual; like *B. carinatus,* but mostly shorter, stiff, erect, rather narrow; spikelets 5–7-fld.; glumes subequal; the upper glume ca. as long as the lowest lemma; lemmas hirsute toward the margin, sometimes with some hairs on the back, the apical teeth 0.7–2 mm. long; $2n = 42$ (Stebbins et al., 1944).—Dry open places, mostly below 2000 ft.; V. Grassland, Foothill Wd., Chaparral, Coastal Sage Scrub, Creosote Bush Scrub; from Yolo and Fresno cos. to San Diego Co.; to Ariz., L. Calif., Santa Barbara Ids. March–May.

7. **B. marginàtus** Nees. Perennial, rather stout, 6–12 dm. high; sheaths pilose; blades ± pilose, flat, 4–8(–12) mm. wide; panicles erect, rather narrow, 1–2 dm. long, the lower branches erect or ± spreading; spikelets 2.5–3.5 cm. long, closely 7–8-fld.; glumes broad, scabrous or scabrous-pubescent, the 1st subacute, 3–5-nerved, 7–9 mm. long, the 2d obtuse, 5–7-nerved, 9–11 mm. long; lemmas subcoriaceous, pubescent, 11–14 mm. long; awns usually 4–7 mm. long; $n = 21$ (Nielsen & Humphrey, 1937).—Dry open places, below 11,000 ft.; many Plant Communities; much of cismontane and montane Calif.; to B.C., S. Dak., New Mex. April–July. Variable and intergrading with *B. carinatus.* Plants with glabrous sheaths have been called var. *seminùdus* Shear and those with pubescent sheaths and large spreading panicles with branches as much as 2 dm. long, var. *làtior* Shear.

8. **B. marítimus** (Piper) Hitchc. [*B. marginatus* ssp. *m.* Piper.] Perennial with stout culms 3–7 dm. tall, ± geniculate at base with numerous basal leafy shoots; sheaths smooth or scaberulous; blades mostly 6–8 mm. wide, scabrous; panicles mostly 1–2 dm. long, strict, the short branches erect; spikelets 3–4 cm. long, 4–6 mm. wide, the awns 2–3 mm. long.—Coastal Strand and sandy places along the coast from San Luis Obispo Co. to s. Ore.; Anacapa and San Miguel ids. April–July.

9. **B. polyánthus** Scribn. [*B. multiflorus* Scribn., not Weigel.] Perennial, with fairly

stout culms 6–12 dm. tall; sheaths smooth, glabrous; blades 5–15 mm. wide, scabrous; panicles commonly 1.5–2.5 dm. long, with ascending branches; spikelets glabrous or scaberulous, ± glossy, rather loose at anthesis, compressed, 3–3.5 cm. long, 7–11-fld.; lemmas 13–15 mm. long; awns 4–6 mm. long.—Montane Coniferous F.; Yosemite National Park; to Wash., Mont., Tex. July–Aug.

10. **B. ciliàtus** L. FRINGED BROME. Perennial, the culms slender, 5–14 dm. tall, glabrous or retrorsely pubescent, the uppermost segms. always pubescent; sheaths pilose to glabrous; blades 5–10 mm. wide, pubescent to pilose above, mostly glabrous beneath; panicle 1–2 dm. long, open, the branches ascending to drooping, up to 1.5 dm. long; spikelets 1.5–2.5 cm. long, 4–9-fld.; glumes glabrous, the 1st 1–3-nerved, the 2d 3-nerved; lemmas 10–12 mm. long, pubescent on margin of lower part, glabrous across the back, 5–7-nerved; awn 3–5 mm. long; $2n = 14$ (Wagnon, 1952).—About meadows and moist places, 7500–9600 ft.; Montane Coniferous F.; San Bernardino Mts., Sierra Nevada (Inyo Co., Tulare Co.); to Alaska, Atlantic Coast, L. Calif. July–Aug.

11. **B. pseudolaèvipes** Wagnon. Perennial, the culms 6–12 dm. tall; sheaths glabrous to pilose; blades 3–9 mm. wide, glabrous or pilose (frequently on margins only); panicle 1–2 dm. long, erect to nodding, open with ascending to spreading branches; spikelets 1.5–3.5 cm. long, 4–10-fld.; glumes mostly pubescent, the 1st 4–6 mm. long, 3-nerved, the 2d mostly 5-nerved, 6.5–8 mm. long; lemmas 10–12 mm. long, pubescent over the back, sometimes on margins only, acute to obtuse; awns 3–5 mm. long; $2n = 14$ (Wagnon, 1952).—Dry often shaded places, below 2500 ft.; Chaparral, Coastal Sage Scrub, Foothill Wd., etc.; Coast Ranges from San Francisco Bay to Kern and San Diego cos.; Santa Cruz and Santa Catalina ids. April–June.

12. **B. vulgàris** (Hook.) Shear. [*B. purgans* var. *v.* Hook. *B. v.* var. *eximius* Shear.] Perennial, the culms slender, 6–12 dm. tall, with pubescent nodes; sheaths pilose to glabrous; blades ± pilose, 6–12 mm. wide, glabrous to pilose above, usually pilose beneath; panicle mostly 1–1.5 dm. long, open, branches ascending to drooping; spikelets 1.5–3 cm. long, 4–9-fld.; glumes glabrous to pilose, narrow, the 1st 6–8 mm. long, mostly 1-nerved, the 2d 9–12 mm. long, 3-nerved; lemma 11–16 mm. long, glabrous to pubescent over the back or sometimes only on the margins, mostly 7-nerved; awn 7–11 mm. long; $2n = 14$ (Wagnon, 1952).—Wooded and shaded places, rocky ravines, etc., below 6000 ft.; many Plant Communities; Coast Ranges and w. base of Sierra Nevada from Monterey and Butte cos. n.; to B.C., Alta., Wyo. May–Aug.

13. **B. laèvipes** Shear. Perennial, the culms 5–15 dm. tall, often with a decumbent base and rooting at lower nodes; sheaths and blades glabrous, the latter 4–10 mm. wide; panicle broad, lax, drooping, 1–2 dm. long; spikelets 2.5–3.5 cm. long, 5–11-fld.; glumes glabrous, smooth, the 1st 6–9 mm. long, 3-nerved, the 2d 5-nerved, 10–12 mm. long; lemmas 12–15 mm. long, densely pubescent on the margins, unevenly sparsely pubescent across the back or just on the lower half, 7-nerved; awn 4–6 mm. long; $2n = 14$ (Wagnon, 1952).—Shaded stream banks and brushy slopes, below 8600 ft.; many Plant Communities; cismontane and montane Calif.; to Wash. May–Aug.

14. **B. Suksdórfii** Vasey. Perennial, the culms 3–10 dm. tall, the 2–3 nodes glabrous; sheaths and blades glabrous except for the scabrous margins of the latter which are 4–12 mm. wide; panicle narrow, rather dense, 7–13 cm. long; spikelets 1.5–3 cm. long, 5–7-fld.; glumes glabrous or sparsely pubescent, the 1st 7–11 mm. long, 1-nerved, the 2d 9–12 mm. long, 3-nerved; awn 2–5 mm. long; $2n = 14$ (Wagnon, 1952).—Rocky slopes, woods and meadows, 4000–11,000 ft.; Yellow Pine F. to Subalpine F.; Sierra Nevada from Tulare Co. n., Coast Ranges from Trinity Co. n.; to Wash. July–Aug.

15. **B. Orcuttiànus** Vasey. Perennial, the culms erect, 7–15 dm. tall, with 2–3 retrorsely pubescent nodes; sheaths pilose or ± velvety; blades glabrous, with finely scabrous margins, 6–12 mm. wide, sometimes glaucous; panicle 1–1.5 dm. long, erect, or pyramidal; spikelets 2–4 cm. long, 5–7(–11)-fld., subterete; glumes glabrous to pubescent, the 1st 5–8 mm. long, 1–3-nerved, the 2d 8–10 mm. long, 3-nerved; lemmas 10–16 mm. long, short-pubescent to glabrous across the back, short-pubescent to scabrous on margins, 3-, 5-, or 7-nerved; awn 5–8 mm. long; $2n = 14$ (Wagnon, 1952).—Meadows, woods and rocky places in brush, 3000–8000 ft.; Yellow Pine F. to Lodgepole F.; mts. from San Diego Co. to s. Wash. June–Sept.

Var. **Hállii** Hitchc. in Jeps. Blades soft-pubescent above and beneath; glumes and lemmas pubescent.—Dry wooded places, 5000–10,000 ft.; Yellow Pine F. to Subalpine F.; mts. from Monterey and Fresno cos. to Riverside Co.

16. **B. grándis** (Shear) Hitchc. in Jeps. [*B. Orcuttianus* var. *g.* Shear. *B. Porteri* var. *assimilis* Davy.] Perennial, with culms 9–14 dm. tall; nodes 3–6, retrorsely pubescent; sheaths retrorsely pubescent; blades 1.5–3 dm. long, 5–12 mm. wide, densely short-pubescent on both surfaces, sometimes not on upper; panicle 1.5–2 dm. long, open, with slender drooping branches; spikelets 2.5–3.5 cm. long, 7–9-fld.; glumes pubescent, the 1st mostly 3-nerved, the 2d 3–5-nerved; lemmas 11–14 mm. long, pubescent over the back and on the margins, 7-nerved; awn 3–6 mm. long; $2n = 14$ (Stebbins & Love, 1941).—Dry open or wooded slopes, below 8000 ft.; Chaparral, Coastal Sage Scrub, Foothill Wd., etc.; San Diego Co. to Santa Clara and Tuolumne cos. April–July.

17. **B. inérmis** Leyss. SMOOTH BROME. Perennial with creeping rhizomes; culms 5–14 dm. tall, smooth; sheaths smooth; blades 1–3 dm. long, glabrous, 5–15 mm. wide; panicle 1–2 dm. long, erect, open; spikelets 2–3 cm. long, 8–10-fld.; glumes glabrous, subulate, the 1st 6–8 mm. long, mostly 1-nerved, the 2d 7–10 mm. long, 3-nerved; lemmas 9–13 mm. long, glabrous or scabrous to sparsely puberulent on margins, mostly 7-nerved, awnless or with an awn to 3 mm. long; $2n = 56$ (Hill & Meyers, 1948).—Occasional in waste places, about meadows, etc. Native of Eurasia. May–Aug.

18. **B. eréctus** Huds. Perennial, 5–10 dm. tall, the nodes glabrous; sheaths mostly glabrous; blades 1–2 dm. long, 2–4.5 mm. wide, often involute or folded; panicle 1–2 dm. long, erect, with erect or ascending branches; spikelets 2–3 cm. long, mostly 5–8-fld.; glumes glabrous, subulate, the 1st 7–9 mm. long, 1-nerved, the 2d 9–11 mm. long, 3-nerved; lemmas 11–13 mm. long, sparsely pubescent on back and margins or glabrous, 5–7-nerved; awn 5–7 mm. long; $2n = 56$ (Elliott, 1948).—Occasionally reported from Calif.; native of Eu.

19. **B. Pórteri** (Coult.) Nash. [*B. Kalmii* var. *P.* Coult. *B. anomalus*, Calif. refs., not Rupr.] Perennial, the culms 3–8 dm. tall, often tufted; nodes glabrous or retrorse-pubescent; sheaths mostly glabrous; blades 1–2.5 dm. long, 2–5 mm. wide, usually erect and glabrous; panicle 7–15 cm. long, drooping, the axis mostly puberulent, branches ascending, arcuate; spikelets 13–15 mm. long, 5–11-fld.; glumes pubescent, the 1st 5–7 mm. long, subulate or with acute tip, 3-nerved, the 2d 6–10 mm. long, 3-nerved; lemmas 8–13 mm. long, pubescent on margins and across back, rarely on margins only, mostly 7-nerved; awn 1.5–3 mm. long; $2n = 14$ (Wagnon, 1952).—Rare, on dry exposed slopes or in woods, mostly 7000–10,500 ft.; Yellow Pine F., Bristle-cone Pine F.; Sierra Nevada (Eldorado Co.), San Bernardino Mts., Inyo-White Mts.; B.C. to Manitoba, Rocky Mts., New Mex., Ariz. July–Aug.

20. **B. Richardsònii** Link. Perennial, the culms 5–9 dm. tall, with glabrous nodes; sheaths glabrous or pilose; blades mostly glabrous, 1–2.5 dm. long, 3–7 mm. wide; panicle 1–2.5 dm. long, open, the branches ascending to spreading or drooping, up to 1.4 dm. long; spikelets 1.7–3.3 cm. long, 6–10-fld.; glumes glabrous, the 1st 1- or rarely 3-nerved, 8–10 mm. long, the 2d 3-nerved, 9–12 mm. long; lemmas 10–14 mm. long, pubescent on margins of lower half, glabrous across the back or ± pubescent on nerves; awn 3–5 mm. long; $2n = 28$ (Elliott, 1949).—Dry open places, 4000–11,000 ft.; largely Pinyon-Juniper Wd., Bristle-cone Pine F.; Inyo and San Bernardino cos.; to B.C., S. Dak., Tex. June–Aug.

21. **B. brizaefórmis** F. & M. RATTLESNAKE GRASS. Annual, the culms 3–6 dm. tall; sheaths and blades pilose-pubescent, the latter mostly 1–3 mm. wide; panicle 0.5–1.5 dm. long, lax, secund, nodding; spikelets 1.5–2.5 cm. long, ca. 10 mm. wide; glumes broad, obtuse, smooth or minutely scabrous, the 1st 3–5-nerved, ca. half as long as the wider 2d which is 5–9-nerved, 6–8 mm. long; lemmas 10 mm. long, broad, obtuse, smooth, scarious-margined, nearly or quite awnless; $n = 7$ (Avdulov, 1928).—Occasional in sandy and waste places, below 6000 ft.; introd. from Eu. June–July.

22. **B. secalìnus** L. CHESS. CHEAT. Annual, the culms 3–6 dm. tall; sheaths smooth; blades mostly glabrous, 5–8 mm. wide; panicle pyramidal, nodding, 7–15 cm. long; spikelets 2–2.5 cm. long, mostly 5–15-fld., lance-ovoid; glumes smooth, the 1st 3–5-nerved, 4–6 mm. long, the 2d 7-nerved, 6–8 mm. long, obtuse; lemmas 7-nerved, 6–8 mm. long, elliptic, obtuse, smooth or scabrous, the margins strongly inrolled in fr.; awn undulate, mostly 3–5 mm. long; florets turgid in fr. and ± distant, so that a side view

of the spikelet allows light to pass through at base of each floret; $n = 7$ (Nielsen, 1939), 14 (Cugnac & Simonet, 1941).—Weed in grain fields and waste places especially in n. Calf. and in much of U.S.; native of Eu. June–July.

23. **B. commutàtus** Schrad. HAIRY CHESS. Resembling *B. secalinus;* sheaths retrorse-pilose; blades ± pubescent; spikelets 15–20 mm. long, 5–8-fld.; glumes narrower, the upper lanceolate; lemmas with an obtuse angle on the margin just above the middle, not so strongly inrolled in fr.; awn straight, commonly longer; the florets imbricate in fr. leaving no spaces at their base; $n = 28$ (Nielsen, 1939).—Weed in fields and .waste places, especially in n. Calif.; native of Eu. May–July.

24. **B. móllis** L. [*B. hordeaceus* Calif. refs.] SOFT CHESS. Annual, softly pubescent throughout; culms 2–8 dm. tall; sheaths retrorse-pubescent; blades mostly 2–5 mm. wide; panicle contracted, erect, 5–10 cm. long or smaller; spikelets 15–20 mm. long; glumes broad, obtuse, coarsely pilose, or scabrous-pubescent, the 1st 3-5-nerved, 4–6 mm. long, the 2d 5-7-nerved, 7–8 mm. long; lemmas broad, 7-nerved, obtuse, coarsely pilose or scabrous-pubescent, bidentate, 8–9 mm. long, hyaline on margin; awn stoutish, 6–9 mm. long; $2n = 28$ (Stebbins & Love, 1941).—Common weed in waste places; native of Eu. April–July.

25. **B. mollifórmis** Lloyd. Near to *B. mollis,* mostly 1–2 dm. tall; lower sheaths felty-pubescent, upper glabrous; blades narrow, with scattered stiff hairs on upper surface; panicle 2–4 cm. long, ovoid, dense, few-fld.; spikelets oblong, compressed, 12–18 mm. long; glumes ca. 6 mm. long, the 2d broader, loosely pilose with spreading hairs; lemmas thinner and narrower, closely imbricate, ca. 8 mm. long, appressed-pilose; awn 5–7 mm. long.—Waste places; reported from s. Calif.; introd. from Eu. April–May.

26. **B. racemòsus** L. Mostly 3–8 dm. high, annual; sheaths and blades pubescent; panicle slender, racemiform, suberect, the branches ascending, mostly shorter than their 1–2 spikelets; spikelets elliptic-lanceolate, 6–10-fld., 1–2 cm. long, glabrous or scabrous, not pubescent; glumes ca. 7–8 mm. long; lemmas 7–9 mm. long, the upper shorter than the lower, faintly 7-nerved; awns straightish, 7–8 mm. long.—Waste places particularly in n. Calif.; introd. from Eu. May–July.

27. **B. scopàrius** L. Resembling *B. molliformis;* culms 2–3 dm. high; sheaths soft-pubescent; blades glabrous, scabrous or sparingly pilose; panicle erect, contracted, 3–7 cm. long; spikelets ca. 1.5 cm. long, 3–4 mm. wide; lemmas ca. 7 mm. long, glabrous, narrow; awn 5–8 mm. long, becoming divaricate.—Reported from Mariposa Co.; native of Eurasia.

28. **B. japónicus** Thunb. JAPANESE CHESS. Annual, the culms erect or geniculate at base, 4–7 dm. tall; sheaths and blades pilose, the latter 2–7 mm. wide; panicle 1–2 dm. long, broadly pyramidal, diffuse, ± drooping; spikelets 2–2.5 cm. long; 1st glume 3-nerved, acute, 4–6 mm. long, the 2d 5-nerved, obtuse, 6–8 mm. long; lemmas glabrous, 7–9 mm. long, firm, obscurely 9-nerved, the margins ± inrolled at maturity; awn 8–10 mm. long, somewhat twisted and strongly divaricate at maturity; $n = 7$ (Avdulov, 1928).—Occasional weed in waste places; introd. from Eu. May–July.

29. **B. arvénsis** L. Resembling *B. japonicus;* spikelets thinner, less turgid, often tinged purple; the hyaline margins of the lemmas ending in prolonged acute teeth; awn straight, 7–10 mm. long; $n = 7$ (Avdulov, 1928).—Reported from Calif.; introd. from Eu.

30. **B. arenàrius** Labill. AUSTRALIAN CHESS. Annual, the culms slender, 2–4 dm. high; sheaths and blades pilose; the latter 3–6 mm. wide; panicle pyramidal, open, nodding, 1–1.5 dm. long, the spreading branches and pedicels sinuously curved; spikelets 1–2 cm. long, ca. 5–9-fld.; glumes densely pilose, acute, scarious-margined, the 1st narrower, 3-nerved, 8 mm. long, the 2d 7-nerved, 10 mm. long; lemmas densely pilose, 7-nerved, 10 mm. long; awn straight, 10–16 mm. long.—Dry places, below 6000 ft.; many Plant Communities; widely scattered over Calif.; introd. from Australia. April–July.

31. **B. rígidus** Roth. [*B. villosus* Forsk. *B. maximus* Desf., not Gilib.] RIPGUT GRASS. Annual, the culms 4–7 dm. tall; sheaths and blades pilose, the latter 2–10 mm. wide; panicle open or rather compact, rather few-fld., 6–12 cm. long, the lower branches mostly 1–2 cm. long; spikelets 3–4 cm. long, usually 5-7-fld.; glumes smooth, acuminate, the 1st 1-nerved, 16–20 mm. long, the 2d 3-nerved, 25–30 mm. long; lemmas 5-nerved,

25–30 mm. long, scabrous or puberulent, 2-toothed, the teeth 3–4 mm. long; awns stout, 3.5–5 cm. long, scabrous; $2n = 28$ (Stebbins & Love, 1941).—Common weed in waste places, fields, etc., at low elevs.; introd. from Eu. April–June.

Var. **Gussònei** (Parl.) Coss. & Durieu. Panicle more open, the lower branches up to 1 dm. long.—With the sp., especially common n. in Calif.; introd. from Eu.

32. **B. stérilis** L. Resembling L. *rigidus,* the culms less robust, 5–10 dm. tall; sheaths pubescent; blades pubescent, mostly 1–3 mm. wide; panicle 1–2 dm. long, with drooping branches; spikelets 2.5–3.5 cm. long, 6–10-fld.; glumes lanceolate-subulate, smooth or scabrous, the 1st 1-nerved, 7–9 mm. long, the 2d 3-nerved, 11–13 mm. long; lemmas linear-lanceolate, 5–7-nerved, 17–20 mm. long, scabrous to scabrous-pubescent, bidentate, the teeth 2 mm. long; awn 2–3 cm. long; $n = 7$ (Staehlin, 1929).—Occasional in waste places, Del Norte and Humboldt cos., occasional farther s.; introd. from Eu. April–June.

33. **B. rùbens** L. Foxtail Chess. Annual with culms 1.5–4 dm. tall, puberulent below the panicle; sheaths and blades pubescent, the latter mostly 1.5–5 mm. wide; panicle erect, ovoid, compact, reddish, 2–7 cm. long; spikelets 7–11-fld., ca. 2.5 cm. long; glumes narrow, acuminate, pubescent or sometimes smooth, the 1st 1-nerved, 7–9 mm. long, the 2d 3-nerved, 10–12 mm. long; lemmas 5-nerved, lanceolate, acute, 12–16 mm. long, ending in 2 long hyaline teeth; awn 18–22 mm. long.—Common, troublesome weed in waste and cult. ground at low elevs., especially in cent. and s. Calif., deserts and cismontane; introd. from s. Eu. March–June.

34. **B. madriténsis** L. Resembling *B. rubens,* but the culms glabrous below the less dense panicles; sheaths mostly pubescent; panicles 5–10 cm. long, oblong-ovoid in outline; glumes 9–12 and 14–16 mm. long; lemmas narrow, linear-lanceolate, 14–18 mm. long, the teeth 2–3 mm. long; awn 16–22 mm. long; $n = 14$ (Cugnac & Simonet, 1941).—Occasional in open ground and waste places; introd. from Eu. April–June.

35. **B. tectòrum** L. Cheat Grass. Downy Cheat. Annual, the slender culms 3–6 dm. tall; sheaths and blades pubescent, the latter mostly 1.5–4 mm. wide; panicle broad, drooping, 5–12 cm. long, with slender reddish branches; spikelets nodding, 1–2 cm. long; glumes villous, the 1st 1-nerved, 4–6 mm. long, the 2d 3-nerved, 8–10 mm. long; lemmas lanceolate, villous, 5-nerved, ca. 1 cm. long, the teeth 2–3 mm. long; awn 12–14 mm. long; $n = 7$ (Cugnac & Simonet, 1941).—Common weed, especially at above 3000 ft. (lower n.), desert and cismontane slopes and mts.; introd. from Eu. May–June.

Var. **glabràtus** Spenner. (Var. *nudus* Klett & Richter.) Spikelets glabrous.—With the sp.

36. **B. Trìnii** Desv. Erect annual, the culms rather slender, 3–6 dm. tall; sheaths and blades ± pilose, the latter mostly 3–8 mm. wide; panicle 1–2 dm. long, rather dense; spikelets 5–7-fld., 1.5–2 cm. long; glumes lanceolate, acuminate, smooth, the 1st mostly 1-nerved, 8–10 mm. long, the 2d broader, mostly 3-nerved, 12–16 mm. long; lemmas sparsely coarse-pubescent, 5-nerved, 12–14 mm. long, acuminate, the teeth narrow, 2 mm. long; awn 1.5–2 cm. long, twisted below, bent below the middle and strongly divaricate at maturity.—Dry slopes and plains, below 5000 ft.; many Plant Communities; from Siskiyou Co. s., especially in the deserts; to Ore., Colo., L. Calif., Chile. March–May.

Var. **excélsus** Shear. Spikelets larger, the lemmas 7-nerved; awns divaricate, but not twisted.—Panamint Mts.; near Lake Mead, Ariz.

2. Brachypòdium Beauv.

Annuals or perennials with erect racemes of subsessile spikelets. Fls. several to many, the rachilla disarticulating above the glumes and between the florets. Glumes unequal, sharp-pointed, 5- and 7-nerved; lemmas firm, rounded or ± flattened on the back, 7-nerved, acuminate, awned or mucronate; palea as long as the body of the lemma, concave, the keels pectinate-ciliate. A small genus of Eurasia, with 2 spp. from Mex. and S. Am. (Greek, *brachys,* short, and *podion,* foot, because of the short pedicels.)

1. **B. distáchyon** (L.) Beauv. [*Bromus distachyos* L.] Branched and geniculate at base, 1.5–3 dm. high, with pubescent nodes; sheaths and blades sparsely pilose to subglabrous; ligule 1.5–2 mm. long, pubescent; blades flat, 3–4 mm. wide; raceme strict,

the segms. of the axis alternately concave; spikelets 1–5, ± imbricate, 2–3.5 cm. long, 5–6 mm. wide; awns slender, erect, 1–2 cm. additional; lemmas scabrous, 8–9 mm. long; $n = 15$ (Avdulov, 1931).—Becoming well established in grassy and waste places of cent. and n. Calif., Santa Catalina Id.; native of Eurasia. April–May.

3. Festùca L. Fescue

Annuals or perennials, with spikelets in narrow or open panicles. Spikelets few- to several-fld., the rachilla disarticulating above the glumes and between the florets, the uppermost floret reduced. Glumes narrow, membranaceous, acute, unequal, the 1st sometimes very small; lemmas rounded on the back, membranaceous or indurate, 5-nerved, the nerves often obscure, acute to obtusish, awned from the tip or rarely from a bifid apex, sometimes awnless. Ca. 100 spp. of temp. and cool regions. Many spp. are important forage grasses. (Ancient Latin name for some grass.)

(Piper, C. V. N. Am. species of Festuca. Contr. U.S. Natl. Herb. 10: 1–48, 1906. St. Yves, A. Contribution a l'étude des Festuca. Candollea 2: 229–316, 1925.)

A. Plants annual; florets often cleistogamous, with 1 or 3 anthers (Section Vulpia).
    B. Spikelets densely 5–12-fld.; lemmas without scarious margin . . . . 1. *F. octoflora*
    BB. Spikelets loosely 1–5-fld.; lemmas with narrow scarious margin.
        C. Infl. narrow, branches ascending or appressed-ascending.
            D. Lemmas conspicuously long-ciliate toward apex . . . . 2. *F. megalura*
            DD. Lemmas not ciliate.
                E. Lower glume ⅔ to ¾ as long as the second . . . . 3. *F. dertonensis*
                EE. Lower glume less than half as long as the second . . . . 4. *F. myuros*
        CC. Infl. broader, the principal branches spreading.
            D. Spikelets glabrous, or ± scabrous, but not with long well developed hairs.
                E. The spikelets usually 3–5-fld.; only the main branches of the infl. divergent
                    5. *F. pacifica*
                EE. The spikelets mostly 1–2-fld.; all branches of infl. divergent or reflexed
                    9. *F. reflexa*
            DD. Spikelets pubescent, with long, well developed hairs.
                E. Pedicels appressed or slightly spreading; lower branches of panicle usually spreading or reflexed.
                    F. Lemmas glabrous; glumes pubescent . . . . 6. *F. confusa*
                    FF. Lemmas pubescent.
                        G. Lemmas hirsute; glumes glabrous or pubescent; lower branches of panicle spreading or reflexed . . . . 7. *F. Grayi*
                        GG. Lemmas woolly-pubescent; glumes glabrous; lower branches of panicle short, scarcely spreading . . . . 8. *F. arida*
                EE. Pedicels and panicle-branches all finally spreading or reflexed.
                    F. Glumes glabrous; lemmas pubescent . . . . 10. *F. microstachys*
                    FF. Glumes pubescent.
                        G. Lemmas pubescent . . . . 11. *F. Eastwoodae*
                        GG. Lemmas glabrous . . . . 12. *F. Tracyi*
AA. Plants perennial; florets opening; anthers mostly 3 (Section Festuca).
    B. Awns scarcely if at all evident, less than 1 mm. long.
        C. Lf.-blades flat, 4–8 mm. wide.
            D. Panicle 1–2 dm. long; lemmas 5–7 mm. long . . . . 16. *F. elatior*
            DD. Panicle 1.5–3.5 dm. long; lemmas 7–10 mm. long . . . . 17. *F. arundinacea*
        CC. Lf.-blades involute, narrower.
            D. Panicle open, the branches mostly in pairs, spreading to ascending; blades soft, green. Subalpine . . . . 19. *F. viridula*
            DD. Panicle narrow, the branches closely ascending; blades stiff, ± glaucous. E. Mojave Desert . . . . 24. *F. arizonica*
    BB. Awns well developed, 2–many mm. long.
        C. Lf.-blades flat, rather soft and lax.
            D. Florets long-stipitate, the rachilla appearing to be jointed a short distance below the floret. N. Calif. . . . . 13. *F. subuliflora*
            DD. Florets not stipitate.
                E. Lemma indistinctly nerved; awn terminal; blades 3–10 mm. wide
                    14. *F. subulata*
                EE. Lemma distinctly 5-nerved; awn from between 2 teeth; bl . . . s 2–4 mm. wide
                    15. *F. Elmeri*
        CC. Lf.-blades narrow or involute, not lax.
            D. Collar and mouth of sheath villous . . . . 18. *F. californica*
            DD. Collar and mouth of sheath not villous.
                E. Awn of lemma longer than the body; ovary pubescent at the summit
                    21. *F. occidentalis*
                EE. Awn of lemma shorter than the body; ovary glabrous at the summit.
                    F. Culms loosely tufted and decumbent at the base; blades smooth; sheaths red-brown with conspicuous pale nerves . . . . 20. *F. rubra*

FF. Culms closely tufted, not decumbent; blades smooth or scabrous; sheaths
    not red-brown at base.
    G. Blades scabrous; plants 3–10 dm. tall . . . . 23. *F. idahoensis*
    GG. Blades smooth; plants 1–2 dm. tall . . . . 22. *F. brachyphylla*

1. **F. octoflòra** Walt. [*Vulpia o.* Rydb.] Six-weeks Fescue. Tufted annual, the stems 0.5–4 dm. tall, glabrous or puberulent; blades involute, 2–8 cm. long; panicle narrow, erect, racemiform, 2–10 cm. long; spikelets 5–10 mm. long, densely 5–13-fld.; glumes subulate-lanceolate, the 1st 1-nerved, 3 mm. long, the 2d 3-nerved, 4 mm. long; lemmas lanceolate, convex, firm, glabrous or scabrous, 4–5 mm. long; awn scabrous, 2–4 mm. long.—Common in dry open places, below 5000 ft.; many Plant Communities; cismontane and desert Calif.; to Wash., N.Y., Fla., Tex. April–June.

Ssp. **hirtélla** Piper. Lemmas hirtellous or pubescent. With the sp.; to B.C., Kans., Fla., Mex.

2. **F. megalùra** Nutt. [*Vulpia m.* Rydb.] Foxtail Fescue. Simple or tufted annual, glabrous, 2–6 dm. tall; sheaths and blades smooth; the latter flat or involute; panicle narrow, 7–20 cm. long, with appressed branches; spikelets 4–5-fld., 8–11 mm. long; glumes glabrous, very unequal, the 1st less than 2 mm. long, the 2d 4–5 mm. long; lemma obscurely 5-nerved, 4–6 mm. long, scabrous and ciliate on upper half; awn ca. 8–15 mm. long, scabrous.—Common in dry open places, below 5500 ft.; many Plant Communities; cismontane Calif.; to B.C., Mont., Ida., L. Calif.; S. Am. April–June.

3. **F. dertonénsis** (All.) Asch. & Graebn. [*Bromus d.* All. *Vulpia d.* Volk. *F. bromoides* auth.] Similar to *F. megalura*, panicle 5–10 cm. long; 1st glume 4 mm. long, 2d 6 mm. long; lemma 7–8 mm. long, not ciliate; awn 10–12 mm. long.—Occasional in cismontane Calif.; introd. from Eu. April–June.

4. **F. Myùros** L. [*Vulpia M. K.* Gmel.] Like *F. megalura*, but panicle commonly somewhat shorter; 1st glume a little shorter; lemmas without cilia; *n* = 7 (Avdulov, 1928), 21 (Tanji, 1925).—Occasional in cismontane Calif.; introd. from Eu. March–May.

5. **F. pacífica** Piper. [*Vulpia p.* Rydb.] Erect or geniculate at base, 2–5 dm. tall, quite glabrous; blades soft, loosely involute; panicle 5–12 cm. long, the lower branches solitary, somewhat distant, spreading, subsecund, 1–3 cm. long; spikelets 3–6-fld., on appressed pedicels; glumes glabrous, the 1st 1-nerved, 4 mm. long, the 2d 3-nerved, 5 mm. long; lemmas lanceolate, mostly scabrous, 6–7 mm. long, with an awn 10–15 mm. long.—Dry open places below 5000 ft.; many Plant Communities; cismontane Calif.; to B.C., L. Calif., Ariz. March–June.

6. **F. confùsa** Piper. Resembling *F. pacifica;* sheaths retrorsely pilose; foliage pubescent; spikelets 2–3-fld.; glumes hirsute; lemmas glabrous.—Occasional on dry hills, below 5000 ft.; many Plant Communities; cismontane Calif. to Wash. March–June.

7. **F. Gràyi** (Abrams) Piper. [*F. microstachys Grayi* Abrams.] Like *F. pacifica*, often somewhat stouter; sheaths and sometimes blades pubescent; glumes glabrous to ± villous; lemmas pubescent, puberulent or ± villous.—Frequent below 6000 ft., in open places; many Plant Communities; cismontane Calif. to Ore. March–June.

8. **F. árida** Elmer. Desert Fescue. Annual with erect or spreading culms, 0.5–1.5 dm. high; sheaths glabrous; blades glabrous, loosely involute, mostly less than 4 cm. long; panicle narrow, 2–5 cm. long, with appressed branches or the lower spreading; glumes subequal, 5–6 mm. long, glabrous; lemmas densely woolly, ca. 5 mm. long, the awn 5–10 mm. long.—Dry often sandy places; reported from ne. Calif.; to e. Wash., Ida., w. Nev. May–July.

9. **F. refléxa** Buckl. [*Vulpia r.* Rydb.] Annual, the culms 2–4 dm. tall; sheaths smooth or pubescent; blades narrow, flat or ± involute, 2–10 cm. long; panicle 5–12 cm. long, branches and spikelets divaricate in maturity; spikelets mostly 1–3-fld., 5–7 mm. long; 1st glume 2–4 mm. long, the 2d 4–5 mm. long; lemmas glabrous or scaberulous, 5–6 mm. long; awns 5–8 mm. long.—Dryish open places, below 5000 ft.; many Plant Communities; cismontane Calif.; to Wash., Utah, Ariz. April–June.

10. **F. micróstachys** Nutt. [*Vulpia m.* Munro ex Benth.] Like *F. reflexa;* the glumes glabrous; lemmas pubescent.—Rare, in open places below 5500 ft.; several Plant Communities; cismontane Calif., Panamint Mts.; to Wash. April–June.

11. **F. Eastwoòdae** Piper. Resembling *F. reflexa;* panicle-axis and branches pubescent; glumes hirsute; lemmas hirsute; awns 3–10 mm. long.—Occasional in open places; Foothill Wd., Chaparral, etc.; San Benito and Monterey cos. n. in the Coast Ranges to Ore. April–June.

12. **F. Tràcyi** Hitchc. Resembling *F. reflexa;* axis of panicle pubescent; glumes villous, the 1st 1.5–2 mm. long, the 2d 3–4 mm. long; lemmas glabrous, ca. 4 mm. long; awn 4–9 mm. long.—Open places, Foothill Wd., etc.; Kern, Colusa, San Benito, Kings, Napa, and Lake cos. April–May.

13. **F. subulíflora** Scribn. in Macoun. Perennial, with erect slender glabrous culms 6–10 dm. tall; blades flat, pubescent, 3–6 mm. wide or loosely involute upon drying; panicle loose, 1–2 dm. long, nodding, the branches drooping, naked at base; spikelets loosely 3–5-fld., the rachilla pubescent or hispid, with internodes to 2 mm. long; florets long-stipitate; glumes narrow, acuminate, the 1st 3–4 mm. long, the 2d 4–5 mm.; lemmas scaberulous toward base, 6–8 mm. long; awn ± flexuous, 10–15 mm. long.— Moist shaded places, below 3000 ft.; Mixed Evergreen F., Redwood F.; Humboldt and Del Norte cos.; to B.C. June–July.

14. **F. subulàta** Trin. in Bong. Perennial with scaberulous culms 4–12 dm. high; sheaths subglabrous; blades thin, lax, 3–10 mm. wide, mostly scabrous; panicle very lax, drooping, 1.5–4 dm. long, the branches mostly in 2's and 3's, naked near base, finally spreading or reflexed, the lower to 1.5 dm. long; spikelets loosely 3–5-fld.; glumes subulate, the 1st ca. 3 mm., the 2d 5 mm. long; lemmas ± keeled, scaberulous toward apex; awn 5–20 mm. long.—Shaded banks and moist places; Redwood F., Montane Coniferous F.; Humboldt Co. to Modoc and Mariposa cos.; to Alaska, Mont., Wyo., Utah. June–Aug.

15. **F. Élmeri** Scribn. & Merr. Perennial, loosely tufted, the culms slender, 4–10 dm. tall; blades flat, scabrous or pubescent on upper surface, 2–4 mm. wide; panicle open, 1–2 dm. long, with slender ± drooping branches naked toward base, the lower to 1 dm. long; spikelets 3–4-fld.; glumes lance-acuminate, the 1st 2–2.5 mm. long, the 2d 3–4 mm. long; lemmas membranaceous, hispidulous, ca. 6 mm. long, minutely 2-toothed at apex; awn 2–8 mm. long.—Wooded slopes, below 5000 ft.; Mixed Evergreen F., Foothill Wd., N. Oak Wd., Yellow Pine F.; Coast Ranges from Monterey Co. n. to Ore. May–June.

Ssp. **luxùrians** Piper. [*F. Jonesii* var. *conferta* Hack.] Spikelets 5–6-fld.; panicles rather congested.—San Francisco Bay region.

16. **F. elàtior** L. MEADOW FESCUE. Perennial, with stout culms 5–12 dm. high; sheaths smooth; blades flat, 4–8 mm. wide, 1–2 dm. long; panicles erect or with nodding summit, subsimple to much-branched, 1–2 dm. long, the branches bearing spikelets nearly to the base; spikelets mostly 6–8-fld., 8–12 mm. long; glumes lanceolate, 3 and 4 mm. long; lemmas oblong-lanceolate, coriaceous, 5–7 mm. long, scarious at apex, rarely short-awned.—Occasional in meadows and waste places as escape from cult.; introd. from Eu. May–July.

17. **F. arundinàcea** Schreb. [*F. elatior* var. *a.* Wimm.] Culms somewhat taller and more robust than in *F. elatior;* blades 2.5–7 dm. long; panicles 1.5–3 dm. long, with more branches and spikelets; spikelets 8–12 mm. long, 4–8-fld., most violet-tinged; lemmas 7–10 mm. long; $n = 14$ (Staehlin, 1929), 21 (Jenkins, 1933).—Occasionally introd. from Eu. May–June.

18. **F. califórnica** Vasey. Perennial with rather stout culms 6–12 dm. high; sheaths ± scabrous, the collar villous; blades flat or involute when dry, firm, scabrous; panicle loose, few-branched, these with few large spikelets toward the ends; spikelets compressed, ca. 5-fld.; glumes lance-oblong, firm, glabrous except for the scabrous keel, 6–8 mm. long; lemmas 8–10 mm. long, convex, acuminate or short-awned; $n = 28$ (Stebbins & Love, 1941).—Wood borders, shaded places, etc., below 6000 ft.; Mixed Evergreen F. to Yellow Pine F.; Coast Ranges from Monterey Co. to Ore. April–July.

Var. **Paríshii** (Piper) Hitchc. [*F. aristulata* P. Piper. *F. P.* Hitchc.] Culms 4–6 dm. tall; sheaths puberulent; blades involute, 1.5–2.5 dm. long; panicle ca. 1 dm. long; awn 3–4 mm. long.—Dry benches, 4000–6500 ft.; Chaparral, Yellow Pine F.; San Bernardino Mts.

19. **F. virídula** Vasey. Loosely tufted perennial, the culms slender, smooth, 6–10 dm. high; sheaths smooth; blades erect, 1–2 mm. wide, flat or loosely involute; panicle open 1–1.5 dm. long, the branches mostly paired, spreading to ascending, slender, rather remote, naked below; spikelets 3–6-fld., 10–12 mm. long; glumes smooth, the lower ca. 5–6 mm. long, the 2d 8–9 mm. long; lemmas firm, keeled toward apex, 6–7 mm. long.—Subalpine meadows, 6000–9000 ft.; Sierra Nevada from Alpine to Sierra cos.; Coast Ranges from Sonoma Co. n.; to B.C., Ida., Colo. July–Aug.

20. **F. rùbra** L. [*F. r.* ssp. *densiuscula* Hack.] RED FESCUE. Perennial, with loosely tufted culms bent or decumbent at the ± reddish base, 4–10 dm. high; lower sheaths brown, thin, fibrillose; blades smooth, soft, ± involute; panicle 0.3–2 dm. long, mostly narrow, with erect branches; spikelets 4–6-fld., pale, often with purplish tinge; mostly 7–8 mm. long; glumes smooth, the lower 3–4, the upper 5–7 mm. long; lemmas 5–7 mm. long, smooth or scabrous upward; awn scabrous, ca. 2–4 mm. long; $n = 7$ (Staehlin, 1929).—Meadows and moist places, from sea level to 8500 ft.; many Plant Communities; from Monterey Co. and San Bernardino Mts. n.; through cooler parts of N. Am. and Eurasia. May–July.

21. **F. occidentàlis** Hook. Perennial, tufted, the culms erect, slender 4–10 dm. tall; sheaths smooth; blades many, largely basal, narrow-involute, ± smooth; panicle 0.7–2 dm. long, often drooping above, the branches in 1's or 2's; spikelets loosely 3–5-fld., 6–10 mm. long, on slender pedicels; lemmas rather thin, 5–6 mm. long, scaberulous upward; awn slender, 6–10 mm. long.—Dry rocky wooded banks and slopes, below 6500 ft.; Redwood F., Mixed Evergreen F. to Red Fir F.; Tulare Co. to Siskiyou Co., San Mateo Co. to Del Norte Co.; to B.C., Mich. May–Aug.

22. **F. brachyphýlla** Schult. [*F. ovina* ssp. *b.* Piper. *F. ovina* ssp. *saximontana* var. *Purpusiana* St. Yves.] Tufted perennial, with erect very slender culms 1–1.5 dm. high; blades smooth, filiform, soft, angled in drying, 2–6 cm. long, numerous; panicle short, narrow, few-fld., 2–5 cm. long; glumes and lemmas broad; spikelets 2–5-fld.; lemmas 3–3.5 mm. long; awn 1–3 mm. long.—Dry ± rocky places, 9400–14,000 ft.; Subalpine F., Bristle-cone Pine F., Alpine Fell-fields; San Bernardino Mts., Sierra Nevada, Inyo-White Mts.; to Arctic region and s. to Colo., Wis., Vt. July–Sept.

23. **F. idahoénsis** Elmer. Perennial with densely tufted, smooth or ± scabrous culms 3–10 dm. tall; lvs. numerous, mostly basal, the blades firm, stiffish, scabrous, 1.5–3 dm. long, involute; panicle narrow, 1–2 dm. long, the branches ascending or appressed, scabrous; spikelets mostly 5–7-fld., scabrous; glumes ca. 1.5–2 and 3 mm. long; lemmas ca. 7 mm. long, subterete, ± purplish; awns mostly 2–4 mm. long.—Dry openings in woods and on rocky slopes, mostly below 5000 ft.; several Plant Communities; Coast Ranges from San Mateo Co. n. to Siskiyou, Lassen, Modoc cos.; to B.C., Alta., Colo. May–July. Plants near this but with spikelets 3–5-fld. and awn not over 1 mm. long, may represent *F. altaica* Trin.

24. **F. arizónica** Vasey. [*F. ovina a.* Hack.] Resembling *F. idahoensis,* the lvs. stiffer, more glaucous; glumes ca. 4 and 5–6 mm. long; lemmas 5–6 mm. long, awnless or nearly so.—Dry slopes at 5500 ft.; Pinyon-Juniper Wd.; Clark Mt., e. San Bernardino Co.; Ariz. to Colo., Tex. June.

4. Sclerópoa Griseb.

Annuals with slightly branched 1-sided panicles. Spikelets several-fld., linear, ± compressed, the rachilla thick, disarticulating above the glumes and between the florets, remaining as a minute stipe to the floret above. Glumes unequal, short, acutish, strongly nerved, the 1st 1-nerved, the 2d 3-nerved. Lemmas subterete, obscurely 5-nerved, obtuse, slightly scarious at apex. A small genus of the Old World. (Greek, *skleros,* hard, and *poa,* grass, because of the stiff panicle.)

1. **S. rígida** (L.) Griseb. [*Poa r.* L.] Culms spreading to erect, 1–3 dm. tall; blades flat, 1–2 mm. wide; panicles stiff, narrow, 0.5–1 dm. long, dense, the branches short, bearing spikelets from near the base; pedicels thick, ± spreading; spikelets 4–10-fld., 5–8 mm. long; glumes ca. 1.5 and 2.5 mm. long; lemmas ca. 2.5 mm. long, glabrous, awnless; $n = 7$ (Avdulov, 1931).—Becoming locally a fairly frequent weed in cent. coastal Calif.; introd. from Eu. May–Sept.

5. Puccinéllia Parl. ALKALI GRASS

Low mostly pale annuals or perennials, tufted. Lf.-sheaths open. Panicles narrow to open. Spikelets several-fld., subterete, the rachilla disarticulating above the glumes and between the florets; glumes shorter than the first lemma, the 1st mostly 1-nerved, the 2d 3-nerved. Lemmas usually firm, rounded on the back, obtuse or acute, rarely acuminate, usually scarious and often erose at apex, 5-nerved, the nerves mostly obscure or indistinct, sometimes conspicuous. Styles absent. Ca. 28 spp., mostly of ± saline or alkaline places; N. Hemis. (Named for B. *Puccinelli,* Italian botanist.)

(Clausen, R. T. Suggestion for assignment of Torreyochloa to Puccinellia. Rhodora 54: 42–45, 1952.)
A. Plants annual, not more than 2 dm. high.
    B. Lemmas obtuse, pubescent on the nerves only. Mojave Desert . . . . 1. *P. Parishii*
    BB. Lemmas acute, ± pubescent over the back. San Joaquin V. . . . . 2. *P. simplex*
AA. Plants perennial, mostly larger.
    B. Lemmas with nerves obscure or indistinct; tufted plants, not having creeping rhizomes.
        C. Lvs. mostly in a short basal tuft; panicle mostly 5–10 cm. long. E. of Sierra Nevada
                3. *P. Lemmoni*
        CC. Lvs. well distributed along stems; panicle 5–20 cm. long.
            D. Lemmas 2–3 mm. long; panicle open, the branches spreading to reflexed.
                E. Lemmas broad, obtuse or truncate, not narrowed above, ca. 2 mm. long;
                    panicle-branches usually reflexed. Introd. . . . . 4. *P. distans*
                EE. Lemmas narrowed into an obtuse apex, ca. 3 mm. long; panicle-branches
                    usually spreading. Native . . . . 5. *P. airoides*
            DD. Lemmas 3–4 mm. long; panicle narrow, mostly with ascending branches. Sea
                beaches . . . . 6. *P. grandis*
    BB. Lemmas with conspicuous nerves; plants with creeping rhizomes.
        C. Panicles 3–8 cm. long; lvs. 3–6 mm. wide. Near timberline.
            D. Spikelets 5–6 mm. long; plants 3–6 dm. high . . . . 7. *P. erecta*
            DD. Spikelets ca. 3 mm. long; plants mostly 1–2 dm. high . . . . 8. *P. californica*
        CC. Panicles 10–20 cm. long; lvs. 6–12 mm. wide; spikelets 4–5 mm. long. Mostly well
        below timberline . . . . 9. *P. pauciflora*

1. **P. Paríshii** Hitchc. Tufted glabrous annual, the culms 3–20 cm. tall; blades flat to subinvolute, scarcely 1 mm. wide; panicle narrow or wider, 1–8 cm. long; spikelets 3–6-fld., 3–5 mm. long; glumes ca. 1.5 and 2 mm. long; lemmas ca. 2 mm. long, obtuse to truncate, scarious and ± erose at apex, pubescent on nerves nearly to the apex; $n = 7$ (Church, 1949).—Alkaline seeps, at 2900 ft.; Joshua Tree Wd.; Rabbit Springs, Mojave Desert; Tuba City, Ariz. April–May.

2. **P. símplex** Scribn. Annual, resembling *P. Parishii;* spikelets 6–8 mm. long; glumes 1 and 2 mm. long; lemmas 2.5 mm. long, acute, ± pubescent over the back; $2n = 56$ (Church, 1949).—Alkaline spots; largely V. Grassland; Cent. V. and interior valleys of Coast Ranges. March–May.

3. **P. Lemmònii** (Vasey) Scribn. [*Poa L.* Vasey. *Puccinellia rubida* Elmer.] Tufted perennial with slender culms 1.5–4 dm. high; blades mostly basal, the slender blades ± glaucous, involute, 4–10 cm. long; panicle pyramidal to rather narrow, 5–10 cm. long, the branches slender, flexuous, fascicled, the lower naked on lower half; spikelets 3–5-fld., 5–6 mm. long; glumes mostly 1-nerved, the first 2 mm., the 2d 3 mm. long; lemmas 3–3.5 mm. long, glabrous, acute; $n = 7$ (Church, 1949).—Moist alkaline places, 4000–7000 ft.; Pinyon-Juniper Wd., Sagebrush Scrub, N. Juniper Wd.; Mono to Modoc and Siskiyou cos.; to Wash., Ida., Nev. May–Aug.

4. **P. dístans** (L.) Parl. [*Poa d. L.*] Perennial with tufted culms, spreading at base or erect, 2–4 dm. tall; blades flat or ± involute, 2–4 mm. wide, 2–10 cm. long; panicle pyramidal, loose, 5–15 cm. long, the branches rather distant, fascicled, the lower spreading to reflexed, naked in their lower half; spikelets 4–6-fld., 4–5 mm. long; glumes 1 and 2 mm. long; lemmas rather thin, obtuse or truncate, ca. 2 mm. long, with few short hairs at base; $n = 7, 21$ (Avdulov, 1931).—Moist, ± alkaline places, as occasional introd. (Baldwin Lake); from Eu. June–July.

5. **P. airoìdes** (Nutt.) Wats. & Coult. [*Poa a.* Nutt. *P. Nuttalliana* Schult.] Tufted perennial, the culms slender, mostly erect, rather stiff and firm at base, 3–6 dm. long; blades 1–3 mm. wide, flat to ± involute; panicle pyramidal, open, mostly 1–2 dm. long, the branches scabrous, distant, spreading, naked near base, to 10 cm. long; spikelets 3–6-fld., 4–7 mm. long, with rather distant florets; glumes 1.5 and 2 mm. long, the 2d 3-nerved; lemmas 2–3 mm. long, narrowed into an obtuse apex; $n = 28$ (Church, 1949).—Alkaline places, below 7000 ft.; Mixed Evergreen F., Sagebrush Scrub, Yellow Pine F., etc.; scattered places in Calif.; to B.C., Wis., Kans., New Mex. June–Sept. Plants with anthers ca. 2 mm. long; lemmas 4–5 mm. long and from subsaline places may be **P. marítima** (Huds.) Parl.

6. **P. grándis** Swall. [*P. nutkaensis* of Calif. refs.] Perennial with densely tufted culms, 3–9 dm. tall; lvs. scattered, smooth, firm, the blades 2–3.5 mm. wide, involute on drying; panicles 1–2 dm. long, pyramidal, scabrous on the branches; spikelets 8–15 mm. long, 5–12-fld., appressed; glumes ca. 2 and 3 mm. long; lemmas 3–4 mm. long,

subacute to obtuse, sparsely pilose at base.—Coastal Salt Marsh and alkaline soils, w. Sacramento Co., San Mateo Co. to Alaska. June–Aug.

7. **P. erécta** (Hitchc.) Munz. [*Glyceria e.* Hitch. *Panicularia e.* Hitch. *Torreyochloa e.* Church.] Perennial, with slender creeping rhizomes and culms mostly 3–6 dm. tall; sheaths smooth, the broad ligule 3–4 mm. long; blades flat, mostly 5–12 cm. long, 4–6 mm. wide; panicle mostly 5–8 cm. long, rather narrow, with ± ascending branches; spikelets 4–6-fld., 5–6 mm. long, often purplish-tinged; lemmas ca. 3 mm. long, scaberulous, ± erose at summit; *n* = 7 (Church, 1949).—Wet places, often in the water, 9000–11,000 ft.; Subalpine F., Alpine Fell-fields; Sierra Nevada from Tulare Co. n., Trinity Co.; Ore., Nev. July–Sept.

8. **P. califórnica** (Beetle) Munz. [*Glyceria c.* Beetle. *Torreyochloa c.* Church.] Questionably distinct from *P. erecta*, supposedly lower; blades to 5 mm. wide; panicle 3–6 cm. long; spikelets ca. 3 mm. long, 3–5-fld.; lemmas barely 2 mm. long.—Wet places, 10,000–11,000 ft.; Subalpine F.; Sierra Nevada, Tulare, and Fresno cos. Aug.–Sept.

9. **P. pauciflòra** (Presl) Munz. [*Glyceria p.* Presl. *Torreyochloa p.* Church.] With running rhizomes, the culms mostly 5–12 dm. high; blades flat, 5–15 mm. wide, 10–15 cm. long; panicle 1–2 dm. long, loose and open or narrow; spikelets mostly 5–6-fld., 4–5 mm. long, often purplish; glumes ca. 1 and 1.5 mm. long, broad, with erose-scarious margins; lemmas oblong, 2–2.5 mm. long, scaberulous on nerves, with scarious rounded ± erose tips; *n* = 7 (Church, 1949).—Wet banks and shallow water, from sea level to 10,000 ft.; many Plant Communities; Sierra Nevada, Coast Ranges from San Mateo Co. n.; to Alaska, S. Dak., New Mex. July–Sept.

6. Glycèria R. Br. MANNA GRASS

Aquatic or marsh perennials with creeping rhizomes and flat blades. Lf.-sheaths connate. Panicles open or contracted. Spikelets few–many-fld., subterete, the rachilla disarticulating above the glumes and between the florets. Glumes unequal, 1-nerved. Lemmas broad, convex on the back, firm, usually obtuse, scarious at apex, 5–9-nerved, the usually prominent nerves parallel. Styles present. Ca. 30 spp. in temp. regions in both hemis. (Greek, *glukeros*, sweet, the seed of some spp. being so.)

Spikelets linear, mostly 1 cm. or longer; panicles narrow, erect.
   Lemmas glabrous between the slightly scabrous nerves, 3–4 mm. long . . . . 1. *G. borealis*
   Lemmas scaberulous or hirtellous between the distinctly scabrous nerves.
      The lemmas ca. 3 mm. long, rounded at the summit . . . . 2. *G. leptostachya*
      The lemmas 4–6 mm. long, ± angled at summit.
         Culms 6–10 dm. tall; blades 3–12 mm. wide. Native . . . . 3. *G. occidentalis*
         Culms 1–7 dm. tall; blades 2–3 mm. wide. Introd. 4. *G. declinata*
Spikelets ovate or oblong, usually not more than 5 mm. long; panicles usually nodding.
   Lf.-blades 2–4(–8) mm. wide, rather firm, ± folded; spikelets 3–4 mm. long . . . . 5. *G. striata*
   Lf.-blades 6–12 mm. wide, flat, rather thin; spikelets 4–6 mm. long . . . . 6. *G. elata*

1. **G. boreàlis** (Nash) Batch. [*Panicularia b.* Nash.] Culms erect or decumbent at base, 5–10 dm. tall; blades flat or folded, mostly 2–4 mm. wide, scabrous above, erect; panicle mostly 2–4 dm. long, narrow, the branches and pedicels appressed; spikelets mostly 6–12-fld., 10–15 mm. long, not purple-tinged; glumes ca. 1.5 and 3 mm. long; lemmas rather thin, obtuse, 3–4 mm. long, strongly 7-nerved, scarious at tip; *n* = 10 (Church, 1949).—Wet places and shallow water, 2750–7000 ft.; Yellow Pine F., Red Fir F.; Mariposa to Siskiyou and Modoc cos.; Alaska, Pa., New Mex., Ariz. June–Aug.

2. **G. leptostàchya** Buckl. [*Panicularia Davyi* Merr.] Culms 1–1.5 m. tall; sheaths ± scaberulous; blades flat, ± scabrous above, 4–7(–10) mm. wide; panicle 2–4 dm. long, with ascending branches in 2's and 3's; spikelets 10–20 mm. long, 8–14-fld., often purplish; glumes 1.5 and 3 mm. long; lemmas firm, broadly rounded, ca. 3 mm. long, scaberulous on the nerves and between; *n* = 20 (Church, 1949).—Freshwater Marsh, Marin and Sonoma cos. and about a lake at 7000 ft., Eldorado Co.; Ore., Wash. May–June.

3. **G. occidentàlis** (Piper) J. C. Nels. [*Panicularia o.* Piper. *G. fluitans* of Calif. refs.] Base decumbent, rooting, 6–12 dm. tall, the culms ± succulent; blades 3–12 mm. wide, ± scabrous above; panicle 3–5 dm. long, narrow, opening somewhat at anthesis; spike-

lets 15–20 mm. long; glumes very unequal, ca. 2 and 4 mm. long, obtuse; lemmas usually purplish at tip, 4–6 mm. long, scabrous, 7–9-nerved; $n = 20$ (Church, 1949).—Swampy places and shallow water; Freshwater Marsh, etc.; Marin Co. n.; to B.C., Ida., Nev. June–Aug.

4. **G. declinàta** Brebiss. [*G. Cookei* Swall.] Erect from a decumbent branching base, 2–7 dm. tall; sheaths keeled, scaberulous; ligule 5–7 mm. long; blades 3–12 cm. long, 2–6 mm. broad; spikelets 15–20 mm. long, appressed; glumes obtuse, the 1st ca. 2 mm., the 2d 3–3.5 mm. long; lemma 4–5 mm. long, scabrous, 7-nerved, obtuse, irregularly dentate; $n = 10$ (Church, 1949).—Moist canyons and meadows, at ca. 3000 ft.; Shasta City, Siskiyou Co.; Nev.; introd. from Eu. May–June.

5. **G. striàta** (Lam.) Hitchc. [*Poa s.* Lam. *Panicularia nervata*, Calif. refs.] Fowl Meadow Grass. Stems slender, loosely to densely tufted, 3–12 dm. high; sheaths scabrous; blades flat 2–6(–10) mm. wide, scabrous above, often folded, fairly firm, pale green; panicle open, ovoid, 1–2 dm. long, nodding, the capillary branches divergent or reflexed in age; spikelets greenish or purplish, 2–4 mm. long, 3–7-fld.; glumes ca. 0.5 and 1 mm. long, obtuse; lemmas sharply 7-nerved, oblong, ca. 2 mm. long, obtuse; $n = 10$ (Church, 1949).—Meadows and wet places, 5000–8000 ft.; Lodgepole F., Red Fir F.; Yolla Bolly Mts., Trinity Co. to Siskiyou and Modoc cos., s. to cent. Sierra Nevada; to B.C., Nfld., Fla., Tex., Ariz., Mex. July–Aug.

6. **G. elàta** (Nash) Hitchc. [*Panicularia e.* Nash.] Near to *G. striata*, the culms 10–18 dm. tall, ± succulent; plants darker green; blades flat, thin, 6–12 mm. wide; panicle diffuse, 1.5–3 dm. long, the branches spreading or the lower deflexed; spikelets 6–8-fld., 3–6 mm. long; glumes ca. 1 and 1.5 mm. long; lemmas 2–3 mm. long; $n = 10$ (Church, 1949).—Wet places, below 8500 ft.; Yellow Pine F. to Lodgepole Pine F., Redwood F.; from San Jacinto and San Bernardino mts., Sierra Nevada to Coast Ranges of Mendocino, Humboldt, and Siskiyou cos.; to B.C., Mont., Colo., New Mex. July–Aug.

G. grándis Wats., with glumes ca. 1.5 and 2 mm. long and pale in contrast to the purple lemmas, may occur at Bridgeport, Mono Co. and n.

7. Pleuropògon R. Br. Semaphore Grass

Soft annuals or perennials, with flat blades and loose racemes. Spikelets rather large, several–many-fld., linear, the rachilla disarticulating above the glumes and between the florets. Glumes mostly unequal, membranaceous or subhyaline, scarious at the ± lacerate tip, the 1st 1-nerved, the 2d obscurely 3-nerved. Lemmas 7-nerved, with a basal round hardened callus, entire to 2-toothed and mucronate or short-awned at apex. Palea 2-keeled, the keels winged. Ca. 5 spp. of w. U.S. and 1 circumboreal. (Greek, *pleura*, side, and *pogon*, beard.)

(Benson, L. Taxonomic Studies. Am. Journ. Bot. 28: 358–360, 1941.)

Lemmas awnless or mucronate, thick, firm, strongly nerved .... 1. *P. Davyi*
Lemmas with awns 1–12 mm. long.
   The lemma 4–6 mm. long; lf.-blades mostly 2–4 mm. wide .... 2. *P. californicus*
   The lemma 6–9 mm. long; lf.-blades mostly 5–12 mm. wide.
      Awn 5–20 mm. long; spikelets reflexed or spreading .... 3. *P. refractus*
      Awn 1–2.5 mm. long; spikelets erect or ascending .... 4. *P. Hooverianus*

1. **P. Dàvyi** L. Benson. Glabrous subpalustrine perennial, with short rhizomes; culms 1–several, erect, 6–10 dm. tall; blades 1–3 dm. long, 6–9 mm. broad; infl. subspicate, 2–3 dm. long; spikelets sessile or subsessile, mostly 10–20-fld., 2–5.5 cm. long; glumes subequal; lemmas 5.5–7.5 mm. long, strongly nerved, obtuse, awnless or mucronate; palea markedly winged.—Wet places, below 2000 ft.; Mixed Evergreen F., N. Oak Wd.; Lake and Mendocino cos. March–May.

2. **P. califórnicus** (Nees) Benth. in Vasey. [*Lophochlaena c.* Nees. *P. Douglasii* Trin.] Glabrous terrestrial perennial with slender elongate rhizomes; culms suberect, 3–7 dm. high; blades 3–15 cm. long, 2–4(–5) mm. wide; infl. 1–2 dm. long, with 5–10 remote short-pedicelled spikelets; these 6–12-fld., 1.2–2.5 cm. long; glumes unequal, the 1st 1–3 mm. long, the 2d 3–6 mm.; lemmas 4–6 mm. long, bifid to subentire at apex, scabrous, obtuse, erose; awn 6–12 mm. long; palea markedly winged, the wings cleft, forming a tooth near the middle.—Wet places at low elevs.; largely Redwood F., Mixed Evergreen F.; Stanislaus Co., near mouth of Sacramento R., Alameda Co. to Humboldt Co. March–May.

3. **P. refráctus** (Gray) Benth. ex Vasey. [*Lophochlaena r.* Gray.] Glabrous sub-palustrine perennial with slender rhizomes; culms 8–12 dm. high, few to many; blades 1–3 dm. long, 5–7(–12) mm. wide; raceme 1.5–3 dm. long, the spikelets 2–6(–8) cm. apart, reflexed at maturity, 6–12-fld., 2–3.5 cm. long; rachilla-joints not glandular-swollen at base; glumes 3.5–5 and 5–7 mm. long, scarious, similar; lemmas 8–9 mm. long, ca. 2 mm. broad, bifid, minutely ciliate-denticulate at rounded apex; awn 5–20 mm. long; palea scarcely or minutely toothed on wings.—Wet meadows, mt. streams, etc., below 5000 ft.; Redwood F., Douglas-Fir F., Yellow Pine F.; from Russian Gulch, Mendocino Co. n. to Wash. May–Aug.

4. **P. Hooveriànus** (L. Benson) J. T. Howell. [*P. refractus* var. *H.* L. Benson.] Similar to *P. r.*, but pedicels erect or ascending at maturity; base of rachilla-joints glandular-swollen; lower glume subentire, acute, upper coarsely and irregularly few-toothed; lemma coarsely few-toothed at apex; awn ca. 1 mm. long; palea-wings with subulate divergent appendages ca. 2 mm. long.—Meadows, Mixed Evergreen F.; Marin and Mendocino cos.

8. Hesperóchloa Rydb.

Densely tufted rhizomatous dioecious perennial. Lvs. firm, narrow, flat or loosely involute. Panicles narrow, erect. Spikelets 3–5-fld., the rachilla disarticulating above the glumes and between the florets. Glumes subequal or unequal, shorter than the 1st lemma, lanceolate, acute, the 1st 1-nerved, the 2d 3-nerved. Lemmas rounded on back, acute or acuminate, awnless, 5-nerved; palea as long as lemma, scabrous-ciliate on the keels; stigmas sessile, long, slender. One sp. (Greek, *esperis*, western, and *chloa*, grass.)

1. **H. Kíngii** (Wats.) Rydb. [*Poa K.* Wats. *Festuca K.* Cassidy. *F. confinis* Vasey.] Clumps large, dense, erect, 5–8 dm. tall; sheaths smooth, striate; blades firm, 3–6 mm. wide, 0.5–3 dm. long; panicle 7–20 cm. long, with short appressed branches; the ♂ infl. denser with somewhat larger spikelets than the ♀; spikelets 7–12 mm. long; glumes thin, the 1st 3–4 mm., the 2d 5–6 mm. long; lemmas 5–8 mm. long, acute to acuminate, scabrous.—Dry slopes, 7000–12,000 ft.; Yellow Pine F., Bristle-cone Pine F.; San Bernardino Mts., Inyo-White Mts., s. Sierra Nevada; Sweetwater Mts., to Ore., Mont., Nebr., Colo. June–Aug.

9. Pòa L. BLUEGRASS
Treatment by David D. Keck

Annuals or usually perennials. Blades relatively narrow, flat, folded or involute, ending in a boat-shaped tip. Panicles open or contracted. Spikelets 2–8-fld., the rachilla disarticulating above the glumes and between the florets, the uppermost floret often rudimentary. Glumes acute, keeled, the 1st 1-nerved, the 2d larger and usually 3-nerved. Lemmas somewhat keeled, acute or sometimes rounded, often scarious apically, awnless, 5-nerved (rarely 3-nerved by reduction of the intermediate nerves). Palea nearly equaling the lemma. Ca. 150 spp. in all temp. and cool regions of both hemis. (Ancient Greek, *poa*, grass or fodder.)

A. Plants annual.
  B. Lemmas pubescent, but not webbed at base. Cismontane .... 1. *P. annua*
  BB. Lemmas glabrous or pubescent, with a tuft of cobwebby hairs at base.
    C. Panicle at length open, the elongated branches not floriferous to base. Cismontane
                                                               2. *P. Bolanderi*
    CC. Panicle narrow, the short appressed branches floriferous to base. Deserts
                                                              3. *P. Bigelovii*
AA. Plants perennial (*P. Kelloggii* sometimes behaving as an annual).
  B. Creeping rhizomes present.
    C. Plants dioecious, the sexes morphologically similar.
      D. Sandbinders of the immediate coast.
        E. Lemmas 6 mm. or more long; lvs. coarse. Cent. and n. coast
                                                        4. *P. Douglasii*
        EE. Lemmas 3–3.5 mm. long; lvs. soft. N. coast .... 5. *P. confinis*
      DD. Woodland spp.
        E. Blades involute, canescent ventrally, the collar puberulent, the ligule very
            short. Del Norte Co. n. .... 6. *P. Piperi*
        EE. Blades flat, essentially glabrous, the collar glabrous or nearly so, the ligule
            3–4 mm. long. Siskiyou Co. .... 7. *P. rhizomata*
    CC. Plants not dioecious, the florets perfect (all ♀ in *P. nervosa*).

D. Panicle open, pyramidal, its elongated lower branches lax and only in outer half floriferous.

    E. Lemmas glabrous or puberulent but not webbed at base; florets all ♀. High montane . . . . 8. *P. nervosa*

    EE. Lemmas with a tuft of cobwebby hairs at base; florets perfect.

        F. Lemmas pilose on keel and marginal nerves; strongly perennial. Ubiquitous . . . . 9. *P. pratensis*

        FF. Lemmas glabrous except for the web; annual or weakly perennial. With the coast redwood . . . . 10. *P. Kelloggii*

DD. Panicle more contracted, oblong, its short lower branches not lax; lemmas not webbed at base (sometimes vestigially webbed in *P. compressa*).

    E. Culms conspicuously flattened, 2-edged . . . . 11. *P. compressa*

    EE. Culms terete or slightly flattened.

        F. Culms leafy only near base; panicle very dense. San Bernardino Mts.
                                      12. *P. atropurpurea*

        FF. Culms leafy well above the middle. N. Calif. . . . . 13. *P. fibrata*

BB. Creeping rhizomes wanting.

    C. Florets mostly converted into dark purple bulblets; culm with bulblike base
                                        14. *P. bulbosa*

    CC. Florets normal, green; culms not bulblike at base.

        D. Lemmas with a tuft of cobwebby hairs at base.

            E. Lemmas glabrous, or the keel slightly pubescent. Nw. Calif. . . 15. *P. trivialis*

            EE. Lemmas pubescent on keel and marginal nerves. E. Calif.

                F. Lower panicle branches in pairs, capillary . . . . 16. *P. leptocoma*

                FF. Lower panicle branches several together, not capillary or elongated
                                        17. *P. palustris*

        DD. Lemmas not webbed at base; all bunchgrasses.

            E. Lemmas glabrous or scabrid, but not pubescent or puberulent on lower part of nerves.

                F. Spikelets compressed, the lemmas keeled.

                    G. Lemmas less than 4 mm. long, usually purple; blades mostly less than 8 cm. long. High alpine dwarfs.

                      H. Culms scarcely exceeding the basal tuft of lvs.

                        I. Panicle 1–2 cm. long; spikelets 3–4 mm. long, usually very purple; glumes longer than 1st floret; lemmas 2–3 mm. long . . . . 18. *P. Lettermanii*

                        II. Panicle 2–3 cm. long; spikelets 4–6 mm. long, purple or greenish; glumes shorter than 1st floret; lemmas 3–5 mm. long . . . . 19. *P. Suksdorfii*

                    HH. Culms well exceeding the basal tuft of lvs., naked above the basal third, often very slender, mostly 10–25 cm. high
                                      20. *P. Hanseni*

                  GG. Lemmas longer.

                      H. Blades filiform to capillary, ± scabrid, the basal mostly 12 cm. or more long; lemmas 4.5–5.5 mm. long, pale. Middle elevs., n. Great Basin . . . . 21. *P. Cusickii*

                      HH. Blades mostly wider, not scabrous dorsally.

                        I. Lower lemmas mostly ca. 6 mm. long, the spikelet pale and shining; lvs. firm, short, often recurved; culms mostly 2 dm. or less tall. Subalpine from Siskiyou Mts. to Donner Pass . . . . 22. *P. Pringlei*

                      II. Lower lemmas 4–5 mm. long, the spikelet purple or green; basal lvs. linear, elongated, culm lvs. ± flattened and wider; culms mostly 3–5 dm. tall. High elevs. throughout the Sierra Nevada . . . . 23. *P. epilis*

              FF. Spikelets little compressed, elongated, the lemmas rounded dorsally, the keel obscure.

                  G. Ligule long, acuminate or sharply acute. Meadows.

                      H. Panicle narrow, elongated. Transmontane . 24. *P. nevadensis*

                      HH. Panicle ovoid-oblong, dense. Napa V. . . . . . . 25. *P. napensis*

                  GG. Ligule short, rounded or obtuse. Drier slopes. Great Basin.

                      H. Blades involute, greenish. Often in alkaline or rather sterile soils . . . . 26. *P. juncifolia*

                      HH. Blades flat, glaucous. In nonalkaline soils . . . . 27. *P. ampla*

        EE. Lemmas ± hairy on back, keel, or nerves at least towards base, sometimes obscurely so (occasionally glabrous in *P. scabrella*).

            F. Spikelets distinctly compressed; lemmas keeled.

                G. Lemmas puberulent towards base only. Coastal . 28. *P. unilateralis*

                GG. Lemmas ± villous on lower half of nerves.

                    H. Spikelets 5–7-flowered, 6–10 mm. long, pale or greenish, the bracts with prominent hyaline margin. Mid-altitudes; trans-Sierran . . . . 29. *P. Fendleriana*

                    HH. Spikelets 2–4-flowered, 2.5–5 mm. long, purplish, the bracts only moderately hyaline-margined. Alpine dwarf; Sierra Nevada . . . . 30. *P. rupicola*

FF. Spikelets little compressed, elongate; lemmas rounded on back, crisp-
puberulent toward base, the keel obscure.
G. Spring-flowering and ± ephemeral, summer dormant. Low to moder-
ate elevs.
H. Panicle usually contracted.
I. Lvs. short, often less than 10 cm.; culms ± capillary,
usually less than 30 cm. high. N. Great Basin
31. *P. Sandbergii*
II. Lvs. longer; culms not capillary, more than 30 cm. high.
Mostly cismontane and Mojave Desert . . 32. *P. scabrella*
HH. Panicle open, the lower branches at right angles to the axis.
Interior foothills . . . . 33. *P. tenerrima*
GG. Summer-flowering and summer-active. Montane to high alpine.
H. Panicle contracted.
I. Culms 4–12 dm. high. Mid-altitudes. Mts. of n. Calif. n.
34. *P. Canbyi*
II. Culms 1.5–3 dm. high. Alpine. Sierra Nevada
35. *P. incurva*
HH. Panicle open, the lower branches at right angles to the axis.
High Sierra Nevada . . . . 36. *P. gracillima*

1. **P. ánnua** L. ANNUAL BLUEGRASS. Tufted winter annual, bright green, glabrous;
culms many, often decumbent at base, 5–30 cm. high, leafy; blades lax, flat, up to 4 mm.
wide; ligule 1.5–2 mm. long, obtuse or truncate; panicle pyramidal, open, 3–8 cm.
long; spikelets 3–6-fld., 3–6 mm. long; glumes unequal; lemmas prominently to obscurely pilose along lower third of the 5 distinct nerves, not webbed at base, the keel
often ciliate almost throughout; anthers 0.5–1 mm. long; $2n = 28$ (Avdulov, 1928).—
Common weed of gardens and lawns, widely distributed at low elevs., abundant in
the coastal districts, rare in the arid interior; across the continent from Calif. to Alaska,
Nfld., Fla.; s. at higher elevs. to trop. Am.; introd. from Eu. Jan.–July.

2. **P. Bolánderi** Vasey. [*P. Howellii* var. *Chandleri* Davy. *P. B.* (ssp.) *C.* Piper.]
Slender annual; culms 2–6 dm. high; sheaths glabrous; blades relatively short and broad,
abruptly acute; spikelets usually 2–3-fld.; glumes relatively broad; lemmas glabrous
except for web, scabridulous on keel and rarely on sides, the intermediate nerves very
faint; $2n = 28$ (Stebbins, 1943).—Frequent in open fir forests, granitic soils, 5000–
10,000 ft.; Red Fir F.; San Jacinto Mts., Sierra Nevada from Kern Co. n., mts. of n.
Calif. w. to e. Humboldt and w. Glenn cos.; n. to Wallowa Mts., Ore., and Blue Mts.,
Wash. July–Aug.

Ssp. **Howéllii** (Vasey & Scribn.) Keck. [*P. H.* Vasey & Scribn. *P. H.* var. *microsperma*
Vasey. *P. B.* var. *H.* Jones.] Culms 2–9 dm. high; sheaths glabrous or sometimes
scaberulous; blades relatively elongated, gradually acuminate; spikelets 2–5-fld.; glumes
narrow; lemmas pubescent on the lower part, sometimes obscurely so, the 5 nerves
all distinct.—Frequent on shaded, rather openly wooded slopes, mostly below 2500 ft.;
N. Coastal Coniferous F., Douglas-Fir F., Foothill Wd.; common in the Coast Ranges,
Siskiyou and Del Norte cos. to San Diego Co., rare in the cent. Sierran foothills; w.
of the Cascades to Vancouver Id. April–June.

3. **P. Bigelòvii** Vasey & Scribn. [*P. annua* var. *stricta* Vasey.] Tufted light green
winter annual, with erect leafy culms 1–3(–5) dm. high; blades soft, flat, rather short,
1–4 mm. wide; ligule 2–4 mm. long, laciniate; panicle narrow, interrupted, 5–15 cm.
long, the branches short, appressed; spikelets 3–5-fld., 4–6 mm. long, broad; glumes
firm, 3-nerved, conspicuously white-margined, the keel scabrous; lemmas 2.5–4 mm.
long, webbed at base, conspicuously white-villous on keel and lateral nerves, in Calif.
material also rather hairy on the internerves and intervals; anthers 0.5–0.7 mm. long.—
Infrequent in desert canyons or open ground, often in the protecting shade of cliffs
or rocks, up to 4000 ft.; Creosote Bush Scrub; w. borders of Colo. Desert, Riverside
and San Diego cos., and Panamint Mts., Inyo Co.; e. to Okla., Tex., n. Mex., the range
not continuous. March–May.

4. **P. Douglásii** Nees. [*Brizopyrum D. H.* & A. *P. californica* Steud.] Dioecious
tufted perennial, spreading by deep-seated rhizomes and aerial leafy runners up to 1
or 2 m. long; culms rather stout, 2–4 dm. high; sheaths scarious, loose; blades firm,
involute, glaucescent, densely puberulent ventrally, often equaling the culm; ligule
mostly less than 1 mm. long, truncate, fimbriate; panicle ovate to ovate-oblong, 2–4(–6)
cm. long, pale and tawny; spikelets 3–9-fld., 6–9 mm. long; glumes broad, chartaceous,
shiny, ciliolate on margin and hispidulous on keel, 5–6.8 mm. long; lemmas 3-nerved,

the nerves pilose below, the base feebly webbed; ripe anthers 2.5–3.5 mm. long; $2n = 28$ (Hartung, 1946).—Coastal dunes, where an important sandbinder, 5–50 ft.; Coastal Strand; Humboldt Co. to Point Sur, Monterey Co., San Miguel and Santa Rosa ids. March–July.

Ssp. **macrántha** (Vasey) Keck. [*P. m.* Vasey. *Melica m.* Beal.] Panicle often less compact and more elongated, 5–12 cm. long; spikelets 9–15 mm. long; glumes and 1st lemma 7–10 mm. long; ripe anthers 3–5 mm. long; $2n = 28$ (Armstrong, 1937).— Stabilized or moving dunes; Coastal Strand; Point Arena, Mendocino Co., n.; to Puget Sound. May–July.

5. **P. confínis** Vasey. Dioecious tufted perennial, spreading by deep-seated slender rhizomes; culms very slender, erect, often geniculate at base, 1–3 dm. high; blades soft, involute, those of the innovations numerous and frequently as long as the culms; ligule 0.3–1.5 mm. long, acute or truncate; panicle rather dense, 1–3 cm. long, the short scabrous branches sharply ascending; spikelets 3–4-fld. with a rudiment, 4–6 mm. long; glumes respectively lance-ovate and broadly ovate; lemmas 2.5–4 mm. long, scaberulous, vestigially webbed at base, faintly 5-nerved; ripe anthers 1.5–2 mm. long; $2n = 42$ (Hartung, 1946).—Sea beaches, dunes and sandy fields in proximity to the ocean; Coastal Strand; Marin Co. n.; to Vancouver Id. April–July.

6. **P. Pìperi** Hitchc. Dioecious; rhizomatous, often densely tufted, pallid or glaucous; culms 2–6 dm. high, the innovations numerous; sheaths of the culm smooth, of the innovations puberulent, especially at throat; blades involute, narrow, canescent ventrally, those of the culm usually few and short, of the innovations 10–30 cm. long, stiffish; ligule of culm lvs. 0.5–2 mm. long, rounded, of the innovations much-reduced; panicle 5–10 cm. long; spikelets 3–7-fld., 5–10 mm. long; glumes broad; lemmas 4–5.5 mm. long, acute, smooth to scaberulous, copiously webbed at base; ripe anthers 2–3.6 mm. long.—Dry or rocky wooded slopes, 500–1500 ft.; Yellow Pine F., Chaparral; Siskiyou Mts. of Del Norte Co.; adjacent Ore. April–May.

7. **P. rhizòmata** Hitchc. Closely related to *P. Piperi*, from which it differs in the following: usually greener, but sometimes ± glaucous; sheaths glabrous or those of the innovations minutely scaberulous; blades flat or folded, glabrous on both sides, soft; ligule of culm lvs. and innovations 3.5–5(–8) mm. long; lemma pubescent on keel and at base of marginal nerves, webbed at base; $2n = 28$ (Hartung, 1946).—In deep soil, moist forest openings, 1500–3000 ft.; Yellow Pine F.; Siskiyou Mts. of cent. Siskiyou Co. April–May.

8. **P. nervòsa** (Hook.) Vasey. [*Festuca n.* Hook. *P. Olneyi* Piper.] Strongly rhizomatous, tufted; culms erect, 3–6 dm. high; lower sheaths usually retrorsely pubescent and purple, the upper smooth or scabrous; culm-blades usually flat, rather short, to 3 mm. wide, smooth or ventrally puberulent, those of the innovations folded, narrower; ligule 0.5–2 mm. long, rounded or truncate, puberulent; panicle open, heavy, 5–10 cm. long, the branches filiform; spikelets rather loosely 3–8-fld., 5–9 mm. long, commonly purplish; florets always ♀ in Calif. material; glumes broad; lemmas 3.5–5 mm. long, scabrous on the nerves or throughout, not webbed; $2n = 56, 63, 70$ (Hartung, 1946).—Gravelly soils of open coniferous woods, grassy slopes, or ridges, 4500–12,500 ft.; Yellow Pine F. to Alpine Fell-fields; abundant in the Sierra Nevada from Nevada Co. s., elsewhere rare, Mt. Pinos, San Bernardino Mts., n. to Modoc and Siskiyou cos.; n. to B.C., e. to Rocky Mts. June–Aug.

9. **P. praténsis** L. KENTUCKY BLUEGRASS. Rhizomatous, forming dense sods; culms tufted, 3–10 dm. high; innovations numerous; sheaths smooth; blades green or glaucescent, flat or folded, mostly 2–3 mm. wide, the basal up to 30 cm. long, soft, smooth, or the margin scabrid; ligule mostly 1 mm. long, truncate; panicle open, pyramidal, 5–15 cm. long; spikelets 3–5-fld., 4–6 mm. long, green or purplish; glumes scabrous on keel; lemmas 2–3 mm. long, obtuse or acute, copiously webbed at base, sericeous on keel and marginal nerves, glabrous between; $2n = 49$ to 84 inclusive in w. Am. (Hartung, 1946).—Meadows, stream banks, open woods and open ground, sea level to 8000 ft. (to 11,300 ft. in Tuolumne Co.); many Plant Communities; widespread except on the deserts and in the Cent. V.; throughout n. N. Am. and Eurasia. May–Aug. Often considered to be wholly a European introd., but many montane races, at least, are native.

10. **P. Kellóggii** Vasey. [*P. Bolanderi* var. *K.* Jones.] Weakly rhizomatous perennial,

sometimes behaving as an annual; culms solitary or few in a tuft, decumbent and often rooting at base, erect, 4–8 dm. high; sheaths smooth; blades green, flat, elongated, 2–4 mm. wide, smooth, or the margin often scabrous; ligule 1.5–2.5 mm. long, obtuse; panicle very open, pyramidal, 10–15 cm. long, the relatively few-spikeleted branches mostly in 2's or 3's; spikelets 4–6 mm. long, loosely 2–3-fld. with a rudiment; lemmas 4–5 mm. long, narrow, tapering to an almost subulate tip, prominently 5-nerved, glabrous except for web at base; anthers 1.5–2 mm. long; $2n = 56$ (Carnegie Lab., unpubl.).—Moist shady places, 50–800 ft.; Redwood F.; near the coast, Humboldt Co. to Santa Cruz Co. April–June. Seldom collected.

11. **P. compréssa** L. CANADIAN BLUEGRASS. Rhizomatous; culms rather scattered, decumbent and ± flattened at base, erect, wiry, leafy, 2–5 dm. high; lvs. bluish-green, short, with flat rather narrow blade and short blunt ligule less than 1 mm. long; panicle narrow, rather dense, 4–7 cm. long; spikelets crowded on the short branches, ovate-oblong, usually 4–6-fld.; glumes 2 mm. long, 3-nerved, acute; lemmas 2.5 mm. long, obtuse, usually purplish near tips, the keel and marginal nerves sericeous toward base, the web scant or wanting, the intermediate nerves faint; $2n = 42$ (Avdulov, 1930); this is the usual number, but races have been found with $2n = 14, 35, 49, 50$, and 56.— In open or partially shaded places, old meadows, etc., 30–6000 ft.; many Plant Communities; Modoc, Siskiyou, and Del Norte cos. s., largely cismontane and infrequent to cent. Calif., rare in s. Calif.; to Atlantic Coast, Alaska; Eurasia. Presumably introd. from Eu., but possibly some races are native. May–July.

12. **P. atropurpurèa** Scribn. Rhizomatous; culms 3–4.5 dm. high, leafy only in lower third; lvs. numerous, mostly basal, ± tightly folded, firm, smooth, the margin finely scabrid, 8–15 cm. long; ligule up to 2.5 mm. long, usually much shorter; panicle spikelike, dense, 2.5–5 cm. long, 1–1.5 cm. wide, purplish; spikelets tightly 3–5-fld., 3.5–5 mm. long; glumes broad, chartaceous, 1.5 and 2 mm. long respectively; lemmas 2.5–3 mm. long, entirely smooth, without web, faintly nerved.—Meadows and grassy slopes, 6000–7000 ft.; Yellow Pine F.; Bear V., San Bernardino Mts. May–June.

13. **P. fibràta** Swall. Loosely tufted, with short rhizomes to 2 cm. long; culms erect, 2–6 dm. high; lvs. pallid, glaucescent, the sheaths firm, striate, the blades flat or usually folded, very firm, those of the innovations up to 17 cm. long, those of the culm much shorter, ± pungent, scabrous; ligule 1–1.5 mm. long, deltoid, obtuse at very tip; panicle 5–15 cm. long, dense, often somewhat interrupted toward base, the short appressed branchlets floriferous to base; spikelets compactly 3–4-fld., 5–7 mm. long; glumes hyaline-margined; lemmas lance-oblong, 3–4.5 mm. long, obscurely to obviously scabrid, nearly smooth to crisp-puberulent on nerves or over the back toward base; anthers 1.9–2.1 mm. long; $2n = 63, 64$ (Hartung, 1946).—Local, in dry alkaline soils, open fields, 2500–5000 ft.; Alkali Sink, Yellow Pine F.; Grenada, Shasta V. (Siskiyou Co.), Duncan Horse Camp (Modoc Co.), Squaw V. (Lassen Co.). June–July.

14. **P. bulbòsa** L. Tufted, 2–5 dm. high, the culms somewhat bulbous at base; sheaths smooth; blades of innovations usually folded, of culms usually flat, 1–2 mm. wide, smooth, or the margin scabrid; ligule 2–3 mm. long; panicle lax but narrow, up to 8 cm. long, the branches ascending and scabrous; spikelets mostly proliferous, the 4–6 florets converted into bulblets with dark purple base and prolonged foliaceous tip to the lemma; unaltered lemmas 2.5 mm. long, sericeous on keel and marginal nerves, webbed at base; $2n = 28$ (Armstrong, 1937) and 42.—Pastures and disturbed land, 100–4500 ft.; Foothill Wd., Yellow Pine F., etc.; Modoc Co. to Del Norte Co., s. to Tuolumne and San Luis Obispo cos.; n. to B.C., e. to the Atlantic; introd. from Eu. April–July. In Calif. a weakly rooted summer-dormant perennial that is slowly spreading, but still rare s. of Shasta Co.

15. **P. triviàlis** L. Culms erect from a decumbent subrhizomatous base rooting at the lowest nodes, often ± geniculate and lax, 4–12 dm. high, somewhat scabrous; sheaths ± scabrous, at least toward the summit; blades flat, lax, bright green, elongated, up to 6 mm. wide, scabrous ventrally; ligule 4–7 mm. long, obtuse; panicle 8–20 cm. long, open but the branches not reflexed; spikelets very many, 2–3-fld., 3–4 mm. long, the bracts often prominently hyaline-margined; glumes very unequal; lemmas 2–3 mm. long, smooth except for a sparse silky pubescence on the sharp keel and the prominent web at base, the nerves strong; $2n = 14$ (Avdulov, 1928).—In moist or swampy often

shaded ground, 20–1500 ft.; Redwood F.; near the coast, from Marin Co. to Del Norte Co.; n. to Alaska; e. U.S.; introd. from Eu. May–July.

16. **P. leptocòma** Trin. Culms loosely tufted, 3–6(–10) dm. high; sheaths smooth, or sometimes scaberulous; blades flat, lax, bright green, short, up to 3 or 4 mm. wide, smooth or scabrid; ligule 2–4 mm. long, obtuse, glabrous; panicle nodding, open, the capillary branches mostly in pairs, naked below; spikelets 2–5-fld., 4–6 mm. long, narrow, glumes very unequal, the lower often almost subulate; lemmas 3–4 mm. long, acuminate, pilose along lower third to half of keel and marginal nerves and smooth between, or nearly glabrous except for web at base, the nerves rather faint; anthers 0.4–0.9 mm. long.—Rare on wet stream banks and boggy meadows, 6000–10,500 ft.; Lodgepole F., Subalpine F.; Warner, Marble, and Sweetwater mts., and Sierra Nevada from Nevada Co. to Tulare Co.; n. to Alaska, e. to Rocky Mts. July–Aug.

17. **P. palústris** L. Fowl Bluegrass. Loosely tufted, the culms decumbent and rooting at the sparingly branched often purplish base, 3–12 dm. high; sheaths smooth or scaberulous; blades 1–3 mm. wide, often scabrous, especially ventrally; ligule 3–5 mm. long, obtuse; panicle slightly nodding, open, to 25 cm. long, the branches in rather distant fascicles of 4–6, naked below; spikelets 2–4-fld., 3–4.5 mm. long; glumes unequal; lemmas bronze-tipped, 2.5–3 mm. long, villous on keel and marginal nerves, webbed at base, the intermediate nerves obscure; $2n = 28$ (Armstrong, 1937).— Moist meadows and ditches, open ground, 5000–6500 ft.; Montane Coniferous F., Sagebrush Scrub; scattered stations, mostly e. flank of the Sierra Nevada, Lassen Co. to Inyo Co., Big Pines (San Gabriel Mts.), Santa Ana R. (San Bernardino Mts.); adjacent Nev., circumpolar, to Alaska, Nfld., Va.; Eurasia. July–Aug.

18. **P. Lettermánii** Vasey. [*Atropis* L. Beal.] Densely tufted dwarf, 3–10 cm. high, the culms not much longer than the crowded basal rosette; blades spreading, folded or flat, mostly 1 mm. wide; ligule 1–2 mm. long, acute; panicle narrow, rather dense, 1–2 cm. long; spikelets purple, 2–4-fld., 3–4 mm. long; glumes subequal, longer than the 1st floret; lemmas 2–3 mm. long, smooth, ovate-oblong, hyaline at apex, obscurely keeled; anthers 0.35–0.5 mm. long.—Rocky slopes, rare, 12,000–13,500 ft.; Alpine Fellfields; crest of Sierra Nevada in Fresno, Inyo, and Tulare cos.; Ruby Range, Nev., Uinta Mts., Utah, Rocky Mts., Colo. to B.C. July–Aug.

19. **P. Suksdòrfii** (Beal) Vasey ex Piper. [*Atropis* S. Beal. *P. Pringlei* in part, of Calif. auth.] Densely cespitose tufts 5–15 cm. high, scarcely exceeded by the leafy culms; sheaths smooth (or puberulous about throat); blades mostly basal, firm, involute or folded, rarely flat, striate, glaucous, up to 10 cm. long, 1–2 mm. wide, ± puberulent ventrally and often scabrous dorsally toward the blunt thickened apex; ligule 2–3 mm. long, triangular, very thin, entire or lacerate; panicle narrow, rather dense, 2–3 cm. long, purple or greenish; spikelets 2–4-fld., narrow, 4–6 mm. long; glumes nearly as long as 1st floret; lemmas 3–5 mm. long, smooth or scaberulous, rather thin, 3-nerved.— Frequent in rock crevices or talus slopes, 11,000–13,700 ft.; Alpine Fell-fields; Sierra Nevada, from Tulare Co. to Mono Co.; Wallowa Mts., Ore., Cascade Range, Wash. July–Sept.

20. **P. Hánseni** Scribn. [*P. Pringlei* var. *H.* Smiley. *P. Leibergii* in part, of Calif. auth.] Small often somewhat lax tufts with capillary culms 10–25 cm. high, naked above the base and much exceeding the basal lvs.; blades often longer and narrower than those of *P. Suksdorfii;* panicle narrow, 2–4 cm. long, the short branches appressed, or sometimes open; florets all perfect or all ♀, rarely all ♂; anthers 1.5–2.5 mm. long; $2n = 42$ (Hartung, 1946).—Common on moist meadowy slopes, forming sods, also in gravels and on rocky ledges, 7000–12,000 ft.; Lodgepole F. to Alpine Fell-fields; Sierra Nevada, from Eldorado Co. to Tulare Co. July–Aug.

21. **P. Cusíckii** Vasey. [*P. filifolia* Vasey. *P. capillarifolia* Scribn. & Will.] Densely tufted, the erect culms 1.5–5 dm. high, with ca. 2 short lvs. near base; blades very numerous, filiform, erect, short or elongate (up to 25 cm.), less than 1 mm. wide, ± scabrous, sometimes obscurely so; ligule 1.5–2.5 mm. long, acute; panicle usually contracted, not very dense, in some forms rather open, usually pallid and shining, 3–7 cm. long; spikelets 2–5-fld., 5–7.5 mm. long in ours; glumes rather broad, unequal, somewhat shorter than 1st floret; lemmas 4–5.5 mm. long, smooth or scabrous, rather thin, often sharply acute; $2n = 42$ (Hartung, 1946).—Meadows, grassy hillsides and

sagebrush plains, in open stands, 5000–8300 ft.; Sagebrush Scrub, N. Juniper Wd.; e. flank of Sierra Nevada and valleys from Alpine Co. n.; to B.C., Alta., Rocky Mts. May–July. Highly variable.

22. **P. Prínglei** Scribn. [*P. argentea* Howell. *Melica a.* Beal. *M. nana* Beal. *Atropis P.* Beal.] Densely tufted, the erect culms 1–2 or 3 dm. high, leafy to the middle; sheaths becoming loose, scarious, shining; blades often recurved, involute, firm, green or glaucescent, narrow but not filiform, 3–10(–15) cm. long, smooth, puberulent ventrally; ligule 2–5 mm. long, decurrent; panicle narrow, not very dense, usually pale and shining, 2–5 cm. long; spikelets 3–5-fld., 7–12 mm. long; glumes broad, thin, equal, smooth, scarcely equaling the 1st floret; lemmas 5–7.5 mm. long, smooth or scabrous, acute or obtusish, 3- or 5-nerved.—Scattered tufts in sand, gravel, or rocks on exposed summits, cliffs, or talus, 6600–10,000 ft.; Red Fir F. to Alpine Fell-fields; Siskiyou Mts. to n. Trinity Co. and Placer Co. (in atypical form in n. Mono Co. at 11,300 ft.); adjacent Ore. July–Sept.

23. **P. épilis** Scribn. [*P. purpurascens* var. *e.* Jones.] Loosely tufted, culms not very dense, erect, 1.5–5 dm. high, leafy below; blades bright green, those of the innovations usually folded or involute, 6–20 cm. long, narrow or even filiform, those of the culms 2 or 3, flat or sometimes folded, 2–6 cm. long, 1.5–3 mm. wide, smooth; ligule 2–4 mm. long; panicle usually condensed, ovoid or oblong, sometimes more open, green or purplish, 2–6 cm. long; spikelets 3–5-fld., 5–7.5 mm. long; glumes unequal; lemmas 4–5.5 mm. long, usually minutely scabrous, acutish, 3- or 5-nerved; $2n = 56, 84$ (Hartung, 1946).—Gravelly or rocky ridges or meadows, 8000–12,000 ft.; Lodgepole F. to Alpine Fell-fields; Sierra Nevada, from Tulare Co. to Nevada Co.; to B.C., Alta., Colo. July–Aug.

24. **P. nevadénsis** Vasey ex Scribn. [*Atropis pauciflora* Thurb. *P. p.* Benth. ex Vasey, not R. & S. *Panicularia Thurberiana* Kuntze. *Poa T.* Vasey. *Poa limosa* Scribn. & Will. *Atropis n.* Beal.] NEVADA BLUEGRASS. Tufted, the many erect culms 5–10 dm. high, leafy throughout; lvs. numerous, usually elongate (up to 30 cm.); sheaths and blades often scabrous, the blades flat or folded or involute, bright green or pale, 1–3.5 mm. wide, often subcapillary and stiffish; ligule of upper lvs. 3–6 mm. long, acute or acuminate; panicle narrow, elongated, up to 25 cm. long, pale, rather loose, the branches appressed; spikelets 2–5(–7)-fld., 4–8 mm. long; glumes medium wide, the 2d ca. equaling the 1st floret; lemmas 3.5–5 mm. long, scabrous apically or throughout, only slightly anthocyanous, usually obtusish; $2n = 63$ (Hartung, 1946).—Moist meadows, 3000–10,000 ft.; Sagebrush Scrub, Yellow Pine F., Red Fir F.; mostly transmontane, Inyo Co. to Modoc Co.; to Wash., Mont., Colo., n. Ariz. June–July.

25. **P. napénsis** Beetle. Tufted, glaucescent, glabrous to scabridulous, the few rather stiff erect culms to 7 dm. high; blades of the innovations to 20 cm. long, of the culms to 15 cm. long, both alike, folded, 1 mm. wide, stiffly erect; ligule 4–6 mm. long, obtuse or acute; lower sheaths very scarious; panicle ovoid-oblong, condensed, pale but somewhat purplish, 10–15 cm. long, 2–5 cm. wide; spikelets crowded in the outer part of the scabrous branchlets, 5–6-fld., 6–7 mm. long; glumes firm, essentially smooth, broadly white-margined; lemmas 3–4 mm. long, glabrous at base, scabrid toward the rather obtuse scarious tip.—Alkaline meadows around hot springs, 300–600 ft.; V. Grassland; upper Napa V., Napa Co. May.

26. **P. juncifòlia** Scribn. [*P. brachyglossa* Piper. *P. Fendleriana* var. *j.* Jones.] Densely tufted, light green or glaucescent, the slender erect culms to 10 dm. high; lvs. numerous, largely basal, the blades mostly tightly involute, 10–20 cm. long, less than 2 mm. wide, smooth or scabrid; ligule mostly 1–2 mm. long, obtuse; panicle narrow, elongated, strict; spikelets 3–6-fld., 7–10 mm. long; glumes broad; lemmas 3–5 mm. long, scabrous apically or throughout, like the glumes not prominently margined, obtuse or rounded; $2n = 63, 78, 84$ (Hartung, 1946).—Not common, moist or dry alkaline meadows and slopes, 2600–8500 ft.; Sagebrush Scrub, Yellow Pine F., Red Fir F.; mostly e. flank of Sierra Nevada, from Tulare Co. to Siskiyou Co.; to Wash., Mont., Colo. June–July.

27. **P. ámpla** Merr. Moderate to large tufts, glaucous or green, the slender erect culms 6–18 dm. high; lvs. many, largely basal, the blades flat, 20–50 cm. long, 1–3 mm. wide; ligule 1–2 mm. long, rounded; panicle narrow, elongated, usually rather dense; spikelets 4–7-fld., 7–10 mm. long; glumes acuminate, scabrous-keeled, the 2d ca.

equaling the 1st floret; lemmas 4–6 mm. long, scabrous apically, green or ± purplish; $2n = 63$ (Hartung, 1946).—Infrequent in prairies, grassy hillsides, or open woods, 3000–9400 ft.; N. Juniper Wd., Yellow Pine F., Red Fir F.; mostly transmontane, Mono Co. to Modoc and Siskiyou cos.; to Wash., Mont., New Mex. May–Aug.

28. **P. unilateràlis** Scribn. ex Vasey. [*Atropis u.* Beal.] Densely tufted, the culms erect, or sometimes decumbent at base, 1–5 dm. high; lvs. numerous, crowded below, green or glaucous, with loose scarious sheaths and flat or folded smooth blades 5–15 cm. long, 1–5 mm. wide; ligule 3–5 mm. long, obtuse to acuminate, broad; panicle dense, spikelike, 3–6(–10) cm. long, 1–2 cm. wide; spikelets 3–8-fld., 6–10 mm. long; glumes broad, glabrous except for the minutely hispidulous keel; lemmas 3–4.5 mm. long, broad, often firm, obscurely to obviously short-hirsute toward base of keel and lateral nerves, not webbed; anthers 1.5–2.5 mm. long; $2n = 42$ (Stebbins, 1945).— In open stands on ocean bluffs, sand dunes, and open grassy slopes close to the coast, 10–400 ft.; Coastal Strand, N. Coastal Scrub, Coastal Sage Scrub; Monterey Co. n.; to Curry Co., Ore. March–June.

29. **P. Fendleriàna** (Steud.) Vasey. [*Eragrostis F.* Steud. *Panicularia F.* Kuntze. *Atropis F.* Beal. *P. longiligula* Scribn. & Will. *P. F.* var. *l.* Gould.] Mutton Grass. Incompletely dioecious; tufted, the many erect culms 2–6 dm. high, scabrid toward the panicle; basal blades flat, folded or involute, glaucescent, stiffish, 10–20(–35) cm. long, those of the culm lvs. mostly very short and appressed or obsolete; ligule extremely variable, from less than 1 mm. long and truncate to 12 mm. long and acuminate, scarcely correlated with the length, flatness or color of blade; panicle oblong, contracted, pale, 3–10 cm. long, 1–2 cm. wide; lemmas 3–5 mm. long, very broad and rounded, villous on keel and marginal nerves, usually smooth between.—Rocky mesas, open hillsides, or in timber, 3200–10,200 ft.; Pinyon-Juniper Wd., Yellow Pine F. to Lodgepole F.; Sierra Nevada, from Plumas Co. to Tulare Co., San Bernardino, San Jacinto, and Santa Rosa mts., mts. of e. Mojave Desert; more common e. to Chihuahua, Okla., Black Hills, and n. to e. Ore., B.C. May–July.

30. **P. rupícola** Nash ex Rydb. [*P. rupestris* Vasey, not With.] Timberline Bluegrass. Densely tufted, the stiffly erect rather crowded culms 1–2 dm. high; lvs. sparse, usually smooth, the blades involute, stiffish, short; ligule 0.7–2 mm. long, obtuse, laciniate; panicle slender, compact, usually purplish, 1.5–5 cm. long, rarely over 1 cm. wide; spikelets 1–4-fld., 3–5 mm. long; glumes broad, scabrid apically, often rather prominently 3-nerved; lemmas silky on midrib and lateral nerves but lacking web at base; anthers 1.1–1.5 mm. long.—Rather frequent on rocky screes and ridges, 11,000–13,000 ft.; Alpine Fell-fields; Sierra Nevada, from Mono Co. to Tulare Co., White Mts.; e. to Rocky Mts., from New Mex. to Alta. July–Aug.

31. **P. Sandbérgii** Vasey. [*P. Buckleyana* var. S. Jones.] Dense often small tufts, the erect or divergent wiry culms much exceeding the short basal lvs., 1.5–4 dm. high; lvs. soft, slender, mostly basal; ligule 2–5 mm. long, acute; panicle slender, elongated, not very dense; spikelets 3–5-fld., 4–7 mm. long; lemmas 3–4 mm. long, usually obviously puberulent over lower half or third; anthers ca. 2 mm. long; $2n = 82$, 84, 86 (Hartung, 1946).—Open plains and hillsides, 3000–8000 ft.; Sagebrush Scrub, N. Juniper Wd., Yellow Pine F., Red Fir F.; Great Basin, from Sierra V. to Modoc Co.; to B.C., Mont., Colo. April–June.

32. **P. scabrélla** (Thurb.) Benth. ex Vasey. [*P. tenuifolia* Nutt. ex Buckl., not A. Rich. *Atropis californica* Munro ex Gray, and *A. c.* Munro ex Thurb. in Wats. *A. s.* Thurb. in Wats. *A. t.* Thurb. in Wats. *P. californica* Scribn., not Steud. *P. Orcuttiana* Vasey. *P. Buckleyana* Nash. *P. capillaris* Scribn. *P. nudata* Scribn.] Small to moderate tufts, green, the slender erect smooth or scabrid culms 4–10 dm. high; lvs. largely basal, soft, slender, the sheaths smooth or scabrid; ligule 3–7 mm. long, acuminate; panicle usually narrow and contracted, sometimes open, 5–15 cm. long; spikelets 3–7-fld., 6–10 mm. long; lemmas 3–5 mm. long, obviously to very obscurely puberulent toward base, or (in the Coast Ranges and Mojave Desert) sometimes glabrous; anthers ca. 2 mm. long; $2n = 63$, 66, 84, 86 (Hartung, 1946).—Common in good soils on open or protected hillsides, sea level to 5000 ft.; many Plant Communities; almost throughout cismontane Calif. but also crossing the n. Sierra Nevada and the Mojave Desert to w. and s. Nev.; n. to Wash., s. to L. Calif. Feb.–June. A highly variable sp.

33. **P. tenérrima** Scribn. Small green tufts, the near-capillary erect culms 2–5 dm.

high, smooth; lvs. crowded at base, those of the innovations filiform, 4–8 cm. long, those of the culms almost as narrow, loosely involute, somewhat shorter; ligule less than 1 mm. long, acute; panicle open, lax, 7–9 cm. long, the capillary branches usually in pairs, floriferous only toward the tips; spikelets loosely 2–3-fld. with a rudiment, 4.5– 6.5 mm. long; lemmas linear-oblong, 3–4.2 mm. long, smooth in outer half, crisp-puberulent only well down toward base; anthers 1.6–2.1 mm. long; $2n = 42$ (Stebbins). —Rare; in thin rocky soil, sometimes serpentine, in open places, 1500–2000 ft.; Foot-hill Wd.; Sierran foothills of Eldorado Co., Black Mt., Ventura Co. A dainty relative of *P. scabrella* with the open panicle of *P. gracillima*.

34. **P. Cánbyi** (Scribn.) Piper. [*Glyceria C.* Scribn. *Atropis C.* Beal.] Densely tufted; culms 4–12 dm. high, naked in upper half; lvs. in a large basal rosette, bright green or sometimes glaucescent, flat or folded, elongated, 0.5–2 mm. wide; ligule 2–5 mm. long; panicle narrow, rather compact, 6–15 cm. long; spikelets 3–6-fld., 5–9 mm. long; lemmas narrowly lanceolate to lance-oblong, obtuse or acute, 3.5–5 mm. long, crisp-puberulent in lower half on nerves or over the rounded back, the intermediate nerves obscure; anthers 1.5–2.5 mm. long; $2n = 72$, 82–85 (Hartung, 1946).—Open rocky slopes, 5000–8600 ft.; N. Juniper Wd., Yellow Pine F., Red Fir F.; mts. of n. Calif., from Lassen and Tehama cos. n.; to B.C., Sask., Colo. June–Aug.

35. **P. incúrva** Scribn. & Will. Densely tufted; culms very slender, 1.5–4 dm. high; lvs. mostly basal, flat or folded, sometimes involute, mostly 1 mm. wide; ligule 2–4 mm. long; panicle narrow, rather dense, 5–8(–10) cm. long; spikelets 2–4-fld., 5–7 mm. long; lemmas as in *Canbyi*, often scabrid apically, the crisp puberulence often limited to very base; anthers 1.5–2.5 mm. long; $2n = 90$, 93, 94, 99, 105–106 (Hartung, 1946).—Common on exposed rocky ridges and slopes, rarely in open woods, 7000– 12,500 ft.; Red Fir F. to Alpine Fell-fields; Marble Mts., Siskiyou Co., Sierra Nevada, from Lassen Co. to Tulare Co., Inyo Mts., Mt. Pinos, Mt. San Gorgonio. July–Aug. Sometimes forming fairy rings of some size through the dying-out of the center.

36. **P. gracíllima** Vasey. [*P. invaginata* Scribn. & Will., in part.] Densely tufted, the slender erect numerous culms 2–6 dm. high; lvs. pallid, glaucescent, mostly basal, the basal sheaths usually purplish, the blades 5–15 cm. long, 0.5–1.5 mm. wide, often filiform, flat or involute; ligule 1.5–4 mm. long, acuminate, often laciniate; panicle 5–10 cm. long, ovate or pyramidal, open, the lower branches horizontal with branchlets again spreading, naked below; spikelets 2–4-fld., 4–7 mm. long; glumes broadly lanceo-late, prominently hyaline-margined; lemmas lance-oblong, acute or obtuse, puberulent or sparsely pilose toward base, especially on keel and marginal nerves; anthers 1.7–2.3 mm. long; $2n = 81$, 84, 86 (Hartung, 1946).—Open rocky slopes or dry alpine meadows, 6000–12,200 ft.; Red Fir F. to Alpine Fell-fields; common in the Sierra Nevada, from Tulare Co. to Butte Co., Trinity and Siskiyou cos.; to B.C., Alta., n. Rocky Mts. July–Aug.

10. Brìza L. QUAKING GRASS

Annuals or perennials, with erect culms, flat blades and open panicles with broad spikelets on capillary pedicels. Spikelets several-fld., often cordate, the florets crowded and horizontally spreading, the rachilla disarticulating above the glumes and between the florets. Glumes subequal, broad, papery, with scarious margins; lemmas papery, several-nerved; palea much shorter than the lemma. Ca. 20 spp., of Eurasia, N. Afr., S. Am. (Greek, *briza,* a kind of grain.)

Plant perennial; sheaths longer than the narrow blades; spikelets ca. 5 mm. long . . . . 1. *B. media*
Plant annual; sheaths shorter than the blades.
    Spikelets 4–5 mm. wide, ca. 3 mm. long; panicle erect . . . . 2. *B. minor*
    Spikelets 10 mm. wide, ca. 12 mm. long; panicle drooping . . . . 3. *B. maxima*

1. **B. mèdia** L. Erect, 2.5–7 dm. high; ligule of the upper lf. ca. 1 mm. long, truncate; blades 2–5 mm. wide; panicle erect, the stiff capillary branches spreading; spikelets nodding, 5–12-fld., orbicular, ca. 5 mm. long, ca. as broad, brown, shining; lemmas naviculate; $n = 7$ (Avdulov, 1931).—Occasionally reported, as from Sonoma Co.; native of Old World.

2. **B. mìnor** L. Annual, 1–5 dm. high; ligule of upper lf. ca. 5 mm. long, acute; blades 2–10 mm. wide; panicle erect, its slender branches spreading; spikelets pendent,

3–6-fld., pale or plum-colored, broadly cordate, 3–4 mm. long; lemmas strongly ventricose below; $n = 5$ (Avdulov, 1931).—Common in waste places and fields, especially in cent. and n. Calif.; introd. from Eu. April–July.

3. **B. máxima** L. Annual, the culms erect or with decumbent base, 3–6 dm. tall; uppermost ligule mostly 3–6 mm. long; blades mostly 3–6 mm. wide; panicle drooping; spikelets rather few, 1–2 cm. long, 1–1.5 cm. broad; glumes and lemmas mostly margined with purple or brown; $n = 7$ (Kattermann, 1933).—Established in oak groves, on roadsides, etc., from San Luis Obispo Co. n., especially abundant in Sonoma Co.; native of Eu. April–July.

11. Eragróstis Beav. LOVE GRASS

Annuals or perennials with loose or dense terminal panicles. Spikelets strongly compressed, few–many-fld., the uppermost florets sterile; rachilla articulated but sometimes not disjointed until after the fall of the glumes and lemmas with the seed. Glumes keeled, somewhat unequal, much shorter than the spikelets, the 1st 1-nerved, the 2d rarely 3-nerved. Lemmas 3-nerved, broad, keeled, acute or acuminate; paleae shorter than their lemmas or ca. as long. Over 100 spp., of trop. and temp. regions. (Greek, *eros*, love, and *agrostis*, a grass.)

A. Plants perennial.
  B. Spikelets usually red-brown, 10–40-fld., subsessile .... 1. *E. oxylepis*
  BB. Spikelets gray-green, 7–11-fld., short-pedicelled .... 14. *E. curvula*
AA. Plants annual.
  B. Plants glandular or warty along the lf.-margins, keel of lemmas, or on panicle branches.
    C. Panicle narrow, rather dense; spikelets 6–10-fld. .... 7. *E. lutescens*
    CC. Panicle open, at least ¼ as wide as long; spikelets 10–40-fld.
      D. Spikelets 2.5–3 mm. wide; lemmas ca. 2.5 mm. long; glands prominent on keels of
        lemmas .... 8. *E. megastachya*
      DD. Spikelets ca. 2 mm. wide; lemmas ca. 2 mm. long; glands mostly on panicle-
        branches and lvs. .... 9. *E. poaeoides*
  BB. Plants not glandular on lemmas, panicle-branches, or lf.-margins, sometimes so on sheaths and
    near nodes.
    C. Spikelets ca. 1 mm. wide, slender, linear.
      D. Spikelets 3–5 mm. long; lemmas 1–1.5 mm. long; pedicels mostly longer than the
        spikelets .... 2. *E. pilosa*
      DD. Spikelets 5–7 mm. long; lemmas ca. 2 mm. long; pedicels mostly shorter than the
        spikelets .... 6. *E. Orcuttiana*
    CC. Spikelets ca. 1.5 mm. wide or wider, linear to ovoid.
      D. Plants creeping, rooting at the nodes to form mats .... 3. *E. hypnoides*
      DD. Plants not creeping and forming mats.
        E. Panicle narrow, with ascending branches bearing spikelets almost to their
          bases; spikelets linear, mostly 12–15-fld. and ca. 1 cm. long . 10. *E. Barrelieri*
        EE. Panicle open, ± diffuse.
          F. Some or all of the spikelets on pedicels 1–5 mm. long; spikelets linear at
            maturity, appressed along primary panicle-branches.
            G. Primary branches of panicle simple or the lower with a branchlet
              bearing 2–3 spikelets; culms slender, mostly less than 3 dm. high
                                4. *E. pectinacea*
            GG. Primary branches of panicle usually bearing branchlets with few to
              several spikelets; culms rather robust, mostly over 3 dm. in height
                                5. *E. diffusa*
          FF. Some or all of the spikelets on pedicels 5–8 mm. long; spikelets ovate
            to linear, not appressed along the primary panicle-branches.
            G. Lf.-blades 5–10 mm. wide; spikelets lanceolate to ovate.
              H. Spikelets narrow-lanceolate, 1.8–2 mm. broad, lead-color,
                mostly 8–12-fld.; panicle with ascending branches
                              11. *E. neomexicana*
              HH. Spikelets ovate-lanceolate, 2–2.5 mm. broad, straw color or
                purplish, 3–9-fld.; panicle open, with long spreading pedicels
                              12. *E. mexicana*
            GG. Lf.-blades mostly 1–3 mm. wide; spikelets linear, 1.6–1.8 mm.
              wide, 8–15-fld.; panicle open, with spreading branches
                              13. *E. arida*

1. **E. oxylèpis** (Torr.) Torr. [*Poa o.* Torr. *E. secundiflora* Calif. refs.] Perennial with stiff culms, erect or decumbent at base, 3–6 dm. tall; sheaths pilose at throat; blades flat, ± involute on drying, 1–4(–5) mm. wide; panicle 0.5–2.5 dm. long, with several to many stiff ascending or spreading branches; spikelets mostly aggregate on very short

branchlets, ± red-brown, compressed, subsessile, linear, 10–40-fld., 8–15 mm. long; glumes 1-nerved, the 2d 2 mm. long; lemmas prominently 3-nerved, closely imbricate, 3 mm. long, abruptly acute, scabrous on keel; *n* = 20 (Brown, 1950).—Reported from San Diego; Colo., Kans. to Fla. and Mex.

2. **E. pilòsa** (L.) Beauv. [*Poa p.* L.] Annual, the culms slender, erect or with decumbent base, 1–5 dm. tall; blades flat, 1–3 mm. wide; panicle open, 0.5–2 dm. long, the capillary branches flexuous, ascending or spreading, bearing spikelets in their distal half or more; spikelets lance-linear, ca. 1 mm. wide, 3–9-fld., 3–5 mm. long, the spreading pedicels mostly longer; glumes glabrous, 0.5 and 1 mm. long; lemmas inconspicuously nerved, 1.2–1.5 mm. long.—Occasional as a waif in waste places; introd. from Eu. July–Sept.

3. **E. hypnoìdes** (Lam.) BSP. [*Poa h.* Lam.] Branching creeping annual, rooting at the nodes and forming mats to several dm. across; lf.-blades scabrous or pubescent on upper side, 1–4 cm. long; panicles simple or nearly so, usually open, ± ellipsoid, 1–6 cm. long; spikelets linear-lanceolate, 10–35-fld., mostly 5–15 mm. long; lemmas thin, acuminate, 1.5–2 mm. long; paleae half as long.—Sand bars and sandy shores; several Plant Communities; Colusa, Stanislaus, and Mariposa cos. n.; to Wash., Atlantic Coast, S. Am. July–Sept.

4. **E. pectinàcea** (Michx.) Nees. [*Poa p.* Michx.] Like *E. pilosa*, the panicle-branches bearing spikelets from near the base; spikelets lance-ovate, 1.3–1.7 mm. wide, mostly more than 5 mm. long, often 10–15-fld.; lemmas mostly more than 1.5 mm. long.—Occasional in Calif.; sandy places, ditches, etc., through most of U.S. July–Oct.

5. **E. diffùsa** Buckl. Annual, with culms erect from a spreading base, 3–5 dm. high, sheaths somewhat pilose above; blades 2–8 cm. long, 1–3 mm. wide, ± involute; panicle diffuse, 0.7–2 dm. long, the primary branches with appressed secondary branchlets bearing few to several spikelets; spikelets 5–many-fld.; glumes 1-nerved, the 2d 1.5 mm. long; lemmas ca. 1.5 mm. long.—Occasional weed in waste places especially in Cent. V. and s. Calif.; to Ida., Wyo., Tex., Mex. June–Sept.

6. **E. Orcuttiàna** Vasey. Annual, the culms ascending from a ± decumbent base, 6–10 dm. tall; sheaths glabrous; blades flat, 2–6 mm. wide; panicle open, 1.5–3 dm. long, the branches, branchlets and pedicels spreading, slender, flexuous, with glabrous axils; spikelets linear, 6–10-fld., 5–7 mm. long, ca. 1 mm. wide; 2d glume 1–1.5 mm. long; lemmas ca. 1.8 mm. long, ± loosely arranged.—Moist fields and waste places; many Plant Communities; cismontane Calif. especially n.; to Ore., Colo., Ariz. May–Oct.

7. **E. lutéscens** Scribn. Freely branched annual, 0.5–2 dm. tall; sheaths and blades with many glandular depressions; blades flat, 2–3 mm. wide, pale; panicle narrow, erect, 0.2–1 dm. long, the ascending to appressed branches with glandular depressions; spikelets 6–10-fld., 5–7 mm. long; glumes 1.5 and 2 mm. long; lemmas ca. 2 mm. long, prominently nerved, acute.—Occasionally reported from sandy damp places, Calif.; to Wash., Ida., Colo., Ariz., Mex. July–Oct.

8. **E. megastàchya** (Koel.) Link. [*Poa m.* Koel. *E. cilianensis* Calif. refs.] Stink Grass. Weedy strong-scented annual with spreading to ascending culms 2–5 dm. long; with a ring of glands below the nodes; lf.-blades 3–8 mm. wide, with wartlike glands on the margins; panicles 0.5–2 dm. long, greenish lead-color, densely-fld.; spikelets 10–40-fld., 5–17 mm. long, 2.5–3 mm. wide, with closely imbricated florets; pedicels and keels of glumes and lemmas ± glandular; lemmas ca. 2.5 mm. long, with prominent lateral nerves; *n* = 10 (Avdulov, 1928).—Waste places, cismontane and sometimes desert Calif.; through most of U.S.; introd. from Eu. June–Oct.

9. **E. poaeoìdes** Beauv. ex R. & S. Resembling *E. megastachya;* smaller, more slender; panicles open; spikelets 1.5–2 mm. wide, 8–20-fld., the florets less densely imbricated, the bases or rachilla-joints visible; lemmas less than 2 mm. long.—Occasional in waste places in s. Calif.; sparingly introd. from Eu. June–Sept.

10. **E. Barreliéri** Daveau. Annual with culms erect or decumbent at base, 2–5 dm. tall, branched below; sheaths pilose at summit; blades flat, 2–4 mm. wide; panicle erect, open but narrow, 0.8–1.5 dm. long, with ascending branches or stiffly spreading, few-fld. and bearing spikelets nearly to the base, glabrous in axils; spikelets linear, mostly 12–15-fld., ca. 1 cm. long, 1.5 mm. wide; lemmas ca. 2 mm. long.—Waste places, Ventura Co.; to Colo., Kans., Tex. June–July.

11. **E. neomexicàna** Vasey. Rather stout annual, darkish-green, 5–10 dm. tall; sheaths

glabrous, with pilose throats; blades flat, 3–8 mm. broad; panicle 2–4 dm. long, with ascending branches and ascending spikelets mostly overlapping or imbricate; spikelets dark gray-green, ± ovate, 8–12-fld., 5–8 mm. long, 1.8–2 mm. broad; lemmas 2–2.3 mm. long.—Waste places, desert and cismontane, s. Calif.; to Atlantic Coast; introd. from Eu. May–Sept.

12. **E. mexicàna** (Hornem.) Link. [*Poa m.* Hornem.] Like *E. neomexicana*, but lower, the culms often simple; lvs. pale, 1–5 mm. broad; panicle 0.4–2 dm. long, open, with spreading branches and panicles; spikelets mostly not more than 7-fld., 2–2.5 mm. broad; lemmas 1.8–2.5 mm. long; $n = 30$ (Avdulov, 1928).—Waste places and road-sides, cismontane and desert areas; Calif. to Tex., Mex. July–Oct.

13. **E. árida** Hitchc. Annual, the culms branching at base, erect or ± decumbent, 2–4 dm. long; sheaths with a dense line of hairs part way along the collar; blades mostly flat, glabrous, 1–2 mm. wide; panicle ⅓–½ of length of entire plant, open, with flexuous spreading branches, branchlets, and pedicels, sparsely pilose in lower axils; spikelets oblong to linear, 8–15-fld., 5–10 mm. long, 1.5–2 mm. wide, ± compressed; glumes acute, 1 and 1.5 mm. long; lemmas 1.6–1.8 mm. long.—Reported from Calif.; to Mo., Tex., Mex.

14. **E. cúrvula** (Schrad.) Nees. [*Poa c.* Schrad.] Densely tufted erect perennial 6–12 dm. tall; sheaths keeled, glabrous or sparsely hispid, the lower hairy toward their base; blades involute, finely pointed, scabrous; panicle 2–3 dm. long, the branches ascending, naked at base; spikelets 7–11-fld., gray-green, 8–10 mm. long; lemmas ca. 2.5 mm. long, with prominent nerves.—Reported from Contra Costa Co.; s. states; introd. from Afr. Aug.–Oct.

12. Cutándia Willk.

Annuals, erect or decumbent, many-stemmed. Lvs. narrow, plane or involute. Infl. a dense panicle, ± secund, with very short branches bearing few spikelets. These narrow, 2–13-fld., the rachilla articulate between the florets, glabrous. Glumes narrow, rigid, unequal, 1–3-nerved. Lemmas prominently 3-nerved, keeled, entire or 2-dentate, mucronulate or rarely aristate. Paleae shorter than the lemmas, narrow, 2-keeled. Caryopsis oblong. Ca. 6 spp. of Medit. region.

1. **C. memphítica** (Spreng.) Richt. [*Dactylis m.* Spreng.] Stems several from base, spreading, 1–1.5 dm. long, glabrous; sheaths and lf.-blades glabrous, the latter 2–4 mm. wide, flat, 4–7 cm. long; panicles few-fld., the branches flat, scabrous, zigzag; spikelets on short pedicels, 2–3-fld., 8–9 mm. long; glumes ca. 6 and 7 mm. long, keeled; lemmas 5–7 mm. long.—Sandy soil, nursery at Devil's Canyon, San Bernardino Mts.; introd. from Eu. May.

13. Dissanthèlium Trin.

Annual or perennial grasses with flat lf.-blades and narrow panicles. Spikelets mostly 2-fld., the florets distant; rachilla slender, disarticulating above the glumes and between the florets. Glumes subequal, acuminate, membranaceous to papery, exceeding the lower floret, the 1st 1-nerved, the 2d 3-nerved. Lemmas strongly compressed, acute, awnless, 3-nerved. Paleae somewhat shorter than lemmas. Two spp., Calif. to S. Am. (Greek, *dissos*, double, and *anthelion*, a small fl.)

1. **D. califórnicum** (Nutt.) Benth. [*Stenochloa c.* Nutt.] Annual, ca. 3 dm. tall, ± decumbent or spreading; blades flat, 1–1.5 dm. long, 2–4 mm. wide; panicle 1–1.5 dm. long, narrow but loose, with fascicles of ascending flexuous branches, some of which are floriferous to their base; glumes 3–4 mm. long, almost equal; lemmas pubescent, ca. 2 mm. long.—Open ground; Coastal Sage Scrub; Santa Catalina and San Clemente ids.; ids. of L. Calif.

14. Monanthóchloe Englem.

Spreading wiry-stemmed perennial with clusters of short subulate lvs. Plants dioecious; spikelets 3–5-fld., the uppermost florets rudimentary, the rachilla disarticulating slowly in ♀ spikelets. Glumes wanting. Lemmas rounded on back, convolute, narrowed above,

3-nerved, membranaceous. Paleae narrow, 2-nerved. One sp. (Greek, *monos*, one, and *anthos*, fl.)

1. **M. littoràlis** Englem. Low, extensively creeping, with short erect branches; lf.-blades 5–10 mm. long, falcate, in ± remote clusters; spikelets 1–few, scarcely evident, partly surrounded by lf.-bases.—Coastal Salt Marsh; Santa Barbara to L. Calif.; Santa Catalina Id.; to Tex., Fla., Mex., Cuba. May–June.

15. Distíchlis Raf. SALT GRASS

Low dioecious perennials with wiry culms ascending from strong creeping or deeply running rhizomes. Ligule short and evenly serrate; lf.-blades 2-ranked, flat or ± involute. Staminate infl. a dense spicate panicle exceeding the blades; ♀ equal to or shorter than blades. Spikelets few–many-fld. Glumes unequal, broad, 3–7-nerved. Lemmas closely to loosely imbricate, 9–11-nerved, coriaceous. Paleae usually a little shorter than lemmas, 2-keeled, serrate on keels, often with a few long hairs on the back. Caryopsis brown. Ca. 5 spp., salt- or alkali-tolerant, of temp. N. and S. Am. and Sudan. (Greek, *distichos*, 2-ranked, in reference to lvs.)

(Beetle, A. A. The N. Am. variations of D. spicata. Bull. Torrey Bot. Club 70: 638–650, 1943.)

1. **D. spicàta** (L) Greene var. **stolonífera** Beetle. Culms 2–3 dm. tall, often prostrate, with a strong tendency to form stolons; blades erect, 1–2 dm. long, the upper exceeding the ♀ panicle and often equaling the ♂; the former green or purplish, club-shaped, 1.5–5 cm. long, often 2 cm. thick, of 8–35 crowded spikelets; these 5–9-fld., ca. 1 cm. long, 4 mm. broad; lower glume 2.5 mm. long, upper 3.5 mm.; lemmas 5 mm. long, faintly nerved; palea broadly winged below, with hyaline margins, serrate on keels above; caryopsis ca. 2 mm. long; ♂ infl. of 6–20 spikelets, these 7–10-fld.; glumes 3 and 3.5 mm. long; lemmas 3.5 mm. long; *n* = 20 (Stebbins & Love, 1941).—Coastal Salt Marsh; Orange Co. to Ore.; Catalina, Santa Cruz. ids. Apr.–July.

KEY TO VARIETIES

Panicle congested, the short pedicels hidden, of uniformly 5–9-fld. spikelets. Coastal Salt Marsh
  var. *stolonifera*
Panicle of approximate but scarcely congested spikelets, the pedicels easily visible; florets 3–14.
  Lvs. divaricate; culms and lvs. rigid. Mojave and Colo. deserts .... var. *divaricata*
  Lvs. mostly ascending; culms and lvs. lax.
    Blades 1–2 dm. long, equally spaced on the culm, often equaling or exceeding the infl.; spikelets 4–6 mm. broad. Away from the coast from Orange and Inyo cos. n. .... var. *stricta*
    Blades seldom 1 dm. long, usually crowded at the base, shorter than the infl.; spikelets 2–4 mm. broad. Interior cismontane Calif. .... var. *nana*

Var. **divaricàta** Beetle. Culms 1–4 dm. tall, stiffly erect; blades rarely more than 5 cm. long, rigid, divaricate, exceeded by both ♂ and ♀ panicles; spikelets 0.5–1.5 cm. long, 5–12-fld., 3–6 mm. broad, the glumes 3 and 3.5 mm. long; lemmas ca. 4 mm. long, not prominently nerved.—Alkali Sink and alkaline flats in Creosote Bush Scrub; Mojave and Colo. deserts; to Son., L. Calif. April–Sept.

Var. **strícta** (Torr.) Beetle. [*Uniola* s. Torr.] Culms 1–3.5 dm. tall, erect or ± decumbent, blades to 2 dm. long, equal to or longer than ♂ panicles, these green, drying straw-brown, 2–7 cm. long; spikelets 2–7 cm. long, 5–20-fld., 4–7 mm. broad; glumes 2–3 and 3–4 mm. long; lemmas 3.5–6 mm. long, firm, with broad hyaline margin; *n* = 20 (Stebbins & Love, 1941).—Alkaline spots in many Plant Communities; San Diego Co. to Siskiyou, Modoc, Lassen, and Mono cos.; to Sask., Dakota, Kans., Tex. May–Aug.

Var. **nàna** Beetle. Culms erect, 1–4 dm. high, the blades up to 6 cm. long, exceeded by both ♂ and ♀ panicles, the latter green and purplish, 1–4 cm. long, of 3–12 spikelets, these 0.5–2 cm. long, 3–4 mm. broad, 3–18-fld.; glumes 3 and 3.5 mm. long; lemmas ca. 3.5 mm. long.—On ± alkaline soils; mostly V. Grassland; San Diego Co. to L. A. Co., Kern to Alameda and Stanislaus cos., Lassen Co.

16. Dáctylis L. ORCHARD GRASS

Coarse tufted perennials with flat blades and glomerate panicles. Spikelets 2–6-fld., compressed, subsessile, strongly overlapping, the florets perfect or the upper ♂.

Glumes unequal, ciliate or scabrous on the sharp keel, acute or mucronate. Lemmas 5-nerved, hispid or ciliate-keeled, with short awn-points; paleae slightly shorter than lemmas. Two or 3 spp. of Old World. (Greek, *dactylos,* a finger.)

1. **D. glomeràta** L. Glaucous, scabrous, 4–15 dm. tall; lvs. 7–30 cm. long, 2–6 mm. wide; panicle 1–2.5 dm. long, with ascending branches that become erect in fr.; lemmas 5–6 mm. long, pointed; $n = 7$ (Felfoeldy, 1949), 14 or 21 (Skovsted, 1939).—Natur. ± sparingly, waste places, lawns, etc.; native of Eu. An important hay grass. May–Aug.

17. Cynosùrus L. DOGTAIL

Annuals or perennials with flat narrow blades and dense terminal spikelike panicles. Spikelets dimorphous, the terminal terete fertile one of each fascicle 2–3-fld., sessile, almost concealed by the modified lower sterile ones which are reduced to a rigid fan-like distichous group of narrow glumes and pointed or awned lemmas. Fertile spikelet with 2 rigid glumes; lemmas longer, terete, 3-keeled, firm, mucronate; paleas with 2 ciliate keels. Ca. 4 spp. of Medit. region. (Greek, *cynos,* of a dog, and *oura,* tail.)

Plants annual; panicles subcapitate; awns evident .... 1. *C. echinatus*
Plants perennial; panicle narrow, spikelike; awns inconspicuous .... 2. *C. cristatus*

1. **C. echinàtus** L. Annual, 2–4 dm. tall; blades short, soft, 1–3 mm. wide; panicle 1–4 cm. long, 2–3 cm. thick, bristly; pairs of spikelets 8–10 mm. long; awns of lemmas 6–12 mm. long; $n = 7$ (Avdulov, 1928).—Common locally in fields, waste places, etc., as introd. from Eu. May–July.

2. **C. cristàtus** L. Erect perennial 2–8 dm. high; basal lvs. long, soft, 1–3 mm. wide; panicle strict, 1-sided, 2–8 cm. long, 0.5–1 cm. thick; pairs of spikelets ca. 5 mm. long; awns of lemmas mostly not more than 1 mm. long; $n = 7$ (Avdulov, 1931).—Occasionally reported as from Los Angeles, San Francisco, and Napa and Humboldt cos.; introd. from Eu.

18. Lamárckia Moench. GOLDENTOP

Low, annual, erect, with flat blades and oblong 1-sided compact panicles of crowded fascicled spikelets, the fertile being hidden (except the awns) by the many sterile ones; fascicles falling entire. Fertile spikelet terminal, 1-fld., the floret on a slender stipe, a rudimentary floret on a long rachilla-joint; both awned, with narrow, acuminate or short-awned glumes; lemma broader, scarcely nerved, with a delicate awn just below the apex. Sterile spikelets linear, 1–3 in a fascicle; glumes 2, narrow, 1-nerved; lemmas many, empty, awnless, obtuse, a reduced spikelet borne on the pedicel with each sterile one. One sp. of s. Eu. (J. B. *Lamarck,* French naturalist.)

1. **L. aúrea** (L.) Moench. [*Cynosurus a.* L. *Achyrodes a.* Kuntze.] Erect or ± decumbent at very base, 1–3.5 dm. tall; lvs. smooth; ligule prominent; panicle 2–7 cm. long, 1–2 cm. wide, shining, golden to purplish; pedicels fascicled, pubescent, drooping or spreading; fertile spikelet ca. 2 mm. long, the sterile 6–8 mm. long; glumes hyaline, 2 mm. long; $n = 7$ (Avdulov, 1931).—Common weed in s. Calif., less so n.; introd. from Medit. region. Feb.–May.

19. Arúndo L. GIANT REED

Tall perennial reeds with broad blades and large plumelike terminal panicles. Spikelets several-fld., the florets successively smaller; rachilla glabrous, disarticulating above the glumes and between the florets; glumes ± unequal, membranaceous, 3-nerved, narrow, slender-pointed, ca. as long as the spikelet; lemmas thin, 3-nerved, long-pilose, narrowed upward, the nerves ending in slender teeth, the middle one becoming an awn. Ca. 6 spp. of warmer parts of Old World. (Ancient Latin name.)

1. **A. Dònax** L. Culms stout, to 6 or 7 m. high; blades to 6 cm. wide; panicle 3–6 dm. long; spikelets ca. 12 mm. long; $n = 55$ (Hunter, 1934).—Moist places like ditches, streams, seeps, desert and cismontane Calif.; introd. from Eu. March–Sept.

20. Cortadèria Stapf. PAMPAS GRASS

Large tussock grasses; lvs. largely near the base; blades narrow, attenuate, usually serrulate on edges; panicles large, terminal, plumelike. Spikelets several-fld.; internodes of rachilla jointed, the lower part glabrous, the upper bearded; florets stipitate. Glumes exceeding lower florets. Plants dioecious, the ♂ spikelets covered with long hairs. Five spp. of S. Am. (Native Argentine name, *cortadera,* cutting, because of lf.-margins.)

1. **C. Selloàna** (Schult.) Asch. & Graebn. [*Arundo S.* Schult.] Perennial, 2–4(–6) m. tall; lvs. 3–9 mm. wide, with tufted hairs at the throat of the sheath; ♀ panicles white to pink, 3–9 dm. long, silky-hairy; spikelets 15–18 mm. long; n = 38 (Hunter, 1934).— Escaped, as along coastal bluffs, Ventura to Monterey cos.; commonly cult.; native of Argentina. Late summer.

21. Phragmìtes Trin. REED

Perennial reeds with stout leafy culms and large terminal panicles. Spikelets 3–7-fld., the rachilla clothed with long silky hairs disarticulating above the glumes and at the base of each segm. between the florets, the lowest floret ♂ or neuter; glumes 3-nerved or the upper 5-nerved, acute, lanceolate, unequal, the second shorter than the florets. Lemmas narrow, long-acuminate, glabrous, 3-nerved, the florets successively smaller, the summits of all subequal. Ca. 3 spp., 1 of Asia, 1 of Argentina, 1 cosmopolitan. (Greek, *phragma,* fence, i.e., hedgelike.)

1. **P. commùnis** Trin. var. **Berlándieri** (Fourn.) Fern. [*P. B.* Fourn.] COMMON REED. Culms stout, 2–4 m. high, from long creeping rhizomes; blades 2–6 dm. long, 1–5 cm. wide; panicle tawny, 1–3 dm. long, densely fld.; spikelets 12–15 mm. long; n = 24 or 48 (Avdulov, 1931).—Forming canelike thickets in wet places, below 5000 ft.; edge of Alkali Sink, Creosote Bush Scrub, deserts; and in scattered localities, many Plant Communities, cismontane Calif.; to Atlantic Coast and Mex. July–Nov.

Ampelodésmos mauritànicus (Poir.) Dur. & Schinz is a robust perennial forming large clumps 2–3 m. tall; blades long, wiry, curved at base; panicle 2–5 dm. long, with slender drooping scabrous branches; spikelets crowded, 2–5-fld., 12–15 mm. long, densely pilose on lemma and rachilla joints.—Reported as escape from cult., as in Napa Co.; from s. Eu.

22. Mélica L. MELIC GRASS

Perennial, the culms frequently bulbous at base; sheaths closed; blades flat. Panicle simple or compound, narrow to spreading; spikelets 1–6-fld., articulation above or below the glumes; terminal floret or florets sterile, similar to the fertile or reduced. Glumes less firm than lemmas, with papery or hyaline margins and apices, not keeled, obtuse or acute, 3–5-nerved. Lemmas firm, not keeled, with hyaline apices and upper margins, usually 7-nerved, awned or not; palea usually ¾ the lemma. Ca. 60 spp., in the cooler parts of the world. (*Melica,* an Italian name for a kind of sorghum, probably from the sweet juice [*mel,* honey]).

(Boyle, W. S. A cyto-taxonomic study of the N. Am. species of Melica. Madroño 8: 1–26, 1945.)
A.   Articulation above the glumes and between the florets; spikelets ascending to erect.
    B.   Lemmas awned.
        C.   Awn short, mostly less than 4 mm. long; lf.-blades mostly 1–3 dm. long.
            D.   Culms not bulbous at base; lemmas obtuse. Widespread . . . . 3. *M. Harfordii*
            DD. Culms bulbous at base; lemmas acute or acuminate. Marin Co.
                                                    2. *M. Geyeri* var. *qristulata*
        CC. Awn 5–12 mm. long; blades of lvs. 0.6–1.4 dm. long . . . . 4. *M. aristata*
    BB. Lemmas not awned.
        C.   Culms bulbous at bases.
            D.   Lemmas tapering-acuminate, mostly ciliate-pubescent on nerves . . 1. *M. subulata*
            DD. Lemmas acute or obtuse, glabrous.
                E.   Rachilla swollen, usually wrinkled in drying; 1st glume ca. 3.5 mm. long
                                                        5. *M. fugax*
                EE. Rachilla normal, not swollen; 1st glume ca. 5–8 mm. long.
                    F.   Panicle open and broad when mature, with long spreading branches; 1st and 2d florets ca. 2.5 mm. long; spikelets ca. 16 mm. long
                                                  2. *M. Geyeri*

FF. Panicle narrow or slightly spreading; 1st and 2d florets mostly less than
2 mm. long; spikelets shorter.
G. Bulb of culm small, globose, the bulb not attached directly to the
rhizome; 1st glume less than half as long as spikelet
6. *M. spectabilis*
GG. Bulb of culm attached directly to rhizome; 1st glume more than
half as long as spikelet.
H. Panicle very narrow, the branches and spikelets appressed.
Dry rocky places.
I. Rudiment (terminal floret) blunt, not exserted; no woody
rhizome present. Mostly below 4000 ft.
7. *M. californica*
II. Rudiment tapering above, usually exserted; woody rhizome
present. Mostly above 4000 ft. . . . . 8. *M. bulbosa*
HH. Panicle ± spreading, the branches stiffly ascending. Meadows
8. *M. bulbosa* var. *inflata*
CC. Culms not bulbous at base.
D. Fertile florets 1–2.
E. Rudiment 1 mm. long, on a stipe 2.5 mm. long; lemma pubescent near tip
9. *M. Torreyana*
EE. Rudiment mostly 2 or more mm. long, on a stipe 0.5 mm. long; lemma
glabrous or scabrous . . . . 10. *M. imperfecta*
DD. Fertile florets more than 2.
E. Lemmas pilose-ciliate on lower margins . . . . 3. *M. Harfordii*
EE. Lemmas glabrous or scabrous.
F. Palea ca. half as long as lemma; spikelets ca. 14 mm. long
11. *M. frutescens*
FF. Palea ¾ as long as lemma; spikelets ca. 1 cm. long.
G. Rudiment mostly blunt, not exserted; no woody rhizome present.
Below 4000 ft. . . . . 7. *M. californica*
GG. Rudiment tapering above, slightly exserted; woody rhizome present.
Above 4000 ft. . . . . 8. *M. bulbosa*
AA. Articulation below the glumes; spikelet falling entire at maturity, reflexed . . . . 12. *M. stricta*

1. **M. subulàta** (Griseb.) Scribn. [*Bromus s.* Griseb. *M. acuminata* Bol.] Culms 6–12
dm. high, bulbous at base and attached to a rhizome; blades 2–10 mm. wide; panicle
1–2.5 dm. long, mostly narrow, the branches rarely longer than 9 cm.; spikelets 1–2.8
cm. long, loosely 2–5-fld.; glumes acutish, the 1st 4–7 mm., the 2d 6–9 mm. long, thin,
purplish or brownish; lemmas narrowed above, pilose-ciliate on backs, the first 8–15 mm.
long; $n = 9$ (Boyle, 1945).—Somewhat shaded woods, banks, etc., below 5000 ft.;
Redwood F., Mixed Evergreen F., Yellow Pine F., etc.; Coast Ranges from Marin Co.
n., Sierra Nevada from Nevada Co. n.; to Alaska, Rocky Mts., Chile. May–July.
2. **M. Gèyeri** Munro ex Bol. [*Bromelica G.* Farw.] Culms 1–2 m. tall, bulbous at base
and attached to rhizomes; blades 2–8 mm. wide, often pubescent above; panicle 1–2.7
dm. long, loose, spreading, the branches to 1.4 dm. long; spikelets 0.8–2.4 cm. long,
loosely 2–6-fld.; glumes subacute, 3.5–7 and 5.5–11 mm. long, tinged bronze or purplish; lemmas subacute, the lowest 8–11 mm. long; $n = 9$ (Boyle, 1945).—Open woods,
below 5000 ft.; Redwood F., Mixed Evergreen F., Yellow Pine F., etc.; Coast Ranges
from Monterey Co. n., rare in cent. Sierra Nevada; to n. Ore. April–July.
Var. **aristulàta** J. T. Howell. Awns on lemmas, 0.5–2 mm. long.—Shaded woods,
Marin Co.
3. **M. Harfórdii** Bol. [*Bromelica H.* Farw. *M. H.* var. *minor* Vasey.] Culms 6–12 dm.
long, not bulbous at base; blades 2–6 mm. wide; panicle 0.6–2.3 dm. long, narrow;
spikelets 0.7–2 cm. long, 2–6-fld.; glumes obtuse to subacute, 4–10 and 5–11 mm. long;
lemmas mostly short-awned, the lowest lemma 6–16 mm. long, emarginate to obtuse;
$n = 9$ (Boyle, 1945).—Dry slopes and open woods, below 7000 ft.; Redwood F., Mixed
Evergreen F., Yellow Pine F., etc.; Coast Ranges from Monterey Co. n., cent. Sierra
Nevada; to B.C. May–July.
4. **M. aristàta** Thurb. ex Bol. [*Bromelica a.* Farw.] Culms 6–12 dm. long, not bulbous
at base; blades 3–6 mm. wide, often pubescent; panicle 1–2.3 dm. long, mostly narrow;
spikelets 1–2 cm. long, 2–3-fld.; glumes obtuse to subacute, 7–11 and 7–12 mm. long;
lemmas awned, the lowest 8–13 mm. long, the awns 5–12 mm.; $n = 9$ (Stebbins & Love,
1941).—Dry open woods, 4500–10,000 ft.; Yellow Pine F. to Subalpine F.; San Bernardino Mts., Sierra Nevada and Humboldt Co. to s. Wash. June–Aug.
5. **M. fùgax** Bol. Culms 3–6.5 dm. long, prominently bulbous at base, usually attached
to a light rhizome; blades 2–4 mm. wide; panicle 0.8–1.8 dm., the branches short,
appressed to spreading; spikelets 0.4–1.7 cm. long, loosely 2–4-fld., with a swollen,

spongy, ± wrinkled rachilla; glumes obtuse, 3–5 and 3.5–7 mm. long; 1st lemma 4–7 mm. long; $n = 9$ (Boyle, 1945).—Dry open places or in woods, mostly 4000–7000 ft.; Yellow Pine F., Red Fir F.; Sierra Nevada to Ore., Ida. May–July. Plants with glabrous culms and lvs. have been called var. *madophýlla* Piper; those with narrow panicles, var. *inexpánsa* Suksd.; and with panicle-branches reflexed, var. *Macbrìdei* (Rowland) Beetle.

6. **M. spectábilis** Scribn. Culms 3–10 dm. long, bulbous at base, each bulb connected to the rhizome by a slender stem; blades 2–5 mm. wide; panicle 0.5–2.5 dm. long, mostly narrow, the branches often flexuous; spikelets 0.7–1.9 cm. long, 3–7-fld., turgid, on capillary often flexuous pedicels; glumes obtuse to subacute, the 1st 3.5–5.5 mm. long, the 2d 5–7 mm.; lemmas obtuse to subacute, the lowest 6–9 mm. long, broad, purplish-tinged below the scarious brownish apex, prominently nerved.—Moist meadows or open woods, mostly above 4000 ft.; Yellow Pine F., Red Fir F.; Coast Ranges from Mendocino Co. n.; to B.C., Wyo., Utah. May–July.

7. **M. califórnica** Scribn. Culms 6–13 dm. long, ± enlarged below, but not definitely bulbous, densely tufted; blades 2–5 mm. wide; ligule 2–5 mm. long; panicle 0.4–3 dm. long, very narrow, mostly dense, often interrupted below; spikelets 0.5–1.5 cm. long, 2–5-fld., chaffy; glumes subequal, blunt, scarcely equal to last floret; lemmas obtuse, acute or emarginate, the lowest 5–9 mm. long; rudiment usually blunt or obovoid, not exserted; $n = 9$ (Stebbins & Love, 1941).—Dry rocky exposed slopes, below 3000 ft.; Foothill Wd., Mixed Evergreen F., Yellow Pine F., etc.; Coast Ranges from Ventura Co. n. April–May.

Var. **nevadénsis** Boyle. Panicle very dense, especially toward tips; spikelets slightly V-shaped; glumes acute and often exceeding last floret; spikelets slightly shorter.— Largely below 4000 ft.; Foothill Wd., Yellow Pine F.; w. base of Sierra Nevada.

8. **M. bulbòsa** Geyer ex Porter & Coult. Culms 2–6 dm. tall, usually bulbous at base, the bulbs attached directly to the woody rhizome; blades 2–5 mm. wide; panicle very narrow, with few to 25 spikelets; these 0.6–2.4 cm. long, 2–5-fld.; glumes obtuse to acute, 5–9 and 6–10 mm. long, shorter than spikelet; lemmas usually obtuse, purple-tinged below the scarious tip, the 1st 6–11 mm. long; rudiment narrow, pointed, exserted; $n = 9$ (Boyle, 1945).—Meadows and moist places, mostly 7000–11,000 ft.; Red Fir F. to Subalpine F.; s. cent. Sierra Nevada n. to Humboldt and Siskiyou cos.; to B.C., Colo., w. Tex. July–Aug.

Var. **inflàta** (Bol.) Boyle. [*M. poaeoides* var. *i.* Bol. *M. i.* Vasey.] Panicle wider, longer, with stiffly ascending branches; spikelets pale, averaging 16 mm. long; lemmas acute, strongly nerved.—Uncommon, ca. 4000–5000 ft.; cent. Sierra Nevada to Wash.

9. **M. Torreyàna** Scribn. Culms weak, slender, 3–10 dm. long; blades 1–4 mm. wide; panicle 0.8–2.5 dm. long, with appressed or occasionally spreading branches; spikelets 0.4–0.7 cm. long, 1- or occasionally 2-fld.; glumes ca. as long as spikelet, mostly acute; lemmas subacute, pubescent on upper dorsal surface; $n = 9$ (Boyle, 1945).—In shady woods, thickets, etc., below 2500 ft.; Mixed Evergreen F., Chaparral, etc.; Coast Ranges from San Luis Obispo Co. to Humboldt Co., Sierra Nevada from Mariposa to Butte cos. March–June.

10. **M. imperfécta** Trin. [*M. colpodioides* Nees. *M. panicoides* & *poaeoides* Nutt. *M. Parishii* Vasey.] Culms erect, 3–11 dm. tall; blades 1–6 mm. wide; panicle 0.5–3.6 dm. long, narrow or spreading, the branches often fascicled; spikelets 0.4–0.7 cm. long, usually 1-, occasionally 2-fld.; glumes obtuse to acutish; lemmas not pubescent above, acute to obtuse, 3–7 mm. long; $n = 9$ (Boyle, 1945).—Common on dry open often rocky slopes, below 4000 ft.; Coastal Sage Scrub, Chaparral, S. Oak Wd., Foothill Wd., etc.; cismontane Calif. from Santa Clara and Eldorado cos. s.; occasional, Creosote Bush Scrub to Pinyon-Juniper Wd., Mojave Desert; to L. Calif. April–May. Plants with lower panicle-branches spreading or reflexed, and pubescent lf.-blades have been called var. *refrácta* Thurb. in Wats.; those with similar panicles and glabrous blades, var. *flexuòsa* Bol.; and those with culms less than 3 dm. high, glabrous, with very narrow blades, var. *mìnor* Scribn.

11. **M. frutéscens** Scribn. Culms 1–2 m. tall, stout; blades 2–4 mm. wide; panicle 1–4 dm. long, narrow, dense, pale and shining to purple-tinged; spikelets 1.2–1.8 cm. long, 3–6-fld.; glumes papery, 7–12 and 9–15 mm. long, prominently 5-nerved; lemmas ± obtuse, papery-scarious above, the first 8–11 mm. long, 7-nerved; $n = 9$ (Boyle,

1945).—Dry slopes, below 5000 ft.; Creosote Bush Scrub to Pinyon-Juniper Wd.; deserts; and Coastal Sage Scrub, Foothill Wd., etc.; interior cismontane Calif. from San Luis Obispo Co. s.; to L. Calif., Ariz. March–May.

12. **M. strícta** Bol. Culms 2–5(–8) dm. high, densely tufted, purplish near base where thickened but not bulbous; blades 2–5 mm. wide; panicle 0.3–2 dm. long, narrow, simple, with appressed branches; spikelets 0.6–2.3 cm. long, 2–5-fld.; glumes acute to emarginate, 6–16 and 6–18 mm. long; lemmas obtuse to acute, the first 8–16 mm. long, anthers 1–2 mm. long; $n = 9$ (Boyle, 1945).—Rocky places, 4000–11,600 ft.; Sagebrush Scrub to Subalpine F., Bristle-cone F.; N. Coast Ranges from Humboldt, Tehama cos., etc., San Bernardino Mts., Sierra Nevada, Inyo-White Mts.; to Ore., Utah. June–Aug. Var. *albicáulis* Boyle, with paler lower sheath; anthers 2–3 mm. long; glumes less acute, ranges in mts. from Ventura Co. to San Bernardino Co.

23. Ectospérma Swall.

Stiff perennial, freely branched, from a long branched thick scaly rhizome with woolly nodes; flowering culms erect or ascending, 2.5–3.5 dm. tall, ridged, glabrous except for the puberulent summit. Lvs. distant, stiff, harsh, the sheaths villous on the margin near the summit; ligule a ring of hairs; blades 5–10 cm. long, 3–5 mm. wide, with pungent apex. Panicles narrow, simple, 0.5–1 dm. long, with pubescent compressed branches. Spikelets several-fld., the glumes and lemmas persistent on the short-jointed rachilla; glumes subequal, broad, 7–11-nerved, 9–14 mm. long, ca. as long as spikelet; lemmas rounded on back, imbricate, thin, 5–7-nerved, 7–9 mm. long, densely villous on margins of lower part; palea equal to lemma. Monotypic. (Greek, *ectos,* free from, and *spermos,* seed.)

1. **E. Alexándrae** Swall. Forming extensive masses on sand dunes, 3000–3500 ft.; Creosote Bush Scrub; s. end of Eureka V. and mouth of Marble Canyon, Inyo Mts., Inyo Co. April–June.

24. Trìdens R. & S.

Tufted perennials, rarely stoloniferous or rhizomatous; blades usually flat. Infl. an open to contracted or capitate panicle. Spikelets several-fld., the rachilla disarticulating above the glumes and between the florets. Glumes membranaceous, often thin, subequal, the 1st sometimes narrower, 1-nerved, the 2d rarely 3–5-nerved. Lemmas broad, rounded on back, emarginate to 2-lobed, 3-nerved, the lateral nerves near the margin, the midrib mostly excurrent between the lobes as a point or short awn, all nerves usually pubescent toward base; palea broad. Perhaps 25 spp. N. and S. Am. (Latin, *tria,* thrice, and *dens,* tooth, referring to the lemma.)

Panicle slender, naked, spikelike; plants 2–5 dm. high .... 1. *T. muticus*
Panicle ovoid or headlike; plants mostly shorter.
   Infl. leafy; plants low, creeping .... 2. *T. pulchellus*
   Infl. not leafy; plants not creeping .... 3. *T. pilosus*

1. **T. mùticus** (Torr.) Nash. [*Tricuspis m.* Torr. *Triodia m.* Scribn.] Culms slender, densely tufted, 2–5 dm. tall; sheaths and blades scaberulous, the former loosely pilose, especially at summit; blades ± involute, 1–3 mm. wide; panicle 4–14 cm. long, interrupted; spikelets 6–8-fld., ca. 1 cm. long, pale to purplish, subterete; glumes scaberulous, 5–6 mm. long, 1-nerved; lemmas ca. 5 mm. long, obtuse, pilose on nerves near their base, thin, ± purplish.—Rocky places, often on limestone, 3000–6000 ft.; Creosote Bush Scrub to Pinyon-Juniper Wd.; deserts from Chuckawalla and Eagle mts., Providence Mts., Clark Mt. to e. Sierra Nevada; to Utah, Texas, Mex. April–May, Oct.–Nov.

2. **T. pulchéllus** (HBK.) Hitchc. [*Triodia p.* HBK. *Tricuspis p.* Torr.] FLUFF GRASS. Low, tufted, mostly 5–12 cm. high; culms slender, puberulent to scabrous, with a long internode bearing a terminal fascicle of narrow lvs. and bending over and rooting at tip to produce other culms; blades involute, sharp-pointed, scabrous, curved, mostly 2–4 cm. long; infl. a capitate panicle in the fascicle, mostly with 1–5 subsessile white woolly spikelets; glumes glabrous, subequal, acuminate, awn-pointed, 6–8 mm. long; lemmas 4 mm. long, long-pilose below, cleft ca. halfway, the awn scarcely exceeding the

obtuse lobes.—Frequent in dry rocky places, below 5000 ft.; Creosote Bush Scrub, Joshua Tree Wd.; both deserts; to Tex., Mex. Feb.–May.

3. **T. pilòsus** (Buckl.) Hitchc. [*Uralepis p.* Buckl.] Densely tufted, 1–3 dm. high; sheaths pilose at throat; blades 1–1.5 mm. wide, flat or folded, ± pilose, in a dense basal cluster; panicle well exserted, ovoid, 1–2 cm. long, pale or purplish; spikelets 3–10, short-pedicelled, 6–12-fld., 1–1.5 cm. long, compressed; glumes ca. ⅔ as long as lower florets; lemmas ca. 6 mm. long, pilose below and on margin, acute, with awn 1–2 mm. long.—Dry slopes, 5000–6000 ft.; Pinyon-Juniper Wd.; Clark Mt. and Mescal Range, e. San Bernardino Co.; to w. Kans., Tex., cent. Mex. May–June.

25. Neostápfia Davy

Low annual with loose sheaths merging into rather broad flat blades without definite junction. Panicles dense, cylindric, the axis prolonged beyond the spikelets into a naked portion or with small bracts. Spikelets few-fld., subsessile, closely imbricate around a simple axis; rachilla disarticulating beyond the florets. Glumes wanting. Lemmas flabellate, prominently many-nerved. One sp. (Named for Otto *Stapf,* English botanist.)

1. **N. colusàna** (Davy) Davy. [*Stapfia c.* Davy. *Anthochloa c.* Scribn. *Davyella c.* Scribn.] Culms from a decumbent base, 0.7–3 dm. long; foliage pale green, blades 5–10 cm. long, 6–12 mm. wide, minutely ciliate with raised sticky glands on margins and nerves; panicles at length exserted, 3–7 cm. long, 8–12 mm. thick; spikelets 6–7 mm. long; lemmas 5 mm. long, very broad, ciliolate-fringed, viscid-glandular on nerves. —About vernal rain pools; V. Grassland; Colusa, Stanislaus, Merced cos. May–July.

26. Orcúttia Vasey

Low annuals, with short blades, spikes or spikelike racemes. Spikelets rather large, subsessile, several-fld., the upper florets reduced, the rachilla continuous. Glumes subequal, shorter than lemmas, broad, irregularly 2–5-toothed, many-nerved, the nerves extending into teeth. Lemmas firm, prominently 13- or 15-nerved, the broad summit toothed; palea broad, as long as lemma. Ca. 4 spp. of Calif., L. Calif. (Named for C. R. *Orcutt,* early California botanist.)

(Hoover, R. F. The genus Orcuttia. Bull. Torrey. Bot. Club 68: 149–156, 1941.)
Lemmas with 7–11 very short teeth. Cent. V. . . . . 1. *O. Greenei*
Lemmas with 5 rather long acuminate or awn-tipped teeth.
    Racemes 2–5 cm. long, often capitate, the spikelets usually crowded toward the summit; teeth of
    lemma unequal; nerves of lemma relatively faint. Cent. V. to L. Calif. . . . . 2. *O. californica*
    Racemes 5–10 cm. long, not capitate, the spikelets rather evenly distributed; teeth of lemma un-
    equal; nerves of lemma prominent.
        Lf.-blades 1–2 mm. wide; spikelets mostly 2–10-fld., glabrous. Shasta and Tehama cos.
                                                              3. *O. tenuis*
        Lf.-blades 2–6 mm. wide; spikelets mostly 10–40-fld., pilose. San Joaquin V. . . . . . . 4. *O. pilosa*

1. **O. Greènei** Vasey. Culms 0.5–3 dm. long, suberect; blades 2–3 cm. long, subinvolute; raceme 3–7 cm. long, pale green; spikelets 1–1.5 cm. long, loosely papillose-pilose; glumes 4–5 mm. long, toothed or subentire; lemmas 5–7 mm. long, obtuse or truncate, the teeth mucronate, but not awned.—Moist open places; V. Grassland; Tehama to Tulare cos. May–June.

2. **O. califórnica** Vasey. Sparingly to moderately pilose, the culms 0.5–1.5 dm. long; sheaths loose; blades 2–4 cm. long; raceme loose below, dense upward, 2–5 cm. long; spikelets alternate on opposite sides of the axis, 8–12 mm. long, densely to sparsely pilose; glumes sharply toothed, 2–4 mm. long; lemmas 4–5 mm. long, deeply toothed at tip.—Drying mud flats; V. Grassland; s. Western Ave., Los Angeles; Murietta Hot Springs, Menifee (w. Riverside Co.); n. L. Calif. May–June.

Var. **inaequàlis** (Hoov.) Hoov. [*O. i.* Hoov.] Infl. shorter, more capitate; lemmas unequally toothed.—V. Grassland; Stanislaus to Tulare cos.

Var. **víscida** Hoov. Plants more viscid; lemma-teeth awned, the head appearing bristly. —V. Grassland, etc.; Sacramento Co.

3. **O. ténuis** Hitchc. Culms slender, tufted, erect, 0.5–1.2 dm. long; lvs. mostly basal, mostly thinly pilose, 1–2 cm. long, strongly nerved; raceme 5–9 cm. long, the lower spikelets distant, the upper approximate; spikelets 6–12 erect, 2–10-fld., mostly sparsely

pilose; glumes 3–5 mm. long, 2–5-toothed at apex; lemmas 4.5–6 mm. long, parted above the middle into 5 equal awn-tipped teeth.—Vernal pools; V. Grassland, Foothill Wd.; Shasta, Tehama, Lake cos. May–July.

4. **O. pilòsa** Hoov. Densely tufted, 0.5–2 dm. tall, erect or decumbent at base, viscid at maturity; sheaths and blades pilose; racemes 5–10 cm. long; spikelets 8–18, 2-ranked, 10–40-fld., appressed or ± spreading, the upper crowded; glumes ca. 3 mm. long, irregularly 3-toothed; lemmas 4–5 mm. long, parted above middle into 5 sub-equal teeth.—Vernal pools; V. Grassland; Stanislaus to Madera cos. May–July.

27. Blepharidáchne Hack.

Low annuals or perennials, with short dense panicles not much exserted from the foliage. Spikelets compressed, rather few, 4-fld., the rachilla disarticulating above the glumes, but not between the florets. Glumes subequal, compressed, 1-nerved, thin, smooth; lemmas 3-nerved, deeply 3-lobed, the nerves extending into awns, conspicuously ciliate; 1st and 2d florets sterile, 3d fertile, 4th reduced to a 3-awned rudiment. A small genus. (Greek, *blepharis*, eyelash, and *achne*, chaff, because of the ciliate lemma.)

1. **B. Kíngii** (Wats.) Hack. [*Eremochloe K.* Wats.] Looking like *Tridens pulchellus*, but not rooting at upper nodes; culms 3–8 cm. long; sheaths with broad hyaline margins; blades involute, curved, stiff, 1–3 cm. long; panicle subcapitate, 1–2 cm. long, often purplish; glumes ca. 7–8 mm. long, acuminate, exceeding florets; sterile lemmas ca. 6 mm. long, long-ciliate on margins, pilose at base, cleft almost to middle, awn-tipped; fertile lemmas similar.—Dry rocky places, 5000–7000 ft.; Pinyon-Juniper Wd.; Panamint, Grapevine, and Inyo-White mts., Inyo Co.; to Utah. May.

28. Enneapògon Desv. ex Beauv.

Slender tufted perennials, with numerous culms and flat to subinvolute lvs. Panicle spikelike, gray green, feathery. Spikelets 3-fld., the 1st floret fertile, the 2d smaller and sterile, the 3d rudimentary. Glumes strongly 7-nerved, longer than the body of the lemmas. Lemmas rounded on the back, firm, the truncate apex with 9 plumose equal awns; palea slightly longer than body of lemma. One sp. (*Ennea*, nine, and *pogon*, beard, because of the 9 awns on the lemma.)

(Chase, A., *in* Madroño 7: 187–189, 1946.)

1. **E. Desvaúxii** Beauv. [*Pappophorum Wrightii* Wats.] Culms slender, 2–4 dm. long, the nodes pubescent; blades ca. 1 mm. wide; panicle 2–5 cm. long; lowest lemma 4–5 mm. long.—Dry limestone slope, 5000–6000 ft.; Pinyon-Juniper Wd.; Clark Mt., e. San Bernardino Co.; to Utah, Tex., Mex., S. Am. Aug.–Sept.

29. Agropỳron Gaertn. Wheat Grass

Mostly perennials, often with creeping rhizomes; the culms usually erect. Spikes usually erect, green or purplish, with mostly solitary spikelets at the nodes. Spikelets several-fld., sessile, placed flatwise at each joint of a continuous rachis, the rachilla articulating above the glume and between the florets. Glumes 2, equal, firm, several-nerved, usually shorter than the 1st lemma, acute or awned, rarely obtuse or notched. Lemma convex on back, rather firm, 5–7-nerved, mostly acute or awned from the apex; palea shorter than lemma. Ca. 60 spp. in temp. regions. (Greek, *agrios*, wild, and *puros*, wheat, the 2 original spp. being weeds in wheat.)

(Gould, F. W. Nomenclatorial changes in Elymus with a key to the Californian species. Madroño 9: 120–128, 1947.)
A. Lemmas awned, the awns 10–30 mm. long; plants typically without creeping rhizomes.
   B. Awns of lemmas curving outward at maturity.
      C. Rachis readily disarticulating at maturity.
         D. Internodes of spike 5–10 mm. long; culms erect at base .... 13. *A. saxicola*
         DD. Internodes of spike 3–5 mm. long; culms prostrate-spreading ... 10. *A. Scribneri*
      CC. Rachis not disarticulating at maturity; internodes of spikes usually 1 cm. or longer.
         D. Culms erect at base, usually 4 or more dm. high; at least some lf.-blades 10 cm. or longer.
            E. The culms slender; blades narrow, usually involute; spikes straight, with closely appressed spikelets. Ne. Calif. .... 11. *A. spicatum*

EE. Culms stout; blades 4–6 or more mm. broad, flat; spikes flexuous, with ±
 spreading spikelets. Sierra Nevada .... 12. A. arizonicum
 DD. Culms usually decumbent at base, mostly 1.5–3.5 dm. long; blades mostly less than
 10 cm. long .... 9. A. Pringlei
 BB. Awns of lemmas straight or undulate, not curving outward.
 C. Rachis readily disarticulating at maturity; glumes mostly attenuate with awns 2–5 cm.
 long .... 15. A. Saundersii
 CC. Rachis not readily disarticulating; glume-awns 1–3 cm. long.
 D. Spikes quite dense, the spikelets overlapping ½–⅔ their length; internodes of
 rachis mostly 4–8 mm. long .... 7. A. subsecundum
 DD. Spikes not dense, the spikelets overlapping the one above on the opposite side of
 the rachis ¼ or less of their length; internodes of rachis largely 10 mm. or more
 long .... 14. A. Parishii var. laeve
 AA. Lemmas awnless or with awns not more than 6 mm. long.
 B. Glumes subulate to narrowly lanceolate, inconspicuously nerved, hard or tough in texture,
 awn-tipped .... 3, A. Smithii
 BB. Glumes broadly lanceolate, strongly 3–9-nerved, thin, or if thickened having an obtuse apex.
 C. Plants with creeping rhizomes.
 D. Internodes of culms 1–3 cm. long; rachis disarticulating at maturity. Seashore
 6. A. junceum
 DD. Internodes of culms mostly over 4 cm. long; rachis not disarticulating.
 E. Lemmas glabrous or scabrous.
 F. Blades flat, thin and lax, bright green, rarely glaucous .... 2. A. repens
 FF. Blades mostly involute, stiff, mostly glaucous .... 5. A. riparium
 EE. Lemmas finely pubescent .... 4. A. dasystachyum
 CC. Plants without creeping rhizomes.
 D. Spikelets much compressed, crowded on the rachis .... 1. A. desertorum
 DD. Spikelets not much compressed.
 E. Nodes of culms glabrous; spikelets 3–5-fld. .... 8. A. trachycaulum
 EE. Nodes of culms pubescent; spikelets mostly 6–8-fld. .... 14. A. Parishii

1. **A. desertòrum** (Fisch.) Schult. [*Triticum d.* Fisch. ex Link.] Densely tufted,
2.5–10 dm. tall, the culms slender, erect or bent at base; sheaths glabrous or the lower
spreading-hirsute; blades 2–4(–5) mm. wide; spike 5–9 cm. long, 7–11 mm. thick, ±
bristly; rachis short-jointed, pubescent; spikelets closely spaced on rachis, 8–12 mm.
long, 5–7-fld., ± spreading; glumes and lemmas firm, glabrous to sparsely ciliate on the
keel, both abruptly narrowed to awns 2–3 mm. long; lemma ca. 6 mm. long, the awn
mostly bent to one side.—Occasional in grainfields, etc.; introd. from Russia.

2. **A. rèpens** (L.) Beauv. [*Triticum r.* L. *Elymus r.* Gould.] QUACK GRASS. Rhizome
yellow-green, creeping, scaly; culms green or glaucous, 5–10 dm. tall, erect or curved
below; blades flat, thin, mostly ± pilose on upper surface, 6–10 mm. wide; spike 5–
10(–15) cm. long; spikelets mostly 4–6-fld., 1–1.5 cm. long, glabrous or scaberulous
on rachilla; glumes 3–7-nerved, awn-pointed, 8–10 mm. long; lemmas 10 mm. long, with
an awn 1–5 mm. long; palea obtuse, almost as long as lemma; $n = 14$, 21 (Avdulov,
1931).—Weed in waste places, etc., San Francisco, Humboldt Co., etc.; more troublesome in e. states; native of Eu. June–Aug.

3. **A. Smíthii** Rydb. [*Elymus S.* Gould. *A. occidentale* Scribn.] W. WHEATGRASS.
Rhizomes creeping, scaly, gray or tawny; culms usually glaucous, erect, 3–10 dm.
tall; sheaths glabrous; blades firm, scabrous, mostly 2–4 mm. wide, involute on drying;
spike erect, 7–15 cm. long, with scabrous rachis; spikelets closely imbricate, 6–10-fld.,
1–2 cm. long, the rachilla scabrous; glumes rigid, rather faintly nerved, short-awned,
10–12 mm. long; lemmas ca. 10 mm. long, glabrous, often pubescent near base; palea
scabrous-pubescent on keels; $n = 14$, 28 (Hartung, 1946).—Dry alkaline places, 5000–
6500 ft.; Sagebrush Scrub, etc.; Mono Co. to Modoc Co.; to e. Wash., Mich., Kans.,
Tex. June–Aug.

4. **A. dasystàchyum** (Hook.) Vasey. [*Triticum repens* var. *d.* Hook. *Elymus subvillosus*
Gould.] Rhizomes creeping; culms 4–8 dm. tall; blades flat to involute, 1–3 mm. wide,
scabrous; spike 6–12 cm. long; spikelets ± loosely imbricate, 4–8-fld., 1–1.5 cm. long,
the rachilla ± pubescent; glumes acute or awn-pointed, scabrous or pubescent, 6–9 mm.
long; lemmas 8–10 mm. long, scabrous- to villous-pubescent, awnless to mucronate;
palea obtuse, ca. 1 cm. long; $n = 14$ (Hartung, 1946).—Dry sandy places, ca. 2000–
4000 ft.; Foothill Wd., Yellow Pine F.; Siskiyou Co., Lassen Co.; to B.C., Mich., Ill.
June–July.

5. **A. ripàrium** Scribn. & Sm. [*Elymus Rydbergii* Gould.] Like *A. dasystachyum*,
with vigorous rhizomes, narrower blades, more imbricate spikelets; lemmas glabrous

or somewhat pubescent on edges of lower part; $n = 14$ (Hartung, 1946).—Dry or moist places, collected in June 1916, near Riverside; to Wash., Alta., N. Dak., Colo.

6. **A. juncèum** (L.) Beauv. [*Triticum j.* L. *Elymus multimodus* Gould.] Rhizomes creeping; culms 3–7 dm. tall; blades pubescent, loosely involute, glaucous; spikelets glabrous, 5–8-fld.; glumes 9-nerved, acutish; $n = 14$ (Peto, 1930).—Coastal dunes, San Francisco; native of Eu. June–Sept.

7. **A. subsecúndum** (Link) Hitchc. [*Triticum s.* Link. *Elymus pauciflorus* ssp. *s.* Gould. *A. caninum* Calif. refs.] Without creeping rhizomes; culms tufted, erect, green or glaucous, 5–10 dm. tall, mostly with glabrous sheaths; blades flat, 3–8 mm. wide; spike 6–15 cm. long, the rachis ± scabrous; spikelets imbricate, few-fld., with villous rachilla; glumes 4–7-nerved, awned; lemmas 5-nerved, with an awn 1–3 cm. long; $n = 14$ (Hartung, 1946).—Moist woods and meadows, below 10,800 ft.; Montane Coniferous F.; Sierra Nevada to Modoc Co.; Alaska, Atlantic Coast. June–Aug.

8. **A. trachycaúlum** (Link) Malte. [*Triticum t.* Link. *Elymus pauciflorus* Gould, not Lam. *A. tenerum* Vasey. *A. pauciflorum* Hitchc.] Much like *A. subsecundum,* the blades mostly 2–4 mm. wide; spike slender, 10–25 cm. long; spikelets remote to imbricate; glumes and lemmas awnless or nearly so; $n = 14$ (Peto, 1929).—Moist to dry and rocky places, below 11,000 ft.; Montane Coniferous F., Chaparral, Mixed Evergreen F., etc.; San Jacinto Mts. n.; through Coast Ranges and Sierra Nevada, White Mts.; to Alaska and Atlantic Coast, Mex. June–Aug.

9. **A. Prínglei** (Scribn. & Sm.) Hitchc. [*A. Gmelini* var. *P.* Scribn. & Sm. *Elymus sierrus* Gould.] Culms 3–5 dm. tall, tufted, curved at base; blades flat or loosely involute, 1–3 mm. wide; spike ± flexuous, 4–7 cm. long, with slender scabrous rachis; spikelets mostly 3–7, rather remote, 3–5-fld.; glumes narrow, 7–8 mm. long, 3-nerved; the awn ca. 5 mm. long; lemmas with a scabrous divergent awn 15–25 mm. long; palea 10–12 mm. long; $n = 14$ (Hartung, 1946).—Rocky slopes, 7500–11,000 ft.; Lodgepole F. to Subalpine F.; Sierra Nevada, from Tulare to Sierra cos. July–Aug.

10. **A. Scríbneri** Vasey. [*Elymus S.* Jones.] Culms tufted, prostrate or decumbent-spreading, 2–4 dm. long; blades flat or loosely involute, 1–3 mm. wide; spike long-exserted, dense, 3–7 cm. long, with glabrous internodes, disarticulating readily; spikelets 3–5-fld.; glumes narrow, the 1 obscurely nerved, the other distinctly 2–3-nerved, awned; lemmas with strong divergent awn 15–25 mm. long; palea slightly exceeding body of lemma, 2-toothed.—Rocky slopes, 10,000–13,600 ft.; Alpine Fell-fields; White Mts., Sweetwater Mts.; to Mont., New Mex. July–Aug.

11. **A. spicàtum** (Pursh) Scribn. & Sm. [*Festuca s.* Pursh. *Elymus s.* Gould. *Agropyron divergens* Nees.] Tufted, erect, green or glaucous, 6–10 dm. tall; sheaths glabrous; blades 1–2(–4) mm. wide, flat to loosely involute, pubescent on upper side, glabrous beneath; spike slender, 8–15 cm. long, with scaberulous rachis and rather long internodes; spikelets mostly 6–8-fld., often shorter than the internodes of the spike; glumes rather narrow, mostly not awned, ca. half as long as spikelet, glabrous or scabrous on the nerves; lemmas ca. 1 cm. long, the strongly divergent awn 10–20 mm. long; $n = 7,$ 14 (Hartung, 1946).—Plains and dry hills, 3000–5000 ft.; Sagebrush Scrub, N. Juniper Wd., etc.; Lassen to Siskiyou and Modoc cos.; to Alaska, Mich., New Mex. June–July.

12. **A. arizónicum** Scribn. & Sm. [*Elymus a.* Gould.] Like *A. spicatum* but taller and coarser; the spike flexuous, 15–30 cm. long; blades 4–6 mm. wide; spikelets somewhat spreading, the awns 15–30 mm. long.—Eel Ridge, nw. Calif.; to Nev., w. Tex., n. Mex. June–July.

13. **A. saxícola** (Scribn. & Sm.) Piper. [*Elymus s.* Scribn. & Sm.] Tufted, erect, 3–8 dm. tall; blades flat to loosely involute, glabrous or ± pubescent, 1–4 mm. wide; spike 5–12 cm. long, the rachis-internodes ± scabrous, 5–10 mm. long; spikelets imbricate, ca. twice as long as internodes, 4–6-fld., with minutely scabrous rachilla; glumes 2-nerved, with a divergent awn 5–20 mm. long; lemmas ca. 8 mm. long, the divergent awn 20–50 mm. long; palea ca. as long as body of lemma, obtuse or truncate.—Dry or rocky slopes, 9000 ft.; Subalpine F.; Mt. Dana, Sierra Nevada; to Wash., Ida. July–Aug.

14. **A. Paríshii** Scribn. & Sm. [*Elymus Stebbinsii* Gould.] Culms 7–12 dm. tall, pubescent at nodes; blades flat or loosely involute, 2–4 mm. wide; spike slender, nodding, 10–25 cm. long, the rachis-internodes 15–25 mm. long; spikelets 4–7-fld., ca. 2 cm.

long, scaberulous on rachillae; glumes 3–5-nerved, 10–15 mm. long, acute; lemmas acute or short-awned (1–8 mm.); palea as long as lemma, obtuse.—Dry slopes below 5000 ft.; Chaparral, Yellow Pine F.; Cuyamaca Mts. to Sierra Nevada and Santa Lucia Mts. Most of our plants have glabrous nodes and are the var. *laève*, Scribn. & Sm.

15. **A. Saúndersii** (Vasey) Hitchc. [*Elymus S.* Vasey.] Erect, 6–10 dm. tall; blades flat or loosely involute; spikes erect, 8–15 cm. long, ± purplish; spikelets 10–15 mm. long (excluding awns), 2–5-fld.; glumes 2–3–5-nerved, scabrous at least on midrib, the awn 20–50 mm. long; lemmas scabrous, with awns 7–30 mm. long.—Dry slopes, 4500–10,500 ft.; Sagebrush Scrub to Subalpine F.; Sweetwater Mts., Sierra Nevada, Tulare and Inyo to Modoc cos.; to Ida., Wyo., Ariz. July–Aug. California material has been described as *Elymus Saundersii* var. *californicus* Hoov.

30. Élymus L. RYE GRASS

Annual or perennial, with flat or rarely convolute blades and cylindric spikes. Spikelets sessile, 2–6-fld., sessile in 2's, or rarely 3's at each node of a continuous rachis, the rachillae disarticulating above the glumes and between the florets. Glumes equal, rigid, narrow, sometimes even awnlike. Lemmas rounded on back or subterete, obscurely 5-nerved, acute or usually awned from tip. Ca. 45 spp. of n. temp. regions. (Greek, *elumos*, ancient name for a grain.)

(Gould, F. W. Nomenclatorial changes in Elymus with a key to the California species. Madroño 9: 120–128, 1947.)
A.   Lemmas awned, the awns mostly 10–30 mm. long; plants typically without creeping rhizomes.
  B.   Plants annual; awns of lemmas 30–80 mm. long. Introd. weed . . . . 1. *E. caput-medusae*
  BB. Plants perennial.
    C.   Awns of lemmas curving outward at maturity.
      D.   Sheaths and blades pubescent; glumes relatively thin, not indurate at base
                                                                        9. *E. glaucus Jepsonii*
      DD. Sheaths and blades mostly glabrous; glumes firm, indurate at base
                                                                        12. *E. canadensis*
    CC. Awns of lemmas ± straight, not curving outward at maturity.
      D.   Rachis not disarticulating at maturity; glumes usually broadly lanceolate, 3–5-nerved; culms usually in small clusters . . . . 9. *E. glaucus*
      DD. Rachis disarticulating at maturity; glumes narrowly lanceolate to subulate, 1–3-nerved; culms usually in dense clumps.
        E.   Spikes ca. 5 mm. broad, body of lemma ca. 6–8 mm. long . 10. *E. Macounii*
        EE. Spikes 8–10 mm. broad; body of lemma 8–10 mm. long . . . 11. *E. aristatus*
AA. Lemmas awnless, or the awns not more than 6 mm. long.
  B.   Glumes broadly lanceolate, strongly 3–9-nerved, thin or if thickened, with an obtuse apex.
    C.   Plants without rhizomes . . . . 9. *E. glaucus* ssp. *virescens*
    CC. Plants with rhizomes . . . . 2. *E. mollis*
  BB. Glumes subulate, or if lanceolate then inconspicuously nerved, hard or tough in texture, awn-tipped or acute.
    C.   Culms finely pubescent below the infl.; plants with slender rhizomes. Seashore
                                                                        3. *E. vancouverensis*
    CC. Culms glabrous below the infl.
      D.   Spikelets 6–40 at a node of the rachis; culms usually 6–10 mm. in diam.; blades 15–35 mm. broad. Coastal . . . . 7. *E. condensatus*
      DD. Spikelets mostly 1–6 at a node; culms mostly less than 6 mm. in diam.; blades 3–15 mm. broad.
        E.   Spikelets mostly more than one at a node.
          F.   Culm-nodes or their vicinity with fine usually dense pubescence; plants mostly without rhizomes . . . . 8. *E. cinereus*
          FF. Culm-nodes glabrous; plants with rhizomes . . . . 4. *E. triticoides*
        EE. Spikelets mostly solitary at the nodes.
          F.   Culms mostly 2.5–10 dm. tall; spikes well exserted.
            G.   Plants with rhizomes . . . . 4. *E. triticoides*
            GG. Plants without rhizomes . . . . 6. *E. salinus*
          FF. Culms 1–2 dm. high; spikes little exserted, often exceeded by blades
                                                                        5. *E. pacificus*

1. **E. cáput-medùsae** L. Annual, the culms branched basally, decumbent to erect, slender, 2–6 dm. tall; blades narrow, 3–6 cm. long; spike 2–5 cm. long without the awns; glumes subulate, with a slender awn 1–2.5 cm. long; lemmas lanceolate, 3-nerved, 6 mm. long, scabrous, with a flat awn 5–10 cm. long.—Occasional as a weed on grassy slopes, etc., particularly in n. Calif.; native of Eu. June–July.

2. **E. móllis** Trin. ex Spreng. Perennial with creeping rhizomes; culms stout, glaucous,

smooth or pubescent above, 6–12 dm. tall; sheaths and blades mostly smooth; blades 7–12 mm. wide, firm, sometimes involute on drying; spike erect, dense, soft, 7–25 cm. long; glumes lanceolate, flat, many-nerved, pubescent or scabrous, 12–25 mm. long, acuminate, ca. equal to spikelet; lemmas scabrous to pubescent, pointed.—Coastal Strand; Monterey Co. n.; to Alaska, Atlantic Coast, Eurasia. June–Aug.

3. **E.** × **vancouverénsis** Vasey (pro sp.) nothomorph **califórnicus** Bowden. Perennial with abundant creeping rhizomes; culms erect, 8–15 dm. tall; blades 5–8 mm. wide, scabrous above; spike ± interrupted, purplish, 2–3 dm. long; glumes linear-lanceolate, firm, gradually acuminate, 1–1.5 cm. long, sparsely long-villous; lemmas firm, 10–15 mm. long, narrowed into a short awn.—Coastal Strand; Marin Co. to B.C. July–Aug.

4. **E. triticoìdes** Buckl. [*E. condensatus* var. *t.* Thurb. *E. Orcuttianus* Vasey.] Rhizomes scaly, extensively creeping; culms mostly glaucous, 2–3.5 mm. in diam., 6–12 dm. tall, mostly in large masses; sheaths smooth or scabrous; blades mostly 2–6 mm. wide, flat or soon involute; spike erect, slender to ± dense, sometimes compound, 1–2 dm. long; spikelets 12–20 mm. long; glumes subulate or narrow, firm, 0–1–3-nerved, 8–14 mm. long, awn-tipped; lemmas 6–10 mm. long, glabrous, brownish to purplish, firm, awn-tipped; *n* = 14, 21 (Gould, 1945).—Moist and alkaline places, below 7500 ft.; many Plant Communities; throughout the state; to Wash., Mont., Tex., L. Calif. June–July. Plants with pubescent sheaths and blades are the var. *pubéscens* Hitchc.

Ssp. **multiflòrus** Gould. Culms 3.5–5 mm. in diam.; blades 6–15 mm. broad; spikelets 17–25 mm. long; *n* = 21 (Gould, 1945).—With the sp.

5. **E. pacíficus** Gould. [*Agropyron arenicola* Davy, not *E. a.* Scribn. & Sm.] Rhizomes slender, creeping; culms ± spreading, 1–2 dm. tall; blades involute, mostly exceeding culms; spike 2–5 cm. long; spikelets solitary, 12–15 mm. long; glumes nerveless, firm, short-awned; lemmas ca. 10 mm. long, obscurely nerved, pointed or awn-tipped, narrowly hyaline on margin.—Coastal Strand; Monterey to Mendocino cos.

6. **E. salìnus** Jones. Tufted, erect, 3–8 dm. tall, purplish-brown at base, sometimes scabrous below nodes and below spike; sheaths scabrous; blades firm, involute, mostly scabrous; spike slender, 5–12 cm. long; spikelets 10–15 mm. long, mostly solitary; glumes subulate, 4–8 mm. long; lemmas ca. 10 mm. long, awnless or nearly so, glabrous or scabrous, rarely sparsely strigose, obscurely nerved.—Rocky slopes, 4500–6500 ft.; Pinyon-Juniper Wd.; Providence, New York, and Clark mts., e. San Bernardino Co.; to Ida., Wyo., Colo. May–June.

7. **E. condensàtus** Presl. Rhizomes short, thick; culms mostly in dense clumps, 1.5–3.5 m. high; blades firm, strongly nerved, flat, 1.5–3 cm. wide, glabrous or pubescent; spikes erect, dense, 1.5–5 dm. long, mostly ± compound; spikelets often in 3's or 5's, 10–15 mm. long, 3–6-fld.; glumes subulate or flat and narrow, usually 1-nerved or nerveless, ca. as long as 1st lemma; lemmas glabrous to ± strigose, short-awned or acute, hyaline-margined; *n* = 14, 28 (Gould, 1945).—Dry places below 5000 ft.; Coastal Sage Scrub, Chaparral, S. Oak Wd., Foothill Wd., etc.; Santa Cruz Co. to L. Calif.; Joshua Tree Wd., etc., mts. of n. Mojave Desert. June–Aug.

8. **E. cinèreus** Scribn. & Merr. [*E. condensatus* var. *pubens* Piper.] Less robust than *E. condensatus*, 0.6–2 m. tall, typically without rhizomes, harsh-puberulent especially at nodes; sheaths and blades pubescent or glabrous, the latter mostly less than 1.5 cm. wide; spikes 1–2.5 dm. long, mostly not branched; glumes and lemmas like those of *E. condensatus*, but the lemmas ± pubescent; *n* = 14, 28 (Gould, 1945).—Dry places below 10,000 ft.; largely Sagebrush Scrub, Pinyon-Juniper Wd., etc.; deserts from San Bernardino Co. n., e. of Sierra Nevada to Siskiyou and Shasta cos.; B.C., Saskatchewan, New Mex. June–Aug.

9. **E. glaúcus** Buckl. [*E. villosus* var. *glabriusculus* Torr. *E. g.* var. *breviaristatus* Davy. *E. g.* var. *maximus* Davy. *E. hispidulus* Davy. *E. angustifolius* Davy. *E. a.* var. *caespitosus* Davy.] Culms tufted, 6–12 dm. tall; sheaths smooth or scabrous; blades flat, mostly 8–15 mm. wide, mostly scabrous, sometimes involute; spike long-exserted, 0.5–2 dm. long; spikelets 10–12 mm. long; glumes ca. as long, strongly 2–5-nerved, acuminate or awn-pointed; lemmas awned, the awn 1–2 times the body-length, erect to spreading; *n* = 14 (Hartung, 1946).—Grassy and wooded places, below 7500 ft.; many Plant Communities; cismontane Calif., occasional on desert; to Alaska, Ontario, Ark., New Mex. June–Aug.

Ssp. **viréscens** (Piper) Gould. [*E. v.* Piper. *E. pubescens* Davy.] Lemmas awnless or the awns to 4 mm. long; $n = 14$ (Hartung, 1946).—Grassy and wooded places, cis-montane Calif., San Diego Co. to Alaska.

Ssp. **Jepsònii** (Davy) Gould. [*E. g.* var. *J.* Davy.] Sheaths and blades ± pubescent.—Largely Montane Coniferous F.; Calif. to B.C., Mont.

10. **E. Macoùnii** Vasey. Densely tufted, the culms slender, 5–10 dm. tall; sheaths mostly glabrous; blades erect, subinvolute, mostly scabrous, 2–5 mm. wide; spike slender, 4–12 cm. long, tardily disarticulating; spikelets imbricate, mostly 2-fld., ca. 10 mm. long without the awns; glumes very narrow, scabrous, ca. 1 cm. long, short-awned; lemmas scabrous toward tip, with slender awns 1–2 cm. long.—Meadows and open places; e. slope of Sierra Nevada from Nevada Co. n., to Siskiyou Co.; B.C., Minn., New Mex. June–Aug. Material referred here is largely sterile and supposed to represent hybrids between *Agropyron* and *Hordeum.*

11. **E. aristàtus** Merr. Tufted, 7–10 dm. tall; sheaths glabrous; blades flat, 5–10 mm. wide; spike 6–14 cm. long, tardily disarticulating; spikelets often in 3's, imbricate, 1–2-fld., ca. 10 mm. long without the awns; glumes subsetaceous, 1–2 cm. long; lemmas sparsely scabrous at least upward, the awn 1–2 cm. long.—Dry places, ca. 7100 ft.; Sagebrush Scrub; Mono Co.; to Wash., Wyo. June–July.

12. **E. canadénsis** L. Erect, tufted, smooth, 1–1.5 m. tall; blades flat, scabrous above, 1.5–3 dm. long, 5–15 mm. wide; spike 1–2 dm. long; glumes ± indurate, ca. 3-nerved, ending in an awn; lemmas scabrous-pubescent, ca. 10 mm. long, the awn 2–3 cm. long and divergent when dry; $n = 14$ (Hartung, 1946).—Reported from Sonoma Co.; n. to Wash., e. to Atlantic Coast. June–Aug.

31. Sitànion Raf. SQUIRRELTAIL

Erect cespitose perennials with bristly spikes. Spikelets sessile, usually 2 at each node of the disarticulating rachis, 2–few-fld. Glumes narrow or setaceous, 1–3-nerved, the nerves prominent, extending into 1 or more awns, the awns equal or unequal. Lemmas firm, convex on back, subterete, 5-nerved, with 1 or more long slender awns. Palea firm, nearly as long as lemma-body, the 2 keels serrulate. Ca. 5–6 spp. of w. U.S. (Greek, *sitos*, grain for food.)

Spike much longer than broad; glumes linear-lanceolate, 2–4 nerved .... 1. *S. Hansenii*
Spike not longer than broad; glumes setaceous, 1- or obscurely 2-nerved.
  Glumes cleft into 3–several fine divisions .... 2. *S. jubatum*
  Glumes entire or 2-cleft .... 3. *S. Hystrix*

1. **S. Hansènii** (Scribn.) J. G. Sm. [*Elymus H.* Scribn. *S. anomalum* J. G. Sm.] Culms smooth, 5–10 dm. high; sheaths glabrous; blades flat, scabrous, 2–8 mm. wide; spike 8–20 cm. long; glumes 2–3-nerved, long-awned; lower lemmas ca. 8 mm. long, with mature awns divergent and 4–5 cm. long; $n = 14$ (Stebbins & Love, 1941).—Dry open often rocky places below 13,000 ft.; many Plant Communities; widely spread in Calif., but particularly in mts. of the Mojave Desert, Sierra Nevada, etc.; to Wash., Wyo. June–Aug. Most plants referred here are apparently sterile hybrids between *Elymus glaucus* and *Sitanion Hystrix* or *S. jubatum.*

2. **S. jubàtum** J. G. Sm. [*S. multisetum* J. G. Sm. *Elymus m.* Davy. *S. polyanthrix* J. G. Sm. *S. breviaristatum* J. G. Sm.] Plants 2–6 dm. high; sheaths smooth, villous or scabrous; blades flat or involute, 1–4 mm. wide; spike dense, 3–10 cm. long, bushy; each lobe of the glumes with an awn 3–8 cm. long; lemmas 8–10 mm. long, smooth or scabrous toward the tip, long-awned; $n = 14$ (Stebbins & Love, 1941).—Rocky or brushy slopes and waste places, below 10,000 ft.; many Plant Communities; cismontane Calif. to White Mts., Inyo Co.; to Wash., Utah, Ariz., L. Calif. May–July.

3. **S. Hýstrix** (Nutt.) J. G. Sm. [*Aegilops H.* Nutt. *S. minus* J. G. Sm. *S. californicum* J. G. Sm. *S. glabrum* J. G. Sm. *S. longifolium* J. G. Sm. *Elymus elymoides* Swezey.] Plants 1–5 dm. high; sheaths glabrous to pubescent; blades flat or involute, glabrous to pubescent, 1–5 mm. wide; spike mostly 2–8 cm. long; glumes narrow, 1–2-nerved, the nerves extending into scabrous awns 2–7 cm. long and sometimes bifid to middle; lemmas smooth to strigose or scabrous, with awns 2–10 cm. long; $n = 14$ (Stebbins & Love, 1941).—Common in dry open places, below 13,000 ft.; many Plant Communities; much of Calif.; to B.C., S. Dak., Tex. Mex. July–Aug.

32. Hýstrix Moench

Erect perennials with flat lvs. and bristly loosely fld. spikes. Spikelets 2–4-fld., sessile, 1–3 at each node of the flattened rachis, divergent. Glumes reduced to short or minute awns, the 1st mostly obsolete or both wanting. Lemmas convex, rigid, long-awned, 5-nerved, the nerves obscure except toward apex. Palea ca. as long as body of lemma. Ca. 4 spp., temp. (Greek, *hustrix,* porcupine.)

1. **H. califórnica** (Bol.) Kuntze. [*Gymnostichium c.* Bol.] Culms stout, 1–2 m. high; sheaths hispid or the upper smooth; blades largely 1–2 cm. wide; spike 1–2.5 dm. long; spikelets usually 3–4 at a node, 1.2–1.5 cm. long, ± ascending; glumes 0; lemmas 5-nerved above, the nerves ciliate-hispid with short stiff hairs, awn straight, stout, rough, ca. 2 cm. long.—Woods and shade; Redwood F., Closed-cone Pine F., Douglas-Fir F., Mixed Evergreen F., etc.; near the coast from Santa Cruz Co. to Sonoma Co. June–Aug.

33. Tríticum L.

Annuals with flat blades and thick spikes. Spikelets 2–5-fld., solitary, placed flatwise at each joint of the rachis, the rachilla often disarticulating above the glumes and between the florets. Glumes rigid, keeled, 3–several-nerved, the apex abruptly mucronate or toothed or awned. Lemmas broad, keeled, many-nerved, pointed or awned. Ca. 15 spp., of Medit. region and w. Asia. (Latin name for wheat.)

1. **T. aestìvum** L. WHEAT. Six–10 dm. tall; blades 1–2 cm. wide; spike mostly 5–12 cm. long; rachis-internodes 3–6 mm. long; spikelets broad; glumes usually keeled toward 1 side.—Commonly cult. and often escaping in waste places; native of Old World. April–July.

34. Aègilops L. GOATGRASS

Annuals with flat blades and usually awned spikes. Spikelets 2–5-fld., solitary, placed flatwise at each joint of the rachis, the spike usually disarticulating near the base at maturity and falling entire. An Old World genus. (Greek name for a grass.)

Spikelets subovate; rachis not disarticulating .... 1. *A. ovata*
Spikelets cylindrical; rachis finally disarticulating.
    Glumes with 1 awn .... 2. *A. cylindrica*
    Glumes with 3 awns .... 3. *A. triuncialis*

1. **A. ovàta** L. Tufted, 1.5–2.5 dm. tall; blades short, sharp-pointed; spike thick, of 2–4 subovate spikelets, the upper sterile; glumes with 4 spreading awns 2–3 cm. long; lemmas mostly with 1 long and 2 short awns; $n = 14$ (Kihara, 1937).—Reported as a weed in Glenn and Mendocino cos.; introd. from Eu. May–July.

2. **A. cylíndrica** Host. Four–6 dm. tall; spike 5–10 cm. long; spikelets 8–10 mm. long; glumes several-nerved, keeled at 1 side, the keel ending in an awn; lemmas of upper spikelets awned, the awns ca. 5 cm. long; $n = 14$ (Kihara, 1937).—Reported as a weed in Siskiyou Co.; introd. from Eu. May–June.

3. **A. triunciàlis** L. Two–4 dm. tall; spike 3–4 cm. long; fertile spikelets 3–5; glumes with 3 strong scabrous awns 4–8 cm. long; lemmas with 3 rigid unequal awns.—Weed on range land, as in Marin, Tuolumne cos.; introd. from Eu. May–July.

35. Secàle L. RYE

Like *Triticum,* but with subulate 1-nerved glumes. Spikelets 2-fld., alternating on a long zigzag rachis. Lemmas keeled and long-awned. Three spp. of Eurasia. (Old Latin name of a grain.)

1. **S. cereàle** L. Tufted annual, 1–1.5 m. tall, blue-green; blades to ca. 12 mm. broad, long-pointed; spike slender, nodding, 7–15 cm. long; spikelets with 2 fertile fls. and sometimes a 3d sterile one; lemmas long, narrow, awned.—Cult. and sometimes escaping in waste places and fields; native sw. Asia. May–Aug.

36. Hórdeum L. BARLEY

Annual and perennial grasses with flat blades and dense bristly spikes, disarticulating at the base of the rachis segms., this remaining as a stipe below the attached 3 spikelets. Spikelets 1-fld., 3 at each node, one or all of spikelets fertile and, according to the number that are fertile, producing spikes with 6, 4, or 2 rows. Glumes very narrow, like an invol. subtending the 3 spikelets. Lemmas awned, rounded on the back. Ca. 20 spp. of temp. regions. (Ancient Latin name for barley.)

(Covas, G. Observations on the N. Am. spp. of Hordeum. Madroño 10: 1–21, 1949.)
Plants perennial.
  Glumes and awns 1.8–8 cm. long .... 3. *H. jubatum*
  Glumes and awns less than 1.8 cm. long.
    Blades pubescent, 1.5–5 mm. wide; anthers usually 1.5–3 mm. long; glumes of cent. spikelet 1.5–
    2.5 times as long as palea .... 1. *H. californicum*
    Blades mostly glabrous, 3–9 mm. wide; anthers mostly 1–1.5 mm. long; glumes of cent. spikelet
    often scarcely longer than palea .... 2. *H. brachyantherum*
Plants annual.
  Glumes of cent. spikelet and the inner ones of the lateral spikelets with ciliate margins.
    Spike with 6–8 spikelets to 1 cm. of rachis; stamens of cent. florets included at anthesis
                                                                        8. *H. Stebbinsii*
    Spike with 3–5 spikelets to 1 cm. of rachis; stamens of cent. florets exserted at anthesis
                                                                        9. *H. leporinum*
  Glumes not ciliate.
    Auricles long; rachis continuous; all 3 spikelets sessile, fertile .... 10. *H. vulgare*
    Auricles obsolete; rachis articulate; lateral spikelets pedicellate, usually not fertile.
      Inner glumes of lateral spikelets strongly broadened, 0.6–1.8 mm. wide ...... 4. *H. pusillum*
      Inner glumes and outer linear-subulate, less than 0.6 mm. wide.
        Spike ovate to ovate-oblong, usually less than 5 cm. long; awns and glumes strongly spread-
        ing at maturity; bases of glumes of lateral spikelets prominent above the pedicel
                                                                        5. *H. Hystrix*
        Spike linear-oblong, usually over 5 cm. long; awns and glumes suberect; bases of glumes of
        lateral spikelets not prominent above the pedicel.
          Cent. spikelet 13–22 mm. long, including awn; pedicels of lateral spikelets almost straight;
          lateral florets with acute but awnless lemmas .... 6. *H. depressum*
          Cent. spikelet 26–32 mm. long including awn; pedicels of lateral spikelets curved; lateral
          florets with acuminate very short awned apex .... 7. *H. arizonicum*

1. **H. califórnicum** Covas & Steb. [*H. nodosum,* Calif. refs. in part.] Perennial, 2–6.5 dm. tall; blades 1.5–5 mm. wide, usually pubescent above and beneath; spike linear-oblong, green or purplish, 2.5–8 cm. long; rachis articulate; cent. spikelet sessile, 12–22 mm. long including awns; glumes setaceous, scabrous, 8–17 mm. long; lemma usually glabrous, scabrous toward apex, tapering into awn 7–15 mm. long; palea 5.5–9.5 mm. long, acuminate; lateral spikelets pedicellate; glumes setaceous, floret mostly neuter; lemma commonly subulate; *n* = 7 (Covas & Stebbins, 1949).—Grassy and brushy places, below 8500 ft.; many Plant Communities; cismontane and montane Calif.; to Ore. April–Aug.

2. **H. brachyántherum** Nevskii. [*H. nodosum* Calif. refs., in part.] Tufted perennial 2–7 dm. tall; blades 3–8 mm. wide, mostly glabrous; spike 8–10 cm. long, sometimes purplish; floret of cent. spikelet 7–10 mm. long, the awn ca. 1 cm., the glumes slightly shorter; glumes of lateral spikelets usually unequal, somewhat shorter, the floret ♂ to reduced and neuter, with awn 2–5 mm. long; *n* = 14 (Covas & Stebbins, 1949).—Moist places, below 11,000 ft.; many Plant Communities; cismontane and montane Calif., White Mts.; to Alaska, Labrador, New Mex. May–Aug.

3. **H. jubàtum** L. FOXTAIL. Tufted perennial 3–6 dm. tall; blades 2–5 mm. wide, scabrous; spike nodding, 5–10 cm. long, ca. as wide, pale, soft; lateral spikelets reduced to 1–3 spreading awns; glumes of perfect spikelet awnlike, 2.5–6 cm. long, spreading; lemma 6–8 mm. long with awn equal to glumes; *n* = 14 (Stebbins & Love, 1941).— Common in open moist and waste places, often a bad weed in ± alkaline pastures and meadows, below 10,000 ft.; many Plant Communities; here and there throughout the state; to Alaska, Atlantic Coast (where introd.), Mex. May–July.

Var. **caespitòsum** (Scribn.) Hitchc. [*H. c.* Scribn.] Awns 1.8–3.5 cm. long.—With the sp.

4. **H. pusíllum** Nutt. Annual, 1–3.5 dm. tall; blades erect, flat; spike erect, 2–7 cm. long; 1st glume of lateral spikelets and both glumes of cent. dilated above the base, with a slender awn 8–15 mm. long; lemma of cent. spikelet awned, smooth, of lateral

spikelets awn-pointed; $n = 7$ (Kihara, 1924).—In open ± alkaline places, below 5000 ft.; cismontane s. Calif.; to Wash., Atlantic Coast, Argentina. April–May.

5. **H. Hýstrix** Roth. [*H. Gussoneanum* Parl.] Annual, the culms spreading at base, 1.5–4 dm. long; sheaths and blades ± pubescent; spike erect, 1.5–3 cm. long, the rachis usually not breaking easily; glumes setaceous, rigid, ca. 12 mm. long; lemma of cent. spikelet 5 mm. long, with a somewhat longer awn than the glumes; fl. of lateral spikelets reduced, short-awned; $n = 7$ (Covas, 1949).—Alkaline or waste places, especially in cent. cismontane Calif.; to B.C., Atlantic Coast; native of Eu. April–June.

6. **H. depréssum** (Scribn. & Sm.) Rydb. [*H. nodosum* var. *d.* Scribn. & Sm.] Annual, the culms geniculate at base, 0.5–4.5 dm. long; upper sheaths often inflated; blades pubescent, 2–4 mm. wide; spike erect, 4–7 cm. long; fl. of cent. spikelet 7–8 mm. long, subterete, the awn ca. 10 mm. long; awns of glumes and glumes of lateral spikelets subequal, ca. 2 cm. long; $n = 14$ (Covas, 1949).—Moist ± alkaline places, at low elevs.; many Plant Communities; to B.C. April–May.

7. **H. arizónicum** Covas. [*H. adscendens* Hitchc., not HBK.] Annual, 2–6 dm. tall, lower sheaths pubescent, upper glabrous; blades 3–5 mm. wide, sparsely pubescent; spike erect, 3–12 cm. long; floret of cent. spikelet 8–9 mm. long, the awn 15–22 mm. long, the glumes slightly shorter; glumes of lateral florets nearly as long, one slightly dilated, the floret reduced to a short-awned lemma; $n = 21$? (Covas, 1949).—Irrigated places, Imperial Co.; to Ariz.

8. **H. Stébbinsi** Covas. [*H. murinum* Calif. refs., in part.] Annual, 1–5 dm. tall; lvs. glaucous with smooth sheaths; blades mostly sparsely pubescent, 2.5–7 mm. wide, auricled at base; spike ovate-oblong, 4–9 cm. long, dense; cent. spikelet 16–36 mm. long (including awn), sessile; glumes linear-lanceolate, 3-nerved, 12–22 mm. long, ciliate on both margins; floret pedicelled; lemma glabrous, the awn 8–25 mm. long; lateral spikelets on a slender pedicel, ciliate inside, the florets usually neuter; $n = 7$ (Covas, 1949).—Weed in waste places and fields; cismontane Calif. to Wash., Okla. Native of Eu. April–May.

9. **H. leporìnum** Link. [*H. murinum*, Calif. refs. in part.] Annual, spreading; sheaths glabrous; blades pilose to glabrous, with well developed auricle; spike 5–9 cm. long, often partly enclosed by the inflated uppermost sheath; glumes of cent. spikelet lanceo-late, 3-nerved, long-ciliate on both margins, the nerves scabrous, the awn 2–2.5 cm. long; floret 1–1.2 cm. long, the awn 3–4 cm. long; lateral spikelets usually ♂, the glumes much shorter, unlike; lemma broad, 10–20 mm. long, the awn 2–4 cm. long; $n = 14$ (Covas, 1949).—Weed in waste places, fields, etc.; native of Eu. April–June.

10. **H. vulgàre** L. COMMON BARLEY. Erect annual, 6–12 dm. tall; blades 5–15 mm. wide; spike 2–10 cm. long excluding awns; the 3 spikelets sessile; glumes divergent at base, narrow, nerveless, stout-awned; awn of lemma straight, erect, mostly 10–15 cm. long; $n = 7$ (Kihara, 1924).—Sometimes found in waste places and fields, as an escape from cult.; native of Old World. BEARDLESS BARLEY (*L. v.* var. *trifurcàtum* [Schlecht.] Alef.) is also met. April–July.

37. Lòlium L. DARNEL. RYEGRASS

Annuals or perennials with flat blades and simple erect stems. Spikes terminal, flat. Spikelets several-fld., solitary, sessile, placed edgewise to the continuous rachis, one edge fitting the alternate notches; rachilla disarticulating above the glumes and be-tween the florets. First glume wanting except on the terminal spikelet, the 2d out-ward, 3–5-nerved, equaling or exceeding the 2d floret. Lemmas convex, 5–7-nerved, obtuse, acute or awned. Ca. 8 spp. Eurasian. (Ancient Latin name.)

Glume shorter than the spikelet; perennial.
  Lemmas nearly or quite awnless; culms subcompressed .... 1. *L. perenne*
  Lemmas, at least the upper, awned; culms cylindric .... 2. *L. multiflorum*
Glume as long or longer than the spikelet; annual.
  Spike 15–25 cm. long, flat; spikelets much wider than the rachis .... 3. *L. temulentum*
  Spike 5–10 cm. long, thickish; spikelets scarcely wider than the rachis .... 4. *L. strictum*

1. **L. perénne** L. PERENNIAL or ENGLISH RYEGRASS. Culms erect or decumbent at the ± reddish base, 3–6 dm. high; auricles at top of sheath obsolete; blades glossy, 2–4 mm. wide, spike mostly 15–25 cm. long; spikelets mostly 6–10-fld.; lemmas 5–7 mm. long,

awnless or almost so; $n = 7$ (Thomas, 1936).—Used in lawns and sometimes natur.; native of Eu. May–Sept.

2. **L. multiflòrum** Lam. Italian Ryegrass. More robust than *L. perenne,* to 1 m. tall, pale or yellowish at base; auricles prominent; spikelets 10–20-fld., 1.5–2.5 cm. long; lemmas 7–8 mm. long, at least the upper awned; $n = 7$ (Peto, 1933).—Common lawn grass and escaping; native of Eu. June–Aug.

3. **L. temuléntum** L. Darnel. Annual, 6–9 dm. tall; blades mostly 3–8 mm. wide; spike stiff; glumes ca. 2.5 cm. long, at least as long as the 5–7-fld. spikelet; florets plump, the lemmas to 8 mm. long, obtuse, with an awn 6–12 mm. long; $n = 7$ (Jenkin & Thomas, 1938).—Common in waste places and fields, cismontane Calif.; introd. from Eu. April–June. Var. *leptochaèton* A. Br. has the lemmas awnless.

4. **L. stríctum** Presl. Annual, with stiff culms and spikes and 1–3 dm. tall; spike thickish, 5–10 cm. long, the rachis thick but flattish and angled; spikelet 3–9-fld., lanceolate; glumes obtuse.—Reported from about Berkeley; native of Eu.

38. Monérma Beauv.

Low annuals with slender cylindric spikes. Spikelets 1-fld., embedded in the hard articulate rachis and falling attached to the joints. First glume wanting except on the last spikelet; second glume even with the surface of the rachis, hardened, nerved, acuminate, longer than the rachis-joint. Lemma with its back to the rachis, hyaline, shorter than the glume, 3-nerved. Palea slightly shorter than lemma, hyaline. One sp. (Greek, *monos,* one, and *erma,* support, in reference to the single spike.)

1. **M. cylíndrica** (Willd.) Coss. & Durieu. [*Rottboellia c.* Willd. *Lepturus c.* Trin.] Branched, spreading, the stems 1–5 dm. long; lvs. narrow; spikes 1–2 dm. long; glume 6 mm. long, acuminate; lemmas 5 mm. long; rachis disarticulating at maturity; $n = 13$ (Avdulov, 1931), 26 (Hunter, 1934).—Coastal Salt Marsh; San Diego to Colusa Co.; introd. from Old World. May–July.

39. Parápholis C. E. Hubb.

Low annuals, with slender cylindric spikes. Spikelets 1–2-fld., embedded in the cylindric articulate rachis and falling attached to the joints. Glumes 2, placed in front of the spikelet and enclosing it, coriaceous, 5-nerved, pointed. Lemma with its back to the rachis, smaller than the glumes, hyaline, 1-nerved. (Greek, *para,* beside, and *pholis,* scale, because of the 2 glumes side by side.)

1. **P. incúrva** (L.) C. E. Hubb. [*Aegilops i.* L. *Pholiurus i.* Schinz & Thell.] Sickle Grass. Tufted, decumbent below, 1–2 dm. high; blades narrow; spike 7–10 cm. long, curved; spikelets 7 mm. long, acuminate; $n = 18$ (Avdulov, 1931).—Coastal Salt Marsh, Coastal Strand; San Diego to Humboldt Co.; Ore., e. Coast; native of Eu. April–June.

40. Scribnèria Hack.

Low annual with short narrow blades and linear cylindrical spikes. Spikelets 1-fld., solitary, laterally compressed, appressed flatwise against the ± thickened continuous rachis, the rachilla disarticulating above the glumes. Glumes equal, firm, narrow, pointed, keeled on outer nerves, the 1st 2-nerved, the 2d 4-nerved. Lemma shorter than glumes, membranaceous, faintly nerved, short-bifid at tip, the midvein ending in a slender awn. Palea ca. as long as lemma. (Named for F. Lamson-*Scribner,* student of grasses.)

1. **S. Bolánderi** (Thurb.) Hack. [*Lepturus B.* Thurb.] Culms 0.5–3 dm. long, tufted, ascending to erect; lvs. few, subfiliform; ligule ca. 3 mm. long; spike 2–11 cm. long, ca. 1 mm. thick; spikelets ca. 7 mm. long; lemmas pubescent near base, the erect awn 2–4 mm. long.—Many habitats below 9000 ft.; many Plant Communities; from San Luis Obispo and Tulare cos. to Siskiyou and Modoc cos. March–June.

41. Schísmus Beauv.

Low tufted annuals with filiform blades and small panicles, the slender pedicels finally disarticulating at base and falling with the spikelet or with the glumes. Spikelets several-fld., the rachilla disarticulating above the glumes and between the florets. Glumes subequal, exceeding the 1st floret, white-margined. Lemmas broad, rounded on back, several-nerved, pilose along lower part of edge, hyaline at tip, 2-toothed. Palea broad, hyaline, nerved at margin. Old World. (Greek, *schismos*, splitting, because of the 2-toothed lemmas.)

Glumes 4–5 mm. long; lemmas ca. 2 mm. long, rounded and emarginate at tip; palea rounded, as long as lemma .... 1. *S. barbatus*
Glumes 5–6 mm. long; lemmas 2.5–3 mm. long, with 2 acute terminal lobes; palea acute, shorter than lemma .... 2. *S. arabicus*

1. **S. barbàtus** (L.) Thell. [*Festuca b.* L.] Culms slender, 0.5–3 dm. long; blades mostly less than 10 cm. long; panicle narrow to oval, 1–5 cm. long, mostly dense, pale or purplish; spikelets ca. 5-fld.; glumes 5–7-nerved, shorter than spikelets, acute; lemmas 9-nerved, the margins appressed-pilose on lower part, teeth minute.—Becoming common in waste places and mud flats, etc.; Imperial Co., Riverside Co., San Bernardino Co., Los Angeles Co., Cent. V., etc.; native of Eurasia, Afr. March–Apr.
2. **S. arábicus** Nees. Like the preceding, but spikelets slightly larger, 5–7-fld.; lemmas 2.5–3 mm. long, more pilose on margins and back, with 2 acute lobes at tip; palea acute.—Open places, Cent. V., w. Mojave Desert; native of sw. Asia. March–May.

42. Koelèria Pers.

Tufted perennials or annuals with narrow blades and shining spikelike panicles. Spikelets 2–4-fld., compressed, the rachilla disarticulating above the glumes and between the florets. Glumes mostly subequal in length, dissimilar in shape, the 1st narrow, 1-nerved, the 2d wider, broadened above the middle, 3–5-nerved; lemmas ± scarious, shining, the lowermost slightly longer than the glumes, faintly 5-nerved, pointed, the awn if present from just below the apex. Ca. 20 spp. of temp. regions. (For G. L. *Koeler*, an early grass student.)

Plants perennial; culms puberulent below panicle .... 1. *K. cristata*
Plants annual; culms glabrous below panicle .... 2. *K. phleoides*

1. **K. cristàta** (L.) Pers. [*Aira c.* L.] JUNEGRASS. Culms erect, 3–6 dm. tall, leafy at base; sheaths pubescent, at least the lower; blades flat or involute, 1–3 mm. wide, glabrous or especially the lower pubescent; panicle erect, dense, often lobed or interrupted below, 4–14 cm. long; spikelets mostly 4–5 mm. long; glumes and lemmas scaberulous, 3–4 mm. long; $n = 14$ (Stebbins & Love, 1941).—Dryish open places below 11,500 ft.; many Plant Communities; cismontane and montane Calif., higher mts. of Mojave Desert; to B.C., Atlantic Coast; Old World. May–July. A taller loosely tufted form with lvs. to 3 dm. long and more open panicles has been called var. *longifòlia* Vasey; open woods in Coast Ranges.
2. **K. phleoides** (Vill.) Pers. [*Festuca p.* Vill.] Annual, 1–3 dm. high; sheaths and blades sparsely pilose; panicle dense, 2–7 cm. long, obtuse; spikelets 2–4 mm. long; glumes acute; lemmas short-awned from a bifid apex.—Becoming established at various points in Calif. (San Luis Obispo and Kern to Shasta and Butte cos.); native of Eu. April–July.

43. Sphenópholis Scribn. WEDGE GRASS

Mostly slender perennials, usually with flat lf.-blades and narrow terminal panicles. Spikelets 2–3-fld., the rachilla prolonged beyond the upper floret. Glumes subequal but unlike in shape, the 1st narrow, 1-nerved, the 2d broader, 3–5-nerved, becoming coriaceous. Lemmas firm, scarcely nerved, awnless or with an awn from just below the apex, the 1st a little shorter or a little longer than the 2d glume. Palea hyaline, exposed.

Ca. 6 spp. of New World. (Greek, *sphen,* wedge, and *pholis,* horny scale, in reference to the hard obovate 2d glume.)

1. **S. obtusàta** (Michx.) Scribn. [*Aira o.* Michx. *S. o.* var. *lobata* Scribn.] Culms tufted, erect, 3–6 dm. tall; sheaths glabrous to pubescent; blades 1–3 dm. long, 2–5 mm. wide, glabrous to pubescent; panicle spikelike to interrupted or lobed, 0.5–2 dm. long; spikelets 2.5–3.5 mm. long, the 2 florets very close together; 2d glume very broad, subcucullate, ± inflated at maturity, 5-nerved, scabrous; lemmas minutely papillose, the 1st ca. 2.5 mm. long.—Damp and open places, as at San Bernardino and in Amador and Fresno cos.; to B.C., Maine, W.I., Mex. April–July.

44. Trisètum Pers.

Tufted perennials with flat blades and open or contracted shining panicles. Spikelets usually 2-fld., sometimes 3–5-fld., the rachilla prolonged beyond the upper floret, usually villous. Glumes ± unequal, acute, the 2d usually exceeding the 1st floret. Lemmas usually short-bearded at base, 2-toothed at apex, the teeth often awned, usually bearing from the back below the apex a straight or bent awn. Ca. 65 spp. of colder and temp. regions. (Latin *tri,* three, and *setum,* bristle, because of the awn and 2 teeth of lemma.)

(Louis-Marie, Father. The genus Trisetum in Am. Rhodora 30: 210–223, 237–245, 1928.)
Awn included within the glumes or wanting .... 1. *T. Wolfii*
Awn exserted.
  Panicle dense, spikelike, sometimes slightly interrupted below; plants densely tufted . 2. *T. spicatum*
  Panicle loose and open to contracted, but not spikelike; plants solitary or in small tufts.
    The panicle pale green, sometimes tinged purple; spikelets usually 2-fld. Native .. 3. *T. cernuum*
    The panicle yellowish; spikelets mostly 3–4-fld. Introd. .... 4. *T. flavescens*

1. **T. Wólfii** Vasey. [*T. subspicatum* var. *muticum* Bol.] Erect, 3–10 dm. tall, loosely tufted; sheaths scabrous, rarely the lower pilose; blades flat, scabrous, rarely pilose, 2–4 mm. wide; panicle erect, rather dense, green or pale, 8–15 cm. long; spikelets 2-(–3)-fld., 5–7 mm. long; glumes subequal, acuminate, ca. 5 mm. long; lemmas obtusish, scaberulous, 4–5 mm. long, awnless or nearly so.—Meadows, 7000–10,000 ft.; Tulare and Inyo cos. n.; to Wash., Alta., Mont., New Mex. July–Aug.

2. **T. spicàtum** (L.) Richt. [*Aira s.* L.] Culms slender, erect, densely tufted, 1–4 dm. high, glabrous or puberulent; sheaths and usually the blades puberulent; panicle dense, almost spikelike, pale or dark purple, 5–15 cm. long; spikelets 4–6 mm. long; 1st glume 1-nerved, the 2d broader, acute, 3-nerved; lemmas scaberulous, 5 mm. long, the first exceeding the glumes, with setaceous teeth; awn attached ca. ⅓ below the tip, 5–6 mm. long, geniculate, exserted.—Alpine slopes and banks, 7200–13,000 ft.; Subalpine F., Alpine Fell-fields; San Jacinto and San Bernardino mts., Sierra Nevada, N. Coast Ranges; to Alaska, Arctic and Antarctic regions. July–Aug. Variable; two forms being: var. *mólle* (Michx.) Beal [*Avena m.* Michx.], with densely pubescent foliage; and var. *Cóngdoni* (Scribn. & Merr.) Hitchc. [*T. C.* Scribn. & Merr.], subglabrous and with spikelets 6–7 mm. long.

3. **T. cérnuum** Trin. Culms rather lax, 6–12 dm. tall; sheaths glabrous to sparsely pilose; blades thin, scabrous, 6–12 mm. wide; panicle lax, drooping, 1.5–3 dm. long, with whorled filiform flexuous branches bearing spikelets near their ends; spikelets 6–12 mm. long, usually 3-fld., the 1st fl. exceeding the 2d glume; 1st glume narrow, acuminate, 1-nerved, 0.5–2 mm. long, the 2d broad, 3-nerved, 3–4 mm. long, occasionally reduced; lemma 5–6 mm. long, the teeth setaceous; awns slender, curved, 5–10 mm. long, attached 1–2 mm. below tip.—Moist woods, below 4000 ft.; Redwood F., Mixed Evergreen F., Douglas-Fir F.; Mendocino Co. n. near the coast; to Alaska, Alta., Mont. May–Aug.

Var. **canéscens** (Buckl.) Beal. [*T. c.* Buckl.] Blades mostly 2–7 mm. wide; spikelets of lowest fascicle nearly sessile; 1st glume 3–4 mm. long, the 2d 5–7 mm. long; $n = 21$ (Stebbins & Love, 1941).—In and about woods, below 5000 ft.; Mixed Evergreen F., Douglas-Fir F., Yellow Pine F., etc.; Coast Ranges from Santa Cruz Co. n.; to B.C., Mont.

Var. **projéctum** (Louis-Marie) Beetle. [*T. p.* Louis-Marie.] Foliage more velvety; blades mostly 1–3 mm. wide and those of the culm mostly less than 1 dm. long;

panicles less interrupted.—Mostly damp banks, 4000–9000 ft.; Montane Coniferous F.; San Jacinto and San Bernardino mts., Sierra Nevada.

4. **T. flavéscens** (L.) Beauv. [*Avena f.* L.] Tall, 5–8 dm. high; blades 2–10 mm. broad; sheaths glabrous or ± pilose; panicle loose, yellowish, 0.5–2 dm. long; spikelets 5–6 mm. long, 2–3-fld.; lemmas smooth or slightly scabrous, 4–6 mm. long; $n = 12$ (Avdulov, 1931), 14 (Nakajima, 1930).—Occasional weed, as in Humboldt Co.; introd. from Eu. June–July.

45. Deschámpsia Beauv. HAIRGRASS

Annual or perennial grasses, with flat or involute lvs. and open or contracted panicles. Spikelets 2-fld., disarticulating above the glumes and between the florets, the hairy rachilla extended beyond the florets or rarely terminated by a ♂ one. Glumes subequal, keeled, acute, membranous, shining. Lemmas thin, truncate, 2–4-toothed at summit, bearded at base, with a slender awn from or below the middle, the awn straight, bent or twisted. Palea narrow, 2-nerved. A genus of cold and temp. areas. (Named for J. C. Loiseleur-*Deslongchamps*, 1774–1849, French botanist.)

(Lawrence, W. E. Some ecotypic relations of D. caespitosa. Am. Journ. Bot. 32: 298–314, 1945.)
Plant annual; spikelets 4–8 mm. long; panicle open .... 1. *D. danthonioides*
Plants perennial.
    Panicle narrow, elongate, ca. ⅓ the length of the stem, with appressed branches; spikelets 4–6 mm. long .... 2. *D. elongata*
    Panicle open or contracted, if narrow, not more than ¼ the length of the stem.
        Blades 4–6 mm. wide, thin; glumes exceeding the florets. Siskiyou Co. ...... 3. *D. atropurpurea*
        Blades 1.5–4 mm. wide, firm; glumes not exceeding the upper floret .... 4. *D. caespitosa*

1. **D. danthonioìdes** (Trin.) Munro ex Benth. [*Aira d.* Trin. *D. calycina* Presl.] Culms rather few, slender, 1–5 dm. tall; blades few, 2–8 cm. long, ca. 1 mm. wide; panicle open, 5–12 cm. long, the branches with few spikelets near the ends; glumes 6–8 mm. long, acuminate, 3-nerved; lemmas smooth, 2–3 mm. long, truncate, the geniculate awns 4–6 mm. long.—Moist places, about meadows, etc., below 9000 ft.; many Plant Communities; montane s. Calif. and cismontane and montane cent. and n. Calif.; to Alaska; Chile. March–Aug. A poorly defined form with glumes ca. 4 mm. long, is var. **grácilis** (Vasey) Munz. [*D. g.* Vasey.] It occurs on mud flats after winter pools; Coastal Sage Scrub, V. Grassland; cismontane s. Calif., s. San Joaquin V.

2. **D. elongàta** (Hook.) Munro ex Benth. [*Aira e.* Hook. *D. e.* vars. *ciliata* and *tenuis* Vasey.] Tufted perennial, the culms slender, erect, 3–10 dm. long; blades flat, soft, 1–1.5 mm. wide; panicle 1–3 dm. long, almost spicate; pedicels short, appressed; glumes 3-nerved, 4–6 mm. long; lemmas 2–3 mm. long, like those of *D. danthonioides*, the awns straighter, to ca. 4 mm. long.—Wet places, 4500–10,100 ft. in s. Calif., down to the coast in cent. and n. Calif.; many Plant Communities; San Diego Co. n.; to Alaska, Wyo., Ariz.; Mex., Chile. May–Aug.

3. **D. atropurpùrea** (Wahl.) Scheele. [*Aira a.* Wahl.] Culms loosely tufted, erect, smooth, purplish at base, 4–8 dm. tall; blades rather soft, flat, 5–10 cm. long, 4–6 mm. wide; panicle open, 5–10 cm. long, the branches few, capillary, naked below; spikelets mostly purplish, broad; glumes ca. 5 mm. long, the 2d 3-nerved, exceeding the florets; lemmas scabrous, ca. 2.5 mm. long, the awn of the 1st included, of the 2d geniculate, exserted.—Moist places, ca. 6000–7500 ft.; Red Fir F.; Mt. Shasta and Marble Mts., Siskiyou Co.; to Alaska, Labrador, Colo. July–Aug.

4. **D. caéspitosa** (L.) Beauv. [*Aira c.* L.] Densely tufted perennial, 6–12 dm. tall; lvs. principally basal, flat or folded, 1.5–4 mm. wide, short or elongate; panicle open, nodding, 1–2 dm. long, the branches capillary, scabrous, the branchlets bearing spikelets toward their tips; spikelets 4–5 mm. long, green to darkly anthocyanous, the florets distant; glumes 1-nerved or the 2d obscurely 3-nerved, acute, ca. as long as the florets; lemmas smooth; awn from near the base, straight or somewhat geniculate, usually somewhat exserted; $n = 13$ (Lawrence, 1945).—Wet meadows, etc., 6000–8500 ft., s. Calif., 3300–12,500 Sierra Nevada; Montane Coniferous F., Alpine Fell-fields; San Bernardino Mts., Sierra Nevada, N. Coast Ranges; to Alaska, Atlantic Coast, Eurasia. July–Aug.

Ssp. **beringénsis** (Hult.) W. E. Lawr. [*D. b.* Hult.] Glumes 5–7 mm. long; awn usually exceeding the lemma; $n = 13$ (Lawrence, 1945).—Marshes and wet places up to 2500 ft.; edge Coastal Salt Marsh, Coastal Prairie, Mixed Evergreen F., etc.; Marin Co. to Alaska. May–July.

Ssp. **holcifórmis** (Presl.) W. E. Lawr. [*D. h.* Presl.] Panicle narrow, condensed, erect; spikelets 6–8 mm. long; awn exceeding lemma, usually ± adnate.—Boggy and marshy places; Coastal Salt March, Freshwater Marsh, N. Coastal Scrub, etc.; along and near the coast from Monterey Co. to cent. Ore. May–July.

46. Aìra L. Hairgrass

Rather delicate annuals with narrow or open panicles of small spikelets. Lf.-blades lax, subfiliform. Spikelets 2-fld., disarticulating above the glumes, the rachilla not prolonged. Glumes boat-shaped, subequal, 1-nerved or obscurely 3-nerved, acute, membranaceous or subscarious. Lemmas firm, rounded on back, tapering into 2 slender teeth and usually with a slender geniculate twisted exserted awn from below the middle of the back. Ca. 5 spp. from warmer parts of Eu. (Old Greek name for a grass.)

Panicle dense, spikelike .... 1. *A. praecox*
Panicle open.
    Lower floret with awn as long as that of upper floret .... 2. *A. caryophyllea*
    Lower floret awnless or nearly so .... 3. *A. elegans*

1. **A. praècox** L. [*Aspris p.* Nash.] Tufted, 1–2 dm. tall; blades setaceous; panicle narrow, dense, 1–3 cm. long; spikelets yellowish, shining, 3.5–4 mm. long; lemmas bidentate at apex, the upper awn 2–4 mm. long; $n = 7$ (Hagerup, 1939).—Sandy places, Marin to Del Norte cos.; to B.C., coastal e. U.S.; introd. from Eu. May–July.

2. **A. caryophyllèa** L. [*Aspris c.* Nash.] Culms 1–few, slender, erect, 1–3 dm. tall; panicle open, the spikelets silvery, shining, 3 mm. long, clustered toward ends of spreading capillary branches; lemma of both florets with geniculate awn 4 mm. long, the apical teeth setaceous; $n = 7$ (Wulff, 1937 b).—Common in open places, below 5000 ft.; many Plant Communities; much of cismontane Calif.; introd. from Eu. April–June.

3. **A. élegans** Willd. ex Gaudin. [*A. capillaris* Calif. refs.] Resembling *A. caryophyllea*; panicle more diffuse; spikelets 2.5 mm. long; lower floret awnless or nearly so, upper with an awn 3 mm. long.—Reported from Marin, Sonoma, and Humboldt cos.; introd. from Eu.

47. Avèna L. Oat

Annual or perennial grasses, low to rather tall, with narrow to open mostly rather few-fld. panicles of rather large spikelets. Spikelets 2–3-fld., the rachilla bearded, disarticulating above the glumes and between the florets. Glumes subequal, membranaceous or papery, several-nerved, longer than the lower floret, usually exceeding the upper floret. Lemmas indurate except toward the summit, 5–9-nerved, bidentate at apex, with a dorsal bent and twisted awn. Ca. 50 spp. of temp. regions. (The ancient Latin name.)

Teeth of lemmas awned or setaceous; pedicels capillary .... 1. *A. barbata*
Teeth of lemmas acute, not setaceous; pedicels stoutish.
    Lemmas pubescent with long brown hairs; spikelets usually 3-fld. .... 2. *A. fatua*
    Lemmas subglabrous.
        Spikelets mostly 3-fld.; awn present, strongly geniculate .... 2. *A. fatua* var. *glabrata*
        Spikelets mostly 2-fld.; awn wanting, or if present, weakly geniculate .... 3. *A. sativa*

1. **A. barbàta** Brot. Slender Wild Oat. Culms slender, 3–6 dm. tall; blades flat, commonly 3–7 mm. wide; panicle lax; spikelets on curved capillary pedicels; lemma with stiff red hairs, the teeth ending in slender setae 4 mm. long; $n = 14$ (Huskins, 1927).—Common weed in waste fields and on open slopes, largely replacing native grasses; native of Old World. March–June.

2. **A. fátua** L. Wild Oat. Culms stout, 3–7 dm. tall; blades 4–8 mm. wide; panicle loose, open, with horizontal branches; spikelets usually 3-fld.; florets readily falling from

glumes; glumes ca. 2.5 cm. long, the rachilla and lower part of lemma with long stiff mostly brownish hairs; lemmas nerved above, ca. 2 cm. long, with acuminate teeth; awn stout, geniculate, twisted below, 3–4 cm. long; $n = 21$ (Philp, 1933).—Waste and cult. places as a common weed; introd. from Eu. April–June. A form with lemmas quite glabrous is var. *glabràta* Peterm.; with the sp.

3. **A. satìva** L. CULTIVATED OAT. Like *A. fatua;* spikelets mostly 2-fld., the florets not separating readily from the glumes; lemma glabrous; awns straight, often wanting; $n = 21$ (Emme, 1930).—Sometimes escaping from cult.; introd. from Eu. April–June.

48. Arrhenátherum Beauv.

Rather tall perennials, with flat blades and narrow panicles. Spikelets 2-fld., the lower floret ♂, the upper perfect, the rachilla disarticulating above the glumes and produced beyond the florets. Glumes rather broad and papery, the 1st 1-nerved, the 2d a little longer than the 1st and ca. as long as the spikelet, 3-nerved. Lemmas 5-nerved, hairy on the callus, the lower awned from near the base with a twisted geniculate exserted awn, the upper with a short straight slender awn from just below the tip. Ca. 6 spp. of temp. parts of Eurasia. (Greek, *arren*, masculine, and *ather*, awn.)

1. **A. elàtius** (L.) Presl. [*Avena e.* L.] TALL OATGRASS. Culms erect, smooth, 1–1.5 m. tall; blades scabrous, 5–10 mm. wide; panicle pale or purplish, shining, 1.5–3 dm. long; spikelets 7–8 mm. long; glumes scaberulous, the 2d ca. as long as florets; lemmas scabrous, the lower awn ca. twice as long as its lemma; $n = 14$ (Avdulov, 1931).—Occasional escape from cult. in cent. and n. Calif.; introd. from Old World. May–July.

49. Hólcus L. VELVET GRASS

Perennials with flat blades and contracted panicles. Spikelets 2-fld., the pedicel disarticulating below the glumes, the rachilla curved and ± elongate below the 1st floret, not prolonged beyond the 2d. Glumes subequal, longer than the 2 florets; 1st floret perfect, the lemma awnless; 2d ♂, the lemma with a short awn on the back. Ca. 8 spp. of Old World. (Old Latin name for a grain.)

Rhizomes not present .... 1. *H. lanatus*
Rhizomes present .... 2. *H. mollis*

1. **H. lanátus** L. [*Notholcus l.* Nash.] Grayish, velvety-pubescent, erect, 3–10 dm. tall; blades 4–8 mm. wide; panicle 8–15 cm. long, contracted, pale, tinged with purple; spikelets 4 mm. long; glumes villous, the 2d broader than the 1st, 3-nerved; lemmas smooth, shining, the 2d with a hooked awn; $n = 7$ (Avdulov, 1928).—Abundantly escaped below 7500 ft., through much of the state, from cult. in meadows and fields; native of Eu. June–Aug.

2. **H. móllis** L. [*Notholcus m.* Hitchc.] CREEPING VELVET GRASS. Glabrous, 5–10 dm. tall, with vigorous slender rhizomes; panicle ovate or oblong, 6–10 cm. long; spikelets 4–5 mm. long; glumes glabrous; awn of 2d floret geniculate, exserted, ca. 3 mm. long; $n = 14$ (Litardière, 1949).—Introd. in Humboldt and Mendocino cos.; native of Eu. June–Aug.

50. Sieglíngia Bernh.

Densely tufted perennial with narrow short blades and narrow simple few-fld. panicle. Spikelets 4–5-fld., the rachilla disarticulating above the glumes and between the florets. Glumes equal, acute, the 1st 1–3-nerved, the 2d 3–5-nerved. Lemmas firm, 7–9-nerved, bifid, the midnerve ending in a short flat mucro, the margins pilose toward the base. (Named for Professor *Siegling* of Germany in 1800.)

1. **S. decúmbens** (L.) Bernh. [*Festuca d.* L.] Erect, 2–5 dm. tall; lvs. crowded near the base, the blades 2–3 mm. wide; panicles 2–7 cm. long, with 3–15 narrowly ovoid spikelets 6–12 mm. long; lemmas 5–6 mm. long; $n = 18$ (Scheerer, 1940), 62 (Maude, 1940).—Occasional escape from cult., as near Berkeley; introd. from Eu. June–July.

51. Danthònia Lam. & DC. Oatgrass

Tufted perennials with few-fld. open or spikelike panicles of rather large spikelets. Spikelets several-fld., the rachilla disarticulating above the glumes and between the florets. Glumes subequal, broad, papery, acute, mostly exceeding the upper floret. Lemmas rounded on back, obscurely several-nerved, bifid at tip, the teeth mostly slender-awned and a stout twisted geniculate awn arising from between the teeth. Ca. 100 spp. of temp. regions. (Named for E. *Danthoine*, French botanist of early 18th century.)

Lemma glabrous on back, pilose on only the margin.
Panicle narrow, with appressed pedicels. High montane .... 1. *D. intermedia*
Panicle open, with spreading or reflexed pedicels.
The panicle usually of 1 spikelet .... 2. *D. unispicata*
The panicle of few to several spikelets .... 3. *D. californica*
Lemma pilose on the back at the base and on the margins. Nw. Calif. .... 4. *D. pilosa*

1. **D. intermèdia** Vasey. Culms 1.5–4 dm. tall; sheaths smooth; blades ± involute, glabrous or sparsely pilose; panicle purplish, narrow, few-fld., 2–5 cm. long, each branch with 1 spikelet; glumes 12–13 mm. long; lemmas 7–8 mm. long, appressed-pilose on margin, scaberulous at tip, the teeth acuminate, aristate-tipped; terminal segm. of awn 6–8 mm. long.—Damp banks, etc., (5200–)9600–11,000 ft.; Subalpine F., Alpine Fellfields; Sierra Nevada from Tulare and Inyos cos. to Eldorado Co., N. Coast Ranges of Humboldt and Trinity cos., Mt. Shasta; to Alaska, Nfld., New Mex. July–Sept.

2. **D. unispicàta** (Thurb.) Munro ex Macoun. [*D. californica* var. *u.* Thurb.] Culms 1–2 dm. long, in spreading tufts; sheaths and blades pilose to glabrous; panicle mostly of 1 spikelet; this ca. 12–15 mm. long (excluding awns); lemma usually glabrous above the hairy callus, 5–7 mm. long, gradually acuminate into the awns; *n* = 18 (Stebbins & Love, 1941).—Rocky hills above 4500 ft.; Sagebrush Scrub to Lodgepole F.; Sierra Nevada, to Modoc Co., Trinity Co., Humboldt Co.; to B.C., Mont., Colo. May–July. Doubtfully specifically distinct from the next sp.

3. **D. califórnica** Bol. Three–10 dm. tall, glabrous; sheaths glabrous, pilose at throat; blades 1–2 dm. long, often flat, glabrous; panicle mostly with 2–5 spikelets, the pedicels 1–2 cm. long; glumes 15–20 mm. long; lemmas (excluding awns) 8–10 mm. long, glabrous on back, the teeth long-aristate; awns having a terminal segm. 5–10 mm. long; *n* = 18 (Stebbins & Love, 1941).—Frequent, dry hills and meadows, below 5000 ft.; many Plant Communities; from Monterey and Tulare cos. n.; to B.C., Colo., New Mex. May–July.

Var. **americàna** (Scribn.) Hitchc. [*D. a.* Scribn.] Tending to be lower and ± spreading, spreading-pilose.—Up to 8000 ft.; with the sp. and s. to San Bernardino and Cuyamaca mts.

4. **D. pilòsa** R. Br. Tufted, 3–6 dm. tall, loosely pilose; panicle narrow, several-fld.; spikelets ca. 6-fld.; glumes 13–14 mm. long; lemma pilose at base and on margin, the teeth with awns 6–8 mm. long, cent. awn 12–15 mm. long; *n* = 24 (Calder, 1937).—Occasionally introd. from Santa Barbara to Humboldt and Siskiyou cos.; native of Australia. May–July.

52. Calamagróstis Adans. Reedgrass

Perennial usually fairly tall grasses, mostly with creeping rhizomes. Spikelets 1-fld., small, in open or more frequently narrow sometimes spikelike panicles. Rachilla disarticulating above the glumes, prolonged beyond the palea as a short often hairy bristle. Glumes subequal, acute or acuminate. Lemma shorter, usually more delicate than glumes, usually 5-nerved, the midvein exserted as an awn, the callus with a tuft of long hairs. Ca. 100 spp. of cool and temp. regions. (Greek, *kalamos*, reed, and *agrostis*, a kind of grass.)

(Stebbins, G. L., Jr. A revision of some N. Am. species of Calamagrostis. Rhodora 32: 35–57, 1930. Nygren, A. Investigations on N. Am. Calamagrostis. Hereditas 40: 377–397, 1954.)
A. Awn longer than the glumes, geniculate.
    B. Panicle open, the branches spreading, naked below.
        C. Blades mostly basal and not more than 2 mm. wide, often involute .... 1. *C. Breweri*
        CC. Blades well distributed, 5–9 mm. broad, flat .... 2. *C. Bolanderi*
    BB. Panicle compact, the branches appressed, floriferous from their base.

C. Glumes ca. 10 mm. long, gradually long-acuminate; awn nearly 1 cm. long above the
bend .... 3. *C. foliosa*
CC. Glumes 6–8 mm. long, abruptly acute or acuminate; awn usually less than 5 mm. long
above the bend.
 D. Panicle anthocyanous. Alpine (Sierra Nevada) .... 4. *C. purpurascens*
 DD. Panicle straw color, serpentine. Marin to Lake cos. .... 5. *C. ophitidis*
AA. Awn not or scarcely longer than the glumes, straight or geniculate.
 B. Awn bent, protruding sidewise from glumes; callus-hairs shorter than lemma.
  C. Sheaths ± pubescent on the collar .... 6. *C. rubescens*
  CC. Sheaths glabrous on the collar.
   D. Culms stout, mostly over 1 m. high; ligule 3–8 mm. long.
    E. Panicles loose, the branches ascending or spreading .... 7. *C. nutkaensis*
    EE. Panicles compact .... 8. *C. densa*
   DD. Culms slender, 4–8 dm. tall; ligule 2–3 mm. long .... 9. *C. koelerioides*
 BB. Awn straight, included; callus-hairs usually not much shorter than lemma.
  C. Panicle rather loose and open.
   D. Callus-hairs copious, ca. as long as lemma; awn delicate, straight
                       10. *C. canadensis*
   DD. Callus-hairs rather scant, ca. half as long as lemma; awn stronger, weakly bent
                        11. *C. lactea*
  CC. Panicle quite contracted.
   D. Lvs. and culms harsh and scabrous; ligule 4–6 mm. long ..... 12. *C. inexpansa*
   DD. Lvs. and culms mostly smooth, the blades sometimes scabrous on upper surface;
ligule 1–3 mm. long .... 13. *C. crassiglumis*

1. **C. Brèweri** Thurb. [*C. Lemmonii* Kearn.] Densely tufted, slender, erect, 1.5–3 dm. high; lvs. mostly basal, usually involute; panicle ovoid, open, purple, 3–8 cm. long, the lower branches slender, spreading, few-fld., 1–2 cm. long; glumes 3–4 mm. long, 1-nerved, smooth, acute; lemma almost as long, cuspidate-toothed, the awn geniculate, borne near the base, twisted below, exserted, ca. 2 mm. long above the bend; $n = 14$, 21 (Nygren, 1954).—Meadows, 6200–12,200 ft.; Subalpine F., Alpine Fell-fields; Sierra Nevada, Trinity Co. July–Sept.

2. **C. Bolánderi** Thurb. Culms 1–1.5 m. tall, from slender rhizomes; sheaths scabrous; blades flat, scattered, smoothish, 5–9 mm. wide; panicle open, 1–2 dm. long, the spreading branches naked below, up to 5 or 10 cm. long; glumes 3–4 mm. long, purple, scabrous, acute; lemma very scabrous, ca. as long as glumes, the awn geniculate, from near the base, exserted, ca. 2 mm. long above the bend; $n = 28$ (Nygren, 1954).—Freshwater Marsh, meadows in Closed-cone Pine F., N. Coastal Scrub; Sonoma to Humboldt cos. June–Aug.

3. **C. foliòsa** Kearn. Tufted, erect, 3–6 dm. tall; lvs. many, crowded toward base, the blades involute, firm, smooth, almost as long as the culm; panicle pale, dense, 5–12 cm. long; glumes ca. 1 cm. long, acuminate; lemma 5–7 mm. long, acuminate, the apex with 4 setaceous teeth, the awn from near the base, bent, ca. 8 mm. long above the bend, callus-hairs 3 mm. long.—Rocky places near the coast; N. Coastal Scrub; Sonoma to Humboldt cos. May–Aug.

4. **C. purpuráscens** R. Br. in Richards. Tufted, erect, 4–6(-10) dm. high; sheaths mostly scabrous; blades 2–4 mm. wide, rather thick, scabrous, ± involute; panicle dense, ± anthocyanous, spikelike, 5–12 cm. long; glumes 6–8 mm. long, scabrous; lemma nearly as long, the apex with 4 setaceous teeth, the awn from near the base, exserted ca. 2 mm.; callus-hairs rather short; $n = 20$, 24, 27, 28, etc. (Nygren, 1954).—Rocky places, 9500–13,000 ft.; Subalpine F., Alpine Fell-fields; Sierra Nevada, White Mts. and Siskiyou Co. (at 4500 ft.); to Alaska, Quebec, Colo. July–Sept.

5. **C. ophítidis** (J. T. Howell) Nygren. [*C. foliosa* var. *o.* Howell.] Near to *C. purpurascens* but more harshly scabrid; blades strongly involute, gray-scabrous; panicle more lax, mostly straw color; $n = 14$ (Nygren, 1954).—On serpentine; Mixed Evergreen F., etc.; Marin to Sonoma and Lake cos. May–June.

6. **C. rubéscens** Buckl. [*C. aleutica* var. *angusta* Vasey. *C. subflexuosa* Kearn. *C. fasciculata* Kearn.] Rhizomes creeping; culms tufted, slender, 6–10 dm. long; sheaths smooth, ± pubescent on collar; blades erect, 2–4 mm. wide, flat or ± involute; panicle spikelike or ± interrupted, pale or purplish, 6–15 cm. long; glumes 4–5 mm. long; lemma pale, thin, ca. 4 mm. long, smooth, obscurely nerved, the geniculate awn from near the base, 1–2 mm. long above the bend, exserted from side of glumes; callus-hairs short, scant; $n = 28$ (Nygren, 1954).—Open banks, rocky places, etc., below 2500 ft.; Chaparral, Mixed Evergreen F., Redwood F., Yellow Pine F., etc.; Santa Cruz Co. to Siskiyou Co., Eldorado Co.; to B.C., Colo. June–Sept.

7. **C. nutkaénsis** (Presl) Steud. [*Deyeuxia n.* Presl. *C. aleutica* var. *patens* Kearn.] Rhizomes short; culms stout, 1–1.5 m. tall; blades flat, 6–12 mm. wide, later involute, scabrous; panicle usually purplish, narrow, rather loose, 15–30 cm. long, with ascending branches; glumes 5–7 mm. long, acuminate; lemma ca. 4 mm. long, indistinctly nerved, the stoutish awn from near the base, slightly geniculate, ca. as long as lemma; callus-hairs shortish; $n = 14$ (Nygren, 1954).—Moist places and swamps, up to 5700 ft.; Closed-cone Pine F., Redwood F., Mixed Evergreen F., Freshwater Swamp, etc.; near coast from Monterey Co. n.; to Alaska. May–Aug.

8. **C. dénsa** Vasey. Rhizomes short, stout; culms stout, densely tufted, 1 m. or more tall; sheaths slightly scabrous; ligule 3–5 mm. long; blades flat or ± involute, scabrous, 1.5–2.5 dm. long; panicle spicate, pale, 10–15 cm. long; glumes 4.5–5 mm. long, scaberulous; lemma 3.5–4 mm. long, the awn bent, ca. as long as lemma, ± exserted at one side; callus-hairs ca. 1 mm. long; $n = 14$ (Nygren, 1954).—Dry hills, ca. 3400–5000 ft.; Chaparral, Yellow Pine F.; e. San Diego Co. June–July.

9. **C. koelerioìdes** Vasey. Like *C. densa*, culms more slender, 4–8 dm. tall; ligule 2–3 mm. long; blades flat, scabrous, 2–3 mm. wide; panicles often purplish; lemma almost as long as glumes; $n = 14$, 15 (Nygren, 1954).—Dry hills and banks, below 7300 ft.; many Plant Communities; Coast Ranges from San Luis Obispo Co. n.; to Wash., Wyo. June–Aug.

10. **C. canadénsis** (Michx.) Beauv. [*Arundo c.* Michx.] Rhizomes long, creeping; culms suberect, tufted, 6–15 dm. tall; sheaths mostly glabrous; blades long, flat, scabrous, 4–8 mm. wide; panicle nodding, dense to open, 1–2.5 dm. long, green to purplish; glumes subequal, 2.8–3.8 mm. long, rounded on back, acute or acuminate; lemma 1.7–3 mm. long, awn inserted near the middle, extending to slightly beyond the tip; callus-hairs ca. as long as lemma; $n = 21$, 28, etc. (Nygren, 1954).—Decayed logs, meadows, etc., at 5000–11,200 ft.; Red Fir F. to Subalpine F.; Sierra Nevada from Tulare Co. n., Coast Ranges from Mendocino and Tehama cos. n.; to B.C., Nfld., Mo., New Mex. July–Sept. A form with spikelets 3.8–4.5 mm. long and more keeled glumes is var. *robústa* Vasey, and has been reported from near Lake Tahoe. Another with spikelets 4.5–6 mm. long and thick opaque glumes scabrous or pubescent on the keel, is var. *Langsdórfii* (Link) Inman and has been reported from Yosemite V.

11. **C. láctea** Beal. Culms weak, 8–15 dm. tall, subgeniculate at nodes; rhizome short, knotty; sheaths scaberulous; blades flat, lax, scabrous, 6–12 mm. wide; panicle pale, narrowly pyramidal; glumes 5–6 mm. long, scabrous, acuminate; lemma shorter than glumes, scabrous, with setaceous tooth at apex, awn from near the base of lemma and ca. its length.—Occasional in mts. and a possible hybrid between a form of *C. canadensis* and *C. nutkaensis*.

12. **C. inexpánsa** Gray. [*C. micrantha* var. *sierrae* Jones. *C. californica* Kearn.] Rhizomes slender; culms tufted, 4–12 dm. high, often scabrous below the panicle; sheaths smooth to scabrous; ligule 4–6 mm. long; blades firm, flat or ± involute, scabrous, 2–4 mm. wide; panicle narrow, dense, 5–15(–20) cm. long; glumes 3–4 mm. long, abruptly acuminate, scaberulous; lemma 3–4 mm. long, scabrous, the awn from ca. the middle, straightish, ca. as long as glumes; callus-hairs ½–¾ as long as lemma; $n = 14$, 28, 42, etc. (Nygren, 1954).—Meadows and moist places, 4500–11,000 ft.; Red Fir F. to Subalpine F.; Sierra Nevada from Tulare to Lassen cos.; Siskiyou Co. June–Aug. A sp.-complex in which California plants vary from having glumes 3–4.5 mm. long (var. *brévior* [Vasey] Stebbins) to 4–4.2 mm. long (*C. califórnica* Kearn.)

13. **C. crassiglùmis** Thurb. Rhizomes short; culms stiff, 1.5–4 dm. high; blades flat or ± involute, smooth, firm, 4–5 mm. wide; panicle narrow, dense, spikelike, 2–5 cm. long, purple; glumes 3–4 mm. long, ovate, acuminate, firm; lemma ca. half as long, broad, obtuse or abruptly pointed, the awn from the middle, ca. as long as lemma; callus-hairs 3 mm. long; $n = 70$ (Nygren, 1954).—Swampy places; Freshwater Marsh, N. Coastal Scrub; Marin and Mendocino cos. to Del Norte Co.; to Alaska. June–July.

53. Ammóphila Host. BEACHGRASS

Coarse tough perennials with scaly tough rhizomes and pale dense spikelike panicles. Spikelets 1-fld., compressed, the rachilla disarticulating above the glumes and produced beyond the palea as a short hairy bristle. Glumes subequal, chartaceous. Lemma similar and a little shorter, with bearded callus. Ca. 3 spp. on sandy coasts of Eu. and N. Am. (Greek, *ammos*, sand, and *philos*, loving).

    1. **A. arenària** (L.) Link. [*Arundo a.* L.] Stout, 6–10 dm. tall; blades very long, soon involute; ligule 10–30 mm. long; panicle 1–3 dm. long; glumes 8–10 mm. long, scabrous; $n = 7$, 14 (Westergaard, 1941).—Introd. in dune areas along the coast; native of Eu. May–Aug.

54. Agróstis L. Bent Grass

    Annual or perennial grasses, delicate to rather coarse, with mostly flattish lvs. and open to contracted panicles of small spikelets. Spikelets 1-fld., disarticulating above the glumes, the rachilla usually not prolonged. Glumes subequal, acute, acuminate or occasionally awned, usually scabrous on the keel, sometimes also on the back. Lemma obtuse, usually shorter and thinner than the glumes, mostly 3-nerved, awnless or dorsally awned. Palea usually shorter than lemma, 2-nerved, or more often small and nerveless. Perhaps 100 spp. in temp. and colder regions, many of importance for forage and lawns. (Greek name of a grass.)

A. Plants annual; lemma with a slender awn.
  B. Spikelets ca. 1.5 mm. long; lemma awned below the tip .... 6. *A. exigua*
  BB. Spikelets at least 2.5 mm. long; lemma awned from the middle.
    C. Apex of lemma subentire or obscurely toothed; lemma 1.7–1.9 mm. long
                                                  7. *A. microphylla*
    CC. Apex of lemma with 2 or 4 awns.
      D. Lemma pilose; glumes 3.5–4 mm. long .... 8. *A. tandilensis*
      DD. Lemma glabrous except on the callus; glumes 5–6 mm. long.
        E. The lemma thin, glabrous, 3 mm. long; palea obsolete
                                               7. *A. microphylla* var. *Hendersoni*
        EE. The lemma firm, scabrous, 3.2–3.5 mm. long; palea ca. ⅓ as long as lemma
                                               9. *A. aristiglumis*
AA. Plants perennial; lemma awned or awnless.
  B. Palea evident, 2-nerved, at least half as long as lemma.
    C. Rachilla prolonged beyond the palea as a minute bristle.
      D. Lemma pubescent; panicle diffuse .... 1. *A. avenacea*
      DD. Lemma glabrous; panicle rather narrow .... 2. *A. Thurberiana*
    CC. Rachilla not so prolonged.
      D. Glumes scabrous on the keel and on the back; panicle contracted, lobed, the short branches densely verticillate .... 3. *A. semiverticillata*
      DD. Glumes scabrous on keel only; panicle open, or if contracted, not lobed or with densely verticillate branches.
        E. Branches of panicle or some of them floriferous from base; ligule 3–6 mm. long .... 4. *A. alba*
        EE. Branches of panicle naked at base; ligule 1–2 mm. long ...... 5. *A. tenuis*
  BB. Palea obsolete or a minute nerveless scale, sometimes up to ca. 0.5 mm. long.
    C. Plants spreading by creeping rhizomes (these short in *A. lepida*).
      D. Hairs at base of lemma 1–2 mm. long; culms 6–9 dm. tall ...... 11. *A. Hallii*
      DD. Hairs at base of lemma minute or none; culms mostly lower.
        E. Rhizomes short; tufted alpine plants .... 12. *A. lepida*
        EE. Rhizomes long and slender; mostly from below 7500 ft.
          F. Panicle spikelike, 5–10 cm. long. N. Coast .... 13. *A. pallens*
          FF. Panicle open, 10–15 cm. long. Widespread .... 14. *A. diegoensis*
    CC. Plants without rhizomes, stolons sometimes developed.
      D. Panicle narrow, contracted, at least some of lower branches bearing spikelets from the base.
        E. Culms slender, mostly 1–2 dm. tall, tufted, with lvs. basal; blades not more than 7 cm. long, less than 2 mm. wide; panicles mostly less than 5 mm. wide.
          F. The culms spreading; panicles strict, greenish; lemma with a minute awn or the midnerve ending below the apex. N. Coast . 15. *A. Blasdalei*
          FF. The culms erect; panicles narrow but loose, purple; lemma awnless, the midnerve reaching the apex. From above 5000 ft. .... 16. *A. variabilis*
        EE. Culms stouter, taller, not in tufts with dense basal lvs.; at least some of blades 8–10 cm. long, 3–5 mm. wide; glumes scabrous on keel.
          F. Lemma acute, not toothed; palea minute.
             G. Panicle loose, the branches whorled, not densely flowered at base; lemma-awn twisted, geniculate. N. Coast Ranges .. 10. *A. ampla*
            GG. Panicle dense to loose, the branches crowded and densely fld. at base; lemma awnless or awned. Widespread .... 17. *A. exarata*
          FF. Lemma minutely 4-toothed; palea ¼–⅓ as long as lemma. N. Coast
                                             18. *A. californica*
      DD. Panicle open, sometimes diffuse; branches very slender, scabrous, the lower branches not bearing spikelets at the base.

E. Lemma awned from near the base. San Luis Obispo and Santa Barbara cos.
                 19. *A. Hooveri*
EE. Lemma awnless or awned from the middle or above.
 F. Panicle very diffuse, the branches capillary, not flexuous, the spikelets
   arranged at the ends .... 20. *A. scabra*
 FF. Panicle open but not diffuse, the branches branching at or below the
   middle.
  G. Lemmas mostly awnless; ligule 1–2 mm. long.
   H. Spikelets ca. 2 mm. long; plants 1–3 dm. high. High mon-
     tane .... 21. *A. idahoensis*
   HH. Spikelets 2.5–3 mm. long; plant 6–9 dm. high. From below
     7000 ft. .... 22. *A. oregonensis*
  GG. Lemmas mostly awned; ligule 4–11 mm. long. Low Coast Ranges
                 23. *A. longiligula*

1. **A. avenàcea** Gmel. [*A. retrofracta* Willd.] Tufted perennial, the culms erect or decumbent below, 2–6 dm. tall; sheaths smooth; ligule of culm-lvs. 3–5 mm. long; blades flat, scabrous, 1–2 mm. wide; panicle diffuse, 1.5–3 dm. long, the branches capillary, reflexed at maturity; glumes acuminate, 3–4 mm. long; lemma ca. half as long as glumes, thin, pubescent, short-bearded on callus and at the middle with a slender geniculate twisted awn exserted ca. half the length of the glumes.—Introd. s. of Stockton; from Polynesia. June–July.

2. **A. Thurberiàna** Hitchc. Tufted perennial, very slender, 2–4 dm. tall; lvs. somewhat crowded near base, the blades ca. 2 mm. wide; panicle rather narrow, ± drooping, rather lax, 5–7 cm. long; spikelets green to purplish, 2 mm. long; lemma almost as long as glumes, the palea ca. ⅔ as long.—Moist places, 4500–11,000· ft.; Red Fir F. to Subalpine F.; Sierra Nevada from Tulare Co. to Plumas Co., N. Coast Ranges, Humboldt and Siskiyou cos.; to B.C. July–Aug.

3. **A. semiverticillàta** (Forsk.) C. Chr. [*Phalaris s.* Forsk. *Polypogon s.* Hylander. *A. verticillata* Vill.] From decumbent at base to creeping and rooting; blades firm, short to elongate; panicle contracted, 3–10 cm. long, densely fld., lobed, with short whorled branches; spikelets usually falling entire; glumes equal, obtuse, 2 mm. long, scabrous; lemma 1 mm. long, awnless, truncate and toothed; palea nearly as long; $n = 14$ (Avdulov, 1931).—Weed along ditches and in moist places, widely distributed in state; introd. from Eu. May–June.

4. **A. álba** L. Redtop. Perennial, forming turf, with elongate stolons and erect leafy sterile shoots; culms 5–12 dm. tall; blades 3–8 mm. wide; panicle ovoid to ellipsoid, purplish to green, rather lax, 1–3 dm. long, the branches spreading at anthesis, more erect in fr.; spikelets 2–3.5 mm. long; glumes subequal, usually scabrous on keel; lemmas rarely awned; $n = 14$ (Sokolovskaja, 1938), 21 (Stuckey & Banfield, 1946), etc.—Escape in moist places, below 7500 ft.; introd. from Eu. June–Sept.

Var. **palústris** (Huds.) Pers. [*A. p.* Huds.] Creeping Bent. Densely matted, often with repent stolons; lvs. 1–5 mm. wide, sometimes involute; panicle straw color to bronze, ± cylindric, 2–18 cm. long, with appressed or ascending branches.—Salt or brackish marshes, Marin Co. n.; to B.C., Atlantic Coast. June–Aug.

5. **A. ténuis** Sibth. Colonial Bent. Tufted, slender, 2–4 dm. tall, with short stolons but not creeping rhizomes; ligule 1–2 mm. long; blades 1–3 mm. wide; panicle 5–10 cm. long, open, delicate, the branches slender, naked below; spikelets not crowded, 2–3 mm. long; lemma nearly equal to glumes, awnless; $n = 14$ (Avdulov, 1931).—Escape from cult. especially in n. part of state; native of Eu. July–Sept. Var. *aristàta* (Parn.) Druce has the lemma awned.

6. **A. exígua** Thurb. Delicate annual, 0.3–1 dm. tall; blades 0.5–2 cm. long, scabrous, subinvolute; panicle 1–5 cm. long, open at maturity; glumes 1.5 mm. long, scaberulous; lemma equal to glumes, with a delicate bent awn 5–6 mm. long from below the tip; palea none.—V. Grassland, etc.; Sacramento V., Napa to Shasta and Butte cos. April–May.

7. **A. microphýlla** Steud. Loosely tufted annual with erect culms 1–3.5 dm. tall, slender, smooth; stem-lvs. 2–3; sheaths smoothish; blades scabrid, 2–5 cm. long, flat or loosely involute, 1.5–3 mm. wide; panicle spikelike, 2–9 cm. long; glumes awned, 3–4 mm. long, aristate, green to purplish, subequal, scabrous on keel; lemma 2 mm. long, broadly oblong, 4-toothed at apex, obscurely 4-nerved, the awn from above the middle, geniculate below, hispidulous, 3.5–4 mm. long; palea none.—Vernal pools, near the

coast, San Diego Co. to Humboldt Co.; several Plant Communities; Ore., L. Calif. May–July.

Var. **intermèdia** Beetle. Awned glumes 4–6 mm. long; lemma 3 mm. long, the awn 7 mm. long.—Vernal pools away from the coast; Lake, Napa, Calaveras, and Merced cos.

Var. **Héndersoni** (Hitchc.) Beetle. [*A. H.* Hitchc.] Awned glumes 7–8 mm. long; lemma 4 mm. long, its awn 10 mm. long.—Vernal pools; V. Grassland, Calaveras and Shasta cos.; to Ore.

8. **A. tandilénsis** (Ktze.) Parodi. [*A. Kennedyàna* Beetle.] Near to *A. microphylla;* panicle dense, 1–4 cm. long; 1st glume 4 mm. long, 2d 3 mm.; lemma deeply bifid, loosely pilose, the body 1.5–2 mm. long, the terminal teeth 1 mm., the awn 5–6 mm. long.—Vernal pools, San Diego, La Jolla and in Solano Co.; Argentine. April.

9. **A. aristiglùmis** Swall. Annual, erect, 0.5–1.5 dm. tall; blades flat, 1.5–3 mm. wide, 3–15 cm. long; panicle mostly 3–6 cm. long, dense; glumes 5–6 mm. long, attenuate into an awn 1–2 mm. long, the 1st glume 1-nerved, the 2d 3-nerved; lemma 3–3.5 mm. long, firm, scabrous, 5-nerved, awned from the back, the awn geniculate, 6–7 mm. long.—Diatomaceous shale, w. of Mt. Vision, Point Reyes, Marin Co. May.

10. **A. ámpla** Hitchc. Tufted perennial, 3–6 dm. tall; blades 3–5 mm. wide, scabrous on margins and minutely so on nerves; panicle generally well exserted, ca. 1 dm. long, the lowest fascicle somewhat remote; glumes subequal, 2.8–3 mm. long including the 1 mm. awn, scaberulous; lemma 1.5–2 mm. long, bifid at apex, the awn 3 mm. long, geniculate near the middle; palea obsolete.—Damp places, below 5000 ft.; many Plant Communities; San Gabriel Mts., Santa Cruz and Tulare cos. to Humboldt Co.; Ore., Wash. June–July.

11. **A. Hállii** Vasey. [*A. Davyi* Scribn.] Perennial, erect, 6–9 dm. tall, with creeping rhizomes; ligule 2–7 mm. long; blades flat, 2–5 mm. wide; panicle 1–2 dm. long, narrow, open, with verticillate branches; glumes ca. 4 mm. long; lemma awnless, 3 mm. long, with basal hair-tuft half as long; palea none; n = 21 (Stebbins & Love, 1941).— Dry woods, below 4000 ft.; Mixed Evergreen F., Yellow Pine F.; Coast Ranges from Santa Barbara Co. to Ore. May–July. A form near the coast of Mendocino Co., with narrower panicles, more involute blades and more straw-colored, is var. *Prínglei* (Scribn.) Hitchc. [*A. P.* Scribn.]

12. **A. lépida** Hitchc. Tufted perennial, 2–3.5 dm. tall, with short rhizomes; ligule to 4 mm. long; lvs. mostly basal, the blades firm, flat or folded, 1–1.5 mm. wide; panicle purple, erect, 1–1.5 dm. long, the lower branches 2–5 cm. long, spreading; glumes 3 mm. long, smooth; lemma 2 mm. long; palea obsolete.—Moist places, 8000–10,700 ft.; Lodgepole F. to Subalpine F.; Sierra Nevada of Tulare and Inyo cos., San Bernardino Mts. July–Aug.

13. **A. pállens** Trin. Rhizomes creeping; culms erect, 2–6 dm. tall; ligule 2–3 mm. long; blades flat or ± involute, 1–4 mm. wide; panicle almost spikelike, 5–10 cm. long; glumes 2.5–3 mm. long; lemma a little shorter, awnless; palea obsolete.—Dunes; Coastal Strand; San Francisco to Del Norte Co.; to Wash. June–Aug.

14. **A. diegoénsis** Vasey. Rhizomes creeping; culms erect, 4–10 dm. tall; blades flat, lax, 2–6 mm. wide; ligule 2–3 mm. long; panicle narrow, open, 10–15 cm. long, with ascending branches, some of which are naked below; spikelets much as in *A. pallens,* awned or awnless; n = 21 (Stebbins & Love, 1941).—Meadows and woods, below 7500 ft.; many Plant Communities; cismontane and montane Calif.; to B.C., Mont., Nev. April–Aug.

15. **A. Blasdàlei** Hitchc. Densely tufted, 1–1.5 dm. tall; blades involute, subfiliform or narrow, 2–4 cm. long; panicle narrow, subspicate, strict, 2–3 cm. long, with appressed branches; spikelets 2.5–3 mm. long; lemma ca. 1.8 mm. long, awnless or very short-awned; palea ca. 0.3 mm. long, nerveless.—Coastal Strand on dunes, N. Coastal Scrub on cliffs; Marin to Mendocino cos. May–July.

16. **A. variábilis** Rydb. [*A. Rossae* Calif. refs.] Densely tufted, 1–2.5 dm. tall; blades flat, to ca. 1 mm. wide; panicle 2–6 cm. long, the branches ascending; spikelets purplish, ca. 2.5 mm. long; lemma 1.5 mm. long, awnless; palea minute.—Rocky slopes, 5000–12,000 ft.; mostly Subalpine F. to Alpine Fell-fields; Sierra Nevada from Tulare Co. n., Coast Ranges from Lake Co. n., to Modoc and Siskiyou cos.; to B.C., Alta., Colo. July–Aug.

17. **A. exaràta** Trin. Mostly tufted, 2–12 dm. tall, slender to stoutish; sheaths smooth to ± scabrous; ligule 4–6 mm. long; blades flat, 2–10 mm. wide, mostly scabrous; panicle narrow, rather open to dense and interrupted, 0.5–3 dm. long; glumes subequal, 2.5–4 mm. long, acuminate to awn-tipped, scabrous on keel; lemma 1.7–2 mm. long, ending in a prickle or short awn; palea minute; $n = 21$ (Stebbins & Love, 1941).—Moist open places, below 10,000 ft.; many Plant Communities; through much of the state; to Alaska, Nebr., Tex., Mex. June–Aug. Variable; among the outstanding forms are: var. *pacífica* Vasey, with an awn longer than glumes; and var. *monolèpis* (Torr.) Hitchc. [*Polypogon monspeliensis* var. *m.* Torr.] with narrow dense panicle; glumes mostly awn-tipped; awn of lemma 1.5–2 mm. longer than glumes.

18. **A. califórnica** Trin. Tufted, stoutish, 1.5–6 dm. tall; ligule truncate, ca. 3–4 mm. long; blades flat, firm, 3–7 mm. wide; panicle dense, spikelike, mostly 2–10 cm. long; spikelets ca. 3 mm. long; glumes acute or acuminate, scabrous on keel and sides; lemma slightly shorter than glumes, awnless or awned; palea ca. as long as lemma.—Coastal Strand and N. Coastal Scrub; Santa Cruz to Del Norte cos. May–Aug.

19. **A. Hoòveri** Swall. Slender, densely tufted, 5–7 dm. tall; ligule ca. 3 mm. long; blades lax, ca. 1 mm. wide; panicle 7–17 cm. long, loose, with ascending branches; spikelets ± purplish, 2–2.5 mm. long, the 2d glume slightly shorter than the 1st; lemma 2 mm. long, 5-nerved, scaberulous, awned from near the base, the awn bent, slightly exceeding glumes; palea obsolete.—Dry sandy places; Foothill Wd.; San Luis Obispo and Santa Barbara cos. June.

20. **A. scàbra** Willd. [*A. hiemalis*, Calif. refs.] Erect, tufted, 3–9 dm. tall; ligule 2–5 mm. long; blades flat, 1–3 mm. wide, 8–20 cm. long, scabrous; panicles 1.5–2.5 dm. long, with distant brittle scabrous spreading or drooping branches; spikelets 2–2.7 mm. long, loosely placed at ends of branches; glumes unequal, acuminate, scabrous on keels; lemma 1.5–1.7 mm. long, awnless.—Moist places, like meadows, etc., 3500–10,000 ft.; Montane Coniferous F.; San Jacinto Mts., n. through Sierra Nevada, N. Coast Ranges; to Alaska, Nfld. July–Sept. A form with short panicle-branches has been called var. *geminàta* (Trin.) Swall. [*A. g.* Trin.]

21. **A. idahoénsis** Nash. [*A. tenuis* Vasey, not Sibth. *A. tenuiculmis* Nash.] Near to *A. scabra*, mostly 1–3 dm. tall, panicle 5–10 cm. long with capillary flexuous branches; spikelets 1.5–2 mm. long; lemma ca. 1.3 mm. long.—Mountain meadows, 5000–11,500 ft.; Montane Coniferous F.; San Jacinto Mts., n. through Sierra Nevada, Coast Ranges of Humboldt and Siskiyou cos.; to Wash., Mont., New Mex. July–Aug.

22. **A. oregonénsis** Vasey. [*A. Hallii* var. *californica* Vasey.] Culms 6–9 dm. tall; blades 2–4 mm. wide; ligule 1–2 mm. long; panicle oblong, 1–3 dm. long, open, the branches whorled, stiffish, ascending, up to 5 or 10 cm. long and branching above middle; glumes 2.5–3 mm. long; lemmas 1.5 mm. long, awnless; palea ca. 0.5 mm. long.—Wet places, below 7000 ft.; many Plant Communities; San Jacinto and San Bernardino mts., Sierra Nevada, N. Coast Ranges; to B.C., Mont., Wyo. June–Aug.

23. **A. longilígula** Hitchc. Culms erect, 6–8 dm. high; ligule 4–6 mm. long; blades 3–4 mm. wide, 10–15 cm. long, scabrous; panicle narrow, but open, bronze-purple, 1–2 dm. long, with scabrous branches; glumes 4 mm. long; lemma 2.5 mm. long, awned from near the middle, the awn bent, exserted; palea ca. 3 mm. long, minute.—Marshy places; Redwood F., Mixed Evergreen F., N. Coastal Coniferous F.; Mendocino to Del Norte cos.; to Ore. May–July.

Var. **austràlis** J. T. Howell. Awn of lemma obsolete to 1 mm. long; ligule 5–11 mm. long.—Marshy places in Closed-cone Pine F.; Marin to Mendocino cos.

55. Cínna L. WOOD REEDGRASS

Tall perennials with flat blades and paniculate infl. Spikelets 1-fld., disarticulating below the glumes, the rachilla forming a stipe below the floret and produced behind the palea as a minute bristle. Glumes subequal, 1–3-nerved. Lemma resembling glumes, almost as long, 3-nerved, with a short straight awn just below the apex. Palea 1-keeled. Three spp., N. Hemis. (*Kinna*, an old Greek name for a grass.)

1. **C. latifòlia** (Trev.) Griseb. [*Agrostis l.* Trev. *C. Bolanderi* Scribn.] Culms 0.5–1.5 m. tall; blades 10–15 mm. wide; panicle 1.5–3 dm. long, with spreading flexuous capillary branches naked at the base; spikelets ca. 4 mm. long; awn of lemma not more than 1

mm. long; palea 2-nerved, the nerves close together.—Moist places, 4500–9500 ft.; Montane Coniferous F.; Sierra Nevada from Tulare Co. n., N. Coast Ranges from Tehama to Siskiyou cos.; to Alaska, Atlantic Coast, Eurasia. July–Aug.

56. Alopécurus L. Foxtail

Low perennials or some taller or some annual, with flat blades and soft dense spike-like panicles. Spikelets 1-fld., disarticulating below the glumes, strongly compressed laterally. Glumes equal, usually united at base, ciliate on keel. Lemma ca. as long as glumes, 5-nerved, obtuse, the margins united at base, with a slender dorsal awn from below the middle which is included or 2–3 times as long as spikelet. Palea wanting. Ca. 25 spp. in temp. regions of N. Hemis. (Greek, *alopex*, fox, and *oura*, tail, referring to cylindrical panicle.)

A. Plants perennial.
   B. Spikelets 5–6 mm. long. Introd. plants . . . . 2. *A. pratensis*
   BB. Spikelets 2–4 mm. long. Native spp.
      C. Awn scarcely exceeding the glumes . . . . 3. *A. aequalis*
      CC. Awn exserted 2–5 mm.
         D. Panicle 3–4 mm. thick; spikelets 2.5 mm. long; awn exserted 2–3 mm.
                                                4. *A. geniculatus*
         DD. Panicle 4–6 mm. thick; spikelets ca. 3 mm. long; awn exserted 3–5 mm.
                                                5. *A. pallescens*
AA. Plants annual.
   B. Spikelets 4–6 mm. long; panicle quite loose.
      C. Panicle 2–4 cm. long . . . . 6. *A. saccatus*
      CC. Panicle 4–10 cm. long . . . . 1. *A. myosuroides*
   BB. Spikelets 2–3.5 mm. long; panicle dense.
      C. The spikelets 2–2.5 mm. long; anthers 0.5 mm. long . . . . 7. *A. carolinianus*
      CC. The spikelets 3–3.5 mm. long; anthers ca. 1 mm. long. . . . . 8. *A. Howellii*

1. **A. myosuroìdes** Huds. [*A. agrestis* L.] Erect or decumbent annual, 2–7 dm. tall, slightly scabrous; blades mostly 2–3 mm. wide; panicle slender, 2–10 cm. long, 3–5 mm. thick; spikelets 6–7 mm. long; glumes glabrous except for the short-ciliate keels, united to near the middle; lemmas slightly longer, the awn exserted 5–7 mm.; anthers 2.5–3 mm. long; *n* = 7 (Tischler, 1934).—Fields and waste places; reported from Calif.; to Wash., Atlantic Coast; native of Eurasia.

2. **A. praténsis** L. Erect perennial, 3–8 dm. tall; blades 2–6 mm. wide; panicle 3–7 cm. long, 7–10 mm. thick; glumes 5 mm. long, villous on keel and pubescent on sides; awn exserted 2–5 mm.; *n* = 14 (Marschal, 1920).—Reported from marshes, Shasta and Humboldt cos.; native of Eurasia. May–June.

3. **A. aequàlis** Sobol. Perennial, usually glaucous, 2–7 dm. tall; blades 1–4 mm. wide; panicle slender, 2–7 cm. long, whitish-drab to mouse-color, 3–5.5 mm. thick; spikelets 2 mm. long; glumes silky with long-ciliate keels; awn of lemma scarcely exserted; anthers 0.6–1 mm. long; *n* = 7 (Avdulov, 1931).—Moist places, 5000–11,500 ft., mts. of s. Calif. and Sierra Nevada, down to sea level in n. coastal region; many Plant Communities, but mostly Montane Coniferous F., Redwood F., etc.; to Alaska, Atlantic Coast, Eurasia. May–July.

4. **A. genículatus** L. Near to *A. aequalis*, but with more decumbent culms rooting at the nodes; awn exserted 2–3 mm.; spikelets 2.5 mm. long, with dark purple tip; anthers ca. 1.5 mm. long; *n* = 14 (Avdulov, 1928).—Wet places with *A. aequalis* in mts. from San Diego Co. to Siskiyou and Modoc cos.; to B.C., Atlantic Coast; Eurasia. June–July.

5. **A. palléscens** Piper. Tufted pale green perennial, 3–5 dm. tall; sheaths somewhat inflated; panicle pale, dense, 4–6 mm. thick; glumes ca. 3 mm. long, ciliate on keel, strigose on sides; lemmas awned near base, the awn exserted 3–5 mm.; anthers ca. 2 mm. long.—Wet places, mud flats, etc., n. Calif. (Lassen Co.) to B.C., Mont. May–June.

6. **A. saccàtus** Vasey. [*A. californicus* Vasey.] Annual, 1–2.5 dm. tall, with inflated upper sheaths; panicle 2–4 cm. long, partly included or short-exserted; spikelets 4–5 mm. long, the awn exserted 5–6 mm.; anthers 1 mm. long.—Mud flats, etc., much of cismontane Calif.; Ore. to Wash. March–May.

7. **A. caroliniànus** Walt. Tufted annual 1–5 dm. tall; panicle simple or branched;

spikelets 2–2.5 mm. long, pale, the awn ca. as long again as the spikelet; anthers 0.5 mm. long.—Weed in vineyard in Fresno Co.; waste places throughout the U.S. May–June.

8. **A. Howéllii** Vasey. [*A. californicus* Vasey.] Annual, 1.5–3 dm. tall, ± geniculate at lower nodes; sheaths inflated; panicle oblong to linear, 2–6 cm. long, 4–7 mm. thick; glumes 3–3.5 mm. long, ciliate on keel, appressed-pilose on lateral nerves; awn attached toward base of lemma, exserted 3–5 mm.; anthers orange, ca. 1 mm. long.—Wet places at low elevs.; Mixed Evergreen F., V. Grassland, Chaparral, Coastal Sage Scrub, etc.; Coast Ranges and Cent. V. from San Diego Co. n.; to Ore. March–June.

57. Polypògon Desf. BEARD GRASS

Annual or perennial grasses, usually decumbent, with flat blades and dense bristly spicate panicles. Spikelets 1-fld., the pedicel articulating below the glumes, leaving a short-pointed callus attached; glumes equal, entire or 2-lobed, awned from the tip or from between the lobes, the awn slender, straight. Lemma much shorter than the glumes, hyaline, usually with a slender straight awn shorter than the awns of the glumes. Ca. 10 spp., temp. (Greek, *polus*, many, and *pogon*, beard, because of the bristly infl.)

Awns of glumes 6–10 mm. long; plants annual.
  Glumes slightly lobed, the lobes not ciliate .... 1. *P. monspeliensis*
  Glumes prominently lobed, the lobes ciliate-fringed .... 2. *P. maritimus*
Awns of glumes 3–5 mm. long; plants perennial.
  Awn of lemma conspicuous, usually exserted beyond the glumes .... 3. *P. interruptus*
  Awn of lemma inconspicuous, included within the glumes .... 4. *P. elongatus*

1. **P. monspeliénsis** (L.) Desf. [*Alopecurus m.* L.] Erect annual or decumbent at base, the culms 1.5–5 dm. long; ligule 5–6 mm. long; blades mostly 4–6 mm. wide; panicle 2–15 cm. long, 1–2 cm. thick, tawny yellow when mature; glumes ca. 2 mm. long, hispidulous, the awns 6–8 mm. long; lemma smooth, shining, ca. half as long as glumes, the awn slightly exceeding them; $n = 14$ (Avdulov, 1931).—Common in low waste places as a weed; introd. from Eu. April–Aug.

2. **P. marítimus** Willd. Upright or spreading annual, the stems 2–3 dm. long; ligule to 6 mm. long; blades 2–4 mm. wide; panicle 2–5 cm. long, 1 cm. thick, cylindric; glumes ca. 3 mm. long, villous, the deep lobes ciliate-fringed, awns 7–10 mm. long; lemma awnless.—Introd. in Napa, Lake, Amador cos., etc.; native of Eu. June.

3. **P. interrúptus** HBK. [*P. lutosus*, Calif. refs.] Tufted perennial, the culms geniculate below, 3–8 dm. tall; ligule 2–5 mm. long; blades mostly 4–6 mm. wide; panicle oblong, 5–15 cm. long, ± interrupted or lobed; glumes 2–3 mm. long, scabrous, the awns 3–5 mm. long; lemma smooth, shining, 1 mm. long, minutely toothed at the truncate apex, its awn ca. 2 mm. long.—Wet and waste places through much of state; natur. from Eu. May–Aug.

4. **P. elongàtus** HBK. Erect or decumbent perennial, to 1 m. tall; blades to 20 cm. long and 1 cm. wide; panicle 1.5–3 dm. long, rather dense; glumes 2–3 mm. long with an awn equally long; lemma 1.5 mm. long, the awn from below the apex, obsolete to 1–2 mm. long.—Coastal Salt Marsh; Contra Costa and San Luis Obispo cos.; native of Cent. and S. Am.

58. Phlèum L. TIMOTHY

Annual or perennial grasses with erect culms, flat blades and dense cylindrical panicles. Spikelets 1-fld., laterally compressed, disarticulating above the glumes. Glumes equal, ciliate on the keels, abruptly mucronate or awned or gradually acute. Lemma shorter than the glumes, hyaline, broadly truncate, 3–5-nerved. Palea narrow, almost as long as lemma. Ca. 10 spp. in temp. regions. (Greek, *phleos*, old name for marsh reed.)

Panicle long-cylindrical; awn of glumes 1 mm. long .... 1. *P. pratense*
Panicle ovoid, not more than twice as long as thick; awn of glumes 2 mm. long .... 2. *P. alpinum*

1. **P. praténse** L. CULTIVATED TIMOTHY. Culms 4–10 dm. high, forming large perennial clumps; blades mostly 5–8 mm. wide; panicles 3–12 cm. long, with crowded spreading spikelets; glumes ca. 3.5 mm. long, truncate with a stout awn 1 mm. long, pectinate-

ciliate on keel; $n = 21$ (Gregor & Sansome, 1930).—Escaping from cult. in waste places, roadsides, etc.; from Eurasia. May–June.

2. **P. alpìnum** L. MOUNTAIN T. Culms. 2–5 dm. high, from a decumbent tufted base; perennial; blades mostly 4–6 mm. wide; panicle mostly 1.5–2.5 cm. long; glumes ca. 5 mm. long, hispid-ciliate on the keel, the awns ca. 2 mm. long; $n = 7$, 14 (Gregor and Sansome, 1930).—Wet meadows, bogs, etc., at 5000–11,500 ft., or lower in N. Coast Ranges; Montane Coniferous F., mostly from Red Fir F. upward, San Jacinto and San Bernardino mts., Sierra Nevada, N. Coast Ranges from Lake Co. n.; to Alaska, Atlantic Coast, Mex., Eurasia, S. Am. July–Aug.

59. Gastrídium Beauv.

Annuals with flat blades and pale shining spicate panicles. Spikelets 1-fld., the rachilla disarticulating above the glumes. Glumes unequal, enlarged or swollen at base; lemma much shorter than glumes, hyaline, broad, truncate, awned or awnless; palea ca. as long as lemma. Two spp., of Medit. region. (Greek, *gastridion*, small pouch, because of the subsaccate glumes.)

1. **G. ventricòsum** (Gouan) Schinz & Thell. [*Agrostis v.* Gouan.] NIT GRASS. Culms 1–5 dm. tall; panicle 3–8 cm. long, dense; lf.-blades 2–12 cm. long, spikelike; glumes 3 mm. long (the 2d ca. ¼ shorter), long-pointed; lemma globular, pubescent, with geniculate awn 5 mm. long; $n = 7$ (Rutland, 1941).—Weed on dry open ground and in waste places, most of cismontane Calif.; introd. from Eu. May–Sept.

60. Lagùrus L.

An annual grass with dense ovoid or oblong heads and flat blades. Spikelets 1-fld., the rachilla disarticulating above the glumes, pilose under the floret. Glumes equal, thin, 1-nerved, tapering into a plumose aristiform point. Lemma shorter than the glumes, thin, glabrous, gradually narrowed into 2 slender naked awns and with a dorsal slender exserted awn. Palea narrow, thin, the 2 keels ending in short awns. One sp., of Medit. region. (Greek, *lagos*, hare, and *oura*, tail.)

1. **L. ovàtus** L. HARE's TAIL. Culms 1–3 dm. tall, slender, pubescent, with ± inflated sheaths; panicle 2–3 cm. long, almost as thick, pale, downy, bristling with dark awns; glumes narrow, 10 mm. long, the lemma-awns much exceeding glumes; $n = 7$ (Avdulov, 1931).—Cult. and occasionally natur., especially in Bay Region and about Monterey. May–July.

61. Muhlenbérgia Schreb.

Perennial or annual, tufted or with scaly rhizomes, the culms simple or much branched. Lvs. flat or involute. Infl. narrow, sometimes spikelike, sometimes an open panicle. Spikelets 1-fld., the rachilla disarticulating above the glumes. Glumes usually shorter than lemma, obtuse to acuminate or awned, keeled or convex on back, the first sometimes obsolete. Lemma firm-membranaceous, 3-nerved, with a very short callus, rarely long-pilose, usually minutely so, the apex acute, awned from the tip or just below it, or from between the short lobes, sometimes mucronate. Ca. 80 spp., mostly in Mex. and sw. U.S. (Named for G. H. E. *Muhlenberg*, 1753–1815, Am. botanist.)

A. Plants annual.
  B. Lemma awned; panicle open, with spreading branches .... 1. *M. microsperma*
  BB. Lemma awnless.
    C. Pedicels capillary, divergent, 3–10 mm. long; glumes pubescent .. 2. *M. minutissima*
    CC. Pedicels appressed, to ca. 1 mm. long; glumes glabrous .... 3. *M. filiformis*
AA. Plants perennial.
  B. Rhizomes present, usually prominent, scaly, creeping, often branching.
    C. Lf.-blades 0.5–2 mm. wide, mostly short and involute.
      D. Panicles open, the pedicels capillary, 2–15 mm. long .... 4. *M. asperifolia*
      DD. Panicles narrow, ± condensed, the pedicels stouter, appressed, shorter.
        E. Lemma and palea glabrous.
          F. Culms smooth, widely creeping, the blades fine, recurved-spreading;
            ligules ca. 1 mm. long .... 5. *M. utilis*
        FF. Culms nodulose-roughened, erect or decumbent at base, sometimes spreading, but not widely creeping; blades ascending; ligules 2–3 mm. long.

G.   Plants mostly 1–5 dm. tall; panicle 2–10 cm. long
                                                                          6. *M. Richardsonis*
    GG.   Plants 0.5–1.5 dm. tall; panicle 1–3 cm. long . . . . 3. *M. filiformis*
EE.   Lemma and palea pilose or villous on lower half.
    F.   Awns 6–10 mm. long; blades 1–3 cm. long . . . . 7. *M. Arsenei*
    FF.   Awns 1–3 mm. long; blades 5–10 cm. long . . . . 8. *M. glauca*
CC.   Lf.-blades flat, at least some of them 3–5 or more mm. wide.
    D.   Hairs at base of floret copious, as long as the lemma body . . . . 9. *M. andina*
    DD.   Hairs at base of floret inconspicuous, not more than half as long as lemma.
        E.   Glumes with stiff scabrous awn-tips much exceeding the awnless lemma;
             spikelets 5–7 mm. long . . . . 10. *M. racemosa*
        EE.   Glumes with soft awn-tips not exceeding the body of the lemma; spikelets
              2–4 mm. long.
            F.   Sheaths glabrous; spikelets 2–3 mm. long . . . . 11. *M. mexicana*
            FF.   Sheaths scabrous; spikelets 3–4 mm. long . . . . 18. *M. californica*
BB.   Rhizomes not developed, the culms tufted.
    C.   Second glume 3-toothed . . . . 13. *M. montana*
    CC.   Second glume not distinctly 3-toothed.
        D.   Panicles 5–10 cm. long.
            E.   The panicle narrow, with ascending branches; glumes ca. 1 mm. long
                                                                          14. *M. Jonesii*
            EE.   The panicle open, with spreading branches; glumes ca. 2 mm. long
                                                                          15. *M. Porteri*
        DD.   Panicle 25–50 cm. long, slender . . . . 16. *M. rigens*

1. **M. microspérma** (DC.) Kunth. [*Trichochloa m.* D.C. *M. purpurea* Nutt.] Branching usually purplish annual with spreading culms 1–3 dm. long; sheaths smooth or scaberulous; ligule 1 mm. long; blades 2–5 cm. long, flat, 1 mm. wide; panicles narrow, lax, 2–10 cm. long; glumes ovate, 1-nerved, unequal, the 2d longer, 1 mm. long; lemma narrow, 3-nerved, 3 mm. long; awn capillary, 10–15 mm. long.—Common in dry open ± disturbed places at low elevs.; V. Grassland, Coastal Sage Scrub, Creosote Bush Scrub, etc.; Butte Co., Monterey and Kern cos. to L. Calif., Nev., Ariz. March–May. Cleistogamous spikelets develop at the base of the lower sheaths.

2. **M. minutíssima** (Steud.) Swall. [*Agrostis m.* Steud. *M. confusa*, Calif. refs. *Sporobolus microspermus* Calif. refs.] Erect to spreading annual, 0.5–3 dm. tall; blades flat, ca. 1 mm. wide; panicle ½–¾ the entire plant, the pedicels slender, ascending; spikelets 1.2–1.5 mm. long, the glumes ½–⅔ as long, minutely pilose; lemma minutely silky-pubescent along edges and midvein.—Moist sandy places, 4000–7500 ft.; Pinyon-Juniper Wd., Sagebrush Scrub, Yellow Pine F.; San Jacinto Mts., San Bernardino Mts., Yosemite V., Mono Lake, Feather R., Butte Co., Trinity Co.; to Wash., Mont., Tex., Mex. July–Oct.

3. **M. filifórmis** (Thurb.) Rydb. [*Vilfa depauperata f.* Thurb. *V. gracillima* Thurb.] Apparently annual or perennial, loosely tufted, with fibrous roots and somewhat spreading base; culms filiform, 0.5–2(–3) dm. high; ligule ca. 2 mm. long; blades flat, 1–3 cm. long; panicles narrow, interrupted, 1–3 cm. long, few-fld.; glumes ovate, 1 mm. long, abruptly narrowed at apex; lemma lanceolate, acute, 2 mm. long, minutely pubescent.—Open moist places, mostly 5000–11,000 ft.; Montane Coniferous F.; San Jacinto and San Bernardino mts., Sierra Nevada, Yolla Bolly Mts. to Siskiyou Co.; to B.C., S. Dak., New Mex. June–Aug.

4. **M. asperifòlia** (Nees & Mey.) Parodi. [*Vilfa a.* Nees & Mey. *Sporobolus a.* Nees & Mey.] SCRATCHGRASS. Pale or glaucous perennial with creeping rhizomes; culms 3–6 dm. long, ascending; sheaths smooth; ligule minute, erose-toothed; blades 2–5 cm. long, 2 mm. wide, scabrous; panicles diffuse, 1–1.5 dm. long, ca. as wide, the branches scabrous; spikelets 1.5 mm. long; glumes slightly unequal, almost as long as spikelet; lemma minutely mucronate.—Occasional in wet muddy often subalkaline places, mostly below 7000 ft.; many Plant Communities; s. and interior cismontane and desert Calif.; to B.C., N.Y., Tex., Mex., s. S. Am. July–Oct.

5. **M. ùtilis** (Torr.) Hitchc. [*Vilfa u.* Torr.] Perennial with widely creeping rhizomes; culms decumbent, widely spreading, with fine lvs., the blades less than 1 mm. wide, soon involute, ca. 2–3 cm. long; panicle narrow, 1–4 cm. long, interrupted; spikelets ca. 2 mm. long, the glumes scarcely half as long; lemma pale, apiculate.—Occasional, wet places where it forms mats; Coastal Sage Scrub, Creosote Bush Scrub, etc.; Ventura, w. San Bernardino and Inyo cos. to Nev., Mex. Oct.–March.

6. **M. Richardsònis** (Trin.) Rydb. [*Vilfa R.* Trin. *V. squarrosa* Trin. *M. s.* Rydb.] Perennial with creeping rhizomes, the culms wiry, 1–5 dm. long, smooth; sheaths

smooth; ligules 1–2 mm. long; blades 2–5 cm. long, flat or involute; panicle narrow, interrupted, or sometimes rather close and spikelike, 2–10 cm. long; spikelets 2–3 mm. long, the glumes ca. half as long; lemma lanceolate, mucronate, smooth; $n = 20$ (Stebbins & Love, 1941).—Dry or open moist ground, 5000–11,000 ft.; Santa Rosa, San Jacinto, San Bernardino, and San Gabriel mts., Sierra Nevada to Modoc Co.; to Wash., Atlantic Coast, New Mex., L. Calif. June–Aug. The lower more rhizomatous, more decumbent form with smaller fls. is *M. squarròsa* and the taller more slender and erect form is *M. Richardsònis.*

7. **M. Arsènei** Hitchc. Tufted perennial, 1–4 dm. high with wiry culms; lvs. near the base, the slender blades involute, 1–3 cm. long; panicle narrow, rather loose, purplish, 2–10 cm. long, with ascending branches; spikelets 4–5 mm. long without the glumes, these shorter, acute, awnless; lemma with a flexuous awn 6–10 mm. long.— Dry limestone slopes, 5000–6000 ft.; Pinyon-Juniper Wd.; Clark Mts., e. Mojave Desert to Utah, New Mex. Aug.–Sept.

8. **M. glaùca** (Nees) Mez. [*Podosaemum g.* Nees. *M. Lemmoni* Scribn.] Perennial, from a slender creeping woody rhizome; culms slender, wiry, erect or ascending, 2–6 dm. tall; blades flat or ± inrolled, 5–10 cm. long, 1–2 mm. wide; panicle narrow, contracted, interrupted, 5–12 cm. long; spikelets 3–4 mm. long, the acuminate glumes almost as long; lemma sparsely pilose below, acuminate into an awn 1–3(–8) mm. long.—Rare, in dry rocky places; Creosote Bush Scrub; e. San Diego Co. to Tex., n. Mex.

9. **M. andìna** (Nutt.) Hitchc. [*Calamagrostis a.* Nutt. *M. comata* Thurb.] Perennial with numerous scaly rhizomes; culms erect or ± spreading, 5–10 dm. tall, pubescent at the nodes; sheaths smooth or scaberulous; ligule 1 mm. long, short-ciliate; blades flat, 2–6 mm. wide, scabrous; panicle narrow, spicate, ± interrupted, 7–15 cm. long, grayish or purplish; glumes narrow, 1-nerved, 3–4 mm. long; lemma 3 mm. long, the awn 4–8 mm. long, the basal hairs almost as long as lemma-body.—Open flats, 6500–10,000 ft.; Montane Coniferous F.; San Bernardino Mts., Panamint Mts., Sierra Nevada, Mt. Shasta; at lower elev., 24 mi. n. of Bitterwater (San Benito Co.); to B.C., Atlantic Coast. July–Sept.

10. **M. racemòsa** (Michx.) BSP. [*Agrostis r.* Michx.] Perennial, from creeping scaly rhizomes; culms rather stout, 3–10 dm. tall, usually ± branched from middle nodes; sheaths loose, keeled; ligule 1–1.5 mm. long; blades flat, 2–7 mm. wide, 4–17 cm. long; panicle narrow, compact, 3–14 cm. long, sometimes lobed; spikelets 5–7 mm. long, the glumes narrow, subequal, stiffly awn-tipped; lemma 2.5–3.5 mm. long, acuminate, rarely short-awned, pilose below; $n = 20$ (Avdulov, 1931).—Moist places; Mixed Evergreen F.?; S. Fork, Eel R., Humboldt Co.; to Wash., Alta., Mich., Okla. July–Sept.

11. **M. mexicàna** (L.) Trin. [*Agrostis m.* L. *M. foliosa,* Calif. refs.] Perennial from creeping scaly rhizomes, the ± scabrous culms 6–10 dm. tall; blades lax, mostly 2–4 mm. wide, often 10–20 cm. long; panicle narrow, mostly long-exserted, 10–15 cm. long; spikelets 2–3 mm. long, the glumes narrow, attenuate, awn-tipped, ca. as long as the pointed or awn-tipped lemma; $n = 20$ (Avdulov, 1931).—Moist thickets and woods; Mixed Evergreen F., Foothill Wd., etc.; Humboldt, Butte cos.; to B.C., Quebec, N. Car. July–Aug. The forma *setiglùmis* (Wats.) Fern. with glumes bearing an awn 1–2 mm. long has been found near Orleans, Humboldt Co.; to Wash., S. Dak.

12. **M. califórnica** Vasey. Pale leafy perennial, ± creeping at base; culms ascending, ± woody below, 3–6 dm. long; sheaths scaberulous; blades flat, 3–6 mm. wide, scabrous, 5–15 cm. long; panicle narrow, dense, interrupted, 7–15 cm. long; glumes narrow, 3–4 mm. long, pointed; lemma 3 mm. long, with an awn 1–2 mm. long.—Occasional in wet places, below 7500 ft.; Coastal Sage Scrub, Chaparral, Yellow Pine F.; cismontane s. Calif., especially about San Bernardino V., to edge of desert. July–Sept.

13. **M. montàna** (Nutt.) Hitchc. [*Calycodon m.* Nutt. *M. gracilis,* Calif. refs.] Densely tufted erect perennial, 1.6–6 dm. high; sheaths glabrous, mostly basal, becoming flat and loose; ligule to 6 mm. long; blades 1–2 mm. wide; panicle narrow, 5–15 cm. long, rather lax but with almost appressed branches; 1st glume acute, 1.5 mm. long, the 2d slightly longer, 3-nerved and -toothed; lemma 3–4 mm. long, pilose below, with a slender awn 10–15 mm. long.—Dry places, 4500–10,500 ft.; Montane Coniferous F.; Sierra Nevada from Tulare to Nevada cos., Trinity Co. to Mont., Tex., Mex. June–Aug.

14. **M. Jònesii** (Vasey) Hitchc. [*Sporobolus J.* Vasey.] Closely tufted perennial 2–4 dm. tall; lvs. mostly basal, the sheaths finally flattened and loose; ligule 2–4 mm. long; blades subfiliform, scabrous, mostly basal; panicle narrow, 5–15 cm. long, lax; spikelets 3–4 mm. long; glumes broad, scabrous-puberulent, ca. ⅓ as long as spikelet, obtuse, often erose; lemma obscurely pubescent below, acuminate to almost awned.—Dry places, 4000–6500 ft.; Montane Coniferous F.; Placer to Modoc and Siskiyou cos. July–Aug.

15. **M. Pórteri** Scribn. Perennial, with culms ± woody at base, widely spreading or ascending through bushes, wiry, freely branched, 3–8 dm. long; sheaths smooth, spreading away from the branches; blades mostly ca. 1 mm. wide, flat, 2–8 cm. long, early deciduous; panicle open, 5–10 cm. long, the branches slender, widely spreading, few-fld.; glumes narrow, acuminate, 2 mm. long; lemma pilose, 3–4 mm. long, with awn 6–12 mm. long.—Dry brushy slopes, 3000–5000 ft.; Shadscale Scrub, Joshua Tree Wd.; w. Colo. Desert, Mojave Desert from Twentynine Palms e.; to w. Tex., Colo. June–Oct.

16. **M. rìgens** (Benth.) Hitchc. [*Epicampes r.* Benth.] Culms tufted, erect, 7–15 dm. tall, rather slender; sheaths smooth or ± scabrous, covering the nodes; blades scabrous, 1.5–2.5 dm. long, involute, with long slender point; panicle slender, mostly spicate, 2–6 dm. long; glumes 2–3 mm. long, scarcely keeled, acute to obtuse; lemma scaberulous, sparsely pilose below, slightly exceeding glumes, 3-nerved upward, awnless; $n = 20$ (Stebbins & Love, 1941).—Dry or damp places, below 7000 ft.; V. Grassland, Chaparral, Yellow Pine F.; foothills of Sierra Nevada from Shasta Co. s., Monterey to San Diego cos. and Little San Bernardino Mts.; L. Calif. June–Sept.

62. Sporóbolus R. Br. DROPSEED

Annuals or perennials with flat or involute lvs. and narrow or spreading panicles. Spikelets 1-fld., the rachilla disarticulating above the glumes. Glumes 1-nerved, usually unequal, the 2d often as long as the spikelet. Lemma membranaceous, 1-nerved, awnless. Palea usually prominent and as long as or longer than the lemma. Caryopsis falling readily from the spikelet at maturity. Ca. 90 spp. in warm regions. (Greek, *spora*, seed, and *ballein*, to throw.)

A. Glumes subequal, much shorter than lemma. Introd. weed, as in lawns . . . . 2. *S. Poiretii*
AA. Glumes unequal, or if equal, as long as lemma. Native.
    B. Panicle spikelike; spikelets 2.5 mm. long . . . . 1. *S. contractus*
    BB. Panicle ± open, not spikelike; spikelets 1.5–2 mm. long.
        C. Sheath with tuft of hairs at throat; glumes scabrous on keel.
            D. Panicle ± included in the sheath, the branches ascending . . . . . . 3. *S. cryptandrus*
            DD. Panicle exserted, the branches spreading . . . . 4. *S. flexuosus*
        CC. Sheath naked or sparingly ciliate at the throat; glumes glabrous . . . . . . 5. *S. airoides*

1. **S. contráctus** Hitchc. [*S. strictus* Merr., not Franch.] Tufted perennial 6–10 dm. high; blades 3–5 mm. wide, flat or involute, 10–15 cm. long; panicles spicate, 1.5–3(–5) dm. long; spikelets 2.5 mm. long; 1st glume ⅓ as long as spikelet; 2d glume, lemma, and palea subequal.—Occasional in dry places, below 6000 ft.; Creosote Bush Scrub to Pinyon-Juniper Wd.; Colo. and e. Mojave deserts; to Colo., w. Tex., Son. Sept.–Oct.

2. **S. Poirètii** (R. & S.) Hitchc. [*Axonopus P.* R. & S.] Erect perennial, ± tufted, 3–10 dm. tall; blades flat to subinvolute, 2–5 mm. wide, fine-pointed; panicle mostly spicate, 1–4 dm. long; spikelets ca. 2 mm. long; glumes subequal, ca. half as long as spikelet; lemma acutish; $n = 18$ (Avdulov, 1931).—Occasional in lawns and near cult.; native from Va. and Tenn. to Fla. and S. Am. July–Sept.

3. **S. cryptándrus** (Torr.) Gray. [*Agrostis c.* Torr.] Tufted perennial 3–8 dm. tall; sheaths with conspicuous tuft of white hairs at summit; blades flat, 2–5 mm. wide, ± involute when dry, fine-pointed, 5–15 cm. long; panicle 5–20 cm. long, the straight branches ascending, the basal part often included; spikelets pale to dark, 2 mm. long; 1st glume ⅓–½ as long, the 2d ca. equal to the acute lemma; $n = 18$ (Brown, 1950), 9 (Nielsen, 1939).—Dry rocky places, 3800–8200 ft.; Joshua Tree Wd., Pinyon-Juniper Wd.; San Bernardino and San Jacinto mts. to N.Y., Clark, and Inyo-White mts.; to Atlantic Coast, La. Occasionally found as weed in San Bernardino V. May–Aug.

4. **S. flexuòsus** (Thurb.) Rydb. [*Vilfa cryptandra* var. *f.* Thurb.] Resembling *S. cryptandrus* but with more open, often elongate panicles with slender spreading or drooping flexuous loosely fld. branches.—Occasional on the deserts, below 4000 ft.;

largely Creosote Bush Scrub; w. Colo. Desert, Mojave Desert from Little San Bernardino Mts. to Death V. area, V. Wells; to Utah, w. Tex., n. Mex. Sept.–Oct.

5. **S. airoìdes** (Torr.) Torr. [*Agrostis a.* Torr.] Perennial with large dense tufts 3–10 dm. tall; sheaths pilose at throat; ligule pilose; blades involute to flat, 1–3 mm. wide, 5–35 cm. long; panicle 1–4 dm. long, diffuse; the spikelets distal on the branches, 1.5–2 mm. long, obtuse; glumes nerved, unequal, acute, glabrous, the 1st ca. half as long as spikelet; *n* = 63 (Stebbins & Love, 1941).—Moist alkaline places, below 5000 ft.; Alkali Sink, Coastal Sage Scrub, etc.; cismontane and desert s. Calif.; to Wash., S. Dak., Tex., Mex. April–Oct.

Var. **Wrìghtii** (Munro ex Scribn.) Gould. [*S. W.* Munro.] Panicle relatively dense, 3–6 dm. long, the many short crowded densely fld. branches floriferous nearly to the base; blades usually flat; *n* = 18 (Brown, 1950).—Alkaline soil, Colo. R. bottom, Whipple Mts.; to Tex., Okla., Mex. April–May.

63. Crýpsis Ait.

Spreading annual with headlike infl. in axils of a pair of broad spathes that are enlarged sheaths with short rigid blades. Spikelets 1-fld., disarticulating below the glumes. Glumes subequal, narrow, acute. Lemma broad, thin, 1-nerved; palea like lemma, splitting between the nerves. Fr. readily falling from the lemma and palea, the seed free from the thin pericarp. One sp., of Medit. region. (Greek, *krupsis,* concealment, because of partly hidden infl.)

1. **C. niliàca** Fig. & DeNot. [*C. aculeata* (L.) Ait.] Prostrate, much-branched, forming mats to 3 dm. in diam.; sheaths tuberculate, bearded at summit; blades flat with involute apex, spreading, 2–5 cm. long, readily deciduous from sheaths; glumes ca. 3 mm. long.—Becoming quite widely spread on mud flats, sand bars, etc.; several Plant Communities; from San Diego to Humboldt and Modoc cos.; native of Old World. June–Sept.

64. Heleóchloa Host ex Roem.

Low spreading tufted annuals with dense oblong spicate panicles, the subtending lvs. with inflated sheaths and reduced blades. Spikelets 1-fld., the rachilla mostly disarticulating above the glumes. Glumes subequal, narrow, acute. Lemma broader, thin, 1-nerved, slightly exceeding glumes; palea nearly as long as lemma, readily splitting between nerves. A small genus, native of Eu. (Greek, *helos,* marsh, and *chloa,* grass.)

1. **H. schoenoìdes** (L.) Host. [*Phleum s.* L.] Tufted, branching, erect to spreading, 1–3 dm. long; sheaths often ± inflated; blades flat, with slender involute tips, 2–4 mm. wide, mostly less than 10 cm. long; panicle pale, 1–4 cm. long, 8–10 mm. thick; spikelets ca. 3 mm. long; pericarp readily separating; *n* = 18 (Avdulov, 1931).—Occasional in waste places, as in Butte, Sacramento, Colusa, Yolo, Napa, and Merced cos.; introd. from s. Eu. June–Oct.

65. Oryzópsis Michx. RICEGRASS

Mostly slender tufted perennials, with flat or involute blades and narrow or open panicles. Spikelets 1-fld., disarticulating above the glumes. Glumes subequal, obtuse to acuminate. Lemma indurate, ca. as long as glumes, broad, subterete, usually pubescent, with a short blunt callus and a short deciduous sometimes bent and twisted awn; palea enclosed by the edges of the lemma. Ca. 20 spp., N. Temp. (Greek, *oruza,* rice, and *opsis,* appearance.)

Lemma smooth, rarely pubescent.
  Blades flat, 5–10 mm. wide; spikelets ca. 3 mm. long .... 1. *O. miliacea*
  Blades ± involute, less than 2 mm. wide; spikelets 3–4 mm. long .... 2. *O. micrantha*
Lemma pubescent.
  Pubescence of lemma short, appressed; glumes 3.5–4 mm. long. Alpine .... 3. *O. Kingii*
  Pubescence of lemma long and silky; glumes 6–10 mm. long. Of lower elevs.
    Branches of panicle and capillary pedicels divaricately spreading .... 4. *O. hymenoides*
    Branches of panicle erect or appressed-ascending.
      Awn 12 mm. long; culms 3–6 dm. tall .... See 4. *O.* x *Bloomeri*
      Awn 6 mm. or less long; culms mostly 1.5–3 dm. tall .... 5. *O. Webberi*

1. **O. miliàcea** (L.) Benth. [*Agrostis m.* L.] Culms erect from a decumbent base, 6–15 dm. long; ligule ca. 2 mm. long; blades flat, 5–10 mm. wide; panicle 1.5–3 dm. long, loose, the branches spreading with numerous short-pedicelled spikelets beyond the middle; glumes acuminate, 3 mm. long; lemmas smooth, 2 mm. long, with a straight awn ca. 4 mm. long; $n = 12$ (Avdulov, 1928).—In widely separated waste places, cismontane Calif.; introd. from Medit. region. April–Sept.

2. **O. micrántha** (Trin. & Rupr.) Thurb. [*Urachne m.* Trin. & Rupr.] Densely tufted, slender, erect, 3–7 dm. tall; ligule ca. 1 mm. long; blades flat or involute, 0.5–2 mm. wide; panicle open, 10–15 cm. long, the branches distant, 2–5 cm. long, spreading to reflexed, with appressed spikelets toward the ends; glumes 3–4 mm. long; lemma elliptic, mostly glabrous, 2–2.5 mm. long, yellow or brown, with a straight awn 5–10 mm. long; $n = 11$ (Johnson, 1945).—Dry limestone crevices, etc., 6000–8800 ft.; Pinyon-Juniper Wd.; Clark, Kingston, and White mts., e. San Bernardino and Inyo cos.; to Sask., N. Dak., New Mex. June–Sept.

3. **O. Kíngii** (Bol.) Beal. [*Stipa K.* Bol.] Culms tufted, slender, 2–4 dm. tall; lvs. mostly basal, filiform, involute, flexuous, 3–15 cm. long; ligule ca. 1 mm. long; panicle narrow, loose, with short slender appressed or ascending few-fld. branches; glumes broad, papery, nerveless, obtuse, purple below, the 1st ca. 3.5 mm., the 2d slightly longer; lemma elliptic, 3–3.5 mm. long, ± strigose, the awn curved, not twisted, ca. 12 mm. long, fairly persistent; $n = 11$ (Johnson, 1945).—Meadow borders and damp banks, 9300–11,500 ft.; Subalpine F., Lodgepole F., etc.; Sierra Nevada from Tulare and Inyo cos. to Tuolumne and Mono cos. July–Aug.

4. **O. hymenoìdes** (R. & S.) Ricker. [*Stipa h.* R. & S. *O. cuspidata* Benth. *O. membranacea* Vasey.] Cespitose, 3–6 dm. tall; sheaths smooth or scaberulous; ligule 5–6 mm. long; blades slender, involute, almost as long as culms; panicle 8–15 cm. long, with slender spreading dichotomous branches and capillary pedicels; glumes ca. 6–7 mm. long, puberulent, 3-nerved, awn-pointed; lemma turgid, 3 mm. long, densely long-pilose with white hairs 3 mm. long, the awn ca. 4 mm. long, readily deciduous; $n = 24$ (Johnson & Rogler, 1943).—Common in dry sandy places, below 10,400 ft.; Creosote Bush Scrub to Bristle-cone Pine F. and Lodgepole F.; both deserts n. to arid slopes of Sierra Nevada and Mt. Shasta; to B.C., Manitoba, Tex., n. Mex. April–July. Apparently hybridizing with a number of spp. of *Stipa*, some of the hybrids having ± appressed branches in the infl. and awns to 12 mm. long; they vary in the indumentum of the awn from scabrous to plumose and of the sheath from glabrous to villous; they are mostly quite sterile. One of them was described as *O. Bloòmeri* (Bol.) Ricker from Mono Co. and has been shown by Johnson (Am. Journ. Bot. 32: 599–608, 1945) to have as one probable parent *Stipa occidentalis;* others have S. *Elmeri,* S. *Thurberiana,* or S. *californica.*

5. **O. Wébberi** (Thurb.) Benth. ex Vasey. [*Eriocoma W.* Thurb.] Erect, cespitose, densely tufted, 1.5–3 dm. high; blades filiform, involute, scabrous; panicle narrow, 2.5–5 cm. long, with appressed branches; glumes ca. 8 mm. long, narrow; lemma 6 mm. long, densely long-pilose, the awn up to 6 mm. long, not twisted.—Rocky slopes, ca. 5000–10,500 ft.; Sagebrush Scrub, Pinyon-Juniper Wd., to Lodgepole F.; e. slope Sierra Nevada, Inyo-White Mts., Inyo Co. to Tuolumne and Lassen cos.; to Ida., Colo., Nev. June–July.

66. **Stìpa** L. Speargrass. Needlegrass

Tufted perennials with involute lvs. and terminal panicles of 1-fld. spikelets which are articulate above the glumes, the articulation oblique, leaving a bearded sharp-pointed callus at the base of the floret. Glumes membranaceous, often papery, narrow, acute to acuminate or aristate. Lemma narrow, terete, firm or indurate, strongly convolute, terminating in a prominent awn, the junction of the body and awn evident, the awn twisted below, geniculate, mostly persistent; palea enclosed in the lemma. Ca. 100 spp. of temp. regions. (Greek, *stupe,* tow, alluding to the feathery awns of the type sp.)

(Dedecca, D. M. Studies on the California species of Stipa. Madroño 12: 129–139, 1954.)

A.   Panicle narrow, slender or contracted; lemma less than 4 times as long as palea.
    B.   Lemma densely appressed-villous, with hairs 3–4 mm. long and rising above the summit like pappus-crown.

C. Lemma ca. 8 mm. long; awn 4–5 cm. long .... 3. *S. coronata*
CC. Lemma up to 6 mm. long; awn ca. 2 cm. long .... 18. *S. pinetorum*
BB. Lemma sparsely strigose to subglabrous, never long-villous.
  C. Awns pubescent, most commonly plumose.
    D. First segm. of the once-geniculate awn strongly plumose, the hairs 5–8 mm. long
                                              1. *S. speciosa*
    DD. First segm. of the awn conspicuously pubescent, the hairs not more than 2 mm. long.
      E. Lemma bilobed at top, the lobes extending into 2 lateral awns 2–3 mm. long on each side of the cent. awn .... 2. *S. Stillmanii*
      EE. Lemma entire or obscurely lobed at top.
        F. Ligule hyaline, 3–6 mm. long .... 8. *S. Thurberiana*
        FF. Ligule opaque, less than 3 mm. long.
          G. Palea 4–5 mm. long .... 10. *S. latiglumis*
          GG. Palea not more than 3 mm. long.
            H. Sheaths glabrous.
              I. Hairs on upper part of lemma longer than those below; awn with rather short hairs .... 12. *S. californica*
              II. Hairs of lemma all short; awn with rather long hairs
                                      11. *S. occidentalis*
            HH. Sheaths pubescent .... 9. *S. Elmeri*
  CC. Awns scabrous to subglabrous, rarely appressed-hispid, not plumose.
    D. Mature lemma pale or later brownish, mostly more than 1 cm. long .. 4. *S. comata*
    DD. Mature lemma yellowish, not more than 8 mm. long.
      E. Lemma broad; 1st glume 5-nerved .... 13. *S. Lemmoni*
      EE. Lemma narrow; 1st glume 3-nerved.
        F. Palea ½–¼ as long as lemma; lemma short-hirsute throughout.
          G. Culms densely pubescent below the nodes; palea 3–4 mm. long
                                    17. *S. diegoensis*
          GG. Culms glabrous below the nodes; palea not more than 2.5 mm. long.
            H. Awn 4–6 cm. long, obscurely geniculate, with flexuous terminal segm.; lemma short-hirsute except near the glabrous top .... 19. *S. arida*
            HH. Awn mostly less than 5 cm. long, if 4 cm. long, twice-geniculate, the terminal segm. essentially straight.
              I. Sheaths, at least the lowermost, pubescent
                                 16. *S. Williamsii*
              II. Sheaths glabrous .... 14. *S. columbiana*
        FF. Palea nearly as long as lemma; hairs at summit of lemma longer than those of the body .... 15. *S. Lettermani*
AA. Panicle broad, open, loose, with spreading branches; lemma at least 4 times as long as palea.
  B. Lemma less than 7 mm. long; awn up to 4 cm. long .... 7. *S. lepida*
  BB. Lemma more than 7 mm. long; awn much more than 4 cm. long.
    C. Lemma slender, cylindrical; middle culm-lvs. 1.2–2.4 mm. broad; foliage usually glaucous; awn slender, flexuous beyond the second bend, mostly 9–12 times as long as lemma .... 6. *S. cernua*
    CC. Lemma fusiform; middle culm-lvs. 2.4–6 mm. broad; foliage usually green; awn stout, stiff, mostly 7–9 times as long as lemma .... 5. *S. pulchra*

1. **S. speciòsa** Trin. & Rupr. Culms many, erect, 3–6 dm. high; sheaths firm, the lowermost pubescent, the throat short-villous; ligule short; blades slender, coriaceous, 2–4 dm. long, involute-filiform, mostly basal; panicles dense, narrow, 1–1.5 dm. long; glumes subequal, pale, papery, ca. 15 mm. long, the 1st 3-nerved, the 2d 5-nerved; lemma ca. 8 mm. long, densely long-pilose on lower half or more, the hairs 6–8 mm. long; awn with 1 sharp bend, the 1st section 1.5–2 cm. long, long-pilose, the upper section scabrous, ca. 2.5 cm. long; n = 30 (Stebbins & Love, 1941).—Dry rocky places, below 6700 ft.; Creosote Bush Scrub, Joshua Tree Wd., etc., on both deserts, occasional in Chaparral, etc. of interior valleys, cismontane s. Calif. n. to San Luis Obispo Co.; to Colo., Ariz., S. Am. April–June.

2. **S. Stillmánii** Bol. Stout, 6–10 dm. tall; sheaths smooth, puberulent at throat and collar; ligule very short; blades scattered, folded or involute, firm, the upper filiform; panicle narrow, dense or interrupted below, 1–2 dm. long; glumes equal, 14–16 mm. long, papery, scaberulous, acuminate into an awn-point, the 1st 3-, the 2d 5-nerved; lemma 9 mm. long, short-pilose, 2-toothed; awn ca. 2.5 cm. long, 1- or indistinctly 2-geniculate, scabrous.—Rare, rocky slopes, 4000–6000 ft.; Montane Coniferous F.; n. Sierra Nevada. June–July.

3. **S. coronàta** Thurb. in Wats. Culms very stout, 1–2 m. tall, 4–6 mm. thick at base, erect, glaucous; sheaths glabrous with pubescent throat; blades flat, 3–6 dm. long, 5–10 mm. wide, involute toward tip; panicle pale or purplish, narrow, ± nodding, 2–5 dm. long; glumes unequal, the 1st 15–20 mm. long, the 2d 13–16 mm. long, both 5-nerved; lemma 9 mm. long, appressed-villous; awn 3.5–5 cm. long, scabrous,

twice-bent, the 1st segm. ca. 1 cm. long, the 2d a little shorter; $n = 20$ (Stebbins & Love, 1941).—Dry slopes, usually below 5000 ft.; Chaparral, Coastal Sage Scrub, Foothill Wd.; Santa Lucia Mts., to San Diego Co.; L. Calif. April–June.

Var. **depauperàta** (Jones) Hitchc. [*S. Parishii* var. *d.* Jones. *S. Parishii* Vasey.] Culms usually 3–6 dm. tall; panicle 1–1.5 dm. long, dense; awns ca. 2.5 cm. long, once-geniculate.—Dry rocky slopes, mostly at 3000–9000 ft.; Pinyon-Juniper Wd., Yellow Pine F., etc.; desert slopes of mts. from San Diego Co. to e. San Bernardino and Inyo cos.; Nev., Utah, Ariz. May–Aug.

4. **S. comàta** Trin. & Rupr. Culms glabrous, 3–6 dm. high; sheaths glabrous, naked at throat; ligule thin, 3–4 mm. long; blades 1–3 dm. long, 1–2 mm. wide, flat or involute; panicle narrow, 1–2 dm. long, commonly included at base; glumes 15–20 mm. long, with ± hyaline attenuate tips; lemma 8–12 mm. long, pale or ± brownish, ± sparsely pubescent; awn 10–15 cm. long, indistinctly bent twice, slender, loosely twisted below, flexuous above, often deciduous.—Dry places, 5500–8600 ft.; largely Yellow Pine F., Lodgepole F., Pinyon-Juniper Wd.; San Bernardino Mts., Argus Mts., White Mts., Masonic Mts., Sweetwater Mts., Sierra Nevada, etc.; to Alaska, Ind., Tex. June–July. A form with the 3d segm. of the awn shorter and straighter and an exserted panicle is the var. *intermèdia* Scribn. & Tweedy; it has been collected in the Sierra Nevada; $n = 22$, 23 (Stebbins & Love, 1941).

5. **S. púlchra** Hitchc. [*S. setigera* Calif. refs.] Culms 6–10 dm. tall; blades flat or involute, 2.5–6 mm. broad, deep green; panicle nodding, loose, ca. 1.5–2 dm. long, with slender spreading branches; lower glume 15–26 mm. long, the 2d slightly shorter, 3–5-nerved; lemma 7.5–13 mm. long, fusiform, pubescent throughout or at base and on nerves to middle or top; awn 6–9 cm. long, short-pubescent to 2d bend, the 1st segm. 1.5–2 cm. long, the 2d shorter, the 3d 4–6 cm.; $n = 32$ (Stebbins & Love, 1941).—Dry slopes, below 5000 ft.; Chaparral, Coastal Sage Scrub, Foothill Wd., etc.; Coast Ranges, Humboldt to San Diego cos., Sierran foothills; Channel Ids.; to L. Calif. March–May.

6. **S. cérnua** Steb. & Love. Culms mostly 6–9 dm. tall, in rather large clumps; basal lvs. many, narrow, glaucous, cauline 1.2–2.4 mm. wide; panicle open with slender flexuous branches; glumes acuminate, the 1st 12–19 mm. long, the 2d shorter; lemma 5–10 mm. long, papillose, pilose below and on nerves; awn 6–11 cm. long, twice bent, the terminal segm. flexuous, scabrous or basally short-pubescent; $n = 35$ (Stebbins & Love, 1941).—Dry slopes, below 4500 ft.; Foothill Wd., Chaparral, Coastal Sage Scrub, etc.; Coast Ranges from Tehama to San Diego cos. and Sierran foothills. April–May.

7. **S. lépida** Hitchc. [*S. eminens,* Calif. refs.] Slender, puberulent below the nodes, 6–10 dm. tall; sheaths smooth, rarely puberulent, ± villous at throat; ligule very short; blades 1–3 dm. long, flat, 2–4 mm. wide, pubescent on basal part of upper surface; panicle rather loose, open, 1.5–2 dm. long, with slender distant branches; glumes 3-nerved, smooth, acuminate, the 1st 6–10 mm. long, the second ca. 4–8 mm.; lemma ca. 6 mm. long, brown, sparingly villous, with hairy tufted tip; awn indistinctly twice bent, 2.5–4 cm. long, scabrous; $n = 17$ (Stebbins & Love, 1941).—Dry slopes, below 4000 ft.; Chaparral, Coastal Sage Scrub, Coastal Prairie, etc.; Coast Ranges from L. Calif. to Humboldt Co.; Channel Ids. March–May.

Var. **Andersònii** (Vasey) Hitchc. [*S. eminens* var. *A.* Vasey.] Culms more slender, blades more slender and involute; panicle narrower, reduced.—With the sp.

8. **S. Thurberiàna** Piper. Culms 1.5–5 dm. tall; sheaths smooth or ± scabrous, mostly basal; ligule 3–6 mm. long, hyaline; blades 1–2.5 dm. long, involute-filiform, scabrous to soft-pubescent, flexuous; panicle 0.8–1.5 dm. long, strigose, the awn 4–5 cm. long, twice bent, the 1st and 2d segms. plumose with hairs 1–2 mm. long; $n = 17$ (Stebbins & Love, 1941).—Dry open woods, 5000–8500 ft.; N. Juniper Wd., Pinyon-Juniper Wd. etc.; Siskiyou and Modoc cos. to White Mts., Inyo Co. June–July.

9. **S. Élmeri** Piper & Brodie ex Scribn. Culms erect, puberulent, 4–8 dm. tall; sheaths pubescent; ligule very short; blades 1.5–3 dm. long, 2–4 mm. wide, flat or later involute, pubescent on upper surface; panicle narrow, 1.5–3.5 dm. long, rather loose; glumes 12–14 mm. long, long-acuminate, hyaline above; lemma ca. 7 mm. long, strigose; awn 4–5 cm. long, twice bent into subequal segms., the 1st and 2d finely plumose; $n = 18$ (Stebbins & Love, 1941).—Dry gravelly and rocky places, 5600–11,200 ft.;

Montane Coniferous F.; San Jacinto and San Bernardino mts., Sierra Nevada; to Wash., Ida., Nev. June–Aug.

10. **S. latiglùmis** Swall. Culms slender, 5–10 dm. tall, strigose below; at least lower sheaths pubescent; blades flat or ± involute, pilose on upper surface; ligule 1–4 mm. long; panicle narrow, 1.5–3 dm. long, loosely fld.; glumes subequal, firm, acute to acuminate, 3-nerved, ± purplish, 1.3–1.5 cm. long; lemma densely pubescent, 8–9 mm. long, the awn twice bent, 3.5–4.5 cm. long, plumose on 1st and 2d segms.; $n = 35$ (Dedecca, 1954).—Dry slopes, 4000–6000 ft.; Yellow Pine F.; San Jacinto Mts., Sierra Nevada n. to Tuolumne Co. June–July.

11. **S. occidentàlis** Thurb. [*S. stricta* var. *sparsiflora* Vasey. *S. o.* var. *montana* Merr. & Davy.] Culms slender, 1.5–4 dm. tall; sheaths glabrous to pubescent; ligule ca. 1 mm. long; blades 1–2 dm. long, 1–2 mm. wide, mostly involute, white-puberulent on upper surface; panicle 1–2 dm. long, lax; glumes ca. 12 mm. long, hyaline at the pointed tips; lemma pale brown, ca. 7 mm. long, sparsely strigose; awn 3–4 cm. long, twice bent, plumose, the hairs on the 2 lower segms. ca. 1 mm. long, on the upper short; $n = 18$ (Stebbins & Love, 1941).—Dry open woods, 5000–11,500 ft.; Montane Coniferous F., Alpine Fell-fields; San Jacinto, San Bernardino, and San Gabriel mts., Sierra Nevada, Panamint Mts., n. to Siskiyou and Modoc cos.; Wash. to Ariz. June–Aug.

12. **S. califórnica** Merr. & Davy. Culms 6–15 dm. high, glabrous or with pubescent nodes; sheaths glabrous or the lower puberulent; ligule firm, 1–2 mm. long; blades 1–4 mm. wide, flat (later ± involute), 1–2 dm. long; panicle 1.5–3 dm. long or more, slender, pale; glumes pale, hyaline, glabrous, 3-nerved, ca. 12 mm. long; lemma 6–8 mm. long, sparsely villous, the uppermost hairs ca. 1.5 mm. long; awn 2.5–3.5 cm. long, twice bent, the 1st and 2d segms. plumose; $n = 18$ (Stebbins & Love, 1941).—Dry open places, 4500–10,400 ft.; Sagebrush Scrub to Lodgepole F.; White Mts. (Inyo Co.), San Jacinto Mts., Sierra Nevada n. to Trinity, Humboldt, Siskiyou and Modoc cos.; to Wash., Ida., Nev. May–Aug.

13. **S. Lemmònii** (Vasey) Scribn. [*S. Pringlei* var. *L.* Vasey. *S. L.* var. *Jonesii* Scribn.] Culms 3–8 dm. tall, scaberulous, usually puberulent below the nodes; ligule 1–3 mm. long; blades 1–2 dm. long, flat or involute, 1–2 mm. wide; panicle 5–12 cm. long, narrow, pale or purplish; glumes 8–10 mm. long, broad, firm, the 1st 5-nerved, the 2d 3-nerved; lemma 6–7 mm. long, pale or brownish, strigose-villous; awn 2–3.5 cm. long, strigose to the 2d bend; $n = 18$ (Stebbins & Love, 1941).—Dry open places, below 7500 ft.; Yellow Pine F., Mixed Evergreen F., etc.; Tehachapi Mts. through Sierra Nevada, Lake Co. n. to B.C., Ida. May–July. The var. *Jònesii* has glumes ca. 8 mm. long; lemma ca. 6 mm. long, awn scarcely 2 cm. long; with the sp. Var. *pubéscens* Crampt. has sheaths and blades pubescent, and occurs at 4000 ft. on serpentine; Chaparral; Tehama Co.

14. **S. columbiàna** Macoun. Near to *S. Williamsii;* 3–6 dm. tall; sheaths naked at throat, glabrous; ligule 1–2 mm. long; blades 1–2.5 dm. long, 1–3 mm. wide; panicle 0.7–2 dm. long, narrow, dense, purplish; glumes ca. 1 cm. long; lemma 6–7 mm. long, pubescent, awn 2–2.5 cm. long, twice bent; $n = 22$ (Nielsen, 1939).—Dry slopes, 4000–10,500 ft.; Sierra Nevada to Siskiyou Co.; to Alaska, S. Dak., Tex. June–Sept.

15. **S. Lettermánii** Vasey. Culms 2–3 dm. tall, in large tufts; blades filiform, strongly involute, 2–3 dm. long; panicle slender, narrow, 1–1.5 dm. long; glumes equal, 7–9 mm. long; lemma 4.5–6 mm. long, the apical hairs 1.3–1.7 mm. long, those on the body 0.8–1.3 mm. long; lemma-lobes 0.8–1 mm. long; palea 3.5–4.5 mm. long; awn glabrous on the 2 lower segms., ± scaberulous on the 3d; $n = 33$ (Stebbins & Love, 1941).—Dry flats, 5000–10,000 ft.; Yellow Pine F.; San Bernardino Mts., e. slope of Sierra Nevada to Modoc Co.; to Mont., New Mex. June–July.

16. **S. Williámsii** Scribn. Erect, 6–10 dm. tall, puberulent especially about the nodes; sheaths puberulent; ligule very short; blades ± puberulent, 1–3 mm. wide, 1–2 dm. long; panicle narrow, 1.5–2 dm. long; glumes thin, subequal, 1 cm. long; lemma ca. 7 mm. long; awn usually 3–5 cm. long.—Dry places; Sagebrush Scrub; Modoc Co. to Wash., Mont., Colo. June–July.

17. **S. diegoénsis** Swall. Culms 7–10 dm. tall, scaberulous, pubescent; ligule 1–2 mm. long; blades 1.5–4 dm. long, 2–4 mm. wide, flat or involute; panicle 1.5–3 dm. long, dense, narrow; glumes acuminate, the 1st 9–10 mm., 1-nerved, the 2d 8–9 mm. long, 3-nerved; lemma ca. 7 mm. long, with summit hairs 1–2 mm. long; awn 2–3 cm. long,

twice bent, scabrous.—Along vernal stream, at 800 ft.; Chaparral; Jamul, e. San Diego Co., n. L. Calif. May–June.

18. **S. pinetòrum** Jones. In large tufts, 3–5 dm. tall; ligule to ca. 0.5 mm. long; lvs. mostly basal, the blades involute-filiform, 5–12 cm. long, slightly scabrous; panicle narrow, 8–10 cm. long; glumes ca. 9 mm. long; lemma narrowly fusiform, ca. 5 mm. long, with a conspicuous tuft of hairs above and with 2 hyaline teeth 1 mm. long at apex; awn ca. 2 cm. long, twice bent, subglabrous.—Dry rocky places, 7000–12,500 ft.; Pinyon-Juniper Wd., Bristle-cone Pine F., Subalpine F.; Panamint, Inyo-White, Sweetwater mts., e. slope of s. Sierra Nevada n. to Eldorado Co.; to Ida., Mont., Colo. June–Aug.

19. **S. árida** Jones. Culms 3–8 dm. tall; blades 1–2 dm. long, 1–2 mm. wide, scabrous, flat or involute; panicle 1–1.5 dm. long, narrow, pale or silvery; glumes 8–12 mm. long; lemma 4–5 mm. long, strigose below, scaberulous upward; awn 4–6 cm. long, loosely twisted for 1–2 cm., flexuous beyond.—Dry probably limestone slopes, 4000–5700 ft.; Pinyon-Juniper Wd., etc.; Funeral Mts., Clark Mt. (e. Mojave Desert); to Colo., Tex. May–June.

67. Arístida L. TRIPLE-AWNED GRASS

Tufted annuals or perennials with narrow blades and narrow or open panicles. Spikelets 1-fld., the rachilla disarticulating obliquely above the glumes. Glumes equal or unequal, narrow, acute to awn-tipped. Lemma indurate, narrow, terete, convolute, with a hard sharp-pointed, usually minutely bearded callus and terminating above in a usually trifid awn. Ca. 150 spp., in warmer regions. (Latin, *arista*, awn.)

(Henrard, J. T. A critical revision of the genus Aristida. Med. van Rijks Herb., Leiden 58: 1–464, 1927.)

A. Neck of fr. jointed at base. Deserts . . . . 1. *A. californica*
AA. Neck of fr. not jointed at base.
  B. Lateral awns wanting or reduced to mere points . . . . 2. *A. Orcuttiana*
  BB. Lateral awns evident.
    C. Plants annual.
      D. Awns 10–15 mm. long . . . . 3. *A. adscensionis*
      DD. Awns 40–70 mm. long . . . . 4. *A. oligantha*
    CC. Plants perennial.
      D. Glumes ca. equal.
        E. Panicle open with spreading branches naked at the base.
          F. Summit of lemma narrowed into a twisted neck 2–5 mm. long
            5. *A. divaricata*
          FF. Summit of lemma somewhat narrowed but not twisted . . 6. *A. hamulosa*
        EE. Panicle narrow, the branches not horizontally spreading . . . . 7. *A. Parishii*
      DD. Glumes definitely unequal.
        E. Lemma tapering into a slender ± twisted beak 5–6 mm. long; awns 1.5–2.5 cm. long, widely spreading . . . . 8. *A. glauca*
        EE. Lemma beakless or only short-beaked.
          F. Panicle rather loose and nodding, with slender flexuous branches
            9. *A. purpurea*
          FF. Panicle erect, stiff with mostly appressed branches.
            G. Panicle mostly more than 1.5 dm. long, the branches several-fld.; awns ca. 2 cm. long . . . . 10. *A. Wrightii*
            GG. Panicle mostly less than 1.5 dm. long, the branches few-fld.; awns 2–several cm. long.
              H. Lemma gradually narrowed above, scaberulous on upper half; lvs. mostly in a short curly cluster at base of plant
                11. *A. Fendleriana*
              HH. Lemma scarcely narrowed above, scaberulous only at tip; lvs. not conspicuously basal . . . . 12. *A. longiseta*

1. **A. califórnica** Thurb. [*A. Jonesii* Vasey. *A. c.* var. *fugitiva* Vasey.] Tufted perennial, the culms wiry, pubescent, 1–3 dm. tall; sheaths pubescent at throat and collar; blades mostly involute, 1–4 cm. long; panicles loose, 2–5 cm. long, few-branched; glumes quite smooth, 1-nerved, awnless, the 1st 8 mm. long, the 2d 12 mm. long; lemma 5–7 mm. long, glabrous below, scaberulous toward summit; neck jointed at base, spirally twisted, 15–20 mm. long; awns equal, 2.5–3 cm. long, spreading horizontally.—Occasional in dry sandy places, low elevs.; Creosote Bush Scrub; both deserts; to Son. April–May.

2. **A. Orcuttiàna** Vasey. [*A. Schiedeana*, Calif. refs.] Erect perennial, the culms

scaberulous, 3–6 dm. tall; sheaths scaberulous, villous at throat; blades flat or involute, 1.5–3 mm. wide, the lower 1–2 dm. long; panicles open, 1–3 dm. long, nodding or drooping; glumes subequal, 10–15 mm. long; lemma 8–10 mm. long, narrowed into a scabrous twisted column 10–17 mm. long; cent. awn divergent, 5–10 mm. long, the lateral awns up to 1 or 2 mm. long.—Reported from near San Diego; to Tex., Mex. Sept.

3. **A. adscensiònis** L. [*A. bromoides* HBK.] Annual, branched at base, often with purplish tinge, erect or spreading, 1–8 dm. tall; blades 1.5–5 cm. long, usually involute; panicle narrow, rather dense, 3–10 cm. long; 1st glume 5–7 mm. long, the 2d 8–10 mm.; lemma 6–9 mm. long, compressed toward the scarcely beaked top; awns ca. equal, mostly 1–1.5 cm. long; $n = 11$ (Avdulov, 1931).—Dry open places and rocky hills, below 3500 ft.; Creosote Bush Scrub, Coastal Sage Scrub, etc.; e. San Luis Obispo Co. to the deserts; to Tex., Kans., Argentina. Feb.–June.

4. **A. oligántha** Michx. Much-branched annual, 3–5 dm. high; blades flattish, to ca. 1 mm. wide; panicle loose, 1–2 dm. long; spikelets short-pedicelled; glumes subequal, 20–30 mm. long, awn-tipped; lemma ca. 20 mm. long, the subequal awns 4–7 cm. long.—Occasional, dry slopes; V. Grassland, Foothill Wd., Coastal Prairie, Sagebrush Scrub; e. side of Sacramento V. s. to Merced Co., Coast Ranges, Lake to Humboldt and Modoc cos.; Ore., Ariz., e. U.S. Aug.–Sept.

5. **A. divaricàta** Humb. & Bonpl. ex Willd. Erect or prostrate-spreading perennial, usually 3–6 dm. long; blades flat or loosely involute, to 3 mm. wide and 15 cm. long; panicle large, diffuse, 1.5–3 dm. long with spreading or reflexed branches naked below; glumes subequal, 10 mm. long; lemma 10 mm. long, the beak 2–5 mm. long, twisted; awns subequal, 1–1.5 cm. long.—Occasional on dry slopes; Coastal Sage Scrub, V. Grassland, etc.; Bakersfield to San Bernardino V. and San Diego; to Tex., Oaxaca. May–July.

6. **A. hamulòsa** Henr. Like *A. divaricata;* lemma somewhat narrowed at summit, not twisted; cent. awn somewhat longer than the lateral ones; $n = 22$ (Stebbins & Love, 1941).—About same range as *A. divaricata;* to w. Tex., Guatemala. May–Nov.

7. **A. Paríshii** Hitchc. Tufted perennial, 2–5 dm. high; blades ± involute, sometimes flat, 1–2.5 dm. long, 1–2 mm. wide; panicle narrow, 1–3 dm. long, with appressed or ascending branches; glumes short-awned, the 1st 12 mm., the 2d 13–14 mm. long; lemma ca. 12 mm. long, with a short ± twisted beak; awn subequal, ca. 2.5 cm. long, divergent.—Dry or rocky slopes, below 4000 ft.; Coastal Sage Scrub, Chaparral, Creosote Bush Scrub; interior cismontane s. Calif., deserts; to Nev., Ariz. April–June.

8. **A. glaùca** (Nees) Walp. [*Chaetaria g.* Nees.] Erect perennial, 2–4 dm. tall; blades involute, curved or flexuous, 5–10 cm. long, ca. 1 mm. thick; panicle narrow, erect, few-fld., 7–15 cm. long; 1st glume 5–8 mm. long, 2d ca. twice as long; lemma 10–12 mm. long, with a slender ± twisted beak 5–6 mm. long; awns equal, widely divergent, 1.5–2.5 cm. long.—Rare in dry places, below 4000 ft.; Creosote Bush Scrub, both deserts; to Utah, Tex., cent. Mex. March–May, Sept.

9. **A. purpùrea** Nutt. Tufted perennial, 3–5 dm. tall; blades usually involute, 1–1.5 mm. wide when unrolled, 5–12 cm. long; panicle narrow, nodding, rather lax, usually purplish, 1–2 dm. long, with capillary branches; 1st glume 6–8 mm. long, 2d ca. 12–15 mm.; lemma ca. 10 mm. long, scarcely beaked, tuberculate-scabrous in lines upward; awns subequal, spreading, 3–5 cm. long.—Occasional; Coastal Sage Scrub, Creosote Bush Scrub; cismontane and desert s. Calif.; to Ark., Kans., Tex., n. Mex. May–July.

10. **A. Wrìghtii** Nash. Erect cespitose perennial, 3–5 dm. tall; sheaths villous at throat and often with hairy line across collar; blades involute, 1–2 dm. long, scabrous; panicle erect, narrow, 1.5–2 dm. long, with appressed branches; glumes unequal, 1-nerved, acuminate, the 1st 6 mm. long, the 2d ca. 12 mm.; lemma 10–12 mm. long, scaberulous above; awns subequal, ca. 2 cm. long, divergent.—Occasional on stony slopes, below 5000 ft.; largely Creosote Bush Scrub, Joshua Tree Wd.; Colo. and Mojave deserts; near Hemet, w. Riverside Co.; to Okla., Tex., cent. Mex. March–June, Sept.

11. **A. Fendleriàna** Steud. Densely cespitose erect perennial, 1.5–3 dm. high; blades crowded at base of plant, forming a curly tuft, 3–5 cm. long; panicle narrow, 5–10 cm. long, with a few loosely arranged appressed spikelets; glumes unequal, smooth, awnless, 1-nerved, the 1st ca. 6 mm., the 2d 10–15 mm. long; lemma ca. 18 mm. long, scaberulous on upper part; awns 2–5 cm. long, spreading.—Rare, dry slopes, 4000–6000 ft.; Pinyon-

Juniper Wd.; e. Mojave Desert (New York Mts., Deep Springs, etc.), Santa Rosa Mts.; to N. Dak., Mont., Tex., Mex. May–July.

12. **A. longisèta** Steud. var. **robústa** Merr. Much like *A. Fendleriana*, 3–5 dm. tall, the blades not in conspicuous basal tufts; panicle stiff, 5–9 cm. long; awns mostly 4–5 cm. long.—Dry slopes; Creosote Bush Scrub; e. San Diego Co., Chuckwalla Mts.; to Wash., Minn. March–May.

68. Hilària HBK. GALLETA

Stiff perennial grasses with solid culms and narrow blades. Spikelets sessile, in groups of 3, appressed to the axis, in terminal spikes. The spikelet-groups falling from the axis entire, the cent. one fertile, mostly 1-fld., the 2 lateral spikelets ♂, 2-fld. Glumes coriaceous, those of the 3 spikelets forming a false invol. sometimes connate at base, ± asymmetric, usually bearing an awn on 1 side from about the middle. Lemma and palea hyaline, subequal. Ca. 5 spp., sw. U.S. to Cent. Am. (Auguste St. *Hilaire*, French naturalist.)

Stems with a felty pubescence. Widespread on deserts .... 1. *H. rigida*
Stems glabrous or slightly puberulent. E. Mojave Desert .... 2. *H. Jamesii*

1. **H. rígida** (Thurb.) Benth. ex Scribn. [*Pleuraphis r.* Thurb.] Forming large open clumps with woody rhizomes; culms rigid, 5–8 dm. tall, leafy; sheaths often glabrate; blades spreading, stiff, involute, 2–6 cm. long; spikes peduncled, 4–8 cm. long; spikelets ca. 8 mm. long, woolly-ciliate, several-awned; lemma 3-nerved, villous on back, enclosing the palea and a rudimentary 2d floret in the cent. spikelet; lateral spikelets with a 2d floret like the 1st.—Common in sandy places, below 4000 ft.; Creosote Bush Scrub, Joshua Tree Wd.; both deserts; to Utah, Ariz., Son. Feb.–June.

2. **H. Jàmesii** (Torr.) Benth. [*Pleuraphis J.* Torr.] Culms erect, 1.5–3 dm. high, slender, from decumbent bases and tough scaly rhizomes; sheaths glabrous or slightly scabrous, sparingly villous about the ligule; blades mostly 2–5 cm. long, 2–4 mm. wide, soon involute; spikelets 6–8 mm. long, villous at base.—Dry slopes, 4000–7500 ft.; mostly Pinyon-Juniper Wd.; Mid Hills near New York Mts., Clark Mt., Death V. region, Argus Mts., White Mts., Inyo Co.; to Wyo., Tex. May–June.

69. Leptóchloa Beauv. SPRANGLETOP

Annuals or perennials, with flat lvs. and simple elongate spikes or racemes borne on a common axis forming a long or sometimes shorter panicle. Spikelets 2- to several-fld., sessile or short-pedicelled, along 1 side of a slender rachis; rachilla disarticulating above the glumes and between the florets. Glumes subequal to unequal, awnless or mucronate, 1-nerved, usually shorter than first lemma. Lemmas obtuse or acute, sometimes 2-toothed and mucronate or short-awned from between the teeth, 3-nerved, the nerves sometimes pubescent. Ca. 20 spp. of warmer parts of world. (Greek, *leptos*, slender, and *chloa*, grass, because of the slender spikes.)

Sheaths papillose-hispid; glumes exceeding 1st floret .... 1. *L. filiformis*
Sheaths smooth; glumes shorter than 1st floret.
  Lemmas awned; spikelets 7–12 mm. long .... 2. *L. fascicularis*
  Lemmas awnless; spikelets 5–7 mm. long .... 3. *L. uninervia*

1. **L. filifórmis** (Lam.) Beauv. [*Festuca f.* Lam.] Annual, often ± reddish or purple; culms 3–10 dm. high; sheaths sparsely papillose-hairy; spikes 20–40, each 3–15 cm. long, spreading-ascending; spikelets 3 mm. long; glumes mucronate, nearly equaling the 3 or 4 small awnless florets; spikelets 1–2 mm. long; *n* = 10 (Brown, 1950).—Moist depressions; Creosote Bush Scrub; Imperial Co.; to Atlantic Coast, trop. Am. Sept.–Dec.

2. **L. fasciculàris** (Lam.) Gray. [*Festuca f.* Lam.] Smooth annual, with culms 3–10 dm. tall; blades flat to loosely involute; panicles ± included, mostly 1–2 dm. long, with numerous racemes 7–12 cm. long; spikelets 3–4 mm. long, the glumes much shorter than the florets; lemmas short-awned.—Usually in moist sometimes alkaline places; V. Grassland, Creosote Bush Scrub, etc.; Imperial V., Owens V. to Lassen Co., Cent. V., etc.; as street weed, San Francisco; to Wash., Atlantic Coast, Tex., etc. June–Oct.

3. **L. uninérvia** (Presl) Hitchc. & Chase. [*Megastachya u.* Presl.] Resembling *L.*

fascicularis; panicle more dense; glumes obtuse; lemmas not awned, merely apiculate.— Moist ± alkaline places; many Plant Communities; desert and cismontane Calif., mostly below 2000 ft., occasional to 7000 ft.; to Ore., N. Car., Mex., etc., S. Am. March–Dec.

70. Eleusíne Gaertn. Goose Grass

Annual grasses with 2 to several stout spikes, digitate at the summit of the culms, sometimes with 1 or 2 a short distance below. Spikelets few- to several-fld., compressed, sessile, closely imbricate, in 2 rows along 1 side of a broad rachis; rachilla disarticulating above the glumes and between the florets. Glumes unequal, broad, acute, 1-nerved, shorter than 1st lemma. Lemmas acute, with 3 strong green nerves close together, forming a keel. Ca. 6 spp. of E. Hemis. (Name from *Eleusis,* a Greek town.)

1. **E. índica** (L.) Gaertn. [*Cynosurus i.* L.] Culms flattened, decumbent, 3–5 dm. long; sheaths loose, overlapping, compressed; blades flat or folded, 3–8 mm. wide; spikes mostly 2–6, rarely 1, 4–15 cm. long; $n = 9$ (Avdulov, 1931), 18 (Moffett & Hurcombe, 1949).—Occasional as street weed, etc.; introd. from Eu. July–Oct.

71. Dactyloctènium Willd.

Annuals or perennials, with flat blades and 2–several short thick spikes, digitate and widely spreading at summit of culms. Spikelets 3–5-fld., compressed, sessile, closely imbricate, in 2 rows along 1 side of narrow flat rachis; rachilla disarticulating above 1st glume and between florets. Glumes ± unequal, broad, 1-nerved, the 1st persistent upon rachis, 2d mucronate or short-awned below tip, deciduous. Lemmas firm, broad, keeled, acuminate or short-awned, 3-nerved, the lateral nerves faint; palea ca. as long as lemma. Small genus of warm regions. (Greek, *daktulos,* finger, and *ktenion,* small comb, because of pectinate arrangement of spikelets.)

1. **D. aegýptium** (L.) Beauv. [*Cynosurus a.* L.] Culms rooting at nodes, compressed, 2–4(–8) dm. long, forming mats; blades flat, ciliate; spikes 1–5 cm. long; $n = 18$ (Moffett & Hurcombe, 1949).—Reported as weed from Calif. and other states; introd. from Old World Trop.

72. Cỳnodon Rich.

Creeping perennials with stolons or rhizomes; blades short; spikes several, slender, digitate at summit of erect culms. Spikelets 1-fld., awnless, sessile in 2 rows along 1 side of a slender continuous rachis and closely appressed; rachilla disarticulating above glumes. Glumes narrow, acuminate, 1-nerved, subequal, shorter than floret. Lemma firm, compressed, pubescent on keel, 3-nerved, the lateral nerves near the margin. Ca. 6 spp. of warmer regions. (Greek, *kuon,* dog, and *odous,* tooth, because of hard scales on rhizomes.)

Lf.-blade with 5 primary nerves; spikes 2–6.
    Flowering culms 1–4 dm. long; lf.-blades glabrous or pilose; spikes 4–5 from one level. Common
                                        1. *C. Dactylon*
    Flowering culms 6–10 dm. long; lf.-blades with stiff spreading hairs; spikes many, not arising at
    one level. Local, near Merced . . . . 2. *C. plectostachyus*
Lf.-blade with 3 primary nerves; spikes usually 2 . . . . 3. *C. transvaalensis*

1. **C. Dáctylon** (L.) Pers. [*Panicum D.* L.] Bermuda Grass. Culms flattened, wiry, glabrous, from tough woody scaly rhizomes, the flowering culms flattened, ± erect, 1–4 dm. long; ligule a conspicuous ring of white hairs; blades flat, glabrous or pilose on upper surface, mostly 1–3 cm. long; spikes usually 4 or 5, 2–5 cm. long; spikelets imbricate, 2 mm. long, the acute lemma boat-shaped; $n = 18$ (Brown, 1950).—Common weed forming very tough sod in waste and low places (fields, lawns, orchards, etc.) through much of state; to Ore., s. states; warm regions of both hemis. June–Aug. Sometimes used as a lawn grass, but mostly unsightly in winter. Also but incorrectly called Devil's Grass.

2. **C. plectostàchyus** (K. Schum.) Pilg. [*Leptochloa p.* K. Schum.] Larger and coarser than no. 1, more richly branched.—Locally introd. about Merced, Livingston; native of E. Afr.

3. **C. transvaálensis** Davy. Extensively creeping, the blades mostly not more than

1 mm. wide; spikes 1–3, the spikelets narrower and glumes shorter than in the preceding sp.; $n = 10$ (Hurcombe, 1947).—Being used as a lawn grass and escaping as at Bard, Imperial Co.; introd. from S. Afr.

73. Beckmánnia Host. SLOUGH GRASS

Rather tall erect grasses with flat lvs. and a terminal elongated narrow nearly simple panicle. Spikelets 1–2-fld., laterally compressed, subcircular, subsessile, closely imbricate, in 2 rows along one side of a continuous slender rachis, disarticulating below the glumes, falling entire. Glumes equal, inflated, obovate, 3-nerved, rounded and apiculate above. Lemma narrow, 5-nerved, acuminate, ca. equal to glumes; palea almost as long. Two spp. of cooler parts of N. Hemis. (J. *Beckmann*, 1739–1811, professor at Goettingen.)

1. **B. Syzigáchne** (Steud.) Fern. [*Panicum S.* Steud. *B. erucaeformis* Calif. refs.] Light green annual 5–10 dm. high; sheaths loose, overlapping; blades 1–2.5 dm. long, 5–8 mm. wide; panicles 1–2.5 dm. long; spikes appressed, 1–2 cm. long; spikelets distended, ca. 3 mm. long; glumes transversely wrinkled, deeply keeled, the acuminate apex of the lemma protruding; $n = 7$ (Avdulov, 1931).—Wet places, below 3500 ft.; several Plant Communities; from San Francisco region n.; to Alaska, Atlantic Coast. May–July.

74. Spartìna Schreb. CORD GRASS

Coarse perennials with strong creeping rootstocks, simple rigid culms, long tough lvs. and 2 to many (in ours) appressed spikes scattered along the main axis. Spikelets 1-fld., much flattened laterally, sessile and usually closely imbricate on one side of a continuous rachis, disarticulating below the glumes. Glumes keeled, 1-nerved, or the 2d with a 2d nerve on 1 side, acute or short-awned, the 1st shorter, the 2d often longer than the lemma. Lemma firm, keeled, the lateral nerves obscure; palea 2-nerved, keeled and flattened. Ca. 14 spp., widely distributed. (Greek, *spartine*, a cord.)

(Mobberley, D. G. Taxonomy and distribution of the genus Spartina. Iowa State College Jour. Sci. 30: 471–574, 1956.)

Blades mostly more than 8 mm. wide; spikes closely approximate forming a cylindric infl. Coastal marshes .... 1. *S. foliosa*
Blades mostly less than 5 mm. wide; spikes distinct, appressed or spreading. Alkaline interior meadows
2. *S. gracilis*

1. **S. foliòsa** Trin. [*S. leiantha* Benth.] Culms stout, 3–12(–15) dm. tall, up to 1 cm. thick at base, usually rooting at lower nodes; blades 8–12 mm. wide at base, smooth; infl. dense, ca. 1.5(–2.5) dm. long; spikes many, approximate, closely appressed, 3–5(–8) cm. long; spikelets very flat, 9–12 mm. long; glumes firm, glabrous or hispid-ciliate on keel, acute, the first narrow, ½–⅔ as long as 2d, smooth, the 2d slightly hispidulous; lemma hispidulous on sides, shorter than 2d glume; palea thin, longer than lemma; $2n = 56$ (Church, 1940).—Coastal Salt Marsh; Del Norte Co. to L. Calif. July–Nov.

2. **S. grácilis** Trin. Culms 6–10 dm. tall, more slender; blades 1.5–2 dm. long, 3–5 mm. wide, involute in age; spikes 4–8, appressed, 2–4 cm. long; spikelets 6–8 mm. long; glumes ciliate on keel, the first ½ as long as 2d, ciliate on keel; palea as long as lemma; $2n = 42$ (Church, 1940).—Alkaline places, 6000–7000 ft.; Sagebrush Scrub; Inyo and Mono cos.; to B.C., Kans., New Mex. June–Aug.

75. Chlòris Sw. FINGER GRASS

Tufted perennials, sometimes annuals, with flat or folded scabrous blades and 2 to several, sometimes showy and feathery spikes aggregated in ± digitate fashion. Spikelets with 1 perfect floret, sessile, in 2 rows along 1 side of a continuous rachis, the rachilla disarticulating above the glumes, produced beyond the perfect floret and with 1 or more empty florets. Glumes ± unequal, the 1st shorter, narrow, acute; lemma keeled, usually broad, 1–5-nerved, often awned from between the short apical teeth; awn slender or reduced to a mucro. Ca. 60 spp. of warmer parts of world. (*Chloris*, goddess of fls.)

Lemmas firm, dark brown, awnless or mucronate .... 1. *C. distichophylla*
Lemmas pale or fuscous, distinctly awned.
  Plant producing long stout stolons; awns 1–5 mm. long .... 2. *C. Gayana*
  Plant not stoloniferous, erect or decumbent and rooting at nodes; awns 5–10 mm. long.
    Lemma conspicuously ciliate-villous, the spikes feathery .... 3. *C. virgata*
    Lemma minutely ciliate, the spikes not feathery .... 4. *C. verticillata*

1. **C. distichophýlla** Lag. Culms 3–9 dm. tall, tufted, leafy; blades mostly 7–17 cm. long, 3–7 mm. wide; ligule a dense ring of hairs less than 0.5 mm. long; spikes ± brownish, commonly 8–15, usually 6–8 cm. long, closely aggregate, ascending to drooping; spikelets 2-fld., ca. 2.3 mm. long; glumes ca. 1 mm. long, minutely scabrous; lemmas awnless, brown, the lower ca. 2 mm. long, lanceolate, villous on margins; $n = 10$ (Krishnaswamy, 1940).—Reported as escape from cult., near San Diego; S. Am. June–Sept.

2. **C. Gayàna** Kunth. RHODES GRASS. Culms 1–1.5 m. tall, with long stout leafy stolons; blades 5–40 cm. long, 3–5 mm. wide; spikes several to many, erect or ascending, 5–10 cm. long; spikelets pale-tawny, crowded; lemma 3 mm. long, hispid on margin near summit, the awn 1–5 mm. long; $n = 10$ (Brown, 1950).—Locally common as a weed, escaping from cult. for forage; introd. from Afr. Aug.–Sept.

3. **C. virgàta** Sw. [*C. elegans* HBK.] Annual, 4–6(–8) dm. tall; blades flat, 7–40 cm. long, 2–7 mm. wide; spikes several, 2–8 cm. long, whitish or tawny, feathery or silky; spikelets crowded, 2.5–4 mm. long; glumes 1-nerved, the 1st 1.5–2 mm., the 2d 3–3.5 mm. long; lemma 3 mm. long, long-ciliate, the awn 5–10 mm. long; $n = 10$ (Moffett & Hurcombe, 1949).—Common in waste places, San Joaquin V., interior s. Calif., less so in Sacramento V.; to Nebr., Tex., and farther e.; trop. Am. April–Sept.

4. **C. verticillàta** Nutt. Tufted, 1–4 dm. tall, erect or decumbent at base; lvs. crowded at base; blades 1–3 mm. wide, obtuse; spikes slender, 7–10(–15) cm. long, in 1–3 whorls, spreading; spikelets ca. 3 mm. long; fertile lemma pubescent on nerves, the awn 5–8 mm. long; $n = 40$ (Brown, 1950).—Escape about Berkeley; introd. from farther e. June–Aug.

76. **Bouteloùa** Lag. GRAMA GRASS

Annuals or perennials, mostly tufted. Spikelets few to many, 1–2-fld., crowded in 2 rows and forming few to many 1-sided, ± curved, sessile spikes; rachis usually conspicuously prolonged beyond the spikelets. Lower fls. perfect, the upper when present ♂; glumes 2, narrow, acute, unequal, keeled. Lemma usually thinner, broader, 3-nerved, the nerves extending into short awns or mucros. Palea 2-nerved, sometimes 2-awned. Ca. 40 spp., mostly N. Am. Valuable range grasses. (C. *Boutelou*, 1774–1842, Spanish horticulturist.)

A. Spikelets not pectinately arranged, the spikes falling entire at maturity.
  B. Plants annual .... 1. *B. aristidoides*
  BB. Plants perennial.
    C. Spikes 35–50, 5–15 mm. long, each with 3–7 spikelets .... 2. *B. curtipendula*
    CC. Spikes mostly 7–12, 20–25 mm. long, each with 8–11 spikelets ...... 3. *B. radicosa*
AA. Spikelets pectinately arranged at maturity, the spikes persistent, the florets falling from the persistent glumes.
  B. Plants annual .... 4. *B. barbata*
  BB. Plants perennial.
    C. The plants decumbent or stoloniferous; culms white-woolly .... 8. *B. eriopoda*
    CC. The plants erect or nearly so; culms not lanate.
      D. Spikes normally less than 4.
        E. Rachis prolonged beyond the spikelets as a naked point; glumes tuberculate
                                     6. *B. hirsuta*
        EE. Rachis not prolonged; glumes not tuberculate .... 7. *B. gracilis*
      DD. Spikes normally 4 or more.
        E. Culms 1.2–1.5 dm. high; spikes 1–2 cm. long; spikelets ca. 12 .. 9. *B. trifida*
        EE. Culms 2.5–5 dm. high; spikes 2.5–3 cm. long; spikelets 40–50
                                     5. *B. Rothrockii*

1. **B. aristidoìdes** (HBK.) Griseb. [*Dinebra a.* HBK.] Erect or spreading annual, the culms slender, 1.5–4 dm. long; sheaths and blades smooth, the latter rather small and few; spikes mostly 8–14, pedunculate, reflexed, readily falling, the base of the rachis forming a sharp bearded point; spikelets 2–4, narrow, appressed; glumes narrow,

acuminate, the 1st ca. 2 mm. long, the 2d almost 4 mm.; lemma strigose on nerves, the lateral nerves ending in awned teeth; rudimentary floret of a pilose stipe and 3 awns.— Occasional in dry sandy places; Creosote Bush Scrub on Colo. Desert, Yellow Pine F. in San Jacinto Mts., and Pinyon-Juniper Wd., Clark Mt., e. Mojave Desert; to Nev., Tex., n. Mex. April–Sept.

2. **B. curtipéndula** (Michx.) Torr. [*Chloris c.* Michx.] Perennial, with scaly rhizomes; culms tufted, 3–8 dm. tall; blades flat or subinvolute, 3–4 mm. wide; spikes 35–50, purplish, 1–2 cm. long, spreading or twisted to 1 side of the slender axis 1.5–2.5 dm. long; spikelets 5–8, appressed or ascending, 6–10 mm. long; glumes unequal, the 1st 4–5 mm., the 2d 7 mm. long; fertile lemma acute, mucronate, 3-nerved, 3-toothed; rudiment with 3 awns; $2n = 28$, 35, 40, 42, 45, 56, 70, 98 (Fults, 1942).—Dry rocky slopes, 4500–6000 ft.; Pinyon-Juniper Wd., etc.; Santa Rosa Mts. (Riverside Co.), Little San Bernardino Mts., New York Mts., Clark Mt. (San Bernardino Co.); to Atlantic Coast, Mex., Argentina. May–Aug.

3. **B. radicòsa** (Fourn.) Griffiths. [*Atheropogon r.* Fourn.] Tufted perennial 6–8 dm. tall; blades 2–3 mm. wide, papillose-ciliate; axis 1–1.5 dm. long with 7–12 oblong spikes 2–3 cm. long; spikelets mostly 8–11; glumes broad, ± unequal, the 2d ca. 6 mm. long; lemmas smooth, 3-awned; rudiment lanceolate, with 3 long awns.—Reported from Calif. (*Orcutt*, probably about Colo. Desert); to New Mex., Mex. Aug.–Sept.

4. **B. barbàta** Lag. Tufted annual with prostrate or spreading culms 1–3 dm. long; sheaths and blades glabrous; blades 1–1.5 mm. wide, 1–4 cm. long; spikes 4–7, 1–1.5 cm. long; spikelets 25–40, 2.5–4 mm. long, almost as wide; the 2d glume twice the 1st; fertile lemma pilose at least along the sides, the awns from minute to as long as body; rudiment ± bearded at base, cleft, awned, a 2d rudiment not awned.—Gravelly and sandy washes, etc., below 5000 ft.; Creosote Bush Scrub to Pinyon-Juniper Wd.; Colo. and e. Mojave deserts; to Colo., Tex., Mex. July–Dec.

5. **B. Rothróckii** Vasey. Perennial, the culms erect or spreading, 3–5 dm. long; sheaths and blades smooth, or the latter with a few long hairs and 2–3 mm. wide; axis 1–2.5 dm. long, with 4–12 spikes 2.5–3 cm. long; spikelets 40–50, ca. 5 mm. long; glumes persistent, unequal, scabrous on keel and back, 2-toothed, the 2d ca. twice as long as 1st; fertile lemma pilose at base, deeply cleft, the awns 1–2 mm. long, spreading; rudiment bearded, cleft, with broad rounded lobes, awned; $n = 11$ (Fults, 1942).—Jamacha, e. San Diego Co.; Ariz., n. Mex. Aug.–Oct.

6. **B. hirsùta** Lag. Erect perennial, 2–4 dm. tall; sheaths smooth; blades papillose-hairy, ca. 2 mm. wide, flat or subinvolute; spikes mostly 1–4, 2.5–3.5 cm. long; spikelets 35–45, ca. 5 mm. long, the 2d glume tuberculate-hirsute, the tubercles black; fertile lemma 3-cleft, awn-tipped; rudiment puberulent to bearded, cleft, with black awns; $2n = 21$, 37, 42 (Fults, 1942).—Reported from Jamacha, e. San Diego Co.; to Wis., Tex., n. Mex. July–Sept.

7. **B. grácilis** (HBK.) Lag. [*Chondrosium g.* HBK.] Densely tufted perennial, 1–4 dm. tall; sheaths and blades glabrous, the latter flat or loosely involute, 1–2 mm. wide; spikes usually 2, 2.5–5 cm. long, curved-spreading; spikelets many (up to 80), ca. 5 mm. long; fertile lemma pilose, slender-awned; rudiment densely bearded at summit of rachilla, cleft to base, with rounded lobes and slender awns; $2n = 20$, 28, 35, 40, 42, 61, 62, 77, 84 (Fults, 1942; Snyder & Harlan, 1953).—Dry places, 5500–8500 ft.; Pinyon-Juniper Wd., Yellow Pine F.; San Bernardino Mts., New York Mts., Clark Mt.; to Wis., Manitoba, Tex., Mex. May–Aug.

8. **B. eriópoda** (Torr.) Torr. [*Chondrosium e.* Torr.] Tufted perennial with swollen bases and wiry white-woolly culms 4–6 dm. long; blades 1–1.5 mm. wide; spikes 3–8, loosely ascending, 2–3 cm. long; spikelets 12–20, 7–10 mm. long, narrow; fertile lemma acuminate, awned; rudiment slender, cleft, the lobes awned; $n = 14$ (Brown, 1950).—Dry slopes, 3500–6000 ft.; Joshua Tree Wd., Pinyon-Juniper Wd.; Clark Mt., etc., e. Mojave Desert; to Okla., Tex., Colo., n. Mex. June–Sept.

9. **B. trífida** Thurb. Tufted perennial, 1–2 dm. tall; blades mostly 1–2 cm. long; spikes 3–7, 1–2 cm. long, ascending; spikelets ca. 12, purplish, 7–10 mm. long; fertile lemma pubescent toward base, cleft over halfway, the awns 5 mm. long, winged toward base; rudiment cleft to base, awned.—Found in Providence Mts. and Death V.; to Nev., Tex., n. Mex. May–Sept.

77. Hieróchloe R. Br. Vanilla Grass

Perennial erect sweet-smelling grasses with small panicles of broad bronze-colored spikelets. These with 1 terminal perfect floret and 2 lateral ♂ florets, disarticulating above the glumes, the ♂ falling attached to the ♀ one. Glumes equal, 3-nerved, broad, thin, smooth, acute; ♂ lemmas ca. as long as glumes, boat-shaped, hairy on margins; fertile lemma ± indurate, nearly smooth, awnless; palea 3-nerved, rounded on back. Ca. 17 spp. of cooler regions. (Greek, *hieros*, sacred, and *chloe*, grass, some spp. being used in church festivals.)

1. **H. occidentàlis** Buckl. [*H. macrophylla* Thurb. *Torresia m.* Hitchc.] Culms few, 6–9 dm. tall, with creeping rhizomes and long lvs.; sheaths scabrous; blades flat, 2–5 dm. long, 8–15 mm. wide; panicle mostly open, 0.7–1.5 dm. long, with subcapillary branches; spikelets 4–5 mm. long, the glumes with pale shining margin; ♂ lemmas almost or quite awnless; fertile lemma strigose toward apex.—Woods below 2000 ft.; Redwood F., Mixed Evergreen F., Closed-cone Pine F.; Monterey Co. n.; to Wash. Jan.–July.

78. Anthoxánthum L. Vernal Grass

Sweet-smelling annual or perennial grasses with flat blades and spikelike panicles. Spikelets with 1 perfect terminal floret and 2 sterile lemmas, these falling away with the fertile floret. Glumes unequal, acute or mucronate; sterile lemmas shorter than the glumes, awned from back; fertile lemma awnless, shorter than the sterile. Palea 1-nerved, rounded on back, enclosed in lemma. Ca. 4 spp. of Eurasia. (Greek, *anthos*, fl., and *xanthos*, yellow.)

Plants perennial .... 1. *A. odoratum*
Plants annual .... 2. *A. aristatum*

1. **A. odoràtum** L. Sweet Vernal Grass. Tufted, erect, 3–6 dm. tall; blades 2–5 mm. wide; panicle brownish-yellow, long-exserted, 2–6 cm. long; spikelets 8–10 mm. long; glumes scabrous; 1st sterile lemma short-awned below apex, 2d with a twisted geniculate awn from near base; fertile lemma ca. 2 mm. long, brown, smooth, shining; $n = 5$, 10 (Oestergren, 1942).—Lawns, pastures, waste places, more abundant n. in the state; escape from cult.; introd. from Eurasia. May–June.

2. **A. aristàtum** Boiss. Annual, 2–3 dm. high; blades 1–2 mm. wide; panicle 2–3 cm. long; spikelets ca. 5–7 mm. long; $n = 5$ (Avdulov, 1928).—Waste places; Marin, Sonoma, Shasta, Del Norte cos., etc.; introd. from Eu. May–June.

79. Ehrhárta Thumb.

Annual or perennial, erect to spreading, with flat blades and narrow panicles. Spikelets laterally compressed, with 1 fertile floret and 2 large sterile lemmas below enclosing the fertile floret; rachilla disarticulating above the glumes, the fertile floret and sterile lemmas falling together. Glumes ovate, obscurely keeled; sterile lemmas indurate, compressed, 3- or 5-nerved; fertile lemma indurate, ovate, obtuse, 5-nerved. A small genus. (Named for F. Ehrhart.)

1. **E. erécta** Lam. Perennial with creeping rootstock, the culms 3–6 dm. tall; blades 5–12 cm. long, 4–9 mm. wide; panicles 0.6–1.5 dm. long, with ascending ± spreading branches; spikelets 3–3.5 mm. long; sterile lemmas awnless, the first smooth, the 2d transversely wrinkled; $n = 12$ (Parthasarathy, 1939).—Natur. on Berkeley campus of the Univ. of Calif.; introd. from S. Afr. March–June.

80. Phálaris L. Canary Grass

Annuals or perennials, with many flat blades and dense spikelike panicles. Spikelets 1-fld., laterally flattened and with 2 sterile lemmas below the fertile floret. Glumes equal, boat-shaped, often winged on the keel. Sterile lemmas reduced to small usually minute scales; fertile lemma coriaceous, shorter than glumes, enclosing the faintly 2-nerved palea. Ca. 20 spp., in temp. Eu. and Am. (An ancient Greek name for a grass.)

A. Spikelets in groups of 7, 1 fertile surrounded by 6 sterile, the group falling entire
1. *P. paradoxa*
AA. Spikelets all alike, not falling entire in groups.
  B. Plants perennial.
    C. Rhizomes lacking; panicle 2–5 cm. long, dense, ovate or short-oblong . 2. *P. californica*
    CC. Rhizomes present; panicle 5–18 cm. long, narrow.
      D. Sterile lemmas both developed, ca. 1 mm. long . . . . 3. *P. arundinacea*
      DD. Sterile lemma solitary, ca. 1.5 mm. long . . . . 4. *P. tuberosa*
  BB. Plants annual.
    C. Glumes broadly winged; panicle ovate to short-oblong.
      D. Sterile lemma solitary; fertile lemma 3 mm. long . . . . 5. *P. minor*
      DD. Sterile lemmas 2; fertile lemma 4–6 mm. long.
        E. Sterile lemmas half as long as the fertile . . . . 6. *P. canariensis*
        EE. Sterile lemmas 0.6 mm. long or less . . . . 7. *P. brachystachys*
    CC. Glumes wingless or nearly so; panicles linear or oblong.
      D. Glumes wingless, acuminate; fertile lemma turgid, the apex smooth, acuminate
8. *P. Lemmonii*
      DD. Glumes narrowly winged toward the summit, abruptly pointed or acute; fertile
        lemma less turgid, villous to the acute apex.
        E. Panicle tapering to each end, mostly 2–6 mm. long . . . . . . 9. *P. caroliniana*
        EE. Panicle subcylindric, mostly 6–15 cm. long . . . . 10. *P. angusta*

1. **P. paradóxa** L. Tufted annual, 3–6 dm. tall; panicle dense, 2–6 cm. long, oblong, with narrowed base where often enclosed in the enlarged upper sheath; spikelets in groups of 6–7, the cent. one fertile, with subulate-acuminate glumes bearing a toothlike wing near the middle of the keel, the others sterile; fertile lemma 3 mm. long, with a few hairs near summit; sterile lemmas obsolete; $n = 7$ (Parthasarathy, 1938).—Occasional in grainfields, waste places, etc.; cismontane Calif.; introd. from Medit. region. May–Aug. A form with outer glumes of all sterile spikelets clavate (instead of only lower part of panicles having such) is var. *praemórsa* (Lam.) Coss. & Durieu, and is also found in Calif.

2. **P. califórnica** H. & A. [*A. amethystina*, Calif. refs.] Perennial, often in dense tufts, 4–12 dm. high; blades 8–15 mm. wide; panicle 2–5 cm. long, 2–2.5 cm. thick, often tinged with purple; glumes 6–8 mm. long, narrow, with smooth keel, sharp but not winged; fertile lemma lance-ovate, ca. 4 mm. long, ± strigose; sterile lemmas ½–⅔ as long; $n = 14$ (Stebbins & Love, 1941).—Canyons and damp places in woods, below 2500 ft.; Redwood F., Closed-cone Pine F., Mixed Evergreen F., etc.; Coast Ranges from San Luis Obispo Co. to Ore. May–Nov.

3. **P. arundinàcea** L. Perennial with creeping rhizomes, glaucous, 6–15 dm. tall; lf.-blades 10–20 mm. wide; panicle 7–18 cm. long, with spreading branches during anthesis; glumes ca. 5 mm. long, narrow, acute, scabrous on keel; fertile lemma lanceolate, 4 mm. long, ± strigose; sterile lemmas villous, 1 mm. long; $n = 7$, 14 (Church, 1929).—Wet banks and moist places below 5000 ft.; V. Grassland to Yellow Pine F.; San Joaquin V. to Siskiyou and Modoc cos.; to Atlantic Coast; Eurasia. May–Aug.

4. **P. tuberòsa** L. var. **stenóptera** (Hack.) Hitchc. [*P. s.* Hack.] HARDING GRASS. Perennial with a loose branching rhizomatous base; culms stout, to 1.5 m. tall; panicle narrow, not branched, 5–15 cm. long; glumes 5–6 mm. long, scabrous on keel; fertile lemma 4 mm. long, ovate-lanceolate, strigose; sterile lemma usually 1, ca. ⅓ as long as fertile; $n = 14$ (Nielsen & Humphrey, 1937).—Occasional escape in moist places, Marin, Colusa, Humboldt cos.; from Old World. June–Sept.

5. **P. mìnor** Retz. Erect annual, 3–9 dm. tall; lvs. 4–15 mm. wide; panicle ovate-oblong, mostly 2–5 cm. long; spikelets narrow, the glumes oblong, 4–6 mm. long, strongly winged on keel, ± green-striped; fertile lemma ovate, acute, villous, ca. 3 mm. long; sterile lemma 1, ca. 1 mm. long; $n = 14$ (Avdulov, 1931).—Frequent weed in waste and disturbed places; many Plant Communities; cismontane Calif.; to Ore., Atlantic Coast; introd. from Medit. region. April–July.

6. **P. canariénsis** L. Erect annual, 3–6 dm. high; the dense broad panicle 1.5–4 cm. long; spikelets broad, pale with green stripes; glumes 7–8 mm. long, abruptly pointed, the green keel prominently pale-winged; fertile lemma 5–6 mm. long, densely strigose; sterile lemmas at least ½ as long as fertile.—Occasional weed in waste places; cismontane Calif. to Atlantic Coast; introd. from Medit. region. April–July.

7. **P. brachystàchys** Link. Like *P. canariensis*, but with smaller spikelets, the glumes ca. 6 mm. long; fertile lemma 4–5 mm. long; sterile lemmas ca. 0.5 mm. long; $n = 6$ (Miege, 1939).—Reported from Butte, Contra Costa, San Luis Obispo cos.; Ore., Tex.; introd. from Medit. region. May–June.

8. **P. Lemmònii** Vasey. Erect annual, 3–9 dm. tall; lvs. 3–9 mm. broad; panicles 5–15 cm. long, subcylindric or ± lobed below, often purplish; glumes ca. 5 mm. long, narrow, scabrous, acuminate, not winged on keel; fertile lemma lance-ovate, acuminate, 3.5–4 mm. long, brown when mature, strigose except at the acuminate tip; sterile lemmas less than ⅓ as long; *n* = 7 (Parthasarathy, 1938).—Moist places below 2000 ft.; Coastal Sage Scrub, V. Grassland, Foothill Wd., Mixed Evergreen F., etc.; cismontane Calif. from San Diego Co. to Butte and Lake cos. April–June.

9. **P. caroliniàna** Walt. Erect annual, 3–6 dm. tall; panicle oblong, 2–6 cm. long, glumes 5–6 mm. long, oblong, abruptly narrowed to an acute apex, scabrous on keel; fertile lemma lanceolate, acute, strigose, 3.5–4 mm. long; sterile lemmas 1–2 mm. long.— Uncommon in old fields and waste places; widely scattered cismontane stations; introd. from e. U.S. April–May.

10. **P. angústa** Nees ex Trin. Annual, 1–1.5 m. tall; panicle subcylindric, 6–15 cm. long, ca. 8 mm. thick; glumes 3.5–4 mm. long, narrow, abruptly pointed, the keel scabrous and narrowly winged upward; fertile lemma ovate-lanceolate, acute; sterile lemmas ca. ⅓ as long; *n* = 7 (Saura, 1943).—Uncommon, wet places at low elevs.; several Plant Communities; Solano and Butte cos. s. to San Diego Co.; to La., S. Am. May–June.

81. Leèrsia Sw. RICE CUTGRASS

Perennial grasses, usually with creeping rhizomes; blades flat, scabrous; panicles mostly open. Spikelets 1-fld., strongly laterally compressed, disarticulating from the pedicel; glumes none. Lemma chartaceous, broad, oblong to oval, boat-shaped, usually 5-nerved, the lateral pair of nerves near the margins, often hispid-ciliate; palea as long as lemma, narrower, usually 3-nerved; stamens 6 or fewer. Ca. 10 spp. of trop. and n. temp. Am.; 1 sp. in Eurasia. (J. D. *Leers*, 1727–1774, German botanist.)

1. **L. oryzoìdes** (L.) Sw. [*Phalaris o.* L. *Homalocenchrus o.* Poll.] Ascending or sprawling, the culms 1–1.5 m. long, terete, from slender elongate rhizomes; lvs. scabrous, 8–10 mm. wide; panicles terminal and axillary, 1–2 dm. long, with flexuous finally spreading branches; spikelets oblong, 4–6 mm. long, 1.5–2 mm. wide; sparsely his-pidulous, with bristly ciliate keels; axillary panicles reduced, partly included in the sheaths and with cleistogamous spikelets; *n* = 24 (Ramanujam, 1938).—Marshes, stream banks, etc.; about Riverside and San Bernardino, Inyo Co., from Napa to Mendocino and Trinity, Siskiyou cos., etc.; to B.C., Maine, Tex., Fla.; Eu. Aug.–Oct.

Orỳza satìva L. RICE. Annual, 1–2 m. tall; panicle rather dense, drooping, 1.5–4 dm. long; spikelets 7–10 mm. long; lemma mucronate to awned.—Cult. and sometimes natur.; Sacramento V.; native of Old World.

82. Digitària Heist. CRAB GRASS

Annual or perennial, erect to prostrate, often weedy. Racemes slender, digitate or approximate on a short axis. Spikelets in 2's or 3's, rarely 1, subsessile or short-pedicelled, alternate in 2 rows on one side of a 3-angled rachis; spikelets lanceolate or elliptic. First glume minute or none; 2d glume equal to or shorter than sterile lemma. Fertile lemma cartilaginous, with pale hyaline margins. Ca. 75 spp. of warmer regions. (Latin, *digitus*, finger, because of the infl.)

Sheaths pilose or villous; spikelets 2.5–3.5 mm. long; fertile lemma pale .... 1. *D. sanguinalis*
Sheaths glabrous; spikelets 2 mm. long; fertile lemma brown .... 2. *D. Ischaemum*

1. **D. sanguinàlis** (L.) Scop. [*Panicum s.* L. *Syntherisma s.* Dulac.] Annual, usually much-branched at base, often purplish, the culms 1–6(–9) dm. long, with ascending flowering shoots; blades 5–10 mm. wide, pubescent to scaberulous; racemes few to several, 5–15 cm. long, digitate; spikelets ca. 3 mm. long; 1st glume minute, 2d ca. 1.5 mm. long, ciliate; sterile lemma nerved, the lateral internerves strigose; fertile lemma pale; *n* = 18 (Avdulov, 1931).—Common weed in lawns, gardens, etc.; native of Eu. June–Sept.

2. **D. Ischaèmum** (Schreb.) Schreb. ex Muhl. [*Panicum I.* Schreb.] Glabrous, the culms 2–4 dm. long, prostrate to ascending; lvs. 3–6 mm. wide; racemes 1–6, commonly purplish, 1–9 cm. long; glume and sterile lemma equal, short-villous between the

nerves, as long as brown lemma; $n = 18$ (Brown, 1948).—Natur. scatteringly, as about Loma Linda, Los Angeles, Ventura, Petaluma, Sacramento; native of Eu. Sept.–Nov.

83. Stenotáphrum Trin. St. Augustine Grass

Creeping stoloniferous perennials with short flowering culms, rather broad and short obtuse blades, and terminal and axillary racemes. Spikelets embedded in 1 side of an enlarged and flattened corky rachis tardily disarticulating toward the tip at maturity, the spikelets remaining attached to the joints. First glume small, 2d glume and sterile lemma subequal, the latter with a palea or a ♂ fl. Fertile lemma chartaceous. Three spp. of warm regions. (Greek, *stenos,* narrow, and *taphros,* trench, because of the cavities in the rachis.)

1. **S. secundátum** (Walt.) Kuntze. [*Ischaemum s.* Walt.] Rather coarse, bright green, the culms compressed, branched, with flowering stems 1–3 dm. tall; blades 5–15 cm. long, 4–10 mm. wide; racemes 5–10 cm. long; spikelets in 1's or 2's, 4–5 mm. long; $n = 10$ (Brown, 1948).—Escape in low waste places from cult. in lawns; native se. U.S. and trop. Am. July–Sept.

84. Erióchloa HBK. Cup Grass

Annual or perennial often-branched grasses. Panicles terminal, of several to many spreading or appressed racemes, usually approximate along a common axis. Spikelets ± pubescent, solitary or in 2's, short-pedicelled or subsessile, in 2 rows on 1 side of a narrow rachis, the back of the fertile lemma turned from the rachis; lower rachilla-joint thickened, forming a ± ringlike callus below the 2d glume, the 1st glume minute and sheathing this. Lemma usually enclosing a hyaline palea or sometimes a ♂ fl. Fertile lemma indurate, minutely papillose-rugose, mucronate or awned, the awns often readily deciduous. Ca. 15 spp. of warmer regions, mostly of Am. (Greek, *erion,* wool, and *chloa,* grass, because of pubescent spikelets.]

Spikelets 4–5 mm. long, the awn not over 1 mm. long .... 1. *E. gracilis*
Spikelets (including awns) 7–10 mm. long, the awn ca. as long as body .... 2. *E. aristata*

1. **E. grácilis** (Fourn.) Hitchc. [*Helopus g.* Fourn. *E. Lemmonii* var. *g.* Gould.] Annual, 4–10 dm. tall; blades flat, smooth, 5–10 mm. wide; racemes several to many, approximate, 2–4 cm. long, ascending to spreading, softly pubescent on axis and rachis; pedicels short-pilose; spikelets 4–5 mm. long, sparsely strigose, acuminate or short-awned; fr. ca. 3 mm. long, apiculate.—Irrigated fields, orchards, etc., Riverside and San Bernardino areas, Imperial V.; to Okla., Tex. Aug.–Sept.

E. contrácta Hitchc., with pubescent lf.-blades and an awn ca. 1 mm. long, is occasional in the state; native from Ariz. to Nebr. and La.

2. **E. aristàta** Vasey. Annual, 3–8 dm. tall; blades 10–12 mm. wide; racemes several, overlapping, 3–4 cm. long, the rachis pilose; spikelets ca. 5 mm. long, the glume and sterile lemma awned; fr. 3.5 mm. long, apiculate.—Colo. R. bottom near Palo Verde, Yuma; Ariz. to Mex. June–Nov.

85. Paspàlum L.

Perennial grasses (ours), with 1–many racemes digitate or racemose at the summit of the culm and branches. Spikelets 1-fld., planoconvex, subsessile, solitary or in 2's, in 2 rows on 1 side of a continuous narrow or dilated rachis. The back of the fertile lemma toward the rachis; 1st glume mostly wanting; 2d glume and sterile lemma subequal; lemma and palea chartaceous-indurated, with inrolled margins. A large genus of warmer parts of world. (Greek, *paspalos,* a kind of millet.)

Racemes a pair at the summit of the culm .... 1. *P. distichum*
Racemes several to many, forming a panicle.
  The racemes laxly spreading; spikelets 3.5 mm. long .... 2. *P. dilatatum*
  The racemes suberect; spikelets 2 mm. long .... 3. *P. Urvillei*

1. **P. dístichum** L. Knot Grass. Creeping and rooting at the nodes, freely branching, the flowering shoots ascending, 2–6 dm. high; blades thin, 4–10 mm. wide, 5–15 cm.

long; panicles terminal, the racemes mostly 2, usually erect, 2–7 cm. long; rachis 2–3 mm. wide; spikelets solitary, ovate, sparsely pubescent, 2.5–4 mm. long; $n = 20$ (Brown, 1948).—Along the coast and interior ditches; V. Grassland, Freshwater Marsh, Coastal Salt Marsh, etc.; cismontane Calif.; to Wash., Atlantic Coast, S. Am. June–Oct.

2. **P. dilatàtum** Poir. Culms stout, 5–15 dm. tall, growing in clumps, glabrous throughout except the spikelets; lvs. 5–10 mm. wide, elongate; racemes 4–10, densely fld., 5–10 cm. long, somewhat spreading; spikelets in pairs, ovate, 3–3.5 mm. long; glume and sterile lemma long-ciliate; $n = 20$ (Brown, 1948).—Roadsides, ditches and waste places, low elevs., cismontane Calif.; to Ore., Atlantic Coast; native of S. Am. May–Nov.

3. **P. Urvíllei** Steud. [*P. Larrangai* Arech.] Culms in large clumps, erect, 1–2 m. tall; lower sheaths hirsute; blades elongate, 3–15 mm. wide; panicle erect, 1–4 dm. long, of ca. 12–20 ascending racemes 7–14 cm. long; spikelets 2.2–2.7 mm. long, ovate, pointed, fringed with long white hairs, the glume silky; $n = 20$ (Burton, 1940).—Rare as introd.; s. Calif. and Butte Co.; native of S. Am. April–July.

86. Pánicum L. Panic Grass

Annual or perennial grasses of various habit. Spikelets ± compressed dorsiventrally, in open or compact panicles, sometimes racemes. Glumes 2, green, nerved, mostly unequal, the 1st often minute, the 2d typically equaling the sterile lemma, the latter simulating a 3d glume, bearing in its axil a membranaceous or hyaline palea. Fertile lemma and palea chartaceous-indurate, mostly obtuse, the nerves obsolete, the margins inrolled over an inclosed palea of the same texture. A very large genus of both hemis., particularly in warmer regions; some spp. in cult. (*Panicum*, an old Latin name for common millet.)

A. Plants perennial.
  B. Spikelets 6–7 mm. long .... 1. *P. Urvilleanum*
  BB. Spikelets less than 4 mm. long.
    C. The spikelets turgid, strongly nerved, sparsely hispid, 3.3 mm. long
                                                       2. *P. Scribnerianum*
    CC. The spikelets not turgid or strongly nerved, pubescent, not over 2.6 mm. long.
      D. Spikelets not more than 2 mm. long.
        E. Blades glabrous on both surfaces .... 3. *P. Lindheimeri*
        EE. Blades pubescent, at least on lower surface.
          F. Plants velvety-pubescent .... 4. *P. thermale*
          FF. Plants ± pubescent, but not velvety.
            G. Culms 3–6 dm. long; vernal blades pubescent above.
              H. Upper surfaces of lf.-blades strigose; autumnal form erect
                  or ascending .... 5. *P. huachucae*
              HH. Upper surfaces of lf.-blades pilose; autumnal form decumbent-spreading .... 6. *P. pacificum*
            GG. Culms 1–2 dm. long; vernal blades glabrous above
                                    7. *P. occidentale*
      DD. Spikelets about 3 mm. long .... 8. *P. shastense*
AA. Plants annual.
  B. Mature fertile lemma transversely strigose; spikelets almost 4 mm. long .. 9. *P. arizonicum*
  BB. Mature fertile lemma smooth.
    C. First glume not more than ¼ as long as spikelet, truncate or broadly triangular; sheaths smooth .... 10. *P. dichotomiflorum*
    CC. First glume as much as ½ as long as spikelet, acute or acuminate; sheaths hispid.
      D. Panicle erect; spikelets not more than 4 mm. long.
        E. The panicle more than half as long as entire plant.
          F. Fr. without scar at base .... 11. *P. capillare*
          FF. Fr. with a lunate scar at base .... 12. *P. Hillmanii*
        EE. The panicle not more than ⅓ the entire length of the plant .. 13. *P. hirticaule*
      DD. Panicle drooping; spikelets 4.5–5 mm. long .... 14. *P. miliaceum*

1. **P. Urvilleànum** Kunth. [*P. U.* var. *longiglume* Scribn.] Robust perennial with creeping rhizomes and solitary or tufted culms 5–10 dm. tall; nodes densely bearded; sheaths overlapping, densely retrorse-villous; blades elongate, 2–5 dm. long, 4–8 mm. wide; panicle 2–3 dm. long, rather narrow; spikelets 6–7 mm. long, densely villous; 1st glume acuminate, ⅔ as long as spikelet; $n = 18$ (Nuñez, 1946).—Occasional in sandy places, Creosote Bush Scrub, both deserts; V. Grassland, San Jacinto; to Ariz., Argentina, Chile. March–May.

2. **P. Scribneriànum** Nash. Vernal form erect, 2–5 dm. tall, glabrous to harshly

puberulent or pilose; sheaths striate, subglabrous to papillose-hispid; blades 5–10 cm. long, 6–12 mm. wide; strigose beneath; panicle 4–8 cm. long; spikelets 3.3 mm. long, obovoid, blunt, ± sparsely pubescent; *n* = 9 (Brown, 1948).—Meadows; Foothill Wd., Mixed Evergreen F.; Shasta to Trinity and Humboldt cos.; to B.C., Maine, Va., Mex. May–Aug.

3. **P. Lindheìmeri** Nash. [*P. Funstoni* Scribn. & Merr.] Vernal culms stiffly ascending or spreading, 3–6 dm. long, glabrous or ± pubescent below; lvs. glabrous except on the ciliate margin of the base of the blades; ligule a ring of hairs 4–5 mm. long; blades 6–8 mm. wide, glabrous; panicle 4–7 cm. long, ca. as wide; spikelets 1.4–1.6 mm. long, obovate; autumnal phase ± spreading to prostrate; *n* = 9 (Brown, 1948).—Rare open moist places; V. Grassland; Cent. V.; New Mex. to Quebec and Fla. July–Aug.

4. **P. thermàle** Bol. [*P. lassenianum* Schmoll.] Vernal phase gray-green, densely tufted, 1–3 dm. tall, the nodes with a ring of hairs; ligule 3 mm. long; blades thick, 3–8 cm. long, 5–12 mm. wide; panicle 3–6 cm. long, with villous axis; spikelets 2 mm. long, pilose.—Wet saline places about hot springs; Sonoma Co., Humboldt Co., Lassen Peak; to Wash., Alta., Utah. June–Aug.

5. **P. huachùcae** Ashe. Vernal form typically stiff, upright, with copious spreading papillose pubescence throughout; nodes bearded; blades firm, erect or ascending, 4–8 cm. long, 6–8 mm. wide, short-pilose on upper surface; panicle 4–7 cm. long, pilose; spikelets 1.6–1.8 mm. long, obovoid, turgid, pubescent; autumnal form erect, with fascicled branches and crowded ascending blades; *n* = 9 (Brown, 1948).—Reported from Lytle Creek Canyon, San Gabriel Mts. and from San Bernardino Mts.; Ariz. to Atlantic Coast. June–Aug.

6. **P. pacíficum** Hitchc. & Chase. Vernal form light green, tufted, spreading or ascending, 3–6 dm. tall; nodes pilose; sheaths papillose-pilose; blades erect or ascending, 5–10 cm. long, 5–8 mm. wide; papillose-pilose on upper surface; panicles 5–10 mm. long; spikelets ca. 2 mm. long, papillose-pubescent; 1st glume ¼ to ⅓ the length of the spikelet, truncate; autumnal phase prostrate-spreading.—Moist places below 4500 ft.; Chaparral, Yellow Pine F., to Redwood F.; mts. of San Diego Co., San Jacinto and San Bernardino Mts., Coast Ranges, n. to Del Norte Co., Sierra Nevada; to B.C., Mont., Ariz. May–July.

7. **P. occidentàle** Scribn. Vernal form yellowish-green, 2.5–5 dm. tall, the culms leafy, ascending to spreading, pilose, with short-bearded nodes; sheaths sparsely papillose-pubescent; ligule 3–4 mm. long; blades 4–8 cm. long, 5–7 mm. wide, strigose beneath; panicle 4–7 cm. long; spikelets 1.8 mm. long, pubescent; autumnal form branching from the lower nodes.—Moist and peaty places, below 8000 ft.; many Plant Communities; scattered stations in much of Calif. June–Aug.

8. **P. shasténse** Scribn. & Merr. Vernal culms 3–5 dm. tall, pilose, with short-bearded nodes; sheaths papillose-pilose, with spreading hairs; ligule 2–3 mm. long; blades 6–8 mm. wide, sparsely pilose above, pilose beneath; panicle 6–8 dm. long; spikelets 2.5 mm. long; autumnal phase spreading with geniculate nodes and elongate arched internodes.—Meadows, Castle Crag, Shasta Co.

9. **P. arizónicum** Scribn. & Merr. Annual, branched from base, 2–5 dm. tall; culms glabrous except below the panicle and sometimes at the nodes; sheaths glabrous to pubescent; blades 5–15 cm. long, 6–12 mm. wide; glabrous or papillose-hispid beneath, ciliate near base; panicle 7–20 cm. long, the branches loosely fld., pubescent and hirsute; spikelets 4 mm. long, obovate-elliptic, hirsute to glabrous.—Reported long ago from Jamacha, e. San Diego Co.; Ariz. to Tex., Mex. July.

10. **P. dichotomiflòrum** Michx. Much-branched annual from a geniculate base, mostly glabrous except for the ring of white hairs at the ligule and the sparse pilosity on upper surface of blades; culms 5–10 dm. long; blades 1–5 dm. long, 3–20 mm. wide; panicles terminal and axillary, mostly included at base, 1–4 dm. long; spikelets 2–3 mm. long.—Moist places as a weed; Loma Linda, Fresno, Modesto; e. U.S. July–Oct.

11. **P. capillàre** L. var. **occidentàle** Rydb. [*P. barbipulvinatum* Nash.] Erect annual, 2–8 dm. tall, papillose-hispid to subglabrous, usually with short flowering branches at base; sheaths hispid; blades 1–2.5 dm. long, 5–15 mm. wide, pubescent; panicles diffuse, densely fld., often half the height of the plant, exserted; spikelets 2.5–4 mm. long, attenuate at tip, subsessile along the ultimate branchlets; fr. 1.7–1.8 mm. long.—Waste places at scattered localities, much of Calif.; to B.C.; Atlantic Coast. July–Sept.

12. **P. Hillmánii** Chase. Resembling *P. capillare*, but without short flowering branches at base, with stouter stiffer culms and panicles, and spikelets on more appressed pedicels; sterile palea more developed; fr. 2 mm. long, darker, with a prominent lunate basal scar.—Reported from Calif.; to Kans., Tex.

13. **P. hirticaùle** Presl. Erect, 2–6 dm. tall, papillose-hispid especially on the sheaths; blades 5–15 cm. long, 4–12 mm. wide; panicle 7–15 cm. long; spikelets ca. 3 mm. long, red-brown; 1st glume ½–¾ as long, acuminate; 2d glume and sterile lemma acuminate.—Open sandy places; Creosote Bush Scrub; Colo. Desert; to Tex., S.A. July–Oct.

14. **P. miliàceum** L. BROOM-CORN MILLET. Culms stout, erect, 2–10 dm. tall; sheaths papillose-hispid; lvs. 1–2.5 dm. long, 1–2.5 cm. wide; panicles dense, drooping at maturity, 1–3 dm. long, rather compact, the branches many, very scabrous, bearing spikelets near their ends; spikelets 4.5–5 mm. long, ovoid, acuminate, many-nerved; fr. 3 mm. long, straw color to red-brown; $n = 18$ (Avdulov, 1931).—Occasional escape from cult.; introd. from Old World. Aug.–Oct.

87. Echinóchloa Beauv.

Coarse annual (in ours) grasses with compressed sheaths, long flat blades and terminal panicles of stout short densely fld. 1-sided racemes. Spikelets 1-fld., sometimes with a ♂ fl. below the terminal perfect one, almost sessile. Glumes unequal, spiny-hispid, mucronate. Sterile lemma similar and awned from apex or mucronate only, inclosing a hyaline palea. Fertile lemma and palea chartaceous, acuminate; margins of the glume inrolled at summit, where the palea is not included. Ca. 15 spp. of warm regions. (Greek, *echinos*, hedgehog, and *chloa*, grass, referring to echinate spikelets.)

Awns usually, at least in part, longer than the spikelets .... 1. *E. crusgalli*
Awns shorter than the spikelets or wanting.
    Spikelets 4 mm. long; branches compound .... 1. *E. crusgalli* var. *zelayensis*
    Spikelets 3 mm. long; spikelets simple .... 2. *E. colonum*

1. **E. crusgálli** (L.) Beauv. [*Panicum c.* L.] BARNYARD GRASS. Culms stout, branching from the base, ascending or erect, 3–10 dm. tall; sheaths and blades glabrous, the latter 5–15 mm. wide; panicle dense, 1–3 dm. long, of many erect or spreading racemes, purple to green, erect or drooping; spikelets 3 mm. long, densely crowded in 3 or 4 rows; nerves strongly tuberculate-hispid; awns usually exceeding spikelet; $n = 18$ (Brown, 1948).—Natur. in waste places and damp cult. ground; introd. from Old World. July–Oct.

Var. **zelayénsis** (HBK.) Hitchc. [*Oplismenus z.* HBK.] Spikelets 4 mm. long, acuminate or short-awned, papillose, less strongly hispid.—Imperial V.; to Mex. Var. *frumentàcea* (Roxb.) Wight. [*Panicum f.* Roxb.] with awnless more turgid spikelets and hispid nerves, not or scarcely tuberculate.—Reported from cent. Calif.

2. **E. colònum** (L.) Link. [*Panicum c.* L.] JUNGLE RICE. Prostrate to erect, 2–4 dm. long; sheaths and blades smooth; blades 3–6 mm. wide; panicle 5–15 cm. long, the racemes several, 1–2 cm. long; spikelets 3 mm. long, crowded, subsessile; glumes and sterile lemma short-pointed, rather soft, faintly nerved, the nerves weakly hispid-scabrous; $n = 36$ (Janaki Ammal, 1945).—Moist places as a weed in s. Calif.; native of Old World. July–Sept.

88. Setària Beauv. BRISTLY FOXTAIL

Annual or perennial grasses with flat lvs. and cylindrical spikelike or looser panicles. Spikelets as in *Panicum,* but surrounded by few or many persistent awnlike branches which spring from the rachis below the articulation of the spikelets. Ca. 65 spp. of warm and trop. regions; some of use for forage. (Latin, *seta,* bristle.)

A. Bristles below each spikelet numerous, at least more than 5.
  B. Plants annual; spikelets 3 mm. long; lower floret ♂, the palea well developed .. 1. *S. glauca*
  BB. Plants perennial; spikelets 2–3 mm. long.
    C. Panicle 4–9 cm. long, the bristles yellow or purple, 3–18 mm. long; fr. not rugose
                                                 2. *S. geniculata*
    CC. Panicle 8–15 cm. long, the bristles orange to purple, 3–6 mm. long; fr. finely rugose
                                                 3. *S. sphacelata*

AA. Bristles below each spikelet 1 or 3.
  B.   The bristles ± retrorsely scabrous.
    C. Panicles cylindric; spikelets 2 mm. long, green .... **4. *S. verticillata***
    CC. Panicles looser; spikelets larger, brown when mature .... **5. *S. Carnei***
  BB. The bristles not retrorsely scabrous .... **6. *S. viridis***

1. **S. glaùca** (L.) Beauv. [*Panicum g.* L. *S. lutescens* (Weigel) Hubb. *Chaetochloa l.* Stuntz.] Annual with culms branching at base, compressed, spreading or erect, 3–6 dm. tall; lvs. flat, with a spiral twist, glaucous, 3–10 mm. wide; panicle 2–8 cm. long, dense, ca. 1 cm. thick; bristles 3–8 mm. long, upwardly scabrous; spikelets 3 mm. long; 1st glume half, 2d ⅔ as long as the striate undulate-rugose fertile lemma; *n* = 18 (Avdulov, 1931).—Weed in fields, waste places, etc.; native of Old World. June–Oct.

2. **S. geniculata** (Lam.) Beauv. [*Panicum g.* Lam. *Chaetochloa g.* Millsp. & Chase.] Perennial, ± cespitose, 3–10 dm. high, the culms slender, compressed, often geniculate at base; sheaths overlapping; blades 1–3 dm. long, 3–7 mm. wide; panicle 4–9 cm. long, 4–8 mm. thick; bristles 5–10, yellowish, 5–10 mm. long, upwardly scabrous; spikelets 2 mm. long; fr. undulate-rugose; *n* = 36 (Kishimoto, 1938).—Dry or moist open places, at low elevs.; V. Grassland, Coastal Sage Scrub, etc.; Cent. V., s. Calif.; to Mass., Fla., Argentina. May–Sept.

3. **S. sphacelàta** (Schumacher.) Stapf & C. E. Hubb. [*Panicum s.* Schumacher.] Tufted perennial, the culms 5–15 dm. tall, flattened; blades 4–10 mm. wide; panicle dense, 8–15 cm. long; bristles 5 or more, 3–6 mm. long.—Escape in wet places, Kern, Stanislaus, and Butte cos.; introd. from Afr.

4. **S. verticillàta** (L.) Beauv. [*Panicum v.* L. *Chaetochloa v.* Scribn.] Annual, 3–6 dm. tall, tufted; lvs. scabrous, 5–10 mm. wide; panicle erect but not stiff, ± cylindric, 5–15 cm. long, 7–15 mm. thick; bristles solitary, downwardly barbed, 3–6 mm. long; spikelets 2–2.5 mm. long; 1st glume ⅓, 2d as long as sterile lemma; fertile lemma obscurely transverse-rugose; *n* = 9 (Krishnaswamy, 1935), 18 (Avdulov, 1931).—Occasional weed, cismontane s. Calif., San Joaquin Co.; introd. from Eu. May–July.

5. **S. Cárnei** Hitchc. Like the preceding sp., but with looser panicles and larger spikelets, brown at maturity.—Weed in vineyards, Fresno Co.; introd. from Australia. July–Sept.

6. **S. víridis** (L.) Beauv. [*Panicum v.* L. *Chaetochloa v.* Scribn.] Annual, branched at base, 1–4 dm. high; lvs. not twisted, scabrous on margins, to ca. 1 cm. wide; panicles 2–5 cm. long, green or purple, densely fld.; bristles 7–12 mm. long; spikelets 2–2.5 mm. long; *n* = 9 (Tateoka, 1954).—Occasional weed; introd. from Eu. June–Aug.

89. Pennisètum Rich.

Annual or perennial grasses, often branched, usually with flat blades and dense spike-like panicles. Spikelets solitary or in groups of 2 or 3, surrounded by an invol. of bristles (sterile branchlets), these not united except at very base, often plumose, deciduous with the spikelets. First glume shorter than the spikelet, sometimes obsolete, 2d glume shorter than or equaling the sterile lemma. Fertile lemma chartaceous, smooth, thin-margined, enclosing the palea. An Old World genus, some spp. like *P. glaucum* (pearl millet) grown for food, others for ornament. (Latin, *penna*, feather, and *seta*, bristle, because of the plumose bristles of some spp.)

Culms extensively creeping; spikelets few, hidden in the upper sheath .... 1. *P. clandestinum*
Culms not creeping; panicle exserted.
  Panicle oval, tawny, 3–10 cm. long .... 2. *P. villosum*
  Panicle elongate, purple or rosy, 15–35 cm. long .... 3. *P. setaceum*

1. **P. clandestìnum** Hochst. ex Chiov. KIKUYU GRASS. Low-growing stoloniferous and rhizomatous perennial, the stolons with short internodes; sheaths broad, inflated; blades narrow; infl. of 2–4 spikelets almost entirely enclosed in the upper sheath of the short culm.—A dangerous weed in orchards and gardens; reported from San Diego and Orange cos.; introd. from Afr.

2. **P. villòsum** R. Br. FEATHERTOP. Tufted perennial, 3–6 dm. tall, pubescent below the panicle; blades 3–5 mm. wide; panicle dense, feathery; spikelets 1–4 in a fascicle; fascicles short-peduncled, with a tuft of white hairs at base of peduncle; bristles many, spreading, the inner very plumose, the longer 4–5 cm. long; 2*n* = 45 (Avdulov, 1931).

—Cult. for ornament and escaping in sandy places, s. Calif.; introd. from Afr. June–Aug.

3. **P. setàceum** (Forsk.) Chiov. [*Phalaris s.* Forsk. *Pennisetum Ruppelii* Steud.] FOUNTAIN GRASS. Tufted perennial, ca. 1 m. tall; blades scabrous, numerous, mostly 2–3 mm. wide; panicle 1.5–3.5 dm. long, mostly pink or purple, the fascicles peduncled, rather loosely arranged, with 1–3 spikelets; bristles plumose toward base, unequal, the longer 3–4 cm. long; $2n = 27$ (Avdulov, 1931).—Escaping from cult. and establishing itself at scattered localities from San Diego to Ventura cos., Alameda Co.; introd. from Afr. July–Oct.

90. Cénchrus L. SANDBUR. BURGRASS

Ours low branching annuals with flat blades and simple racemes of spiny burs terminating the culm and branches. Spikelets 1-fld., few in a cluster, acuminate, subtended by a short-pedicelled ovoid or globular invol. of rigid connate spines which is deciduous at maturity. Glumes shorter than the lemma. Ca. 25 spp. of warmer regions. (Greek, *kegchros*, a kind of millet.)

Invol. with a ring of slender bristles at base; spikelets usually 4 in each bur .... 1. *C. echinatus*
Invol. with no ring of slender bristles at base; spikelets usually 2 in each bur .... 2. *C. pauciflorus*

1. **C. echinàtus** L. Culms compressed, usually geniculate, 2.5–6 dm. long; blades 3–8 mm. wide, pilose on upper surface near base; raceme 3–10 cm. long; burs 4–7 mm. long, equally broad or broader, pubescent, the lobes of the invol. bent inward but not interlocking; $n = 17$ (Avdulov, 1931).—Occasional weed as in San Diego Co.; to S. Car. and S. Am. Oct.

2. **C. pauciflòrus** Benth. Sometimes forming large mats, the spreading culms 2–9 dm. long, rather stout; blades 2–7 mm. wide; raceme usually 3–8 cm. long, the burs somewhat crowded, mostly 4–6 mm. long and wide, pubescent; spines many, spreading or reflexed, flat, some of the upper 4–5 mm. long, usually villous at base; $n = 18$ (Brown, 1948).—Sandy places as a weed; Cent. V., s. Calif.; to Ore., Atlantic Coast, S. Am. July–Sept.

91. Imperàta Cyrill.

Erect rather coarse perennials with leafy stems and narrow silky terminal panicles. Spikelets all alike, paired, awnless, unequally pedicellate on a slender continuous rachis, surrounded by long silky hairs. Glumes subequal, membranaceous. Sterile lemma, fertile lemma and palea thin and hyaline. Ca. 5 spp. of warm regions. (F. *Imperato*, Italian naturalist.)

1. **I. brevifòlia** Vasey. [*I. Hookeri* Rupr.] SATINTAIL. Culms 1–1.5 m. tall from scaly rhizomes; ligule villous; lvs. 1–5 dm. long, 8–15 mm. wide, glabrous; uppermost lvs. reduced; panicle spicate, 1.5–2.5 dm. long, somewhat tawny or pinkish, soft-silky; spikelets 3 mm. long, with hairs twice as long.—Rare, moist places; Chaparral, Coastal Sage Scrub, Creosote Bush Scrub, etc.; scattered locations mostly in s. Calif.; to Utah, Mex. Sept.–May.

Erianthus ravénnae (L.) Beauv. RAVENNA GRASS. Culms stout, to 4 m. high; panicle to 6 dm. long, ± silvery; spikelets awnless or nearly so.—Reported from Imperial Co.; native of Eu.

92. Andropògon L. BEARD GRASS

Rather coarse grasses (ours perennials) with solid leafy stems. Infl. a panicle or compound infl. of racemes with numerous spikelets. Spikelets in pairs at each node of the jointed rachis, 1 sessile and perfect, the other pedicellate and ♂ or much reduced. Glumes of fertile spikelet narrow, awnless, several-nerved, flat or concave. Sterile lemmas hyaline, empty. Fertile lemma hyaline, usually with a bent and twisted awn. Palea hyaline, minute or wanting. Ca. 150 spp. of warmer regions. (Greek, *aner* (andr-) man, and *pogon*, beard, because of the hairy pedicels of the ♂ or sterile spikelets.)

Racemes solitary on each peduncle; rachis-joints oblique and hollow at summit ...... 1. *A. cirratus*
Racemes 2–many on each peduncle.
    The racemes 2–several on each peduncle, digitate; joints of rachis slender, sometimes with a shallow groove on 1 side.
        Infl. decompound, the profuse pairs of racemes aggregate in an elongate or corymbose mass; spathes rarely more than 2 mm. wide .... 2. *A. glomeratus*
        Infl. not decompound or dense, the 2–4 racemes well separated from the other groups; spathes much inflated .... 3. *A. virginicus*
    The racemes several to many; joints of rachis flat, the margins thick and ciliate, the center very thin
                                                                    4. *A. barbinodis*

1. **A. cirràtus** Hack. Plants pale, glaucous to purplish, 4–7 dm. high, slender-stemmed; blades flat, 1–4 mm. wide, mostly scabrous; raceme exserted, 3–6 cm. long; sessile spikelet 8–9 mm. long, the awn 5–10 mm. long; pedicellate spikelet scarcely reduced, awnless, the pedicel stiffly ciliate on one side near the top; $n = 10$ (Brown, 1950).—Reported from Jamacha, e. San Diego Co.; Ariz. to Tex., n. Mex.

2. **A. glomeràtus** (Walt.) BSP. [*Cinna g.* Walt.] Tufted, leafy, 5–10 dm. tall, the stems corymbosely branched above; sheaths and blades scabrous; ligules villous; blades 1–5 dm. long, 3–6 mm. wide; infl. compound, 1–3 dm. long, the racemes paired; sessile spikelet 5 mm. long; awn 1–2 cm. long, not geniculate; $n = 10$ (Brown, 1950).—Occasional in wet places; Coastal Sage Scrub, Chaparral, Creosote Bush Scrub, etc.; s. Calif., introd. in cent. Calif.; to Ky., W.I., Cent. Am. Sept.–March.

3. **A. virgínicus** L. Culms 5–10 dm. tall, branched above; lower sheaths equitant, compressed, keeled; blades 2–5 mm. wide, flat or folded; infl. elongate, the racemes in 2's or 4's, partly included and shorter than the brownish spathes; sessile spikelet ca. 3 mm. long; awn 1–2 cm. long; $n = 10$ (Church, 1936).—Occasional weed in waste places; Yuba, Sonoma, Butte cos.; to Atlantic Coast, Cent. Am. Sept.–Jan.

4. **A. barbinòdis** Lag. [*A. saccharoides* Calif. refs.] Tufted, 6–13 dm. high, often branching below, the nodes glabrous to hispid; blades commonly glaucous, subglabrous, 3–6 mm. wide; panicle long-exserted, silvery white, silky, dense, oblong, 7–15 cm. long; racemes 2–4 cm. long; rachis-joints and pedicels silky; spikelets 4 mm. long, the awn geniculate, twisted below, 1–1.5 cm. long; pedicellate spikelet reduced; $n = 90$ (Gould, 1953).—Dry slopes below 4000 ft.; Coastal Sage Scrub, Joshua Tree Wd., etc.; s. Calif.; to Okla., Tex., Mex. Feb.–Sept.

93. Sórghum Moench.

Tallish annuals or perennials, with flat blades and terminal panicles of 1- to 5-jointed tardily disarticulating racemes. Spikelets in pairs, 1 sessile and fertile, the other pedicellate, sterile but well developed, usually ♂. Ca. 20 spp., mostly of Old World. (*Sorgho*, the Italian name.)

Plants perennial; panicle openly branched .... 1. *S. halepense*
Plants annual; panicle compactly branched .... 2. *S. bicolor*

1. **S. halepénse** (L.) Pers. [*Holcus h.* L.] JOHNSON GRASS. Rhizomes heavy, scaly, extensively creeping; stems erect, glabrous, coarse, 5–15 dm. tall; ligules ciliate; blades 1–5 dm. long, 1–2 cm. wide; panicles 1–3 dm. long, purplish; fertile spikelets 5 mm. long, with pubescent glumes; awn 1 cm. long, deciduous; $n = 10, 20$ (Janaki Ammal, 1945).—Common in low wet and waste places, especially in s. Calif. and in interior valleys, less so near the cent. and n. coast; native of Old World. May–Aug.

2. **S. bícolor** (L.) Moench. [*S. vulgare* Pers.] SORGHUM. Annual, more robust; panicle more compact; $n = 10$ (Kuwada, 1915).—Occasional escape from cult. Known in many forms in agriculture.

ABBREVIATIONS OF AUTHORS' NAMES

Compiled by David D. Keck

Abrams.   Abrams, LeRoy, 1874–1956, professor of botany, Stanford; author of "Flora of Los Angeles and Vicinity" (1917), "Illustrated Flora of the Pacific States" (vol. I, 1923, vol. II, 1944, vol. III, 1947, vol. IV, in preparation).

Achey.   Achey, Daisy Bird (*later* Marshall), 1906——, Pomona College.

Adams.   Adams, Joseph Edison, 1903——, professor of botany, North Carolina; *Arctostaphylos.*

Adans.   Adanson, Michel, 1727–1806, France; author of "Familles des Plantes" (1763); originated some 1600 generic names.

Aellen.   Aellen, Paul, 1896——, professor in Basel; student of Chenopodiaceae.

Ag.   Agardh, Carl Adolf, Bishop of Carlstadt, 1785–1859, professor in Lund; noted algologist.

Ag., J. G.   Agardh, Jacob Georg, the son, 1813–1901, professor of botany, Lund; algologist and student of *Lupinus.*

Ait.   Aiton, William, 1731–1793, famous gardener at Kew; issued "Hortus Kewensis" (1789).

Ait. f.   Aiton, William Townsend, the son, 1766–1849, director of Kew, 1793–1841; issued "Hortus Kewensis," ed. 2 (1810–1813).

Alef.   Alefeld, Friedrich Georg Christoph, 1820–1872, Germany.

All.   Allioni, Carlo, 1725–1804, professor of botany, Turin; author of "Flora Pedemontana" (1785).

Ames.   Ames, Oakes, 1874–1950, director of the Botanical Museum, Harvard; orchidologist.

Anderss.   Andersson, Nils Johan, 1821–1880, director of the Botanical Museum, Stockholm; collected in California in 1852; *Salix, Andropogon.*

Andr.   Andrews, Henry C., English botanical artist and engraver of the early 19th century; monographer of *Geranium, Erica,* and *Rosa.*

André.   André, Édouard François, 1840–1911, French explorer and collector in South America; first editor of L'Illustration Horticole, later editor-in-chief of Revue Horticole.

Andrz.   Andrzejowski, Antoni Lukianowicz, 1784–1868, professor of botany, Vilna.

Anon.   Anonymous authority.

Antoine.   Antoine, Franz, 1815–1886, director of the royal gardens at Schönbrunn, Vienna; Coniferae.

Appleg.   Applegate, Elmer Ivan, 1867–1949, student of the Oregon flora; *Erythronium.*

Arcang.   Arcangeli, Giovanni, 1840–1921, Italy.

Ard.   Arduino, Pietro, 1728–1805, Padua, Italy.

Arech.   Arechavaleta, José, 1838–1912, Spanish-born Uruguayan botanist.

Arènes.   Arènes, Jean, contemporary French botanist.

Armstr. & Thornb.   Armstrong, Margaret (Neilson), 1867——, New York authoress, and John James Thornber, 1872——, professor of botany, Arizona.

Arn.   Arnott, George Arnold Walker, 1799–1868, Edinburgh and Glasgow botanist.

Arthur.   Arthur, Joseph Charles, 1850–1942, authority on plant rusts, Purdue; editor of Botanical Gazette.

Arv.-Touv.   Arvet-Touvet, Jean Maurice Casimir, 1841–1913, France.

Asch.   Ascherson, Paul Friedrich August, 1834–1913, professor of botany, Berlin.

Asch. & Graebn.   Ascherson, P. F. A., and K. O. R. P. P. Graebner, authors of "Synopsis der Mitteleuropaischen Flora" (1896–1917).

Asch. & Magnus.   Ascherson, P. F. A., and Paul Wilhelm Magnus, 1844–1914, professor of botany, Berlin.

Ashe.   Ashe, William Willard, 1872–1932, botanist and forester, U.S. Forest Service.

Aubl.   Aublet, Jean Baptiste Christophe Fusée, 1720–1778; French botanical collector and author of a French Guiana flora.

Aud.   Audubon, John James Laforest, 1780–1851, eminent American ornithologist and artist.

Auth.   Authors, referring to usage by various writers.

Avé-Lall.   Avé-Lallemant, Julius Leopold Eduard, 1803–1867, German associate of the botanic garden in St. Petersburg.

Bab.   Babington, Charles Cardale, 1808–1895, professor of botany, Cambridge, England.

Babc.   Babcock, Ernest Brown, 1877–1954, professor of genetics, California; monographer of *Crepis.*

Babc. & Hall.   Babcock, E. B., and H. M. Hall.

Babc. & Steb.   Babcock E. B., and G. L. Stebbins, Jr.

Bacig.   Bacigalupi, Rimo Charles, 1901——, curator of the Jepson Herbarium; Saxifragaceae.

Bailey.   Bailey, Liberty Hyde, 1858–1954, Cornell University; eminent horticulturist and author, "Standard Cyclopedia of Horticulture" (1914–1917), "Manual of Cultivated Plants" (1925), "Hortus" (1930); founder of Gentes Herbarum; student of *Carex, Rubus,* palms.

Bailey, V.   Bailey, Virginia Edith (Long), 1908——, assistant to W. L. Jepson.

Baill.   Baillon, Henri Ernest, 1827–1895, Paris physician; author of "Histoire des Plantes" (1867–1895) and many other works.

Baker.   Baker, John Gilbert, 1834–1920, keeper of the herbarium, Kew, 1890–1899; student of ferns, Amaryllidaceae, Bromeliaceae, Iridaceae, Liliaceae, Compositae, and the flora of tropical Africa.

Baker, E. G.   Baker, Edmund Gilbert, the son, 1864——, British Museum, London.
Baker, M. S.   Baker, Milo Samuel, 1868——, professor in Santa Rosa Junior College; authority on *Viola*.
Baker & Clausen.   Baker, M. S., and J. C. Clausen.
Balbis.   Balbis, Giovanni Battista, 1765–1831, professor of botany, Turin.
Balf.   Balfour, John Hutton, 1808–1884, professor of botany and director of the botanic garden, Edinburgh.
Ball.   Ball, Carleton Roy, 1873–1958, American authority on *Salix*.
Ball & Bracelin.   Ball, C. R., and Nina Floy (Burfield) Bracelin, 1890——, assistant in the herbarium, Univ. California.
Ball, J.   Ball, John, 1818–1889, Irish traveler in both North and South America.
Banks.   Banks, Sir Joseph, 1743–1820, accompanied Capt. Cook in his first voyage of circumnavigation in 1768; early director of Kew; president of the Royal Society.
Barb.   Barbey, William, 1842–1914, botanist of Geneva; *Epilobium*.
Barkl.   Barkley, Fred Alexander, 1908——, Warner-Chilcott Laboratories, New Jersey; Anacardiaceae.
Barneby.   Barneby, Rupert Charles, 1911——, student of the Great Basin flora; *Astragalus*.
Barnh.   Barnhart, John Hendley, 1871–1949, bibliographer, New York Botanical Garden; Lentibulariaceae.
Barr.   Barratt, Joseph, 1796–1882, Connecticut; geologist.
Bartl.   Bartling, Friedrich Gottlieb, 1798–1875, professor of botany, Göttingen.
Bartlett.   Bartlett, Harley Harris, 1886——, professor of botany, Michigan.
Barton.   Barton, William Paul Crillon, 1786–1856, professor of botany, Pennsylvania; author of "A Flora of North America" (1821–1823).
Batch.   Batchelder, Frederick William, 1838–1911, New Hampshire amateur botanist and ornithologist.
Batsch.   Batsch, August Johann Georg Karl, 1761–1802, German horticultural writer.
Bauh.   Bauhin, Caspar, 1560–1624, professor of medicine, Basel; author of the Pinax; maker of the first distinct diagnoses of species.
Baumann.   Baumann, Émile Napoléon, 1835–1910, Alsace; horticulturist.
Baxter.   Baxter, Edgar Martin, author of "California Cactus" (1935).
Bdg.   Brandegee, Townshend Stith, 1843–1925, civil engineer, and his wife gave their valuable herbarium and library to the University of California in 1906.
Bdg., K.   Brandegee, Mary Katherine (Layne) (Curran), 1844–1920, physician and botanist, California.
Beal.   Beal, William James, 1833–1924, Michigan agrostologist.
Beane.   Beane, Lawrence, 1901——, Fresno, Calif., musician.
Beane & Vollmer.   Beane, L., and Albert Michael Vollmer, retired San Francisco physician.
Beauv.   Palisot de Beauvois, Baron Ambroise Marie François Joseph, 1752–1820, French naturalist; Gramineae.
Beauverd.   Beauverd, Gustave, 1867–1942, Herbier Boissier, Geneva.
Bebb.   Bebb, Michael Schuck, 1833–1895, Illinois; *Salix*.
Beck, G.   Beck-Mannagetta, Günther (Günther Ritter Beck von Mannagetta und Larchenau), 1856–1931, professor of botany, Prague; author of "Flora von NiederÖsterreich" (1890–1893); Orobanchaceae.
Beetle.   Beetle, Alan Ackerman, 1913——, professor of agronomy, Wyoming; Gramineae, Cyperaceae.
Beetle & Tofsrud.   Beetle, A. A., and Robert B. Tofsrud, California (Davis).
Beetle, D. E.   Beetle, Dorothy Erna (Schoof), 1916——, University of California; *Fritillaria*.
Bég. & Bel.   Béguinot, Augusto, 1875–1940, professor of botany, Genoa, and Nicola Belosersky.
Behr.   Behr, Hans Hermann, 1818–1904, emtomologist and botanist of California Academy of Sciences.
Behr & Kell.   Behr, H. H., and A. Kellogg.
Beissn.   Beissner, Ludwig, 1843–1927, Germany.
Bell.   Bell, Clyde Ritchie, 1921——, professor of botany, North Carolina.
Benn.   Bennett, Arthur, 1943–1939, English builder and amateur botanist.
Benson, G. T.   Benson, Gilbert Thereon, 1896–1928, librarian, Dudley Herbarium; author of "The Trees and Shrubs of Western Oregon" (1930).
Benson, L.   Benson, Lyman David, 1909——, professor of botany, Pomona College; author (with Robert A. Darrow) of "A Manual of Southwestern Desert Trees and Shrubs" (1945), "Plant Classification" (1957); *Ranunculus*, Cactaceae.
Benth.   Bentham, George, 1800–1884, long-time president of the Linnaean Society; outstanding English taxonomist; author of "Plantas Hartwegianas . . . enumerat" (1839); "Handbook of the British Flora" (1858, with later editions); "Flora Hongkongensis" (1861); "Flora Australiensis" (1863–1878); and monographic works on Leguminosae, Labiatae, Scrophulariaceae, etc.
Benth. & Hook.   Bentham, G., and Sir J. D. Hooker, authors of "Genera Plantarum" (1862–1883).
Benth. & Muell.   Bentham, G., and F. J. H. von Mueller.
Berckmans.   Berckmans, Prosper Jules Alphonse, 1830–1910, Belgian-born horticulturist and entomologist of Georgia.
Berg.   Bergius, Peter Jonas, 1730–1790, Swedish botanist.
Berger.   Berger, Alwin, 1871–1931, curator of the garden at La Mortola, Italy; student of Cactaceae and other succulents.
Berl.   Berlandier, Jean Louis, 1805–1851, Switzerland; first collector in Texas; Grossulariaceae.
Bernh.   Bernhardi, Johann Jacob, 1774–1850, professor of botany, Erfurt.
Bertol.   Bertoloni, Antonio, 1775–1869, professor of botany, Bologna; author of "Flora Italica" (1833–54).

Bess.   Besser, Wilibald Swibert Joseph Gottlieb von, 1784–1842, Austria and Poland; professor in the Wolhynien Lyceum (Poland); student of the flora of Galicia and southwest Russia.
Bessey.   Bessey, Charles Edwin, 1845–1915, professor of botany, Nebraska; author of a system of phylogeny.
Beyr.   Beyrich, Heinrich Carl, 1796–1834, Prussian botanist who collected in Georgia, South Carolina, and Texas.
Bickn.   Bicknell, Eugene Pintard, 1859–1925, New York banker and amateur botanist.
Bieb.   Marschall von Bieberstein, Baron Friedrich August, 1768–1826, Stuttgart; author of works on the flora of southern Russia and the Caucasus.
Bigel.   Bigelow, Jacob, 1787–1879, professor of botany, Boston.
Bigel. & Engelm.   Bigelow, John Milton, 1804–1878, surgeon attached to the Mexican Boundary Commission and the Pacific Railroad Survey under Lt. Whipple; and G. Engelmann.
Bioletti.   Bioletti, Frederic Theodore, 1865–1939, professor of viticulture, California.
Bisch.   Bischoff, Gottlieb Wilhelm, 1797–1854, professor of botany, Heidelberg.
Bitter.   Bitter, Friedrich August Georg, 1873–1927, professor in Göttingen; Ophioglossaceae.
Biv.   Bivona-Bernardi, Baron Antonio, 1778–1837, Messina, Sicily.
Blake.   Blake, Sidney Fay, 1892——, U.S.D.A., Beltsville; authority on Compositae.
Blanch.   Blanchard, William Henry, 1850–1922, New England teacher.
Blank.   Blankinship, Joseph William, 1862–1938, plant pathologist, Montana.
Blasd.   Blasdale, Walter Charles, 1871——, professor of chemistry, California.
Blume.   Blume, Carl Ludwig von, 1796–1862, German director of the botanic garden at Batavia and writer on the flora of Java.
Boeck.   Boeckeler, Johann Otto, 1803–1899, apothecary-botanist of Oldenburg; Cyperaceae.
Boenn.   Boenninghausen, Clemens Maria Friedrich von, 1785–1864, German botanist.
Bogin.   Bogin, Clifford, 1920——, New York Botanical Garden; Sagittaria.
Boiss.   Boissier, Edmond Pierre, 1810–1885, Geneva; one of the outstanding systematists of the 19th century; author of the monumental "Flora Orientalis" (1867–1884); monographer of Plumbaginaceae and Euphorbia.
Boiss. & Reut.   Boissier, E. P., and Georges François Reuter, 1805–1872, Switzerland; conservator of the Herbarium Boissier.
Boivin.   Boivin, Joseph Robert Bernard, 1916——, Department of Agriculture, Ottawa; Thalictrum.
Bol.   Bolander, Henry Nicholas, 1831–1897, made extensive collections in California from 1863–1875, mostly under the State Geological Survey.
Bolus, L.   Bolus, Harriet Margaret Louisa (Kensit), 1877——, curator of the Bolus Herbarium, Kirstenbosch, South Africa.
Bong.   Bongard, August Heinrich Gustav, 1786–1839, professor of botany, St. Petersburg; monographer of Brazilian plants; describer of Mertens' Alaskan collection.
Booth.   Booth, William Beattie, 1804–1874, English gardener and horticulturist.
Boothman.   Boothman, H. Stuart, of Wisley; Aquilegia.
Boott.   Boott, Francis, 1792–1863, American caricologist and physician who settled in London.
Boott, W.   Boott, William, 1805–1887, Boston; Carex.
Borb.   Borbás, Vinczé von, 1844–1905, Hungarian botanist.
Boreau.   Boreau, Alexandre, 1803–1875, director of the botanic garden, Angers.
Borkh.   Borkhausen, Moritz Balthasar, 1760–1806, Germany.
Bory.   Bory de Saint-Vincent, Jean Baptiste George Marcellin, Baron de, 1778–1846, French traveler and naturalist.
Borzi.   Borzi, Antonino, 1852–1921, director of the botanic garden, Palermo.
Bosse.   Bosse, Julius Friedrich Wilhelm, 1788–1864, Oldenburg.
Bowden.   Bowden, Wray Merrill, 1914——, Department of Agriculture, Canada; cytogenetics.
Boyle.   Boyle, William Sidney, 1915——, professor of botany, Utah State Agricultural College; Melica.
Br., A.   Braun, Alexander Carl Heinrich, 1805–1877, professor of botany and director of the botanical garden, Berlin; Characeae, Selaginella.
Br., A. and Bouché.   Braun, A. C. H., and Carl David Bouché, 1809–1881, inspector in Berlin botanical garden.
Br., N. E.   Brown, Nicholas Edward, 1849–1934, Kew; authority on the South African flora.
Br., P.   Browne, Patrick, 1720?–1790, Irish physician who wrote on the plants of Jamaica.
Br., R.   Brown, Robert, 1773–1858, librarian and first keeper of botany, British Museum; a chief exponent of the Natural System; noted morphologist and cytologist.
Brack.   Brackenridge, William Dunlop, 1810–1893, Scotsman; member of the U.S. exploring expedition under Captain Wilkes.
Brads.   Bradshaw, Robert Vernon, 1896——, Stanford; Lathyrus.
Brainerd.   Brainerd, Ezra, 1844–1924, president, Middlebury College; Viola.
Brand.   Brand, August, 1863–1930, German student of Polemoniaceae, Hydrophyllaceae and Boraginaceae.
Brébiss.   Brébisson, Louis Alphonse de, 1798–1872, French author of a flora of Normandy.
Brenckle.   Brenckle, Jacob Frederic, 1875——, South Dakota physician.
Brew.   Brewer, William Henry, 1828–1910, geologist and botanist, leader of the field parties in the California State Geological Survey.
Brew. & Wats.   Brewer, W. H., and S. Watson, authors of "Botany of California" (1876–1880).
Brign.   Brignoli di Brunnhoff, Giovanni de, 1774–1857, professor of botany, Modena.
Briq.   Briquet, John Isaac, 1870–1931, director of the Conservatoire botanique, Geneva; Labiatae, Umbelliferae, Compositae; noted for his work to advance modern nomenclature through the Botanical Congresses.
Britt. & Br.   Britton, N. L., and Addison Brown, 1830–1913, New York amateur botanist and patron of science; authors of "Illustrated Flora of the Northern States and Canada" (1913).

Britt. & Rend.   Britten, J., and A. B. Rendle.
Britt. & Rose   Britton, N. L., and J. N. Rose, authors of an elaborate monograph of the Cactaceae
(1919–1923).
Britt. & Rusby.   Britton, N. L., and H. H. Rusby.
Britten.   Britten, James, 1846–1924, botanist, British Museum; editor of the Journal of Botany,
British and Foreign for nearly 45 years.
Britton.   Britton, Nathaniel Lord, 1859–1934, director-in-chief, New York Botanical Garden, 1896–
1930; flora of North America, West Indies, Bolivia.
Brongn.   Brongniart, Adolphe Théodore, 1801–1876, noted paleobotanist and systematist of Paris.
Brot.   Brotero, Felix da Silva Avellar, 1744–1828, professor of botany, Coimbra, Portugal; author
of "Flora Lusitanica" (1804).
Broun.   Broun, Maurice, 1906–——, Massachusetts ornithologist.
Brouss.   Broussonet, Pierre Marie Auguste, 1761–1807, botanist of Montpellier.
Brown, S. W.   Brown, Spencer Wharton, 1918–——, professor of genetics, California.
Brühl.   Brühl, Paul Johannes, 1855–——, professor of botany, Calcutta.
Brumhard.   Brumhard, Philipp, 1879–——, Germany.
BSP.   Britton, N. L.; Emerson Ellick Sterns, 1846–1926; and Justus Ferdinand Poggenburg 1840–
1893.
Bubani.   Bubani, Pietro, 1806–1888, Italy; author of "Flora Pyrenaea" (1897–1901).
Buch.   Buchenau, Franz Georg Philipp, 1831–1906, professor in Bremen; Alismataceae, Juncaceae.
Buch. & Fern.   Buchenau, F. G. P., and M. L. Fernald.
Buchh.   Buchholz, John Theodore, 1888–1951, professor of botany, Illinois; authority on the embry-
ology of the Coniferae.
Buch.-Ham.   Buchanan-Hamilton, Francis, 1762–1829, Scotch surgeon in India; superintendent of
the botanic garden, Calcutta; author of "Prodromus Florae Nepalensis" (1825).
Buckl.   Buckley, Samuel Botsford, 1809–1884, state geologist of Texas; collected plants, shells, and
insects from Georgia to Texas.
Buist.   Buist, Robert, 1805–1880, Philadelphia florist and writer on horticulture.
Bunge.   Bunge, Alexander Andrejewitsch von, 1803–1890, professor of botany, Dorpat; student of
the flora of Russia and central Asia; *Astragalus*.
Burgsd.   Burgsdorf, Friedrich August Ludwig von, 1747–1802, German forester.
Burm. f.   Burman, Nikolaus Laurens, 1734–1793, Amsterdam.
Burnat.   Burnat, Émile, 1828–1920, Geneva botanist who amassed a private herbarium of 220,000
specimens.
Burnham.   Burnham, Stewart Henry, 1870–1943, New York Botanical Garden and Cornell.
Bush.   Bush, Benjamin Franklin, 1858–1937, postmaster, Independence, Missouri.
Butters.   Butters, Frederic King, 1878–1945, professor of botany, Minnesota.
Butters & Abbe.   Butters, F. K., and Ernst Cleveland Abbe, 1905–——, professor of botany, Min-
nesota.
Calloni.   Calloni, Silvio, 1851–1931, assistant in the institute of zoology, University of Pavia.
Camb.   Cambessèdes, Jacques, 1799–1863, France.
Camp.   Camp, Wendell Holmes, 1904–——, professor of botany, Connecticut; *Vaccinium*.
Campb.   Campbell, Gloria Rae (*later* Day), 1924–——, Rancho Santa Ana Botanic Garden.
Campd.   Campdera, François, 18th–19th century, France; author of "Monographie des Rumex"
(1819).
Campst., R. Br.   Brown, Robert, 1842–1895, British botanist born in Campster, hence the ab-
breviation.
Canby.   Canby, William Marriott, 1831–1904, Delaware businessman; accumulator of a large
herbarium.
Canby, M.   Canby, Margaret Leslie (*later* Ries, *later* Funai), 1904–——, Pomona College; *Core-
throgyne*.
Canby & Rose.   Canby, W. M., and J. N. Rose.
Card.   Card, Hamilton Hye, 1877–——, schoolteacher in Fillmore, Ill.
Carr.   Carrière, Élie Abel, 1818–1896, Muséum d'Histoire Naturelle, Paris; editor of Revue Horticole;
Coniferae.
Carter.   Carter, Annetta Mary, 1907–——, assistant curator, Herbarium of the University of Cali-
fornia.
Carter, W. R.   Carter, William R., author of a flora of Vancouver and Queen Charlotte Islands
(1921).
Caruel.   Caruel, Teodoro, 1830–1898, Italy.
Casp.   Caspary, Johann Xaver Robert, 1818–1887, professor of botany, Königsberg.
Cass.   Cassini, Count Alexandre Henri Gabriel de, 1781–1832, France; author of "Dictionnaire des
Sciences naturelles"; Compositae.
Cassidy.   Cassidy, James, 1844?–1889, professor of botany, Colorado Agricultural College.
Cav.   Cavanilles, Antonio José, 1745–1804, professor of botany and director of the botanic gar-
dens, Madrid; author of "Icones et descriptiones plantarum" (1791–1801).
Čelak.   Čelakovsky, Ladislav Josef, 1834–1902, professor of botany, Prague; author of "Prodromus
der Flora von Böhmen."
Cerv.   Cervantes, Vincente de, 1759?–1829, professor of botany and director of the botanic garden,
Mexico City.
Chaix.   Chaix, Dominique, Abbé, 1730–1799, student of the flora of the French Alps.
Cham.   Chamisso, Adelbert Ludwig von (*formerly* Louis Charles Adelaide Chamisso de Boncourt),
1781–1838, poet-naturalist, Berlin; botanist on the ship *Rurik*, which visited California in 1816.
Cham. & Schlecht.   Chamisso, A. von, and D. F. L. von Schlechtendal.
Chamb.   Chambers, Kenton Lee, 1929–——, Stanford and Yale.

Chandl.   Chandler, Harley Pierce, 1875–1918, California school teacher.
Chapm.   Chapman, Alvan Wentworth, 1809–1899, Florida; author of "Flora of the Southern United States" 1860, 2d ed. 1883, 3d ed. 1897).
Chase.   Chase, Mary Agnes (Merrill), 1869——, Custodian of Grasses, U.S. National Herbarium; eminent agrostologist.
Chât.   Châtelain, Jean Jacques, 18th century. France; *Corallorhiza*.
Chev.   Chevallier, François Fulgis, 1796–1840, France.
Chiov.   Chiovenda, Emilio, 1871——, director of the Botanical Institute of the University, Modena.
Choisy.   Choisy, Jacques Denys, 1799–1859, professor in Geneva; collaborator in de Candolle's Prodromus; Guttiferae, Convolvulaceae, Nyctaginaceae.
Chr., C.   Christensen, Carl Frederik Albert, 1872–1942, Danish student of ferns.
Christ.   Christ, Konrad Hermann Heinrich, 1833–1933, Swiss authority on ferns; Coniferae, *Carex*, *Rosa*; author of the classic "Pflanzenleben der Schweiz" (1879).
Church.   Church, George Lyle, 1903——, professor of botany, Brown; cytotaxonomy of grasses.
Citerne.   Citerne, Paul Emile Charles, 1857——, French student of Berberidaceae.
Ckll.   Cockerell, Theodore Dru Alison, 1866–1948, professor of zoology, Colorado.
Clairv.   Clairville, Joseph Philippe de, 1742–1830, Switzerland.
Clarke.   Clarke, Charles Baron, 1832–1906, superintendent of the Royal Botanic Gardens, Calcutta; student of the flora of India; Cyperaceae.
Clausen.   Clausen, Robert Theodore, 1911——, professor of botany, Cornell; Ophioglossaceae, *Sedum*.
Clausen & Keck.   Clausen, Jens Christian, 1891——, experimental taxonomist, Carnegie Institution, and D. D. Keck.
Clausen & Uhl.   Clausen, R. T., and Charles Harrison Uhl, 1918——, professor of botany, Cornell.
Clayt.   Clayton, John, 1685–1773, physician in Virginia; collector for Gronovius.
Clem. & Clem.   Clements, Frederic Edward, 1874–1945, plant ecologist and climatologist, Carnegie Institution, and Edith Gertrude (Schwartz) Clements, 1877——.
Clokey.   Clokey, Ira Waddell, 1878–1950, Colorado and California; accumulator of a large herbarium now at Berkeley; flora of the Charleston Mts., Nevada.
Clus.   Clusius, Carolus (Charles de l'Ecluse), 1525–1609, professor of botany, Leiden.
Clute.   Clute, Willard Nelson, 1869–1950, professor of botany, Butler; editor of the Fern Bulletin and the American Botanist.
Cockayne.   Cockayne, Leonard, 1855–1934, New Zealand forester.
Cockayne & Allan.   Cockayne, L., and Harry Howard Barton Allan, 1882——, New Zealand ecologist and taxonomist.
Cogn.   Cogniaux, Célestin Alfred, 1841–1916, Verviers, Belgium; Cucurbitaceae; Melastomaceae, Orchidaceae.
Cole.   Cole, Donald, 1925——, Claremont Graduate School; *Rosa*.
Coleman.   Coleman, Nathan, 1825–1887, Michigan.
Comm.   Commerson, Philibert, 1727–1773, France; naturalist of the Bouganville expedition, 1767–1769, South America and Africa.
Compton.   Compton, Gladys Ruth (*later* Hutchison), 1909——, Pomona College.
Congd.   Congdon, Joseph Whipple, 1834–1910, California lawyer and amateur botanist; collected particularly in Sonoma and Mariposa counties; his herbarium, mounted on oversized sheets, is at Minnesota.
Const.   Constance, Lincoln, 1909——, professor of botany, California; student of Umbelliferae and Hydrophyllaceae.
Const. & Roll.   Constance, L., and R. C. Rollins.
Cooper.   Cooper, James Graham, 1830–1902, American physician and ornithologist.
Copel.   Copeland, Edwin Bingham, 1873——, research associate in botany, California; authority on ferns.
Copel. f.   Copeland, Herbert Faulkner, the son, 1902——, professor, Sacramento Junior College; student of phylogeny, plant anatomy, Ericaceae.
Correll.   Correll, Donovan Stewart, 1908——, Texas Research Foundation, Renner.
Cory.   Cory, Victor Louis, 1880——, Texas.
Coss.   Cosson, Ernest Saint-Charles, 1819–1889, France.
Coss. & Durieu.   Cosson, E. St.-C., and M. C. Durieu de Maisonneuve, authors of "Flore d'Algérie" (1854–1867).
Coss. & Germ.   Cosson, E. St.-C., and Jacques Nicolaus Ernest Germain de Saint-Pierre, 1815–1882, Paris physician; authors of a flora of Paris (1845).
Coult.   Coulter, John Merle, 1851–1928, professor of botany, Chicago; founder and long-time editor of the Botanical Gazette; author of "Manual of the Botany of the Rocky Mountain Region" (1885).
Coult. & Evans.   Coulter, J. M., and Walter Harrison Evans, 1863–1941, U.S.D.A.
Coult. & Fish.   Coulter, J. M., and Elmon McLean Fisher, 1861——.
Coult. & Nels.   Coulter, J. M., and Aven Nelson.
Coult. & Rose.   Coulter, J. M., and J. N. Rose.
Cov.   Coville, Frederick Vernon, 1867–1937, curator, U.S. National Herbarium, 1893–1937; botanist of the Death Valley Expedition in 1891.
Cov. & Britt.   Coville, F. V., and N. L. Britton.
Cov. & Gilman.   Coville, F. V., and Marshall French Gilman, 1871–1944, California collector.
Cov. & Grant.   Coville, F. V., and A. L. Grant.
Cov. & Leib.   Coville, F. V., and J. B. Leiberg.
Cov. & Mort.   Coville, F. V., and C. V. Morton.
Cov. & Rose.   Coville, F. V., and J. N. Rose.
Covas.   Covas, Guillermo, 1915——, Argentine agrostologist and geneticist.

Covas & Steb.   Covas, G., and G. L. Stebbins, Jr.
Craig.   Craig, Thomas Theodore, 1907——, Pomona College; iris breeder; *Gilia* (*Hugelia*).
Crampt.   Crampton, Beecher, 1918——, curator, Agronomy Herbarium, Davis.
Crantz.   Crantz, Heinrich Johann Nepomuk von, 1722–1799, professor of medicine, Vienna; author of a flora of Austria; Cruciferae, Umbelliferae.
Crép.   Crépin, François, 1830–1903, director of the botanic garden, Brussels.
Cronq.   Cronquist, Arthur John, 1919——, curator, New York Botanical Garden; Compositae.
Cronq. & Keck.   Cronquist, A. J., and D. D. Keck.
Crum.   Crum, Ethel Katherine, 1886–1943, assistant curator, Herbarium of the University of California; *Potentilla, Monolopia.*
Cunn., A.   Cunningham, Allan, 1791–1839, Kew Collector and Colonial Botanist in New South Wales.
Cunn., A. M.   Cunningham, Alida Mabel, 1868——, Purdue.
Cunn., R.   Cunningham, Richard, 1793–1835, brother of Allan; employed by W. T. Aiton and prepared second edition of "Hortus Kewensis"; killed by Australian natives.
Curran.   Curran, Mary Katharine (Layne) (*later* Brandegee), 1844–1920, physician and botanist, California.
Curt.   Curtis, William, 1746–1799, England, founder of the Botanical Magazine; author of "Flora Londinensis" (1772–1798).
Cutler.   Cutler, Hugh Carson, 1912——, Missouri Botanical Garden; *Ephedra*, economic botany.
Cyrill.   Cirillo, Domenico, 1739–1799, professor of botany, Naples.
Dalla-Torre.   Dalla-Torre, Karl Wilhelm von, 1850–1928, Germany; author (with H. Harms) of the important index "Genera Siphonogamarum" (1900–1907).
Daniels.   Daniels, Francis Potter, 1869–1947, Georgia State College for Women.
Danser.   Danser, Benedictus Hubertus, 1891–1943, Netherlands.
Darl., J.   Darlington, Josephine, 1905——, Missouri Botanical Garden; *Mentzelia.*
Daveau.   Daveau, Jules Alexandre, 1852–1929, director, Botanic Garden, Lisbon.
Davenp.   Davenport, George Edward, 1833–1907, Massachusetts businessman; student of ferns.
Davids., A.   Davidson, Anstruther, 1860–1932, Los Angeles physician.
Davids., J. F.   Davidson, John Fraser, 1911——, professor of botany, Nebraska.
Davids. & Mox.   Davidson, A., and G. L. Moxley, authors of "Flora of Southern California" (1923).
Davis.   Davis, Kary Cadmus, 1867–1936, Minnesota and Tennessee.
Davy.   Burtt-Davy, Joseph, 1870–1940, Imperial Forestry Institute, Oxford; at the University of California, 1892–1902, as an agrostologist.
Davy & Merr.   Davy, J. B., and E. D. Merrill.
DC.   Candolle (also Décandolle), Augustin Pyramus de, 1778–1841, professor of botany, Geneva; first in an illustrious line of systematists; founder of the Prodromus, a fundamental work in the development of the modern phylogenetic system.
DC., A.   Candolle, Alphonse Louis Pierre Pyramus de, the son, 1806–1893, Geneva; author of the last 10 vols. of the Prodromus; founder of Monographiae Phanerogamarum, a continuation and revision of the Prodromus.
DC. & Duby.   Candolle, A. P. de, and J. É. Duby.
Dcne.   Decaisne, Joseph, 1807–1882, director, Jardin des Plantes, Paris; Asclepiadaceae, Plantaginaceae.
Delpino.   Delpino, Giacomo Giuseppe Federico, 1833–1905, Italian taxonomist.
Dempst.   Dempster, Lauramay (Tinsley), 1905——, research assistant to W. L. Jepson; Jepson Herbarium.
Desf.   Desfontaines, Réné Louiche, 1750–1833, professor in the Jardin des Plantes, Paris; author of "Flora Atlantica" (1798–1800).
Desr.   Desrousseaux, Louis Auguste Joseph, 1753–1838, French cloth manufacturer; contributer to Lamarck's Encyclopedia.
Desv.   Desvaux, Augustin Nicaise, 1784–1856, professor of botany, Angers; editor of Journal de Botanique.
Detl.   Detling, LeRoy Ellsworth, 1898——, professor of botany, Oregon; *Dentaria, Descurainia.*
de Vries.   Vries, Hugo de, 1848–1935, brilliant Dutch botanist; one of the rediscoverers of Mendel's principles (1900) and exponent of the Mutation Theory.
Dewey.   Dewey, Rev. Chester, 1784–1867, professor, University of Rochester; specialist in *Carex.*
Dickson.   Dickson, James, 1738–1822, Scotch nurseryman; writer on cryptogams.
Dieck.   Dieck, Dr. G., 19th century, proprietor of a large German estate.
Diels.   Diels, Friedrich Ludwig Emil, 1874–1945, director of the botanical garden and museum, Berlin; student of the flora of China; Droseraceae, Menispermaceae.
Dietr.   Dietrich, Friedrich Gottlieb, 1768–1850, garden director at Eisenach.
Dietr., A.   Dietrich, Albert, 1795–1856, Berlin.
Dietr., D.   Dietrich, David Nathanael Friedrich, 1800–1888, Jena.
Dill.   Dillenius (Dillen), Johann Jacob, 1684–1747, professor of botany, Oxford; especially known as a bryologist.
Dippel.   Dippel, Leopold, 1827–1914, Germany.
Dod.   Dodoens, Rembert (Rembertus Dodonaeus), 1517–1585, professor in Leiden; published a work on plants illustrated by 1340 copper engravings.
Dole.   Dole, Eleazer Johnson, 1888——, professor of botany, Vermont.
Domb.   Dombey, Joseph, 1742–1796, French botanist and traveler in Peru and Chile.
Domin.   Domin, Karel, 1882–1954, professor of botany and director of the Botanical Institute and Gardens, Prague.
Don, D.   Don, David, brother of George, 1799–1841, professor in King's College, London; librarian to the Linnaean Society.

Don, G.   Don, George, 1798–1856, Scotch collector for the Horticultural Society in Brazil, West Indies, and Africa.
Donn.   Donn, James, 1758–1813, under Aiton at Kew, later curator of Cambridge Garden; author of "Hortus Cantabrigiensis" (1796), which went through 13 editions.
Dougl.   Douglas, David, 1798–1834, ardent Scotch collector in northwestern America, taking nearly 500 species in California alone for the Royal Horticultural Society; collected extensively along the Columbia River: ". . . his cost of maintenance during three years came to £66."
Drew, E.   Drew, Elmer Reginald, 1865–1930, professor of physics, Stanford.
Drew, W.   Drew, William Brooks, 1908——, professor of botany, Michigan State.
Druce.   Druce, George Claridge, 1850–1932, professor at Oxford.
Drum. & Hutch.   Drummond, James Ramsay, 1851–1921, grandson of Thomas Drummond, the famous collector, and J. Hutchinson, both of Kew.
Dryand.   Dryander, Jonas Carlsson, 1748–1810, librarian for Sir Joseph Banks, and later for the Linnaean Society.
Duby.   Duby, Jean Étienne, 1798–1885, Geneva cleric; Primulaceae.
Duchartre.   Duchartre, Pierre Étienne Simon, 1811–1894, French botanist.
Duchn.   Duchesne, Antoine Nicolas, 1747–1827, France; author of a work on useful plants, especially strawberries.
Dudl.   Dudley, William Russel, 1849–1911, professor of botany, Stanford.
Dufour.   Dufour, Jean Marie Léon, 1780–1865, France.
Duhamel.   Duhamel de Monceau, Henri Louis, 1700–1781, France.
Dulac.   Dulac, Joseph, Abbé, author of a flora of the Department of the High Pyrenees (1867).
Dum.-Cours.   Dumont de Courset, Baron Georges Louis Marie, 1746–1824, French horticultural writer.
Dumort.   Dumortier, Count Barthélemy Charles Joseph, 1797–1878, president of the Belgian Chamber of Deputies.
Dunal.   Dunal, Michel Félix, 1789–1856, professor of botany, Montpellier; Vacciniaceae, Solanaceae.
Dundas.   Dundas, Frederic Winn, 1911——, Pomona College.
Dunkle.   Dunkle, Meryl Byron, 1888——, junior college teacher, Santa Catalina Island and Long Beach.
Dunn, D.   Dunn, David Baxter, 1917——, California (Los Angeles), professor of botany, Missouri; Lupinus.
Dunn, S. T.   Dunn, Stephen Troyte, 1868–1938, official guide, Kew.
Durand.   Durand, Elias Magloire, 1794–1873, Philadelphia pharmacist.
Dur., T.   Durand, Théophile Alexis, 1855–1912, Brussels.
Dur. & Hilg.   Durand, E. M., and Theodore Charles Hilgard, 1828–1875, Philadelphia physician; wrote a description of plants collected in California by Lt. Williamson's party of the Pacific Railroad Survey.
Dur. & Jacks.   Durand, T. A., and Benjamin Daydon Jackson, 1846–1927, botanical secretary of the Linnaean Society, 1880–1902, general secretary, 1902–1926; compiler of Index Kewensis.
Dur. & Schinz.   Durand, T. A. and H. Schinz.
Durazz.   Durazzo, Ippolito, 1750–1818, Italy.
Durieu.   Durieu de Maisonneuve, Michel Charles, 1796–1878, director of the botanic gardens, Bordeaux; student of the flora of southern France, Spain, and especially Algeria.
Dusén.   Dusén, Per Karl Hjalmar, 1855–1926, Swedish civil engineer who made extensive botanical trips to various countries including Brazil, where he became amanuensis of the Museum of Rio de Janeiro.
Dyal.   Dyal, Sarah Creecie (later Nielsen), 1907——, Cornell; Plectritis.
Dykes.   Dykes, William Rickatson, 1877–1925, secretary, Royal Horticultural Society; author of "The Genus Iris."
E. & P.   Engler, H. G. A., and K. A. E. Prantl, authors of "Die natürlichen Pflanzenfamilien" (1887–1915) (2d ed. 1924–1958 incomplete).
Eastw.   Eastwood, Alice, 1859–1953, curator of botany, California Academy of Sciences, 1892–1950; tireless student of the West American flora and of cultivated plants.
Eastw. & Mox.   Eastwood, A., and G. L. Moxley.
Eat.   Eaton, Amos, 1776–1842, New York; produced first manual in America with descriptions in English.
Eat., A. A.   Eaton, Alvah Augustus, 1865–1908, New England; student of ferns.
Eat., D. C.   Eaton, Daniel Cady, grandson of Amos, 1834–1895, professor of botany, Yale; authority on ferns.
Eat. & Wright.   Eaton, Amos, and John Wright, 1811–1846, professor of physiology, Rensselaer Institute.
Ehrend.   Ehrendorfer, Friedrich, 1927——, Botanical Institute, University of Vienna.
Ehrh.   Ehrhart, J. Friedrich, 1742–1795, Switzerland; pupil of Linnaeus; advocate of monomial nomenclature.
Eichler.   Eichler, August Wilhelm, 1839–1887, Berlin; important editor of and contributor to Flora Brasiliensis.
Ell.   Elliott, Stephen, 1771–1830, professor in Charleston; author of "A sketch of the botany of South Carolina and Georgia."
Elmer.   Elmer, Adolph Daniel Edward, 1870–1942, collector in California, Washington, and the Philippine Islands.
Elwes.   Elwes, Henry John, 1846–1922, English world traveler and patron of science; Lilium.
Endl.   Endlicher, Stephen Friedrich Ladislaus, 1804–1849, professor of botany and director of the botanic garden, Vienna; author of "Genera Plantarum" (1836–1840), and many other big works; student of Coniferae.

Engelm. Engelmann, George, 1809–1884, physician in St. Louis and eminent botanist; painstaking student of North American *Cuscuta, Juncus,* Euphorbiaceae, *Isoetes, Yucca,* Cactaceae, *Pinus, Abies, Juniperus, Agave,* and *Vitis.*

Engelm. & Bigel. Engelmann, G., and J. M. Bigelow.

Engl. Engler, Adolf (Heinrich Gustaf Adolf), 1844–1930, director of the botanic garden and museum, Berlin; founder and editor of Botanische Jahrbücher, Die Vegetation der Erde, and Das Pflanzenreich; outstanding systematist and tireless worker; Araceae, monographer of *Saxifraga.*

Engl. & Irm. Engler, H. G. A., and Edgar Irmscher, 1887——, curator of the herbarium, Hamburg.

Ensign. Ensign, Margaret Ruth (*later* Lewis), 1919——, Pomona College; *Forsellesia.*

Epl. Epling, Carl Clawson, 1894——, professor of botany, California (Los Angeles); student of Labiatae.

Eschs. Eschscholtz, Johann Friedrich, 1793–1831, zoologist in Dorpat; surgeon and naturalist on the ship *Rurik,* of the Russian Romanzoff Expedition under Kotzebue, which visited California in 1816, and on the *Predpriaetie,* also under Kotzebue, which reached San Francisco and the Sacramento River in the fall of 1824.

Estes. Estes, Frederick Earle, 1902——, Pomona College; *Malvastrum.*

Ewan. Ewan, Joseph Andorfer, 1909——, professor of botany, Tulane; *Delphinium.*

F. & M. Fischer, F. E. L., and C. A. Meyer.

Fabr. Fabricius, Philipp Conrad, 1714–1774, professor in Helmstädt.

Farw. Farwell, Oliver Atkins, 1867–1944, consulting botanist for Parke, Davis & Co., Detroit.

Fassett. Fassett, Norman Carter, 1900–1954, professor of botany, Wisconsin.

Fedde. Fedde, Friedrich Karl Georg, 1873–1942, professor and editor in Berlin; Papaveraceae.

Fée. Fée, Antoine Laurent Apollinaire, 1789–1874, professor of botany, Strassburg; noted student of cryptogams.

Fenley. Fenley, Kittie Lucille (*later* Parker), 1910——, California and Arizona.

Fenzi. Fenzi, Emanuele Orazio (*later* in California known as Francesco Franceschi), 1843–1924, Italian horticulturist who lived in California from 1892–1913 and published "Santa Barbara Exotic Flora."

Fenzl. Fenzl, Eduard, 1808–1879, director, Botanical Garden, Vienna.

Ferg. Ferguson, Alexander McGowen, 1874——, Texas plant breeder.

Fern. Fernald, Merritt Lyndon, 1873–1950, director of Gray Herbarium, 1937–1947; a founder of Rhodora (associate editor, 1899–1928, editor, 1929–1950); noted plant geographer and systematist; *Potamogeton;* author of "Gray's Manual of Botany, Eighth Edition."

Fern. & Brack. Fernald, M. L., and Amelia Ellen Brackett, 1896–1926, Gray Herbarium.

Fern. & Griscom. Fernald, M. L., and Ludlow Griscom, 1890——, Harvard research ornithologist.

Fern. & Macbr. Fernald M. L., and J. F. Macbride.

Fern. & Weath. Fernald, M. L., and C. A. Weatherby.

Fern. & Wieg. Fernald, M. L., and K. M. Wiegand.

Ferris. Ferris, Roxana Judkins (Stinchfield), 1895——, Dudley Herbarium; editor of "Illustrated Flora of the Pacific States, vol. IV.

Fieb. Fieber, Franz Xaver, 1807–1872, Hungary.

Fig. & De Not. Figari, Antonio, 1804–1870, and Giuseppe De Notaris, 1805–1877, Italian students of the Egyptian flora.

Fisch. Fischer, Friedrich Ernst Ludwig von, 1782–1854, director of the botanic garden, St. Petersburg, 1823–1850.

Fisch. & Avé-Lall. Fischer, F. E. L., and J. L. E. Avé-Lallemant.

Fisch. & Trautv. Fischer, F. E. L., and E. R. Trautvetter.

Fisch., G. Fischer, Gustav, 1889——, Berlin.

Florin. Florin, Carl Rudolf, 1894——, director of Hortus Bergianus, Stockholm, 1944——; Gymnosperms, Paleobotany.

Flous. Flous, Mlle F., Toulouse, France; Coniferae.

Focke. Focke, Wilhelm Olbers, 1834–1922, physician in Berne; student of *Rubus.*

Forbes. Forbes, James, 1773–1861, English gardener, Woburn Abbey; author of several books on garden plants.

Forsk. Forsskål, Petter (*also* Pehr Forskål), 1732–1763, Finnish student of Linnaeus, who traveled to Arabia and wrote a flora of Egypt and Arabia; died on the desert of starvation and exposure after repeated encounters with bandits.

Forst. & Forst. f. Forster, Johann Reinhold, 1729–1798, Halle, and his son, J. G. A. Forster, who traveled together to Russia and England and on Captain Cook's Second Voyage.

Forst. f. Forster, Johann Georg Adam (*also* George Forster), 1754–1794, professor of natural history, Cassel.

Fosb. Fosberg, Francis Raymond, 1908——, U.S. Geological Survey, Washington; student of the California flora, later of the South American and Polynesian floras.

Foster. Foster, Robert Crichton, 1904——, Gray Herbarium; Iridaceae.

Foucaud. Foucaud, Julien, 1847–1904, France.

Foug. Fougeroux de Bondaroy, Auguste Denis, 1732–1789, France.

Fourn. Fournier, Eugène Pierre Nicolas, 1834–1884, physician in Paris; Asclepiadaceae.

Fourr. Fourreau, Pierre Jules, 1844–1871, France; student of the flora of the Rhone Valley.

Franco. Franco, João Manuel Antonio Paes do Amaral, 1921——, Lisbon, Portugal.

Fraser. Fraser, John, 1750–1811, nurseryman of Chelsea; zealous collector of North American plants.

Frém. Frémont, John Charles, 1813–1890, soldier and explorer; first botanical collector in the Sierra Nevada and first presidential candidate of the Republican Party.

Freyn. Freyn, Josef Franz, 1845–1903, Prague; distinguished student of the flora of Austria-Hungary and northern Asia, specialist in Liliaceae and Ranunculaceae.

Friedl. Friedland, Solomon, 1912——, New York Botanical Garden; *Hemicarpha.*

Fries.   Fries, Elias Magnus, 1794–1878, professor of botany, Uppsala; noted mycologist and student of *Hieracium*.

Fries & Broberg.   Fries, E. M., and Sv. P. Broberg.

Fries, Th.   Fries, Thore Magnus, the son, 1832–1913, professor of botany, Uppsala; lichenologist and student of *Hieracium*.

Frye.   Frye, Theodore Christian, 1869——, professor of botany, University of Washington.

Frye & Rigg.   Frye, T. C., and George Burton Rigg, 1872——, professor of botany, University of Washington; authors of "Northwest Flora" (1912).

Gaertn.   Gaertner, Joseph, 1732–1791, physician near Stuttgart; writer on the structure of fruit and seeds.

Gaertn. f.   Gaertner, Carl Friderich von, 1772–1850, son of Joseph; Stuttgart.

Gaertn., Mey. & Scherb.   Gaertner, Philipp Gottfried, Bernhard Meyer, and Johannes Scherbius, 18th-century authors of a flora of Wetterau.

Gale.   Gale, Shirley (*later* Cross), Gray Herbarium; Rhynchospora.

Gand.   Gandoger, Michel, 1850–1926, French abbé; author of "Flora Europae" (27 vols.); voluminous writer; amasser of a huge herbarium now at Lyon; a "splitter" who named thousands of unacceptable species.

Garcke.   Garcke, Friedrich August, 1819–1904, curator of the herbarium, Berlin; author of "Flora von Nord- und Mittel-Deutschland" (1849) that went through 20 editions.

Garden.   Garden, Alexander, 1730–1791, physician and amateur botanist of Charleston, South Carolina, for whom Linnaeus named the genus *Gardenia*.

Garrett.   Garrett, Albert Osmun, 1870–1948, high-school teacher in Salt Lake City.

Gars.   Garsault, François Alexandre Pierre de, 1691–1778, France.

Gates.   Gates, Reginald Ruggles, 1882——, English geneticist.

Gaudin.   Gaudin, Jean François Gottlieb Philippe, 1766–1833, Swiss agrostologist.

Gay.   Gay, Claude, 1800–1873, French author of a history of Chile, which included eight volumes on the botany.

Gay, J.   Gay, Jacques Étienne, 1786–1864, French student of the flora of Switzerland and the Pyrenees; systematist and morphologist.

Gay & Durieu.   Gay, J. E., and M. C. Durieu de Maisonneuve.

Gentry.   Gentry, Howard Scott, 1903——, U.S.D.A., Beltsville; made important botanical explorations in Sonora, Mexico.

Gesn.   Gesner, Conrad, 1516–1565, Zürich; physician, philosopher, and noted botanist.

Geyer.   Geyer, Carl Andreas, 1809–1853, German botanist with Nicollet's expedition; collected in northern Idaho and Washington in 1844.

Gilbert.   Gilbert, Benjamin Davis, 1835–1907, American fern student.

Gilib.   Gilibert, Jean Emmanuel, 1741–1814, professor in Wilno, later in Lyon; author of "Flora Lithuanica" (1785).

Gill.   Gill, Lake Shore, 1900——, American forest pathologist.

Gill. & Hook.   Gillies, J., and W. J. Hooker.

Gillett.   Gillett, George Willson, 1917——, professor of botany, Michigan State Univ.

Gillett, J.   Gillett, John Montague, 1918——, Department of Agriculture, Canada. Gentianaceae.

Gillies.   Gillies, John, 1747–1836, Scotch physician who resided some years in Argentina and collected in Chile.

Gilly.   Gilly, Charles Louis, 1911——, Michigan State.

Gleason.   Gleason, Henry Allan, 1882——, assistant director and head curator, New York Botanical Garden; author of "New Britton & Brown Illustrated Flora" (1952).

Gleditsch.   Gleditsch, Johann Gottlieb, 1714–1786, director, Botanic Garden, Berlin.

Gmel.   Gmelin, Johann Georg, 1709–1755, traveled in Siberia and Kamchatka, 1733–1743, and gave first account of their floras; author of the classic "Flora Sibirica" (1747–1769); later professor in Tübingen.

Gmel., C. C.   Gmelin, Carl Christian, 1762–1837, physician in Karlsruhe; author of "Flora Badensis Alsatica" (1805–1826).

Gmel., J. F.   Gmelin, Johann Friedrich, nephew of J. G., 1748–1804, professor in Tübingen, then in Göttingen; editor of the 13th edition of Linne's Systema Naturae (1788–1793).

Gmel., S. G.   Gmelin, Samuel Gottlieb, nephew of J. G., cousin of J. F., 1743–1774, traveled with Pallas in southeastern Russia; collaborated with his uncle in producing "Flora Sibirica."

Godron.   Godron, Dominique Alexandre, 1807–1880, botanist of Nancy, France.

Goldb.   Goldbach, Karl Ludwig, 1793–1824, Moscow.

Goldie.   Goldie, John, 1793–1886, Scotsman who traveled through eastern Canada and New England in 1819.

Good.   Goodenough, Samuel, Bishop of Carlisle, 1743–1827, England; one of the three founders of the Linnaean Society of London; *Carex*.

Goodd.   Goodding, Leslie Newton, 1880——, botanist, Soil Conservation Service, Arizona.

Goodd., C.   Goodding, Charlotte Olive (*later* Reeder), the daughter, 1916——, Yale; *Bouteloua*.

Goodm.   Goodman, George Jones, 1904——, professor of botany, Oklahoma; West American genera of Polygonaceae.

Goodm. & L. Benson.   Goodman, G. J., and L. D. Benson.

Gord.   Gordon, George, 1806–1879, superintendent of the Horticultural Gardens, Chiswick, near London.

Gord. &' Glend.   Gordon, G., and Robert Glendinning, fl. 1844–1858, of the Chiswick Nursery, authors of "The Pinetum" (1858, ed. 2, 1875).

Gouan.   Gouan, Antoine, 1733–1821, professor of botany, Montpellier; correspondent and friend of Linnaeus; first Frenchman to accept binary nomenclature for plants.

Goujon.   Goujon, J., fl. 1872, French horticultural writer.

Gould.   Gould, Frank W, 1913——, professor of range and forestry, A. & M. College, Texas.

Graebn.   Graebner, Karl Otto Robert Peter Paul, 1871–1933, professor of botany, Berlin.
Grah.   Graham, Robert, 1786–1845, professor of botany, Edinburgh.
Grant.   Grant, Adele (Lewis), 1881———, California; monographer of *Mimulus.*
Grant, A.   Grant, Alva (Day) (Hansen), 1920———, California, Rancho Santa Ana Botanic Garden; Polemoniaceae.
Grant, A. & V.   Grant, A. D. H., and V. E. Grant.
Grant, G. B.   Grant, George Barnard, 1849–1917, his private herbarium, mostly of southern California plants, is deposited at Stanford.
Grant, V.   Grant, Verne Edwin, 1917———, cytogeneticist, Rancho Santa Ana Botanic Garden; Polemoniaceae.
Grant, V. & A.   Grant, V. E., and A. D. H. Grant.
Gray.   Gray, Asa, 1810–1888, professor of botany, Harvard; preëminent American systematist; author of "Manual of the Botany of the Northern United States" (1848, now through 8 editions), [Gamopetalae of California] in Brewer and Watson's "Botany of California" (1876), "Synoptical Flora of North America" (1878–1897), botanical textbooks, numerous reports of collections and revisions.
Gray, S. F.   Gray, Samuel Frederick, 1766–1836, England; author of "Natural Arrangement of British Plants" (1821), a work much in advance of its time.
Greene.   Greene, Edward Lee, 1843–1915, professor of botany, California, 1885–1895, then at Catholic University of America and Smithsonian Institution; editor of Pittonia, and Leaflets of Botanical Observation and Criticism; believer in absolute priority in nomenclature.
Greenm.   Greenman, Jesse More, 1867–1951, curator of the herbarium, Missouri Botanical Garden, 1913–1948; *Senecio.*
Greenman & Roush.   Greenman, J. M., and E. M. F. Roush.
Gren. & Godr.   Grenier, Jean Charles Marie, 1808–1875, professor of botany, Besançon, France, and Dominique Alexandre Godron, 1807–1880, professor in Nancy; authors of the noted "Flore de France" (1848–1856).
Grev. & Balf.   Greville, Robert Kaye, 1794–1866, professor in Edinburgh, and J. H. Balfour.
Grev. & Hook.   Greville, R. K., and W. J. Hooker.
Griffiths.   Griffiths, David, 1867–1935, botanist, U.S.D.A.
Griffiths & Hare.   Griffiths, D., and Raleigh Frederick Hare, 1870–1934, plant chemist, New Mexico College of Agriculture.
Griseb.   Grisebach, August Heinrich Rudolf, 1814–1879, professor of botany, Göttingen; noted for his "Vegetation der Erde" (1872, ed. 2, 1884), and studies on the West Indian flora; Gentianaceae.
Groenl.   Groenland, Johannes, 1824–1891, an editor of Revue Horticole, Paris.
Gron.   Gronovius, Johannes Fridericus, 1690–1762, senator in Leiden, friend of Linnaeus, author of "Flora Virginica" (1743, ed. 2, 1762).
Gross.   Grosser, Wilhelm Carl Heinrich, 1869———, director of the agricultural research station, Breslau; Cistaceae.
Gross, H.   Gross, Hugo, 19th century, East Prussia.
Grub.   Grubov, Valery Ivanovich, 1917———, Russian monographer of *Rhamnus* and *Frangula.*
Guill.   Guillemin, Jean Baptiste Antoine, 1796–1842, French botanist who traveled in Brazil.
Gürke & Harms.   Gürke, Robert Louis August Max, 1854–1911, Germany, and Hermann August Theodor Harms, 1870–1942, botanical museum, Berlin-Dahlem.
Guss.   Gussone, Giovanni, 1787–1866, professor of botany, Naples; author of valuable works on the flora of southern Italy and Sicily.
H. & A.   Hooker, W. J., and G. A. W. Arnott, authors of "The Botany of Captain Beechey's Voyage" (1830–1841), in which many California species are described.
Hack.   Hackel, Eduard, 1850–1926, noted Austrian agrostologist.
Haenke.   Haenke, Thaddaeus, 1761–1817, Bohemia; phytographer for King of Spain; first botanist, with Luis Née, in California, visiting San Diego and Monterey with the Malaspina Expedition in 1791. His collections were described by K. B. Presl, to whom refer.
Hagström.   Hagström, Johan Oskar, 1860–1922, Sweden; *Potamogeton, Ruppia.*
Haines.   Haines, Adelbert Lee, 1915———, California; *Yucca.*
Hall.   Hall, Harvey Monroe, 1874–1932, professor of botany, California, then staff member Carnegie Institution of Washington; pioneer in experimental taxonomy; student of the Compositae; *Haplopappus,* Madiinae.
Hall, C. C.   Hall, Carlotta (Case), the wife, 1880–1949, California; student of ferns.
Hall & Clem.   Hall, H. M., and F. E. Clements, authors of "The Phylogenetic Method in Taxonomy" (1923), in which North American *Artemisia, Chrysothamnus,* and *Atriplex* are monographed.
Hall & Hall.   Hall, H. M., and C. C. Hall, authors of "A Yosemite Flora" (1912).
Haller.   Haller, Albrecht (Albert) von, 1708–1777, professor in Göttingen, later in Berne; Swiss botanist, physician, poet, and statesman.
Haller f.   Haller, Albrecht von, the son, 1758–1823, Berne.
Hallier f. Hallier, Hans Gottfried, 1868–1932, Dutch botanist of Buitenzorg and Leiden; student of phylogeny.
Ham., A.   Hamilton, Arthur, of Geneva, Switzerland, author of a monograph on *Scutellaria* in 1832.
Hanks.   Hanks, Lenda Tracy, 1879–1944, New York schoolteacher.
Hanson.   Hanson, Peter, 1824–1887, artist and amateur lily grower of Brooklyn, New York.
Hara.   Hara, Hiroshi, contemporary Japanese taxonomist.
Hartw.   Hartweg, Karl Theodor, 1812–1871, German collector sent to Mexico by the London Horticultural Society, and later to California in 1846 and 1847.
Harv.   Harvey, William Henry, 1811–1866, professor of botany and keeper of the herbarium, Trinity College, Dublin; made known the California collections of Thomas Coulter; author (with O. W. Sonder) of "Flora Capensis" (1859–1865).

Harv. & Gray.   Harvey, W. H., and A. Gray.

Harvey, M.   Harvey, Margaret (*later* Dildine), 1919——, Pomona College; *Fremontia.*

Hassk.   Hasskarl, Justus Carl, 1811–1894, German botanist at Buitenzorg, Java.

Haum.   Hauman, Lucien (*formerly* Hauman-Merck), 19th century, German botanist attached to the natural history museum, Buenos Aires; later in Brussels.

Hausskn.   Haussknecht, Heinrich Karl, 1838–1903, professor in Weimar; monographer of *Epilobium.*

Haw.   Haworth, Adrian Hardy, 1767–1833, English entomologist; student of succulents; *Mesembryanthemum.*

Hay.   Hayek, August von, 1871–1928, plant geographer, University of Vienna; author of a flora of Steiermark.

Hayne.   Hayne, Friedrich Gottlob, 1763–1832, professor of botany, Berlin.

HBK.   Humboldt, Baron Friedrich Wilhelm Heinrich Alexander von, 1769–1859, German zoologist; Aimé Jacques Alexandre Bonpland, 1773–1858, French botanist, the two forming the most famous scientific expedition to tropical America; and Carl Sigismund Kunth, professor of botany, Berlin, who wrote the text of their descriptive work "Nova genera et species Plantarum" (1815–1825).

Heckard.   Heckard, Lawrence Ray, 1923——, professor of botany, Illinois.

Hegelm.   Hegelmaier, Christof Friedrich, 1833–1906, professor of botany, Tübingen; Lemnaceae, *Callitriche.*

Heimerl.   Heimerl, Anton, 1857——, professor in Vienna; Nyctaginaceae, *Achillea.*

Heiser.   Heiser, Charles Bixler, Jr., 1920——, professor of botany, Indiana; Compositae.

Heist.   Heister, Lorenz, 1683–1758, professor in Helmstädt.

Heller.   Heller, Amos Arthur, 1867–1944, diligent collector in California; editor of Muhlenbergia.

Hemsl.   Hemsley, William Botting, 1843–1924, keeper of the herbarium, Kew, 1899–1908; author of the 5 vol. work on the botany in "Biologia Centrali-Americana" (1879–1888).

Henders.   Henderson, Louis Forniquet, 1853–1942, professor of botany, Oregon.

Henr.   Henrard, Jan Theodoor, 1881——, Conservator, Rijksherbarium, Leiden; *Aristida.*

Henr. & Thell.   Henrard, J. T., and A. Thellung.

Henry.   Henry, Joseph Kaye, 1866–1930, professor of English, University of British Columbia; collector and flora writer.

Henshaw.   Henshaw, Julia Willmothe (Henderson), American authoress of a popular book on mountain and wild flowers (1906).

Herb.   Herbert, Hon. William, 1778–1847, English politician, then a churchman, becoming Dean of Manchester; Amaryllidaceae.

Herder.   Herder, Ferdinand Godfried Theobald Maximilian von, 1828–1896, Russia; student of the flora of Siberia and Russian North America.

Herm., F. J.   Hermann, Frederick Joseph, 1906——, National Arboretum Herbarium, Beltsville; *Carex.*

Hester.   Hester, J. Pinckney, contemporary southern California writer on succulents.

Heynh.   Heynhold, Gustav, Saxony, wrote in the 1840's; *Hieracium.*

Hiern.   Hiern, William Philip, 1839–1925, British Museum; student of the flora of tropical Africa and India.

Hieron.   Hieronymous, Georg Hans Emo Wolfgang, 1846–1921, professor in Berlin; student of the Argentine and Andean flora.

Hilend.   Hilend, Martha Luella (*later* Kinsey), 1902——, Pomona College and California (Los Angeles).

Hilend & Howell.   Hilend, M., and J. T. Howell, *Galium.*

Hill.   Hill, Sir John, 1716–1775, London physician; author of herbals and nature books; produced the first flora of England on the Linnaean system.

Hill, E. J.   Hill, Ellsworth Jerome, 1833–1917, Illinois high-school teacher; student of mosses and ferns.

Himmelb.   Himmelbaur, Wolfgang, 1886–1937, professor of systematic botany, Vienna; Berberidaceae.

Hitchc.   Hitchcock, Albert Spear, 1865–1935, U.S. National Herbarium; leading American agrostologist; author of "Manual of the Grasses of the United States."

Hitchc. & Chase.   Hitchcock, A. S., and M. A. Chase.

Hitchc., C. L.   Hitchcock, Charles Leo, 1902——, professor of botany, University of Washington; *Lycium, Draba, Lepidium, Lathyrus, Sidalcea,* etc.

Hitchc., E.   Hitchcock, Edward, 1793–1864, professor of chemistry and natural history, and president, Amherst.

Hitchc. & Maguire.   Hitchcock, C. L., and B. Maguire.

Hitchc. & Sharsm.   Hitchcock, C. L., and C. W. Sharsmith.

Hochr.   Hochreutiner, Bénédict Pierre Georges, 1873——, director, Conservatoire de Botanique, Geneva.

Hochst.   Hochstetter, Christian Ferdinand, 1787–1860, botanist of Stuttgart.

Hoeck.   Hoeck, Fernando, 1858–1915, Germany.

Hoffm.   Hoffmann, Georg Franz, 1760–1826, professor of botany, Göttingen, later in Moscow; student of lichens, *Salix,* and Umbelliferae.

Hoffm., R.   Hoffman, Ralph, 1870–1932, botanist and ornithologist; director, Santa Barbara Museum of Natural History; student of the botany of the islands of the Santa Barbara Channel.

Hoffmsg.   Hoffmannsegg, Count Johann Centurius von, 1766–1849, Dresden.

Hoffmsg. & Link.   Hoffmannsegg, Count J. C. von, and J. H. F. Link, authors of "Flore Portugaise" (1809–1820).

Holl & Heynh.   Holl, Friedrich, and G. Heynhold, authors of "Flora von Sachsen" (1842).

Hollick & Britt.   Hollick, Charles Arthur, 1857–1933, paleobotanist, New York Botanical Garden, and N. L. Britton.

Holm.   Holm, Herman Theodor, 1854–1932, botanist of Denmark, Greenland, and the United States.
Holm, R.   Holm, Richard William, 1925——, curator of Dudley Herbarium, Stanford.
Holz.   Holzinger, John Michael, 1853–1929, teacher in Minnesota.
Hook.   Hooker, Sir William Jackson, 1785–1865, director of Kew, 1841–1865; author of "Flora boreali-americana" and many other illustrious works; founder and editor of Journal of Botany and Icones Plantarum; editor of Botanical Magazine.
Hook. f.   Hooker, Sir Joseph Dalton, the son, 1817–1911, director of Kew, 1865–1885; talented editor and student of the New Zealand, Himalayan, and Indian floras.
Hook. & Grev.   Hooker, W. J., and R. K. Greville, authors of "Icones Filicum" (1831).
Hook. & Harv.   Hooker, W. J., and W. H. Harvey.
Hoopes.   Hoopes, Josiah, 1832–1904, Pennsylvania nurseryman; student of the Coniferae.
Hoov.   Hoover, Robert Francis, 1913——, professor, California State Polytechnic College, San Luis Obispo.
Hopk., L. S.   Hopkins, Lewis Sylvester, 1872——, dean, Culver-Stockton College, Missouri.
Hopk., M.   Hopkins, Milton, 1906——, professor of botany, Oklahoma, 1936–1945; editor, Henry Holt & Co.; *Arabis.*
Hoppe.   Hoppe, David Heinrich, 1760–1846, professor in Regensburg; editor of Flora.
Horkel.   Horkel, Johann, 1769–1846, Germany.
Hornem.   Hornemann, Jens Wilken, 1770–1841, professor of botany, Copenhagen.
Hort.   Hortorum, *of gardens,* used with plants of garden or unknown origin.
Host.   Host, Nicolaus Thomas, 1761–1834, imperial physician in Vienna; author of "Flora Austriaca" (1827–1831).
House.   House, Homer Doliver, 1878–1949, New York state botanist, 1914–1948; author of "Wild Flowers of New York."
Houst.   Houstoun, William, 1695–1733, Scotch collector in the West Indies.
Houtt.   Houttuyn, Martin, 1720–1794, Dutch physician and naturalist.
Howard & Manton.   Howard, Harold W., and Irène Manton, professor of botany, Leeds; English cytogeneticists.
Howell.   Howell, Thomas Jefferson, 1842–1912, Portland, Oregon; author of "A Flora of Northwest America" (1897–1903).
Howell, J. T.   Howell, John Thomas, 1903——, California Academy of Sciences; editor of Leaflets of Western Botany; diligent collector and student of many California genera.
Hu.   Hu, Shiu-ying, 1910——, Arnold Arboretum, student of the Chinese flora.
Hubb., C. E.   Hubbard, Charles Edward, 1900——, agrostologist, Kew.
Hubb., F. T.   Hubbard, Frederick Tracy, 1875——, Massachusetts.
Huds.   Hudson, William, 1730–1793, London apothecary; author of "Flora Anglica" (1762, first of 3 editions).
Hult.   Hultén, Oskar Eric Gunnar, 1894——, professor of botany, Stockholm; student of American Arctic floras; author of "Flora of Kamtchatka" (1927–1929), "Flora of the Aleutian Islands" (1937), "Flora of Alaska and Yukon" (1941–1950).
Hult. & St. John.   Hultén, O. E. G., and H. St. John.
Humb.   Humboldt, Baron Friedrich Wilhelm Heinrich Alexander von, 1769–1859, German zoologist, and Aimé Jacques Alexandre Bonpland, 1773–1858, French botanist; authors of the classic "Voyage aux Régions Équinoxiales du Nouveau Continent" (1805–1837).
Hutch.   Hutchinson, John, 1884——, Kew; author of a new system of phylogeny (1926, 1934).
Huth.   Huth, Ernst, 1845–1897, Germany.
Hylander.   Hylander, Nils, 1904——, Institution for Systematic Botany, Uppsala.
Iltis.   Iltis, Hugh Hellmut, 1925——, professor of botany, Wisconsin; Capparidaceae.
Ingram.   Ingram, John William, Jr., 1924——, Southern California, California (Berkeley), and Bailey Hortorium; Euphorbiaceae.
Inman.   Inman, Ondess Lamar, 1890–1942, professor of botany, Idaho.
Ives.   Ives, Eli, 1779–1861, professor at Yale.
Jacq.   Jacquin, Nicolaus Joseph Baron von, 1727–1817, professor of botany and director of the botanic garden, Vienna; noted systematist.
Jacq. f.   Jacquin, Joseph Franz Baron von, the son, 1766–1839, professor of botany, Schemnitz.
James.   James, Edwin, 1797–1861, first botanist in Colorado, with Major Long's Expedition to the Rocky Mountains (1819–1820).
Jancz.   Janczewski, Eduard, Ritter von Glinka, 1846–1918, professor in Cracow; monographer of *Ribes.*
Jeps.   Jepson, Willis Linn, 1867–1946, professor of botany, California; author of "A Flora of California," "A Manual of the Flowering Plants of California," and other works; founder of the California Botanical Society.
Jeps. & Ames.   Jepson, W. L., and Alma Union Ames (*later* Weigart), 189?——.
Jeps. & Bail.   Jepson, W. L., and V. L. Bailey.
Jeps. & Hoov.   Jepson, W. L., and R. F. Hoover.
Jeps. & Mason.   Jepson, W. L., and H. L. Mason.
Jeps. & Rydb.   Jepson, W. L., and P. A. Rydberg.
Jeps. & Tracy.   Jepson, W. L., and J. P. Tracy.
Jeps. & Wies.   Jepson, W. L., and A. E. Wieslander.
Johansen.   Johansen, Donald Alexander, 1901——, California cytologist and embryologist.
Johnson.   Johnson, Arthur Monrad, 1878–1943, professor of botany, California (Los Angeles); student of the Saxifragaceae and phylogeny.
Johnst., J. R.   Johnston, John Robert, 1880——, American plant pathologist.
Jones.   Jones, Marcus Eugene, 1852–1934, Utah mining consultant; assembled very extensive herbarium of Great Basin plants now at Pomona College; published his botanical observations in a

private journal, "Contributions to Western Botany," that is marked by its cutting criticism of almost all contemporaries.

Jones, G. N.   Jones, George Neville, 1904——, professor of botany, Illinois; author of floras on the Olympic Peninsula (1936), Mt. Rainier (1938), and Illinois (1945).

Jones & Jones.   Jones, G. N., and Florence Freeman Jones, 1913——, the wife.

Jones, Q.   Jones, Quentin, 1920——, U.S.D.A., Beltsville.

Jord.   Jordan, Alexis, 1814–1897, Lyon, France: proved the existence of many genetically distinct races (which he elevated to specific rank) in such complexes as *Erophila verna;* such microspecies are now often called jordanons.

Jtn.   Johnston, Ivan Murray, 1898——, professor of botany, Harvard; authority on the Boraginaceae.

Juss.   Jussieu, Antoine Laurent de, nephew of Bernard, 1748–1836, professor in the Jardin des Plantes, Paris; first characterizer of natural families; expounder of his uncle's new system.

Juss., A.   Jussieu, Adrien Henri Laurent de, son of Antoine, 1797–1853, professor in the Jardin des Plantes.

Juss., B.   Jussieu, Bernard de, 1699–1776, arranger of the garden of La Trianon at Versailles after a new system of classification, which was an outgrowth of the teaching of Tournefort.

Just.   Just, Johann Leopold, 1841–1891, Karlsruhe; founder of Just's Botanischer Jahresbericht.

Kalm.   Kalm, Pehr (Peter), 1715–1779, professor of natural science, Åbo, Finland; pupil of Linnaeus who traveled in eastern North America (1747–1749), and published in 1765 the first part of "Flora Fennica."

Karst.   Karsten, Gustav Karl Wilhelm Hermann, 1817–1908, professor in Vienna; author of "Flora von Deutschland, Oesterreich und der Schweiz" (1895).

Kaulf.   Kaulfuss, Georg Friedrich, 1786–1830, professor in Halle; student of ferns.

Kearn.   Kearney, Thomas Henry, 1874–1956, U.S.D.A. and California Academy of Sciences; taxonomist and cotton breeder; Malvaceae.

Kearn. & Peeb.   Kearney, T. H., and R. H. Peebles, authors of "Flowering Plants and Ferns of Arizona" (1942), "Arizona Flora" (1951).

Keck.   Keck, David Daniels, 1903——, assistant director, New York Botanical Garden; experimental taxonomy; *Penstemon,* Madiinae, flora of California.

Kell.   Kellogg, Albert, 1813–1887, San Francisco physician and botanist; a founder of the California Academy of Sciences.

Kell. & Behr.   Kellogg, A., and H. H. Behr.

Keller.   Keller, Johann Christoph, 1737–1796, German artist and engraver.

Keller, A. C.   Keller, Allan Charles, 1914——, Claremont Graduate School; *Acer.*

Kenn.   Kennedy, Patrick Beveridge, 1874–1930, professor of agronomy, California; *Trifolium.*

Kenn. & McDer.   Kennedy, P. B., and L. F. McDermott.

Ker.   Ker, John Bellenden, or John Ker Bellenden, or, before 1804, John Gawler, 1764–1842, England; first editor of Edwards' Botanical Register.

Kern.   Kerner von Marilaun, Ritter Anton Josef, 1831–1898, professor of botany, Innsbruck, later Vienna; made the first scientific transplant experiments using climatically unlike gardens; author of "Pflanzenleben" (1887–1891).

Keys.   Keyserling, Count Alexander Friedrich Michael Leberecht Arthur von, 1815–1891, St. Petersburg; paleontologist and student of ferns.

Kindb.   Kindberg, Nils Conrad, 1832–1910, professor in Linköping, Sweden.

Kit.   Kitaibel, Paul, 1757–1817, professor of botany and chemistry, and director of the botanic garden, Budapest.

Klatt.   Klatt, Friedrich Wilhelm, 1825–1897, Hamburg; Iridaceae.

Kleeb.   Kleeberger, George Reinard, schoolteacher at Weaverville and in San Jose Normal School (1885–1886).

Klett & Richter.   Klett, Gustav Theodor, 1808?–1882, and Hermann Eberhard Friedrich Richter, 1808–1876, authors of a flora of Leipzig (1830).

Klotzsch.   Klotzsch, Johann Friedrich, 1805–1860, curator of the herbarium, Berlin; monographer of Begoniaceae.

Kl. & Gke.   Klotzsch, J. F., and F. A. Garcke.

Knuth.   Knuth, Reinhard Gustav Paul, 1874–1957, German student of Primulaceae, Geraniaceae, etc.

Knuth, F. M.   Knuth-Knuthenborg, Frederik Marcus, 1904——, German student of Cacti.

Koch.   Koch, Wilhelm Daniel Joseph, 1771–1849, professor of botany, Erlangen; author or editor of three floras of Germany and Switzerland; Umbelliferae.

Koch, K.   Koch, Karl Heinrich Emil, 1809–1879, professor in Berlin; author of "Hortus Dendrologicus" (1853).

Koehne.   Koehne, Bernhard Adalbert Emil, 1848–1918, professor in Berlin; Lythraceae.

Koel.   Koeler, Georg Ludwig, 1765–1807, professor in Mainz; author of a treatise on the grasses of Germany and France.

Kom.   Komarov, Vladimir Leontievich, 1869–1945, president, Academy of Sciences of the U.S.S.R.; student of the flora of Kamchatka, organizer and editor of "Flora of the U.S.S.R."

Konig.   Koenig, Carl Dietrich Eberhard, 1774–1851, German geologist at the British Museum; while in England (most of his life), he was known as "Charles Konig."

Kost.   Kosteletzky, Vincenz Franz, 1801–1887, writer on medical botany, Prague.

K.-Pol.   Kozo-Poliansky, Boris Mikhailovič, 1890——, Russia; Umbelliferae.

Krause.   Krause, Ernst Hans Ludwig, 1859——, marine staff doctor, Kiel; *Rubus.*

Krautter.   Krautter, Louis, 1880–1910, Pennsylvania; *Penstemon.*

Kudo.   Kudo, Yushûn, 1887–1932, director, Botanical Garden, Taihoku, Formosa.

Kuhn.   Kuhn, Friedrich Adalbert Maximilian, 1842–1894, German student of ferns.

Kükenth.   Kükenthal, Georg, 1864——, clergyman of Coburg, Germany; authority on Cyperaceae.

Kunth.   Kunth, Carl Sigismund, 1788–1850, professor of botany, Berlin; excellent systematist and
   voluminous writer.
Kuntze.   Kuntze, Carl Ernst Otto, 1843–1907, German advocate of strict priority in nomenclature;
   author of "Revisio Generum Plantarum" (1891), in which the names of over 30,000 species were
   changed.
Kunze.   Kunze, Gustav, 1793–1851, botanist and physician of Leipzig; student of ferns.
Kurtz.   Kurtz, Fritz (Federico), 1854–1920, German botanist who moved to Argentina in 1884
   as professor of botany, Córdoba.
Kusnez.   Kusnezov, Nicolai Ivanovitch, 1864–1932, Russia.
L.   Linnaeus, Carolus (afterwards Carl von Linné), 1707–1778, Sweden; founder of binomial
   nomenclature and the Sexual System of classification; the "Father of Botany."
L. f.   Linné, Carl von, the son, 1741–1783, successor to his father in the professorship of botany
   in Uppsala.
Labill.   Labillardière, Jacques Julien Houtton de, 1755–1834, France; botanist and traveler.
Laest.   Laestadius, Lars Levi, 1800–1861, pastor of parochial schools in Lapland.
Lag.   Lagasca y Segura, Mariano, 1776–1839, professor and director of the botanic garden, Madrid;
   his collections were destroyed by a mob, and he was exiled to England.
Lag. & Rodr.   Lagasca y Segura, M., and José Demetrio Rodriguez, 1780–1846, director of the
   botanic garden, Madrid.
Lagrèze-Fossat.   Lagrèze-Fossat, Adrien Rose Arnaud, 1814–1874, French advocate who wrote
   on botany from 1838–1847.
Laicharding.   Laicharding, Johann Nepomuk von, 1754–1797, Innsbruck.
Lam.   Lamarck, Jean Baptiste Pierre Antoine de Monet de, 1744–1829, famous French botanist
   and zoologist who propounded a theory of evolution by the transmission of acquired characters;
   first to use dichotomous keys in natural history in his "Flore Françoise" (1778).
Lam. & DC.   Lamarck, J. B. P. A. de M. de, and A. P. De Candolle.
Lamb.   Lambert, Aylmer Bourke, 1761–1842, vice-president of the Linnaean Society in Lon-
   don; patron of botany; author of "A Description of the Genus Pinus" (1803–1842, in 5 editions).
Lange.   Lange, Johan Martin Christian, 1818–1898, professor of botany, Copenhagen.
Larisey.   Larisey, Mary Maxine, 1909——, professor, School of Pharmacy, Medical College of
   South Carolina.
Lasch.   Lasch, Wilhelm Gottfried, 1786–1863, German apothecary.
Lawr., G. H. M.   Lawrence, George Hill Mathewson, 1910——, director, Bailey Hortorium, Cor-
   nell.
Lawr., W. E.   Lawrence, William Evans, 1883–?1950; professor of ecology, Oregon State;
   Deschampsia.
Lawson.   Lawson, George, 1827–1895, professor of natural history, Kingston, Ontario, Edin-
   burgh; Nymphaeaceae.
Lawson, P. & C.   Lawson, Peter, d. 1820, and Sir Charles Lawson, the son(?), 1794–1873, Edin-
   burgh nurserymen.
Lec. and Lam.   Lecoq, Henri, 1802–1871, and Martial Lamotte, 1820–1883, authors of a catalog
   of plants of central France (1848).
Ledeb.   Ledebour, Carl Friedrich von, 1785–1851, professor in Dorpat; author of "Flora Altaica"
   (1829–1883) and "Flora Rossica" (1842–1853).
Lehm.   Lehmann, Johann Georg Christian, 1792–1860, director of the botanic garden, Hamburg;
   authority on *Potentilla* and other genera.
Leib.   Leiberg, John Bernhard, 1853–1913, botanical collector for the U.S.D.A.
Leichtl.   Leichtlin, Max, 1831–1910, founder of the famous garden at Baden-Baden; introducer
   of numerous ornamental species.
Lejeune.   Lejeune, Alexandre Louis Simon, 1779–1858, Belgian physician.
Lem.   Lemaire, Charles Antoine, 1801–1871, professor in Ghent; student of Cactaceae and culti-
   vated plants; editor of Flore des Serres (1845–1855) and L'Illustration Horticole (1854–1869).
Lemmon.   Lemmon, John Gill, 1832–1908, pioneer California botanist and schoolteacher in
   Sierra Valley; Coniferae.
Lenz.   Lenz, Lee Wayne, 1915——, geneticist at Rancho Santa Ana Botanic Garden; *Iris*.
Leonard.   Leonard, Emery Clarence, 1892——, U.S. National Herbarium.
Lepechin.   Lepechin, Ivan, 1737–1802, Petrograd.
Less.   Lessing, Christian Friedrich, 1809–1862, German physician; student of Compositae.
Lév.   Léveillé, Augustin Abel Hector, 1863–1918, Le Mans, France; monographer of *Oenothera*.
Lewis.   Lewis, Frank Harlan, 1919——, professor of botany, California (Los Angeles); *Trichostema,
   Delphinium, Clarkia*.
Lewis & Epl.   Lewis, F. H., and C. C. Epling.
Lewis & Ernst.   Lewis, F. H., and Wallace Roy Ernst, 1928——, California (Los Angeles).
Lewis & Lewis.   Lewis, F. H., and Margaret (Ensign) Lewis. See Ensign.
Lewis & Vasek.   Lewis, F. H., and Frank Charles Vasek, 1927——, professor at California (River-
   side).
Ley.   Ley, Frances Arline (*later* Fitch), 1919——, Claremont Graduate School; *Holodiscus*.
Leyss.   Leysser, Friedrich Wilhelm von, 1731–1815, author of a flora of Halle.
L'Hér.   L'Héritier de Brutelle, Charles Louis, 1746–1800, celebrated French botanist.
Liebm.   Liebmann, Friedrik Michael, 1813–1856, Denmark; collected extensively in Mexico.
Lightf.   Lightfoot, John, 1735–1788, English clergyman; author of "Flora Scotica" (1777).
Lilja.   Lilja, Nils, 1808–1870, Sweden; author of "Skånes Flora" (1838).
Liljebl.   Liljeblad, Samuel, 1761–1815, Sweden.
Lindb.   Lindberg, Sextus Otto, 1835–1889, Finland; noted student of mosses.
Lindb. f.   Lindberg, Harald, the son, 1871——, custodian of the botanical museum of the uni-
   versity, Helsinki.

Lindbl. Lindblom, Alexis Eduard, 1807–1853, botanist of Lund.
Linden. Linden, Jean Jules, 1817–1898, director of the botanic garden in Brussels; orchidologist.
Lindl. Lindley, John, 1799–1865, professor of botany, London; editor of Edwards's Botanical Register (1829–1847), horticulturist and textbook writer.
Lindl. & Gord. Lindley, J., and G. Gordon.
Lindl. & Paxt. Lindley, J., and J. Paxton.
Lindsay. Lindsay, George Edmund, 1916——, director, Museum of Natural History, San Diego.
Lingelsh. Lingelsheim, Alexander von, 1874–1937, professor in Breslau; Oleaceae.
Link. Link, Johann Heinrich Friedrich, 1767–1851, professor of natural science and director of the botanic garden, Berlin.
Link & Otto. Link, J. H. F., and C. F. Otto.
Lint & Epl. Lint, Harold L., 1917——, California State Polytechnic College, San Dimas, and C. C. Epling.
Little. Little, Elbert Luther, Jr., 1907——, U.S. Forest Service, Washington, D.C.
Lloyd. Lloyd, James, 1810–1896, London and Nantes; student of the flora of western France.
Lloyd, F. Lloyd, Francis Ernest, 1868–1947, professor of botany, McGill; author of works on guayule and carnivorous plants.
Lloyd & Underw. Lloyd, F. E., and L. M. Underwood.
Lodd. Loddiges, Conrad, 1738–1826, English nurseryman.
Loefl. Loefling, Pehr (Peter), 1729–1756, Swedish student of Linnaeus, who traveled to Venezuela for him and died there.
Lois. Loiseleur-Deslongchamps, Jean Louis Augusta, 1774–1849, French physician; author of "Flora Gallica" (1806–1807, 2d ed. 1828).
Loja. Lojacono-Pojero, Michele, docent of botany, Palermo; published from 1878–1909.
Loud. Loudon, John Claudius, 1783–1843, English horticulturist; prolific author and editor of garden books.
Louis-Marie. Lalonde, Louis (Père Louis-Marie), 1896——, professor of botany, Institute Agronomique d'Oka, La Trappe, Quebec.
Löve, A. Löve, Askell, 1916——, professor of botany, University of Montreal; cytotaxonomist.
Lowe. Lowe, Rev. Richard Thomas, 1802–1874, Englishman who wrote on the flora of Madeira.
Ludw. Ludwig, Christian Gottlieb, 1709–1773, professor in Leipzig.
Luerss. Luerssen, Christian, 1843–1916, professor of botany and director of the botanic garden, Königsberg.
Lunell. Lunell, Joel, 1851–1920, North Dakota advocate of strict priority in nomenclature.
Lynch. Lynch, Richard Irwin, 1850–1924, curator of the Botanic Garden, Cambridge, England.
Lyon. Lyon, Harold Lloyd, 1879–1957, pathologist, Hawaiian Sugar Planters' Association.
M. & J. Munz, P. A., and I. M. Johnston.
Macbr. Macbride, James Francis, 1892——, Chicago Natural History Museum; student of the West American flora; author of "Flora of Peru."
Macbr. & Nels. Macbride, J. F., and A. Nelson.
Macbr. & Pays. Macbride, J. F., and E. B. Payson.
MacM. MacMillan, Conway, 1867–1929, state botanist, Minnesota.
McClat. McClatchie, Alfred James, died in 1906, American student of *Eucalyptus.*
McClint. & Epl. McClintock, Elizabeth May, 1912——, botanist, California Academy of Sciences, and C. C. Epling.
McDer. McDermott, Laura Frances, 1882–1923, teacher in San Diego high school; *Trifolium.*
McGreg. McGregor, Ernest Alexander, 1880——, California entomologist and botanist.
McKelvey. McKelvey, Susan Adams (Delano), 1883——, Arnold Arboretum; *Yucca.*
McMinn. McMinn, Howard Earnest, 1892——, professor of botany, Mills; author of "An Illustrated Manual of California Shrubs" (1939) and other works; *Ceanothus, Diplacus.*
McMinn, Babcock & Righter. McMinn, H. E., E. B. Babcock, and Francis Irving Righter, 1897——, U.S. Forest Service, Berkeley.
McNair. McNair, James Birtley, 1889——, California phytochemist.
Macoun. Macoun, John, 1832–1920, Government Naturalist of Canada.
Macoun, J. M. Macoun, James Melville, the son, 1862–1920, curator of the National Herbarium of Canada.
McVaugh. McVaugh, Rogers, 1909——, professor of botany, Michigan; Campanulaceae.
Maguire. Maguire, Bassett, 1904——, curator, New York Botanical Garden; flora of the Great Basin and of northeastern South America.
Maguire & Holmgren. Maguire, B., and Arthur Herman Holmgren, 1912——, professor of botany, Utah State Agricultural College.
Maire, Weiller & Wilcz. Maire, René Charles Joseph Ernest, 1878–1949, professor of botany, Alger; Marc Weiller, 19th century, French artillery officer; and Ernst Wilczek, 1867——, professor in the University, Lausanne.
Malte. Malte, Malte Oscar, 1880–1933, Chief Botanist, National Herbarium of Canada, 1921–1933.
Mansf. Mansfeld, Rudolf, 1901——, assistant in the botanical museum, Berlin.
Marsh. Marshall, Humphrey, 1722–1801, Pennsylvania; the father of American dendrology; author of "Arbustum Americanum" (1785).
Marshall & Bock. Marshall, William Taylor, cactus collector and grower, and Thor Methven Bock, artist, California authors of "Cactaceae" (1941).
Mart. Martius, Karl (Carl) Friedrich Philipp von, 1794–1868, professor in Munich; founder of the classic "Flora Brasiliensis" (1840–1906) and prolific writer on systematic botany and zoology.
Martin, F. L. Martin, Floyd L., 1909——, Claremont Graduate School, *Cercocarpus.*
Martin. Martin, James Stillman, 1914——, Washington, *Trifolium.*
Masf. Masferrer y Arquimbau, Ramón, 1850–1884, Spain.

Mason.   Mason, Herbert Louis, 1896——, professor of botany, California; Polemoniaceae and fossil floras of California.
Mason & Bacig.   Mason, H. L., and R. C. Bacigalupi.
Mason & A. Grant.   Mason, H. L., and A. D. H. Grant.
Mason, C. T.   Mason, Charles Thomas, nephew of H. L., 1918——, professor of botany, Arizona.
Mason, S. C.   Mason, Silas Cheever, 1857–1935, senior horticulturist, U.S.D.A., Riverside; expert on date culture.
Mast.   Masters, Maxwell Tylden, 1833–1907, England; editor of The Gardeners' Chronicle; contributor to Martius's "Flora Brasiliensis" and Oliver's "Flora of Tropical Africa."
Math.   Mathias, Mildred Esther (*later* Hassler), 1906——, professor of botany, California (Los Angeles); Umbelliferae.
Math. & Const.   Mathias, M. E., and L. Constance.
Mattf.   Mattfeld, Johannes, 1895–1951, curator in the herbarium, Berlin; Compositae, Caryophyllaceae, Cyperaceae.
Maxim.   Maximowicz, Carl Johann (Karl Ivanovich Maksimovich), 1827–1891, director of the botanic garden, St. Petersburg.
Maxon.   Maxon, William Ralph, 1877–1948, Curator of Plants, U.S. National Herbarium; ferns.
Mayr.   Mayr, Heinrich, 1856–1911, German forester; author of a work on the forests of North America (1890).
Medic.   Medicus, Friedrich Casimir, 1736–1808, Germany.
Meigen.   Meigen, Johann Wilhelm, 1764–1845, Germany; author of a German flora, but published principally as an entomologist.
Meinsh.   Meinshausen, Karl Friedrich, 1819–1899, Russia.
Meissn.   Meisner (*also* Meissner), Carl Friedrich, 1800–1874, professor of botany, Basel; Polygonaceae, Lauraceae, Ericaceae, Convolvulaceae.
Menz.   Menzies, Archibald, 1754–1842, Scotland; first botanist to visit Pacific Coast of North America; collected in California on three successive years, 1792–1794, as surgeon on the English ship *Discovery;* presented his plants to Banks.
Merr.   Merrill, Elmer Drew, 1876–1956, administrator of the botanical collections, Harvard; prolific contributor on the flora of the Philippine Islands, China, and Indo-Malaysia; at first a student of American grasses.
Merr. & Davy.   Merrill, E. D., and J. B. Davy.
Merriam.   Merriam, Clinton Hart, 1855–1942, founder and chief of U.S. Bureau of Biological Survey.
Mert. & Koch.   Mertens, Franz Carl, 1764–1831, professor in Bremen, and W. D. J. Koch.
Mett.   Mettenius, Georg Heinrich, 1823–1866, professor of botany, Leipzig; *Salvinia.*
Mey., C. A.   Meyer, Carl Anton von, 1795–1855, director of the botanic garden, St. Petersburg.
Mey., E.   Meyer, Ernst Heinrich Friedrich, 1791–1858, professor of botany, Königsberg; Juncaceae.
Mey., F. G.   Meyer, Frederick G., 20th century, Missouri Botanical Garden; *Valeriana.*
Mey., G. F. W.   Meyer, Georg Friedrich Wilhelm, 1782–1856, professor in Göttingen; author of Flora of Hanover.
Meyen & Walp.   Meyen, Franz Julius Ferdinand, 1804–1840, German botanical artist who collected in Brazil and Peru; and W. G. Walpers.
Mez.   Mez, Carl Christian, 1866–1944, professor in Königsberg; Bromeliaceae.
Micheli.   Micheli, Pietro Antonio, 1679–1737, director of the gardens in Florence; Onagraceae.
Michx.   Michaux, André, 1746–1802, France and United States; author of "Flora Boreali-Americana" (1803); *Quercus.*
Michx., f.   Michaux, François André, the son, 1770–1855, France; author of "The North American Sylva" (1817–1819, first English ed.).
Miers.   Miers, John, 1789–1879, London; student of the South American flora.
Milde.   Milde, Carl August Julius, 1824–1871, German student of Pteridophyta.
Mill.   Miller, Philip, 1691–1771, England; author of "The Gardeners Dictionary" (1731), which went through eight editions.
Miller, G. N.   Miller, Gertrude N., 20th century, Cornell; *Fraxinus.*
Millsp.   Millspaugh, Charles Frederick, 1854–1923, curator, Department of Botany, Field Museum, 1894–1923.
Millsp. & Chase.   Millspaugh, C. F., and M. A. Chase.
Millsp. & Nutt.   Millspaugh, C. F., and Lawrence William Nuttall, 1857–1933, authors of "Flora of Santa Catalina Island" (1923).
Millsp. & Sherff.   Millspaugh, C. F., and E. E. Sherff.
Miq.   Miquel, Frederik Anton Willem, 1811–1871, professor of botany, Utrecht; Urticaceae, Primulaceae.
Mirb.   Mirbel, Charles François Brisseau de, 1776–1854, Paris.
Mitch.   Mitchell, John, 1676–1768, English-born Virginia physician.
Mkze.   Mackenzie, Kenneth Kent, 1877–1934, New York lawyer; authority on American *Carex.*
Mkze. & Bush.   Mackenzie, K. K., and B. F. Bush.
Mlkn.   Milliken, Jessie (*later* Brown), 1877——, California; Polemoniaceae.
Moç.   Moçino, José Mariano, 1757–1820, Mexican physician; collected in California in 1792.
Moç. & Ses.   Moçiño, J. M., and M. de Sessé y Lacasta; their works on the flora of New Spain and Mexico were published in the latter part of the 19th century.
Moehr.   Moehring, Paul Heinrich Gerhard, 1710–1792, physician, botanist and ornithologist of Oldenburg.
Moench.   Moench, Conrad, 1744–1805, professor of botany, Marburg.
Mol.   Molina, Juan Ignacio (Giovanni Ignazio), 1740–1829, Chilean Jesuit; author of a natural history of Chile (1782).
Mold.   Moldenke, Harold Norman, 1909——, New York Botanical Garden; Verbenaceae.

Moore, H. E.   Moore, Harold Emery, 1917——, professor of botany, Bailey Hortorium, Cornell.
Moore, T.   Moore, Thomas, 1821–1887, curator, Chelsea Botanic Garden, England; student of ferns and orchids.
Moq.   Moquin-Tandon, Christian Horace Benedict Alfred, 1804–1863, France; Chenopodiaceae, Amaranthaceae.
Moran.   Moran, Reid Venable, 1916——, Stanford and Cornell; Crassulaceae.
Moretti.   Moretti, Giuseppe, 1782–1853, professor of botany and director of the botanic garden, Pavia.
Moric.   Moricand, Moïse Étienne, 1779–1854, Geneva, Switzerland.
Morière.   Morière, Jules (Pierre Gilles Morière), 1817–1888, Normandy.
Moris.   Moris, Giuseppe Giacinto, 1796–1869, professor of botany, Turin.
Morong.   Morong, Rev. Thomas, 1827–1894, Massachusetts amateur botanist; Naiadaceae.
Morr. & Dec.   Morren, Charles François Antoine, 1807–1858, professor of botany, Ghent, and J. Decaisne.
Morris.   Morris, Edward Lyman, 1870–1913, high-school teacher, Washington, D.C.; Plantaginaceae.
Morrison.   Morrison, John Laurence, 1911——, California.
Mort.   Morton, Conrad Vernon, 1905——, curator, U.S. National Herbarium.
Mox.   Moxley, George Loucks, 1871——, student of the southern California flora from 1898–1923.
Muell.-Arg.   Mueller, Jean (Argoviensis, i.e., of Aargau), 1828–1896, Switzerland; Euphorbiaceae, Buxaceae, Resedaceae.
Muell., C.   Mueller, Karl, 1820–1889, German horticulturist.
Muell., F.   Mueller, Baron Ferdinand Jacob Heinrich von, 1825–1896, director, Melbourne Botanic Garden; eminent student of the Australian flora.
Muell., P. J.   Mueller, Philipp Jakob, 1832–1889, Alsatian botanist; student of European *Rubus*.
Muhl.   Muhlenberg, Henry (*formerly* Gotthilf Heinrich Ernst Muehlenberg, also Heinrich Ludwig Muehlenberg), 1753–1815, Lutheran minister and pioneer botanist of Pennsylvania.
Mull., C. H.   Muller (formerly Mueller), Cornelius Herman, 1909——, professor at California (Santa Barbara).
Munro.   Munro, William, 1818–1880, English general and agrostologist.
Munson.   Munson, Thomas Volney, 1843–1913, grape breeder of Dennison, Texas.
Munz.   Munz, Philip Alexander, 1892——, professor of botany, Pomona, later director, Rancho Santa Ana Botanic Garden; author of "Manual of Southern California Botany" (1935); Onagraceae.
Munz. & Hitchc.   Munz, P. A., and C. L. Hitchcock.
Munz. & Keck.   Munz, P. A., and D. D. Keck.
Munz & McBurn.   Munz, P. A., and Jean (Pitzer) McBurney, 1909——, Pomona College.
Munz & Roos.   Munz, P. A., and John Christian Roos, 1918——, College of Medical Evangelists, Loma Linda, California.
Munz & West.   Munz, P. A., and Louise West (*later* Voss, *later* Ashcroft), 1910——, Pomona College.
Murbeck.   Murbeck, Svante Samuel, 1859–1946, professor of botany, Lund.
Murr, J.   Murr, Josef, 1864–1932, Switzerland.
Murr.   Murray, Johann Anders, 1740–1791, Swedish professor of medicine and botany, Göttingen; student of Linnaeus and editor of some of his later editions.
Murr., A.   Murray, Andrew, 1812–1878, Edinburgh entomologist; student of conifers.
Mutis.   Mutis, José Celestino, 1732–1811, Spanish botanical explorer in Colombia; correspondent of Linnaeus.
Nakai.   Nakai, Takenoshin, 1882–1952, professor of botany, Tokyo.
Nash.   Nash, George Valentine, 1864–1921, head gardener, New York Botanical Garden; agrostologist.
Naud.   Naudin, Charles Victor, 1815–1899, director of the botanic garden, Villa Thuret, Antibes; student of the flora of the Pyrenees, of the Solanaceae, Melastomaceae, Cucurbitaceae, and *Eucalyptus*.
Neck.   Necker, Noel Joseph de, 1729–1793, Mannheim.
Neé.   Neé, Luis, 18th–19th century, botanist with T. Haenke on the Malaspina Expedition, visiting California in 1791; his collections are preserved at Madrid.
Nees.   Nees von Esenbeck, Christian Gottfried Daniel, 1776–1858, professor of botany, Breslau; prolific botanical writer; Acanthaceae, Cyperaceae, Gramineae.
Nees & Arn.   Nees von Esenbeck, C. G. D., and G. A. W. Arnott.
Nees & Mey.   Nees von Esenbeck, C. G. D., and F. J. F. Meyen.
Nels., A.   Nelson, Aven, 1859–1952, professor of botany and president, Wyoming; student of the Rocky Mountain flora; reviser of Coulter's "New Manual of Botany of the Central Rocky Mountains" (1909).
Nels., E.   Nelson, Elias Emanuel, 1876——, U.S.D.A. Experimental Farm, Bend, Oregon; horticulture, forage plants for arid regions.
Nels., J. C.   Nelson, James Carlton, 1867–1944, high-school teacher in Salem, Oregon.
Nels. & Kenn.   Nelson, A., and P. B. Kennedy.
Nels. & Macbr.   Nelson, A., and J. F. Macbride.
Nestl.   Nestler, Christian Gottfried, 1778–1832, professor of botany, Strassburg; monographer of *Potentilla* (1816).
Nevskii.   Nevskiĭ, Sergeĭ Arsenjevič, 1908–1938, senior agrostologist at Botanical Institute of Academy of Science in U.S.S.R., Leningrad.
Newb.   Newberry, John Strong, 1822–1892, physician and botanist on the Pacific Railroad Survey under Lt. Williamson; also with Capt. Ives on the Colorado River in 1857–58; later noted as a paleobotanist and geologist at Columbia School of Mines.
Newm.   Newman, Edward, 1801–1876, English naturalist, entomologist, ornithologist, student of ferns.

Newsom.   Newsom, Vesta Marie (*later* Griffith), 1902–1958, Pomona College; *Collinsia.*
Nichols.   Nicholson, George, 1874–1908, curator at Kew; author of "Illustrated Dictionary of Gardening," 1885–1889.
Niedz.   Niedenzu, Franz Josef, 1857–1937, Germany; Malpighiaceae.
Nieuwl.   Nieuwland, Julius Aloysius Arthur, 1878–1936, professor of botany (1904–1918) and professor of organic chemistry (1918–1936), Notre Dame; founder and first editor of The American Midland Naturalist.
Nort.   Norton, John Bitting Smith, 1872——, plant pathologist, University of Maryland.
Nutt.   Nuttall, Thomas, 1786–1859, Philadelphia; made important collections in California in 1835 from San Francisco to San Diego; author of "The Genera of North American Plants" (1818), "The North American Sylva" (1842); also a noted ornithologist.
Nygren.   Nygren, Axel, 1912——, cytologist of Uppsala.
Nym.   Nyman, Carl Fredrik, 1820–1893, curator of the herbarium, Stockholm.
Oakes.   Oakes, William, 1799–1849, Massachusetts; student of the Vermont flora.
Oeder.   Oeder, Georg Christian von, 1728–1791, professor of botany, Copenhagen; first editor of "Flora Danica" (1761–1771).
Oerst.   Oersted, Anders Sandøe, 1816–1872, Danish botanist who collected in Costa Rica and Colombia.
Ogden.   Ogden, Eugene Cecil, 1905——, state botanist, New York.
Olney.   Olney, Col. Stephen Thayer, 1812–1878, Providence, Rhode Island; student of *Carex;* bequeathed his rich herbarium, together with an endowment, to Brown University.
O'Neill.   O'Neill, Hugh Thomas, 1894——, curator of Langlois Herbarium, Catholic University, Washington; Cyperaceae.
Onno.   Onno, Max, 1903——, German botanist working in Vienna.
Opiz.   Opiz, Philipp Maximilian, 1787–1858, zealous Bohemian botanist who named great numbers of "species" of little worth.
Orcutt.   Orcutt, Charles Russell, 1864–1929, San Diego collector.
Ort.   Gómez Ortega, Casimiro, 1740–1818, director of the botanic garden, Madrid.
Ortgies.   Ortgies, Karl Eduard, 1829–1916, plant collector connected with the Zürich botanical garden.
Osterh.   Osterhout, George Everett, 1858–1937, Colorado amateur botanist.
Ottley.   Ottley, Alice Maria, 1882——, professor of botany, Wellesley; *Lotus.*
Otto.   Otto, Christoph Friedrich, 1783–1856, garden director of Schöneberg, near Berlin.
Otto & Dietr.   Otto, C. F., and A. Dietrich.
Ownbey.   Ownbey, Francis Marion, 1910——, professor of botany, State College of Washington; *Calochortus, Allium.*
Ownbey & Aase.   Ownbey, F. M., and Hannah Caroline Aase, 1883——, professor of botany, State College of Washington.
Ownbey, G.   Ownbey, Gerald Bruce, 1916——, brother of Marion, professor of botany, Minnesota.
Pall.   Pallas, Peter Simon, 1741–1811, German student of the Russian and Siberian flora; early monographer of *Astragalus* (1800); also eminent as a zoologist.
Palla.   Palla, Eduard, 1864–1922, professor of botany, Graz; Cyperaceae.
Palmer,   Palmer, Ernest Jesse, 1875——, field collector for Missouri Botanical Garden and Arnold Arboretum, 1913–1948.
Parish.   Parish, Samuel Bonsall, 1838–1928, pioneer botanist of San Bernardino who made known much of the southern California flora.
Parker.   Parker, Kittie Lucille (Fenley). *See* Fenley.
Parl.   Parlatore, Filippo, 1816–1877, professor of botany, Florence; Gnetaceae, Coniferae.
Parn.   Parnell, Richard, 1810–1882, Scotch ink manufacturer and agrostologist.
Parodi.   Parodi, Lorenzo Raimundo, 1895——, agrostologist of Buenos Aires.
Parry.   Parry, Charles Christopher, 1823–1890, Colorado and Iowa botanist with the Mexican Boundary Survey.
Parry & Gray.   Parry, C. C., and A. Gray.
Pax.   Pax, Ferdinand Albin, 1858–1942, professor of botany, Breslau; Aceraceae, Primulaceae.
Pax & K. Hoffm.   Pax, F. A., and Kaethe Hoffmann, professor in Breslau; made the huge contribution to Das Pflanzenreich on the Euphorbiaceae.
Paxt.   Paxton, Sir Joseph, 1801–1865, England; editor of Paxton's Magazine of Botany and author of other horticultural works.
Pays.   Payson, Edwin Blake, 1893–1927, professor of botany, Wyoming; Cruciferae, *Cryptanthe.*
Pays. & St. John.   Payson, E. B., and H. St. John.
Peck.   Peck, Morton Eaton, 1871——, professor of biology, Willamette; author of "A Manual of the Higher Plants of Oregon" (1941).
Peck & Appleg.   Peck, M. E., and E. I. Applegate.
Peeb.   Peebles, Robert Hibbs, 1900–1956, cotton agronomist, U.S.D.A.
Penl.   Penland, Charles William Theodore, 1899——, professor of botany, Colorado College.
Penn.   Pennell, Francis Whittier, 1886–1952, curator of botany, Academy of Natural Sciences of Philadelphia; authority on Scrophulariaceae.
Penn. & Keck.   Pennell, F. W., and D. D. Keck.
Perdue.   Perdue, Robert Edward, Jr., 1924——, botanist, U.S.D.A., Beltsville.
Perry, L. M.   Perry, Lily May, 1895——, botanist, Arnold Arboretum; *Verbena.*
Pers.   Persoon, Christian Hendrik, 1761–1836, bizarre individual and brilliant mycologist, author of "Synopsis Plantarum" (1805–1807), and other valuable botanical works.
Peterm.   Petermann, Wilhelm Ludwig, 1806–1855, Leipzig.
Petr.   Petrak, Franz, 1886——, Vienna mycologist and monographer of *Cirsium.*
Petrovic.   Petrović, Sava, 1839–1889, Serbian botanist.

Pfeiffer. Pfeiffer, Norman Etta, 1889——, Boyce Thompson Institute for Plant Research; *Isoetes.*
Phil. Philippi, Rudolf Amandus, 1808–1904, director, Museo Nacional and professor of botany, Santiago; student of the botany, zoology, and paleontology of Chile.
Phillips. Phillips, Lyle L., 1923——, Claremont Graduate School and University of Washington; *Lupinus.*
Pick. Pickering, Charles, 1805–1878, physician and botanist on the Wilkes Expedition; plant geographer.
Pierce & Fosb. Pierce, Wright, Claremont, California, photographer and ornithologist, and F. R. Fosberg.
Pilg. Pilger, Robert Knud Friedrich, 1876–1953, director of the botanic garden and museum, Berlin; Plantaginaceae, Coniferae.
Piper. Piper, Charles Vancouver, 1867–1926, agrostologist, U.S.D.A.; author of "Flora of the State of Washington" (1906).
Piper & Brodie. Piper, C. V., and David Arthur Brodie, 1868——, U.S.D.A.
Planch. Planchon, Jules Émile, 1823–1888, professor of botany, Montpellier; Ulmaceae.
Plum. Plumier, Charles, 1646–1704, French Franciscan monk who wrote on the flowering plants and ferns of tropical America.
Poe. Poe, Ione, 1901——, Pomona College; *Plantago.*
Poelln. Poellnitz, Dr. Karl von, contemporary farmer and amateur botanist of Thüringen.
Poepp. Poeppig, Eduard Friedrich, 1798–1868, professor of zoology, Leipzig, who published on plants collected during a trip to Chile, Peru, and the Amazon.
Pohl. Pohl, Richard Walter, 1916——, professor of botany, Iowa State.
Poir. Poiret, Jean Louis Marie, 1755–1834, French botanist, who completed Lamarck's "Encyclopédie Méthodique Botanique."
Poll. Pollich, Johann Adam, 1740–1780, author of a flora of the Palatinate.
Pollard. Pollard, Charles Louis, 1872–1945, Vermont librarian and collector.
Porter. Porter, Thomas Conrad, 1822–1901, professor of botany, Lafayette College, Pennsylvania; student of the Colorado flora.
Porter & Coult. Porter, T. C., and J. M. Coulter.
Porter, C. L. Porter, Cedric Lambert, 1905——, professor of botany, Wyoming.
Pospischal. Pospischal, Alfred, *Convolvulus.*
Post. Post, Douglas Manners, 1920——, California, Illinois; *Frasera.*
Praeger. Praeger, Robert Lloyd, 1865–1953, librarian, National Library, Dublin; Crassulaceae.
Prain. Prain, Sir David, 1857–1944, director of Kew, 1905–1922; an editor of Hooker's Icones Plantarum and Curtis's Botanical Magazine; student of the flora of India.
Prantl. Prantl, Karl Anton Eugen, 1849–1893, professor of botany, Breslau.
Prescott. Prescott, John D., died in 1837 in Petrograd; his herbarium of 28,000 sheets, rich in collections of Douglas, Nuttall, and Scouler, is now at Oxford.
Presl. Presl, Karel Bořiwog, 1794–1852, professor of natural history, Prague; in "Reliquiae Haenkeanae" he described the collections made along the western side of the American continent by Thaddaeus Haenke, the first botanist to visit California.
Presl, J. Presl, Jan Swatopluk, the brother, 1791–1849, professor in Prague.
Presl, J. & C. Presl, J. S., and K. B. Presl, authors of "Flora Čechica" (1819).
Pritz. Pritzel, Georg August, 1815–1874, bibliographer, Academy of Sciences, Berlin; Lycopodiaceae, *Anemone;* author of the invaluable "Thesaurus Literaturae Botanicae" (1851, 2d ed. 1872–[1877]).
Prov. Provancher, Léon, Abbé, 1820–1892, Quebec naturalist; founder of Le Naturaliste Canadien, author of "Flore Canadienne" (1862).
Purdy. Purdy, Carlton Elmer (*also* Carl Purdy), 1861–1945, Ukiah, California nurseryman specializing in native bulbs.
Purpus. Purpus, Joseph Anton, 1860–1932, botanical collector in southwestern United States and Mexico.
Pursh. Pursh, Frederick Traugott, 1774–1820, born in Saxony, settled in Philadelphia; author of "Flora Americae Septentrionalis" (1814).
Quick. Quick, Clarence Roy, 1902——, U.S. Forest Service, Berkeley; *Ribes.*
R. & P. Ruiz Lopez, Hipólito, 1754–1815, and José Antonio Pavon, 175(?)–1844, Spanish botanical explorers and authors of a flora of Peru and Chile (1794 and 1798–1802).
R. & S. Roemer, J. J., and J. A. Schultes, produced an edition of Linnaeus's "Systema Vegetabilium" (1817–1830).
Rabenh. Rabenhorst, Gottlob Ludwig, 1806–1881, German author of cryptogamic floras.
Raeusch. Raeuschel, Ernst Adolf, 18th–19th century, Germany.
Raf. Rafinesque (*or* Rafinesque-Schmaltz), Constantine Samuel, 1783–1840, Kentucky; brilliant, eccentric pioneer naturalist; profligate author of binomials, with many "species" quite untraceable.
Raim. Raimann, Rudolf, 1863–1896, Vienna.
Raffenau-Delile. Delile, Alire Raffeneau, 1778–1850, professor in Montpellier, France.
Rapaices. Rapaics von Ruhmwerth, Raymund, 1885——, professor in the Agricultural Academy, Budapest.
Rattan. Rattan, Volney, 1840–1915, gifted teacher of botany in California schools; author of "A Popular California Flora" (1879) that went through 8 editions.
Raup. Raup, Hugh Miller, 1901——, research associate, Arnold Arboretum; botanical explorer in northwestern Canada.
Ray. Ray (*or* Wray), Rev. John, 1627–1705, England; author of the classical "Historia Plantarum" (1686–1704).
Rchb. Reichenbach, Heinrich Gottlieb Ludwig, 1793–1879, professor in Dresden; first editor of Icones Florae Germanicae et Helveticae and author of many other extensive works on plants and animals.

Rchb. f.   Reichenbach, Heinrich Gustav, the son, 1823–1889, professor of botany and director of the botanic garden, Hamburg; orchidologist.

Rchb. f. & Baker.   Reichenbach, H. G., and J. G. Baker.

Rech. f.   Rechinger, Karl Heinz, 1906——, Natural History Museum, Vienna; flora of the Mediterranean and the Near East; *Rumex*.

Reed.   Reed, Clyde Franklin, 20th-century; Baltimore, Maryland; student of ferns.

Regel.   Regel, Eduard August von, 1815–1892, director of the botanic garden, St. Petersburg; editor of Gartenflora; Betulaceae.

Rehd.   Rehder, Alfred, 1863–1949, curator of the herbarium, Arnold Arboretum; author of "Manual of Cultivated Trees and Shrubs Hardy in North America" (1927, 2d ed. 1940).

Reichard.   Reichard, Johann Jacob, 1743–1782, Frankfurt am Main.

Reiche.   Reiche, Karl Friedrich (*later* Carlos Federico Reiche), 1860–1929, German author of "Flora de Chile" (1896–1911).

Remy.   Remy, Ezechiel Jules, 1826–1893, French student of the Andean flora.

Rendle.   Rendle, Alfred Barton, 1865–1938, keeper of botany, British Museum.

Retz.   Retzius, Anders Johan, 1742–1821, professor in Lund; prolific writer on botany and zoology.

Reuthe.   Reuthe, G., German horticultural writer of the 1880's.

Rich.   Richard, Louis Claude Marie, 1754–1821, French collector in South America and the West Indies.

Rich., A.   Richard, Achille, 1794–1852, the son, physician and botanical demonstrator to the faculty of medicine, Paris.

Richards.   Richardson, Sir John, 1787–1865, Scotch botanist and zoologist attached to Capt. Sir John Franklin's expedition to arctic America.

Richt.   Richter, Karl (Carl), 1855–1891, Vienna; collector of a large herbarium.

Ricker.   Ricker, Percy Leroy, 1878——, U.S.D.A., Washington; Gramineae, Leguminosae.

Riehl.   Riehl, Nicholas, 1808–1852, St. Louis, Missouri; Nurseryman.

Riv.   Rivinus, August Quirinus (Bachmann), 1652–1723, professor of botany, Leipzig.

Rob.   Robinson, Benjamin Lincoln, 1864–1935, curator of Gray Herbarium, 1892–1935; student of Compositae; an editor of Gray's "Synoptical Flora of North America."

Rob. & Greenm.   Robinson, B. L., and J. M. Greenman.

Robbins.   Robbins, Guy Thomas, 1916——, Jepson Herbarium, California.

Robbins, J. W.   Robbins, James Watson, 1801–1879, Massachusetts.

Rock.   Rock, Howard Francis Leonard, 1925——, professor of botany, Tennessee.

Roehl.   Roehling, Johann Christoph, 1757–1813, German clergyman; author of "Deutschlands Flora" (1796), with later editions.

Roem.   Roemer, Johann Jacob, 1763–1819, professor of botany, Zürich; active editor.

Roem., M.   Roemer, Max J., Germany; published in the first half of the 19th century.

Roezl & Leichtl.   Roezl, Benito (Benedict), 1824–1885, Austrian seedsman and collector, who collected near San Diego in 1869 and 1870 and was near Lake Tahoe before 1880, also in Mexico and Central America, and M. Leichtlin.

Rohrb.   Rohrbach, Paul, 1847–1871, Berlin; Portulacaceae, *Silene*.

Rolfe.   Rolfe, Robert Allen, 1855–1921, Kew gardener and orchid specialist.

Roll.   Rollins, Reed Clark, 1911——, director of Gray Herbarium; Cruciferae; *Parthenium*.

Rose.   Rose, Joseph Nelson, 1862–1928, associate botanist, U.S. National Herbarium; Cactaceae, Umbelliferae, and Crassulaceae.

Rose & Davids.   Rose, J. N., and A. Davidson.

Rose & Jtn.   Rose, J. N., and I. M. Johnston.

Rose & Painter.   Rose, J. N., and Joseph Hannum Painter, 1879–1908, U.S. National Museum.

Rose & Standl.   Rose, J. N., and P. C. Standley.

Rosend.   Rosendahl, Carl Otto, 1875–1956, professor of botany, Minnesota; Saxifragaceae.

Rosend. & Butt.   Rosendahl, C. O., and F. K. Butters.

Rosend., Butt. & Lak.   Rosendahl, C. O., F. K. Butters, and Olga Lakela, 1890——, professor of botany, Minnesota (Duluth Branch).

Rosend. & Rydb.   Rosendahl, C. O., and P. A. Rydberg.

Rossb., G.   Rossbach, George Bowyer, 1910——, professor, West Virginia Wesleyan College; *Erysimum*.

Rossb., R. P.   Rossbach, Ruth (Peabody) (*later* Berendsen), 1915?——, Gray Herbarium; *Spergularia*.

Rostk. & Schmidt.   Rostkovius, Friedrich Wilhelm Gottlieb, 1770–1848, physician in Stettin, and Wilhelm Ludwig Ewald Schmidt, 1804–1843, Stettin, authors of "Flora Sedinensis" (1824).

Roth.   Roth, Albrecht Wilhelm, 1757–1834, German physician.

Rothm.   Rothmaler, Werner, contemporary botanist of Griefswald, Germany.

Rothr.   Rothrock, Joseph Trimble, 1839–1922, professor of botany, Pennsylvania; surgeon and botanist on Lt. Wheeler's Survey.

Rottb.   Rottboell, Christen Friis, 1727–1797, professor of botany and director of the botanical garden, Copenhagen.

Rouleau.   Rouleau, Joseph Albert Ernest, 1916——, Montreal.

Roush.   Roush, Eva Myrtelle (Fling), 1890——, Arnold Arboretum; *Sidalcea*.

Rouy.   Rouy, Georges C. Ch., 1851–1924, senior author of "Flore de France" (1893–1913).

Rowland.   Rowland, Verner Hawsbrook, 1883——, high-school principal and school superintendent in Wyoming and Nebraska.

Rowlee.   Rowlee, Willard Winfield, 1861–1923, American student of willows.

Roxb.   Roxburgh, William, 1751–1815, Scotch physician and director of the Royal Botanic Gardens, Calcutta; author of "Flora Indica" (1820–1824).

Ruempl.   Ruempler, Theodor, 1817–1891, Germany.

Rupp. Ruppius (Rupp), Heinrich Bernhard, 1688–1719, author of a flora of Jena (1718) with later editions.

Rupr. Ruprecht, Franz Joseph, 1814–1870, curator of the herbarium of the Academy of Science, St. Petersburg; Gramineae, Umbelliferae, *Botrychium*, algae.

Rusby. Rusby, Henry Hurd, 1855–1940, dean of the New York College of Pharmacy; an active collector in South America.

Russell. Russell, Norman Hudson, 1921——, professor of botany, Grinnell College; *Viola*.

Rydb. Rydberg, Per Axel, 1860–1931, curator, New York Botanical Garden; author of "Flora of Colorado" (1906), "Flora of the Rocky Mountains and Adjacent Plains" (1917), "Flora of the Prairies and Plains of Central North America" (1932), etc.

Rylands. Rylands, Thomas Glazebrook, 1818–1900, English wire manufacturer, diatomist, and fern student.

Sab. Sabine, Joseph, 1770–1837, secretary of the Horticultural Society of London.

Salisb. Salisbury, Richard Anthony (*born* Markham), 1761–1829, England; early proponent of the natural system of classification.

Salm-Dyck. Salm-Reifferscheid-Dyck, Prince Joseph Franz Maria Anton Hubert Ignaz, 1773–1861, Germany; owner of a fine living collection of succulents; Cactaceae.

Salzm. Salzmann, Philipp, 1781–1851, born in Germany, botanized in Brazil, Spain, North Africa, south France.

Samuels. Samuelsson, Gunnar, 1885–1944, professor of botany, Uppsala.

Sanson. Sanson, M., Petrograd botanist of the 1830's.

Sarg. Sargent, Charles Sprague, 1841–1927, creator of Arnold Arboretum; author of "The Silva of North America" (1891–1902), "Manual of the Trees of North America" (1905).

Savi. Savi, Gaetano, 1769–1844, Italian author of a flora of Pisa.

Schaffn. Schaffner, Wilhelm, died in 1882, German collector in Mexico.

Schaffn., J. H. Schaffner, John Henry, 1866–1939, professor of botany, Ohio State; *Equisetum*; phylogeny.

Schauer. Schauer, Johann Conrad, 1813–1848, professor at Griefswald; Verbenaceae.

Sch. Bip. Schultz, Carl Heinrich, *Bipontinus* (i.e., of Zweibrücken), 1805–1867, Germany; Compositae.

Scheele. Scheele, Georg Heinrich Adolf, 1808–1864, German botanist who described plants from Texas (1849).

Scheutz. Scheutz, Nils Johan Wilhelm, 1836–1889, Sweden.

Schinz & Keller. Schinz, Hans, 1858–1941, director, botanic garden and museum, Zürich, and editor of numerous works on the flora of subtropical Africa, and Robert Keller, 19th–20th century, rector of the gymnasium, Winterthur, authors of "Flora der Schweiz" (1900, cf. next entry).

Schinz & Thell. Schinz, H., and A. Thellung, authors of the 4th edition of Schinz and Keller's "Flora der Schweiz" (1923).

Schk. Schkuhr, Christian, 1741–1811, university mechanic in Wittenberg; student of the German flora.

Schlecht. Schlechtendal, Diederich Franz Leonhard von, 1794–1866, professor of botany, Halle; Elaeagnaceae.

Schlecht. & Cham. Schlechtendal, D. F. L. von, and A. von Chamisso.

Schlechter. Schlechter, Friedrich Reichardt Rudolf, 1872–1925, Berlin; famous botanical collector.

Schleich. Schleicher, Johann Christoph, 1768–1834, Switzerland.

Schleid. Schleiden, Matthias, Jakob, 1804–1881, Germany; author of botanical handbooks.

Schmidt, F. Schmidt, Friedrich, 1832–1908, St. Petersburg paleontologist who made some studies of the eastern Siberian flora.

Schmidt, F. W. Schmidt, Franz Wilibald, 1764–1796, professor of botany, Prague.

Schmoll. Schmoll, Hazel Marguerite, 1890——, Chicago Natural History Museum.

Schneid., C. K. Schneider, Camillo Karl (*formerly* Carl Camillo), 1876–1951, Austria and Germany; dendrologist.

Schoenl. Schoenlein, Johann Lucas, 1793–1864, Bamberg, Bavaria.

Schott. Schott, Heinrich Wilhelm, 1794–1865, director of the royal garden in Schönbrunn, Vienna; Araceae.

Schrad. Schrader, Heinrich Adolph, 1767–1836, professor of botany, Göttingen; monographer of *Verbascum*.

Schrank. Schrank, Franz von Paula von, 1747–1835, professor of botany, Munich; author of floras of Bavaria, Salisburg, and Monaco.

Schreb. Schreber, Johann Christian Daniel von, 1739–1810, professor in Erlangen; editor of the 8th edition of Linnaeus's "Genera Plantarum" (1789).

Schreib. Schreiber, Beryl Olive (*later* Jesperson), 1911——, U.S. Forest Service; later with Sunset Magazine.

Schrenk. Schrenk, Alexander Gustav von, 1816–1876, of the botanic garden, St. Petersburg.

Schub. Schubert, Bernice Giduz, 1913——, U.S.D.A., Beltsville.

Schult. Schultes, Joseph August, 1773–1831, professor of botany, Vienna, Cracow, and Landeshut.

Schult. f. Schultes, Julius Hermann, the son, 1804–1840, Vienna.

Schultz, F. Schultz, Friedrich Wilhelm, 1804–1876, German physician, brother of Karl Heinrich (Schultz-Bipontinus); student of the flora of the upper Rhine.

Schulz, E. D. Schulz, Ellen Dorothy (*later* Quillin), 1892——, San Antonio, Texas; popular botanical author.

Schulz, O. E. Schulz, Otto Eugen, 1874–1936, German taxonomist; Cruciferae.

Schum. Schumacher, Heinrich Christian Friederich, 1757–1830, Danish botanist.

Schum., K. Schumann, Karl Moritz, 1851–1904, curator of the herbarium, Berlin; Cactaceae.

Schw. Schweinitz, Lewis David von, 1780–1834, Pennsylvania clergyman; noted student of fungi.

Schwant. Schwantes, G., fl. 1928–1932, professor in Kiel; Mesembryanthemeae.

Schwarz.   Schwarz, O., Vienna.
Scop.   Scopoli, Johann Anton (Giovanni Antonio), 1723–1788, physician and professor of natural
   history, Pavia.
Scribn.   Scribner, Frank Lamson, 1851–1938, agrostologist, U.S.D.A., Washington.
Scribn. & Ball.   Scribner, F. L., and C. R. Ball.
Scribn. & Merr.   Scribner, F. L., and E. D. Merrill.
Scribn. & Sm.   Scribner, F. L., and J. G. Smith.
Scribn. & Tweedy.   Scribner, F. L., and Frank Tweedy, 1854–1937, topographic engineer, U.S.
   Geological Survey; collected in Yellowstone Park and the Pacific Northwest.
Scribn. & Will.   Scribner, F. L., and Thomas Albert Williams, 1865–1900, agrostologist, U.S.D.A.
Seem.   Seemann, Berthold Carl, 1825–1871, German naturalist and world traveler living in Eng-
   land; author of "Botany of the Voyage of H.M.S. Herald" (1852–1857).
von Seem.   Seemen, Karl Otto von, 1838–1910, German botanist.
Sendt.   Sendtner, Otto, 1813–1859, professor of botany, Monaco.
Ser.   Seringe, Nicolas Charles, 1776–1858, professor in Lyon; important collaborator in De
   Candolle's Prodromus; Caryophyllaceae, Rosaceae, Cucurbitaceae, *Salix, Aconitum.*
Ses.   Sessé y Lacasta, Martin de, 175?–1809, director of the botanic garden in Mexico City.
Ses. & Moc.   Sessé y Lacasta, M. de, and J. M. Moçiño.
Seub.   Seubert, Moritz, 1818–1878, professor in Karlsruhe; student of monocots, Amaranthaceae.
Shaf.   Shafer, John Adolf, 1863–1918, custodian of the museum, New York Botanical Garden;
   made the most extensive collections of Cuban plants.
Shan. & Const.   Shan, Ren Hwa, contemporary Shanghai botanist, and L. Constance.
Sharp.   Sharp, Ward McClintic, 1904——, U.S. Fish and Wildlife Service, State College, Pa.
Sharsm., C. W.   Sharsmith, Carl William, 1903——, professor of botany, San Jose State; alpine
   flora of the Sierra Nevada.
Sharsm., H. K.   Sharsmith, Helen Katherine (Meyers), 1905——, herbarium botanist, California;
   flora of Mt. Hamilton Range.
Shear.   Shear, Cornelius Lott, 1865–1956, plant pathologist and agrostologist, U.S.D.A., Wash-
   ington.
Sheld.   Sheldon, Edmund Perry, 1869——, resident first in Minnesota, later in Portland, Oregon;
   forestry.
Sherff.   Sherff, Earl Edward, 1886——, botanist, Chicago Teachers College; student of *Bidens*
   and the Hawaiian flora.
Shinners.   Shinners, Lloyd Herbert, 1918——, professor of botany, Southern Methodist.
Shuttl.   Shuttleworth, Robert James, 1810–1874, English botanist and conchologist who resided
   most of his life in Berne; amassed an herbarium of 170,000 specimens now in British Museum.
Sibth.   Sibthorp, John, 1758–1796, professor of botany, Oxford; author of "Flora Graeca" (1806–
   1840).
Sieb. & Zucc.   Siebold, Philipp Franz von, 1796–1866, Germany, made several trips to study the
   botany, agriculture, and ethnography of Japan, and J. G. Zuccarini, authors of a flora of Japan
   (1843–1846).
Sieber.   Sieber, Franz Wilhelm, 1789–1844, Bohemian botanical traveler and collector in the
   Alps, Africa, and West Indies.
Sims.   Sims, John, 1749–1831, England; for 25 years editor of Curtis's Botanical Magazine.
Sm.   Smith, Sir James Edward, 1759–1828, England; founder and for 40 years president of the
   Linnaean Society; purchaser of the Linnaean herbarium and library, now at the Linnaean Society.
Sm., C. P.   Smith, Charles Piper, 1877–1955, high-school teacher, San Jose; *Lupinus.*
Sm., J.   Smith, John, 1798–1888, gardener at Kew; student of ferns.
Sm., J. G.   Smith, Jared Gage, 1866——, agrostologist, U.S.D.A., Washington, later in Hawaii.
Small.   Small, John Kunkel, 1869–1938, head curator, New York Botanical Garden; author
   of "Flora of the Southeastern United States" (1903); and "Manual of the Southeastern Flora"
   (1933).
Small & Rydb.   Small, J. K., and P. A. Rydberg.
Smiley.   Smiley, Frank Jason, 1880——, professor of botany and geology, Occidental; author of
   "A Report upon the Boreal Flora of the Sierra Nevada of California" (1921).
Smyth.   Smyth, Bernard Bryan, 1843–1913, librarian, Kansas Academy of Science, Topeka.
Sobol.   Sobolevski, Gregory Fedorovitch, 1741–1807, author of "Flora Petropolitana" (1799).
Soland.   Solander, Daniel Carl, 1736–1782, England; gifted Swedish student of Linnaeus; accom-
   panied Banks on Capt. Cook's first voyage of circumnavigation.
Solms.   Solms-Laubach, Count Hermann Maximilian Carl Ludwig Friedrich, 1842–1915, professor
   of botany, Strassburg; Lennoaceae.
Souster.   Souster, J. E. S., 20th century, assistant curator of the gardens, Kew.
Spach.   Spach, Édouard, 1801–1879, France; author of "Histoire naturelle des Végétaux Phanéro-
   games" (1834–1848).
Spenner.   Spenner, Fridolin Karl Leopold, 1798–1841, professor in Freiburg.
Sprague.   Sprague, Thomas Archibald, 1877——, Scotch taxonomist and botanical nomenclaturist
   on the staff at Kew for 45 years.
Sprague, E.   Sprague, Elizabeth F., 1915?——, Claremont Graduate School.
Spreng.   Sprengel, Kurt Polykarp Joachim, 1766–1833, professor of medicine and botany, Halle;
   author of works on the flora of Halle, on the Umbelliferae, on the history of botany, and editor of
   the 18th edition of Linnaeus's "Systema Vegetabilium" (1825–1828).
Spring.   Spring, Frédéric Antoine, 1814–1872, professor in Lüttich, Belgium; Lycopodiaceae.
Stacey.   Stacey, John William, 1871–1943, San Francisco publisher; *Carex.*
Standl.   Standley, Paul Carpenter, 1884——, curator, U.S. National Herbarium, 1909–1928,
   Chicago Natural History Museum, 1928–1950; student of the Mexican and Central American
   floras.

Stanf.   Stanford, Ernest Elwood, 1888——, professor of botany, College of the Pacific.
Stapf.   Stapf, Otto, 1857–1933, Austrian botanist on the staff at Kew from 1890, as keeper from 1909–1922, contributor to Harvey and Sonder's "Flora Capensis" and Oliver's "Flora of Tropical Africa."
Stapf & C. E. Hubb.   Stapf, O., and C. E. Hubbard.
Steb.   Stebbins, George Ledyard, Jr., 1906——, professor of genetics, California; cytogeneticist and cytotaxonomist; *Crepis, Antennaria,* Gramineae.
Steb. & Love.   Stebbins, G. L., Jr., and Robert Merton Love, 1909——, professor of agronomy, California (Davis).
Stearn.   Stearn, William Thomas, 1911——, botanist at British Museum.
Steinh.   Steinheil, Adelphe, 1810–1839, Alsace.
Stephan.   Stephan, Christian Friedrich, 1757–1814, professor in Moscow.
Steud.   Steudel, Ernst Gottlieb, 1783–1856, German physician and botanical bibliographer; agrostologist.
Stev.   Steven, Christian von, 1781–1863, Russian state councilor; Crimean flora.
Stewart, M. G.   Stewart, Margaret Gaylord (*later* Grover), 1911——, student at Pomona College.
Stewart, S. R.   Stewart, Sara R. (*later* Hinckley), 20th century; Gray Herbarium.
Stewart, W. S.   Stewart, William Sheldon, 1914——, director, Los Angeles State and County Arboretum.
Steyerm.   Steyermark, Julian Alfred, 1909——, curator, Chicago Natural History Museum; botanical explorer and student of Central and South American floras; *Grindelia.*
St. Hil.   Saint-Hilaire, Auguste de (Augustin François Cesar Prouvencal de), 1779–1853, France; collected 7,000 species of plants during extensive travels in Brazil and Paraguay; published many works.
St. John.   St. John, Harold, 1892——, professor of botany, Hawaii, formerly at Pullman; student of the flora of eastern Washington.
St. John & Const.   St. John, H., and L. Constance.
St. John & Warren.   St. John, H., and Fred Adelbert Warren, 1902——, State College of Washington.
Stocking.   Stocking, Kenneth Morgan, 1911——, College of the Pacific; *Marah, Echinocystis.*
Stockw.   Stockwell, William Palmer, 1898–1950, director of the Institute of Forest Genetics, Placerville and Berkeley.
Stockw. & Right.   Stockwell, W. P., and Francis Irving Righter, 1897——, U.S. Forest Service, Berkeley.
Stoker.   Stoker, Fred, died in 1943, English physician and horticulturist.
Stokes.   Stokes, Jonathan, 1755–1831, English friend of the younger Linnaeus.
Stokes, S.   Stokes, Susan Gabriella, 1868–1954, high-school teacher in San Diego; student of *Eriogonum.*
Stuntz.   Stuntz, Stephen Conrad, 1875–1918, United States.
St. Yves.   Saint-Yves, Alfred, 1855–1933, France; grasses.
Sudw.   Sudworth, George Bishop, 1864–1927, chief dendrologist, U.S. Forest Service, Washington; author of "Forest Trees of the Pacific Slopes" (1908).
Suksd.   Suksdorf, Wilhelm Nikolaus, 1850–1932, amateur botanist, Bingen, Washington, whose valuable herbarium was bequeathed to State College of Washington.
Svens.   Svenson, Henry Knute, 1897——, Brooklyn Botanic Garden, American Museum of Natural History, U.S. Geological Survey; *Eleocharis.*
Sw.   Swartz, Olof Peter, 1760–1818, professor in Stockholm; student of Linnaeus, organizer of the botany of the West Indies.
Swall.   Swallen, Jason Richard, 1903——, head curator, U.S. National Herbarium; Gramineae.
Sweet.   Sweet, Robert, 1783–1835, England; monographer of the Geraniaceae (1820–1830); horticulturist and ornithologist.
Swezey.   Swezey, Goodwin Deloss, 1851–1934, professor of astronomy, Nebraska.
Swingle.   Swingle, Walter Tennyson, 1871–1952, U.S.D.A., Washington; authority on *Citrus;* introduced commercial date palm into California in 1900, and the fig insect, Acala cotton, etc.
T. & G.   Torrey, J., and A. Gray, authors of "A Flora of North America (1838–1840[-1843])."
Tausch.   Tausch, Ignaz Friedrich, 1793–1848, Bohemian botanist known for his studies in *Hieracium.*
Tayl., G.   Taylor, George, 1904——, keeper of botany, British Museum (Natural History), 1928–1956; director, Royal Botanic Gardens, Kew, 1956——.
Templeton.   Templeton, Bonnie Carolyn, 1906——, curator, botany department, Los Angeles Museum.
Ten.   Tenore, Michele, 1780–1861, professor of botany, Naples.
Thed.   Thedenius, Knut Fredrik, 1814–1894, Sweden.
Thell.   Thellung, Albert, 1881–1928, keeper in the Botanical Institute of the University of Zürich; Cruciferae.
Thomas.   Thomas, John Hunter, 1928——, professor of biology, Occidental College.
Thomps., H. J.   Thompson, Henry Joseph, 1921——, professor of botany, California (Los Angeles); *Dodecatheon.*
Thomps., W.   Thompson, William, 1823–1903, English amateur gardener.
Thornb.   Thornber, John James, 1872——, professor of botany, Arizona.
Thouars.   Du Petit-Thouars, Louis Marie Aubert, 1758–1831, French taxonomist and botanical writer.
Thuill.   Thuillier, Jean Louis, 1757–1822, professor in Paris; author of a flora of Paris (1790).
Thunb.   Thunberg, Carl Pehr, 1743–1828, professor of botany, Upsala; student of Linnaeus, collected in South Africa, Japan, Ceylon; prolific author of notable works including "Flora Japonica" (1784) and "Flora Capensis" (1820).

Thurb.   Thurber, George, 1821–1890, New York; botanist with the Mexican Boundary Survey, editor of the American Agriculturist.
Tides.   Tidestrom, Ivar T., 1864–1956, professor of botany, Catholic University; author of "Flora of Utah and Nevada" (1925) and "A Flora of Arizona and New Mexico" (1941).
Todaro.   Todaro, Agostino, 1818–1892, director of the botanic garden, Palermo.
Torr.   Torrey, John, 1796–1873, physician; professor of chemistry and botany, College of Physicians and Surgeons, New York; collected in California in 1865 and described the plants from many of the early exploring expeditions to the West Coast; an outstanding systematist.
Torr. & Frém.   Torrey, J., and J. C. Frémont.
Torr. & Hook.   Torrey, J., and Sir W. J. Hooker.
Toumey.   Toumey, James William, 1864–1932, professor of forestry, Yale.
Tourn.   Tournefort, Joseph Pitton de, 1656–1708, professor of botany in the royal garden, Paris; first clear characterizer of genera; his botanical system was the highest pre-Linnean development.
Tracy.   Tracy, Joseph Prince, 1879–1953, title examiner of Eureka and amasser of a fine herbarium, now deposited at Berkeley, of the North Coast Ranges.
Traub.   Traub, Hamilton Paul, 1890——, U.S.D.A.; horticulturist.
Trautv.   Trautvetter, Ernst Rudolph von, 1809–1889, Russia.
Trel.   Trelease, William, 1857–1945, director, Missouri Botanical Garden, 1885–1912, then professor of botany, Illinois, 1913–1926; *Yucca, Agave, Epilobium, Quercus, Piper*.
Trev.   Treviranus, Ludolph Christian, 1779–1864, professor of botany, Bonn.
Trevisan.   Trevisan di San Leon, Count Vittore Benedetto Antonio, 1818–1897, Milan; proprietor of a very rich herbarium, especially cryptogams.
Trew.   Trew, Christoph Jakob, 1695–1769, physician and naturalist of Nürnberg, Germany.
Trin.   Trinius, Carl Bernhard von, 1778–1844, Russian court physician, also poet and noted agrostologist.
Trin. & Rupr.   Trinius, C. B., and F. J. Ruprecht.
Tryon.   Tryon, Rolla Milton, 1916——, curator, Missouri Botanical Garden; pteridophytes.
Tucker.   Tucker, John Maurice, 1916——, professor of botany, California (Davis).
Tuckerm.   Tuckerman, Edward, 1817–1886, professor at Amherst; lichenologist.
Tulasne.   Tulasne, Louis René, 1815–1885, French student of cryptogams.
Turcz.   Turczaninow, Nicolaus (Nikilai Stepanovich Turchaninov), 1796–1864, Russia; author of "Flora baicalensidahurica."
Turner.   Turner, Billie Lee, 1925——, professor of botany, Texas.
Turp.   Turpin, Pierre Jean François, 1775–1840, French botanical artist and naturalist.
Uline & Bray.   Uline, Edwin Burton, 1867–1933, New York high-school principal, and William L. Bray, 1865–1953, botanist and dean of the graduate school, Syracuse.
Underw.   Underwood, Lucien Marcus, 1853–1907, professor of botany, Columbia; student of ferns.
Urb.   Urban, Ignatz, 1848–1931, German authority on the flora of tropical America, monographer of *Medicago*, Loasaceae.
Urb. & Gilg.   Urban, I., and Ernst Friedrich Gilg, 1867–1933, both of the Botanical Museum, Berlin.
Urv.   Dumont d'Urville, Jules Sébastian César, 1790–1842, French naval officer and botanist, who commanded an expedition of circumnavigation.
Vahl.   Vahl, Martin Hendriksen, 1749–1804, professor of botany, Copenhagen; student of Linnaeus and one of the editors of "Flora Danica."
Vail.   Vail, Anna Murray, 1863——, librarian, New York Botanical Garden.
Vaill.   Vaillant, Sébastien, 1669–1722, pupil of Tournefort who wrote a flora of Paris; the first to detect that the pollen-grain is the male sex-cell.
Van Es.   Van Eseltine, Glen Parker, 1888–1938, U.S.D.A.
Van Houtte.   Van Houtte, Louis, 1810–1876, horticulturist of Ghent; editor of Flore des Serres (1845–1855), continued as Journal général d'Horticulture (1856–1865), continued as Annales générales d'Horticulture (1865–1883).
Van Melle.   Van Melle, Peter Jacobus, 1891–1953, nurseryman of Poughkeepsie, N.Y.
Vasey.   Vasey, George, 1822–1893, eminent agrostologist, U.S.D.A., Washington.
Vasey & Rose.   Vasey, G., and J. N. Rose.
Vasey & Scribn.   Vasey, G., and F. L. Scribner.
Vatke.   Vatke, Georg Carl Wilhelm, 1849–1889, assistant in the botanical garden, Berlin.
Vell.   Velloso, José Marianno da Conceição, 1742–1811, Rio de Janiero; writer on the flora of Brazil.
Vent.   Ventenat, Étienne Pierre, 1757–1808, professor of botany, Paris; horticulturist.
Vest.   Vest, Lorenz Chrysanth von, 1776–1840, professor in Graz, Austria.
Victorin.   Marie-Victorin, Frère (*formerly* Conrad Kirouac), 1885–1944, director of the botanical institute, Montreal; author of "Flore Laurentienne" (1935), the first flora in which chromosome numbers appeared.
Vill.   Villars, Dominique, 1745–1814, physician and professor in Grenoble, finally in Strassburg; author of a basic work on the flora of the West Alps.
Viv.   Viviani, Domenico, 1772–1840, Italy.
Vog.   Vogel, Julius Rudolph Theodor, 1812–1841, German explorer in Africa.
Volk.   Volkart, Albert, 1873——, Switzerland.
Voss.   Voss, Andreas, 1857–1924, German student of conifers.
Voss, J.   Voss, John William, 1907——, Pomona College; school principal; *Phacelia*.
Wagnon.   Wagnon, Harvey Keith, 1916——, State Department of Agriculture, Sacramento.
Wahl, H. A.   Wahl, Herbert Alexander, 1900——, professor of botany, Pennsylvania State; *Chenopodium*.

Wahl.   Wahlenberg, Göran (Georg), 1780–1851, professor of botany, Upsala; plant geographer, author of "Flora Lapponica" (1812), "De Vegetatione et Climate Helvetiae" (1813); "Flora Carpatorum Principalium" (1814); "Flora Upsaliensis" (1820); "Flora Suecica" (1824–1826).

Waldst. & Kit.   Waldstein-Wartemberg, Count Franz Adam von, 1759–1823, Austria, and P. Kitaibel, authors of "Descriptiones et Icones Plantarum rariorum Hungariae" (1802–1812).

Wall.   Wallich, Nathanael (*formerly* Nathan Wolff), 1786–1854, Danish physician; superintendent of the Royal Botanic Garden, Calcutta, 1815–1841.

Wallr.   Wallroth, Carl Friedrich Wilhelm, 1792–1857, German physician and botanist.

Walp.   Walpers, Wilhelm Gerhard, 1816–1853, German author of "Repertorium Botanices Systematicae" (1842–1853).

Walt.   Walter, Thomas, 1740–1788, Charleston planter; author of "Flora Caroliniana" (1788).

Walth., E.   Walther, Edward Eric, 1892——, director, Strybing Arboretum, Golden Gate Park, San Francisco.

Walton.   Walton, F. A., Birmingham, England, grower of succulents; editor of The Cactus Journal.

Wang.   Wangenheim, Friedrich Adam Julius von, 1747–1800, German forester.

Wanger.   Wangerin, Walther Leonhard, 1884–1938, professor in Danzig; Cornaceae.

Ward.   Ward, George Henry, 1916——, professor of botany, Knox College; *Artemisia.*

Wats.   Watson, Sereno, 1826–1892, assistant to Asa Gray, curator of Gray Herbarium, 1888–1892, critical student of West American plants.

Wats. & Coult.   Watson, S., and J. M. Coulter.

Wats. & Rothr.   Watson, S., and J. T. Rothrock.

Watt.   Watt, David Allan Poe, 1830–1917, Montreal; student of ferns.

Waugh.   Waugh, Frank Albert, 1869–1947, professor of horticulture, Amherst.

Weath.   Weatherby, Charles Alfred, 1875–1949, Gray Herbarium; student of ferns; bibliographer.

Webb.   Webb, Philip Barker, 1793–1854, England; botanical traveler and accumulator of a rich herbarium and botanical library, which are now in Florence.

Webb & Berthel.   Webb, P. B., and Sabin Berthelot, 1794–1880, French consul on Teneriffe; authors of "Histoire naturelle des Îles Canaries" (1835–1844).

Webber, J. M.   Webber, John Milton, 1897——, Agricultural Research Service, Berkeley, California; cotton breeder.

Weber.   Weber, Georg Heinrich, 1752–1828, professor at Kiel.

Weber, A.   Weber, A., 19th–20th century, Germany; French army doctor who traveled in Mexico; Cacti.

Weber, W.   Weber, William Alfred, 1918——, professor of botany, Colorado.

Weberb.   Weberbauer, August, 1871–1948, Royal Botanic Gardens, Breslau; collected in Africa and the Andes; Rhamnaceae.

Weigel.   Weigel, Christian Ehrenfried, 1748–1831, professor at Griefswald.

Weinm.   Weinmann, Johann Anton, 1782–1858, director, botanic garden, St. Petersburg.

Weiss.   Weiss (*also* Weis), Friedrich Wilhelm, born in 1744, privatdocent in Göttingen.

Wendl.   Wendland, Hermann, 1823–1903, director of the royal garden at Herrenhausen, Hannover; Palmae.

Wesmael.   Wesmael, Alfred, 1832–1905, Belgium.

Wettst.   Wettstein, Richard Ritter von Westersheim, 1863–1931, director of the botanic garden, Vienna; outstanding systematist, proposer of an important phylogenetic system.

Wheeler.   Wheeler, Louis Cutter, 1910——, professor of botany, Southern California; *Euphorbia.*

Wheelock.   Wheelock, William Efner, 1852–1927, New York physician.

Wherry.   Wherry, Edgar Theodore, 1885——, professor of botany, Pennsylvania; Polemoniaceae.

White.   White, Theodore Greeley, 1872–1901, student at New York Botanical Garden.

Wibel.   Wibel, August Wilhelm Eberhard Christoph, 1775–1814, physician in Wertheim.

Wieg.   Wiegand, Karl McKay, 1873–1942, professor of botany, Cornell.

Wies. & Schreib.   Wieslander, Albert Everett, 1890——, U.S. Forest Service, Berkeley, and B. O. Schreiber.

Wiggers.   Wiggers, Fredericus Henricus, author of a flora of Holstein (1780).

Wiggins.   Wiggins, Ira Loren, 1899——, professor of botany and director of the natural history museum, Stanford; student of the flora of the Sonoran Desert; Malvaceae, ferns.

Wiggins & Stockw.   Wiggins, I. L., and W. P. Stockwell.

Wiggins & Wolf.   Wiggins, I. L., and C. B. Wolf.

Wight.   Wight, Robert, 1797–1872, English student of the flora of India.

Wight, W.   Wight, William Franklin, 1874–1954, U.S.D.A., Palo Alto.

Wikstr.   Wikström, Johann Emanuel, 1789–1856, director of the Botanical Museum, Stockholm.

Willd.   Willdenow, Carl Ludwig, 1765–1812, director of the Berlin Botanical Garden (1801–1812); produced the fourth edition of Linnaeus' "Species Plantarum" (1797–1810).

Williams, F. N.   Williams, Frederic Newton, 1862–1923, England.

Williams, L.   Williams, Louis Otho, 1908——, American botanist, United Fruit Company, Honduras; later Beltsville.

Willk.   Willkomm, Heinrich Moritz, 1821–1895, professor of botany at Tharandt, Dorpat, and Prague; author, with J. M. C. Lange, of a noted Spanish flora.

Wilmott.   Wilmott, Alfred James, 1888–1950, keeper of botany, British Museum.

Wimmer.   Wimmer, Christian Friedrich Heinrich, 1803–1868, school official in Breslau.

With.   Withering, William, 1741–1799, author of a British flora.

Wittr.   Wittrock, Veit Brecher, 1839–1914, director, Hortus Bergianus, Stockholm.

Wolf, C. B.   Wolf, Carl Brandt, 1905——, botanist, Rancho Santa Ana Botanic Garden, 1930–1945.

Wolf, T.   Wolf, Franz Theodor, 1841–1924, German geologist in Ecuador until 1891, then a teacher in Dresden; monographer of *Potentilla* (1908).

Wolff, H.   Wolff, Hermann (Karl Friedrich August Hermann), 1866–1929, German veterinary surgeon and botanist; student of Umbelliferae.

Wood.   Wood, Alphonso, 1810–1881, principal, Brooklyn Female Academy; author of "Class-Book of Botany" (1845, 2d ed. 1861), in which dichotomous keys were first employed in America; collected extensively in California in 1866.

Woodson.   Woodson, Robert Everard, 1904——, senior taxonomist, Missouri Botanical Garden; authority on Apocynaceae.

Woot.   Wooton, Elmer Ottis, 1865–1945, professor of biology, New Mexico State College, 1890–1911, U.S.D.A., Washington, 1911–1935; student of the New Mexican flora.

Woot. & Standl.   Wooton, E. O., and P. C. Standley, authors of "Flora of New Mexico" (1915).

Wormsk.   Wormskjöld, Morten, 1783–1845, Danish lieutenant; collected in Greenland, also took part in Kotzebue's first voyage on the "Rurik," leaving the expedition at Kamchatka, where he remained for two years.

Wulf.   Wulfen, Franz Xaver (Freiherr) von, 1728–1805, professor at Klagenfurt, Hungary.

Wynd.   Wynd, Frederick Lyle, 1904——, owner and director, Maple Leaf Soil Laboratory, Canada.

Yunck.   Yuncker, Truman George, 1891——, professor of botany, De Pauw; monographer of *Cuscuta, Piper,* and *Peperomia.*

Zabel.   Zabel, Hermann, 1832–1912, German dendrologist.

Zahn.   Zahn, Karl Hermann, 1865——, teacher in Karlsruhe; *Hieracium.*

Zeile.   Zeile, Elsie May (*later* Lovegrove), assistant to W. L. Jepson.

Zeissold.   Zeissold, H., fl. 1895, German student of cacti.

Zinn.   Zinn, Johann Gottfried, 1727–1759, professor of medicine, Göttingen.

Zucc.   Zuccarini, Joseph Gerhard, 1797–1848, professor of botany, Munich.

GLOSSARY

Compiled by David D. Keck

Abaxial. Located on the side away from the axis.

Abortive. Imperfect or barren.

Abrupt. Terminating suddenly; not tapering.

Acaulescent. Stemless or essentially so. (Cf. *Caulescent.*)

Accessory. Additional to the usual number of organs.

Accrescent. Increasing in size with age, as often with the calyx after flowering.

Accumbent. Lying against something, as cotyledons against the radicle.

Acerose. Needle-shaped, as the leaves of pines.

Acicular. Needlelike.

Actinomorphic. Exhibiting radial symmetry, as a regular flower.

Acumen. A tapering point.

Acuminate. Gradually tapering to a short point. (Cf. *Acute, Attenuate.*)

Acute. Sharp-pointed, but less tapering than acuminate.

Adaxial. Located on the side nearest the axis.

Adnate. Grown together with an unlike part, as the calyx-tube with an inferior ovary, or an anther by its whole length with the filament.

Adventitious, adventive. Out of the usual place; introduced but not yet naturalized.

Aestivation. The arrangement of the parts in a flower bud.

Aggregate. Collected into dense clusters or tufts. *Aggregate fruit:* one formed by the clustering together of pistils that were distinct in the flower, as blackberry.

Akene (achene). A small, dry, hard, indehiscent, 1-seeded fruit.

Alliaceous. Onionlike, usually in respect to odor.

Alpine. Strictly applicable to plants growing above timber line.

Alternate. Any arrangement of parts along the axis other than opposite or whorled; situated regularly between other organs, as stamens alternate with petals.

Alveolate. Honeycombed; with deep angular cavities (alveoli) separated by thin partitions; faveolate.

Ament. Catkin.

Amplexicaul. Clasping the stem, as the base of certain leaves.

Ampliate. Enlarged; dilated.

Anastomosing. Netted; particularly applied to veins so connected by cross veins as to form a network.

Anatropous. An inverted and straight ovule, with the micropyle next to the hilum.

Ancipital. Two-edged, as certain flattened stems.

Androecium. The whole set of stamens.

Androgynous. Having staminate and pistillate flowers in the same inflorescence, or in *Carex* in the same spikelet, the former above the latter.

Anemophilous. Wind-pollinated.

Annual. Of one year's or season's duration from seed to maturity and death. *Winter annual:* a plant from autumn-germinating seed that fruits in the following spring.

Annular. Circular; in the form of a ring, or marked transversely by rings.

Annulus. A ring-shaped part or organ, such as surrounds the sporangium in some ferns.

Anterior. On the front side; in the flower the side away from the axis and toward the subtending bract.

Anther. The pollen-bearing part of the stamen.

Antheridium. The male sexual organ of ferns, etc., analogous to the anther.

Antheriferous. Anther-bearing.

Anthesis. Strictly, the time of expansion of the flower, but also used for the period during which the flower is open and functional.

Anthocarp. A structure in which the fruit proper is united with the perianth or receptacle.

Anthocyanous. Showing anthocyanin in the herbage, a class of soluble glucoside pigments producing reddish or purplish coloring.

Antipetalous. Opposite the petals. *Antisepalous:* opposite the sepals.

Antrorse. Directed upward or forward.

Apetalous. Without petals.

Aphyllopodic. Without leaves at the base.

Aphyllous. Leafless.

Apical. Situated at the tip.

Apiculate. Terminated abruptly in a little point.

Apomictic. Producing seed without any form of fertilization or sexual union.

Appendage. Any attached supplementary or secondary part.

Appressed. Pressed flat against another organ.

Approximate. Near together.

Arachnoid. Cobwebby; of soft and slender entangled hairs.

Arborescent. Treelike in tendency.

Archegonium. The female sexual organ of ferns, etc., analogous to the pistil.

Arcuate. Moderately curved, as if bent like a bow.

Areola (-ae) A little area defined on a surface, as the angular space between vein reticulations. In Compositae the circle at the summit of the akene where sat the corolla. *Areolate:* with areolae; reticulate.

Aril. A process of the placenta adhering about the hilum of a seed. *Arillate:* with an aril.

Aristate. Awn- or bristle-tipped, as the floral bracts of barley. *Aristulate:* bearing a short awn.

Articulate. Jointed; having a place for natural separation with a clean-cut scar.

Ascending. Rising obliquely or curving upward.

Asepalous. Without sepals.

Asperous. Rough to the touch; scabrid.

Assurgent. Ascending, rising.

Attenuate. Slenderly tapering or prolonged; more gradual than acuminate.

Auricle. An ear-shaped appendage. *Auriculate:* bearing auricles.

Awn. A terminal slender bristle on an organ.

Axil. Upper angle formed by a leaf or branch with the stem.

Axile. Belonging to, or situated in, the axis.

Axillary. In an axil.

Axis. The central stem along which parts or organs are arranged; the central line of any organ.

Baccate. Berrylike and pulpy.

Banner. Upper petal of a papilionaceous flower.

Barbate. Bearded with long stiff hairs.

Barbed. Bearing sharp rigid reflexed points like the barb of a fish-hook.

Barbellate. With short, usually stiff hairs. *Barbellulate:* the diminutive.

Barbulate. Finely bearded.

Basal. Relating to, or situated at, the base.

Basifixed. Attached by the base.

Beak. A prolonged firm tip, particularly of a seed or fruit. *Beaked:* ending in a beak.

Bearded. Bearing long stiff hairs.

Berry. A pulpy indehiscent fruit with no true stone, as the tomato.

Bi- or *Bis-.* Latin prefix signifying two, twice, or doubly.

Bidentate. Having two teeth.

Biennial. Of two years' duration from seed to maturity and death.

Bifid. Two-cleft to about the middle.

Bifurcate. Two-forked or -pronged.

Bilabiate. Two-lipped (calyx or corolla).

Bilocular. Two-celled.

Binate. In pairs.

Bipartite. Divided into two parts almost to the base; two-parted.

Bipinnate. Doubly or twice pinnate.

Bipinnatifid. Twice pinnately cleft.

Bisected. Completely divided into two parts.

Bladdery. Thin and inflated.

Blade. The expanded part of a leaf or petal.

Bole. The trunk or stem of a tree.

Bract. A reduced leaf subtending a flower, usually associated with an inflorescence. *Bracteate:* with bracts.

Bractlet. A secondary bract borne on a pedicel instead of subtending it; sepaloid organs subtending the sepals in many Rosaceae. *Bracteolate:* with bractlets.

Bristly. Bearing stiff hairs.

Bud. An undeveloped stem, leaf, or flower. Buds are often enclosed by reduced or specialized leaves termed bud-scales.

Bulb. An underground leaf-bud with thickened scales or coats like the onion.

Bulbel. Daughter bulbs arising around the mother bulb.

Bulblet. A small bulb, especially one borne aerially as in a leaf axil or in the inflorescence.

Bullate. Blistered or puckered.

Caducous. Falling off very early or prematurely.

Calcarate. Spurred.

Callosity. A hardened thickening.

Callus. A callosity; the thickened extension at the base of the lemma in some grasses. *Callose:* bearing callosities.

Calyculate. Bearing bracts around the calyx imitating an outer calyx; said of the short bracts imitating an outer involucre in some Compositae.

Calyptra. A lid or hood.

Calyx. The external, usually green, whorl of a flower, contrasted with the inner showy corolla.

Calyx-lobe. In a gamosepalous calyx, the free projecting parts.

Campanulate. Bell-shaped.

Campylotropous. (ovule or seed). So curved as to bring apex and base nearly together.

Canaliculate. Longitudinally channeled or grooved.

Canescent. Covered with grayish-white or hoary fine hairs.

Capillary. Hairlike; exceedingly slender.

Capitate. Head-shaped; aggregated into very dense clusters or heads.

Capitulum. See *Head*.

Capsule. A dry dehiscent fruit composed of more than one carpel.

Carinate. Keeled; with a sharp longitudinal ridge.

Carpel. A simple pistil, or one of the modified leaves forming a compound pistil.

Carpophore. A prolongation of the floral axis between the carpels, as that which supports the pendulous fruit of the Umbelliferae.

Cartilaginous. Like cartilage in texture; tough.

Caruncle. An excrescense or outgrowth at or near the hilum of certain seeds.

Caryopsis. The grain or fruit of grasses.

Castaneous. Chestnut-colored; dark brown.

Catkin. A scaly deciduous spike; ament.

Caudate. Bearing a tail or slender taillike appendage.

Caudex. The woody base of an otherwise herbaceous perennial.

Caulescent. With an obvious leafy stem; plants with radical leaves and flowers on a scape are called acaulescent.

Cauline. Belonging to the stem.

Cell. A cavity of an anther containing the pollen, or of an ovary containing the ovules. *Cellular:* made up of cells or marked off so as to resemble cells.

Cenospecies. All the ecospecies so related that they may exchange genes among themselves to a limited extent through hybridization.

Centripetal. Growing from without toward the center; an indeterminate inflorescence.

Cernuous. Nodding; drooping.

Cespitose. In little tufts or dense clumps; said of low plants of turfy habit.

Chaff. Thin dry scales.

Channeled. Deeply grooved longitudinally, like a gutter.

Chaparral. A xerophytic formation of dense, impenetrable thickets composed of stiff or thorny, mostly small-leaved, evergreen shrubs.

Chartaceous. With the texture of writing paper.

Ciliate. Fringed with hairs on the margin. *Ciliolate:* the diminutive.

Cincinnus (*-i*). A curl; used in the plural here for the branches of a unilateral scorpioid cyme.

Cinereous. Ash-colored; light gray.

Circinate. Coiled from the top downward with the apex as a center.

Circumpolar. Occurring around the pole, as of arctic plants mostly confined to far northern latitudes.

Circumscissile. Dehiscing by a transverse line around the fruit or anther, the top falling as a lid.

Cismontane. This side of the mountains, or west of the main Sierran crest, as opposed to the deserts.

Clasping. Leaf partly or wholly surrounding the stem; amplexicaul.

Clavate. Club-shaped; gradually thickened toward the apex from a slender base. *Clavellate:* the diminutive.

Claw. The narrow petiolelike base of some petals and sepals.

Cleft. Cut about halfway to the midrib.

Cleistogamous. Small flowers self-fertilizing without opening; usually additional to the ordinary flower and inconspicuous, as in some violets.

Coalescent. Said of organs of one kind that have grown together.

Cochleate. Coiled like a snail shell.

Coherent. Congenitally united with another organ of the same kind (*coalescent*), or of another kind (*adnate*).

Collar. Outer side of the grass leaf at the junction of sheath and blade.

Columella. The persistent axis of certain capsules.

Column. Body formed by union of stamens and pistil in orchids, or of stamens in mallows and milkweeds.

Coma. A tuft of hairs, particularly on a seed. *Comose:* furnished with a coma.

Commissure. The face by which two carpels cohere, as in Umbelliferae.

Complanate. Flattened.

Complete. Having all the parts belonging to it, as a flower with sepals, petals, stamens, and pistils.

Complicate. Folded together.

Compound. Having two or more similar parts in one organ. *Compound leaf:* one with two or more separate leaflets. *Compound pistil:* having two or more carpels united.

Compressed. Flattened laterally.

Concave. Hollow.

Concolor. Of uniform color.

Conduplicate. Folded together lengthwise, as the leaves of many grasses.

Confluent. Blending of one part into another.

Congested. Crowded together.

Conglomerate. Densely clustered.

Conic. Cone-shaped, with the point of attachment at the broad base.

Conjugate. Joined in pairs.

Connate. Congenitally united, as similar organs joined as one. *Connate-perfoliate:* united at base in pairs around the supporting axis.

Connective. Portion of the filament connecting the two cells of an anther.

Connivent. Converging or coming together, but not organically united.

Constricted. Tightened or drawn together.

Continuous. Not articulated (grass rhachis) or interrupted.

Contorted. Twisted, bent, or distorted.

Contracted. Narrowed in a particular place, or shortened; the opposite of open or spreading (inflorescence).

Convex. Rounded on the surface.

Convolute. Rolled up longitudinally. Said of blades or floral envelopes in the bud when one edge is outside and the other inside.

Coralloid. Corallike.

Cordate. Heart-shaped with the notch at the base and ovate in general outline.

Coriaceous. Leathery in texture; tough.

Corm. A short, bulblike, underground stem, as the "bulb" of gladiolus.

Corneous. Of the texture of horn.

Corniculate. Bearing little horns or hornlike processes.

Corolla. The inner perianth of a flower, composed of colored petals, which may be almost wholly united.

Corona. A crown; the crownlike cup found at the orifice of the corolla-tube in *Narcissus.*

Coroniform. Crown-shaped.

Corrugated. Wrinkled; folded.

Cortex. Rind or bark.

Corymb. A flat-topped or convex racemose flower-cluster, the lower or outer pedicels longer, their flowers opening first. *Corymbose:* in corymbs.

Costa. A rib. *Costate:* ribbed; having longitudinal elevations.

Cotyledon. The primary leaf or leaves of the embryo.

Crateriform. Shallowly cup-shaped.

Creeping. Spreading over or beneath the ground and rooting at the nodes.

Crenate. Having the margin cut with rounded teeth; scalloped. *Crenulate:* the diminutive.

Crested. Having a crest, elevated appendage, or ridge on the summit of an organ.

Crinite. Bearded with long and weak hairs.

Crispate. Crisped or curled.

Crisped. Irregularly curled (said of hairs or leaf-margins).

Cristate. Crested or tufted.

Crown. The persistent base of an herbaceous perennial; the top of a tree; a circle of appendages on the throat of a corolla, etc.

Cruciferous. Cross-bearing; a flower with four petals placed opposite each other at right angles.

Crustaceous. Of brittle texture.

Cucullate. Hooded, or hood-shaped.

Cucullus. A hoodlike process on some seeds; cf. caruncle.

Culm. The type of hollow or pithy slender stem found in grasses and sedges.

Cuneate, cuneiform. Wedge-shaped; triangular, with the narrow part at point of attachment.

Cupulate. Cup-shaped, as the cup (involucre) of the acorn.

Cuspidate. Tipped with a cusp, or sharp, short, rigid point.

Cylindraceous. Somewhat or nearly cylindrical.

Cyme. A flat-topped or convex paniculate flower-cluster, with central flowers opening first. *Cymose:* arranged in cymes. *Cymule:* a small or few-flowered cyme.

Deciduous. Falling off, as petals fall after flowering, or leaves of nonevergreen trees in autumn.

Declined. Curved downward.

Decompound. More than once divided or compounded.

Decumbent. Lying down, but with the tip ascending.

Decurrent. Extending down the stem below the insertion; said of leaves or ligules.

Decussate. Opposite pairs (usually leaves) alternating at right angles with those above or below.

Deflexed. Turned abruptly downward.

Dehiscent. Opening spontaneously when ripe to discharge the contents, as an anther or seed vessel.

Deliquescent. Dissolving or melting away.

Deltoid. Equilaterally triangular.

Dendritic. Treelike, as the branching hairs of some Cruciferae.

Dentate. Having the margin cut with sharp salient teeth not directed forward.

Denticulate: slightly and finely toothed.

Depauperate. Dwarf, starved.

Depressed. Low, as if flattened from above.

Dextrorse. Turned to the right.

Diadelphous. Stamens united by their filaments into two sets.

Diandrous. Having two stamens.

Diaphanous. Transparent.

Dicarpellary. Having two carpels.

Dichotomous. Repeatedly forking in pairs.

Didymous. Twin; found in pairs.

Didynamous. With four stamens in two pairs of unequal length, as in most Labiatae.

Diffuse. Scattered; widely spread.

Digitate. Fingered; shaped as an open hand; compound with the members arising from one point.

Dilated. Flattened and broadened, as an expanded filament.

Dimorphic, dimorphous. Having two forms, as flowers with short stamens and long styles or long stamens and short styles.

Dioecious. Having staminate and pistillate flowers on different plants.

Diploid. Having two basic chromosome sets (twice the number in normal, haploid gametes).

Disarticulating. Separating joint from joint at maturity.

Discoid. Disklike. In the Compositae, a head without ray-florets.

Discrete. Separate; not coalescent.

Disk. A fleshy development of the receptacle about the base of the ovary. In Compositae, the tubular flowers (disk-florets) of the head as distinct from the ray.

Dissected. Deeply divided into numerous fine segments.

Distal. Opposite the point of attachment; apical; away from the axis.

Distichous. In two vertical rows or ranks.

Distinct. Separate; not united with parts in the same circle. Cf. *Free.*

Divaricate. Widely divergent.

Divergent. Extending away from each other by degrees.

Divided. Separated to the base.

Dolabriform. Axe-shaped or hatchet-shaped.

Dorsal. Pertaining to the back; the surface turned away from the axis.

Dorsifixed. Attached to the back.

Dorsiventral. Having distinct dorsal and ventral surfaces.

Downy. Closely covered with very short and weak soft hairs.

Drooping. Erectish at base but bending downward above, as the branches of a grass panicle.

Drupe. A fleshy one-seeded indehiscent fruit containing a stone with a kernel; a stone-fruit such as a plum.

E-, Ex-. Latin prefix meaning without, out of, from.

Ebracteate. Without bracts.

Echinate. Prickly, a hedgehog.

Ecospecies. All individuals so related that they are able to exchange genes freely without loss of fertility or vigor in the offspring.

Ecotype. Those individuals that are fitted to survive in only one kind of environment occupied by the species.

Edaphic. Pertaining to, or influenced by, soil conditions.

Ellipsoid. An elliptic solid.

Elliptic. In the form of a flattened circle more than twice as long as broad.

Emarginate. With a small notch at the apex.

Embryo. The incipient plantlet in the seed.

Emersed, emergent. Raised above the water.

Enation. An outgrowth on the surface of an organ.

Endocarp. The inner layer of the pericarp.

Endogenous. Forming new tissue within.

Endosperm. The nutritive tissue surrounding the embryo of a seed and formed within the embryo-sac.

Ensiform. Sword-shaped, as the leaves of *Iris.*

Entire. Undivided; the margin continuous, not incised or toothed.

Entomophilous. Insect-pollinated.

Epappose. Without pappus.

Ephemeral. Lasting for a day or less.

Epicarp. The outer layer of the pericarp.

Epigynous. Borne on the ovary; said of floral parts when the ovary is wholly inferior.

Equitant. Astride, as if riding, such as the leaves of an Iris.

Erect. Upright in relation to the ground, or sometimes perpendicular to the surface of attachment. A lip of a corolla is erect when in line with the tube.

Erose. Irregularly toothed as if gnawed.

Estipulate. Without stipules.

Evanescent. Quickly disappearing.

Evergreen. Remaining green through the winter.

Exalbuminous. Without endosperm, referring to seeds.

Excurrent. Projecting beyond the edge, as the midrib of a mucronate leaf.

Exocarp. The outer layer of the pericarp.

Exogenous. Forming new tissue outside the old.

Explanate. Spread out flat.

Exserted. Protruding, as stamens projecting beyond the corolla; not included.

Extrorse. Facing outward from the axis, as the dehiscence of an anther.

Falcate. Sickle-shaped.

Farinaceous. Containing starch; mealy in texture.

Farinose. Covered with a meallike powder.

Fascicle. A close cluster or bundle of flowers, leaves, stems, or roots. *Fasciculate:* in a fascicle.

Fastigiate. Clustered, parallel, erect branches.

Faveolate, favose. Honeycombed, as the receptacle in many Compositate; alveolate.

Fenestrate. With transparent areas or windowlike openings.

Ferruginous. Rust-colored.

Fertile. Said of pollen-bearing stamens and seed-bearing fruits.

Fetid. Disagreeably odorous.

Fibrillose. Furnished with little fibers.

-fid. A suffix meaning deeply cut.

Filament. A thread, especially the stalk of an anther.

Filiform. Threadlike.

Fimbriate. Fringed (with longer or coarser hairs as compared with ciliate). *Fimbrillate:* the diminutive.

Fistular, fistulous. Hollow throughout, as the leaf of an onion.

Flabellate, flabelliform. Fan-shaped; broadly wedge-shaped.

Flaccid. Weak, limp, soft or flabby.

Flagellate. With very slender runners.

Fleshy. Thick and juicy; succulent.

Flexuous. Zigzag.

Floccose. With *flocs* or tufts of soft woolly hair. *Flocculent:* the diminutive.

Floral tube. A more or less elongate tube consisting of perianth or other floral parts.

Floret. The individual flower of the Compositae and Gramineae; a small flower of a dense cluster.

Floriferous. Bearing or producing flowers.

Foliaceous. Leaflike; said especially of sepals or bracts that in texture or appearance resemble leaves.

Foliolate. Having leaflets.

Foliose. Having numerous leaves.

Follicle. A dry, monocarpellary fruit, opening only on the ventral suture, as in the larkspur.

Foveate. Pitted. *Foveolate:* the diminutive.

Free. Not joined to other organs; the reverse of adnate.

Frond. Leaf of a fern.

Fruit. The ripened pistil with all its accessory parts.

Frutescent, fruticose. Shrubby or bushy in the sense of being woody. *Fruticulose:* applied to a little shrub.

Fugacious. Perishing very early.

Fulvous. Tawny; dull yellow.

Funnelform. Gradually widening upwards, like a funnel.

Furcate. Forked.

Furfuraceous. Covered with branlike scales; scurfy.

Fuscous. Grayish-brown.

Fusiform. Spindle-shaped; thickest near the middle and tapering toward each end.

Galea. The helmetlike upper lip in certain bilabiate corollas. *Galeate:* having a galea.

Gametophyte. The sexual form of the plant (as in ferns) contrasted with the sporophyte or asexual form.

Gamopetalous. Corolla with petals united. Same as sympetalous and monopetalous.

Gamophyllous. Composed of coalescent leaves or leaflike organs.

Gamosepalous. Calyx with sepals united.

Geminate. In pairs, twin, binate.

Genetic. That which is inherited.

Geniculate. Bent abruptly, as a knee.

Gibbous. Swollen on one side; ventricose.

Glabrous. Without hairs; incorrectly used

in the sense of smooth, the antonym of rough. *Glabrate:* Almost glabrous; tending to be glabrous. *Glabrescent:* becoming glabrous.

Gland. A depression, protuberance, or appendage on the surface of an organ, which secretes a usually sticky fluid. *Glandular:* bearing glands or glandlike.

Glaucous. Covered or whitened with a bloom, as a cabbage leaf. *Glaucescent:* becoming glaucous.

Globose. Spherical or rounded. *Globular:* somewhat or nearly globose.

Glochid. A barbed hair or bristle. *Glochidiate:* barbed at the tip, as a bristle.

Glomerate. Densely compacted in clusters or heads. *Glomerulate:* arranged in small clusters.

Glomerule. A compact capitate cyme.

Glumes. The pair of bracts at the base of a grass spikelet. *Glumaceous:* resembling glumes.

Glutinous. With a gluey exudation.

Graduate. Marked with small regular distances.

Granular, granulose. Covered with very small grains or granules; minutely mealy.

Gregarious. Growing in groups or colonies.

Gynaecandrous. Having staminate and pistillate flowers in the same spikelet, the latter above the former.

Gynandrous. Stamens adnate to the pistil.

Gynobase. An elongation or dilation of the receptacle to support the carpels or nutlets, as in many borages.

Gynoecium. The pistils collectively of a flower.

Gynophore. The prolonged stipe of a pistil, as in *Cleome.*

Habit. General appearance of a plant.

Habitat. The normal situation in which a plant lives.

Halophyte. A plant of salty or alkaline soils.

Hastate. Halberd-shaped; of the shape of an arrowhead but with the basal lobes turned outward.

Haustoria. The suckerlike attachment organs of parasites like *Cuscuta.*

Head. A dense globular cluster of sessile or subsessile flowers arising essentially from the same point on the peduncle; capitulum.

Hemispheric. Shaped like half a sphere.

Herb. A plant without persistent woody stem, at least above ground.

Herbaceous. Pertaining to an herb; opposed to woody; having the texture or color of a foliage leaf; dying to the ground each year.

Herbage. Collectively, the green parts of a plant.

Heterogamous. Producing two or more kinds of flowers.

Heteromorphic. Of more than one kind of form.

Heterosporous. Having spores of two sizes or shapes.

Hexamerous. Having the floral whorls composed of six members.

Hexaploid. Having six basic chromosome sets (six times the gametic number).

Hilum. The scar at the point of attachment of an ovule or seed.

Hirsute. Rough with coarse or shaggy hairs. *Hirsutulous:* the diminutive.

Hirtellous. Minutely hirsute.

Hispid. Rough with stiff or bristly hairs. *Hispidulous:* the diminutive.

Hoary. Covered with white down.

Holosericeous. Covered with fine and silky hairs.

Holotype. The one specimen on which a species or other taxon is based.

Homogamous. A head or cluster with flowers alike throughout.

Homonym. In nomenclature, a name rejected because it duplicates a name previously and validly published for a group of the same rank based on a different type.

Homosporous. With spores all alike in size and shape.

Host. A plant which nourishes a parasite.

Humistrate. Spread over the surface of the ground.

Hyaline. Colorless or translucent, transparent.

Hybrid. A cross between two species.

Hydrophyte. Partially or wholly immersed water plant.

Hypanthium. A cup-shaped enlargement of the receptacle on which the calyx, corolla, and often the stamens are inserted; in perigyny the "calyx-tube."

Hypogynous. Borne on the receptacle below or free from the pistil; said of petals or stamens.

Imbricate. Overlapping as shingles on a roof.

Imparipinnate. Odd-pinnate, having a terminal leaflet.

Immersed. Growing under water.

Incised. Cut rather deeply and sharply.

Included. Not protruding beyond the surrounding organ or envelope.

Incumbent. Lying upon anything; said of cotyledons when the back of one rests against the stalk of the embryo.

Incurved. Bending inwards.

Indefinite. Relating to number, inconstant or very numerous.

Indehiscent. Not splitting open, as an akene.

Indeterminate. Not terminated absolutely, as a raceme.

Indigenous. Native to the country.

Indument. Any hairy covering or pubescence.

Induplicate. Valvate aestivation in which the margins of the leaves are bent or folded inward.

Indurate. Hard, hardened.

Indusium. In ferns, the epidermal outgrowth that covers or invests the sorus.

Inequilateral. Unequal-sided.

Inferior. Lower or beneath. *Inferior ovary:* one that is adnate to the hypanthium and situated below the calyx-lobes.

Inflated. Blown up; bladdery.

Inflexed. Turned abruptly inward.

Inflorescence. The flower-cluster of a plant, or, more correctly, the disposition of the flowers on an axis.

Infundibuliform. Funnelform.

Innate. Borne on the apex of the support; in an anther the antithesis of adnate.

Innocuous. Harmless, hence unarmed or spineless.

Innovation. In Gramineae a new sterile basal shoot; an offshoot from a stem.

Insectivorous. Consuming insects, i.e. by dissolving out the organic parts.

Inserted. Attached to or growing upon.

Insertion. The place or mode of attachment of an organ to its support.

Intercostal. Situated between ribs or costae.

Internerves. Spaces between the nerves.

Internode. The portion of stem between two nodes.

Interrupted. Not continuous or regular.

Interval. Space between ridges.

Introrse. Turned inward, towards the axis.

Intruded. Projecting inward or forward.

Inverted. Upside down; turned over.

Involucel. A secondary involucre, as the bracts subtending the secondary umbels in the Umbelliferae.

Involucrate. Having an involucre.

Involucre. A whorl of bracts (phyllaries) subtending a flower cluster, as in the heads of Compositae.

Involute. With the edges rolled inward, i.e., toward the upper side.

Irregular. Showing a lack of uniformity; asymmetric, as a zygomorphic flower.

Isotype. A specimen of the type collection other than the holotype.

Joint. The node of a grass culm; an articulation.

Keel. A prominent dorsal ridge, analogous to the keel of a boat; the two lower and united petals of a papilionaceous corolla.

Labellum. A lip; the odd petal of an orchid.

Labiate. Lipped; a member of the Labiatae.

Lacerate. Appearing irregularly torn or cleft.

Laciniate. Cut into narrow lobes or segments.

Lamellate. Made up of thin plates.

Lamina. The blade or expanded part of a leaf, petal, etc.

Lanate, lanose. Woolly; densely clothed with long entangled hairs. *Lanulose:* short-woolly.

Lanceolate. Lance-shaped; much longer than broad, tapering from below the middle to the apex and (more abruptly) to the base.

Lanuginous. Cottony or woolly; lanate.

Lateral. At or on the side.

Lax. Loose, distant.

Leaflet. A segment of a compound leaf.

Legume. A superior 1-celled fruit of a simple pistil usually dehiscent into two valves, having the seeds attached along the ventral suture; a leguminous plant.

Lemma. In grasses, the lower of the two bracts immediately enclosing the floret.

Lenticels. Corky spots on young bark, arising in relation to epidermal stomates.

Lenticular. Lens-shaped.

Lepidote. Covered with small scurfy scales; scurfy; furfuraceous.

Ligneous. Woody.

Ligule. The strap-shaped part of a ray corolla in Compositae; the thin, collarlike appendage on the inside of the blade at the junction with the sheath in grasses. *Ligulate:* provided with a ligule; strap-shaped or tongue-shaped.

Limb. A border; in particular, the expanded portion of a gamopetalous corolla or a gamosepalous calyx.

Linear. Resembling a line; long and narrow, of uniform width, as the leaf-blade of grasses.

Lineate. Marked with parallel lines.

Lingulate. Tongue-shaped.

Lip. One of the two divisions of a bilabiate corolla or calyx, hence an upper lip and a lower lip, although one lip may be wanting; the upper lip of orchids, by a twist of the ovary, usually appears to be the lower.

Littoral. Of a shore, particularly of the seashore.

Lobe. A division or segment of an organ, usually rounded or obtuse; cut less than halfway to the midrib (of a leaf). *Lobed:* bearing lobes.

Locular. Having cells or loculi, as a bilocular pistil or anther is two-celled.

Loculicidal. Dehiscent longitudinally through the middle of the back of a pericarp, between the partitions into the cavity.

Lodicules. The 2 or 3 minute hyaline scales at the base of the stamens in grasses, representing the perianth.

Loment. A legume made up of 1-seeded joints.

Lunate. Crescent-shaped.

Lyrate. Lyre-shaped; pinnatifid, with the terminal lobe considerably larger than the others.

Macro-. Greek prefix meaning large or, more properly, long.

Macrosporangium. The organ in which macrospores are produced.

Macrospore. The larger of the two kinds of spores in Selaginellaceae, etc.

Macrosporophyll. The modified leaf that bears the macrosporangium.

Maculate. Spotted or blotched.

Malpighiaceous hairs. Straight appressed hairs attached by the middle and tapering to the free tips.

Mammillate. Having nipples.

Marcescent. When withered not falling off, especially leaves and corollas.

Marginate. Distinctly margined.

Maritime. Of the seacoast.

Medial. Of the middle.

Medullary. Pertaining to the pith.

Megaspore. A synonym of macrospore.

Membranaceous, membranous. Of the nature of a membrane; thin, soft, and pliable.

-merous. Greek suffix, having parts, as pentamerous or 5-merous, having 5 parts.

Mesocarp. The middle layer or coat of a fruit.

Mesophyte. A plant that grows under medium moisture conditions.

Micro-. Greek prefix meaning small.

Micropyle. The minute orifice in the integuments of an ovule through which the pollen tube enters the seed cavity.

Microsporangium. The organ in which microspores are produced.

Microspore. The smaller of the two kinds of spores in Selaginellaceae, etc.

Microsporophyll. The modified leaf that bears the microsporangium.

Midrib. The central rib of a leaf or other organ.

Monadelphous. Stamens united by their filaments into a tube surrounding the gynoecium, as in malvaceous flowers.

Moniliform. Resembling a string of beads.

Monocephalous. Bearing only one head.

Monoecious. Having staminate and pistillate flowers on the same plant but not perfect ones.

Monotypic. Having a single type or representative, as a genus with only one species.

Montane. Pertaining to mountains.

Mucro. A small and short abrupt tip of an organ, as the projection of the midrib of a leaf. *Mucronate:* with a mucro. *Mucronulate:* minutely mucronate.

Multi-. Latin prefix for many.

Multicipital. Descriptive of a crown of roots or a caudex from which several stems arise.

Multifid. Cleft into many narrow lobes or segments.

Muricate. Rough with short and firm sharp excrescences. *Muriculate:* the diminutive.

Muticous. Pointless, blunt, awnless.

Naked. With a usual covering wanting, as a flower destitute of a perianth.

Napiform. Turnip-shaped.

Nascent. In the act of being formed.

Navicular. Cymbiform; boat-shaped; shaped like the bow of a canoe.

Nectariferous. Nectar-bearing; having a *nectary,* or an organ which secretes nectar.

Nerve. A simple vein or slender rib of a leaf or bract. *Nervelet:* an ultimate branch of a nerve.

Neutral. Devoid of functional stamens or gynoecium.

Node. The joint of a stem; the point of insertion of a leaf or leaves.

Nodding. Hanging down. Cf. *Pendent.*

Nodose. Furnished with knots or knobby nodes. *Nodulose:* the diminutive.

Nut. A hard-shelled and one-seeded indehiscent fruit derived from a simple or a compound ovary.

Nutlet. Diminutive of nut; applied to any small and dry nutlike fruit or seed. Thicker-walled than an akene.

Ob-. Latin prefix signifying the reverse or contrariwise.

Obcompressed. Flattened the other way, antero-posteriorly instead of laterally, opposite of the usual way.

Obconic. Inversely conical, with the point of attachment at the small end.

Obcordate. Inversely cordate, the notch at the apex.

Oblanceolate. Inversely lanceolate.

Oblique. Of unequal sides (as in leaves), slanting.

Oblong. Much longer than broad with nearly parallel sides.

Obovate. Inversely ovate.

Obovoid. Inversely ovoid.

Obsolete. Rudimentary or not evident; applied to an organ that is almost entirely suppressed: vestigial.

Obtuse. Blunt or rounded at the end.

Ochroleucous. Yellowish-white.

Ocrea. A sheath around the stem derived from the leaf-stipules; used chiefly in the Polygonaceae.

Offsets. Short basal lateral shoots from which new plants can develop.

Operculum. A lid. *Operculate:* furnished with a lid.

Opposite. Set against, as leaves when two at a node; one part before another, as a stamen in front of a petal.

Orbicular, orbiculate. Approximately circular in outline.

Orifice. The mouthlike opening of a tubular corolla at the junction of limb and throat or tube.

Orthotropous (ovule or seed). Erect, straight, with the micropyle at the apex and the hilum at the base.

Oval. Broadly elliptic.

Ovary. The part of the pistil that contains the ovules.

Ovate. With the outline of a hen's egg in longitudinal section, the broader end downward.

Ovoid. Solid ovate or solid oval.

Ovule. The megasporangium of a seed plant; the body in the ovary which becomes a seed. *Ovulate, ovuliferous:* bearing ovules.

Palate. In personate corollas, the projecting part of the lower lip, which closes the throat, as in the snapdragon.

Palea. One of the chafflike scales on the receptacle of many Compositae; the in-

ner bract of a grass floret, often partly invested by the lemma. *Paleaceous:* chaffy; composed of small membranaceous scales.

Palmate. Hand-shaped with the fingers spread; in a leaf, having the lobes or divisions radiating from a common point.

Palmatifid. Cleft so as to resemble the outstretched fingers of the hand. *Palmatisect:* palmately divided.

Palustrine. Of or pertaining to marshes.

Panduriform. Fiddle-shaped; obovate and with a contraction on each side.

Panicle. A compound racemose inflorescence. *Paniculate:* borne in a panicle.

Pannose. Feltlike in texture or appearance.

Papilionaceous. Applied to the butterfly-like corolla of the pea, with banner, wings, and keel.

Papillate, papillose. Bearing minute conical processes or papillae.

Pappose. Pappus-bearing.

Pappus. The modified calyx-limb in Compositae, consisting of a crown of bristles or scales on the summit of the akene.

Parietal. Attached to the wall of the ovary, instead of to the axis; said of ovules or a placenta.

Paripinnate. See Pinnate.

Parted. Deeply cleft nearly to the base.

Pectinate. With narrow closely set divisions like the teeth of a comb.

Pedate. Palmate, with the lateral lobes 2-cleft; said of leaves.

Pedicel. The stalk of a single flower in a flower-cluster or of a spikelet in grasses. *Pedicellate:* having a pedicel, as opposed to sessile.

Peduncle. The general term for the stalk of a flower or a cluster of flowers. *Pedunculate:* having a peduncle.

Pellicle. A thin skin or filmy covering.

Pellucid. Transparent, clear.

Peltate. Shield-shaped; a flat body having the stalk attached to the lower surface instead of at the base or margin.

Pendent, pendulous. Suspended or hanging, as an ovule that hangs from the side of the locule; nodding.

Penicillate. Ending in a tuft of fine hairs.

Pepo. A gourd fruit, with hard rind, one-loculed, many seeded.

Perennial. Lasting from year to year.

Perfect. A flower having both stamens and pistils.

Perfoliate. With the leaf entirely surrounding the stem.

Perianth. The floral envelopes collectively; usually used when calyx and corolla are not clearly differentiated.

Pericarp. The ripened walls of the ovary, referring to a fruit.

Perigynium. The inflated saclike organ surrounding the pistil in *Carex.*

Perigynous. Borne around the ovary in contrast to beneath it, as the stamens and corolla are inserted on the floraltube.

Perisperm. The nutritive tissue surrounding the embryo of the seed and formed outside the embryo-sac.

Persistent. Remaining attached, as a calyx on the fruit.

Personate. A bilabiate corolla having the throat nearly closed by a prominent palate.

Petal. One of the leaves of a corolla, usually colored. *Petaloid:* having the aspect of or colored as petals.

Petiole. A leaf-stalk. *Petiolate:* having a petiole.

Petiolule. The stalk of a leaflet.

Phyllary. An individual bract of the involucre of a Composite.

Phyllotaxy. Leaf arrangement.

Phyllode. A dilated petiole serving as a leaf-blade.

Phylogeny. The race history of a plant or natural group; relationship by descent.

Phyllopodic. With a leafy base.

Phylum. A primary division of the plant kingdom.

Pilose. Bearing soft and straight spreading hairs. *Pilosulous:* the diminutive.

Pinna. A leaflet or primary division of a pinnate leaf.

Pinnate. A compound leaf, having the leaflets arranged on each side of a common petiole; featherlike. *Odd-pinnate:* pinnate with a single terminal leaflet (imparipinnate). *Abruptly pinnate:* pinnate without an odd terminal leaflet (paripinnate).

Pinnatifid. Pinnately cleft into narrow lobes not reaching to the midrib.

Pinnatisect. Pinnately divided to the midrib.

Pinnule. A division of a pinna.

Pistil. The ovule-bearing organ of a flower, consisting of stigma and ovary, usually with a style between; gynoecium.

Pistillate. Provided with pistils and without stamens; female.

Pitted. Having little depressions or pits: foveate.

Placenta. The ovule-bearing surface in the ovary.

Placentation. The arrangement or orientation of the placentas.

Plane. Surface flat and even, not curved.

Plicate. Plaited; folded as a fan.

Plumose. Feathery; having fine hairs on each side as a plume.

Pod. Any dry dehiscent fruit; specifically a legume.

Pollen. The male fecundating spores found in the anther.

Pollinia. The pollen-masses of the orchids and milkweeds.

Polygamous. Bearing unisexual and bisexual flowers on the same plant.

Polymorphous. Of various forms.

Polyploid. Having a chromosome number that is a multiple of a basic number for a group of forms. *Polyploid series:* examples—$2n = 8$ (diploid), 16 (tetraploid), 24 (hexaploid), 32 (octoploid), etc. *Polyploid complex:* intimately related members of a polyploid series.

Polystichous. In several vertical rows or ranks.

Pome. An applelike fruit.

Porrect. Directed outward and forward; vertical to the substratum.

Posterior. On the side toward the axis; the upper side of the flower.

Praemorse. As it were, bitten off; said of roots.

Precocious. Flowering before the appearance of the leaves.

Prickle. Sharp outgrowth of the bark or epidermis. *Prickly:* armed with prickles, as the rose.

Procumbent. Trailing on the ground, but not rooting.

Proliferous. Bearing offsets, bulbils, or other vegetative progeny; abnormal or redundant development, as when a leafy shoot develops from a flower part.

Prostrate. Lying flat upon the ground.

Proterandrous. Shedding the pollen before the stigma of the flower is receptive. *Proterogynous:* having the stigma receptive before the stamens of the flower mature.

Proximal. Nearest the axis or base, as contrasted with distal.

Pruinose. Covered with a coarse waxy powder, more pronounced than when glaucous.

Puberulent. Minutely pubescent.

Pubescent. Covered with short soft hairs; downy.

Pulverulent. Dusted as with fine powder.

Pulvinate. Cushion-shaped.

Pulvinus. A swelling close under the insertion of a leaf; the swollen base of a petiole.

Punctate. Dotted with punctures or with translucent pitted glands or with colored dots. *Puncticulate:* the diminutive.

Pungent. Ending in a rigid, sharp point or prickle; acrid (to the taste or smell).

Pustulate. Bearing irregular blisterlike swellings or *pustules,* mostly at the bases of hairs.

Pyriform. Pear-shaped.

Quadrate. Square.

Quadri-. Latin prefix signifying four.

Raceme. A simple, elongated, indeterminate inflorescence with each flower subequally pedicelled. *Racemose:* of the nature of a raceme or in racemes.

Rachilla. A small rachis, specifically the axis of a grass spikelet.

Rachis. The axis of a spike or raceme, or of a compound leaf.

Radiate. Spreading from a common center; bearing rays.

Radical. Belonging to or proceeding from the root.

Radicle. That portion of the embryo below the cotyledons.

Rameal. Belonging to a branch.

Ramose. Branching or branchy. *Ramulose:* with many branchlets.

Raphe. The ridge connecting the two ends of an anatropous ovule.

Ray. A primary branch of an umbel; the ligule of a ray-floret in Compositae, the ray-florets being marginal and differentiated from the disk-florets.

Receptacle. That portion of the floral axis upon which the flower parts are borne, or, in Compositae, that which bears the florets in the head.

Reclined, reclinate. Turned downward, with the tip resting on the ground.

Recurved. Bent backwards.

Reflexed. Abruptly bent downward.

Regular. Said of a flower having radial symmetry, with the parts in each series alike.

Relict. A localized plant left over from an earlier geological period.

Remote. Distantly spaced.

Reniform. Kidney-shaped.

Repand. With an undulating margin, less strongly wavy than sinuate.

Repent. Creeping (prostrate and rooting).

Replicate. Folded backward.

Replum. The septum of certain pods that persist after the valves have fallen, as in the fruit of Cruciferae.

Resiniferous. Producing resin.

Resupinate. Upside down; inverted by the twisting of the pedicel, as the flowers of orchids.

Reticulate. With a network; net-veined.

Retrorse. Bent backward or downward.

Retuse. Notched shallowly at a rounded apex.

Revolute. Rolled backward from both margins, i.e., toward the underside.

Rhizome. An underground stem or rootstock, with scales at the nodes and producing leafy shoots on the upper side and roots on the lower side. *Rhizomatous:* having a rhizome.

Rhombic. Somewhat diamond-shaped. *Rhomboidal.* A solid with a rhombic outline.

Rib. The primary vein of a leaf, or a ridge on a fruit. *Ribbed:* with prominent ribs.

Ringent. Gaping, as the mouth of an open-throated bilabiate corolla.

Rootstock. See *Rhizome.*

Rosette. A crowded cluster of radiating leaves appearing to rise from the ground.

Rostrate. Beaked. *Rostellate:* the diminutive.

Rosulate. In the form of a rosette.

Rotate. Wheel-shaped; said of a sympetalous corolla with obsolete tube and with a flat and circular limb.

Rotund. Rounded in outline.

Rubellous. Reddish. *Rubescent:* turning red.

Ruderal. Weedy; growing in waste places.

Rudimentary. Imperfectly developed; vestigial.

Rufous. Reddish-brown.

Ruga(-e). A wrinkle or fold.

Rugose. Wrinkled. *Rugulose:* the diminutive.

Runcinate. Sharply pinnatifid or incised, the lobes pointing downward.

Runner. A slender trailing stem rooting at the nodes or end; a very slender stolon.

Sac. The cavity of an anther.

Saccate. Furnished with a sac or pouch.

Sagittate. Shaped as an arrowhead, with the basal lobes turned downward. Cf. *Hastate.*

Salverform. A corolla with slender tube abruptly expanding into a flat limb.

Samara. An indehiscent winged fruit.

Saprophyte. A plant living on dead organic matter and hence without chlorophyll.

Sarmentose. Bearing long slender prostrate runners.

Scabrous. Rough to the touch, owing to the structure of the epidermis or to the presence of short stiff hairs. *Scabrid:* somewhat rough. *Scaberulous:* minutely roughened.

Scalariform. Ladderlike.

Scale. Any thin scarious bract; usually a vestigial leaf. *Scaly:* squamose; scarious.

Scandent. Climbing.

Scape. A leafless peduncle rising from the ground in acaulescent plants. *Scapiform, scapose:* resembling or bearing a scape.

Scarious. Thin, dry, and membranaceous, not green.

Scorpioid. Said of a unilateral inflorescence circinately coiled in the bud.

Scrobiculate. Marked by minute or shallow depressions.

Scurfy. Clothed with small branlike scales; furfuraceous.

Secund. Arranged on one side only; unilateral.

Seed. The ripened ovule.

Seep(s). A moist spot where underground water comes to or near the surface.

Segment. A division or part of a leaf or other organ that is cleft or divided but not truly compound.

Sepal. A leaf or segment of the calyx. *Sepaloid:* sepallike.

Septicidal. Dehiscence of a capsule through the septa and between the locules.

Septum. A partition between cavities. *Septate:* divided by partitions or septa.

Seriate. Disposed in series or rows.

Sericeous. Silky; clothed with appressed fine and straight hairs.

Serrate. Saw-toothed, the sharp teeth pointing forward. *Serrulate:* finely serrate.

Sessile. Attached directly by the base; not stalked, as a leaf without a petiole.

Seta. A bristle, or a rigid, sharp-pointed, bristlelike organ. *Setaceous:* bristly or bristlelike. *Setigerous:* bristle-bearing.

Setose. Clothed with bristles. *Setulose:* bearing minute bristles.

Several. Fewer than *many*, perhaps 6 to 8 or 10.

Sheath. The tubular basal part of the leaf that encloses the stem, as in grasses and sedges. *Sheathing:* enclosed as by a sheath, vaginate.

Shrub. A woody plant of smaller proportions than a tree, which usually produces several branches from the base.

Sigmoid. Doubly curved, like the letter S.

Silique. A narrow many-seeded capsule of the Cruciferae, with 2 valves splitting from the bottom and leaving the placentae with the false partition (replum) between them. *Silicle:* a short silique, not much longer than wide.

Silky. See sericeous.

Simple. Unbranched, as a stem or hair; uncompounded, as a leaf; single, as a pistil of one carpel.

Sinistrorse. Directed toward the left.

Sinuate. With a strongly wavy margin. Cf. *Repand.*

Sinus. The cleft or recess between two lobes of an expanded organ such as a leaf.

Smooth. Not rough to the touch. Cf. *glabrous*, without hairs, which may be either smooth or scabrous.

Solitary. Borne singly.

Sordid. Of a dull or dirty hue.

Sorus (plural, *sori*). A fruit-dot, or -cluster on the back of the fronds of ferns.

Spadix. A spike on a succulent axis enveloped in a spathe.

Spathe. A broad sheathing bract enclosing a spadix, as in the calla. *Spathaceous:* resembling or having a spathe.

Spatulate. Like a spatula, a knife rounded above and gradually narrowed to the base.

Spicate. Having the form of or arranged in a spike. *Spiciform:* spikelike.

Spike. An elongated rachis of sessile flowers or spikelets.

Spikelet. A secondary spike; the ultimate flower-cluster in grasses, consisting of two glumes and one or more florets, and in sedges.

Spine. A sharp-pointed, stiff, woody body, arising from below the epidermis; commonly the counterpart of a leaf or stipule. Cf. *Prickle. Spinescent:* more or less spiny, spine-tipped. *Spinose:* bearing spines. *Spinulose:* bearing diminutive spines.

Sporangium. A spore case or sac.

Spore. The reproductive body of pteridophytes and lower plants, analogous to the seed.

Sporocarp. A receptacle containing sporangia or spores.

Sporophyll. A spore-bearing leaf.

Sporophyte. The asexual or diploid generation of ferns and their allies, the fern plant itself.

Spreading. Diverging almost to the horizontal; nearly prostrate. *Spreading hairs:* not at all appressed, but erect. *Spread-*

ing lower lip: diverging from the main axis of the flower.

Spur. A slender, saclike, nectariferous process from a petal or sepal.

Squama. A scale, usually the homologue of a leaf. *Squamella:* Diminutive squama, applied to some types of pappus in Compositae. *Squamellate:* like a little scale. *Squamose, squamate:* covered with scales; scaly. *Squamulose:* provided with small scales.

Squarrose. Spreading rigidly at right angles or more, as the tips of bracts.

Stamen. The male organ of the flower, which bears the pollen.

Staminate. Having stamens but not pistils; said of a flower or plant that is male, hence not seed-bearing.

Staminiferous. Stamen-bearing.

Staminode. A sterile stamen (lacking an anther), or what corresponds to a stamen.

Stellate. Star-shaped. *Stellulate:* resembling a little star or stars.

Sterile. Infertile or barren, as a stamen lacking an anther, a flower wanting a pistil, a seed without an embryo, etc.

Stigma. The receptive part of the pistil on which the pollen germinates. *Stigmatic:* pertaining to the stigma.

Stipe. The leaf-stalk of a fern; the stalk beneath an ovary. *Stipitate:* having a stipe or stalk, as an elevated gland.

Stipule. One of the pair of usually foliaceous appendages found at the base of the petiole in many plants. *Stipulate:* possessing stipules.

Stolon. A modified stem bending over and rooting at the tip; or creeping and rooting at the nodes; or a horizontal stem that gives rise to a new plant at its tip. Cf. *Runner,* a very slender stolon, and *Rhizome,* a subterranean stem. *Stoloniferous:* having stolons.

Stomate. A breathing pore or aperture in the epidermis.

Stomium. Line of dehiscence in a fern sporangium.

Stone. The bony endocarp of a drupe.

Stramineous. Straw-colored.

Striate. Marked with fine longitudinal lines or furrows.

Strict. Very straight and upright, not at all lax or spreading.

Strigose. Clothed with sharp and stiff appressed straight hairs. *Strigillose* or *strigulose:* minutely strigose.

Strobilus. Conelike aggregation of sporophylls.

Strophiole. An appendage at the hilum of certain seeds.

Style. The contracted portion of the pistil between the ovary and the stigma. *Stylebranches* may be only in part stigmatic, the remainder then being *appendage.*

Stylopodium. An enlargement or disklike expansion at base of the style, as in Umbelliferae.

Sub-. Latin prefix meaning somewhat, almost, of inferior rank, beneath.

Subtend. To be below and close to, as the leaf subtends the shoot borne in its axil.

Subulate. Awl-shaped.

Succulent. Juicy; fleshy and soft.

Suffrutescent. Obscurely shrubby; very little woody, but not necessarily low.

Suffruticose. Woody but very low; diminutively shrubby. Cf. *Fruticulose.*

Sulcate. Longitudinally grooved, furrowed, or channeled.

Sulcus (-*i*). A furrow or groove.

Superior. Growing above, as an ovary that is free from the other floral organs.

Suture. The line of dehiscence of fruits or anthers; the line of a natural union or division between coherent parts.

Swale. A moist meadowy area.

Symmetrical. Said of a flower having the same number of parts in each circle.

Sympatric. Growing together with, or having the same range as.

Sympetalous. With petals united in a one-piece corolla; gamopetalous.

Sympodium. An apparent main axis formed of successive secondary axes, each of which represents one fork of a dichotomy, the other being much weaker or entirely suppressed. *Sympodial:* of the nature of a sympodium.

Syn-. Greek prefix meaning united.

Synonym. A systematic name, as for a species, that was superfluous when published, or for some other reason is rejected in favor of another.

Taproot. A primary stout vertical root giving off small laterals but not dividing.

Taxon (plural, *taxa*). Any taxonomic unit, as an order, genus, variety, etc.

Tawny. Dull brownish-yellow; fulvous.

Tendril. A slender, coiling or twining organ by which a climbing plant grasps its support.

Terete. Cylindrical; round in cross section.

Ternary. Consisting of threes; trimerous.

Ternate. In threes, as a leaf consisting of three leaflets.

Tesselated. The surface marked by checkered work, either as depressions or color patterns.

Testa. The outer seed-coat.

Tetradynamous. Having four long and two short stamens.

Tetragonal, tetragonous. Four-angled.

Tetramerous. Having the floral organs in fours or multiples of four.

Tetraploid. Having four basic chromosome sets (four times the gametic number).

Thorn. See *Spine,* but technically a sharp-pointed stiff woody body derived from a modified branch.

Throat. The orifice of a gamopetalous corolla; the expanded portion between the limb and the tube proper.

Thyrse, thyrsus. A compact, ovate panicle; strictly, with main axis indeterminate, but with other axes cymose. *Thyrsoid:* like a thyrse.

Tomentose. With tomentum; covered with a rather short, densely matted, soft white wool. *Tomentulose:* the diminutive. *Tomentum:* a covering of such densely matted woolly hairs.

Tooth. Any small marginal lobe. *Toothed:* dentate.

Torose. Cylindrical with alternate swellings and constrictions, as a rhizome. *Torulose:* the diminutive.

Tortile. Twisted or twining.

Tortuous. Bent or twisted in different directions.

Torus. See *Receptacle.*

Tri-. A Greek and Latin prefix signifying three, thrice, or triply.

Triandrous. Having three stamens.

Trichotomous. Three-forked.

Trifid. Three-cleft to about the middle.

Trigonous. Three-angled, the faces between plane.

Trimerous. Having the parts in threes.

Tripartite. Three-parted.

Triquetrous. Three-edged, the faces between concave.

Triternate. Thrice ternate.

Truncate. As if cut off squarely at the end.

Tube. The narrow basal portion of a gamopetalous corolla or a gamosepalous calyx.

Tuber. A thickened, solid, and short underground stem, with many buds. *Tuberous:* bearing a tuber; resembling a tuber.

Tubercle. A small tuberlike prominence or nodule; the persistent base of the style in some Cyperaceae. *Tubercled, tuberculate:* beset with tubercles or warty excrescences; verrucose.

Tuberiferous. Bearing tubers.

Tubular. Shaped like a hollow cylinder.

Tunicate, tunicated. Having coats or tunics, as a bulb.

Turbinate. Top-shaped; inversely conical.

Turgid. Swollen; inflated.

Umbel. A flat or convex flower-cluster in which the pedicels arise from a common point, like rays of an umbrella. *Umbellate:* borne in an umbel. *Umbellet:* a secondary umbel. *Umbelliferous:* bearing umbels. *Umbelliform:* umbel-shaped.

Umbilicus. A navel; the hilum of a seed. *Umbilicate:* depressed in the center.

Umbonate. Bearing an umbo or boss or conical projection in the center, as the scale of a pine cone.

Uncinate. Hooked at the tip.

Undershrub. A very low shrub.

Undulate. Wavy; repand; with less pronounced "waves" than sinuate.

Unguiculate. Contracted at base into a claw.

Unilateral. One-sided, or turned to one side of an axis: secund.

Unilocular. Having one locule or cell.

Uniseriate. Arranged in one horizontal row.

Unisexual. Flowers having only stamens or pistils; of one sex.

Urceolate. Urn-shaped or pitcherlike, contracted at the mouth.

Utricle. A small, bladdery one-seeded fruit. *Utricular:* having little bladders; inflated.

Vaginate. Loosely surrounded by a sheath.

Valve. One of the segments into which a dehiscent capsule or legume separates. *Valvate:* opening as if by valves, as most capsules and some anthers; in aestivation, meeting at the edges without overlapping.

Vein. A vascular bundle of a leaf or other flat organ. Cf. *Nerve. Veinlet:* one of the ultimate branches of a vein.

Velum. The membranous indusium in *Isoetes.*

Velutinous. Velvety; covered with a fine and dense silky pubescence.

Venation. The arrangement of the veins of a leaf; nervation; veining. *Venose:* veiny; abounding in veins. *Venulose:* abounding in veinlets.

Venter. Belly; under part.

Ventral. Relating to the inward face of an organ, in relation to the axis; anterior; front; opposed to dorsal.

Ventricose. Inflated or swelling out on one side or unequally; gibbous.

Vermicular, vermiform. Worm-shaped or wormlike.

Vermiculate. Marked with tortuous impressions, as if worm-eaten.

Vernation. The arrangement of foliage leaves within the bud.

Vernicose. As if varnished.

Verrucose. Warty; covered with wartlike excrescenses; tuberculate.

Versatile. An anther attached near the middle and capable of swinging freely on the filament.

Verticil. A whorl, or circular arrangement of similar parts about the same point on an axis. *Verticillate:* whorled.

Verticillaster. A false whorl, composed of a pair of nearly sessile cymes in the axils of opposite leaves, as in many mints.

Vesicle. A little bladder or air cavity. *Vesicular:* pertaining to, or having the form of, a vesicle.

Vespertine. Blossoming in the evening.

Vestigial. Reduced to a vestige or trace of a part or organ once more perfectly developed.

Villous. Bearing long and soft and not matted hairs; shaggy.

Virgate. Wand-shaped; slender, straight and erect.

Viscid, viscous. Sticky; glutinous. *Viscidulous:* slightly viscid.

Whorl. A ring of similar organs radiating from a node; verticil.

Wing. A thin and usually dry extension bordering an organ; a lateral petal of a papilionaceous flower. *Winged:* bearing a wing; alate.

Woolly. Having long, soft, entangled hairs; lanate. Cf. *Tomentose.*

Xerophyte. A drought-resistant or desert plant.

Zygomorphic, zygomorphous. Bilaterally symmetrical; that which can be bisected by only one plane into similar halves.

INDEX

Common names that are hyphenated or consist of more than one word are to be sought under the last word. For example, oaks, such as Blue Oak, Canyon Oak, and Poison-Oak, are under *Oak* in alphabetical order. Compound names spelled as a single word, such as Fireweed and Watercress, are listed under the initial letter.

1593

SUPPLEMENT
to
A CALIFORNIA
FLORA

PHILIP A. MUNZ

Introduction

A CALIFORNIA FLORA by Munz and Keck appeared in 1959, and ten years have now passed since the book was actually written. Very active floristic work has been underway in California since then and as many revisions and monographs have appeared for groups having species in the state, an attempt should now be made to present as many changes and corrections as possible. Furthermore, with the critical work being done at present on the flora of Europe, reexamination of many types has meant new appraisal of many Linnaean and other older specific names, and changes therefore have to be made for some of the Old World species which have become established here.

Since 1959 I have attempted to record such changes and corrections as I have found, but many others have been made available to me through the kindness of friends to whom I am most grateful. I refer particularly to John Thomas Howell and Thomas C. Fuller, as well as Rimo Bacigalupi, Clare B. Hardham, Al Hobart, Beatrice Howitt, L. L. Kiefer, Peter H. Raven, James L. Reveal, Robert F. Thorne, Ernest C. Twisselmann, H. L. Wedberg, L. C. Wheeler and Louis B. Ziegler. Their names and those of others who have helped appear in italics for individual bits of information in the present treatise. In some cases names represent the collector on whose specimen the record is based; in others the person who published or made known the information in other ways.

In organizing this *Supplement*, page numbers refer of course to the page in the FLORA. In discussing a plant the genus will ordinarily be given, then the number that the species bears in the FLORA:

p. 631.   PENSTEMON
      4. P. heterodoxus Gray.

The above two lines indicate that *Penstemon heterodoxus* Gray is treated on page 631 in the FLORA and that it is the fourth species discussed in the genus. Thus the reference

1

is easily made to the FLORA and the information presented in the *Supplement* can be correlated with that in the FLORA. Abbreviations used in the *Supplement* are the same as those in the FLORA.

It is hoped that persons possessing a copy of the FLORA can insert in the margins notes to treatment in the *Supplement*. Since some corrections were made in the second and subsequent printings of the FLORA, occasional references are given for the original 1959 printing and apply only to it, but in such cases this fact is indicated.

p. 14.   Sixth paragraph beginning "Deep pervious soil," insert "Cascade Mountains and" before "Sierra Nevada."

p. 17.   Fifth paragraph from top of page, beginning "Foothills," insert after "inner Coast Ranges" the words "and some eastern slopes of outer Coast Ranges."
Line 3 from bottom of page, change San Luis Obispo County to Monterey County.

p. 18.   Paragraph 1 of "26. Alpine Fell-fields." Change *Astrágalus tegetàrius* to *A. Kentrophyta.*
In paragraph 2 of "27. Northern Juniper Woodland" insert "Cascade Mountains and the" before "Sierra Nevada."

p. 19.   Change for the 1959 printing:
Line 11, Arthur R. Cronquist to Arthur J. Cronquist.
Line 15, Edward R. Balls to Edward K. Balls.
Line 18, Steven S. Tillett to Stephen S. Tillett.

p. 22.   For 1959 printing, under LYCOPODIUM add:
1. L. clavàtum L. $2n = 68$ (Löve & Löve, 1958); $2n = 34$ (Mehra & Verma, 1957).
2. L. inundàtum L. at Humboldt Bay, Humboldt Co.; to Canada and New England. In it the sporophylls scarcely differ in general appearance from the foliage lvs.

p. 24.   SELAGINELLA
7. S. densa Rydb. var. scopulòrum (Maxon) Tryon. Change to S. Engelmannii Hieron. var. s. (Maxon) Reed.
10. S. asprélla Maxon. Reported from Kern R. Canyon, Tulare Co. at 2800 ft., *J. T. Howell*; also in Kern Co., *Twisselmann.*

p. 25.   ISOETES
For 1959 printing transpose lines 4 and 5 of generic description.
3. I. muricàta Durieu var.hespéria Reed. Add to synonymy: *I. echinosperma* Dur. var. *hesperia* (Reed) Löve.

p. 27.   EQUISÈTUM
1. and 2. *E. Funstònii* and *E. kansànum* are considered to be the same and are reduced to synonymy under 3. E. laevigàtum by R. L. Hauke (Beihefte zur Nova Hedwigia 8: 1–123. 1963).
4. E. hyemàle var. *califórnicum* Milde and var. *robústum* (A. Br.) A. A. Eat. are put in synonymy under var. affine (Engelm.) A. A. Eat. by Hauke.
4a. E. × Ferríssii Clute [*E. hyemale* var. *affine* × *E. laevigatum*]. It approaches complete intermediacy between its parents and is said to occur in Calif., as in n. Monterey Co., San Gabriel Mts., etc.

p. 28.   In Key to Orders, change first line as follows:

A.   Plants mostly fernlike, terrestrial or erect on mud, not floating or prostrate on mud; producing one kind of spore in sporangia.

p. 29.   BOTRYCHIUM
1. B. multífidum ssp. silaifòlium (Presl) Clausen. In paragraph 2 transfer sentence "More common in Calif." etc. to 2. B. símplex var. compósitum (Lasch) Milde as a second paragraph.
2. B. símplex E. Hitchc. $n = 5$ (Wagner, 1955). The sp. has been collected at elevs. up to 11,000 ft., *Howell.*
2a. B. pumícola Cov.; an Oregon sp. has possibly been collected on Mt. Shasta by *W. B. Cooke* near the s. wall of Diller Canyon. It differs from *B. simplex* as follows:

Sterile blade simple or pinnate, sometimes subternately divided, stalked, inserted at various heights. . . . . *B. simplex*
Sterile blade pinnately divided, with the basal divisions again divided, giving a ternate appearance; the blade sessile, inserted above the middle of the plant. . . . . *B. pumicola*

p. 30.   B. Lunària var. minganénse (Victorin) Dole is treated as a subsp. by Calder and Taylor.

p. 31.   PTERIDÀCEAE
    In Key to Genera, delete "1. *Pteridium*" at end of "C" and insert after "C":

> D.   Fronds usually 3 times pinnate in lower part. Common native. ...... 1. *Pteridium*
> DD. Fronds once pinnate. Natur. in canyons on s. face of San Gabriel Mts... 1a. *Pteris*

p. 32.   PTERIDIUM
    1. **P. aquilìnum** (L.) Kuhn var. **pubescens** Underwood is the correct name.

1a. Ptèris L.

    A large genus of warmer parts of the world, the plants of medium or large size. Sori on a narrow receptacle connected in a marginal line under a simple indusium formed of the revolute margin of the frond, mostly connecting the ends of the free veins. Our sp. with once-pinnate lvs.
    1. **P. vittàta** L. LADDER-BRAKE. Rootstock stout; lvs. dark green, erect or nearly so, 2–5 dm. long, clustered; petioles green, scaly; lfts. firm, lanceolate, ± acuminate; *n* = 58 (Kurita, 1963).—Natur. from cult. at ca. 1500 ft., Eaton Canyon, *Kiefer*, San Dimas Canyon, *Beach*, Big Dalton Canyon, *Beach*, San Gabriel Canyon, *Hutt* (all in San Gabriel Mts.). Native of China.

p. 32.   CHEILANTHES
p. 33.   6. **C. Coóperae** D. C. Eat. ranges n. in Coast Ranges to Santa Cruz and San Luis Obispo cos. and in Sierra Nevada is in Fresno and Tulare cos.
    8. **C. Paríshii** Davenp. near Quail Springs, Little San Bernardino Mts. and in Anza-Borrego State Park, San Diego Co., *L. L. Kiefer.*
p. 34.   12. **C. intertéxta** (Maxon) Maxon. Near Lake Arrowhead and Big Bear City, *L. L. Kiefer*; Cienega Seca Creek, *Munz*; Bluff Lake, *Johnston*; all in San Bernardino Mts.
    13. **C. Covíllei** Maxon in Marin Co., *Howell, Raven*; Mendocino Co., *Kiefer.*
    15. **C. Carlótta-Hálliae** Wagner & Gilbert. Tufted, 6–27 cm. tall; rhizome 1–5 cm. long, clothed with old petioles; lvs. wiry, coriaceous, deltoid, 4 times divided, the ultimate segms. linear- to lance-acuminate, 2–8 mm. long; petioles dark brown.—Dry rocky places, Marin, Monterey, San Benito and San Luis Obispo cos. Howell (Am. Fern J. 5 : 19. 1960) finds it growing with *Cheilanthes siliquosa* Maxon (*Onychium densum* of the FLORA) and *C. californica* Mett. (*Aspidotis c.* of the FLORA) and feels that it is probably a hybrid between them.
    ALEURITÓPTERIS
    1. **A. cretàcea** (Liebm.) Fourn. It has been pointed out that this name is not applicable to our California plants (Howell, Am. Fern J. 50: 22. 1960). Rather than make a new combination in this genus, it is suggested that they be included under *Notholaèna* with N. califórnica D. C. Eat. and ssp. nigréscens Ewan as the names.
p. 35.   PELLAÈA
    In Key to Species, *P. compácta* ranges n. to Placer Co. and in addition to that correction there can be inserted after "Fronds 2–3 times pinnate" etc.:

> Pinnules with greenish or undifferentiated borders. .... 3. *P. mucronata*
> Pinnules with opaque whitish borders. .... 3a. *P. longimucronata*

p. 36.   2. **P. brachýptera** (T. Moore) Baker; occurs in Placer Co., *Howell.*
    3. **P. longimucronàta** Hook. Resembling *P. mucronata*, but with broader pinnules with opaque whitish borders (instead of greenish or undifferentiated borders) and by longer sporangium-stalks.—In the Providence and New York mts., e. San Bernardino Co., *L. Kiefer*; to Colo., New Mex.
    3. **P. mucronàta** (D. C. Eat.) D. C. Eat. Delete the paragraph on var. *californica* (Lemmon) M. & J. and place it in synonymy under species no. 1. **P. compacta.**
    3. **P. mucronàta** and 5. **P. Bridgèsii.** Kiefer feels that these hybridize in the Sierra Nevada.
    4. **P. andromedifòlia** (Kaulf.) Fée. *n* = 29, 87 (Tryon, 1965). Var. **pubéscens**

Baker. L. Kiefer writes that this var. is quite distinct in its narrow blade, dense pubescence, heavy stipe, large glossy green (not blue-green pinnae which are never in threes. He reports it from Point Mugu (Ventura Co.) and the Channel Ids.

6. **P. Brèweri** D. C. Eat. $n = 29$ (Tryon & Britton, 1958).

p. 37. CRYPTOGRÁMMA

1. **C. acrostichoìdes** R. Br. $2n = 60$ (Löve & Löve, 1965).

PITYROGRÁMMA

Alt and Grant (Brittonia 12: 153–170. 1960) recognize **P. triangulàris** (Kaulf.) Maxon with $n = 30$, 45 or 60 chromosomes and **P. t.** var. **Maxònii** Weath. They treat as species: *P. pállida* (Weath.) Alt & Grant with $n = 30$ and *P. viscòsa* (D. C. Eat.) Maxon with $n = 30$.

J. T. Howell (Leafl. W. Bot. 9: 223. 1962) describes **P. triangulàris** var. **semipállida** from Butte Co. as resembling var. *pállida*, but the fronds green and glandless above and the stipes dark green.

p. 38. Hoover (Am. Fern J. 56: 19. 1966) proposes **P. triangulàris** var. **víridis** as lacking evident waxy powder on the lvs. so as to be green underneath.—Santa Cruz Id.; San Luis Obispo, Santa Cruz, Lake and Tuolumne cos.

ADIÁNTUM

1. After **A. pedàtum** L. var. **aleùticum** Rupr. [*A. p. L.* ssp. *a.* Calder and Taylor.] $2n = 58$ (Löve, 1964), insert paragraph:

A. Tràcyi C. C. Hall ex Wagner. Near *A. pedatum*, but with 3 major divisions to each frond instead of 5–9 and with a broader frond than in *A. Jordanii.*—Range, Humboldt to Marin cos.

2. **A. Jórdanii** K. Mull., not C. Muell.

p. 39. Top of page, change Polystichum to Adiantum.

ASPIDIÀCEAE

In Key to Genera, drop "2. *Polystichum*" from end of "B" and after "B" insert:

```
C.  Veins free. .......................................... 2.  Polystichum
CC. Veins anastomosing. ............................... 2a.  Cyrtomium
```

WOÓDSIA

3. **W. Plúmmerae** Lemmon. Differs from *W. oregana* in the indusium having non-ciliated instead of ciliated lobes. Reported by Dodge (Nova Hedwigia 16: 107. 1964) from "Colorado Desert," San Diego Co., *Orcutt.* Since Dodge cites a number of Orcutt collections from Cantillas Mts., etc. in L. Calif., one wonders whether this label is faulty.

p. 40. POLYSTICHUM

2. **P. munìtum** (Kaulf.) Presl. $n = 41$ (Löve, 1964).

Var. **ímbricans** (D. C. Eat.) Maxon. Kiefer feels that this is distinct enough to rank as a sp. He reports finding hybrids between *P. munitum* and *P. scopulinum* at White Mt., n. Siskiyou Co.

3. **P. Lemmònii** Underw. Kruckeberg (Am. Fern J. 54: 123. 1964) questions its occurrence on granite, believing it a plant of serpentine.

4a. **P. Kruckebérgii** Wagner. Like *P. scopulinum,* but lvs. smaller (8–30 cm. long); pinnae largely triangular (5–15 mm. long) and deeply and sharply toothed, the teeth smaller than in *scopulinum;* primary veins fewer (ca. 3–9 pairs); stipe mostly shorter.—Tuolumne and Siskiyou cos.; to B.C., Utah.

5. **P. califórnicum** (D. C. Eat.) Underw. Wagner (Am. Fern J. 53: 7. 1963) writes that this is probably hybrid between *P. Dudleyi* and *P. munitum.* In the description of *P. californicum* in the FLORA transpose lines 6 and 7.

p. 41. **2a. Cyrtòmium Presl.**

Near to *Polystichum,* but differing in its anastomosing veins. Several spp., Old World. (Name Greek, a *bow*).

1. **C. falcàtum** (L.f.) Presl. [*Polypodium f.* L. f.] HOLLY-FERN. Stiff, erect,

the stipes shaggy; fronds dark green, 3–6 dm. long, 10–20 cm. wide, pinnate; pinnae alternate, 7–10 cm. long, ± ovate; $n = 82$ (Mitui, 1965).—Reported as natur. at La Jolla, San Diego, Big Dalton Canyon (Los Angeles Co.), *Kiefer*. Native of E. Asia, S. Afr., Polynesia.

DRYÓPTERIS

1. D. dilatàta (Hoffm.) Gray changed to D. austrìaca (Jacq.) Woynar by Morton. $n = 41$ (Löve, 1964).

p. 42.   2. D. argùta (Kaulf.) Watt. $n = 41$ (Löve, 1964).

3. D. Filix-más (L.) Schott. $n = 82$ (Löve, 1964).

LASTRÈA

Morton (Am. Fern J. 48: 136–141. 1958) uses Thelýpteris and

1. T. pubérula (Baker) Morton. Reported as in Santa Monica Mts., *Kiefer*.

2. T. nevadénsis (Baker) Clute. Apparently the Tuolumne Co. reference in the last line of the description in the FLORA refers to *Athyrium Filix-femina*, not this sp.

p. 44. BLÉCHNUM

1. B. Spicant (L.) Roth, not With. Löve and Löve (Bot. Tidsskr. 62: 94. 1966) refer Pacific N. Am. material to ssp. nippónicum (Kunze) Löve & Löve.

WOODWÁRDIA

1. W. fimbriàta Sm. in Rees. $n = 34$ (Manton & Sledge, 1954).

ASPLÈNIUM

Change Key to Species to:

Fronds fernlike, the blade with 5–30 pairs of pinnae.
  Stipe and rachis dark chestnut or purplish brown throughout.
    S. California. . . . . 1. *A. vespertinum*
  Stipe red-brown, the upper part and rachis green. . . . . 2. *A. viride*
Fronds grasslike, the blade consisting of 2–3 alternate linear segms. Rare, Tulare Co. in Sierra
Nevada. . . . . 3. *A. septentrionale*

p. 45.   1. A. vespertìnum Maxon. At Sherwood Lake, Santa Monica Mts., *B. Joe*.

2. A. víride Huds. $2n = 72$. (Meyer, 1958).

3. A. septentrionàle (L.) Hoffm. [*Acrostichum s. L.*] Fronds grasslike, the stipe much longer than the blade, the latter consisting of 2–3 alternate linear segms.—Columbine Lake above Sawtooth Pass, Tulare Co., *Howell*; to S. Dak., Okla., New Mex.; Eurasia.

POLYPÒDIUM

1. P. Scoùleri Harv. & Grev. $n = 37$ (Manton, 1951).

2. P. hespérium Maxon the preferable name. Kiefer (in letter) reports hybrids between *P. h.* and *P. californicum* and between *P. c.* and *P. Scouleri*.

3. P. califórnicum Kaulf. Lloyd and Lang report 2 races; one diploid ($n = 37$) from San Francisco s., the other tetraploid ($n = 74$) from Monterey Co. to Humboldt Co. and in Sierran foothills.

p. 46.   4. P. Glycyrrhìza D. C. Eat. Reported from Yankee Hill, Butte Co., *Howell*.

p. 47. MARSÍLEA

1. M. vestìta auth., not Hook. & Grev. should be changed to M. mucronàta R. Br.

PILULÀRIA

1. P. americàna A. Br., not R. Br.

p. 49. ÀBIES

1. A. bracteata D. Don ex Poiteau not (D. Don) Nutt.

2a. A. amábilis (Dougl.) Forbes. LOVELY FIR. [*Picea amabilis Dougl.*] Like *A. grandis* in having the lvs. dark green above and with stomates only on the silvery white under surface, but differs in having the bracts gradually narrowed into a slender tip instead of abruptly so; the cones oblong and purple instead of cylindric and green; lvs. erect on the branches instead of in flatter sprays.— Ranging from the Crater Lake region in Ore. to Alaska, but occurs above Diamond Lake, Marble Mts., Siskiyou Co., *Philip A. Lewis*.

p. 51. PÌNUS

In Key to Species AA, BB, CC, D, change EE. to:

> EE. Lvs. dull gray-green, 12–28 cm. long; branchlets glaucous; cones 15–35 cm. long. .... 16. *P. Jeffreyi*
> EEE. Lvs. dull gray-green, 10–15 cm. long; cones 10–12 cm. long.
> 16a. *P. washoensis*

p. 53. **11. P. remoràta** Mason. Reported from Marin Co., *Howell.*

13. P. Murrayàna Grev. & Balf. is considered to be a ssp. of *P. contorta Dougl.* by Critchfield (Maria Moors Cabot Foundation Publ. 3: 1–118. 1957).

15. P. ponderòsa Lawson according to Little.

p. 54. **16a. P. washoénsis** Mason & Stockwell. Lvs. gray-green, 10–15 cm. long; cones 10–12 cm. long, reddish purple to purplish black.—Warner Mts., Modoc Co.; Mt. Rose, w. Nev., Blue Mts., ne. Ore. *Haller* (Madroño 16: 126–132) and Critchfield & Allenbaugh (Madroño 18: 63–64. 1965).

17. P. radiàta D. Don. line 3, insert "mostly" before the last word "in." Becoming established in Marin Co., as seedlings from cult. trees, *Howell.*

18. P. attcnuàta Lemmon. In n. Tulare Co. on road to Mineral King, *G. A. Sanger.*

p. 55. **19. P. Sabiniàna** Dougl. At 5200 ft. in Yosemite.

PÌCEA

In description of genus given as lvs. sometimes stomatiferous on upper side; this may be misleading since lvs. may twist and the upper side turns down.

1. P. Engelmánnii Parry ex Engelm. treated as *P. glauca* ssp. *E. T. M. C. Taylor* (Madroño 15: 144. 1959).

p. 56. TSUGA

1. T. Mertensiàna (Bong.) Carr. s. to Silliman Lake, Tulare Co., at 10,000 ft., *Kaune.*

p. 57. PSEUDOTSÙGA

1. P. Menzièsii (Mirb.) Franco. Reported from near Lompoc, Santa Barbara Co., *Howell.*

p. 59. LIBOCÈDRUS

Mostly being broken up into genera for n. and s. hemispheres and for our sp. the best name is probably **Calocèdrus decúrrens** (Torr.) Florin (Cf. Taxon 5: 192. 1956). Calocèdrus Kurz is a genus of the n. hemis. and has 3 spp.

p. 61. CUPRÉSSUS

1. C. Macnabiàna A. Murr. reported by W. W. Wagener from e. of Amador City in Sect. 20, T. 7N, R. 11E on Sutter Creek Quadrangle. A review of data on distribution is by Griffin & Stone (Madroño 19: 19–27. 1967).

2. C. Bàkeri Jeps. In Plumas Co., *Wagener & Quick* (Aliso 5: 351. 1963).

3. C. nevadénsis Abrams. Reported from e. slope of Greenhorn Range and n. slope of Breckenridge Mt., Kern Co., *Twisselmann.* E. L. Little, Jr. makes new combinations: *C. arizónica* Greene var. *nevadénsis* (Abrams) Little and var. *Stephensònii* (Wolf) Little (Madroño 18: 164. 1966).

6. C. macrocárpa Hartw. ex Gord. $2n = 22$ (Hunziker, 1958). Occurs at Point Lobos, Monterey Co. and colonizing in Marin Co., *Howell.*

p. 63. JUNÍPERUS

> Lvs. usually in 3's, etc. .... 2. *J. californica*
> Lvs. usually in 2's, etc. .... 3. *J. osteosperma*

3. J. osteospérma (Torr.) Little reported from n. side of San Gabriel Mts., *Vasek.*

p. 64. **4. J. occidentàlis** Hook. F. C. Vasek (Brittonia 18: 350–372. 1967) restricts ssp. occidentàlis to area from Susanville, Lassen Co. n. to Wash., Ida. Submonoecious with brownish bark; lvs. in 2's or 3's; cotyledons usually 2 in young seedlings. Subsp. austràlis Vasek occurs from Lassen Co. and the Yolla Bolly Mts. s. to San Bernardino Co. A larger tree, mostly dioecious, with reddish brown bark; lvs. usually in 3's; cotyledons 2–4 in seedlings.

p. 65.   TÓRREYA
        1. **T. califórnica** Torr. ranges s. to Fremont Peak, Monterey and San Benito cos., *P. Arnaud.*

p. 67.   EPHEDRA
        6. **E. víridis** Cov. $n = 14$ (Raven, Kyhos & Hill, 1965).

p. 68.   For 1959 printing, in the upper key, invert the page numbers after Groups 3, 4, 5.

p. 69.   For 1959 printing, in AA, B, C move DD down 3 lines to be after the first EE. Line 11 from the bottom of the page and following lines, change to:

    EE. Fls. hypogynous, the ovary not so inclosed.
      F.   Style and stigma single.
        G.   Calyx not tubular.
          H.   Lvs. subulate, squarrose-spreading, 3–6 mm. long.
                          Caryophyllaceae (Loeflingia) p. 285
          HH. Lvs. not subulate.
            I.   Plants perennial; lvs. oblong, 8–18 mm. long. Saline marshes.
                            Gláux p. 404
            II.   Plants annual; lvs. ovate.
               J.   Calyx 5–6-parted; stamens 6–7 . . *Eremocarpus* p. 162
               JJ.   Calyx largely 4-parted; stamens. . .4. *Urticaceae* p. 920
      GG. Calyx tubular, corolla-like, subtended by bracts often forming a calyx-
          like invol. . . . . *Nyctaginaceae* p. 388

p. 71.   In CC, D at top of page, after "Sepals 2," insert "sometimes 3."
        In CC, DD, etc., below "II. Lvs. alternate" insert:

          I′.   Lvs. reduced to small scales; petals 4; thorny shrubs.
                        Koeberliniaceae p. 174
          I′I′.   Lvs. well developed; plants not thorny shrubs.

Then proceed to "J. Lf.-blades compound," etc.
About middle of page, after CC, DD, EE, FF, GG, H, II, JJ change K to:

          K.   Ovary 3–5-loculed.
             L.   Styles 3, bilobed; stamens united in ranks
                 of 5. *Ditaxis* p. 162
             LL. Style 1, with 3 stigmas; stamens not so
                 united. . . . . *Helianthemum* p. 173
          KK. Ovary or ovaries 1-loculed, etc. as in text.

p. 72.   About middle of page, after "J. Lvs. for most part alternate," change to:

          K.   Sepals 2, or if more, the lvs. fleshy.
             L.   Sepals 2; plants not large vines; stamens 1–3.
                          Portulacaceae p. 295
             LL. Sepals 5; plants large vines; stamens 5.
                          Basellaceae p. 306
          KK. Sepals 3-flowered, etc.

About middle of page, after "M. Fls. regular," etc. insert:

          N.   Ovary 1-celled
                 Caryophyllaceae p. 273
          NN. Ovary 2–5-celled. . *Linaceae* p. 152

Insert on lower part of page, after CC, D. "Ovary inferior":

    D′.   Plant scabrous with short barbed hairs; fr. dry. . . . . *Petalonyx* p. 177
    D′D′. Plants not scabrous with short barbed hairs.

p. 73.   Line 6 from top of page, change to read:

        G.   Lvs. compound, consisting of 2 or more lfts. (separately deciduous)

and toward middle of page change GG. to:

        GG. Lvs. simple, but sometimes simply divided, looking like lfts., but not
          separately deciduous.

About middle of page, insert after "II. The fr. not a legume":

Ia. Lvs. pellucid-punctate (gland-dotted); petals 4, white; lvs. linear; stamens 8. ... *Rutaceae* (Cneoridium) p. 992

IaIa. Lvs. not as above; stamens not 8; petals not 4.

then proceed to "J. Trees with large palmately" etc. and change:

JJ. Trees or shrubs, the lvs. not large and palmately lobed.

J'. Lvs. small, scalelike, linear, caducous; branchlets rigid, spine-tipped. Colo. Desert.
Koeberliniaceae p. 17

J'J'. Lvs. large, pinnately veined.

K. Stamens 4–5, etc.

p. 74. Near top of page, change II to "Lvs. simple and entire or divided, but not compound (including *Geraniaceae* and *Cruciferae* with divided lvs.)."

In middle of page under "II. The fls. not papilionaceous; fr. not a legume," change to:

J. Lvs. not peltate.

K. Fr. a 1-loculed, 3-valved caps.; plants not over 3 dm. high. .... *Violaceae* p. 183

KK. Fr. a 5-loculed, elastically dehiscent, explosive caps.; plants to 1 m. high. *Balsaminaceae* p. 970

JJ. Lvs. peltate; fr. 3-lobed, 3-loculed, each locule 1-seeded. .... *Tropaeolaceae* p. 151

p. 75. In middle of page, "JJ. Stamens 4–7; fr. not opening by a lid."

A few lines farther down, change to:

NN. Ovary 2-loculed; stigma usually 1.

O. Lvs. opposite; fls. in dense axillary clusters, these forming interrupted leafy spikes.
Loganiaceae p. 59?

OO. Lvs. alternate; infl. never spicate. ..... *Solanaceae* p. 590

p. 76. Change G and GG as follows:

G. Lvs. opposite and evidently stipulate, or whorled and not stipulate.
Rubiaceae p. 1037

GG. Lvs. opposite or perfoliate, rarely stipulate and when so, the stipules minute .... *Caprifoliaceae* p. 1046

p. 77. CALYCÁNTHUS

At end of generic description insert as reference "Nicely, K. A. A monographic study of the Calycanthaceae." (Castanea 30: 38–80. 1965.) Two spp. of *Calycanthus* are recognized.

p. 79. RANUNCULÀCEAE

In Key to Genera, after "Petals present" and "Sepals not spurred," add

Petals with a nectariferous pit or scale at base; ovule and seed erect or ascending.
6. *Ranunculus*

Petals unappendaged; ovule and seed suspended. .... 6a. *Adonis*

CÁLTHA

2. C. palústris L. Lf.-blades 5–20 cm. wide, crenate or dentate; sepals yellow, 10–15 mm. long.—Marsh near Forestville, Sonoma Co., *Rubtzoff*. A sp. of the e. U.S. and Eu.

p. 83. DELPHÍNIUM

Key to Species G, H, II, J, change to:

KK. Petioles glabrous to puberulent or strigose.

L. Follicles 14–28 mm. long, etc.
31. *D. Andersonii*

LL. Follicles 6–15 mm. long.

M. The follicles puberulent, 10–15 mm. long; fls. lavender to blue-purple. Monterey and San Luis Obispo cos.
28. *D. umbraculorum*

MM. The follicles sparsely hairy, 6–10 mm.
                          long; fls. dark blue-purple. W. edge of
                          Colo. Desert.
                                                24. *D. Parishii* ɛsp. *subglobosum*

Key to Species GG, H, II, JJ, delete "24. *D. Parishii*" and insert:

K. Fls. mostly bluish; sepals 8–12 mm. long; folli-
                     cles 8–14 mm. long. . . . . 24. *D. Parishii*
                 KK. Fls. white; sepals 6–7 mm. long; follicles 7–9
                     mm. long. . . . . 25. *D. inopinum*

In last line of key, change 29 to 22, for the 1959 printing.

1. **D. Ajàcis** L. It is becoming prevalent to separate the annual spp. of *Delphinium* in which the 2 upper petals fuse, the 2 laterals disappear, and the carpel is 1, and recognize the genus **Consólida** (DC) S. F. Gray. The present sp. is the only one treated in this FLORA to which this would apply and it should bear the name **Consólida ambígua** (L) Ball & Heywood.

p. 84.   2. **D. Purpùsii** Bdg. In Tulare Co., *Thorne*.

4. **D. decòrum** F. & M. is reported from as far s. as Monterey Co., *Howitt & Howell*.

p. 89.   26. **D. Párryi** Gray reported from Monterey Co., *Howitt & Howell*.

p. 90.   29. **D. variegàtum** T. & G. Plants from Carmel and Monterey constitute **D. Hutchinsònae** Ewan. Stems fistulose, 4.5–6 dm. tall, ± pubescent; lvs. with 5 broadly cuneate-obovate segms., each of which is trifid; fls. dark purple; sepals 15–20 mm. long; spur 10–12 mm. long; petals dark, the laminae ca. 6 mm. long; follicles purple-veined, strigulose; seeds winged on angles.—Monterey coast.

p. 91.   ACONÌTUM

1e. **A. columbiànum** Nutt. in Greenhorn Range, Kern Co., *Twisselmann*.

p. 92.   RANÚNCULUS

In Key, AA, BB, C, DD, EE, after line F insert a line:

F′.   Plants with stems; perennials.

and proceed to G. and GG. After the line GG, insert the line:

F′F′.   Plants stemless annuals. Ne. Calif. . . . . 27a. *R. testiculatus*

p. 94.   3. **R. àcris** L. 2*n* = 14 (Jørgensen et al., 1958). In Nevada Co., *True & Howell*.

p. 96.   6. **R. cànus** Benth. Reported from as far south as Monterey Co., *Howitt & Howell*.

8. **R. Macoùnii** Britton. 2*n* = 42 (Löve & Kapoor, 1968).

p. 98.   13. **R. muricàtus** L. Reported from Madera and Fresno cos., *Weiler*.

15. **R. Eschscholtzii** Schlecht. 2*n* = ca. 56 (Löve & Kapoor, 1968).

p. 99.   17. **R. alismifòlius** Geyer ex Benth. In line 1 of Key, correct to read:

"Receptacle 3–5 mm. long in fr." 2*n* = 16 (Löve & Kapoor, 1968).

p. 100.   23. Change to **R. bonariénsis** Poir var. **trisépalus** (Gill.) Lourteig. [*R. t.* Gill. *R. alveolatus* Carter.] Add to range: Chile and Argentina.

p. 101.   26. **R. Cymbalària** Pursh var. **saximontànus** Fern.; at elevs. up to 10,500 ft. in White Mts., *Blakley & Muller*.

27a. **R. testiculàtus** Crantz. [*Ceratocephalus t.* A. Kern.] On p. 92 it keys out with spp. 7 and 8, but is easily recognized as a stemless annual, 4–7 cm. tall, the lvs. basal, divided into linear parts; petals 3–5 mm. long; stamens 10–20; each ak. with a beak twice as long as the body.—Reported from Lassen and Modoc cos., *Fuller*. Native of Old World.

p. 102.   31. **R. subrígidus** W. Drew. Reported from Fulmor Lake, San Jacinto Mts., *Raven*.

6a. Adònis L. PHEASANT'S EYE

A small Eurasian genus like *Ranunculus*, but lacking a nectariferous scale or pit at the base of each petal. Cauline lvs. dissected into numerous linear segms. Fls. solitary at ends of stem or branches.

1. **A. aestivàlis** L. Annual, 2–5 dm. high; fls. solitary, 15–35 mm. across;

petals 6–8, yellow to reddish, 10–17 mm. long.—Reported from Canby, Modoc Co., *Fuller*, and Lassen Co., *Sweeney*. Native of Eurasia.

ANEMÒNE

p. 103.	6. A. quinquefòlia L. var. Gràyii (Behr. & Kell.) Jeps. occurs in Monterey Co., *Howitt & Howell.*

p. 104.	ISOPÝRUM

1. I. occidentàle H. & A. occurs in the San Emigdio Range, Kern Co., *Twisselmann.*

p. 106.	THALÍCTRUM

The literature reference at the end of the generic description should read Contr. Gray Herb. 152.

p. 107.	4. T. polycárpon (Torr.) Wats. $2n = 14$ pairs (Raven, Kyhos & Hill, 1965). At end of generic description cite Ahrendt, L. W. A. Berberis and Mahonia. J. Linn. Soc. 57: 1–408. 1961.

p. 110.	8. B. Dictyòta Jeps. Ahrendt separates *B. D.* and *B. californica* Jeps. on basis of former having lvs. lustrous above and lfts. with 4–8 teeth on each side, the latter with lvs. dull above and the lfts. with 8–12 teeth on each side.

9. B. ampléctens (Eastw.) Wheeler. $2n = 14$ pairs (Raven, Kyhos & Hill, 1965).

12. B. Hígginsiae Munz. Add synonym *Mahonia H.* (Munz) Ahrendt.

13. B. Nevínii Gray. Found in San Francisquito Canyon, n. Los Angeles Co., *Mrs. Thompson.*

p. 113.	BRASÈNIA

1. B. Schrèberi J. F. Gmel. $2n = 80$ (Löve, 1964).

p. 114.	CERATOPHÝLLUM

1. C. demérsum L. Insert synonym *C. apiculatum* Cham.

ANEMÓPSIS

1. A. califórnica Hook. $2n = 22$ pairs (Raven, Kyhos & Hill, 1965).

FREMONTODÉNDRON Cov.

This is the correct name for *Fremontia* Torr., since the latter name has not been conserved, although it had been proposed for conservation when the FLORA was being written. Names to be used under *Fremontodendron* are:

p. 115.	1. F. mexicànum A. Davids.

2. F. califórnicum Cov.

p. 116.	Ssp. napénse (Eastw.) Munz. [*F. n.* Lloyd].

Ssp. obispoénse (Eastw.) Munz.

Ssp. crassifòlium (Eastw.) J. H. Thomas.

Ssp. decúmbens (Lloyd) Munz, comb. nov. [*F. d.* Lloyd, Brittonia 17: 382–384. 1965.] Plant 2–4 m. broad, to 1 m. tall; fls. orange to red-brown, 3–3.6 cm. diam. Eldorado Co.

AYÈNIA

1. *A. californica* Jeps. is placed in synonymy under *A.* compácta Rose in a revision by Cristòbal (Opera Lilloana 4:1–230. 1960).

p. 117.	ABÙTILON

1. A. Theophrásti Medic. $2n = 42$ (E. B. Smith, 1965).—Reported from Monterey Co., *Howitt & Howell.*

4. A. críspum (L.) Sweet is often treated as **Bogenhárdia c.** Kearn., because of the numerous inflated carpels and the fr. not umbilicate apically.

p. 119.	SPHAERÁLCEA

1. S. Orcúttii Rose. $2n = 10$ (Krapovickas, 1957).

3. S. Émoryi ssp. variábilis (Ckll.) Kearn. Add as synonym *S.E.* var. *californica* Shinners. $2n = 20$ (Krapovickas, 1957).

4. S. ambígua Gray. $n = 5$ (Bates, 1967).

Ssp. montícola Kearn. $n = 15$ (Bates, 1967).

Ssp. rugòsa Kearn. $n = 10$ (Bates, 1967).

p. 122.	**Eremálche** Greene

Use instead of *Malvastrum* for California species and change generic description to

Low annual herbs. Lvs. orbicular or palmately parted, stellate-pubescent. Fls.

solitary or in pairs in the upper lf.-axils. Involucellate bractlets 3, distinct, persistent. Sepals somewhat united at base. Petals white to rose-purple, hairy along the margins of the claws. Stamineal column simple, glabrous. Style-branches from one and one-half to two times as long as the stamineal column, filiform, as many as the carpels. Stigmas capitate. Carpels 10–40, indehiscent, 1-ovulate, glabrous, reticulate or transversely ridged on the back and angles. Embryo of the solitary seed forming an incomplete circle; endosperm scanty. (Greek, referring to the desert habitat.) A genus comprised of the following four species.

1. **E. rotundifòlia** (Gray) Greene. [*Malvastrum r.* Gray. *Sphaeralcea r.* Jeps.] Description as in the FLORA. $n = 10$ (Bates, 1967).

2. **E. éxilis** (Gray) Greene. [*Malvastrum e.* Gray. *Sphaeralcea e.* Jeps.] Description in the FLORA. $n = 10$ (Bates, 1967).

3. **E. kernénsis** Wolf. [*Malvastrum k.* Munz.] $n = 10$ (Bates, 1967).

4. **E. Párryi** (Greene) Greene. [*Malvastrum P.* Greene. *Sphaeralcea P.* Jeps.] Description as in the FLORA. $n = 10$ (Bates, 1967).

p. 123.   MALACOTHÁMNUS

Line 4 from top of page, insert (Greek, *malakos,* soft, and *thamnos,* shrub).

p. 124.   3. **M. níveus** (Eastw.) Kearn. in Monterey Co., *Howitt & Howell.*

p. 125.   6. **M. orbiculàtus** (Greene) Greene. In Temblor Range, Kern Co., *Twisselmann.*

p. 126.   16. **M. Jònesii** (Munz) Kearn. In Monterey Co., *Howitt & Howell.*

p. 127.   18. **M. fascículatus** (Nutt. in T. & G.) Greene ssp. **catalinénsis** (Eastw.) Thorne. New synonym under var. **catalinensis.**

p. 128.   LAVATERA

3. **L. crètica** L. Tracy, San Joaquin Co., *Fuller;* Monterey and Ventura cos., *Howell.*

p. 130.   SÌDA

1. **S. hederàcea** Torr. I. D. Clement (Contr. Gray Herb. 180: 52. 1957) uses **S. lepròsa** (Ort.) K. Schum. var. **hederàcea** K. Schum.

2. **S. rhombifòlia** L. at Orland, Glenn Co., *Fuller.*

p. 132.   SIDALCEA

5. **S. calycòsa** ssp. **rhizómata** (Jeps.) Munz in Mendocino Co., *Howell.*

p. 133.   In Key, AA, BB, CC, DD, EE, FF, G, HH, insert "Sierra Nevada" before "Humboldt."

p. 135.   13. **S. oregàna** (Nutt.) Gray ssp. **spicàta** (Regel) C. L. Hitchc. not Greene.

p. 136.   17. **S. Hickmánii** ssp. **Paríshii** (Rob.) C. L. Hitchc. $n = 10$ (Bates, 1967).

p. 137.   HIBÍSCUS

1. **H. triònum** L. in Imperial Co., *Fuller,* and Plumas Co., *Fuller.*

2. **H. califórnicus** Kell. is a perennial in the Botanic Garden.

p. 140.   GERÀNIUM

In Key to Species, after A, B, C insert "C′":

C′.   Peduncles 2-fld.; petals pink to red; root a taproot; caudex branched.
D.   Stems with retrorse-spreading stiff shining hairs; petals scarcely longer than sepals; seeds coarsely reticulate. . . . . 1. *G. pilosum*
DD. Stems with retrorse appressed dull hairs; petals ca. twice as long as sepals; seeds finely reticulate. . . . . 2. *G. retrorsum*
C′C′. Peduncles 1-fld.; petals white, with faint pink edging; roots almost tuberous; caudex not branched. . . . . 1a. *G. microphyllum*

In Key to Species after A, B, CC, insert "C′":

C′.   Carpel-bodies deciduous from the styles at maturity, each with 2 fibrous appendages near the top. . . . . 2a. *G. Robertianum*
C′C′. Carpel-bodies permanently attached to the styles, unappendaged.

Then go to D. in the old key.

p. 141.   1a. **G. microphýllum** Hook. f. Stems slender, with long white and shorter hairs; lvs. dark green; fls. ca. 8 mm. in diam.; carpels smooth, strigose.—Olema, etc., Marin Co. Native of New Zealand.

2a. **G. Robertiànum** L. Annual, commonly branched at base, the branches ± decumbent, 1–5 dm. long, glandular-pubescent; lvs. with ovate divisions 1.5–6 cm. long; sepals 6–8.5 mm. long; petals red-purple, 8–11 mm. long; style-column 1–1.5 cm. long, excluding the slender subulate beak; carpel-bodies 2.5 mm. long, wrinkled.—Established in San Francisco, *Howell.* Native of Eurasia.

    3. **G. Bicknéllii** var. **lóngipes** (Wats.) Fern. in Monterey Co., *Hardham.*

p. 144.   ERÒDIUM

    1. **E. texànum** Gray. Occurs in the El Paso Range, Kern Co., *Twisselmann.*

p. 146.   PELARGÒNIUM

    6. **P. grossularioìdes** (L.) Ait. Reported from Cambria, San Luis Obispo Co. and from San Mateo Co.

p. 147.   ÓXALIS

In Key to Species, under "A. Petals yellow," change to:

> B.   Plants from underground bulblets; petals ca. 2 cm. long.
> C.   The plants acaulescent; lfts. 10–25 mm. long. . . . . 1. *O. Pes-caprae*
> CC. The plants caulescent; lfts. ca. 10 mm. long. . . . . 1a. *O. incarnata*

Under AA, BB, CC, DD omit "11. *O. Martiana*" and add:

> E.   Lfts. suborbicular, 2.5–6 cm. wide; petals 12–18 mm. long.
>   11. *O. Martiana*
> EE. Lfts. narrowed toward base, ca. 1 cm. wide; petals 20–23 mm. long.
>   12. *O. purpurea*

After 1. **O. pes-cáprae**, insert:

    1a. **O. incarnàta** L. Perennial from a bulb; stems 15–30 cm. high; lfts. glabrous, 10 mm. long, 8 mm. wide; sepals 7 mm. long; petals yellow, 17–20 mm. long.—Natur. in San Francisco, *Howell et al.* Native of S. Afr.

p. 148.   2. **O. láxa** H. & A. Pacific Grove, Monterey Co., *Howitt & Howell.*

    3. **O. corniculàta** L. $2n = 24, 36, 42, 48$ (Eiten, 1963).

    5. and 6. **O. pilòsa** Nutt. and **O. califórnica** (Abrams) Knuth. Eiten (Am. Midl. Nat. 69: 303. 1963) uses *O. álbicans* ssp. *pilosa* (Nutt.) Eiten and *O. a.* ssp. *califórnica* (Abrams) Eiten, with **O. álbicans** ssp. **álbicans** e. and s. of Calif.

    7. **O. hírta** L. Occurs at Elk, Mendocino Co., *Fuller.*

p. 149.   11. **O. Martiàna** Zucc. Found in Monterey Co., *Howitt & Howell.*

    12. **O. purpùrea** Thunb. Acaulescent from a rounded bulb; lvs. rosulate, mostly 3–8; lfts. 3, glabrous and green above, impressed-punctate and ± violet beneath, densely ciliate, to ca. 1 cm. long and wide; petioles 1–5 cm. long; peduncles few, 1.5–2 times as long as petioles; sepals 5–7 mm. long, lanceolate; corolla 2–2.3 cm. long, purple on the limb, yellow in throat.—Reported as escape in Santa Cruz, Monterey, Santa Barbara cos. *Howell.* From S. Afr.

p. 149.   LIMNANTHÀCEAE

In family description, change "Carpels 3 or 5" to "Carpels 3 to 5."

p. 150.   LIMNÁNTHES

    3. **L. Doúglasii** R. Br. var. **ròsea** (Hartw.) in Benth. C. T. Mason. Ranges s. to Fresno Co., *Weiler.*

    5. **L. montàna** Jeps. Reported from the Greenhorn Mts., Kern Co., *Twisselmann.*

p. 152.   LINÀCEAE

    C. Marvin Rogers (Madroño 18: 181–184. 1966) recognizes a new genus **Sclerolìnon** for **S. dígynum** and Helen K. Sharsmith (Univ. of Calif. Publ. Bot. 32: 235–314. 1961) separates the other species treated as *Linum* in the FLORA into **Lìnum** and **Hesperolìnon**, the former including spp. 1–4 and 6, the latter 7–15.

LÌNUM

In Key to Species change AA to:

> A.   Petals blue or red, 1–20 mm. long; styles 5; caps. 5–10 mm. long.
> AA. Petals white, rose or yellow, 2–8 mm. long (–15 in *puberulum*); styles often 2–3,
>   sometimes quite united; caps. 2–4 mm. long.

AA, B, C for sp. 5 "Coast Ranges and Sierra Nevada."

p. 153.   2. Change *L. angustifòlium* Huds. to **L. biénne** Mill.

4. **L. perénne** L. ssp. **Lewísii** (Pursh.) Hult. H. G. Baker (Huntiana 2: 141–161. 1965) reports strong crossing barriers between Old World *Linum perenne* L. and New World L. Lewísii Pursh and considers them separate spp.

5. **L. dígynum** Gray [*Cathartolinum d.* Small.] is treated as:

2. Sclerolìnon Rogers.

Glabrous annual. Lvs. opposite, oblong, lacking stipular glands. Sepals lance-oblong, obtuse, lacerate-denticulate and glandular on the margins. Petals yellow. Teeth between stamens none; styles 2, free; stigmas capitate; caps. 4-loculed. (Greek, *skleros*, hard, and *linos*, flax.)

Sclerolìnon dígynum (Gray) Rogers. [*Linum d.* Gray.]

p. 154.   Beginning with sp. no. 7, *Linum drymarioides* Curran, insert:

Hesperolìnon Small

Slender-stemmed annual herbs, ± glaucous, essentially glabrous, or with some puberulence above the axils. Lvs. in whorls of 4 at basal nodes, becoming irregularly whorled at upper nodes and opposite or alternate on ultimate branches, usually early caducous, entire, linear or lanceolate to oblong or almost round, sessile, fleshy, progressively reduced up the stems into small bracts. Fls. in cymes; pedicels filiform, pseudocleistogamous. Sepals 5, united below. Petals 5, erect to spreading, white to lavender or pink or yellow, caducous, clawed. Stamens 5; fils. filiform; anthers versatile. Ovary of 2 or 3 carpels; ovules 4 or 6. Styles as many as carpels, with minute stigmas. Seeds triangular, plump, shining, dark tan to brown. (Greek, *hesperos*, western, and *linos*, flax.)

Petal claw pouches in margins of lamella; petal attachment to apex of cup in sinuses between fils.; carpels 2–3; styles 2–3; stigmas minute.

A. Lvs. lanceolate to rounded ± clasping at base, margined with stipitate glands.
   B. Petals yellow; lvs. irregularly whorled to mainly alternate. ... 8. *H. adenophyllum*
   BB. Petals white to lavender or pink; lvs. in whorls of 4. .... 9. *H. drymarioides*
AA. Lvs. linear to narrow-oblong, with narrow, non-clasping base and not prominently glandular on margin.
   B. Main axis of stem usually long, the primary branches little-spreading; pedicels 0.5–2 mm. long in anthesis.
      C. Sepals glabrous; styles 5–7 mm. long.
         D. Petals yellow; dehisced antlers yellow. ..../........... 10. *H. Breweri*
         DD. Petals white or pink; dehisced anthers white or pink. 11. *H. californicum*
      CC. Sepals pubescent; styles 4–4.5 mm. long. .... 12. *H. congestum*
   BB. Main axis of stem usually short, the primary branches widely spreading; pedicels 0.5–25 mm. long at anthesis.
      C. Carpels and styles 2.
         D. Petals and dehisced anthers yellow. .... 4. *H. bicarpellatum*
         DD. Petals and dehisced anthers whitish. .... 7. *H. didymocarpon*
      CC. Carpels and styles 3.
         D. Petals white to pinkish.
            E. Petals 1.5–3.5 mm. long, not or little spreading; fils. included.
                     1. *H. micranthum*
            EE. Petals 4–6 mm. long, widely spreading; fils. exserted.
               F. Pedicels 1–5 mm. long in fl., not strongly reflexed, the buds not pendent. .... 6. *H. disjunctum*
               FF. Pedicels 5–15 mm. long in fl., early deflexed from branch, the buds pendent. .... 3. *H. spergulinum*
         DD. Petals yellow.
            E. Pedicels 2–25 mm. long; petals 1.5–2.5 mm. long, not or scarcely spreading. .... 2. *H. Clevelandii*
            EE. Pedicels 0.5–3 mm. long; petals 4–5 mm. long, widely spreading.
                     5. *H. tehamense*

1. **H. micránthum** (Gray) Small [*Linum micranthum* Gray.] No. 13 under *Linum* in the FLORA.

2. **H. Clevelándii** (Greene) Small. [*Linum C.* Greene.] No. 14 under *Linum* in the FLORA.

3. **H. spergulìnum** (Gray) Small. [*Linum s.* Gray.] No. 12 under *Linum* in the FLORA.

4. **H. bicarpellàtum** (H. K. Sharsm.) H. K. Sharsm. [*Linum b.* H. K. Sharsm.] No. 15 under *Linum* in the FLORA.

5. **H. tehaménse** H. K. Sharsm. Close to *H. bicarpellatum,* but tricarpellate; petals 4–5 mm. long instead of 3–3.5 mm.; plant ± hoary, not relatively glabrous.—Tehama and Glenn cos.

6. **H. disjúnctum** H. K. Sharsm. Near to *H. micranthum,* but with larger fls. and more spreading perianth; fils. and style exserted.—Serpentine, Inner Coast Ranges, Tehama Co. to Fresno and Monterey cos.

7. **H. didymocárpum** H. K. Sharsm. Near *H. bicarpellatum,* but with petals white to pinkish; dehisced anthers, fils. and styles white.—Serpentine w. of Big Canyon Creek, Lake Co.

8. **H. adenophÿllum** (Gray) Small. [*Linum a.* Gray.] No. 8 under *Linum* in the FLORA. Humboldt to Lake cos.

9. **H. drymarioìdes** (Curran) Small. [*Linum d.* Curran.] No. 7 under *Linum* in the FLORA.

10. **H. Brèweri** (Gray) Small. [*Linum B.* Gray.] No. 9 under *Linum* in the FLORA.

11. **H. califórnicum** (Benth.) Small. [*Linum c.* Benth.] No. 11 under *Linum* in the FLORA.

12. **H. congéstum** (Gray) Small. [*Linum c.* Gray.] No. 10 under *Linum* in the FLORA.

p. 156.   POLÝGALA

2. **P. subspinòsa** Wats. var. **heterorhÿncha** Barneby, $2n = 38$ (Raven, Kyhos & Hill, 1965).

p. 157.   4. **P. cornùta** Kell. var. **Físhiae** Jeps. $2n = 9$ pairs (Raven *et al.,* 1965).

p. 158.   FAGÒNIA

1. **F. laèvis** Standl. Treatment by D. M. Porter (Contr. Gray Herb. 192: 119. 1963). Branches essentially glabrous; pedicels and sepals stipitate-glandular; stipules subulate, reflexed to spreading, 1–6 mm. long.

2. **F. pachyacántha** Rydb. in Vail & Rydb. [*F. californica* var. *glutinosa* Pringle ex Vail.] Ultimate branches densely stipitate-glandular or with subsessile glands; pedicels and sepals glandular to glabrate; stipules linear-subulate, spreading to slightly reflexed, 3–16 mm. long.

LÁRREA

1. **L. tridentata** Sessé & Moçiño for N. Am. plants according to Porter (Contr. Gray Herb. 192: 110–113. 1963). Lft.-veins usually dark, not lined with hairs (lined with white hairs 1–2 mm. long in S. Am.); lfts. obliquely lanceolate to falcate (not obovate to elliptic); stipules obovate, acute to short-acuminate, 1–4 mm. long, free from stem (not broadly ovate, rounded to obtuse at apex, 1–2 mm. long, clasping the stem).

ZYGOPHÝLLUM

1. **Z. Fabàgo** L. var. **brachycárpum** Boiss. Reported from 12 mi. w. of Tipton, Tulare Co., *Fuller.* In the last line the record from Hamlin should read Patterson, Stanislaus Co., collected by *Hamlin.*

p. 159.   KALLSTROÈMIA

1. **K. califórnica** (Wats.) Vail in Calif. and also var. **brachystÿlis** (Vail) Kearn. & Peebles, according to Porter (Contr. Gray Herb. 192: 131. 1963). The sp. has 3–6 pairs of lfts. that are to 3 mm. wide and 7 mm. long; petals 1–2 mm. wide, 3–5 mm. long; beak of fr. 1–2 mm. long; nutlets sharply tuberculate.

Var. **brachystÿlis** has lfts. 2–4 pairs, 3–10 mm. wide, 6–21 mm. long; petals 2–3 mm. wide, 4–6 mm. long; beak of fr. 2–3 mm. long; nutlets rounded-tuberculate.

p. 161.   TETRACÓCCUS

2. **T. ilicifòlius** Cov. & Gilman. Type locality at 3200 ft., *H. T. Harvey.*

3. **T. Hállii** Bdg. ranges into L. Calif. (Sierra San Pedro Martir), *Jaeger.*

p. 162.   CRÒTON

2. **C. califórnicus** Muell-Arg. $n = 14$ (Szweykowski, 1965).

p. 163. DITÁXIS

5. **D. lanceolàta** (Benth.) Pax & K. Hoffm. The last word in line 1 should be *Argythamnia* (in the 1959 printing).

p. 166. EUPHÓRBIA

Key A, BB, C, change to:

D. Floral lvs. broad at base; caps. with tubercles.
  E. Plant glabrous annual. . . . . 2. *E. spathulata*
  EE. Plant pubescent perennial. . . . . 2a. *E. oblongata*
DD. Floral lvs. narrowed at base; caps. smooth.
  E. Lvs. obovate; glands 4, round to elliptical . . . . 3. *E. Helioscopia*
  EE. Lvs. linear to ovate; gland 1, flattened-obconic, tangentially bilabiate.
                                                                            3a. *E. dentata*

p. 167. Key C, change D. to:

D. Stipules united into a broad white membranous scale.
  E. Plant perennial; ♂ fls. 12–20. 23.. *E. albomarginata*
  EE. Plant annual; ♂ fls. 5–10. . . . . 23a. *E. serpens*

p. 168. 2a. **E. oblongàta** Griseb. Like *E. spathulata,* but pubescent, perennial with rhizomes; lvs. sessile, oblong-lanceolate, the floral ovate, mucronate; upper stem-lvs. ca. 4–6 cm. long; caps. warty.—Contra Costa and San Joaquin cos., Santa Clara–San Mateo county line, *Thorne & Raven.* Native of e. Medit.

3a. **E. dentàta** Michx. Like *E. Helioscopia,* but stem not tipped by an umbel; stipules glandlike; each invol. with only 1 gland.—San Joaquin Co., *Fuller.* Native annual from e. of Rocky Mts.

4. **E. Ésula** L. $2n = 60$ (R. J. Moore, 1958).

7. **E. crenulàta** Engelm. ascends to 6000 ft., Lassen Park, *Howell.*

p. 170. 16. **E. maculàta** L. Burch (Rhodora 68: 163. 1966) concludes that E. nutáns Lag. is the correct name for our plant. Wheeler in 1960 reported it from near Lodi, San Joaquin Co.

p. 171. 23a. **E. sérpens** HBK. Like *E. albomarginata* in having stipules united, but is a prostrate annual with 5–10 ♂ fls. instead of 12 or more; lf.-blades 2–7 mm. long.—Hunter's Point, Salton Sea, and San Francisco, *Howell.* Widely distributed in N. and S. Am.

p. 172. 32. **E. prostràta** Ait. Alameda Co., *Howell.*

p. 173. HELIÁNTHEMUM

1. **H. Greènei** Rob. Add *Crocanthemum occidentale* Janchen to synonymy.

p. 174. 4. **H. guttàtum** (L.) Mill. Differs from the 3 spp. treated in the FLORA in being annual, slender, stellate-pubescent and pilose, 6–30 cm. high; lvs. opposite, elliptic-lanceolate. Native of Eu.

CÍSTUS

1. **C. villòsus** L. occurs in cult. in a number of forms, more than one of which has become natur. Var. **taúricus** Grosser is sparsely hairy, not glandular; var. **undulàtus** Grosser is glandular.

Insert: 21a. **Koeberliniàceae.** JUNCO FAMILY

Almost leafless trees or shrubs with pale green, spine-tipped stiff branchlets. Lvs. alternate, minute, scalelike, soon deciduous. Fls. small, in small axillary racemes. Sepals 4–5. Petals 4–5, deciduous. Stamens hypogynous. Ovary of 2–5 united carpels, superior.

1. Koeberlínia Zucc.

Branches spinose, interlocking. Sepals 4. Petals 4, greenish white. Stamens 8. Fr. a rounded berry. (*C. L. Köberlin,* German clergyman.)

1. **K. spinòsa** Zucc. Shrubby or arborescent, with thin smooth, green or yellowish bark on the young branches; lvs. 1.5–2 mm. long; racemes 1–6 mm. long; berry 4–5 mm. diam., grayish.—Reported from Chocolate Mts., Imperial Co., *Wiggins;* to Texas, Mex.

p. 174.   TAMARÍX

(M. Zohary of Israel in the 1965 Annual Research Report for the U.S.D.A. treats the following spp. as in Calif.) B. R. Baum (Baileya 15: 19–25. 1967) agrees with this listing.

A. Fls. 5-merous.
    B. Lvs. vaginate; bracts not vaginate, but somewhat clasping. . . . . *T. aphylla*
    BB. Lvs. sometimes auriculate, but not vaginate or amplexicaul.
      C. Staminate fls. with fils. inserted below the disc.
        D. Petals obovate, widened distally; bracts ovate, subobtuse.
          In more saline places. (Includes *T. pentandra* auth.)
                                          T. ramosissima Ledeb.
        DD. Petals oblong-ovate, narrowed distally. Non-saline habitats.
                                            T. chinensis Lour.
      CC. Staminate fls. with fils. inserted around the disc.
        D. Petals more than 2.25 mm. long, subpersistent, narrowed slightly toward the apex; racemes 5–10 mm. broad. . . . . *T. africana* Poir.
        DD. Petals not more than 2 mm. long, soon deciduous; racemes 4–5 mm. broad.
          E. The petals elliptical, equally wide in both halves; disc with 5 rounded lobes, each between 2 stamens. . . . . *T. aralensis* Bunge
          EE. The petals elliptic-ovate, definitely wider in lower half; disc not lobed, the fils. confluent at base. Rare. . . . . *T. gallica* L.
AA. Fls. 4-merous; petals oblong, scarcely 2 mm. long; bracts diaphanous.
    (Includes *T. tetranda* auth.) . . . . *T. parviflora* DC.

p. 176.   FOUQUIÈRIA

1. **F. spléndens** Engelm. $2n = 12$ (Raven, Kyhos & Hill, 1965).

p. 178.   PETALÓNYX

After generic description insert (Davis, W. S. and H. J. Thompson. A revision of Petalonyx, etc. Madroño 19: 1–18. 1967).

1. **P. nítidus** Wats. $n = 23$ (Davis & Thompson, 1967).
2. **P. lineàris** Greene. $n = 23$ (Davis & Thompson, 1967).
3. **P. Gilmánii** Munz reduced to **P. Thúrberi** ssp. **Gilmánii** Davis & Thompson. $n = 23$.
4. **P. Thúrberi** Gray. $n = 23$ (Davis & Thompson, 1967).

p. 180.   MENTZÈLIA

9. **M. nìtens** Greene. var. **Jònesii** (Urban & Gilg) Darlington. Petals 8–12 mm. long. Owens V., *Wiggins*.

p. 181.   12. **M. Líndleyi** T. & G. in Monterey Co., *Howitt & Howell*.

Ssp. **cròcea** (Kell.) C. B. Wolf maintained as a sp. by H. J. Thompson (Brittonia 12: 81–92. 1960).

p. 183.   EÙCNIDE

Thompson, H. J. & W. R. Ernst. Fl. biol. & systematics. (Jour. Arn. Arb. 48: 56–88. 1967.)

1. **E. ùrens** (Gray) Parry reported from the El Paso Range, Kern Co., *Twisselmann*. $n = 21$ (Thompson & Ernst, 1967).

p. 184.   VIÒLA

Change A of Key to:

A. Stipules almost as large as lvs., leaflike with large terminal lobe. Introd. annuals.
    B. Petals scarcely longer than sepals; spur ca. equal to appendages. . . 23. *V. arvensis*
    BB. Petals longer than sepals; spur rather longer than appendages. . . . . 24. *V. tricolor*

p. 185.   1. **V. glabélla** Nutt. In Greenhorn Range, Kern Co., *Twisselmann*.
p. 186.   5. **V. Bàkeri** ssp. **shasténsis** M. S. Baker differs from *V. Bakeri* in having pubescent caps. and a few short appressed hairs on the faces of the sepals. Amador, Lassen, Tehama, and Trinity cos.; to Ore. Clausen (Madroño 17: 175) does not maintain this ssp.

7. **V. áurea** Kell. J. Clausen says **V. a.** and ssp. **mohavénsis** Baker & Clausen should be sspp. of **V. purpurea** Kell. (Madroño 17: 175 & 295. 1964).

Ssp. **mohavénsis** Baker & Clausen reported from Monterey Co., *Howitt & Howell*.

p. 188.   8. **V. purpùrea** ssp. **geophỳta** Baker & Clausen occurs in volcanic ash at 4000–6000 ft.

10. **V. pedunculàta** ssp. **tenuifòlia** Baker & Clausen reported from Santa
Lucia Mts., Monterey Co., *Howitt & Howell.*

p. 189.   13. **V. Sheltònii** Torr. found in Greenhorn Range, Kern Co., *Twisselmann.*

p. 190.   22. **V. adúnca** Sm. var. **oxýceras** (Wats.) Jeps. is a separate sp. to M. S.
Baker and a ssp. to J. Clausen as *V. a.* ssp. *o.* (Wats.) Piper. It has thinner lvs.
never cordate at base, sometimes wider than long (not ovate); fls. smaller, with
upper 4 petals mostly in 1 plane, but the lower petal in another plane (instead
of all in the same plane); seeds with ratio 1.85 to 1 for length to width
(*adunca* almost 2:1).

p. 191.   24. **V. trìcolor** L. WILD PANSY. Corolla 15–25 mm. long, blue-violet or yellow
or combination of these; spur up to twice as long as appendages.—Established
in counties about San Francisco Bay; native to Eu.

HYPÈRICUM

In Key after line beginning "Sepals ovate to obovate," etc. change to:

Lvs. 1–4 cm. long; herbs to 7 dm. tall.
    The lvs. oblong to ovate, mostly flat; sepals mostly without black dots. . . . **4.** *H. formosum*
    The lvs. linear to lanceolate, mostly folded; sepals with many black marginal dots.
                                                                 5. *H. concinnum*
Lvs. 5–7 cm. long; shrubs to 5 m. tall. . . . . **6.** *H. canariense*

p. 192.   3. **H. perforàtum** L. At Kernville, Kern Co., *Twisselmann.*

5. **H. concínnum** Benth. $2n = 8$ pairs (Raven, Kyhos & Hill, 1965).

6. **H. canariénse** L. Shrub to 5 m. tall; lvs. oblong-lanceolate, narrowed at
base, 5–7 cm. long; fls. 2.5–3 cm. diam., in panicles; sepals ovate, acute, ciliate.
—Established in colonies along creeks at Montecito and on hill roads n. of Santa
Barbara, *Pollard;* native of Old World.

PAPAVERÀCEAE

In Key to Genera, after A, delete B and BB and change to:

    B. Basal lvs. broadly linear without petiole; plants pubescent.
        C. Carpels more than 3. . . . . **1.** *Platystemon*
        CC. Carpels 3 . . . . **1a.** *Hesperomecon*
    BB. Basal lvs. spatulate, narrowed at base; plants essentially glabrous. . . **2.** *Meconella*

p. 193.   PLATYSTÈMON

After generic description insert as a reference: Ernst, W. R. Floral morphology and systematics of Platystemon and its allies Hesperomecon and Meconella.
Univ. Kans. Sci. Bull. 47: 25–70. 1967.

Change generic description of Platystemon to:

1. Platystèmon Benth. CREAM CUPS

Low villous annuals with lvs. not narrowed at base, ± alternate below, opposite or whorled above. Fls. terminal on long peduncles. Sepals 3. Petals 6, white
to yellowish. Stamens hypogynous, many. Carpels more than 3, each forming a
locule around a central chamber; carpels at first united, separate in fr., each
several-ovuled, breaking transversely into indehiscent 1-seeded joints. Rest of
old description valid, but seeds with wall of fr. adherent.

    1. **P. califórnicus** Benth.

1a. Hesperomècon Greene
(*Platystigma* Benth., not R. Br.).

Plants villous. Lvs. broadly linear, not narrowed at the base. Stamens usually
many. Carpels 3, the ovary forming 1 locule. Seeds free from wall of fr. (Greek,
hesperos, western, and *mekon,* poppy).

p. 194.   1. **H. lineàris** (Benth.) Greene. [*Platystigma l.* Benth. *Platystemon l.* Curran.
Meconella l. Nels & Macbr.] Additional synonymy in FLORA at top of p. 194.
Description as in FLORA, for *Meconella linearis.*

2. Meconélla Nutt. in T. & G.

Plants glabrous or with a few short hairs on sepals. Basal lvs. spatulate, distinctly narrowed at base, the blades ± deltoid to orbicular, upper lvs. ± linear. Stamens 4–6 in 1 series or ca. 12 in 2 series. Carpels 3, the ovary a single locule. Fr. frequently spirally twisted. Seeds lustrous black, free. Three spp. (Greek, *mekon*, poppy, and *ella*, diminutive.)

A.   Receptacle ca. as broad as long, without rim; stamens 6, the anthers frequently as long as or longer than the fils. .... *M. denticulata*

AA. Receptacle broader than long, with small rim beneath insertion of sepals; anthers very much shorter than fils.

B.   Stamens mostly ca. 12, biseriate or unequal. .... 2. *M. californica*

BB. Stamens 4–6, in 1 series, ca. equal in length. .... 3. *M. oregana*

    1. **M. denticulàta** Greene. [*Platystemon d.* (Greene) Greene. *M. kakoethes* Fedde. *M. oregana* var. *denticulata* of the FLORA.]

    2. **M. califórnica** Torr. & Frém. of the FLORA.

    3. **M. oregàna** Nutt. in T. & G. of the FLORA.

p. 195. ARCTOMÈCON

    1. **A. Merriàmii** Cov. *n* = 12 (Ernst, 1958).

DENDROMÈCON

    1. **D. rígida** Benth. *n* = 28 (Ernst, 1958). Add Piute Mts., Kern Co. as extension of range.

    2. **D. Harfórdii** Kell. treated by Raven (Aliso 5: 321, 1963) as ssp. of **D. rígida** Benth. with var. **rhamnoides** as a synonym of *D. rigida.* Thorne has *D. rigida* ssp. *rhamnoìdes* (Greene) Thorne.

p. 196. ESCHSCHÓLZIA

W. Ernst (Madroño 17: 280–294. 1964) shows that certain corrections are necessary: In Key to Species, change A, B to:

B.   Buds nodding, usually with some hairs; herbage ± canescent-hairy.

C.   Sepals 11–21 mm. long; petals 15–32 mm. long. Inner S. Coast Ranges from w. Merced Co. to Ventura Co. and w. San Joaquin Co. ...... 5. *E. Lemmonii*

CC. Sepals 5–13 mm. long; petals 5–16 mm. long. Outer S. Coast Ranges, San Luis Obispo and Monterey cos. .... 5a. *E. hypecoides*

Insert after A, BB, CC, DD, E:

F.   Sepals 7–8 mm. long; petals 10–38 mm. long. 1. *E. caespitosa*

FF. Sepals 3–4 mm. long; petals ca. 4 mm. long. 1a. *E. rhombipetala*

    1. **E. caespitòsa** Benth. Delete *E. rhombipetala* and *E. hypecoides* as synonyms. *n* = 6 (Ernst, 1959).

p. 197.   1a. **E. rhombipétala** Greene. Sepals 3–4 mm. long (not 7–18 mm. as in *E. caespitosa*); petals ca. 4 mm. long (not 10–38 mm.); buds erect.—It has been found in Contra Costa, San Joaquin, Alameda, Stanislaus and San Luis Obispo cos.

    2. **E. élegans** Greene. *n* = 17 (Ernst, 1959).

    3. **E. minutiflòra** Wats. *n* = 18 (Ernst, 1959).

E. minutiflora Wats. var. *darwinensis* Jones is treated as a sp., **E. Covíllei** Greene by Mosquin (Madroño 16: 91–96. 1961) on the basis of fl.-size, more northern distribution and chromosome number (*n* = 18 in *E. minutiflora* and 12 in *E. Covillei*).

    4. **E. Paríshii** Greene. 2*n* = 6 pairs (Raven, Kyhos & Hill, 1965).

    5. **E. Lemmònii** Greene. *n* = 6 (Ernst, 1959).

    5a. **E. hypecoìdes** Benth. [*E. eximia* Greene. *E. alcicornis* Greene. *E. delitescens* Greene ex Fedde. *E. asprella* Greene?]. Agreeing with *E. Lemmonii* in having nodding buds; differing in sepals 5–13 (not 11–21) mm. long; petals 5–16 (not 15–32) mm. long, bright yellow, often with a diffuse orange spot near the base (not mostly dark orange or red orange).—Outer S. Coast Ranges of San Benito, Monterey, w. Fresno and n. San Luis Obispo cos. (while *E. Lem-*

monii ranges from w. Merced Co. to Ventura Co., in the Coast Ranges and adjacent margin of the San Joaquin V.).
6. **E. Lóbbii** Greene. In the Greenhorn Range, Kern Co. ,*Twisselmann.*

p. 198.   8. **E. califórnica** Cham. $2n = 6$ pairs (Raven, Kyhos & Hill, 1965). Twisselmann recognizes an especially large plant from sand flats n. of Kernville as **E. prócera** Greene.
GLAÙCIUM
1. **G. flàvum** Crantz. $2n = 14$ (Larsen, 1954).

p. 199.   **Argemòne**
1. **A. corymbòsa** Greene. $n = 14$ (Ernst, 1959).
2. **A. muníta** Dur. & Hilg. In the Temblor Range, Kern Co., *Twisselmann*; Monterey Co., *Howitt & Howell.*

p. 200.   STYLOMÈCON
1. **S. heterophýlla** (Benth.) G. Tayl. $2n = 56$ (Ernst, 1958).

p. 201.   PAPÀVER
2. **P. Rhoèas** L. $2n = 14, 15, 21$ (Koopmans, 1955).
3. **P. califórnicum** Gray. $2n = 28$ (Ernst, 1958).
After 3. **P. califórnicum** Gray insert:
A plant superficially resembling *P. californicum* Gray but with caps. to 2.5 cm. long, obovoid-oblong, more than twice as long as wide, growing spontaneously near Pacific Highlands, Monterey Co., has been identified by Dr. P. F. Yeo as **P. dùbium** L., a European sp.
5. **P. ápulum** Ten. var. **micránthum** (Boreau) Fedde reported from Temblor Range, Kern Co., *Twisselmann.*
6. **P. hýbridum** L. Annual, 2–5 dm. tall, stiff-strigose; lvs. 2–3 times pinnately lobed into bristle-pointed narrow segms.; fls. 2–4 cm. in diam.; sepals bristly; petals round-obovate, crimson with a blackish basal spot; caps. ± globose, 1–1.5 cm. in diam.—A heavy infestation reported in a vineyard 1 mi. w. of Madera, Madera Co., *T. C. Fuller*; w. Kern and e. San Luis Obispo cos., *Twisselmann.* Native of Eurasia.
CÁNBYA
1. **C. cándida** Parry. $n = 8$ (Ernst, 1958).

p. 203.   DICÉNTRA
Chromosome counts reported by Ernst are $n = 12$ for *D. chrysántha* (H. & A.) Walp., $n = 8$ for *D. formòsa* (Andr.) Walp., and $n = 16$ for *D. ochroleùca* Engelm.
3. **D. formòsa** (Andr.) Walp. s. to Monterey Co., *Howitt & Howell.*
Ssp. **nevadénsis** (Eastw.) Munz is treated as a sp. by K. R. Stern (Brittonia 13: 1–57. 1961).

p. 203.   5. **D. pauciflòra** Wats. reported from the Greenhorn Range, Kern Co., *Twisselmann.*

p. 204.   FUMÀRIA
1. **F. officinalis** L. $2n = 32$ (Löve & Löve, 1956).

p. 206.   Capparidaceae spelled "Capparàceae" by Ernst (Jour. Arn. Arb. 44: 81–95. 1963).

p. 207.   ISÓMERIS
1. **I. arbòrea** Nutt. $2n = 20$ pairs (Raven, Kyhos & Hill, 1965).
OXÝSTILIS
1. **O. lùtea** Torr. & Frém. $2n = 20$ pairs (Raven, Kyhos & Hill, 1965).

p. 208.   WISLIZÈNIA
1. **W. rcfrácta** Engelm. Lvs. trifoliolate, the lfts. mostly broadly obovate; caps. valves at most faintly tuberculate.
Var. **Pálmeri** (Gray) Jtn. Lvs. mostly simple, or if bifoliolate, the lfts. narrower; caps. valves strongly tuberculate to sometimes horned.—From near mouth of Colo. R. to Son.
CLEÒME
1. **C. serrulàta** Pursh. $n = 17$ (Raven, Kyhos & Hill, 1965).
2. **C. lùtea** Hook. $2n = 17$ pairs (Raven, Kyhos & Hill, 1965).

p. 210.   CRUCÍFERAE
In Key to Genera, under A, change B to:

    B.   Silicles not winged; stems with branched hairs.
         C.   Fruiting pedicels slender, 2–6 mm. long, recurved; style shorter than the fr.
                     41. *Athysanus*
         CC.   Fruiting pedicels stout, ca. 1 mm. long, not recurved; style almost as long as
             the fr. .... 41a. *Euclidium*

p. 211.   In Key AA, BB, C, DD, EE, FF, GG, HH, I, JJ, change K to:

                 K.   Lvs. simple, entire or finely serrate.
                     L.   The lvs. auriculate-clasping; fls. yellow.
                           53. *Conringia*
                     LL.   The lvs. not auriculate-clasping; fls. purple
                       to whitish. .... 54. *Hesperis*

p. 215.   THELYPÒDIUM
         5. **T. flexuòsum** Rob. in Gray. $n = 13$ (Rollins, 1966).
         7. **T. laciniàtum** var. **millefòlium** (A. Nels.) Pays. $n =$ ca. 14 (Rollins, 1966).
         9. **T. flavéscens** (Hook.) Wats. Rollins puts this in the genus *Caulanthus*;
$n = 14$ (Rollins, 1966).

p. 216.   10. **T. Lemmònii** Greene. $n = 14$ (Rollins, 1966).
         11. **T. lasiophýllum** (H. & A.) Greene. Rollins has in *Caulanthus*; $n = 14$
(Rollins, 1966).

p. 217.   STREPTÁNTHUS
In Key after A, B, change:

         CC.   Annual or biennial; sepals usually lacking short stiff terminal hairs; upper
pair of fls. connate.
           D.   Siliques ascending to recurved-spreading, 2.5–7 cm. long; sepals about
equally broad.
              E.   Lower lvs. saliently lobed, spatulate-obovate, long-petioled; plants
0.5–1.8 dm. high. Marin Co. .... 9. *S. batrachopus*
              EE.   Lower lvs. toothed or entire, short-petioled; plants mostly 3–7 dm.
high.
                  F.   Basal lvs. oblanceolate to ovate or obovate; petals in dissimilar
pairs.
                     G.   Sepals alike; petals 7–10 mm. long.
                         H.   Plants annual, 3–6 dm. tall; sepals largely purple, not
setose; petals white or with purple veins. Glenn Co.
to San Benito Co. .... 10. *S. Breweri*
                       HH. Plants biennial, to 10 dm. tall; sepals greenish-yellow
to red-purple, often setose.
                           I.   Petals creamy to light salmon with brownish or
orange veins. Sonoma and Lake cos.
                               10a. *S. Morrisonii*
                          II.   Petals white. Sonoma Co. .. 10b. *S. brachiatus*

GG, FF, and DD as in the FLORA.
After "BB. Infl. with some conspicuous broad leafy bracts among lower fls."
change to:

         C.   Middle cauline lvs. oblong to obovate.
           D.   Siliques 1.5–3 mm. broad.
              E.   The siliques arcuate-spreading; basal lvs. entire or toothed.
                             7. *S. tortuosus*
              EE.   The siliques ascending; basal lvs. pinnatifid. 18b. *S. Farnsworthianus*
           DD. Siliques 1 mm. broad, erect. .... 8. *S. gracilis*
         CC.   Middle cauline lvs. linear or pinnate; siliques deflexed.
           D.   Fls. generally yellow, rarely with purplish tinge; lvs. entire and linear
or with linear lobes. .... 18. *S. diversifolius*
           DD. Fls. violet or purplish; lvs. not with linear segms.; siliques 1.5–1.7 mm.
broad. Tehipite V., Fresno Co. .... 18a. *S. fenestratus*

AA, etc. remains as in the FLORA.

p. 218.   3. **S. cordàtus** Nutt. Insert:
Var. **piuténsis** J. T. Howell. Siliques flattened, broad, but stems taller, to 1 m.,
and with woodier base.—Piute Mts., Kern Co.
         7. **S. tortuòsus** Kell. $n = 14$ (Rollins, 1966).

p. 219.   10. **S. Brèweri** Gray. $n = 14$ (Rollins, 1966).

10a. **S. Morrisònii** F. W. Hoffm. Glabrous glaucous biennial, to 10 dm. tall; upper and lower surfaces of juvenile and of lower lvs. usually uniformly green; upper stem-lvs. auriculate-spatulate to -ovate, clasping, entire or few-toothed; calyx greenish-yellow, becoming golden-yellow in age, glabrous or with a few scattered hairs, to 8 mm. long; petals creamy white to light salmon with brownish or orange veins; lower petals 1 cm. long; upper connate fils. orange; siliques erect or divergent, 2–7 cm. long, torulose.—Serpentine, Big and East Austin Creeks, Sonoma Co.

Ssp. **elàtus** F. W. Hoffm. Upper surface of juvenile and of lower lvs. heavily mottled with purple-brown, lower surface uniformly purplish; calyx greenish-yellow to golden-yellow; upper connate fils. yellow.—Serpentine, head of St. Helena and Bucksnort creeks, Lake Co.

Ssp. **hirtiflòrus** F. W. Hoffm. Upper surface of juvenile lvs. mottled with purple-brown, lower surface uniformly purple; calyx red-purple, densely hirsute; upper fils. orange with 2 purple stripes.—Serpentine, head of East Austin Creek, Sonoma Co.

10b. **S. brachiàtus** F. W. Hoffm. More or less woody biennial to 4.5 dm. tall, glabrous, glaucous below; stem lvs. crisped, auriculate, to 5.5 cm. long, 2.5 cm. wide, entire to coarsely serrate; fls. 8 mm. long; calyx purplish, glabrous, usually reticulate; upper fils. orange, with 2 purplish lines; siliques erect, torulose, purplish, to 6.5 cm. long.—Serpentine, east of Pine Flat, Sonoma Co.

p. 220.   12. **S. glandulòsus** Hook. A. R. Kruckeberg (Madroño 14: 217–226. 1958) recognizes in the *S. glandulosus* complex:

(1) **S. glandulòsus** Hook. Plants ± pubescent below, 3–7 dm. tall; infl. secund; fls. lilac-lavender to purple or more often purplish-black, rarely rose; $n = 14$.—Serpentine, San Luis Obispo Co. to Tehama Co.

Ssp. **pulchéllus** (Greene) Kruckeberg. Plants often dwarfish, 1–4 dm. tall; fls. reddish-purple, usually secund, crowded on the short simple to branched racemes; siliques divaricate or ascending, 4–6 cm. long; $n = 14$.—Serpentine, Marin Co.

Ssp. **secúndus** (Greene) Kruckeberg. Fls. in open or crowded secund racemes; siliques usually arcuate, 5–6 cm. long. Var. **secúndus** (Greene) Kruckeberg. Fls. greenish-yellow, tinged with rose or purple as blotches at the base of the petal laminae; $n = 14$.—Marin Co. Var. **sonoménsis** Kruckeberg. Fls. yellow, white or greenish-white; $n = 14$.—Sonoma Co. Var. **Hoffmánii** Kruckeberg. Fls. rose to rose-purplish; $n = 14$.—East Austin Creek, Sonoma Co.

(2) **S. álbidus** Greene. Plants ± pubescent below, usually 6–10 dm. high; infl. not secund; fls. greenish-white.—S. of San Jose, Santa Clara Co.

Ssp. **peramoènus** (Greene) Kruckeberg. Fls. lilac-lavender.—Alameda, Contra Costa, Santa Clara cos.

(3) **S. nìger** Greene. Plants glabrous throughout; infl. zigzag in outline; fls. purplish-black; $n = 14$.—Tiburon Peninsula, Marin Co.

13. **S. insígnis** Jeps. $n = 14$ (Rollins, 1966).

p. 221.   17. **S. polygaloìdes** Gray. Basal lvs. 4–8 cm. long, pinnately divided to the linear rachis with the divisions ca. 3–8 on each side, spreading, 2–15 mm. long, mostly linear to linear-oblong.

18. **S. diversifòlius** Wats. Lower and middle cauline lvs. entire and linear or pinnately divided with linear lobes; fls. generally yellow, rarely tinged purplish; siliques 1–1.25 mm. broad; seeds brown, 1.5 mm. long; $n = 14$ (Rollins, 1966).—Foothill Wd., Yellow Pine F., below 5000 ft.; Amador and Butte cos. to Tulare Co.

18a. **S. fenestràtus** (Greene) J. T. Howell. [*Pleiocardia f.* Greene.] Lvs. deeply divided, the segms. usually oblongish or broader, not linear-filiform; fls. violet or purplish; siliques 1.5–1.7 mm. broad; seeds blackish-brown, 1.5–2 mm. long.—Yellow Pine F., Tehipite V., Fresno Co.

18b. **S. Farnsworthiànus** J. T. Howell. Basal lvs. pinnatifid, midcauline lvs. deeply pinnately lobed to subentire; pedicels to 5 mm. long; fls. 10–15 mm. long, the petals whitish with purple nerves; siliques ascending, straight or curved, 7–9 (–12) cm. long, 3 mm. wide.—At ca. 3000–4000 ft., Fresno and Kern cos., Sierra Nevada to Glennville.

p. 222.   CAULÁNTHUS

1. **C. amplexicaùlis** Wats. var. **barbárae** (J. T. Howell) Munz comb. nov. [*Streptanthus a.* var. *barbarae* J. T. Howell, Leafl. W. Bot. 9: 223. 1962.] Differing from *C. a.* Wats. in having sepals ochroleucous to yellowish; siliques to 15 cm. long.—San Rafael Mts., Santa Barbara Co.

2. **C. Coòperi** (Wats.) Pays. $2n = 14$ pairs (Raven, Kyhos & Hill, 1965).

p. 223.   4. **C. Coùlteri** Wats. $n = 14$ (Rollins, 1966).

Var. **Lemmònii** (Wats.) Munz. Rollins treats this as of specific rank, 1966. $n = 14$ (Rollins, 1966).

5. **C. inflàtus** Wats. $n = 10$ (Rollins, 1966).

p. 224.   STREPTANTHELLA

1. **S. longiróstris** (Wats.) Rydb. $n = 14$ (Rollins, 1966).

Var. **derelícta** J. T. Howell. $n = 14$ (Rollins, 1966).

p. 225.   SUBULÀRIA

1. **S. aquática** L. ssp. **americàna** Mulligan & Calder for American plants which have more persistent sepals than the European; mature silicles more elliptic; lowest pedicel-axil 30°–50° rather than 50°–90°.

CARDÀRIA

G. A. Mulligan & C. Frankton (Can. Jour. Bot. 40: 1411–1425. 1962) recognize for this genus:

1. **C. Dràba** (L.) Desv. Silicles and sepals glabrous; silicles cordate; $n = 32$.

2. **C. chalepénsis** (L.) Handel-Mazzetti [*C. Draba* var. *repens* O. E. Schulz.] Silicles and sepals mostly glabrous; silicles subreniform to obovoid; $n = 40$.

3. **C. pubéscens** (C. A. Mey.) Jarmolenko. [*C. p.* var. *elongata* Rollins.] Silicles and sepals with short simple hairs; silicles strongly inflated, ovoid or subglobose; $n = 8$.

p. 226.   LEPÍDIUM

In Key, in the E. leading to "10. *L. lasiocarpum*," insert "sometimes" after "Petals."

1. **L. campéstre** (L.) R. Br. $2n = 16$ (Mulligan, 1957).—Reported from Monterey Co., *Howitt & Howell.*

p. 227.   3. **L. latifòlium** L. In Eldorado Co. and n. along w. slope of Sierra Nevada, *Fuller;* in Yolo, Sonoma, Monterey cos., *Howell.*

4. **L. stríctum** (Wats.) Rattan. $n = $ ca. 16 (Rollins, 1966).

7. **L. densiflòrum** Schrad. In Greenhorn Mts., Kern Co., *Twisselmann.*

Var. **pubicárpum** (Nels.) Thell. $2n = 16$ pairs (Raven, Kyhos & Hill, 1965).

8. **L. pinnatifidum** Ledeb. In Glenn Co., Santa Barbara Co., *Fuller;* in Monterey and San Francisco cos., *Howell et al.*

p. 228.   9. **L. virgínicum** var. **pubéscens** (Greene) Thell., not C. L. Hitchc.

11. **L. nítidum** Nutt. is sometimes used for plants with caps. 3–4 mm. long, while var. **insígne** Greene has caps. 6 mm. long, the var. ranging in e. Monterey and San Luis Obispo cos. and Tehachapi Mts.

Var. **oregànum** C. L. Hitchc. is reported from Goose Lake, Kern Co., *Twisselmann.*

p. 230.   18. **L. Jarédii** Bdg. $n = 8$ (Rollins, 1966).

CORÓNOPUS

2. Use **C. squamàtus** (Forskål) Ascherson, not *C. procumbens* Gilib.

THLÁSPI

Substitute the following for the Key in the FLORA:

Plants annual; silicles roundel, deeply notched; seeds with concentric ridges. 1. *T. arvense*
Plants perennial; silicles cuneate-obovate, scarcely or not notched; seeds smooth.
    Silicles conspicuously acute when mature; styles ca. 3 mm. long. At low elevs., Mixed
      Evergreen F., Humboldt Co. to sw. Ore. .... 3. *T. californicum*
    Silicles obovate or obcordate, largely emarginate; styles 1.5–2 mm. long. At 3000–6500
      ft., Yellow Pine F. and above; Siskiyou, Trinity and Modoc cos.; to Wash., Ida.
                                              2. *T. Fendleri*

p. 231.   2. **T. Féndleri** Gray var. **hespérium** (Pays.) C. L. Hitchc. [*T. glaucum* var. *h.* Pays.] Delete *T. californicum* from synonymy.
            3. **T. califórnicum** Wats. Stems 2–3 dm. high, reddish; lvs. glaucous, the basal petioled, toothed, elliptic-obovate, the cauline oblong-ovate, sessile, clasping, at least some of them as long as the internodes; pedicels spreading, 7–10 mm. long, rather stout; sepals 3 mm. long, white-margined; petals white, 5–6 mm. long; silicles 7–10 mm. long, usually acute; styles 2–3 mm. long.—Serpentine, 400–1200 ft., Mixed Evergreen F., Humboldt Co. to Josephine Co., Ore.

sisÝmbrium

            3. **S. orientàle** L. *n* = 7 (Rollins, 1966). Temblor Range and Tehachapi Mts., Kern Co., *Twisselmann*.
            4. **S. Ìrio** L. *n* = 7, 14, 21 28 (Koshov, 1955). Reported from many counties in Central V. of Calif., *Weiler*.

p. 232.   arabidópsis

            1. **A. Thaliàna** (L.) Heynh. For 1959 printing, the name Gray in the synonymy should be Gay.

p. 234.   cakìle

            2. **C. marítima** Scop. can be reported from Santa Cruz Id., Santa Rosa Id., Point Mugu and near El Segundo, Los Angeles Co.

p. 235.   isàtis

            1. **I. tinctòria** L. In Sierra and Nevada cos., *Fuller*.

p. 236.   brássica

In Key change line 1 of this page:

> D. Petals ca. 3 mm. wide; beak of silique 2–10 mm. long, the apex narrower than the stigma.
>   E.   Plants annual; petals 7–8 mm. long; beak 5–8 mm. long. . . . . . . 6. *B. juncea*
>   EE. Plants perennial; petals 9–10 mm. long; beak 9–10 mm. long. 6a. *B. fruticulosa*

            5. **B. geniculàta** (Desf.) J. Ball, without a parenthesis, in 1959 printing.
            6a. **B. fruticulòsa** Cyrillo. Lvs. not auriculate, all lyrate or pinnately lobed; sepals shorter than pedicels; beak often 1–2-seeded.—Dunes in Sunset district, San Francisco, *Rubtzoff*.

p. 238.   barbarèa

Substitute "Scop." for "R. Br." as authority for genus.
            2. **B. vérna** (Mill.) Asch. Reported from Monterey Co., *Howitt & Howell*.
            3. **B. orthóceras** Ledeb. Found in Monterey Co., *Howitt & Howell. n* = 8 (Rollins, 1966).

p. 239.   roríppa

            2. **R. sinuàta** (Nutt.) Hitchc. *n* = 8 (Rollins, 1966).
            4. **R. subumbellàta** Roll. *n* = 5 (Rollins, 1966).
            5. **R. curvisilíqua** (Hook.) Bessey. *n* = 8 (Rollins, 1966).

p. 240.   nastúrtium

Peter S. Green (Rhodora 64: 32–43. 1962) uses:
            1. **Roríppa Nastúrtium-aquáticum** (L.) Britt. & Rendle instead of *Nasturtium officinale*.
            2. **Roríppa microphÝlla** (Boenn.) Hyland. instead of *Nasturtium m.* It is doubtful that this sp. occurs in Calif., the Jonesville record cited in the flora being *Cardámine Brèweri*.

p. 242.   cardámine

            3. **C. Brèweri** Wats. *n* = 42–48 (Rollins, 1966).

p. 243.   dentària

In last line of Key at top of page, insert "or simple" after "3–7-foliolate."
            2. **D. califórnica** Nutt. var. *integrifòlia* (Nutt.) Detl. Rollins (1966) has D. integrifòlia Nutt. *n* = 16, and var. **califórnica** (Nutt.) Jeps., *n* = 8 or 16.

p. 244.   idahòa

            1. **I. scapígera** (Hook.) Nels. & Macbr. 2*n* = 16 (Raven, Kyhos & Hill, 1965). Occurs in Kern Co., at Glennville, *Twisselmann*.

p. 245.   lyrocárpa

            1. **L. Còulteri** var. **Pálmeri** (Wats.) Roll. 2*n* = 20 (Rollins, 1941).

DITHÝREA

Change *D. califórnica* var. *marítima* to **D. marítima** A. Davids. Perennial from heavy cordlike underground rhizomes; lvs. to 1 dm. long, the blades rounded, 2–5 cm. in diam., fleshy, subentire; silicles 14–15 mm. broad (as opposed to 10 mm. in *D. californica*); *n* = 40 (Thompson orally for count by Miss Bartholomew) as against 10 for *californica* (Rollins, 1966).—Coastal Strand, Los Angeles Co. to San Luis Obispo Co., San Nicolas Id.

p. 246.   PHYSÀRIA

1. **P. Chàmbersii** Roll. $2n = 10$ (Rollins, 1966).

LESQUERÉLLA

1. **L. Pálmeri** Wats. $n = 5$ (Rollins, 1966).
3. **L. occidentàlis** Wats. Ascends to 9700 ft., Lassen Park.

p. 247.   PHOENICAÙLIS

In line 1 of generic description change "scapes" to "stems" in the 1959 printing.

CAPSÉLLA

1. **C. Búrsa-pastòris** (L.) Medic. $2n = 32$ (Löve & Löve, 1956).

p. 248.   DRÀBA

In Key to Species, AA, BB, C, insert "mostly" before "unforked hairs" in the 1959 printing.

Change end of Key to:

FF. Lvs. 2–7 mm. wide.
    G. Silicles lanceolate, 1.3–3.5 mm. wide; seeds not winged, 1–1.4 mm. long.
        H. Petals yellow, 5–6 mm. long. . . . . 21. *D. cruciata*
        HH. Petals white, 2.5–5 mm. long. . . . . 21a. *D. nivalis*
    GG. Silicles ovate, 3–6 mm. wide; seeds winged, ca. 2 mm. long.
                                 22. *D. asterophora*

p. 249.   1. **D. vérna** L. In Monterey Co., *Howitt & Howell*; in Greenhorn Range, Kern Co., *Twisselmann*; Fresno and Kern cos., *Weiler*.

p. 250.   3. **D. réptans** (Lam.) Fern. $n = 15$ (Mulligan, 1956).
5. **D. nemoròsa** L. $n = 8$ (Mulligan, 1966).

p. 251.   10. **D. Lemmònii** Wats. and 12. **D. Brèweri** Wats. Both spp. have been taken at elevations of 14,200 ft.
13. **D. praeálta** Greene. $n = 28$ (Mulligan, 1966).

p. 252.   17. **D. crassifòlia** Grah. $n = 20$ (Mulligan, 1966).
18. **D. oligospérma** Hook. $2n =$ ca. 60 (Rollins, 1966).

p. 253.   21a. **D. nivàlis** Liljebl. var. **elongàta** Wats. Near *D. cruciata* Pays., but has white petals 2.5–5 mm. long; lvs. cinereous with stalked, irregularly branched and stellate trichomes; stems glabrous to cinereous; silicles glabrous to stellate. —Found at Convict Lake Basin, Sierra Nevada, Mono Co., at 10,800 ft., *Major & Bamberg*; Rocky Mts.

41a. Euclídium R. Br.

A monotypic genus of Eurasia.
1. **E. syriàcum** (L.) R. Br. Annual, divaricately branched, 1–3 dm. tall, stellate or with branched hairs; lvs. lanceolate to oblanceolate, entire to subrepand, 1–3 cm. long, with short winged petiole; fls. loosely spicate; pedicels 0.8–1 mm. long, erect; petals 1–1.2 mm. long; silicle ovoid, hispid, 2–3 mm. long.—Reported as weed near Adin, Lassen Co., *McCaskill*; known from Wash., Ida.; native in Eurasia.

p. 254.   THYSANOCÁRPUS

2. **T. cúrvipes** Hook. $n = 7$ (Rollins, 1966).
Var. **elegáns** (F. & M.) Rob. Rollins considers this to be a species. $2n = 28$ (Rollins, 1966).

p. 255.   3. **T. laciniàtus** Nutt. var. **crenàtus** (Nutt.) Brew. reported from as far n. as Tehama Co., *Wagnon*.

p. 257.   ÁRABIS

In Key BB, CC, D, EE, F, G, H delete "Cauline lvs. auricled" and "Cauline lvs. not auricled" from I and II respectively.

p. 258. In Key to Arabis, for *A. Lemmonii,* change to "cauline lvs. oblong-lanceolate to ovate, mostly glabrous."

p. 259. 3. **A. blepharoyphýlla** H. & A. Occurs in Monterey Co., *Howitt & Howell.*

4. **A. modésta** Roll. On line between Napa and Yolo cos. *Hemphill.*

p. 260. 9. **A. Drummóndii** Gray. $2n = 14$ (Rollins, 1966).

10. **A. divaricárpa** A. Nels. $2n = 14, 22$ (Rollins, 1966).

12. **A. Lemmònii** Wats. $2n = 14$ (Rollins, 1966).

p. 262. 15. **A. sparsiflòra** var. **califórnica** Roll. $2n = 22$ (Raven, Kyhos & Hill, 1965).

p. 263. 21. **A. Holboéllii** var. **pinetòrum** (Tides.) Roll. $2n = 21$ (Rollins, 1966).

Var. **pendulocárpa** (A. Nels.) Roll. $2n = 14$ (Rollins, 1966).

p. 264. 28. **A. repánda** Wats. Occurs in the Greenhorn Range, Kern Co., *Twisselmann.*

p. 265. 34. **A. platyspérma** Gray. Found in the Greenhorn Range, *Twisselmann.*

p. 266. SÍBARA

1. **S. virgínica** (L.) Roll. $2n = 16$ (Rollins, 1966). Occurs in Fresno and Kern cos., *Weiler.*

2. **S. filifòlia** (Greene) Greene. Collected on Catalina Id., *Trask* in 1901.

p. 268. ERÝSIMUM

1. **E. cheiranthoìdes** L. A garden weed in Fresno, *Weiler.*

2. **E. repándum** L. $n = 8$ (Mulligan, 1966).

3. **E. perénne** (Wats. ex Cov.) Abrams. On Kern R. Plateau and in Piute Mts., Kern Co., *Twisselmann.*

4. **E. argillòsum** (Greene) Rydb. At 9725 ft., White Mts., *Blakley & Muller.*

5. **E. capitàtum** (Dougl.) Greene. $2n = 36$ (Mulligan, 1966).

Twisselmann recognizes a var. **stellàtum** with orange to maroon fls. and published by J. T. Howell as *E. asperum* var. *stellatum.* He also separates E. moníliforme Eastw. (Temblor, San Emigdio ranges, etc., Kern Co.) with pale fls., entire lvs. and slender pods constricted between the seeds.

p. 269. 7. **E. ammóphilum** Heller. $n = 18$ (Mulligan, 1966).

8. **E. concínnum** Eastw. $n = $ ca. 18 (Rollins, 1966).

p. 270. ALÝSSUM

1. **A. alyssoìdes** L. Reported from Oakland, Alameda Co.; Warner Mts., Modoc Co., *Fuller.*

p. 271. ## 54. Hésperis L. ROCKET

Annual to perennial, caulescent. Lvs. lanceolate or lance-ovate, serrulate, acuminate to acute, sessile or short-petioled, gradually reduced up the stem. Infl. racemose or paniculate, particularly in fr. Fls. purple to white, shortpediceled. Siliques subcylindric, very slender; stigma lobed, erect. Seeds in 1 row in each locule, oblong, marginless. Cotyledons incumbent. (Greek, *hesperos, evening* or *evening-star,* because of the evening fragrance of the fls.)

1. **H. matronàlis** L. Stems 3–8 dm. tall, with simple spreading hairs; lvs. 3–12 cm. long, 1–3.5 cm. wide, mostly on short petioles; calyx ca. 7 mm. long; petals 15–20 mm. long; siliques 5–14 cm. long.—Sparingly natur. as a weed, as at Trinity Center, Trinity Co., and Quincy, Plumas Co., *Howell.* Native of Old World.

p. 271. Illustration for *Elátine* at top of page, the seed drawn is of the *brachyspermaobovata* group.

CHORÍSPORA

1. **C. tenélla** (Pall.) DC. Collected in Lassen Co., *Anderson in 1963.*

MATTHÌOLA

2. **M. bicórnis** DC. Roadside near Tehachapi, Kern Co., *Fuller.* Siliques 8–30 cm. long, branchlike, with 2 long slender hornlike processes from the backs of the stigma-lobes. From the E. Medit.

p. 273. BÉRGIA

At end of generic description, dates for *Bergius* should be 1730–1790.

p. 274.   CARYOPHYLLÀCEAE
Key under AA, B, C, DD insert

E. Petals bifid or bilobed. Common. . . . . .2. *Cerastium*
EE. Petals entire. Rare. . . . . .2a. *Holosteum*

Key, under AA, BB, CC, D, E, F, insert

G. Calyx-tube lacking white scarious seam between the teeth; plant annual; fls. 10–20 mm. diam. . . . . 12. *Vaccaria*
GG. Calyx-tube with white scarious seam between the teeth; plant perennial; fls. 4–5 mm. diam. 12a. *Gypsophila*

At end of Key, change the last word *"Tunica"* to *Kaulrauschia."*

p. 275.   STELLÀRIA
1. S. mèdia (L.) Vill. not Cyrill as in 1959 printing.
4. S. Jamesiàna Torr. is treated as *Arenària J.* (Torr.) Shinners in Sida 1: 50. 1962.
5. S. gramínea L. reported from Stanford University campus and San Francisco. $2n = 26$ (Löve & Löve).
6. S. longipès Goldie. $2n = 104$ (Jørgensen et al., 1958).

p. 276.   8. S. sitchàna var. Bongardiàna (Fern.) Hult. in Greenhorn Range, Kern Co., *Twisselmann.*
10. S. críspa C. & S. Kern Co., *Twisselmann.*
11. S. obtùsa Engelm. In Nevada Co. and on Lassen Peak, *Howell.*

CERÁSTIUM
Key, change last line to "Annual without persistent basal sterile offsets" and add:

Caps.-teeth revolute; sepals 4–6 mm. long; caps. 5–9 mm. long. . . . . 3. *C. glomeratum*
Caps.-teeth plane on the margin; sepals 8–10 mm. long; caps. 18 mm. long.
3a. *C. dichotomum*

p. 277.   2. C. vulgàtum L. $2n = 72, 126, 144, 180$ (Blackburn & Morton, 1956).
3. Change *C. viscòsum* to C. glomeràtum Thuill. [*C. v.* many auth.] $n = 36$ (Huynk., 1965).
3a. C. dichótomum L. Annual, differing from *C. viscosum* in having sepals 8–10 mm. long; caps. 18 mm. long, the caps.-teeth plane on the margin (not revolute).—Reported from 5 mi. e. of Montague, Siskiyou Co., *Fuller;* native Medit. to Iran.

p. 277.   ## 2a. Holósteum L. JAGGED CHICKWEED

Annuals or biennials, with several fls. borne in an umbel on a long terminal peduncle. Sepals 5. Petals 5, usually jagged or denticulate to the point. Stamens mostly 3–5. Styles mostly 3. Pod ovoid, 1-celled, many-seeded.
1. H. umbellàtum L. Glaucous, glandular-pubescent annual, 5–20 cm. tall; lvs. oblong; fls. white or pink, small; styles 3; caps. cylindric, deeply 6-toothed. —Grenada, Siskiyou Co., *Howell;* from Old World.

p. 278.   ARENÀRIA
In Key after A, BB, under C. delete "3. *A. californica*" and insert:

D. Plant 2–5 cm. high; branches decumbent; seeds smooth. . . 4. *A. pusilla* var. *diffusa*
DD. Plant 3–10 cm. high; branches erect; seeds rough. . . . . 3. *A. californica*

p. 279.   4. A. pusílla Wats. In Greenhorn Range, Kern Co., *Twisselmann* and in San Luis Obispo Co., *Hardham.*
p. 280.   8. A. obtusilòba (Rydb.) Fern. $n = 13$ (Wiens & Halleck, 1963).
p. 282.   16. A. macradènia Wats. Kern Plateau, Kern Co., *Twisselmann.*
Var. Parishiòrum Rob. In Tehachapi Mts., Kern Co., *Twisselmann.*
18. A. macrophýlla Hook. In Santa Lucia Mts., *Howitt & Howell.*
p. 283.   SPERGULÀRIA
2. S. atrospérma R. P. Rossb. in Plumas Co. and sw. Kern Co., *Howell.*
p. 284.   4. S. Boccònii (Scheele) Foucaud. $2n = 36$ (Blackburn & Morton, 1957).
p. 285.   **Polycárpon**

1. **P. tetraphýllum** (L.) L. $2n = 54$ (Blackburn & Morton, 1957).

LOEFLÍNGIA

1. **L. squarròsa** Nutt. At 4000 ft. in Plumas Co., *Howell.*

SILÈNE

Fls. solitary or more often cymose, some spp. (*S. californica, S. Lemmonii* and *S. Sargentii*) are diurnal, others (*S. montana, S. Grayi, S. verecunda*) vespertine (letter from *H. L. Buckalew*).

p. 287.   Key under OO, change to:

> P. Calyx usually plainly constricted below the ovary, 10–12 mm. long; appendages of petals 1–2 mm. long; infl. tending to have more than 1 fl. in a cymule. Lake and Mono cos. s. .... 22. *S. verecunda*
> PP. Calyx slightly constricted below the ovary, ca. 13 mm. long; appendages of petals scarcely 1 mm. long; infl. with a single fl. in a cymule. Siskiyou Co. 22a. *S. marmorensis*

2. **S. conoìdea** L. Collected at Greenville, Plumas Co., *Budaj.*

p. 288.   4. **S. gállica** L. $2n = 24$ (Blackburn & Morton, 1957).

7. **S. califórnica** Durand. $2n = 72$ (Kruckeberg, 1960).

p. 289.   10. **S. invìsa** Hitchc. & Maguire. $2n = 48$ (Kruckeberg, 1960). Reported from Lassen Park.

11. **S. campanulàta** Wats. $2n = 48$ (Kruckeberg, 1960).

12. **S. apérta** Greene. $2n = 48$ (Kruckeberg, 1960).

15. **S. montàna** Wats. G. Bocquet (Candollea 20: 49–50, 1965) shows that *Silene montana* Wats. is antedated by *S. m.* Arrondeau. The next name available is **S. bernardìna** Wats., of which ssp. **bernardìna** is *S. montana* var. *b.* of the FLORA; ssp. **Maguìrei** Bocquet is *S. montana* of the FLORA; and ssp. **Maguìrei var. siérrae** (Hitchc. & Maguire) Bocquet is *S. montana* var. *sierrae* of the FLORA. For *S. bernardina* ssp. *Maguirei* $2n = 48$ (Kruckeberg, 1960). Both *S. bernardina* and var. *sierrae* have been reported from the Greenhorn Mts., Kern Co., *Twisselmann.*

p. 290.   17. **S. Lemmònii** Wats. $2n = 48$ (Kruckeberg, 1960).

18. **S. occidentàlis** Wats. $2n = 48$ (Kruckeberg, 1960).

19. **S. Gràyii** Wats. $2n = 48$ (Kruckeberg, 1960).

p. 291.   22. **S. verecúnda** Wats. $2n = 48$ (Kruckeberg, 1960).

Ssp. **Andersònii** (Clokey) Hitchc. & Maguire. $2n = 48$ (Kruckeberg, 1960).

22a. **S. marmorénsis** Kruckeberg. Stems slender, 2.5–4 dm. long, simple, retrorsely glandular-pubescent above; cauline lvs. 5–7 pairs, lanceolate, 3–4.5 cm. long, 3–5 mm. wide; infl. 1–2 dm. long; pedicels 7–10 mm. long, glandular; calyx ca. 13 mm. long, glandular, the lobes 3 mm. long; petals pale pink above, 12–16 mm. long, the blade 4–6 mm. long, bilobed over half its length, the lobes oblong, appendages 2, oblong, scarcely 1 mm. long.—Near Somes Bar, Siskiyou Co., in Yellow Pine F.

p. 292.   LÝCHNIS

2. **L. álba** Mill. $2n = 24$ (Mulligan, 1957).

p. 293.   ## 12a. Gypsóphila L.

Herbs, branched or diffuse, glaucous, scanty-leafy when in bloom; fls. small, many, cymose-paniculate; calyx 5-toothed, scarious between the nerves; petals 5, entire or emarginate; stamens 10; styles 2 (3). Over 100 spp. in Old World.

1. **G. paniculàta** L. Perennial, 6–9 dm. high; lvs. lanceolate; fls. numerous,

white, 4–5 mm. diam., in crowded corymbs in panicles.—Near Janesville, Lassen Co. and Macdoel and Weed, Siskiyou Co., *Fuller;* Eurasia.

DIÁNTHUS

Insert Key:

Fls. in heads or clusters surrounded by invol.-like bracts.
Tufted perennial; involucral bracts glabrous. . . . . 1. *D. barbatus*
Annual or biennial; involucral bracts hairy. . . . . 2. *D. Armeria*
Fls. 1-2-3, not in heads. . . . . 3. *D. deltoides*

2. **D. Armèria** L. Hairy involucres; plant annual or biennial in duration; lvs. linear; fls. bright red with pale dots.—Magalla, Butte Co., *Howell;* introd. from Eu.

3. **D. deltoìdes** L. Fls. 1, rarely 2–3, not in involucral heads; petals rose or white with pale spots and a dark basal band.—Reported from se. Siskiyou Co. and Huntington Lake, Fresno Co., *Howell.* Introd. from Eu.

TÙNICA

1. Change *T. prolífera* (L.) Scop. to **Kohlráuschia velùtina** (Guss.) Reichb. because of misdetermination and add Shasta and Sacramento cos. for occurrence.

p. 295. SCLERÁNTHUS

1. **S. ánnuus** L. $2n = 48$ (Blackburn & Morton, 1957).

p. 295.   **35a. Basellàceae.** BASÉLLA FAMILY

Climbing fleshy perennial herbs with ca. 20 spp., mostly native in trop. Am. Rootstocks tuberous; lvs. alternate, usually petioled, entire, mostly fleshy, broad, glabrous. Fls. bisexual, regular, racemose, small, with 2 bracts, 2 sepals, 5 persistent petals remaining closed. Stamens 5, opposite the petals. Ovary superior, 1-loculed, 1-ovuled; styles usually 3, with cleft or entire stigmas. Fr. indehiscent, fleshy, inclosed by the persistent corolla.

1. Boussingaúltia HBK.

Stems much branched. Fls. in axillary and terminal spikelike racemes. Sepals nearly flat, not winged. Ca. 14 spp.

1. **B. grácilis** Miers var. **pseùdo-baselloìdes** Bailey. MADEIRA VINE. MIGNON-ETTE VINE. A twining vine 3–6 m. tall, producing little tubercles in the lf.-axils by means of which propagation occurs; lvs. ovate, 2.5–7.5 cm. long, subcordate at base, short-petioled; racemes to 3 dm. long, many-fld.; fls. white, aging black, fragrant.—Late summer. Occasional escape from cult., as in San Francisco.

p. 296. PORTULÁCA

1. **P. oleràcea** L. $2n = 36$ (Sharma & Bhatt, 1956).

p. 297. LEWÍSIA

2. and 5. **L. Leàna** and **L. Cotylèdon.** Tucker et al. (Cactus & Succ. Soc. Am., March, 1964) report a hybrid between these species.

3. **L. Congdònii** (Rydb.) J. T. Howell. In Foothill Wd., Yellow Pine F., at 2000–7000 ft., *J. T. Howell.*

p. 298. 7. **L. pygmaèa** (Gray) Rob. in Gray. $n = $ ca. 33 (Wiens & Halleck, 1962).
13. **L. redivìva** Pursh. At sea level in San Francisco Bay area.

Var. **mìnor** (Rydb.) Munz. In Greenhorn Range, Kern Co., *Twisselmann;* and 1.5 mi. n. of Kenworthy Ranger Station, San Jacinto Mts., Riverside Co., *Ziegler.*

p. 300. CLAYTÒNIA

R. J. Davis (The N. Am. perennial spp. of Claytonia, Brittonia 18: 285–300. 1967) recognizes the taxa I have in the FLORA and includes also in *Claytonia,* with the same specific names: **Móntia parvifòlia** and **M. Chamíssoi.**

1. **C. umbellàta** Wats. $2n = 16$ (Davis, 1967).
2. **C. lanceolàta** Pursh. $2n = 16$ (Davis, 1967); $n = 8$ (Wiens & Halleck, 1962).

3. and 4. **C. bellidifòlia** Rydb. and **C. nevadénsis** Wats. reported by Buckalew (letter) as having lvs. entirely red when they push out of the ground and
remaining so for some time, turning green on upper surface by anthesis but
remaining red beneath and on a vein around the perimeter.

4. **C. nevadénsis** Wats. Occurs in Shasta Co., Calif. and Harney Co., Ore.,
Chambers.

p. 301.   MÓNTIA

1. **M. parvifòlia** (Moç. ex DC.) Greene. $2n = 20$ (Nilsson, 1966).

4. **M. lineáris** (Dougl. ex Hook.) Greene, not (Dougl.) Greene.

p. 302.   6, 7, and 8. D. Moore (Bot. Notiser 116: 16. 1963) for the *M. fontana*
group gives 3 spp. for California:

6. **M. mìnor** Gmel. [*M. verna* Neck.] Seeds (0.8) 1–1.2 (–1.4) mm. diam.,
dull, entirely covered with rather broad tubercles.—Monterey Co. to s. Ore.

7. **M. Hállii** (Gray) Greene. Seeds 0.6–1.2 mm. diam., rather shiny with
7–11 rows of slender tubercles around the keel.—L. Calif. to B.C.

8. **M. Funstònii** Rydb. Seeds 0.7–1.2 mm. diam., somewhat shiny, with keel
tubercles generally low, sometimes none.—Scattered locations (in Calif. at
above 6000 ft.), n. L. Calif. to B.C.

p. 303.   12. **M. perfoliàta** (Donn) Howell. $n = 6$, 12, 18 (Raven, 1962); 18 (W. H.
Lewis, 1963).

13. **M. spathulàta** (Dougl.) Howell. In Temblor Range, Kern Co., *Twisselmann.* $2n = 48$ (Nilsson, 1966).

p. 304.   Var. víridis A. Davids. Also in Greenhorn Range, *Twisselmann.*

14. **M. gypsophiloìdes** (F. & M.) Howell. $2n = 16$ (Nilsson, 1966). In Temblor Range, *Twisselmann.*

15. **M. sibírica** (L.) Howell. $n = 12$ (Raven, 1962; W. H. Lewis, 1963);
$2n = 18$ pairs (D. E. Anderson, 1963).

p. 305.   CALYPTRÍDIUM

2. **C. Párryi** var. **Hésseae** Thomas. In the Santa Lucia Mts., *Howitt &
Howell.*

p. 306.   6. **C. umbellàtum** var. **caudicíferum** (Gray) Jeps. at elevs. up to 14,200 ft.,
Mr. Whitney, *Raven.*

p. 307.   GLÌNUS

1. **G. lotoìdes** L. In San Luis Obispo Co., *Hardham*; Kern Co., *Twisselmann.*

CYPSELÈA

1. **C. humifùsa** Turp. Northeast of Graton, Sonoma Co., *Rubtzoff*; Clear
Lake, Lake Co., *M. S. Baker, Mason.*

TRIÁNTHEMA

1. **T. Portulacástrum** L. Occurs in Kern and Tulare cos., *Twisselmann.*

p. 308.   TETRAGÒNIA

1. Change *T. expansa* to **T. tetragonioìdes** (Pall.) O. Kuntze.

p. 309.   MESEMBRYÁNTHEMUM

1. **M. nodiflòrum** L. $n = 18$ (Reese, 1957).

4. **M. cordifòlium** L. f. In Monterey Co., *Howitt & Howell.*

p. 311.   OPÚNTIA
In Key, change last line by deleting "19. *O. occidentalis.*" Add:

> E.   Joints ± elongate; petals 1.5 times as long as wide; stigma often longer than
> wide; fr. not with a deeply impressed umbilicus. . . . . 19. *O. littoralis*
> EE. Joints nearly round; petals narrow, twice as long as wide; stigma often wider
> than long; fr. rounded, with deeply depressed umbilicus. . . . . . . 20. *O. oricola*

p. 313.   11. **O. Wrightiàna** (Baxter) Peeb. extends into L. Calif.

p. 315.   19. **O. occidentàlis** Engelm. & Bigel. Benson & Walkington (Ann. Mo. Bot.
Gard. 52: 262–273. 1965) discuss the *O. occidentalis* group and report that:

(1) *O. occidentalis* was based on a type from Cucamonga, San Bernardino
Co. which proves to be the Mission Cactus, **O. megacántha** Salm-Dyck and
not a native California plant.

(2) **O. littoràlis** (Engelm.) Ckll. is the specific name to be used and consists of several vars. which can be keyed out as follows:

A. Spines none or a few along the top of the joint; spines 6–12 (–20) mm. long; fls. magenta. From Glendora, Los Angeles Co. to Riverside Co., at low elevs. (mostly below 2000 ft.) .... var. *austrocalifórnica*
AA. Spines usually on most of the joint; spines 25–69 mm. long; fls. yellow or the center reddish.
  B. Joints green, not glaucous; spines 5–11 per areole. Near the coast from Santa Barbara Co. to L. Calif. .... var. *littoràlis*
  BB. Joints moderately glaucous; spines 1–4 (–6) per areole. Away from the coast.
    C. Spines brown or dark gray. Newhall, Los Angeles Co., to San Bernardino and Riverside cos. at mostly below 2000 ft. .... var. *Vàseyi*
    CC. Spines reddish to gray.
      D. The spines reddish with yellow-white tips; fls. yellow. Largely at 3000–7000 ft., San Gabriel, San Bernardino and San Jacinto mts. var. *Pièrcei*
      DD. The spines red and yellow to gray; fls. often with a reddish center. Mts. of e. Mojave Desert. .... var. *Martiniàna*

In the above, O. littoràlis (Engelm.) Ckll. var. littoràlis excludes O. orícola Philbrick.

O. l. var. Vàseyi (Coulter) Benson & Walkington includes O. occidentalis var. V. and var. Covillei of p. 316 of the FLORA.

O. l. var. austrocalifórnica Benson & Walkington. Low, the joints elongate-obovate, ca. 12.5–20 cm. long; petals pale purple.

O. l. var. Pièrcei (Fosberg) Benson & Walkington much as O. o. var. P. in the FLORA.

O. l. var. Martiniàna (L. Benson) L. Benson. At 2000–8000 ft., e. San Bernardino Co.; to Nev., Ariz.

p. 316.   20. O. orícola Philbrick. A coastal sp. separable from O. *littoralis* in having yellow, subhooked spines to 2 cm. long (instead of white with red-brown base and 1.5–3.5 cm. long); joints nearly round (not elongate); petals narrow, twice as long as wide (not 1.5 times); stigma often wider than long (not longer than wide); and fr. rounded, with deeply depressed umbilicus.—Ranging from Santa Barbara to n. L. Calif.

p. 320.   POLYGONÀCEAE

In description of family, insert after "stipules," the words "sometimes obsolete."

p. 322.   CHORIZÁNTHE

In Key change F. to "Calyx-lobes entire or nearly so."

p. 323.   Line 16 from top, change to:

    FF. All the calyx-lobes not entire.
      G. Both outer and inner calyx-lobes deeply bilobed, the divisions sharply acute; fls. white, 5.5–6 mm. long. Sierra Madre, Santa Barbara Co.
          25d. *C. Blakleyi*
      GG. Not both outer and inner calyx-lobes deeply bilobed.
        H. Not all the calyx-lobes erose.
          I. The outer and inner calyx-lobes fimbriate. 24. *C. fimbriata*
          II. The outer calyx-lobes entire to bilobed.
            J. Outer calyx-lobes entire, the inner fimbriate.
              K. Outer calyx-lobes roundish, erect, purplish; inner oblong, erect; fls. 3.5–5 mm. long. Serpentine, Monterey Co. to Santa Barbara Co. 25. *C. Palmeri*
              KK. Outer calyx-lobes obovate, flaring, white; inner oblong, erect; fls. 4–4.5 mm. long. San Benito and Monterey cos. to Santa Barbara Co.
                  25c. *C. obovata*
            JJ. The outer calyx-lobes bilobed or erose; inner shallowly bilobed.
              K. Outer calyx-lobes erose; inner shallowly bilobed; fls. 4–4.5 mm. long. E. Monterey Co., San Benito Co., w. Fresno Co., San Luis Obispo Co.
                  25b. *C. ventricosa*
              KK. Outer calyx-lobes bilobed; inner fimbriate; fls. 5–6 mm. long. San Benito and Monterey cos. to Santa Barbara Co. .... 25a. *C. biloba*
        HH. All the calyx-lobes ± erose.
          I. Involucral teeth straight. .... 16. *C. valida*
          II. Involucral teeth uncinate. .... 28. *C. Parryi*
    EE. Involucral teeth very unequal, the anterior one usually longer than involucral tube, the others relatively short.
      F. Elongate anterior involucral tooth straight.

G. The outer calyx-lobes obovate, shallowly bilobed, well exserted; the
inner lobes half as long, fimbriate; fls. ca. 3.5 mm. long, whitish;
stamens 9. San Luis Obispo and Monterey cos. ... 26a. *C. rectispina*
GG. The outer calyx-lobes linear-oblong, obscurely erose, almost included; the inner minutely erose; fls. white, almost 3 mm. long.
San Benito, w. Fresno, Monterey, San Luis Obispo and Kern cos.
26. *C. uniaristata*
FF. Elongate anterior involucral tooth uncinate; outer calyx-lobes ovate, minutely erose; inner erose; fls. 3.5 mm. long, pink. Inner Coast Ranges,
Mendocino and Lake cos. to Ventura Co., Sierran foothills of Tulare and
Kern cos. .... 27. *C. Clevelandii*

p. 324.   5. **C. polygonoìdes** T. & G. in Monterey Co., *Howitt & Howell.*
p. 328.   25a. **C. bilòba** Goodm. [*C. Palmeri* Wats. var. *biloba* Munz.]. Annual, to 3.5
dm. tall, appressed curly-pubescent and commonly with some longer spreading
hairs toward the infl.; lvs. basal, elliptic and sessile to oblanceolate and petioled,
1–5 cm. long, strigose; bracts at lower branches similar to lvs., but awn-pointed;
upper bracts to ca. 8 mm. long; invols. 5–7 mm. long, gray-pubescent and with
coarser hairs on ribs, the tube 4–5 mm. long; fls. partly exserted, 5–6 mm. long,
the lobes 2 mm. long, the outer obovate, obcordate to bilobed, inner oblong,
obtuse, fimbriate in upper third.—San Benito and Monterey cos., to Santa Barbara Co.
25b. **C. ventricòsa** Goodman. [*C. Palmeri* var. *v.* Munz.] Diffuse, 1–3 dm.
tall, spreading-pubescent; lvs. basal, oblanceolate, long- or short-petiolate, the
blades to 4.5 cm. long, hirsute beneath, less so above; lower bracts like the lvs.,
the upper subulate; invols. ventricose, the tube ca. 3.5 mm. long, sparsely
pubescent except on the ribs (with short ascending hairs), 5 teeth uncinate,
the 6th elongate, ca. 2 mm. long, straight or uncinate; perianth partly exserted,
4–4.5 mm. long, the outer lobes broadly obcordate, subentire or erose, ca. 1.5
mm. long, the inner squarish, emarginate, ca. 1 mm. long, fimbriate in distal
half.—E. Monterey Co., San Benito Co., w. Fresno Co., San Luis Obispo Co.
25c. **C. obovàta** Goodm. Erect, 1–3 dm. tall, subappressed or spreading
pubescent; lvs. oblanceolate, long petioled, the blade 1–2 cm. long, densely
soft-hirsute beneath, sparsely so above; lower bracts foliose, strigose, mucronate
to awn-pointed, the upper subulate; invols. urceolate, 4.5–5 mm. long, grayish
with ascending hairs, sometimes less pubescent, the tube 3–4 mm. long, the
teeth divergent, 5 uncinate, the anterior 2 mm. long, straight or curved downward; calyx 4–4.5 mm. long, glabrous, the tube slightly longer than the outer
lobes, these obovate, the inner truncate, finely fimbriate.—Santa Barbara Co. to
San Benito and Monterey cos.
25d. **C. Blàkleyi** Hardham. Erect, 0.5–2 dm. tall, bright yellow-green, with
long spreading hairs throughout; lvs. basal, long-petioled, the petiole and blade
ca. 2 cm. long; invols. urceolate-cylindric, the tube 3.5–4 mm. long, glabrous
except for long spreading hairs on the ribs of the older invols. and short upcurled hairs on the younger; 5 teeth short, uncinate, the anterior 1.5–2 mm.
long, slightly recurved or nearly straight; calyx white, 5.5–6 mm. long, the inner
lobes deeply bilobed, with remotely dentate margins, ca. 0.5 mm. long, the
outer slightly longer, narrow, similarly bilobed.—Sierra Madre, Santa Barbara
Co.
26a. **C. rectispìna** Goodm. Spreading to decumbent, the stems 1–2 dm. long,
grayish strigose; lvs. oblanceolate to spatulate, 1.5–3 cm. long, obtuse, villous-
hirsute; bracts foliose, 0.5–1 cm. long, awn-tipped; invols. urceolate-cylindric,
gray-pubescent, the tube 2–2.5 mm. long, 5 teeth, short, uncinate, widely spreading, the anterior 1 as long as or longer than the tube, straight, divergent; calyx
partly exserted, ca. 3.5 mm. long, strigulose on outer surface, segms. very unequal, the outer 3 nearly as long as calyx-tooth, broadly obovate to suborbicular,
truncate, subentire, the inner 3 half as long, oblong, erose to finely fimbriate,
mostly obtuse.—San Luis Obispo and Santa Barbara cos.
p. 329.   30. **C. califórnica** (Benth.) Gray. Monterey Co., *Howitt & Howell.*
p. 331.   OXYTHÈCA
6. **O. caryophylloìdes** Parry. Pine Mt., Ventura Co., *Pollard, Blakley, Twisselmann.*

7. Eriógonum Michx. WILD BUCKWHEAT
This section represents pages 332 through 354 of the FLORA.

Annual or perennial herbs or shrubs with basal or cauline, alternate lvs. and often with alternate or more commonly whorled scalelike or foliaceous bracts, entire and estipitate. Fls. perfect or sometimes also imperfect, borne in invols. Invols. campanulate to turbinate or cylindric, 4–10-lobed or -toothed, awnless, few- to many-fld., sessile or peduncled. Pedicels ± exserted, intermixed with setaceous bractlets and jointed at summit with the base of the perianth or with a slender and stalklike, stipitate base. Perianth commonly called "calyx," 6-parted or -cleft, petaloid, with 2 series of 3 segms. each. Stamens 9, the fils. filiform, often pilose at the base, inserted at base of perianth. Ovary 1-celled, 3-angled or -winged; styles 3; stigmas capitate. Aks. mostly 3-angled, sometimes lenticular. A N. Am. genus of ca. 205 spp., mostly w.; some of importance as bee-plants, others with some horticultural possibility. (Greek, *erion*, wool, and *gonu*, knee or joint, the type of the genus, *E. tomentosum* Michx. being hairy at the nodes.)

(Stokes, Susan G., The genus Eriogonum, 1–132. 1936.)

James L. Reveal, a graduate student at Brigham Young University, Provo, Utah, has spent several years studying the genus *Eriogonum* and has kindly prepared a manuscript with a key to the California species as he now sees them (August, 1967). He has written up various changes in name and status and inserted many new taxa, as well as chromosome counts. Many of the species proposed since January 1, 1958 (the date at which the FLORA had to be closed to changes and new species) were included in the *Supplement* manuscript which I submitted to the University of California Press in June, 1966, however some of these have since received different status in Mr. Reveal's treatment.

I have decided, on looking over his many suggestions and corrections for this complex genus, to include the text from the FLORA for the species he has not changed in order to provide a continuous and usable treatment. Although it is therefore difficult to separate our contributions my work mostly consisted of filling in gaps from the FLORA and dividing up *Eriogonum latifolium* as here treated. I wish to make it clear that the new interpretations, corrections arising from the study of types, and many other important suggestions are by Mr. Reveal and he should receive the credit for the improvement over the FLORA. Since he has used the varietal rank instead of the subspecific rank employed by Miss Stokes in her revision, for the sake of uniformity I have followed him in regard to the species which he has not written up.

Philip A. Munz

Mr. Reveal is grateful to the United States National Herbarium and the Smithsonian Institution which sponsored his Predoctoral Internship in Washington, D.C. (from September 1966 to February 1967) where his part of the paper was basically prepared. His field work and herbaria visits were largely supported by an NSF grant to Dr. Arthur Cronquist for studies on the Intermountain Flora through a cooperative program between the New York Botanical Garden and Utah State University. His contributions were submitted to the Department of Botany, Brigham Young University, as partial fulfillment of the requirements for Doctoral Research credit given the Fall Semester of 1966–1967.

A. Calyx stipelike at the attenuated base (see also *E. saxatile* and *E. crocatum*); bracts leafy, indefinite in number (2–several). (Subgenus *Oligogonum* Nutt.)
  B. Invols. with lobes at least half as long as tube and usually reflexed or spreading.
    C. Calyx pubescent externally.
      D. Flowering stems or scapes without subtending bracts and with solitary terminal invols. Inyo Co. to Modoc Co., e. of the Sierra Nevada crest. . . . . 1. *E. caespitosum*
      DD. Flowering stems with whorled subtending bracts at the base of the umbel or near the middle of the stems.
        E. Invols. solitary, terminal, not immediately subtended by leafy bracts, the flowering stems with whorled bracts near the middle.
          F. Invol. lobes oblong, as long or longer than the tube; calyx densely pilose externally; lvs. mostly tomentose above; aks. densely hairy above the middle. E. slope of Sierra Nevada, Nevada Co. north, ne. Calif. below 8000 ft. . . . . 2. *E. Douglasii*
          FF. Invol. lobes broadly triangular, shorter than the tube; calyx sparsely pilose externally; lvs. subglabrous and green above; aks. glabrous or with few scattered hairs above the middle. Tulare Co. . . . 3. *E. Twisselmannii*

EE. Invols. more than 1, umbellate, subtended by 2–several leafy bracts, stems
without a whorl of bracts.
    F.   Lvs. ± glabrate above; calyx, including the stipe 7–9 mm. long. Above
3000 ft. . . . . 4. *E. sphaerocephalum*
    FF. Lvs. densely white-tomentose above and below; calyx, including the stipe,
5–7 mm. long. Below 2000 ft. . . . . 5. *E. tripodum*
CC. Calyx glabrous externally.
    D. Invols. solitary, without subtending bracts immediately below invols., flowering stem
with whorled leafy bracts near the middle: fls. yellow.
        E. Lvs. densely tomentose above and below, occasionally glabrate above, (5–)
10–15 mm. long. W. slopes of Sierra Nevada, Nevada Co. to Fresno Co., 3000–
5000 ft. . . . . 6. *E. Prattenianum*
        EE. Lvs. glabrate and green above, ± oval, mostly less than 8 mm. long. Scott
Mts., Siskiyou and Trinity cos., 7000–9000 ft. . . . . 7. *E. siskiyouense*
    DD. Invols. clustered and subtended by 2–several bracts.
        E. Flowering stems with a whorl of leafy bracts at about midlength; lvs. mostly
linear-oblanceolate to oblanceolate. Modoc Co. . . . . 8. *E. heracleoides*
        EE. Flowering stems without a whorl of leafy bracts.
            F. Scapes erect or nearly so; invols. up to 10 mm. long.
               G. Lvs. less than 2 cm. long, not cordate at the base; invols. 3–5 mm.
long. Throughout Calif. . . . . 9. *E. umbellatum*
               GG. Lvs. 2–10 (–20) cm. long, usually cordate at the base; invols. 6–10
mm. long. N. Coast Ranges. . . . . 10. *E. compositum*
            FF. Scapes usually flat on the ground; invols. 9–12 mm. long; lvs. 2–4 cm.
long. N. Coast Ranges and n. Sierra Nevada. . . . . 11. *E. Lobbii*
BB. Invols. with lobes much shorter than tube, toothlike and suberect.
    C. Calyx pubescent externally.
        D. Infl. capitate, subtended by 5 membranaceous bracts; scapes 8–30 cm. long, erect;
calyx yellowish.
            E. Lvs. subglabrous to short-pilose, 1–3 cm. long, calyx cream-colored, 5–6 mm.
long. Inyo and Mono cos. . . . . 12. *E. latens*
            EE. Lvs. hirtellous to glabrescent, 0.5–1.5 cm. long; calyx bright yellow, 3 mm.
long. Del Norte and Siskiyou cos. . . . . 13. *E. hirtellum*
        DD. Infl. subcapitate to umbellate, subtended by 2 membranaceous bracts; scapes 4–8
(–15) cm. long, suberect to nearly prostrate; calyx white to rose. Shasta and Siski-
you cos. . . . . 14. *E. pyrolifolium*
    CC. Calyx glabrous externally.
        D. Bracts in a whorl near middle of flowering stem which bears a single invol.
            E. Calyx yellow; lvs. densely white-tomentose on both surfaces.
                F. Lvs. elliptic to ovate, 5–15 mm. long; scapes 1–3 dm. long; calyx 5–7
mm. long in fruit. W. slope of Sierra Nevada, Nevada Co. to Fresno Co.,
3000–5000 ft. . . . . 6. *E. Prattenianum*
               FF. Lvs. rounded, 1–3 cm. long; scapes 4–6 cm. long; calyx 3–5 mm. long in
fruit. Siskiyou Co., 7500–9000 ft. . . . . 15. *E. alpinum*
            EE. Calyx white or pink; lvs. oblanceolate to spatulate, silky pubescent below, less
so above. Mendocino Co. . . . . 16. *E. Kelloggii*
        DD. Bracts subtending the umbel or head of several invols.
            E. Styles less than 1 mm. long; fils. pilose.
                F. Calyx white to pink with a reddish midrib; style ca. 1 mm. long. Mostly
Tulare Co. . . . . 17. *E. polypodum*
               FF. Calyx yellow; styles 0.5 mm. long.
                   G. Lvs. glabrate above, mostly rounded at the base; infl. mostly open.
Tuolumne Co. n. to cent. Ore. . . . . 18. *E. marifolium*
                   GG. Lvs. densely white-tomentose on both surfaces, mostly acute at the
base; infl. mostly congested. Alpine Co. to Tulare Co. 19. *E. incanum*
            EE. Styles 2–4 mm. long; fils. densely woolly.
                F. Calyx ochroleucous; lvs. broadly ovate, rounded or subcordate at base.
N. Sierra Nevada. . . . . 20. *E. ursinum*
               FF. Calyx yellow; lvs. obovate to spatulate, attenuated at the base. N. Coast
Ranges. . . . . 21. *E. ternatum*
AA. Calyx not stipelike at the base; bracts not leafy, regularly 3 in number.
    B. Stems not jointed internally.
        C. Invols. campanulate to turbinate, not angled or ribbed, 4-or 5-toothed or -lobed, rarely
obscurely nerved at the base; mostly peduncled. (Subgenus *Ganysma* [S. Wats.] Greene.)
            D. Lvs. basal and also on the lower nodes, tomentose or floccose except in *E. sperguli-
num*.
               E. Basal lvs. oblanceolate to oblong-obovate, not revolute for most part, tomentose
or floccose.
                  F. Invols. ± glandular-puberulent to pubescent externally.
                    G. Calyx-segms. dissimilar, the outer segms. ovate, elliptic, or roundish,
the inner segms. narrowly lanceolate or oblong, and longer.
                       H. Outer segms. obovate to elliptic, not obviously inflated, or if so,
only near the base, the inner segms. spatulate; stamens conspicu-
ously exserted. . . . . 22. *E. angulosum*
                     HH. Outer segms. elliptic to roundish or obovate, obviously inflated

at maturity, the inner segms. narrowly lanceolate; stamens included.

I.   Outer segms. inflated at the base and middle, the sides of
segms. incurved below, the inner segms. obtuse to acute;
peduncles and invols. glandular-puberulent, with non-capi-
tate hairs. . . . . 23. *E. maculatum*

II.   Outer segms. inflated above the middle, the apex curved
inward, the inner segms. acute to acuminate; peduncles and
invols. with capitate-glandular hairs. . . . 24. *E. viridescens*

GG. Calyx-segms. similar or nearly so, the outer segms. oblong, not inflated.

H.   Fls. not concealed by cottonlike tomentum inside the invols.;
invols. 2 mm. long. . . . . 25. *E. gracillimum*

HH. Fls. concealed by tufts of cottonlike tomentum inside the invols.;
invols. 3 mm. long. . . . . 26. *E. gossypinum*

FF. Invols. glabrous or densely tomentose, not glandular.

G.   Invols. glabrous.

H.   Calyx hispid externally; lvs. floccose to glabrous, 2–8 cm. long;
plants up to 7 dm. tall. . . . . 32. *E. Ordii*

HH. Calyx glabrous externally; lvs. floccose below, less so above, 1.5–
2.5 cm. long; plants up to 3 dm. tall. . . . . . . 51. *E. argillosum*

GG. Invols. tomentose.

H.   Lvs. mostly basal or subbasal, rarely axillary; calyx smooth, not
papillose; ak. beaks granular. W. Kern Co., ne. San Luis Obispo
and se. Monterey cos. . . . . 50. *E. temblorense*

HH. Lvs. mostly cauline and axillary; calyx papillose; ak. beaks papillose. San Benito Co., w. Fresno Co., and sw. Merced Co.

52. *E. vestitum*

EE. Basal lvs. linear, revolute, pilose; invols. 0.5–1 mm. long. 27. *E. spergulinum*

DD. Lvs. strictly basal, pilose or tomentose at least below.

E.   Invols. 4-lobed or -toothed.

F.   Calyx pubescent with hooked hairs externally.

G.   Invols. 2-fld.; aks. exserted; calyx ca. 1 mm. long. 28. *E. hirtiflorum*

GG. Invols. 4–6-fld.; aks. not exserted; calyx ca. 1.5 mm. long.

29. *E. inerme*

FF. Calyx puberulent or hispidulous externally, the hairs not hooked.

G.   Calyx white, 1.5–2 mm. long, apex notched or apiculate; lvs. spatulate, ciliate or pilose. . . . . 30. *E. apiculatum*

GG. Calyx pink or yellow, 0.5–1 mm. long, the segms. not apiculate.

H.   Calyx pink, 0.5–0.7 mm. long; lvs. spatulate, hirsute; invols.
0.5–0.7 mm. long. Montane. . . . . 31. *E. Parishii*

HH. Calyx yellow, 1–2 mm. long; invols. 0.7–2 mm. long. Deserts.

I.   Lvs. short-hirsute, suborbicular, 1–2.5 cm. long; invols.
0.7–1 mm. long. . . . . 34. *E. trichopes*

II.   Lvs. floccose or glabrous, obovate to oblanceolate, 2–8 cm.
long; invols. 1–2 mm. long. . . . . 32. *E. Ordii*

EE. Invols. 5-lobed or -toothed.

F.   Calyx pubescent or puberulent externally.

G.   Invols. glabrous externally.

H.   Outer segms. not saccate-dilated.

I.   Lf.-blades short-hirsute; invols. turbinate.

J.   Calyx pink or whitish with reddish midveins; lvs. hirsutulous and slightly glandular. . . 33. *E. glandulosum*

JJ.   Calyx yellow.

K.   Plants strictly annual; branchlets numerous and
whorled at each node; invols. usually 4-lobed,
occasionally some 5-lobed. . . . . 34. *E. trichopes*

KK. Plants perennial but flowering the first year;
branchlets few at each node, not in whorls.

35. *E. inflatum*

II.   Lf.-blades woolly; calyx 1 mm. long; invols. broadly campanulate. . . . . 40. *E. reniforme*

HH. Outer calyx-segms. saccate-dilated at each side of the cordate
base at maturity, sparsely puberulent at the base of the perianth tube, yellow maturing reddish or reddish with white lobes.

37. *E. Thomasii*

GG. Invols. glandular-puberulent externally.

H.   Calyx white to red, the outer segms. rounded and narrowed
abruptly to a narrow clawlike base, slightly glandular externally
at the base and with a white tuft of hairs within. 38. *E. Thurberi*

HH. Calyx yellow, the outer segms. obovate, smooth, glandular on
entire outer surface, glabrous within. . . . . 39. *E. pusillum*

FF. Calyx glabrous externally.

G.   Lvs. pilose-hispid, not woolly. Inyo and Mono cos. 36. *E. esmeraldense*

GG. Lvs. tomentose below.

   H. Stems glabrous or glandular, not woolly.
    I. Outer calyx-segms. panduriform, crisped, not cordate; peduncles slender, 5–25 mm. long. . . . . 41. *E. cernuum*
    II. Outer calyx-segms. not panduriform or crisped.
     J. Outer calyx-segms. cordate at the bases; peduncles stoutish.
      K. Peduncles deflexed.
       L. Stems glabrous.
        M. Invols. narrowly-turbinate to turbinate, peduncles 0–15 mm. long; calyx white, oblong.
         N. Invols. 1.5–3 mm. long; plants variously branched; calyx 1–2.5 mm. long, not gibbous at maturity. . . . . 43. *E. deflexum*
         NN. Invols. 1–1.5 mm. long; plants branching in a series of flat-topped layers, pagoda-like; calyx 1.5 mm. long, gibbous at the base at maturity. . . . . 44. *E. Rixfordii*
        MM. Invols. hemispheric, sessile; calyx yellow to reddish-yellow, suborbicular.
                45. *E. Hookeri*
       LL. Stems glandular, stems short and stout, crowns flat-topped. . . . 46. *E. brachypodum*
      KK. Peduncles erect; plants 3–10 dm. high; branches erect and whiplike with peduncles 3–5 mm. long; calyx 1.5–2 mm. long. . . . . 47. *E. insigne*
     JJ. Outer calyx-segms. obtuse at the base.
      K. Plants glandular; invols. deflexed, sessile or subsessile above, peduncled to 10 mm. below; calyx white becoming reddish. . . . . . . 48. *E. eremicola*
      KK. Plants glabrous.
       L. Peduncles short, straight and erect, less than 1 mm. long; calyx white to reddish. Inyo Co. 42. *E. Hoffmannii*
       LL. Peduncles long, curving upwards, 1–3 (–5) cm. long; calyx white to yellowish. Nevada Co. to Lassen Co. . . . . 49. *E. collinum*
  HH. Stems woolly.
    I. Lvs. basal or subbasal, oblong to elliptic; styles 0.7–1 mm. long; invols. 2–2.5 mm. long; stamens 2–2.5 mm. long. W. Kern Co., se. Monterey Co. and ne. San Luis Obispo Co. . . . . 53. *E. temblorense*
    II. Lvs. strictly basal, roundish; styles 0.1–0.3 mm. long; invols. 2 mm. long; stamens 1–1.5 mm. long. E. Monterey Co. and sw. Fresno Co. . . . . 55. *E. Eastwoodianum*
CC. Invols. cylindric or cylindric-turbinate to turbinate or prismatic, often 5–6 nerved, angled, or ribbed, mostly sessile, solitary or congested into heads, the teeth usually short.
  D. Plants annual; lvs mostly in basal rosettes. (Subgenus *Oregonium* [S. Wats.] Greene.)
    E. Flowering branches with short branchlets, usually of a single internode; invols. axillary and terminal.
     F. Calyx 1–1.5 mm. long, yellow to cream-white. Deserts.
      G. Calyx yellow, 1 mm. long, the segms. connate only at the base, calyx tube acutish; branches capillary, spreading and open infl. Mojave Desert, Inyo Co. to San Bernardino Co. . . . . 50. *E. mohavense*
      GG. Calyx whitish to cream-colored, 1–1.5 mm. long, the segms. connate about half the length of the calyx, calyx tube green or reddish, campanulate; branches stouter, erect and narrow infl. Mono. Co.
                51. *E. ampullaceum*
    FF. Calyx 1.5–2.5 mm. long, white to rose or red. Coast Ranges.
     G. Stems tomentose.
      H. Invols. glabrous, 2.5 mm. long; calyx 1.5–2.5 mm. long. Santa Clara Co. to San Benito and Monterey cos. . . 52. *E. argillosum*
     HH. Invols. tomentose.
      I. Invols. 1.5–2 mm. long.
       J. Lvs. basal and cauline, oblong to elliptic; styles 0.7–1 mm. long; stamens 1.5–2.5 mm. long.
        K. Lvs. mostly basal or subbasal, rarely axillary; calyx smooth; ak. beaks granular. W. Kern Co., ne. San Luis Obispo and se. Monterey cos.
               53. *E. temblorense*
        KK. Lvs. mostly cauline and axillary; calyx and ak. beaks papillose. San Benito Co., w. Fresno and sw. Merced cos. . . . . 54. *E. vestitum*

JJ. Lvs. strictly basal, roundish; styles 0.1–0.3 mm. long; stamens 1–1.5 mm. long. E. Monterey Co. and sw. Fresno Co. .... 55. *E. Eastwoodianum*

    II. Invols. 3–4 mm. long; calyx 2 mm. long. Contra Costa Co.
56. *E. truncatum*

GG. Stems glabrous.
  H. Lvs. round to reniform.
    I. Invols. 2.5 mm. long, 5-lobed; calyx 2–2.5 mm. long. Alameda Co. to Kern Co. .... 57. *E. Covilleanum*
    II. Invols. 3–4 mm. long, 8-lobed with short teeth; calyx 1.5 mm. long. The Pinnacles, Monterey and San Benito cos.
58. *E. Nortonii*
  HH. Lvs. oblong to oblong-ovate.
    I. Invols. 3–4 mm. long, distinctly 5-lobed; calyx 1.5–2.5 mm. long. Marin and Contra Costa cos. ...... 59. *E. caninum*
    II. Invols. 2.5–3 mm. long, obscurely 5-lobed; calyx 1.5 mm. long. Santa Clara Co. to San Benito and Monterey cos.
52. *E. argillosum*

EE. Flowering branches elongate, virgate, and bearing invols. at the nodes, the lateral ones appressed.
  F. Calyx glabrous or minutely puberulent externally.
    G. Invols. 2–5 mm. long.
      H. Lvs. oblong-obovate to oblanceolate; stems tomentose.
        I. Invols. 4–5 mm. long, cylindric, with minute teeth; outer calyx segms. narrowly obovate; aks. 2 mm. long.
60. *E. roseum*
        II. Invols. 2–3 mm. long, turbinate, with prominent teeth; outer calyx-segms. broadly obovate; aks. ca. 1 mm. long.
61. *E. gracile*
      HH. Lvs. rounded or nearly so; stems usually glabrous or only sparsely floccose.
        I. Stems simple below, glabrous except at the base; outer calyx-segms. more than twice as long as wide. Mts. of s. Calif. .... 62. *E. Davidsonii*
        II. Stems mostly branched from the base, glabrous or floccose; outer calyx-segms. less than twice as long as wide. Cent. and n. Calif. .... 63. *E. vimineum*
    GG. Invols. 1–1.5 mm. long.
      H. Outer calyx-segms. fan-shaped, their sides incurved below the broad truncate apices, yellowish to reddish; branches usually incurved at the summit in age; invols. few fld. San Bernardino Co. to Mono Co. .... 64. *E. nidularium*
      HH. Outer calyx-segms. not as above.
        I. Outer calyx-segms. not hastate, glabrous or glandular, 1–2 mm. long, white or yellow; lvs. strictly basal.
          J. Calyx-segms. 1.5–2 mm. long.
            K. Stems densely tomentose; fls. 3–10 per invol.; plants densely branched and spreading. San Bernardino Co. to Inyo Co. ... 65. *E. Palmerianum*
            KK. Stems glabrous to sparsely floccose; fls. more than 10 per invol.; plants usually sparsely branched and erect. San Bernardino Co. n. along the e. side of the Sierra Nevada. .... 66. *E. Baileyi*
          JJ. Calyx-segms. 1–1.5 mm. long.
            K. Calyx yellow. Deserts, San Bernardino Co. to Mono Co. .... 67. *E. brachyanthum*
            KK. Calyx white. S. Coast Ranges, Santa Clara Co. to San Luis Obispo Co. .... 68. *E. elegans*
        II. Calyx-segms. hastate at the base when mature, 0.8–1.2 mm. long, white to pink; lvs. basal or also cauline; stems tomentose. San Bernardino Mts. south. ... 69. *E. foliosum*
  FF. Calyx densely hairy externally. Inner N. Coast Ranges.
70. *E. dasyanthemum*

DD. Plants perennial; lvs. often cauline as well as basal. (Subgenus *Eucycla* [Nutt.] Kuntze.)
  E. Invols. solitary at the nodes, the lateral ones appressed to the branches.
    F. Calyx stipitate with long, winged attenuated bases, 5–7 mm. long.
      G. Infl. 1–1.5 dm. across. open and lax; calyx narrowed to a 3-angled base, pinkish to white or yellowish; aks. ca. 2 mm. long. Mts. of s. Calif. .... 71. *E. saxatile*
      GG. Infl. 0.3–0.8 dm. across, dense; calyx narrowed to a tubular base, sulphur yellow; aks. ca. 3 mm. long. N. base of Santa Monica Mts.
72. *E. crocatum*
    FF. Calyx astipitate, 2–5 mm. long.
      G. Invols. 2–6 mm. long.

H. Infl. with invols. in cymes or panicles.
   I. Lvs. more than 2 cm. long. . . . . 94. *E. nudum*
   II. Lvs. less than 2 cm. long.
      J. Calyx glabrous.
         K. Lvs. rotund-ovate, 3–5 (–10) mm. long and wide; plants herbaceous, 8–20 cm. high, glabrous. Amador Co. . . . . 73. *E. apricum*
         KK. Lvs. lanceolate, oblong, or elliptic, 8–20 mm. long; plants subshrubs or shrubs.
            L. Infl. a compact terminal cyme; invols. tomentose or glabrous; outer calyx-segs. subcordate at the base. . . . 74. *E. microthecum*
            LL. Infl. a divaricately branched panicle; invols. glabrous.
               M. Outer calyx-segs. round, subcordate at the base; branches green, dichotomous, ascending. . . 75. *E. Heermannii*
               MM. Outer calyx-segs. obovate, narrowed at the base; branches grayish, mostly horizontal, tiered. . . 76. *E. Plumatella*
      JJ. Calyx silky-villous, yellowish; plants shrubby, 6–15 dm. high. Imperial Co. . . . . 77. *E. deserticola*
HH. Infl. with invols. placed racemosely along the branches; invols. tomentose.
   I. Plants shrubby, up to 15 dm. high; lvs. elliptic to oblong, 2–3 cm. long; invols. racemosely arranged on the ends of the fragile branches. . . . . 78. *E. Kearneyi*
   II. Plants low subshrubs less than 3 dm. high.
      J. Plants suffrutescent and much branched at the base or densely cespitose; lvs. many, lance-elliptic to oblanceolate, ± revolute.
         K. Invols. 5–6 mm. long; calyx 4–5 mm. long, ochroleucous with red midribs. Santa Lucia Mts.
                              79. *E. Butterworthianum*
         KK. Invols. 2–3 mm. long; calyx 2–4 mm. long, white to pink. Sierra Nevada and mts. of s. Calif.
                              80. *E. Wrightii*
      JJ. Plants not suffrutescent at the base or densely cespitose; basal lvs. few, rounded to broadly ovate.
         K. Basal lvs. roundish to broadly ovate, 1.5–4 cm. long, on petioles 1–5 cm. long; plants from a highly branched, woody, spreading caudex, suberect; invols. and calyx 3–5 mm. long.
                              81. *E. panamintense*
         KK. Basal lvs. oblong, 3–5 cm. long, on petioles 3–5 cm. long; plants arising from a single, woody, little branched caudex, erect; invols. 2–3.5 mm. long; calyx 2.5–3 mm. long. . . 82. *E. racemosum*
   GG. Invols. 6–7 mm. long; loosely branched whitish tomentulose herbs 8–18 dm. high. S. Coast Ranges. . . . . 83. *E. elongatum*
EE. Invols. mostly clustered or in heads.
   F. Calyx-segs. dissimilar, the outer segms. often twice as wide as the inner segms., lobes dividing the calyx to the swollen basal joint.
      G. Infl. capitate.
         H. Outer calyx-segs. plane, not inflated; lvs. 6–20 (–60) mm. long. Common in s. and e. Calif. . . . . 84. *E. ovalifolium*
         HH. Outer calyx-segs. rounded and inflated; lvs. 2–4 mm. long. Rare, Panamint Mts., Inyo Co. . . . . 85. *E. Gilmanii*
      GG. Infl. cymose-umbellate with divaricate rays usually more than 2 mm. long. N. Calif. . . . . 86. *E. strictum*
FF. Calyx-segs. similar or nearly so.
   G. Plants cespitose, matted, herbaceous, the caudex much branched and woody; flowering stems scapelike.
      H. Calyx villous externally and internally; ovary sparsely pilose; lvs. lanate-tomentose. Inyo Co. . . . . 87. *E. Shockleyi*
      HH. Calyx and ovary not as above.
         I. Invol. a distinct rigid tube, 3–4 mm. long.
            J. Infl. compactly cymose-umbellate, the rays up to 5 mm. long; calyx whitish becoming reddish, finely glandular-hairy. Kern and Tulare cos. . . . . . . 88. *E. Breedlovei*
            JJ. Infl. tightly capitate.
               K. Calyx yellow; lf.-blades 5–35 mm. long.
                   L. Flowering stems 1–2.5 dm. tall, floccose; lf.-blades 2–3.5 cm. long. Below 6000 ft., Nevada Co. north. . . . . 89. *E. ochrocephalum*

LL. Flowering stems less than 1 dm. tall, glabrous or glandular; lf.-blades 4–15 mm. long.
　M. 　Stems and invols. glandular. Above 9000 ft., Inyo and Fresno cos. north to Placer Co. ... 90. *E. anemophilum*
　MM. 　Stems and invols. glabrous. Below 6000 ft., Modoc Co. north.
91. *E. chrysops*
　KK. Calyx white; lf.-blades mostly less than 5 mm. long, densely white-tomentose on both surfaces; stems glabrous to floccose. Mono Co. south.
92. *E. Kennedyi*
II. Invols. membranaceous and indistinctly forming a tube, 2–3 mm. long; calyx rose to red. White Mts... 93. *E. gracilipes*
GG. Plants shrubby, or if herbaceous, then not cespitose.
　H. Plants essentially herbaceous, only the base woody.
　　I. 　Lvs. spreading, oblong-ovate to ovate, obtuse, mostly 2–6 cm. long.
　　　J. 　Invols. 2–5 mm. long. Mainland.
　　　　K. 　Invols. and flowering stems tomentose; heads 1–few, mostly 1.5–3 cm. across. Immediate seacoast.
94. *E. latifolium*
　　　　KK. Invols. and flowering stems glabrous, or if tomentose then plants with heads many and plants from interior of n. Calif. .... 95. *E. nudum*
　　　JJ. Invols. 5–6 mm. long. Insular. ...... 96. *E. grande*
　　II. 　Lvs. erect, ± lanceolate, acute.
　　　J. 　Flowering stems glabrous or villous, 4–8 dm. long; lf.-blades 4–15 cm. long. ....97. *E. elatum*
　　　JJ. Flowering stems tomentose, 2–5 dm. long; lf.-blades 2–5 cm. long. .... 98. *E. pendulum*
　HH. Plants definitely shrubby.
　　I. 　Lvs. narrowly linear or nearly so, the blades less than 2 cm. long, strongly fascicled; shrubs with terminal cymose or subumbellate infl. .... 99. *E. fasciculatum*
　　II. 　Lvs. linear-oblong to orbicular, or if linear, more than 2 cm. long.
　　　J. 　Heads in dense compound cymes. Insular.
　　　　K. 　Lvs. linear or narrowly oblong, ± revolute, 2–3 cm. long. .... 100. *E. arborescens*
　　　　KK. Lvs. oblong-ovate, plane, 3–10 cm. long.
101. *E. giganteum*
　　　JJ. Heads terminal on 2-forked peduncles or scattered along the stems. Mostly mainland.
　　　　K. 　Calyx glabrous externally; lf.-blades 5–15 mm. long. .... 102. *E. parvifolium*
　　　　KK. Calyx white-villous; lf.-blades 15–30 mm. long.
103. *E. cinereum*
BB. Stems internally jointed. Death Valley. (Subgenus *Clastomyelon* Cov. & Mort.)
104. *E. intrafractum*

Subgenus **Oligogònum** Nutt.

1. **E. caespitòsum** Nutt. [*E. sericoleucum* Greene ex Tidestr. *E. sphaerocephalum* var. *s.* S. Stokes.] Low compact matted perennial from much-branched woody caudices; lvs. elliptic to oblong-spatulate, densely white-tomentose, 5–15 mm. long, short-petioled, 1–3 mm. wide, crowded on the tips of the short branches, ± revolute; flowering stems scapelike, bractless, slender, 3–8 cm. high, somewhat loosely tomentose; invols. solitary, terminal, the tube turbinate, ca. 3 mm. long, with somewhat longer linear lobes that become reflexed; calyx yellow, 2.5–4 mm. long in anthesis, later reddish and 4–6 (–10) mm. long in fruit, pubescent especially toward the stipelike base, the segms. similar, oblance-oblong; fils. pilose basally; aks. lanceolate in outline, somewhat 3-angled, ca. 3 mm. long, often pubescent at the apex.—Dry gravelly slopes and flats, 5000–8600 ft.; Sagebrush Scrub, N. Juniper Wd., Yellow Pine F.; White Mts., Inyo Co. to Modoc Co.; to Ida., Mont., Colo. May–July.

2. **E. Douglásii** Benth. [*E. caespitosum* var. *D.* Jones. *E. c.* ssp. *D.* S. Stokes.] Rather loosely matted perennials from much-branched woody caudices, the plants to 3–4 dm. across; lvs. mostly linear to linear-spatulate, tomentose on both surfaces, 5–20 mm. long, the petioles often making up ⅓ of this; flowering stems loosely tomentose, 4–12 cm. long,

with a whorl of oblanceolate leafy bracts near the middle and a single terminal invol.; involucral tube turbinate, 3 mm. long, with reflexed oblong lobes ca. as long; calyx yellow, later often reddish, 5–8 mm. long, villous-pubescent on midribs and base, segms. narrowly obovate; fils. pilose basally; aks. lanceolate in outline, 3-angled, pubescent on upper half, ca. 3 mm. long.—Infrequent, dry rocky places, 4500–8000 ft.; Sagebrush Scrub to Yellow Pine F.; e. slope of Sierra Nevada from Nevada Co. to Modoc Co.; s. to Siskiyou Co.; to Wash., Ida., Nev. May–July.

Considerable work is necessary on this species. The type specimen is distinctive and it appears that much of what has passed in California and adjacent southern Oregon as *Eriogonum Douglasii* is actually a capitate form of *E. sphaerocephalum*. The distinguishing feature which separates the two is that *E. Douglasii* has densely tomentose lvs. on both surfaces while the *E. sphaerocephalum* form has lvs. that are subglabrous above and generally larger. In addition the involucre characteristics used in the key may be used to separate the two species.

3. **E. Twisselmánnii** (J. T. Howell) Reveal, stat. & comb. nov. [*E. Douglasii* var. *T. J. T. Howell*, Leafl. West. Bot. 10: 13. 1963.] Loose matted perennials from a much-branched spreading woody caudex, the plants 1–2 dm. high; lvs. oblanceolate or elliptic, tomentose below, subglabrous and green above, 6–10 mm. long, the petioles 2–4 mm. long; flowering stems loosely tomentose, 5–12 cm. long, with a whorl of oblanceolate leafy bracts near the middle and a single terminal invol.; invols. campanulate-cupulate, ca. 5 mm. long, the base of the tube subtruncate to broadly turbinate, with 6–9 triangular lobes, reflexed, about as long as the tube; calyx ochroleuceous, brownish, or yellowish, 5–6 mm. long, with sparse long hairs and numerous minute glandular hairs externally, and numerous minute glandular hairs or long non-glandular hairs internally, segms. subequal, the outer segms. obtuse or emarginate, 5 mm. long and 3.5 mm. wide, the inner segms. broadly oblanceolate, 6 mm. long and 3 mm. wide, the calyx tube about 1.5 mm. long, stipe 1 mm. long; aks. 5.5 mm. long, lanceolate-ovoid in outline, smooth, glabrous or with few scattered hairs.—Dry rocky outcrops, 7900–8200 ft.; Yellow Pine F., near The Needles, s. Tulare Co. July–Sept.

Eriogonum Twisselmannii is closely related to *E. Douglasii* where J. T. Howell placed it, but the latter is found nearly 300 miles to the north and differs in several morphological aspects, growth habit, pubescence, as well as the distinct geographical separation.

4. **E. sphaerocéphalum** Dougl. ex Benth. [*E. s.* var. *brevifolium* S. Stokes ex Jones.] Caudices much branched, woody, with decumbent leafy branches 5–12 cm. long; lvs. mostly oblanceolate, in whorls at the upper nodes, ± glabrate above, grayish-lanate below, 1–3 cm. long including the slender petioles; flowering stems ascending to erect, 5–15 cm. long, with a whorl of leafy oblanceolate bracts at the middle, capitate, simple, or umbellate above, each peduncle bearing 1 invol. with broadly turbinate tubes 3–4 mm. long, the lobes ca. as long; calyx bright sulphur-yellow, villous-tomentose, 7–8 mm. long, the stipitate base slender and ca. 2 mm. long, the segms. oblong-ovate; aks. lanceolate-ovoid in outline, 3-angled and pubescent above, ca. 3 mm. long.—Dry rocky places, 3000–7000 ft.; Sagebrush Scrub, N. Juniper Wd., Yellow Pine F.; Lassen Co.; to Wash., Ida., Nev. May–July.

Var. **halimioìdes** (Gand.) S. Stokes. [*E. h.* Gand.] Calyx whitish to pale yellow or pinkish; *n* = 20 (Reveal, 1965).—Dry rocky places, 3000–7500 ft.; Sagebrush Scrub, N. Juniper Wd.; Siskiyou, Lassen, and Modoc cos.; E. Ore., w. Nev. to Ida.

The common form of *Eriogonum sphaerocephalum* in California is the var. *halimioides*, and in many cases, this form has been recognized as *E. Douglasii* rather than *E. sphaerocephalum*. As noted under *E. Douglasii*, the two forms may be separated by the pubescence of the lvs., and when the type of *E. Douglasii* is considered, it is possible that to a great degree most of the material called *E. Douglasii* in both California and Oregon is actually a capitate form of *E. sphaerocephalum* var. *halimioides*.

5. **E. tripòdum** Greene. Caudices woody, loosely branched; lvs. in whorls at the tips of branches, narrowly oblanceolate, 1.5–2.5 cm. long, white-tomentose on both surfaces, revolute; flowering stems slender, 2–3 dm. high, bearing a whorl of foliaceous bracts near the middle at the base of a 3-rayed umbel, the rays naked or bracted, glabrous; invols. solitary and tomentose, the spreading or reflexed lobes shorter than the tube; calyx yellow, 4–5 mm. long, villous-tomentose, the stipelike base ca. 2 mm. long; fils. pilose basally; aks. narrow-ovoid, strongly angled, pubescent at the apex, ca. 2 mm. long.—Gravelly

slopes, often on serpentine, below 2000 ft.; Foothill Wd.; inner Coast Ranges and Sierran foothills, Tehama and Lake cos. and Tuolumne and Mariposa cos. June–July.

6. **E. Pratteniànum** Durand. [*E. umbellatum* ssp. *serratum* S. Stokes, as to type, not as to concept.] Low tufted perennials from a branched woody caudex, 10–15 cm. across; lvs. elliptical to ovate, lf.-blades acutish, 5–15 mm. long, 2–7 mm. wide, densely white-felty tomentose on both surfaces, occasionally becoming subglabrous above at maturity, short petioled, 1–5 mm. long, tomentose, densely crowded at the tips of the branches with 5–15 lvs.; scapes slender, erect, (5–) 10–30 cm. high, with a whorl of leafy bracts near the middle, narrowly elliptic, 7–10 mm. long, 2–4 mm. wide, similar to the lvs.; invols. solitary, without bracts at the base, campanulate, the tubes 3–4 mm. long, arachnoid-tomentose externally, glabrous internally, 8–10-lobed, the lobes 1.5–2.5 mm. long, erect or reflexed at maturity, the bractlets linear, 2–3 mm. long, whitish with several long marginal cells, the pedicels glabrous, (3–) 5–7 mm. long; calyx yellow, glabrous, 3–4 mm. long including the stipe in anthesis, the stipe 1–1.5 mm. long, the calyx-segms. similar or nearly so, obovate to spatulate, 2.5–3 mm. long, 1–1.5 mm. wide; stamens exserted, fils. 3–4 mm. long, sparsely pilose basally; ovary glabrous, styles (0.5–) 1–1.5 mm. long; calyx 6–7 mm. long in fruit; aks. lance-ovate, 4–5 mm. long, light yellowish-brown to rustic, smooth and glabrous, gradually tapering to the apex, mostly indistinctly 3-angled, embryo straight.—Dry rocky ridges and outcrops, 3000–5000 ft.; Yellow Pine F.; w. slope of the Sierra Nevada from Nevada Co. south to Fresno Co. May–July.

Eriogonum Prattenianum has been totally neglected in all monographs and revisions of the genus, as well as all floristic treatments in California, since the species was published in the *Journ. Acad. Nat. Sci. Phila.* ser. II, 3: 100. 1855. The holotype at the Academy of Natural Sciences in Philadelphia consists only of a single stem broken off the plant slightly below the whorl of foliaceous bracts near the middle, yet the species is so distinct that even from this fragmentary material it is possible to match it with the present day collections. The original collection was made by Henry Pratten in 1851 from near Nevada City, Nevada Co., California. Although it has been recollected and named *E. umbellatum* ssp. *serratum,* the above description is the first prepared for the species as defined by the type.

Eriogonum Prattenianum has been known for several years to California botanists as the Sierra Nevada form of *E. siskiyouense.* The Siskiyou Buckwheat, however, is known only from its type locality on Mt. Eddy in the Scott Mountains where it occurs on rocky ridges from 7000 to 9000 feet elevation. The Sierran Buckwheat grows on rocky open places from 3000 to 5000 feet in the Yellow Pine Forest belt. The two species also differ in their flowering times. Actually, *E. Prattenianum* seems to be more closely related to *E. tripodum* which occurs in the same general area but always at a lower elevation.

Stokes based her name, *Eriogonum umbellatum* ssp. *serratum,* on A. A. *Heller 13208* from near Grass Valley, Nevada Co., but her resulting concept of the name was applied to specimens of *E. umbellatum* var. *polyanthum.* The specimens of *E. Prattenianum,* from Tehipite Valley, Fresno Co., differ from the others in that the lvs. are subglabrate above and may represent a distinct variety, but the lack of adequate material dictates that it remain unnamed at this time.

7. **E. siskiyouénse** Small. [*E. ursinum* var. *s.* S. Stokes.] Low matted or tufted perennials from a woody base, with short compact branches 3–7 cm. long; lvs. oval to spatulate, crowded at ends of branches, 5–8 mm. long, acutish, short-petioled, glabrate above, tomentose below; scapes slender, erect, 3–8 cm. high, with a whorl of leafy bracts near the middle; invol. solitary, without bracts at its base, campanulate, somewhat arachnoid-tomentose, the tube 3.5–4 mm. long, the reflexed lobes ca. as long; calyx yellow, glabrous, 5 mm. long, short-stipitate, the outer segms. oblong, rounded at the apex, the inner segms. somewhat narrower; fils. pilose basally; aks. narrow-ovoid, glabrous, 3-angled.—Rocky ridges, 7000–9000 ft., in Subalpine F.; Mt. Eddy, Siskiyou and Trinity cos. Aug.–Sept.

8. **E. heracleoìdes** Nutt. var. **angustifòlium** (Nutt.) T. & G. [*E. h.* of most authors. *E. a.* Nutt.] Loosely tufted plants from branched woody caudices; lvs. linear, 2–5 cm. long, 2–6 mm. wide, white-tomentose, especially beneath, short-petioled; flowering stems 2–4 dm. high, tomentose, usually with a whorl of leafy bracts near the middle of the stems and another at the base of the umbel; umbels simple or compound, the rays 2–5 cm. long; invols. solitary, turbinate, woolly-tomentose, the tubes 2.5–3 mm. long, the lobes 3–4 mm. long, spreading or reflexed; calyx glabrous, 4.5–6 mm. long including the stipitate base which is tubular and 1.5–2 mm. long, the segms. oblong-ovate; fils. pilose basally; aks. light brown, narrow, somewhat 3-angled, ca. equally pointed at both ends, pubescent at the apex, ca. 2 mm. long.—Occasional in dry places, 6000–7500 ft.; Sage-

brush Scrub, N. Juniper Wd., Yellow Pine F.; Warner Mts., Modoc Co.; to B.C., Mont., Utah. June–Aug. The var. *heracleoides* occurs chiefly in se. Wash., the Wallowa Mts., Ore. and w. Ida.

9. **E. umbellàtum** Torr. Caudex open or depressed, the plants cespitose to subshrubby; lvs. spatulate to suborbicular, mostly less than 2 cm. long and not more than 3–4 times as long as broad, tomentose to glabrate, the petioles short to long; flowering stems tomentose to glabrous, erect; infl. usually umbellately or cymosely divided, or in reduced forms, often simple or capitate; invols. usually deeply lobed, the lobes erect; calyx cream to yellow, maturing to a rose or red in most, the midribs often large and obvious, up to 10 mm. long; aks. usually sparsely pubescent at the apex, up to 4 mm. long.—In numerous situations throughout much of California.

In the *Eriogonum umbellatum* complex there are about 20 distinct infraspecific populations now currently recognized, and certainly more exist. The treatment given here is an attempt to summarize the species as best as possible at the present time, but no doubt several errors are made or continued here. While not all of the types have been seen, a majority have been and it is hoped from these studies that the nomenclature proposed here is more stable than in the past. Likewise it is hoped that the definition of the various varieties is clearer and more exact than previously presented in any treatment to this date. However, it must be quickly mentioned, that we are perpetuating the past handling of var. *stellatum* which is here defined to include at least two and possibly three distinct types. The holotype of var. *stellatum*, however, has not been seen, and it is not known at this time which part of this overall complex, found throughout much of the western United States, is represented by the type and which which remain to be described or recognized. In California, the following varieties may be tentatively recognized.

KEY TO VARIETIES

A. Primary rays of umbels simple, not branched or bracteate in the middle.
  B. Umbels subcapitate or rays few and scarcely more than 1 cm. long.
    C. Calyx bright yellow, 4–7 mm. long.
      D. Lvs. slightly tomentose below, not dense or matted, subglabrous to glabrous above; scapes slender, 1–3 dm. high. Common Sierra Nevada form mostly below 10,000 ft.
                                         (a) var. *umbellatum*
      DD. Lvs. densely and thickly matted white-tomentose below, subglabrous to glabrous above; scapes stout, 2–4 dm. high. Common form in n. Calif., Siskiyou and Trinity cos. e. to Modoc Co. .... (g) var. *polyanthum*
    CC. Calyx mostly whitish to red, 3–8 mm. long, occasionally pale yellow, or if yellow, then plants mainly above 9000 ft.; scapes 0.4–1.2 dm. high. Mostly montane.
      D. Lvs. green and glabrate above; calyx mostly whitish to yellow or reddish to purplish.
        E. Calyx yellow or pale yellow, 2–4 mm. long. Sierra Nevada above 10,000 ft. to Mt. Shasta and Warner Range down to 9000 ft. .... (b) var. *Covillei*
        EE. Calyx reddish to purplish or cream-colored, 3–8 mm. long.
          F. Calyx cream-colored to pale yellow with a tannish midrib. E. slope of the Sierra Nevada to San Bernardino Mts.; Nev. .... (c) var. *dicrocephalum*
          FF. Calyx reddish-brown to pink with a large reddish or purplish midrib. Inyo, Panamint, and Grapevine mts.; s. Nev. .... (d) var. *versicolor*
      DD. Lvs. densely white-woolly on both surfaces; calyx deep red, 4–5 mm. long. San Gabriel Mts. .... (e) var. *minus*
  BB. Umbels open, the rays mostly 2.5 or more cm. long.
    C. Calyx whitish, cream-colored, or pale yellow; lvs. tomentose (at least sparsely so) on both surfaces. E. slope of the Sierra Nevada to San Bernardino Mts., Nev.
                                         (c) var. *dicrocephalum*
    CC. Calyx bright yellow, often becoming tinged with red.
      D. Lvs. tomentose, at least below.
        E. Lvs. slightly tomentose below, not dense or matted, subglabrous to glabrous above; scapes slender, 1–3 dm. high. Common form in the Sierra Nevada mostly below 10,000 ft. .... (a) var. *umbellatum*
        EE. Lvs. densely and thickly matted white-tomentose below, subglabrous to glabrous above; scapes stout, 2–4 dm. high. Common form in n. Calif., Siskiyou and Trinity cos. e. to Modoc Co. .... (g) var. *polyanthum*
      DD. Lvs. totally glabrous on both surfaces; calyx 7–10 mm. long. Placer Co. to Modoc Co. .... (f) var. *Torreyanum*
AA. Primary rays of umbels usually branched, or if not, then with bracts near the middle.
  B. Lvs. totally glabrous on both surfaces.
    C. Plants 1.5–3 dm. high, woody only at the base; calyx 7–10 mm. long; infl. with bracts near the middle, rarely divided; lvs. elliptic. Placer Co. to Modoc Co.
                                         (f) var. *Torreyanum*
    CC. Plants 4–12 dm. high, woody about half the height of the plants; calyx 3–6 mm. long; infl. compoundly divided; lvs. narrowly lanceolate. Mono, Inyo, and Tulare cos.
                                        (k) var. *chlorothamnus*
  BB. Lvs. pubescent, at least below.

C. Plants 4–12 dm. high, woody about half the height of the plants; calyx 3–6 mm. long; lvs. narrowly lanceolate, subglabrate below. Mono, Inyo, and Tulare cos.

(k) var. *chlorothamnus*

CC. Plants up to 5 dm. high, woody only at the base; lvs. mostly elliptic, tomentose, at least below.

D. Lvs. sparsely tomentose above and below, the pubescence even on both surfaces. Inyo-White Mts., Panamint Mts. .... (j) var. *subaridum*

DD. Lvs. densely tomentose, at least below.

E. Calyx 3–7 mm. long, including the stipe.

F. Lvs. densely tomentose below, subglabrous to glabrous above. Coast Ranges, w. slope of the Sierra Nevada, mts. of s. Calif. (i) var. *stellatum*

FF. Lvs. densely tomentose on both surfaces. Central Coast Ranges, Glenn Co. to San Benito Co. .... (l) var. *bahiiforme*

EE. Calyx 7–10 (–12) mm. long, including the stipe. N. Coast Ranges.

(h) var. *speciosum*

(a) Var. **umbellàtum**. [*E. polyanthum* of California authors, not Benth., in large part. *E. reclinatum* Greene.] Plants 2–6 dm. high; lvs. tomentose below, less so to glabrous above, lf.-blades elliptic to ovate, 1–2 cm. long, petioles slender; scapes slender, mostly 1–3 dm. long, the umbels few rayed, bractless; invol. tubes 2–3.5 mm. long with reflexed lobes as long; calyx yellow, 4–7 mm. long; $n = 40$ (Reveal 1965, 1968).—Rather common on dry slopes and ridges, 2500–10,000 ft.; Sagebrush Scrub to Yellow Pine F. and Subalpine F.; Coast Ranges from Tehama Co. to Humboldt and Siskiyou cos., Sierra Nevada from Tulare Co. n.; to Wash., and e. to Colo. June–Sept.

This variety has been called ssp. *polyanthum* for several years and considered the common California form although the type of *polyanthum* is a different plant entirely. The distinguishing features of the common California form do differ slightly from the Rocky Mountain plants which are true var. *umbellatum*, and especially in the large subshrubby size of the plants found along the eastern slope of the Sierra Nevada which Greene named *E. reclinatum*. In considering the entire range of var. *umbellatum*, however, the California plant gradually blends into the low Rocky Mountain form through northern Nevada and Idaho.

(b) Var. **Covíllei** (Small) Munz & Reveal comb. nov. [*E. C. Small, Bull. Torr. Bot. Club* 25: 42. 1898. *E. ursinum* var. *C. S.* Stokes in name. *E. umbellatum* var. *polypodum* of S. Stokes, not *E. p.* Small. *E. umbellatum* ssp. *C.* Munz.] Plants less than 1.5 dm. high; lvs. lightly tomentose on both surfaces, lf.-blades narrowly elliptic, 0.5–1 cm. long, the petioles as long; scapes slender, less than 1 dm. long, the umbels capitate or nearly so; invols. 1.5–3 mm. long with reflexed lobes as long; calyx yellow, 2–4 mm. long.—Dry gravelly rocky soil, above 10,000 ft.; Alpine Fell-fields of Sierra Nevada, Subalpine F., Mt. Shasta and Warner Range above 9000 ft. July–Sept.

The var. *Covillei* is the high alpine form of *Eriogonum umbellatum* which occurs in the Sierra Nevada. It is closely related to var. *Hausknechtii* (Dammer) Jones which occurs in similar habitats in the high mountains of Oregon. These two varieties seem to have evolved from var. *umbellatum*, while the var. *Porteri* (Small) S. Stokes which is found in the high mountains of northeastern Nevada and Utah was derived from the glabrous-leaved form of var. *umbellatum*, the var. *intectum* A. Nels. All of the alpine varieties are characterized by their capitate or subcapitate inflorescences, but differ from each other in leaf pubescence and geographical distribution.

(c) Var. **dicrocephalum** Gand. [*E. aridum* Greene. *E. azaleastrum* Greene. *E. u.* ssp. *aridum* S. Stokes. *E. u.* var. *aridum* C. L. Hitchc.] Lf.-blades tomentose on both surfaces, or subglabrous to glabrous above, 1.5–2 cm. long; scapes 1–2.5 dm. high; rays few, (0.5–) 1–3 cm. long; calyx whitish, cream-colored, or pale yellow, often with a large tannish midrib, 4–8 mm. long including the stipe.—Occasional, below 10,000 ft.; Pinyon-Juniper Wd., Lodgepole F.; e. slope of Sierra Nevada, White Mts., Mono Co. s. to San Gabriel Mts., San Bernardino Co.; n. Nev., w. Utah. June–Aug.

A recent examination of the type of var. *dicrocephalum* shows that this is an earlier name for what has been known as *aridum*.

(d) Var. **versicólor** S. Stokes. Low, matted perennials less than 1.5 dm. high; lvs. lightly tomentose on both surfaces, the lf.-blades elliptic, 0.5–1.5 cm. long, the petioles ca. as long; scapes slender, less than 1.5 dm. long, the umbel rays few, less than 1.5 cm. long; calyx reddish-brown to rose or pink with a large reddish or purplish midrib, 3–6 mm. long.—Occasional, below 9000 ft.; Pinyon-Juniper Wd.; Panamint, Inyo, and Grapevine mts., s. Nev.

The var. *versicolor* is closely related to var. *dicrocephalum* and occasionally the two are difficult to distinguish. The California plants are simply branched in the inflorescence, but in the Charleston Mts. of southern Nevada, biumbellate forms are not infrequently found.

(e) Var. **mìnus** Jtn. [*E. m.* Ewan.] Low and densely matted; lf.-blades round-ovate, 4–10 mm. long, permanently densely white-woolly on both surfaces; scapes 3–12 cm. long; rays of umbels 1–3, 5–20 mm. long; calyx 4–5 mm. long, often deep red.—Dry stony slopes, 8000–10,000 ft.; Lodgepole F., Subalpine F.; San Gabriel Mts. July–Sept.

(f) Var. **Torreyànum** (Gray) Jones. [*E. T.* Gray.] Lvs. glabrous, elliptic, 3–6 cm. long; infl. compound, bracteate near the middle of the rays; calyx 7–10 mm. long, bright yellow, numerous in each invol. the heads up to 5 cm. in diam.—Dry gravelly or stony places at ca. 6000 ft.; Placer Co., reported as far n. as Modoc Co. July–Aug.

The var. *Torreyanum* is definitely known only from the Donner Pass area although it has been reported as far north as Modoc Co. This variety is exceedingly distinct and no other forms are known to be near it. The plants from eastern Oregon which have been called var. *Torreyanum* actually represent another variety that may be called var. **glabérrimum** (Gand.) Reveal [based on *E. g.* Gand., *Bull. Soc. Bot. Belg.* 42: 197. 1906]. This variety differs mainly in that the Oregon plant has smaller flowers in equally smaller heads in distinct compoundly umbellate inflorescences.

(g) Var. **polyánthum** (Benth.) Jones. [*E. p.* Benth. *E. dumosum* Greene. *E. modocense* Greene. *E. u.* sspp. *p.* and *d.* S. Stokes. *E. u.* var. *m.* S. Stokes.] Plants 3–10 dm. high; lvs. densely matted and felty white-tomentose below, ovate to elliptic, 5–30 mm. long, 5–20 mm. wide, short-petioled; scapes stoutish, 1–4 dm. long with large bracts 1–4 cm. long; invols. 4–6 mm. long, 4–10 mm. wide; calyx yellow, 6–7 mm. long.—Dry sandy or gravelly places below 5000 ft.; Sagebrush Scrub, N. Juniper Wd. up to Red Fir F.; Placer Co. n. to Siskiyou and Modoc cos.; s. Ore. July–Aug.

The recognition of var. *polyanthum* as here described differs considerably from the treatments given before in the various state floras. The type of *polyanthum* was collected by John C. Frémont near Mt. Shasta, and is the large, bright yellow-flowered subshrub so often seen on the loose sandy soils of that area. Nearly all of the references to *polyanthum* in California should actually be applied to var. *umbellatum*.

(h) Var. **speciòsum** (Drew) S. Stokes. [*E. s.* Drew.] Plants 5–10 dm. high; similar to var. *polyanthum* but with compound infl. and larger fls., 7–10 (–12) mm. long.—Dry rocky places mostly below 3000 ft.; Yellow Pine F., Foothill Wd.; N. Coast Ranges from Del Norte Co. to Siskiyou Co. July–Aug.

(i) Var. **stellàtum** (Benth.) Jones. [*E. s.* Benth. *E. u.* ssp. *s.* S. Stokes.] Basal lvs. spatulate-obovate, glabrate above, densely tomentose below, the petioles often exceeding the blades, the lf.-blades 1.5–2.5 cm. long; rays of the umbels often compoundly branched with foliaceous bracts at the base of each division; calyx usually bright yellow, sometimes with a reddish tinge, mostly 6–8 mm. long.—Dry rocky slopes, 3000–7500 ft.; Yellow Pine F. to Red Fir F.; Trinity to Siskiyou and Del Norte cos., along the w. slope of the Sierra Nevada to Merced Co., mts. of s. Calif.; to Wash., Mont., e. to Utah. July–Sept.

As noted at the beginning of this discussion, the var. *stellatum* is not yet clearly understood. In California, the variety, as here defined, includes not only the northern California element which is probably true var. *stellatum*, but also an element from near Yosemite Valley which differs in its pubescent leaf and shorter and more compact stature, and the plants from the mountains of southern California. This last population is apparently unnamed. It may be distinguished by its densely tomentose stems and leaves and wide spreading low growth. However, in many respects, the southern California plant approaches var. *bahiiforme* of the Coast Ranges, and until this relationship can be investigated, it would be unwise to propose a new variety for this seemingly distinct population.

(j) Var. **subáridum** S. Stokes. [*E. u.* ssp. *s.* Munz.] Lvs. very finely and closely tomentose on both surfaces, not densely tomentose below, lf.-blades 1–1.5 cm. long; infl. compound; calyx mostly yellow, occasionally pale yellow, 6–7 mm. long.—Dry rocky slopes, 5000–9000 ft.; Pinyon-Juniper Wd.; White, Inyo, Argus, and Panamint mts., across the Mojave Desert to s. Nev., Utah, and sw. Colo. July–Aug.

(k) Var. **chlorothámnus** Reveal, var. nov. A var. *stellato* differt fruticosis, 5–10 dm. altis, foliis fere subglabris vel glabris, ellipticis vel oblanceolatis, 5–20 mm. longis et 4–6 mm. latis. Plants 5–10 dm. high; lvs. subglabrous or glabrous on both surfaces, narrowly elliptic or oblanceolate, 5–20 mm. long, 4–6 mm. wide, short-petioled; scapes slender,

1–2.5 dm. long; invols. 3–6 mm. long; calyx yellow, 3–6 mm. long.—Dry gravelly places; Sagebrush Scrub, 7000–9000 ft.; Owens V., Inyo and Mono cos., Little Kern River, Tulare Co. July–Sept. TYPE: Summit of Sherwin Grade, 4 miles s. of Tom's Place, on a dirt road 0.5 mile e. of U.S. Hwy. 395, Mono Co., Calif., 23 July 1966, *N. H. Holmgren & J. L. Reveal 2938*. Holotype deposited at the Intermountain Herbarium, Utah State University.

For several years this plant has gone under the name of var. *bahiiforme*, as a syntype collected by Horn is representative of this plant. By typifying var. *bahiiforme* (see below), it becomes necessary to recognize this distinct variety. A part of the overall complex associated with var. *stellatum*, the var. *chlorothamnus* is actually more closely related to var. *subaridum* which occurs nearby on the White Mts.

(1) Var. **bahiifórme** (T. & G.) Jeps. [*E. polyanthum* var. *b.* T. & G. *E. stellatum* var. *b.* Wats. *E. trichotomum* Small. *E. Smallianum* Heller. *E. u.* var. S. S. Stokes. *E. u.* ssp. *b.* Munz.] Lvs. ± felty, white-tomentose on both surfaces, the petioles scarcely as long as the blades, the lf.-blades 1–1.5 cm. long; infl. compound, or at least with bracts on the primary rays; calyx 5–8 mm. long, yellow becoming tinged with red.—Dry rocky places, mostly above 1000 ft.; Yellow Pine F., Foothill Wd.; Coast Ranges from Lake and Glenn cos. s. to San Benito Co. July–Sept. LECTOTYPE: New Idria, San Benito Co., Calif., *Brewer 771* (GH).

By typifying the var. *bahiiforme* with the New Idria collection this allows for the exclusion of the Owens Valley plants which are now named var. *chlorothamnus*, and it also allows for the exclusion of the plants from the mountains of southern California which have, on occasion, been called var. *bahiiforme*. As noted under var. *stellatum*, these southern California plants are presently placed there although it is likely that they represent a distinct group.

10. **E. compósitum** Dougl. ex Benth. [*E. c.* var. *citrinum* S. Stokes.] Perennials with ± branched woody caudices; lvs. basal, ovate to lance-ovate, usually truncate or more commonly cordate at the base, densely white-tomentose below, usually glabrate and greenish above, the lf.-blades 2–10 (–20) cm. long, the petioles as long or longer; scapes stout, 2–4 (–5) dm. long, subglabrous, ascending to erect; infl. simple to compound, umbellate, subtended by narrow foliaceous bracts; invols. campanulate, 6–10 mm. long, mostly pilose-tomentose externally, with 5 linear, finally reflexed lobes; calyx glabrous, ochroleucous to pale yellow, becoming 5–6 mm. long, the stipe 1–1.5 mm. long; calyx-segms. similar, ± oblong-ovate; fils. pilose basally; aks. narrowly triangular-ovoid, pubescent above, light brown, 5–6 mm. long; $n = 20$ (Reveal, 1968).—Dry rocky hills and slopes or ledges, below 7500 ft.; Yellow Pine F., Red Fir F.; Lake Co. to Del Norte and Siskiyou cos.; to Wash., Ida. May–July.

11. **E. Lóbbii** T. & G. [*E. L.* var. *minus* T. & G.] Few-branched from stout woody caudices, covered with hairy bases of dead lvs.; lvs. in tufted rosettes, mostly round-oval, plane, densely tomentose especially below, the lf.-blades 1–4 cm. long, abruptly narrowed into petioles as long or longer; scapes flat on the ground or decumbent, tomentose, 0.5–2 dm. long, with foliaceous bracts subtending the 2- to several-rayed umbels; rays 1–3 cm. long, woolly-hirsute; invols. campanulate, 8–12 mm. long, the lobes reflexed; calyx white to rose (especially in age), 5–7 mm. long, the base scarcely stipitate, the stipe mostly less than 1 mm. long, the calyx-segms. oblong-obovate; fils. hairy basally; aks. lance-ovoid, glabrous, shining, olive-green, 4.5–6 mm. long, 3-angled at the apex.— Gravelly slopes and ridges, 5500–8000 ft.; Red Fir F. to Alpine Fell-fields; Coast Ranges, Lake Co. to Humboldt and Siskiyou cos. and 5500–12,000 ft., Sierra Nevada, from Inyo and Mariposa cos. to Plumas Co.; w. Nev. June–Aug.

In western Nev., var. *robustius* (Greene) Jones [*E. r.* Greene.] occurs and may be distinguished by its larger flowers, leaves, and higher and more robust stature when seen in the field. It is to be expected at low elevations in the foothills of the Sierra Nevada in eastern Placer, Nevada, or Sierra cos.

12. **E. làtens** Jeps. [*E. monticola* S. Stokes.] Caudex woody, with short branches; lvs. basal, round-ovate to elliptic-obovate, subglabrous to short-pilose, paniculate, the lf.-blades 1–3 cm. long, abruptly narrowed to longer petioles; scapes naked, 1.5–3 dm. long; infl. capitate, 2–3.5 cm. in diam., subtended by membranous rose-colored bracts; invols. few, campanulate, sparsely pilose, the lobes oblong-ovate, becoming recurved; calyx cream-colored to pale yellow, strigose near base, 5–6 mm. long at low elevs., 3–4 mm. long in alpine situations, not obviously stipitate at base; fils. pubescent basally; aks. lance-

ovoid, glabrous, 3–4 mm. long.—Dry stony slopes and ridges, 6500–11,000 ft.; Pinyon-Juniper Wd., Red Fir F. to Alpine Fell-fields; e. slope of the Sierra Nevada and White Mts., Inyo and Mono cos.; w. Nev. July–Aug.

13. **E. hirtéllum** Howell & Bacig. Perennials with woody, long-creeping rhizomes covered with old lf.-bases and with small clusters of lvs.; lvs. basal, ovate, elliptic, or broadly oblanceolate, 0.5–1.5 cm. long, 0.3–0.8 cm. wide, petioles 0.3–2 cm. long, the petioles and lf.-blades hirtellous or the blades sometimes glabrescent; scapes naked, erect and slender, 8–25 cm. long; infl. capitate, subtended by 5 narrowly oblong hirtellous bracts; invols. turbinate, 5–6 mm. long, 5–6 toothed, the teeth 1 mm. long, erect, hirtellous externally; calyx bright yellow, densely white-pilose externally near the base, glabrous within, the calyx-segms. dissimilar, the outer segms. 3 mm. long, much broadened and subrounded above a clawlike base, to 3 mm. wide, the inner segms. 2.5 mm. long, 1 mm. wide, not obviously stipitate at the base; fils. hirsutulous basally; aks. not known.—Dry serpentine ridge-top, Klamath Mts., Del Norte and Siskiyou cos., 5300 ft. July–Sept.

This new species was recently proposed by Howell & Bacigalupi in *Leafl. West. Bot.* 9: 174. 1961, and is most closely related to *Eriogonum latens* Jeps. which is known only from Inyo and Mono counties. These species, along with *E. Lobbii* and *E. pyrolifolium* have indistinctly stipitated bases, but their overall morphological similarities with the other members in the subgenus *Oligogonum* indicate that they fit here rather than elsewhere.

14. **E. pyroliifòlium** Hook. Caudex woody, stout, few-branched from a strong taproot; lvs. basal, ovate to rounded, the lf.-blades glabrous, 1.5–2 cm. long, abruptly narrowed to the equally long, villous petioles; scapes naked, 4–8 (–15) cm. long; infl. capitate or umbellate, subtended by 2 lanceolate bracts; rays simple, up to 6 mm. long; invols. campanulate, loosely woolly externally, with short, erect, teeth; calyx whitish to rose, loosely villous, 5–6 mm. long, obscurely stipitate, somewhat sparsely pubescent and glandular-puberulent at the base and along the midribs; fils. pilose basally; aks. lance-ovoid, pubescent above, ca. 5 mm. long.—Dry gravelly and sandy slopes, 8500–10,500 ft.; mostly Alpine Fell-fields; Mt. Lassen, Little Mt. Hoffmann, and Mt. Shasta; n. in the typical and in a more common form, var. *coryphaèum* T. & G. which has densely tomentose lvs. below; to Wash. and Mont. July–Aug.

15. **E. alpìnum** Engelm. Caudices rather slender, with elongate underground branches; lvs. basal, rounded, densely white-tomentose on both surfaces, the lf.-blades 1–3 cm. long; scapes 4–6 cm. long, densely white-tomentose, bracted well above the middle and each bearing a single campanulate invol. with 5–6 or more short erect teeth; calyx yellow, glabrous, 3–5 mm. long, short-stipitate; fils. somewhat pubescent at the base; aks. glabrous.—Loose slopes and ridges, 7500–9000 ft.; Subalpine F., Alpine Fell-fields; Mt. Eddy and Scott Mtn., Siskiyou Co. Aug.–Sept.

16. **E. Kellóggii** Gray. Caudex cespitose, branched and forming loose mats, the stems branched, loosely tomentose, with rosettes of lvs. at the tips; lvs. oblanceolate to spatulate, silky-tomentose beneath and often less so above, 4–10 mm. long, short-petioled; scapes slender, 4–7 cm. long, with a whorl of leafy bracts near the middle; invols. solitary, turbinate, tomentose, with short erect teeth; calyx whitish or pinkish, glabrous, 5–7 mm. long, stipitate at base; fils. pilose basally; aks. angled-conical, glabrous except for the sparsely pubescent apices, ca. 5 mm. long.—Dry ridges, ca. 4000 ft.; Yellow Pine F.; known definitely only from Red Mt., n. Mendocino Co. Possibly also on Black Butte Mts., Glenn Co.

17. **E. polypòdum** Small. [*E. umbellatum* var. *p.* S. Stokes, as to name and type. *E. ursinum* var. *Covillei* S. Stokes as to application, not *E. umbellatum* var. *C. E. ursinum* var. *venosum* S. Stokes.] Low spreading perennials from branched woody caudices, forming loose mats; lvs. ovate with rounded or cordate bases, revolute, the lf.-blades less than 1 cm. long, the petioles stout, shorter than blades; scapes erect, 5–15 cm. tall, slender; infl. capitate or few-rayed umbels; invols. turbinate, 3–4 mm. long; calyx chalky white with reddish midribs, 2.5–3 mm. long, calyx-segms. oblong; plants polygamo-dioecious; fils. pilose basally; aks. ca. 3 mm. long, with a wrinkled cellular-papillose adherent basal part which is over half their entire length and free from the seed.—Dry sandy to gravelly flats and slopes often among boulders, ca. 8000–10,500 ft.; Subalpine F.; s. Sierra Nevada, mostly in Tulare Co. July–Aug.

18. **E. marifòlium** T. & G. [*E. m.* var. *apertum* S. Stokes. *E. cupulatum* S. Stokes.]

Caudex loosely and much branched, forming mats with tomentose branchlets with terminal tufts of lvs.; lvs. ovate to oval, densely white-tomentose below, ± glabrate and green above, 5–15 mm. long, the petioles ca. as long; scapes slender, usually sparsely tomentose, 0.5–2 (–4) dm. long; infl. capitate or more frequently open umbels with subtending linear bracts. the central ray often shorter than the lateral rays, the lateral rays up to 2–3 cm. long; invols. 2–3 mm. long, sparsely pubescent externally, with short erect teeth; plants polygamo-dioecious; calyx yellowish, often tinged with red along the midribs, glabrous, the ♀ fls. 4–5 mm. long with stipes ca. 0.5 mm. long, the ♂ fls. mostly less than 3 mm. long; fils. pilose basally; aks. glabrous except sometimes at the very tip, angled-conic with narrowed bases, greenish, 3.5–4 mm. long; styles ca.' 0.4 mm. long; $2n = 32$ (Stokes & Stebbins, 1955).—Dry gravelly or sandy soil on flats or slopes, 3500–11,000 ft.; Red Fir F., Subalpine F.; Sierra Nevada mostly n. of Fresno Co. to Siskiyou Co. and cent. Ore., w. Nev. July–Aug.

19. **E. incànum** T. & G. [*E. marifolium* var. *i.* Jones. *E. rosulatum* Small. *E. ursinum* var. *r.* S. Stokes.] Caudex densely cespitose, forming branched mats 1–3 dm. across; lvs. oblong-ovate, lf.-blades 5–15 mm. long, 3–7 mm. wide, densely tomentose on both surfaces, narrowing to tomentose petioles 5–10 mm. long; scapes tomentose, 0.1–2 dm. high; infl. capitate during early anthesis, subtended by a whorl of 3–6 lanceolate, tomentose bracts, 4–5 mm. long, the infl. becoming umbellate after fertilization with a central sessile invol. and rays 2–20 mm. long especially on the ♀ plants, ♂ infl. remaining subcapitate; invols. broadly turbinate, 2.5–3 mm. long with 5–8 broadly triangular teeth; calyx pale yellow to yellow, glabrous, 2.5–3 mm. long, the ♀ fls. becoming 4–6 mm. long in fruit, polygamo-dioecious; fils. pilose basally; aks. tomentulose at the apex to nearly or quite glabrous, ca. 3 mm. long.—Gravelly and rocky slopes and ridges, 7000–12,000 ft.; Red Fir F. to Alpine Fell-fields; Tulare and Inyo cos. to Tuolumne and Alpine cos. July–Sept.

The recognition of this species which is often associated with the related *Eriogonum marifolium* comes after an intensive study of both in the herbarium and in the field. In the Tioga Pass area the two are distinct even though they can be seen growing together. The male plants of *E. marifolium* are often confused with the unrelated *E. umbellatum*, but this is often due to lack of both sexes being collected or observed.

20. **E. ursìnum** Wats. [*E. ovatum* Greene.] Caudex woody, branched, forming loose mats; lvs. tufted at ends of branchlets, ovate, the lf.-blades 8–14 (–25) mm. long, densely white-tomentose below, often glabrate above, the petiole short; scapes 2–4 dm. long, villous-tomentulose to subglabrate; infls. of compact umbels, the rays often 1–3 cm. long, the infl. subtended by narrow lfy. bracts, the rays often bearing smaller bracts; invols. woolly-villous, subcampanulate, with short broad teeth; calyx ochroleucous, glabrous, 5–6 mm. long, with a stipe 1 mm. long; fils. woolly basally; aks. subconic, subglabrous, ca. 4 mm. long, narrowed at the base; styles ca. 2.5 mm. long.—Dry open gravelly places, 3500–8000 ft.; Yellow Pine F., Red Fir F.; Sierra Nevada from Placer Co. to Butte and Shasta cos. May–Aug.

Var. **nervulòsum** S. Stokes. Plants rhizomatous; scapes erect, 4–6 cm. long; infl. congested, subcapitate.—Stony places; Red Fir F.; Snow Mt., Lake Co. Aug.–Sept.

21. **E. ternàtum** Howell. [*E. ursinum* var. *confine* S. Stokes.] Caudex woody, much-branched, forming mats; lvs. in terminal tufts on the branchlets, obovate to oblong, glabrate to tomentose above, densely tomentose below, the lf.-blades 10–15 mm. long, the petioles ca. as long; scapes 1–3 dm. high, bracted at the summit and also on the rays, the rays 1–2 cm. long; invols. turbinate, 5–8 mm. long, woolly-tomentose externally, the teeth ca. 2 mm. long; calyx sulphur-yellow, glabrous, 3–5 mm. long, stipitate at the base; fils. woolly basally; aks. pilose at the apex; styles 3 mm. long.—Rocky places, 2000–6000 ft.; Yellow Pine F., Red Fir F.; Del Norte and Siskiyou cos.; sw. Ore. June–Aug.

Var. **Congdònii** (S. Stokes) J. T. Howell [*E. ursinum* var. *C.* S. Stokes.] Lvs. revolute, narrowly elliptic or ± oblong; branches of infl. ebracteolate.—At 5000–7000 ft.; Red Fir F.; Trinity and Siskiyou cos. July–Aug.

Subgenus **Ganýsma** (S. Wats.) Greene.

22. **E. angulòsum** Benth. Annuals, erect, with spreading dichotomous ± angled stems 1–4 (–9) dm. long, whitish-tomentose to glabrate; basal lvs. oblanceolate to oblance-

oblong, the lf.-blades 1–3 cm. long, short-petioled, tomentose below, glabrate above, revolute and crisped on the edges; stem-lvs. well distributed, sessile, 0.5–2 cm. long, lanceolate to oblanceolate; peduncles arising from most axils, slender, 1–2 cm. long, glabrous or sparsely tomentose; invols. open-turbinate, 1.5–2.5 mm. long, ± puberulent, with broad rounded lobes; calyx rose, tipped with white, ca. 1.5 mm. long, minutely glandular-puberulent, the outer calyx-segms. obovate to elliptical, deeply concave, sometimes with an inflated area near the base, the inner segms. narrowly spatulate, longer; stamens 2–3 mm. long, usually well exserted; aks. ca. 1 mm. long, grayish, the body subovoid with a sharply angled triangular beak.—Dry open places, mostly below 2500 ft.; V. Grassland, Foothill Wd., Joshua Tree Wd., Pinyon-Juniper Wd.; S. Coast Ranges from Contra Costa Co. to San Diego Co. (rare in the s.), San Joaquin V. and foothills of the Sierra Nevada to w. Mojave Desert. May–Nov.

23. **E. maculàtum** Heller. [*E. angulosum* vars. *rectipes, pauciflorum, flabellatum,* and *patens* Gand. *E. a.* var. *m.* Jeps. *E. a.* ssp. *m.* S. Stokes.] Annuals with 1–several branches from the caudices, 1–2 (–3) dm. high, tomentose almost throughout; basal lvs. lanceolate to obovate, 1–3 (–4) cm. long, 1–1.5 (–2) cm. wide, narrowing to short petioles, tomentose below, glabrate to sparsely pubescent above, occasionally revolute and crisped on the margins; cauline lvs. sessile, 0.5–2 cm. long, 3–10 mm. wide, lanceolate to oblanceolate, tomentose below, glabrate above, becoming reduced above, subtended by scalelike bracts, 1–3 mm. long; branches widely spreading to give an open subglobose crown; peduncles filiform, (5–) 10–30 mm. long, arising axillary, often glandular-puberulent; invols. campanulate, 1–1.5 (–2) mm. long, 1.5–3 (–3.5) mm. wide, glandular-puberulent externally, glabrous to woolly within, with 5 rounded lobes; calyx white to yellow, pink or red with conspicuous rose-purple spots on the outer inflated segms., 1–2.5 mm. long, glandular-puberulent, calyx-segms. dissimilar, the outer segms. elliptic to roundish or obovate, with an inflated area at the base and middle with the sides of the blades incurved below, the inner segms. obtuse to lanceolate, longer; stamens 1–1.5 (–2) mm. long; aks. 1–1.5 mm. long, grayish, the ovoid bases tapering to long, 3-angled beaks; *n* = 20 (Reveal, 1965).—Dry, often sandy or gravelly places, below 7000 ft.; Creosote Bush Scrub, Joshua Tree Wd., Sagebrush Scrub, Pinyon-Juniper Wd.; deserts, e. Wash. to San Diego Co.; to Utah, Ariz., L. Calif. April–Nov.

24. **E. viridéscens** Heller. [*E. angulosum* var. *v.* Jones. *E. bidentatum* Jeps. *E. a.* sspp. *v.* and *b.* S. Stokes.] Annuals with 1–several stems from the caudex, 1–2 (–3) dm. high, tomentose almost throughout; basal lvs. lanceolate to obovate, 2–4 cm. long, 1.5–2 cm. wide, narrowing to petioles 1–2 (–3) cm. long, tomentose below, glabrate to sparsely pubescent above; cauline lvs. sessile, 0.5–2 cm. long, 3–10 mm. wide, similar to the basal lvs. only more reduced above; branches widely spreading to an open subglobose crown; peduncles filiform, 1–2 cm. long, arising axillary, sparsely finely pubescent with capitate hairs; invols. campanulate, 2–3 mm. long, 2–4 mm. wide, finely pubescent externally, tomentose internally, with 5 rounded lobes; calyx white, pink, or red, often with reddish spots on the outer segms., 1–2.5 mm. long, calyx-segms. dissimilar, the outer segms. broadly expanded above the middle of the segms., obovate or spatulate, the apices truncate, the inner segms. acute to acuminate, longer; stamens included; aks. 1–1.5 mm. long, grayish to brownish, ovoid with long 3-angled beaks.—Dry plains and hills about upper San Joaquin V. from inner Monterey Co. and Merced Co. to w. Mojave Desert. May–Oct.

25. **E. gracíllimum** Wats. [*E. angulosum* var. *g.* Jones. *E. a.* ssp. *g.* S. Stokes. *E. variabile* Heller. *E. a.* var. *v.* Parish. *E. a.* var. *victorense* Jones. *E. a.* ssp. *victorense* S. Stokes.] Annuals, freely branched at or just above the base, 1–4 dm. high, thinly woolly nearly throughout; basal lvs. oblong to oblanceolate, 2–4 cm. long, short-petioled, densely tomentose below, less so above, with revolute, crisped edges; cauline lvs. lance-oblong, well distributed, similar to the basal lvs.; peduncles filiform, 8–25 mm. long, glabrous; invols. subcampanulate, angled, 2 mm. long, glandular-puberulent, shallowly 5-lobed; calyx rose, tipped with white, 2–2.5 mm. long, the segms. similar, oblong to elliptic, frequently crenulate; aks. shining black, ca. 1 mm. long, shortly beaked; *n* = 20 (Reveal, 1968).— Common on sandy plains, below 3500 ft.; Grassland, Foothill Wd., Joshua Tree Wd., Coastal Sage Scrub; inner S. Coast Ranges from Merced and Monterey cos. to w. Mojave Desert and interior s. Calif. April–Sept.

26. **E. gossýpinum** Curran. Diffusely dichotomous, erect, slender-stemmed annuals, 0.5–2 dm. high, tomentose; basal lvs. broadly oblanceolate, 1.5–4 cm. long; stem lvs. lanceolate, somewhat smaller; peduncles 2–15 mm. long; invols. turbinate, 3 mm. long, deeply lobed and filled with dense cottony tomentum; calyx white, 1.5 mm. long, concealed in the tomentum of the invol. so as to appear pubescent, the calyx-segms. similar, linear-oblong; aks. 1 mm. long, brownish; $n = 20$ (Reveal, 1967).—Uncommon, sandy places, below 3000 ft.; V. Grassland; about the head of the San Joaquin V., Kings and Kern cos. to se. San Luis Obispo Co. April–Sept.

This group of five species which centers around *Eriogonum angulosum* has been treated in several ways. All except *E. gossypinum* have been reduced to an infraspecific rank under *E. angulosum* using a variety of names. Several years ago, J. T. Howell (1944) wrote an excellent review of these species which is followed here with only minor changes and additions in the ranges and chromosome numbers of some of the species.

27. **E. spergulìnum** Gray. [*Oxytheca s.* Greene.] Erect, slender-stemmed annuals, forking freely above with widely spreading branches, 1–5 dm. high, the internodes generally with capitate glands; basal lvs. linear, 2–3 cm. long, hispid, short-petioled; cauline lvs. linear, whorled, sessile, becoming bracteate above; peduncles filiform, 5–12 mm. long; invols. solitary, glabrous, 0.5–1 mm. long, deeply 4-lobed; calyx white with rose midribs, glabrous or sparsely pubescent, 2.5–3.5 mm. long, the segms. oblong; stamens usually exserted, the anthers 0.35–0.5 mm. long, oblong to elliptic; aks. brownish, ca. 1.5 mm. long, lanceolate in outline, narrowed gradually into narrow beaks.—Dry gravelly flats and gentle slopes, 5000–10,000 ft.; Montane Coniferous F.; Sierra Nevada from Eldorado Co. to Tulare Co. June–Aug.

Var. **Reddingiànum** (Jones) J. T. Howell. [*Oxytheca R.* Jones.] Internodes generally stipitate-glandular; calyx 1.5–2.5 mm. long; anthers 0.2–0.33 mm. long, roundish.—More common, 4000–11,000 ft.; Montane Coniferous F.; N. Coast Ranges from Lake Co. n., Sierra Nevada s. to Mt. Pinos, Ventura Co.; n. to Ore., Ida., Nev. June–Sept.

Var. **praténse** (S. Stokes) J. T. Howell. [*E. p.* S. Stokes.] Internodes usually glandless; calyx ± hirsutulous, ca. 2 mm. long.—Similar places, 8000–11,500 ft.; Sierra Nevada in Tulare and Inyo cos. July–Aug.

28. **E. hirtiflòrum** Gray ex Wats. [*Oxytheca h.* Greene.] Annuals, 5–15 cm. high, repeatedly dichotomously branched, glandular-puberulent; lvs. basal and at the lower nodes, obovate to spatulate, 1–2.5 cm. long, ciliate, narrowed into winged petioles; invols. sessile in the forks and along the branches, or short-peduncled, narrow, 2-fld., ca. 1 mm. long; calyx reddish, ca. 1 mm. long, hirsutulous with hooked hairs, the segms. oblong; aks. narrow, ca. 1 mm. long, exceeding the calyx, with broad obtuse angled beaks.—Occasional, dry gravelly places, below 6000 ft.; Chaparral, Foothill Wd., Yellow Pine F.; N. Tujunga Creek, Los Angeles Co.; Council Rock, Ventura Co.; The Pinnacles, San Benito Co.; Greenhorn Range, Kern Co.; N. Coast Ranges and Sierra Nevada. June–Oct.

29. **E. inérme** (Wats.) Jeps. [*Oxytheca i.* Wats. *E. vagans* Wats.] Annuals, dichotomously or trichotomously forked just above or at the base, then repeatedly dichotomous, 5–30 cm. high, sparingly stipitate-glandular; lvs. basal, spatulate, 1–2 cm. long, sessile, ciliate, otherwise glabrous; bracts 3-lobed; invols. on pedicellike peduncles, 4-lobed nearly to the base, subglabrous, 1.5 mm. long; calyx rose, 1.5 mm. long, hispid with hooked hairs, the segms. oblong, the inner segms. retuse and smaller than the outer; aks. scarcely if at all longer than the calyx.—Uncommon, dry barren soils or along moist edges of meadows, below 7000 ft.; Foothill Wd., Chaparral, Yellow Pine F.; Coast Ranges from Lake Co. to Kern Co. June–July.

Var. hispídulum Goodm. Invols. hispidulous.—From 3000–6000 ft.; Yellow Pine F.; San Bernardino Mts., Greenhorn Range; Sierra Nevada from Tulare Co. to Tuolumne Co. June–Aug.

30. **E. apiculàtum** Wats. [*E. a.* var. *subvirgatum* S. Stokes.] Erect annuals, usually simple at the base, dichotomously or trichotomously branched above, spreading, 2–9 dm. high, somewhat glandular-pubescent in lower portions of the internodes and peduncles, with ultimately very slender branchlets; lvs. strictly basal, oblanceolate to obovate, the lf.-blades 1.5–4 cm. long, pilose, glandular, the petioles ca. as long; bracts 1–2 mm. long; peduncles in forks and scattered along the branchlets, filiform, 2–35 mm. long, often de-

flexed; invols. 1.5 mm. long, glabrous, 4-lobed about half their length; calyx white, 1.5–2 mm. long, puberulent, the segms. oblong-obovate, notched to apiculate; aks. ca. 1.5 mm. long, not distinctly beaked.—Dry open places in disintegrated granite, 3600–8000 ft.; Joshua Tree Wd., Pinyon-Juniper Wd., Yellow Pine F.; Joshua Tree National Monument, San Jacinto, Santa Rosa, Palomar, and Cuyamaca mts. July–Aug.

31. **E. Parishii** Wats. Annuals with 1–3 erect stems 1–3 dm. high, diffusely branched so as to form a dense rounded mass of very slender ultimate branchlets, glaucous, glabrous except for short-stipitate glands above the nodes; lvs. basal, spatulate, 2–6 cm. long, hirsute; peduncles capillary but rigid, 4–12 mm. long; invols. solitary, 5-lobed, ca. 0.6 mm. long; calyx pinkish, 1–2 per invol. minutely puberulent, ca. 0.6 mm. long, the outer segms. ovate, the inner segms. oblong-spatulate; aks. ca. 1 mm. long, dark brown, the subglobose base tapering to a stout, somewhat angled, beak; $2n = 40$ (Stokes & Stebbins, 1955), $n = 20$ (Reveal, 1967).—Dry gravelly places, 4000–9000 ft.; Pinyon-Juniper Wd., Yellow Pine F., Lodgepole F.; s. Sierra Nevada, Tulare and Inyo cos., south through the San Gabriel Mts. to n. L. Calif. July–Sept.

32. **E. Órdii** Wats. [*E. tenuissimum* Eastw.] Annuals, diffusely paniculate with many capillary ultimate branches, floccose near the base, glabrous above, 4–7 dm. high; basal lvs. oblong-obovate to oblong-oblanceolate, the lf.-blades 2–8 cm. long, on equally long petioles, floccose-tomentose to ± woolly, especially below; bracts of the lower nodes often in foliaceous whorls and similar to the basal lvs., the upper bracts reduced, subulate; peduncles capillary, 1–2 cm. long; invols. solitary, turbinate, 1 mm. long, 4-toothed; calyx white, tinged with pink or pale yellowish, 1.5 mm. long, densely pubescent, 1–3 fls. per invol., the calyx-segms. oblong-ovate; aks. shining, olive-green, ca. 1.5 mm. long, ovoid, 3-angled on the stout beak.—Dry disturbed and barren places, below 3000 ft.; Foothill Wd.; inner Coast Ranges of Monterey and San Benito cos., n. base of Tehachapi Mts. and near Oildale, Kern Co.; also Creosote Bush Scrub and Pinyon-Juniper Wd., in n. Los Angeles Co., and at Split Mt., Colo. Desert; w. Ariz. March–June.

33. **E. cárneum** (J. T. Howell) Reveal, stat. nov. [*E. glandulosum* var. *c.* J. T. Howell, *Leafl. West. Bot.* 8: 38. 1956.] Annuals with one, rarely more, stems from the caudex, erect, 1–2.5 dm. high, glandular with small tack-shaped glands, numerous on the lower nodes becoming only somewhat less numerous above; lvs. basal, broadly elliptic, 6–12 mm. long, 5–10 mm. wide, pilose-hirsutulous and slightly glandular, on petioles 3–12 mm. long, stiffly hispid; bracts sparsely hirsutulous within and at the acute apices in some, glandular along the upper margins, 1 mm. long; branches numerous, spreading, trichotomous at the first node with 1–several branchlets at this node, dichotomous above with ever gradually shortening branches above; peduncles slender, one or two at each node, straight, deflexed or nearly so, 2–3 mm. long below, gradually becoming subsessile above, glandular nearly the entire length; invols, narrowly turbinate, 0.8–1.2 mm. long, glabrous, 5-lobed, few-fld.; calyx white, maturing pinkish, with red midribs, 1–1.8 mm. long, densely pilose externally with numerous long thin white hairs, the calyx-segms. narrowly lanceolate, 0.3–0.4 mm. wide; stamens included, the fils. white or red, the anthers roundish, red; aks. black, shining, 1–1.3 mm. long, the large globose base abruptly tapering to a short beak less than 0.4 mm. long.—Dry sandy soil, 3500–4500 ft.; Sagebrush Scrub, Pinyon-Juniper Wd.; extreme e. Inyo and ne. San Bernardino cos.; sw. Nev. June–Aug.

Eriogonum glandulosum (Nutt.) Nutt. ex Benth. is an exceedingly rare species known only from seven collections. The type of *E. glandulosum* was collected by Gambel along the Old Spanish Trail in southwestern Utah in 1841. Subsequent collections have essentially come from Lincoln Co. north to Elko Co., Nevada, and the adjacent counties in Utah. When J. T. Howell named the var. *carneum* he based the description on a small series of specimens. Recently, a relatively large collection from near Mercury, Nevada (*Beatley 4997*) closed the geographical distance between the two varieties yet strengthened the morphological differences between them.

The inflorescence of *Eriogonum glandulosum* is somewhat ascending with few branches. The peduncles, which are usually glandular about half their 5–15 mm. long length, are ascending or curving upwards in a fashion similar to *E. collinum*. The involucres are broadly turbinate and 1.5–2 mm. long. The flowers, which are yellow, are wider in fruit than those of *E. carneum*. The flower pubescence of *E. glandulosum* is composed of thick white stiff hairs which are less numerous and shorter than in *E. carneum*.

Eriogonum carneum differs in several ways. The inflorescence tends to be more flat-topped, and with the numerous branches, it is more dense. The peduncles are straight, deflexed, and less than 4 mm. long. The glands cover nearly the entire length of the peduncles. The narrowly turbinate involucres are 0.8–1.2 mm. long. The white flowers are exceedingly narrow in fruit, and when compared with *E. glandulosum*, the pilose hairs are more dense, longer and slender, and somewhat less stiff.

When taking into consideration the above morphological differences and the fact that even in Nevada the two remain perfectly distinct, it seems best to elevate the western population to the specific rank.

34. **E. tríchopes** Torr. [*E. trichopodum* Torr. ex Benth. *E. clavatum* Small. *E. t.* ssp. *c.* S. Stokes. *E. cordatum* Torr. ?, nom. dubium. *E. t.* ssp. *cordatum* S. Stokes.] Annuals with 1–several stems from the base, trichotomous at the first node and dichotomous or trichotomous at the upper nodes, with few to many whorled branchlets at each node especially on the lower nodes, glabrous and glaucous nearly throughout, infrequently inflated at the lower internodes; lvs. in a basal rosette, round to somewhat round-oblong, ± cordate at the base, hirsute, often somewhat crinkled, the lf.-blades 1–2 cm. long, on petioles as long to twice as long; peduncles capillary, 8–15 mm. long; invols. turbinate, scarcely 1 mm. long, glabrous, 2–few-fld., essentially 4-lobed; calyx yellow to green, 1 mm. long in anthesis, 1.5 mm. long in fr., white-strigulose, the segms. ovate; aks. shining, brown, 1.5 mm. long, narrow-ovoid, 3-angled with stout beaks; $n = 16$ (Reveal, 1965).—Common in washes and on mesas, below 5500 ft.; mostly Creosote Bush Scrub, Joshua Tree Wd.; Mojave and Colo. deserts, inner S. Coast Ranges; to Utah and New Mex., Son. and L. Calif. April–Aug.

The oldest name for this species is very likely *E. cordatum*, but as early as 1870, Torrey & Gray reported that the specimens had been lost, and attempts to find any collections in both American and European herbaria have failed. At present, as the description is not adequate enough for definite identification, it seems best to consider the name to be a nomen dubium. One name, long associated with *E. trichopes*, is *E. trichopes* ssp. *minor* (Benth.) S. Stokes, based on a series of specimens collected in New Mexico. These plants do not differ from typical *E. trichopes*, and the name should be considered a synonym.

35. **E. inflàtum** Torr. & Frém. DESERT TRUMPET. Perennials, but flowering the first year, or in the northern part of its overall range, a strict annual, 2–10 dm. high, glabrous and glaucous nearly throughout or somewhat hirsute at the very base, with 1–several stems which are simple below and dichotomous above to form diffuse panicles, the stems usually conspicuously inflated in the upper portion of the lower internodes; lvs. oblong-ovate to rounded or even subreniform, usually cordate or truncate at the base, 1–2.5 (–3) cm. long, short-hirsute, green, somewhat crisped, on petioles 2–5 cm. long; peduncles capillary, in forks and racemosely along the branchlets, 5–20 mm. long; invols. glabrous, 1–1.5 mm. long, turbinate, 5-lobed, the lobes occasionally with stipitate glands, several-fld.; calyx yellow often with red-brown midribs, conspicuously and densely pubescent, 2–2.5 mm. long, the segms. lance-ovate; aks. brown, ca. 2 mm. long, sharply 3-angled, lance-ovoid, narrowed gradually toward the apices; $n = 16$ (Stone & Raven, 1958; Reveal, 1965).—Common in washes and along mesas, below 6000 ft.; Creosote Bush Scrub, Joshua Tree Wd., Sagebrush Scrub, Pinyon-Juniper Wd.; Mojave and Colo. deserts, n. to Mono Co.; to Colo., Ariz., and L. Calif. March–Oct.

Var. **deflàtum** Jtn. [*E. glaucum* Small.] Stems not inflated, plants up to 15 dm. high; $n = 16$ (Reveal, 1967). Largely on the Colo. Desert, also in Death V.; L. Calif.

Eriogonum inflatum is often easily confused with *E. trichopes*, yet this confusion is really not necessary. While both species can be annuals, the lower nodes of *E. trichopes* often bear several small branchlets arranged in whorls and the inflorescences often have more numerous branches so that the plants appear filmy when viewed from a distance. As to *E. inflatum*, its branches are usually stouter and fewer in number, and the lower nodes never have whorls of branchlets. The perennial phase of *E. inflatum* is distinct enough, however, it is the first-year flowering plants and the strict annual plants that present problems. Throughout much of eastern Utah and adjacent western Colorado only the strict annual phase is found, although in the southern parts of these states the perennial phase may occur. The biology of the two phases is rather distinct, as is the morphology to some degree, but how to consistently separate the first-year annuals from the strict annuals has not been determined except by means of their respective geographical distribution. At present, therefore, no new taxa are proposed, and only the problem is presented.

36. **E. esmeraldénse** Wats. Glabrous annuals, 1- to few-branched, then repeatedly branched, 1–3 dm. high, the ultimate branches very slender; lvs. basal, somewhat round-obovate, 6–15 mm. long, pilose-hispid, on petioles ca. as long; peduncles filiform, 5–15 mm. long; invols. narrow-turbinate, 1 mm. long, 5-lobed, few-fld.; calyx white to pink, glabrous, the segms. ± oblong, obtuse or retuse; aks. narrow-ovoid, brown, shining, ca.

1 mm. long, gradually narrowed to a stout beak.—Dry gravelly places, 6000–9800 ft.; Pinyon-Juniper Wd.; foothills and low mts. of Inyo and Mono cos., w. Nev. July–Sept.

37. **E. Thomàsii** Torr. [*E. minutiflorum* Wats.] Annuals, 1–several stemmed at the base, 1–2.5 dm. high, glabrous and glaucous nearly throughout, repeatedly trichotomous; lvs. basal, round to round-reniform, 1–2 cm. wide, often glabrate above, densely white-woolly below, on petioles 2–5 cm. long; peduncles filiform, 5–15 mm. long; invols. ca. 1 mm. long, deeply 5-lobed, several-fld.; calyx at first yellow, later white to rose, ca. 1 mm. long, hispidulous externally, the outer segms. ovate, with a saclike dilation on each side of the cordate base when mature, the inner segms. spatulate; aks. ca. 1 mm. long, dark brown, shining, round-ovoid with 3-angled beaks; *n* = 20 (Raven et al., 1965; Reveal, 1967).— Common in dry sandy places, below 5000 ft.; Creosote Bush Scrub, Joshua Tree Wd.; Mojave and Colo. deserts n. to Inyo Co.; to sw. Utah, Ariz., L. Calif. March–June.

38. **E. Thúrberi** Torr. [*E. cernuum* ssp. *T.* S. Stokes. *E. c.* ssp. *viscosum* S. Stokes. *E. T.* var. *Parishii* Gand.] Annuals, simple or several-stemmed from the base, diffusely and trichotomously branched, 1–3 dm. high, floccose at least in the lower parts; lvs. basal, oblong-ovate, 1–3 cm. long, densely white-woolly below, glabrate above, on petioles 1–3 cm. long; peduncles capillary, 5–25 mm. long, ± glandular-puberulent; invols. broadly turbinate, glandular-puberulent, 2 mm. long, 5-lobed to near the middle; calyx rose to whitish, 1–1.5 mm. long, glandular-puberulent near the base, the outer segms. roundish or broadly ovate, abruptly narrowed to a clawlike base, with a tuft of white-cottony tomentum within, the inner segms. narrowly lanceolate; aks. almost black, shining, 0.6– 0.8 mm. long, round-ovoid with a short sharp beak.—Sandy places below 5000 ft.; Coastal Sage Scrub, Chaparral; Los Angeles region to San Diego Co.; Creosote Bush Scrub, Joshua Tree Wd., Colo. Desert and occasional on Mojave Desert; Ariz., L. Calif. April–July.

39. **S. pusíllum** T. & G. [*E. reniforme* ssp. *p.* S. Stokes. *E. comosum* var. *playanum* Jones. *E. r.* var. *asarifolium* Gand.] Annuals, erect, simple or branched at the base, trichotomously branched above, with glabrous and glaucous stems 1–3 dm. high; lvs. basal, the lf.-blades rounded to oblong-ovate, 1–3 cm. long, densely white-woolly on both surfaces, somewhat greenish above in some, on petioles ca. as long; peduncles very slender, 1–3 cm. long; invols. broadly turbinate to campanulate, glandular-puberulent, 1.5 mm. long, with 5 broad rounded lobes, several-fld.; calyx 1.5 mm. long, glandular puberulent, yellow, later with reddish midribs, the outer segms. obovate, the inner segms. oblong, both elongating to ca. 3 mm. in fr.; aks. dark, shining, 0.6–0.8 mm. long, round-ovoid, with short stout beaks; *n* = 16 (Reveal, 1965).—Common on plains and mesas, mostly 2500–6500 ft.; Creosote Bush Scrub, Joshua Tree Wd., Pinyon-Juniper Wd.; deserts from Mono Co. s. to Palm Springs region; Nev., Utah. March–July.

40. **E. renifórme** Torr. & Frém. [*E. r.* var. *comosum* Jones. *E. c.* Jones.] Annuals with 1 to several stems, ± floccose below, glabrous above, divergently trichotomously branched, 0.5–2.5 dm. high; lvs. basal, round-reniform or rounded, mostly 1–2 cm. wide, mostly white-woolly on both surfaces, or nearly glabrous above, with somewhat crisped margins; peduncles 4–15 mm. long, capillary; invols. glabrous, subcampanulate, almost 2 mm. long, shallowly and broadly 5-lobed, several fld.; calyx pale yellow, 1–1.5 mm. long, glandular-puberulent, the outer segms. elliptic-ovate, the inner segms. narrower; aks. brown, shining, round-lenticular, scarcely beaked, ca. 1 mm. long; *n* = 16 (Reveal, 1965).—Common in sandy places below 4500 ft.; Creosote Bush Scrub, Joshua Tree Wd.; Mojave and Colo. deserts, Inyo Co., s. Nev., L. Calif. March–June.

41. **E. cérnuum** Nutt. [*E. c.* var. *tenue* T. & G.] Annuals, glabrous and glaucous, diffusely branched from or above the base, 1–3 dm. high, less than 5 cm. high at high elevs.; lvs. basal, rounded, (0.5–) 1–2 cm. long, densely white-woolly below, subglabrate above, the petioles up to 4 cm. long; peduncles capillary, deflexed, straight or curved, 5–25 mm. long; invols. turbinate, 1.5–2 mm. long, 5-lobed; calyx white, often becoming rose tinged, glabrous, 1–1.5 mm. long, attenuate at the base, the segms. oblong-obovate, panduriform, undulate and often emarginate; aks. slender, ca. 1.5 mm. long.—At 7000–10,000 ft.; Panamint and White mts., e. slope of Sierra Nevada; to Ore., Rocky Mts. and Great Plains. June–Sept.

Var. **viminàle** (S. Stokes) Reveal, stat. nov. [*E. c.* ssp. *v.* S. Stokes, *Gen. Eriog.* 41. 1936.] Invols. sessile or peduncled less than 2 mm.—Lassen Co., Great Basin of Nev. and w. Utah.

42. **E. Hoffmánnii** S. Stokes. Annuals; scapes usually one, rarely more, 1–5 dm. high; lvs. basal, 1–4 cm. long, 2–4 cm. wide, suborbicular to subcordate, densely white-tomentose below, less so to glabrous and green above; stems less than 5 cm. long, glabrous; with spreading, glabrous branches; peduncles lacking; invols. erect, turbinate, 1–2 mm. long, 5-lobed; calyx 1.5 mm. long, white with greenish midribs, glabrous, the segms. spatulate; aks. brown, 2 mm. long, the globose bases tapering to roughened 3-angled beaks; *n* = 20 (Reveal, 1968).—Dry talus slopes, 4000–5000 ft.; Pinyon-Juniper Wd.; Wild Rose Spring and Townes Pass, Panamint Mts., Inyo Co. July–Sept.

Var. **robústius** S. Stokes. Plants up to 10 dm. high with stems up to 4 dm. long; lvs. basal or sheathing up the stems 1–3 (–5) cm., the lvs. 2–5 cm. long, 3–8 cm. wide, crisped; infl. erect and strict, with long whiplike branches, the secondary branches at right angles to the main branches; peduncles, when present, less than 1 mm. long, erect; calyx white or reddish with reddish midribs, the segms. ovate; aks. not known.—Dry sandy washes, 1000–2000 ft.; Creosote Bush Scrub; Black and Funeral mts., Death V., Inyo Co. Aug.–Nov.

43. **E. defléxum** Torr. Annuals with 1–several stems from the base, glabrous and glaucous nearly throughout, up to 7 dm. high; lvs. basal, cordate, reniform to nearly orbicular, 1–4 cm. long, 2–5 cm. wide, densely white-tomentose below, less so to subglabrous and green above; infl. mostly spreading, open to diffuse; peduncles deflexed, up to 3 mm. long, usually sessile; invols. 1.5–2 mm. long, turbinate, 5-lobed; calyx white to pinkish, 1–2 mm. long, the outer segms. ovate to ovate-elliptic, or oblong, usually cordate at the bases, the inner segms. lanceolate to ovate; aks. reddish-brown to dark brown, 2–3 mm. long, the subglobose bases tapering to stout 3-angled beaks; *n* = 20 (Reveal, 1964; 1968). —Common in sandy to gravelly washes and slopes, mostly below 6000 ft.; Creosote Bush Scrub to Pinyon-Juniper Wd.; Mojave and Colo. deserts, Inyo Co. to San Diego and Imperial cos.; to Utah, Ariz., L. Calif. May–Oct.

Var. **barátum** (Elmer) Reveal. [*E. b.* Elmer. *E. Watsonii* of all Calif. references, not *E. W.* Torr. & Gray.] Plants up to 10 dm. high, slender or stout, stems and branches often inflated, forming erect, strict crowns, the branches few, often elongated and whiplike; peduncles mostly 5–15 mm. long; invols. 2.5–3 mm. long, narrowly turbinate; calyx 2–2.5 mm. long; *n* = 20 (Reveal, 1968).—Common on talus, gravelly slopes of mountains and ridges or passes up to 9500 ft.; Montane Coniferous F.; s. Mono Co. and Inyo Co. s. to Ventura and Los Angeles cos.; w. Nev. June–Oct.

44. **E. Rixfórdii** S. Stokes. [*E. deflexum* ssp. *R.* Munz.] Annuals with several stems 2–4 dm. high; lvs. basal, 1–3 cm. long and wide, cordate to orbicular, margins often crispate, densely white-tomentose below, less so to subglabrous and green above; stems erect, the central main stem stout and 3–20 cm. long, dividing repeatedly at the first node, the outer stems slender and 4–12 cm. long, glabrous; branches widely spreading to nearly horizontal in position, diffuse, the few to many branches of varying lengths, the main center branch the longest, dichotomous, with angles above the second to fourth node near to 90°, intricately and divaricately branched so as to form a nearly globose crown suggesting a pagoda by the varying layers of the secondary flat-topped branches, the internodes of the upper branches short, rarely over 5 mm. long, glabrous; invols. turbinate-campanulate, 1–1.5 mm. long, 1.5–2 mm. wide, 5-lobed, glabrous; calyx white with greenish midribs, 1.5 mm. long, glabrous, the outer segms. oblong-oval, the bases cordate and gibbous at maturity, the inner whorl narrowly lanceolate, longer than the outer segms.; aks. dark brown, 1.5 mm. long, lenticular; *n* = 20 (Reveal, 1968).—Dry sandy to gravelly soils, below 5000 ft.; Creosote Bush Scrub to Pinyon-Juniper Wd.; Panamint Mts., Death V., Inyo Co. July–Oct.

45. **E. Hoòkeri** Wats. [*E. deflexum* ssp. *H.* S. Stokes.] Annuals with only a single stem, rarely more, from the caudex, 1–4 (–6) dm. high; lvs. basal, 2–5 cm. long, 2–6 cm. wide, cordate to subreniform, margins often upward rolled on the edges, densely white felty-tomentose below, white-tomentose above; stems glabrous, erect; branches spreading, the crown subglobose to flat-topped with the outer edge lower than the center so as to give an "umbrella" appearance, glabrous; peduncles lacking; invols. deflexed, hemispheric, 1–2 mm. long, 1.5–3 mm. wide, 5-lobed; calyx yellow, becoming reddish-yellow at maturity, 1.5–2 mm. long, glabrous, the segms. dissimilar, the outer segms. orbicular, cordate or hastate at the base, the inner segms. oblong-ovate and shorter than the outer, the outer expanding in fr., becoming nearly isodiametric, 2–2.5 mm. long and

wide; aks. light brown, 2–2.5 mm. long, the globose bases tapering to 3-angled beaks; *n* = 20 (Reveal, 1968).—Rare in Pinyon-Juniper Wd., from 5000–6000 ft.; s. Mono Co. and n. Inyo Co.; e. to Colo., Wyo., and n. Ariz. July–Oct.

46. **E. brachypòdum** T. & G. [*E. Parryi* Gray. *E. deflexum* var. *b.* Munz. *E. d.* sspp. *b.* and *P. S.* Stokes.] Annuals with 1–several stems, 5–40 cm. high; lvs. basal, 2–4 cm. long, 2–5 cm. wide, orbicular to cordate, densely white-tomentose below, less so to subglabrous and green above; stems stout, erect, 2–7 cm. long, glandular; branches horizontal in a low flat-topped crown or spreading and forming a subglobose one, trichotomous at the first node, usually dichotomous above, the ultimate branches often with alternating secondary branches which too have alternating branches of varying lengths, becoming shortest in the last of the secondary branches and toward the tips of the main ones, glandular throughout; invols. peduncled up to 15 mm. long, slender to stoutish, deflexed; invols. turbinate to campanulate, 1–2.5 mm. long, 1.5–2.5 mm. wide, 5-lobed; calyx white becoming reddish at maturity, 1–2.5 mm. long, glabrous, the outer segms. ovate to oblong, the base cordate to auriculate, the inner segms. narrower and shorter; aks. brown to nearly black, 1.5–2 mm. long, the subglobose bases tapering to 3-angled beaks; *n* = 20 (Reveal, 1968).—Creosote Bush Scrub to Pinyon-Juniper Wd., below 7500 ft.; Kern and San Bernardino cos., across Inyo Co. to Nev., s. Utah, nw. Ariz. March–Oct.

47. **E. insìgne** Wats. [*E. deflexum* var. *i.* Jones. *E. d.* ssp. *i.* S. Stokes.] Annuals with usually only a single stem from the caudex, up to 10 dm. high; lvs. basal, up to 8 cm. long and wide, subcordate to orbicular, the base subcordate, densely white-tomentose below, less so to subglabrous and green above; stems erect, stout, up to 20 cm. long, glabrous; branches erect and strict, dichotomous or trichotomous, often 4–5 times taller than wide, the long whiplike branches with alternating right-angled secondary ones which are also with alternating branches, the tips of the main and secondary branches ending in a raceme of 2–8 invols., glabrous throughout; invols. erect on peduncles up to 3 mm. long; invols. turbinate, 2–2.5 (–3) mm. long, 1.5–2.5 mm. wide, 5-lobed; calyx 1.5–2 mm. long, white, glabrous, the outer segms. oblong, the base cordate, the inner segms. narrower and shorter, rarely over 1 mm. wide; aks. dark brown to grayish-black, 2–2.5 mm. long, the globose base tapering to long 3-angled beaks; *n* = 20 (Reveal, 1968).—Sandy soils mostly below 4000 ft.; Creosote Bush Scrub to Sagebrush Scrub; San Diego Co. n. to s. Inyo Co.; s. Nev., sw. Utah, nw. Ariz. May–Nov.

48. **E. eremìcola** Howell & Reveal. Annuals with one or occasionally several stems from the caudex, 8–25 cm. high; lvs. basal, 1–2.5 cm. long and wide, rounded, subcordate at the base, densely white-tomentose below, less so to glabrous above, on petioles up to 3 cm. long; stems slender, 3–10 cm. long, capitate-glandular; branches spreading so as to form a subglobose crown, ± open, trichotomous at the first node, dichotomous or trichotomous above, glandular; peduncles slender, deflexed, sessile or subsessile above, peduncled to 10 mm. below, glandular; invols. turbinate-campanulate, 1.8–2 mm. long, 1–1.5 mm. wide, 5-lobed, sparsely glandular; calyx whitish becoming reddish at maturity, 2–2.5 mm. long, glabrous, the calyx-segms. ovate-oblong, the bases obtuse to subcordate; aks. brown, 2 mm. long, with a stout 3-angled beak.—Sandy to gravelly soils from 7500 to 10,000 ft.; Pinyon-Juniper Wd., Yellow Pine F.; Telescope Peak, Panamint Mts., Inyo Co. June–Sept.

The Panamint Mountains harbour several interesting buckwheats, and *Eriogonum eremicola* is certainly one of these. Since the species was described (*Leafl. West. Bot.* 10: 174. 1965) little additional information has been obtained upon it. It is placed following those species centering around *E. deflexum* only because it does not seem to fit better anywhere else, and it approaches *E. Watsonii* closer than *E. Hoffmannii* to which this plant was associated for several years.

49. **E. collìnum** S. Stokes ex Jones. [*E. nutans* of Calif. authors, not T. & G.] Annuals with 1–several stems from the caudex, up to 7 dm. high, floccose at the base becoming glabrous nearly throughout at maturity; lvs. basal, 1–3 cm. long, 1–3.5 cm. wide, round, cordate, elliptic to obovate, the base subcordate to reniform, sparsely hirsute to densely white-tomentose below, subglabrous to glabrate or glabrous above, on petioles 1–5 cm. long; stems trichotomous at the first node, with rather open and divaricated dichotomous branches; peduncles curving or ascending upwards, mostly slender, 1–3 (–5) cm. long; invols. turbinate, (1.5–) 2–3 mm. long, 1.5–2.5 mm. wide, 5-lobed; calyx white with a yellowish cast, to pinkish-yellow, or yellow, glabrous except for the pustulose or rarely

hirsutulous basal perianth tube, the calyx-segms. ovate, lanceolate to spatulate, crispate in some; aks. brownish, 2–2.5 mm. long, fusiform; $n = 18$ (Reveal, 1966).—Dry sandy to heavy clay soils, mostly below 6000 ft.; Sagebrush Scrub, N. Juniper Wd.; Nevada Co. n. to Lassen Co.; w. Nev., s. Ida. June–Sept.

For several years this species has been confused with *Eriogonum nutans* in California and with *E. reniforme* in Nevada, even though it is related to neither. The recent article by Reveal (*Madroño* 18: 167–173. 1966) has reviewed the species and its relationships.

Subgenus **Oregònium** (S. Wats.) Greene.

50. E. mohavense Wats. [*E. delicatulum* Wats.] Erect annuals, diffusely and repeatedly dichotomously or trichotomously branched at or above the base, 1–3 dm. high, glabrous and green throughout except at the nodes and lvs., the ultimate branchlets capillary; lvs. basal, rounded or broadly oblong, closely white-woolly, 0.6–2 cm. long, on petioles up to twice as long; invols. sessile in the forks and often terminal on the branchlets, hence in subcymose infls.; glabrous except at the throat, turbinate, 1.7–2 mm. long; calyx yellow, glabrous, ca. 1 mm. long, the outer segms. oblance-oblong to subelliptic, the inner narrower; aks. dark brown to nearly black, ca. 1 mm. long, the stout beaks 3-angled and muriculate.—Dry sandy and gravelly places, mostly ca. 2000–4000 ft.; Creosote Bush Scrub, Joshua Tree Wd.; Mojave Desert from e. base of the San Bernardino Mts. to Owens V. May–Aug.

51. E. ampullàceum J. T. Howell. [*E. mohavense* ssp. *a.* S. Stokes.] Erect annuals 1–3 dm. high, glabrous and glaucous nearly throughout; lvs. basal, rotund to subcordate, 0.5–1.5 cm. long and wide, white- tomentose, on petioles 0.5–4 cm. long; infl. dichotomous or trichotomous, mostly narrow and strict, erect; invols. sessile, turbinate-campanulate, 1.5–2 mm. long and wide, 5-lobed; calyx 1 mm. long, the lobes whitish, obtuse, the perianth tube broadly campanulate; stamens as long as to slightly shorter than the calyx; calyx 1.5 mm. long in fruit; aks. 1 mm. long, with globose bases tapering to long 3-angled sharp beaks.—Dry sandy soil, 6500–7000 ft.; Sagebrush Scrub; Mono Co. July–Sept.

52. E. argillòsum J. T. Howell. Erect annuals 1–3 dm. high, with 1–several stems, glabrous to sparingly tomentose; basal lvs. oblong, 1.5–2.5 cm. long, white-woolly below, less so above, slightly revolute, on petioles as long as or longer than the blades; cauline lvs. at first node somewhat smaller, becoming still more reduced above; peduncles filiform; invols. turbinate, 2.5–3 mm. long, obscurely 5-lobed, scarious below the sinuses, glabrous; calyx white or rose with dark midribs, 1.5 mm. long, the outer segms. oblong, somewhat broader than the inner; aks. 2–2.5 mm. long.—Clay and serpentine; Foothill Wd.; inner Coast Ranges, Santa Clara Co. to San Benito and Monterey cos. March–June.

53. E. temblorénse Howell & Twisselmann. Erect annuals 1–8 dm. high, with 1–several stems, densely white-tomentose almost throughout; lvs. mostly basal or subbasal, rarely axillary, lvs. elliptical, 2–2.5 cm. long, 1–1.5 cm. wide, white-woolly-tomentose on both surfaces, on petioles about as long; lower stem lvs., when present, similar but gradually reduced upwardly; lower peduncles up to 2 cm. long, erect, becoming sessile above, axillary or racemosely arranged; invols. turbinate, 2–2.5 mm. long, 5-lobed, with narrowly scarious sinuses; calyx white with green midribs, 1.5 mm. long in anthesis, becoming reddish and 2–2.5 mm. long in fr., segms. ± similar, oblong-ovate, rarely smooth, with cells longitudinally striate-lineate, ± alveolate at the base; ovary granular, styles 0.7–1 mm. long; aks. 2–2.8 mm. long with a slender granular or papillose beak; $n = 17$ (Reveal, 1968).—Dry slopes; V. Grassland; inner Coast Ranges; se. Monterey, ne. San Luis Obispo, and w. Kern cos. July–Sept.

54. E. vestìtum J. T. Howell. Erect annuals, 1–4 dm. high, simple below, branched above, densely white-tomentose nearly throughout, leafy; basal lvs. elliptic to elliptic-oblong, 1–3 cm. long, tomentose on both surfaces, on petioles ca. as long; lower cauline lvs. similar but gradually reduced above; peduncles 1–6 cm. long or almost lacking, axillary or racemosely arranged; invols. turbinate-campanulate, 2 mm. long, 5-lobed, with narrow scarious sinuses; calyx white with red midribs, the segms. oblong-ovate, 1.5–2 mm. long, papillose; aks. 2.5 mm. long, with slender papillose beaks; $n = 17$ (Reveal,

1967).—Dry slopes, V. Grassland; inner Coast Ranges, San Benito and w. Fresno cos. May–June.

55. **E. Eastwoodiànum** J. T. Howell. [*E. truncatum* var. *adsurgens* Jeps. *E. Covilleanum* ssp. *a. Abrams. E. vimineum* ssp. *a.* S. Stokes.] Erect annuals, branched from the base, 2–5 dm. high, floccose-tomentose nearly throughout; lvs. basal, suborbicular, 1–3 cm. long and wide, woolly below, subglabrate above, on petioles 3–8 cm. long; upper branches forming a cymose infl.; invols. peduncled in the forks, sessile on branchlets or terminal, turbinate, 2 mm. long, distinctly 5-toothed; calyx white, 2 mm. long, glabrous, the outer segms. elliptic or oblong-obovate to obtuse, the inner segms. oblong, smaller; aks. 2 mm. long, brownish; $n = 17$ (Reveal, 1967). Stokes & Stebbins, 1955, reported the taxon *adsurgens* to be $2n = 22$, but as no vouchers were noted, it is impossible to check this count.—Diatomaceous shale; Foothill Wd.; mts. of w. Fresno and e. Monterey cos. June–July.

The type of var. *adsurgens* deposited in the Jepson Herbarium at the University of California is the same kind of plant later named by J. T. Howell as *Eriogonum Eastwoodianum,* and it is therefore reduced to synonymy.

The series of species, *Eriogonum Eastwoodianum, E. vestitum,* and *E. temblorense* all occur in the inner Coast Ranges and they are not easily distinguished. The seasonal variation within each species tends to approach features of other species at different times of the year, so that *E. temblorense* is similar to late season forms of *E. vestitum,* and *E. Eastwoodianum* often resembles young plants of *E. temblorense.* All three have the same chromosome number, $n = 17$, and approach the definition of the subgenus *Ganysma* in several respects. Nevertheless, these species are morphologically and cytologically closer to *E. argillosum, E. truncatum,* and *E. Covilleanum,* good members of the subgenus *Oregonium,* rather than to any species in *Ganysma.*

56. **E. truncàtum** T. & G. Erect annuals, with 1–several stems, 1–3 dm. high, floccose-tomentose nearly throughout; lvs. basal and at the lower nodes, oblong-oblanceolate to obovate, tomentose below, less so and greenish above, 2–5 cm. long, attenuate to petioles ca. as long; infl. open, dichotomously or trichotomously branched, subcymose; invols. subsessile, 1–few in the forks and at the ends of branchlets, tomentose, 2.5–3 mm. long, turbinate, shallowly and broadly 5-toothed, the sinuses almost filled with membranes; calyx light rose-colored, ca. 2 mm. long, glabrous, the outer segms. elliptic-obovate, the inner segms. slightly narrower; aks. dark brown, ca. 2 mm. long, narrow-ovoid, with broad 3-angled beaks.—Dry slopes, 1000–1500 ft.; edge of Chaparral, e. base of Mt. Diablo, Contra Costa Co. April–June.

57. **E. Covilleànum** Eastw. [*E. vimineum* var. *C.* S. Stokes.] Erect annuals, simple below or branched from the base, glabrous, the slender stems 1–4 dm. long, dichotomously or trichotomously branched above in cymose fashion; lvs. basal, suborbicular, densely white-tomentose below, subglabrous above, 0.5–1.5 cm. long, on longer slender petioles; invols. solitary, ca. 2.5 mm. long, sessile at the forks and nodes or terminal, narrowly turbinate, glabrous without, 5-veined, the margins subentire and ciliate; calyx 2–2.5 mm. long, rose or white with red midveins, the segms. elliptic, puberulent toward the base without; aks. ca. 2 mm. long, ovoid, with prominent muriculate beaks; $n = 17$ (Reveal, 1967).—Shale and serpentine talus, 1200–2000 ft.; Chaparral, Foothill Wd.; inner Coast Ranges, Alameda Co. to the Temblor Range, Kern Co. April–June.

58. **E. Nortònii** Greene. [*E. vimineum* ssp. *N.* S. Stokes.] Erect annuals, 0.5–2 dm. high, with reddish stems, glabrous nearly throughout; lvs. basal and at the lower nodes, round to reniform, deeply emarginate, densely white-tomentose below, subglabrous to glabrous and green above, 5–15 mm. long, on longer petioles; invols. in the forks and at the tips of short slender branchlets, solitary, broadly turbinate, 3–4 mm. long, with ca. 8 short blunt teeth; calyx white to rose, 1.5 mm. long, the segms. obovate, glabrous; aks. ca. 1 mm. long, ovoid, with stout 3-angled beaks.—Dry rocky slopes, 1500–4000 ft.; Chaparral; The Pinnacles, inner Coast Ranges, Monterey and San Benito cos. May–June.

59. **E. canìnum** (Greene) Munz. [*E. vimineum* var. *c.* Greene. *E. v.* var. *californicum* Gand.] Widely spreading annuals, with several glabrous and reddish stems from the base, 1.5–3.5 dm. long, repeatedly dichotomously or trichotomously branched; lvs. basal and at the lower nodes oblong-ovate, densely white-tomentose below, subglabrous to glabrous and green above, subcuneate at the base, 0.5–3 cm. long, on petioles 2–3 times longer; invols. glabrous, narrowly turbinate, 5-ribbed and -toothed, 3–4 mm. long; calyx rose-red, 1.5–2.5 mm. long, the segms. obovate, glabrous; aks. lance-ovoid, reddish, tapering gradu-

ally to the 3-angled beaks; $n = 12$ (Reveal, 1968).—Dry rocky slopes on shale and serpentine, 1000–2000 ft.; Coastal Prairie; Marin Co. and the Oakland Hills, Contra Costa Co. June–Sept.

The plant in the N. Coast Ranges and on the western slope of the Sierra Nevada, which resembles this species, seems to be unnamed. This plant is taller, more erect, and with larger involucres and flowers than found in *Eriogonum caninum.* The type of *E. pedunculatum* S. Stokes from Calaveras Co. seems to be this kind of plant, but its relationships with *E. vimineum* must be studied before any nomenclatural changes can be logically made.

60. **E. ròseum** Dur. & Hilg. [*E. virgatum* Benth. *E. v.* var. *r.* T. & G. *E. v.* var. *rubidum* Jeps. ex Bauer. *E. vimineum* ssp. *v.* S. Stokes.] Erect annuals, simple or with few ascending virgate branches, floccose-tomentose nearly throughout, 1–8 dm. high; lvs. at the base and lower nodes, the basal lvs. oblong-oblanceolate, 1–3 cm. long, with equally long petioles; invols. cylindric, 4–5 mm. long, 5-toothed, sessile and rather remote, tomentose; calyx yellow to pink or white, 2 mm. long, the outer segms. narrowly obovate, the inner oblong, glabrous; aks. almost 2 mm. long, narrowly ovoid with broad 3-angled scaberulous beaks; $2n = 18$ (Stokes & Stebbins, 1955), $n = 9$ (Reveal, 1967).—Dry, often sandy or gravelly to rocky places, below 5000 ft.; Chaparral, Foothill Wd., Yellow Pine F., N. Oak Wd., etc.; away from the immediate coast, s. Ore. to n. Ventura and Los Angeles cos., especially in the ranges bordering the Great V. June–Oct.

61. **E. grácile** Benth. [*E. vimineum* ssp. *g.* S. Stokes. *E. acetoselloides* Torr. ex Benth. *E. leucocladon* Benth. *E. roseum* var. *l.* Hoover. *E. verticillatum* Nutt.] Erect annuals, strictly or rather diffusely branched from base, 2–5 dm. high, thinly tomentose to floccose-tomentose throughout, the branchlets slender, ascending; lvs. mostly basal, oblanceolate to oblong, the lf.-blades 1–3 (–4) cm. long, tomentose, especially below, on petioles ca. as long; cauline-lvs. becoming strongly reduced above; invols. 1.8–2 (–3) mm. long, turbinate, subglabrous, the 5 teeth conspicuous, rigid; calyx white, pinkish, or yellowish, 1.5–2 mm. long, the outer segms. broadly obovate, the inner segms. oblong, glabrous; aks. ca. 1 mm. long, ovoid, with prominent 3-angled beaks; $2n = 22$ (Stokes & Stebbins, 1955).—Common in dry cismontane washes, on mesas, etc., below 3500 (5000) ft.; Coastal Sage Scrub, Chaparral, Foothill Wd., S. Oak Wd., etc.; inner Coast Ranges and Great V. from Vaca Mts. s. through s. Calif.; L. Calif. July–Oct.

Var. **citharifórme** (Wats.) Munz. [*E. c.* Wats *E. vimineum* var. *c.* S. Stokes. *E. agninum* Greene.] Lvs. practically all basal, crisped on the margins and with conspicuously winged petioles.—San Luis Obispo, Santa Barbara and Ventura cos. May–Aug.

Var. **polygonoìdes** (S. Stokes) Munz. [*E. vimineum* ssp. *p.* S. Stokes.] Stems glabrous and glaucous; lvs. elliptic to ovate-lanceolate, with narrow petioles.—Grain fields and dry slopes; San Luis Obispo to Santa Ynez, Santa Barbara Co. Sept.–Nov.

This species is exceedingly variable and needs critical study in order to understand the inner relationships of the various taxa. As the species is presented here, it is believed to include all of the forms referrable to it. The type of *E. verticillatum* is a non-flowering, exceedingly immature specimen which is believed to represent *E. gracile.*

62. **E. Davidsònii** Greene. [*E. vimineum* var. *D.* S. Stokes. *E. molestum* of authors, not Wats. *E. m.* var. *D.* Jeps. *E. v.* var. *aviculare* S. Stokes. *E. v.* var. *glabrum* S. Stokes.] Erect annuals, few-branched or usually simple at the base, glabrous and glaucous except on the lvs., 1–5 dm. high; lvs. basal, rounded to reniform, 1–2 (–4) cm. long and wide, densely white-tomentose especially below, crisped or undulate, on longer petioles; invols. cylindric-turbinate, 3–5 mm. long, glabrous, scarious between the ribs, sessile and remote, few on a branch; calyx white or tinged with pink to rose-red, 1.5–2 mm. long, the outer segms. oblong-obovate, the inner segms. slightly narrower, glabrous; fils. pilose basally; aks. brown, narrow-ovoid, shining, ca. 2 mm. long, narrowed slightly to stout muriculate, 3-angled beaks.—Occasional, dry places under pines, 3000–7000 ft.; Chaparral, Pinyon-Juniper Wd., Joshua Tree Wd., Yellow Pine F.; Monterey and Tulare cos. s. to San Diego Co.; Ariz. and L. Calif. June–Sept.

For years this plant has been known as *Eriogonum molestum.* However, with the critical examination of types in the Gray Herbarium, Harvard University, this taxon was found to represent not an annual, but the perennial *E. nudum* var. *pauciflorum* (see no. 95). When Watson proposed the species, he based *E. molestum* on a series of collections. In the Gray Herbarium are all those mentioned, Palmer, Cleveland, Parish Brothers, and Nevin. Of these, all represent *E. nudum* except the Palmer collection which is *E. Davidsonii*. Nevertheless, as the majority of the collections represent the perennial, and the entire description is based on the perennial, and in particular the Parish Brothers collection, it is here selected as the lectotype: San Jacinto Mts., San Diego Co., California, July 1881, *Parish & Parish 972* GH! Isotypes: MO! PH! US!

With the typification of *Eriogonum molestum,* the name *E. Davidsonii* must now be used. In the past, *E. Davidsonii* has been recognized as a variety of *E. molestum* based on the length of the involucres. As this separation was due to the fact that there was some confusion with the perennial *E. nudum*, no such separation will be made here.

63. E. vimíneum Dougl. ex Benth. [*E. luteolum* Greene. *E. v.* var. *l.* S. Stokes. ?*E. pedunculatum* S. Stokes.] Erect annuals, with several stems arising from the base, branched above, 1–3 dm. high, ± floccose-tomentose below, becoming glabrous above, with slender, ± greenish and glaucous stems; lvs. generally all basal, round to round-ovate, white-woolly below, glabrate above, 1–2 cm. long, on longer petioles; invols. usually sessile along the branches, narrow-cylindric, 2–3.5 (–4) mm. long, with short blunt teeth, glabrous; calyx white to rose or yellowish, 2 mm. long, the outer segms. obovate with rounded apices, the inner narrower, glabrous; fils. glabrous or pilose basally; aks. 2 mm. long, red-brown, ovoid with prominent muriculate, 3-angled beaks; $2n = 24$ (Stokes & Stebbins, 1955).—Common in dry rocky and sandy places, below 6000 ft.; Chaparral, Foothill Wd., Yellow Pine F., etc.; Coast Ranges and Sierran foothills, Monterey, Santa Clara and Mariposa cos. n. to Wash., Ida., e. to Utah and Ariz. June–Sept.

64. E. nidulàrium Cov. [*E. vimineum* ssp. *n.* S. Stokes. *E. n.* var. *luciense* Jones.] Erect annuals, repeatedly forked from near the base, 5–15 (–20) cm. high, floccose-tomentose almost throughout, forming dense masses of numerous branches with short internodes and in age the tips of the branches becoming curved inward; lvs. basal, rounded, sometimes cordate at the base, 1–2 cm. broad, on much longer petioles; invols. cylindrical-turbinate, 1 mm. long, sessile in all the forks and along the branches, few-fld.; calyx red, white, or yellow, 1.5–2 (–3) mm. long, the outer segms. obovate, dilated and truncate at the apices, fan-shaped, the inner narrower, glabrous; fils. glabrous; aks. ca. 1 mm. long, narrow-ovoid, with long scaberulous beaks.—Common in dry gravelly and rocky places, mostly below 7000 .ft.; Creosote Bush Scrub, Joshua Tree Wd., Pinyon-Juniper Wd.; deserts from Mono Co. to San Bernardino Co.; Ore., Nev., w. Utah, and Ariz. April–Oct.

65. E. Palmeriànum Reveal, nom. nov. [based on *E. Plumatella* var. *Palmeri* T. & G., Proc. Amer. Acad. 8: 180. 1870, not *E. Palmeri* Wats. *E. Baileyi* var. *tomentosum* Wats.] Densely branched and spreading annuals, 1–3 dm. high, forming broad, often flat-topped crowns up to 3 dm. across, densely tomentose nearly throughout; lvs. basal, suborbicular to cordate, 0.5–1.5 cm. long, 0.5–2 cm. wide, densely tomentose below, less so and often green above, on petioles 1.5–3 cm. long; invols. campanulate, 1–1.5 mm. long, sparsely tomentose with ciliated margins, sessile at the nodes and along the subvirgate branchlets, few-fld.; calyx white to pink with reddish-brown midribs, 1.5–2 mm. long, the outer segms. oblong to fan-shaped, somewhat constricted near the middle and flaring above, the inner segms. narrower; aks. 1.5–1.8 mm. long, the scabrous 3-angled beaks ca. 1 mm. long.—Dry sandy or gravelly flats mostly below 7000 ft.; Creosote Bush Scrub, Joshua Tree Wd., Pinyon-Juniper Wd.; e. San Bernardino Co. to n. Inyo Co.; s. Nev. and n. Ariz. to extreme sw. Colo. June–Oct.

66. E. Báileyi Wats. [*E. vimineum* ssp. *B.* S. Stokes. *E. v.* var. *B.* R. J. Davis. *E. v.* var. *multiradiatum* S. Stokes. *E. gracile* var. *effusum* T. & G. *E. restioides* Gand.] Erect annuals, diffusely branched from the base, 1–4 dm. high, forming broad round-topped crowns, glabrous except at the white-woolly base of the stems; lvs. basal, suborbicular, 5–20 mm. wide, densely white-tomentose on both surfaces, on somewhat longer petioles; invols. tubular-campanulate, 1–1.5 mm. long, glabrous except for the ciliated margins, sessile at the nodes of the subvirgate branchlets, several-fld.; calyx white to pink, ca. 1.5 mm. long, the outer segms. oblong or oblong-obovate, somewhat constricted near the middle and flaring above, the inner narrower, glabrous or glandular; aks. dark brown, ca. 1 mm. long, the globose bases gradually narrowing into stout muriculate 3-angled beaks.—Dry sandy or gravelly flats and banks, mostly 2500–7500 ft.; Creosote Bush Scrub, Joshua Tree Wd. to Yellow Pine F.; e. end of San Bernardino Mts. across the Mojave Desert along the e. side of the Sierra Nevada; e. Ore., w. Nev. May–Sept.

Var. divaricàtum (Gand.) Reveal, comb. nov. [*E. praebens* Gand. *E. p.* var. *d.* Gand., Bull. Soc. Bot. Belg. 42: 196. 1906. *E. commixtum* Greene ex Tidestr. *E. vimineum* var. *c.*

S. Stokes.] Stems sparsely floccose-tomentose; calyx glandular.—Dry sandy soil, mostly 4500–6000 ft.; Pinyon-Juniper Wd.; Carson City, Nev. n. to Placer and Nevada cos.; w. Nev.

For several years the name *Eriogonum Baileyi* var. *tomentosum* has been used as the name for the tomentose form of this species. However, in reviewing the types involved, a large number of problems arose. First, Watson did not select a type for var. *tomentosum*, and a recent investigation at the Gray Herbarium revealed only the Arizona specimen cited (a Palmer collection), but those from the other states mentioned were not found. For this reason, no lectotype is selected. The Palmer collection, however, was named first as *E. Plumatella* var. *Palmeri*. It must be recalled that to Torrey & Gray, the name *E. Plumatella* was the same plant that is now named *E. nidularium*. This plant represents what has been called in Stokes' monograph and in the Arizona Flora by Kearney & Peebles, *E. densum* Greene. Nevertheless, the type of *E. densum* from New Mexico is a grazed form of *E. polycladon* Benth. In field studies in Nevada by N. H. Holmgren and Reveal, it has been possible to study both *E. nidularium* and the plant that has been called *E. densum*, and the two are consistently distinguishable. Therefore, with this in mind, the name *E. Palmerianum* is proposed as a new name. The new name also replaces *E. Baileyi* var. *tomentosum* to a large degree and it makes it necessary to propose a name for the true tomentose form of *E. Baileyi*.

In studying the *E. Baileyi* complex, the tomentose plants have been found only in the western Nevada area from about Carson City north to north of Reno. This form just enters California in extreme eastern Placer and Nevada counties. The earliest varietal name for this plant is *E. praebens* var. *divaricatum*, and thus a new combination is made.

67. **E. brachýanthum** Cov. [*E. Baileyi* var. *b.* Jeps. *E. vimineum* var. *b.* S. Stokes.] Erect annuals, simple or diffusely branched from the base, 1–3 dm. high, forming broad round-topped crowns, glabrous except at the white-woolly base of the stems; lvs. basal, rounded to broadly ovate, 5–20 mm. wide, densely white-tomentose on both surfaces or rarely somewhat glabrate above, on petioles somewhat longer; invols. turbinate, 1 mm. long, glabrous, closely appressed to the branchlets, few-fld.; calyx yellow, 0.6–0.8 mm. long, the outer segms. oblong to oblong-obovate, the inner slightly narrower, glabrous; aks. dark brown, ca. 1 mm. long, the globose bases gradually narrowing into long exserted 3-angled beaks.—Dry sandy to gravelly slopes, mostly 2500–7500 ft.; Creosote Bush Scrub, Joshua Tree Wd., Pinyon-Juniper Wd.; Mojave Desert n. to Mono Co.; w. Nev. May–Aug.

68. **E. elegáns** Greene. [*E. vimineum* var. *e.* Jeps. *E. Baileyi* ssp. *e.* Munz.] Erect annuals, usually simple at the base, 1–4 dm. high, forming an open crown of gray to reddish branches, repeatedly dichotomously branched above, glabrous nearly throughout except for the white-woolly base of the stems; lvs. rounded to subcordate or oblong, basal, 3–15 mm. long and wide, white-woolly on both surfaces, undulate on the margins, on somewhat longer petioles; invols. turbinate, 1–1.5 mm. long, glabrous except for the ciliated margins, sessile at the nodes of the branches, few-fld.; calyx rose to pink with reddish midribs, 1–1.5 mm. long, the segms. oblong-obovate, glandular-puberulent externally; aks. brown, ca. 1 mm. long, the globose bases narrowing to 3-angled beaks.—Dry sandy flats and washes; V. Grassland, Foothill Wd.; S. Coast Ranges, Santa Clara Co. to San Luis Obispo Co. May–Aug.

69. **E. foliòsum** Wats. [*E. Baileyi* var. *tomentosum*, in part.] Spreading annuals with imperfectly dichotomously branched tomentose branches, 1–3 dm. long, these often becoming glabrate toward their tips at maturity; basal lvs. ovate to oblong, 0.5–1 cm. long, 0.3–1 cm. wide, densely white-tomentose below, less so to glabrous above, on tomentose petioles up to 2 cm. long; cauline lvs., when present, in pairs at the nodes, narrowly ovate, sessile or nearly so; invols. turbinate, 0.8–1.2 mm. long, solitary, sessile, 5-lobed; calyx white with pink midribs, becoming pink or rose at maturity, 0.8–1.2 mm. long, the outer segms. broadly hastate at the base when mature, obtuse at the apices, the margins ± crispate, the inner segms. narrower, oblong, slightly longer than the outer; fils. minutely pubescent basally; aks. bronze to brown, 1–1.2 mm. long, with long, sharp, 3-angled beaks.—Rare in sandy to gravelly places below 4000 ft.; Pinyon-Juniper Wd.; San Bernardino Mts., s. to extreme n. L. Calif. March–Aug.

This species has not been included in the California flora as it has been called *Eriogonum Baileyi* var. *tomentosum* in southern California. Although Stokes mentioned its presence in her monograph, her citation has been ignored. Her basis, no doubt, is the same as ours, *Abrams 2894* from Bear Valley, San Bernardino Mts. The type of *E. foliosum* was collected either in extreme southern San Diego Co. or adjacent northern Baja California.

Our plant is var. *foliosum* which has been somewhat confused with the var. **hastàtum** (Wiggins)

Reveal, stat. & comb. nov. (based on *Eriogonum hastatum* Wiggins, Contr. Dudley Herb. 1: 165, pl. 12, fig. 2A–B. 1933) of northern Baja California. The var. *hastatum* differs from the var. *foliosum* in having longer, spreading to decumbent branches up to 5 dm. long, basal lvs. 1–4 cm. long, invols. 2–3 mm. long which are only shallowly divided by the 5 lobes, a calyx 1.5–2.5 mm. long, and achenes 1.8–2 mm. long.

70. **E. dasyánthemum** T. & G. [*E. vimineum* var. *eriocladon* Benth. *E. d.* var. *Jepsonii* Greene.] Erect annuals, branched from the base or above, 2–6 dm. high, floccose-tomentose nearly throughout, sometimes glabrate; lvs. basal, roundish, white-woolly below, glabrate above, 1–2 cm. wide, on petioles ca. as long; lower nodes also with some lvs.; invols. subcylindric, ca. 4 mm. long, 5-ribbed and -toothed, sessile, scattered along the branchlets, tomentose between the ribs; calyx white or rose, 2 mm. long, pubescent externally, the segms. oblong-obovate; aks. 1.5–2 mm. long, with scabrellous 3-angled beaks; $2n = 24$ (Stokes & Stebbins, 1955).—Dry slopes; Foothill Wd., Chaparral, V. Grassland; inner Coast Ranges from Lake Co. to Tehama Co. Aug.–Oct.

Subgenus **Eucỳcla** (Nutt.) Kuntze.

71. **E. saxátile** Wats. [*E. Bloomeri* Parish. *E. Stokesae* Jones. *E. s.* var. *S.* Jones.] Perennial herbs with few-branched caudices clothed with the crowded closely white-felted lvs.; lvs. basal, rounded or broadly obovate, the lf.-blades 1–2.5 cm. long, on petioles ca. as long or longer; flowering stems ascending, rather slender, 1–3 (–4.5) dm. high, closely tomentose or floccose, forking above with ascending or spreading branches; invols. turbinate, 3–4 mm. long, solitary at the nodes, scattered along the branches, tomentulose, many-fld.; calyx white, pinkish, or yellowish, 5–7 mm. long, glabrous, narrowed to a sharply triangular narrow base, the outer segms. oblanceolate, the inner segms. obovate and larger; aks. brownish, glabrous, ca. 2 mm. long, ± winged and 3-angled, narrowly elliptic in outline, not beaked; $n = 20$ (Reveal, 1967).—Dry rocky slopes and ridges, mostly 4000–11,000 ft.; Joshua Tree Wd. to Subalpine F.; San Jacinto and Little San Bernardino mts., w. and n. to San Gabriel, Argus and Panamint mts., s. Sierra Nevada to Fresno Co., Santa Lucia Mts. Nev. May–July.

72. **E. crocàtum** A. Davids. [*E. saxatile* var. *c.* Munz.] Perennial subshrubs, the caudices loosely branched and clothed with old lvs.; lvs. sheathing up the stems, lf.-blades broadly ovate, 1.5–3.5 cm. long, the petioles shorter to ca. as long; flowering stems terminal, 1–2 (–3) dm. high, 1–2 forked at right angles, forming rather dense cymes 3–8 cm. across; invols. broadly campanulate, 3–4 mm. long, white-woolly; calyx sulphur-yellow, 5–6 mm. long, narrowed into a tubular stipe-like base, the outer calyx-segms. oblance-oblong, the inner segms. wider than the outer segms., glabrous; aks. brownish, lance-ovoid, ca. 3 mm. long, glabrous, somewhat 3-angled; $2n = 40$ (Stokes & Stebbins, 1955). —Rocky slopes, at ca. 500 ft.; Coastal Sage Scrub; Conejo Grade, n. base of Santa Monica Mts., Ventura Co. April–July.

73. **E. àpricum** J. T. Howell. Compact perennial herbs with short woody branching caudices covered with old persistent lvs.; lvs. basal, round-ovate, 3–5 (–10) mm. long and wide, densely and persistently tomentose below, glabrate above at maturity, the bases cordate or round, the apices obtuse or acutish, on petioles 3–10 (–25) mm. long; stems dying back each year, 8–20 cm. long, erect, slender, glabrous; infl. dichotomous or trichotomous; invols. campanulate, 2–2.5 mm. long, solitary, sessile and terminal, rarely spicate, glabrous without, sparsely pubescent within, 5-lobed; calyx white with reddish midribs, 2–3 mm. long, segms. oblong, glabrous without, sparsely long-pilose below the middle within; aks. 2.5–3 mm. long, the subglobose bases tapering to roughened 3-angled beaks; $n = 20$ (Stebbins, per. comm. to J. T. Howell and to J. L. Reveal).—Clay hills at ca. 300 ft.; Chaparral; near Ione and Buena Vista, Amador Co. June–Sept.

This unusually odd addition to the California buckwheats is not easily explained, and from our present knowledge, not understood. Howell suggested that its affinities might lie with the annual *E. vimineum*, but this now seems unlikely in view of the chromosomal differences. He also suggested that it may have originated from the same ancestral stock that gave rise to *E. Batemanii*, but this species belongs to a species complex that is almost entirely restricted to the Green-Colorado river basin of western Colorado and eastern Utah. The new California species, published in *Leafl West Bot.* (7: 237. 1955), is placed before *E. microthecum* only as a matter of convenience and does not reflect an opinion of its relationship although it is probably closer to this species than to any other group.

74. **E. microthècum** Nutt. Low bushy half-shrubs 1–4 dm. high from freely branched woody bases; lvs. linear to linear-oblanceolate, or narrowly obovate to elliptic, 1–3 cm. long, grayish white-tomentose below, glabrous and greenish above, not revolute, narrowing gradually to short petioles; flowering stems herbaceous, leafy from slightly less to considerably more than half their length, 5–20 cm. long, usually glabrous; infl. freely branching, open, cymose, usually flat-topped, 2–20 cm. long; invols. narrowly turbinate, (2–) 2.5–3 mm. long, sparsely floccose or more commonly glabrous except for the ciliated, rounded 5-lobes, solitary, mostly pedunculate becoming sessile above; calyx yellow, 2–3 mm. long, the outer segms. round-oblong to obovate, with subcordate bases, the inner narrower, elliptic, glabrous; aks. narrow, 1.5–2 mm. long.—Dry places mostly below 7000 ft.; N. Juniper Wd., Yellow Pine F.; Lassen and Modoc cos.; e. Ore., w. Ida., and n. Nev.

It should be noted in the beginning that the treatment of this species complex for California is not complete, and the only reason for this is that Reveal would prefer to publish the new varieties that have been found in a separate paper. To date, there are at least three additional new varieties for the state. One, which will be a new combination, is common in Arizona and just enters the state. Two undescribed varieties have been discovered in the San Bernardino Mts. and the Cuyamaca Mts. A possibly third undescribed variety may occur in the alpine regions of the Sierra Nevada. This entire species complex will be the subject of the sixth paper in the series of *Notes on Eriogonum* by Reveal.

The concept of the var. *microthecum* differs from that presented in California floras in the past, and actually it was not until the *Eriogonum* treatment by Hitchcock in the Pacific Northwest Flora series (1964) that its exact nature was noted. The type which was collected by Nuttall in northeastern Oregon represents that part of the species as outlined above which enters California only in Lassen and Modoc cos. Because of the close relationship of var. *microthecum* to another yellow-flowered variety, it is imperative that the following variety be proposed at this time.

Var. **ambíguum** (Jones) Reveal, comb. nov. [based on *E. aureum* Jones var. *a.* Jones, *Proc. Calif. Acad.* ser. 2, 5: 720. 1895. *E. m.* var. *expansum* S. Stokes.] Lvs. floccose above, usually not revolute; stems usually floccose-tomentose above; invols. turbinate, 2–2.5 mm. long; calyx yellow.—Dry rocky and gravelly places, Pinyon-Juniper Wd., Yellow Pine F., 5000–10,000 ft.; e. slopes of the Sierra Nevada from Inyo Co. n. to Placer Co., and White Mts.; w. Nev. July–Sept.

Var. **laxiflòrum** Hook. [*E. confertiflorum* Benth. *E. m.* var. *c.* T. & G. *E. m.* sspp. *c.* and *l.* S. Stokes. *E. m.* var. *panamintense* S. Stokes. *E. effusum* var. *limbatum* S. Stokes. *E. tenellum* var. *erianthum* Gand.] Plants (0.5–) 2–4 dm. high; lvs. usually revolute, narrow; calyx white.—Dry rocky places, mostly 5000–10,000; Inyo Co. n. to Wash.; Rocky Mts. July–Oct. The common form in California.

It should be noted that the author citation of var. *laxiflorum* differs from that usually presented, but Hooker proposed the name in 1854, while Bentham, who is usually cited, proposed the same name in 1856. A name long associated with the California flora is *E. effusum* var. *rosamarinoides* Benth. In reviewing the types of this complex, it was discovered that the var. *rosamarinoides* is an excellent variety of *E. effusum* which was collected by J. C. Frémont in Kansas, though how it came to be cited by Bentham as coming from California is unknown.

75. **E. Heermánnii** Dur. & Hilg. [*E. H.* ssp. *occidentale* S. Stokes.] Woody and branched shrubs, the stems erect, woody and floccose in lower portions, glabrous and light green above, 3–7 (–15) dm. high; lvs. lance-oblong to oblanceolate, 1–2 (–2.5) cm. long, green above, floccose below, somewhat undulate, short-petioled; infl. a cymose panicle of dichotomously branched rigid branchlets, almost or quite smooth, subterete, the lower internodes 2–4 (–6) cm. long; invols. solitary in the forks or terminal, broadly turbinate, 2 mm. long, rather deeply lobed, glabrous; calyx yellowish-white, 3–4 mm. long, the outer segms. orbicular, subcordate at the bases, the inner oblong, glabrous; aks. narrow, 3-angled, 2–2.5 mm. long.—Dry slopes and ridges, 2000–7000 ft.; Pinyon-Juniper Wd., Foothill Wd., Joshua Tree Wd.; borders of San Joaquin V. (San Benito and Kern cos.) and w. Mojave Desert (Little San Bernardino Mts. to Inyo Co.); w. Nev. July–Sept.

Var. **floccòsum** Munz. Lower internodes of infl. 2–3 cm. long, the branchlets floccose-tomentose; fls. 2–3 mm. long.—Largely in Pinyon-Juniper Wd.; mts. of e. Mojave Desert, San Bernardino Co. Aug.–Oct.

Var. **argénse** (Jones) Munz. [*E. sulcatum* var. *a.* Jones. *E. Howellii* S. Stokes.] Low subshrubs, 1–2 dm. high; stems numerous, slender and rather delicate; infl. very compact and intricate, the lower internodes 0.5–1.2 cm. long, the branchlets glabrous and scabrous under a lens; calyx ca. 2 mm. long—Dry rocky slopes and ridges, 5000–8000 ft.; Pinyon-Juniper Wd.; Inyo and Mono cos.; Nev. July–Oct.

Var. **sulcàtum** (Wats.) Munz & Reveal, comb. nov. [*E. sulcatum* Wats., *Proc. Am. Acad.* 14: 296. 1879. *E. H.* ssp. *s.* Munz.] Intricately branched rounded bushes 1.5–4 dm. high; internodes short, strongly angled and grooved, but not scabrellous; fls. as in the typical form of the sp.—At 6300–6800 ft.; Pinyon-Juniper Wd.; Kingston Mts., e. Mojave Desert; to sw. Utah, nw. Ariz. July–Oct.

76. **E. Plumatélla** Dur. & Hilg. [*E. Palmeri* Wats.] Rather woody shrubs at the base, with several erect stems 3–6 dm. high, leafy on lower portions, white-tomentulose almost throughout, forked above; lvs. oblanceolate to oblong-lanceolate, 8–15 mm. long, revolute, acute, hoary-tomentose, with short slender petioles; invols. borne on mostly horizontal branches which spread in tiers to one side of the main axis and form an intricate mass ending in short-noded branchlets; invols. solitary but close together, glabrous, sessile, turbinate-cylindric, 2.5 mm. long; calyx glabrous, white, 2 mm. long, the outer segms. obovate, the inner narrower; fls. pilose basally; aks. narrow, brown, slightly angled, scaberulous above.—Dry stony places, below 4500 ft.; Creosote Bush Scrub, Joshua Tree Wd., Shadscale Scrub, Pinyon-Juniper Wd.; Kern R. region to Walker Pass, Mojave Desert; apparently to Utah, Ariz. Aug.–Oct.

Var. **Jaègeri** (M. & J.) S. Stokes ex Munz. [*E. nodosum* var. *J. M. & J.*] Infl. green and glabrous.—Mojave Desert; to w. Ariz.

77. **E. desertícola** Wats. Erect shrubs 6–12 (–15) dm. tall, much branched, the ultimate branchlets white-tomentose and leafy when young, becoming glabrous, green, and leafless in age; lvs. oblong-ovate to round-oblong, 5–15 mm. long, sometimes wider, white-woolly, on petioles 5–12 mm. long; invols. solitary, terminal, woolly, 1.5 mm. long, with 4 rounded teeth, short-peduncled; calyx yellow with green or reddish midribs, silky-villous, the segms. oblong-obovate, 2–3 mm. long; fils. pubescent basally; aks. dark, lance-ovoid, strigose, scabrellous above, 3 mm. long.—Locally common along sandy washes, dunes, etc.; Creosote Bush Scrub; Imperial Co. from the Salton Sink to dunes w. of Yuma. Sept.–Dec.

78. **E. Keárneyi** Tidestr. [*E. nodosum* var. *K. S. Stokes.*] Subshrubs with few to several scraggly, fragile stems arising from a woody taproot, usually densely tomentose nearly throughout, 3–8 (–10) dm. high; lvs. on the lower third of the plant, broadly oblanceolate to elliptic, the lf.-blades 1.5–2.5 (–3) cm. long, 5–10 (–12) mm. wide, densely tomentose on both surfaces, petioles up to 1 cm. long; infl. making up more than half the height of the plant; invols. racemosely disposed, scattered and appressed to the branches, turbinate, 2–2.5 mm. long; calyx white with reddish midribs, glabrous, 1.5–2 mm. long, connate nearly ½ the length and forming a campanulate perianth tube, this maturing reddish, the segms. obovate, nearly similar; calyx ca. 2.5 mm. long in fruit; aks. 2 mm. long, the subglobose bases tapering to 3-angled beaks.—Dry sandy soil, 4500–7000 ft.; Sagebrush Scrub, Pinyon-Juniper Wd.; Mono Lake Basin, Mono Co.; Nev., w. Utah, nw. Ariz. July–Oct.

Var. **monoénse** (S. Stokes) Reveal. [*E. nodosum* of California authors, in part. *E. n.* ssp. *m.* S. Stokes.] Plants 8–15 dm. high; lvs. 2–3 cm. long; invols. 2.5–3 mm. long, clustering at the ends of the branches; calyx 2.5–3 mm. long.—Dry gravelly places, 6000–8500 ft.; Sagebrush Scrub, Pinyon-Juniper Wd.; Mono and Inyo cos. July–Oct.

The var. *Kearneyi* is known in California only from the northeast shore of Mono Lake where it is common and somewhat unusual. Equally scattered in this population are glabrous stemmed forms along with the normal densely tomentose plants. Our plants are disjunct from the other known populations, the closest being in Washoe Co., and from near Tonopah, Nye Co., Nevada. A similar Tonopah-Mono Lake disjunct is *Astragalus pseudiodanthus* Barneby. The var. *monoense*, proposed in *Leafl. West. Bot.* 10: 334. 1966, is distinct from the Mono Lake plants, but somewhat similar to the high desert forms in western Utah.

79. **E. Butterworthiànum** J. T. Howell. Caudex low, branched, woody, plants 1–1.5 dm. high; lvs. linear or narrowly elliptical, ± revolute, (0.5–) 1–2 cm. long, 1–4 mm. wide, tomentose on both surfaces, gradually tapering to short petioles; flowering stems 1–3 cm. long, few, simple or divided 1 or 2 times, densely tomentose; invols. mostly solitary, sessile, tomentose, turbinate, 5–6 mm. long, 5-lobed, the teeth 1–2 mm. long,

deposed racemosely along the stems; calyx ochroleucous with reddish midribs, 4–5 mm. long, glabrous, the segms. obovate; fils. ciliated basally; aks. narrow, ca. 3 mm. long.— Crevices of sandstone, 2200 ft.; Mixed Evergreen F.; Santa Lucia Range, Monterey Co. June–Aug.

This species is closest to *Eriogonum Wrightii*, but how close remains to be determined by cytological studies. Named in 1961 (*Leafl. West. Bot.* 9: 153.), the plant has not been found beyond the type area which is near The Indians in the Santa Lucia Range.

80. **E. Wrìghtii** Torr. ex Benth. Low, highly branched perennial subshrubs from branched woody caudices, 1.5–4 dm. high; lvs. on the lower half of the plants, crowded, oblanceolate to elliptic, 0.5–1.5 cm. long, 2–5 (–7) mm. wide, mostly entire, densely white-tomentose above and below, or rarely subglabrous and green above, the petioles 2–5 mm. long; flowering stems several, tomentose or rarely glabrous, up to 25 cm. long; infl. racemose, 5–30 cm. long, once or twice dichotomous or trichotomous; invols. solitary, turbinate, 2–2.5 mm. long, ± tomentose; calyx whitish or pink, 2.5–3.5 mm. long, glabrous, the outer segms. broadly obovate, the inner less so; flls. ± pilose basally; aks. narrow, 2.5–3 mm. long, 3-angled, somewhat scaberulous above; $2n = 34$ (Stokes & Stebbins, 1955).—Gravelly and rocky places, mostly below 5000 ft.; Pinyon-Juniper Wd.; Mojave and Colo. deserts e. to w. Tex. Aug.–Oct.

Var. **trachygònum** (Torr. ex Benth.) Jeps. [*E. W.* ssp. *t.* S. Stokes. *E. t.* Torr. ex Benth.] Low, woody, subshrubs, 2–4 dm. high; lvs. broadly elliptic to elliptic, 1.5–3 cm. long, (3–) 5–10 mm. wide, densely tomentose below, usually less so above, not revolute; invols. 2–3 mm. long, ± tomentose; calyx white to pink, 3–4 mm. long.—Exposed and open rocky places, mostly below 6000 ft.; Foothill Wd., Yellow Pine F.; inner Coast Ranges and w. base of Sierra Nevada, Shasta Co. to n. Los Angeles Co. Aug.–Oct.

Var. **membranàceum** S. Stokes ex Jeps. [*E. W.* ssp. *m.* S. Stokes.] Woody, branched subshrubs, 2–4 dm. high and up to 6 dm. across; petiole bases dilated into glabrate, brownish sheaths which clasp the stems; lvs. strongly involute, 3–6 (–10) mm. long, 1–3 (–4) mm. wide; invols. 2–3 mm. long, often glabrous; calyx white to pink, 3–4 mm. long.— Dry stony places below 6000 ft.; Chaparral, Joshua Tree Wd., Pinyon-Juniper Wd.; Little San Bernardino and Santa Monica mts. to San Diego Co.; L. Calif. Aug.–Oct.

Var. **nodòsum** (Small) Reveal. [*E. n.* Small.] Woody, branched, densely white-lanate shrubs up to 6 dm. high and 15 dm. across; petiole bases dilated but not extending completely around the stems, lvs. 8–12 mm. long, 2.5–4 mm. wide, tomentose, not revolute; invols. 1.5–2.5 mm. long; calyx white, 3–4 mm. long.—Dry stony places below 3000 ft.; Creosote Bush Scrub, Pinyon-Juniper Wd.; w. edge of Colo. Desert e. to Twentynine Palms; L. Calif. Aug.–Nov. The var. *Prínglei* (Coult. & Fish.) Reveal, a form with more numerous branches, shorter invols. and flls., occurs in southwestern Arizona and may be expected in southeastern California.

Var. **subscapòsum** Wats. [*E. W.* ssp. *s.* S. Stokes. *E. junceum* Greene. *E. curvatum* Small. *E. W.* var. *c.* Munz. *E. Kennedyi* ssp. *pinorum* S. Stokes.] Low and loosely matted, 1–2.5 dm. high; lvs. crowded, 5–12 mm. long, grayish- or brownish-white; flowering stems slender, up to 1.5 dm. long, glabrous to floccose; infl. 5–15 cm. long; invols. 1.5–3 mm. long.—Rocky and gravelly places, 5000–11,000 ft.; Montane Coniferous F.; Sierra Nevada to San Jacinto and San Bernardino mts. July–Oct.

Var. **olanchénse** (J. T. Howell) Reveal, comb. nov. [*E. Kennedyi* var. *o.* J. T. Howell, *Leafl. West. Bot.* 6: 151. 1951.] Low and densely matted cespitose perennials, less than 3 cm. high; lvs. crowded, 1–2.5 mm. long, densely white-tomentose; flowering stems slender, up to 1 cm. long, subglabrous; infl. 1–1.5 cm. long; invols. 0.8–1.3 mm. long.—Dry granitic places, 11,500–11,800 ft.; Olancha Pk., Tulare Co. July–Aug.

The *Eriogonum Wrightii* complex is composed of several morphologically distinct, but closely related and often sympatric varieties that occur in the dry places throughout much of southern California. The species has suffered considerable confusion in the herbarium with *E. Kennedyi*, and it is hoped that the present key will alleviate most of the problems encountered in the past.

81. **E. panamintènse** Morton. [*E. reliquum* S. Stokes. *E. racemosum* var. *desertorum* S. Stokes.] Caudex low and matted, branched, woody, often up to 4 dm. across; lvs. basal and along the lower nodes, not crowded, densely white-tomentose below, less so above, elliptic, ovate or obovate, 15–40 mm. long, 10–25 mm. wide, on tomentose peti-

oles 1–5 cm. long; flowering stems several to many, white-tomentose, 1.5–3 cm. high, 1–3 times dichotomous; cauline lvs. verticillate, ± orbicular, similar to the basal lvs. but reduced, short petioled; invols. solitary, sessile, scattered racemosely along the branches and in the forks, 3–5 mm. long, densely white-tomentose; calyx white to whitish-brown, 3.5–5 mm. long, glabrous, the segms. ± similar, oblanceolate; fils. pilose basally; aks. narrow, 3-angled, ca. 3 mm. long.—Dry rocky slopes, 5000–9000 ft.; Pinyon-Juniper Wd., Yellow Pine F.; Inyo-White to New York and Clark mts.; sw. Nev. May–Oct.

Var. mensícola (S. Stokes) Reveal, comb. & stat. nov. [*E. m.* S. Stokes, *Leafl. West. Bot.* 3: 16. 1941.] Lvs. strictly basal, the lf.-blades 1–1.5 cm. long and wide, rotund, densely tomentose on both surfaces; bracts not leafy, lanceolate to lance-ovate, mostly 2–6 mm. long; invols. few, scattered along the stems, 2–4 mm. long; calyx 3–4 mm. long. —Dry rocky slopes, 5000–8000 ft.; Pinyon-Juniper Wd.; Inyo-White and Panamint mts.; Sheep Range, Nev. July–Sept.

The recognition of *Eriogonum panamintense* came after Reveal had spent considerable time studying *E. racemosum*, *Wrightii*, and *panamintense* in the field and in the herbarium. *Eriogonum panamintense*, for several years, has been associated with *E. racemosum*, a form of which just enters California, although *E. panamintense* is more closely related to *E. Wrightii*. The growth habit of *E. panamintense* and *E. racemosum* is totally different, with the latter having a single, or at most, three stems arising from a short, little branched caudex, while the former has several branches arising from a highly branched, spreading matted woody caudex.

82. E. racemòsum Nutt. Caudex woody, few-branched, compact; lvs. basal, oblong to oblong-ovate, the lf.-blades mostly 2.5–3.5 cm. long, glabrate above and closely white-tomentose below, on petioles as long or longer; flowering stems slender, mostly 1.5–3 dm. high, tomentose, trichotomous once or twice, usually not leafy bracted; invols. solitary and arranged racemosely along the upper branches, tubular-campanulate, 3–4 mm. long, tomentose; calyx cream-white, 2.5–3 mm. long, the segms. oblong-oblanceolate; fils. pilose basally; aks. lance-ovoid, brownish, ca. 2 mm. long, 3-angled; $n = 18$ (Reveal, 1968).—At ca. 6000–7000 ft.; Pinyon-Juniper Wd.; White and Cottonwood mts., Inyo Co.; to Colo., New Mex. July–Aug.

The plant simply called *Eriogonum racemosum* in this treatment extends from the White Mountains eastward to the Toquima Range of central Nevada. This part of the species is somewhat similar to the kind of plant found in northern New Mexico and adjacent Colorado, but differs considerably from the kind of plant found in eastern Nevada and most of Utah. The western phase of the species seems to be distinct, but until the entire species is studied in some detail, no new taxa are proposed.

83. E. elongàtum Benth. [*E. denudatum* Nutt.] Perennial herbs, mostly loosely branched at the base, whitish-tomentulose nearly throughout, leafy in lower portion, passing into elongate, leafless paniculately forked infls. above, 6–12 (–18) dm. high; lvs. lance-oblong to narrowly ovate, crisped-undulate, somewhat glabrate above, white-tomentose below, 3–5 cm. long, cuneate at base, short-petioled; invols. remotely scattered, oblong-cylindric, 6–7 mm. long, tomentose, truncate, obscurely 5-toothed; calyx white or pinkish, glabrous, 2.5–3 mm. long, the segms. obovate, the inner slightly longer than the outer, somewhat pubescent within; fils. glabrous; aks. dark, narrow, 2–2.5 mm. long, glabrous, somewhat 3-angled; $2n = 34$ (Stokes & Stebbins, 1955).—Dry rocky places, below 6000 ft.; Coastal Sage Scrub, Chaparral, Foothill Wd.; Coast Ranges from Monterey and San Benito cos. to n. L. Calif. Aug.–Nov.

84. E. ovalifòlium Nutt. Cespitose perennial with closely branched woody caudices thickly beset with lvs., densely white-tomentose; lvs. basal, round to obovate, the lf.-blades 5–15 (–20) mm. long, short-petioled; flowering stems scapose, slender, white-woolly, 1–2 dm. high; infl. capitate, 1.5–2.5 cm. in diam.; invols. several, white-woolly, commonly 4–5 mm. long, narrowly cylindric; calyx whitish or rarely yellow to ochro-leucous, glabrous, 4–5 mm. long, the outer lobes elliptic, subcordate at the base, the inner segms. spatulate, exserted; fils. pilose basally; aks. glabrous, 2–2.5 mm. long, 3-angled; $2n = 40$ (Stokes & Stebbins, 1955), $n = 20$ (Reveal, 1965).—Dry slopes and flats, mostly 5000–7000 ft.; Sagebrush Scrub, N. Juniper Wd., Pinyon-Juniper Wd.; e. slope of Sierra Nevada, n. and e. to Alta., Rocky Mts. May–July.

Var. multiscàpum Gand. [*E. orthocaulon* Small. *E. o.* Nutt. var. *o.* C. L. Hitchc. *E. o.* var. *celsum.* A. Nels.] Lf.-blades oblong to obovate, 3–6 cm. long; flowering stems ca. 1–3 dm. high, fls. mostly yellow; $n = 20$ (Reveal, 1965).—Dry slopes and flats, Lassen and Modoc cos., 4500–6000 ft.; Sagebrush Scrub, N. Juniper Wd.; s. Ore., Nev. and Ida. May–June.

Var. **vinèum** (Small) Nels. [*E. v.* Small. *E. o.* ssp. *v.* S. Stokes.] Lf.-blades round-ovate, 7–12 mm. long, white, felty-tomentose; flowering stems ca. 1 dm. high; infl. capitate, up to 4 cm. in diam.; calyx 5–7 mm. long, pale.—Desert slopes, San Bernardino Mts., 5000–8000 ft. May–June.

As presently defined, the var. *vineum* is restricted to the San Bernardino Mts. and is the large flowered form of *E. ovalifolium*. The application of this name in the various floras and revisions has been mainly to var. *nivale*.

Var. **nivàle** (Canby) Jones. [*E. n.* Canby. *E. rhodanthum* Nels. & Kenn. *E. eximium* Tidestr. *E. o.* ssp. *e.* S. Stokes.] Lvs. nearly round, mostly less than 1 cm. in diam., white to rusty tomentose; flowering stems less than 1 dm. high; calyx white with reddish midribs to rose, 2–3 mm. long; *n* = 20 (Reveal, 1965).—Dry flats and ridges, 7000–12,000 ft.; Montane Coniferous F., Alpine Fell-fields; Sierra Nevada to Ore., w. Utah. July–Aug.

The var. *nivale* is the high alpine form of the species which occurs in the mountains of California, Nevada, and western Utah, but is gradually replaced in the north by the var. *depressum* Blank. The species, *E. rhodanthum* and *E. eximium,* are essentially the same kind of plant, the first coming from the higher elevations of Mt. Rose, Nevada, and the second from the lower elevations. However, while the population is distinctive, it too blends into adjacent forms of var. *nivale* and for this reason, the two species are reduced to synonyms.

85. **E. Gilmánii** S. Stokes. Low compact perennial from an elongated woody root, the caudex covered with old lvs.; lvs. crowded in a basal rosette, few to ca. 14, suberect, densely white-tomentose, the lf.-blades 2–4 mm. long, elliptic, with petioles margined and ca. as long; scapes solitary, 1–2 cm. high and with 2–3 cymosely arranged invols.; invols. turbinate, 1.5 mm. long; calyx reddish, glabrous, the outer segms. inflated, rounded on the back and orbicular in shape so as to form a globose segm. 3–4 mm. in diam., the inner segms. narrower, longer, slightly exserted; aks. 2.5 mm. long, acutely 3-angled.—At 6200 ft., Pinyon Mesa, Panamint Mts., Inyo Co. Aug.–Sept.

86. **E. stríctum** Benth. ssp. **prolíferum** (T. & G.) S. Stokes. [*E. ovalifolium* var. *p.* Wats. *E. p.* T. & G. *E. Greenei* Gray. *E. niveum* var. *G.* S. Stokes.] Caudex compactly branched, the plants white-tomentose; lvs. basal, the blades ovate, obtuse, sometimes bicolored, 1–2 cm. long, on slender long petioles; flowering stems erect, naked, slender, floccose, 1.5–2.5 dm. high; infl. cymose-umbellate, the rays divaricate, to 4 cm. long; invols. largely solitary in the forks and terminal, oblong-turbinate, tomentose, 5–6 mm. long, 5-toothed; calyx cream to white, 4–5 mm. long, the outer segms. broadly elliptic with obcordate bases, the inner narrower, obovate, somewhat exserted; fils. pilose basally; aks. glabrous, narrow, 3-angled.—Dry open places, 3500–6000 ft.; Sagebrush Scrub, N. Juniper Wd., Yellow Pine F.; Siskiyou Co. and Plumas Co. n.; to B.C. and Ida. June–Aug.

Var. **anserìnum** (Greene) R. J. Davis. [*E. a.* Greene. *E. strictum* ssp. *a.* S. Stokes. *E. flavissimum* Gand. *E. ovalifolium* ssp. *f.* S. Stokes. *E. proliferum* ssp. *a.* Munz. *E. s.* var. *f.* C. L. Hitchc.] Invols. 1–3 in clusters on rays up to 5 cm. long; calyx yellow.— Similar situations, e. Siskiyou Co., Modoc Co.; adjacent Nev. and Ore. June–July.

The var. *proliferum* and var. *anserinum* have suffered much misunderstanding as the list of synonymy shows. *Eriogonum strictum* occurs to the north, with two forms coming into California, both of which are often mistaken for *E. ovalifolium,* a related, but distinct species. The cymose-umbellate inflorescences of *E. strictum,* sen. lat., quickly distinguish it from the capitate inflorescences of *E. ovalifolium* even though both may have similar flower color, size of plants, and leaf color and shape.

87. **E. Shóckleyi** Wats. [*E. acaule* of Intermountain authors, not Nutt.] Caudex pulvinate, branched, low, the plants with a dense brownish-white tomentum; lvs. crowded, oblanceolate to elliptic or spatulate, plane, the lf.-blades 3–8 mm. long, 2–4 mm. wide, with petioles ca. as long or slightly longer, densely tomentose on both surfaces; flowering stems scapelike, lacking to 2 cm. long, slender, tomentose; infl. capitate, with 4–10 invols.; invols. campanulate, ± membranaceous but forming distinct tubes, 2–2.5 mm. long, the 5–7 lobes dividing the tube to near the middle; calyx brownish with tan to rusty midribs, densely pubescent without, slightly so within, 2.5–3 mm. long, the segms. obovate, similar; fils. pilose basally; aks. lance-ovoid, ca. 2.5 mm. long, sparsely pubescent. —Dr. rocky slopes and clay hills, 6000–8000 ft.; Sagebrush Scrub, Pinyon-Juniper Wd.; Last Chance Mts., Inyo Co.; e. to Utah, w. Colo., n. Ariz. May–July.

This previously unreported species for California is to be expected along the slopes of the desert ranges adjacent to Nevada in Inyo and possibly Mono cos. Our plant is var. *Shockleyi.*

88. **E. Breedlóvei** (J. T. Howell) Reveal. [*E. ochrocephalum* var. *B.* J. T. Howell.] Caudex cespitose, branched, low, the plants densely olive-green to whitish tomentose with glandular hairs nearly throughout; lvs. crowded, broadly elliptic, plane or with margins slightly revolute, lf.-blades 2–8 (–10) mm. long, 2–4 (–6) mm. wide, densely tomentose below, less so and green above, petioles up to 1 cm. long, petiole bases broadened, hairy on both surfaces; flowering stems scapelike, erect, 1.5–6 cm. tall, densely glandular-hairy, the 3 triangular bracts ca. 2 mm. long; infl. compactly cymose-umbellate with short rays up to 3 (–5) mm. long, the branches often bracteate; invols. 4–6, turbinate-campanulate, 3.5–4 mm. long and wide, glandular-puberulent, 7–9 lobed, the triangular to acute lobes 1–1.5 mm. long, forming a distinct tube; calyx whitish becoming reddish, 2.5–3.5 (–4) mm. long, the segms. cuneate, somewhat dissimilar, the outer segms. wider than the inner, finely glandular-hairy within and without; fils. ± pilose basally; aks. lance-ovoid, 2–3 mm. long, sparsely and minutely hairy above the middle.— Dry rocky outcrops, 7800–8200 ft.; Red Fir F.; Piute Mts., Kern Co., Baker Point, s. Tulare Co. June–Aug.

Eriogonum Breedlovei is an interesting addition to the California buckwheats, and especially to the section *Capitata* T. & G. It is similar to *E. Cusickii* Jones and *E. exilifolium* Reveal in that the inflorescence is not tightly capitate, but rather a compact cymose-umbellate head. However, this single feature seems to be a parallel characteristic and does not reflect any relationship between these species. The Piute Buckwheat, as stated by Howell when he proposed the variety, is most closely related to what is here called *E. anemophilum.* The variety was proposed in *Leafl. West. Bot.* 10: 14, 1963, but following additional field work, it was raised to the species level in *Leafl. West. Bot.* 10: 335. 1966.

89. **E. ochrocéphalum** Wats. Caudex ± cespitose, branched, forming loosely branching mats, the plants with dense whitish tomentum; lvs. crowded, ovate to obovate, or lanceolate to oblong, mostly 2–3.5 cm. long, equally grayish-tomentose on both surfaces, with petioles nearly twice as long; flowering stems scapelike, 1–2.5 dm. long, slender, glabrous and greenish to floccose; infl. capitate, 10–15 mm. across; invols. several, turbinate-campanulate, tubular and rigid, (3–) 3.5–5 mm. long, with 6–8 teeth; fls. yellowish, glabrous, 2–3 mm. long, the segms. ovate-oblong, essentially alike; fils. pilose basally; aks. lance-ovoid, 1.5–2 mm. long.—Dry loose, often volcanic soil, or gumbo clay hills; 4000–6000 ft.; Pinyon-Juniper Wd.; Nevada Co. n.; Ore., Nev., and Ida. May–June.

90. **E. anemóphilum** Greene. [*E. rosense* Nels. & Kenn. *E. ochrocephalum* var. *agnellum* Jeps. *E. o.* sspp. *a.* and *a.* S. Stokes.] Caudex densely cespitose, branched, low, the plants with dense olive-green to whitish tomentum; lvs. crowded, oblanceolate, 4–15 mm. long, 2.5–5 mm. wide, densely white-tomentose below, greenish-tomentose above, the petioles from shorter to as long as the lvs.; flowering stems scapelike, 1–9 cm. tall, slender, glandular; infl. capitate, 6–15 mm. across; invols. few, turbinate, tubular and rigid, 3–3.5 mm. long, with 6–8 teeth; calyx yellow and becoming reddish, glabrous or more frequently glandular, 2–3 mm. long, the segms. obovate, essentially alike; fils. pilose basally; aks. lance-ovoid, ca. 1.5 mm. long; $n = 20$ (Reveal, 1967).—Dry granitic and volcanic soils, 9000–12,000 ft.; Yellow Pine F., Red Fir F., Alpine Fell-fields; Inyo and Fresno cos. to Placer Co.; Nev. July–Aug.

The introduction of the name *Eriogonum anemophilum* comes as a result of two studies, one in the field and the other in the herbarium. The type of *E. ochrocephalum* came from the clay hills north of Reno, Nevada, and extends from the Reno area northward into southern Oregon where it is largely replaced by the var. *calcareum* (S. Stokes) Peck. In the high mountains of Oregon eastward to the Idaho-Montana state line is a related species, *E. chrysops* Rydb. The relationship between *E. chrysops* and *E. ochrocephalum* is the same as that found between *E. anemophilum* and *E. ochrocephalum.* Both species occur in the high mountains, with *E. chrysops* ranging from 7000 to 11,000 feet elevation and *E. anemophilum* occurring from 9000 feet to 12,000 feet elevation. While the distinction between *E. ochrocephalum* and the alpine species is distinct, the relationship between the two alpine species is in need of critical study.

Reveal has seen only an isotype of *Eriogonum anemophilum,* and the holotypes of *E. rosense* and *E. ochrocephalum* var. *agnellum* compare rather favorably with it. The isotype is slightly larger than either of the other specimens, but this condition falls well within the variation seen in the Sierra Nevada populations, although *E. anemophilum* was collected in the West Humboldt Range of western Nevada. The type of *E. rosense* was taken from the summit of Mt. Rose in extreme western Nevada.

91. **E. chrýsops** Rydb. [*E. ochrocephalum* ssp. *c.* S. Stokes.] Caudex densely cespitose, branched, low, the plants with dense grayish-white tomentum; lvs. crowded, spatulate

to oblanceolate, 4–6 mm. long, 2–3 mm. wide, densely tomentose on both surfaces, the petioles shorter than the blades; flowering stems scapelike, 2–5 cm. tall, slender, glabrous; infl. capitate, ca. 1 cm. across; invols. 3–7, sessile, campanulate, 2–2.5 mm. long, with 5 lobes, glabrous or nearly so; calyx yellow, glabrous, 2.5–3 mm. long, the segms. oblong to oblong-obovate, essentially alike; fils. pilose basally; aks. lance-ovoid, 2–2.5 mm. long. —Dry volcanic soils, at 5500 ft.; N. Juniper Wd.; Modoc Co. n. se. Ore. to Mont. May– June.

Eriogonum chrysops is included in the California flora based upon *Austin,* without number, and *Balls 14780,* both of which differ in a few minor respects from the typical form in Oregon and Idaho.

92. **E. Kénnedyi** Porter ex Wats. [*E. K.* var. *austromontanum* of authors, in part.] Caudex branched, woody, forming dense leafy mats with numerous lvs.; lf.-blades elliptic to oblong, grayish- to brownish-white tomentose, 2–4 (–5) mm. long, 0.5–1.5 (–2) mm. wide, ± revolute in some, subsessile; flowering stems scapelike, wiry, glabrous, 4–12 cm. long; infl. capitate, 4–8 mm. across; invols. few, sparsely tomentose to glabrous, tur- binate, angled, 1.5–2.5 mm. long; calyx glabrous, white with reddish midribs, 1.5–2.5 mm. long, the segms. broadly elliptical, somewhat rounded at the base; fils. subglabrous; aks. ca. 2 mm. long, papillose-puberulent, lance-ovoid, 3-angled.—Dry stony to gravelly slopes and ridges, 5000–7000 ft.; San Bernardino Mts. and Mt. Pinos. April–June.

The var. *Kennedyi* is totally redefined here as a result of an investigation of the original type collec- tion. In the past, the concept of var. *Kennedyi* has been appplied to what is here called var. *Purpusii.* This misapplication has been the result of the inability to exactly determine the type locality. Originally cited as simply "Kern County" it has been assumed that this meant Inyo Co. However this is not the case. At the suggestion of E. C. Twisselmann who is the author of the Kern County Flora, a careful comparison was made with the type and other collections. The original Kennedy collection was found to compare most favorably with specimens from the lower slopes of Mt. Pinos which is just south of the Kern Co. line.

Var. austromontànum M. & J. [*E. K.* ssp. *a.* S. Stokes.] Loosely matted; lf.-blades oblanceolate, (4–) 6–10 (–12) mm. long, 1–2 mm. wide, the lvs. often sheathing up the stems; stems floccose, 8–15 cm. long; invols. turbinate, tomentose, 2.5–4 mm. long; calyx white with reddish-brown midribs, 2–3 mm. long, oblong-obovate, gradually contracted into a cuneate base; aks. 3.5–4 mm. long.—Dry stony slopes, 6300–6500 ft.; Yellow Pine F.; Bear Valley, San Bernardino Mts. July–Aug.

The var. *austromontanum* has been defined in the past to include what is here called var. *Kennedyi* and var. *austromontanum.* By removing the smaller and more compact, early flowering, var. *Kennedyi,* the var. *austromontanum* becomes a distinct taxon.

Var. alpígenum M. & J. [*E. K.* ssp. *a.* Munz.] Mats very dense and woody; lvs. 2–4 mm. long, 0.7–1.5 mm. wide; stems less than 2 cm. long, densely white-tomentose; infl. capitate, 4–8 mm. wide; invols. turbinate, 1.5–2 mm. long, tomentose; calyx white to reddish with reddish-brown midribs, 1.5–2.5 mm. long; aks. 2 mm. long.—Dry granitic gravel slopes and ridges, 10,000–11,500 ft.; Alpine Fell-fields; San Gorgonio Peak, San Bernardino Mts., somewhat lower to 8750 ft. in the San Gabriel Mts. and on Mt. Pinos. July–Aug.

The var. *alpigenum* as treated here is the same as treated previously, the only changes being in the distribution.

Var. Purpùsii (Bdg.) Reveal, stat. & comb. nov. [*E. P.* Bdg., *Bot. Gaz.* 27: 457. 1899.] Lvs. white-tomentose, oblong, (2.5–) 3–6 mm. long, 1.5–3.5 mm. wide; stems thin and wiry, glabrous or rarely tomentose, 4–10 cm. high; infl. capitate, 8–15 mm. wide; invols. turbinate-campanulate, glabrous to sparsely tomentose, 1.5–2 mm. long; calyx white with greenish midribs, 2–2.5 mm. long; aks. 3 mm. long.—Dry granitic flats and slopes, 5000– 8000 ft.; Sagebrush Scrub, Pinyon-Juniper Wd.; e. slope of Sierra Nevada from Mono Co. s. to Argus and Coso mts., Inyo Co. May–June.

To most California authors, this form of the overall species, *E. Kennedyi,* has been regarded as the typical form of the species. It is interesting to note that in the Stokes monograph, she had correctly determined what part of the species was typical *E. Kennedyi* and had recognized the distinctiveness of var. *Purpusii.* The fact that the taxon *Purpusii* is reduced under *E. Kennedyi* is mainly due to the following variety, and to the near intermediacy of some southern forms of var. *Purpusii* with the Bear Valley populations of var. *Kennedyi.*

Var. pinícola Reveal, var. nov. A var. *Purpusii* differt foliis 3–5 mm. longis, 1–4 mm. latis, scapis 5–13 cm. altis; involucris 2.5–3.5 mm. longis, perianthiis albis vel rufis, 2.5–3.5 mm. longis. Lvs. grayish- to rusty-white tomentose, oblong, 3–5 mm. long, 1–4 mm. wide; stems thin and wiry, glabrous, reddish, 5–13 cm. high; infl. capitate, ca. 1 cm. across; invols. turbinate-campanulate, sparsely tomentose, 2.5–3.5 mm. long; calyx white to reddish with green to reddish-brown midribs, 2.5–3.5 mm. long; aks. 3 mm. long.— Dry exposed ridgetops, 4900–5600 ft.; Pinyon Wd., Jeffrey Pine F.; known at present only from Sweetwater Ridge and Pine Tree Canyon, Kern Co. May–June. *Type:* Sweetwater Ridge, south of Cache Peak, Kern Co., California, 9 June 1966, *E. C. Twisselmann 12360.* Holotype at the California Academy of Sciences.

This new variety is most closely related to the var. *Purpusii* which occurs on the east side of the Sierra Nevada, and differs in several technical characteristics. This form is also somewhat similar to var. *austromontanum.*

93. **E. gracílipes** Wats. [*E. Kennedyi* ssp. *g.* S. Stokes. *E. ochrocephalum* var. *g.* J. T. Howell.] Caudex cespitose, branched, low, the plants with a dense whitish tomentum; lvs. crowded, oblanceolate to elliptic, plane, the lf.-blades 1–2 cm. long, with petioles ca. as long, tomentose and glandular below, less so above; flowering stems scapelike, 3–8 cm. long, slender, glandular; infl. capitate, 5–7 invols.; invols. campanulate, membranaceous, not rigid, 2–3 mm. long, flaring at the throat, the 5 lobes deeply divided to near the base; calyx white with reddish midribs becoming rose at maturity, glabrous, 2–3 mm. long, the segms. obovate, similar; fils. pilose basally; aks. lance-ovoid, ca. 2 mm. long.—Dry rocky slopes and ridges, 10,000–13,000 ft.; Bristlecone Pine F.; White Mts., Mono and Inyo cos., w. Nev. July–Sept.

The species in the section *Capitata,* which include numbers 87 through 93 are often narrowly endemic and isolated on high mountain peaks. *Eriogonum gracilipes* is most closely related to *E. Holmgrenii* Reveal of the Snake Range in eastern Nevada, which in turn seems to be close to *E. Kingii* Torr. & Gray of the Ruby Mts., northeastern Nevada. Rarely are the species in this section found growing together, and thus the various forms tend to be closely related yet very distinct.

94. **E. latifòlium** Sm. in Rees. [*E. arachnoideum* H. & A.] Caudex low, woody, densely leafy, often much-branched; lf.-blades persistent, ovate to almost oblong, cordate or rounded at the bases, obtuse or acute, often crisped on the margins, lanate or somewhat glabrous above, densely white-lanate below, 2.5–5 cm. long, the petioles shorter or longer, woolly, expanded at base; flowering stems stout, leafless, tomentulose, 2–6 dm. long, simple or rarely 2–4 forked, the forks simple or rarely forked again; infl. of capitate clusters 1.5–3 cm. across, terminal and also sessile in the forks; invols. tomentose, numerous, 3.5–4 mm. long, shallowly 5-toothed; calyx glabrous or with a few scattered hairs near the base in some, white to rose, 3 mm. long, the segms. obovate, subequal; fils. villous basally; aks. glabrous, brown, lance-ovoid, 3-angled, ca. 4 mm. long; $2n = 40$ (Stokes & Stebbins, 1955), $n = 20$ (Reveal, 1968).—Cliffs and sandy places along the immediate coast, Coastal Strand and N. Coastal Scrub; San Luis Obispo Co. n. to Ore. June–Sept.

95. **E. nùdum** Dougl. ex Benth. [*E. latifolium* ssp. *n.* S. Stokes. *E. oblongifolium* Benth. *E. oblanceolatum* Greene. *E. l.* var. *parvulum* S. Stokes. *E. longulum* Greene.] Caudex short, simple or few-branched, the leaf-bearing area not elongated; lf.-blades oblong to oblanceolate or broadly elliptic-ovate, 1–6 cm. long, rounded at the apex, glabrate above, white-lanate below, ± undulate-crisped, the petioles often much longer; flowering stems commonly 1–few, erect, slender, 3–10 dm. high, glabrous or nearly so, glaucous, usually forking or trichotomous near the middle, then branching again; invols. usually in clusters, subcylindric, 3–5 mm. long, glabrous or slightly woolly; calyx 2–2.5 mm. long, mostly white with some pink, or sometimes yellow, usually glabrous without; fils. pilose basally; aks. 1.5–3 mm. long; $2n = 40$ (Stokes & Stebbins, 1955), $n = 20$ (Reveal, 1968).—Dry, usually rocky places, up to ca. 8000 ft.; many Plant Communities; Coast Ranges from about San Francisco Bay n., Sierra Nevada from Tulare Co. n.; to Wash., Nev. June–Nov.

Exceedingly variable and intergrading with the varieties treated below. Subalpine plants from the Sierra Nevada, 7000–11,000 ft., tend to have several stems from the base, 2–3 dm. high and very slender, with lf.-blades 1–2 cm. long, and with branched infls. with largely solitary invols. which may be known as var. dedúctum (Greene) Jeps. [*E. d.* Greene.] The extreme in such reduction, with simple stems 1–2 dm. long and ending in single heads of several invols., occurs at elevations from ca. 10,000–12,000 ft. in s. Sierra Nevada and may be known as var. scapígerum (Eastw.) Jeps. [*E. s.* Eastw. *E. latifolium* var. *s.* S. Stokes.]

Key to Varieties of E. nudum

A. Invols. and flowering stems ± tomentose; heads usually several to many; lvs. in basal rosettes. Interior of n. Calif. . . . . var. *oblongifolium*

AA. Invols. and flowering stems mostly glabrous or if tomentose, then lvs. scattered along the woody caudices.

    B. Lvs. scattered along the woody caudices, the lf.-blades very strongly undulate-crisped, 3–7 cm. long. Coast Ranges of cent. Calif.

        C. Stems not strongly inflated. Coast Ranges, Sonoma Co. to Monterey Co. var. *auriculatum*

        CC. Stems strongly inflated. Inner Coast Ranges, Merced Co. to Kern Co. . . . . var. *indictum*

    BB. Lvs. in basal rosettes, the lf.-blades plane or slightly crisped, mostly 1–5 cm. long.

        C. Invols. 2–6 in a cluster.

            D. Fls. mostly glabrous externally, white to pink, rarely yellow.

                E. Flowering stems branched above. Lower elevations. . . . . var. *nudum*

                EE. Flowering stems scapose, short, ending in a solitary head. Subalpine.

                                                var. *scapigerum*

            DD. Fls. pubescent externally, yellow to white. Desert edges and borders of San Joaquin and Salinas valleys. . . . . var. *pubiflorum*

        CC. Invols. solitary, rarely in pairs. Mostly of pine belt.

            D. Branches several from the base. High Sierra Nevada. . . . . var. *deductum*

            DD. Branches 1–few from the base. Mts. of s. Calif. . . . . var. *pauciflorum*

Var. oblongifòlium Wats. [*E. sulphureum* Greene. *E. latifolium* ssp. *s.* S. Stokes. *E. nudum* var. *s.* Jeps. *S. Harfordii* Small. *E. capitatum* Heller.] Lvs. basal, the lf.-blades largely oblong-spatulate, 2–4 cm. long; flowering stems 5–10 dm. high, white-tomentose, dichotomously branched; invols. mostly 3–6 in a cluster, tomentose, 3–5 mm. long; calyx white or rose or yellowish, 3–4 mm. long, pubescent without near the base; $n = 20$ (Reveal, 1965).—Dry slopes mostly below 4000 ft.; N. Oak Wd., Yellow Pine F., Foothill Wd.; Napa Co. to Humboldt and Siskiyou cos.; up to 7000 ft. from Nevada Co. to Modoc Co.; adjacent Ore., Nev. May–Aug.

Var. auriculàtum (Benth.) Tracy ex Jeps. [*E. a.* Benth. *E. latifolium* ssp. *a.* S. Stokes. *E. l.* ssp. *decurrens* S. Stokes. *E. l.* var. *alternans* S. Stokes.] Stems 2–10 (–20) dm. high, caudexlike at the bases, with lvs. on lower part; lf.-blades oblong to elliptic, obtuse at apex, truncate or subcordate at base, green above, white-tomentose below, 3–7 cm. long; flowering stems glabrous and glaucous, or rarely tomentose, often fistulose; invols. solitary or in pairs, 3–4 mm. long; calyx usually cream to pink, sometimes yellowish, mostly glabrous, in heads ca. 1 cm. across; $2n = 80$ (Stokes & Stebbins, 1955).—Dry, often stony places; Coastal Strand, Chaparral, V. Grassland; Coast Ranges from Sonoma Co. to Monterey Co. July–Sept.

A form with unusually strong inflated flowering stems, very robust, and with fls. yellow to whitish, is var. indíctum (Jeps.) Reveal, comb. nov. [*E. i.* Jeps., Fl. Calif. 2: 421. 1913. *E. latifolium* var. *i.* S. Stokes.] from Merced Co. to Kern Co.; $2n = 80$ (Stokes & Stebbins. 1955).

Var. pubiflòrum Benth. [*E. saxicola* Heller. *E. gramineum* S. Stokes. *E. latifolium* ssp. *s.* S. Stokes. *E. l.* ssp. *Westonii* S. Stokes.] Flowering stems glabrous, glaucous, 3–6 dm. high, cymose above; invols. clustered, rarely solitary, subcampanulate; calyx yellow to white, pubescent without; $n = 20$ (Reveal, 1965).—Dry hot places below 6000 ft.; largely Foothill Wd., Joshua Tree Wd., Pinyon-Juniper Wd., Yellow Pine F.; Santa Ana Mts., w. Mojave Desert from San Gabriel Mts. n. to Modoc Co., Siskiyou and Humboldt cos. and along the inner Coast Ranges to Santa Lucia Mts. June–Sept.

Var. pauciflòrum Wats. [*E. molestum* Wats. *E. vimineum* ssp. *m.* S. Stokes. *E. latifolium* ssp. *p.* S. Stokes. *E. n.* var. *perturbum* Jones.] Caudex rather simple with lvs. crowded; lf.-blades oblong-ovate, 1.5 cm. long, green or glabrate above, white-woolly below, on petioles as long or longer; flowering stems 3–8 dm. high, slender, glabrous, glaucous, forked several times; invols. 1, rarely 2, at a place, rather few on a branch, 5–7 mm. long; calyx whitish, glabrous, 2 mm. long.—Dry slopes, 5000–9000 ft.; Yellow Pine F., Red Fir F.; Cuyamaca Mts., to Santa Rosa and San Bernardino mts. Aug.–Oct.

96. E. grànde Greene. [*E. nudum* var. *g.* Jeps. *E. latifolium* ssp. *g.* S. Stokes.] Caudex woody at the base, few-branched with elongated leaf-bearing areas 2–3 dm. long at the base; lf.-blades oblong-ovate, 3–10 cm. long, greenish above, closely white-woolly below, strongly undulate-crisped, the petioles much longer; flowering stems 8–15 dm. long, glabrous, glaucous, forking above; invols. in clusters of 1–3, turbinate, 5–6 mm. long, sub-

glabrous without; calyx whitish, ca. 3 mm. long, the segms. oblong-obovate, spreading; fils. pilose basally; aks. 2.5–3 mm. long; $2n = 40$ (Stokes & Stebbins, 1955).—Bluffs and cliffs, Coastal Sage Scrub, Chaparral; Santa Cruz, Santa Catalina, Anacapa and San Clemente ids. June–Oct.

A form from San Miguel and the w. end of Santa Cruz ids., is lower and more decumbent with a tendency toward subcapitate cymes with red fls. and may be known as var. **rubéscens** (Greene) Munz. [*E. r.* Greene. *E. latifolium* ssp. *r.* S. Stokes.]

97. **E. elàtum** Dougl. ex Benth. Caudex woody, branched or simple; lvs. basal, erect, the lf.-blades lanceolate to lance-ovate, 4–15 cm. long, acutish, green and glabrate above, somewhat tomentose below but not hoary, the petioles ca. as long, villous; flowering stems 4–8 dm. long, glabrous, glaucous, repeatedly trichotomous above, somewhat inflated in some; invols. in terminal clusters of 2–4, solitary in some, glabrous or somewhat pubescent, turbinate, ca. 4 mm. long, 5-toothed; calyx white or pinkish, 2.5 mm. long, pubescent without, the segms. obovate; fils. glabrous except at very base; aks. brownish, ca. 4 mm. long, subovoid with a rather prominently 3-angled beak; $n = 20$ (Hitchcock, 1964; Reveal, 1967).—Dry rocky slopes, 4000–9500 ft.; Sagebrush Scrub, N. Juniper Wd., Yellow Pine F.; Kern Plateau, Kern Co., Mono and Eldorado cos. to Modoc, Siskiyou and Trinity cos.; to Wash., Ida., Nev. June–Sept.

Var. **villòsum** Jeps. [*E. e.* var. *incurvum* Jeps. *E. e.* ssp. *glabrescens* S. Stokes.] Flowering stems villous-pubescent.—Panamint Mts. to Siskiyou Co.; w. Nev.

98. **E. péndulum** Wats. Base woody, few-branched, ascending or decumbent, 1–2.5 dm. high below the flowering branches; lvs. crowded near tips of basal branches, lf.-blades lance-oblong, (1.5–) 2–4 (–5) cm. long, obtusish, thinly floccose above, densely white-tomentose beneath, short-petioled (less than 5 mm.); flowering stems white-tomentose, 2–5 dm. high, leafy-bracted at nodes; invols. solitary, bractless, sessile or peduncled, the peduncles stout, to 1 dm. long, invols. turbinate-campanulate, white-tomentose, 3.5–5 mm. high, with 6–8 shallow lobes; calyx densely villous, 3–6 mm. long, the segms. narrow-oblong; stamens exserted, the fils. densely pilose; aks. villous, 3–5 mm. long.—Dry slopes below 3000 ft.; Mixed Evergreen F.; Del Norte Co.; adjacent Ore. Aug.–Sept.

99. **E. fascículatum** Benth. [*E. f.* var. *maritimum* Parish. *E. f.* var. *oleifolium* Gand. *E. aspalathoides* Gand. *E. f.* ssp. *a.* S. Stokes. *E. rosmarinifolium* Nutt.] CALIFORNIA BUCKWHEAT. Low spreading shrubs, the stems ± decumbent, 6–12 dm. long, branched, leafy; branchlets loosely pubescent to subglabrous, ending in leafless peduncles 3–10 (–15) cm. long, bearing ± open cymose infl. with many capitate clusters at the tips; lvs. numerous, fascicled, oblong-linear to linear-oblanceolate, green and glabrate above, white-woolly beneath, 6–15 mm. long, strongly revolute; invols. prismatic, 3–4 mm. high, glabrous, with 5 short acute teeth; calyx white or pinkish, ca. 3 mm. long, nearly or quite glabrous without, the outer segms. broadly elliptic, the inner obovate; fils. subglabrous basally; aks. lance-ovoid, light brown, angled, shining, ca. 2 mm. long; $2n = 40$ (Stokes & Stebbins, 1955), $n = 20$ (Reveal, 1967).—Dry slopes and canyons near the immediate coast; Coastal Sage Scrub; Santa Barbara to n. L. Calif.

The definition of var. *fasciculatum* as presented here is that which has been used in floras for many years. When Bentham described *E. fasciculatum,* he based his description on two collections, one by Menzies and one by Douglas. The Douglas specimens (BR, GH, K, MO, NY) which Reveal has seen represent what is here called var. *foliolosum.* The only Menzies collection that he has seen is deposited at Glasgow (GL) and does represent what is here called var. *fasciculatum;* there is no record of this collection at Kew. At present it is not known whether Bentham saw the Glasgow specimen or not, but it seems that in order to retain the definition of this species as understood for the last hundred years it will be necessary to typify the name on the Menzies collection. However, until all the English herbaria are investigated, such a step is not yet taken.

Var. **foliolòsum** (Nutt.) S. Stokes ex Jones. [*E. rosmarinifolium* var. *f.* Nutt. *E. fasciculatum* ssp. *foliolosum* S. Stokes. *E. f.* var. *obtusiflorum* S. Stokes.] Upper surface of lvs., outer surface of calyx, invols., etc. pubescent; peduncles 1–2 dm. long; $2n = 80$ (Stebbins, 1942).—Common on interior cismontane slopes and mesas, below 3000 ft.; Chaparral, Coastal Sage Scrub; Monterey Co. and San Benito Co. to n. L. Calif. Exceedingly variable. March–Oct.

Var. polifòlium (Benth.) T. & G. [*E. f.* ssp. *p.* S. Stokes. *E. p.* Benth.] Plants commonly 2–5 (–8) dm. tall; lvs. densely canescent to hoary above, commonly less revolute; invols. and calyx pubescent; heads solitary or in reduced cymes; 2n = 40 (Stebbins, 1942).—Common on dry slopes, below 7000 ft.; Sagebrush Scrub to Pinyon-Juniper Wd.; both deserts to San Joaquin V. and Inyo Co. and interior of s. Calif.; to Utah and Ariz., L. Calif. April–Nov.

Var. flavovíride M. & J. [*E. f.* ssp. *f.* S. Stokes.] Low, 2–3 dm. tall; lvs. yellow-green, subglabrous above, strongly revolute; peduncles glabrous; invols. and calyx subglabrous, the latter quite reddish.—Rocky places, below 4000 ft.; Creosote Bush Scrub; Eagle Mts., e. Riverside Co. to Little San Bernardino Mts. and Sheephole Mts., San Bernardino Co. March–May.

100. **E. arboréscens** Greene. Loosely branched shrubs 6–15 (–20) dm. tall, the stems to 1 dm. thick, with shreddy bark; branchlets tomentose when young, later glabrate, purplish and glaucous; lvs. in crowded terminal tufts, linear to oblong, revolute, 2–3 cm. long, densely white-tomentose below, glabrate above; infl. dense terminal leafy-bracted cymes, 5–15 cm. across; invols. tomentose, 3 mm. long, turbinate, with obtuse oval teeth; calyx whitish to pink, 2 mm. long, villous at the base; fils. glabrous; aks. lance-ovoid, shining, angled, ca. 2.5 mm. long; n = 20 (Reveal, 1968).—Rocky slopes and canyon walls; Coastal Sage Scrub, Chaparral; Santa Cruz, Santa Rosa and Anacapa ids. April–Sept.

101. **E. gigantèum** Wats. St. Catherine's Lace. Coarse rounded branching shrubs, open, 3–20 (–35) dm. high, the cent. trunk to 1 dm. thick, the younger branches tomentose, then glabrate and dark, with lvs. toward the tips; lf.-blades leathery, oblong-ovate to ovate, 3–7 (–10) cm. long, closely white-tomentose below, cinereous and somewhat glabrate above, on stout petioles 1–3 cm. long; peduncles stout, 1–3 dm. long, tomentose, later glabrate, bearing large 2–3-forked horizontal cymes often several dm. across and with leafy bracts at the forks; invols. crowded, campanulate; 3–4 mm. long, tomentose, with short obtuse teeth, subsessile or on short slender peduncles; calyx white, becoming rusty in age, ca. 2 mm. long, white-hairy, the segms. obovate; fils. hairy; aks. brown, shining, narrow-ovoid, angled above, ca. 2 mm. long; n = 20 (Reveal, 1968).—Dry slopes; Chaparral, Coastal Sage Scrub; Santa Catalina Id. May–Aug.

Var. compáctum Dunkle. Lvs. oblong; tomentum on young growth looser; involucral peduncles very stout.—Santa Barbara Id.

Var. formòsum K. Bdg. [*E. f.* K. Bdg. *E. g.* ssp. *f.* Raven.] Lvs. oblong-lanceolate.—San Clemente Id.

102. **E. parvifòlium** Sm. in Rees. [*E. p.* var. *commune* Benth. *E. p.* var. *crassifolium* Benth.] Shrubs with loosely branched decumbent or prostrate stems 3–10 dm. long, thinly floccose, densely leafy to the summit; lvs. fascicled, round-ovate to lance-oblong, thickish, revolute, sometimes cordate at the base, 5–15 mm. long, on shorter petioles, the lf.-blades green and glabrate above, densely white-tomentose below; flowering stems few, mostly 2–5 cm. long, simple or forked, bearing compact heads 1–2 cm. in diam.; invols. glabrate or somewhat woolly, turbinate-campanulate, 3–4 mm. long; calyx white or tinged rose, glabrous, ca. 3 mm. long, the segms. obovate; fils. pilose basally; aks. ovoid-deltoid, shining, brown, 2.5 mm. long; 2n = 40 (Stokes & Stebbins, 1955).—Common on bluffs and dunes along the coast; Coastal Strand, Coastal Sage Scrub; Monterey Co. to San Diego Co. Mostly summer, but with some fls. throughout the year.

Two ill-defined forms are: (1) with greenish-yellow fls., Point Lobos, Monterey Co., var. lùcidum (J. T. Howell ex S. Stokes) Reveal, stat. nov. [*E. p.* ssp. *lucidum* J. T. Howell ex S. Stokes, Gen. Eriog. 87. 1936.] and (2) diffusely branched, with lanceolate lvs. 15–30 mm. long and white fls. in heads scarcely 1 cm. in diam. and broad infl. 1–2 dm. in diam.; from Santa Paula Canyon, Ventura Co., var. Pàynei (C. B. Wolf ex Munz) Reveal, stat. nov. [*E. p.* ssp. *P.* C. B. Wolf ex Munz, *Aliso* 2: 80. 1949.]

103. **E. cinèreum** Benth. Freely branched shrubs, 6–15 dm. high, tomentulose, leafy below the infl.; lvs. ovate, 1.5–3 cm. long, obtuse, cuneate at the base, greenish-cinereous above, white-tomentulose below, crisped-undulate; short-petioled; flowering stems elongated, dichotomous, with scattered heads; invols. cylindric-turbinate, tomentulose, 3–4 mm. long, somewhat angled, 5-toothed; calyx densely white-villous, ca. 3 mm. long, the segms. narrow-obovate, whitish to pinkish; fils. subglabrous; aks. brown, deltoid-ovoid, sharply angled, somewhat roughened, ca. 2 mm. long; n = 40 (Reveal, 1968).—Beaches

and bluffs near the coast; Coastal Strand, Coastal Sage Scrub; Santa Barbara to San Pedro, Santa Rosa Id. June–Dec.

Subgenus Clastomỳelon Cov. & Mort.

104. **E. intrafráctum** Cov. & Mort. Perennial, woody at the base, from distinct taproots; lvs. basal, oblong-ovate, somewhat whitish-pilose, the lf.-blades 2.5–7 cm. long, the petioles somewhat longer; flowering stems usually solitary, simple below, sometimes branched in infl., rather stout, glabrous, glaucous, transversely jointed into hollow ring-like segms., each segm. 3–10 mm. long, becoming easily fractured, 6–12 dm. high; infl. usually of 2–3 virgate branches 2–4 dm. long and sometimes with shorter secondary branches; invols. usually in whorls of 3 at each node, usually 1 in the axil of each of 3 bracts, 5-parted into oblong lobes which become more divided with expanding fls., short-pilose; calyx yellow, tinged with red, pubescent, ca. 2 mm. long, the lobes oblong-lanceolate, subequal; aks. flask-shaped, brownish, almost 2 mm. long, 3-ridged on lower part, then abruptly narrowed into triangular beaks.—Local and rare, limestone crevices, 2000–5000 ft.; Creosote Bush Scrub; Grapevine and Panamint mts., Inyo Co. May–Oct.

p. 355.   RUMEX
          Key AA, B, CC, DD, E, FF, change to:

> **G.** Valves in fruit deltoid, acute, one valve with a large ovate callosity; panicle branches often ascending.
> **7. *R. salicifolius***
> **GG.** Valves in fruit ovate to ovate-lanceolate, generally all with prominent callosities; panicle-branches curved-spreading.
> **8. *R. transitorius***

p. 356.   3. **R. paucifòlius** Nutt. ex Wats. Various chromosome counts have been reported: $2n = 28$ (Löve & Sarkar, 1956); $2n = 14, 28$ (B. W. Smith, 1958). In 1967 Löve and Evenson (Taxon 16: 423–425) gave $2n = 14$ for the Rocky Mountain plant and $2n = 28$ for the Sierran. They referred the former to *Acetosa paucifolia* (Nutt.) Löve and the latter to *A. gracilescens* (Rech. f.) Löve & Evenson.
          4. **R. venòsus** Pursh. $2n = 40$ (Sarkar, 1958).
p. 357.   5. **R. califórnicus** Rech. f. $2n = 20$ (Sarkar, 1958).
          R. utahénsis Rech. f. For 1959 printing transpose lines 2 and 3.
          6. **R. crássus** Rech. f. $2n = 20$ (Sarkar, 1958).
          7. **R. salicifòlius** Weinm. $2n = 20$ (Sarkar, 1958).
          8. **R. transitòrius** Rech. f. $2n = 20$ (Sarkar, 1958).
          10. **R. trianguliválvis** (Danser) Rech. f. California plants are var. **oreolápathum** Rech. f.
p. 358.   11. **R. hymenosèpalus** Torr. ranges n. to Monterey Co., *Howitt & Howell.* $2n = 40$ (Löve & Patil, 1967).
          12. **R. fenestràtus** Greene. $2n = $ ca. 200 (Löve, 1967).
          13. **R. occidentàlis** Wats. $2n = $ ca. 140 (Wellington, 1957).
p. 359.   19. **R. stenophýllus** Ledeb. $2n = 60$ (Löve, 1967).
          23. **R. fuegìnus** Phil, $2n = 40$ (Löve & Löve, 1967).
          24. **R. persicarioìdes** L. $2n = 40$ (Löve & Löve, 1967).
p. 360.   POLÝGONUM
          In Key to Species, after A, B, C, D, EE, F insert:

> **G.** Branch-lvs. much smaller than stem-lvs.; perianth divided almost to base; fr. trigonous with 3 concave sides.
> **4. *P. aviculare***
> **GG.** Branch-lvs. not much smaller than stem-lvs.; perianth divided ca. half its length; fr. with 2 sides convex, 1 concave.
> **4a. *P. arenastrum***

p. 360.   In Key A, B, CC, D, EE, FF, change to:

> **GG.** Upper lvs. not reduced to bracts; calyx with yellowish margins.

H. Pedicels exserted from the ocreae; fls. clustered at apex of stems; lvs. lacking conspicuous lateral veins.

7. *P. ramosissimum*

HH. Pedicels included within the ocreae; fls. not markedly clustered at apex of stems; lvs. rugulose-veiny when dry. . . . . 7a. *P. prolificum*

p. 361. In key in line 2 at top of page, insert "heads or" between "or" and "in." In Key AA, change BB to:

BB. Infl. of heads or open panicles and the lvs. broad, or the infl. of small axillary clusters and terminal spike with the lvs. cordate-sagittate.

B′. Fls. in dense globular heads on long peduncles; lvs. ovate to elliptic; stems prostrate, rooting. . . . . 36. *P. capitatum*

B′B′. Fls. in open panicles or in small axillary clusters and terminal spikes.

C. Stems twining; plant annual, etc., as in FLORA.

3. Change **P. Fòwleri** Rob. to **P. marinénse** Martens & Raven with shining aks., whereas they are dull in *P. Fowleri.* **P. marinense** ranges in saline marshes of Marin Co., but *P. Fowleri* is from Puget Sound n. and e.

p. 362. 4a. **P. arenástrum** Bor. Forming dense prostrate mats 1–16 dm. across; lvs. to 20 mm. long, 5 mm. wide; infl. 2–3-fld.; calyx greenish-white or pink; fr. 1.5–2.5 mm. long, brown to black; $2n = 40$. San Clemente Id., *Raven.* From Eu.

6. **P. pátulum** Bieb. $2n = 20$ (Löve & Löve, 1956).

7a. **P. prolíficum** (Small) Rob. [*P. ramosissimum* var. *p.* Small.] Differing from *P. ramosissimum* in its pedicels (included within, rather than exserted from the ocreae) and less shiny frs.; fls. not markedly clustered at apex of stems.—Found at Cutting's Wharf on Napa R., Napa Co., *Howell;* central and e. N. Am.

8. **P. Douglásii** Greene. $2n = 40$ (Löve & Löve, 1956).

Var. **latifòlium** (Engelm.) Greene reported from the Greenhorn Range, Kern Co., *Twisselmann.*

p. 363. 12. **P. Kellóggii** Greene. Add to the synonymy [*P. imbricatum* auth.]

16. **P. califórnicum** Meissn. S. in Sierra Nevada to Fresno Co., *Weiler.*

p. 366. 35. **P. sachalinénse** F. Schmidt ex Maxim. Now reported from Calif. as in Humboldt, Mendocino, and Siskiyou cos., *Fuller;* in Grass V., Nevada Co., and Camiso, El Dorado Co., *Fuller;* Atascadero, San Luis Obispo Co., *Fuller.*

36. **P. capitàtum** Ham. in Don. With many stems or branches creeping from a woody rootstock, leafy, glandular-hirsute; lvs. 1–3 cm. long, elliptical, acute, short-petioled, the petiole 2-auricled at its base; stipules short-cupular; heads 1–3, 6–18 mm. diam.; peduncles glabrous or glandular hipid; calyx pink, 5-cleft, the segms. obtuse.—Garden escape at Pacific Grove, Monterey Co., *Howitt & Howell.*

p. 368. CHENOPÒDIUM

In Key under A, BB, C, DD, add:

E. Plants annual, prostrate; seeds 0.9–1.1 mm. broad; pericarp gray-striped or mottled. Coastal sand, San Luis Obispo and Santa Barbara cos. . . . . 3a. *C. carnosulum*

EE. Plants mostly perennial, ascending or erect; seeds 0.7 mm. broad; pericarp thin, deciduous, gland-dotted. Widespread.

3. *C. ambrosioides*

In Key AA, B, CC, DD, E, F, G, delete "18. *C. album*" at end of line and add:

H. Blades at least 1½ times as long as wide; sepals not united to broadest part of fruit, variously keeled.

18. *C. album*

HH. Blades scarcely if at all longer than broad, basal lobes often bipartite; sepals united to or above broadest part of fruit, usually strongly keeled. 18a. *C. opulifolium*

p. 369. In Key under AA, B, CC, DD, EE:

F. Lf.-blades linear to narrow-lanceolate or narrow-oblong, short-petiolate, the blades mostly 1–3-nerved.

G. Lvs. entire, 1-nerved, mostly 2–3 mm. wide.

11. *C. leptophyllum*

GG. Lvs. narrow-lanceolate or broader, the lower 4–18 mm.
 wide. . . . . 12. *C. desiccatum*
FF. Lf.-blades lance-ovate to ovate or broader, long-petioled, pin-
 nately veined.
 G. Main lf.-blades definitely longer than broad.
  H. Pericarp separable; lvs. mostly entire; seed 1 mm.
   broad.
   I. Lvs. oblong or oval, 0.8–2 cm. long, mostly less
    than ⅓ as broad. . . . . 12. *C. desiccatum*
   II. Lvs. ovate to triangular-oblong, 1.5–3 cm. long,
    mostly more than ⅓ as broad. . . 15. *C. atrovirens*
  HH. Pericarp attached; lvs. sometimes toothed.
   I. Seeds 1–1.5 mm. broad; plants mostly 3–12 dm.
    high, openly branched.
    J. Lvs. thin, ovate-lanceolate or broader; seeds
     1.2–1.5 mm. broad. . . . 14. *C. incognitum*
    JJ. Lvs. firm, ovate-lanceolate or narrower;
     seeds 1.0 mm. broad. . . . . . 12a. *C. hians*
   II. Seeds 0.7–0.8 mm. wide; plants 2–3 dm. high,
    bushy-branched . . . . 16. *C. nevadense*
GG. Main lf.-blades scarcely if at all, etc. as in the FLORA.

Key under AA, BB, CC, D, change EE to:

  EE. Fls. in large spicate glomerules; calyx fleshy and bright red in fr.
   F. Lvs. truncate to cordate-hastate at base, the margins usually
    strongly toothed; principal glomerules in well developed plants
    usually 6–10 mm. in diam.; stigma 0.3–0.4 mm. long, flexuous.
               6. *C. capitatum*
   FF. Lvs. tapering or truncate-hastate at base, the margins some-
    what toothed or entire; glomerules usually smaller; stigmas
    chiefly 0.1–0.2 mm. long, squarrose. . . . . 6a. *C. Overi*

p. 370. 3a. **C. carnosùlum** Moq. var. **patagónicum** (Phil.) Wahl. [*C. p.* Phil.] Prostrate annual, the branches 1–3 dm. long; primary lvs. narrow-ovate, to 9 mm.
wide, the base cuneate.—Sand near coast, San Luis Obispo and Santa Barbara
cos.; Chile.
  4. **C. Bòtrys** L. $2n = 18$ (Mulligan, 1961).
  6. **C. capitàtum** (L). Asch. $2n = 18$ (Mulligan, 1957).
  6a. **C. Òveri** Aellen. [*Blitum hastatum* Rydb., not *C. h.* Phil.] Stem slender,
2–4 dm. tall; lvs. very thin, the blades 3–7 cm. long, ovate to lance-ovate, the
upper smaller, not hastate; fls. in the upper axils and in slender, interrupted,
terminal spikes.—Lassen Co.; to Rocky Mts.
  8. **C. chenopodioìdes** (L.) Aellen, not *C. rubrum* L.
p. 371. 12. **C. desiccàtum**, not *dessicatum*.
  Var. **leptophylloìdes** (J. Murr.) Wahl. Reported from Monterey Co., *Howitt*
and *Howell*.
  12a. **C. hìans** Standl. Ill-scented annual, 4–8 dm. high, sparsely branched,
copiously farinose; petioles stout, up to half as long as the blades; lf.-blades
elliptic-oblong to narrowly lance-oblong, 1.2–3 dm. long, green and glabrate
above, white-farinose beneath; glomerules large, in stout dense erect spikes;
calyx farinose; pericarp closely adherent; seed 0.8–1 mm. broad, nearly smooth.
—Inyo and Mono cos. to Siskiyou Co.; to New Mex. and n.w. U.S. and adjacent
Canada.
  15. **C. atròvirens** Rydb. [*C. Fremontii* var. *a.* Fosberg.]
  17. **C. Vulvària** L. reported as in Monterey Co., *Howitt & Howell*.
p. 372. 18. **C. álbum** L. Wahl has identified a *Howell* specimen from San Bernardino
as **C. missouriénse** Aellen, differing from *C. album* in having seeds 0.9–1.2 mm.
broad, not 1.1–1.5 mm. Native of cent. U.S.
  18a. **C. opulifòlium** Schrad. Near to *C. album* (annual, 3–8 dm. tall, farinose,
glaucous) but with thicker lvs. with basal lobes so that the blade is subtrilobed,
the median lobe is short and the blade as a whole is scarcely longer than wide.
—An Eurasian plant sparsely introd. in this country, exemplified from California
by: King City, Monterey Co., *K. Esau* in 1927 and by 5 miles nw. of College
City, Colusa Co., *J. H. Thomas* in 1960.
  20. **C. Berlandièri** Moq. var. **sinuàtum** (J. Murr.) Wahl. [Add to possible

synonymy *C. B.* var. *californicum* Aellen.] This var. is supposed to have thin membranous lvs. Most Calif. material heretofore referred to *C. album* seems to belong here. It may be distinguished with difficulty from

Var. **Zscháckei** (Murr.) Murr. [*C. Z.* Murr.] Lvs. larger, thin to coriaceous; sepals usually strongly keeled; seeds 1.2–1.5 mm. in diam.—Reported by *Howell* from Marin and Monterey cos., by *Twisselmann* as a widespread summer weed in Kern Co. It seems to occur in alkaline spots from San Bernardino Co. to Lassen Co.

CYCLOLÒMA

1. **C. atriplícifolium** (Spreng.) Coult. n. to Tehama Co., *Howell.*

p. 373. MONOLÈPIS

1. **M. Nuttalliàna** (Schult.) Greene ascends to 10,000 ft. in White Mts.

p. 374. ATRIPLÉX

Key, AA, B, delete "32. *A. canescens*" and insert:

C.   Lvs. narrowly spatulate to narrowly oblong, 1.5–5 cm. long; fruiting bracts stalked. Common. .... 32. *A. canescens*
CC.   Lvs. oblong to subobovate, 1.2–2 cm. long; fruiting bracts hardly stalked. Rare weed. .... 33. *A. Vesicaria*

p 375. 2. **A. pátula** ssp. **hastàta** (L.) Hall & Clements. $2n = 18$ (Löve, 1964).
3. **A. ròsea** L. $2n = 18$ (Mulligan, 1957).

p. 376. 9. **A. Serenàna** A. Nels. reported from interior Monterey Co., *Howitt & Howell.*

p. 378. 19. **A. semibaccàta** R. Br. Found as far n. as Monterey Co., *Howitt & Howell.*

p. 379. 26. **A. lentifórmis** (Torr.) Wats. ssp. **Breweri** (Wats.) Hall & Clem. Found near Maricopa, Kern Co., *Twisselmann.*

p. 380. 33. **A. Vesicària** Heward in Hook. f. Bushy shrub with a scaly white tomentum; lvs. oblong to subobovate, 12–20 mm. long, short-petioled; ♂ fls. in small clusters forming dense leafless spikes 12–25 mm. long; ♀ fls. few together, in axillary clusters; fr. bracts suborbicular, 6–10 mm. in diam., entire, flat, but each with a membranous inflated appendage on the disk nearly as large as the bract itself.—Weed in the Northridge area, Los Angeles Co., *Fuller*; native of Australia.

p. 381. KÒCHIA

3. **K. scopària** (L.) Schrad. var. **subvillòsa** Moq. A very hairy form reported from Santa Barbara and San Francisco.

Var. **cúlta** Farwell. With dense ovoid to globular habit, very narrow lvs. mostly with long hairs particularly toward the base, purple-red in autumn. Escape in Fresno, *Fuller.*

p. 383. SUAÈDA

1. **S. depréssa** (Pursh) Wats. $2n = 36$ (Mulligan, 1965).

p. 384. SALSÒLA

Dr. T. C. Fuller has sent specimens to two European botanists for study: Dr. P. Aellen of Basel and Dr. V. Botschantzev of Leningrad. Apparently our common sp., introduced from Russia, should be called **S. pestífera** A. Nels. [*S. kali* ssp. *ruthenica* (Ilgin) Soó.] In it the fruiting calyx is 3–6 mm. broad.

A new record for Calif. is **S. Paulsènii** Litv. Plant 1–5 dm. tall, ca. 5–6 dm. across, glabrous or sparsely papillose; lvs. 1.5–3 cm. long, semicylindrical, mucronate, yellow; bracts ovate at base, with a linear spinose apex; bracteoles partly connate with the solitary fls.; perianth with a short tube and small stiffly erect spinose tips to the segms.; wings membranous, veined, the fruiting calyx 8–9 mm. wide.—Common in disturbed ground 2 mi. e. of Barstow, San Bernardino Co., *T. C. Fuller*; and at 6000 ft. in San Bernardino Mts.; native of Russia.

HALOGÈTON

1. **H. glomeràtus** (Bieb.) C. A. Mey. in Led. Reported from e. Mojave Desert (Halloran Summit, etc.), *Fuller.*

p. 385. AMARANTHÀCEAE

Key to Genera:

A. Lvs. alternate; anthers 4-celled; plants nearly or quite glabrous. ..... 1. *Amaranthus*
AA. Lvs. largely opposite; anthers 2-celled; plants white stellate-woolly or villous.
    B. Fls. glomerate, with an invol. of upper lvs.
        C. Stamens perigynous. Mostly deserts. .... 2. *Tidestromia*
        CC. Stamens hypogynous. Santa Barbara Co. .... 2a. *Brayulinia*
    BB. Fls. in axillary headlike spikes, without invol. Mostly near beaches. 3. *Alternanthera*

AMARÁNTHUS
Key to Species, change to:

A. Sepals of the ♀ fls. broadened upward, the calyx ± urceolate.
    B. Fls. dioecious; sepals of ♀ fls. not fimbriate.
        C. Bracts 2–3 times as long as the pistillate calyx.
            D. Infl. not leafy; bracts rigid and spinose; plants not viscid-pubescent.
                                    1. *A. Palmeri*
            DD. Infl. leafy at least below; bracts not spinose; plants viscid-pubescent.
                                    2. *A. Watsonii*
        CC. Bracts not longer than pistillate calyx. .... 2a. *A. arenicola*
    BB. Fls. monoecious; sepals of ♀ flls. fimbriate. .... 3. *A. fimbriatus*

AA. etc. as in the FLORA.
In the Key AA, BB, CC, D substitute:

    E. Style-branches recurved; lateral branches of infl. few to none.
        F. Base of style-branches slender, forming shallow saddle; midrib
            of bract very slender, rather long excurrent; sepals obovate to
            spatulate, obtuse or emarginate, recurved; lateral spikes of infl.
            few to none. .... 6a. *A. caudatus*
        FF. Bases of style-branches stout, forming cleft at summit of broad
            tower; midrib of bract very thick, excurrent; sepals oblong,
            acute, straight; lateral spikes of infl. long, few, widely spaced.
                        6. *A. Powellii*
    EE. Style-branches erect; lateral spikes of infl. numerous, crowded.
        F. Sepals oblong, acute, straight; midrib of bract long, excurrent;
            style-branches with slender bases.
            G. Lateral spikes of infl. long; sepals very short, 1.5 mm.
                long; bracts to 1.5 times as long as sepals, usually red or
                purple. .... 7a. *A. cruentus*
            GG. Lateral spikes of infl. short; sepals moderately long, 1.5–2
                mm.; bracts twice as long as sepals, usually green or pale
                reddish. .... 7. *A. hybridus*
        FF. Sepals narrowly obovate, emarginate, recurved; midrib of bracts
        barely excurrent; style-branches with moderately stout bases.
                        5. *A. retroflexus*

DD. "Infl. wholly of axillary glomerules," etc. as in the FLORA.

p. 386.   1. **A. Pálmeri** Wats. $2n = 45$ (Grant, 1958). In the San Joaquin V., *Twisselmann.*

2a. **A. arenícola** Jtn. Differing from *A. Palmeri* and *A. Watsonii* (both of which have bracts 2–3 times as long as the ♀ calyx) in the bracts not exceeding the ♀ calyx which is 2.5–3 mm. long—Adventive in Monterey and Santa Barbara cos., *Howell*; native Ida. to Colo.

4. **A. defléxus** L. $2n = 34$ (Grant, 1959).

A sp. resembling *A. deflexus*, but with wrinkled seeds is **A. gracilis** Desf. Reported from San Francisco.

5. **A. retrofléxus** L. $2n = 34$ (Grant, 1959).

6. **A. Powéllii** Wats. $2n = 34$ (Grant, 1959).

6a. **A. caudàtus** L. Infl. thick and pendulous, terminal spike extremely long, laterals few and short or absent; bract short or medium, with slender, rather long-excurrent midrib; sepals recurved, broadly obovate or spatulate, obtuse to emarginate, 1.5–2 mm. long.—Widely introd. in N. Am. Native of Cent. and S. Am.

7. **A. hýbridus** L. $2n = 32$ (Grant, 1959).

7a. **A. cruéntus** L. Infl. lax, the terminal spike short, the laterals long, very numerous and crowded; bracts extremely short, not much longer than the sepals, with long-excurrent midrib; sepals straight, 1.5 mm. long, oblong, acute, Widely introd. in N. Am. Native in Cent. and S. Am.

8. Apparently the correct name for this sp. is **A. blitoìdes** Wats. for Am. plants and *A. graecìzans* L. for European. $2n = 32$ (Grant, 1959).

Tucker and Sauer (Madroño 14: 252–261. 1958) in discussing weedy populations of *Amaranthus* from the Sacramento-San Joaquin Delta presented evidence to show that several spp. were involved, which added 6a and 7a as here presented to those already treated in the FLORA.

11. **A. spinòsus** L. $2n = 34$ (Grant, 1959).

p. 387.   TIDESTRÒMIA

1. **T. oblongifòlia** (Wats.) Small. Wiggins recognizes ssp. **cryptántha** (Wats.) Wiggins on basis of smaller lvs. (2–10 mm. long); invol. deeper (3–4 mm.) and about the Salton Sea.

2a. Brayulínea Small

Prostrate to decumbent perennial herbs; stems branched at base, often zigzag. Fls. perfect, subtended by bracts and in axillary clusters. Sepals 5, pubescent. Stamens 5, perigynous; fils. broad; anthers 1-celled. Ovary flattened, 1-celled; style short; stigma notched. Utricle membranous, indehiscent.

1. **B. dénsa** (Willd.) Small. [*Illecebrum d.* Willd.] Prostrate perennial, densely lanate; cauline lvs. wing-petioled, elliptic to broadly oval, 3–15 mm. long; fls. densely glomerate; bracts ovate, scarious, white; calyx 2–2.5 mm. long. —Reported from Lompoc, Santa Barbara Co., *Howell.* Native from Ariz. to Texas and S. Am.

p. 388.   ALTERNÁNTHERA

1. *A. rèpens* (L.) Kuntze. Change to **A. pungens** HBK, since the comb. *A. repens* had earlier been made for another sp.

2. *A. philoxeroìdes* (Mart.) Griseb. A bad weed in irrigation canals, as at Visalia, Tulare Co., *Fuller.*

PHYTOLÁCCA

2. **P. heterotépala** H. Walter. In San Francisco, *Howell.* Differs from *P. americana* in having sepals unequal; stamens 13–20, while *P. a.* has sepals equal; stamens 10.

p. 389.   In caption under Fig. 40 change "inferior" ovary to "superior."
p. 392.   OXÝBAPHUS

4. **O. pùmilus** (Standl.) Standl. Found at Kenworthy, Hemet V., San Jacinto Mts., *Ziegler.*

p. 393.   MIRÁBILIS

4. **M. laèvis** (Benth.) Curran. At 6 mi. se. of Friant Dam on San Joaquin R., Fresno Co. and at other spots in Sierran foothills, *Quibell.*

p. 394.   ABRONIA

Key, in line 3 from bottom, omit word "annuals."

S. S. Tillett (Brittonia 19: 299–327. 1967) has a study of our maritime spp. and proposes:

A. latifòlia and *A. maritíma* are perennial, with perianth limb reflexed and without central eyespot. *A. grácilis* and *A. umbellàta* are annual, with the perianth limb plane and with evident central white eyespot. He separates *A. gracilis* Benth. from *A. umbellata* Lam. as follows:

Lvs. very thin, oval, deeply crenate to sinuately lobed; open sandy areas of scrub at low to middle elevations. Baja Calif. and possibly Imperial and San Diego cos. . . *A. gracilis* Benth.
Lvs. thicker, oval to elliptic or rhomboidal, asymmetrical, the margin entire to somewhat irregular; strand and disturbed areas in coastal scrub bordering strand. . . *A. umbellata* Lam.

p. 395.   6. **A. umbellàta** Lam. Tillett keys out 3 sspp. of which 2 occur in Calif.:

Perianth light to dark magenta, displayed in a nearly hemispheric umbel, the tube 9–13 mm. long, the limb 7–16 mm. in diam.; wings of anthocarp very well developed, broadly rounded; Sonoma Co. to Baja Calif. . . . . ssp. *umbellata*
Perianth light magenta, somewhat yellowed, displayed in a poorly opened umbel, the tube

6.5–10 mm. long, the limb 6–8.5 mm. diam.; wings less well developed, angled above; Marin Co. to s. Ore. . . . . ssp. *breviflora*

Tillett finds F₁ hybrids between *A. maritima* and *A. umbellata* generally perennial, the pubescence of the lvs. as in *A. m.* Lf. shape and thickness intermediate. Umbel angle as in *A. u.* Perianth tube short, the throat wide as in *A. m.* The corolla limb like *A. u.* in size, reflexed as in *A. m.* Eyespot present.

A. álba Eastw., *A. insuláris* Standl., *A. neurophylla* Standl., *A. platyphýlla* Standl. and *A. variábilis* Standl. represent plants with introgression between *A. umbellata* and *A. maritima*.

A. mìnor Standl. is based on plants of *A. umbellata* introgressed by *A. latifòlia.* Lvs. broad, perianth somewhat reflexed, throat less constricted, anthocarp wings smaller and not extending above the apex, yellow color on the tube and underside of the perianth limb.

p. 396.   6. **A. umbellata** Lam. ssp. **platyphylla** (Standl.) Ferris, not Munz.

BATIDÀCEAE

For 1959 printing, in line 2 from bottom of page, change "exstipulate" to "stipulate."

p. 398.   STŸRAX

1. **S. officinalis** var. **fulvescens** (Eastw.) Munz & Jtn. $2n = 8$ pairs (Raven, Kyhos & Hill, 1965).

p. 403.   DODECÁTHEON

9. **D. pulchéllum** (Raf.) Merr. In the synonymy *"Eximie"* should be *"Exinia."*

p. 404.   CENTÚNCULUS

The genus is best reduced to synonymy under **Anágallis** and the usable binomial is **Anágallis mínima** (L.) Krause.

p. 406.   PLANTÀGO

2. **P. heteróphylla** group. I. J. Bassett (Can. J. Bot. 44: 467–479. 1966) in a study of the spp. referred in the FLORA to *P. heterophylla* and *P. Bigelovii*, recognizes the following for California; which can be keyed out under A, BB, C as follows:

> D.   Caps. 10–25-seeded; seeds 0.5–0.8 mm. long. . . . . . . 2. *P. heterophylla*
> DD. Caps. 4–9-seeded; seeds 0.75–2.5 mm. long.
>   E.   Corolla lobes mostly erect in age, forming a beak; seeds 4, ca. 0.75–
>   1.8 mm. long. . . . . 2a. *P. pusilla*
>   EE. Corolla lobes spreading or reflexed in age, not forming a beak; seeds
>   4–9, 1.5–2.5 mm. long.
>   F.   Scape and lvs. mostly erect; plants 5–15 cm. high; seeds mostly
>   4–5, roughly or finely rugose-pitted, dark brown, elliptic oblong,
>   1.75–2.5 mm. long. . . . . 3a. *P. elongata*
>   FF. Scape and lvs. mostly decumbent to semierect; plants 1.5–8 cm.
>   high; seeds 4–9, irregularly and coarsely pitted, dark brown to
>   black, slightly angled in outline, 1.5–2 mm. long.
>   3. *P. Bigelovii*

2. **P. heteróphylla** Nutt. Occurs only in the e. U.S.

2a. **P. pusílla** Nutt. Plant pubescent to glabrous, erect or strongly ascending, 2–10 cm. high; lvs. ⅓ to ¾ as high as the scape, all basal; scapes several to many; spikes 1.5–6 cm. long; bracts triangular-ovate, scarious-margined, slightly shorter than to equaling the calyx, 1.5–2 mm. long; sepals obovate; corolla lobes 0.5 mm. long, mostly erect and forming a beak over the caps.; caps. ovoid, circumscissile below the middle, ca. 2 mm. long; seeds 4, dark brown, pitted, ca. ⅓ as wide as long, 0.75–1.25 (–1.8) mm. long; $2n = 12$.—Cited from San Diego; Ore., Wash., e. U.S.

p. 407.   3. **P. Bigelòvii** Gray. Plant ± pubescent, decumbent to semierect, 3–5 cm. high; lvs. entire, linear to subfiliform, 1–7 cm. long; bracts ovate, 2 mm. long, hyaline-margined; sepals broadly obovate, ca. 2 mm. long; corolla lobes spreading to sharply reflexed in fr., 0.5–1 mm. long; caps. oblong-ovoid, 2–3 mm. long, circumscissile just below the middle; seeds mostly 4–5, dark brown to black, oblong, slightly angled, irregularly and coarsely pitted, 1.5–2 mm. long; $2n = 20$.—Well distributed in cismontane Calif., mostly near the coast; to B.C.

Ssp. califórnica (Greene) Bassett. [*P. c.* Greene.] Plants 4–8 cm. tall; lvs. often with a few teeth; seeds mostly 6–9.—Mostly inland; central Calif. to L. Calif., Son.

3a. **P. elongàta** Pursh. Plants appressed-pubescent with septate hairs, mostly erect, 5–15 cm. high; lvs. linear to subfiliform, ⅓–¾ as high as the scape; scapes few to many; spikes mostly lax, 2.5–8 cm. long; bracts ovate, 2–2.5 mm. long, hyaline-margined; sepals ovate, 2–2.5 mm. long; corolla-lobes to 1 mm. long, rarely closing; caps. ovoid, circumscissile just below the middle, 2.5–3.5 mm. long; seeds mostly 4, uniform in size, mostly 4–5 mm. long, dark brown, roughly or finely rugose-pitted, 1.75–2.5 mm. long; $2n = 12$.—Alkaline areas, w. central Calif.; to s. Can. and w. Miss. Valley.

Ssp. **pentaspérma** Bassett. Lvs. and scapes usually subglabrous; spikes generally dense; seeds mostly 5, irregular in shape, one smaller than the others; $2n = 36, 12$.—Saline and alkaline places; central Calif. to B.C.

4. **P. eríopoda** Torr. $2n = 24$ (Bassett, 1967).

7. **P. virgínica** L. $2n = 24$ (Fujiwara, 1956).

p. 408.   13. **P. Púrshii** R. & S. var. **pícta** (Morris) Pilg. should be changed to var. **oblónga** (Morris) Shinners, *P. picta* Morris being antedated by *P. p.* Colenso.

14. **P. erécta** Morris. $2n = 20$ and ssp. **rigídior** Pilg. $2n = 42$ (Moore, 1962).

15. **P. insulàris** Eastw. $2n = 8$ (Moore, 1962); (Raven, Kyhos & Hill, 1965).

p. 409.   16. **P. índica** L. $2n = 12$ (Fujiwara, 1956).

p. 410.   LIMÒNIUM

In Key at top of page, after line 3 insert 4th line:

Fls. ca. 4 mm. long, white. Rare escape, San Francisco. . . . . 2a. *L. perfoliatum*

2a. **L. perfoliàtum** (Karelin ex Boiss.) Kuntze. [*Statice p.* Karelin.] Perennial, to 6 dm. high; lvs. entire, oblong-spatulate, 3–8 cm. long; fls. white, ca. 4 mm. long.—Escape from cult., salt marsh, Islair Creek, San Francisco; from Caspian Sea.

2. **L. Perèzii** F. T. Hubb. reported from 4 mi. n. of Paso Robles, San Luis Obispo Co., *Twisselmann.*

p. 411.   ERICÀCEAE

Key to Genera, A, B, CC, DD, change to:

DD. Corolla urn-shaped or tubular-campanulate.
    E.   Corolla urn-shaped; anthers mucronate; lvs. broad, petioled. Native alpine. . . . . **7.** *Leucothoe*
    EE. Corolla tubular-campanulate; anthers with 2 hairy awns at the base. Escape in n. Calif. . . . . **7a.** *Erica*

p. 412.   LÈDUM

1. **L. glandulòsum** Nutt. ssp. **columbiànum** (Piper) C. L. Hitchc. occurs in Monterey Co., *Howitt & Howell.*

p. 413.   RHODODÉNDRON

2. **R. occidentàle** (T. & G.) Gray. Found in Gabilan Range, Monterey Co., *Howitt & Howell* and on Fremont Peak, San Benito Co., *Arnaud.*

KÁLMIA

1. Add **K. polifòlia** Wang. [*K. p.* ssp. *occidentalis* (Small) Abrams. *K. o.* Small.] Differing from var. *microphýlla* (Hook.) Hall in usually being 2–4 dm. tall; lvs. mostly 2–4 cm. long, less than half as broad; fls. 12–18 mm. broad.—Siskiyou and Modoc cos. to Can. and ne. U.S.

p. 414.

7a. Érica L. HEATH

Large genus of shrubs and subshrubs native in S. Afr. and Medit. region. Lvs. usually in whorls of 3–6, small, needle-like. Fls. 1–many, usually nodding; calyx short, 4-parted; corolla withering-persistent, ± cylindrical, with 4 small lobes; stamens usually 8; ovary 4- or 8-celled, with 2–many ovules in each cell; caps. loculicidal.

1. **E. lusitánica** Rudolph. SPANISH HEATH. Erect, dense shrub, 2–3 m. tall;

young stems with short simple hairs; lvs. irregularly arranged or 3–5 in a whorl, glabrous; fls. very many, along entire length of branches, white or pink, ca. 4 mm. long, tubular-campanulate.—On cleared land near Eureka, Humboldt Co.; native sw. Eu.

p. 415.   GAULTHÈRIA

    2. **G. humifùsa** (Grah.) Rydb. In Eldorado Co.

    3. **G. ovatifòlia** Gray. In Sierra, Butte and Eldorado cos., *Howell.*

ARBÙTUS

    1. **A. Menziesii** Pursh. $2n = 26$ (Stebbins & Major, 1965).

p. 417.   ARCTOSTÁPHYLOS

In Key, A, BB, change C to:

> C.   Ovary with short stiff hairs.
>   D.   Young branchlets bristly hairy; fr. splitting open and falling early. Amador Co. to Calaveras Co. . . . . 2. *A. myrtifolia*
>   DD. Young branchlets with a fine, quickly deciduous pubescence; fr. not splitting as above. San Mateo Co. . . . . 7a. *A. pacifica*

p. 418.   In Key AA, BB, C, change D to:

> D.   Lvs. pale green or gray-green; plants without a basal burl.
>   E.   Bark rough and shreddy; erect shrub. . . . . 24. *A. morroensis*
>   EE. Bark smooth; prostrate or decumbent shrub. . . . . . 37a. *A. cruzensis*

p. 419.   In Key, AA, BB, CC, D, EE, FF insert:

> GGG.   Branchlets gray-tomentulose; bracts canescent. San Luis Obispo and Monterey cos. . . . . 29. *A. obispoensis*

In Key, AA, BB, CC, DD, E, FF, G, HH, change to:

> I.   Lvs. cordate or auriculate at base.
>   J.   Shrub 6–12 dm. high; corolla pinkish-white. Santa Cruz Mts. 30. *A. glutinosa*
>   JJ. Shrub 20–40 dm. high; corolla white. S. Santa Lucia Mts. . . . . . 35a. *A. Hooveri*

p. 420.   At end of Key change to:

> GG. Pubescence on branchlets glandular.
>   H.   Lvs. oblong-lanceolate to ovate-lanceolate; berries viscid. Marin Co. . . . . 34. *A. virgata*
>   HH. Lvs. elliptic or ovate to broadly oval; berries hairy, but not viscid. Monterey Co. 33a. *A. montereyensis*

    2. **A. myrtifòlia** Parry. Ranges quite widely in Amador and Calaveras cos., *Gankin.*

    3. **A. nissenàna** Merriam. Mostly low and sprawling; corolla white to pink, *Knight.* A study by Schmid, Mallory & Tucker (Brittonia 20: 34–43. 1968) indicates that *A. nissenana* and *A. viscida* may hybridize freely.

    4. **A. Ùva-úrsi** (L.) Spreng. var. **coáctilis** Fern. & Macbr. at Point Sur, Monterey Co., Sonora Pass and Convict Lake Basin in Sierra Nevada. Wiens & Halleck give $n = 13$ for a collection from Colo.

    5. **A. Edmúndsii** J. T. Howell seems likely to be a hybrid. J. B. Roof proposes var. **parvifòlia** with lvs. ca. 7 mm. long; frs. 6 mm. diam.; Little Sur R., Monterey Co.

p. 421.   7a. **A. pacífica** Roof. Near the *A. Hookeri* complex, forming carpets of pastel-green hue, on a sandstone outcrop at ca. 1100 ft., ne. slope of San Bruno Mt., San Mateo Co. Lvs. rounded to lanceolate, 10–18 mm. long, reticulate above and beneath, finely serrulate; infls. few, the bracts and rachises minutely puberulent; pedicels glabrous, to 1.5 mm. long; corolla white, slender, 4 mm. long; mature fr. flattened, 6 mm. broad; nutlets separable.

    7b. **A. Hearstiòrum** Hoov. & Roof is proposed as a new sp. near to *A. Hookeri* G. Don from near the Arroyo de la Cruz, Hearst Ranch, nw. San Luis Obispo Co. Resembling *A. Hookeri*, but more prostrate, forming mats; petioles ca. 1 mm. long; lf.-blades 10–18 mm. long; infl. small, mostly 3–6-fld.

p. 422.   13. **A. nevadénsis** Gray. In the Greenhorn Range, Kern Co., *Twisselmann.*

p. 423.   14. **A. púngens** var. **montàna** (Eastw.) Munz. $2n = 52$ (Stebbins & Major, 1965).

17. **A. pátula** Greene. Ascends to 11,000 ft. in the Sierra Nevada.

p. 425.   27. **P. canéscens** Eastw. $2n = 26$ (Stebbins & Major, 1965).

p. 426.   29. **A. obispoénsis** Eastw. Add as synonym *A. luciana* P. V. Wells.

33a. **A. montereyénsis** Hoov. Near to *A. columbiana* Piper. Erect, 1.25 m. tall; branchlets with spreading gland-tipped hairs; lvs. elliptic to oval, 15–34 mm. long, mostly obtuse, sparsely hairy on midrib; panicle glandular-hairy; bracts lanceolate, ca. as long as flowering pedicels; corolla tinged with some pink; fr. sparsely hairy, not viscid.—Sand at s. edge of Monterey Airport, Monterey Co.

p. 427.   35a. **A. Hoòveri** P. V. Wells. Tall, to over 4 m.; branchlets densely short-pubescent and glandular-hispid; lvs. ovate to oblong, 3–5 cm. long, cordate, on petioles 2–10 mm. long, gray-green, glandular-hairy on both surfaces; infl. large, open, leafy-bracted, the bracts to 15 mm. long; pedicels to 10 mm. long; corolla 7–8 mm. long; fr. glandular-viscid; nutlets separable.—Near summit of Nascimiento Pass, Santa Lucia Mts., s. Monterey Co. (Near *A. Andersonii* and *A. glandulosa*.)

p. 427.   Mr. James B. Roof has gone into the matter of *A. imbricata* Eastw. in some detail, proposes a new sp., **A. montaraénsis** Roof, and discusses in general the genus on Montara Mt. and San Bruno Mt. (The Four Seasons 2 (3): 6–16. 1967). He feels that *A. imbricata* is not a variety of *A. Andersonii* Gray, as it is treated in the FLORA, but a definitely distinct species. It consists of 5 separate colonies occupying "clean and otherwise bare sandstone outcroppings" on San Bruno Mt., at 1000–1300 ft. He feels that it is of hybrid origin (*A. montaraensis* × *A. Uva-ursi*).

His *A. montaraensis* is compared with *A. pallida* Eastw. (*A. Andersonii* var. *pallida* in the FLORA). It is erect, 1–5 m. tall, twice as high as *pallida*, which is divaricately spreading; has vivid bright green lvs. 3–5 cm. long (not glaucous); lvs. ciliate on margin when young (not so in *pallida*); heavily ciliate bracts (bracts not ciliate in *pallida*); long-hairy glandular ovary (short-hairy, less glandular in *pallida*). It grows on granitic sand and sandstone, at 500–1500 ft., in w. San Mateo Co., from near Lake Pilarcitos to Scarper Peak and Montara Mt.

p. 428.   37a. **A. cruzénsis** Roof is proposed as near *A. pajaroensis* and *A. pechoensis*. Decumbent, spreading to 3 m. across; lvs. sessile, pale dull green, oblong, 2–3 cm. long; infl. compact, subcapitate, closely fine-pubescent; bracts to 10 mm. long; corolla 6 mm. long; fr. depressed-globose, to 10 mm. wide, finely hirsutulous.—Arroyo de la Cruz Creek, on Highway 1, nw. San Luis Obispo Co.

38. **A. pechoénsis** var. **viridíssima** Eastw. Delete the insular references.

p. 428.   38a. **A. refugióensis** Gankin. (The Four Seasons 2 (2): 13. 1967). Erect, 2.5–4 m. tall, 2–3.5 m. wide; without basal burl; young growth with both short and long gland-tipped setose hairs; lvs. mostly sessile, cordate, clasping, entire to serrulate, imbricate, 2.5–4.5 cm. long, 2–3 cm. wide, with equal numbers of stomates on both surfaces; infl. branched, the rachis fine-pubescent and glandular-setose; bracts foliaceous, 5–10 mm. long; pedicels 6–9 mm. long; calyx-segms. 2 mm. long; corolla white to pinkish; ovary glabrous; fr. globose, 1–1.5 cm. in diam., the nutlets coalesced.—At 2250 feet, Refugio Pass and region, Santa Barbara Co. Near *A. pechoensis*, but fr. not depressed. At Rancho Santa Ana Bot. Gard., the herbarium has 3 sheets from Refugio Pass which do not support the distinctness of the proposed sp.

39. **A. glandulòsa** Eastw. var. **Cushingiàna** (Eastw.) Adams ex McMinn. Reported from Monterey Co., *Howitt and Howell*.

p. 429.   41. **A. crustàcea** Eastw. var. **Ròsei** (Eastw.) McMinn. J. B. Roof (The Four Seasons 1: 1–15. 1964) presents data for recognition of **A. Ròsei** Eastw. as a sp. Burls more irregular and pitted; bark terracotta red, exfoliating in long strips (not purple-black and smooth); lvs. often truncate at the base, brighter green, rarely toothed; new branchlets slightly pubescent (not bristly); fls. largely before March 15 (not on into April). He would extend its range from San Francisco to colonies in Monterey Co. (Ft. Ord region, Garrapata Creek, Rocky Creek, Bixby Creek, and Plaskett Creek).

p. 430.   43. A. subcordàta Eastw. On Santa Catalina Id., *Thorne.*

p. 432.   EMPÈTRUM

1. American material passing as *E. nigrum* L. should probably be called E. hermaphrodìtum (Lange) Hagerup, being tetraploid, with bisexual fls. and larger pollen grains. [*E. Eamesii* ssp. *h.* D. Löve.]

p. 433.   PÝROLA

1. B. Krisa (Bot. Jahrb. 85: 612–637. 1966) recognizes Pýrola califórnica Krisa [*P. asarifolia* Am. auth., *P. a.* var. *incarnata* Fern. & var. *purpurea* (Bunge) Fern.] for plants with lance-ovate bracts, 4.5–7.5 mm. long, usually as long as the pedicels; calyx-lobes 3–3.5 mm. long; anthers 2.5–3 mm. long, short-pointed, and

P. bracteàta Hook. [*P. asarifolia* var. *bracteata* (Hook.) Jeps. *P. rotundifolia* var. *b.* Gray.] Bracts linear-lanceolate, 7.5–8.5 mm. long, usually twice as long as the pedicels; calyx-lobes lanceolate and in upper third prolonged into a long point 3.9–4.4 mm. long.

p. 434.   MONÈSES

1. M. uniflòra (L.) Gray ssp. reticulàta (Nutt.) Calder & Taylor has been published.

p. 435.   CHIMÁPHILA

1. C. umbellàta var. occidentàlis (Rydb.) Blake. $2n = 26$ (Raven, Kyhos & Hill, 1965).

ALLÓTROPA

1. A. virgàta T. & G. ex Gray. $2n = 26$ (D. E. Anderson, 1965).

p. 436.   PTERÓSPORA

1. P. andromedèa Nutt. is said by Bakshi to be parasitic on root fungi.

p. 437.   PITYÒPUS

1. P. califórnicus (Eastw.) Copel. f. is reported also as in Lake and Colusa cos.

p. 438.   MICROCÀLA

1. M. quadrángularis (Lam.) Griseb. D. M. Post (Madroño 19: 134. 1967) takes up the name *Cicendia quadrangularis* (Lam.) Griseb. "The more familiar generic name *Microcala* is apparently invalid due to its being superfluous when published." $n = 13$ (Post, 1967).

p. 440.   EUSTÒMA

1. E. exaltàtum (L.) Salisb. is older as a comb. than (L.) Griseb.

p. 443.   GENTIÀNA

10. G. tenélla Rottb. is placed in *Comastoma* as *C. tenellum* Toyokuni.

11. *Gentianópsis simplex* (Gray) Iltis is proposed. Its range can be extended in the Sierra Nevada s. into the Greenhorn Range, Kern Co., *Twisselmann.*

12. *Gentianópsis holopetala* (Gray) Iltis is proposed.

p. 446.   MENYÁNTHES

1. M. trifoliàta L. Once reported as in San Francisco Co.; also in Mendocino Co. Wade, 1956, gave $2n = 54, 108$.

p. 447.   FRAXINUS

1. F. dipétala H. & A. $2n = 23$ pairs (Raven, Kyhos & Hill, 1965).

p. 452.   ASCLEPIADÀCEAE

In Key to Genera, in line 3, change to:

Fls. borne in umbels or racemes (then add:)
    The fls. not white. Native. . . . . 2. *Sarcostemma*
    The fls. white. Escape from cult. . . . . 2a. *Araujia*

CYNÁNCHUM L. not *Cynanchium* as in 1959 printing.

p. 453.   SARCOSTÉMMA

1. S. cynanchoìdes Dcne. ssp. Hartwégii (Vail) R. Holm. Reported from Ojai V., Ventura Co., *Pollard.*

2. **S. hirtéllum** (Gray) R. Holm. Corolla-lobes subpilose without, glabrous to glabrate within (fide *Bacigalupi*).

2a. Araújia Brot.

Corolla tube inflated at the base; lobes 5, overlapping in the bud; crown with 5 scales attached at or below the middle of the tube; stigma often 2-beaked at apex.

1. **A. sericófera** Brot. BLADDER-FLOWER. Vigorous climber; stem covered with pale down when young; lvs. ovate-oblong, 5–10 cm. long, pale green, minutely pitted beneath; fls. white, salverform, 2–3 cm. across, the tube 12 mm. long; pod grooved, 12 cm. long, 5–7 cm. wide.—Escape from cult., Riverside to Placer cos.; native of S. Am.

p. 454. ASCLÈPIAS

In description of genus, under "Follicles," insert "mostly" before "acuminate."

p. 456. 7. **A. subulàta** Dcne. in A. DC. $2n = 22$ (W. H. Lewis, 1961).

p. 457. 14. **A. curassávica** L. $2n = 22$ (Huynh, 1965).

p. 458. CONVOLVULÀCEAE

In Key to Genera, from last line delete "4. *Convolvulus*," add two lines:

Stigma oblong, ± cylindrical, the stigmatic area and style distinct; ovary with an incomplete septum . . . . 5. *Calystegia*
Stigma linear, the stigmatic area and the style ± continuous; ovary with a complete septum. . . . . 4. *Convolvulus*

DICHÓNDRA

1. Change *D. rèpens* Forst. & Forst. f. to **D. Donnelliàna** Tharp & Johnston.

p. 459. IPOMOÈA

Change Key to Species to:

Sepals ca. 7–12 mm. long
  Corolla 5–6 cm. long . . . . 1. *I. purpurea*
  Corolla 1.5–2 cm. long . . . . 1a. *I. triloba*
Sepals ca. 20–30 mm. long
  Corolla 2–4 cm. long; lvs. not canescent . . . . 2. *I. nil*
  Corolla 6–8 cm. long; lvs. silvery-canescent . . . . 3. *I. mutabilis*

1. **I. purpùrea** (L.) Roth. $n = 15$ (A. Jones, 1964).

1a. **I. trilòba** L. Climbing annual herb; stems glabrous; lvs. cordate, entire or 3-lobed, 3–6 cm. long, 2–5 cm. broad, glabrous, subacuminate; petioles 3–5 cm. long; infl. axillary, peduncled, 1–5-fld., umbellate; calyx 8 mm. long, the sepals ciliate, oblong-acuminate; corolla ca. 15 mm. long.—Escape in Imperial and Riverside cos.; native in trop. Am.

2. **I. hederàcea** (L.) Jacq. A. Jones (Jour. Heredity 55: 216–219. 1964) uses **I. níl** (L.) Roth.

3. **I. mutábilis** Ker.-Gawl. Weedy pubescent perennial; lvs. to 1 dm., ca. equally long and wide, cordate, whitish-tomentose beneath; calyx 2–3 cm. long; corolla vivid ultramarine blue, to ca. 8 cm. across.—Escaping as a weed in waste places, as in Santa Barbara and Ventura cos.; native of Mex.

CONVÓLVULUS

Amend the description as follows:

Pollen ± elongate. Stigmas 2, linear, ± applanate, acutate at apices, the stigmatic area and the style ± continuous; ovary 2-locular, the septum complete. A fairly large genus as amended.

A. Plants annual; corolla ca. 6 mm. long, deeply cleft. . . . . 1. *C. simulans*
AA. Plants perennial; corolla 2–6 cm. long, not cleft.
  B. Corolla purple to rose, 2.5–3 cm. long; stems climbing. . . . . 2. *C. althaeoides*
  BB. Corolla white or with some pink, 1.5–2 cm. long; stems largely prostrate.
                                            3. *C. arvensis*

1. **C. símulans** L. M. Perry is no. 16 in the FLORA.
2. **C. althaeoìdes** L. is no. 17 in the FLORA.
3. **C. arvénsis** L. is no. 15 in the FLORA. $n = 24$ (Khoshoo & Sachideva, 1961).

5. Calystègia R. Br. MORNING-GLORY

Resembling *Convolvulus*, but with pollen sphaeroidal. Stigmas oblong, ±
cylindrical with blunt apices, the stigmatic area and style distinct. Caps. 1-locu-
lar with an incomplete septum. A fairly large genus, of which the spp. have
largely been referred to *Convolvulus* (Greek, *kalux*, cup, and *stegos*, a covering).

A. Calyx enclosed or closely subtended by a pair of large sepallike bracts.
 B. Corolla purple to rose.
  C. Lvs. reniform, obtuse, fleshy, 2–5 cm. broad; prostrate seaside herbs.
                           1. *C. Soldanella*
  CC. Lvs. ovate-hastate, thin, acute to acuminate, 6–10 cm. long; climbing swamp
    plant. . . . . 2. *C. sepium*
 BB. Corolla white to cream, sometimes pinkish in age.
  C. Stems mostly over 1 m. long, twining or trailing.
   D. Plants of swamps and marshes, the stems entirely herbaceous. 2. *C. sepium*
   DD. Plants of dry places, the stems ± woody at base. . . . . 3. *C. macrostegia*
  CC. Stems mostly 1–5 dm. long, erect to prostrate.
   D. Herbage glabrous; corolla 4–5 cm. long. N. Calif. . . . 4. *C. atriplicifolia*
   DD. Herbage pubescent to tomentose.
    E. Corolla 4–5 cm. long; plant stemless or nearly so. Central Coast
     Ranges.
     F. Plant pilose-pubescent; pedicels much shorter than petioles.
                         5. *C. subacaulis*
     F. Plant ± tomentose; pedicels ca. as long as pedicels. 6. *C. collina*
    EE. Corolla 2.5–3.5 cm. long; plants usually with stems 1–4 dm. long,
     variously pubescent. . . . . 7. *C. malacophylla*
AA. Calyx subtended by more remote bracts that are not much like sepals.
 B. Bracts hastately lobed, like the upper lvs. Interior Calif. from Shasta Co. to San
  Diego Co. . . . . 8. *C. fulcrata*
 BB. Bracts entire.
  C. Plants puberulent to pubescent.
   D. Stems trailing or erect, 3–6 dm. long; outer sepals rounded at apex.
                      9. *C. polymorpha*
   DD. Stems climbing, taller; outer sepals acuminate. . . . . 10. *C. occidentalis*
  CC. Plants glabrous.
   D. Bracts broadly oblong to oval, attached near the base of the calyx. Desert
    slopes of the San Gabriel Mts. . . . . 11. *C. Peirsonii*
   DD. Bracts subulate to narrowly lanceolate, usually well below the calyx.
    E. Plants climbing; basal lobes of lvs. usually broad and toothed.
                      12. *C. purpurata*
    EE. Plants not climbing; basal lobes of lvs. linear, entire. 13. *C. longipes*

p. 460. 1. **C. Soldanélla** (L.) R. Br. is *Convolvulus* sp. no. 1 of the FLORA.

  2. **C. sèpium** (L.) R. Br. is *Convolvulus* sp. no. 2 of the FLORA. It is European with the basal lf.-lobes having conspicuous subacute angles. It is questionable whether typical *sepium* is in Calif. $2n = 20$ (Smith, 1965).

  Ssp. **americàna** (Sims) Brummitt. [*Convolvulus s.* var. *a.* Sims.] Lf.-lobes
more sharply angled. Perhaps natur., as about San Bernardino; e. U.S.

  Ssp. **limnóphila** (Greene) Brummitt. [*Convolvulus l.* Greene. *C. s.* var. *repens*
auth.] Plants mostly only 6–9 dm. high, herbage glabrous or very slightly pubescent; lvs. narrow, sagittate, the basal lobes ¼–⅓ the length of the body; bracts
unequal, large, the lower partly enfolding the truncate upper one; corolla large,
pinkish.—Tidal marshes, San Francisco Bay region.

  Ssp. **Binghàmiae** (Greene) Brummitt. [*Convolvulus B.* Greene. *Convolvulus
sepium* var. *dumetorum* Pospichal.] Plant glabrous throughout, 1–2 mm. high;
lvs. mostly obtuse at apex; bracts ca. 8–10 mm. long, ca. half as long as the
sepals.—Coastal marshes, Santa Barbara Co. to Orange Co.

  3. **C. macrostègia** (Greene) Brummitt. [*Convolvulus m.* Greene. *Convolvulus
occidentalis* var. *m.* Munz. *Volvulus m.* Farwell.] Is *Convolvulus* sp. no. 3 in
the FLORA.

A. Bracts largely 2–3 cm. long; corolla 5–6 cm. long. Insular. . . . . *C. macrostegia*
AA. Bracts 1–1.5 cm. long; corolla 2–4.5 cm. long. Mostly mainland.
 B. The bracts mostly subcordate at base, membranous and purplish. Near the coast,
  Monterey Co. s. . . . . Ssp. *cyclostegia*
 BB. The bracts mostly rounded at the base, firm and greenish. Largely away from the
  coast.

C. Lvs. and stems cinereous with dense tomentulose puberulence. Interior s.
Calif. . . . . Ssp. *arida*
CC. Lvs. and stems glabrous or nearly so.
    D. Middle lobe of lvs. narrowly to deltoid-lanceolate; corolla 3–3.5 cm. long.
      E. Basal lobes of lvs. less than half as long as middle lobe, not strongly
      divergent. Ventura Co. to Orange Co. and Catalina Id. Ssp. *intermedia*
      EE. Basal lobes of lvs. at least half as long as middle lobe, strongly
      divergent. Dry hills about San Diego and to w. Riverside and e.
      Orange cos. . . . . Ssp. *longiloba*
    DD. Middle lobe of lvs. narrowly linear; corolla 2–2.5 cm. long. W. River-
    side to n. L. Calif. . . . . Ssp. *tenuifolia*

p. 461.   Ssp. **cyclostègia** (House) Brummitt. [*Convolvulus c.* House.] This is *Convolvulus* sp. no. 4 of the FLORA.

Ssp. **árida** (Greene) Brummitt. [*Convolvulus a.* Greene.] This is *Convolvulus* sp. no. 5 of the FLORA.

Ssp. **intermèdia** (Abrams) Brummitt. [*Convolvulus aridus* ssp. *i.* Abrams.]

Ssp. **longilòba** (Abrams) Brummitt. [*Convolvulus aridus* ssp. *l.* Abrams.]

Ssp. **tenuifòlia** (Abrams) Brummitt. [*Convolvulus aridus* ssp. *t.* Abrams.]

4. **C. atriplícifolia** Hallier f. [*Convolvulus nyctagineus* Greene.] This is *Convolvulus* sp. no. 6 in the FLORA.

5. **C. subacaùlis** H. & A. [*Convolvulus s.* Greene. and var. *dolosus* Jeps. *Convolvulus californicus* Choisy.] This is *Convolvulus* sp. no. 7 of the FLORA.

6. **C. collìna** (Greene) Brummitt. [*Convolvulus c.* Greene. *Convolvulus malacophyllus* ssp. *c.* Abrams.]

Ssp. **tridactylòsa** (Eastw.) Brummitt. [*Convolvulus t.* Eastw.] Prostrate and trailing, gray-tomentose throughout; lvs. 3-parted, cuneate at base, the divisions widely spreading, the middle from ovate-triangular to narrower, ca. 2 cm. long, 4–10 mm. wide, mucronate, the lateral divisions oblong, obtuse, 1–2 cm. long, 5–10 mm. wide; petioles flexuous, the lowest 5 cm. long; pedicels shorter than petioles; bracts lanceolate, acute, equaling or shorter than the elliptical mucronate sepals, these tomentose; fils. shorter than the style.—Mts. near Covelo, Mendocino Co.

7. **C. malacophýlla** (Greene) Munz, comb. nov. [*Convolvulus m.* Greene, Pittonia 3: 326. 1898. *Calystegia fulcrata* ssp. *m.* Brummitt. *Calystegia villosa* Kell., not Raf.] Plants densely gray-tomentose; lf.-blades triangular-hastate; bracts 10–15 mm. long.—Sierra Nevada from Tulare Co. n., to Trinity and Siskiyou cos. This is *Convolvulus* sp. no. 8 of the FLORA.

p. 462.   Ssp. **pedicellàta** (Jeps.) Munz, comb. nov. [*Convolvulus villosus* var. *pedicellatus* Jeps., Man. Fl. Pl. Calif., 777. 1925. *Convolvulus malacophyllus* ssp. *p.* Abrams. *Calystegia fulcrata* ssp. *p.* Brummitt.] Lf.-blades narrowly lanceolate; plants densely gray-tomentose; bracts 10–15 mm. long.—Coast Ranges, Alameda Co. to Ventura Co.

Ssp. **tomentélla** (Greene) Munz, comb. nov. [*Convolvulus tomentellus* Greene. Pittonia 3: 327. 1898. *Calystegia fulcrata* ssp. *t.* Brummitt.] This is sp. no. 9 under *Convolvulus* in the FLORA.

Var. **deltoìdea** (Greene) Munz, comb. nov. [*Convolvulus deltoideus* Greene, Pittonia 3: 331. 1898. *Calystegia fulcrata* ssp. *tomentella* var. d. Brummitt.] Lvs. 1.2–1.5 cm. long, deltoid, somewhat broader; herbage short-canescent. At 3000–5000 ft.; Foothill Wd., Yellow Pine F.; Tehachapi Mts., Mt. Pinos.

8. **C. fulcràta** (Gray) Brummitt. [*Convolvulus f.* (Gray) Greene.] This is *Convolvulus* sp. no. 10 in the FLORA.

9. **C. polymórpha** (Greene) Munz, comb. nov. [*Convolvulus polymorphus* Greene, Pittonia 3: 331. 1898.] This is *Convolvulus* sp. no. 11 in the FLORA.

10. **C. occidentàlis** (Gray) Brummitt. [*Convolvulus o.* Gray.] This is sp. no. 12 of *Convolvulus* in the FLORA.

p. 463.   11. **C. Peirsònii** (Abrams) Brummitt. [*Convolvulus P.* Abrams.] This is *Convolvulus* sp. no. 13 in the FLORA.

12. **C. purpuràta** (Greene) Brummitt. [*Convolvulus luteolus* var. *p.* Greene.] This is *Convolvulus occidentalis* var. *purpuratus* of the FLORA.

Ssp. **solanénsis** (Jeps.) Brummitt. [*Convolvulus luteolus* var. *s.* Jeps.] This is the *Convolvulus occidentalis* var. *solanensis* in the FLORA, with ochroleucous fls. and from Solano Co.

Ssp. **saxícola** (Eastw.) Brummitt. [*Convolvulus s.* Eastw.] This is the *Convolvulus occidentalis* var. *saxicola* of the FLORA. Lvs. small, round-ovate. S. Sonoma Co. and adjacent Marin Co.

13. **C. lóngipes** (Wats.) Brummitt. [*Convolvulus l.* Wats.] This is sp. no. 14 under *Convolvulus* in the FLORA.

p. 465.   CUSCÙTA

In Key after BB, C, DD, EE, delete "13. *C. salina*" at end of line and add:

> F. Fls. 2–3 mm. long; corolla-lobes ovate-lanceolate; scales attached to the corolla-tube most of their length; anthers oval, the fils. well developed. 13. *C. salina*
> FF. Fls. 3–4 mm. long; corolla-lobes lanceolate; scales commonly free; anthers oval-oblong. . . . . 13a. *C. nevadensis*

p. 466.   10. **C. indécora** Choisy. $n = 15$ (Raven, Kyhos & Hill, 1965).

p. 467.   11. Change name from *C. subinclusa* Dur. & Hilg. to **C. Ceanòthi** Behr.

13a. **C. nevadénsis** Jtn. [*C. Veatchii apoda* Yuncker. *C. salina a.* Yuncker.] Fls. 3–4 mm. long, the lobes of the calyx and corolla more lanceolate than in *C. salina* and also somewhat longer; scales mostly broader and commonly free; anthers oval-oblong, subsessile.—On *Atriplex*, etc., Towne's Pass, Panamint Mts., *Eastwood & Howell*; Nev.

14. **C. denticulàta** Engelm. $2n = 15$ pairs (Raven, Kyhos & Hill, 1965).

p. 469.   POLEMÒNIUM

In Key, last line, for 1959 printing of the FLORA, change 69 to 6.

p. 470.   4. **P. pulchérrimum** Hook.

Wherry (Aliso 6: 99. 1967) recognizes:

> A. Herbage sparingly pubescent; habit lax; corolla normally violet. From n. Calif. northward. . . . . var. *pulcherrimum*
>   [*P. Berryi* Eastw., a synonym]
> AA. Herbage copiously pubescent; habit compact; corolla white or nearly so. Mt. Shasta to Mt. Rainier. . . . . var. *pilosum* (Greenm.) Brand
>   [*P. shastense* Baker ex Eastw.]

p. 475.   PHLÓX

In Key, change last FF to:

> FF. Lvs. 3–10 mm. long, often ± arched and spreading; plants pulvinate, densely tomentose throughout.
>   G. Lvs. 5–10 mm. long, plane, subulate, not closely imbricated or concealing the stem. . . . . 12. *P. Hoodii*
>   GG. Lvs. 3–5 mm. long, concave, oblong-elliptic, imbricated, completely concealing the stem. . . . . 13. *P. bryoides*

p. 477.   10. **P. diffùsa** Benth. ssp. **subcarinàta** Wherry. Reported from Piute Mt., Kern Co., *Twisselmann*.

13. **P. bryoìdes** Nutt. Very compactly pulvinate, 5–10 cm. broad; lvs. closely imbricated, completely concealing the stem, 3–5 mm. long; 3-ribbed on lower surface; fls. solitary, sessile; calyx ca. 5 mm. long; corolla white to lilac, 7–10 mm. long.—Mt. Lassen Park at 8000 ft.; Ore. to Wyo., Nev.

MICRÓSTERIS

1. **M. grácilis** (Hook.) Greene. [*Gilia g.* Hook. Omit "Dougl."]

p. 479.   GÍLIA

In Key, under AA, B, CC, D, E, FF, drop "23. *G. ophthalmoides*" from end of line and insert on next line:

> H. Corolla 7–12 mm. long, the tube well exserted from the calyx. . . . . 23. *G. ophthalmoides*
> HH. Corolla 3.5–5 mm. long, the tube included in the calyx. . . . . 23a. *G. Clokeyi*

In Key, under AA, B, CC, DD, E, F drop "28. *G. brecciarum*" and add:

> G. Corolla with broad throat, white to violet. Deserts and e. of Sierra Nevada. . . . . 28. *G. brecciarum*

GG. Corolla slender in form, deep violet with purple tube.
Inner S. Coast Range, Kern, Ventura and Santa Barbara
cos. .... 28a. *G. jacens*

In Key, AA, BB, C, insert "mostly" between "glomerules" and "of."

p. 480. In Key, AA, BB, C, DD, EE, FF, GG, insert "Largely" after "pollen mostly blue."

p. 482. 1. **G. capitàta** ssp. **abrótanifòlia** (Nutt. ex Greene) V. Grant. Reported from Santa Lucia Mts., *Howitt & Howell.*

p. 485. 15. **G. ochroleùca** Jones ssp. **bizonàta** A. & V. Grant. Reported from Monterey Co., *Howitt & Howell.*

p. 486. 17. **G. leptántha** Parish ssp. **vívida** A. & V. Grant transferred to 15. **G. ochroleuca** ssp. vivida A. & V. Grant.

p. 487. 19. **G. tenuiflòra** Benth. is reported from the Temblor Range, Kern Co., *Twisselmann.*

p. 488. 21. Under **G. intèrior** (Mason & Grant) A. Grant is mentioned **G. austro-occidentalis** (A. & V. Grant) A. & V. Grant which can be considered a sp.
23a. **G. Clòkeyi** Mason. Differing from **G. ophthalmoìdes** Brand in having corollas 3.5–5 mm. long, the tube included in the calyx, the throat pale yellow below, white above; $n = 9$.—E. San Bernardino and Inyo cos.; to New Mex.

p. 490. 28a. **G. jàcens** A. & V. Grant. Resembling *G. brecciarum* in habit of branching and fl.-size, *G. tenuiflora* in fl.-shape and color, *G. leptantha* in lf.-dissection. From *G. brecciarum* and *G. leptantha* it differs by the slender form and deep violet to purple corollas; from *G. tenuiflora* by smaller corollas (5–7 mm. long) and the spreading habit of branching. S. Coast Ranges, Kern Co. to n. Santa Barbara and Ventura cos.

p. 493. IPOMÓPSIS
2. **I. aggregàta** ssp. **Bridgèsii** V. & A. Grant. Reported from Kern Plateau, Kern Co., *Twisselmann.*
3. **I. congésta** (Hook.) V. Grant. [Add *Gilia congesta* ssp. *palmifrons* Brand to synonyms.]

p. 495. ERIÁSTRUM
1. **E. densifòlium** ssp. **elongàtum** (Benth.) Mason occurs in Monterey Co., *Howitt & Howell.*
3. **E. sapphirìnum** ssp. **dasyánthum** (Brand) Mason reported from Monterey Co., *Howitt & Howell.*

p. 501. NAVARRÈTIA
4. **N. plieántha** Mason. Loch Lomond, Bennett Mt., Lake Co., *Rubtzoff.*

p. 502. 11. **N. cotulifòlia** (Benth.) H. & A. Outer Coast Ranges, Sonoma Co., *Rubtzoff.*
12. **N. nigellifórmis** Greene. In the Greenhorn Range, Kern Co., *Twisselmann.*

p. 503. 17. **N. mitracárpa** Greene in the Greenhorn Range, Kern Co., *Twisselmann.*
Ssp. Jaredii (Eastw.) Mason in Monterey Co., *Howitt & Howell.*
20. **N. divaricàta** (Torr.) Greene in the Greenhorn Mts., *Twisselmann.*

p. 506. LEPTODÁCTYLON
1. **L. púngens** (Torr.) Rydb. ssp. **Hállii** (Parish) Mason ascends to 10,500 ft., in the White Mts.

p. 509. LINÁNTHUS
2. **L. pygmaèus** (Brand) J. T. Howell ssp. **continentàlis** Raven is proposed for plants from the mainland, restricting insular plants as *L. pygmaeus.* On Guadalupe and San Clemente ids., the corolla is lavender-blue, 4.5–6 mm. long, hence surpassing the calyx by 1.5–2.2 mm.; on the mainland it is white, 3.8–6 mm. long, therefore 0–1.4 mm. longer than the calyx.

p. 513. 26. **L. nudàtus** Greene. Synonym *Gilia n.* Greene, not Brand.

p. 515. 32. **L. bìcolor** (Nutt.) Greene ranges on the mainland s. to Point Sal, Santa Barbara Co., *C. F. Smith.*

p. 519. EUCRÝPTA
2. **E. micrántha** (Torr.) Heller. $n = 6$ or 12 (Cave & Constance, 1963).

p. 522. NEMÓPHILA

7. **N. heterophýlla** F. & M. in the San Emigdio Range, Kern Co., *Twisselmann.*

p. 523.   PHACÈLIA
          In Key, change 7 to 57, at end of line 3.

p. 525.   In Key AA, B, CC, D, E, FF, GG, H, I, go to:

> J.   Plant glandular-hirsute, 20–60 cm. high;
>   corolla whitish, 5–6 mm. long; lvs. entire
>   to dentate-lobed, the petiole ca. as long as
>   the blade; calyx-lobes unequal. San Luis
>   Obispo and Monterey cos. ... 55. *P. grisea*
> JJ. Plant glandular-villous, 10–40 cm. tall; co-
>   rolla lavender to violet, 6–7 mm. long; lvs.
>   entire or with salient teeth, the petioles al-
>   most as long as blades; calyx-lobes unequal.
>   W. base of Sierra Nevada and to Modoc Co.
>   56. *P. Purpusii*
> JJJ. Plant glandular-pubescent, 2–12 cm. tall;
>   corolla lavender, 5–7 mm. long; lvs. crenate,
>   the petiole ½ to ⅓ as long as the blade;
>   calyx-lobes subequal. Ventura Co.
>   55a. *P. Hardhamiae*

p. 527.   In Key AA, BB, CC, DD, EE, change F to:

> F.   Corolla yellow, 2.5–4 mm. long; stamens 1.5–2 mm. long; fls.
>   5-merous.
> G.   Plant finely glandular throughout. Inyo and Mono cos.
>   86. *P. inyoensis*
> GG. Plant not glandular. Mono Co. n. ..... 86a. *P. scopulina*

3. **P. serícea** (Grah.) Gray ssp. **ciliòsa** (Rydb.) Gillett is the name taken up for our plants which range from Modoc Co. to adjacent Ore. and to nw. Ariz. *P. sericea* ssp. *sericea* is ne. of Calif. and has campanulate rather than ± urceolate corollas.

p. 528.   4. **P. Bolánderi** Gray. $n = 11$ (Cave & Constance, 1963).

5. **P. ramosíssima** Dougl. ex Lehm. is in Monterey Co., *Howitt & Howell.*
     Var. **válida** Peck is reported from the Piute Mts., Kern Co., *Twisselmann.*

p. 529.   7. **P. Lyònii** Gray. $n = 11$ (Cave & Constance, 1963).

8. **P. floribúnda** Greene. $n = 11$ (Cave & Constance, 1959).

9. After **P. dístans** Benth. insert **P. umbròsa** Greene. Resembling *P. distans,* but with slender weak stems to ca. 3 dm. high; calyx-lobes linear; corolla tubular-campanulate, 3–6 mm. long.—W. edge of Colo. Desert, from Santa Rosa Mts. to L. Calif.

p. 530.   12. **P. cicutària** Greene var. **híspida** (Gray) J. T. Howell is reported as in Monterey Co., *Howitt & Howell.*

13. **P. cryptántha** Greene occurs in King's R. region, Fresno Co., *Howell.*

14. **P. vállis-mórtae** var. **helióphila** J. Voss. $n = 11$ (Cave & Constance, 1963).

p. 531.   16. **P. Rattánii** Gray is in Ventura Co., *Hardham.*

19. **P. crenulàta** Torr. $n = 11$ (Cave & Constance, 1963).

20. **P. minutiflòra** J. Voss. $n = 11$ (Cave & Constance, 1963).

p. 532.   22. **P. pedicellàta** Gray. $n = 11$ (Cave & Constance, 1963).

25. **P. imbricàta** Greene. Heckard adds **P. oreópola** Heckard which keys out:

Corolla white; stems usually more than 5 dm. long, erect to ascending; Coast Ranges, w. Sierra Nevada and cismontane s. Calif. below 5000 ft. .... *P. imbricata*
Corolla largely lavender to pale pink; stems 2.5–5 dm. long, decumbent to ascending. Above 5000 ft.
  Plants decumbent; corolla narrowly cylindrical with incurved lobes; lobes of rosette lvs. 3–5 pairs. San Gabriel Mts. .... *P. oreopola* Heckard
  Plants ascending; corolla urceolate; lobes of rosette lvs. 1–2 pairs or lvs. entire. San Bernardino Mts. .... *P. oreopola* ssp. *simulans* Heckard

p. 533.   25. **P. imbricàta** ssp. **bernardìna** (Greene) Heckard. $n = 11$ (Heckard, 1960).

26. **P. califórnica** Cham. Add. *P. Biolettii* Greene as synonym. $n = 22$ (Heckard, 1960).

27. **P. egèna** (Greene ex Brand) Const. $n = 22$ (Cave & Constance, as *bernardina*). Occurs in Monterey Co., *Howitt & Howell*.

29. **P. heterophýlla** Pursh. Heckard (Univ. Calif. Publ. Bot. 32: 68–76. 1960) has 2 sspp. under *P. heterophylla*: the ssp. **heterophýlla** ranging from Wash. and cent. Ore. to Mont. and Mex. Calif. material is referred to ssp. **virgàta** (Greene) Heckard, a biennial; infl. virgate; lower stem and rosette lvs. lack gland-tipped hairs; infl. and calyx-lobes usually with dull, whitish hairs.— Range Wash. to nw. Wyo., n. Calif., e. Nev. [Synonyms are *P. californica* var. *rubacea* Jeps. and *P. Peirsoniae* Williams.]

p. 534.   30. **P. hastàta** Dougl. ex Lehm. ssp. **compácta** (Greene) Heckard. $n = 22$ (Cave & Constance, 1947, as *frigida*).

31. **P. nemoràlis** Greene. Delete *P. Bioletti* Greene as a synonym. The typical form occurs in cent. Calif.

Ssp. **oregonénsis** Heckard, with stouter stems, 2 or more pairs of lobes on rosette lvs. (instead of 1 pair) and upper surface of lvs. sparsely appressed hispid or strigose instead of spreading hispid.—Nw. Calif. to Wash. $n = 22$ Cave & Constance, 1942 as *nemoralis*).

32. **P. mutábilis** Greene. Dele *P. californica* var. *rubacea* Jeps. $n = 11$, 22 (Cave & Constance, 1942, 1958).

33. **P. corymbòsa** Jeps. $n = 11$, 22 (Cave & Constance, 1958; Kruckeberg, 1956).

34. **P. frígida** Greene. Heckard recognizes 2 sspp.:

Ssp. **frígida**. Lvs. ovate or lance-ovate to lenticular in outline with cuneate to obtuse base; upper lf.-surface smooth-strigose, grayish to whitish; calyx-lobes linear-lanceolate; corolla white, broadly campanulate; $n = 33$ (Heckard, 1960). —Modoc Co. and Mt. Shasta to n. Ore.

Ssp. **dasyphýlla** (Macbr.) Heckard. [*P. d.* Greene.] Lvs. lanceolate or lenticular to narrow-lanceolate with attenuate base; upper lf.-surface shaggy-strigose or with hairs little appressed, green to yellowish-green; calyx-lobes linear or linear-lanceolate; corolla white to lavender, tubular-campanulate; $n = 22$, 33 (Heckard, 1960).—Sierra Nevada and White Mts.

p. 535.   37. **P. Leònis** J. T. Howell. $n = 11$ (Cave & Constance, 1963).

40. **P. orógenes** Brand. $n = 9$ (Cave & Constance, 1959).

p. 536.   43. **P. marcéscens** Eastw. ex Macbr. $n = 8$ (Cave & Constance, 1963).

47. **P. austromontàna** J. T. Howell. $n = 9$ (Cave & Constance, 1959).

p. 537.   52. Howitt & Howell use **P. Davidsònii** Gray instead of *P. curvipes* var. *macrantha* (Parish) Munz and report it for Monterey Co.

p. 538.   53. **P. Douglásii** var. **cryptántha** Brand. $n = 11$ (Cave & Constance, 1959), as *P. stellaris*.

55. **P. grísea** Gray. $n = 9$ (Cave & Constance, 1959). Found near San Marcos Pass, Santa Barbara Co., *Raven*.

55a. **P. Hárdhamiae** Munz. Resembles *P. grisea* and *P. Purpusii* in its small promptly deciduous, pelviform corolla, deeply parted style, glandular pubescence, lax, scorpioid infl., dentate rather than pinnate lvs., but differs from the former in smaller stature, 2–12 cm. tall, more constantly crenate lvs.; longer petioles with reference to blade length (½ to ⅓), more lax, few-fld. cymes, lavender rather than white corollas, subequal rather unequal calyx-lobes, more southern occurrence. From *P. Purpusii* it differs in smaller, more delicate habit, rounder lvs. with more shallow, less pointed teeth, more coastal range. It was found at Rose Lake, Ventura Co., at 3600 ft.

p. 539.   57. **P. perityloìdes** Cov. $n = 11$ (Cave & Constance, 1959).

58. **P. suaveòlens** Greene. In Monterey Co., *Howitt & Howell* and San Luis Obispo Co., *Hardham*.

60. **P. rotundifòlia** Torr. ex Wats. $n = 12$ (Cave & Constance, 1959).

61. **P. mustelìna** Cov. $n = 12$ (Cave & Constance, 1959).

62. **P. Peirsoniàna** J. T. Howell. $n = 12$ (Cave & Constance, 1959).

p. 540.   68. **P. Nashiàna** Jeps. $n = 11$ (Cave & Constance, 1963).

p. 541.   74. **P. pachyphýlla** Gray. $n = 11$ (Cave & Constance, 1959).

p. 542.   75. **P. calthifòlia** Brand. $n = 12$ (Cave & Constance, 1959).

76. **P. neglécta** Jones. $n = 11$ (Cave & Constance, 1959).

77. **P. Ivesiàna** Torr. reported as having $n = 11$, *Cave & Constance.*

77a. **P. pediculoìdes** (J. T. Howell) Const. with $n = 23$. [*P. Ivesiana* var. *p.* J. T. Howell.]

79. **P. affìnis** Gray. $n = 12$ (Cave & Constance, 1959).

p. 544.　86. **P. inyoénsis** (Macbr.) J. T. Howell. $n = 12$ (Cave & Constance, 1959).

86a. **P. scopulìna** (A. Nels.) J. T. Howell. [*Emmenanthe s.* A. Nels. *Miltitzia s.* Rydb.] Diffuse annual, with short spreading hairs, not glandular; corolla yellow, 2.5–4 mm. long at anthesis, ca. equalling the calyx, to 5 mm. in fruit; style 1–2.5 mm. long; longer fils. usually surpassing sinuses; $n = 12$ (Cave & Constance, 1959).—Bridgeport, Mono Co.; to Mont., Wash.

87. **P. tetrámera** J. T. Howell. $2n = 22$ (Cave & Constance, 1963).

p. 545.　NÀMA

In Key, in line leading to *N. aretioides*, change to:

Corolla 7–15 mm. long, 7–12 mm. broad. . . . . *N. aretioides*

1. **N. Rothróckii** Gray. $n = 17$ (Cave & Constance, 1959).

p. 546.　5. **N. aretioìdes** (H. & A.) Brand. Add to synonyms *N. a.* f. *californicum* Brand.

p. 547.　8. **N. pusíllum** Lemmon. $n = 7$ (Cave & Constance, 1959).

9. **N. depréssum** Lemmon ex Gray. $n = 7$ (Cave & Constance, 1959).

p. 548.　ERIODÍCTYON

In Key, change line 1 at top of page to:

Calyx 2–4 mm. long; cyme open; corolla sparsely pubescent.
　Calyx-lobes 3–4 mm. long, sparsely hirsutulous; corolla 5–6 mm. long, with white
　limb; seeds almost 1 mm. long. E. Mojave Desert. . . . . 1. *E. angustifolium*
　Calyx-lobes 2–3 mm. long, ciliate; corolla 11–15 mm. long, with lavender limb; seeds
　0.4 mm. long. San Luis Obispo Co. . . . . 1a. *E. altissimum*

1a. **E. altíssimum** P. V. Wells. Viscid shrub 2–4 m. tall; lvs. linear, 6–9 cm. long, 2–4 mm. wide, revolute; infl. with branches 4–9 cm. long, the fls. secund; calyx-segms. narrow-lanceolate, 2–3 mm. long, ciliate; corollas purplish, 11–15 mm. long, villous without.—At 880 ft., Indian Knob, 4 mi. n. of Pismo, San Luis Obispo Co.

4. **E. trichocàlyx** var. **lanàtum** Jeps. $n = 14$ (Cave & Constance, 1959).

5. **E. tomentòsum** Benth. $n = 14$ (Cave & Constance, 1959). Reported from Temblor Range, Kern Co., *Twisselmann.*

p. 549.　6. **E. Tráskiae** ssp. **Smíthii** Munz. subsp. nov. Shrub 0.5–2 m. tall; lvs. elliptic to ovate, the blades 3–7 cm. wide, to 10 cm. long; calyx 4.5–5 mm. long; corolla 6–7 mm. long. (Folia elliptica vel ovata, laminis 3–7 cm. latis, ad 10 cm. longis; calyx 4.5–5 mm. longus; corolla 6–7 mm. longa.) Type, San Marcos Pass, Santa Barbara Co., *Clifton F. Smith 1621,* July 4, 1950 (Pomona College Herbarium). It is a pleasure to dedicate this subspecies to Clifton Smith of Santa Barbara, who has done so much to make known the flora of his county. The proposed subspecies is found in Chaparral up to 2000 ft., on the mainland from San Luis Obispo Co. to Ventura Co.

In *E. Tráskiae* Eastw. ssp. **Tráskiae** of Catalina Id. the lf.-blades are mostly narrow-elliptic, ca. 1.5–2.5 (–3) cm. wide and to 6 or 7 cm. long; the calyx is 6–6.5 mm. long; corolla 7.5–8 mm. long.

7. **E. crássifolium** Benth. var. **crássifolium.** $n = 14$ (Cave & Constance, 1959).

Var. **denudàtum** Abrams. Reported from the San Emigdio Range, Kern Co., *Twisselmann.*

p. 549.　HESPEROCHÌRON

1. **H. pùmilus** (Griseb.) Porter. $n = 8$ (Cave & Constance, 1959).

p. 550.　TRICÁRDIA

1. **T. Watsònii** Torr. ex Wats. $n = 8$ (Cave & Constance, 1959).

p. 552.　AMMOBRÒMA

1. **A. sonòrae** Torr. ex Gray. $2n = 18$ (Moore, 1962).

p. 552.　BORAGINÀCEAE

In Key to Genera, after line 4 from the bottom of the page "Nutlets erect," etc., add:

Corolla rotate; anthers connivent itno a cone. . . . . 5a. *Borago*
Corolla tubular, funnelform or salverform; anthers not forming a cone.
Attachment of nutlet surrounded by an annular rim, strongly convex and leaving a pit upon the low receptacle.
Scales or appendages in throat of corolla linear and acute.
Scales or appendages broad and blunt. . . . . 6. *Symphytum*
Lvs. oblong or lanceolate, obscurely veined. 6a. *Anchusa*
Lvs. ovate, netted-veined. . . . . 6b. *Pentaglottis*

p. 553. In Key at top of page after "Nutlets attached ± laterally" etc. and after "The nutlets not armed with conspicuous prickles," insert:

Calyx at maturity conspicuously expanded and net-veined; fls. on short recurved pedicels.
13a. *Asperugo*
Calyx not so expanded.
"Corolla bright blue" etc. as in the FLORA.

p. 554. HELIOTRÒPIUM
Change Key to:

Plants perennial.
The plant glabrous, succulent; corolla white or pale. Native. . . . . 1. *H. curassavicum*
The plant soft-hairy; corolla purple. Escape from cult. . . . . 3. *H. amplexicaule*
Plants annual.
Corolla 8–14 mm. broad. Desert native. . . . . 2. *H. convolvulaceum*
Corolla 4 mm. broad. Escape in n. Calif. . . . . 4. *H. europaeum*

Add 4. **H. europaèum** L. Erect annual, 1.5–8 dm. high, hoary-pubescent; lvs. oval; fls. in scorpioid spikes; corolla white or bluish, to ca. 4 mm. broad; nutlets 4, tuberculate.—Tehama and Butte cos. and Modoc-Siskiyou area, *Howell*; native of Eu.

p. 555. PECTOCARYA
1. **P. lineàris** DC. var. **feróocula** Jtn. $2n = 24$ pairs (Raven, Kyhos & Hill, 1965).
2. **P. recurvàta** Jtn. $n = 12$ (Di Fulvio, 1965).
3. **P. platycárpa** (M. & J.) M. & J. Add as synonym *P. linearis* var. *platycarpa* Cronq.

p. 556. 4. **P. penicillàta** (H. & A.) A. DC. $2n = 12$ pairs (Raven, Kyhos & Hill, 1965). Add as synonym *P. linearis* var. *penicillata* Jones.
CYNOGLÓSSUM

p. 557. 3. **C. officinàle** L. Biennial, soft short-hairy, leafy; upper lvs. lanceolate, closely sessile; infl. paniculate, of nearly bractless racemes; corolla red-purple; nutlets flat on the broad upper face, 5–7 mm. long, overtopped by the beaklike style.—Established on hillside 6 mi. s. of McCloud (hence in n. Shasta Co.) and 0.3 mi. s. of McCloud, Siskiyou Co., *T. C. Fuller*. Introd. from Eurasia.

5a. Boràgo L. BORAGE

Erect, strigose-hispid herbs. Lvs. alternate. Fls. blue, in loose leafy cymes and on long pedicels. Calyx with 5 linear segms. Corolla rotate, with 5 acute lobes, the throat with scales or hairy crests. Stigma subentire. Nutlets attached by the base. Old World.
1. **B. officinàlis** L. Coarse hairy annual, 4–6 dm. high; lvs. oblong or ovate, to 15 cm. long; corolla 2 cm. wide; stamens exserted, forming a cone 6 mm. high; nutlets verrucose.—Escape from cult., San Francisco. From Medit. region.

6a. Anchùsa L. ALKANET. BUGLOSS

Hispid or villous herbs with panicled leafy-bracteate scorpioid cymes or racemes with elongate branches; fls. blue, violet or white; calyx-lobes narrow.

Corolla trumpet-shaped, the throat closed by scales. Nutlets basally attached. Eurasia.

1. **A. prócera** Bess. Stem to 18 dm. high, simple or branched at base; lvs. oblong-lanceolate, the cauline clasping; infl. of crowded branches; bracts broadly triangular; calyx subsessile; fls. blue.—Hayfork, Trinity Co., *Howell*. From Se. Eu.

6b. Pentaglóttis Tausch

Like *Anchusa*, but with ovate, netted-veined lvs. and nutlets with a stalked small attachment rather than sessile and broad.

1. **P. sempervìrens** (L.) Tausch. [*Anchusa s.* L.] Hairy, 3–9 dm. tall; lvs. to 3 dm. long; cymes very hispid in long-peduncled subcapitate pairs; fls. subsessile; corolla ca. 10 mm. across, bright blue.—Locally established, San Francisco; from Eu.

p. 558. MYOSÒTIS

2. **M. láxa** Lehm. At Kernville, etc., Kern Co., *Twisselmann*.

4. **M. latifòlia** Poir., not *M. sylvatica* (cf. Johnston, Wrightia 2: 16–17. 1959).—Occurs in Monterey Co., *Howitt & Howell*, as well as in the range given in the FLORA. $2n = 18$ (Merxmuller & Grau, 1963).

p. 559. LITHOSPÉRMUM

Change Key to:

A.   Plants annual; nutlets tubercled, dull. . . . . 1. *L. arvense*
AA. Plants perennial; nutlets smooth, shining.
   B.   Upper lvs. crowded, lance-linear, with attenuate apex.
      C.   Corolla pale yellowish, often greenish-tinted, the tube 4–6 mm. long, the limb 7–13 mm. wide, the lobes entire or nearly so. . . . . 2. *L. ruderale*
      CC. Corolla bright yellow, the tube 12–30 mm. long, the limb 12–20 mm. wide, the lobes erose. . . . . 2a. *L. incisum*
   BB. Upper lvs. scattered, lance-ovate, acute to obtuse; fls. golden-yellow.
                                                       3. *L. californicum*

2a. **L. incìsum** Lehm. Agreeing with *L. ruderale* in the narrow lvs. with attenuate apex, but with bright yellow corolla having a tube 12–20 mm. long and a limb 7–13 mm. wide and with corolla-lobes erose.—New York Mts., e. San Bernardino Co., *Balls & Everett*; to Mont. and B.C.

p. 560. ÈCHIUM

Plant herbaceous, ± hispid with stiffish white hairs; infl. a loose panicle of coiled racemes.
                                                  1. *E. plantagineum*
Plant shrubby, not at all hispid; infl. cylindric and spikelike, of numerous densely flowered 1-sided spikelets. . . . . 2. *E. fastuosum*

2. **E. fastuòsum** Ait. Shrubby, 1–2 m. high, soft, grayish-hirsute; lvs. lanceolate, acuminate; infl. cylindric, spikelike, ca. 15 cm. long, consisting of many densely-fld., coiled, 1-sided spikelets; fls. purple or dark blue, 12 mm. diam.—Escaping from cult. as in San Francisco and Marin cos.; from the Canary Ids.

p. 563. HACKÈLIA

12. **H. velùtina** (Piper) Jtn. Extends e. to the Greenhorn Range, Kern Co., *Twisselmann*.

13a. Asperùgo L. MADWORT

Annual herb, with much branched stems, spreading or procumbent. Fls. axillary, few, small, blue. Calyx lobed to about the middle. Corolla salverform, the wide tube nearly closed by the appendages. Stigma capitellate. Calyx in fr. greatly enlarged, strongly anastomose-veined. Nutlets flattened, minutely verrucose.

1. **A. procúmbens** L. Three to 7 dm. high, with stout recurved prickles; lvs. oblanceolate, 3–6 cm. long; corolla 2–3 mm. long.—Adventive in Modoc Co., *Heiser*; from Eurasia.

p. 564. CRYPTÁNTHA

In Key, A, B, delete the *C. circumscissa* at the end of the line and insert:

C.　Corolla 1–4 mm. in diam.; pollen grains 7–9 long, oblong. 1. *C. circumscissa*
CC.　Corolla 4–6 mm. in diam.; pollen grains 5.5–6.5 long. . . . . 1a. *C. similis*

p. 565.　In Key, line 6 from bottom of page, the calyx may be 2–4 mm. in *C. incana.*

p. 567.　　1. **C. circumscíssa** (H. & A). Jtn. Corolla 1–4 mm. broad; pollen grains 7–9 μ long; $n = 12$.

Var. rosulàta J. T. Howell can now become ssp. **rosulàta** Mathew & Raven.

1a. **C. símilis** Mathew & Raven. Differing from **C. circumscissa** in having the corolla 4–6 mm. broad; pollen grains 5.5–6.5 μ long; $n = 6$. Mojave Desert.

2. **C. micrántha** (Torr.) Jtn. and ssp. **lépida** (Gray) Mathew & Raven. Both have $n = 12$.

p. 568.　　7. **C. utahénsis** (Gray) Greene occurs in the Greenhorn Mts., Kern Co., *Twisselmann.*

p. 569.　　18. **C. intermèdia** (Gray) Greene. $n = 12$ (Di Fulvio, 1965).

p. 579.　PLAGIOBÓTHRYS

2. **P. Jònesii** Gray. In the El Paso Range, Kern Co., *Twisselmann.*

p. 580.　　8. **P. infectìvus** Jtn. In the Temblor Range, Kern Co., *Twisselmann.*

p. 582.　　14. **P. califórnicus** var. **fulvescens** Jtn. In Monterey Co., *Howitt & Howell.*

p. 583.　　22. **P. Austíniae** (Greene) Jtn. Ranges n. to Jackson Co., Ore.

p. 586.　　36. **P. hispídulus** (Greene) Jtn. *Allocarya h.* Greene, not Jtn.

p. 587.　　39. **P. reticulàtus** var. **Rossianòrum** Jtn. As far s. as Monterey Co., *Howitt & Howell.*

p. 589.　AMSÍNCKIA

8. **A. Menzièsii** (Lehm.) Nels & Macbr. Cronquist (Vasc. Pl. Pac. N. W. 4: 181. 1959) keys out the following:

A.　Stem spreading-hispid and also evidently puberulent or strigose with shorter and softer, ± retrorse hairs; pubescence of the lvs. tending to be ascending instead of widely spreading; corolla 5–8 mm. long. S. Calif. to B.C., Ida., Utah. . . . . *A. retrorsa* Suksd.
AA.　Stem spreading-hispid, nearly or quite without shorter and softer hairs below the infl.; hairs of the lvs. often widely spreading.
　　B.　Corolla 7–10 mm. long, the tube well exserted. . . . . *A. intermedia* F. & M.
　　BB.　Corolla 4–17 mm. long, the tube scarcely exserted.
A. Menziesii (Lehm.) Nels & Macbr.

p. 590.　　　　　　　　61a. **Loganiàceae.** LOGANIA FAMILY

Herbs to vines or trees. Lvs. simple, usually opposite, stipulate. Infl. of leafy interrupted spikes to cymose. Fls. regular, usually perfect, 4–5-merous. Calyx-lobes imbricate. Corolla sympetalous, the lobes valvate, imbricate or contorted. Stamens as many as the corolla-lobes, alternate with them. Ovary superior, 2-loculed. Style usually simple; stigma capitate or 2-lobed. Fr. in ours a caps. A family of ca. 600 spp., largely tropical.

Fls. in headlike clusters in an interrupted spike; lvs. densely woolly; plant 2–3 dm. high. E. Mojave Desert. . . . . 1. *Buddleja*
Fls. in paniculate cymes; lvs. glabrous above; plant 4–5 m. tall. Escape from cult.
　　　　　　　　　　　　　　　　　　　　　　　　　　　　　　　　　　　2. *Chilianthus*

1. **Búddleja** L.

Shrubs to trees; lvs. simple, entire to dentate. Fls. mostly 4-merous. Calyx campanulate. Corolla salverform or rotate-campanulate, the lobes ovate or rounded. Anthers subsessile on throat or tube of corolla. Fr. a septicidal caps.; valves 2-cleft at apex. Seeds many. Ca. 70 spp. of warm N. & S. Am., Asia, S. Afr.

1. **B. utahénsis** Cov. Much-branched shrub 2–3 dm. high, densely lanate-tomentose; lvs. subsessile, linear, with revolute margins, 1–3 cm. long; axils usually with fascicles of very small lvs.; fls. in glomerules forming 2–4 heads in an interrupted spike, ca. 10–15 mm. thick; corolla creamy-yellow to purple, 4–5 mm. long, the lobes rounded, 1 mm. long, the tube tomentulose.—Dry rocky slopes, 3500–5500 ft., Joshua Tree Wd., Pinyon-Juniper Wd.; Kingston and Panamint mts., San Bernardino and Inyo cos.; to Utah. May–Oct.

2. **B. Davídii** Franchet. SUMMER LILAC. To 5 m. tall, deciduous; lvs. 5–30 cm. long, white-felted beneath; panicles 25–40 cm. long; fls. lilac to purple. Escape from cult., as in San Francisco; native of China.

2. Chiliánthus Burchell

Arborescent shrubs. Lvs. entire to toothed or lobed. Fls. in terminal pyramidal or subspherical cymes. Fls. small, 4-merous. Corolla tube short, scarcely exserted. Stamens inserted near the base of the corolla-tube, elongate, exserted. Caps. 2-loculed. Seeds small. A S. Afr. genus of ca. 4 spp.

1. **C. oleàceus** Burchell. [*C. arboreus* Benth.] To 4 or 6 m. tall; stems 4-angled; lvs. lanceolate, 7–10 cm. long, rusty-scurfy beneath, but soon glabrate, ± revolute; fls. fragrant, creamy-white, 2.5 mm. long, many, in panicles to 1 dm. across.—Escaping from cult., Saddle Peak, Santa Monica Mts., Los Angeles Co., *Raven & Thompson*.

p. 590. SOLANÀCEAE

In Key to Genera, after AA, B, CC, DD delete word *Physalis* at end of line and insert:

> E.   Ovary 2-loculed; fruiting calyx 5-toothed. . . . . **4. Physalis**
> EE. Ovary 3–5-loculed; fruiting calyx 5-parted. . . . . **4a. Nicandra**
> BB. Corolla not rotate.
>   C.   Fr. a berry.
>   D.   Corolla urceolate, white. . . . . **5. Salpichroa**
>   DD. Corolla tubular, red. . . . . **5a. Cestrum**
>   CC. Fr. a caps.; etc. as in FLORA.

p. 592. LÝCIUM

7. Use **L. bárbarum** L. instead of *L. halimifolium* Mill.

p. 593. PHÝSALIS

After generic description, insert Waterfall, U.T. Taxonomic study of the genus *Physalis* in N. Am. n. of Mex. Rhodora 60: 106–114, 128–141, 152–173. 1958.

In Key, AA, BB, CC, D, after EE, add:

> EEE. Corolla yellow, broader; lvs. mostly entire; perennial.   11a. *P. subglabrata*

p. 594. 2. Change *P. Fendleri* Gray to **P. hederifòlia** var. **cordifòlia** (Gray) Waterfall.

5. Under **P. crassifòlia** Benth. Waterfall recognizes var. **versícolor** (Rydb.) Waterfall. [*P. v.* Rydb.] Corolla often bluish on drying; flowering calyx 3–4 mm. long (instead of 4–6 mm.), on peduncles 5–10 times as long (not 6–7). Colo. Desert; to Nev., Ariz.

p. 595. 6. Waterfall restricts **P. pubéscens** L. to the se. states for the U.S. and places Calif. material in var. **intcgrifòlia** (Dunal) Waterfall. [*P. hirsuta* var. i. Dunal.] Lvs. with fewer teeth or entire and mostly flaccid and translucent.

10. Change *P. Wrightii* Gray to **P. acutifòlia** (Miers) Sandwith. [*Saracha acutifolia* Miers.] Kern Co. s. to Imperial Co.; to Tex., Mex.

11. Waterfall reduces *P. lanceifolia* Nees to **P. angulàta** L. var. **lanceifòlia** (Nees) Waterfall. $n = 12$ (Gottschalk, 1954).

11a. **P. subglabràta** Mack. & Bush. [*P. virginiana* Mill. var. *s.* Waterfall.] Plants nearly glabrous; lvs. ovate to lance-ovate, mostly entire, sometimes slightly sinuate-dentate; corolla usually 15–20 mm. long; anthers bluish; fruiting calyces mostly 25–35 mm. long.—In a field near Montague, Siskiyou Co., *Fuller*; native of cent. U.S.

12. **P. ixocárpa** Brot. Howitt & Howell report from Monterey Co.

4a. Nicándra Adans.

Annual herbs differing from *Physalis* in the 3–5-loculed ovary and deeply parted calyx.

1. **N. Physalòdes** Caertn. [*Atropa P.* L.] To 1 m., much branched; lvs. ovate

to oblong, 10–15 mm. long, sinuate-toothed; fls. blue, 2.5–4 cm. long and broad, enclosed in the enlarged green calyx 2.5–4 cm. long. Escape, as at Santa Barbara, *Pollard*; native of Peru.

SALPICHRÒA

1. Change S. *rhomboidea* (Gill. & Hook.) Miers to **S. origanifòlia** (Lam.) Thell.

p. 596.   **5a. Céstrum L.**

Shrubs with alternate entire lvs. Fls. tubular, in axillary or terminal cymes. Corolla with a long tube, 5-lobed. Stamens 5. Fr. a berry. A large genus of warm regions.

1. C. **fascículatum** Miers. Shrub 1–2 m. tall; lvs. ovate-acuminate; cymes capitate; corolla deep rose-red.—Natur. in pasture near Hydesville, Humboldt Co.; native of Mex.

SOLÀNUM

Species no. 1 in the Key should be spelled *sarrachoides*, not as in 1959 printing. In Key the first GG after A, B, etc. should be changed in the 1959 printing to:

> G.   Infl. umbelliform; calyx-lobes distinct, reflexed at maturity, etc.

In Key A, BB, CC, DD, E add:

> E′.   Low native shrub to ca. 1 m. high; berry whitish. Common in Coast Ranges. . . . . 10. S. *umbelliferum*
> E′E′. Tall climbing introduced shrub; berry greenish yellow.
> 10a. S. *Gayanum*

In Key after A, BB, CC, DD, EE, FF, under G and GG delete distribution.
In Key change AA to "Plants prickly."

p. 597.   1. S. sarrachoìdes Sendt. ex. Mart., correct spelling for 1959 printing.

6. S. triflòrum Nutt. Reported from Greenhorn Mts., Kern Co., *Twisselmann*.

p. 598.   8. S. *aviculare* Forst. f. should probably be called S. laciniàtum Ait. $n = 24$ (Gottschalk, 1954).

10a. S. Gayànum (Remy) Phil. f. Vigorous climbing shrub, covered with stellate hairs; lvs. petioled, oval-oblong to elliptic, 5–8 cm. long; fls. terminal, lavender-violet, on branched peduncles; corolla stellate-pubescent without; berries greenish-yellow.—Natur. about San Francisco; native of Chile.

p. 599.   12. S. Xánti var. montànum Munz. Collected in Riverside Co. at 4800 ft. upper end of Hemet V., San Jacinto Mts., *Ziegler*, and at 6500 ft., Santa Rosa Mts., *Weatherby*. Varying in glandulosity, but on the whole matching well material from the San Bernardino Mts.

14. S. sisymbriifòlium Lam. $n = 12$ (Gottschalk, 1954).

15. S. rostràtum Dunal. $n = 12$ (De Lisle, 1965).

p. 600.   18. S. lanceolàtum Cav. Reported from Contra Costa Co., *Fuller*; Monterey Co., *Howitt & Howell*; and Ventura Co., *Fuller*.

19. S. marginàtum L. f. Found 4 mi. e. of Pala, San Diego Co., *Fuller*.

20. S. carolinénse L. $n = 12$ (Heiser, 1953). Reported as occurring in El Dorado Co., *Fuller*; Sonoma Co., *Fuller*.

DATURA

1. D. meteloìdes A. DC. There is a question as to the proper name for the species that has been so widely called D. *meteloides*. Ewan (Rhodora 40: 317–323. 1944) felt that there are inconsistencies in the drawing on which the name was based in 1852 and the applicability uncertain. He proposed the use of the later name D. Wrìghtii Regel (1881). Barclay (Bot. Mus. Leaflets, Harvard Univ. 8: 245–272. 1959) considered D. *meteloides* identical with D. inóxia Miller, 1786. The latter was based on material cultivated in England from material from Vera Cruz, Mex. If our common southwestern plant is the same as this more southern one, the name inoxia has long priority.

p. 601.   NICOTIÀNA

In Key, Change B to:

> B. Lvs. ± auriculate-clasping; fls. open during the day. Perennial or biennial.
> C. Calyx campanulate; corolla 18–22 mm. long. Desert native.   2. *N. trigonophylla*
> CC. Calyx oblong or subglobose; corolla 65–85 mm. long. Coastal introduction.
> 2a. *N. sylvestris*

p. 602. 2a. **N. sylvéstris** Spegazzini & Comes. Viscid robust perennial 1–2 m. tall; lvs. 20–50 cm. long, the cauline sessile; flowering calyx 10–18 mm. long; corolla white, 6.5–8.5 cm. long exclusive of the limb; caps. ovoid, 15–18 mm. long.— Natur. Pacific Grove, Monterey Co.; native of S. Am.

p. 603. SCROPHULARIÀCEAE
In family decription, line 1, change last word "or" to "to." (1959 prtg.)

p. 605. LINDÉRNIA
2. **L. anagallídea** (Michx.) Penn. occurs in Kern Co., *Twisselmann*.

p. 608. MÍMULUS
In Key, middle of page, "28. *M. purpureus*" should be "29" in 1959 printing.

p. 610. 1. **M. Lewísii** Pursh. n = 8 (Vickery et al., 1958).
3. **M. moschàtus** Dougl. ex Lindl. n = 16 (Vickery et al., 1958).

p. 611. 5. **M. primuloìdes** Benth. Occurs in Greenhorn Range, Kern Co., *Twisselmann*.

p. 612. 12. **M. discólor** Grant is in the opinion of Dr. Bacigalupi a synonym of **M. montioìdes** Gray.
13. **M. Pálmeri** Gray. Occurs in Monterey Co., *Hardham*.

p. 613. 15. **M. floribúndus** Dougl. ex Lindl. n = 16 (Mukherjee & Vickery, 1961).
p. 614. 21. **M. Paríshii** Greene. Reported from the Piute Mts., Kern Co. and Kern Plateau, Tulare Co., *Twisselmann*.

p. 615. 27. **M. barbàtus** Greene. Kern Plateau, Kern Co., *Twisselmann*.
30. **M. diffùsus** Grant. Apparently occurs in Monterey Co., *Hardham* collections.

p. 616. 34. **M. guttàtus** Fisch. ex DC. Mukherjee & Vickery (Madroño 16: 141–154. 1962) report chromosome counts as follows: *M. guttatus* ssp. *guttatus* n = 14; *M. guttatus* ssp. *litoralis*, n = 14; *M. guttatus* var. *puberulus*, n = 14. In Madroño 17: 156–160. 1964, they report *guttatus* as having 14, 16, 28. Vickery (Evolution 18: 52–70. 1964) recognizes **M. platycàlyx** Penn., which can be separated from *M. guttatus* in having the fruiting calyx as wide as long (not ⅔ as wide); corolla 15–20 mm. long (not 18–45) and plant very slender. It occurs in the Sierra Nevada from Mariposa Co. to Tulare Co.
Twisselmann feels that **M. microphýllus** Benth. is quite distinct in being depauperate and occurring in moist cracks in granite, Kern Co.
M. guttàtus ssp. **arenícola** Penn. Accepted by J. T. Howell as a recognizable ssp. Plant low, 5–20 cm. tall, much branched, the infl. not very pubescent; lf.-blades 1–2 cm. long; corolla 25–35 mm. long. Coastal Strand; Monterey Co.

p. 617. 36. **M. nasùtus** Greene. Delete *M. platycalyx* as a synonym. n = 14 or 13 Mukherjee & Vickery, 1962).
37. **M. Tilíngii** Régel. Mukherjee and Vickery in 1962 report n = 14 for *M. T.*, but for var. *corallìnus* (Greene) Grant n = 24.

p. 618. 46. **M. Whítneyi** Gray can be reported from Kern Co. (Greenhorn Range), *Twisselmann*.
48. **M. Fremóntii** (Benth.) Gray occurs in Monterey Co., *Howitt & Howell*, and in w. Kern Co., *Twisselmann*.
50. **M. víscidus** Congd. In Greenhorn Range, Kern Co., *Twisselmann*.

p. 620. 54. **M. Rattánii** Gray. In Monterey Co., *Howitt & Howell*, and in San Luis Obispo Co., *Hardham*.

p. 621. 57. **M. brévipes** Benth. n = 8 (Mukherjee & Vickery, 1962).
58. **M. mohavénsis** Lemmon. n = 7 (Carlquist, 1953).
59. **M. pygmaèus** Grant. On flats w. of Lake Almanor, Plumas Co., *Vesta Holt*.
61. **M. Douglásii** (Benth. in DC.) Gray. In Greenhorn Mts., *Twisselmann*.

p. 622.   68. **M. Congdònii** Rob. In the Greenhorn Mts., Kern Co., *Twisselmann*.
p. 624.   73. **M. aurantìacus** Curt. $n = 10$ (Mukherjee & Vickery, 1961).
p. 625.   GRATÌOLA
   In Key, change B. and BB. to:

   B.   Lvs. attenuate; sepals 7–11 mm. long; plants often copiously glandular-puberulent
        above; corollas with yellowish throat and white limb. . . . . 2. *G. ebractcata*
   BB. Lvs. blunt; sepals 4–6 mm. long, 3 of them fused almost half way; plants essen-
        tially glabrous; corolla yellow, only the 3 lower lobes white. . . 3. *G. heterosepala*

p. 626.   2. **G. ebracteàta** Benth. ascends to 6800 ft., Lassen Nat. Park, *Gillett*.
   3. **G. heterosèpala** Mason & Bacig. Also in Sacramento & Madera cos., *Baci-galupi*.
   LIMOSÉLLA
   1. **L. aquática** L. in Monterey Co., *Howitt & Howell*.
p. 627.   3. **L. subulàta** Ives. $2n = 20$ (Löve & Löve, 1958).
   VERBÁSCUM
   3. **V. virgàtum** Stokes ex With. Delete "s." in last line. Reported from Fresno
   Co., *Weiler*.
p. 631.   PÉNSTEMON
   3. **P. oreócharis** Greene. Cronquist (Vasc. Pl. Pac. N.W. 4: 402) reduces to
   synonymy under **P. Rydbérgii** A. Nels.
p. 633.   12. **P. albomarginàtus** Jones. $2n = 8$ pairs (Raven, Kyhos & Hill, 1965).
p. 637.   38. **P. caèsius** Gray. Greenhorn & San Emigdio ranges, Kern Co., *Twissel-mann*.
p. 640.   48. **P. Newbérryi** Gray found on the Kern Plateau, Kern Co., *Twisselmann*.
   51. **P. nemoròsus** (Dougl. ex Lindl.) Trautv. is transferred to *Nothochelone*
   (Gray) Straw as *N. nemorosa* (Dougl. ex Lindl.) Straw. See Brittonia 18: 85.
   1966. Seeds winged all around.
pp. 640–642.
   52–58. The following were transferred by Straw (Brittonia 18: 87–88. 1966)
   to the genus *Kéckia* Straw. Later (Brittonia 19: 203–204. 1967) he made the
   following combinations under *Keckiella* Straw:
   52. *K. Rothróckii* (Gray) Straw and subsp. *jacinténsis* (Abrams) Straw.
   53. *K. breviflòra* (Lindl.) Straw and ssp. *glabrisèpala* (Keck) Straw.
   54. *K. Lemmònii* (Gray) Straw.
   55. *K. antirrhinoìdes* (Benth.) Straw and ssp. *microphýlla* (Gray) Straw.
   56. *K. cordifòlia* (Benth.) Straw.
   57. *K. corymbòsa* (Benth.) Straw.
   58. *K. ternàta* (Torr.) Straw and ssp. *septentrionàlis* (Munz & Jtn.) Straw.
p. 642.   SCROPHULÀRIA
   After the generic description insert: Shaw, R. J. The biosystematics of *Scrophu-laria* in w. N. Am. Aliso 5: 147–178. 1962.
   Change the Key to:

   A.   Infl. villous, the hairs tipped with small glands; plants shrubby in age; sterile fil. absent
        or rudimentary. Catalina and San Clement ids. . . . . 2. *S. villosa*
   AA. Infl. puberulent or short-pubescent; stems herbaceous; sterile fil. well developed.
        B.   Sterile fil. clavate to obovate, sometimes with an acute apex.
             C.   Corolla dark maroon, the upper half blackish; tube urceolate with a con-
                  stricted orifice; sterile fil. with an acute apex. Santa Barbara Co. 3. *S. atrata*
             CC. Corolla dark maroon or garnet-brown; orifice not constricted.
                  D.   The corolla distinctly bicolored; lf.-blades cuneate. From the Sierra Ne-
                       vada to w. Nev. . . . . 1a. *S. desertorum*
                  DD. The corolla not distinctly bicolored; lf.-bases truncate to cordate. W. of
                       the Sierra Nevada and s. mts. . . . . 1. *S. californica*
        BB. Sterile fil. flabellate, wider than long. . . . . 4. *S. lanceolata*

   1. **S. califórnica** Cham. & Schlecht. and
   Ssp. **floribúnda** (Greene) Shaw. [*S. c.* var. *f.* Greene.]
   1a. **S. desertòrum** (Munz) Shaw. [*S. californica* var. *d.* Munz.]
p. 643.   COLLÍNSIA

In Key, change "14. *C. bartsiaefolia*" to "4. *C. bartsiifolia*," 1959 printing.
Change C. at bottom of page to:

    C. Calyx-lobes obtuse or obtusish.
        D. Corolla 7–10 mm. long; upper fils. bearded. Ventura Co. to San Bernardino Co. .... 7. *C. Parryi*
        DD. Corolla 6–7 mm. long; fils. glabrous. Monterey Co. .... 7a. *C. antonina*

p. 644.   In Key, CC, DD, change "Upper fils. glabrous" to "Upper fils bearded" in the 1959 printing.
    3. Use **C. multicólor** Lindl. & Paxt. and put *C. franciscana* Bioletti in synonymy.

p. 645.   6. **C. tinctòria** Hartw. ex Benth. occurs in the Greenhorn Mts., Kern Co., *Twisselmann*.

    7a. **C. antonìna** Hardham. With few branches, 1–8 (–14) cm. high; petioles and pedicels puberulent and sparsely glandular; lf.-blades pubescent above, glabrous beneath, crenulate, oblong, 4–8 mm. long; fls. 1–3 at each node; pedicels to 1 cm. long in fr.; calyx 5–6 mm. long, puberulent; corolla 6–7 mm. long, white with red spots at base of lobes; $n = 7$.—San Antonio Hills, near Jolon-Bradley road, Monterey Co.

    Ssp. purpùrea Hardham. Corolla purple, white at base of upper lobes and with red spots. Same region.

p. 648.   GALVÈZIA
    1. **G. speciòsa** (Nutt.) Gray. $2n = 15$ pairs (Raven, Kyhos & Hill, 1965).

p. 649.   MOHÁVEA
    1. **M. confertiflòra** Heller. $2n = 15$ pairs (Raven, Kyhos & Hill, 1965).
    2. **M. breviflòra** Cov. in El Paso Range, Kern Co., *Twisselmann*.

p. 650.   ANTIRRHÌNUM
    4. **A. multiflòrum** Penn. $2n = 16$ pairs(Raven, Kyhos & Hill, 1965).

p. 651.   7. **A. Nuttaliànum** Benth. in DC. $2n = 16$ pairs (Raven, Kyhos & Hill, 1965).
p. 652.   12. **A. ovàtum** Eastw. in Temblor Range, Kern Co., *Twisselmann*.
p. 653.   LINÀRIA
    2. **L. dalmática** (L.) Mill. Reported from a number of counties: Siskiyou, Modoc, Lassen, Butte, Alpine, Sierra, Monterey, Shasta and Sacramento.
    5. **L. canadénsis** var. **texàna** Penn. $2n = 12$ (W. H. Lewis, 1958).
    7. **L. reticulàta** (SM.) Desf. may be an older name for **L. pinifolia** (Poir.) Thell. Recently reported for Santa Barbara Co., *Fuller*.

p. 654.   KÍCKXIA
    2. **K. Elátine** (L.) Dumort. caused abandonment of ca. 1500 acres of barley stubble sw. of Dayton, Butte Co., *Fuller*.

    DIGITÀLIS
    1. **D. purpùrea** L. Spontaneous in Nevada Co., *Howell & True*.

p. 655.   VERÓNICA
    Key, A, B, change CC to:

        CC. Caps. wider than long, deeply notched; corolla pubescent in the tube.
           D. Fls. in racemes, the upper lvs. bractlike. Common. ... 4. *V. serpyllifolia*
           DD. Fls. solitary in the axils of lvs. much like the cauline lvs. Lawns, Golden Gate Park, San Francisco .... 4a. *V. filiformis*

    Key A, BB, change to:

        BB. Plant annual; fls. from most axils.
           C. Corolla 2–2.5 mm. wide.
               C′. Pedicels 1–2 mm. long; lvs. not coarsely toothed.
                  D. Lvs. linear-oblong to spatulate; corolla white. ...... 5. *V. peregrina*
                  DD. Lvs. rounded to oval; corolla bright blue. .... 6. *V. arvensis*
               C′C′. Pedicels 6–12 mm. long; lvs. with 2 or 3 large teeth on each side near the base. Sacramento Co. .... 6a. *V. hederifolia*

    Key, under AA, B, add:

        CCC. Lvs. triangular-ovate, sessile or very short-petioled, not clasping; corolla ca. 10 mm. across, deep bright blue with white eye. ...... 9a. *V. Chamaedrys*

4a. **V. filifórmis** Sm. Prostrate pubescent perennial, forming mats; lvs. to 6 mm. long, round to ovate, subcordate, short-petioled; fls. to 8 mm. in diam.— Natur. in San Francisco; native of Asia Minor.

p. 656.　　6a. **V. hederifòlia** L. Annual, the stems 5–25 cm. long, pubescent; lvs. petioled, cordate, 3–5-lobed; fls. long-pedicelled, solitary in the axils; corolla rotate, 2–2.5 mm. broad, lilac or blue; seeds 2.5–3 mm. broad.—Natur. in Sacramento Co., *Crampton*; from Eu.

9a. **V. Chamaèdrys** L. Near *V. americana* in its lvs. not clasping and fls. to 10 or 12 mm. across; plant erect or nearly so; lvs. sessile, rounded at base, pubescent.—Lawns, San Francisco; native of Eu.

12. **V. comòsa** Richt. Reported from Monterey Co., *Howitt & Howell*.

p. 660.　PEDÍCULARIS

10. **P. racemòsa** Dougl. ex Hook. Corolla pink to purplish.

Ssp. **álba** Penn. Corolla white or ochroleucous; lvs. rather narrow.—E. of the Sierra Nevada and Cascade Mts.

p. 661.　ORTHOCÁRPUS

1. **O. campéstris** Benth. Found at 6600 ft., Mt. Lassen Park, *Gillett*.

p. 662.　　6. **O. castillejoìdes** var. **humboldtiénsis** Keck. $2n = 12$ pairs (D. E. Anderson, 1965).

p. 665.　CASTILLÈJA

Key, change AA, B, C and CC to (following Bacigalupi & Heckard, Leafl. W. Bot. 10: 285. 1966):

    C. Floral bracts oblong to linear-oblong, either entire or with 3 relatively short, very blunt lobes toward their tips; foliage pubescent, the lvs. not involute; axillary shoots well developed; upper surface of galea ± densely and shaggily pubescent.

    D. Dorsal portion of galea thin, pale orange to yellow; distal portions of the calyx and of the often terminally lobed floral bracts scarlet; lvs. dark green, lance-oblong. . . . . 30 & 31. *C. subinclusa* and *C. franciscana*

    DD. Dorsal portion of galea thicker, dark green; floral bracts mostly entire and usually green and not suffused by any other color; distal portion of calyx rose-pink; lvs. gray-green, mostly linear-oblong. Inner S. Coast Ranges from San Benito Co. to nw. Los Angeles Co. and to Kern River V., also in San Diego Co.; L. Calif. . . . . 31a. *C. Jepsonii*

    CC. Floral bracts cuneate in outline, deeply and digitately 3-cleft or 3-parted, the spreading lobes narrow and acute, green to rose-pink; foliage and stems (except infl. and very base of stem) glabrous and glaucous, the narrowly linear lvs. involute; axillary shoots mostly weakly developed or lacking; dorsal surface of galea glabrate to sparsely and finely pubescent. . . 29. *C. linariifolia*

Key, middle of page, change J. and JJ. to:

    J. Lvs. lanceolate; lower corolla-lip not exserted. North Coast Ranges and Sierra Nevada. . . . . 17. *C. Applegatei*

    JJ. Lvs. almost linear; lower corolla-lip exserted. Sierra Nevada. . . . . 18. *C. disticha*

Key, CC, D, EE, change to:

    EE. Plant not glandular-pubescent below the infl.

    E'. Plant whitened by an arachnoid-lanose coat of long flexuous hairs; lvs. linear; sepals of each side united to a roundish tip. Channel Ids. . . . . 27. *C. hololeuca*

    E'E'. Plant not grayish-woolly.

    F. Calyx-lobes mostly less than 2 mm., etc. as in the FLORA.

p. 667.　　3. Change to **C. longispìca** A. Nels. and put *C. psittacina* in synonymy.

p. 668.　　4. **C. pilòsa** (Wats.) Rydb. $n = 12$ (Heckard, 1958).

5. Add as synonym under **C. nàna** Eastw., *C. rubida* Piper var. *monoensis* (Jeps.) Edwin.

9. **C. brevilobàta** Piper. $n = 12$ (Heckard, 1958).

10. **C. Brèweri** Fern. Occurs in the Piute Mts., Kern Co., *Twisselmann*.

14. **C. Wìghtii** Elmer. $n = 12$ (Gillett, 1954).

14a. **C. inflàta** Penn. $n = 36$ (Heckard, 1958).

14b. **C. litoràlis** Penn. Ownbey, in Pacific N.W. Fl. 4: 312, treats as a sp.

15. **C. Ròseana** Eastw. *n* = 12 (Heckard, 1958).

p. 670.   18. **C. dísticha** Eastw. Found in Piute Mts., Kern Co., *Twisselmann*.

p. 671.   22. **C. affìnis** H. & A. is used by Bacigalupi (Leafl. W. Bot. 10: 286. 1966) for plants with the lvs. having slender hairs, while **C. affìnis** var. **contentiòsa** (J. F. Macbr.) Bacig. [*C. Douglasii* var. *c.* J. F. Macbr.] has vitreous, thick-based, cellular trichomes on the lf.-margins.—Coast Ranges, Monterey Co. to Santa Monica Mts., Anacapa and Santa Rosa ids.

24. **C. chromòsa** A. Nels. *n* = 12 (Heckard, 1958).

25. **C. Gleasònii** Elmer [*C. pruinosa* ssp. *G.* Munz.] treated as a distinct sp., from the San Gabriel Mts. by Bacigalupi.

p. 672.   28. **C. plagiótoma** Gray. Reported from Piute Mts., Kern Co., *Twisselmann* and from San Luis Obispo and Fresno cos., *Eastwood*.

29. **C. linariifòlia** Benth. *n* = 12 (Heckard, 1958).

31a. **C. Jepsònii** Bacig. & Heckard. Pilose and puberulous leafy, gray-green perennial, the usually many strict stems 6–12 dm. tall, striate, with sterile axillary, quite leafy shoots; cauline lvs. mostly linear-oblong, mostly entire, with long and short hairs, the lvs. 3–8 cm. long, 2–6 mm. wide; infl. eventually as long as 4 dm., with some gland-tipped hairs often among the septate pilosity; upper bracts often tridentate and distally rose-pink; calyx narrow, striate, 2.5–3 cm. long; corolla conspicuously exserted and curved outward, 2.5–5 cm. long, the galea ca. 2 cm. long, the upper surface deep green and densely covered with short, thickish, often gland-tipped hairs, its apex obtuse, the narrow pink margins thin and glabrous; lower lip rudimentary, 1–2.5 mm. long, dark red-purple; *n* = 12.—At 1500–7500 ft.; inner South Coast Ranges from ne. San Benito Co. to Los Angeles Co. and Piute Mts., Kern Co., also in se. San Diego Co.; L. Calif.

32. **C. stenántha** Gray. *n* = 12 (Heckard, 1958).

p. 673.   CORDYLÁNTHUS

In Key, Corolla in *C. Nevinii* is not truly glabrous, *Bacigalupi*.

p. 675.   9. **C. Nevínii** Gray. On Piute Mt. and Mt. Pinos, *Twisselmann*.

p. 676.   1. **C. bernardìnus** Munz, collected in the Nelson Range, Inyo Co., near the head of Grapevine Canyon, *Roos*.

13. **C. Ferrisiànus** Penn. Reported from Greenhorn Mts. and Piute Mts., Kern Co., *Twisselmann*.

16. **C. filifòlius** Nutt. ex Benth. in the Santa Lucia Mts., *Hardham*.

p. 677.   17. **C. pilòsus** Gray. Change to "Plants 5–7 dm. tall."

p. 679.   MARTYNIÀCEAE

In description of the family insert after "Caps." and after "endocarp" the word "often," since in *Ibicella* the caps. is not crested but echinate.

PROBOSCÍDEA

Change description to "Coarse viscid-pubescent plants."

p. 681.   PINGUÌCULA

S. J. Casper (Fedde Repert. 66: 1–148. 1962) refers our plants to **P. macró-ceras** Link with corolla-lobes of lower lip obovate; calyx deeply divided; corolla, including spur, 20–30 mm. long; spur 6–11 mm. long, while in *P. vulgàris* L. the corolla-lobes of the lower lip are more oblong; the calyx is less deeply divided; the corolla, including spur, is 15–22 mm. long; spur 3–6 mm. long. **P. macróceras** is transcontinental and in ne. Asia. *P. vulgàris* ranges from Greenland to cent. Siberia and the Caucasus. Calder & Taylor make the comb. *P. vulgaris* ssp. *macroceras* (Link).

p. 682.   OROBÁNCHE

1. Change name of **O. uniflòra** L. var. **minùta** (Suksd.) Achey to **O. u.** ssp. **occidentàlis** (Greene) Abrams ex Ferris and have as synonyms: [*Aphyllon u.* var. *o.* Gray and *A. minutum* Suksd. and *O. u.* var. *minuta* Achey.]

p. 863.   2. **O. fascículata** Nutt. 2*n* = 24 pairs (Raven, Kyhos & Hill, 1965).

p. 684.   5. **O. Grayàna** var. **Feùdgei** Munz. Reported from Temblor Range, Kern Co., *Twisselmann*.

Var. **Nelsonii** Munz. $2n = 24$ pairs (D. E. Anderson, 1963).

6. **O. califórnica** Cham. & Schlecht. var. **corymbosa** (Rydb.) Munz. Add to synonymy *O. corymbosa* Ferris. Reported from Piute Mts., Kern Co., *Twisselmann.*

7. **O. ludoviciana** var. **Coòperi** G. Beck. $2n = 24$ pairs (Raven, Kyhos and Hill, 1965).

p. 685.   10. **O. ramòsa** L. Reported as in Sacramento Co., *Brown* and in San Joaquin Co., *Nichols.*

p. 686.   BELOPERÒNE

1. **B. califórnica** Benth. $2n = 28$ (Grant, 1956).

VERBENÀCEAE

Change Key to:

A.   Fls. sessile in spikes which are simple or clustered; lvs. simple.
     B.   Nutlets 4; fls. in terminal spikes. . . . . 1. *Verbena*
     BB. Nutlets 2; fls. in dense spikes or heads.
          C.   Spikes globose or cylindrical; fr. dry. . . . . 2. *Lippia*
          CC. Spikes flat-topped; fr. drupaceous. . . . . 3. *Lantana*
AA. Fls. pedicelled; lvs. palmately compound. . . . . 4. *Vitex*

VERBÈNA

In Key, insert "mostly" after "Fls." in A.

In Key, A, B, C, change DD to:

     DD. Lvs. short-petioled, not auriculate-clasping.
         E.   Infl. lax, elongate; fls. distant; pubescence on rachis, bractlets, and
              calyx very minute, closely appressed. . . . . 2. *V. litoralis*
         EE. Infl. dense, contracted; fls. mostly congested; pubescent on rachis,
              bractlets and calyx spreading. . . . . 2a. *V. brasiliensis*

p. 687.   In Key, add at very end:

     BBB. The ultimate lf.-divisions linear; corolla ca. 10 mm. long. . . . . 12. *V. tenuisecta*

2. Moldenke (Phytologia 8: 313. 1962) apparently takes up the name **V. brasiliénsis** Vell. for my *V. litoralis,* but **V. litoràlis** HBK is all right for plants from Amador and San Joaquin cos. (Phytologia 10: 67). $2n = 56$ (Huynh, 1965).

p. 687.   2a. **V. brasiliénsis** Vell. Confused with **V. litoràlis** HBK, but the infl. more condensed, the pubescence of rachis, bracts and calyx more spreading.—Cited in Calif. from Amador, Butte, Eldorado, Nevada, Sacramento, San Joaquin, Solano, Stanislaus, Sutter and Yuba cos. Native of S. Am.

3. After **V. hastàta** L. insert:

Var. **scàbra** Moldenke. Lf.-bases more rigid; lvs. conspicuously scabrous on upper surface, often ± conspicuously pubescent beneath.—Cited from Modoc, San Joaquin and Shasta cos.; to B.C., Mont.

6a. **V. Clemensòrum** Moldenke described as possible hybrid between *V. officinalis* and *V. robusta,* from Jackson, Amador Co. Coarse herb with glabrous stems, stiff ovate incised lvs. 2.5–8 cm. long; infl. spicate, compound, ± puberulent, elongate; corolla 2 mm. wide.

7. **V. lasiostàchys** Link. $2n = 7$ pairs (Raven, Kyhos & Hill, 1965).

p. 688.   8. **V. robústa** Greene. Is reported from Mendocino and Amador cos.

9. **V. bracteàta** Lag. & Rodr. [*V. bracteosa* Michx.] Moldenke has described **V. califórnica** from Keystone, Tuolumne Co. which keys out to **V. bracteàta**, but is said to have a fruiting calyx 3.5–4 mm. long; corolla not known.

10. **V. Gooddíngii** Briq. Add:

Var. **nepetifòlia** Tidestr. [*V. bipinnatifida* var. *n.* Jeps.] Lvs. shallowly lobed or usually more coarsely toothed.—Lanfair, e. Mojave Desert; to Ariz., L. Calif., Nuevo Leon.

11. **V. ténera** Spreng. and **V. tenuisécta** Briq. are confused and both are to be looked for in Calif. According to one author, *V. tenera* has the calyx-hairs appressed and *V. tenuisecta* ascending-spreading. The latter is described as having lvs. 2–4 cm. long, segms. mostly ca. 1 mm. wide; corolla-limb ca. 10 mm.

wide.—Reported from Bakersfield region, Kern Co., *Twisselmann*; native of
S. Am. All California material which I have seen seems to be *V. tenuisecta.*

p. 689   3. Lantàna L.

Shrubs or herbs; scabrous-hirsute, pubescent or tomentose. Lvs. opposite,
dentate, often rugose. Fls. small, sessile in the axils of bracts, forming dense
spikes or heads, which are terminal or axillary. Calyx very small. Corolla somewhat irregular, but not bilabiate; tube slender; lobes 4–5. Stamens 4, included.
Ovary 2-loculed, forming a fleshy drupe with 2 bony nutlets. (Old Latin name.)
    1. L. montevidénsis Briq. Stems weak, vinelike, ca. 1 m. long; lvs. ovate,
2–5 cm. long; fls. rose-lilac, in heads 2.5–3 cm. across.—Occasionally spontane-
ous, as at Claremont; native of S. Am.

                                          4. Vìtex L.

Trees or shrubs. Lvs. persistent or deciduous, opposite, digitate, mostly with
3–7 lfts. Fls. white, blue or yellow, in few to many-fld. cymes which are often
panicled; calyx campanulate, usually 5-toothed; corolla tubular-funnelform, the
limb slightly 2-lipped; lobes 5. Stamens 4, often exserted. Fr. a small drupe,
with a 4-celled stone. A rather large genus of warmer regions. (Ancient Latin
name.)
    1. V. Ágnus-cástus L. Deciduous shrubs to 3 m. high, aromatic, gray-pubes-
cent; lvs. opposite, digitately 5–7-foliolate, the lfts. linear-lanceolate, 5–15 cm.
long, gray-felted beneath; fls. pale violet, in terminal racemes 7–20 cm. long;
corolla tubular, 8–9 mm. long.—S. Eu. Found in alkali sink e. of Weedpatch,
Kern Co., *Twisselmann.*

p. 690.   LABIÀTAE
Key AA, BB, C, DD, E, F, G, HH, II, JJ, KK, change "lvs." to "fls." in 1959
printing.

p. 692.   TRICHOSTÈMA
    1. T. oblóngum Benth. s. to Greenhorn Mts., Kern Co., *Twisselmann.*
    3. T. simulàtum Jeps. in Alpine Co., *Lewis.*
    4. T. rubisépalum Elmer. $n = 7$ (H. Lewis, 1960).

p. 693.   6. T. micránthum Gray. $n = 7$ (H. Lewis, 1960).
          SALAZÀRIA
    1. S. mexicàna Torr. $2n = 50$ pairs (Raven, Kyhos & Hill, 1965).

p. 694.   SCUTELLÀRIA. .
In last part of Key, S. *antirrhinoides* has definite petioles, while they are quite
lacking in S. *siphocampyloides* and S. *Austiniae.*
    2. S. tuberòsa Benth. Reported from near Woody, Kern Co., *Twisselmann.*
    Put ssp. *austràlis* Epl. into synonymy under S. tuberosa. Add
    Var. símilis Jeps. Calyx very densely villous. Range of the sp.
    4. S. Bolánderi Gray. Found in Greenhorn Mts., *Twisselmann.*

p. 697.   GLECHÒMA instead of *Glecoma* for genus no. 9.
          PRUNÉLLA
    1. P. vulgàris L. var. atropurpùrea Fern. Bracts often purplish and short-
ciliate; corollas dark purple.—Along the coast, from San Mateo Co. to Sonoma
Co.
    Var. parviflòra (Poir.) DC. is a lawn plant in Golden Gate Park, San Fran-
cisco; introd. from Eu. and has corollas scarcely exceeding the bracts.
          PHLÒMIS
    1. P. fruticòsa L. $2n = 20$ (Strid, 1965).

p. 699.   STÀCHYS
    4. S. Emersònii Piper should be called S. mexicàna Benth. . .

p. 700.   8. S. pycnántha Benth. is apparently a serpentine plant. The Tehama Co.
reference in the FLORA is a misidentification for S. *rigida* ssp. *rivularis* Epl.

p. 701.   ACANTHOMÍNTHA

2. **A. obovàta** Jeps. is reported from Monterey Co., *Howitt & Howell. 2n =* 19 pairs (Raven, Kyhos & Hill, 1965).

p. 702.   SÁLVIA

In Key under A, B, CC, change D to:

> D. Plants introd. perennials.
>> E. Corolla blue, 10–15 mm. long. . . . . 2. *S. verbenacea*
>> EE. Corolla purple, ca. 25 mm. long. . . . . 2a. *S. pratensis*

Under A, BB, insert:

> CCC. Calyx tubular; lvs. oval, obtuse, corolla red. Introd. . . . . . . 6a. *S. Grahamii*

Under AA, B, insert:

> CCC. Fls. 30 mm. long, scarlet; lvs. 7–14 cm. long; plant ca. 4 m. tall. Introd.
> 8a. *S. longistyla*

p. 703.   2a. **S. praténsis** L. Perennial, to 7 dm. high, erect, pubescent; lvs. largely basal, oblong-ovate, those of infl. cordate-ovate; racemes glandular, subsimple, the whorls remote, 6-fld.; corolla bright blue, sometimes red or white, 2.5 cm. long.—Found as escape near Yreka, Siskiyou Co., *Fuller*; native of Eu.

3. **S. carduàcea** Benth. *n* = 16 (Epling, Lewis & Raven, 1962).

4. **S. Columbàriae** Benth. var. **Ziègleri** Munz. var. nov. Plants much like the typical form, but persistent until fall or early winter; lvs. rather coarsely sub-sinuately pinnatifid into broad oblong divisions with low rounded teeth; calyx ca. 7 mm. long, the 2 upper lobes united for ca. 2 mm., then with 2 divergent spines almost 2 mm. long; lower lip 2-lobed, the lobes ca. 0.5 mm. long, each ending in a spine almost 2 mm. long; corolla ca. 8 mm. long, deep blue, in general like that figured by Epling (Ann. Mo. Bot. Gard. 25: pl. 16. 1938), with shorter stamens; these fertile.

Folia crasse subsinuate pinnatifida, lobis oblongis; calyce ca. 2 mm. longo, labio superiore de 2 spinis divergentibus et ca. 2 mm. longis, inferiore de lobis 0.5 mm. longis, spinosis; corolla ca. 8 mm. longa.

TYPE: from disturbed soil along a road about one mile north of Kenworthy Ranger Station, Hemet Valley, San Jacinto Mts., Riverside County, growing at about 5,000 ft. elev., Sept. 8, 1964, *Louis B. Ziegler* (RSA). Another collection from the same region is *L. B. Ziegler* Nov. 10, 1965.

The proposed variety is for a population extending for a mile or more and remaining green and floriferous many months after the general population of *S. Columbariae* has dried up in the late spring. Mr. Ziegler has had it under observation for more than one season, the plants ripening seeds and sending out new branches that go on into flowering. The coarse lvs. and smaller fls. can be matched by specimens from other parts of the range of the sp., but the very robust and long continued growth seem unique. It is a pleasure to dedicate this plant to Mr. Louis B. Ziegler who has made a number of other interesting botanical discoveries in the same area.

5. **S. funèrea** Jones. *n* = c. 32 (Epling, Lewis & Raven, 1960).

6. **S. Greàtae** Bdg. *2n* = c. 30 (Epling, Lewis & Raven, 1960).

p. 704.   6a. **S. Gràhamii** Benth. Shrub, to ca. 1 m. tall; lvs. oval, obtuse, rounded or cuneate at base, crenate; racemes to more than 3 dm. long; floral whorls 2-fld.; corolla ca. 2.5 cm. long, crimson, purplish in age, the midlobe of the lower lip obcordate, large, with 2 small white spots.—Escape from cult, as in Marin, Sonoma and Santa Barbara cos.; native of Mex. The correct name may be **S. microphýlla** Benth. in H. & B.

8a. **S. longistỳla** Benth. Herb, 4–4.5 m. tall, tomentose-villous; lvs. petioled, broad-ovate, 7–14 cm. long; racemes 5–8 dm. long; corolla scarlet, ca. 3 cm. long, with long-exserted stamens and style.—Escape from cult., Big Sur, Monterey Co., *Howell*; native of Mex.

7. **S. spathàcea** Greene. *2n* = 30 (Epling, Lewis & Raven, 1960).

8. **S. sonoménsis** Greene. *2n* = 30 (Epling, Lewis & Raven, 1960).

p. 705. 10. **S. pachyphýlla** Epl. ex Munz. $2n = 30$ (Epling, Lewis & Raven, 1960).
11. **S. mellífera** Greene. $2n = 30$ (Epling, Lewis & Raven, 1960).
12. **S. Múnzii** Epl. $2n = 30$ (Epling, Lewis & Raven, 1960).
13. **S. Brandègei** Munz. $n = 15$ (Epling, Lewis & Raven, 1960).
14. **S. eremostàchya** Jeps. $2n = 30$ (Epling, Lewis & Raven, 1960).
15. **S. mohavénsis** Greene. $n = 15$ (Epling, Lewis & Raven, 1960).

p. 706. 16. **S. Clevelándii** (Gray) Greene. $2n = 30$ (Epling, Lewis & Raven, 1960).
17. **S. leucophýlla** Greene. $2n = 30$ (Epling, Lewis & Raven, 1960). Found
in Monterey Co., *Howitt & Howell*.
18. **S. Vàseyi** (Porter) Parish. $n = 15$ (Epling, Lewis & Raven, 1960).
19. **S. apiàna** Jeps. $n = 15$ (Epling, Lewis & Raven, 1960).

p. 707. LEPECHÍNIA
1. **L. calcyìna** (Benth.) Epl. in Munz. $n = 16$ (Epling, 1948); $2n = 17$
pairs (Raven, Kyhos & Hill, 1965).
3. **L. frágrans** (Greene) Epl. $n = 16$ (Epling, 1948).
4. **L. Gánderi** Epl. $n = 16$ (Epling, 1948).

p. 710. POGÓGYNE
4. **P. serpylloìdes** (Torr.) Gray. $2n = 19$ pairs (Raven, Kyhos & Hill, 1965).
5. **P. zizyphoroìdes** Benth. $2n = 19$ pairs (Raven, Kyhos & Hill, 1965).

p. 712. MONARDÉLLA
5. **M. villòsa** Benth. Mrs. Clare Hardham has proposed a number of new spp.
in the *M. villosa* complex and the following key is presented in place of the one
in the FLORA:

A.   Upper parts of the plant and lower surfaces of lvs. pubescent to villous.
  B.   Lvs. ovate to roundish, mostly green above.
    C.   The lvs. almost glabrous. Near the coast, San Luis Obispo Co. to Marin Co.
                                                 5a. *M. subglabra*
    CC. The lvs. ± villous-pubescent to tomentose beneath.
      D.   Outer leafy bracts 3 pairs; lvs. ± woolly-pubescent beneath. Santa Bar-
         bara Co. to Marin Co. 5. *M. v.* var. *franciscana*
      DD. Outer leafy bracts 1–2 pairs.
        E.   Lvs. densely white-tomentose beneath with branched hairs. W. San
           Luis Obispo and Monterey cos. . . . . 5. *M. v.* var. *obispoensis*
        EE. Lvs. ± villous-pubescent beneath with simple hairs. San Luis
           Obispo to Humboldt cos. . . . . 5. *M. villosa*
  BB. Lvs. lanceolate to ovate-lanceolate, grayish above.
    C.   The lvs. almost entire, 20–25 mm. long, soft villous-pubescent, especially be-
      neath. Monterey Co. to w. Ore., possibly Sierra Nevada. 5. *M.v.* ssp. *subserrata*
    CC. The lvs. evidently serrate.
      D.   Hairs rather long, curved, multicellular; lvs. 7–17 mm. long. Diablo
         Range, San Benito Co. and e. Monterey Co. . . . . 5b. *M. benitensis*
      DD. Hairs minute and glandular. San Antonio Hills, Santa Lucia Range.
                                               5c. *M. antonina*
AA. Upper parts of plant and lower lf.-surfaces puberulent at most.
  B.   Bracts leaflike, not markedly ciliate and not purplish. Sierra Nevada and Siskiyou
    Mts. . . . . 5. *M. v.* ssp. *Sheltonii*
  BB. Bracts membranous, ciliate, the inner purple. San Mateo Co. to Sonoma Co.
                                                5. *M. v.* ssp. *neglecta*

p. 713. 5a. **M. subglàbra** (Hoover) Hardham. [*M. villosa* var. *s.* Hoover.] Rhizomatous perennial; stems with short down-curved hairs; lvs. broadly ovate, 7–16
mm. long, 4–9 mm. wide; petioles 2.5–6 mm. long; blades with 6–10 prominent
veins, thickened serrate margins, subglabrous except for dense glandular pubescence beneath; bracts in 3 series: *outer* of 1 or 2 pairs, ± remote, petioled or
not, leafy at tip, membranous at base; *middle* of 3 pairs with leafy tips and
membranous bases; the *inner* of a few pairs of membranous bracts; leafy bracts
often reflexed in age; calyx 7–9 mm. long, 13-veined and with sparsely pubescent teeth 1.3 mm. long; $n = 20$.—San Luis Obispo Co. to Marin Co., below
2000 ft., on exposed or woody places.
5b. **M. beniténsis** Hardham. Woody at base, somewhat rhizomatous, forming small clumps 3–5 dm. tall; stems many-branched from above the base; lvs.
lanceolate to ovate-lanceolate, 7–17 mm. long, 6–10 mm. wide, usually 8-veined;
petiole ca. 3 mm. long, margined; lf.-margins remotely serrate, the teeth gland-

tipped; pubescence of lvs., stems and foliar bracts of rather long curved multi-cellular hairs and of microscopic glandular hairs beneath the longer ones; fl.-head usually with 1 (–2) pairs of leafy bracts; next 3 pairs less leaflike; inner bracts of several pairs, linear, leathery-membranous; calyx 13-nerved, 6.5–8 mm. long; corolla 13–14.5 mm. long; calyx-teeth 1.66 mm. long, long bristly-hairy; $n = 21$.—Serpentine along Clear Creek, Diablo Range, San Benito Co. and possibly Griswold Hills and Priest Valley.

5c. **M. antonìna** Hardham. Perennial, woody at base, from small clumps to large patches; stems often many-branched; lvs. lance-ovate, 10–26 mm. long, 3–15 mm. wide, cuneate at base; petioles 2–8 mm. long; lf.-margins remotely serrate, usually 7–8-veined; pubescence of numerous minute gland-tipped hairs and short down-curved white hairs; leafy bracts 1 pair, ca. 8 mm. long, 3 mm. wide, 7–8-veined; middle bracts of 3–4 pairs, leafy-tipped; inner bracts of a few pairs, narrow, membranous; calyx 6–9 mm. long, ca. 13-nerved, with teeth 1.3 mm. long; corolla bluish purple, 11–16 mm. long; $n = 21$.—On silicious shale, Foothill Wd., San Antonio Hills of Santa Lucia Range, between Bradley and King City, Monterey Co.

p. 714. 9. **M. odoratíssima** Benth. ssp. **parvifòlia** (Greene) Epl. in the Greenhorn Range, Kern Co., *Twisselmann*; White Mts., *Howell*.

10. **M. Robisònii** Epl. in Munz.; Granite Mts. n. of Amboy, San Bernardino Co., *Haller*; $2n = 21$ pairs (Raven, Kyhos & Hill, 1965).

p. 715. 13. **M. undulàta** Benth. $2n = 21$ pairs (Raven, Kyhos & Hill, 1965).

18. **M. éxilis** (Gray) Greene. Ascends to 6200 ft., S. Fork of Kern R., Kern Co., *Griesel.*

p. 716. LÝCOPUS

After generic description insert: "Henderson, N.C. A taxonomic revision of the genus Lycopus. Am. Midl. Nat. 68: 95–138. 1962."

1. **L. americànus** Muhl. In Kern R. region, Kern Co., *Twisselmann*; Lake, San Francisco and Inyo cos., *Rubtzoff.*

p. 717. MÉNTHA
In Key add a last line:

Lvs. lanceolate to lance-ovate, 5–12 cm. long, pubescent to tomentose above, white-tomentose beneath. . . . . 7. *M. longifolia*

p. 718. 5. **M. citràta** Ehrh. Fourth word should be "glabrous."

7. **M. longifòlia** Huds. [*M. silvestris* L.] Stoloniferous; stems erect, puberulent to tomentose, 4–12 dm. high; lvs. nearly sessile, lanceolate to lance-ovate, 5–12 cm. long, sharply serrate, pubescent to tomentose above, white-tomentose beneath; spikes thickish, ± dense, especially above; corolla purplish, 3 mm. long. —Escaped, as at Santa Barbara, *C. Smith*; Ojai Valley, Ventura Co., *Pollard*; from Eurasia.

p. 719. TILLAÈA
Raven and others use *Crassula* instead of *Tillaea*.
2. **T. mucòsa** L. Reported from Calaveras Co., *Gankin*.

p. 720. PARVISÈDUM
4. **P. Congdònii** (Eastw.) Clausen. In the Greenhorn Range, Kern Co., *Twisselmann.*

p. 721. DÚDLEYA
In Key, A, B, CC, DD, EE, FF, G, HH, delete "Insular. 7. *D. Greenei*" from II and insert:

J. Caudex 2–5 cm. thick; rosette-lvs. 1.5–3 cm. wide. Insular. . . . . 7. *D. Greenei*
JJ. Caudex 1–1.8 cm. thick; rosette-lvs. 3–7 mm. wide. Mainland. . . . . 7a. *D. Bettinae*

p. 723. 4a. Delete *Cotyledon Palmeri* Wats. from synonymy under 4. **D. farinosa** and insert **D. Pálmeri** (Wats.) Britton & Rose. Caudex 2–4 cm. thick, to 2 dm. long, loosely branching; rosettes 5–20 cm. in diam.; lvs. 15–25, green or reddish, oblong-lanceolate, acute to acuminate, 5–20 cm. long, 1.5–5 cm. wide; floral stems 2–6 dm. tall, 5–11 mm. thick, with ca. 3 simple or forked branches; cymes circinnate, 5–8 cm. long, 5–14-fld.; pedicels erect, 2–10 mm. long; sepals deltoid-ovate, acute, 3–5 mm. long; petals yellow with red, elliptic, acute, 11–16 mm. long, connate for 1.5–2 mm.; $2n = 68, 85, 119$.—San Luis Obispo and Monterey cos. May–June.

p. 724.   7a. **D. Bettìnae** Hoov. Much like *D. parva,* but branches below the rosettes 10–18 mm. in diam.; lvs. evergreen, terete to semiterete, 2–7 cm. long, 3–7 mm. wide after drying; fl.-stems 1.5–2.5 dm. tall, few-branched to simple; pedicels erect, 1–4 mm. long; calyx-lobes ovate, 3–5 mm. long; petals straw-color, often purplish-tinged toward apex, erect or slightly curved outward at the tips.— Serpentine, 1 mi. s. of Cayucos and 1 mi. w. of Cerro Romauldo, San Luis Obispo Co. Hoover separates this proposed sp. from its 2 associates as follows:

> Plants eventually developing a dense cluster of numerous rosettes; lvs. terete to semiterete; petals cream-color or straw-color, often with midrib purple-tinged toward apex.
>> Branches of caudex just below the rosettes 4–8 mm. in diam.; lvs. shriveling in summer, in dried specimens less than 2 mm. wide at the middle. . . . . 3. *D. parva*
>> Branches of caudex just below the rosettes 10–18 mm. in diam.; lvs. evergreen after drying, 3–7 mm. wide at middle. . . . . 7a. *D. Bettinae*
> Plants usually with a single rosette, rarely more than 4 or 5 even when old; lvs. flat, comparatively thin; petals with purple midrib and purple striations on either side.
>> 8. *D. Abramsii* ssp. *murina*

p. 725.   15a. Delete *Stylophyllum Ḥassei* Rose from synonymy under 15. **D. virens** and insert **D. Hássei** (Rose) Moran. [*S. H.* Rose. *Cotyledon H.* Fedde. *Echeveria H.* Berger.] Caudex 1–3 cm. thick, to 3 dm. long, much branched; rosette 6–10 cm. diam.; lvs. 15–30, farinose, linear-lanceolate, obtuse, 3–10 cm. long, 5–15 mm. wide, 2–4 mm. thick; floral stems 1–3 dm. long, with 2–4 simple or forked branches; cymes 2–8 cm. long, 4–14-fld.; pedicels erect, 1–5 mm. long; sepals deltoid-ovate, acute, 2–3 mm. long; petals white, 8–10 mm. long; connate for 1.5–2 mm., spreading from middle; $n = 34$.—Catalina Id.; L. Calif. May–June.

p. 727.   SÈDUM

In Key, change last DD to:

> DD. Fls. white.
>> E.   Lvs. narrowed toward base; petals ca. 8 mm. long. San Bernardino Mts. . . . . 10. *S. niveum*
>> EE. Lvs. not narrowed toward base; petals ca. 4 mm. long. Escape from cult. . . . . 10a. *S. album*

p. 728.   7. **S. Ròsea** (L.) Scop. ssp. **integrifòlium** (Raf.) Hult. $n = 16$ (Wiens & Halleck, 1962).

p. 729.   10a. **S. álbum** L. Glabrous, creeping, evergreen, forming large mats; fl.-stems 7–15 cm. high; lvs. alternate, linear-oblong to obovate or even globular, 3–15 mm. long, terete or flattened above, sessile; fls. ca. 9 mm. in diam.; petals white, obtuse, ca. equaling the stamens; fr. erect, white, streaked red. Fls. in early summer.—Reported as becoming natur., especially near the coast; native Eurasia.

p. 729.   CONGDÒNIA

1. **C. pinetòrum** (Bdg.) Jeps. R. Moran (Leafl. W. Bot. 6: 62–63. 1950) expressed doubt as to whether this plant originally came from California. The only collection known is the type at the University of California, for which evidence was presented that it may have come from Mexico, not California. Furthermore, the name *Congdonia* was used earlier for a genus of *Rubiaceae.*

p. 730.   COTYLÈDON

1. **C. orbiculàta** L. Reported as highly poisonous to sheep and goats, *Fuller.*

p. 731.   SAXIFRAGACEAE

In Key, A, BB, CC, DD, EE, FF, GG, HH, II, change J. to:

> J.   Ovary 2-celled.
>> K.   Placentae axile; petals obovate or spatulate. Widespread. . . . . . 4. *Boykinia*
>> KK. Placentae parietal; petals filiform. In extreme n. Calif. . . . 11a. *Bensoniella*
> JJ.   Ovary 1-celled, etc. as in the FLORA.

p. 734. SAXÍFRAGA

In Key, AA, BB, change C to:

    C.   Lower stems with ± horizontal perennial branches densely covered with elongate lvs.

        D.   Lvs. entire, linear; petals oblanceolate, 4–5 mm. long. From Tulare Co. n.

                                                    5. *S. Tolmiei*

        DD. Lvs. mostly 3-lobed, ± spatulate; petals obovate to oblong-obovate. Marble Mts., Siskiyou Co. . . . . 5a. *S. caespitosa*

1. **S. débilis** Engelm. Raven (Leafl. W. Bot. 10: 142) says this sp. grows at 10,000–12,400 ft., Madera, Mono, Fresno, Tulare, Inyo cos.

3. Change name to **S. odontolòma** Piper, fide Bacigalupi. [*S. arguta* auth, not D. Don.]

p. 735.   4. **S. Mertensiàna** Bong. $2n = $ ca. 48 (Beamish, 1961).

5. **S. Tólmiei** T. & G. $n = 15$ (Beamish, 1961).

5a. **S. caespitòsa** L. Dense tufted perennial from a woody, often branching rootstock, the stems depressed, 3–6 cm. long; lvs. mostly cuneate with 3 obtuse apical lobes, glandular-puberulent and villous-ciliate, 5–10 mm. long; flowering stems 4–10 cm. high, few-fld., glandular-pubescent; sepals ca. 3 mm. long; petals 4–5 mm. long.—Damp rocky places at 6500 ft., Marble Mts., Siskiyou Co., *Gilbert Muth.* Previously known from Lincoln Co., Ore. n. and e. through the Rocky Mts. to n. Ariz. Identification of the Muth specimen by R. Bacigalupi.

p. 736.   LITHOPHRÁGMA

After generic description, insert: (Taylor, R. L. The genus Lithophragma. Univ. Calif. Publ. Bot. 37: 1–122. 1965).

p. 737.   Change Key to:

A.   Basal lvs. truly compound, trifoliolate; stems stout, 4–6 dm. high. San Clemente Id.

                                                   10. *L. maximu* ᴀ

AA. Basal lvs. not truly compound, but merely lobed; stems slender.

    B.   The basal lvs. not lobed to near their bases.

        C.   Petals entire or shallowly toothed.

            D.   Fl.-tube with a ± rounded, not definitely acute base; pedicels 1–3 mm. long; stem-lvs. alternate.

                E.   Base of petal blade not involute or toothed; fl.-tube campanulate with a truncate base. . . . . 1. *L. heterophyllum*

                EE. Base of petal blade somewhat involute, minutely toothed or laciniate; fl.-tube obtuse or rounded at base. . . . . 2. *L. Bolanderi*

            DD. Fl.-tube with an acute base; pedicels 5–10 mm. long; stem lvs. mostly opposite. . . . . 4. *L. Cymbalaria*

        CC. Petals deeply parted.

            D.   Corolla spreading widely; petals not very lacerate on margins.

                E.   Pedicels 1–2 mm. long; petals 4–7 mm. long; fl.-tube truncate at base. . . . . 1. *L. heterophyllum*

                EE. Pedicels 2–8 mm. long; petals 6–10 mm. long; fl.-tube obconic at base. . . . . 5. *L. affine*

            DD. Corolla campanulate, the petals with very lacerate margins. N. Calif.

                                          3. *L. campanulatum*

    BB. The basal lvs. lobed almost to their base.

        C.   Petals mostly 3–5 mm. long; fl.-tube ± rounded at base.

            D.   Fls. 3–8; fl.-tube scarcely striate.

                E.   Stem lvs. not bearing axillary bulblets; stems 2–3.5 mm. tall.

                                          6. *L. tenellum*

                EE. Stem lvs. often bearing axillary bulblets which replace the fls.; stems 1–2 dm. tall. . . . . 7. *L. glabrum*

            DD. Fls. 8–20; fl. tube ± striate; stems 2–5 dm. tall. . . . . 6. *L. tenellum*

        CC. Petals mostly 6–10 mm. long; fl.-tube ± acute at the base.

            D.   Base of fl.-tube acutish, but not obconic. From San Luis Obispo Co. south.

                                   5. *L. affine* ssp. *mixtum*

            DD. Base of fl.-tube obconic. From Kern and San Benito cos. north.

                E.   Fl.-tube 3 times as long as broad; petals with 3 broadly oblong lobes.

                                        8. *L. trifoliatum*

                EE. Fl.-tube twice as long as broad; petals with 3–7 linear-oblong lobes.

                                      9. *L. parviflorum*

1. **L. heterophýllum** (H. & A.) T. & G. [*Tellima h.* H. & A. *L. trilobum* Rydb.] Slender perennial 2–4 dm. tall, the flowering stalks 1–several; herbage

glandular-pubescent or ± hirsutulose; basal lvs. round-reniform, 1.5–4 cm. wide, crenately shallowly lobed; cauline lvs. alternate, usually much reduced, mostly deeply 3-cleft; fls. mostly 3–9; pedicels shorter than fl.-tube which is campanulate with a truncate base; corolla widely spreading; petals 5–12 mm. long, usually 3- (sometimes 5- or 7-) lobed; seeds tuberculate; $2n = 14$ (Taylor, 1965).—Mostly in S. Oak and Foothill Wd., below 4500 ft.; Los Angeles Co. to Humboldt Co.

2. **L. Bolánderi** Gray. [*Tellima heterophylla* var. *B*. Jeps. *T. scabrella* Greene. *L. s.* Greene. *L. heterophylla* var. *s.* Jeps. *L. s.* var. *Peirsonii* Jeps.] Flowering stalks 2.5–8.5 dm. tall, several; herbage pubescent; basal lvs. orbicular, 3–5-lobed; cauline lvs. 2–3, much reduced; fls. 3–5-many; pedicels not longer than fl.-tube which is campanulate, obtuse at base; corolla wide-spreading, the petals ovate-elliptic, 4–7 mm. long, mostly entire or with small serrations near base; seeds tuberculate; $2n = 14, 28, 35, 42$ (Taylor).—Foothill Wd., Yellow Pine F., etc.; Los Angeles and Kern cos. n. to Shasta Co.; San Francisco Bay to Mendocino Co.

3. **L. campanulàtum** Howell. [*L. laciniata* Eastw. ex Rydb.] Slender to rather robust, 2.5–4.5 dm. tall; herbage moderately pubescent; basal lvs. round, trilobed, the lobes apiculately toothed; cauline lvs. 1–2, trilobed or trifoliate; fls. 2–11, very short-pedicelled; fl.-tube broadly campanulate; corolla campanulate, the white petals ovate-elliptic, 3–7 mm. long, palmately lobed with lacerate margin, ligulate; seeds large, the tubercles in distinct rows.—Semi-open slopes, N. Oak Wd., Yellow Pine F.; below 7500 ft.; Lake and Colusa cos. to sw. Ore.

4. **L. Cymbalària** T. & G. [*Tellima C*. Walp.] Slender, 2–4 dm. tall, sparingly pubescent; basal lvs. reniform, weakly 3-lobed; fls. 2–5 (–8), the pedicels to twice the turbinate fl.-tube; corolla wide-spreading, bowl-shaped, the petals 4–5 mm. long, entire, spatulate; seeds tuberculate; $2n = 14 + 1$ (Taylor).—Shaded woods, etc. below 3000 ft.; Palomar Mts. (San Diego Co.), Ventura and Santa Barbara cos. to Stanislaus Co.; Santa Cruz and Santa Rosa ids.

5. **L. affine** Gray. Robust, 2–5 dm. tall, variously pubescent; basal lvs. palmately 3–5-lobed, each lobe sub-lobed into cuspidate lobules; cauline lvs. 1–3, more dissected; fls. 5–9 (–15); fl.-tube widely inflated, obconic below, the pedicels shorter than or equalling it; petals 6–13 mm. long, always 3-lobed; seeds not tuberculate; $2n = 14, 21,$ or 28 (Taylor).—Grassy banks below 3500 ft., many Plant Communities; below 7000 ft.; Santa Barbara Co. to Humboldt Co.

Subsp. **míxtum** R. L. Taylor. [*Tellima tripartita* Greene. *L. t.* Greene.] Basal lvs. 3-lobed, each lobe often subdivided into spreading lobules; petals 4–9 mm. long, 3-lobed; $2n = 28, 35$ (Taylor).—Below 6500 ft.; San Luis Obispo Co. to L. Calif.

6. **L. tenéllum** Nutt. in T. & G. [*Tellima t.* Walp. *L. rupicola* Greene. *L. australis* Rydb. *L. breviloba* Rydb.] Stems 1.5–3 dm. tall; herbage light green and sparsely pubescent; basal lvs. round, simple and irregularly 3–5-lobed or digitately compound; cauline lvs. 2, pinnatifid; fls. 3–12; pedicels not longer than fl.-tube which is campanulate or hemispheric; petals mostly pink, 3–7 mm. long, palmately 5-parted; seeds smooth or wrinkled; $2n = 14$ and 35 (Taylor).—Occasional, 2000–7000 ft. or higher; various Communities; San Gabriel and San Bernardino mts., Butte, Plumas and Modoc cos.; to Ida., Rocky Mts.

p. 738.   7. **L. glàbrum** Nutt. in T. & G. [*Tellima g*. Walp. *L. bulbifera* Rydb. *L. g.* var. *b*. Jeps.] Slender, 1–3.5 dm. tall, subglabrous to sparingly pubescent; basal lvs. orbicular, usually trifoliate, the segms. many-cleft or round-lobed; cauline lvs. 2–4, much reduced; often with bulbils in the axils; fls. 1–5-7, the petals often exceeding the fl.-tubes which are campanulate with an acute or hemispheric base; petals pink or rarely white, ovate, 3.5–6.5 mm. long, palmately 5-parted; seeds tuberculate; $2n = 14, 28$ (Taylor).—In Calif., in dry open places, Sagebrush Scrub, Montane Coniferous F., 4500–11,000 ft.; Sierra Nevada to Siskiyou Co.; to B.C., Rocky Mts.

8. **L. trifoliàtum** Eastw. [*L. parviflora* var. *t*. Jeps.] Slender, 2–5 dm. tall, densely pubescent; basal lvs. round, digitately trifoliate, the segms. many-lobed;

cauline lvs. 2–3, deeply lobed; fls. 4–8, the pedicels not exceeding the length of the fl.-tube which is elongate-obconic; petals pink, obovate-rhombic, 9–11 mm. long, always 3-cleft; seeds smooth or wrinkled; $2n = 28$ (Taylor).—Largely Foothill Wd., below 2000 ft.; w. slope of Sierra Nevada from Tehama Co. to Placer Co.

9. **L. parviflòrum** (Hook.) T. & G. [*Tellima p.* Hook. *Pleurendotria p.* Raf. *L. austromontana* Heller.] Slender, 2–5 dm. tall, nearly glabrous to densely pubescent; basal lvs. orbicular, 3-parted or digitately trifoliate; cauline lvs. 2–3, much like the basal; fls. 4–14, the pedicels not longer than the fl.-tubes, which are oblong-obconic; petals white or pink, obovate-rhombic, 7–16 mm. long, always 3-cleft; seeds smooth or wrinkled; $2n = 14, 21, 28, 35$ (Taylor).—Open slopes 2000–6000 ft.; San Diego Co., Tehachapi Mts. to San Benito Co. and N. Coast Ranges, Sierra Nevada to Modoc Co., to B.C. and Rocky Mts.

10. **L. máximum** Bacig. Stout perennial, the basal lvs. truly compound, the 3 lfts. rhombic, cuneate; petioles to 15 cm. long; fl.-stems stout, 4–6 dm. tall; fl.-tube campanulate, ca. 6 mm. long at anthesis, including the sepals, 8 mm. in fruit; petals 4 mm. long, digitately incised.—San Clemente Id.

p. 738.   CHRYSOSPLÈNIUM

1. **C. glechomifòlium** Nutt. in T. & G. $2n = 9$ pairs (Anderson in 1965).

p. 739.   TIARÉLLA

1. **T. trifoliàta** L. ssp. **unifoliàta** (Hook.) Kern in Madroño 18: 159. 1966.

p. 739.   11a. **Bensoniélla** Morton.

Perennial with slender branching scaly rootstocks and simple scapiform flowering branches. Lvs. basal, cordate, petioled. Fl.-tube campanulate, free from the ovary. Sepals 5, three approximate, the other 2 more distant, all 3-nerved. Petals 5, filiform, entire. Stamens 5, opposite the sepals. Pistil 2-valved at apex. Carpels subcompressed, sharply angled on the back. Ovules many. (G. T. *Benson*, former botanist at Stanford.)

1. **B. oregòna** (Abrams & Bacig.) Morton. [*Bensonia o.* Abrams & Bacig.] Petioles slender, 3–7 cm. long, with elongate brownish hairs; lf.-blades 2.5–4.5 cm. long, crenately 7-lobed; scape 2 dm. high, sparsely pilose; sepals 2 mm. long, stipitate-glandular; petals ca. 3 mm. long.—Between 4000 and 5000 ft., Siskiyou Mts., nw. Calif. and sw. Ore.

TÉLLIMA

1. **T. grandiflòra** (Pursh.) Dougl. ranges s. into Eldorado Co.

p. 742.   HEÙCHERA

6. **H. rubéscens** Torr. var. **pachypòda** (Greene) Rosend. Dr. Bacigalupi informs me that a study of the type specimen of var. *alpicola* as used in the FLORA shows that the correct name is **pachypòda**. This ranges s. into the Piute Mts., Kern Co., *Twisselmann*.

Var. *alpícola* Jeps. is the correct name for var. *Rydbergiana* of the FLORA and that name goes into synonymy.

p. 744.   PHILADÉLPHUS

The citation of the Hitchcock reference to Madroño in the 1959 printing should be to vol. 7.

2. **P. Lewísii** Pursh ssp. **califórnicus** (Benth.) Munz, if treated as a var., would have as its oldest name *P. L.* var. *parvifòlius* Torr.

p. 745.   CARPENTÈRIA

1. **C. califórnica** Torr. $n = 10$ (Ernst, 1964); $2n = 20$ (Raven, Kyhos & Hill, 1965).

p. 747.   RÌBES

4. **R. laxiflòrum** Pursh. $2n = 8$ pairs (D. E. Anderson, 1963).

p. 748.   7. **R. viscosíssimum** Pursh, Twisselmann reports from Kern Co. (Sunday Peak).

p. 749.   10. **R. sanguíneum** var. **glutinòsum** Loud. $2n = 8$ pairs (D. E. Anderson, 1963).

11. **R. malvàceum** Sm. occurs in the Tehachapi Mts., Kern Co., *Twisselmann.*

p. 751.   18. **R. quercetòrum** Greene. is reported from 26 mi. n. of Essex, Mojave Desert, San Bernardino Co., at 4400 ft., *Haller.*

19. **R. velutìnum** Greene. In line 3, change "modal" to "nodal."

p. 753.   27. **R. califórnicum** H. & A. is common on the e. side of the Santa Lucia Mts., San Luis Obispo Co., *Hardham*; also in the Temblor Range, Kern Co., *Twisselmann.*

p. 756.   ROSÀCEAE

In Key, in the E and EE just above AA, for the 1959 printing, change to:

> E.   Pistil 1; fls. bisexual; lvs. mostly serrate. . . . . .30. *Prunus*
> EE. Pistils usually 5; fls. unisexual or bisexual; lvs. entire. 31. *Osmaronia*

In the Key, under AA, BB, change CC to:

> CC. Plants evergreen, with persistent lvs.
>   D.   Carpels with leathery walls at maturity; styles 2–3; fls. in large corymbose panicles; lvs. sharply toothed. Native. . . . . 37. *Heteromeles*
>   DD. Carpels bony at maturity with 2-seeded nutlets; lvs. entire or crenate. Introduced.
>   E.   Lvs. entire; shrubs spineless; styles 2–5. . . . . 37a. *Cotoneaster*
>   EE. Lvs. crenate; shrubs thorny; styles 5. . . . . 37b. *Pyracantha*

LYONOTHÁMNUS

1. **L. floribúndus** Gray 2n = c. 48 (Stebbins & Major, 1965). Var. *asplenifòlius* (Greene) Bdg. Raven makes the comb. ssp. *a.* (Greene) Raven. 2n = 27 pairs (Raven, Kyhos and Hill, 1965).

p. 757.   SPIRAÈA

In Key, after the word "Fls." delete "white" and "pink" in the 1959 printing.

p. 758.   ARÚNCUS

1. **A. vulgàris** Raf. 2n = 18 (Löve, 1964).

p. 759.   HOLODÍSCUS

1. **H. díscolor** (Pursh) Maxim. has been called "Indian Arrow-wood" in Mendocino Co.

p. 761.   HORKÈLIA

In Key, top of page, under FF, change GG to:

> GG. Cymes few-fld.; lfts. 6–12 pairs.
>   H.   Cymes open; lfts. 3.5–5 mm. long. White Mts.
>   11. *H. hispidula*
>   HH. Cymes congested; lfts. smaller. Kern Plateau.
>   11a. *H. tularensis*

p. 761.   3. **H. califórnica** Cham. & Schlecht. Change in line 2 of 1959 printing the word *carmelina* to *carmeliana*. The sp. has been reported from Monterey Co., *Howitt & Howell.*

p. 763.   11a. **H. tularénsis** (J. T. Howell) Munz, comb. nov. [*Potentilla tularensis* J. T. Howell, Leaflets W. Bot. 10: 254–255. 1966.] Odorless compact cespitose or loosely pulvinate herb, with multicipital caudex, pale cinereous with appressed and spreading hairs; stems erect, 3–10 cm. tall; basal lvs. rosulate, 2–4 cm. long, the petioles 0.5–1.5 cm. long; lfts. in 6–10 pairs, the lower ± petiolulate, the upper crowded, 3–5-palmatifid, the segms. oblong to obovate; cyme laxly few-fld.; fls. 4–5 mm. long; fl.-tube cupulate, 1–1.5 mm. high, pilose within; bracteoles 1–2 mm. long; sepals 2–3.5 mm. long; petals white, linear-oblanceolate, 2–3.5 mm. long; fils. 10, subulate-dilate; styles ca. 11; aks. 2.5 mm. long.—At 9430 ft., Bald Mt., Kern Plateau, Tulare Co., *Twisselmann.*

p. 764.   15. **H. tridentàta** Torr.; line 7 from bottom of page, for 1959 printing change "petals" to "sepals."

p. 765.   IVÈSIA

In Key, AA, BB, C, DD, EE, change F to:

> F.   Fl.-tube campanulate or turbinate to saucer-shaped; sepals 3.5–5.5 mm. long; bractlets thin; herbage densely villous or tomentose.

G. The fl.-tube glabrous within, 2–2.5 mm. deep; pistils 4–7; lfts. 20–35 pairs, 4–15 mm. long. Plumas Co. to Sierra Co.
H. Petals white, much exceeding the sepals.
   11. *I. sericoleuca*
   HH. Petals yellow, not exceeding sepals. . . 11a. *I. aperta*
GG. Fl.-tube saucer-shaped; sepals 2.5–3 mm. long; bractlets thickened; lfts. 25–50 pairs, 2–5 mm. long. Siskiyou to Trinity Co. . . . . 12. *I. Pickeringii*

p. 767.   9. **I. purpuráscens** (Wats.) Keck. Reported from Piute Mts., Kern Co., *Twisselmann*.

11a. **I. apérta** (J. T. Howell) Munz, comb. nov. [*Potentilla aperta* J. T. Howell, Leaflets W. Bot. 9: 239, 1962.] Like *I. sericoleuca* (Rydb.) Rydb. in habit and pubescence and like *I. Kingii* Wats. in its open fls. with a saucer-shaped floral tube. If *I. sericoleuca* is restricted to plants with turbinate fl.-tube and white petals and much exceeding the sepals, the name *I. aperta* can be applied to those with yellow petals not exceeding the sepals and with open shallow, starlike fls.—Sierra Valley, Plumas and Sierra cos.

p. 772.   POTENTÍLLA

14. **P. saxòsa** Lemmon ex Greene ssp. **siérrae** Munz is reported from Weldon, Kern Co., *Twisselmann*.

p. 773.   17. **P. grácilis** Dougl. ex Hook. ssp. **Nuttállii** (Lehm.) Keck has been collected at elevs. up to 11,000 ft.

21. **P. récta** L. has been found wild in San Francisco.

p. 774.   23. **P. anserìna** L. $2n = 28$ (Taylor, 1967).

26. **P. glandulòsa** Lindl. Line 8 of description, the second word "petals" should be changed to "sepals" in the 1959 printing.

p. 775.   Ssp. **Hansènii** (Greene) Keck is reported from the Piute Mts., Kern Co., *Twisselmann*.

FRAGÀRIA

As an additional reference for the genus cite "Staudt, G. Taxonomic studies in the genus Fragaria." Canad. J. Bot. 40: 869–886, 1962.

p. 776.   1. Staudt would consider our w. material of **F. chiloénsis** (L.) Duchn. to be ssp. **pacífica** Staudt.

3. **F. califórnica** Cham. & Schlecht. is **F. vésca** L. ssp. **califórnica** Staudt.

4. *F. platypétala* Rydb. is **F. virginiàna** L. ssp. **platypétala** Staudt.

p. 777.   SANGUISÓRBA

1. For S. *occidentàlis* Nutt. use **S. ánnua** (Nutt. ex Hook.) T. & G. according to Nordborg (Opera Bot. 11: 64, 1966).

2. **S. mìnor** Scop. $2n = 28, 56$ (Nordborg, 1958).

3. **S. microcéphala** Presl. Add to synonymy S. *officinalis* ssp. *m.* Calder & Taylor.

p. 778.   ACAÈNA

2. **A. anserinifòlia** (J. R. & G. Forst.) Druce. [*A. Sanguisorba* Vahl.] Prostrate, creeping, much-branched undershrub with short leafy erect stems 2–15 cm. high; lfts. 3–4 pairs, oblong, crenate-serrate; fls. in globose heads, purplish. —Growing with *Pinus radiata*, *Ceanothus griseus* and *Heteromeles* on Peter Pan road, just off Highway 1, s. of Wild Cat Creek, Carmel Highlands, Monterey Co., *E. K. Balls*. Native of Australia, New Zealand.

AGRIMÒNIA

1. **A. gryposèpala** Wallr. Reported from Eldorado Co., *Fuller*.

ADENÓSTOMA

1. **A. fasciculàtum** H. & A. $2n = 9$ pairs (Raven, Kyhos & Hill, 1965).

2. **A. sparsifòlium** Torr. $2n = 9$ pairs (Raven, Kyhos & Hill, 1965).

p. 779.   GÈUM

1. **G. macrophýllum** Willd. Occurs in the Piute Mts., Kern Co., *Twisselmann*.

p. 780.   PÚRSHIA

1. **P. tridentàta** (Pursh) DC. $2n = 18$ (Cave, 1956). South to Kern Plateau, Kern Co., *Twisselmann*.

p. 781.   CHAMAEBÀTIA

1. **C. foliolòsa** Benth. Found in the Greenhorn Mts., Kern Co., *Twisselmann*.

p. 782. **CERCOCÁRPUS**

C. betuloìdes var. *Tráskiae* (Eastw.) Dunkle. Dr. Thorne has recently found several individuals in an arroyo in the Salta Verde on Catalina Id. The very heavy coriaceous large lvs., impressed veins, felty tomentum, thick pedicels, etc. suggest species rank, **C. Tráskiae** Eastw.

3. **C. ledifòlius** Nutt. reported as on St. John Mt., sw. Glenn Co., and in nw. Tehama Co., *Tucker*.

4. **C. intricàtus** Wats. is common in the King's River region.

p. 783. **RÙBUS**

In Key, change A to:

A. Plants ± herbaceous, creeping; stipules broad, almost or quite free.
  B. Stems not prickly; sepals 6–7 mm. long; petals white. Humboldt and Siskiyou cos.
   1. *R. lasiococcus*
  BB. Stems with curved prickles; sepals 7–9 mm. long; petals dull purple. Del Norte Co.
   1a. *R. nivalis*

In Key, AA, BB, C, DD, change and insert:

D. Fls. borne in long terminal racemes or large panicles; prickles stout, broad-based or none. Escapes from cult.
  E. Infl. long, racemose or corymbose.
   F. Lvs. of primocanes non-glandular and velvety above; infl. 7–12-fld. .... 8a. *R. pensilvanicus*
   FF. Lvs. of primocanes glandular and pubescent above; infl. with few to many fls. in a long cluster. ...... 8b. *R. alleghaniensis*
  EE. Infl. a large panicle.
   F. Lfts. deeply cut or dissected, not whitish-tomentose or canescent beneath. .... 6. *R. laciniatus*
   FF. Lfts. not deeply cut, whitish-tomentose or gray-canescent beneath.
    G. Canes and infl. quite unarmed; lf.-margins finely serrate.
     7. *R. ulmifolius*
    GG. Canes and infl. armed; lf.-margins coarsely unequally serrate or toothed. .... 8. *R. procerus*

1a. **R. nivàlis** Dougl. Creeping stems 3–12 dm. long, sparingly armed; lvs. simple to ± distinctly 3-lobed; fls. usually 1; sepals 7–9 mm. long; petals dull purple.—At 4100 ft., upper East Fork of Illinois River, Del Norte Co., *Hobart*; to B.C., Ida.

p. 785. 8a. **R. pensilvánicus** Poir. in Lam. Upright but usually diffuse bramble; lvs. soft-pubescent beneath; floral lvs. and lfts. narrow and acuminate; pedicels nearly or quite unarmed; infl. ± corymbiform.—Reported as natur. in Humboldt Co., *Pollard*; Butte Co., *Howell*; Monterey Co., *Howitt & Howell*; native of e. N. Am.

8b. **R. alleghaniénsis** Porter ex Bailey. Large prickly highbush blackberry; lfts. narrow-acuminate, mostly pubescent and glandular; pedicels glandular; infl. long-racemiform.—Plumas Co.; native of e. N. Am.

10. **R. glaucifòlius** Kell. S. to Piute Mts., Kern Co., *Twisselmann*.

p. 786. **RÒSA**

In Key AA, BB, change the rest of the Key to:

C. Sepals glandular-hispid on back.
  D. Lvs. simply serrate, without gland-tipped teeth; fls. mostly corymbose.
   3. *R. pisocarpa*
  DD. Lvs. doubly serrate with gland-tipped teeth; fls. mostly solitary.
   7. *R. pinetorum*
CC. Sepals not glandular-hispid on back.
  D. The sepals and styles deciduous in fr.; pistils few; pedicels 1–3 cm. long, ± reflexed in fr.; fr. 4–8 mm. thick in maturity. .... 9. *R. gymnocarpa*
  DD. The sepals and styles persistent in fr.; pedicels usually less than 2 cm. long, not reflexed in fr.; fr. more than 7 mm. thick at maturity.
   E. Stems armed with stout flattened recurved prickles; pedicels villous; fl.-tube often externally pilose when young; sepals ± pubescent on backs. .... 6. *R. californica*
   EE. Stems armed with straight or ascending weak, slender prickles; pedicels not villous; fl.-tube glabrous; sepals not pubescent. 8. *R. Woodsii*

p. 787.   7. **R. pinetòrum** Heller in Greenhorn Mts., Kern Co., *Twisselmann*.
p. 788.   8. **R. Woòdsii** Lindl. var. **ultramontàna** (Wats.) Jeps. In Kern Co. (Kernville and Mt. Pinos region), *Twisselmann*.
p. 789.   PRÙNUS
            2. **P. subcordàta** Benth. in Monterey Co., *Howitt & Howell*.

p. 795.   38. Cotoneáster Medic.

Shrubs, sometimes arborescent, evergreen or deciduous, not thorny. Lvs. numerous, alternate, short-petioled, simple and entire, stipulate. Fls. white or pink, small but many, solitary or in cymose clusters terminating lateral spurs, appearing after the lvs. are out; fl.-tube adnate to ovary. Sepals 5, small, persistent. Petals 5. Stamens ca. 20. Pistil 1; ovary 2–5-celled; styles 2–5, distinct. Fr. a red or dark pome. Ca. 50 spp. of the Old World. (Latin, *quince-like,* from the lvs. of some spp.)
            1. **C. pannòsa** Franch. Arching evergreen shrub to 3 m. tall; lvs. elliptic- to ovate-oblong, ± glabrous above, white-tomentose beneath, 1.5–4 cm. long; infl. corymbose; petals white, spreading, roundish; calyx tomentose; fr. red, globose-ovoid, 8 mm. in diam.—Escape from cult., in Marin Co., Ventura Co., etc. From China.
            2. **C. Franchétii** Boiss. Fls. pinkish; lvs. yellowish-tomentose beneath, to 3 cm. long.—Occasional escape from cult., as in San Francisco region. Native to China.

                          39. Pyracántha Roem. FIRETHORN

Near to *Cotoneaster,* but with heavy thorns; lvs. usually crenate or serrate; fls. in corymbs; pistil 1 (actually of 5 pistils ventrally connate or coherent along basal half and 5-celled basally, separating at maturity); fr. with 5 nutlets. Ca. half a dozen spp. of Medit. region and Asia. (Greek, *fire* and *thorn,* from red frs. and thorns.)
            1. **P. angustifòlia** (Franch.) Schneid. [*Cotoneaster a.* Franch.] Evergreen shrub to 3 or 4 m. tall; lvs. narrow-oblong, roundish at ends, 2.5–5 cm. long, gray-tomentose beneath; infl. corymbose, few-fld.; calyx tomentose; petals white; fr. orange-yellow, 6–8 mm. in diam. flattish-globose.—Occasional escape from cult. as in Marin Co., *Howell*; native of China.
p. 795.   CROSSOSÒMA
            1. **C. califórnicum** Nutt. $2n = 12$ (Raven, Kyhos & Hill, 1965).
p. 798.   ALBÍZIA
            1. **A. distàchya** (Vent.) Macbr. $2n = 26$ (Frahm-Leliveld, 1957).
            PROSÒPIS
            Add as reference: Johnston, M. C. The N. Am. mesquites. Prosopis sect. Algarobia. Brittonia 14: 72–90. 1962.
            1. **P. glandulòsa** Torr. var. **Torreyàna** (L. Benson) M. C. Jtn. is used for our common Mesquite, while **P. juliflòra** (Sw.) DC., which name has also been applied to Calif. plants, is restricted to the area s. of the U.S. If our plants are to be considered a distinct sp., apparently they should be called **P. odoràta** Torr. & Frém. The reference to "Grantland" School under var. **velutina** should read "Grantville" School, *Norland*.
            2. **P. velùtina** Woot., which can be separated from **P. glandulosa** by having lfts. less than 5 times as long as broad, is native e. of Calif. and becomes natur. occasionally in the state; San Diego Co., Orange Co., Santa Barbara Co., Mariposa Co.
p. 799.   CÉRCIS
            1. **C. occidentàlis** Torr. ex Gray. $2n = 14$ (Taylor, 1967). Found at Onyx, Kern Co., *Twisselmann*.
p. 800.   CÁSSIA
            1. **C. armàta** Wats. $2n = 14$ pairs (Raven, Kyhos & Hill, 1965).

3. **C. tomentòsa** L. f. Shrub 3–4 m. high; twigs and lower surface of lvs. tomentose; lfts. 6–8 pairs, oblong, 2–4 cm. long, each pair with a gland at base; fls. deep yellow, 2.5 cm. in diam.; fertile stamens 7; pod 12 cm. long.—Established in Marin Co. and at Santa Barbara; native of Mex., S. Am.

CERCÍDIUM

1. **C. flóridum** Benth. 2n = 14 pairs (Raven, Kyhos & Hill, 1965).

p. 802.   PAPILIONOÌDEAE

In Key to Genera, after CC, change D to "Rachis of lf." and after EE, FF, delete "35. *Pisum*" and insert:

> G.   Calyx-lobes leafy; pod many-seeded. . . . . 35. *Pisum*
> GG. Calyx-lobes long-subulate; pod 1–2 seeded. . . . . 36. *Lens*

p. 803.   THERMÓPSIS

3. **T. macrophỳlla** H. & A. Add:
Var. agnìna J. T. Howell. Plants to 2 m. tall; lfts. ± villous-pubescent on both sides; lower calyx-lobes deltoid, 3.5–4 mm. long.—Santa Ynez Mts., Santa Barbara Co.

p. 804.   PICKERÍNGIA

1. **P. montàna** Nutt. 2n = 14 pairs (Raven, Kyhos & Hill, 1965).

p. 804.   LUPÌNUS

In Key delete word *fragrantissimus* in line 15 from bottom of page, in the 1959 printing.
In Key, the last word on the page should be *polycarpus*, not *micranthus*.

p. 805.   In Key, under JJ. about two-thirds of way from top of page, change "19. *L. Moranii*" to "19. *L. guadalupensis.*"

p. 806.   In Key, after BB, C, D, EE insert:

> EEE. Stems and lvs. subglabrous or slightly strigose; corolla yellow, 10
>   mm. long. Se. Shasta Co. . . . . . . 71. *L. Andersonii* var. *Christinae*

Just below the above, "74. *L. latifolius*" should be "76. *L. latifolius,*" in the 1959 printing.

p. 807.   About one-third way down the page "56. *L. Grayii*" should be "58. *L. Grayii.*"
p. 809.   2. **L. rùber** Heller occurs in Monterey Co., *Howitt & Howell.*
p. 811.   5. Walter Knight (Four Seasons 1(3): 8–9. 1965) thinks that *Lupinus* **Mìlo-Bàkeri** C. P. Sm. is a distinct sp. from *L. luteolus* Kell.: (1) *L. M.-B.* not yet in full bloom on June 27, while *L. l.* is past; (2) *L. M.-B.* to ca. 165 cm. tall, *L. l.* to 75 cm.; (3) lfts. on *L. M.-B.* oblanceolate, keeled to 45°, on *L. l.* largely obovate, keeled to 90°; (4) fls. on *L. M.-B.* blue, yellowish in age only, on *L. l.* yellow throughout; (5) stem on *L. M.-B.* with pith until old and in seed, on *L. l.* generally hollow from early. *L. M.-B.* he finds to be a local endemic in Round Valley at Covelo, Mendocino Co.

p. 812.   12. **L. concínnus** J. G. Agardh. Dunn, Christian and Dziekanowski after several years of work on the *L. concinnus* complex, make the following changes (Cf. Aliso 6: 45–50. 1966):
L. concínnus Agardh.
L. concínnus ssp. **optàtus** (C. P. Sm.) Dunn
L. concínnus ssp. **Orcúttii** (Wats.) Dunn
L. Agardhiànus Heller
L. pállidus Bdg.
L. brévior (Jeps.) Christian & Dunn.
They find *L. pallidus* and *L. Agardhianus* intersterile with the rest of the complex, but *L. pallidus* interfertile with *L. sparsiflorus* Benth. (no. 18 of the FLORA). **L. brévior** is an obligate selfer and ranges from Imperial Co. n.
13. **L. Stìversii** Kell. ascends to 5500 ft. in King's River Canyon.

p. 813.   18. **L. sparsiflòrus** Benth. was also studied by Dunn, Christian and Dziekanowski who report as follows:
L. sparsiflòrus Benth. is of cismontane S. Calif.

Ssp. **inopinàtus** (C. P. Sm.) Dziekanowski & Dunn is of the San Diego region and adjacent L. Calif.

Ssp. **mohavénsis** Dziekanowski & Dunn occurs in the Mojave Desert and into Ariz. and Sonora. Generally smaller than cismontane plants and in its fls. They give the following Key:

A. Keel glabrous.
  B. Lfts. linear spatulate, mostly subglabrous above; plants sparsely spreading-pilose; fls. bluish-purple. . . . . *L. Agardhianus*
  BB. Lfts. spatulate-oblanceolate to obovate, amply pubescent above; plants abundantly spreading-pilose; fls. pinkish-white to lavender. . . . . *L. concinnus*
AA. Keel ciliate above and/or below the claws.
  B. Fls. off-white to bluish or purplish, 5–7 mm. long or less; plants usually less than 15 cm. tall.
    C. Plants silky to strigose, often decumbent or prostrate; lfts. pubescent on both sides. . . . . *L. pallidus*
    CC. Plants glabrous or sparsely pilose; lfts. glabrous above. . . . . *L. brevior*
  BB. Fls. blue to pink or purple, 7–13 mm. long; plants generally more than 15 cm. tall.
    C. The fls. pink to magenta with a yellow center spot; lfts. obovate to oblanceolate with the tip rounded, fleshy in texture; plants appearing succulent.
        L. arizonicus
    CC. The fls. blue or purplish, with a yellowish-white center spot; lfts. linear to filiform, not fleshy; plants not appearing succulent.
      D. Fls. 10–13 mm. long. Cismontane, Ventura to Riverside cos. *L. sparsiflorus*
      DD. Fls. 7–11 mm. long.
        E. Banner oblong-oval, longer than wide. San Diego Co. and n. L. Calif.
            L. sparsiflorus inopinatus
        EE. Banner orbicular, emarginate. Mojave Desert to Ariz. ond Son.
            L. sparsiflorus mohavensis

(I am very grateful to Dr. David Dunn for making the above information available to me.)

p. 814. 19. *L. Moranii* Dunkle should be changed to **L. guadalupénsis** Greene according to *Dunn*.

22. *L.* **nànus** Dougl. in Benth. In the Key, change "Largest lfts. 1.5–5 mm." to "5–7.5 mm." for the 1959 printing.

p. 816. 28. Change name to **L. polycárpus** Greene. [*L. micranthus* Dougl. in Lindl., not Gussone, fide *Dunn*.]

p. 817. 32. *L.* **séllulus** Kell. var. **ártulus** (Jeps.) Eastw. has a range from Plumas Co. to Modoc Co.

p. 818. 35. *L.* **Culbertsònii** Greene. has been collected in Mono Co., *Hardham*.

p. 823. 61. *L.* **arbòreus** Sims. Add as synonym: *L. macrocarpus* H. & A.

p. 824. 62. *L.* **albifróns** Benth. is apparently on Santa Cruz and Santa Catalina ids. $2n = 24$ pairs (Raven, Kyhos & Hill, 1965).

p. 827. 76. *L.* **latifòlius** Agardh. Add to synonymy *L. lasiotropis* Greene ex Eastw.

p. 828. 78. *L.* **polyphýllus** Lindl. ssp. **superbus** (Heller) Munz occurs in the Piute and Greenhorn mts., Kern Co., *Twisselmann*.

p. 829. 82. *L.* **magníficus** Jones var. **glarécola** Jones, not *glareola*.

p. 830. CÝTISUS
Change Key to:

A. Fls. white; banner hairy on the back. . . . . 1. *C. proliferus*
AA. Fls. yellow; banner mostly glabrous.
  B. Stems sharply angled, leafless or nearly so; pods hairy along margins only.
        2. *C. scoparius*
  BB. Stems obtusely angled or ridged, leafy; pods hairy all over.
    C. Racemes nearly capitate, 3–9-fld., at the ends of short lateral branchlets.
      D. Shrub 10–30 dm. high; lfts. flat, obovate; fls. 10–12 mm. long.
          3. *C. monspessulanus*
      DD. Shrub 2–8 dm. high; lfts. revolute, linear-oblong; fls. 12–14 mm. long.
          3a. *C. linifolius*
    CC. Racemes ± elongate, 6–many-fld., secund, terminal and lateral; fls. 12–14 mm. long.
      D. Lfts. less than 2 cm. long, usually obovate; petioles to 8 mm. long.
        E. The racemes dense, 2–5 cm. long; lfts. glabrous above, densely silky-villous beneath, often only 3–6 mm. long; calyx-lobes ± ovate, rather abruptly narrowed to an acuminate apex. . . 4. *C. canariensis*
        EE. The racemes lax, 5–10 cm. long; lfts. silky-pubescent above and be-

neath, 8–18 mm. long; calyx-lobes lanceolate, gradually narrowed to
the acuminate tip. . . . . 4a. *C.* × *racemosus*
    DD. Lfts. 1.5–5 cm. long, oblong-obovate; petioles often 12 or more mm.
long; fls. 13–15 mm. long. . . . . 5. *C. stenopetalus*

2. **C. scopàrius** (L.) Link. Specimens of a shrub spreading on a road bank
ca. 3 miles down Page Mill road from intersection with Moody road, 4 mi. s. of
Palo Alto, Santa Clara Co., with sharply angled stems and solitary yellow fls.
with deep red tips to the wings, has been distributed as **C.** × **Dallimorei** Rolfe.
[*C. multiflorus* × *C. scoparius* var. *Andreanus*.] In some ways it more closely
resembles descriptions and plates of *C. scoparius* var. *Andreanus* (Puissant), a
shrub originating apparently in Normandy.

3. **C. monspessulànus** L. Abundant in San Luis Obispo Co.; near Placerville,
Eldorado Co., *Thorne*; Santa Catalina Id., *Wolf.*

3a. **C. linifòlius** (L.) Lam. [*Genista l.* L.] Low shrub 2–8 dm. high, with
erect, appressed-silky branches; lfts. linear or linear-lanceolate, revolute at margin, nearly glabrous and shining above, silvery-pubescent beneath, 12–25 mm.
long; fls. 12–14 mm. long; pod torulose.—Adventive in Santa Barbara Co. and
on Santa Catalina Id., *Fuller.* From Medit. region.

4. **C. canariénsis** (L.) Kuntze. Amend to: lfts. glabrous above, densely silky-
villous beneath, often only 3–6 mm. long; calyx-lobes lance-ovate, abruptly narrowed to the acuminate tip.

4a. **C.** × **racemòsus** Nichol. Near to *C. canariensis* in appearance, but with
longer more lax racemes; lfts. silky-pubescent above and beneath, largely 8–18
mm. long; calyx-lobes lanceolate, gradually attenuate.—Monterey and Santa
Clara cos., *Fuller.* Probably of garden origin.

5. *C. maderénsis* Masf. Change to **C. stenopétalus** (Webb) Christ. and amend
description to: lfts. to 3.5 cm. long; petioles to 15 mm. long; racemes to 10 cm.
long.—Found also at Pacific Grove, Monterey Co., *Fuller.* Native of Canary Ids.

p. 831.   MEDICÁGO
Insert after generic description: Heyn, C. C. The annual spp. of Medicago.
Scripta Hierosolymitana 12: 1–154. 1963.

2. **M. lupulìna** L. 2*n* = 16 (Mulligan, 1957).

4. Change *M. hispida* Gaertn. to **M. polymórpha** L. and var. *confinis* (Koch)
Burnat to **M. p.** var. **brevispìna** (Benth.) Heyn.

p. 832.   MELILÒTUS. The noun is feminine, not masculine as in the FLORA.

2. **M. índica** (L.) All. 2*n* = 8 pairs (Raven, Kyhos & Hill, 1965).

TRIFÒLIUM
In Key, after A, B, C, and D, insert:

        D'.   **Fls. few in a head, becoming reflexed and surrounded by later sterile fls.**
             forming a bur. . . . . 1a. *T. subterraneum*
      D'D'. Fls. more numerous and not forming a bur.
          E. "Stipules lance-oblong," etc. as in FLORA.

p. 833.   In Key, line 12 from bottom of page, *T. oreganum* should be no. 16, not 6.
p. 834.   In Key after CC. "Invol. flat, rotate," D, EE, insert:

        F.   Invol. 12–15 mm. broad; calyx not inflated and vesiculose in
          fr. Native. . . . . 39. *T. Wormskjoldii*
      FF. Invol. 5–6 cm. broad; calyx conspicuously inflated and vesicu-
          lose in fr. Introd. . . . . 39a. *T. fragiferum*

In Key, after CC, DD, EE, FF insert:

        G.   Calyx teeth triangular, abruptly spine-pointed; fls. pur-
          plish; corolla 4 mm. long. . . . . . . 44a. *T. glomeratum*
      GG. Calyx teeth not abruptly spine-pointed, dilated.
          H. Corolla 12–15 mm. long; plants erect to ± de-
            cumbent. Widely distributed.   43. *T. tridentatum*
         HH. Corolla 8–10 mm. long; plants subprostrate. Mon-
            terey Peninsula. . . . . 42. *T. polyodon*
      GGG. Calyx teeth not dilated, etc. replacing GG. in the
          FLORA.

1a. **T. subterràneum** L. Hairy annual; prostrate; lfts. obovate, apically notched, to 1 cm. long; fertile fls. ca. 2–5 in a head without an invol., 8–12 mm. long, cream-yellow, conspicuously reflexed after anthesis and then surrounded by sterile fls. forming a bur-like cluster.—Introd. in Humboldt, Sonoma and Santa Cruz cos., *Howell*; from Eu.

2. **T. procúmbens** L. In Monterey Co., *Howitt & Howell. n* = 7 (Larsen, 1955).

4. **T. bìfidum** Gray. As far s. as Monterey Co., *Howitt & Howell.* 2*n* = 16 (Mosquin & Gillett, 1965).

p. 835.   6. **T. Pálmeri** Wats. 2*n* = 16 (Gillett & Mosquin, 1967).

7. **T. ciliolàtum** Benth. *n* = 8 (Gillett & Mosquin); 2*n* = 16 (Mosquin & Gillett, 1965).

8. **T. Brèweri** Wats. 2*n* = 16 (Mosquin & Gillett, 1965); *n* = 8 (G. & M., 1967).

11. **T. Beckwíthii** Brew. ex Wats. 2*n* = ca. 48 (Gillett & Mosquin, 1967).

p. 836.   14. **T. prodúctum** Greene. *n* = 8 (Gillett & Mosquin, 1967).

15. **T. eriocéphalum** Nutt. *n* = 8 (Gillett & Mosquin, 1967).

17. **T. lóngipes** Nutt. *n* = 16, 24 (Mosquin & Gillett, 1965); *n* = 8, 16, 24 (Gillett & Mosquin, 1967).

p. 837.   18. **T. macrocéphalum** (Pursh) Poir. *n* = 16, ca. 84 (Gillett & Mosquin, 1967).

19. **T. Andersònii** Gray. *n* = 8 (Gillett & Mosquin, 1967).

20. **T. monoénse** Greene. 2*n* = 16 (Gillett & Mosquin, 1967).

25. **T. arvénse** L. Reported from Monterey Peninsula, *Howitt & Howell.*

26. **T. incarnàtum** L. Reported from Monterey Co., *Howitt & Howell* and from several cos. from Fresno Co. and Monterey Co. north, *Weiler.*

p. 838.   27. **T. hírtum** All. In the Santa Ynez Mts., Santa Barbara Co., *Raven*; Santa Lucia Mts., Monterey Co., *Howitt & Howell*; Mariposa Co., *Howell*; Fresno and Madera cos., *Weiler*; Kern Co., *Twisselmann.*

28. **T. Macràei** H. & A. 2*n* = 16 (Gillett & Mosquin, 1967).

30. **T. dichótomum** H. & A. In San Luis Obispo Co., *Hardham.*

33. **T. cyathíferum** Lindl. *n* = 8 (Gillett & Mosquin, 1967).

p. 839.   35. **T. Gràyii** Loja. At Atascadero, San Luis Obispo Co., *Eastwood.*

36. **T. mìcrodon** H. & A. 2*n* = 16 (Mosquin & Gillett, 1965).

37. **T. microcéphalum** Pursh. *n* = 8 (Gillett & Mosquin, 1967).

38. **T. monánthum** Gray. 2*n* = 16 (Mosquin & Gillett, 1965; G. & M. 1967).

p. 840.   39. **T. Wormskíoldii** Lehm. *n* = 16 (Mosquin & Gillett, 1965; G. & M. 1967).

39a. **T. fragíferum** L. Perennial with creeping, rooting stems; lfts. cuneate-obovate, 6–12 mm. long; fls. pink to white, in dense globose heads ca. 12 mm. in diam., on very long peduncles; calyx inflated and vesiculose in fr.—Found at King City, Monterey Co., *Howell*; Davis, Yolo Co., *Skaggs*; North Highland, Sacramento Co., *Fuller*; from Medit. region.

44. should be **T. variegàtum** Nutt. in T. & G.

p. 841.   44a. **T. glomeràtum** L. Glabrous annual; lfts. 5–8 mm. long, obovate-cuneate, sharply serrate; stipules usually ovate with long points; heads sessile, globular; fls. purplish; calyx strongly ribbed, whitish, the teeth triangular, spinescent; corolla 4 mm. long.—Natur. in Yuba Co., *Crampton*; from Eu.

45. **T. tridentàtum** Lindl. 2*n* = 16 (Mosquin & Gillett, 1965).

p. 842.   49. **T. depauperàtum** Desv. 2*n* = 16 (Mosquin & Gillett, 1965). Reported from Temblor Range, Kern Co., *Twisselmann* and from Madera and Fresno cos., *Weiler.*

LOTUS

In 1959 printing of the FLORA, in the Key, under A, BB, CC, D, line F belongs after EE, not before it.

p. 845.   10. **L. grandiflòrus** (Benth.) Greene. 2*n* = 7 pairs (Raven, Kyhos & Hill, 1965); *n* = 7 (Grant & Sidhu,, 1967).

11. **L. rígidus** (Benth.) Greene. 2*n* = 7 pairs (Raven, Kyhos & Hill, 1965).

p. 846.   13. **L. strigòsus** (Nutt.) Greene. 2*n* = 7 pairs (Raven, Kyhos & Hill, 1965).

15. **L. salsuginòsus** Greene. $2n = 7$ pairs (Raven, Kyhos & Hill, 1965); $n = 7$ (Grant & Sidhu, 1967). The sp. reported as in the San Emigdio Range, Kern Co., *Twisselmann*.

Var. **brevivexíllus** Ottley. $2n = 7$ pairs (Raven, Kyhos & Hill, 1965). Has been found in the El Paso Range, Kern Co., *Twisselmann*.

16. **L. denticulàtus** (E. Drew) Greene. $n = 6$ (Grant & Sidhu, 1967).

p. 847.   17. **L. subpinnàtus** Lag. $n = 6$ (Grant & Sidhu, 1967).

18. **L. humistràtus** Greene. $n = 6$ (Grant & Sidhu, 1967).

21. **L. Purshiànus** (Benth.) Clem & Clem. $n = 7$ (Grant & Sidhu, 1967).

p. 848.   24. **L. nevadénsis** Greene. $n = 7$ (Grant & Sidhu, 1967).

25. **L. Douglásii** Greene. $2n = 14$ (Grant, 1965).

27. **L. argophýllus** (Gray) Greene. $2n = 7$ pairs (Raven, Kyhos & Hill, 1965); $n = 7$ (Grant & Sidhu, 1967).

Ssp. **adsúrgens** (Dunkle) Raven. [*L. a.* var. *adsurgens* Dunkle.] A beautiful silvery suffrutescent plant with densely crowded ascending lvs. San Clemente Id.

Ssp. **ornithòpus** (Greene) Raven. [*Hosackia o.* Greene.] $2n = 7$ pairs (Raven, Kyhos & Hill, 1965).

p. 849.   29. **L. hamàtus** Greene. Occurs in San Luis Obispo Co., *Hardham*.

30. **L. Benthàmii** Greene. [*L. cytisoides* Benth., not *cystoides*.] $2n = 7$ pairs (Raven, Kyhos & Hill, 1965).

31. **L. scopàrius** (Nutt. in T. & G.) Ottley. $n = 7$ (Grant & Sidhu, 1967).

Var. **dendroìdeus** (Greene) Ottley. $n = 7$ (Grant & Sidhu, 1967). Add:

Ssp. **Tráskiae** (Eastw. ex Abrams) Raven. [*L. T.* Eastw. ex Abrams.] Pods mostly 3–5 cm. long and 4–8-seeded.—San Clemente Id.

p. 850.   34. **L. corniculàtus** L. $n = 12$ (Grant & Sidhu, 1967).

35. **L. uliginòsus** Schk. Add to synonymy *L. trifoliolatus* Eastw. *L. pedunculatus* auth., not Cav.

PSORÀLEA

In Key, delete "6. *P. rigida*" and on next line insert:

Fls. racemose, calyx short-pubescent; plant 3–6 dm. high. . . . . 6. *P. rigida*
Fls. subcapitate or subumbellate; calyx conspicuously white-hairy; plant to 20 dm. high. Introd. . . . . 6a. *P. bituminosa*

p. 851.   1. **P. califórnica** Wats. $2n = 11$ pairs (Raven, Kyhos & Hill, 1965).
p. 851.   3. **P. lanceolàta** ssp. **scabra** (Nutt.) Piper. $2n = 11$ pairs (Raven, Kyhos and Hill, 1965).—At 6500 ft., 5.5 mi. w. of state line, along Highway 31, Mono Co., *Reveal*.

5. **P. physòdes** Dougl. $n = 11$ (Raven, Kyhos & Hill, 1965).

6a. **P. bituminòsa** L. Half shrub to 2 m. tall; lvs. 3-foliolate; lfts. lanceolate, 2–5 cm. long; stipules lance-subulate, 5–7 mm. long; fls. lilac with an almost white keel and banner somewhat red in age.—Millard Canyon, San Gabriel Mts., Los Angeles Co., *Griesel*. Possibly established here and at Pleasanton, Alameda Co., after being tested as a fire-resistant plant by forest personnel, *Fuller*. From Old World.

7. **P. macrostàchya** DC. $2n = 11$ pairs (Raven, Kyhos & Hill, 1965).

p. 853.   DÀLEA

In Key, AA, B, C and CC, change "lfts." to "lvs."

3. **D. Párryi** T. & G. $2n = 10$ pairs (Raven, Kyhos & Hill, 1965).

5. **D. Schóttii** var. **puberula** (Parish) Munz. $n = 10$ (Turner, 1963); $2n = 10$ pairs (Raven, Kyhos & Hill, 1965).

6. **D. polyadènia** Torr. ex Wats. $2n = 10$ pairs (Raven, Kyhos & Hill, 1965).

p. 854.   PETALOSTEMON, not *Petalostemum* in 1959 printing. Cf. Taxon 8: 293. 1959.
p. 855.   SESBANIA. Authority for genus is Adans., corr. Scop.
p. 856.   ASTRÁGALUS

Before the Key, insert Barneby, R. C. Atlas of N. Am. Astragalus 2 vols., 1188 pages, 1964.

In Key, change *A. Hornii* from 81 to no. 82.

p. 864.   In Key, after S at top of page, delete "84. *A. iodanthus*" and insert:

T. Stems decumbent, not zigzag; lfts. 5–15 mm.
 long; petals ochroleucous with purple keel-tip;
 pods 2–4 cm. long . . . . 84. *A. iodanthus*
TT. Stems prostrate, abruptly zigzag; lfts. 5–9 mm.
 long; petals red-violet; pods 2 cm. long.
 84a. *A. pseudiodanthus*

1. **A. Gambeliànus** Sheld. Reported from Kern Co., *Twisselmann*.

2. **A. didymocárpus** H. & A. $2n = 12$ pairs (Raven, Kyhos & Hill, 1965).

p. 865.   Var. **obispénsis** (Rydb.) Jeps. $2n = 13$ pairs (Raven, Kyhos & Hill, 1965).

3. **A. Brèweri** Gray. $n = $ ca. 12 (Raven, Kyhos & Hill, 1965).

p. 866.   9. **A. Nuttalliànus** DC. var. **imperféctus**, not no. 3 as in 1959 printing.

p. 867.   12. **A. calycòsus** Torr. $2n = 22$ (Ledingham & Fahselt, 1964).

p. 868.   17. **A. inyoénsis** Sheld. $2n = 22$ (Ledingham & Fahselt, 1964).

p. 870.   25. **A. Serènoi** (Kuntze) Sheld. $2n = 24$? (Ledingham & Fahselt, 1964).

p. 871.   29. **A. dasyglóttis** Fisch. ex DC., not *A. agrestis* Dougl. Cf. Barneby, Leaflets W. Bot. 9: 51. 1959.

p. 873.   40. **A. coccíneus** Bdg. $2n = 22$ (Ledingham & Fahselt, 1964).

41. **A. funèreus** Jones [*Xylophacos f.* Rydb.] was omitted from the FLORA. Cespitose perennial with stems to ca. 1 dm. long, the plant densely villous with tangled hairs; lfts. 13–17, oval to obovate, obtuse, 5–8 mm. long, silky- or cottony-tomentose with tangled hairs; racemes subcapitate, 3–10-fld.; calyx-tube 7–8 mm. long, black-hairy, the subulate teeth ca. 3 mm. long; corolla rose-purple, the keel 21–26 mm. long, the wing-petals equaling or shorter; pod 2.5–4 cm. long, densely hairy.—At 4000–5000 ft., Death Valley region. March–Apr.

p. 873.   42. **A. Púrshii** Dougl. In 1964 Barneby reduced var. *longilobus* Jones to synonymy under typical **A. Purshii** and included the concept named *longilobus* in the FLORA in var. **tínctus** Jones.

p. 874.   Var. **léctulus** Jones. $2n = 22$ (Ledingham & Fahselt, 1964).

46. **A. Johánnis-Howéllii** Barneby $2n = 22$ (Ledingham & Fahselt, 1964).

p. 875.   49. **A. Ravènii** (Barneby. $2n = 24$ (Ledingham & Fahselt, 1964); $2n = 22$ (Raven, Kyhos & Hill, 1965).

p. 877.   58. **A. miguelénsis** Greene. $2n = 22$ (Ledingham & Fahselt, 1964).

p. 878.   62. **A. asymmétricus** Sheld. Occurs in the Temblor Range, Kern Co., *Twisselmann*.

p. 879.   65, 66, 69. Barneby called in 1964 sp. 66 of the FLORA **A. trichópodus** (Nutt.) Gray, with *A. capíllipes* Jones as a synonym. Sp. 65 of the FLORA became **A. trichópodus** var. **lónchus** (Jones) Barneby with *Phaca canescens* Nutt. and *P. encenadae* Rydb. as synonyms. $2n = 11$ pairs (Raven, Kyhos & Hill, 1965). Sp. 67 was called **A. trichópodus** var. **phóxus** (Jones) Barneby. [*A. Antiselli* var. *phoxus* Jones; *A. Antiselli* Gray; *Homalobus Antiselli* Rydb.; *A. trichopodus* var. *A.* Jeps.; *A. Hasseanthus* Sheld., *A. gaviotus* Elmer; and *A. Antiselli* var. *gaviotus* Munz.]

p. 880.   71. **A. Vaseyi** Wats. changed by Barneby (1964) to **A. Pálmeri** Gray. [*A. Vaseyi* Wats.; *A. metanus* Jones; *A. Vaseyi* var. *Johnstonii* Munz & McBurn.]

p. 881.   74. **A. oóphorus** Wats. $2n = 24$ (Ledingham & Fahselt, 1964).

p. 883.   83. *A. lentiginosus* var. *carinatus* Jones reduced to synonymy under typical **A. lentiginòsus** Dougl. by Barneby (1964).

p. 884.   83. **A. lentiginòsus** var. **Fremóntii** (Gray) Wats. $2n = 22$ (Ledingham & Fahselt, 1964).

p. 884.   **A. lentiginòsus** var. **coachéllae** Barneby in Shreve & Wiggins used for our material referred in the FLORA to var. *Coulteri*.

84a. **A. pseudiodánthus** Barneby. Near to *A. iodanthus* Wats., but with prostrate, abruptly zigzag stems; villous-pubescent; lfts. 5–9 mm. long; fls. red-violet, 8–9 mm. long; pods 2 cm. long.—Old beach, n. side of Mono Lake, Mono Co., *Reveal*; w. Nev. June.

p. 886.   92. **A. Whítneyi** Gray. $2n = 22$ (Ledingham & Fahselt, 1964).

p. 888.   ORNITHÒPUS

2. **O. pinnàtus** (Mill.) Druce. Differs from *O. roseus* by having lvs. glabrous; fls. yellow; pods strongly bent. Taken at same station as *O. roseus*. Native of Eu.

ALHÀGI

1. **A. camelòrum** Fisch. $n = 8$ (Baquar et al., 1965).—Reported from Contra Costa Co. and Afton Canyon, Mojave Desert, San Bernardino Co., *Fuller.*

p. 889.   LÁTHYRUS

In Key, AA, B, C, D, E, delete "2. *L. sphericus*" at end of line and add:

F.   Stems winged; fls. 10–13 mm. long. . . . . 2a. *L. Cicera*
FF. Stems angled, not winged; fls. 10 mm. long. . . 2. *L. sphericus*

p. 890.   2a. **L. Cicèra** L. Annual, 2–6 dm. high, glabrous; stems winged; lfts. 2, lanceolate to linear; fls. reddish, 10–13 mm. long, solitary; pods 3–4 cm. long, 8–10 mm. wide.—Reported from San Mateo Co., *Howell*, and Amador Co., *Fuller.* From Eurasia.

p. 891.   4. **L. tingitànus** L. Lfts. 2, not 4.

p. 892.   11. **L. vestìtus** Nutt. ex T. & G. and ssp. **Bolánderi** (Wats.) C. L. Hitchc. are both in Monterey Co., *Howitt & Howell.*

13. **L. laetiflòrus** Greene. $2n = 7$ pairs (Raven, Kyhos & Hill, 1965).—Found in the San Emigdio Range, Kern Co., *Twisselmann.*

p. 896.   VÍCIA

8. **V. benghalénsis** L. $2n = 12$ (Srivastaoa, 1963).—Reported from Monterey Co., *Howitt & Howell.*

11. **V. dasycárpa** Ten.—Reported from Catalina Id., *Thorne*; Monterey Co., *Howitt & Howell.*

16. **V. angustifòlia** Reichard (Not *angusifolia* as in 1959 printing of the FLORA). $2n = 6$ pairs (Raven, Kyhos & Hill, 1965).

p. 897.   36. **Léns** Moench. LENTIL

Like *Pisum*, but the fls. small, inconspicuous, whitish; calyx-lobes very narrow; seeds 1–2. Eurasia.

1. **L. culinàris** Medic. [*Ervum Lens* L.] Annual, to 4 dm. tall; lfts. 4–7 pairs; fls. 1–3, to ca. 6 mm. long; pod to 2 cm. long, almost as broad.—Occasional escape; reported from Ventura and San Francisco cos.

PLÁTANUS

1. **P. racemòsa** Nutt. $2n = 21$ pairs (Raven, Kyhos & Hill, 1965).

p. 898.   KRAMÈRIA

1. **K. Gràyii** Rose & Painter. $n = 6$ (Turner, 1959).

p. 900.   ÁLNUS

3. **A. oregòna** Nutt.—Reported from as far s. as Monterey Co., *Howitt & Howell.*

p. 901.   1. **Chrysolèpsis** Hjelmquist. CHINQUAPIN

Trees or shrubs, evergreen, the buds with imbricated scales. Lvs. simple. Catkins staminate or androgynous; fls. 3–7 (–11), fasciculate, staminate always bracteolate; calyx 5–6-parted; stamens several. Pistillate fls. at base of staminate, 3 in a cupule of 7 free valves (5 outer and 2 inner, the latter separating the 3 trigonous fruits from one another); styles 3. Fr. maturing in the second season, the spiny invol. inclosing the nuts, these angled. (Greek, *chrysos*, gold, and *lepis*, scale). Two spp. of w. N. Am. (Hjelmquist, Bot. Notiser, Suppl. 2: 117. 1948; 113: 377. 1960).

1. **C. chrysophýlla** (Dougl. ex Hook.) Hjelmquist. [*Castanopsis c.* A. DC. *Castanea c.* Dougl. ex Hook.] See description in the FLORA under *Castanopsis.*— Extends s. to Marin Co., possibly Santa Cruz Mts.

Var. **mìnor** (Benth.) Munz, comb. nov. [*Castanea chrysophylla* var. *minor* Benth., Pl. Hartweg., 337. 1857. *Castanopsis c.* var. *m.* A. DC.]

p. 902.   2. **C. sempervìrens** (Kell.) Hjelmquist. [*Castanopsis s.* Dudl. *Castanea s.* Kell.] For description see *Castanopsis* in the FLORA.

LITHOCARPUS
1. **L. densiflorus** (H. & A.) Rehd. at De Sabla, Butte Co., *Vesta Holt.*

p. 904.   QUÉRCUS
3. **Q. agrifòlia** Neé. $2n = 12$ pairs (Raven, Kyhos & Hill, 1965).
Var. **frutéscens** Engelm., not Jeps. as in 1959 printing.

p. 905.   6. **Q. Douglásii** H. & A. A supposed hybrid with *Q. lobata* Neé has been described as **Q.** × **jolonensis** Sarg., from Monterey Co.
9. **Q. dumòsa** Nutt. A hybrid supposedly with *Q. lobata* Neé is **Q.** × **Townei** Palmer, once described from near Pasadena.

p. 907.   16. **Q. Dúnnii** Kell. seems to have priority as a sp. name over *Q. Palmeri* Engelm.

p. 909.   JÙGLANS
1. **J. califórnica** Wats. is in Monterey Co., *Howitt & Howell,* where it grows in the Santa Lucia Mts.

p. 910.   PÓPULUS
2. **P. trichocárpa** T. & G. is sometimes treated as a ssp.; *P. balsamifera* L. ssp. *t.* Brayshaw. (Cf. Brayshaw, Can. Field-Naturalist 79: 95. 1965).
P. acuminàta Rydb. is reported by *Twisselmann* from the upper Kern R. and by *DeDecker* from Lone Pine Creek. Lvs. much longer than wide, not toothed near apex. A sp. from e. of Calif. in Rocky Mt. region. Sometimes considered a hybrid.

p. 912.   SÀLIX
In Key, line 9 from top of page, change HH to:

> HH. Shrub to ca. 1 m. high, with gray-tomentose twigs; lvs. obovate to elliptic.
> I. Lvs. mostly 2–3 cm. long; fils. mostly free; style 0.5–0.8 mm. long. E. slope of Sierra Nevada.
> 29a. *S. brachycarpa*
> II. Lvs. 1.5–8 cm. long; fils. ± united; style ca. 1 mm. long. Del Norte Co. ... 29. *S. delnortensis*

In Key, after AA, B, CC, D, EE, F, GG, insert:

> H. Margins of lf.-blades and petioles near the apex with conspicuous yellowish glands; branchlets and upper surface of lvs. shiny. .... 1. *S. lasiandra*
> HH. Margins of lf.-blades and petioles not or not conspicuously glandular; branchlets and upper surface of lvs. not shiny. .... 4. *S. laevigata*

In Key, after AA, BB, C, insert:

> C′. Plants with "weeping" habit, having long pendulous branches; lvs. largely 10–15 cm. long, very long-acuminate. Escape from cult. .. 3. *S. babylonica*
> C′C′. Plants not conspicuous in the pendulous branches. Native.
> D. Stamens 4–6; etc. as in the FLORA.

3. **S. Goòddingii** Ball. $2n = 19$ pairs (Raven, Kyhos & Hill, 1965).

p. 193.   3a. **S. babylónica** L. WEEPING WILLOW. Broad-headed large tree with long flexible hanging branches; lvs. 8–15 cm. long, long-acuminate, finely serrulate; stipules rarely developed.—Escape from cult., as at Santa Barbara, *Pollard.* Native of Cuba.

p. 914.   7. **S. melanópsis** Nutt. is referred to S. *exigua* ssp. *m.* Cronq. in vol. 2 of Vasc. Pls. Pac. N. W. 1964. *Twisselmann* reports from along Kern R. above Lake Isabella and from Tehachapi Mts.
9. **S. lùtea** Nutt. is made a synonym of S. *rigida* Muhl. by Cronquist.
Var. **Watsònii** (Bebb) Jeps. on Kern Plateau, *Twisselmann.*

p. 915.   12. **S. pseudocordàta** Anderss. is referred to S. *myrtillifolia* Anderss. by Cronquist.
13. **S. lasiólepis** Benth. $2n = 38$ pairs (Raven, Kyhos & Hill, 1965).

p. 916.   20. **S. anglòrum** Cham. var. **antiplásta** C. K. Schneid. is referred to S. *arctica* Pall. by Cronquist.

22. **S. planifòlia** Pursh var. **mònica** (Bebb) C. K. Schneid. is referred to *S. phyllicifolia* var. *monica* Jeps.

p. 918.   29a. **S. brachycárpa** Nutt. Erect shrub 2–10 dm. tall; young twigs dark or reddish under the villous tomentum; lvs. entire, obovate to elliptic, hairy when young, mostly 2–3 cm. long; ament scales mostly brown; stamens 2; caps. hairy, 3–5 mm. long.—Convict Creek, Mono Co. at 9900–10,600 ft., *Major and Bamberg*; Ore. to Alaska, Rocky Mts.; Quebec.

31. **S. Geyeriàna** Anderss. var. **argéntea** C. K. Schneid. found at Pine Flat, Kern Co., *Twisselmann*.

ÚLMUS
Change Key to Species to:

A.   Lvs. doubly serrate, unequal at base.
B.   Lvs. with scattered pubescence, scabrous above; branchlets pubescent until the second year. . . . . 1. *U. procera*
BB.   Lvs. with axillary tufts beneath, smooth above; branchlets subglabrous.
2. *U. carpinifolia*
AA.   Lvs. usually simply serrate; subequal at base.
B.   Branchlets soon glabrous; fls. in spring before the lvs.; samara ca. 12 mm. long.
3. *U. pumila*
BB.   Branchlets pubescent; fls. in late summer; samara ca. 8 mm. long.   4. *U. parvifolia*

p. 919.   3. **U. pùmila** L. SIBERIAN ELM. Small tree; branchlets soon glabrous; lvs. elliptic to oblong-lanceolate, 2–3.5 cm. long, short-acuminate; fls. in the spring. —Cult. in interior and desert areas; native of Siberia.

4. **U. parvifòlia** Jacq. [*U. chinensis* Pers.] CHINESE ELM. Partially evergreen in mild climate; branchlets pubescent; lvs. elliptic to ovate, 2–3.5 cm. long, acute; fls. in late summer.—Seeding itself as at Pasadena, *Howell*, Santa Barbara, *Pollard*; native of e. Asia.

CÉLTIS
1. **C. reticulàta** Torr. is probably the better name for *C. Douglasii* Planch.

MORÀCEAE
In Key to Genera, add a third line:

Plant a tree; lvs. lobed, alternate. . . . . 3. *Morus*

p. 920.   HÙMULUS
1. **H. Lùpulus** L. HOPS. $2n = 20$ (Jacobsen, 1957).

3. Mòrus L. MULBERRY

Trees with milky juice, alternate lobed lvs.; monoecious or dioecious; fls. in small cylindrical catkin-like spikes, the ♂ soon falling, the ♀ ripening into a blackberry-like juicy cluster; perianth 4-parted; stamens 4; each ovary becoming a drupelet inclosed in the enlarged fleshy perianth. Ca. a dozen spp. (*Morus*, the classical name.)

1. **M. álba** L. WHITE MULBERRY. Tree to 15 m.; lvs. broad-ovate, the blade 5–15 cm. long, largely scallop-toothed, often irregularly lobed; fr. 2–5 cm. long, whitish to purple, sweet.—Kern R., near Stockdale Country Club, Kern Co., *Twisselmann*. Natur. from China.

Ficus carìca L., the COMMON FIG, sometimes escapes, as on Catalina Id., *Thorne*.

URTICÀCEAE
In Key to Genera, delete "3. *Parietaria*" from end of third line and add:

Fls. in axillary glomerate clusters. . . . . 3. *Parietaria*
Fls. solitary. . . . . 4. *Helxine*

ÚRTICA
1. **U. holosericea** Nutt. Hitchcock (Vasc. Pl. Pac. N.W. 2: 91. 1964) uses *U. dioica* L. ssp. *gracilis* vars. *holosericea, Lyallii* and *californica*.

p. 921. HESPEROCNIDE
1. **H. tenélla** Torr. reported from Temblor Range and Greenhorn Range, Kern Co., *Twisselmann.*
PARIETARIA
3. **P. pensylvánica** Muhl. $n = 8$ (E. B. Smith, 1963).

4. Helxìne Req. BABY'S-TEARS

Delicate creeping herb; lvs. alternate; monoecious; fls. minute, solitary in axils; ♂ fls. with 4-parted calyx and 4 stamens; ♀ calyx tubular, contracted at mouth; ♂ fls. with a 3-lvd. invol., ♀ with a 3-lobed invol.; ak. ovoid, included in the invol. One sp., from Corsica, Sardinia.
1. **H. Soleiròlii** Req. Lvs. mostly less than 6 mm. long, unequal-sided.— Natur. at Cambria, San Luis Obispo Co., *Hardham,* and at other stations from Lake Co. to Santa Barbara Co., *Howell.*

p. 922. AMMÁNNIA
1. **A. coccínea** Rottb. Reported from Marin Co., *Howell.*

p. 923. PÉPLIS
1. **P. Pórtula** L. [*Lythrum P.* (L.) D. A. Webb.]
LYTHRUM
In Key to Species, under "Fls. sessile or nearly so," insert:

Petals 1.5–2 mm. long; plant annual. . . . . 1. *L. Hyssopifolia*
Petals 7–10 mm. long; plant perennial. . . . . 1a. *L. Salicaria*

1a. **L. Salicària** L. PURPLE LOOSESTRIFE. Stout erect perennial 6–12 dm. tall; lvs. opposite or whorled, sessile, ± lanceolate, 3–10 cm. long; spikes 1–4 dm. long; petals red-purple, 7–10 mm. long.—Natur. in e. U.S. and adventive 0.5 mi. e. of Grass V., Nevada Co., *Fuller* and 2.5 mi. s. of Oroville, Butte Co., *Heinrichs.* Native of Eu.
2. **L. tribracteàtum** Salzm. ex Ten. Reported from Lake Co., *Fuller.*
ONAGRÀCEAE
In Key at bottom of page, change to

Sepals persistent. . . . . 1. *Ludwigia*
Sepals deciduous after anthesis.
  Fls. 4–merous normally, etc. as in the FLORA.

pp. 924–925. Combine *Jussiaea* and *Ludwigia* as follows:

1. Ludwígia L.

Our spp. herbaceous, mostly of wet places, sometimes floating in open water, sometimes with basal vegetative shoots creeping and rooting at the nodes. Underwater parts often swollen and spongy. Lvs. alternate or opposite, rarely whorled, mostly simple, membranaceous or rarely coriaceous. Stipules present, at least in upper part of plant. Fls. yellow or white, solitary and axillary, or in terminal spikes or heads. Fl.-tube not prolonged beyond the ovary, usually with 2 bracteoles at the base of the ovary or summit of the pedicel. Fls. diurnal, regular, 3–7-merous, but mostly 4-merous. Sepals persistent after anthesis. Petals 0–7, caducous. Stamens in 1 or 2 series, each series usually as many as the sepals; anthers usually versatile, basifixed in very small fls. Ovary with as many locules as sepals; style simple, ± produced above the disc; stigma capitate or hemispheric, often slightly lobed. Ovary cylindrical to obconic, many-ovuled, the ovules pluriseriate to uniseriate in each locule. Pluriseriate seeds naked and with evident raphe, the uniseriate surrounded by an endocarp. Caps. dehiscing by a terminal pore or by flaps separating from the valvelike top or more irregularly.

A. Stamens in 2 series, mostly 8 or 10 in number; petals 10–20 mm. long.
    B. Flowering stems usually floating or creeping; lvs. oblong, 1–10 cm. long; bracteoles at base of ovary deltoid; caps. mostly 2–3 mm. thick. Well distributed.
                                                                    1. *L. peploides*

BB. Flowering stems usually erect; lvs. ± lanceolate, mostly 5–10 cm. long; bracteoles
lanceolate; caps. 3–4 mm. thick. Local. 2. *L. uruguayensis*
AA. Stamens in 1 series, 4–5 in number; petals none or small and quickly shed.
    B. Ovary with 4 evident longitudinal green bands; basal bracteoles from not evident
       to ca. 1 mm. long; petals none. . . . . 3. *L. palustris*
    BB. Ovary lacking green bands; bracteoles above the base and 1–5 mm. long; petals
       present, but easily shed. . . . . 4. *L. repens*

p. 925.   1. **L. peploìdes** (HBK) Raven. [*Jussiaea p.* HBK. *J. repens* var. *p.* Griseb.
J. r. var. *californica* Wats. *J. c.* Jeps.] Description as in the FLORA.
    Ssp. **montevidénsis** (Spreng.) Raven. [*J. m.* Spreng.] In Eldorado Co., *Fuller.*
    2. **L. uruguayénsis** (Camb.) Hara. [*Jussiaea u.* Camb. in St. Hil. *J. grandi-flora* Michx. *J. Michauxiana* Fern.] Description as on p. 925.
    3. **L. palústris** (L.) Ell. [*Isnardia p.* Ell. *L. p.* var. *americana* Fern. & Gris-com.] Description as for var. *americana* on p. 925.
    Var. **pacífica** Fern. & Griscom, as on p. 925.

p. 926.   4. **L. rèpens** Forster var. **stipitàta** (Fern. & Griscom) Munz. [*L. natans* Ell.
var. *s.* Fern. & Griscom.] Use description on p. 926.

ZAUSCHNÈRIA

In Key, change Z. *californica* to Z. *c. mexicana* and Z. *c. angustifolia* to
Z. *californica.*
    3. **Z. califórnica** Presl. [Z. *c.* ssp. *angustifolia* Keck.] Amend description to:
Suffrutescent at base, often much branched, the stems 3–7 dm. long, tomentose-canescent, ± glandular; lvs. linear, densely tomentose-canescent, the lower
opposite or subopposite, lateral veins usually not evident; fls. 3–4 cm. long; fl.-tube largely 2–3 cm. long; sepals erect, lanceolate, 8–10 mm. long, scarlet;
petals 2-cleft, scarlet, 8–15 mm. long; stamens well exserted; style and stigma
surpassing stamens; caps. sessile or nearly so, linear, 4-angled, 8-nerved, often
curved, with a short beak and 1–2 cm. long; $n = 30$.—Dry slopes below 2000
ft., Coastal Sage Scrub, Chaparral, etc., Coast Ranges from Monterey Co. to
San Diego Co.; Catalina Id. Aug.–Oct.
    Ssp. **mexicàna** (Presl) Raven. [Z. *m.* Presl. Z. *villosa* Greene. Z. *californica*
var. *v.* Jeps. Z. *Eastwoodae* Moxley. Z. *velutina* Eastw. ex Moxley.] Suffrutes-cent at base, the stems to 9 dm. long; lvs. lanceolate to linear-lanceolate or
oblong-lanceolate, mostly 3–5 mm. wide, green to gray-pilose; $n = 30$.—Dry,
mostly stony or gravelly places below 3800 ft., Sonoma and Lake cos. to L.
Calif. Variable, the var. *villosa* having been used to designate plants from the
Santa Barbara Ids. with long soft hairs.

p. 927.   Use Ssp. **latifòlia** and Z. **càna** as they are in the FLORA.

EPILÒBIUM

In Key, line 4 from bottom of page, change to "Stems 3–20 dm. tall."

p. 928.   In Key, after DD, E, insert:

    E′. Petals yellow, 14–18 mm. long; stems mostly 3–7 dm. tall. Siski-
      you Mts. . . . . 7b. *E. luteum*
    E′E′. Petals pink to white, 2–10 mm. long.
      F. Petals 5–10 mm. long, etc. as in the FLORA.

    1. **E. angustifòlium** L. Mosquin (Brittonia 18: 167–187. 1966) proposes:

A. Abaxial lf. midribs glabrous; lvs. (3)–10–(30) mm. wide, (35)–85–(170) mm. long;
   pollen grains commonly less than 85μ diam. Circumpolar, s. to ca. the s. limits of the
   boreal forest and in Rocky Mts. to Wyo. . . . . Ssp. *angustifolium*
AA. Abaxial lf. midribs glabrous to pubescent; lvs. (5)–20–(40) mm. wide, (60)–110–
   (220) mm. long; pollen grains over 85μ diam. S. Canada and U.S., as well as Eurasia.
                                          Ssp. *circumvagum*

    Subsp. **angustifòlium** includes var. *intermedium* of the FLORA. $n = 18$ (Mos-quin).
    Subsp. **circumvàgum** Mosquin includes var. *macrophyllum* of the FLORA. $n = 36$ (Mosquin).

p. 929.   3. **E. obcordàtum** Gray. Transpose lines 8 and 9 of 1959 printing. Lvs. glau-cous, 6–16 mm. long; fl. tube 2–4 mm.—Sierra Nevada to Ida.

Ssp. **siskiyouénse** Munz. [*E. o.* var. *laxum* (Hausskn.) Dempster in Jeps.] Plants more greenish; lvs. mostly 10–22 mm. long, subacute; fl.-tube 2–2.5 mm. long; longer stamens ca. half as long as petals.—Siskiyou and Trinity cos.; to Jackson Co., Ore.

p. 930.   7. **E. minùtum** Lindl. ex Hook. *n* = 13. Reported from as far s. as Ventura Co., *Hardham.*

7a. **E. foliòsum** (Nutt. ex T. & G.) Suksd. [*E. minutum* var. *f.* T. & G. *E. m.* var. *Biolettii* Greene.] Resembling *E. minutum*, but lvs. linear to lance-linear, sharply pointed, somewhat toothed, with more tendency to axillary fascicles and narrow petioles; fls. smaller; sepals ca. 1 mm. long; petals scarcely 2 mm. long; caps. short-pedicelled; *n* = 16. Ranging with *E. minutum*, particularly northward.

7b. **E. lùteum** Pursh. Perennial with creeping underground rootstock and well developed turions; stems subsimple, erect, 2–7 dm. tall; lvs. ± ovate, 2–7 cm. long, very slightly reduced up the stem; fls. few, nodding in the bud; petals yellow, obcordate, 14–18 mm. long.—Moist places, Siskiyou Co., *Bacigalupi*; to B.C., Alta.

p. 931.   11. **E. Pringleànum** Hausskn. In line 2 of description delete "slender" from middle of line.

12. **E. Halleànum** Hausskn. Reported from Marin and Santa Cruz cos.

p. 932.   19. **E. adenocaúlon** Hausskn. var. **occidentàle** Trel. in Monterey Co., *Howell.*

p. 933.   BOISDUVÁLIA

New reference to literature: Raven, P. H. and D. M. Moore. A revision of Boisduvalia. Brittonia 17: 238–253. 1965.

p. 934.   3. **B. glabélla** (Nutt.) Walp. *n* = 15 (Kurabayashi, 1962).

4. **B. strícta** (Gray) Greene occurs in the Greenhorn Mts., Kern Co., *Twisselmann*; in Monterey Co., *Howitt & Howell*; San Bernardino Mts., *Charlotte Bringle.*

6. Reduce *B. pállida* Eastw. to synonymy under 5. **B. macrántha** Heller.

p. 935.   CLÁRKIA

In Key, after AA, B, C, change to:

> D.   Petals 6–12 mm. long, 3–7 mm. broad.
> D'.   Pollen dull yellow-gray to dark blue-gray; petals 6–12 mm. long, lanceolate to rhombic, with or without dark spots. Widely distributed.
>   27. *C. rhomboidea*
> D'D'. Pollen yellow, petals 6–7.5 mm. long, obovate, not spotted. Plumas and Yuba cos. .... 28a. *C. stellata*

In Key, after AA, BB, CC, D, E, change to:

> F.   Some long hairs present on ovary and calyx; lf.-width ¼–⅗ the length; petal-width ⅛–¾ the length; style usually well exserted. Lake and Plumas cos. to San Diego Co.
>   24. *C. unguiculata*
> FF. Long hairs absent; lf.-width less than ⅛ the length; petal-width usually less than ½ the length; leafy bracts width usually less than ¼ the length.
>   G.   Style usually well exserted; sepals dark red-purple; petals usually with a large dark red-purple spot at base of blade; often with 5–6 fls. open on one stem at same time. Tulare Co. .... 25a. *C. springvillensis*
>   GG. Style seldom well exserted; sepals mostly green or only slightly reddish; petal spot, if present, small and well defined or, if large, not sharply defined; usually 1–3 fls. open on stem at the same time.
>   H.   The style equaling or only slightly exceeding the anthers; lvs. usually bright green; petals pink, with or without a purple spot at the base of the blade, or white. S. Tulare and n. Kern cos. .... 25. *C. exilis*
>   HH. The style equaling the anthers to well exserted; lvs. usually gray-green; petals pink, sometimes with a darker blotch at base of blade, or petals reduced, sepal-like, unexpanded and wrinkled. Inner Coast Ranges, Alameda Co. to w. Kern Co.
>   25b. *C. tembloriensis*

In Key, AA, BB, CC, D, EE, F, G, HH, I, insert at end of line 2:
"5. *C. gracilis*" (for 1959 printing).

p. 936.   In Key, AA, BB, CC, D, EE, FF, G, H, I, petals should be 10–30, not 10–13
mm. long (for 1959 printing).
In Key, after DD, E, F, G, delete "2. *C. rubicunda*" and insert:

> H. Fl.-tube 4–10 mm. long; petals 10–30 mm. long,
> obovate to fan-shaped. Monterey Co. to Marin Co.
> 2. *C. rubicunda*
> HH. Fl.-tube 1–3 mm. long; petals 5–13 mm. long,
> wedge-shaped. San Francisco. .. 2a. *C. franciscana*

p. 938.   2a. **C. franciscàna** Lewis & Raven. Short description in the FLORA.
5. **C. grácilis** (Piper) Nels. & Macbr. in the synonymy *G. a.* var. *pygmaea*
Jeps. should be forma (1959 printing).

p. 939.   10. **C. Dàvyi** (Jeps.) Lewis & Lewis In Santa Cruz Co.; on Santa Rosa Id.,
Raven.

p. 940.   12. **C. purpùrea** ssp. **quadrivúlnera** (Dougl.) Lewis & Lewis. In the synonymy
var. *capitata* Jeps. and var. *flagellata* Jeps. should read forma. (1959 printing.)

p. 941.   16. **C. bilòba** (Durand) Nels. & Macbr. ssp. **Brandègeae** (Jeps.) Lewis &
Lewis. In the synonymy, for 1959 printing change "var." to "f."

p. 942.   21. **C. símilis** Lewis & Ernst. In Monterey Co., *Howitt & Howell*.
24. **C. unguiculàta** Lindl. Line 4 from bottom of description, change "short-
petioled" to "short-pediceled."
25a. **C. springvillénsis** Vasek. Near to *C. unguiculata*; lf.-blades bright green,
2–9 cm. long, 5–20 mm. wide; fl.-tube 3–4 mm. long; sepals 12–16 mm. long,
puberulent, usually dark red; petals 13–16 mm. long, including a narrow red
claw 7–9 mm. long and a limb 6–8 mm. long, 7–10 mm. broad, lavender-pink
and usually with a dark purplish spot at base of limb; style exceeding stamens;
n = 9.—Foothill Wd., near Springville, Ranger Station, Tulare Co.
25b. **C. tembloriénsis** Vasek. Near to *C. unguiculata*, erect, to 8 dm. tall; lf.-
blades gray-green, 2–7 cm. long, 3–13 mm. wide; fl.-tube 2(3) mm. long;
sepals 9–16 mm. long, 2–3 mm. wide, puberulent; petals 13–17 mm.
long, including a narrow claw 5–11 mm. long 1(2) mm. wide, the limb scarcely
wider than the base; style often not exceeding the stamens; *n* = 9.—Inner
Coast Ranges from e. Alameda and w. San Joaquin cos. to w. Kern and e. San
Luis Obispo cos.

p. 943.   28a. **C. stellàta** Mosquin. Erect, to 1 m. tall, simple or branched, subglabrous
to strigulose; lf.-blades lanceolate to elliptic or ovate, 1–5 cm. long, 5–20 mm.
broad, on petioles 5–30 mm. long; infl.-rachis recurved in bud; fl.-tube 1.5–2
mm. long; sepals 5–7 mm. long; petals obovate, 6–7.5 mm. long, 3–4 mm.
broad, shallowly 3-lobed, the central lobe ca. 1 mm. longer than the lateral,
lavender-purple, not flecked, red-purple at the base; anthers pale; pollen yellow;
style shorter than the stamens; mature caps. 2–2.5 cm. long, 2–3 mm. broad,
dry and quadrangular, often with red streaks, straight or slightly curved; *n* = 7.
—Plumas and Yuba cos.

p. 944.   OENOTHÈRA
References to literature: Raven, Brittonia 16: 276–299. 1964 and Munz, N.
Am. Flora, Series II, Part 5: 79–177. 1965.
In Key, A, B, CC, change to:

> CC. Fls. white to rose, the buds often nodding.
> D.   Caps. sterile and slender in lower part, thicker, fertile and ± winged in
> upper part; seeds in more than 2 rows in each locule; fl.-tube 0.4–2 cm.
> long.
> E.   Petals white to pink, 2.5–4 cm. long; plants with running under-
> ground rootstocks; buds nodding. .... 7. *Oe. speciosa*
> EE. Petals rose to red-violet, 0.5–1 cm. long; plants from a ± woody
> caudex; buds erect. .... 7a. *Oe. rosea*
> DD. Caps. cylindric, sessile, not sterile in lower part; seeds in 1 row in a
> locule; fl.-tube 2–4 cm. long.
> E.   Plants annual or surviving longer, from a deep taproot; basal lvs.
> ± rhombic, the blades 2–10 cm. long; caps. usually woody, with
> exfoliating epidermis. .... 8. *Oe. deltoides*

EE. Plants perennial, largely from running underground rootstocks; basal lvs. tending to be smaller and more narrow; caps. usually not woody.

F. Plants greenish and subglabrous to strigose; caps. 2.5 mm. thick at base. S. Calif., mostly west of the deserts. 9. *Oe. californica*

FF. Plants canescent to hoary, usually with some spreading hairs especially in the upper parts.

G. Stems 4–8 dm. long; lvs. runcinate-pinnatifid, 3–12 cm. long; free sepal-tips 1–3 mm. long. Sand dunes about Antioch. . . . . 8. *Oe. deltoides Howellii*

GG. Stems 1–5 dm. long; lvs. subentire to deeply sinuate-dentate, 1–6 cm. long; free sepal-tips lacking or 1 mm. long. Deserts. . . . . 9a. *Oe. avita*

BB. Caps. crested etc., as in the FLORA.

p. 945. In Key, AA, BB, CC, D, E, FF, GG, delete H and HH. and end GG. in 27. *Oe. Boothii.*

In Key, near bottom of page, keep F, G and GG, omitting H and HH and change to:

H. Fls. small, the petals 2–5 mm. long.

I. The plants low, commonly less than 15 cm. tall, with stems less than 1 mm. in diam. and glabrous or finely pubescent; lvs. less than 2 mm. wide.

J. Sepals 1.5–2 mm. long; petals 1.5–2.5 mm. long. Mojave Desert to e. Wash.

29. *Oe. contorta*

JJ. Sepals 2–3 mm. long; petals 3.5–5 mm. long. Central Valley and adjacent areas to the west. . . . . 29a. *Oe. cruciata*

II. The plants taller or coarser, with stems more than 1 mm. in diam.

J. Lvs. rather broad, the principal 2–5 mm. broad, often coarsely and sharply serrate; petals 3–4 mm. long; plant with an abundant spreading pubescence. E. middle Calif.

29b. *E. pubens*

JJ. Lvs. mostly 1–2 mm. broad, scarcely or not toothed; petals 2–3 mm. long; plant subglabrous to strigulose or finely spreading-pubescent. W. of the Cordillera, from L. Calif. to s. Ore. . . . . 29c. *Oe. dentata*

HH. Fls. larger, the petals usually 5–14 mm. long.

I. Caps. sessile; petals mostly 5–8 mm. long; lvs. narrow, mostly 1–2.5 mm. broad. Contra Costa and Butte cos. to Mojave Desert and s. Calif.

30a. *Oe. campestris*

II. Caps. tending to be pediceled and often somewhat clavate, the fls. larger, with petals 10–14 mm. long; lvs. mostly wider, denticulate. Mojave Desert and borders. . . . . 31. *Oe. kernensis*

p. 946. In Key, change G and GG at top of page to:

G. Mature caps. oblong-pyramidal, 12–15 mm. long, almost straight. San Clemente Id. . . . . . . 32. *Oe. guadalupensis*

GG. Mature caps. curved or contorted, 15–40 mm. long.

H. Plants pallid with a closely appressed pubescence, mostly prostrate or nearly so. Deserts.

I. Sepals 3–4 mm. long; petals 3–6 mm. long. From Inyo and e. San Diego cos. to Ariz.

33a. *Oe. Abramsii*

II. Sepals 7–9 mm. long; petals 8–12 mm. long. W. edge of Colo. Desert. . . . . . 34a. *Oe. Hallii*

HH. Plants green, not pallid (or if so, growing on Coastal Strand), the pubescence largely spreading, sometimes almost lacking.

I. Fls. small, the petals 1.5–7 mm. long.

J. Petals 2–4 mm. long; stems prostrate to ascending or erect, not reddish; cauline lvs. sessile; foliage hirsutulous or villous. Mendocino and Glenn cos. to L. Calif.

K. Stems semiprostrate; cauline lvs. oblong-

lanceolate, obtuse, sessile but not
clasping. . . . . 33. *Oe. micrantha*
KK. Stems erect or ascending; cauline lvs.
oblong-ovate to broadly ovate, acute
with subcordate clasping base.
33a. *Oe. hirtella*
JJ. Petals 5–7 mm. long; stems nearly erect,
slender, reddish, subglabrous or nearly so.
Central Calif. to L. Calif. 33c. *Oe. ignota*
II. Fls. larger, the petals mostly 8–22 mm. long.
J. Plants of sea-bluffs and inland, greenish;
greenish cauline lvs. lanceolate to lance-
ovate, acute, wavy-margined, thin.
34. *Oe. bistorta*
JJ. Plants of sea-beaches, usually grayish to
silvery; cauline lvs. lance-oblong to orbicular-ovate, obtuse, not wavy-margined, thick
in texture. . . . . 35. *Oe. cheiranthifolia*

In Key, DD, EE, change to:

F. Lvs. all cauline, well distributed, simple, mostly cordate-orbicu-
lar; caps. short-pediceled, coarse-cylindrical; pollen shed in
tetrads.
G. Fl.-tube 5–14 mm. long; style 8–23 mm. long.
37. *Oe. cardiophylla*
GG. Fl.-tube 18–40 mm. long; style 30–58 mm. long.
37a. *Oe. arenaria*
FF. Lvs. mostly basal, often pinnatifid, largely ovate, oblong or
lanceolate; caps. usually prominently pediceled; pollen grains
shed individually.
G. Caps. linear, usually over 2 cm. long and less than 2 mm.
in diam.
H. Stigma surrounded by anthers at anthesis; petals less
than 6 mm. long; style less than 6 mm. long; infl.
erect in bud. . . . . 41. *Oe. Walkeri*
HH. Stigma elevated above anthers at maturity; petals
usually more than 6 mm. long; style over 6 mm. long.
38. *Oe. brevipes*
GG. Caps. distinctly clavate, usually more than 2 mm. in diam.
H. Corolla lavender in anthesis; flowering mostly in late
summer and fall. . . . . 42. *Oe. heterochroma*
HH. Corolla yellow or white at anthesis; flowering mostly
in spring and early summer.
I. Mature pedicels and caps. ascending or spreading; corolla yellow or white. Widespread on des-
erts. 39. *Oe. claviformis*
II. Mature pedicels and caps. sharply deflexed;
corollas bright yellow. Region s. of Death Valley.
39a. *Oe. Munzii*

p. 948.   5. **Oe. strícta** Ledeb. ex Link. Now reported from Santa Barbara, Monterey
and Santa Cruz cos.

7a. **Oe. ròsea** L'Hérit. ex Ait. [*Hartmannia r.* G. Don.] Perennial, blooming
the first year, from a somewhat woody caudex, freely branched, to 1 m. or more
tall, ± strigulose throughout; lvs. not crowded, the lower oblanceolate or wider,
subentire to pinnatifid, 2–5 cm. long, the cauline gradually reduced up the stem;
fls. in simple slender erect bracteate racemes; fl.-tube 4–8 mm. long; sepals
5–8 mm. long; petals rose to red-violet, 5–10 mm. long; caps. proper obovoid,
8–10 mm. long, 3–4 mm. thick, with 4 wings to ca. 1 mm. wide, caps. passing
at base into a hollow ribbed pedicel 5–20 mm. long; $n = 7$.—Texas to S. Am.
Increasingly found as an adventive in Calif.

p. 949.   8. **Oe. deltoìdes** Torr. & Frém. The varieties *cognata*, *Piperi*, *Howellii* have
been raised to subspecies, with W. Klein as the authority. Var. *cinerácea* (Jeps.)
Munz is reduced to synonymy under *Oe. deltoides*.

Ssp. *eurekénsis* becomes 9a. **Oe. ávita** W. Klein ssp. *eurekénsis* (Munz) Klein.

9. **Oe. califórnica** Wats. is retained for the tetraploid plants ($n = 14$) of cismontane s. Calif. and

9a. **Oe. ávita** W. Klein should be used for the desert plants of e. Calif. and
w. Ariz. with $n = 7$. These tend to be shaggier, with more long spreading hairs
especially in the upper parts.

p. 950.   10. *Oe. caespitòsa* Nutt. var. *longiflora* (Heller) Munz can be reduced to synonymy under **Oe. caespitòsa** var. **marginàta** (Nutt.) Munz. *n* = 7 (Kurabayashi et al., 1962).

     12. **Oe. primivèris** Gray can be keyed out as follows:

> Plants practically acaulescent; fl.-tube 3–5 cm. long. Deserts, mostly above 1500 ft.
> Petals 10–22 mm. long; sepals 10–20 mm. long; caps. 17–30 mm. long. St. George, Utah to extreme e. San Bernardino Co., then se. to Texas, Son. . . . . *Oe. primivèris*
> Petals 25–40 mm. long; sepals 20–25 mm. long; caps. 25–50 mm. long. Sw. Nev. and Inyo Co., Calif. to Riverside Co. . . . . Ssp. *bufonis*
> Plants with stems 1–4 dm. long; fl.-tube 4–6 cm. long; petals mostly 3–4 cm. long. From below 900 ft., Imperial Co., Calif. to Yuma and Pima cos., Ariz. . . . . Ssp. *caulescens*

Synonymy is as follows:

> **Oe. primivèris** Gray. [*Lavauxia p.* Small. *Oe. Johnsoni* Parry.]
> **Oe primivèris** Gray ssp. **bufònis** (M. E. Jones) Munz. [*Oe. b.* Jones. *Lavauxia lobata* A. Nels.]
> **Oe. primivèris** ssp. **caulèscens** (Munz) Munz. [*Oe. p.* var. *c.* Munz.]

     15. **Oe. leptocárpa** Greene. Add as synonym *Camissonia californica* Raven.

p. 951.   16. **Oe. Pálmeri** Wats. Add as synonym *Camissonia P.* Raven.

     17. **Oe. graciliflòra** H. & A. Add as synonym *Camissonia g.* Raven.

     18. *Oe. heteràntha* Nutt. Change name to **Oe. subacaúlis** (Pursh) Garrett with synonyms: *Jussiaea s.* Pursh. *Camissonia s.* Raven. *Oe. heterantha* Nutt.

     19. **Oe. ovàta** Nutt. in T. & G. Additional synonym *Camissonia o.* Raven.

     20. **Oe. tanacetifòlia** T. & G. Additional synonym *Camissonia t.* Raven.

p. 952.   21. **Oe. breviflòra** T. & G. Additional synonym *Camissonia b.* Raven.

     22. **Oe. refrácta** Wats. Additional synonym *Camissonia r.* Raven.

     23. **Oe. chamaenerioìdes** Gray. Additional synonym *Camissonia c.* Raven.

     24. **Oe. mìnor** (A. Nels.) Munz. Additional synonym *Camissonia m.* Raven.

     25., 26., and 27. combined under **Oe. Boòthii** for which the following key is presented:

> A. Mature caps. merely curved or bent, not distinctly coiled or contorted, subfusiform or linear in shape.
>   B. Lvs. well distributed, glandular-pubescent to -villous; stem epidermis exfoliating tardily if at all; caps. 10–15 mm. long.
>   C. Stems largely 1.5–4 dm. tall, the central one usually more prominent than the lateral; basal lvs. ovate to lance-ovate; fl.-tube 4–8 mm. long; petals 3.5–9 mm. long. Northeast and east of Calif. . . . . 25. *Oe. Boothii*
>   CC. Stems largely 1–2 dm. tall, the branches sometimes as prominent as the central stem; basal lvs. lanceolate to lance-ovate; fl.-tube 3–5 mm. long; petals 4–4.5 mm. long. Mono and Inyo cos. and adjacent Nev. . . . . Ssp. *intermedia*
>   BB. Lvs. largely near the base of the plant, subglabrous to strigulose, lance-ovate to oblanceolate; stem epidermis exfoliating promptly; caps. 15–25 mm. long.
>   C. The caps. not more than 2 mm. thick at base, not conspicuously quadrangular or thickened and indurated at the angles, scarcely woody; plant slender, 2–5 dm. tall.
>   D. Caps. with simple curve about ⅓ the way from the base, so that the tip spreads away from the stem axis.
>   E. Base of mature caps. ca. 2 mm. thick, the body curved and with spreading tips.
>   F. Exfoliating epidermis of stem straw- or flesh-color; petals 4.5–5 mm. long, white except possibly when aging. Monterey and Stanislaus cos. to Kern and Los Angeles cos. . . Ssp. *decorticans*
>   FF. Exfoliating epidermis of stems white to reddish; petals 3.5–4 mm. long, red. Mts. about the w. end of the Mojave Desert. Ssp. *rutila*
>   EE. Base of mature caps. ca. 1 mm. thick, the body straight or curved; petals 3–3.5 mm. long. Inyo Mts., Inyo Co. . . . . Ssp. *inyoensis*
>   DD. Caps. often contorted so that the tip points down; base of caps. 1–1.5 mm. thick; epidermis of stems white. W. Mojave Desert. Ssp. *desertorum*
>   CC. The caps. 2.5–3 mm. thick at base, conspicuously quadrangular and much thickened and indurated at the angles, quite woody; plants low and coarse, rarely more than 2 dm. high. Deserts from e. Inyo Co. to Imperial and San Diego cos., thence to Utah, Ariz. . . . . Ssp. *condensata*
> A. Mature caps. usually distinctly coiled or contorted, not merely bent and curved, not subfusiform in shape; lateral stems often prominent; plants finely pubescent to short-villous. E. Ore. to San Bernardino Co. and Ariz. and Utah. . . . . Ssp. *alyssoides*

pp. 952.–   27. Synonymy for the sspp. of **Oe. Boòthii** Dougl. ex Lehm. in Hook.
953.   [*Sphaerostigma B.* Walp. *Camissonia B.* Raven. *Sphaerostigma senex* A. Nels.
S. Lemmonii A. Nels.]
    Ssp. **intermèdia** Munz.
    Ssp. **decórticans** (H. & A.) Munz. [*Gaura d.* H. & A. *Oe. d.* Greene. *Sphae-rostigma d.* Small. *Camissonia Boothii* ssp. *d.* Raven.]
    Ssp. **rùtila** (Davidson) Munz. [*Oe. r.* Davidson. *Oe. decorticans* var. *r.* Munz.]
    Ssp. **inyoénsis** Munz.
    Ssp. **desertòrum** (Munz) Munz. [*Oe. decorticans* var. *d.* Munz. *Camissonia Boothii* ssp. *desertorum* Raven.]
    Ssp. **condensàta** (Munz) Munz. [*Oe. decorticans* var. *c.* Munz. *Camissonia Boothii* ssp. *c.* Raven.]
    Ssp. **alyssoìdes** (H. & A.) Munz. [*Oe. a.* H. & A. *Oe. a.* var. *villosa* Wats. *Camissonia Boothii* ssp. *a.* Raven.]
    28. **Oe. andìna** Nutt. Add to synonymy *Camissonia a.* Raven.
    29. **Oe. contórta** Dougl. ex Hook. Use following key:

Caps. sessile, curved or straight, 25–35 mm. long, ending in a definite beak; *n* = 7. Siski-you and Lassen cos. to B.C., Ida. . . . . *Oe. contorta*
Caps. definitely pediceled, not attenuate into a beak, 17–25 (–30) mm. long, frequently curved into a half circle; *n* = 7. Mojave Desert to Wyo., Utah, e. Wash. . . . . . var. *flexuosa*

p. 954.   Add to synonymy under Var. **flexuòsa**: *Camissonia parvula* (Nutt.) Raven.
    29a. **Oe. cruciàta** (Wats.) Munz. [*Oe. dentata* var. *c.* Wats. *Oe. campestris* var. *c.* Greene. *Sphaerostigma campestre* var. *minus* Small.] Bushy, slender-stemmed annual, 5–15 cm. tall, subglabrous or with some spreading hairs, rather leafy; lvs. linear to oblong-linear, 10–25 mm. long, 1–2.5 mm. wide, subentire to denticulate, with appressed or spreading hairs; fls. few, solitary in axils of foliose bracts; fl.-tube 1–2 mm. long; sepals 2–3 mm. long; petals 3.5–5 mm. long, yellow, aging orange; caps. linear, 1.5–2.5 cm. long, sessile, short-beaked, straight or somewhat contorted.—Open places below 3500 ft., Central Valley from Butte Co. to Kern Co. and w. to Lake Co.
    29b. **Oe. pùbens** (Wats.) Munz. *n* = 14. Use instead of *Oe. contorta* var. *pubens* (Wats.) Cov.
    29c. Instead of *Oe. contorta* var. *epilobioides* (Greene) Munz use **Oe. den-tàta** Cav. [*Sphaerostigma d.* Walp. *Camissonia d.* Reiche. *Oe. contorta* vars. *strigulosa* and *epilobioides* Munz.] Much like *Oe. contorta*, the stems commonly 1.5–3 dm. tall, subglabrous to strigulose or spreading-pubescent, mostly not over 1 mm. in diam., freely branched, often glandular in the infl.; lvs. well distributed, linear, 1–2 mm. wide, 2–3 cm. long; caps. 1.5–2.5 cm. long, sessile, beaked or not; *n* = 14, 21.—Dry disturbed places mostly below 5000 ft., L. Calif. to s. Ore., w. of the Cordillera; Chile.
    30. **Oe. campéstris** Greene. [*Oe. dentata* var. *c.* Jeps. *Camissonia c.* Raven.]— From Butte and Contra Costa cos. s. to Kern and Santa Barbara cos.
    Ssp. **Paríshii** (Abrams) Munz. [*Oe. dentata* var. *P.* Munz.] Stem subglabrous or with short appressed hairs; petals 5–8 mm. long; *n* = 7.—Santa Barbara to Los Angeles Co. and from San Bernardino to w. Riverside Co.
    Oe. dentata var. *Johnstonii* Munz is an uncertain quantity and approaches Var. *Gilmanii* which is now transferred to
p. 955.   31. **Oe. kernénsis** Munz. This is keyed as follows:

A.   Plants canescent with short spreading nonglandular hairs in the lower parts; caps. often pediceled, cylindric-clavate, not beaked. Walker Pass region, Kern Co. . . Ssp. *kernensis*
AA.   Plants finely glandular-pubescent almost throughout or somewhat strigulose below; caps. sessile or pediceled, linear, ± beaked.
    B.   Central stem tending to be the most prominent; plants usually strongly glandular throughout; lvs. plane, mostly ca. 2–3 mm. wide. S. Death Valley region, Inyo Co.
                                                                                    Ssp. *Gilmanii*
    BB.   Central stem not more prominent than the principal branches; plant weakly glandu-lar; lvs. ± crisped on margins, often 3–4 mm. wide. Mojave Desert from Pilot Knob to Kelso. . . . . Ssp. *mojavensis*

Synonymy as follows:
Oe. kernénsis Munz. [*Camissonia k.* Raven.] $n = 7$.
Oe. kernénsis ssp. Gilmánii (Munz) Munz. [*Oe. dentata* var. *G.* Munz.] $n = 7$.
Oe. kernénsis Munz ssp. mojavénsis Munz. $n = 7$.

32. **Oe. guadalupénsis** Wats. ssp. **clementina** Raven. [*Camissonia g.* ssp. *c.* Raven.] Plants from San Clemente Id. differ from the Guadalupe Id. ones in being more pubescent.

33. **Oe. micrántha** Hornem. Additional synonym *Camissonia m.* Raven.
33a. **Oe. Abrámsii** Macbr. [*Oe. micrantha* var. *exfoliata* (A. Nels.) Munz. *Camissonia pallida* Raven.]
33b. **Oe. hirtélla** Greene. [*Oe. micrantha* var. *h.* Jeps. *Oe. m.* var. *Jonesii* Munz. *Camissonia h.* Raven. *Oe. m.* var. *Reedii* Jeps.] $n = 7$.
33c. **Oe. ignòta** (Jeps.) Munz. [*Oe. micrantha* var. *i.* Jeps. *Oe. hirta* var. *i.* Munz. *Camissonia i.* Raven.]

p. 956.   34. **Oe. bistórta** Nutt. ex T. & G. Add to synonymy *Camissonia b.* Raven.
34a. **Oe. Hállii** (Davidson) Munz. [*Sphaerostigma H.* Davidson. *Oe. bistorta* var. *H.* Jeps. *Camissonia H.* Raven.]

35. **Oe. cheiranthifòlia** Hornem. ex Spreng. ssp. **suffruticòsa** (Wats.) Munz instead of var. *s.* Wats. Add to synonyms *Camissonia c.* ssp. *s.* Raven. $n = 7$ (Kurabayashi et al., 1962).

p. 957.   37. **Oe cardiophýlla** Torr. Add synonym *Camissonia c.* Raven. Plants rather slender; pubescence mostly villous.—San Bernardino Co. southward and eastward.
Ssp. **robústa** Raven. [*Camissonia cardiophylla r.* Raven.] Coarse, the pubescence mostly glandular.—Inyo Co.
37a. **Oe. arenària** (A. Nels.) Raven. [*Chylismia a.* A. Nels. *Oe. cardiophylla* var. *splendens* Munz & Jtn. *Oe. c.* var. *longituba* Jeps. *Camissonia arenaria* Raven.] Fl.-tube 18–40 mm. long; style 30–58 mm. long.—From Riverside Co. to sw. Ariz. and adjacent Son.

38. **Oe. brévipes** Gray. Insert the following key:

Buds not individually pendulous before opening; petals not fading red, usually more than 6 mm. long.
    Plants villous, stout-stemmed; sepals villous as well as glandular and with subapical free tips; caps. usually 2–3 mm. in diam. Deserts of se. Calif. to sw. Utah and Ariz.   *Oe. brevipes*
    Plants strigose, rather villous below; sepals glandular-pubescent to canescent, without free caudate tips; caps. mostly 1–1.5 mm. in diam. Inyo and Riverside cos. to Ariz. and Utah.
                            ssp. *pallidula*
Buds individually pendulous before opening; petals often fading red, less than 6 mm. long.
Imperial Valley to Yuma Co., Ariz. .... ssp. *arizonica*

Add following synonymy.
Oe. brévipes Gray. [*Camissonia b.* Raven. *Oe. divaricata* Greene.]
Ssp. pallídula (Munz) Raven. [*Oe. b.* var. *p.* Munz. *Oe. p.* Munz. *Camissonia brevipes* ssp. *p.* Raven.]
Ssp. arizónica Raven. [*Camissonia brevipes* ssp. *a.* Raven.]

39. **Oe. clavifórmis** Torr. & Frém. Use the following key:

A. Lower parts of plant villous with spreading pubescence; sepals with free caudate tips arising below the apices; petals usually yellow and not changing color when fading. San Diego and Imperial cos. .... ssp. *Peirsonii*
AA. Lower parts of plant strigose or glabrous, but not with spreading hairs; sepals with or without free tips.
    B. Petals white, the fl.-tube orange-brown.
        C. Plants variously pubescent on stems and in infl.
            D. Lateral lfts. reduced in number, the lvs. nearly simple; basal rosette compact. E. central Calif. to s. Ore. .... ssp. *integrior*
           DD. Lateral lfts. generally well developed and numerous, the basal rosette not compact.
               E. Sepals often with free subapical tips; terminal lfts. often large and nearly cordate; buds and infl. often silky-strigose. Death Valley region. .... ssp. *funerea*
               EE. Sepals usually lacking free tips; terminal lfts. usually inconspicuous; buds and lfts. not silky. Well distributed. .... ssp. *aurantiaca*

CC. Plants usually glabrous in infl. and on buds; lateral lfts. usually well developed. W. deserts of Calif. . . . . *Oe. claviformis*
BB. Petals yellow, the fl.-tube yellow or orange-brown.
C. Plants strigose above and in the infl.; sepals sometimes with free tips. Se. Imperial Co. to Son. . . . . ssp. *yumae*
CC. Plants usually glabrous above; sepals without free tips.
D. Lvs. lanceolate, narrow, evenly dentate, often almost without lateral lfts.; fl.-tube dark. Mono and Inyo cos. . . . . ssp. *lancifolia*
DD. Lvs. pinnate, the terminal lfts. ovate, often blunt, the lateral lfts. often well developed. Ne. Calif. to Ore., Ida. . . . . ssp. *cruciformis*

Add to synonymy of *Oe. claviformis*:

Oe. clavifórmis Torr. & Frém. [*Chylismia scapoidea c.* Small. *C. c.* Heller. *Camissonia c.* Raven.]—Largely of the w. Mojave Desert.

Ssp. aurantiàca (Wats.) Raven. [*Oe. scapoidea* var. *a.* Wats. *Chylismia s.* var. *a.* Wats. ex Davids. & Moxley. *Oe. claviformis* var. *a.* Munz. *Camissonia c.* ssp. *a.* Raven. *Chylismia a.* Johansen.]—Lincoln Co., Nev. to L. Calif.

Ssp. yùmae Raven. [*Camissonia claviformis* ssp. *y.* Raven.]—Se. Imperial Co., to adjacent Ariz. and Son.

Ssp. Peirsònii (Munz) Raven. [*Oe c.* var. *P.* Munz. *Chylismia P.* Johansen. *Camissonia c.* var. *P.* Raven.]—Colo. Desert e. of Salton Sea, to n. L. Calif.

Ssp. intégrior Raven. [*Camissonia c.* ssp. *i.* Raven. *Oe. scapoidea* var. *purpurascens* Wats.]—E. central Calif. to w. Nev., se. Ore.

Ssp. crucifórmis (Kell.) Raven. [*Oe. cruciformis* Kell. *Camissonia claviformis* ssp. *cruciformis* Raven. *Oe. c.* ssp. *citrina* Raven.]—Ne. Calif. to se. Ore., Ida., w. Nev.

Ssp. lancifòlia (Heller) Raven. [*Chylismia l.* Heller. *Camissonia c.* ssp. *l.* Raven.]

39a. **Oe. Múnzii** Raven. [*Camissonia M.* Raven.] Like *Oe. claviformis*, with numerous branches at base and above, ± strigose; lvs. pinnate, with well developed lfts.; fls. yellow; mature pedicels and caps. sharply deflexed.—Region s. of Death V., Inyo Co.

p. 958.
41. Change to *Oe.* Wálkeri (A. Nels.) Raven ssp. tórtilis (Jeps.) Raven. [*Oe. scapoidea* var. *t.* Jeps. *Camissonia W.* ssp. *t.* Raven.] Annual to short-lived perennial, with well developed basal rosette, etc.—E. cent. Calif. to w. Utah and Nev.

42. **Oe. heterochròma** Wats. Add synonym *Camissonia h.* Raven.

Ssp. monoénsis (Munz) Raven. [*Oe. h.* var. *m.* Munz. *Camissonia h.* ssp. *m.* Raven.] $n = 7$ (Kurabayashi et al., 1962).

GAYOPHÝTUM
Cite: Lewis, H. and J. Szeykowski, The genus Gayophytum. Brittonia 16: 343–391. 1964.
Substitute the following key:

A. Plants flowering from near the base, the first fls. at 1–4 nodes above the cotyledons.
B. The plants branched only in their lower portion, secondary branches few or none, the branching not dichotomous.
C. Caps. with dorsal and ventral valves remaining attached to the septum at maturity, the lateral valves free; seeds usually obliquely placed in the caps, and often ca. 15–18 in each locule. . . . . 1. *G. humile*
CC. Caps. with all 4 valves free from the septum at maturity, the seeds usually vertically placed in the caps. and 5–10 in each locule. . . . . 2. *G. racemosum*
BB. The plants branched throughout, the secondary branches evident.
C. Branches mostly separated by 2–8 nodes; seeds in even rows, not staggered; caps. not torulose. . . . . 3. *G. decipiens*
CC. Branches mostly at every node or every other node; seeds often plainly staggered, those in one row alternate with those in the other; caps. often plainly torulose. . . . . 7. *G. diffusum parviflorum*
AA. Plants not flowering near the base, but only several to many nodes above the cotyledons.
B. Caps. irregularly lumpy by failure of part of the ovules to develop; a high percentage of pollen grains empty. Wash. to Nev. and Calif. . . . . 4. *G. heterozygum*
BB. Caps. ± regular in outline, entire or torulose, all of the ovules developing; pollen grains almost all good.
C. Petals 0.5–1 mm. long; caps. 2–5 mm. long, mostly shorter than the pedicels. Wash. to cent. Calif., Mont., Colo. . . . . 5. *G. ramosissimum*

CC. Petals mostly 1.5–7 mm. long; caps. 3–12 mm. long, equaling or longer than the pedicels.
    D.   Seeds 1–5 (–6) in a caps., each ca. 1.5 mm. long; caps. ca. as long as pedicels; petals mostly 1.5–2 mm. long. Mts. of s. Calif.
                                            6. *G. oligospermum*
   DD. Seeds usually 6–10 or more in a caps.; each ca. 1–1.2 mm. long; petals mostly 1.5–7 mm. long.
    E.   Petals generally 1.5–2.5 mm. long; style not surpassing stamens. B.C. to Mont., S. Dak., New Mex., L. Calif.
                                    7. *G. diffusum parviflorum*
    EE. Petals 3–7 mm. long; style surpassing stamens.
      F.   The petals largely 3–4.5 mm. long; sepals 2–3 mm. long. Wash. to Plumas Co., Calif. and to e. Mont. and Wyo., also in San Bernardino Mts. .... 7. *G. diffusum*
      FF. The petals 4–7 mm. long; sepals 3–4 mm. long. Sierra Nevada and Greenhorn Mts. .... 8. *G. eriospermum*

1. **G. hùmile** Juss. [*G. pumilum* Wats. *G. Nuttallii* T. & G. as to lectotype.] Use description on p. 959. Change range as s. to the San Bernardino Mts.

2. **G. racemòsum** T. & G. [*G. caesium* T. & G. *G. rasemosum* var. *c.* Munz. *G. Helleri* Rydb. *G. ramosissimum* var. *pygmeum* Jeps. *G. Helleri* var. *glabrum* Munz. *G. humile* var. *hirtellum* Munz.] Use description on p. 959, changing range as from Ventura Co. n.

3. **G. decípiens** Lewis & Szweykowski. Plants 1–3 dm. tall, simple or branched at base and ± throughout, the branches mostly separated by 2–8 nodes, subglabrous throughout, or finely strigulose especially above, or occasionally with short spreading hairs; lvs. well distributed, the lower 2–3 cm. long, the uppermost 0.5–1.2 cm. long and bractlike; fls. from near the base of the plant or beginning higher; sepals ca. 1 mm. long; petals mostly 1–1.2 mm. long; lower stamens ca. ⅔ the length of the petals; pistil slightly shorter than the petals; caps. erect or recurved, somewhat flattened, not strongly constricted between the seeds; seeds evenly spaced, not staggered, usually more than 10 in a caps.; $n = 7$.—Occasional in dry sandy and gravelly places, Sagebrush Scrub and Pine F., San Bernardino and San Gabriel mts., through the Sierra Nevada and desert ranges to Wash., Ida., Utah, n. Ariz.

4. **G. heterozýgum** Lewis and Szweykowski. [*G. diffusum* var. *villosum* Munz.] Mostly 1.5–5 dm. tall, freely and repeatedly branched, usually in the upper half, the successive branches often at succeeding internodes, the central axis more prominent than the lateral branches, often ± zigzag, subglabrous throughout or strigulose or spreading-pubescent especially toward the tips; lower lvs. to 6 or 7 cm. long and 4–6 mm. wide; cauline half as long, the uppermost are linear bracts; pedicels mostly 2–7 mm. long; sepals largely 1.5–2.5 mm. long; petals 2–4 mm. long; longer stamens ca. equaling the petals, with many of the pollen grains empty; caps. 5–10 mm. long, irregularly lumpy; seeds 1–1.5 mm. long; $n = 7$.—Local in Chaparral and Montane F., San Jacinto Mts. n. through the Coast Ranges and Sierra Nevada; to Wash., w. Nev.

5. **G. ramosíssimum** T. & G. Use description on p. 958. S. to Inyo Co.

6. **G. oligospérmum** Lewis and Szweykowski. Plants 2–8 dm. tall, repeatedly branched in upper parts, sometimes also below, erect, with the principal axis ± zigzag; lower lvs. early deciduous, cauline linear, 2–5 cm. long; pedicels filiform, 3–6 mm. long, arising only in the most distal axils, ascending to reflexed; sepals often strigulose, 1–1.5 mm. long; petals 1.5–2 mm. long; longer stamens ca. ⅔ as long as petals; style slightly exceeding stamens; caps. ca. equaling pedicels, mostly strigulose, conspicuously constricted between the seeds; seeds 1–5 (–6), those in one locule alternating with those in the other, ca. 1.5 mm. long; $n = 7$.—Dry ridges and slopes, 4500–7500 ft., Montane Coniferous F., San Gabriel and San Bernardino mts. to Cuyamaca Mts.

7. **G. diffùsum** T. & G. Erect, simple or usually branched and mostly above and repeatedly, largely at successive nodes, 1–5 or more dm. tall, strigulose or with spreading hairs above; lower lvs. lance-linear to linear, petioled, largely 2–4 cm. long, the cauline gradually reduced up the stem; pedicels 3–8 mm. long, ascending to erect; sepals 2–3 mm. long; petals 3–4 (–5) mm. long;

longer stamens ca. equaling petals; style somewhat surpassing longer stamens; caps. 5–12 mm. long, glabrous or pubescent; seeds 1–1.2 mm. long; $n = 14$.— Coniferous F., Plumas Co. to Wash., Mont., Wyo.; also in San Bernardino Mts. Ssp. **parviflòrum** Lewis & Szweykowski. [*G. Nuttallii* auth., not T. & G. *G. ramosissimum* var. *strictipes* Hook. *G. lasiospermum* var. *Hoffmannii* Munz. *G. Helleri* var. *rosulatum* Jeps. *G. Nuttallii* vars. *Abramsii* & *intermedium* Munz. *G. i.* Rydb.] Pedicels 1.5–5 mm. long, erect to spreading or deflexed; petals 1–2.5 mm. long; longer stamens not surpassing petals; style nearly as long as petals; caps. with or sometimes without constrictions between seeds, largely 4–12 mm. long; seeds mostly 3–10, often staggered, those in one locule alternating with those in the other, ca. 1 mm. long; $n = 14$.—Montane Coniferous F., B.C. to L. Calif., New Mex., S. Dak.

　　　8. **G. eriospérmum** Cov. [*G. lasiospermum* var. *e.* Munz.] Closely resembling *G. diffusum* as to stature, habit and foliage; fls. larger; sepals 3–4 mm. long; petals 4–7 mm. long; $n = 7$.—Coniferous F., Sierra Nevada and Greenhorn Range.

p. 960.　　**GÁURA**

　　　4. **G. odoràta** Ses. ex Lag. In Santa Cruz Co., *Fuller.*

　　　5. **G. coccínea** Pursh. $2n = 14$ (Taylor, 1967).

p. 961.　　**HETEROGÁURA**

　　　1. **H. heterándra** (Torr.) Cov. Also in S. Coast Ranges, from Santa Lucia Mts. and Diablo Range south.

CIRCAÈA

　　　1. **C. alpìna** L. ssp. **pacífica** (Asch. & Magnus) Raven.

HALORAGÀCEAE

Change Key to Genera to:

A.　Plants aquatic, with some submerged lvs.
　　B.　Submerged lvs. pinnatifid; stamens 4 or 8 .... 1. *Myriophyllum*
　　BB.　Submerged lvs. simple, entire; stamen 1. .... 2. *Hippuris*
AA.　Plants terrestrial.
　　B.　Lvs. large, round to subreniform. .... 3. *Gunnera*
　　BB.　Lvs. medium in size, lance-ovate. .... 4. *Haloragis*

p. 962.　　**MYRIOPHÝLLUM**

　　　1. **M. brasiliénse** Camb.—In San Francisco and Marin cos., *Howell.*

　　　4. **M. hippuroìdes** Nutt. ex T. & G.—At 7200 ft. in Fresno Co., *Quibell.*

p. 963.　　　　　　　　　　　　　　　　**3. Gúnnera L.**

Perennial herbs with creeping rhizomes. Lvs. radical, petioled, large, round to subreniform, entire to lobed. Infl. scapose, the upper fls. ♂, the lower ♀, with bisexual between. Calyx-lobes 2–3. Petals 2, hooded, or 0. Ca. 25 spp. of the S. Hemis. (J. E. *Gunner*, 1718–1773, Norwegian botanist).

　　　1. **G. chilénsis** Lam. Rhizome short, thick; petiole large, fleshy; laminae to 2 m. diam., palmately lobed and incised; infl. a large spike to 1 m. tall; fls. apetalous; fr. red.—Escaped in Marin Co., *Howell*; native of Chile.

　　　　　　　　　　　　　　　　4. Haloràgis Forst.

Herbs or shrubs. Lvs. entire to serrate. Fls. small, pedicelled to subsessile. Calyx-lobes 2–4. Petals 2–4, concave. Stamens 4–8. Ovary 2–4-loculed; styles 2–4. A rather large genus of the S. Hemis. (Greek, *halos*, round, and *rhagos*, berry.)

　　　1. **H. erécta** (Murr.) Schindler. [*Cercodia e.* Murr.] Erect low leafy shrub; lvs. opposite, lance-ovate, serrate; fls. small, nodding; petals carinate, ca. 3 mm. long.—Reported from San Francisco as natur. plants.

p. 963.　　**MYRTÀCEAE**

Fr. a dehiscent caps.; calyx-lobes and petals united to form a lid or cap which dehisces transversely. .... 1. *Eucalyptus*
Fr. fleshy, black, sweet; perianth not forming a lid. .... 2. *Eugenia*

EUCALÝPTUS

In Key to Species, delete "3. *E. tereticornis*" at end of last line and add:

Lid 2–4 times longer than the calyx-tube. . . . . 3. *E. tereticornis*
Lid ca. as long as calyx-tube. . . . . 4. *E. camaldulensis*

4. **E. camaldulénsis** Dehnhardt. [*E. rostrata* Schlecht., not Cav.] Tall tree, with smooth deciduous bark; lvs. narrowly lanceolate; fls. 6–12 mm. across, 4–8 together in a stalked umbel; lid conical not beaked; fr. subglobular.—Reported as natur. at Santa Barbara, *Pollard* and El Paso Creek, Kern Co., *Twisselmann.*

2. Eugènia L.

Evergreen trees or shrubs closely related to *Myrtus*. Fls. solitary or clustered in the axils. Calyx 4–5-lobed. Stamens many. Ovary 2–3-loculed with many to few ovules. Fr. a berry crowned by the calyx. A large genus in warmer regions. (Prince *Eugene* of Savoy, 1663–1736, patron of hort.)

1. **E. apiculàta** DC. Shrub or small tree, finely pubescent; lvs. 1–2.5 cm. long, ovate, sharply pointed; fls. white, solitary, ca. 2 cm. wide; stamens crowded in a ring; fr. fleshy, black, sweet.—Natur. in Marin Co., *Howell.* From Chile.

p. 964. CALLÍTRICHE

3. The comb. **C. heterophýlla** Pursh ssp. **Bolánderi** Calder & Taylor has been published.

p. 966. ARISTOLÒCHIA

1. **A. califórnica** Torr. has been collected at Millerton Lake, Madera Co., 22 mi. ne. of Friant, *Barbara Brock.* $2n = 28$ (Gregory, 1956); $2n = 16$ pairs (Raven, 1965).

p. 969. VITÀCEAE

Key to Genera:

Lvs. simple, palmately lobed. . . . . 1. *Vitis*
Lvs. palmately compound. . . . . 2. *Parthenocissus*

p. 970.   ## 2. Parthenocíssus Planch.

Climbers with tendrils, which are often disk-like at their tips. Fr. a small 1–4-seeded berry. Ca. one doz. spp. of N. Am. and Asia. Gr., *virgin-ivy.*)

1. **P. insérta** (Kerner) Fritsch. [*P. vitacea* (Knorr) Hitchc.] A high climbing woody vine with long-petioled palmately compound lvs. and few-branched tendrils; fls. in terminal umbellate clusters; berries almost black, ca. 6 mm. diam. —Reported from lower Kern Canyon, Kern Co., *Twisselmann.* Native in e. N. Am.

94A. Balsaminàceae. JEWEL-WEED FAMILY

Succulent herbs. Lvs. alternate, simple. Fls. irregular, showy or the later fls. small and apetalous. Sepals 3, the 2 lateral small, greenish, the posterior sepal large, petaloid, saccate and spurred. Petals 3 or 5, with 2 cleft into unequal lobes. Stamens 5, short; fils. with scalelike appendages on inner side and ± united; anthers connivent or coherent. Ovary oblong, 5-loculed; style short or obsolete; stigma 5-lobed. Ovules several in each locule. Caps. slender, elastically dehiscent into 5 coiled valves. Two genera and ca. 200 spp., largely of trop. Asia.

1. Impàtiens L.

Characters of the family:

Fls. pale yellow. Humboldt Co. . . . . 1. *I. occidentalis*
Fls. pink-purple. San Bernardino Co. . . . . 2. *I. Balfouri*

1. **I. occidentàlis** Rydb. Annual, ca. 1 m. high; lvs. oval, 2–10 cm. long, coarsely few-toothed (serrate-dentate); infl. 3–5-fld.; lateral sepals 6 mm. long; spur ca. 1.5 cm. long; anterior petal pale yellow, ca. 7 mm. long, 10 mm. wide; caps. 15–20 mm. long.—Apparently once reported from Humboldt Co.; W. Wash. to Alaska.

2. **I. Balfoùrii** Hook. f. Perennial, to 1 m. tall; lvs. 7–12 cm. long, with many sharp recurved teeth; infl. 6–8-fld.; lateral sepals ca. 6–7 mm. long; spur ca. 2 cm. long; anterior petal pink-purple.—Along stream near Mentone, San Bernardino Co., *Roos*; Felton and Santa Cruz, Santa Cruz Co., *Hasse*; Humboldt Co., *Pollard*. Himalayan.

p. 971. CONDÀLIA

Under **Condalia** add as synonyms *Condaliopsis lycioides* (Gray) Suesseng. and *Condaliopsis Parryi* (Torr.) Suesseng.

p. 972. RHÁMNUS

2. **R. crocèa** ssp. **ilicifòlia** C. B. Wolf. $2n = 12$ pairs (Raven, Kyhos & Hill, 1965).

p. 975. CEANÒTHUS

In Key A, BB, CC, D, EE, change "lvs. white" to "fls. white," in the 1959 printing.

p. 976. 1. **C. velutìnus** Dougl. ex Hook. var. *laevigàtus* (Hook) T. & G.; change to var. **Hoòkeri** M. C. Johnston.

p. 977. 3. **C. integérrimus** var. **califórnicus** (Kell.) G. T. Benson. Kern Co., (Greenhorn Pass, Tejon Canyon), *Twisselmann*.

p. 978. 8. **C. incànus** T. & G. occurs in Monterey Co., *Howitt & Howell*.

C. × Van Rensselaeri Roof. [*C. incanus* × *C. thyrsiflorus*.]Ca. 5 m. tall, evergreen; some branches sometimes streaked with red-brown; lvs. prominently 3-veined from base, dark green, dull, lighter underneath, 1–3.5 cm. long, 1–2 cm. wide; fls. white, in panicles.—Sandy slope near Lake Merced, San Francisco.

11. **C. oligánthus** Nutt. in T. & G. is reported from Monterey Co., *Howitt & Howell*.

p. 980. 19a. **C. Hearstiòrum** Hoov. & Roof. Proposed for small colonies of plants from near Arroyo de la Cruz, Hearst Ranch, nw. San Luis Obispo Co. Reported as near to *C. dentatus,* but prostrate and matted; branchlets villous when young; lvs. oblong, green and glabrous above, white-tomentose beneath, glandular-dentate, 10–17 mm. long, on petioles 1–2 mm. long; infl. 5–9 mm. long; fls. deep vivid blue; caps. ca. 4 mm. diam.

p. 981. 25. The combination *C. megacarpus* ssp. *insularis* has been made by Raven.

p. 983. 33. **C. rígidus**; add:

Var. **álbus** Roof. Procumbent, with low arcuate branches; lvs. bright green; fls. white.—Yankee Point, Carmel Highlands, Monterey Co.

p. 984. 41. **C. pùmilus** Greene. Plants taken in Eldorado Co. at 2000 ft. elev., on n. slope of Pine Hill, 3–4 mi. east-northeast of Rescue, by *Thorne*, have lvs. like *C. pumilus* in size and shining upper surface, but with broader blunter teeth.

p. 985. SIMMÓNDSIA

1. **S. chinénsis** C. K. Schneid. $2n = 26$ pairs (Raven, Kyhos & Hill, 1965).

p. 986. STAPHYLÈA

1. **S. Bolánderi** Gray. $2n = 13$ pairs (Raven, Kyhos & Hill, 1965).

p. 987. DÍRCA

1. **D. occidentàlis** Gray. $2n = 38$ (Stebbins & Major, 1965).

ELAEAGNÀCEAE should be OLEASTER FAMILY in 1959 printing.

p. 988. COMÁNDRA

1. **C. pállida** DC. Piehl (Mem. Torrey Bot. Club 22: 65. 1965) uses C. umbellàta (L.) Nutt. ssp. califórnica (Rydb.) Piehl. [*C. californica* Eastw. ex Rydb. *C. nudiflora* A. Davids.] Ranging from Vancouver Id. through Wash., w. Ore., to Kern Co., Calif. and in Ariz. The name *pallida* is restricted as C. umbellàta ssp. pallida (A. DC.) Piehl, to n. Ariz. and the n. Rocky Mts. and Cascades. It has grayer lvs. than ssp. califórnica.

p. 989.   LORANTHÀCEAE
          In Key to Genera add third line:

Berry subsessile, rounded; ♀ sepals 4; anthers several-celled. . . . .   3. *Viscum*

          ARCEUTHÒBIUM
          In Key to Species, for hosts add *Tsuga* for *A. campylopodum.*
p. 990.   1. **A. americànum** Nutt. ex Engelm. in Gray. *n* = 14 (Wiens, 1964).—Rang-
          ing from Tulare Co. n., *Kujet.*
           2. **A. Douglásii** Engelm.—From Siskiyou and Shasta cos.
           3. **A. campylopòdum** Engelm. in Gray. *n* = 14 (Wiens, 1964).
          PHORADÉNDRON
          (Wiens, D. Revision of acatophyllous species of Phoradendron. Brittonia 16:
          11–54. 1964.)
           1. **P. califórnicum** Nutt. *n* = 14 (Wiens, 1964).
           2. **P. juniperìnum** Engelm. [*P. ligatum* Trel.] Internodes usually less than 1
          cm. long; plant usually erect; *n* = 14. Parasitic on *Juniperus.*
p. 991.   Ssp. **Libocèdri** (Engelm.) Wiens. [*P. j.* var. *L.* Engelm. *P. L.* Howell.] Inter-
          nodes usually over 1 cm. long; plant often pendulous in age; *n* = 14. Parasitic
          on *Libocedrus,* now *Calocedrus.*
           3. Change **P. Bolleànum** (Seem.) Eichler var. *densum* (Torr.) Fosb. to ssp.
          dénsum (Torr.) Wiens, with *n* = 14, 27 and
           Var. *pauciflòrum* (Torr.) Fosb. to ssp. **pauciflòrum** (Torr.) Wiens, with
          n = 14.
           4. Change *P. flavescens* (Pursh) Nutt. var. *macrophyllum* Engelm. to **P. to-
          mentòsum** (DC.) Engelm. ex Gray ssp. **macrophýllum** (Engelm.) Wiens, with
          n = 14, and change
           Var. *villòsum* (Nutt.) Engelm. to **P. villòsum** (Nutt.) Nutt., with *n* = 14.

                                      3. Víscum L.

           Fls. unisexual. Sepals much reduced; petals sepaloid, usually 4. Stamens ses-
          sile, opening by pores. Berries white, viscous. Genus of Old World. (Latin,
          mistletoe.)
           1. **V. álbum** L. Woody evergreen; lvs. narrowly obovate, leathery, 5–8 cm.
          long; fls. 3–5, subsessile; berry 1 cm. diam.—Reported as on apple trees and
          maple, ca. 1 mi. n. of Sebastopol, Sonoma Co., *Howell.* Native, Eurasia.
p. 992.   CNEORÍDIUM
           1. **C. dumòsum** Hook. f. 2*n* = 18 pairs (Raven, Kyhos & Hill, 1965).
p. 994.   AÉSCULUS
           1. **A. califórnica** Nutt. *n* = 20 (Ornduff & Lloyd, 1965).
p. 996.   ÀCER
           3. **A. macrophýllum** Pursh. 2*n* = 26 (Wright, 1957).
p. 998.   RHÙS
           2. **R. diversilòba** T. & G. 2*n* = 15 pairs (Raven, Kyhos & Hill, 1965). Add to
          synonymy: *Toxicodendron radicans* L. ssp. *diversiloba* (T. & G.) Thorne.
p. 1000.   Fig. 99 illustrates *Angelica* sp., not *Sphenosciadium capitellatum.*
p. 1003. HYDROCÓTYLE
           H. sibthorpioìdes Lam. Native of Asia, Afr., not of Eu.
p. 1004. BÒWLESIA
           After the generic description insert: Mathias, M. E. & L. Constance. A revi-
          sion of the genus Bowlesia Ruiz & Pavon and its relatives. Univ. Calif. Publ.
          Bot. 38: 1–73. 1965.
           1. **B. incàna** R. & P. *n* = 16 (Bell & Constance, 1960).
p. 1005. SANÍCULA
           5a. **S. símulans** Hoov. Proposed for plants from Monterey Co. to Santa Bar-
          bara Co., said to differ from *S. arguta* Greene in the bractlets not being cus-
          pidate, but ovate to oblong, 1.5–2 mm. in maximum width and the fls. a paler
          yellow.

p. 1007.   TÓRILIS
           Use the following key:

A.   Umbels sessile or short-pedunculate, capitate, opposite the lvs. . . . . 1. *T. nodosa*
AA. Umbels usually long-pedunculate, spreading, terminal and lateral.
     B.   Invol. of several bracts, 1 to each ray; bristles incurved-ascending, shorter than
          the width of the fr. . . . . .2. *T. japonica*
     BB. Invol. none or of one bract.
          C.   Plant appressed-hispid throughout; rays 2–10; both carpels uncinate-bristly.
                                                                                   3. *T. arvensis*
          CC. Plant spreading-pubescent; rays 2–3, one carpel with a glochidiate append-
               age. . . . . 4. *T. heterophylla*

           2. **T. japónica** (Houtt.) DC. is probably not in Calif.
           3. **T. arvénsis** (Huds.) Link has little or no invol., but has 2–10 rays and
           both carpels are uncinate-bristly.
           4. **T. heterophýlla** Guss. Annual, spreading-pubescent, 2–6 dm. high; lvs.
           bipinnatisect, the uppermost much different from the lower, undivided or 3-
           parted; umbels long-pedunculate, terminal; rays 2–3; invol. none; fls. white or
           pink; carpels usually dimorphic, usually only one with glochidiate appendage.
           —Reported from Butte, Sonoma and Humboldt cos., *Howell.* From the Medit.
           region.
           ANTHRÍSCUS
           1. Change *A. scandicina* (Weber) Mansf. to **A. neglécta** Boiss. & Reuter
           var. **Scándix** (Scop.) Hylander. [*A. scandicina* Calif. auth.]

p. 1008.   OSMORHÌZA
           3. **O. depauperàta** Phil. $n = 11$ (Bell & Constance, 1960).

p. 1011.   SÌUM
           1. **S. suáve** Walt. $n = 6$ (Bell & Constance, 1960).

p. 1012.   PERIDERÍDIA
           2. Insert "P." before "Kellóggii."—Occurs in Monterey Co., *Howitt &
           Howell.*

p. 1013.   3. **P. Gáirdneri** (H. & A.) Math. $n = 17$ (Bell & Constance, 1960).
           4. **P. oregàna** (Wats.) Math. $n = 18$ (Bell & Constance, 1960).
           5. **P. Paríshii** Nels. & Macbr. $n = 17, 18, 19$ (Bell & Constance, 1960).
           6. **P. Bolánderi** Nels. & Macbr. $n = 18, 19, 20$ (Bell & Constance, 1960).
           Reported from Greenhorn Range, Kern Co., *Twisselmann.*
           7. **P. Prínglei** Nels. & Macbr. Reported from Monterey Co., *Howitt &
           Howell.*

p. 1014.   LIGÚSTICUM
           1. **L. apiifòlium** (Nutt.) Gray. $n = 11$ (Bell & Constance, 1960).
           3. **L. Gràyii** Coult. & Rose. $n = 11$ (Bell & Constance, 1960).

p. 1015.   CICÙTA
           2. **C. Douglásii** Coult. & Rose. $n = 12$ (Bell & Constance, 1960).
           OREONANA
           1. **O. Cleméntis** (Jones) Jeps. Flowers from late May to Aug., *Buckalew.*

p. 1018.   LOMÀTIUM
           Additional reference: Theobald, W. L. Lomatium dasycarpum-mohavense
           complex. Brittonia 18: 1–18. 1966.

p. 1019.   In Key AA, BB, CC, D, E, FF, GG, HH, I, JJ, change to:

           JJ.   Fls. white.
                 K.   Lf.-blades ± oblong, 4–10 cm. long,
                      the ultimate segms. 1–4 mm. long,
                      not crowded. At ca. 5000 ft., e. Las-
                      sen Co. . . . . 21a. *L. Ravenii*
                 KK. Lf.-blades narrow-ovate in outline,
                      1.5–5 cm. long, the ultimate segms.
                      1–2 mm. long, crowded. At 10,000
                      ft. or higher, Inyo Mts.
                                              22. *L. inyoense*

p. 1012.   1. **L. lùcidum** (Nutt.) Jeps. $n = 11$ (Bell & Constance, 1960).
           4. **L. parvifòlium** var. **pállidum** Jeps. $n = 22$ (Bell & Constance, 1960).
           9. **L. leptocárpum** Coult. & Rose. $n = 11$ (Bell & Constance, 1960).

p. 1023.   16. **L. caruifòlium** (H. & A.) Coult. & Rose. Occurs in the Greenhorn Range, Kern Co., *Twisselmann*.

    19. **L. nevádense** (Wats.) Coult. & Rose. $n = 11$ (Bell & Constance, 1960).

p. 1024.   21. Theobald uses **L. foeniculàceum** (Nutt.) Coult. & Rose ssp. **fimbriàtum** Theobald for Calif. plants heretofore treated as *L. MacDougallii* and reduces *L. inyoense* Math. & Const. to **L. foeniculàceum** ssp. **inyoénse** Theobald. He separates the two sspp. as follows:

Petals glabrous. .... ssp. *inyoense*
Petals pubescent along the margins. .... ssp. *fimbriatum*

    21a. **L. Ravènii** Math. & Const. Acaulescent perennial, 1.5–4 dm. high, villous; lvs. ± oblong, 4–10 cm. long, ternate-bipinnate, the ultimate divisions linear, 1–4 mm. long; petioles 2–10 cm. long; peduncles 1–3 dm. long; fertile rays of umbel 5–18, unequal, 2–7 cm. long; involucel of several linear, ± scarious, distinct, villous or ciliolulate bractlets 2–4 mm. long; petals white; anthers purple; fr. oval, 6–8 mm. long, with narrow thin wings.—Near Ravendale, Lassen Co.

    23. **L. dasycárpum** (T. & G.) Coult. & Rose is maintained by Theobald with ssp. **tomentòsum** (Benth.) Theobald, which is sp. no. 26 in the FLORA.

    24. **L. mohavénse** (Coult. & Rose) Coult. & Rose. Theobald distinguishes:

Ultimate divisions of lvs. obovate, rarely more than 2–3 times longer than broad; petals purple or yellow; fr. pubescent. Mojave Desert and higher elevs., of Ventura and Santa Barbara cos. .... ssp. *mohavense*
Ultimate divisions of lvs. oblong to oblong-obovate, usually more than 3 times longer than broad; petals usually yellow; fr. pubescent to glabrate. Higher elevs., s. margin of Mojave Desert, w. and nw. Colo. Desert, into L. Calif. .... ssp. *longilobum* Theobald

p. 1025.   26. **L. tomentòsum** (Benth.) Coult. & Rose. $n = 11$ (Bell & Constance, 1960).

    30. **L. disséctum** (Nutt.) Math. & Constance. $n = 11$ (Bell & Constance, 1960).

    Var. **multífidum** Math. & Constance. $n = 11$ (Bell & Constance, 1960).

p. 1026.   33. **L. triternàtum** var. **macrocárpum** (Coult. & Rose) Math. $n = 11$ (Bell & Constance, 1960).

    35. **L. nudicáule** (Pursh) Coult. & Rose. $n = 11$ (Bell & Constance, 1960).

p. 1027.   ANGÉLICA

    2. **A. linearilòba** Gray var. **Culbertsònii** Jeps. occurs in the Greenhorn Range, Kern Co., *Twisselmann*. The variety is a taxon of questionable validity based on a collection from Little Kern R., with lf.-segms. 8–9 mm. wide.

p. 1028.   5. **A. tomentòsa** Wats. $n = 11$ (Bell & Constance, 1960).

p. 1029.   SPHENOSCIÀDIUM

    1. **S. capitellàtum** Gray. $n = 11$ (Bell & Constance, 1960).

p. 1030.   PTERÝXIA

    1. **P. terebinthìna** (Hook.) Coult. & Rose. var. **califórnica** Math. $n = 11$ (Bell & Constance, 1960).

p. 1031.   CYMÓPTERUS

    2. **C. desertícola** Bdg. $n = 11$ (Bell & Constance, 1960).

    5. **C. panaminténsis** Coult. & Rose. Reported from the El Paso Range, Kern Co., *Twisselmann*.

    Var. **acutifòlius** (Coult. & Rose) Munz. $n = 11$ (Bell & Constance, 1960).

p. 1032.   ERÝNGIUM

    3. **E. pinnatiséctum** Jeps. $n = 16$ (Bell & Constance, 1960).

p. 1033.   4. **E. alismifòlium** Greene. $n = 16$ (Bell & Constance, 1960).

    6. **E. aristulàtum** Jeps. $n = 16, 32$ (Bell & Constance, 1960).

    7. **E. Vàseyi** var. **castrénse** (Jeps.) Hoov. $n = 32$ (Bell & Constance, 1960).

    Var. **globòsum** (Jeps.) Hoov. [*E. spinulosum* Math.] Found in the Greenhorn Range, Kern Co., *Twisselmann*.

p. 1034.   CÓRNUS

    2. **C. stolonífera** Michx. $2n = 22$ (Taylor, 1967).

p. 1035.   5. **C. Nuttallii** Aud. has been found in the Greenhorn Range, Kern Co., *Twisselmann*.

p. 1036.   GÁRRYA

3. **G. flavéscens** Wats. var. **pállida** (Eastw.) Bacig. *ex* Ewan. Reported from the Tehachapi and Greenhorn ranges, Kern Co., *Twisselmann*.

p. 1037.   4. **G. ellíptica** Dougl. 2n = 22 (Van Horn, 1963).

RUBIÀCEAE

Substitute the following Key to Genera:

A. Lvs. in whorls mostly of 4 or more; plants herbs or low shrubs.
    B. Fls. solitary or in cymes, pedicelled; fls. lacking 3 basal bracts.
        C. Corolla rotate.
            D. The corolla with 4 free lobes; fr. dry or fleshy, of 2 1-seeded mericarps.
                                                                    1. *Galium*
            DD. The corolla with 5 free lobes; fr. berry, 1-seeded and derived from 1
                mericarp. . . . . 1a. *Rubia*
        CC. Corolla funnelform.
            D. Calyx of 4(–6) distinct teeth, persistent in fr.; corolla pinkish-lilac.
                                                                    2. *Sherardia*
            DD. Calyx an inconspicuous ridge; corolla whitish or bright blue.
                                                                    2a. *Asperula*
    BB. Fls. in elongate spikes, sessile, each fl. with 3 basal bracts. . . . . 2b. *Crucianella*
AA. Lvs. opposite, or if whorled, on large shrubs.
    B. Plants low perennial herbs; fls. in cymes. 3. *Kelloggia*
    BB. Plants large shrubs.
        C. Fls. in heads. Plants native. . . . . 4. *Cephalanthus*
        CC. Fls. in compound clusters. Escape from gardens. . . . . 5. *Coprosma*

p. 1038.   GÀLIUM

Insert the following references to literature:

Ehrendorfer, F. Evolution of the G. multiflorum complex in w. N. Am. I. Diploids and polyploids in this diverse group. Madroño 16: 109–122. 1961. Dempster, L. T. New names and combs. in the genus Galium. Brittonia 10: 181–192. 1958. A re-evaluation of G. multiflorum and related taxa. Brittonia 11: 105–122. 1959. Dempster, L. T. & G. L. Stebbins. The fleshy-fruited Galium spp. of Calif. I. Cytological findings and some taxonomic conclusions. Madroño 18: 105–112. 1965.

In Key, under AA, B, CC, change to:

CC. Plants perennial; fr. often glabrous.
    C'.   Lvs. 15–50 mm. long, in 4's, elliptic to elliptic-ovate; fr. with hooked
          bristles. Del Norte Co. . . . . 8c. *G. oreganum*
    C'C'. Lvs. less than 20 mm. long; fr. glabrous.
        D. The lvs. 2–4 mm. long; fls. greenish or yellow. Santa Lucia Mts.
                                                                    16a. *G. Hardhamiae*
        DD. The lvs. 5–20 mm. long, 4–6 at a node.
            E. Fls. 1.5–2 mm. broad, 1–3 in upper axils or on bractlets;
                pedicels strongly arcuate in age. Widespread. 9. *G. trifidum*
            EE. Fls. 2.5–3.5 mm. broad, in small cymes; peduncles often
                curved, but pedicels straight. Nw. Calif. . . 10. *G. cymosum*

p. 1039.   In Key at top of page, FF, GG, HH, II, delete "Sierra Nevada."
In Key under CC, change D as follows:

D. Corolla-lobes glabrous or minutely pubescent.
    E. Hairs of fr. ascending and subappressed. At 7000 ft. or more, San
        Gabriel and San Bernardino mts. . . . . 29. *G. Jepsonii*
    EE. Hairs on fr. spreading.
        F. Corolla reddish, with lance-acuminate lobes; plants often
            polygamous. E. Mojave Desert.   . . . . 26. *G. Wrightii*
        FF. Corolla greenish-white, the lobes mostly acute.
            G. Lvs. linear to lanceolate.
                H. Infl. merely bracteate, the fls. abundant; pedicels
                    short to none. Cismontane.   27. *G. angustifolium*
                HH. Infl. leafy, the fls. rather few. Modoc and Siskiyou
                    cos. . . . . 38. *G. serpenticum*
            GG. Lvs. ovate to orbicular.
                H. Herbage glabrous and shiny; lvs. acuminate, ± ar-
                    cuate; infl. divaricately branched. Modoc Co. to
                    Mono Co. . . . . 33. *G. multiflorum*

HH. Herbage minutely pubescent or glabrous, dull; lvs.
merely acute, not arcuate; infl. divaricately branched.
  I. Lvs. fusiform or ovate, commonly thin. E. Mono
    Co. .... 38. *G. Watsonii*
  II. Lvs. round to broadly ovate, usually thickish.
    J. Frs., including hairs, 5–9 mm. wide, in-
      cluded or scarcely exserted from the lvs.
      Lake and Placer cos. n. 36. *G. Grayanum*
    JJ. Frs., including hairs, 3–4 mm. wide, ex-
      serted well beyond the lvs. Mts. of e. cen-
      tral Calif. .... 37. *G. hypotrichium*
DD. Corolla-lobes mostly hairy to hispid.
  E. Lvs. acerose-acute or -acuminate, the midveins prominent, lateral
    veins mostly lacking, margin often strongly revolute. Deserts,
    mostly below 4500 ft. .... 30. *G. stellatum*
  EE. Lvs. acute, but not acerose-acuminate, the lateral veins usually
    present, sometimes as prominent as the midvein.
    F. Fl.-clusters drooping; mature frs., including hairs, 4–5 mm.
      in diam.; corolla-lobes silky-hairy without. Mts. about w. end
      of Mojave Desert. .... 31. *G. Hallii*
    FF. Fl.-clusters not drooping; frs. mostly smaller; corolla-lobes
      mostly bristly hairy.
      G. Lvs. linear-oblong to oblong, 1–2 mm. wide; plants
        tufted, 0.5–2 dm. tall, cinereous-pubescent. San Gabriel
        Mts. to San Jacinto Mts. .... 28. *G. gabrielense*
      GG. Lvs. lance-ovate to roundish, mostly wider.
        H. Herbage glabrous, except uppermost bracts.
          I. Lvs. 2–8 mm. long. Kern and San Bernardino
            cos. to Inyo Co. .... 34. *G. Matthewsii*
          II. Lvs. 6–17 mm. long. Inyo Co. to Utah.
            32a. *G. magnifolium*
        HH. Herbage hispid.
          I. Lvs. of a given whorl unequal, 3–7 mm. long;
            fls. ca. 1.5 mm. in diam., borne in sessile glom-
            erules on a virgate or sparsely branched infl.
            San Gabriel Mts. to Santa Rosa and Kingston
            mts. .... 35. *G. Parishii*
          II. Lvs. of a given whorl subequal, 5–12 mm.
            long.
            J. The lvs. elliptical or ovate, tapering at
              both ends, not arcuate, generally 1-nerved.
              32. *G. Munzii*
            JJ. The lvs. broadly ovate, mostly round at
              the base, abruptly narrowed to a sharp
              apex, ± arcuate, the broader pair often
              3-nerved. .... 33. *G. multiflorum*

p. 1040.   **G. parisiénse** L. in Santa Barbara Co., *Raven*; Monterey Co., *Howitt and Howell*; Kern Co., *Twisselmann*.

3. Change *G. tricorne* Stokes to **G. tricornùtum** Dandy.

4. **G. Aparìne** L. $2n =$ ca. 66 (Löve & Löve, 1956); 44 (Kliphius, 1962).

5. **G. spùrium** L., the smooth-fruited form reported from Marin Co., *Howell*.

6. **G. Mollùgo** L. $2n = 44$ (Kliphuis, 1962).

8. Change *G. asperrimum* Gray to **G. mexicànum** HBK. var. **aspérulum** (Gray) Dempster.

8a. **G. oregànum** Britt. Perennial with slender creeping rootstocks; stems arising singly, erect, 1–4 dm. tall, glabrous; lvs. in 4's, hispid-ciliate on margins, elliptic to elliptic-ovate, 1.5–5 cm. long; infl. branched; corolla 3 mm. diam.; fr. 2 mm. high, with many hooked bristles.—Del Norte Co.; to Wash.

p. 1041.   13. **G. muràle** (L.) All. Sierran foothills, *Howell*.

14. **G. Andrèwsii** Gray. $2n = 22$. Add:

Var. **gaténse** Dempster. Strikingly hairy; $2n = 88$.—Los Gatos Creek, w. Fresno Co. and San Carlos Peak above New Idria, San Benito Co.

15. **G. ambíguum** Wight. $2n = 22$.

Var. **siskiyouénse** Ferris. $2n = 66$.

16a. **G. Hardhàmiae** Dempster. Dioecious perennial herb with hispid internodes 0.5–2.5 cm. long; lvs. 6 at a node, ovate, acute, sparsely hispid, 2–4 mm. long; infl. narrow, with short branches; corolla rotate, 2 mm. across, yellow or green; fr. fleshy, glabrous, didymous; $2n = 22$.—Santa Lucia Mts. of San Luis Obispo and Monterey cos.

p. 1042.   17. **G. califórnicum** H. & A. $2n = 88$ or 132. Add:
                  Ssp. **luciénse** Dempster & Stebbins. Plants to 1.5 dm. tall, soft-pubescent,
                  pale green; lvs. 4–6 (–10) mm. long, not armed at apex; ovary densely pubes-
                  cent; fr. white, fleshy; $2n = 44.$—Region of Cone Peak, Santa Lucia Mts.,
                  Monterey Co.
                  18. **G. muricàtum** Wight. $2n = 22, 44.$
                  19. **G. Bolánderi** Gray. [*G. pubens* Gray, sp. no. 21 in the FLORA.] $2n = $
                  66. *G. pubens* proves to be hexaploid forms of *G. Bolanderi.*
                  20. **G. sparsiflòrum** Wight. $2n = 22.$
                  21a. **G. gránde** McClatchie treated as a sp. by Dempster. $n = $ more than
                  110.
                  22. Under **G. Nuttállii** Gray ($2n = 22$) insert the following key:

A.   Lvs. acute at apex, tipped with a stoutish hair or bristle.
     B.   The lvs. linear or lanceolate to narrowly ovate, 3–8 mm. long, narrowed gradu-
          ally to the apex. Near the coast, San Diego Co. to San Benito Co. Channel Ids.
          (Include ssp. *insulare* Ferris as synonym). . . . . *G. Nuttallii*
AA. Lvs. obtuse or rounded at the apex, tipped with a weak hair or none.
     B.   The lvs. oval or oblong. Coast Ranges, San Diego Co. to Ore.
                                                                 Var. *ovalifolium* Dempster
     BB. The lvs. linear, Sierra Nevada foothills and inner Coast Ranges.
                                                                 Var. *tenue* Dempster

                  22a. **G. Cliftonsmíthii** (Dempst.) Dempst. & Stebbins. [*G. Nuttallii* var. *C.*
                  Dempster.] Principal stems acutely angled; internodes 3–10 cm. long; lvs.
                  6–14 mm. long, long-acuminate; secondary herbage in large tufts; fls. very
                  few, green or pale yellow, with long-acuminate lobes; $2n = $ ca. 182–189.—
                  Mts. below 4000 ft., Santa Barbara Co. to Monterey Co.
p. 1043.   24. **G. buxifòlium** Greene is treated by Dempster as *G. catalinense* var. *b.*
                  Dempster.
                  27. **G. angustifòlium** Nutt. Insert:
                  Var. **onycénse** Dempster. Low tufted masses, gray in color, 12–25 cm. tall;
                  fr. 2 mm. diam.—Onyx, Kern Co.
p. 1044.   30. **G. stellàtum** ssp. **erèmicum** Ehrend. $2n = 22$ (Ehrendorfer, 1961).
                  32. **G. Múnzii** Hilend & Howell. Delete *G. Matthewsii* var. *scabridum* Jeps.
                  from synonymy. Add:
                  Var. **kingstonénse** Dempster. Corolla campanulate, not rotate as in *G.
                  Munzii,* larger, clear pink; $2n = 44.$—Kingston Mts., Mojave Desert.
                  Var. **cárneum** Hilend & Howell. A variable series, often tall and wiry, pos-
                  sibly representing hybrids between *G. Munzii* and *G. Matthewsii.*—Panamint
                  Mts.
                  32a. **G. magnifòlium** (Dempster) Dempster. [*G. Matthewsii* var. *m.* Demp-
                  ster. *G. Munzii* f. *glabrum* Ehrend.] To 4 dm. tall, wiry, suffrutescent lvs.
                  glabrous, acute or mostly acuminate, lanceolate to ovate, 6–17 mm. long; infl.
                  much branched, divaricate; corollas and ultimate bracts hispid.—Between
                  2700 and 7300 ft., s. Inyo Co. to se. Utah and n. central Ariz.
                  33. **G. multiflòrum** Kell. [*G. Bloomeri* Gray and var. *hirsutum* Gray. *G.
                  Matthewsii* var. *scabridum* Jeps.] $2n = 22$ (Ehrendorfer, 1961).
                  34. **G.Matthèwsii** Gray. $2n = 22$ (Ehrendorfer, 1961).—Reported from the
                  Walker Pass region, Kern Co., *Twisselmann.*
                  35. **G. Paríshii** Hilend & Howell, $2n = 22$ (Ehrendorfer, 1961).
p. 1045.   36. **G. Grayànum** Ehrend. $2n = 22, 44$ (Ehrendorfer, 1961). [*G. G.* ssp.
                  glabrescens Ehrend.]
                  37. **G. hypotríchium** Gray. $2n = 22.$ Ssp. **subalpìnum** Ehrend. $2n = 44.$
                  Intermediate between *G. hypotrichium* and *G. Munzii.*
                  38. **G. Watsonii** (Gray) Heller. [*G. multiflorum* var. *W.* Gray.] Plants mod-
                  erately low, not at all stiff, usually lax; lvs. generally fusiform, thin, 12–24 mm.
                  long, glabrous or with scant microscopic pubescence, not shiny; infl. little
                  branched, not divaricate; corollas glabrous or nearly so. $2n = 22$ (Ehrendorfer,
                  1961).—E. Mono Co., between 7500 and 10,000 ft.; to Utah.
                  39. **G. serpénticum** Dempster. [*G. Watsonii* (Gray) Heller f. *scabridum*

Ehrend.] Cespitose perennial with erect branches 12–28 cm. tall; lvs. linear or lanceolate, 5–20 mm. long, glabrous to hispidulous, acute; panicles leafy, the straight branches 2.5 cm. long; corolla rotate, glabrous, ± yellowish to greenish; fr. with long brown hairs, thus 6 mm. diam.; $2n = 22$ (Ehrendorfer, 1961).—Modoc and Siskiyou cos.; to Wash., Ida.

1a. Rùbia L.

Perennial herbs, frequently rather stiff, hispid or prickly. Lvs. in whorls of 4–8, rarely opposite. Fls. small, in cymes, 5-merous. Invol. none. Calyx-tube ovoid or globose, without limb. Corolla rotate or slightly campanulate. Ovary 2-loculed with 1 ovule in each locule; fr. fleshy. Ca. 40 spp. of Old and New Worlds. (Latin *red*, a dye extracted from the root.)

1. **R. tinctòrum** L. MADDER. To ca. 1 m. high, glabrous; lvs. lanceolate, with a conspicuous network of lateral veins beneath; fls. yellow, in axillary and terminal cymes; fr. a red-brown berry.—Escape at Niles, Alameda Co., *Howell.* From Medit. region.

SHERÁRDIA

1. **S. arvénsis** L. $2n = 22$ (Löve & Löve, 1956).

p. 1046. ## 5. Coprósma J. R. & G. Forst.

Evergreen shrubs or small trees. Lvs. shining, commonly opposite, obtuse or notched at apex. Fls. small, solitary or fascicled, imperfect, white or greenish. Corolla funnelform or campanulate, 4–5-lobed. Fr. an ovoid or globose usually 2-celled drupe. Ca. 60 spp., largely of S. Hemis. (Greek name, referring to the usually foetid odor of the plants.)

1. **C. rèpens** A. Rich. Shrub or to 8 m. tall; lvs. ± fleshy, broad-oblong, 6–8 cm. long, with glandlike pits on undersurface at base of lateral veins; fls. ♂ and ♀, small; drupe orange-red, ca. 10 mm. long. Occasional escape from cult., near the coast.—From New Zealand. Cult. in Calif. as *C. Baueri.*

CAPRIFOLIÀCEAE

Line 1 of family description, insert "sometimes herbs" at end of first sentence.

p. 1047. SAMBÙCUS

1. Originally spelled S. cerùlea Raf.

2. **S. mexicàna** Presl. $2n = $ ca. 36 (Raven, Kyhos & Hill, 1965).

3. **S. callicárpa** Greene. Add as synonym S. *racemosa* var. *arborescens* Gray.

4. **S. microbòtrys** Rydb. Add as synonym, S. *racemosa* var. *m.* Kearney & Peebles.

p. 1049. SYMPHORICÁRPOS

1. **S. rivularis** Suksd. Add as synonym, S. *albus* var. *laevigatus* (Fern.) Blake. Reported from Greenhorn Range, Kern Co., *Twisselmann.*

3. **S. hespérius** G. N. Jones. Insert as synonym, S. *mollis* Nutt. ssp. *h.* Abrams ex Ferris.

4. **S. acùtus** (Gray) Dieck. Reported from Greenhorn and Piute mts., Kern Co., *Twisselmann.*

6. **S. Paríshii** Rydb. Extends through the Tehachapi and Piute mts. and to Mt. Pinos, *Twisselmann.*

p. 1050. LONÍCERA

In Key, AA, B, change to:

B. Fls. mostly in 1 whorl, sometimes 2–3; lvs. glabrous except for the ciliate margins.
  C. Corollas with a short, nearly regular limb; lvs. deciduous. . . . . 5. *L. ciliosa*
  CC. Corollas conspicuously 2-lipped, the limb ca. as long as the tube; lvs. ± evergreen. . . . . 5a. *L. etrusca*

p. 1051. 4. **L. utahénsis** Wats. Probably not in Calif.; the old collection referred to it probably was wrongly identified and was a specimen of **L. tatárica** L., a deciduous shrub to 3 m. tall, with ovate lvs. 2–6 cm. long, and with pink or white to crimson bilabiate fls. 2–2.5 cm. long, borne in pairs on slender axillary peduncles.—From Eurasia.

5a. **L. etrúsca** Santi. Vigorous evergreen vine; lvs. 4–9 cm. long, not connate except just below the infl.; infl. stipitate-glandular; bracts subtending fls. round; fls. 3–5 cm. long, yellowish white tinged with purplish red.—Reported from Del Norte and Humboldt cos., w. Ore. Introd. from Eu.

6. **L. subspicàta** H. & A. in Monterey Co., *Howitt & Howell.*

p. 1054.   VALERIANÉLLA

1. Change *V. olitòria* (L.) Poll. to *V.* Locústa (L.) Betcke.

PLECTRÌTIS

To literature citations add: Dempster, L. T. Dimorphism in the fruits of Plectritis, and its taxonomic implications. Brittonia 10: 14–27. 1958. Morey, D. H. Changes in nomenclature in the genus Plectritis. Contr. Dudley Herb. 5: 119–121. 1959.

Change the treatment in the FLORA to:

A. Corolla bilabiate, funnelform, pale to medium pink, the spur obsolete to usually less than ⅓ the length of the corolla; fr. keeled, the keel rarely grooved, acutely angled or smoothly rounded, winged or wingless, the wings thin, pubescent, the hairs flexible, evenly tapered, ± obtuse. . . . . 1. *P. congesta*
AA. Corolla regular, white to bilabiate, red, tubular or ± funnelform, the spur usually more than ⅓ the length of the corolla; fr. keeled, the keel often with a dorsal groove, obtusely angled, winged or wingless, the wings stiff and often thick, the margins thickened, often grooved, pubescent, the hairs slightly clavate, cylindrical or long and curly.
   B. Corolla essentially regular, white or light pink, usually lacking red spots at the base of the ventral lip, stout and with a clavate spur; keel of the fr. without 2 brush-like rows of hairs. . . . . 2. *P. macrocera*
   BB. Corolla strongly bilabiate, pink to light red, usually with 2 red spots at the base of the ventral lip, slender and with a slender spur; keel of the fr. often with 2 brush-like rows of hairs. . . . . 3. *P. ciliosa*

1. **P. congésta** (Lindl.) DC. [*Valerianella congesta* Lindl.] Erect, 1.5–6 dm. tall; lvs. obovate to ovate, 1–5 cm. long; corolla pale pink to pink, subcampanulate to funnelform, 4.5–9.5 mm. long, strongly bilabiate, the spur short to obsolete, usually with expanded tip; fr. pale yellow to brown, 2–4.5 mm. long, the keel sharply angled, acute, very slightly indented dorsally, wings broad to obsolescent, with thin subscarious margins that are basally connivent and spreading above.—Shaded places at low elevs., near the coast, San Luis Obispo Co. n.; to B.C. Apr.–May.

Ssp. **nítida** (Heller) Morey. [*P. n.* Heller.] Frs. shining, the keel smoothly rounded, the wings, when present, connivent at base and apex.—Shaded, ± wet places, coastal lowlands, Monterey Co. to Mendocino Co.

Ssp. **brachystèmon** (F. & M.) Morey. [*P. b.* F. & M. *Betckea major* F. & M. *P. samolifolia* Calif. refs. *Valerianella anomala* Gray. *V. aphanoptera* Gray. *V. magna* Greene. *P. involuta* Suksd.] Fls. 1–3.5 mm. long, often spurless, the corolla tubular-funnelform.—Brushy montane areas, Monterey Co. n.; to B.C.

2. **P. macrócera** T. & G. [*P. Jepsonii* (Suksd.) Davy. *P. glabra* Jeps. *P. Eichleriana* (Suksd.) Heller. *P. collina* Heller.] Plants slender, 1–6.5 dm. tall; lvs. obovate to lance-ovate; corolla white to pale pink, 2–3.5 mm. long, stout, subcampanulate to broadly tubular, regular to weakly bilabiate, the spur stout, ca. twice as long as broad, 1–1½ times as long as the tube, not or slightly expanded at the tip; fr. pale straw-yellow to red-brown, 2–4 mm. long, ± pubescent; keel rounded to rather angular, obtusely angled, often with a dorsal groove; wings thick, expanded to obsolete, with a usually grooved marginal thickening. $n = 18$ (Raven, Kyhos & Hill, 1965).—Usually shaded places below 4000 ft., Foothill Wd., Oak Wd., Yellow Pine F., etc., S. Calif. to s. Wash. Apr.–May.

Ssp. **Gràyii** (Suksd.) Morey. [*Aligera G.* Suksd. *A. mamillata* Suksd.] Expanded wings thin, with a narrow marginal thickening scarcely grooved; fr. with a median ridge on ventral surface usually bearing a multiseriate row of bristles.—Open or shaded places, s. Calif. to Wash., Mont., Utah.

3. **P. cilìòsa** (Greene) Jeps. [*Valerianella c.* Greene. *Aligera macroptera* Suksd. *A. californica* Suksd.] Slender, 1–5.5 dm. tall; lvs. obovate to oblong-fusiform; corolla deep pink with dark spots at base of ventral lobes, 5.5–8.5 mm. long, slender, bilabiate, the spur slender, usually exceeding ovary; fr. pale straw-yellow to brown, 3–4 mm. long, ± pubescent; keel rounded, obtusely angled, with a deep dorsal groove; wings variously expanded, bounded with a usually grooved, marginal thickening.—Open sunny places below 6000 ft., many Plant Communities, San Benito and Monterey cos. to Mendocino Co. and w. base of Sierra Nevada. Apr.–May.

Ssp. **insígnis** (Suksd.) Morey. [*Aligera i.* Suksd. *A. rubens* Suksd. *A. patelliformis* Suksd. *P. Davyana* Jeps.] Corolla 1.5–3.5 mm. long.—Yellow Pine F. and below, inland coastal valleys, San Diego Co. to Napa Co.; sparingly to Wash.; n. L. Calif.

p. 1056.   CENTRÁNTHUS

Change spelling to **Kentránthus** Neck.

p. 1057.   DÍPSACUS

1. **D. sylvéstris** Huds. has been incorrectly named in books and should be called **D. fullònum** L.

2. **D. satìvus** (L.) Honckeny is the correct name for the cult. or FULLERS' TEASEL.

p. 1058.   CUCURBITÀCEAE

In Key to Genera, A, change BB by deleting "4. *Citrullus*" and inserting:

    C. Anther-connective without terminal appendage; seeds with obtuse margin.
                                                                 4. *Citrullus*
    CC. Anther-connective terminated by an appendage; seeds not margined.
                                                                 5. *Cucumis*

p. 1060.   BRYÒNIA

After the genus *Marah* add **Bryònia dioìca** Jacq. BRYONY. Like *Marah,* but with small red globular berries. In cultivation and an escape in Golden Gate Park, San Francisco. From the Old World.

CUCÚRBITA

1. **C. foetidíssima** HBK. Occurs in Monterey Co., *Howitt & Howell.*

p. 1061.   3. **C. palmàta** Wats. Occurs in Monterey Co., *Howitt & Howell.*

CITRÚLLUS

1. Change *C. vulgaris* Schrad. var. *citroìdes* Bailey to **C. lanàtus** (Thunb.) Mansf. var. **citroìdes** (Bailey) Mansf. as the name for CITRON. For the WATERMELON the correct name is **C. lanàtus** (Thunb.) Mansf.

5. Cucùmis L.

Herbaceous scabrous plants; tendrils simple; lvs. 3–7-lobed, rounded-obtuse; pepo globose. (Greek, *kykyon,* cucumber.) An Afr. genus of ca. 40 spp.

1. **C. myriocárpus** Naud. Green trailing annual with angulate-striate, rough-hairy branches; lvs. 4–5 cm. long, long-petioled, the prominent lobes and sinus rotundate, the middle lobe large; fls. very small; pepo ca. 2 cm. in diam., subglobose, marked with darker green bands and beset with weak deciduous prickles; $2n = 24$ (Shimotsuma, 1965).—Native of S. Afr. An infestation of some size was reported from near Ballard, Santa Barbara Co., *Fuller.*

2. **C. Mèlo** L. var. **Dudaím** (L.) Dunal. [*C. D. L. C. odoratissimus* Moench.] Lvs. scarcely lobed; fls. relatively large; fr. medium orange, smooth, 5–6 cm. in diam., very fragrant.—Weed in Asparagus fields w. of El Centro, Imperial Co., *Fuller.*

p. 1062.   CAMPÁNULA

In Key AA, B, C, add "n. to Ore." for *C. prenanthoides.*

p. 1063.   3. **C. prenanthoìdes** Durand. Add synonym *Asyneuma p.* McVaugh; $2n = 34$ (Gadella, 1963).

4. **C. Scoùleri** Hook. For the 1959 printing of the FLORA, transpose line 6 and lines 7 and 8.

7. **C. califórnica** (Kell.) Heller. Reported from the Santa Cruz Mts., *Howell*.

p. 1064. TRIODÀNIS

1. **T. biflòra** (R. & P.) Greene. $2n = 28$ (Löve & Solbrig., 1065).

p. 1065. GITHÓPSIS

3. **G. specularioìdes** Nutt. reported from the Greenhorn Mts., Kern Co., *Twisselmann*; Madera and Fresno cos., *Weiler*.

p. 1066. NEMÁCLADUS

In Key after first DD insert:

> D′. Plants 0.5–1 cm. tall; pedicels exceeded by the subtending bract; no glands evident on the ovary. . . . . 6a. *N. Twisselmannii*
> D′D′. Plants 3–12 cm. tall; pedicels not exceeded by the subtending bract; ovary-glands evident.
> E, EE, etc. for spp. 4, 5, 6.

1. **N. longiflòrus** Gray. In line 4 of description, delete "fls. 4–7 mm. long."

p. 1067. 6a. **N. Twisselmánnii** J. T. Howell. Rosulate, 0.5–1 cm. tall, grayish, hirsutulose; lvs. entire, the basal spatulate, 2–3 mm. long, the cauline oblong, 3 mm. long; fls. 2–3 mm. long at anthesis, 4–5 mm. in fr.; corolla 2–3 mm. long, the tube 1 mm., the lobes hirsutulose; glands of ovary obsolete; seeds 0.75 mm. long, with rows of ca. 8 pits.—Sw. of Pine Flat, Kern Plateau, Kern Co., at 7350 ft.

p. 1068. 10. Appendages apparently are present in **N. glandulíferus** Jeps.

11. **N. capillàris** Greene occurs in Ventura Co., *Hardham*; and in Monterey Co., *Howitt & Howell*.

p. 1069. DOWNÍNGIA

In the Key, under A, BB insert:

> C. Corolla bright blue, the lower lip with a central bilobed white spot; lobes of upper lip narrow, usually parallel or crossed over each other. 2. *D. elegans*
> CÇ. Corolla lavender-blue, the lower lip with a central white area containing 2 bright orange-yellow spots; lobes of upper lip broader, widely divergent.
>   2a. *D. Bacigalupii*

p. 1070. 1. **D. insígnis** Greene. $n = 11$ (Wood, 1961).

2a. **D, Bacigalùpii** Weiler. Plants 5–30 cm. tall; corolla lavender-blue, usually with prominent, more deeply colored veins especially on lower corolla-lobes except in the central white area and orange-yellow spots; lobes of lower lip rounded, abruptly pointed.—Sierra Co., Calif. to sw. Ore., sw. Ida. In *D. elegans* $n = 10$; in *D. Bacigalupii* $n = 12$.

3. **D. pusílla** (G. Don) Torr. $n = 11$ (Wood, 1962).

5. **D. bicornùta** Gray. $n = 11$ (Wood, 1961).

Var. **pícta** Hoov. $n = 11$ (Wood, 1961).

6. **D. pulchélla** (Lindl.) Torr. $n = 11$ (Wood, 1961).

p. 1071. 7. **D. ornatíssima** Greene. $n = 11$ (Wood, 1961).

8. **D. cuspidàta** (Greene) Greene. $n = 11$ (Wood, 1961).—In Fresno and Kern cos., *Weiler*.

9. **D. bélla** Hoov. occurs on Mt. Abel Road, Ventura Co., *Twisselmann*.

10. **D. concólor** Greene. $n = 8, 9$ (Wood, 1961).

p. 1073. Family number for COMPOSITAE is 119.

p. 1074. In Key to Tribes of Compositae, under G insert "usually" after "style-branches."

In Key, A, B, C, D, E, FF, GG, H, *Anthemideae*, change to p. 1227 and in HH, II the third word is "aks." not "ask."

p. 1075. In Artificial Key, under A, "Anthers tailed or sagittate at their base" under BB add:

> C. Style without ring of hairs or distinct thickened ring below the branches; plants not prickly; receptacle naked. . . . . *Inuleae* p. 1257
> CC. Style with a ring of hairs or a thickened ring below the papillate branches, the branches generally connate near the tip; plants usually prickly; receptacle usually densely bristly. . . . . *Cynareae* p. 1271

In Key under **Group A,** after AA, B, C, and D, insert:

> E. Phyllaries 1–3; head usually 1-fld. Introd. . . . . . . 50a. *Flaveria*
> EE. Phyllaries several; head several-fld. Native. . . . . 51. *Baeria*

p. 1077. In Key under **Group B,** after AA, BB, etc. EE, FF, GG, HH, insert:

> H'. Plants rather succulent, glabrous except for a microscopic glandular pubescence and very sparse flocs of wool in lf.-axils and below the heads.
> 65. *Blennosperma*
> H'H'. Plants not as above.
> I. Herbage not white-woolly; annual herbs.
> J. Lvs. 1–3 times ternately divided; plants 3–6 dm. high. San Bernardino and San Jacinto mts. . . . . 54. *Bahia*
> JJ. Lvs. coarsely sinuate-dentate; plants 1–3 dm. high. Garden escape.
> 121a. *Dimorphotheca*

In Key to **Group C,** follow through A, BB, CC, DD "Awns or teeth retrorsely barbed. Herbaceous"; delete "12. *Bidens*" and add:

> E. Inner phyllaries free essentially to the base. . . . . 12. *Bidens*
> EE. Inner phyllaries connate to the middle or higher.
> 12a. *Thelesperma*

p. 1079. In Key under **Group D,** after AA, BB, CC, D, EE, FF, G, change to:

> H. Plants 1–4 dm. tall.
> I. Branches filiform; lvs. linear, entire.
> 86. *Tracyina*
> II. Branches coarser; lvs. broader, dentate to pinnatifid. 111. *Senecio*
> HH. Plants taller, coarse; lvs. dentate to lobed.
> 117. *Erechtites*

p. 1083. In Key, FF, G insert "usually" after "pappus."
p. 1084. WYÈTHIA
 1. **W. ovàta** T. & G. Greenhorn Mts., Kern Co., *Twisselmann.*
p. 1086. BALSAMORHÌZA
 4. **B. macrolèpis** Sharp ssp. **platylèpis** (Sharp) Ferris. [*B. p.* Sharp.] Pubescence unusually dense, silvery; pinnae of lvs. not or scarcely lobed; outer phyllaries shorter than to equaling the disk.—Nevada Co. to Modoc and Siskiyou cos.; sw. Ore., w. Nev. (Delete *B. platylepis* from synonymy under **B. Hookeri**).
p. 1087. VIGUIÈRA
 1. **V. deltoìdea** var. **Paríshii** (Greene) Vasey & Rose. *n* = 18 (Heiser, 1960).
p. 1089. HELIÁNTHUS
 4. **H. ciliàris** DC. At Hayward, Alameda Co., *Fuller* and in Orange Co., *Fuller.*
p. 1090. RUDBÉCKIA
 2. **R. califórnica** Gray. In line 4 change "peduncles" to "petioles" in the 1959 printing.
p. 1092. ENCÈLIA
 1. **E. farinòsa** Gray var. **ràdians** Bdg. ex Blake reported from se. Calif. by Wiggins. Lvs. soon glabrate; invol. essentially glabrous; disk fls. purplish.
 2. **E. frutéscens** (Gray) Gray. *n* = 17 (Jackson, 1960).
p. 1093. VERBESÌNA
The combination **V. encelioìdes** ssp. **auriculàta** (Rob. & Greenm.) Coleman has been published.
p. 1094. ECLÍPTA
 1. **E. álba** (L.) Hassk. *n* = 12 (Mehra et al., 1965).—Found in moist and cult. places, Kern Co., *Twisselmann.*
BÌDENS
In Key, change first line to:

"Aks. (at least the central ones) linear-tetragonal" and add:
Lvs. pinnate with 3–5 lfts. Common in s. Calif. . . . . 1. *B. pilosa*
Lvs. mostly entire. Rare, San Joaquin Co. . . . . 1a. *B. connata*

1. **B. pilòsa** L. *n* = 12, 14 (Powell & Turner, 1963).—Occurs in Santa Cruz Co., *Howell.*
1a. **B. connàta** Muhl. var. **petiolàta** (Nutt.) Farwell. Lvs. mostly unlobed, tapering to slender or narrowly margined petioles; inner aks. to 8 mm. long, the awns retrorsely barbed.—Stanislaus River, Caswell Memorial State Park, San Joaquin Co., *Rubtzoff.* Native of e. U.S.
2. **B. laèvis** (L.) BSP. *n* = 11 (Torres, 1958).—In Marin and adjacent counties, *Howell.*
3. **B. cérnua** L. At Greenfield, Kern Co., *Twisselmann.*
p. 1095.   4. **B. frondòsa** L. 2*n* = 48 (Löve, 1964).

12a. Thelespérma Less.

Mostly perennial, smooth, glabrous. Lvs. opposite, usually finely dissected. Heads pedunculate, invol. double, the outer phyllaries often narrow, the inner broader, connate in at least the lower half. Rays yellow, if present; disk fls. mostly yellow; aks. papillose, with 2 narrow, retrorsely barbed teeth. A genus of several spp., largely of sw. N. Am. (Greek, *thele*, nipple, and *sperma*, seed, referring to the papillose aks.)

1. **T. megapotàmicum** (Spreng.) Kuntze. [*Bidens m.* Spreng. *T. gracile* (Torr.) Gray.] Three to 6 dm. tall; lvs. twice 3–5-nately parted into linear lobes; peduncles commonly 1–2 dm. long; heads normally discoid, 1–1.5 cm. thick; outer phyllaries free, 2–3 mm. long; inner much longer, connate; aks. dark, subtended by scarious receptacular bracts.—Once established on Catalina Id. Ranging from Ariz. to Utah, Wyo., Nebr., Tex., Mex.; s. to S. Am.
p. 1096.   COREÓPSIS
2a. **C. Atkinsoniàna** Dougl. reported from Escalon, San Joaquin Co., *Fuller.*
p. 1097.   9. **C. Bigelòvii** (Gray) Hall. Taken on road to Cedar Grove, Fresno Co., *Howell.*
p. 1098.   GUIZÒTIA
1. **G. abyssínica** (L. f.) Cass. Reported from San Francisco, *Howell.*
GALINSÒGA
1. **G. parviflòra** Cav. *n* = 8, 16 (Turner, Powell & King, 1962). Recorded as occasional in San Francisco Bay area, *Howell.*
p. 1099.   EASTWOÒDIA
1. **E. élegans** Bdg. *n* = 9 (Solbrig et al., 1964).
MELAMPÒDIUM
1. **M. perfoliàtum** HBK. *n* = 12 (Turner & King, 1962).
p. 1100.   OXYTÈNIA
1. **O.** *acerosa* Nutt. is combined with Iva as **I. aceròsa** (Nutt.) Jackson in R. C. Jackson, A revision of the genus Iva. Univ. Kans. Sci. Bull. 41: 793–876. 1960. *n* = 18 (Jackson, 1960).
p. 1101.   ÌVA
2. **I. axillàris** Pursh. 2*n* = 36, 54 (Mulligan, 1961).
Ssp. **robústior** (Hook.) Bassett. [*I. a.* var. *r.* Hook. *I. a.* var. *pubescens* Gray. *I. a.* Calif. refs.] *n* = 18, 27 (Payne et al., 1964).—Ranging from Calif. to B.C., S. Dak. and New Mex. Ssp. **axillàris** is east of the Continental Divide. Our Calif. plant is reported from Monterey Co., *Howitt & Howell.*
DICÒRIA
1. **D. canéscens** T. & G. *n* = 18 (Payne et al., 1964).
p. 1102.   HYMENOCLÈA
1. **H. Salsòla** T. & G. *n* = 18 (Payne et al., 1964).
2. **H. monogỳra** T. & G. *n* = 18 (Payne et al., 1964).
pp. 1102–
  1105.   AMBRÒSIA and FRANSÈRIA

Shinners (Field & Lab. 17: 170–176. 1949) unites *Ambrosia* and *Franseria.*
W. W. Payne (A reëvaluation of the genus Ambrosia. Jour. Arn. Arb. 45:
401–438. 1964) also unites them and for California we have the following
spp.:

1. **A. acanthicárpa** Hook. *n* = 18. Reported from Monterey Co., *Howitt &
Howell.*

2. **A. ambrosioìdes** (Cav.) Payne. [*Franseria a.* Cav.] *n* = 18.

3. **A. artemisiifòlia** L. 2*n* = 36 (Mulligan, 1965).

4. **A. Chamissònis** (Less.) Greene. *n* = 18. It occurs on the mainland s. to
Point Sal, Santa Barbara Co., *C. F. Smith.* Payne does not recognize ssp. *bi-
pinnatisecta* of the FLORA.

5. **A. chenopodiifòlia** (Benth.) Payne. [*Franseria c.* Benth. in Hinds.]
n = 36.

6. **A. confertiflòra** DC. *n* = 36, 54. Reported from Monterey Co., *Howitt &
Howell.*

7. **A. dumòsa** (Gray) Payne. [*Franseria d.* Gray in Torr. & Frém. in Frém.]
n = 18, 36, 54, 63, (72?).

8. **A. eriocéntra** (Gray) Payne. [*Franseria e.* Gray.] *n* = 18.

9. **A. ilicifòlia** (Gray) Payne. [*Franseria i.* Gray.] *n* = 18.

10. **A. psilostàchya** DC. var. **califórnica** (Rydb.) Blake in Tidestr. differs
from var. **psilostàchya** in having spreading instead of appressed or ascending
hairs on the stems. Both varieties may occur in Calif.

11. **A. trífida** L. 2*n* = 24 (Mulligan, 1957). Reported from Byron, San Joaquin Co., *Fuller.*

p. 1111.　LÀYIA

9a. **L. Ziègleri** Munz. sp. nov. Annual, divaricately branched from base,
1–3 dm. high, with spreading, both stiff and soft hairs, and others ± appressed, leafy especially in the lower parts; basal leaves 2–4 cm. long, 3–4 mm.
wide, winged-petiolate, then slightly wider and dentate in distal part; cauline
lvs. largely turned to 1 side, gradually reduced upward, entire, subsessile,
lance-oblong or narrower, the lower ca. 2.5 cm. long, the upper less than 1
cm. long, mostly 2–3 mm. wide, blunt, short-hairy, not glandular; heads hemispheric, solitary at ends of branches, on naked peduncles 1–8 cm. long and
± black-stipitate-glandular toward tip as are the receptacle and phyllaries;
phyllaries ca. 13–14, ± uneven in length, ca. 7–8 mm. long, oblong, obtuse,
with short stiff hairs and the stipitate glands; rays golden, ca. 8 mm. long,
rather sharply 3-lobed; ray-aks. black, epappose, glabrous, ca. 3 mm. long,
completely enclosed in the subtending phyllaries; disk-fls. separated from the
rays by a double row of green acute stiff lance-oblong bracts 6–7 mm. long,
the fls. ca. 35 in number, yellow, the corolla 4–5 mm. long, the pappus bristles
ca. 30, stiff, ca. 3 mm. long, scarcely plumose or flattened, without woolly
hairs near the base; disk-aks. dark, ca. 3 mm. long, strigose with short white
hairs.

Planta annua, base divaricate ramosa, 1–3 dm. alta, ± hirsuta et strigosa,
foliosa; foliis basalibus oblongo-lanceolatis, 2–4 cm. longis, 3–4 mm. latis,
pauce serratis, foliis caulium integris, lanceolato-oblongis, 1–2.5 cm. longis,
non glandulosis; caputibus hemisphericis, solitariis; pedunculis 1–8 cm. longis,
superne stipitato-glandulosis; phyllariis glandulosis, ca. 13–14, inaequalibus,
7–8 mm. longis, oblongis, acute 3-lobatis; akeniis nigris, glabris, obcompressis,
ca. 3 mm. longis; bracteis inter flores ligulatos et flores disci in 2 seriebus,
viridibus, acutis, 6–7 mm. longis; floribus disci ca. 35, aureis, corolla 4–5 mm.
longa, setis pappi ca. 30, teretibus, vix plumosis, base non lanosis; akeniis disci
ca. 3 mm. longis, strigosis.

Type from grassy-meadowy slope between Mountain Center and Keen Camp
Summit, San Jacinto Mts., Riverside Co., at 4750 feet, *Louis B. Ziegler,* ca.
June 1, 1966 (RSA). Isotype material being distributed to other herbaria.

This proposed sp. seems nearest to *Layia glandulosa* (Hook.) H. & A. ssp.
lutea Keck and to *L. pentachaeta* Gray. It resembles the former in its basal

dentate leaves, entire cauline leaves, stipitate-glandular peduncles and involucres, ring of bracts between ray- and disk-fls., yellow anthers, setaceous pappus, etc., but differs in the more numerous subterete pappus-bristles which are scarcely plumose and not wholly at the base. From *L. pentachaeta* (which it resembles in lack of wool on the terete pappus-bristles) it differs in its involucre not being pustulate-hirsute and the less plumose pappus. From both taxa it is well removed geographically.

p. 1120.   HEMIZÒNIA
              7. **H. Hallìàna** Keck. Occurs at Cholame, San Luis Obispo Co., *Twisselmann*.
p. 1121.   12. **H. fasciculàta** (DC.) T. & G. Reported from Monterey Co., *Howitt & Howell*.
              16. **H. púngens** (H. & A.) T. & G. was found in Marin Co., *Howell*.
p. 1122.   20. **H. Fítchii** Gray. Reported from the Greenhorn Mts., Kern Co., *Twisselmann*.
p. 1124.   HOLOCÁRPHA
              1. **H. macradènia** (DC.) Greene. Collected ca. 1961 at Watsonville, Santa Cruz Co., *Robbins*.
              2. **H. virgàta** (Gray) Keck; in Marin Co., *Howell*.
              3. **H. obcónica** (Clausen & Keck) Keck. Reported as common on low foothills on e. side of San Joaquin Valley s. to mesas e. of Bakersfield., *Twisselmann*.
p. 1125.   CALYCADÈNIA
              In Key under AA, B, C change "more" to "none" for 1959 printing.
p. 1127.   4. **C. villòsa** DC. Between Brites Valley and Tehachapi Valley, Kern Co., *Twisselmann*.
p. 1128.   11. **C. spicàta** (Greene) Greene. In Greenhorn Range, Kern Co., *Twisselmann*.
p. 1130.   HELÈNIEAE
              Key under last F at bottom of page, insert:

> F′.   Phyllaries 1–3; head usually 1-fld. Introd. . . 50a. *Flaveria*
> F′F′. Phyllaries several; head several-fld. Native. . . . . 51. *Baeria*

p. 1131.   Genera 99. *Hymenopappus* and 100. *Hymenothrix* belong to the *Anthemideae*, p. 1227, and are keyed out there also.
p. 1132.   PSILÓSTROPHE
              1. **P. Coòperi** (Gray) Greene. $n = 16$ (Jackson, 1960; Raven & Kyhos, 1961.
              BÀILEYA
              1. **B. pauciràdiata** Harv. & Gray. $n = 16$ (Raven & Kyhos, 1961).
              2. **B. pleniràdiata** Harv. & Gray. $n = 16$ (Raven & Kyhos, 1961).
              3. **B. multiràdiata** Harv. & Gray. $n = 16$ (Raven & Kyhos, 1961).
p. 1133.   WHÍTNEYA
              1. **W. dealbàta** Gray. $n = 8, 14$ (Stebbins & Major, 1965).
              VENEGÀSIA
              1. **V. carpesioìdes** DC. $n = 19$ (Raven & Kyhos, 1961).
              JAÚMEA
              1. **J. carnòsa** (Less.) Gray. $n = 19$ (Raven & Kyhos).
p. 1134.   EATONÉLLA
              2. **E. Congdònii** Gray. $n = 10$ (Raven & Kyhos, 1961).
              PERÍTYLE
              1. **P. Emòryi** Torr. $n = 53–57$ (Raven & Kyhos, 1961). Reported from El Paso Range, Kern Co., *Twisselmann*.
p. 1135.   LAPHÀMIA and PERÍTYLE
              Shinners (Southwestern Naturalist 4: 204–205. 1959) unites *Laphamia* and *Perityle* under the latter name and makes the combs. *Perityle villosa* (Blake) Shinners for **Laphàmia villòsa** Blake and *P. intricata* (Bdg.) Shinners for **Laphàmia intricàta** Bdg.
              Add **Laphàmia inyóensis** Ferris, differing from **L. villòsa** Blake by the lvs.

being triangular-toothed, mostly truncate at the base; longer hairs to 1–1.5 mm. instead of 0.6 mm.—Inyo Mts., Inyo Co.

p. 1136. HÚLSEA

3a. **H. inyóensis** (Keck) Munz. stat. nov. [*H. californica* T. & G. ex Gray ssp. *inyoensis* Keck, Aliso 4: 101. 1958.] For the most part a much less robust plant than *H. californica* which is an annual and with stems to 7 or 8 mm. thick when well developed. *H. inyoensis* is perennial and often with several stems from an underground base, 3–4 dm. tall, 3–4 mm. in diam. Lvs. green and glandular, the lower tending to be more coarsely and deeply dentate than in *H. californica*. Phyllaries broadly oblong-lanceolate, acuminate to acute, less woolly than in *H. californica*. Aks. 7–7.5 mm. long instead of 5 mm., the pappus-paleae subequal, 1 mm. long, instead of 2–3 mm. long and unequal.

Known from the Inyo and Panamint ranges of Inyo Co., while *H. californica* occurs in the mountains of San Diego Co. To *H. inyoensis* can be referred *Mary DeDecker 356* from junction of Al Rose and Mazourka roads, west slope of Inyo Mts., at 6700 feet, June 10, 1956, and *Mary DeDecker 1371* from west of Goldbelt Springs, Panamint Mts., at 5400 feet, April 30, 1961.

5. **H. vestìta** Gray. $n = 19$ (Raven & Kyhos, 1961).

p. 1137. 6. **H. nàna** Gray. $n = 19$ (Raven & Kyhos, 1961).

GAILLÁRDIA

1. **G. pulchélla** Foug. $n = 17, 18$ (Biddulph, 1944).

2. Add **G. aristàta** Pursh. BLANKETFLOWER. Reported from 2 mi. west of Tehachapi, Kern Co., 7 mi. s. of Grass V., Nevada Co., and 2.4 mi. n. of Hallelujah Junction, Lassen Co., *Fuller*. It differs from *G. pulchella* Foug. in being perennial, with lvs. 5–15 cm. long; ligules 25–30 mm. long, purple at base, the remainder yellow. Native from Ore. to B.C. and N. Dak.

p. 1138. HYMENÓXYS

2. Under **H. Lemmònii** (Greene) Ckll. the Ariz. reference pertains to *H. helenioides* (Rydb.) Ckll., not to *H. Lemmonii*.

4. **H. odoràta** DC. $n = 11$ (Raven & Kyhos, 1961).

p. 1139. HELÈNIUM

1. **H. amárum** (Raf.) Rock. $n = 15$ (Turner & Ellison, 1960).

2. **H. Hoopèsii** Gray. $n = 15$ (Raven & Kyhos, 1961).

4. **H. pubérulum** DC. $n = 29$ (Raven & Kyhos, 1961).

5. **H. Bigelòvii** Gray. $n = 16$ (Raven & Kyhos, 1961). Reported from Greenhorn Mts., Kern Co., *Twisselmann*.

p. 1140. <center>50a. Flavèria Juss.</center>

Glabrous to pubescent annuals, ± succulent. Lvs. opposite, entire or toothed, sessile, sometimes connate. Heads individually inconspicuous. Invol. narrow, prismatic; phyllaries 1–8, subequal. Ray mostly 1 and yellow, or 0. Disk-fls. 1–15. Ak. narrow, 8–10-ribbed. Pappus wanting or rarely of 2–4 scales. Ca. 10 spp., mostly American. (Latin, *flavus*, yellow.)

1. **F. trinérvia** (Spreng.) C. Mohr. Stem 2–12 dm. tall, widely branched, subglabrous; lvs. linear to linear-elliptic, 3–10 cm. long, serrate, 3-ribbed; heads usually 1-fld., in axillary or involucrate clusters; corolla of ♀ fls. 1.5 mm. long, the ligule oblique; corolla of perfect fl. 2 mm. long; ak. 2 mm. long.—Ala. to Ariz. & S. Am. Natur. at Calimesa, Riverside Co., *Fuller*.

p. 1140. LASTHÈNIA

Dr. Ornduff (A biosystematic survey of the goldfield genus Lasthenia. Univ. Calif. Publ. Bot. 40: 1–92. 1966) combines the three genera of the FLORA: *Baeria, Lasthenia* and *Crockeria* under **Lasthènia**. Herewith is given his key for determination of spp.:

A. Phyllaries free.
B. Receptacle subulate; phyllaries usually 3–6; anther tips subulate or deltoid with wartlike glands.
C. Rays very short or apparently absent; invol. cylindrical; disk florets mostly with tetramerous corollas. . . . . 5. *L. microglossa*

CC. Rays longer, over 2 mm. long; invol. turbinate to campanulate; disk florets mostly with pentamerous corollas.
D. Stems fine and wiry; lvs. linear and entire; anther tips subulate.
4. *L. leptalea*
DD. Stems coarser; lvs. with occasional very short lateral teeth; anther tips deltoid with wartlike glands. .... 3. *L. debilis*
BB. Receptacle conic to subglobose; phyllaries usually more than 6; anther tips not as above.
C. Lvs., especially the middle ones, usually pinnately lobed or cleft; corollas remaining yellow in dilute aqueous alkali.
D. Pappus always present, monomorphic, of paleae tapering to an awn; pubescence eglandular; invol. turbinate; phyllaries persistent after ripening of aks.; anther tips deltoid. .... 6. *L. platycarpha*
DD. Pappus present or absent, monomorphic or dimorphic; invol. ± hemispheric; phyllaries deciduous upon ripening of aks.; anther tips linear to ovate.
E. Pappus, when present, strongly dimorphic (except in rare forms of *L. Fremontii*), the longer member awns; herbage eglandular.
F. Ak. bodies over 1.5 mm. long. .... 15. *L. minor*
FF. Ak bodies under 1.5 mm. long.
G. Pappus usually of 2 or more long awns alternating with very short scales (rarely of long awns only or missing).
11. *L. Fremontii*
GG. Pappus usually of 1 long awn and several very short scales. .... 13. *L. Burkei*
EE. Pappus, when present, ± monomorphic, or if dimorphic the longer members not awns; herbage usually glandular. .. 14. *L. coronaria*
CC. Lvs. all essentially entire; corollas turning deep red in dilute aequous alkali.
D. Plants biennial or short-lived perennials; coastal .... 2. *L. macrantha*
DD. Plants annual; coastal or inland. .... 1. *L. chrysostoma*
AA. Phyllaries united into a partial cup, the tips free.
B. Phyllaries united into a cup over ⅔ their length; aks., if epappose, over 1.8 mm. long.
C. Aks. epappose.
D. Aks. obovate to oblong, strongly flattened, with a conspicuous marginal fringe of stiff blunt hairs. .... 8. *L. chrysantha*
DD. Aks. ± clavate, not strongly flattened; ak. pubescence, if any, not restricted to margins.
E. Aks. glabrous or pubescent with rusty or yellowish wartlike papillae. .... 9. *L. glabrata*
EE. Aks. pubescent with short curved hairs. .... 10. *L. Ferrisiae*
CC. Aks. pappose. .... 7. *L. glaberrima*
BB. Phyllaries united ⅛–½ their length; epappose aks. under 1.5 mm. long.
12. *L. conjugens*

A list of the spp. is given as recognized by Ornduff, with chromosome numbers and synonymy:

1. **L. chrysóstoma** (F. & M.) Greene. $n = 8, 16$. Equals No. 3 under *Baeria* in the FLORA with the two sspp. there recognized. [*Baeria chrysostoma* F. & M. *B. gracilis* DC. *Burrielia tenerrima* DC. *B. hirsuta* Nutt. *B. longifolia* Nutt. *B. parviflora* Nutt. *Baeria Clevelandii* Gray. *B. curta* Gray. *B. Palmeri* var. *clementina* Gray. *Lasthenia hirsutula* Greene. *Baeria h.* Greene. *B. chrysostoma* ssp. *h.* Ferris. *B. c.* var. *gracilis* formae *crassa* and *nuda* Hall.]

2. **L. macrántha** (Gray) Greene. $n = 16, 24$. Equals No. 2 under *Baeria* in the FLORA. [*Burrielia chrysostoma* var. *m.* Gray. *Baeria m.* var. *pauci-aristata* Gray. *B. m.* var. *littoralis* Jeps. *B. m.* var. *thalassophila* J. T. Howell.]

3. **L. débilis** (Greene ex Gray) Ornduff. $n = 4$. Equals No. 8 of *Baeria* in the FLORA. [*Baeria d.* Greene ex Gray.]

4. **L. leptàlea** (Gray) Ornduff. $n = 8$. Equals No. 7 under *Baeria* in the FLORA [*B. l.* (Gray) Gray. *Burrielia l.* Gray.]

5. **L. microglóssa** (DC.) Greene. $n = 12$. This is No. 9 of *Baeria* in the FLORA. [*Baeria m.* Greene.]

6. **L. platycárpha** (Gray) Greene. $n = 4$. This is No. 4 under *Baeria* in the FLORA. [*Burrielia p.* Gray. *Baeria p.* Gray. *B. carnosa* Greene. *Lasthenia c.* Greene.]

7. **L. glabérrima** DC. $n = 5$. This is No. 1 under *Lasthenia* in the FLORA. [*Rancagua g.* Endl. ex Walp.]

8. **L. chrysántha** (Greene ex Gray) Greene. $n = 7$. This is No. 1 under *Crockeria* in the FLORA. [*Crockeria* Greene ex Gray.]

9. **L. glabràta** Lindl. $n = 7$. This is No. 2 under *Lasthenia* in the FLORA. [*L. californica* DC. in part. *Hologymne c.* Bartl. *H. glabrata* Bartl. *Monolopia californica* Fisch., Mey & Avé-Lall. *M. glabrata* Fisch., Mey & Avé-Lall.]

10. **L. Ferrísiae** Ornduff. $n = 7$. Aks. intermediate between those of *L. glabrata* ssp. *Coulteri* and *L. chrysantha*; somewhat flattened, obovate to oblong, sparingly to densely clothed on faces and margins with short, whitish or straw-colored curved hairs and papillae.—Common in the San Joaquin Valley from Contra Costa Co. to n. Kern and San Luis Obispo cos.; less common in Butte and Colusa cos. Feb.–May.

11. **L. Fremóntii** (Torr. ex Gray) Greene. $n = 6$. This is No. 6 under *Baeria* in the FLORA. [*Dichaeta F.* Torr. ex Gray. *Burrielia F.* Benth. *Baeria F.* Gray. *B. F.* var. *heterochaeta* Hoov.]

12. **L. cónjugens** Greene. $n = 6$. In the FLORA this is in synonymy under *Baeria Fremontii.* [*B. F.* var. *c.* Ferris.] Description as for *L. Fremontii,* but differing by its phyllaries which are united ¼–⅓ their length; aks. always epappose, glabrous, shining; receptacle hemispheric, densely pubescent.— Vernal pools, formerly from Santa Barbara Co. to Mendocino Co. at various stations.

13. **L. Búrkei** (Greene) Greene. $n = 6$. In the FLORA this was cited in synonymy under *Baeria Fremontii.* [*B. Burkei* Greene.] Like *L. Fremontii* in most respects, but the pappus consisting of 1 long awn and many very short scales.—From vernal pools and wet meadows from s. Mendocino Co. to central Sonoma Co. and in s. Lake Co.

14. **L. coronària** (Nutt.) Ornduff. $n = 5$, or 4. This is No. 1 under *Baeria* in the FLORA as *B. californica.* [*Hymenoxys c.* Hook. *Ptilomeris anthemoides* Nutt. *Burrielia a.* Gray. *Actinolepis a.* Gray. *Baeria a.* Gray. *B. coronaria* f. *a.* Voss. *B. aristata* f. *a.* Hall. *Ptilomeris a.* Nutt. *B. a.* Cov. *P. coronaria* Nutt. *Hymenoxys californica* β *c.* T. & G. *Actinolepis c.* Gray. *B. c.* Gray. *Ptilomeris mutica* Nutt. *Hymenoxys m.* T. & G. *Actinolepis m.* Gray. *Baeria m.* Gray. *B. coronaria* f. *m.* Voss. *B. aristata* f. *m.* Hall. *Ptilomeris affinis* Nutt. *Actinolepis a.* Benth. & Hook. *Baeria a.* Gray. *B. aristata* var. *a.* Hall. *Ptilomeris tenella* Nutt. *Actinolepis t.* Gray. *Baeria t.* Gray. *B. aristata* var. *affinis* f. *truncata* Hall. *B. Parishii* Wats. *B. aristata* var. *P.* Hall. *B. a.* var. *P. f. quadrata* Hall. *B. a.* var. *P. f. varia* Hall.]

15. **L. mìnor** (DC.) Ornduff. $n = 4$. This is No. 5 under *Baeria* in the FLORA. [*Monolepis m.* DC. *Eriophyllum m.* Rydb. *Baeria m.* Ferris. *Dichaeta tenella* Nutt. *Baeria uliginosa* var. *t.* Gray. *B. u.* var. *tenera* Gray. *Lasthenia tenella* Greene. *Baeria tenella* Greene. *Dichaeta uliginosa* Nutt. *Baeria u.* Gray. *Lasthenia u.* Greene.]

Ssp. **marítima** (Gray) Ornduff. [*Burrielia m.* Gray. *Baeria m.* Gray. *Baeria minor* ssp. *m.* Ferris.]

p. 1145. MONOLÒPIA

3. **M. lanceolàta** Nutt. $n = 10$ (Raven & Kyhos, 1961).

p. 1146. ERIOPHÝLLUM

1. **E. lanàtum** (Pursh) Forbes. J. S. Mooring (Madroño 18: 236–239. 1966) reports the following: var. **lanàtum** $2n = 16$; var. **achillaeoìdes** $2n = 16$, 32, 48; var. **aphanáctis** $2n = 32$; var. **arachnoìdeum,** $2n = $ ca. 16; var. **crocèum,** $2n =32$; var. **cuneàtum,** $2n = $ ca. 32; var. **grandiflorum,** $2n = 16$, 32; var. **integrifòlium,** $2n = 16$; var. **lanceolàtum,** $2n = 16$; var. **leucophýllum,** $2n = 32$.

p. 1147. 1. **E. lanàtum** (Pursh) Forbes var. **achillaeoìdes** (DC.) Jeps. occurs in Monterey Co., *Howitt & Howell* and in San Luis Obispo Co., *Hardham.*

p. 1149. 4. **E. confertiflòrum** (DC.) Gray var. **laxiflòrum** Gray seems worth recognition. The upper stems are very slender; lvs. with narrow divisions; infl. open, few-fld.; peduncles 1–3 cm. long; heads. small.—Santa Clara Co. to Tehachapi Mts., Kern Co., thence n. along Sierran foothills to Mariposa Co.

p. 1150. 8. **E. ambíguum** (Gray) Gray var. **paleàceum** (Bdg.) Ferris. Invols. 5–7 mm. high (4.5–5 in typical *ambiguum*); phyllaries broadly acute, not indurate except for carinate ridge; disk-fls. 2–3 mm. long (as against 1.3–2 mm.)—

Mono Co. to Death Valley and to Riverside Co.; Nev. *E. ambiguum* ranges
from Ft. Tejon and the Tehachapi Mts. to the Greenhorn Mts., Kern Co.

p. 1152. RIGIOPÁPPUS
   1. R. leptoclàdus Gray. $n = 9$, not 8 as reported in 1959 printing.

p. 1153. TRICHOPTÍLIUM
   1. T. incìsum (Gray) Gray. $n = 13$ (Ravens & Kyhos, 1961).

p. 1154. CHAENÁCTIS
   Spp. 1, 2 and 4 of the FLORA have $n = 6$ (Raven & Kyhos, 1961).
   Abrams & Ferris (Ill. Fl. Pac. States 4: 240. 1960) recognize under
   7. C. Douglásii (Hook.) H. & A. the following taxa in Calif.:

   A. Annual to biennial, floccose-canescent, to 6 dm. high; fls. white to pinkish; phyllaries
      moderately glandular-puberulent and loosely canescent; pappus-paleae not over half
      the length of the corollas. Ne. Calif. to B.C., Mont., Ariz. . . . . Var. *achilleifolia*
   AA. Perennial.
      B. To 1 dm. high; pappus-paleae to half the corolla length; heads few. From 7000–
         9000 ft. Desert ranges and Sierra Nevada to Wash., Mont. . . . . . . Var. *montana*
      BB. From 1.5–3 dm. high; pappus-paleae more than half the corolla length. From
         3700–10,000 ft., e. side of the Sierra Nevada, and from Tulare Co. to Nevada Co.
                                                                               Var. *rubricaulis*

   3. C. suffrutéscens Gray. $n = 6$ (Mooring, 1965).
   6. C. alpìna (Gray) Jones. $n = 6$ (Mooring, 1965).
   7. C. Douglásii (Hook.) H. & A. In the above treatment var. Douglásii is
   n. of Calif. with $n = 12$ (Gillett, 1954; Raven & Kyhos, 1961); $2n = 12, 24,$
   25 (Mooring, 1965).
   Var. achilleifòlia (H. & A.) A. Nels. [C. a. H. & A.] Mooring (Brittonia 17:
   25. 1965) suggests that Calif. plants are var. *achilleifolia* with $n = 6$ (Raven
   & Kyhos, 1961).
   Var. montàna Jones. [C. *panamintensis* Stockwell.] $n = 6$ (Raven & Kyhos,
   1961).
   Var. rubricaúlis (Rydb.) Ferris. [C. *r.* Rydb.] $n = 6$ (Raven & Kyhos,
   1961).
   8. C. glabriúscula DC. Raven & Kyhos, 1961, give $n = 6$ for vars. lanòsa,
p. 1156. denudàta, cúrta, gracilénta, tenuifòlia. Abrams & Ferris (Ill. Fl. Pac. States 4:
   242. 1960) recognizes C. tanacetifòlia Gray as distinct from C. glabriúscula in
   having 8 pappus-paleae in 2 series instead of ca. 4 in 1 series. They say that
   the former is 6–15 cm. high, few- to several-stemmed and has lvs. with the
   pinnate divisions crowded and has a var. gracilénta with plant 15–25 cm. high,
   more erect; lvs. with pinnae more remote.—C. tanacetifòlia Gray occurs in the
   inner Coast Ranges of Solano, Lake, Napa, and Yolo cos., Santa Clara Co. to
   San Benito. Var. gracilénta (Greene) Stockwell [C. *g.* Greene. C. *g.* var. *fili-*
   folia Jeps. C. *glabriuscula* var. *g.* Keck], Yolo, Lake, and Napa cos.
   9. C. macrántha D. C. Eat. $n = 6$ (Raven & Kyhos, 1961).
   10. C. carphoclínia Gray. $n = 8$ (Raven & Kyhos, 1961).
   Var. attenuàta (Gray) Jones. $n = 8$ (Raven & Kyhos, 1961).
   11. C. Fremóntii Gray. $n = 5$ (Raven & Kyhos, 1961).
p. 1157. 12. C. Xantiàna Gray. $n = 7$ (Raven & Kyhos, 1961).
   13. C. stevioìdes H. & A. $n = 5$ (Raven & Kyhos, 1961).
   Var. brachypáppa (Gray) Hall. $n = 5$ (Raven & Kyhos, 1961).
   14. C. artemisiifòlia (Harv. & Gray) Gray. $n = 8$ (Raven & Kyhos, 1961).
p. 1158. PALAFÓXIA
   1. P. lineàris (Cav.) Lag. $n = 12$ (Raven & Kyhos, 1961); $n = 10$ (Ball,
   1965).
   Var. gigantèa Jones. $n = 12$ (Raven & Kyhos, 1961).
   AMBLYOPÁPPUS
   1. A. pusíllus H. & A. $n = 8$ (Raven & Kyhos, 1961).
p. 1159. BLENNOSPÉRMA
   1. B. nànum (Hook.) Blake var. robústum J. T. Howell. $n = 7$ (Ornduff,
   1960).

2. **B. Bàkeri** Heiser. Reported from 4 mi. s. of El Verano, s. of Sonoma, *W. Roderick.*

PÉCTIS

1. **P. pappòsa** Harv. & Gray ex Gray. $n = 12$ (Raven & Kyhos, 1961).

p. 1160. NICOLLÈTIA

1. **N. occidentális** Gray. $n = 10$ (Raven & Kyhos, 1961).

TAGÈTES

1. **T. minúta** L. Reported from 8 mi. n. of Visalia, Tulare Co., *Fuller*; Exeter, Tulare Co., *Haworth.*

p. 1161. DYSSÒDIA

1. **D. porophylloìdes** Gray. $n = 13$ (Raven & Kyhos, 1961).

2. **D. Coóperi** Gray. $n = 13$ (Raven & Kyhos, 1961).

4. **D. Thúrberi** (Gray) Rob. Hitherto known from mts. of e. Mojave Desert, San Bernardino Co. Taken by *L. B. Ziegler* at Pinyon Flats, San Jacinto Mts., Riverside Co. in a spot ca. 175 ft. in diam. and containing several hundred plants.

POROPHÝLLUM

1. **P. grácile** Benth. $n = 24$ (Raven & Kyhos, 1961).

p. 1163. ASTERÈAE

In Key to Genera, under A, B, CC, DD, E, change F to:

> F. Phyllaries in ± distinct vertical ranks.
> G. Upright stems annual; lvs. rigid, ± persistent.
> 81. *Petradoria*
> GG. Upright stems perennial; lvs. not rigid or persistent.
> 82. *Chrysothamnus*

p. 1164. GRINDÈLIA

1. **G. hùmilis** H. & A. $n = 6$ (Raven et al., 1960).

2. **G. hirsùtula** H. & A. $n = 12$ (Raven et al., 1960).

p. 1165. 3. **G. prócera** Greene. $n = 12$ (Dunford, 1964). Delete *G. camporum* var. *parviflora* as a synonym. The two above taxa can apparently be distinguished:

> A. Rays mostly 32–44, the laminae 8–10 mm. long; aks. 2–4 mm. long, usually dull to fuscous-brown.—San Joaquin V. from Sacramento Co. to Kern Co. and to n. base of San Gabriel Mts. . . . . *G. procera*
> AA. Rays 16–18, the laminae 7–8 mm. long; aks. 4–5.2 mm. long, strawcolor to light brown. About San Francisco Bay. . . . . *G. camporum* var. *parviflora*

4. **G. marítima** (Greene) Steyerm. $n = 12$ (Raven et al., 1960).

5. **G. strícta** DC. ssp. **venulòsa** (Jeps.) Keck. $2n = 24$ (De Jong & Montgomery, 1963). Reported from Monterey Co., *Howitt & Howell. G. arenicola* Steyerm.; $n = 12$ (Raven et al., 1960).

6. **G. latifòlia** Kell.; $2n = 24$ (De Jong & Montgomery, 1963); $n = 12$ (Raven et al., 1960).

Ssp. **platyphýlla** (Greene) Keck. [*G. rubricaulis* var. *permixta* Steyerm.; $n = 12$ (Raven et al., 1964).]

7. **G. squarròsa** (Pursh) Dunal; $n = 6$ (Raven et al., 1960).

p. 1166. 8. **G. Hállii** Steyerm.; $n = 6$ (Raven et al., 1960).

9. **G. campòrum** Greene; $n = 12$ (Raven et al., 1960). [*G. robusta* var. *Davyi* Jeps.; $n = 6$ (Raven et al., 1960).] Insert:

Var. **parviflòra** Steyerm. Rays 16–18, the laminae 7–8 mm. long; aks. 4–5.2 mm. long, stramineous or light brown.—About San Francisco Bay.

11. **G. robústa** Nutt.; $n = 12$ (Raven et al., 1960). [*G. rubricaulis* var. *elata* Steyerm., $n = 6$ (Raven et al., 1960).]

GUTIERRÈZIA

Solbrig, O. T. Cytotaxonomic and evolutionary studies in the N. Am. spp. of Gutierrezia. Contr. Gray Herb. 188: 1–63. 1960. The Calif. spp. of Gutierrezia. Madroño 18: 75–84. 1965.

> A. Heads with only 2–3 fls.; invol. very narrow, to 1.5 mm. wide; aks. of disk fls. aborted. From desert fringes. . . . . 4. *G. microcephala*
> AA. Heads with more than 4 fls.; invol. turbinate, 1.5 or more mm. wide; aks. of disk fls. fertile.

B.   The heads clustered at ends of branchlets; fls. 5–10; invol. less than 5 mm. wide.
     S. Calif. . . . . 3. *G. Sarothrae*
BB. The heads mostly solitary at the ends of branchlets; fls. usually more than 10;
     invol. sometimes more than 5 mm. wide.
     C.   Infl. loosely corymbose; heads 6–10 mm. high, 4–6 mm. wide; open, little-
          branched shrub. San Francisco Bay area. . . . . 1. *G. californica*
     CC. Infl. paniculate; heads 4–7 mm. high, 2–5 mm. wide; a globose, much-
          branched shrub. Coast Ranges and S. Calif. . . . . 2. *G. bracteata*

p. 1167.   1. **G. califórnica** (DC.) T. & G. [*Brachyris c.* DC. *Xanthocephalum c.*
        Greene.] Invol. 6–10 mm. high, 4–6 mm. broad, turbinate to campanulate,
        the phyllaries ca. 20, in 3 overlapping series, lanceolate to ovate; ligulate fls.
        ca. 9, the ligules narrowly lanceolate, 3–4 mm. long; tubular fls. ca. 11, gla-
        brous, 3–4 mm. high. On serpentine, about San Francisco Bay (Angel Id.,
        Oakland, Point Bonita).
                2. **G. bracteàta** Abrams. [*G. californica* var. *b.* Hall.] Invol. 5–6.5 mm. high,
        conical to turbinate, the phyllaries in ca. 3 rows, narrow, elongate, carinate or
        strongly convex, to 4 mm. long and 2 mm. wide; ligulate fls. usually 5 (3–6),
        6–8 mm. long; tubular fls. usually 4 (2–6), 4–6 mm. long. $n = 8$ (Solbrig et
        al., 1964).—Inner Coast Ranges, Yolo Co. to Riverside Co. and L. Calif.
                3. **G. Saròthrae** (Pursh) Britt. & Rusby. As in the FLORA. $n = 12$ (Solbrig
        et al., 1964).
                4. **G. microcéphala** (DC.) Gray. As in the FLORA.
        AMPHIPÁPPUS
                10. **A. Fremóntii** T. & G.; $n = 9$ (Raven et al., 1960).
p. 1168.   ACAMPTOPÁPPUS
                1. **A. sphaerocéphalus** (Harv. & Gray) Gray; $n = 9$ (Raven et al., 1960).
                2. **A. Shóckleyi** Gray; $n = 9$ (Raven et al., 1960).
        CHRYSÓPSIS
                1. **C. villòsa** (Pursh) Nutt. var. **Bolánderi** (Gray) Gray ex Jeps. $n = 9$
        (Raven et al., 1960).
p. 1169.   Var. **fastigiàta** (Greene) Hall; $n = 9$ (Raven et al., 1960).
                Var. **híspida** (Hook.) Gray ex D. C. Eat. $n = 9$ (Raven et al.)
p. 1170.   2. **C. oregòna** (Nutt.) Gray var. **scabérrima** Gray; occurs in the Greenhorn
        Mts., Kern Co., *Twisselmann.*
                3. **C. Brèweri** Gray also in the Greenhorn Mts.
        HETEROTHÈCA
                Shinners in 1951, Wagenknecht in 1960, and Harms in 1965 agree that
        Chrysopsis and *Heterotheca* should be combined under *Heterotheca*, but the
        taxa in *Chrysopsis* are still so imperfectly understood that not all the neces-
        sary combinations have been made.
p. 1171.   1. **H. grandiflòra** Nutt.; $n = 9$ (Raven et al., 1960).
                2. **H. subaxillàris** (Lam.) Britt. & Rusby was found in Ventura Co., *Pollard.*
        $n = 9$ (Smith, 1965).
p. 1172.   CHAETOPÁPPA
                2. **C. aúrea** (Nutt.) Keck; $n = 9$ (Solbrig et al., 1964).
                3. **C. frágilis** (Bdg.) Keck; $n = 9$ (Solbrig et al., 1964).
                4. **C. bellidiflòra** (Greene) Keck; $n = 9$ (Solbrig et al., 1964). Occurs in
        the Santa Lucia Mts., Monterey Co., *Howitt & Howell.*
                5. **C. éxilis** (Gray) Keck; $n = 9$ (Solbrig et al., 1964).
                6. **C. alsinoìdes** (Greene) Keck; $n = 9$ (Solbrig et al., 1964).
p. 1173.   HAPLOPÁPPUS
        In Key under AA and before B insert:

        A′.   Pappus deciduous, ± in a ring; tall annual or biennial herbs with spiny-dentate lvs.
              (Section Prionopsis). . . . . 4a. *H. ciliatus*
        A′A′. Pappus persistent, if plant herbaceous (Section Blepharodon).

p. 1175.   1. **H. grácilis** (Nutt.) Gray. $n = 2, 3$ (Jackson, 1965).
                1a. **H. Ravènii** Jackson near to *H. gracilis* (Nutt.) Gray, but with shorter
        pappus bristles and shorter aks. and with the bases of the larger bristles ca.

twice as wide as in *H. gracilis*. In *H. Ravenii* the invol. is always hirsute, in *H. gracilis* usually strigose. In H. R. $n = 4$; in *H. g. n = 2* or 3. Some plants previously referred to *H. gracilis* belong to *H. Ravenii*.

4a. **H. ciliàtus** (Nutt.) DC. [*Donia c.* Nutt. *Aster c.* O. Kuntze.] Erect, annual or biennial, 5–15 dm. high; stems very leafy to top, glabrous; lvs. oval to oblong, dentate with spine-tipped teeth, very obtuse, 3–8 cm. long; heads few, in open cymes; invol. 12–18 mm. high, the phyllaries in several series, the outer ± squarrose; rays many, the ligules 12–18 mm. long.—Mo. to Tex., Okla., New Mex. Introd. in Ventura and San Francisco cos., *Howell*.

5. **H. carthamoìdes** ssp. **Cusíckii** (Gray) Hall has been found in Plumas Co., *Howell*.

p. 1176.   8. **H. apargioìdes** Gray; $n = 6$ (Solbrig et al., 1964).

p. 1178.   12. **H. Whítneyi** Gray. Occurs in the Greenhorn Mts., Kern Co., *Twisselmann*.

17. **H. Macronèma** Gray. $n = 9$ (Solbrig et al., 1964). Reported from the White Mts., *Howell*.

p. 1180.   24. **H. linearifòlius** DC.; $n = 9$ (Raven et al., 1960).

Var. *intèrior* Jones; $n = 9$ (Solbrig et al., 1964).

25. **H venètus** (HBK.) Blake ssp. **vernonioìdes** (Nutt.) Hall; $n = 6$ (Raven et al., 1960).

p. 1181.   26. **H. acradènius** (Greene) Blake; $n = 12$ (Raven et al., 1960).

Ssp. **eremóphilus** (Greene) Hall; $2n = 12$ (Raven et al., 1960).

27. **H. cànus** (Gray) Blake. $n = 5$ (Raven et al., 1960).

28. **H. squarròsus** H. & A.; $n = 5$ (Raven et al., 1960).

p. 1182.   29. **H. pinifòlius** Gray. Sends out long underground shoots to form new plants. Occurs in Monterey Co., *Howitt & Howell*.

30. **H. ericoìdes** (Less.) H. & A. $2n = 18$ (De Jong & Montgomery, 1963).

p. 1183.   32. **H. Pálmeri** Gray ssp. **pachylèpis** Hall; $n = 9$ (Raven et al., 1960). Reported from San Emigdio Range, Kern Co., *Twisselmann*.

35. **H. Coóperi** (Gray); $n = 9$ (Raven et al., 1960).

36. **H. arboréscens** (Gray) Hall; $n = 9$ (Raven et al., 1960). Reported from the Greenhorn and Piute ranges, Kern Co., *Twisselmann*.

37. **H. Paríshii** (Greene) Blake; $2n = 18$ (DeJong & Montgomery, 1963).

p. 1184.   38. **H. cuneàtus** Gray; $n = 9$ (Solbrig et al., 1964). Occurs in Monterey Co., *Howitt & Howell*.

p. 1185.   SOLIDÀGO

1. **S. occidentàlis** (Nutt.) T. & G.; $n = 9$ (Raven et al., 1960).

2. **S. califórnica** Nutt; $n = 9$ (Raven et al., 1960).

p. 1186.   6. **S. confìnis** Gray; $n = 9$ (Raven et al., 1960; Solbrig et al., 1964).

7. **S. spectábilis** (D. C. Eat.) Gray. At Kernville, Kern Co., *Twisselmann*.

8. **S. spathulàta** DC.; $n = 9$ (Raven et al., 1960).

PETRADÒRIA

Change generic description to:

Suffrutescent herbs with woody caudex; stems several, annual, leafy to the
· infl., resinous, prominently striate; lvs. linear to lanceolate or oblanceolate, 3–5-parallel-veined, coriaceous, entire; infl. open, racemose to corymbose; invol. cylindric, the phyllaries in ± vertical ranks, keelless or nearly so, ovate-oblong; fls. 4–7, yellow, the rays present or absent; aks. somewhat compressed, glabrous; pappus brownish, the bristles somewhat unequal, capillary, finely twisted.

A.   Ray-fls. present; invol. 5–9 mm. high. . . . . 1. *P. pumila*
AA. Ray-fls. absent; invol. 11 or more mm. high. . . . . 2. *P. discoidea*

p. 1187.   2. **P. discoìdea** L. C. Anderson. [*Chrysothamnus gramineus* Hall.] See description of sp. no. 7 on p. 1189. Cf. Anderson, L. R. Studies on Petradoria. Trans. Kans. Acad. Sci. 66: 632–684. 1964.

CHRYSOTHÁMNUS

1. **C. panículàtus** (Gray) Hall. $2n = 18$ (De Jong & Montgomery, 1963).

p. 1188.　　4. **C. viscidiflòrus** (Hook.) Nutt. ssp. **pùmilus** (Nutt.) Hall & Clem. occurs in the Piute Mts. and on Mt. Pinos, Kern Co., *Twisselmann*. L. C. Anderson (Madroño 17: 223–4. 1964) recognizes ssp. **hùmilis** (Greene) Hall & Clem. on the basis of long narrow invols., few fls. in a head; style branches sparsely or not exserted; stigmatic appendages long. Hence separate from ssp. **pubérulus** (D. C. Eat.) Hall & Clem.

　　　　　5. **C. axillàris** Keck. L. C. Anderson makes this a ssp. of *C. viscidiflorus*.

p. 1189.　　7. **C. gramíneus** Hall. See **Petradoria discoidea** L. C. Anderson.

　　　　　8. **C.** × **Bolánderi** (Gray) Greene. Considered to be a hybrid between *C. nauseosus* and *Haplopappus Macronema* by Anderson & Reveal, Madroño 18: 225–232. 1966.

p. 1190.　　8. **C. Párryi** (Gray) Greene ssp. **vulcánicus** (Greene) Hall & Clem. is found in the Greenhorn Mts., Kern Co., *Twisselmann*.

p. 1191.　　9. **C. nauseòsus** (Pall.) Britton ssp. **consímilis** (Greene) Hall & Clem.; $n =$ 9 (Raven et al., 1960).

　　　　　Ssp. **mohavénsis** (Greene) Hall & Clem.; $n = 9$ (Raven et al., 1960).

p. 1192.　MONÓPTILON

　　　　　2. **M. bellioìdes** (Gray) Hall; $2n = 8$ (Raven et al., 1960).

p. 1194.　PSILÁCTIS

　　　　　1. **P. Còulteri** Gray. $n = 5$ (Solbrig et al., 1964).—Turner and Horne (Brittonia 16: 316–331. 1964) include *Psilactis* in *Machaeranthera* and have the combination *M. Coulteri* (Gray) Turner & Horne.

p. 1196.　ÁSTER

　　　　　2. **A. oregonénsis** (Nutt.) Cronq. ssp. **califórnicus** (Durand) Keck. Add to the synonymy *Sericocarpus o*. ssp. *c*. Ferris.

p. 1197.　　9. **A. subspicàtus** Nees; $n = 25$ (Raven et al., 1960).

p. 1198.　　10. **A. Eatònii** (Gray) Howell occurs in the Piute Mts., Kern Co., *Twisselmann*.

　　　　　13. **A. occidentàlis** (Nutt.) T. & G. Ferris (Madroño 15: 128. 1959) recognizes under *A. occidentalis* (which she restricts to the Sierra Nevada and n.):

　　　　　Var. **Paríshii** (Gray) Ferris. [*A. Fremontii* var. *P*. Gray.] Ca. 3 dm. tall; lvs. slightly clasping, linear-lanceolate, the basal oblanceolate; heads in a short cymose panicle (instead of 1 to few); invol. 5–8 mm. high (instead of 6 mm.) —Mts. of s. Calif. to L. Calif.

　　　　　Var. **delectábilis** (Hall) Ferris. [*A. d*. Hall.] Lvs. definitely clasping; heads 1–few; invol. 8–10 mm. high. $n = 16$ (Raven et al., 1960). Fresno and Tulare cos. in the Sierra Nevada.

p. 1199.　　16. **A. adscéndens** Lindl. in Hook. Found in Piute Mts., Kern Co., *Twisselmann*.

p. 1200.　　17. **A. pauciflòrus** Nutt. occurs at Isabella, Kern Co., *Twisselmann*.

　　　　　18. **A. alpígenus** ssp. **Andersònii** (Gray) Onno; $n = 18$ (Raven et al., 1960).

　　　　　21. **A. spinòsus** Benth; $n = 9$ (Raven et al,. 1960).

p. 1201.　　23. **A. éxilis** Ell.; $2n = 10$ (Huziwara, 1958); $n = 5$ (Solbrig et al., 1964).

p. 1202.　MACHAERÁNTHERA ·

　　　　　2. **M. leucanthemifòlia** (Greene) Greene; $n = 5$ (Raven et al., 1960).

p. 1203.　　4. **M. shasténsis** Gray var. **glossophýlla** (Piper) Cronq. & Keck. Next to last word in paragraph should be "*M*." not "*A*." *canescens*.

p. 1204.　　5. **M. canéscens** (Pursh) Gray; $n = 4$ (Raven et al., 1960; Solbrig et al., 1964).

　　　　　6. **M. tortifòlia** (Gray) Cronq. & Keck; $n = 6$ (Raven et al., 1960).

p. 1205.　CORETHRÓGYNE

　　　　　1. **C. califórnica** var. **obovàta** (Benth.) Kuntze; $n = 5$ (Raven et al., 1960).

　　　　　3. **C. filaginifòlia** (H. & A.) Nutt.; $2n = 10$ (De Jong & Montgomery, 1963).

p. 1206.　　Var. **pinetòrum** Jtn.; $n = 5$ (Raven et al., 1960).

　　　　　Var. **brevícula** (Greene) Canby; $n = 5$ (Raven et al., 1960).

　　　　　Var. **bernardìna** (Abrams) Hall; $n = 5$ (Raven et al., 1960).

　　　　　Var. **virgàta** (Benth.) Gray; $n = 5$ (Raven et al., 1960).

p. 1207.   Var. linifòlia Hall is treated as a sp. *C. linifolia* (Hall) Ferris in Abrams and Ferris, Ill. Fl. Pac. States 4: 238. 1960.

p. 1210.   ERÍGERON

    In Key after CC, D insert:

> D′.   Lvs. to 2.5 cm. long, in part 3-toothed or -lobed at apex; heads on solitary peduncles, to 2 cm. across; rays many, pink or whitish or purplish. Garden escape. .... 39a. *E. Karwinskianus*
> D′D′. Lvs. not much toothed or lobed at apex. Native.
> E.   Root-crown, etc. as in the FLORA.

p. 1211.   4. **E. philadélphicus** L.; $2n = 18$ (Mulligan, 1957).

    5. **E. glaùcus** Ker. $n = 9$ (Pagni, 1954; Raven et al., 1960).

p. 1214.   21. **E. aphanáctis** (Gray) Greene; $n = 9$ (Solbrig, et al., 1964).

    23. **E. Bloòmeri** Gray; $n = 9$ (Solbrig et al., 1964).

p. 1217.   38. **E. Brèweri** var. **porphyreticus** (Jones) Cronq.; $2n = 18$ (Montgomery & Yang, 1960).

p. 1218.   39. **E. foliòsus** Nutt. in Tehachapi Mts., Kern Co., *Twisselmann.*

    Var. **Hartwégii** (Greene) Jeps. at Tejon Pass, Kern Co., *Twisselmann.*

    Var. **stenophýllus** (Nutt.) Gray; $n = 9$ (Raven et al., 1960). Reported from Monterey Co., *Howitt & Howell.*

p. 1220.   44. **E. strigòsus** Muhl. $n = 35–36$ (Taylor, 1967).

p. 1221.   LESSÍNGIA

    1. Ferris (Contr. Dudley Herb. 5: 102. 1959) transferred **L. germanòrum** Cham. var. **Peirsònii** J. T. Howell to *L. Lemmonii* Gray var. *Peirsonii*. She recognized

p. 1222.   *L. ramulosíssima* A. Nels as *L. Lemmonii* var. *r.* (A. Nels.) Ferris, with narrower invols. than in *L. Lemmonii*. She had *L. glandulifera* Gray $n = 5$ (Raven et al., 1960) as a sp. with var. *tomentosa* (Greene) Ferris and var. *pectinata* (Greene) Jeps.

    3. **L. nemáclada** Greene; $n = 5$ (Solbrig et al., 1964) [*L. mendocina* Greene; $n = 5, 6$ (Raven et al., 1960).]

p. 1223.   6. **L. ramulòsa** Gray in Benth.; $n = 5$ (Raven et al., 1960).

    Ferris had *L. micradenia* Greene as a sp. with var. *glabrata* (Keck) Ferris. and var. *arachnoidea* (Greene) Ferris.

    7. **L. hololeùca** Greene; $n = 5$ (Solbrig et al., 1964).

p. 1224.   CONÝZA

    The authority for the genus should be Less., not L., for the 1959 printing.

    1a. After *C. canadensis* insert **C. Bilboàna** Remy. Like *C. canadensis* in appearance, but the corollas of the outer fls. obliquely tubular, not ligulate.— A weed in the San Francisco region; native of S. Am.

p. 1226.   BÁCCHARIS

    3. **B. Douglásii** DC.; $n = 9$ (Solbrig et al., 1964).

    4. **B. glutinòsa** Pers.; $n = 9$ (Turner et al., 1961).

    5. **B. vimínea** DC.; $n = 9$ (Solbrig et al., 1964).

    6. **B. Emòryi** Gray; $2n = 18$ (De Jong & Montgomery, 1963).

p. 1227.   8. **B. sergiloìdes** Gray. $2n = 18$ (De Jong & Montgomery, 1963).

    9. **B. sarothroìdes** Gray; $2n = 18$ (De Jong & Montgomery, 1963).

p. 1228.   ÁNTHEMIS

    Key out *Anthemis* as follows:

> A.   Plants annual; rays white.
> B.   Phyllaries brown, the summit largely membranaceous, dark; stems very slender, ± reddish. Sonoma Co. .... 1. *A. fuscata*
> BB. Phyllaries not brown, scarious-tipped or -margined.
> C.   Plant glabrous or slightly hairy, foetid; aks. tubercled; receptacle-scales linear-acute. .... 2. *A. Cotula*
> CC. Plant pubescent or woolly, aromatic; aks. ribbed, not tubercled; receptacle-scales lanceolate-cuspidate. .... 3. *A. arvensis*
> BB. Plant perennial; rays yellow. .... 4. *A. tinctoria*

    3. **A. arvénsis** L. One–6 dm. high; lvs. 3–5 cm. long, bipinnatifid; phyllaries viscid-tomentose, narrowly hyaline-margined; rays 15–20; pappus a minute crown or none.—Garberville, Humboldt Co.; native of Eu.

4. A. tinctòria L. Perennial, 3–9 dm. high, whitish-villous; lvs. evenly pinnate, the pinnae incised; rays 20–30, yellow; pappus a short crown.—Escape in Plumas and Alameda cos.; from Eu.

ACHILLÈA

1. A. Millefòlium L. reported from 2 mi. out of Magalia along Pentz road, Butte Co., *Howell.*

p. 1230. HYMENÒTHRIX

Cite as reference: Turner, B. L. Taxonomy ˉ Hymenothrix. Brittonia 14: 101–119. 1962.

2. H. Loomísii Blake. Recorded as e. of Weimar, Placer Co., *Fuller.*

p. 1231. CHRYSÁNTHEMUM

1. C. carinàtum L. $n = 9$ (Rana, 1965). Reported from Oceano. San Luis Obispo Co., *Fuller.*

4. C. Leucánthemum L. $2n = 18, 36$ (Böcher & Larsen, 1957).

5. C. anethifòlium Webb. & Barth. $n = 9$ (Linder & Lambert, 1965).

6. **C. Parthènium** (L.) Bernh. Delete "Pers."

p. 1232. TANACETUM

2. T. Douglásii DC. $n = 27$ (Raven). Raven calls attention to the fact that this sp. forms large mats, while *T. camphoràtum* Less., $n = 27$ (Raven) has more individual plants.

3. T. camphoràtum Less. is recorded by *Raven* as from Samoa, Humboldt Co.

p. 1233. MATRICÀRIA

Ferris in Abrams Ill. Fl. Pac. States 4: 399. 1960 recognizes:

A. Pappus-crown minute, entire, the aks. with 2 glandular lines which extend the length of the ak. but not into the minute pappus-crown; mature heads usually 6–9 mm. high. . . . . 1. *M. matricarioides*
AA. Pappus-crown evident, with 2 short lobes or teeth, these bearing oblanceolate or elliptic glands which scarcely extend onto the body of the ak.; mature heads usually 10–11 mm. high. . . . . 2. *M. occidentalis*

1. M. matricarioìdes (Less.) Porter.

2. M. occidentàlis Greene. [*Chamomilla o.* Rydb.] Much like the preceding in its general appearance, but differing in the above technical characters. Apparently quite widely distributed in Calif.

p. 1234. ARTEMÍSIA

To be added to literature citations: Beetle, A. A. New names within the section Tridentatae. Rhodora 61: 82–85. 1959; and A study of Sagebrush. Univ. Wyo. Bull. 368. 1960:

In Key, change DD to:

DD. Fls. of center of disk sterile, the outer ♀ and fertile.
   D′.   Plants from a taproot; lvs. largely basal, the divisions ca. 1 mm. wide. Marble Mts. . . . . 9a. *A. campestris*
   D′D′. Plants from horizontal rhizomes; lvs. well distributed, they or their divisions mostly more than 3 mm. wide.
      E.   Principal lvs. narrow, 1 cm. or less wide exclusive of lobes when present, tomentose on both sides or green above; stems rarely more than 1 m. tall.
      E′.   Lvs. entire to bipinnatifid with entire lobes, these usually rather broad. . . . . 10. *A. ludoviciana*
      E′E′. Lvs. finely divided, at least some of them bipinnatifid, some of the segms. again toothed.
                        10a. *A. Michauxiana*
    EE. Principal lvs. 1–5 cm. wide, etc. as in FLORA.

2. A. tridentàta Nutt. should be reported from the inner S. Coast Ranges: on San Carlos Creek 1 mi. n. of New Idria, San Benito Creek at 2400 ft. and 0.5 mi. below New Idria at 3700 ft., *Howbecker & Quibell*; and from 12 mi. nw. of Coalinga, Fresno Co., in the n. half Sect. 1, Township 19S, Range 14E, Mt. Diablo Base and Meridian, *H. W. Wolfram.* Mr. Wolfram reports the latter station as covering over 400 acres.

3. **A. arbúscula** Nutt. ssp. **nòva** (A. Nels.) Ward occurs n. of Kenworthy Ranger Station, San Jacinto Mts., *L. B. Ziegler.* This material frequently layers.

p. 123ϵ.   7. **A. califórnica** Less. var. **insulàris** (Rydb.) Munz. Raven (Aliso 5: 341. 1963) proposes a new name *A. nesiótica* Raven. It differs from *A. californica* in the lvs. not or little revolute, less lobed, the segms. 1–3 dm. wide; ray-fls. to 15; disk-fls. to 40 (as opposed to 6–10 and 15–30 respectively in *A. californica*). San Clemente, Santa Barbara and San Nicolas ids.

9a. **A. campéstris** L. ssp. **boreàlis** (Pallas) Hall & Clem. var. **Wormskióldii** (Besser) Cronq. A perennial 1.5–3 dm. high, with a reddish, ± canescent stem and crow?ᵈ rather narrow panicle; lvs. minutely gray-silky, mostly basal, the blades 1.5–2 cm. long, the divisions linear, scarcely 1 mm. wide; heads with invols. ca. 4 mm. high, ± villous.—On dry wind-swept bluff at 7200 ft., Marble Rim, Marble Mts., Siskiyou Co., July 22, 1966, *Gilbert Muth*. Heretofore reported from n. Ore. n. and e. The specimen seen is the basis of the above description and, so far as I can determine, falls into the above classification.

p. 1237.   10a. **A. Michauxiàna** Bess. Perennial herb 2–4 dm. high; lvs. bipinnatifid with the secondary lobes again toothed, the lobes linear, widely spreading and acute; infl. narrowly paniculiform, with heads nodding, at least at first; invol. 3.5–4 mm. high, glabrous or sparingly tomentose.—Occasional in n. Calif.; to B.C., Alta., Utah. May–Aug.

13. **A. Dracúnculus** L. ascends to 10,300 ft. in White Mts.

p. 1238.   SENECIÒNEAE

In Key, line 5 from bottom of page, after "perennial" add "or annual."

p. 1239.   ADENOCÁULON

1. **A. bìcolor** Hook.; $n = 23$ (Ornduff et al., 1963).

DIMERÈSIA

1. **D. Howéllii** Gray; $n = 7$ (Ornduff et al., 1963).

p. 1240.   ÁRNICA

1. **A. Chamissònis** Less. ssp. **foliòsa** (Nutt.) Maguire; $n = 53–54$ (Ornduff et al., 1963).

p. 1241.   3. **A. amplexicáulis** Nutt.; $n = 33–34$ (Ornduff et al., 1963).

4. **A. móllis** Hook. is found in the White Mts.

5. **A. diversifòlia** Greene. $n = 54–57$ (Taylor, 1967).

6. **A. Párryi** Gray; $n = $ ca. 36 (Ornduff et al., 1963).

p. 1242.   10. **A. latifòlia** Bong.; $n = 19$ (Ornduff et al., 1963).
p. 1243.   15. **A. discoìdea** Benth.; $n = 38$ (Ornduff et al., 1963).
p. 1244.   RAILLARDÉLLA

1. **R. Mùirii** Gray. Reported from Tulare Co., *Howell*.

p. 1245.   SENÈCIO

Add to literature citations: Barkley, T. M. A revision of Senecio aureus L. and allied spp. Trans. Kans. Acad. Sci. 65: 318–408. 1962.

In Key after A, BB, CC, DD, E, insert:

> E'.   Fls. white, blue, pink or purple-red; lvs. large, cordate-ovate to -triangular. Garden escape. .... 9a. *S. cruentus*
> E'E'.   Fls. yellow to cream. Native.
> F.   Herbage ± villous, etc. as in the FLORA.

p. 1246.   1. **S. Lyònii** Gray. $n = 20$ (Ornduff, et al., 1963).
p. 1247.   2. **S. Douglásii** DC. var. **tularénsis** Munz was found in the Greenhorn Range, Kern Co., *Twisselmann*.

Var. **monoénsis** (Greene) Jeps.; $n = 20$ (Ornduff et al., 1963).

3. **S. Blochmániae** Greene. $n = 20$ (Ornduff et al., 1963).

4. **S. spartioìdes** T. & G. $n = 20$ (Ornduff et al., 1963).

5. **S. Fremóntii** T. & G. and var. **occidentàlis** Gray. $n = 20$ (Ornduff et al., 1963).

7. **S. Clarkiànus** Gray occurs in Nevada Co., *True & Howell*, and in the Greenhorn Mts., Kern Co., *Twisselmann*.

p. 1248.   8. **S. triangulàris** Hook. $n = 20, 40$ (Ornduff et al., 1963).

9. **S. sérra** Hook. $n = 20$ (Ornduff et al., 1963).—Reported from the Piute Mts., Kern Co., *Twisselmann*.

9a. **S. cruéntus** DC. The cult. form (Florists' *Cineraria*) with large cordate-ovate to triangular lvs., white tomentose beneath, and with large heads to 7 cm. across and fls. in shades of blue, pink, purple to white.—An occasional escape from gardens, as in the San Francisco area. Native of Canary Ids.

10. **S. integérrimus** Nutt. var. **màjor** (Gray) Cronq.; $n = 40$ (Ornduff et al.). Var. **exaltàtus** (Nutt.) Cronq.; $n = 20, 40$ (Ornduff et al., 1963).

11. **S. aronicoìdes** DC. [*S. exaltatus* var. *uniflosculosus* Gray.] $n = 20$ (Ornduff et al., 1963). The sp. has been collected 10 mi. sw. of Salinas, Monterey Co., *Hardham*.

12. **S. foètidus** Howell. $n = 20$ (Ornduff et al., 1963).

p. 1249.   13. **S. hydróphilus** Nutt. [*S. foetidus* var. *hydrophiloides* T. M. Barkley.] $n = 20$ (Ornduff et al., 1963).

14. **S. astéphanus** Greene. The range extends into Monterey Co., *Howitt & Howell.*

16. **S. cànus** Hook. $n = 23, 46$ (Ornduff et al., 1963). $n = 69$ (Taylor, 1967).

18. **S. bernardìnus** Greene. $n = 23$ (Ornduff et al., 1963).

19. **S. ionophýllus** Greene. $n = 23$ (Ornduff et al., 1963). It occurs in the Tehachapi Mts. and on Piute Creek, Kern Co., *Twisselmann.*

p. 1250.   20. **S. Greènei** Gray. $n = 20–23$ (Ornduff et al., 1963).

21. **S. werneriifòlius** Gray. $n = 22$ (Wiens & Halleck, 1962).

23. **S. pauciflòrus** Pursh. Barkley recognizes **S. pseudáureus** Rydb. for *S. pauciflorus* var. *fallax* Greenm. ex Jeps. and **S. indécorus** Greene for *S. pauciflorus* var. *jucundulus* Jeps., separating them as follows:

A.   Basal lvs. subcordate to truncate at base, lanceolate; heads 5–20; florets yellow.
                                                                    S. pseudaureus

AA. Basal lvs. cuneate to rounded at base, broader.
     B.   Lvs. thickish, suborbicular; heads mostly 1–6; florets orange. .... *S. pauciflorus*
     BB. Lvs. thin, membranaceous, elliptic-ovate to oblong or subreniform; heads usually
          8–20; florets yellow. .... *S. indecorus*

24. **S. cymbalarioìdes** Nutt. Change name to **S. streptanthifòlius** *Greene,* since there was an earlier use of the name S. c. by Buck. $n = 23$ (Ornduff et al., 1963).

25. **S. Clevelándii** Gray. $n = 23$ (Ornduff et al., 1963).

p. 1251.   27. **S. Brèweri** Davy. $n = 23$ (Ornduff et al., 1963).

28. **S. eurycéphalus** T. & G. $n = 23$ (Ornduff et al., 1963).

29. **S. multilobàtus** T. & G. $n = 23$ (Ornduff et al., 1963).

31. **S. Jacobaèa** L. $2n = 32, 40$ (Böcher & Larsen, 1955). Reported from Humboldt and Del Norte cos., *Fuller.*

p. 1252.   34. **S. aphanáctis** Greene. $n = 20$ (Ornduff, 1963).

35. **S. sylváticus** L. $n = 20$ (Ornduff et al., 1963).

36. **S. vulgàris** L. $n = 20$ (Ornduff et al., 1963).

37. **S. mohavénsis** Gray. $n = 20$ (Ornduff et al., 1963).

38. **S. mikanioìdes** Otto. $n = 10$ (Ornduff et al., 1963).

CROCÍDIUM

1. **C. multicáule** Hook. Found at 1600 ft., 2 mi. w. of jct. of Watts V. Road with Maxon Road, Fresno Co., *Weiler;* and 10.2 mi. e. of jct. of Pine Flat Road and Trimmer Springs Road, *Weiler;* also in Kern Co., *Twisselmann* and San Luis Obispo Co., *Hardham.* $n = 9$ (Ornduff et al., 1963).

p. 1253.   PETASÌTES

1. **P. palmàtus** (Ait.) Gray. $2n = 30$ pairs (D. E. Anderson, 1963).

PSATHYRÒTES

1. **P. ramosíssima** (Torr.) Gray. $n = 17$ (Ornduff et al., 1963).

2. **P. ánnua** (Nutt.) Gray. $n = 17$ (Ornduff et al., 1963).

p. 1254.   LUÌNA

1. **L. hypoleùca** Benth. $n = 30$ (Ornduff et al., 1963).

ERECHTITES

1. **E. argùta** (A. Rich.) DC. Add as synonym *Senecio glomeratus* Desf. The sp. has been recorded as in San Luis Obispo Co., *Hardham,* and in Monterey

Co., *Howitt & Howell.*

2. **E. prenanthoìdes** (A. Rich.) DC. Add as synonym *Senecio minimus* (Poir.) DC. The sp. occurs in Monterey Co., *Howitt & Howell.*

PEUCEPHÝLLUM

1. **P. Schóttii** (Gray) Gray. *n* = 20 (Ornduff et al., 1963).

p. 1255.   TETRADÝMIA

1. **T. axillàris** A. Nels. *n* = 30 (Ornduff et al., 1963).

p. 1256.   5. **T. glabràta** Gray. *n* = 62 (Ornduff et al., 1963).

LEPIDOSPÁRTUM

1. **L. squamàtum** (Gray) Gray. *n* = 30, ca. 45 (Ornduff et al., 1963).

2. **L. latisquàmum** Wats. *n* = 30 (Ornduff et al., 1963).

p. 1257.   CALENDÙLEAE

Change Key to:

A.   Aks. strongly incurved; lvs. entire or nearly so. . . . . 121. *Calendula*
AA.   Aks. straight; lvs. toothed or lobed. . . . . 121a. *Dimorphotheca*

121a. Dimorphothèca Vaill. Cape-Marigold

Herbs or undershrubs; lvs. alternate, entire to pinnatifid. Heads solitary on terminal peduncles; invol. broadly campanulate; phyllaries in 1 series, with scarious margins; receptacle naked. Ray-fls. ♀, fertile in 1 row, the style divided into 2 long stigmatic branches. Disk-fls. bisexual, fertile; style with 2 short branches. Aks. of ray-fls. 3-angled to subterete, usually wrinkled or tuberculate, or disk-fls. smooth and with thickened margins. Ca. 7 spp. of S. Afr.

1. **D. sinuàta** DC. Annual, 1–3 dm. high, loosely branched, glandular-pubescent; lvs. oblong-lanceolate, to 9 cm. long, coarsely sinuate-dentate; heads 3.5 cm. across, the rays orange-yellow.—Reported from San Diego, Riverside, San Bernardino, Ventura, Santa Barbara and Kern cos.

2. **D. Ecklònis** DC. Shrubby perennial with narrow toothed lvs.; peduncles bearing large solitary daisy-like heads having rays white above, purplish beneath; disk blackish-blue; head closing at night.—Occasional escape from cult., as at Santa Barbara, *Pollard.*

p. 1257.   INULÈAE

In Key at bottom of Page, after A, insert:

A.   Pappus of 5–many equally long bristles. . . . . 122. *Inula*
AA.   Pappus of an outer row of short scales and an inner of long bristles. 122a. *Pulicaria*

p. 1258.   ### 122a. Pulicària Gaertn.

Distinguished from *Inula* by the 2-ranked pappus, the outer row with short scales, the inner with longer bristles. (Latin, *pulicarius,* flea-like.)

1. **P. hispánica** (Boiss.) Boiss. Annual to perennial herb, to ca. 1 m. tall, branched, short-villous throughout; lvs. alternate, oblong to narrow-oblanceolate, clasping at base, 1–6 cm. long, entire; heads many, 7–10 mm. diam.; phyllaries linear, attenuate, 3–4.8 mm. long, villous; rays yellow, the ligules 1.5–2 mm. long; aks. cylindrical, 1 mm. long, brownish.—Adventive along streams, etc., from San Luis Obispo, Orange, Riverside cos. From the Medit. region. (Cf. Raven, Aliso 5: 251–4. 1963).

PLÙCHEA

2. **P. serícea** (Nutt.) Cov. In Kern R. region, *Twisselmann.* *n* = 10 (Turner, Powell and King, 1962).

p. 1259.   GNAPHÀLIUM

In Key, after AA, BB, C, D, EE change to:

EE.   Infl. paniculate.
     F.   Phyllaries dingy, straw-colored to pale brownish; ♀ fls. 21–
         32; hermaphrodite fls. 11–14; heads campanulate-subglo-
         bose. Trinity and Plumas cos. n. . . . . 7a. *G. Macounii*

FF. Phyllaries pink at least when young; ♀ fls. 45–58; hermaphrodite fls. 6; heads turbinate to narrow-campanulate. Humboldt Co. to Orange Co. . . . . 7. *G. ramosissimum*

In Key, AA, BB, C, DD, *G. bicolor* should be no. 9; *G. leucocephalum* no. 8.
1. **G. purpùreum** L. 2*n* = 28 (Huynh, 1965).
3. **G. japónicum** Thunb. In Del Norte, Napa and San Joaquin cos., *Fuller*.

p. 1260. 7a. **G. Macóunii** Greene. [*G. decurrens* Ives, not L. *G. Ivesii* Nels & Macbr.]
Annual or biennial, 4–9 dm. tall, corymbose above, glandular-pilose, increasingly tomentose above; lvs. thin, lanceolate to linear, 3–10 cm. long, sessile,
shortly decurrent; panicle short, to 1.5 dm. across; heads 5–6 mm. high; phyllaries imbricate, dingy, thin-scarious and shining, woolly at base; corollas yellowish; pappus-bristles falling separately.—Yellow Pine F., Trinity and Plumas
cos.; to B.C. and Quebec. July–Oct.
9. **G. bicolor** Bioletti. Add Sierran foothills, from Madera Co. to Tulare
Co., *Ferris*.
11. **G. beneòlens** A. Davids. Occurs in the Greenhorn Range, Kern Co.,
Twisselmann; in Monterey Co., *Howitt & Howell*.

p. 1261. 13. **G. lùteo-álbum** L. The first word in line 3 should be "cm." not "dm."
Occurs at Kernville, etc., Kern Co., *Twisselmann*; on San Clemente Id., *Raven*.

p. 1262. ANTENNÀRIA
4. Ferris in Abrams, Ill. Fl. Pac. States 4: 476. 1960, separates **A. luzuloìdes** T. & G. (with heads in a close corymbiform cyme; lvs. mostly 3-nerved,
the lower 3–10 cm. long, the cauline little reduced upward) and occurring
from Lasssen Co. n. and **A. microcéphala** Gray (heads in a loose or close
panicle; lvs. 1-nerved or the midrib obscure; lower lvs. 2–4.5 cm. long; cauline
lvs. abruptly reduced upward) and from Lassen, Glenn and Trinity cos. n.
6. **A. ròsea** Greene. Ferris recognizes also **A. marginàta** Greene [*A. dioica*
var. *m.* Jeps.] from the San Bernardino Mts., as having basal lvs. glabrate on
upper surface, 1–3 cm. long; plant cespitose, 2–20 cm. high; invols. campanulate, woolly at base, 6–8 mm. high, the phyllaries with pale green base, often
brown or purplish submedial spot.

p. 1263. ANÁPHALIS
1. Change **A. margaritàcea** to (L.) Benth. ex C. B. Clarke.

p. 1264. FILÀGO
F. vulgàris Lam. [*F. germanica* L.] has been reported from Mendocino Co.
It differs from the spp. enumerated in the FLORA by branching proliferately
from below the lowest capitate cluster of heads (instead of these being terminal or axillary) and by the outer receptacular bracts being cuspidate with the
receptacle subulate. A native of Eu.
Chrtek & Holub (Preslia 35: 9. 1963) place *F. gallica*, *F. californica*, *F. arizonica*, and *F. depressa* in a genus *Oglifa*.

p. 1265. PSILOCÁRPHUS
1. **P. brevíssimus** Nutt. Small poolbeds, Greenhorn foothills, Kern Co.,
Twisselmann.
3. **P. tenéllus** Nutt. var. **ténuis** (Eastw.) Cronq. is reported in Marin Co.
and extending to the coast in Monterey Co., *Howell*.

p. 1266. EVÁX
1. **E. cauléscens** (Benth.) Gray is reported from a vernal pool 6 mi. e. of
Paso Robles on the road to Creston, San Luis Obispo Co., *Hardham*.
Var. **hùmilis** (Greene) Jeps. in the Greenhorn Range, Kern Co., *Twisselmann*.
2. **E. sparsiflòra** (Gray) Jeps. var. **brevifòlia** (Gray) Jeps. is in the San
Marcos Pass, Santa Barbara Co., *Raven*.
3. **E. acáulis** (Kell.) Greene in the Santa Lucia Mts., Monterey Co., *Howitt
& Howell*.

p. 1267. TRICHOCORÒNIS
1. **T. Wrìghtii** (T. & G.) Gray. *n* = 15 (Turner, Powell & King, 1962).

p. 1267. HOFMEISTÈRIA

In Phytologia 12: 469 King & Robinson make the combination **Pleuro-corònis plurisèta** (Gray) King & Robinson.

p. 1269.   BRICKÉLLIA

2. **B. multiflòra** Kell. has been found in Jawbone Canyon, w. Mojave Desert, Kern Co., *Twisselmann.*

5. **B. microphýlla** (Nutt.) Gray collected by *Hardham* at Caliente Creek, Kern Co.

p. 1271.   CYNARÈAE

In Key, under AA, B, CC insert:

> C′.   Receptacle densely pitted, the pits membranous-bordered, not densely setose. .... 139a. *Onopordum*
> C′C′.   Receptacle densely setose.
> D.   Fils. united below, as in FLORA.

p. 1272.

139a. Onopórdum L.

Tall herbs with erect stems and sinuate or pinnatifid spiny lvs., the cauline lvs. decurrent and the stems conspicuously spiny-winged. Heads homogamous; invol. of many rows of conspicuous spiny tegules. Receptacle naked, deeply pitted, the pits with toothed membranous borders. Florets all tubular, hermaphrodite, usually red-purple; anthers with terminal subulate appendages and short basal tails. Aks. obovoid, compressed or 4-angled; pappus of many rows of rough hairs united into a basal ring. Ca. 20 spp., Eurasian. (Greek name for Cotton Thistle.)

> A.   Invol. glandular-pubescent; mature lvs. green, glabrescent. .... 1. *O. tauricum*
> AA.   Invol. arachnoid; lvs. gray or whitish beneath. .... 2. *O. Acanthium*

1. **O. táuricum** Willd. Biennial, 3–5 dm. tall, branched; invol. sub-globose, the phyllaries lanceolate, spine-tipped; fls. purplish, glabrous.—Medit. region. Adventive in Siskiyou Co., sw. of Dorris, *Fuller.*

2. **O. Acánthium** L. SCOTCH THISTLE. Biennial, 5 to 20 dm. tall, with a close arachnoid tomentum; invol. subglobose, 3–5 cm. diam.; phyllaries lanceolate; fls. reddish purple. 2*n* = 34 (Moore & Frankton, 1962).—Reported from Siskiyou, Modoc, Lake, Lassen, Monterey, San Benito, Tulare and Sierra cos., *Fuller.*

p. 1273.   SÍLYBUM

1. **S. Mariànum** (L.) Gaertn. 2*n* = 34 (Moore & Frankton, 1962).

CARTHÀMUS

1. **C. lanàtus** L. Now known from Humboldt, Napa, Sonoma and s. Mendocino cos.

2. **C. baèticus** (Boiss. & Reut.) Nym. [*C. lanatus* ssp. *creticus* (L.) Holmb.] In Calaveras Co., *Fuller.*

3. **C. tinctòrius** L. In the Temblor Range, at Arvin, etc., Kern Co., *Twisselmann.*

p. 1275.   CÍRSIUM

In Key after D, EE, FF, GG insert:

> G′.   Plants annual, sometimes biennial, glabrous to ± arachnoid. San Luis Obispo and Santa Barbara cos.
> 12a. *C. loncholepis*
> G′G′.   Plants perennial, if subglabrous, then with rhizome-like roots, if arachnoid, then densely so.
> H.   Heads solitary, etc. as in the FLORA.

p. 1276.   In Key after last HH, add:

> I.   Heads often broader than long, hemispheric; phyllaries with spines 5–10 mm. long, the outer phyllaries often reflexed.
> 29. *C. neomexicanum*
> II.   Heads subglobose; phyllaries with spines mostly 3–7 mm. long, the outer phyllaries spreading or ascending. ... 29a. *C. utahense*

5. **C. remotifòlium** (Hook.) DC. From the synonymy delete *C. r.* ssp. *pseudocarlinoides* Petr.

p. 1277.   6. **C. callilèpis** (Greene) Jeps. Howell recognizes **C. c.** var. **pseudocarlinoìdes** (Petrak) J. T. Howell as having more loosely ascending, not appressed phyllaries and less graduated.—Marin Co. to Ore.

7. **C. quercetòrum** (Gray) Jeps. In Sacramento and Solano cos., *Fuller*.

10. **C. crassicáule** (Gray) Jeps. Lost Hills, Kern Co., *Twisselmann*.

12. **C. foliòsum** (Hook.) DC. $2n = 34$ (Ownbey & Hsi, 1963).

p. 1278.   12a. **C. loncholèpis** Petr. Near *C. foliosum* (Hook.) DC. Annual or biennial, with a taproot, nearly or quite glabrous; heads solitary or clustered at ends of branches; phyllaries loosely imbricate, mostly linear- or ovate-lanceolate.—San Luis Obispo and Santa Barbara cos.

13. **C. Drummóndii** T. & G. $2n = 34$ (Ownbey & Hsi, 1963).

14. **C. tiogànum** (Congd.) Petr. [*Cnicus t.* Congd.]

15. **C. undulàtum** (Nutt.) Spreng. $2n = 26$ (Frankton & Moore, 1961). Collected 3.5 mi. ne. of Livermore, at Escondido, San Diego Co. and in Lassen Co., Alameda Co., *Fuller*. Once reported also from Catalina Id.

17. **C. Brèweri** (Gray) Jeps. J. T. Howell in Abrams, Ill. Fl. Pac. States 4: 520. 1960, uses **C. Douglásii** DC. [*C. Breweri* var. *Wrangelii* Petr.] for plants to 1.8 m. tall; lvs. more spiny and with short decurrent bases; heads 2.5–3.5 cm. long, bowl-shaped.—Monterey Co. to Mendocino Co.

For **C. Brèweri** (Gray) Jeps. [*C. B.* vars. *canescens* and *lanosissimum* Petr.] Howell uses **C. Douglásii** var. **canéscens** (Petr.) J. T. Howell. To 2.5 m. tall; lvs. less spiny and with longer decurrent bases; heads 2–3 cm. long.—Lake Co. to Siskiyou Co. and in the Sierra Nevada from Lake Tahoe north.

18. **C. Vàseyi** (Gray) Jeps. is **C. hydrophìlum** (Greene) Jeps. var. **Vàseyi** (Gray) J. T. Howell.

p. 1279.   20. **C. ochrocéntrum** Gray. $2n = 30, 31,$ `32 (Ownbey & Hsi, 1963). Reported from Lassen, Plumas, Los Angeles & San Diego cos.

21. **C. Andersònii** (Gray) Petr. in the Piute Mts., Kern Co., *Twisselmann*.

24. **C. cymòsum** (Greene) J. T. Howell. In Tehachapi Mts., *Twisselmann*.

25. Change *C. Còulteri* Harv. & Gray to **C. proteànum** J. T. Howell.

p. 1280.   26. **C. occidentàle** (Nutt.) Jeps. Add as synonyms *C. Coulteri* Harv. & Gray. *C. o.* var. *C.* Jeps.

30. **C. arvénse** (L.) Scop. $2n = 34$ (Moore & Frankton, 1962).

Var. **mìte** Wimmer & Grabowski. Lvs. nearly glabrous, undulate-margined or very shallowly lobed.—In Modoc Co., *Fuller*.

31. **C. utahénse** Petr. Differing from *C. neomexicanum* Gray in having subglobose (not hemispheric) heads; phyllaries with spines 3–7 mm. long, the outer phyllaries spreading or ascending (not 5–10 mm. or reflexed).—E. of the Sierra Nevada; to the Rocky Mts.

p. 1281.   CÁRDUUS

1. **C. pycnocéphalus** L. $2n = 54$ (Moore & Frankton, 1962). Known now from Butte, Mariposa, Eldorado, Nevada, and Fuller cos., *Fuller*.

2. **C. tenuiflòrus** Curt. Additional reports: Santa Barbara Co., Sierra Nevada foothills in Tulare Co., *Fuller*, Riverside Co., *Howell*.

3. **C. nùtans** L. Near Victorville, San Bernardino Co., *Fuller*; 2 mi. s. of Bordertown, Nev. in Nevada Co., and Long Valley nw. part of Sierra Co., *Fuller*; 4 mi. s. of Castella, Shasta Co.

CENTÁUREA

In Key, after AA, B, CC, change to:

> D. Invol. cylindric, not more than 1 cm. long.
> E. Invol. ca. 7 mm. long; fls. usually pink or lavender; pappus bristles slender, to 2.5 mm. long. . . . . 9. *C. virgata*
> EE. Invol. ca. 1 cm. long; fls. usually white; pappus none or of scales less than 1 mm. long. . . . . 9a. *C. diffusa*

p. 1282.   2. **C. Cinerària** L. $2n = 18$ (Gori, 1954; Larsen, 1956).

3. **C. maculòsa** Lam. Now in Lassen, Nevada, Placer, Siskiyou, and Trinity cos., *Fuller*.

9. **C. virgàta** Lam. var. **squarròsa** (Willd.) Boiss. Add Shasta Co., *Fuller*.

9a. **C. diffùsa** Lam. Differing from *C. virgata* in having white rather than pink or lavender fls.; invol. longer.—Reported from Lassen, Siskiyou, and Modoc cos., *Fuller;* native of e. Medit. region.

10. **C. dilùta** Ait. Vista, San Diego Co., *Fuller.*

p. 1283.   11. **C. ibèrica** Trev. Carmel R., Monterey Co., *Howitt & Howell;* e. of Livermore, Alameda Co., *Marsh.*

13. **C. sulphùrea** Willd. Reported from Sacramento, Eldorado and San Joaquin cos., *Fuller.*

15. **C. meliténsis** L. $2n = 36$ (Chiappini, 1955).

CNÌCUS

1. **C. benedíctus** L. $n = 11$ (Moore & Frankton, 1962).

p. 1285.   TRÍXIS

1. **T. califórnica** Kell. $n = 27$ (Turner, Powell & King, 1962).

p. 1287.   CICHORIÈAE

In Key after AA, BB, CC, D, E, change to:

     F.   Plants nonscapose, usually much branched.
         F'.   Plants shrubs 1–2 m. high; lvs. 6–13 cm. long, in tufts at ends of branches, irregularly sinuate-toothed. San Clemente Id. .... 163a. *Munzothamnus*
         F'F'.   Plants herbaceous, or at least lower in height.
             G.   Pappus not plumose.
                 G'.   Plants perennial or annual; heads relatively slender.
                     H.   The pappus of 5 rigid tapering awns.
                         159. *Chaetadelphia*
                     HH. The pappus of many capillary bristles.
                         160 *Lygodesmia*
                 G'G'.   Plants annual; heads subglobose.
                       163. *Malacothrix*
             GG. Pappus plumose, etc. as in the FLORA.

In Key after AA, BB, CC, D, change EE to:

     EE. Aks. epappose.
         F.   Florets white; glaucous scapose native with broad spinulose-denticulate lvs. Deserts. .... 167. *Atrichoseris*
         FF. Florets yellow; stems leafy. Adventive. ... 167a. *Rhagadiolus*

In Key, after DD, EE, F, GG, change to:

     HH. Aks. beaked.
         H'.   Invol. of 1 row of 7–8 phyllaries.
                 170a. *Urospermum*
         H'H'. Invol. of 2 or more rows of phyllaries.
             I.   Stems leafy.
                 J.   Outer phyllaries broader than the inner; plants rough-bristly throughout. .... 172. *Picris*
                 JJ. Outer and inner phyllaries equal in a single subcylindric series and with small green scales at base; plants hirsute at base only.
                       172a. *Chondrilla*
             II. Stems scapose; outer phyllaries very small.
                   173. *Leontodon*

Toward end of Key, change H to:

     H. Pappus present.
         H'.   Phyllaries usually 2-seriate; pappus of many bristles.
             I, II, J and JJ. as in the FLORA.
         H'H'. Phyllaries in 1–2 series; pappus of 4–10 scales. .... 180. *Tolpis*

p. 1288.   CICHÒRIUM

2. **C. Endìva** L. ENDIVE. Bracts subtending heads leafy, commonly longer than the heads; fls. purple, in heads ca. 4 cm. across.—Cult. as a salad plant and escaped in the Salinas Valley, Monterey Co., *Howell.* From Eurasia.

p. 1289.   MICRÓSERIS

In Key, AA, BB, CC, DD, E should be Alameda not Amador Co., in 1959 printing.

In Key after A, B, CC "Pappus-bristles not plumose, 6–10" insert:

> D.   Invol. 25–75-fld., 10–25 mm. high. Sonoma and Lassen cos. to Wash.
>   1. *M. laciniata*
> DD. Invol. 15–25-fld., 8–18 mm. high. Siskiyou Mts. . . . . 1a. *M. Howellii*

p. 1290.   1a. **M. Howéllii** Gray. [*Scorzonella H.* Greene.] Perennial with fleshy fusi-form taproot; stem slender, erect, 1.5–5 dm. tall; lvs. chiefly basal, entire to laciniate-pinnatifid with slender lobes; heads 15–25-fld.; invol. 8–18 mm. high; florets pale yellow; pappus 6–12 mm. long, the paleae 6–10 in number, 3–6 mm. long, awned.—Siskiyou Mts.

       6. Change *M. Lindleyi* (DC.) Gray to **M. linearifòlia** (Nutt.) Sch-Bip.

p. 1291.   9. **M. acumìnata** Greene. In last line change Amador to Alameda, in 1959 printing.

p. 1293.   AGOSÉRIS

      4. **A. apargioìdes** (Less.) Greene var. **Eastwoodiae** (Fedde) Munz, not Q. Jones. [*A. Eastwoodae* Fedde in Just, Bot. Jahresb. 31: 808. 1904.]

p. 1294.   PHALACRÓSERIS

      1. **P. Bolánderi** Gray. In Nevada Co., *True*.

p. 1295.   STEPHANOMÈRIA

Insert paragraph after 1. **S. virgàta** Benth.:

      S. carotífera Hoov., a perennial from near Morro Bay, San Luis Obispo Co. was described by Hoover for plants differing from *S. virgata*, an annual, which it resembles in heads, fls., aks. The proposed sp., however, has a fleshy root and divaricate branching.

p. 1297.   MALACÒTHRIX

Insert literature refs.: Williams, E. W. The genus Malacothrix. Am. Midl. Natur. 58: 494–512. 1957. Davis, Wm. S. and P. H. Raven. Three new spp. related to M. Clevelandii, Madroño 16: 258–266. 1962.

p. 1298.   Insert in Key after FF, (lines 1 and 2 at top of page):

> G.   Aks. less than 1.7 mm. long, fusiform, with 5 of the 18 ribs more prominent than the others; invol. less than 8 mm. high.
> H.   Aks. brown or straw-colored; cauline lvs. often toothed; plants usually unbranched below. Tehama Co. and Glenn Co. to L. Calif. . . 6. *M. Clevelandii*
> HH. Aks. dark purplish-brown, rarely paler; cauline lvs. entire; plants often well-branched from base. Santa Cruz Id., Hueneme Beach, Ventura Co. 6a. *M. similis*
> GG. Aks. more than 1.7 mm. long, subcylindrical, gray-brown to straw-colored, with 15 equally prominent ribs; invol. 7–10 mm. high. Deserts, Inyo Co. to e. San Diego Co.
>   6b. *M. Stebbinsii*

      1. **M. Còulteri** Harv. & Gray var. **cognàta** Jeps. has stem lvs. parted nearly to the midrib into linear divisions. Santa Cruz and Santa Rosa ids. San Pedro Hills, Los Angeles Co.

      4. **M. sonchoìdes** (Nutt.) T. & G. The report in the FLORA of $2n = 18$ is an error.

      6. **M. Clevelándii** Gray. Stems usually 1, sometimes several from base; basal lvs. linear to linear-lanceolate, dentate, pinnatifid or lobed; cauline lvs. often toothed; heads cylindrical to narrow-campanulate, 4–8 mm. high, 19–67-fld.; ligules yellow; pollen grains ca. 25μ in diam.; aks. truncate-fusiform, 1.4–1.8 mm. long, slightly curved, 5 of the 15 ribs more prominent than the others, apex of the ak. flared, bordered by a ring of 14–17 white-scarious teeth; persistent seta on ak. 1; $n = 7$.—From Tehama and Glenn cos. to L. Calif.

      6a. **M. símilis** Davis & Raven. Usually branched from the base; basal lvs. linear-lanceolate, entire to pinnatifid; cauline lvs. subentire; heads narrow-campanulate, 6–10 mm. high, 32–73-fld.; pollen-grains ca. 30μ in diam.; aks.

truncate-fusiform, 1.4–1.7 mm. long, slightly curved, dark purplish-brown, 5 of the 15 ribs prominent; ak.-apex with ca. 18 white-scarious teeth, persistent seta 1; *n* = 14.—Santa Cruz Id.; Hueneme Beach in Ventura Co.; L. Calif.

6b. **M. Stebbínsii** Davis & Raven. Usually simple at the base; basal lvs. lanceolate to oblanceolate, dentate, rarely pinnatifid; heads campanulate, 7–10 mm. high, 19–70-fld.; aks. 1.7–2.3 mm. long, rarely curved, with 15 equally prominent ribs, apex bordered by a ring of 14–17 white-scarious teeth, persistent seta usually 1.—Deserts, Inyo Co. to e. San Diego Co.; Nev., Ariz., Son.

p. 1299.   8. **M. indécora** Greene and 9. **M. squálida** Greene are both treated as vars. of M. foliòsa Gray by Williams.

10. **M. incàna** (Nutt.) T. & G. Williams makes the combination **M. i.** var. **succulénta** (Elmer) Williams.

11. **M. saxátilis** (Nutt.) T. & G. Ferris recognizes var. **altíssima** (Greene) Ferris [*M. a.* Greene.] as distinct from var. **tenuifòlia** in having the stems partly subterranean, arising singly from branches of a deep-seated root, instead of at the surface of the ground; lvs. 1–2 dm. long, deeply laciniate-lobed.— Tehachapi Mts., Mt. Pinos and mts. of Santa Barbara Co. s. to Santa Monica Mts., Los Angeles Co.

p. 1300.   12. **M. Blàirii** (M. & J.) M. & J. Transfer to:

163a. Munzothámnus Raven

Shrub 1–1.5 m. tall; lvs. in tufts at ends of branches, large, coarsely sinuate or lobulate. Invols. 9–12-fld., narrow. Florets purple. Pappus setae thick, few, plumose. One sp. from San Clemente Id. (*Munz* and *shrub*).

1. **M. Blàirii** (M. & J.) Raven. [*Stephanomeria B.* M. & J. *Malacothrix B.* M. & J.] Lvs. obovate to oblong-obovate, 5–15 cm. long; infl. paniculate, leafless, 1–2 dm. long; aks. 3–3.5 mm. long; pappus-bristles many, 4 mm. long.— Rocky canyon-walls; Coastal Sage Scrub; San Clemente Id. July–Sept.

p. 1301.   ### 167a. Rhagadìolus Juss.

Annual, divergently branched above. Lvs. largely basal, much reduced upward, the lower oblanceolate or broader, petioled, toothed to pinnatifid. Heads rather small, long-peduncled; invol. double, the outer phyllaries minute, the inner 5–8, much enlarged in 'fr., keeled. Fls. yellow. Aks. linear-subulate, without pappus, the outer rather persistent with the invol., the inner soon deciduous. Medit.

1. **R. édulis** Willd. Low annual; lower lvs. broadly oblong-oblanceolate, dentate; heads ca. 1 cm. high; inner phyllaries linear and to 2.5 cm. in fr.— Adventive near Napa, Napa Co., *Raven.*

p. 1302.   TRAGOPÒGON

3. **T. praténsis** L. Reported from San Luis Obispo Co., *Frey;* Kern Co., *Twisselmann.*

170a. Urospérmum Scop.

Annual to biennial, few-branched plants. Heads solitary on long peduncles, medium to large; phyllaries 7–8 in 1 series, connate at base. Florets yellow. Aks. ending in a hollow beak separated from the true ovary-cavity by a septum. Pappus-bristles plumose, in 2 series hanging together at the base in a ring and falling simultaneously. Two spp. Medit. (Greek, *oura*, tail, and *sperma*, seed.)

1. **U. picroìdes** (L.) Schmidt. Three–4 dm. tall, hispid, even subspinose below; lvs. pubescent-hispid, the lower oblong-obovate, pinnatifid or dentate, the upper lanceolate, clasping; peduncles long, naked; heads medium-large; aks. flat, tuberculate.—Adventive on Univ. of Calif. campus at Berkeley, **Carter.**

HYPOCHOÈRIS
In Key, line 1, insert "more or less" after "annual."

p. 1303.   172a. Chondrílla L.

Caulescent, branched, hispid below. Lower lvs. sinuate or runcinate, the upper entire. Heads small, subsessile, 7–12-fld., yellow. Invol. of linear subequal phyllaries in 1 series. Aks. oblong, slender-beaked. (Greek, *chondrile*, a kind of endive or chicory.)

1. C. juncèa L. SKELETON WEED. Biennial, 4–10 dm. tall.—Reported as weed from San Luis Obispo, Placer, Nevada, Sacramento and Eldorado cos.; from the Medit. basin.

SÓNCHUS

1. S. arvénsis L. *n* = 9 (Mehra et al., 1965).—In Siskiyou Co., *Fuller*; near Guadalupe, Santa Barbara Co., *Fuller*.

p. 1304.   4. S. ásper (L.) Garsault. 2*n* = 18 (various workers). Said to have, as a distinctive character, thin papery aks. rather than thickish ones.

p. 1305.   LACTÙCA

5. L. salígna L. Reported from Monterey Co., *Howitt & Howell*.

7. L. ludoviciàna (Nutt.) DC. [*Sonchus l.* Nutt.] Near *L. canadensis* in its beaked aks. but the fruiting invols. 15–22 mm. long (not 10–15) and pappus 7–12 mm. long (not 5–7); florets 20–56 (not 13–22).—Reported from San Bernardino, *Ferris*; native of central U.S.

HIERÀCIUM
Change line 3 of Key to:

Ligules yellow or orange; stem and invol. pubescent to hirsute.

Change line 4 to:

Heads yellow.

Add as last line of Key:

Heads orange-red. .... 8. *H. aurantiacum*

4. H. argùtum Nutt. var. Paríshii (Gray) Jeps. in San Luis Obispo Co., *Hardham*.

p. 1306.   6. H. grácile Hook. Add to synonymy *H. triste* ssp. *g.* Calder & Taylor.

8. H. aurantiàcum L. Long-hirsute, rank perennial with many-leaved basal rosettes and coarse rooting stolons; lvs. 5–20 cm. long; scapes 2–7 dm. tall, naked or with 1–2 lvs.; heads orange-red, ca. 2 cm. across, in corymbs.—In lawns, Grass Valley, Nevada Co., *Fuller*; native of Europe.

p. 1307.   CRÈPIS
In Key change AA to:

AA. Introduced annuals or biennials, or perennial in *C. bursifolia.*
   B.   Invols. 5–8 mm. long; aks. 2–3.5 mm. long. .... 10. *C. capillaris*
   BB. Invols. 8–12 mm. long; aks. 3.5–8 mm. long.
      C.   Aks. 6–8 mm. long; invols. not usually strongly setose.
         D.   Annual or biennial; beak of ak. not filiform, scarcely longer than the
            body. .... 11. *C. vesicaria*
         DD. Perennial; beak of the ak. filiform, ca. twice the length of the ak.-body.
                                          13. *C. bursifolia*
      CC. Aks. 3–5 mm. long; invols. strongly setose. .... 12. *C. setosa*

p. 1309.   In the 1959 printing the lines for 10, 11, 12 were badly jumbled. Change as follows:

10. C. capillaris (L.) Wallr. First 3 lines OK. Transpose lines 3, 4, and 5 from 12. *C. setòsa*. The last 2 lines are OK, and are followed by "June–Aug." from the end of *setosa*.

11. C. vesicària L. ssp. taraxacifòlia (Thuill.) Thell. First 4 lines OK, followed by the last 3 lines from Ssp. *Hallii* above, beginning "9–13."

12. **C. setòsa** Haller f. Combine the first 2 lines of *C. setosa* and the last 3 lines of *C. vesicaria* ssp. *taraxacifolia*. Add:

13. **C. bursifòlia** L. Perennial, 1–3.5 dm. tall, with several stems from a woody caudex, tomentulose, cymosely branched above, each with 2–14 heads; lvs. mostly basal, these 5–15 cm. long, oblanceolate in outline, lyrately pinnatifid; invols. 9–11 mm. long, cylindric, canescent, farinose, sometimes somewhat setulose; aks. 6–7.5 mm. long, 10-ribbed, the filamentous beak ca. twice as long as the body; pappus 3–4 mm. long.—Adventive in region of San Francisco Bay. Native of Italy. May–June.

p. 1310.

180. Tólpis Gaertn.

Low herbs with oblong to oblong-spathulate, entire to pinnatifid lvs. or the upper lvs. linear. Heads long-peduncled, yellow. Phyllaries in 1–2 series; receptacle naked. Aks. oblong, 6–8-costate; pappus of 4–10 seta-like scales. (Greek, *tolupe*, ball, referring to the form of the invol.)

1. **T. barbàta** Willd. Annual, 2–4 dm. tall, the stem glabrous, branched; lvs. toothed, the lower petioled; heads 1.5–3 cm. in diam.; pappus-bristles 4–5. —Escape at Pacific Grove, Monterey Co.; native of Medit. region.

2. **T. umbellàta** Bertol. Heads 1–1.5 cm. diam.—Escape nw. of Ukiah, Mendocino Co.; native of Medit. region.

p. 1311.   MONOCOTYLEDÒNEAE

In Key, after AA, B, CC, D, EE, FF, add:

> GGG. Lvs. clustered at the nodes or on short branches each with a sheath and a narrow blade. . . *Halodule* p. 1324

In Key, after AA, BB, change to:

> C.   Carpels ± free, 1-loculed.
>   D.   Parts of perianth usually 2 (1–3), petaloid. *Aponogetonaceae* p. 131.?
>
>   DD. Parts of perianth 6, in 2 series, the inner petaloid.
>   *Alismataceae* p. 1312

p. 1312.   Caption for Fig. 121 should read *Echinodorus Berteroi*.

p. 1313.   MACHAEROCÁRPUS

1. **M. califórnicus** (Torr.) Small. In Bear V., Kern Co., *Twisselmann*.

1a. Aponogetonàceae. APONOGETON FAMILY

Aquatic perennial herbs with tuberous rhizome and floating or submerged lvs.

1. Aponogèton L. f.

Lvs. long-petioled, linear-oblong, with many parallel and transverse veins. Infl. a simple or forked spike. Fls. mostly bisexual; perianth parts mostly 2, petaloid. Stamens 6 or more, in 2 whorls. Pistils 3–4, free. Ovules 2–8, basal. (Origin of name uncertain.)

1. **A. distàchyus** L. f. CAPE POND-WEED. Fls. white, fragrant, with purplish anthers.—Escape from cult., San Mateo Co., *Rubtzoff*. From S. Afr.

p. 1313.   SAGITTÀRIA

4. **S. cuneàta** Sheld. Reported from near sea level, Marin Co., *Howell*.

p. 1314.   ALÍSMA

Add as citations to literature: Hendricks, A. J. A revision of the genus Alisma. Am. Midl. Nat. 58: 470–493. 1957. Pogan, E. Taxonomic value of A. triviale Pursh and A. subcordatum Raf. Can. J. Bot. 41: 1011–1013. 1963.

1. **A. triviàle** Pursh. $2n = 28$.

2. Change *A. Geyeri* Torr. to **A. subcordàtum** Raf. with sepals in anthesis 2–2.5 mm. long; petals white, 1–2 mm. long; aks. 1.5–2 mm. long; $2n = 14$. —Yosemite Valley, Mariposa Co., Modoc Co.; to N.Y., Quebec.

3. **A. lanceolàtum** With. Lvs. narrow-lanceolate, 0.5–2 dm. long; petals

rose, ± pointed apically; $2n = 26$.—Natur. in Sonoma Co., Marin Co., Colusa Co., Placer Co.; from Eurasia.

p. 1317. POTAMOGÈTON
4. **P. Robbínsii** Oakes. $n = 26$ (Stern, 1961).
5. **P. críspus** L. Found in Sonoma Co., *Rubtzoff*.
7. **P. foliòsus** Raf. $2n = 28$ (Stern, 1961).
8. **P. pusíllus** L. $2n = 26$ (Harada, 1956).

p. 1318. 10. **P. diversifòlius** Raf. Occurs in Sonoma, Lake and Modoc cos., *Rubtzoff*.
11. **P. epihýdrus** ssp. **Nuttállii** (Cham. & Schlecht.) Calder & Taylor.
13. **P. amplifòlius** Tuckerm. $n = 26$ (Stern, 1961).
15. **P. nàtans** L. $2n = 52$ (Löve & Löve, 1956); $n = 26$ (Stern, 1961).

p. 1319. 16. **P. gramíneus** L. $n = 26$ (Stern, 1961).
17. **P. illinoénsis** Morong. $n = 52$ (Stern, 1961).
18. **P. praelóngus** Wulf. $2n = 52$ (Löve & Löve, 1956).
19. **P. Richardsònii** (Benn.) Rydb. $n = 26$ (Stern, 1961). Reported from Lake Co., *Rubtzoff*.

RÙPPIA
In Key, "Lvs. obtusish" goes with **R. spiràlis** ($2n = 10$) and "lvs. acute" goes with **R. marítima** ($2n = 20$).

p. 1320. TRIGLÓCHIN
Used by Linnaeus as a neuter noun. Our species should be:
1. **T. striàtum** R. & P.

p. 1321. 2. **T. palústre** L. $2n = 24$ (Larsen, K., 1966).
3. **T. marítimum** L. $2n = 12$ to 144 (Löve & Löve, 1958). Found in Monterey Co., *Howitt & Howell*.
4. **T. concínnum** Davy. $2n = 24$ (Larsen, 1966); 48 (Löve & Löve, 1958). Löve & Löve recognize **T. débile** (Jones) Löve & Löve, with $2n = 96$.

LILAÈA
1. **L. scilloìdes** (Poir.) Haum. $2n = 12$ (Larsen, 1966).

p. 1324. ZANNICHÉLLIA
1. **Z. palústris** L. $2n = 12$ (Reese, 1957).

HALÓDULE
Under Zannichelliaceae add 2. **Halódule** Endl.
Dioecious; style shorter than the stigma (instead of longer); pollen filiform (not globose); carpels 2 and united, or 1. Fr. a nutlet.
1. **H. Wrìghtii** Aschers. [*Diplanthera* W. Aschers.] Lvs. bicuspidate at the semilunate apex; anthers ca. 6 mm. long; mature fr. black.—Introd. intentionally into the Salton Sea from Texas and now well established. Native in se. states and W. Indies.

NÀJAS
2. **N. flexilis** (Willd.) Rostk. & Schmidt is apparently not in the California flora and plants so identified in the past are largely *N. guadalupensis* fide *Thorne*.

p. 1325. ELODÈA
Cite as a reference: St. John, H. Monograph of the genus Elodea. Part 1. Research studies Wash. State Univ. 30: 19–44. 1962.

p. 1326. 1. **E. dénsa** Plancha. Add as a synonym *Egeria densa* St. John.
2. **E. canadénsis** Michx. $n = 24$ (Harada, 1956).

2a. **E. Brandègeae** St. John. Differing from *E. canadensis* by having upper and middle lvs. 5–8 mm. long, rather than 6–13; fls. perfect, not dioecious; sepals 3.8 mm. long, as against 2.2 mm.; stigmas 0.3 mm. long, entire, as opposed to 4 mm., bifid.—Truckee, Sierra Nevada.

p. 1327. In Key AA, B, CC, DD, EE, change to "Anthers often seemingly basifixed." In Key, under A, BB, C, D, change to:

    E.   Lvs. not equitant.
       F.   The lvs. 5–10 dm. long, dry; pedicels 2–5 cm. long.
                        1. *Xerophyllum*

FF. The lvs. shorter, fleshy; pedicels ca. 1 cm. long.

1a. *Asphodelus*

EE. Lvs. equitant.

F. and FF. as in the FLORA.

p. 1328.

1a. Asphodèlus L. ASPHODEL

Near to *Narthecium,* but lvs. spirally arranged, not equitant. Fls. white to pinkish, with jointed pedicels; perianth-segms. with a single colored midvein; stamens dilated at base and covering the ovary. Several spp. Old World.

1. A. fistulòsus L. Root system of several tubers; stems 2–6 dm. high, glabrous; lvs. straight, semiterete; fls. in open panicles; perianth 8–12 mm. long; caps. 4–6 mm. long, subglobose.—Adventive in San Diego and Santa Barbara cos. Native of Medit. region.

NARTHÈCIUM

1. N. califórnicum Baker. $n = 13$ (Cave, 1966). Howell reports this as far s. as Tulare Co.

TOFIÈLDIA

1. T. glutinòsa Pers. ssp. occidentàlis (Wats.) C. L. Hitchc. $n = 15$ (Cave, 1966).

p. 1329.

SCHOENOLÍRION

1. S. álbum Durand. $n = 26$ (Cave, 1966).

CHLORÓGALUM

1. C. pomeridiànum (DC.) Kunth $n = 15$ (Cave, 1966).

Var. divaricàtum (Lindl.) Hoov. extends as far s. as Santa Barbara Co.

p. 1330.

5. C. purpùreum Bdg. var. redúctum Hoov. Described as 1–2 dm. tall.—La Panza road, San Luis Obispo Co., apparently on serpentine soil.

p. 1331.

SMILACÌNA

1. S. racemòsa (L.) Desf. $2n = 36$ (Therman, 1956).

MAIÁNTHEMUM

1. M. dilatàtum (Wood) Nels. & Macbr. $n = 18$ (Therman, 1956). Occurs in San Mateo Co., *Howell.*

p. 1332.

DÍSPORUM

3. D. Hoòkeri (Torr.) Nichols. In San Luis Obispo Co., *Hardham.*

p. 1333.

STENÁNTHIUM

1. S. occidentàle Gray. $n = 8$ (Cave, 1966).

p. 1335.

ZIGADÈNUS

6. Z. exaltàtus Eastw. In Piute and Greenhorn mts., Kern Co., *Twisselmann.*

VERÀTRUM

1, 2, 3. V. insolìtum Jeps., V. fimbriàtum Gray, V. califórnicum Durand. $n = 16$ (Cave, 1966).

p. 1336.

ASPÁRAGUS

1. A. officinàlis L. is largely dioecious.

p. 1337.

ERYTHRÒNIUM

3. E. tuolumnénse Appleg. $n = 12$ (Cave, 1966).

4. E. grandiflòrum Pursh. Add ssp. Pusatèrii Munz & Howell. Lvs. green, not mottled, ± crisped-undulate, 2–5 dm. long, the blade 1.5–2.5 dm. long; infl. 1–3 dm. tall, 1–3-fld.; perianth-segms. 2–3 cm. long, yellowish orange on lower half or third, cream color in distal part; anthers yellow.—Rocky soil below Hockett Lakes, at ca. 8000 ft., Tulare Co.

p. 1340.

FRITILLÀRIA

6. F. agréstis Greene. Temblor Range, Kern Co., *Twisselmann.*

8a. F. Roderíckii Knight. Proposed as a new sp. near to *B. biflora* in having the lvs. mostly near the ground, but differing in the possession of some rice-grain bulblets; perianth-segms. creamy-greenish on lower third and the inner surface with raised veins; fils. in 1, not 2 ranks, equal in length; $n = 12$.—On clay, Mendocino Co.

p. 1341.

15. F. atropurpùrea Nutt. $n = 12$ (Bottino, 1965).

p. 1347.   CALOCHÓRTUS

3. **C. pulchéllus** Dougl. reported from Upper Cache Creek, w. Colusa Co., *Holt.*

p. 1351.   23a. **C. símulans** (Hoov.) Munz. comb. nov. [*Mariposa simulans,* Hoov. Leafl. W. Bot. 4: 4. 1944.] Differs from *V. venustus* Dougl. ex Benth. in petals not having margins above lower third, and with a red spot surrounding the glandular area and frequently a small red spot immediately above it; petals cuneate, with straight sides.—Central San Luis Obispo Co. In *C. venustus* the petals have a conspicuous dark spot and frequently also a lighter spot near the apex, petals inwardly curved so as to be clawlike near the base.

25. **C. Véstae** Purdy. Reported from Greenhorn Range, Kern Co., *Twisselmann.*

p. 1352.   30. **C. invenústus** Greene. $2n = 7$ pairs (Raven, Kyhos & Hill, 1965). Found in Kern Co., *Twisselmann.*

p. 1353.   34. **C. clavàtus** Wats. Insert: *C. clavatus* in its typical form is largely on soil of serpentine origin, has deep yellow petals and uniformly deep purple anthers.—From San Luis Obispo Co. to Santa Barbara Co.

Ssp. **pállidus** (Hoov.) Munz, comb. nov. [*Mariposa clavata* (Wats.) Hoov. var. *pallida* Hoov. Leafl. W. Bot. 10: 126. 1964.] Differs from var. *clavatus* by having petals light yellow, the hairs gradually enlarged toward the apex, distinctly less than 0.1 mm. in dried specimens, fungoid processes of gland smaller and less branched; anthers yellow to pale or medium purple.—S. Coast Ranges from San Joaquin and Stanislaus cos. to n. Los Angeles Co. and Kern Co.

Ssp. **recurvifòlius** (Hoov.) Munz, comb. nov. [*Mariposa clavata* var. *recurvifolia* Hoov. Leafl. W. Bot. 10: 126–127. 1964.] Plants dwarf, 9–12 cm. tall, the internodes to 2 cm. long; lvs. strongly recurved; fls. of typical *clavata.*— N. of Arroyo de la Cruz, San Luis Obispo Co., on ocean bluff.

Var. **àvius** Jeps. Sepals equaling or exceeding petals; gland in a deeper pocket.—Base of Sierra Nevada, Eldorado Co. to Mariposa Co.

36. **C. Weèdii** Wood extends s. into L. Calif.

p. 1357.   EICHHORNIA

1. **E. crássipes** (Mart.) Solms. The first word in the last line of the description should be "serious" not "series" in the 1959 printing.—Reported from a pond, San Ysidro, San Diego Co., *Fuller.*

ARÀCEAE

Add to the Key:

Spathe greenish without, whitish within; lvs. sagittate. . . . . 4. *Arum*

p. 1358.   ## 4. Árum L.

Low simple herbs with underground tubers, hastate or sagittate lvs. and scape bearing a spathe, that withers after anthesis, and a spadix of minute naked fls. Several spp. (Greek, *aron,* ancient name.)

1. **A. itálicum** Mill. To 1 m. high; lvs. surpassing the spathe which is long-pointed, green without, whitish within.—Established on Mendocino Prairie and Albion River, Mendocino Co., *Marie Kelley.* From the Medit. region.

LYSICHÌTON

1. **L. americànum** Hult. & St. John. $2n = 28$ (Löve & Kawano, 1961).

p. 1358.   LEMNÀCEAE

At end of family description insert: Daubs, E. H. A monograph of Lemnaceae. Ill. Biol. Mon. 34: 1–118. 1965.

SPIRODÈLA

2. **S. oligorhìza** (Kurz) Hegelm. A definite collection cited from Calif. is *Heckard & Bacigalupi 7693.*

LÉMNA

Use the following key:

A. Fronds usually submerged, long-stipitate, many remaining attached, forming long chains. . . . . 1. *L. trisulca*
AA. Fronds usually floating, short-stipitate or sessile, mostly 2–5 attached.
    B. Dorsal surface flat, smooth, with no prominent protuberances, 1-veined or veinless.
        C. Plants narrow-elliptical, often 8–10 attached; base asymmetrical.
                6. *L. valdiviana*
        CC. Plants oval, symmetrical, seldom more than 2 remaining attached.
                2. *L. minima*
    BB. Dorsal surface with ± prominent protuberances, indistinctly to prominently 3-veined.
        C. Root sheath with definite wings or appendages. . . . . 4. *L. perpusilla*
        CC. Root sheath without definite wings or appendages.
            D. Ventral surface of frond flat to slightly convex, but not inflated.
                E. Dorsal surface dark green, apex symmetrical; air spaces not prominent. . . . . 3. *L. minor*
                EE. Dorsal surface mottled yellow-green, apex asymmetrical; air spaces not prominent. . . . . 5. *L. gibba*
           DD. Ventral surface of frond noticeably convex, the air spaces inflated; both surfaces showing red-purple coloring.
                E. Air spaces strongly inflated, gibbous; apex asymmetrical.
                      5. *L. gibba*
                EE. Air spaces slightly inflated, the apex symmetrical. . . 7. *L. obscura*

p. 1359.   7. **L. obscùra** (Austin) Daubs. [*L. minor* var. *o.* Austin.] Plants solitary or 2–3 attached, elliptic-orbicular to obovate, slightly asymmetrical; ventral surface strongly red-purple, slightly inflated.—Occasional in Calif.; e. and s. U.S.; Mex.

WOLFFIÉLLA

    1. **W. oblónga** (Phil.) Hegelm. Daubs does not have this sp. as occurring in Calif.

    2. **W. lingulàta** (Hegelm.) Hegelm. Reported now as from Marin, Sonoma and Butte cos.

WÓLFFIA

    1. **W. columbiàna** Karst. according to Daubs does not have brown pigment. He reports as from Calif. a plant that does have brown pigment cells making it punctate, namely **W. punctàta** Griseb.

p. 1362.   YÚCCA

    4. **Y. Whípplei** Torr. ssp. **caespitòsa** (Jones) Haines ranges n. to Tehipite Valley, Fresno Co.

NOLÌNA

    1. **N. interràta** Gentry. In the first line of the description it should read "2–3 m. long" not "mm."; in the 1959 printing.

    2. **N. Párryi** ssp. **Wólfii** Munz. Kern Plateau, Kern Co., *Twisselmann.*

p. 1363.   3. **N. Bigelòvii** (Torr.) Wats. *n* = 19 (Cave, 1964).
p. 1364.   WASHINGTÒNIA

    1. **W. filífera** (Lindl.) Wendl. is natur. in Kern Co., *Twisselmann.*

p. 1365.   SPARGÀNIUM

    4. **S. multipedunculàtum** (Morong) Rydb. reported as in Sonoma Co., *Rubtzoff.*

p. 1366.   TÝPHA

    S. Galen Smith (Am. Midl. Nat. 78: 257–287. 1967) recognizes three species: **T. latifòlia** L., **T. angustifòlia** L. and **T. domingénsis** Pers. **T.** × **glaùca** Godron represents hybrids between *T. angustifolia* and *T. latifolia*, *T. domingensis* and *T. latifolia*, and trihybrids. It occurs rather widely in central California.

p. 1367.   AMARYLLIDÀCEAE

In Key at bottom of page, after line 2 "Fils. not appendaged" etc., insert:

Perianth-segms. connate at base or to the middle. . . . . 1a. *Nothoscordum*
Perianth-segms. distinct or barely united at the base into a ring.
    Pedicels not subtended etc. as in the FLORA.

p. 1369.   ÁLLIUM

In Key, A, B, CC, DD, E, change F:

F. Perianth-segms. 8–12 mm. long; bracts 15–20 mm. long.
  G. Perianth-segms. lanceolate, bright rose, the inner serru-
      late at tip. Native in n. Calif. .... 36. *A. acuminatum*
  GG. Perianth-segms. rounded at apex, white, not serrulate.
      Garden escape. .... 36a. *A. neapolitanum*
FF. Perianth-segms. 5–8 mm. long, not serrulate; bracts 8–12 mm.
    long. Middle and s. Calif.
  G. and GG. as in the FLORA.

p. 1373.

16. **A. unifòlium** Kell. s. into San Luis Obispo Co., *Hardham.*
18. **A. praècox** Bdg. *n* = 14 (Lenz, 1966).

p. 1378.

36a. **A. neapolitànum** Cyr. Bulb coats with quadrate reticulations having very heavy thick walls; scape 3–6 dm. high, subterete; lvs. 2–3 or more, lance-linear, loose-spreading, shorter than the scape, 1–3 cm. broad; pedicels 3–4 cm. long; fls. pure white, the perianth-segms. ovate, obtuse, 10–12 mm. long; stamens included; ovary not crested; stigma subentire.—Garden escape reported from Butte, Yolo, Sacramento, Napa, Sonoma, Marin, San Francisco, and Orange cos. From the Medit. region.

36b. **A. triquètrum** L. Near *A. neapolitanum,* but stem of the scape sharply 3-angled; lvs. 4–10 mm. broad; perianth-segms. oblong-lanceolate, 12–18 mm. long; stigma trifid.—Said by *Howell* to be a garden escape in Calif. Also from the Medit. region.

38. **A. Davísiae** Jones can be reported from above Kenworthy Ranger Station, San Jacinto Mts., Riverside Co. at 5300 ft., *L. B. Ziegler.*

1a. Nothoscórdum Kunth.

Stock a tunicated bulb. Plant without onion smell. Perianth campanulate, the segms. joined at the base into a short tube. Ovules 4–12 for each locule.

1. **N. inòdorum** (Ait.) Nichols. Lvs. basal, 2.5–3 dm. long; scape 2–4 dm. long; spathe 2-valved; fls. many, scented; perianth-segms. 8–14 mm. long, dull white, with greenish base and reddish midrib outside.—Reported as adventive in Marin and Fresno cos., *Fuller.* From se. U.S.

p. 1378.

MUÍLLA

1. **M. marítima** (Torr.) Wats. 2*n* = 20 (Lenz, 1966).

p. 1379.

BLOOMÈRIA

Hoover (Herbertia 11: 21. 1955) described *B. humilis* from San Luis Obispo Co., which he separated from *B. crocea*:

Corm never with offsets; scape rarely less than 15 cm. tall; basal lf. always solitary; perianth-segms. abruptly spreading from base; lower portion of fils. papillose. .. 1. *B. crocea*
Corm often with offsets; scape rarely more than 8 cm. tall; basal lvs. 1 or 2; perianth-segms. approximate toward base, gradually curving outward; lower portion of fils. often smooth. .... 1a. *B. humilis*

1. **B. cròcea** (Torr.) Cov. 2*n* = 18 (Lenz, 1966).
2. **B. Cleveléndii** Wats. is transferred to *Muilla Clevelandii* (Wats.) Hoov. because of having more lvs., which are not keeled or channeled and in not having the lower portion of the fils. terminate in a cuplike insertion for the upper portion.

p. 1380.

BRODIAÈA

In Key insert after A and B:

B′. Spathe-valves 2; fls. solitary. .... 1a. *B. uniflora*
B′B′. Spathe-valves 3 or more; fls. several.
  C. Perianth-tube obtuse, etc. as in the FLORA.

p. 1381.

In Key, AA, B, CC, DD, E, FF, GG, change HH to:

HH. The fils. 1 mm. long.
  I. Fils. broadly triangular; staminodia with in-
      curved apex. Monterey Co. to San Diego Co.,
      Santa Cruz Id. .... 16. *B. jolonensis*
  II. Fils. linear; staminodia erect. San Clement Id.
      22a. *B. kinkiensis*

1a. **B. uniflòra** (Lindl.) Engler. [*Triteleia u.* Lindl. *Ipheion u.* Raf.] Spring bloomer with onionlike odor; from small deep-seated bulbs; lvs. nearly flat; scape 1.5–2 dm. high, bracted about midway; fl. white with bluish tinge, 3–3.5 cm. across.—Escape from cult., as at Santa Barbara, *Pollard*. Originally from Argentina. Work now being done on what the FLORA has as one genus, *Brodiaea*, indicates that there should be a number of genera recognized.

3. **B. clementìna** (Hoov.) Munz. $2n = 16$ (Niehaus, 1965).

4. **B. láxa** (Benth.) Wats. $2n = 16, 32, 48$ (Lenz, 1966).

p. 1382. 6. **B. Dúdleyi** (Hoov.) Munz. $n = 8$ (Niehaus, 1965).

7. **B. lùgens** (Greene) Baker. Reported from Pinnacles, San Benito Co., *Hoover*.

8. **B. lùtea** (Lindl.) Mort. var. **Coòkii** (Hoov.) Munz, comb. nov. [*Triteleia ixiodes* var. *Cookii* Hoov., Plant Life 11: 19. 1955.] Described from the Santa Lucia Mts., above San Simeon, San Luis Obispo Co. for plant with perianth white, purple-tinged without, the segms. reflexed.

[*B. ixioides* $2n = 14$ (Lenz, 1966).]

Var. **scàbra** (Greene) Munz. Add *Triteleia scabra* Hoov. as syn.

Var. **analìna** (Greene) Munz. Add as synonym *Triteleia l.* var. *a.* Hoov. $n = 20$ (Niehaus, 1965); $2n = 10$ (Lenz, 1966). Occurs in Piute Mts., Kern Co., *Twisselmann*.

p. 1383. 10. **B. hyacinthìna** (Lindl.) Baker. The word in synonymy should be spelled *Hesperoscordum*. The sp. *hyacinthina* is reported from San Luis Obispo Co., *Hardham*. $n = 28, 35$ (Niehaus, 1965); $2n = 28$ (Lenz, 1966).

Var. **Greènei** (Hoov.) Munz. $2n = 16$ (Niehaus, 1965).

11. **B. gracilis** Wats. $n = 8$ (Niehaus, 1965).

14. **B. élegans** Hoov. $n = 8$ (Niehaus, 1965). Add:

Var. **austràlis** Hoov. Staminodia longer than the stamens, purple-tinged, obtuse, with slightly involute margin.—Tulare Co.

p. 1384. 15. **B. coronària** (Salisb.) Engler. $2n = 24$ (Niehaus, 1965).

Var. **macropòda** (Torr.) Hoov. $2n = 36$ (Niehaus, 1965).

16. **B. jolonénsis** Eastw. $2n = 12$ (Niehaus, 1965).

18. **B. pállida** Hoov. $n = 6$ (Niehaus, 1965).

p. 1385. 20. **B. califórnica** Lindl. var. **leptándra** (Greene) Hoov. $2n = 12$ (Niehaus, 1965).

21. **B. appendiculàta** Hoov. $n = 6$ (Niehaus, 1965).

22. **B. filifòlia** Wats. $2n = 32$ (Niehaus, 1965).

22a. **B. kinkiénsis** Niehaus. Corm with heavy fibrous outer coat; lvs. linear, 2–4 dm. long; scape 2–3 dm. tall; pedicels 3–8 cm. long; perianth-tube whitish with brown-purple midribs, rounded at base, 12 mm. long, 4–5 mm. wide, not splitting as caps. matures; perianth-segms. violet, 13–17 mm. long, spreading, the outer oblong, the inner obovate; staminodia erect, 7 mm. long, 3 mm. wide, cuspidate; fils. 1 mm. long; anthers 4–5 mm. long, retuse; $2n = 32$ (Niehaus).—San Clemente Id.

23. **B. Orcúttii** (Greene) Baker. $2n = 24$ (Lenz, 1966).

24. **B. pulchélla** (Salisb.) Greene. $2n = 18$ pairs (Raven, Kyhos & Hill, 1965); $2n = 18, 36, 45, 54, 72$ (Lenz, 1966).

Var. **pauciflòra** (Torr.) Mort. $2n = 36$ (Lenz, 1966).

25. **B. multiflòra** Benth. $2n = 18, 45$ (Lenz, 1966).

p. 1386. 27. **B. volùbilis** (Morière) Baker. $2n = 18$ (Lenz, 1966).

p. 1387. IRIDÀCEAE

Change the Key to:

A. Spathes more than 1-fld.
    B. Infl. umbeliate; the perianth not more than 2 cm. long.
        C. Perianth-segms. all alike. .... 2. *Sisyrinchium*
        CC. Perianth-segms. not alike, the 3 inner longer than the 3 outer. 2a. *Libertia*
    BB. Infl. not umbellate; perianth-segms. not all alike, or, if so, more than 2 cm. long.
        C. Perianth with 3 erect and 3 spreading or drooping segms.; fls. not in an elongate spikelike infl. .... 1. *Iris*
        CC. Perianth not as above; infl. spikelike.

D. The perianth irregular, the 3 upper segms. larger than the 3 lower.
  E. Style branches simple, not bifid.; fls. 3–7 cm. long.
    F. Perianth tube constricted near or below the middle into a
      narrow cylindrical or filiform basal part. ... 5. *Chasmanthe*
    FF. Perianth tube tapering gradually from base to throat, curved.
                                        3. *Gladiolus*
  EE. Style branches bifid; fls. mostly less than 3.5 cm. long. 6. *Freesia*
DD. The perianth regular or essentially so.
  E. Lvs. 2.5–6 cm. wide. .... 4. *Watsonia*
  EE. Lvs. setaceous. .... 7. *Romulea*
AA. Spathes 1–3-fld., often appearing calyx-like.
  B. Lvs. setaceous; infl. 1–3-fld.; perianth with red-lilac limb. .... 7. *Romulea*
  BB. Lvs. 1 cm. or more broad; infl. a lax panicle of several many-fld. spikes; perianth
    with orange-red limb. .... 8. *Crocosmia*

p. 1388.   ìRIS

Line 1 at top of page. Change "Studies" to "Revision" in 1959 printing.
In Key, change to:

AA. Rhizome 20–40 mm. thick.
  B. Lvs. linear, generally less than 10 mm. wide.
    C. Lvs. light green, mostly 3–6 mm. wide; spathes largely scarious; fls. usually
      2–3 on a stem. Mostly in Montane Coniferous F. ..... 12. *I. missouriensis*
    CC. Lvs. dark green, ca. 6–10 mm. wide; spathe largely herbaceous; fls. up to 8
      on a stem. Plants from near the coast. .... 13. *I. longipetala*
  BB. Lvs. ensiform, 15–25 mm. wide; fls. 2–3 on a stem, yellow. Escape from cult.
                                       14. *I. Pseudoacorus*

p. 1390.   4. **I. chrysophýlla** Howell. Add as synonym *I. tenax* ssp. *c.* Clarkson.

    8. **I. bracteàta** Wats. Insert as a synonym *I. tenax* ssp. *b.* Clarkson.

p. 1391.   10. **I. innominàta** Henders. Add as synonym *I. tenax* ssp. *i.* Clarkson.

    11. **I. Douglasiàna** Herb. Insert as synonyms *I. tenax* ssp. *D.* Clarkson and
I. tenax ssp. *Thompsonii* Clarkson.

p. 1392.   14. **I. Pseudàcorus** L. Erect, glabrous, rather glaucous, 4–15 dm. high;
rhizome often 3–4 cm. diam.; lvs. 15–25 mm. wide, ca. as long as the compressed terete scape; fls. 8–10 cm. diam., yellow, the outer segms. often
purple-veined with an orange spot near the base; caps. elliptic, apiculate.—
Garden-escape in wet places, Sonoma Co., *Rubtzoff* and near Mettler, Kern
Co., *Twisselmann*.

SISYRÍNCHIUM

Change description of genus to "Tufted annuals or usually perennials from a
short rootstock."
Insert before the present Key:

Plants annual. .... 7. *S. minus*
Plants perennial.
  Fls. mostly blue, etc. as in present Key.

p. 1393.   4. **S. halóphilum** Greene. In Piute and Tehachapi mts., Kern Co., *Twisselmann*.

    7. **S. mìnus** Engelm. & Gray. Small annual, the fls. lavender-pink to purple-
rose, white with yellow eye, or all yellow.—Collected in grassy field on Sepul-
veda Blvd., Los Angeles, *F. W. Gould in 1944*; native of Tex., La. Apr.–May.

2a. Libértia Spreng.

Perennial herbs with short creeping rhizome; lvs. linear, equitant; perianth
without any tube above the ovary, the segms. obovate, the 3 outer shorter
than the 3 inner. Several spp. in the S. Hemis. (Marie *Libert*, 1782–1865,
Belgian student of liverworts.)

    1. **L. formòsa** Grah. Lvs. 3–4.5 dm. long, rigid; outer perianth-segms.
brown, the inner white.—Natur. in San Francisco. Native of Chile.

p. 1394.   WATSÒNIA

    1. Change to **W. bulbillifera** Matthews & L. Bolus. [*W. angusta* Calif. auth.,
not Ker.] With clusters of bulbils at lower nodes of infl.—Known from Mendocino and Sonoma cos.

p. 1394.   **5. Chasmánthe N. E. Br.**

Differing from *Gladiolus* by having the perianth tube constricted near or below the middle into a narrow or basal part. Ca. 9 spp.; African. (Greek, *chasme*, gaping, and *anthe*, flower.)

1. **C. aethiòpica** (L.) N. E. Br. [*Antholyza a.* L.] Stems ca. 1 m. high; fls. red-yellow, 3–6 cm. long, the cylindrical part ca. ½ the whole.—Cult. and escaped, as in Santa Barbara, *Pollard.* Point Lobos State Reserve, Monterey Co., *Fuller.* Native of S. Afr.

6. Freèsia Klatt

Cormous plants with plane narrow lvs. below and showy fls. in loose secund spikes at top of the slender stem; perianth tubular, funnel-shaped, the segms. ± unequal. One or more spp. S. Afr. (E. M. *Fries,* 1795?–1876, Swedish botanist.)

1. **F. refrácta** Klatt. Two–4 dm. tall; basal lvs. 1–1.5 dm. long; fls. solitary in the short spathes, usually ± yellow, to 3.5 cm. long, fragrant.—Common garden plant occasionally becoming established as at San Luis Obispo and San Francisco. Native of Afr.

7. Romulèa Maratti

Corm tunicated. Foliage lvs. tufted, slender, linear. Scape simple or branched. Fls. 1–3 in a spathe, long-peduncled. Perianth-segms. in 2 similar series; tube short. Style-branches linear, bifid. Caps. 3-lobed. Ca. 70 spp. of Medit. region and S. Afr. (*Romulus,* one of the founders of Rome.)

1. **R. ròsea** Eckl. Corm globose, 8–12 mm. thick; lvs. 1.5–3 dm. long, setaceous; peduncle to 1.5 dm. long, 1–3-fld.; outer spathe 2 cm. long; perianth with yellow throat and red-lilac limb.—Adventive at Carmel Highlands, Monterey Co., *Fuller*; native of S. Afr.

8. Crocósmia Planch.

Corm with reticulated tunics; lvs. many, equitant; infl. a panicle of several spikes; spathe-valves calyxlike, notched or cut; fl. 1 in each spathe; perianth with tube somewhat dilated above; segms. subequal, oblong to obovate; stamens 3, inserted at base of funnel. S. Afr. (Greek, *crocus,* saffron, and *osme,* smell, because of odor of dried fls. immersed in water.)

1. **C. crocosmiflòra** N. E. Br. MONTBRETIA. Sts. to 1 m., branching; lvs. 2–4 dm. long; fls. orange-crimson, 3–5 cm. diam.—Garden hybrid, widely cult., sometimes establishing itself as a wild plant.

p. 1396.   CYPRIPÈDIUM
3. **C. califórnicum** Gray. Reported from Belden, Plumas Co.

HABENÀRIA
2. **H. élegans** (Lindl.) Boland. $2n = 21$ pairs (Raven, Kyhos & Hill, 1965). Calder & Taylor have as new combs.: *H. unalascensis* ssp. *maritima* and *elata.*
3. **H. dilatàta** var. leucostàchys (Lindl.) Ames. $n = 21$ (Raven, Kyhos and Hill, 1965).

p. 1397.   SPIRÁNTHES
2. **S. porrifòlia** Lindl. Reported from Monterey Co., *Howitt & Howell.*

p. 1398.   LÍSTERA
1. **L. cordàta** (L.) R. Br. in Ait. has greenish or purplish fls., the 2 forms usually growing intermingled. $2n = 36$–38 (Löve & Löve, 1956). $n = 19$ (Taylor, 1967).—Collected in Del Norte Co., *Munz.* The greenish form has been called *L. nephrophylla* Rydb., [*L. c.* var. *n* Hult., *L. c.* ssp. *n.* Löve & Löve.]

EPIPÁCTIS
Insert after generic description

The lip distinctly 3-lobed; lateral lobes erect and forming a sac which is papillose within;
mid-lobe usually linear-oblanceolate; sepals 1.2–1.3 cm. long. . . . . 1. *E. gigantea*
The lip not 3-lobed; sac not papillose within; the apical part of the lip usually triangular-
ovate; sepals 1–1.2 cm. long. . . . . 2. *E. Helleborine*

p. 1399   1. **E. gigantèa** Dougl. ex Hook. $n = 20$ (Raven, Kyhos & Hill, 1965).
             2. **E. Helleborìne** (L.) Crantz. [*Serapias H. L.*] Differing from *E.
gigantea* particularly in its smaller fls. and non-lobed lip.—Reported from the counties
about San Francisco Bay, *Howell*; found also in the e. U.S.; native of Old
World.

GOODYÈRA
             1. **G. oblongifòlia** Raf. reported from Santa Cruz Mts., *Howell*.

CALÝPSO
             1. **C. bulbòsa** (L.) Oakes was found in the Santa Cruz Mts., *Crandall*.
Calder & Taylor distinguish western plants as *C.* bulbòsa ssp. occidentàlis
(Holz) Calder & Taylor.

p. 1400.   CORALLORHÌZA
             1. **C. striàta** Lindl. reported from Fresno Co., *Collett*.
             3. Calder & Taylor use the comb. *C.* maculàta Raf. ssp. *Mertensiàna*
(Dougl.) Calder & Taylor.

p. 1402.   JÚNCUS
             In Key, AA, B, CC, change D to:

             D. Perennials, usually with flat lf.-blades.
                 D′. Anthers 3, red to dark purple; perianth 2–3.5 mm. long. Nevada
                      Co. . . . . 21a. *J. marginatus*
                 D′D′. Anthers 6.
                 E. Lf.-sheath passing gradually, etc. as in the FLORA.

p. 1403.   In Key change the last DD to:

             DD. Anthers usually shorter than fils.; style short.

p. 1404.   2. **J. Párryi** Engelm. Found in the White Mts. at 11,800 ft.
             5. **J. mexicànus** Willd. In the White Mts. at 11,900 ft.
p. 1405.   6. **J. bálticus** Willd. $2n = 40$ (Löve & Löve, 1956).
             11. **J. bufònis** L. var. **Congdònii** (Wats.) J. T. Howell. [*J. C.* Wats.] Seeds
nearly smooth, shining, translucent.—Merced Co.
p. 1406.   12. **J. sphaerocárpus** Nees. $2n = 36$ (Snogerup, 1958).—Rare in the S.
Coast Ranges.
p. 1407.   17. **J. Covíllei** Piper. Ranges s. to Marin Co., *Howell*.
             18. **J. orthophýllus** Cov. Found in Kern Co., *Twisselmann*.
             20. **J. macrophýllus** Cov. occurs in the Greenhorn Range, Kern Co., *Twis-
selmann* and in Monterey Co., *Howell*.
             21a. **J. marginàtus** Rostk. Cespitose from a short thick and often knotty
rhizome; stems slender, 2–8 dm. high; lvs. green, flat, soft, the basal 4–20
cm. long, 1–4 mm. broad; infl. 1–10 cm. long, with ca. 2–30 heads of 2–12
fls. each; perianth 3–3.5 mm. long, reddish brown; stamens 3, with reddish
anthers; caps. rounded, beakless; seeds brown, many-ribbed, 0.5 mm. long.—
Ca. 7 mi. e. of Nevada City, Nevada Co., at 3100 ft.; Ariz. and Rocky Mts. to
New England.
p. 1408.   26. **J. Kellóggii** Engelm. in Monterey Co., *Howitt & Howell*.
             27. **J. capillàris** F. J. Herm. Report for Monterey Co. erroneous.
             28. **J. bryoìdes** F. J. Herm. is in Monterey Co. at The Indians, having 2
bracts instead of the usual 1.
p. 1409.   30. **J. unciàlis** Greene. is in Monterey Co.
             34. **J. acuminàtus** Michx. F. J. Hermann uses f. **sphaerocéphalus** Herm. for
the Calif. plant.
             35. **J. rugulòsus** Engelm. has been found n. into Monterey Co., *Howitt &
Howell*; Kern Co., *Twisselmann*.
p. 1410.   37. **J. articulàtus** L. $2n = 80$ (Löve & Löve, 1956).

41. **J. Mertensiànus** Bong. F. J. Hermann (Leafl. W. Bot. 10: 81–86. 1964) revises this complex as follows: (combining spp. 41, 42, 43 of the FLORA)

Heads usually solitary, sometimes 2, many-fld. (12- or more-).
 Perianth-segms. purplish-black, flaccid, narrow and exposing much of the mature caps.; bracts spathaceous; anthers usually much shorter than the fils.; auricles rounded, opaque, 1–2 mm. long. Calif. to Rocky Mts. . . . . *J. Mertensianus*
 Perianth-segms. brown, stiffish; bract narrow; anthers and fils. usually subequal; auricles rounded to acute, translucent. Mts. of s. Calif. . . . . Var. *Duranii*
 (*J.M.* ssp. *M.* var. *Duranii* F. J. Herm.)
Heads usually several to many, few-fld. (12- or fewer-), usually dark brown; perianth-segms. usually stiffish; bracts not spathaceous; anthers longer than the fils. Mts., Calif. to Wash., Wyo. . . . . Ssp. *gracilis*
 (*J.M.* ssp. *gracilis* (Engelm.) F. J. Herm. *J. phaeocephalus* var. *gracilis* Engelm. *J. nevadensis* Wats.)

Rubtzoff (Leafl. W. Bot. 10: 168. 1965) reports **J. Mertensiànus** ssp. **grácilis** from Boggs Lake, s. Lake Co. The nearest previously known station was Trinity Co., *Mason*.

49. **J. ensifòlius** Wikstr. Greenhorn Mts., Kern Co., *Twisselmann*.

p. 1412. **LÚZULA**

3. **L. subcongésta** (Wats.) Jeps. not Buch.

p. 1413. 5. **L. spicàta** (L.) DC. occurs in the White Mts. Add to synonymy *L. s.* ssp. *saximontana* Löve & Löve.

6. **L. comòsa** E. Mey. Add to the synonyms *L. multiflora* (Retz.) Lejeune var. *comosa* (E. Mey.) St. John. Change "anthers" etc. to "anthers equal to or longer than fils."

p. 1414. **CYPERÀCEAE**

In legend under Fig. 133, the next to the last word in line 2 should be "spikelets" not "spikes."

p. 1415. At end of Key to Genera, change to:

AA. Fls. all unisexual, ♂ and ♀ in separate spikes or separate parts of the same spike.
 B. Female fls. enclosed in perigynia, which often end in a beak. Many spp. Common.
                              12. *Carex*
 BB. Female fls. without perigynia, but closely enfolded by an inner glume. One sp., Convict Lake Basin, Sierra Nevada. . . . . 13. *Kobresia*

SCÍRPUS

Following in part changes suggested by T. Koyama (Can. J. Bot. 40: 913–937. 1962 and 41: 1117–1122. 1963) and by others, I submit the following key for California species:

A. Bristles much exserted, upwardly barbed. . . . . 1. *S. criniger*
AA. Bristles, where present, included within scales and usually downwardly barbed.
 B. Infl. subtended by involucral lvs.
  C. Involucral lvs. 2–5, usually exceeding the infl.
   D. Spikelets small, 0.3–0.6 cm. long.
    E. Aks. lenticular; style 2-fid; stamens 2. . . . . 2. *S. microcarpus*
    EE. Aks. triangular; style 3-fid; stamens 3.
     F. Perianth bristles 2–4 mm. long, the teeth antrorse, rather scattered; infl. mostly with primary rays and less diffuse branching.
                          3. *S. Congdonii*
     FF. Perianth bristles 1–2.5 mm. long, the teeth retrorse, close together; infl. open, diffuse. . . . . 3a. *S. diffusus*
   DD. Spikelets larger, 1–4 cm. long.
    E. Aks. rhomboid-obovoid, 3-angled with pale, slightly concave sides; perianth bristles 6, strongly scabrous with spinules, persistent on mature aks.; lvs. evenly distributed the full length of the culms.
                          4. *S. fluviatilis*
    EE. Aks. obovate to broadly so, lenticular or compressed-triangular with convex sides; perianth bristles fewer than 5, scaberulous with minute hairlike appressed spinules, deciduous; lvs. mostly basal but sometimes with a few on the culm.
     F. Floral scales rufescent, chartaceous, not translucent, very tightly appressed, the awn short and abruptly recurved.
                          5. *S. robustus*
     FF. Floral scales pale- to chestnut-brown but not rufescent, thin-membranaceous, semitranslucent particularly on the hyaline margin, ± loose, the awn long and gradually recurved.

    G. Spikelets ovoid, usually more than 6 mm. thick, solitary or clustered in groups of 2 or 3 on umbel rays, sometimes all congested in a large head; aks. broadly ovate, more frequently digynous. **6. *S. maritimus* var. *paludosus***
    GG. Spikelets lanceolate to linear lanceolate, usually less than 6 mm. thick, clustered in groups of 2–5 on well-elongated umbel rays; aks. oblong-obovate, more frequently trigynous. . . . . **6. *S. maritimus* var. *tuberosus***
  CC. Involucral lf. solitary, often appearing as a continuation of the culm.
    D. Culm leafy, triangular or subterete; spikelets few, 1–12.
      E. Perennials with rhizomes; culms not filiform; bristles present.
        F. Culm subterete; scales not awned.
          G. Spikelets 2–8; style 2-cleft; lvs. ca. 2 mm. wide. Mono Co. to Modoc Co. . . . . **8. *S. nevadensis***
          GG. Spikelet 1; style 3-cleft; lvs. 0.5–1 mm. wide. Del Norte Co., Nevada Co. . . . . **8a. *S. subterminalis***
        FF. Culm sharply triangular; scales short-awned.
          G. Infl. with the 2nd and 3rd bracts scalelike and to as long as the lowest spikelet; lf.-blades convolute; plant to ca. 1 m. tall.
            H. Floral scales rusty brown or yellow brown, thinly coriaceous along the midvein, the long excurrent mucro exceeding the acute teeth at the scale apex. **7. *S. americanus* var. *longispicatus***
            HH. Floral scales purple-fuscous or purplish-sanguineous, membranaceous along the midvein, the short upright mucro shorter than or equalling the rounded small teeth at the scale apex. **7. *S. americanus* var. *monophyllus***
          GG. Infl. without second and third bracts; lf.-blades flat.
            H. Aks. light brown or gray, the surface minutely pitted, common; native. . . . . **9. *S. Olneyi***
            HH. Aks. dark brown, the surface horizontally rugose; rare weed of rice fields. . . . . . . **10. *S. mucronatus***
      EE. Annuals with fibrous roots, or rarely perennials; culms filiform; bristles absent.
        F. Annuals; aks. punctate or transversely corrugate.
          G. Scales sharply keeled, acute or acuminate; involucral lf. to 2.5 cm. long. . . . . **13. *S. koilepis***
          GG. Scales only slightly keeled at tip.
            H. The scales obtuse; aks. punctate. Native. **12. *S. cernuus***
            HH. The scales acuminate; aks. transversely corrugate. Rare adventive. . . . . **12a. *S. saximontanus***
        FF. Perennials; aks. longitudinally ribbed. . . . . . **11. *S. setaceus***
    DD. Lvs. of culm reduced to basal sheaths with short blades, to 8 cm. long; culm stout and terete; spikelets numerous in umbels.
      E. Bristles filiform, barbed retrorsely.
        F. Scales well exceeding the ak., pale brown and sanguineous-tinged, with dark red gummy spots at least on the upper half; spikelets on relatively short rigid rays. . . . . . . **14. *S. acutus***
        FF. Scales equalling or only very slightly exceeding the ak., rusty brown, smooth or with a few dark red gummy spots on the upper midvein only; spikelets ± nodding on elongate slender rays. . . . . **15. *S. validus***
      EE. Bristles broad, plumose with dense soft hairs. . . **16. *S. californicus***
  BB. Infl. subtended by the long-awned lowermost scale of the spikelet; true involucral lvs. absent.
    C. Plants not stoloniferous; bristles 6; ak. strongly trigonous. Tulare and Inyo cos. to Tuolumne and Mono cos. at high altitudes. . . . . . **17. *S. Clementis***
    CC. Plant stoloniferous, with slender scaly stolons and filiform rhizomes; bristles 0; ak. compressed. Convict Lake, Mono Co. . . . . **18. *S. Rollandii***

p. 1416.   **1. S. crìniger** Gray. At elevs. as low as 150 ft., Del Norte Co.

    3a. S. diffùsus Schuyler. Resembling *S. Congdonii*, the infl. more diffuse with primary rays mostly again divided; perianth bristles shorter, their teeth retrorse, quite crowded, rather than antrorse and scattered.—Mostly below 6500 ft., Humboldt Co. to Lake Co. and Tulare Co.

    4. S. fluviátilis (Torr.) Gray reported from Santa Cruz Mts., *Thomas*; Sonoma Co., *Baker*; and Napa Co., *Jussel*.

    5. S. robústus Pursh as understood by Koyama and keyed out above is found in Coastal Salt Marsh of Calif., Mex., E. N. Am. and in S. Am.

p. 1417.   **6.** Koyama recognizes **S. marítimus** L. var. **paludòsus** (A. Nels.) Kükenth.

[*S. paludosus* A. Nels. was in synonymy under *S. robustus* in the FLORA. *S. pacificus* Britt. ex Parish] and S: m. var. tuberòsus (Desf.) R. & S. [*S. tuberosus* Desf. in the FLORA] as varieties in the **S. marítimus** complex.—The first named is native in N. Am. and common at low elevs. in Freshwater Marsh, etc., while the second is an Old World plant natur. in Calif. and Quebec. An additional record for it in Calif. is Kern Co., *Twisselmann*.

7. **S. americànus** Pers. For Calif. Koyama recognizes two vars. as shown in the new key: Var. **longispicàtus** Britt. [*S. a.* var. *polyphyllus* (Böckeler) Beetle, in part] which he gives as from inland Calif. to inland B.C., Mex., W. Indies, S. Am.

Var. **monophýllus** (Presl) Koyama [*S. m.* Presl.] which he ascribes to coastal Calif.; n. to coastal B.C. and s. to S. Am.

8a. **S. subterminàlis** Torr. Aquatic perennial, with slender nodulose culms 3–10 dm. long; lvs. slender, channeled, 2–5 dm. long, 0.5–1 mm. wide; spikelets solitary, terminal, oblong-cylindrical, 6–10 mm. long, subtended by an erect subulate involucral lf.; fls. 6–10; scales light brown with green midveins; stamens 3; style 3-cleft to middle.—N. Del Norte Co., *Hobart*, Nevada Co., *Howell & True*; to B.C., Atlantic Coast. July–Aug.

12a. **S. saximontànus** Fern. [*S. supinus* L. var. *s.* Koyama]. Tufted annual; culms slender, terete, simple, unequal, to 4 dm. tall; basal sheaths mostly bladeless; elongate cauline blade occasional; invol. erect; spikelets 1–7, becoming cylindrical, 0.5–1.5 cm. long; scales cuspidate-acuminate, with green keel; styles 3-cleft; ak. strongly 3-angled, the subequal faces slightly convex.—Damp shores, Colo. to S. Dak., Kans., Tex., etc. Reported as in Ventura, Kern, Colusa, Glenn, and Butte cos.

p. 1418.   14. **S. acùtus** Muhl. Add as syn. *S. lacustris* ssp. *glaucus* (Smith) Hartman. So treated by Koyama, Can. J. Bot. 40: 926. 1962.

15. **S. válidus** Vahl. [Add as syn., *S. lacustris* L. ssp. *validus* (Vahl) T. Koyama, Can. J. Bot. 40: 927. 1962.]—Reported from as far south as San Francisco, *Rubtzoff* and Santa Barbara Co., *C. Smith*.

16. **S. califórnicus** (C. A. Mey.) Steud. in Napa and Sonoma cos., *Rubtzoff*.

17. **S. Cleméntis** Jones. In third line from end of description, "1–5 mm." should read "1.5 mm."

18. **S. Rollándii** Fern. [*S. pumilus* auth., not Vahl. *S. p.* ssp. *R.* Raymond.] Stoloniferous with slender scaly stolons and filiform rhizomes; culms in tufts, 5–15 cm. high; spikelet solitary, 3–4 mm. long; scales ovate, brownish; bristles 0; ak. blackish.—Calcareous places at 10,200 to 10,600 ft., Convict Lake Basin, Mono Co.; Colo. to Alta., e. Can.

HELEÓCHARIS.   The spelling **Eleócharis** is preferable.

In Key after A, BB, CC:

    D.   Scales spir 'y arranged; tubercle not 3–lobed.
       D'.   Tubercle confluent with the ak., merely conic. and not forming a distinct caps. .... 2. *H. pauciflora*
       D'D'. Tubercle obviously differentiated from the ak., forming a distinct apical caps.
       E.   Tubercle long-subulate, etc. as in the FLORA.

p. 1419.   1. **H. aciculàris** (L.) R. & S. var. **rádicans** (Poir.) Britton. In Sonoma Co., *Rubtzoff*, and in Marin and Mendocino cos.

2. **H. pauciflòra** (Lightf.) Link. Reported from Sonoma and Monterey cos. and the White Mts. Clapham, Tutin & Warburg in the Fl. Brit. Isles use **H. quinqueflòra** (F. X. Hartmann) Schwarz instead of *pauciflora,* but I have seen no discussion as to the reason.

p. 1420.   3. **H. párvula** (R. & S.) Link. In Napa Co., *Rubtzoff* and in Marin Co.

5. **H. montevidénsis** var. **Paríshii** Grant. $n = 5$ (Raven et al., 1965).

p. 1421.   11. J. T. Howell (Wasmann J. Biol. 22: 163.) uses **H. macrostàchya** Britton. instead of **H. palústris** (L.) R. & S. $2n = 38$ (Strandhede, 1967).

p. 1423.   HEMICÁRPHA

2. **H. occidentàlis** Gray. Reported from Greenhorn Mts., Kern Co. and Kern Plateau, Tulare Co., *Hardham*.

p. 1424.   CYPÈRUS

In Key change AA, B, C, DD, EE to:

EE.   Scales ovate to rounded, not recurved at tip.
    F.   Spikes dense, often lobate; scales ovate-orbicular, not mucro-
         nate. . . . . 8. *C. diffusus*
    FF. Spikes loose; scales ovate, mucronate. . . . . 8a. *C. fuscus*

1. C. nìger var. capitàtus (Britton) O'Neill ranges to Humboldt and Shasta cos., *Rubtzoff*. Ascends to 5700 ft. in Lassen Nat. Park.

Var. rivulàris (Kunth) V. Grant. Correct name is var. castàneus (Pursh) Kükenthal.

p. 1425.   8. C. diffórmis L. Rubtzoff (Leafl. W. Bot. 10: 68) sums up records from Coast Ranges: Sonoma, Napa, Marin, San Francisco cos.

8a. C. fúscus L. Tufted; invol. of 2–4 divergent lvs.; umbel condensed or rayed, the spikes subcapitate; spikelets purple-brown, 3–12 mm. long, the scales ca. 1 mm. long.—Stanislaus R. at Caswell Memorial State Park; San Joaquin Co., *Rubtzoff*. Found in e. U.S.; native of Old World.

p. 1427.   KYLLÍNGA

1. K. brevifòlia Rottb. List as synonym *Cyperus brevifolius* Rottb.

SCHOÈNUS

1. S. nigricáns L. $n = 22$ (Davies, 1956).

p. 1428.   RHYNCHÓSPORA

2. Change name to R. glomeràta (L.) Vahl var. mìnor Britton. [*R. g.* var. *capitellata* Kükenthal.] Reported from Plumas Co. and region, *Rubtzoff*.

4. R. califórnica Gale in Marin Co. at Point Reyes.

p. 1429.   CÀREX

At end of generic description add: (*Carex*, the classical Latin name, of obscure derivation).

p. 1433.   In Key after FF, G, H, II insert:

J.   Perigynia lance-ovate, tapering gradually
     into a beak ca. ¼ the length of the body.
     At above 4000 ft., mts. of Calif.
                                    46. *C. mariposana*
JJ. Perigynia broadly ovate, tapering abruptly
     into a beak almost ½ the body length.
                                    46a. *C. molesta*

p. 1434.   In Key under **Firmiculmes** change to:

A.   Culms terete or obtusely triangular, usually smooth. . . . . 86. *C. multicaulis*
AA. Culms sharply triangular, scabrous above.
    B.   Rootstocks prolonged; spikelet solitary; pistillate scales short awn-tipped or awn-
         less, not foliaceous. . . . . 87. *C. Geyeri*
    BB. Rootstocks not prolonged, the plants cespitose; spikelets sometimes 2 or 3; pis-
         tillate scales ± foliaceous. . . . . 87a. *C. Tompkinsii*

p. 1436.   1. C. capitàta L. $2n = 50$ (Löve & Löve, 1956).
p. 1437.   3. C. Brèweri Boott. Occurs in the White Mts., Inyo Co.
p. 1438.   9. C. pánsa Bailey. Add as synonym *C. arenicola* ssp. *pansa* Koyama & Calder.
             10. C. Eléocharis Bailey. To 13,500 ft. in the White Mts.
             11. C. simulàta Mkze. s. to the Piute Mts., Kern Co., *Twisselmann*.
             12. C. vallícola Dewey taken at Monitor Pass, Alpine Co.
p. 1439.   17. C. tumulícola Mkze. On San Clemente Id., *Raven*.
             21. C. vicària Bailey. In Greenhorn Range, Kern Co., *Twisselmann*.
p. 1440.   23. C. Dúdleyi Mkze. in San Luis Obispo Co., *Hardham*.
             24. C. diandra Schrank. $2n = 60$ (Löve & Löve, 1956).
p. 1441.   26. C. Jònesii Bailey. In Greenhorn Range, Kern Co., *Twisselmann*.
             30. C. dìsperma Dewey. $2n = 70$ (Löve & Löve, 1965).
             32. C. canéscens L. $2n = 56$ (Löve & Löve, 1956).
p. 1442.   34. C. laeviculmis Meinsh. In Nevada Co., *True & Howell*.
p. 1443.   38. C. angústior Mkze. In the Greenhorn Range, Kern Co., *Twisselmann*.

39. **C. leptopòda** Mkze. [*C. Deweyana* ssp. *l.* (Mkze.) Calder & Taylor.] In Monterey Co., *Howitt & Howell*; and San Luis Obispo Co., *Hoover*.

40. **C. Bolánderi** Olney in Monterey Co., *Howitt & Howell*.

p. 1444.   46. **C. mariposàna** Bailey. In the Greenhorn Mts., Kern Co., *Twisselmann*.

46a. **C. molésta** Mkze. Cespitose, the culms 3–10 dm. high, roughened above, brownish-black at the base; lf.-blades 1–3 dm. long, 2–3 mm. wide; spikes 4–8, gynaecandrous, in a head 2–3 cm. long, the spikes subglobose, 6–9 mm. long; scales ovate, yellowish-brown with 3-nerved green center and hyaline margins; perigynia ovate, 4.5 mm. long, rounded at the base, ± nerved, tapering abruptly into a beak almost half the length of the body; beak flat, serrulate, brownish-tipped, shallowly bidentate.—Santa Barbara, *Pollard*. From the e. and central U.S.

p. 1445.   52. **C. subfúsca** W. Boott. In the White Mts.

p. 1446.   55. **C. gracílior** Mkze. Greenhorn Range, Kern Co., *Twisselmann*.

p. 1447.   61. **C. phaeocéphala** Piper f. **Eastwoodiàna** (Stacey) F. J. Herm. [*C. E. Stacey.*] Perigynium typically broadest at or below the middle, not above. Mono Co.; to Wash., Wyo.

p. 1449.   73. **C. leptàlea** Wahl. $2n = 52$ (Löve & Löve, 1965).

p. 1450.   77. **C. globòsa** Boott. On Santa Cruz Id.

78. **C. Braìnerdii** Mkze. in the Greenhorn Range, Kern Co., *Twisselmann*; ascending to 9100 ft., Lassen Nat. Park.

80. **C. Róssii** Boott. In the White Mts.

p. 1452.   87a. **C. Tompkínsii** J. T. Howell Cespitose, the stems many, erect, acutely triangular, scabrous above, 1–4 dm. tall; lvs. of the season blade-bearing, generally 2 or 3, near base of stems or above, the sheaths cylindric, hyaline, brown-tinged at mouth, blades flat or caniculate, to 4 dm. long, 1.5–3 mm. wide; spikelets bractless, androgynous, solitary and terminal or with 1 or 2 lateral; stigmas 3; scales green with hyaline margins, awn-tipped or semi-foliaceous; perigynia 5–6 mm. long, 2-ribbed, greenish, short-beaked; ak. 4–5 mm. long.—Kings River, Fresno Co. at 3200–5500 ft.

p. 1454.   98. **C. mendocinénsis** Olney in Monterey Co., *Howitt & Howell*, and San Luis Obispo Co., *Hardham*.

101. **C. Lemmònii** W. Boott in the Greenhorn Range, Kern Co., *Twisselmann*.

p. 1455.   104. **C. luzulìna** Olney in San Luis Obispo Co., *Hardham*.

108. **C. lanuginòsa** Michx. at various stations in Kern Co., *Twisselmann*.

p. 1456.   108a. **C. lasiocárpa** Ehrh. edges of pond, west of Lake Center Public Campground, s. Plumas Co., at 6700 ft., *E. K. Balls*. Differs from *C. lanuginosa* by having lvs. filiform-convolute, 0.5–2 mm. wide (not flat with revolute margins and 2–5 mm. wide).—Wash. to Atlantic Coast; Eurasia.

111. **C. amplifòlia** Boott. Collected in Kern Co., *Twisselmann*.

112. **C. limòsa** L. $2n = 62$ (Löve & Löve, 1956); $2n = 64$ (Löve & Löve, 1965).

p. 1458.   123. **C. gymnóclada** Holm. Hermann proposes *C. scopulorum* var. *bracteosa* (Bailey) F. J. Herm. for *C. gymnoclada*.

p. 1459.   125. **C. Kellóggii** W. Boott in the Greenhorn Range, Kern Co., *Twisselmann*.

129. **C. aquátilis** Wahl. $n = 38$ (Davies, 1956); $2n = 76$ (Löve & Löve, 1956).

p. 1460.   133. **C. nudàta** W. Boott. The lowest bract is much shorter than the infl., not the culms.

p. 1461.   137. **C. spíssa** Bailey as far n. as Monterey Co., *Howitt & Howell*.

142. **C. vesicària** L. $2n = 74$ (Löve & Löve, 1965).

p. 1462.   144. **C. rostràta** Stokes $2n = 76$ (Löve & Löve, 1956)

13. Kobrèsia Willd.

Slender arctic and mountain sedges with erect culms leafy below and with spikelets few-fld., variously grouped. Scales of spikelets 1-fld., the lower fls.

usually ♀, the upper ♂. Stamens 3. Lacking perigynia and perianth-bristles, but ♀ fl. closely enfolded by the inner glume. Ovary oblong, narrowed into a short style; stigmas 2–3, linear. Ak. sessile, obtusely angled. (Von *Kobres*, a German naturalist.) Ca. 30 spp.

1. **K. myosuroìdes** (Vill.) Fiori & Paol. [*K. Bellardii* (All.) Degland]. Culms very slender, 1–4.5 dm. tall; lvs. shorter, narrow, the margins ± revolute; old sheaths fibrillose, brown; spike subtended by a short bract or bractless, usually densely fld., 1.5–3 cm. long, 3–4 mm. diam.; aks. scarcely 2 mm. long, 1 mm. thick, appressed.—At 9700 to 10,600 ft., moist places, Convict Basin, Mono Co.; to Arctic Am., Eurasia.

p. 1465.   FESTÙCEAE

In Key HH, II, J, change K:

> K. Spikelets in racemes.
> L. Racemes short, dense, over-topped by the lvs.; spikelets awnless. . . . . . 6a. *Sclerochloa*
> LL. Racemes elongate, loose, ex-serted; spikelets awned or mu-cronate. . . . . . . 7. *Pleuropogon*

HÓRDEAE

In B in the 8th line from the bottom of the page, the word "latter" should be "lateral" in the 1959 printing.

p. 1466.   AGROSTÍDEAE

In Key AA, BB, insert before C:

> B′. Lemma firm, bearing a long straight delicate awn just below the tip; palea ca. as long as lemma. . . . . 53a. *Apera*
> B′B′. Lemma thin or membranous.
> C. Glumes mostly longer, etc. as in the FLORA.

p. 1467.   PANICÈAE

In Key after AA, BB, change to:

> C. Glumes awned or mucronate; apex of palea not inclosed by the lemma.
> D. Infl. paniculate; spikelets silky. . . . . 86a. *Rhynchelytrum*
> DD. Infl. of unilateral racemes along a common axis; spikelets not silky. 87. *Echinochloa*
> CC. Glumes awnless, etc. as in the FLORA.

Key after AA, BB, CC, DD, EE, change to:

> F. Racemes slender, 3–12.
> G. Fr. flexible; 1st glume reduced, but present. 82. *Digitaria*
> GG. Fr. rigid; 1st glume wanting. . . . . 82a. *Axonopus*
> FF. Racemes stout, in pairs. . . . . 85. *Paspalum*

ANDROPOGÒNEAE

> A. Spikelets all alike, fertile, surrounded by copious soft hairs; infl. a narrow panicle. 91. *Imperata*
> AA. Spikelets unlike, the sessile perfect, the pedicellate ♂ or neuter.
> B. Racemes of several joints.
> C. Fertile spikelet with a hairy-pointed callus, formed of the attached support-ing rachis joint or pedicel; awns strong, brown. . . . . 92b. *Heteropogon*
> CC. Fertile spikelet without a callus, the rachis disarticulating below the spike-let; awns slender.
> D. Lower pair of spikelets like the others of the raceme. 92. *Andropogon*
> DD. Lower pair of spikelets sterile, awnless; racemes in pairs on slender flexuous peduncles. . . . . 92a. *Hyparrhenia*

p. 1468.   BB. Racemes reduced to 1 or few joints, these in a compound panicle. 93. *Sorghum*

p. 1469.   BRÒMUS

1. *B. cathárticus* Vahl should be changed to **B. Willdenòvii** Kunth. (cf. Raven, Brittonia 12: 221. 1961).

p. 1470.   2. Change *B. Haenkeànus* (Presl) Kunth to **B. unioloìdes** HBK. The sp. has recently been reported from Cantil, Kern Co., *Twisselmann*, and Monterey Co., *Howitt & Howell.* 2n = 42 (Schulz-Schaeff. & Mark, 1957).

8. **B. marítimus** Hitchc. 2n = 56 (Schulz-Schaeff. & Mark, 1957).

p. 1471.   9. **B. polyánthus** Scribn. $2n = 56$ (Schulz-Schaeff. & Mark, 1957).
              12. **B. vulgàris** (Hook.) Shear. Santa Cruz. Id.
p. 1472.   18. **B. eréctus** Huds. $n = 21 + 4$ (Schulz-Schaeff., 1956); $2n = 42, 56$, 70, 112 (Hill, 1965).
              19. **B. Pórteri** (Coult.) Nash in the Sierra Nevada of Tulare and Inyo cos., *Raven.*
              20. **B. Richardsònii** Link. $2n = 28$ (Mitchell & Wilton, 1965).
p. 1473.   26. **B. racemòsus** L. $n = 14$ (Schulz-Schaeffer, 1956).
              29. **B. arvénsis** L. Reported from Madera Co., *Raven.*
              31. *B. rígidus* Roth should be changed to **B. diándrus** Roth. $n = 21$ (Schulz-Schaeffer, 1956).
p. 1474.   32. **B. stérilis** L. In Monterey Co., *Howitt & Howell.*
              33. **B. rùbens** L. $2n =$ ca. 28 (Reese, 1957).
              36. **B. Trìnii** E. Desv. in Gay.
p. 1476.   FESTÙCA
              10. **F. microstàchys** var. **símulans** (Hoov.) Hoov. has spikelets 3–6-fld. instead of 1–3 and is common on hills and plains of Kern Co. and into interior San Luis Obispo Co.
              11. **F. Eastwoòdiae** Piper in the Temblor Range, Kern Co., *Twisselmann.*
p. 1477.   12. **F. Tràcyi** Hitchc. In Tuolumne Co., *Raven.*
              16. **F. elátior** L. $2n = 14$ (Bowden, 1960). Change to **F. praténsis** Huds. Cf. Terrell (Brittonia 19: 129. 1967).
              18. **F. califórnica** Vasey. In San Luis Obispo Co., *Hoover* and in Sierra Nevada of Eldorado Co., *Crampton.*
p. 1478.   20. **F. rùbra** L. In San Luis Obispo Co. $2n = 42$ (Jorgensen et al., 1958).
              21. **F. occidentàlis** Hook. In the Greenhorn Range, Kern Co., *Twisselmann,* and in Santa Barbara Co., *Pollard* and Monterey Co., *Howitt & Howell.*
p. 1480.   PUCCINÉLLIA Twisselmann reports both **P. erécta** and **P. pauciflòra** as in Kern Co.
              GLYCÈRIA
              2. **G. leptostàchya** Buckl. At San Francisco, *Howell & Raven.*
p. 1481.   3. **G. occidentàlis** (Piper) J. C. Nels. In San Mateo Co., *Thomas.*
              4. **G. declinàta** Brebiss. In Calaveras and Stanislaus cos., *Crampton.*
              6. **G. elàta** (Nash) Hitchc. In Sonoma and Marin cos., *Rubtzoff.*

6a. Sclerochlòa Beauv.

Low tufted annual with broad upper sheaths, folded blades and dense spikelike racemes. Spikelets subsessile, imbricate in 2 rows on 1 side of the broad thick rachis, 3-fld., the upper floret sessile. Glumes broad, the first 3-nerved, the second 7-nerved; lemmas rounded on back, with 5 prominent parallel nerves and hyaline margins. (Gr., *skleros,* hard, and *chloa,* grass).

1. **S. dura** (L.) Beauv. [*Cynosurus d.* L.] Two to 7 cm. tall; lf. blades 7–18 mm. long, 1–3 mm. wide; raceme 1–2 cm. long; spikelets 6–7 mm. long.—Adventive in Shasta V., Siskiyou Co., *Fuller.* From s. Eu.

PLEUROPÒGON
              2. **P. califórnicus** (Nees) Benth. in Vasey. San Luis Obispo Co., *Hoover.*
p. 1482.   3. **P. refráctus** (Gray) Benth. ex Vasey. $2n = 18$ pairs (D. E. Anderson, 1965).
p. 1483.   PÒA
              In Key after BB, CC, D, EE, FF, add:

> G. Glumes lanceolate, acute, shorter than the first lemma; ligules of culm lvs. 3–5 mm. long. .... 17. *P. palustris*
> GG. Glumes narrower, acuminate, ca. as long as the first lemma; ligules very short. .... 17a. *P. nemoralis*

p. 1484.   1. **P. ánnua** L. A much used common name in Calif. is WINTERGRASS. $2n = 14$ (Hovin, 1958).
              3. **P. Bigelòvii** Vasey & Scribn. In Red Rock Canyon, w. Mojave Desert, *Twisselmann.*

p. 1486.   14. **P. bulbòsa** L. ranges s. to Fresno Co. in the Sierra Nevada, *Raven;* several localities in Kern Co., *Twisselmann;* Santa Barbara Co., *Fuller;* w. Riverside Co., *Lathrop.*
                15. **P. triviàlis** L. In San Francisco.

p. 1487.   17a. **P. nemoràlis** L. Culms tufted, 3–7 dm. tall; ligule very short; blades ca. 2 mm. wide; panicle 4–10 cm. long, the branches spreading; spikelets 2–5-fld., 3–5 mm. long; glumes narrow, sharply acuminate, ca. as long as the first floret; lemmas 2–3 mm. long, sparsely webbed at base, pubescent on keel and marginal nerves.—Natur. in Golden Gate Park, San Francisco. From Eu.

p. 1488.   23. **P. épilis** Scribn. From 5500 to 12,000 ft.
                24. **P. nevadénsis** Vasey ex Scribn. at Isabella, Kern Co., *Twisselmann.*

                BRIZA
                2. **B. mìnor** L. $n = 7$ (Gould, 1958).

p. 1491.   ERAGRÓSTIS
                In Key, AA, BB, C, change to:

>   D.   Spikelets 3–5 mm. long; lemmas 1–1.5 mm. long; pedicels mostly longer than the spikelets; surface of the grain smooth; side of grain opposite the embryo rounded. . . . . 2. *E. pilosa*
>   DD. Spikelets 5–7 mm. long; lemmas ca. 2 mm. long; pedicels mostly shorter than the spikelets; surface of grain reticulate, side of grain opposite the embryo flat or grooved. . . . . 6. *E. Orcuttiana*

p. 1492.   2. **E. pilòsa** (L.) Beauv. $2n = 60$ (Tateoka, 1965).
                3. **E. hypnoìdes** (Lam.) BSP. Santa Cruz Co., Sonoma and Lake cos. to Siskiyou Co., *Rubtzoff.*
                4. **E. pectinàcea** (Michx.) Nees. $n = 20, 30$ (Gould, 1958).
                5. **E. diffùsa** Buckl. in Sonoma Co., *Rubtzoff;* in Marin and San Francisco cos., *Howell.* $2n = 60$ (Gould, 1965).
                8. *E. megastàchya* (Koel.) Link. Probably the preferable name is **E. cilianensis** (All.) E. Mosher.
                9. **E. poaeoìdes** Beauv. ex R. & S. In Sonoma Co., *Rubtzoff.*

p. 1493.   14. **E. cúrvula** (Schrad.) Nees. In San Diego, Contra Costa and Solano cos., *Fuller;* in Yolo Co., *Crampton.* $2n = 14$ (Reese, 1957).

                DISSANTHÈLIUM
                Swallen and Tovar (Phytologia 11: 361–376. 1965) recognize 17 spp. most of them from S. Am.

p. 1496.   CORTADÈRIA
                1. **C. Selloàna** (Schult.) Asch. & Graebn. Natur. at San Francisco and in the North Coast Ranges, *Howell;* a heavy infestation of 1100 acres, e. side of Big Lagoon, Humboldt Co., *Fuller.*

p. 1497.   MÉLICA
                2. **M. Gèyeri** var. **aristulàta** J. T. Howell in Tehama Co., *Crampton.*
                3. **M. Harfórdii** Bol. In San Luis Obispo Co., *Hoover.*
                4. **M. aristàta** Thurb. ex Bol. in the Santa Lucia Mts., *Hardham.*

p. 1499.   ECTOSPÉRMA Swall., not Vaucher.
                Renamed **Swallènea** Soderstrom & Decker, with **S. Alexándrae** (Swallen) Soderstrom & Decker.

                TRÌDENS
                Tateoka (Am. J. Bot. 48: 565–573. 1961) recognizes two genera *Tridens* R. & S. with **T. mùticus** (Torr.) Nash [$2n = 40$ (Tateoka)] and *Erioneuron* Nash with **E. pulchéllum** (HBK) Tateoka and **E. pilosum** (Buckl.) Nash. [$2n = 16$ (Tateoka).]

p. 1500.   NEOSTÁPFIA
                1. **N. colusàna** (Davy) Davy. $2n = 40$ (Stebbins & Major, 1965).

                ORCÚTTIA
                Begin the Key for the 4 spp. treated in the FLORA with:

>   A.   Lemmas toothed at the apex; florets 10–40 in a spikelet; lodicules obsolete. (Use the key in the FLORA for spp. 1–4.)

AA. Lemmas not toothed at the apex, but erose on the margin and with a terminal mucro; florets 5–10 in a spikelet; lodicules 2, fused to the palea. .... 5. *O. mucronata*

p. 1501.

5. **O. mucronàta** Crampton. Culms decumbent, 2.5–12 cm. long; lvs. 1–4 cm. long, viscid; infl. 1.5–6 cm. long; spikelets 7–19, spirally arranged, 7–13 mm. long, 5–10-fld.; glumes 4–7 mm. long, unequal; lemmas coriaceous, 5–7 mm. long.—Dry lake, 12 mi. sw. of Dixon, Solano Co.

p. 1502.

AGROPÝRON

In Key, AA, BB, C, DD, change EE to:

EE. Lemmas finely pubescent.
F. Glumes acute or awn-pointed; lvs. flat to involute.
4. *A. dasystachyum*
FF. Glumes truncate; lvs. flat. .... 4a. *A. trichophorum*

In Key, AA, BB, CC, change D to:

D. Spikelets much compressed.
E. Spikelets crowded on rachis; glumes narrowed to short awns.
1. *A. desertorum*
EE. Spikelets shorter than the internodes; glumes obtuse or truncate.
6a. *A. elongatum*

1. **A. desertòrum** (Fisch.) Schult. $2n = 28$ (Sarkar, 1956). In Plumas and Mono cos., *Raven.*

4a. **A. trichóphorum** (Link) Richt. [*Triticum t.* Link.] Plants with creeping rhizomes; culms tallish; lvs. flat; spikelets pubescent, awnless; glumes several-nerved, truncate.—Reported as probably established after being seeded as TOPAR WHEATGRASS on brush-burns, Greenhorn Range, Kern Co., *Twisselmann*; Siskiyou Co., *Fuller.*

5. **A. ripàrium** Scribn. & Sm. $2n = 42$ (Tateoka, 1956).

p. 1503.

6a. **A. elongàtum** (Host) Beauv. [*Triticum e.* Host.] TALL WHEATGRASS. Not creeping, 3–10 dm. tall, glabrous; lvs. glaucous, inrolled, stiff; spike elongate, very lax; spikelets spaced, compressed, oval, 4–8-fld.; glumes obtuse or truncate, 7–9-nerved.—Established in Plumas, Lassen, and Kern cos. From the Medit. region.

7. **A. subsecúndum** (Link) Hitchc. In Tehachapi Mts., ne. of Lebec, *Twisselmann.*

10. **A. Scríbneri** Vasey. $2n = 28$ (Tateoka, 1956).

13. **A. saxícola** (Scribn. & Sm.) Piper is considered to be a hybrid by F. Douglas Wilson.

p. 1504.

15. **A. × Sáundersii** (Vasey) Hitchc. is also a hybrid, between *A. trachycaulum* (Link) Malte and *Sitanion Hystrix* (Nutt.) J. G. Smith. It is reported by *Raven* from the Sierra Nevada of Alpine, Tuolumne, Madera, Fresno, Tulare, Mono, and Inyo cos.

ÉLYMUS

1. **E. cáput-medùsae** L. is recorded from Solano, Alameda, Sacramento, San Joaquin, Tulare, and Fresno cos. McKell, Robison and Major use the name *Taeniatherum asperum* (Simonkai) Nevski for plants referred to E. cáput-medùsae by Am. auth.

p. 1505.

2. **E. móllis** (Trin ex Spreng. $2n = 18$ (Löve & Löve, 1956). Ranges s. to San Luis Obispo Co., *Hoover.*

3. **E. móllis** Trin. ex Spreng. × **E. triticoìdes** Buckl. for *E. vancouverensis* Vasey nothomorph *californicus* Bowden.

5. **E. pacíficus** Gould in San Luis Obispo Co., *Hoover.*

p. 1506.

SITÀNION

Insert literature citation: Wilson, F. D. Revision of Sitanion. Brittonia 15: 303–323. 1963.

Use this Key:

A. Spikelets 3 at each node of the rachis; florets of central spikelet fertile, those of lateral spikelets reduced, rudimentary. Siskiyou Co. .... 5. *S. hordeoides*
AA. Spikelets usually 2 at each node of the rachis, if 3, at least some florets of lateral spikelets fertile.

B. Lowermost floret of 1 or both spikelets at each rachis-node sterile and reduced to a subulate or lanceolate structure, giving the appearance of extra glume segms.
C. Glumes entire or bifid. . . . . 3. *S. Hystrix*
CC. Glumes 3–many-cleft; awns of the lemmas exceeding those of the glumes.
2. *S. jubatum*
BB. Lowermost floret fertile, not reduced.
C. Glumes subulate, entire; awns of the glumes exceeding those of the lemmas.
4. *S. longifolium*
CC. Glumes usually lanceolate, entire or 2–several-cleft; awns of the lemmas exceeding those of the glumes. . . . . 1. *"S." × Hansenii*

3. **S. Hystrix** (Nutt.) J. G. Smith. Delete *S. californicum* and *S. longifolium* from the synonyms. At least 1 glume of each node of the rachis 2-cleft; awns of the glumes exceeding those of the lemmas.—E. of the Sierra Nevada, deserts from Modoc Co. to Riverside Co.; to B.C., S. Dak.

Var. **califórnicum** (J. G. Smith) F. D. Wilson. [*S. c.* J. G. Smith. *S. minus* var. *c.* Jtn.] Glumes entire; awns of the lemmas exceeding those of the glumes.—Mts., s. Calif. to B.C., Mont., Utah.

4. **S. longifòlium** J. G. Smith. Plants 2.5–6 dm. tall, usually loosely cespitose; culms slender to robust, erect to spreading; blades 2–5 mm. wide, usually glabrous above; spikes 7–15 cm. long; spikelets mostly 2 at a node, few–several-fld.; glumes entire, usually 1-nerved, subulate, with spreading setaceous awns 5–12 cm. long; lemmas 7–12 mm. long, the central nerve extending into a stout spreading, setaceous awn 5–10 cm. long.—At 2000–10,000 ft., s. Calif. to S. Dak., Tex., Mex.

5. **S. hordeoìdes** Suksd. Plants 1–2 dm. tall, loosely cespitose; sheaths puberulent to villous; blades 1–4 mm. wide; spikes 3–6 cm. long, dense; spikelets 3 at a node, the central spikelets usually with 2 sterile lateral florets and 1 fertile terminal one; glumes subulate to narrowly lanceolate, extending into slender scabrous awns 1.5–5 cm. long; lemmas obscurely 5-nerved, ca. 10 mm. long.—Dry rocky places, Siskiyou Co., to Wash., Ida.

p. 1507.   AÈGILOPS
1. **A. ovàta** L. The Glenn Co. reference in the FLORA should have been Willits, Mendocino Co., according to *Fuller*.
3. **A. triunciàlis** L. The rachis does not disarticulate.

p. 1509.   HÓRDEUM
5. *H. Hýstrix* Roth should be changed to **H. geniculàtum** Allioni.
8. *H. Stébbinsii* Covas should be changed to **H. gláucum** Steud.
10. In **H. vulgàre** L. all spikelets produce large seed. In a cult. Barley **H. dístichon** L. which is reported as adventive in Monterey Co.; the spike is 2-rowed, not 4- or 6-rowed. In it the lateral spikelets are sterile.

p. 1510.   LÒLIUM
2. **L. multiflòrum** Lam. var. **mùticum** (DC.) Volkart reported from Monterey Co., *Howitt & Howell* and from San Francisco, *Raven*. It differs from the sp. in having all lemmas awnless, instead of at least the upper awned. Other San Francisco collections, with branched panicles, are var. **ràmosum** Guss., *Howell*.

PARÁPHOLIS
1. **P. incúrva** (L.) C. E. Hubb. $n = 21$ (Gould, 1958).

SCRIBNÈRIA
1. **S. Bolánderi** (Rhurb.) Hack. $2n = 26$ (Stebbins & Major, 1965).—Reported from Santa Barbara Co., *Raven*; Kern Co., *Twisselmann*.

p. 1511.   SCHÍSMUS
S. barbàtus (L.) Thell. and **S. arábicus** Nees are both recorded from Monterey Co., *Howitt & Howell*. For both spp., $n = 6$ (Gould, 1958).

KOELÈRIA
1. **K. cristàta** (L.) Pers. is an illegitimate name. Apparently **K. macrántha** (Ledeb.) Spreng. may be used for our plant (Voss, Rhodora 68: 441. 1966). $n = 7$ (Tateoka, 1955); $2n = 14$ (Bowden, 1960).

p. 1512.   SPHENÓPHOLIS

Citation: Erdman, K. S. Taxonomy of the genus Sphenopholis. Iowa State Jour. Sci. 39: 289–336. 1965.

1. **S. obtusàta** (Michx.) Scribn. $2n = 14$ (Erdman, 1965).

TRISÈTUM

2. **T. spicàtum** (L.) Richt. $2n = 28$ (Tateoka, 1954).

3. **T. cérnuum** ssp. **canéscens** (Buckl.) Calder & Taylor.—Found in the Greenhorn Mts., Kern Co., *Twisselmann*, and in San Luis Obispo Co., *Hoover*.

p. 1513.   DESCHÁMPSIA

2. **D. elongàta** (Hook.) Munro ex Benth. $2n = 26$ (Bowden, 1960).

4. **D. cespitòsa** (L.) Beauv. for original spelling. $2n = 26, 27, 28$ (Bowden, 1960). Reported from Kern Plateau, Kern Co., *Twisselmann*; Mt. Pinos, *Hoffmann*.

p. 1514.   Ssp. **holicifórmis** (Presl) W. E. Lawr. extends s. to San Luis Obispo Co., *Hoover*.

AÌRA

2. **A. caryophyllèa** L. $2n = 14, 28$ (Böcher & Larsen, 1958).

3. **A. élegans** Willd. ex Gaudin, $2n = 14$ (Böcher & Larsen, 1958).

AVÈNA

1. **A. barbàta** Brot. $2n = 28$ (Martinoli, 1955).

2. **A. fátua** L. Grows to a height of 1.9 m.

p. 1515.   HÓLCUS

1. **H. lanàtus** L. To 2 m. tall.

2. **H. móllis** L. $2n = 28, 35, 42, 49$ (Jones, 1958).

p. 1516.   DANTHÒNIA

1. **D. intermèdia** Vasey. $2n \doteq$ ca. 98 (Taylor, 1967).

3. **D. califórnica** Bol. ranges s. to San Luis Obispo Co., *Hoover*.

p. 1517.   CALAMAGRÓSTIS

5. **C. ophítidis** (J. T. Howell) Nygren. The synonym should be *C. purpurascens* var. *ophitidis* J. T. Howell.

6. **C. rubéscens** Buckl. ranges s. to Monterey Co., *Howitt & Howell*, and to San Luis Obispo Co., *Hoover*. Santa Cruz Id., *Blakely & Muller*.

p. 1518.   7. **C. nutkaénsis** (Presl.) Steud. has been reported from San Luis Obispo Co., *Hoover*.

10. Change the name of *C. canadensis* var. *Langsdorfii* (Link) Inman to **C. c.** var. **scàbra** (Presl) Hitchc. and add Mt. Lassen area for its distribution.

p. 1519.   **53a. Apèra** Adans.

Near *Agrostis*. Annual, with compound panicle, the lemma chartaceous, terete, shortly bifid, with a well developed awn from the sinus. Three spp. Eurasia.

1. **A. interrúpta** (L.) Beauv. Low, tufted, 1.5–6 dm. tall; lvs. ± convolute, smooth, narrow, short; ligule to 5 mm. long, truncate; panicle 3–18 cm. long, narrow, ± interrupted; spikelets 1.5–2 mm. long.—Found 2.6 mi. w. of Alturas, Modoc Co.; introd. from Eu.

AGRÓSTIS

In Key, under AA, BB, CC, D, E, change to:

> F.   Lemmas 2 mm. long or less.
>   G.   The culms spreading; panicles strict, greenish; lemmas with a minute awn or the midnerve ending below the apex. Coastal. (See also var. *marinensis*.)
>   15. *A. Blasdalei*
>   GG. The culms erect; panicles narrow but loose, purple; lemmas awnless, the midnerve reaching the apex. Montane.
>   16. *A. variabilis*
> FF. Lemmas 2.5 mm. long or more; awn ca. 2 mm. long. N. Coastal. . . . . 16a. *A. clivicola*

p. 1520.   1. **A. avenàcea** Gmel. *Raven* records it from Eldorado Co.

4. **A. stolonífera** L. var. **màjor** (Gaud.) Farwell. [*A. alba* auth. *A. gigantea* Roth.] for REDTOP and var. **palústris** (Huds.) Farwell for CREEPING BENT.

5. **A. ténuis** Sibth. $2n = 28, 32, 34$ (Bowden, 1960).

p. 1521.   12. **A. lépida** Hitchc.—Occurs in n. Fresno Co., *Raven.*

15. **A. Blasdàlei** Hitchc. var. **marinénsis** Crampton. Glumes 3–4 mm. long; lemmas 2.5–2.8 mm. long.—Coast of Marin Co.

16a. **A. clivícola** Crampton (Brittonia 19: 174. 1967). Perennial; culms tufted, glabrous, 1–3.5 dm. long, more or less prostrate; panicle dense, 2–7 cm. long; glumes 2.8–4 mm. long, the awns to 0.5 mm. long; lemmas 2.5–3 mm. long, scabrous over the back; $2n = 42$.—Coastal bluffs, Mendocino and Sonoma cos.

Var. **púnta-reyesénsis** Crampton. Culms erect; lf.-blades thinnish, pointed; $2n = 42$.—Marin and Sonoma cos.

p. 1522.   18. *A. califórnica* Trin. Change to **A. densiflora** Vasey (cf. Chambers, Madroño 18: 251. 1966).

20. **A. scàbra** Willd. $2n = 42$ (Bowden, 1960).

CÍNNA

1. **C. latifòlia** (Trev.) Griseb. $2n = 28$ (Tateoka, 1954).

p. 1523.   ALOPÉCURUS

3. **A. aequàlis** Sobol. $2n = 7$ pairs (D. E. Anderson, 1965).

Var. **sonoménsis** Rubtzoff. More robust and erect than the sp.; lf. blades to 7.5 mm. wide; panicle 2.5–9 cm. long, 4–8 mm. wide; awn exserted 1–2.5 mm. —Moist places, Marin and Sonoma cos.

7. **A. carolinianus** Walt. In Merced and Madera cos., *Rubtzoff.*

p. 1524.   POLYPÒGON

2. **P. marítimus** Willd. *Rubtzoff* finds this quite widespread in Calif.

3. **P. interrúptus** HBK. Rubtzoff apparently refers most Calif. material formerly identified as this sp. to **P. austràlis** Brongn. It differs from *P. i.* in a shorter ligule, to 2 mm. long, shorter blades, lax more purplish panicle and slender ± tangled hairs.—Originally from S. Am. It has been found in Orange, Riverside, San Bernardino, Inyo cos. and from Stanislaus to Butte cos. and from Sonoma to Humboldt cos. Also in Monterey Co., *Howitt* and *Howell,* and Santa Clara and Ventura cos.

p. 1525.   PHLÈUM

2. **P. alpìnum** L. Add *P. commutatum* Gaud. as a synonym. In the North Coast Ranges, the sp. comes s. as far as San Francisco Bay.

p. 1526.   MUHLENBÉRGIA

In Key after BB, insert:

> B′.   Culms decumbent and rooting at the nodes; glumes minute, the first glume
>   sometimes wanting. .... 12a. *M. Schreberi*
> B′B′. Culms erect or spreading, but not rooting at the nodes;
>   C.   Second glume 3-toothed, etc. as in the FLORA.

p. 1526.   4. **M. asperifòlia** (Nees & Mey.) Parodi. $n = 10$ (Pohl & Mitchell, 1965).

6. **M. Richardsònis** (Trin.) Rydb. Reported by Twisselmann from Mt. Pinos and Piute Mts., Kern Co.

p. 1527.   9. **M. andìna** (Nutt.) Hitchc. in Napa Co., *Howell.* $n = 10$ (Pohl & Mitchell, 1965).

11. **M. mexicàna** (L.) Trin. f. **ambígua** (Torr.) Fern. with conspicuously awned lemmas, should be used instead of f. *setiglumis* (Wats.) Fern. and occurs in Mendocino, Humboldt, Butte, and Plumas cos., *Rubtzoff.*

12. **M. califórnica** Vasey. $n = 40$ (Pohl & Mitchell, 1965).

12a. **M. Schrèberi** Gmel. Branches 1–3 dm. long; blades flat, to 6 cm. long, 2–4 mm. wide; panicles slender, lax, nodding, 5–15 cm. long; glumes minute, the first often obsolete, the second 0.1–0.2 mm. long; lemma 2 mm. long, awn 2–5 mm. long. $n = 20$ (Pohl & Mitchell, 1965).—Introd. on property of Felix Gillet Nursery, Nevada City, Nevada Co., *Fuller.* From e. N. Am.

p. 1528. SPORÓBOLUS

Insert at beginning of Key:

0. Plants annual. . . . . 6. *S. vaginiflorus*
00. Plants perennial.
    A. Glumes subequal, much shorter etc. as in the FLORA. Spp. 1–5.

1. S. contráctus Hitchc. In Stanislaus and San Joaquin cos., *Fuller.*

3. S. cryptándrus (Torr.) Gray. at Andrews Camp, Bishop Creek, Inyo Co., at 9000 ft. and 5 mi. s. of Coleville, Mono Co., *Raven;* in Yolo and Solano cos., *Crampton.* n = 19, 36 (Gould, 1958); 2n = 36 (Bowden, 1960).

p. 1529. 6. S. vaginiflòrus (Torr.) Wood. Annual, 2–4 dm. high; blades slender, subinvolute; panicles mostly not more than 3 cm. long; glumes acute, subequal, 3–5 mm. long; lemmas as long as glumes or longer, acute to acuminate, sparsely pubescent.—Reported from Nevada and Shasta cos., *Raven.* Native of e. U.S.

HELEÓCHLOA

1. H. schoenoìdes (L.) Host. Additional records are: Kern Co. (Lake Isabella) and San Luis Obispo Co., *Twisselmann;* Los Angeles Co. (Bouquet Canyon) *Raven;* Monterey Co., *Howitt & Howell;* Tulare Co., *Twisselmann.*

p. 1530. ORYZÓPSIS

B. L. Johnson (Am. J. Bot. 47: 736–742. 1960) discusses a number of hybrids between *Oryzopsis hymenoides* (R. & S.) Ricker and *Stipa speciosa* Trin. and (in Am. J. Bot. 50: 228–234. 1963) a hybrid between *O. hymenoides* and *S. pinetorum* Jones from Inyo Mts.

p. 1531. STÌPA

In Key after A, BB, C, DD, EE, FF, GG, H, change I to:

    I. Hairs on upper part of lemma longer than those below; awn with rather short hairs.
    J. Lemma more than 2.3 times as long as the palea; culms glabrous below the nodes; sheaths sparsely villous, sometimes glabrous at the throat; awn-hairs 0.2–1 mm. long. . . . . . 12. *S. californica*
    JJ. Lemma less than 2.3 times as long as the palea; culms scabrous below the nodes; sheaths glabrous at the throat; awn-hairs 0.5–1.5 mm. long. . . . 12a. *S. nevadensis*

In Key after AA, BB, CC, change D to:

    D. Mature lemma brownish.
    E. Lemma 8–12 mm. long, ± sparsely pubescent; awn 10–15 cm. long. Common native. . . . . 4. *S. comata*
    EE. Lemma 3.5–6 mm. long, pubescent in lines; awn 1.1–1.8 cm. long. Rare introduction. . . . . 4a. *S. brachychaeta*

1. S. speciòsa Trin. & Rupr. N. to Fresno Co. in the Sierra Nevada.

2. S. Stillmánii Bol. *Howell* reports as in Shasta, Tehama, Plumas, and Nevada cos.

p. 1532. 3. S. coronàta Thurb. in Wats. In Napa Co., *Rubtzoff.*

4. S. comàta Trin. & Rupr. To 11,300 ft. on the e. side of the Sierra Nevada, *Raven.*

4a. S. brachychaèta Godr. Densely cespitose perennial to 1 m. tall; blades firm, flat or loosely involute; panicle narrow, open, the few spikelets on slender pedicels; glumes 6–8 mm. long; lemma 3.5–6 mm. long, brown, pubescent in lines; awn 1.1–1.8 cm. long.—Reported from s. edge of Fresno and near Camarillo, Ventura Co., *Fuller.* Native of Argentina.

8. S. Thurberiàna Piper. Reported from Cuyama Valley, Ventura Co., *Clifton Smith;* ranges n. to Wash., Ida.

9. S. Élmeri Piper & Brodie; 10. S. occidentalis Thurb.; etc. J. Maze (Leafl. W. Bot. 10: 159–160. 1965) presents a different key by which to distinguish spp. in this complex:

A. Awn pubescent, at least at the base, plumose or subplumose.
  B. Hairs of the first segm. of the awn subequal to or exceeding those at the tip of
       the lemma; transition from pubescence of lemma to that of awn gradual.
    C. Plant more than 4 dm. tall. .... 9. *S. Elmeri*
    CC. Plant less than 4 dm. tall. .... 11. *S. occidentalis*
  BB. Hairs of the first segm. of the awn shorter than those at the tip of the lemma;
       transition from pubescence of lemma to that of awn abrupt.
    C. Palea ca. half as long as lemma; floret less than 2.2 times as long as palea,
        the hairs at tip of palea mostly ca. 1 mm. long; glabrous area on the inside
        curve of the callus obtuse to acute and usually not well extended toward the
        lemma, callus tip shorter and more acute than that of *S. californica*, callus
        less than 1 mm. long. .... 12a. *S. nevadensis*
    CC. Palea mostly less than half as long as lemma; floret more than 2.2 times as
        long as palea; hairs at tip of palea mostly less than 1 mm. long, or glabrous
        area on the inside curve of the callus more acuminate and well extended
        toward the lemma, callus tip usually longer and more acuminate than that
        of *S. nevadensis*; callus ca. 1 mm. long. .... 12. *S. californica*
AA. Awn scabridulous to hirtellous.
  B. Lemma with few, if any, longer hairs at the tip, the hairs not spreading, or gla-
       brous area on the inside curve of the callus short-acute and not well extended
       toward the lemma. .... 14. *S. columbiana*
  BB. Lemma with many distinctly longer hairs at the tip, the hairs usually spreading;
       the glabrous area on the inside curve of the callus long-acute and well extended
       toward the lemma; glabrous tip of the callus acuminate. ..... 12. *S. californica*

p. 1533.   12. **S. califórnica** Merr. & Davy. Shirley Meadows, Kern Co., *Twisselmann*.

 12a. **S. nevadénsis** B. L. Johnson. Culms often in tufts, 4–8 dm. tall, scabrous below the nodes; sheaths glabrous or nearly so, glabrous at throat; ligule
ca. 0.5 mm. long; cauline lvs. 1–2.5 dm. long, 1–3 mm. wide, becoming involute, ± puberulent above, glabrous below; panicle narrow, 1–2.5 dm. long;
glumes 7–10 mm. long; lemma 5–6.5 mm. long, 1.7–2.3 times as long as the
palea, sparsely villous, the summit hairs ca. 1.5 mm. long; awn 2–3.5 cm. long,
twice geniculate, plumose on first and second segms., awn hairs 0.5–1.5 mm.
long; $2n = 68$.—E. slope of the Sierra Nevada, Modoc Co. to Kern Co., to
Ida., Nev.

 14. **S. columbiàna** Macoun. $n = 18$ (Johnson, 1962).

 15. **S. Lettermánii** Vasey restricted by Johnson to San Bernardino Mts. $n =$
16 (Johnson, 1962).

p. 1534.   18. **S. pinetòrum** Jones. Eldorado to Tulare and Inyo cos., *Raven*.

p. 1535.   ARÍSTIDA

 3. **A adscensiònis** L. Known also from San Luis Obispo Co., Santa Barbara
Co., San Diego Co.

 5. **A. divaricàta** Humb. & Bonpl. ex Willd. $n = 11$ (Gould, 1958).

 9. **A. purpùrea** Nutt. $n = 11$ (Gould, 1958).

 11. **A. Fendleriàna** Steud. $n = 33$ (Gould, 1958).

p. 1536.   HILÀRIA

 2. **H. Jàmesii** (Torr.) Benth. $2n = 18$ pairs.

 3. **H. Belángeri** (Steud.) Nash. A slender stoloniferous plant 1–3 dm. tall,
with bearded nodes; spikes 2–3 cm. long.—Reported as se. Calif.; to Tex.,
n. Mex. *Wiggins.*

LEPTOCHLÒA

 2. **L. fasciculàris** (Lam.) Gray in Monterey Co., *Howitt & Howell* and in
Sonoma Co., *Rubtzoff,* as well as Santa Barbara Co.

 3. **L. uninérvia** Hitchc. & Chase. $n = 10$ (Gould, 1958).

p. 1537.   4. **L. víscida** (Scribn.) Beal. Differing from the 3 spp. treated in the FLORA
in having lemmas 2 mm. long, viscid on back; panicle usually less than 10 cm.
long, tinged purple; sheaths scabrous.—Reported from Kern Co., *Twisselmann;*
ranging to Tex., n. Mex.

ELEUSÌNE

 2. **E. tristàchya** (Lam.) Lam. [*Cynosurus t.* Lam.] Differing from *E. indica*
by the fewer and shorter spikes (1–3 in number, 1–2.5 cm. long, 8–10 mm.
thick).—Reported from 2.5 mi. n. of Clements, San Joaquin Co., *Fuller.* Introd. from Africa.

DACTYLOCLÈNIUM

1. **D. aegýptium** (L.) Beauv. 2n = 40 (Tateoka, 1965). Taken at Bonsall, San Diego Co., *Dixon in 1965*.

CỲNODON

1. **C. dáctylon** not Dactylon. 2n = 40 (Tateoka, 1954).

p. 1538.　SPARTÌNA

Change Key to:

A. Blades mostly more than 8 mm. wide; spikes closely approximate forming a cylindric
    infl. Coastal marshes and dunes. .... 1. *S. foliosa*
AA. Blades less than 5 mm. wide; spikes distinct, appressed or spreading.
    B. Blades usually flat; glumes conspicuously hispid-ciliate on the keels; spikes sev-
        eral, appressed. Interior alkaline meadows. .... 2. *S. gracilis*
    BB. Blades usually involute; glumes scabrous on the keels; spikes few, ascending or
        spreading. Southampton Bay. .... 3. *S. patens*

3. **S. pàtens** (Ait.) Muhl. SALTMEADOW CORDGRASS. Culms slender, usually less than 1 m. tall, with long slender rhizomes; blades mostly involute, less than 3 mm. wide; spikes 2 to several, 2–5 cm. long, rather remote on the axis; spikelets 8–12 mm. long; first glume ca. half as long as the floret, the second longer than the lemma; lemma 5–7 mm. long, emarginate.—Reported from Southampton Bay in a marsh, nw. of Benicia, Solano Co., *Mall*; e. U.S.

p. 1539.　CHLÒRIS

1. **C. distichophýlla** Lag. 2n = 40 (Huynh, 1965).

2. **C. Gayàna** Kunth. 2n = 40 (Tateoka, 1965).

3. **C. virgàta** Sw. Reported from near Hilts, Siskiyou Co., *Fuller*.

p. 1540.　BOUTELOÙA

2. **B. curtipéndula** (Michx.) Torr. n = 10 (Gould, 1958). Gould and Kapadia (Brittonia 16: 203. 1965) refer California material to var. **caéspitosa** Gould & Kapadia as more stiffly erect than in the sp. Crampton reports the sp. as from Yolo Co.

3. **B. radicòsa** (Fourn.) Griffiths. 2n = 60 (Gould, 1965).

4. **B. barbàta** Lag. 2n = 20 (Gould, 1965).

6. **B. hirsùta** Lag. n = 10, 23 (Gould, 1958); 2n = 46, ca. 52 (Gould, 1965).

7. **B. grácilis** (HBK) Lag. Reported from the Goleta campus of the University of California, *Ernst*. 2n = 20, 40, 60 (Gould, 1965).

9. **B. trífida** Thurb. 2n = 20 (Gould, 1965).

p. 1541.　HIEROCHLÒE

1. **H. occidentàlis** Buckl. 2n = 21 pairs (D. E. Anderson, 1965).

ANTHOXÁNTHUM

1. **A. odoràtum** L. 2n = 28 (Tateoka, 1954); 2n = 20 (Gray, 1965).

EHRHÁRTA

2. Add **E. calycìna** Sm. Spikelets 7–8 mm. long; sterile lemmas thinly silky-villous; fertile lemma silky on the nerves.—Reported from San Luis Obispo Co., *Twisselmann*, and Ventura Co., *Pollard*. Introd. from S. Afr.

PHÁLARIS

Introduce after generic description: Anderson, D. E. Taxonomy and distribution of the genus Phalaris. Iowa State Jour. Sci. 36: 1–96. 1961.

p. 1542.　In next to last line of Key, "2–6 mm." should read "2–6 cm."

1. **P. paradóxa** L. 2n = 14 (Ambastha, 1956).

2. **P. califórnica** H. & A. In synonymy *P. amethystina* not *A. amethystina*. 2n = 28 (Ambastha, 1956).

3. **P. arundinàcea** L. 2n = 28 (Ambastha, 1956).

4. *P. tuberòsa* L. var. *stenoptera* (Hack.) Hitchc. Anderson uses **P. aquática** L. Howell (Leafl. W. Bot. 10: 40–41. 1963) uses *P. stenóptera* Hack. and reports it as widespread in Calif.

6. **P. canariénsis** L. 2n = 12 (Löve & Löve, 1956).

p. 1543.　9. **P. caroliniàna** Walt. 2n = 14 (Ambastha, 1956).

10. **P. angústa** Nees ex Trin. 2n = 14 (Ambastha, 1956). Taken in Sonoma Co., *M. Baker*.

DIGITÀRIA

Swallen (Wiggins, Fl. Sonoran Desert 1: 282. 1964) recognizes **D. ad-scéndens** (HBK) Henr. as a weed and distinguishes it from **D. sanguinàlis** (L.) Scop. by having the second glume two-thirds as long as the spikelet; fr. pale (instead of 2nd glume half as long as spikelet; fr. lead-colored). He has the former ranging w. to Calif., the latter w. to Texas. *D. adscendens* grows to 1.5 m. tall, not 1–6 (–9) dm., as in the text.

p. 1544.   **82a. Axonòpus** Beauv.

Stoloniferous or tufted perennials, rarely annuals. Blades usually flat or folded. Racemes slender, spikelike, digitate or racemose along the main axis. Spikelets depressed-biconvex, oblong, usually obtuse, solitary, subsessile, alternate, in 2 rows on one side of a 3-angled rachis. First glume wanting; second glume and sterile lemma equal; fertile lemma and palea indurate, the lemma oblong-elliptic, usually obtuse. (Gr. *axon*, axis, and *pous*, foot.)

1. **A. compréssus** (Sw.) Beauv. Stoloniferous; culms compressed, 1.5–5 dm. long; blades 8–25 cm. long, 8–12 mm. wide; raceme-spikes 2–5, mostly 4–8 cm. long; spikelets 2.2–2.5 (–2.8) mm. long, pilose.—Reported as occasional lawn weed, although sometimes planted for lawns. Se. U.S. to S. Am.

ERIOCHLÒA

1. **E. grácilis** (Fourn.) Hitchc. In Kern Co., *Twisselmann*.

2. **E. aristàta** Vasey. 2n = 36 (Gould, 1965).

p. 1545.   PASPÀLUM

1. **P. dístichum** L. 2n = 60 (de Wet, 1958).

2. **P. dilatàtum** Poir. 2n = 50 (Tateoka, 1955).

PÁNICUM

In Key, after A, BB, CC, D, change E to:

    E. Blades glabrous on both surfaces.
          F. Ligule a ring of hairs 4–5 mm. long. .... 3. *P. Lindheimeri*
          FF. Ligule obsolete. .... 3a. *P. agrostoides*
    EE. Blades pubescent, at least on lower surface.
          F. Plants velvety-pubescent. .... 4. *P. thermale*
          FF. Plants ± pubescent, but not velvety.
              G. Vernal blades pubescent above.
                H and HH as in the FLORA.
             GG. Vernal blades glabrous above. .... 7. *P. occidentale*

p. 1546.   3a. **P. agrostoìdes** Spreng. In dense clumps 5–10 dm. tall; lf. blades 2–5 dm. long, 5–12 mm. wide, flat; panicles terminal and axillary, 1–3 dm. long; pedicels 1- to several-haired near summit; spikelets reddish, ca. 2 mm. long.—Reported from Calif.; e. U.S.; central Am.

p. 1546.   9. **P. arizónicum** Scribn. & Merr. In Monterey Co., *Howitt & Howell*.

10. **P. dichotomiflòrum** Michx. n = 27 (Gould, 1958). Additional records are Monterey Co., *Howitt & Howell*; Nevada Co., Stanislaus Co., *Fuller*. Widely distributed in the state, *Rubtzoff*.

11. **P. capillàre** L. is probably in California, especially as a weed in San Francisco. It differs from the var. **occidentale** Rydb. in having longer, more pubescent blades; panicles less exserted and narrower; spikelets 2–2.5 mm. long.

p. 1547.   12. **P. Hillmánii** Chase. In Yolo and Solano cos., *Crampton*.

86a. Rhynchelỳtrum Nees

Perennials or annuals, with rather open panicles of silky spikelets, these on short capillary pedicels. First glume minute, villous; second glume and sterile lemma equal, gibbous below, raised on a stipe above the first glume, emarginate, short-awned, covered, except toward the slightly spreading apex, with long silky hairs, the palea well developed. Lemma cartilaginous, boat-shaped.

A genus of several spp. (Greek, *rhynchos,* beak, and *elytron,* scale, referring to the beaked second glume and sterile lemma.)

1. **R. ròseum** (Nees) Stapf and Hubb. [*Tricholaena rosea* Nees.] NATAL GRASS. Perennial, ca. 1 m. tall; blade flat, 2–5 mm. wide; panicle rosy to pink, 1–1.5 dm. long, with slender ascending branches; spikelets 5 mm. long. Abundant weed at La Mesa, San Diego Co., *T. C. Fuller;* natur. from S. Afr.

ECHINOCHLÒA

1. **E. crusgálli** (L.) Beauv. $n = 36$ (Gould, 1958).

2. **E. colònum** (L.) Link. $2n = 54$ (Tateoka, 1965).—Reported from Palo Alto, Santa Clara Co.

3. **E. oryzícola** (Vasinger) Vasinger var. **mùtica** Vasinger. Spikelets 5 mm. long, awnless.—Rice fields near Biggs, Butte Co.; originally from Eurasia.

SETÀRIA

After generic description cite: Rominger, J. M. Taxonomy of Setaria in North America. Ill. Biol. Mon. 29: 1–132. 1962.

p. 1548.   In Key to **Setaria,** after AA, BB add:

> C. Upper surface of lvs. scabrous; spikelets 1.8–2.2 mm. long; panicles at maturity nodding from apex. .... 6. *S. viridis*
> CC. Upper surface of lf. blades pilose or strigose; spikelets 2.5–3 mm. long; panicles at maturity nodding from near the base. .... 7. *S. Faberi*

1. Change *S. gláuca* (L.) Beauv. to **S. lutéscens** (Weigel) Hubb.

2. **S. geniculàta** (Lam.) Beauv. $2n = 36, 72$ (Gould, 1965).

3. **S. sphacelàta** (Schumacher) Stapf & Hubb. $2n = 36$ (de Wet, 1954); $2n = 18$ (de Wet, 1958).

4. *S. Cárnei* Hitchc. Rominger says that the Fresno specimen so identified is S. verticillàta (L.) Beauv.

7. **S. Fàberi** Herrm. Annual, 5–20 dm. tall; lf. blades scabrous and soft hairy on upper surface; panicles arching and drooping from near the base, 6–20 cm. long; spikelets 2.5–3 mm. long, subtended by 3 (1–6) bristles, each ca. 1 cm. long.—Reported as adventive in Marin Co., *Howell;* Solano Co., *Crampton;* and Los Angeles Co., *Fuller.*

PENNISÈTUM

1. **P. clandestìnum** Hochst. ex Chiov. $2n = 36$ (Narayan, 1955). Reported from San Francisco, Alameda, and Shasta cos.

2. **P. villòsum** R. Br. Renamed as *Cenchrus longisetus* M. C. Johnston.

p. 1549.   3. **P. setàceum** (Forsk.) Chiov. Occurs in Monterey Co., *Howitt & Howell.*

CÉNCHRUS

Cite as a reference: De Lisle, D. G. Taxonomy and distribution of the genus Cenchrus. Iowa State Jour. Sci. 37: 259–351. 1963.

In Key change *C. pauciflorus* to *C. incertus* and add *C. longispinus:*

> A.   Invol. with a ring of slender bristles at base; spikelets usually 4 in each bur.
>   1. *C. echinatus*
> AA. Invol. with no ring of slender bristles at base; spikelets usually 2 in each bur.
>   B.   Spines broader at base, less than 45 in number, 2–5 mm. long. . . 2. *C. incertus*
>   BB. Spines slender, usually more than 50 in number, 3.5–7 mm. long.
>   3. *C. longispinus*

1. **C. echinàtus** L. $2n = 68$ (Tateoka, 1955). $2n = 70$ (Gould, 1965). In Imperial Co., *DeLisle;* Solano Co., *Fuller.*

2. **C. incértus** M. A. Curtis [*C. pauciflorus* Benth.] $2n = 34$ (Tateoka, 1955). Solano Co., *Crampton;* Daggett, San Bernardino Co., *Fuller.*

3. **C. longispìnus** (Hack. in Kneucker) Fern. [*C. echinatus* f. *l.* Hack. in Kneucker.] Forming large clumps with many branches; culms terete, 1–9 dm. tall; sheaths pilose on margins and at throat; ligule a rim of ciliate hairs 0.7–1.7 mm. long; blades 6–18 cm. long; infl. compact, 4–10 cm. long; burs ± globose, 8–12 mm. long; spines slender, 3.5–7 mm. long; spikelets 2–3 in a bur, 6–8 mm. long; $n = 17$.—Reported from Merced, Riverside, Solano, Yolo, Lassen, and Monterey cos. Native of e. U.S.

IMPERÀTA
1. **I. brevifòlia** Vasey. Found as far n. as Centerville, Fresno Co., *Fuller*.

p. 1550.   ANDROPÒGON
3. **A. virgínicus** L. from Shasta, Butte and Yuba cos., s. to Fresno Co., *Fuller*.
5. **A. saccharoìdes** Sw. is reported from s. Calif. eastward by Swallen in Wiggins, Fl. Sonoran Desert 1: 299. 1964, differing from *A. barbinodis* by having numerous racemes on a relatively long axis (not few on a short axis), panicle long-exserted; nodes glabrous or appressed-hispid (not densely bearded); spikelets sessile, 4 mm. long (not 5–6 mm.).
Var. **Torreyànus** (Steud.) Hack.—Near Fairfield, Solano Co., *Crampton*.

92a. Hyparrhènia Anderss. ex Stapf

Tall perennials, the pairs of racemes and their spathes ± crowded, forming a large elongate infl. Spikelets in pairs, those of the lower pairs alike, sterile and awnless; fertile spikelets 1–few in each raceme, terete or flattened on the back, the base usually elongate into a sharp callus. Fertile lemma with a strong geniculate awn.
1. **H. hírta** (L.) Stapf. [*Andropogon h.* L.] To ca. 1 m. tall; blades to 3 mm. wide, ± involute, flexuous; racemes whitish or grayish, silky-villous.—Reported from Los Angeles, *Raven*. From the Old World.

92b. Heteropògon Pers.

Annual or perennial, with flat or folded blades and usually solitary terminal racemes. Lower few pairs of spikelets alike, ♂, awnless; remaining sessile spikelets fertile, long-awned; pedicellate spikelets ♂ like lower ones; rachis continuous below, bearing fertile spikelets above, disarticulating at base of each joint, the joint forming a sharp barbed callus below fertile spikelet; glumes of fertile spikelet dark brown, the 1st enclosing the 2nd; lemmas hyaline, fertile one with a long stout twisted geniculate awn. (Greek, *heteros*, different, and *pogon*, beard, referring to the awnless ♂ and awned ♀ spikelets.)
1. **H. contórtus** (L.) Beauv. ex Roem. & Schult. [*Andropogon c.* L.] Perennial, tufted, 2–8 dm. tall, glabrous, with a few flowering branches at upper nodes; blades 5–15 cm. long, 3–7 mm. wide, scabrous; raceme usually long-exserted, 4–7 cm. long; 1st glume hirsute with spreading hairs; awns 5–12 cm. long, hirsute; spikelet ca. 1 cm. long.—Dehesa School, San Diego Co., *Gander*; n. Imperial Co., *Wheeler*; Ariz. to Texas and south.

SÓRGHUM
Fuller collected 3 additional spp. of *Sorghum* from near the Experimental Farm near Bard, Imperial Co. One of these, **S. sudanénse** (L.) Pers., has been reported from the w. side of the San Joaquin Valley in Kern Co., *Twisselmann*. It is an annual, 2–3 m. tall; lf.-blades 1.5–3 dm. long, 8–12 mm. wide; panicle erect, loose, 1.5–3 dm. long, the branches subverticillate. Escaped from cult. as a hay and pasture grass.
Another, **S. virgàtum** (Hack.) Stapf was taken by *Fuller* 1.5 mi. sw. of Bard, also n. of Indio, Riverside Co. It is a tall annual with a narrow slender open panicle and narrowly lanceolate green finely awned spikelets.
The third, **S. lanceolàtum** Stapf was also from 1.5 mi. sw. of Bard, Imperial Co. It is a robust annual to 1.5 m. tall; blades 3–6 dm. long, 2–3.5 cm. wide; panicle 2.5–4 dm. long with ascending branches; rachis joints and pedicels ciliate; spikelets ca. 6 mm. long, silky-pubescent; awn ca. 1 cm. long. From trop. Afr.

CORRECTIONS TO THE INDEX OF THE FLORA

p. 1594. Aesculus, 994

p. 1595. Allium
lacunosum, 1378

p. 1599. Arbutus
Menziesii, not
italicized.

p. 1600. Artemisia
1238 after vari-
ous spp. should
be 1236
Artemiastrum
should be
Artemisiastrum,
1236

p. 1603. Barbarea
orthoceras, not
orthoceras

p. 1604. *Bermudiana*
should precede
Bernardia

p. 1605. Change order to:
Brachyris
Bracken
Brake
Brandegea
Brasenia
Brassica
Braya
Brevoortia
Breweria
Brewerina
Brickellia
Bride, Mourning-
Brier, Sweet-
Briza
Brizopyrum
Brodiaea
Bryanthus,
on p. 1605
after Brush

p. 1608. Campanula
prenanthoides,
1063
Cardionema,
not italicized

p. 1610. Ceanothus
cordulatus, 978

p. 1612. Chicory, 1288

p. 1614. Clarkia
deflexa, 942

p. 1615. Coleogyne
ramosissima, 782

p. 1617. Crypsis
niliaca,
not niliacea
Cryptantha
pterocarya, 567
Purpusii, 568

p. 1621. Eastwoodia, 1099

p. 1622. Ephedra
viridis, 67

p. 1627. Eupatorium
occidentale, 1268

p. 1629. Gastridium, 1525

p. 1632. Gnaphalium
decurrens, 1260

p. 1633. Gymnosteris,
add:
minuscula, 474
nudicaulis, 474

p. 1634. Heleochloa
schoenoides,
1529

p. 1637. Hymenoclea
monogyra, 1102
Salsola, 1102

p. 1639. Koeberlinia
spinosa, 174

p. 1641. Liguliflorae, 1286
Ligusticum
Grayii, 1014
Pringlei, 1014

p. 1643. Lonicera
involucrata, 1050
flavescens, 1051
Ledebourii,
1051

p. 1644. Lotus
junceus,
not italicized
Lunaria
annua, 241
Lupinus
Benthamii,
not italicized

p. 1647. *Mammillaria*,
not italicized

p. 1649. *Microcala*,
not italicized

p. 1652. Nemophila
macrocarpa and
maculata should
be above
Menziesii
Neostapfia
colusana
misspelled

p. 1653. Oenothera
Jepsonii, p. 946

p. 1659. Phragmites
should precede
Phyla

p. 1661. Poa bulbosa,
1486

p. 1670. Scrophularia
californica,
not california

p. 1671. Senecio
petrocallis, 1250

p. 1672. Snow Plant, 436
Solomon's-Seal,
False, 1330

p. 1676. Thelesperma,
1095
Thlaspi should
precede Thorn,
Box-Thorn
Cotton- through
Thysanocarpus
(p. 1677) should
precede Tillaea.

p. 1677. Trifolium
californicum,
838

p. 1678. Macraei, 838
Trisetum
spicatum
not italicized

p. 1679. Vancouveria
to come after
Valerianella

p. 1680. Vicia
californica, 895
madrensis, 895

p. 1681. Zigadenus
Fremontii, 1334
inezianus, 1334

Index to the Supplement

Names in italics are largely those in italics in the text. Common names that are hyphenated or consist of more than one word, such as Poison-Oak and Blue Oak, are to be sought under the second word, in this case Oak. Compound names spelled as a single word, like Fireweed and Watermelon, are listed under the initial letter.

List of Families for A CALIFORNIA FLORA